D0772461

CRC Handbook of Chemistry and Physics

91st Edition

EDITORIAL ADVISORY BOARD

CRC Handbook of Chemistry and Physics

A Ready-Reference Book of Chemical and Physical Data

Editor-in-Chief

W. M. Haynes, Ph.D.
Scientist Emeritus
National Institute of Standards and Technology

Associate Editor

David R. Lide, Ph.D.
Former Director, Standard Reference Data
National Institute of Standards and Technology

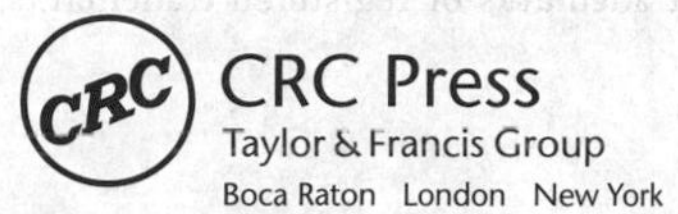

CRC Press is an imprint of the
Taylor & Francis Group, an **informa** business

CRC Press
Taylor & Francis Group
6000 Broken Sound Parkway NW, Suite 300
Boca Raton, FL 33487-2742

Printed in the United States of America on acid-free paper
10 9 8 7 6 5 4 3 2 1

International Standard Book Number: 978-1-4398-2077-3 (Hardback)

Visit the Taylor & Francis Web site at
http://www.taylorandfrancis.com

and the CRC Press Web site at
http://www.crcpress.com

FOREWORD

I enjoyed being a student in the '1950s,' when my university Physics Department had two copies of the *CRC Handbook of Chemistry and Physics*, but a single subscription to the *Physical Review*. Professors and students alike had used these bound volumes for years and years. What a joy it was to search for some fact in these pages. In the *Handbook* the actual needed data could usually be recovered efficiently via the indexing, but the real joy was in some moments "lost" in reading some other tables of data that suddenly had become interesting. Both the *Handbook* and the *Physical Review* exhibited the spontaneous "snap-open" character imposed by generations of faculty and young researchers wanting to find the specific heat of bismuth, or the definition of the pascal, or to read the discovery announcement of the deuteron, or nuclear fission or … . A few hours spent browsing the *CRC Handbook of Chemistry and Physics* for me was the more rewarding, in view of its broader domain and focus on the properties of real stuff. And hence the continuing market for well-used copies, even 15 years old.

But what about all the tables of numbers in these *CRC Handbooks*? Do we now really need a printed book, when we have the Internet available? To me the answer is completely and transparently "YES" we do indeed need *books* of *carefully reviewed data*. The locations of frequently used tables are instantly visible, their organization permanent and unchanging. The carefully reviewed attribute also deserves further comment: Science is a special domain, where we interested little workers are curious to understand how Nature works, and perhaps sometimes can use a new insight to design some new process, or reaction, or chain of equations. For these data users, while it is helpful to have the data conveniently available, having the *best* data is critical. In many cases the decisive point will come much later: After all our calculations, is this derived quantity just slightly positive or slightly negative? Or even more interestingly, zero within the uncertainty of our evaluation? Eventually we should be able to rid our work of mathematical errors, but uncertainties in data are intrinsic to the measurement process. And in these days, the progress in some measurement fields pushes against the accuracy/reproducibility limits of the standards with which we quantify our experimental results. Indeed the team of this (now-retired) writer has had this experience — the improved measurement techniques led to challenge the accuracy of realization of the definition of an SI unit, the meter, and ultimately to its replacement definition as a derived quantity, via an adopted value for the speed of light (299,792,458 m/s).

Sometimes we find "quick" and "approximate" results also to be empowering. In "real" life we often just need to know some realistic numbers for preliminary design of some apparatus or process. Or one recent evening after looking at the heating bill, I just really want to have some realistic numbers to try to account for the thermal (in-)efficiency of my new house. Whatever the technical topic, I'll wager reliable information comes faster from the well-known *Handbook* than from the Web. I *do* also love Google and the Web, but the *Handbook* data are created by — and scrutinized by — the expert user community, with any detected errors fully traceable and normally corrected in the next printing. And there is no software start-up time needed!

Another case where serendipity, and unexpected curiosity count is the accidental discovery of interesting "stuff" far from the search topic. For example, in the 90th edition it is wonderful also to have a credible table of the mineral contents – as well as caloric content – of many foods.

In this 91st edition, the traditional (and huge) editing effort and good critical judgment have been brought to this task by Dr. W. M. "Mickey" Haynes of the NIST, who has taken over the helm from long-time editor Dr. David R. Lide. Enjoy!

John L. Hall, PhD
Nobel Laureate Physics, 2005
JILA, University of Colorado and
National Institute of Standards and Technology
Boulder, Colorado
March 2010

PREFACE

I am pleased to serve as the fifth Editor-in-Chief of the *CRC Handbook of Chemistry and Physics*. The *Handbook* has been a staple in my library and life during my career, and I am very appreciative to the previous editors for their efforts to ensure the reliability of its information. I thank the publisher for giving me this opportunity, and I hope that my background in properties research will continue to contribute to the level of excellence established for the *Handbook* and to ensure that it remains the first source for data for physical and chemical properties and related topics.

I have been working with David Lide, Editor-in-Chief of the *CRC Handbook of Chemistry and Physics* from 1989–2009, for the past two years during a transition period. I am much indebted to him for his collegial and expert guidance during this period. My main goals, as I take on this project, are to maintain the high standards of quality that have been a hallmark for the *Handbook*, to provide respected sources of documentation for the information published in the *Handbook*, and to expand the coverage of the diverse subject matter in the *Handbook* consistent with advances in science and technology. One cannot overestimate the invaluable contributions that David Lide made to the *Handbook* during his tenure. I would be remiss without noting the most significant of them.

The numerous major changes made to the *Handbook* during the tenure of David Lide were carried out to make it more user friendly, to expand its coverage, and to emphasize quality control, especially in terms of the reliability of the information and its documentation. These major changes include: (a) development of Internet and CD-ROM versions of the *Handbook*; (b) reorganization of the content into 16 sections that bring together related topics; (c) creation of a more systematic index; (d) consistent use of chemical conventions according to IUPAC and ISO recommendations and standards; (e) timely updates of such basic information as fundamental constants, atomic weights, SI units, etc.; (f) major reorganization and expansion of comprehensive, widely used tables for the properties of organic and inorganic compounds, while emphasizing compounds of greatest industrial and research interest; (g) expanded coverage of biochemistry, geophysics, astronomy, and environmental science; and (h) better communication with users through focus groups and other feedback mechanisms.

The 91st Edition of the *Handbook* includes new tables and major expansions and updates, especially in the following areas:

Section 6: Fluid Properties
- New tables on thermophysical properties of selected fluids at saturation and the dependence of liquid density on temperature and pressure
- Major updates for tables on the density of water and properties of ice and D_2O
- Major update and expansion of the table on critical constants of organic compounds

Section 8: Analytical Chemistry
- Major updates for tables on the ionization constants of water and heavy water

Section 9: Molecular Structure and Spectroscopy
- Updates for tables on atomic radii of the elements, bond dissociation energies, and spectroscopic constants of diatomic molecules

Section 10: Atomic, Molecular Structure and Spectroscopy
- Major update for the table on atomic transition probabilities (added new elements) and updates for tables on electron affinities and atomic and molecular polarizabilities

Section 12: Properties of Solids
- New table on electron stopping powers of elements

Section 13: Polymer Properties
- New tables on abbreviations in polymer science and on physical properties of polymers

In order to maintain a manageable number of pages and allow space for growth of the *Handbook*, the indexes for molecular formulae and CAS registry numbers for the Physical Constants of Organic Compounds table and the index for CAS registry numbers for the Physical Constants of Inorganic Compounds table have been removed from the hard-copy edition of the *Handbook*. However, they are available in the electronic versions of the *Handbook* and by email request to the Editor-in-Chief: william.haynes@taylorandfrancis.com

The assistance and cooperation of the Thermodynamics Research Center (TRC) of the National Institute of Standards and Technology (NIST) in updating, expanding, and validating thermophysical properties of organic compounds in the table on critical constants should be recognized. A considerable effort was devoted to this activity for this edition and will be expanded to other tables in the future. TRC specializes in the collection, evaluation, and correlation of thermophysical and thermochemical property data for organic compounds.

The success of the *Handbook* is very dependent on feedback from its users. The Editor-in-Chief will appreciate any suggestions from readers on proposed new topics for the *Handbook* or comments on how the usefulness of the *Handbook* may be improved in future editions. Please send your comments to the Editor-in-Chief: william.haynes@taylorandfrancis.com

Numerous international experts make key contributions to the *Handbook*. These contributors are listed on the pages immediately following the Preface. Their efforts play a significant role in contributing to the wide range of material covered in the *Handbook*. I should also acknowledge the contributions of the Editorial Advisory Board of the *Handbook*, whose members are listed in the front matter. I am much indebted to Fiona Macdonald, Publisher (Chemical & Life Sciences), CRC Press/Taylor & Francis Group for her support and assistance in the preparation of the *Handbook*. Thanks also to Glen Butler, Pam Morrell, Mimi Williams, Theresa Delfron, and James Yanchak for their careful and cooperative work in the production of the *Handbook*.

W. M. Haynes
March 2010

The 91st Edition of the *CRC Handbook of Chemistry and Physics* is dedicated in memory of my parents, William Paul and Nellie Young Haynes

Note on the Ordering of Chemical Compounds: Several different ordering schemes for lists of chemical compounds are used in this book. The long tables, Physical Constants of Organic Compounds and Physical Constants of Inorganic Compounds, are ordered by name (generally the systematic name), but indexes to synonyms, formulas, and CAS Registry Numbers are available. If the table is very short and includes only familiar substances, the listing is usually alphabetical by name or common formula. Many tables of intermediate length are ordered by molecular formula using a modification of the Hill convention. In this convention the molecular formula is written with C first, H second, and then all other elements in alphabetical order of their chemical symbols. For tables with organic compounds only, the sequence of entries is determined by the alphabetical order of elements in the molecular formula and the number of atoms of each element, in ascending order, e.g., C_3H_7Cl, C_3H_7N, C_3H_7NO, $C_3H_7NO_2$, etc. (For organic compounds, a quick way to look up the molecular formula is to use the Physical Constants of Organic Compounds table, which starts on Page **3**-1, and its synonym index on Page **3**-524.) In tables containing non-carbon compounds, those are usually listed first, followed by a separate listing of compounds that do contain carbon. This is a departure from the strict Hill convention as followed by the Chemical Abstracts Service, where the molecular formulas beginning with A and B precede the formulas for carbon-containing compounds, while those beginning with D … Z follow. For tabular displays, as opposed to an index, it appears more convenient to the user if the non-carbon compounds are listed as a block, rather than being split by the longer list of carbon compounds.

CURRENT CONTRIBUTORS

Lev. I. Berger
California Institute of Electronics and Materials Science
2115 Flame Tree Way
Hemet, California 92545

Charles E. Carraher
Department of Chemistry and Biochemistry
Florida Atlantic University
Boca Raton, Florida 33431

Robert D. Chirico
Thermodynamics Research Center
Thermophysical Properties Division
National Institute of Standards and Technology
Boulder, Colorado 80305

Ivan Cibulka
Department of Physical Chemistry
Institute of Chemical Technology
CZ-166 28 Prague, Czech Republic

Arthur K. Covington
Department of Chemistry
University of Newcastle
Newcastle upon Tyne NE1 7RU
England

Christopher J. Cramer
Department of Chemistry
University of Minnesota
Minneapolis, Minnesota 55455

Vladimir Diky
Thermodynamics Research Center
Thermophysical Properties Division
National Institute of Standards and Technology
Boulder, Colorado 80305

Michael Frenkel
Thermodynamics Research Center
Thermophysical Properties Division
National Institute of Standards and Technology
Boulder, Colorado 80305

Jeffrey R. Fuhr
Atomic Physics Division
National Institute of Standards and Technology
Gaithersburg, Maryland 20899

Jürgen Gmehling
Universität Oldenburg
Falkutät V, Technische Chemie
D-26111 Oldenburg, Germany

Robert N. Goldberg
Biotechnology Division
National Institute of Standards and Technology
Gaithersburg, Maryland 20899

Allan H. Harvey
Thermophysical Properties Division
National Institute of Standards and Technology
Boulder, Colorado 80305

Steven R. Heller
Chemical and Biochemical Reference Data Division
National Institute of Standards and Technology
Gaithersburg, Maryland 20899

Norman E. Holden
National Nuclear Data Center
Brookhaven National Laboratory
Upton, New York 11973

Andrei Kazakov
Thermodynamics Research Center
Thermophysical Properties Division
National Institute of Standards and Technology
Boulder, Colorado 80305

Henry V. Kehiaian (deceased)

Daniel E. Kelleher
Atomic Physics Division
National Institute of Standards and Technology
Gaithersburg, Maryland 20899

Carolyn A. Koh
Center for Hydrate Research
Colorado School of Mines
1600 Illinois Street
Golden, Colorado 80401

William H. Koppenol
Dept. CHAB
Lab. f. Anorg. Chemie, HC1 H211
Wolfgang-Pauli-Strasse 10
ETH Hönggerberg
CH-8093 Zürich, Switzerland

Eric W. Lemmon
Thermophysical Properties Division
National Institute of Standards and Technology
Boulder, Colorado 80305

Frank J. Lovas
8616 Melwood Rd.
Bethesda, Maryland 20817

Yu-Ran Luo
School of Chemistry and Material Science
University of Science and Technology of China
Hefei 230026, China

Serguei N. Lvov
Department of Energy and Mineral Engineering
Pennsylvania State University
University Park, Pennsylvania 16802

Manjeera Mantina
Department of Chemistry
University of Minnesota
Minneapolis, Minnesota 55455

William C. Martin
Atomic Physics Division
National Institute of Standards and Technology
Gaithersburg, Maryland 20899

Alan D. McNaught
8 Cavendish Avenue
Cambridge CB1 7US
England

Thomas M. Miller
Air Force Research Laboratory/VSBP
29 Randolph Rd.
Hanscom AFB, Massachusetts 01731-3010

Nasser Moazzen-Ahmadi
Department of Physics and Astronomy
University of Calgary
2500 University Drive NW
Calgary, Alberta T2N 1N4, Canada

Peter J. Mohr
Physics Laboratory
National Institute of Standards and Technology
Gaithersburg, Maryland 20899

Chris D. Muzny
Thermodynamics Research Center
Thermophysical Properties Division
National Institute of Standards and Technology
Boulder, Colorado 80305

Irving Ozier
Department of Physics and Astronomy
University of British Columbia
6224 Agricultural Road
Vancouver, British Columbia V6T 1Z1, Canada

Larissa I. Podobedova
Atomic Physics Division
National Institute of Standards and Technology
Gaithersburg, Maryland 20899

Cedric J. Powell
Surface and Microanalysis Science Division
National Institute of Standards and Technology
Gaithersburg, Maryland 20899

Joseph Reader
Atomics Physics Division
National Institute of Standards and Technology
Gaithersburg, Maryland 20899

E. Dendy Sloan
Center for Hydrate Research
Colorado School of Mines
1600 Illinois Street
Golden, Colorado 80401

Lewis E. Synder
Astronomy Department
University of Illinois
Urbana, Illinois 61801

Barry N. Taylor
Physics Laboratory
National Institute of Standards and Technology
Gaithersburg, Maryland 20899

Donald G. Truhlar
Department of Chemistry
University of Minnesota
Minneapolis, Minnesota 55455

Rosendo Valero
Chemistry Department
University of Coimbra
Coimbra, Portugal

Wolfgang L. Wiese
Physics Laboratory
National Institute of Standards and Technology
Gaithersburg, Maryland 20899

Christian Wohlfarth
Martin Luther University
Institute of Physical Chemistry
Mühlpforte 1
06108 Halle (Saale), Germany

Daniel Zwillinger
Mathematics Department
Rensselaer Polytechnic Institute
Troy, New York 12180

TABLE OF CONTENTS

SECTION 1: BASIC CONSTANTS, UNITS, AND CONVERSION FACTORS

CODATA Recommended Values of the Fundamental Physical Constants ... 1-1
Standard Atomic Weights (2007) ... 1-12
Atomic Masses and Abundances ... 1-14
Electron Configuration and Ionization Energy of Neutral Atoms in the Ground State ... 1-18
International Temperature Scale of 1990 (ITS-90) ... 1-20
Conversion of Temperatures from the 1948 and 1968 Scales to ITS-90 ... 1-21
International System of Units (SI) ... 1-23
Units for Magnetic Properties ... 1-27
Conversion Factors ... 1-28
Conversion of Temperatures ... 1-38
Conversion Factors for Energy Units ... 1-39
Conversion Factors for Pressure Units ... 1-40
Conversion Factors for Thermal Conductivity Units ... 1-41
Conversion Factors for Electrical Resistivity Units ... 1-42
Conversion Factors for Chemical Kinetics ... 1-43
Conversion Factors for Ionizing Radiation ... 1-44
Values of the Gas Constant in Different Unit Systems ... 1-46

SECTION 2: SYMBOLS, TERMINOLOGY, AND NOMENCLATURE

Symbols and Terminology for Physical and Chemical Quantities ... 2-1
Expression of Uncertainty of Measurements ... 2-13
Nomenclature for Chemical Compounds ... 2-15
Nomenclature for Inorganic Ions and Ligands ... 2-16
Organic Substituent Groups and Ring Systems ... 2-23
Representation of Chemical Structures with the IUPAC International Chemical Identifier (InChI) ... 2-27
Scientific Abbreviations, Acronyms, and Symbols ... 2-29
Greek, Russian, and Hebrew Alphabets ... 2-43
Definitions of Scientific Terms ... 2-44
Thermodynamic Functions and Relations ... 2-69
Nobel Laureates in Chemistry and Physics ... 2-70

SECTION 3: PHYSICAL CONSTANTS OF ORGANIC COMPOUNDS

Physical Constants of Organic Compounds ... 3-1
Synonym Index of Organic Compounds ... 3-524
Diamagnetic Susceptibility of Selected Organic Compounds ... 3-549

SECTION 4: PROPERTIES OF THE ELEMENTS AND INORGANIC COMPOUNDS

The Elements ... 4-1
Physical Constants of Inorganic Compounds ... 4-43
Formula Index of Inorganic Compounds ... 4-102
Physical Properties of the Rare Earth Metals ... 4-115
Melting, Boiling, Triple, and Critical Point Temperatures of the Elements ... 4-121
Heat Capacity of the Elements at 25 °C ... 4-123
Vapor Pressure of the Metallic Elements — Equations ... 4-124
Vapor Pressure of the Metallic Elements — Data ... 4-126
Density of Molten Elements and Representative Salts ... 4-127
Magnetic Susceptibility of the Elements and Inorganic Compounds ... 4-130
Index of Refraction of Inorganic Liquids ... 4-136
Physical and Optical Properties of Minerals ... 4-137
Crystallographic Data on Minerals ... 4-144

SECTION 5: THERMOCHEMISTRY, ELECTROCHEMISTRY, AND KINETICS

CODATA Key Values for Thermodynamics ... 5-1
Standard Thermodynamic Properties of Chemical Substances ... 5-4
Thermodynamic Properties as a Function of Temperature ... 5-43
Thermodynamic Properties of Aqueous Ions ... 5-66
Heat of Combustion ... 5-68
Energy Content of Fuels ... 5-69

Electrical Conductivity of Water 5-70
Electrical Conductivity of Aqueous Solutions 5-71
Standard KCl Solutions for Calibrating Conductivity Cells 5-72
Molar Conductivity of Aqueous HF, HCl, HBr, and HI 5-73
Equivalent Conductivity of Electrolytes in Aqueous Solution 5-74
Ionic Conductivity and Diffusion at Infinite Dilution 5-75
Activity Coefficients of Acids, Bases, and Salts 5-78
Mean Activity Coefficients of Electrolytes as a Function of Concentration 5-80
Enthalpy of Dilution of Acids 5-84
Enthalpy of Solution of Electrolytes 5-85
Enthalpy of Hydration of Gases 5-86
Chemical Kinetic Data for Stratospheric Modeling 5-90

SECTION 6: FLUID PROPERTIES

Thermophysical Properties of Water and Steam 6-1
Vapor Pressure and Other Saturation Properties of Water 6-5
Standard Density of Water 6-7
Fixed-Point Properties of H_2O and D_2O 6-9
Properties of Saturated Liquid D_2O 6-10
Properties of Ice and Supercooled Water 6-12
Vapor Pressure of Ice 6-13
Melting Point of Ice as a Function of Pressure 6-13
Permittivity (Dielectric Constant) of Water at Various Frequencies 6-14
Thermophysical Properties of Air 6-15
Thermophysical Properties of Fluids 6-21
Thermophysical Properties of Selected Fluids at Saturation 6-38
Virial Coefficients of Selected Gases 6-46
Van Der Waals Constants for Gases 6-55
Mean Free Path and Related Properties of Gases 6-56
Influence of Pressure on Freezing Points 6-57
Critical Constants of Organic Compounds 6-58
Critical Constants of Inorganic Compounds 6-82
Sublimation Pressure of Solids 6-85
Vapor Pressure 6-87
Vapor Pressure of Fluids at Temperatures Below 300 K 6-117
Vapor Pressure of Saturated Salt Solutions 6-125
IUPAC Recommended Data for Vapor Pressure Calibration 6-126
Enthalpy of Vaporization 6-127
Enthalpy of Fusion 6-145
Compressibility and Expansion Coefficients of Liquids 6-155
Temperature and Pressure Dependence of Liquid Density 6-157
Volumetric Properties of Aqueous Sodium Chloride Solutions 6-162
Properties of Cryogenic Fluids 6-163
Properties of Liquid Helium 6-164
Properties of Refrigerants 6-165
Properties of Gas Clathrate Hydrates 6-168
Ionic Liquids 6-173
Density and Specific Volume of Mercury 6-177
Thermal Properties of Mercury 6-178
Melting Curve of Mercury 6-179
Vapor Pressure of Mercury 6-180
Surface Tension of Common Liquids 6-181
Surface Tension of Aqueous Mixtures 6-185
Permittivity (Dielectric Constant) of Liquids 6-186
Permittivity (Dielectric Constant) of Gases 6-208
Azeotropic Data for Binary Mixtures 6-209
Viscosity of Gases 6-228
Viscosity of Liquids 6-229
Viscosity of Carbon Dioxide Along the Saturation Line 6-234
Viscosity and Density of Aqueous Hydroxide Solutions 6-235
Viscosity of Liquid Metals 6-236
Thermal Conductivity of Gases 6-238
Thermal Conductivity of Liquids 6-240

Diffusion in Gases 6-245
Diffusion of Gases in Water 6-247
Diffusion Coefficients in Liquids at Infinite Dilution 6-248

SECTION 7: BIOCHEMISTRY

Properties of Amino Acids 7-1
Structures of Common Amino Acids 7-3
Properties of Purine and Pyrimidine Bases 7-5
The Genetic Code 7-6
Properties of Fatty Acids and Their Methyl Esters 7-7
Composition and Properties of Common Oils And Fats 7-9
Carbohydrate Names and Symbols 7-14
Standard Transformed Gibbs Energies of Formation for Biochemical Reactants 7-16
Apparent Equilibrium Constants for Enzyme-Catalyzed Reactions 7-19
Thermodynamic Quantities for the Ionization Reactions of Buffers in Water 7-23
Biological Buffers 7-26
Typical pH Values of Biological Materials and Foods 7-27
Structure and Functions of Some Common Drugs 7-28
Chemical Constituents of Human Blood 7-45
Chemical Composition of the Human Body 7-48
Nutrient Values of Foods 7-49

SECTION 8: ANALYTICAL CHEMISTRY

Preparation of Special Analytical Reagents 8-1
Standard Solutions of Acids, Bases, and Salts 8-5
Standard Solutions of Oxidation and Reduction Reagents 8-7
Organic Analytical Reagents for the Determination of Inorganic Substances 8-8
Flame and Bead Tests 8-13
Acid-Base Indicators 8-15
Fluorescent Indicators 8-18
Conversion Formulas for Concentration of Solutions 8-19
Electrochemical Series 8-20
Reduction and Oxidation Potentials for Certain Ion Radicals 8-30
Practical pH Measurements on Natural Waters 8-37
Buffer Solutions Giving Round Values of pH at 25 °C 8-39
Dissociation Constants of Inorganic Acids and Bases 8-40
Dissociation Constants of Organic Acids and Bases 8-42
Concentrative Properties of Aqueous Solutions: Density, Refractive Index, Freezing Point Depression, and Viscosity 8-52
Ionization Constant of Water 8-78
Ionization Constant of Normal and Heavy Water 8-79
Solubility of Selected Gases in Water 8-80
Solubility of Carbon Dioxide in Water at Various Temperatures and Pressures 8-84
Aqueous Solubility and Henry's Law Constants of Organic Compounds 8-85
Aqueous Solubility of Inorganic Compounds at Various Temperatures 8-121
Solubility Product Constants 8-127
Solubility of Common Salts at Ambient Temperatures 8-130
Solubility of Hydrocarbons in Seawater 8-131
Solubility of Organic Compounds in Pressurized Hot Water 8-133
Solubility Chart 8-136
Reduction of Weighings in Air to Vacuo 8-138
Volume of One Gram of Water 8-139
Properties of Carrier Gases for Gas Chromatography 8-140
Solvents for Ultraviolet Spectrophotometry 8-141
^{13}C Chemical Shifts of Useful Nmr Solvents 8-142
Mass Spectral Peaks of Common Organic Solvents 8-143
Proton NMR Shifts of Common Organic Solvents 8-150

SECTION 9: MOLECULAR STRUCTURE AND SPECTROSCOPY

Bond Lengths in Crystalline Organic Compounds 9-1
Bond Lengths in Organometallic Compounds 9-17
Structure of Free Molecules in the Gas Phase 9-19
Characteristic Bond Lengths in Free Molecules 9-48

Atomic Radii of the Elements 9-49
Dipole Moments 9-51
Hindered Internal Rotation 9-60
Bond Dissociation Energies 9-65
Electronegativity 9-99
Force Constants for Bond Stretching 9-100
Fundamental Vibrational Frequencies of Small Molecules 9-101
Spectroscopic Constants of Diatomic Molecules 9-104
Infrared Correlation Charts 9-110
Nuclear Spins, Moments, and Other Data Related to NMR Spectroscopy 9-115
Proton NMR Chemical Shifts for Characteristic Organic Structures 9-118
^{13}C-NMR Absorptions of Major Functional Groups 9-119

SECTION 10: ATOMIC, MOLECULAR, AND OPTICAL PHYSICS

Line Spectra of the Elements 10-1
Atomic Transition Probabilities 10-93
Electron Affinities 10-147
Proton Affinities 10-167
Atomic and Molecular Polarizabilities 10-186
Ionization Energies of Atoms and Atomic Ions 10-196
Ionization Energies of Gas-Phase Molecules 10-199
X-Ray Atomic Energy Levels 10-217
Electron Binding Energies of the Elements 10-221
Natural Width of X-Ray Lines 10-227
Photon Attenuation Coefficients 10-228
Classification of Electromagnetic Radiation 10-233
Sensitivity of the Human Eye to Light of Different Wavelengths 10-235
Black Body Radiation 10-236
Characteristics of Infrared Detectors 10-238
Index of Refraction of Inorganic Crystals 10-239
Refractive Index and Transmittance of Representative Glasses 10-243
Index of Refraction of Water 10-244
Index of Refraction of Liquids for Calibration Purposes 10-245
Index of Refraction of Air 10-246
Index of Refraction of Gases 10-247
Characteristics of Laser Sources 10-248
Infrared Laser Frequencies 10-254
Infrared and Far-Infrared Absorption Frequency Standards 10-261

SECTION 11: NUCLEAR AND PARTICLE PHYSICS

Summary Tables of Particle Properties 11-1
Table of the Isotopes 11-2
Neutron Scattering and Absorption Properties 11-170
Cosmic Radiation 11-183

SECTION 12: PROPERTIES OF SOLIDS

Techniques for Materials Characterization: Experimental Techniques Used to Determine the Composition, Structure, and Energy States of Solids and Liquids 12-1
Symmetry of Crystals 12-5
Ionic Radii in Crystals 12-11
Polarizabilities of Atoms and Ions in Solids 12-13
Crystal Structures and Lattice Parameters of Allotropes of the Elements 12-15
Phase Transitions in the Solid Elements at Atmospheric Pressure 12-19
Lattice Energies 12-21
The Madelung Constant and Crystal Lattice Energy 12-34
Elastic Constants of Single Crystals 12-35
Electrical Resistivity of Pure Metals 12-41
Electrical Resistivity of Selected Alloys 12-43
Electrical Resistivity of Graphite Materials 12-46
Permittivity (Dielectric Constant) of Inorganic Solids 12-47
Curie Temperature of Selected Ferroelectric Crystals 12-56
Properties of Antiferroelectric Crystals 12-57
Dielectric Constants of Glasses 12-58

Properties of Superconductors 12-59
High-Temperature Superconductors 12-75
Organic Superconductors 12-77
Properties of Semiconductors 12-80
Selected Properties of Semiconductor Solid Solutions 12-93
Properties of Organic Semiconductors 12-95
Diffusion Data for Semiconductors 12-99
Properties of Magnetic Materials 12-107
Organic Magnets 12-116
Electron Inelastic Mean Free Paths 12-119
Electron Stopping Powers 12-121
Electron Work Function of the Elements 12-123
Secondary Electron Emission 12-124
Optical Properties of Selected Elements 12-125
Optical Properties of Selected Inorganic and Organic Solids 12-150
Elasto-Optic, Electro-Optic, and Magneto-Optic Constants 12-169
Nonlinear Optical Constants 12-183
Phase Diagrams 12-186
Heat Capacity of Selected Solids 12-204
Thermal and Physical Properties of Pure Metals 12-205
Thermophysical Properties of Stainless Steel 310 12-207
Thermal Conductivity of Metals and Semiconductors as a Function of Temperature 12-208
Thermal Conductivity of Alloys as a Function of Temperature 12-210
Thermal Conductivity of Crystalline Dielectrics 12-211
Thermal Conductivity of Ceramics and Other Insulating Materials 12-213
Thermal Conductivity of Glasses 12-215
Thermoelectric Properties of Metals and Semiconductors 12-219
Fermi Energy and Related Properties of Metals 12-221
Properties of Commercial Metals and Alloys 12-223
Hardness of Minerals and Ceramics 12-224

SECTION 13: POLYMER PROPERTIES

Abbreviations Used in Polymer Science and Technology 13-1
Physical Properties of Selected Polymers 13-3
Nomenclature for Organic Polymers 13-5
Solvents for Common Polymers 13-9
Glass Transition Temperature for Selected Polymers 13-10
Dielectric Constant of Selected Polymers 13-17
Pressure–Volume–Temperature Relationships for Polymer Melts 13-18
Upper Critical (UCST) and Lower Critical (LCST) Solution Temperatures of Binary Polymer Solutions 13-23
Vapor Pressures (Solvent Activities) for Binary Polymer Solutions 13-41
Specific Enthalpies of Solution of Polymers and Copolymers 13-46
Solubility Parameters of Selected Polymers 13-74

SECTION 14: GEOPHYSICS, ASTRONOMY, AND ACOUSTICS

Astronomical Constants 14-1
Properties of the Solar System 14-2
Satellites of the Planets 14-4
Interstellar Molecules 14-7
Mass, Dimensions, and Other Parameters of the Earth 14-10
Geological Time Scale 14-12
Acceleration Due to Gravity 14-13
Density, Pressure, and Gravity as a Function of Depth within the Earth 14-14
Ocean Pressure as a Function of Depth and Latitude 14-15
Properties of Seawater 14-16
Abundance of Elements in the Earth's Crust and in the Sea 14-18
Solar Irradiance at the Earth 14-19
U.S. Standard Atmosphere (1976) 14-20
Geographical and Seasonal Variations in Solar Radiation 14-26
Major World Earthquakes 14-27
Weather-Related Scales 14-31
Infrared Absorption by the Earth's Atmosphere 14-33
Atmospheric Concentration of Carbon Dioxide, 1958–2008 14-34

Mean Temperatures in the United States, 1900–1992 14-36
Global Temperature Trend, 1880–2009 14-38
Global Warming Potential of Greenhouse Gases 14-39
Atmospheric Electricity 14-41
Speed of Sound in Various Media 14-48
Attenuation and Speed of Sound in Air as a Function of Humidity and Frequency 14-50
Speed of Sound in Dry Air 14-51
Musical Scales 14-52
Characteristics of Human Hearing 14-53

SECTION 15: PRACTICAL LABORATORY DATA

Standard ITS-90 Thermocouple Tables 15-1
Secondary Reference Points on the ITS-90 Temperature Scale 15-10
Relative Sensitivity of Bayard-Alpert Ionization Gauges to Various Gases 15-12
Laboratory Solvents and Other Liquid Reagents 15-13
Miscibility of Organic Solvents 15-23
Density of Solvents as a Function of Temperature 15-25
Dependence of Boiling Point on Pressure 15-26
Ebullioscopic Constants for Calculation of Boiling Point Elevation 15-27
Cryoscopic Constants for Calculation of Freezing Point Depression 15-28
Freezing Point Lowering by Electrolytes in Aqueous Solution 15-29
Correction of Barometer Readings to 0 °C Temperature 15-30
Determination of Relative Humidity from Dew Point 15-31
Determination of Relative Humidity from Wet and Dry Bulb Temperatures 15-32
Constant Humidity Solutions 15-33
Standard Salt Solutions for Humidity Calibration 15-34
Low-Temperature Baths for Maintaining Constant Temperature 15-35
Metals and Alloys with Low Melting Temperature 15-36
Wire Tables 15-37
Characteristics of Particles and Particle Dispersoids 15-38
Density of Various Solids 15-39
Density of Sulfuric Acid 15-40
Density of Ethanol–Water Mixtures 15-41
Dielectric Strength of Insulating Materials 15-42
Coefficient of Friction 15-47
Flame Temperatures 15-49
Allocation of Frequencies in the Radio Spectrum 15-50

SECTION 16: HEALTH AND SAFETY INFORMATION

Handling and Disposal of Chemicals in Laboratories 16-1
Flammability of Chemical Substances 16-13
Threshold Limits for Airborne Contaminants 16-29
Octanol–Water Partition Coefficients 16-43
Protection against Ionizing Radiation 16-48
Annual Limits on Intakes of Radionuclides 16-49
Chemical Carcinogens 16-53

APPENDIX A: MATHEMATICAL TABLES

Miscellaneous Mathematical Constants A-1
Decimal Equivalents of Common Fractions A-2
Quadratic Formula A-2
Exponential and Hyperbolic Functions and Their Common Logarithms A-3
Natural Trigonometric Functions to Four Places A-6
Relation of Angular Functions in Terms of One Another A-8
Derivatives A-9
Integration A-11
Integrals A-15
Differential Equations A-46
Fourier Series A-57
Fourier Expansions for Basic Periodic Functions A-59
The Fourier Transforms A-61
Series Expansion A-65
Vector Analysis A-68

Orthogonal Curvilinear Coordinates A-75
Transformation of Integrals A-77
Bessel Functions A-78
The Factorial Function A-80
The Gamma Function A-81
The Beta Function A-82
The Error Function A-83
Orthogonal Polynomials A-83
Tables of Orthogonal Polynomials A-86
Clebsch–Gordan Coefficients A-87
Normal Probability Function A-88
Percentage Points, Student's *t*-Distribution A-91
Percentage Points, Chi-Square Distribution A-91
Percentage Points, *F*-Distribution A-93
Moment of Inertia for Various Bodies of Mass A-97

APPENDIX B: SOURCES OF PHYSICAL AND CHEMICAL DATA B-1

INDEX I-1

Units

Section 1
Basic Constants, Units, and Conversion Factors

CODATA Recommended Values of the Fundamental Physical Constants 1-1
Standard Atomic Weights (2007) 1-12
Atomic Masses and Abundances 1-14
Electron Configuration and Ionization Energy of Neutral Atoms in the Ground State 1-18
International Temperature Scale of 1990 (ITS-90) 1-20
Conversion of Temperatures from the 1948 and 1968 Scales to ITS-90 1-21
International System of Units (SI) 1-23
Units for Magnetic Properties 1-27
Conversion Factors 1-28
Conversion of Temperatures 1-38
Conversion Factors for Energy Units 1-39
Conversion Factors for Pressure Units 1-40
Conversion Factors for Thermal Conductivity Units 1-41
Conversion Factors for Electrical Resistivity Units 1-42
Conversion Factors for Chemical Kinetics 1-43
Conversion Factors for Ionizing Radiation 1-44
Values of the Gas Constant in Different Unit Systems 1-46

CODATA RECOMMENDED VALUES OF THE FUNDAMENTAL PHYSICAL CONSTANTS: 2006

Peter J. Mohr, Barry N. Taylor, and David B. Newell

These tables give the 2006 self-consistent set of values of the basic constants and conversion factors of physics and chemistry recommended by the Committee on Data for Science and Technology (CODATA) for international use. The 2006 adjustment takes into account the data considered in the 2002 adjustment as well as the data that became available between 31 December 2002, the closing date of that adjustment, and 31 December 2006, the closing date of the new adjustment. The new data has led to a significant reduction in the uncertainties of many recommended values. The 2006 set replaces the previously recommended 2002 CODATA set and may also be found on the World Wide Web at physics.nist.gov/constants.

This report was prepared by the authors under the auspices of the CODATA Task Group on Fundamental Constants. The members of the task group are:

F. Cabiati, Istituto Nazionale di Ricerca Metrologica, Italy

K. Fujii, National Metrology Institute of Japan, Japan

S. G. Karshenboim, D. I. Mendeleyev All-Russian Research Institute for Metrology, Russian Federation

I. Lindgren, Chalmers University of Technology and Göteborg University, Sweden

B. A. Mamyrin (deceased), A. F. Ioffe Physical-Technical Institute, Russian Federation

W. Martienssen, Johann Wolfgang Goethe-Universität, Germany

P. J. Mohr, National Institute of Standards and Technology, United States of America

D. B. Newell, National Institute of Standards and Technology, United States of America

F. Nez, Laboratoire Kastler-Brossel, France

B. W. Petley, National Physical Laboratory, United Kingdom

T. J. Quinn, Bureau international des poids et mesures

B. N. Taylor, National Institute of Standards and Technology, United States of America

W. Wöger, Physikalisch-Technische Bundesanstalt, Germany

B. M. Wood, National Research Council, Canada

Z. Zhang, National Institute of Metrology, China (People's Republic of)

References

1. Mohr, P. J., Taylor, B. N., and Newell, D. B., "CODATA recommended values of the fundamental physical constants: 2006," *J. Phys. Chem. Ref. Data* 37, 1187, 2008; also appears in *Rev. Mod. Phys.* 80, 633, 2008, and online at <physics.nist.gov/constants>.
2. Yao, W. M., et al., *J. Phys. G* 33, 1, 2006.

TABLE I: An abbreviated list of the CODATA recommended values of the fundamental constants of physics and chemistry based on the 2006 adjustment.

Quantity	Symbol	Numerical value	Unit	Relative std. uncert. u_r
speed of light in vacuum	c, c_0	299 792 458	m s^{-1}	(exact)
magnetic constant	μ_0	$4\pi \times 10^{-7}$	N A^{-2}	
		$= 12.566\,370\,614... \times 10^{-7}$	N A^{-2}	(exact)
electric constant $1/\mu_0 c^2$	ϵ_0	$8.854\,187\,817... \times 10^{-12}$	F m^{-1}	(exact)
Newtonian constant of gravitation	G	$6.674\,28(67) \times 10^{-11}$	m^3 kg^{-1} s^{-2}	1.0×10^{-4}
Planck constant	h	$6.626\,068\,96(33) \times 10^{-34}$	J s	5.0×10^{-8}
$h/2\pi$	$\hbar$	$1.054\,571\,628(53) \times 10^{-34}$	J s	5.0×10^{-8}
elementary charge	e	$1.602\,176\,487(40) \times 10^{-19}$	C	2.5×10^{-8}
magnetic flux quantum $h/2e$	Φ_0	$2.067\,833\,667(52) \times 10^{-15}$	Wb	2.5×10^{-8}
conductance quantum $2e^2/h$	G_0	$7.748\,091\,7004(53) \times 10^{-5}$	S	6.8×10^{-10}
electron mass	m_e	$9.109\,382\,15(45) \times 10^{-31}$	kg	5.0×10^{-8}
proton mass	m_p	$1.672\,621\,637(83) \times 10^{-27}$	kg	5.0×10^{-8}
proton-electron mass ratio	m_p/m_e	$1836.152\,672\,47(80)$		4.3×10^{-10}
fine-structure constant $e^2/4\pi\epsilon_0\hbar c$	α	$7.297\,352\,5376(50) \times 10^{-3}$		6.8×10^{-10}
inverse fine-structure constant	α^{-1}	$137.035\,999\,679(94)$		6.8×10^{-10}
Rydberg constant $\alpha^2 m_e c/2h$	R_∞	$10\,973\,731.568\,527(73)$	m^{-1}	6.6×10^{-12}
Avogadro constant	N_A, L	$6.022\,141\,79(30) \times 10^{23}$	mol^{-1}	5.0×10^{-8}

TABLE I: (Continued.)

Quantity	Symbol	Numerical value	Unit	Relative std. uncert. u_r
Faraday constant $N_A e$	F	96 485.3399(24)	C mol^{-1}	2.5×10^{-8}
molar gas constant	R	8.314 472(15)	J mol^{-1} K^{-1}	1.7×10^{-6}
Boltzmann constant R/N_A	k	$1.380\,6504(24) \times 10^{-23}$	J K^{-1}	1.7×10^{-6}
Stefan-Boltzmann constant $(\pi^2/60)k^4/\hbar^3 c^2$	σ	$5.670\,400(40) \times 10^{-8}$	W m^{-2} K^{-4}	7.0×10^{-6}
Non-SI units accepted for use with the SI				
electron volt: (e/C) J	eV	$1.602\,176\,487(40) \times 10^{-19}$	J	2.5×10^{-8}
(unified) atomic mass unit $1\ \mathrm{u} = m_u = \frac{1}{12}m(^{12}\mathrm{C}) = 10^{-3}\ \mathrm{kg\ mol^{-1}}/N_A$	u	$1.660\,538\,782(83) \times 10^{-27}$	kg	5.0×10^{-8}

TABLE II: The CODATA recommended values of the fundamental constants of physics and chemistry based on the 2006 adjustment.

Quantity	Symbol	Numerical value	Unit	Relative std. uncert. u_r
		UNIVERSAL		
speed of light in vacuum	c, c_0	299 792 458	m s^{-1}	(exact)
magnetic constant	μ_0	$4\pi \times 10^{-7}$	N A^{-2}	
		$= 12.566\,370\,614... \times 10^{-7}$	N A^{-2}	(exact)
electric constant $1/\mu_0 c^2$	ϵ_0	$8.854\,187\,817... \times 10^{-12}$	F m^{-1}	(exact)
characteristic impedance of vacuum $\sqrt{\mu_0/\epsilon_0} = \mu_0 c$	Z_0	376.730 313 461...	Ω	(exact)
Newtonian constant of gravitation	G	$6.674\,28(67) \times 10^{-11}$	m^3 kg^{-1} s^{-2}	1.0×10^{-4}
	$G/\hbar c$	$6.708\,81(67) \times 10^{-39}$	(GeV/c^2)$^{-2}$	1.0×10^{-4}
Planck constant	h	$6.626\,068\,96(33) \times 10^{-34}$	J s	5.0×10^{-8}
in eV s		$4.135\,667\,33(10) \times 10^{-15}$	eV s	2.5×10^{-8}
$h/2\pi$	$\hbar$	$1.054\,571\,628(53) \times 10^{-34}$	J s	5.0×10^{-8}
in eV s		$6.582\,118\,99(16) \times 10^{-16}$	eV s	2.5×10^{-8}
$\hbar c$ in MeV fm		197.326 9631(49)	MeV fm	2.5×10^{-8}
Planck mass $(\hbar c/G)^{1/2}$	m_P	$2.176\,44(11) \times 10^{-8}$	kg	5.0×10^{-5}
energy equivalent in GeV	$m_P c^2$	$1.220\,892(61) \times 10^{19}$	GeV	5.0×10^{-5}
Planck temperature $(\hbar c^5/G)^{1/2}/k$	T_P	$1.416\,785(71) \times 10^{32}$	K	5.0×10^{-5}
Planck length $\hbar/m_P c = (\hbar G/c^3)^{1/2}$	l_P	$1.616\,252(81) \times 10^{-35}$	m	5.0×10^{-5}
Planck time $l_P/c = (\hbar G/c^5)^{1/2}$	t_P	$5.391\,24(27) \times 10^{-44}$	s	5.0×10^{-5}
		ELECTROMAGNETIC		
elementary charge	e	$1.602\,176\,487(40) \times 10^{-19}$	C	2.5×10^{-8}
	e/h	$2.417\,989\,454(60) \times 10^{14}$	A J^{-1}	2.5×10^{-8}
magnetic flux quantum $h/2e$	Φ_0	$2.067\,833\,667(52) \times 10^{-15}$	Wb	2.5×10^{-8}
conductance quantum $2e^2/h$	G_0	$7.748\,091\,7004(53) \times 10^{-5}$	S	6.8×10^{-10}
inverse of conductance quantum	G_0^{-1}	12 906.403 7787(88)	Ω	6.8×10^{-10}
Josephson constant[1] $2e/h$	K_J	$483\,597.891(12) \times 10^9$	Hz V^{-1}	2.5×10^{-8}
von Klitzing constant[2] $h/e^2 = \mu_0 c/2\alpha$	R_K	25 812.807 557(18)	Ω	6.8×10^{-10}
Bohr magneton $e\hbar/2m_e$	μ_B	$927.400\,915(23) \times 10^{-26}$	J T^{-1}	2.5×10^{-8}
in eV T^{-1}		$5.788\,381\,7555(79) \times 10^{-5}$	eV T^{-1}	1.4×10^{-9}
	μ_B/h	$13.996\,246\,04(35) \times 10^9$	Hz T^{-1}	2.5×10^{-8}

[1]See Table IV for the conventional value adopted internationally for realizing representations of the volt using the Josephson effect.
[2]See Table IV for the conventional value adopted internationally for realizing representations of the ohm using the quantum Hall effect.

TABLE II: (Continued).

Quantity	Symbol	Numerical value	Unit	Relative std. uncert. u_r
	μ_B/hc	$46.686\,4515(12)$	$\mathrm{m^{-1}\,T^{-1}}$	2.5×10^{-8}
	μ_B/k	$0.671\,7131(12)$	$\mathrm{K\,T^{-1}}$	1.7×10^{-6}
nuclear magneton $e\hbar/2m_p$	μ_N	$5.050\,783\,24(13) \times 10^{-27}$	$\mathrm{J\,T^{-1}}$	2.5×10^{-8}
in eV $\mathrm{T^{-1}}$		$3.152\,451\,2326(45) \times 10^{-8}$	$\mathrm{eV\,T^{-1}}$	1.4×10^{-9}
	μ_N/h	$7.622\,593\,84(19)$	$\mathrm{MHz\,T^{-1}}$	2.5×10^{-8}
	μ_N/hc	$2.542\,623\,616(64) \times 10^{-2}$	$\mathrm{m^{-1}\,T^{-1}}$	2.5×10^{-8}
	μ_N/k	$3.658\,2637(64) \times 10^{-4}$	$\mathrm{K\,T^{-1}}$	1.7×10^{-6}
ATOMIC AND NUCLEAR				
General				
fine-structure constant $e^2/4\pi\epsilon_0\hbar c$	α	$7.297\,352\,5376(50) \times 10^{-3}$		6.8×10^{-10}
inverse fine-structure constant	α^{-1}	$137.035\,999\,679(94)$		6.8×10^{-10}
Rydberg constant $\alpha^2 m_e c/2h$	R_∞	$10\,973\,731.568\,527(73)$	$\mathrm{m^{-1}}$	6.6×10^{-12}
	$R_\infty c$	$3.289\,841\,960\,361(22) \times 10^{15}$	Hz	6.6×10^{-12}
	$R_\infty hc$	$2.179\,871\,97(11) \times 10^{-18}$	J	5.0×10^{-8}
$R_\infty hc$ in eV		$13.605\,691\,93(34)$	eV	2.5×10^{-8}
Bohr radius $\alpha/4\pi R_\infty = 4\pi\epsilon_0\hbar^2/m_e e^2$	a_0	$0.529\,177\,208\,59(36) \times 10^{-10}$	m	6.8×10^{-10}
Hartree energy $e^2/4\pi\epsilon_0 a_0 = 2R_\infty hc$ $= \alpha^2 m_e c^2$	E_h	$4.359\,743\,94(22) \times 10^{-18}$	J	5.0×10^{-8}
in eV		$27.211\,383\,86(68)$	eV	2.5×10^{-8}
quantum of circulation	$h/2m_e$	$3.636\,947\,5199(50) \times 10^{-4}$	$\mathrm{m^2\,s^{-1}}$	1.4×10^{-9}
	h/m_e	$7.273\,895\,040(10) \times 10^{-4}$	$\mathrm{m^2\,s^{-1}}$	1.4×10^{-9}
Electroweak				
Fermi coupling constant[3]	$G_F/(\hbar c)^3$	$1.166\,37(1) \times 10^{-5}$	$\mathrm{GeV^{-2}}$	8.6×10^{-6}
weak mixing angle[4] θ_W (on-shell scheme) $\sin^2\theta_W = s_W^2 \equiv 1 - (m_W/m_Z)^2$	$\sin^2\theta_W$	$0.222\,55(56)$		2.5×10^{-3}
Electron, e^-				
electron mass	m_e	$9.109\,382\,15(45) \times 10^{-31}$	kg	5.0×10^{-8}
in u, $m_e = A_r(e)$ u (electron relative atomic mass times u)		$5.485\,799\,0943(23) \times 10^{-4}$	u	4.2×10^{-10}
energy equivalent	$m_e c^2$	$8.187\,104\,38(41) \times 10^{-14}$	J	5.0×10^{-8}
in MeV		$0.510\,998\,910(13)$	MeV	2.5×10^{-8}
electron-muon mass ratio	m_e/m_μ	$4.836\,331\,71(12) \times 10^{-3}$		2.5×10^{-8}
electron-tau mass ratio	m_e/m_τ	$2.875\,64(47) \times 10^{-4}$		1.6×10^{-4}
electron-proton mass ratio	m_e/m_p	$5.446\,170\,2177(24) \times 10^{-4}$		4.3×10^{-10}
electron-neutron mass ratio	m_e/m_n	$5.438\,673\,4459(33) \times 10^{-4}$		6.0×10^{-10}
electron-deuteron mass ratio	m_e/m_d	$2.724\,437\,1093(12) \times 10^{-4}$		4.3×10^{-10}
electron to alpha particle mass ratio	m_e/m_α	$1.370\,933\,555\,70(58) \times 10^{-4}$		4.2×10^{-10}
electron charge to mass quotient	$-e/m_e$	$-1.758\,820\,150(44) \times 10^{11}$	$\mathrm{C\,kg^{-1}}$	2.5×10^{-8}
electron molar mass $N_A m_e$	$M(e)$, M_e	$5.485\,799\,0943(23) \times 10^{-7}$	$\mathrm{kg\,mol^{-1}}$	4.2×10^{-10}
Compton wavelength $h/m_e c$	λ_C	$2.426\,310\,2175(33) \times 10^{-12}$	m	1.4×10^{-9}
$\lambda_C/2\pi = \alpha a_0 = \alpha^2/4\pi R_\infty$	$\bar{\lambda}_C$	$386.159\,264\,59(53) \times 10^{-15}$	m	1.4×10^{-9}
classical electron radius $\alpha^2 a_0$	r_e	$2.817\,940\,2894(58) \times 10^{-15}$	m	2.1×10^{-9}
Thomson cross section $(8\pi/3)r_e^2$	σ_e	$0.665\,245\,8558(27) \times 10^{-28}$	$\mathrm{m^2}$	4.1×10^{-9}

[3]Value recommended by the Particle Data Group (Yao *et al.*, 2006).

[4]Based on the ratio of the masses of the W and Z bosons m_W/m_Z recommended by the Particle Data Group (Yao *et al.*, 2006). The value for $\sin^2\theta_W$ they recommend, which is based on a particular variant of the modified minimal subtraction ($\overline{\mathrm{MS}}$) scheme, is $\sin^2\hat{\theta}_W(M_Z) = 0.231\,22(15)$.

TABLE II: (Continued).

Quantity	Symbol	Numerical value	Unit	Relative std. uncert. u_r
electron magnetic moment	μ_e	$-928.476\,377(23) \times 10^{-26}$	J T^{-1}	2.5×10^{-8}
to Bohr magneton ratio	μ_e/μ_B	$-1.001\,159\,652\,181\,11(74)$		7.4×10^{-13}
to nuclear magneton ratio	μ_e/μ_N	$-1838.281\,970\,92(80)$		4.3×10^{-10}
electron magnetic moment anomaly $\|\mu_e\|/\mu_B - 1$	a_e	$1.159\,652\,181\,11(74) \times 10^{-3}$		6.4×10^{-10}
electron g-factor $-2(1+a_e)$	g_e	$-2.002\,319\,304\,3622(15)$		7.4×10^{-13}
electron-muon magnetic moment ratio	μ_e/μ_μ	$206.766\,9877(52)$		2.5×10^{-8}
electron-proton magnetic moment ratio	μ_e/μ_p	$-658.210\,6848(54)$		8.1×10^{-9}
electron to shielded proton magnetic moment ratio (H_2O, sphere, 25°C)	μ_e/μ'_p	$-658.227\,5971(72)$		1.1×10^{-8}
electron-neutron magnetic moment ratio	μ_e/μ_n	$960.920\,50(23)$		2.4×10^{-7}
electron-deuteron magnetic moment ratio	μ_e/μ_d	$-2143.923\,498(18)$		8.4×10^{-9}
electron to shielded helion magnetic moment ratio (gas, sphere, 25°C)	μ_e/μ'_h	$864.058\,257(10)$		1.2×10^{-8}
electron gyromagnetic ratio $2\|\mu_e\|/\hbar$	γ_e	$1.760\,859\,770(44) \times 10^{11}$	s^{-1} T^{-1}	2.5×10^{-8}
	$\gamma_e/2\pi$	$28\,024.953\,64(70)$	MHz T^{-1}	2.5×10^{-8}
Muon, μ^-				
muon mass	m_μ	$1.883\,531\,30(11) \times 10^{-28}$	kg	5.6×10^{-8}
in u, $m_\mu = A_r(\mu)$ u (muon relative atomic mass times u)		$0.113\,428\,9256(29)$	u	2.5×10^{-8}
energy equivalent	$m_\mu c^2$	$1.692\,833\,510(95) \times 10^{-11}$	J	5.6×10^{-8}
in MeV		$105.658\,3668(38)$	MeV	3.6×10^{-8}
muon-electron mass ratio	m_μ/m_e	$206.768\,2823(52)$		2.5×10^{-8}
muon-tau mass ratio	m_μ/m_τ	$5.945\,92(97) \times 10^{-2}$		1.6×10^{-4}
muon-proton mass ratio	m_μ/m_p	$0.112\,609\,5261(29)$		2.5×10^{-8}
muon-neutron mass ratio	m_μ/m_n	$0.112\,454\,5167(29)$		2.5×10^{-8}
muon molar mass $N_A m_\mu$	$M(\mu)$, M_μ	$0.113\,428\,9256(29) \times 10^{-3}$	kg mol^{-1}	2.5×10^{-8}
muon Compton wavelength $h/m_\mu c$	$\lambda_{C,\mu}$	$11.734\,441\,04(30) \times 10^{-15}$	m	2.5×10^{-8}
$\lambda_{C,\mu}/2\pi$	$\bar{\lambda}_{C,\mu}$	$1.867\,594\,295(47) \times 10^{-15}$	m	2.5×10^{-8}
muon magnetic moment	μ_μ	$-4.490\,447\,86(16) \times 10^{-26}$	J T^{-1}	3.6×10^{-8}
to Bohr magneton ratio	μ_μ/μ_B	$-4.841\,970\,49(12) \times 10^{-3}$		2.5×10^{-8}
to nuclear magneton ratio	μ_μ/μ_N	$-8.890\,597\,05(23)$		2.5×10^{-8}
muon magnetic moment anomaly $\|\mu_\mu\|/(e\hbar/2m_\mu) - 1$	a_μ	$1.165\,920\,69(60) \times 10^{-3}$		5.2×10^{-7}
muon g-factor $-2(1+a_\mu)$	g_μ	$-2.002\,331\,8414(12)$		6.0×10^{-10}
muon-proton magnetic moment ratio	μ_μ/μ_p	$-3.183\,345\,137(85)$		2.7×10^{-8}
Tau, τ^-				
tau mass[5]	m_τ	$3.167\,77(52) \times 10^{-27}$	kg	1.6×10^{-4}
in u, $m_\tau = A_r(\tau)$ u (tau relative atomic mass times u)		$1.907\,68(31)$	u	1.6×10^{-4}

[5]This and all other values involving m_τ are based on the value of $m_\tau c^2$ in MeV recommended by the Particle Data Group (Yao *et al.*, 2006), but with a standard uncertainty of 0.29 MeV rather than the quoted uncertainty of −0.26 MeV, +0.29 MeV.

TABLE II: (Continued).

Quantity	Symbol	Numerical value	Unit	Relative std. uncert. u_r
energy equivalent	$m_\tau c^2$	$2.847\,05(46) \times 10^{-10}$	J	1.6×10^{-4}
in MeV		1776.99(29)	MeV	1.6×10^{-4}
tau-electron mass ratio	m_τ/m_e	3477.48(57)		1.6×10^{-4}
tau-muon mass ratio	m_τ/m_μ	16.8183(27)		1.6×10^{-4}
tau-proton mass ratio	m_τ/m_p	1.893 90(31)		1.6×10^{-4}
tau-neutron mass ratio	m_τ/m_n	1.891 29(31)		1.6×10^{-4}
tau molar mass $N_A m_\tau$	$M(\tau)$, M_τ	$1.907\,68(31) \times 10^{-3}$	kg mol^{-1}	1.6×10^{-4}
tau Compton wavelength $h/m_\tau c$	$\lambda_{C,\tau}$	$0.697\,72(11) \times 10^{-15}$	m	1.6×10^{-4}
$\lambda_{C,\tau}/2\pi$	$\bar{\lambda}_{C,\tau}$	$0.111\,046(18) \times 10^{-15}$	m	1.6×10^{-4}
Proton, p				
proton mass	m_p	$1.672\,621\,637(83) \times 10^{-27}$	kg	5.0×10^{-8}
in u, $m_p = A_r(p)$ u (proton relative atomic mass times u)		1.007 276 466 77(10)	u	1.0×10^{-10}
energy equivalent	$m_p c^2$	$1.503\,277\,359(75) \times 10^{-10}$	J	5.0×10^{-8}
in MeV		938.272 013(23)	MeV	2.5×10^{-8}
proton-electron mass ratio	m_p/m_e	1836.152 672 47(80)		4.3×10^{-10}
proton-muon mass ratio	m_p/m_μ	8.880 243 39(23)		2.5×10^{-8}
proton-tau mass ratio	m_p/m_τ	0.528 012(86)		1.6×10^{-4}
proton-neutron mass ratio	m_p/m_n	0.998 623 478 24(46)		4.6×10^{-10}
proton charge to mass quotient	e/m_p	$9.578\,833\,92(24) \times 10^{7}$	C kg^{-1}	2.5×10^{-8}
proton molar mass $N_A m_p$	$M(p)$, M_p	$1.007\,276\,466\,77(10) \times 10^{-3}$	kg mol^{-1}	1.0×10^{-10}
proton Compton wavelength $h/m_p c$	$\lambda_{C,p}$	$1.321\,409\,8446(19) \times 10^{-15}$	m	1.4×10^{-9}
$\lambda_{C,p}/2\pi$	$\bar{\lambda}_{C,p}$	$0.210\,308\,908\,61(30) \times 10^{-15}$	m	1.4×10^{-9}
proton rms charge radius	R_p	$0.8768(69) \times 10^{-15}$	m	7.8×10^{-3}
proton magnetic moment	μ_p	$1.410\,606\,662(37) \times 10^{-26}$	J T^{-1}	2.6×10^{-8}
to Bohr magneton ratio	μ_p/μ_B	$1.521\,032\,209(12) \times 10^{-3}$		8.1×10^{-9}
to nuclear magneton ratio	μ_p/μ_N	2.792 847 356(23)		8.2×10^{-9}
proton g-factor $2\mu_p/\mu_N$	g_p	5.585 694 713(46)		8.2×10^{-9}
proton-neutron magnetic moment ratio	μ_p/μ_n	−1.459 898 06(34)		2.4×10^{-7}
shielded proton magnetic moment (H_2O, sphere, 25°C)	μ'_p	$1.410\,570\,419(38) \times 10^{-26}$	J T^{-1}	2.7×10^{-8}
to Bohr magneton ratio	μ'_p/μ_B	$1.520\,993\,128(17) \times 10^{-3}$		1.1×10^{-8}
to nuclear magneton ratio	μ'_p/μ_N	2.792 775 598(30)		1.1×10^{-8}
proton magnetic shielding correction $1 - \mu'_p/\mu_p$ (H_2O, sphere, 25°C)	σ'_p	$25.694(14) \times 10^{-6}$		5.3×10^{-4}
proton gyromagnetic ratio $2\mu_p/\hbar$	γ_p	$2.675\,222\,099(70) \times 10^{8}$	s^{-1} T^{-1}	2.6×10^{-8}
	$\gamma_p/2\pi$	42.577 4821(11)	MHz T^{-1}	2.6×10^{-8}
shielded proton gyromagnetic ratio $2\mu'_p/\hbar$ (H_2O, sphere, 25°C)	γ'_p	$2.675\,153\,362(73) \times 10^{8}$	s^{-1} T^{-1}	2.7×10^{-8}
	$\gamma'_p/2\pi$	42.576 3881(12)	MHz T^{-1}	2.7×10^{-8}
Neutron, n				
neutron mass	m_n	$1.674\,927\,211(84) \times 10^{-27}$	kg	5.0×10^{-8}
in u, $m_n = A_r(n)$ u (neutron relative atomic mass times u)		1.008 664 915 97(43)	u	4.3×10^{-10}
energy equivalent	$m_n c^2$	$1.505\,349\,505(75) \times 10^{-10}$	J	5.0×10^{-8}
in MeV		939.565 346(23)	MeV	2.5×10^{-8}
neutron-electron mass ratio	m_n/m_e	1838.683 6605(11)		6.0×10^{-10}

TABLE II: (Continued).

Quantity	Symbol	Numerical value	Unit	Relative std. uncert. u_r
neutron-muon mass ratio	m_n/m_μ	8.892 484 09(23)		2.5×10^{-8}
neutron-tau mass ratio	m_n/m_τ	0.528 740(86)		1.6×10^{-4}
neutron-proton mass ratio	m_n/m_p	1.001 378 419 18(46)		4.6×10^{-10}
neutron molar mass $N_A m_n$	$M(n)$, M_n	$1.008\,664\,915\,97(43) \times 10^{-3}$	kg mol^{-1}	4.3×10^{-10}
neutron Compton wavelength $h/m_n c$	$\lambda_{C,n}$	$1.319\,590\,8951(20) \times 10^{-15}$	m	1.5×10^{-9}
$\lambda_{C,n}/2\pi$	$\lambdabar_{C,n}$	$0.210\,019\,413\,82(31) \times 10^{-15}$	m	1.5×10^{-9}
neutron magnetic moment	μ_n	$-0.966\,236\,41(23) \times 10^{-26}$	J T^{-1}	2.4×10^{-7}
to Bohr magneton ratio	μ_n/μ_B	$-1.041\,875\,63(25) \times 10^{-3}$		2.4×10^{-7}
to nuclear magneton ratio	μ_n/μ_N	$-1.913\,042\,73(45)$		2.4×10^{-7}
neutron g-factor $2\mu_n/\mu_N$	g_n	$-3.826\,085\,45(90)$		2.4×10^{-7}
neutron-electron magnetic moment ratio	μ_n/μ_e	$1.040\,668\,82(25) \times 10^{-3}$		2.4×10^{-7}
neutron-proton magnetic moment ratio	μ_n/μ_p	$-0.684\,979\,34(16)$		2.4×10^{-7}
neutron to shielded proton magnetic moment ratio (H_2O, sphere, 25°C)	μ_n/μ'_p	$-0.684\,996\,94(16)$		2.4×10^{-7}
neutron gyromagnetic ratio $2\lvert\mu_n\rvert/\hbar$	γ_n	$1.832\,471\,85(43) \times 10^{8}$	s^{-1} T^{-1}	2.4×10^{-7}
	$\gamma_n/2\pi$	29.164 6954(69)	MHz T^{-1}	2.4×10^{-7}
Deuteron, d				
deuteron mass	m_d	$3.343\,583\,20(17) \times 10^{-27}$	kg	5.0×10^{-8}
in u, $m_d = A_r(d)$ u (deuteron relative atomic mass times u)		2.013 553 212 724(78)	u	3.9×10^{-11}
energy equivalent	$m_d c^2$	$3.005\,062\,72(15) \times 10^{-10}$	J	5.0×10^{-8}
in MeV		1875.612 793(47)	MeV	2.5×10^{-8}
deuteron-electron mass ratio	m_d/m_e	3670.482 9654(16)		4.3×10^{-10}
deuteron-proton mass ratio	m_d/m_p	1.999 007 501 08(22)		1.1×10^{-10}
deuteron molar mass $N_A m_d$	$M(d)$, M_d	$2.013\,553\,212\,724(78) \times 10^{-3}$	kg mol^{-1}	3.9×10^{-11}
deuteron rms charge radius	R_d	$2.1402(28) \times 10^{-15}$	m	1.3×10^{-3}
deuteron magnetic moment	μ_d	$0.433\,073\,465(11) \times 10^{-26}$	J T^{-1}	2.6×10^{-8}
to Bohr magneton ratio	μ_d/μ_B	$0.466\,975\,4556(39) \times 10^{-3}$		8.4×10^{-9}
to nuclear magneton ratio	μ_d/μ_N	0.857 438 2308(72)		8.4×10^{-9}
deuteron g-factor μ_d/μ_N	g_d	0.857 438 2308(72)		8.4×10^{-9}
deuteron-electron magnetic moment ratio	μ_d/μ_e	$-4.664\,345\,537(39) \times 10^{-4}$		8.4×10^{-9}
deuteron-proton magnetic moment ratio	μ_d/μ_p	0.307 012 2070(24)		7.7×10^{-9}
deuteron-neutron magnetic moment ratio	μ_d/μ_n	$-0.448\,206\,52(11)$		2.4×10^{-7}
Triton, t				
triton mass	m_t	$5.007\,355\,88(25) \times 10^{-27}$	kg	5.0×10^{-8}
in u, $m_t = A_r(t)$ u (triton relative atomic mass times u)		3.015 500 7134(25)	u	8.3×10^{-10}
energy equivalent	$m_t c^2$	$4.500\,387\,03(22) \times 10^{-10}$	J	5.0×10^{-8}
in MeV		2808.920 906(70)	MeV	2.5×10^{-8}
triton-electron mass ratio	m_t/m_e	5496.921 5269(51)		9.3×10^{-10}
triton-proton mass ratio	m_t/m_p	2.993 717 0309(25)		8.4×10^{-10}

TABLE II: (Continued).

Quantity	Symbol	Numerical value	Unit	Relative std. uncert. u_r
triton molar mass $N_A m_t$	$M(t)$, M_t	$3.015\,500\,7134(25) \times 10^{-3}$	kg mol^{-1}	8.3×10^{-10}
triton magnetic moment	μ_t	$1.504\,609\,361(42) \times 10^{-26}$	J T^{-1}	2.8×10^{-8}
to Bohr magneton ratio	μ_t/μ_B	$1.622\,393\,657(21) \times 10^{-3}$		1.3×10^{-8}
to nuclear magneton ratio	μ_t/μ_N	$2.978\,962\,448(38)$		1.3×10^{-8}
triton g-factor $2\mu_t/\mu_N$	g_t	$5.957\,924\,896(76)$		1.3×10^{-8}
triton-electron magnetic moment ratio	μ_t/μ_e	$-1.620\,514\,423(21) \times 10^{-3}$		1.3×10^{-8}
triton-proton magnetic moment ratio	μ_t/μ_p	$1.066\,639\,908(10)$		9.8×10^{-9}
triton-neutron magnetic moment ratio	μ_t/μ_n	$-1.557\,185\,53(37)$		2.4×10^{-7}
Helion, h				
helion (^{3}He nucleus) mass	m_h	$5.006\,411\,92(25) \times 10^{-27}$	kg	5.0×10^{-8}
in u, $m_h = A_r(h)$ u (helion relative atomic mass times u)		$3.014\,932\,2473(26)$	u	8.6×10^{-10}
energy equivalent	$m_h c^2$	$4.499\,538\,64(22) \times 10^{-10}$	J	5.0×10^{-8}
in MeV		$2808.391\,383(70)$	MeV	2.5×10^{-8}
helion-electron mass ratio	m_h/m_e	$5495.885\,2765(52)$		9.5×10^{-10}
helion-proton mass ratio	m_h/m_p	$2.993\,152\,6713(26)$		8.7×10^{-10}
helion molar mass $N_A m_h$	$M(h)$, M_h	$3.014\,932\,2473(26) \times 10^{-3}$	kg mol^{-1}	8.6×10^{-10}
shielded helion magnetic moment (gas, sphere, 25°C)	μ'_h	$-1.074\,552\,982(30) \times 10^{-26}$	J T^{-1}	2.8×10^{-8}
to Bohr magneton ratio	μ'_h/μ_B	$-1.158\,671\,471(14) \times 10^{-3}$		1.2×10^{-8}
to nuclear magneton ratio	μ'_h/μ_N	$-2.127\,497\,718(25)$		1.2×10^{-8}
shielded helion to proton magnetic moment ratio (gas, sphere, 25°C)	μ'_h/μ_p	$-0.761\,766\,558(11)$		1.4×10^{-8}
shielded helion to shielded proton magnetic moment ratio (gas/H_2O, spheres, 25°C)	μ'_h/μ'_p	$-0.761\,786\,1313(33)$		4.3×10^{-9}
shielded helion gyromagnetic ratio $2\lvert\mu'_h\rvert/\hbar$ (gas, sphere, 25°C)	γ'_h	$2.037\,894\,730(56) \times 10^{8}$	s^{-1} T^{-1}	2.8×10^{-8}
	$\gamma'_h/2\pi$	$32.434\,101\,98(90)$	MHz T^{-1}	2.8×10^{-8}
Alpha particle, α				
alpha particle mass	m_α	$6.644\,656\,20(33) \times 10^{-27}$	kg	5.0×10^{-8}
in u, $m_\alpha = A_r(\alpha)$ u (alpha particle relative atomic mass times u)		$4.001\,506\,179\,127(62)$	u	1.5×10^{-11}
energy equivalent	$m_\alpha c^2$	$5.971\,919\,17(30) \times 10^{-10}$	J	5.0×10^{-8}
in MeV		$3727.379\,109(93)$	MeV	2.5×10^{-8}
alpha particle to electron mass ratio	m_α/m_e	$7294.299\,5365(31)$		4.2×10^{-10}
alpha particle to proton mass ratio	m_α/m_p	$3.972\,599\,689\,51(41)$		1.0×10^{-10}
alpha particle molar mass $N_A m_\alpha$	$M(\alpha)$, M_α	$4.001\,506\,179\,127(62) \times 10^{-3}$	kg mol^{-1}	1.5×10^{-11}
PHYSICOCHEMICAL				
Avogadro constant	N_A, L	$6.022\,141\,79(30) \times 10^{23}$	mol^{-1}	5.0×10^{-8}
atomic mass constant $m_u = \frac{1}{12}m(^{12}C) = 1$ u	m_u	$1.660\,538\,782(83) \times 10^{-27}$	kg	5.0×10^{-8}

TABLE II: (Continued).

Quantity	Symbol	Numerical value	Unit	Relative std. uncert. u_r
$= 10^{-3}$ kg mol$^{-1}/N_A$				
energy equivalent	$m_u c^2$	$1.492\,417\,830(74) \times 10^{-10}$	J	5.0×10^{-8}
in MeV		$931.494\,028(23)$	MeV	2.5×10^{-8}
Faraday constant[6] $N_A e$	F	$96\,485.3399(24)$	C mol^{-1}	2.5×10^{-8}
molar Planck constant	$N_A h$	$3.990\,312\,6821(57) \times 10^{-10}$	J s mol^{-1}	1.4×10^{-9}
	$N_A hc$	$0.119\,626\,564\,72(17)$	J m mol^{-1}	1.4×10^{-9}
molar gas constant	R	$8.314\,472(15)$	J mol^{-1} K^{-1}	1.7×10^{-6}
Boltzmann constant R/N_A	k	$1.380\,6504(24) \times 10^{-23}$	J K^{-1}	1.7×10^{-6}
in eV K^{-1}		$8.617\,343(15) \times 10^{-5}$	eV K^{-1}	1.7×10^{-6}
	k/h	$2.083\,6644(36) \times 10^{10}$	Hz K^{-1}	1.7×10^{-6}
	k/hc	$69.503\,56(12)$	m^{-1} K^{-1}	1.7×10^{-6}
molar volume of ideal gas RT/p				
$T = 273.15$ K, $p = 101.325$ kPa	V_m	$22.413\,996(39) \times 10^{-3}$	m^3 mol^{-1}	1.7×10^{-6}
Loschmidt constant N_A/V_m	n_0	$2.686\,7774(47) \times 10^{25}$	m^{-3}	1.7×10^{-6}
$T = 273.15$ K, $p = 100$ kPa	V_m	$22.710\,981(40) \times 10^{-3}$	m^3 mol^{-1}	1.7×10^{-6}
Sackur-Tetrode constant				
(absolute entropy constant)[7]				
$\frac{5}{2} + \ln[(2\pi m_u k T_1/h^2)^{3/2} k T_1/p_0]$				
$T_1 = 1$ K, $p_0 = 100$ kPa	S_0/R	$-1.151\,7047(44)$		3.8×10^{-6}
$T_1 = 1$ K, $p_0 = 101.325$ kPa		$-1.164\,8677(44)$		3.8×10^{-6}
Stefan-Boltzmann constant				
$(\pi^2/60)k^4/\hbar^3 c^2$	σ	$5.670\,400(40) \times 10^{-8}$	W m^{-2} K^{-4}	7.0×10^{-6}
first radiation constant $2\pi hc^2$	c_1	$3.741\,771\,18(19) \times 10^{-16}$	W m^2	5.0×10^{-8}
first radiation constant for spectral radiance $2hc^2$	c_{1L}	$1.191\,042\,759(59) \times 10^{-16}$	W m^2 sr^{-1}	5.0×10^{-8}
second radiation constant hc/k	c_2	$1.438\,7752(25) \times 10^{-2}$	m K	1.7×10^{-6}
Wien displacement law constants				
$b = \lambda_{max} T = c_2/4.965\,114\,231...$	b	$2.897\,7685(51) \times 10^{-3}$	m K	1.7×10^{-6}
$b' = \nu_{max}/T = 2.821\,439\,372... c/c_2$	b'	$5.878\,933(10) \times 10^{10}$	Hz K^{-1}	1.7×10^{-6}

TABLE III: The variances, covariances, and correlation coefficients of the values of a selected group of constants based on the 2006 CODATA adjustment. The numbers in bold above the main diagonal are 10^{16} times the numerical values of the relative covariances; the numbers in bold on the main diagonal are 10^{16} times the numerical values of the relative variances; and the numbers in italics below the main diagonal are the correlation coefficients.[1]

	α	h	e	m_e	N_A	m_e/m_μ	F
α	**0.0047**	**0.0002**	**0.0024**	**−0.0092**	**0.0092**	**−0.0092**	**0.0116**
h	*0.0005*	**24.8614**	**12.4308**	**24.8611**	**−24.8610**	**−0.0003**	**−12.4302**
e	*0.0142*	*0.9999*	**6.2166**	**12.4259**	**−12.4259**	**−0.0048**	**−6.2093**
m_e	*−0.0269*	*0.9996*	*0.9992*	**24.8795**	**−24.8794**	**0.0180**	**−12.4535**
N_A	*0.0269*	*−0.9996*	*−0.9991*	*−1.0000*	**24.8811**	**−0.0180**	**12.4552**
m_e/m_μ	*−0.0528*	*0.0000*	*−0.0008*	*0.0014*	*−0.0014*	**6.4296**	**−0.0227**
F	*0.0679*	*−0.9975*	*−0.9965*	*−0.9990*	*0.9991*	*−0.0036*	**6.2459**

[1] The relative covariance is $u_r(x_i, x_j) = u(x_i, x_j)/(x_i x_j)$, where $u(x_i, x_j)$ is the covariance of x_i and x_j; the relative variance is $u_r^2(x_i) = u_r(x_i, x_i)$; and the correlation coefficient is $r(x_i, x_j) = u(x_i, x_j)/[u(x_i)u(x_j)]$.

---

[6] The numerical value of F to be used in coulometric chemical measurements is 96 485.3401(48) [5.0×10^{-8}] when the relevant current is measured in terms of representations of the volt and ohm based on the Josephson and quantum Hall effects and the internationally adopted conventional values of the Josephson and von Klitzing constants K_{J-90} and R_{K-90} given in Table IV.

[7] The entropy of an ideal monoatomic gas of relative atomic mass A_r is given by $S = S_0 + \frac{3}{2} R \ln A_r - R \ln(p/p_0) + \frac{5}{2} R \ln(T/\text{K})$.

TABLE IV: Internationally adopted values of various quantities.

Quantity	Symbol	Numerical value	Unit	Relative std. uncert. u_r
relative atomic mass[1] of ^{12}C	$A_r(^{12}\mathrm{C})$	12		(exact)
molar mass constant	M_u	1×10^{-3}	kg mol^{-1}	(exact)
molar mass of ^{12}C	$M(^{12}\mathrm{C})$	12×10^{-3}	kg mol^{-1}	(exact)
conventional value of Josephson constant[2]	K_{J-90}	483 597.9	GHz V^{-1}	(exact)
conventional value of von Klitzing constant[3]	R_{K-90}	25 812.807	Ω	(exact)
standard atmosphere		101 325	Pa	(exact)

[1] The relative atomic mass $A_r(X)$ of particle X with mass $m(X)$ is defined by $A_r(X) = m(X)/m_u$, where $m_u = m(^{12}\mathrm{C})/12 = M_u/N_A =$ 1 u is the atomic mass constant, M_u is the molar mass constant, N_A is the Avogadro constant, and u is the unified atomic mass unit. Thus the mass of particle X is $m(X) = A_r(X)$ u and the molar mass of X is $M(X) = A_r(X)M_u$.
[2] This is the value adopted internationally for realizing representations of the volt using the Josephson effect.
[3] This is the value adopted internationally for realizing representations of the ohm using the quantum Hall effect.

TABLE V: Values of some x-ray-related quantities based on the 2006 CODATA adjustment of the values of the constants.

Quantity	Symbol	Numerical value	Unit	Relative std. uncert. u_r
Cu x unit: $\lambda(\mathrm{CuK}\alpha_1)/1\,537.400$	xu(CuKα_1)	$1.002\,076\,99(28) \times 10^{-13}$	m	2.8×10^{-7}
Mo x unit: $\lambda(\mathrm{MoK}\alpha_1)/707.831$	xu(MoKα_1)	$1.002\,099\,55(53) \times 10^{-13}$	m	5.3×10^{-7}
ångstrom star: $\lambda(\mathrm{WK}\alpha_1)/0.209\,010\,0$	Å*	$1.000\,014\,98(90) \times 10^{-10}$	m	9.0×10^{-7}
lattice parameter[1] of Si (in vacuum, 22.5°C)	a	$543.102\,064(14) \times 10^{-12}$	m	2.6×10^{-8}
{220} lattice spacing of Si $a/\sqrt{8}$ (in vacuum, 22.5°C)	d_{220}	$192.015\,5762(50) \times 10^{-12}$	m	2.6×10^{-8}
molar volume of Si $M(\mathrm{Si})/\rho(\mathrm{Si}) = N_A a^3/8$ (in vacuum, 22.5°C)	$V_m(\mathrm{Si})$	$12.058\,8349(11) \times 10^{-6}$	m^3 mol^{-1}	9.1×10^{-8}

[1] This is the lattice parameter (unit cell edge length) of an ideal single crystal of naturally occurring Si free of impurities and imperfections, and is deduced from measurements on extremely pure and nearly perfect single crystals of Si by correcting for the effects of impurities.

TABLE VI: The values in SI units of some non-SI units based on the 2006 CODATA adjustment of the values of the constants.

Quantity	Symbol	Numerical value	Unit	Relative std. uncert. u_r
	Non-SI units accepted for use with the SI			
electron volt: (e/C) J	eV	$1.602\,176\,487(40) \times 10^{-19}$	J	2.5×10^{-8}
(unified) atomic mass unit: $1\,\mathrm{u} = m_u = \frac{1}{12}m(^{12}\mathrm{C}) = 10^{-3}$ kg mol$^{-1}/N_A$	u	$1.660\,538\,782(83) \times 10^{-27}$	kg	5.0×10^{-8}
	Natural units (n.u.)			
n.u. of velocity: speed of light in vacuum	c, c_0	299 792 458	m s^{-1}	(exact)
n.u. of action: reduced Planck constant $(h/2\pi)$	$\hbar$	$1.054\,571\,628(53) \times 10^{-34}$	J s	5.0×10^{-8}
in eV s		$6.582\,118\,99(16) \times 10^{-16}$	eV s	2.5×10^{-8}
in MeV fm	$\hbar c$	$197.326\,9631(49)$	MeV fm	2.5×10^{-8}

TABLE VI: (Continued.)

Quantity	Symbol	Numerical value	Unit	Relative std. uncert. u_r
n.u. of mass:				
electron mass	m_e	$9.109\,382\,15(45) \times 10^{-31}$	kg	5.0×10^{-8}
n.u. of energy	$m_e c^2$	$8.187\,104\,38(41) \times 10^{-14}$	J	5.0×10^{-8}
in MeV		$0.510\,998\,910(13)$	MeV	2.5×10^{-8}
n.u. of momentum	$m_e c$	$2.730\,924\,06(14) \times 10^{-22}$	kg m s^{-1}	5.0×10^{-8}
in MeV/c		$0.510\,998\,910(13)$	MeV/c	2.5×10^{-8}
n.u. of length ($\hbar/m_e c$)	$\lambda\!\!\!^{-}_C$	$386.159\,264\,59(53) \times 10^{-15}$	m	1.4×10^{-9}
n.u. of time	$\hbar/m_e c^2$	$1.288\,088\,6570(18) \times 10^{-21}$	s	1.4×10^{-9}
Atomic units (a.u.)				
a.u. of charge:				
elementary charge	e	$1.602\,176\,487(40) \times 10^{-19}$	C	2.5×10^{-8}
a.u. of mass:				
electron mass	m_e	$9.109\,382\,15(45) \times 10^{-31}$	kg	5.0×10^{-8}
a.u. of action:				
reduced Planck constant ($h/2\pi$)	$\hbar$	$1.054\,571\,628(53) \times 10^{-34}$	J s	5.0×10^{-8}
a.u. of length:				
Bohr radius (bohr) ($\alpha/4\pi R_\infty$)	a_0	$0.529\,177\,208\,59(36) \times 10^{-10}$	m	6.8×10^{-10}
a.u. of energy:				
Hartree energy (hartree) ($e^2/4\pi\epsilon_0 a_0 = 2R_\infty hc = \alpha^2 m_e c^2$)	E_h	$4.359\,743\,94(22) \times 10^{-18}$	J	5.0×10^{-8}
a.u. of time	$\hbar/E_h$	$2.418\,884\,326\,505(16) \times 10^{-17}$	s	6.6×10^{-12}
a.u. of force	E_h/a_0	$8.238\,722\,06(41) \times 10^{-8}$	N	5.0×10^{-8}
a.u. of velocity (αc)	$a_0 E_h/\hbar$	$2.187\,691\,2541(15) \times 10^{6}$	m s^{-1}	6.8×10^{-10}
a.u. of momentum	$\hbar/a_0$	$1.992\,851\,565(99) \times 10^{-24}$	kg m s^{-1}	5.0×10^{-8}
a.u. of current	$eE_h/\hbar$	$6.623\,617\,63(17) \times 10^{-3}$	A	2.5×10^{-8}
a.u. of charge density	e/a_0^3	$1.081\,202\,300(27) \times 10^{12}$	C m^{-3}	2.5×10^{-8}
a.u. of electric potential	E_h/e	$27.211\,383\,86(68)$	V	2.5×10^{-8}
a.u. of electric field	E_h/ea_0	$5.142\,206\,32(13) \times 10^{11}$	V m^{-1}	2.5×10^{-8}
a.u. of electric field gradient	E_h/ea_0^2	$9.717\,361\,66(24) \times 10^{21}$	V m^{-2}	2.5×10^{-8}
a.u. of electric dipole moment	ea_0	$8.478\,352\,81(21) \times 10^{-30}$	C m	2.5×10^{-8}
a.u. of electric quadrupole moment	ea_0^2	$4.486\,551\,07(11) \times 10^{-40}$	C m^2	2.5×10^{-8}
a.u. of electric polarizability	$e^2a_0^2/E_h$	$1.648\,777\,2536(34) \times 10^{-41}$	C^2 m^2 J^{-1}	2.1×10^{-9}
a.u. of 1st hyperpolarizability	$e^3a_0^3/E_h^2$	$3.206\,361\,533(81) \times 10^{-53}$	C^3 m^3 J^{-2}	2.5×10^{-8}
a.u. of 2nd hyperpolarizability	$e^4a_0^4/E_h^3$	$6.235\,380\,95(31) \times 10^{-65}$	C^4 m^4 J^{-3}	5.0×10^{-8}
a.u. of magnetic flux density	$\hbar/ea_0^2$	$2.350\,517\,382(59) \times 10^{5}$	T	2.5×10^{-8}
a.u. of magnetic dipole moment ($2\mu_B$)	$\hbar e/m_e$	$1.854\,801\,830(46) \times 10^{-23}$	J T^{-1}	2.5×10^{-8}
a.u. of magnetizability	$e^2a_0^2/m_e$	$7.891\,036\,433(27) \times 10^{-29}$	J T^{-2}	3.4×10^{-9}
a.u. of permittivity ($10^7/c^2$)	$e^2/a_0 E_h$	$1.112\,650\,056\ldots \times 10^{-10}$	F m^{-1}	(exact)

TABLE VII: The values of some energy equivalents derived from the relations $E = mc^2 = hc/\lambda = h\nu = kT$, and based on the 2006 CODATA adjustment of the values of the constants; 1 eV = (e/C) J, 1 u = $m_u = \frac{1}{12}m(^{12}\mathrm{C}) = 10^{-3}$ kg mol$^{-1}/N_A$, and $E_h = 2R_\infty hc = \alpha^2 m_e c^2$ is the Hartree energy (hartree).

Relevant unit

	J	kg	m^{-1}	Hz
1 J	(1 J) = 1 J	(1 J)$/c^2$ = 1.112 650 056... × 10^{-17} kg	(1 J)$/hc$ = 5.034 117 47(25) × 10^{24} m^{-1}	(1 J)$/h$ = 1.509 190 450(75) × 10^{33} Hz
1 kg	(1 kg)c^2 = 8.987 551 787... × 10^{16} J	(1 kg) = 1 kg	(1 kg)c/h = 4.524 439 15(23) × 10^{41} m^{-1}	(1 kg)c^2/h = 1.356 392 733(68) × 10^{50} Hz
1 m^{-1}	(1 m^{-1})hc = 1.986 445 501(99) × 10^{-25} J	(1 m^{-1})h/c = 2.210 218 70(11) × 10^{-42} kg	(1 m^{-1}) = 1 m^{-1}	(1 m^{-1})c = 299 792 458 Hz
1 Hz	(1 Hz)h = 6.626 068 96(33) × 10^{-34} J	(1 Hz)h/c^2 = 7.372 496 00(37) × 10^{-51} kg	(1 Hz)$/c$ = 3.335 640 951... × 10^{-9} m^{-1}	(1 Hz) = 1 Hz
1 K	(1 K)k = 1.380 6504(24) × 10^{-23} J	(1 K)k/c^2 = 1.536 1807(27) × 10^{-40} kg	(1 K)k/hc = 69.503 56(12) m^{-1}	(1 K)k/h = 2.083 6644(36) × 10^{10} Hz
1 eV	(1 eV) = 1.602 176 487(40) × 10^{-19} J	(1 eV)$/c^2$ = 1.782 661 758(44) × 10^{-36} kg	(1 eV)$/hc$ = 8.065 544 65(20) × 10^{5} m^{-1}	(1 eV)$/h$ = 2.417 989 454(60) × 10^{14} Hz
1 u	(1 u)c^2 = 1.492 417 830(74) × 10^{-10} J	(1 u) = 1.660 538 782(83) × 10^{-27} kg	(1 u)c/h = 7.513 006 671(11) × 10^{14} m^{-1}	(1 u)c^2/h = 2.252 342 7369(32) × 10^{23} Hz
1 E_h	(1 E_h) = 4.359 743 94(22) × 10^{-18} J	(1 E_h)$/c^2$ = 4.850 869 34(24) × 10^{-35} kg	(1 E_h)$/hc$ = 2.194 746 313 705(15) × 10^{7} m^{-1}	(1 E_h)$/h$ = 6.579 683 920 722(44) × 10^{15} Hz

	K	eV	u	E_h
1 J	(1 J)$/k$ = 7.242 963(13) × 10^{22} K	(1 J) = 6.241 509 65(16) × 10^{18} eV	(1 J)$/c^2$ = 6.700 536 41(33) × 10^{9} u	(1 J) = 2.293 712 69(11) × 10^{17} E_h
1 kg	(1 kg)c^2/k = 6.509 651(11) × 10^{39} K	(1 kg)c^2 = 5.609 589 12(14) × 10^{35} eV	(1 kg) = 6.022 141 79(30) × 10^{26} u	(1 kg)c^2 = 2.061 486 16(10) × 10^{34} E_h
1 m^{-1}	(1 m^{-1})hc/k = 1.438 7752(25) × 10^{-2} K	(1 m^{-1})hc = 1.239 841 875(31) × 10^{-6} eV	(1 m^{-1})h/c = 1.331 025 0394(19) × 10^{-15} u	(1 m^{-1})hc = 4.556 335 252 760(30) × 10^{-8} E_h
1 Hz	(1 Hz)h/k = 4.799 2374(84) × 10^{-11} K	(1 Hz)h = 4.135 667 33(10) × 10^{-15} eV	(1 Hz)h/c^2 = 4.439 821 6294(64) × 10^{-24} u	(1 Hz)h = 1.519 829 846 006(10) × 10^{-16} E_h
1 K	(1 K) = 1 K	(1 K)k = 8.617 343(15) × 10^{-5} eV	(1 K)k/c^2 = 9.251 098(16) × 10^{-14} u	(1 K)k = 3.166 8153(55) × 10^{-6} E_h
1 eV	(1 eV)$/k$ = 1.160 4505(20) × 10^{4} K	(1 eV) = 1 eV	(1 eV)$/c^2$ = 1.073 544 188(27) × 10^{-9} u	(1 eV) = 3.674 932 540(92) × 10^{-2} E_h
1 u	(1 u)c^2/k = 1.080 9527(19) × 10^{13} K	(1 u)c^2 = 931.494 028(23) × 10^{6} eV	(1 u) = 1 u	(1 u)c^2 = 3.423 177 7149(49) × 10^{7} E_h
1 E_h	(1 E_h)$/k$ = 3.157 7465(55) × 10^{5} K	(1 E_h) = 27.211 383 86(68) eV	(1 E_h)$/c^2$ = 2.921 262 2986(42) × 10^{-8} u	(1 E_h) = 1 E_h

STANDARD ATOMIC WEIGHTS (2007)

This table of atomic weights includes the changes made in 2007 by the IUPAC Commission on Isotopic Abundances and Atomic Weights. Those changes affected the following elements: Lu, Mo, Ni, Yb, and Zn.

The Standard Atomic Weights apply to the elements as they exist naturally on Earth, and the uncertainties take into account the isotopic variation found in most laboratory samples. Further comments on the variability are given in the footnotes.

The number in parentheses following the atomic weight value gives the uncertainty in the last digit. An atomic weight entry in brackets indicates that the element that has no stable isotopes; the value given is the atomic mass in u (or the mass number, if the mass is not accurately known) for the isotope of longest half-life. Thorium, protactinium, and uranium have no stable isotopes, but the terrestrial isotopic composition is sufficiently uniform to permit a standard atomic weight to be specified.

Reference

Wieser, M. E., and Berglund, M., "Atomic Weights of the Elements 2007," *Pure Appl. Chem.* 81, 2131, 2009.

Name	Symbol	Atomic no.	Atomic weight	Footnotes
Actinium	Ac	89	[227.0277]	a
Aluminum	Al	13	26.9815386(8)	
Americium	Am	95	[243.0614]	a
Antimony	Sb	51	121.760(1)	g
Argon	Ar	18	39.948(1)	g r
Arsenic	As	33	74.92160(2)	
Astatine	At	85	[209.9871]	a
Barium	Ba	56	137.327(7)	
Berkelium	Bk	97	[247.0703]	a
Beryllium	Be	4	9.012182(3)	
Bismuth	Bi	83	208.98040(1)	
Bohrium	Bh	107	[272]	a
Boron	B	5	10.811(7)	g m r
Bromine	Br	35	79.904(1)	
Cadmium	Cd	48	112.411(8)	g
Calcium	Ca	20	40.078(4)	g
Californium	Cf	98	[251.0796]	a
Carbon	C	6	12.0107(8)	g r
Cerium	Ce	58	140.116(1)	g
Cesium	Cs	55	132.9054519(2)	
Chlorine	Cl	17	35.453(2)	g m r
Chromium	Cr	24	51.9961(6)	
Cobalt	Co	27	58.933195(5)	
Copernicium	Cn	112	[285]	a
Copper	Cu	29	63.546(3)	r
Curium	Cm	96	[247.0704]	a
Darmstadtium	Ds	110	[281]	a
Dubnium	Db	105	[262.1141]	a
Dysprosium	Dy	66	162.500(1)	g
Einsteinium	Es	99	[252.0830]	a
Erbium	Er	68	167.259(3)	g
Europium	Eu	63	151.964(1)	g
Fermium	Fm	100	[257.0951]	a
Fluorine	F	9	18.9984032(5)	
Francium	Fr	87	[223.0197]	a
Gadolinium	Gd	64	157.25(3)	g
Gallium	Ga	31	69.723(1)	
Germanium	Ge	32	72.64(1)	
Gold	Au	79	196.966569(4)	
Hafnium	Hf	72	178.49(2)	
Hassium	Hs	108	[277]	a
Helium	He	2	4.002602(2)	g r
Holmium	Ho	67	164.93032(2)	
Hydrogen	H	1	1.00794(7)	g m r
Indium	In	49	114.818(3)	
Iodine	I	53	126.90447(3)	
Iridium	Ir	77	192.217(3)	
Iron	Fe	26	55.845(2)	
Krypton	Kr	36	83.798(2)	g m
Lanthanum	La	57	138.90547(7)	g
Lawrencium	Lr	103	[262.1097]	a
Lead	Pb	82	207.2(1)	g r
Lithium	Li	3	6.941(2)	b g m r
Lutetium	Lu	71	174.9668(1)	g
Magnesium	Mg	12	24.3050(6)	
Manganese	Mn	25	54.938045(5)	
Meitnerium	Mt	109	[276]	a
Mendelevium	Md	101	[258.0984]	a
Mercury	Hg	80	200.59(2)	
Molybdenum	Mo	42	95.96(2)	g
Neodymium	Nd	60	144.242(3)	g
Neon	Ne	10	20.1797(6)	g m
Neptunium	Np	93	[237.0482]	a
Nickel	Ni	28	58.6934(4)	
Niobium	Nb	41	92.90638(2)	
Nitrogen	N	7	14.0067(2)	g r
Nobelium	No	102	[259.1010]	a
Osmium	Os	76	190.23(3)	g
Oxygen	O	8	15.9994(3)	g r
Palladium	Pd	46	106.42(1)	g
Phosphorus	P	15	30.973762(2)	
Platinum	Pt	78	195.084(9)	
Plutonium	Pu	94	[244.0642]	a
Polonium	Po	84	[208.9824]	a
Potassium	K	19	39.0983(1)	g
Praseodymium	Pr	59	140.90765(2)	
Promethium	Pm	61	[144.9127]	a
Protactinium	Pa	91	231.03588(2)	
Radium	Ra	88	[226.0254]	a
Radon	Rn	86	[222.0176]	a
Rhenium	Re	75	186.207(1)	
Rhodium	Rh	45	102.90550(2)	
Roentgenium	Rg	111	[280]	a
Rubidium	Rb	37	85.4678(3)	g
Ruthenium	Ru	44	101.07(2)	g
Rutherfordium	Rf	104	[261.1088]	a
Samarium	Sm	62	150.36(2)	g
Scandium	Sc	21	44.955912(6)	

Name	Symbol	Atomic no.	Atomic weight	Footnotes	Name	Symbol	Atomic no.	Atomic weight	Footnotes
Seaborgium	Sg	106	[266.1219]	a	Thulium	Tm	69	168.93421(2)	
Selenium	Se	34	78.96(3)	r	Tin	Sn	50	118.710(7)	g
Silicon	Si	14	28.0855(3)	r	Titanium	Ti	22	47.867(1)	
Silver	Ag	47	107.8682(2)	g	Tungsten	W	74	183.84(1)	
Sodium	Na	11	22.98976928(2)		Ununhexium	Uuh	116	[292]	a
Strontium	Sr	38	87.62(1)	g r	Ununquadium	Uuq	114	[289]	a
Sulfur	S	16	32.065(5)	g r	Uranium	U	92	238.02891(3)	g m
Tantalum	Ta	73	180.94788(2)		Vanadium	V	23	50.9415(1)	
Technetium	Tc	43	[97.9072]	a	Xenon	Xe	54	131.293(6)	g m
Tellurium	Te	52	127.60(3)	g	Ytterbium	Yb	70	173.054(5)	g
Terbium	Tb	65	158.92535(2)		Yttrium	Y	39	88.90585(2)	
Thallium	Tl	81	204.3833(2)		Zinc	Zn	30	65.38(2)	
Thorium	Th	90	232.03806(2)	g	Zirconium	Zr	40	91.224(2)	g

[a] No stable isotope exists. The atomic mass in u (or the mass number, if the mass is not accurately known) is given in brackets for the isotope of longest half-life.

[b] Commercially available Li materials have atomic weights that range between 6.939 and 6.996; if a more accurate value is required, it must be determined for the specific material.

[g] Geological specimens are known in which the element has an isotopic composition outside the limits for the normal material. The difference between the atomic weight of the element in such specimens and that given in the table may exceed the stated uncertainty.

[m] Modified isotopic compositions may be found in commercially available material because it has been subject to an undisclosed or inadvertent isotopic fractionation. Substantial deviations in atomic weight of the element from that given in the table can occur.

[r] Range in isotopic composition of normal terrestrial material prevents a more precise atomic weight being given; the tabulated value should be applicable to any normal material.

ATOMIC MASSES AND ABUNDANCES

This table lists the mass (in atomic mass units, symbol u) and the natural abundance (in percent) of the stable nuclides and a few important radioactive nuclides. A complete table of all nuclides may be found in Section 11 ("Table of the Isotopes").

The atomic masses were taken from the 2003 evaluation of Audi, Wapstra, and Thibault (References 2, 3). The number in parentheses following the mass value is the uncertainty in the last digit(s) given. An asterisk * after an entry indicates the mass value was derived not purely from experimental data, but at least partly from systematic trends.

Natural abundance values were taken from the IUPAC Technical Report "Atomic Weight of the Elements: Review 2000" (Reference 4); these entries are also followed by uncertainties in the last digit(s) of the stated values. This uncertainty includes both the estimated measurement uncertainty and the reported range of variation in different terrestrial sources of the element (see Reference 4 for full details and caveats regarding elements whose abundance is variable). The absence of an entry in the Abundance column indicates a radioactive nuclide not present in nature or an element whose isotopic composition varies so widely that a meaningful natural abundance cannot be defined.

References

1. Holden, N. E., "Table of the Isotopes", in Lide, D. R., Ed., *CRC Handbook of Chemistry and Physics, 86th Ed.*, CRC Press, Boca Raton FL, 2005.
2. Audi, G., Wapstra, A. H., and Thibault, *Nucl. Phys.*, A729, 336, 2003.
3. Audi, G., and Wapstra, A. H., Atomic Mass Data Center, World Wide Web site, <http://www.nndc.bnl.gov/amdc/index.html>.
4. de Laeter, J. R., Böhlke, J. K., De Bièvre, P., Hidaka, H., Peiser, H. S., Rosman, K. J. R., and Taylor, P. D. P., *Pure Appl. Chem.* 75, 683, 2003.

Z	Isotope	Mass in u	Abundance in %
1	^{1}H	1.00782503207(10)	99.9885(70)
	^{2}H	2.0141017778(4)	0.0115(70)
	^{3}H	3.0160492777(25)	
2	^{3}He	3.0160293191(26)	0.000134(3)
	^{4}He	4.00260325415(6)	99.999866(3)
3	^{6}Li	6.015122795(16)	7.59(4)
	^{7}Li	7.01600455(8)	92.41(4)
4	^{9}Be	9.0121822(4)	100
5	^{10}B	10.0129370(4)	19.9(7)
	^{11}B	11.0093054(4)	80.1(7)
6	^{11}C	11.0114336(10)	
	^{12}C	12.0000000(0)	98.93(8)
	^{13}C	13.0033548378(10)	1.07(8)
	^{14}C	14.003241989(4)	
7	^{14}N	14.0030740048(6)	99.636(7)
	^{15}N	15.0001088982(7)	0.364(7)
8	^{16}O	15.99491461956(16)	99.757(16)
	^{17}O	16.99913170(12)	0.038(1)
	^{18}O	17.9991610(7)	0.205(14)
9	^{18}F	18.0009380(6)	
	^{19}F	18.99840322(7)	100
10	^{20}Ne	19.9924401754(19)	90.48(3)
	^{21}Ne	20.99384668(4)	0.27(1)
	^{22}Ne	21.991385114(19)	9.25(3)
11	^{22}Na	21.9944364(4)	
	^{23}Na	22.9897692809(29)	100
	^{24}Na	23.99096278(8)	
12	^{24}Mg	23.985041700(14)	78.99(4)
	^{25}Mg	24.98583692(3)	10.00(1)
	^{26}Mg	25.982592929(30)	11.01(3)
13	^{27}Al	26.98153863(12)	100
14	^{28}Si	27.9769265325(19)	92.223(19)
	^{29}Si	28.976494700(22)	4.685(8)
	^{30}Si	29.97377017(3)	3.092(11)
15	^{31}P	30.97376163(20)	100
	^{32}P	31.97390727(20)	
16	^{32}S	31.97207100(15)	94.99(26)
	^{33}S	32.97145876(15)	0.75(2)
	^{34}S	33.96786690(12)	4.25(24)
	^{35}S	34.96903216(11)	
	^{36}S	35.96708076(20)	0.01(1)
17	^{35}Cl	34.96885268(4)	75.76(10)
	^{37}Cl	36.96590259(5)	24.24(10)
18	^{36}Ar	35.967545106(29)	0.3365(30)
	^{38}Ar	37.9627324(4)	0.0632(5)
	^{40}Ar	39.9623831225(29)	99.6003(30)
19	^{39}K	38.96370668(20)	93.2581(44)
	^{40}K	39.96399848(21)	0.0117(1)
	^{41}K	40.96182576(21)	6.7302(44)
	^{42}K	41.96240281(24)	
	^{43}K	42.960716(10)	
20	^{40}Ca	39.96259098(22)	96.941(156)
	^{42}Ca	41.95861801(27)	0.647(23)
	^{43}Ca	42.9587666(3)	0.135(10)
	^{44}Ca	43.9554818(4)	2.086(110)
	^{45}Ca	44.9561866(4)	
	^{46}Ca	45.9536926(24)	0.004(3)
	^{47}Ca	46.9545460(24)	
	^{48}Ca	47.952534(4)	0.187(21)
21	^{45}Sc	44.9559119(9)	100
22	^{46}Ti	45.9526316(9)	8.25(3)
	^{47}Ti	46.9517631(9)	7.44(2)
	^{48}Ti	47.9479463(9)	73.72(3)
	^{49}Ti	48.9478700(9)	5.41(2)
	^{50}Ti	49.9447912(9)	5.18(2)
23	^{50}V	49.9471585(11)	0.250(4)
	^{51}V	50.9439595(11)	99.750(4)
24	^{50}Cr	49.9460442(11)	4.345(13)
	^{51}Cr	50.9447674(11)	
	^{52}Cr	51.9405075(8)	83.789(18)
	^{53}Cr	52.9406494(8)	9.501(17)
	^{54}Cr	53.9388804(8)	2.365(7)
25	^{54}Mn	53.9403589(14)	
	^{55}Mn	54.9380451(7)	100
26	^{52}Fe	51.948114(7)	
	^{54}Fe	53.9396105(7)	5.845(35)
	^{55}Fe	54.9382934(7)	
	^{56}Fe	55.9349375(7)	91.754(36)
	^{57}Fe	56.9353940(7)	2.119(10)
	^{58}Fe	57.9332756(8)	0.282(4)
	^{59}Fe	58.9348755(8)	
27	^{57}Co	56.9362914(8)	

Z	Isotope	Mass in u	Abundance in %
	^{58}Co	57.9357528(13)	
	^{59}Co	58.9331950(7)	100
	^{60}Co	59.9338171(7)	
28	^{58}Ni	57.9353429(7)	68.0769(89)
	^{59}Ni	58.9343467(7)	
	^{60}Ni	59.9307864(7)	26.2231(77)
	^{61}Ni	60.9310560(7)	1.1399(6)
	^{62}Ni	61.9283451(6)	3.6345(17)
	^{63}Ni	62.9296694(6)	
	^{64}Ni	63.9279660(7)	0.9256(9)
29	^{63}Cu	62.9295975(6)	69.15(3)
	^{64}Cu	63.9297642(6)	
	^{65}Cu	64.9277895(7)	30.85(3)
30	^{64}Zn	63.9291422(7)	48.268(321)
	^{65}Zn	64.9292410(7)	
	^{66}Zn	65.9260334(10)	27.975(77)
	^{67}Zn	66.9271273(10)	4.102(21)
	^{68}Zn	67.9248442(10)	19.024(123)
	^{70}Zn	69.9253193(21)	0.631(9)
31	^{67}Ga	66.9282017(14)	
	^{68}Ga	67.9279801(16)	
	^{69}Ga	68.9255736(13)	60.108(9)
	^{71}Ga	70.9247013(11)	39.892(9)
32	^{68}Ge	67.928094(7)	
	^{70}Ge	69.9242474(11)	20.38(18)
	^{72}Ge	71.9220758(18)	27.31(26)
	^{73}Ge	72.9234589(18)	7.76(8)
	^{74}Ge	73.9211778(18)	36.72(15)
	^{76}Ge	75.9214026(18)	7.83(7)
33	^{75}As	74.9215965(20)	100
34	^{74}Se	73.9224764(18)	0.89(4)
	^{75}Se	74.9225234(18)	
	^{76}Se	75.9192136(18)	9.37(29)
	^{77}Se	76.9199140(18)	7.63(16)
	^{78}Se	77.9173091(18)	23.77(28)
	^{79}Se	78.9184991(18)	
	^{80}Se	79.9165213(21)	49.61(41)
	^{82}Se	81.9166994(22)	8.73(22)
35	^{79}Br	78.9183371(22)	50.69(7)
	^{81}Br	80.9162906(21)	49.31(7)
36	^{78}Kr	77.9203648(12)	0.355(3)
	^{80}Kr	79.9163790(16)	2.286(10)
	^{82}Kr	81.9134836(19)	11.593(31)
	^{83}Kr	82.914136(3)	11.500(19)
	^{84}Kr	83.911507(3)	56.987(15)
	^{86}Kr	85.91061073(11)	17.279(41)
37	^{85}Rb	84.911789738(12)	72.17(2)
	^{86}Rb	85.91116742(21)	
	^{87}Rb	86.909180527(13)	27.83(2)
38	^{84}Sr	83.913425(3)	0.56(1)
	^{85}Sr	84.912933(3)	
	^{86}Sr	85.9092602(12)	9.86(1)
	^{87}Sr	86.9088771(12)	7.00(1)
	^{88}Sr	87.9056121(12)	82.58(1)
	^{89}Sr	88.9074507(12)	
	^{90}Sr	89.907738(3)	
39	^{89}Y	88.9058483(27)	100
40	^{90}Zr	89.9047044(25)	51.45(40)
	^{91}Zr	90.9056458(25)	11.22(5)
	^{92}Zr	91.9050408(25)	17.15(8)
	^{94}Zr	93.9063152(26)	17.38(28)
	^{96}Zr	95.9082734(30)	2.80(9)
41	^{93}Nb	92.9063781(26)	100
42	^{92}Mo	91.906811(4)	14.77(31)
	^{94}Mo	93.9050883(21)	9.23(10)
	^{95}Mo	94.9058421(21)	15.90(9)
	^{96}Mo	95.9046795(21)	16.68(1)
	^{97}Mo	96.9060215(21)	9.56(5)
	^{98}Mo	97.9054082(21)	24.19(26)
	^{99}Mo	98.9077119(21)	
	^{100}Mo	99.907477(6)	9.67(20)
43	^{97}Tc	96.906365(5)	
	^{98}Tc	97.907216(4)	
	^{99}Tc	98.9062547(21)	
44	^{96}Ru	95.907598(8)	5.54(14)
	^{98}Ru	97.905287(7)	1.87(3)
	^{99}Ru	98.9059393(22)	12.76(14)
	^{100}Ru	99.9042195(22)	12.60(7)
	^{101}Ru	100.9055821(22)	17.06(2)
	^{102}Ru	101.9043493(22)	31.55(14)
	^{104}Ru	103.905433(3)	18.62(27)
	^{106}Ru	105.907329(8)	
45	^{103}Rh	102.905504(3)	100
46	^{102}Pd	101.905609(3)	1.02(1)
	^{104}Pd	103.904036(4)	11.14(8)
	^{105}Pd	104.905085(4)	22.33(8)
	^{106}Pd	105.903486(4)	27.33(3)
	^{108}Pd	107.903892(4)	26.46(9)
	^{110}Pd	109.905153(12)	11.72(9)
47	^{107}Ag	106.905097(5)	51.839(8)
	^{109}Ag	108.904752(3)	48.161(8)
48	^{106}Cd	105.906459(6)	1.25(6)
	^{108}Cd	107.904184(6)	0.89(3)
	^{110}Cd	109.9030021(29)	12.49(18)
	^{111}Cd	110.9041781(29)	12.80(12)
	^{112}Cd	111.9027578(29)	24.13(21)
	^{113}Cd	112.9044017(29)	12.22(12)
	^{114}Cd	113.9033585(29)	28.73(42)
	^{116}Cd	115.904756(3)	7.49(18)
49	^{111}In	110.905103(5)	
	^{113}In	112.904058(3)	4.29(5)
	^{115}In	114.903878(5)	95.71(5)
50	^{112}Sn	111.904818(5)	0.97(1)
	^{113}Sn	112.905171(4)	
	^{114}Sn	113.902779(3)	0.66(1)
	^{115}Sn	114.903342(3)	0.34(1)
	^{116}Sn	115.901741(3)	14.54(9)
	^{117}Sn	116.902952(3)	7.68(7)
	^{118}Sn	117.901603(3)	24.22(9)
	^{119}Sn	118.903308(3)	8.59(4)
	^{120}Sn	119.9021947(27)	32.58(9)
	^{122}Sn	121.9034390(29)	4.63(3)
	^{124}Sn	123.9052739(15)	5.79(5)
51	^{121}Sb	120.9038157(24)	57.21(5)
	^{123}Sb	122.9042140(22)	42.79(5)
52	^{120}Te	119.904020(10)	0.09(1)
	^{122}Te	121.9030439(16)	2.55(12)
	^{123}Te	122.9042700(16)	0.89(3)
	^{124}Te	123.9028179(16)	4.74(14)
	^{125}Te	124.9044307(16)	7.07(15)
	^{126}Te	125.9033117(16)	18.84(25)
	^{128}Te	127.9044631(19)	31.74(8)

Z	Isotope	Mass in u	Abundance in %
	^{130}Te	129.9062244(21)	34.08(62)
53	^{123}I	122.905589(4)	
	^{125}I	124.9046302(16)	
	^{127}I	126.904473(4)	100
	^{129}I	128.904988(3)	
	^{131}I	130.9061246(12)	
54	^{124}Xe	123.9058930(20)	0.0952(3)
	^{126}Xe	125.904274(7)	0.0890(2)
	^{128}Xe	127.9035313(15)	1.9102(8)
	^{129}Xe	128.9047794(8)	26.4006(82)
	^{130}Xe	129.9035080(8)	4.0710(13)
	^{131}Xe	130.9050824(10)	21.2324(30)
	^{132}Xe	131.9041535(10)	26.9086(33)
	^{134}Xe	133.9053945(9)	10.4357(21)
	^{136}Xe	135.907219(8)	8.8573(44)
55	^{129}Cs	128.906064(5)	
	^{133}Cs	132.905451933(24)	100
	^{134}Cs	133.906718475(28)	
	^{136}Cs	135.9073116(20)	
	^{137}Cs	136.9070895(5)	
56	^{130}Ba	129.9063208(30)	0.106(1)
	^{132}Ba	131.9050613(11)	0.101(1)
	^{133}Ba	132.9060075(11)	
	^{134}Ba	133.9045084(4)	2.417(18)
	^{135}Ba	134.9056886(4)	6.592(12)
	^{136}Ba	135.9045759(4)	7.854(24)
	^{137}Ba	136.9058274(5)	11.232(24)
	^{138}Ba	137.9052472(5)	71.698(42)
	^{140}Ba	139.910605(9)	
57	^{138}La	137.907112(4)	0.090(1)
	^{139}La	138.9063533(26)	99.910(1)
58	^{136}Ce	135.907172(14)	0.185(2)
	^{138}Ce	137.905991(11)	0.251(2)
	^{140}Ce	139.9054387(26)	88.450(51)
	^{141}Ce	140.9082763(26)	
	^{142}Ce	141.909244(3)	11.114(51)
	^{144}Ce	143.913647(4)	
59	^{141}Pr	140.9076528(26)	100
60	^{142}Nd	141.9077233(25)	27.2(5)
	^{143}Nd	142.9098143(25)	12.2(2)
	^{144}Nd	143.9100873(25)	23.8(3)
	^{145}Nd	144.9125736(25)	8.3(1)
	^{146}Nd	145.9131169(25)	17.2(3)
	^{148}Nd	147.916893(3)	5.7(1)
	^{150}Nd	149.920891(3)	5.6(2)
61	^{145}Pm	144.912749(3)	
	^{147}Pm	146.9151385(26)	
62	^{144}Sm	143.911999(3)	3.07(7)
	^{147}Sm	146.9148979(26)	14.99(18)
	^{148}Sm	147.9148227(26)	11.24(10)
	^{149}Sm	148.9171847(26)	13.82(7)
	^{150}Sm	149.9172755(26)	7.38(1)
	^{152}Sm	151.9197324(27)	26.75(16)
	^{154}Sm	153.9222093(27)	22.75(29)
63	^{151}Eu	150.9198502(26)	47.81(6)
	^{153}Eu	152.9212303(26)	52.19(6)
64	^{152}Gd	151.9197910(27)	0.20(1)
	^{154}Gd	153.9208656(27)	2.18(3)
	^{155}Gd	154.9226220(27)	14.80(12)
	^{156}Gd	155.9221227(27)	20.47(9)
	^{157}Gd	156.9239601(27)	15.65(2)
	^{158}Gd	157.9241039(27)	24.84(7)
	^{160}Gd	159.9270541(27)	21.86(19)
65	^{159}Tb	158.9253468(27)	100
66	^{156}Dy	155.924283(7)	0.056(3)
	^{158}Dy	157.924409(4)	0.095(3)
	^{160}Dy	159.9251975(27)	2.329(18)
	^{161}Dy	160.9269334(27)	18.889(42)
	^{162}Dy	161.9267984(27)	25.475(36)
	^{163}Dy	162.9287312(27)	24.896(42)
	^{164}Dy	163.9291748(27)	28.260(54)
67	^{165}Ho	164.9303221(27)	100
68	^{162}Er	161.928778(4)	0.139(5)
	^{164}Er	163.929200(3)	1.601(3)
	^{166}Er	165.9302931(27)	33.503(36)
	^{167}Er	166.9320482(27)	22.869(9)
	^{168}Er	167.9323702(27)	26.978(18)
	^{170}Er	169.9354643(30)	14.910(36)
69	^{169}Tm	168.9342133(27)	100
70	^{168}Yb	167.933897(5)	0.13(1)
	^{169}Yb	168.935190(5)	
	^{170}Yb	169.9347618(26)	3.04(15)
	^{171}Yb	170.9363258(26)	14.28(57)
	^{172}Yb	171.9363815(26)	21.83(67)
	^{173}Yb	172.9382108(26)	16.13(27)
	^{174}Yb	173.9388621(26)	31.83(92)
	^{176}Yb	175.9425717(28)	12.76(41)
71	^{175}Lu	174.9407718(23)	97.41(2)
	^{176}Lu	175.9426863(23)	2.59(2)
72	^{174}Hf	173.940046(3)	0.16(1)
	^{176}Hf	175.9414086(24)	5.26(7)
	^{177}Hf	176.9432207(23)	18.60(9)
	^{178}Hf	177.9436988(23)	27.28(7)
	^{179}Hf	178.9458161(23)	13.62(2)
	^{180}Hf	179.9465500(23)	35.08(16)
73	^{180}Ta	179.9474648(24)	0.012(2)
	^{181}Ta	180.9479958(19)	99.988(2)
74	^{180}W	179.946704(4)	0.12(1)
	^{182}W	181.9482042(9)	26.50(16)
	^{183}W	182.9502230(9)	14.31(4)
	^{184}W	183.9509312(9)	30.64(2)
	^{186}W	185.9543641(19)	28.43(19)
75	^{185}Re	184.9529550(13)	37.40(2)
	^{187}Re	186.9557531(15)	62.60(2)
76	^{184}Os	183.9524891(14)	0.02(1)
	^{186}Os	185.9538382(15)	1.59(3)
	^{187}Os	186.9557505(15)	1.96(2)
	^{188}Os	187.9558382(15)	13.24(8)
	^{189}Os	188.9581475(16)	16.15(5)
	^{190}Os	189.9584470(16)	26.26(2)
	^{192}Os	191.9614807(27)	40.78(19)
77	^{191}Ir	190.9605940(18)	37.3(2)
	^{193}Ir	192.9629264(18)	62.7(2)
78	^{190}Pt	189.959932(6)	0.014(1)
	^{192}Pt	191.9610380(27)	0.782(7)
	^{194}Pt	193.9626803(9)	32.967(99)
	^{195}Pt	194.9647911(9)	33.832(10)
	^{196}Pt	195.9649515(9)	25.242(41)
	^{198}Pt	197.967893(3)	7.163(55)
79	^{197}Au	196.9665687(6)	100
	^{198}Au	197.9682423(6)	
80	^{196}Hg	195.965833(3)	0.15(1)

Z	Isotope	Mass in u	Abundance in %
	^{197}Hg	196.967213(3)	
	^{198}Hg	197.9667690(4)	9.97(20)
	^{199}Hg	198.9682799(4)	16.87(22)
	^{200}Hg	199.9683260(4)	23.10(19)
	^{201}Hg	200.9703023(6)	13.18(9)
	^{202}Hg	201.9706430(6)	29.86(26)
	^{203}Hg	202.9728725(18)	
	^{204}Hg	203.9734939(4)	6.87(15)
81	^{201}Tl	200.970819(16)	
	^{203}Tl	202.9723442(14)	29.52(1)
	^{205}Tl	204.9744275(14)	70.48(1)
82	^{204}Pb	203.9730436(13)	1.4(1)
	^{206}Pb	205.9744653(13)	24.1(1)
	^{207}Pb	206.9758969(13)	22.1(1)
	^{208}Pb	207.9766521(13)	52.4(1)
	^{210}Pb	209.9841885(16)	
83	^{207}Bi	206.9784707(26)	
	^{209}Bi	208.9803987(16)	100
84	^{209}Po	208.9824304(20)	
	^{210}Po	209.9828737(13)	
85	^{210}At	209.987148(8)	
	^{211}At	210.9874963(30)	
86	^{211}Rn	210.990601(7)	
	^{220}Rn	220.0113940(24)	
	^{222}Rn	222.0175777(25)	
87	^{223}Fr	223.0197359(26)	
88	^{223}Ra	223.0185022(27)	
	^{224}Ra	224.0202118(24)	
	^{226}Ra	226.0254098(25)	
	^{228}Ra	228.0310703(26)	
89	^{227}Ac	227.0277521(26)	
90	^{228}Th	228.0287411(24)	
	^{230}Th	230.0331338(19)	
	^{232}Th	232.0380553(21)	100
91	^{231}Pa	231.0358840(24)	100
92	^{233}U	233.0396352(29)	
	^{234}U	234.0409521(20)	0.0054(5)
	^{235}U	235.0439299(20)	0.7204(6)
	^{236}U	236.0455680(20)	
	^{238}U	238.0507882(20)	99.2742(10)
93	^{237}Np	237.0481734(20)	
	^{239}Np	239.0529390(22)	
94	^{238}Pu	238.0495599(20)	
	^{239}Pu	239.0521634(20)	
	^{240}Pu	240.0538135(20)	
	^{241}Pu	241.0568515(20)	
	^{242}Pu	242.0587426(20)	
	^{244}Pu	244.064204(5)	
95	^{241}Am	241.0568291(20)	
	^{243}Am	243.0613811(25)	
96	^{243}Cm	243.0613891(22)	
	^{244}Cm	244.0627526(20)	
	^{245}Cm	245.0654912(22)	
	^{246}Cm	246.0672237(22)	
	^{247}Cm	247.070354(5)	
	^{248}Cm	248.072349(5)	
97	^{247}Bk	247.070307(6)	
	^{249}Bk	249.0749867(28)	
98	^{249}Cf	249.0748535(24)	
	^{250}Cf	250.0764061(22)	
	^{251}Cf	251.079587(5)	
	^{252}Cf	252.081626(5)	
99	^{252}Es	252.082980(50)	
100	^{257}Fm	257.095105(7)	
101	^{256}Md	256.094060(60)	
	^{258}Md	258.098431(5)	
102	^{259}No	259.10103(11)*	
103	^{262}Lr	262.10963(22)*	
104	^{261}Rf	261.108770(30)*	
105	^{262}Db	262.11408(20)*	
106	^{263}Sg	263.11832(13)*	
107	^{264}Bh	264.12460(30)*	
108	^{265}Hs	265.13009(15)*	
109	^{268}Mt	268.13873(34)*	
110	^{281}Ds	281.16206(78)*	
111	^{272}Rg	273.15362(36)*	

ELECTRON CONFIGURATION AND IONIZATION ENERGY OF NEUTRAL ATOMS IN THE GROUND STATE

William C. Martin

The ground state electron configuration, ground level, and ionization energy of the elements hydrogen through rutherfordium are listed in this table. The electron configurations of elements heavier than neon are shortened by using rare-gas element symbols in brackets to represent the corresponding electrons. See the references for details of the notation for Pa, U, and Np. Ionization energies to higher states (and more precise values of the first ionization energy for certain elements) may be found in the table "Ionization Energies of Atoms and Atomic Ions" in Section 10 of this *Handbook*.

References

1. Martin, W. C., Musgrove, A., Kotochigova, S., and Sansonetti, J. E., NIST Physical Reference Data Web Site, <http://physics.nist.gov/PhysRefData/IonEnergy/ionEnergy.html>, October 2004.
2. Martin, W. C., and Wiese, W. L., "Atomic Spectroscopy", in *Atomic, Molecular, & Optical Physics Handbook*, ed. by G.W.F. Drake (AIP, Woodbury, NY, 1996) Chapter 10, pp. 135-153.

Z		Element	Ground-state configuration	Ground level	Ionization energy (eV)
1	H	Hydrogen	$1s$	$^2S_{1/2}$	13.5984
2	He	Helium	$1s^2$	$^1S_0$	24.5874
3	Li	Lithium	$1s^2\,2s$	$^2S_{1/2}$	5.3917
4	Be	Beryllium	$1s^2\,2s^2$	$^1S_0$	9.3227
5	B	Boron	$1s^2\,2s^2\,2p$	$^2P^o_{1/2}$	8.2980
6	C	Carbon	$1s^2\,2s^2\,2p^2$	$^3P_0$	11.2603
7	N	Nitrogen	$1s^2\,2s^2\,2p^3$	$^4S^o_{3/2}$	14.5341
8	O	Oxygen	$1s^2\,2s^2\,2p^4$	$^3P_2$	13.6181
9	F	Fluorine	$1s^2\,2s^2\,2p^5$	$^2P^o_{3/2}$	17.4228
10	Ne	Neon	$1s^2\,2s^2\,2p^6$	$^1S_0$	21.5645
11	Na	Sodium	[Ne] $3s$	$^2S_{1/2}$	5.1391
12	Mg	Magnesium	[Ne] $3s^2$	$^1S_0$	7.6462
13	Al	Aluminum	[Ne] $3s^2\,3p$	$^2P^o_{1/2}$	5.9858
14	Si	Silicon	[Ne] $3s^2\,3p^2$	$^3P_0$	8.1517
15	P	Phosphorus	[Ne] $3s^2\,3p^3$	$^4S^o_{3/2}$	10.4867
16	S	Sulfur	[Ne] $3s^2\,3p^4$	$^3P_2$	10.3600
17	Cl	Chlorine	[Ne] $3s^2\,3p^5$	$^2P^o_{3/2}$	12.9676
18	Ar	Argon	[Ne] $3s^2\,3p^6$	$^1S_0$	15.7596
19	K	Potassium	[Ar] $4s$	$^2S_{1/2}$	4.3407
20	Ca	Calcium	[Ar] $4s^2$	$^1S_0$	6.1132
21	Sc	Scandium	[Ar] $3d\,4s^2$	$^2D_{3/2}$	6.5615
22	Ti	Titanium	[Ar] $3d^2\,4s^2$	$^3F_2$	6.8281
23	V	Vanadium	[Ar] $3d^3\,4s^2$	$^4F_{3/2}$	6.7462
24	Cr	Chromium	[Ar] $3d^5\,4s$	$^7S_3$	6.7665
25	Mn	Manganese	[Ar] $3d^5\,4s^2$	$^6S_{5/2}$	7.4340
26	Fe	Iron	[Ar] $3d^6\,4s^2$	$^5D_4$	7.9024
27	Co	Cobalt	[Ar] $3d^7\,4s^2$	$^4F_{9/2}$	7.8810
28	Ni	Nickel	[Ar] $3d^8\,4s^2$	$^3F_4$	7.6398
29	Cu	Copper	[Ar] $3d^{10}\,4s$	$^2S_{1/2}$	7.7264
30	Zn	Zinc	[Ar] $3d^{10}\,4s^2$	$^1S_0$	9.3942
31	Ga	Gallium	[Ar] $3d^{10}\,4s^2\,4p$	$^2P^o_{1/2}$	5.9993
32	Ge	Germanium	[Ar] $3d^{10}\,4s^2\,4p^2$	$^3P_0$	7.8994
33	As	Arsenic	[Ar] $3d^{10}\,4s^2\,4p^3$	$^4S^o_{3/2}$	9.7886
34	Se	Selenium	[Ar] $3d^{10}\,4s^2\,4p^4$	$^3P_2$	9.7524
35	Br	Bromine	[Ar] $3d^{10}\,4s^2\,4p^5$	$^2P^o_{3/2}$	11.8138
36	Kr	Krypton	[Ar] $3d^{10}\,4s^2\,4p^6$	$^1S_0$	13.9996
37	Rb	Rubidium	[Kr] $5s$	$^2S_{1/2}$	4.1771
38	Sr	Strontium	[Kr] $5s^2$	$^1S_0$	5.6949
39	Y	Yttrium	[Kr] $4d\,5s^2$	$^2D_{3/2}$	6.2173
40	Zr	Zirconium	[Kr] $4d^2\,5s^2$	$^3F_2$	6.6339
41	Nb	Niobium	[Kr] $4d^4\,5s$	$^6D_{1/2}$	6.7589
42	Mo	Molybdenum	[Kr] $4d^5\,5s$	$^7S_3$	7.0924
43	Tc	Technetium	[Kr] $4d^5\,5s^2$	$^6S_{5/2}$	7.28
44	Ru	Ruthenium	[Kr] $4d^7\,5s$	$^5F_5$	7.3605

Z		Element	Ground-state configuration	Ground level	Ionization energy (eV)
45	Rh	Rhodium	[Kr] $4d^8\ 5s$	$^4F_{9/2}$	7.4589
46	Pd	Palladium	[Kr] $4d^{10}$	$^1S_0$	8.3369
47	Ag	Silver	[Kr] $4d^{10}\ 5s$	$^2S_{1/2}$	7.5762
48	Cd	Cadmium	[Kr] $4d^{10}\ 5s^2$	$^1S_0$	8.9938
49	In	Indium	[Kr] $4d^{10}\ 5s^2\ 5p$	$^2P^o_{1/2}$	5.7864
50	Sn	Tin	[Kr] $4d^{10}\ 5s^2\ 5p^2$	$^3P_0$	7.3439
51	Sb	Antimony	[Kr] $4d^{10}\ 5s^2\ 5p^3$	$^4S^o_{3/2}$	8.6084
52	Te	Tellurium	[Kr] $4d^{10}\ 5s^2\ 5p^4$	$^3P_2$	9.0096
53	I	Iodine	[Kr] $4d^{10}\ 5s^2\ 5p^5$	$^2P^o_{3/2}$	10.4513
54	Xe	Xenon	[Kr] $4d^{10}\ 5s^2\ 5p^6$	$^1S_0$	12.1298
55	Cs	Cesium	[Xe] $6s$	$^2S_{1/2}$	3.8939
56	Ba	Barium	[Xe] $6s^2$	$^1S_0$	5.2117
57	La	Lanthanum	[Xe] $5d\ 6s^2$	$^2D_{3/2}$	5.5769
58	Ce	Cerium	[Xe] $4f\ 5d\ 6s^2$	$^1G^o_4$	5.5387
59	Pr	Praseodymium	[Xe] $4f^3\ 6s^2$	$^4I^o_{9/2}$	5.473
60	Nd	Neodymium	[Xe] $4f^4\ 6s^2$	$^5I_4$	5.5250
61	Pm	Promethium	[Xe] $4f^5\ 6s^2$	$^6H^o_{5/2}$	5.582
62	Sm	Samarium	[Xe] $4f^6\ 6s^2$	$^7F_0$	5.6437
63	Eu	Europium	[Xe] $4f^7\ 6s^2$	$^8S^o_{7/2}$	5.6704
64	Gd	Gadolinium	[Xe] $4f^7\ 5d\ 6s^2$	$^9D^o_2$	6.1498
65	Tb	Terbium	[Xe] $4f^9\ 6s^2$	$^6H^o_{15/2}$	5.8638
66	Dy	Dysprosium	[Xe] $4f^{10}\ 6s^2$	$^5I_8$	5.9389
67	Ho	Holmium	[Xe] $4f^{11}\ 6s^2$	$^4I^o_{15/2}$	6.0215
68	Er	Erbium	[Xe] $4f^{12}\ 6s^2$	$^3H_6$	6.1077
69	Tm	Thulium	[Xe] $4f^{13}\ 6s^2$	$^2F^o_{7/2}$	6.1843
70	Yb	Ytterbium	[Xe] $4f^{14}\ 6s^2$	$^1S_0$	6.2542
71	Lu	Lutetium	[Xe] $4f^{14}\ 5d\ 6s^2$	$^2D_{3/2}$	5.4259
72	Hf	Hafnium	[Xe] $4f^{14}\ 5d^2\ 6s^2$	$^3F_2$	6.8251
73	Ta	Tantalum	[Xe] $4f^{14}\ 5d^3\ 6s^2$	$^4F_{3/2}$	7.5496
74	W	Tungsten	[Xe] $4f^{14}\ 5d^4\ 6s^2$	$^5D_0$	7.8640
75	Re	Rhenium	[Xe] $4f^{14}\ 5d^5\ 6s^2$	$^6S_{5/2}$	7.8335
76	Os	Osmium	[Xe] $4f^{14}\ 5d^6\ 6s^2$	$^5D_4$	8.4382
77	Ir	Iridium	[Xe] $4f^{14}\ 5d^7\ 6s^2$	$^4F_{9/2}$	8.9670
78	Pt	Platinum	[Xe] $4f^{14}\ 5d^9\ 6s$	$^3D_3$	8.9588
79	Au	Gold	[Xe] $4f^{14}\ 5d^{10}\ 6s$	$^2S_{1/2}$	9.2255
80	Hg	Mercury	[Xe] $4f^{14}\ 5d^{10}\ 6s^2$	$^1S_0$	10.4375
81	Tl	Thallium	[Xe] $4f^{14}\ 5d^{10}\ 6s^2\ 6p$	$^2P^o_{1/2}$	6.1082
82	Pb	Lead	[Xe] $4f^{14}\ 5d^{10}\ 6s^2\ 6p^2$	$^3P_0$	7.4167
83	Bi	Bismuth	[Xe] $4f^{14}\ 5d^{10}\ 6s^2\ 6p^3$	$^4S^o_{3/2}$	7.2855
84	Po	Polonium	[Xe] $4f^{14}\ 5d^{10}\ 6s^2\ 6p^4$	$^3P_2$	8.414
85	At	Astatine	[Xe] $4f^{14}\ 5d^{10}\ 6s^2\ 6p^5$	$^2P^o_{3/2}$	
86	Rn	Radon	[Xe] $4f^{14}\ 5d^{10}\ 6s^2\ 6p^6$	$^1S_0$	10.7485
87	Fr	Francium	[Rn] $7s$	$^2S_{1/2}$	4.0727
88	Ra	Radium	[Rn] $7s^2$	$^1S_0$	5.2784
89	Ac	Actinium	[Rn] $6d\ 7s^2$	$^2D_{3/2}$	5.17
90	Th	Thorium	[Rn] $6d^2\ 7s^2$	$^3F_2$	6.3067
91	Pa	Protactinium	[Rn] $5f^2(^3H_4)\ 6d\ 7s^2$	$(4,3/2)_{11/2}$	5.89
92	U	Uranium	[Rn] $5f^3(^4I^o_{9/2})\ 6d\ 7s^2$	$(9/2,3/2)^o_6$	6.1941
93	Np	Neptunium	[Rn] $5f^4(^5I_4)\ 6d\ 7s^2$	$(4,3/2)_{11/2}$	6.2657
94	Pu	Plutonium	[Rn] $5f^6\ 7s^2$	$^7F_0$	6.0260
95	Am	Americium	[Rn] $5f^7\ 7s^2$	$^8S^o_{7/2}$	5.9738
96	Cm	Curium	[Rn] $5f^7\ 6d\ 7s^2$	$^9D^o_2$	5.9914
97	Bk	Berkelium	[Rn] $5f^9\ 7s^2$	$^6H^o_{15/2}$	6.1979
98	Cf	Californium	[Rn] $5f^{10}\ 7s^2$	$^5I_8$	6.2817
99	Es	Einsteinium	[Rn] $5f^{11}\ 7s^2$	$^4I^o_{15/2}$	6.42
100	Fm	Fermium	[Rn] $5f^{12}\ 7s^2$	$^3H_6$	6.50
101	Md	Mendelevium	[Rn] $5f^{13}\ 7s^2$	$^2F^o_{7/2}$	6.58
102	No	Nobelium	[Rn] $5f^{14}\ 7s^2$	$^1S_0$	6.65
103	Lr	Lawrencium	[Rn] $5f^{14}\ 7s^2\ 7p$?	$^2P^o_{1/2}$?	4.9?
104	Rf	Rutherfordium	[Rn] $5f^{14}\ 6d^2\ 7s^2$?	$^3F_2$?	6.0?

INTERNATIONAL TEMPERATURE SCALE OF 1990 (ITS-90)

B. W. Mangum

A new temperature scale, the International Temperature Scale of 1990 (ITS-90), was officially adopted by the Comité International des Poids et Mesures (CIPM), meeting 26—28 September 1989 at the Bureau International des Poids et Mesures (BIPM). The ITS-90 was recommended to the CIPM for its adoption following the completion of the final details of the new scale by the Comité Consultatif de Thermométrie (CCT), meeting 12—14 September 1989 at the BIPM in its 17th Session. The ITS-90 became the official international temperature scale on 1 January 1990. The ITS-90 supersedes the present scales, the International Practical Temperature Scale of 1968 (IPTS-68) and the 1976 Provisional 0.5 to 30 K Temperature Scale (EPT-76).

The ITS-90 extends upward from 0.65 K, and temperatures on this scale are in much better agreement with thermodynamic values that are those on the IPTS-68 and the EPT-76. The new scale has subranges and alternative definitions in certain ranges that greatly facilitate its use. Furthermore, its continuity, precision, and reproducibility throughout its ranges are much improved over that of the present scales. The replacement of the thermocouple with the platinum resistance thermometer at temperatures below 961.78 °C resulted in the biggest improvement in reproducibility.

The ITS-90 is divided into four primary ranges:

1. Between 0.65 and 3.2 K, the ITS-90 is defined by the vapor pressure-temperature relation of ^{3}He, and between 1.25 and 2.1768 K (the λ point) and between 2.1768 and 5.0 K by the vapor pressure–temperature relations of ^{4}He. T_{90} is defined by the vapor pressure equations of the form:

$$T_{90}/\mathrm{K} = A_0 + \sum_{i=1}^{9} A_i\left[(\ln(p/\mathrm{Pa}) - B)/C\right]^i$$

The values of the coefficients A_i, and of the constants A_o, B, and C of the equations are given below.

2. Between 3.0 and 24.5561 K, the ITS-90 is defined in terms of a ^{3}He or ^{4}He constant volume gas thermometer (CVGT). The thermometer is calibrated at three temperatures — at the triple point of neon (24.5561 K), at the triple point of equilibrium hydrogen (13.8033 K), and at a temperature between 3.0 and 5.0 K, the value of which is determined by using either ^{3}He or ^{4}He vapor pressure thermometry.
3. Between 13.8033 K (−259.3467 °C) and 1234.93 K (961.78 °C), the ITS-90 is defined in terms of the specified fixed points given below, by resistance ratios of platinum resistance thermometers obtained by calibration at specified sets of the fixed points, and by reference functions and deviation functions of resistance ratios which relate to T_{90} between the fixed points.
4. Above 1234.93 K, the ITS-90 is defined in terms of Planck's radiation law, using the freezing-point temperature of either silver, gold, or copper as the reference temperature.

Full details of the calibration procedures and reference functions for various subranges are given in:

The International Temperature Scale of 1990, *Metrologia,* 27, 3, 1990; errata in *Metrologia,* 27, 107, 1990.

Defining Fixed Points of the ITS-90

Material[a]	Equilibrium state[b]	Temperature	
		T_{90} (K)	t_{90} (°C)
He	VP	3 to 5	−270.15 to −268.15
e-H_2	TP	13.8033	−259.3467
e-H_2 (or He)	VP (or CVGT)	≈17	≈ −256.15
e-H_2 (or He)	VP (or CVGT)	≈20.3	≈ −252.85
Ne[c]	TP	24.5561	−248.5939
O_2	TP	54.3584	−218.7916
Ar	TP	83.8058	−189.3442
Hg[c]	TP	234.3156	−38.8344
H_2O	TP	273.16	0.01
Ga[c]	MP	302.9146	29.7646
In[c]	FP	429.7485	156.5985
Sn	FP	505.078	231.928
Zn	FP	692.677	419.527
Al[c]	FP	933.473	660.323
Ag	FP	1234.93	961.78
Au	FP	1337.33	1064.18
Cu[c]	FP	1357.77	1084.62

[a] e-H_2 indicates equilibrium hydrogen, that is, hydrogen with the equilibrium distribution of its ortho and para states. Normal hydrogen at room temperature contains 25% para hydrogen and 75% ortho hydrogen.

[b] VP indicates vapor pressure point; CVGT indicates constant volume gas thermometer point; TP indicates triple point (equilibrium temperature at which the solid, liquid, and vapor phases coexist); FP indicates freezing point, and MP indicates melting point (the equilibrium temperatures at which the solid and liquid phases coexist under a pressure of 101 325 Pa, one standard atmosphere). The isotopic composition is that naturally occurring.

[c] Previously, these were secondary fixed points.

Values of Coefficients in the Vapor Pressure Equations for Helium

Coef. or constant	^{3}He 0.65—3.2 K	^{4}He 1.25—2.1768 K	^{4}He 2.1768—5.0 K
A_0	1.053 447	1.392 408	3.146 631
A_1	0.980 106	0.527 153	1.357 655
A_2	0.676 380	0.166 756	0.413 923
A_3	0.372 692	0.050 988	0.091 159
A_4	0.151 656	0.026 514	0.016 349
A_5	−0.002 263	0.001 975	0.001 826
A_6	0.006 596	−0.017 976	−0.004 325
A_7	0.088 966	0.005 409	−0.004 973
A_8	−0.004 770	0.013 259	0
A_9	−0.054 943	0	0
B	7.3	5.6	10.3
C	4.3	2.9	1.9

CONVERSION OF TEMPERATURES FROM THE 1948 AND 1968 SCALES TO ITS-90

This table gives temperature corrections from older scales to the current International Temperature Scale of 1990 (see the preceding table for details on ITS-90). The first part of the table may be used for converting Celsius temperatures in the range −180 to 4000 °C from IPTS-68 or IPTS-48 to ITS-90. Within the accuracy of the corrections, the temperature in the first column may be identified with either t_{68}, t_{48}, or t_{90}. The second part of the table is designed for use at lower temperatures to convert values expressed in kelvins from EPT-76 or IPTS-68 to ITS-90.

The references give analytical equations for expressing these relations. Note that Reference 1 supersedes Reference 2 with respect to corrections in the 630 to 1064 °C range.

References

1. Burns, G. W. et al., in *Temperature: Its Measurement and Control in Science and Industry*, Vol. 6, Schooley, J. F., Ed., American Institute of Physics, New York, 1993.
2. Goldberg, R. N. and Weir, R. D., *Pure and Appl. Chem.*, 64, 1545, 1992.

t/°C	$t_{90}-t_{68}$	$t_{90}-t_{48}$
−180	0.008	0.020
−170	0.010	0.017
−160	0.012	0.007
−150	0.013	0.000
−140	0.014	0.001
−130	0.014	0.008
−120	0.014	0.017
−110	0.013	0.026
−100	0.013	0.035
−90	0.012	0.041
−80	0.012	0.045
−70	0.011	0.045
−60	0.010	0.042
−50	0.009	0.038
−40	0.008	0.032
−30	0.006	0.024
−20	0.004	0.016
−10	0.002	0.008
0	0.000	0.000
10	−0.002	−0.006
20	−0.005	−0.012
30	−0.007	−0.016
40	−0.010	−0.020
50	−0.013	−0.023
60	−0.016	−0.026
70	−0.018	−0.026
80	−0.021	−0.027
90	−0.024	−0.027
100	−0.026	−0.026
110	−0.028	−0.024
120	−0.030	−0.023
130	−0.032	−0.020
140	−0.034	−0.018
150	−0.036	−0.016
160	−0.037	−0.012
170	−0.038	−0.009
180	−0.039	−0.005
190	−0.039	−0.001
200	−0.040	0.003
210	−0.040	0.007
220	−0.040	0.011
230	−0.040	0.014
240	−0.040	0.018
250	−0.040	0.021
260	−0.040	0.024
270	−0.039	0.028
280	−0.039	0.030

t/°C	$t_{90}-t_{68}$	$t_{90}-t_{48}$
290	−0.039	0.032
300	−0.039	0.034
310	−0.039	0.035
320	−0.039	0.036
330	−0.040	0.036
340	−0.040	0.037
350	−0.041	0.036
360	−0.042	0.035
370	−0.043	0.034
380	−0.045	0.032
390	−0.046	0.030
400	−0.048	0.028
410	−0.051	0.024
420	−0.053	0.022
430	−0.056	0.019
440	−0.059	0.015
450	−0.062	0.012
460	−0.065	0.009
470	−0.068	0.007
480	−0.072	0.004
490	−0.075	0.002
500	−0.079	0.000
510	−0.083	−0.001
520	−0.087	−0.002
530	−0.090	−0.001
540	−0.094	0.000
550	−0.098	0.002
560	−0.101	0.007
570	−0.105	0.011
580	−0.108	0.018
590	−0.112	0.025
600	−0.115	0.035
610	−0.118	0.047
620	−0.122	0.060
630	−0.125	0.075
640	−0.11	0.12
650	−0.10	0.15
660	−0.09	0.19
670	−0.07	0.24
680	−0.05	0.29
690	−0.04	0.32
700	−0.02	0.37
710	−0.01	0.41
720	0.00	0.45
730	0.02	0.49
740	0.03	0.53
750	0.03	0.56

t/°C	$t_{90}-t_{68}$	$t_{90}-t_{48}$
760	0.04	0.60
770	0.05	0.63
780	0.05	0.66
790	0.05	0.69
800	0.05	0.72
810	0.05	0.75
820	0.04	0.76
830	0.04	0.79
840	0.03	0.81
850	0.02	0.83
860	0.01	0.85
870	0.00	0.87
880	−0.02	0.87
890	−0.03	0.89
900	−0.05	0.90
910	−0.06	0.92
920	−0.08	0.93
930	−0.10	0.94
940	−0.11	0.96
950	−0.13	0.97
960	−0.15	0.97
970	−0.16	0.99
980	−0.18	1.00
990	−0.19	1.02
1000	−0.20	1.04
1010	−0.22	1.05
1020	−0.23	1.07
1030	−0.23	1.10
1040	−0.24	1.12
1050	−0.25	1.14
1060	−0.25	1.17
1070	−0.25	1.19
1080	−0.26	1.20
1090	−0.26	1.20
1100	−0.26	1.2
1200	−0.30	1.4
1300	−0.35	1.5
1400	−0.39	1.6
1500	−0.44	1.8
1600	−0.49	1.9
1700	−0.54	2.1
1800	−0.60	2.2
1900	−0.66	2.3
2000	−0.72	2.5
2100	−0.79	2.7
2200	−0.85	2.9
2300	−0.93	3.1

t/°C	$t_{90}-t_{68}$	$t_{90}-t_{48}$
2400	−1.00	3.2
2500	−1.07	3.4
2600	−1.15	3.7
2700	−1.24	3.8
2800	−1.32	4.0
2900	−1.41	4.2
3000	−1.50	4.4
3100	−1.59	4.6
3200	−1.69	4.8
3300	−1.78	5.1
3400	−1.89	5.3
3500	−1.99	5.5
3600	−2.10	5.8
3700	−2.21	6.0
3800	−2.32	6.3
3900	−2.43	6.6
4000	−2.55	6.8

T/K	$T_{90}-T_{76}$	$T_{90}-T_{68}$
5	−0.0001	
6	−0.0002	
7	−0.0003	
8	−0.0004	
9	−0.0005	
10	−0.0006	
11	−0.0007	
12	−0.0008	
13	−0.0010	
14	−0.0011	−0.006
15	−0.0013	−0.003
16	−0.0014	−0.004
17	−0.0016	−0.006
18	−0.0018	−0.008
19	−0.0020	−0.009
20	−0.0022	−0.009
21	−0.0025	−0.008
22	−0.0027	−0.007
23	−0.0030	−0.007
24	−0.0032	−0.006
25	−0.0035	−0.005
26	−0.0038	−0.004
27	−0.0041	−0.004
28		−0.005
29		−0.006
30		−0.006
31		−0.007
32		−0.008

T/K	$T_{90}-T_{76}$	$T_{90}-T_{68}$	T/K	$T_{90}-T_{76}$	$T_{90}-T_{68}$	T/K	$T_{90}-T_{76}$	$T_{90}-T_{68}$	T/K	$T_{90}-T_{76}$	$T_{90}-T_{68}$
33		−0.008	57		0.000	81		0.008	150		0.014
34		−0.008	58		0.001	82		0.008	160		0.014
35		−0.007	59		0.002	83		0.008	170		0.013
36		−0.007	60		0.003	84		0.008	180		0.012
37		−0.007	61		0.003	85		0.008	190		0.012
38		−0.006	62		0.004	86		0.008	200		0.011
39		−0.006	63		0.004	87		0.008	210		0.010
40		−0.006	64		0.005	88		0.008	220		0.009
41		−0.006	65		0.005	89		0.008	230		0.008
42		−0.006	66		0.006	90		0.008	240		0.007
43		−0.006	67		0.006	91		0.008	250		0.005
44		−0.006	68		0.007	92		0.008	260		0.003
45		−0.007	69		0.007	93		0.008	270		0.001
46		−0.007	70		0.007	94		0.008	273.16		0.000
47		−0.007	71		0.007	95		0.008	300		−0.006
48		−0.006	72		0.007	96		0.008	400		−0.031
49		−0.006	73		0.007	97		0.009	500		−0.040
50		−0.006	74		0.007	98		0.009	600		−0.040
51		−0.005	75		0.008	99		0.009	700		−0.055
52		−0.005	76		0.008	100		0.009	800		−0.089
53		−0.004	77		0.008	110		0.011	900		−0.124
54		−0.003	78		0.008	120		0.013			
55		−0.002	79		0.008	130		0.014			
56		−0.001	80		0.008	140		0.014			

INTERNATIONAL SYSTEM OF UNITS (SI)

The International System of Units, abbreviated as SI (from the French name *Le Système International d'Unités*), was established in 1960 by the 11th General Conference on Weights and Measures (CGPM) as the modern metric system of measurement. The core of the SI is the seven base units for the physical quantities length, mass, time, electric current, thermodynamic temperature, amount of substance, and luminous intensity. These base units are:

Base quantity	SI base unit	
	Name	Symbol
length	meter	m
mass	kilogram	kg
time	second	s
electric current	ampere	A
thermodynamic temperature	kelvin	K
amount of substance	mole	mol
luminous intensity	candela	cd

The SI base units are defined as follows:

ampere: The ampere is that constant current which, if maintained in two straight parallel conductors of infinite length, of negligible circular cross-section, and placed 1 meter apart in vacuum, would produce between these conductors a force equal to $2 \cdot 10^{-7}$ newton per meter of length.

candela: The candela is the luminous intensity, in a given direction, of a source that emits monochromatic radiation of frequency $540 \cdot 10^{12}$ hertz and that has a radiant intensity in that direction of 1/683 watt per steradian.

kelvin: The kelvin, unit of thermodynamic temperature, is the fraction 1/273.16 of the thermodynamic temperature of the triple point of water.

kilogram: The kilogram is the unit of mass; it is equal to the mass of the international prototype of the kilogram.

meter: The meter is the length of the path travelled by light in vacuum during a time interval of 1/299 792 458 of a second.

mole: The mole is the amount of substance of a system which contains as many elementary entities as there are atoms in 0.012 kilogram of carbon 12. When the mole is used, the elementary entities must be specified and may be atoms, molecules, ions, electrons, other particles, or specified groups of such particles.

second: The second is the duration of 9 192 631 770 periods of the radiation corresponding to the transition between the two hyperfine levels of the ground state of the cesium 133 atom.

SI derived units

Derived units are units which may be expressed in terms of base units by means of the mathematical symbols of multiplication and division (and, in the case of °C, subtraction). Certain derived units have been given special names and symbols, and these special names and symbols may themselves be used in combination with those for base and other derived units to express the units of other quantities. The next table lists some examples of derived units expressed directly in terms of base units:

Physical quantity	SI derived unit	
	Name	Symbol
area	square meter	m^2
volume	cubic meter	m^3
speed, velocity	meter per second	m/s
acceleration	meter per second squared	m/s^2
wave number	reciprocal meter	m^{-1}
density, mass density	kilogram per cubic meter	kg/m^3
specific volume	cubic meter per kilogram	m^3/kg
current density	ampere per square meter	A/m^2
magnetic field strength	ampere per meter	A/m
concentration (of amount of substance)	mole per cubic meter	mol/m^3
luminance	candela per square meter	cd/m^2
refractive index	(the number) one	1[a]

[a] The symbol "1" is generally omitted in combination with a numerical value.

For convenience, certain derived units, which are listed in the next table, have been given special names and symbols. These names and symbols may themselves be used to express other derived units. The special names and symbols are a compact form for the expression of units that are used frequently. The final column shows how the SI units concerned may be expressed in terms of SI base units. In this column, factors such as m^0, kg^0 ..., which are all equal to 1, are not shown explicitly.

Physical quantity	Name	Symbol	SI derived unit expressed in terms of: Other SI units	
			Other SI units	SI base units
plane angle	radian[a]	rad	$m \cdot m^{-1} = 1$[b]	
solid angle	steradian[a]	sr[c]	$m^2 \cdot m^{-2} = 1$[b]	
frequency	hertz	Hz	s^{-1}	
force	newton	N	$m \cdot kg \cdot s^{-2}$	
pressure, stress	pascal	Pa	N/m^2	$m^{-1} \cdot kg \cdot s^{-2}$
energy, work, quantity of heat	joule	J	$N \cdot m$	$m^2 \cdot kg \cdot s^{-2}$
power, radiant flux	watt	W	J/s	$m^2 \cdot kg \cdot s^{-3}$
electric charge, quantity of electricity	coulomb	C	$s \cdot A$	
electric potential difference, electromotive force	volt	V	W/A	$m^2 \cdot kg \cdot s^{-3} \cdot A^{-1}$
capacitance	farad	F	C/V	$m^{-2} \cdot kg^{-1} \cdot s^4 \cdot A^2$
electric resistance	ohm	Ω	V/A	$m^2 \cdot kg \cdot s^{-3} \cdot A^{-2}$
electric conductance	siemens	S	A/V	$m^{-2} \cdot kg^{-1} \cdot s^3 \cdot A^2$
magnetic flux	weber	Wb	$V \cdot s$	$m^2 \cdot kg \cdot s^{-2} \cdot A^{-1}$

Physical quantity	Name	Symbol	SI derived unit expressed in terms of: Other SI units	SI base units
magnetic flux density	tesla	T	Wb/m^2	kg · s^{-2} · A^{-1}
inductance	henry	H	Wb/A	m^2 · kg · s^{-2} · A^{-2}
Celsius temperature	degree Celsius[d]	°C		K
luminous flux	lumen	lm	cd · sr[c]	m^2 · m^{-2} · cd = cd
illuminance	lux	lx	lm/m^2	m^2 · m^{-4} · cd = m^{-2} · cd
activity (of a radionuclide)	becquerel	Bq		s^{-1}
absorbed dose, specific energy (imparted), kerma	gray	Gy	J/kg	m^2 · s^{-2}
dose equivalent, ambient dose equivalent, directional dose equivalent, personal dose equivalent, organ equivalent dose	sievert	Sv	J/kg	m^2 · s^{-2}
catalytic activity	katal	kat		s^{-1} · mol

[a] The radian and steradian may be used with advantage in expressions for derived units to distinguish between quantities of different nature but the same dimension. Some examples of their use in forming derived units are given in the next table.

[b] In practice, the symbols rad and sr are used where appropriate, but the derived unit "1" is generally omitted in combination with a numerical value.

[c] In photometry, the name steradian and the symbol sr are usually retained in expressions for units.

[d] It is common practice to express a thermodynamic temperature, symbol T, in terms of its difference from the reference temperature T_0 = 273.15 K. The numerical value of a Celsius temperature t expressed in degrees Celsius is given by t/°C = T/K-273.15. The unit °C may be used in combination with SI prefixes, e.g., millidegree Celsius, m°C. Note that there should never be a space between the ° sign and the letter C, and that the symbol for kelvin is K, not °K.

The SI derived units with special names may be used in combinations to provide a convenient way to express more complex physical quantities. Examples are given in the next table:

Physical Quantity	SI derived unit: Name	Symbol	As SI base units
dynamic viscosity	pascal second	Pa · s	m^{-1} · kg · s^{-1}
moment of force	newton meter	N · m	m^2 · kg · s^{-2}
surface tension	newton per meter	N/m	kg · s^{-2}
angular velocity	radian per second	rad/s	m · m^{-1} · s^{-1} = s^{-1}
angular acceleration	radian per second squared	rad/s^2	m · m^{-1} · s^{-2} = s^{-2}
heat flux density, irradiance	watt per square meter	W/m^2	kg · s^{-3}
heat capacity, entropy	joule per kelvin	J/K	m^{-3} · kg · s^{-2} · K^{-1}
specific heat capacity, specific entropy	joule per kilogram kelvin	J/(kg · K)	m^2 · s^{-2} · K^{-1}
specific energy	joule per kilogram	J/kg	m^2 · s^{-2}
thermal conductivity	watt per meter kelvin	W/(m · K)	m · kg · s^{-3} · K^{-1}
energy density	joule per cubic meter	J/m^3	m^{-1} · kg · s^{-2}
electric field strength	volt per meter	V/m	m · kg · s^{-3} · A^{-1}
electric charge density	coulomb per cubic meter	C/m^3	m^{-3} · s · A
electric flux density	coulomb per square meter	C/m^2	m^{-2} · s · A
permittivity	farad per meter	F/m	m^{-3} · kg^{-1} · s^4 · A^2
permeability	henry per meter	H/m	m · kg · s^{-2} · A^{-2}
molar energy	joule per mole	J/mol	m^2 · kg · s^{-2} · mol^{-1}
molar entropy, molar heat capacity	joule per mole kelvin	J/(mol · K)	m^2 · kg · s^{-2} · K^{-1} · mol^{-1}
exposure (x and γ rays)	coulomb per kilogram	C/kg	kg^{-1} · s · A
absorbed dose rate	gray per second	Gy/s	m^2 · s^{-3}
radiant intensity	watt per steradian	W/sr	m^4 · m^{-2} · kg · s^{-3} = m^2 · kg · s^{-3}
radiance	watt per square meter steradian	W/(m^2 · sr)	m^2 · m^{-2} · kg · s^{-3} = kg · s^{-3}
catalytic (activity) concentration	katal per cubic meter	kat/m^3	m^{-3} · s^{-1} · mol

In practice, with certain quantities preference is given to the use of certain special unit names, or combinations of unit names, in order to facilitate the distinction between different quantities having the same dimension. For example, the SI unit of frequency is designated the hertz, rather than the reciprocal second, and the SI unit of angular velocity is designated the radian per second rather than the reciprocal second (in this case retaining the word radian emphasizes that angular velocity is equal to 2π times the rotational frequency). Similarly the SI unit of moment of force is designated the newton meter rather than the joule.

In the field of ionizing radiation, the SI unit of activity is designated the becquerel rather than the reciprocal second, and the SI units of absorbed dose and dose equivalent the gray and sievert, respectively, rather than the joule per kilogram. In the field of catalysis, the SI unit of catalytic activity is designated the katal rather than the mole per second. The special names becquerel, gray, sievert, and katal were specifically introduced because of the dangers to human health which might arise from mistakes involving the units reciprocal second, joule per kilogram and mole per second.

Units for dimensionless quantities, quantities of dimension one

Certain quantities are defined as the ratios of two quantities of the same kind, and thus have a dimension which may be expressed by the number one. The unit of such quantities is necessarily a derived unit coherent with the other units of the SI and, since it is formed as the ratio of two identical SI units, the unit also may be expressed by the number one. Thus the SI unit of all quantities having the dimensional product one is the number one. Examples of such quantities are refractive index, relative permeability, and friction factor. Other quantities having the unit 1 include "characteristic numbers" like the Prandtl number and numbers which represent a count, such as a number of molecules, degeneracy (number of energy levels), and partition function in statistical thermodynamics. All of these quantities are described as being dimensionless, or of dimension one, and have the coherent SI unit 1. Their values are simply expressed as numbers and, in general, the unit 1 is not explicitly shown. In a few cases, however, a special name is given to this unit, mainly to avoid confusion between some compound derived units. This is the case for the radian, steradian and neper.

SI prefixes

The following prefixes have been approved by the CGPM for use with SI units. Only one prefix may be used before a unit. Thus 10^{-12} farad should be designated pF, not μμF.

Factor	Name	Symbol	Factor	Name	Symbol
10^{24}	yotta	Y	10^{-1}	deci	d
10^{21}	zetta	Z	10^{-2}	centi	c
10^{18}	exa	E	10^{-3}	milli	m
10^{15}	peta	P	10^{-6}	micro	μ
10^{12}	tera	T	10^{-9}	nano	n
10^{9}	giga	G	10^{-12}	pico	p
10^{6}	mega	M	10^{-15}	femto	f
10^{3}	kilo	k	10^{-18}	atto	a
10^{2}	hecto	h	10^{-21}	zepto	z
10^{1}	deka	da	10^{-24}	yocto	y

The kilogram

Among the base units of the International System, the unit of mass is the only one whose name, for historical reasons, contains a prefix. Names and symbols for decimal multiples and submultiples of the unit of mass are formed by attaching prefix names to the unit name "gram" and prefix symbols to the unit symbol "g".

Example : 10^{-6} kg = 1 mg (1 milligram) *but not* 1 μkg (1 microkilogram).

Units used with the SI

Many units that are not part of the SI are important and widely used in everyday life. The CGPM has adopted a classification of non-SI units: (1) units accepted for use with the SI (such as the traditional units of time and of angle); (2) units accepted for use with the SI whose values are obtained experimentally; and (3) other units currently accepted for use with the SI to satisfy the needs of special interests.

(1) Non-SI units accepted for use with the International System

Name	Symbol	Value in SI units
minute	min	1 min = 60 s
hour	h	1 h= 60 min = 3600 s
day	d	1 d = 24 h = 86 400 s
degree	°	$1° = (\pi/180)$ rad
minute	′	$1' = (1/60)° = (\pi/10\,800)$ rad
second	″	$1'' = (1/60)' = (\pi/648\,000)$ rad
liter	l, L	$1\text{L} = 1\ \text{dm}^3 = 10^{-3}\ \text{m}^3$
metric ton	t	$1\ \text{t} = 10^3$ kg
neper[a]	Np	1 Np = 1
bel[b]	B	1 B = (1/2) ln 10 Np

[a] The neper is used to express values of such logarithmic quantities as field level, power level, sound pressure level, and logarithmic decrement. Natural logarithms are used to obtain the numerical values of quantities expressed in nepers. The neper is coherent with the SI, but is not yet adopted by the CGPM as an SI unit. In using the neper, it is important to specify the quantity.

[b] The bel is used to express values of such logarithmic quantities as field level, power level, sound-pressure level, and attenuation. Logarithms to base ten are used to obtain the numerical values of quantities expressed in bels. The submultiple decibel, dB, is commonly used.

(2) Non-SI units accepted for use with the International system, whose values in SI units are obtained experimentally

Name	Symbol	Value in SI Units
electronvolt[b]	eV	$1\ \text{eV} = 1.602\ 176\ 53(14) \cdot 10^{-19}$ J[a]
dalton[c]	Da	$1\ \text{Da} = 1.660\ 538\ 86(28) \cdot 10^{-27}$ kg[a]
unified atomic mass unit[c]	u	1 u = 1 Da
astronomical unit[d]	ua	$1\ \text{ua} = 1.495\ 978\ 706\ 91(06) \cdot 10^{11}$ m[a]

[a] For the electronvolt and the dalton (unified atomic mass unit), values are quoted from the 2002 CODATA set of the Fundamental Physical Constants (p. 1-1 of this Handbook). The value given for the astronomical unit is quoted from the IERS Conventions 2003 (D.D. McCarthy and G. Petit, eds., IERS Technical Note 32, Frankfurt am Main: Verlag des Bundesamts für Kartographie und Geodäsie, 200). The value of ua in meters comes from the JPL ephemerides DE403 (Standish E.M. 1995, "Report of the IAU WGAS Sub-Group on Numerical Standards", in "Highlights of Astronomy", Appenlzer ed., pp 180-184, Kluwer Academic Publishers, Dordrecht). It has been determined in "TDB" units using Barycentric Dynamical Time TDB as a time coordinate for the barycentric system.

[b] The electronvolt is the kinetic energy acquired by an electron in passing through a potential difference of 1 V in vacuum.

[c] The Dalton and unified atomic mass unit are alternative names for the same unit, equal to 1/12 of the mass of an unbound atom of the nuclide ^{12}C, at rest and in its ground state. The dalton may be combined with SI prefixes to express the masses of large molecules in kilodalton, kDa, or megadalton, MDa.

[d] The astronomical unit is a unit of length approximately equal to the mean Earth-Sun distance. It is the radius of an unperturbed circular Newtonian orbit about the Sun of a particle having infinitesimal mass, moving with a mean motion of 0.017 202 098 95 radians/day (known as the Gaussian constant).

(3) Other non-SI units currently accepted for use with the International System

Name	Symbol	Value in SI Units
nautical mile		1 nautical mile = 1852 m
knot		1 nautical mile per hour = (1852/3600) m/s
are		$1\ \text{a} = 1\ \text{dam}^2 = 10^2\ \text{m}^2$
hectare	ha	$1\ \text{ha} = 1\ \text{hm}^2 = 10^4\ \text{m}^2$
bar	bar	$1\ \text{bar} = 0.1\ \text{MPa} = 100\ \text{kPa} = 10^5\ \text{Pa}$
ångström	Å	$1\ \text{Å} = 0.1\ \text{nm} = 10^{-10}\ \text{m}$
barn	b	$1\ \text{b} = 100\ \text{fm}^2 = 10^{-28}\ \text{m}^2$

Other non-SI units

The SI does not encourage the use of cgs units, but these are frequently found in old scientific texts. The following table gives the relation of some common cgs units to SI units.

Name	Symbol	Value in SI units
erg	erg	$1\ \text{erg} = 10^{-7}$ J
dyne	dyn	$1\ \text{dyn} = 10^{-5}$ N
poise	P	$1\text{P} = 1\text{dyn} \cdot \text{s/cm}^2 = 0.1\ \text{Pa} \cdot \text{s}$
stokes	St	$1\ \text{St} = 1\ \text{cm}^2/\text{s} = 10^{-4}\ \text{m}^2/\text{s}$
gauss	G	$1\text{G} \triangleq 10^{-4}$ T
oersted	Oe	$1\ \text{Oe} \triangleq (1000/4\pi)$ A/m
maxwell	Mx	$1\text{Mx} \triangleq 10^{-8}$ Wb
stilb	sb	$1\ \text{sb} = 1\ \text{cd/cm}^2 = 10^4\ \text{cd/m}^2$
phot	ph	$1\ \text{ph} = 10^4$ lx
gal	Gal	$1\ \text{Gal} = 1\ \text{cm/s}^2 = 10^{-2}\ \text{m/s}^2$

Note: The symbol ≙ should be read as "corresponds to"; these units cannot strictly be equated because of the different dimensions of the electromagnetic cgs and the SI.

Examples of other non-SI units found in the older literature and their relation to the SI are given below. Use of these units in current texts is discouraged.

Name	Symbol	Value in SI units
curie	Ci	1 Ci = $3.7 \cdot 10^{10}$ Bq
roentgen	R	1 R = $2.58 \cdot 10^{-4}$ C/kg
rad	rad	1 rad = 1 cGy = 10^{-2} Gy
rem	rem	1 rem = 1 cSv = 10^{-2} Sv
X unit		1 X unit ≈ $1.002 \cdot 10^{-4}$ nm
gamma	γ	1 γ = 1 nT = 10^{-9} T
jansky	Jy	1 Jy = 10^{-26} W · m^{-2} · Hz^{-1}
fermi		1 fermi = 1 fm = 10^{-15} m
metric carat		1 metric carat = 200 mg = $2 \cdot 10^{-4}$ kg
torr	Torr	1 Torr = (101325/760) Pa
standard atmosphere	atm	1 atm = 101325 Pa
calorie[(a)]	cal	1 cal = 4.184 J
micron	µ	1 µ = 1 µm = 10^{-6} m

[(a)] Several types of calorie have been used; the value given here is the so-called "thermochemical calorie".

Prefixes for binary multiples

In December 1998 the International Electrotechnical Commission (IEC), the leading international organization for worldwide standardization in electrotechnology, approved as an IEC International Standard names and symbols for prefixes for binary multiples for use in the fields of data processing and data transmission. The prefixes are as follows:

Prefixes for binary multiples

Factor	Name	Symbol	Origin	Derivation
2^{10}	kibi	Ki	kilobinary: $(2^{10})^1$	kilo: $(10^3)^1$
2^{20}	mebi	Mi	megabinary: $(2^{10})^2$	mega: $(10^3)^2$
2^{30}	gibi	Gi	gigabinary: $(2^{10})^3$	giga: $(10^3)^3$
2^{40}	tebi	Ti	terabinary: $(2^{10})^4$	tera: $(10^3)^4$
2^{50}	pebi	Pi	petabinary: $(2^{10})^5$	peta: $(10^3)^5$
2^{60}	exbi	Ei	exabinary: $(2^{10})^6$	exa: $(10^3)^6$

Examples and comparisons with SI prefixes

one **kibibit**	1 Kibit = 2^{10} bit = **1024 bit**
one **kilobit**	1 kbit = 10^3 bit = **1000 bit**
one **mebibyte**	1 MiB = 2^{20} B = **1 048 576 B**
one **megabyte**	1 MB = 10^6 B = **1 000 000 B**
one **gibibyte**	1 GiB = 2^{30} B = **1 073 741 824 B**
one **gigabyte**	1 GB = 10^9 B = **1 000 000 000 B**

It is suggested that in English, the first syllable of the name of the binary-multiple prefix should be pronounced in the same way as the first syllable of the name of the corresponding SI prefix, and that the second syllable should be pronounced as "bee."

It is important to recognize that the new prefixes for binary multiples are not part of the International System of Units (SI), the modern metric system. However, for ease of understanding and recall, they were derived from the SI prefixes for positive powers of ten. As can be seen from the above table, the name of each new prefix is derived from the name of the corresponding SI prefix by retaining the first two letters of the name of the SI prefix and adding the letters "bi," which recalls the word "binary." Similarly, the symbol of each new prefix is derived from the symbol of the corresponding SI prefix by adding the letter "i," which again recalls the word "binary." (For consistency with the other prefixes for binary multiples, the symbol Ki is used for 2^{10} rather than ki.)

References

1. Taylor, B. N., and Thompson, A., *The International System of Unit (SI)*, NIST Special Publication 330, National Institute of Standards and Technology, Gaithersburg, MD, 2008.
2. Bureau International des Poids et Mesures, *Le Système International d'Unités (SI)*, 8th French and English Edition, BIPM, Sèvres, France, 2006.
3. Thompson, A., and Taylor, B. N., *Guide for the Use of the International System of Unit (SI)*, NIST Special Publication 811, National Institute of Standards and Technology, Gaithersburg, MD, 2008.
4. NIST Physical Reference Data web site, http://physics.nist.gov/cuu/Units/index.html, October 2004.
5. Amendment 2 to IEC International Standard IEC 60027-2, 1999-01, Letter symbols to be used in electrical technology – Part 2: Telecommunications and electronics.
6. IEC 60027-2, Second edition, 2000-11, Letter symbols to be used in electrical technology - Part 2: Telecommunications and electronics.
7. Barrow, B., "A Lesson in Megabytes," *IEEE Stand. Bearer*, January 1997, p. 5.

UNITS FOR MAGNETIC PROPERTIES

Quantity	Symbol	Gaussian & cgs emu [a]	Conversion factor, C [b]	SI & rationalized mks [c]
Magnetic flux density, magnetic induction	B	gauss (G) [d]	10^{-4}	tesla (T), Wb/m^2
Magnetic flux	Φ	maxwell (Mx), G · cm^2	10^{-8}	weber (Wb), volt second (V · s)
Magnetic potential difference, magnetomotive force	U, F	gilbert (Gb)	$10/4\pi$	ampere (A)
Magnetic field strength, magnetizing force	H	oersted (Oe),[e] Gb/cm	$10^3/4\pi$	A/m[f]
(Volume) magnetization [g]	M	emu/cm^3 [h]	10^3	A/m
(Volume) magnetization	$4\pi M$	G	$10^3/4\pi$	A/m
Magnetic polarization, intensity of magnetization	J, I	emu/cm^3	$4\pi \times 10^{-4}$	T, Wb/m^2 [i]
(Mass) magnetization	σ, M	emu/g	1	A·m^2/kg
			$4\pi \times 10^{-7}$	Wb·m/kg
Magnetic moment	m	emu, erg/G	10^{-3}	A·m^2, joule per tesla (J/T)
Magnetic dipole moment	j	emu, erg/G	$4\pi \times 10^{-10}$	Wb·m[i]
(Volume) susceptibility	χ, κ	dimensionless, emu/cm^3	4π	dimensionless
			$(4\pi)^2 \times 10^{-7}$	henry per meter (H/m), Wb/(A·m)
(Mass) susceptibility	χ_ρ, κ_ρ	cm^3/g, emu/g	$4\pi \times 10^{-3}$	m^3/kg
			$(4\pi)^2 \times 10^{-10}$	H·m^2/kg
(Molar) susceptibility	χ_{mol}, κ_{mol}	cm^3/mol, emu/mol	$4\pi \times 10^{-6}$	m^3/mol
			$(4\pi)^2 \times 10^{-13}$	H·m^2/mol
Permeability	μ	dimensionless	$4\pi \times 10^{-7}$	H/m, Wb/(A·m)
Relative permeability[j]	μ_r	not defined		dimensionless
(Volume) energy density, energy product [k]	W	erg/cm^3	10^{-1}	J/m^3
Demagnetization factor	D, N	dimensionless	$1/4\pi$	dimensionless

[a] Gaussian units and cgs emu are the same for magnetic properties. The defining relation is $B = H + 4\pi M$.
[b] Multiply a number in Gaussian units by C to convert it to SI (e.g., $1\ \mathrm{G} \times 10^{-4}\ \mathrm{T/G} = 10^{-4}\ \mathrm{T}$).
[c] SI (*Système International d'Unités*) has been adopted by the National Bureau of Standards. Where two conversion factors are given, the upper one is recognized under, or consistent with, SI and is based on the definition $B = \mu_0(H + M)$, where $\mu_0 = 4\pi \times 10^{-7}$ H/m. The lower one is not recognized under SI and is based on the definition $B = \mu_0 H + J$, where the symbol I is often used in place of J.
[d] 1 gauss = 10^5 gamma (γ).
[e] Both oersted and gauss are expressed as $\mathrm{cm}^{-1/2} \cdot \mathrm{g}^{1/2} \cdot \mathrm{s}^{-1}$ in terms of base units.
[f] A/m was often expressed as "ampere–turn per meter" when used for magnetic field strength.
[g] Magnetic moment per unit volume.
[h] The designation "emu" is not a unit.
[i] Recognized under SI, even though based on the definition $B = \mu_0 H + J$. See footnote c.
[j] $\mu_r = \mu/\mu_0 = 1 + \chi$, all in SI. μ_r is equal to Gaussian μ.
[k] $B \cdot H$ and $\mu_0 M \cdot H$ have SI units J/m^3; $M \cdot H$ and $B \cdot H/4\pi$ have Gaussian units erg/cm^3.

Reference

R. B. Goldfarb and F. R. Fickett, U.S. Department of Commerce, National Bureau of Standards, Boulder, Colorado 80303, March 1985, NBS Special Publication 696. Superintendent of Documents, U.S. Government Printing Office, Washington, DC 20402, 1985.

CONVERSION FACTORS

The following table gives conversion factors from various units of measure to SI units. It is reproduced from NIST Special Publication 811, *Guide for the Use of the International System of Units (SI)*. The table gives the factor by which a quantity expressed in a non-SI unit should be multiplied in order to calculate its value in the SI. The SI values are expressed in terms of the base, supplementary, and derived units of SI in order to provide a coherent presentation of the conversion factors and facilitate computations (see the table "International System of Units" in this section). If desired, powers of ten can be avoided by using SI prefixes and shifting the decimal point if necessary.

Conversion from a non-SI unit to a different non-SI unit may be carried out by using this table in two stages, e.g.,

$$1 \text{ cal}_{th} = 4.184 \text{ J}$$

$$1 \text{ Btu}_{IT} = 1.055056 \text{ E+03 J}$$

Thus,

$$1 \text{ Btu}_{IT} = (1.055056 \text{ E+03} \div 4.184) \text{ cal}_{th} = 252.164 \text{ cal}_{th}$$

Conversion factors are presented for ready adaptation to computer readout and electronic data transmission. The factors are written as a number equal to or greater than one and less than ten with six or fewer decimal places. This number is followed by the letter E (for exponent), a plus or a minus sign, and two digits that indicate the power of 10 by which the number must be multiplied to obtain the correct value. For example:

$$3.523\,907 \text{ E-02 is } 3.523\,907 \times 10^{-2}$$

or

$$0.035\,239\,07$$

Similarly:

$$3.386\,389 \text{ E+03 is } 3.386\,389 \times 10^{3}$$

or

$$3\,386.389$$

A factor in boldface is exact; i.e., all subsequent digits are zero. All other conversion factors have been rounded to the figures given in accordance with accepted practice. Where less than six digits after the decimal point are shown, more precision is not warranted.

It is often desirable to round a number obtained from a conversion of units in order to retain information on the precision of the value. The following rounding rules may be followed:

1. If the digits to be discarded begin with a digit less than 5, the digit preceding the first discarded digit is not changed.

   Example: 6.974 951 5 rounded to 3 digits is 6.97

2. If the digits to be discarded begin with a digit greater than 5, the digit preceding the first discarded digit is increased by one.

   Example: 6.974 951 5 rounded to 4 digits is 6.975

3. If the digits to be discarded begin with a 5 and at least one of the following digits is greater than 0, the digit preceding the 5 is increased by 1.

   Example: 6.974 851 rounded to 5 digits is 6.974 9

4. If the digits to be discarded begin with a 5 and all of the following digits are 0, the digit preceding the 5 is unchanged if it is even and increased by one if it is odd. (Note that this means that the final digit is always even.)

   Examples:
   6.974 951 5 rounded to 7 digits is 6.974 952
   6.974 950 5 rounded to 7 digits is 6.974 950

Reference

Thompson, A., and Taylor, B. N., *Guide for the Use of the International System of Units (SI)*, NIST Special Publication 811, 2008 Edition, Superintendent of Documents, U.S. Government Printing Office, Washington, DC 20402, 2008.

Factors in **boldface** are exact

To convert from	to	Multiply by	
abampere	ampere (A)	**1.0**	**E+01**
abcoulomb	coulomb (C)	**1.0**	**E+01**
abfarad	farad (F)	**1.0**	**E+09**
abhenry	henry (H)	**1.0**	**E–09**
abmho	siemens (S)	**1.0**	**E+09**
abohm	ohm (Ω)	**1.0**	**E–09**
abvolt	volt (V)	**1.0**	**E–08**
acceleration of free fall, standard (g_n)	meter per second squared (m/s^2)	**9.806 65**	**E+00**
acre (based on U.S. survey foot)[a]	square meter (m^2)	4.046 873	E+03
acre foot (based on U.S. survey foot)[a]	cubic meter (m^3)	1.233 489	E+03
ampere hour (A · h)	coulomb (C)	**3.6**	**E+03**
ångström (Å)	meter (m)	**1.0**	**E–10**
ångström (Å)	nanometer (nm)	**1.0**	**E–01**
apostilb (asb)	candela per meter squared (cd/m^2)	3.183 098	E–01
are (a)	square meter (m^2)	**1.0**	**E+02**
astronomical unit (ua or AU)	meter (m)	1.495 979	E+11
atmosphere, standard (atm)	pascal (Pa)	**1.013 25**	**E+05**
atmosphere, standard (atm)	kilopascal (kPa)	**1.013 25**	**E+02**
atmosphere, technical (at)[b]	pascal (Pa)	**9.806 65**	**E+04**
atmosphere, technical (at)[b]	kilopascal (kPa)	**9.806 65**	**E+01**

[a] The U.S. survey foot equals (1200/3937) m. 1 international foot = 0.999998 survey foot.

[b] One technical atmosphere equals one kilogram-force per square centimeter (1 at = 1 kgf/cm^2).

To convert from	to	Multiply by	
bar (bar)	pascal (Pa)	**1.0**	**E+05**
bar (bar)	kilopascal (kPa)	**1.0**	**E+02**
barn (b)	square meter (m^2)	**1.0**	**E−28**
barrel [for petroleum, 42 gallons (U.S.)](bbl)	cubic meter (m^3)	1.589 873	E−01
barrel [for petroleum, 42 gallons (U.S.)](bbl)	liter (L)	1.589 873	E+02
biot (Bi)	ampere (A)	**1.0**	**E+01**
British thermal unit$_{IT}$ (Btu$_{IT}$)[c]	joule (J)	1.055 056	E+03
British thermal unit$_{th}$ (Btu$_{th}$)[c]	joule (J)	1.054 350	E+03
British thermal unit (mean) (Btu)	joule (J)	1.055 87	E+03
British thermal unit (39 °F) (Btu)	joule (J)	1.059 67	E+03
British thermal unit (59 °F) (Btu)	joule (J)	1.054 80	E+03
British thermal unit (60 °F) (Btu)	joule (J)	1.054 68	E+03
British thermal unit$_{IT}$ foot per hour square foot degree Fahrenheit [Btu$_{IT}$ · ft/(h · ft^2 · °F)]	watt per meter kelvin [W/(m · K)]	1.730 735	E+00
British thermal unit$_{th}$ foot per hour square foot degree Fahrenheit [Btu$_{th}$ · ft/(h · ft^2 · °F)]	watt per meter kelvin [W/(m · K)]	1.729 577	E+00
British thermal unit$_{IT}$ inch per hour square foot degree Fahrenheit [Btu$_{IT}$ · in/(h · ft^2 · °F)]	watt per meter kelvin [W/(m · K)]	1.442 279	E−01
British thermal unit$_{th}$ inch per hour square foot degree Fahrenheit [Btu$_{th}$ · in/(h · ft^2 · °F)]	watt per meter kelvin [W/(m · K)]	1.441 314	E−01
British thermal unit$_{IT}$ inch per second square foot degree Fahrenheit [Btu$_{IT}$ · in/(s · ft^2 · °F)]	watt per meter kelvin [W/(m · K)]	5.192 204	E+02
British thermal unit$_{th}$ inch per second square foot degree Fahrenheit [Btu$_{th}$ · in/(s · ft^2 · °F)]	watt per meter kelvin [W/(m · K)]	5.188 732	E+02
British thermal unit$_{IT}$ per cubic foot (Btu$_{IT}$/ft^3)	joule per cubic meter (J/m^3)	3.725 895	E+04
British thermal unit$_{th}$ per cubic foot (Btu$_{th}$/ft^3)	joule per cubic meter (J/m^3)	3.723 403	E+04
British thermal unit$_{IT}$ per degree Fahrenheit (Btu$_{IT}$/°F)	joule per kelvin (J/k)	1.899 101	E+03
British thermal unit$_{th}$ per degree Fahrenheit (Btu$_{th}$/°F)	joule per kelvin (J/k)	1.897 830	E+03
British thermal unit$_{IT}$ per degree Rankine (Btu$_{IT}$/°R)	joule per kelvin (J/k)	1.899 101	E+03
British thermal unit$_{th}$ per degree Rankine (Btu$_{th}$/°R)	joule per kelvin (J/k)	1.897 830	E+03
British thermal unit$_{IT}$ per hour (Btu$_{IT}$/h)	watt (W)	2.930 711	E−01
British thermal unit$_{th}$ per hour (Btu$_{th}$/h)	watt (W)	2.928 751	E−01
British thermal unit$_{IT}$ per hour square foot degree Fahrenheit [Btu$_{IT}$/(h · ft^2 · °F)]	watt per square meter kelvin [W/(m^2 · K)]	5.678 263	E+00
British thermal unit$_{th}$ per hour square foot degree Fahrenheit [Btu$_{th}$/(h · ft^2 · °F)]	watt per square meter kelvin [W/(m^2 · K)]	5.674 466	E+00
British thermal unit$_{th}$ per minute (Btu$_{th}$/min)	watt (W)	1.757 250	E+01
British thermal unit$_{IT}$ per pound (Btu$_{IT}$/lb)	joule per kilogram (J/kg)	**2.326**	**E+03**
British thermal unit$_{th}$ per pound (Btu$_{th}$/lb)	joule per kilogram (J/kg)	2.324 444	E+03
British thermal unit$_{IT}$ per pound degree Fahrenheit [Btu$_{IT}$/(lb · °F)]	joule per kilogram kelvin (J/(kg · K)]	**4.1868**	**E+03**
British thermal unit$_{th}$ per pound degree Fahrenheit [Btu$_{th}$/(lb · °F)]	joule per kilogram kelvin [J/(kg · K)]	**4.184**	**E+03**
British thermal unit$_{IT}$ per pound degree Rankine [Btu$_{IT}$/(lb · °R)]	joule per kilogram kelvin [J/(kg · K)]	**4.1868**	**E+03**
British thermal unit$_{th}$ per pound degree Rankine [Btu$_{th}$/(lb · °R)]	joule per kilogram kelvin [J/(kg · K)]	**4.184**	**E+03**
British thermal unit$_{IT}$ per second (Btu$_{IT}$/s)	watt (W)	1.055 056	E+03
British thermal unit$_{th}$ per second (Btu$_{th}$/s)	watt (W)	1.054 350	E+03

[c] The Fifth International Conference on the Properties of Steam (London, July 1956) defined the International Table calorie as 4.1868 J. Therefore the exact conversion factor for the International Table Btu is 1.055 055 852 62 kJ. Note that the notation for the International Table used in this listing is subscript "IT". Similarly, the notation for thermochemical is subscript "th." Further, the thermochemical Btu, Btu$_{th}$, is based on the thermochemical calorie, cal$_{th}$, where cal$_{th}$ = 4.184 J exactly.

To convert from	to	Multiply by	
British thermal unit$_{IT}$ per second square foot degree Fahrenheit [$Btu_{IT}/(s \cdot ft^2 \cdot °F)$]	watt per square meter kelvin [$W/(m^2 \cdot K)$]	2.044 175	E+04
British thermal unit$_{th}$ per second square foot degree Fahrenheit [$Btu_{th}/(s \cdot ft^2 \cdot °F)$]	watt per square meter kelvin [$W/(m^2 \cdot K)$]	2.042 808	E+04
British thermal unit$_{IT}$ per square foot (Btu_{IT}/ft^2)	joule per square meter (J/m^2)	1.135 653	E+04
British thermal unit$_{th}$ per square foot (Btu_{th}/ft^2)	joule per square meter (J/m^2)	1.134 893	E+04
British thermal unit$_{IT}$ per square foot hour [($Btu_{IT}/(ft^2 \cdot h)$]	watt per square meter (W/m^2)	3.154 591	E+00
British thermal unit$_{th}$ per square foot hour [$Btu_{th}/(ft^2 \cdot h)$]	watt per square meter (W/m^2)	3.152 481	E+00
British thermal unit$_{th}$ per square foot minute [$Btu_{th}/(ft^2 \cdot min)$]	watt per square meter (W/m^2)	1.891 489	E+02
British thermal unit$_{IT}$ per square foot second [($Btu_{IT}/(ft^2 \cdot s)$]	watt per square meter (W/m^2)	1.135 653	E+04
British thermal unit$_{th}$ per square foot second [$Btu_{th}/(ft^2 \cdot s)$]	watt per square meter (W/m^2)	1.134 893	E+04
British thermal unit$_{th}$ per square inch second [$Btu_{th}/(in^2 \cdot s)$]	watt per square meter (W/m^2)	1.634 246	E+06
bushel (U.S.) (bu)	cubic meter (m^3)	3.523 907	E–02
bushel (U.S.) (bu)	liter (L)	3.523 907	E+01
calorie$_{IT}$ (cal_{IT})[c]	joule (J)	**4.1868**	**E+00**
calorie$_{th}$ (cal_{th})[c]	joule (J)	**4.184**	**E+00**
calorie (cal) (mean)	joule (J)	4.190 02	E+00
calorie (15 °C) (cal_{15})	joule (J)	4.185 80	E+00
calorie (20 °C) (cal_{20})	joule (J)	4.181 90	E+00
calorie$_{IT}$, kilogram (nutrition)[d]	joule (J)	**4.1868**	**E+03**
calorie$_{th}$, kilogram (nutrition)[d]	joule (J)	**4.184**	**E+03**
calorie (mean), kilogram (nutrition)[d]	joule (J)	4.190 02	E+03
calorie$_{th}$ per centimeter second degree Celsius [$cal_{th}/(cm \cdot s \cdot °C)$]	watt per meter kelvin [$W/(m \cdot K)$]	**4.184**	**E+02**
calorie$_{IT}$ per gram (cal_{IT}/g)	joule per kilogram (J/kg)	**4.1868**	**E+03**
calorie$_{th}$ per gram (cal_{th}/g)	joule per kilogram (J/kg)	**4.184**	**E+03**
calorie$_{IT}$ per gram degree Celsius [$cal_{IT}/(g \cdot °C)$]	joule per kilogram kelvin [$J/(kg \cdot K)$]	**4.1868**	**E+03**
calorie$_{th}$ per gram degree Celsius [$cal_{th}/(g \cdot °C)$]	joule per kilogram kelvin [$J/(kg \cdot K)$]	**4.184**	**E+03**
calorie$_{IT}$ per gram kelvin [$cal_{IT}/(g \cdot K)$]	joule per kilogram kelvin [$J/(kg \cdot K)$]	**4.1868**	**E+03**
calorie$_{th}$ per gram kelvin [$cal_{th}/(g \cdot K)$]	joule per kilogram kelvin [$J/(kg \cdot K)$]	**4.184**	**E+03**
calorie$_{th}$ per minute (cal_{th}/min)	watt (W)	6.973 333	E–02
calorie$_{th}$ per second (cal_{th}/s)	watt (W)	**4.184**	**E+00**
calorie$_{th}$ per square centimeter (cal_{th}/cm^2)	joule per square meter (J/m^2)	**4.184**	**E+04**
calorie$_{th}$ per square centimeter minute [$cal_{th}/(cm^2 \cdot min)$]	watt per square meter (W/m^2)	6.973 333	E+02
calorie$_{th}$ per square centimeter second [$cal_{th}/(cm^2 \cdot s)$]	watt per square meter (W/m^2)	**4.184**	**E+04**
candela per square inch (cd/in^2)	candela per square meter (cd/m^2)	1.550 003	E+03
carat, metric	kilogram (kg)	**2.0**	**E–04**
carat, metric	gram (g)	**2.0**	**E–01**
centimeter of mercury (0 °C)[e]	pascal (Pa)	1.333 22	E+03
centimeter of mercury (0 °C)[e]	kilopascal (kPa)	1.333 22	E+00
centimeter of mercury, conventional (cmHg)[e]	pascal (Pa)	1.333 224	E+03

[d] The kilogram calorie or "large calorie" is an obsolete term used for the kilocalorie, which is the calorie used to express the energy content of foods. However, in practice, the prefix "kilo" is usually omitted.

[e] Conversion factors for mercury manometer pressure units are calculated using the standard value for the acceleration of gravity and the density of mercury at the stated temperature. Additional digits are not justified because the definitions of the units do not take into account the compressibility of mercury or the change in density caused by the revised practical temperature scale, ITS-90. Similar comments also apply to water manometer pressure units. Conversion factors for conventional mercury and water manometer pressure units are based on ISO 31-3.

To convert from	to	Multiply by	
centimeter of mercury, conventional (cmHg)[e]	kilopascal (kPa)	1.333 224	E+00
centimeter of water (4 °C)[e]	pascal (Pa)	9.806 38	E+01
centimeter of water, conventional (cmH_2O)[e]	pascal (Pa)	**9.806 65**	**E+01**
centipoise (cP)	pascal second (Pa · s)	**1.0**	**E–03**
centistokes (cSt)	meter squared per second (m^2/s)	**1.0**	**E–06**
chain (based on U.S. survey foot) (ch)[a]	meter (m)	2.011 684	E+01
circular mil	square meter (m^2)	5.067 075	E–10
circular mil	square millimeter (mm^2)	5.067 075	E–04
clo	square meter kelvin per watt ($m^2 \cdot K/W$)	1.55	E–01
cord (128 ft^3)	cubic meter (m^3)	3.624 556	E+00
cubic foot (ft^3)	cubic meter (m^3)	2.831 685	E–02
cubic foot per minute (ft^3/min)	cubic meter per second (m^3/s)	4.719 474	E–04
cubic foot per minute (ft^3/min)	liter per second (L/s)	4.719 474	E–01
cubic foot per second (ft^3/s)	cubic meter per second (m^3/s)	2.831 685	E–02
cubic inch (in^3)[f]	cubic meter (m^3)	1.638 706	E–05
cubic inch per minute (in^3/min)	cubic meter per second (m^3/s)	2.731 177	E–07
cubic mile (mi^3)	cubic meter (m^3)	4.168 182	E+09
cubic yard (yd^3)	cubic meter (m^3)	7.645 549	E–01
cubic yard per minute (yd^3/min)	cubic meter per second (m^3/s)	1.274 258	E–02
cup (U.S.)	cubic meter (m^3)	2.365 882	E–04
cup (U.S.)	liter (L)	2.365 882	E–01
cup (U.S.)	milliliter (mL)	2.365 882	E+02
curic (Ci)	becquerel (Bq)	**3.7**	**E+10**
darcy[g]	meter squared (m^2)	9.869 233	E–13
day (d)	second (s)	**8.64**	**E+04**
day (sidereal)	second (s)	8.616 409	E+04
debye (D)	coulomb meter (C · m)	3.335 641	E–30
degree (angle) (°)	radian (rad)	1.745 329	E–02
degree Celsius (temperature) (°C)	kelvin (K)	$T/K = t/°C$**+273.15**	
degree Celsius (temperature interval) (°C)	kelvin (K)	**1.0**	**E+00**
degree centigrade (temperature)[h]	degree Celsius (°C)	$t/°C \approx t/\text{deg.cent.}$	
degree centigrade (temperature interval)[h]	degree Celsius (°C)	1.0	E+00
degree Fahrenheit (temperature) (°F)	degree Celsius (°C)	$t/°C = (t/°F$ **–32)/1.8**	
degree Fahrenheit (temperature) (°F)	kelvin (K)	$T/K = (t/°F$ **+459.67)/1.8**	
degree Fahrenheit (temperature interval)(°F)	degree Celsius (°C)	5.555 556	E–01
degree Fahrenheit (temperature interval) (°F)	kelvin (K)	5.555 556	E–01
degree Fahrenheit hour per British thermal unit$_{IT}$ (°F · h/Btu$_{IT}$)	kelvin per watt (K/W)	1.895 634	E+00
degree Fahrenheit hour per British thermal unit$_{th}$ (°F · h/Btu$_{th}$)	kelvin per watt (K/W)	1.896 903	E+00
degree Fahrenheit hour square foot per British thermal unit$_{IT}$ (°F · h · ft^2/Btu$_{IT}$)	square meter kelvin per watt ($m^2 \cdot K/W$)	1.761 102	E–01
degree Fahrenheit hour square foot per British thermal unit$_{th}$ (°F · h · ft^2/Btu$_{th}$)	square meter kelvin per watt ($m^2 \cdot K/W$)	1.762 280	E–01
degree Fahrenheit hour square foot per British thermal unit$_{IT}$ inch [°F · h · ft^2/(Btu$_{IT}$ · in)]	meter kelvin per watt (m · K/W)	6.933 472	E+00
degree Fahrenheit hour square foot per British thermal unit$_{th}$ inch [°F · h · ft^2/(Btu$_{th}$ · in)]	meter kelvin per watt (m · K/W)	6.938 112	E+00
degree Fahrenheit second per British thermal unit$_{IT}$ (°F · s/Btu$_{IT}$)	kelvin per watt (K/W)	5.265 651	E–04
degree Fahrenheit second per British thermal unit$_{th}$ (°F · s/Btu$_{th}$)	kelvin per watt (K/W)	5.269 175	E–04
degree Rankine (°R)	kelvin (K)	$T/K = (T/°R)$**/1.8**	
degree Rankine (temperature interval) (°R)	kelvin (K)	5.555 556	E–01
denier	kilogram per meter (kg/m)	1.111 111	E–07
denier	gram per meter (g/m)	1.111 111	E–04
dyne (dyn)	newton (N)	**1.0**	**E–05**
dyne centimeter (dyn · cm)	newton meter (N · m)	**1.0**	**E–07**
dyne per square centimeter (dyn/cm^2)	pascal (Pa)	**1.0**	**E–01**

[f] The exact conversion factor is 1.638 706 4 E–05.

[g] The darcy is a unit for expressing the permeability of porous solids, not area.

[h] The centigrade temperature scale is obsolete; the degree centigrade is only approximately equal to the degree Celsius.

To convert from	to	Multiply by	
electronvolt (eV)	joule (J)	1.602 177	E−19
EMU of capacitance (abfarad)	farad (F)	**1.0**	**E+09**
EMU of current (abampere)	ampere (A)	**1.0**	**E+01**
EMU of electric potential (abvolt)	volt (V)	**1.0**	**E−08**
EMU of inductance (abhenry)	henry (H)	**1.0**	**E−09**
EMU of resistance (abohm)	ohm (Ω)	**1.0**	**E−09**
erg (erg)	joule (J)	**1.0**	**E−07**
erg per second (erg/s)	watt (W)	**1.0**	**E−07**
erg per square centimeter second [erg/(cm^2 · s)]	watt per square meter (W/m^2)	**1.0**	**E−03**
ESU of capacitance (statfarad)	farad (F)	1.112 650	E−12
ESU of current (statampere)	ampere (A)	3.335 641	E−10
ESU of electric potential (statvolt)	volt (V)	2.997 925	E+02
ESU of inductance (stathenry)	henry (H)	8.987 552	E+11
ESU of resistance (statohm)	ohm (Ω)	8.987 552	E+11
faraday (based on carbon 12)	coulomb (C)	9.648 531	E+04
fathom (based on U.S survey foot)[a]	meter (m)	1.828 804	E+00
fermi	meter (m)	**1.0**	**E−15**
fermi	femtometer (fm)	**1.0**	**E+00**
fluid ounce (U.S.) (fl oz)	cubic meter (m^3)	2.957 353	E−05
fluid ounce (U.S.) (fl oz)	milliliter (mL)	2.957 353	E+01
foot (ft)	meter (m)	**3.048**	**E−01**
foot (U.S. survey ft)[a]	meter (m)	3.048 006	E−01
footcandle	lux (lx)	1.076 391	E+01
footlambert	candela per square meter (cd/m^2)	3.426 259	E+00
foot of mercury, conventional (ftHg)[e]	pascal (Pa)	4.063 666	E+04
foot of mercury, conventional (ftHg)[e]	kilopascal (kPa)	4.063 666	E+01
foot of water (39.2 °F)[e]	pascal (Pa)	2.988 98	E+03
foot of water (39.2 °F)[e]	kilopascal (kPa)	2.988 98	E+00
foot of water, conventional (ftH$_2$O)[e]	pascal (Pa)	2.989 067	E+03
foot of water, conventional (ftH$_2$O)[e]	kilopascal (kPa)	2.989 067	E+00
foot per hour (ft/h)	meter per second (m/s)	8.466 667	E−05
foot per minute (ft/min)	meter per second (m/s)	**5.08**	**E−03**
foot per second (ft/s)	meter per second (m/s)	**3.048**	**E−01**
foot per second squared (ft/s^2)	meter per second squared (m/s^2)	**3.048**	**E−01**
foot poundal	joule (J)	4.214 011	E−02
foot pound-force (ft · lbf)	joule (J)	1.355 818	E+00
foot pound-force per hour (ft · lbf/h)	watt (W)	3.766 161	E−04
foot pound-force per minute (ft · lbf/min)	watt (W)	2.259 697	E−02
foot pound-force per second (ft · lbf/s)	watt (W)	1.355 818	E+00
foot to the fourth power (ft^4)[i]	meter to the fourth power (m^4)	8.630 975	E−03
franklin (Fr)	coulomb (C)	3.335 641	E−10
gal (Gal)	meter per second squared (m/s^2)	**1.0**	**E−02**
gallon [Canadian and U.K. (Imperial)] (gal)	cubic meter (m^3)	**4.546 09**	**E−03**
gallon [Canadian and U.K. (Imperial)] (gal)	liter (L)	**4.546 09**	**E+00**
gallon (U.S.) (gal)	cubic meter (m^3)	3.785 412	E−03
gallon (U.S.) (gal)	liter (L)	3.785 412	E+00
gallon (U.S.) per day (gal/d)	cubic meter per second (m^3/s)	4.381 264	E−08
gallon (U.S.) per day (gal/d)	liter per second (L/s)	4.381 264	E−05
gallon (U.S.) per horsepower hour [gal/(hp · h)]	cubic meter per joule (m^3/J)	1.410 089	E−09
gallon (U.S.) per horsepower hour [gal/(hp · h)]	liter per joule (L/J)	1.410 089	E−06
gallon (U.S.) per minute (gpm)(gal/min)	cubic meter per second (m^3/s)	6.309 020	E−05
gallon (U.S.) per minute (gpm)(gal/min)	liter per second (L/s)	6.309 020	E−02
gamma (γ)	tesla (T)	**1.0**	**E−09**
gauss (Gs, G)	tesla (T)	**1.0**	**E−04**
gilbert (Gi)	ampere (A)	7.957 747	E−01

[i] This is a unit for the quantity second moment of area, which is sometimes called the "moment of section" or "area moment of inertia" of a plane section about a specified axis.

To convert from	to	Multiply by	
gill [Canadian and U.K. (Imperial)] (gi)	cubic meter (m^3)	1.420 653	E–04
gill [Canadian and U.K. (Imperial)] (gi)	liter (L)	1.420 653	E–01
gill (U.S.) (gi)	cubic meter (m^3)	1.182 941	E–04
gill (U.S.) (gi)	liter (L)	1.182 941	E–01
gon (also called grade) (gon)	radian (rad)	1.570 796	E–02
gon (also called grade) (gon)	degree (angle) (°)	**9.0**	**E–01**
grain (gr)	kilogram (kg)	**6.479 891**	**E–05**
grain (gr)	milligram (mg)	**6.479 891**	**E+01**
grain per gallon (U.S.) (gr/gal)	kilogram per cubic meter (kg/m^3)	1.711 806	E–02
grain per gallon (U.S.) (gr/gal)	milligram per liter (mg/L)	1.711 806	E+01
gram-force per square centimeter (gf/cm^2)	pascal (Pa)	**9.806 65**	**E+01**
gram per cubic centimeter (g/cm^3)	kilogram per cubic meter (kg/m^3)	**1.0**	**E+03**
hectare (ha)	square meter (m^2)	**1.0**	**E+04**
horsepower (550 ft · lbf/s) (hp)	watt (W)	7.456 999	E+02
horsepower (boiler)	watt (W)	9.809 50	E+03
horsepower (electric)	watt (W)	**7.46**	**E+02**
horsepower (metric)	watt (W)	7.354 988	E+02
horsepower (U.K.)	watt (W)	7.4570	E+02
horsepower (water)	watt (W)	7.460 43	E+02
hour (h)	second (s)	**3.6**	**E+03**
hour (sidereal)	second (s)	3.590 170	E+03
hundredweight (long, 112 lb)	kilogram (kg)	5.080 235	E+01
hundredweight (short, 100 lb)	kilogram (kg)	4.535 924	E+01
inch (in)	meter (m)	**2.54**	**E–02**
inch (in)	centimeter (cm)	**2.54**	**E+00**
inch of mercury (32 °F)[e]	pascal (Pa)	3.386 38	E+03
inch of mercury (32 °F)[e]	kilopascal (kPa)	3.386 38	E+00
inch of mercury (60 °F)[e]	pascal (Pa)	3.376 85	E+03
inch of mercury (60 °F)[e]	kilopascal (kPa)	3.376 85	E+00
inch of mercury, conventional (inHg)[e]	pascal (Pa)	3.386 389	E+03
inch of mercury, conventional (inHg)[e]	kilopascal (kPa)	3.386 389	E+00
inch of water (39.2 °F)[e]	pascal (Pa)	2.490 82	E+02
inch of water (60 °F)[e]	pascal (Pa)	2.4884	E+02
inch of water, conventional (inH_2O)[e]	pascal (Pa)	2.490 889	E+02
inch per second (in/s)	meter per second (m/s)	**2.54**	**E–02**
inch per second squared (in/s^2)	meter per second squared (m/s^2)	**2.54**	**E–02**
inch to the fourth power (in^4)[i]	meter to the fourth power (m^4)	4.162 314	E–07
kayser (K)	reciprocal meter (m^{-1})	**1.0**	**E+02**
kelvin (K)	degree Celsius (°C)	$t/°C = T/K - 273.15$	
kilocalorie$_{IT}$ (kcal$_{IT}$)	joule (J)	**4.1868**	**E+03**
kilocalorie$_{th}$ (kcal$_{th}$)	joule (J)	**4.184**	**E+03**
kilocalorie (mean) (kcal)	joule (J)	4.190 02	E+03
kilocalorie$_{th}$ per minute (kcal$_{th}$/min)	watt (W)	6.973 333	E+01
kilocalorie$_{th}$ per second (kcal$_{th}$/s)	watt (W)	**4.184**	**E+03**
kilogram-force (kgf)	newton (N)	**9.806 65**	**E+00**
kilogram-force meter (kgf · m)	newton meter (N · m)	**9.806 65**	**E+00**
kilogram-force per square centimeter (kgf/cm^2)	pascal (Pa)	**9.806 65**	**E+04**
kilogram-force per square centimeter (kgf/cm^2)	kilopascal (kPa)	**9.806 65**	**E+01**
kilogram-force per square meter (kgf/m^2)	pascal (Pa)	**9.806 65**	**E+00**
kilogram-force per square millimeter (kgf/mm^2)	pascal (Pa)	**9.806 65**	**E+06**
kilogram-force per square millimeter (kgf/mm^2)	megapascal (MPa)	**9.806 65**	**E+00**
kilogram-force second squared per meter ($kgf \cdot s^2/m$)	kilogram (kg)	**9.806 65**	**E+00**
kilometer per hour (km/h)	meter per second (m/s)	2.777 778	E–01
kilopond (kilogram-force) (kp)	newton (N)	**9.806 65**	**E+00**
kilowatt hour (kW · h)	joule (J)	**3.6**	**E+06**
kilowatt hour (kW · h)	megajoule (MJ)	**3.6**	**E+00**

To convert from	to	Multiply by	
kip (1 kip=1000 lbf)	newton (N)	4.448 222	E+03
kip (1 kip=1000 lbf)	kilonewton (kN)	4.448 222	E+00
kip per square inch (ksi) (kip/in²)	pascal (Pa)	6.894 757	E+06
kip per square inch (ksi) (kip/in²)	kilopascal (kPa)	6.894 757	E+03
knot (nautical mile per hour)	meter per second (m/s)	5.144 444	E–01
lambert[j]	candela per square meter (cd/m²)	3.183 099	E+03
langley (cal_{th}/cm²)	joule per square meter (J/m²)	**4.184**	**E+04**
light year (l.y.)[k]	meter (m)	9.460 73	E+15
liter (L)[l]	cubic meter (m³)	**1.0**	**E–03**
lumen per square foot (lm/ft²)	lux (lx)	1.076 391	E+01
maxwell (Mx)	weber (Wb)	**1.0**	**E–08**
mho	siemens (S)	**1.0**	**E+00**
microinch	meter (m)	**2.54**	**E–08**
microinch	micrometer (μm)	**2.54**	**E–02**
micron (μ)	meter (m)	**1.0**	**E–06**
micron (μ)	micrometer (μm)	**1.0**	**E+00**
mil (0.001 in)	meter (m)	**2.54**	**E–05**
mil (0.001 in)	millimeter (mm)	**2.54**	**E–02**
mil (angle)	radian (rad)	9.817 477	E–04
mil (angle)	degree (°)	**5.625**	**E–02**
mile (mi)	meter (m)	**1.609 344**	**E+03**
mile (mi)	kilometer (km)	**1.609 344**	**E+00**
mile (based on U.S. survey foot) (mi)[a]	meter (m)	1.609 347	E+03
mile (based on U.S. survey foot) (mi)[a]	kilometer (km)	1.609 347	E+00
mile, nautical[m]	meter (m)	**1.852**	**E+03**
mile per gallon (U.S.) (mpg) (mi/gal)	meter per cubic meter (m/m³)	4.251 437	E+05
mile per gallon (U.S.) (mpg) (mi/gal)	kilometer per liter (km/L)	4.251 437	E–01
mile per gallon (U.S.) (mpg) (mi/gal)[n]	liter per 100 kilometer (L/100 km)	divide 235.215 by number of miles per gallon	
mile per hour (mi/h)	meter per second (m/s)	**4.4704**	**E–01**
mile per hour (mi/h)	kilometer per hour (km/h)	**1.609 344**	**E+00**
mile per minute (mi/min)	meter per second (m/s)	**2.682 24**	**E+01**
mile per second (mi/s)	meter per second (m/s)	**1.609 344**	**E+03**
millibar (mbar)	pascal (Pa)	**1.0**	**E+02**
millibar (mbar)	kilopascal (kPa)	**1.0**	**E–01**
millimeter of mercury, conventional (mmHg)[e]	pascal (Pa)	1.333 224	E+02
millimeter of water, conventional (mmH_2O)[e]	pascal (Pa)	**9.806 65**	**E+00**
minute (angle) (′)	radian (rad)	2.908 882	E–04
minute (min)	second (s)	**6.0**	**E+01**
minute (sidereal)	second (s)	5.983 617	E+01
nit	candela per meter squared (cd/m²)	**1.0**	**E+00**
nox	lux (lx)	**1.0**	**E–03**
oersted (Oe)	ampere per meter (A/m)	7.957 747	E+01
ohm centimeter (Ω · cm)	ohm meter (Ω · m)	**1.0**	**E–02**
ohm circular-mil per foot	ohm meter (Ω · m)	1.662 426	E–09
ohm circular-mil per foot	ohm square millimeter per meter (Ω · mm²/m)	1.662 426	E–03
ounce (avoirdupois) (oz)	kilogram (kg)	2.834 952	E–02
ounce (avoirdupois) (oz)	gram (g)	2.834 952	E+01
ounce (troy or apothecary) (oz)	kilogram (kg)	3.110 348	E–02
ounce (troy or apothecary) (oz)	gram (g)	3.110 348	E+01
ounce [Canadian and U.K. fluid (Imperial)] (fl oz)	cubic meter (m³)	2.841 306	E–05

[j] The exact conversion factor is $10^4/\pi$.

[k] This conversion factor is based on 1 d = 86 400 s; and 1 Julian century = 36 525 d. (See *The Astronomical Almanac for the Year 1995*, page K6, U.S. Government Printing Office, Washington, DC, 1994.)

[l] In 1964 the General Conference on Weights and Measures reestablished the name "liter" as a special name for the cubic decimeter. Between 1901 and 1964 the liter was slightly larger (1.000 028 dm³); when one uses high-accuracy volume data of that time, this fact must be kept in mind.

[m] The value of this unit, 1 nautical mile = 1852 m, was adopted by the First International Extraordinary Hydrographic Conference, Monaco, 1929, under the name "International nautical mile."

[n] For converting fuel economy, as used in the U.S., to fuel consumption.

To convert from	to	Multiply by	
ounce [Canadian and U.K. fluid (Imperial)] (fl oz)	milliliter (mL)	2.841 306	E+01
ounce (U.S. fluid) (fl oz)	cubic meter (m^3)	2.957 353	E–05
ounce (U.S. fluid) (fl oz)	milliliter (mL)	2.957 353	E+01
ounce (avoirdupois)-force (ozf)	newton (N)	2.780 139	E–01
ounce (avoirdupois)-force inch (ozf · in)	newton meter (N · m)	7.061 552	E–03
ounce (avoirdupois)-force inch (ozf · in)	millinewton meter (mN · m)	7.061 552	E+00
ounce (avoirdupois) per cubic inch (oz/in^3)	kilogram per cubic meter (kg/m^3)	1.729 994	E+03
ounce (avoirdupois) per gallon [Canadian and U.K. (Imperial)] (oz/gal)	kilogram per cubic meter (kg/m^3)	6.236 023	E+00
ounce (avoirdupois) per gallon [Canadian and U.K. (Imperial)] (oz/gal)	gram per liter (g/L)	6.236 023	E+00
ounce (avoirdupois) per gallon (U.S.)(oz/gal)	kilogram per cubic meter (kg/m^3)	7.489 152	E+00
ounce (avoirdupois) per gallon (U.S.)(oz/gal)	gram per liter (g/L)	7.489 152	E+00
ounce (avoirdupois) per square foot (oz/ft^2)	kilogram per square meter (kg/m^2)	3.051 517	E–01
ounce (avoirdupois) per square inch (oz/in^2)	kilogram per square meter (kg/m^2)	4.394 185	E+01
ounce (avoirdupois) per square yard(oz/yd^2)	kilogram per square meter (kg/m^2)	3.390 575	E–02
parsec (pc)	meter (m)	3.085 678	E+16
peck (U.S.) (pk)	cubic meter (m^3)	8.809 768	E–03
peck (U.S.) (pk)	liter (L)	8.809 768	E+00
pennyweight (dwt)	kilogram (kg)	1.555 174	E–03
pennyweight (dwt)	gram (g)	1.555 174	E+00
perm (0 °C)	kilogram per pascal second square meter [kg/(Pa · s · m^2)]	5.721 35	E–11
perm (23 °C)	kilogram per pascal second square meter [kg/(Pa · s · m^2)]	5.745 25	E–11
perm inch (0 °C)	kilogram per pascal second meter [kg/(Pa · s · m)]	1.453 22	E–12
perm inch (23 °C)	kilogram per pascal second meter [kg/(Pa · s · m)]	1.459 29	E–12
phot (ph)	lux (lx)	**1.0**	**E+04**
pica (computer) (1/6 in)	meter (m)	4.233 333	E–03
pica (computer) (1/6 in)	millimeter (mm)	4.233 333	E+00
pica (printer's)	meter (m)	4.217 518	E–03
pica (printer's)	millimeter (mm)	4.217 518	E+00
pint (U.S. dry) (dry pt)	cubic meter (m^3)	5.506 105	E–04
pint (U.S. dry) (dry pt)	liter (L)	5.506 105	E–01
pint (U.S. liquid) (liq pt)	cubic meter (m^3)	4.731 765	E–04
pint (U.S. liquid) (liq pt)	liter (L)	4.731 765	E–01
point (computer) (1/72 in)	meter (m)	3.527 778	E–04
point (computer) (1/72 in)	millimeter (mm)	3.527 778	E–01
point (printer's)	meter (m)	3.514 598	E–04
point (printer's)	millimeter (mm)	3.514 598	E–01
poise (P)	pascal second (Pa · s)	**1.0**	**E–01**
pound (avoirdupois) (lb)[o]	kilogram (kg)	4.535 924	E–01
pound (troy or apothecary) (lb)	kilogram (kg)	3.732 417	E–01
poundal	newton (N)	1.382 550	E–01
poundal per square foot	pascal (Pa)	1.488 164	E+00
poundal second per square foot	pascal second (Pa · s)	1.488 164	E+00
pound foot squared (lb · ft^2)	kilogram meter squared (kg · m^2)	4.214 011	E–02
pound-force (lbf)[p]	newton (N)	4.448 222	E+00
pound-force foot (lbf · ft)	newton meter (N · m)	1.355 818	E+00
pound-force foot per inch (lbf · ft/in)	newton meter per meter (N · m/m)	5.337 866	E+01
pound-force inch (lbf · in)	newton meter (N · m)	1.129 848	E–01
pound-force inch per inch (lbf · in/in)	newton meter per meter (N · m/m)	4.448 222	E+00
pound-force per foot (lbf/ft)	newton per meter (N/m)	1.459 390	E+01
pound-force per inch (lbf/in)	newton per meter (N/m)	1.751 268	E+02
pound-force per pound (lbf/lb) (thrust to mass ratio)	newton per kilogram (N/kg)	**9.806 65**	**E+00**

[o] The exact conversion factor is 4.535 923 7 E–01. All units that contain the pound refer to the avoirdupois pound unless otherwise specified.

[p] If the local value of the acceleration of free fall is taken as g_n=9.806 65 m/ s^2 (the standard value), the exact conversion factor is 4.448 221 615 260 5 E+00.

To convert from	to	Multiply by	
pound-force per square foot (lbf/ft²)	pascal (Pa)	4.788 026	E+01
pound-force per square inch (psi) (lbf/in²)	pascal (Pa)	6.894 757	E+03
pound-force per square inch (psi) (lbf/in²)	kilopascal (kPa)	6.894 757	E+00
pound-force second per square foot (lbf · s/ft²)	pascal second (Pa · s)	4.788 026	E+01
pound-force second per square inch (lbf · s/in²)	pascal second (Pa · s)	6.894 757	E+03
pound inch squared (lb · in²)	kilogram meter squared (kg · m²)	2.926 397	E−04
pound per cubic foot (lb/ft³)	kilogram per cubic meter (kg/m³)	1.601 846	E+01
pound per cubic inch (lb/in³)	kilogram per cubic meter (kg/m³)	2.767 990	E+04
pound per cubic yard (lb/yd³)	kilogram per cubic meter (kg/m³)	5.932 764	E−01
pound per foot (lb/ft)	kilogram per meter (kg/m)	1.488 164	E+00
pound per foot hour [lb/(ft · h)]	pascal second (Pa · s)	4.133 789	E−04
pound per foot second [lb/(ft · s)]	pascal second (Pa · s)	1.488 164	E+00
pound per gallon [Canadian and U.K. (Imperial)] (lb/gal)	kilogram per cubic meter (kg/m³)	9.977 637	E+01
pound per gallon [Canadian and U.K. (Imperial)] (lb/gal)	kilogram per liter (kg/L)	9.977 637	E−02
pound per gallon (U.S.) (lb/gal)	kilogram per cubic meter (kg/m³)	1.198 264	E+02
pound per gallon (U.S.) (lb/gal)	kilogram per liter (kg/L)	1.198 264	E−01
pound per horsepower hour [lb/(hp · h)]	kilogram per joule (kg/J)	1.689 659	E−07
pound per hour (lb/h)	kilogram per second (kg/s)	1.259 979	E−04
pound per inch (lb/in)	kilogram per meter (kg/m)	1.785 797	E+01
pound per minute (lb/min)	kilogram per second (kg/s)	7.559 873	E−03
pound per second (lb/s)	kilogram per second (kg/s)	4.535 924	E−01
pound per square foot (lb/ft²)	kilogram per square meter (kg/m²)	4.882 428	E+00
pound per square inch (*not* pound-force) (lb/in²)	kilogram per square meter (kg/m²)	7.030 696	E+02
pound per yard (lb/yd)	kilogram per meter (kg/m)	4.960 546	E−01
psi (pound-force per square inch) (lbf/in²)	pascal (Pa)	6.894 757	E+03
psi (pound-force per square inch) (lbf/in²)	kilopascal (kPa)	6.894 757	E+00
quad (10^{15} Btu_{IT})[c]	joule (J)	1.055 056	E+18
quart (U.S. dry) (dry qt)	cubic meter (m³)	1.101 221	E−03
quart (U.S. dry) (dry qt)	liter (L)	1.101 221	E+00
quart (U.S. liquid) (liq qt)	cubic meter (m³)	9.463 529	E−04
quart (U.S. liquid) (liq qt)	liter (L)	9.463 529	E−01
rad (absorbed dose) (rad)	gray (Gy)	**1.0**	**E−02**
rem (rem)	sievert (Sv)	**1.0**	**E−02**
revolution (r)	radian (rad)	6.283 185	E+00
revolution per minute (rpm) (r/min)	radian per second (rad/s)	1.047 198	E−01
rhe	reciprocal pascal second [(Pa · s)⁻¹]	**1.0**	**E+01**
rod (based on U.S. survey foot) (rd)[a]	meter (m)	5.029 210	E+00
roentgen (R)	coulomb per kilogram (C/kg)	**2.58**	**E−04**
rpm (revolution per minute) (r/min)	radian per second (rad/s)	1.047 198	E−01
second (angle) (″)	radian (rad)	4.848 137	E−06
second (sidereal)	second (s)	9.972 696	E−01
shake	second (s)	**1.0**	**E−08**
shake	nanosecond (ns)	**1.0**	**E+01**
skot	candela per meter squared (cd/m²)	3.183 098	E−04
slug (slug)	kilogram (kg)	1.459 390	E+01
slug per cubic foot (slug/ft³)	kilogram per cubic meter (kg/m³)	5.153 788	E+02
slug per foot second [slug/(ft · s)]	pascal second (Pa · s)	4.788 026	E+01
square foot (ft²)	square meter (m²)	**9.290 304**	**E−02**
square foot per hour (ft²/h)	square meter per second (m²/s)	**2.580 64**	**E−05**
square foot per second (ft²/s)	square meter per second (m²/s)	**9.290 304**	**E−02**
square inch (in²)	square meter (m²)	**6.4516**	**E−04**
square inch (in²)	square centimeter (cm²)	**6.4516**	**E+00**
square mile (mi²)	square meter (m²)	2.589 988	E+06
square mile (mi²)	square kilometer (km²)	2.589 988	E+00

To convert from	to	Multiply by	
square mile (based on U.S. survey foot) (mi^2)[a]	square meter (m^2)	2.589 998	E+06
square mile (based on U.S. survey foot) (mi^2)[a]	square kilometer (km^2)	2.589 998	E+00
square yard (yd^2)	square meter (m^2)	8.361 274	E−01
statampere	ampere (A)	3.335 641	E−10
statcoulomb	coulomb (C)	3.335 641	E−10
statfarad	farad (F)	1.112 650	E−12
stathenry	henry (H)	8.987 552	E+11
statmho	siemens (S)	1.112 650	E−12
statohm	ohm (Ω)	8.987 552	E+11
statvolt	volt (V)	2.997 925	E+02
stere (st)	cubic meter (m^3)	**1.0**	**E+00**
stilb (sb)	candela per square meter (cd/m^2)	**1.0**	**E+04**
stokes (St)	meter squared per second (m^2/s)	**1.0**	**E−04**
tablespoon	cubic meter (m^3)	1.478 676	E−05
tablespoon	milliliter (mL)	1.478 676	E+01
teaspoon	cubic meter (m^3)	4.928 922	E−06
teaspoon	milliliter (mL)	4.928 922	E+00
tex	kilogram per meter (kg/m)	**1.0**	**E−06**
therm (EC)[q]	joule (J)	**1.055 06**	**E+08**
therm (U.S.)[q]	joule (J)	**1.054 804**	**E+08**
ton, assay (AT)	kilogram (kg)	2.916 667	E−02
ton, assay (AT)	gram (g)	2.916 667	E+01
ton-force (2000 lbf)	newton (N)	8.896 443	E+03
ton-force (2000 lbf)	kilonewton (kN)	8.896 443	E+00
ton, long (2240 lb)	kilogram (kg)	1.016 047	E+03
ton, long, per cubic yard	kilogram per cubic meter (kg/m^3)	1.328 939	E+03
ton, metric (t)	kilogram (kg)	**1.0**	**E+03**
tonne (called "metric ton" in U.S.) (t)	kilogram (kg)	**1.0**	**E+03**
ton of refrigeration (12 000 Btu_{IT}/h)	watt (W)	3.516 853	E+03
ton of TNT (energy equivalent)[r]	joule (J)	**4.184**	**E+09**
ton, register	cubic meter (m^3)	2.831 685	E+00
ton, short (2000 lb)	kilogram (kg)	9.071 847	E+02
ton, short, per cubic yard	kilogram per cubic meter (kg/m^3)	1.186 553	E+03
ton, short, per hour	kilogram per second (kg/s)	2.519 958	E−01
torr (Torr)	pascal (Pa)	1.333 224	E+02
unit pole	weber (Wb)	1.256 637	E−07
watt hour (W · h)	joule (J)	**3.6**	**E+03**
watt per square centimeter (W/cm^2)	watt per square meter (W/m^2)	**1.0**	**E+04**
watt per square inch (W/in^2)	watt per square meter (W/m^2)	1.550 003	E+03
watt second (W · s)	joule (J)	**1.0**	**E+00**
yard (yd)	meter (m)	**9.144**	**E−01**
year (365 days)	second (s)	**3.1536**	**E+07**
year (sidereal)	second (s)	3.155 815	E+07
year (tropical)	second (s)	3.155 693	E+07

[q] The therm (EC) is legally defined in the Council Directive of 20 December 1979, Council of the European Communities (now the European Union, EU). The therm (U.S.) is legally defined in the Federal Register of July 27, 1968. Although the therm (EC), which is based on the International Table Btu, is frequently used by engineers in the United States, the therm (U.S.) is the legal unit used by the U.S natural gas industry.

[r] Defined (not measured) value.

CONVERSION OF TEMPERATURES

From	To	
Celsius	Fahrenheit	$t_F/°F = (9/5)\, t/°C + 32$
	Kelvin	$T/K = t/°C + 273.15$
	Rankine	$T/°R = (9/5)\,(t/°C + 273.15)$
Fahrenheit	Celsius	$t/°C = (5/9)\,[(t_F/°F) - 32]$
	Kelvin	$T/K = (5/9)\,[(t_F/°F) - 32] + 273.15$
	Rankine	$T/°R = t_F/°F + 459.67$
Kelvin	Celsius	$t/°C = T/K - 273.15$
	Rankine	$T/°R = (9/5)\, T/K$
Rankine	Fahrenheit	$t_F/°F = T/°R - 459.67$
	Kelvin	$T/K = (5/9)\, T/°R$

Definition of symbols:

T = thermodynamic (absolute) temperature

t = Celsius temperature (the symbol θ is also used for Celsius temperature)

t_F = Fahrenheit temperature

Designation of Large Numbers

	U.S.A.	Other countries
10^6	million	million
10^9	billion	milliard
10^{12}	trillion	billion
10^{15}	quadrillion	billiard
10^{18}	quintillion	trillion
100^{100}	googol	
10^{googol}	googolplex	

CONVERSION FACTORS FOR ENERGY UNITS

If greater accuracy is required, use the Energy Equivalents section of the Fundamental Physical Constants table.

	Wavenumber $\bar{\nu}$ cm^{-1}	Frequency ν MHz	Energy E aJ	Energy E eV	Energy E E_h	Molar energy E_m kJ/mol	Molar energy E_m kcal/mol	Temperature T K
$\bar{\nu}$: 1 cm^{-1}	≙ 1	2.997925×10^4	1.986447×10^{-5}	1.239842×10^{-4}	4.556335×10^{-6}	11.96266×10^{-3}	2.85914×10^{-3}	1.438769
ν: 1 MHz	≙ 3.33564×10^{-5}	1	6.626076×10^{-10}	4.135669×10^{-9}	1.519830×10^{-10}	3.990313×10^{-7}	9.53708×10^{-8}	4.79922×10^{-5}
1 aJ	≙ 50341.1	1.509189×10^9	1	6.241506	0.2293710	602.2137	143.9325	7.24292×10^4
E: 1 eV	≙ 8065.54	2.417988×10^8	0.1602177	1	3.674931×10^{-2}	96.4853	23.0605	1.16045×10^4
E_h	≙ 219474.63	6.579684×10^9	4.359748	27.2114	1	2625.500	627.510	3.15773×10^5
E_m: 1 kJ/mol	≙ 83.5935	2.506069×10^6	1.660540×10^{-3}	1.036427×10^{-2}	3.808798×10^{-4}	1	0.239006	120.272
1 kcal/mol	≙ 349.755	1.048539×10^7	6.947700×10^{-3}	4.336411×10^{-2}	1.593601×10^{-3}	4.184	1	503.217
T: 1 K	≙ 0.695039	2.08367×10^4	1.380658×10^{-5}	8.61738×10^{-5}	3.16683×10^{-6}	8.31451×10^{-3}	1.98722×10^{-3}	1

Examples of the use of this table:

1 aJ ≙ 50341 cm^{-1}

1 eV ≙ 96.4853 kJ mol^{-1}

The symbol ≙ should be read as meaning corresponds to or is equivalent to.

$E = h\nu = hc\bar{\nu} = kT$; $E_m = N_A E$; E_h is the Hartree energy.

CONVERSION FACTORS FOR PRESSURE UNITS

	Pa	kPa	MPa	bar	atm	Torr	μmHg	psi
Pa	1	0.001	0.000001	0.00001	9.8692×10^{-6}	0.0075006	7.5006	0.0001450377
kPa	1000	1	0.001	0.01	0.0098692	7.5006	7500.6	0.1450377
MPa	1000000	1000	1	10	9.8692	7500.6	7500600	145.0377
bar	100000	100	0.1	1	0.98692	750.06	750060	14.50377
atm	101325	101.325	0.101325	1.01325	1	760	760000	14.69594
Torr	133.322	0.133322	0.000133322	0.00133322	0.00131579	1	1000	0.01933672
μmHg	0.133322	0.000133322	1.33322×10^{-7}	1.33322×10^{-6}	1.31579×10^{-6}	0.001	1	1.933672×10^{-5}
psi	6894.757	6.894757	0.006894757	0.06894757	0.068046	51.7151	51715.1	1

To convert a pressure value from a unit in the left-hand column to a new unit, multiply the value by the factor appearing in the column for the new unit. For example:

1 kPa = 9.8692×10^{-3} atm
1 Torr = 1.33322×10^{-4} MPa

Notes: μmHg is often referred to as "micron"
Torr is essentially identical to mmHg
psi is an abbreviation for the unit pound–force per square inch
psia (as a term for a physical quantity) implies the true (absolute) pressure
psig implies the true pressure minus the local atmospheric pressure

CONVERSION FACTORS FOR THERMAL CONDUCTIVITY UNITS

MULTIPLY ↓ by appropriate factor to OBTAIN→	Btu_{IT} h^{-1} ft^{-1} °F^{-1}	Btu_{IT} in. h^{-1} ft^{-2} °F^{-1}	Btu_{th} h^{-1} ft^{-1} °F^{-1}	Btu_{th} in h^{-1} ft^{-2} °F^{-1}	cal_{IT} s^{-1} cm^{-1} °C^{-1}	cal_{th} s^{-1} cm^{-1} °C^{-1}	$kcal_{th}$ h^{-1} m^{-1} °C^{-1}	J s^{-1} cm^{-1} K^{-1}	W cm^{-1} K^{-1}	W m^{-1} K^{-1}	mW cm^{-1} K^{-1}
Btu_{IT} h^{-1} ft^{-1} °F^{-1}	1	12	1.00067	12.0080	4.13379×10^{-3}	4.13656×10^{-3}	1.48916	1.73073×10^{-2}	1.73073×10^{-2}	1.73073	17.3073
Btu_{IT} in h^{-1} ft^{-2} °F^{-1}	8.33333×10^{-2}	1	8.33891×10^{-2}	1.00067	3.44482×10^{-4}	3.44713×10^{-4}	0.124097	1.44228×10^{-3}	1.44228×10^{-3}	0.144228	1.44228
Btu_{th} h^{-1} ft^{-1} °F^{-1}	0.999331	11.9920	1	12	4.13102×10^{-3}	4.13379×10^{-3}	1.48816	1.72958×10^{-2}	1.72958×10^{-2}	1.72958	17.2958
Btu_{th} in. h^{-1} ft^{-2} °F^{-1}	8.32776×10^{-2}	0.999331	8.33333×10^{-2}	1	3.44252×10^{-4}	3.44482×10^{-4}	0.124014	1.44131×10^{-3}	1.44131×10^{-3}	0.144131	1.44131
cal_{IT} s^{-1} cm^{-1} °C^{-1}	2.41909×10^{2}	2.90291×10^{3}	2.42071×10^{2}	2.90485×10^{3}	1	1.00067	3.60241×10^{2}	4.1868	4.1868	4.1868×10^{2}	4.1868×10^{3}
cal_{th} s^{-1} cm^{-1} °C^{-1}	2.41747×10^{2}	2.90096×10^{3}	2.41909×10^{2}	2.90291×10^{3}	0.999331	1	3.6×10^{2}	4.184	4.184	4.184×10^{2}	4.184×10^{3}
$kcal_{th}$ h^{-1} m^{-1} °C^{-1}	0.671520	8.05824	0.671969	8.06363	2.77592×10^{-3}	2.77778×10^{-3}	1	1.16222×10^{-2}	1.16222×10^{-2}	1.16222	11.6222
J s^{-1} cm^{-1} K^{-1}	57.7789	6.93347×10^{2}	57.8176	6.93811×10^{2}	0.238846	0.239006	86.0421	1	1	1×10^{2}	1×10^{3}
W cm^{-1} K^{-1}	57.7789	6.93347×10^{2}	57.8176	6.93811×10^{2}	0.238846	0.239006	86.0421	1	1	1×10^{2}	1×10^{3}
W m^{-1} K^{-1}	0.577789	6.93347	0.578176	6.93811	2.38846×10^{-3}	2.39006×10^{-3}	0.860421	1×10^{-2}	1×10^{-2}	1	10
mW cm^{-1} K^{-1}	5.77789×10^{-2}	0.693347	5.78176×10^{-2}	0.693811	2.38846×10^{-4}	2.39006×10^{-4}	8.60421×10^{-2}	1×10^{-3}	1×10^{-3}	0.1	1

CONVERSION FACTORS FOR ELECTRICAL RESISTIVITY UNITS

To convert FROM ↓ multiply by appropriate factor to OBTAIN →	abΩ cm	μΩ cm	Ω cm	StatΩ cm	Ω m	Ω cir. mil ft^{-1}	Ω in.	Ω ft
abohm centimeter	1	1×10^{-3}	10^{-9}	1.113×10^{-21}	10^{-11}	6.015×10^{-3}	3.937×10^{-10}	3.281×10^{-11}
microohm centimeter	10^{3}	1	10^{-6}	1.113×10^{-18}	10^{-8}	6.015	3.937×10^{-7}	3.281×10^{-6}
ohm centimeter	10^{8}	10^{6}	1	1.113×10^{-12}	1×10^{-2}	6.015×10^{6}	3.937×10^{-1}	3.281×10^{-2}
statohm centimeter (esu)	8.987×10^{20}	8.987×10^{17}	8.987×10^{11}	1	8.987×10^{9}	5.406×10^{18}	3.538×10^{11}	2.949×10^{10}
ohm meter	10^{11}	10^{8}	10^{2}	1.113×10^{-10}	1	6.015×10^{8}	3.937×10^{1}	3.281
ohm circular mil per foot	1.662×10^{2}	1.662×10^{-1}	1.662×10^{-7}	1.850×10^{-19}	1.662×10^{-9}	1	6.54×10^{-6}	5.45×10^{-9}
ohm inch	2.54×10^{9}	2.54×10^{6}	2.54	2.827×10^{-12}	2.54×10^{-2}	1.528×10^{7}	1	8.3×10^{-2}
ohm foot	3.048×10^{10}	3.048×10^{7}	3.048×10^{-1}	3.3924×10^{-11}	3.048×10^{-1}	1.833×10^{8}	12	1

CONVERSION FACTORS FOR CHEMICAL KINETICS

Equivalent Second Order Rate Constants

A \ B	$cm^3\ mol^{-1}s^{-1}$	$dm^3\ mol^{-1}s^{-1}$	$m^3\ mol^{-1}s^{-1}$	cm^3 $molecule^{-1}s^{-1}$	$(mmHg)^{-1}s^{-1}$	$atm^{-1}s^{-1}$	$ppm^{-1}min^{-1}$	$m^2kN^{-1}s^{-1}$
$1\ cm^3\ mol^{-1}\ s^{-1}$ =	1	10^{-3}	10^{-6}	1.66×10^{-24}	$1.604 \times 10^{-5}\ T^{-1}$	$1.219 \times 10^{-2}\ T^{-1}$	2.453×10^{-9}	$1.203 \times 10^{-4}\ T^{-1}$
$1\ dm^3\ mol^{-1}\ s^{-1}$ =	10^3	1	10^{-3}	1.66×10^{-21}	$1.604 \times 10^{-2}\ T^{-1}$	$12.19\ T^{-1}$	2.453×10^{-6}	$1.203 \times 10^{-1}\ T^{-1}$
$1\ m^3\ mol^{-1}\ s^{-1}$ =	10^6	10^3	1	1.66×10^{-18}	$16.04\ T^{-1}$	$1.219 \times 10^4\ T^{-1}$	2.453×10^{-3}	$120.3\ T^{-1}$
$1 cm^3\ molecule^{-1}s^{-1}$ =	6.023×10^{23}	6.023×10^{20}	6.023×10^{17}	1	$9.658 \times 10^{18} T^{-1}$	$7.34 \times 10^{21}\ T^{-1}$	1.478×10^{15}	$7.244 \times 10^{19} T^{-1}$
$1\ (mmHg)^{-1}\ s^{-1}$ =	$6.236 \times 10^4\ T$	$62.36\ T$	$6.236 \times 10^{-2}\ T$	$1.035 \times 10^{-19}\ T$	1	760	4.56×10^{-2}	7.500
$1\ atm^{-1}\ s^{-1}$	$82.06\ T$	$8.206 \times 10^{-2}\ T$	$8.206 \times 10^{-5}\ T$	$1.362 \times 10^{-22}\ T$	1.316×10^{-3}	1	6×10^{-5}	9.869×10^{-3}
$1\ ppm^{-1}\ min^{-1}$ = at 298 K, 1 atm total pressure	4.077×10^8	4.077×10^5	407.7	6.76×10^{-16}	21.93	1.667×10^4	1	164.5
$1\ m^2kN^{-1}s^{-1}$ =	$8314\ T$	$8.314\ T$	$8.314 \times 10^{-3}\ T$	$1.38 \times 10^{-20}\ T$	0.1333	101.325	6.079×10^{-3}	1

To convert a rate constant from one set of units A to a new set B find the conversion factor for the row A under column B and multiply the old value by it, e.g.. to convert $cm^3\ molecule^{-1}\ s^{-1}$ to $m^3\ mol^{-1}\ s^{-1}$ multiply by 6.023×10^{17}.

Table adapted from High Temperature Reaction Rate Data No. 5, The University, Leeds (1970).

Equivalent Third Order Rate Constants

A \ B	$cm^6\ mol^{-2}s^{-1}$	$dm^6\ mol^{-1}s^{-1}$	$m^6\ mol^{-2}\ s^{-1}$	cm^6 $molecule^{-2}\ s^{-1}$	$(mmHg)^{-2}\ s^{-1}$	$atm^{-2}\ s^{-1}$	$ppm^{-2}\ min^{-1}$	$m^4\ kN^{-2}s^{-1}$
$1\ cm^6\ mol^{-2}\ s^{-1}$=	1	10^{-6}	10^{-12}	2.76×10^{-48}	$2.57 \times 10^{-10} T^{-2}$	$1.48 \times 10^{-4}\ T^{-2}$	1.003×10^{-19}	$1.477 \times 10^{-8} T^{-2}$
$1\ dm^6\ mol^{-2}\ s^{-1}$ =	10^6	1	10^{-6}	2.76×10^{-42}	$2.57 \times 10^{-4} T^{-2}$	$148\ T^{-2}$	1.003×10^{13}	$1.477 \times 10^{-2} T^{-2}$
$1\ m^6\ mol^{-2}\ s^{-1}$=	10^{12}	10^6	1	2.76×10^{-36}	$257\ T^{-2}$	$1.48 \times 10^8 T^{-2}$	1.003×10^{-7}	$1.477 \times 10^4 T^{-2}$
$1cm^6\ molecule^{-2}s^{-1}$=	3.628×10^{47}	3.628×10^{41}	3.628×10^{35}	1	$9.328 \times 10^{37} T^{-2}$	$5.388 \times 10^{43} T^{-2}$	3.64×10^{28}	$5.248 \times 10^{39} T^{-2}$
$1\ (mmHg)^{-2}\ s^{-1}$ =	$3.89 \times 10^9\ T^2$	$3.89 \times 10^3\ T^2$	$3.89 \times 10^{-3}\ T^2$	$1.07 \times 10^{-38} T^2$	1	5.776×10^5	3.46×10^{-5}	56.25
$1\ atm^{-2}\ s^{-1}$ =	$6.733 \times 10^3\ T^2$	$6.733 \times 10^{3}\ T^2$	$6.733 \times 10^{-9}\ T^2$	$1.86 \times 10^{-44} T^2$	1.73×10^{-6}	1	6×10^{-11}	9.74×10^{-5}
$1\ ppm^{-2}\ min^{-1}$= at 298K, 1 atm total pressure	9.97×10^{18}	9.97×10^{12}	9.97×10^6	2.75×10^{-29}	2.89×10^4	1.667×10^{10}	1	1.623×10^6
$1\ m^4 kN^{-2}s^{-1}$=	$6.91 \times 10^7\ T^2$	$6.91\ T^2$	$69.1 \times 10^{-5}\ T^2$	$1.904 \times 10^{-40}\ T^2$	0.0178	1.027×10^4	6.16×10^{-7}	1

From *J. Phys. Chem. Ref. Data*, 9, 470, 1980, by permission of the authors and the copyright owner, the American Institute of Physics.

CONVERSION FACTORS FOR IONIZING RADIATION

Conversion between SI and Other Units

Quantity	Symbol for quantity	Expression in SI units	Expression in symbols for SI units	Special name for SI units	Symbols using special names	Conventional units	Symbol for conventional unit	Value of conventional unit in SI units
Activity	A	1 per second	s^{-1}	becquerel	Bq	curie	Ci	3.7×10^{10} Bq
Absorbed dose	D	joule per kilogram	$J\ kg^{-1}$	gray	Gy	rad	rad	0.01 Gy
Absorbed dose rate	$\dot{D}$	joule per kilogram second	$J\ kg^{-1}\ s^{-1}$		$Gy\ s^{-1}$	rad	$rad\ s^{-1}$	$0.01\ Gy\ s^{-1}$
Average energy per ion pair	W	joule	J			electronvolt	eV	1.602×10^{-19} J
Dose equivalent	H	joule per kilogram	$J\ kg^{-1}$	sievert	Sv	rem	rem	0.01 Sv
Dose equivalent rate	$\dot{H}$	joule per kilogram second	$J\ kg^{-1}\ s^{-1}$		$Sv\ s^{-1}$	rem per second	$rem\ s^{-1}$	$0.01\ Sv\ s^{-1}$
Electric current	I	ampere	A			ampere	A	1.0 A
Electric potential difference	U, V	watt per ampere	$W\ A^{-1}$	volt	V	volt	V	1.0 V
Exposure	$\dot{X}$	coulomb per kilogram	$C\ kg^{-1}$			roentgen	R	$2.58 \times 10^{-4}\ C\ kg^{-1}$
Exposure rate	X	coulomb per kilogram second	$C\ kg^{-1}\ s^{-1}$			roentgen per second	$R\ s^{-1}$	$2.58 \times 10^{-4}\ C\ kg^{-1}\ s^{-1}$
Fluence	ϕ	1 per meter squared	m^{-2}			1 per centimeter squared	cm^{-2}	$1.0 \times 10^{4}\ m^{-2}$
Fluence rate	Φ	1 per meter squared second	$m^{-2}\ s^{-1}$			1 per centimeter squared second	$cm^{-2}\ s^{-1}$	$1.0 \times 10^{4}\ m^{-2}\ s^{-1}$
Kerma	K	joule per kilogram	$J\ kg^{-1}$	gray	Gy	rad	rad	0.01 Gy
Kerma rate	$\dot{K}$	joule per kilogram second	$J\ kg^{-1}\ s^{-1}$		$Gy\ s^{-1}$	rad per second	$rad\ s^{-1}$	$0.01\ Gy\ s^{-1}$
Lineal energy	y	joule per meter	$J\ m^{-1}$			kiloelectron volt per micrometer	$keV\ \mu m^{-1}$	$1.602 \times 10^{-10}\ J\ m^{-1}$
Linear energy transfer	L	joule per meter	$J\ m^{-1}$			kiloelectron volt per micrometer	$keV\ \mu m^{-1}$	$1.602 \times 10^{-10}\ J\ m^{-1}$
Mass attenuation coefficient	μ/ρ	meter squared per kilogram	$m^2\ kg^{-1}$			centimeter squared per gram	$cm^2\ g^{-1}$	$0.1\ m^2\ kg^{-1}$
Mass energy transfer coefficient	μ_{tr}/ρ	meter squared per kilogram	$m^2\ kg^{-1}$			centimeter squared per gram	$cm^2\ g^{-1}$	$0.1\ m^2\ kg^{-1}$
Mass energy absorption coefficient	μ_{en}/ρ	meter squared per kilogram	$m^2\ kg^{-1}$			centimeter squared per gram	$cm^2\ g^{-1}$	$0.1\ m^2\ kg^{-1}$
Mass stopping power	S/ρ	joule meter squared per kilogram	$J\ m^2\ kg^{-1}$			MeV centimeter squared per gram	$MeV\ cm^2\ g^{-1}$	$1.602 \times 10^{-14}\ J\ m^2\ kg^{-1}$
Power	P	joule per second	$J\ s^{-1}$	watt	W	watt	W	1.0 W
Pressure	p	newton per meter squared	$N\ m^{-2}$	pascal	Pa	torr	torr	(101325/760)Pa
Radiation chemical yield	G	mole per joule	$mol\ J^{-1}$			molecules per 100 electron volts	molecules $(100\ eV)^{-1}$	$1.04 \times 10^{-7}\ mol\ J^{-1}$
Specific energy	z	joule per kilogram	$J\ kg^{-1}$	gray	Gy	rad	rad	0.01 Gy

Conversion of Radioactivity Units from MBq to mCi and μCi

MBq	mCi	MBq	mCi	MBq	mCi	MBq	mCi	MBq	mCi
7000	189.	700	18.9	70	1.89	7	189	0.7	18.9
6000	162.	600	16.2	60	1.62	6	162	0.6	16.2
5000	135.	500	13.5	50	1.35	5	135	0.5	13.5
4000	108.	400	10.8	40	1.08	4	108	0.4	10.8
3000	81.	300	8.1	30	810	3	81	0.3	8.1
2000	54.	200	5.4	20	540	2	54	0.2	5.4
1000	27.	100	2.7	10	270	1	27	0.1	2.7
900	24.	90	2.4	9	240	0.9	24		
800	21.6	80	2.16	8	220	0.8	21.6		

Conversion of Radioactivity Units from mCi and μCi to MBq

mCi	MBq	mCi	MBq	mCi	MBq	μCi	MBq	μCi	MBq	μCi	MBq
200	7400	40	1480	5	185	1000	37.0	200	7.4	30	1.11
150	5550	30	1110	4	148	900	33.3	100	3.7	20	0.74
100	3700	20	740	3	111	800	29.6	90	3.33	10	0.37
90	3330	10	370	2	74.0	700	25.9	80	2.96	5	0.185
80	2960	9	333	1	37.0	600	22.2	70	2.59	2	0.074
70	2590	8	296			500	18.5	60	2.22	1	0.037
60	2220	7	259			400	14.8	50	1.85		
50	1850	6	222			300	11.1	40	1.48		

Conversion of Radioactivity Units

100 TBq (10^{14} Bq)	=	2.7 kCi (2.7×10^3 Ci)	100 kBq (10^5 Bq)	=	2.7 μCi (2.7×10^{-6}Ci)
10 TBq (10^{13} Bq)	=	270 Ci (2.7×10^2 Ci)	10 kBq (10^4 Bq)	=	270 nCi (2.7×10^{-7} Ci)
1 TBq (10^{12} Bq)	=	27 Ci (2.7×10^1 Ci)	1 kBq (10^3 Bq)	=	27 nCi (2.7×10^{-8} Ci)
100 GBq (10^{11} Bq)	=	2.7 Ci (2.7×10^0 Ci)	100 Bq (10^2 Bq)	=	2.7 nCi (2.7×10^{-9} Ci)
10 GBq (10^{10} Bq)	=	270 mCi (2.7×10^{-1} Ci)	10 Bq (10^1 Bq)	=	270 pCi (2.7×10^{-10} Ci)
1 GBq (10^9 Bq)	=	27 mCi (2.7×10^{-2} Ci)	1 Bq (10^0 Bq)	=	27 pCi (2.7×10^{-11} Ci)
100 MBq (10^8 Bq)	=	2.7 mCi (2.7×10^{-3} Ci)	100 mBq (10^{-1} Bq)	=	2.7 pCi (2.7×10^{-12} Ci)
10 MBq (10^7 Bq)	=	270 μCi (2.7×10^{-4} Ci)	10 mBq (10^{-2} Bq)	=	270 fCi (2.7×10^{-13} Ci)
1 MBq (10^6 Bq)	=	27 μCi (2.7×10^{-5} Ci)	1 mBq (10^{-3} Bq)	=	27 fCi (2.7×10^{-14} Ci)

Conversion of Absorbed Dose Units

SI Units		Conventional	SI Units		Conventional
100 Gy (10^2 Gy)	=	10,000 rad (10^4 rad)	100 μGy (10^{-4} Gy)	=	10 mrad (10^{-2} rad)
10 Gy (10^1 Gy)	=	1,000 rad (10^3 rad)	10 μGy (10^{-5} Gy)	=	1 mrad (10^{-3} rad)
1 Gy (10^0 Gy)	=	100 rad (10^2 rad)	1 μGy (10^{-6} Gy)	=	100 μrad (10^{-4} rad)
100 mGy (10^{-1} Gy)	=	10 rad (10^1 rad)	100 nGy (10^{-7} Gy)	=	10 μrad (10^{-5} rad)
10 mGy (10^{-2} Gy)	=	1 rad (10^0 rad)	10 nGy (10^{-8} Gy)	=	1 μrad (10^{-6} rad)
1 mGy (10^{-3} Gy)	=	100 mrad (10^{-1} rad)	1 nGy (10^{-9} Gy)	=	100 nrad (10^{-7} rad)

Conversion of Dose Equivalent Units

100 Sv (10^2 Sv)	=	10,000 rem (10^4 rem)	100 μSv (10^{-4} Sv)	=	10 mrem (10^{-2} rem)
10 Sv (10^1 Sv)	=	1,000 rem (10^3 rem)	10 μSv (10^{-5} Sv)	=	1 mrem (10^{-3} rem)
1 Sv (10^0 Sv)	=	100 rem (10^2 rem)	1 μSv (10^{-6} Sv)	=	100 μrem (10^{-4} rem)
100 mSv (10^{-1} Sv)	=	10 rem (10^1 rem)	100 nSv (10^{-7} Sv)	=	10 μrem (10^{-5} rem)
10 mSv (10^{-2} Sv)	=	1 rem (10^0 rem)	10 nSv (10^{-8} Sv)	=	1 μrem (10^{-6} rem)
1 mSv (10^{-3} Sv)	=	100 mrem (10^{-1} rem)	1 nSv (10^{-9} Sv)	=	100 nrem (10^{-7} rem)

VALUES OF THE GAS CONSTANT IN DIFFERENT UNIT SYSTEMS

In SI units the value of the gas constant, R, is:

R = 8.314472 Pa m³ K⁻¹ mol⁻¹
= 8314.472 Pa L K⁻¹ mol⁻¹
= 0.08314472 bar L K⁻¹ mol⁻¹

This table gives the appropriate value of R for use in the ideal gas equation, $PV = nRT$, when the variables are expressed in other units. The following conversion factors for pressure units were used in generating the table:

1 atm = 101325 Pa
1 psi = 6894.757 Pa
1 torr (mmHg) = 133.322 Pa [at 0 °C]
1 in Hg = 3386.38 Pa [at 0 °C]
1 in H_2O = 249.082 Pa [at 4 °C]
1 ft H_2O = 2988.98 Pa [at 4 °C]

Reference

Mohr, P. J., Taylor, B. N., and Newell, D. B., "CODATA recommended values of the fundamental physical constants: 2006", *J. Phys. Chem. Ref. Data* 37, 1187, 2008.

Units of V, T, n			Units of P						
V	T	n	kPa	atm	psi	mmHg	in Hg	in H_2O	ft H_2O
ft³	K	mol	0.2936228	0.00289784	0.0425864	2.20236	0.0867070	1.17881	0.0982351
		lb·mol	133.1851	1.31443	19.3168	998.973	39.3296	534.704	44.5587
	°R	mol	0.1631238	0.00160990	0.0236591	1.22353	0.0481706	0.654900	0.0545751
		lb·mol	73.99170	0.730242	10.7316	554.984	21.8498	297.058	24.7548
cm³	K	mol	8314.472	82.0574	1205.91	62363.8	2455.27	33380.4	2781.71
		lb·mol	3771381	37220.6	546993	282878000	1113690	15141100	1261760
	°R	mol	4619.151	45.5875	669.951	34646.5	1364.03	18544.7	1545.39
		lb·mol	2095211	20678.1	303885	15715400	618717	8411730	700979
L	K	mol	8.314472	0.0820574	1.20591	62.3638	2.45527	33.3804	2.78171
		lb·mol	3771.381	37.2206	546.993	28287.8	1113.69	15141.1	1261.76
	°R	mol	4.619151	0.0455875	0.669951	34.6465	1.36403	18.5447	1.54539
		lb·mol	2095.211	20.6781	303.885	15715.4	618.717	8411.73	700.979
m³	K	mol	0.008314472	0.0000820574	0.00120591	0.0623638	0.00245527	0.0333804	0.00278171
		lb·mol	3.771381	0.0372206	0.546993	28.2878	1.11369	15.1411	1.26176
	°R	mol	0.004619151	0.0000455875	0.000669951	0.0346465	0.00136403	0.0185447	0.00154539
		lb·mol	2.095211	0.0206781	0.303885	15.7154	0.618717	8.41173	0.700979

Section 2
Symbols, Terminology, and Nomenclature

Symbols and Terminology for Physical and Chemical Quantities 2-1
Expression of Uncertainty of Measurements 2-13
Nomenclature for Chemical Compounds 2-15
Nomenclature for Inorganic Ions and Ligands 2-16
Organic Substituent Groups and Ring Systems 2-23
Representation of Chemical Structures with the IUPAC International Chemical Identifier (InChI) 2-27
Scientific Abbreviations, Acronyms, and Symbols 2-29
Greek, Russian, and Hebrew Alphabets 2-43
Definitions of Scientific Terms 2-44
Thermodynamic Functions and Relations 2-69
Nobel Laureates in Chemistry and Physics 2-70

SYMBOLS AND TERMINOLOGY FOR PHYSICAL AND CHEMICAL QUANTITIES

The International Organization for Standardization (ISO), International Union of Pure and Applied Chemistry (IUPAC), and the International Union of Pure and Applied Physics (IUPAP) have jointly developed a set of recommended symbols for physical and chemical quantities. Consistent use of these recommended symbols helps assure unambiguous scientific communication. The list below is reprinted from Reference 1 with permission from IUPAC. Full details may be found in the following references:

1. Ian Mills, Ed., *Quantities, Units, and Symbols in Physical Chemistry,* Blackwell Scientific Publications, Oxford, 1988. Third Edition: RSC Publishing, Cambridge, UK, 2007.
2. E. R. Cohen and P. Giacomo, *Symbols, Units, Nomenclature, and Fundamental Constants in Physics,* Document IUPAP–25, 1987; also published in *Physica* 146A, 1–68, 1987.
3. *ISO Standards Handbook 2: Units of Measurement,* International Organization of Standardization, Geneva, 1982.

GENERAL RULES

The value of a physical quantity is expressed as the product of a numerical value and a unit, e.g.:

$T = 300$ K
$V = 26.2$ cm^3
$C_p = 45.3$ J mol^{-1} K^{-1}

The symbol for a physical quantity is always given in italic (sloping) type, while symbols for units are given in roman type. Column headings in tables and axis labels on graphs may conveniently be written as the physical quantity symbol divided by the unit symbol, e.g.:

T/K
V/cm^3
C_p/J mol^{-1} K^{-1}

The values in the table or graph axis are then pure numbers. Subscripts to symbols for physical quantities should be italic if the subscript refers to another physical quantity or to a number, e.g.:

C_p – heat capacity at constant pressure
B_n – nth virial coefficient

Subscripts that have other meanings should be in roman type:

m_p – mass of the proton
E_k – kinetic energy

The following tables give the recommended symbols for the major classes of physical and chemical quantities. The expression in the Definition column is given as an aid in identifying the quantity but is not necessarily the complete or unique definition. The SI Unit gives one (not necessarily unique) expression for the coherent SI unit for the quantity. Other equivalent unit expressions, including those that involve SI prefixes, may be used.

Name	Symbol	Definition	SI unit
Space and Time			
cartesian space coordinates	x, y, z		m
spherical polar coordinates	r, θ, ϕ		m, 1, 1
generalized coordinate	q, q_i		(varies)
position vector	$\boldsymbol{r}$	$\boldsymbol{r} = x\boldsymbol{i} + y\boldsymbol{j} + z\boldsymbol{k}$	m
length	l		m
special symbols:			
height	h		
breadth	b		
thickness	d, δ		
distance	d		
radius	r		
diameter	d		
path length	s		
length of arc	s		
area	A, A_s, S		m^2
volume	$V, (v)$		m^3
plane angle	$\alpha, \beta, \gamma, \theta, \phi$...	$\alpha = s/r$	rad, 1
solid angle	ω, Ω	$\omega = A/r^2$	sr, 1
time	t		s
period	T	$T = t/N$	s
frequency	ν, f	$\nu = 1/T$	Hz
circular frequency, angular frequency	ω	$\omega = 2\pi\nu$	rad s^{-1}, s^{-1}
characteristic time interval, relaxation time, time constant	τ, T	$\tau = \lvert \mathrm{d}t/\mathrm{dln}x \rvert$	s
angular velocity	ω	$\omega = \mathrm{d}\phi/\mathrm{d}t$	rad s^{-1}, s^{-1}
velocity	$\boldsymbol{v}, \boldsymbol{u}, \boldsymbol{w}, \boldsymbol{c}, \dot{\boldsymbol{r}}$	$\boldsymbol{v} = \mathrm{d}\boldsymbol{r}/\mathrm{d}t$	m s^{-1}

Name	Symbol	Definition	SI unit
speed	v, u, w, c	$v = \|\mathbf{v}\|$	m s^{-1}
acceleration	$\mathbf{a}, (g)$	$\mathbf{a} = \mathrm{d}\mathbf{v}/\mathrm{d}t$	m s^{-2}
Classical Mechanics			
mass	m		kg
reduced mass	μ	$\mu = m_1 m_2/(m_1 + m_2)$	kg
density, mass density	ρ	$\rho = m/V$	kg m^{-3}
relative density	d	$d = \rho/\rho$	1
surface density	ρ_A, ρ_S	$\rho_A = m/A$	kg m^{-2}
specific volume	v	$v = V/m = 1/\rho$	m^3 kg^{-1}
momentum	$\mathbf{p}$	$\mathbf{p} = m\mathbf{v}$	kg m s^{-1}
angular momentum, action	$\mathbf{L}$	$\mathbf{L} = \mathbf{r} \times \mathbf{p}$	J s
moment of inertia	I, J	$I = \Sigma m_i r_i^2$	kg m^2
force	$\mathbf{F}$	$\mathbf{F} = \mathrm{d}\mathbf{p}/\mathrm{d}t = m\mathbf{a}$	N
torque, moment of a force	$\mathbf{T}, (\mathbf{M})$	$\mathbf{T} = \mathbf{r} \times \mathbf{F}$	N m
energy	E		J
potential energy	E_p, V, Φ	$E_p = \int \mathbf{F} \cdot ds$	J
kinetic energy	E_k, T, K	$E_k = 1/2mv^2$	J
work	W, w	$W = \int \mathbf{F} \cdot ds$	J
Hamilton function	H	$H(q, p) = T(q, p) + V(q)$	J
Lagrange function	L	$L(q, \dot{q}) = T(q, \dot{q}) - V(q)$	J
pressure	p, P	$p = F/A$	Pa, N m^{-2}
surface tension	γ, σ	$\gamma = \mathrm{d}W/\mathrm{d}A$	N m^{-1}, J m^{-2}
weight	$G, (W, P)$	$G = mg$	N
gravitational constant	G	$F = Gm_1 m_2/r^2$	N m^2 kg^{-2}
normal stress	σ	$\sigma = F/A$	Pa
shear stress	τ	$\tau = F/A$	Pa
linear strain, relative elongation	ε, e	$\varepsilon = \Delta l/l$	1
modulus of elasticity, Young's modulus	E	$E = \sigma/\varepsilon$	Pa
shear strain	γ	$\gamma = \Delta x/d$	1
shear modulus	G	$G = \tau/\gamma$	Pa
volume strain, bulk strain	θ	$\theta = \Delta V/V_0$	1
bulk modulus, compression modulus	K	$K = -V_0(\mathrm{d}p/\mathrm{d}V)$	Pa
viscosity, dynamic viscosity	η, μ	$\tau_{x,z} = \eta(\mathrm{d}v_x/\mathrm{d}z)$	Pa s
fluidity	ϕ	$\phi = 1/\eta$	m kg^{-1} s
kinematic viscosity	ν	$\nu = \eta/\rho$	m^2 s^{-1}
friction coefficient	$\mu, (f)$	$F_{\mathrm{frict}} = \mu F_{\mathrm{norm}}$	1
power	P	$P = \mathrm{d}W/\mathrm{d}t$	W
sound energy flux	P, P_a	$P = \mathrm{d}E/\mathrm{d}t$	W
acoustic factors			
reflection factor	ρ	$\rho = P_r/P_0$	1
acoustic absorption factor	$\alpha_a, (\alpha)$	$\alpha_a = 1 - \rho$	1
transmission factor	τ	$\tau = P_{tr}/P_0$	1
dissipation factor	δ	$\delta = \alpha_a - \tau$	1
Electricity and Magnetism			
quantity of electricity, electric charge	Q		C
charge density	ρ	$\rho = Q/V$	C m^{-3}
surface charge density	σ	$\sigma = Q/A$	C m^{-2}
electric potential	V, ϕ	$V = \mathrm{d}W/\mathrm{d}Q$	V, J C^{-1}
electric potential difference	$U, \Delta V, \Delta\phi$	$U = V_2 - V_1$	V
electromotive force	E	$E = \int(\mathbf{F}/Q) \cdot ds$	V
electric field strength	$\mathbf{E}$	$\mathbf{E} = \mathbf{F}/Q = -\mathrm{grad}\ V$	V m^{-1}
electric flux	Ψ	$\Psi = \int \mathbf{D} \cdot \mathrm{d}A$	C
electric displacement	$\mathbf{D}$	$\mathbf{D} = \varepsilon \mathbf{E}$	C m^{-2}
capacitance	C	$C = Q/U$	F, C V^{-1}
permittivity	ε	$\mathbf{D} = \varepsilon \mathbf{E}$	F m^{-1}
permittivity of vacuum	ε_0	$\varepsilon_0 = \mu_0^{-1} c_0^{-2}$	F m^{-1}
relative permittivity	ε_r	$\varepsilon_r = \varepsilon/\varepsilon_0$	1
dielectric polarization (dipole moment per volume)	$\mathbf{P}$	$\mathbf{P} = \mathbf{D} - \varepsilon_0 \mathbf{E}$	C m^{-2}
electric susceptibility	χ_e	$\chi_e = \varepsilon r - 1$	1
electric dipole moment	$\mathbf{p}, \boldsymbol{\mu}$	$\mathbf{p} = Q\mathbf{r}$	C m

Name	Symbol	Definition	SI unit
electric current	I	$I = \mathrm{d}Q/\mathrm{d}t$	A
electric current density	$\boldsymbol{j}, \boldsymbol{J}$	$I = \int \boldsymbol{j} \cdot \mathrm{d}\boldsymbol{A}$	$\mathrm{A\ m^{-2}}$
magnetic flux density, magnetic induction	$\boldsymbol{B}$	$\boldsymbol{F} = Q\boldsymbol{v} \times \boldsymbol{B}$	T
magnetic flux	Φ	$\Phi = \int \boldsymbol{B} \cdot \mathrm{d}\boldsymbol{A}$	$\mathrm{A\ m^{-2}}$
magnetic field strength	$\boldsymbol{H}$	$\boldsymbol{B} = \mu\boldsymbol{H}$	$\mathrm{A\ m^{-2}}$
permeability	μ	$\boldsymbol{B} = \mu\boldsymbol{H}$	$\mathrm{N\ A^{-2}}$, $\mathrm{H\ m^{-1}}$
permeability of vacuum	μ_0		$\mathrm{H\ m^{-1}}$
relative permeability	μ_r	$\mu_r = \mu/\mu_0$	1
magnetization (magnetic dipole moment per volume)	$\boldsymbol{M}$	$\boldsymbol{M} = \boldsymbol{B}/\mu_0 - \boldsymbol{H}$	$\mathrm{A\ m^{-1}}$
magnetic susceptibility	$\chi, \kappa, (\chi_m)$	$\chi = \mu_r - 1$	1
molar magnetic susceptibility	χ_m	$\chi_m = V_m\chi$	$\mathrm{m^3\ mol^{-1}}$
magnetic dipole moment	$\boldsymbol{m}, \boldsymbol{\mu}$	$E_p = -\boldsymbol{m} \cdot \boldsymbol{B}$	$\mathrm{A\ m^2}$, $\mathrm{J\ T^{-1}}$
electrical resistance	R	$R = U/I$	Ω
conductance	G	$G = 1/R$	S
loss angle	δ	$\delta = (\pi/2) + \phi_I - \phi_U$	1, rad
reactance	X	$X = (U/I)\sin\delta$	Ω
impedance (complex impedance)	Z	$Z = R + iX$	Ω
admittance (complex admittance)	Y	$Y = 1/Z$	S
susceptance	B	$Y = G + iB$	S
resistivity	ρ	$\rho - E/j$	Ω m
conductivity	κ, γ, σ	$\kappa = 1/\rho$	$\mathrm{S\ m^{-1}}$
self-inductance	L	$E = -L(\mathrm{d}I/\mathrm{d}t)$	H
mutual inductance	M, L_{12}	$E_1 = L_{12}(\mathrm{d}I_2/\mathrm{d}t)$	H
magnetic vector potential	$\boldsymbol{A}$	$\boldsymbol{B} = \nabla \times \boldsymbol{A}$	$\mathrm{Wb\ m^{-1}}$
Poynting vector	$\boldsymbol{S}$	$\boldsymbol{S} = \boldsymbol{E} \times \boldsymbol{H}$	$\mathrm{W\ m^{-2}}$
Quantum Mechanics			
momentum operator	$\dot{\boldsymbol{p}}$	$\dot{\boldsymbol{p}} = -ih\nabla$	$\mathrm{m^{-1}\ J\ s}$
kinetic energy operator	$\hat{T}$	$\hat{T} = -(h^2/2m)\nabla^2$	J
Hamiltonian operator	$\hat{H}$	$\hat{H} = \hat{T} + V$	J
wavefunction, state function	Ψ, ψ, ϕ	$\hat{H}\psi = E\psi$	$(\mathrm{m^{-3/2}})$
probability density	P	$P = \psi^*\psi$	$(\mathrm{m^{-3}})$
charge density of electrons	ρ	$\rho = -eP$	$(\mathrm{C\ m^{-3}})$
probability current density	$\boldsymbol{S}$	$\boldsymbol{S} = -i\hbar(\psi^*\nabla\psi - \psi\nabla\psi^*)/2m_e$	$(\mathrm{m^{-2}\ s^{-1}})$
electric current density of electrons	$\boldsymbol{j}$	$\boldsymbol{j} = -e\boldsymbol{S}$	$(\mathrm{A\ m^{-2}})$
matrix element of operator Â	$A_{ij}, \langle i\vert\hat{A}\vert j\rangle$	$A_{ij} = \int\psi_i{}^*\hat{A}\psi_j\mathrm{d}\tau$	(varies)
expectation value of operator Â	$\langle A\rangle, \bar{A}$	$\langle A\rangle = \int\psi^*\hat{A}\Psi\mathrm{d}\tau$	(varies)
hermitian conjugate of Â	$\hat{A}^\dagger$	$(\hat{A}^\dagger)_{ij} = (A_{ji})^*$	(varies)
commutator of Â and $\hat{B}$	$[\hat{A}, \hat{B}], [\hat{A}, \hat{B}]_-$	$[\hat{A}\hat{B}] = \hat{A}\hat{B} - \hat{B}\hat{A}$	(varies)
anticommutator	$[\hat{A}, \hat{B}]_+$	$[\hat{A}, \hat{B}]_+ = \hat{A}\hat{B} + \hat{B}\ddot{A}$	(varies)
spin wavefunction	$\alpha; \beta$		1
coulomb integral	H_{AA}	$H_{AA} = \int\psi_A{}^*\hat{H}\psi_A\mathrm{d}\tau$	J
resonance integral	H_{AB}	$H_{AB} = \int\psi_A{}^*\hat{H}\psi_B\mathrm{d}\tau$	J
overlap integral	S_{AB}	$S_{AB} = \int\psi_A{}^*\psi_B\mathrm{d}\tau$	1
Atoms and Molecules			
nucleon number, mass number	A		1
proton number, atomic number	Z		1
neutron number	N	$N = A - Z$	1
electron rest mass	m_e		kg
mass of atom, atomic mass	m_a, m		kg
atomic mass constant	m_u	$m_u = m_a(^{12}\mathrm{C})/12$	kg
mass excess	Δ	$\Delta = m_a - Am_u$	kg
elementary charge, proton charge	e		C
Planck constant	h		J s
Planck constant/2π	$\hbar$	$\hbar = h/2\pi$	J s
Bohr radius	a_0	$a_0 = 4\pi\varepsilon_0\hbar^2/m_e e^2$	m
Hartree energy	E_h	$E_h = \hbar^2/m_e a_0{}^2$	J
Rydberg constant	R_∞	$R_\infty = E_h/2hc$	$\mathrm{m^{-1}}$
fine structure constant	α	$\alpha = e^2/4\pi\varepsilon_0\hbar c$	1

Name	Symbol	Definition	SI unit
ionization energy	E_i		J
electron affinity	E_{ea}		J
dissociation energy	E_d, D		J
from the ground state	D_0		J
from the potential minimum	D_e		J
principal quantum number (H atom)	n	$E = -hcR/n^2$	1
angular momentum quantum numbers	see under Spectroscopy		
magnetic dipole moment of a molecule	$\boldsymbol{m}, \boldsymbol{\mu}$	$E_p = -\boldsymbol{m} \cdot \boldsymbol{B}$	J T^{-1}
magnetizability of a molecule	ξ	$\boldsymbol{m} = \xi B$	J T^{-2}
Bohr magneton	μ_B	$\mu_B = e\hbar/2m_e$	J T^{-1}
nuclear magneton	μ_N	$\mu_N = (m_e/m_p)\mu_B$	J T^{-1}
magnetogyric ratio (gyromagnetic ratio)	γ	$\gamma = \mu/L$	C kg^{-1}
g factor	g		1
Larmor circular frequency	ω_L	$\omega_L = (e/2m)B$	s^{-1}
Larmor frequency	ν_L	$\nu_L = \omega_L/2\pi$	Hz
longitudinal relaxation time	T_1		s
transverse relaxation time	T_2		s
electric dipole moment of a molecule	$\boldsymbol{p}, \boldsymbol{\mu}$	$E_p = -\boldsymbol{p} \cdot \boldsymbol{E}$	C m
quadrupole moment of a molecule	$\boldsymbol{Q}; \boldsymbol{\Theta}$	$E_p = 1/2Q: V'' = 1/3\Theta: V''$	C m^2
quadrupole moment of a nucleus	eQ	$eQ = 2 \cdot \langle \Theta_{ZZ} \rangle$	C m^2
electric field gradient tensor	$\boldsymbol{q}$	$q_{\alpha\beta} = -\partial^2 V/\partial\alpha\partial\beta$	V m^{-2}
quadrupole interaction energy tensor	χ	$\chi_{\alpha\beta} = eQq_{\alpha\beta}$	J
electric polarizability of a molecule	α	p (induced) $= \alpha E$	C m^2 V^{-1}
activity (of a radioactive substance)	A	$A = -\mathrm{d}N_B/\mathrm{d}t$	Bq
decay (rate) constant, disintegration (rate) constant	λ	$A = \gamma N_B$	s^{-1}
half life	$t_{1/2}$, $T_{1/2}$		s
mean life	τ		s
level width	Γ	$\Gamma = \hbar/\tau$	J
disintegration energy	Q		J
cross section (of a nuclear reaction)	σ		m^2
Spectroscopy			
total term	T	$T = E_{tot}/hc$	m^{-1}
transition wavenumber	$\tilde{\nu}$, (ν)	$\tilde{\nu} = T' - T''$	m^{-1}
transition frequency	ν	$\nu = (E' - E'')/h$	Hz
electronic term	T_e	$T_e = E_e/hc$	m^{-1}
vibrational term	G	$G = E_{vib}/hc$	m^{-1}
rotational term	F	$F = E_{rot}/hc$	m^{-1}
spin orbit coupling constant	A	$T_{s.o.} = A\langle \hat{\boldsymbol{L}} \cdot \hat{\boldsymbol{S}} \rangle$	m^{-1}
principal moments of inertia	$I_A; I_B; I_C$	$I_A \le I_B \le I_C$	kg m^2
rotational constants,			
in wavenumber	$\tilde{A}; \tilde{B}; \tilde{C}$	$\tilde{A} = h/8\pi^2 cI_A$	m^{-1}
in frequency	$A; B; C$	$A = h/8\pi^2 I_A$	Hz
inertial defect	Δ	$\Delta = I_C - I_A - I_B$	kg m^2
asymmetry parameter	κ	$\kappa = \dfrac{(2B - A - C)}{(A - C)}$	1
centrifugal distortion constants,			
S reduction	$D_J; D_{JK}; D_K; d_1; d_2$		m^{-1}
A reduction	$\Delta_J; \Delta_{JK}; \Delta_K; \delta_J; \delta_K$		m^{-1}
harmonic vibration wavenumber	$\omega_e; \omega_r$		m^{-1}
vibrational anharmonicity constant	$\omega_e x_e; x_{rs}; g_{u'}$		m^{-1}
vibrational quantum numbers	$v_r; l_t$		1
Coriolis zeta constant	ζ_{rs}^{α}		1
angular momentum quantum numbers	see additional information below		
degeneracy, statistical weight	g, d, β		1
electric dipole moment of a molecule	$\boldsymbol{p}, \boldsymbol{\mu}$	$E_p = -\boldsymbol{p} \cdot \boldsymbol{E}$	C m
transition dipole moment of a molecule	$\boldsymbol{M}, \boldsymbol{R}$	$\boldsymbol{M} = \int \psi' \boldsymbol{p} \psi'' \mathrm{d}\tau$	C m
molecular geometry, interatomic distances,			
equilibrium distance	r_e		m
zero–point average distance	r_z		m

Name	Symbol	Definition	SI unit
ground state distance	r_0		m
substitution structure distance	r_s		m
vibrational coordinates,			
internal coordinates	R_i, r_i, θ_j, etc.		(varies)
symmetry coordinates	S_i		(varies)
normal coordinates			
mass adjusted	Q_r		$kg^{1/2}$ m
dimensionless	q_r		1
vibrational force constants,			
diatomic	$f, (k)$	$f = \partial^2 V/\partial r^2$	J m^{-2}
polyatomic,			
internal coordinates	f_{ij}	$f_{ij} = \partial^2 V/\partial r_i \partial r_j$	(varies)
symmetry coordinates	F_{ij}	$F_{ij} = \partial^2 V/\partial S_i \partial S_j$	(varies)
dimensionless normal coordinates	$\phi_{rst...}, k_{rst...}$		m^{-1}
nuclear magnetic resonance (NMR),			
magnetogyric ratio	γ	$\gamma = \mu/I\hbar$	C kg^{-1}
shielding constant	σ_A	$B_A = (1 - \sigma_A)B$	1
chemical shift, δ scale	δ	$\delta = 10^6(\nu - \nu_0)/\nu_0$	1
(indirect) spin–spin coupling constant	J_{AB}	$\hat{H}/h = J_{AB}\hat{I}_A \cdot \hat{I}_B$	Hz
direct (dipolar) coupling constant	D_{AB}		Hz
longitudinal relaxation time	T_1		s
transverse relaxation time	T_2		s
electron spin resonance, electron paramagnetic resonance (ESR, EPR),			
magnetogyric ratio	γ	$\gamma = \mu/s\hbar$	C kg^{-1}
g factor	g	$h\nu = g\mu_B B$	1
hyperfine coupling constant,			
in liquids	a, A	$\hat{H}_{hfs}/h = a\hat{S} \cdot \hat{I}$	Hz
in solids	T	$\hat{H}_{hfs}/h = \hat{S} \cdot T \cdot \hat{I}$	Hz

Angular momentum	Operator symbol	Quantum number symbol: Total	*Z*–axis	*z*-axis
electron orbital	$\hat{L}$	L	M_L	Λ
one electron only	$\hat{l}$	l	m_l	λ
electron spin	$\hat{S}$	S	M_S	Σ
one electron only	$\hat{s}$	s	m_s	σ
electron orbital + spin	$\hat{L} + \hat{S}$			$\Omega = \Lambda + \Sigma$
nuclear orbital (rotational)	$\hat{R}$	R		K_R, k_R
nuclear spin	$\hat{I}$	I	M_I	
internal vibrational				
spherical top	$\hat{l}$	$l(l\zeta)$		K_l
other	$\hat{j}, \hat{\pi}$			$l(l\zeta)$
sum of $R + L(+j)$	$\hat{N}$	N		K, k
sum of $N + S$	$\hat{J}$	J	M_J	K, k
sum of $J + I$	$\hat{F}$	F	M_F	

Electromagnetic Radiation

Name	Symbol	Definition	SI unit
wavelength	λ		m
speed of light			
in vacuum	c_0		m s^{-1}
in a medium	c	$c = c_0/n$	m s^{-1}
wavenumber in vacuum	$\tilde{\nu}$	$\tilde{\nu} = \nu/c_0 = 1/n\lambda$	m^{-1}
wavenumber (in a medium)	σ	$\sigma = 1/\lambda$	m^{-1}
frequency	ν	$\nu = c/\lambda$	Hz
circular frequency, pulsatance	ω	$\omega = 2\pi\nu$	s^{-1}, rad s^{-1}
refractive index	n	$n = c_0/c$	1
Planck constant	h		J s

Name	Symbol	Definition	SI unit
Planck constant/2π	$\hbar$	$\hbar = h/2\pi$	J s
radiant energy	Q, W		J
radiant energy density	ρ, w	$\rho = Q/V$	J m^{-3}
spectral radiant energy density			
in terms of frequency	ρ_ν, w_ν	$\rho = d\rho/d\nu$	J m^{-3} Hz^{-1}
in terms of wavenumber	$\rho_{\tilde{\nu}}, w_{\tilde{\nu}}$	$\rho_{\tilde{\nu}} = d\rho / d\tilde{\nu}$	J m^{-2}
in terms of wavelength	ρ_λ, w_λ	$\rho_\lambda = d\rho/d\lambda$	J m^{-4}
Einstein transition probabilities			
spontaneous emission	A_{nm}	$dN_n/dt = -A_{nm}N_n$	s^{-1}
stimulated emission	B_{nm}	$dN_n/dt = -\rho_{\tilde{\nu}}(\tilde{\nu}_{nm}) \times B_{nm}N_n$	s kg^{-1}
stimulated absorption	B_{mn}	$dN_n/dt = -\rho_{\tilde{\nu}}(\tilde{\nu}_{nm}) B_{mn}N_m$	s kg^{-1}
radiant power, radiant energy per time	Φ, P	$\Phi = dQ/dt$	W
radiant intensity	I	$I = d\Phi/d\Omega$	W sr^{-1}
radiant exitance (emitted radiant flux)	M	$M = d\Phi/dA_{source}$	W m^{-2}
irradiance, (radiant flux received)	$E, (I)$	$E = d\Phi/dA$	W m^{-2}
emittance	ε	$\varepsilon = M/M_{bb}$	1
Stefan–Boltzmann constant	σ	$M_{bb} = \sigma T^4$	W m^{-2} K^{-4}
first radiation constant	c_1	$c_1 = 2\pi hc_0^2$	W m^2
second radiation constant	c_2	$c_2 = hc_0/k$	K m
transmittance, transmission factor	τ, T	$\tau = \Phi_{tr}/\Phi_0$	1
absorptance, absorption factor	α	$\alpha = \Phi_{abs}/\Phi_0$	1
reflectance, reflection factor	ρ	$\rho = \Phi_{refl}/\Phi_0$	1
(decadic) absorbance	A	$A = -\lg(1-\alpha_i)$	1
napierian absorbance	B	$B = -\ln(1-\alpha_i)$	1
absorption coefficient			
(linear) decadic	a, K	$a = A/l$	m^{-1}
(linear) napierian	α	$\alpha = B/l$	m^{-1}
molar (decadic)	ε	$\varepsilon = a/c = A/cl$	m^2 mol^{-1}
molar napierian	κ	$\kappa = \alpha/c = B/cl$	m^2 mol^{-1}
absorption index	k	$k = \alpha/4\pi\tilde{\nu}$	1
complex refractive index	$\hat{n}$	$\hat{n} = n + ik$	1
molar refraction	R, R_m	$R = \dfrac{(n^2-1)}{(n^2+2)}V_m$	m^3 mol^{-1}
angle of optical rotation	α		1, rad
Solid State			
lattice vector	$\boldsymbol{R}, \boldsymbol{R}_0$		m
fundamental translation vectors for the crystal lattice	$\boldsymbol{a}_1; \boldsymbol{a}_2; \boldsymbol{a}_3, \boldsymbol{a}; \boldsymbol{b}; \boldsymbol{c}$	$\boldsymbol{R} = n_1\boldsymbol{a}_1 + n_2\boldsymbol{a}_2 + n_3\boldsymbol{a}_3$	m
(circular) reciprocal lattice vector	$\boldsymbol{G}$	$\boldsymbol{G} \cdot \boldsymbol{R} = 2\pi m$	m^{-1}
(circular) fundamental translation vectors for the reciprocal lattice	$\boldsymbol{b}_1; \boldsymbol{b}_2; \boldsymbol{b}_3, \boldsymbol{a}^*; \boldsymbol{b}^*; \boldsymbol{c}^*$	$\boldsymbol{a}_i \cdot \boldsymbol{b}_k = 2\pi\delta_{ik}$	m^{-1}
lattice plane spacing	d		m
Bragg angle	θ	$n\lambda = 2d \sin\theta$	1, rad
order of reflection	n		1
order parameters			
short range	σ		1
long range	s		1
Burgers vector	$\boldsymbol{b}$		m
particle position vector	$\boldsymbol{r}, \boldsymbol{R}_j$		m
equilibrium position vector of an ion	$\boldsymbol{R}_0$		m
displacement vector of an ion	$\boldsymbol{u}$	$\boldsymbol{u} = \boldsymbol{R} - \boldsymbol{R}_0$	m
Debye–Waller factor	B, D		1
Debye circular wavenumber	q_D		m^{-1}
Debye circular frequency	ω_D		s^{-1}
Grüneisen parameter	γ, Γ	$\gamma = \alpha V/\kappa C_v$	1
Madelung constant	$\alpha, \mathcal{M}$	$E_{coul} = \dfrac{\alpha N_A z_+ z_- e^2}{4\pi\varepsilon_0 R_0}$	1
density of states	N_E	$N_E = dN(E)/dE$	J^{-1} m^{-3}
(spectral) density of vibrational modes	N_ω, g	$N_\omega = dN(\omega)/d\omega$	s m^{-3}

Name	Symbol	Definition	SI unit
resistivity tensor	ρ_{ik}	$\boldsymbol{E} = \boldsymbol{\rho} \cdot \boldsymbol{j}$	Ω m
conductivity tensor	σ_{ik}	$\sigma = \rho^{-1}$	S m^{-1}
thermal conductivity tensor	λ_{ik}	$J_q = -\boldsymbol{\lambda} \cdot \text{grad } T$	W m^{-1} K^{-1}
residual resistivity	ρ_R		Ω m
relaxation time	τ	$\tau = l/v_F$	s
Lorenz coefficient	L	$L = \lambda/\sigma T$	V^2 K^{-2}
Hall coefficient	A_H, $\boldsymbol{R}_H$	$\boldsymbol{E} = \boldsymbol{\rho} \cdot \boldsymbol{j} + R_H(\boldsymbol{B} \times \boldsymbol{j})$	m^3 C^{-1}
thermoelectric force	E		V
Peltier coefficient	Π		V
Thomson coefficient	μ, (τ)		V K^{-1}
work function	Φ	$\Phi = E_\infty - E_F$	J
number density, number concentration	n, (p)		m^{-3}
gap energy	E_g		J
donor ionization energy	E_d		J
acceptor ionization energy	E_a		J
Fermi energy	E_F, ε_F		J
circular wave vector, propagation vector	k, $\boldsymbol{q}$	$k = 2\pi/\lambda$	m^{-1}
Bloch function	$u_k(\boldsymbol{r})$	$\psi(\boldsymbol{r}) = u_k(\boldsymbol{r}) \exp(\mathrm{i}\boldsymbol{k} \cdot \boldsymbol{r})$	m$^{-3/2}$
charge density of electrons	ρ	$\rho(\boldsymbol{r}) = -e\psi^*(\boldsymbol{r})\psi(\boldsymbol{r})$	C m^{-3}
effective mass	m^*		kg
mobility	μ	$\mu = v_{drift}/E$	m^2 V^{-1} s^{-1}
mobility ratio	b	$b = \mu_n/\mu_p$	1
diffusion coefficient	D	$\mathrm{d}N/\mathrm{d}t = -DA(\mathrm{d}n/\mathrm{d}x)$	m^2 s^{-1}
diffusion length	L	$L = \sqrt{D\tau}$	m
characteristic (Weiss) temperature	θ, θ_w		K
Curie temperature	T_C		K
Néel temperature	T_N		K
Statistical Thermodynamics			
number of entities	N		1
number density of entities, number concentration	n, C	$n = N/V$	m^{-3}
Avogadro constant	L, N_A		mol^{-1}
Boltzmann constant	k, k_B		J K^{-1}
gas constant (molar)	R	$R = Lk$	J K^{-1} mol^{-1}
molecular position vector	$\boldsymbol{r}$ (x, y, z)		m
molecular velocity vector	$\boldsymbol{c}(c_x, c_y, c_z)$, $\boldsymbol{u}(u_x, u_y, u_z)$	$\boldsymbol{c} = \mathrm{d}\boldsymbol{r}/\mathrm{d}t$	m s^{-1}
molecular momentum vector	$\boldsymbol{p}(p_x, p_y, p_z)$	$\boldsymbol{p} = m\boldsymbol{c}$	kg m s^{-1}
velocity distribution function (Maxwell)	$f(c_x)$	$f(c_x) = (m/2\pi kT)^{1/2} \times \exp(-mc_x^2/2kT)$	m^{-1} s
speed distribution function (Maxwell–Boltzmann)	$F(c)$	$F(c) = (m/2\pi kT)^{3/2} \times 4\pi c^2 \exp(-mc^2/2kT)$	m^{-1} s
average speed	$\bar{c}$, $\bar{u}$, $\langle c \rangle$, $\langle u \rangle$	$\bar{c} = \int cF(c)\mathrm{d}c$	m s^{-1}
generalized coordinate	q		(m)
generalized momentum	p	$p = \partial L/\partial \dot{q}$	(kg m s^{-1})
volume in phase space	Ω	$\Omega = (1/h)\int p\mathrm{d}q$	1
probability	P		1
statistical weight, degeneracy	g, d, W, ω, β		1
density of states	$\rho(E)$	$\rho(E) = \mathrm{d}N/\mathrm{d}E$	J^{-1}
partition function, sum over states, for a single molecule	q, z	$q = \sum_i g_i \exp(-\varepsilon_i/kT)$	1
for a canonical ensemble (system, or assembly)	Q, Z		1
microcanonical ensemble	Ω		1
grand (canonical ensemble)	Ξ		1
symmetry number	σ, s		1
reciprocal temperature parameter	β	$\beta = 1/kT$	J^{-1}
characteristic temperature	Θ		K

Name	Symbol	Definition	SI unit
General Chemistry			
number of entities (e.g. molecules, atoms, ions, formula units)	N		1
amount (of substance)	n	$n_B = N_B/L$	mol
Avogadro constant	L, N_A		mol^{-1}
mass of atom, atomic mass	m_a, m		kg
mass of entity (molecule, or formula unit)	m_f, m		kg
atomic mass constant	m_u	$m_u = m_a(^{12}C)/12$	kg
molar mass	M	$M_B = m/n_B$	$kg\ mol^{-1}$
relative molecular mass (relative molar mass, molecular weight)	M_r	$M_{r,B} = m_B/m_u$	1
molar volume	V_m	$V_{m,B} = V/n_B$	$m^3\ mol^{-1}$
mass fraction	w	$w_B = m_B/\Sigma m_i$	1
volume fraction	ϕ	$\phi_B = V_B/\Sigma V_i$	1
mole fraction, amount fraction, number fraction	x, y	$x_B = n_B/\Sigma n_i$	1
(total) pressure	p, P		Pa
partial pressure	p_B	$p_B = y_B p$	Pa
mass concentration (mass density)	γ, ρ	$\gamma_B = m_B/V$	$kg\ m^{-3}$
number concentration, number density of entities	C, n	$C_B = N_B/V$	m^{-3}
amount concentration, concentration	c	$c_B = n_B/V$	$mol\ m^{-3}$
solubility	s	$s_B = c_B$ (saturated solution)	$mol\ m^{-3}$
molality (of a solute)	$m, (b)$	$m_B = n_B/m_A$	$mol\ kg^{-1}$
surface concentration	Γ	$\Gamma_B = n_B/A$	$mol\ m^{-2}$
stoichiometric number	ν		1
extent of reaction, advancement	ξ	$\Delta\xi = \Delta n_B/\nu_B$	mol
degree of dissociation	α		1
Chemical Thermodynamics			
heat	q, Q		J
work	w, W		J
internal energy	U	$\Delta U = q + w$	J
enthalpy	H	$H = U + pV$	J
thermodynamic temperature	T		K
Celsius temperature	θ, t	$\theta/°C = T/K - 273.15$	°C
entropy	S	$dS \geq dq/T$	$J\ K^{-1}$
Helmholtz energy (Helmholtz function)	A	$A = U - TS$	J
Gibbs energy (Gibbs function)	G	$G = H - TS$	J
Massieu function	J	$J = -A/T$	$J\ K^{-1}$
Planck function	Y	$Y = -G/T$	$J\ K^{-1}$
surface tension	γ, σ	$\gamma = (\partial G/\partial A_s)_{T,p}$	$J\ m^{-2}$, $N\ m^{-1}$
molar quantity X	X_m	$X_m = X/n$	(varies)
specific quantity X	x	$x = X/m$	(varies)
pressure coefficient	β	$\beta = (\partial p/\partial T)v$	$Pa\ K^{-1}$
relative pressure coefficient	α_p	$\alpha_p = (1/p)(\partial p/\partial T)_V$	K^{-1}
compressibility,			
isothermal	κ_T	$\kappa_T = -(1/V)(\partial V/\partial p)_T$	Pa^{-1}
isentropic	κ_S	$\kappa_S = -(1/V)(\partial V/\partial p)_S$	Pa^{-1}
linear expansion coefficient	α_l	$\alpha_l = (1/l)(\partial l/\partial T)$	K^{-1}
cubic expansion coefficient	α, α_V, γ	$\alpha = (1/V)(\partial V/\partial T)_p$	K^{-1}
heat capacity,			
at constant pressure	C_p	$C_p = (\partial H/\partial T)_p$	$J\ K^{-1}$
at constant volume	C_V	$C_V = (\partial U/\partial T)_V$	$J\ K^{-1}$
ratio of heat capacities	$\gamma, (\kappa)$	$\gamma = C_p/C_V$	1
Joule–Thomson coefficient	μ, μ_{JT}	$\mu = (\partial T/\partial p)_H$	$K\ Pa^{-1}$
second virial coefficient	B	$pV_m = RT(1 + B/V_m + \ldots)$	$m^3\ mol^{-1}$
compression factor (compressibility factor)	Z	$Z = pV_m/RT$	1
partial molar quantity X	$X_B, (X'_B)$	$X_B = (\partial X/\partial n_B)_{T,p,n_j \neq B}$	(varies)
chemical potential (partial molar Gibbs energy)	μ	$\mu_B = (\partial G/\partial n_B)_{T,p,n_j \neq B}$	$J\ mol^{-1}$
absolute activity	λ	$\lambda_B = \exp(\mu_B/RT)$	1

Name	Symbol	Definition	SI unit
standard chemical potential	$\mu^{\ominus}, \mu^{o}$		J mol^{-1}
standard partial molar enthalpy	$H_B^{\ominus}$	$H_B^{\ominus} = \mu_B^{\ominus} + TS_B^{\ominus}$	J mol^{-1}
standard partial molar entropy	$S_B^{\ominus}$	$S_B^{\ominus} = -(\partial\mu_B^{\ominus}/\partial T)_p$	J mol^{-1} K^{-1}
standard reaction Gibbs energy (function)	$\Delta_r G^{\ominus}$	$\Delta_r G^{\ominus} = \sum_B \nu_B \mu_B^{\ominus}$	J mol^{-1}
affinity of reaction	$A, (\mathcal{A})$	$A = -(\partial G/\partial\xi)_{p,T} = -\sum_B \nu_B \mu_B$	J mol^{-1}
standard reaction enthalpy	$\Delta_r H^{\ominus}$	$\Delta_r H^{\ominus} = \sum_B \nu_B H_B^{\ominus}$	J mol^{-1}
standard reaction entropy	$\Delta_r S^{\ominus}$	$\Delta_r S^{\ominus} = \sum_B \nu_B S_B^{\ominus}$	J mol^{-1} K^{-1}
equilibrium constant	$K^{\ominus}, K$	$K^{\ominus} = \exp(-\Delta_r G^{\ominus}/RT)$	1
equilibrium constant,			
pressure basis	K_p	$K_p = \prod_B p_B^{\nu_B}$	Pa$^{\Sigma\nu}$
concentration basis	K_c	$K_c = \prod_B c_B^{\nu_B}$	(mol m^{-3})$^{\Sigma\nu}$
molality basis	K_m	$K_m = \prod_D m_B^{\nu_B}$	(mol kg^{-1})$^{\Sigma\nu}$
fugacity	$f, \tilde{p}$	$f_D = \lambda_B \lim_{p\to 0}(p_B/\lambda_B)_T$	Pa
fugacity coefficient	ϕ	$\phi_B = f_B/p_B$	1
activity and activity coefficient referenced to Raoult's law, (relative) activity	a	$a_B = \exp\left[\frac{\mu_B - \mu_B^*}{RT}\right]$	1
activity coefficient	f	$f_B = a_B/x_B$	1
activities and activity coefficients referenced to Henry's law, (relative) activity,			
molality basis	a_m	$a_{m,B} = \exp\left[\frac{\mu_B - \mu_B^{\ominus}}{RT}\right]$	1
concentration basis	a_c	$a_{c,B} = \exp\left[\frac{\mu_B - \mu_B^{\ominus}}{RT}\right]$	1
mole fraction basis	a_x	$a_{x,B} = \exp\left[\frac{\mu_B - \mu_B^{\ominus}}{RT}\right]$	1
activity coefficient,			
molality basis	γ_m	$a_{m,B} = \gamma_{m,B} m_B/m^{\ominus}$	1
concentration basis	γ_c	$a_{c,B} = \gamma_{c,B} c_B/c^{\ominus}$	1
mole fraction basis	γ_x	$a_{x,B} = \gamma_{x,B} x_B$	1
ionic strength,			
molality basis	I_m, I	$I_m = ½ \Sigma m_B z_B^2$	mol kg^{-1}
concentration basis	I_c, I	$I_c = ½ \Sigma c_B z_B^2$	mol m^{-3}
osmotic coefficient,			
molality basis	ϕ_m	$\phi_m = (\mu_A^* - \mu_A)/(RTM_A \Sigma m_B)$	1
mole fraction basis	ϕ_x	$\phi_x = (\mu_A - \mu_A^*)/(RT \ln x_A)$	1
osmotic pressure	Π	$\Pi = c_B RT$ (ideal dilute solution)	Pa

(i) Symbols used as subscripts to denote a chemical process or reaction

These symbols should be printed in roman (upright) type, without a full stop (period).

vaporization, evaporation (liquid → gas)	vap
sublimation (solid → gas)	sub
melting, fusion (solid → liquid)	fus
transition (between two phases)	trs
mixing of fluids	mix
solution (of solute in solvent)	sol
dilution (of a solution)	dil
adsorption	ads
displacement	dpl
immersion	imm

	reaction in general	r
	atomization	at
	combustion reaction	c
	formation reaction	f
(ii) Recommended superscripts		
	standard	$\ominus$, °
	pure substance	*
	infinite dilution	∞
	ideal	id
	activated complex, transition state	‡
	excess quantity	E

Name	Symbol	Definition	SI unit
Chemical Kinetics			
rate of change of quantity X	$\dot{X}$	$\dot{X} = \mathrm{d}X/dt$	(varies)
rate of conversion	$\dot{\xi}$	$\dot{\xi} = \mathrm{d}\xi/dt$	mol s^{-1}
rate of concentration change (due to chemical reaction)	r_B, v_B	$r_B = \mathrm{d}c_B/dt$	mol m^{-3} s^{-1}
rate of reaction (based on amount concentration)	v	$v = \dot{\xi}/V = v_B^{-1}\mathrm{d}c_B/dt$	mol m^{-3} s^{-1}
partial order of reaction	n_B	$v = k\Pi c_B^{n_B}$	1
overall order of reaction	n	$n = \Sigma n_B$	1
rate constant, rate coefficient	k	$v = k\Pi c_B^{n_B}$	(mol^{-1} m^3)$^{n-1}$ s^{-1}
Boltzmann constant	k, k_B		J K^{-1}
half life	$t_{1/2}$	$c(t_{1/2}) = c_0/2$	s
relaxation time	τ	$\tau = 1/(k_1 + k_{-1})$	s
energy of activation, activation energy	E_a, E	$E_a = RT^2\,\mathrm{d}\ln k/\mathrm{d}T$	J mol^{-1}
pre-exponential factor	A	$k = A\exp(-E_a/RT)$	(mol^{-1} m^3)$^{n-1}$ s^{-1}
volume of activation	$\Delta^{\ddagger}V$	$\Delta^{\ddagger}V = -RT \times (\partial \ln k/\partial p)_T$	m^3 mol^{-1}
collision diameter	d	$d_{AB} = r_A + r_B$	m
collision cross-section	σ	$\sigma_{AB} = \pi d_{AB}^2$	m^2
collision frequency	Z_A		s^{-1}
collision number	Z_{AB}, Z_{AA}		m^{-3} s^{-1}
collision frequency factor	z_{AB}, z_{AA}	$z_{AB} = Z_{AB}/Lc_Ac_B$	m^3 mol^{-1} s^{-1}
standard enthalpy of activation	$\Delta^{\ddagger}H^{\ominus}, \Delta H^{\ddagger}$		J mol^{-1}
standard entropy of activation	$\Delta^{\ddagger}S^{\ominus}, \Delta S^{\ddagger}$		J mol^{-1} K^{-1}
standard Gibbs energy of activation	$\Delta^{\ddagger}G^{\ominus}, \Delta G^{\ddagger}$		J mol^{-1}
quantum yield, photochemical yield	ϕ		1
Electrochemistry			
elementary charge (proton charge)	e		C
Faraday constant	F	$F=eL$	C mol^{-1}
charge number of an ion	z	$z_B = Q_B/e$	1
ionic strength	I_c, I	$I_c = ½\,\Sigma c_i z_i^2$	mol m^{-3}
mean ionic activity	$a_\pm$	$a_\pm = m_\pm\gamma_\pm/m^{\ominus}$	1
mean ionic molality	$m_\pm$	$m_\pm^{(\nu_+ + \nu_-)} = m_+^{\nu+}m_-^{\nu-}$	mol kg^{-1}
mean ionic activity coefficient	$\gamma_\pm$	$\gamma_\pm^{(\nu_+ + \nu_-)} = \gamma_+^{\nu+}\gamma_-^{\nu-}$	1
charge number of electrochemical cell reaction	$n, (z)$		1
electric potential difference (of a galvanic cell)	$\Delta V, E, U$	$\Delta V = V_R - V_L$	V
emf, electromotive force	E	$E = \lim_{I\to 0} \Delta V$	V
standard emf, standard potential of the electrochemical cell reaction	$E^{\ominus}$	$E^{\ominus} = -\Delta_r G^{\ominus}/nF = (RT/nF)\ln K^{\ominus}$	V
standard electrode potential	$E^{\ominus}$		V
emf of the cell, potential of the electrochemical cell reaction	E	$E = E^{\ominus} - (RT/nF) \times \Sigma\nu_i \ln a_i$	V
pH	pH	$\mathrm{pH} \approx -\lg\left[\frac{c(\mathrm{H}^+)}{\mathrm{mol\ dm}^{-3}}\right]$	1
inner electric potential	ϕ	$\nabla\phi = -\boldsymbol{E}$	V
outer electric potential	ψ	$\psi = Q/4\pi\varepsilon_0 r$	V

Name	Symbol	Definition	SI unit
surface electric potential	χ	$\chi = \phi - \psi$	V
Galvani potential difference	$\Delta\phi$	$\Delta_\alpha^\beta\phi = \phi^\beta - \phi^\alpha$	V
volta potential difference	$\Delta\psi$	$\Delta_\alpha^\beta\psi = \psi^\beta - \psi^\alpha$	V
electrochemical potential	$\tilde{\mu}$	$\tilde{\mu}_B^{\ \alpha} = (\partial G/\partial n_B^{\ \alpha})$	J mol^{-1}
electric current	I	$I = \mathrm{d}Q/\mathrm{d}t$	A
(electric) current density	j	$j = I/A$	A m^{-2}
(surface) charge density	σ	$\sigma = Q/A$	C m^{-2}
electrode reaction rate constant	k	$k_{ox} = I_a/(nFA\prod_i c_i^{n_i})$	(varies)
mass transfer coefficient, diffusion rate constant	k_d	$k_{d,B} = \lvert\nu_B\rvert I_{l,B}/nFcA$	m s^{-1}
thickness of diffusion layer	δ	$\delta_B = D_B/k_{d,B}$	m
transfer coefficient (electrochemical)	α	$\alpha_c = \dfrac{-\lvert\nu\rvert RT\partial\ln\lvert I_c\rvert}{nF\,\partial E}$	1
overpotential	η	$\eta = E_I - E_{I=0} - IR_u$	V
electrokinetic potential (zeta potential)	ζ		V
conductivity	κ, (σ)	$\kappa = j/E$	S m^{-1}
conductivity cell constant	K_{cell}	$K_{cell} = \kappa R$	m^{-1}
molar conductivity (of an electrolyte)	Λ	$\Lambda_B = \kappa/c_B$	S m^2 mol^{-1}
ionic conductivity, molar conductivity of an ion	λ	$\lambda_B = \lvert z_B\rvert Fu_B$	S m^2 mol^{-1}
electric mobility	u, (μ)	$u_B = v_B/E$	m^2 V^{-1} s^{-1}
transport number	t	$t_B = j_B/\Sigma j_i$	1
reciprocal radius of ionic atmosphere	κ	$\kappa = (2F^2I/\varepsilon RT)^{1/2}$	m^{-1}
Colloid and Surface Chemistry			
specific surface area	a, a_s, s	$a = A/m$	m^2 kg^{-1}
surface amount of B, adsorbed amount of B	$n_B^{\ s}$, $n_B^{\ a}$		mol
surface excess of B	$n_B^{\ \sigma}$		mol
surface excess concentration of B	Γ_B, ($\Gamma_B^{\ \sigma}$)	$\Gamma_B = n_B^{\ \sigma}/A$	mol m^{-2}
total surface excess concentration	Γ, (Γ^σ)	$\Gamma = \sum_i \Gamma_i$	mol m^{-2}
area per molecule	a, σ	$a_B = A/N_B^{\ \sigma}$	m^2
area per molecule in a filled monolayer	a_m, σ_m	$a_{m,B} = A/N_{m,B}$	m^2
surface coverage	θ	$\theta = N_B^{\ \sigma}/N_{m,B}$	1
contact angle	θ		1, rad
film thickness	t, h, δ		m
thickness of (surface or interfacial) layer	τ, δ, t		m
surface tension, interfacial tension	γ, σ	$\gamma = (\partial G/\partial A_s)_{T,p}$	N m^{-1}, J m^{-2}
film tension	Σ_f	$\Sigma_f = 2\gamma_f$	N m^{-1}
reciprocal thickness of the double layer	κ	$\kappa = [2F^2I_c/\varepsilon RT]^{1/2}$	m^{-1}
average molar masses			
number–average	M_n	$M_n = \Sigma n_iM_i/\Sigma n_i$	kg mol^{-1}
mass–average	M_m	$M_m = \Sigma n_iM_i^2/\Sigma n_iM_i$	kg mol^{-1}
Z–average	M_Z	$M_Z = \Sigma n_iM_i^3/\Sigma n_iM_i^2$	kg mol^{-1}
sedimentation coefficient	s	$s = v/a$	s
van der Waals constant	λ		J
retarded van der Waals constant	β, B		J
van der Waals–Hamaker constant	A_H		J
surface pressure	π^s, π	$\pi^s = \gamma^0 - \gamma$	N m^{-1}
Transport Properties			
flux (of a quantity X)	J_X, J	$J_X = A^{-1}\,\mathrm{d}X/\mathrm{d}t$	(varies)
volume flow rate	q_V, $\dot{V}$	$q_v = \mathrm{d}V/\mathrm{d}t$	m^3 s^{-1}
mass flow rate	q_m, $\dot{m}$	$q_m = \mathrm{d}m/\mathrm{d}t$	kg s^{-1}
mass transfer coefficient	k_d		m s^{-1}
heat flow rate	ϕ	$\phi = \mathrm{d}q/\mathrm{d}t$	W
heat flux	J_q	$J_q = \phi/A$	W m^{-2}
thermal conductance	G	$G = \phi/\Delta T$	W K^{-1}
thermal resistance	R	$R = 1/G$	K W^{-1}
thermal conductivity	λ, k	$\lambda = J_q/(\mathrm{d}T/\mathrm{d}l)$	W m^{-1} K^{-1}

Name	Symbol	Definition	SI unit
coefficient of heat transfer	$h, (k, K, \alpha)$	$h = J_q/\Delta T$	W m^{-2} K^{-1}
thermal diffusivity	a	$a = \lambda/\rho c_p$	m^2 s^{-1}
diffusion coefficient	D	$D = J_n/(\mathrm{d}c/\mathrm{d}l)$	m^2 s^{-1}

The following symbols are used in the definitions of the dimensionless quantities: mass (m), time (t), volume (V), area (A), density (ρ), speed (v), length (l), viscosity (η), pressure (p), acceleration of free fall (g), cubic expansion coefficient (α), temperature (T), surface tension (γ), speed of sound (c), mean free path (λ), frequency (f), thermal diffusivity (a), coefficient of heat transfer (h), thermal conductivity (k), specific heat capacity at constant pressure (c_p), diffusion coefficient (D), mole fraction (x), mass transfer coefficient (k_d), permeability (μ), electric conductivity (κ), and magnetic flux density (B).

Name	Symbol	Definition	SI unit
Reynolds number	Re	$Re = \rho vl/\eta$	1
Euler number	Eu	$Eu = \Delta p/\rho v^2$	1
Froude number	Fr	$Fr = v/(lg)^{1/2}$	1
Grashof number	Gr	$Gr = l^3 g\alpha\Delta T\rho^2/\eta^2$	1
Weber number	We	$We = \rho v^2 l/\gamma$	1
Mach number	Ma	$Ma = v/c$	1
Knudsen number	Kn	$Kn = \lambda/l$	1
Strouhal number	Sr	$Sr = lf/v$	1
Fourier number	Fo	$Fo = at/l^2$	1
Péclet number	Pe	$Pe = vl/a$	1
Rayleigh number	Ra	$Ra = l^3 g\alpha\Delta T\rho/\eta a$	1
Nusselt number	Nu	$Nu = hl/k$	1
Stanton number	St	$St = h/\rho vc_p$	1
Fourier number for mass transfer	Fo^*	$Fo^* = Dt/l^2$	1
Péclet number for mass transfer	Pe^*	$Pe^* = vl/D$	1
Grashof number for mass transfer	Gr^*	$Gr^* = l^3 g\left(\frac{\partial \rho}{\partial x}\right)_{T,p}\left(\frac{\Delta x\rho}{\eta}\right)$	1
Nusselt number for mass transfer	Nu^*	$Nu^* = k_d l/D$	1
Stanton number for mass transfer	St^*	$St^* = k_d/v$	1
Prandtl number	Pr	$Pr = \eta/\rho a$	1
Schmidt number	Sc	$Sc = \eta/\rho D$	1
Lewis number	Le	$Le = a/D$	1
magnetic Reynolds number	Rm, Re_m	$Rm = v\mu\kappa l$	1
Alfvén number	Al	$Al = v(\rho\mu)^{1/2}/B$	1
Hartmann number	Ha	$Ha = Bl\,(\kappa/\eta)^{1/2}$	1
Cowling number	Co	$Co = B^2/\mu\rho v^2$	1

EXPRESSION OF UNCERTAINTY OF MEASUREMENTS

In general, the result of a measurement is only an approximation or estimate of the true value of the quantity subject to measurement, and thus the result is of limited value unless accompanied by a statement of its uncertainty. Much (but not all) of the scientific data appearing in the literature does include some indication of the uncertainty, but this may be stated in many different ways and is often explained poorly. In an effort to encourage consistency in uncertainty statements, the International Committee for Weights and Measures (CIPM) initiated a project, in collaboration with several other international organizations, to prepare a set of guidelines expressing international consensus on the recommended method of stating uncertainties. This project resulted in the publication of the *Guide to the Expression of Uncertainty in Measurement* (Reference 1), which is often referred to as *GUM*. The recommendations of *GUM* have been summarized by the National Institute of Standards and Technology in *NIST Technical Note 1297, Guidelines for Evaluating the Uncertainty of NIST Measurement Results* (Reference 2).

In the notation of *GUM*, we are concerned with the **measurand**, i.e., the quantity that is being measured. In physics and chemistry this is usually called a **physical quantity** and represents some inherent characteristic of a material, system, or process that can be expressed in numerical terms — specifically as the product of a number and a reference, commonly called a **unit.** Thus the density of water at room temperature is (approximately) 0.998 g/mL (grams per milliliter) or, alternatively 998 kg m^{-3} (kilograms per meter cubed). This statement gives the most likely value of the measurand, to this level of precision, but gives no information on how much the stated value might differ from the true value.

It is important to differentiate between the terms **error** and **uncertainty**. The error in a measurement is the difference between the measured value and the true value; the error can be stated if the true value is known (to some level of accuracy). The uncertainty is an estimate of the maximum reasonable extent to which the measured value is believed to deviate from the true value, in a situation where the true value is not known (most often the case). The result of a measurement can unknowably be very close to the true value, and thus have negligible error, even though its uncertainty is large.

The uncertainty of the result of a measurement generally consists of several components, which may be grouped in two types according to the method used to estimate their numerical values:

Type A. Those which are evaluated by statistical methods
Type B. Those which are evaluated by other means

The terms "random uncertainty" and "systematic uncertainty" are often used, but these terms do not always correspond in a simple way to the A and B categories. This is because the nature of an uncertainty component is conditioned by how the quantity appears in the mathematical model that describes the current measurement process. An uncertainty component arising from a systematic effect may in some cases be evaluated by methods of Type A while in other cases by methods of Type B.

In the *GUM* formulation, each component of uncertainty, whether in the A or B category, is represented by an estimated standard deviation, termed **standard uncertainty**, symbol u_i, and equal to the positive square root of the estimated variance u_i^2.

For an uncertainty component of Type A, $u_i = s_i$, where s_i is the statistically estimated standard deviation, as determined from a series of observations by appropriate statistical analysis. Any valid statistical method may be used. Examples are calculating the standard deviation of the mean of a series of independent observations; using the method of least squares to fit a curve to data in order to estimate parameters of the curve and their standard deviations; and carrying out an analysis of variance (ANOVA) in order to identify and quantify random effects in certain types of measurements. Details of statistical analysis are given in References 4–7 and many other places.

In a similar manner, each uncertainty component of Type B is represented by a quantity u_j, which is obtained from an assumed probability distribution based on all the available information about the measurement process. Since u_j is treated like a standard deviation, the standard uncertainty in each Type B component is simply u_j. The evaluation of u_j is usually based on scientific judgment using all the relevant information available, which may include

- previous measurement data
- experience with, or general knowledge of, the behavior and properties of relevant materials and instruments
- manufacturer's specifications
- data provided in calibrations and other reports
- uncertainties assigned to reference data taken from handbooks.

The specific approach to evaluating the standard uncertainty u_j of a Type B uncertainty will depend on the detailed model of the measurement process. The following are examples of steps that may be used:

1. Convert a quoted uncertainty (for example, in a calibration factor) that is a stated multiple of an estimated standard deviation to a standard uncertainty by dividing the quoted uncertainty by the multiplier.
2. Convert a quoted uncertainty that defines a "confidence interval" having a stated level of confidence, such as 95% or 99%, to a standard uncertainty by treating the quoted uncertainty as if a normal distribution had been used to calculate it (unless otherwise indicated) and dividing it by the appropriate factor for such a distribution. These factors are 1.960 and 2.576 for the two levels of confidence given.
3. Model knowledge of the quantity in question by a normal distribution and estimate lower and upper limits a_- and a_+ such that the best estimated value of the quantity is $(a_+ + a_-)/2$ (i.e., the midpoint of the limits) and there is 1 chance out of 2 (i.e., a 50 percent probability) that the value of the quantity lies in the interval a_- to a_+. Then $u_j \approx 1.48\, a$, where $a = (a_+ - a_-)/2$ is the half-width of the interval.
4. Model knowledge of the quantity in question by a normal distribution and estimate lower and upper limits a_- and a_+ such that the best estimated value of the quantity is $(a_+ + a_-)/2$ and there is about a 2 out of 3 chance (i.e., a 67 percent probability) that the value of the quantity lies in the interval a_- to a_+. Then $u_j \approx a$, where $a = (a_+ - a_-)/2$.
5. Estimate lower and upper limits a_- and a_+ for the value of the quantity in question such that the probability that the value lies in the interval a_- to a_+ is, for all practical purposes, 100 percent. Provided that there is no contradictory information, treat the quantity as if it is equally probable for its value to lie anywhere within the interval a_- to a_+;

that is, model it by a uniform or rectangular probability distribution. The best estimate of the value of the quantity is then $(a_+ + a_-)/2$ with $u_j = a/\sqrt{3}$ where $a = (a_+ - a_-)/2$. If the distribution used to model the quantity is triangular rather than rectangular, then $u_j = a/\sqrt{6}$. The rectangular distribution is a reasonable default model in the absence of any other information. But if it is known that values of the quantity in question near the center of the limits are more likely than values close to the limits, a triangular or a normal distribution may be a better model.

When all the standard uncertainties of Type A and Type B have been determined in this way, they should be combined to produce the **combined standard uncertainty** (suggested symbol u_c), which may be regarded as the estimated standard deviation of the measurement result. This process, often called the *law of propagation of uncertainty* or "root-sum-of-squares," involves taking the square root of the sum of the squares of all the u_i. In many practical measurement situations, the probability distribution characterized by the measurement result y and its combined standard uncertainty $u_c(y)$ is approximately normal (Gaussian). When this is the case, $u_c(y)$ defines an interval $y - u_c(y)$ to $y + u_c(y)$ about the measurement result y within which the value of the measurand Y estimated by y is believed to lie with a level of confidence of approximately 68 percent. That is, it is believed with an approximate level of confidence of 68 percent that $y - u_c(y) \leq Y \leq y + u_c(y)$, which is commonly written as $Y = y \pm u_c(y)$.

In fundamental metrological research (involving physical constants, calibration standards, and the like) the combined standard uncertainty u_c is normally used as the statement of uncertainty in a measurement. In most cases, however, it is desirable to use a measure of uncertainty that defines an interval about the measurement result y within which the value of the measurand Y is confidently believed to lie. The measure of uncertainty intended to meet this requirement is termed **expanded uncertainty,** suggested symbol U, and is obtained by multiplying $u_c(y)$ by a **coverage factor,** suggested symbol k. Thus $U = ku_c(y)$ and it is believed with high confidence that $y - U \leq Y \leq y + U$, which is commonly written as $Y = y \pm U$. The value of the coverage factor k is chosen on the basis of the desired level of confidence to be associated with the interval defined by $U = ku_c$. Typically, k is in the range 2 to 3. When the normal distribution applies, $U = 2u_c$ (i.e., $k = 2$) defines an interval having a level of confidence of approximately 95 percent, and $U = 3u_c$ defines an interval having a confidence level greater than 99 percent. In current international practice it is most common to use $k = 2$, corresponding to about 95 percent confidence, but the value of k should be stated in each case to avoid confusion. See References 1 and 2 for methods of calculating k when a value other than $k = 2$ is needed for a specific requirement.

Summary of Key Steps

- Group the uncertainty components into Type A (can be evaluated by statistical methods) and Type B (must be evaluated by other means).
- Determine the standard uncertainty for each component of Type A by statistical methods and for each component of Type B by other suitable methods, based on modeling the measurement process.
- Take the square root of the sum of the squares of all the standard uncertainties to get the combined standard uncertainty u_c.
- Specify a coverage factor k which, when multiplied by u_c, gives the expanded uncertainty U. In fundamental metrological research $k = 1$ is usually chosen; in other cases, $k = 2$ (corresponding to a confidence level of about 95%) is the most common choice.

References

1. ISO, *Guide to the Expression of Uncertainty in Measurement*, International Organization for Standardization, Geneva, Switzerland, 1993. Several supplements have been published; see Bich, W., Cox, M. C., and Harris, P. M., "Evolution of the *Guide to the Expression of Uncertainty in Measurement*," *Metrologia* 43, S161, 2006.
2. Taylor, B. N., and Kuyatt, C. E., *Guidelines for Evaluating and Expressing the Uncertainty of NIST Measurement Results*, NIST Technical Note 1297, National Institute of Standards and Technology, Gaithersburg, MD, 1994; available for free download at <physics.nist.gov/cuu/Uncertainty/bibliography.html>.
3. Bell, S., *A Beginner's Guide to Uncertainty of Measurement*, National Physical Laboratory, Teddington, Middlesex, UK, 2001; available on the Internet through <www.npl.co.uk/server.php?show=ConWebDoc.1785>.
4. Eisenhart, C., "Realistic Evaluation of the Precision and Accuracy of Instrument Calibration Systems," *J. Res. Natl. Bur. Stand.* (U.S.) 67C, 161, 1963.
5. Mandel, J., *The Statistical Analysis of Experimental Data*, Dover Publishers, New York, 1984.
6. Nantrella, M. G., *Experimental Statistics*, NBS Handbook 91, U.S. Government Printing Office, Washington, DC, 1966.
7. Box, G. E. P., Hunter, J. S., and Hunter, W. G., *Statistics for Experimenters: Design, Innovation, and Discovery, 2nd Edition*, John Wiley & Sons, Hoboken, NJ, 2005.

NOMENCLATURE FOR CHEMICAL COMPOUNDS

The International Union of Pure and Applied Chemistry (IUPAC) maintains several commissions that deal with the naming of chemical substances. In general, the approach of IUPAC is to present rules for arriving at names in a systematic manner, rather than recommending a unique name for each compound. Thus there are often several alternative "IUPAC names," depending on which nomenclature system is used, each of which may have advantages in specific applications. However, each of these names will be unambiguous.

Organizations such as the Chemical Abstacts Service and the Beilstein Institute that prepare indexes to the chemical literature must adopt a system for selecting unique names in order to avoid excessive cross referencing. Chemical Abstracts Service uses a system which groups together compounds derived from a single parent compound. Thus most index names are inverted (e.g., Benzene, bromo rather than bromobenzene; Acetic acid, sodium salt rather than sodium acetate).

Recommended names for the most common substituent groups, ligands, ions, and organic rings are given in the two following tables, "Nomenclature for Inorganic Ions and Ligands" and "Organic Substituent Groups and Ring Systems." For the basics of macromolecular nomenclature, see "Nomenclature for Organic Polymers" in Section 13.

Some of the most useful recent guides to chemical nomenclature, prepared by IUPAC and other organizations such as the International Union of Biochemistry and Molecular Biology (IUBMB) and the American Chemical Society are listed below. These books contain citations to the more detailed nomenclature documents in each area. Two very useful web sites providing links to nomenclature documents are:

www.iupac.org/publications/index.html
www.chem.qmul.ac.uk/iupac/

Inorganic Chemistry

Block, B. P., Powell, W. H., and Fernelius, W. C., *Inorganic Chemical Nomenclature, Principles and Practice*, American Chemical Society, Washington, 1990.

Nomenclature of Inorganic Chemistry - IUPAC Recommendations 2005. Connelly, N.G., Damhus, T., Hartshorn, R. M., and Hutton, A. T., The Royal Society of Chemistry, 2005.

Organic Chemistry

International Union of Pure and Applied Chemistry, *A Guide to IUPAC Nomenclature of Organic Compounds, Recommendations 1993,* Panico, R., Powell, W. H., and Richer, J.-C., Eds., Blackwell Scientific Publications, Oxford, 1993.

International Union of Pure and Applied Chemistry, *Glossary of Class Names of Organic Compounds and Reactive Intermediates Based on Structure*, Moss, G. P., Smith, P. A. S., and Tavernier, D., Eds., *Pure & Appl. Chem.* 67, 1307, 1995.

International Union of Pure and Applied Chemistry, *Basic Terminology of Stereochemistry*, Moss, G. P., Ed., *Pure & Appl. Chem.* 68, 2193, 1996.

Rhodes, P. H., *The Organic Chemist's Desk Reference*, Chapman & Hall, London, 1995.

Macromolecular Chemistry

International Union of Pure and Applied Chemistry, *Compendium of Macromolecular Nomenclature*, Metanomski, W. V., Ed., Blackwell Scientific Publications, Oxford, 1991.

International Union of Pure and Applied Chemistry, *Glossary of Basic Terms in Polymer Science*, Jenkins, A.D., Kratochvil, P., Stepto, R. F. T., and Suter, U. W., Eds., *Pure & Appl. Chem.* 68, 2287, 1996.

Biochemistry

International Union of Biochemistry and Molecular Biology, *Biochemical Nomenclature and Related Documents, 2nd Edition, 1992,* Portland Press, London, 1993; includes recommendations of the IUPAC-IUBMB Joint Commission on Biochemical Nomenclature.

International Union of Biochemistry and Molecular Biology, *Enzyme Nomenclature, 1992,* Academic Press, Orlando, FL, 1992.

IUPAC-IUBMB Joint Commission on Biochemical Nomenclature, *Nomenclature of Carbohydrates, Recommendations 1996,* McNaught, A. D., Ed., *Pure & Appl. Chem.* 68, 1919, 1996.

General

Chemical Abstracts Service, *Naming and Indexing Chemical Substances for Chemical Abstracts, Appendix IV, Chemical Abstracts 1994 Index Guide.*

Principles of Chemical Nomenclature: a Guide to IUPAC Recommendations, Leigh, G. J., Favre, H. A., and Metanomski, W. V., Blackwell Science, 1998.

NOMENCLATURE FOR INORGANIC IONS AND LIGANDS

Willem H. Koppenol

The entries below were selected from Table IX of Connelly, N. G., Damhus, T., Hartshorn, R. M. and Hutton, A. T., Eds., *Nomenclature of Inorganic Chemistry. IUPAC Recommendations 2005*, The Royal Society of Chemistry, 2005. Two changes were made: in the case of the hypohalides, the oxidohalogenate names are listed, not the new halooxygenate names. Thus, for BrO^- the still acceptable name "oxidobromate(1−)" is listed, not the more correct, but less palatable, "bromooxygenate(1−)." Similarly, and for reasons of consistency, $ClO^{\bullet}$ is not named oxygen (mono)chloride, but chlorine mono(o) oxide. The symbol '⊃' is used for dividing names when this is made necessary by a line break. When the name is reconstructed from the name given in the table, this symbol should be omitted. Thus, all *hyphens* in the table are true parts of the names. The symbols '>' and '<' placed next to an element symbol both denote two single bonds connecting the atom in question to two other atoms. For a given compound, the various systematic names, if applicable, are given in the order: stoichiometric names, substitutive names, additive names and hydrogen names. Acceptable names that are not entirely systematic (or not formed according to any of the systems mentioned above) are given at the end after a semicolon. No order of preference is implied by the order in which formulae and names are listed. Reprinted by permission of IUPAC.

Formula for uncharged atom or group	**Name**			
	Uncharged atoms or molecules (including zwitterions and radicals) or substituent groups[a]	***Cations (including cation radicals) or cationic substituent groups***[a]	***Anions (including anion radicals) or anionic substituent groups***[b]	***Ligands***[c]
H	hydrogen $H^{\bullet}$, hydrogen(•), monohydrogen (natural or unspecified isotopic composition) $^{1}H^{\bullet}$, protium(•), monoprotium $^{2}H^{\bullet} = D^{\bullet}$, deuterium(•), monodeuterium $^{3}H^{\bullet} = T^{\bullet}$, tritium(•), monotritium	hydrogen (general) H^+, hydrogen(1+), hydron (natural or unspecified isotopic composition) $^{1}H^+$, protium(1+), proton $^{2}H^+ = D^+$, deuterium(1+), deuteron $^{3}H^+ = T^+$, tritium(1+), triton	hydride (general) H^-, hydride (natural or unspecified isotopic composition) $^{1}H^-$, protide $^{2}H^- = D^-$, deuteride $^{3}H^- = T^-$, tritide	hydrido protido deuterido tritido
H_2	H_2, dihydrogen D_2, dideuterium T_2, ditritium	$H_2^{\bullet+}$, dihydrogen(•1+) $^{1}H_2^{\bullet+}$, diprotium(•1+) $D_2^{\bullet+}$, dideuterium(•1+) $T_2^{\bullet+}$, ditritium(•1+)		
D, see H				
D_2, see H_2				
T, see H				
T_2, see H_2				
F	fluorine $F^{\bullet}$, fluorine(•), monofluorine –F, fluoro	fluorine (general) F^+, fluorine(1+)	fluoride (general) F^-, fluoride(1−); fluoride	fluorido (general) F^-, fluorido(1−); fluorido
F_2	F_2, difluorine	$F_2^{\bullet+}$, difluorine(•1+)	$F_2^{\bullet-}$, difluoride(•1−)	F_2, difluorine
Cl	chlorine (general) $Cl^{\bullet}$, chlorine(•), monochlorine –Cl, chloro	chlorine (general) Cl^+, chlorine(1+)	chloride (general) Cl^-, chloride(1−); chloride	chlorido (general) Cl^-, chlorido(1−); chlorido
Cl_2	Cl_2, dichlorine	$Cl_2^{\bullet+}$, dichlorine(•1+)	$Cl_2^{\bullet-}$, dichloride(•1−)	Cl_2, dichlorine $Cl_2^{\bullet-}$, dichlorido(•1−)
Br	bromine (general) $Br^{\bullet}$, bromine(•), monobromine –Br, bromo	bromine (general) Br^+, bromine(1+)	bromide (general) Br^-, bromide(1−); bromide	bromido (general) Br^-, bromido(1−); bromido
Br_2	Br_2, dibromine	$Br_2^{\bullet+}$, dibromine(•1+)	$Br_2^{\bullet-}$, dibromide(•1−)	Br_2, dibromine
I	iodine (general) $I^{\bullet}$, iodine(•), monoiodine –I, iodo	iodine (general) I^+, iodine(1+)	iodide (general) I^-, iodide(1−); iodide	iodido (general) I^-, iodido(1−); iodido
I_2	I_2, diiodine	$I_2^{\bullet+}$, diiodine(•1+)	$I_2^{\bullet-}$, diiodide(•1−)	I_2, diiodine

ClO	ClO, chlorine mon(o)oxide $ClO^{\bullet}$, oxidochlorine(•); chlorosyl –ClO, oxo-λ^3-chloranyl; chlorosyl –OCl, chlorooxy		ClO^{-}, oxidochlorate(1–); hypochlorite	ClO^{-}, oxidochlorato(1–); hypochlorito
ClO_2	ClO_2, chlorine dioxide $ClO_2^{\bullet}$, dioxidochlorine(•) $ClOO^{\bullet}$, chloridodioxygen ⁐ (*O*–*O*) (•), –ClO_2, dioxo-λ^5-chloranyl; chloryl –OClO, oxo-λ^3-chloranyloxy	ClO_2^{+}, dioxidochlorine(1+) (*not* chloryl)	ClO_2^{-}, dioxidochlorate(1–); chlorite	ClO_2^{-}, dioxidochlorato(1–); chlorito
ClO_3	ClO_3, chlorine trioxide $ClO_3^{\bullet}$, trioxidochlorine(•) –ClO_3, trioxo-λ^7-chloranyl; perchloryl –$OClO_2$, dioxo-λ^5-chloranyloxy	ClO_3^{+}, trioxidochlorine(1+) (*not* perchloryl)	ClO_3^{-}, trioxidochlorate(1–); chlorate	ClO_3^{-}, trioxidochlorato(1–); chlorato
ClO_4	ClO_4, chlorine tetraoxide $ClO_4^{\bullet}$, tetraoxidochlorine(•) –$OClO_3$, trioxo-λ^7-chloranyloxy		ClO_4^{-}, tetraoxidochlorate(1–); perchlorate	ClO_4^{-}, tetraoxidochlorato(1–); perchlorato
IO	IO, iodine mon(o)oxide $IO^{\bullet}$, oxidoiodine(•); iodosy l –IO, oxo-λ^3-iodanyl; iodosyl –OI, iodooxy	IO^{+}, oxidoiodine(1+) (*not* iodosyl)	IO^{-}, oxidoiodate(1–); hypoiodite $IO^{\bullet 2-}$, oxidoiodate(•2–)	IO^{-}, oxidoiodato(1–); hypoiodito
IO_2	IO_2, iodine dioxide $IO_2^{\bullet}$, dioxidoiodine(•) –IO_2, dioxo-λ^5-iodanyl; iodyl –OIO, oxo-λ^3-iodanyloxy	IO_2^{+}, dioxidoiodine(1+) (*not* iodyl)	IO_2^{-}, dioxidoiodate(1–); iodite	IO_2^{-}, dioxidoiodato(1–); iodito
IO_3	IO_3, iodine trioxide $IO_3^{\bullet}$, trioxidoiodine(•) –IO^3, trioxo-λ^7-iodanyl; periodyl –OIO_2, dioxo-λ^5-iodanyloxy	IO_3^{+}, trioxidoiodine(1+) (*not* periodyl)	IO_3^{-}, trioxidoiodate(1–); iodate	IO_3^{-}, trioxidoiodato(1–); iodato
IO_4	IO_4, iodine tetraoxide $IO_4^{\bullet}$, tetraoxidoiodine(•) –OIO_3, trioxo-λ^7-iodanyloxy		IO_4^{-}, tetraoxidoiodate(1–); periodate	IO_4^{-}, tetraoxidoiodato(1–); periodato
O	oxygen (general) O, monooxygen $O^{2\bullet}$, oxidanylidene, monooxygen(2•) >O, oxy, epoxy (in rings) =O, oxo	oxygen (general) $O^{\bullet +}$, oxygen(•1+)	oxide (general) $O^{\bullet -}$, oxidanidyl, oxide(•1–) O^{2-}, oxide(2–); oxide –O^{-}, oxido	O^{2-}, oxido
O_2	O_2, dioxygen $O_2^{2\bullet}$, dioxidanediyl, dioxygen(2•) –OO–, dioxidanediyl; peroxy	$O_2^{\bullet +}$, dioxidanyliumyl, dioxygen(•1+) O_2^{2+}, dioxidanebis(ylium), dioxygen(2+)	$O_2^{\bullet -}$, dioxidanidyl, dioxide(•1–); superoxide (*not* hyperoxide) O_2^{2-}, dioxidanediide, dioxide(2–); peroxide	dioxido (general) O_2, dioxygen $O_2^{\bullet -}$, dioxido(•1–); superoxido O_2^{2-}, dioxidanediido, dioxido(2–); peroxido
O_3	O_3, trioxygen; ozone –OOO–, trioxidanediyl		$O_3^{\bullet -}$, trioxidanidyl, trioxide(•1–); ozonide	O_3, trioxygen; ozone $O_3^{\bullet -}$, trioxido(•1–); ozonido
HO	$HO^{\bullet}$, oxidanyl, hydridooxygen(•); hydroxyl –OH, oxidanyl; hydroxy	HO^{+}, oxidanylium, hydridooxygen(1+); hydroxylium	HO^{-}, oxidanide, hydroxide	HO^{-}, oxidanido; hydroxido
HO_2	$HO_2^{\bullet}$, dioxidanyl, hydridodioxygen(•) hydrogen dioxide –OOH, dioxidanyl; hydroperoxy	HO_2^{+}, dioxidanylium, hydridodioxygen(1+)	HO_2^{-}, dioxidanide, hydrogen(peroxide)(1–)	HO_2^{-}, dioxidanido, hydrogen(peroxido)(1–)
S	sulfur (general) S, monosulfur =S, sulfanylidene; thioxo –S–, sulfanediyl	sulfur (general) S^{+}, sulfur(1+)	sulfide (general) $S^{\bullet -}$, sulfanidyl, sulfide(•1–) S^{2-}, sulfanediide, sulfide(2–); sulfide –S^{-}, sulfido	sulfido (general) $S^{\bullet -}$, sulfanidyl, sulfido(•1–) S^{2-}, sulfanediido, sulfido(2–)

S_2	S_2, disulfur –SS–, disulfanediyl >S=S, sulfanylidene-λ[4]-sulfanediyl; sulfinothioyl	$S_2^{\bullet+}$, disulfur(•1+)	$S_2^{\bullet-}$, disulfanidyl, disulfide(•1–) S_2^{2-}, disulfide(2–), disulfanediide –SS⁻, disulfanidyl	S_2^{2-}, disulfido(2–), disulfanediido
HS	$HS^{\bullet}$, sulfanyl, hydridosulfur(•) –SH, sulfanyl	HS^+, sulfanylium, hydridosulfur(1+)	HS^-, sulfanide, hydrogen(sulfide)(1–)	HS^-, sulfanido, hydrogen(sulfido)(1–)
SO	SO, sulfur mon(o)oxide [SO], oxidosulfur >SO, oxo-λ[4]-sulfanediyl; sulfinyl	$SO^{\bullet+}$, oxidosulfur(•1+) (*not* sulfinyl or thionyl)	$SO^{\bullet-}$, oxidosulfate(•1–)	[SO], oxidosulfur
SO_2	SO_2, sulfur dioxide [SO_2], dioxidosulfur >SO_2, dioxo-λ[6]-sulfanediyl; sulfuryl, sulfonyl		$SO_2^{\bullet-}$, dioxidosulfate(•1–) SO_2^{2-}, dioxidosulfate(2–), sulfanediolate	[SO_2], dioxidosulfur SO_2^{2-}, dioxidosulfato(2–), sulfanediolato
SO_3	SO_3, sulfur trioxide		$SO_3^{\bullet-}$, trioxidosulfate(•1–) SO_3^{2-}, trioxidosulfate(2–); sulfite $–S(O)_2(O^-)$, oxidodioxo-λ[6]-sulfanyl; sulfonato	SO_3^{2-}, trioxidosulfato(2–); sulfito
SO_4	$–OS(O)_2O–$, sulfonylbis(oxy)		$SO_4^{\bullet-}$, tetraoxidosulfate(•1–) SO_4^{2-}, tetraoxidosulfate(2–); sulfate	SO_4^{2-}, tetraoxidosulfato(2–); sulfato
S_2O_3			$S_2O_3^{\bullet-} = SO_3S^{\bullet-}$, trioxido-1κ[3]*O*-disulfate(*S–S*)(•1–), trioxidosulfidosulfate(•1–) $S_2O_3^{2-} = SO_3S^{2-}$, trioxido-1κ[3]*O*-disulfate(*S–S*)(2–), trioxidosulfidosulfate(2–); thiosulfate, sulfurothioate	$S_2O_3^{2-} = SO_3S^{2-}$, trioxido-1κ[3]*O*-disulfato(*S–S*)(2–), trioxidosulfidosulfato(2–); thiosulfato, sulfurothioato
Se	Se (general) Se, monoselenium >Se, selanediyl =Se, selanylidene; selenoxo	selenium	selenide (general) $Se^{\bullet-}$, selanidyl, selenide(•1–) Se^{2-}, selanediide, selenide(2–); selenide	selenido (general) $Se^{\bullet-}$, selanidyl, selenido(•1–) Se^{2-}, selanediido, selenido(2–)
SeO	SeO, selenium mon(o)oxide [SeO], oxidoselenium >SeO, seleninyl			[SeO], oxidoselenium
SeO_2	SeO_2, selenium dioxide [SeO_2], dioxidoselenium >SeO_2, selenonyl		SeO_2^{2-}, dioxidoselenate(2–)	[SeO_2], dioxidoselenium SeO_2^{2-}, dioxidoselenato(2–)
SeO_3	SeO_3, selenium trioxide		$SeO_3^{\bullet-}$, trioxidoselenate(•1–) SeO_3^{2-}, trioxidoselenate(2–); selenite	SeO_3^{2-}, trioxidoselenato(2–); selenito
SeO_4			SeO_4^{2-}, tetraoxidoselenate(2–); selenate	SeO_4^{2-}, tetraoxidoselenato(2–); selenato
Te	tellurium >Te, tellanediyl =Te, tellanylidene; telluroxo	tellurium	telluride (general) $Te^{\bullet-}$, tellanidyl, telluride(•1–) Te^{2-}, tellanediide, telluride(2–); telluride	tellurido (general) $Te^{\bullet-}$, tellanidyl, tellurido(•1–) Te^{2-}, tellanediido, tellurido(2–)
CrO_2	CrO_2, chromium dioxide, chromium(IV) oxide			
UO_2	UO_2, uranium dioxide	UO_2^{+}, dioxidouranium(1+) [*not* uranyl(1+)] UO_2^{2+}, dioxidouranium(2+) [*not* uranyl(2+)]		

NpO_2	NpO_2, neptunium dioxide	NpO_2^+, dioxidoneptunium(1+) [*not* neptunyl(1+)] NpO_2^{2+}, dioxidoneptunium(2+) [*not* neptunyl(2+)]		
PuO_2	PuO_2, plutonium dioxide	PuO_2^+, dioxidoplutonium(1+) [not plutonyl(1+)] PuO_2^{2+}, dioxidoplutonium(2+) [not plutonyl (2+)]		
N	nitrogen $N^{\bullet}$, nitrogen(•), mononitrogen –N<, azanetriyl; nitrilo –N=, azanylylidene ≡N, azanylidyne	nitrogen (general) N^+, nitrogen(1+)	nitride (general) N^{3-}, nitride(3–), azanetriide; nitride $=N^-$, azanidylidene; amidylidene $-N^{2-}$, azanediidyl	N^{3-}, nitrido(3–), azanetriido
N_2	N_2, dinitrogen $=N^+=N^-$, (azanidylidene) azaniumylidene; diazo –N=N–, diazane-1,2-diylidene; hydrazinediylidene –NN=, diazene 1,2 diyl; azo	$N_2^{\bullet+}$, dinitrogen(1+) N_2^{2+}, dinitrogen(2+) $-N^+{\equiv}N$, diazyn-1-ium-1-yl	N_2^{2-}, dinitride(2–) N_2^{4-}, dinitride(4–), diazanetetraide; hydrazinetetraide	N_2, dinitrogen N_2^{2-}, dinitrido(2–) N_2^{4-}, dinitrido(4–), diazanetetraido; hydrazinetetraido
N_3	$N_3^{\bullet}$, trinitrogen(•) $-N=N^+=N^-$, azido		N_3^-, trinitride(1–); azide	N_3^-, trinitrido(1–); azido
NH	$NH^{2\bullet}$, azanylidene, hydridonitrogen(2•); nitrene >NH, azanediyl =NH, azanylidene; imino	NH^+, azanyliumdiyl, hydridonitrogen(1+) NH^{2+}, azanebis(ylium), hydridonitrogen(2+)	NH^-, azanidyl, hydridonitrate(1–) NH^{2-}, azanediide, hydridonitrate(2–); imide $-NH^-$, azanidyl; amidyl	NH^{2-}, azanediido, hydridonitrato(2–); imido
NH_2	$NH_2^{\bullet}$, azanyl, dihydridonitrogen(•); aminyl $-NH_2$, azanyl; amino	NH_2^+, azanylium, dihydridonitrogen(1+)	NH_2^-, azanide, dihydridonitrate(1–); amide	NH_2^-, azanido, dihydridonitrato(1–), amido
NH_3	NH_3, azane (parent hydride name), amine (parent name for certain organic derivatives), trihydridonitrogen; ammonia	$NH_3^{\bullet+}$, azaniumyl, trihydridonitrogen(•1+) $-NH_3^+$, azaniumyl; ammonio	$NH_3^{\bullet-}$, azanuidyl, trihydridonitrate(•1–)	NH_3, ammine
NH_4	$NH_4^{\bullet}$, λ^5-azanyl, tetrahydridonitrogen(•)	NH_4^+, azanium; ammonium		
H_2NO	$H_2NO^{\bullet}$, aminooxidanyl, dihydridooxidonitrogen(•); aminoxyl $HONH^{\bullet}$, hydroxyazanyl, hydridohydroxidonitrogen(•) –NH(OH), hydroxyazanyl, hydroxyamino $-ONH_2$, aminooxy $-NH_2(O)$, oxo-λ^5-azanyl; azinoyl		$HONH^-$, hydroxyazanide, hydridohydroxidonitrate(1–) H_2NO^-, azanolate, aminooxidanide, dihydridooxidonitrate(1–)	$NHOH^-$, hydroxyazanido, hydridohydroxidonitrato(1–) H_2NO^-, azanolato, aminooxidanido, dihydridooxidonitrato(1–)
N_2H_2	HN=NH, diazene $^-N=NH_2^+$, diazen-2-ium-1-ide $H_2NN^{2\bullet}$, diazanylidene, hydrazinylidene $=NNH_2$, diazanylidene; hydrazinylidene $^{\bullet}HNNH^{\bullet}$, diazane-1,2-diyl; hydrazine-1,2-diyl –HNNH–, diazane-1,2-diyl; hydrazine-1,2-diyl	$HNNH^{2+}$, diazynediium	$HNNH^{2-}$, diazane-1,2-diide, hydrazine-1,2-diide H_2NN^{2-}, diazane-1,1-diide, hydrazine-1,1-diide	HN=NH, diazene $^-N=NH_2^+$, diazen-2-ium-1-ido $HNNH^{2-}$, diazane-1,2-diido, hydrazine-1,2-diido H_2NN^{2-}, diazane-1,1-diido, hydrazine-1,1-diido

N_2H_3	$H_2NNH^{\bullet}$, diazanyl, trihydridodinitrogen(*N–N*)(•); hydrazinyl $-NHNH_2$, diazanyl; hydrazinyl $^{2-}NNH_3^+$, diazan-2-ium-1,1-diide	$H_2N{=}NH^+$, diazenium	H_2NNH^-, diazanide, hydrazinide	$^{2-}NNH_3^+$, diazan-2-ium-1,1-diido H_2NNH^-, diazanido, hydrazinido
N_2H_4	H_2NNH_2, diazane (parent hydride name), hydrazine (parent name for organic derivatives) $^-NHNH_3^+$, diazan-2-ium-1-ide	$H_2NNH_2^{\bullet+}$, diazaniumyl, bis(dihydridonitrogen)(*N–N*)(•1+); hydraziniumyl $H_2N{=}NH_2^{2+}$, diazenediium		H_2NNH_2, diazane, hydrazine $^-NHNH_3^+$, diazan-2-ium-1-ido
NO	NO, nitrogen mon(o)oxide (*not* nitric oxide) $NO^{\bullet}$, oxoazanyl, oxidonitrogen(•); nitrosyl –N=O, oxoazanyl; nitroso >$N(O)^-$, oxo-λ^5-azanyl; azoryl =$N(O)^-$, oxo-λ^5-azanylidene; azorylidene ≡N(O), oxo-λ^5-azanylidyne; azorylidyne $-O^+{=}N^-$, azanidylideneoxidaniumyl	NO^+, oxidonitrogen(1+) (*not* nitrosyl) $NO^{\bullet 2+}$, oxidonitrogen(2+)	NO^-, oxidonitrate(1–) $NO^{(2\bullet)-}$, oxidonitrate(2•1–)	NO, oxidonitrogen (general); nitrosyl = oxidonitrogen-κ*N* (general) NO^+, oxidonitrogen(1+) NO^-, oxidonitrato(1–)
NO_2	NO_2, nitrogen dioxide $NO_2^{\bullet}$ = $ONO^{\bullet}$, nitrosooxidanyl, dioxidonitrogen(•); nitryl $-NO_2$, nitro –ONO, nitrosooxy	NO_2^+, dioxidonitrogen(1+) (*not* nitryl)	NO_2^-, dioxidonitrate(1–); nitrite $NO_2^{\bullet 2-}$, dioxidonitrate(•2–)	NO_2^-, dioxidonitrato(1–); nitrito $NO_2^{\bullet 2-}$, dioxidonitrato(•2–)
NO_3	NO_3, nitrogen trioxide $NO_3^{\bullet}$ = $O_2NO^{\bullet}$, nitrooxidanyl, trioxidonitrogen(•) $ONOO^{\bullet}$, nitrosodioxidanyl, (dioxido)oxidonitrogen(•) $-ONO_2$, nitrooxy		NO_3^-, trioxidonitrate(1–); nitrate $NO_3^{\bullet 2-}$, trioxidonitrate(•2–) $[NO(OO)]^-$, (dioxido)oxidonitrate(1–); peroxynitrite	NO_3^-, trioxidonitrato(1–); nitrato $NO_3^{\bullet 2-}$, trioxidonitrato(•2–) $[NO(OO)]^-$, oxidoperoxidonitrato(1–); peroxynitrito
N_2O	N_2O, dinitrogen oxide (*not* nitrous oxide) NNO, oxidodinitrogen(*N—N*) –N(O)=N–, azoxy		$N_2O^{\bullet -}$, oxidodinitrate(•1–)	N_2O, dinitrogen oxide (general) NNO, oxidodinitrogen(*N—N*) $N_2O^{\bullet -}$, oxidodinitrato(•1–)
N_2O_3	N_2O_3, dinitrogen trioxide O_2NNO, trioxido-1κ^2*O*,2κ*O*-dinitrogen(*N–N*) $NO^+NO_2^-$, oxidonitrogen(1+) dioxidonitrate(1–) ONONO, dinitrosooxidane, μ-oxidobis(oxidonitrogen)		$N_2O_3^{2-}$ = $[O_2NNO]^{2-}$, trioxido-1κ$_2$*O*,2κ*O*-dinitrate(*N–N*)(2–)	
N_2O_4	N_2O_4, dinitrogen tetraoxide O_2NNO_2, bis(dioxidonitrogen)(*N–N*) ONOONO, 1,2-dinitrosodioxidane, 2,5-diazy-1,3,4,6-tetraoxy-[6] catena $NO^+NO_3^-$, oxidonitrogen(1+) trioxidonitrate(1–)			
N_2O_5	N_2O_5, dinitrogen pentaoxide O_2NONO_2, dinitrooxidane, $NO_2^+NO_3^-$, dioxidonitrogen(1+) trioxidonitrate(1–)			
NS	NS, nitrogen monosulfide $NS^{\bullet}$, sulfidonitrogen(•) –N=S, sulfanylideneazanyl; thionitroso	NS^+, sulfidonitrogen(1+) (*not* thionitrosyl)	NS^-, sulfidonitrate(1–)	NS, sulfidonitrogen, sulfidonitrato, thionitrosyl (general) NS^+, sulfidonitrogen(1+) NS^-, sulfidonitrato(1–)

P	phosphorus (general) $P^{\bullet}$, phosphorus(•), monophosphorus >P–, phosphanetriyl	phosphorus (general) P^{+}, phosphorus(1+)	phosphide (general) P^{-}, phosphide(1–) P^{3-}, phosphide(3–), phosphanetriide; phosphide	P^{3-}, phosphido, phosphanetriido
PO	$PO^{\bullet}$, oxophosphanyl, oxidophosphorus(•), phosphorus mon(o)oxide; phosphoryl >P(O)–, oxo-λ^5-phosphanetriyl; phosphoryl =P(O)–, oxo-λ^5-phosphanylidene; phosphorylidene ≡P(O), oxo-λ^5-phosphanylidyne; phosphorylidyne	PO^{+}, oxidophosphorus(1+) (*not* phosphoryl)	PO^{-}, oxidophosphate(1–)	
PO_2	–P(O)$_2$, dioxo-λ^5-phosphanyl		PO_2^{-}, dioxidophosphate(1–)	PO_2^{-}, dioxidophosphato(1–)
PO_3			PO_3^{-}, trioxidophosphate(1–) $PO_3^{\bullet 2-}$, trioxidophosphate(•2–) PO_3^{3-}, trioxidophosphate(3–); phosphite $(PO_3^{-})_n$ = $(P(O)_2O)_n^{n-}$, *catena*-poly[(dioxidophosphate-μ-oxido)(1–)]; metaphosphate –P(O)(O$^-$)$_2$, dioxidooxo-λ^5-phosphanyl; phosphonato	PO_3^{-}, trioxidophosphato(1–) $PO_3^{\bullet 2-}$, trioxidophosphato(•2–) PO_3^{3-}, trioxidophosphato(3–); phosphito
PO_4			$PO_4^{\bullet 2-}$, tetraoxidophosphate(•2–) PO_4^{3-}, tetraoxidophosphate(3–); phosphate	PO_4^{3-}, tetraoxidophosphato(3–); phosphato
PS	$PS^{\bullet}$, sulfidophosphorus(•); –PS, thiophosphoryl	PS^{+}, sulfidophosphorus(1+) (*not* thiophosphoryl)		
AsO_3			AsO_3^{3-}, trioxidoarsenate(3–); arsenite, arsorite –As(=O)(O$^-$)$_2$, dioxidooxo-λ^5-arsanyl; arsonato	AsO_3^{3-}, trioxidoarsenato(3–); arsenito, arsorito
AsO_4			AsO_4^{3-}, tetraoxidoarsenate(3–); arsenate, arsorate	AsO_4^{3-}, tetraoxidoarsenato(3–); arsenato, arsorato
VO	VO, vanadium(II) oxide, vanadium mon(o)oxide	VO^{2+}, oxidovanadium(2+) (*not* vanadyl)		
CO	CO, carbon mon(o)oxide >C=O, carbonyl =C=O, carbonylidene	$CO^{\bullet+}$, oxidocarbon(•1+) CO^{2+}, oxidocarbon(2+)	$CO^{\bullet-}$, oxidocarbonate(•1–)	CO, oxidocarbon, oxidocarbonato (general); carbonyl = oxidocarbon-κ*C* (general) $CO^{\bullet+}$, oxidocarbon(•1+) $CO^{\bullet-}$, oxidocarbonato(•1–)
CO_2	CO_2, carbon dioxide, dioxidocarbon		$CO_2^{\bullet-}$, oxidooxomethyl, dioxidocarbonate(•1–)	CO_2, dioxidocarbon $CO_2^{\bullet-}$, oxidooxomethyl, dioxidocarbonato(•1–)
CO_3			$CO_3^{\bullet-}$, trioxidocarbonate(•1–), $OCOO^{\bullet-}$, (dioxido)oxidocarbonate(•1–), oxidoperoxidocarbonate(•1–) CO_3^{2-}, trioxidocarbonate(2–); carbonate	CO_3^{2-}, trioxidocarbonato(2–); carbonato
CS	carbon monosulfide >C=S, carbonothioyl; thiocarbonyl =C=S, carbonothioylidene	$CS^{\bullet+}$, sulfidocarbon(•1+)	$CS^{\bullet-}$, sulfidocarbonate(•1–)	CS, sulfidocarbon, sulfidocarbonato, thiocarbonyl (general); $CS^{\bullet+}$, sulfidocarbon(•1+) $CS^{\bullet-}$, sulfidocarbonato(•1–)
CS_2	CS_2, disulfidocarbon, carbon disulfide		$CS_2^{\bullet-}$, sulfidothioxomethyl, disulfidocarbonate(•1–)	CS_2, disulfidocarbon $CS_2^{\bullet-}$, sulfidothioxomethyl, disulfidocarbonato(•1–)

CN	$CN^{\bullet}$, nitridocarbon(•); cyanyl –CN, cyano –NC, isocyano	CN^+, azanylidynemethylium, nitridocarbon(1+)	CN^-, nitridocarbonate(1−); cyanide	nitridocarbonato (general) CN^-, nitridocarbonato(1−); cyanido = [nitridocarbonato(1−)-κC]
CNO	$OCN^{\bullet}$, nitridooxidocarbon(•) –OCN, cyanato –NCO, isocyanato –ONC, λ[2]-methylidene⊃ azanylylideneoxy –CNO, (oxo-λ[5]- azanylidynemethyl		OCN^-, nitridooxidocarbonate(1−); cyanate ONC^-, carbidooxidonitrate(1−); fulminate $OCN^{\bullet 2-}$, nitridooxidocarbonate(•2−)	OCN^-, nitridooxidocarbonato(1−); cyanato ONC^-, carbidooxidonitrato(1−); fulminato
CNS	$SCN^{\bullet}$, nitridosulfidocarbon(•) –SCN, thiocyanato –NCS, isothiocyanato –SNC, λ[2]-methylidene⊃ azanylylidenesulfanediyl –CNS, (sulfanylidene-λ[5]- azanylidynemethyl		SCN^-, nitridosulfidocarbonate(1−); thiocyanate SNC^-, carbidosulfidonitrate(1−)	SCN^-, nitridosulfidocarbonato(1−); thiocyanato SNC^-, carbidosulfidonitrato(1−)
CNSe	$SeCN^{\bullet}$, nitridoselenidocarbon(•) –SeCN, selenocyanato –NCSe, isoselenocyanato –SeNC, λ[2]-methylidene⊃ azanylylideneselanediyl –CNSe, (selanylidene-λ[5]- azanylidynemethyl		$SeCN^-$, nitridoselenidocarbonate(1−); selenocyanate $SeNC^-$, carbidoselenidonitrate(1−)	$SeCN^-$, nitridoselenidocarbonato(1−); selenocyanato $SeNC^-$, carbidoselenidonitrato(1−)

[a] Where an element symbol occurs in the first column, the unmodified element name is listed in the second and third columns. The unmodified name is generally used when the element appears as an electropositive constituent in the construction of a stoichiometric name (Sections IR-5.2 and IR-5.4). Names of homoatomic cations consisting of the element are also constructed using the element name, adding multiplicative prefixes and charge numbers as applicable (Sections IR-5.3.2.1 to IR-5.3.2.3). The sections mentioned refer to parts of *Nomenclature of Inorganic Chemistry. IUPAC Recommendations 2005*, see above.

[b] Where an element symbol occurs in the first column, the fourth column gives the element name appropriately modified with the ending 'ide' (hydride, nitride, etc.). The 'ide' form of the element name is generally used when the element appears as an electronegative constituent in the construction of a stoichiometric name (Sections IR-5.2 and IR-5.4). Names of homoatomic anions consisting of the element in question are also constructed using this modified form, adding multiplicative prefixes and charge numbers as applicable (Sections IR-5.3.3.1 to IR-5.3.3.3). Examples are given in the Table of names of some specific anions, e.g. chloride(1−), oxide(2−), dioxide(2−). In certain cases, a particular anion has the 'ide' form itself as an accepted short name, e.g., chloride, oxide. If specific anions are named, the 'ide' form of the element name with no further modification is given as the first entry in the fourth column, with the qualifier '(general)'. The sections mentioned refer to parts of *Nomenclature of Inorganic Chemistry. IUPAC Recommendations 2005*, see above.

[c] Ligand names must be placed within enclosing marks whenever necessary to avoid ambiguity, cf. Section IR-9.2.2.3. Some ligand names must always be enclosed. For example, if 'dioxido' is cited as is, it must be enclosed so as to distinguish it from two 'oxido' ligands; if combined with a multiplicative prefix it must be enclosed because it starts with a multiplicative prefix itself. A ligand name such as 'nitridocarbonato' must always be enclosed to avoid interpreting it as two separate ligand names, 'nitrido' and 'carbonato'. In this table, however, these enclosing marks are omitted for the sake of clarity. Note that the ligand names given here with a charge number can generally also be used without if it is not desired to make any implication regarding the charge of the ligand. For example, the ligand name '[dioxido(•1−)]' may be used if one wishes explicitly to consider the ligand to be the species dioxide(•1−), whereas the ligand name '(dioxido)' can be used if no such implications are desirable. The section mentioned refer to parts of *Nomenclature of Inorganic Chemistry. IUPAC Recommendations 2005*, see above.

ORGANIC SUBSTITUENT GROUPS AND RING SYSTEMS

The first part of this table lists substituent groups and their line formulas. A substituent group is defined by IUPAC as a group that replaces one or more hydrogen atoms attached to a parent structure. Such groups are sometimes called radicals, but IUPAC now reserves the term radical for a free molecular species with unpaired electrons. IUPAC does not recommend some of these names, which are marked here with asterisks (e.g., amyl*), but they are included in this list because they are often encountered in the older literature. Substituent group names that are formed by systematic rules (e.g., methyl from methane, ethyl from ethane, etc.) are included here only for the first few members of a homologous series.

In the second part of the table a number of common organic ring compounds are shown, with the conventional numbering of the ring positions indicated.

The help of Warren H. Powell in preparing this table is greatly appreciated. Pertinent references may be found in the table "Nomenclature of Chemical Compounds."

Substituent Groups

Name	Formula
acetamido (acetylamino)	CH_3CONH-
acetoacetyl	CH_3COCH_2CO-
acetonyl	CH_3COCH_2-
acetyl	CH_3CO-
acryloyl* (1-oxo-2-propenyl)	$CH_2=CHCO-$
alanyl (from alanine)	$CH_3CH(NH_2)CO-$
β-alanyl	$H_2N(CH_2)_2CO-$
allyl (2-propenyl)	$CH_2=CHCH_2-$
allylidene (2-propenylidene)	$CH_2=CHCH=$
amidino (aminoiminomethyl)	$H_2NC(=NH)-$
amino	H_2N-
amyl* (pentyl)	$CH_3(CH_2)_4-$
anilino (phenylamino)	C_6H_5NH-
anisidino	$CH_3OC_6H_4NH-$
anthranoyl (2-aminobenzoyl)	$2\text{-}H_2NC_6H_4CO-$
arsino	AsH_2-
azelaoyl (from azelaic acid)	$-OC(CH_2)_7CO-$
azido	N_3-
azino	$=N-N=$
azo	$-N=N-$
azoxy	$-N(O)=N-$
benzal* (benzylidene)	$C_6H_5CH=$
benzamido (benzoylamino)	C_6H_5CONH-
benzhydryl (diphenylmethyl)	$(C_6H_5)_2CH-$
benzoxy* (benzoyloxy)	C_6H_5COO-
benzoyl	C_6H_5CO-
benzyl	$C_6H_5CH_2-$
benzylidene	$C_6H_5CH=$
benzylidyne	$C_6H_5C=$
biphenylyl	$C_6H_5C_6H_4-$
biphenylene	$-C_6H_4-C_6H_4-$
butoxy	C_4H_9O-
sec-butoxy (1-methylpropoxy)	$C_2H_5CH(CH_3)O-$
tert-butoxy (1,1-dimethylethoxy)	$(CH_3)_3CO-$
butyl	$CH_3(CH_2)_3-$
sec-butyl (1-methylpropyl)	$CH_3CH_2CH(CH_3)-$
tert-butyl (1,1-dimethylethyl)	$(CH_3)_3C-$
butyryl (1-oxobutyl)	$CH_3(CH_2)_2CO-$
caproyl* (hexanoyl)	$CH_3(CH_2)_4CO-$
capryl* (decanoyl)	$CH_3(CH_2)_8CO-$
caprylоyl* (octanoyl)	$CH_3(CH_2)_6CO-$
carbamido (carbamoylamino)	$H_2NCONH-$
carbamoyl (aminocarbonyl)	H_2NCO-
carbamyl (aminocarbonyl)	H_2NCO-
carbazoyl (hydrazinocarbonyl)	$H_2NNHCO-$
carbethoxy (ethoxycarbonyl)	C_2H_5OCO-
carbonyl	$=C=O$
carboxy	$HOOC-$
cetyl* (hexadecyl)	$CH_3(CH_2)_{15}-$
chloroformyl (chlorcarbonyl)	$ClCO-$
cinnamoyl	$C_6H_5CH=CHCO-$
cinnamyl (3-phenyl-2-propenyl)	$C_6H_5CH=CHCH_2-$
cinnamylidene	$C_6H_5CH=CHCH=$
cresyl* (hydroxymethylphenyl)	$HO(CH_3)C_6H_4-$
crotonoyl	$CH_3CH=CHCO-$
crotyl (2-butenyl)	$CH_3CH=CHCH_2-$
cyanamido (cyanoamino)	$NCNH-$
cyanato	$NCO-$
cyano	$NC-$
decanedioyl	$-OC(CH_2)_8CO-$
decanoyl	$CH_3(CH_2)_8CO-$
diazo	$N_2=$
diazoamino	$-NHN=N-$
disilanyl	H_3SiSiH_2-
disiloxanyloxy	$H_3SiOSiH_2O$
disulfinyl	$-S(O)S(O)-$
dithio	$-SS-$
enanthoyl* (heptanoyl)	$CH_3(CH_2)_5CO-$
epoxy	$-O-$
ethenyl (vinyl)	$CH_2=CH-$
ethynyl	$HC\equiv C-$
ethoxy	C_2H_5O-
ethyl	CH_3CH_2-
ethylene	$-CH_2CH_2-$
ethylidene	$CH_3CH=$
ethylthio	C_2H_5S-
formamido (formylamino)	$HCONH-$
formyl	$HCO-$
furmaroyl (from fumaric acid)	$-OCCH=CHCO-$
furfuryl (2-furanylmethyl)	$OC_4H_3CH_2-$
furfurylidene (2-furanylmethylene)	$OC_4H_3CH=$
glutamoyl (from glutamic acid)	$-OC(CH_2)_2CH(NH_2)CO-$
glutaryl (from glutaric acid)	$-OC(CH_2)_3CO-$
glycylamino	H_2NCH_2CONH-
glycoloyl; glycolyl (hydroxyacetyl)	$HOCH_2CO-$
glycyl (aminoacetyl)	H_2NCH_2CO-
glyoxyloyl; glyoxylyl (oxoacetyl)	$HCOCO-$
guanidino	$H_2NC(=NH)NH-$
guanyl (aminoiminomethyl)	$H_2NC(=NH)-$
heptadecanoyl	$CH_3(CH_2)_{15}CO-$
heptanamido	$CH_3(CH_2)_5CONH-$
heptanedioyl	$-OC(CH_2)_5CO-$
heptanoyl	$CH_3(CH_2)_5CO-$
hexadecanoyl	$CH_3(CH_2)_{14}CO-$
hexamethylene (1,6-hexanediyl)	$-(CH_2)_6-$
hexanedioyl	$-OC(CH_2)_4CO-$
hippuryl (N-benzoylglycyl)	$C_6H_5CONHCH_2CO-$
hydrazino	H_2NNH-
hydrazo	$-HNNH-$
hydrocinnamoyl	$C_6H_5(CH_2)_2CO-$

Substituent	Formula
hydroperoxy	HOO-
hydroxyamino	HONH-
hydroxy	HO-
imino	HN=
iodoso* (iodosyl)	OI-
iodyl	O_2I-
isoamyl* (isopentyl; 3-methylbutyl)	$(CH_3)_2CH(CH_2)_2$-
isobutenyl (2-methyl-1-propenyl)	$(CH_3)_2C{=}CH$-
isobutoxy (2-methylpropoxy)	$(CH_3)_2CHCH_2O$-
isobutyl (2-methylpropyl)	$(CH_3)_2CHCH_2$-
isobutylidene (3-methylpropylidene)	$(CH_3)_2CHCH{=}$
isobutyryl (2-methyl-1-oxopropyl)	$(CH_3)_2CHCO$-
isocyanato	OCN-
isocyano	CN-
isohexyl (4-methylpentyl)	$(CH_3)_2CH(CH_2)_3$-
isoleucyl (from isoleucine)	$C_2H_5CH(CH_3)CH(NH_2)CO$-
isonitroso* (hydroxyamino)	HON=
isopentyl (3-methylbutyl)	$(CH_3)_2CH(CH_2)_2$-
isopentylidene (3-methylbutylidene)	$(CH_3)_2CHCH_2CH{=}$
isopropenyl (1-methylethenyl)	$CH_2{=}C(CH_3)$-
isopropoxy (1-methylethoxy)	$(CH_3)_2CHO$-
isopropyl (1-methylethyl)	$(CH_3)_2CH$-
isopropylidene (1-methylethylidene)	$(CH_3)_2C{=}$
isothiocyanato (isothiocyano)	SCN-
isovaleryl* (3-methyl-1-oxobutyl)	$(CH_3)_2CHCH_2CO$-
lactoyl (from lactic acid)	$CH_3CH(OH)CO$-
lauroyl (from lauric acid)	$CH_3(CH_2)_{10}CO$-
lauryl (dodecyl)	$CH_3(CH_2)_{11}$-
leucyl (from leucine)	$(CH_3)_2CHCH_2CH(NH_2)CO$-
levulinoyl (from levulinic acid)	$CH_3CO(CH_2)_2CO$-
malonyl (from malonic acid)	$-OCCH_2CO$-
mandeloyl (from mandelic acid)	$C_6H_5CH(OH)CO$-
mercapto	HS-
mesityl	2,4,6-$(CH_3)_3C_6H_2$-
methacryloyl (from methacrylic acid)	$CH_2{=}C(CH_3)CO$-
methallyl (2-methyl-2-propenyl)	$CH_2{=}C(CH_3)CH_2$-
methionyl (from methionine)	$CH_3SCH_2CH_2CH(NH_2)CO$-
methoxy	CH_3O-
methyl	H_3C-
methylene	$H_2C{=}$
methylthio	CH_3S-
myristoyl (from myristic acid)	$CH_3(CH_2)_{12}CO$-
myristyl (tetradecyl)	$CH_3(CH_2)_{13}$-
naphthyl	$(C_{10}H_7)$-
naphthylene	$-(C_{10}H_6)$-
neopentyl (2,2-dimethylpropyl)	$(CH_3)_3CCH_2$-
nitramino (nitroamino)	O_2NNH-
nitro	O_2N-
nitrosamino (nitrosoamino)	ONNH-
nitrosimino (nitrosoimino)	ONN=
nitroso	ON-
nonanoyl (from nonanoic acid)	$CH_3(CH_2)_7CO$-
oleoyl (from oleic acid)	$CH_3(CH_2)_7CH{=}CH(CH_2)_7CO$-
oxalyl (from oxalic acid)	-OCCO-
oxo	O=
palmitoyl (from palmitic acid)	$CH_3(CH_2)_{14}CO$-
pentamethylene (1,5-pentanediyl)	$-(CH_2)_5$-
pentyl	$CH_3(CH_2)_4$-
tert-pentyl	$CH_3CH_2C(CH_3)_2$-
phenacyl	$C_6H_5COCH_2$-
phenacylidene	$C_6H_5COCH{=}$
phenethyl (2-phenylethyl)	$C_6H_5CH_2CH_2$-
phenoxy	C_6H_5O-
phenyl	C_6H_5-
phenylene (benzenediyl)	$-C_6H_4$-
phosphino* (phosphanyl)	H_2P-
phosphinyl* (phosphinoyl)	$H_2P(O)$-
phospho	O_2P-
phosphono	$(HO)_2P(O)$-
phthaloyl (from phthalic acid)	1,2-$C_6H_4(CO\text{-})_2$
picryl (2,4,6-trinitrophenyl)	2,4,6-$(NO_2)_3C_6H_2$-
pimeloyl (from pimelic acid)	$-OC(CH_2)_5CO$-
piperidino (1-piperidinyl)	$C_5H_{10}N$-
pivaloyl (from pivalic acid)	$(CH_3)_3CCO$-
prenyl (3-methyl-2-butenyl)	$(CH_3)_2C{=}CHCH_2$-
propargyl (2-propynyl)	$HC'CCH_2$-
1-propenyl	$-CH{=}CHCH_3$
2-propenyl (allyl)	$CH_2{=}CHCH_2$-
propionyl* (propanyl)	CH_3CH_2CO-
propoxy	$CH_3CH_2CH_2O$-
propyl	$CH_3CH_2CH_2$-
propylidene	$CH_3CH_2CH{=}$
pyrryl (pyrrolyl)	C_4H_4N-
salicyloyl (2-hydroxybenzoyl)	2-HOC_6H_4CO-
selenyl* (selanyl; hydroseleno)	HSe-
seryl (from serine)	$HOCH_2CH(NH_2)CO$-
siloxy	H_3SiO-
silyl	H_3Si-
silylene	$H_2Si{=}$
sorboyl (from sorbic acid)	$CH_3CH{=}CHCH{=}CHCO$-
stearoyl (from stearic acid)	$CH_3(CH_2)_{16}CO$-
stearyl (octadecyl)	$CH_3(CH_2)_{17}$-
styryl (2-phenylethenyl)	$C_6H_5CH{=}CH$-
suberoyl (from suberic acid)	$-OC(CH_2)_6CO$-
succinyl (from succinic acid)	$-OCCH_2CH_2CO$-
sulfamino (sulfoamino)	$HOSO_2NH$-
sulfamoyl (sulfamyl)	H_2NSO_2-
sulfanilyl [(4-aminophenyl)sulfonyl]	4-$H_2NC_6H_4SO_2$-
sulfeno	HOS-
sulfhydryl (mercapto)	HS-
sulfinyl	OS=
sulfo	HO_3S-
sulfonyl (sulfuryl)	$-SO_2$-
terephthaloyl	1,4-$C_6H_4(CO\text{-})_2$
tetramethylene	$-(CH_2)_4$-
thienyl (from thiophene)	(C_4H_3S)-
thiocarbonyl (carbothionyl)	=CS
thiocarboxy	HOSC-
thiocyanato (thiocyano)	NCS-
thionyl* (sulfinyl)	-SO-
threonyl (from threonine)	$CH_3CH(OH)CH(NH_2)CO$-
toluidino [(methylphenyl)amino]	$CH_3C_6H_4NH$-
toluoyl (methylbenzoyl)	$CH_3C_6H_4CO$-
tolyl (methylphenyl)	$CH_3C_6H_4$-
α-tolyl (benzyl)	$C_6H_5CH_2$-
tolylene (methylphenylene)	$-(CH_3C_6H_3)$-
tosyl [(4-methylphenyl) sulfonyl)]	4-$CH_3C_6H_4SO_2$-
triazano	H_2NNHNH-
trimethylene (1,3-propanediyl)	$-(CH_2)_3$-
trityl (triphenylmethyl)	$(C_6H_5)_3C$-
valeryl* (pentanoyl)	$CH_3(CH_2)_3CO$-
valyl (from valine)	$(CH_3)_2CHCH(NH_2)CO$-
vinyl (ethenyl)	$CH_2{=}CH$-
vinylidene (ethenylidene)	$CH_2{=}C{=}$
xylidino [(dimethylphenyl)amino]	$(CH_3)_2C_6H_3NH$-
xylyl (dimethylphenyl)	$(CH_3)_2C_6H_3$-
xylylene [phenelenebis(methylene)]	$-CH_2C_6H_4CH_2$-

Organic Ring Compounds

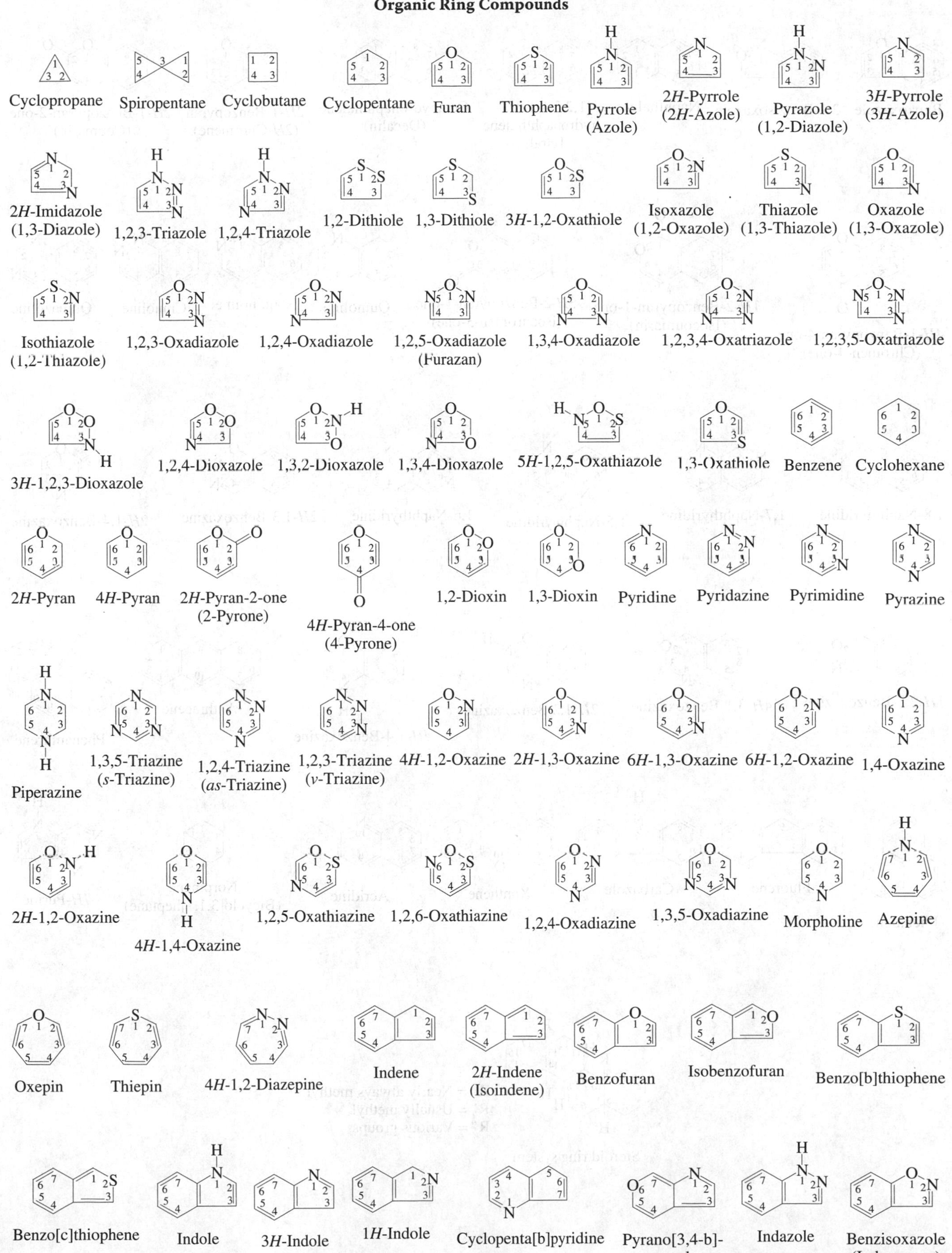
Cyclopropane
Spiropentane
Cyclobutane
Cyclopentane
Furan
Thiophene
Pyrrole (Azole)
2H-Pyrrole (2H-Azole)
Pyrazole (1,2-Diazole)
3H-Pyrrole (3H-Azole)
2H-Imidazole (1,3-Diazole)
1,2,3-Triazole
1,2,4-Triazole
1,2-Dithiole
1,3-Dithiole
3H-1,2-Oxathiole
Isoxazole (1,2-Oxazole)
Thiazole (1,3-Thiazole)
Oxazole (1,3-Oxazole)
Isothiazole (1,2-Thiazole)
1,2,3-Oxadiazole
1,2,4-Oxadiazole
1,2,5-Oxadiazole (Furazan)
1,3,4-Oxadiazole
1,2,3,4-Oxatriazole
1,2,3,5-Oxatriazole
3H-1,2,3-Dioxazole
1,2,4-Dioxazole
1,3,2-Dioxazole
1,3,4-Dioxazole
5H-1,2,5-Oxathiazole
1,3-Oxathiole
Benzene
Cyclohexane
2H-Pyran
4H-Pyran
2H-Pyran-2-one (2-Pyrone)
4H-Pyran-4-one (4-Pyrone)
1,2-Dioxin
1,3-Dioxin
Pyridine
Pyridazine
Pyrimidine
Pyrazine
Piperazine
1,3,5-Triazine (s-Triazine)
1,2,4-Triazine (as-Triazine)
1,2,3-Triazine (v-Triazine)
4H-1,2-Oxazine
2H-1,3-Oxazine
6H-1,3-Oxazine
6H-1,2-Oxazine
1,4-Oxazine
2H-1,2-Oxazine
4H-1,4-Oxazine
1,2,5-Oxathiazine
1,2,6-Oxathiazine
1,2,4-Oxadiazine
1,3,5-Oxadiazine
Morpholine
Azepine
Oxepin
Thiepin
4H-1,2-Diazepine
Indene
2H-Indene (Isoindene)
Benzofuran
Isobenzofuran
Benzo[b]thiophene
Benzo[c]thiophene
Indole
3H-Indole
1H-Indole
Cyclopenta[b]pyridine
Pyrano[3,4-b]-pyrrole
Indazole
Benzisoxazole (Indoxazene)

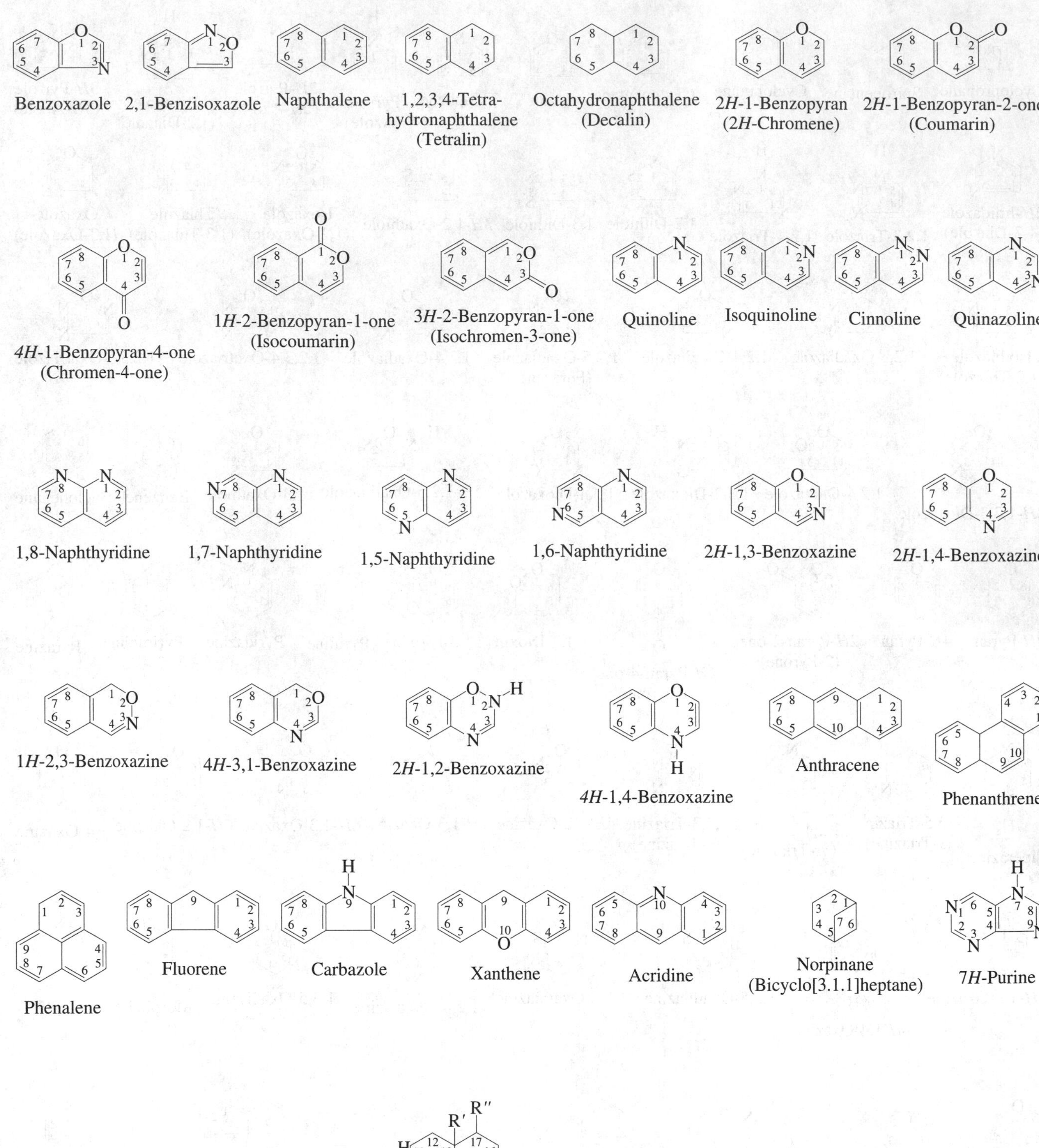
Benzoxazole
2,1-Benzisoxazole
Naphthalene
1,2,3,4-Tetrahydronaphthalene (Tetralin)
Octahydronaphthalene (Decalin)
2H-1-Benzopyran (2H-Chromene)
2H-1-Benzopyran-2-one (Coumarin)
4H-1-Benzopyran-4-one (Chromen-4-one)
1H-2-Benzopyran-1-one (Isocoumarin)
3H-2-Benzopyran-1-one (Isochromen-3-one)
Quinoline
Isoquinoline
Cinnoline
Quinazoline
1,8-Naphthyridine
1,7-Naphthyridine
1,5-Naphthyridine
1,6-Naphthyridine
2H-1,3-Benzoxazine
2H-1,4-Benzoxazine
1H-2,3-Benzoxazine
4H-3,1-Benzoxazine
2H-1,2-Benzoxazine
4H-1,4-Benzoxazine
Anthracene
Phenanthrene
Phenalene
Fluorene
Carbazole
Xanthene
Acridine
Norpinane (Bicyclo[3.1.1]heptane)
7H-Purine
R = Nearly always methyl
R′ = Usually methyl
R″ = Various groups
Steroid ring system

REPRESENTATION OF CHEMICAL STRUCTURES WITH THE IUPAC INTERNATIONAL CHEMICAL IDENTIFIER (INCHI)

Stephen R. Heller and Alan D. McNaught

The IUPAC International Chemical Identifier (InChI) is a freely available, non-proprietary identifier for chemical substances that can be used in both printed and electronic data sources. It is generated from a computerized representation of a molecular structure diagram, which can be produced by chemical structure-drawing software. Its use enables linking of diverse data compilations and unambiguous identification of chemical substances. A full description of the Identifier and software for its generation are available from the IUPAC Web site (Ref. 1), and a helpful compilation of answers to frequently asked questions has been put together at the Unilever Centre for Molecular Science Informatics (Ref. 2). Commercial structure-drawing software that will generate the Identifier is available from several organizations, listed on the IUPAC Web site.

The conversion of structural information to the Identifier is based on a set of IUPAC structure conventions, and rules for normalization and canonicalization (conversion to a single, predictable sequence) of an input structure representation. The resulting InChI is simply a series of characters that serve to uniquely identify the structure from which it was derived. The InChI uses a layered format to represent all available structural information relevant to compound identity. InChI layers are listed below. Each layer in an InChI representation contains a specific type of structural information. These layers, automatically extracted from the input structure, are designed so that each successive layer adds additional detail to the Identifier. The specific layers generated depend on the level of structural detail available and whether or not allowance is made for tautomerism. Of course, any ambiguities or uncertainties in the original structure will remain in the InChI.

This layered structure design offers a number of advantages. If two structures for the same substance are drawn at different levels of detail, the one with the lower level of detail will, in effect, be contained within the other. Specifically, if one substance is drawn with stereo-bonds and the other without, the layers in the latter will be a subset of the former. The same will hold for compounds treated by one author as tautomers and by another as exact structures with all H-atoms fixed. This can work at a finer level. For example, if one author includes double bond and tetrahedral stereochemistry, but another omits stereochemistry, the latter InChI will be contained in the former.

The InChI layers are

1. Formula
2. Connectivity (no formal bond orders)
    a. disconnected metals
    b. connected metals
3. Isotopes
4. Stereochemistry
    a. double bond (*Z/E*)
    b. tetrahedral (sp^3)
5. Tautomers (on or off)

Charges are not part of the basic InChI, but rather are added at the end of the InChI string.

Two examples of InChI representations are given below. It is important to recognize, however, that InChI strings are intended for use by computers and end users need not understand any of their details. In fact, the open nature of InChI and its flexibility of representation, after implementation into software systems, may allow chemists to be even less concerned with the details of structure representation by computers.

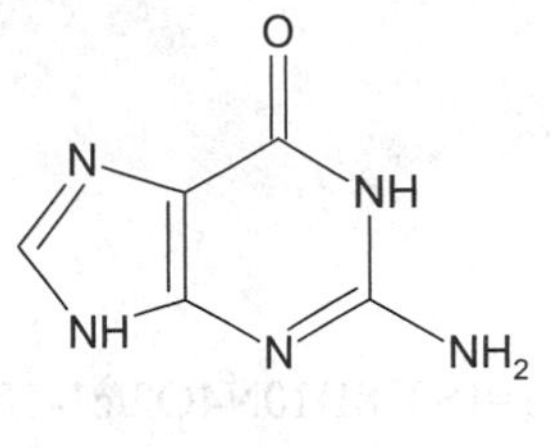

guanine

InChI=1/C5H5N5O/c6-5-9-3-2(4(11)10-5)7-1-8-3/h1H,(H4,6,7,8,9,10,11)/f/h8,10H,6H2

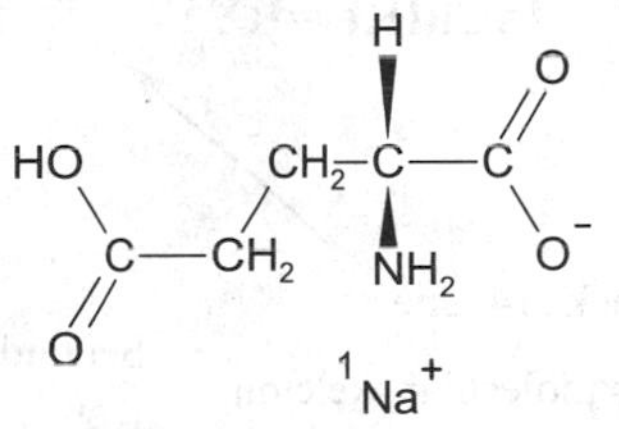

monosodium glutamate

InChI=1/C5H9NO4.Na/c6-3(5(9)10)1-2-4(7)8;/h3H,1-2,6H2,(H,7,8)(H,9,10);/q;+1/p-1/t3-;/m1./s1/fC5H8NO4.Na/h7H;/q-1;m

The layers in the InChI string are separated by the '/' character followed by a lowercase letter (except for the first layer, the chemical formula), with the layers arranged in predefined order. In the examples the following segments are included

- InChI version number
- /- chemical formula
- /c connectivity-1.1 (excluding terminal H)
- /h connectivity-1.2 (locations of terminal H, including mobile H attachment points)
- /q charge
- /p proton balance
- /t sp^3 (tetrahedral) parity
- /m parity inverted to obtain relative stereo (1 = inverted, 0 = not inverted)
- /s stereo type (1 = absolute, 2 = relative, 3 = racemic)
- /f chemical formula of the fixed-H structure if it is different
- /h connectivity-2 (locations of fixed mobile H)
- /q charge
- /t sp^3 (tetrahedral) parity
- /m parity inverted to obtain relative stereo (1 = inverted, 0 = not inverted, . = inversion does not affect the parity)
- /s stereo type (1 = absolute, 2 = relative, 3 = racemic)

One of the most important applications of InChI is the facility to locate mention of a chemical substance using Internet-based search engines. This is made easier by using a shorter (compressed) form of InChI, known as InChIKey. The InChIKey is a 27-character representation that, because it is compressed, cannot be reconverted into the original structure, but it is not subject to the undesirable and unpredictable breaking of longer character strings by some search engines. The usefulness of the InChIKey as a search tool is enhanced by its derivation from a "standard" InChI. i.e., an InChI produced with standard option settings for features such as tautomerism and stereochemistry. An example is shown below; the "standard" InChI is denoted by the letter "S" after the version number.

InChI=1S/C8H10N4O2/c1-10-4-9-6-5(10)7(13)12(3)8(14)11(6)2/h4H,1-3H3 (caffeine)

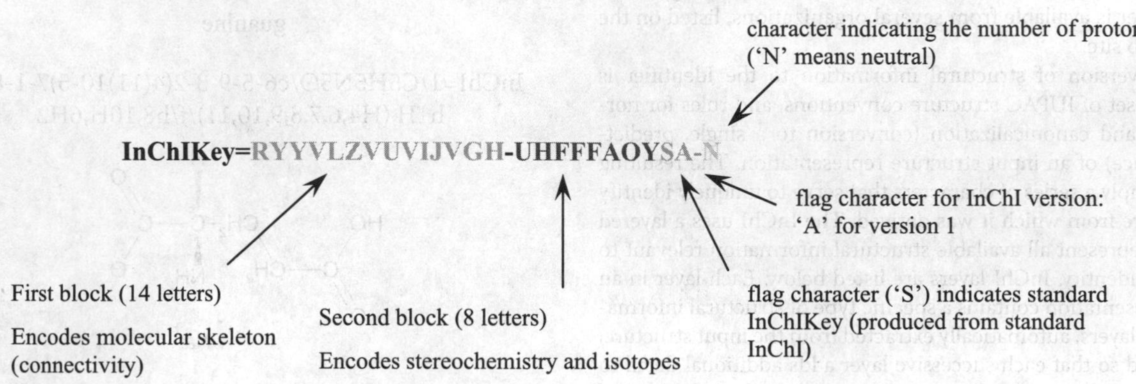

Use of InChIKey also allows searches based solely on atomic connectivity (first 14 characters). Software for generating InChIKey is available from the IUPAC Web site (Ref. 1).

The enormous databases compiled by organizations such as PubChem (Ref. 4), the U.S. National Cancer Institute (NCI), and ChemSpider (Ref. 5) contain millions of InChIs and InChIKeys, which allow sophisticated searching of these collections. PubChem provides InChI-based structure-search facilities for both identical and similar structures (Ref. 6), and ChemSpider offers both search facilities and Web services enabling a variety of InChI and InChIKey conversions (Ref. 7). The NCI Chemical Structure Lookup Service (Ref. 8) provides InChI-based search access to over 39 million chemical structures from over 80 different public and commercial data sources.

References

1. http://www.iupac.org/inchi
2. http://wwmm.ch.cam.ac.uk/inchifaq/
3. *Pure Appl. Chem.*, in preparation.
4. http://pubchem.ncbi.nlm.nih.gov
5. http://www.chemspider.com
6. http://pubchem.ncbi.nlm.nih.gov/search
7. http://www.chemspider.com/InChI.asmx
8. http://cholla.chemnavigator.com/cgi-bin/lookup/new/search

SCIENTIFIC ABBREVIATIONS, ACRONYMS, AND SYMBOLS

This table lists some abbreviations, acronyms, and symbols encountered in the physical sciences. Most entries in italic type are symbols for physical quantities; for more details on these, see the table "Symbols and Terminology for Physical and Chemical Quantities" in this section. Additional information on units may be found in the table "International System of Units (SI)" in Section 1. Many of the terms to which these abbreviations refer are included in the tables "Definitions of Scientific Terms" in Section 2 and "Techniques for Materials Characterization" in Section 12. Useful references for further information are given below.

Publication practices vary with regard to the use of capital or lower case letters for many abbreviations. An effort has been made to follow the most common practices in this table, but much variation is found in the literature. Likewise, policies on the use of periods in an abbreviation vary considerably. Periods are generally omitted in this table unless they are necessary for clarity. Periods should never appear in SI units. The SI prefixes (m, k, M, etc.) are included here, but they should never be used alone. Selected combinations of these prefixes with SI units (e.g., mg, kV, MW) are also included.

Abbreviations are listed in alphabetical order without regard to case. Entries beginning with Greek letters fall at the end of the table.

References

1. *Quantities, Units, and Symbols in Physical Chemistry, Third Edition*, IUPAC 2007, RSC Publishing, 2007.
2. Kotyk, A., *Quantities, Symbols, Units, and Abbreviations in the Life Sciences*, Humana Press, Totawa, NJ, 1999.
3. Rhodes, P. H., *The Organic Chemist's Desk Reference*, Chapman & Hall, London, 1995.
4. Minkin, V., Glossary of Terms used in Theoretical Organic Chemistry, *Pure Appl. Chem.* 71, 1919–1981, 1999.
5. Brown, R. D., Ed., Acronyms Used in Theoretical Chemistry, *Pure Appl. Chem.* 68, 387–456, 1996.
6. *Quantities and Units, ISO Standards Handbook, Third Edition*, International Organization for Standardization, Geneva, 1993.
7. Cohen, E. R., and Giacomo, P., Symbols, Units, Nomenclature, and Fundamental Constants in Physics, *Physica* 146A, 1–68, 1987.
8. *Chemical Acronyms Database*, Indiana University, <www.oscar.chem.indiana.edu/cfdocs/ libchem/acronyms/ acronymsearch.html>.
9. *Acronyms and Symbols*, <www3.interscience.wiley.com/stasa/>.
10. *IUPAC Compendium of Chemical Terminology* (Gold Book), <goldbook.iupac.org>.
11. IUPAC-IUB Joint Commission on Biochemical Nomenclature, *Pure & Appl. Chem.* 56, 595, 1984.

A	ampere; alanine; adenine (in genetic code)
Å	ångström
A	absorbance; area; Helmholtz energy; mass number
A_H	Hall coefficient
A_r	atomic weight (relative atomic mass)
a	atto (SI prefix for 10^{-18})
a	absorption coefficient; acceleration; activity; van der Waals constant
a_0	Bohr radius
AAA	acetoacetanilide
Aad	2-aminoadipic acid
AAF	2-(acetylamino)fluorene
AAN	aminoacetonitrile
AAO	acetaldehyde oxime
AAS	atomic absorption spectroscopy
ABA	abscisic acid; acrylonitrile-butadiene acrylate
Abe	abequose
ABL	α-acetylbutyrolactone
ABS	acrylonitrile-butadiene-styrene copolymer
abs	absolute
Abu	2-aminobutanoic acid
Ac	acetyl; acetate
ac, AC	alternating current
ACAC	acetylacetone
Aces	2-[(2-amino-2-oxoethyl)amino]ethanesulfonic acid
ACS	acrylonitrile-chlorinated polyethylene-styrene copolymer
ACT	activated complex theory
ACTH	adrenocorticotropic hormone
A/D	analog to digital
Ad	adamantyl
Ada	[(carbamoylmethyl)imino]diacetic acid
Ade	adenine
ADI	acceptable daily intake
Ado	adenosine
ADP	adenosine diphosphate; ammonium dihydrogen phosphate
ads	adsorption
AE	appearance energy
ae	eon (10^9 years)
AEP	1-(2-aminoethyl)piperazine
AEPD	2-amino-2-ethyl-1,3-propanediol
AES	atomic emission spectroscopy; Auger electron spectroscopy
AF	audio frequency
AFM	atomic force microscopy
Ahx	2-aminohexanoic acid
AI	artificial intelligence
AIBN	2,2'-azobis[isobutyronitrile]
AICA	5-amino-1*H*-imidazole-4-carboxamide
AIM	atoms in molecules (method)
AIP	aluminum isopropoxide
Al	Alfén number
Ala	alanine
alc	alcohol
ALE	atomic layer epitaxy
aliph.	aliphatic
alk.	alkaline
All	allose
Alt	altrose
AM	amplitude modulation
Am	amyl
am	amorphous solid
AMA	acrylate maleic anhydride terpolymer
AMMA	acrylate-methyl methacrylate copolymer
AMP	adenosine monophosphate
AMPD	2-amino-2-methyl-1,3-propanediol
AMS	accelerator mass spectrometry

AMTCS	amyltrichlorosilane [trichloropentylsilane]
amu	atomic mass unit (recommended symbol is u)
AN	acetonitrile; acrylonitrile
anh, anhyd	anhydrous
ANOVA	analysis of variance
antilog	antilogarithm
ANTU	1-naphthalenylthiourea
AO	atomic orbital
AOM	angular overlap model
AP	ethylene-propylene copolymer
APAD	3-acetylpyridine adenine dinucleotide
APAP	acetyl *p*-amino phenol (acetaminophen)
Ape	2-aminopentanoic acid
API	atmospheric pressure ionization
Api	apiose
APM	atomic probe microanalysis
Apm	2-aminopimelic acid
APO	amorphous polyolefin
APPI	atmospheric pressure photoionization
APS	appearance potential spectroscopy; adenosine phosphosulfate
APW	augmented plane wave
aq	aqueous
Ar	aryl
Ara	arabinose
Ara-ol	arabinitol
Arg	arginine
ARPES	angular resolved photoelectron spectroscopy
AS	acrylonitrile styrene copolymer
ASA	acetylsalicylic acid; acrylonitrile-styrene-acrylonitrile block copolymer
ASC	4-(acetylamino)benzenesulfonyl chloride
ASCII	American National Standard Code for Information Interchange
ASE	aromatic stabilization model
Asn	asparagine
Asp	aspartic acid
at	atomization
ATCP	4-amino-3,5,6-trichloro-2-pyridinecarboxlic acid
ATEE	*N*-acetyl-*L*-tyrosine ethyl ester
ATLC	adsorption thin layer chromatography
atm	standard atmosphere
ATP	adenosine triphosphate
ATR	attenuated total internal reflection
at.wt.	atomic weight
AU	astronomical unit (ua is also used); polyurethane
AUC	area under the time-concentration curve
av	average
avdp	avoirdupois
B	bel; asparagine or aspartic acid (unspecified)
B	magnetic flux density; second virial coefficient; susceptance
b	barn
b	van der Waals constant; molality
BA	benzyladenine
BAL	British anti-Lewisite [2,3-dimercapto-1-propanol]
BAP, BaP	benzo[a]pyrene
bar	bar (pressure unit)
bbl	barrel
BBP	benzyl butyl phthalate
BCB	bromocresol blue
bcc	body centered cubic
BCF	bioconcentration factor
BCG	bromocresol green
BCME	bis(chloromethyl) ether
BCNU	*N,N′*-bis(2-chloroethyl)-*N*-nitrosourea
BCP	bromocresol purple
BCPB	bromochlorophenol blue
BCPE	1,1-bis(4-chlorophenyl)ethanol
BCS	Bardeen-Cooper-Schrieffer (theory)
BDE	bond dissociation energy
BDEA	butyldiethanolamime
BDMA	benzyldimethylamine
Bé	Baumé
BEBO	bond energy bond order (method)
BEI	biological exposure index
BEM	biological effect monitoring
BEP	2-butyl-2-ethyl-1,3-propanediol
Bes	2-[bis(2-hydroxyethyl)amino]ethanesulfonic acid
BET	Brunauer-Emmett-Teller (isotherm)
BeV	billion electronvolt
BGE	butyl glycidyl ether
BHA	*tert*-butyl-4-hydroxyanisole
BHC	benzene hexachloride [hexachlorobenzene]
Bhn	Brinell hardness number
BHT	butylated hydroxytoluene [2,6-di-*tert*-butyl-4-methylphenol]
Bi	biot
Bicine	*N,N*-bis(2-hydroxyethyl)glycine
BIRD	blackbody infrared radiative dissociation
Bistris	2-[bis(2-hydroxyethyl)amino]-2-(hydroxymethyl) propane-1,3-diol
Bistris-propane	1,3-bis[tris(hydroxymethyl)methylamino]propane
BLO	γ-butyrolactone
BN	bond number; benzonitrile
BNS	nuclear backscattering spectroscopy
BO	Born-Oppenheimer (approximation); bond order
BOD	biochemical oxygen demand
BON	β-hydroxynaphthoic acid
BP	base peak (in mass spectrometry); benzo[a]pyrene
bp	boiling point; base pair
BPB	bromophenol blue
BPG	2,3-bis(phospho)-*D*-glycerate
BPL	β-propiolactone
BPO	benzoyl peroxide
bpy	2,2′-bipyridine
Bq	becquerel
Br	butyryl
BRE	bond resonance energy
BrUrd	5-bromouridine
BS	Birge-Sponer extrapolation
BSE	back scattered electron(s)
BSSE	basis set superposition error
BTMSA	1,2-bis(trimethylsilyl)acetylene
Btu	British thermal unit
BTX	benzene, toluene, and xylene

Bu	butyl
bu	bushel
BVE	butyl vinyl ether
Bz	benzoyl
Bzl	benzyl
C	coulomb; cysteine; cytosine (in genetic code)
°C	degree Celsius
C	capacitance; heat capacity; number concentration
c	centi (SI prefix for 10^{-2}); combustion reaction
c	amount concentration; specific heat; velocity
c_0	speed of light in vacuum
CA	collisional activation; cellulose acetate
ca.	approximately
CAB	cellulose acetate butyrate
CADD	computer-assisted drug design
cal	calorie
calc	calculated
cAMP	adenosine cyclic 3',5'-(hydrogen phosphate)
CAN	ceric ammonium nitrate
CAR	carbon fiber
CARS	coherent anti-Stokes Raman spectroscopy
CAS	complete active space
CASRN	Chemical Abstracts Service Registry Number
CAT	computerized axial tomography; clear air turbulence
CBE	chemical beam epitaxy
CBS	complete basis set (of orbitals)
CC	coupled cluster; combustion calorimetry
cc	cubic centimeter
CCD	charge-coupled device
CD	circular dichroism
cd	candela; condensed (phase)
CDAA	2-chloro-*N,N*-diallylacetamide
CDNO	complete neglect of differential overlap
CDP	cytidine 5'-diphosphate
CDT	1,5,9-cyclododecatriene
CDTA	(1,2-cyclohexylenedinitrilo)tetraacetic acid monohydrate
CDW	charge density waves
CED	cohesive energy density
CEM	channel electron multiplier
CEP	counter electrophoresis
CEPA	coupled electron-pair approximation
cf.	compare
CFC	chlorofluorocarbon compound
cfm	cubic feet per minute
CFRP	carbon reinforced plastics
cgs	centimeter-gram-second system
Chaps	3-[3-(cholamidopropyl)dimethylammonio]-1-propanesulfonic acid
Ches	2-(*N*-cyclohexylamino)ethanesulfonic acid
CHF	coupled Hartree-Fock (method)
Chl	chlorophyll
Cho	choline
CHT	1,3,5-cycloheptatriene
Ci	curie
CI	configuration interaction; chemical ionization; color index
CID	charge-injection device; collision-induced dissociation
CIDEP	chemically induced dynamic electron polarization
CIDNP	chemically induced dynamic nuclear polarization
CIE	countercurrent immunoelectrophoresis
cir	circular
CKFF	Cotton-Kraihanzel force field
CL	cathode luminescence (spectroscopy)
CLT	central limit theorem
cm	centimeter
c.m.	center of mass
CMC	carboxymethylcellulose
c.m.c.	critical micelle concentration
CMO	canonical molecular orbital
CMP	cytidine 5'-monophosphate; chemical measurement process
CN	coordination number; cellulose nitrate
CNDO	complete neglect of differential overlap
Co	Cowling number
COC	cycloolefin copolymer
COD	chemical oxygen demand; 1,4-cyclooctadiene
conc	concentrated; concentration
const	constant
COOP	crystal orbital overlap population
cos	cosine
cosh	hyperbolic cosine
COSY	correlation spectroscopy
COT	1,3,5,7-cyclooctatetraene
cot	cotangent
coth	hyperbolic cotangent
CP	chemically pure
Cp	cyclopentadienyl
Cp*	pentamethylcyclopentadienyl
cP	centipoise
cp	candle power
CPA	coherent potential approximation
CPC	centrifugal partition chromatography
cpd	contact potential difference
CPE	chlorinated polyethylene
CPL	circular polarization of luminescence
CPR	chlorophenol red
cps	cycles per second
CPT	charge conjugation/space inversion/time inversion (theorem)
CPU	central processing unit
CPVC	chlorinated poly(vinyl chloride)
CR	chloroprene rubber (neoprene)
cr, cryst	crystalline (phase)
CRF	charge remote fragmentation
CRU	constitutional repeating unit (in polymer nomenclature)
CSA	camphorsulfonic acid
csc	cosecant
CSR	charge stripping reaction
CT	charge transfer
ct	carat
CTA	cellulose triacetate
CTEM	conventional transmission electron microscopy
CTFE	chlorotrifluoroethylene
CTP	cytidine 5'-triphosphate
CTR	controlled thermonuclear reaction
cu	cubic

CV cyclic voltammetry
CVD chemical vapor deposition
cw continuous wave
cwt hundredweight (112 pounds)
Cy cyclohexyl
Cya cysteic acid
Cyd cytidine
cyl cylinder
Cys cysteine
Cyt cytosine
D debye unit; aspartic acid
D diffusion coefficient; dissociation energy; electric displacement
d day; deuteron; deci (SI prefix for 10^{-1})
d distance; density; dextrorotatory
2,4-D 2,4-dichlorophenoxyacetic acid
D/A digital to analog
Da dalton
DA donor-acceptor (complex)
da deka (SI prefix for 10^{1})
DAA diacetone alcohol
DAB 4-(dimethylamino)azobenzene
Dab 2,4-diaminobutanoic acid
DACH *trans*-1,2-diaminocyclohexane
DAIP diallyl isophthalate plasticizer
DAP diammonium phosphate
DART direct analysis in real time mass spectrometry
dB decibel
DBA dibenz[a,h]anthracene; dibenzylamine
DBCP 1,2-dibromo-3-chloropropane
DBED dibenzyl ethylene diamine
DBM dibutyl maleate
DBMC 2,4-di-*tert*-butyl-5-methylphenol
DBMS database management system
DBP dibutyl phthalate; 2,3-dibromo-1-propanol
DBPC 2,6-di-*tert*-butyl-*p*-cresol
dc, DC direct current
DCB dicyanobenzene
DCBP 4,4'-dichlorobenzophenone
DCEE dichloroethyl ether
DCHA dicyclohexylamine
DCM dichloromethane
DCNP 2,6-dichloro-4-nitrophenol
DCP 2,4-dichlorophenol
DCPD dicyclopentadiene
DDM 4,4'-diaminodiphenylmethane
DDT dichlorodiphenyltrichloroethane
DE delocalization energy; delayed extraction
DEA *N,N*-diethylaniline; diethanolamine
Dec decyl
dec decomposes
DEET diethyltoluamide [*N,N*-diethyl-3-methylbenzamide]
deg degree
DEK diethyl ketone
den density
DEP 2,2-diethyl-1,3-propanediol
DES diethyl sulfate
DESI desorption electrospray ionization (in mass spectrometry)
det determinant
dev deviation
DFT density functional theory
dGlc 2-deoxyglucose
DHBA 2,3-dihydroxybenzoic acid
DHH dehydroheliotridine
DHR dehydroretronecine
DHU dihydrouridine
DI desorption ionization
diam diameter
DIBA diisobutyl adipate
DIBK diisobutyl ketone
dil dilute; dilution
DIM diatomics in molecules (method); digital imaging microscopy
DIPA diisopropanolamine
dm decimeter
DMA *N,N*-dimethylaniline
DMAB 4-(dimethylamino)azobenzene
DMAC *N,N*-dimethylacetamide
DMAE *N,N*-dimethylethanolamine
DMBA 7,12-dimethylbenz[a]anthracene
DME 1,2-dimethoxyethane
DMF *N,N*-dimethylformamide
DMP dimethyl phthalate
DMS dimethyl sulfide
DMSO dimethyl sulfoxide
DMT dimethyl terephthalate; dimethyl tartrate
DN donor number
DNA deoxyribonucleic acid
DNase deoxyribonuclease
DNB 1,3-dinitrobenzene
DNMR dynamic NMR spectroscopy
DNP dinitropyrene
Dod dodecyl
DOP dioctyl phthalate
DOS density of states; digital operating system; dioctyl sebacate
doz dozen
DP, d.p. degree of polymerization
DPA diphenylamine
DPG *N,N'*-diphenylguanidine
dpl displacement
Dpm 2,6-diaminopimelic acid
dpm disintegrations per minute
dps disintegrations per second
DPU *N,N'*-diphenylurea
dr dram
DRE Dewar resonance energy
dRib 2-deoxyribose
DRIFT diffuse reflectance infrared Fourier transform
DRP dynamic reaction path
DRS diffuse reflectance spectroscopy
DS degree of substitution
DSC differential scanning calorimetry
DTA differential thermal analysis
DTBP di-*tert*-butyl peroxide

DVB	divinylbenzene
dyn	dyne
DZ	double-zeta (type of basis set)
E	exa (SI prefix for 10^{18}); glutamic acid
E	electric field strength; electromotive force; energy; Young's modulus of elasticity; entgegen (*trans* configuration)
E_h	Hartree energy
e	electron; base of natural logarithms
e	elementary charge; linear strain
EA	electron affinity
EAA	ethylene acrylic acid copolymer; ethyl acetoacetate
EAK	ethyl amyl ketone (3-octanone)
EAN	effective atomic number
EC	ethyl cellulose
ECD	electron capture dissociation
ECP	effective core potential
ECR	electron cyclotron resonance
ECTFE	ethylene-chlorotrifluoroethylene copolymer
ED	electron diffraction
EDAX	energy dispersive analysis by x-rays
EDB	ethylene dibromide [1,2-dibromoethane]
EDC	ethylene dichloride [1,2-dichloroethane]
EDI	estimated daily intake
EDS	energy-dispersive x-ray spectroscopy
EDTA	ethylenediaminetetraacetic acid
EEA	ethylene-ethyl acetate copolymer
EEDQ	ethyl 2-ethoxy 1(2*H*)-quinolinecarboxylate
EEL	environmental exposure level
EELS	electron energy loss spectroscopy
EES	excitation emission spectrum
EFF	empirical force field
EFFF	energy factored force field
EG	equilibrium in the gas phase
EGA	evolved gas analysis
EGG	Einstein-Guth-Gold equation
EHMO, EHT	extended Hückel molecular orbital (theory)
EIMS	electron impact mass spectrometry
EIS	electron impact spectroscopy; electrochemical impedance spectroscopy
ELISA	enzyme-linked immunosorbent assay
ELS	energy loss spectroscopy
EM	extended molarity; electron microscopy
EMAC	ethylene-methyl acrylate copolymer
emf	electromotive force
EMPA, EMA	electron probe microanalysis
emu	electromagnetic unit system
en	ethylenediamine
ENDOR	electron-nuclear double resonance
EOS	equation of state
EP	epoxy resin
EPDS	electron photodetachment spectroscopy
EPM	ethylene-propylene copolymer
EPR	electron paramagnetic resonance; ethylene propylene rubber
EPS	expanded polystyrene
EPT-76	provisional low temperature scale of 1976
EPTC	dipropylcarbamothioic acid, *S*-ethyl ester
EPXMA	electron probe x-ray microanalysis
eq, eqn	equation
eqQ	quadrupole coupling constant
erf	error function
erg	erg (energy unit)
ES	equilibrium in solution
ESA	electrostatic energy analyzer
ESCA	electron spectroscopy for chemical analysis
ESD	electron stimulated desorption
e.s.d.	estimated standard deviation
ESI	electrospray ionization
ESR	electron spin resonance
est	estimated
esu	electrostatic unit system
ET	ephemeris time; electron transfer
Et	ethyl
ETA	electrothermal analysis
ETFE	ethylene tetrafluoroethylene polymer
Etn	ethanolamine
ETO	ethylene oxide
ETS	electron tunneling spectroscopy
ETU	ethylene thiourea
EU	polyether polyurethane
Eu	Euler number
e.u.	entropy unit
eV	electronvolt
EVA	ethylene-vinyl acetate copolymer
EVE	ethyl vinyl ether
EXAFS	extended x-ray absorption fine structure (spectroscopy)
EXELFS	extended energy loss fine structure
exp	exponential function
expt	experimental
ext	external
F	farad; phenylalanine
°F	degree Fahrenheit
F	Faraday constant; force; angular momentum
f	formation reaction; femto (SI prefix for 10^{-15})
f	activity coefficient; aperture ratio; focal length; force constant; frequency; fugacity
FAB	fast atom bombardment
FAD	flavine adenine dinucleotide
FAIMS	high-field asymmetric waveform ion mobility spectrometry
FA-SIFT	flowing afterglow – selected ion-flow tube
fcc	face centered cubic
FD	field desorption
FEL	free electron laser
FEM	field emission microscopy
FEMO	free electron molecular orbital
FEP	fluorinated ethylene propylene
FET	field effect transistor
FI	field ionization
fid	free induction decay
FIM	field ion microscopy
FIR	far infrared
fl	fluid (phase)
FM	frequency modulation
Fo	Fourier number

fp	freezing point
fpm	feet per minute
fps	feet per second; foot-pound-second system
Fr	franklin
Fr	Froude number
FRP	fibrous glass reinforced polyester; fiber reinforced plastic
Fru	fructose
FSGO	floating spherical gaussian orbitals
FT	Fourier transform
ft	foot
ft-lb	foot pound
FTIR	Fourier transform infrared spectroscopy
FTMS	Fourier transform mass spectrometry
FTNMR	Fourier transform nuclear magnetic resonance
fus	fusion (melting)
FVP	flash vacuum pyrolysis
FWHM	full width at half maximum
G	gauss; guanine (in genetic code); giga (SI prefix for 10^9); glycine
G	electrical conductance; Gibbs energy; gravitational constant; sheer modulus
g	gram; gas (phase)
g	acceleration due to gravity; degeneracy; Landé g-factor; statistical weight
GABA	γ-aminobutyric acid
Gal	gal; galactose
gal	gallon
GalN	galactosamine
GB	gas-phase basicity
GC	gas chromatography
GC-MS	gas chromatography-mass spectroscopy
GDMS	glow discharge mass spectroscopy
GDP	guanosine 5'-diphosphate
gem	geminal (on the same carbon atom)
GeV	gigaelectronvolt
GF	glass reinforced
GIAO	gauge invariant atomic orbital
GIBMS	guided ion beam mass spectrometry
gl	glacial
Gla	4-carboxyglutamic acid
GLC	gas-liquid chromatography
Glc	glucose
GlcA	gluconic acid
GlcN	glucosamine
GlcNAc	*N*-acetylglucosamine
GlcU	glucuronic acid
Gln	glutamine
GLP	good laboratory practice
Glu	glutamic acid
Glx	glutamine or glutamic acid (unspecified)
Gly	glycine
GMP	guanosine 5'-monophosphate
GMT	Greenwich mean time
GPC	gel-permeation chromatography
gpm	gallons per minute
gps	gallon per second
Gr	Grashof number
gr	grain
Gra	glyceraldehyde
Gri	glyceric acid
Grn	glycerone [dihydroxyacetone]
Gro	glycerol
GTO	gaussian-type orbital
GTP	guanosine 5'-triphosphate
Gua	guanine
Gul	gulose
Guo	guanosine
GUT	grand unified theory
GVB	generalized valence bond (method)
GWS	Glashow-Weinberg-Salam (theory)
Gy	gray; gigayear
H	henry; histidine
H	enthalpy; Hamiltonian function; magnetic field
H_0	Hubble constant
h	helion; hour; hecto (SI prefix for 10^2)
h	Planck constant
Ha	Hartmann number
ha	hectare
HAM	hydrogenic atoms in molecules
hav	haversine
Hb	hemoglobin
HCA	heterocyclic amine; hexachloroacetone
HCB	hexachlorobenzene
hcp	hexagonal closed packed
Hcy	homocysteine
HCZ, HCTZ	hydrochlorothiazide
HDL	high-density lipoprotein
HDPE	high-density polyethylene
HDS	hydrodesulfurization
HEIS	high-energy ion scattering
HEP	high energy physics
Hepes	4-(2-hydroxyethyl)-1-piperazineethanesulfonic acid
Hepps	4-(2-hydroxyethyl)-1-piperazinepropanesulfonic acid
HF	high frequency; Hartree-Fock (method)
HFA	hexafluoroacetone
HFO	Hartree-Fock orbital
hfs	hyperfine structure
HHPA	hexahydrophthalic anhydride
HIPS	high-impact polystyrene
His	histidine
HMC	high strength molding compound
HMDA	hexamethylenediamine
HMO	Hückel molecular orbital
HMT	hexamethylenetetramine
HMX	cyclotetramethylenetetranitramine
HN1	2-chloro-*N*-(2-chloroethyl)-*N*-ethylethanamine
HOAc	acetic acid
HOC	halogenated organic compound(s)
HOMAS	harmonic oscillator model of aromatic stabilization
HOMO	highest occupied molecular orbital
HOSE	harmonic oscillator stabilization energy
Hp	heptyl
hp	horsepower
HPLC	high-performance liquid chromatography

HPMS	high pressure mass spectrometry
HQ	*p*-hydroquinone
hr	hour
HRE	Hückel resonance energy
HREELS	high resolution electron energy loss spectroscopy
HREM	high resolution electron microscopy
HSAB	hard-soft acid-base (theory)
HSE	homodesmotic stabilization energy
Hse	homoserine
HVA	homovanillic acid
Hx	hexyl
Hyl	5-hydroxylysine
Hyp	hypoxanthine; 4-hydroxyproline
Hz	hertz
I	isoleucine; inositol; ionomer
I	electric current; ionic strength; moment of inertia; nuclear spin angular momentum; radiant intensity
i	square root of minus one
i	electric current
I/O	input/output
IAT	international atomic time
IC	integrated circuit
ICD	induced circular dichroism
ICP	inductive-coupled plasma
ICR	ion cyclotron resonance
ICVTST	improved canonical variational transition-state theory
ID	inside diameter
id	ideal (solution)
Ido	iodose
IdoA	iduronic acid
IDP	inosine 5'-diphosphate
IE	ionization energy
i.e.p.	isoelectric point
IEPA	independent electron pair approximation
IF	intermediate frequency
IGLO	individual gauge for localized orbitals
IIR	isobutylene-isoprene rubber (butyl rubber)
IKES	ion kinetic energy spectrometry
Ile	isoleucine
Im	imaginary part
IMFP	inelastic mean free path (of electrons)
imm	immersion
IMP	inosine 5'-monophosphate
IMPATT	impact ionization avalanche transit time
IMS	ion mobility spectrometry
in.	inch
InChI	IUPAC International Chemical Identifier
INDO	immediate neglect of differential overlap
Ino	inosine
INS	inelastic neutron scattering; ion neutralization spectroscopy
Ins	*myo*-inositol
int	internal
IP	ionization potential
IPA	isopropyl alcohol
IPMA	ion probe microanalysis
IPN	interpenetrating polymer network
IPR	isotope perturbation of resonance
IPTS	International Practical Temperature Scale
IQ	2-amino-3-methyl-3*H*-imidazo(4,5-f)quinoline
IR	infrared
IRAS	infrared reflection-absorption spectroscopy
IRC	intrinsic reaction coordinate
IRMPD	infrared multiphoton dissociation
IRMS	isotope ratio mass spectrometry
IRS	infrared spectroscopy
isc	intersystem crossing
ISE	ion-selective electrode; isodesmic stabilization energy
ISS	ion scattering spectroscopy
IT	ion trap; information technology
ITP	inosine 5'-triphosphate
ITS	International Temperature Scale (1990)
IU	international unit
IVE	isobutyl vinyl ether
J	joule; leucine or isoleucine (unspecified)
J	angular momentum; electric current density; flux; Massieu function
j	angular momentum; electric current density
JT	Jahn-Teller (effect)
K	kelvin; lysine
K	absorption coefficient; bulk modulus; equilibrium constant; kinetic energy
k	kilo (SI prefix for 10^3)
k	absorption index; Boltzmann constant; rate constant; thermal conductivity; wave vector
kat	katal (unit of catalytic activity)
kb	kilobar; kilobases (DNA or RNA)
KC-MS	Knudson cell mass spectrometry
kcal	kilocalorie
KDP	potassium dihydrogen phosphate
KE	kinetic energy
KERD	kinetic energy release distributions
keV	kiloelectronvolt
KG	kinetics in the gas phase
kg	kilogram
kgf	kilogram force
KIE	kinetic isotope effect
kJ	kilojoule
km	kilometer
Kn	Knudsen number
kPa	kilopascal
KS	kinetics in solution
kt	karat
KTP	potassium titanium phosphate
kV	kilovolt
kva	kilovolt ampere
kW	kilowatt
kwh	kilowatt hour
L	liter; lambert; leucine
L	Avogadro constant; inductance; Lagrange function; angular momentum
l	liter; liquid (phase)
l	angular momentum; length; mean free path; levorotatory
Lac	lactose
LAH	lithium aluminum hydride
lat.	latitude
lb	pound

lbf	pound force
LC	liquid chromatography; liquid crystal
LC-MS	liquid chromatography-mass spectrometry
lc	liquid crystal (phase)
LCAO	linear combination of atomic orbitals
LD	lethal dose; laser desorption
LDA	local density approximation; lithium diisopropylamide
LDL	low-density lipoprotein
LDPE	low-density polyethylene
LDV	laser-Doppler velocimetry
Le	Lewis function
LE	localization energy
LEC	liquid exchange chromatography
LED	light emitting diode
LEED	low-energy electron diffraction
LEIS	low-energy ion scattering
Leu	leucine
LFER	linear free energy relationships
LFL	lower flammable limit
LI	laser ionization
lim	limit
LIMS	laser ionization mass spectroscopy; laboratory information management system
liq	liquid
LIT	linear ion trap
LLCT	ligand to ligand charge transfer
lm	lumen
LMCT	ligand to metal charge transfer
LMMS	laser microprobe mass spectrometry
LMO	localized molecular orbital
LMR	laser magnetic resonance
ln	logarithm (natural)
LNDO	local neglect of differential overlap
log	logarithm (common)
LOMO	lowest occupied molecular orbital
long.	longitude
LPE	linear polyethylene
LPG	liquid petroleum gas
LPHP	laser-powered homogeneous pyrolysis
LPU	law of propagation of uncertainty
LSFE	linear field stabilization energy
LSI	liquid secondary ionization
LST	local sidereal time
LT	local time
LTE	local thermodynamic equilibrium
LUMO	lowest unoccupied molecular orbital
lx	lux
ly	langley
l.y.	light year
Lys	lysine
Lyx	lyxose
M	molar (as in 0.1 M solution); mega (SI prefix for 10^6); methionine
M	magnetization; molar mass; mutual inductance; torque; angular momentum component; median
M_r	molecular weight (relative molar mass)
m	meter; molal (as in 0.1 m solution); metastable (isotope); milli (SI prefix for 10^{-3})
m	magnetic dipole moment; mass; molality; angular momentum component; *meta* (locant on aromatic ring)
Ma	Mach number
MA	maleic anhydride
MAAc	methyl amyl acetate
Mal	maltose
Man	mannose
MASNMR	magic angle spinning nuclear magnetic resonance
max	maximum
Mb	myoglobin
MBE	molecular beam epitaxy
MBER	molecular beam electron resonance
MBK	methyl butyl ketone
MBOCA	4,4′-methylenebis[2-chloroaniline]
MBPT	many body perturbation theory
MBS	methyl methacrylate butadiene styrene terpolymer
MC	Monte Carlo (method)
MCAA	monochloroacetic acid
MCD	magnetic circular dichroism
MCP	microchannel plate
MCPA	(4-chloro-2-methylphenoxy)acetic acid
MCPF	modified coupled pair functional
MCS	Monte Carlo simulation
MCSCF	multiconfigurational self-consistent field (approximation)
MD	molecular dynamics (method)
MDI	methylene diphenylisocyanate
MDPE	medium density polyethylene
Me	methyl
MeCCNU	1-(2-chloroethyl)-3-(4-methylcyclohexyl)-1-nitrosourea
MeIQ	2-amino-3,4-dimethylimidazo[4,5-f]quinoline
MeIQx	2-amino-3,8-dimethylimidazo[4,5-f]quinoxaline
MEK	methyl ethyl ketone
MEP	molecular electrostatic potential
MERP	minimum energy reaction path
Mes	4-morpholineethanesulfonic acid
MESFET	metal-semiconductor field-effect transistor
Met	methionine
MeV	megaelectronvolt
meV	millielectronvolt
MF	molecular formula; melamine-formaldehyde resin
mg	milligram
MHD	magnetohydrodynamics
mi	mile
MIAK	methyl isoamyl ketone
MIBK	methyl isobutyl ketone
MIC	methyl isocyanate
MIK	methyl isobutyl ketone
MIKES	mass-analyzed ion kinetic energy spectrometry
min	minimum; minute
MINDO	modified INDO (method)
MIPK	methyl isopropyl ketone
MIR	mid infrared
misc	miscible
MKS	meter-kilogram-second system
MKSA	meter-kilogram-second-ampere system
mL, ml	milliliter
MM	molecular mechanics

mm	millimeter
MMDR	microwave-microwave double resonance
mmf	magnetomotive force
mmHg	millimeter of mercury
MNA	*m*-nitroaniline
MNDO	modified neglect of diatomic overlap
MNT	*m*-nitrotoluene
MNU	*N*-methyl-*N*-nitrosourea
MO	molecular orbital; methyl orange
MODR	microwave-optical double resonance
mol	mole
mol.wt.	molecular weight
mon	monomeric form
Mops	4-morpholinepropanesulfonic acid
MOS	metal-oxide semiconductor
MOSFET	metal-oxide semiconductor field-effect transistor
mp	melting point
MPa	megapascal
MPA	Mulliken population analysis
Mpc	megaparsec
MPD	2-methyl-2,4-pentanediol
MPI	multiphoton ionization
MPTP	1,2,3,6-tetrahydro-1-methyl-4-phenylpyridine
MR	methyl red
MRD	multireference double substitution (method)
MRI	magnetic resonance imaging
mRNA	messenger RNA
MS	mass spectroscopy
ms	millisecond
MSA	methanesulfonic acid
MSDS	Material Safety Data Sheet
MSF	methanesulfonyl fluoride
MS-K	mass spectroscopy – kinetic method
MSL	mean sea level
MTBE	methyl *tert*-butyl ether
MTD	maximum tolerable dose
Mur	muramic acid
mV	millivolt
MVK	methyl vinyl ketone
MW	megawatt; microwave; molecular weight
mW	milliwatt
MWD	molecular weight distribution
Mx	maxwell
N	newton; asparagine
N	angular momentum; neutron number; number density
N_A	Avogadro constant
n	neutron; nano (SI prefix for 10^{-9})
n	amount of substance; number density; principal quantum number; refractive index; normal (in chemical formulas)
NAA	nuclear activation analysis; 1-naphthaleneacetic acid
NAAD	nicotinic acid adenine dinucleotide
NAD	nicotinamide adenine dinucleotide
NADH	reduced NAD
NADP	NAD phosphate
NANA	*N*-acetylneuraminic acid
NAO	natural atomic orbital
NBO	natural bond orbital
nbp	normal boiling point
NBR	nitrile butadiene rubber [poly(butadiene-co-acrylonitrile)]
NDELA	*N*-nitrosodiethanolamine
NEDOR	nuclear electron double resonance
NEM	*N*-ethylmorpholine
Neu	neuraminic acid
NEXAFS	near-edge x-ray absorption fine structure
ng	nanogram
NHO	natural hybrid orbital
NHOMO	next-to-highest occupied molecular orbital
NICI	negative ion chemical ionization
NICS	nuclear independent chemical shift
NIR	near infrared; ribosylnicotinamide
nm	nanometer
NMN	β-nicotinamide mononucleotide
NMR	nuclear magnetic resonance
Nn	nonyl
NNDO	neglect of nonbonded differential overlap
NO	natural orbital
NOE	nuclear Overhauser effect
NOEL	no observed effect level
NOx	nitrogen oxides
NP	nitropyrene
NPA	natural population analysis
NQR	nuclear quadrupole resonance
NR	natural rubber
NRA	nuclear reaction analysis
ns	nanosecond
NSE	neutron spin echo
NTA	nitrilotriacetic acid
NTP	normal temperature and pressure
Nu	nucleophile
Nu	Nusselt number
o	*ortho* (locant on aromatic ring)
OAA	oxaloacetic acid
obs, obsd	observed
Oc	octyl
OD	optical density; outside diameter
ODMR	optically detected magnetic resonance
Oe	oersted
OFGF	outer valence Green's function (method)
ONA	*o*-nitroaniline
ORD	optical rotatory dispersion
Oro	orotate; orotidine
oz	ounce
P	poise; peta (SI prefix for 10^{15}); proline
P	power; pressure; probability; sound energy flux
p	proton; pico (SI prefix for 10^{-12})
p	dielectric polarization; electric dipole moment; momentum; pressure; bond order; *para* (as aromatic ring locant)
Pa	pascal
PA	proton affinity; pyrrolizidine alkaloid; polyamide (nylon)
PAA	poly(acrylic acid)
PABA	*p*-aminobenzoic acid
PABS	*p*-aminobenzenesulfonamide
PAC	photoacoustic calorimetry

PAH polycyclic aromatic hydrocarbon(s)
PAI polyamide-imide
PAL polyaniline
PAM polyacrylamide
PAN 1-(2-pyridylazo)-2-naphthol; polyacrylonitrile
PAR 4-(2'-pyridylazo)resorcinol
PARA polyaryl amide
PAS photoacoustic spectroscopy; polyarylsulfone
PB polybutylene
PBA poly(butyl acrylate)
PBAN polybutylene-acrylonitrile copolymer
PBB polybrominated biphenyl
PBD poly(1,3-butadiene)
PBI polybenzimidazole
PBMA poly(butyl methacrylate)
PBS polybutadiene-styrene copolymer
PBT poly(butylene terephthalate)
PC paper chromatography; photocalorimetry; polycarbonate
pc parsec
PCB polychlorinated biphenyl
PCHO paraldehyde (2,4,6-trimethyl-1,3,5-trioxane)
PCL polycaprolactone
PCM polarizable continuum model
PCNB pentachloronitrobenzene
PCP pentachlorophenol
PCR polymerase chain reaction
PCT poly(cyclohexylene terephthalate)
PCTFE polymonochlorotrifluoroethylene
PD potential difference
PDB *p*-dichlorobenzene
pdl poundal
PDMS poly(dimethylsiloxane)
PE polyethylene
Pe pentyl
Pe Péclet number
pe probable error
PEA poly(ethyl acrylate)
PEEK poly(ether ether ketone)
PEG poly(ethylene glycol)
PEI polyetherimide
PEK polyetherketone
PEL permissible exposure limit
PEO poly(ethylene oxide)
PES photoelectron spectroscopy; potential energy surface; polyethersulfone
PET positron emission tomography; poly(ethylene terephthalate); pentaerythritol tetranitrate
peth petroleum ether
PEX crosslinked polyethylene
PF phenol-formaldehyde resin
pf power factor
PFOA perfluorooctanoic acid
pg picogram
Ph phenyl
pH negative log of hydrogen ion concentration
Phe phenylalanine
PhIP 2-amino-1-methyl-6-phenylimidazo[4,5-b]pyridine
PHPMS pulsed high pressure mass spectrometry
PI polyimide
pI isoelectric point
PIB polyisobutylene
PIMS photoionization mass spectrometry
PIN p-intrinsic-n (diode)
Pipes 1,4-piperazinediethanesulfonic acid
PIV particle-image velocimetry
PIXE particle induced x-ray emission
p*K* negative log of ionization constant
PLM principle of least motion
PLOT porous-layer open-tabular (column)
PLS partial least squares
pm picometer
PMA poly(methyl acrylate)
PMAC phenylmercuric acetate
PMMA poly(methyl methacrylate)
PMO perturbation MO (theory)
PMP polymethylpentene
PMS polymethylstyrene; *p*-methylstyrene
PNA *p*-nitroaniline
PNDO partial neglect of differential overlap
PNO pair natural orbitals
PNRA prompt nuclear reaction analysis
PNT *p*-nitrotoluene
PO polyolefin
POAV π-orbital axis vector
pol polymeric form
POM polyoxymethylene
POx phosphorus oxides
PP polypropylene
ppb parts per billion
PPC chlorinated polypropylene
PPE poly(phenylene ether)
ppm parts per million
PPO poly(phenylene oxide)
PPOX polypropylene oxide
PPP Pariser-Parr-Pople (method)
PPS poly(phenylene sulfide)
PPSU poly(phenylene sulfone)
PPT poly(propylene terephthalate)
ppt parts per thousand; precipitate
Pr propyl
Pr Prandtl number
PRDDO partial retention of diatomic differential overlap
Pro proline
PS photoelectron spectroscopy; polystyrene
ps picosecond
PSD photon stimulated desorption
psi pounds per square inch
psia pounds per square inch absolute
psig pounds per square inch gage
PT perturbation theory
pt pint
PTFE poly(tetrafluoroethylene)
PTME poly(tetramethylene terephthalate)
PTMS propyltrimethoxysilane
PTP *p*-terphenyl

PTU phenylthiourea
PU polyurethane
Pu purine
PVA poly(vinyl alcohol)
PVAc poly(vinyl acetate)
PVC poly(vinyl chloride)
PVD physical vapor deposition
PVDC poly(vinylidene chloride)
PVDF poly(vinylidene fluoride)
PVF poly(vinyl fluoride)
PVK poly(vinyl carbazole)
PVME poly(methyl vinyl ether)
PVOH poly(vinyl alcohol)
PVP poly(vinyl pyrrolidone)
PVT pressure-volume-temperature
Py pyrimidine
PyMS pyrolysis mass spectrometry
p.z.c. point of zero charge
Q electric charge; heat; partition function; quadrupole moment; radiant energy; vibrational normal coordinate; glutamine
q electric field gradient; flow rate; heat; wave vector (phonons)
QCD quantum chromodynamics
QCI quadratic configuration interaction
QCT quasi-classical trajectory (method)
QED quantum electrodynamics
Q.E.D. quod erat demonstrandum (which was to be proved)
QIT quadrupole ion trap
QMRE quantum mechanical resonance energy
QMS quadrupole mass spectrometry
QSAR quantitative structure-activity relations
QSO quasi-stellar object
qt quart
quad quadrillion BTU ($=1.055 \cdot 10^{18}$ joules)
Qui quinovose
q.v. quod vide (which you should see)
R roentgen; arginine; alkyl radical (in chemical formulas)
°R degree Rankine
R electrical resistance; gas constant; molar refraction; Rydberg constant; coefficient of multiple correlation
r reaction (as in $\Delta_r H$)
r position vector; radius
RA right ascension
rad radian
RAIRS reflection-absorption infrared spectroscopy
RAM random access memory
RBS Rutherford back scattering
Rbu, Rul ribulose
RCI ring current index
RDA rubidium dihydrogen arsenate
RDS rate determining step
RDX Royal Demolition Explosive (hexahydro-1,3,5-trinitro-1,3,5-triazine)
Re real part
RE resonance energy
RED radial electron distribution
REELS reflection electron energy loss spectroscopy
REM reflection electron microscopy
rem roentgen equivalent man
REMPI resonance-enhanced multiphoton ionization
REPE resonance energy per electron
RF radiofrequency
RGA residual gas analyzer
Rha rhamnose
RHEED reflection high-energy electron diffraction
RHF restricted Hartree-Fock (theory)
RI resonance ionization
RIA radioimmunoassay
Rib ribose
Ribulo ribulose
rms root-mean-square
RNA ribonucleic acid
RNase ribonuclease
ROHF restricted open shell Hartree-Fock
ROM read only memory
ROMP ring opening metathesis polymerization
ROP ring opening polymerization
RPA random phase approximation
RPH reaction path Hamiltonian
RPLC reversed-phase liquid chromatography
rpm revolutions per minute
rps revolutions per second
RRK Rice-Ramsperger-Kassel (theory)
RRKM Rice-Ramsperger-Kassel-Marcus (theory)
rRNA ribosomal RNA
RRS resonance Raman spectroscopy
RS Raman spectroscopy
RSC reaction-solution calorimetry
Ry rydberg
S siemens; serine
S area; entropy; probability current density; Poynting vector; symmetry coordinate; spin angular momentum
s second; solid (phase)
s path length; spin angular momentum; symmetry number; sedimentation coefficient; solubility; symmetrical (as stereochemical descriptor)
SAED selected area electron diffraction
SALC symmetry adapted linear combinations
SALI surface analysis by laser ionization
SAM scanning Auger microscopy
SAMS self-assembled monolayers
SANS small angle neutron scattering
SAR structure-activity relationship
Sar sarcosine
sat, satd saturated
SAXS small angle x-ray scattering
SB styrene butadiene copolymer
SBS styrene butadiene styrene block copolymer
Sc Schmidt number
SC spin-coupled (method)
SCD state correlation diagram
SCE saturated calomel electrode
SCF self-consistent field (method); supercritical fluid
SCP single cell protein
SCR silicon-controlled rectifier
SCRF self-consistent reaction field (method)

sd	standard deviation
SDA	sulfadiazine
SDW	spin density wave
SE	strain energy
SEBS	styrene ethylene butylene styrene block copolymer
SEC	size exclusion chromatography
sec	secant; second
sec	secondary (in chemical name)
SECSY	spin-echo correlated spectroscopy
Sed	sedoheptulose
SEELFS	surface extended energy loss fine structure
SEM	scanning electron microscopy; standard error of the mean
sepn	separation
Ser	serine
SERS	surface-enhanced Raman spectroscopy
SET	single electron transfer
SEXAF	surface extended x-ray absorption fine structure
SFC	supercritical fluid chromatography
Sh	Sherwood number
Shy	thiohypoxanthine
SI	International System of Units; surface ionization
SID	surface-induced dissociation
SILAR	successive ionic layer adsorption and reaction
SIM	selected ion monitoring
SIMS	secondary-ion spectroscopy
sin	sine
sinh	hyperbolic sine
SIPN	semi-interpenetrating polymer network
SIS	styrene isoprene styrene block copolymer
SLAM	scanning laser acoustic microscopy
SLUMO	second lowest unoccupied molecular orbital
SMILES	simplified molecular input line entry system
SMMA	styrene methyl methacrylate copolymer
SMO	semiempirical molecular orbital
SMOW	Standard Mean Ocean Water (Vienna)
SNMS	sputtered neutral mass spectroscopy
Sno	thiouridine
SNU	solar neutrino unit
SOJT	second-order Jahn-Teller (effect)
sol	soluble; solution
soln, sln	solution
SOMO	singly occupied molecular orbital
Sor	sorbose
sp gr	specific gravity
SPM	scanned probe microscopy
SPST	single-pulse shock tubes
sq	square
Sr	Strouhal number
sr	steradian
Srd	6-thioinosine
SSMS	source spark mass spectroscopy
St	stoke
St	Stanton number
std, stnd	standard (state)
STEL	short-term exposure limit
STEM	scanning transmission electron microscope
STM	scanning tunneling microscopy
STO	Slater-type orbital
STP	standard temperature and pressure
sub, subl	sublimes; sublimation
Suc, Sac	sucrose
Sur	thiouracil
Sv	sievert
SWIFT	stored waveform inverse Fourier transform
T	tesla; tera (SI prefix for 10^{12}); threonine
T	kinetic energy; period; term value; temperature (thermodynamic); torque; transmittance
t	metric tonne; triton
t	Celsius temperature; thickness; time; transport number
TAC	time-to-amplitude converter
TAI	International Atomic Time
Tal	talose
tan	tangent
tanh	hyperbolic tangent
Taps	3-{[2-hydroxy-1,1-bis(hydroxymethyl)ethyl]amino}-1-propanesulfonic acid
TBE	1,1,2,2-tetrabromoethane
TBP	tributyl phosphate
TC	titration calorimetry
TCA	trichloroacetic acid
TCB, TCBA	2,3,6-trichlorobenzoic acid
TCE	trichloroethylene
TCG	Geocentric Coordinated Time
TCNE	tetracyanoethylene
TCNQ	tetracyanoquinodimethane
TCP	tricresyl phosphate
TCSCF	two configuration self-consistent field
TDA	toluene-2,4-diamine
TDI	toluene diisocyanate
tDNA	transfer DNA
TE	transverse electric
TEA	triethanolamine; triethylamine
TED	transferred electron device; transmission electron diffraction
TEDA	triethylenediamine
TEELS	transmission electron energy loss spectroscopy
TEM	transverse electromagnetic; transmission electron microscope
temp	temperature
TEO	thermoplastic elastic olefin
TEPP	tetraethyl pyrophosphate
tert	tertiary (in chemical name)
Tes	2-{[2-hydroxy-1,1-bis(hydroxymethyl)ethyl]amino}-1-propanesulfonic acid
TFD	Thomas-Fermi-Dirac (method)
TFE	tetrafluoroethylene
TGA	thermogravimetric analysis
Thd	ribosylthymine
THEED	transmission high energy electron diffraction
theor	theoretical
thf, THF	tetrahydrofuran
THQ	1,2,3,4-tetrahydroquinoline
Thr	threonine
Thy	thymine
TI	thermal ionization

TIPA	triisopropanolamine
TL	thermoluminescence
TLC	thin-layer chromatography
TLV	threshold limit value
TM	transverse magnetic
TMAB	tetrabutylammonium bromide
TMAO	trimethylamine oxide
TMCP	tri-*m*-cresyl phosphate
TMEDA	*N,N,N',N'*-tetramethyl-1,2-ethanediamine
TMMV	threshold molecular weight value
TMS	tetramethylsilane
TNA	2,4,6-trinitroaniline
TNB	1,3,5-trinitrobenzene
TNM	tetranitromethane
TNT	2,4,6-trinitrotoluene
TOCP	tri-*o*-cresyl phosphate
TOF	turnover frequency
TOF-MS	time-of-flight mass spectrometer
tol	tolyl
TON	turnover number
TOPO	trioctylphosphine oxide
Torr	torr (pressure unit)
TOTP	tri-*o*-tolyl phosphate
TPE	thermoplastic elastomer
TPTA	triphenyltin acetate
TPTC	triphenyltin chloride
TRE	topological resonance energy
Tre	trehalose
Tricine	*N*-[2-hydroxy-1,1-bis(hydroxymethyl)ethyl]glycine
Tris	2-amino-2-(hydroxymethyl)-1,3-propanediol
TRMC	time-resolved microwave conductivity
tRNA	transfer RNA
Trp	tryptophan
trs	transition
TS	transition state
TSS	transition state spectroscopy
TST	generalized transition-state theory
TTF	tetrathiofulvalene
Tyr	tyrosine
U	uracil (in genetic code)
U	electric potential difference; internal energy
u	unified atomic mass unit
u	Bloch function; electric mobility; velocity
ua	astronomical unit (AU is also used)
UBFF	Urey-Bradley force field
UDMH	1,1-dimethylhydrazine
UDP	uridine 5'-diphosphate
UHF	ultrahigh frequency; unrestricted Hartree-Fock (method)
UHMWPE	ultrahigh molecular weight polyethylene
ULDPE	ultra low density polyethylene
ULPE	ultra linear polyethylene
UMP	uridine 5'-monophosphate
uns, unsym	unsymmetrical (as chemical descriptor)
UPS, UPES	ultraviolet photoelectron spectroscopy
Ura	uracil
Urd	uridine
USP	United States Pharmacopeia
UT	universal time
UTC	coordinated universal time
UTP	uridine 5'-triphosphate
UV	ultraviolet
V	volt; valine
V	electric potential; potential energy; volume
v	reaction rate; specific volume; velocity; vibrational quantum number; vicinal (as chemical descriptor)
v/v	volume per volume (volume of solute divided by volume of solution, expressed as percent)
VA	vinyl acetate, vanillic acid
Val	valine
vap	vaporization
VAT	vibration assisted tunneling
VB	valence band; valence bond (theory)
VCD	vibrational circular dichroism
VDW	van der Waals interaction
VHF	very high frequency
vic	vicinal (on adjacent carbon atom)
VIS	visible region of the spectrum
vit	vitreous (phase)
VLDPE	very low density polyethylene
VLPP	very low pressure pyrolysis
VMA	vanilmandelic acid
VOC	volatile organic compound(s)
VOFF	valence orbital force field
VPC	vapor phase chromatography
VSEPR	valence shell electron-pair repulsion (method)
VSIP	valence state ionization potential
VSLI	very large scale integrated (circuit)
VSMOW	Vienna Standard Mean Ocean Water
VTCS	vinyltrichlorosilane
VUV	vacuum ultraviolet
W	watt; tryptophan
W	radiant energy; statistical weight; work
w	energy density; mass fraction; velocity; work
w/v	weight per volume (mass of solute divided by volume of solution, usually expressed as g/100 mL)
w/w	weight per weight (mass of solute divided by mass of solution, expressed as percent)
WAXS	wide angle x-ray scattering
Wb	weber
We	Weber number
WKB	Wentzel-Kramers-Brillouin (approximation)
WLF	Williams-Landel-Ferry (equation)
WLN	Wiswesser line notation
wt	weight
X	X unit; halogen (in chemical formula)
X	reactance
x	mole fraction
X, Xaa	unspecified amino acid
XAFS	x-ray absorption fine structure
Xan	xanthine
XANES	x-ray absorption near-edge structure
Xao	xanthosine
Xle	leucine or isoleucine (unspecified)
XLPE	crosslinked polyethylene
Xlu, Xul	xylulose
XPS, XPES	x-ray photoelectron spectroscopy

XRD x-ray diffraction
XRF x-ray fluorescence
XRS x-ray spectroscopy
Xyl xylose
Y yotta (SI prefix for 10^{24}); tyrosine
Y admittance; Planck function; Young's modulus
y yocto (SI prefix for 10^{-24})
y mole fraction for gas (when *x* refers to liquid phase)
y, yr year
YAG yttrium aluminum garnet
yd yard
YIG yttrium iron garnet
Z zetta (SI prefix for 10^{21}); glutamine or glutamic acid (unspecified)
Z atomic number; compression factor; collision number; impedance; partition function; zusammen (*cis*-configuration)
z zepto (SI prefix for 10^{-21})
z charge number (of an ion); collision frequency factor
ZDO zero differential overlap
ZINDO Zerner's INDO method
ZPE, ZPVE zero point vibrational energy
ZULU Greenwich mean time
α alpha particle
α absorption coefficient; degree of dissociation; electric polarizability; expansion coefficient; fine structure constant
β beta particle
β reciprocal temperature parameter (= $1/kT$)
γ photon; gamma (obsolete mass unit = μg)
γ activity coefficient; conductivity; magnetogyric ratio; mass concentration; ratio of heat capacities; surface tension
Γ Grüneisen parameter; level width; surface concentration
Δ inertial defect; mass excess
δ chemical shift; Dirac delta function; Kronecker delta; loss angle
ε emittance; Levi-Civita symbol; linear strain; molar absorption coefficient; permittivity
ζ Coriolis coupling constant; electrokinetic potential
η overpotential; viscosity
κ compressibility; conductivity; magnetic susceptibility; molar absorption coefficient
λ absolute activity; radioactive decay constant; thermal conductivity; wavelength
Λ angular momentum; ionic conductivity
μ muon; micro (SI prefix for 10^{-6})
μ chemical potential; electric dipole moment; electric mobility; friction coefficient; Joule-Thompson coefficient; magnetic dipole moment; mobility; permeability
μF microfarad
μg microgram
μm micrometer
μs microsecond
ν frequency; kinematic velocity; stoichiometric number
ν_e neutrino
$\tilde{\nu}$ wavenumber
Π osmotic pressure; Peltier coefficient
π pion
ρ density; reflectance; resistivity
σ electrical conductivity; cross section; normal stress; shielding constant (NMR); Stefan-Boltzmann constant; surface tension; standard deviation
τ transmittance; chemical shift; shear stress; relaxation time
Φ magnetic flux; potential energy; radiant power; work function
ϕ electrical potential; fugacity coefficient; osmotic coefficient; quantum yield; wavefunction
χ magnetic susceptibility
χ_e electric susceptibility
ψ wavefunction
Ω ohm
Ω axial angular momentum; solid angle
ω circular frequency; angular velocity; harmonic vibration wavenumber; statistical weight

GREEK, RUSSIAN, AND HEBREW ALPHABETS

The following table presents the Hebrew, Greek, and Russian alphabets, their letters, the names of the letters, and the English equivalents.

Hebrew[1,3]

Letter	Name	Equivalent
א	aleph	’[2]
ב	beth	b, bh
ג	gimel	g, gh
ד	daleth	d, dh
ה	he	h
ו	waw	w
ז	zayin	z
ח	heth	ḥ
ט	teth	ṭ
י	yodh	y
כ ך	kaph	k, kh
ל	lamedh	l
מ ם	mem	m
נ ן	nun	n
ס	samekh	s
ע	ayin	
פ ף	pe	p, ph
צ ץ	sadhe	ṣ
ק	qoph	q
ר	resh	r
שׂ	sin	ś
שׁ	shin	sh
ת	taw	t, th

Greek[4]

Letter	Name	Equivalent
Α α	alpha	a
Β β	beta	b
Γ γ	gamma	g, n
Δ δ	delta	d
Ε ε	epsilon	e
Ζ ζ	zeta	z
Η η	eta	ē
Θ θ	theta	th
Ι ι	iota	i
Κ κ	kappa	k
Λ λ	lambda	l
Μ μ	mu	m
Ν ν	nu	n
Ξ ξ	xi	x
Ο ο	omicron	o
Π π	pi	p
Ρ ρ	rho	r, rh
Σ σ ς	sigma	s
Τ τ	tau	t
Υ υ	upsilon	y, u
Φ φ	phi	ph
Χ χ	chi	ch
Ψ ψ	psi	ps
Ω ω	omega	ō

Russian

Letter	Equivalent
А а	a
Б б	b
В в	v
Г г	g
Д д	d
Е е	e
Ж ж	zh
З з	z
И и Й й	i, ĭ
К к	k
Л л	l
М м	m
Н н	n
О о	o
П п	p
Р р	r
С с	s
Т т	t
У у	u
Ф ф	f
Х х	kh
Ц ц	ts
Ч ч	ch
Ш ш	sh
Щ щ	shch
Ъ ъ[5]	”
Ы ы	y
Ь ь[6]	’
Э э	e
Ю ю	yu
Я я	ya

[1] Where two forms of a letter are given, the second one is the form used at the end of a word.
[2] Not represented in transliteration when initial.
[3] The Hebrew letters are primarily consonants; a few of them are also used secondarily to represent certain vowels, when provided at all, by means of a system of dots or strokes adjacent to the consonated characters.
[4] The letter gamma is transliterated "n" only before velars; the letter upsilon is transliterated "u" only as the final element in diphthongs.
[5] This sign indicates that the immediately preceding consonant is not palatized even though immediately followed by a palatized vowel.
[6] This sign indicates that the immediately preceding consonant is palatized even though not immediately followed by a palatized vowel.

DEFINITIONS OF SCIENTIFIC TERMS

Brief definitions of selected terms of importance in chemistry, physics, and related fields of science are given in this section. The selection process emphasizes the following types of terms:

- Physical quantities
- Units of measure
- Classes of chemical compounds and materials
- Important theories, laws, and basic concepts.

Individual chemical compounds are not included.

Definitions have taken wherever possible from the recommendations of international or national bodies, especially the International Union of Pure and Applied Chemistry (IUPAC) and International Organization for Standardization (ISO). For physical quantities and units, the recommended symbol is also given. The source of such definitions is indicated by the reference number in brackets following the definition. In many cases these official definitions have been edited in the interest of stylistic consistency and economy of space. The user is referred to the original source for further details.

An asterisk (*) following a term indicates that further information can be found by consulting the index of this handbook under the entry for that term.

References

1. *ISO Standards Handbook 2, Units of Measurement*, International Organization for Standardization, Geneva, 1992.
2. *Quantities, Units, and Symbols in Physical Chemistry, Second Edition*, International Union of Pure and Applied Chemistry, Blackwell Scientific Publications, Oxford, 1993.
3. *Compendium of Chemical Terminology*, International Union of Pure and Applied Chemistry, Blackwell Scientific Publications, Oxford, 1987.
4. *A Guide to IUPAC Nomenclature of Organic Compounds*, International Union of Pure and Applied Chemistry, Blackwell Scientific Publications, Oxford, 1993.
5. *Glossary of Class Names of Organic Compounds and Reactive Intermediates Based on Structure, Pure and Applied Chemistry*, 67, 1307, 1995.
6. *Compendium of Analytical Nomenclature*, International Union of Pure and Applied Chemistry, Blackwell Scientific Publications, Oxford, 1987.
7. *Nomenclature of Inorganic Chemistry*, International Union of Pure and Applied Chemistry, Blackwell Scientific Publications, Oxford, 1990.
8. *Glossary of Basic Terms in Polymer Science, Pure and Applied Chemistry*, 68, 2287, 1996.
9. *The International Temperature Scale of 1990, Metrologia*, 27, 107, 1990.
10. *Compilation of ASTM Standard Definitions*, American Society of Testing and Materials, Philadelphia, 1990.
11. *ASM Metals Reference Book*, American Society for Metals, Metals Park, OH, 1983.

***Ab initio* method** - An approach to quantum-mechanical calculations on molecules which starts with the Schrödinger equation and carries out a complete integration, without introducing empirical factors derived from experimental measurement.

Absorbance (*A*) - Defined as $-\log(1-\alpha) = \log(1/\tau)$, where α is the absorptance and τ the transmittance of a medium through which a light beam passes. [2]

Absorbed dose (*D*) - For any ionizing radiation, the mean energy imparted to an element of irradiated matter divided by the mass of that element. [1]

Absorptance (α) - Ratio of the radiant or luminous flux in a given spectral interval absorbed in a medium to that of the incident radiation. Also called absorption factor. [1]

Absorption coefficient (*a*) - The relative decrease in the intensity of a collimated beam of electromagnetic radiation, as a result of absorption by a medium, during traversal of an infinitesimal layer of the medium, divided by the length traversed. [1]

Absorption coefficient, molar (ε) - Absorption coefficient divided by amount-of-substance concentration of the absorbing material in the sample solution ($\varepsilon = a/c$). The SI unit is m^2/mol. Also called extinction coefficient, but usually in units of $mol^{-1}dm^3cm^{-1}$. [2]

Acceleration - Rate of change of velocity with respect to time.

Acceleration due to gravity (*g*)* - The standard value (9.80665 m/s^2) of the acceleration experienced by a body in the earth's gravitational field. [1]

Acenes - Polycyclic aromatic hydrocarbons consisting of fused benzene rings in a rectilinear arrangement. [5]

Acid - Historically, a substance that yields an H^+ ion when it dissociates in solution, resulting in a pH<7. In the Brönsted definition, an acid is a substance that donates a proton in any type of reaction. The most general definition, due to G.N. Lewis, classifies any chemical species capable of accepting an electron pair as an acid.

Acid dissociation constant (K_a)* - The equilibrium constant for the dissociation of an acid HA through the reaction $HA + H_2O \rightleftharpoons A^- + H_3O^+$. The quantity $pK_a = -\log K_a$ is often used to express the acid dissociation constant.

Actinides - The elements of atomic number 89 through 103, e.g., Ac, Th, Pa, U, Np, Pu, Am, Cm, Bk, Cf, Es, Fm, Md, No, Lr. [7]

Activation energy* - In general, the energy that must be added to a system in order for a process to occur, even though the process may already be thermodynamically possible. In chemical kinetics, the activation energy is the height of the potential barrier separating the products and reactants. It determines the temperature dependence of the reaction rate.

Activity - For a mixture of substances, the absolute activity λ of substance B is defined as $\lambda_B = \exp(\mu_B/RT)$, where μ_B is the chemical potential of substance B, R the gas constant, and T the thermodynamic temperature. The relative activity a is defined as $a_B = \exp[(\mu_B-\mu_B^\circ)/RT]$, where μ_B° designates the chemical potential in the standard state. [2]

Activity coefficient (γ)* - Ratio of the activity a_B of component B of a mixture to the concentration of that component. The value of γ depends on the method of stating the composition. For mole fraction x_B, the relation is $a_B = \gamma_B x_B$; for molarity c_B, it is $a_B = \gamma_B c_B/c^\circ$, where c° is the standard state composition (typically chosen as 1 mol/L); for molality m_B, it is $a_B = \gamma_B m_B/m^\circ$, where m° is the standard state molality (typically 1 mol/kg). [2]

Activity, of radioactive substance (*A*) - The average number of spontaneous nuclear transitions from a particular energy state occurring in an amount of a radionuclide in a small time interval divided by that interval. [1]

Acyl groups - Groups formed by removing the hydroxy groups from oxoacids that have the general structure RC(=O)(OH) and replacement analogues of such acyl groups. [5]

Adiabatic process - A thermodynamic process in which no heat enters or leaves the system.

Admittance (*Y*) - Reciprocal of impedance. $Y = G + iB$, where G is conductance and B is susceptance. [1]

Adsorption - A process in which molecules of gas, of dissolved substances in liquids, or of liquids adhere in an extremely thin layer to surfaces of solid bodies with which they are in contact. [10]

Albedo* - The ratio of the light reflected or scattered from a surface to the intensity of incident light. The term is often used in reference to specific types of terrain or to entire planets.

Alcohols - Compounds in which a hydroxy group, -OH, is attached to a saturated carbon atom. [5]

Aldehydes - Compounds RC(=O)H, in which a carbonyl group is bonded to one hydrogen atom and to one R group. [5]

Aldoses - Aldehydic parent sugars (polyhydroxyaldehydes $H[CH(OH)]_nC(=O)H$, $n>1$) and their intramolecular hemiacetals. [5]

Aldoximes - Oximes of aldehydes: RCH=NOH. [5]

Alfvén number (*Al*) - A dimensionless quantity used in plasma physics, defined by $Al = v(\rho\mu)^{1/2}/B$, where ρ is density, v is velocity, μ is permeability, and B is magnetic flux density. [2]

Alfvén waves - Very low frequency waves which can exist in a plasma in the presence of a uniform magnetic field. Also called magnetohydrodynamic waves.

Alicyclic compounds - Aliphatic compounds having a carbocyclic ring structure which may be saturated or unsaturated, but may not be a benzenoid or other aromatic system. [5]

Aliphatic compounds - Acyclic or cyclic, saturated or unsaturated carbon compounds, excluding aromatic compounds. [5]

Alkali metals - The elements lithium, sodium, potassium, rubidium, cesium, and francium.

Alkaline earth metals - The elements calcium, strontium, barium, and radium. [7]

Alkaloids - Basic nitrogen compounds (mostly heterocyclic) occurring mostly in the plant kingdom (but not excluding those of animal origin). Amino acids, peptides, proteins, nucleotides, nucleic acids, and amino sugars are not normally regarded as alkaloids. [5]

Alkanes - Acyclic branched or unbranched hydrocarbons having the general formula C_nH_{2n+2}, and therefore consisting entirely of hydrogen atoms and saturated carbon atoms. [5]

Alkenes - Acyclic branched or unbranched hydrocarbons having one carbon-carbon double bond and the general formula C_nH_{2n}. Acyclic branched or unbranched hydrocarbons having more than one double bond are alkadienes, alkatrienes, etc. [5]

Alkoxides - Compounds, ROM, derivatives of alcohols, ROH, in which R is saturated at the site of its attachment to oxygen and M is a metal or other cationic species. [5]

Alkyl groups - Univalent groups derived from alkanes by removal of a hydrogen atom from any carbon atom: C_nH_{2n+1}-. The groups derived by removal of a hydrogen atom from a terminal carbon atom of unbranched alkanes form a subclass of normal alkyl (*n*-alkyl) groups. The groups RCH_2-, R_2CH-, and R_3C- (R not equal to H) are primary, secondary, and tertiary alkyl groups, respectively. [5]

Alkynes - Acyclic branched or unbranched hydrocarbons having a carbon-carbon triple bond and the general formula C_nH_{2n-2}, RC≡CR′. Acyclic branched or unbranched hydrocarbons having more than one triple bond are known as alkadiynes, alkatriynes, etc. [5]

Allotropy - The occurrence of an element in two or more crystalline forms.

Allylic groups - The group CH_2=$CHCH_2$- (allyl) and derivatives formed by substitution. The term 'allylic position' or 'allylic site' refers to the saturated carbon atom. A group, such as -OH, attached at an allylic site is sometimes described as "allylic". [5]

Amagat volume unit - A non-SI unit previously used in high pressure science. It is defined as the molar volume of a real gas at one atmosphere pressure and 273.15 K. The approximate value is 22.4 L/mol.

Amides - Derivatives of oxoacids R(C=O)(OH) in which the hydroxy group has been replaced by an amino or substituted amino group. [5]

Amine oxides - Compounds derived from tertiary amines by the attachment of one oxygen atom to the nitrogen atom: R_3N^+-O^-. By extension the term includes the analogous derivatives of primary and secondary amines. [5]

Amines - Compounds formally derived from ammonia by replacing one, two, or three hydrogen atoms by hydrocarbyl groups, and having the general structures RNH_2 (primary amines), R_2NH (secondary amines), R_3N (tertiary amines). [5]

Amino acids* - Compounds containing both a carboxylic acid group (-COOH) and an amino group ($-NH_2$). The most important are the α-amino acids, in which the $-NH_2$ group in attached to the C atom adjacent to the -COOH group. In the β-amino acids, there is an intervening carbon atom. [4]

Ampere (A)* - The SI base unit of electric current. [1]

Ampere's law - The defining equation for the magnetic induction $\boldsymbol{B}$, viz., $d\boldsymbol{F} = I d\boldsymbol{l} \times \boldsymbol{B}$, where $d\boldsymbol{F}$ is the force produced by a current I flowing in an element of the conductor $d\boldsymbol{l}$ pointing in the direction of the current.

Ångström (Å) - A unit of length used in spectroscopy, crystallography, and molecular structure, equal to 10^{-10} m.

Angular momentum (*L*) - The angular momentum of a particle about a point is the vector product of the radius vector from this point to the particle and the momentum of the particle; i.e., $\boldsymbol{L} = \boldsymbol{r} \times \boldsymbol{p}$. [1]

Angular velocity (ω) - The angle through which a body rotates per unit time.

Anilides - Compounds derived from oxoacids R(C=O)(OH) by replacing the -OH group by the -NHPh group or derivative formed by ring substitution. Also used for salts formed by replacement of a nitrogen-bound hydrogen of aniline by a metal. [5]

Anion - A negatively charged atomic or molecular particle.

Antiferroelectricity* - An effect analogous to antiferromagnetism in which electric dipoles in a crystal are ordered in two sublattices that are polarized in opposite directions, leading to zero net polarization. The effect vanishes above a critical temperature.

Antiferromagnetism* - A type of magnetism in which the magnetic moments of atoms in a solid are ordered into two antiparallel aligned sublattices. Antiferromagnets are characterized by a zero or small positive magnetic susceptibility. The susceptibility increases with temperature up to a critical value, the Néel temperature, above which the material becomes paramagnetic.

Antiparticle - A particle having the same mass as a given elementary particle and a charge equal in magnitude but opposite in sign.

Appearance potential* - The lowest energy which must be imparted to the parent molecule to cause it to produce a particular specified parent ion. This energy, usually stated in eV, may be imparted by electron impact, photon impact, or in other ways. More properly called appearance energy. [3]

Appearance potential spectroscopy (APS) - See Techniques for Materials Characterization, page 12-1.

Are (a) - A unit of area equal to 100 m^2. [1]

Arenes - Monocyclic and polycyclic aromatic hydrocarbons. See aromatic compounds. [5]

Aromatic compounds - Compounds whose structure includes a cyclic delocalized π-electron system. Historical use of the term implies a ring containing only carbon (e.g., benzene, naphthalene), but it is often generalized to include heterocyclic structures such as pyridine and thiophene. [5]

Arrhenius equation - A key equation in chemical kinetics which expresses the rate constant k as $k = A\exp(-E_a/RT)$, where E_a is the activation energy, R the molar gas constant, and T the temperature. A is called the preexponential factor and, for simple gas phase reactions, may be identified with the collision frequency.

Arsines - AsH_3 and compounds derived from it by substituting one, two or three hydrogen atoms by hydrocarbyl groups. $RAsH_2$, R_2AsH, R_3As (R not equal to H) are called primary, secondary and tertiary arsines, respectively. [5]

Aryl groups - Groups derived from arenes by removal of a hydrogen atom from a ring carbon atom. Groups similarly derived from heteroarenes are sometimes subsumed in this definition. [5]

Astronomical unit (AU)* - The mean distance of the earth from the sun, equal to 1.49597870 × 10^{11} m.

Atomic absorption spectroscopy (AAS) - See Techniques for Materials Characterization, page **12**-1.

Atomic emission spectroscopy (AES) - See Techniques for Materials Characterization, page **12**-1.

Atomic force microscopy (AFM) - See Techniques for Materials Characterization, page 12-1.

Atomic mass* - The mass of a nuclide, normally expressed in unified atomic mass units (u).

Atomic mass unit (u)* - A unit of mass used in atomic, molecular, and nuclear science, defined as the mass of one atom of ^{12}C divided by 12. Its approximate value is 1.66054 × 10^{-27} kg. Also called the unified atomic mass unit. [1]

Atomic number (*Z*) - A characteristic property of an element, equal to the number of protons in the nucleus.

Atomic weight (A_r)* - The ratio of the average mass per atom of an element to 1/12 of the mass of nuclide ^{12}C. An atomic weight can be defined for a sample of any given isotopic composition. The standard atomic weight refers to a sample of normal terrestrial isotopic composition. The term relative atomic mass is synonymous with atomic weight. [2]

Attenuated total reflection (ATR) - See Techniques for Materials Characterization, page 12-1.

Auger effect - An atomic process in which an electron from a higher energy level fills a vacancy in an inner shell, transferring the released energy to another electron which is ejected.

Aurora - An atmospheric phenomenon in which streamers of light are produced when electrons from the sun are guided into the thermosphere by the earth's magnetic field. It occurs in the polar regions at altitudes of 95—300 km.

Avogadro constant (N_A)* - The number of elementary entities in one mole of a substance.

Azeotrope - A liquid mixture in a state where the variation of vapor pressure with composition at constant temperature (or, alternatively, the variation of normal boiling point with composition) shows either a maximum or a minimum. Thus when an azeotrope boils the vapor has the same composition as the liquid.

Azides - Compounds bearing the group $-N_3$, viz. $-N=N^+=N^-$; usually attached to carbon, e.g. PhN_3, phenyl azide or azidobenzene. Also used for salts of hydrazoic acid, HN_3, e.g. NaN_3, sodium azide. [5]

Azines - Condensation products, $R_2C=NN=CR_2$, of two moles of a carbonyl compound with one mole of hydrazine. [5]

Azo compounds - Derivatives of diazene (diimide), HN=NH, wherein both hydrogens are substituted by hydrocarbyl groups, e.g., PhN=NPh, azobenzene or diphenyldiazene. [5]

Balmer series - The series of lines in the spectrum of the hydrogen atom which corresponds to transitions between the state with principal quantum number n = 2 and successive higher states. The wavelengths are given by $1/\lambda = R_H(1/4 - 1/n^2)$, where n = 3,4,... and R_H is the Rydberg constant for hydrogen. The first member of the series ($n = 2 \rightleftharpoons 3$), which is often called the H_α line, falls at a wavelength of 6563 Å.

Bar (bar) - A unit of pressure equal to 10^5 Pa.

Bardeen-Cooper-Schrieffer (BCS) theory - A theory of superconductivity which is based upon the formation of electron pairs as a result of an electron-lattice interaction. The theory relates the superconducting transition temperature to the density of states and the Debye temperature.

Barn (b) - A unit used for expressing cross sections of nuclear processes, equal to 10^{-28} m^2.

Barrel - A unit of volume equal to 158.9873 L.

Baryon - Any elementary particle built up from three quarks. Examples are the proton, neutron, and various short-lived hyperons. Baryons have odd half-integer spins.

Base - Historically, a substance that yields an OH^- ion when it dissociates in solution, resulting in a pH>7. In the Brönsted definition, a base is a substance capable of accepting a proton in any type of reaction. The more general definition, due to G.N. Lewis, classifies any chemical species capable of donating an electron pair as a base.

Becquerel (Bq)* - The SI unit of radioactivity (disintegrations per unit time), equal to s^{-1}. [1]

Beer's law - An approximate expression for the change in intensity of a light beam that passes through an absorbing medium, viz., $\log(I/I_0) = -\varepsilon cl$, where I_0 is the incident intensity, I is the final intensity, ε is the molar (decadic) absorption coefficient, c is the molar concentration of the absorbing substance, and l is the path length. Also called the Beer-Lambert law

Binding energy* - A generic term for the energy required to decompose a system into two or more of its constituent parts. In nuclear physics, the binding energy is the energy difference between a nucleus and the separated nucleons of which it is composed (the energy equivalent of the mass defect). In atomic physics, it is the energy required to remove an electron from an atom.

Biot (Bi) - A name sometimes used for the unit of current in the emu system.

Birefringence - A property of certain crystals in which two refracted rays result from a single incident light ray. One, the ordinary ray, follows the normal laws of refraction, while the other, the extraordinary ray, exhibits a variable refractive index which depends on the direction in the crystal.

Black body radiation* - The radiation emitted by a perfect black body, i.e., a body which absorbs all radiation incident on it and reflects none. The wavelength dependence of the radiated energy density ρ (energy per unit volume per unit wavelength range) is given by the Planck formula

$$\rho = \frac{8\pi hc}{\lambda^5 (e^{hc/\lambda kt} - 1)}$$

where λ is the wavelength, h is Planck's constant, c is the speed of light, k is the Boltzmann constant, and T is the temperature.

Black hole - A very dense object, formed in a supernova explosion, whose gravitational field is so large that no matter or radiation can escape from the object.

Bloch wave function - A solution of the Schrödinger equation for an electron moving in a spatially periodic potential; used in the band theory of solids.

Bohr magneton (μ_B)* - The atomic unit of magnetic moment, defined as $eh/4\pi m_e$, where h is Planck's constant, m_e the electron mass, and e the elementary charge. It is the moment associated with a single electron spin.

Bohr, bohr radius (a_0)* - The radius of the lowest orbit in the Bohr model of the hydrogen atom, defined as $\varepsilon_0 h^2/\pi m_e e^2$, where ε_0 is the permittivity of a vacuum, h is Planck's constant, m_e the electron mass, and e the elementary charge. It is customarily taken as the unit of length when using atomic units.

Boiling point - The temperature at which the liquid and gas phases of a substance are in equilibrium at a specified pressure. The normal boiling point is the boiling point at normal atmospheric pressure (101.325 kPa).

Boltzmann constant (k)* - The molar gas constant R divided by Avogadro's constant.

Boltzmann distribution - An expression for the equilibrium distribution of molecules as a function of their energy, in which the number of molecules in a state of energy E is proportional to $\exp(-E/kT)$, where k is the Boltzmann constant and T is the temperature.

Bond strength - See Dissociation energy.

Born-Haber cycle* - A thermodynamic cycle in which a crystalline solid is converted to gaseous ions and then reconverted to the solid. The cycle permits calculation of the lattice energy of the crystal.

Bose-Einstein distribution - A modification of the Boltzmann distribution which applies to a system of particles that are bosons. The number of particles of energy E is proportional to $[e^{(E-\mu)/kT}-1]^{-1}$, where μ is a normalization constant, k is the Boltzmann constant, and T is the temperature.

Boson - A particle that obeys Bose-Einstein Statistics; specifically, any particle with spin equal to zero or an integer. This includes the photon, pion, deuteron, and all nuclei of even mass number.

Boyle's law - The empirical law, exact only for an ideal gas, which states that the volume of a gas is inversely proportional to its pressure at constant temperature.

Bragg angle (θ) - Defined by the equation $n\lambda = 2d\sin\theta$, which relates the angle θ between a crystal plane and the diffracted x-ray beam, the wavelength λ of the x-rays, the crystal plane spacing d, and the diffraction order n (any integer).

Bravais lattices* - The 14 distinct crystal lattices that can exist in three dimensions. They include three in the cubic crystal system, two in the tetragonal, four in the orthorhombic, two in the monoclinic, and one each in the triclinic, hexagonal, and trigonal systems.

Breakdown voltage - The potential difference at which an insulating substance undergoes a physical or chemical change that causes it to become a conductor, thus allowing current to flow through the sample.

Bremsstrahlung - Electromagnetic radiation generated when the velocity of a charged particle is reduced (literally, "braking radiation"). An example is the x-ray continuum resulting from collisions of electrons with the target in an x-ray tube.

Brewster angle - The angle of incidence for which the maximum degree of plane polarization occurs when a beam of unpolarized light is incident on the surface of a medium of refractive index n. At this angle, the angle between the reflected and refracted beams is 90°. The value of the Brewster angle is $\tan^{-1}n$.

Brillouin scattering - The scattering of light by acoustic phonons in a solid or liquid.

Brillouin zone - A region of allowed wave vectors and energy levels in a crystalline solid, which plays a part in the propagation of waves through the lattice.

British thermal unit (Btu) - A non-SI unit of energy, equal to approximately 1055 J. Several values of the Btu, defined in slightly different ways, have been used.

Brownian motion - The random movements of small particles suspended in a fluid, which arise from collisions with the fluid molecules.

Brunauer-Emmett-Teller method (BET) - See Techniques for Materials Characterization, page 12-1.

Buffer* - A solution designed to maintain a constant pH when small amounts of a strong acid or base are added. Buffers usually consist of a fairly weak acid and its salt with a strong base. Suitable concentrations are chosen so that the pH of the solution remains close to the pK_a of the weak acid.

Calorie (cal) - A non-SI unit of energy, originally defined as the heat required to raise the temperature of 1 g of water by 1 °C. Several calories of slightly different values have been used. The thermochemical calorie is now defined as 4.184 J.

Candela (cd)* - The SI base unit of luminous intensity. [1]

Capacitance (C) - Ratio of the charge acquired by a body to the change in potential. [1]

Carbamates - Salts or esters of carbamic acid, $H_2NC(=O)OH$, or of N-substituted carbamic acids: $R_2NC(=O)OR'$, (R′ = hydrocarbyl or a cation). The esters are often called urethanes or urethans, a usage that is strictly correct only for the ethyl esters. [5]

Carbenes - The electrically neutral species H_2C: and its derivatives, in which the carbon is covalently bonded to two univalent groups of any kind or a divalent group and bears two non-bonding electrons, which may be spin-paired (singlet state) or spin-non-paired (triplet state). [5]

Carbinols - An obsolete term for substituted methanols, in which the name carbinol is synonymous with methanol. [5]

Carbohydrates - Originally, compounds such as aldoses and ketoses, having the stoichiometric formula $C_n(H_2O)_n$ (hence "hydrates of carbon"). The generic term carbohydrate now includes mono-, oligo-, and polysaccharides, as well as their reaction products and derivatives. [5]

Carboranes - A contraction of carbaboranes. Compounds in which a boron atom in a polyboron hydride is replaced by a carbon atom with maintenance of the skeletal structure. [5]

Carboxylic acids - Oxoacids having the structure RC(=O)OH. The term is used as a suffix in systematic name formation to denote the -C(=O)OH group including its carbon atom. [5]

Carnot cycle - A sequence of reversible changes in a heat engine using a perfect gas as the working substance, which is used to demonstrate that entropy is a state function. The Carnot cycle also provides a means to calculate the efficiency of a heat engine.

Catalyst - A substance that participates in a particular chemical reaction and thereby increases its rate but without a net change in the amount of that substance in the system. [3]

Catenanes, catena compounds - Hydrocarbons having two or more rings connected in the manner of links of a chain, without a covalent bond. More generally, the class catena compounds embraces functional derivatives and hetero analogues. [5]

Cation - A positively charged atomic or molecular particle.

Centipoise (cP) - A common non-SI unit of viscosity, equal to mPa s.

Centrifugal distortion - An effect in molecular spectroscopy in which rotational levels are lowered in energy, relative to the values of a rigid rotor, as the rotational angular momentum increases. The effect may be understood classically as a stretching of the bonds in the molecule as it rotates faster, thus increasing the moment of inertia.

Ceramic - A nonmetallic material of very high melting point.

Cerenkov radiation - Light emitted when a beam of charged particles travels through a medium at a speed greater than the speed of light in the medium. It is typically blue in color.

Cgs system of units - A system of units based upon the centimeter, gram, and second. The cgs system has been supplanted by the International System (SI).

Chalcogens - The Group VIA elements (oxygen, sulfur, selenium, tellurium, and polonium). Compounds of these elements are called chalcogenides. [7]

Chaotic system - A complex system whose behavior is governed by deterministic laws but whose evolution can vary drastically when small changes are made in the initial conditions.

Charge - See Electric charge.

Charles' law - The empirical law, exact only for an ideal gas, which states that the volume of a gas is directly proportional to its temperature at constant pressure.

Charm - A quantum number introduced in particle physics to account for certain properties of elementary particles and their reactions.

Chelate - A compound characterized by the presence of bonds from two or more bonding sites within the same ligand to a central metal atom. [3]

Chemical potential - For a mixture of substances, the chemical potential of constituent B is defined as the partial derivative of the Gibbs energy G with respect to the amount (number of moles) of B, with temperature, pressure, and amounts of all other constituents held constant. Also called partial molar Gibbs energy. [2]

Chemical shift* - A small change in the energy levels (and hence in the spectra associated with these levels) resulting from the effects of chemical binding in a molecule. The term is used in fields such as NMR, Mössbauer, and photoelectron spectroscopy, where the energy levels are determined primarily by nuclear or atomic effects.

Chiral molecule - A molecule which cannot be superimposed on its mirror image. A common example is an organic molecule containing a carbon atom to which four different atoms or groups are attached. Such molecules exhibit optical activity, i.e., they rotate the plane of a polarized light beam.

Chlorocarbons - Compounds consisting solely of chlorine and carbon. [5]

Chromatography* - A method for separation of the components of a sample in which the components are distributed between two phases, one of which is stationary while the other moves. In gas chromatography the gas moves over a liquid or solid stationary phase. In liquid chromatography the liquid mixture moves through another liquid, a solid, or a gel. The mechanism of separation of components may be adsorption, differential solubility, ion-exchange, permeation, or other mechanisms. [6]

Clapeyron equation - A relation between pressure and temperature of two phases of a pure substance that are in equilibrium, viz., $dp/dT = \Delta_{trs}S/\Delta_{trs}V$, where $\Delta_{trs}S$ is the difference in entropy between the phases and $\Delta_{trs}V$ the corresponding difference in volume.

Clathrates - Inclusion compounds in which the guest molecule is in a cage formed by the host molecule or by a lattice of host molecules. [5]

Clausius (Cl) - A non-SI unit of entropy or heat capacity defined as cal/K = 4.184 J/K. [2]

Clausius-Clapeyron equation - An approximation to the Clapeyron equation applicable to liquid-gas and solid-gas equilibrium, in which one assumes an ideal gas with volume much greater than the condensed phase volume. For the liquid-gas case, it takes the form $d(\ln p)/dT = \Delta_{vap}H/RT^2$, where R is the molar gas constant and $\Delta_{vap}H$ is the molar enthalpy of vaporization. For the solid-gas case, $\Delta_{vap}H$ is replaced by the molar enthalpy of sublimation, $\Delta_{sub}H$.

Clausius-Mosotti equation - A relation between the dielectric constant ε_r at optical frequencies and the polarizability α:

$$\frac{\varepsilon_r - 1}{\varepsilon_r + 2} = \frac{\rho N_A \alpha}{3M\varepsilon_0}$$

where ρ is density, N_A is Avogadro's number, M is molar mass, and ε_0 is the permittivity of a vacuum.

Clebsch-Gordon coefficients - A set of coefficients used to describe the vector coupling of angular momenta in atomic and nuclear physics.

Codon - A set of three bases, chosen from the four primary bases found in the DNA molecule (uracil, cytosine, adenine, and guanine), which specifies the production of a particular amino acid or carries some other genetic instruction. For example, the codon UCA specifies the amino acid serine, CAG specifies glutamine, etc. There are a total of 64 codons.

Coercive force - The magnetizing force at which the magnetic flux density is equal to zero. [10]

Coercivity* - The maximum value of coercive force that can be attained when a magnetic material is symmetrically magnetized to saturation induction. [10]

Coherent anti-Stokes Raman spectroscopy (CARS) - See Techniques for Materials Characterization, page 12-1.

Colloid - Molecules or polymolecular particles dispersed in a medium that have, at least in one direction, a dimension roughly between 1 nm and 1 μm. [3]

Color center - A defect in a crystal that gives rise to optical absorption, thus changing the color of the material. A common type is the F-center, which results when an electron occupies the site of a negative ion.

Compressibility (κ)* - The fractional change of volume as pressure is increased, viz., $\kappa = -(1/V)(dV/dp)$. [1]

Compton wavelength (λ_C)* - In the scattering of electromagnetic radiation by a free particle (e.g., electron, proton), $\lambda_C = h/mc$ is the increase in wavelength, at a 90° scattering angle, corresponding to the transfer of energy from radiation to particle. Here h is Planck's constant, c the speed of light, and m the mass of the particle.

Conductance (G)* - For direct current, the reciprocal of resistance. More generally, the real part of admittance. [1]

Conductivity, electrical (σ)* - The reciprocal of the resistivity. [1]

Conductivity, thermal - See Thermal conductivity.

Congruent transformation - A phase transition (melting, vaporization, etc.) in which the substance preserves its exact chemical composition.

Constitutional repeating unit (CRU) - In polymer science, the smallest constitutional unit, the repetition of which constitutes a regular macromolecule, i.e., a macromolecule with all units connected identically with respect to directional sense. [8]

Copolymer - A polymer derived from more than one species of monomer. [8]

Coriolis effect - The deviation from simple trajectories when a mechanical system is described in a rotating coordinate system. It affects the motion of projectiles on the earth and in molecular spectroscopy leads to an important interaction between the rotational and vibrational motions. The effect may be described by an additional term in the equations of motion, called the Coriolis force.

Cosmic rays* - High energy nuclear particles, electrons, and photons, originating mostly outside the solar system, which continually bombard the earth's atmosphere.

Coulomb (C)* - The SI unit of electric charge, equal to A s. [1]

Coulomb's law - The statement that the force F between two electrical charges q_1 and q_2 separated by a distance r is $F = (4\pi\varepsilon_0)^{-1} q_1q_2/r^2$, where ε_0 is the permittivity of a vacuum.

Covalent bond - A chemical bond between two atoms whose stability results from the sharing of two electrons, one from each atom.

Cowling number (*Co*) - A dimensionless quantity used in plasma physics, defined by $Co = B^2/\mu\rho v^2$, where ρ is density, v is velocity, μ is permeability, and B is magnetic flux density. [2]

CPT theorem - A theorem in particle physics which states that any local Lagrangian theory that is invariant under proper Lorentz transformations is also invariant under the combined operations of charge conjugation, C, space inversion, P, and time reversal, T, taken in any order.

Critical point* - In general, the point on the phase diagram of a two-phase system at which the two coexisting phases have identical properties and therefore represent a single phase. At the liquid-gas critical point of a pure substance, the distinction between liquid and gas vanishes, and the vapor pressure curve ends. The coordinates of this point are called the critical temperature and critical pressure. Above the critical temperature, it is not possible to liquefy the substance.

Cross section (σ)* - A measure of the probability of collision (or other interaction) between a beam of particles and a target which it encounters. In rough terms it is the effective area the target particles present to the incident ones; however, the precise definition depends on the nature of the interaction. A general definition of σ is the number of encounters per unit time divided by nv, where n is the concentration of incident particles and v their velocity.

Crosslink - In polymer science, a small region in a macromolecule from which at least four chains emanate, and formed by reactions involving sites or groups on existing macromolecules or by interactions between existing macromolecules. [8]

Crown compounds - Macrocyclic polydentate compounds, usually uncharged, in which three or more coordinating ring atoms (usually oxygen or nitrogen) are or may become suitably close for easy formation of chelate complexes with metal ions or other cationic species. [5]

Crust* - The outer layer of the solid earth, above the Mohorovicic discontinuity. Its thickness averages about 35 km on the continents and about 7 km below the ocean floor.

Cryoscopic constant (E_f)* - The constant that expresses the amount by which the freezing point T_f of a solvent is lowered by a non-dissociating solute, through the relation $\Delta T_f = E_f\, m$, where m is the molality of the solute.

Curie (Ci) - A non-SI unit of radioactivity (disintegrations per unit time), equal to 3.7×10^{10} s^{-1}.

Curie temperature (T_C)* - For a ferromagnetic material, the critical temperature above which the material becomes paramagnetic. Also applied to the temperature at which the spontaneous polarization disappears in a ferroelectric solid. [1]

Cyanohydrins - Alcohols substituted by a cyano group, most commonly, but not limited to, examples having a CN and an OH group attached to the same carbon atom. They are formally derived from aldehydes or ketones by the addition of hydrogen cyanide. [5]

Cycloalkanes - Saturated monocyclic hydrocarbons (with or without side chains). See alicyclic compounds. Unsaturated monocyclic hydrocarbons having one endocyclic double or one triple bond are called cycloalkenes and cycloalkynes, respectively. [5]

Cyclotron resonance - The resonant absorption of energy from a system in which electrons or ions that are orbiting in a uniform magnetic field are subjected to radiofrequency or microwave radiation. The resonance frequency is given by $\nu = eH/2\pi m^*c$, where e is the elementary charge, H is the magnetic field strength, m^* is the effective mass of the charged particle, and c is the speed of light. The effect occurs in both solids (involving electrons or holes) and in low pressure gasses (involving ions)

Dalton (Da) - A name sometimes used in biochemistry for the unified atomic mass unit (u).

De Broglie wavelength - The wavelength associated with the wave representation of a moving particle, given by h/mv, where h is Planck's constant, m the particle mass, and v the velocity.

De Haas-Van Alphen effect - An effect observed in certain metals and semiconductors at low temperatures and high magnetic fields, characterized by a periodic variation of magnetic susceptibility with field strength.

Debye equation* - The relation between the relative permittivity (dielectric constant) ε_r, polarizability α, and permanent dipole moment μ in a dielectric material whose molecules are free to rotate. It takes the form

$$\frac{\varepsilon_r - 1}{\varepsilon_r + 2} = \frac{\rho N_A}{3M\varepsilon_0}\left(\alpha + \frac{\mu^2}{3kT}\right)$$

where ρ is density, N_A is Avogadro's number, M is molar mass, and ε_0 is the permittivity of a vacuum.

Debye length - In the Debye-Hückel theory of ionic solutions, the effective thickness of the cloud of ions of opposite charge which surrounds each given ion and shields the Coulomb potential produced by that ion.

Debye temperature (θ_D)* - In the Debye model of the heat capacity of a crystalline solid, $\theta_D = h\nu_D/k$, where h is Planck's constant, k is the Boltzmann constant, and ν_D is the maximum vibrational frequency the crystal can support. For $T << \theta_D$, the heat capacity is proportional to T^3.

Debye unit (D) - A non-SI unit of electric dipole moment used in molecular physics, equal to 3.335641×10^{-30} C m.

Debye-Waller factor (*D*) - The factor by which the intensity of a diffraction line is reduced because of lattice vibrations. [1]

Defect - Any departure from the regular structure of a crystal lattice. A Frenkel defect results when an atom or ion moves to an interstitial position and leaves behind a vacancy. A Schottky defect involves either a vacancy where the atom has moved to the surface or a structure where a surface atom has moved to an interstitial position.

Degree of polymerization - The number of monomeric units in a macromolecule or an oligomer molecule. [8]

Dendrite - A tree-like crystalline pattern often observed, for example, in ice crystals and alloys in which the crystal growth branches repeatedly.

Density (ρ)* - In the most common usage, mass density or mass per unit volume. More generally, the amount of some quantity (mass, charge, energy, etc.) divided by a length, area, or volume.

Density of states (N_E, ρ) - The number of one-electron states in an infinitesimal interval of energy, divided by the range of that interval and by volume. [1]

Dew point* - The temperature at which liquid begins to condense as the temperature of a gas mixture is lowered. In meteorology, it is the temperature at which moisture begins to condense on a surface in contact with the air.

Diamagnetism - A type of magnetism characterized by a negative magnetic susceptibility, so that the material, when placed in an external magnetic field, becomes weakly magnetized in the direction opposite to the field. This magnetization is independent of temperature.

Diazo compounds - Compounds having the divalent diazo group, $=N^+=N^-$, attached to a carbon atom, e.g., $CH_2=N_2$ diazomethane. [5]

Dielectric constant (ε)* - Ratio of the electric displacement in a medium to the electric field strength. Also called permittivity. [1]

Dienes - Compounds that contain two fixed double bonds (usually assumed to be between carbon atoms). Dienes in which the two double-bond units are linked by one single bond are termed conjugated. [5]

Differential scanning calorimetry (DSC) - See Techniques for Materials Characterization, page 12-1.

Differential thermal analysis (DTA) - See Techniques for Materials Characterization, page 12-1.

Diffusion* - The migration of atoms, molecules, ions, or other particles as a result of some type of gradient (concentration, temperature, etc.).

Diopter - A unit used in optics, formally equal to m^{-1}. It is used in expressing dioptic power, which is the reciprocal of the focal length of a lens.

Dipole moment, electric (*p*,μ)* - For a distribution of equal positive and negative charge, the magnitude of the dipole moment vector is the positive charge multiplied by the distance between the centers of positive and negative charge distribution. The direction is given by the line from the center of negative charge to the center of positive charge.

Dipole moment, magnetic (*m*,μ) - Formally defined in electromagnetic theory as a vector quantity whose vector product with the magnetic flux density equals the torque. The magnetic dipole generated by a current I flowing in a small loop of area A has a magnetic moment of magnitude IA. In atomic and nuclear physics, a magnetic moment is associated with the angular momentum of a particle; e.g., an electron with orbital angular momentum $\boldsymbol{l}$ exhibits a magnetic moment of $-e\boldsymbol{l}/2m_e$ where e is the elementary charge and m_e the mass of the electron. [1]

Disaccharides - Compounds in which two monosaccharides are joined by a glycosidic bond. [5]

Dislocation - An extended displacement of a crystal from a regular lattice. An edge dislocation results when one portion of the crystal has partially slipped with respect to the other, resulting in an extra plane of atoms extending through part of the crystal. A screw dislocation transforms successive atomic planes into the surface of a helix.

Dispersion - Splitting of a beam of light (or other electromagnetic radiation) of mixed wavelengths into the constituent wavelengths as a result of the variation of refractive index of the medium with wavelength.

Dissociation constant* - The equilibrium constant for a chemical reaction in which a compound dissociates into its constituent parts.

Dissociation energy (D_e)* - For a diatomic molecule, the difference between the energies of the free atoms at rest and the minimum in the potential energy curve. The term bond dissociation energy (D_0), which can be applied to polyatomic molecules as well, is used for the difference between the energies of the fragments resulting when a bond is broken and the energy of the original molecule in its lowest energy state. The term bond strength implies differences in enthalpy rather than energy.

Domain - A small region of a solid in which the magnetic or electric moments of the individual units (atoms, molecules, or ions) are aligned in the same direction.

Domain wall - The transition region between adjacent ferromagnetic domains, generally a layer with a thickness of a few hundred ångström units. Also called Bloch wall.

Doppler effect - The change in the apparent frequency of a wave (sound, light, or other) when the source of the wave is moving relative to the observer.

Dose equivalent (*H*) - The product of the absorbed dose of radiation at a point of interest in tissue and various modifying factors which depend on the type of tissue and radiation. [1]

Drift velocity - The velocity of charge carriers (electrons, ions, etc.) moving under the influence of an electric field in a medium which subjects the carriers to some frictional force.

Dyne (dyn) - A non-SI (cgs) unit of force, equal to 10^{-5} N.

Ebullioscopic constant (E_b)* - The constant that expresses the amount by which the boiling point T_b of a solvent is raised by a non-dissociating solute, through the relation $\Delta T_b = E_b m$, where m is the molality of the solute.

Eddy currents - Circulating currents set up in conducting bulk materials or sheets by varying magnetic fields.

Effinghausen effect - The appearance of a temperature gradient in a current carrying conductor that is placed in a transverse magnetic field. The direction of the gradient is perpendicular to the current and the field.

Eigenvalue - An allowed value of the constant a in the equation $Au = au$, where A is an operator acting on a function u (which is called an eigenfunction). In quantum mechanics, the outcome of any observation is an eigenvalue of the corresponding operator. Also called characteristic value.

Einstein - A non-SI unit used in photochemistry, equal to one mole of photons.

Einstein temperature (θ_V) - In the Einstein theory of the heat capacity of a crystalline solid, $\theta_V = h\nu/k$, where h is Planck's constant, k is the Boltzmann constant, and ν is the vibrational frequency of the crystal.

Einstein transition probability - A constant in the Einstein relation $A_{ij} + B_{ij}\rho$ for the probability of a transition between two energy levels i and j in a radiation field of energy density ρ. The A_{ij} coefficient describes the probability of spontaneous emission, while B_{ij} and B_{ji} govern the probability of stimulated emission and absorption, respectively ($B_{ij} = B_{ji}$).

Elastic limit - The greatest stress which a material is capable of sustaining without any permanent strain remaining after complete release of the stress. [10]

Elastic modulus - See Young's modulus.

Electric charge (*Q*) - The quantity of electricity; i.e., the property that controls interactions between bodies through electrical forces.

Electric current (*I*) - The charge passing through a circuit per unit time. [1]

Electric displacement (*D*) - A vector quantity whose magnitude equals the electric field strength multiplied by the permittivity of the medium and whose direction is the same as that of the field strength.

Electric field strength (*E*) - The force exerted by an electric field on a point charge divided by the electric charge. [1]

Electric potential (*V*) - A scalar quantity whose gradient is equal to the negative of the electric field strength.

Electrical conductance - See Conductance.

Electrical resistance - See Resistance.

Electrical resistivity - See Resistivity.

Electrochemical series* - An arrangement of reactions which produce or consume electrons in an order based on standard electrode potentials. A common arrangement places metals in decreasing order of their tendency to give up electrons.

Electrode potential* - The electromotive force of a cell in which the electrode on the left is the standard hydrogen electrode and that on the right is the electrode in question. [2]

Electrolysis - The decomposition of a substance as a result of passing an electric current between two electrodes immersed in the sample.

Electromotive force (emf) - The energy supplied by a source divided by the charge transported through the source. [1]

Electron* - An elementary particle in the family of leptons, with negative charge and spin of 1/2.

Electron affinity* - The energy difference between the ground state of a gas-phase atom or molecule and the lowest state of the corresponding negative ion.

Electron cyclotron resonance (ECR) - See Techniques for Materials Characterization, page 12-1.

Electron energy loss spectroscopy (EELS) - See Techniques for Materials Characterization, page 12-1.

Electron nuclear double resonance (ENDOR) - See Techniques for Materials Characterization, page 12-1.

Electron paramagnetic resonance (EPR) - See Techniques for Materials Characterization, page 12-1.

Electron probe microanalysis (EPMA) - See Techniques for Materials Characterization, page 12-1.

Electron spectroscopy for chemical analysis (ESCA) - See Techniques for Materials Characterization, page 12-1.

Electron spin (*s*) - The quantum number, equal to 1/2, that specifies the intrinsic angular momentum of the electron.

Electron stimulated desorption (ESD) - See Techniques for Materials Characterization, page 12-1.

Electron volt (eV)* - A non-SI unit of energy used in atomic and nuclear physics, equal to approximately 1.602177×10^{-19} J. The electron volt is defined as the kinetic energy acquired by an electron upon acceleration through a potential difference of 1 V. [1]

Electronegativity* - A parameter originally introduced by Pauling which describes, on a relative basis, the power of an atom or group of atoms to attract electrons from the same molecular entity. [3]

Electrophoresis - The motion of macromolecules or colloidal particles in an electric field. [3]

Emissivity (ε)* - Ratio of the radiant flux emitted per unit area to that of an ideal black body at the same temperature. Also called emittance. [1]

Emu - The electromagnetic system of units, based upon the cm, g, and s plus the emu of current (sometimes called the abampere).

Enantiomers - A chiral molecule and its non-superposable mirror image. The two forms rotate the plane of polarized light by equal amounts in opposite directions. Also called optical isomers.

Energy (*E,U*)* - The characteristic of a system that enables it to do work.

Energy gap* - In the theory of solids, the region between two energy bands, in which no bound states can occur.

Enols, alkenols - The term refers specifically to vinylic alcohols, which have the structure HOCR′=CR_2. Enols are tautomeric with aldehydes (R′ = H) or ketones (R′ not equal to H). [5]

Enthalpy (*H*)* - A thermodynamic function, especially useful when dealing with constant-pressure processes, defined by $H = E + PV$, where E is energy, P pressure, and V volume. [1]

Enthalpy of combustion* - The enthalpy change in a combustion reaction. Its negative is the heat released in combustion.

Enthalpy of formation, standard* - The enthalpy change for the reaction in which a substance is formed from its constituent elements, each in its standard reference state (normally refers to 1 mol, sometimes to 1 g, of the substance).

Enthalpy of fusion* - The enthalpy change in the transition from solid to liquid state.

Enthalpy of sublimation - The enthalpy change in the transition from solid to gas state.

Enthalpy of vaporization* - The enthalpy change in the transition from liquid to gas state.

Entropy (*S*)* - A thermodynamic function defined such that when a small quantity of heat dQ is received by a system at temperature T, the entropy of the system is increased by dQ/T, provided that no irreversible change takes place in the system. [1]

Entropy unit (e.u.) - A non-SI unit of entropy, equal to 4.184 J/K mol.

Ephemeris time - Time measured in tropical years from January 1, 1900.

Epoxy compounds - Compounds in which an oxygen atom is directly attached to two adjacent or non-adjacent carbon atoms of a carbon chain or ring system; thus cyclic ethers. [5]

Equation of continuity - Any of a class of equations that express the fact that some quantity (mass, charge, energy, etc.) cannot be created or destroyed. Such equations typically specify that the rate of increase of the quantity in a given region of space equals the net current of the quantity flowing into the region.

Equation of state* - An equation relating the pressure, volume, and temperature of a substance or system.

Equilibrium constant (*K*)* - For a chemical reaction aA + bB ⇌ cC + dD, the equilibrium constant is defined by:

$$K = \frac{a_C^{\ c} \cdot a_D^{\ d}}{a_A^{\ a} \cdot a_B^{\ b}}$$

where a_i is the activity of component i. To a certain approximation, the activities can be replaced by concentrations. The equilibrium constant is related to $\Delta_r G°$, the standard Gibbs energy change in the reaction, by $RT \ln K = -\Delta_r G°$.

Equivalent conductance - See Conductivity, electrical

Erg (erg) - A non-SI (cgs) unit of energy, equal to 10^{-7} J.

Esters - Compounds formally derived from an oxoacid RC(=O)(OH) and an alcohol, phenol, heteroarenol, or enol by linking, with formal loss of water from an acidic hydroxy group of the former and a hydroxy group of the latter. [5]

Esu - The electrostatic system of units, based upon the cm, g, and s plus the esu of charge (sometimes called the statcoulomb or franklin).

Ethers - Compounds with formula ROR, where R is not equal to H. [5]

Euler number (*Eu*) - A dimensionless quantity used in fluid mechanics, defined by $Eu = \Delta p/\rho v^2$, where p is pressure, ρ is density, and v is velocity. [2]

Eutectic - The point on a two-component solid-liquid phase diagram which represents the lowest melting point of any possible mixture. A liquid having the eutectic composition will freeze at a single temperature without change of composition.

Excitance (*M*) - Radiant energy flux leaving an element of a surface divided by the area of that element. [1]

Exciton - A localized excited state consisting of a bound electron-hole pair in a molecular or ionic crystal. The exciton can propagate through the crystal.

Exosphere - The outermost part of the earth's atmosphere, beginning at about 500 to 1000 km above the surface. It is characterized by densities so low that air molecules can escape into outer space.

Expansion coefficient - See thermal expansion coefficient.

Extended electron energy loss fine structure (EXELFS) - See Techniques for Materials Characterization, page 12-1.

Extended x-ray absorption fine structure (EXAFS) - See Techniques for Materials Characterization, page 12-1.

Extinction coefficient - See Absorption coefficient, molar.

F-Center - See Color center.

Fahrenheit temperature (°F) - The temperature scale based on the assignment of 32°F = 0 °C and a temperature interval of °F =(5/9)°C; i.e., t/°F = (9/5)t/°C + 32.

Farad (F)* - The SI unit of electric capacitance, equal to C/V. [1]

Faraday constant (*F*)* - The electric charge of 1 mol of singly charged positive ions; i.e., $F = N_A e$, where N_A is Avogadro's constant and e is the elementary charge. [1]

Faraday effect* - The rotation of the plane of plane-polarized light by a medium placed in a magnetic field parallel to the direction of the light beam. The effect can be observed in solids, liquids, and gasses.

Fatty acids - Aliphatic monocarboxylic acids derived from or contained in esterified form in an animal or vegetable fat, oil, or wax. Natural fatty acids commonly have a chain of 4 to 28 carbons (usually unbranched and even-numbered), which may be saturated or unsaturated. By extension, the term is sometimes used to embrace all acyclic aliphatic carboxylic acids. [5]

Fermat's principle - The law that a ray of light traversing one or more media will follow a path which minimizes the time required to pass between two given points.

Fermi (f) - Name sometimes used in nuclear physics for the femtometer.

Fermi level - The highest energy of occupied states in a solid at zero temperature. Sometimes called Fermi energy. The Fermi surface is the surface in momentum space formed by electrons occupying the Fermi level.

Fermi resonance - An effect observed in vibrational spectroscopy when an overtone of one fundamental vibration closely coincides in energy with another fundamental of the same symmetry species. It leads to a splitting of vibrational bands.

Fermi-Dirac distribution - A modification of the Boltzmann distribution which takes into account the Pauli exclusion principle. The number of particles of energy E is proportional to $[e^{(E-\mu)/kT}+1]^{-1}$, where μ is a normalization constant, k the Boltzmann constant, and T the temperature. The distribution is applicable to a system of fermions.

Fermion - A particle that obeys Fermi-Dirac statistics. Specifically, any particle with spin equal to an odd multiple of 1/2. Examples are the electron, proton, neutron, muon, etc.

Ferrimagnetism* - A type of magnetism in which the magnetic moments of atoms in a solid are ordered into two nonequivalent sublattices with unequal magnetic moments, leading to a nonzero magnetic susceptibility.

Ferrite - A ferrimagnetic material of nominal formula MFe_2O_4, where M is a divalent metal; widely used in microwave switches and other solid state devices.

Ferroelectricity* - The retention of electric polarization by certain materials after the external field that produced the polarization has been removed.

Ferromagnetism* - A type of magnetism in which the magnetic moments of atoms in a solid are aligned within domains which can in turn be aligned with each other by a weak magnetic field. Some ferromagnetic materials can retain their magnetization when the external field is removed, as long as the temperature is below a critical value, the Curie temperature. They are characterized by a large positive magnetic susceptibility.

Fick's law - The statement that the flux J of a diffusing substance is proportional to the concentration gradient, i.e., $J = -D(dc/dx)$, where D is called the diffusion coefficient.

Field - A mathematical construct which describes the interaction between particles resulting from gravity, electromagnetism, or other physical phenomena. In classical physics a field is described by equations. Quantum field theory introduces operators to represent the physical observables.

Field emission microscopy (FEM) - See Techniques for Materials Characterization, page 12-1.

Field ion microscopy (FIM) - See Techniques for Materials Characterization, page 12-1.

Fine structure - The splitting in spectral lines that results from interactions of the electron spin with the orbital angular momentum.

Fine structure constant (α)* - Defined as $e^2/2hc\varepsilon_0$, where e is the elementary charge, h Planck's constant, c the speed of light, and ε_0 the permittivity of a vacuum. It is a measure of the strength of the electromagnetic interaction between particles.

First radiation constant (c_1)* - Constant ($= 2\pi hc^2$) in the equation for the radiant excitance M_λ of a black body:

$$M_\lambda = \frac{c_1\lambda^{-5}\Delta\lambda}{e^{c_2/\lambda T} - 1}$$

where λ is the wavelength, T is the temperature, and $c_2 = hc/k$ is the second radiation constant.

Flash point - The lowest temperature at which vapors above a volatile combustible substance will ignite in air when exposed to a flame. [10]

Fluence (*F*) - Term used in photochemistry to specify the energy per unit area delivered in a given time interval, for example by a laser pulse. [2]

Fluorocarbons - Compounds consisting solely of fluorine and carbon. [5]

Fluxoid - The quantum of magnetic flux in superconductivity theory, equal to $hc/2e$, where h is Planck's constant, c the velocity of light, and e the elementary charge.

Force (*F*) - The rate of change of momentum with time. [1]

Force constants (*f, k*)* - In molecular vibrations, the coefficients in the expression of the potential energy in terms of atom displacements from their equilibrium positions. In a diatomic molecule, $f = d^2V/dr^2$, where $V(r)$ is the potential energy and r is the interatomic distance. [2]

Fourier number (*Fo*) - A dimensionless quantity used in fluid mechanics, defined by $Fo = at/l^2$, where a is thermal diffusivity, t is time, and l is length. [2]

Fourier transform infrared spectroscopy (FTIR) - A technique for obtaining an infrared spectrum by use of an interferometer in which the path length of one of the beams is varied. A Fourier transformation of the resulting interferogram yields the actual spectrum. The technique is also used for NMR and other types of spectroscopy.

Fractals - Geometrical objects that are self-similar under a change of scale; i.e., they appear similar at all levels of magnification. They can be considered to have fractional dimensionality. Examples occur in diverse fields such as geography (rivers and shorelines), biology (trees), and solid state physics (amorphous materials).

Franck-Condon principle - An important principle in molecular spectroscopy which states that the nuclei in a molecule remain essentially stationary while an electronic transition is taking place. The physical interpretation rests on the fact that the electrons move much more rapidly than the nuclei because of their much smaller mass.

Franklin (Fr) - Name sometimes given to the unit of charge in the esu system.

Fraunhofer diffraction - Diffraction of light in situations where the source and observation point are so far removed that the wave surfaces may be considered planar.

Fraunhofer lines - Sharp absorption lines in the spectrum of sunlight, caused by absorption of the solar blackbody radiation by atoms near the sun's surface.

Free radical - See Radicals. The term "free radical" is often used more broadly for molecules that have a paramagnetic ground state (e.g., O_2) and sometimes for any transient or highly reactive molecular species.

Freezing point - See Melting point.

Frequency (ν)* - Number of cycles of a periodic phenomenon divided by time. [1]

Fresnel diffraction - Diffraction of light in a situation where the source and observation point are sufficiently close together that the curvature of the wave surfaces must be taken into account.

Froude number (*Fr*) - A dimensionless quantity used in fluid mechanics, defined by $Fr = v/(lg)^{1/2}$, where v is velocity, l is length, and g is acceleration due to gravity. [2]

Fugacity (f_B) - For a gas mixture, the fugacity of component B is defined as the absolute activity λ_B times the limit, as the pressure p approaches zero at constant temperature, of p_B/λ_B. [2]

Fullerenes - Compounds composed solely of an even number of carbon atoms, which form a cage-like fused-ring polycyclic system with twelve five-membered rings and the rest six-membered rings. The archetypal example is [60]fullerene, where the atoms and bonds delineate a truncated icosahedron. The term has been broadened to include any closed cage structure consisting entirely of three-coordinate carbon atoms. [5]

Fulvalenes - The hydrocarbon fulvalene and its derivatives formed by substitution (and by extension, analogues formed by replacement of one or more carbon atoms of the fulvalene skeleton by a heteroatom). [5]

Fulvenes - The hydrocarbon fulvene and its derivatives formed by substitution (and by extension, analogues formed by replacement of one or more carbon atoms of the fulvene skeleton by a heteroatom). [5]

Fundamental vibrational frequencies* - In molecular spectroscopy, the characteristic vibrational frequencies obtained when the vibrational energy is expressed in normal coordinates. They determine the primary features of the infrared and Raman spectra of the molecule.

γ - Name sometimes used for microgram.

γ-rays* - Electromagnetic radiation (photons) with energy greater than about 0.1 MeV (wavelength less than about 1 pm).

g*-Factor of the electron - The proportionality factor in the equation relating the magnetic moment μ of an electron to its total angular momentum quantum number J, i.e., $\mu = -g\mu_B J$, where μ_B is the Bohr magneton. Also called Landé factor.

Gal - A non-SI unit of acceleration, equal to 0.01 m/s. Also called galileo.

Gallon (US) - A unit of volume equal to 3.785412 L.

Gallon (UK, Imperial) - A unit of volume equal to 4.546090 L.

Gauss (G) - A non-SI unit of magnetic flux density (B) equal to 10^{-4} T.

Gaussian system of units - A hybrid system used in electromagnetic theory, which combines features of both the esu and emu systems.

Gel - A colloidal system with a finite, but usually rather small, yield stress (the sheer stress at which yielding starts abruptly). [3]

Genetic code* - The set of relations between each of the 64 codons of DNA and a specific amino acid (or other genetic instruction).

Gibbs energy (*G*)* - An important function in chemical thermodynamics, defined by $G = H\text{-}TS$, where H is the enthalpy, S the entropy, and T the thermodynamic temperature. Sometimes called Gibbs free energy and, in older literature, simply "free energy". [2]

Gibbs phase rule - The relation $F = C - P + 2$, where C is the number of components in a mixture, P is the number of phases, and F is the degrees of freedom, i.e., the number of intensive variables that can be changed independently without affecting the number of phases.

Glass transition temperature* - The temperature at which an amorphous polymer is transformed, in a reversible way, from a viscous or rubbery condition to a hard and relatively brittle one. [10]

Glow discharge mass spectroscopy (GDMS) - See Techniques for Materials Characterization, page 12-1.

Gluon - A hypothetical particle postulated to take part in the binding of quarks, in analogy to the role of the photon in electromagnetic interactions.

Glycerides - Esters of glycerol (propane-1,2,3-triol) with fatty acids, widely distributed in nature. They are by long-established custom subdivided into triglycerides, 1,2- or 1,3-diglycerides, and 1- or 2-monoglycerides, according to the number and positions of acyl groups. [5]

Glycols - Dihydric alcohols in which two hydroxy groups are on different carbon atoms, usually but not necessarily adjacent. Also called diols. [5]

Grain (gr) - A non-SI unit of mass, equal to 64.79891 mg.

Grain boundary - The interface between two regions of different crystal orientation.

Grashof number (*Gr*) - A dimensionless quantity used in fluid mechanics, defined by $Gr = l^3 g\alpha\Delta T\rho^2/\eta^2$, where T is temperature, ρ is density, l is length, η is viscosity, α is cubic expansion coefficient, and g is acceleration of gravity. [2]

Gravitational constant (*G*)* - The universal constant in the equation for the gravitational force between two particles, $F = Gm_1m_2/r^2$, where r is the distance between the particles and m_1 and m_2 are their masses. [1]

Gray (Gy)* - The SI unit of absorbed dose of radiation, equal to J/kg. [1]

Gregorian calendar - The modification of the Julian calendar introduced in 1582 by Pope Gregory XII which specified that a year divisible by 100 is a leap year only if divisible by 400.

Grignard reagents - Organomagnesium halides, RMgX, having a carbon–magnesium bond (or their equilibrium mixtures in solution with $R_2Mg + MgX_2$). [5]

Gruneisen parameter (γ) - Defined by $\gamma = \alpha_V/\kappa\, c_V\, \rho$, where α_V is the cubic thermal expansion coefficient, κ is the isothermal compressibility, c_V is the specific heat capacity at constant volume, and ρ is the mass density. γ is independent of temperature for most crystalline solids. [1]

Gyromagnetic ratio (γ) - Ratio of the magnetic moment of a particle to its angular momentum. Also called magnetogyric ratio.

Hadron - Any elementary particle that can take part in the strong interaction. Hadrons are subdivided into baryons, with odd half integer spins, and mesons, which have zero or integral spin.

Hall effect* - The development of a transverse potential difference V in a conducting material when subjected to a magnetic field H perpendicular to the direction of the current. The potential difference is given by $V = R_H BJt$, where B is the magnetic induction, J the current density, t the thickness of the specimen in the direction of the potential difference, and R_H is called the Hall coefficient.

Halocarbon - A compound containing no elements other than carbon, hydrogen, and one or more halogens. In common practice, the term is used mainly for compounds of no more than four or five carbon atoms.

Halogens - The elements F, Cl, Br, I, and At. Compounds of these elements are called halogenides or halides. [7]

Hamiltonian (*H*) - An expression for the total energy of a mechanical system in terms of the momenta and positions of constituent particles. In quantum mechanics, the Hamiltonian operator appears in the eigenvalue equation $H\psi = E\psi$, where E is an energy eigenvalue and ψ the corresponding eigenfunction.

Hardness* - The resistance of a material to deformation, indentation, or scratching. Hardness is measured on various scales, such as Mohs, Brinell, Knoop, Rockwell, and Vickers. [10]

Hartmann number (*Ha*) - A dimensionless quantity used in plasma physics, defined by $Ha = Bl(\kappa/\eta)^{1/2}$, where B is magnetic flux density, l is length, κ is electric conductivity, and η is viscosity. [2]

Hartree (E_h)* - An energy unit used in atomic and molecular science, equal to approximately $4.3597482 \times 10^{-18}$ J.

Hartree-Fock method - A iterative procedure for solving the Schrödinger equation for an atom or molecule in which the equation is solved for each electron in an initial assumed potential from all the other electrons. The new potential that results is used to repeat the calculation and the procedure continued until convergence is reached. Also called self-consistent field (SCF) method.

Heat capacity* - Defined in general as dQ/dT, where dQ is the amount of heat that must be added to a system to increase its temperature by a small amount dT. The heat capacity at constant pressure is $C_p = (\partial H/\partial T)_p$; that at constant volume is $C_V = (\partial E/\partial T)_V$, where H is enthalpy, E is internal energy, p is pressure, V is volume, and T is temperature. An upper case C normally indicates the molar heat capacity, while a lower case c is used for the specific (per unit mass) heat capacity. [1]

Heat of formation, vaporization, etc. - See corresponding terms under Enthalpy.

Hectare (ha) - A unit of area equal to 10^4 m^2. [1]

Heisenberg uncertainty principle - The statement that two observable properties of a system that are complementary, in the sense that their quantum-mechanical operators do not commute, cannot be specified simultaneously with absolute precision. An example is the position and momentum of a particle; according to this principle, the uncertainties in position Δq and momentum Δp must satisfy the relation $\Delta p \Delta q \geq h/4\pi$, where h is Planck's constant.

Heitler-London model - An early quantum-mechanical model of the hydrogen atom which introduced the concept of the exchange interaction between electrons as the primary reason for stability of the chemical bond.

Helicon - A low-frequency wave generated when a metal at low temperature is exposed to a uniform magnetic field and a circularly polarized electric field.

Helmholz energy (*A*) - A thermodynamic function defined by $A = E - TS$, where E is the energy, S the entropy, and T the thermodynamic temperature. [2]

Hemiacetals - Compounds having the general formula $R_2C(OH)OR'$ (R' not equal to H). [5]

Henry (H)* - The SI unit of inductance, equal to Wb/A. [1]

Henry's law * - An expression which applies to an ideal dilute solution in which one or more gasses are dissolved, viz., $p_i = H_i x_i$, where p_i is the partial pressure of component i above the solution, x_i is its mole fraction in the solution, and H_i is the Henry's law constant (a characteristic of the given gas and solvent, as well as the temperature).

Hermitian operator - An operator A that satisfies the relation $\int u_m{}^*Au_n dx = (\int u_n{}^*Au_m\, dx)^*$, where * indicates the complex conjugate. The eigenvalues of Hermitian operators are real, and eigenfunctions belonging to different eigenvalues are orthogonal.

Hertz (Hz) - The SI unit of frequency, equal to s^{-1}. [1]

Heterocyclic compounds - Cyclic compounds having as ring members atoms of at least two different elements, e.g., quinoline, 1,2-thiazole, bicyclo[3.3.1]tetrasiloxane. [5]

Heusler alloys - Alloys of manganese, copper, aluminum, nickel, and sometimes other metals which find important uses as permanent magnets.

Holography - A technique for creating a three-dimensional image of a object by recording the interference pattern between a light beam diffracted from the object and a reference beam. The image can be reconstructed from this pattern by a suitable optical system.

Homopolymer - A polymer derived from one species of (real, implicit, or hypothetical) monomer. [8]

Hooke's law - The statement that the ratio of stress to strain is a constant in a totally elastic medium.

Horse power - A non-SI unit of energy, equal to approximately 746 W.

Hubble constant - The ratio of the recessional velocity of an extragalactic object to the distance of that object. Its value is about 2×10^{-18} s^{-1}.

Huckel theory - A simple approximation for calculating the energy of conjugated molecules in which only the resonance integrals between neighboring bonds are considered. Also called CNDO method (complete neglect of differential overlap).

Hume-Rothery rules - A set of empirical rules for predicting the occurrence of solid solutions in metallic systems. The rules involve size, crystal structure, and electronegativity.

Hund's rules - A series of rules for predicting the sequence of energy states in atoms and molecules. One of the important results is that when two electrons exist in different orbitals, the state with their spins parallel (triplet state) lies at lower energy than the state with antiparallel spins (singlet).

Hydrazines - Hydrazine (diazane), H_2NNH_2, and its hydrocarbyl derivatives. When one or more substituents are acyl groups, the compound is a hydrazide. [5]

Hydrocarbon - A compound containing only carbon and hydrogen. [5]

Hydrolysis - A reaction occurring in water in which a chemical bond is cleaved and a new bond formed with the oxygen atom of water.

Hyperfine structure - Splitting of energy levels and spectral lines into several closely spaced components as a result of interaction of nuclear spin angular momentum with other angular momenta in the atom or molecule.

Hysteresis* - An irreversible response of a system (parameter A) as a function of an external force (parameter F), usually symmetric with respect to the origin of the A vs. F graph after the initial application of the force. A common example is magnetic induction vs. magnetic field strength in a ferromagnet.

Ideal gas law - The equation of state $pV = RT$, which defines an ideal gas, where p is pressure, V molar volume, T temperature, and R the molar gas constant.

Ideal solution - A solution in which solvent-solvent and solvent-solute interactions are identical, so that properties such as volume and enthalpy are exactly additive. Ideal solutions follow Raoult's law, which states that the vapor pressure p_i of component i is $p_i = xp_i{}^*$, where x_i is the mole fraction of component i and $p_i{}^*$ the vapor pressure of the pure substance i.

Ignition temperature* - The lowest temperature at which combustion of a material will occur spontaneously under specified conditions. Sometimes called autoignition temperature, kindling point. [10]

Imides - Diacyl derivatives of ammonia or primary amines, especially those cyclic compounds derived from diacids. Also used for salts having the anion $RN_2{}^-$. [5]

Impedance (*Z*) - The complex representation of potential difference divided by the complex representation of current. In terms of reactance X and resistance R, the impedance is given by $Z = R + iX$. [1]

Index of refraction (*n*)* - For a non-absorbing medium, the ratio of the velocity of electromagnetic radiation *in vacuo* to the phase velocity of radiation of a specified frequency in the medium. [1]

Inductance - The ratio of the electromagnetic force induced in a coil by a current to the rate of change of the current.

Inductive coupled plasma mass spectroscopy (ICPMS) - See Techniques for Materials Characterization, page 12-1.

Inertial defect - In molecular spectroscopy, the quantity I_c-I_a-I_b for a molecule whose equilibrium configuration is planar, where I_a, I_b, and I_c are the effective principal moments of inertia. The inertial defect for a rigid planar molecule would be zero, but vibration–rotation interactions in a real molecule lead to a positive inertial defect.

Insulator - A material in which the highest occupied energy band (valence band) is completely filled with electrons, while the next higher band (conduction band) is empty. Solids with an energy gap of 5 eV or more are generally considered as insulators at room temperature. Their conductivity is less than 10^{-6} S/m and increases with temperature.

Intercalation compounds - Compounds resulting from reversible inclusion, without covalent bonding, of one kind of molecule in a solid matrix of another compound, which has a laminar structure. The host compound, a solid, may be macromolecular, crystalline, or amorphous. [5]

International System of Units (SI)* - The unit system adopted by the General Conference on Weights and Measures in 1960. It consists of seven base units (meter, kilogram, second, ampere, kelvin, mole, candela), plus derived units and prefixes. [1]

International Temperature Scale (ITS-90)* - The official international temperature scale adopted in 1990. It consists of a set of fixed points and equations which enable the thermodynamic temperature to be determined from operational measurements. [9]

Ion - An atomic or molecular particle having a net electric charge. [3]

Ion exchange - A process involving the adsorption of one or several ionic species accompanied by the simultaneous desorption (displacement) of one or more other ionic species. [3]

Ion neutralization spectroscopy (INS) - See Techniques for Materials Characterization, page 12-1.

Ionic strength (*I*) - A measure of the total concentration of ions in a solution, defined by $I = 1/2\Sigma_i z_i^2 m_i$, where z_i is the charge of ionic species i and m_i is its molality. For a 1-1 electrolyte at molality m, $I = m$.

Ionization constant* - The equilibrium constant for a reaction in which a substance in solution dissociates into ions.

Ionization potential* - The minimum energy required to remove an electron from an isolated atom or molecule (in its vibrational ground state) in the gaseous phase. More properly called ionization energy. [3]

Irradiance (*E*) - The radiant energy flux incident on an element of a surface, divided by the area of that element. [1]

Isentropic process - A thermodynamic process in which the entropy of the system does not change.

Ising model - A model describing the coupling between two atoms in a ferromagnetic lattice, in which the interaction energy is proportional to the negative of the product of the spin components along a specified axis.

Isobar - A line connecting points of equal pressure on a graphical representation of a physical system.

Isochore - A line or surface of constant volume on a graphical representation of a physical system.

Isoelectric point* - The pH of a solution or dispersion at which the net charge on the macromolecules or colloidal particles is zero. In electrophoresis there is no motion of the particles in an electric field at the isoelectric point.

Isomers - In chemistry, compounds that have identical molecular formulas but differ in the nature or sequence of bonding of their atoms or in the arrangement of their atoms in space. In physics, nuclei of the same atomic number Z and mass number A but in different energy states. [3]

Isomorphs - Substances of different chemical nature but having the same crystal structure.

Isotactic macromolecule - A tactic macromolecule, essentially comprising only one species of repeating unit which has chiral or prochiral atoms in the main chain in a unique arrangement with respect to its adjacent constitutional units. [8]

Isotherm - A line connecting points of equal temperature on a graphical representation of a physical system.

Isothermal process - A thermodynamic process in which the temperature of the system does not change.

Isotones - Nuclides having the same neutron number N but different atomic number Z. [3]

Isotopes - Two or more nuclides with the same atomic number Z but different mass number A. The term is sometimes used synonymously with nuclide, but it is preferable to reserve the word nuclide for a species of specific Z and A. [3]

Jahn-Teller effect - An interaction of vibrational and electronic motions in a nonlinear molecule which removes the degeneracy of certain electronic energy levels. It can influence the spectrum, crystal structure, and magnetic properties of the substance.

Johnson noise - Electrical noise generated by random thermal motion of electrons in a conductor or semiconductor. Also called thermal noise.

Josephson effect - The tunneling of electron pairs through a thin insulating layer which separates two superconductors. When a potential difference is applied to the superconductors, an alternating current is generated whose frequency is precisely proportional to the potential difference. This effect has important applications in metrology and determination of fundamental physical constants.

Joule (J)* - The SI unit of energy, equal to N m. [1]

Joule-Thomson coefficient (μ) - A parameter which describes the temperature change when a gas expands adiabatically through a nozzle from a high pressure to a low pressure region. It is defined by $\mu = (\partial T/\partial p)_H$, where H is enthalpy.

Julian calendar - The calendar introduced by Julius Caesar in 46 B.C. which divided the year into 365 days with a leap year of 366 days every fourth year.

Julian date (JD) - The number of days elapsed since noon Greenwich Mean Time on January 1, 4713 B.C. Thus January 1, 2000, 0h (midnight) will be JD 2,451,543.5. This dating system was introduced by Joseph Scaliger in 1582.

Kaon - One of the elementary particles in the family of mesons. Kaons have a spin of zero and may be neutral or charged.

Kelvin (K)* - The SI base unit of thermodynamic temperature. [1]

Kepler's laws - The three laws of planetary motion, which established the elliptical shape of planetary orbits and the relation between orbital dimensions and the period of rotation.

Kerr effect* - An electrooptical effect in which birefringence is induced in a liquid or gas when a strong electric field is applied perpendicular to the direction of an incident light beam. The Kerr constant k is given by n_1-n_2 = $k\lambda E^2$, where λ is the wavelength, E is the electric field strength, and n_1 and n_2 are the indices of refraction of the ordinary and extraordinary rays, respectively.

Ketenes - Compounds in which a carbonyl group is connected by a double bond to an alkylidene group: $R_2C=C=O$. [5]

Ketones - Compounds in which a carbonyl group is bonded to two carbon atoms: $R_1R_2C=O$ (neither R may be H). [5]

Kilogram (kg)* - The SI base unit of mass. [1]

Kinetic energy (E_k, T) - The energy associated with the motion of a system of particles in a specified reference frame. For a single particle of mass m moving at velocity v, $E_k = 1/2mv^2$.

Kirchhoff's laws - Basic rules for electric circuits, which state (a) the algebraic sum of the currents at a network node is zero and (b) the algebraic sum of the voltage drops around a closed path is zero.

Klein-Gordon equation - A relativistic extension of the Schrödinger equation.

Klein-Nishima formula - An expression for the scattering cross section of a photon by an unbound electron, based upon the Dirac electron theory.

Knight shift - The change in magnetic resonance frequency of a nucleus in a metal relative to the same nucleus in a diamagnetic solid. The effect is due to the polarization of the conduction electrons in the metal.

Knudsen number (*Kn*) - A dimensionless quantity used in fluid mechanics, defined by $Kn = \lambda/l$, where λ is mean free path and l is length. [2]

Kondo effect - A large increase in electrical resistance observed at low temperatures in certain dilute alloys of a magnetic metal in a nonmagnetic material.

Kramers-Kronig relation - A set of equations relating the real and imaginary parts of the index of refraction of a medium

Lactams - Cyclic amides of amino carboxylic acids, having a 1-azacycloalkan-2-one structure, or analogues having unsaturation or heteroatoms replacing one or more carbon atoms of the ring. [5]

Lactones - Cyclic esters of hydroxy carboxylic acids, containing a 1-oxacycloalkan-2-one structure, or analogues having unsaturation or heteroatoms replacing one or more carbon atoms of the ring. [5]

Lagrangian function (*L*) - A function used in classical mechanics, defined as the kinetic energy minus the potential energy for a system of particles.

Lamb shift - The small energy difference between the $^2S_{1/2}$ and $^2P_{1/2}$ levels in the hydrogen atom, which results from interactions between the electron and the radiation field.

Laminar flow - Smooth, uniform, non-turbulent flow of a gas or liquid in parallel layers, with little mixing between layers. It is characterized by small values of the Reynolds number.

Landé *g*-factor - See *g*-Factor of the electron

Langevin function - The mathematical function $L(x) = (e^x+e^{-x})/(e^x-e^{-x})-1/x$, which occurs in the expression for the average dipole moment of a group of rotating polar molecules in an electric field: $\mu_{av} = \mu L(\mu E/kT)$, where μ is the electric dipole moment of a single molecule, E is the electric field strength, k is the Boltzmann constant, and T is the temperature.

Lanthanides - The elements of atomic number 57 through 71, which share common chemical properties: La, Ce, Pr, Nd, Pm, Sm, Eu, Gd, Tb, Dy, Ho, Er, Tm, Yb, Lu. [7]

Larmor frequency (ν_L) - The precession frequency of a magnetic dipole in an applied magnetic field. In particular, a nucleus in a magnetic field of strength B has a Larmor frequency of $\gamma B/2\pi$, where γ is the magnetogyric ratio of the nucleus.

Laser* - A device in which an optical cavity is filled with a medium where a population inversion can be produced by some means. When the resonant frequency of the cavity bears the proper relation to the separation of the inverted energy levels, stimulated emission occurs, producing a highly monochromatic, coherent beam of light.

Laser ionization mass spectroscopy (LIMS) - See Techniques for Materials Characterization, page 12-1.

Lattice constants* - Parameters specifying the dimensions of a unit cell in a crystal lattice, specifically the lengths of the cell edges and the angles between them.

Lattice energy* - The energy per ion pair required to separate completely the ions in a crystal lattice at a temperature of absolute zero.

Laue diagram - A diffraction pattern produced when an x-ray beam passes through a thin slice of a crystal and impinges on a detector behind the crystal.

Lenz's law - The statement that the current induced in a circuit by a change in magnetic flux is so directed as to oppose the change in flux

Leonard-Jones potential - A simple but useful function for approximating the interaction between two neutral atoms or molecules separated by a distance r by writing the potential energy as $U(r) = 4\varepsilon\{(r_0/r)^{12} - (r_0/r)^6\}$, where ε and r_0 are adjustable parameters. In this form the depth of the potential well is ε and the minimum occurs at $2^{1/6}r_0$. The $(1/r)^{12}$ term is often replaced by other powers of $1/r$.

Lepton - One of the class of elementary particles that do not take part in the strong interaction. Included are the electron, muon, and neutrino. All leptons have a spin of 1/2.

Lewis number (*Le*) - A dimensionless quantity used in fluid mechanics, defined by $Le = a/D$, where a is thermal diffusivity and D is diffusion coefficient. [2]

Ligand field theory - A description of the structure of crystals containing a transition metal ion surrounded by nonmetallic ions (ligands). It is based on construction of molecular orbitals involving the *d*-orbitals of the central metal ion and combinations of atomic orbitals of the ligands.

Light year (l.y.) - A unit of distance used in astronomy, defined as the distance light travels in one year in a vacuum. Its approximate value is 9.46073×10^{15} m.

Lignins - Macromolecular constituents of wood related to lignans, composed of phenolic propylbenzene skeletal units, linked at various sites and apparently randomly. [5]

Ligroin - The petroleum fraction consisting mostly of C_7 and C_8 hydrocarbons and boiling in the range 90-140 °C; commonly used as a laboratory solvent.

Lipids - A loosely defined term for substances of biological origin that are soluble in nonpolar solvents. They consist of saponifiable lipids, such as glycerides (fats and oils) and phospholipids, as well as nonsaponifiable lipids, principally steroids. [5]

Lipoproteins - Clathrate complexes consisting of a lipid enwrapped in a protein host without covalent binding, in such a way that the complex has a hydrophilic outer surface consisting of all the protein and the polar ends of any phospholipids. [5]

Liter (L)* - A synonym for cubic decimeter. [1]

Lithosphere* - The outer layer of the solid earth, extending from the base of the mantle to the surface of the crust.

Lorentz contraction - The reduction in length of a moving body in the direction of motion, given by the factor $(1-v^2/c^2)^{1/2}$, where v is the velocity of the body and c the velocity of light. Also known as the FitzGerald-Lorentz contraction.

Lorentz force - The force exerted on a point charge Q moving at velocity $\boldsymbol{v}$ in the presence of external fields $\boldsymbol{E}$ and $\boldsymbol{B}$. It is given (in SI units) by $\boldsymbol{F} = Q(\boldsymbol{E} + \boldsymbol{v} \times \boldsymbol{B})$.

Loss angle (δ) - For a dielectric material in an alternating electromagnetic field, δ is the phase difference between the current and the potential difference. The function $\tan \delta$ is a measure of the ratio of the power dissipated in the dielectric to the power stored.

Low energy electron diffraction (LEED) - See Techniques for Materials Characterization, page 12-1.

Lumen (lm)* - The SI unit of luminous flux, equal to cd sr. [1]

Luminous flux (Φ) - The intensity of light from a source multiplied by the solid angle. The SI unit is lumen. [1]

Lux (lx)* - The SI unit of illuminance, equal to cd sr m^{-2}. [1]

Lyddane-Sachs-Teller relation - A relation between the phonon frequencies and dielectric constants of an ionic crystal which states that $(\omega_T/\omega_L)^2 = \varepsilon(\infty)/\varepsilon(0)$, where ω_T is the angular frequency of transverse optical phonons, ω_L that of longitudinal optical phonons, $\varepsilon(0)$ is the static dielectric constant, and $\varepsilon(\infty)$ the dielectric constant at optical frequencies.

Lyman series - The series of lines in the spectrum of the hydrogen atom which corresponds to transitions between the ground state (principal quantum number $n = 1$) and successive excited states. The wavelengths are given by $1/\lambda = R_H(1-1/n^2)$, where $n = 2,3,4,...$ and R_H is the Rydberg constant for hydrogen. The first member of the series ($n = 1 \leftrightarrow 2$), which is often called the Lyman-α line, falls at a wavelength of 1216 Å, and the series converges at 912 Å, the ionization limit of hydrogen.

Mach number (*Ma*) - A dimensionless quantity used in fluid mechanics, defined by $Ma = v/c$, where v is velocity and c is the speed of sound. [2]

Macromolecule - A molecule of high relative molecular mass (molecular weight), the structure of which essentially comprises the multiple repetition of units derived, actually or conceptually, from molecules of low relative molecular mass. [8]

Madelung constant* - A constant characteristic of a particular crystalline material which gives a measure of the electrostatic energy binding the ions in the crystal.

Magnetic field strength (*H*) - An axial vector quantity, the curl of which is equal to the current density, including the displacement current. [1]

Magnetic induction (*B*) - An axial vector quantity such that the force exerted on an element of current is equal to the vector product of this element and the magnetic induction. [1]

Magnetic moment - See Dipole moment, magnetic.

Magnetic susceptibility (χ_m, κ)* - Defined by $\chi_m = (\mu-\mu_0)/\mu_0$, where μ is the permeability of the medium and μ_0 the permeability of a vacuum. [1]

Magnetization (*M*) - Defined by $M = (B/\mu_0)-H$, where B is magnetic induction, H magnetic field strength, and μ_0 the permeability of a vacuum. [1]

Magnetogyric ratio (γ) - Ratio of the magnetic moment of a particle to its angular momentum. Also called gyromagnetic ratio.

Magneton - See Bohr magneton, Nuclear magneton.

Magnetostriction* - The change in dimensions of a solid sample when it is placed in a magnetic field.

Magnon - A quantum of magnetic energy associated with a spin wave in a ferromagnetic or antiferromagnetic crystal.

Mantle - The layer of the earth between the crust and the liquid outer core, which begins about 2900 km below the Earth's surface.

Maser - A device in which a microwave cavity is filled with a medium where a population inversion can be produced by some means. When the resonant frequency of the cavity bears the proper relation to the separation of the inverted energy levels, the device can serve as an amplifier or oscillator at that frequency.

Mass (*m*)* - Quantity of matter. Mass can also be defined as "resistance to acceleration".

Mass defect (*B*) - Defined by $B = Zm(^1\mathrm{H}) + Nm_n - m_a$, where Z is the atomic number, $m(^1\mathrm{H})$ is the mass of the hydrogen atom, N is the neutron number, m_n is the rest mass of the neutron, and m_a is the mass of the atom in question. Thus Bc^2 can be equated to the binding energy of the nucleus if the binding energy of atomic electrons is neglected. [1]

Mass excess (Δ) - Defined by $\Delta = m_a - Am_u$, where m_a is the mass of the atom, A the number of nucleons, and m_u the unified atomic mass constant ($m_u = 1$ u). [1]

Mass fraction (w_B) - The ratio of the mass of substance B to the total mass of a mixture. [1]

Mass number (*A*) - A characteristic property of a specific isotope of an element, equal to the sum of the number of protons and neutrons in the nucleus.

Mass spectrometry - An analytical technique in which ions are separated according to the mass/charge ratio and detected by a suitable detector. The ions may be produced by electron impact on a gas, a chemical reaction, energetic vaporization of a solid, etc. [6]

Massieu function - A thermodynamic function defined by $J = -A/T$, where A is the Helmholz energy and T the thermodynamic temperature. [2]

Matthiessen's rule - The statement that the electrical resistivity ρ of a metal can be written as $\rho = \rho_L+\rho_i$, where ρ_L is due to scattering of conduction electrons by lattice vibrations and ρ_i to scattering by impurities and imperfections. If the impurity concentration is small, ρ_i is temperature independent.

Maxwell (Mx)* - A non-SI unit of magnetic field strength (H) equal to 10^{-8} Wb. [1]

Maxwell's equations - The fundamental equations of electromagnetism. In a form appropriate to SI units, they are:

$$\text{curl}\, \boldsymbol{H} = \partial \boldsymbol{D}/\partial t + \boldsymbol{j}$$
$$\text{div}\, \boldsymbol{B} = 0$$
$$\text{curl}\, \boldsymbol{E} = -\partial \boldsymbol{B}/\partial t$$
$$\text{div}\, \boldsymbol{D} = \rho$$

where $\boldsymbol{H}$ is the magnetic field strength, $\boldsymbol{B}$ the magnetic induction, $\boldsymbol{E}$ the electric field strength, $\boldsymbol{D}$ the electric displacement, $\boldsymbol{j}$ the current density, ρ the charge density, and t is time.

Maxwell-Boltzmann distribution - An expression for the fraction of molecules $f(v)$ in a gas that have velocity v within a specified interval. It takes the form

$$f(v) = 4\pi(M/2\pi RT)^{3/2}\, v^2 e^{-Mv^2/2RT}$$

where M is the molar mass, R the molar gas constant, and T the temperature.

Mean free path* - The average distance a gas molecule travels between collisions.

Meissner effect - The complete exclusion of magnetic induction from the interior of a superconductor.

Melting point* - The temperature at which the solid and liquid phases of a substance are in equilibrium at a specified pressure (normally taken to be atmospheric unless stated otherwise).

Mercaptans - A traditional term abandoned by IUPAC, synonymous with thiols. This term is still widely used. [5]

Meson - Any elementary particle that has zero or integral spin. Mesons are responsible for the forces between protons and neutrons in the nucleus.

Mesosphere - The part of the Earth's atmosphere extending from the top of the stratosphere (about 50 km above the surface) to 80–90 km. It is characterized by a decrease in temperature with increasing altitude.

Metal - A material in which the highest occupied energy band (conduction band) is only partially filled with electrons. The electrical conductivity of metals generally decreases with temperature.

Metallocenes - Organometallic coordination compounds in which one atom of a transition metal such as iron, ruthenium or osmium is bonded to and only to the face of two cyclopentadienyl ligands which lie in parallel planes. [5]

Meter (m)* - The SI base unit of length. [1]

Methine group - In organic compounds, the -C= group. [5]

Mho - An archaic name for the SI unit siemens (reciprocal ohm).

Micelle - A particle formed by the aggregation of surfactant molecules (typically, 10 to 100 molecules) in solution. For aqueous solutions, the hydrophilic end of the molecule is on the surface of the micelle, while the hydrophobic end (often a hydrocarbon chain) points toward the center. At the critical micelle concentration (cmc) the previously dissolved molecules aggregate into a micelle.

Micron (μ) - An obsolete name for micrometer.

Mie scattering - The scattering of light by spherical dielectric particles whose diameter is comparable to the wavelength of the light.

Milky way - The band of light in the night sky resulting from the stars in the galactic plane. The term is also used to denote the galaxy in which the sun is located.

Miller indices (*hkl*) - A set of indices used to label planes in a crystal lattice. [2]

Millimeter of mercury (mmHg) - A non-SI unit of pressure, equal to 133.322 Pa. The name is generally considered interchangeable with torr.

Mobility (μ)* - In solid state physics, the drift velocity of electrons or holes in a solid divided by the applied electric field strength. The term is used in a similar sense in other fields.

Molality (*m*) - A measure of concentration of a solution in which one states the amount of substance (i.e., number of moles) of solute per kilogram of solvent. Thus a 0.1 molal solution (often written as 0.1 m) has $m = 0.1$ mol/kg.

Molar mass - The mass of one mole of a substance. It is normally expressed in units of g/mol, in which case its numerical value is identical with the molecular weight (relative molecular mass). [1]

Molar quantity - It is often convenient to express an extensive quantity (e.g., volume, enthalpy, heat capacity, etc.) as the actual value divided by amount of substance (number of moles). The resulting quantity is called molar volume, molar enthalpy, etc.

Molar refraction (*R*) - A property of a dielectric defined by the equation $R = V_m[(n^2-1)/(n^2+2)]$, where n is the index of refraction of the medium (at optical wavelengths) and V_m the molar volume. It is related to the polarizability α of the molecules that make up the medium by the Lorenz-Lorentz equation, $R = N_A\alpha/3\varepsilon_0$, where N_A is Avogadro's constant and ε_0 is the permittivity of a vacuum.

Molarity (*c*) - A measure of concentration of a solution in which one states the amount of substance (i.e., number of moles) of solute per liter of solution. Thus a 0.1 molar solution (often referred to as 0.1 M) has a concentration $c = 0.1$ mol/L.

Mole (mol)* - The SI base unit of amount of substance. [1]

Mole fraction (x_B) - The ratio of the amount of substance (number of moles) of substance B to the total amount of substance in a mixture. [1]

Molecular orbital - See Orbital.

Molecular weight (M_r)* - The ratio of the average mass per molecule or specified entity of a substance to 1/12 of the mass of nuclide ^{12}C. Also called relative molar (or molecular) mass. [1]

Moment of inertia (*I*) - The moment of inertia of a body about an axis is the sum (or integral) of the products of its elements of mass and the squares of their distances from the axis. [1]

Momentum (*p*) - The product of mass and velocity. [1]

Monomer - A substance consisting of molecules which can undergo polymerization, thereby contributing constitutional units to the essential structure of a macromolecule. [8]

Monosaccharides - A term which includes aldoses, ketoses, and a wide variety of derivatives. [5]

Mössbauer effect - The recoilless emission of γ-rays from nuclei bound in a crystal under conditions where the recoil energy associated with the γ emission is taken up by the crystal as a whole. This results in a very narrow line width, which can be exploited in various types of precise measurements.

Muon* - An unstable elementary particle of spin 1/2 and mass about 200 times that of the electron.

Naphtha - The petroleum fraction consisting mostly of C_6 to C_8 hydrocarbons and boiling in the range 80–120 °C. Solvents derived from this fraction include ligroin and petroleum ether.

Nautical mile - A non-SI unit of length, equal to exactly 1852 m.

Navier-Stokes equations - A set of complex equations for the motion of a viscous fluid subject to external forces.

Néel temperature (T_N)* - The critical temperature above which an antiferromagnetic substance becomes paramagnetic. [1]

Nernst effect - The production of an electric field in a conductor subject to an applied magnetic field and containing a transverse temperature gradient. The electric field is perpendicular to the magnetic field and the temperature gradient.

Network - In polymer science, a highly ramified macromolecule in which essentially each constitutional unit is connected to each other constitutional unit and to the macroscopic phase boundary by many permanent paths through the macromolecule, the number of such paths increasing with the number of intervening bonds. The paths must on the average be coextensive with the macromolecule. [8]

Neutrino - A stable elementary particle in the lepton family. Neutrinos have zero (or at least near-zero) rest mass and spin 1/2.

Neutron* - An elementary particle on spin 1/2 and zero charge. The free neutron has a mean lifetime of 887 seconds. Neutrons and protons, which are collectively called nucleons, are the constituents of the nucleus.

Neutron activation analysis (NAA) - See Techniques for Materials Characterization, page 12-1.

Neutron number (*N*) - A characteristic property of a specific isotope of an element, equal to the number of neutrons in the nucleus.

Newton (N)* - The SI unit of force, equal to m kg s^{-2}. [1]

Nitriles - Compounds having the structure RC≡N; thus C-substituted derivatives of hydrocyanic acid, HC≡N. [5]

Nitrosamines - N-Nitroso amines: compounds of the structure R_2NNO. Compounds RNHNO are not ordinarily isolatable, but they, too, are nitrosamines. The name is a contraction of N-nitrosoamine and, as such, does not require the N locant. [5]

Nuclear magnetic resonance (NMR)* - A widely used technique in which the resonant absorption of radiofrequency radiation by magnetic nuclei in a magnetic field is measured. The results give important information on the local environment of each nucleus.

Nuclear magneton (μ_N)* - The unit of nuclear magnetic moment, defined as $eh/4\pi m_p$, where h is Planck's constant, m_p the proton mass, and e the elementary charge.

Nuclear quadrupole resonance (NQR) - See Techniques for Materials Characterization, page 12-1.

Nuclear reaction analysis (NRA) - See Techniques for Materials Characterization, page 12-1.

Nuclear spin (*I*) - The quantum number that specifies the intrinsic angular momentum of a particular nucleus. The magnitude of the angular momentum is given by $[I(I+1)]^{1/2}\,h/2\pi$, where h is Planck's constant.

Nucleic acids* - Macromolecules, the major organic matter of the nuclei of biological cells, made up of nucleotide units, and hydrolyzable into certain pyrimidine or purine bases (usually adenine, cytosine, guanine, thymine, uracil), D-ribose or 2-deoxy-D-ribose. [5]

Nucleon - A collective term for the proton and neutron.

Nucleosides - Ribosyl or deoxyribosyl derivatives (rarely, other glycosyl derivatives) of certain pyrimidine or purine bases. They are thus glycosylamines or N-glycosides related to nucleotides by the lack of phosphorylation. [5]

Nucleotides - Compounds formally obtained by esterification of the 3′ or 5′ hydroxy group of nucleosides with phosphoric acid. They are the monomers of nucleic acids and are formed from them by hydrolytic cleavage. [5]

Nuclide - A species of atoms in which each atom has identical atomic number Z and identical mass number A. [3]

Nusselt number (*Nu*) - A dimensionless quantity used in fluid mechanics, defined by $Nu = hl/k$, where h is coefficient of heat transfer, l is length, and k is thermal conductivity. [2]

Nyquist theorem - An expression for the mean square thermal noise voltage across a resistor, given by $4RkT\Delta f$ where R is the resistance, k the Boltzmann constant, T the temperature, and Δf the frequency band within which the voltage is measured.

Octanol-water partition coefficient (*P*)* - A measure of the way in which a compound will partition itself between the octanol and water phases in the two-phase octanol-water system, and thus an indicator of certain types of biological activity. Specifically, P is the ratio of the concentration (in moles per liter) of the compound in the octanol phase to that in the water phase at infinite dilution. The quantity normally reported is log P.

Oersted (Oe) - A non-SI unit of magnetic field (H), equal to 79.57747 A/m.

Ohm (Ω)* - The SI unit of electric resistance, equal to V/A. [1]

Ohm's law - A relation among electric current I, potential difference V, and resistance R, viz., $I = V/R$. At constant temperature the resistance for many materials is constant to high precision.

Olefins - Acyclic and cyclic hydrocarbons having one or more carbon-carbon double bonds, apart from the formal ones in aromatic compounds. The class olefins subsumes alkenes and cycloalkenes and the corresponding polyenes. [5]

Oligomer - A substance consisting of molecules of intermediate relative molecular mass (molecular weight), the structure of which essentially comprises the multiple repetition of units derived, actually or conceptually, from molecules of low relative molecular mass. In contrast to a polymer, the properties of an oligomer can vary significantly with the removal of one or a few of its units. [8]

Oligopeptides - Peptides containing from three to nine amino groups. [5]

Onsager relations - An important set of equations in the thermodynamics of irreversible processes. They express the symmetry between the transport coefficients describing reciprocal processes in systems with a linear dependence of flux on driving forces.

Optical rotary power - Angle by which the plane of polarization of a light beam is rotated by an optically active medium, divided by path length and by concentration of the active constituent. Depending on whether mass or molar concentration is used, the modifier "specific" or "molar" is attached. [2]

Orbital - A one-electron wavefunction. Atomic orbitals are classified as *s*-, *p*-, *d*-, or *f*-orbitals according to whether the angular momentum quantum number $l = 0, 1, 2$, or 3. Molecular orbitals, which are usually constructed as linear combinations of atomic orbitals, describe the distribution of electrons over the entire molecule.

Oscillator strength (*f*) - A measure of the intensity of a spectroscopic transition, defined by

$$f = \frac{8\pi^2 m_e \nu}{3he^2}\left|\mu_{ij}\right|^2$$

where ν is the frequency, μ_{ij} the transition dipole moment, m_e the mass of the electron, e the elementary charge, and h Planck's constant.

Osmosis - The flow of a solvent in a system in which two solutions of different concentration are separated by a semipermeable membrane which cannot pass solute molecules. The solvent will flow from the side of lower concentration to that of higher concentration, thus tending to equalize the concentrations. The pressure that must be applied to the more concentrated side to stop the flow is called the osmotic pressure.

Osmotic coefficient (ϕ) - Defined by $\phi = \ln a_A/(M_A\Sigma m_B)$, where M_A is the molar mass of substance A (normally the solvent), a_A is its activity, and the m_B are molalities of the solutes. [1]

Osmotic pressure (Π) - The excess pressure necessary to maintain osmotic equilibrium between a solution and the pure solvent separated by a membrane permeable only to the solvent. In an ideal dilute solution $\Pi = c_B RT$, where c_B is the amount-of-substance concentration of the solute, R is the molar gas constant, and T the temperature. [1,2]

Ostwald dilution law - A relation for the concentration dependence of the molar conductivity Λ of an electrolyte solution, viz.,

$$\frac{1}{\Lambda} = \frac{1}{\Lambda^\circ} + \frac{\Lambda c}{K(\Lambda^\circ)^2}$$

where c is the solute concentration, K is the equilibrium constant for dissociation of the solute, and $\Lambda°$ is the conductivity at $c\Lambda = 0$.

Ounce (oz) - A non-SI unit of mass. The avoirdupois ounce equals 28.34952 g, while the troy ounce equals 31.10348 g.

Overpotential (η) - In an electrochemical cell, the difference between the potential of an electrode and its zero-current value.

Oximes - Compounds of structure $R_2C{=}NOH$ derived from condensation of aldehydes or ketones with hydroxylamine. Oximes from aldehydes may be called aldoximes; those from ketones may be called ketoximes. [5]

Oxo compounds - Compounds containing an oxygen atom, =O, doubly bonded to carbon or another element. The term thus embraces aldehydes, carboxylic acids, ketones, sulfonic acids, amides and esters. [5]

Ozonides - The 1,2,4-trioxolanes formed by the reaction of ozone at a carbon-carbon double bond, or the analogous compounds derived from acetylenic compounds. [5]

Pair production - A process in which a photon is converted into a particle and its antiparticle (e.g., an electron and positron) in the electromagnetic field of a nucleus.

Paraffins - Obsolescent term for saturated hydrocarbons, commonly but not necessarily acyclic. Still widely used in the petrochemical industry, where the term designates acyclic saturated hydrocarbons, and stands in contradistinction to naphthenes. [5]

Paramagnetism* - A type of magnetism characterized by a positive magnetic susceptibility, so that the material becomes weakly magnetized in the direction of an external field. The magnetization disappears when the field in removed. In the simplest approximation (Curie's law) the susceptibility is inversely proportional to temperature.

Parity - The property of a quantum-mechanical wave function that describes its behavior under the symmetry operation of coordinate inversion. A parity of +1 (or even) is assigned if the wave function does not change sign when the signs of all the coordinates are changed; the parity is −1 (or odd) if the wave function changes sign under this operation.

Parsec (pc) - A unit of distance defined as the distance at which 1 astronomical unit (AU) subtends an angle of 1 second of arc. It is equal to 206264.806 AU or 3.085678×10^{16} m.

Particle induced x-ray emission (PIXE) - See Techniques for Materials Characterization, page 12-1.

Partition function (*q*, *z*) - For a single molecule, $q = \Sigma_i g_i \exp(\varepsilon_i/kT)$, where ε_i is an energy level of degeneracy g_i, k the Boltzmann constant, and T the absolute temperature; the summation extends over all energy states. For a system of N non-interacting molecules which are indistinguishable, as in an ideal gas, the canonical partition function $Q = qN/N!$.

Pascal (Pa)* - The SI unit of pressure, equal to N/m^2. [1]

Paschen series - The series of lines in the spectrum of the hydrogen atom which corresponds to transitions between the state with principal quantum number $n = 3$ and successive higher states. The wavelengths are given by $1/\lambda = R_H(1/9\text{-}1/n^2)$, where $n = 4,5,6,...$ and R_H is the Rydberg constant. The first member of the series ($n = 3 \leftrightarrow 4$), which is often called the P_α line, falls in the infrared at a wavelength of 1.875 μm.

Paschen-Back effect - In atomic spectroscopy, the decoupling of electron spin from orbital angular momentum as the strength of an external magnetic field is increased.

Pauli exclusion principle - The statement that two electrons in an atom cannot have identical quantum numbers; thus if there are two electrons in the same orbital, their spin quantum numbers must be of opposite sign.

Pearson symbol - A code for designating crystallographic information, including the crystal system, the lattice type, and the number of atoms per unit cell.

Péclet number (*Pe*) - A dimensionless quantity used in fluid mechanics, defined by $Pe = vl/a$, where v is velocity, l is length, and a is thermal diffusivity. [2]

Peltier effect - The absorption or generation of heat (depending on the current direction) which occurs when an electric current is passed through a junction between two materials.

Peptides - Amides derived from two or more amino carboxylic acid molecules (the same or different) by formation of a covalent bond from the carbonyl carbon of one to the nitrogen atom of another with formal loss of water. [5]

Permeability (μ) - Magnetic induction divided by magnetic field strength; i.e. $\mu = B/H$. The relative permeability $\mu_r = \mu/\mu_0$, where μ_0 is the permeability of a vacuum. [1]

Permittivity (ε) - Ratio of the electric displacement in a medium to the electric field strength. Also called dielectric constant. [1]

Peroxides - Compounds of structure ROOR in which R may be any organic group. In inorganic chemistry, salts of the anion O_2^{-2} [5]

Peroxy acids - Acids in which an acidic -OH group has been replaced by an -OOH group; e.g., $CH_3C(=O)OOH$ peroxyacetic acid, $PhS(=O)_2OOH$ benzeneperoxysulfonic acid. [5]

Petroleum ether - The petroleum fraction consisting of C_5 and C_6 hydrocarbons and boiling in the range 35–60 °C; commonly used as a laboratory solvent.

pH* - A convenient measure of the acid-base character of a solution, usually defined by $\text{pH} = -\log[c(H^+)/\text{mol L}^{-1}]$, where $c(H^+)$ is the concentration of hydrogen ions. The more precise definition is in terms af activity rather than concentration. [2]

Phenols - Compounds having one or more hydroxy groups attached to a benzene or other arene ring. [5]

Phonon - A quantum of energy associated with a vibrational mode of a crystal lattice.

Phosphines - PH_3 and compounds derived from it by substituting one, two or three hydrogen atoms by hydrocarbyl groups.

RPH_2, R_2PH and R_3P (R not equal to H) are called primary, secondary and tertiary phosphines, respectively. [5]

Phosphonium compounds - Salts (and hydroxides) $[R_4P]^+X^-$ containing tetracoordinate phosphonium ion and the associated anion. [5]

Phosphonium ylides - Compounds having the structure $R_3P^+\text{-}C^-R_2 \rightleftharpoons R_3P{=}CR_2$. Also known as Wittig reagents. [5]

Phosphorescence - The process by which a molecule is excited by light to a higher electronic state and then undergoes a radiationless transition to a state of different multiplicity from which it decays, after some delay, to the ground state. The emitted light is normally of longer wavelength than the exciting light because vibrational energy has been dissipated.

Photoelectric effect - The complete absorption of a photon by a solid with the emission of an electron.

Photon - An elementary particle of zero mass and spin 1/2. The photon is involved in electromagnetic interactions and is the quantum of electromagnetic radiation.

Photon stimulated desorption (PSD) - See Techniques for Materials Characterization, page 12-1.

Pinacols - Tetra(hydrocarbyl)ethane-1,2-diols, $R_2C(OH)C(OH)R_2$, of which the tetramethyl example is the simplest one and is itself commonly known as pinacol. [5]

Pion - An elementary particle in the family of mesons. Pions have zero spin and may be neutral or charged. They participate in the strong interaction which holds the nucleus together.

p*K** - The negative logarithm (base 10) of an equilibrium constant *K*. For pK_a, see Acid dissociation constant.

Planck constant (*h*)* - The elementary quantum of action, which relates energy to frequency through the equation $E = h\nu$.

Planck distribution - See Black body radiation

Planck function - A thermodynamic function defined by $Y = -G/T$, where *G* is Gibbs energy and *T* thermodynamic temperature. [2]

Plasma - A highly ionized gas in which the charge of the electrons is balanced by the charge of the positive ions, so that the system as a whole is electrically neutral.

Plasmon - A quantum associated with a plasma oscillation in the electron gas of a solid.

Point group* - A group of symmetry operations (rotations, reflections, etc.) that leave a molecule invariant. Every molecular conformation can be assigned to a specific point group, which plays a major role in determining the spectrum of the molecule.

Poise (P) - A non-SI unit of viscosity, equal to 0.1 Pa s.

Poiseuille's equation - A formula for the rate of flow of a viscous fluid through a tube:

$$\frac{dV}{dt} = \frac{(p_1^{\,2} - p_2^{\,2})\pi r^4}{16 l \eta p_0}$$

where *V* is the volume as measured at pressure p_0; p_1 and p_2 are the pressures at each end of the tube; *r* is the radius and *l* the length of the tube; and η is the viscosity.

Poisson ratio (μ) - The absolute value of the ratio of the transverse strain to the corresponding axial strain resulting from uniformly distributed axial stress below the proportional limit (i.e., where Hooke's law is valid). [10]

Polariton - A quantum associated with the coupled modes of photons and optical phonons in an ionic crystal.

Polarizability (α)* - The change in dipole moment of a molecule produced by an external electric field; specifically, $\alpha_{ab} = \partial p_a/\partial E_b$, where p_a is the dipole moment component on the *a* axis and E_b is the component of the electric field strength along the *b* axis. [2]

Polymer - A substance composed of molecules of high relative molecular mass (molecular weight), the structure of which essentially comprises the multiple repetition of units derived, actually or conceptually, from molecules of low relative molecular mass. A single molecule of a polymer is called a macromolecule. [8]

Polypeptides - Peptides containing 10 or more amino acid residues. See also Peptides. [5]

Polysaccharides - Compounds consisting of a large number of monosaccharides linked glycosidically. This term is commonly used only for those containing more than ten monosaccharide residues. Also called glycans. [5]

Porphyrins - Natural pigments containing a fundamental skeleton of four pyrrole nuclei united through the α-positions by four methine groups to form a macrocyclic structure (porphyrin is designated porphine in Chemical Abstracts indexes). [5]

Positron - The antiparticle of the electron. It has the same mass and spin as an electron, and an equal but opposite charge.

Positronium - The hydrogen-like "atom" formed from a positron nucleus and an electron. Its lifetime is very short because of annihilation of the positron and electron.

Potential - See Electric potential.

Potential energy (E_p, *V*, *U*) - The portion of the energy of a system that is associated with its position in a force field.

Pound (lb) - A non-SI unit of mass, equal to 0.4535924 kg.

Power (*P*) - Rate of energy transfer. For electrical circuits, this is equal to the product of current and potential difference, $P = IV$. [1]

Poynting vector (*S*) - For electromagnetic radiation, the vector product of the electric field strength and the magnetic field strength. [1]

Prandtl number (*Pr*) - A dimensionless quantity used in fluid mechanics, defined by $Pr = \eta/\rho a$, where η is viscosity, ρ is density, and *a* is thermal diffusivity. [2]

Pressure* - Force divided by area. [1]

Proteins - Naturally occurring and synthetic polypeptides having molecular weights greater than about 10,000 (the limit is not precise). See also Peptides. [5]

Proton* - A stable elementary particle of unit positive charge and spin 1/2. Protons and neutrons, which are collectively called nucleons, are the constituents of the nucleus.

Pulsar - A neutron star which rotates rapidly and emits electromagnetic radiation in regular pulses at a frequency related to the rotation period.

Purine bases* - Purine and its substitution derivatives, especially naturally occurring examples. [5]

Pyrimidine bases* - Pyrimidine and its substitution derivatives, especially naturally occurring examples. [5]

Q-switching - A technique for obtaining very high power from a laser by keeping the *Q* factor of the laser cavity low while the population inversion builds up, then suddenly increasing the *Q* to initiate the stimulated emission.

Quad - A unit of energy defined as 10^{15} Btu, equal to approximately 1.055056×10^{18} J.

Quadrupole moment - A coefficient of the third term (after monopole and dipole) in the power series expansion of the electric potential of an array of charges. A nucleus of spin greater than 1/2 has a non-vanishing nuclear quadrupole moment which can interact with the electric field gradient of the surrounding electrons. Molecular quadrupole moments have an influence on intermolecular forces.

Quality factor (*Q*) - The ratio of the absolute value of the reactance of an electrical system to the resistance; thus a measure of the energy stored per cycle relative to the energy dissipated.

Quantum yield - In photochemistry, the number of moles transformed in a specific process, either physically (e.g., by emission of photons) or chemically, per mole of photons absorbed by the system. [3]

Quark - An elementary entity which has not been directly observed but is considered a constituent of protons, neutrons, and other hadrons.

Quasar - An extragalactic object emitting electromagnetic radiation at a very high power level and showing a very large red shift, thus indicating that the object is receding at a speed approaching the speed of light.

Quasicrystal - A solid having conventional crystalline properties but whose lattice does not display translational periodicity.

Quaternary ammonium compounds - Derivatives of ammonium compounds, $NH_4^+ Y^-$, in which all four of the hydrogens bonded to nitrogen have been replaced with hydrocarbyl groups. Compounds having a carbon-nitrogen double bond (i.e. $R_2C{=}N^+R_2Y^-$) are more accurately called iminium compounds. [5]

Quinones - Compounds having a fully conjugated cyclic dione structure, such as that of benzoquinones, derived from aromatic compounds by conversion of an even number of -CH= groups into -C(=O)- groups with any necessary rearrangement of double bonds. [5]

Racemic mixture - A mixture of equal amounts of a pair of enantiomers (optical isomers); such a mixture is not optically active.

Rad - A non-SI unit of absorbed dose of radiation, equal to 0.01 Gy.

Radiance (*L*) - The radiant intensity in a given direction from an element of a surface, divided by the area of the orthogonal projection of this element on a plane perpendicular to the given direction. [1]

Radiant intensity (*I*) - The radiant energy flux leaving an element of a source within an element of solid angle, divided by that element of solid angle. [1]

Radicals - Molecular entities possessing an unpaired electron, such as $\cdot CH_3$, $\cdot SnH_3$, $\cdot Cl$. (In these formulas the dot, symbolizing the unpaired electron, should be placed so as to indicate the atom of highest spin density, if this is possible). [5]

Raman effect - The inelastic scattering of light by a molecule, in which the incident photon either gives up to, or receives energy from, one of the internal vibrational modes of the molecule. The scattered light thus has either a lower frequency (Stokes radiation) or higher frequency (anti-Stokes radiation) than the incident light. These shifts provide a measure of the normal vibrational frequencies of the molecule.

Rankine cycle - A thermodynamic cycle which can be used to calculate the ideal performance of a heat engine that uses a condensable vapor as the working fluid (e.g., a steam engine or a heat pump).

Rankine temperature - A thermodynamic temperature scale based on a temperature interval °R = (5/9) K ; i.e., $T/°R = (9/5)T/K = t/°F + 459.67$.

Raoult's law - The expression for the vapor pressure p_i of component i in an ideal solution, viz., $p_i = x_i p_{i0}$, where x_i is the mole fraction of component i and p_{i0} the vapor pressure of the pure substance i.

Rare earth elements - The elements Sc, Y, and the lanthanides (La, Ce, Pr, Nd, Pm, Sm, Eu, Gd, Tb, Dy, Ho, Er, Tm, Yb, Lu). [7]

Rayleigh number (*Ra*) - A dimensionless quantity used in fluid mechanics, defined by $Ra = l^3 g\alpha\Delta T\rho/\eta a$, where l is length, g is acceleration of gravity, α is cubic expansion coefficient, T is temperature, ρ is density, η is viscosity, and a is thermal diffusivity. [2]

Rayleigh scattering - The scattering of light by particles which are much smaller than the wavelength of the light. It is characterized by a scattered intensity which varies as the inverse fourth power of the wavelength.

Rayleigh wave - A guided elastic wave along the surface of a solid; also called surface acoustic wave.

Reactance (*X*) - The imaginary part of impedance. For an inductive reactance L and a capacitive reactance C in series, the reactance is $X = L\omega - 1/(C\omega)$, where ω is 2π times the frequency of the current. [1]

Red shift - A displacement of a spectral line toward longer wavelengths. This can occur through the Doppler effect (e.g., in the light from receding galaxies) or, in the general theory of relativity, from the effects of a star's gravitational field.

Reflectance (ρ) - Ratio of the radiant or luminous flux at a given wavelength that is reflected to that of the incident radiation. Also called reflection factor. [1]

Reflection high energy electron diffraction (RHEED) - See Techniques for Materials Characterization, page 12-1.

Relative humidity* - The ratio of the partial pressure of water vapor in air to the saturation vapor pressure of water at the same temperature, expressed as a percentage. [10]

Relative molar mass - See Molecular weight.

Rem - A non-SI unit of dose equivalent, equal to 0.01 Sv.

Resistance (*R*) - Electric potential difference divided by current when there is no electromotive force in the conductor. This definition applies to direct current. More generally, resistance is defined as the real part of impedance. [1]

Resistivity (ρ) - Electric field strength divided by current density when there is no electromotive force in the conductor. Resistivity is an intrinsic property of a material. For a conductor of uniform cross section with area A and length L, and whose resistance is R, the resistivity is given by $\rho = RA/L$. [1]

Reynolds number (*Re*) - A dimensionless quantity used in fluid mechanics, defined by $Re = \rho v l/\eta$, where ρ is density, v is velocity, l is length, and η is viscosity. [2]

Rheology - The study of the flow of liquids and deformation of solids. Rheology addresses such phenomena as creep, stress relaxation, anelasticity, nonlinear stress deformation, and viscosity.

Ribonucleic acids (RNA) - Naturally occurring polyribonucleotides. See also nucleic acids, nucleosides, nucleotides, ribonucleotides. [5]

Ribonucleotides - Nucleotides in which the glycosyl group is a ribosyl group. See also nucleotides. [5]

Roentgen (R) - A unit used for expressing the charge (positive or negative) liberated by x-ray or γ radiation in air, divided by the mass of air. A roentgen is defined as 2.58×10^{-4} C/kg.

Rotational constants - In molecular spectroscopy, the constants appearing in the expression for the rotational energy levels as a function of the angular momentum quantum numbers. These constants are proportional to the reciprocals of the principal moments of inertia, averaged over the vibrational motion.

Rutherford back scattering (RBS) - See Techniques for Materials Characterization, page 12-1.

Rydberg constant (R_∞)* - The fundamental constant which appears in the equation for the energy levels of hydrogen-like atoms; i.e., $E_n = hcR_\infty Z^2\mu/n^2$, where h is Planck's constant, c the speed of light, Z the atomic number, μ the reduced mass of nucleus and electron, and n the principal quantum number ($n = 1, 2, ...$).

Rydberg series - A regular series of lines in the spectrum of an atom or molecule, with the spacing between successive lines becoming smaller as the frequency increases (wavelength decreases). The series eventually converges to a limit which usually corresponds to the complete removal of an electron from the atom or molecule.

Sackur-Tetrode equation* - An equation for the molar entropy S_m of an ideal monatomic gas: $S_m = R\ln(e^{5/2}\, V/N_A\Lambda^3)$, where R is the molar gas constant, V is the volume, and N_A is Avogadro's number. The constant Λ is given by $\Lambda = h/(2\pi mkT)^{1/2}$, where h is Planck's constant, m the atomic mass, k the Boltzmann constant, and T the temperature.

Salinity (S)* - A parameter used in oceanography to describe the concentration of dissolved salts in seawater. It is defined in terms of electrical conductivity relative to a standard solution of KCl. When expressed in units of parts per thousand, S may be roughly equated to the concentration of dissolved material in grams per kilogram of seawater.

Salt - An ionic compound formed by the reaction of an acid and a base.

Scanned probe microscopy (SPM) - See Techniques for Materials Characterization, page 12-1.

Scanning electron microscopy (SEM) - See Techniques for Materials Characterization, page 12-1.

Scanning laser acoustic microscopy (SLAM) - See Techniques for Materials Characterization, page 12-1.

Scanning transmission electron microscopy (STEM) - See Techniques for Materials Characterization, page 12-1.

Scanning tunneling microscopy (STM) - See Techniques for Materials Characterization, page 12-1.

Schiff bases - Imines bearing a hydrocarbyl group on the nitrogen atom: $R_2C{=}NR'$ (R' not equal to H). Considered by many to be synonymous with azomethines. [5]

Schmidt number (Sc) - A dimensionless quantity used in fluid mechanics, defined by $Sc = \eta/\rho D$, where η is viscosity, ρ is density, and D is diffusion coefficient. [2]

Schottky barrier - A potential barrier associated with a metal-semiconductor contact. It forms the basis for the rectifying device known as the Schottly diode.

Schrödinger equation - The basic equation of wave mechanics which, for systems not dependent on time, takes the form:

$$-(\hbar/2m)\nabla^2\psi + V\psi = E\psi$$

where ψ is the wavefunction, V is the potential energy expressed as a function of the spatial coordinates, E is an energy eigenvalue, ∇^2 is the Laplacian operator, $\hbar$ is Planck's constant divided by 2π, and m is the mass.

Second (s)* - The SI base unit of time. [1]

Second radiation constant (c_2)* - See First radiation constant.

Secondary ion mass spectroscopy (SIMS) - See Techniques for Materials Characterization, page 12-1.

Seebeck effect - The development of a potential difference in a circuit where two different metals or semiconductors are joined and their junctions maintained at different temperatures. It is the basis of the thermocouple.

Selenides - Compounds having the structure RSeR (R not equal to H). They are thus selenium analogues of ethers. Also used for metal salts of H_2Se. [5]

Semicarbazones - Compounds having the structure $R_2C{=}NNHC({=}O)NH_2$, formally derived by condensation of aldehydes or ketones with semicarbazide [$NH_2NHC({=}O)NH_2$]. [5]

Semiconductor - A material in which the highest occupied energy band (valence band) is completely filled with electrons at T = 0 K, and the energy gap to the next highest band (conduction band) ranges from 0 to 4 or 5 eV. With increasing temperature electrons are excited into the conduction band, leading to an increase in the electrical conductivity.

Semiquinones - Radical anions having the structure -O-Z-O· where Z is an ortho- or para-arylene group or analogous heteroarylene group; they are formally generated by the addition of an electron to a quinone. [5]

SI units* - The International System of Units adopted in 1960 and recommended for use in all scientific and technical fields. [1]

Siemens (S)* - The SI unit of electric conductance, equal to Ω^{-1}. [1]

Sievert (Sv)* - The SI unit of dose equivalent (of radiation), equal to J/kg. [1]

Silanes - Saturated silicon hydrides, analogues of the alkanes; i.e., compounds of the general formula Si_nH_{2n+2}. Silanes may be subdivided into silane, oligosilanes, and polysilanes. Hydrocarbyl derivatives are often referred to loosely as silanes. [5]

Silicones - Polymeric or oligomeric siloxanes, usually considered unbranched, of general formula $[\text{-OSiR}_2\text{-}]_n$ (R not equal to H). [5]

Siloxanes - Saturated silicon-oxygen hydrides with unbranched or branched chains of alternating silicon and oxygen atoms (each silicon atom is separated from its nearest silicon neighbors by single oxygen atoms). [5]

Skin effect - The concentration of high frequency alternating currents near the surface of a conductor.

Slater orbital - A particular mathematical expression for the radial part of the wave function of a single electron, which is used in quantum-mechanical calculations of the energy and other properties of atoms and molecules.

Small angle neutron scattering (SANS) - See Techniques for Materials Characterization, page 12-1.

Snell's law - The relation between the angle of incidence i and the angle of refraction r of a light beam which passes from a medium of refractive index n_0 to a medium of index n_1, viz., $\sin i/\sin r = n_1/n_0$.

Solar constant* - The mean radiant energy flux from the sun on a unit surface normal to the direction of the rays at the mean distance of the earth from the sun. The value is approximately 1373 W/m^2.

Solar wind - The stream of high velocity hydrogen and helium ions emitted by the sun which flows through the solar system and beyond.

Soliton - A spatially localized wave in a solid or liquid that can interact strongly with other solitons but will afterwards regain its original form.

Solubility* - A quantity expressing the maximum concentration of some material (the solute) that can exist in another liquid or solid material (the solvent) at thermodynamic equilibrium at specified temperature and pressure. Common measures of solubility include the mass of solute per unit mass of solution (mass fraction), mole fraction of solute, molality, molarity, and others.

Solubility product constant (K_{sp})* - The equilibrium constant for the dissolution of a sparsely soluble salt into its constituent ions.

Space group* - A group of symmetry operations (reflections, rotations, etc.) that leave a crystal invariant. A total of 230 space groups have been identified.

Spark source mass spectroscopy (SSMS) - See Techniques for Materials Characterization, page 12-1.

Specific gravity - Ratio of the mass density of a material to that of water. Since one must specify the temperature of both the sample and the water to have a precisely defined quantity, the use of this term is now discouraged.

Specific heat - Heat capacity divided by mass. See Heat capacity.

Specific quantity - It is often convenient to express an extensive quantity (e.g., volume, enthalpy, heat capacity, etc.) as the actual value divided by mass. The resulting quantity is called specific volume, specific enthalpy, etc.

Specific rotation $[\alpha]^\theta_\lambda$ - For an optically active substance, defined by $[\alpha]^\theta_\lambda = \alpha/\gamma l$, where α is the angle through which plane polarized light is rotated by a solution of mass concentration γ and path length l. Here θ is the Celsius temperature and λ the wavelength of the light at which the measurement is carried out. Also called specific optical rotatory power. [2]

Spin (s, I)* - A measure of the intrinsic angular momentum of a particle, which it possesses independent of its orbital motion. The symbol s is used for the spin quantum number of an electron, while I is generally used for nuclear spin.

Spiro compounds - Compounds having one atom (usually a quaternary carbon) as the only common member of two rings. [5]

Stacking fault - An error in the normal sequence of layer growth in a crystal.

Standard mean ocean water (SMOW) - A standard sample of pure water of accurately known isotopic composition which is maintained by the International Atomic Energy Agency. It is used for precise calibration of density and isotopic composition measurements.

Standard reduction potential (E°) - The zero-current potential of a cell in which the specified reduction reaction occurs at the right-hand electrode and the left-hand electrode is the standard hydrogen electrode. Also called Standard electrode potential.

Standard state - A defined state (specified temperature, pressure, concentration, etc.) for tabulating thermodynamic functions and carrying out thermodynamic calculations. The standard state pressure is usually taken as 100,000 Pa (1 bar), but various standard state temperatures are used. [2]

Stanton number (St) - A dimensionless quantity used in fluid mechanics, defined by $St = h/\rho v c_p$, where h is coefficient of heat transfer, ρ is density, v is velocity, and c_p is specific heat capacity at constant pressure. [2]

Stark effect - The splitting of an energy level of an atom or molecule, and hence a splitting of spectral lines arising from that level, as a result of the application of an external electric field.

Statistical weight (g) - The number of distinct states corresponding to the same energy level. Also called degeneracy.

Stefan-Boltzmann constant (σ)* - Constant in the equation for the radiant exitance M (radiant energy flux per unit area) from a black body at thermodynamic temperature T, viz. $M = \sigma T^4$. [1]

Stibines - SbH_3 and compounds derived from it by substituting one, two or three hydrogen atoms by hydrocarbyl groups: R_3Sb. $RSbH_2$, R_2SbH, and R_3Sb (R not equal to H) are called primary, secondary and tertiary stibines, respectively. [5]

Stochastic process - A process which involves random variables and whose outcome can thus be described only in terms of probabilities.

Stoichiometric number (ν) - The number appearing before the symbol for each compound in the equation for a chemical reaction. By convention, it is negative for reactants and positive for products. [2]

Stokes (St) - A non-SI unit of kinematic viscosity, equal to 10^{-4} m^2/s.

Stokes' law - The statement, valid under certain conditions, that the viscous force F experienced by a sphere of radius a moving at velocity v in a medium of viscosity η is given by $F = -6\pi\eta a v$.

Strain - The deformation of a body that results from an applied stress.

Stratosphere - The part of the earth's atmosphere extending from the top of the troposphere (typically 10 to 15 km above the surface) to about 50 km. It is characterized by an increase in temperature with increasing altitude.

Stress - Force per unit area (pressure) applied to a body. Tensile stress tends to stretch or compress the body in the direction of the applied force. Sheer stress results from a tangential force which tends to twist the body.

Strong interaction - The short range (order of 1 fm) attractive forces between protons, neutrons, and other hadrons which are responsible for the stability of the nucleus.

Strouhal number (Sr) - A dimensionless quantity used in fluid mechanics, defined by $Sr = lf/v$, where l is length, f is frequency, and v is velocity. [2]

Structure factor - In x-ray crystallography, the sum of the scattering factors of all the atoms in a unit cell, weighted by an appropriate phase factor. The intensity of a given reflection is proportional to the square of the structure factor.

Sublimation pressure - The pressure of a gas in equilibrium with a solid at a specified temperature.

Sulfides - Compounds having the structure RSR (R not equal to H). Such compounds were once called thioethers. In an inorganic sense, salts or other derivatives of hydrogen sulfide. [5]

Sulfones - Compounds having the structure, $RS(=O)_2R$ (R not equal to H), e.g. $C_2H_5S(=O)_2CH_3$, ethyl methyl sulfone. [5]

Sulfonic acids - $HS(=O)_2OH$, sulfonic acid, and its S-hydrocarbyl derivatives. [5]

Sulfoxides - Compounds having the structure $R_2S=O$ (R not equal to H), e.g., $Ph_2S=O$, diphenyl sulfoxide. [5]

Superconductor - A material that experiences a nearly total loss of electrical resistivity below a critical temperature T_c. The effect can occur in pure metals, alloys, semiconductors, organic compounds, and certain inorganic solids.

Superfluid - A fluid with near-zero viscosity and extremely high thermal conductivity. Liquid helium exhibits these properties below 2.186 K (the λ point).

Supernova - A star in the process of exploding because of instabilities which follow the exhaustion of its nuclear fuel.

Surface analysis by laser ionization (SALI) - See Techniques for Materials Characterization, page 12-1.

Surface tension (γ,σ)* - The force per unit length in the plane of the interface between a liquid and a gas, which resists an increase in the area of that surface. It can also be equated to the surface Gibbs energy per unit area.

Surfactant - A substance which lowers the surface tension of the medium in which it is dissolved, and/or the interfacial tension with other phases, and accordingly is positively adsorbed at the liquid-vapor or other interfaces. [3]

Susceptance (*B*) - Imaginary part of admittance. [1]

Svedberg - A non-SI unit of time, used to express sedimentation coefficients, equal to 10^{-13} s.

Syndiotactic macromolecule - A tactic macromolecule, essentially comprising alternating enantiomeric configurational base units which have chiral or prochiral atoms in the main chain in a unique arrangement with respect to their adjacent constitutional units. In this case the repeating unit consists of two configurational base units that are enantiomeric. [8]

Tacticity - The orderliness of the succession of configurational repeating units of a macromolecule or oligomer molecule. In a tactic macromolecule essentially all the configurational repeating units are identical with respect to directional sense. See Configurational repeating unit, Isotactic, Syndiotactic. [8]

Tautomerism - Isomerism of the general form G-X-Y=Z ⇌ X=Y-Z-G, where the isomers (called tautomers) are readily interconvertible; the atoms connecting the groups X, Y, Z are typically any of C, H, O, or S, and G is a group which becomes an electrofuge (i.e., a group that does not carry away the bonding electron pair when it leaves its position in the molecule) or nucleofuge (a group that does carry away the bonding electrons when leaving) during isomerization. The commonest case, when the electrofuge is H^+, is also known as prototropy. A common example, written so as to illustrate the general pattern given above, is keto-enol tautomerism, such as

$H\text{-}O\text{-}C(CH_3)=CH\text{-}CO_2Et$ (enol) ⇌ $(CH_3)C(=O)\text{-}CH_2\text{-}CO_2Et$ (keto)

In some cases the interconversion rate between tautomers is slow enough to permit isolation of the separate keto and enol forms. [5]

Tensile strength* - In tensile testing, the ratio of maximum load a body can bear before breaking to original cross-sectional area. Also called ultimate strength. [11]

Terpenes - Hydrocarbons of biological origin having carbon skeletons formally derived from isoprene $[CH_2=C(CH_3)CH=CH_2]$. [5]

Terpenoids - Natural products and related compounds formally derived from isoprene units. They contain oxygen in various functional groups. The skeleton of terpenoids may differ from strict additivity of isoprene units by the loss or shift of a methyl (or other) group. [5]

Tesla (T)* - The SI unit of magnetic flux density (*B*), equal to V s/m². [1]

Thermal conductivity* - Rate of heat flow divided by area and by temperature gradient. [1]

Thermal diffusivity - Thermal conductivity divided by density and by specific heat capacity at constant pressure. [1]

Thermal expansion coefficient (α)* - The linear expansion coefficient is defined by $\alpha_l = (1/l)(dl/dT)$; the volume expansion coefficient by $\alpha_V = (1/V)(dV/dT)$. [1]

Thermionic emission - The emission of electrons from a solid as a result of heat. The effect requires a high enough temperature to impart sufficient kinetic energy to the electrons to exceed the work function of the solid.

Thermodynamic laws - The foundation of the science of thermodynamics:

First law: The internal energy of an isolated system is constant; if energy is supplied to the system in the form of heat dq and work dw, then the change in energy $dU = dq + dw$.

Second law: No process is possible in which the only result is the transfer of heat from a reservoir and its complete conversion to work.

Third law: The entropy of a perfect crystal approaches zero as the thermodynamic temperature approaches zero.

Thermoelectric power - For a bar of a pure material whose ends are at different temperatures, the potential difference divided by the difference in temperature of the ends. See also Seeback effect.

Thermogravimetric analysis (TGA) - See Techniques for Materials Characterization, page 12-1.

Thermosphere - The layer of the Earth's atmosphere extending from the top of the mesosphere (typically 80–90 km above the surface) to about 500 km. It is characterized by a rapid increase in temperature with increasing altitude up to about 200 km, followed by a leveling off in the 300–500 km region.

Thiols - Compounds having the structure RSH (R not equal to H). Also known by the term mercaptans (abandoned by IUPAC); e.g., CH_3CH_2SH, ethanethiol. [5]

Thomson coefficient (μ, τ) - The heat power developed in the Thomson effect (whereby heat is evolved in a conductor when a current is flowing in the presence of a temperature gradient), divided by the current and the temperature difference. [1]

Tonne (t) - An alternative name for megagram (1000 kg). [1]

Torque (*T*) - For a force ***F*** that produces a torsional motion, $\boldsymbol{T} = \boldsymbol{r} \times \boldsymbol{F}$, where ***r*** is a vector from some reference point to the point of application of the force.

Torr - A non-SI unit of pressure, equal to 133.322 Pa. The name is generally considered interchangeable with millimeter of mercury.

Townsend coefficient - In a radiation counter, the number of ionizing collisions by an electron per unit path length in the direction of an applied electric field.

Transducer - Any device that converts a signal from acoustical, optical, or some other form of energy into an electrical signal (or vice versa) while preserving the information content of the original signal.

Transistor - A voltage amplifier using controlled electron currents inside a semiconductor.

Transition metals - Elements characterized by a partially filled *d* subshell. The First Transition Series comprises Sc, Ti, V, Cr, Mn, Fe, Co, Ni, Cu. The Second and Third Transition Series include the lanthanides and actinides, respectively. [7]

Transition probability* - See Einstein transition probability.

Transmittance (τ) - Ratio of the radiant or luminous flux at a given wavelength that is transmitted to that of the incident radiation. Also called transmission factor. [1]

Tribology - The study of frictional forces between solid surfaces.

Triple point* - The point in p,T space where the solid, liquid, and gas phases of a substance are in thermodynamic equilibrium. The corresponding temperature and pressure are called the triple point temperature and triple point pressure.

Troposphere - The lowest part of the earth's atmosphere, extending to 10–15 km above the surface. It is characterized by a decrease in temperature with increasing altitude. The exact height varies with latitude and season.

Tunnel diode - A device involving a p-n junction in which both sides are so heavily doped that the Fermi level on the p-side lies in the valence band and on the n-side in the conduction band. This leads to a current-voltage curve with a maximum, so that the device exhibits a negative resistance in some regions.

Ultraviolet photoelectron spectroscopy (UPS) - See Techniques for Materials Characterization, page 12-1.

Umklapp process - A process involving the interaction of three or more waves (lattice or electron) in a solid in which the sum of the wave vectors does not equal zero.

Unified atomic mass unit (u)* - A unit of mass used in atomic, molecular, and nuclear science, defined as the mass of one atom of ^{12}C divided by 12. Its approximate value is 1.66054×10^{-27} kg. [1]

Universal time (t_U, UT) - Mean solar time counted from midnight at the Greenwich meridian. Also called Greenwich mean time (GMT). The interval of mean solar time is based on the average, over one year, of the time between successive transits of the sun across the observer's meridian.

Vacancy - A missing atom or ion in a crystal lattice.

Van Allen belts - Two toroidal regions above the earth's atmosphere containing protons and electrons. The outer belt at about 25,000 km above the surface is probably of solar origin. The inner belt at about 3000 km contains more energetic particles from outside the solar system.

Van der Waals' equation* - An equation of state for fluids which takes the form:

$$pV_m = RT\left(\frac{1}{V_m - b} - \frac{a}{V_m^{\ 2}}\right)$$

where p is pressure, V_m is molar volume, T is temperature, R is the molar gas constant, and a and b are characteristic parameters of the substance which describe the effect of attractive and repulsive intermolecular forces, respectively.

Van der Waals' force - The weak attractive force between two molecules which arises from electric dipole interactions. It can lead to the formation of stable but weakly bound dimer molecules or clusters.

Van't Hoff equation - The equation expressing the temperature dependence of the equilibrium constant K of a chemical reaction:

$$\frac{d\ln K}{dT} = \frac{\Delta_r H^\circ}{RT^2}$$

where $\Delta_r H^\circ$ is the standard enthalpy of reaction, R the molar gas constant, and T the temperature. Also called van't Hoff isochore.

Vapor pressure* - The pressure of a gas in equilibrium with a liquid (or, in some usage, a solid) at a specified temperature.

Varistor - A device that utilizes the properties of certain metal oxides with small amounts of impurities, which show abrupt nonlinearities at specific voltages where the material changes from a semiconductor to an insulator.

Velocity (v) - Rate of change of distance with time.

Verdet constants (V)* - Angle of rotation of a plane polarized light beam passing through a medium in a magnetic field, divided by the field strength and by the path length.

Virial equation of state* - An equation relating the pressure p, molar volume V_m, and temperature T of a real gas in the form of an expansion in powers of the molar volume, viz., $pV_m = RT(1+BV_m^{-1}+CV_m^{-2}+ ...)$, where R is the molar gas constant. B is called the second virial coefficient, C the third virial coefficient, etc. The virial coefficients are functions of temperature.

Viscosity (η)* - The proportionality factor between sheer rate and sheer stress, defined through the equation $F = \eta\, A(dv/dx)$, where F is the tangential force required to move a planar surface of area A at velocity v relative to a parallel surface separated from the first by a distance x. Sometimes called dynamic or absolute viscosity. The term kinematic viscosity (symbol ν) is defined as η divided by the mass density.

Volt (V)* - The SI unit of electric potential, equal to W/A. [1]

Volume fraction (ϕ_j) - Defined as $V_j/\Sigma_i V_i$, where V_j is the volume of the specified component and the V_i are the volumes of all the components of a mixture prior to mixing. [2]

Watt (W)* - The SI unit of power, equal to J/s. [1]

Wave function - A function of the coordinates of all the particles in a quantum mechanical system (and, in general, of time) which fully describes the state of the system. The product of the wave function and its complex conjugate is proportional to the probability of finding a particle at a particular point in space.

Weak interaction - The weak forces (order of 10^{-12} of the strong interaction) between elementary particles which are responsible for beta decay and other nuclear effects.

Weber (Wb)* - The SI unit of magnetic flux, equal to V s. [1]

Weber number (*We*) - A dimensionless quantity used in fluid mechanics, defined by $We = \rho v^2 l/\gamma$, where ρ is density, v is velocity, l is length, and γ is surface tension. [2]

Weight - That force which, when applied to a body, would give it an acceleration equal to the local acceleration of gravity. [1]

Wiedeman-Franz law - The law stating that the thermal conductivity k and electrical conductivity σ of a pure metal are related by $k = L\sigma T$, where T is the temperature and L (called the Lorenz ratio) has the approximate value 2.45×10^{-8} V²/K².

Wien displacement law - The relation, which can be derived from the Planck formula for black body radiation, that

$\lambda_{max}T = 0.0028978$ m K, where λ_{max} is the wavelength of maximum radiance at temperature T.

Wigner-Seitz method - A method of calculating electron energy levels in a solid using a model in which each electron is subject to a spherically symmetric potential.

Wittig reagents - See phosphonium ylides.

Work (*W*) - Force multiplied by the displacement in the direction of the force. [1]

Work function (Φ)* - The energy difference between an electron at rest at infinity and an electron at the Fermi level in the interior of a substance. It is thus the minimum energy required to remove an electron from the interior of a solid to a point just outside the surface. [1]

X unit (X) - A unit of length used in x-ray crystallography, equal to approximately 1.002×10^{-13} m.

X-ray photoelectron spectroscopy (XPS) - See Techniques for Materials Characterization, page 12-1.

Yield strength - The stress at which a material exhibits a specified deviation (often chosen as 0.2% for metals) from proportionality of stress and strain. [11]

Young's modulus (*E*) - In tension or compression of a body below its elastic limit, the ratio of stress to corresponding strain. Since strain is normally expressed on a fractional basis, Young's modulus has dimensions of pressure. Also called elastic modulus. [11]

Zeeman effect - The splitting of an energy level of an atom or molecule, and hence a splitting of spectral lines arising from that level, as a result of the application of an external magnetic field.

Zener diode - A control device utilizing a p-n junction with a well defined reverse-bias avalanche breakdown voltage.

Zeotrope - A liquid mixture that shows no maximum or minimum when vapor pressure is plotted against composition at constant temperature. See Azeotrope.

Zero-point energy - The energy possessed by a quantum mechanical system as a result of the uncertainty principle even when it is in its lowest energy state; e.g., the difference between the lowest energy level of a harmonic oscillator and the minimum in the potential well.

Zeta potential (ζ) - The electric potential at the surface of a colloidal particle relative to the potential in the bulk medium at a long distance. Also called electrokinetic potential.

Zwitterions - Neutral compounds having formal unit electrical charges of opposite sign. Some chemists restrict the term to compounds with the charges on non-adjacent atoms. Sometimes referred to as inner salts, dipolar ions (a misnomer). [5]

THERMODYNAMIC FUNCTIONS AND RELATIONS

p = pressure V = volume T = temperature
n_i = amount of substance i
$x_i = n_i/\Sigma_j n_j$ = mole fraction of substance i

Energy	U
Entropy	S
Enthalpy	$H = U + pV$
Helmholtz energy	$A = U - TS$
Gibbs energy	$G = U + pV - TS$
Isobaric heat capacity	$C_p = (\partial H/\partial T)_p$
Isochoric heat capacity	$C_V = (\partial U/\partial T)_V$
Isobaric expansivity	$\alpha = V^{-1}(\partial V/\partial T)_p$
Isothermal compressibility	$\kappa_T = \ V^{-1}(\partial V/\partial p)_T$
Isentropic compressibility	$\kappa_S = -V^{-1}(\partial V/\partial p)_S$
	$\kappa_T - \kappa_S = T\alpha^2 V/C_p$
	$C_p \quad C_V = T\alpha^2 V/\kappa_T$
Gibbs-Helmholtz equation	$H = G - T(\partial G/\partial T)_p$
Maxwell relations	$(\partial S/\partial p)_T = -(\partial V/\partial T)_p$
	$(\partial S/\partial V)_T = -(\partial p/\partial T)_V$
Joule-Thomson expansion	$\mu_{JT} = (\partial T/\partial p)_H = -\{V - T(\partial V/\partial T)_p\}/C_p$
	$\phi_{JT} - (\partial H/\partial p)_T = V - T(\partial V/\partial T)_p$
Partial molar quantity	$X_i = (\partial X/\partial n_i)_{T,p,nj\neq i}$
Chemical potential	$\mu_i = (\partial G/\partial n_i)_{T,p,nj\neq i}$
Perfect gas [symbol ^{pg}]	$pV = (\Sigma_i n_i)RT$
	$\mu_i^{pg} = \mu_i^\theta + RT\ln(x_i p/p^\theta)$
Fugacity	$f_i = (x_i p)\exp\{(\mu_i - \mu_i^{pg})/RT\}$
Activity coefficient	$\gamma_i = f_i/(x_i f_i^\theta)$
Gibbs-Duhem relation	$0 = S\mathrm{d}T - V\mathrm{d}p + \Sigma_i n_i \mathrm{d}\mu_i$

[Superscript θ in above equations indicates standard state]

Notation for chemical and physical changes ($X = H, S, G$, etc.):

Chemical reaction	$\Delta_r X$
Formation from elements	$\Delta_f X$
Combustion	$\Delta_c X$
Fusion (cry→liq)	$\Delta_{fus} X$
Vaporization (liq→gas)	$\Delta_{vap} X$
Sublimation (cry→gas)	$\Delta_{sub} X$
Phase transition	$\Delta_{trs} X$
Solution	$\Delta_{sol} X$
Mixing	$\Delta_{mix} X$
Dilution	$\Delta_{dil} X$

NOBEL LAUREATES IN CHEMISTRY AND PHYSICS

Full details on nationality and basis of the awards can be found at <nobelprize.org/>.

Chemistry

2009 Venkatraman Ramakrishnan, Thomas A. Steitz, Ada E. Yonath
2008 Martin Chalfie, Osamu Shimomura, Roger Y. Tsien
2007 Gerhard Ertl
2006 Roger D. Kornberg
2005 Yves Chauvin, Robert H. Grubbs, Richard R. Schrock
2004 Aaron Ciechanover, Avram Hershko, Irwin Rose
2003 Peter Agre, Roderick MacKinnon
2002 John B. Fenn, Koichi Tanaka, Kurt Wüthrich
2001 William S. Knowles, Ryoji Noyori, K. Barry Sharpless
2000 Alan Heeger, Alan G. MacDiarmid, Hideki Shirakawa
1999 Ahmed Zewail
1998 Walter Kohn, John Pople
1997 Paul D. Boyer, John E. Walker, Jens C. Skou
1996 Robert F. Curl Jr., Sir Harold Kroto, Richard E. Smalley
1995 Paul J. Crutzen, Mario J. Molina, F. Sherwood Rowland
1994 George A. Olah
1993 Kary B. Mullis, Michael Smith
1992 Rudolph A. Marcus
1991 Richard R. Ernst
1990 Elias James Corey
1989 Sidney Altman, Thomas R. Cech
1988 Johann Deisenhofer, Robert Huber, Hartmut Michel
1987 Donald J. Cram, Jean-Marie Lehn, Charles J. Pedersen
1986 Dudley R. Herschbach, Yuan T. Lee, John C. Polanyi
1985 Herbert A. Hauptman, Jerome Karle
1984 Bruce Merrifield
1983 Henry Taube
1982 Aaron Klug
1981 Kenichi Fukui, Roald Hoffmann
1980 Paul Berg, Walter Gilbert, Frederick Sanger
1979 Herbert C. Brown, Georg Wittig
1978 Peter Mitchell
1977 Ilya Prigogine
1976 William Lipscomb
1975 John Cornforth, Vladimir Prelog
1974 Paul J. Flory
1973 Ernst Otto Fischer, Geoffrey Wilkinson
1972 Christian Anfinsen, Stanford Moore, William H. Stein
1971 Gerhard Herzberg
1970 Luis Leloir
1969 Derek Barton, Odd Hassel
1968 Lars Onsager
1967 Manfred Eigen, Ronald G.W. Norrish, George Porter
1966 Robert S. Mulliken
1965 Robert B. Woodward
1964 Dorothy Crowfoot Hodgkin
1963 Karl Ziegler, Giulio Natta
1962 Max F. Perutz, John C. Kendrew
1961 Melvin Calvin
1960 Willard F. Libby
1959 Jaroslav Heyrovsky
1958 Frederick Sanger
1957 Lord Todd
1956 Sir Cyril Hinshelwood, Nikolay Semenov
1955 Vincent du Vigneaud
1954 Linus Pauling
1953 Hermann Staudinger
1952 Archer J.P. Martin, Richard L.M. Synge
1951 Edwin M. McMillan, Glenn T. Seaborg
1950 Otto Diels, Kurt Alder
1949 William F. Giauque
1948 Arne Tiselius
1947 Sir Robert Robinson
1946 James B. Sumner, John H. Northrop, Wendell M. Stanley
1945 Artturi Virtanen
1944 Otto Hahn
1943 George de Hevesy
1942 *No prize awarded*
1941 *No prize awarded*
1940 *No prize awarded*
1939 Adolf Butenandt, Leopold Ruzicka
1938 Richard Kuhn
1937 Norman Haworth, Paul Karrer
1936 Peter Debye
1935 Frédéric Joliot, Irène Joliot-Curie
1934 Harold C. Urey
1933 *No prize awarded*
1932 Irving Langmuir
1931 Carl Bosch, Friedrich Bergius
1930 Hans Fischer
1929 Arthur Harden, Hans von Euler-Chelpin
1928 Adolf Windaus
1927 Heinrich Wieland
1926 The Svedberg
1925 Richard Zsigmondy
1924 *No prize awarded*
1923 Fritz Pregl
1922 Francis W. Aston
1921 Frederick Soddy
1920 Walther Nernst
1919 *No prize awarded*
1918 Fritz Haber
1917 *No prize awarded*
1916 *No prize awarded*
1915 Richard Willstätter
1914 Theodore W. Richards
1913 Alfred Werner
1912 Victor Grignard, Paul Sabatier
1911 Marie Curie
1910 Otto Wallach
1909 Wilhelm Ostwald
1908 Ernest Rutherford
1907 Eduard Buchner
1906 Henri Moissan
1905 Adolf von Baeyer
1904 Sir William Ramsay
1903 Svante Arrhenius
1902 Emil Fischer
1901 Jacobus H. van't Hoff

Physics

2009 Charles K. Kao, Willard S. Boyle, George E. Smith
2008 Makoto Kobayashi, Toshihide Maskawa, Yoichiro Nambu
2007 Albert Fert, Peter Grünberg
2006 John C. Mather, George F. Smoot
2005 Roy J. Glauber, John L. Hall, Theodor W. Hänsch
2004 David J. Gross, H. David Politzer, Frank Wilczek
2003 Alexei A. Abrikosov, Vitaly L. Ginzburg, Anthony J. Leggett
2002 Raymond Davis Jr., Masatoshi Koshiba, Riccardo Giacconi
2001 Eric A. Cornell, Wolfgang Ketterle, Carl E. Wieman
2000 Zhores I. Alferov, Herbert Kroemer, Jack S. Kilby
1999 Gerardus't Hooft, Martinus J.G. Veltman
1998 Robert B. Laughlin, Horst L. Störmer, Daniel C. Tsui
1997 Steven Chu, Claude Cohen-Tannoudji, William D. Phillips
1996 David M. Lee, Douglas D. Osheroff, Robert C. Richardson
1995 Martin L. Perl, Frederick Reines
1994 Bertram N. Brockhouse, Clifford G. Shull
1993 Russell A. Hulse, Joseph H. Taylor Jr.
1992 Georges Charpak
1991 Pierre-Gilles de Gennes
1990 Jerome I. Friedman, Henry W. Kendall, Richard E. Taylor
1989 Norman F. Ramsey, Hans G. Dehmelt, Wolfgang Paul
1988 Leon M. Lederman, Melvin Schwartz, Jack Steinberger
1987 J. Georg Bednorz, K. Alex Müller
1986 Ernst Ruska, Gerd Binnig, Heinrich Rohrer
1985 Klaus von Klitzing
1984 Carlo Rubbia, Simon van der Meer
1983 Subramanyan Chandrasekhar, William A. Fowler
1982 Kenneth G. Wilson
1981 Nicolaas Bloembergen, Arthur L. Schawlow, Kai M. Siegbahn
1980 James Cronin, Val Fitch
1979 Sheldon Glashow, Abdus Salam, Steven Weinberg
1978 Pyotr Kapitsa, Arno Penzias, Robert Woodrow Wilson
1977 Philip W. Anderson, Sir Nevill F. Mott, John H. van Vleck
1976 Burton Richter, Samuel C.C. Ting
1975 Aage N. Bohr, Ben R. Mottelson, James Rainwater
1974 Martin Ryle, Antony Hewish
1973 Leo Esaki, Ivar Giaever, Brian D. Josephson
1972 John Bardeen, Leon N. Cooper, Robert Schrieffer
1971 Dennis Gabor
1970 Hannes Alfvén, Louis Néel
1969 Murray Gell-Mann
1968 Luis Alvarez
1967 Hans Bethe
1966 Alfred Kastler
1965 Sin-Itiro Tomonaga, Julian Schwinger, Richard P. Feynman
1964 Charles H. Townes, Nicolay G. Basov, Aleksandr M. Prokhorov
1963 Eugene Wigner, Maria Goeppert-Mayer, J. Hans D. Jensen
1962 Lev Landau
1961 Robert Hofstadter, Rudolf Mössbauer
1960 Donald A. Glaser
1959 Emilio Segrè, Owen Chamberlain
1958 Pavel A. Cherenkov, Il'ja M. Frank, Igor Y. Tamm
1957 Chen Ning Yang, Tsung-Dao Lee
1956 William B. Shockley, John Bardeen, Walter H. Brattain
1955 Willis E. Lamb, Polykarp Kusch
1954 Max Born, Walther Bothe
1953 Frits Zernike
1952 Felix Bloch, E. M. Purcell
1951 John Cockcroft, Ernest T.S. Walton
1950 Cecil Powell
1949 Hideki Yukawa
1948 Patrick M.S. Blackett
1947 Edward V. Appleton
1946 Percy W. Bridgman
1945 Wolfgang Pauli
1944 Isidor Isaac Rabi
1943 Otto Stern
1942 *No prize awarded*
1941 *No prize awarded*
1940 *No prize awarded*
1939 Ernest Lawrence
1938 Enrico Fermi
1937 Clinton Davisson, George Paget Thomson
1936 Victor F. Hess, Carl D. Anderson
1935 James Chadwick
1934 *No prize awarded*
1933 Erwin Schrödinger, Paul A.M. Dirac
1932 Werner Heisenberg
1931 *No prize awarded*
1930 Sir Venkata Raman
1929 Louis de Broglie
1928 Owen Willans Richardson
1927 Arthur H. Compton, C.T.R. Wilson
1926 Jean Baptiste Perrin
1925 James Franck, Gustav Hertz
1924 Manne Siegbahn
1923 Robert A. Millikan
1922 Niels Bohr
1921 Albert Einstein
1920 Charles Edouard Guillaume
1919 Johannes Stark
1918 Max Planck
1917 Charles Glover Barkla
1916 *No prize awarded*
1915 William Bragg, Lawrence Bragg
1914 Max von Laue
1913 Heike Kamerlingh Onnes
1912 Gustaf Dalén
1911 Wilhelm Wien
1910 Johannes Diderik van der Waals
1909 Guglielmo Marconi, Ferdinand Braun
1908 Gabriel Lippmann
1907 Albert A. Michelson
1906 J.J. Thomson
1905 Philipp Lenard
1904 Lord Rayleigh
1903 Henri Becquerel, Pierre Curie, Marie Curie
1902 Hendrik A. Lorentz, Pieter Zeeman
1901 Wilhelm Conrad Röntgen

Physics

2009 Charles K. Kao, Willard S. Boyle, George E. Smith
2008 Makoto Kobayashi, Toshihide Maskawa, Yoichiro Nambu
2007 Albert Fert, Peter Grünberg
2006 John C. Mather, George F. Smoot
2005 Roy J. Glauber, John L. Hall, Theodor W. Hänsch
2004 David J. Gross, H. David Politzer, Frank Wilczek
2003 Alexei A. Abrikosov, Vitaly L. Ginzburg,
Anthony J. Leggett
2002 Raymond Davis Jr., Masatoshi Koshiba,
Riccardo Giacconi
2001 Eric A. Cornell, Wolfgang Ketterle, Carl E. Wieman
2000 Zhores I. Alferov, Herbert Kroemer, Jack S. Kilby
1999 Gerardus 't Hooft, Martinus J.G. Veltman
1998 Robert B. Laughlin, Horst L. Störmer, Daniel C. Tsui
1997 Steven Chu, Claude Cohen-Tannoudji,
William D. Phillips
1996 David M. Lee, Douglas D. Osheroff,
Robert C. Richardson
1995 Martin L. Perl, Frederick Reines
1994 Bertram N. Brockhouse, Clifford G. Shull
1993 Russell A. Hulse, Joseph H. Taylor Jr.
1992 Georges Charpak
1991 Pierre-Gilles de Gennes
1990 Jerome I. Friedman, Henry W. Kendall, Richard E. Taylor
1989 Norman F. Ramsey, Hans G. Dehmelt, Wolfgang Paul
1988 Leon M. Lederman, Melvin Schwartz, Jack Steinberger
1987 J. Georg Bednorz, K. Alex Müller
1986 Ernst Ruska, Gerd Binnig, Heinrich Rohrer
1985 Klaus von Klitzing
1984 Carlo Rubbia, Simon van der Meer
1983 Subramanyan Chandrasekhar, William A. Fowler
1982 Kenneth G. Wilson
1981 Nicolaas Bloembergen, Arthur L. Schawlow,
Kai M. Siegbahn
1980 James Cronin, Val Fitch
1979 Sheldon Glashow, Abdus Salam, Steven Weinberg
1978 Pyotr Kapitsa, Arno Penzias, Robert Woodrow Wilson
1977 Philip W. Anderson, Sir Nevill F. Mott, John H. van Vleck
1976 Burton Richter, Samuel C.C. Ting
1975 Aage N. Bohr, Ben R. Mottelson, James Rainwater
1974 Martin Ryle, Antony Hewish
1973 Leo Esaki, Ivar Giaever, Brian D. Josephson
1972 John Bardeen, Leon N. Cooper, Robert Schrieffer
1971 Dennis Gabor
1970 Hannes Alfvén, Louis Néel
1969 Murray Gell-Mann
1968 Luis Alvarez
1967 Hans Bethe
1966 Alfred Kastler
1965 Sin-Itiro Tomonaga, Julian Schwinger,
Richard P. Feynman
1964 Charles H. Townes, Nicolay G. Basov,
Aleksandr M. Prokhorov
1963 Eugene Wigner, Maria Goeppert-Mayer, J. Hans D. Jensen
1962 Lev Landau
1961 Robert Hofstadter, Rudolf Mössbauer
1960 Donald A. Glaser
1959 Emilio Segrè, Owen Chamberlain
1958 Pavel A. Cherenkov, Il'ja M. Frank, Igor Y. Tamm
1957 Chen Ning Yang, Tsung-Dao Lee
1956 William B. Shockley, John Bardeen, Walter H. Brattain
1955 Willis E. Lamb, Polykarp Kusch
1954 Max Born, Walther Bothe
1953 Frits Zernike
1952 Felix Bloch, E. M. Purcell
1951 John Cockcroft, Ernest T.S. Walton
1950 Cecil Powell
1949 Hideki Yukawa
1948 Patrick M.S. Blackett
1947 Edward V. Appleton
1946 Percy W. Bridgman
1945 Wolfgang Pauli
1944 Isidor Isaac Rabi
1943 Otto Stern
1942 No prize awarded
1941 No prize awarded
1940 No prize awarded
1939 Ernest Lawrence
1938 Enrico Fermi
1937 Clinton Davisson, George Paget Thomson
1936 Victor F. Hess, Carl D. Anderson
1935 James Chadwick
1934 No prize awarded
1933 Erwin Schrödinger, Paul A.M. Dirac
1932 Werner Heisenberg
1931 No prize awarded
1930 Sir Venkata Raman
1929 Louis de Broglie
1928 Owen Willans Richardson
1927 Arthur H. Compton, C.T.R. Wilson
1926 Jean Baptiste Perrin
1925 James Franck, Gustav Hertz
1924 Manne Siegbahn
1923 Robert A. Millikan
1922 Niels Bohr
1921 Albert Einstein
1920 Charles Edouard Guillaume
1919 Johannes Stark
1918 Max Planck
1917 Charles Glover Barkla
1916 No prize awarded
1915 William Bragg, Lawrence Bragg
1914 Max von Laue
1913 Heike Kamerlingh Onnes
1912 Gustaf Dalén
1911 Wilhelm Wien
1910 Johannes Diderik van der Waals
1909 Guglielmo Marconi, Ferdinand Braun
1908 Gabriel Lippmann
1907 Albert A. Michelson
1906 J.J. Thomson
1905 Philipp Lenard
1904 Lord Rayleigh
1903 Henri Becquerel, Pierre Curie, Marie Curie
1902 Hendrik A. Lorentz, Pieter Zeeman
1901 Wilhelm Conrad Röntgen

Section 3
Physical Constants of Organic Compounds

Physical Constants of Organic Compounds 3-1
Synonym Index of Organic Compounds 3-524
Diamagnetic Susceptibility of Selected Organic Compounds 3-549

PHYSICAL CONSTANTS OF ORGANIC COMPOUNDS

The basic physical constants and structure diagrams for about 10,900 organic compounds are presented in this table. An effort has been made to include the compounds most frequently encountered in the laboratory, the workplace, and the environment. Particular emphasis has been given to substances that are considered environmental or human health hazards. In making the selection of compounds for the table, added weight was assigned to the appearance of a compound in various lists or reference sources such as:

- Laboratory reagent lists, e.g., the ACS *Reagent Chemicals* volume (Ref. 1)
- The DIPPR list of industrially important compounds (Ref. 2) and the (much larger) TSCA inventory of chemicals used in commerce
- The *Hazardous Substances Data Bank* (Ref. 3)
- The UNEP list *Persistent Organic Pollutants* (Ref. 4)
- Chemicals on Reporting Rules (CORR), a database of about 7500 regulated compounds prepared by the Environmental Protection Agency (Ref. 5)
- The EPA Integrated Risk Information System (IRIS), a database of human health effects of exposure to chemicals in the environment (Ref. 6)
- Compendia of chemicals of biochemical or medical importance, such as *The Merck Index* (Ref. 10)
- Specialized tables in this *Handbook*

It should be noted that the above lists vary widely in their choice of chemical names, and even in the use of Chemical Abstracts Registry Numbers. To the extent possible, we have attempted to systematize the names and registry numbers for this table.

Clearly, criteria of this type are somewhat subjective, and compounds considered important by some users have undoubtedly been omitted. Suggestions for additional compounds or other improvements are welcomed.

The data in the table have been derived from many sources, including both the primary literature and evaluated compilations. The *Handbook of Data on Organic Compounds, Third Edition* (Ref. 7) and the *Chapman & Hall/CRC Combined Chemical Dictionary* (Ref. 8) were important sources. Other useful compilations of physical property data for organic compounds are listed in Refs. 9–19. Many boiling point values (and some melting point and density values) were taken from recent physical chemistry literature dealing with fluid properties. Where conflicts were found, the value deemed most reliable was chosen.

The table is arranged alphabetically by substance name, which generally is either an IUPAC systematic name or, in the case of pesticides, pharmaceuticals, and other complex compounds, a simple trivial name. Names in ubiquitous use, such as acetic acid and formaldehyde, are adopted rather than their systematic equivalents. Synonyms are given in the column following the primary name, and structure diagrams are given on the page facing the data listing. The explanation of the data columns follows:

- **No.**: An identification number used in the indexes.
- **Name**: Primary name of the substance.
- **Synonym**: A synonym in common use. When the primary name is non-systematic, a systematic name may appear here.
- **Mol. Form.**: The molecular formula written in the Hill convention.
- **CAS RN**: The Chemical Abstracts Service Registry Number for the compound.
- **Mol. Wt.**: Molecular weight (relative molar mass) as calculated with the 2001 IUPAC Standard Atomic Weights.
- **Physical Form**: A notation of the physical phase, color, crystal type, or other features of the compound at ambient temperature. Abbreviations are given below.
- **mp**: Normal melting point in °C. A value is sometimes followed by "dec," indicating decomposition is observed at the stated temperature (so that it is probably not a true melting point). The notation "tp" indicates a triple point, where solid, liquid, and gas are in equilibrium.
- **bp**: Normal boiling point in °C, if it is available. This is the temperature at which the liquid phase is in equilibrium with the vapor at a pressure of 760 mmHg (101.325 kPa). A notation "sp" following the value indicates a sublimation point, where the vapor pressure of the solid phase reaches 760 mmHg. When a notation such as "dec" or "exp" (explodes) follows the value, the temperature may not be a true boiling point. A simply entry "sub" indicates the solid has a significant sublimation pressure at ambient temperatures. The boiling point at reduced pressure is listed in some cases, with or without the normal boiling point. Here the superscript indicates the pressure in mmHg.
- **den**: Density (mass per unit volume) in g/cm^3. The temperature in °C is indicated by a superscript. Values refer to the liquid or solid phase, and all values are true densities, not specific gravities. The number of decimal places gives a rough estimate of the accuracy of the value.
- n_D: Refractive index, at the temperature in °C indicated by the superscript. Unless otherwise indicated, all values refer to a wavelength of 589 nm (sodium D line). Values are given only for liquids and solids.
- **Solubility**: Qualitative indication of solubility in common solvents. Abbreviations are:

i	insoluble
sl	slightly soluble
s	soluble
vs	very soluble
msc	miscible
dec	decomposes

Abbreviations for solvents are given below.

In order to facilitate the location of compounds in the table, an index to synonyms is provided. Indexes to Molecular Formulas and CAS Registry Numbers are available in the electronic versions of the *Handbook* or by request via e-mail (william.haynes@taylorandfrancis.com).

- **Synonym Index**: Includes common synonyms, but not the primary name by which the table is arranged.

The assistance of Fiona Macdonald in checking names and formulas is gratefully acknowledged, as well as the efforts of Janice Shackleton, Trupti Desai, Nazila Kamaly, Matt Griffiths, and Lawrence Braschi in preparing the structure diagrams.

List of Abbreviations

Ac	acetyl
Ac_2O	acetic anhydride
AcOEt	ethyl acetate
ac	acid
ace	acetone
al	alcohol (ethanol)
alk	alkali
amor	amorphous
anh	anhydrous
aq	aqueous
bipym	bipyramidal
bl	blue
blk	black
bp	boiling point
br	brown
bt	bright
Bu	butyl
BuOH	1-butanol
bz	benzene
chl	chloroform
col	colorless
con, conc	concentrated
cry	crystals
ctc	carbon tetrachloride
cy, cyhex	cyclohexane
dec	decomposes
den	density
dil	dilute
diox	dioxane
dk	dark
DMF	dimethylformamide
DMSO	dimethyl sulfoxide
efflor	efflorescent
Et	ethyl
EtOH	ethanol
eth	diethyl ether
exp	explodes
fl	flakes
flr	fluorescent
fum	fumes, fuming
gl	glacial
gr	gray
gran	granular
grn	green
hex	hexagonal
HOAc	acetic acid
hp	heptane
hx	hexane
hyd	hydrate
hyg	hygroscopic
i	insoluble
i-	iso-
iso	isooctane
lf	leaves
lig	ligroin
liq	liquid
lo	long
mcl	monoclinic
Me	methyl
MeCN	acetonitrile
MeOH	methanol
misc, msc	miscible
mp	melting point
n	refractive index
nd	needles
oct	octahedra, octahedral
oran	orange
orth	orthorhombic
os	organic solvents
pa	pale
peth	petroleum ether
Ph	phenyl
PhCl	chlorobenzene
$PhNH_2$	aniline
$PhNO_2$	nitrobenzene
pl	plates
pow	powder
Pr	propyl
PrOH	1-propanol
pr	prisms
purp	purple
py	pyridine
pym	pyramids, pyramidal
reac	reacts
rhom	rhombic
s	soluble
sat	saturated
sc	scales
sl	slightly soluble
soln	solution
sp	sublimation point
stab	stable
sub	sublimes
sulf	sulfuric acid
syr	syrup
tab	tablets
tcl	triclinic
tetr	tetragonal
tfa	trifluoroacetic acid
thf, THF	tetrahydrofuran
tol	toluene
tp	triple point
trg	trigonal
unstab	unstable
vap	vapor
viol	violet
visc	viscous
vol	volatile
vs	very soluble
w	water
wh	white
xyl	xylene
ye	yellow

References

1. American Chemical Society, *Reagent Chemicals, Ninth Edition*, Oxford University Press, New York, 2000.
2. Daubert, T. E., Danner, R. P., Sibul, H. M., and Stebbins, C. C., *Physical and Thermodynamic Properties of Pure Compounds: Data Compilation*, extant 2002 (core with supplements), Taylor & Francis, Bristol, PA.
3. National Library of Medicine, *Hazardous Substances Data Bank*, <http://toxnet.nlm.nih.gov/cgi-bin/sis/htmlgen?HSDB>.
4. United Nations Environmental Program, *Persistent Organic Pollutants*, <http://www.chem.unep.ch/pops/>.

5. Environmental Protection Agency, *Chemicals on Reporting Rules*, <http://www.epa.gov/opptintr/CORR>.
6. Environmental Protection Agency, *Integrated Risk Information System*, <http://www.epa.gov/iris/index.html>.
7. Lide, D. R., and Milne, G. W. A., Editors, *Handbook of Data on Organic Compounds, Third Edition*, CRC Press, Boca Raton, FL, 1993.
8. Macdonald, F., Editor, *Chapman & Hall/CRC Combined Chemical Dictionary*, <http://www.chemnetbase.com/scripts/ccdweb.exe>.
9. Linstrom, P. J., and Mallard, W. G., Editors, *NIST Chemistry WebBook*, NIST Standard Reference Database No. 69, July 2001, National Institute of Standards and Technology, Gaithersburg, MD 20899, <http://webbook.nist.gov>.
10. Thermodynamics Research Center, National Institute of Standards and Technology, *TRC Thermodynamic Tables*, <http:/trc.nist.gov>.
11. O'Neil, M. J., Editor, *The Merck Index, Thirteenth Edition*, Merck & Co., Rahway, NJ, 2001.
12. Stevenson, R. M., and Malanowski, S., *Handbook of the Thermodynamics of Organic Compounds*, Elsevier, New York, 1987.
13. Riddick, J. A., Bunger, W. B., and Sakano, T. K., *Organic Solvents*, Fourth Edition, John Wiley & Sons, New York, 1986.
14. *Physical Constants of Hydrocarbon and Non-Hydrocarbon Compounds*, ASTM Data Series DS 4B, ASTM, Philadelphia, 1988.
15. *Beilstein Database*, <http://www.mdli.com/products/xfirebeilstein.html>.
16. *Landolt-Börnstein Numerical Data and Functional Relationships in Science and Technology*, <http://www.landolt-boernstein.com>.
17. Vargaftik, N.B., Vinogradov, Y. K., and Yargin, V. S., *Handbook of Physical Properties of Liquids and Gases, Third Edition*, Begell House, New York, 1996
18. Lide, D. R., and Kehiaian, H. V., *Handbook of Thermophysical and Thermochemical Data*, CRC Press, Boca Raton, FL, 1994.
19. Lide, D. R., Editor, *Properties of Organic Compounds*, <http://www.chemnetbase.com/scripts/pocweb.exe>.

No.	Name	Synonym	Mol. Form.	CAS RN	Mol. Wt.	Physical Form	mp/°C	bp/°C	den/ g cm-3	n_D	Solubility
1	Abate	Temephos	$C_{16}H_{20}O_6P_2S_3$	3383-96-8	466.469		30		1.32		
2	Abietic acid		$C_{20}H_{30}O_2$	514-10-3	302.451	mcl pl (al-w)	173.5	250^{9}			vs ace, bz, eth, EtOH
3	Abscisic acid		$C_{15}H_{20}O_4$	21293-29-8	264.318	cry (chl-peth)	160	sub 120			vs ace, eth, chl
4	Acacetin	5,7-Dihydroxy-2-(4-methoxyphenyl)-4*H*-1-benzopyran-4-one	$C_{16}H_{12}O_5$	480-44-4	284.263	ye nd (95% al)	263				vs EtOH
5	Acebutolol, (±)		$C_{18}H_{28}N_2O_4$	37517-30-9	336.426	cry	121				
6	Acedapsone		$C_{16}H_{16}N_2O_4S$	77-46-3	332.374	pa ye nd (eth) lf (dil al)	290				sl H_2O
7	Acenaphthene	1,2-Dihydroacenaphthylene	$C_{12}H_{10}$	83-32-9	154.207		93.4	279	1.222^{20}	1.6048^{95}	i H_2O; sl EtOH, chl; vs bz; s HOAc
8	Acenaphthylene	Acenaphthalene	$C_{12}H_8$	208-96-8	152.192		91.8	280	0.8987^{16}		i H_2O; vs EtOH, eth, bz; sl chl
9	1,2-Acenaphthylenedione		$C_{12}H_6O_2$	82-86-0	182.175	ye nd (HOAc)	261	sub	1.4800^{20}		i H_2O; sl EtOH, bz, HOAc; s lig
10	Acenocoumarol	Nicoumalone	$C_{19}H_{15}NO_6$	152-72-7	353.325	cry (ace aq)	198				i H_2O
11	Acephate	Phosphoramidothioic acid, acetyl-, *O,S*-dimethyl ester	$C_4H_{10}NO_3PS$	30560-19-1	183.166		88		1.35^{20}		
12	Acepromazine		$C_{19}H_{22}N_2OS$	61-00-7	326.455	oran oil		$230^{0.5}$			
13	Acesulfame		$C_4H_5NO_4S$	33665-90-6	163.153	nd (bz)	123.2				s bz, chl
14	Acetaldehyde	Ethanal	C_2H_4O	75-07-0	44.052	vol liq or gas	-123.37	20.1	0.7834^{18}	1.3316^{20}	msc H_2O, EtOH, eth, bz; sl chl
15	Acetaldehyde phenylhydrazone		$C_8H_{10}N_2$	935-07-9	134.178		99.5	150^{40}, 135^{21}			vs EtOH
16	Acetaldoxime	Acetaldehyde oxime	C_2H_5NO	107-29-9	59.067	nd	45	115	0.9656^{20}	1.4264^{20}	s H_2O, chl; msc EtOH, eth
17	Acetamide	Ethanamide	C_2H_5NO	60-35-5	59.067	trg mcl (al-eth)	80.16	222.0	0.9986^{85}	1.4278	vs H_2O, EtOH
18	Acetanilide	*N*-Phenylacetamide	C_8H_9NO	103-84-4	135.163		114.3	304	1.2190^{15}		sl H_2O; vs EtOH, ace; s eth, s bz, tol
19	Acetazolamide	*N*-[5-(Aminosulfonyl)-1,3,4-thiadiazol-2-yl]acetamide	$C_4H_6N_4O_3S_2$	59-66-5	222.246		260.5				sl H_2O
20	Acethion		$C_8H_{17}O_4PS_2$	919-54-0	272.322	liq		$137^{1.5}$	1.18^{20}		
21	Acetic acid	Ethanoic acid	$C_2H_4O_2$	64-19-7	60.052		16.64	117.9	1.0446^{25}	1.3720^{20}	msc H_2O, EtOH, eth, ace, bz; s chl, CS_2
22	Acetic acid, 2-phenylhydrazide		$C_8H_{10}N_2O$	114-83-0	150.177	hex pr (eth)	130.0				vs H_2O, EtOH; sl eth, chl, tfa; s bz
23	Acetic anhydride		$C_4H_6O_3$	108-24-7	102.089	liq	-74.1	139.5	1.082^{20}	1.3901^{20}	vs H_2O; s EtOH, bz; msc eth; sl ctc
24	Acetoacetanilide		$C_{10}H_{11}NO_2$	102-01-2	177.200	pr or nd (bz or lig)	86				sl H_2O; s EtOH, eth, bz, chl, acid, lig
25	Acetoacetic acid		$C_4H_6O_3$	541-50-4	102.089	cry (eth)	36.5	dec 100			vs H_2O, eth, EtOH
26	2-Acetoacetoxyethyl methacrylate	2-(Methacryloyloxy)ethyl acetoacetate	$C_{10}H_{14}O_5$	21282-97-3	214.215	liq		$100^{0.8}$	1.122	1.4560^{20}	
27	Acetochlor		$C_{14}H_{20}ClNO_2$	34256-82-1	269.768	ye liq		$134^{0.4}$		1.5272^{20}	sl H_2O
28	Acetohexamide		$C_{15}H_{20}N_2O_4S$	968-81-0	324.396	cry (EtOH aq)	188				i H_2O, eth; sl EtOH, chl; s py
29	Acetohydrazide		$C_2H_6N_2O$	1068-57-1	74.081		67	137^{25}			s H_2O, EtOH; sl eth
30	Acetohydroxamic acid	*N*-Hydroxyacetamide	$C_2H_5NO_2$	546-88-3	75.067	hyg cry	90				
31	1-Acetonaphthone		$C_{12}H_{10}O$	941-98-0	170.206		34	297	1.1171^{21}	1.6280^{22}	i H_2O; s EtOH, eth, ace, chl
32	2-Acetonaphthone		$C_{12}H_{10}O$	93-08-3	170.206	nd (lig, dil al)	56	302			sl EtOH, ctc
33	Acetone	2-Propanone	C_3H_6O	67-64-1	58.079	liq	-94.7	56.05	0.7845^{25}	1.3588^{20}	msc H_2O, EtOH, eth, ace, bz, chl
34	Acetone cyanohydrin		C_4H_7NO	75-86-5	85.105		-19	82^{23}	0.932^{19}	1.3992^{20}	vs H_2O, EtOH, eth; s ace, bz, chl; i peth
35	Acetone (2,4-dinitrophenyl) hydrazone		$C_9H_{10}N_4O_4$	1567-89-1	238.200	ye nd or pl (al)	128				i H_2O; s EtOH, eth, bz, chl, AcOEt
36	Acetone (1-methylethylidene) hydrazone	Dimethyl ketazine	$C_6H_{12}N_2$	627-70-3	112.172	liq	-12.5	133	0.8390^{20}	1.4535^{20}	msc H_2O, EtOH, eth; s ace
37	Acetone thiosemicarbazide		$C_4H_9N_3S$	1752-30-3	131.199	ye cry	176				s ace

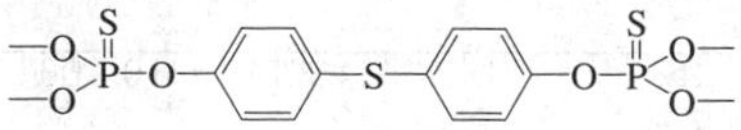

Abate

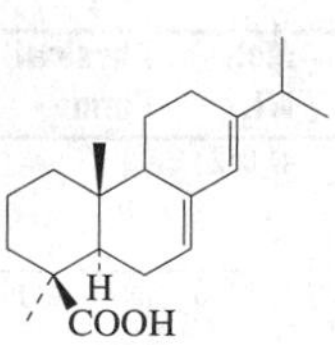

Abietic acid

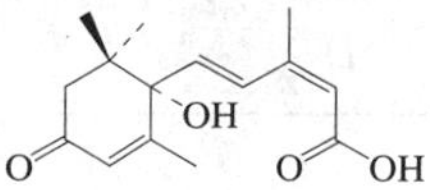

Abscisic acid

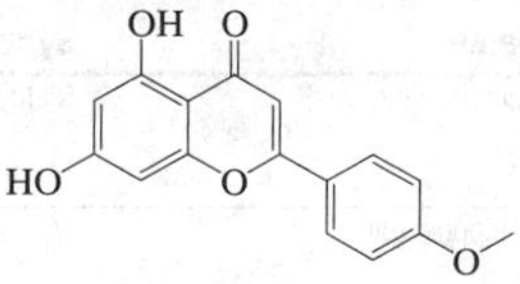

Acacetin

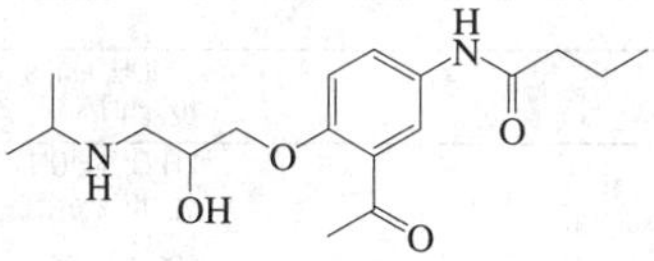

Acebutolol, (±)

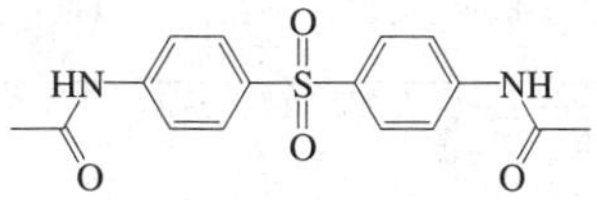

Acedapsone

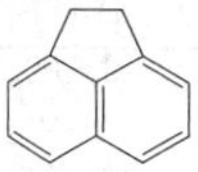

Acenaphthene

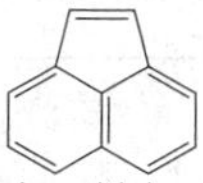

Acenaphthylene

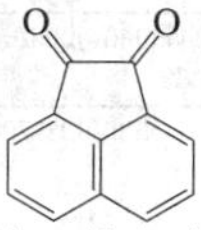

1,2-Acenaphthylenedione

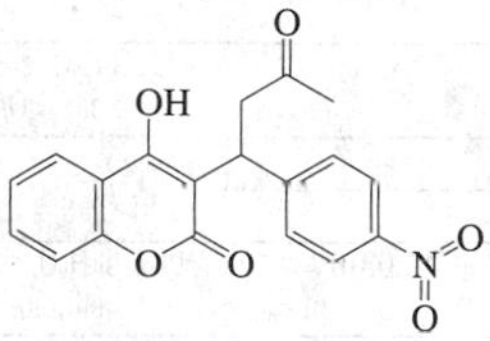

Acenocoumarol

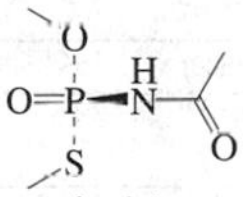

Acephate

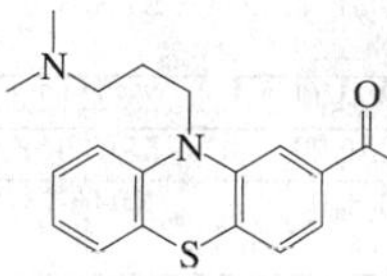

Acepromazine

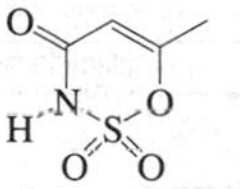

Acesulfame

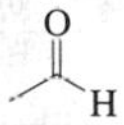

Acetaldehyde

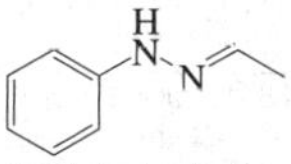

Acetaldehyde phenylhydrazone

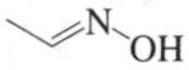

Acetaldoxime

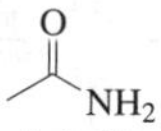

Acetamide

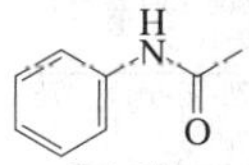

Acetanilide

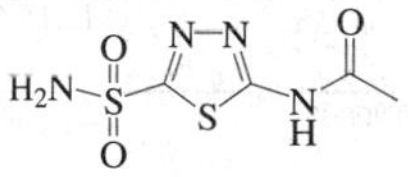

Acetazolamide

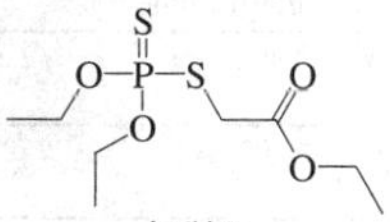

Acethion

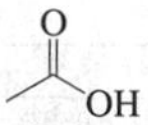

Acetic acid

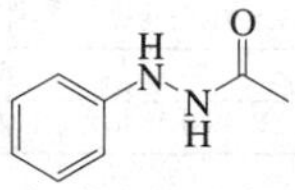

Acetic acid, 2-phenylhydrazide

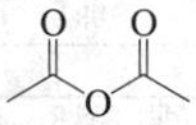

Acetic anhydride

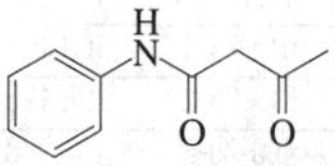

Acetoacetanilide

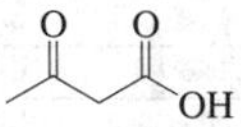

Acetoacetic acid

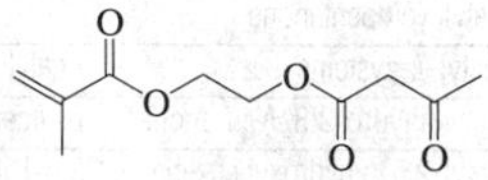

2-Acetoacetoxyethyl methacrylate

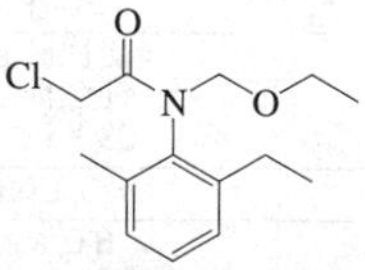

Acetochlor

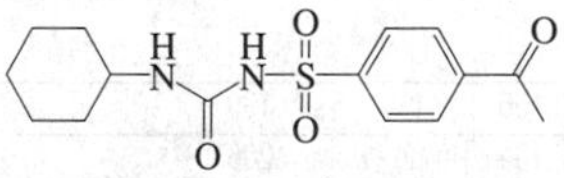

Acetohexamide

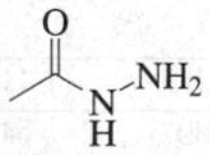

Acetohydrazide

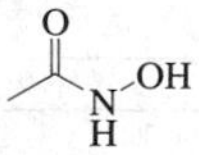

Acetohydroxamic acid

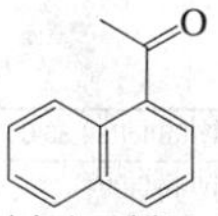

1-Acetonaphthone

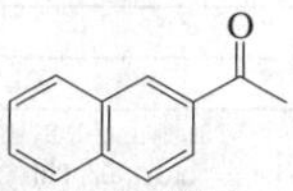

2-Acetonaphthone

Acetone

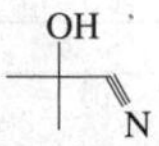

Acetone cyanohydrin

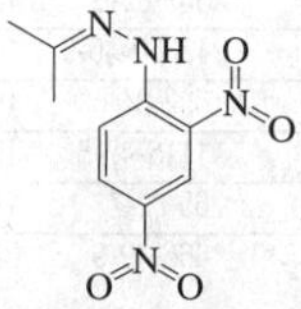

Acetone (2,4-dinitrophenyl)hydrazone

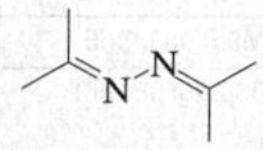

Acetone (1-methylethylidene)hydrazone

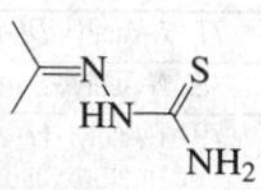

Acetone thiosemicarbazide

No.	Name	Synonym	Mol. Form.	CAS RN	Mol. Wt.	Physical Form	mp/°C	bp/°C	den/ g cm-3	n_D	Solubility
38	Acetonitrile	Methyl cyanide	C_2H_3N	75-05-8	41.052	liq	-43.82	81.65	0.7857^{20}	1.3442^{30}	msc H_2O, EtOH, eth, ace, bz, ctc
39	Acetophenone	Methyl phenyl ketone	C_8H_8O	98-86-2	120.149	mcl pr or pl	20.5	202	1.0281^{20}	1.5372^{20}	sl H_2O; s EtOH, eth, ace, bz, con sulf, chl
40	Acetophenone azine	Methylphenyl ketazine	$C_{16}H_{16}N_2$	729-43-1	236.311		120				
41	Acetoxon	Acetophos	$C_8H_{17}O_5PS$	2425-25-4	256.257	liq		$73^{0.005}$			
42	*N*-Acetylacetamide		$C_4H_7NO_2$	625-77-4	101.105	nd (eth)	79	223.5			s H_2O, EtOH, eth, chl, lig
43	*N*-Acetyl-*L*-alanine		$C_5H_9NO_3$	97-69-8	131.130		125				
44	4-(Acetylamino)benzenesulfonyl chloride	Acetylsulfanilyl chloride	$C_8H_8ClNO_3S$	121-60-8	233.673	nd (bz), pr (bz-chl)	149				vs EtOH, eth; s bz, chl
45	2-(Acetylamino)benzoic acid		$C_9H_9NO_3$	89-52-1	179.172	nd (HOAc)	187.5				sl H_2O; s EtOH; vs eth, ace, bz, HOAc
46	4-(Acetylamino)benzoic acid		$C_9H_9NO_3$	556-08-1	179.172	nd (HOAc)	256.5				i H_2O; s EtOH; sl eth, tfa
47	2-(Acetylamino)-2-deoxy-*D*-glucose	*N*-Acetyl-*D*-glucosamine	$C_8H_{15}NO_6$	7512-17-6	221.208		205				
48	2-(Acetylamino)-2-deoxy-*D*-mannose	*N*-Acetyl-*D*-mannosamine	$C_8H_{15}NO_6$	3615-17-6	221.208	cry (ace aq)	128				dec alk
49	2-(Acetylamino)fluorene		$C_{15}H_{13}NO$	53-96-3	223.270	cry (dil al)	193				i H_2O; s EtOH, eth, HOAc
50	4-(Acetylamino)fluorene		$C_{15}H_{13}NO$	28322-02-3	223.270	br cry (bz)	200				
51	6-(Acetylamino)hexanoic acid	ε-Acetamidocaproic acid	$C_8H_{15}NO_3$	57-08-9	173.210	cry (ace)	104.5				
52	4-Acetylanisole		$C_9H_{10}O_2$	100-06-1	150.174	pl (peth)	38.5	258	1.0818^{41}	1.547^{41}	sl H_2O; s EtOH, eth, ace, chl
53	2-Acetylbenzoic acid		$C_9H_8O_3$	577-56-0	164.158	nd (w), pr (bz)	114.5	111^{2}			vs H_2O, eth, EtOH
54	3-Acetylbenzoic acid		$C_9H_8O_3$	586-42-5	164.158		172	111^{2}			s H_2O; msc EtOH
55	4-Acetylbenzoic acid		$C_9H_8O_3$	586-89-0	164.158	nd (w)	208	sub			vs H_2O
56	Acetyl benzoylperoxide	Acetozone	$C_9H_8O_4$	644-31-5	180.158	wh nd (lig)	37	130^{19}			vs eth
57	Acetyl bromide	Ethanoyl bromide	C_2H_3BrO	506-96-7	122.948	liq	-96	76	1.6625^{16}	1.4486^{20}	msc eth, bz, chl; s ace
58	Acetyl chloride	Ethanoyl chloride	C_2H_3ClO	75-36-5	78.497	liq	-112.8	50.7	1.1051^{20}	1.3886^{20}	msc eth, ace, bz, chl; s ctc
59	Acetylcholine bromide		$C_7H_{16}BrNO_2$	66-23-9	226.112	hyg cry	146				vs H_2O
60	Acetylcholine chloride		$C_7H_{16}ClNO_2$	60-31-1	181.661		150				s H_2O, EtOH; i eth
61	Acetylcholine iodide		$C_7H_{16}INO_2$	2260-50-6	273.112	hyg	163				
62	2-Acetylcyclohexanone		$C_8H_{12}O_2$	874-23-7	140.180		-11	112^{18}, 101^{14}	1.0782^{25}	1.5138^{20}	s ctc
63	2-Acetylcyclopentanone		$C_7H_{10}O_2$	1670-46-8	126.153			73^{20}	1.0431^{25}	1.4906^{20}	
64	*N*-Acetyl-*L*-cysteine	Acetylcysteine	$C_5H_9NO_3S$	616-91-1	163.195	cry (w)	109.5				
65	3-Acetyldihydro-2(3*H*)-furanone	α-Acetylbutyrolactone	$C_6H_8O_3$	517-23-7	128.126			107^{5}	1.1846^{20}	1.4585^{20}	vs H_2O
66	1-Acetyl-2,5-dihydroxybenzene	2,5-Dihydroxyacetophenone	$C_8H_8O_3$	490-78-8	152.148	ye grn nd (dil al or w)	205.3				sl H_2O, eth, bz; s EtOH
67	Acetylene	Ethyne	C_2H_2	74-86-2	26.037	col gas	-80.7 (triple point)	-84.7 sp	0.377^{25} (p>1 atm)		sl H_2O, EtOH, CS_2; s ace, bz, chl
68	*N*-Acetylethanolamine		$C_4H_9NO_2$	142-26-7	103.120		63.5	166^{8}	1.1079^{25}	1.4674^{20}	msc H_2O; s ace; sl bz, lig
69	Acetyl fluoride	Ethanoyl fluoride	C_2H_3FO	557-99-3	62.042	vol liq or gas	-84	20.8	1.032^{25}		msc EtOH, eth; s bz, chl; sl CS_2
70	*N*-Acetylglutamic acid		$C_7H_{11}NO_5$	1188-37-0	189.166	pr (w)	199				s H_2O, EtOH
71	*N*-Acetylglycine	Aceturic acid	$C_4H_7NO_3$	543-24-8	117.104	lo nd (w, MeOH)	206				vs H_2O, ace, EtOH
72	*trans*-1-Acetyl-4-hydroxy-*L*-proline	Oxaceprol	$C_7H_{11}NO_4$	33996-33-7	173.167	cry (Ac)	132				vs H_2O, MeOH
73	1-Acetyl-1*H*-imidazole		$C_5H_6N_2O$	2466-76-4	110.114		104.5				sl H_2O; s EtOH, eth, chl, THF
74	Acetyl iodide	Ethanoyl iodide	C_2H_3IO	507-02-8	169.948			108	2.0673^{20}	1.5491^{20}	vs eth
75	Acetyl isothiocyanate		C_3H_3NOS	13250-46-9	101.127			132.5	1.1523^{13}	1.5231^{18}	s eth, CS_2
76	*N6*-Acetyl-*L*-lysine		$C_8H_{16}N_2O_3$	692-04-6	188.224		265 dec				
77	*N*-Acetyl-*DL*-methionine		$C_7H_{13}NO_3S$	1115-47-5	191.248		114.5				
78	*N*-Acetyl-*L*-methionine	Methionamine	$C_7H_{13}NO_3S$	65-82-7	191.248		105.5				
79	1-Acetyl-17-methoxyaspidospermidine	Aspidospermine	$C_{22}H_{30}N_2O_2$	466-49-9	354.485	nd or pr (al) nd (peth)	208	220^{2}			sl H_2O, eth; s EtOH, bz, chl
80	*N*-Acetyl-*N*-methylacetamide		$C_5H_9NO_2$	1113-68-4	115.131	liq	-25	195; 114.5^{61}	1.0663^{25}	1.4502^{25}	msc H_2O; i eth
81	1-Acetyl-3-methylpiperidine		$C_8H_{15}NO$	4593-16-2	141.211	liq	-13.6	239	0.9684^{25}	1.4731^{25}	vs H_2O
82	3-Acetyl-6-methyl-2*H*-pyran-2,4(3*H*)-dione	Dehydroacetic acid	$C_8H_8O_4$	520-45-6	168.148		109	270			vs H_2O, eth; sl EtOH, chl

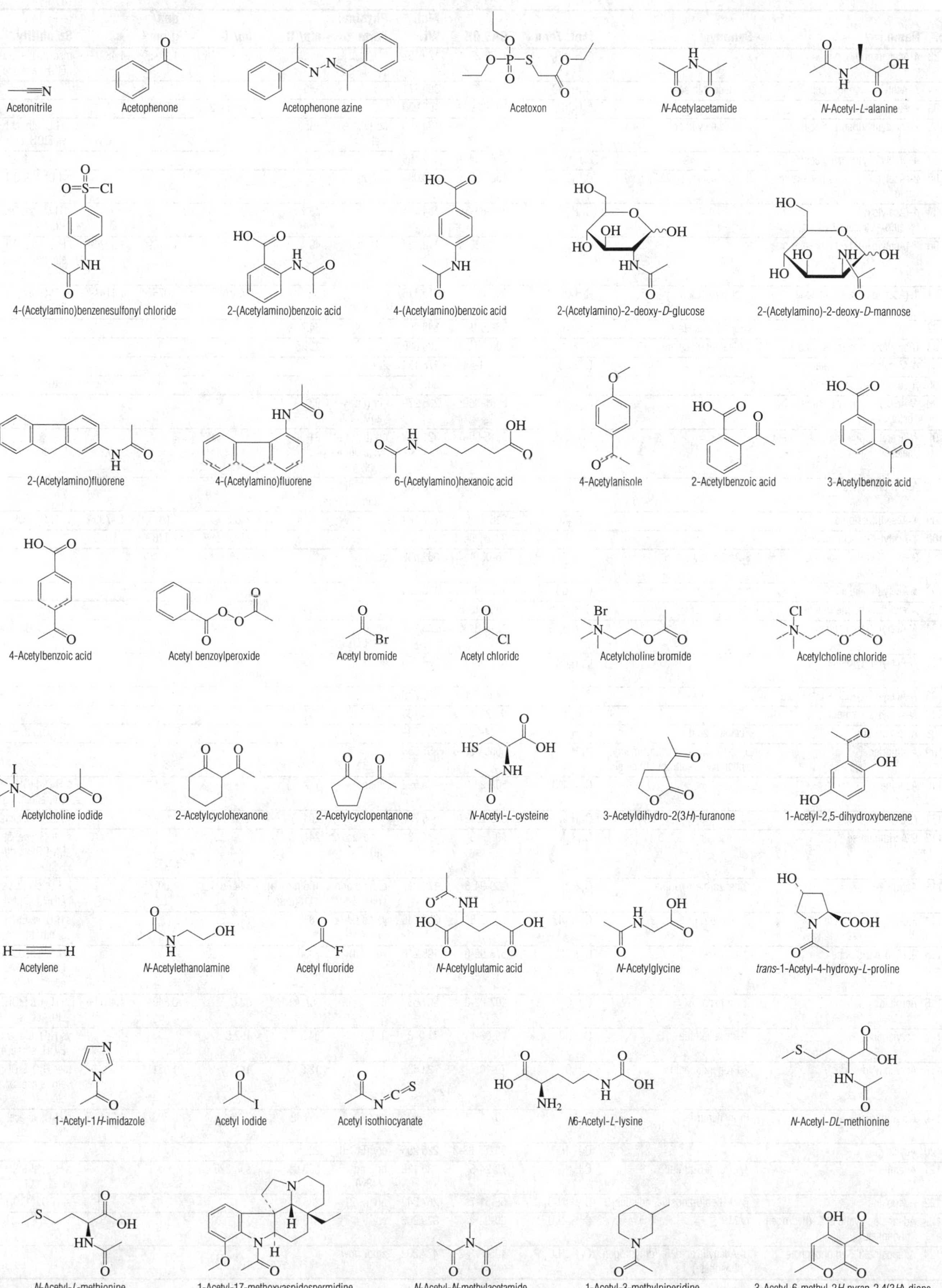

Acetonitrile
Acetophenone
Acetophenone azine
Acetoxon
N-Acetylacetamide
N-Acetyl-L-alanine
4-(Acetylamino)benzenesulfonyl chloride
2-(Acetylamino)benzoic acid
4-(Acetylamino)benzoic acid
2-(Acetylamino)-2-deoxy-D-glucose
2-(Acetylamino)-2-deoxy-D-mannose
2-(Acetylamino)fluorene
4-(Acetylamino)fluorene
6-(Acetylamino)hexanoic acid
4-Acetylanisole
2-Acetylbenzoic acid
3-Acetylbenzoic acid
4-Acetylbenzoic acid
Acetyl benzoylperoxide
Acetyl bromide
Acetyl chloride
Acetylcholine bromide
Acetylcholine chloride
Acetylcholine iodide
2-Acetylcyclohexanone
2-Acetylcyclopentanone
N-Acetyl-L-cysteine
3-Acetyldihydro-2(3H)-furanone
1-Acetyl-2,5-dihydroxybenzene
Acetylene
N-Acetylethanolamine
Acetyl fluoride
N-Acetylglutamic acid
N-Acetylglycine
trans-1-Acetyl-4-hydroxy-L-proline
1-Acetyl-1H-imidazole
Acetyl iodide
Acetyl isothiocyanate
N6-Acetyl-L-lysine
N-Acetyl-DL-methionine
N-Acetyl-L-methionine
1-Acetyl-17-methoxyaspidospermidine
N-Acetyl-N-methylacetamide
1-Acetyl-3-methylpiperidine
3-Acetyl-6-methyl-2H-pyran-2,4(3H)-dione

No.	Name	Synonym	Mol. Form.	CAS RN	Mol. Wt.	Physical Form	mp/°C	bp/°C	den/g cm^{-3}	n_D	Solubility
83	4-Acetylmorpholine		$C_6H_{11}NO_2$	1696-20-4	129.157		14.5	152^{50}, 118^{12}	1.1145^{20}	1.4827^{20}	msc H_2O; s EtOH, ace, ctc
84	*N*-Acetylneuraminic acid	Aceneuramic acid	$C_{11}H_{19}NO_9$	131-48-6	309.271		186				
85	Acetyl nitrate		$C_2H_3NO_4$	591-09-3	105.050			exp 60; 22^{70}	1.24^{15}		
86	2-(Acetyloxy)benzoic acid	Acetylsalicylic acid	$C_9H_8O_4$	50-78-2	180.158	nd (w), mcl tab (w)	135				s H_2O, eth, chl; vs EtOH; sl bz
87	4-(Acetyloxy)benzoic acid		$C_9H_8O_4$	2345-34-8	180.158		188.5				
88	2-(Acetyloxy)-5-bromobenzoic acid	5-Bromoacetylsalicylic acid	$C_9H_7BrO_4$	1503-53-3	259.054	nd (al)	60				i H_2O; vs EtOH, eth
89	4-(Acetyloxy)-3-methoxybenzaldehyde		$C_{10}H_{10}O_4$	881-68-5	194.184		78				sl H_2O; vs EtOH, eth
90	2-(Acetyloxy)-1-phenylethanone		$C_{10}H_{10}O_3$	2243-35-8	178.184	orth pl	49	270	1.1169^{65}	1.5036^{65}	i H_2O; vs EtOH, eth, chl; sl bz, lig
91	1-(Acetyloxy)-2-propanone	Acetoxyacetone	$C_5H_8O_3$	592-20-1	116.116			171; 63^{11}	1.0757^{20}	1.4141^{20}	vs H_2O, eth, EtOH
92	(Acetyloxy)tributylstannane	Tributyltin acetate	$C_{14}H_{30}O_2Sn$	56-36-0	349.097		84.7				
93	(Acetyloxy)triphenylstannane	Triphenyltin acetate	$C_{20}H_{18}O_2Sn$	900-95-8	409.066		121.5				
94	4-Acetylphenyl acetate		$C_{10}H_{10}O_3$	13031-43-1	178.184						s ctc, CS_2
95	*N*-Acetyl-*L*-phenylalanine		$C_{11}H_{13}NO_3$	2018-61-3	207.226		173.5				s EtOH
96	*N*-Acetyl-*L*-phenylalanine, ethyl ester		$C_{13}H_{17}NO_3$	2361-96-8	235.279	cry (EtOH aq)	93				
97	*N*-Acetyl-*L*-phenylalanine, methyl ester		$C_{12}H_{15}NO_3$	3618-96-0	221.252	nd (peth) or visc oil (chl)	91				
98	Acetyl phosphate		$C_2H_5O_5P$	590-54-5	140.032	unstab in soln					
99	1-Acetylpiperidine		$C_7H_{13}NO$	618-42-8	127.184	liq	-13.4	226.5	1.011^{9}	1.4790^{25}	vs H_2O, EtOH
100	1-Acetyl-4-piperidinone		$C_7H_{11}NO_2$	32161-06-1	141.168			218; $124^{0.2}$	1.146^{25}	1.5026^{20}	
101	3-Acetylpyridine adenine dinucleotide	3-Acetyl NAD	$C_{22}H_{28}N_6O_{14}P_2$	86-08-8	662.436	solid					
102	4-Acetylthioanisole		$C_9H_{10}OS$	1778-09-2	166.239		81.5				
103	Acetyl thiocholine iodide		$C_7H_{16}INOS$	1866-15-5	289.177		205				
104	*N*-Acetyl-*L*-tryptophan		$C_{13}H_{14}N_2O_3$	1218-34-4	246.261	nd (dil MeOH)	189.5				s H_2O, EtOH, alk
105	*N*-Acetyl-*L*-tyrosine		$C_{11}H_{13}NO_4$	537-55-3	223.226	cry (w); pl (diox)	153				
106	*N*-Acetyl-*L*-tyrosine ethyl ester		$C_{13}H_{17}NO_4$	840-97-1	251.279		80.5				
107	*N*-Acetyl-*L*-valine		$C_7H_{13}NO_3$	96-81-1	159.183		164				
108	Acid Fuchsin	Fuchsin, acid	$C_{20}H_{17}N_3Na_2O_9S_3$	3244-88-0	585.539						sl H_2O, EtOH
109	Acifluorfen	5-[2-Chloro-4-(trifluoromethyl) phenoxy]-2-nitrobenzoic acid	$C_{14}H_7ClF_3NO_5$	50594-66-6	361.658		150				
110	Aconine		$C_{25}H_{41}NO_9$	509-20-6	499.596	amor	132				s H_2O, EtOH, chl; sl eth, lig
111	Aconitine		$C_{34}H_{47}NO_{11}$	302-27-2	645.737	orth lf	204				vs bz, EtOH, chl
112	9-Acridinamine	Aminacrine	$C_{13}H_{10}N_2$	90-45-9	194.231	ye nd (ace or al)	241				s EtOH, ace; sl DMSO; vs dil HCl
113	Acridine	Dibenzo[b,e]pyridine	$C_{13}H_9N$	260-94-6	179.217	orth nd or pr (al)	106(form a); 110(form b)	344.86	1.005^{20}		i H_2O; sl ctc; vs EtOH, eth, bz
114	3,6-Acridinediamine	Proflavine	$C_{13}H_{11}N_3$	92-62-6	209.246	ye nd (al or w)	285				s H_2O; vs EtOH; sl eth, bz
115	9(10*H*)-Acridinone		$C_{13}H_9NO$	578-95-0	195.216	ye lf (al)	>300				i H_2O, eth, bz; sl EtOH; s HOAc, alk
116	Acrolein	2-Propenal	C_3H_4O	107-02-8	56.063	liq	-87.7	52.6	0.840^{20}	1.4017^{20}	vs H_2O; s EtOH, eth, ace; sl chl
117	Acrylamide	2-Propenamide	C_3H_5NO	79-06-1	71.078	lf (bz)	84.5	192.6			vs H_2O, chl; s EtOH, eth, ace
118	Acrylic acid	2-Propenoic acid	$C_3H_4O_2$	79-10-7	72.063		12.5	141	1.0511^{20}	1.4224^{20}	msc H_2O, EtOH, eth; s ace, bz, ctc
119	Acrylonitrile	Propenenitrile	C_3H_3N	107-13-1	53.063	liq	-83.48	77.3	0.8007^{25}	1.3911^{20}	s H_2O; vs ace, bz, eth, EtOH
120	Acyclovir		$C_8H_{11}N_5O_3$	59277-89-3	225.205	cry (EtOH)	225				
121	Adenine	1*H*-Purin-6-amine	$C_5H_5N_5$	73-24-5	135.128	orth nd (+3w)	360 dec	sub 220			s H_2O; sl EtOH; i eth, chl
122	Adenosine	β-*D*-Ribofuranoside, adenine-9	$C_{10}H_{13}N_5O_4$	58-61-7	267.242	n(w+3/2)	235.5				sl H_2O; i EtOH
123	Adenosine cyclic 3′,5′-(hydrogen phosphate)	cAMP	$C_{10}H_{12}N_5O_6P$	60-92-4	329.206	cry	219				
124	Adenosine 3′,5′-diphosphate	3′-Adenylic acid, 5′-(dihydrogen phosphate)	$C_{10}H_{15}N_5O_{10}P_2$	1053-73-2	427.202	amor pow					

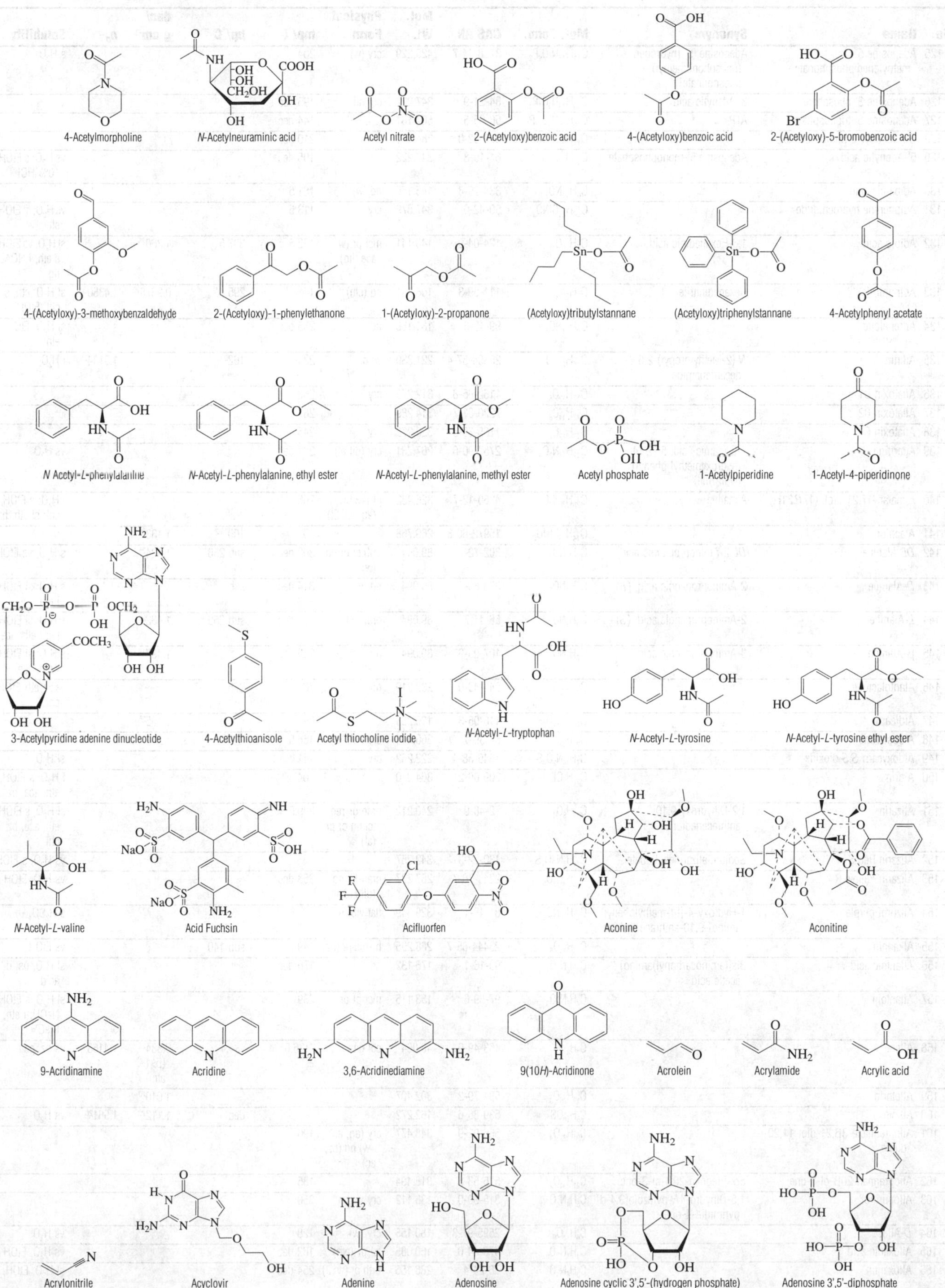
4-Acetylmorpholine
N-Acetylneuraminic acid
Acetyl nitrate
2-(Acetyloxy)benzoic acid
4-(Acetyloxy)benzoic acid
2-(Acetyloxy)-5-bromobenzoic acid
4-(Acetyloxy)-3-methoxybenzaldehyde
2-(Acetyloxy)-1-phenylethanone
1-(Acetyloxy)-2-propanone
(Acetyloxy)tributylstannane
(Acetyloxy)triphenylstannane
4-Acetylphenyl acetate
N-Acetyl-L-phenylalanine
N-Acetyl-L-phenylalanine, ethyl ester
N-Acetyl-L-phenylalanine, methyl ester
Acetyl phosphate
1-Acetylpiperidine
1-Acetyl-4-piperidinone
3-Acetylpyridine adenine dinucleotide
4-Acetylthioanisole
Acetyl thiocholine iodide
N-Acetyl-L-tryptophan
N-Acetyl-L-tyrosine
N-Acetyl-L-tyrosine ethyl ester
N-Acetyl-L-valine
Acid Fuchsin
Acifluorfen
Aconine
Aconitine
9-Acridinamine
Acridine
3,6-Acridinediamine
9(10H)-Acridinone
Acrolein
Acrylamide
Acrylic acid
Acrylonitrile
Acyclovir
Adenine
Adenosine
Adenosine cyclic 3′,5′-(hydrogen phosphate)
Adenosine 3′,5′-diphosphate

No.	Name	Synonym	Mol. Form.	CAS RN	Mol. Wt.	Physical Form	mp/°C	bp/°C	den/ g cm-3	n_D	Solubility
125	Adenosine 5'-methylenediphosphonate	Adenosine, 5'-[hydrogen (phosphonomethyl) phosphonate]	$C_{11}H_{17}N_5O_9P_2$	3768-14-7	425.229	cry (w)	204				s H_2O
126	Adenosine 3'-phosphate	3'-Adenylic acid	$C_{10}H_{14}N_5O_7P$	84-21-9	347.222	col nd	195 dec				
127	Adenosine 5'-triphosphate	ATP	$C_{10}H_{16}N_5O_{13}P_3$	56-65-5	507.181		144 dec				
128	*S*-Adenosyl-*L*-homocysteine		$C_{14}H_{20}N_6O_5S$	979-92-0	384.411		210 dec				
129	5'-Adenylic acid	Adenosine 5'-monophosphate	$C_{10}H_{14}N_5O_7P$	61-19-8	347.222		195 dec				vs H_2O; s EtOH, 10% HCl
130	Adipamic acid		$C_6H_{11}NO_3$	334-25-8	145.156	nd (w)	161.5				
131	Adiphenine hydrochloride		$C_{20}H_{26}ClNO_2$	50-42-0	347.879	cry	113.5				vs H_2O; sl EtOH, eth
132	Adipic acid	1,6-Hexanedioic acid	$C_6H_{10}O_4$	124-04-9	146.141	mcl pr (w, ace, lig)	152.5	337.5	1.360^{25}		sl H_2O; vs EtOH; s eth; i HOAc, lig
133	Adiponitrile	Hexanedinitrile	$C_6H_8N_2$	111-69-3	108.141	nd (eth)	1	295	0.9676^{20}	1.4380^{20}	sl H_2O, eth; s chl, EtOH
134	Adrenalone		$C_9H_{11}NO_3$	99-45-6	181.188	nd	235 dec				sl H_2O, EtOH, eth
135	Affinin	*N*-(2-Methylpropyl)-2,6,8-decatrienamide	$C_{14}H_{23}NO$	25394-57-4	221.339	ye oil	23	$162^{0.5}$		1.5134^{25}	i H_2O
136	Aflatoxin B1		$C_{17}H_{12}O_6$	1162-65-8	312.273	cry	268				
137	Aflatoxin B2		$C_{17}H_{14}O_6$	7220-81-7	314.289		287.5				
138	Aflatoxin G1		$C_{17}H_{12}O_7$	1165-39-5	328.273	cry	245				
139	Agaritine	*L*-Glutamic acid, 5-[2-[4-(hydroxymethyl)phenyl] hydrazide]	$C_{12}H_{17}N_3O_4$	2757-90-6	267.281	cry (dil al)	207 dec				vs H_2O
140	Ajmalan-17,21-diol, (17*R*,21α)	Ajmaline	$C_{20}H_{26}N_2O_2$	4360-12-7	326.432	pl (+3.5w) (aq AcOEt)	206				i H_2O; s EtOH, chl; sl eth, bz
141	Alachlor		$C_{14}H_{20}ClNO_2$	15972-60-8	269.768		40	$100^{0.02}$	1.133^{25}		
142	*DL*-Alanine	*DL*-2-Aminopropanoic acid	$C_3H_7NO_2$	302-72-7	89.094	orth pr or nd (w)	300 dec	sub 250	1.424^{25}		s H_2O; vs EtOH
143	*D*-Alanine	2-Aminopropanoic acid, (*R*)	$C_3H_7NO_2$	338-69-2	89.094	nd (w, al)	314 dec	sub			s H_2O; sl EtOH; i eth
144	*L*-Alanine	2-Aminopropanoic acid, (*S*)	$C_3H_7NO_2$	56-41-7	89.094	orth (w)	297 dec	sub 250	1.432^{22}		s H_2O; sl EtOH, py; i eth, ace
145	β-Alanine	3-Aminopropanoic acid	$C_3H_7NO_2$	107-95-9	89.094	nd, orth pr (al)	200 dec		1.437^{19}		s H_2O; sl EtOH; i eth, ace
146	Alantolactone		$C_{15}H_{20}O_2$	546-43-0	232.319	nd	76	275			vs bz, eth, EtOH, chl
147	Aldicarb		$C_7H_{14}N_2O_2S$	116-06-3	190.263		99		1.195^{25}		
148	Aldosterone		$C_{21}H_{28}O_5$	52-39-1	360.444	cry (HOAc)	166.5				
149	Aldoxycarb *S*,*S*-dioxide		$C_7H_{14}N_2O_4S$	1646-88-4	222.262	cry	141				sl H_2O
150	Aldrin		$C_{12}H_8Cl_6$	309-00-2	364.910		104				i H_2O; s EtOH, eth, ace, bz
151	Alizarin	1,2-Dihydroxy-9,10-anthracenedione	$C_{14}H_8O_4$	72-48-0	240.212	oran or red tcl nd or pr (al)	289.5				sl H_2O; s EtOH, eth, ace, bz; i chl
152	Alizarin Red S	Sodium alizarinesulfonate	$C_{14}H_7NaO_7S$	130-22-3	342.257						vs H_2O; s EtOH
153	Alizarin Yellow R		$C_{13}H_9N_3O_5$	2243-76-7	287.227	oran-br nd (dil HOAc)	253 dec				vs H_2O, EtOH
154	Alizurol purple	1-Hydroxy-4-[(4-methylphenyl) amino]-9,10-anthracenedione	$C_{21}H_{15}NO_3$	81-48-1	329.349	flat viol nd					s H_2SO_4
155	Alkannin		$C_{16}H_{16}O_5$	23444-65-7	288.295	br-red pr (bz)	149	sub 140			vs EtOH
156	Allantoic acid	Bis[(aminocarbonyl)amino] acetic acid	$C_4H_8N_4O_4$	99-16-1	176.132	nd	170 dec				sl H_2O, os, dil acid
157	Allantoin		$C_4H_6N_4O_3$	97-59-6	158.116	mcl pl or	239				sl H_2O; s EtOH, NaOH; i eth, MeOH
158	Allene		C_3H_4	463-49-0	40.064	col gas	-136.6	-34.4	0.584^{25} (p>1 atm)	1.4168	vs bz, peth
159	Allethrin		$C_{19}H_{26}O_3$	584-79-2	302.407				1.010^{20}		
160	Allicin		$C_6H_{10}OS_2$	539-86-6	162.272			dec	1.112^{20}	1.561^{20}	vs H_2O
161	Allopregnane-3β,21-diol-11,20-dione		$C_{21}H_{32}O_4$	566-02-9	348.477	cry (aq, ac, +w) nd (bz, ac)	190				
162	Allopregnan-20β-ol-3-one	5α-Pregnan-20β-ol-3-one	$C_{21}H_{34}O_2$	516-58-5	318.494		185				
163	Allopurinol	1,5-Dihydro-4*H*-pyrazolo[3,4-d] pyrimidin-4-one	$C_5H_4N_4O$	315-30-0	136.112	cry	350				
164	*D*-Allose		$C_6H_{12}O_6$	2595-97-3	180.155	cry (w)	128				vs H_2O
165	Alloxanic acid		$C_4H_4N_2O_5$	470-44-0	160.085	tcl pr (eth)	162 dec				vs H_2O, EtOH
166	Alloxantin		$C_8H_6N_4O_8$	76-24-4	286.156	orth pr (w+2)	254 dec				sl H_2O, EtOH, eth

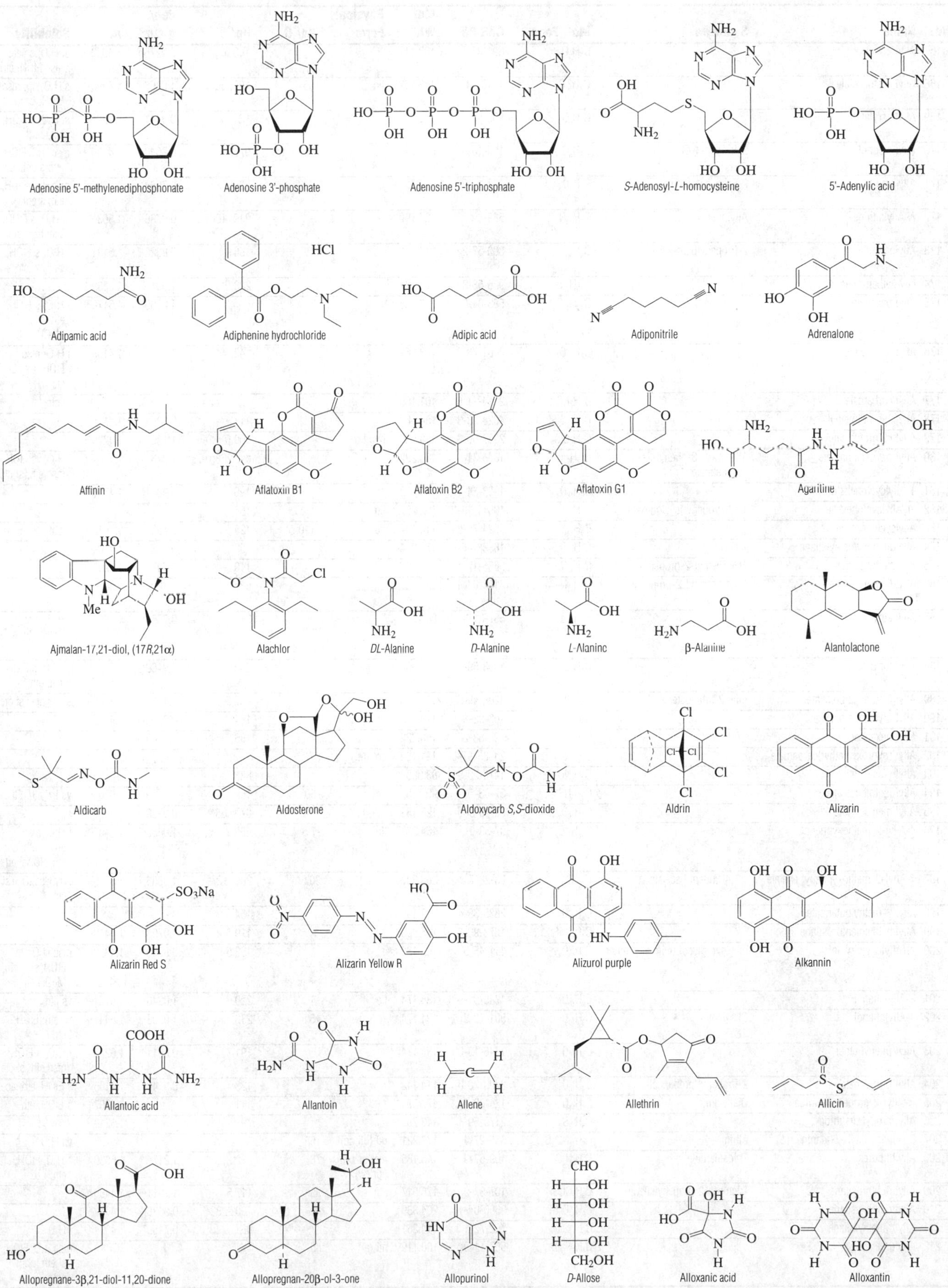

Adenosine 5'-methylenediphosphonate
Adenosine 3'-phosphate
Adenosine 5'-triphosphate
S-Adenosyl-L-homocysteine
5'-Adenylic acid
Adipamic acid
Adiphenine hydrochloride
Adipic acid
Adiponitrile
Adrenalone
Affinin
Aflatoxin B1
Aflatoxin B2
Aflatoxin G1
Agaritine
Ajmalan-17,21-diol, (17R,21α)
Alachlor
DL-Alanine
D-Alanine
L-Alanine
β-Alanine
Alantolactone
Aldicarb
Aldosterone
Aldoxycarb S,S-dioxide
Aldrin
Alizarin
Alizarin Red S
Alizarin Yellow R
Alizurol purple
Alkannin
Allantoic acid
Allantoin
Allene
Allethrin
Allicin
Allopregnane-3β,21-diol-11,20-dione
Allopregnan-20β-ol-3-one
Allopurinol
D-Allose
Alloxanic acid
Alloxantin

No.	Name	Synonym	Mol. Form.	CAS RN	Mol. Wt.	Physical Form	mp/°C	bp/°C	den/ g cm^{-3}	n_D	Solubility
167	Allyl acetate		$C_5H_8O_2$	591-87-7	100.117			103.5	0.9275^{20}	1.4049^{20}	sl H_2O; s ace; msc EtOH, eth
168	Allyl acetoacetate		$C_7H_{10}O_3$	1118-84-9	142.152	liq	-85	196; 66.5^{14}	1.0366^{20}	1.4398^{20}	s H_2O, lig; msc EtOH, bz
169	Allyl acrylate		$C_6H_8O_2$	999-55-3	112.127			121	0.9441^{20}	1.4320^{20}	sl H_2O; s EtOH, eth, acid
170	Allyl alcohol	2-Propen-1-ol	C_3H_6O	107-18-6	58.079	liq	-129	97.4	0.8540^{20}	1.4135^{20}	msc H_2O, EtOH, eth; s chl
171	Allylamine	2-Propen-1-amine	C_3H_7N	107-11-9	57.095	liq	-88.2	53.3	0.758^{20}	1.4205^{20}	msc H_2O, EtOH, eth; s chl
172	*N*-Allylaniline	Allylphenylamine	$C_9H_{11}N$	589-09-3	133.190			219; 106^{12}	0.9736^{25}	1.563^{20}	sl H_2O; s EtOH, ace; msc eth
173	Allylbenzene	2-Propenylbenzene	C_9H_{10}	300-57-2	118.175	liq	-40	156	0.8920^{20}	1.5131^{20}	i H_2O; s EtOH, eth, bz, ctc
174	α-Allylbenzenemethanol		$C_{10}H_{12}O$	936-58-3	148.201			228.5	1.004^{18}	1.5289^{21}	
175	Allyl benzoate		$C_{10}H_{10}O_2$	583-04-0	162.185				1.0569^{15}	1.5178^{20}	i H_2O; s EtOH, eth, ace, MeOH
176	Allyl butanoate		$C_7H_{12}O_2$	2051-78-7	128.169			142; 44.5^{15}	0.9017^{20}	1.4158^{20}	i H_2O; msc EtOH, eth; sl ctc
177	Allyl carbamate		$C_4H_7NO_2$	2114-11-6	101.105						sl ctc
178	Allylchlorodimethylsilane		$C_5H_{11}ClSi$	4028-23-3	134.680			111	0.8964^{20}	1.4195^{20}	
179	Allyl chloroformate		$C_4H_5ClO_2$	2937-50-0	120.535	hyg liq		109.5	1.136	1.4220^{20}	
180	Allyl *trans*-cinnamate	Allyl *trans*-3-phenyl-2-propenoate	$C_{12}H_{12}O_2$	1866-31-5	188.222			dec 268; 163^{17}	1.048^{23}	1.530^{20}	i H_2O; vs EtOH; msc eth; sl ctc
181	1-Allylcyclohexanol		$C_9H_{16}O$	1123-34-8	140.222			190	0.9341^{22}	1.4756^{22}	
182	1-Allylcyclohexene	1-(2-Propenyl)cyclohexene	C_9H_{14}	13511-13-2	122.207	liq		156			
183	Allylcyclopentane		C_8H_{14}	3524-75-2	110.197	liq	-110.7	125	0.793^{25}	1.4412^{20}	s chl
184	Allyldiethoxymethylsilane		$C_8H_{18}O_2Si$	18388-45-9	174.314			155	0.8572^{25}	1.4104^{20}	
185	Allyldiethylamine	*N,N*-Diethyl-2-propen-1-amine	$C_7H_{15}N$	5666-17-1	113.201			110	0.7477^{25}	1.4209^{20}	
186	Allyldimethylamine	*N,N*-Dimethyl-2-propen-1-amine	$C_5H_{11}N$	2155-94-4	85.148			63.5	0.7094^{25}	1.4010^{20}	
187	Allyl ethyl ether		$C_5H_{10}O$	557-31-3	86.132			67.6	0.7651^{20}	1.3881^{20}	i H_2O; msc EtOH, eth; s ace
188	Allyl formate		$C_4H_6O_2$	1838-59-1	86.090			83.6	0.9460^{20}		sl H_2O; s EtOH; msc eth
189	Allyl 2-furancarboxylate	Allyl 2-furanoate	$C_8H_8O_3$	4208-49-5	152.148			207.5	1.115^{25}	1.4945^{20}	s eth, ace; sl ctc
190	Allyl glycidyl ether		$C_6H_{10}O_2$	106-92-3	114.142			154	0.9698^{20}	1.4332^{20}	
191	Allyl hexanoate		$C_9H_{16}O_2$	123-68-2	156.222			186	0.8869^{20}		
192	Allyl (hydroxymethyl)carbamate		$C_5H_9NO_3$	24935-97-5	131.130	cry (tol)	57				
193	Allyl isocyanate		C_4H_5NO	1476-23-9	83.089			88			
194	Allyl isothiocyanate		C_4H_5NS	57-06-7	99.155	liq	-80	152	1.0126^{20}	1.5306^{20}	vs bz, eth, EtOH
195	Allyl methacrylate		$C_7H_{10}O_2$	96-05-9	126.153			67^{50}, 55^{30}	0.9335^{20}	1.4360^{20}	
196	4-Allyl-2-methoxyphenol	Eugenol	$C_{10}H_{12}O_2$	97-53-0	164.201	liq	-7.5	253.2	1.0652^{20}	1.5405^{20}	i H_2O; msc EtOH, eth; s chl, HOAc, oils
197	4-Allyl-2-methoxyphenyl acetate	1,3,4-Eugenol acetate	$C_{12}H_{14}O_3$	93-28-7	206.237	pr (al)	30.5	281; 127^6	1.0806^{20}	1.5205^{20}	i H_2O; s EtOH; sl ctc
198	Allyl 3-methylbutanoate		$C_8H_{14}O_2$	2835-39-4	142.196			154			
199	Allylmethyldichlorosilane		$C_4H_8Cl_2Si$	1873-92-3	155.099			119.5	1.0758^{20}	1.4419^{20}	
200	2-(Allyloxy)ethanol	Ethylene glycol monoallyl ether	$C_5H_{10}O_2$	111-45-5	102.132			158.5	0.9580^{20}	1.4358^{20}	msc H_2O; vs EtOH; s bz, ctc, MeOH
201	2-Allylphenol		$C_9H_{10}O$	1745-81-9	134.174	liq	-6	220	1.0246^{15}	1.5181^{20}	vs eth
202	4-Allylphenol	Chavicol	$C_9H_{10}O$	501-92-8	134.174		15.8	238	1.0203^{15}	1.5441^{18}	vs eth, EtOH, chl
203	Allyl phenyl ether		$C_9H_{10}O$	1746-13-0	134.174			191.7	0.9811^{20}	1.5223^{20}	i H_2O; s EtOH; msc eth; sl ctc
204	Allyl propanoate	2-Propenyl propanoate	$C_6H_{10}O_2$	2408-20-0	114.142			123	0.9140^{20}	1.4105^{20}	s EtOH, eth, ace
205	*N*-Allyl-2-propen-1-amine	Diallylamine	$C_6H_{11}N$	124-02-7	97.158			111		1.4387^{20}	s EtOH, eth
206	Allyl propyl disulfide		$C_6H_{12}S_2$	2179-59-1	148.289			79^{13}		1.5219^{20}	
207	3-(Allylsulfinyl)-*L*-alanine, (*S*)	Alliin	$C_6H_{11}NO_3S$	556-27-4	177.221	nd (dil ac)	165				vs H_2O
208	Allylthiourea	Thiosinamine	$C_4H_8N_2S$	109-57-9	116.185	mcl or orth pr (w)	78		1.217^{20}	1.5936^{78}	s H_2O, EtOH; sl eth; i bz
209	Allyltrichlorosilane	Trichloro-2-propenylsilane	$C_3H_5Cl_3Si$	107-37-9	175.517		35	117.5	1.2011^{20}	1.4460^{20}	
210	Allyltriethoxysilane		$C_9H_{20}O_3Si$	2550-04-1	204.339			100^{50}, 82^{28}	0.9030^{20}	1.4072^{20}	
211	Allyltrimethylsilane		$C_6H_{14}Si$	762-72-1	114.261			85	0.7158^{25}	1.4074^{20}	i H_2O
212	Allylurea		$C_4H_8N_2O$	557-11-9	100.119	nd (al)	85				msc H_2O, EtOH; sl eth, chl; i peth

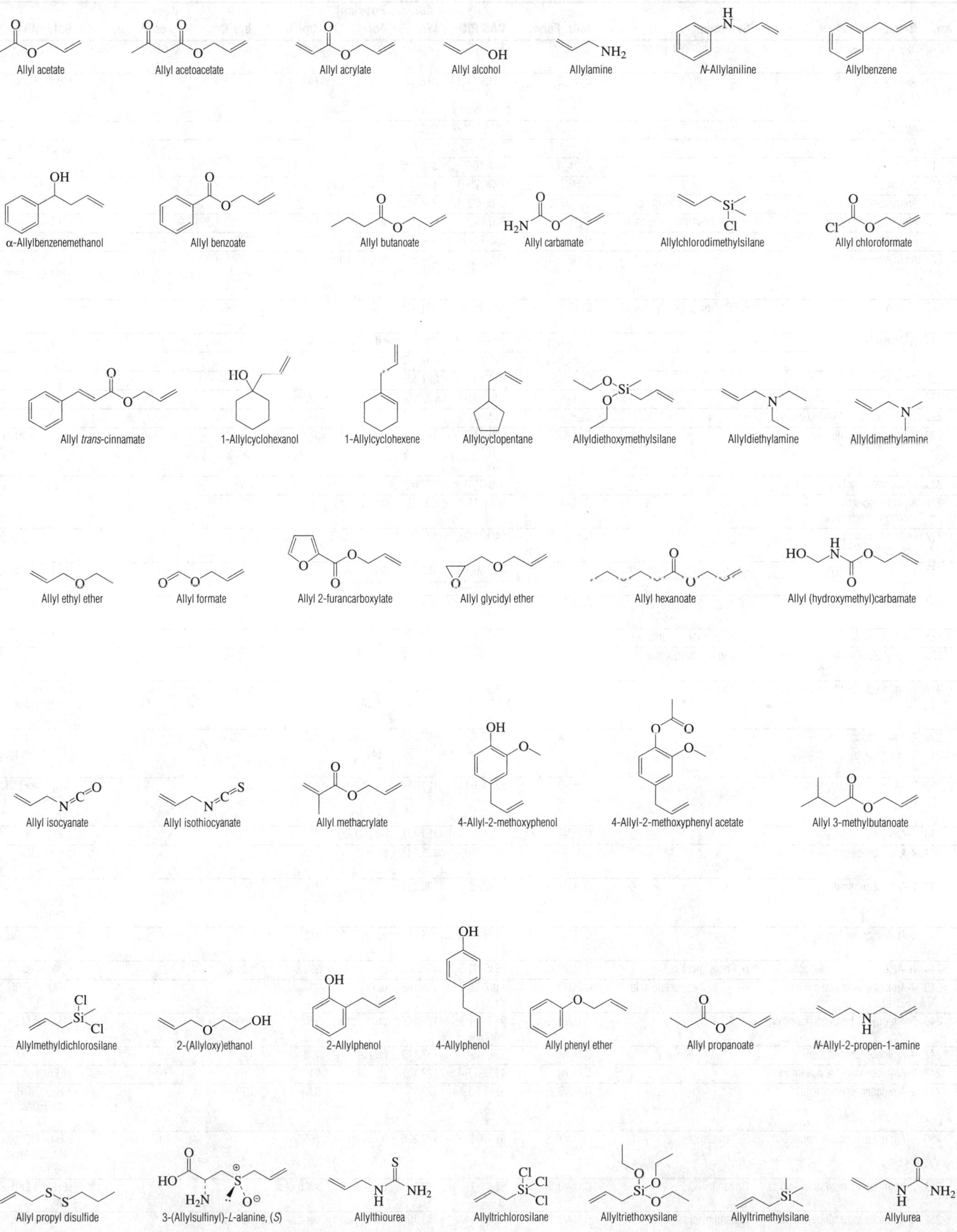

Allyl acetate
Allyl acetoacetate
Allyl acrylate
Allyl alcohol
Allylamine
N-Allylaniline
Allylbenzene
α-Allylbenzenemethanol
Allyl benzoate
Allyl butanoate
Allyl carbamate
Allylchlorodimethylsilane
Allyl chloroformate
Allyl trans-cinnamate
1-Allylcyclohexanol
1-Allylcyclohexene
Allylcyclopentane
Allyldiethoxymethylsilane
Allyldiethylamine
Allyldimethylamine
Allyl ethyl ether
Allyl formate
Allyl 2-furancarboxylate
Allyl glycidyl ether
Allyl hexanoate
Allyl (hydroxymethyl)carbamate
Allyl isocyanate
Allyl isothiocyanate
Allyl methacrylate
4-Allyl-2-methoxyphenol
4-Allyl-2-methoxyphenyl acetate
Allyl 3-methylbutanoate
Allylmethyldichlorosilane
2-(Allyloxy)ethanol
2-Allylphenol
4-Allylphenol
Allyl phenyl ether
Allyl propanoate
N-Allyl-2-propen-1-amine
Allyl propyl disulfide
3-(Allylsulfinyl)-L-alanine, (S)
Allylthiourea
Allyltrichlorosilane
Allyltriethoxysilane
Allyltrimethylsilane
Allylurea

No.	Name	Synonym	Mol. Form.	CAS RN	Mol. Wt.	Physical Form	mp/°C	bp/°C	den/ g cm⁻³	n_D	Solubility
213	Allyl vinyl ether	3-(Ethenyloxy)-1-propene	C_5H_8O	3917-15-5	84.117			66	0.7900^{20}	1.4062^{20}	i H_2O; s eth, ace, chl
214	Aloin A		$C_{21}H_{22}O_9$	1415-73-2	418.395		149.3				s H_2O, EtOH, ace; sl eth, bz; i chl
215	Alphaprodine		$C_{16}H_{23}NO_2$	15867-21-7	261.360	cry	103				
216	Alstonidine		$C_{22}H_{24}N_2O_4$	25394-75-6	380.437	cry (eth)	189				vs ace, EtOH
217	Alstonine		$C_{21}H_{20}N_2O_3$	642-18-2	348.395	ye nd (ace)	207 dec				
218	*D*-Altrose		$C_6H_{12}O_6$	1990-29-0	180.155	pr (MeOH,al)	103.5				vs H_2O
219	Aluminum 2-butoxide	2-Butanol, aluminum salt	$C_{12}H_{27}AlO_3$	2269-22-9	246.322			197^{20}			
220	Aluminum distearate	Hydroxyaluminum distearate	$C_{36}H_{71}AlO_5$	300-92-5	610.928	wh pow	145				i H_2O
221	Aluminum ethanolate	Aluminum ethoxide	$C_6H_{15}AlO_3$	555-75-9	162.163	liq/wh solid	140	200^7			dec H_2O; sl xyl
222	Aluminum isopropoxide		$C_9H_{21}AlO_3$	555-31-7	204.243	hyg wh solid	119	135^{10}, $94^{0.5}$			reac H_2O; s EtOH, bz, peth, chl
223	Alverine	*N*-Ethyl-bis(3-phenylpropyl) amine	$C_{20}H_{27}N$	150-59-4	281.435	oil		$166^{0.3}$			
224	α-Amanitin		$C_{39}H_{54}N_{10}O_{14}S$	23109-05-9	918.970	nd	254 dec				
225	Amaranth dye		$C_{20}H_{11}N_2Na_3O_{10}S_3$	915-67-3	604.472	dk red pow					s H_2O
226	Ametryn		$C_9H_{17}N_5S$	834-12-8	227.330		88				
227	Amminetrimethylboron		$C_3H_{12}BN$	1830-95-1	72.945		73.5				
228	19-Amino-8,11,13-abietatriene		$C_{20}H_{31}N$	1446-61-3	285.467	cry	44.5				
229	2-Aminoacetamide		$C_2H_6N_2O$	598-41-4	74.081	hyg nd (chl)	67.5				vs H_2O, EtOH; sl eth, bz; s ace, chl
230	Aminoacetonitrile		$C_2H_4N_2$	540-61-4	56.066			58^{15}			vs EtOH
231	Aminoacetonitrile monohydrochloride		$C_2H_5ClN_2$	6011-14-9	92.527	hyg cry (al)	165 dec				
232	α-Aminoacetophenone hydrochloride		$C_8H_{10}ClNO$	5468-37-1	171.624		194 dec				
233	1-Aminoadamantane hydrochloride	Adamantanamine hydrochloride	$C_{10}H_{18}ClN$	665-66-7	187.710	cry (al-eth)	360 dec				vs H_2O, EtOH
234	2-Aminoadipic acid		$C_6H_{11}NO_4$	626-71-1	161.156	pl (w)	207.0				sl H_2O, EtOH, eth
235	3-Aminoalanine	2,3-Diaminopropionic acid	$C_3H_8N_2O_2$	515-94-6	104.108	hyg rosettes	110				vs H_2O
236	1-Amino-9,10-anthracenedione	1-Aminoanthraquinone	$C_{14}H_9NO_2$	82-45-1	223.227	red nd (al)	253.5	sub			vs ace, bz, EtOH, chl
237	2-Amino-9,10-anthracenedione	2-Aminoanthraquinone	$C_{14}H_9NO_2$	117-79-3	223.227	red nd (al, HOAc)	304.5	sub			i H_2O, eth; sl EtOH; s ace, bz, chl
238	4-Aminoazobenzene		$C_{12}H_{11}N_3$	60-09-3	197.235	oran mcl nd (al)	127	>360			sl H_2O, lig; s EtOH, eth, bz, chl
239	2-Aminobenzaldehyde		C_7H_7NO	529-23-7	121.137	silv lf	40.5	80^2			sl H_2O; vs EtOH, eth; s bz, chl; i lig
240	3-Aminobenzaldehyde		C_7H_7NO	1709-44-0	121.137	nd (AcOEt)	29				s eth, acid
241	4-Aminobenzaldehyde		C_7H_7NO	556-18-3	121.137	pl (w)	71.5				s H_2O, EtOH, eth, acid
242	2-Aminobenzamide		$C_7H_8N_2O$	88-68-6	136.151		110.5 dec				s H_2O, EtOH; sl eth, bz; vs AcOEt
243	4-Aminobenzamide		$C_7H_8N_2O$	2835-68-9	136.151	ye cry (+1/4w)	183				sl H_2O; s EtOH, eth
244	α-Aminobenzeneacetic acid, (±)	α-Phenylglycine	$C_8H_9NO_2$	2835-06-5	151.163	pl	292 dec	sub 255			s alk; sl os
245	4-Aminobenzeneacetic acid	*p*-Aminophenylacetic acid	$C_8H_9NO_2$	1197-55-3	151.163	pl (w)	200 dec				i H_2O; sl EtOH, DMSO
246	5-Amino-1,3-benzenedicarboxylic acid		$C_8H_7NO_4$	99-31-0	181.147	pr(al), pl(w)	360	sub			i H_2O; sl EtOH
247	4-Aminobenzeneethanol		$C_8H_{11}NO$	104-10-9	137.179	nd (al)	108				
248	2-Aminobenzenemethanamine		$C_7H_{10}N_2$	4403-69-4	122.167		61	269			vs EtOH
249	2-Aminobenzenemethanol		C_7H_9NO	5344-90-1	123.152		83.5	273			s H_2O, EtOH, eth, HOAc; vs bz, chl
250	4-Aminobenzenesulfonamide	Sulfanilamide	$C_6H_8N_2O_2S$	63-74-1	172.205	lf (dil al)	165.5		1.08^{25}		s H_2O, EtOH, eth, ace; sl chl, peth
251	2-Aminobenzenesulfonic acid	Orthanilic acid	$C_6H_7NO_3S$	88-21-1	173.190	pr (+ 1/2w)	>320 dec				sl H_2O; i EtOH, eth
252	3-Aminobenzenesulfonic acid	Metanilic acid	$C_6H_7NO_3S$	121-47-1	173.190	nd, pr (w +1)	dec				sl H_2O, EtOH; i eth
253	4-Aminobenzenesulfonic acid	Sulfanilic acid	$C_6H_7NO_3S$	121-57-3	173.190	orth pl or mcl (w+2)	288		1.485^{25}		sl H_2O; i EtOH, eth
254	4-Aminobenzenesulfonyl fluoride	*p*-Sulfanilyl fluoride	$C_6H_6FNO_2S$	98-62-4	175.181		68.5				

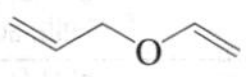

Allyl vinyl ether

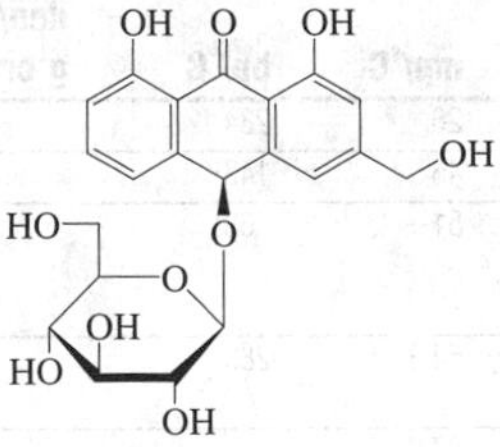

Aloin A

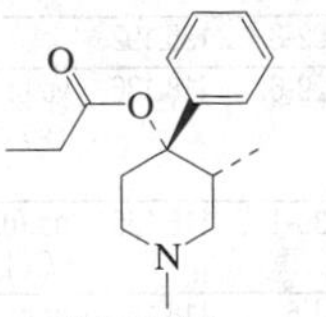

Alphaprodine

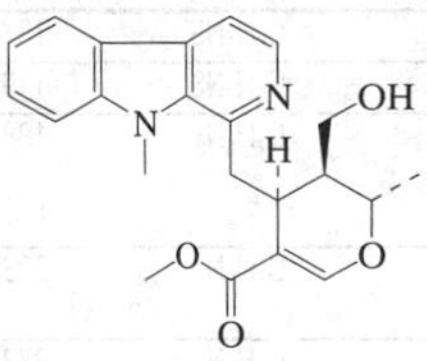

Alstonidine

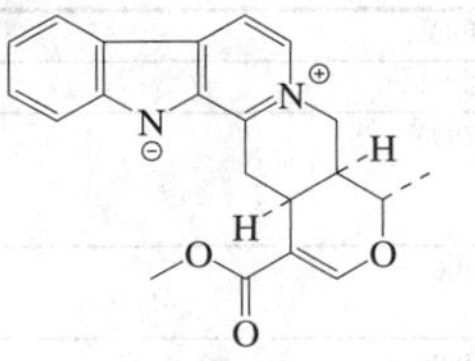

Alstonine

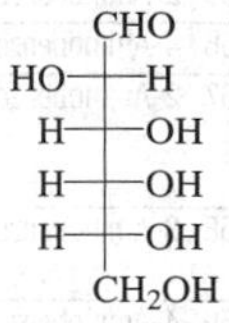

D-Altrose

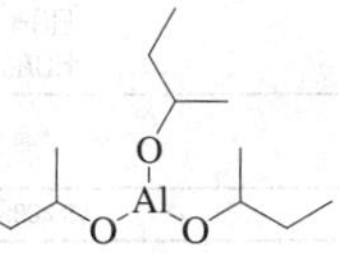

Aluminum 2-butoxide

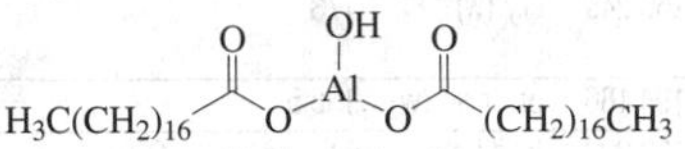

Aluminum distearate

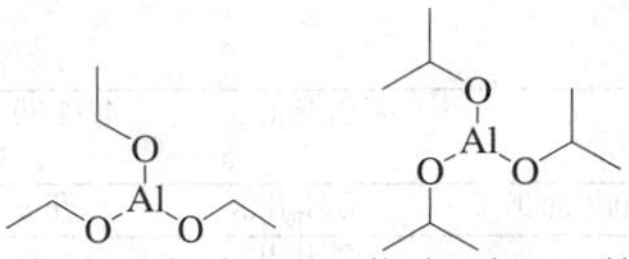

Aluminum ethanolate

Aluminum isopropoxide

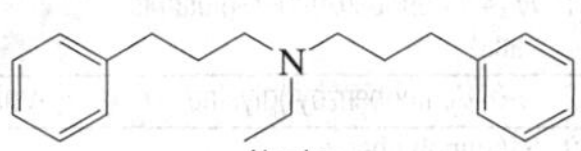

Alverine

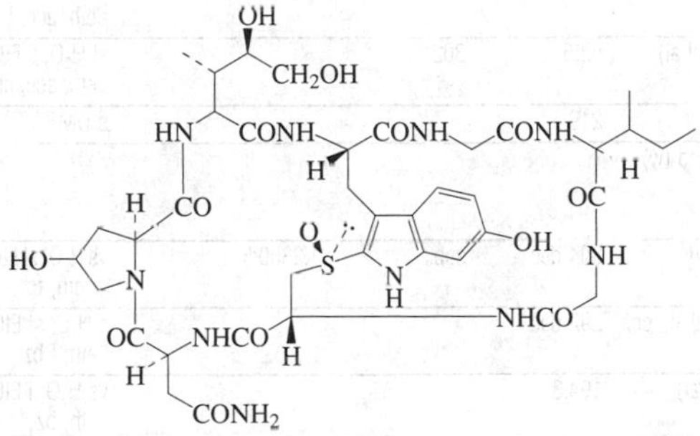

α-Amanitin

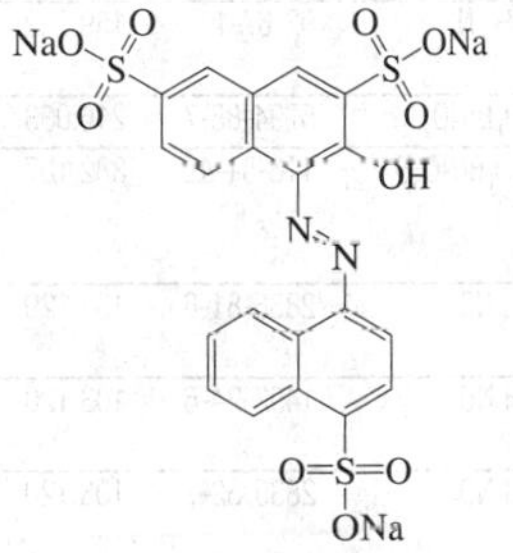

Amaranth dye

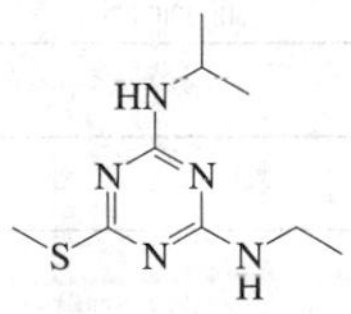

Ametryn

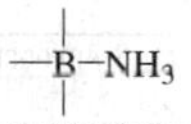

Amminetrimethylboron

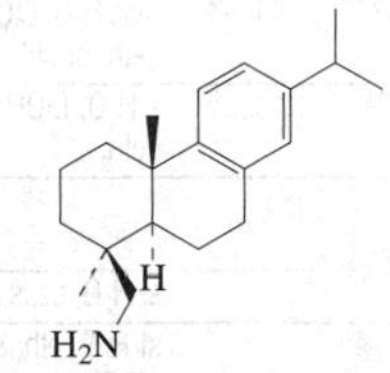

19-Amino-8,11,13-abietatriene

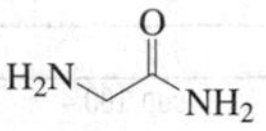

2-Aminoacetamide

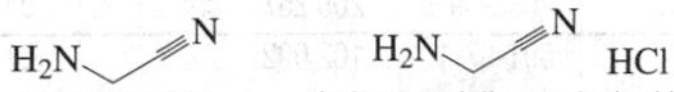

Aminoacetonitrile

Aminoacetonitrile monohydrochloride

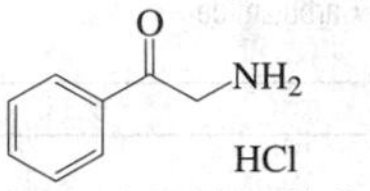

α-Aminoacetophenone hydrochloride

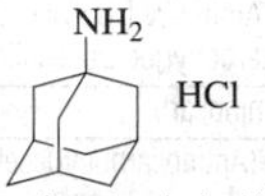

1-Aminoadamantane hydrochloride

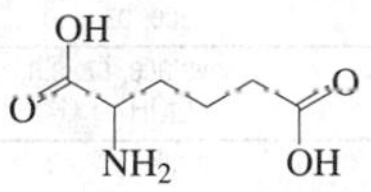

2-Aminoadipic acid

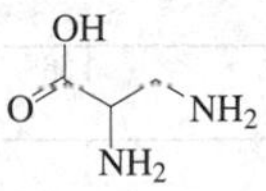

3-Aminoalanine

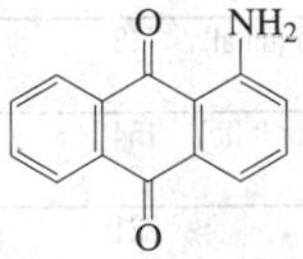

1-Amino-9,10-anthracenedione

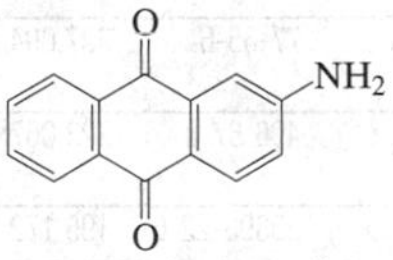

2-Amino-9,10-anthracenedione

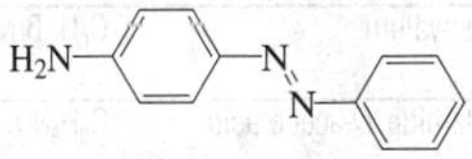

4-Aminoazobenzene

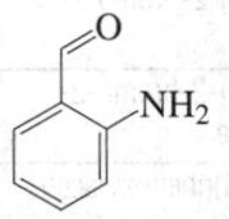

2-Aminobenzaldehyde

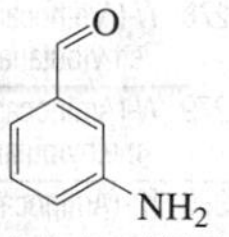

3-Aminobenzaldehyde

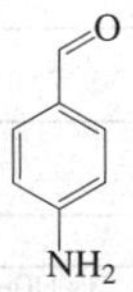

4-Aminobenzaldehyde

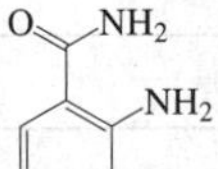

2-Aminobenzamide

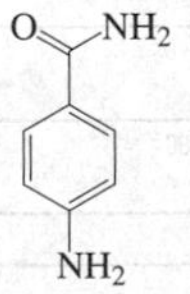

4-Aminobenzamide

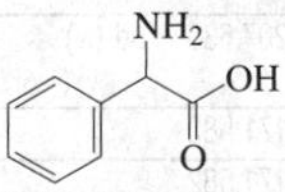

α-Aminobenzeneacetic acid, (±)

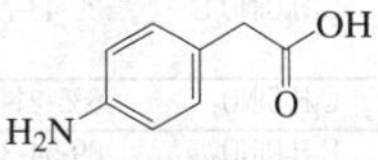

4-Aminobenzeneacetic acid

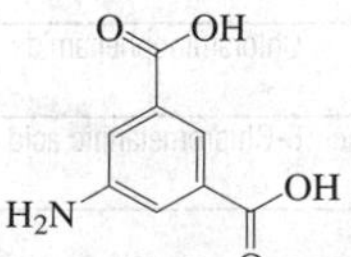

5-Amino-1,3-benzenedicarboxylic acid

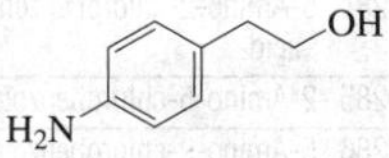

4-Aminobenzeneethanol

2-Aminobenzenemethanamine

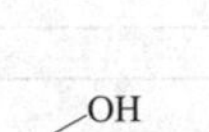

2-Aminobenzenemethanol

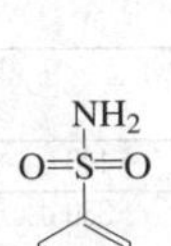

4-Aminobenzenesulfonamide

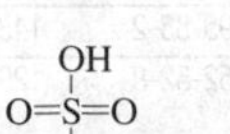

2-Aminobenzenesulfonic acid

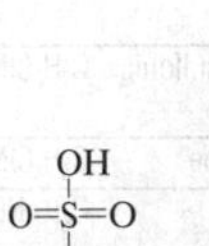

3-Aminobenzenesulfonic acid

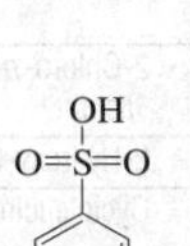

4-Aminobenzenesulfonic acid

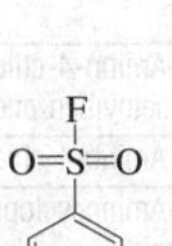

4-Aminobenzenesulfonyl fluoride

No.	Name	Synonym	Mol. Form.	CAS RN	Mol. Wt.	Physical Form	mp/°C	bp/°C	den/ g cm^{-3}	n_D	Solubility
255	2-Aminobenzenethiol		C_6H_7NS	137-07-5	125.192		26	234		1.4606^{20}	s EtOH, eth
256	4-Aminobenzenethiol		C_6H_7NS	1193-02-8	125.192		46	143^{17}			s H_2O, EtOH
257	2-Aminobenzonitrile		$C_7H_6N_2$	1885-29-6	118.136	ye pr (CS_2) nd (peth)	51	263			sl H_2O; vs EtOH, eth, ace, bz; i peth
258	3-Aminobenzonitrile		$C_7H_6N_2$	2237-30-1	118.136	nd (dil al or CCl_4)	54.3	289			sl H_2O; vs EtOH, eth, ace, chl
259	4-Aminobenzonitrile		$C_7H_6N_2$	873-74-5	118.136	pr or pl (w)	87.0				sl H_2O, ctc; vs EtOH, eth, ace, bz
260	4-Aminobenzophenone		$C_{13}H_{11}NO$	1137-41-3	197.232	lf (dil al)	124	246^{13}			sl H_2O, tfa; s EtOH, eth, HOAc
261	*N*-(4-Aminobenzoyl)-*L*-glutamic acid		$C_{12}H_{14}N_2O_5$	4271-30-1	266.249	cry (w)	173				
262	*N*-(4-Aminobenzoyl)glycine	*p*-Aminohippuric acid	$C_9H_{10}N_2O_3$	61-78-9	194.186	pr or nd (w)	198.5				vs ace, bz, EtOH
263	2-Aminobiphenyl		$C_{12}H_{11}N$	90-41-5	169.222	lf (dil al)	51	299			i H_2O; s EtOH, eth, bz; sl DMSO, peth
264	3-Aminobiphenyl		$C_{12}H_{11}N$	2243-47-2	169.222	nd	31.5				sl H_2O; s EtOH, eth, ace, bz
265	4-Aminobiphenyl	*p*-Biphenylamine	$C_{12}H_{11}N$	92-67-1	169.222	lf (dil al)	53.5	302			sl H_2O; s EtOH, eth, ace, chl
266	2-Amino-5-bromobenzoic acid	5-Bromoanthranilic acid	$C_7H_6BrNO_2$	5794-88-7	216.033	nd	219.5				s DMSO
267	1-Amino-4-bromo-9,10-dihydro-9,10-dioxo-2-anthracenesulfonic acid	1-Amino-4-bromoanthraquinone-2-sulfonic acid	$C_{14}H_8BrNO_5S$	116-81-4	382.187	red nd (w)					
268	*DL*-2-Aminobutanoic acid		$C_4H_9NO_2$	2835-81-6	103.120	lf (w)	304 dec	sub	1.2300^{20}		vs H_2O; sl EtOH; i eth, bz
269	*L*-2-Aminobutanoic acid		$C_4H_9NO_2$	1492-24-6	103.120	lf (dil al), cry (al)	292 dec				s H_2O; sl EtOH, eth; i bz
270	*DL*-3-Aminobutanoic acid		$C_4H_9NO_2$	2835-82-7	103.120	nd (al)	194.3				vs H_2O; i EtOH, eth, bz
271	4-Aminobutanoic acid	γ-Aminobutyric acid	$C_4H_9NO_2$	56-12-2	103.120	pr or nd (al) lf (MeOH-eth)	203 dec				vs H_2O; sl EtOH, ace; i eth, bz
272	2-Amino-1-butanol, (±)		$C_4H_{11}NO$	13054-87-0	89.136	liq	-1.0	178	0.9162^{20}	1.4489^{25}	msc H_2O, EtOH, eth; sl chl
273	4-Amino-1-butanol		$C_4H_{11}NO$	13325-10-5	89.136			205; 125^{34}	0.967^{12}	1.4625^{20}	s H_2O, EtOH; i eth
274	4-Amino-*N*-[(butylamino) carbonyl]benzenesulfonamide	Carbutamide	$C_{11}H_{17}N_3O_3S$	339-43-5	271.336		144.5				
275	Aminocarb		$C_{11}H_{16}N_2O_2$	2032-59-9	208.257	cry	94				sl H_2O, bz; s ace
276	*N*-(Aminocarbonyl)acetamide		$C_3H_6N_2O_2$	591-07-1	102.092		218	sub 180			sl H_2O, eth; s EtOH
277	[4-[(Aminocarbonyl)amino] phenyl]arsonic acid	Carbarsone	$C_7H_9AsN_2O_4$	121-59-5	260.079	nd (w)	174				sl H_2O, DMSO, EtOH; i eth, chl; s alk
278	*N*-(Aminocarbonyl)-2-bromo-2-ethylbutanamide	Carbromal	$C_7H_{13}BrN_2O_2$	77-65-6	237.094	orth (dil al)	118		1.544^{25}		sl H_2O, chl; s ace, bz
279	*N*-(Aminocarbonyl)-2-bromo-3-methylbutanamide	Bromisovalum	$C_6H_{11}BrN_2O_2$	496-67-3	223.067	nd or lf (to)	154	sub	1.56^{15}		vs ace, bz, eth, EtOH
280	[2-(Aminocarbonyl)phenoxy]acetic acid	Salicylamide *O*-acetic acid	$C_9H_9NO_4$	25395-22-6	195.172		221				s alk
281	7-Aminocephalosporanic acid		$C_{10}H_{12}N_2O_5S$	957-68-6	272.277	cry					
282	1-Amino-5-chloro-9,10-anthracenedione	1-Amino-5-chloroanthraquinone	$C_{14}H_8ClNO_2$	117-11-3	257.673		212				
283	4-Amino-6-chloro-1,3-benzenedisulfonamide	Chloraminophenamide	$C_6H_8ClN_3O_4S_2$	121-30-2	285.729		254.5				
284	5-Amino-2-chlorobenzenesulfonic acid	6-Chlorometanilic acid	$C_6H_6ClNO_3S$	88-43-7	207.635	nd (w)	280 dec				
285	2-Amino-5-chlorobenzoic acid		$C_7H_6ClNO_2$	635-21-2	171.582		211				
286	5-Amino-2-chlorobenzoic acid		$C_7H_6ClNO_2$	89-54-3	171.582		188		1.519^{15}		vs EtOH
287	2-Amino-5-chlorobenzophenone	2-Benzoyl-4-chloroaniline	$C_{13}H_{10}ClNO$	719-59-5	231.677	ye nd	100.5				vs H_2O, EtOH, peth, chl
288	2-Amino-4-chloro-5-methylbenzenesulfonic acid	2-Chloro-*p*-toluidine-5-sulfonic acid	$C_7H_8ClNO_3S$	88-51-7	221.662	short nd (w)					
289	2-Amino-4-chlorophenol	2-Hydroxy-5-chloroaniline	C_6H_6ClNO	95-85-2	143.571		140				sl DMSO
290	1-Aminocyclopentanecarboxylic acid	Cycloleucine	$C_6H_{11}NO_2$	52-52-8	129.157	cry (al-w)	330 dec				
291	7-Aminodeacetoxycephalosporanic acid		$C_8H_{10}N_2O_3S$	22252-43-3	214.241		241 dec				
292	1-Amino-1-deoxy-*D*-glucitol	Glucamine	$C_6H_{15}NO_5$	488-43-7	181.187	cry (MeOH)	127				vs H_2O, EtOH
293	2-Amino-2-deoxy-*D*-glucose	*D*-Glucosamine	$C_6H_{13}NO_5$	3416-24-8	179.171						vs H_2O

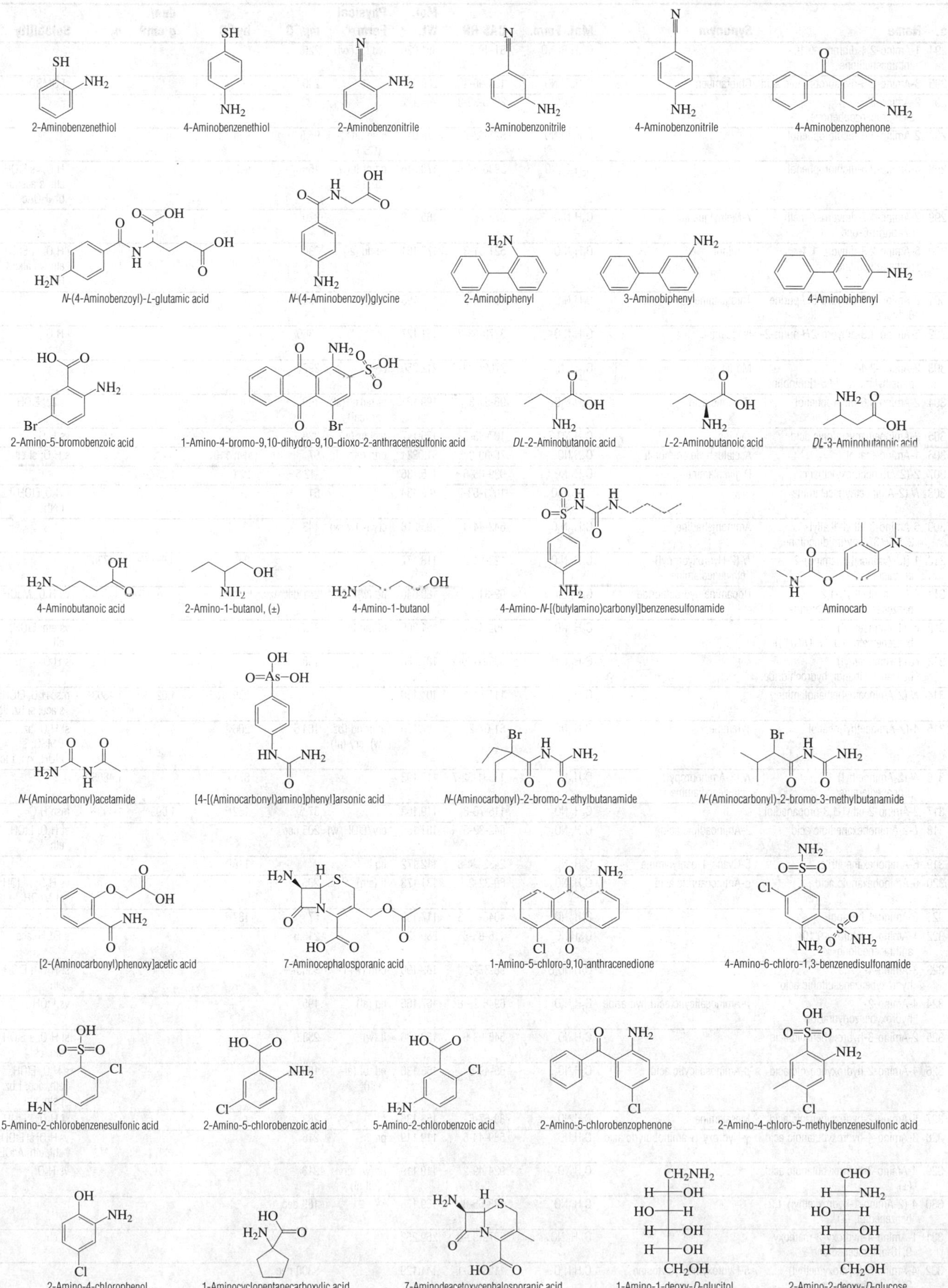

2-Aminobenzenethiol
4-Aminobenzenethiol
2-Aminobenzonitrile
3-Aminobenzonitrile
4-Aminobenzonitrile
4-Aminobenzophenone
N-(4-Aminobenzoyl)-L-glutamic acid
N-(4-Aminobenzoyl)glycine
2-Aminobiphenyl
3-Aminobiphenyl
4-Aminobiphenyl
2-Amino-5-bromobenzoic acid
1-Amino-4-bromo-9,10-dihydro-9,10-dioxo-2-anthracenesulfonic acid
DL-2-Aminobutanoic acid
L-2-Aminobutanoic acid
DL-3-Aminobutanoic acid
4-Aminobutanoic acid
2-Amino-1-butanol, (±)
4-Amino-1-butanol
4-Amino-N-[(butylamino)carbonyl]benzenesulfonamide
Aminocarb
N-(Aminocarbonyl)acetamide
[4-[(Aminocarbonyl)amino]phenyl]arsonic acid
N-(Aminocarbonyl)-2-bromo-2-ethylbutanamide
N-(Aminocarbonyl)-2-bromo-3-methylbutanamide
[2-(Aminocarbonyl)phenoxy]acetic acid
7-Aminocephalosporanic acid
1-Amino-5-chloro-9,10-anthracenedione
4-Amino-6-chloro-1,3-benzenedisulfonamide
5-Amino-2-chlorobenzenesulfonic acid
2-Amino-5-chlorobenzoic acid
5-Amino-2-chlorobenzoic acid
2-Amino-5-chlorobenzophenone
2-Amino-4-chloro-5-methylbenzenesulfonic acid
2-Amino-4-chlorophenol
1-Aminocyclopentanecarboxylic acid
7-Aminodeacetoxycephalosporanic acid
1-Amino-1-deoxy-D-glucitol
2-Amino-2-deoxy-D-glucose

No.	Name	Synonym	Mol. Form.	CAS RN	Mol. Wt.	Physical Form	mp/°C	bp/°C	den/ g cm^{-3}	n_D	Solubility
294	1-Amino-2,4-dibromo-9,10-anthracenedione		$C_{14}H_7Br_2NO_2$	81-49-2	381.020	red nd (xyl)	226				
295	3-Amino-2,5-dichlorobenzoic acid	Chloramben	$C_7H_5Cl_2NO_2$	133-90-4	206.027		200				sl DMSO
296	2-Amino-2',5-dichlorobenzophenone		$C_{13}H_9Cl_2NO$	2958-36-3	266.122		≈80				
297	2-Amino-4,6-dichlorophenol		$C_6H_5Cl_2NO$	527-62-8	178.016	long nd (CS_2)	95.5	sub 70			
298	4-Amino-2,6-dichlorophenol		$C_6H_5Cl_2NO$	5930-28-9	178.016	nd or lf (w, bz)	168	sub			i H_2O; vs EtOH, eth; s ace; sl bz, HOAc
299	2-Amino-1,7-dihydro-7-methyl-6*H*-purin-6-one	7-Methylguanine	$C_6H_7N_5O$	578-76-7	165.153		370				
300	5-Amino-2,3-dihydro-1,4-phthalazinedione	Luminol	$C_8H_7N_3O_2$	521-31-3	177.161	ye nd (al)	330.5				i H_2O; sl EtOH, eth; vs alk; s HOAc
301	2-Amino-1,7-dihydro-6*H*-purine-6-thione	Thioguanine	$C_5H_5N_5S$	154-42-7	167.193		>360				
302	6-Amino-1,3-dihydro-2*H*-purin-2-one	Isoguanine	$C_5H_5N_5O$	3373-53-3	151.127		>360				i H_2O
303	2-Amino-3,4-dimethylimidazo[4,5-f]quinoline	Me-IQ	$C_{12}H_{12}N_4$	77094-11-2	212.250	cry	297				
304	2-Amino-4,6-dinitrophenol	Picramic acid	$C_6H_5N_3O_5$	96-91-3	199.121	dk red nd (al) pr (chl)	169				vs bz, EtOH
305	2-Aminoethanesulfonic acid	Taurine	$C_2H_7NO_3S$	107-35-7	125.147	mcl pr (w)	328				vs H_2O
306	1-Aminoethanol	Acetaldehyde ammonia	C_2H_7NO	75-39-8	61.083	orth (eth-al)	97	dec 110			s H_2O; sl eth
307	2-(2-Aminoethoxy)ethanol	Diglycolamine	$C_4H_{11}NO_2$	929-06-6	105.136		-12.5	221	1.0572^{20}		
308	*N*-(2-Aminoethyl)acetamide		$C_4H_{10}N_2O$	1001-53-2	102.134		51				s H_2O, EtOH, bz; i eth
309	6-Amino-3-ethyl-1-allyl-2,4(1*H*,3*H*)-pyrimidinedione	Aminometradine	$C_9H_{13}N_3O_2$	642-44-4	195.218	cry (+1w, w)	143				
310	1-[(2-Aminoethyl)amino]-2-propanol	*N*-(2-Hydroxypropyl) ethylenediamine	$C_5H_{14}N_2O$	123-84-2	118.177			94^{3}	0.9837^{25}	1.4738^{20}	
311	4-(2-Aminoethyl)-1,2-benzenediol, hydrochloride	Dopamine hydrochloride	$C_8H_{12}ClNO_2$	62-31-7	189.640	nd (w)	241 dec				vs H_2O, MeOH
312	α-(1-Aminoethyl) benzenemethanol, [*S*-(*R**,*R**)]-		$C_9H_{13}NO$	492-39-7	151.205	pl(MeOH)	77.5				vs eth, EtOH, chl
313	α-(1-Aminoethyl) benzenemethanol, hydrochloride		$C_9H_{14}ClNO$	53631-70-2	187.666		198.5				s H_2O
314	*N*-(2-Aminoethyl)ethanolamine		$C_4H_{12}N_2O$	111-41-1	104.150			239; 105^{10}	1.0286^{20}	1.4863^{20}	msc H_2O, EtOH; s ace; sl bz, lig
315	4-(2-Aminoethyl)phenol	Tyramine	$C_8H_{11}NO$	51-67-2	137.179	pl or nd (bz, w), cry (al)	164.5	206^{25}			sl H_2O, bz, DMSO; s EtOH, xyl; i tol
316	*N*-(2-Aminoethyl)-1,3-propanediamine	*N*-(3-Aminopropyl) ethylenediamine	$C_5H_{15}N_3$	13531-52-7	117.193			87^{3}		1.4805^{25}	
317	2-Amino-2-ethyl-1,3-propanediol		$C_5H_{13}NO_2$	115-70-8	119.163		37.5	152^{10}	1.099^{20}	1.490^{20}	msc H_2O
318	*L*-2-Aminohexanedioic acid	2-Aminoadipic acid	$C_6H_{11}NO_4$	542-32-5	161.156	cry (EtOH, w)	205 dec				sl H_2O, EtOH, eth
319	6-Aminohexanenitrile	5-Cyano-1-pentylamine	$C_6H_{12}N_2$	2432-74-8	112.172	liq		118^{16}			
320	6-Aminohexanoic acid	ε-Aminocaproic acid	$C_6H_{13}NO_2$	60-32-2	131.173	lf (eth)	205				vs H_2O; i EtOH; sl MeOH
321	6-Amino-1-hexanol		$C_6H_{15}NO$	4048-33-3	117.189		57	137^{30}			
322	1-Amino-4-hydroxy-9,10-anthracenedione		$C_{14}H_9NO_3$	116-85-8	239.226		216.5				s EtOH, ace
323	3-Amino-4-hydroxybenzenesulfonic acid		$C_6H_7NO_4S$	98-37-3	189.190	orth (w+1)	>300				sl H_2O; i EtOH, eth
324	4-Amino-2-hydroxybenzohydrazide	*p*-Aminosalicylic acid hydrazide	$C_7H_9N_3O_2$	6946-29-8	167.165	nd (al)	195				vs EtOH
325	2-Amino-3-hydroxybenzoic acid		$C_7H_7NO_3$	548-93-6	153.136	lf (w)	253.5				sl H_2O; s EtOH, eth, chl
326	4-Amino-2-hydroxybenzoic acid	*p*-Aminosalicylic acid	$C_7H_7NO_3$	65-49-6	153.136	nd, pl (al-eth)	150 dec				s H_2O, EtOH, eth, ace; i bz, peth, chl
327	5-Amino-2-hydroxybenzoic acid	Mesalamine	$C_7H_7NO_3$	89-57-6	153.136		283				sl H_2O; i EtOH
328	3-Amino-4-hydroxybutanoic acid	γ-Hydroxy-β-aminobutyric acid	$C_4H_9NO_3$	589-44-6	119.119	pr	216				vs H_2O; sl EtOH, chl, eth, AcOEt
329	4-Amino-3-hydroxybutanoic acid, (±)		$C_4H_9NO_3$	924-49-2	119.119	pr (w), cry (dil al)	218				vs H_2O
330	4-(2-Amino-1-hydroxyethyl)-1,2-benzenediol, (±)		$C_8H_{11}NO_3$	138-65-8	169.178		189 dec				
331	1-Amino-4-hydroxy-2-methoxy-9,10-anthracenedione		$C_{15}H_{11}NO_4$	2379-90-0	269.253						sl chl
332	4-Amino-5-(hydroxymethyl)-2(1*H*)-pyrimidinone	5-Hydroxymethylcytosine	$C_5H_7N_3O_2$	1123-95-1	141.129		>300 dec				

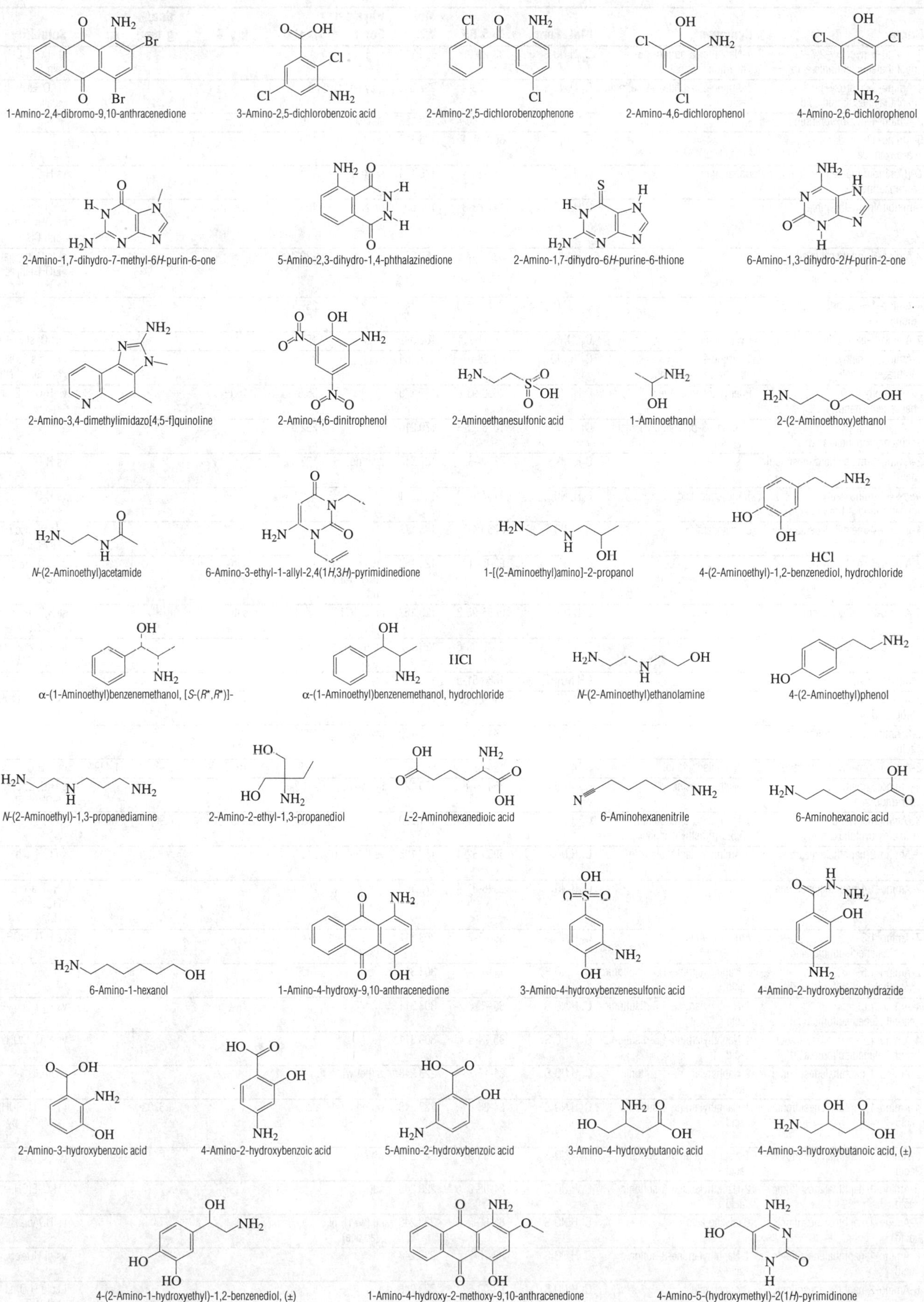
1-Amino-2,4-dibromo-9,10-anthracenedione
3-Amino-2,5-dichlorobenzoic acid
2-Amino-2',5-dichlorobenzophenone
2-Amino-4,6-dichlorophenol
4-Amino-2,6-dichlorophenol
2-Amino-1,7-dihydro-7-methyl-6H-purin-6-one
5-Amino-2,3-dihydro-1,4-phthalazinedione
2-Amino-1,7-dihydro-6H-purine-6-thione
6-Amino-1,3-dihydro-2H-purin-2-one
2-Amino-3,4-dimethylimidazo[4,5-f]quinoline
2-Amino-4,6-dinitrophenol
2-Aminoethanesulfonic acid
1-Aminoethanol
2-(2-Aminoethoxy)ethanol
N-(2-Aminoethyl)acetamide
6-Amino-3-ethyl-1-allyl-2,4(1H,3H)-pyrimidinedione
1-[(2-Aminoethyl)amino]-2-propanol
4-(2-Aminoethyl)-1,2-benzenediol, hydrochloride
α-(1-Aminoethyl)benzenemethanol, [S-(R*,R*)]-
α-(1-Aminoethyl)benzenemethanol, hydrochloride
N-(2-Aminoethyl)ethanolamine
4-(2-Aminoethyl)phenol
N-(2-Aminoethyl)-1,3-propanediamine
2-Amino-2-ethyl-1,3-propanediol
L-2-Aminohexanedioic acid
6-Aminohexanenitrile
6-Aminohexanoic acid
6-Amino-1-hexanol
1-Amino-4-hydroxy-9,10-anthracenedione
3-Amino-4-hydroxybenzenesulfonic acid
4-Amino-2-hydroxybenzohydrazide
2-Amino-3-hydroxybenzoic acid
4-Amino-2-hydroxybenzoic acid
5-Amino-2-hydroxybenzoic acid
3-Amino-4-hydroxybutanoic acid
4-Amino-3-hydroxybutanoic acid, (±)
4-(2-Amino-1-hydroxyethyl)-1,2-benzenediol, (±)
1-Amino-4-hydroxy-2-methoxy-9,10-anthracenedione
4-Amino-5-(hydroxymethyl)-2(1H)-pyrimidinone

No.	Name	Synonym	Mol. Form.	CAS RN	Mol. Wt.	Physical Form	mp/°C	bp/°C	den/ g cm^{-3}	n_D	Solubility
333	4-Amino-5-hydroxy-2,7-naphthalenedisulfonic acid	1-Naphthol-8-amino-3,6-disulfonic acid	$C_{10}H_9NO_7S_2$	90-20-0	319.311						sl H_2O, EtOH, eth
334	4-Amino-3-hydroxy-1-naphthalenesulfonic acid	1-Amino-2-naphthol-4-sulfonic acid	$C_{10}H_9NO_4S$	116-63-2	239.248	gray nd					i H_2O, EtOH, bz; s alk
335	2-Amino-4-hydroxypteridine		$C_6H_5N_5O$	2236-60-4	163.137	ye cry	>360				
336	5-Amino-1*H*-imidazole-4-carboxamide		$C_4H_6N_4O$	360-97-4	126.117	cry (EtOH)	170				
337	*O*-[(Aminoiminomethyl)amino]-*L*-homoserine	Canavanine	$C_5H_{12}N_4O_3$	543-38-4	176.174	cry (al)					vs H_2O
338	(Aminoiminomethyl)urea		$C_2H_6N_4O$	141-83-3	102.095	pr	105	dec 160			s H_2O, py; sl EtOH; i eth, bz, chl, CS_2
339	2-Amino-5-iodobenzoic acid		$C_7H_6INO_2$	5326-47-6	263.033		220 dec				sl H_2O, tfa; vs EtOH, eth, ace; s bz
340	4-Amino-1*H*-isoindole-1,3(2*H*)-dione		$C_8H_6N_2O_2$	2518-24-3	162.146		269.5				
341	4-Amino-3-isoxazolidinone, (*R*)	Cycloserine	$C_3H_6N_2O_2$	68-41-7	102.092		155 dec				s H_2O; sl MeOH
342	1-Amino-2-methyl-9,10-anthracenedione	1-Amino-2-methylanthraquinone	$C_{15}H_{11}NO_2$	82-28-0	237.254		205.5				i H_2O; s EtOH, bz, chl; sl eth
343	α-(Aminomethyl) benzenemethanol	Phenylethanolamine	$C_8H_{11}NO$	7568-93-6	137.179		56.5	160^{17}			vs H_2O; s EtOH
344	β-(Aminomethyl) benzenepropanoic acid	4-Amino-3-phenylbutyric acid	$C_{10}H_{13}NO_2$	1078-21-3	179.216		252 dec				
345	2-Amino-5-methylbenzenesulfonic acid		$C_7H_9NO_3S$	88-44-8	187.216	lt ye nd	132 dec				vs H_2O
346	*trans*-4-(Aminomethyl) cyclohexanecarboxylic acid	Tranexamic acid	$C_8H_{15}NO_2$	1197-18-8	157.211		>300				vs H_2O
347	4-Amino-4-methyl-2-pentanone	Diacetonamine	$C_6H_{13}NO$	625-04-7	115.173			$25^{0.14}$			s H_2O; msc EtOH, eth
348	2-Amino-4-methylphenol		C_7H_9NO	95-84-1	123.152	cry (w), orth (bz), lf or nd	136	sub			sl H_2O, bz; s EtOH, eth, chl; i lig
349	4-Amino-2-methylphenol		C_7H_9NO	2835-96-3	123.152	nd or lf (bz)	176.5	sub			sl H_2O, bz; s EtOH, eth
350	4-Amino-3-methylphenol		C_7H_9NO	2835-99-6	123.152	pr (dil al) cry (bz)	179				sl H_2O; vs EtOH, eth; s DMSO
351	(Aminomethyl)phosphonic acid		CH_6NO_3P	1066-51-9	111.038	cry	309				
352	2-Amino-2-methyl-1,3-propanediol		$C_4H_{11}NO_2$	115-69-5	105.136		110	151^{10}			vs H_2O; s EtOH
353	*L*-3-Amino-2-methylpropanoic acid		$C_4H_9NO_2$	144-90-1	103.120	cry (w)	182				
354	2-Amino-2-methyl-1-propanol	2-Aminoisobutanol	$C_4H_{11}NO$	124-68-5	89.136		25.5	165.5	0.934^{20}	1.449^{20}	msc H_2O; s ctc
355	4-Amino-5-methyl-2(1*H*)-pyrimidinone	5-Methylcytosine	$C_5H_7N_3O$	554-01-8	125.129	pr (w+1/2)	270 dec				s H_2O, acid; sl EtOH; i eth
356	3-(Aminomethyl)-3,5,5-trimethylcyclohexanol	1-Hydroxy-3-aminomethyl-3,5,5-trimethylcyclohexane	$C_{10}H_{21}NO$	15647-11-7	171.280		45.5	265	0.969^{25}	1.4904^{20}	
357	3-Amino-2-naphthalenecarboxylic acid	3-Amino-2-naphthoic acid	$C_{11}H_9NO_2$	5959-52-4	187.195	ye lf (dil al)	216.5				s EtOH, eth
358	2-Amino-1,4-naphthalenedione		$C_{10}H_7NO_2$	2348-81-4	173.169		207				i H_2O, alk; s EtOH, eth, HOAc
359	7-Amino-1,3-naphthalenedisulfonic acid	Amido-G-Acid	$C_{10}H_9NO_6S_2$	86-65-7	303.311	mcl pr or nd (w+4)	274				vs H_2O, EtOH
360	2-Amino-1,5-naphthalenedisulfonic acid	2-Naphthylamine-1,5-disulfonic acid	$C_{10}H_9NO_6S_2$	117-62-4	303.311		>300				
361	4-Amino-1,6-naphthalenedisulfonic acid	1-Naphthylamine-4,7-disulfonic acid	$C_{10}H_9NO_6S_2$	85-75-6	303.311						vs H_2O
362	4-Amino-1,7-naphthalenedisulfonic acid	1-Naphthylamine-4,6-disulfonic acid	$C_{10}H_9NO_6S_2$	85-74-5	303.311						vs H_2O, EtOH
363	2-Amino-1-naphthalenesulfonic acid	2-Naphthylamine-1-sulfonic acid	$C_{10}H_9NO_3S$	81-16-3	223.248	sc(hot w)					s DMSO
364	4-Amino-1-naphthalenesulfonic acid	1-Naphthylamine-4-sulfonic acid	$C_{10}H_9NO_3S$	84-86-6	223.248	wh nd (w+1/2) red-br cry	dec		1.6703^{25}		i H_2O; sl EtOH; s MeOH, py
365	5-Amino-1-naphthalenesulfonic acid	1-Naphthylamine-5-sulfonic acid	$C_{10}H_9NO_3S$	84-89-9	223.248	wh cry					s H_2O; i eth
366	6-Amino-1-naphthalenesulfonic acid	2-Naphthylamine-5-sulfonic acid	$C_{10}H_9NO_3S$	81-05-0	223.248	nd(w)					i H_2O, EtOH, eth
367	7-Amino-1-naphthalenesulfonic acid	Badische acid	$C_{10}H_9NO_3S$	86-60-2	223.248	nd (w+1), pl (aq ace)					vs HOAc
368	8-Amino-1-naphthalenesulfonic acid	1-Naphthylamine-8-sulfonic acid	$C_{10}H_9NO_3S$	82-75-7	223.248	nd					vs gl HOAc
369	6-Amino-2-naphthalenesulfonic acid	Bronner acid	$C_{10}H_9NO_3S$	93-00-5	223.248	lf					i cold H_2O; sl hot H_2O

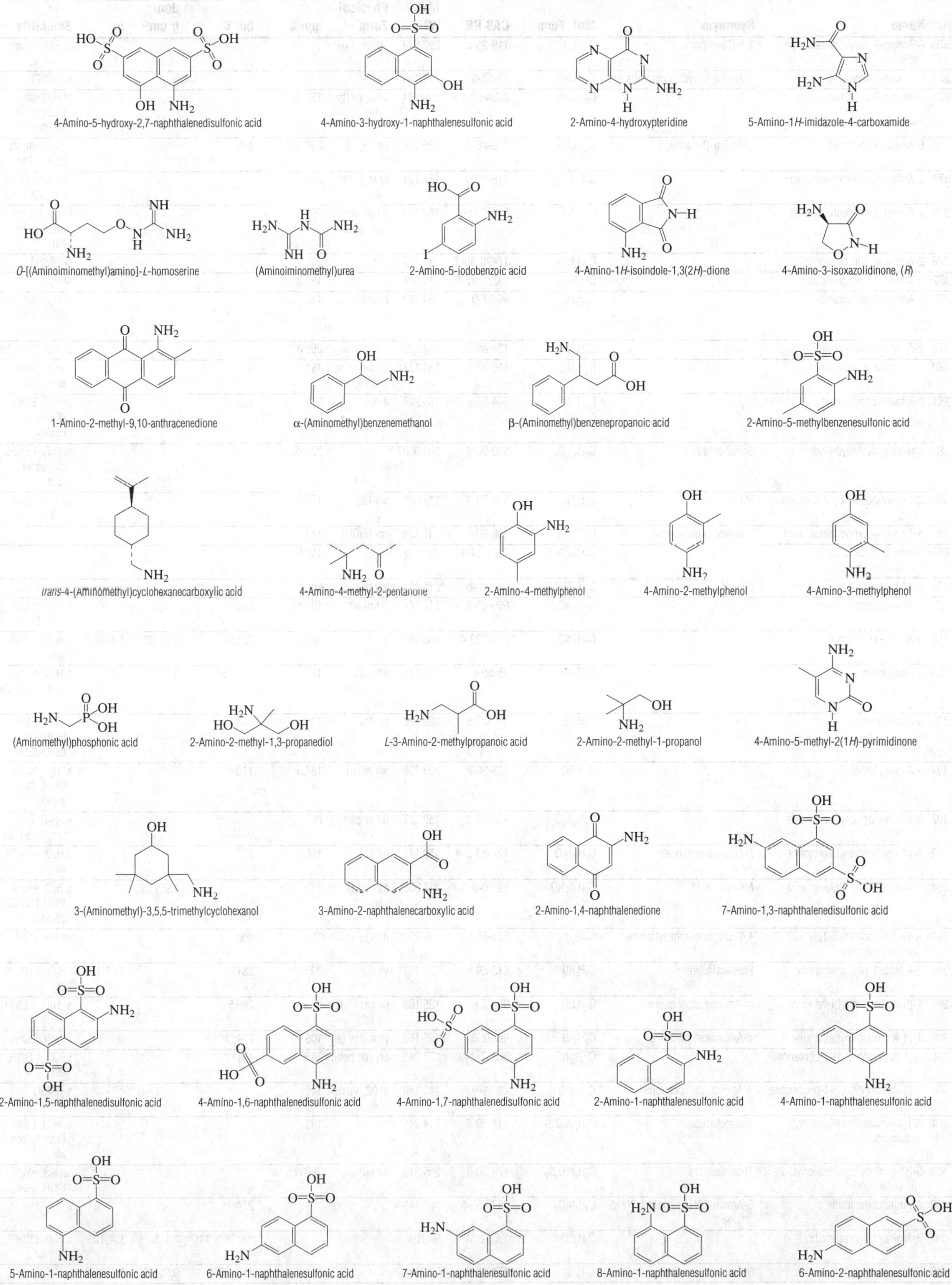
4-Amino-5-hydroxy-2,7-naphthalenedisulfonic acid
4-Amino-3-hydroxy-1-naphthalenesulfonic acid
2-Amino-4-hydroxypteridine
5-Amino-1H-imidazole-4-carboxamide
O-[(Aminoiminomethyl)amino]-L-homoserine
(Aminoiminomethyl)urea
2-Amino-5-iodobenzoic acid
4-Amino-1H-isoindole-1,3(2H)-dione
4-Amino-3-isoxazolidinone, (R)
1-Amino-2-methyl-9,10-anthracenedione
α-(Aminomethyl)benzenemethanol
β-(Aminomethyl)benzenepropanoic acid
2-Amino-5-methylbenzenesulfonic acid
trans-4-(Aminomethyl)cyclohexanecarboxylic acid
4-Amino-4-methyl-2-pentanone
2-Amino-4-methylphenol
4-Amino-2-methylphenol
4-Amino-3-methylphenol
(Aminomethyl)phosphonic acid
2-Amino-2-methyl-1,3-propanediol
L-3-Amino-2-methylpropanoic acid
2-Amino-2-methyl-1-propanol
4-Amino-5-methyl-2(1H)-pyrimidinone
3-(Aminomethyl)-3,5,5-trimethylcyclohexanol
3-Amino-2-naphthalenecarboxylic acid
2-Amino-1,4-naphthalenedione
7-Amino-1,3-naphthalenedisulfonic acid
2-Amino-1,5-naphthalenedisulfonic acid
4-Amino-1,6-naphthalenedisulfonic acid
4-Amino-1,7-naphthalenedisulfonic acid
2-Amino-1-naphthalenesulfonic acid
4-Amino-1-naphthalenesulfonic acid
5-Amino-1-naphthalenesulfonic acid
6-Amino-1-naphthalenesulfonic acid
7-Amino-1-naphthalenesulfonic acid
8-Amino-1-naphthalenesulfonic acid
6-Amino-2-naphthalenesulfonic acid

No.	Name	Synonym	Mol. Form.	CAS RN	Mol. Wt.	Physical Form	mp/°C	bp/°C	den/ g cm^{-3}	n_D	Solubility
370	8-Amino-2-naphthalenesulfonic acid	1,7-Cleve's acid	$C_{10}H_9NO_3S$	119-28-8	223.248	nd or pr (w)					sl EtOH; s eth
371	5-Amino-1-naphthol	1-Amino-6-hydroxynaphthalene	$C_{10}H_9NO$	83-55-6	159.184		170				sl DMSO
372	1-Amino-2-naphthol		$C_{10}H_9NO$	2834-92-6	159.184	silvery lf (bz, eth)	150 dec				sl H_2O, eth; s EtOH; vs dil alk, acid
373	8-Amino-2-naphthol	8-Amino-β-naphthol	$C_{10}H_9NO$	118-46-7	159.184	nd (w, al)	206	sub			s H_2O, eth; vs EtOH; sl bz, lig
374	2-Amino-4-nitrobenzoic acid		$C_7H_6N_2O_4$	619-17-0	182.134	oran pr (dil al)	269				i H_2O; vs EtOH, eth, ace; s xyl
375	2-Amino-5-nitrobenzoic acid		$C_7H_6N_2O_4$	616-79-5	182.134	lf (al), ye nd (w, dil al)	269				i H_2O, bz, chl, xyl; s EtOH, eth
376	2-Amino-5-nitrobenzonitrile		$C_7H_5N_3O_2$	17420-30-3	163.134		203.5				sl DMSO
377	3-Amino-1-nitroguanidine		$CH_5N_5O_2$	18264-75-0	119.084		187.8				sl H_2O
378	2-Amino-4-nitrophenol		$C_6H_6N_2O_3$	99-57-0	154.123	oran pr (+w)	146				sl H_2O, ace; vs EtOH; s eth, bz, HOAc
379	2-Amino-5-nitrophenol		$C_6H_6N_2O_3$	121-88-0	154.123		205.8				s H_2O, EtOH, bz
380	4-Amino-2-nitrophenol		$C_6H_6N_2O_3$	119-34-6	154.123	dk red pl or nd (w, al)	131	110^{12}			s H_2O, EtOH, eth; sl DMSO
381	2-Aminooctanoic acid, (±)		$C_8H_{17}NO_2$	644-90-6	159.227	lf (w)	270	sub			sl H_2O, EtOH, eth, bz; s HOAc
382	Aminooxoacetohydrazide	Semioxamazide	$C_2H_5N_3O_2$	515-96-8	103.080		221 dec				sl H_2O; i EtOH, eth; vs alk, acid
383	*cis*-4-Amino-4-oxo-2-butenoic acid	Maleamic acid	$C_4H_5NO_3$	557-24-4	115.088	cry (al)	172.5				vs H_2O, EtOH
384	5-Amino-4-oxopentanoic acid	5-Aminolevulinic acid	$C_5H_9NO_3$	106-60-5	131.130	cry (EtOH)	118				
385	(Aminooxy)acetic acid, hydrochloride (2:1)		$C_4H_{11}ClN_2O_6$	2921-14-4	218.592		152.5				
386	6-Aminopenicillanic acid	Penicin	$C_8H_{12}N_2O_3S$	551-16-6	216.257	cry (w)	208				
387	5-Aminopentanoic acid		$C_5H_{11}NO_2$	660-88-8	117.147	lf (dil al)	157 dec	dec			s H_2O; sl EtOH; i eth, bz, lig
388	5-Amino-1-pentanol		$C_5H_{13}NO$	2508-29-4	103.163		38.5	221.5	0.9488^{17}	1.4618^{17}	msc H_2O, EtOH, ace
389	2-Aminophenol		C_6H_7NO	95-55-6	109.126	wh orth bipym nd (bz)	174	sub 153	1.328^{25}		s H_2O, eth; vs EtOH; sl bz, tfa
390	3-Aminophenol		C_6H_7NO	591-27-5	109.126	pr (to)	123	164^{11}			s H_2O, tol; vs EtOH, eth; sl bz, DMSO
391	4-Aminophenol		C_6H_7NO	123-30-8	109.126	wh pl (w)	187.5	$110^{0.3}$			sl H_2O, tfa; vs EtOH; i bz, chl; s alk
392	*N*-(3-Aminophenyl)acetamide		$C_8H_{10}N_2O$	102-28-3	150.177	nd or pl (bz)	88				vs H_2O, EtOH, ace; sl eth, bz
393	*N*-(4-Aminophenyl)acetamide	*p*-Aminoacetanilide	$C_8H_{10}N_2O$	122-80-5	150.177	nd (w)	166.5	267			s H_2O; vs EtOH, eth
394	(4-Aminophenyl)arsonic acid	Arsanilic acid	$C_6H_8AsNO_3$	98-50-0	217.055	mcl nd (w, al)	232		1.9571^{10}		s H_2O, eth; sl EtOH, DMSO; i ace, bz
395	*N*-(4-Aminophenyl)-1,4-benzenediamine	4,4'-Diaminodiphenylamine	$C_{12}H_{13}N_3$	537-65-5	199.251	lf (w)	158	dec			vs eth, EtOH
396	2-Amino-1-phenylethanone	Phenacylamine	C_8H_9NO	613-89-8	135.163	ye cry	20	251		1.6160^{20}	i H_2O; s eth; sl ctc
397	1-(3-Aminophenyl)ethanone	*m*-Aminoacetophenone	C_8H_9NO	99-03-6	135.163	pa ye pl (al), lf (eth)	98.5	289.5			sl H_2O; s EtOH
398	1-(4-Aminophenyl)ethanone	*p*-Aminoacetophenone	C_8H_9NO	99-92-3	135.163	ye mcl pr (al)	106	294; 195^{15}			vs eth, EtOH
399	1-(4-Aminophenyl)-1-pentanone		$C_{11}H_{15}NO$	38237-74-0	177.243	cry (bz-peth)	74.5	161^{3}			i H_2O; s EtOH, eth
400	1-(4-Aminophenyl)-1-propanone	*p*-Aminopropiophenone	$C_9H_{11}NO$	70-69-9	149.189	pl (al, w), nd (w)	140				s DMSO
401	*N*-[(4-Aminophenyl)sulfonyl] acetamide	Sulfacetamide	$C_8H_{10}N_2O_3S$	144-80-9	214.241		183				sl H_2O; s EtOH; i eth; vs ace, alk
402	5-[(4-Aminophenyl)sulfonyl]-2-thiazolamine	Thiazolsulfone	$C_9H_9N_3O_2S_2$	473-30-3	255.316	nd (al)	220 dec				vs ace, eth, EtOH, diox
403	4-Aminophthalimide	5-Amino-1*H*-isoindole-1,3(2*H*)-dione	$C_8H_6N_2O_2$	3676-85-5	162.146			$224^{0.5}$			
404	3-Amino-1,2-propanediol, (±)		$C_3H_9NO_2$	13552-31-3	91.109			dec 265; 145^{9}	1.1752^{20}	1.4910^{25}	s H_2O, EtOH; i eth, bz
405	3-Aminopropanenitrile	3-Aminopropionitrile	$C_3H_6N_2$	151-18-8	70.093			185; 88^{20}	0.9584^{20}	1.4396^{20}	
406	2-Amino-1-propanol, (±)		C_3H_9NO	6168-72-5	75.109			174.5		1.4502^{20}	vs H_2O, EtOH, eth; sl chl

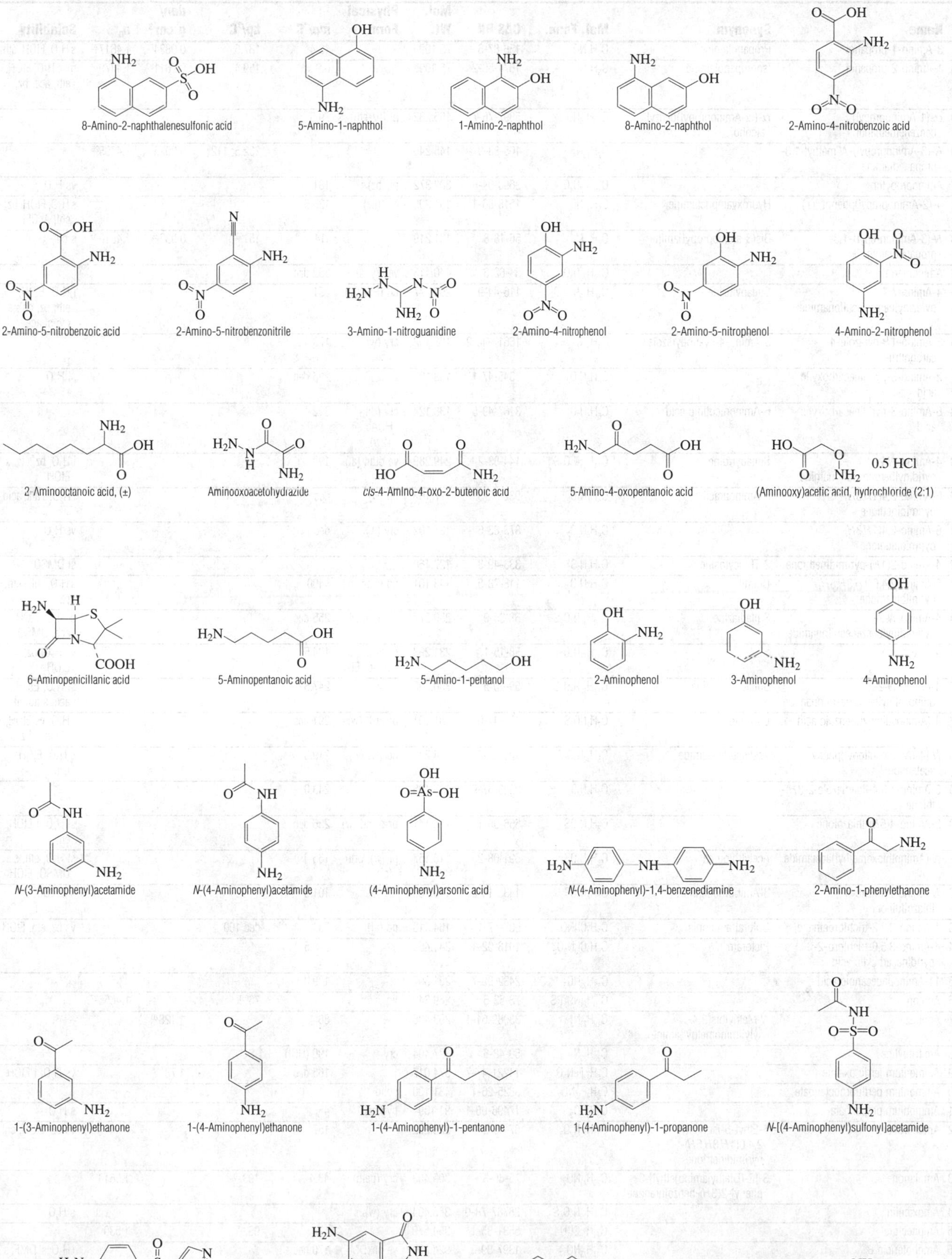

8-Amino-2-naphthalenesulfonic acid
5-Amino-1-naphthol
1-Amino-2-naphthol
8-Amino-2-naphthol
2-Amino-4-nitrobenzoic acid
2-Amino-5-nitrobenzoic acid
2-Amino-5-nitrobenzonitrile
3-Amino-1-nitroguanidine
2-Amino-4-nitrophenol
2-Amino-5-nitrophenol
4-Amino-2-nitrophenol
2-Aminooctanoic acid, (±)
Aminooxoacetohydrazide
cis-4-Amino-4-oxo-2-butenoic acid
5-Amino-4-oxopentanoic acid
(Aminooxy)acetic acid, hydrochloride (2:1)
6-Aminopenicillanic acid
5-Aminopentanoic acid
5-Amino-1-pentanol
2-Aminophenol
3-Aminophenol
4-Aminophenol
N-(3-Aminophenyl)acetamide
N-(4-Aminophenyl)acetamide
(4-Aminophenyl)arsonic acid
N-(4-Aminophenyl)-1,4-benzenediamine
2-Amino-1-phenylethanone
1-(3-Aminophenyl)ethanone
1-(4-Aminophenyl)ethanone
1-(4-Aminophenyl)-1-pentanone
1-(4-Aminophenyl)-1-propanone
N-[(4-Aminophenyl)sulfonyl]acetamide
5-[(4-Aminophenyl)sulfonyl]-2-thiazolamine
4-Aminophthalimide
3-Amino-1,2-propanediol, (±)
3-Aminopropanenitrile
2-Amino-1-propanol, (±)

No.	Name	Synonym	Mol. Form.	CAS RN	Mol. Wt.	Physical Form	mp/°C	bp/°C	den/ g cm^{-3}	n_D	Solubility
407	3-Amino-1-propanol	Propanolamine	C_3H_9NO	156-87-6	75.109		12.4	187.5	0.9824^{26}	1.4617^{20}	s H_2O, EtOH, eth
408	1-Amino-2-propanol	Isopropanolamine	C_3H_9NO	1674-56-2	75.109		0.9	159.4	0.9611^{20}	1.4479^{20}	msc H_2O, EtOH, eth, ace, bz, ctc
409	α-(1-Aminopropyl) benzenemethanol	α-(α-Aminopropyl)benzyl alcohol	$C_{10}H_{15}NO$	5897-76-7	165.232	pl (bz-eth)	79.5				
410	*N*-(3-Aminopropyl)-*N*-methyl-1,3-propanediamine		$C_7H_{19}N_3$	105-83-9	145.246			232.5; 112^{6}	0.9023^{20}	1.4705^{25}	
411	Aminopropylon		$C_{16}H_{22}N_4O_2$	3690-04-8	302.372	pr (bz)	181				vs H_2O
412	4-(2-Aminopropyl)phenol, (±)	Hydroxyamphetamine	$C_9H_{13}NO$	1518-86-1	151.205	cry (bz)	125.5				s H_2O, EtOH, bz, chl, AcOEt
413	*N*-(3-Aminopropyl)-1,3-propanediamine	Bis(3-aminopropyl)amine	$C_6H_{17}N_3$	56-18-8	131.219		-14	151^{50}	0.938^{25}	1.4810^{20}	s chl
414	Aminopterin		$C_{19}H_{20}N_8O_5$	54-62-6	440.413	ye cry	262 dec				
415	4-Amino-*N*-pyrazinylbenzenesulfonamide	Sulfapyrazine	$C_{10}H_{10}N_4O_2S$	116-44-9	250.277	nd ($PhNO_2$)	251				i H_2O, EtOH, eth, bz, chl; s py; sl ace
416	3-Amino-1*H*-pyrazole-4-carbonitrile	3-Amino-4-cyanopyrazole	$C_4H_4N_4$	16617-46-2	108.102	cry (w)	173				
417	2-Amino-3-pyridinecarboxylic acid		$C_6H_6N_2O_2$	5345-47-1	138.124		296 dec				sl H_2O
418	6-Amino-3-pyridinecarboxylic acid	6-Aminonicotinic acid	$C_6H_6N_2O_2$	3167-49-5	138.124	cry (dil HOAc, +2w)	312				
419	4-Amino-*N*-2-pyridinylbenzenesulfonamide	Sulfapyridine	$C_{11}H_{11}N_3O_2S$	144-83-2	249.289	ye oran (al)	192				i H_2O, bz, ctc; s EtOH
420	5-Amino-2,4(1*H*,3*H*)-pyrimidinedione	5-Aminouracil	$C_4H_5N_3O_2$	932-52-5	127.102	nd (w)	dec				i H_2O; s alk, acid
421	6-Amino-2,4(1*H*,3*H*)-pyrimidinedione		$C_4H_5N_3O_2$	873-83-6	127.102	cry (w)	dec				vs H_2O
422	4-Amino-2(1*H*)-pyrimidinethione	2-Thiocytosine	$C_4H_5N_3S$	333-49-3	127.168						sl DMSO
423	5-Amino-2,4,6(1*H*,3*H*,5*H*)-pyrimidinetrione	Uramil	$C_4H_5N_3O_3$	118-78-5	143.101	nd or pl (w)	>400				s H_2O, chl; i eth, bz
424	4-Amino-*N*-2-pyrimidinylbenzenesulfonamide	Sulfadiazine	$C_{10}H_{10}N_4O_2S$	68-35-9	250.277	cry (w), wh pow	255 dec				sl H_2O, EtOH, ace, DMSO
425	Aminopyrine		$C_{13}H_{17}N_3O$	58-15-1	231.293	pr or pl (lig or AcOEt)	134.5				vs H_2O, bz, EtOH
426	4-Amino-*N*-2-quinoxalinylbenzenesulfonamide	Sulfaquinoxaline	$C_{14}H_{12}N_4O_2S$	59-40-5	300.336		247.5				sl H_2O, EtOH, ace; s aq alk
427	4-(Aminosulfonyl)benzoic acid	Carzenide	$C_7H_7NO_4S$	138-41-0	201.201	pr or lf (w)	291 dec				i H_2O; vs EtOH; sl eth; i bz
428	*N*-[4-(Aminosulfonyl)phenyl] acetamide	Acetylsulfanilamide	$C_8H_{10}N_2O_3S$	121-61-9	214.241	nd (HOAc)	219.5				s H_2O, EtOH, ace
429	5-Amino-1,3,4-thiadiazole-2(3*H*)-thione		$C_2H_3N_3S_2$	2349-67-9	133.195		243.0				
430	2-Amino-4(5*H*)-thiazolone		$C_3H_4N_2OS$	556-90-1	116.141	pr or nd (w)	256 dec				sl H_2O; i EtOH, eth
431	*N*-(Aminothioxomethyl)acetamide	Acetylthiourea	$C_3H_6N_2OS$	591-08-2	118.157	pr (w), orth (al)	165				sl H_2O, eth; s DMSO, EtOH
432	*N*-Amino-2-thioxo-4-thiazolidinone	3-Aminorhodanine	$C_3H_4N_2OS_2$	1438-16-0	148.206		101.5				s DMSO
433	1-Amino-2,2,2-trichloroethanol	Chloral ammonia	$C_2H_4Cl_3NO$	507-47-1	164.418	nd (al)	73	dec 100			vs bz, eth, EtOH
434	4-Amino-3,5,6-trichloro-2-pyridinecarboxlic acid	Picloram	$C_6H_3Cl_3N_2O_2$	1918-02-1	241.459		218.5				
435	11-Aminoundecanoic acid		$C_{11}H_{23}NO_2$	2432-99-7	201.307		189.0				
436	Amiton		$C_{10}H_{24}NO_3PS$	78-53-5	269.342	liq		$76^{0.01}$		1.4655^{27}	
437	Amitraz	*N*-Methylbis(2,4-xylyliminomethyl)amine	$C_{19}H_{23}N_3$	33089-61-1	293.406		86		1.128^{20}		
438	Amitriptyline		$C_{20}H_{23}N$	50-48-6	277.404	cry	196 (HCl)				
439	Ammonium ferric oxalate		$C_6H_{12}FeN_3O_{12}$	14221-47-7	374.017		165 dec		$1.78^{17.5}$		vs H_2O; i EtOH
440	Ammonium perfluorooctanoate		$C_8H_4F_{15}NO_2$	3825-26-1	431.100	solid					
441	Ammonium propanoate		$C_3H_9NO_2$	17496-08-1	91.109	hyg cry	45				s H_2O
442	Amobarbital	5-Ethyl-5-isopentyl-2,4,6(1*H*,3*H*,5*H*)-pyrimidinetrione	$C_{11}H_{18}N_2O_3$	57-43-2	226.272		157				vs bz, EtOH, chl
443	Amolanone	3-[2-(Diethylamino)ethyl]-3-phenyl-2(3*H*)-benzofuranone	$C_{20}H_{23}NO_2$	76-65-3	309.403	cry (peth)	43.4	$193^{2.0}$		1.5614^{25}	
444	Amoxicillin		$C_{16}H_{19}N_3O_5S$	26787-78-0	365.404	cry (w)					s H_2O
445	Amphecloral		$C_{11}H_{12}Cl_3N$	5581-35-1	264.579			$96.^{0.5}$		1.530	
446	Amphotericin B		$C_{47}H_{73}NO_{17}$	1397-89-3	924.080	ye pr (DMF)	170 dec				i H_2O; sl DMF; s DMSO
447	Ampicillin		$C_{16}H_{19}N_3O_4S$	69-53-4	349.405	cry	200 dec				sl H_2O
448	Ampyrone		$C_{11}H_{13}N_3O$	83-07-8	203.240	pa ye cry (bz)	109				s H_2O, EtOH, bz, chl; sl eth

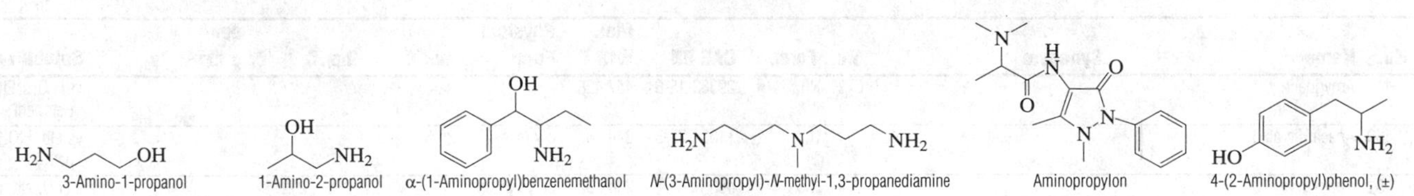

3-Amino-1-propanol

1-Amino-2-propanol

α-(1-Aminopropyl)benzenemethanol

N-(3-Aminopropyl)-*N*-methyl-1,3-propanediamine

Aminopropylon

4-(2-Aminopropyl)phenol, (±)

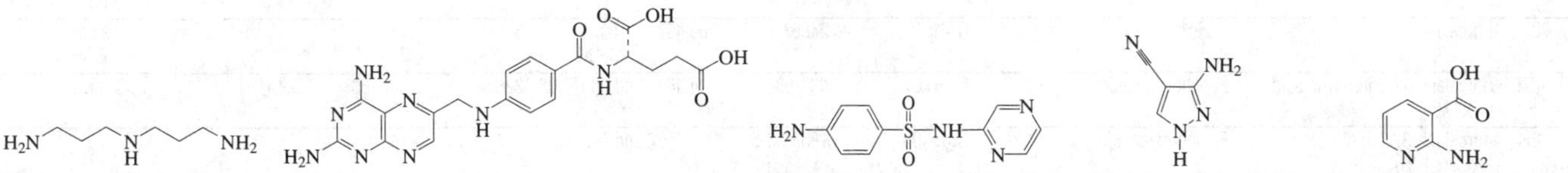

N-(3-Aminopropyl)-1,3-propanediamine

Aminopterin

4-Amino-*N*-pyrazinylbenzenesulfonamide

3-Amino-1*H*-pyrazole-4-carbonitrile

2-Amino-3-pyridinecarboxylic acid

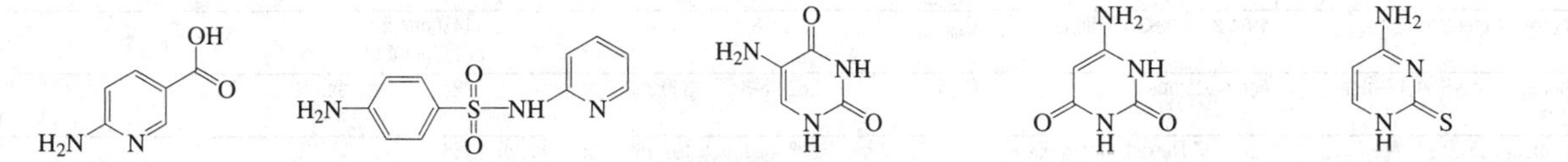

6-Amino-3-pyridinecarboxylic acid

4-Amino-*N*-2-pyridinylbenzenesulfonamide

5-Amino-2,4(1*H*,3*H*)-pyrimidinedione

6-Amino-2,4(1*H*,3*H*)-pyrimidinedione

4-Amino-2(1*H*)-pyrimidinethione

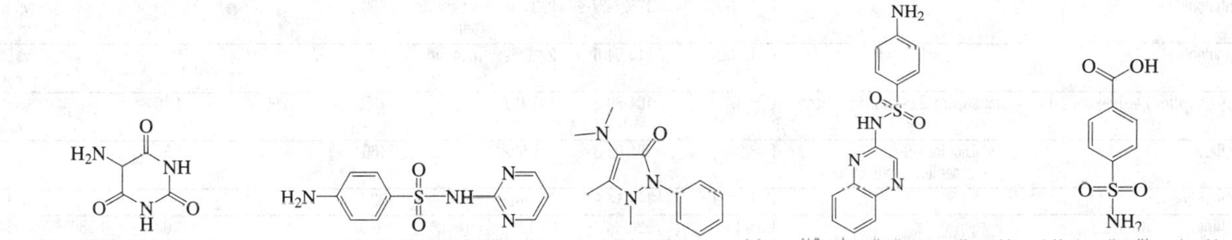

5-Amino-2,4,6(1*H*,3*H*,5*H*)-pyrimidinetrione

4-Amino-*N*-2-pyrimidinylbenzenesulfonamide

Aminopyrine

4-Amino-*N*-2-quinoxalinylbenzenesulfonamide

4-(Aminosulfonyl)benzoic acid

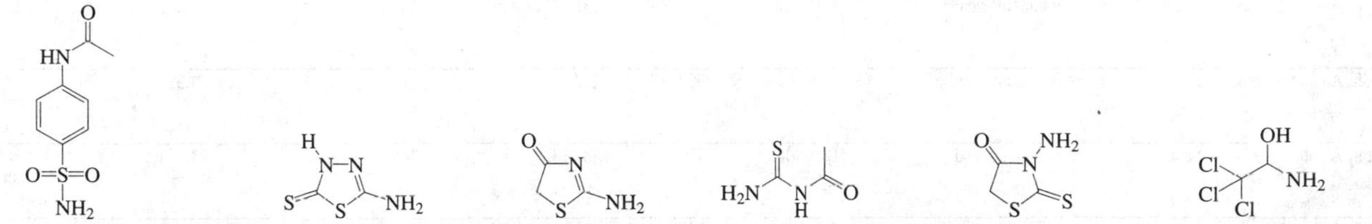

N-[4-(Aminosulfonyl)phenyl]acetamide

5-Amino-1,3,4-thiadiazole-2(3*H*)-thione

2-Amino-4(5*H*)-thiazolone

N-(Aminothioxomethyl)acetamide

N-Amino-2-thioxo-4-thiazolidinone

1-Amino-2,2,2-trichloroethanol

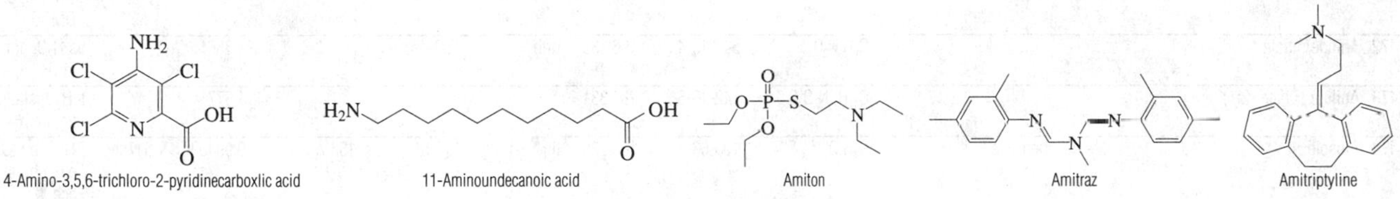

4-Amino-3,5,6-trichloro-2-pyridinecarboxlic acid

11-Aminoundecanoic acid

Amiton

Amitraz

Amitriptyline

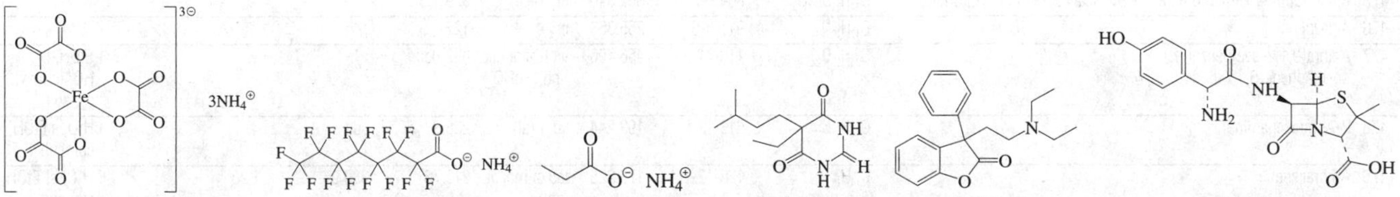

Ammonium ferric oxalate

Ammonium perfluorooctanoate

Ammonium propanoate

Amobarbital

Amolanone

Amoxicillin

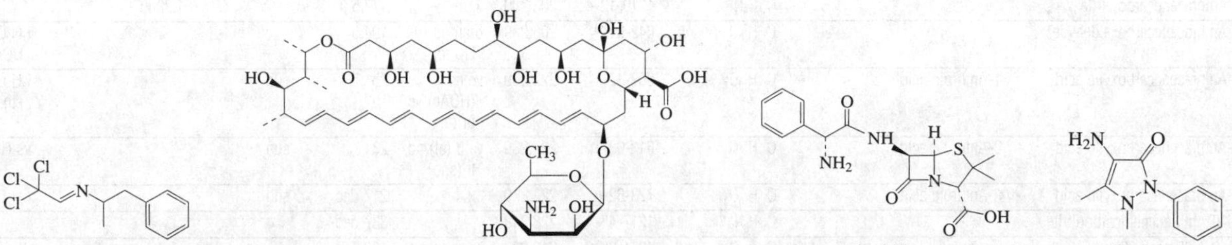

Amphecloral

Amphotericin B

Ampicillin

Ampyrone

No.	Name	Synonym	Mol. Form.	CAS RN	Mol. Wt.	Physical Form	mp/°C	bp/°C	den/ g cm^{-3}	n_D	Solubility
449	Amygdalin		$C_{20}H_{27}NO_{11}$	29883-15-6	457.428		224.5				vs H_2O; sl EtOH; i eth, chl
450	Anacardic acid		$C_{22}H_{32}O_3$	11034-77-8	344.487	cry (ace)	35.5				vs eth, EtOH, peth
451	Anagyrine		$C_{15}H_{20}N_2O$	486-89-5	244.332	pe ye glass		265^{12}, 212^{4}			s H_2O, eth, bz; vs EtOH, chl; i lig
452	Androstane		$C_{19}H_{32}$	24887-75-0	260.457	lf (ace-MeOH)	50	$60^{0.003}$			vs ace, eth, EtOH, peth
453	Androstane-17-carboxylic acid, (5β,17β)	Etiocholanic acid	$C_{20}H_{32}O_2$	438-08-4	304.467	nd (gl HOAc)	228.5	sub 160			
454	Androstane-3,17-diol, (3α,5α,17β)	Epiandrostanediol	$C_{19}H_{32}O_2$	1852-53-5	292.456	nd (ace aq)	223				
455	5α-Androstane-3,17-dione		$C_{19}H_{28}O_2$	846-46-8	288.424	cry (MeOH)	135				
456	5β-Androstane-3,17-dione		$C_{19}H_{28}O_2$	1229-12-5	288.424	cry (ace-hx)	135				
457	Androst-4-ene-3,17-dione	4-Androstene-3,17-dione	$C_{19}H_{26}O_2$	63-05-8	286.408		143(form a); 173(form b)				
458	Androst-4-ene-3,11,17-trione	Adrenosterone	$C_{19}H_{24}O_3$	382-45-6	300.392	nd (al)	222	sub			sl H_2O; s EtOH, eth, ace, chl
459	Anemonin	*trans*-1,7-Dioxadispiro[4.0.4.2] dodeca-3,9-diene-2,8-dione	$C_{10}H_8O_4$	508-44-1	192.169	orth pl (chl) nd (al or bz)	158				vs chl
460	Anhalamine		$C_{11}H_{15}NO_3$	643-60-7	209.242	nd (al)	187.5				vs eth, EtOH
461	Anhalonidine		$C_{12}H_{17}NO_3$	17627-77-9	223.268	oct cry (bz, eth)	160.5				vs H_2O, EtOH
462	Anhalonine		$C_{12}H_{15}NO_3$	519-04-0	221.252	rhom nd	86	$140^{0.02}$			vs EtOH, bz, chl, eth, peth
463	2,5-Anhydro-3,4-dideoxyhexitol	Tetrahydro-2,5-furandimethanol	$C_6H_{12}O_3$	104-80-3	132.157		<-50	265	1.154^{20}		vs H_2O, ace, bz, EtOH
464	Anilazine	2,4-Dichloro-6-(*o*-chloroanilino)-*s*-triazine	$C_9H_5Cl_3N_4$	101-05-3	275.522		160		1.8^{20}		
465	Anileridine		$C_{22}H_{28}N_2O_2$	144-14-9	352.469	cry	83				s H_2O
466	Aniline	Benzenamine	C_6H_7N	62-53-3	93.127	oily liq	-6.02	184.17	1.0217^{20}	1.5863^{20}	s H_2O, ctc, lig; msc EtOH, eth, ace, bz
467	Aniline-2-carboxylic acid	*o*-Anthranilic acid	$C_7H_7NO_2$	118-92-3	137.137	lf (al)	146.5	sub	1.412^{20}		s H_2O, EtOH, eth; sl bz, tfa; vs chl, py
468	Aniline-3-carboxylic acid	*m*-Anthranilic acid	$C_7H_7NO_2$	99-05-8	137.137		173		1.51^{25}		sl H_2O, EtOH; s eth, tfa; vs ace; i bz
469	Aniline-4-carboxylic acid	*p*-Anthranilic acid	$C_7H_7NO_2$	150-13-0	137.137	mcl pr (w)	188.2		1.374^{20}		s H_2O, EtOH, eth; sl ace; i bz, chl
470	Aniline hydrobromide		C_6H_8BrN	542-11-0	174.039		286				
471	Aniline hydrochloride	Benzenamine hydrochloride	C_6H_8ClN	142-04-1	129.588	lf or nd	198		1.2215^{4}		vs H_2O, EtOH; i eth, chl; sl DMSO
472	Aniline nitrate		$C_6H_8N_2O_3$	542-15-4	156.139	orth	190 dec		1.356^{4}		vs H_2O, eth, EtOH
473	Aniline sulfate (2:1)		$C_{12}H_{16}N_2O_4S$	542-16-5	284.331				1.377^{4}		s H_2O; sl EtOH, tfa; i eth
474	Anisole	Methoxybenzene	C_7H_8O	100-66-3	108.138	liq	-37.13	153.7	0.9940^{20}	1.5174^{20}	i H_2O; s EtOH, eth, chl; vs ace, bz
475	Anisotropine methylbromide	Octatropine methylbromide	$C_{17}H_{32}BrNO_2$	80-50-2	362.346	cry (ace)	329				
476	Antazoline		$C_{17}H_{19}N_3$	91-75-8	265.353	cry	122				
477	Anthra[9,1,2-cde]benzo[rst] pentaphene-5,10-dione		$C_{34}H_{16}O_2$	116-71-2	456.490	viol-bl or blk nd ($PhNO_2$)	492 dec				i EtOH, bz, HOAc; s xyl, py, sulf
478	2-Anthracenamine		$C_{14}H_{11}N$	613-13-8	193.244	ye lf (al)	238.8	sub			i H_2O; s EtOH; i con sulf
479	Anthracene		$C_{14}H_{10}$	120-12-7	178.229	tab or mcl pr (al)	215.76	339.9	1.28^{25}		i H_2O; sl EtOH, eth, ace, bz, chl, ctc
480	9-Anthracenecarbonitrile		$C_{15}H_9N$	1210-12-4	203.239		177.5		1.3000^{20}		
481	9-Anthracenecarboxaldehyde		$C_{15}H_{10}O$	642-31-9	206.239	oran nd (dil HOAc)	104.5				i H_2O; s bz, HOAc
482	1-Anthracenecarboxylic acid	1-Anthroic acid	$C_{15}H_{10}O_2$	607-42-1	222.239	ye nd (HOAc) ye pr (al)	251.5	sub			i H_2O; s EtOH, eth; sl bz, chl
483	2-Anthracenecarboxylic acid	2-Anthroic acid	$C_{15}H_{10}O_2$	613-08-1	222.239	ye lf (al) nd, lf (sub)	281	sub			vs HOAc
484	9-Anthracenecarboxylic acid	9-Anthroic acid	$C_{15}H_{10}O_2$	723-62-6	222.239		217 dec	sub			i H_2O; s EtOH
485	9,10-Anthracenedicarbonitrile		$C_{16}H_8N_2$	1217-45-4	228.248		337 dec				
486	9,10-Anthracenediol		$C_{14}H_{10}O_2$	4981-66-2	210.228	br or ye nd	180				vs eth, EtOH

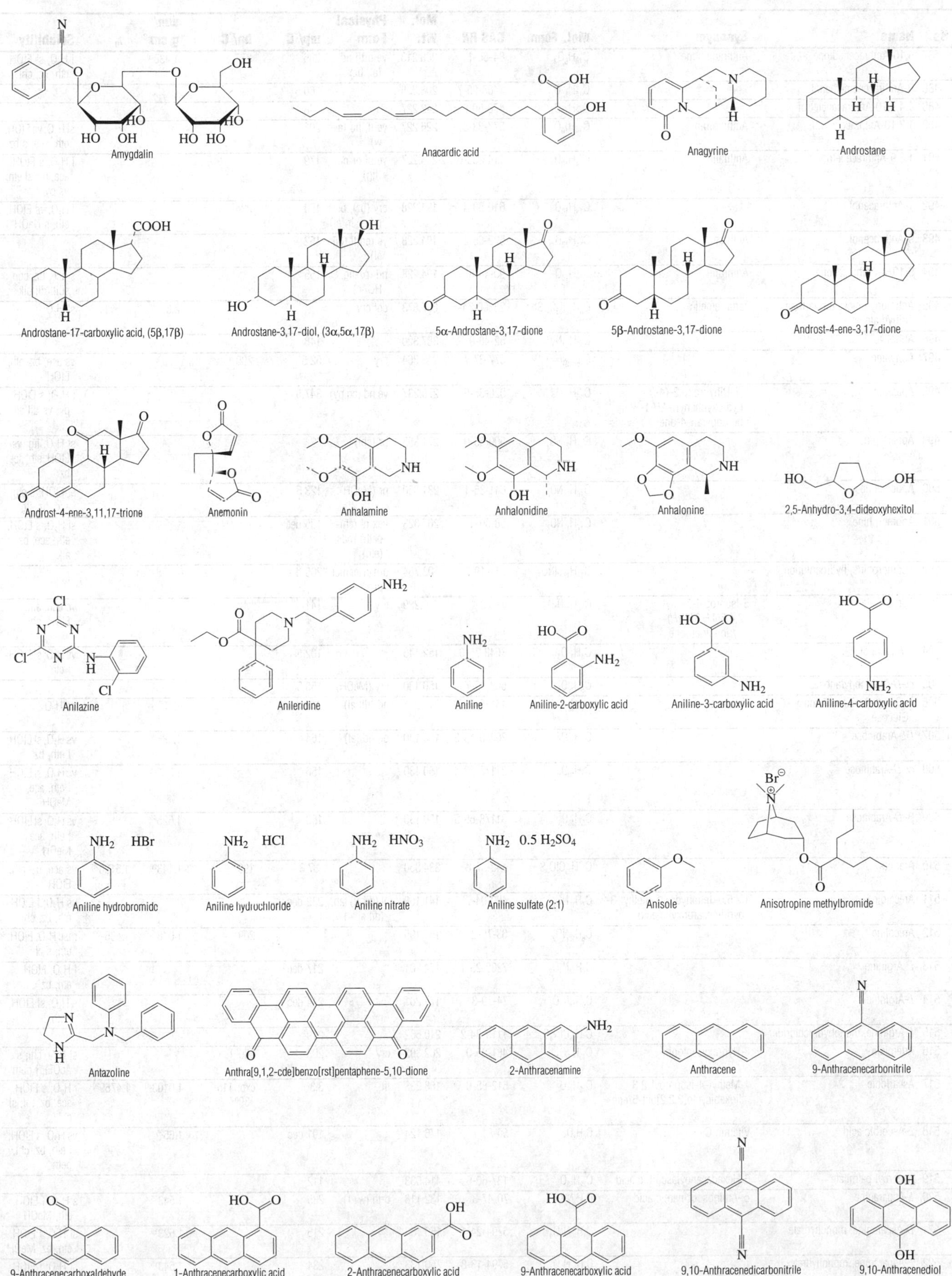
Amygdalin
Anacardic acid
Anagyrine
Androstane
Androstane-17-carboxylic acid, (5β,17β)
Androstane-3,17-diol, (3α,5α,17β)
5α-Androstane-3,17-dione
5β-Androstane-3,17-dione
Androst-4-ene-3,17-dione
Androst-4-ene-3,11,17-trione
Anemonin
Anhalamine
Anhalonidine
Anhalonine
2,5-Anhydro-3,4-dideoxyhexitol
Anilazine
Anileridine
Aniline
Aniline-2-carboxylic acid
Aniline-3-carboxylic acid
Aniline-4-carboxylic acid
Aniline hydrobromide
Aniline hydrochloride
Aniline nitrate
Aniline sulfate (2:1)
Anisole
Anisotropine methylbromide
Antazoline
Anthra[9,1,2-cde]benzo[rst]pentaphene-5,10-dione
2-Anthracenamine
Anthracene
9-Anthracenecarbonitrile
9-Anthracenecarboxaldehyde
1-Anthracenecarboxylic acid
2-Anthracenecarboxylic acid
9-Anthracenecarboxylic acid
9,10-Anthracenedicarbonitrile
9,10-Anthracenediol

No.	Name	Synonym	Mol. Form.	CAS RN	Mol. Wt.	Physical Form	mp/°C	bp/°C	den/g cm^{-3}	n_D	Solubility
487	9,10-Anthracenedione	Anthraquinone	$C_{14}H_8O_2$	84-65-1	208.213	ye orth nd (al, bz)	286	377	1.438^{20}		i H_2O; sl EtOH, eth, bz, chl
488	9-Anthracenemethanol		$C_{15}H_{12}O$	1468-95-7	208.255		160.5				
489	1,4,9,10-Anthracenetetrol		$C_{14}H_{10}O_4$	476-60-8	242.227		148				
490	1,2,10-Anthracenetriol	Anthrarobin	$C_{14}H_{10}O_3$	577-33-3	226.227	ye lf, nd (al-w)	208				sl H_2O; vs EtOH, eth, ace; s bz
491	1,8,9-Anthracenetriol	Anthralin	$C_{14}H_{10}O_3$	1143-38-0	226.227	ye pl or nd (lig)	179				i H_2O; s EtOH, ace, bz; sl eth; vs py
492	1-Anthracenol		$C_{14}H_{10}O$	610-50-4	194.228	cry (bz), br nd or lf (al)	158	234^{13}			i H_2O; vs EtOH, eth; s NaOH
493	9-Anthracenol	Anthranol	$C_{14}H_{10}O$	529-86-2	194.228	ye red lf (dil al)	152				
494	9(10*H*)-Anthracenone	Anthrone	$C_{14}H_{10}O$	90-44-8	194.228	nd (bz-lig, HOAc)	155				s ace, bz, con sulf, dil alk
495	Antimony potassium tartrate trihydrate	Tartar emetic	$C_8H_{10}K_2O_{15}Sb_2$	28300-74-5	667.873	col cry			2.6		sl H_2O
496	Apholate		$C_{12}H_{24}N_9P_3$	52-46-0	387.300		148				
497	Aphylline		$C_{15}H_{24}N_2O$	577-37-7	248.364	cry	52.5	200^4			vs ace, bz, eth, EtOH
498	Apigenin	5,7-Dihydroxy-2-(4-hydroxyphenyl)-4*H*-1-benzopyran-4-one	$C_{15}H_{10}O_5$	520-36-5	270.237	ye nd (aq py)	347.5				i H_2O; s EtOH, py; vs dil alk
499	Apoatropine		$C_{17}H_{21}NO_2$	500-55-0	271.355	pr (chl)	62				sl H_2O, lig; vs EtOH, eth, ace, bz
500	Apocodeine		$C_{18}H_{19}NO_2$	641-36-1	281.350	pr (MeOH)	123.5				sl EtOH; s eth, ace, bz, lig
501	Apomorphine		$C_{17}H_{17}NO_2$	58-00-4	267.323	hex pl (chl-peth) rods (eth)	195 dec				sl H_2O; s EtOH, eth, ace, bz, alk
502	Apomorphine, hydrochloride		$C_{17}H_{18}ClNO_2$	314-19-2	303.784	grn in air mcl pr	205 dec				
503	Aprobarbital	5-Isopropyl-5-allyl-2,4,6(1*H*,3*H*,5*H*)-pyrimidinetrione	$C_{10}H_{14}N_2O_3$	77-02-1	210.229	cry	141				vs ace, eth, EtOH, chl
504	*L*-Arabinitol		$C_5H_{12}O_5$	7643-75-6	152.146		102.5				vs H_2O; sl EtOH; i eth
505	α-*D*-Arabinopyranose		$C_5H_{10}O_5$	608-45-7	150.130	cry (MeOH)	155.5		1.585^{25}		
506	6-*O*-α-*L*-Arabinopyranosyl-*D*-Glucose	Vicianose	$C_{11}H_{20}O_{10}$	14116-69-9	312.271	nd (dil al)	210 dec				vs H_2O
507	*DL*-Arabinose		$C_5H_{10}O_5$	20235-19-2	150.130	pr, nd (al)	164.5		1.585^{20}		vs H_2O; sl EtOH; i eth, bz
508	α-*D*-Arabinose		$C_5H_{10}O_5$	31178-68-4	150.130		156		1.585^{25}		vs H_2O; sl EtOH; i eth, ace, MeOH
509	β-*D*-Arabinose		$C_5H_{10}O_5$	31178-69-5	150.130		156		1.625^{25}		vs H_2O; sl EtOH; i eth, ace, MeOH
510	Aramite		$C_{15}H_{23}ClO_4S$	140-57-8	334.860		-37.3	195^2	1.143^{20}	1.5100^{20}	vs ace, bz, eth, EtOH
511	Arecaidine	1,2,5,6-Tetrahydro-1-methyl-3-pyridinecarboxylic acid	$C_7H_{11}NO_2$	499-04-7	141.168	pl (dil al) tab (dil al +1w)	232 dec				vs H_2O; i EtOH, eth, bz, chl
512	Arecoline		$C_8H_{13}NO_2$	63-75-2	155.195			209	1.0485^{20}	1.486-^{20}	msc H_2O, EtOH, eth; s chl
513	*D*-Arginine		$C_6H_{14}N_4O_2$	7200-25-1	174.201		217 dec				i H_2O, EtOH, eth, bz
514	*L*-Arginine		$C_6H_{14}N_4O_2$	74-79-3	174.201		244 dec				s H_2O; sl EtOH; i eth
515	*L*-Arginine, monohydrochloride		$C_6H_{15}ClN_4O_2$	1119-34-2	210.662		219				
516	Artemisin	8-Hydroxysantonin	$C_{15}H_{18}O_4$	481-05-0	262.302	cry	203	$260^{0.1}$			sl H_2O, chl; s AcOEt; i peth
517	Ascaridole	1-Methyl-4-isopropyl-2,3-dioxabicyclo[2.2.2]oct-5-ene	$C_{10}H_{16}O_2$	512-85-6	168.233	liq	3.3	exp; 115^{15}, $39^{0.2}$	1.0103^{20}	1.4769^{20}	i H_2O; s EtOH, ace, bz, tol; sl chl
518	*L*-Ascorbic acid	Vitamin C	$C_6H_8O_6$	50-81-7	176.124		191 dec		1.65^{25}		vs H_2O; s EtOH; i eth, bz, chl, peth
519	Ascorbyl palmitate	6-Hexadecanoylascorbic acid	$C_{22}H_{38}O_7$	137-66-6	414.533		112				
520	*L*-Asparagine	α-Aminosuccinamic acid	$C_4H_8N_2O_3$	70-47-3	132.118	orth (w+1)	235		1.543^{15}		s H_2O; i EtOH, eth, MeOH
521	*D*-Asparagine, monohydrate		$C_4H_{10}N_2O_4$	5794-24-1	150.133		215		1.523^{15}		sl H_2O; i EtOH, eth, bz, MeOH
522	*L*-Asparagine, monohydrate		$C_4H_{10}N_2O_4$	5794-13-8	150.133		234		1.543^{15}		sl H_2O; i EtOH, eth, bz, MeOH

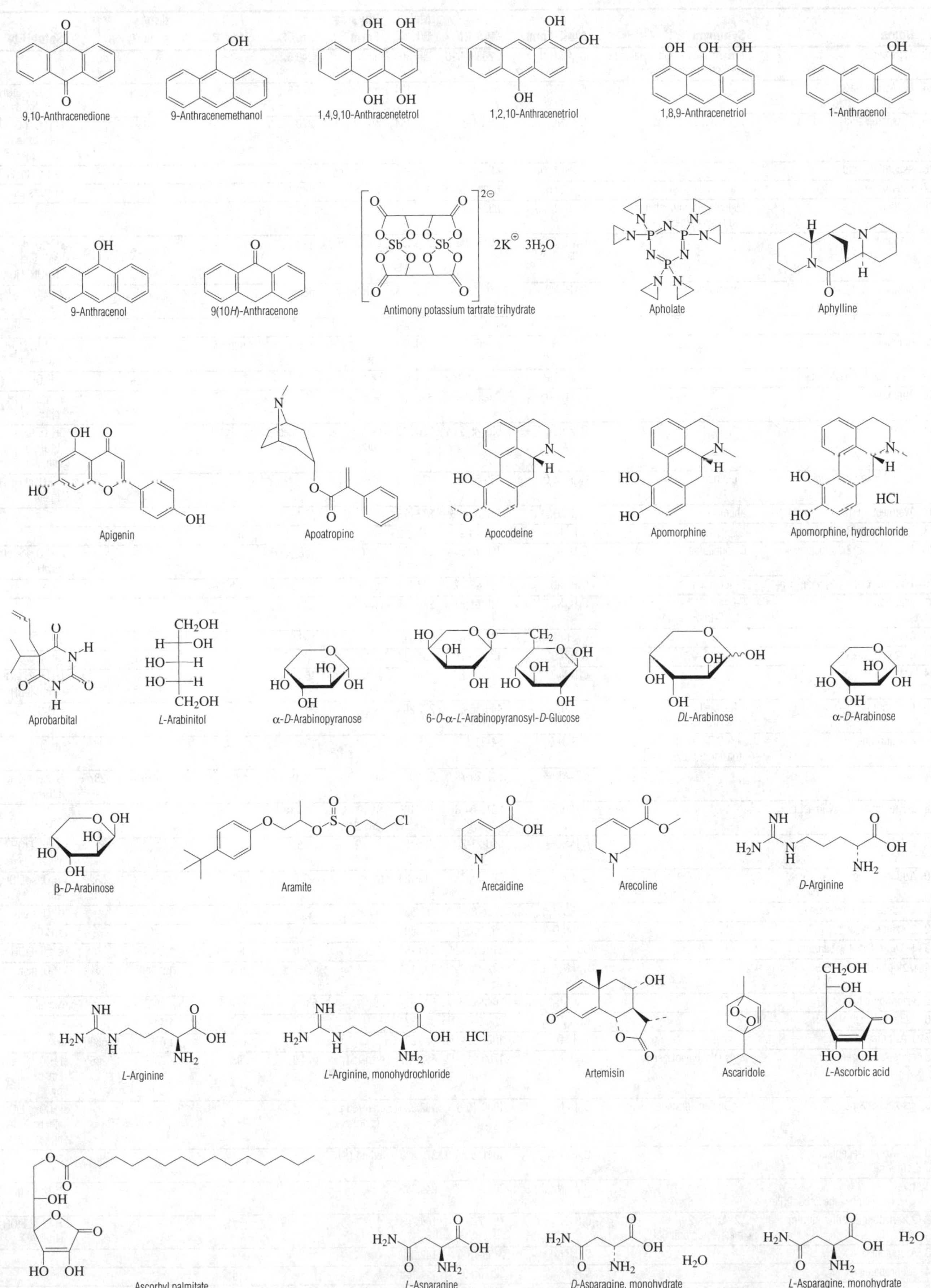
9,10-Anthracenedione
9-Anthracenemethanol
1,4,9,10-Anthracenetetrol
1,2,10-Anthracenetriol
1,8,9-Anthracenetriol
1-Anthracenol
9-Anthracenol
9(10H)-Anthracenone
Antimony potassium tartrate trihydrate
Apholate
Aphylline
Apigenin
Apoatropine
Apocodeine
Apomorphine
Apomorphine, hydrochloride
Aprobarbital
L-Arabinitol
α-D-Arabinopyranose
6-O-α-L-Arabinopyranosyl-D-Glucose
DL-Arabinose
α-D-Arabinose
β-D-Arabinose
Aramite
Arecaidine
Arecoline
D-Arginine
L-Arginine
L-Arginine, monohydrochloride
Artemisin
Ascaridole
L-Ascorbic acid
Ascorbyl palmitate
L-Asparagine
D-Asparagine, monohydrate
L-Asparagine, monohydrate

No.	Name	Synonym	Mol. Form.	CAS RN	Mol. Wt.	Physical Form	mp/°C	bp/°C	den/ g cm^{-3}	n_D	Solubility
523	Aspartame	*L*-α-Aspartyl-*L*-phenylalanine, 2-methyl ester	$C_{14}H_{18}N_2O_5$	22839-47-0	294.303	nd (w)	246.5				
524	*DL*-Aspartic acid		$C_4H_7NO_4$	617-45-8	133.104	mcl pr (w)	277.5		1.6622^{13}		sl H_2O; i EtOH, eth, bz, py
525	*L*-Aspartic acid	*L*-Aminosuccinic acid	$C_4H_7NO_4$	56-84-8	133.104	orth lf (w)	270		1.6603^{13}		sl H_2O; i EtOH, eth, bz; s dil HCl, py
526	Aspergillic acid		$C_{12}H_{20}N_2O_2$	490-02-8	224.299	pa ye rods	98				vs bz, eth, EtOH
527	Astemizole		$C_{28}H_{31}FN_4O$	68844-77-9	458.570	wh cry	149.1				i H_2O; s os
528	Asulam	Methyl [(4-aminophenyl) sulfonyl]carbamate	$C_8H_{10}N_2O_4S$	3337-71-1	230.241		144				
529	Atenolol		$C_{14}H_{22}N_2O_3$	29122-68-7	266.336	cry (AcOEt)	147				sl H_2O, diox, ace; i chl; s MeOH, HOAc
530	Atisine	Anthorine	$C_{22}H_{33}NO_2$	466-43-3	343.503	orth bipym	58.5				vs eth, EtOH, chl
531	Atrazine		$C_8H_{14}ClN_5$	1912-24-9	215.684		173				
532	Atropine		$C_{17}H_{23}NO_3$	51-55-8	289.370	orth nd (dil al)	118.5	sub 95			vs H_2O, EtOH; i eth; sl chl
533	Auramine hydrochloride		$C_{17}H_{22}ClN_3$	2465-27-2	303.83	ye nd (w)	267				sl H_2O
534	Aureothin		$C_{22}H_{23}NO_6$	2825-00-5	397.421	ye pr	158				vs ace, EtOH, chl
535	Aurin		$C_{19}H_{14}O_3$	603-45-2	290.312	dk red lf or orth	309 dec				i H_2O, bz; s EtOH, alk; sl eth, chl
536	Aurin tricarboxylic acid, triammonium salt	Aluminon	$C_{22}H_{23}N_3O_9$	569-58-4	473.433	red-br pow					s H_2O; sl EtOH; i peth
537	Avermectin B1a	Abamectin	$C_{48}H_{72}O_{14}$	71751-41-2	873.078		152				
538	3-Azabicyclo[3.2.2]nonane		$C_8H_{15}N$	283-24-9	125.212			166^{500}			
539	1-Azabicyclo[2.2.2]octane	Quinuclidine	$C_7H_{13}N$	100-76-5	111.185	cry (eth)	158				vs H_2O, ace, eth, EtOH
540	1-Azabicyclo[2.2.2]octan-3-ol	3-Quinuclidinol	$C_7H_{13}NO$	1619-34-7	127.184	cry (bz)	221	sub 120			s ace
541	Azacitidine	4-Amino-1-β-*D*-ribofuranosyl-1,3,5-triazine-2(1*H*)-one	$C_8H_{12}N_4O_5$	320-67-2	244.205	cry	229				
542	Azacyclotridecan-2-one		$C_{12}H_{23}NO$	947-04-6	197.317		152.5				
543	8-Azaguanine		$C_4H_4N_6O$	134-58-7	152.114		300				
544	Azaserine		$C_5H_7N_3O_4$	115-02-6	173.128	ye-grn orth cry	150 dec				vs H_2O; sl EtOH, ace, MeOH
545	Azathioprine	6-[(1-Methyl-4-nitro-1*H*-imidazol-5-yl)thio]-1*H*-purine	$C_9H_7N_7O_2S$	446-86-6	277.263	ye cry	243 dec				sl H_2O, EtOH, chl
546	6-Azauridine	2-β-*D*-Ribofuranosyl-1,2,4-triazine-3,5(2*H*,4*H*)-dione	$C_8H_{11}N_3O_6$	54-25-1	245.189		158				s H_2O
547	Azetidine		C_3H_7N	503-29-7	57.095	liq	-70.0	63	0.8436^{20}	1.4287^{25}	vs ace, bz, eth, EtOH
548	2-Azetidinecarboxylic acid		$C_4H_7NO_2$	2517-04-6	101.105	cry (95% MeOH)	217 dec				
549	2-Azetidinone		C_3H_5NO	930-21-2	71.078		73.5	106^{15}			vs eth, EtOH, chl
550	Azidobenzene		$C_6H_5N_3$	622-37-7	119.124	pa ye oil	-27.5	70^{11}	1.0860^{20}	1.5589^{25}	i H_2O; sl EtOH, eth
551	1-Azido-4-chlorobenzene		$C_6H_4ClN_3$	3296-05-7	153.569		20	96^{20}	1.2634^{25}		i H_2O; s eth
552	2-Azidoethanol		$C_2H_5N_3O$	1517-05-1	87.080			75^{40}	1.146^{24}		vs H_2O
553	1-Azido-4-methylbenzene		$C_7H_7N_3$	2101-86-2	133.151		-29.0	dec 180; 80^{10}	1.0527^{23}		vs eth, EtOH
554	(Azidomethyl)benzene		$C_7H_7N_3$	622-79-7	133.151			108^{23}, 78^{12}	1.0730^{19}	1.5341^{25}	i H_2O; msc EtOH, eth
555	Azinphos ethyl		$C_{12}H_{16}N_3O_3PS_2$	2642-71-9	345.377	nd	53	$111^{0.001}$	1.284^{20}		reac alk
556	Azinphos-methyl		$C_{10}H_{12}N_3O_3PS_2$	86-50-0	317.324		73		1.44^{20}		
557	1-Aziridineethanol		C_4H_9NO	1072-52-2	87.120			168	1.088^{25}	1.4560^{20}	
558	*trans*-Azobenzene	*trans*-Diphenyldiazene	$C_{12}H_{10}N_2$	17082-12-1	182.220	oran-red mcl lf (al)	67.88	293	1.203^{20}	1.6266^{78}	sl H_2O; s EtOH, eth, bz, chl; vs py
559	*cis*-Azobenzene	*cis*-Diphenyldiazene	$C_{12}H_{10}N_2$	1080-16-6	182.220	oran-red pl (peth)	71				sl H_2O; s EtOH, eth, bz, HOAc, lig
560	3,3'-Azobenzenedisulfonyl chloride		$C_{12}H_8Cl_2N_2O_4S_2$	104115-88-0	379.239	red nd (eth)	166.5				vs eth
561	1,1'-Azobiscyclohexanecarbonitrile		$C_{14}H_{20}N_4$	2094-98-6	244.336		100				i H_2O; s lig
562	2,2'-Azobis[isobutyronitrile]	2,2'-Azobis[2-methylpropionitrile]	$C_8H_{12}N_4$	78-67-1	164.208						i H_2O; sl EtOH, eth
563	Azobutane		$C_8H_{18}N_2$	2159-75-3	142.242			60^{18}			
564	Azopropane		$C_6H_{14}N_2$	821-67-0	114.188			114			
565	*cis*-Azoxybenzene	Diphenyldiazene 1-oxide, (*E*)	$C_{12}H_{10}N_2O$	21650-65-7	198.219		87		1.166^{20}	1.633^{20}	

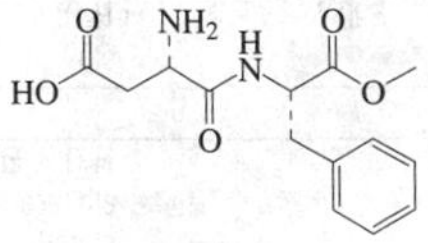

Aspartame

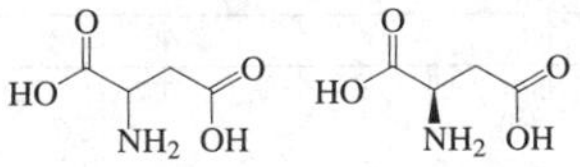

DL-Aspartic acid

L-Aspartic acid

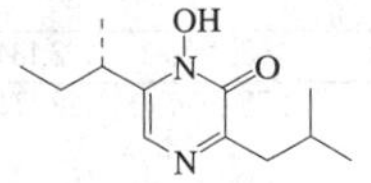

Aspergillic acid

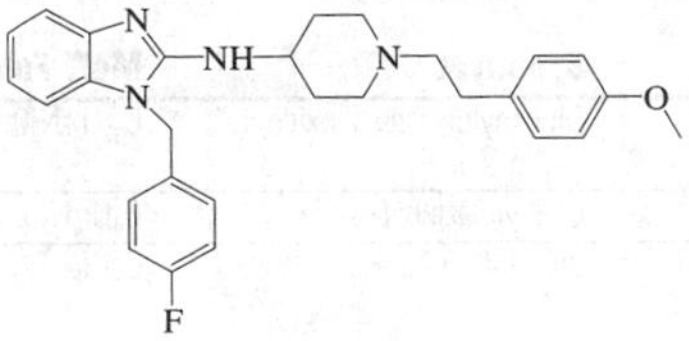

Astemizole

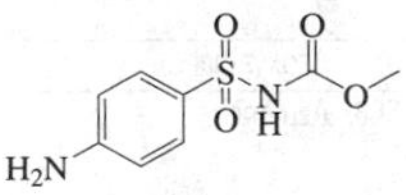

Asulam

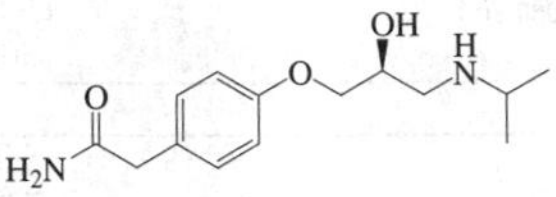

Atenolol

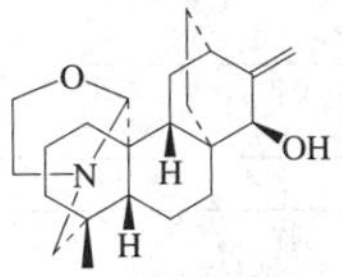

Atisine

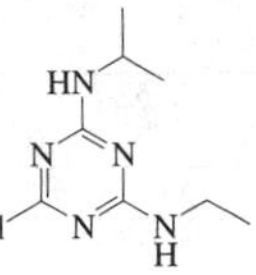

Atrazine

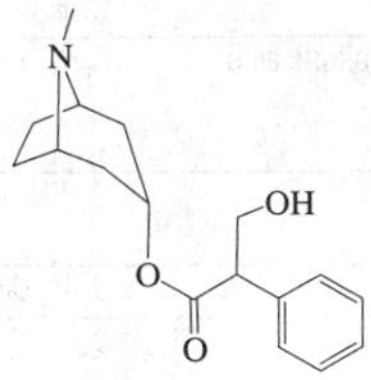

Atropine

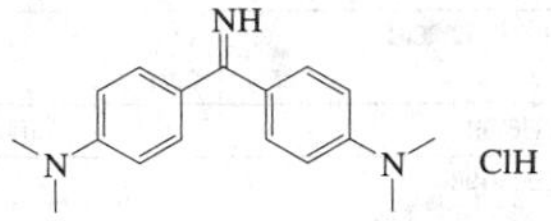

Auramine hydrochloride

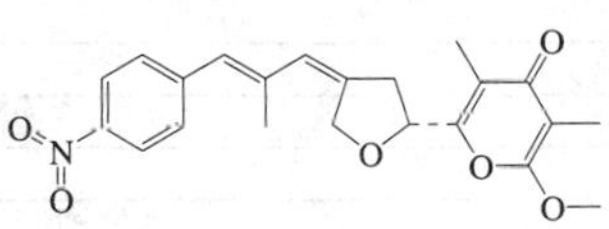

Aureothin

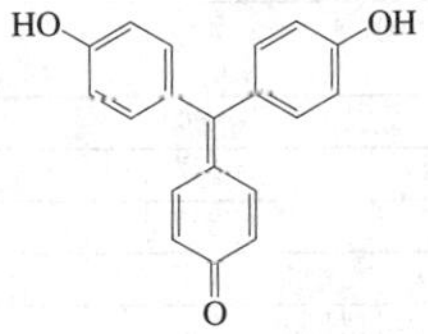

Aurin

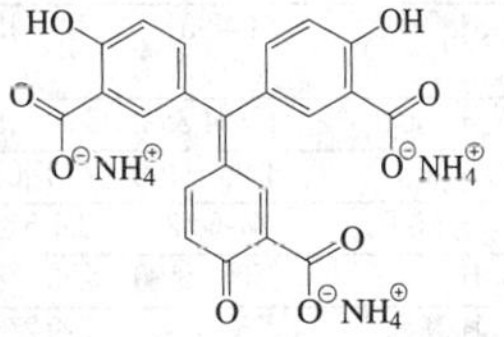

Aurin tricarboxylic acid, triammonium salt

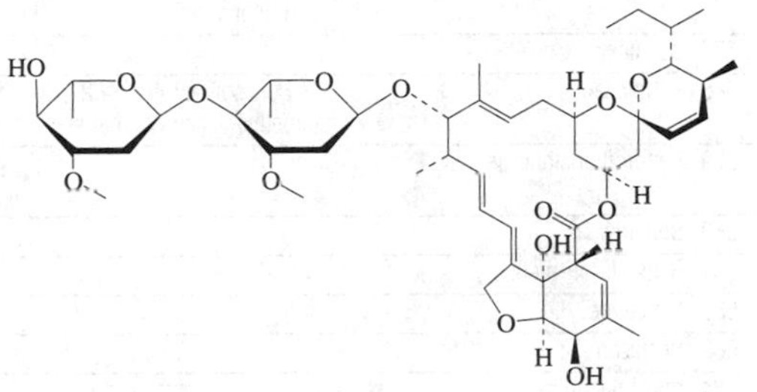

Avermectin B1a

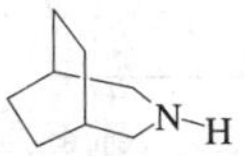

3-Azabicyclo[3.2.2]nonane

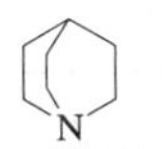

1-Azabicyclo[2.2.2]octane

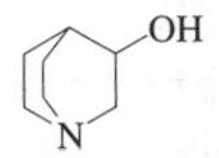

1-Azabicyclo[2.2.2]octan-3-ol

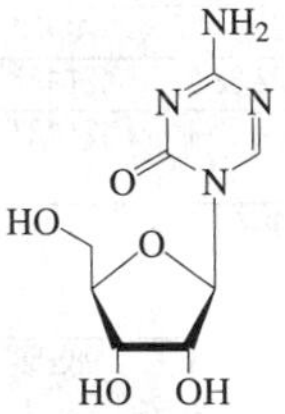

Azacitidine

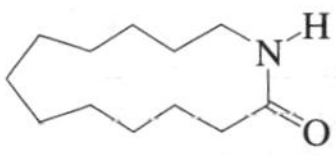

Azacyclotridecan-2-one

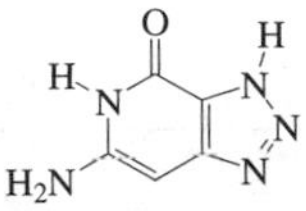

8-Azaguanine

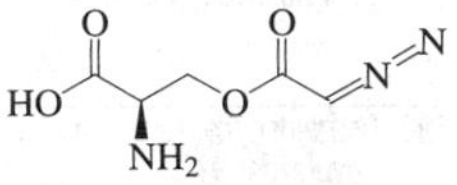

Azaserine

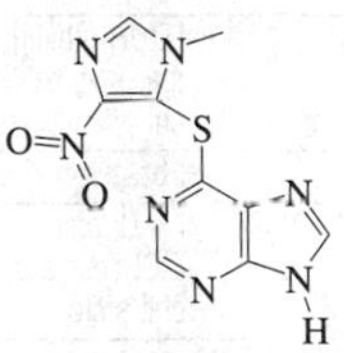

Azathioprine

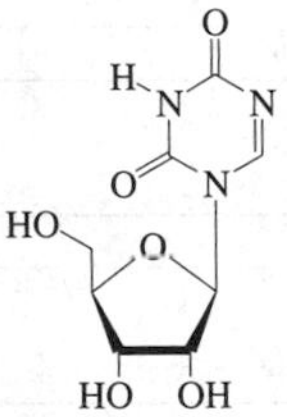

6-Azauridine

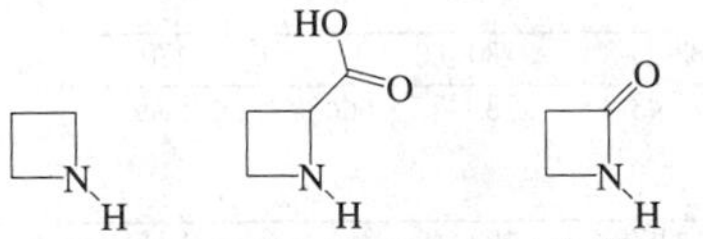

Azetidine

2-Azetidinecarboxylic acid

2-Azetidinone

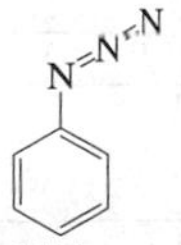

Azidobenzene

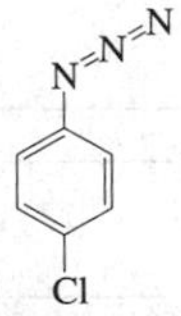

1-Azido-4-chlorobenzene

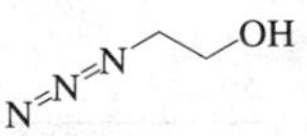

2-Azidoethanol

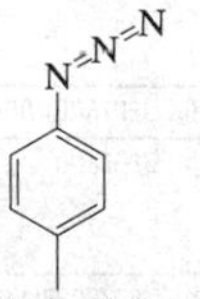

1-Azido-4-methylbenzene

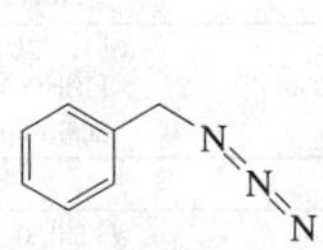

(Azidomethyl)benzene

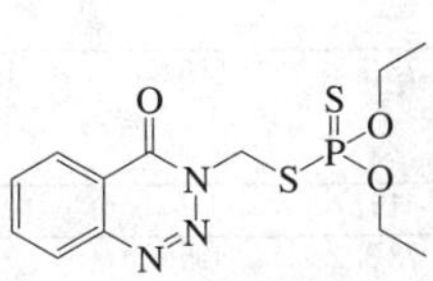

Azinphos ethyl

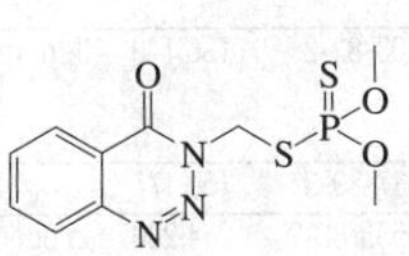

Azinphos-methyl

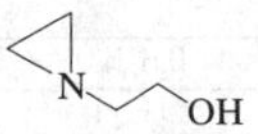

1-Aziridineethanol

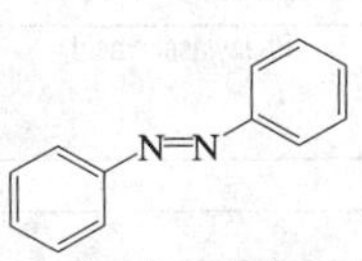

trans-Azobenzene

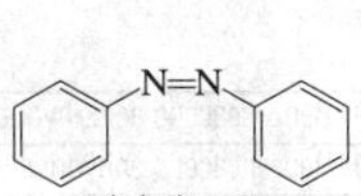

cis-Azobenzene

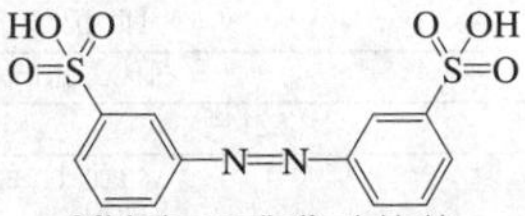

3,3'-Azobenzenedisulfonyl chloride

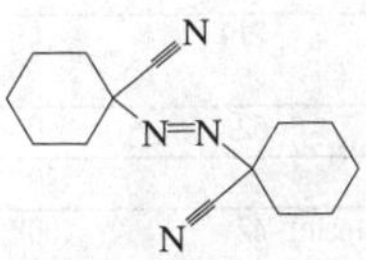

1,1'-Azobiscyclohexanecarbonitrile

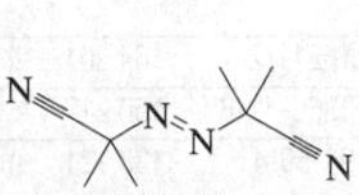

2,2'-Azobis[isobutyronitrile]

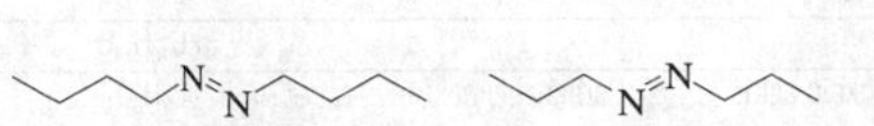

Azobutane

Azopropane

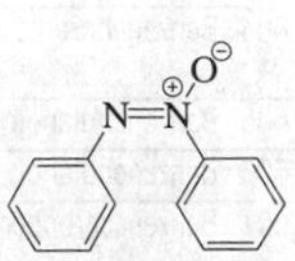

cis-Azoxybenzene

No.	Name	Synonym	Mol. Form.	CAS RN	Mol. Wt.	Physical Form	mp/°C	bp/°C	den/ g cm-3	n_D	Solubility
566	*trans*-Azoxybenzene	Diphenyldiazene 1-oxide, (*Z*)	$C_{12}H_{10}N_2O$	20972-43-4	198.219		34.6		1.1590^{26}		i H_2O; s EtOH, eth
567	Azoxyethane	Diethyldiazine 1-oxide	$C_4H_{10}N_2O$	16301-26-1	102.134	liq		46			
568	Azulene	Bicyclo[5.3.0]decapentaene	$C_{10}H_8$	275-51-4	128.171	bl or gr-blk lf (al)	99	dec 270; 125^{10}			i H_2O; s EtOH, eth, ace, acid; sl chl
569	Balan	*N*-Butyl-*N*-ethyl-2,6-dinitro-4-(trifluoromethyl)aniline	$C_{13}H_{16}F_3N_3O_4$	1861-40-1	335.279		66	$121^{0.5}$, 148^7			
570	Barban		$C_{11}H_9Cl_2NO_2$	101-27-9	258.101		75				
571	Barbital	5,5-Diethylbarbituric acid	$C_8H_{12}N_2O_3$	57-44-3	184.192	nd (w)	190		1.220^{25}		sl H_2O; s EtOH, eth, ace, chl, lig, tfa
572	Barbituric acid		$C_4H_4N_2O_3$	67-52-7	128.086	orth pr (w +2)	248	dec 260			s H_2O, eth; sl EtOH
573	Bayleton	Triadimefon	$C_{14}H_{16}ClN_3O_2$	43121-43-3	293.749		82		1.22^{20}		
574	Bebeerine		$C_{36}H_{38}N_2O_6$	477-60-1	594.696	cry (bz, eth, chl-MeOH)	221				s EtOH, MeOH, eth; vs ace, chl
575	Benactyzine	2-(Diethylamino)ethyl benzilate	$C_{20}H_{25}NO_3$	302-40-9	327.418	cry	51				
576	Benactyzine hydrochloride	2-Diethylaminoethyl benzilate hydrochloride	$C_{20}H_{26}ClNO_3$	57-37-4	363.878		177.5				s H_2O; i eth
577	Benalaxyl		$C_{20}H_{23}NO_3$	71626-11-4	325.402		79		1.27^{25}		
578	Bendiocarb	1,3-Benzodioxol-4-ol, 2,2-dimethyl-, methylcarbamate	$C_{11}H_{13}NO_4$	22781-23-3	223.226		130		1.25^{20}		
579	Bendroflumethiazide		$C_{15}H_{14}F_3N_3O_4S_2$	73-48-3	421.415	cry	225				i H_2O, bz, eth; s EtOH, ace
580	Benomyl		$C_{14}H_{18}N_4O_3$	17804-35-2	290.318		dec				
581	Bensulfuron-methyl		$C_{16}H_{18}N_4O_7S$	83055-99-6	410.402		187				
582	Bensulide		$C_{14}H_{24}NO_4PS_3$	741-58-2	397.514		34.4		1.224^{20}		
583	Bentazon		$C_{10}H_{12}N_2O_3S$	25057-89-0	240.278		138				
584	Benz[c]acridine	12-Azabenz[a]anthracene	$C_{17}H_{11}N$	225-51-4	229.276	nd (dil al)	132				vs bz, eth, EtOH
585	Benzaldehyde	Benzenecarboxaldehyde	C_7H_6O	100-52-7	106.122	liq	-57.1	178.8	1.0401^{25}	1.5463^{20}	sl H_2O; msc EtOH, eth; vs ace, bz
586	Benzaldehyde hydrazone	Benzylidene hydrazine	$C_7H_8N_2$	5281-18-5	120.152	lf	16	140^{14}			s EtOH
587	*cis*-Benzaldehyde oxime		C_7H_7NO	622-32-2	121.137	pr	36.5	200	1.1111^{20}	1.5908^{20}	vs bz, eth, EtOH
588	*trans*-Benzaldehyde oxime		C_7H_7NO	622-31-1	121.137	nd (eth)	35	119^{10}	1.145^{20}		s H_2O; vs EtOH, eth
589	Benzaldehyde, phenylhydrazone		$C_{13}H_{12}N_2$	588-64-7	196.247	nd (lig), pr	157.0				sl EtOH, eth; s ace, bz, liq NH_3
590	Benzaldehyde, (phenylmethylene) hydrazone		$C_{14}H_{12}N_2$	588-68-1	208.258	ye pr (al)	93				i H_2O; s EtOH, eth, ace, bz, chl; sl ctc
591	Benzamide	Benzoic acid amide	C_7H_7NO	55-21-0	121.137	mcl pr or pl (w)	127.3	290	1.0792^{130}		sl H_2O, eth, bz; vs EtOH, ctc, CS_2
592	Benz[a]anthracene	1,2-Benzanthracene	$C_{18}H_{12}$	56-55-3	228.288	lf (al)	160.5	438			i H_2O; vs EtOH
593	Benz[a]anthracene-7,12-dione		$C_{18}H_{10}O_2$	2498-66-0	258.271		170.5				sl EtOH, eth, lig; s ace; vs bz, chl
594	Benzanthrone		$C_{17}H_{10}O$	82-05-3	230.260		170				sl bz
595	Benzene	[6]Annulene	C_6H_6	71-43-2	78.112	orth pr or liq	5.49	80.09	0.8765^{20}	1.5011^{20}	sl H_2O; msc EtOH, eth, ace, chl; s ctc
596	Benzeneacetaldehyde	Phenylacetaldehyde	C_8H_8O	122-78-1	120.149		33.5	195	1.0272^{20}	1.5255^{20}	sl H_2O; s ace; msc EtOH, eth
597	Benzeneacetamide	α-Phenylacetamide	C_8H_9NO	103-81-1	135.163		157				sl H_2O, eth, bz; s EtOH
598	Benzeneacetic acid	Phenylacetic acid	$C_8H_8O_2$	103-82-2	136.149	lf, pl (peth)	76.5	265.5	1.228^6		sl H_2O, chl; vs EtOH, eth; s ace; i lig
599	Benzeneacetic acid, hydrazide		$C_8H_{10}N_2O$	937-39-3	150.177		115.5				
600	Benzeneacetic anhydride		$C_{16}H_{14}O_3$	1555-80-2	254.280	pr or nd (eth)	73.3	195^{12}			vs eth, chl
601	Benzeneacetonitrile	Benzyl cyanide	C_8H_7N	140-29-4	117.149	liq	-23.8	233.5	1.0205^{15}	1.5211^{25}	
602	Benzeneacetyl chloride	Phenylacetyl chloride	C_8H_7ClO	103-80-0	154.594			170^{250}, 105^{24}	1.1682^{20}	1.5325^{20}	vs eth
603	Benzenearsonic acid		$C_6H_7AsO_3$	98-05-5	202.040	cry (w)	158 dec				vs H_2O, EtOH
604	Benzeneboronic acid		$C_6H_7BO_2$	98-80-6	121.930		219				sl H_2O; s EtOH, eth, bz
605	Benzenebutanoic acid		$C_{10}H_{12}O_2$	1821-12-1	164.201	lf (w)	52	290			s H_2O, EtOH, eth
606	Benzenebutanol		$C_{10}H_{14}O$	3360-41-6	150.217			140^{14}		1.5214^{20}	
607	Benzenecarboperoxoic acid	Perbenzoic acid	$C_7H_6O_3$	93-59-4	138.121	mcl pl (peth)	42	100^{14}			vs ace, bz, eth, EtOH
608	Benzenecarbothioamide		C_7H_7NS	2227-79-4	137.203		117				

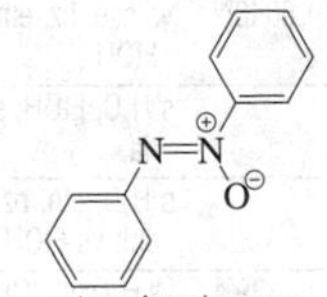

trans-Azoxybenzene

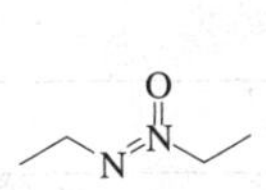

Azoxyethane

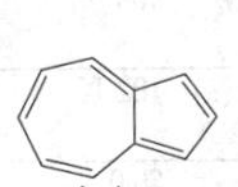

Azulene

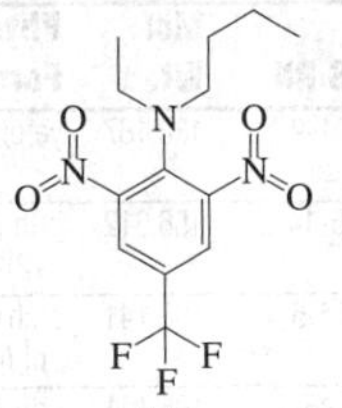

Balan

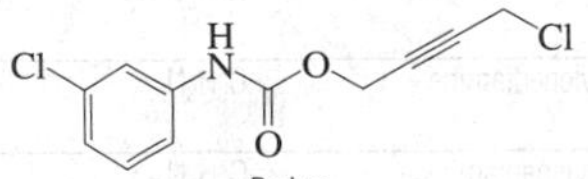

Barban

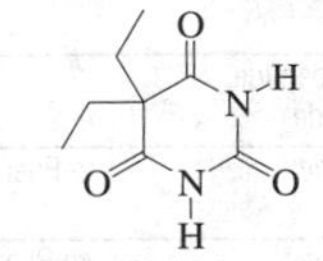

Barbital

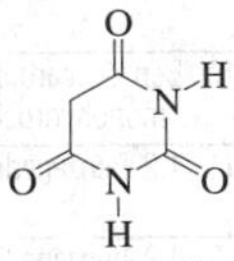

Barbituric acid

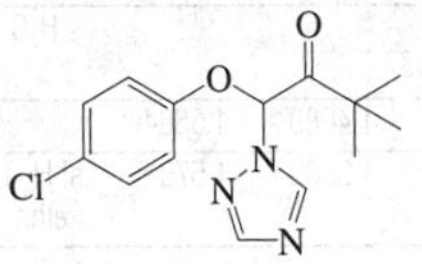

Bayleton

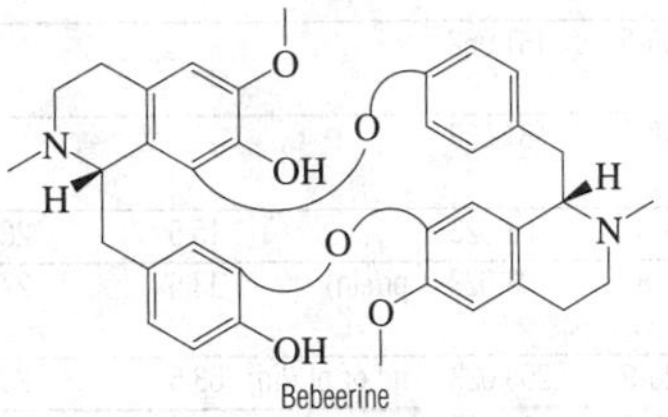

Bebeerine

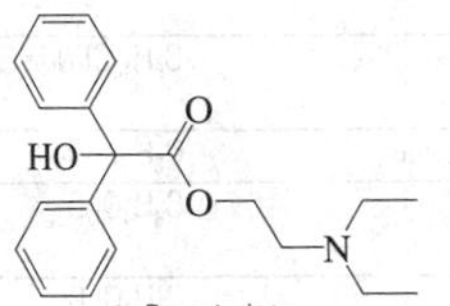

Benactyzine

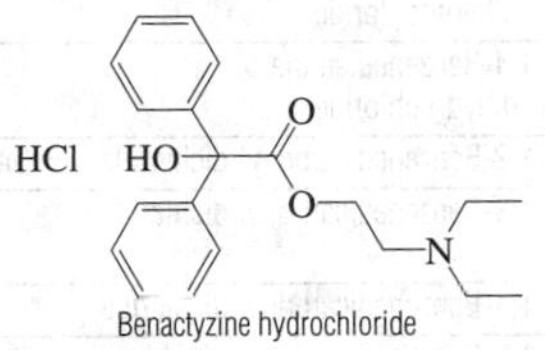

Benactyzine hydrochloride

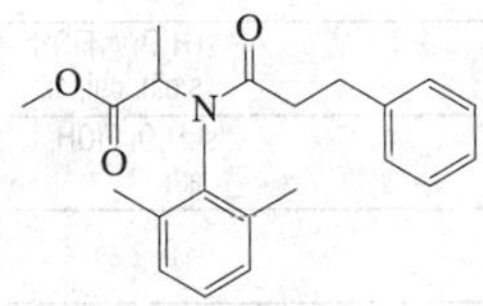

Benalaxyl

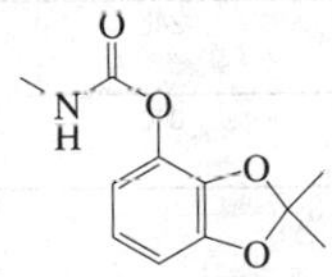

Bendiocarb

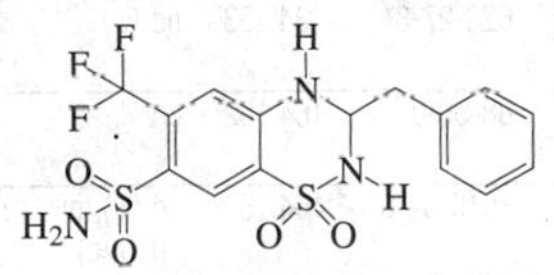

Bendroflumethiazide

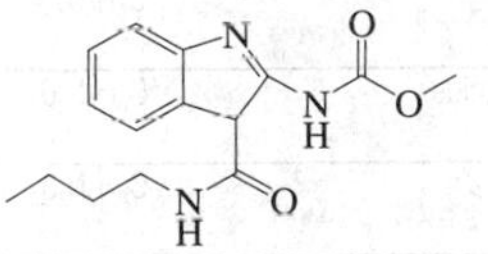

Benomyl

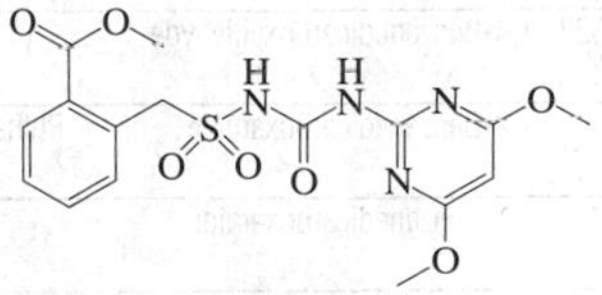

Bensulfuron-methyl

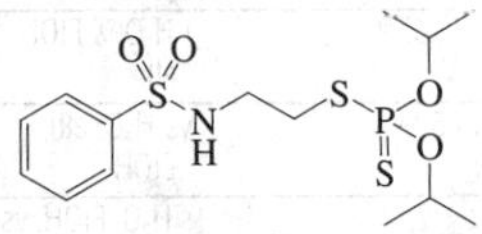

Bensulide

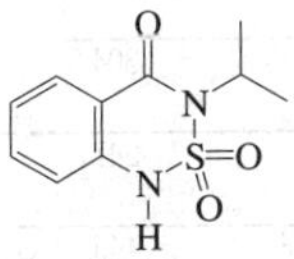

Bentazon

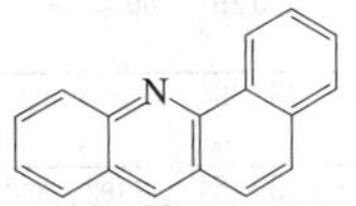

Benz[c]acridine

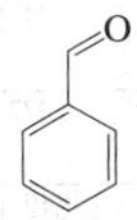

Benzaldehyde

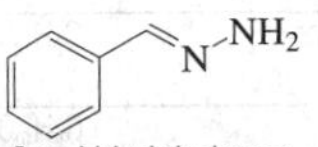

Benzaldehyde hydrazone

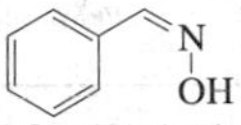

cis-Benzaldehyde oxime

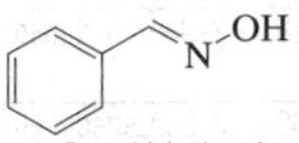

trans-Benzaldehyde oxime

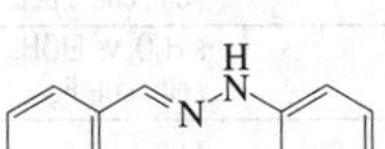

Benzaldehyde, phenylhydrazone

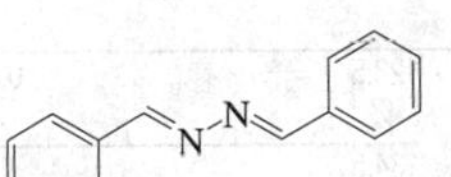

Benzaldehyde, (phenylmethylene)hydrazone

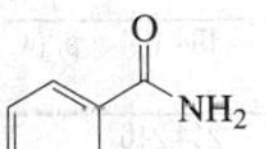

Benzamide

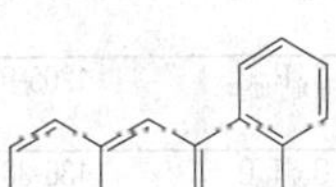

Benz[a]anthracene

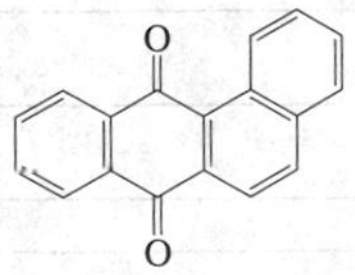

Benz[a]anthracene-7,12-dione

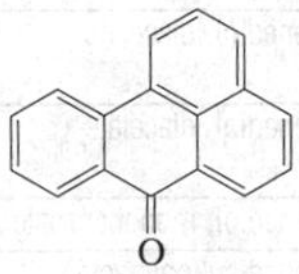

Benzanthrone

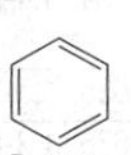

Benzene

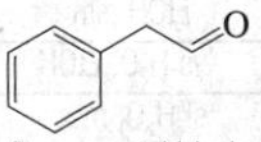

Benzeneacetaldehyde

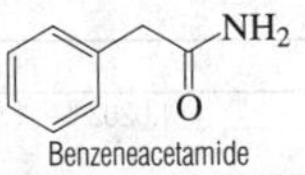

Benzeneacetamide

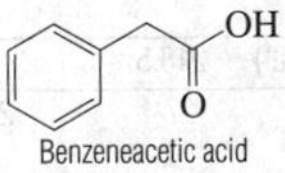

Benzeneacetic acid

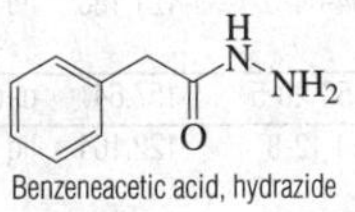

Benzeneacetic acid, hydrazide

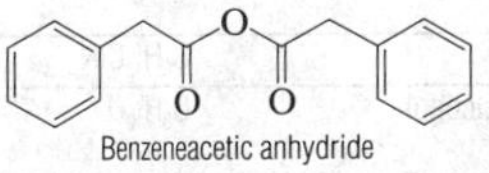

Benzeneacetic anhydride

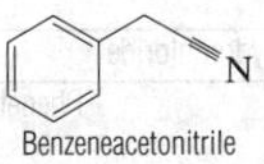

Benzeneacetonitrile

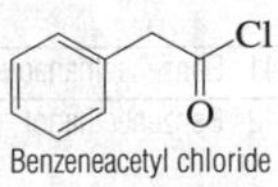

Benzeneacetyl chloride

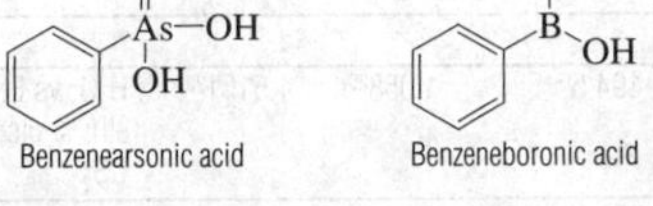

Benzenearsonic acid

Benzeneboronic acid

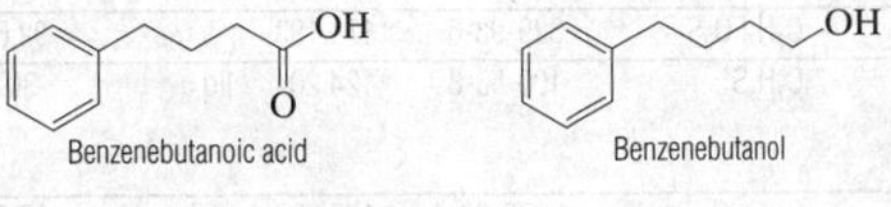

Benzenebutanoic acid

Benzenebutanol

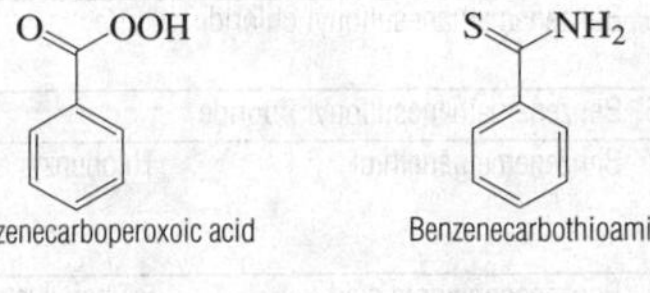

Benzenecarboperoxoic acid

Benzenecarbothioamide

No.	Name	Synonym	Mol. Form.	CAS RN	Mol. Wt.	Physical Form	mp/°C	bp/°C	den/ g cm^{-3}	n_D	Solubility
609	Benzenecarbothioic acid		C_7H_6OS	98-91-9	138.187	ye pl (HOAc)	24	86^{10}	1.28^{20}	1.6040^{20}	vs ace, bz, eth, EtOH
610	Benzenecarboximidamide, monohydrochloride		$C_7H_9ClN_2$	1670-14-0	156.612	orth pr (w +2)	169				s H_2O, EtOH; sl tfa
611	1,2-Benzenediamine	*o*-Phenylenediamine	$C_6H_8N_2$	95-54-5	108.141	brsh ye lf (w) pl (chl)	102.1	257			s H_2O, eth, bz, chl; vs EtOH
612	1,3-Benzenediamine	*m*-Phenylenediamine	$C_6H_8N_2$	108-45-2	108.141	orth (al)	66.0	285	1.0096^{58}	1.6339^{58}	vs H_2O; s EtOH, eth, bz
613	1,4-Benzenediamine	*p*-Phenylenediamine	$C_6H_8N_2$	106-50-3	108.141	wh pl (bz, eth)	141.1	267			sl H_2O; s EtOH, eth, bz, chl
614	1,2-Benzenediamine, dihydrochloride		$C_6H_{10}Cl_2N_2$	615-28-1	181.062		250 dec				
615	1,3-Benzenediamine, dihydrochloride		$C_6H_{10}Cl_2N_2$	541-69-5	181.062						s H_2O
616	1,4-Benzenediamine, dihydrochloride		$C_6H_{10}Cl_2N_2$	624-18-0	181.062						s H_2O
617	1,2-Benzenedicarbonyl dichloride	Phthaloyl chloride	$C_8H_4Cl_2O_2$	88-95-9	203.023		15.5	281.1	1.4089^{20}	1.5684^{20}	
618	1,3-Benzenedicarbonyl dichloride		$C_8H_4Cl_2O_2$	99-63-8	203.023	pr(eth)	43.5	276	1.3880^{17}	1.570^{47}	sl H_2O, EtOH; s eth
619	1,4-Benzenedicarbonyl dichloride		$C_8H_4Cl_2O_2$	100-20-9	203.023	nd or pl (lig)	83.5	258; 125^9			s eth
620	1,2-Benzenedicarboxaldehyde		$C_8H_6O_2$	643-79-8	134.133	ye cry or nd (lig)	55.8	$83^{0.8}$			vs eth, EtOH
621	1,3-Benzenedicarboxaldehyde		$C_8H_6O_2$	626-19-7	134.133	nd (dil al)	89.5	246; 136^{13}			sl H_2O, eth, chl; vs EtOH; s ace, bz
622	1,4-Benzenedicarboxaldehyde		$C_8H_6O_2$	623-27-8	134.133	nd (w)	117	246			sl H_2O; vs EtOH; s eth, chl, alk
623	1,2-Benzenedicarboxamide	Phthalamide	$C_8H_8N_2O_2$	88-96-0	164.162	cry	222	dec			sl H_2O, EtOH; i eth
624	1,4-Benzenedicarboxamide		$C_8H_8N_2O_2$	3010-82-0	164.162	nd (w), pl (HOAc)	322.3				
625	1,2-Benzenedicarboxylic acid, bis(2-butoxyethyl) ester	Bis(2-butoxyethyl) phthalate	$C_{20}H_{30}O_6$	117-83-9	366.448			270			
626	1,2-Benzenedicarboxylic acid, bis(2-methoxyethyl) ester	Bis(2-methoxyethyl) phthalate	$C_{14}H_{18}O_6$	117-82-8	282.289		-60.0	230^{10}	1.1596^{20}		
627	1,2-Benzenedicarboxylic acid, diallyl ester	Diallyl phthalate	$C_{14}H_{14}O_4$	131-17-9	246.259			161^4			
628	1,2-Benzenedicarboxylic acid, dipropyl ester	Dipropyl phthalate	$C_{14}H_{18}O_4$	131-16-8	250.291	liq	-31.0	304.5	1.0767^{20}		i H_2O; s EtOH, eth
629	1,3-Benzenedimethanamine	*m*-Xylene diamine	$C_8H_{12}N_2$	1477-55-0	136.194			247	1.052^{20}		vs H_2O, eth, EtOH
630	1,2-Benzenedimethanol		$C_8H_{10}O_2$	612-14-6	138.164	pl (eth, peth)	64.8	145^3			s H_2O, EtOH; vs eth; sl bz
631	1,3-Benzenedimethanol		$C_8H_{10}O_2$	626-18-6	138.164	nd (bz)	57	156^{13}	1.1610^{18}		vs H_2O, eth, EtOH
632	1,4-Benzenedimethanol		$C_8H_{10}O_2$	589-29-7	138.164	nd (w)	117.5	140^1			vs H_2O, ace, eth, EtOH
633	1,2-Benzenediol, diacetate		$C_{10}H_{10}O_4$	635-67-6	194.184	nd (al)	64.5	142^9			i H_2O; vs EtOH, eth, chl; s peth
634	1,4-Benzenediol, diacetate		$C_{10}H_{10}O_4$	1205-91-0	194.184	pl (w, al)	123.5		0.8731^{25}		s H_2O; vs EtOH, eth, chl, lig
635	1,3-Benzenediol, monobenzoate		$C_{13}H_{10}O_3$	136-36-7	214.216		134.5				
636	1,3-Benzenedisulfonic acid		$C_6H_6O_6S_2$	98-48-6	238.238	hyg cry					
637	1,3-Benzenedisulfonyl dichloride		$C_6H_4Cl_2O_4S_2$	585-47-7	275.130		61.8	$195^{10.5}$			
638	1,2-Benzenedithiol		$C_6H_6S_2$	17534-15-5	142.242		28.5	238.5			vs EtOH, eth, bz; s AcOEt
639	1,3-Benzenedithiol		$C_6H_6S_2$	626-04-0	142.242	lf	27	245			vs bz, eth, EtOH
640	Benzeneethanamine	1-Amino-2-phenylethane	$C_8H_{11}N$	64-04-0	121.180	liq	<0	195	0.9640^{25}	1.5290^{25}	s H_2O, ctc; vs EtOH, eth
641	Benzeneethanamine, hydrochloride		$C_8H_{12}ClN$	156-28-5	157.641	pl or lf (al)	218.5				vs H_2O, EtOH
642	Benzeneethanol	Phenethyl alcohol	$C_8H_{10}O$	60-12-8	122.164	liq	-27	218.8	1.0202^{20}	1.5325^{20}	sl H_2O; msc EtOH, eth
643	Benzenehexacarboxylic acid	Mellitic acid	$C_{12}H_6O_{12}$	517-60-2	342.169	nd (al)	287 dec				vs H_2O; s EtOH, sulf
644	Benzenemethanamine, hydrochloride		$C_7H_{10}ClN$	3287-99-8	143.614		258.3				vs H_2O, EtOH
645	Benzenemethanesulfonyl chloride		$C_7H_7ClO_2S$	1939-99-7	190.648	pr (eth), nd (bz)	93				vs eth, bz
646	Benzenemethanesulfonyl fluoride		$C_7H_7FO_2S$	329-98-6	174.193		92.0				
647	Benzenemethanethiol	Thiobenzyl alcohol	C_7H_8S	100-53-8	124.204	liq	-30	194.5	1.058^{20}	1.5151^{20}	i H_2O; vs EtOH, eth; sl ctc; s CS_2
648	Benzenepentanoic acid	5-Phenylvaleric acid	$C_{11}H_{14}O_2$	2270-20-4	178.228	pl (w), pr (peth)	57.5	190^{30}			sl H_2O; vs EtOH; s os
649	Benzenepentanol		$C_{11}H_{16}O$	10521-91-2	164.244			155^{20}, 150^{18}	0.9725^{20}	1.5156^{20}	vs eth, EtOH

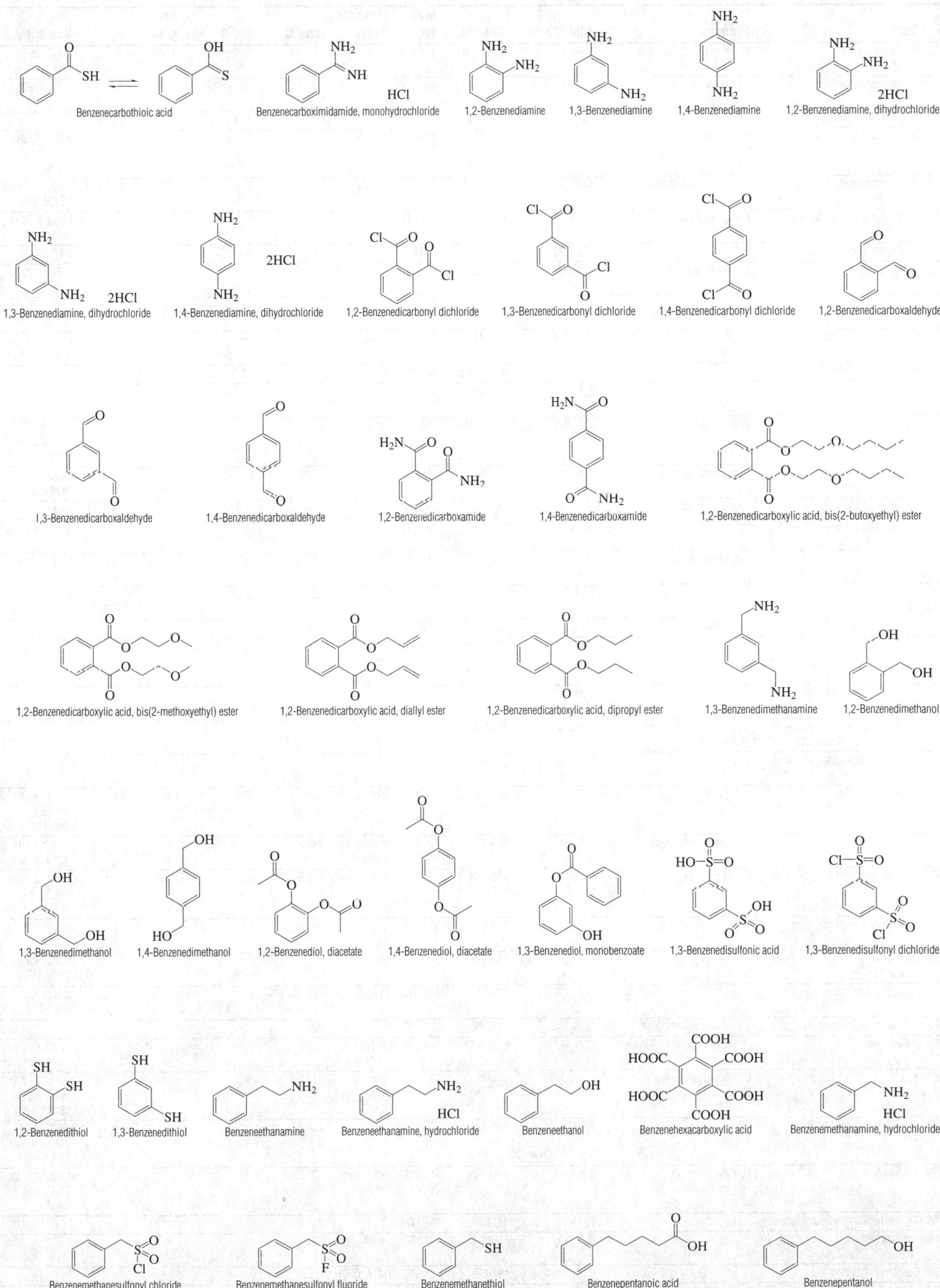
Benzenecarbothioic acid
Benzenecarboximidamide, monohydrochloride
1,2-Benzenediamine
1,3-Benzenediamine
1,4-Benzenediamine
1,2-Benzenediamine, dihydrochloride
1,3-Benzenediamine, dihydrochloride
1,4-Benzenediamine, dihydrochloride
1,2-Benzenedicarbonyl dichloride
1,3-Benzenedicarbonyl dichloride
1,4-Benzenedicarbonyl dichloride
1,2-Benzenedicarboxaldehyde
1,3-Benzenedicarboxaldehyde
1,4-Benzenedicarboxaldehyde
1,2-Benzenedicarboxamide
1,4-Benzenedicarboxamide
1,2-Benzenedicarboxylic acid, bis(2-butoxyethyl) ester
1,2-Benzenedicarboxylic acid, bis(2-methoxyethyl) ester
1,2-Benzenedicarboxylic acid, diallyl ester
1,2-Benzenedicarboxylic acid, dipropyl ester
1,3-Benzenedimethanamine
1,2-Benzenedimethanol
1,3-Benzenedimethanol
1,4-Benzenedimethanol
1,2-Benzenediol, diacetate
1,4-Benzenediol, diacetate
1,3-Benzenediol, monobenzoate
1,3-Benzenedisulfonic acid
1,3-Benzenedisulfonyl dichloride
1,2-Benzenedithiol
1,3-Benzenedithiol
Benzeneethanamine
Benzeneethanamine, hydrochloride
Benzeneethanol
Benzenehexacarboxylic acid
Benzenemethanamine, hydrochloride
Benzenemethanesulfonyl chloride
Benzenemethanesulfonyl fluoride
Benzenemethanethiol
Benzenepentanoic acid
Benzenepentanol

No.	Name	Synonym	Mol. Form.	CAS RN	Mol. Wt.	Physical Form	mp/°C	bp/°C	den/ g cm^{-3}	n_D	Solubility
650	Benzenepropanal	Hydrocinnamic aldehyde	$C_9H_{10}O$	104-53-0	134.174	mcl	47	224; 117^{28}	1.0190^{20}		i H_2O; vs EtOH; msc eth
651	Benzenepropanenitrile	Hydrocinnamonitrile	C_9H_9N	645-59-0	131.174	liq	-1	261; 141^{25}	1.0016^{20}	1.5266^{28}	s EtOH, eth; sl chl
652	Benzenepropanethiol		$C_9H_{12}S$	24734-68-7	152.256			121^{23}, 109^{10}	1.01^{25}	1.5494^{20}	
653	Benzenepropanoic acid	Hydrocinnamic acid	$C_9H_{10}O_2$	501-52-0	150.174	nd (w)	48	279.8	1.0712^{49}		s H_2O, EtOH, eth, ctc, CS_2; vs bz
654	Benzenepropanol	Hydrocinnamyl alcohol	$C_9H_{12}O$	122-97-4	136.190		<-18	235	0.995^{25}	1.5357^{25}	s H_2O, ctc; msc EtOH, eth
655	Benzenepropanol carbamate	Phenprobamate	$C_{10}H_{13}NO_2$	673-31-4	179.216		102				i H_2O; s EtOH, chl
656	Benzenepropanoyl chloride		C_9H_9ClO	645-45-4	168.619			dec 225; 105^{10}	1.135^{21}		s eth, CS_2
657	Benzeneseleninic acid	Phenylseleninic acid	$C_6H_6O_2Se$	6996-92-5	189.07		124.5		1.93^{20}		sl H_2O; i bz; vs alk
658	Benzeneselenol		C_6H_6Se	645-96-5	157.07			183.6; 84^{25}	1.4865^{15}		i H_2O; s EtOH; vs eth, ctc
659	Benzenesulfinic acid		$C_6H_6O_2S$	618-41-7	142.176	pr (w)	84	dec			sl H_2O; s EtOH, eth, bz; i peth
660	Benzenesulfinyl chloride		C_6H_5ClOS	4972-29-6	160.621	pl (peth)	38	$71^{1.5}$	1.3469^{25}	1.3470^{25}	s eth, chl
661	Benzenesulfonamide		$C_6H_7NO_2S$	98-10-2	157.191	lf, nd (w)	156				sl H_2O, tfa; s EtOH, eth
662	Benzenesulfonic acid	Besylic acid	$C_6H_6O_3S$	98-11-3	158.175	nd (bz)	65				vs H_2O, EtOH; i eth; sl bz; s HOAc
663	Benzenesulfonyl chloride	Phenylsulfonyl chloride	$C_6H_5ClO_2S$	98-09-9	176.621		14.5	dec 251	1.3470^{15}		i H_2O; vs EtOH; s eth, ctc
664	Benzenesulfonyl fluoride	Phenylsulfonyl fluoride	$C_6H_5FO_2S$	368-43-4	160.166			203.5	1.3286^{20}	1.4932^{18}	s EtOH, eth
665	1,2,4,5-Benzenetetracarboxylic acid	Pyromellitic acid	$C_{10}H_6O_8$	89-05-4	254.150	tcl pr (w+2)	276				sl H_2O; s EtOH
666	Benzenethiol	Phenyl mercaptan	C_6H_6S	108-98-5	110.177	liq	-14.93	169.1	1.0775^{20}	1.5893^{20}	i H_2O; s EtOH, eth, bz; sl ctc
667	1,3,5-Benzenetricarbonyl trichloride		$C_9H_3Cl_3O_3$	4422-95-1	265.477		36.3	180^{16}			s chl
668	1,2,3-Benzenetricarboxylic acid	Hemimellitic acid	$C_9H_6O_6$	569-51-7	210.140	pr (al)	200		1.546^{20}		vs eth, EtOH
669	1,2,4-Benzenetricarboxylic acid	Trimellitic acid	$C_9H_6O_6$	528-44-9	210.140	nd (w) cry (al) cry (HOAc)	219				vs H_2O, eth, EtOH
670	1,3,5-Benzenetricarboxylic acid		$C_9H_6O_6$	554-95-0	210.140	pr or nd (w+1)	380				sl H_2O; vs EtOH, eth
671	1,2,4-Benzenetricarboxylic acid 1,2-anhydride, 4-chloride	4-(Chloroformyl)phthalic anhydride	$C_9H_3ClO_4$	1204-28-0	210.571		66				
672	1,2,4-Benzenetricarboxylic acid, triallyl ester		$C_{18}H_{18}O_6$	2694-54-4	330.332		<-30		1.164^{20}		
673	1,2,3-Benzenetriol	Pyrogallol	$C_6H_6O_3$	87-66-1	126.110	lf or nd (bz)	133	309	1.453^{4}	1.561^{134}	vs H_2O, EtOH, eth, NH_3; s ace; i bz
674	1,2,4-Benzenetriol	Hydroxyhydroquinone	$C_6H_6O_3$	533-73-3	126.110	pl (eth), lf or pl (w)	140.5				vs H_2O, EtOH, eth; i bz, chl
675	1,3,5-Benzenetriol	Phloroglucinol	$C_6H_6O_3$	108-73-6	126.110	lf or pl (w +2)	218.5	sub	1.46^{25}		sl H_2O; vs EtOH, eth, bz, py; s ace
676	1,2,4-Benzenetriol triacetate		$C_{12}H_{12}O_6$	613-03-6	252.219		99	300			s EtOH, chl, MeOH
677	Benzestrol		$C_{20}H_{26}O_2$	85-95-0	298.419	cry (al)	164				vs ace, eth, EtOH, HOAc
678	Benzethonium chloride		$C_{27}H_{42}ClNO_2$	121-54-0	448.081	pl (chl/eth)	165 (hyd)				vs H_2O; s ace, chl, EtOH
679	Benzidene-3,3'-dicarboxylic acid	3,3'-Dicarboxybenzidine	$C_{14}H_{12}N_2O_4$	2130-56-5	272.256	nd	300 dec				
680	*p*-Benzidine	[1,1'-Biphenyl]-4,4'-diamine	$C_{12}H_{12}N_2$	92-87-5	184.236	nd (w)	120	401			sl H_2O, eth, DMSO; s EtOH
681	Benzil	Diphenylethanedione	$C_{14}H_{10}O_2$	134-81-6	210.228	ye pr (al)	94.87	347	1.084^{102}		i H_2O; vs EtOH, eth; s ace; sl ctc
682	1*H*-Benzimidazol-2-amine		$C_7H_7N_3$	934-32-7	133.151	pl (w)	224				s H_2O, EtOH, ace; sl eth, bz, DMSO
683	1*H*-Benzimidazole	*N,N'*-Methenyl-*o*-phenylenediamine	$C_7H_6N_2$	51-17-2	118.136	orth bipym pl (w)	170.5	>360			sl H_2O, eth; vs EtOH; i bz; s dil alk
684	1*H*-Benzimidazole-2-acetonitrile		$C_9H_7N_3$	4414-88-4	157.172		208.4				
685	1*H*-Benz[de]isoquinoline-1,3(2*H*)-dione		$C_{12}H_7NO_2$	81-83-4	197.190	nd (chl-al)	300				
686	Benzo[c]chrysene		$C_{22}H_{14}$	194-69-4	278.346	nd (AcOH)	126.5				
687	Benzo[g]chrysene	Benzo[a]triphenylene	$C_{22}H_{14}$	196-78-1	278.346	nd (AcOH)	114.5				

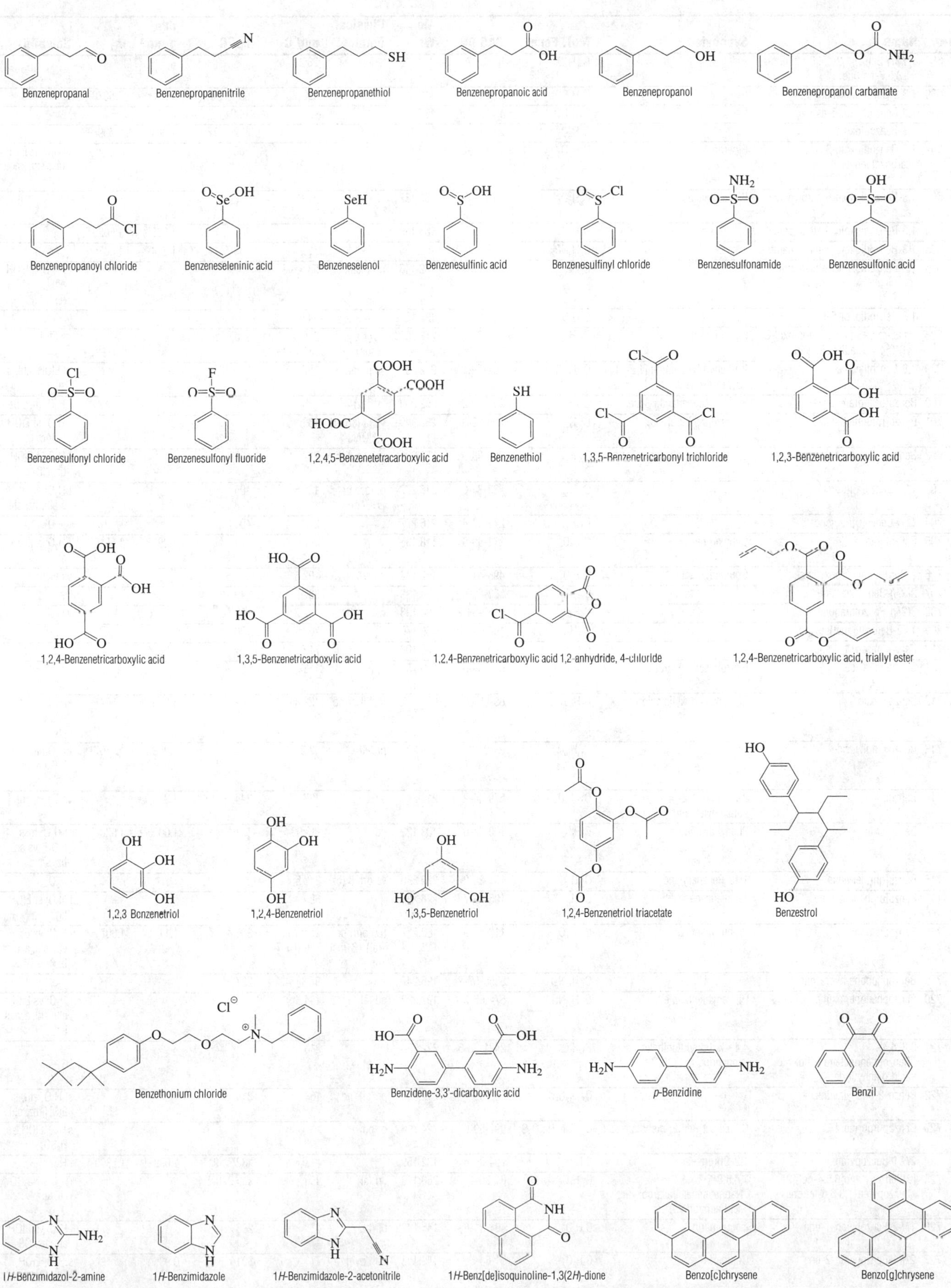

Benzenepropanal
Benzenepropanenitrile
Benzenepropanethiol
Benzenepropanoic acid
Benzenepropanol
Benzenepropanol carbamate
Benzenepropanoyl chloride
Benzeneseleninic acid
Benzeneselenol
Benzenesulfinic acid
Benzenesulfinyl chloride
Benzenesulfonamide
Benzenesulfonic acid
Benzenesulfonyl chloride
Benzenesulfonyl fluoride
1,2,4,5-Benzenetetracarboxylic acid
Benzenethiol
1,3,5-Benzenetricarbonyl trichloride
1,2,3-Benzenetricarboxylic acid
1,2,4-Benzenetricarboxylic acid
1,3,5-Benzenetricarboxylic acid
1,2,4-Benzenetricarboxylic acid 1,2-anhydride, 4-chloride
1,2,4-Benzenetricarboxylic acid, triallyl ester
1,2,3 Benzenetriol
1,2,4-Benzenetriol
1,3,5-Benzenetriol
1,2,4-Benzenetriol triacetate
Benzestrol
Benzethonium chloride
Benzidene-3,3'-dicarboxylic acid
p-Benzidine
Benzil
1H-Benzimidazol-2-amine
1H-Benzimidazole
1H-Benzimidazole-2-acetonitrile
1H-Benz[de]isoquinoline-1,3(2H)-dione
Benzo[c]chrysene
Benzo[g]chrysene

No.	Name	Synonym	Mol. Form.	CAS RN	Mol. Wt.	Physical Form	mp/°C	bp/°C	den/ g cm⁻³	n_D	Solubility
688	1*H*,3*H*-Benzo[1,2-c:4,5-c′]difuran-1,3,5,7-tetrone		$C_{10}H_2O_6$	89-32-7	218.119		285.3				
689	1,3,2-Benzodioxaborole		$C_6H_5BO_2$	274-07-7	119.914		12	88^{156}, 50^{50}	1.2700^{20}	1.5070^{20}	
690	1,3-Benzodioxol-5-amine		$C_7H_7NO_2$	14268-66-7	137.137		42	144^{16}			
691	1,3-Benzodioxole		$C_7H_6O_2$	274-09-9	122.122			172.5; 77^{27}	1.064^{25}	1.5398^{20}	
692	1,3-Benzodioxole-5-carboxaldehyde	Piperonal	$C_8H_6O_3$	120-57-0	150.132		37	263			sl H_2O; vs EtOH; msc eth; s ace, chl
693	1,3-Benzodioxole-5-carboxylic acid	Piperonylic acid	$C_8H_6O_4$	94-53-1	166.132		229				
694	1,3-Benzodioxole-5-ethanamine		$C_9H_{11}NO_2$	1484-85-1	165.189			166^{20}, 101^{1}	1.225^{20}	1.5620^{20}	
695	1,3-Benzodioxole-5-methanamine		$C_8H_9NO_2$	2620-50-0	151.163			139^{13}, $100^{0.07}$	1.214^{25}	1.5635^{20}	
696	1,3-Benzodioxole-5-methanol		$C_8H_8O_3$	495-76-1	152.148	nd (peth)	58	157^{16}			sl H_2O; s EtOH, eth, bz, chl; i lig
697	1,3-Benzodioxol-5-ol		$C_7H_6O_3$	533-31-3	138.121		64.9				
698	*trans,trans*-5-(1,3-Benzodioxol-5-yl)-2,4-pentadienoic acid	Piperinic acid	$C_{12}H_{10}O_4$	136-72-1	218.205	nd (al), ye nd (sub)	215.8	sub			vs EtOH
699	7,8-Benzoflavone	2-Phenyl-4*H*-naphtho[1,2-b]pyran-4-one	$C_{19}H_{12}O_2$	604-59-1	272.297	ye pl (al)	157				sl EtOH, chl; s sulf
700	Benzo[b]fluoranthene	Benz[e]acephenanthrylene	$C_{20}H_{12}$	205-99-2	252.309	nd (bz)	168				i H_2O; msc bz
701	Benzo[j]fluoranthene	Dibenzo[a,jk]fluorene	$C_{20}H_{12}$	205-82-3	252.309	ye pl (al) nd (HOAc)	166				i H_2O; sl EtOH, HOAc
702	Benzo[k]fluoranthene	2,3,1′,8′-Binaphthylene	$C_{20}H_{12}$	207-08-9	252.309	pa ye nd (bz)	217	480			i H_2O; s EtOH, bz, HOAc
703	11*H*-Benzo[a]fluorene		$C_{17}H_{12}$	238-84-6	216.277	pl (ace or HOAc)	189.5	405			i H_2O; sl EtOH; s eth, bz, chl
704	11*H*-Benzo[b]fluorene		$C_{17}H_{12}$	243-17-4	216.277		212	401			i H_2O
705	Benzofuran	Coumarone	C_8H_6O	271-89-6	118.133		<-18	174	1.0913^{25}	1.5615^{17}	i H_2O; s EtOH, eth
706	2-Benzofurancarboxylic acid	Coumarilic acid	$C_9H_6O_3$	496-41-3	162.142	nd (w)	192.5	312.5			vs EtOH
707	2(3*H*)-Benzofuranone		$C_8H_6O_2$	553-86-6	134.133		50	249	1.2236^{14}		
708	3(2*H*)-Benzofuranone		$C_8H_6O_2$	7169-34-8	134.133	red nd (al)	102.5	152^{15}			vs bz
709	1-(2-Benzofuranyl)ethanone		$C_{10}H_8O_2$	1646-26-0	160.170		76	126^{11}			s H_2O
710	Benzofurazan, 1-oxide		$C_6H_4N_2O_2$	480-96-6	136.108		71.5		1.280^{80}		
711	Benzohydrazide	Benzoic acid, hydrazide	$C_7H_8N_2O$	613-94-5	136.151	pl (w)	115	dec 267			s H_2O, EtOH; sl eth, ace, chl
712	Benzoic acid	Benzenecarboxylic acid	$C_7H_6O_2$	65-85-0	122.122	mcl lf or nd	122.35	249.2	1.2659^{15}	1.504^{132}	sl H_2O; vs EtOH, eth; s ace, bz, chl
713	Benzoic anhydride		$C_{14}H_{10}O_3$	93-97-0	226.227	pr (eth)	42.5	360	1.989^{15}	1.5767^{15}	i H_2O, lig; s EtOH, eth; sl chl
714	Benzoin	2-Hydroxy-1,2-diphenylethanone, (±)	$C_{14}H_{12}O_2$	579-44-2	212.244		137	344; 194^{12}	1.310^{20}		vs EtOH, chl
715	Benzonitrile	Phenyl cyanide	C_7H_5N	100-47-0	103.122	liq	-13.99	191.1	1.0093^{15}	1.5289^{20}	sl H_2O; msc EtOH; vs ace, bz; s ctc
716	Benzo[ghi]perylene	1,12-Benzperylene	$C_{22}H_{12}$	191-24-2	276.330	ye-grn lf (bz)	272.5				i H_2O
717	Benzo[c]phenanthrene	Tetrahelicene	$C_{18}H_{12}$	195-19-7	228.288		68				i H_2O; sl EtOH, lig
718	Benzophenone	Diphenyl ketone	$C_{13}H_{10}O$	119-61-9	182.217	(α) orth pr (al); (β) mcl pr	47.9 (α); 26 (β)	305.4	1.111^{18}	1.6077^{19}	i H_2O; vs EtOH, eth, chl, ace; s bz
719	Benzophenone hydrazone		$C_{13}H_{12}N_2$	5350-57-2	196.247		97.3	227^{55}			
720	Benzophenone, oxime	Diphenyl ketoxime	$C_{13}H_{11}NO$	574-66-3	197.232	nd (al)	144				i H_2O; vs EtOH, eth, chl, ace; s bz
721	3,3′,4,4′-Benzophenonetetracarboxylic acid dianhydride	4,4′-Carbonyldiphthalic anhydride	$C_{17}H_6O_7$	2421-28-5	322.226		216				
722	Benzo-2-phenylhydrazide		$C_{13}H_{12}N_2O$	532-96-7	212.246	pr (al), nd (w)	168	314			sl H_2O, eth; s EtOH, bz, chl
723	Benzopurpurine 4B	C.I. Direct Red 2, disodium salt	$C_{34}H_{26}N_6Na_2O_6S_2$	992-59-6	724.716	br pow					s H_2O, EtOH, ac, H_2SO_4
724	2*H*-1-Benzopyran	1,2-Chromene	C_9H_8O	254-04-6	132.159			132^{102}, 91^{13}	1.0993^{16}	1.5869^{24}	i H_2O
725	[2]Benzopyrano[6,5,4-def][2]benzopyran-1,3,6,8-tetrone	1,4,5,8-Naphthalenetetracarboxylic acid anhydride	$C_{14}H_4O_6$	81-30-1	268.178	nd (al)	450	sub 320			i H_2O; s Na_2CO_3, HOAc
726	1*H*-2-Benzopyran-1-one	Isocoumarin	$C_9H_6O_2$	491-31-6	146.143	pl (bz)	47	286			i H_2O; vs EtOH, eth, bz, CS_2
727	2*H*-1-Benzopyran-2-one	Coumarin	$C_9H_6O_2$	91-64-5	146.143	orth pym (eth)	71	301.7	0.935^{20}		s H_2O, EtOH, alk; vs eth, chl, py

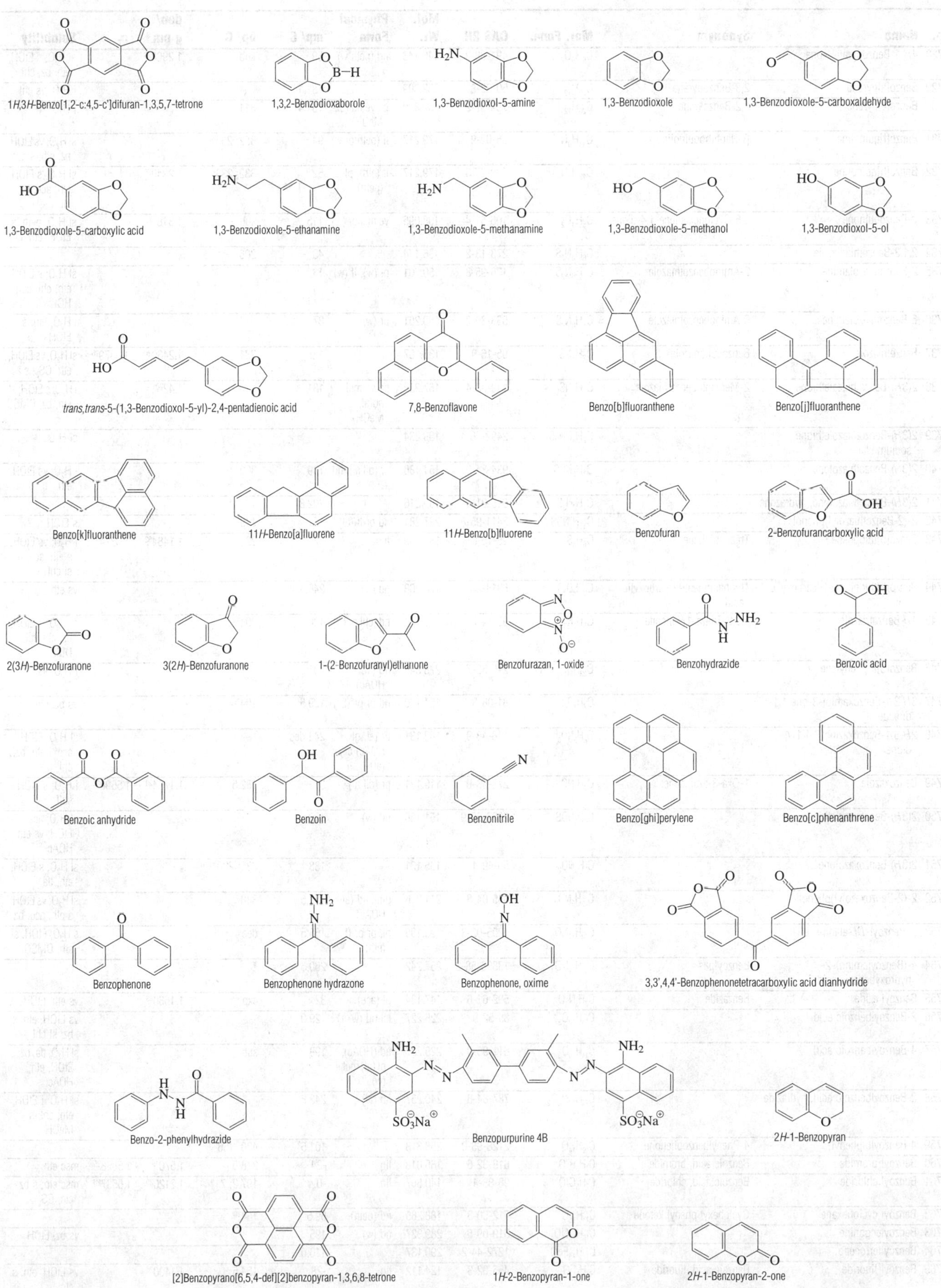
1H,3H-Benzo[1,2-c:4,5-c']difuran-1,3,5,7-tetrone
1,3,2-Benzodioxaborole
1,3-Benzodioxol-5-amine
1,3-Benzodioxole
1,3-Benzodioxole-5-carboxaldehyde
1,3-Benzodioxole-5-carboxylic acid
1,3-Benzodioxole-5-ethanamine
1,3-Benzodioxole-5-methanamine
1,3-Benzodioxole-5-methanol
1,3-Benzodioxol-5-ol
trans,trans-5-(1,3-Benzodioxol-5-yl)-2,4-pentadienoic acid
7,8-Benzoflavone
Benzo[b]fluoranthene
Benzo[j]fluoranthene
Benzo[k]fluoranthene
11H-Benzo[a]fluorene
11H-Benzo[b]fluorene
Benzofuran
2-Benzofurancarboxylic acid
2(3H)-Benzofuranone
3(2H)-Benzofuranone
1-(2-Benzofuranyl)ethanone
Benzofurazan, 1-oxide
Benzohydrazide
Benzoic acid
Benzoic anhydride
Benzoin
Benzonitrile
Benzo[ghi]perylene
Benzo[c]phenanthrene
Benzophenone
Benzophenone hydrazone
Benzophenone, oxime
3,3',4,4'-Benzophenonetetracarboxylic acid dianhydride
Benzo-2-phenylhydrazide
Benzopurpurine 4B
2H-1-Benzopyran
[2]Benzopyrano[6,5,4-def][2]benzopyran-1,3,6,8-tetrone
1H-2-Benzopyran-1-one
2H-1-Benzopyran-2-one

No.	Name	Synonym	Mol. Form.	CAS RN	Mol. Wt.	Physical Form	mp/°C	bp/°C	den/ g cm^{-3}	n_D	Solubility
728	4*H*-1-Benzopyran-4-one		$C_9H_6O_2$	491-38-3	146.143	nd (peth w)	59	sub	1.2900^{20}		sl H_2O; s EtOH, eth, bz, chl
729	Benzo[a]pyrene	2,3-Benzopyrene	$C_{20}H_{12}$	50-32-8	252.309		181.1				i H_2O; vs chl
730	Benzo[e]pyrene	1,2-Benzpyrene	$C_{20}H_{12}$	192-97-2	252.309	pa ye nd (bz-MeOH)	181.4	311			i H_2O
731	Benzo[f]quinoline	β-Naphthoquinoline	$C_{13}H_9N$	85-02-9	179.217	lf (peth or w)	94	352; 203^{8}			sl H_2O; vs EtOH, bz, eth; s ace
732	Benzo[h]quinoline		$C_{13}H_9N$	230-27-3	179.217	lf (eth), pl (peth)	52	339; 233^{47}	1.2340^{20}		sl H_2O; s EtOH, eth, ace, bz, ctc
733	*p*-Benzoquinone	2,5-Cyclohexadiene-1,4-dione	$C_6H_4O_2$	106-51-4	108.095	ye mcl pr (w)	115	sub	1.318^{20}		sl H_2O, peth; s EtOH, eth, chl
734	2,1,3-Benzothiadiazole		$C_6H_4N_2S$	273-13-2	136.174		43	206			
735	2-Benzothiazolamine	2-Aminobenzothiazole	$C_7H_6N_2S$	136-95-8	150.201	pl (w), lf (w)	132				sl H_2O; s EtOH, eth, chl, con HCl
736	6-Benzothiazolamine	6-Aminobenzothiazole	$C_7H_6N_2S$	533-30-2	150.201	pr (w)	87				i H_2O, eth; s EtOH
737	Benzothiazole	Benzosulfonazole	C_7H_5NS	95-16-9	135.187		1.0	231	1.2460^{20}	1.6379^{20}	sl H_2O; vs EtOH, eth, CS_2; s ace
738	2(3*H*)-Benzothiazolethione	2-Mercaptobenzothiazole	$C_7H_5NS_2$	149-30-4	167.252	pa ye mcl nd(al, MeOH)	181		1.42^{20}		i H_2O; s EtOH; sl eth, bz, DMSO
739	2(3*H*)-Benzothiazolethione, sodium salt		$C_7H_4NNaS_2$	2492-26-4	189.234						sl H_2O
740	2(3*H*)-Benzothiazolone		C_7H_5NOS	934-34-9	151.186	pr (dil al), nd	139	360			i H_2O; vs EtOH, eth
741	2(3*H*)-Benzothiazolone, hydrazone		$C_7H_7N_3S$	615-21-4	165.216		202.8				
742	2-(2-Benzothiazolyl)phenol		$C_{13}H_9NOS$	3411-95-8	227.281	nd or lf (al)	131	179^{3}			s EtOH
743	Benzo[b]thiophene	Thianaphthene	C_8H_6S	95-15-8	134.199	lf	32	221	1.1484^{32}	1.6374^{37}	i H_2O; vs EtOH; s eth, ace, bz; sl chl
744	Benzo[b]thiophene-2-carboxylic acid	Thionaphthene-2-carboxylic acid	$C_9H_6O_2S$	6314-28-9	178.208	nd (w)	240.5				vs eth
745	1*H*-Benzotriazole	1,2,3-Triaza-1*H*-indene	$C_6H_5N_3$	95-14-7	119.124	nd (chl or bz)	100	204^{15}			sl H_2O; s EtOH, bz, chl, tol, DMF
746	Benzo[b]triphenylene		$C_{22}H_{14}$	215-58-7	278.346	nd (al, HOAc)	205				i H_2O; vs bz
747	3*H*-2,1-Benzoxathiol-3-one 1,1-dioxide		$C_7H_4O_4S$	81-08-3	184.170	nd or pr (bz)	129.5	184^{18}			vs bz, chl
748	2*H*-3,1-Benzoxazine-2,4(1*H*)-dione		$C_8H_5NO_3$	118-48-9	163.131	pr (al, gl HOAc) cry (al)	243 dec				sl H_2O, EtOH, ace; i eth, bz, chl
749	Benzoxazole	1-Oxa-3-azaindene	C_7H_5NO	273-53-0	119.121	pr (dil al)	31	182.5	1.1754^{20}	1.5594^{20}	i H_2O; s EtOH, sulf
750	2(3*H*)-Benzoxazolethione		C_7H_5NOS	2382-96-9	151.186	nd (w)	196				sl H_2O, ace, EtOH; vs eth, HOAc
751	2(3*H*)-Benzoxazolone		$C_7H_5NO_2$	59-49-4	135.121		138	335; 230^{30}			sl H_2O; s EtOH, eth, tfa
752	2-(2-Benzoxazolyl)phenol		$C_{13}H_9NO_2$	835-64-3	211.216	pink nd (al, HOAc)	123.5	338			sl H_2O; vs EtOH; s eth, ace, bz
753	*N*-Benzoyl-*DL*-alanine		$C_{10}H_{11}NO_3$	1205-02-3	193.199	pl, pr or lf (eth)	165.5	dec			s H_2O, EtOH; sl eth, DMSO
754	4-(Benzoylamino)-2-hydroxybenzoic acid	Benzoylpas	$C_{14}H_{11}NO_4$	13898-58-3	257.242		260.5				
755	Benzoyl azide	Benzazide	$C_7H_5N_3O$	582-61-6	147.134	pl (ace)	32	exp	1.1680^{35}		vs eth, EtOH
756	2-Benzoylbenzoic acid		$C_{14}H_{10}O_3$	85-52-9	226.227	tcl nd (w+1)	129.0				vs EtOH, eth; s bz; sl chl
757	4-Benzoylbenzoic acid		$C_{14}H_{10}O_3$	611-95-0	226.227	nd (HOAc), pl (al) mcl lf (w)	199	sub			sl H_2O, tfa, bz; s EtOH, eth, HOAc
758	2-Benzoylbenzoic acid, hydrazide		$C_{14}H_{12}N_2O_2$	787-84-8	240.257	nd (al)	242.3				sl H_2O; i EtOH, eth, chl; s MeOH
759	4-Benzoylbiphenyl	4-Phenylbenzophenone	$C_{19}H_{14}O$	2128-93-0	258.313		101.5	420; $156^{0.1}$			
760	Benzoyl bromide	Benzoic acid, bromide	C_7H_5BrO	618-32-6	185.018	liq	-24	218.5	1.570^{15}	1.5868^{25}	msc eth
761	Benzoyl chloride	Benzoic acid, chloride	C_7H_5ClO	98-88-4	140.567	liq	-0.4	197.2; 71^{9}	1.2120^{20}	1.5537^{20}	msc eth; s bz, ctc, CS_2
762	Benzoyl cyclohexane	Cyclohexyl phenyl ketone	$C_{13}H_{16}O$	712-50-5	188.265	nd (peth)	59.5	164^{18}			
763	Benzoylecgonine		$C_{16}H_{19}NO_4$	519-09-5	289.327	nd (w)	195				vs bz, EtOH
764	Benzoylferrocene		$C_{17}H_{14}FeO$	1272-44-2	290.137		110.0				
765	Benzoyl fluoride	Benzoic acid, fluoride	C_7H_5FO	455-32-3	124.112	liq	-28	154.5	1.1400^{20}		vs EtOH, eth; s ctc

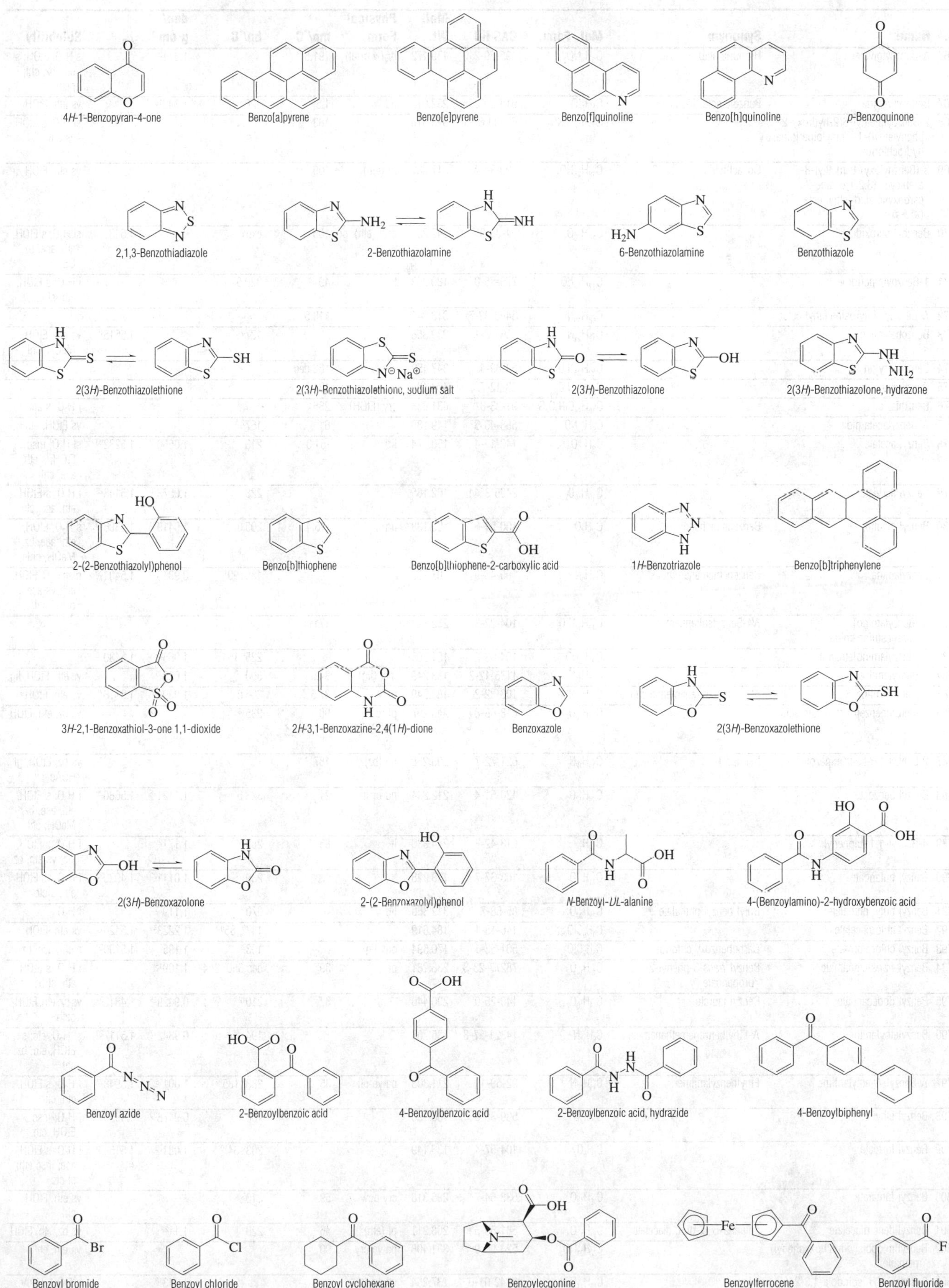
4H-1-Benzopyran-4-one
Benzo[a]pyrene
Benzo[e]pyrene
Benzo[f]quinoline
Benzo[h]quinoline
p-Benzoquinone
2,1,3-Benzothiadiazole
2-Benzothiazolamine
6-Benzothiazolamine
Benzothiazole
2(3H)-Benzothiazolethione
2(3H)-Benzothiazolethione, sodium salt
2(3H)-Benzothiazolone
2(3H)-Benzothiazolone, hydrazone
2-(2-Benzothiazolyl)phenol
Benzo[b]thiophene
Benzo[b]thiophene-2-carboxylic acid
1H-Benzotriazole
Benzo[b]triphenylene
3H-2,1-Benzoxathiol-3-one 1,1-dioxide
2H-3,1-Benzoxazine-2,4(1H)-dione
Benzoxazole
2(3H)-Benzoxazolethione
2(3H)-Benzoxazolone
2-(2-Benzoxazolyl)phenol
N-Benzoyl-DL-alanine
4-(Benzoylamino)-2-hydroxybenzoic acid
Benzoyl azide
2-Benzoylbenzoic acid
4-Benzoylbenzoic acid
2-Benzoylbenzoic acid, hydrazide
4-Benzoylbiphenyl
Benzoyl bromide
Benzoyl chloride
Benzoyl cyclohexane
Benzoylecgonine
Benzoylferrocene
Benzoyl fluoride

No.	Name	Synonym	Mol. Form.	CAS RN	Mol. Wt.	Physical Form	mp/°C	bp/°C	den/ g cm^{-3}	n_D	Solubility
766	*N*-Benzoylglycine	Hippuric acid	$C_9H_9NO_3$	495-69-2	179.172	pr (w or al)	191.5		1.371^{20}		s H_2O, EtOH; sl eth, bz, chl; i peth
767	Benzoyl iodide	Benzoic acid, iodide	C_7H_5IO	618-38-2	232.018	nd	1.5	128^{20}	1.746^{18}		vs eth, EtOH
768	2-Benzoylmethyl-6(2-hydroxy-2-phenylethyl)-1-methylpiperidine, hydrochloride		$C_{22}H_{28}ClNO_2$	63990-84-1	373.916		183.5				sl H_2O; s EtOH; vs chl
769	3-(Benzoyloxy)-8-methyl-8-azabicyclo[3.2.1]octane-2-carboxylic acid, ethyl ester, [1*R*-(*exo,exo*)]	Cocaethylene	$C_{18}H_{23}NO_4$	529-38-4	317.381	pr (eth)	109				vs eth, EtOH
770	Benzoyl peroxide		$C_{14}H_{10}O_4$	94-36-0	242.227	orth (eth), pr	105	exp		1.543	sl H_2O; s EtOH, eth, ace, bz, CS_2
771	1-Benzoylpiperidine		$C_{12}H_{15}NO$	776-75-0	189.253	tcl	49	320.5			i H_2O; s EtOH, eth; sl ctc
772	*N*-Benzoyl-*L*-tyrosine ethyl ester		$C_{18}H_{19}NO_4$	3483-82-7	313.349		119.5				
773	Benzphetamine		$C_{17}H_{21}N$	156-08-1	239.356			$127^{0.02}$		1.5515^{19}	vs eth, EtOH, MeOH, chl
774	Benzpiperylon		$C_{22}H_{25}N_3O$	53-89-4	347.453	cry (al)	182 dec				
775	Benzquinamide		$C_{22}H_{32}N_2O_5$	63-12-7	404.499	cry	131				
776	Benzthiazide		$C_{15}H_{14}ClN_3O_4S_3$	91-33-8	431.938	cry (EtOH)	236				i H_2O; s alk
777	*N*-Benzylacetamide		$C_9H_{11}NO$	588-46-5	149.189		61	157^{2}			vs EtOH, eth
778	Benzyl acetate		$C_9H_{10}O_2$	140-11-4	150.174	liq	-51.3	213	1.0550^{20}	1.5232^{20}	sl H_2O; msc EtOH; s eth, ace, chl
779	Benzyl acrylate		$C_{10}H_{10}O_2$	2495-35-4	162.185			228	1.0573^{20}	1.5143^{20}	i H_2O; s EtOH, eth, ace, ctc
780	Benzyl alcohol	Benzenemethanol	C_7H_8O	100-51-6	108.138	liq	-15.4	205.31	1.0419^{24}	1.5396^{20}	s H_2O, EtOH, eth, ace, bz, MeOH, chl
781	Benzylamine	Benzenemethanamine	C_7H_9N	100-46-9	107.153			185; 90^{12}	0.9813^{20}	1.5401^{20}	msc H_2O, EtOH, eth; vs ace; s bz; sl chl
782	4-(Benzylamino) benzenesulfonamide	*N*4-Benzylsulfanilamide	$C_{13}H_{14}N_2O_2S$	104-22-3	262.327		171				
783	2-[Benzylamino]ethanol		$C_9H_{13}NO$	104-63-2	151.205			225; 154^{12}	1.065^{25}	1.5430^{20}	
784	4-Benzylaniline		$C_{13}H_{13}N$	1135-12-2	183.249	mcl (lig)	34.5	300	1.038^{25}		vs eth, EtOH, lig
785	*N*-Benzylaniline	*N*-Phenylbenzenemethanamine	$C_{13}H_{13}N$	103-32-2	183.249	pr	37.5	306.5	1.0298^{65}	1.6118^{25}	vs eth, EtOH
786	α-Benzylbenzenepropanoic acid		$C_{16}H_{16}O_2$	618-68-8	240.297	pl (peth HOAc) nd (w)	90	235^{18}			vs bz, eth, EtOH
787	2-Benzyl-1*H*-benzimidazole	Bendazol	$C_{14}H_{12}N_2$	621-72-7	208.258	nd (bz)	187				vs bz, EtOH, gl HOAc
788	Benzyl benzoate		$C_{14}H_{12}O_2$	120-51-4	212.244	nd or lf	21	323.5	1.1121^{25}	1.5680^{20}	i H_2O; s EtOH, eth, ace, bz, MeOH, chl
789	4-Benzyl-1,1'-biphenyl		$C_{19}H_{16}$	613-42-3	244.330	lf	85	285^{110}	1.171^{0}		i H_2O; s EtOH, ctc; vs eth, bz
790	Benzyl butanoate		$C_{11}H_{14}O_2$	103-37-7	178.228			239	1.0111^{20}	1.4920^{20}	i H_2O; vs EtOH, eth; s ctc
791	Benzyl butyl phthalate	Butyl benzyl phthalate	$C_{19}H_{20}O_4$	85-68-7	312.360	liq		370	1.119^{25}		i H_2O
792	Benzyl chloroacetate		$C_9H_9ClO_2$	140-18-1	184.619			147^{9}, $85^{0.4}$	1.2223^{4}	1.5426^{18}	vs eth, EtOH
793	Benzyl chloroformate	Carbobenzoxy chloride	$C_8H_7ClO_2$	501-53-1	170.594	oily liq		103^{20}	1.195^{25}	1.5190^{20}	s eth, ace, bz
794	Benzyl *trans*-cinnamate	Benzyl *trans*-3-phenyl-2-propenoate	$C_{16}H_{14}O_2$	78277-23-3	238.281	pr	39	dec 350; 244^{5}	1.109^{15}		i H_2O; s EtOH, eth; sl bz
795	Benzyl dodecanoate	Benzyl laurate	$C_{19}H_{30}O_2$	140-25-0	290.440		8.5	210^{12}	0.9429^{25}	1.4812^{24}	vs bz, eth, EtOH, peth
796	Benzylethylamine	*N*-Ethylbenzenemethanamine	$C_9H_{13}N$	14321-27-8	135.206			194	0.9342^{17}	1.5117^{20}	sl H_2O, ctc; s EtOH, eth, bz, chl
797	*N*-Benzyl-*N*-ethylaniline	Ethylbenzylaniline	$C_{15}H_{17}N$	92-59-1	211.303	pa ye oil	35	288; 185^{22}	1.001^{55}	1.5943^{23}	i H_2O; s EtOH, eth, chl
798	Benzyl ethyl ether		$C_9H_{12}O$	539-30-0	136.190			186	0.9478^{20}	1.4955^{20}	i H_2O; msc EtOH, eth
799	Benzyl formate		$C_8H_8O_2$	104-57-4	136.149			203; 84^{10}	1.081^{20}	1.5154^{20}	i H_2O; s EtOH, ace; msc eth; sl ctc
800	Benzyl fumarate		$C_{18}H_{16}O_4$	538-64-7	296.318	cry pow	59	210^{5}			vs eth, EtOH, chl
801	Benzylidene diacetate	Toluene-α,α-diol, diacetate	$C_{11}H_{12}O_4$	581-55-5	208.211	pl (eth)	46	220	1.11^{20}		vs bz, eth, EtOH
802	Benzylimidobis(*p*-methoxyphenyl) methane		$C_{22}H_{21}NO_2$	524-96-9	331.408	pa ye cry	90				vs eth, chl
803	2-Benzyl-1*H*-isoindole-1,3(2*H*)-dione		$C_{15}H_{11}NO_2$	2142-01-0	237.254	ye nd (al)	116		1.343^{18}		s EtOH, HOAc; sl DMSO

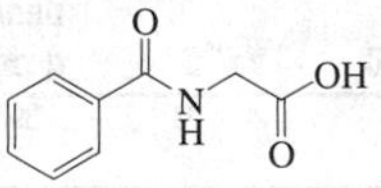

N-Benzoylglycine

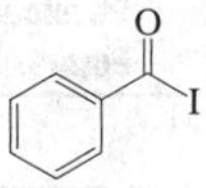

Benzoyl iodide

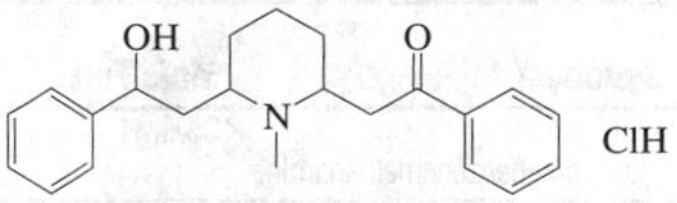

2-Benzoylmethyl-6(2-hydroxy-2-phenylethyl)-1-methylpiperidine, hydrochloride

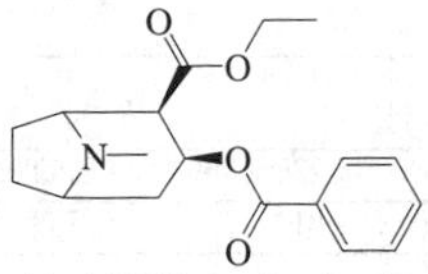

3-(Benzoyloxy)-8-methyl-8-azabicyclo[3.2.1]octane-2-carboxylic acid, ethyl ester, [1R-(exo,exo)]

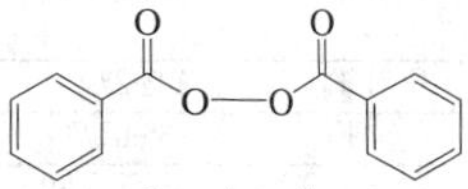

Benzoyl peroxide

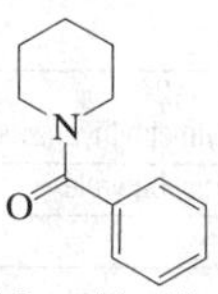

1-Benzoylpiperidine

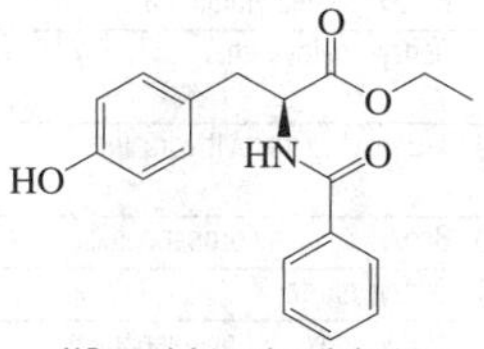

N-Benzoyl-L-tyrosine ethyl ester

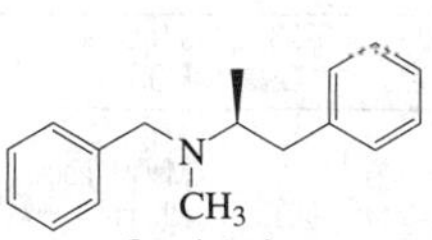

Benzphetamine

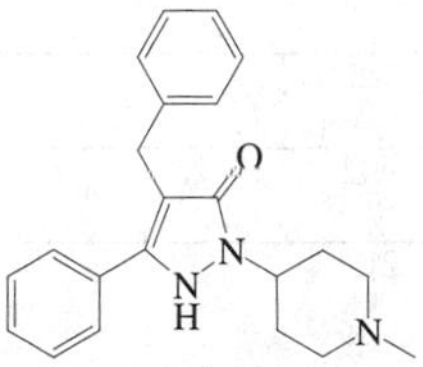

Benzpiperylon

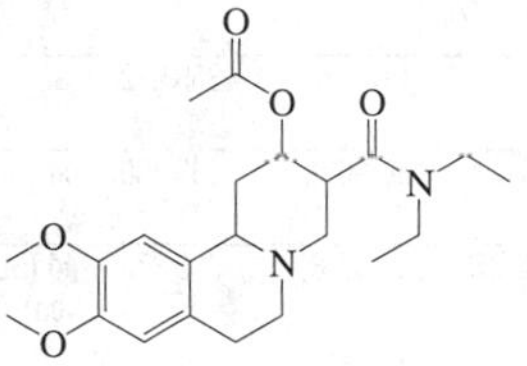

Benzquinamide

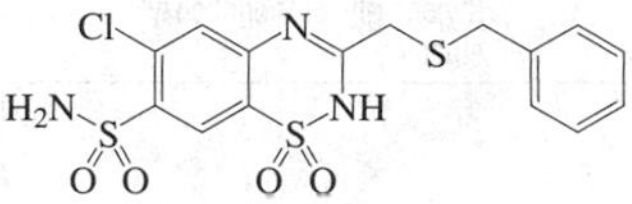

Benzthiazide

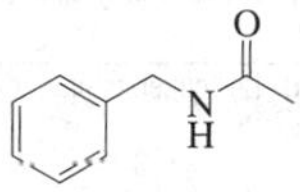

N-Benzylacetamide

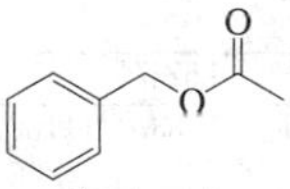

Benzyl acetate

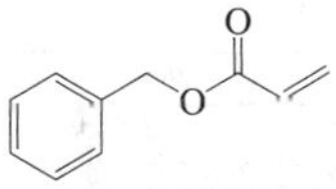

Benzyl acrylate

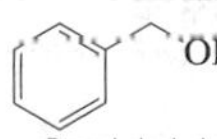

Benzyl alcohol

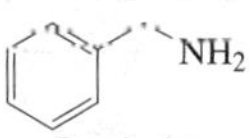

Benzylamine

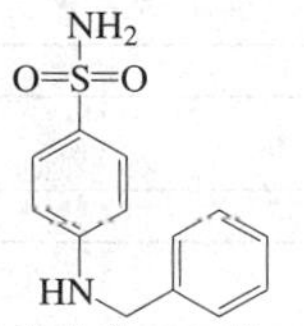

4-(Benzylamino)benzenesulfonamide

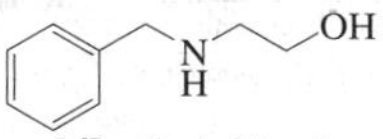

2-[Benzylamino]ethanol

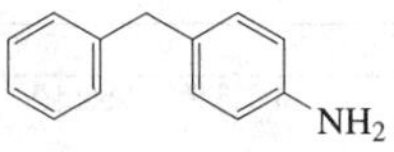

4-Benzylaniline

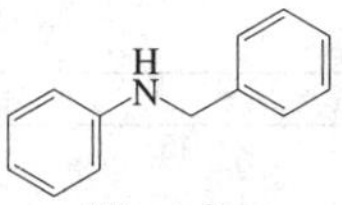

N-Benzylaniline

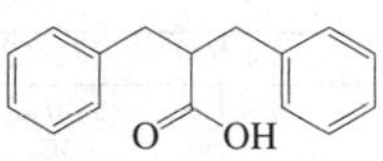

α-Benzylbenzenepropanoic acid

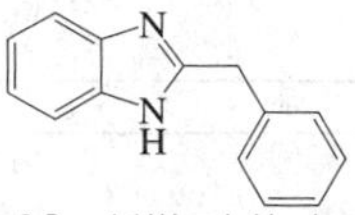

2-Benzyl-1H-benzimidazole

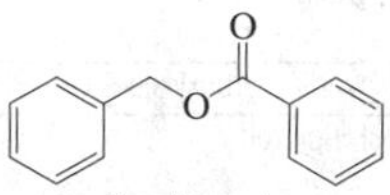

Benzyl benzoate

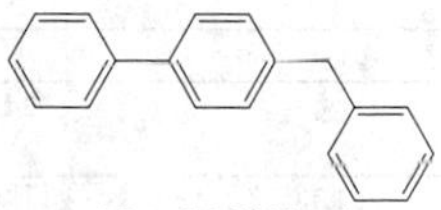

4-Benzyl-1,1'-biphenyl

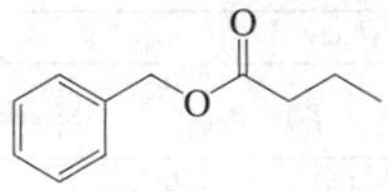

Benzyl butanoate

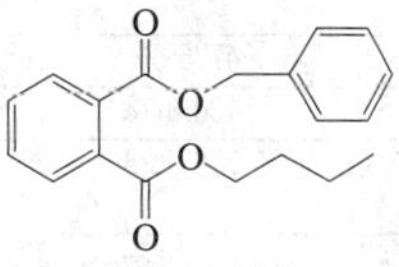

Benzyl butyl phthalate

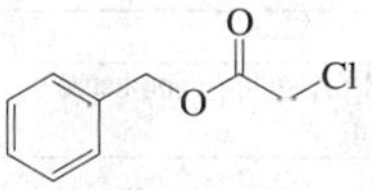

Benzyl chloroacetate

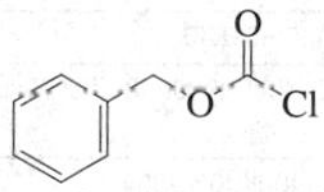

Benzyl chloroformate

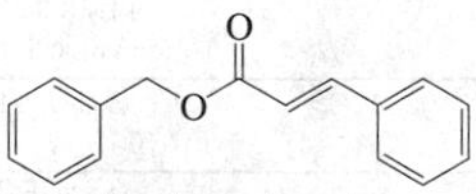

Benzyl trans-cinnamate

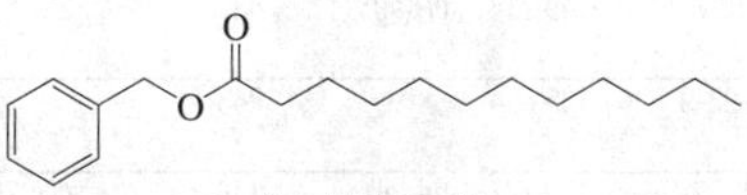

Benzyl dodecanoate

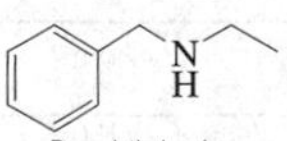

Benzylethylamine

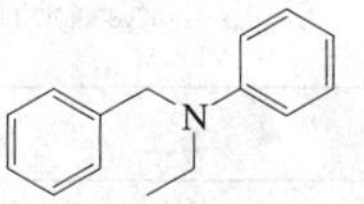

N-Benzyl-N-ethylaniline

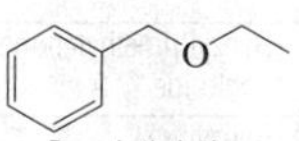

Benzyl ethyl ether

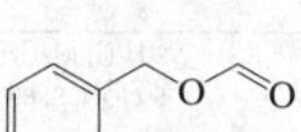

Benzyl formate

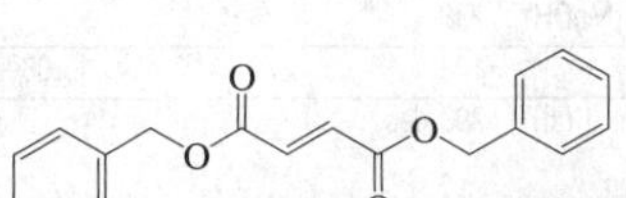

Benzyl fumarate

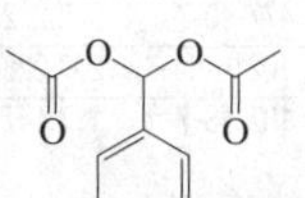

Benzylidene diacetate

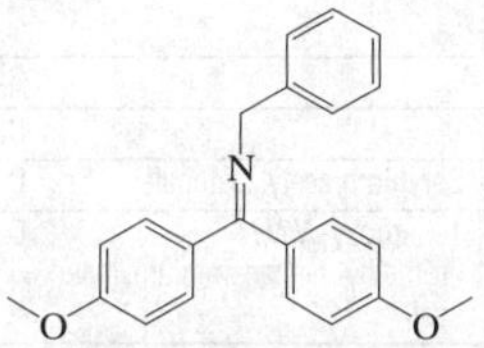

Benzylimidobis(p-methoxyphenyl)methane

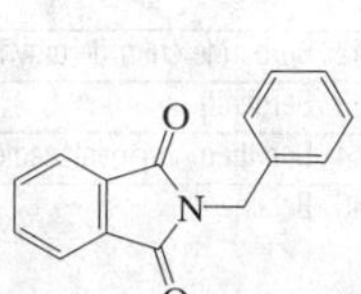

2-Benzyl-1H-isoindole-1,3(2H)-dione

No.	Name	Synonym	Mol. Form.	CAS RN	Mol. Wt.	Physical Form	mp/°C	bp/°C	den/ g cm^{-3}	n_D	Solubility
804	Benzylisopropylamine	*N*-Isopropylbenzenemethanamine	$C_{10}H_{15}N$	102-97-6	149.233			200; 93^{10}	0.892^{25}	1.5025^{20}	
805	Benzyl isothiocyanate	(Isothiocyanatomethyl)benzene	C_8H_7NS	622-78-6	149.214	ye oil		243	1.1246^{16}	1.6049^{15}	i H_2O; msc EtOH; s eth
806	Benzyl methacrylate		$C_{11}H_{12}O_2$	2495-37-6	176.212			144^{50}			
807	Benzyl 3-methylbutanoate		$C_{12}H_{16}O_2$	103-38-8	192.254			245; 136^{25}	0.9983^{15}	1.4884^{20}	
808	Benzyl methyl ether		$C_8H_{10}O$	538-86-3	122.164	liq	-52.6	170	0.9634^{20}	1.5008^{20}	i H_2O, lig; vs EtOH, eth; s bz
809	1-Benzyl-2-methylhydrazine	1-Methyl-2-phenylmethylhydrazine	$C_8H_{12}N_2$	10309-79-2	136.194	liq		117^{20}			
810	Benzyl 2-methylpropanoate	Benzyl isobutyrate	$C_{11}H_{14}O_2$	103-28-6	178.228			228; 114^{20}	1.0159^{18}	1.4883^{20}	
811	Benzyl nitrite		$C_7H_7NO_2$	935-05-7	137.137	oil		81^{35}	1.075^{25}	1.4989^{25}	
812	*N*-Benzyloxycarbonylaspartame		$C_{22}H_{24}N_2O_7$	33605-72-0	428.435	cry	122				
813	Benzyloxycarbonyl-*L*-glutamine		$C_{13}H_{16}N_2O_5$	2650-64-8	280.276		134.5				s DMSO
814	Benzyloxycarbonylglycine		$C_{10}H_{11}NO_4$	1138-80-3	209.199		121				s ace
815	Benzyloxycarbonylglycyl-*L*-leucine		$C_{16}H_{22}N_2O_5$	1421-69-8	322.356		100				
816	Benzyloxycarbonylglycyl-*L*-phenylalanine		$C_{19}H_{20}N_2O_5$	1170-76-9	356.372		126				
817	2-(Benzyloxy)ethanol	Ethylene glycol monobenzyl ether	$C_9H_{12}O_2$	622-08-2	152.190	oil	<-75	256	1.0640^{20}	1.5233^{20}	vs H_2O, eth, EtOH
818	Benzylpenicillin sodium		$C_{16}H_{17}N_2NaO_4S$	69-57-8	356.372	nd (BuOH aq)	215		1.41		vs H_2O; s MeOH; i ace, eth, chl
819	2-Benzylphenol	*o*-Benzylphenol	$C_{13}H_{12}O$	28994-41-4	184.233		21	312		1.5994^{20}	vs ace, bz, EtOH
820	4-Benzylphenol	*p*-Benzylphenol	$C_{13}H_{12}O$	101-53-1	184.233		84	322			s H_2O, EtOH, eth, bz, ctc, HOAc, chl
821	Benzyl phenyl ether		$C_{13}H_{12}O$	946-80-5	184.233	lf (al)	40	286.5			
822	1-Benzylpiperazine		$C_{11}H_{16}N_2$	2759-28-6	176.258			146^{12}		1.5430^{28}	s H_2O, EtOH, eth; sl chl
823	1-Benzylpiperidine		$C_{12}H_{17}N$	2905-56-8	175.270			245	0.9625^{16}	1.5227^{20}	
824	4-Benzylpiperidine		$C_{12}H_{17}N$	31252-42-3	175.270		16.8	270; 150^{17}	0.9970^{20}	1.5337^{25}	i H_2O; s EtOH, eth
825	Benzyl propanoate		$C_{10}H_{12}O_2$	122-63-4	164.201			221	1.0335^{20}		
826	2-Benzylpyridine		$C_{12}H_{11}N$	101-82-6	169.222	nd	12.5	277; 149^{16}	1.067^{0}	1.5785^{20}	i H_2O; s EtOH, eth, chl
827	4-Benzylpyridine		$C_{12}H_{11}N$	2116-65-6	169.222		12.4	288; 180^{31}	1.0612^{20}	1.5818^{20}	i H_2O; s EtOH, ctc; vs eth
828	Benzyl 3-pyridinecarboxylate	Benzyl nicotinate	$C_{13}H_{11}NO_2$	94-44-0	213.232			170^{3}			
829	1-Benzyl-1*H*-pyrrole		$C_{11}H_{11}N$	2051-97-0	157.212		15	247	1.0183^{20}	1.5655^{24}	i H_2O; vs EtOH, eth
830	Benzyl 1,2-pyrrolidinedicarboxylate, (*S*)	*N*-(Benzyloxycarbonyl)-*L*-proline	$C_{13}H_{15}NO_4$	1148-11-4	249.263		78.5			1.5310^{20}	sl chl
831	Benzyl salicylate		$C_{14}H_{12}O_3$	118-58-1	228.243			320	1.1799^{20}	1.5805^{20}	sl H_2O; s EtOH, eth, ctc
832	*O*-Benzyl-*L*-serine	3-(Benzyloxy)-*L*-alanine	$C_{10}H_{13}NO_3$	4726-96-9	195.215		218 dec				
833	Benzylsulfonic acid		$C_7H_8O_3S$	100-87-8	172.202	hyg cry					
834	4-[(Benzylsulfonyl)amino]benzoic acid	*p*-(Benzylsulfonamido)benzoic acid	$C_{14}H_{13}NO_4S$	536-95-8	291.323		229.5				vs EtOH
835	(Benzylsulfonyl)benzene		$C_{13}H_{12}O_2S$	3112-88-7	232.298	nd (al)	146		1.1261^{153}		i H_2O; sl EtOH, eth, bz
836	(Benzylthio)benzene		$C_{13}H_{12}S$	831-91-4	200.299	lf (al)	43.5	197^{27}			i H_2O; s EtOH, eth, con sulf
837	Benzyl thiocyanate	α-Thiocyanatotoluene	C_8H_7NS	3012-37-1	149.214	pr (al)	43	232			i H_2O; s EtOH, eth, chl, CS_2
838	Benzyltrimethylammonium chloride		$C_{10}H_{16}ClN$	56-93-9	185.694		243				vs H_2O; s ace
839	Benzylurea		$C_8H_{10}N_2O$	538-32-9	150.177	nd (al)	148	dec 200			vs ace, EtOH
840	Bephenium chloride		$C_{17}H_{22}ClNO$	13928-81-9	291.816	cry (ace)	135				
841	Berberine		$C_{20}H_{19}NO_5$	2086-83-1	353.369	red-ye nd (w+6) cry (chl)	145				vs eth, EtOH
842	Berberine chloride dihydrate		$C_{20}H_{22}ClNO_6$	5956-60-5	407.845	ye cry					
843	Bergenin		$C_{14}H_{16}O_9$	477-90-7	328.272	cry (MeOH)	238				vs H_2O, EtOH
844	Beryllium 2,4-pentanedioate	Beryllium acetylacetonate	$C_{10}H_{14}BeO_4$	10210-64-7	207.228		108	270	1.168^{20}		
845	Betaine	1-Carboxy-*N,N,N*-trimethylmethanaminium, inner salt	$C_5H_{11}NO_2$	107-43-7	117.147	pr or lf (al)	293 dec				vs H_2O, MeOH; s EtOH; sl eth, chl
846	Betaine, hydrochloride		$C_5H_{12}ClNO_2$	590-46-5	153.608	mcl cry (al)	227.5				vs H_2O
847	Betamethasone		$C_{22}H_{29}FO_5$	378-44-9	392.460	cry (AcOMe)	232 dec				
848	Bethanidine		$C_{10}H_{15}N_3$	55-73-2	177.246	cry (aq MeOH)	196				

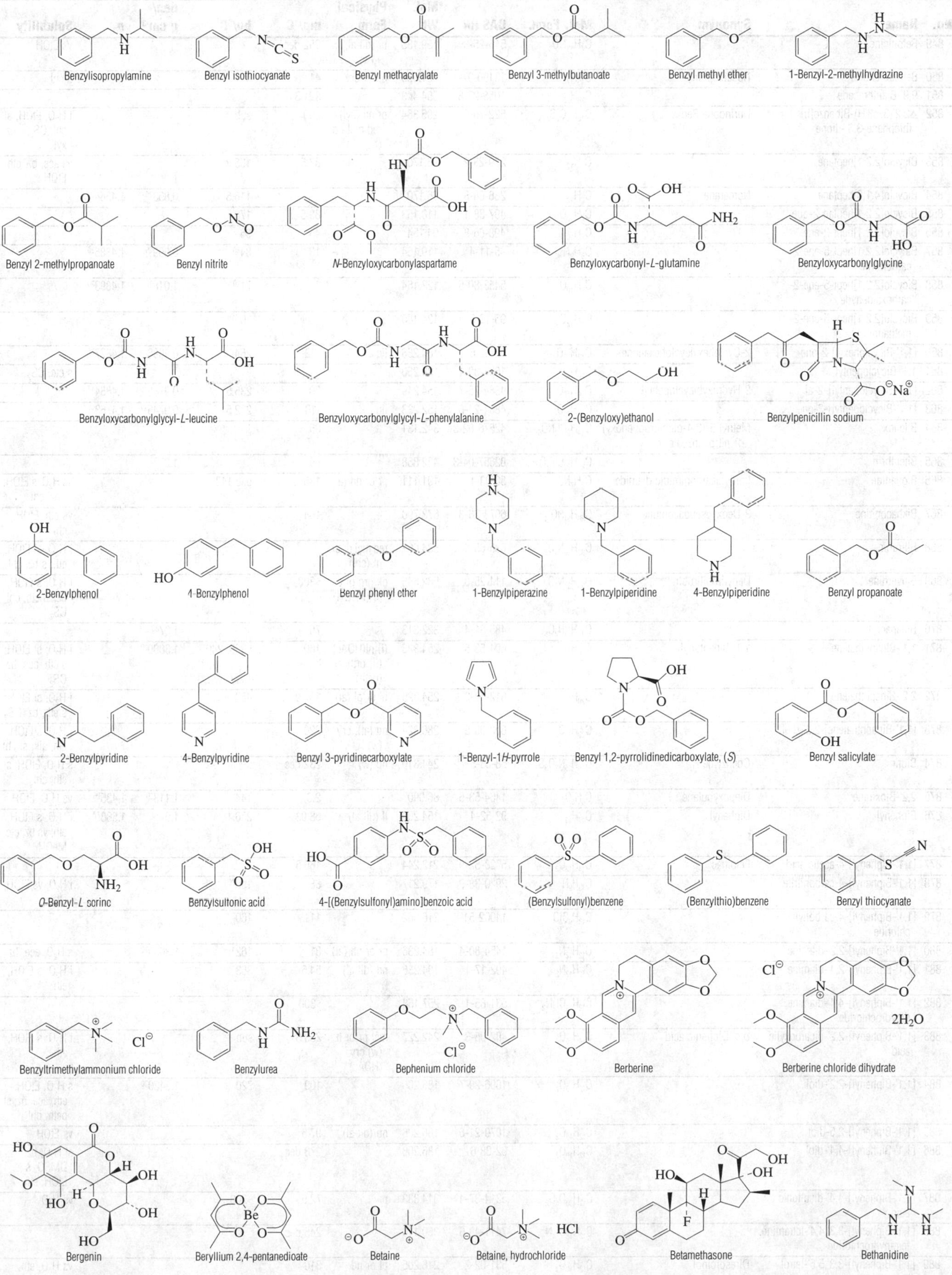

Benzylisopropylamine
Benzyl isothiocyanate
Benzyl methacrylate
Benzyl 3-methylbutanoate
Benzyl methyl ether
1-Benzyl-2-methylhydrazine
Benzyl 2-methylpropanoate
Benzyl nitrite
N-Benzyloxycarbonylaspartame
Benzyloxycarbonyl-L-glutamine
Benzyloxycarbonylglycine
Benzyloxycarbonylglycyl-L-leucine
Benzyloxycarbonylglycyl-L-phenylalanine
2-(Benzyloxy)ethanol
Benzylpenicillin sodium
2-Benzylphenol
4-Benzylphenol
Benzyl phenyl ether
1-Benzylpiperazine
1-Benzylpiperidine
4-Benzylpiperidine
Benzyl propanoate
2-Benzylpyridine
4-Benzylpyridine
Benzyl 3-pyridinecarboxylate
1-Benzyl-1H-pyrrole
Benzyl 1,2-pyrrolidinedicarboxylate, (S)
Benzyl salicylate
O-Benzyl-L-serine
Benzylsulfonic acid
4-[(Benzylsulfonyl)amino]benzoic acid
(Benzylsulfonyl)benzene
(Benzylthio)benzene
Benzyl thiocyanate
Benzyltrimethylammonium chloride
Benzylurea
Bephenium chloride
Berberine
Berberine chloride dihydrate
Bergenin
Beryllium 2,4-pentanedioate
Betaine
Betaine, hydrochloride
Betamethasone
Bethanidine

No.	Name	Synonym	Mol. Form.	CAS RN	Mol. Wt.	Physical Form	mp/°C	bp/°C	den/ g cm^{-3}	n_D	Solubility
849	Betonicine		$C_7H_{13}NO_3$	515-25-3	159.183	pr (dil al, +1w)	252 dec				vs EtOH
850	Betulaprenol 9	Nonaisoprenol	$C_{45}H_{74}O$	13190-97-1	631.069	oil or cry	41				s chl
851	9,9'-Bianthracene		$C_{28}H_{18}$	1055-23-8	354.443		321.3				
852	Δ2,2'(3*H*,3'H)-Bibenzo[b] thiophene-3,3'-dione	Durindone Red	$C_{16}H_8O_2S_2$	522-75-8	296.364	br nd (xyl) red mcl nd (bz)	359	sub			i H_2O, EtOH; sl chl, CS_2; s bz, xyl
853	Bicyclo[2.2.1]heptane		C_7H_{12}	279-23-2	96.170		87.5	105.3			vs ace, bz, eth, EtOH
854	Bicyclo[4.1.0]heptane	Norcarane	C_7H_{12}	286-08-8	96.170			116.5	0.853^{25}	1.4564^{20}	
855	Bicyclo[2.2.1]heptan-2-one		$C_7H_{10}O$	497-38-1	110.153		89.5	170			
856	Bicyclo[2.2.1]hept-2-ene		C_7H_{10}	498-66-8	94.154		45	96			
857	Bicyclo[2.2.1]hept-5-ene-2-carbonitrile		C_8H_9N	95-11-4	119.164		13	84^{10}	0.999^{25}	1.4885^{20}	
858	Bicyclo[2.2.1]hept-5-ene-2-carboxaldehyde		$C_8H_{10}O$	5453-80-5	122.164			71^{20}	1.018^{25}	1.4893^{20}	
859	Bicyclo[2.2.1]hept-5-ene-2-methanol		$C_8H_{12}O$	95-12-5	124.180			103^{20}			
860	[1,1'-Bicyclohexyl]-2-one	2-Cyclohexylcyclohexanone	$C_{12}H_{20}O$	90-42-6	180.286	liq	-32	264	0.9696^{25}	1.4877^{25}	
861	1,1'-Bicyclopentyl		$C_{10}H_{18}$	1636-39-1	138.250						s ctc, CS_2
862	[1,1'-Bicyclopentyl]-2-ol	2-Hydroxybicyclopentyl	$C_{10}H_{18}O$	4884-25-7	154.249		20	235.5	0.9785^{15}	1.4884^{17}	
863	[1,1'-Bicyclopentyl]-2-one		$C_{10}H_{16}O$	4884-24-6	152.233	liq	-13	232.5	0.9745^{21}	1.4763	
864	Bifenox	Methyl 5-(2,4-dichlorophenoxy) -2-nitrobenzoate	$C_{14}H_9Cl_2NO_5$	42576-02-3	342.131		85				
865	Bifenthrin		$C_{23}H_{22}ClF_3O_2$	82657-04-3	422.868		69		1.2^{125}		
866	Biguanide	Imidodicarbonimidic diamide	$C_2H_7N_5$	56-03-1	101.111	pr or nd (al)	136	dec 142			vs H_2O; s EtOH; i bz, chl
867	Bikhaconitine	3-Deoxypseudaconitine	$C_{36}H_{51}NO_{11}$	6078-26-8	673.790		164				vs eth, EtOH, chl
868	Bilirubin		$C_{33}H_{36}N_4O_6$	635-65-4	584.662	red mcl pr or pl (chl)					i H_2O; sl EtOH, eth; s bz, chl
869	Biliverdine	Dehydrobilirubin	$C_{33}H_{34}N_4O_6$	114-25-0	582.646	dk grn pl or pr (MeOH)	>300				i H_2O; s EtOH, bz; sl eth, chl, CS_2
870	Binapacryl		$C_{15}H_{18}N_2O_6$	485-31-4	322.313		70		1.27^{20}		
871	1,1'-Binaphthalene	1,1'-Binaphthyl	$C_{20}H_{14}$	604-53-5	254.325	(i) pl(HOAc) (ii) orth (peth)	160	>360; 240^{12}	1.3000^{20}		i H_2O; sl EtOH; s eth, ace, bz, CS_2
872	2,2'-Binaphthalene		$C_{20}H_{14}$	612-78-2	254.325	bl flr pl (al)	187.9	452			i H_2O; sl EtOH; s eth, bz, CS_2
873	[1,1'-Binaphthalene]-2,2'-diol		$C_{20}H_{14}O_2$	602-09-5	286.324	nd (al), cry (w)	220				i H_2O; s EtOH, eth, alk; sl chl
874	Biotin	Coenzyme R	$C_{10}H_{16}N_2O_3S$	58-85-5	244.310	nd (w)	232 dec				s H_2O, EtOH; sl eth, chl
875	2,2'-Bioxirane	Diepoxybutane	$C_4H_6O_2$	1464-53-5	86.090		2.0	144	1.113^{20}	1.435^{20}	vs H_2O, EtOH
876	Biphenyl	Diphenyl	$C_{12}H_{10}$	92-52-4	154.207	lf (dil al)	68.93	256.1	1.04^{20}	1.588^{77}	i H_2O; s EtOH, eth; vs bz, ctc, MeOH
877	[1,1'-Biphenyl]-4-acetic acid	Felbinac	$C_{14}H_{12}O_2$	5728-52-9	212.244		160.5				
878	[1,1'-Biphenyl]-4-carbonitrile		$C_{13}H_9N$	2920-38-9	179.217		88	190^{20}			i H_2O; vs EtOH, eth
879	[1,1'-Biphenyl]-4-carbonyl chloride		$C_{13}H_9ClO$	14002-51-8	216.662		111	160^{2}			
880	[1,1'-Biphenyl]-2,2'-diamine		$C_{12}H_{12}N_2$	1454-80-4	184.236	pr or nd (al)	81	162^{4}	1.3090^{20}		s H_2O, ace, bz
881	[1,1'-Biphenyl]-2,4'-diamine		$C_{12}H_{12}N_2$	492-17-1	184.236	nd (dil al)	54.5	363			i H_2O; s EtOH, eth
882	[1,1'-Biphenyl]-4,4'-diamine, dihydrochloride		$C_{12}H_{14}Cl_2N_2$	531-85-1	257.158		>300				
883	[1,1'-Biphenyl]-2,2'-dicarboxylic acid	*o,o'*-Diphenic acid	$C_{14}H_{10}O_4$	482-05-3	242.227	mcl pr or lf (w) cry (HOAc)	233.5	sub			i H_2O; s EtOH, eth
884	[1,1'-Biphenyl]-2,2'-diol		$C_{12}H_{10}O_2$	1806-29-7	186.206		109	320	1.3420^{20}		s H_2O, EtOH, eth, ace, bz; sl peth, chl
885	[1,1'-Biphenyl]-2,5-diol		$C_{12}H_{10}O_2$	1079-21-6	186.206	nd (dil al)	97.5				vs EtOH
886	[1,1'-Biphenyl]-4,4'-diol		$C_{12}H_{10}O_2$	92-88-6	186.206		278 dec				sl H_2O, bz, DMSO; s EtOH, eth
887	[1,1'-Biphenyl]-4,4'-disulfonic acid		$C_{12}H_{10}O_6S_2$	5314-37-4	314.333	pr	72.5	>200			vs H_2O
888	[1,1'-Biphenyl]-3,3',4,4'-tetramine, tetrahydrochloride		$C_{12}H_{18}Cl_4N_4$	7411-49-6	360.110		245 dec				
889	[1,1'-Biphenyl]-3,3',5,5'-tetrol	Diresorcinol	$C_{12}H_{10}O_4$	531-02-2	218.205	pl or nd (w+2)	310				vs H_2O, eth, EtOH

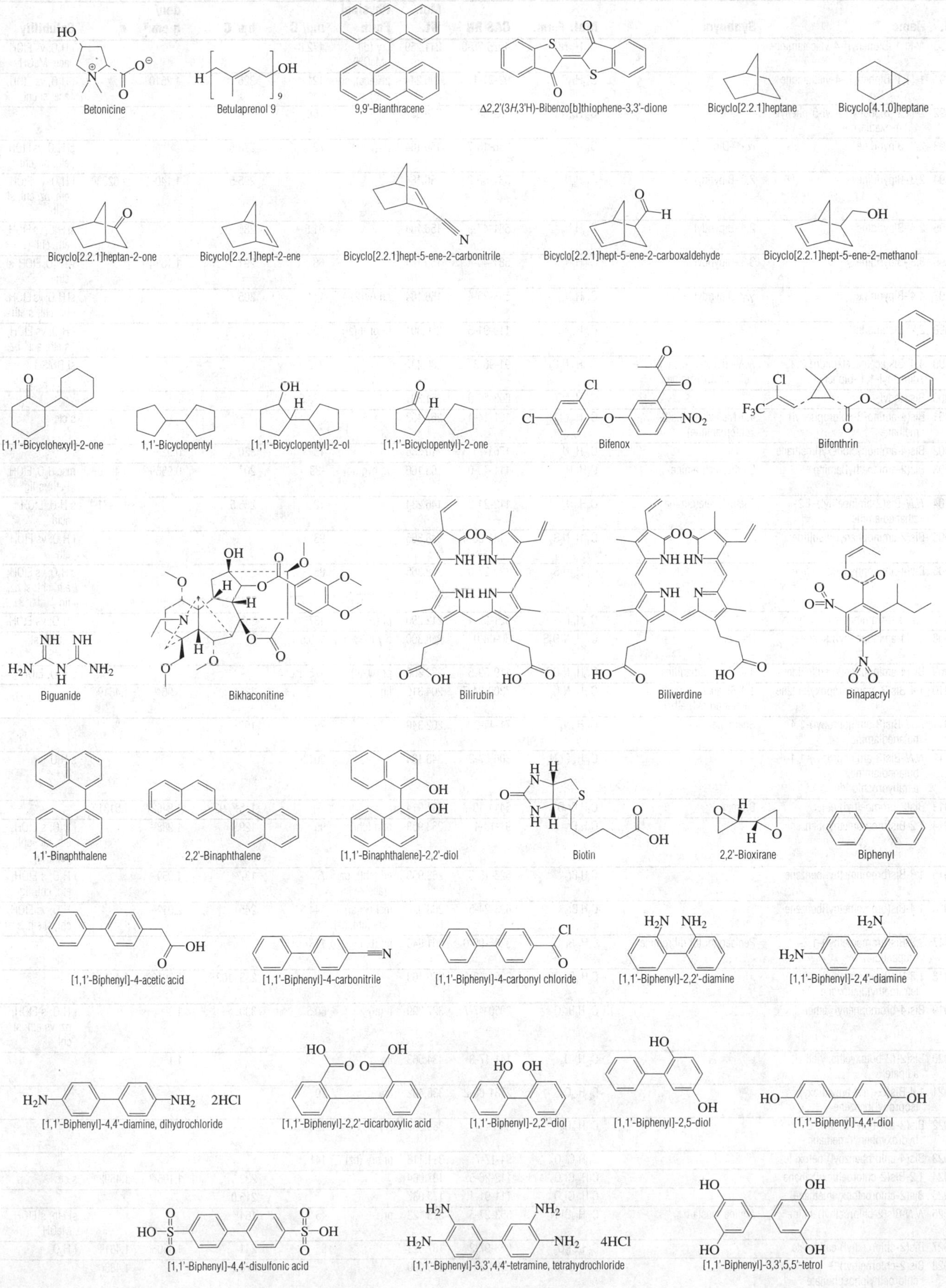
Betonicine
Betulaprenol 9
9,9'-Bianthracene
Δ2,2'(3H,3'H)-Bibenzo[b]thiophene-3,3'-dione
Bicyclo[2.2.1]heptane
Bicyclo[4.1.0]heptane
Bicyclo[2.2.1]heptan-2-one
Bicyclo[2.2.1]hept-2-ene
Bicyclo[2.2.1]hept-5-ene-2-carbonitrile
Bicyclo[2.2.1]hept-5-ene-2-carboxaldehyde
Bicyclo[2.2.1]hept-5-ene-2-methanol
[1,1'-Bicyclohexyl]-2-one
1,1'-Bicyclopentyl
[1,1'-Bicyclopentyl]-2-ol
[1,1'-Bicyclopentyl]-2-one
Bifenox
Bifenthrin
Biguanide
Bikhaconitine
Bilirubin
Biliverdine
Binapacryl
1,1'-Binaphthalene
2,2'-Binaphthalene
[1,1'-Binaphthalene]-2,2'-diol
Biotin
2,2'-Bioxirane
Biphenyl
[1,1'-Biphenyl]-4-acetic acid
[1,1'-Biphenyl]-4-carbonitrile
[1,1'-Biphenyl]-4-carbonyl chloride
[1,1'-Biphenyl]-2,2'-diamine
[1,1'-Biphenyl]-2,4'-diamine
[1,1'-Biphenyl]-4,4'-diamine, dihydrochloride
[1,1'-Biphenyl]-2,2'-dicarboxylic acid
[1,1'-Biphenyl]-2,2'-diol
[1,1'-Biphenyl]-2,5-diol
[1,1'-Biphenyl]-4,4'-diol
[1,1'-Biphenyl]-4,4'-disulfonic acid
[1,1'-Biphenyl]-3,3',4,4'-tetramine, tetrahydrochloride
[1,1'-Biphenyl]-3,3',5,5'-tetrol

No.	Name	Synonym	Mol. Form.	CAS RN	Mol. Wt.	Physical Form	mp/°C	bp/°C	den/ g cm^{-3}	n_D	Solubility
890	*N*-[1,1'-Biphenyl]-4-ylacetamide		$C_{14}H_{13}NO$	4075-79-0	211.259	cry (dil MeOH)	172.8				i H_2O; vs EtOH, ace, MeOH
891	1-[1,1'-Biphenyl]-4-ylethanone		$C_{14}H_{12}O$	92-91-1	196.244	pr (ace), cry (al)	121	326	1.2510^{0}		i H_2O; vs EtOH, ace; sl chl
892	2-[1,1'-Biphenyl]-4-yl-5-phenyl-1,3,4-oxadiazole		$C_{20}H_{14}N_2O$	852-38-0	298.337		168				
893	2,2'-Bipyridine	α,α'-Dipyridyl	$C_{10}H_8N_2$	366-18-7	156.184	pr (peth)	72	273.5			sl H_2O; vs EtOH, eth, bz, chl
894	2,3'-Bipyridine	2,3'-Bipyridyl	$C_{10}H_8N_2$	581-50-0	156.184			295.5	1.140^{20}	1.6223^{20}	i H_2O; vs EtOH, eth, bz, chl; sl peth
895	2,4'-Bipyridine	2,4'-Bipyridyl	$C_{10}H_8N_2$	581-47-5	156.184		61.5	281			sl H_2O; vs EtOH, eth, chl
896	3,3'-Bipyridine	3,3'-Bipyridyl	$C_{10}H_8N_2$	581-46-4	156.184		68	291.5	1.1614^{20}		vs H_2O, EtOH; sl eth
897	4,4'-Bipyridine	γ,γ'-Dipyridyl	$C_{10}H_8N_2$	553-26-4	156.184	nd (w+2)	114	305			sl H_2O; vs EtOH, bz, chl; s eth
898	2,2'-Biquinoline		$C_{18}H_{12}N_2$	119-91-5	256.301	pl or lf (al)	196				i H_2O; vs EtOH; s eth, ace, bz
899	4,4'-Bis(acetoacetamido)-3,3'-dimethyl-1,1'-biphenyl	*N,N*-Bis(acetoacetyl)-3,3'-dimethylbenzidine	$C_{22}H_{24}N_2O_4$	91-96-3	380.437		212				sl DMSO
900	Bisacodyl		$C_{22}H_{19}NO_4$	603-50-9	361.391		133.5				
901	Bis(4-amino-3-chlorophenyl) methane	4,4-Methylene-bis(2-chloroaniline)	$C_{13}H_{12}Cl_2N_2$	101-14-4	267.153						s ctc
902	Bis(4-aminocyclohexyl)methane		$C_{13}H_{26}N_2$	1761-71-3	210.358		15	320	0.92^{75}		
903	Bis(2-aminoethyl)amine	Diethylenetriamine	$C_4H_{13}N_3$	111-40-0	103.166	ye hyg liq	-39	207	0.9569^{20}	1.4810^{25}	msc H_2O, EtOH; i eth; s lig
904	*N,N'*-Bis(2-aminoethyl)-1,2-ethanediamine	Triethylenetetramine	$C_6H_{18}N_4$	112-24-3	146.234		12	266.5		1.4971^{20}	s H_2O, EtOH, acid
905	Bis(2-aminophenyl)disulfide		$C_{12}H_{12}N_2S_2$	1141-88-4	248.366		93				i H_2O; vs EtOH, eth
906	Bis(4-aminophenyl)disulfide		$C_{12}H_{12}N_2S_2$	722-27-0	248.366		85				s H_2O; vs EtOH, eth, chl; sl bz, lig
907	1,2-Bis(4-aminophenyl)ethane		$C_{14}H_{16}N_2$	621-95-4	212.290	pl (w)	137	sub			i H_2O; vs EtOH
908	Bis(4-aminophenyl) sulfone	Dapsone	$C_{12}H_{12}N_2O_2S$	80-08-0	248.300	cry (95% al)	175.5				s EtOH; sl DMSO
909	Bis(4-aminophenyl) sulfoxide	4,4'-Sulfinyldianiline	$C_{12}H_{12}N_2OS$	119-59-5	232.300	pr (w, al)	175 dec				s H_2O, EtOH
910	1,4-Bis(3-aminopropoxy)butane	1,4-Butanediol bis(3-aminopropyl) ether	$C_{10}H_{24}N_2O_2$	7300-34-7	204.310	liq		135^{3}	0.96^{20}	1.4619^{20}	
911	*N,N'*-Bis(3-aminopropyl)-1,4-butanediamine	Spermine	$C_{10}H_{26}N_4$	71-44-3	202.340		29	150^{5}			
912	*N,N'*-Bis(3-aminopropyl)-1,4-butanediamine, tetrahydrochloride		$C_{10}H_{30}Cl_4N_4$	306-67-2	348.184		301.5				s H_2O
913	Bis(2-bromoethyl) ether	Bromex	$C_4H_8Br_2O$	5414-19-7	231.914			115^{32}, 92^{12}	1.8452^{20}	1.5131^{27}	
914	1,2-Bis(bromomethyl)benzene		$C_8H_8Br_2$	91-13-4	263.958	orth (chl)	95	$129^{4.5}$	1.988^{25}		i H_2O; s EtOH, eth, ctc, chl, peth, lig
915	1,3-Bis(bromomethyl)benzene		$C_8H_8Br_2$	626-15-3	263.958	nd (chl), pr (ace)	77	137^{20}	1.959^{25}		i H_2O; s EtOH, eth, chl, lig
916	1,4-Bis(bromomethyl)benzene		$C_8H_8Br_2$	623-24-5	263.958	mcl pr (al), cry (chl, bz)	144.5	245	2.012^{25}		i H_2O; vs EtOH, chl; sl eth; s bz
917	2,2-Bis(bromomethyl)-1,3-propanediol	Pentaerythritol dibromide	$C_5H_{10}Br_2O_2$	3296-90-0	261.940	nd (bz)	113				
918	1,3-Bis(bromomethyl) tetramethyldisiloxane		$C_6H_{16}Br_2OSi_2$	2351-13-5	320.169			233; 103^{15}	1.3918^{25}	1.4719^{25}	
919	Bis(4-bromophenyl) ether		$C_{12}H_8Br_2O$	2050-47-7	327.999	lf (al)	60.5	339	1.8^{25}		i H_2O; s EtOH, bz; vs eth; sl chl
920	Bis(2-(2-butoxyethoxy)ethyl) adipate		$C_{22}H_{42}O_8$	141-17-3	434.563	liq			1.1^{25}		
921	1,4-Bis(α-(*tert*-butyldioxy) isopropyl)benzene		$C_{20}H_{34}O_4$	2781-00-2	338.482	cry	79				
922	Bis(3-*tert*-butyl-5-ethyl-2-hydroxyphenyl)methane		$C_{25}H_{36}O_2$	88-24-4	368.553	cry	123				
923	Bis(4-chlorobenzoyl) peroxide		$C_{14}H_8Cl_2O_4$	94-17-7	311.118	pr cry (bz)	141				
924	1,2-Bis(2-chloroethoxy)ethane		$C_6H_{12}Cl_2O_2$	112-26-5	187.064			232	1.195^{20}	1.4592^{25}	s ctc
925	Bis(2-chloroethoxy)methane		$C_5H_{10}Cl_2O_2$	111-91-1	173.037			215.0			
926	*N,N*-Bis(2-chloroethyl)aniline	Aniline mustard	$C_{10}H_{13}Cl_2N$	553-27-5	218.123	pr	45	164^{14}			sl eth; s EtOH, MeOH
927	Bis(2-chloroethyl) carbonate		$C_5H_8Cl_2O_3$	623-97-2	187.021		8	241	1.3506^{20}	1.461^{20}	i H_2O
928	Bis(2-chloroethyl) 2-chloroethylphosphonate		$C_6H_{12}Cl_3O_3P$	6294-34-4	269.490			170.2^{5}		1.488^{25}	

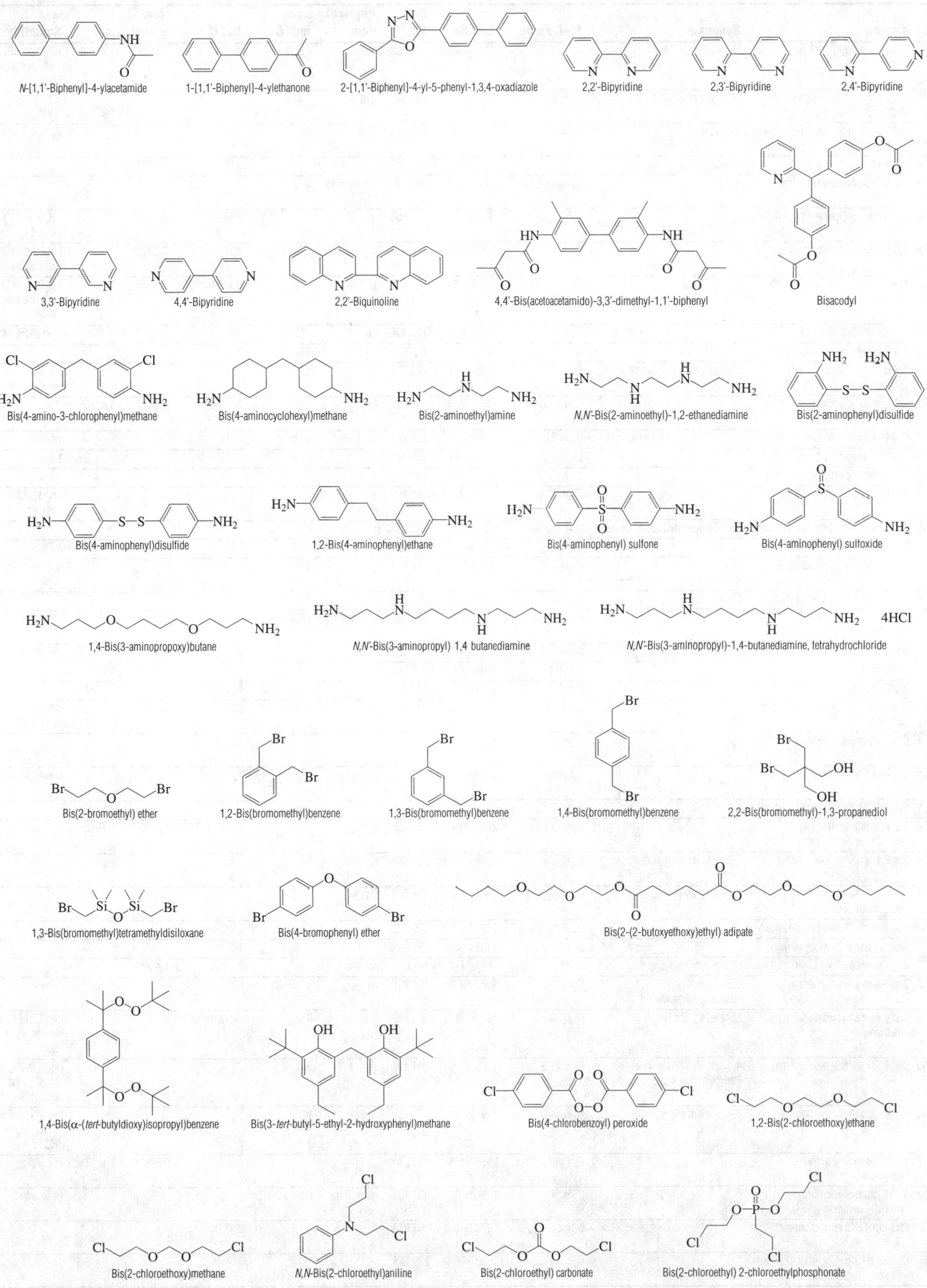

N-[1,1'-Biphenyl]-4-ylacetamide
1-[1,1'-Biphenyl]-4-ylethanone
2-[1,1'-Biphenyl]-4-yl-5-phenyl-1,3,4-oxadiazole
2,2'-Bipyridine
2,3'-Bipyridine
2,4'-Bipyridine
3,3'-Bipyridine
4,4'-Bipyridine
2,2'-Biquinoline
4,4'-Bis(acetoacetamido)-3,3'-dimethyl-1,1'-biphenyl
Bisacodyl
Bis(4-amino-3-chlorophenyl)methane
Bis(4-aminocyclohexyl)methane
Bis(2-aminoethyl)amine
N,N'-Bis(2-aminoethyl)-1,2-ethanediamine
Bis(2-aminophenyl)disulfide
Bis(4-aminophenyl)disulfide
1,2-Bis(4-aminophenyl)ethane
Bis(4-aminophenyl) sulfone
Bis(4-aminophenyl) sulfoxide
1,4-Bis(3-aminopropoxy)butane
N,N'-Bis(3-aminopropyl) 1,4 butanediamine
N,N'-Bis(3-aminopropyl)-1,4-butanediamine, tetrahydrochloride
4HCl
Bis(2-bromoethyl) ether
1,2-Bis(bromomethyl)benzene
1,3-Bis(bromomethyl)benzene
1,4-Bis(bromomethyl)benzene
2,2-Bis(bromomethyl)-1,3-propanediol
1,3-Bis(bromomethyl)tetramethyldisiloxane
Bis(4-bromophenyl) ether
Bis(2-(2-butoxyethoxy)ethyl) adipate
1,4-Bis(α-(tert-butyldioxy)isopropyl)benzene
Bis(3-tert-butyl-5-ethyl-2-hydroxyphenyl)methane
Bis(4-chlorobenzoyl) peroxide
1,2-Bis(2-chloroethoxy)ethane
Bis(2-chloroethoxy)methane
N,N-Bis(2-chloroethyl)aniline
Bis(2-chloroethyl) carbonate
Bis(2-chloroethyl) 2-chloroethylphosphonate

No.	Name	Synonym	Mol. Form.	CAS RN	Mol. Wt.	Physical Form	mp/°C	bp/°C	den/ g cm^{-3}	n_D	Solubility
929	Bis(2-chloroethyl) ether	Dichloroethyl ether	$C_4H_8Cl_2O$	111-44-4	143.012	liq	-51.9	178.5	1.22^{20}	1.451^{20}	i H_2O; s EtOH, eth, ace; msc bz
930	Bis(2-chloroethyl)methylamine hydrochloride	Nitrogen mustard hydrochloride	$C_5H_{12}Cl_3N$	55-86-7	192.515	hyg nd	111.5				
931	*N,N'*-Bis(2-chloroethyl)-*N*-nitrosourea	Carmustine	$C_5H_9Cl_2N_3O_2$	154-93-8	214.049	lt ye pow	31				vs H_2O, EtOH
932	Bis(2-chloroethyl) sulfide	Mustard gas	$C_4H_8Cl_2S$	505-60-2	159.078		13.5	216	1.2741^{20}	1.5313^{20}	
933	1,2-Bis(2-chloroethylsulfonyl) ethane		$C_6H_{12}Cl_2O_4S_2$	3944-87-4	283.193	cry (MeOH/ HOAc)	205				
934	1,2-Bis(chloromethyl)benzene		$C_8H_8Cl_2$	612-12-4	175.056	mcl (lig)	55	239.5	1.393^{25}		i H_2O; vs EtOH, eth, chl; s ctc
935	1,3-Bis(chloromethyl)benzene		$C_8H_8Cl_2$	626-16-4	175.056	cry	34.2	251.5	1.302^{20}		i H_2O; vs EtOH, eth; sl chl
936	1,4-Bis(chloromethyl)benzene		$C_8H_8Cl_2$	623-25-6	175.056	pl (al)	100	dec 245; 135^{16}	1.417^{25}		i H_2O; vs EtOH, eth, ace, chl; sl HOAc
937	Bis(chloromethyl) ether		$C_2H_4Cl_2O$	542-88-1	114.958	liq	-41.5	106	1.323^{15}	1.435^{21}	msc EtOH, eth
938	3,3-Bis(chloromethyl)oxetane		$C_5H_8Cl_2O$	78-71-7	155.022	liq	18.7	101^{27}	1.295^{25}		
939	2,2-Bis(chloromethyl)-1,3-propanediol	Pentaerythritol dichlorohydrin	$C_5H_{10}Cl_2O_2$	2209-86-1	173.037	cry	83	159^{12}			
940	1,3-Bis(chloromethyl) tetramethyldisiloxane		$C_6H_{16}Cl_2OSi_2$	2362-10-9	231.267	liq	-90	204; 92^{21}	1.045^{20}	1.4398^{20}	
941	Bis(4-chlorophenoxy)methane	Di(4-chlorophenoxy)methane	$C_{13}H_{10}Cl_2O_2$	555-89-5	269.123	cry (peth)	70.5	191^{6}			vs ace, bz
942	Bis(4-chlorophenyl) disulfide		$C_{12}H_8Cl_2S_2$	1142-19-4	287.228		72.8				s chl
943	Bis(4-chlorophenyl)ethanedione		$C_{14}H_8Cl_2O_2$	3457-46-3	279.119		197.8				
944	1,1-Bis(4-chlorophenyl)ethanol		$C_{14}H_{12}Cl_2O$	80-06-8	267.150		70				i H_2O, EtOH; s eth, bz
945	1,2-Bis(2-chlorophenyl)-hydrazine	2,2'-Dichlorohydrazobenzene	$C_{12}H_{10}Cl_2N_2$	782-74-1	253.126		87				
946	Bis(4-chlorophenyl)methane		$C_{13}H_{10}Cl_2$	101-76-8	237.124		55.5	188^{18}	1.365^{17}		s EtOH
947	Bis(4-chlorophenyl) sulfone		$C_{12}H_8Cl_2O_2S$	80-07-9	287.162		147.9	250^{10}			sl H_2O; s EtOH, chl
948	*N,N'*-Bis(4-chlorophenyl)thiourea	Di(*p*-chlorophenyl)thiourea	$C_{13}H_{10}Cl_2N_2S$	1220-00-4	297.202	nd	176				
949	1,1-Bis(4-chlorophenyl)-2,2,2-trichloroethanol		$C_{14}H_9Cl_5O$	115-32-2	370.485	cry (petr)	77.5	$180^{0.1}$			i H_2O, os
950	Bis(3-chloropropyl) ether	3-Chloropropyl ether	$C_6H_{12}Cl_2O$	629-36-7	171.064			216; 90.5^{11}	1.136^{20}	1.4158^{20}	s EtOH, eth
951	Bis(2-cyanoethyl) ether		$C_6H_8N_2O$	1656-48-0	124.140			161^{5}, $111^{0.5}$	1.0504^{20}	1.4405^{20}	
952	Bis(2-cyanoethyl) sulfide		$C_6H_8N_2S$	111-97-7	140.206			$163^{1.5}$		1.5047^{20}	
953	Bis(η-cyclopentadienyl)titanium chloride		$C_{10}H_{10}Cl_2Ti$	1271-19-8	248.959	red cry	289	258^{10}	1.60		sl H_2O, bz; s chl, EtOH, tol
954	Bis(η-cyclopentadienyl)zirconium chloride		$C_{10}H_{10}Cl_2Zr$	1291-32-3	292.316			$180^{0.5}$			
955	1,2-Bis(dibromomethyl)benzene		$C_8H_6Br_4$	13209-15-9	421.750	mcl	116.5				sl H_2O; vs chl; i lig
956	Bis(2,4-dichlorobenzoyl) peroxide		$C_{14}H_6Cl_4O_4$	133-14-2	380.008		106				
957	1,3-Bis(dichloromethyl) tetramethyldisiloxane		$C_6H_{14}Cl_4OSi_2$	2943-70-6	300.157			149^{50}, 117^{11}	1.2213^{20}	1.4660^{20}	
958	Bis(2,4-dichlorophenyl)ether	2,2',4,4'-Tetrachlorodiphenyl ether	$C_{12}H_6Cl_4O$	28076-73-5	307.987	cry (eth)	71				
959	4,4'-Bis(diethylamino) benzophenone	Michler's ethyl ketone	$C_{21}H_{28}N_2O$	90-93-7	324.459	lf (al)	95.3				
960	Bis(diethyldithiocarbamate)nickel		$C_{10}H_{20}N_2NiS_4$	14267-17-5	355.232			$202^{0.02}$			
961	Bis(diethyldithiocarbamate)zinc		$C_{10}H_{20}N_2S_4Zn$	14324-55-1	361.948			$178^{0.05}$			
962	Bis(difluoromethyl) ether	Difluoromethyl ether	$C_2H_2F_4O$	1691-17-4	118.030	col gas		2	1.43^{20}		
963	Bis(2-dimethylaminoethyl) ether	2,2'-Oxybis[*N,N*-dimethylethanamine]	$C_8H_{20}N_2O$	3033-62-3	160.257	liq		80^{15}			
964	Bis[4-(dimethylamino)phenyl] methane	Michler's Base	$C_{17}H_{22}N_2$	101-61-1	254.370	pl or tab (al, lig)	91.5	dec 390; 183^{3}			i H_2O; sl EtOH; vs eth, bz; s acid
965	Bis[4-(dimethylamino)phenyl] methanethione	4,4'-Bis(dimethylamino) thiobenzophenone	$C_{17}H_{20}N_2S$	1226-46-6	284.419	pl	204				i H_2O, EtOH, lig; sl eth; s bz, chl, HOAc
966	Bis(4-dimethylaminophenyl) methanol	4,4'-Bis(dimethylamino) benzhydrol	$C_{17}H_{22}N_2O$	119-58-4	270.369		102.0				i H_2O; vs EtOH; s eth, bz, HOAc
967	1,3-Bis(dimethylamino)-2-propanol		$C_7H_{18}N_2O$	5966-51-8	146.230			181.5	0.8788^{20}	1.4418^{20}	vs H_2O
968	4,4'-Bis(dimethylamino) triphenylmethane		$C_{23}H_{26}N_2$	129-73-7	330.465	nd or lf (al, bz)	102				vs bz, eth
969	Bis(dimethyldithiocarbamate) copper		$C_6H_{12}CuN_2S_4$	137-29-1	303.978			$206^{0.01}$			
970	Bis(dimethyldithiocarbamate) nickel		$C_6H_{12}N_2NiS_4$	15521-65-0	299.125			$208^{0.002}$			

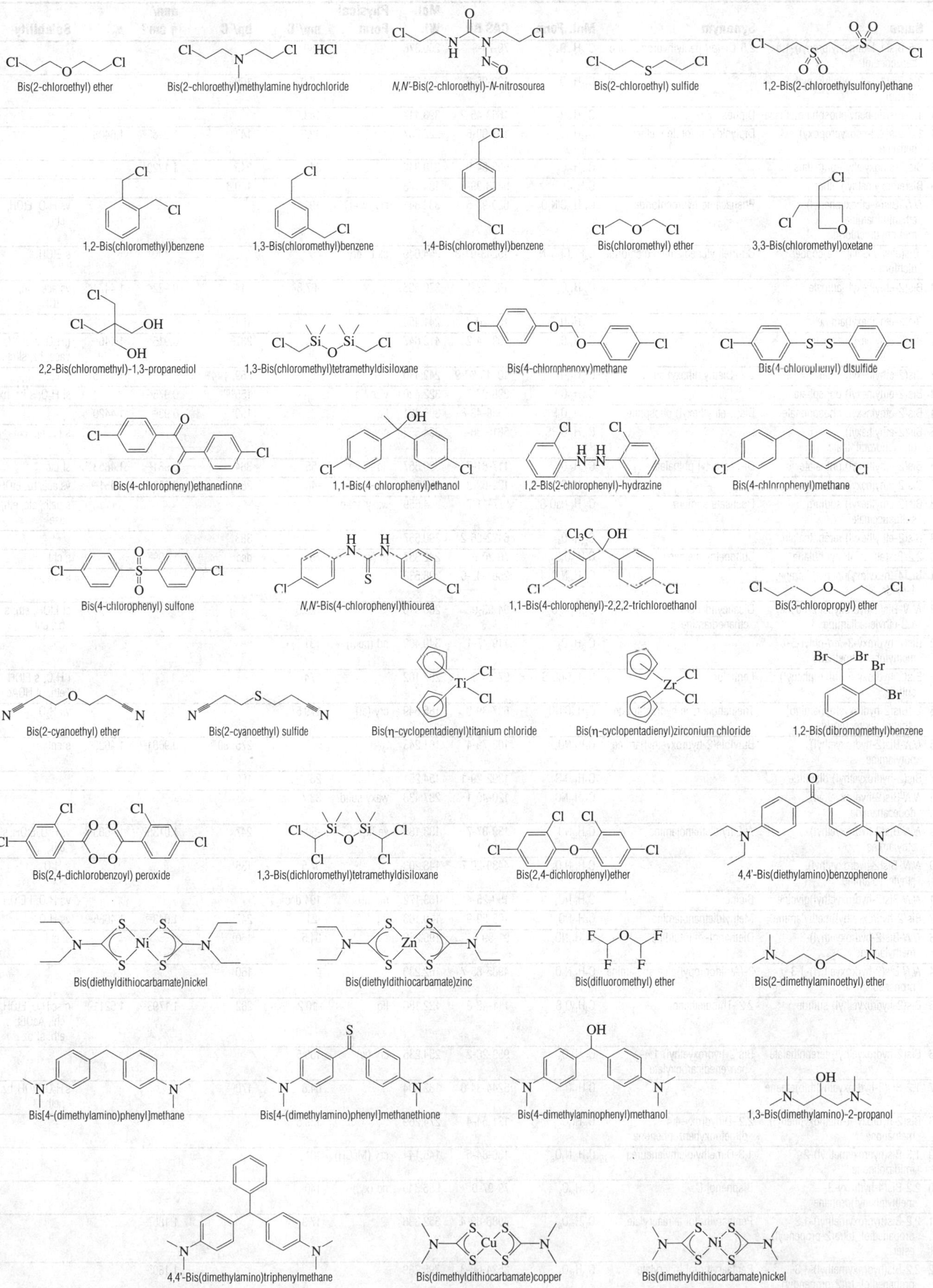
Bis(2-chloroethyl) ether
Bis(2-chloroethyl)methylamine hydrochloride
N,N'-Bis(2-chloroethyl)-N-nitrosourea
Bis(2-chloroethyl) sulfide
1,2-Bis(2-chloroethylsulfonyl)ethane
1,2-Bis(chloromethyl)benzene
1,3-Bis(chloromethyl)benzene
1,4-Bis(chloromethyl)benzene
Bis(chloromethyl) ether
3,3-Bis(chloromethyl)oxetane
2,2-Bis(chloromethyl)-1,3-propanediol
1,3-Bis(chloromethyl)tetramethyldisiloxane
Bis(4-chlorophenoxy)methane
Bis(4-chlorophenyl) disulfide
Bis(4-chlorophenyl)ethanedione
1,1-Bis(4 chlorophenyl)ethanol
1,2-Bis(2-chlorophenyl)-hydrazine
Bis(4-chlorophenyl)methane
Bis(4-chlorophenyl) sulfone
N,N'-Bis(4-chlorophenyl)thiourea
1,1-Bis(4-chlorophenyl)-2,2,2-trichloroethanol
Bis(3-chloropropyl) ether
Bis(2-cyanoethyl) ether
Bis(2-cyanoethyl) sulfide
Bis(η-cyclopentadienyl)titanium chloride
Bis(η-cyclopentadienyl)zirconium chloride
1,2-Bis(dibromomethyl)benzene
Bis(2,4-dichlorobenzoyl) peroxide
1,3-Bis(dichloromethyl)tetramethyldisiloxane
Bis(2,4-dichlorophenyl)ether
4,4'-Bis(diethylamino)benzophenone
Bis(diethyldithiocarbamate)nickel
Bis(diethyldithiocarbamate)zinc
Bis(difluoromethyl) ether
Bis(2-dimethylaminoethyl) ether
Bis[4-(dimethylamino)phenyl]methane
Bis[4-(dimethylamino)phenyl]methanethione
Bis(4-dimethylaminophenyl)methanol
1,3-Bis(dimethylamino)-2-propanol
4,4'-Bis(dimethylamino)triphenylmethane
Bis(dimethyldithiocarbamate)copper
Bis(dimethyldithiocarbamate)nickel

No.	Name	Synonym	Mol. Form.	CAS RN	Mol. Wt.	Physical Form	mp/°C	bp/°C	den/ g cm⁻³	n_D	Solubility
971	2,5-Bis(1,1-dimethylpropyl)-1,4-benzenediol	2,5-Di-*tert*-pentylhydroquinone	$C_{16}H_{26}O_2$	79-74-3	250.376		180				
972	2,4-Bis(1,1-dimethylpropyl) phenol		$C_{16}H_{26}O$	120-95-6	234.376		26.0	169^{22}			
973	1,2-Bis(diphenylphosphino)ethane	Diphos	$C_{26}H_{24}P_2$	1663-45-2	398.417		143.5				
974	1,3-Bis(2,3-epoxypropoxy) benzene	Diglycidyl resorcinol ether	$C_{12}H_{14}O_4$	101-90-6	222.237		42.5	$147^{0.4}$	1.2183^{30}	1.5408^{20}	
975	Bis(2-ethoxyethyl) phthalate		$C_{16}H_{22}O_6$	605-54-9	310.342		34	345	1.1229^{21}		
976	Bis(ethoxymethyl) ether		$C_6H_{14}O_3$	5648-29-3	134.173			140.6			
977	*N,N'*-Bis(4-ethoxyphenyl) ethanimidamide monohydrochloride	Phenacaine hydrochloride	$C_{18}H_{23}ClN_2O_2$	620-99-5	334.841	cry (w+1)	191				vs H_2O, EtOH, chl
978	Bis(ethylenediamine)copper dichloride	Cupriethylenediamine dichloride	$C_4H_{16}Cl_2CuN_4$	15243-01-3	254.649	dk bl cry					s EtOH
979	Bis(2-ethylhexyl) adipate		$C_{22}H_{42}O_4$	103-23-1	370.566		-67.8	214^{5}	0.922^{25}	1.4474^{20}	vs ace, eth, EtOH
980	Bis(2-ethylhexyl)amine		$C_{16}H_{35}N$	106-20-7	241.456			161^{21}			
981	Bis(2-ethylhexyl) azelate		$C_{25}H_{48}O_4$	103-24-2	412.647		-78	237^{5}	0.915^{25}	1.446^{25}	i H_2O; s EtOH, ace, bz; sl ctc
982	Bis(2-ethylhexyl) ether	2,2'-Diethyldihexyl ether	$C_{16}H_{34}O$	10143-60-9	242.440			269; 144^{13}		1.4325^{20}	sl ctc
983	Bis(2-ethylhexyl) phosphate		$C_{16}H_{35}O_4P$	298-07-7	322.420	visc liq		$155^{0.015}$	0.975^{25}		sl H_2O; s bz, hx
984	Bis(2-ethylhexyl) phosphonate	Bis(2-ethylhexyl) phosphite	$C_{16}H_{35}O_3P$	3658-48-8	306.421	liq		150^{1}	0.93^{25}	1.4420^{20}	
985	Bis(2-ethylhexyl) phosphorodithioate		$C_{16}H_{35}O_2PS_2$	5810-88-8	354.552	cry					s bz, hp, chl
986	Bis(2-ethylhexyl) phthalate	Di-*sec*-octyl phthalate	$C_{24}H_{38}O_4$	117-81-7	390.557	liq	-55	384	0.981^{25}	1.4853^{20}	sl ctc
987	Bis(2-ethylhexyl) sebacate		$C_{26}H_{50}O_4$	122-62-3	426.673		-48	256^{5}	0.912^{25}	1.451^{25}	vs ace, bz, EtOH
988	Bis(2-ethylhexyl) sodium sulfosuccinate	Docusate sodium	$C_{20}H_{37}NaO_7S$	577-11-7	444.559	waxy solid					s peth, ctc, eth, ace
989	Bis(2-ethylhexyl) terephthalate		$C_{24}H_{38}O_4$	6422-86-2	390.557			383			
990	2,2-Bis(ethylsulfonyl)butane	Sulfonethylmethane	$C_8H_{18}O_4S_2$	76-20-0	242.357	pl (w)	76	dec	1.199^{85}		s chl
991	Bis[4-(hexyloxy)phenyl]diazene, 1-oxide		$C_{24}H_{34}N_2O_3$	2587-42-0	398.538						s chl
992	*N,N'*-Bis(2-hydroxybenzylidene)-1,2-ethylenediamine	Disalicylidene-1,2-ethanediamine	$C_{16}H_{16}N_2O_2$	94-93-9	268.310		125.5				sl EtOH, eth; s bz, chl
993	Bis(2-hydroxy-3-*tert*-butyl-5-methylphenyl)methane		$C_{23}H_{32}O_2$	119-47-1	340.499	nd (peth)	131				
994	Bis(2-hydroxy-5-chlorophenyl) sulfide	Fenticlor	$C_{12}H_8Cl_2O_2S$	97-24-5	287.162		174				i H_2O; s EtOH, eth, gl HOAc
995	2-[Bis(2-hydroxyethyl)amino] ethanol hydrochloride	Triethanolamine hydrochloride	$C_6H_{16}ClNO_3$	637-39-8	185.649	cry (al)	179.5				vs H_2O
996	*N,N*-Bis(2-hydroxyethyl) butylamine	Butylbis(2-hydroxyethyl)amine	$C_8H_{19}NO_2$	102-79-4	161.243			275; 80^{35}	0.9681^{20}	1.4625^{20}	s chl
997	Bis(2-hydroxyethyl) disulfide		$C_4H_{10}O_2S_2$	1892-29-1	154.251		26	$160^{3.5}$			
998	*N,N*-Bis(2-hydroxyethyl) dodecanamide		$C_{16}H_{33}NO_3$	120-40-1	287.438	waxy solid	38.7				
999	*N,N*-Bis(2-hydroxyethyl) ethylamine	*N*-Ethyldiethanolamine	$C_6H_{15}NO_2$	139-87-7	133.189	ye liq	-50	247	1.0135^{20}	1.4663^{20}	vs H_2O, EtOH; sl eth
1000	*N,N'*-Bis(2-hydroxyethyl) ethylenediamine		$C_6H_{16}N_2O_2$	4439-20-7	148.203		97.5	136^{1}			s H_2O
1001	*N,N*-Bis(2-hydroxyethyl)glycine	Bicine	$C_6H_{13}NO_4$	150-25-4	163.172	nd (al)	194 dec				vs H_2O; i EtOH
1002	Bis(2-hydroxyethyl)methylamine	Methyldiethanolamine	$C_5H_{13}NO_2$	105-59-9	119.163	liq	-21	247	1.043^{25}	1.4685^{20}	vs H_2O
1003	*N,N*-Bis(2-hydroxyethyl)-3-methylaniline	Diethanol-*m*-toluidine	$C_{11}H_{17}NO_2$	91-99-6	195.259		64.5	160^{1}			sl chl
1004	*N,N*-Bis(2-hydroxyethyl)-1,3-propanediamine	3-(Aminopropyl)diethanolamine	$C_7H_{18}N_2O_2$	4985-85-7	162.230			160^{1}			
1005	Bis(2-hydroxyethyl) sulfide	2,2'-Thiodiethanol	$C_4H_{10}O_2S$	111-48-8	122.186	liq	-10.2	282	1.1793^{25}	1.5211^{20}	msc H_2O, EtOH, chl, AcOEt; s eth; sl bz
1006	Bis(2-hydroxyethyl) terephthalate	Bis(2-hydroxyethyl) 1,4-benzenedicarboxylate	$C_{12}H_{14}O_6$	959-26-2	254.235	cry (w)	109.5				
1007	1,2-Bis(2-hydroxyethylthio)ethane		$C_6H_{14}O_2S_2$	5244-34-8	182.304		64.8	$170^{0.5}$			s H_2O, EtOH, bz, peth
1008	Bis(2-hydroxy-4-methoxyphenyl) methanone	2,2'-Dihydroxy-4,4'-dimethoxybenzophenone	$C_{15}H_{14}O_5$	131-54-4	274.269		139.5				
1009	1,3-Bis(hydroxymethyl)-2-imidazolidone	1,3-Dimethylolethyleneurea	$C_5H_{10}N_2O_3$	136-84-5	146.144	cry (MeOH)	101				
1010	2,2-Bis(4-hydroxy-3-methylphenyl)propane	Bisphenol C	$C_{17}H_{20}O_2$	79-97-0	256.340	nd (xyl)	140				
1011	2,2-Bis(hydroxymethyl)-1,3-propanediol, tetra(2-propenoyl) ester	Pentaerythritol tetraacrylate	$C_{17}H_{20}O_8$	4986-89-4	352.336		17.3		1.185^{25}		
1012	2,2-Bis(hydroxymethyl)-1,3-propanediol, tri(2-propenoyl) ester	Pentaerythritol triacrylate	$C_{14}H_{18}O_7$	3524-68-3	298.289				1.180^{20}		

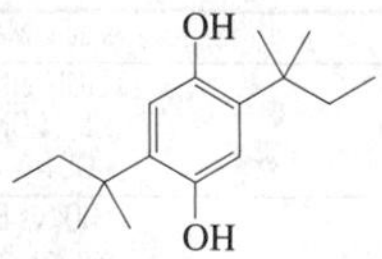

2,5-Bis(1,1-dimethylpropyl)-1,4-benzenediol
2,4-Bis(1,1-dimethylpropyl)phenol 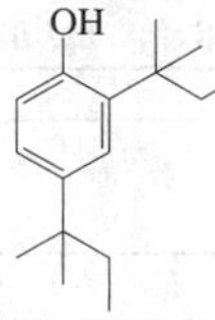
1,2-Bis(diphenylphosphino)ethane 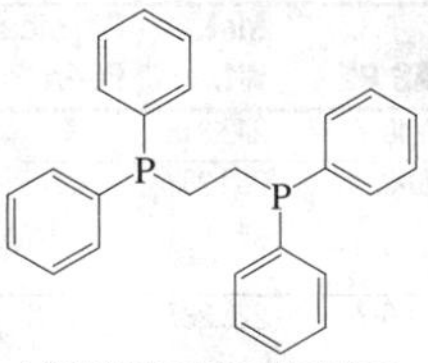
1,3-Bis(2,3-epoxypropoxy)benzene 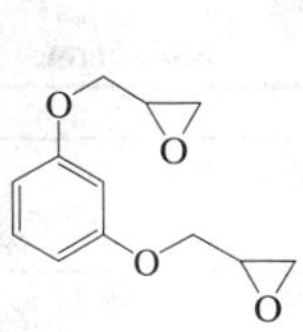
Bis(2-ethoxyethyl) phthalate 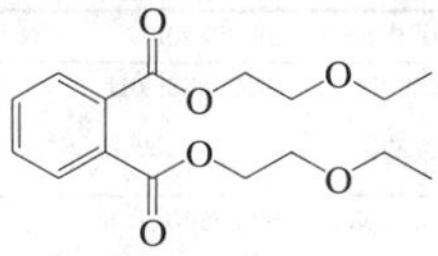
Bis(ethoxymethyl) ether 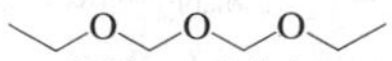

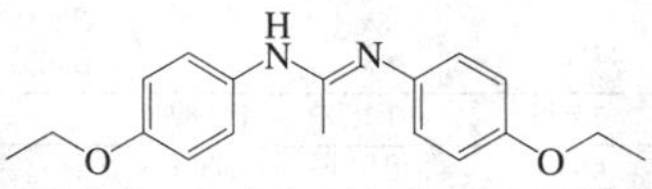

N,N'-Bis(4-ethoxyphenyl)ethanimidamide monohydrochloride
Bis(ethylenediamine)copper dichloride 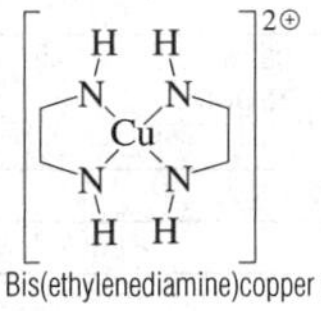
Bis(2-ethylhexyl) adipate 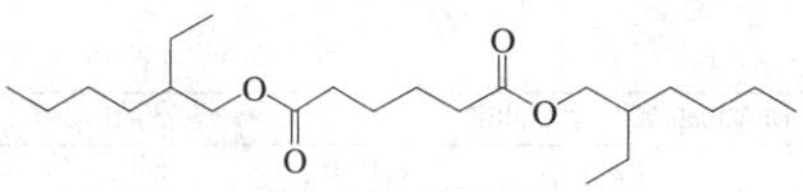
Bis(2-ethylhexyl)amine 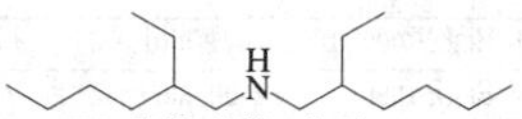

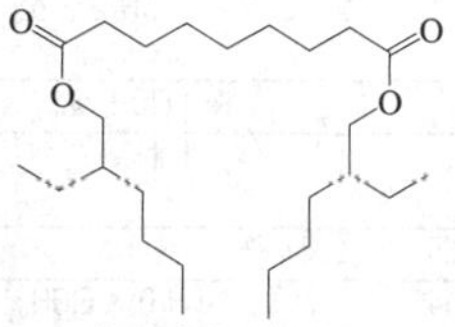

Bis(2-ethylhexyl) azelate
Bis(2-ethylhexyl) ether 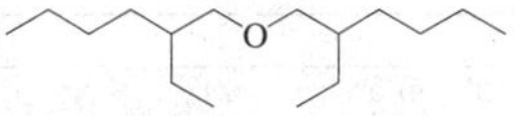
Bis(2-ethylhexyl) phosphate 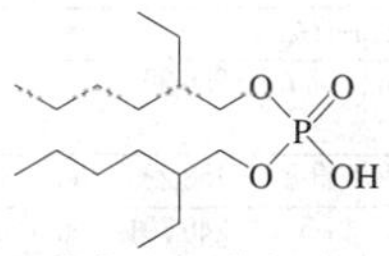
Bis(2-ethylhexyl) phosphonate 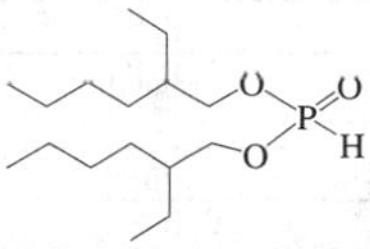
Bis(2-ethylhexyl) phosphorodithioate 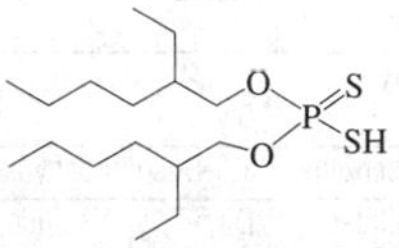
Bis(2-ethylhexyl) phthalate 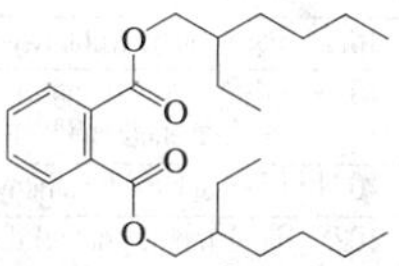

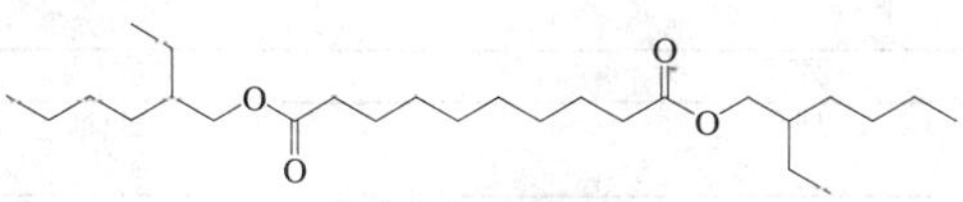

Bis(2-ethylhexyl) sebacate
Bis(2-ethylhexyl) sodium sulfosuccinate 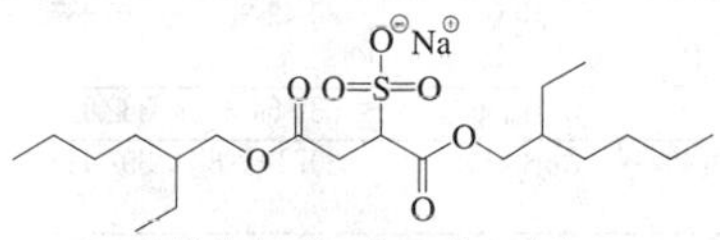
Bis(2-ethylhexyl) terephthalate 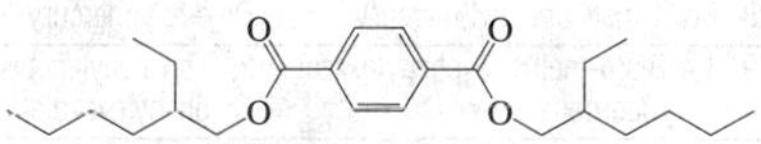

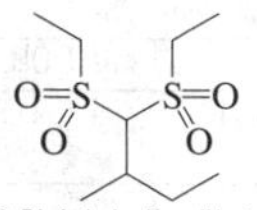

2,2-Bis(ethylsulfonyl)butane
Bis[4-(hexyloxy)phenyl]diazene, 1-oxide 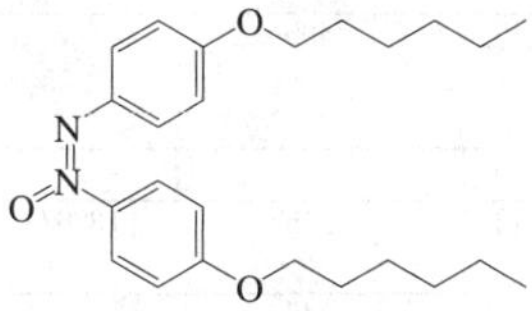
N,N'-Bis(2-hydroxybenzylidene)-1,2-ethylenediamine 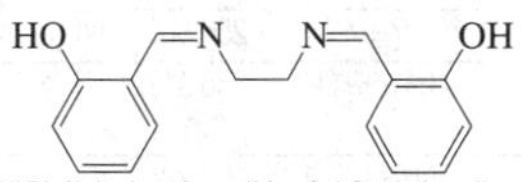
Bis(2-hydroxy-3-*tert*-butyl-5-methylphenyl)methane 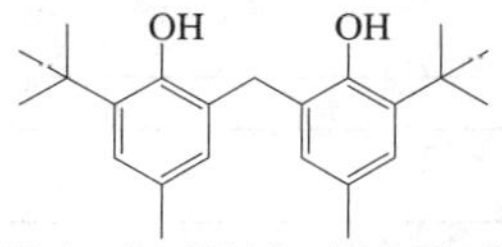
Bis(2-hydroxy-5-chlorophenyl) sulfide 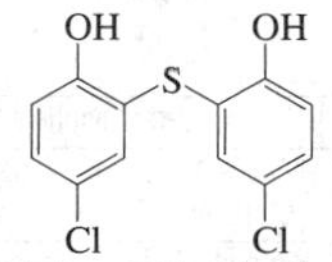

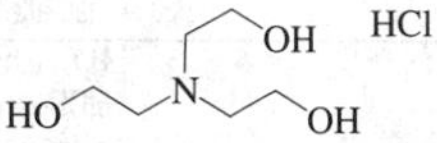

2-[Bis(2-hydroxyethyl)amino]ethanol hydrochloride
N,N-Bis(2-hydroxyethyl)butylamine 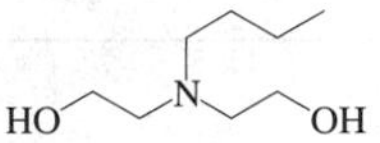
Bis(2-hydroxyethyl) disulfide 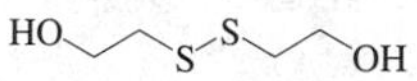
N,N-Bis(2-hydroxyethyl)dodecanamide 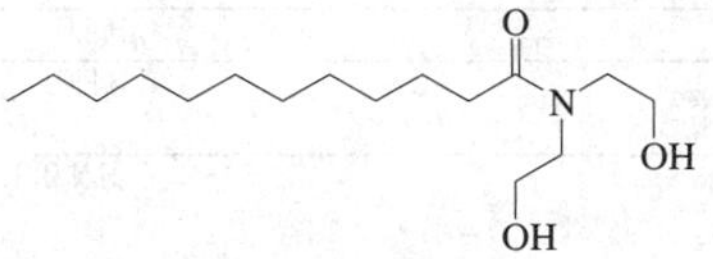
N,N-Bis(2-hydroxyethyl)ethylamine 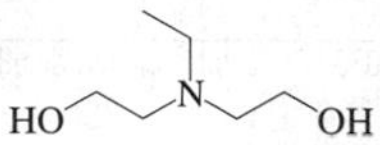

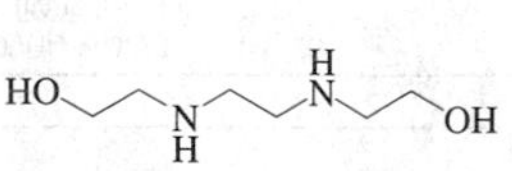

N,N'-Bis(2-hydroxyethyl)ethylenediamine
N,N-Bis(2-hydroxyethyl)glycine 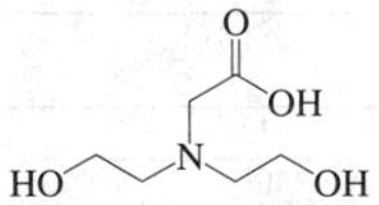
Bis(2-hydroxyethyl)methylamine 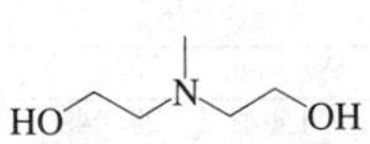
N,N-Bis(2-hydroxyethyl)-3-methylaniline 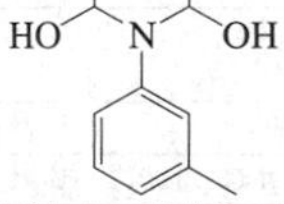
N,N-Bis(2-hydroxyethyl)-1,3-propanediamine 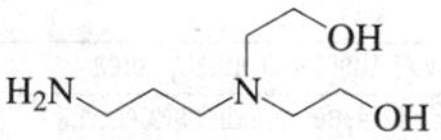

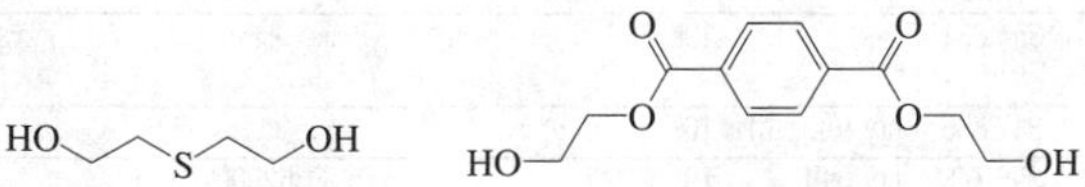

Bis(2-hydroxyethyl) sulfide
Bis(2-hydroxyethyl) terephthalate
1,2-Bis(2-hydroxyethylthio)ethane 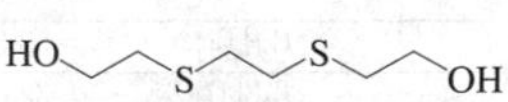
Bis(2-hydroxy-4-methoxyphenyl)methanone 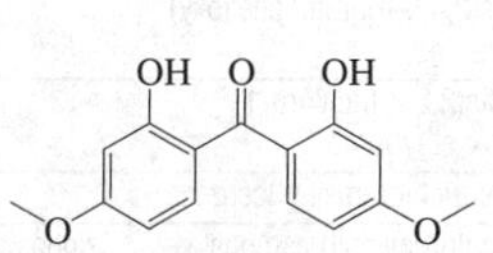

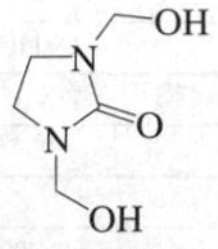

1,3-Bis(hydroxymethyl)-2-imidazolidone
2,2-Bis(4-hydroxy-3-methylphenyl)propane 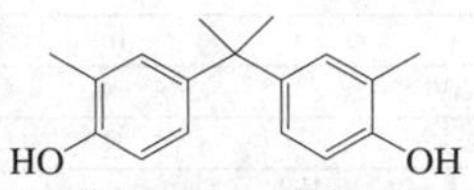
2,2-Bis(hydroxymethyl)-1,3-propanediol, tetra(2-propenoyl) ester 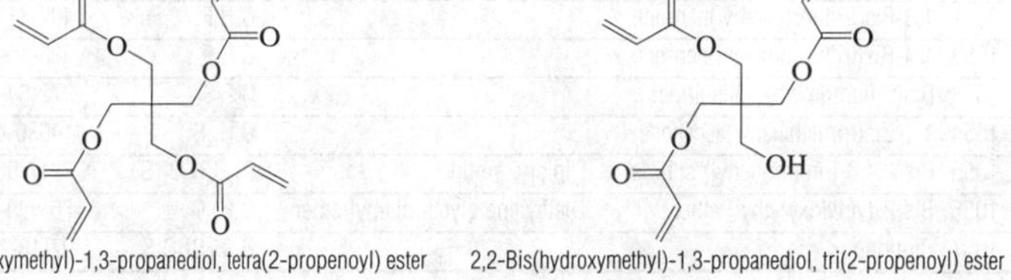
2,2-Bis(hydroxymethyl)-1,3-propanediol, tri(2-propenoyl) ester

No.	Name	Synonym	Mol. Form.	CAS RN	Mol. Wt.	Physical Form	mp/°C	bp/°C	den/ g cm^{-3}	n_D	Solubility
1013	2,2-Bis(4-hydroxyphenyl)butane	Bisphenol B	$C_{16}H_{18}O_2$	77-40-7	242.313		120.5				vs ace, MeOH
1014	Bis(4-hydroxyphenyl)methane	Bisphenol AD	$C_{13}H_{12}O_2$	620-92-8	200.233		162.5	sub			s EtOH, eth, chl, alk; sl DMSO; i CS_2
1015	2,2-Bis(4-hydroxyphenyl)propane	Bisphenol A	$C_{15}H_{16}O_2$	80-05-7	228.287	cry or fl	153	220^{4}, 222^{3}			i H_2O; vs EtOH, eth, bz, alk; s HOAc
1016	2,2-Bis(4-hydroxyphenyl)propane dimethacrylate	Bisphenol A dimethacrylate	$C_{23}H_{24}O_4$	3253-39-2	364.435		73				
1017	Bis(4-hydroxyphenyl) sulfone	Bisphenol S	$C_{12}H_{10}O_4S$	80-09-1	250.270	nd (w), orth bipym	240.5		1.3663^{15}		i H_2O; s EtOH, eth; sl bz, DMSO
1018	Bis(2-mercaptoethyl) sulfide	2,2'-Dimercaptodiethyl sulfide	$C_4H_{10}S_3$	3570-55-6	154.317		-11	135^{18}	1.183^{25}	1.5982^{20}	
1019	Bis(2-methallyl) carbonate		$C_9H_{14}O_3$	64057-79-0	170.205		201.3	66^{3}	0.943^{25}	1.4371^{20}	
1020	Bis(2-methoxyethyl)amine	2-Methoxy-*N*-(2-methoxyethyl) ethanamine	$C_6H_{15}NO_2$	111-95-5	133.189						s ctc
1021	Bis(4-methoxyphenyl)diazene, 1-oxide		$C_{14}H_{14}N_2O_3$	1562-94-3	258.272	ye nd (al)			1.1711^{11}		s EtOH, ace, bz; sl chl
1022	Bis(4-methoxyphenyl)ethanedione		$C_{16}H_{14}O_4$	1226-42-2	270.280		133				sl EtOH, chl
1023	1,4-Bis(methylamino)-9,10-anthracenedione		$C_{16}H_{14}N_2O_2$	2475-44-7	266.294						sl chl
1024	1,3-Bis(1-methylethenyl)benzene	1,3-Diisopropenylbenzene	$C_{12}H_{14}$	3748-13-8	158.239	liq		231	0.925	1.5570^{20}	
1025	Bis(4-methylphenyl) disulfide	Di-*p*-Tolyl disulfide	$C_{14}H_{14}S_2$	103-19-5	246.391	nd or lf (al)	47.5	212^{20}	1.114^{51}		i H_2O; s EtOH, ace; vs eth
1026	Bis(4-methylphenyl) ether	*p*-Tolyl ether	$C_{14}H_{14}O$	1579-40-4	198.260		51	285			vs bz, eth, EtOH
1027	Bis(1-methyl-1-phenylethyl) peroxide	Dicumyl peroxide	$C_{18}H_{22}O_2$	80-43-3	270.367	cry (EtOH)	40	$100^{0.2}$			
1028	Bis(4-methylphenyl)mercury	Di-*p*-tolylmercury	$C_{14}H_{14}Hg$	537-64-4	382.85		245.7				
1029	1,4-Bis(4-methyl-5-phenyloxazol-2-yl)benzene	2,2'-*p*-Phenylenebis(4-methyl-5-phenyloxazole)	$C_{26}H_{20}N_2O_2$	3073-87-8	392.449		232				sl chl
1030	Bis(4-methylphenyl) sulfide	Di-*p*-tolyl sulfide	$C_{14}H_{14}S$	620-94-0	214.326	nd (al)	57.3	>300; 175^{16}			i H_2O; s EtOH, ace, bz, HOAc; sl chl
1031	Bis(4-methylphenyl) sulfone	Di-*p*-tolyl sulfone	$C_{14}H_{14}O_2S$	599-66-6	246.325	pr(bz), nd(w,al)	159	406			sl H_2O, eth; s EtOH, bz, chl, CS_2
1032	*N,N'*-Bis(2-methylphenyl)thiourea		$C_{15}H_{16}N_2S$	137-97-3	256.366	nd (al, sub)					vs bz, EtOH, chl
1033	1,3-Bis(1-methyl-4-piperidyl) propane		$C_{15}H_{30}N_2$	64168-11-2	238.412		13.7	215^{50}	0.8962^{25}	1.4804^{25}	
1034	Bis(methylthio)methane		$C_3H_8S_2$	1618-26-4	108.226			148			
1035	1,2-Bis(*N*-morpholino)ethane		$C_{10}H_{20}N_2O_2$	1723-94-0	200.278	wh-ye (eth,lig)	75	285; 160^{25}			vs H_2O, ace, bz, EtOH
1036	Bismuth acetate		$C_6H_9BiO_6$	22306-37-2	386.111	col tablets	250				i H_2O
1037	Bismuth subsalicylate		$C_7H_5BiO_4$	14882-18-9	362.093	pr					i H_2O, EtOH; reac alk
1038	Bis(2-nitrophenyl) disulfide		$C_{12}H_8N_2O_4S_2$	1155-00-6	308.333		198.5				i H_2O, eth; sl EtOH, ace, bz, HOAc
1039	Bis(3-nitrophenyl) disulfide	Nitrophenide	$C_{12}H_8N_2O_4S_2$	537-91-7	308.333		84				sl EtOH, chl; s eth
1040	Bis(4-nitrophenyl) disulfide		$C_{12}H_8N_2O_4S_2$	100-32-3	308.333		182	$255^{0.1}$			sl EtOH, HOAc
1041	1,2-Bis(4-nitrophenyl)ethane	4,4'-Dinitrobibenzyl	$C_{14}H_{12}N_2O_4$	736-30-1	272.256	ye nd (al,bz)	181.8				i EtOH; sl eth, bz, chl, HOAc
1042	*N,N'*-Bis(4-nitrophenyl)urea	4,4'-Dinitrocarbanilide	$C_{13}H_{10}N_4O_5$	587-90-6	302.242		312 dec				
1043	Bis(2,4-pentanedionato)cobalt	Cobalt(II) bis(acetylacetonate)	$C_{10}H_{14}CoO_4$	14024-48-7	257.149	bl-viol cry	167				
1044	Bis(1-phenylethyl)amine		$C_{16}H_{19}N$	10024-74-5	225.329			296.5	1.018^{15}	1.573	
1045	1,2-Bis(2,4,6-tribromophenoxy) ethane		$C_{14}H_8Br_6O_2$	37853-59-1	687.637	nd (bz/EtOH)	222				
1046	*N,N'*-Bis(2,2,2-trichloro-1-hydroxyethyl)urea		$C_5H_6Cl_6N_2O_3$	116-52-9	354.831		196				vs ace, EtOH
1047	1,4-Bis(trichloromethyl)benzene		$C_8H_4Cl_6$	68-36-0	312.836	cry (bz, eth)	109				s chl
1048	Bis(trichloromethyl) carbonate	Triphosgene	$C_3Cl_6O_3$	32315-10-9	296.748	cry (eth, peth)	79	203	1.6290^{80}		
1049	Bis(tridecyl) thiodipropanoate	Ditridecyl thiodipropionate	$C_{32}H_{62}O_4S$	10595-72-9	542.897			$265^{0.25}$			vs EtOH
1050	3,5-Bis(trifluoromethyl)aniline		$C_8H_5F_6N$	328-74-5	229.123			85^{15}, 76^{10}	1.487^{25}	1.4335^{20}	
1051	1,3-Bis(trifluoromethyl)benzene		$C_8H_4F_6$	402-31-3	214.108			116	1.3790^{25}	1.3916^{25}	i H_2O
1052	1,4-Bis(trifluoromethyl)benzene		$C_8H_4F_6$	433-19-2	214.108	liq		115			
1053	Bis(trifluoromethyl) disulfide		$C_2F_6S_2$	372-64-5	202.141			34.6			vs EtOH, peth
1054	1,2-Bis(trimethylsilyl)acetylene		$C_8H_{18}Si_2$	14630-40-1	170.400		26	134	0.770^{20}	1.413^{20}	
1055	Bis(2,4,6-trinitrophenyl) sulfide	Dipicryl sulfide	$C_{12}H_4N_6O_{12}S$	2217-06-3	456.258	ye cry	230	exp			
1056	Bis[2-(vinyloxy)ethyl] ether	Diethylene glycol divinyl ether	$C_8H_{14}O_3$	764-99-8	158.195			81^{10}			
1057	Bithionol		$C_{12}H_6Cl_4O_2S$	97-18-7	356.052		188		1.73^{25}		vs ace

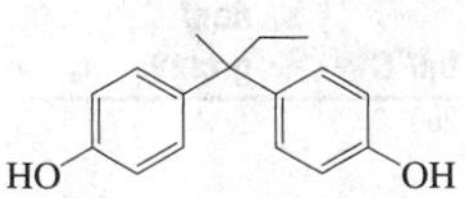

2,2-Bis(4-hydroxyphenyl)butane
Bis(4-hydroxyphenyl)methane 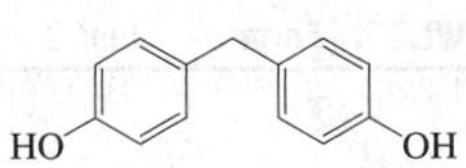
2,2-Bis(4-hydroxyphenyl)propane 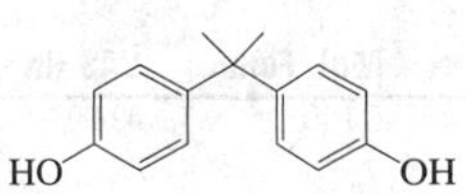
2,2-Bis(4-hydroxyphenyl)propane dimethacrylate 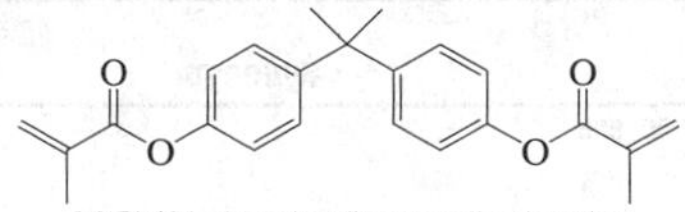

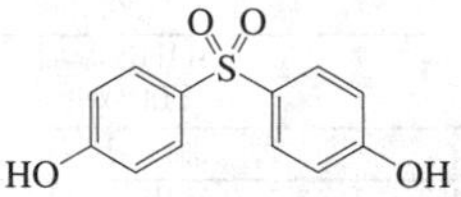

Bis(4-hydroxyphenyl) sulfone
Bis(2-mercaptoethyl) sulfide 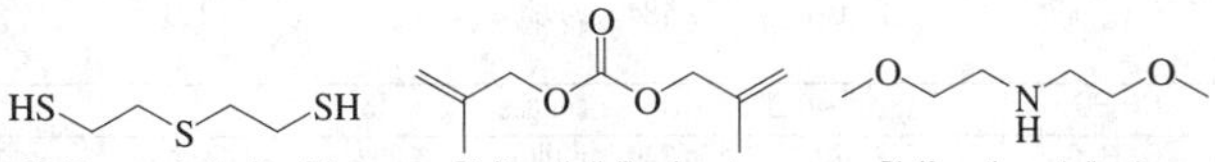
Bis(2-methallyl) carbonate
Bis(2-methoxyethyl)amine
Bis(4-methoxyphenyl)diazene, 1-oxide 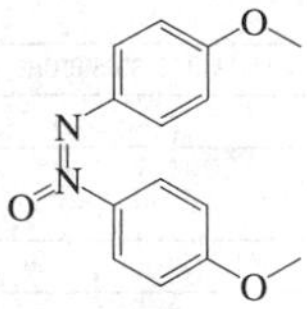
Bis(4-methoxyphenyl)ethanedione 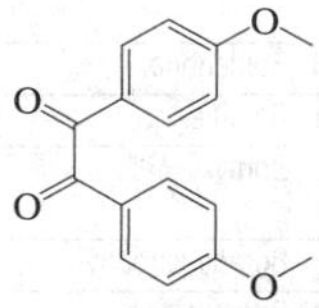

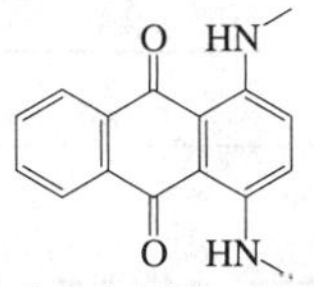

1,4-Bis(methylamino)-9,10-anthracenedione
1,3-Bis(1-methylethenyl)benzene 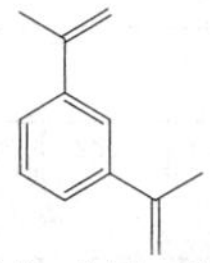
Bis(4-methylphenyl) disulfide 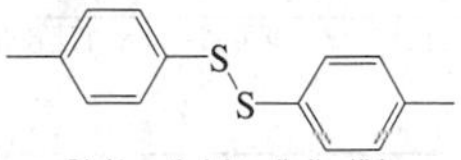
Bis(4-methylphenyl) ether 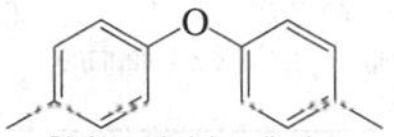
Bis(1-methyl-1-phenylethyl)peroxide 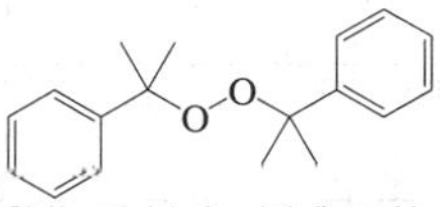

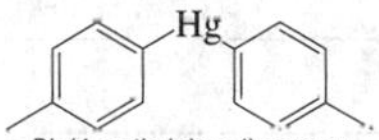

Bis(4-methylphenyl)mercury
1,4-Bis(4-methyl-5-phenyloxazol-2-yl)benzene 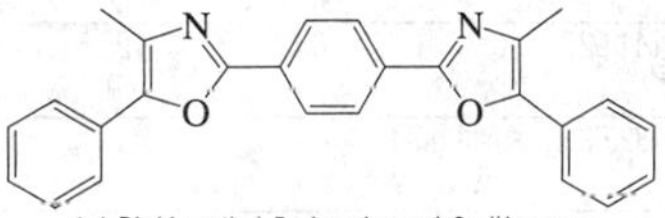
Bis(4-methylphenyl) sulfide 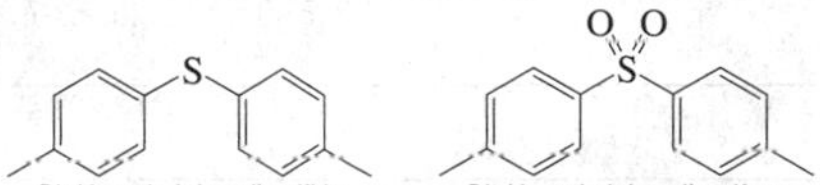
Bis(4-methylphenyl) sulfone
N,N'-Bis(2-methylphenyl)thiourca 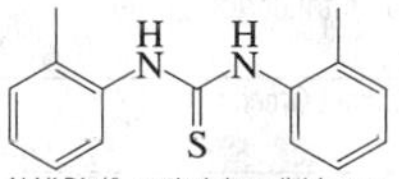

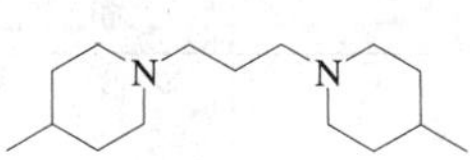

1,3-Bis(1-methyl-4-piperidyl)propane
Bis(methylthio)methane 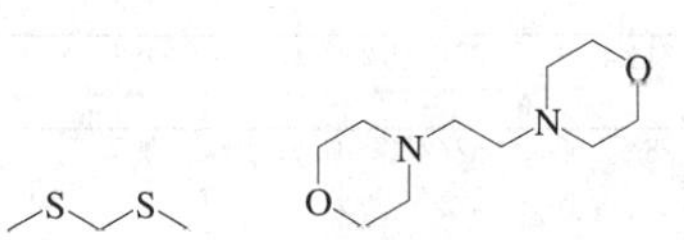
1,2-Bis(*N*-morpholino)ethane
Bismuth acetate 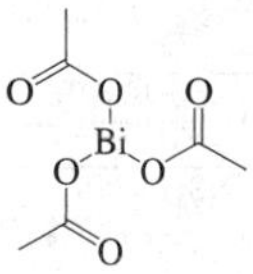
Bismuth subsalicylate 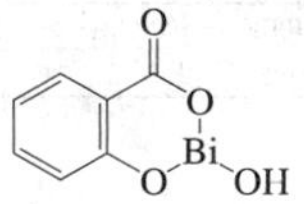
Bis(2-nitrophenyl) disulfide 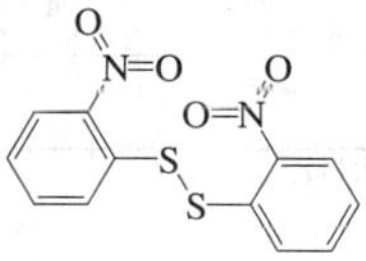

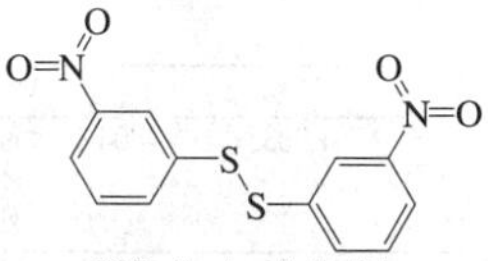

Bis(3-nitrophenyl) disulfide
Bis(4-nitrophenyl) disulfide 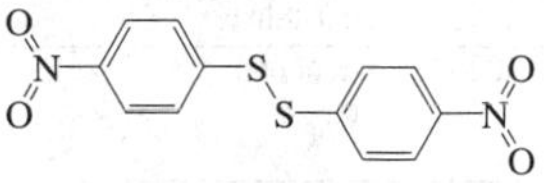
1,2-Bis(4-nitrophenyl)ethane 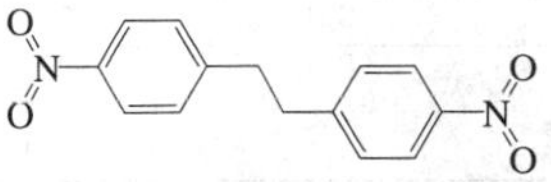
N,N'-Bis(4-nitrophenyl)urea 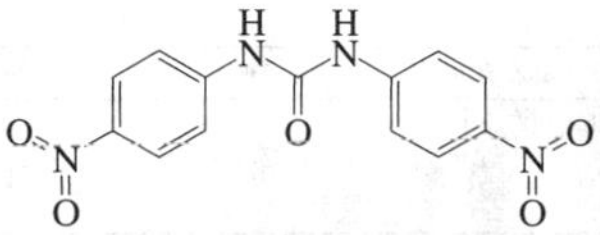

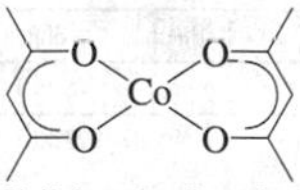

Bis(2,4-pentanedionato)cobalt
Bis(1-phenylethyl)amine 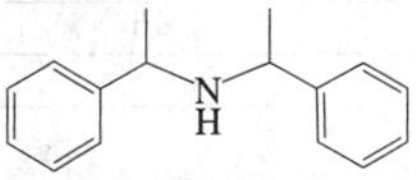
1,2-Bis(2,4,6-tribromophenoxy)ethane 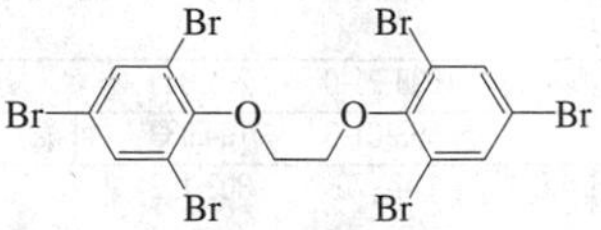
N,N'-Bis(2,2,2-trichloro-1-hydroxyethyl)urea 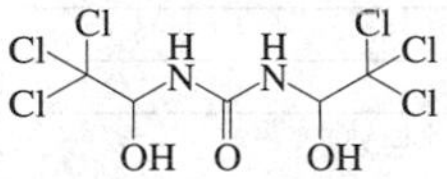
1,4-Bis(trichloromethyl)benzene 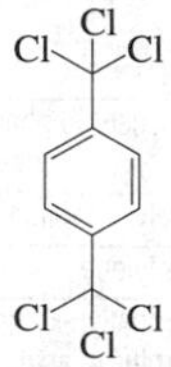

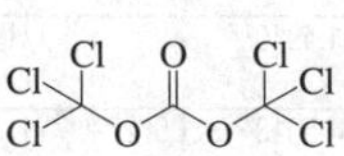

Bis(trichloromethyl) carbonate
Bis(tridecyl) thiodipropanoate 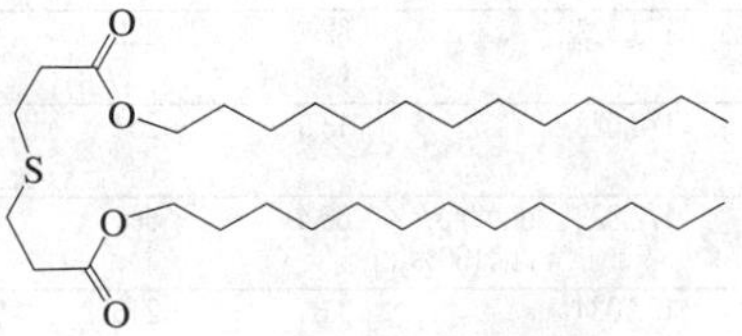
3,5-Bis(trifluoromethyl)aniline 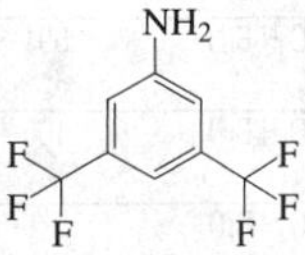
1,3-Bis(trifluoromethyl)benzene 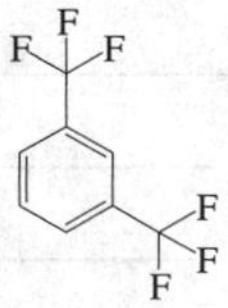
1,4-Bis(trifluoromethyl)benzene 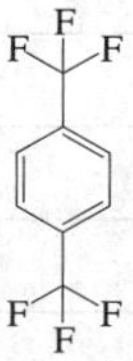

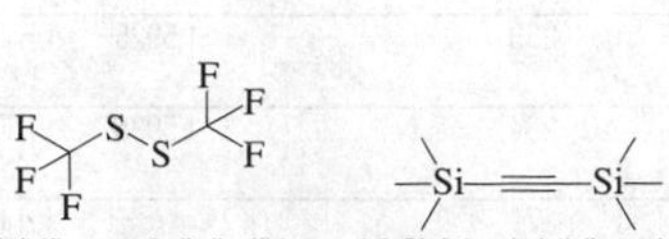

Bis(trifluoromethyl) disulfide
1,2-Bis(trimethylsilyl)acetylene
Bis(2,4,6-trinitrophenyl) sulfide 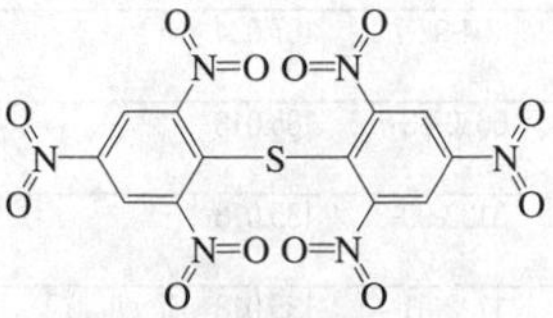
Bis[2-(vinyloxy)ethyl] ether 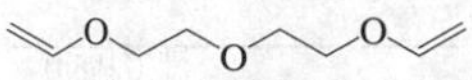
Bithionol

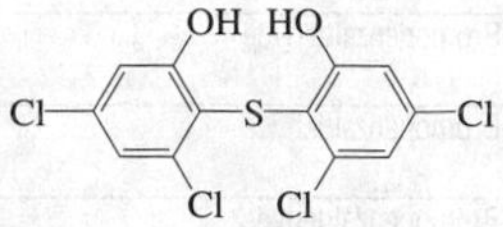

No.	Name	Synonym	Mol. Form.	CAS RN	Mol. Wt.	Physical Form	mp/°C	bp/°C	den/ g cm^{-3}	n_D	Solubility
1058	2,2′-Bithiophene		$C_8H_6S_2$	492-97-7	166.264		33	260			i H_2O; vs EtOH; s eth, ctc, HOAc
1059	Bixin		$C_{25}H_{30}O_4$	6983-79-5	394.504	viol pr (ace)	198				i H_2O; s EtOH, ace; sl eth, bz, HOAc
1060	Boldenone	Dehydrotestosterone	$C_{19}H_{26}O_2$	846-48-0	286.408		165				
1061	Boldine		$C_{19}H_{21}NO_4$	476-70-0	327.375	cry (eth)	163				vs EtOH, chl
1062	Bomyl		$C_9H_{15}O_8P$	122-10-1	282.184	ye oil		160^{17}			sl H_2O; vs ace, EtOH, xyl
1063	Borane carbonyl		CH_3BO	13205-44-2	41.845	col gas	-137	-64			dec H_2O
1064	Borneol, (±)		$C_{10}H_{18}O$	6627-72-1	154.249	lf (lig)	208	sub	1.011^{20}		i H_2O; vs EtOH, eth, bz
1065	*l*-Bornyl acetate		$C_{12}H_{20}O_2$	5655-61-8	196.286		27	223.5	0.982^{25}	1.4626^{20}	sl H_2O; s EtOH, eth
1066	Bornylamine		$C_{10}H_{19}N$	32511-34-5	153.265		163				vs ace, bz, eth, EtOH
1067	Bornyl chloride	2-Chloro-1,7,7-trimethylbicyclo[2.2.1]heptane, *endo*	$C_{10}H_{17}Cl$	464-41-5	172.695	nd	132	207.5			vs bz, eth, EtOH, peth
1068	Bornyl 3-methylbutanoate, (1*R*)	*d*-Bornyl isovalerate	$C_{15}H_{26}O_2$	53022-14-3	238.366			257.5	0.955^{25}		vs eth, EtOH
1069	Boron trifluoride - dimethyl ether complex		$C_2H_6BF_3O$	353-42-4	113.874		-14	dec 127	1.2410^{20}	1.302^{20}	
1070	Boron trifluoride etherate		$C_4H_{10}BF_3O$	109-63-7	141.927	liq	-60.4	125.5	1.125^{25}	1.348^{20}	dec H_2O; vs eth, EtOH
1071	Brilliant Green		$C_{27}H_{34}N_2O_4S$	633-03-4	482.635	small gold cry					vs H_2O, EtOH
1072	Brilliant Yellow		$C_{26}H_{20}N_4Na_2O_8S_2$	3051-11-4	626.569	ye cry (w)					s H_2O, EtOH; sl ace
1073	Brodifacoum		$C_{31}H_{23}BrO_3$	56073-10-0	523.417	off-wh pow	230				i H_2O; sl EtOH, bz; s ace, chl
1074	Bromacil	5-Bromo-3-*sec*-butyl-6-methyluracil	$C_9H_{13}BrN_2O_2$	314-40-9	261.115		158		1.55^{25}		
1075	Bromadiolone		$C_{30}H_{23}BrO_4$	28772-56-7	527.406	ye-wh pow	205				vs DMF; sl ace, chl, EtOH, eth; i hx
1076	Bromal hydrate		$C_2H_3Br_3O_2$	507-42-6	298.756	mcl pr (w+1)	53.5	dec	2.5661^{40}		vs eth, EtOH
1077	Bromdian	Tetrabromobisphenol A	$C_{15}H_{12}Br_4O_2$	79-94-7	543.871		179				s EtOH, eth, bz, chl
1078	*N*-Bromoacetamide		C_2H_4BrNO	79-15-2	137.963	nd (chl-hx)	103.5				vs eth
1079	Bromoacetic acid		$C_2H_3BrO_2$	79-08-3	138.948	hex or orth cry	50	208	1.9335^{50}	1.4804^{50}	msc H_2O, EtOH, eth; s ace, bz; sl chl
1080	Bromoacetone		C_3H_5BrO	598-31-2	136.975	liq	-36.5	138; 31.5^{8}	1.634^{23}	1.4697^{15}	sl H_2O; s EtOH, eth, ace
1081	α-Bromoacetophenone	ω-Bromoacetophenone	C_8H_7BrO	70-11-1	199.045	nd (al) orth pr (al) pl(peth)	50.5	135^{18}	1.647^{20}		i H_2O; s EtOH, peth; vs eth, bz, chl
1082	4-(Bromoacetyl)biphenyl	2-Bromo-4′-phenylacetophenone	$C_{14}H_{11}BrO$	135-73-9	275.140	nd (95% al)	127				
1083	Bromoacetyl bromide		$C_2H_2Br_2O$	598-21-0	201.844			148.5	2.312^{22}	1.5449^{20}	s ace, ctc
1084	Bromoacetylene		C_2HBr	593-61-3	104.933	col gas		4.7			vs eth
1085	5-(2-Bromoallyl)-5-*sec*-butylbarbituric acid	Butallylonal	$C_{11}H_{15}BrN_2O_3$	1142-70-7	303.152		131.5				vs eth, EtOH
1086	5-(2-Bromoallyl)-5-isopropylbarbituric acid	Propallylonal	$C_{10}H_{13}BrN_2O_3$	545-93-7	289.125	cry (dil HOAc, dil al)	181				sl H_2O, eth, bz; vs EtOH, ace, HOAc
1087	2-Bromoaniline		C_6H_6BrN	615-36-1	172.023		32	229	1.578^{20}	1.6113^{20}	i H_2O; s EtOH, eth
1088	3-Bromoaniline		C_6H_6BrN	591-19-5	172.023		18.5	251	1.5793^{20}	1.6260^{20}	sl H_2O; s EtOH, eth
1089	4-Bromoaniline		C_6H_6BrN	106-40-1	172.023	orth bipym nd (60% al)	66.4	dec	1.4970^{100}		i H_2O; s EtOH, eth; sl chl
1090	2-Bromoanisole		C_7H_7BrO	578-57-4	187.034		1.3	216	1.5018^{20}	1.5727^{20}	i H_2O; vs EtOH, eth
1091	3-Bromoanisole		C_7H_7BrO	2398-37-0	187.034			211; 105^{16}		1.5635^{20}	i H_2O; s EtOH, eth, bz, CS_2
1092	4-Bromoanisole		C_7H_7BrO	104-92-7	187.034		13.5	215	1.4564^{20}	1.5642^{20}	sl H_2O; vs EtOH, eth, chl; s ctc
1093	2-Bromobenzaldehyde		C_7H_5BrO	6630-33-7	185.018		21.5	230		1.5925^{20}	i H_2O; vs EtOH, bz; sl ctc
1094	3-Bromobenzaldehyde		C_7H_5BrO	3132-99-8	185.018			234		1.5935^{20}	i H_2O; vs EtOH, eth; sl ctc
1095	4-Bromobenzaldehyde		C_7H_5BrO	1122-91-4	185.018	lf (dil al)	58	67^{2}			i H_2O; vs EtOH, bz; sl chl

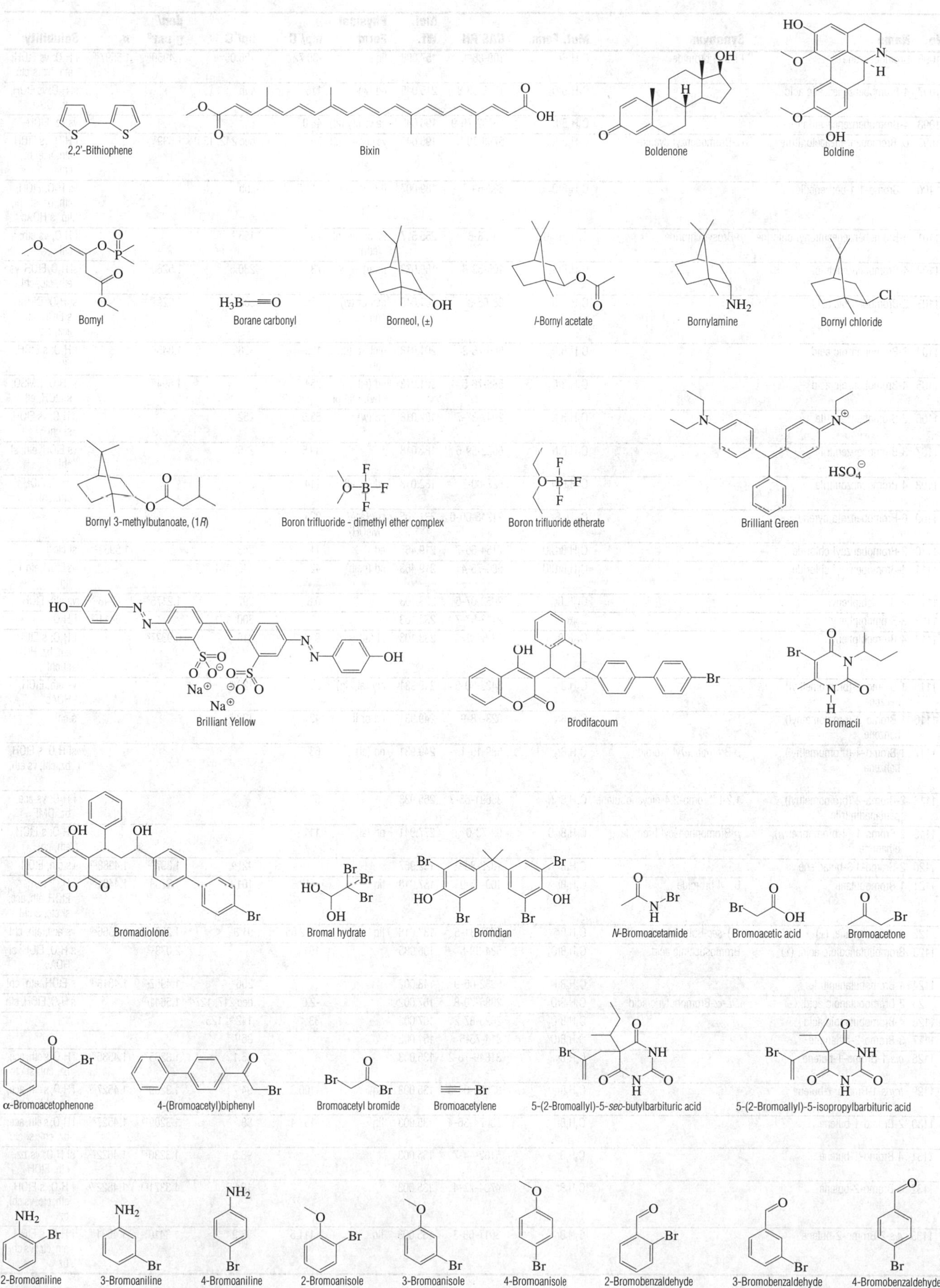
2,2'-Bithiophene
Bixin
Boldenone
Boldine
Bomyl
Borane carbonyl
Borneol, (±)
l-Bornyl acetate
Bornylamine
Bornyl chloride
Bornyl 3-methylbutanoate, (1*R*)
Boron trifluoride - dimethyl ether complex
Boron trifluoride etherate
Brilliant Green
Brilliant Yellow
Brodifacoum
Bromacil
Bromadiolone
Bromal hydrate
Bromdian
N-Bromoacetamide
Bromoacetic acid
Bromoacetone
α-Bromoacetophenone
4-(Bromoacetyl)biphenyl
Bromoacetyl bromide
Bromoacetylene
5-(2-Bromoallyl)-5-*sec*-butylbarbituric acid
5-(2-Bromoallyl)-5-isopropylbarbituric acid
2-Bromoaniline
3-Bromoaniline
4-Bromoaniline
2-Bromoanisole
3-Bromoanisole
4-Bromoanisole
2-Bromobenzaldehyde
3-Bromobenzaldehyde
4-Bromobenzaldehyde

No.	Name	Synonym	Mol. Form.	CAS RN	Mol. Wt.	Physical Form	mp/°C	bp/°C	den/ g cm^{-3}	n_D	Solubility
1096	Bromobenzene	Phenyl bromide	C_6H_5Br	108-86-1	157.008	liq	-30.72	156.06	1.4950^{20}	1.5597^{20}	i H_2O; vs EtOH, eth, bz; s ctc
1097	4-Bromobenzeneacetic acid		$C_8H_7BrO_2$	1878-68-8	215.045	nd (w)	116	sub			sl H_2O; vs EtOH, eth, CS_2
1098	4-Bromobenzeneacetonitrile		C_8H_6BrN	16532-79-9	196.045	pa ye cry (al)	48.0				vs bz, EtOH
1099	α-Bromobenzeneacetonitrile	α-Bromobenzyl cyanide	C_8H_6BrN	5798-79-8	196.045	ye cry (dil al)	29	dec 242; 133^{12}	1.539^{29}		i H_2O; vs EtOH, eth, ace, bz, chl
1100	2-Bromo-1,4-benzenediol		$C_6H_5BrO_2$	583-69-7	189.007	lf (lig), cry (chl)	111.5	sub			vs H_2O, EtOH, eth, bz; sl chl, lig; s HOAc
1101	4-Bromobenzenesulfonyl chloride	*p*-Brosyl chloride	$C_6H_4BrClO_2S$	98-58-8	255.517	tcl or mcl pl (eth)	76	153^{15}			i H_2O; vs eth; s chl
1102	4-Bromobenzenethiol		C_6H_5BrS	106-53-6	189.073	lf (al)	73	230.5	1.5260^{83}		sl H_2O, EtOH; vs eth, ctc, chl
1103	2-Bromobenzoic acid		$C_7H_5BrO_2$	88-65-3	201.018	mcl pr (w), nd	150	sub	1.929^{25}		sl H_2O, DMSO; s EtOH, eth, ace, chl
1104	3-Bromobenzoic acid		$C_7H_5BrO_2$	585-76-2	201.018	mcl nd (dil al)	155	>280	1.845^{20}		i H_2O; s EtOH, eth
1105	4-Bromobenzoic acid		$C_7H_5BrO_2$	586-76-5	201.018	nd (eth), lf (w), mcl pr	254.5		1.894^{20}		sl H_2O, DMSO; s EtOH, eth
1106	2-Bromobenzonitrile		C_7H_4BrN	2042-37-7	182.018	nd (w)	55.5	252			s H_2O; vs EtOH; sl chl
1107	3-Bromobenzonitrile		C_7H_4BrN	6952-59-6	182.018		39.5	225			vs EtOH, eth; sl chl
1108	4-Bromobenzonitrile		C_7H_4BrN	623-00-7	182.018	nd (w, al)	114	236			s H_2O, EtOH, eth, chl
1109	6-Bromobenzo[a]pyrene		$C_{20}H_{11}Br$	21248-00-0	331.205	cry (ace/ MeOH)	223				
1110	2-Bromobenzoyl chloride		C_7H_4BrClO	7154-66-7	219.463	nd	11	243		1.5963^{20}	sl ctc
1111	4-Bromobenzoyl chloride		C_7H_4BrClO	586-75-4	219.463	nd (peth)	42	246; 181^{125}			vs EtOH, eth, bz, lig
1112	2-Bromobiphenyl		$C_{12}H_9Br$	2052-07-5	233.103		0.8	297	1.2175^{26}	1.6248^{25}	vs eth, EtOH
1113	3-Bromobiphenyl		$C_{12}H_9Br$	2113-57-7	233.103			300; 171^{17}		1.6411^{20}	i H_2O
1114	4-Bromobiphenyl		$C_{12}H_9Br$	92-66-0	233.103	pl (al)	91.5	310	0.9327^{25}		i H_2O; s EtOH, eth, bz, HOAc; sl chl
1115	1-Bromo-2-(bromomethyl) benzene		$C_7H_6Br_2$	3433-80-5	249.931	cry (al, lig)	31	129^{19}			vs eth, EtOH, HOAc
1116	1-Bromo-3-(bromomethyl) benzene		$C_7H_6Br_2$	823-78-9	249.931	nd or lf	42	122^{12}			s chl
1117	1-Bromo-4-(bromomethyl) benzene	*p*-Bromobenzyl bromide	$C_7H_6Br_2$	589-15-1	249.931	nd (al)	63				sl H_2O; s EtOH, bz, chl; vs eth, CS_2
1118	2-Bromo-2-(bromomethyl) pentanedinitrile	1,2-Dibromo-2,4-dicyanobutane	$C_6H_6Br_2N_2$	35691-65-7	265.933		52				i H_2O; vs ace, bz, DMF
1119	2-Bromo-1-(4-bromophenyl) ethanone	*p*-Bromophenacyl bromide	$C_8H_6Br_2O$	99-73-0	277.941	nd (al)	111				i H_2O; s EtOH, eth, chl
1120	2-Bromo-1,3-butadiene		C_4H_5Br	1822-86-2	132.987			42^{165}	1.397^{20}	1.4988^{20}	vs eth, EtOH
1121	1-Bromobutane	Butyl bromide	C_4H_9Br	109-65-9	137.018	liq	-112.6	101.6	1.2758^{20}	1.4401^{20}	i H_2O; msc EtOH, eth, ace; sl ctc; s chl
1122	2-Bromobutane, (±)	(±)-*sec*-Butyl bromide	C_4H_9Br	5787-31-5	137.018	liq	-112.65	91.3	1.2585^{20}	1.4366^{20}	vs ace, eth, chl
1123	Bromobutanedioic acid, (±)	Bromosuccinic acid	$C_4H_5BrO_4$	584-98-5	196.985		161		2.073^{25}		s H_2O, EtOH; sl HOAc
1124	4-Bromobutanenitrile		C_4H_6BrN	5332-06-9	148.002			206	1.4967^{20}	1.4818^{20}	s EtOH, eth, chl
1125	2-Bromobutanoic acid, (±)	*DL*-α-Bromobutyric acid	$C_4H_7BrO_2$	2385-70-8	167.002		-2.0	dec 217; 127^{25}	1.5641^{20}		s H_2O, EtOH, eth
1126	4-Bromobutanoic acid		$C_4H_7BrO_2$	2623-87-2	167.002		33	142^{25}, 125^{7}			
1127	3-Bromo-2-butanone		C_4H_7BrO	814-75-5	151.002			36^{11}			
1128	*cis*-1-Bromo-1-butene		C_4H_7Br	31849-78-2	135.003			86.1	1.3265^{15}	1.4536^{20}	i H_2O; s eth, ace, bz, chl; sl ctc
1129	*trans*-1-Bromo-1-butene		C_4H_7Br	32620-08-9	135.003	liq	-100.3	94.7	1.3209^{15}	1.4527^{20}	i H_2O; s eth, ace, bz, chl; sl ctc
1130	2-Bromo-1-butene		C_4H_7Br	23074-36-4	135.003	liq	-133.4	88	1.3209^{15}	1.4527^{20}	i H_2O; s eth, ace, bz, chl; sl ctc
1131	4-Bromo-1-butene		C_4H_7Br	5162-44-7	135.003			98.5	1.3230^{20}	1.4622^{20}	sl H_2O; vs bz, eth, EtOH
1132	1-Bromo-2-butene		C_4H_7Br	4784-77-4	135.003			104.5	1.3371^{25}	1.4822^{20}	i H_2O; s EtOH, eth, ctc; vs chl, bz
1133	*cis*-2-Bromo-2-butene		C_4H_7Br	3017-68-3	135.003	liq	-111.5	93.9	1.3416^{15}	1.4631^{19}	i H_2O; s EtOH, eth, ctc; vs chl, bz

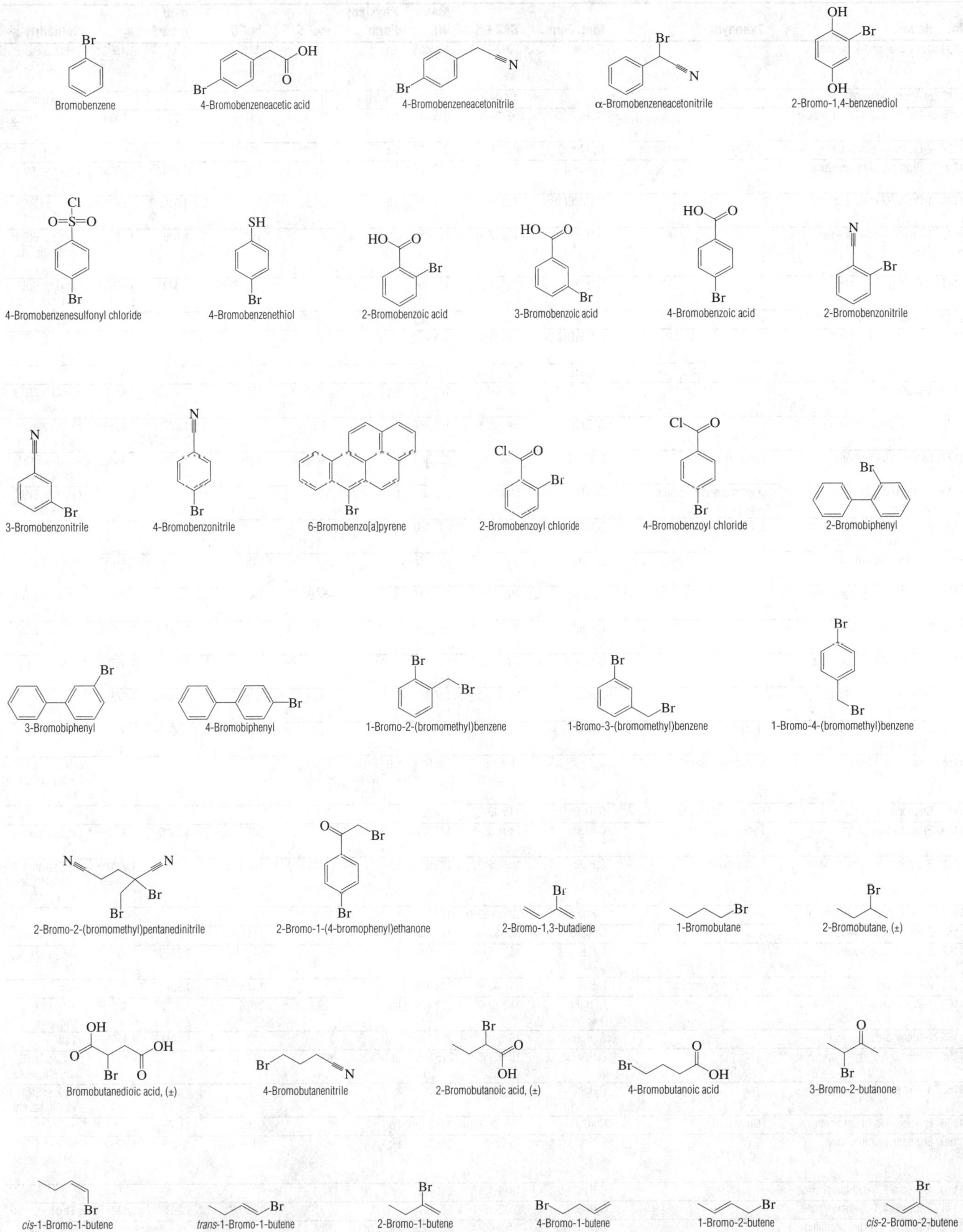
Br
Bromobenzene
OH
O
Br
4-Bromobenzeneacetic acid
N
Br
4-Bromobenzeneacetonitrile
Br
N
α-Bromobenzeneacetonitrile
OH
Br
OH
2-Bromo-1,4-benzenediol
Cl
O=S=O
Br
4-Bromobenzenesulfonyl chloride
SH
Br
4-Bromobenzenethiol
HO O
Br
2-Bromobenzoic acid
HO O
Br
3-Bromobenzoic acid
HO O
Br
4-Bromobenzoic acid
N
Br
2-Bromobenzonitrile
N
Br
3-Bromobenzonitrile
N
Br
4-Bromobenzonitrile
Br
6-Bromobenzo[a]pyrene
Cl O
Br
2-Bromobenzoyl chloride
Cl O
Br
4-Bromobenzoyl chloride
Br
2-Bromobiphenyl
Br
3-Bromobiphenyl
Br
4-Bromobiphenyl
Br
Br
1-Bromo-2-(bromomethyl)benzene
Br
Br
1-Bromo-3-(bromomethyl)benzene
Br
Br
1-Bromo-4-(bromomethyl)benzene
N
N
Br
Br
2-Bromo-2-(bromomethyl)pentanedinitrile
O
Br
Br
2-Bromo-1-(4-bromophenyl)ethanone
Br
2-Bromo-1,3-butadiene
Br
1-Bromobutane
Br
2-Bromobutane, (±)
OH
O
OH
Br O
Bromobutanedioic acid, (±)
Br
N
4-Bromobutanenitrile
Br
O
OH
2-Bromobutanoic acid, (±)
O
Br
OH
4-Bromobutanoic acid
O
Br
3-Bromo-2-butanone
Br
cis-1-Bromo-1-butene
Br
trans-1-Bromo-1-butene
Br
2-Bromo-1-butene
Br
4-Bromo-1-butene
Br
1-Bromo-2-butene
Br
cis-2-Bromo-2-butene

No.	Name	Synonym	Mol. Form.	CAS RN	Mol. Wt.	Physical Form	mp/°C	bp/°C	den/ g cm^{-3}	n_D	Solubility
1134	*trans*-2-Bromo-2-butene		C_4H_7Br	3017-71-8	135.003	liq	-114.6	85.6	1.3323^{15}	1.4602^{16}	i H_2O; s EtOH, eth, ctc; vs chl, bz
1135	(4-Bromobutoxy)benzene		$C_{10}H_{13}BrO$	1200-03-9	229.113	cry (al)	41	154^{18}			sl EtOH, ctc
1136	1-Bromo-4-*tert*-butylbenzene		$C_{10}H_{13}Br$	3972-65-4	213.114		19	231.5	1.2286^{20}	1.5436^{20}	i H_2O; s eth, bz, chl
1137	2-Bromo-3'-chloroacetophenone	3-Chlorophenacyl bromide	C_8H_6BrClO	41011-01-2	233.490	nd	40	397.5			vs EtOH
1138	1-Bromo-2-chlorobenzene		C_6H_4BrCl	694-80-4	191.453	liq	-12.3	204	1.6387^{25}	1.5809^{20}	i H_2O; vs bz; sl ctc
1139	1-Bromo-3-chlorobenzene		C_6H_4BrCl	108-37-2	191.453	liq	-21.5	196	1.6302^{20}	1.5771^{20}	i H_2O; vs EtOH, eth
1140	1-Bromo-4-chlorobenzene		C_6H_4BrCl	106-39-8	191.453	nd or pl (al, eth)	68	196	1.576^{71}	1.5531^{70}	i H_2O; sl EtOH; s eth, bz, ctc, chl
1141	1-Bromo-4-chlorobutane		C_4H_8BrCl	6940-78-9	171.464			175; 63^{10}	1.489^{20}	1.4885^{20}	i H_2O; s EtOH, eth, chl; sl ctc
1142	Bromochlorodifluoromethane	Halon 1211	$CBrClF_2$	353-59-3	165.365	col gas	-159.5	-3.7			
1143	3-Bromo-1-chloro-5,5-dimethylhydantoin		$C_5H_6BrClN_2O_2$	126-06-7	241.471		162				
1144	1-Bromo-1-chloroethane		C_2H_4BrCl	593-96-4	143.410			83	1.667^{10}	1.4660^{20}	
1145	1-Bromo-2-chloroethane		C_2H_4BrCl	107-04-0	143.410	liq	-16.7	107	1.7392^{20}	1.4908^{20}	sl H_2O; s EtOH, eth, chl
1146	Bromochlorofluoromethane		CHBrClF	593-98-6	147.374	liq	-115	36	1.9771^{0}	1.4144^{25}	i H_2O; s eth, ace, chl
1147	Bromochloromethane	Halon 1011	CH_2BrCl	74-97-5	129.384	liq	-87.9	68.0	1.9344^{20}	1.4838^{20}	i H_2O; s EtOH, eth, ace, bz
1148	1-Bromo-4-(chloromethyl) benzene	*p*-Bromobenzyl chloride	C_7H_6BrCl	589-17-3	205.480	nd (al, peth)	50	236			i H_2O; vs EtOH, eth; s peth
1149	2-Bromo-1-(4-chlorophenyl) ethanone	*p*-Chlorophenacyl bromide	C_8H_6BrClO	536-38-9	233.490	nd	96.5				
1150	1-Bromo-2-chloropropane		C_3H_6BrCl	3017-96-7	157.437			118	1.531^{20}	1.4745^{20}	vs ace, bz, eth, EtOH
1151	1-Bromo-3-chloropropane		C_3H_6BrCl	109-70-6	157.437	liq	-58.9	143.3	1.5969^{20}	1.4864^{20}	i H_2O; vs EtOH, eth, chl
1152	2-Bromo-1-chloropropane		C_3H_6BrCl	3017-95-6	157.437			117	1.537^{20}	1.4795^{20}	i H_2O; vs EtOH, eth; s ace, bz
1153	2-Bromo-2-chloropropane		C_3H_6BrCl	2310-98-7	157.437			95	1.495^{20}	1.4575^{20}	vs ace, bz, eth, EtOH
1154	1-Bromo-2-chloro-1,1,2-trifluoroethane		$C_2HBrClF_3$	354-06-3	197.381			52.5	1.8574^{25}	1.3738^{20}	
1155	2-Bromo-2-chloro-1,1,1-trifluoroethane	Halothane	$C_2HBrClF_3$	151-67-7	197.381			50.2; 20^{243}	1.8563^{25}	1.3697^{0}	sl H_2O; s peth
1156	Bromocresol Green	Bromcresol Green	$C_{21}H_{14}Br_4O_5S$	76-60-8	698.014	wh or red (+7w) ye (HOAc)	218.5				sl H_2O; vs EtOH, eth, AcOEt; s bz
1157	Bromocresol Purple	Bromcresol Purple	$C_{21}H_{16}Br_2O_5S$	115-40-2	540.222		241.5				
1158	Bromocycloheptane	Cycloheptyl bromide	$C_7H_{13}Br$	2404-35-5	177.082			101^{40}, 75^{12}	1.3080^{20}	1.4996^{20}	i H_2O; vs eth, chl
1159	Bromocyclohexane	Cyclohexyl bromide	$C_6H_{11}Br$	108-85-0	163.055	liq	-56.5	166.2	1.3359^{20}	1.4957^{20}	i H_2O; msc EtOH, eth, ace, bz, lig, ctc
1160	*trans*-4-Bromocyclohexanol		$C_6H_{11}BrO$	32388-22-0	179.054	pl (hx)	81.5				
1161	2-Bromocyclohexanone		C_6H_9BrO	822-85-5	177.038			114^{32}, 90^{14}	1.340^{25}	1.5085^{25}	
1162	3-Bromocyclohexene		C_6H_9Br	1521-51-3	161.039			81^{40}, 56^{11}	1.3890^{20}	1.5320^{20}	i H_2O; s eth, bz, chl
1163	Bromocyclopentane	Cyclopentyl bromide	C_5H_9Br	137-43-9	149.029			137.5	1.3873^{20}	1.4886^{20}	sl ctc
1164	1-Bromodecane		$C_{10}H_{21}Br$	112-29-8	221.178	liq	-29.2	240.6	1.0702^{20}	1.4557^{20}	i H_2O; vs eth, chl; s ctc
1165	2-Bromodecanoic acid		$C_{10}H_{19}BrO_2$	2623-95-2	251.161		2.0	140^{2}	1.1912^{24}	1.4595^{24}	vs eth
1166	1-Bromo-3,5-dichlorobenzene		$C_6H_3BrCl_2$	19752-55-7	225.898	pr (al)	83	232			i H_2O; s EtOH, eth, chl; vs bz
1167	4-Bromo-1,2-dichlorobenzene		$C_6H_3BrCl_2$	18282-59-2	225.898	pr	25	237			i H_2O; sl EtOH; vs eth, bz, chl
1168	Bromodichlorofluoromethane	Halon 1121	$CBrCl_2F$	353-58-2	181.819	liq		52.8	1.95^{22}		
1169	Bromodichloromethane		$CHBrCl_2$	75-27-4	163.829	liq	-57	90	1.980^{20}	1.4964^{20}	i H_2O; vs EtOH, eth, ace, bz; sl ctc
1170	4-Bromo-2,5-dichlorophenol		$C_6H_3BrCl_2O$	1940-42-7	241.897	nd	71.5				
1171	2-Bromo-1,1-diethoxyethane		$C_6H_{13}BrO_2$	2032-35-1	197.070			170; 66^{18}	1.283^{20}	1.4387^{20}	s EtOH, eth
1172	4-Bromo-*N,N*-diethylaniline		$C_{10}H_{14}BrN$	2052-06-4	228.129	nd or pr	38	270			i H_2O; vs EtOH, eth
1173	Bromodifluoromethane		$CHBrF_2$	1511-62-2	130.920		-145	-14.6	1.55^{16}		s H_2O; vs EtOH
1174	3-Bromo-4,5-dihydro-2(3*H*)-furanone	α-Bromo-γ-butyrolactone	$C_4H_5BrO_2$	5061-21-2	164.986			130^{20}	1.8^{20}	1.5059^{20}	

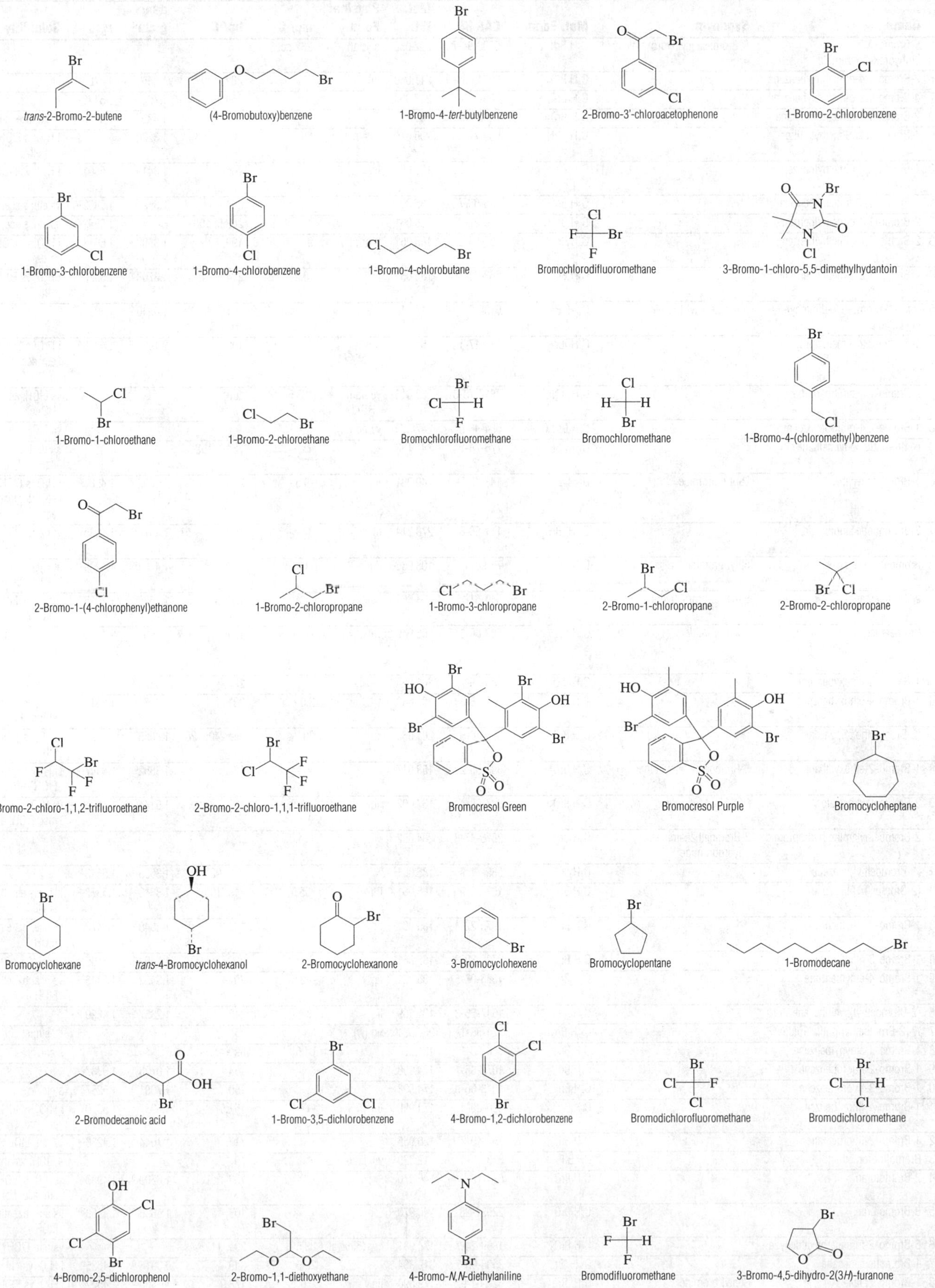

trans-2-Bromo-2-butene
(4-Bromobutoxy)benzene
1-Bromo-4-tert-butylbenzene
2-Bromo-3′-chloroacetophenone
1-Bromo-2-chlorobenzene
1-Bromo-3-chlorobenzene
1-Bromo-4-chlorobenzene
1-Bromo-4-chlorobutane
Bromochlorodifluoromethane
3-Bromo-1-chloro-5,5-dimethylhydantoin
1-Bromo-1-chloroethane
1-Bromo-2-chloroethane
Bromochlorofluoromethane
Bromochloromethane
1-Bromo-4-(chloromethyl)benzene
2-Bromo-1-(4-chlorophenyl)ethanone
1-Bromo-2-chloropropane
1-Bromo-3-chloropropane
2-Bromo-1-chloropropane
2-Bromo-2-chloropropane
1-Bromo-2-chloro-1,1,2-trifluoroethane
2-Bromo-2-chloro-1,1,1-trifluoroethane
Bromocresol Green
Bromocresol Purple
Bromocycloheptane
Bromocyclohexane
trans-4-Bromocyclohexanol
2-Bromocyclohexanone
3-Bromocyclohexene
Bromocyclopentane
1-Bromodecane
2-Bromodecanoic acid
1-Bromo-3,5-dichlorobenzene
4-Bromo-1,2-dichlorobenzene
Bromodichlorofluoromethane
Bromodichloromethane
4-Bromo-2,5-dichlorophenol
2-Bromo-1,1-diethoxyethane
4-Bromo-N,N-diethylaniline
Bromodifluoromethane
3-Bromo-4,5-dihydro-2(3H)-furanone

No.	Name	Synonym	Mol. Form.	CAS RN	Mol. Wt.	Physical Form	mp/°C	bp/°C	den/ g cm⁻³	n_D	Solubility
1175	5-Bromo-*N*,2-dihydroxybenzamide	5-Bromosalicylhydroxamic acid	$C_7H_6BrNO_3$	5798-94-7	232.032	cry (al)	232 dec				
1176	2-Bromo-1,4-dimethoxybenzene		$C_8H_9BrO_2$	25245-34-5	217.060	oil		262; 130^{10}	1.445	1.5700^{20}	
1177	4-Bromo-1,2-dimethoxybenzene		$C_8H_9BrO_2$	2859-78-1	217.060			254.5	1.702^{25}	1.5743^{20}	
1178	2-Bromo-1,1-dimethoxyethane		$C_4H_9BrO_2$	7252-83-7	169.017			149	1.430^{20}	1.4450^{20}	s eth, ace, chl
1179	4-Bromo-*N*,*N*-dimethylaniline		$C_8H_{10}BrN$	586-77-6	200.076		55	264	1.3220^{100}		i H_2O; s EtOH; vs eth
1180	1-Bromo-2,4-dimethylbenzene		C_8H_9Br	583-70-0	185.061	liq	-17	205	1.3419^{20}	1.5501^{20}	i H_2O; vs EtOH, eth, ace
1181	1-Bromo-3,5-dimethylbenzene		C_8H_9Br	556-96-7	185.061			204	1.362^{20}	1.5462^{22}	vs eth; s ace, bz
1182	2-Bromo-1,3-dimethylbenzene		C_8H_9Br	576-22-7	185.061			203.5; 100^{20}		1.5552^{20}	vs eth; s ace, bz
1183	2-Bromo-1,4-dimethylbenzene		C_8H_9Br	553-94-6	185.061	lf or pl	9	199; 88^{13}	1.3582^{18}	1.5514^{18}	i H_2O; vs EtOH; s bz
1184	4-Bromo-1,2-dimethylbenzene		C_8H_9Br	583-71-1	185.061	liq	-0.2	214.5	1.3708^{20}	1.5530^{20}	i H_2O; vs EtOH, eth
1185	*trans*-1-Bromo-3,7-dimethyl-2,6-octadiene	*trans*-Geranyl bromide	$C_{10}H_{17}Br$	6138-90-5	217.146			101^{12}, $47^{0.005}$	1.0940^{22}	1.5027^{20}	
1186	1-Bromo-2,2-dimethylpropane		$C_5H_{11}Br$	630-17-1	151.045			106	1.1997^{20}	1.4370^{20}	i H_2O; s EtOH, eth, ace, bz; vs chl
1187	2-Bromo-4,6-dinitroaniline		$C_6H_4BrN_3O_4$	1817-73-8	262.018	ye nd (al or HOAc)	153.5	sub			vs EtOH, ace; s HOAc
1188	1-Bromo-2,4-dinitrobenzene		$C_6H_3BrN_2O_4$	584-48-5	247.003	ye nd (al)	75				vs EtOH
1189	α-Bromodiphenylmethane		$C_{13}H_{11}Br$	776-74-9	247.130		45	184^{20}, 152^{2}			s EtOH, chl; vs bz
1190	1-Bromododecane	Lauryl bromide	$C_{12}H_{25}Br$	143-15-7	249.231	liq	-9.5	276	1.0399^{20}	1.4583^{20}	i H_2O; s EtOH, eth, ctc; msc ace
1191	2-Bromododecanoic acid		$C_{12}H_{23}BrO_2$	111-56-8	279.214	pl	32	158^{2}	1.1474^{74}	1.4585^{24}	vs bz, eth, EtOH, lig
1192	Bromoethane	Ethyl bromide	C_2H_5Br	74-96-4	108.965	liq	-118.6	38.5	1.4604^{20}	1.4239^{20}	sl H_2O; msc EtOH, eth, chl
1193	2-Bromoethanol	Ethylene bromohydrin	C_2H_5BrO	540-51-2	124.964			150; 51^{4}	1.7629^{20}	1.4915^{20}	msc H_2O, EtOH, eth; sl lig
1194	Bromoethene	Vinyl bromide	C_2H_3Br	593-60-2	106.949	vol liq or gas	-139.54	15.8	1.4933^{20}	1.4380^{20}	i H_2O; s EtOH, eth, ace, bz, chl
1195	1-Bromo-2-ethoxybenzene		C_8H_9BrO	583-19-7	201.060			223	1.4223^{20}		vs eth, EtOH
1196	1-Bromo-4-ethoxybenzene		C_8H_9BrO	588-96-5	201.060		2.0	231	1.4071^{25}	1.5517^{20}	i H_2O; vs EtOH, eth; s chl
1197	(2-Bromoethoxy)benzene		C_8H_9BrO	589-10-6	201.060		39	dec 240; 128^{20}	1.3555^{20}		i H_2O; vs EtOH, eth
1198	1-Bromo-2-ethoxyethane	2-Bromoethyl ethyl ether	C_4H_9BrO	592-55-2	153.017			127.5	1.3852^{0}	1.4447^{20}	sl H_2O; msc EtOH, eth
1199	2-Bromoethyl acetate		$C_4H_7BrO_2$	927-68-4	167.002	liq	-13.8	162.5	1.514^{20}	1.457^{23}	vs H_2O, chl; msc EtOH, eth
1200	2-Bromoethylamine hydrobromide	2-Bromoethanamine hydrobromide	$C_2H_7Br_2N$	2576-47-8	204.892		174.0				
1201	(1-Bromoethyl)benzene		C_8H_9Br	585-71-7	185.061			219; 92^{11}	1.3535^{25}	1.5543^{25}	
1202	(2-Bromoethyl)benzene		C_8H_9Br	103-63-9	185.061	liq	-55.9	219; 105^{18}	1.3643^{20}	1.5372^{20}	i H_2O; s eth, bz; sl ctc
1203	1-Bromo-2-ethylbenzene		C_8H_9Br	1973-22-4	185.061	liq	-67.9	199.3	1.3548^{20}	1.5472^{20}	vs ace, bz, eth, EtOH
1204	1-Bromo-3-ethylbenzene		C_8H_9Br	2725-82-8	185.061			202	1.3493^{20}	1.5465^{20}	
1205	1-Bromo-4-ethylbenzene		C_8H_9Br	1585-07-5	185.061	liq	-43.5	204	1.3423^{20}	1.5445^{20}	vs ace, bz, eth, EtOH
1206	(2-Bromoethyl)cyclohexane		$C_8H_{15}Br$	1647-26-3	191.109	liq	-57	212	1.2357^{20}	1.4899^{20}	
1207	*N*-(2-Bromoethyl)phthalimide		$C_{10}H_8BrNO_2$	574-98-1	254.081	nd (w)	83				vs eth; sl chl
1208	1-Bromo-4-ethynylbenzene		C_8H_5Br	766-96-1	181.030		64.5	89^{16}			s chl
1209	1-Bromo-2-fluorobenzene		C_6H_4BrF	1072-85-1	174.998			154	1.0738^{21}	1.5337^{20}	
1210	1-Bromo-3-fluorobenzene		C_6H_4BrF	1073-06-9	174.998			150	1.7081^{20}	1.5257^{20}	s ctc
1211	1-Bromo-4-fluorobenzene		C_6H_4BrF	460-00-4	174.998	liq	-17.4	151.5	1.593^{15}	1.5310^{15}	i H_2O; s EtOH, eth, chl
1212	1-Bromo-2-fluoroethane		C_2H_4BrF	762-49-2	126.955			71.5	1.7044^{25}	1.4236^{20}	vs eth, EtOH
1213	Bromofluoromethane		CH_2BrF	373-52-4	112.929	vol liq or gas		19			s EtOH; vs chl
1214	2-Bromofuran		C_4H_3BrO	584-12-3	146.970			103	1.6500^{20}	1.4980^{20}	sl H_2O; s EtOH, eth, ace, bz
1215	3-Bromofuran		C_4H_3BrO	22037-28-1	146.970			103	1.6606^{20}	1.4958^{20}	vs ace, bz, eth, EtOH
1216	5-Bromo-2-furancarboxaldehyde		$C_5H_3BrO_2$	1899-24-7	174.981	cry (50% al)	83.5	201; 112^{16}			vs eth, EtOH
1217	1-Bromoheptadecane		$C_{17}H_{35}Br$	3508-00-7	319.364		29.6	349	0.9916^{20}	1.4625^{20}	i H_2O; vs chl
1218	1-Bromoheptane	Heptyl bromide	$C_7H_{15}Br$	629-04-9	179.098	liq	-56.1	178.9	1.1400^{20}	1.4502^{20}	i H_2O; vs EtOH, eth; sl ctc; s chl

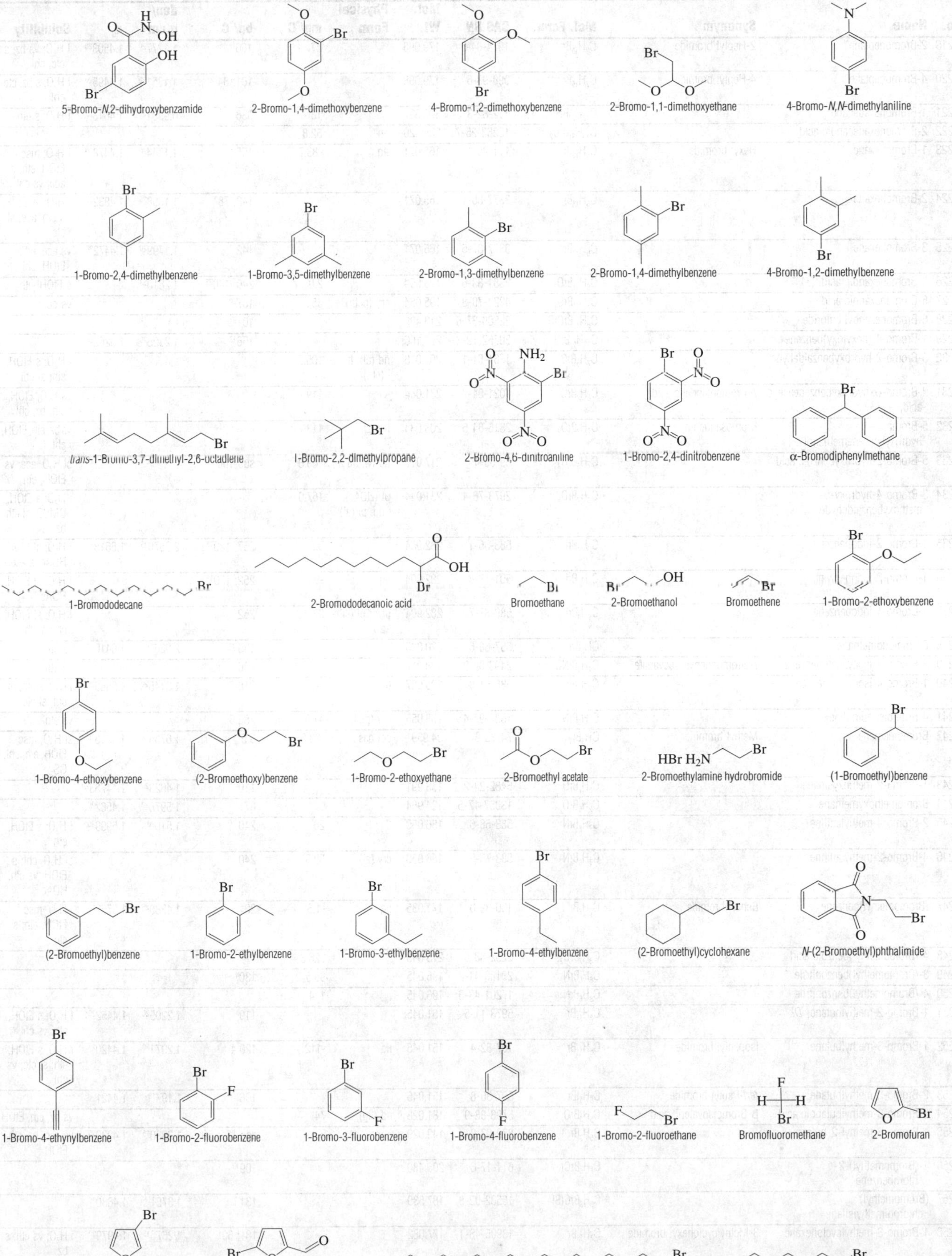

5-Bromo-N,2-dihydroxybenzamide
2-Bromo-1,4-dimethoxybenzene
4-Bromo-1,2-dimethoxybenzene
2-Bromo-1,1-dimethoxyethane
4-Bromo-N,N-dimethylaniline
1-Bromo-2,4-dimethylbenzene
1-Bromo-3,5-dimethylbenzene
2-Bromo-1,3-dimethylbenzene
2-Bromo-1,4-dimethylbenzene
4-Bromo-1,2-dimethylbenzene
trans-1-Bromo-3,7-dimethyl-2,6-octadiene
1-Bromo-2,2-dimethylpropane
2-Bromo-4,6-dinitroaniline
1-Bromo-2,4-dinitrobenzene
α-Bromodiphenylmethane
1-Bromododecane
2-Bromododecanoic acid
Bromoethane
2-Bromoethanol
Bromoethene
1-Bromo-2-ethoxybenzene
1-Bromo-4-ethoxybenzene
(2-Bromoethoxy)benzene
1-Bromo-2-ethoxyethane
2-Bromoethyl acetate
2-Bromoethylamine hydrobromide
(1-Bromoethyl)benzene
(2-Bromoethyl)benzene
1-Bromo-2-ethylbenzene
1-Bromo-3-ethylbenzene
1-Bromo-4-ethylbenzene
(2-Bromoethyl)cyclohexane
N-(2-Bromoethyl)phthalimide
1-Bromo-4-ethynylbenzene
1-Bromo-2-fluorobenzene
1-Bromo-3-fluorobenzene
1-Bromo-4-fluorobenzene
1-Bromo-2-fluoroethane
Bromofluoromethane
2-Bromofuran
3-Bromofuran
5-Bromo-2-furancarboxaldehyde
1-Bromoheptadecane
1-Bromoheptane

No.	Name	Synonym	Mol. Form.	CAS RN	Mol. Wt.	Physical Form	mp/°C	bp/°C	den/ g cm^{-3}	n_D	Solubility
1219	2-Bromoheptane	2-Heptyl bromide	$C_7H_{15}Br$	1974-04-5	179.098		47	166	1.1277^{20}	1.4503^{20}	i H_2O; vs bz; s ctc, chl
1220	4-Bromoheptane	4-Heptyl bromide	$C_7H_{15}Br$	998-93-6	179.098			161; 84^{72}	1.1351^{20}	1.4495^{20}	i H_2O; s bz, ctc, chl
1221	1-Bromohexadecane		$C_{16}H_{33}Br$	112-82-3	305.337		18	336	0.9991^{20}	1.4618^{25}	i H_2O; s eth
1222	2-Bromohexadecanoic acid		$C_{16}H_{31}BrO_2$	18263-25-7	335.320		52.8				
1223	1-Bromohexane	Hexyl bromide	$C_6H_{13}Br$	111-25-1	165.071	liq	-83.7	155.3	1.1744^{20}	1.4478^{20}	i H_2O; msc EtOH, eth; s ace; vs chl
1224	2-Bromohexane		$C_6H_{13}Br$	3377-86-4	165.071			143; 78^{90}	1.1658^{20}	1.4832^{25}	i H_2O; vs EtOH; s eth, ace; sl ctc
1225	3-Bromohexane		$C_6H_{13}Br$	3377-87-5	165.071			142	1.1799^{20}	1.4472^{20}	vs ace, eth, EtOH, chl
1226	2-Bromohexanoic acid, (±)		$C_6H_{11}BrO_2$	2681-83-6	195.054		2.0	240; 140^{23}	1.2810^{33}		s EtOH, eth
1227	6-Bromohexanoic acid		$C_6H_{11}BrO_2$	4224-70-8	195.054	cry (peth)	35	167^{20}			vs peth
1228	6-Bromohexanoyl chloride		$C_6H_{10}BrClO$	22809-37-6	213.499			101^{6}			
1229	1-Bromo-4-(hexyloxy)benzene		$C_{12}H_{17}BrO$	30752-19-3	257.166			156^{13}	1.2306^{20}	1.5262^{20}	
1230	5-Bromo-2-hydroxybenzaldehyde		$C_7H_5BrO_2$	1761-61-1	201.018	nd (al), lf (eth)	105.5				i H_2O; s EtOH, eth; sl chl
1231	4-Bromo-α-hydroxybenzeneacetic acid, (±)	*p*-Bromomandelic acid	$C_8H_7BrO_3$	7021-04-7	231.044		119				vs H_2O, EtOH, eth, bz, chl
1232	5-Bromo-2-hydroxybenzenemethanol	Bromosaligenin	$C_7H_7BrO_2$	2316-64-5	203.034	lf (bz)	113				vs bz, eth, EtOH, chl
1233	5-Bromo-2-hydroxybenzoic acid		$C_7H_5BrO_3$	89-55-4	217.017	nd (w, dil al)	169.8	sub 100			sl H_2O, ace; vs EtOH, eth
1234	3-Bromo-4-hydroxy-5-methoxybenzaldehyde		$C_8H_7BrO_3$	2973-76-4	231.044	pl (HOAc), nd, pl (al)	167.0				i H_2O; s EtOH, DMSO; sl eth, bz
1235	1-Bromo-2-iodobenzene		C_6H_4BrI	583-55-1	282.904		9.5	257; 120^{15}	2.2570^{25}	1.6618^{25}	i H_2O; sl EtOH, HOAc; s ace
1236	1-Bromo-3-iodobenzene		C_6H_4BrI	591-18-4	282.904	liq	-9.3	252; 120^{18}			i H_2O; sl EtOH, HOAc
1237	1-Bromo-4-iodobenzene		C_6H_4BrI	589-87-7	282.904	pr or pl (eth-al)	92	252			i H_2O; sl EtOH, chl; s eth
1238	Bromoiodomethane		CH_2BrI	557-68-6	220.835			139.5	2.926^{17}	1.6410^{20}	vs chl
1239	1-Bromo-4-isocyanatobenzene	*p*-Bromophenyl isocyanate	C_7H_4BrNO	2493-02-9	198.017	nd		226			vs eth
1240	1-Bromo-4-isopropylbenzene		$C_9H_{11}Br$	586-61-8	199.087	liq	-22.5	218.7	1.3145^{20}	1.5569^{20}	i H_2O; s eth, bz, chl; sl ctc
1241	4-Bromoisoquinoline		C_9H_6BrN	1532-97-4	208.055	cry (peth)	41.5	282.5			vs eth
1242	Bromomethane	Methyl bromide	CH_3Br	74-83-9	94.939	col gas	-93.68	3.5	1.6755^{20}	1.4218^{20}	sl H_2O; msc EtOH, eth, chl, CS_2
1243	1-Bromo-2-methoxyethane		C_3H_7BrO	6482-24-2	138.991			110	1.4623^{20}	1.44753^{20}	
1244	Bromomethoxymethane		C_2H_5BrO	13057-17-5	124.964			87	1.5976^{20}	1.4562^{20}	
1245	2-Bromo-4-methylaniline		C_7H_8BrN	583-68-6	186.050	lf	26	240	1.510^{20}	1.5999^{20}	i H_2O; s EtOH, eth
1246	4-Bromo-2-methylaniline		C_7H_8BrN	583-75-5	186.050	cry (al)	59.5	240			sl H_2O, chl; s EtOH; vs eth, HOAc
1247	(Bromomethyl)benzene	Benzyl bromide	C_7H_7Br	100-39-0	171.035	liq	-1.5	201	1.4380^{25}	1.5752^{20}	i H_2O; msc EtOH, eth; s ctc
1248	4-(Bromomethyl)benzoic acid		$C_8H_7BrO_2$	6232-88-8	215.045		226.3				
1249	3-(Bromomethyl)benzonitrile		C_8H_6BrN	28188-41-2	196.045		96.5	130^{4}			
1250	4-(Bromomethyl)benzonitrile		C_8H_6BrN	17201-43-3	196.045		114				
1251	1-Bromo-2-methylbutane, *DL*		$C_5H_{11}Br$	5973-11-5	151.045			119	1.2205^{20}	1.4452^{20}	i H_2O; s EtOH, eth; vs chl
1252	1-Bromo-3-methylbutane	Isopentyl bromide	$C_5H_{11}Br$	107-82-4	151.045	liq	-112	120.4	1.2071^{20}	1.4420^{20}	i H_2O; s EtOH, eth; sl ctc; vs chl
1253	2-Bromo-2-methylbutane	*tert*-Pentyl bromide	$C_5H_{11}Br$	507-36-8	151.045			108	1.197^{18}	1.4421	
1254	3-Bromo-3-methylbutanoic acid	β-Bromoisovaleric acid	$C_5H_9BrO_2$	5798-88-9	181.028	nd (lig)	74				vs bz, eth, EtOH
1255	1-Bromo-3-methyl-2-butene		C_5H_9Br	870-63-3	149.029			dec 131; 50^{40}	1.2930^{15}	1.4930^{15}	vs ace, bz, eth, EtOH
1256	1-(Bromomethyl)-2-chlorobenzene		C_7H_6BrCl	611-17-6	205.480			109^{10}			
1257	(Bromomethyl) chlorodimethylsilane		$C_3H_8BrClSi$	16532-02-8	187.539			131	1.375^{25}	1.4630^{25}	
1258	1-Bromo-3-methylcyclohexane	3-Methylcyclohexyl bromide	$C_7H_{13}Br$	13905-48-1	177.082			181; 60^{11}	1.2676^{15}	1.4979^{20}	i H_2O; vs eth; s bz
1259	(Bromomethyl)cyclohexane		$C_7H_{13}Br$	2550-36-9	177.082			76^{26}	1.283^{20}	1.4907^{30}	vs bz, eth, chl
1260	1-(Bromomethyl)-3-fluorobenzene		C_7H_6BrF	456-41-7	189.025			88^{20}		1.5474^{20}	
1261	3-(Bromomethyl)heptane		$C_8H_{17}Br$	18908-66-2	193.125			67^{10}			

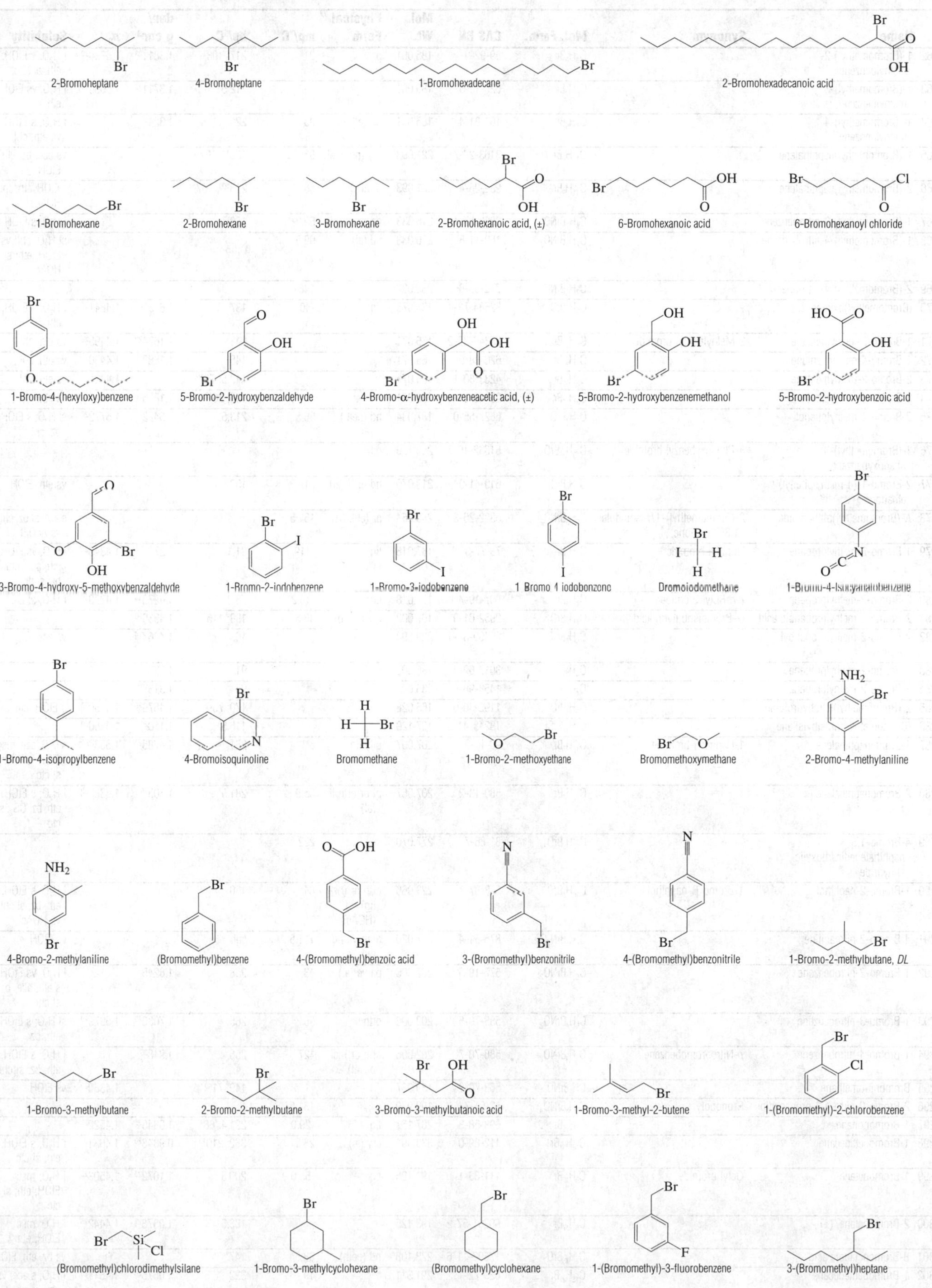
2-Bromoheptane
4-Bromoheptane
1-Bromohexadecane
2-Bromohexadecanoic acid
1-Bromohexane
2-Bromohexane
3-Bromohexane
2-Bromohexanoic acid, (±)
6-Bromohexanoic acid
6-Bromohexanoyl chloride
1-Bromo-4-(hexyloxy)benzene
5-Bromo-2-hydroxybenzaldehyde
4-Bromo-α-hydroxybenzeneacetic acid, (±)
5-Bromo-2-hydroxybenzenemethanol
5-Bromo-2-hydroxybenzoic acid
3-Bromo-4-hydroxy-5-methoxybenzaldehyde
1-Bromo-2-iodobenzene
1-Bromo-3-iodobenzene
1-Bromo-4-iodobenzene
Bromoiodomethane
1-Bromo-4-isocyanatobenzene
1-Bromo-4-isopropylbenzene
4-Bromoisoquinoline
Bromomethane
1-Bromo-2-methoxyethane
Bromomethoxymethane
2-Bromo-4-methylaniline
4-Bromo-2-methylaniline
(Bromomethyl)benzene
4-(Bromomethyl)benzoic acid
3-(Bromomethyl)benzonitrile
4-(Bromomethyl)benzonitrile
1-Bromo-2-methylbutane, DL
1-Bromo-3-methylbutane
2-Bromo-2-methylbutane
3-Bromo-3-methylbutanoic acid
1-Bromo-3-methyl-2-butene
1-(Bromomethyl)-2-chlorobenzene
(Bromomethyl)chlorodimethylsilane
1-Bromo-3-methylcyclohexane
(Bromomethyl)cyclohexane
1-(Bromomethyl)-3-fluorobenzene
3-(Bromomethyl)heptane

No.	Name	Synonym	Mol. Form.	CAS RN	Mol. Wt.	Physical Form	mp/°C	bp/°C	den/ g cm^{-3}	n_D	Solubility
1262	1-(Bromomethyl)-2-methylbenzene		C_8H_9Br	89-92-9	185.061	pr	21	217; 108^{16}	1.3811^{23}	1.5730^{20}	i H_2O; s EtOH, eth, ace, bz
1263	1-(Bromomethyl)-3-methylbenzene		C_8H_9Br	620-13-3	185.061			212.5	1.3711^{23}	1.5660^{20}	i H_2O; vs EtOH, eth
1264	1-(Bromomethyl)-4-methylbenzene		C_8H_9Br	104-81-4	185.061	nd (al)	35	220	1.324^{25}		i H_2O; s EtOH; vs eth, chl
1265	1-(Bromomethyl)naphthalene		$C_{11}H_9Br$	3163-27-7	221.093	cry (peth, al)	56	183^{18}, 167^{10}			vs ace, bz, eth, EtOH
1266	2-(Bromomethyl)naphthalene		$C_{11}H_9Br$	939-26-4	221.093	lf (al)	56	213^{100}, 167^{14}			s EtOH, eth, chl, HOAc
1267	1-(Bromomethyl)-3-nitrobenzene		$C_7H_6BrNO_2$	3958-57-4	216.033	nd or pl (al)	59.3	162^{13}			i H_2O; s EtOH
1268	1-(Bromomethyl)-4-nitrobenzene		$C_7H_6BrNO_2$	100-11-8	216.033	nd (al)	99.5				sl H_2O, chl; vs EtOH, eth; s HOAc
1269	2-(Bromomethyl)-4-nitrophenol		$C_7H_6BrNO_3$	772-33-8	232.032		148				
1270	(Bromomethyl)oxirane, (±)		C_3H_5BrO	82584-73-4	136.975	liq	-40	137	1.615^{14}	1.4841^{20}	i H_2O; s EtOH, eth, bz, chl
1271	1-Bromo-2-methylpentane	2-Methylpentyl bromide	$C_6H_{13}Br$	25346-33-2	165.071			141	1.1624^{20}	1.4495^{20}	vs eth, chl
1272	1-Bromo-4-methylpentane		$C_6H_{13}Br$	626-88-0	165.071			145	1.1683^{20}	1.4490	vs eth, chl
1273	2-Bromo-2-methylpentane		$C_6H_{13}Br$	4283-80-1	165.071			142.5; 70^{100}		1.442^{23}	vs eth, chl
1274	3-Bromo-3-methylpentane		$C_6H_{13}Br$	25346-31-0	165.071			130; 76^{100}	1.1835^{20}	1.4525^{20}	vs eth, chl
1275	2-Bromo-4-methylphenol		C_7H_7BrO	6627-55-0	187.034	nd (peth)	56.5	213.5	1.5422^{25}	1.5772^{20}	sl H_2O; s EtOH, bz, chl
1276	1-(Bromomethyl)-3-phenoxybenzene	3-Phenoxybenzyl bromide	$C_{13}H_{11}BrO$	51632-16-7	263.129	oil					
1277	2-Bromo-1-(4-methylphenyl) ethanone		C_9H_9BrO	619-41-0	213.070	nd or lf (al)	51	157^{14}			vs eth, EtOH
1278	*N*-(Bromomethyl)phthalimide	2-(Bromomethyl)-1*H*-isoindole-1,3(2*H*)-dione	$C_9H_6BrNO_2$	5332-26-3	240.054	pr (chl, bz)	151.5				s ace; sl bz, chl; vs AcOEt
1279	1-Bromo-2-methylpropane	Isobutyl bromide	C_4H_9Br	78-77-3	137.018	liq	-119	91.1	1.272^{15}	1.4348^{20}	i H_2O; vs EtOH, eth, ace, chl, bz; s ctc
1280	2-Bromo-2-methylpropane	*tert*-Butyl bromide	C_4H_9Br	507-19-7	137.018	liq	-16.2	73.3	1.4278^{20}	1.4278^{20}	i H_2O; sl ctc
1281	2-Bromo-2-methylpropanoic acid	α-Bromoisobutyric acid	$C_4H_7BrO_2$	2052-01-9	167.002	cry (peth)	48.5	199; 115^{24}	1.4969^{60}		
1282	2-Bromo-2-methylpropanoyl bromide		$C_4H_6Br_2O$	20769-85-1	229.898			163	1.4067^{14}		vs ace, CS_2
1283	1-Bromo-2-methylpropene		C_4H_7Br	3017-69-4	135.003			91	1.336^{20}		
1284	3-Bromo-2-methylpropene		C_4H_7Br	1458-98-6	135.003			95	1.313^{20}		
1285	2-(Bromomethyl)tetrahydrofuran		C_5H_9BrO	1192-30-9	165.028			170; 70^{22}	1.4679^{20}	1.4850^{20}	s EtOH, eth
1286	(Bromomethyl)trimethylsilane		$C_4H_{11}BrSi$	18243-41-9	167.120			116.5	1.170^{25}	1.4460^{20}	
1287	1-Bromonaphthalene	1-Naphthyl bromide	$C_{10}H_7Br$	90-11-9	207.067	oily liq	6.1	281	1.4785^{20}	1.658^{20}	s H_2O, ace; msc EtOH, eth, bz; sl ctc
1288	2-Bromonaphthalene		$C_{10}H_7Br$	580-13-2	207.067	pl or orth lf (al)	55.9	281.5	1.605^{25}	1.6382^{60}	i H_2O; s EtOH, eth, bz, CS_2; sl ctc
1289	4-Bromo-1,8-naphthalenedicarboxylic anhydride		$C_{12}H_5BrO_3$	81-86-7	277.070		222				
1290	1-Bromo-2-naphthol	1-Bromo-β-naphthol	$C_{10}H_7BrO$	573-97-7	223.066	orth pr (bz-lig) nd (HOAc)	84	130			i H_2O; s EtOH, eth, bz; sl chl; vs HOAc
1291	4-Bromo-2-nitroaniline		$C_6H_5BrN_2O_2$	875-51-4	217.020	oran-ye nd (w)	111.5	sub			vs EtOH
1292	1-Bromo-2-nitrobenzene		$C_6H_4BrNO_2$	577-19-5	202.006	pa ye (al)	43	258	1.6245^{80}		i H_2O; vs EtOH; s eth, ace, bz; sl chl
1293	1-Bromo-3-nitrobenzene		$C_6H_4BrNO_2$	585-79-5	202.006	orth	56	265	1.7036^{20}	1.5979^{20}	sl H_2O; s EtOH, eth, bz
1294	1-Bromo-4-nitrobenzene	*p*-Nitrobromobenzene	$C_6H_4BrNO_2$	586-78-7	202.006	orth or mcl pr (al)	127	256	1.948^{25}		i H_2O; s EtOH, eth, bz; sl chl
1295	Bromonitromethane		CH_2BrNO_2	563-70-2	139.937			149; 71^{40}		1.4880^{20}	vs EtOH
1296	2-Bromo-2-nitro-1,3-propanediol	Bronopol	$C_3H_6BrNO_4$	52-51-7	199.989		131.5				
1297	1-Bromononane		$C_9H_{19}Br$	693-58-3	207.151	liq	-29.0	221.4; 88^4	1.0845^{25}	1.4522^{25}	
1298	1-Bromooctadecane		$C_{18}H_{37}Br$	112-89-0	333.391	cry (al)	28.2	362; 210^{10}	0.9848^{20}	1.4631^{20}	i H_2O; s EtOH, eth; sl ctc
1299	1-Bromooctane	Octyl bromide	$C_8H_{17}Br$	111-83-1	193.125	liq	-55.0	200.8	1.1072^{25}	1.4503^{25}	i H_2O; msc EtOH, eth; sl ctc
1300	2-Bromooctane, (±)		$C_8H_{17}Br$	60251-57-2	193.125			188.5	1.0878^{25}	1.4442^{25}	i H_2O; msc EtOH, eth
1301	8-Bromooctanoic acid		$C_8H_{15}BrO_2$	17696-11-6	223.108	nd (peth)	38.5	147^2			vs bz, eth, EtOH
1302	1-Bromopentadecane		$C_{15}H_{31}Br$	629-72-1	291.311		19	322	1.0675^{20}	1.4611^{20}	i H_2O; s ace; vs chl

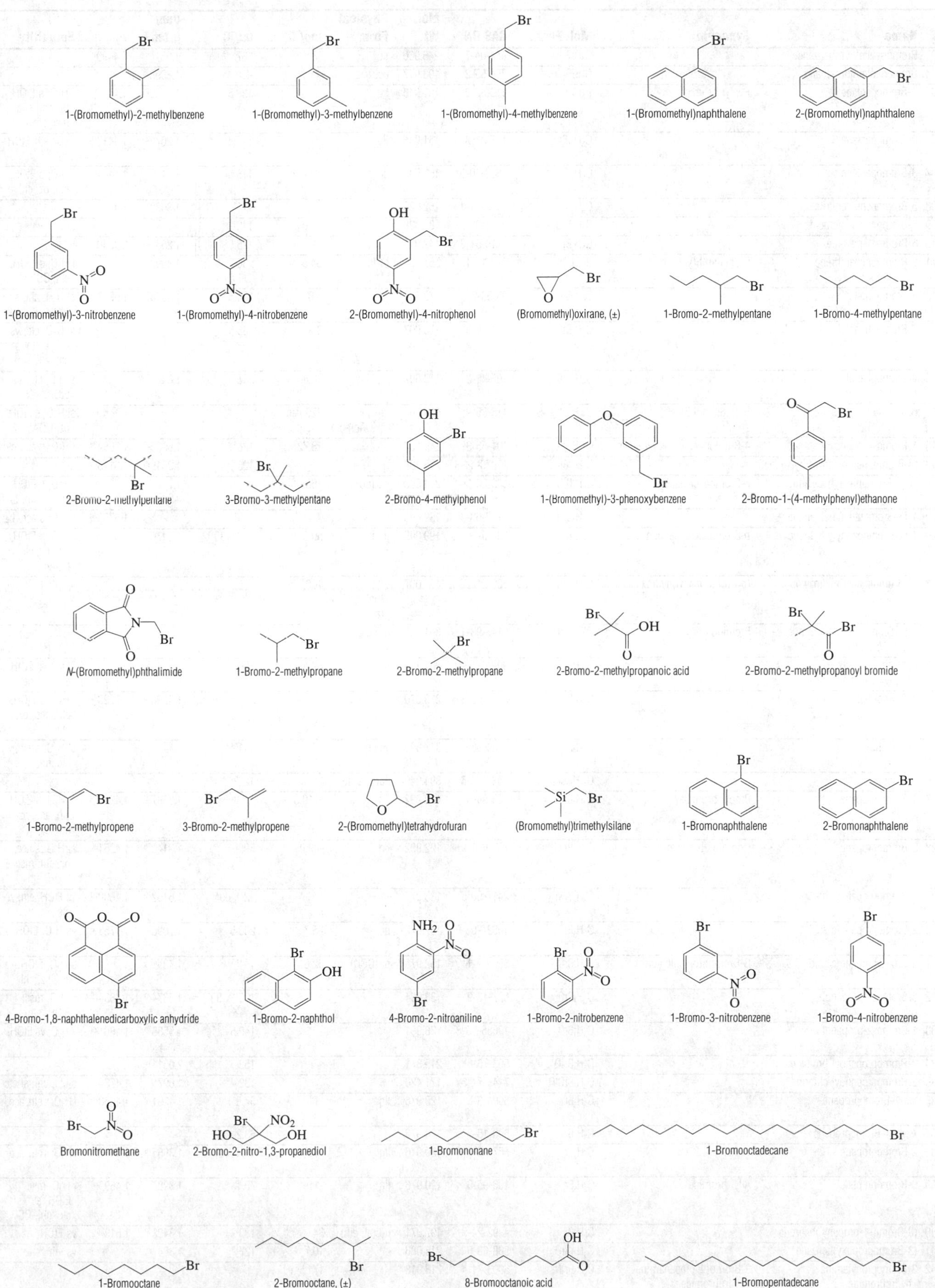
1-(Bromomethyl)-2-methylbenzene
1-(Bromomethyl)-3-methylbenzene
1-(Bromomethyl)-4-methylbenzene
1-(Bromomethyl)naphthalene
2-(Bromomethyl)naphthalene
1-(Bromomethyl)-3-nitrobenzene
1-(Bromomethyl)-4-nitrobenzene
2-(Bromomethyl)-4-nitrophenol
(Bromomethyl)oxirane, (±)
1-Bromo-2-methylpentane
1-Bromo-4-methylpentane
2-Bromo-2-methylpentane
3-Bromo-3-methylpentane
2-Bromo-4-methylphenol
1-(Bromomethyl)-3-phenoxybenzene
2-Bromo-1-(4-methylphenyl)ethanone
N-(Bromomethyl)phthalimide
1-Bromo-2-methylpropane
2-Bromo-2-methylpropane
2-Bromo-2-methylpropanoic acid
2-Bromo-2-methylpropanoyl bromide
1-Bromo-2-methylpropene
3-Bromo-2-methylpropene
2-(Bromomethyl)tetrahydrofuran
(Bromomethyl)trimethylsilane
1-Bromonaphthalene
2-Bromonaphthalene
4-Bromo-1,8-naphthalenedicarboxylic anhydride
1-Bromo-2-naphthol
4-Bromo-2-nitroaniline
1-Bromo-2-nitrobenzene
1-Bromo-3-nitrobenzene
1-Bromo-4-nitrobenzene
Bromonitromethane
2-Bromo-2-nitro-1,3-propanediol
1-Bromononane
1-Bromooctadecane
1-Bromooctane
2-Bromooctane, (±)
8-Bromooctanoic acid
1-Bromopentadecane

No.	Name	Synonym	Mol. Form.	CAS RN	Mol. Wt.	Physical Form	mp/°C	bp/°C	den/ g cm^{-3}	n_D	Solubility
1303	Bromopentafluorobenzene		C_6BrF_5	344-04-7	246.960	liq	-31	137	1.981^{25}	1.4490^{20}	
1304	Bromopentafluoroethane		C_2BrF_5	354-55-2	198.917	col gas		-21	1.8098^{25}		
1305	1-Bromopentane	Pentyl bromide	$C_5H_{11}Br$	110-53-2	151.045	liq	-88.0	129.8	1.2182^{20}	1.4447^{20}	i H_2O; s EtOH, bz, chl; sl ctc; msc eth
1306	2-Bromopentane		$C_5H_{11}Br$	107-81-3	151.045	liq	-95.5	117.4	1.2075^{20}	1.4413^{20}	vs bz, eth, EtOH, chl
1307	3-Bromopentane		$C_5H_{11}Br$	1809-10-5	151.045	liq	-126.2	118.6	1.214^{20}	1.4441^{20}	i H_2O; s EtOH, eth, bz, chl
1308	5-Bromopentanenitrile		C_5H_8BrN	5414-21-1	162.029			111^{12}, 103^{10}	1.3989^{20}	1.4780^{20}	
1309	5-Bromopentanoic acid		$C_5H_9BrO_2$	2067-33-6	181.028		40.0	142^{13}			s chl
1310	5-Bromo-1-pentene		C_5H_9Br	1119-51-3	149.029			125.5	1.2581^{20}	1.4640^{20}	
1311	9-Bromophenanthrene	9-Phenanthryl bromide	$C_{14}H_9Br$	573-17-1	257.125	pr (al)	64.5	>360	1.4093^{10}		i H_2O; s EtOH, eth, CS_2; sl chl
1312	2-Bromophenol		C_6H_5BrO	95-56-7	173.007		5.6	194.5	1.4924^{20}	1.589^{20}	sl H_2O, chl; s EtOH, eth, alk
1313	3-Bromophenol		C_6H_5BrO	591-20-8	173.007		33	236.5			sl H_2O, ctc; vs EtOH, eth; s chl, alk
1314	4-Bromophenol		C_6H_5BrO	106-41-2	173.007		66.4	238	1.840^{15}		s H_2O, chl; vs EtOH, eth
1315	Bromophenol Blue	Bromphenol Blue	$C_{19}H_{10}Br_4O_5S$	115-39-9	669.960	hex pr (HOAc-ace)	279 dec				sl H_2O; s EtOH, bz, HOAc
1316	1-Bromo-4-phenoxybenzene	4-Bromophenyl phenyl ether	$C_{12}H_9BrO$	101-55-3	249.102		18.72	$126^{3.5}$	1.6088^{20}	1.6084^{20}	i H_2O; s eth, ctc
1317	(4-Bromophenoxy)trimethylsilane		$C_9H_{13}BrOSi$	17878-44-3	245.188			126^{25}	1.2619^{20}	1.5145^{20}	
1318	*N*-(4-Bromophenyl)acetamide	*p*-Bromoacetanilide	C_8H_8BrNO	103-88-8	214.060	nd (60% al)	168		1.717^{25}		i H_2O; s EtOH, chl; sl eth, bz
1319	1-(3-Bromophenyl)ethanone		C_8H_7BrO	2142-63-4	199.045		7.5	133^{19}		1.5755^{20}	i H_2O; s ace, bz
1320	1-(4-Bromophenyl)ethanone	*p*-Bromoacetophenone	C_8H_7BrO	99-90-1	199.045	lf (al)	50.5	257; 130^{11}	1.647^{25}	1.647	i H_2O; s EtOH, eth, bz, ctc, HOAc
1321	(4-Bromophenyl)hydrazine	(*p*-Bromophenyl)hydrazine	$C_6H_7BrN_2$	589-21-9	187.037	nd (w), lf (lig), cry (al)	108				vs eth, EtOH, lig
1322	2-(4-Bromophenyl)-1*H*-indene-1,3(2*H*)-dione	Bromindione	$C_{15}H_9BrO_2$	1146-98-1	301.135	cry (lig)	138				
1323	(4-Bromophenyl) phenylmethanone		$C_{13}H_9BrO$	90-90-4	261.113	lf (al)	82.5	350			i H_2O; sl EtOH, eth, bz, peth
1324	2-Bromo-1-phenyl-1-propanone		C_9H_9BrO	2114-00-3	213.070			247.5	1.4298^{20}	1.5720^{20}	i H_2O; s EtOH, eth, ace, bz, ctc
1325	Bromophos		$C_8H_8BrCl_2PS$	2104-96-3	317.999	ye cry	54	$141^{0.01}$			sl H_2O; s eth, ctc, tol
1326	Bromophos-ethyl		$C_{10}H_{12}BrCl_2O_3PS$	4824-78-6	394.049	pale-ye liq		$122^{0.004}$			
1327	1-Bromopropane	Propyl bromide	C_3H_7Br	106-94-5	122.992	liq	-110.3	71.1	1.3537^{20}	1.4343^{20}	sl H_2O; s EtOH, eth, ace, bz, chl, ctc
1328	2-Bromopropane	Isopropyl bromide	C_3H_7Br	75-26-3	122.992	liq	-89.0	59.5	1.3140^{20}	1.4251^{20}	sl H_2O; s ace, bz, chl; msc EtOH, eth
1329	3-Bromopropanenitrile		C_3H_4BrN	2417-90-5	133.975			92^{25}, 69^{7}	1.6152^{20}	1.4800^{20}	vs EtOH, eth; sl ctc
1330	2-Bromopropanoic acid, (±)		$C_3H_5BrO_2$	10327-08-9	152.975	pr	25.7	203.5	1.7000^{20}	1.4753^{20}	vs H_2O, EtOH, eth; sl chl
1331	3-Bromopropanoic acid	β-Bromopropionic acid	$C_3H_5BrO_2$	590-92-1	152.975	pl (CCl_4)	62.5	141^{45}	1.48^{25}		s H_2O, EtOH, eth, bz, chl
1332	3-Bromo-1-propanol		C_3H_7BrO	627-18-9	138.991			105^{185}, 80^{22}	1.5374^{20}	1.4834^{25}	s H_2O; msc EtOH, eth
1333	1-Bromo-2-propanol		C_3H_7BrO	19686-73-8	138.991			146.5	1.5585^{30}	1.4801^{20}	s H_2O; vs EtOH, eth
1334	2-Bromopropanoyl bromide		$C_3H_4Br_2O$	563-76-8	215.871			153	2.0611^{16}		
1335	2-Bromopropanoyl chloride		C_3H_4BrClO	7148-74-5	171.420			132	1.697^{11}	1.4780^{20}	s eth, chl; sl ctc
1336	*cis*-1-Bromopropene		C_3H_5Br	590-13-6	120.976	liq	-113	57.8	1.4291^{20}	1.4560^{20}	i H_2O; s eth, ace, chl
1337	*trans*-1-Bromopropene		C_3H_5Br	590-15-8	120.976			63.2			
1338	2-Bromopropene		C_3H_5Br	557-93-7	120.976	liq	-126	48.4	1.3965^{16}	1.4467^{16}	i H_2O; s eth, ace, chl
1339	3-Bromopropene	Allyl bromide	C_3H_5Br	106-95-6	120.976	liq	-119	70.1	1.398^{20}	1.4697^{20}	i H_2O; msc EtOH, eth; s ctc, chl, CS_2
1340	(3-Bromo-1-propenyl)benzene		C_9H_9Br	4392-24-9	197.071	nd (al, eth)	34	130^{10}	1.3428^{30}	1.613^{20}	vs EtOH
1341	(3-Bromopropoxy)benzene		$C_9H_{11}BrO$	588-63-6	215.086		10.7	127^{18}	1.364^{16}		vs eth
1342	3-Bromopropylamine hydrobromide	3-Bromo-1-propanamine hydrobromide	$C_3H_9Br_2N$	5003-71-4	218.918		171.5				

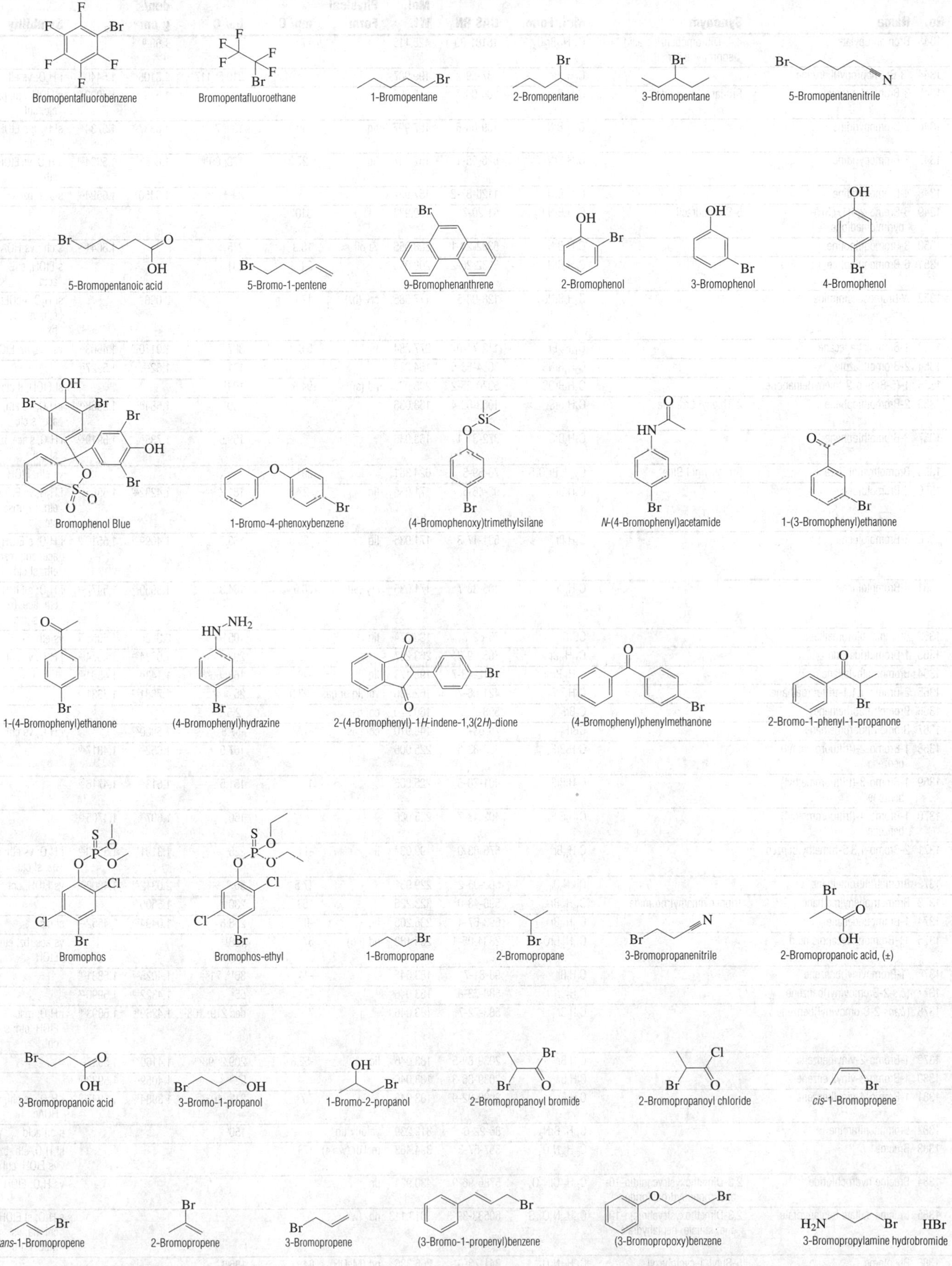

Bromopentafluorobenzene
Bromopentafluoroethane
1-Bromopentane
2-Bromopentane
3-Bromopentane
5-Bromopentanenitrile
5-Bromopentanoic acid
5-Bromo-1-pentene
9-Bromophenanthrene
2-Bromophenol
3-Bromophenol
4-Bromophenol
Bromophenol Blue
1-Bromo-4-phenoxybenzene
(4-Bromophenoxy)trimethylsilane
N-(4-Bromophenyl)acetamide
1-(3-Bromophenyl)ethanone
1-(4-Bromophenyl)ethanone
(4-Bromophenyl)hydrazine
2-(4-Bromophenyl)-1H-indene-1,3(2H)-dione
(4-Bromophenyl)phenylmethanone
2-Bromo-1-phenyl-1-propanone
Bromophos
Bromophos-ethyl
1-Bromopropane
2-Bromopropane
3-Bromopropanenitrile
2-Bromopropanoic acid, (±)
3-Bromopropanoic acid
3-Bromo-1-propanol
1-Bromo-2-propanol
2-Bromopropanoyl bromide
2-Bromopropanoyl chloride
cis-1-Bromopropene
trans-1-Bromopropene
2-Bromopropene
3-Bromopropene
(3-Bromo-1-propenyl)benzene
(3-Bromopropoxy)benzene
3-Bromopropylamine hydrobromide

No.	Name	Synonym	Mol. Form.	CAS RN	Mol. Wt.	Physical Form	mp/°C	bp/°C	den/ g cm⁻³	n_D	Solubility
1343	Bromopropylate	4,4'-Dibromobenzilic acid isopropyl ester	$C_{17}H_{16}Br_2O_3$	18181-80-1	428.115		77		1.59^{20}		
1344	(3-Bromopropyl)benzene		$C_9H_{11}Br$	637-59-2	199.087			219.5; 117^{25}	1.3106^{25}	1.5440^{25}	i H_2O; vs eth
1345	3-Bromo-1-propyne	Propargyl bromide	C_3H_3Br	106-96-7	118.960			89	1.579^{19}	1.4922^{20}	s EtOH, eth, bz, ctc, chl
1346	2-Bromopyridine		C_5H_4BrN	109-04-6	157.997	liq	-40.1	193; 75^{13}	1.6337^{20}	1.5734^{20}	sl H_2O; s EtOH, eth, ctc
1347	3-Bromopyridine		C_5H_4BrN	626-55-1	157.997	liq	-27.3	173; 69^{18}	1.645^{0}	1.5694^{20}	s H_2O; vs EtOH, eth
1348	4-Bromopyridine		C_5H_4BrN	1120-87-2	157.997		0.5	$29^{0.4}$	1.6450^{0}	1.5694^{20}	s ace, bz
1349	5-Bromo-2,4(1*H*,3*H*)-pyrimidinedione	5-Bromouracil	$C_4H_3BrN_2O_2$	51-20-7	190.983		310				
1350	3-Bromoquinoline		C_9H_6BrN	5332-24-1	208.055	ye oil	13.3	275		1.6641^{20}	s chl; vs HOAc
1351	6-Bromoquinoline		C_9H_6BrN	5332-25-2	208.055		24	281			s EtOH, eth, acid
1352	*N*-Bromosuccinimide		$C_4H_4BrNO_2$	128-08-5	177.985	cry (bz)	174		2.098^{25}		sl H_2O, AcOEt, eth; vs ace; i hx
1353	1-Bromotetradecane		$C_{14}H_{29}Br$	112-71-0	277.284		5.6	307	1.0170^{20}	1.4603^{20}	vs ace, bz, EtOH
1354	2-Bromothiazole		C_3H_2BrNS	3034-53-5	164.024			171	1.82^{25}	1.5927^{20}	
1355	1-(5-Bromo-2-thienyl)ethanone		C_6H_5BrOS	5370-25-2	205.072	nd (al)	94.5	103^{4}			sl EtOH; s ctc
1356	2-Bromothiophene	2-Thienyl bromide	C_4H_3BrS	1003-09-4	163.036			150	1.684^{20}	1.5868^{20}	i H_2O; vs eth, ace; s ctc
1357	3-Bromothiophene		C_4H_3BrS	872-31-1	163.036			159.5	1.735^{20}	1.5919^{20}	i H_2O; s ace, bz; sl chl
1358	Bromothymol Blue	Bromthymol Blue	$C_{27}H_{28}Br_2O_5S$	76-59-5	624.381		201				vs eth, EtOH
1359	2-Bromotoluene		C_7H_7Br	95-46-5	171.035	liq	-27.8	181.7	1.4232^{20}	1.5565^{20}	i H_2O; vs EtOH, eth, bz; msc ctc
1360	3-Bromotoluene		C_7H_7Br	591-17-3	171.035	liq	-39.8	183.7	1.4099^{20}	1.5510^{20}	i H_2O; s EtOH, ace, chl; msc eth; sl ctc
1361	4-Bromotoluene		C_7H_7Br	106-38-7	171.035	cry (al)	28.5	184.3	1.3959^{35}	1.5477^{20}	i H_2O; s EtOH, eth, ace, bz, chl; sl ctc
1362	Bromotrichloromethane		$CBrCl_3$	75-62-7	198.274	liq	-5.65	105	2.012^{25}	1.5065^{20}	vs eth, EtOH
1363	1-Bromotridecane		$C_{13}H_{27}Br$	765-09-3	263.257		6.2	292	1.0234^{25}	1.4574^{25}	i H_2O; vs chl
1364	Bromotriethylsilane		$C_6H_{15}BrSi$	1112-48-7	195.173	liq	-49.3	163; 66^{24}	1.143^{20}	1.4561^{20}	
1365	2-Bromo-1,1,1-trifluoroethane		$C_2H_2BrF_3$	421-06-7	162.936	vol liq or gas	-93.9	26	1.7881^{20}	1.3331^{20}	
1366	Bromotrifluoroethene		C_2BrF_3	598-73-2	160.920	col gas		-2.5			
1367	Bromotrifluoromethane		$CBrF_3$	75-63-8	148.910	col gas	-172	-57.8	1.5800^{20}		i H_2O; vs chl
1368	1-Bromo-2-(trifluoromethyl) benzene		$C_7H_4BrF_3$	392-83-6	225.006			167.5	1.652^{25}	1.4817^{20}	
1369	1-Bromo-3-(trifluoromethyl) benzene		$C_7H_4BrF_3$	401-78-5	225.006		1	151.5	1.613^{25}	1.4716^{20}	
1370	1-Bromo-4-(trifluoromethyl) benzene		$C_7H_4BrF_3$	402-43-7	225.006			160	1.607^{25}	1.4705^{25}	
1371	2-Bromo-1,3,5-trimethylbenzene		$C_9H_{11}Br$	576-83-0	199.087	liq	-1	225	1.3191^{10}	1.5510^{20}	i H_2O; vs eth; s bz; sl ctc
1372	Bromotrinitromethane		$CBrN_3O_6$	560-95-2	229.931		17.5	56^{10}	2.0312^{20}	1.4808^{20}	vs EtOH, chl
1373	Bromotriphenylmethane	Triphenylmethyl bromide	$C_{19}H_{15}Br$	596-43-0	323.226		153	230^{15}	1.5500^{20}		
1374	1-Bromoundecane		$C_{11}H_{23}Br$	693-67-4	235.205	liq	-9.7	258.8	1.0494^{25}	1.4552^{25}	sl ctc
1375	11-Bromoundecanoic acid		$C_{11}H_{21}BrO_2$	2834-05-1	265.188	nd (liq)	57	188^{18}			vs ace, bz, eth, EtOH
1376	(1-Bromovinyl)benzene		C_8H_7Br	98-81-7	183.046		-44	86^{14}, 71^{3}	1.4025^{23}	1.5881^{20}	
1377	(*cis*-2-Bromovinyl)benzene		C_8H_7Br	588-73-8	183.046		-7	55^{2}	1.4322^{10}	1.5990^{22}	
1378	(*trans*-2-Bromovinyl)benzene		C_8H_7Br	588-72-7	183.046		7	dec 219; 108^{20}	1.4269^{16}	1.6093^{20}	i H_2O; msc EtOH, eth; s chl
1379	1-Bromo-2-vinylbenzene		C_8H_7Br	2039-88-5	183.046	liq	-52.8	209.2; 98^{20}	1.4160^{20}	1.5927^{20}	
1380	1-Bromo-3-vinylbenzene		C_8H_7Br	2039-86-3	183.046			92^{20}	1.4059^{20}	1.5933^{20}	
1381	1-Bromo-4-vinylbenzene		C_8H_7Br	2039-82-9	183.046		7.7	212; 103^{20}	1.3984^{20}	1.5947^{20}	i H_2O; vs chl; s HOAc
1382	Brompheniramine		$C_{16}H_{19}BrN_2$	86-22-6	319.239	ye oily liq		$150^{0.5}$			s dil acid
1383	Brucine		$C_{23}H_{26}N_2O_4$	357-57-3	394.463	mcl pr (w +4)	178				sl H_2O, eth, bz; vs EtOH, chl
1384	Brucine hydrochloride	2,3-Dimethoxystrychnidin-10-one, monohydrochloride	$C_{23}H_{27}ClN_2O_4$	5786-96-9	430.924	pr					vs H_2O, EtOH
1385	Brucine sulfate heptahydrate	2,3-Dimethoxystrychnidin-10-one, sulfate, heptahydrate	$C_{46}H_{68}N_4O_{19}S$	60583-39-3	1013.113	nd (w)					s H_2O; sl EtOH, chl, tfa; vs MeOH; i bz
1386	Bucolome	5-Butyl-1-cyclohexyl-2,4,6(1*H*,3*H*,5*H*)-pyrimidinetrione	$C_{14}H_{22}N_2O_3$	841-73-6	266.336	nd (MeOH)	84	$186^{0.8}$			

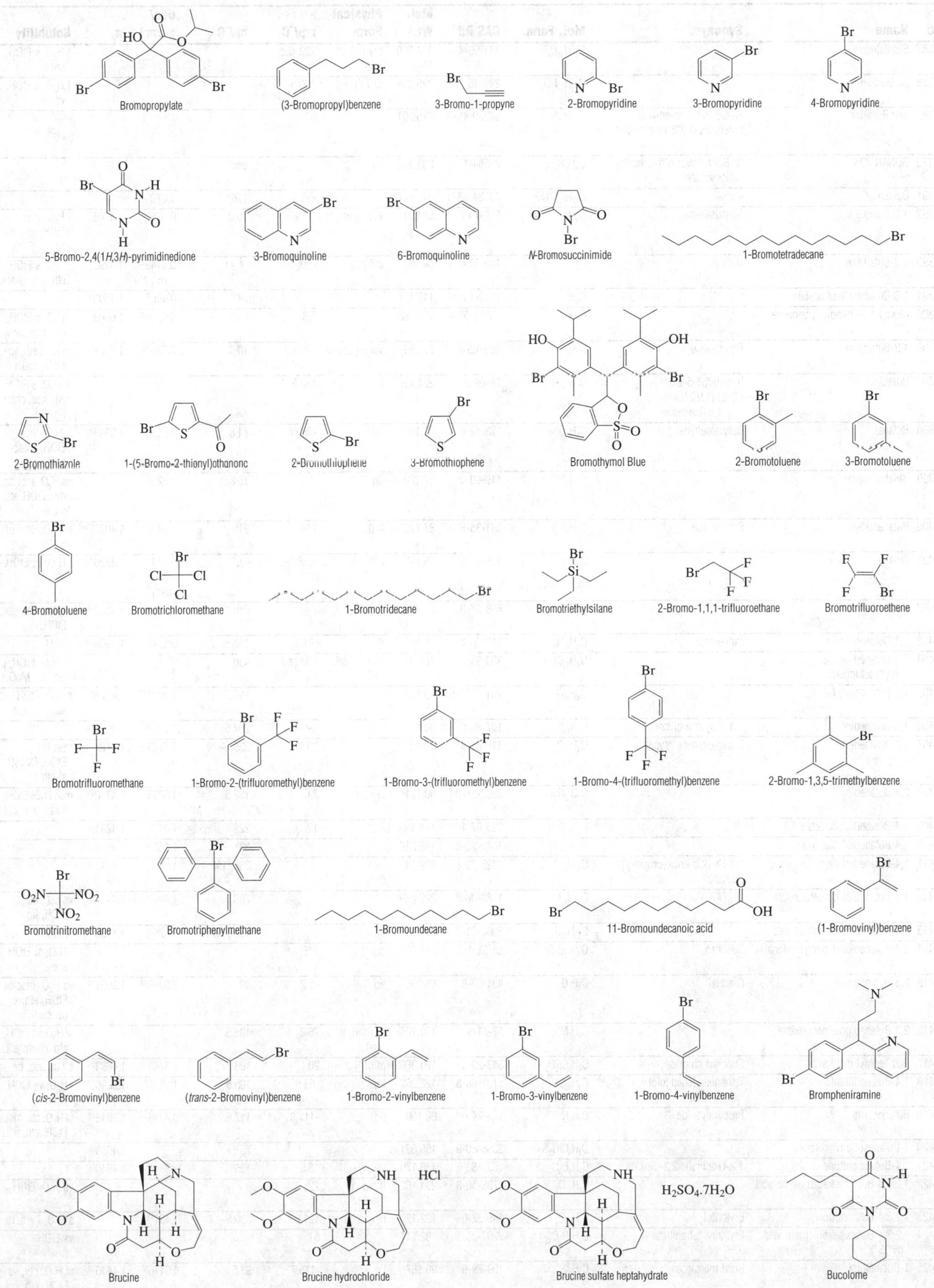
Bromopropylate
(3-Bromopropyl)benzene
3-Bromo-1-propyne
2-Bromopyridine
3-Bromopyridine
4-Bromopyridine
5-Bromo-2,4(1H,3H)-pyrimidinedione
3-Bromoquinoline
6-Bromoquinoline
N-Bromosuccinimide
1-Bromotetradecane
Bromothymol Blue
2-Bromothiazole
1-(5-Bromo-2-thienyl)ethanone
2-Bromothiophene
3-Bromothiophene
2-Bromotoluene
3-Bromotoluene
4-Bromotoluene
Bromotrichloromethane
1-Bromotridecane
Bromotriethylsilane
2-Bromo-1,1,1-trifluoroethane
Bromotrifluoroethene
Bromotrifluoromethane
1-Bromo-2-(trifluoromethyl)benzene
1-Bromo-3-(trifluoromethyl)benzene
1-Bromo-4-(trifluoromethyl)benzene
2-Bromo-1,3,5-trimethylbenzene
Bromotrinitromethane
Bromotriphenylmethane
1-Bromoundecane
11-Bromoundecanoic acid
(1-Bromovinyl)benzene
(cis-2-Bromovinyl)benzene
(trans-2-Bromovinyl)benzene
1-Bromo-2-vinylbenzene
1-Bromo-3-vinylbenzene
1-Bromo-4-vinylbenzene
Brompheniramine
Brucine
HCl
Brucine hydrochloride
H2SO4.7H2O
Brucine sulfate heptahydrate
Bucolome

No.	Name	Synonym	Mol. Form.	CAS RN	Mol. Wt.	Physical Form	mp/°C	bp/°C	den/ g cm^{-3}	n_D	Solubility
1387	Bufotalin		$C_{26}H_{36}O_6$	471-95-4	444.560	cry (+1 al)	223 dec				i H_2O; s EtOH, chl
1388	Bulbocapnine		$C_{19}H_{19}NO_4$	298-45-3	325.359	pr (al)	199.5				i H_2O; s EtOH; vs chl
1389	*sec*-Bumeton	*N*-*sec*-Butyl-*N'*-ethyl-6-methoxy-1,3,5-triazine-2,4-diamine	$C_{10}H_{19}N_5O$	26259-45-0	225.291		87				
1390	BUSAN 72A	(2-Benzothiazolylthio)methyl thiocyanate	$C_9H_6N_2S_3$	21564-17-0	238.352	liq		dec			
1391	Butachlor		$C_{17}H_{26}ClNO_2$	23184-66-9	311.847		<-5	$156^{.5}$	1.070^{25}		
1392	1,2-Butadiene	Methylallene	C_4H_6	590-19-2	54.091	vol liq or gas	-136.2	10.9	0.676^{0}	1.4205^{1}	i H_2O; msc EtOH, eth; vs bz
1393	1,3-Butadiene	Divinyl	C_4H_6	106-99-0	54.091	col gas	-108.91	-4.41	0.6149^{25} (p>1 atm	1.4292^{-25}	i H_2O; s EtOH, eth, bz; vs ace
1394	1,3-Butadien-1-ol acetate		$C_6H_8O_2$	1515-76-0	112.127			58^{40}	0.945^{25}	1.4690^{20}	
1395	(*trans*)-1,3-Butadienylbenzene		$C_{10}H_{10}$	16939-57-4	130.186		2.3	76^{11}	0.9286^{20}	1.6089^{25}	i H_2O; s EtOH, eth, ace, bz
1396	1,3-Butadiyne	Diacetylene	C_4H_2	460-12-8	50.059	vol liq or gas	-36.4	10.3	0.7364^{0}	1.4189^{5}	vs H_2O, eth, ace; s chl, EtOH
1397	Butalbital	5-Isobutyl-5-allyl-2,4,6(1*H*,3*H*,5*H*)-pyrimidinetrione	$C_{11}H_{16}N_2O_3$	77-26-9	224.256	pr	138.5				sl H_2O; s EtOH, eth, ace, chl; i lig
1398	Butanal	Butyraldehyde	C_4H_8O	123-72-8	72.106	liq	-96.86	74.8	0.8016^{20}	1.3843^{20}	s H_2O; msc EtOH; vs ace, bz; sl chl
1399	Butanal oxime		C_4H_9NO	110-69-0	87.120	liq	-29.5	154	0.923^{20}		vs H_2O, ace, bz; msc EtOH, eth; s chl
1400	Butanamide	Butyramide	C_4H_9NO	541-35-5	87.120	lf (bz)	114.8	216	0.8850^{120}	1.4087^{130}	sl H_2O, eth; i bz; s EtOH
1401	Butane		C_4H_{10}	106-97-8	58.122	col gas	-138.3	-0.5	0.573^{25} (p>1 atm)	1.3326^{20}	i H_2O; vs EtOH, eth, chl
1402	Butanedial		$C_4H_6O_2$	638-37-9	86.090			dec 170; 58^{9}	1.065^{20}	1.4262^{18}	vs H_2O, ace, eth, EtOH
1403	1,4-Butanediamine	Putrescine	$C_4H_{12}N_2$	110-60-1	88.151	lf	21.91	158.5	0.877^{25}	1.4969^{20}	s H_2O
1404	1,4-Butanediamine dihydrochloride		$C_4H_{14}Cl_2N_2$	333-93-7	161.073	nd or lf (al, w)	280 dec	sub			vs H_2O, EtOH; i eth, bz, MeOH
1405	1,2-Butanediol, (±)		$C_4H_{10}O_2$	26171-83-5	90.121			190.5	1.0024^{20}	1.4378^{20}	s H_2O, EtOH, ace
1406	1,3-Butanediol	1,3-Butylene glycol	$C_4H_{10}O_2$	107-88-0	90.121		<-50	207.5	1.0053^{20}	1.4401^{20}	
1407	1,4-Butanediol	Tetramethylene glycol	$C_4H_{10}O_2$	110-63-4	90.121		20.4	235	1.0171^{20}	1.4460^{20}	msc H_2O; s EtOH, DMSO; sl eth
1408	2,3-Butanediol		$C_4H_{10}O_2$	6982-25-8	90.121	cry (eth)	7.6	182.5	1.0033^{20}	1.4310^{25}	msc H_2O, EtOH; s eth, ace, chl
1409	1,4-Butanediol diacetate		$C_8H_{14}O_4$	628-67-1	174.195		12	229	1.0479^{15}	1.4251^{15}	
1410	1,4-Butanediol diacrylate		$C_{10}H_{14}O_4$	1070-70-8	198.216			$83^{0.3}$	1.105^{25}		
1411	1,4-Butanediol diglycidyl ether	1,4-Bis(2,3-epoxypropoxy) butane	$C_{10}H_{18}O_4$	2425-79-8	202.248			266; 155^{11}	1.1^{25}	1.4611^{20}	
1412	1,3-Butanediol dimethacrylate		$C_{12}H_{18}O_4$	1189-08-8	226.269			290		1.4495^{25}	vs ace, eth, EtOH, lig
1413	1,4-Butanediol dimethacrylate		$C_{12}H_{18}O_4$	2082-81-7	226.269	liq		133^{4}, $76^{0.027}$	1.025^{20}	1.4560^{20}	sl H_2O
1414	1,4-Butanediol dimethylsulfonate	Busulfan	$C_6H_{14}O_6S_2$	55-98-1	246.301	cry	116				i H_2O; sl EtOH, ace
1415	2,3-Butanedione	Diacetyl	$C_4H_6O_2$	431-03-8	86.090	liq	-1.2	88	0.9808^{18}	1.3951^{20}	vs H_2O; msc EtOH, eth; s bz, ctc
1416	2,3-Butanedione monooxime		$C_4H_7NO_2$	57-71-6	101.105	pr (chl), lf (w)	76.8	185.5			sl H_2O; vs EtOH, eth, chl; s alk
1417	Butanedioyl dichloride	Succinyl chloride	$C_4H_4Cl_2O_2$	543-20-4	154.980	pl or lf	20	193.3	1.3748^{20}	1.4683^{20}	s eth, ace, bz
1418	1,4-Butanedithiol	Tetramethylenedithiol	$C_4H_{10}S_2$	1191-08-8	122.252	liq	-53.9	195.5	1.0021^{0}	1.5290^{20}	i H_2O; vs EtOH; sl ctc
1419	Butanenitrile	Propyl cyanide	C_4H_7N	109-74-0	69.106	liq	-111.9	117.6	0.7936^{20}	1.3842^{20}	sl H_2O, ctc; msc EtOH, eth; s bz
1420	1-Butanesulfonyl chloride		$C_4H_9ClO_2S$	2386-60-9	156.631			75^{10}		1.4559^{20}	
1421	1,4-Butane sultone	1,2-Oxathiane 2,2-dioxide	$C_4H_8O_3S$	1633-83-6	136.170	liq	13.5	135^{4}	1.331^{20}	1.4640^{20}	
1422	1,2,3,4-Butanetetracarboxylic acid		$C_8H_{10}O_8$	1703-58-8	234.160	lf (w) cry (ace)	236.5				vs H_2O, EtOH
1423	1,2,3,4-Butanetetrol	Erythritol	$C_4H_{10}O_4$	149-32-6	122.120	bipym tetr pr	121.5	330.5	1.451^{20}		s H_2O; i eth, bz
1424	1,2,3,4-Butanetetrol tetranitrate, (*R**,*S**)	Erythrityl tetranitrate	$C_4H_6N_4O_{12}$	7297-25-8	302.111		61				vs EtOH
1425	1-Butanethiol	Butyl mercaptan	$C_4H_{10}S$	109-79-5	90.187	liq	-115.7	98.5	0.8416^{20}	1.4440^{20}	sl H_2O, chl; vs EtOH, eth

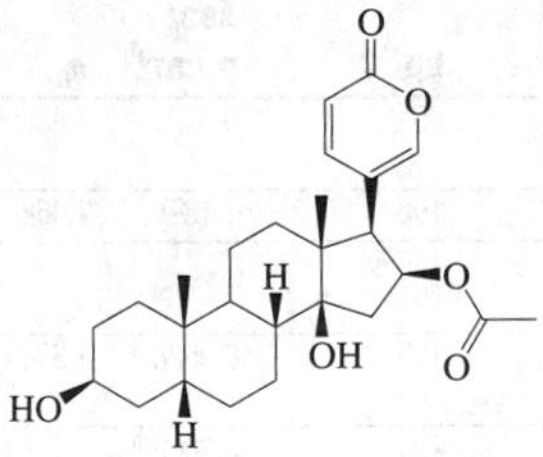

Bufotalin

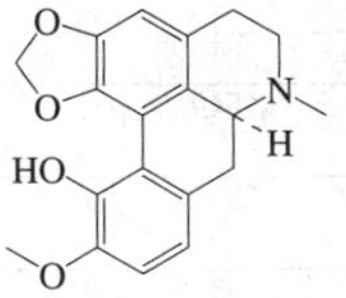

Bulbocapnine

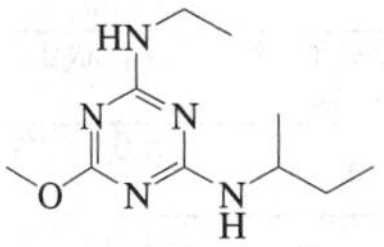

sec-Bumeton

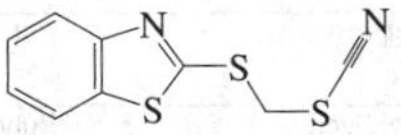

BUSAN 72A

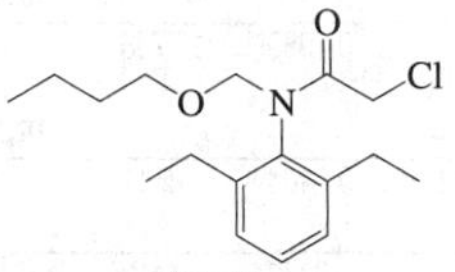

Butachlor

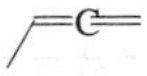

1,2-Butadiene

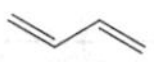

1,3-Butadiene

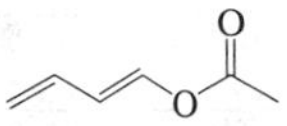

1,3-Butadien-1-ol acetate

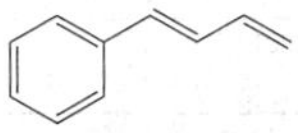

(*trans*)-1,3-Butadienylbenzene

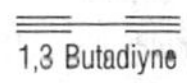

1,3 Butadiyne

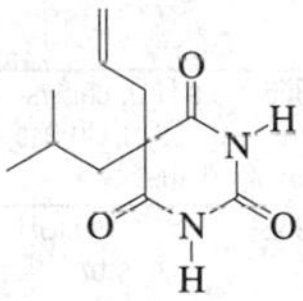

Butalbital

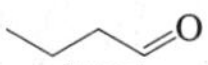

Butanal

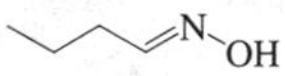

Butanal oxime

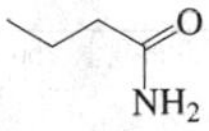

Butanamide

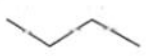

Butane

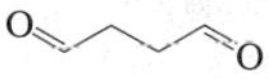

Butanedial

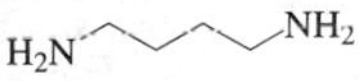

1,4-Butanediamine

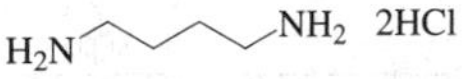

1,4-Butanediamine dihydrochloride

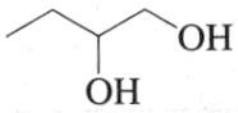

1,2-Butanediol, (±)

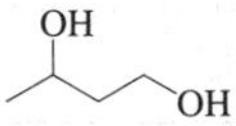

1,3-Butanediol

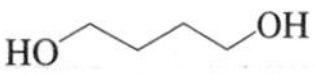

1,4-Butanediol

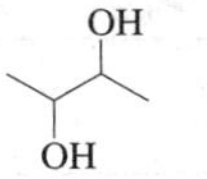

2,3-Butanediol

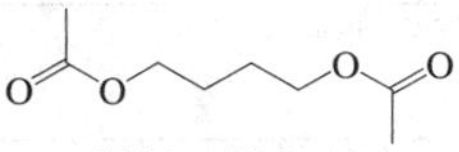

1,4-Butanediol diacetate

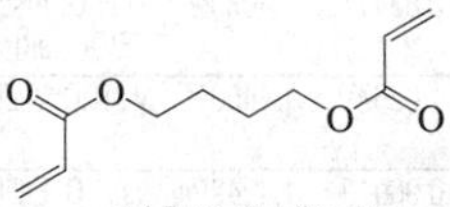

1,4-Butanediol diacrylate

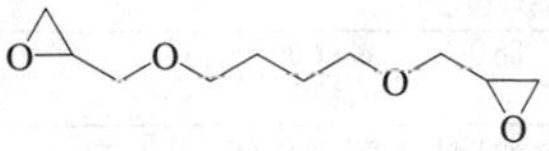

1,4-Butanediol diglycidyl ether

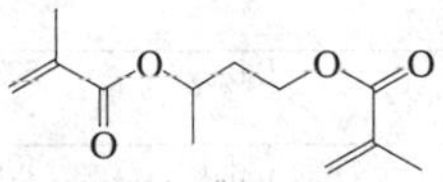

1,3-Butanediol dimethacrylate

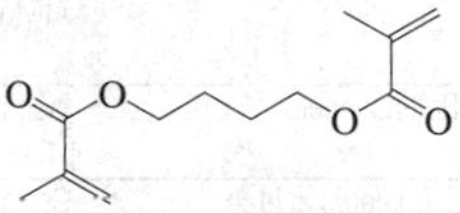

1,4-Butanediol dimethacrylate

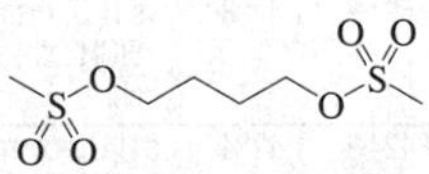

1,4-Butanediol dimethylsulfonate

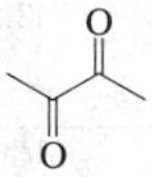

2,3-Butanedione

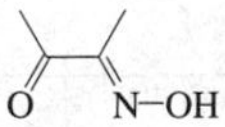

2,3-Butanedione monooxime

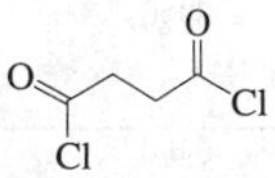

Butanedioyl dichloride

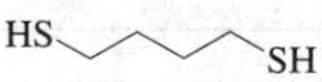

1,4-Butanedithiol

Butanenitrile

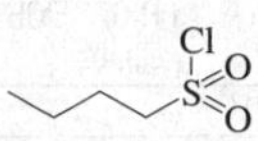

1-Butanesulfonyl chloride

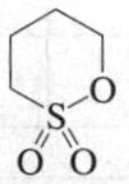

1,4-Butane sultone

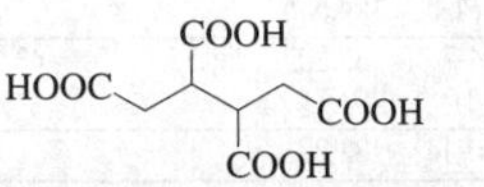

1,2,3,4-Butanetetracarboxylic acid

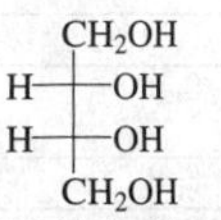

1,2,3,4-Butanetetrol

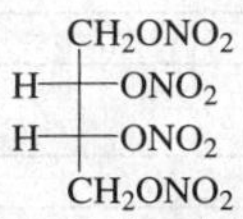

1,2,3,4-Butanetetrol tetranitrate, (R^*,S^*)

1-Butanethiol

No.	Name	Synonym	Mol. Form.	CAS RN	Mol. Wt.	Physical Form	mp/°C	bp/°C	den/ g cm^{-3}	n_D	Solubility
1426	2-Butanethiol	*sec*-Butyl mercaptan	$C_4H_{10}S$	91840-99-2	90.187	liq	-165	85.0	0.8295^{20}	1.4366^{20}	s EtOH, eth, bz, peth; sl ctc
1427	1,2,4-Butanetriol		$C_4H_{10}O_3$	3068-00-6	106.120			190^{18}, 172^{12}	1.18^{20}	1.4688^{20}	vs H_2O, EtOH
1428	Butanilicaine	2-(Butylamino)-*N*-(2-chloro-6-methylphenyl)acetamide	$C_{13}H_{19}ClN_2O$	3785-21-5	254.755	cry	46	145$^{0.001}$			
1429	Butanoic acid	Butyric acid	$C_4H_8O_2$	107-92-6	88.106	liq	-5.1	163.75	0.9528^{25}	1.3980^{20}	msc H_2O, EtOH, eth; sl ctc
1430	Butanoic anhydride	Butyric anhydride	$C_8H_{14}O_3$	106-31-0	158.195	liq	-75	200	0.9668^{20}	1.4070^{20}	s eth; sl ctc
1431	1-Butanol	Butyl alcohol	$C_4H_{10}O$	71-36-3	74.121	liq	-88.6	117.73	0.8095^{20}	1.3988^{20}	s H_2O, bz; msc EtOH, eth; vs ace
1432	2-Butanol	*sec*-Butyl alcohol	$C_4H_{10}O$	78-92-2	74.121	liq	-88.5	99.51	0.8063^{20}	1.3978^{20}	vs H_2O; msc EtOH, eth; s bz, ctc
1433	2-Butanone	Methyl ethyl ketone	C_4H_8O	78-93-3	72.106	liq	-86.64	79.59	0.7999^{25}	1.3788^{20}	vs H_2O; msc EtOH, eth, ace, bz; s chl
1434	2-Butanone (1-methylpropylidene) hydrazone		$C_8H_{16}N_2$	5921-54-0	140.226			171.5	0.8404^{20}	1.4511^{20}	
1435	2-Butanone oxime		C_4H_9NO	96-29-7	87.120	liq	-29.5	152.5	0.9232^{20}	1.4410^{20}	s H_2O, chl; msc EtOH, eth
1436	2-Butanone peroxide	Methyl ethyl ketone peroxide	$C_8H_{16}O_4$	1338-23-4	176.211	col liq		exp 110			sl H_2O; misc os
1437	Butanoyl chloride	*n*-Butyryl chloride	C_4H_7ClO	141-75-3	106.551	liq	-89	102	1.0277^{20}	1.4121^{20}	msc eth
1438	Butaperazine		$C_{24}H_{31}N_3OS$	653-03-2	409.587			275$^{0.05}$			
1439	Butazolamide	*N*-[5-(Aminosulfonyl)-1,3,4-thiadiazol-2-yl]butanamide	$C_6H_{10}N_4O_3S_2$	16790-49-1	250.298	cry	261 dec				
1440	*trans*-2-Butenal	*trans*-Crotonaldehyde	C_4H_6O	123-73-9	70.090	liq	-76	102.2	0.8516^{20}	1.4366^{20}	s H_2O, chl; vs EtOH, eth, ace; msc bz
1441	1-Butene	1-Butylene	C_4H_8	106-98-9	56.107	col gas	-185.34	-6.26	0.588^{25} (p>1 atm)	1.3962^{20}	i H_2O; vs EtOH, eth; s bz
1442	*cis*-2-Butene		C_4H_8	590-18-1	56.107	col gas	-138.88	3.71	0.616^{25} (p>1 atm)	1.3931^{-25}	i H_2O; vs EtOH, eth; s bz
1443	*trans*-2-Butene		C_4H_8	624-64-6	56.107	col gas	-105.52	0.88	0.599^{25} (p>1 atm)	1.3848^{-25}	s bz
1444	*trans*-2-Butenedinitrile		$C_4H_2N_2$	764-42-1	78.072	nd (bz-peth)	96.8	186	0.9416^{111}	1.4349^{111}	s H_2O, EtOH, eth, ace, bz, chl; sl peth
1445	*cis*-2-Butene-1,4-diol		$C_4H_8O_2$	6117-80-2	88.106		2.0	235	1.0698^{20}	1.4782^{20}	s H_2O; vs EtOH
1446	*trans*-2-Butene-1,4-diol		$C_4H_8O_2$	821-11-4	88.106		25	131^{13}	1.0700^{20}	1.4755^{20}	vs H_2O, EtOH
1447	*trans*-2-Butenedioyl dichloride	Fumaric acid dichloride	$C_4H_2Cl_2O_2$	627-63-4	152.964	pa ye lig		159	1.408^{20}	1.5004^{18}	
1448	*cis*-2-Butenenitrile	Isocrotononitrile	C_4H_5N	1190-76-7	67.090	liq		107.4			
1449	*trans*-2-Butenenitrile	Crotononitrile	C_4H_5N	627-26-9	67.090	liq	-51.5	120	0.8239^{20}	1.4225^{20}	s eth, ace
1450	3-Butenenitrile	Allyl cyanide	C_4H_5N	109-75-1	67.090	liq	-87	119	0.8341^{20}	1.4060^{20}	sl H_2O; msc EtOH, eth
1451	*cis*-2-Butenoic acid	Isocrotonic acid	$C_4H_6O_2$	503-64-0	86.090	nd or pr (peth)	15	169	1.0267^{20}	1.4450^{20}	vs H_2O; s EtOH
1452	*trans*-2-Butenoic acid	Crotonic acid	$C_4H_6O_2$	107-93-7	86.090	mcl pr or nd (w, lig)	71.5	184.7	0.9604^{77}	1.4249^{77}	vs H_2O, EtOH; s eth, ace, lig
1453	3-Butenoic acid		$C_4H_6O_2$	625-38-7	86.090	liq	-35	169	1.0091^{20}	1.4239^{20}	s H_2O; msc EtOH, eth
1454	2-Butenoic anhydride	Crotonic acid anhydride	$C_8H_{10}O_3$	623-68-7	154.163			247; 129^{19}	1.0397^{20}	1.4745^{20}	vs eth
1455	*cis*-2-Buten-1-ol	*cis*-Crotyl alcohol	C_4H_8O	4088-60-2	72.106			123	0.8662^{20}	1.4342^{25}	s H_2O
1456	*trans*-2-Buten-1-ol	*trans*-Crotyl alcohol	C_4H_8O	504-61-0	72.106		<-30	121.2	0.8521^{20}	1.4288^{20}	vs H_2O; msc EtOH, eth; s chl
1457	3-Buten-1-ol		C_4H_8O	627-27-0	72.106			113.5	0.8424^{20}	1.4224^{20}	s H_2O, ace; msc EtOH, eth; sl chl
1458	3-Buten-2-ol		C_4H_8O	598-32-3	72.106			97			
1459	3-Buten-2-one	Methyl vinyl ketone	C_4H_6O	78-94-4	70.090			81.4	0.864^{20}	1.4081^{20}	s H_2O, EtOH, bz; vs eth, ace; sl ctc
1460	2-Butenoyl chloride		C_4H_5ClO	10487-71-5	104.535			124.5	1.0905^{20}	1.460^{18}	vs ace
1461	(*trans*-1-Butenyl)benzene		$C_{10}H_{12}$	1005-64-7	132.202	liq	-43.1	198.7	0.9019^{20}	1.5420^{20}	i H_2O; s EtOH, eth, bz, ctc
1462	2-Butenylbenzene		$C_{10}H_{12}$	1560-06-1	132.202			176	0.8831^{20}	1.5101^{20}	
1463	3-Butenylbenzene		$C_{10}H_{12}$	768-56-9	132.202	liq	-70	177	0.8831^{20}	1.5059^{20}	i H_2O; s eth, bz
1464	1-Buten-3-yne	Vinylacetylene	C_4H_4	689-97-4	52.075	col gas		5.1	0.7094^{0}	1.4161^{1}	i H_2O; s bz
1465	Butethamine hydrochloride	2-Isobutylaminoethyl 4-aminobenzoate	$C_{13}H_{21}ClN_2O_2$	553-68-4	272.771	cry	194				s H_2O; sl EtOH, bz, chl; i eth

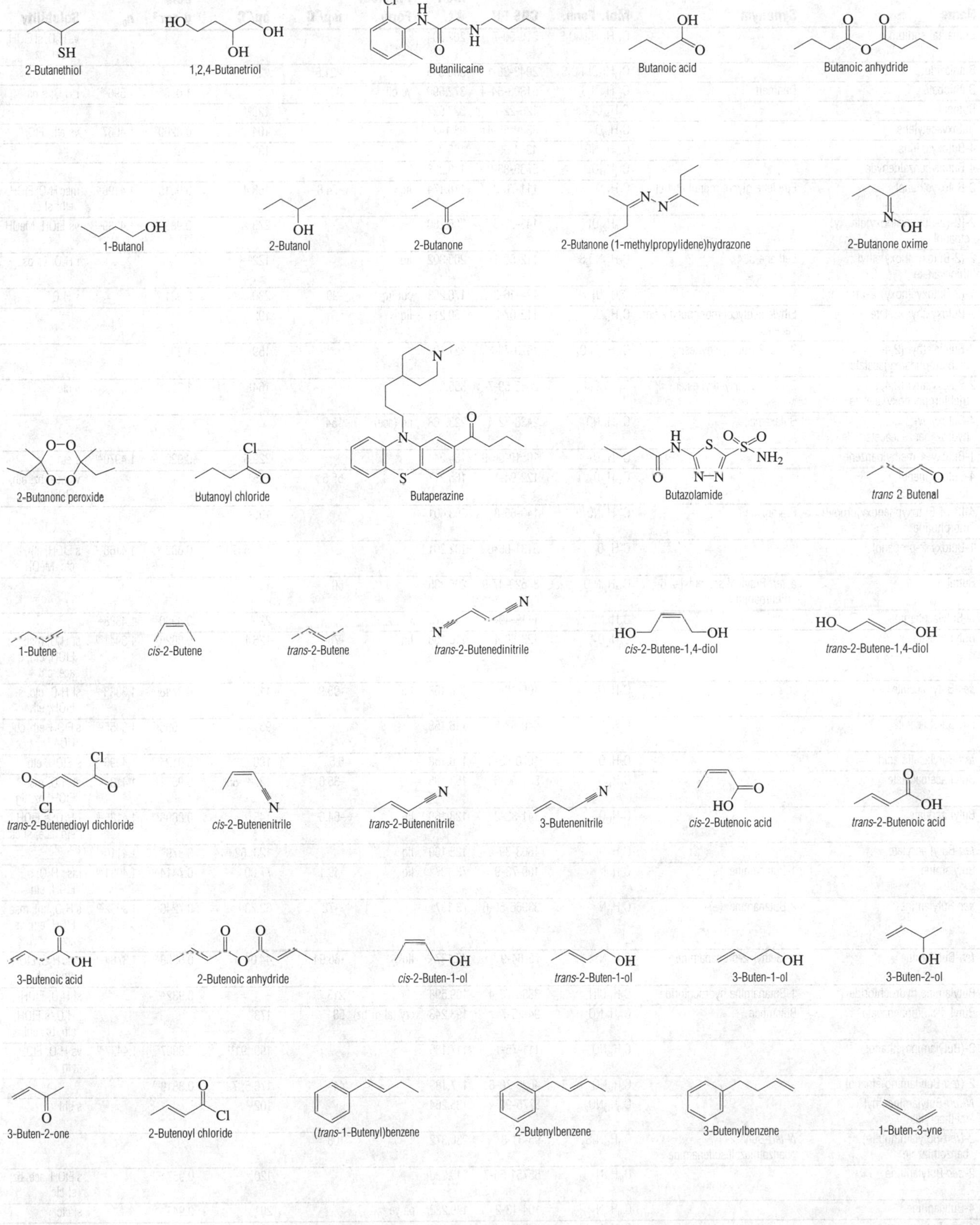

2-Butanethiol
1,2,4-Butanetriol
Butanilicaine
Butanoic acid
Butanoic anhydride
1-Butanol
2-Butanol
2-Butanone
2-Butanone (1-methylpropylidene)hydrazone
2-Butanone oxime
2-Butanone peroxide
Butanoyl chloride
Butaperazine
Butazolamide
trans-2-Butenal
1-Butene
cis-2-Butene
trans-2-Butene
trans-2-Butenedinitrile
cis-2-Butene-1,4-diol
trans-2-Butene-1,4-diol
trans-2-Butenedioyl dichloride
cis-2-Butenenitrile
trans-2-Butenenitrile
3-Butenenitrile
cis-2-Butenoic acid
trans-2-Butenoic acid
3-Butenoic acid
2-Butenoic anhydride
cis-2-Buten-1-ol
trans-2-Buten-1-ol
3-Buten-1-ol
3-Buten-2-ol
3-Buten-2-one
2-Butenoyl chloride
(*trans*-1-Butenyl)benzene
2-Butenylbenzene
3-Butenylbenzene
1-Buten-3-yne
Butethamine hydrochloride

No.	Name	Synonym	Mol. Form.	CAS RN	Mol. Wt.	Physical Form	mp/°C	bp/°C	den/ g cm^{-3}	n_D	Solubility
1466	Buthalital sodium		$C_{11}H_{15}N_2NaO_2S$	510-90-7	262.304						vs H_2O; sl EtOH; i eth, bz
1467	Buthiazide		$C_{11}H_{16}ClN_3O_4S_2$	2043-38-1	353.846		221.5				
1468	Buthiobate	Denmert	$C_{21}H_{28}N_2S_2$	51308-54-4	372.590	ye oil	32		1.0865^{25}	1.596^{26}	i H_2O; s os
1469	Butonate		$C_8H_{14}Cl_3O_5P$	126-22-7	327.527			$129^{0.5}$			
1470	Butoxyacetylene		$C_6H_{10}O$	3329-56-4	98.142			104	0.8200^{20}	1.4067	vs eth, EtOH
1471	4-Butoxyaniline		$C_{10}H_{15}NO$	4344-55-2	165.232			132^{4}			
1472	4-Butoxybenzaldehyde		$C_{11}H_{14}O_2$	5736-88-9	178.228			148^{10}			
1473	2-Butoxyethanol	Ethylene glycol monobutyl ether	$C_6H_{14}O_2$	111-76-2	118.174	liq	-74.8	168.4	0.9015^{20}	1.4198^{20}	msc H_2O, EtOH, eth; sl ctc
1474	2-[2-(2-Butoxyethoxy)ethoxy] ethanol		$C_{10}H_{22}O_4$	143-22-6	206.280			278	0.9890^{20}	1.4389^{20}	vs EtOH, MeOH
1475	2-(2-Butoxyethoxy)ethyl thiocyanate	Lethane 384	$C_9H_{17}NO_2S$	112-56-1	203.302	liq		$122^{0.25}$			i H_2O; vs os
1476	1-(2-Butoxyethoxy)-2-propanol		$C_9H_{20}O_3$	124-16-3	176.253	col liq	-90	230	0.931^{20}		s H_2O
1477	2-Butoxyethyl acetate	Ethylene glycol monobutyl ether acetate	$C_8H_{16}O_3$	112-07-2	160.211	liq		192			
1478	2-Butoxyethyl (2,4-dichlorophenoxy)acetate	2,4-D 2-Butoxyethyl ester	$C_{14}H_{18}Cl_2O_4$	1929-73-3	321.197			159^{1}	1.232^{20}		
1479	2-Butoxyethyl (2,4,5-trichlorophenoxy)acetate	2,4,5-T Butoxyethyl ester	$C_{14}H_{17}Cl_3O_4$	2545-59-7	355.642			164^{1}	1.280^{20}		s ctc
1480	4-Butoxy-*N*-hydroxybenzeneacetamide	Bufexamac	$C_{12}H_{17}NO_3$	2438-72-4	223.268	nd (ace)	154				
1481	1-Butoxy-4-methylbenzene		$C_{11}H_{16}O$	10519-06-9	164.244			229.5	0.9205^{25}	1.4970^{20}	s eth
1482	4-Butoxyphenol		$C_{10}H_{14}O_2$	122-94-1	166.217		65.5	125^{4}			vs ace, bz, eth, EtOH
1483	4-[3-(4-Butoxyphenoxy)propyl] morpholine	Pramoxine	$C_{17}H_{27}NO_3$	140-65-8	293.401			196^{6}			
1484	1-Butoxy-2-propanol		$C_7H_{16}O_2$	5131-66-8	132.201			171.5; 71^{20}	0.882^{20}	1.4168^{20}	s EtOH, eth, bz, ctc, MeOH
1485	Butralin	4-*tert*-Butyl-*N*-*sec*-butyl-2,6-dinitroaniline	$C_{14}H_{21}N_3O_4$	33629-47-9	295.335		60	$135^{0.5}$			
1486	*N*-Butylacetamide		$C_6H_{13}NO$	1119-49-9	115.173			229	0.8960^{25}	1.4388^{25}	
1487	Butyl acetate		$C_6H_{12}O_2$	123-86-4	116.158	liq	-78	126.1	0.8825^{20}	1.3941^{20}	sl H_2O; msc EtOH, eth; s ace, chl
1488	*sec*-Butyl acetate		$C_6H_{12}O_2$	105-46-4	116.158	liq	-98.9	112	0.8748^{20}	1.3888^{20}	sl H_2O, ctc; s EtOH, eth
1489	*tert*-Butyl acetate		$C_6H_{12}O_2$	540-88-5	116.158			95.1	0.8665^{20}	1.3855^{20}	s EtOH, eth, chl, HOAc
1490	*tert*-Butylacetic acid		$C_6H_{12}O_2$	1070-83-3	116.158		6.5	190	0.9124^{20}	1.4096^{20}	s EtOH, eth
1491	Butyl acetoacetate		$C_8H_{14}O_3$	591-60-6	158.195		-35.6	127^{50}, 85^{8}	0.9671^{25}	1.4137^{20}	sl H_2O; msc EtOH, bz, lig
1492	Butyl acrylate		$C_7H_{12}O_2$	141-32-2	128.169	liq	-64.6	145	0.8898^{20}	1.4185^{20}	i H_2O; s EtOH, eth, ace; sl ctc
1493	*tert*-Butyl acrylate		$C_7H_{12}O_2$	1663-39-4	128.169	liq		120; 62^{60}	0.879^{25}	1.4110^{20}	
1494	Butylamine	1-Butanamine	$C_4H_{11}N$	109-73-9	73.137	liq	-49.1	77.00	0.7414^{20}	1.4031^{20}	msc H_2O; s EtOH, eth
1495	*sec*-Butylamine	2-Butanamine, (±)-	$C_4H_{11}N$	33966-50-6	73.137		<-72	62.73	0.7246^{20}	1.3932^{20}	s H_2O, chl; msc EtOH, eth; vs ace
1496	*tert*-Butylamine	2-Methyl-2-propanamine	$C_4H_{11}N$	75-64-9	73.137	liq	-66.94	44.04	0.6958^{20}	1.3784^{20}	msc H_2O, EtOH, eth; s chl
1497	Butylamine hydrochloride	1-Butanamine hydrochloride	$C_4H_{12}ClN$	3858-78-4	109.598		213		0.982^{20}		sl H_2O, EtOH
1498	Butyl 4-aminobenzoate	Butamben	$C_{11}H_{15}NO_2$	94-25-7	193.243	cry (al or bz)	58	173^{8}			i H_2O; s EtOH, eth, bz, chl
1499	2-(Butylamino)ethanol		$C_6H_{15}NO$	111-75-1	117.189			199; 91^{11}	0.8907^{20}	1.4437^{20}	vs H_2O, EtOH, eth
1500	2-(*tert*-Butylamino)ethanol		$C_6H_{15}NO$	4620-70-6	117.189		44	176.5; 72^{14}	0.8818^{20}		
1501	*N*-*tert*-Butylaminoethyl methacrylate		$C_{10}H_{19}NO_2$	3775-90-4	185.264			102^{12}			s chl
1502	2-(*tert*-Butylaminothio) benzothiazole	*N*-*tert*-Butyl-2-benzothiazolesulfenamide	$C_{11}H_{14}N_2S_2$	95-31-8	238.372		108				
1503	2-*sec*-Butylaniline		$C_{10}H_{15}N$	55751-54-7	149.233			120^{16}	0.9574^{20}		s EtOH, ace, bz; sl ctc
1504	4-Butylaniline		$C_{10}H_{15}N$	104-13-2	149.233	pa ye		261	0.945^{20}		sl ctc
1505	4-*sec*-Butylaniline		$C_{10}H_{15}N$	30273-11-1	149.233			238; 118^{15}	0.949^{15}	1.5360^{29}	vs bz, eth
1506	4-*tert*-Butylaniline		$C_{10}H_{15}N$	769-92-6	149.233	ye rd (peth)	17	241	0.9525^{15}	1.5380^{20}	sl H_2O; msc EtOH, eth; vs bz; s ctc
1507	*N*-Butylaniline		$C_{10}H_{15}N$	1126-78-9	149.233	liq	-14.4	243.5	0.9323^{20}	1.5341^{20}	vs eth, EtOH
1508	*N*-*tert*-Butylaniline		$C_{10}H_{15}N$	937-33-7	149.233			215; 95^{19}		1.5270^{20}	s EtOH; vs ace, bz, chl
1509	2-*tert*-Butyl-9,10-anthracenedione		$C_{18}H_{16}O_2$	84-47-9	264.319		99				s ctc, CS_2

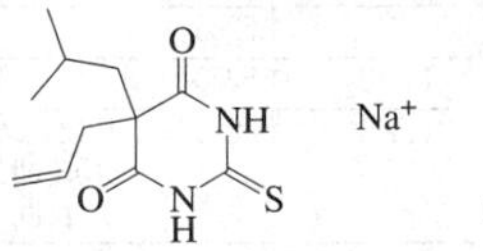

Buthalital sodium

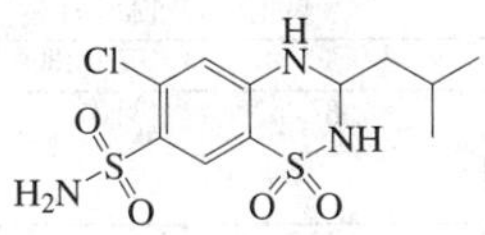

Buthiazide

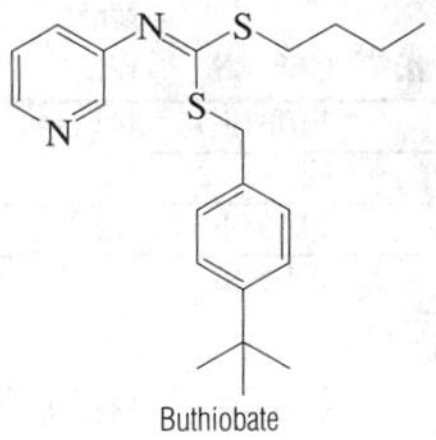

Buthiobate

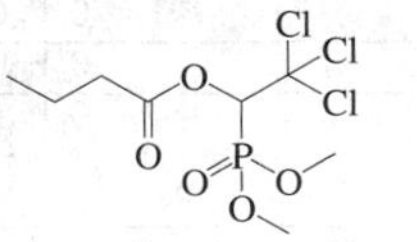

Butonate

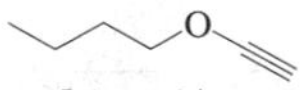

Butoxyacetylene

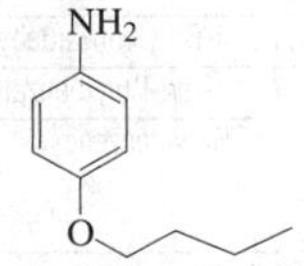

4-Butoxyaniline

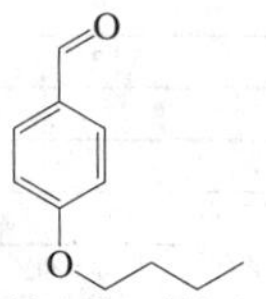

4-Butoxybenzaldehyde

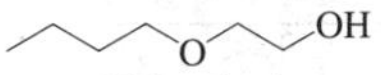

2-Butoxyethanol

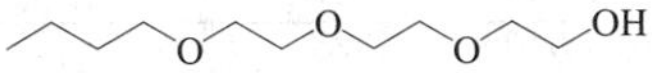

2-[2-(2-Butoxyethoxy)ethoxy]ethanol

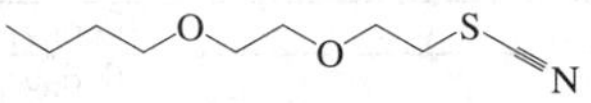

2-(2-Butoxyethoxy)ethyl thiocyanate

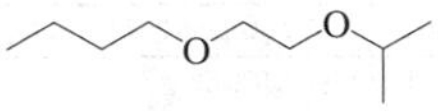

1-(2-Butoxyethoxy)-2-propanol

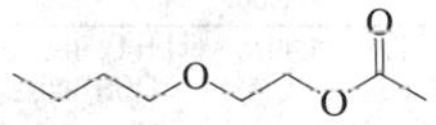

2-Butoxyethyl acetate

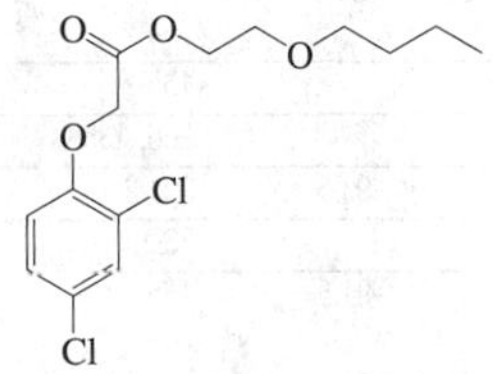

2-Butoxyethyl (2,4-dichlorophenoxy)acetate

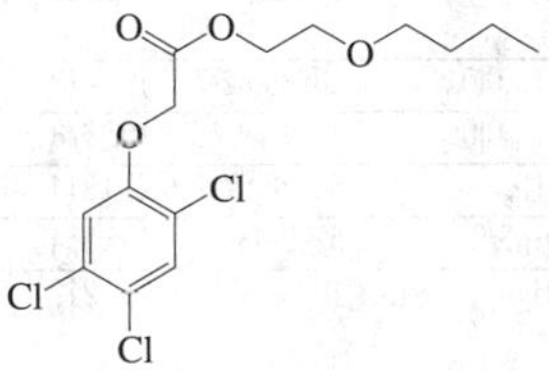

2-Butoxyethyl (2,4,5-trichlorophenoxy)acetate

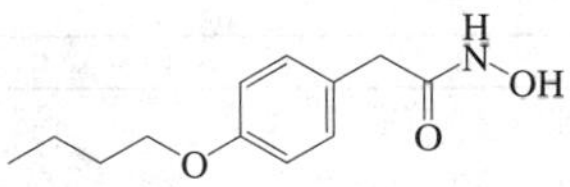

4 Butoxy-*N*-hydroxybenzeneacetamide

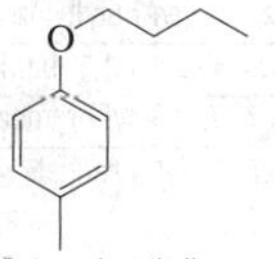

1-Butoxy-4-methylbenzene

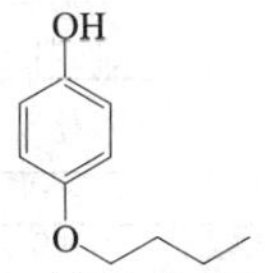

4-Butoxyphenol

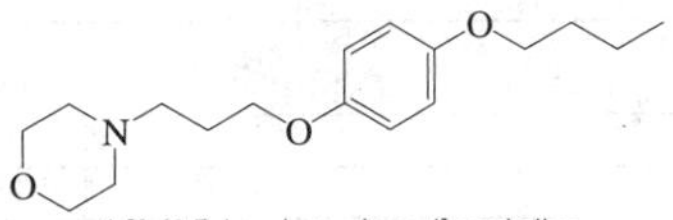

4-[3-(4-Butoxyphenoxy)propyl]morpholine

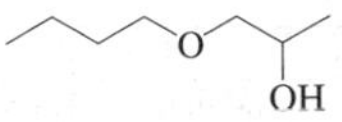

1-Butoxy-2-propanol

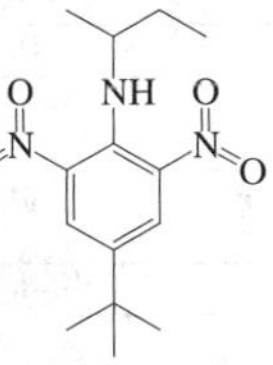

Butralin

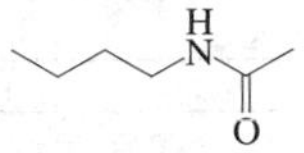

N-Butylacetamide

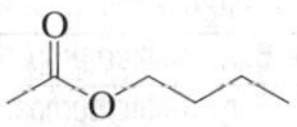

Butyl acetate

sec-Butyl acetate

tert-Butyl acetate

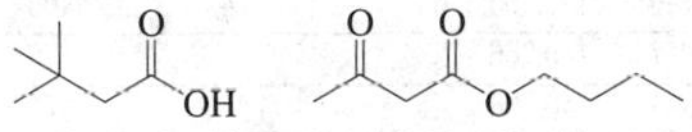

tert-Butylacetic acid

Butyl acetoacetate

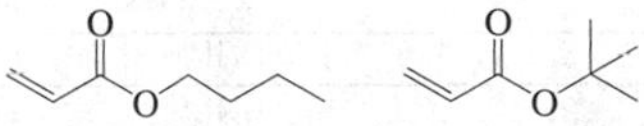

Butyl acrylate

tert-Butyl acrylate

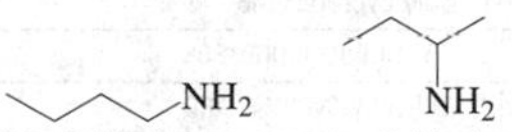

Butylamine

sec-Butylamine

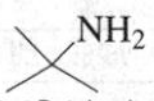

tert-Butylamine

Butylamine hydrochloride

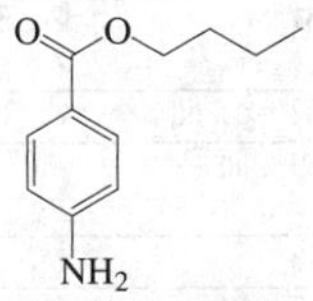

Butyl 4-aminobenzoate

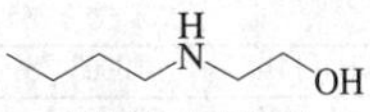

2-(Butylamino)ethanol

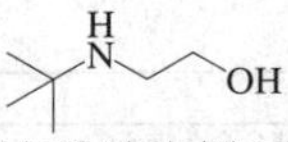

2-(*tert*-Butylamino)ethanol

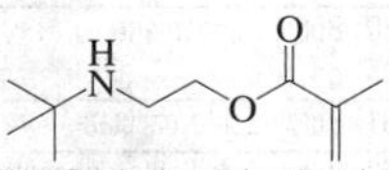

N-*tert*-Butylaminoethyl methacrylate

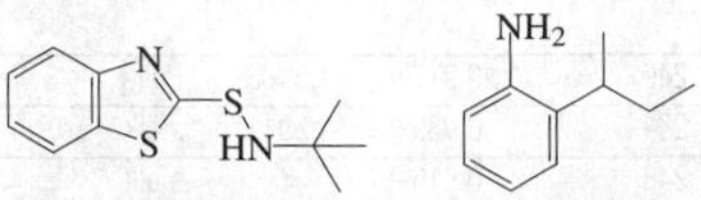

2-(*tert*-Butylaminothio)benzothiazole

2-*sec*-Butylaniline

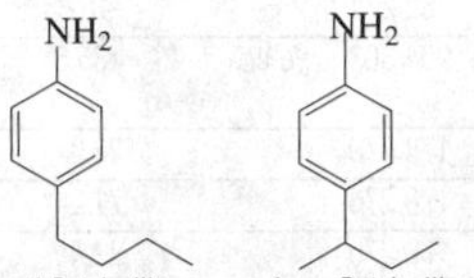

4-Butylaniline

4-*sec*-Butylaniline

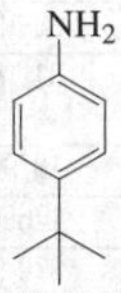

4-*tert*-Butylaniline

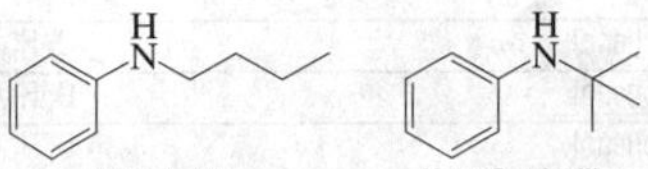

N-Butylaniline

N-*tert*-Butylaniline

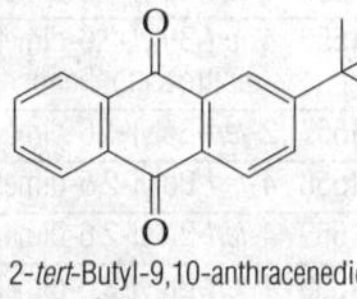

2-*tert*-Butyl-9,10-anthracenedione

No.	Name	Synonym	Mol. Form.	CAS RN	Mol. Wt.	Physical Form	mp/°C	bp/°C	den/ g cm^{-3}	n_D	Solubility
1510	*tert*-Butyl azidoformate	*tert*-Butyl carbonazidate	$C_5H_9N_3O_2$	1070-19-5	143.144	unstab >80		73^{70}			
1511	4-Butylbenzaldehyde		$C_{11}H_{14}O$	1200-14-2	162.228			123^{7}		1.5265	
1512	4-*tert*-Butylbenzaldehyde		$C_{11}H_{14}O$	939-97-9	162.228	liq		107^{11}, 130^{25}	0.970	1.5270^{20}	
1513	Butylbenzene		$C_{10}H_{14}$	104-51-8	134.218	liq	-87.85	183.31	0.8601^{20}	1.4898^{20}	i H_2O; msc EtOH, eth, ace, bz, peth, ctc
1514	*sec*-Butylbenzene, (±)	2-Phenylbutane	$C_{10}H_{14}$	36383-15-0	134.218	liq	-82.7	173.3	0.8621^{20}	1.4902^{20}	i H_2O; msc EtOH, eth, ace, bz, peth, ctc
1515	*tert*-Butylbenzene		$C_{10}H_{14}$	98-06-6	134.218	liq	-57.8	169.1	0.8665^{20}	1.4927^{20}	i H_2O; vs EtOH, eth; msc ace, bz
1516	4-*tert*-Butyl-1,2-benzenediol		$C_{10}H_{14}O_2$	98-29-3	166.217		54.3	285; 160^{22}			s tfa
1517	2-*tert*-Butyl-1,4-benzenediol		$C_{10}H_{14}O_2$	1948-33-0	166.217		128				
1518	*N*-*tert*-Butylbenzenemethanamine		$C_{11}H_{17}N$	3378-72-1	163.260			75^{5}		1.4951^{25}	
1519	4-*tert*-Butylbenzenemethanol		$C_{11}H_{16}O$	877-65-6	164.244			236; 140^{20}	0.928^{25}	1.5179^{20}	
1520	Butyl benzoate		$C_{11}H_{14}O_2$	136-60-7	178.228	liq	-22.4	250.3	1.000^{20}	1.4940^{25}	i H_2O; msc EtOH, eth; s ace; sl ctc
1521	2-*tert*-Butylbenzoic acid		$C_{11}H_{14}O_2$	1077-58-3	178.228	pl (dil al)	80.5				vs EtOH
1522	3-*tert*-Butylbenzoic acid		$C_{11}H_{14}O_2$	7498-54-6	178.228	nd (peth)	128.8				vs EtOH, peth
1523	4-*tert*-Butylbenzoic acid		$C_{11}H_{14}O_2$	98-73-7	178.228	nd (dil al)	164.5				i H_2O; vs EtOH, bz; s chl
1524	4-Butylbenzoyl chloride		$C_{11}H_{13}ClO$	28788-62-7	196.673			155^{26}	1.051^{25}	1.5351^{20}	
1525	4-*tert*-Butylbenzoyl chloride		$C_{11}H_{13}ClO$	1710-98-1	196.673			266; 135^{20}	1.007^{25}	1.5364^{20}	
1526	2-Butyl-1,1′-biphenyl		$C_{16}H_{18}$	54532-97-7	210.314	liq	-9.65	291.2	0.9676^{20}	1.5604^{20}	
1527	*tert*-Butyl bromoacetate		$C_6H_{11}BrO_2$	5292-43-3	195.054			73^{25}		1.4430^{20}	vs eth, EtOH
1528	Butyl butanoate		$C_8H_{16}O_2$	109-21-7	144.212	liq	-91.5	166	0.8700^{20}	1.4075^{20}	i H_2O; msc EtOH, eth; s ctc
1529	Butyl *cis*-2-butenedioate	Monobutyl maleate	$C_8H_{12}O_4$	925-21-3	172.179	oil			1.09^{25}		
1530	Butyl carbamate		$C_5H_{11}NO_2$	592-35-8	117.147	pr	53	dec 204; 108^{14}			vs EtOH; sl chl
1531	Butyl chloroacetate		$C_6H_{11}ClO_2$	590-02-3	150.603			183	1.0704^{20}	1.4297^{20}	vs eth, EtOH
1532	*tert*-Butyl chloroacetate		$C_6H_{11}ClO_2$	107-59-5	150.603			150; 50^{10}		1.4260^{20}	dec H_2O
1533	Butylchlorodimethylsilane		$C_6H_{15}ClSi$	1000-50-6	150.722			139	0.876^{20}	1.5145^{20}	
1534	Butyl chloroformate		$C_5H_9ClO_2$	592-34-7	136.577			142	1.074^{25}	1.4114^{20}	msc eth; s ace; sl ctc
1535	*N*-Butyl-4-chloro-2-hydroxybenzamide	Buclosamide	$C_{11}H_{14}ClNO_2$	575-74-6	227.688		91.5				
1536	Butyl 2-chloropropanoate		$C_7H_{13}ClO_2$	54819-86-2	164.630			184	1.0253^{20}	1.4263^{20}	vs eth
1537	Butyl 3-chloropropanoate		$C_7H_{13}ClO_2$	27387-79-7	164.630			104^{22}, 92^{6}	1.0370^{20}	1.4321^{20}	vs H_2O, eth
1538	*tert*-Butyl chromate		$C_8H_{18}CrO_4$	1189-85-1	230.223	red cry (peth)	-5				reac H_2O
1539	Butyl citrate		$C_{18}H_{32}O_7$	77-94-1	360.443		-20	233^{22}	1.043^{20}	1.4460^{20}	
1540	Butyl cyanoacetate		$C_7H_{11}NO_2$	5459-58-5	141.168			231; 115^{15}	1.0010^{20}	1.4200^{20}	
1541	Butylcyclohexane		$C_{10}H_{20}$	1678-93-9	140.266	liq	-74.73	180	0.7902^{20}	1.4408^{20}	i H_2O
1542	*sec*-Butylcyclohexane		$C_{10}H_{20}$	7058-01-7	140.266			179.3	0.8131^{20}	1.4467^{20}	i H_2O; s ace
1543	*tert*-Butylcyclohexane		$C_{10}H_{20}$	3178-22-1	140.266	liq	-41.2	171.5	0.8127^{20}	1.4469^{20}	i H_2O
1544	2-*tert*-Butylcyclohexanol		$C_{10}H_{20}O$	13491-79-7	156.265		45	139^{95}	0.902^{25}		
1545	*cis*-4-*tert*-Butylcyclohexanol		$C_{10}H_{20}O$	937-05-3	156.265		83	112^{15}			
1546	*trans*-4-*tert*-Butylcyclohexanol		$C_{10}H_{20}O$	21862-63-5	156.265		83	112^{15}			
1547	4-*tert*-Butylcyclohexanone		$C_{10}H_{18}O$	98-53-3	154.249		48	90^{9}			
1548	Butylcyclohexylamine	*N*-Butylcyclohexanamine	$C_{10}H_{21}N$	10108-56-2	155.281		208.3				sl H_2O, ctc; vs EtOH, eth
1549	Butyl cyclohexyl phthalate		$C_{18}H_{24}O_4$	84-64-0	304.382	col liq		≈205^{5}	1.076^{25}		sl H_2O; misc os
1550	Butylcyclopentane		C_9H_{18}	2040-95-1	126.239	liq	-108	156.6	0.7846^{20}	1.4316^{20}	vs ace, bz, eth, EtOH
1551	Butyl dichloroacetate		$C_6H_{10}Cl_2O_2$	29003-73-4	185.048			193.5	1.1820^{20}	1.4420^{20}	vs eth, EtOH
1552	Butyl (2,4-dichlorophenoxy) acetate	2,4-D Butyl ester	$C_{12}H_{14}Cl_2O_3$	94-80-4	277.143		9	133^{1}			
1553	5-Butyldihydro-2(3*H*)-furanone		$C_8H_{14}O_2$	104-50-7	142.196			132^{20}	0.9796^{19}	1.4451^{19}	s EtOH; sl ctc
1554	Butyldimethylamine	*N*,*N*-Dimethyl-1-butanamine	$C_6H_{15}N$	927-62-8	101.190			95	0.7206^{20}	1.3970^{20}	msc H_2O, EtOH, eth, ace, bz
1555	1-*tert*-Butyl-3,5-dimethylbenzene		$C_{12}H_{18}$	98-19-1	162.271	liq	-18	207	0.8668^{20}		s ctc
1556	4-*tert*-Butyl-2,6-dimethyl-3,5-dinitroacetophenone	Musk ketone	$C_{14}H_{18}N_2O_5$	81-14-1	294.303	ye cry	135.5				vs chl
1557	2-*tert*-Butyl-4,6-dimethylphenol		$C_{12}H_{18}O$	1879-09-0	178.270		22.3	249	0.917^{80}	1.5183^{20}	i alk
1558	4-*tert*-Butyl-2,5-dimethylphenol		$C_{12}H_{18}O$	17696-37-6	178.270		71.2	264	0.939^{80}	1.5311^{20}	s alk
1559	4-*tert*-Butyl-2,6-dimethylphenol		$C_{12}H_{18}O$	879-97-0	178.270		82.4	248	0.916^{80}		s alk
1560	1-*tert*-Butyl-3,5-dimethyl-2,4,6-trinitrobenzene		$C_{12}H_{15}N_3O_6$	81-15-2	297.263	pl, nd (al)	110				i H_2O; sl EtOH; s eth, chl
1561	2-*tert*-Butyl-4,6-dinitrophenol		$C_{10}H_{12}N_2O_5$	1420-07-1	240.212	ye solid	126				

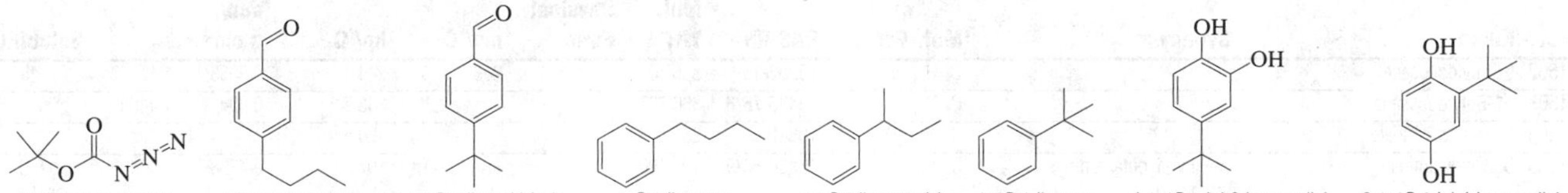

tert-Butyl azidoformate
4-Butylbenzaldehyde
4-*tert*-Butylbenzaldehyde
Butylbenzene
sec-Butylbenzene, (±)
tert-Butylbenzene
4-*tert*-Butyl-1,2-benzenediol
2-*tert*-Butyl-1,4-benzenediol

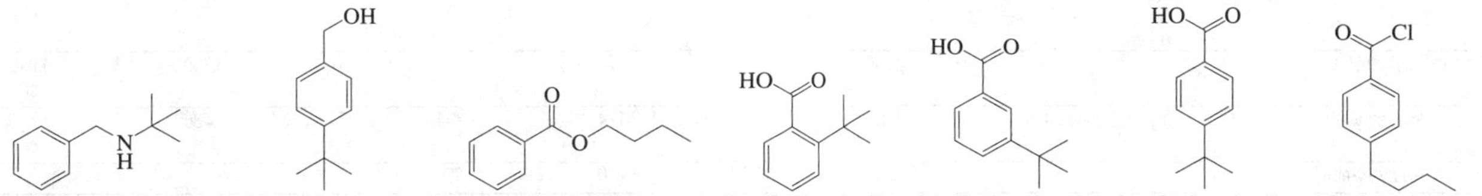

N-*tert*-Butylbenzenemethanamine
4-*tert*-Butylbenzenemethanol
Butyl benzoate
2-*tert*-Butylbenzoic acid
3-*tert*-Butylbenzoic acid
4-*tert*-Butylbenzoic acid
4-Butylbenzoyl chloride

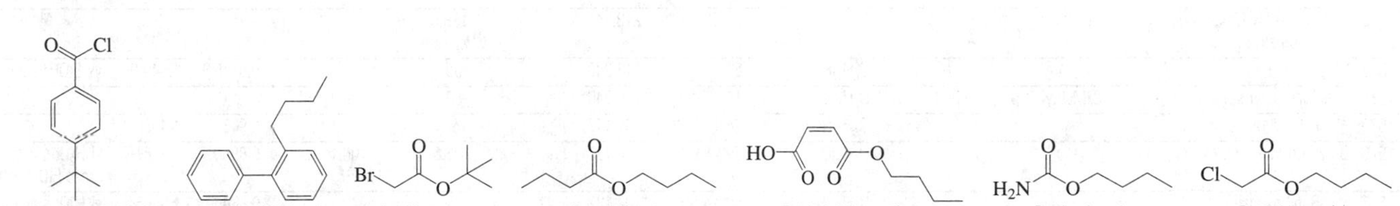

4-*tert*-Butylbenzoyl chloride
2-Butyl-1,1'-biphenyl
tert-Butyl bromoacetate
Butyl butanoate
Butyl *cis*-2-butenedioate
Butyl carbamate
Butyl chloroacetate

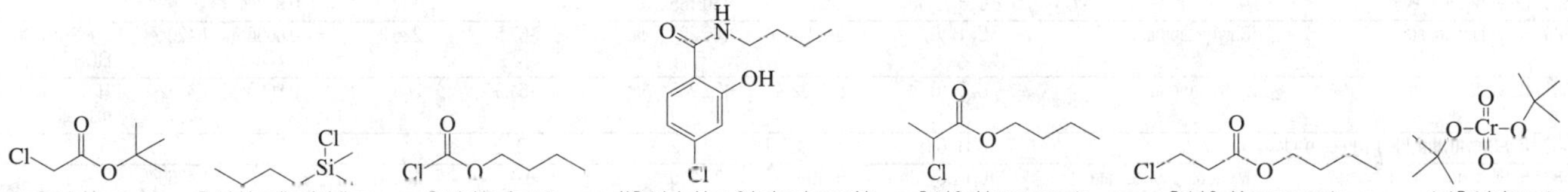

tert-Butyl chloroacetate
Butylchlorodimethylsilane
Butyl chloroformate
N-Butyl-4-chloro-2-hydroxybenzamide
Butyl 2-chloropropanoate
Butyl 3-chloropropanoate
tert-Butyl chromate

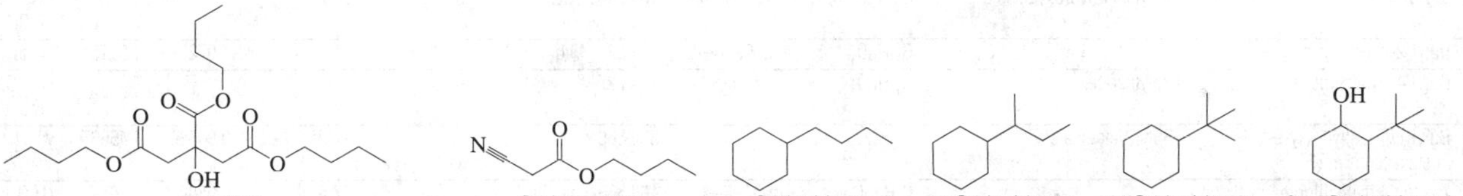

Butyl citrate
Butyl cyanoacetate
Butylcyclohexane
sec-Butylcyclohexane
tert-Butylcyclohexane
2-*tert*-Butylcyclohexanol

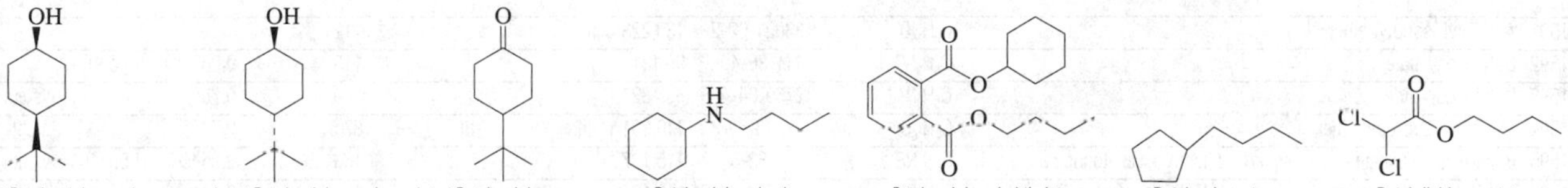

cis-4-*tert*-Butylcyclohexanol
trans-4-*tert*-Butylcyclohexanol
4-*tert*-Butylcyclohexanone
Butylcyclohexylamine
Butyl cyclohexyl phthalate
Butylcyclopentane
Butyl dichloroacetate

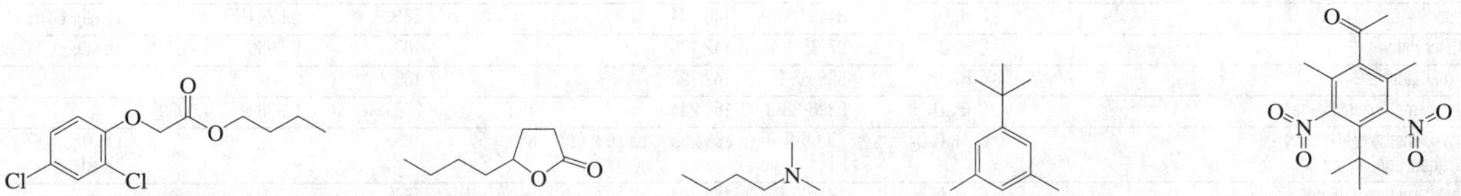

Butyl (2,4-dichlorophenoxy)acetate
5-Butyldihydro-2(3*H*)-furanone
Butyldimethylamine
1-*tert*-Butyl-3,5-dimethylbenzene
4-*tert*-Butyl-2,6-dimethyl-3,5-dinitroacetophenone

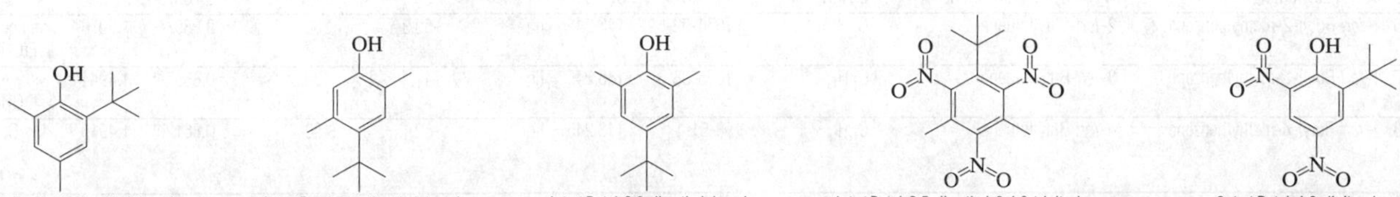

2-*tert*-Butyl-4,6-dimethylphenol
4-*tert*-Butyl-2,5-dimethylphenol
4-*tert*-Butyl-2,6-dimethylphenol
1-*tert*-Butyl-3,5-dimethyl-2,4,6-trinitrobenzene
2-*tert*-Butyl-4,6-dinitrophenol

No.	Name	Synonym	Mol. Form.	CAS RN	Mol. Wt.	Physical Form	mp/°C	bp/°C	den/ g cm^{-3}	n_D	Solubility
1562	5-Butyldocosane		$C_{26}H_{54}$	55282-16-1	366.707		208	244^{10}	0.8058^{20}	1.4503^{20}	
1563	11-Butyldocosane		$C_{26}H_{54}$	13475-76-8	366.707			242.5^{10}	0.8041^{20}	1.4499^{20}	
1564	Butyl dodecanoate		$C_{16}H_{32}O_2$	106-18-3	256.424			180^{18}			
1565	Butylethylamine	*N*-Ethyl-1-butanamine	$C_6H_{15}N$	13360-63-9	101.190			107.5	0.7398^{20}	1.4040^{20}	msc EtOH, eth, ace, bz
1566	1-*tert*-Butyl-4-ethylbenzene		$C_{12}H_{18}$	7364-19-4	162.271	liq	-38.4	211	0.8641^{20}		
1567	Butyl ethyl ether		$C_6H_{14}O$	628-81-9	102.174	liq	-124	92.3	0.7495^{20}	1.3818^{20}	i H_2O; msc EtOH, eth; vs ace
1568	*sec*-Butyl ethyl ether		$C_6H_{14}O$	2679-87-0	102.174			81	0.7503^{20}	1.3802^{20}	i H_2O; vs EtOH, eth
1569	*tert*-Butyl ethyl ether	Ethyl *tert*-butyl ether	$C_6H_{14}O$	637-92-3	102.174	liq	-94	72.6	0.736^{25}	1.3756^{20}	i H_2O; vs EtOH, eth
1570	2-*tert*-Butyl-4-ethylphenol		$C_{12}H_{18}O$	96-70-8	178.270		23	250			
1571	2-Butyl-2-ethyl-1,3-propanediol		$C_9H_{20}O_2$	115-84-4	160.254	wh cry	43.8	262	0.927^{50}	1.4587^{25}	sl H_2O, ace; s EtOH
1572	5-Butyl-5-ethyl-2,4,6(1*H*,3*H*,5*H*)-pyrimidinetrione	Butethal	$C_{10}H_{16}N_2O_3$	77-28-1	212.245		128.5				
1573	Butyl ethyl sulfide		$C_6H_{14}S$	638-46-0	118.240	liq	-95.1	144.3	0.8376^{20}	1.4492^{10}	vs EtOH; s chl
1574	*tert*-Butyl ethyl sulfide	2-Methyl-2-propanethiol	$C_6H_{14}S$	14290-92-7	118.240	liq	-88.9	120.4; 56^{109}			
1575	*N*-*tert*-Butylformamide		$C_5H_{11}NO$	2425-74-3	101.147	liq	16	202	0.903	1.4330^{20}	
1576	Butyl formate		$C_5H_{10}O_2$	592-84-7	102.132	liq	-91.5	106.1	0.8958^{20}	1.3887^{20}	sl H_2O; s ace; msc EtOH, eth
1577	*sec*-Butyl formate		$C_5H_{10}O_2$	589-40-2	102.132			97	0.8846^{20}	1.3865^{20}	sl H_2O; s ace; msc EtOH, eth
1578	*tert*-Butyl formate	1,1-Dimethylethyl formate	$C_5H_{10}O_2$	762-75-4	102.132	liq		82	0.872	1.3790^{20}	
1579	Butyl glycidyl ether		$C_7H_{14}O_2$	2426-08-6	130.185			169; 75^{26}	0.918^{20}		
1580	Butyl heptanoate	Butyl enanthate	$C_{11}H_{22}O_2$	5454-28-4	186.292	liq	-67.5	226.2	0.8638^{20}	1.4204^{20}	vs ace, bz, eth, EtOH
1581	Butyl hexanoate	Butyl caproate	$C_{10}H_{20}O_2$	626-82-4	172.265	liq	-64.3	208	0.8653^{20}	1.4152^{20}	i H_2O; s EtOH; msc eth
1582	*tert*-Butylhydrazine hydrochloride		$C_4H_{13}ClN_2$	7400-27-3	124.612		192.5				
1583	Butyl hydrogen succinate	Monobutyl succinate	$C_8H_{14}O_4$	5150-93-6	174.195		8.6	136.5^{3}	1.0732^{20}	1.4360^{20}	
1584	*tert*-Butyl hydroperoxide		$C_4H_{10}O_2$	75-91-2	90.121		6	dec 89; 36^{17}	0.8960^{20}	1.4015^{20}	s H_2O, EtOH, eth, ctc, chl
1585	*tert*-Butyl-4-hydroxyanisole	Butylated hydroxyanisole	$C_{11}H_{16}O_2$	25013-16-5	180.244	wax	51	268			i H_2O; s peth, EtOH
1586	Butyl 2-hydroxybenzoate		$C_{11}H_{14}O_3$	2052-14-4	194.227	liq	-5.9	271	1.0728^{20}	1.5115^{20}	sl ctc
1587	Butyl 4-hydroxybenzoate	Butylparaben	$C_{11}H_{14}O_3$	94-26-8	194.227		68.5				sl H_2O, ctc; s EtOH
1588	Butyl *cis*-12-hydroxy-9-octadecenoate, (*R*)	Butyl ricinoleate	$C_{22}H_{42}O_3$	151-13-3	354.566			275^{13}	0.9058^{22}	1.4566^{22}	vs eth
1589	*tert*-Butyl hypochlorite		C_4H_9ClO	507-40-4	108.566	ye liq		77.5	0.9583^{18}	1.403^{20}	i H_2O; vs eth, bz; s ace
1590	Butyl isobutyl ether		$C_8H_{18}O$	17071-47-5	130.228	liq		151	0.763^{15}	1.4077^{21}	vs ace, eth, EtOH
1591	*tert*-Butyl isobutyl ether		$C_8H_{18}O$	33021-02-2	130.228	liq		112.0			
1592	Butyl isocyanate		C_5H_9NO	111-36-4	99.131			115	0.880^{20}	1.4060^{20}	
1593	Butyl isocyanide		C_5H_9N	2769-64-4	83.132			120	0.78^{20}		vs eth, EtOH
1594	*tert*-Butyl isopropyl ether		$C_7H_{16}O$	17348-59-3	116.201	liq	-88	87.6	0.7365^{25}		s chl
1595	Butyl isothiocyanate	1-Isothiocyanatobutane	C_5H_9NS	592-82-5	115.197			168	0.9546^{20}	1.501^{20}	vs eth, EtOH
1596	*sec*-Butyl isothiocyanate, (±)	2-Isothiocyanatobutane, (±)	C_5H_9NS	116724-11-9	115.197			159.5	0.944^{12}		vs eth, EtOH
1597	*tert*-Butyl isothiocyanate	2-Isothiocyanato-2-methylpropane	C_5H_9NS	590-42-1	115.197		10.5	140	0.9187^{10}		
1598	Butyl lactate		$C_7H_{14}O_3$	34451-18-8	146.184			77^{10}	0.9744^{27}		vs eth, EtOH
1599	Butyl methacrylate		$C_8H_{14}O_2$	97-88-1	142.196			160	0.8936^{20}	1.4240^{20}	vs eth, EtOH
1600	*tert*-Butyl methacrylate		$C_8H_{14}O_2$	585-07-9	142.196			135.2			
1601	1-*tert*-Butyl-4-methoxybenzene		$C_{11}H_{16}O$	5396-38-3	164.244		19.0	238	0.9383^{20}	1.5039^{20}	
1602	1-*tert*-Butyl-2-methoxy-4-methyl-3,5-dinitrobenzene		$C_{12}H_{16}N_2O_5$	83-66-9	268.265	pa ye lf (al)	85	185^{16}			i H_2O; sl EtOH; s eth, chl
1603	2-*tert*-Butyl-4-methoxyphenol		$C_{11}H_{16}O_2$	121-00-6	180.244			184^{50}			
1604	3-*tert*-Butyl-4-methoxyphenol		$C_{11}H_{16}O_2$	88-32-4	180.244		65				
1605	Butylmethylamine	*N*-Methyl-1-butanamine	$C_5H_{13}N$	110-68-9	87.164			91	0.7637^{15}		
1606	1-*tert*-Butyl-2-methylbenzene	2-*tert*-Butyltoluene	$C_{11}H_{16}$	1074-92-6	148.245	liq	-50.3	200.4	0.8897^{20}	1.5076^{20}	vs ace, bz, eth, EtOH
1607	1-*tert*-Butyl-3-methylbenzene	3-*tert*-Butyltoluene	$C_{11}H_{16}$	1075-38-3	148.245	liq	-41.4	189.3	0.8657^{20}	1.4944^{20}	vs ace, bz, eth, EtOH
1608	1-*tert*-Butyl-4-methylbenzene	4-*tert*-Butyltoluene	$C_{11}H_{16}$	98-51-1	148.245	liq	-52	190	0.8612^{20}	1.4918^{20}	i H_2O; sl EtOH; vs eth, chl; s ace, bz
1609	Butyl 2-methylbutanoate	Butyl *o*-toluate	$C_9H_{18}O_2$	15706-73-7	158.238			179	0.8620^{20}	1.4135^{20}	
1610	Butyl 3-methylbutanoate	Butyl *p*-toluate	$C_9H_{18}O_2$	109-19-3	158.238					1.4058^{25}	

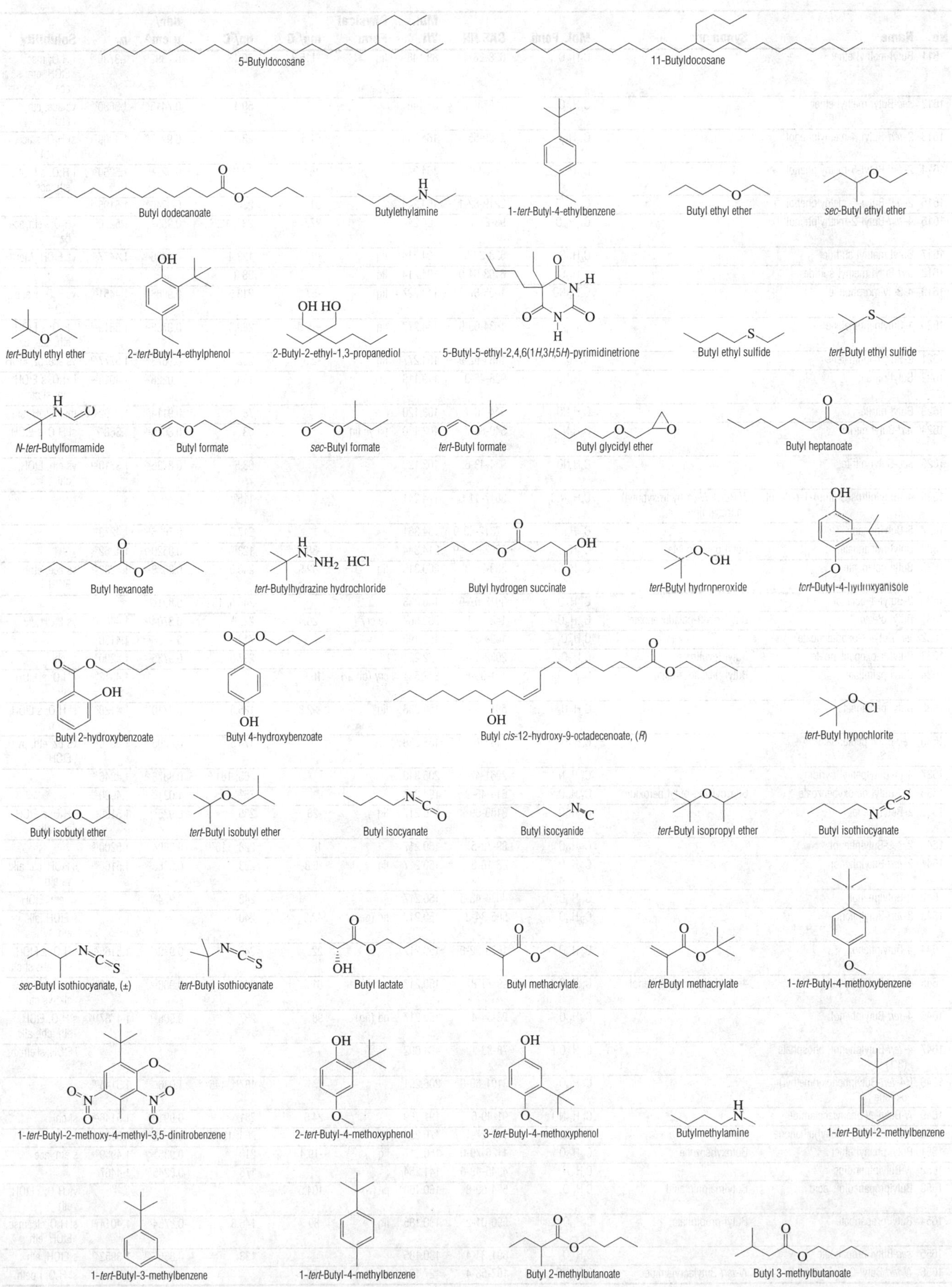

5-Butyldocosane
11-Butyldocosane
Butyl dodecanoate
Butylethylamine
1-tert-Butyl-4-ethylbenzene
Butyl ethyl ether
sec-Butyl ethyl ether
tert-Butyl ethyl ether
2-tert-Butyl-4-ethylphenol
2-Butyl-2-ethyl-1,3-propanediol
5-Butyl-5-ethyl-2,4,6(1H,3H,5H)-pyrimidinetrione
Butyl ethyl sulfide
tert-Butyl ethyl sulfide
N-tert-Butylformamide
Butyl formate
sec-Butyl formate
tert-Butyl formate
Butyl glycidyl ether
Butyl heptanoate
Butyl hexanoate
tert-Butylhydrazine hydrochloride
Butyl hydrogen succinate
tert-Butyl hydroperoxide
tert-Butyl-4-hydroxyanisole
Butyl 2-hydroxybenzoate
Butyl 4-hydroxybenzoate
Butyl cis-12-hydroxy-9-octadecenoate, (R)
tert-Butyl hypochlorite
Butyl isobutyl ether
tert-Butyl isobutyl ether
Butyl isocyanate
Butyl isocyanide
tert-Butyl isopropyl ether
Butyl isothiocyanate
sec-Butyl isothiocyanate, (±)
tert-Butyl isothiocyanate
Butyl lactate
Butyl methacrylate
tert-Butyl methacrylate
1-tert-Butyl-4-methoxybenzene
1-tert-Butyl-2-methoxy-4-methyl-3,5-dinitrobenzene
2-tert-Butyl-4-methoxyphenol
3-tert-Butyl-4-methoxyphenol
Butylmethylamine
1-tert-Butyl-2-methylbenzene
1-tert-Butyl-3-methylbenzene
1-tert-Butyl-4-methylbenzene
Butyl 2-methylbutanoate
Butyl 3-methylbutanoate

No.	Name	Synonym	Mol. Form.	CAS RN	Mol. Wt.	Physical Form	mp/°C	bp/°C	den/ g cm^{-3}	n_D	Solubility
1611	Butyl methyl ether		$C_5H_{12}O$	628-28-4	88.148	liq	-115.7	70.16	0.7392^{25}	1.3736^{20}	i H_2O; msc EtOH, eth; s ace
1612	*sec*-Butyl methyl ether		$C_5H_{12}O$	116783-23-4	88.148			59.1	0.7415^{20}	1.3680^{25}	vs ace, eth, EtOH
1613	2-*tert*-Butyl-4-methylphenol		$C_{11}H_{16}O$	2409-55-4	164.244		51.5	237	0.9247^{75}	1.4969^{75}	sl H_2O; s ace, bz, chl
1614	2-*tert*-Butyl-5-methylphenol		$C_{11}H_{16}O$	88-60-8	164.244		46.5	127^{11}	0.922^{80}	1.5250^{20}	i H_2O; s EtOH, eth, ace
1615	2-*tert*-Butyl-6-methylphenol		$C_{11}H_{16}O$	2219-82-1	164.244		31	230	0.9240^{80}	1.5195^{20}	
1616	4-*tert*-Butyl-2-methylphenol		$C_{11}H_{16}O$	98-27-1	164.244		27.5	237; 132^{20}	0.965^{20}	1.5230^{20}	i H_2O; s eth, ace, bz
1617	Butyl methyl sulfide		$C_5H_{12}S$	628-29-5	104.214	liq	-97.8	123.4	0.8426^{20}	1.4477^{20}	vs EtOH, MeOH
1618	*tert*-Butyl methyl sulfide		$C_5H_{12}S$	6163-64-0	104.214	liq		98.9			
1619	4-Butylmorpholine		$C_8H_{17}NO$	1005-67-0	143.227	liq	-57.1	213.5	0.9068^{20}	1.4451^{20}	vs H_2O, ace, bz, EtOH
1620	1-Butylnaphthalene		$C_{14}H_{16}$	1634-09-9	184.277	liq	-19.8	289.3	0.9738^{20}	1.5819^{20}	i H_2O; s EtOH, eth, ace, bz
1621	2-Butylnaphthalene		$C_{14}H_{16}$	1134-62-9	184.277	liq	-2.5	292	0.9673^{20}	1.5777^{20}	vs ace, bz, EtOH
1622	Butyl nitrate		$C_4H_9NO_3$	928-45-0	119.119			133	1.0228^{30}	1.4013^{23}	i H_2O; s EtOH, eth; sl ctc
1623	Butyl nitrite		$C_4H_9NO_2$	544-16-1	103.120			78	0.9114^{25}	1.3762^{20}	msc EtOH, eth
1624	*tert*-Butyl nitrite		$C_4H_9NO_2$	540-80-7	103.120	pa ye liq		63	0.8670^{20}	1.368^{20}	sl H_2O; s EtOH, eth, chl, CS_2
1625	*sec*-Butyl nitrite		$C_4H_9NO_2$	924-43-6	103.120			68.5	0.8726^{20}	1.3710^{20}	vs eth, EtOH, chl
1626	4-(Butylnitrosoamino)-1-butanol	*N*-Butyl-*N*-(4-hydroxybutyl) nitrosamine	$C_8H_{18}N_2O_2$	3817-11-6	174.241			$115^{0.01}$			
1627	5-Butylnonane		$C_{13}H_{28}$	17312-63-9	184.361			217.5	0.7635^{18}	1.4273^{18}	
1628	Butyl nonanoate	Butyl pelargonate	$C_{13}H_{26}O_2$	50623-57-9	214.344		-38	123^{20}	0.8520^{25}	1.4262^{25}	
1629	Butyl octanoate		$C_{12}H_{24}O_2$	589-75-3	200.318	liq	-42.9	240.5	0.8628^{20}	1.4232^{25}	vs ace, eth, EtOH
1630	2-Butyl-1-octanol		$C_{12}H_{26}O$	3913-02-8	186.333			246.5; 132^{15}	0.891^{20}		
1631	Butyl oleate	Butyl *cis*-9-octadecenoate	$C_{22}H_{42}O_2$	142-77-8	338.567	ye cry	-26.4	227^{15}	0.8704^{15}	1.4480^{25}	vs EtOH
1632	*tert*-Butyl 3-oxobutanoate		$C_8H_{14}O_3$	1694-31-1	158.195			71.5^{11}	0.9756^{20}	1.4180^{20}	
1633	Butyl 4-oxopentanoate	Butyl levulinate	$C_9H_{16}O_3$	2052-15-5	172.221			237.5	0.9735^{20}	1.4290^{20}	sl chl
1634	Butyl palmitate	Butyl hexadecanoate	$C_{20}H_{40}O_2$	111-06-8	312.531	cry (dil al)	16.9			1.4312^{50}	i H_2O; s EtOH, eth
1635	Butyl pentanoate		$C_9H_{18}O_2$	591-68-4	158.238	liq	-92.8	185.8	0.8710^{15}	1.4128^{20}	sl H_2O; s EtOH, eth
1636	*sec*-Butyl pentanoate		$C_9H_{18}O_2$	116836-32-9	158.238			174.5	0.8605^{20}	1.4070^{20}	vs bz, eth, py, EtOH
1637	4-(1-Butylpentyl)pyridine		$C_{14}H_{23}N$	2961-47-9	205.340			265; 181^{50}	0.8878^{25}	1.4846^{25}	
1638	*tert*-Butyl peroxybenzoate	Benzoyl *tert*-butyl peroxide	$C_{11}H_{14}O_3$	614-45-9	194.227			$75^{0.2}$	1.021^{25}	1.4990^{20}	
1639	2-Butylphenol		$C_{10}H_{14}O$	3180-09-4	150.217	liq	-20	235	0.975^{20}	1.5180^{25}	i H_2O; s EtOH, eth, alk
1640	2-*sec*-Butylphenol		$C_{10}H_{14}O$	89-72-5	150.217		16	228; 116^{21}	0.9804^{25}	1.5200^{25}	
1641	2-*tert*-Butylphenol		$C_{10}H_{14}O$	88-18-6	150.217	liq	-6.8	223	0.9783^{20}	1.5160^{20}	s EtOH, ctc, alk; vs eth
1642	3-Butylphenol		$C_{10}H_{14}O$	4074-43-5	150.217			248	0.974^{20}		vs eth, EtOH
1643	3-*tert*-Butylphenol		$C_{10}H_{14}O$	585-34-2	150.217	nd (peth)	42.3	240			s EtOH, alk; vs eth
1644	4-Butylphenol		$C_{10}H_{14}O$	1638-22-8	150.217		22	248	0.976^{22}	1.5165^{25}	i H_2O; s EtOH, eth, alk; sl ctc
1645	4-*sec*-Butylphenol	4-(1-Methylpropyl)phenol	$C_{10}H_{14}O$	99-71-8	150.217		61.5	241	0.986^{20}	1.5182^{21}	i H_2O; s EtOH, alk; vs eth
1646	4-*tert*-Butylphenol		$C_{10}H_{14}O$	98-54-4	150.217	nd (lig)	98	237	0.908^{80}	1.4787^{114}	s H_2O, EtOH, eth, chl, alk
1647	4-*tert*-Butylphenol, phosphate (3:1)		$C_{30}H_{39}O_4P$	78-33-1	494.602						i EtOH; sl eth, bz
1648	[(4-*tert*-Butylphenoxy)methyl] oxirane		$C_{13}H_{18}O_2$	3101-60-8	206.281			167^{14}, $145^{0.5}$	1.036^{25}	1.5145^{20}	
1649	*N*-Butyl-*N*-phenylacetamide		$C_{12}H_{17}NO$	91-49-6	191.269		24.5	281	0.9912^{20}	1.5146^{20}	sl chl
1650	1-(4-*tert*-Butylphenyl)ethanone		$C_{12}H_{16}O$	943-27-1	176.254		17.7	263; 137^{20}	0.9635^{20}	1.518^{15}	
1651	Butyl phenyl ether	Butoxybenzene	$C_{10}H_{14}O$	1126-79-0	150.217	liq	-19.4	210	0.9351^{20}	1.4969^{20}	s eth, ace
1652	*N*-Butylpiperidine		$C_9H_{19}N$	4945-48-6	141.254			176	0.8245^{20}	1.4467^{20}	
1653	Butylpropanedioic acid	Butylmalonic acid	$C_7H_{12}O_4$	534-59-8	160.168	pr (w)	104.5				vs H_2O; s EtOH, eth
1654	Butyl propanoate	Butyl propionate	$C_7H_{14}O_2$	590-01-2	130.185	liq	-89	146.8	0.8754^{20}	1.4014^{20}	sl H_2O, ctc; msc EtOH, eth
1655	*sec*-Butyl propanoate		$C_7H_{14}O_2$	591-34-4	130.185			133	0.8657^{20}	1.3952^{20}	s EtOH, eth
1656	*N*-*tert*-Butyl-2-propenamide	*N*-*tert*-Butylacrylamide	$C_7H_{13}NO$	107-58-4	127.184	cry (bz)	128				sl H_2O; i peth

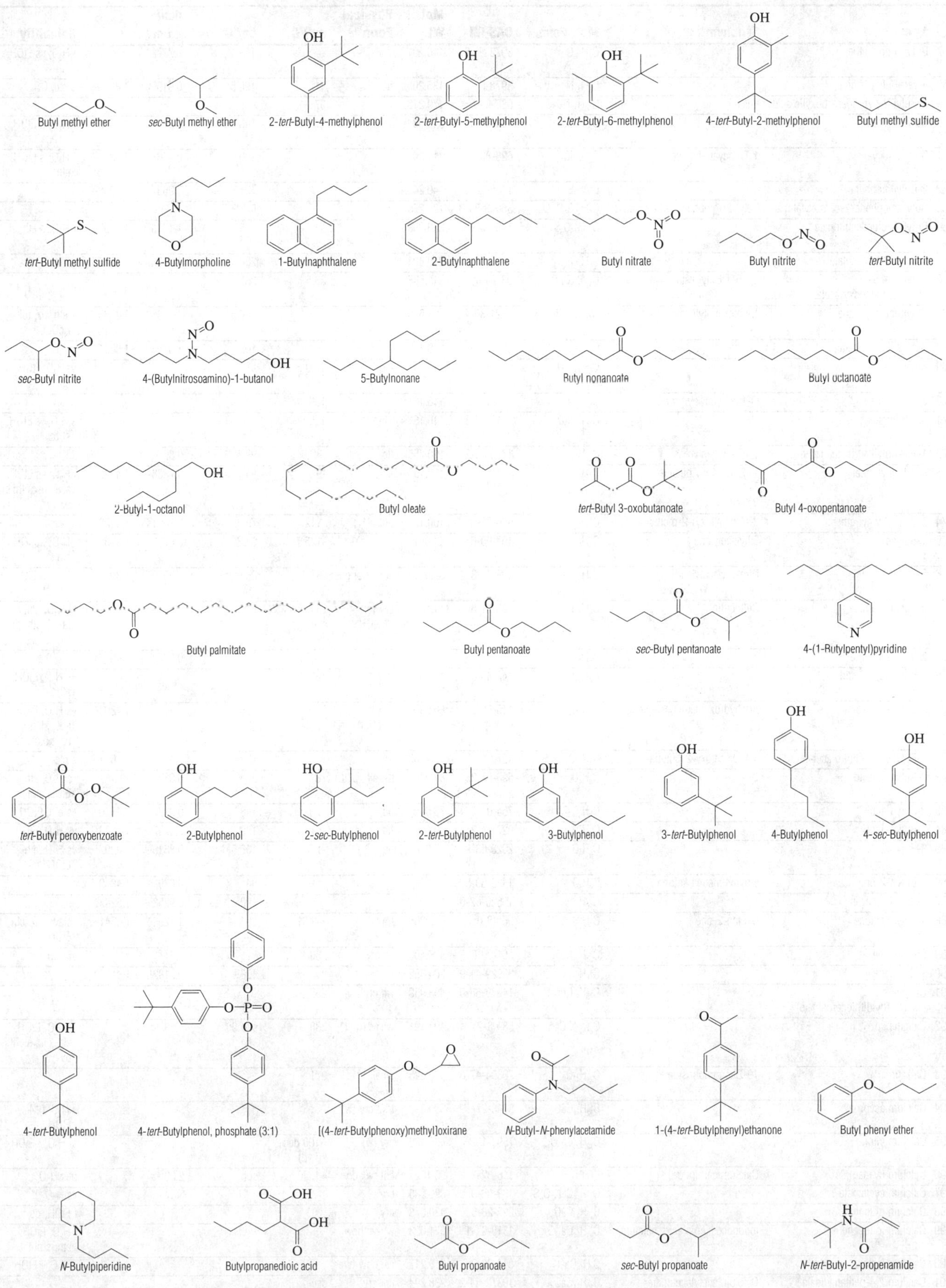

Butyl methyl ether
sec-Butyl methyl ether
2-*tert*-Butyl-4-methylphenol
2-*tert*-Butyl-5-methylphenol
2-*tert*-Butyl-6-methylphenol
4-*tert*-Butyl-2-methylphenol
Butyl methyl sulfide
tert-Butyl methyl sulfide
4-Butylmorpholine
1-Butylnaphthalene
2-Butylnaphthalene
Butyl nitrate
Butyl nitrite
tert-Butyl nitrite
sec-Butyl nitrite
4-(Butylnitrosoamino)-1-butanol
5-Butylnonane
Butyl nonanoate
Butyl octanoate
2-Butyl-1-octanol
Butyl oleate
tert-Butyl 3-oxobutanoate
Butyl 4-oxopentanoate
Butyl palmitate
Butyl pentanoate
sec-Butyl pentanoate
4-(1-Butylpentyl)pyridine
tert-Butyl peroxybenzoate
2-Butylphenol
2-*sec*-Butylphenol
2-*tert*-Butylphenol
3-Butylphenol
3-*tert*-Butylphenol
4-Butylphenol
4-*sec*-Butylphenol
4-*tert*-Butylphenol
4-*tert*-Butylphenol, phosphate (3:1)
[(4-*tert*-Butylphenoxy)methyl]oxirane
N-Butyl-*N*-phenylacetamide
1-(4-*tert*-Butylphenyl)ethanone
Butyl phenyl ether
N-Butylpiperidine
Butylpropanedioic acid
Butyl propanoate
sec-Butyl propanoate
N-*tert*-Butyl-2-propenamide

No.	Name	Synonym	Mol. Form.	CAS RN	Mol. Wt.	Physical Form	mp/°C	bp/°C	den/ g cm^{-3}	n_D	Solubility
1657	Butyl propyl ether		$C_7H_{16}O$	3073-92-5	116.201			118.1	0.7772^{0}		i H_2O; vs EtOH, eth
1658	4-*tert*-Butylpyridine		$C_9H_{13}N$	3978-81-2	135.206	liq	-41	196.5	0.915^{25}	1.4958^{20}	s ctc, CS_2
1659	5-Butyl-2-pyridinecarboxylic acid	Fusaric acid	$C_{10}H_{13}NO_2$	536-69-6	179.216		97				
1660	Butyl stearate		$C_{22}H_{44}O_2$	123-95-5	340.583		27	343	0.854^{25}	1.4328^{50}	i H_2O; s EtOH; vs ace
1661	Butyl thiocyanate	1-Thiocyanobutane	C_5H_9NS	628-83-1	115.197			186	0.9563^{15}	1.4360^{20}	i H_2O; s EtOH, eth
1662	2-Butylthiophene		$C_8H_{12}S$	1455-20-5	140.246			181.5	0.9537^{20}	1.5090^{20}	
1663	Butyl thiophene-2-carboxylate	Butyl 2-thiophenecarboxylate	$C_9H_{12}O_2S$	56053-84-0	184.255			58$^{0.15}$			
1664	Butyl 4-toluenesulfonate		$C_{11}H_{16}O_3S$	778-28-9	228.308			165^{6}	1.1319^{20}	1.5050^{20}	i H_2O; s eth; sl ctc
1665	Butyl trichloroacetate		$C_6H_9Cl_3O_2$	3657-07-6	219.493			204	1.2778^{20}	1.4525^{25}	s ctc
1666	Butyl (2,4,5-trichlorophenoxy)acetate	2,4,5-T Butyl ester	$C_{12}H_{13}Cl_3O_3$	93-79-8	311.588		28.5	337			
1667	Butyltrichlorosilane	Trichlorobutylsilane	$C_4H_9Cl_3Si$	7521-80-4	191.559			148.5	1.1606^{20}	1.4363^{20}	s eth, bz, tol, AcOEt
1668	Butyl trifluoroacetate		$C_6H_9F_3O_2$	367-64-6	170.129			102	1.0268^{22}	1.353^{22}	s chl
1669	Butylurea		$C_5H_{12}N_2O$	592-31-4	116.161	tab (w), nd (bz)	97.0				vs H_2O, EtOH; sl chl
1670	*sec*-Butylurea	(1-Methylpropyl)urea	$C_5H_{12}N_2O$	689-11-2	116.161	pr (w)	169				
1671	*tert*-Butylurea		$C_5H_{12}N_2O$	1118-12-3	116.161		176 dec				s H_2O; vs EtOH; sl bz
1672	1-*tert*-Butyl-4-vinylbenzene	*p-tert*-Butylstyrene	$C_{12}H_{16}$	1746-23-2	160.255	liq	-36.9	99^{14}	0.89^{20}		
1673	Butyl vinyl ether	1-(Ethenyloxy)butane	$C_6H_{12}O$	111-34-2	100.158	liq	-92	94	0.7888^{20}	1.4026^{20}	i H_2O; vs EtOH, ace; msc eth; s bz
1674	*tert*-Butyl vinyl ether	2-(Ethenyloxy)-2-methylpropane	$C_6H_{12}O$	926-02-3	100.158	liq	-112	75	0.7691^{20}	1.3922^{20}	
1675	1-Butyne	Ethylacetylene	C_4H_6	107-00-6	54.091	col gas	-125.7	8.08	0.6783^{0}	1.3962^{20}	i H_2O; s EtOH, eth
1676	2-Butyne	Dimethylacetylene	C_4H_6	503-17-3	54.091	vol liq or gas	-32.2	26.9	0.6910^{20}	1.3921^{20}	i H_2O; s EtOH, eth, ctc
1677	2-Butynediamide	Cellocidin	$C_4H_4N_2O_2$	543-21-5	112.087	cry (dil MeOH)	217 dec				sl H_2O, chl, EtOH, eth, gl HOAc
1678	2-Butynedinitrile		C_4N_2	1071-98-3	76.056		20.5	76.5	0.9708^{25}	1.4647^{25}	
1679	2-Butynedioic acid		$C_4H_2O_4$	142-45-0	114.057		183 dec				vs H_2O, EtOH, eth
1680	2-Butyne-1,4-diol	Bis(hydroxymethyl)acetylene	$C_4H_6O_2$	110-65-6	86.090	pl (bz, AcOEt)	50	238		1.4804^{20}	vs H_2O, EtOH, ace; sl eth; i bz, peth
1681	2-Butyne-1,4-diol diacetate	1,4-Diacetoxy-2-butyne	$C_8H_{10}O_4$	1573-17-7	170.163			122^{10}		1.4611^{20}	s ctc
1682	2-Butynoic acid		$C_4H_4O_2$	590-93-2	84.074	pl (eth, peth)	78	203	0.9641^{20}		vs H_2O, eth, EtOH, chl
1683	2-Butyn-1-ol		C_4H_6O	764-01-2	70.090	liq	-1.1	148	0.9370^{20}	1.4530^{20}	vs eth, EtOH
1684	3-Butyn-1-ol		C_4H_6O	927-74-2	70.090	liq	-63.6	129	0.9257^{20}	1.4409^{20}	vs H_2O, EtOH
1685	3-Butyn-2-ol		C_4H_6O	2028-63-9	70.090	liq	-1.5	106.5	0.8618^{20}	1.4207^{20}	vs H_2O, eth, EtOH
1686	3-Butyn-2-one	Ethynyl methyl ketone	C_4H_4O	1423-60-5	68.074			84	0.8793^{20}	1.4070^{20}	
1687	3-Butynylbenzene		$C_{10}H_{10}$	16520-62-0	130.186			190	0.9258^{20}	1.5208^{20}	
1688	γ-Butyrolactone	Oxolan-2-one	$C_4H_6O_2$	96-48-0	86.090	liq	-43.61	204	1.1296^{20}	1.4341^{20}	vs ace, bz, eth, EtOH
1689	Cacotheline		$C_{21}H_{21}N_3O_7$	561-20-6	427.408	ye cry	>300				sl H_2O
1690	γ-Cadinene		$C_{15}H_{24}$	39029-41-9	204.352			126^{12}	0.9182^{15}	1.3166^{20}	
1691	Cadmium bis(diethyldithiocarbamate)		$C_{10}H_{20}CdN_2S_4$	14239-68-0	408.950	wh cry	255				
1692	Caffeine		$C_8H_{10}N_4O_2$	58-08-2	194.191	wh nd (w+1), hex pr (sub)	238	sub 90	1.23^{19}		sl H_2O, EtOH; i eth, ctc; s chl, py
1693	Calactin	19-Oxogomphoside	$C_{29}H_{40}O_9$	20304-47-6	532.623	small pr (ace)	271				
1694	Calcium ascorbate		$C_{12}H_{14}CaO_{12}$	5743-27-1	390.310	tricl cry (w)					s H_2O; i MeOH, EtOH
1695	Calcium citrate		$C_{12}H_{10}Ca_3O_{14}$	7693-13-2	498.433	cry (w)	≈100 dec (hyd)				sl H_2O; i EtOH
1696	Calcium cyanamide	Calcium carbimide	$CCaN_2$	156-62-7	80.102	col hex cry	≈1340	sub	2.29		dec H_2O
1697	Calcium cyclamate		$C_{12}H_{24}CaN_2O_6S_2$	139-06-0	396.535	cry					vs H_2O
1698	Calcium gluconate		$C_{12}H_{22}CaO_{14}$	299-28-5	430.373	cry					i EtOH, os
1699	Calcium iodobehenate	Iododocosanoic acid, calcium salt	$C_{44}H_{84}CaI_2O_4$	1319-91-1	971.023	wh-ye pow					i H_2O, EtOH, eth; s chl
1700	Calcium lactate		$C_6H_{10}CaO_6$	814-80-2	218.217	wh pow (w)					s H_2O; i EtOH
1701	Calcium 2,4-pentanedioate	Calcium acetylacetonate	$C_{10}H_{14}CaO_4$	19372-44-2	238.294	col cry (MeOH)	dec				

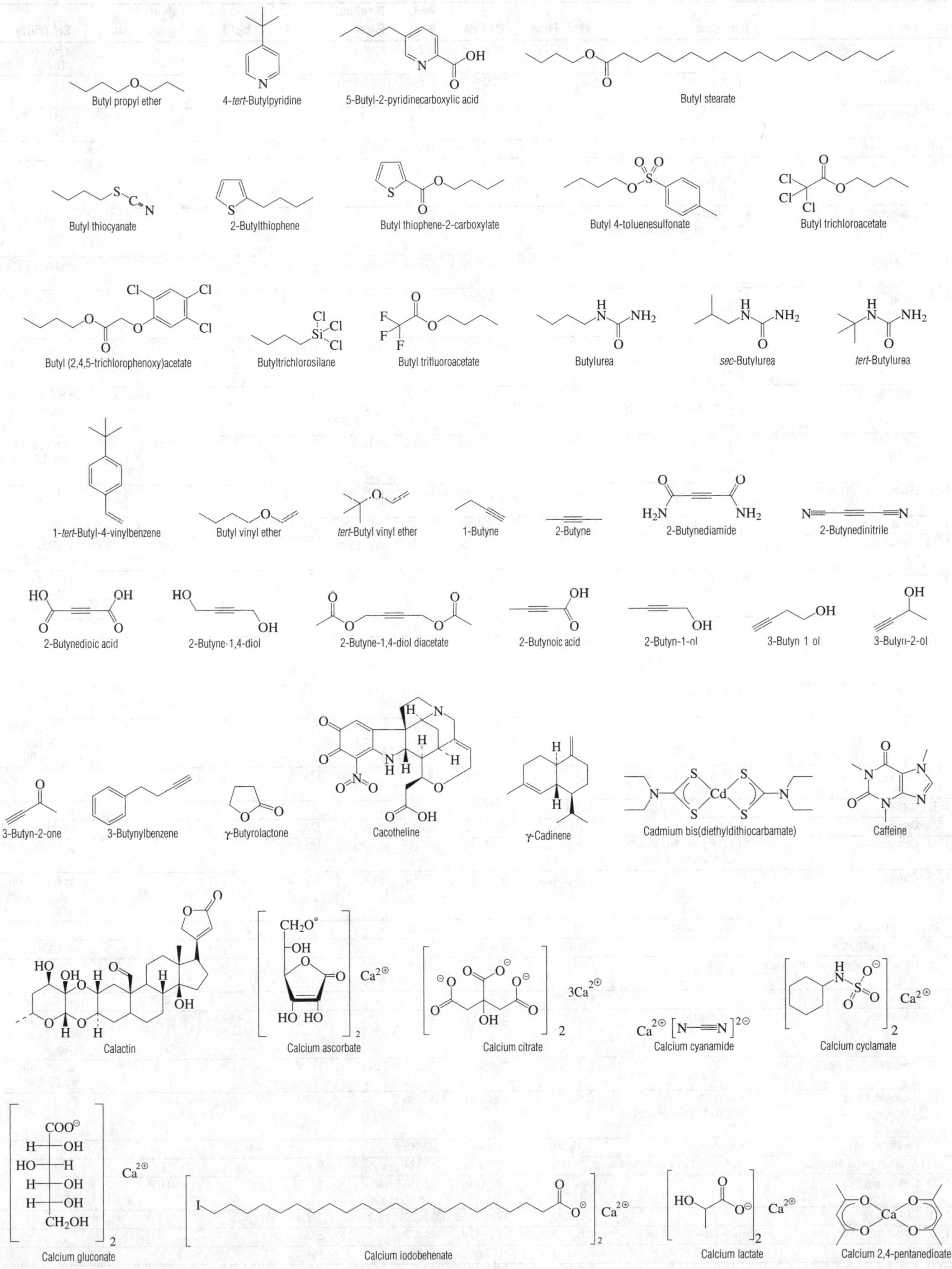

Butyl propyl ether
4-*tert*-Butylpyridine
5-Butyl-2-pyridinecarboxylic acid
Butyl stearate
Butyl thiocyanate
2-Butylthiophene
Butyl thiophene-2-carboxylate
Butyl 4-toluenesulfonate
Butyl trichloroacetate
Butyl (2,4,5-trichlorophenoxy)acetate
Butyltrichlorosilane
Butyl trifluoroacetate
Butylurea
sec-Butylurea
tert-Butylurea
1-*tert*-Butyl-4-vinylbenzene
Butyl vinyl ether
tert-Butyl vinyl ether
1-Butyne
2-Butyne
2-Butynediamide
2-Butynedinitrile
2-Butynedioic acid
2-Butyne-1,4-diol
2-Butyne-1,4-diol diacetate
2-Butynoic acid
2-Butyn-1-ol
3-Butyn-1-ol
3-Butyn-2-ol
3-Butyn-2-one
3-Butynylbenzene
γ-Butyrolactone
Cacotheline
γ-Cadinene
Cadmium bis(diethyldithiocarbamate)
Caffeine
Calactin
Calcium ascorbate
Calcium citrate
Calcium cyanamide
Calcium cyclamate
Calcium gluconate
Calcium iodobehenate
Calcium lactate
Calcium 2,4-pentanedioate

No.	Name	Synonym	Mol. Form.	CAS RN	Mol. Wt.	Physical Form	mp/°C	bp/°C	den/ g cm^{-3}	n_D	Solubility
1702	Calcium thioglycollate		$C_4H_6CaO_4S_2$	814-71-1	222.297	pr (w)	220 dec				s H_2O, chl; sl EtOH; i eth, bz
1703	Calotoxin	4'β-Hydroxy-19-oxogomphoside	$C_{29}H_{40}O_{10}$	20304-49-8	548.622	cry (EtOH)	268				
1704	Calotropin		$C_{29}H_{40}O_9$	1986-70-5	532.623	pl (EtOH)	221				s H_2O, EtOH; i eth
1705	Calusterone		$C_{21}H_{32}O_2$	17021-26-0	316.483	cry (ace)	157.5				
1706	Camphene, (+)	2,2-Dimethyl-3-methylenebicyclo[2.2.1]heptane, (1R)-	$C_{10}H_{16}$	5794-03-6	136.234	nd	52	161	0.8950^{50}	1.4570^{25}	vs eth
1707	Camphene, (-)	2,2-Dimethyl-3-methylenebicyclo[2.2.1]heptane, (1*S*)-	$C_{10}H_{16}$	5794-04-7	136.234		52	158	0.8446^{50}	1.4564^{54}	vs eth
1708	*d*-Camphocarboxylic acid		$C_{11}H_{16}O_3$	18530-30-8	196.243	pr (eth, 50% al)	127.5				vs bz, eth, EtOH
1709	Camphor, (±)	1,7,7-Trimethylbicyclo[2.2.1]heptan-2-one, (±)	$C_{10}H_{16}O$	21368-68-3	152.233	wh rhom cry (EtOH)	178.3	sub			i H_2O; vs EtOH, eth; s ace, bz, ctc
1710	Camphor, (+)	1,7,7-Trimethylbicyclo[2.2.1]heptan-2-one, (1*R*)	$C_{10}H_{16}O$	464-49-3	152.233	pl	178.8	207.4	0.990^{25}	1.5462	i H_2O; vs EtOH, eth; s ace, bz
1711	Camphor, (-)	1,7,7-Trimethylbicyclo[2.2.1]heptan-2-one, (1*S*)	$C_{10}H_{16}O$	464-48-2	152.233		178.6		0.9853^{18}		i H_2O; vs EtOH, eth, HOAc; s ace, bz
1712	Camphoric acid, (±)	1,2,2-Trimethyl-1,3-cyclopentanedicarboxylic acid, (1*RS*, 3*SR*)	$C_{10}H_{16}O_4$	5394-83-2	200.232	pr, lf	202		1.186		sl H_2O; s chl, eth, EtOH
1713	*d*-Camphorsulfonic acid		$C_{10}H_{16}O_4S$	3144-16-9	232.297	pr (HOAc)	195 dec				vs H_2O; i eth; sl HOAc
1714	Canadine, (±)	*DL*-Tetrahydroberberine	$C_{20}H_{21}NO_4$	29074-38-2	339.386	mcl nd (al)	134				vs EtOH, chl
1715	Cannabidiol		$C_{21}H_{30}O_2$	13956-29-1	314.462	rods (peth)	67	188^{2}	1.040^{40}	1.5404^{20}	i H_2O; s EtOH, eth, bz, chl
1716	Cannabinol	6,6,9-Trimethyl-3-pentyl-6*H*-dibenzo[b,d]pyran-1-ol	$C_{21}H_{26}O_2$	521-35-7	310.430	pl, lf (peth)	77	$185^{0.05}$			i H_2O; s EtOH, eth, ace, bz, peth, alk
1717	Canrenone		$C_{22}H_{28}O_3$	976-71-6	340.455	cry (AcOEt)	150				
1718	Cantharidin		$C_{10}H_{12}O_4$	56-25-7	196.200	orth pl	218	sub 84			i H_2O; sl EtOH, eth, ace, bz; s HOAc
1719	Caprolactam	6-Hexanelactam	$C_6H_{11}NO$	105-60-2	113.157	lf (lig)	69.3	270			vs H_2O, bz, EtOH, chl
1720	Capsaicin		$C_{18}H_{27}NO_3$	404-86-4	305.412	mcl pl or sc (peth)	65	$215^{0.01}$			i H_2O; vs EtOH; s eth, bz, peth; sl con HCl
1721	Capsanthin	3,3'-Dihydroxy-β,κ-caroten-6'-one, (3*R*,3'*S*,5'*R*)	$C_{40}H_{56}O_3$	465-42-9	584.871		176				
1722	Captafol		$C_{10}H_9Cl_4NO_2S$	2425-06-1	349.061	cry	161				
1723	Captan		$C_9H_8Cl_3NO_2S$	133-06-2	300.590	cry (CCl_4)	172.5		1.74^{25}		vs chl
1724	Captopril	1-(3-Mercapto-2-methyl-1-oxypropyl)proline	$C_9H_{15}NO_3S$	62571-86-2	217.285	cry (AcOEt)	105				s H_2O, EtOH, chl
1725	Carbachol		$C_6H_{15}ClN_2O_2$	51-83-2	182.648		210 dec				vs H_2O, MeOH; sl EtOH; i eth, chl
1726	Carbamic chloride	Carbamyl chloride	CH_2ClNO	463-72-9	79.486			dec 62			
1727	Carbamodithioic acid		CH_3NS_2	594-07-0	93.172						vs EtOH, eth
1728	Carbamoyl dihydrogen phosphate		CH_4NO_5P	590-55-6	141.021	unstab in soln					
1729	Carbaryl		$C_{12}H_{11}NO_2$	63-25-2	201.221		145		1.228^{25}		vs ace, DMF
1730	Carbazole	Dibenzopyrolle	$C_{12}H_9N$	86-74-8	167.206	pl or lf	246.3	354.69			i H_2O; sl EtOH, eth, bz, chl; s ace
1731	9*H*-Carbazole-9-acetic acid		$C_{14}H_{11}NO_2$	524-80-1	225.243	lf (AcOEt)	215				vs eth, EtOH, chl, HOAc
1732	Carbendazim	Carbamic acid, 1*H*-benzimidazol-2-yl-, methyl ester	$C_9H_9N_3O_2$	10605-21-7	191.186		300 dec		1.45		
1733	Carbetapentane	Pentoxyverine	$C_{20}H_{31}NO_3$	77-23-6	333.465			$165^{0.01}$			
1734	*N*-Carbethoxyphthalimide	*N*-(Ethoxycarbonyl)phthalimide	$C_{11}H_9NO_4$	22509-74-6	219.194		91				
1735	Carbic anhydride		$C_9H_8O_3$	129-64-6	164.158	orth cry (peth)	164.5		1.417^{25}		vs ace, bz, EtOH, chl
1736	Carbimazole		$C_7H_{10}N_2O_2S$	22232-54-8	186.231	cry, pow	123.5				vs ace, chl
1737	Carbobenzoxyhydrazine	Benzyl carbazate	$C_8H_{10}N_2O_2$	5331-43-1	166.177		69.5				
1738	Carbofuran		$C_{12}H_{15}NO_3$	1563-66-2	221.252		151		1.18		
1739	Carboimidic difluoride		CHF_2N	2712-98-3	65.023	gas	-90	-13 dec			

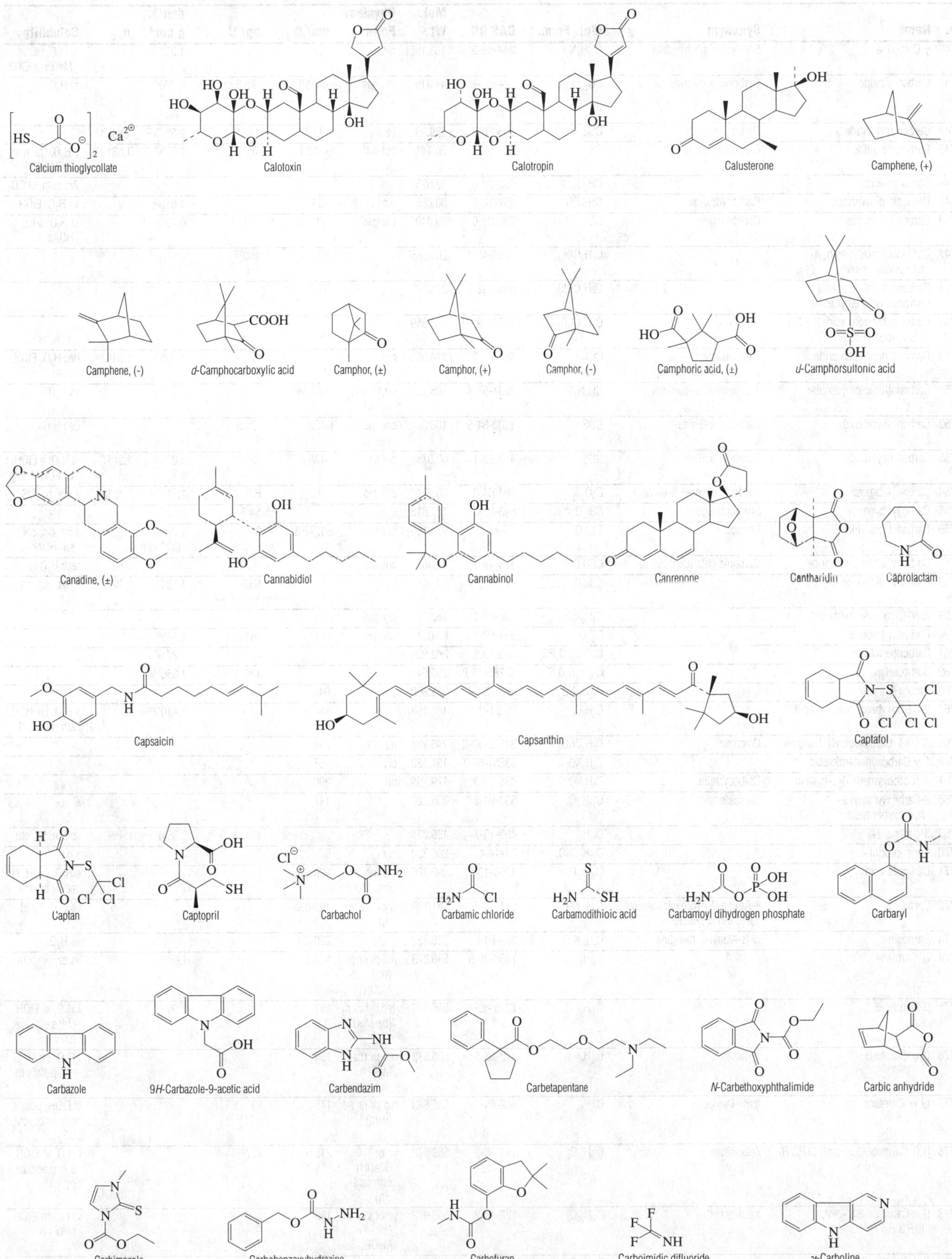
Calcium thioglycollate
Calotoxin
Calotropin
Calusterone
Camphene, (+)
Camphene, (-)
d-Camphocarboxylic acid
Camphor, (±)
Camphor, (+)
Camphor, (-)
Camphoric acid, (±)
d-Camphorsulfonic acid
Canadine, (±)
Cannabidiol
Cannabinol
Canrenone
Cantharidin
Caprolactam
Capsaicin
Capsanthin
Captafol
Captan
Captopril
Carbachol
Carbamic chloride
Carbamodithioic acid
Carbamoyl dihydrogen phosphate
Carbaryl
Carbazole
9H-Carbazole-9-acetic acid
Carbendazim
Carbetapentane
N-Carbethoxyphthalimide
Carbic anhydride
Carbimazole
Carbobenzoxyhydrazine
Carbofuran
Carboimidic difluoride
γ-Carboline

No.	Name	Synonym	Mol. Form.	CAS RN	Mol. Wt.	Physical Form	mp/°C	bp/°C	den/ g cm^{-3}	n_D	Solubility
1740	γ-Carboline	5*H*-Pyrido[4,3-b]indole	$C_{11}H_8N_2$	244-69-9	168.195	nd	225		1.352		sl H_2O, bz; vs MeOH; s EtOH
1741	Carbon dioxide	Carbonic anhydride	CO_2	124-38-9	44.010	col gas	-56.56 tp	-78.5 sp	0.720^{25} (p>1 atm)		sl H_2O
1742	Carbon diselenide	Carbon selenide	CSe_2	506-80-9	169.93	ye liq	-43.7	125.5	2.6823^{20}	1.8454^{20}	
1743	Carbon disulfide	Carbon bisulfide	CS_2	75-15-0	76.141	col liq	-112.1	46	1.2632^{20}	1.6319^{20}	s H_2O, chl; msc EtOH, eth
1744	Carbonic acid		CH_2O_3	463-79-6	62.025						Aq. soln. of CO_2
1745	Carbonic dihydrazide	Carbohydrazide	CH_6N_4O	497-18-7	90.085	nd (dil al)	154		1.616^{20}		vs H_2O, EtOH
1746	Carbon monoxide	Carbon oxide	CO	630-08-0	28.010	col gas	-205.02	-191.5	0.7909^{-19}		sl H_2O; s bz, HOAc
1747	Carbonochloridic acid, 4-nitrophenyl ester		$C_7H_4ClNO_4$	7693-46-1	201.565		80	160^{19}			
1748	Carbonochloridic acid, (4-nitrophenyl)methyl ester		$C_8H_6ClNO_4$	4457-32-3	215.592		32.8				
1749	Carbonochloridic acid, 2,2,2-trichloroethyl ester		$C_3H_2Cl_4O_2$	17341-93-4	211.859			63^{11}			
1750	Carbonothioic dichloride	Thiophosgene	CCl_2S	463-71-8	114.982	red liq		73	1.508^{15}	1.5442^{20}	dec H_2O, EtOH; s eth
1751	Carbonothioic dihydrazide	1,3-Diamino-2-thiourea	CH_6N_4S	2231-57-4	106.151	nd, pl (w) nd, pl (w)	170 dec				vs H_2O
1752	Carbon oxyselenide	Carbonyl selenide	COSe	1603-84-5	106.97	col gas; unstab	-122	-21.5			dec H_2O
1753	Carbon oxysulfide	Carbonyl sulfide	COS	463-58-1	60.075	col gas	-138.8	-50	1.028^{17}	1.24^{-87}	sl H_2O; s EtOH; vs KOH
1754	Carbon suboxide	1,2-Propadiene-1,3-dione	C_3O_2	504-64-3	68.031	col gas	-107	6.8	1.114^{0}	1.4538^{0}	s eth, bz, CS_2
1755	Carbonyl bromide	Bromophosgene	CBr_2O	593-95-3	187.818			64.5	2.52^{15}		
1756	Carbonyl chloride	Phosgene	CCl_2O	75-44-5	98.916	col gas	-127.78	8	1.3719^{25} (p>1 atm		s bz, ctc, chl, tol, HOAc
1757	Carbonyl chloride fluoride	Carbonic chloride fluoride	CClFO	353-49-1	82.461	col gas	-148	-47.2			reac H_2O
1758	Carbonyl dicyanide		C_3N_2O	1115-12-4	80.044	liq	-36	65.5	1.124^{20}	1.3919^{20}	s eth, ace, ctc, chl
1759	*N,N'*-Carbonyldiimidazole		$C_7H_6N_4O$	530-62-1	162.149	cry (bz)	119				
1760	Carbonyl fluoride		CF_2O	353-50-4	66.007	col gas	-111.2	-84.5	1.139^{25}		
1761	Carbophenothion		$C_{11}H_{16}ClO_2PS_3$	786-19-6	342.866			$82^{0.01}$	1.271^{20}		
1762	Carbosulfan		$C_{20}H_{32}N_2O_3S$	55285-14-8	380.544			126	1.056^{20}		
1763	Carboxin		$C_{12}H_{13}NO_2S$	5234-68-4	235.302		94				
1764	2-Carboxybenzeneacetic acid		$C_9H_8O_4$	89-51-0	180.158		184.5		1.4100^{20}		s H_2O, EtOH; sl eth; i bz, chl
1765	*N*-(*D*-1-Carboxyethyl)-*L*-arginine	Octopine	$C_9H_{18}N_4O_4$	34522-32-2	246.264	nd (w)	281				
1766	*L*-γ-Carboxyglutamic acid		$C_6H_9NO_6$	53861-57-7	191.138	cry	167				
1767	*S*-(Carboxymethyl)-*L*-cysteine	Carbocysteine	$C_5H_9NO_4S$	638-23-3	179.195	nd	206				
1768	2-Carboxyphenyl 2-hydroxybenzoate	Salsalate	$C_{14}H_{10}O_5$	552-94-3	258.226		147				sl ace
1769	3-Carene, (+)		$C_{10}H_{16}$	498-15-7	136.234			171; 123^{200}	0.8549^{30}	1.469^{3}	vs ace, bz, eth
1770	Carisoprodol		$C_{12}H_{24}N_2O_4$	78-44-4	260.330	cry	92				s os
1771	Carminic acid		$C_{22}H_{20}O_{13}$	1260-17-9	492.386	red mcl pr (aq, MeOH)	136 dec				s H_2O, EtOH; sl eth; i bz, chl
1772	Carnitine	4-Amino-3-hydroxybutanoic acid trimethylbetaine	$C_7H_{15}NO_3$	541-15-1	161.199	cry (al-ace), hyg	197 dec				vs H_2O, EtOH
1773	Carnosine	*N*-β-Alanyl-*L*-histidine	$C_9H_{14}N_4O_3$	305-84-0	226.232		260				vs H_2O
1774	α-Carotene		$C_{40}H_{56}$	7488-99-5	536.873	red pl or pr (peth, bz-MeOH)	187.5		1.00^{20}		vs bz, eth, chl
1775	β-Carotene		$C_{40}H_{56}$	7235-40-7	536.873	red br hex pr (bz-MeOH)	183		1.00^{20}		i H_2O; sl EtOH, chl; s eth, ace, bz
1776	β,ψ-Carotene	γ-Carotene	$C_{40}H_{56}$	472-93-5	536.873	red pr (bz-MeOH), viol pr (eth)	153				i H_2O, EtOH; sl eth, peth; s bz, chl
1777	ψ,ψ-Carotene	*trans*-Lycopene	$C_{40}H_{56}$	502-65-8	536.873	red pr or nd (peth)	175				sl EtOH, peth; s eth; vs bz, chl, CS_2
1778	β,β-Carotene-3,3'-diol, (3*R*,3'*R*)	Zeaxanthin	$C_{40}H_{56}O_2$	144-68-3	568.872	ye pr (MeOH) orth (chl-eth)	215.5	$227^{0.06}$			i H_2O; sl EtOH; s eth, ace, bz, py, chl
1779	β,ε-Carotene-3,3'-diol, (3*R*,3'*R*,6'*R*)	Xanthophyll	$C_{40}H_{56}O_2$	127-40-2	568.872	ye or viol pr (eth-MeOH)	196				vs bz, eth, EtOH, peth
1780	β,β-Caroten-3-ol, (3*R*)	Cryptoxanthin	$C_{40}H_{56}O$	472-70-8	552.872	garnet red pr (bz-MeOH)	160				vs bz, chl

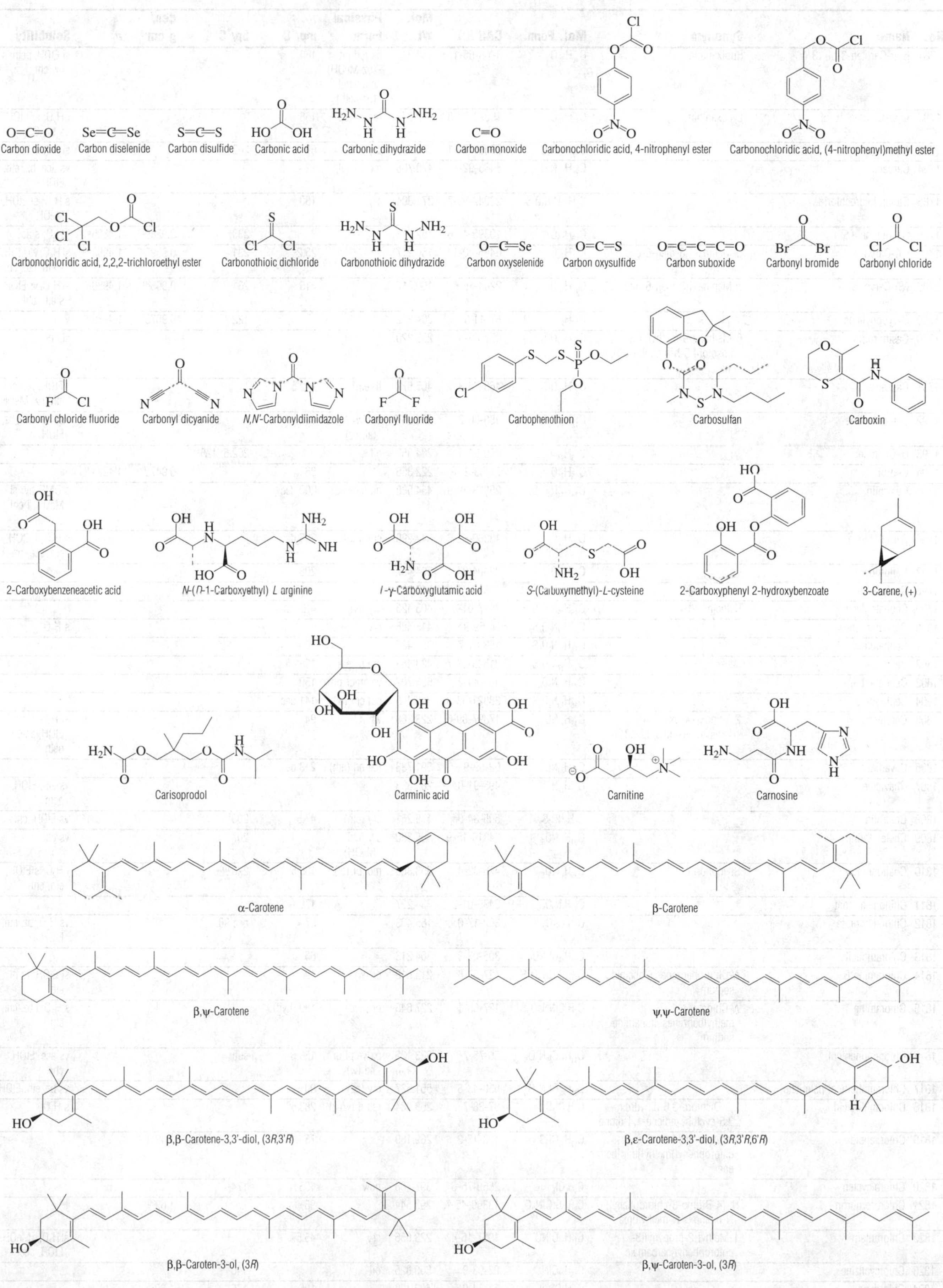

O=C=O
Carbon dioxide
Se=C=Se
Carbon diselenide
S=C=S
Carbon disulfide
Carbonic acid
Carbonic dihydrazide
C≡O
Carbon monoxide
Carbonochloridic acid, 4-nitrophenyl ester
Carbonochloridic acid, (4-nitrophenyl)methyl ester
Carbonochloridic acid, 2,2,2-trichloroethyl ester
Carbonothioic dichloride
Carbonothioic dihydrazide
O=C=Se
Carbon oxyselenide
O=C=S
Carbon oxysulfide
O=C=C=C=O
Carbon suboxide
Carbonyl bromide
Carbonyl chloride
Carbonyl chloride fluoride
Carbonyl dicyanide
N,N'-Carbonyldiimidazole
Carbonyl fluoride
Carbophenothion
Carbosulfan
Carboxin
2-Carboxybenzeneacetic acid
N-(D-1-Carboxyethyl) L-arginine
L-γ-Carboxyglutamic acid
S-(Carboxymethyl)-L-cysteine
2-Carboxyphenyl 2-hydroxybenzoate
3-Carene, (+)
Carisoprodol
Carminic acid
Carnitine
Carnosine
α-Carotene
β-Carotene
β,ψ-Carotene
ψ,ψ-Carotene
β,β-Carotene-3,3'-diol, (3R,3'R)
β,ε-Carotene-3,3'-diol, (3R,3'R,6'R)
β,β-Caroten-3-ol, (3R)
β,ψ-Caroten-3-ol, (3R)

No.	Name	Synonym	Mol. Form.	CAS RN	Mol. Wt.	Physical Form	mp/°C	bp/°C	den/ g cm^{-3}	n_D	Solubility
1781	β,ψ-Caroten-3-ol, (3*R*)	Rubixanthin	$C_{40}H_{56}O$	3763-55-1	552.872	dk red nd (bz-MeOH) oran-red (bz-peth)	160				sl EtOH, peth; s bz, chl
1782	ψ,ψ-Caroten-16-ol	Lycoxanthin	$C_{40}H_{56}O$	19891-74-8	552.872	red pl (bz-MeOH)	168				i H_2O; sl EtOH; s bz, CS_2
1783	Caroverine		$C_{22}H_{27}N_3O_2$	23465-76-1	365.468	cry	69	$202^{0.01}$			sl i-PrOH
1784	Carpaine		$C_{28}H_{50}N_2O_4$	3463-92-1	478.708	mcl pr (al, ace)	121				vs ace, bz, eth, EtOH
1785	Cartap hydrochloride		$C_7H_{16}ClN_3O_2S_2$	22042-59-7	273.804	cry	180				s H_2O; sl EtOH, MeOH
1786	Carvenone, (*S*)		$C_{10}H_{16}O$	10395-45-6	152.233			233	0.9289^{20}	1.4805^{20}	i H_2O; s ace
1787	(*R*)-Carvone	*p*-Mentha-1,8-dien-6-one, (*R*)	$C_{10}H_{14}O$	6485-40-1	150.217		25.2	231	0.9593^{20}	1.4988^{20}	sl H_2O; vs EtOH; s eth, ctc, chl
1788	(*S*)-Carvone	*p*-Mentha-1,8-dien-6-one, (*S*)	$C_{10}H_{14}O$	2244-16-8	150.217		<15	231	0.965^{20}	1.4989^{20}	sl H_2O; vs EtOH; s eth, chl
1789	Caryophyllene		$C_{15}H_{24}$	87-44-5	204.352			$122^{13.5}$	0.9075^{20}	1.4986^{20}	vs bz
1790	Casimiroin	6-Methoxy-9-methyl-1,3-dioxolo[4,5-h]quinolin-8(9*H*)-one	$C_{12}H_{11}NO_4$	477-89-4	233.220						sl chl
1791	Cassaine		$C_{24}H_{39}NO_4$	468-76-8	405.572	fl (eth)	142.5				s EtOH, ace, chl, eth, bz, MeOH
1792	Caulophylline		$C_{12}H_{16}N_2O$	486-86-2	204.267	cry (w+2), nd (al, bz)	137				vs H_2O, ace, bz, EtOH
1793	α-Cedrene		$C_{15}H_{24}$	469-61-4	204.352	oil		262.5; 125^{12}			
1794	Cedrol		$C_{15}H_{26}O$	77-53-2	222.366		86		0.9479^{90}	1.4824^{90}	
1795	Cefazolin		$C_{14}H_{14}N_8O_4S_3$	25953-19-9	454.508	nd (ace aq)	200 dec				s DMF, py; sl MeOH; i chl, bz, eth
1796	β-Cellobiose		$C_{12}H_{22}O_{11}$	13360-52-6	342.296	cry (dil al)	225 dec				s H_2O; i EtOH, eth, ace, bz
1797	Cellotriose		$C_{18}H_{32}O_{16}$	33404-34-1	504.437		208				
1798	Cephalexin		$C_{16}H_{17}N_3O_4S$	15686-71-2	347.389	cry					
1799	Cephaloglycin	Kafocin	$C_{18}H_{19}N_3O_6S$	3577-01-3	405.425	cry (w)	≈220 dec				
1800	Cephaloridine		$C_{19}H_{17}N_3O_4S_2$	50-59-9	415.486	cry					s H_2O
1801	Cephalothin		$C_{16}H_{16}N_2O_6S_2$	153-61-7	396.437		160				
1802	Cephapirin		$C_{17}H_{17}N_3O_6S_2$	21593-23-7	423.463	cry (ace aq)	155				
1803	Cepharanthine		$C_{37}H_{38}N_2O_6$	481-49-2	606.707	ye amor pow	150				
1804	Cephradine		$C_{16}H_{19}N_3O_4S$	38821-53-3	349.405	col cry (w)	141 dec				
1805	Cerulenin	2,3-Epoxy-4-oxo-7,10-dodecadienamide, (2*R*,3*S*)-	$C_{12}H_{17}NO_3$	17397-89-6	223.268	wh nd	94				sl H_2O; s bz, EtOH, ace; i peth
1806	Cevadine		$C_{32}H_{49}NO_9$	62-59-9	591.733	flat nd (eth)	213 dec				
1807	Chavicine		$C_{17}H_{19}NO_3$	495-91-0	285.338						vs eth, EtOH, peth
1808	Cheirolin		$C_5H_9NO_2S_2$	505-34-0	179.261	cry (eth)	47.5	200^3			vs EtOH, chl
1809	Chelerythrine		$C_{21}H_{19}NO_5$	34316-15-9	365.380	cry (chl-MeOH)		207			vs chl
1810	Chelidonine	Stylophorine	$C_{20}H_{19}NO_5$	476-32-4	353.369	mcl pr (al)	135.5	$220^{0.002}$			i H_2O; s EtOH, eth, chl
1811	Chinomethionat		$C_{10}H_6N_2OS_2$	2439-01-2	234.297		170				
1812	Chloral hydrate		$C_2H_3Cl_3O_2$	302-17-0	165.403		57	dec 96	1.9081^{20}		vs H_2O, bz, eth, EtOH
1813	Chlorambucil		$C_{14}H_{19}Cl_2NO_2$	305-03-3	304.213		65				
1814	Chloramine B	*N*-Chlorobenzenesulfonamide sodium	$C_6H_5ClNNaO_2S$	127-52-6	213.618	pr (w)	190				sl EtOH; i chl, eth
1815	Chloramine T	*N*-Chloro-4-methylbenzenesulfonamide sodium	$C_7H_7ClNNaO_2S$	127-65-1	227.645	pr (hyd)	180 (hyd)				s H_2O; i bz, chl, eth
1816	Chloramphenicol		$C_{11}H_{12}Cl_2N_2O_5$	56-75-7	323.129	pa ye pl or nd (w)	150.5	sub			vs ace, EtOH, chl
1817	Chloramphenicol palmitate		$C_{27}H_{42}Cl_2N_2O_6$	530-43-8	561.537	cry (bz)	90				vs bz, eth, EtOH
1818	Chloranilic acid	2,5-Dichloro-3,6-dihydroxy-2,5-cyclohexadiene-1,4-dione	$C_6H_2Cl_2O_4$	87-88-7	208.984	red lf (w+2)	283.5				s H_2O
1819	Chlorbenside	1-Chloro-4[[(4-chlorophenyl)methyl]thio]benzene	$C_{13}H_{10}Cl_2S$	103-17-3	269.189		75		1.4210^{20}		
1820	Chlorbicyclen		$C_9H_6Cl_8$	2550-75-6	397.768	pow	105	174^2			
1821	Chlorbromuron	N'-(4-Bromo-3-chlorophenyl)-N-methoxy-N-methylurea	$C_9H_{10}BrClN_2O_2$	13360-45-7	293.544		96		1.69^{20}		
1822	Chlorbufam	1-Methyl-2-propynyl(3-chlorophenyl)carbamate	$C_{11}H_{10}ClNO_2$	1967-16-4	223.656	cry	45.5				sl H_2O; s MeOH, EtOH, ace
1823	Chlorcyclizine		$C_{18}H_{21}ClN_2$	82-93-9	300.826	oil		$140^{0.12}$			
1824	Chlordane		$C_{10}H_6Cl_8$	57-74-9	409.779		106	175^1	1.60^{25}		

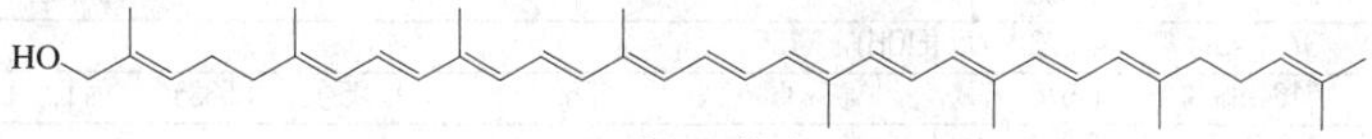

ψ,ψ-Caroten-16-ol

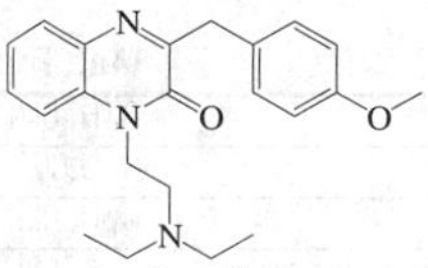

Caroverine

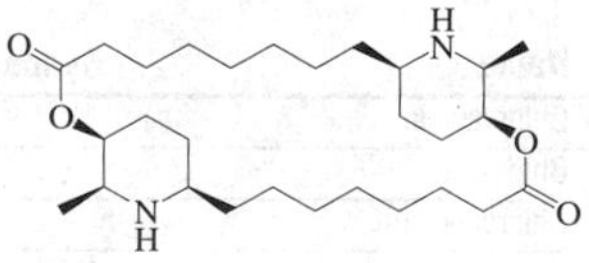

Carpaine

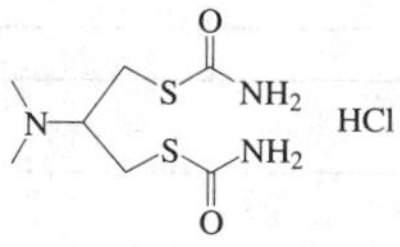

Cartap hydrochloride

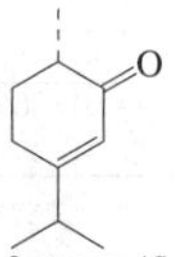

Carvenone, (*S*)

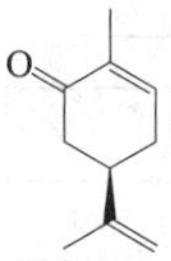

(*R*)-Carvone

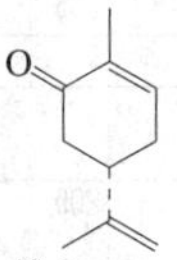

(*S*)-Carvone

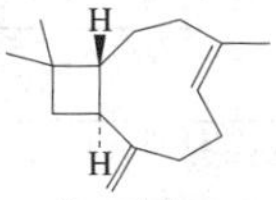

Caryophyllene

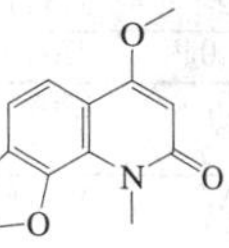

Casimiroin

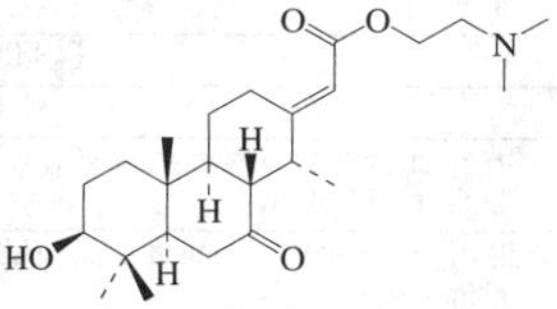

Cassaine

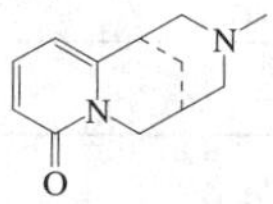

Caulophylline

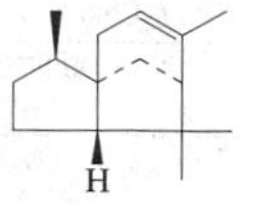

α-Cedrene

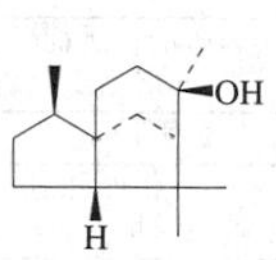

Cedrol

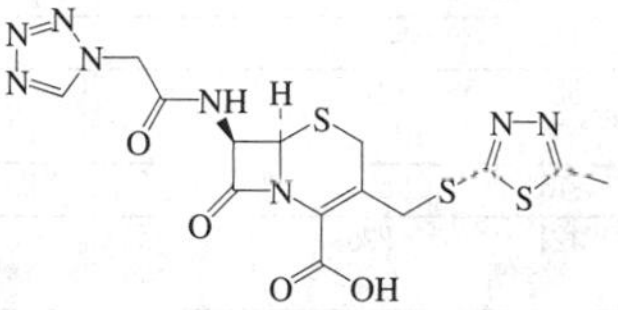

Cefazolin

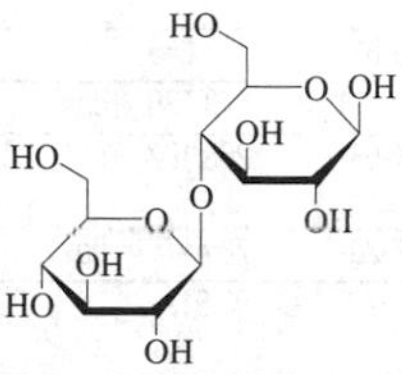

β-Cellobiose

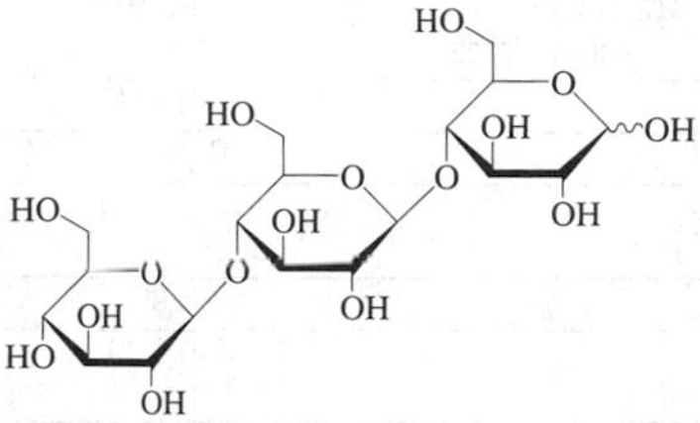

Cellotriose

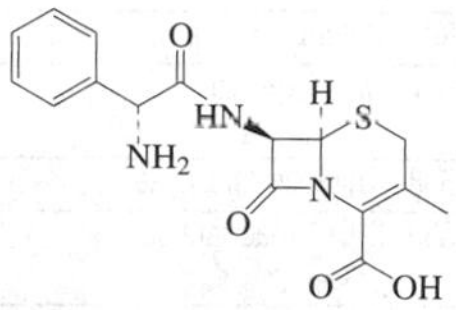

Cephalexin

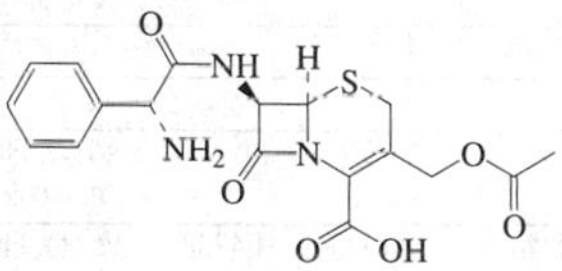

Cephaloglycin

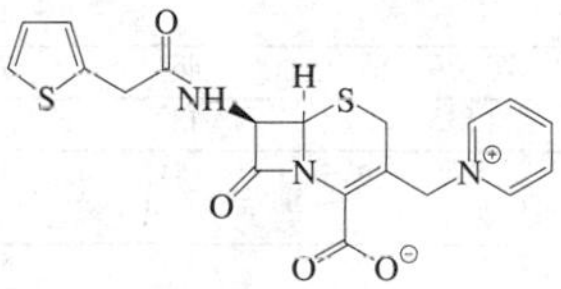

Cephaloridine

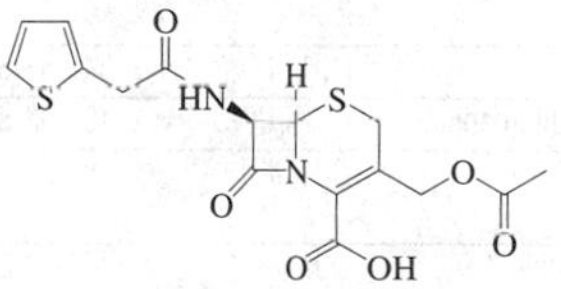

Cephalothin

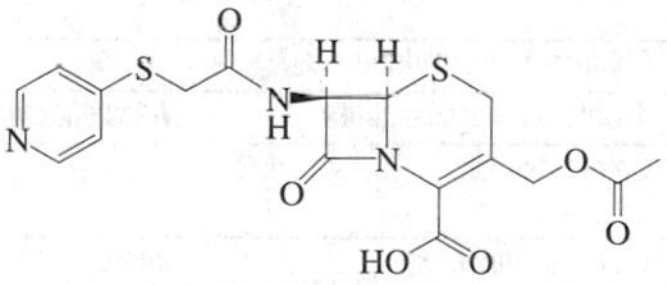

Cephapirin

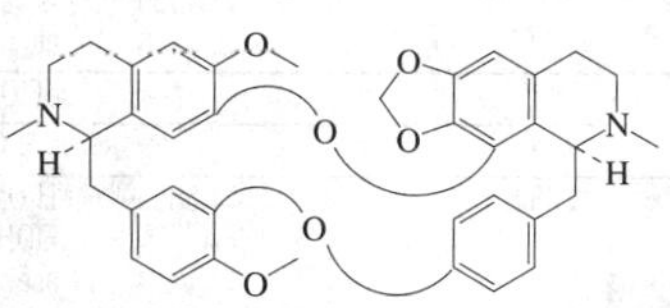

Cepharanthine

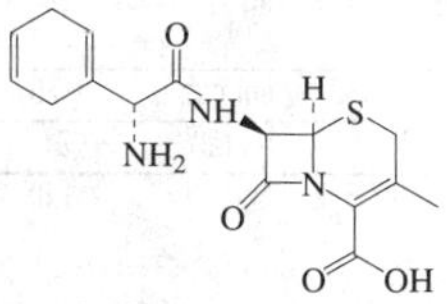

Cephradine

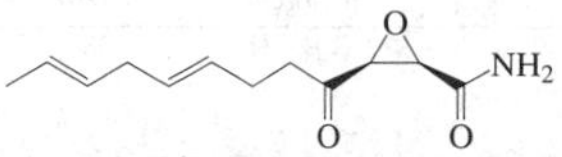

Cerulenin

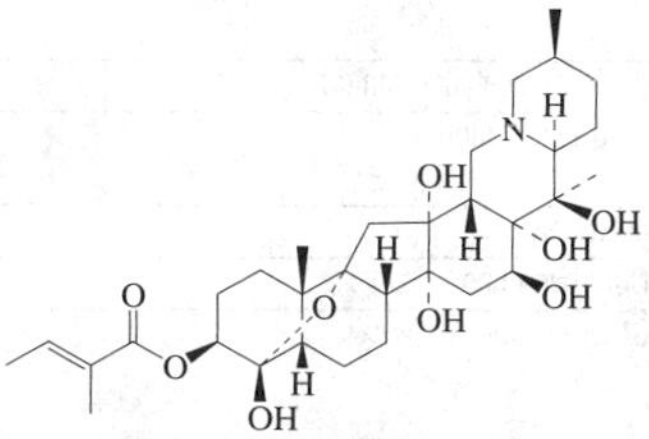

Cevadine

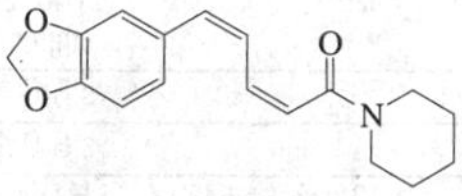

Chavicine

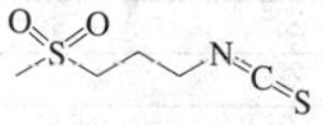

Cheirolin

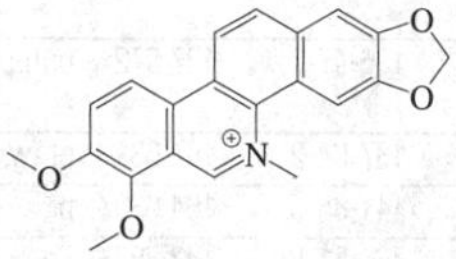

Chelerythrine

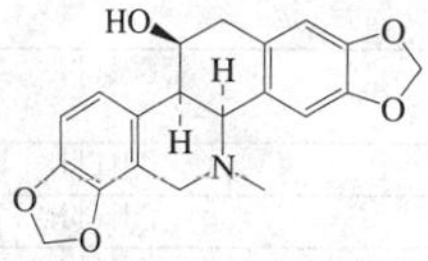

Chelidonine

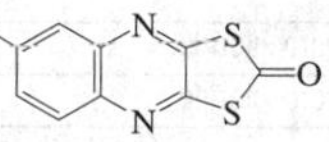

Chinomethionat

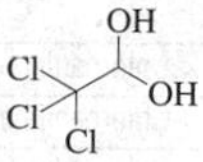

Chloral hydrate

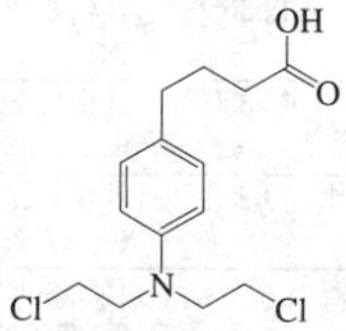

Chlorambucil

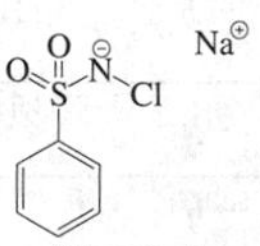

Chloramine B

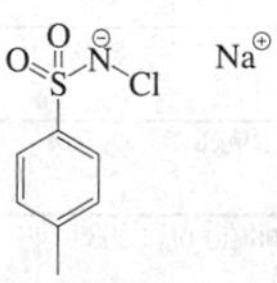

Chloramine T

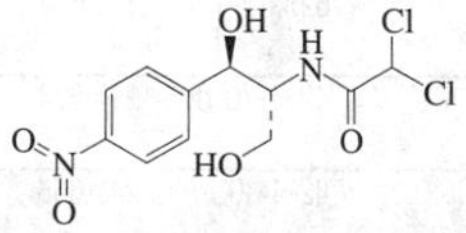

Chloramphenicol

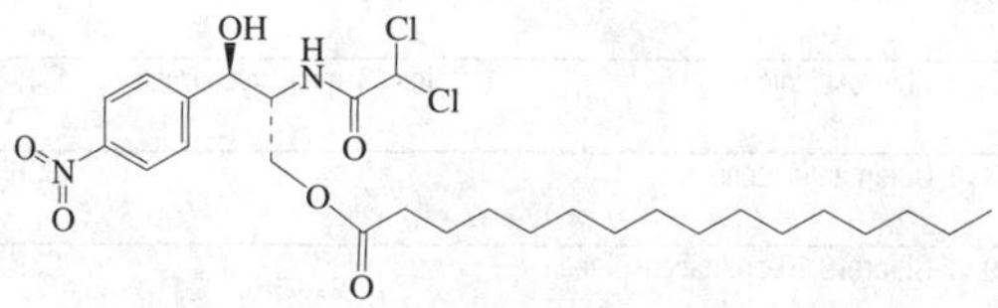

Chloramphenicol palmitate

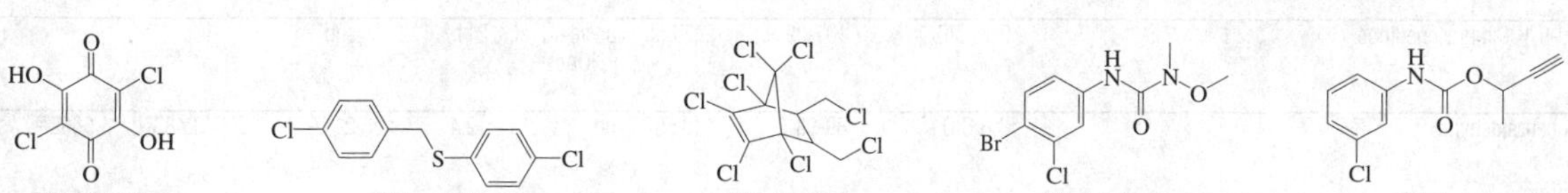

Chloranilic acid

Chlorbenside

Chlorbicyclen

Chlorbromuron

Chlorbufam

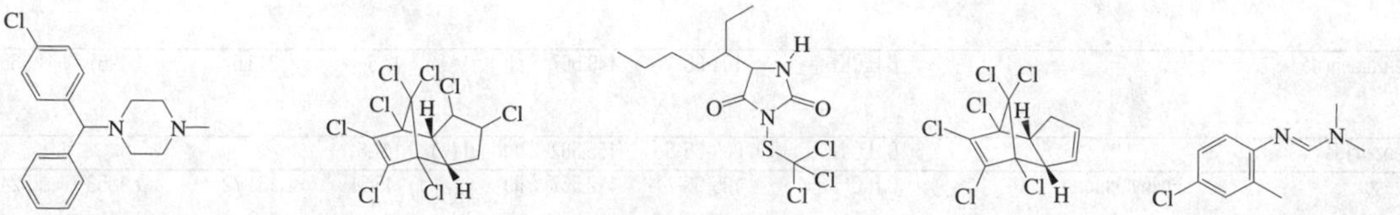

Chlorcyclizine

Chlordane

Chlordantoin

Chlordene

Chlordimeform

No.	Name	Synonym	Mol. Form.	CAS RN	Mol. Wt.	Physical Form	mp/°C	bp/°C	den/ g cm^{-3}	n_D	Solubility
1825	Chlordantoin		$C_{11}H_{17}Cl_3N_2O_2S$	5588-20-5	347.689						s CS_2
1826	Chlordene		$C_{10}H_6Cl_6$	3734-48-3	338.873	cry (EtOH)	155				
1827	Chlordimeform		$C_{10}H_{13}ClN_2$	6164-98-3	196.676		35	$156^{0.4}$	1.105^{25}	1.5885^{25}	vs bz, eth, EtOH
1828	Chlorendic acid	1,4,5,6,7,7-Hexachloro-5-norbornene-2,3-dicarboxylic acid	$C_9H_4Cl_6O_4$	115-28-6	388.844	cry (w)	232				
1829	Chlorendic anhydride		$C_9H_2Cl_6O_3$	115-27-5	370.828		235				
1830	Chlorfenvinphos		$C_{12}H_{14}Cl_3O_4P$	470-90-6	359.569			$170^{0.05}$			
1831	Chlorflurecol	9*H*-Fluorene-9-carboxylic acid, 2-chloro-9-hydroxy-	$C_{14}H_9ClO_3$	2464-37-1	260.672				1.496^{20}		
1832	Chloridazon	3(2*H*)-Pyridazinone, 5-amino-4-chloro-2-phenyl-	$C_{10}H_8ClN_3O$	1698-60-8	221.643		205				
1833	Chlorimuron-ethyl		$C_{15}H_{15}ClN_4O_6S$	90982-32-4	414.821		186				
1834	Chlormephos	Chloromethyl *O,O*-diethyl dithiophosphate	$C_5H_{12}ClO_2PS_2$	24934-91-6	234.705	oil		$83^{0.1}$		1.5244	sl H_2O; misc os
1835	Chlormequat chloride		$C_5H_{13}Cl_2N$	999-81-5	158.069		239 dec				
1836	Chlormezanone		$C_{11}H_{12}ClNO_3S$	80-77-3	273.736	cry	117				sl EtOH
1837	Chlornaphazine		$C_{14}H_{15}Cl_2N$	494-03-1	268.182	pl (peth)	55	210^{5}			vs ace, bz, eth, EtOH
1838	Chloroacetaldehyde		C_2H_3ClO	107-20-0	78.497	liq	-16.3	85.5	1.19		s eth
1839	2-Chloroacetamide		C_2H_4ClNO	79-07-2	93.512		121	225			s H_2O; vs EtOH; sl eth
1840	Chloroacetic acid		$C_2H_3ClO_2$	79-11-8	94.497	mcl pl	63	189.3	1.4043^{40}	1.4351^{55}	vs H_2O; s EtOH, eth, bz, chl; sl ctc
1841	Chloroacetic anhydride		$C_4H_4Cl_2O_3$	541-88-8	170.979	pr (bz)	46	203	1.5497^{20}		
1842	4-Chloroacetoacetanilide	*N*-Acetoacetyl-4-chloroaniline	$C_{10}H_{10}ClNO_2$	101-92-8	211.645		132				
1843	Chloroacetone		C_3H_5ClO	78-95-5	92.524	liq	-44.5	119	1.15^{20}		s H_2O, EtOH, eth, chl
1844	Chloroacetonitrile	Chloromethyl cyanide	C_2H_2ClN	107-14-2	75.497			126.5	1.1930^{20}	1.4202^{25}	vs eth, EtOH
1845	α-Chloroacetophenone	ω-Chloroacetophenone	C_8H_7ClO	532-27-4	154.594	pl(dil al), rhom, lf (peth)	56.5	247	1.324^{15}		i H_2O; vs EtOH, eth, bz; s ace, peth
1846	4-(2-Chloroacetyl)acetanilide		$C_{10}H_{10}ClNO_2$	140-49-8	211.645		218				
1847	Chloroacetyl chloride		$C_2H_2Cl_2O$	79-04-9	112.942	liq	-22	106	1.4202^{20}	1.4530^{20}	msc eth; s ace, ctc
1848	Chloroacetylene		C_2HCl	593-63-5	60.482	col gas	-126	-30			sl EtOH
1849	9-Chloroacridine		$C_{13}H_8ClN$	1207-69-8	213.663	nd (al)	121	sub			vs H_2O, EtOH
1850	2-Chloroaniline		C_6H_6ClN	95-51-2	127.572	liq	-1.9	208.8		1.5895^{20}	i H_2O; msc EtOH; s eth, ace
1851	3-Chloroaniline		C_6H_6ClN	108-42-9	127.572	liq	-10.28	230.5	1.2161^{20}	1.5941^{20}	i H_2O; msc EtOH, eth, ace, bz; s chl
1852	4-Chloroaniline		C_6H_6ClN	106-47-8	127.572	orth pr	70.5	232	1.429^{19}	1.5546^{87}	s H_2O, EtOH, eth, chl
1853	2-Chloroaniline hydrochloride		$C_6H_7Cl_2N$	137-04-2	164.033	pl (w, aq al)	235		1.505^{18}		vs H_2O
1854	3-Chloroaniline hydrochloride		$C_6H_7Cl_2N$	141-85-5	164.033	pl	222				vs H_2O, EtOH
1855	2-Chloroanisole	1-Chloro-2-methoxybenzene	C_7H_7ClO	766-51-8	142.583	liq	-26.8	198.5	1.1911^{20}	1.5480^{20}	i H_2O; s EtOH, eth; sl chl
1856	3-Chloroanisole	1-Chloro-3-methoxybenzene	C_7H_7ClO	2845-89-8	142.583			193.5	1.1759^{12}	1.5365^{20}	i H_2O; s EtOH, eth
1857	4-Chloroanisole	1-Chloro-4-methoxybenzene	C_7H_7ClO	623-12-1	142.583		<-18	197.5	1.201^{20}	1.5390^{20}	i H_2O; vs EtOH, eth, chl; s ctc
1858	1-Chloroanthracene		$C_{14}H_9Cl$	4985-70-0	212.674	lf (HOAc)	83.5		1.1707^{100}	1.6959^{100}	i H_2O; s EtOH, eth, bz, ctc
1859	1-Chloro-9,10-anthracenedione		$C_{14}H_7ClO_2$	82-44-0	242.658	ye nd (to or al)	163	sub			i H_2O; sl EtOH, ctc; msc eth; s bz
1860	2-Chloro-9,10-anthracenedione		$C_{14}H_7ClO_2$	131-09-9	242.658	pa ye nd (al, HOAc)	211	sub			i H_2O, eth; sl EtOH, bz; vs tol; s $PhNO_2$
1861	2-Chlorobenzaldehyde		C_7H_5ClO	89-98-5	140.567	nd	12.4	211.9	1.2483^{20}	1.5662^{20}	sl H_2O; s EtOH, eth, ace, bz, ctc
1862	3-Chlorobenzaldehyde		C_7H_5ClO	587-04-2	140.567	pr	17.5	213.5	1.2410^{20}	1.5650^{20}	sl H_2O, chl; s EtOH, eth, ace, bz
1863	4-Chlorobenzaldehyde		C_7H_5ClO	104-88-1	140.567	pl	47.5	213.5	1.196^{61}	1.555^{61}	s H_2O, ace, chl; vs EtOH, eth, bz
1864	2-Chlorobenzamide		C_7H_6ClNO	609-66-5	155.582	orth nd (w)	141.8				s H_2O, EtOH, eth
1865	Chlorobenzene	Phenyl chloride	C_6H_5Cl	108-90-7	112.557	liq	-45.31	131.72	1.1058^{20}	1.5241^{20}	i H_2O; msc EtOH, eth; vs bz, ctc

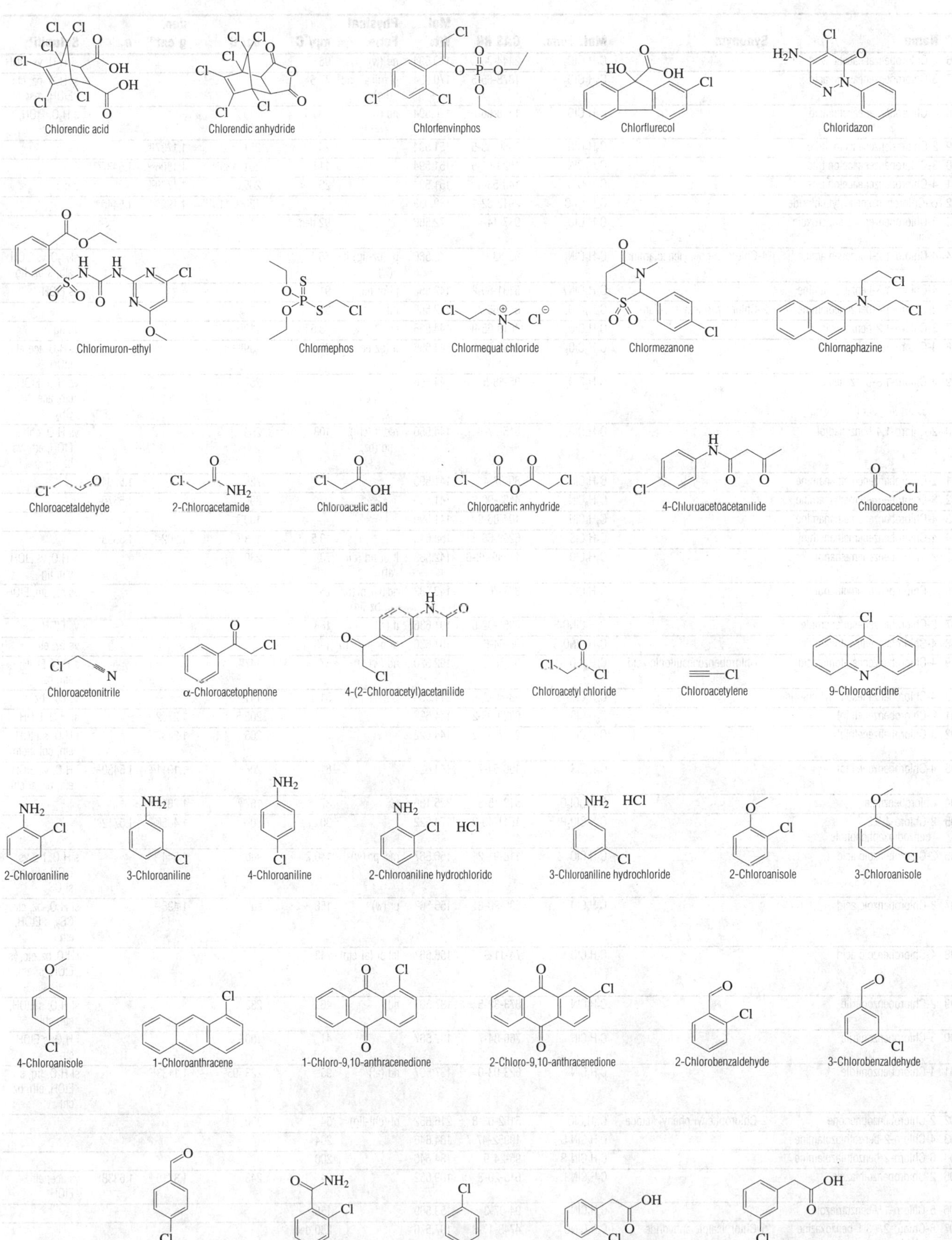
Chlorendic acid
Chlorendic anhydride
Chlorfenvinphos
Chlorflurecol
Chloridazon
Chlorimuron-ethyl
Chlormephos
Chlormequat chloride
Chlormezanone
Chlornaphazine
Chloroacetaldehyde
2-Chloroacetamide
Chloroacetic acid
Chloroacetic anhydride
4-Chloroacetoacetanilide
Chloroacetone
Chloroacetonitrile
α-Chloroacetophenone
4-(2-Chloroacetyl)acetanilide
Chloroacetyl chloride
Chloroacetylene
9-Chloroacridine
2-Chloroaniline
3-Chloroaniline
4-Chloroaniline
2-Chloroaniline hydrochloride
3-Chloroaniline hydrochloride
2-Chloroanisole
3-Chloroanisole
4-Chloroanisole
1-Chloroanthracene
1-Chloro-9,10-anthracenedione
2-Chloro-9,10-anthracenedione
2-Chlorobenzaldehyde
3-Chlorobenzaldehyde
4-Chlorobenzaldehyde
2-Chlorobenzamide
Chlorobenzene
2-Chlorobenzeneacetic acid
3-Chlorobenzeneacetic acid

No.	Name	Synonym	Mol. Form.	CAS RN	Mol. Wt.	Physical Form	mp/°C	bp/°C	den/ g cm^{-3}	n_D	Solubility
1866	2-Chlorobenzeneacetic acid		$C_8H_7ClO_2$	2444-36-2	170.594	nd (w)	96				sl H_2O; vs EtOH
1867	3-Chlorobenzeneacetic acid		$C_8H_7ClO_2$	1878-65-5	170.594	pl (dil al), nd (lig)	77.5				sl H_2O, bz, ctc, EtOH; msc eth
1868	4-Chlorobenzeneacetic acid		$C_8H_7ClO_2$	1878-66-6	170.594	nd (w)	105.5				s H_2O, EtOH, eth, bz
1869	2-Chlorobenzeneacetonitrile		C_8H_6ClN	2856-63-5	151.594		24	251	1.1737^{18}		
1870	3-Chlorobenzeneacetonitrile		C_8H_6ClN	1529-41-5	151.594		11.5	261; 135^{10}	1.1806^{30}	1.5437^{20}	
1871	4-Chlorobenzeneacetonitrile		C_8H_6ClN	140-53-4	151.594		29	265.0	1.1778^{30}		s ctc
1872	α-Chlorobenzeneacetyl chloride		$C_8H_6Cl_2O$	2912-62-1	189.039			120^{23}, 110^{14}	1.196^{25}	1.5440^{20}	
1873	3-Chlorobenzenecarboperoxoic acid		$C_7H_5ClO_3$	937-14-4	172.566		92 dec				
1874	4-Chloro-1,2-benzenediamine	4-Chloro-*o*-phenylenediamine	$C_6H_7ClN_2$	95-83-0	142.586	pl (bz-lig) lf (w)	76				sl H_2O; vs EtOH, eth; s bz, lig
1875	4-Chloro-1,3-benzenediamine		$C_6H_7ClN_2$	5131-60-2	142.586	pl or nd	91				vs EtOH
1876	2-Chloro-1,4-benzenediamine	2-Chloro-*p*-phenylenediamine	$C_6H_7ClN_2$	615-66-7	142.586	nd	64				
1877	3-Chloro-1,2-benzenediol		$C_6H_5ClO_2$	4018-65-9	144.556	cry (lig)	48.5	110^{11}			vs lig
1878	4-Chloro-1,2-benzenediol		$C_6H_5ClO_2$	2138-22-9	144.556	lf (bz-peth)	90.5	$139^{10.5}$			vs H_2O, ace, eth, EtOH
1879	4-Chloro-1,3-benzenediol		$C_6H_5ClO_2$	95-88-5	144.556			257			vs H_2O, EtOH, eth, ace, bz, CS_2
1880	2-Chloro-1,4-benzenediol		$C_6H_5ClO_2$	615-67-8	144.556	red lf (chl), nd (bz)	108	263			vs H_2O, chl; s EtOH, eth; vs bz
1881	2-Chlorobenzenemethanamine		C_7H_8ClN	89-97-4	141.599			72^{2}		1.5594^{25}	
1882	3-Chlorobenzenemethanamine		C_7H_8ClN	4152-90-3	141.599			89^{2}		1.5570^{25}	
1883	4-Chlorobenzenemethanamine		C_7H_8ClN	104-86-9	141.599			109^{13}		1.5566^{25}	
1884	4-Chlorobenzenemethanethiol		C_7H_7ClS	6258-66-8	158.649		19.5	113^{17}	1.202^{25}	1.5893^{20}	
1885	2-Chlorobenzenemethanol		C_7H_7ClO	17849-38-6	142.583	lf or nd (dil al)	73	230			sl H_2O; vs EtOH, eth, lig
1886	4-Chlorobenzenemethanol		C_7H_7ClO	873-76-7	142.583	nd (w), pl (bz or bz-lig)	75	235			vs bz, eth, EtOH
1887	2-Chlorobenzenesulfonamide		$C_6H_6ClNO_2S$	6961-82-6	191.636	lf (al)	188				vs EtOH
1888	4-Chlorobenzenesulfonamide		$C_6H_6ClNO_2S$	98-64-6	191.636	pr or pl (eth)	146				vs bz, eth
1889	4-Chlorobenzenesulfonic acid	*p*-Chlorobenzenesulfonic acid	$C_6H_5ClO_3S$	98-66-8	192.620	nd (w+1)	67	147^{25}			s H_2O, EtOH; i eth, bz
1890	4-Chlorobenzenesulfonyl chloride		$C_6H_4Cl_2O_2S$	98-60-2	211.066		51	141^{15}			vs eth, bz
1891	2-Chlorobenzenethiol		C_6H_5ClS	6320-03-2	144.622			205.5	1.2752^{10}		sl H_2O, EtOH
1892	3-Chlorobenzenethiol		C_6H_5ClS	2037-31-2	144.622			206	1.2637^{13}		i H_2O; s EtOH, eth, chl, peth
1893	4-Chlorobenzenethiol		C_6H_5ClS	106-54-7	144.622		61	206	1.1911^{20}	1.5480^{20}	i H_2O; vs EtOH, eth, bz; sl chl
1894	Chlorobenzilate		$C_{16}H_{14}Cl_2O_3$	510-15-6	325.186		37	$157^{0.07}$	1.2816^{20}		
1895	2-Chloro-1,3,2-benzodioxaphosphole		$C_6H_4ClO_2P$	1641-40-3	174.522		30	80^{20}	1.4650^{20}	1.5712^{20}	
1896	2-Chlorobenzoic acid		$C_7H_5ClO_2$	118-91-2	156.567	mcl pr (w)	140.2	sub	1.544^{20}		s H_2O, bz; vs EtOH, eth, ace; sl CS_2
1897	3-Chlorobenzoic acid		$C_7H_5ClO_2$	535-80-8	156.567	pr (w)	158	sub	1.496^{25}		sl H_2O, bz, ctc, CS_2; s EtOH, eth
1898	4-Chlorobenzoic acid		$C_7H_5ClO_2$	74-11-3	156.567	tcl pr (al-eth)	243				i H_2O, bz, ctc; vs EtOH; sl eth, ace
1899	2-Chlorobenzonitrile		C_7H_4ClN	873-32-5	137.567	nd	46.3	232			sl H_2O; s EtOH, eth, chl
1900	3-Chlorobenzonitrile		C_7H_4ClN	766-84-7	137.567		41	100^{15}			i H_2O; s EtOH, eth
1901	4-Chlorobenzonitrile		C_7H_4ClN	623-03-0	137.567	nd (al)	95	223; 95^{5}	1.1133^{17}		sl H_2O, lig; s EtOH, eth, bz, chl
1902	2-Chlorobenzophenone	2-Chlorophenyl phenyl ketone	$C_{13}H_9ClO$	5162-03-8	216.662	pl (chl-lig)	54	330			
1903	4-Chloro-2-benzothiazolamine		$C_7H_5ClN_2S$	19952-47-7	184.646		204				
1904	6-Chloro-2-benzothiazolamine		$C_7H_5ClN_2S$	95-24-9	184.646		200				
1905	2-Chlorobenzothiazole		C_7H_4ClNS	615-20-3	169.632		24	248	1.3715^{10}	1.6338^{10}	vs ace, eth, EtOH
1906	5-Chloro-1*H*-benzotriazole		$C_6H_4ClN_3$	94-97-3	153.569		158				
1907	6-Chloro-2*H*-3,1-benzoxazine-2,4(1*H*)-dione	5-Chloroisatoic anhydride	$C_8H_4ClNO_3$	4743-17-3	197.576		280 dec				
1908	5-Chloro-2-benzoxazolamine	Zoxazolamine	$C_7H_5ClN_2O$	61-80-3	168.580	pl (bz)	184.5				vs EtOH
1909	2-Chlorobenzoxazole		C_7H_4ClNO	615-18-9	153.566		7	201.5	1.3453^{18}	1.5678^{20}	
1910	5-Chloro-2(3*H*)-benzoxazolone	Chlorzoxazone	$C_7H_4ClNO_2$	95-25-0	169.566	cry (ace)	191.5				vs EtOH, MeOH
1911	2-Chlorobenzoyl chloride		$C_7H_4Cl_2O$	609-65-4	175.012	liq	-4	238		1.5726^{16}	s ctc

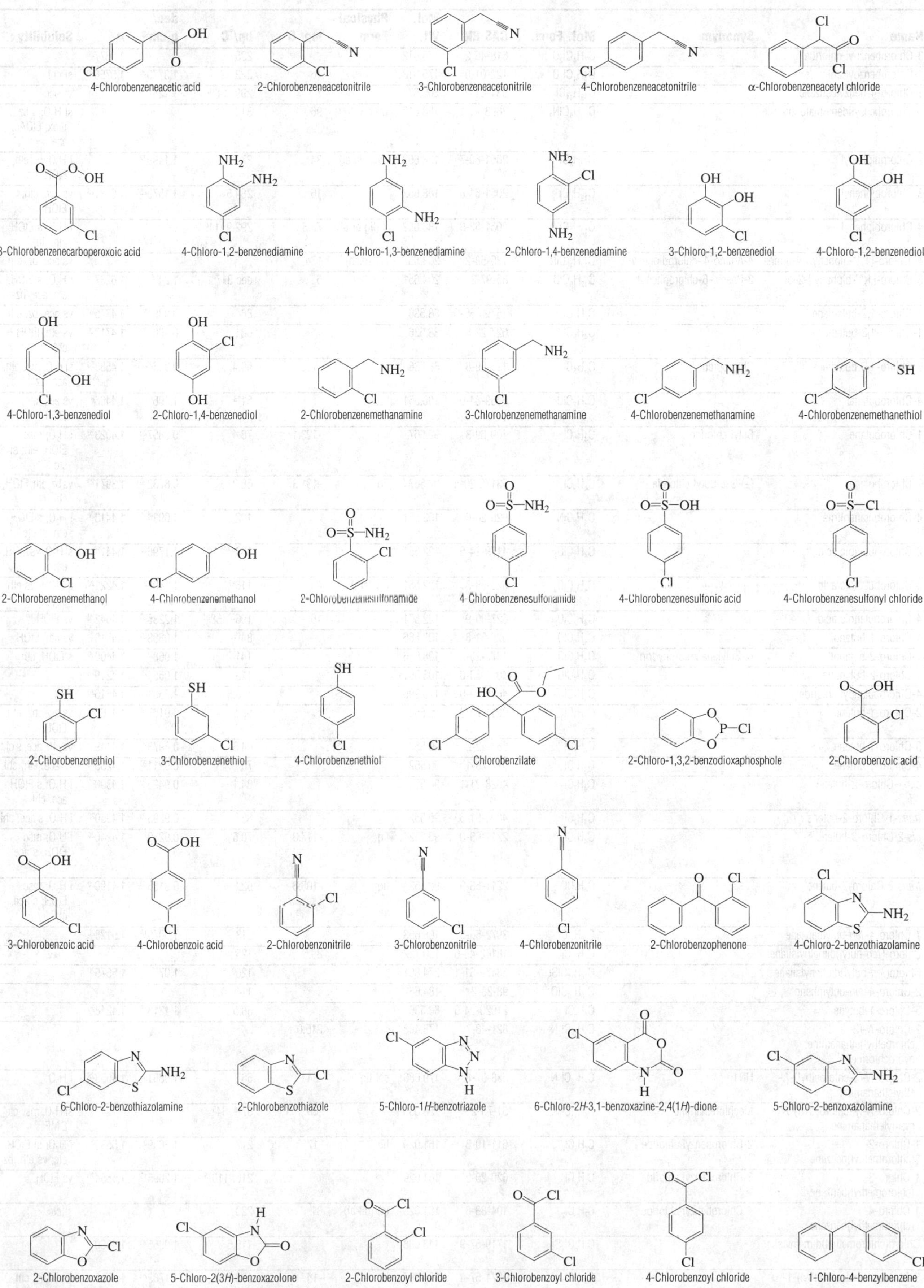

4-Chlorobenzeneacetic acid
2-Chlorobenzeneacetonitrile
3-Chlorobenzeneacetonitrile
4-Chlorobenzeneacetonitrile
α-Chlorobenzeneacetyl chloride
3-Chlorobenzenecarboperoxoic acid
4-Chloro-1,2-benzenediamine
4-Chloro-1,3-benzenediamine
2-Chloro-1,4-benzenediamine
3-Chloro-1,2-benzenediol
4-Chloro-1,2-benzenediol
4-Chloro-1,3-benzenediol
2-Chloro-1,4-benzenediol
2-Chlorobenzenemethanamine
3-Chlorobenzenemethanamine
4-Chlorobenzenemethanamine
4-Chlorobenzenemethanethiol
2-Chlorobenzenemethanol
4-Chlorobenzenemethanol
2-Chlorobenzenesulfonamide
4-Chlorobenzenesulfonamide
4-Chlorobenzenesulfonic acid
4-Chlorobenzenesulfonyl chloride
2-Chlorobenzenethiol
3-Chlorobenzenethiol
4-Chlorobenzenethiol
Chlorobenzilate
2-Chloro-1,3,2-benzodioxaphosphole
2-Chlorobenzoic acid
3-Chlorobenzoic acid
4-Chlorobenzoic acid
2-Chlorobenzonitrile
3-Chlorobenzonitrile
4-Chlorobenzonitrile
2-Chlorobenzophenone
4-Chloro-2-benzothiazolamine
6-Chloro-2-benzothiazolamine
2-Chlorobenzothiazole
5-Chloro-1H-benzotriazole
6-Chloro-2H-3,1-benzoxazine-2,4(1H)-dione
5-Chloro-2-benzoxazolamine
2-Chlorobenzoxazole
5-Chloro-2(3H)-benzoxazolone
2-Chlorobenzoyl chloride
3-Chlorobenzoyl chloride
4-Chlorobenzoyl chloride
1-Chloro-4-benzylbenzene

No.	Name	Synonym	Mol. Form.	CAS RN	Mol. Wt.	Physical Form	mp/°C	bp/°C	den/ g cm^{-3}	n_D	Solubility
1912	3-Chlorobenzoyl chloride		$C_7H_4Cl_2O$	618-46-2	175.012			225		1.5677^{20}	
1913	4-Chlorobenzoyl chloride		$C_7H_4Cl_2O$	122-01-0	175.012		16	222	1.3770^{20}	1.5756^{20}	sl chl
1914	1-Chloro-4-benzylbenzene		$C_{13}H_{11}Cl$	831-81-2	202.679		7.5	299; 147^{8}	1.1247^{20}		vs ace
1915	*o*-Chlorobenzylidene malononitrile		$C_{10}H_5ClN_2$	2698-41-1	188.613	wh cry	96	312			sl H_2O; s bz, diox, EtOAc, ace
1916	2-Chlorobiphenyl		$C_{12}H_9Cl$	2051-60-7	188.652	mcl (dil al)	34	274	1.1499^{32}		i H_2O; vs eth, EtOH, lig
1917	3-Chlorobiphenyl		$C_{12}H_9Cl$	2051-61-8	188.652		16	284.5	1.1579^{25}	1.6181^{25}	vs ace, eth, EtOH
1918	4-Chlorobiphenyl		$C_{12}H_9Cl$	2051-62-9	188.652	lf (lig or al)	78.8	292.9; 146^{10}			i H_2O; s EtOH, eth, lig
1919	4'-Chloro-[1,1'-biphenyl]-4-amine	4-Amino-4'-chlorodiphenyl	$C_{12}H_{10}ClN$	135-68-2	203.667	cry (peth)	134				vs ace, bz, eth
1920	3-Chloro-[1,1'-biphenyl]-2-ol	2-Phenyl-6-chlorophenol	$C_{12}H_9ClO$	85-97-2	204.651		6	dec 317	1.24^{25}	1.6237^{30}	i H_2O; s EtOH, eth, ace, bz
1921	4-Chloro-1,2-butadiene		C_4H_5Cl	25790-55-0	88.536			88	0.9891^{20}	1.4775^{20}	vs ace, bz, eth
1922	1-Chloro-1,3-butadiene		C_4H_5Cl	627-22-5	88.536			68	0.9606^{20}	1.4712^{20}	vs eth, EtOH, chl
1923	2-Chloro-1,3-butadiene	Chloroprene	C_4H_5Cl	126-99-8	88.536	liq	-130	59.4	0.956^{20}	1.4583^{20}	sl H_2O; msc eth, ace, bz
1924	4-Chlorobutanal		C_4H_7ClO	6139-84-0	106.551			51^{13}	1.106^{8}	1.4466^{8}	vs ace, eth, EtOH
1925	1-Chlorobutane	Butyl chloride	C_4H_9Cl	109-69-3	92.567	liq	-123.1	78.4	0.8857^{20}	1.4023^{20}	i H_2O; msc EtOH, eth; sl ctc
1926	2-Chlorobutane	(±)-*sec*-Butyl chloride	C_4H_9Cl	53178-20-4	92.567	liq	-131.3	68.2	0.8732^{20}	1.3971^{20}	vs bz, eth, EtOH, chl
1927	4-Chlorobutanenitrile		C_4H_6ClN	628-20-6	103.551			192	1.0934^{15}	1.4413^{20}	i H_2O; s EtOH, eth; sl ctc
1928	2-Chlorobutanoic acid		$C_4H_7ClO_2$	4170-24-5	122.551			189^{627}, 101^{15}	1.1796^{20}	1.441^{20}	sl H_2O; vs EtOH, eth
1929	3-Chlorobutanoic acid		$C_4H_7ClO_2$	625-68-3	122.551	cry (eth)	16	116^{22}	1.1898^{20}	1.4221^{20}	s EtOH; vs eth; sl ctc
1930	4-Chlorobutanoic acid		$C_4H_7ClO_2$	627-00-9	122.551		16	196^{22}, $68^{0.2}$	1.2236^{20}	1.4642^{20}	vs EtOH
1931	4-Chloro-1-butanol		C_4H_9ClO	928-51-8	108.566			84^{16}	1.0883^{20}	1.4518^{20}	vs eth, EtOH
1932	1-Chloro-2-butanol	α-Butylene chlorohydrin	C_4H_9ClO	1873-25-2	108.566			141	1.068^{25}	1.4400^{20}	s EtOH, eth
1933	3-Chloro-2-butanone		C_4H_7ClO	4091-39-8	106.551			115	1.0554^{25}	1.4219^{20}	
1934	4-Chlorobutanoyl chloride		$C_4H_6Cl_2O$	4635-59-0	140.996			173.5	1.2581^{20}	1.4616^{20}	s eth
1935	2-Chloro-1-butene		C_4H_7Cl	2211-70-3	90.552			58.5	0.9107^{15}	1.4165^{21}	vs ace, bz, eth, EtOH
1936	3-Chloro-1-butene		C_4H_7Cl	563-52-0	90.552			64.5	0.8978^{20}	1.4149^{20}	vs eth, ace; s chl
1937	4-Chloro-1-butene		C_4H_7Cl	927-73-1	90.552			75	0.9211^{20}	1.4233^{20}	vs ace, eth, chl
1938	*cis*-1-Chloro-2-butene		C_4H_7Cl	4628-21-1	90.552			84.1	0.9426^{20}	1.4390^{20}	i H_2O; s EtOH, ace, chl
1939	*trans*-1-Chloro-2-butene		C_4H_7Cl	4894-61-5	90.552			85	0.9295^{20}	1.4350^{20}	i H_2O; s ace, chl
1940	*cis*-2-Chloro-2-butene		C_4H_7Cl	2211-69-0	90.552	liq	-117.3	70.6	0.9239^{20}	1.4240^{20}	i H_2O; msc EtOH; s ace, chl
1941	*trans*-2-Chloro-2-butene		C_4H_7Cl	2211-68-9	90.552	liq	-105.8	62.8	0.9138^{20}	1.4190^{20}	i H_2O; msc EtOH; s ace, chl
1942	1-Chloro-4-*tert*-butylbenzene		$C_{10}H_{13}Cl$	3972-56-3	168.663			213	1.0075^{18}	1.5123^{20}	
1943	Chloro-(*tert*-butyl)dimethylsilane		$C_6H_{15}ClSi$	18162-48-6	150.722		89.5	125			
1944	Chloro(*tert*-butyl)diphenylsilane		$C_{16}H_{19}ClSi$	58479-61-1	274.861			$120^{0.06}$	1.07^{20}	1.5675^{20}	
1945	2-Chloro-4-*tert*-butylphenol		$C_{10}H_{13}ClO$	98-28-2	184.662			114^{8}			
1946	3-Chloro-1-butyne		C_4H_5Cl	21020-24-6	88.536			68.5	1.4218^{25}	1.4218^{25}	
1947	2-Chloro-*N*-(2-chloroethyl)ethanamine, hydrochloride		$C_4H_{10}Cl_3N$	821-48-7	178.488		215.0				
1948	2-Chloro-*N*-(2-chloroethyl)-*N*-ethylethanamine	HN1	$C_6H_{13}Cl_2N$	538-07-8	170.080	col liq	-34	66^{12}	1.0861^{23}	1.4653^{25}	i H_2O
1949	2-Chloro-*N*-(2-chloroethyl)-*N*-methylethanamine	Mechlorethamine	$C_5H_{11}Cl_2N$	51-75-2	156.053		-60	87^{18}, 64^{5}			sl H_2O; msc ctc, DMF
1950	1-Chloro-2-(chloromethyl)benzene	2-Chlorobenzyl chloride	$C_7H_6Cl_2$	611-19-8	161.029	liq	-17	217	1.2699^{0}	1.5530^{20}	i H_2O; sl EtOH, ctc; vs eth, bz
1951	1-Chloro-3-(chloromethyl)benzene	3-Chlorobenzyl chloride	$C_7H_6Cl_2$	620-20-2	161.029			216; 110^{25}	1.2695^{15}	1.5554^{20}	vs EtOH
1952	1-Chloro-4-(chloromethyl)benzene	4-Chlorobenzyl chloride	$C_7H_6Cl_2$	104-83-6	161.029	nd (dil al)	31	223			sl ctc
1953	Chloro(chloromethyl)dimethylsilane		$C_3H_8Cl_2Si$	1719-57-9	143.088			115.5	1.0865^{20}	1.4360^{20}	
1954	3-Chloro-2-(chloromethyl)-1-propene		$C_4H_6Cl_2$	1871-57-4	124.997	liq	-14	138	1.1782^{20}	1.4753	vs EtOH, chl

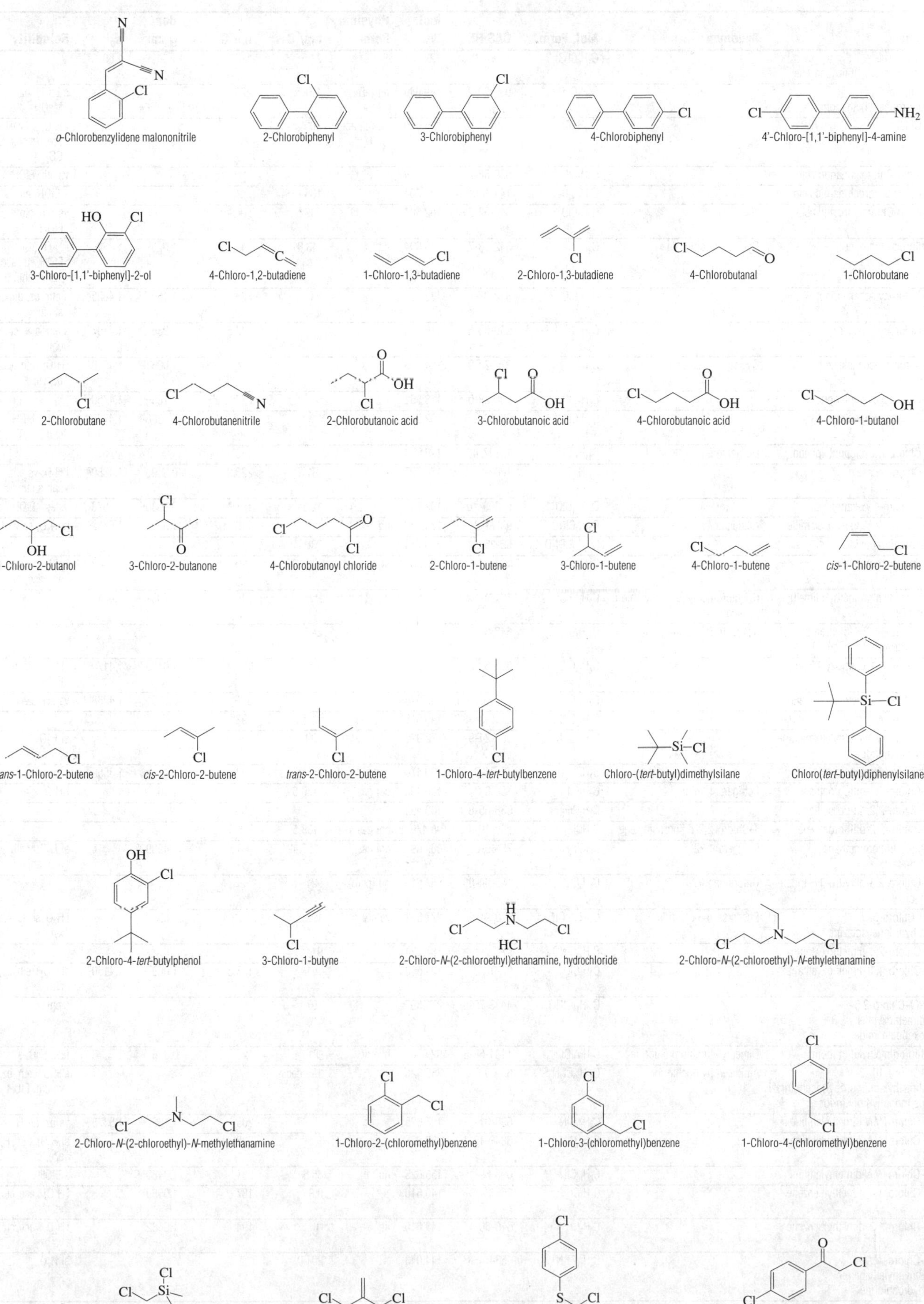

o-Chlorobenzylidene malononitrile
2-Chlorobiphenyl
3-Chlorobiphenyl
4-Chlorobiphenyl
4'-Chloro-[1,1'-biphenyl]-4-amine
3-Chloro-[1,1'-biphenyl]-2-ol
4-Chloro-1,2-butadiene
1-Chloro-1,3-butadiene
2-Chloro-1,3-butadiene
4-Chlorobutanal
1-Chlorobutane
2-Chlorobutane
4-Chlorobutanenitrile
2-Chlorobutanoic acid
3-Chlorobutanoic acid
4-Chlorobutanoic acid
4-Chloro-1-butanol
1-Chloro-2-butanol
3-Chloro-2-butanone
4-Chlorobutanoyl chloride
2-Chloro-1-butene
3-Chloro-1-butene
4-Chloro-1-butene
cis-1-Chloro-2-butene
trans-1-Chloro-2-butene
cis-2-Chloro-2-butene
trans-2-Chloro-2-butene
1-Chloro-4-tert-butylbenzene
Chloro-(tert-butyl)dimethylsilane
Chloro(tert-butyl)diphenylsilane
2-Chloro-4-tert-butylphenol
3-Chloro-1-butyne
2-Chloro-N-(2-chloroethyl)ethanamine, hydrochloride
2-Chloro-N-(2-chloroethyl)-N-ethylethanamine
2-Chloro-N-(2-chloroethyl)-N-methylethanamine
1-Chloro-2-(chloromethyl)benzene
1-Chloro-3-(chloromethyl)benzene
1-Chloro-4-(chloromethyl)benzene
Chloro(chloromethyl)dimethylsilane
3-Chloro-2-(chloromethyl)-1-propene
1-Chloro-4-[(chloromethyl)thio]benzene
2-Chloro-1-(4-chlorophenyl)ethanone

No.	Name	Synonym	Mol. Form.	CAS RN	Mol. Wt.	Physical Form	mp/°C	bp/°C	den/ g cm^{-3}	n_D	Solubility
1955	1-Chloro-4-[(chloromethyl)thio]benzene		$C_7H_6Cl_2S$	7205-90-5	193.094		21.5	128^{12}	1.346^{25}	1.6055^{20}	
1956	2-Chloro-1-(4-chlorophenyl)ethanone		$C_8H_6Cl_2O$	937-20-2	189.039	nd (al)	101.5	270			s EtOH, bz, MeOH
1957	3-Chlorocholest-5-ene, (3β)		$C_{27}H_{45}Cl$	910-31-6	405.099	nd (al, ace)	96				i H_2O; s EtOH, ace, bz, chl; vs CS_2
1958	*trans-o*-Chlorocinnamic acid		$C_9H_7ClO_2$	939-58-2	182.604		212				vs eth, EtOH
1959	*trans-m*-Chlorocinnamic acid		$C_9H_7ClO_2$	14473-90-6	182.604		165				s EtOH, eth
1960	*trans-p*-Chlorocinnamic acid		$C_9H_7ClO_2$	940-62-5	182.604		249.5				vs ace, eth, EtOH
1961	Chlorocyclohexane	Cyclohexyl chloride	$C_6H_{11}Cl$	542-18-7	118.604	liq	-43.81	142	1.000^{20}	1.4626^{20}	i H_2O; msc EtOH, eth, ace, bz; vs chl
1962	2-Chlorocyclohexanone		C_6H_9ClO	822-87-7	132.587		23	82^{15}	1.160^{20}	1.4825^{20}	s eth, bz, diox; sl ctc
1963	1-Chlorocyclohexene		C_6H_9Cl	930-66-5	116.588			142.5	1.0361^{19}	1.4797^{20}	s eth, ace, ctc, chl
1964	Chlorocyclopentane	Cyclopentyl chloride	C_5H_9Cl	930-28-9	104.578			114	1.0051^{20}	1.4510^{20}	i H_2O; s eth, ace, bz, ctc
1965	2-Chlorocyclopentanone		C_5H_7ClO	694-28-0	118.562			87^{19}, 73^{12}	1.185^{25}	1.4750^{20}	
1966	3-Chlorocyclopentene		C_5H_7Cl	96-40-2	102.563			40^{40}, 27^{30}	1.0388^{25}	1.4708^{26}	vs eth, EtOH, chl
1967	4-Chloro-2-cyclopentylphenol	Dowicide 9	$C_{11}H_{13}ClO$	13347-42-7	196.673			183^{18}			
1968	1-Chlorodecane		$C_{10}H_{21}Cl$	1002-69-3	176.727	liq	-31.3	225.9	0.8696^{20}	1.4380^{20}	i H_2O; vs eth, chl; s ctc
1969	10-Chloro-1-decanol		$C_{10}H_{21}ClO$	51309-10-5	192.726		12.5	187^{15}	0.9630^{25}	1.4578^{20}	vs eth, EtOH
1970	2-Chloro-*N,N*-diallylacetamide	Allidochlor	$C_8H_{12}ClNO$	93-71-0	173.640	liq		116^{1}, $92^{0.7}$	1.088^{25}	1.4932^{25}	sl H_2O; s EtOH
1971	Chlordiazepoxide		$C_{16}H_{14}ClN_3O$	58-25-3	299.754		236.2				
1972	Chlorodibromomethane		$CHBr_2Cl$	124-48-1	208.280	liq	-20	120	2.451^{20}	1.5482^{20}	i H_2O; s EtOH, eth, ace, bz
1973	Chloro(dichloromethyl)dimethylsilane	(Dichloromethyl)dimethylchlorosilane	$C_3H_7Cl_3Si$	18171-59-0	177.533	liq	-48	149	1.2369^{20}	1.461^{20}	
1974	5-Chloro-*N*-(3,4-dichlorophenyl)-2-hydroxybenzamide	3',4',5-Trichlorosalicylanilide	$C_{13}H_8Cl_3NO_2$	642-84-2	316.568		247				
1975	2-Chloro-1,1-diethoxyethane		$C_6H_{13}ClO_2$	621-62-5	152.619			157.4	1.0180^{20}	1.4170^{20}	sl H_2O, ctc; msc EtOH, eth
1976	3-Chloro-1,1-diethoxypropane		$C_7H_{15}ClO_2$	35573-93-4	166.646			84^{25}	0.9951^{19}	1.4268^{20}	vs ace, bz
1977	2-Chloro-*N,N*-diethylacetamide		$C_6H_{12}ClNO$	2315-36-8	149.618			192^{25}			
1978	2-Chloro-*N,N*-diethylethanamine, hydrochloride		$C_6H_{15}Cl_2N$	869-24-9	172.096		200				sl H_2O
1979	Chlorodifluoroacetic acid		$C_2HClF_2O_2$	76-04-0	130.478	hyg	25	122		1.3559^{20}	s chl
1980	1-Chloro-1,1-difluoroethane	Refrigerant 142b	$C_2H_3ClF_2$	75-68-3	100.495	col gas	-130.8	-9.1	1.107^{25}		i H_2O; s bz
1981	1-Chloro-2,2-difluoroethane		$C_2H_3ClF_2$	338-65-8	100.495			35.1			
1982	1-Chloro-2,2-difluoroethene	1-Chloro-2,2-difluoroethylene	C_2HClF_2	359-10-4	98.479	col gas	-138.5	-18.5			
1983	Chlorodifluoromethane	Refrigerant 22	$CHClF_2$	75-45-6	86.469	col gas	-157.42	-40.7	1.4909^{-69}		sl H_2O; s eth, ace, chl
1984	7-Chloro-2,3-dihydro-1*H*-inden-4-ol	Chlorindanol	C_9H_9ClO	145-94-8	168.619	nd (peth)	92				
1985	10-Chloro-5,10-dihydrophenarsazine	Phenarsazine chloride	$C_{12}H_9AsClN$	578-94-9	277.581	ye cry	195		1.65		i H_2O; sl ctc, bz, xyl
1986	5-Chloro-2,4-dimethoxyaniline		$C_8H_{10}ClNO_2$	97-50-7	187.624		91				
1987	2-Chloro-1,1-dimethoxyethane		$C_4H_9ClO_2$	97-97-2	124.566			127.5	1.068^{20}	1.4150^{20}	sl EtOH, eth, bz, ctc
1988	*N*-(4-Chloro-2,5-dimethoxyphenyl)-3-oxobutanamide		$C_{12}H_{14}ClNO_4$	4433-79-8	271.697		107				s chl
1989	Chlorodimethylaluminum	Dimethylaluminum chloride	C_2H_6AlCl	1184-58-3	92.504	hyg liq	-21	126	0.996		reac H_2O; s hx
1990	2-Chloro-10-(3-dimethylaminopropyl)phenothiazine monohydrochloride	Aminazin hydrochloride	$C_{17}H_{20}Cl_2N_2S$	69-09-0	355.325		195 dec				s H_2O; i eth, bz; vs chl, EtOH
1991	2-Chloro-*N,N*-dimethylaniline		$C_8H_{10}ClN$	698-01-1	155.625			205	1.1067^{20}	1.5578^{20}	vs bz, EtOH
1992	3-Chloro-*N,N*-dimethylaniline		$C_8H_{10}ClN$	6848-13-1	155.625			232			sl H_2O; s EtOH, ace, bz
1993	4-Chloro-*N,N*-dimethylaniline		$C_8H_{10}ClN$	698-69-1	155.625	nd (al)	35.5	231	1.0480^{100}		s EtOH
1994	2-Chloro-1,4-dimethylbenzene		C_8H_9Cl	95-72-7	140.610		0.8	187	1.0589^{15}		i H_2O; s ace, ctc; vs bz
1995	4-Chloro-1,2-dimethylbenzene		C_8H_9Cl	615-60-1	140.610	liq	-6	194	1.0682^{15}		i H_2O; s ace, ctc; vs bz
1996	2-Chloro-*N,N*-dimethylethanamine, hydrochloride		$C_4H_{11}Cl_2N$	4584-46-7	144.043		201.0				sl H_2O
1997	(2-Chloro-1,1-dimethylethyl)benzene	Neophyl chloride	$C_{10}H_{13}Cl$	515-40-2	168.663			223; 105^{18}	1.047^{20}	1.5247^{20}	vs ace, bz, eth, EtOH

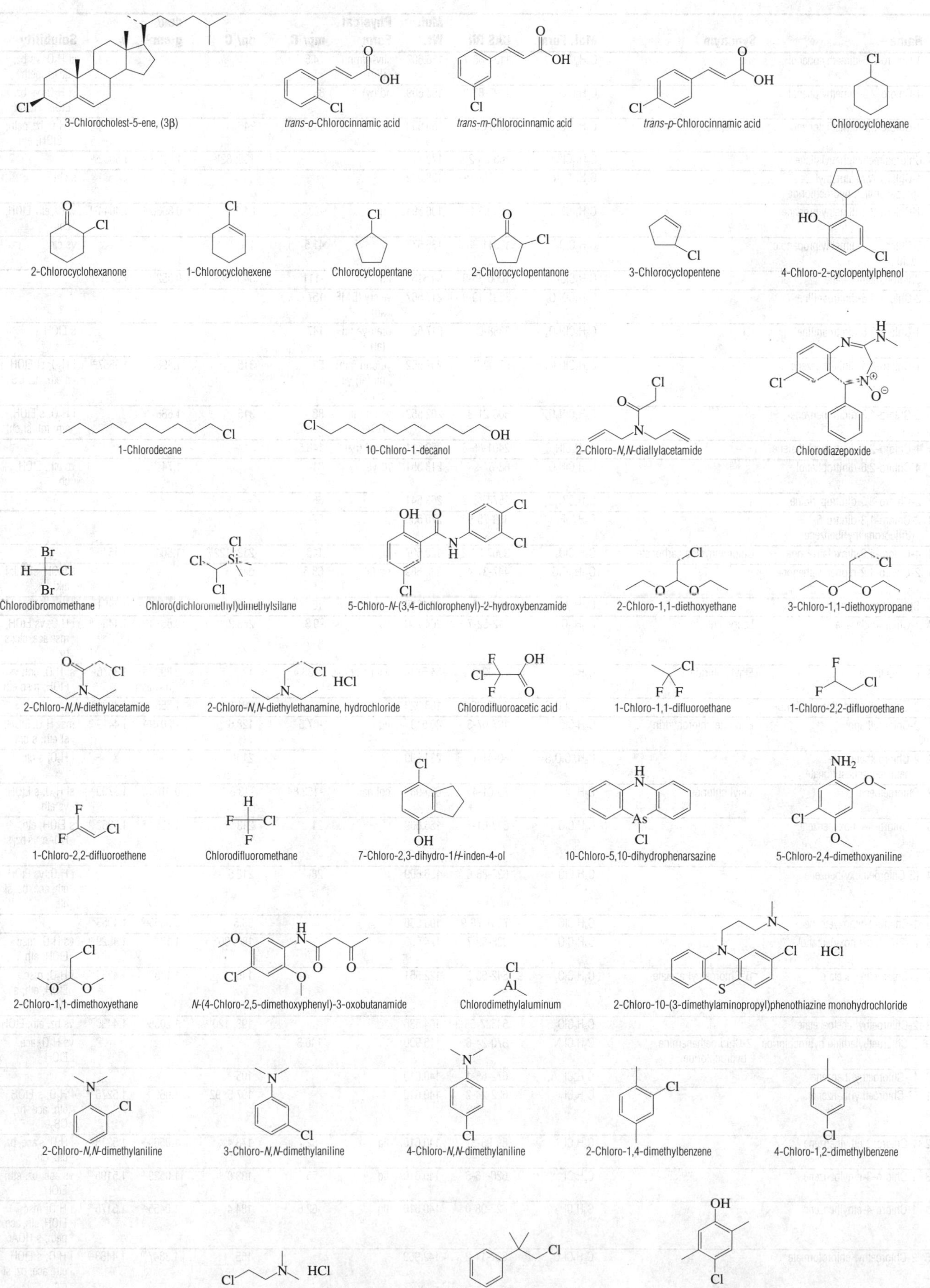

3-Chlorocholest-5-ene, (3β)
trans-o-Chlorocinnamic acid
trans-m-Chlorocinnamic acid
trans-p-Chlorocinnamic acid
Chlorocyclohexane
2-Chlorocyclohexanone
1-Chlorocyclohexene
Chlorocyclopentane
2-Chlorocyclopentanone
3-Chlorocyclopentene
4-Chloro-2-cyclopentylphenol
1-Chlorodecane
10-Chloro-1-decanol
2-Chloro-N,N-diallylacetamide
Chlordiazepoxide
Chlorodibromomethane
Chloro(dichloromethyl)dimethylsilane
5-Chloro-N-(3,4-dichlorophenyl)-2-hydroxybenzamide
2-Chloro-1,1-diethoxyethane
3-Chloro-1,1-diethoxypropane
2-Chloro-N,N-diethylacetamide
2-Chloro-N,N-diethylethanamine, hydrochloride
Chlorodifluoroacetic acid
1-Chloro-1,1-difluoroethane
1-Chloro-2,2-difluoroethane
1-Chloro-2,2-difluoroethene
Chlorodifluoromethane
7-Chloro-2,3-dihydro-1H-inden-4-ol
10-Chloro-5,10-dihydrophenarsazine
5-Chloro-2,4-dimethoxyaniline
2-Chloro-1,1-dimethoxyethane
N-(4-Chloro-2,5-dimethoxyphenyl)-3-oxobutanamide
Chlorodimethylaluminum
2-Chloro-10-(3-dimethylaminopropyl)phenothiazine monohydrochloride
2-Chloro-N,N-dimethylaniline
3-Chloro-N,N-dimethylaniline
4-Chloro-N,N-dimethylaniline
2-Chloro-1,4-dimethylbenzene
4-Chloro-1,2-dimethylbenzene
2-Chloro-N,N-dimethylethanamine, hydrochloride
(2-Chloro-1,1-dimethylethyl)benzene
4-Chloro-2,5-dimethylphenol

No.	Name	Synonym	Mol. Form.	CAS RN	Mol. Wt.	Physical Form	mp/°C	bp/°C	den/ g cm-3	n_D	Solubility
1998	4-Chloro-2,5-dimethylphenol		C_8H_9ClO	1124-06-7	156.609	silv-grn nd (lig)	74.5				sl H_2O; vs bz, EtOH, peth
1999	4-Chloro-2,6-dimethylphenol		C_8H_9ClO	1123-63-3	156.609	nd (w)	83				sl H_2O; vs bz, EtOH, HOAc
2000	4-Chloro-3,5-dimethylphenol	Chloroxylenol	C_8H_9ClO	88-04-0	156.609		115	246			sl H_2O, bz, peth; s EtOH, eth
2001	Chlorodimethylphenylsilane		$C_8H_{11}ClSi$	768-33-2	170.712			195; 82^{16}	1.032^{20}	1.5082^{20}	
2002	1-Chloro-*N*,*N*-dimethyl-2-propanamine, hydrochloride		$C_5H_{13}Cl_2N$	17256-39-2	158.069						s chl
2003	1-Chloro-2,2-dimethylpropane		$C_5H_{11}Cl$	753-89-9	106.594	liq	-20	84.3	0.8660^{20}	1.4044^{20}	vs bz, eth, EtOH, chl
2004	3-Chloro-2,2-dimethylpropanoic acid		$C_5H_9ClO_2$	13511-38-1	136.577		41.5	110^{10}			vs ctc
2005	Chlorodimethylsilane		C_2H_7ClSi	1066-35-9	94.616	liq	-111	34.7	0.852	1.3830^{20}	
2006	2-Chloro-4,6-dinitroaniline		$C_6H_4ClN_3O_4$	3531-19-9	217.567	ye cry (DMF aq)	157				
2007	4-Chloro-2,6-dinitroaniline		$C_6H_4ClN_3O_4$	5388-62-5	217.567	oran-ye nd (al)	147				s EtOH
2008	1-Chloro-2,4-dinitrobenzene		$C_6H_3ClN_2O_4$	97-00-7	202.552	ye orth (eth) nd (al) ye cry	53	315	1.4982^{75}	1.5857^{60}	i H_2O; sl EtOH; s eth, bz, CS_2
2009	2-Chloro-1,3-dinitrobenzene		$C_6H_3ClN_2O_4$	606-21-3	202.552	ye nd (al, HOAc)	88	315	1.6867^{16}		i H_2O; s EtOH, eth, tol; sl chl
2010	1-Chloro-2,4-dinitronaphthalene		$C_{10}H_5ClN_2O_4$	2401-85-6	252.611	ye nd (bz)	146.5				
2011	4-Chloro-2,6-dinitrophenol		$C_6H_3ClN_2O_5$	88-87-9	218.551	pa ye cry	81		1.74^{22}		vs eth, EtOH, chl
2012	2-Chloro-3,5-dinitropyridine		$C_5H_2ClN_3O_4$	2578-45-2	203.541		66.5				
2013	2-Chloro-1,3-dinitro-5-(trifluoromethyl)benzene		$C_7H_2ClF_3N_2O_4$	393-75-9	270.550		57				
2014	4-Chloro-1,3-dioxolan-2-one	Chloroethylene carbonate	$C_3H_3ClO_3$	3967-54-2	122.507	liq	110	213; 122^{18}	1.504	1.4540^{20}	
2015	2-Chloro-1,2-diphenylethanone		$C_{14}H_{11}ClO$	447-31-4	230.689	nd (al)	68.5	dec			s EtOH; sl chl; i alk
2016	Chlorodiphenylmethane		$C_{13}H_{11}Cl$	90-99-3	202.679		16	140^{3}	1.140^{25}	1.5951^{20}	s chl
2017	1-Chlorododecane	Lauryl chloride	$C_{12}H_{25}Cl$	112-52-7	204.780	liq	-9.3	263.2	0.8673^{20}	1.4434^{20}	i H_2O; vs EtOH; msc ace, ctc; s bz
2018	Chloroethane	Ethyl chloride	C_2H_5Cl	75-00-3	64.514	vol liq or gas	-138.4	12.3	0.8902^{25} (p>1 atm	1.3676^{20}	sl H_2O, chl; vs EtOH; msc eth
2019	2-Chloroethanesulfonyl chloride		$C_2H_4Cl_2O_2S$	1622-32-8	163.023			201.5	1.555^{20}	1.4920^{20}	
2020	2-Chloroethanol	Ethylene chlorohydrin	C_2H_5ClO	107-07-3	80.513	liq	-67.5	128.6	1.2019^{20}	1.4419^{20}	msc H_2O, EtOH; sl eth; s chl
2021	2-Chloroethanol, 4-methylbenzenesulfonate		$C_9H_{11}ClO_3S$	80-41-1	234.699			210^{21}			i H_2O; s ctc
2022	Chloroethene	Vinyl chloride	C_2H_3Cl	75-01-4	62.498	col gas	-153.84	-13.8	0.9106^{20}	1.3700^{20}	sl H_2O; s EtOH; vs eth
2023	1-Chloro-4-ethoxybenzene		C_8H_9ClO	622-61-7	156.609		21	213	1.1254^{20}	1.5252^{20}	s EtOH, eth, HOAc; vs bz; sl ctc
2024	(2-Chloroethoxy)benzene		C_8H_9ClO	622-86-6	156.609		28	218.5			i H_2O; vs EtOH, eth, ace, bz; sl ctc
2025	1-Chloro-1-ethoxyethane		C_4H_9ClO	7081-78-9	108.566			93.5	0.9655^{20}	1.4053^{20}	
2026	2-(2-Chloroethoxy)ethanol		$C_4H_9ClO_2$	628-89-7	124.566			180; 80^{5}	1.18^{25}	1.4529^{20}	vs H_2O; msc EtOH, eth
2027	2-Chloroethyl acetate	β-Chloroethyl acetate	$C_4H_7ClO_2$	542-58-5	122.551			145	1.178^{20}	1.4234^{20}	i H_2O; msc EtOH, eth; s ctc
2028	2-Chloroethyl acetoacetate		$C_6H_9ClO_3$	54527-68-3	164.586			198; 120^{19}	1.2055^{21}	1.4430^{20}	vs bz, eth, EtOH
2029	2-Chloroethylamine hydrochloride	2-Chloroethanamine hydrochloride	$C_2H_7Cl_2N$	870-24-6	115.990		146.3				vs H_2O, ace, EtOH
2030	(1-Chloroethyl)benzene		C_8H_9Cl	672-65-1	140.610			105^{50}			
2031	(2-Chloroethyl)benzene		C_8H_9Cl	622-24-2	140.610			197.5; 92^{20}	1.069^{25}	1.5276^{20}	i H_2O; s EtOH, eth, ace, bz, CS_2
2032	1-Chloro-2-ethylbenzene		C_8H_9Cl	89-96-3	140.610	liq	-82.7	178.4	1.0569^{20}	1.5218^{20}	i H_2O; s ace, bz, ctc, chl
2033	1-Chloro-3-ethylbenzene		C_8H_9Cl	620-16-6	140.610	liq	-55	183.8	1.0529^{20}	1.5195^{20}	vs ace, bz, eth, EtOH
2034	1-Chloro-4-ethylbenzene		C_8H_9Cl	622-98-0	140.610	liq	-62.6	184.4	1.0455^{20}	1.5175^{20}	i H_2O; msc EtOH, eth, ace, peth; s HOAc
2035	2-Chloroethyl chloroformate		$C_3H_4Cl_2O_2$	627-11-2	142.969			155	1.3847^{20}	1.4483^{20}	i H_2O; s EtOH, eth, ace, bz; sl ctc

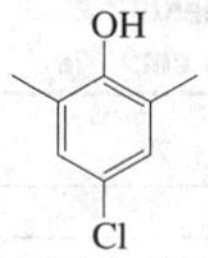

OH Cl
4-Chloro-2,6-dimethylphenol

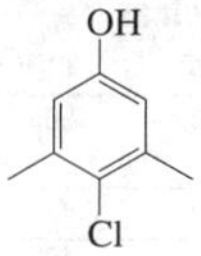

OH Cl
4-Chloro-3,5-dimethylphenol

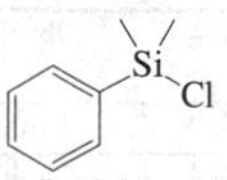

Si Cl
Chlorodimethylphenylsilane

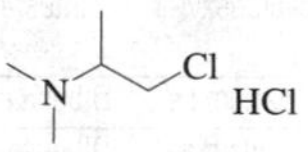

N Cl HCl
1-Chloro-*N*,*N*-dimethyl-2-propanamine, hydrochloride

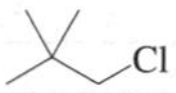

Cl
1-Chloro-2,2-dimethylpropane

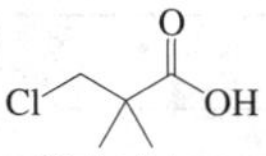

O Cl OH
3-Chloro-2,2-dimethylpropanoic acid

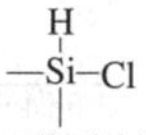

H —Si—Cl
Chlorodimethylsilane

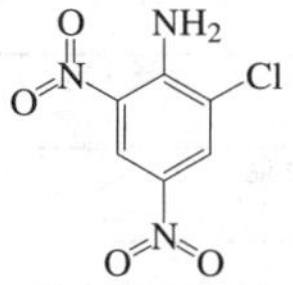

O NH2 N Cl O N O
2-Chloro-4,6-dinitroaniline

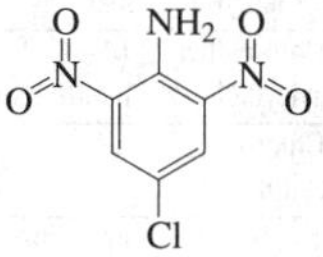

O NH2 O N N O Cl
4-Chloro-2,6-dinitroaniline

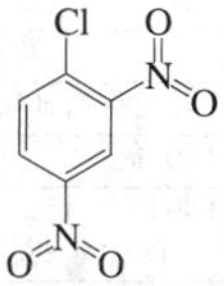

Cl O N O O N O
1-Chloro-2,4-dinitrobenzene

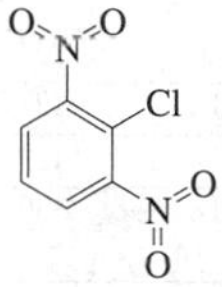

O N O Cl N O O
2-Chloro-1,3-dinitrobenzene

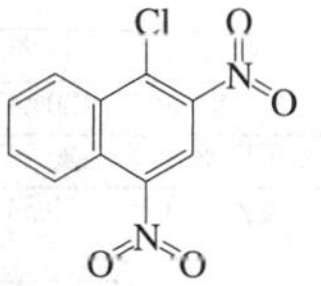

Cl O N O O N O
1-Chloro-2,4-dinitronaphthalene

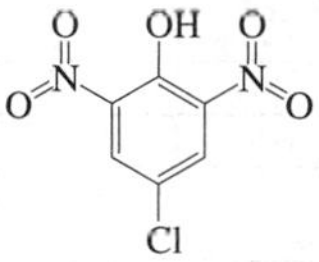

O OH O N N O Cl
4-Chloro-2,6-dinitrophenol

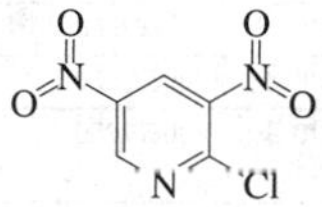

O O N N O N Cl
2-Chloro-3,5-dinitropyridine

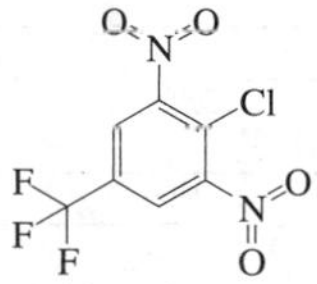

O N O Cl F F F N O O
2-Chloro-1,3-dinitro-5-(trifluoromethyl)benzene

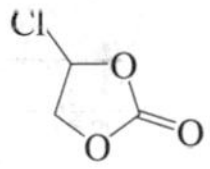

Cl O O O
4-Chloro-1,3-dioxolan-2-one

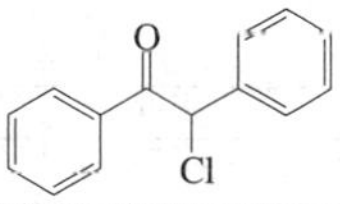

O Cl
2-Chloro-1,2-diphenylethanone

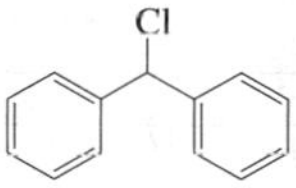

Cl
Chlorodiphenylmethane

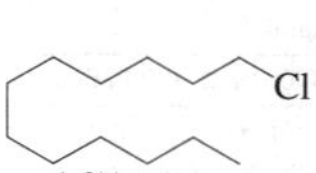

Cl
1-Chlorododecane

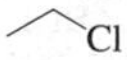

Cl
Chloroethane

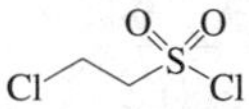

O O Cl S Cl
2-Chloroethanesulfonyl chloride

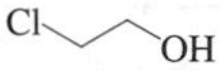

Cl OH
2-Chloroethanol

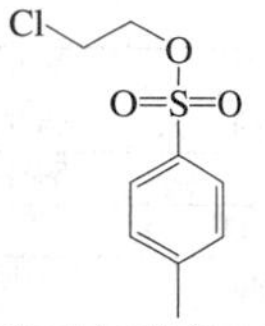

Cl O O=S=O
2-Chloroethanol, 4-methylbenzenesulfonate

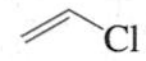

Cl
Chloroethene

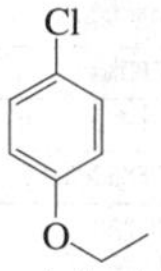

Cl O
1-Chloro-4-ethoxybenzene

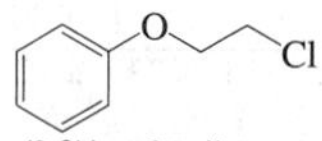

O Cl
(2-Chloroethoxy)benzene

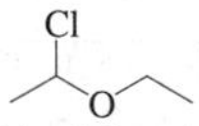

Cl O
1-Chloro-1-ethoxyethane

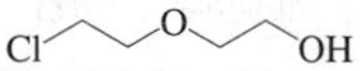

Cl O OH
2-(2-Chloroethoxy)ethanol

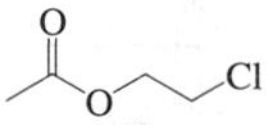

O O Cl
2-Chloroethyl acetate

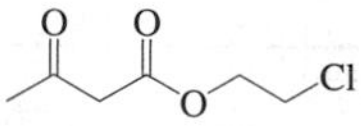

O O O Cl
2-Chloroethyl acetoacetate

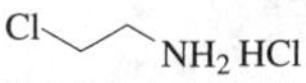

Cl NH2 HCl
2-Chloroethylamine hydrochloride

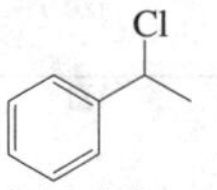

Cl
(1-Chloroethyl)benzene

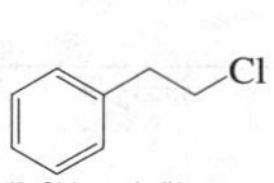

Cl
(2-Chloroethyl)benzene

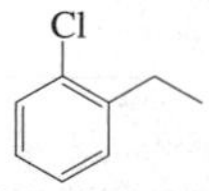

Cl
1-Chloro-2-ethylbenzene

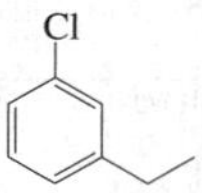

Cl
1-Chloro-3-ethylbenzene

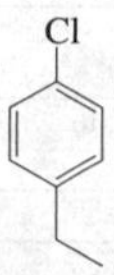

Cl
1-Chloro-4-ethylbenzene

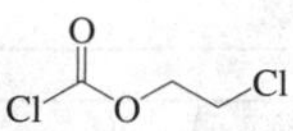

O Cl O Cl
2-Chloroethyl chloroformate

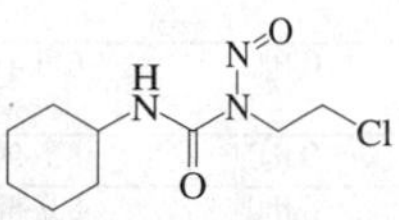

N O H N N Cl O
1-(2-Chloroethyl)-3-cyclohexyl-1-nitrosourea

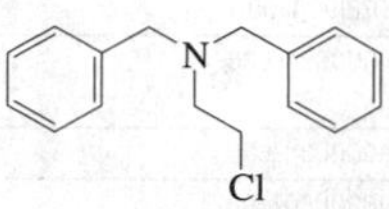

N Cl
N-(2-Chloroethyl)dibenzylamine

No.	Name	Synonym	Mol. Form.	CAS RN	Mol. Wt.	Physical Form	mp/°C	bp/°C	den/ g cm^{-3}	n_D	Solubility
2036	1-(2-Chloroethyl)-3-cyclohexyl-1-nitrosourea	Lomustine	$C_9H_{16}ClN_3O_2$	13010-47-4	233.695	ye pow	90				i H_2O; s EtOH
2037	*N*-(2-Chloroethyl)dibenzylamine	Dibenamine	$C_{16}H_{18}ClN$	51-50-3	259.774	oily liq		169^{3}			
2038	*N*-(2-Chloroethyl)dibenzylamine hydrochloride	Dibenamine hydrochloride	$C_{16}H_{19}Cl_2N$	55-43-6	296.235	cry	194				i H_2O; s EtOH, dil acid
2039	Chloroethyldimethylsilane		$C_4H_{11}ClSi$	6917-76-6	122.669			89.5	0.8675^{20}	1.4105^{20}	
2040	2-Chloroethyl ethyl ether		C_4H_9ClO	628-34-2	108.566			107.5	0.9895^{20}	1.4113^{20}	sl H_2O; msc eth; s chl
2041	2-Chloroethyl isocyanate		C_3H_4ClNO	1943-83-5	105.523			44^{17}			
2042	1-(2-Chloroethyl)-3-(4-methylcyclohexyl)-1-nitrosourea	Semustine	$C_{10}H_{18}ClN_3O_2$	13909-09-6	247.722	cry	64 dec				
2043	5-(2-Chloroethyl)-4-methylthiazole	Clomethiazole	C_6H_8ClNS	533-45-9	161.653	oil		92^{7}	1.233^{25}		
2044	*N*-(2-Chloroethyl)morpholine		$C_6H_{12}ClNO$	3240-94-6	149.618			42^{1}			
2045	4-(2-Chloroethyl)morpholine, hydrochloride		$C_6H_{13}Cl_2NO$	3647-69-6	186.079		185				
2046	1-Chloro-2-(ethylthio)ethane		C_4H_9ClS	693-07-2	124.632			157	1.0663^{25}		
2047	2-Chloroethyl vinyl ether		C_4H_7ClO	110-75-8	106.551	liq	-70	108	1.0495^{20}	1.4378^{20}	vs EtOH, eth; sl chl
2048	3-Chloro-4-fluoroaniline		C_6H_5ClFN	367-21-5	145.562		45.0	227.0			
2049	1-Chloro-2-fluorobenzene		C_6H_4ClF	348-51-6	130.547	liq	-43	137.6	1.2233^{30}	1.4918^{30}	i H_2O; s ace, bz
2050	1-Chloro-3-fluorobenzene		C_6H_4ClF	625-98-9	130.547			127.6	1.221^{25}	1.4911	
2051	1-Chloro-4-fluorobenzene		C_6H_4ClF	352-33-0	130.547	liq	-26.8	130	1.4990^{15}	1.4990^{15}	i H_2O; s EtOH, eth, bz
2052	1-Chloro-1-fluoroethane		C_2H_4ClF	1615-75-4	82.504	vol liq or gas		16.2			
2053	1-Chloro-2-fluoroethane		C_2H_4ClF	762-50-5	82.504			52.8	1.1747^{20}	1.3775^{20}	vs eth, EtOH
2054	Chlorofluoromethane		CH_2ClF	593-70-4	68.478	col gas	-135.1	-9.1			sl H_2O; vs chl
2055	1-Chloro-3-fluoro-2-methylbenzene		C_7H_6ClF	443-83-4	144.574			154	1.191^{25}	1.5026^{20}	
2056	2-Chloro-1-fluoro-4-nitrobenzene	3-Chloro-4-fluoronitrobenzene	$C_6H_3ClFNO_2$	350-30-1	175.545		41.5	229.5			
2057	4-Chloro-1-(4-fluorophenyl)-1-butanone		$C_{10}H_{10}ClFO$	3874-54-2	200.636			136^{6}	1.22^{25}	1.5255^{20}	
2058	3-Chloro-2,5-furandione		C_4HClO_3	96-02-6	132.502		33	196	1.5375^{25}	1.4980^{20}	
2059	1-Chloro-1,2,2,3,3,4,4-heptafluorocyclobutane	Refrigerant C317	C_4ClF_7	377-41-3	216.485	liq or gas	-39.1	25	1.602^{15}		
2060	1-Chloroheptane	Heptyl chloride	$C_7H_{15}Cl$	629-06-1	134.647	liq	-69.5	160.4	0.8762^{20}	1.4264^{20}	i H_2O; msc EtOH, eth; sl ctc; s chl
2061	2-Chloroheptane		$C_7H_{15}Cl$	1001-89-4	134.647			61^{32}, 46^{19}	0.8672^{20}	1.4221^{20}	i H_2O; vs eth; s bz, chl, HOAc
2062	3-Chloroheptane		$C_7H_{15}Cl$	999-52-0	134.647			144; 48^{20}	0.8690^{20}	1.4228^{20}	vs bz, eth
2063	4-Chloroheptane		$C_7H_{15}Cl$	998-95-8	134.647			144	0.8710^{20}	1.4237^{20}	vs bz, eth
2064	7-Chloro-1-heptanol	Heptamethylene chlorohydrin	$C_7H_{15}ClO$	55944-70-2	150.646	cry (peth, bz)	11	150^{20}	0.9998^{15}	1.4537^{25}	vs EtOH, peth
2065	1-Chlorohexadecane		$C_{16}H_{33}Cl$	4860-03-1	260.886		17.9	326.6	0.8635^{20}	1.4503^{20}	i H_2O
2066	1-Chlorohexane	Hexyl chloride	$C_6H_{13}Cl$	544-10-5	120.620	liq	-94.0	135.1	0.8738^{25}	1.4200^{20}	i H_2O; s EtOH, eth, ace, bz; vs chl; sl ctc
2067	2-Chlorohexane	2-Hexyl chloride	$C_6H_{13}Cl$	638-28-8	120.620			122.5	0.8694^{21}	1.4142^{22}	vs ace, bz, eth, EtOH
2068	3-Chlorohexane	3-Hexyl chloride	$C_6H_{13}Cl$	2346-81-8	120.620			123	0.8684^{20}	1.4163^{20}	vs ace, bz, eth, EtOH
2069	6-Chloro-1-hexanol		$C_6H_{13}ClO$	2009-83-8	136.619			107^{12}	1.0241^{20}	1.4550^{20}	sl H_2O; vs EtOH, eth
2070	4-Chloro-17-hydroxyandrost-4-en-3-one, (17β)	Clostebol	$C_{19}H_{27}ClO_2$	1093-58-9	322.869		189				
2071	5-Chloro-2-hydroxybenzaldehyde		$C_7H_5ClO_2$	635-93-8	156.567	pl (al)	100.3	105^{12}			i H_2O; vs EtOH; s eth, alk
2072	4-Chloro-α-hydroxybenzeneacetic acid		$C_8H_7ClO_3$	492-86-4	186.593		120.3				vs bz, EtOH
2073	3-Chloro-4-hydroxybenzoic acid		$C_7H_5ClO_3$	3964-58-7	172.566	nd (w)	171	sub			sl H_2O, bz, chl; vs EtOH, eth, ace
2074	5-Chloro-2-hydroxybenzoic acid		$C_7H_5ClO_3$	321-14-2	172.566	nd (w, al)	174.8				s H_2O, eth; vs EtOH, bz; sl ace
2075	2-Chloro-5-hydroxybenzophenone		$C_{13}H_9ClO_2$	85-19-8	232.662		95.3				i H_2O
2076	3-Chloro-4-hydroxy-5-methoxybenzaldehyde		$C_8H_7ClO_3$	19463-48-0	186.593	tetr	165				i H_2O; s EtOH, HOAc
2077	1-Chloro-2-iodobenzene		C_6H_4ClI	615-41-8	238.453		0.7	234.5	1.9515^{25}	1.6331^{25}	i H_2O; s ace; sl ctc
2078	1-Chloro-3-iodobenzene		C_6H_4ClI	625-99-0	238.453			230	1.9255^{20}		i H_2O; s ace
2079	1-Chloro-4-iodobenzene		C_6H_4ClI	637-87-6	238.453	lf (ace, al)	57	227	1.886^{27}		i H_2O; s EtOH, $PhNO_2$; sl chl
2080	1-Chloro-4-iodobutane		C_4H_8ClI	10297-05-9	218.464	liq		116; 89^{19}	1.785	1.5400^{20}	
2081	Chloroiodomethane		CH_2ClI	593-71-5	176.384			109	2.422^{20}	1.5822^{20}	vs ace, bz, eth, EtOH

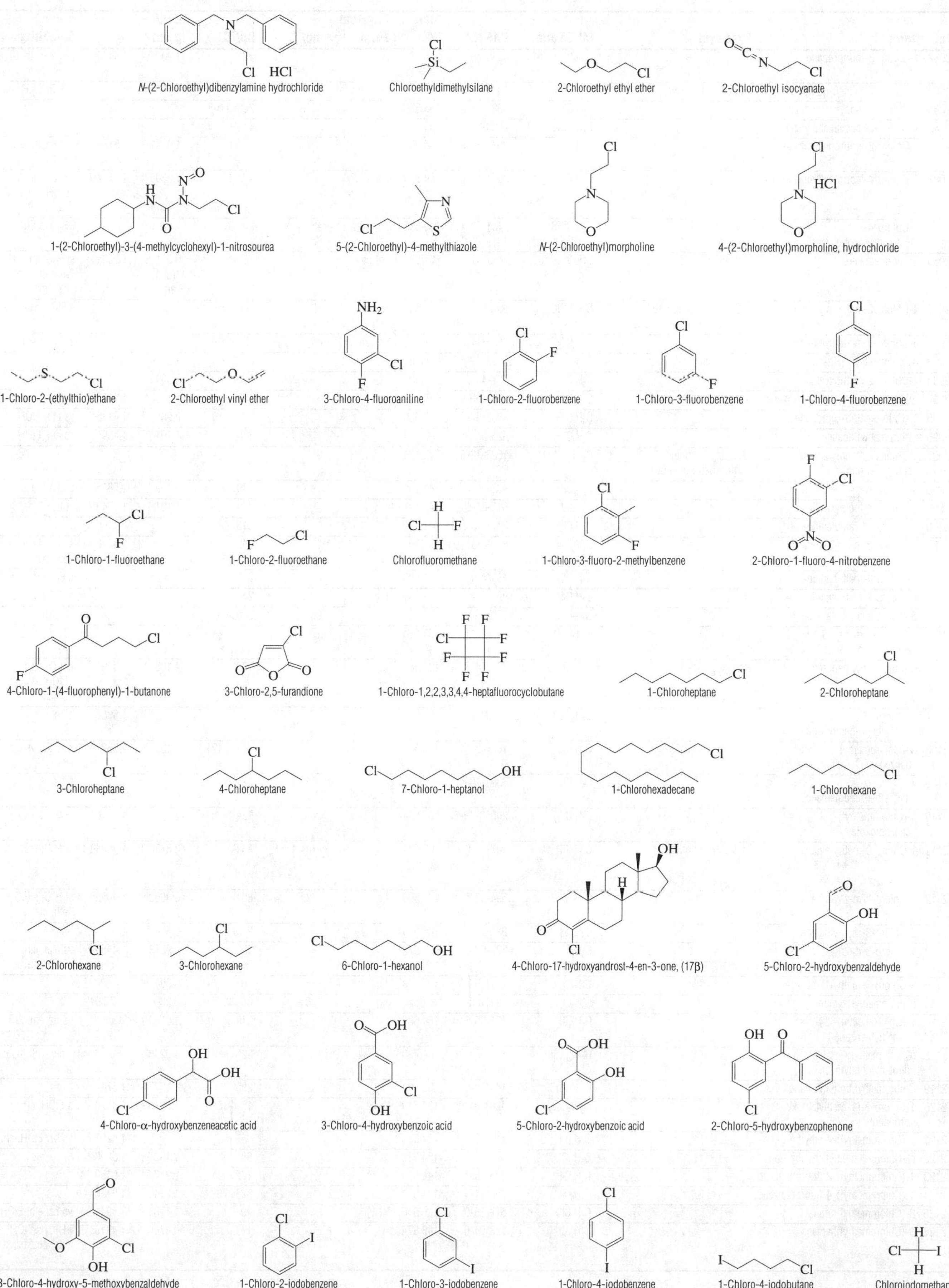

N-(2-Chloroethyl)dibenzylamine hydrochloride
Chloroethyldimethylsilane
2-Chloroethyl ethyl ether
2-Chloroethyl isocyanate
1-(2-Chloroethyl)-3-(4-methylcyclohexyl)-1-nitrosourea
5-(2-Chloroethyl)-4-methylthiazole
N-(2-Chloroethyl)morpholine
4-(2-Chloroethyl)morpholine, hydrochloride
1-Chloro-2-(ethylthio)ethane
2-Chloroethyl vinyl ether
3-Chloro-4-fluoroaniline
1-Chloro-2-fluorobenzene
1-Chloro-3-fluorobenzene
1-Chloro-4-fluorobenzene
1-Chloro-1-fluoroethane
1-Chloro-2-fluoroethane
Chlorofluoromethane
1-Chloro-3-fluoro-2-methylbenzene
2-Chloro-1-fluoro-4-nitrobenzene
4-Chloro-1-(4-fluorophenyl)-1-butanone
3-Chloro-2,5-furandione
1-Chloro-1,2,2,3,3,4,4-heptafluorocyclobutane
1-Chloroheptane
2-Chloroheptane
3-Chloroheptane
4-Chloroheptane
7-Chloro-1-heptanol
1-Chlorohexadecane
1-Chlorohexane
2-Chlorohexane
3-Chlorohexane
6-Chloro-1-hexanol
4-Chloro-17-hydroxyandrost-4-en-3-one, (17β)
5-Chloro-2-hydroxybenzaldehyde
4-Chloro-α-hydroxybenzeneacetic acid
3-Chloro-4-hydroxybenzoic acid
5-Chloro-2-hydroxybenzoic acid
2-Chloro-5-hydroxybenzophenone
3-Chloro-4-hydroxy-5-methoxybenzaldehyde
1-Chloro-2-iodobenzene
1-Chloro-3-iodobenzene
1-Chloro-4-iodobenzene
1-Chloro-4-iodobutane
Chloroiodomethane

No.	Name	Synonym	Mol. Form.	CAS RN	Mol. Wt.	Physical Form	mp/°C	bp/°C	den/ g cm⁻³	n_D	Solubility
2082	1-Chloro-3-iodopropane		C_3H_6ClI	6940-76-7	204.437			171	1.904^{20}	1.5472^{20}	i H_2O; s eth, bz, chl; sl ctc
2083	5-Chloro-7-iodo-8-quinolinol	Iodochlorhydroxyquin	C_9H_5ClINO	130-26-7	305.499	ye br nd (al)	178.5				sl EtOH; s HOAc
2084	1-Chloro-2-isocyanatobenzene		C_7H_4ClNO	3320-83-0	153.566		30.5	200; 115^{43}			sl ctc
2085	1-Chloro-3-isocyanatobenzene		C_7H_4ClNO	2909-38-8	153.566			113^{43}			sl chl
2086	1-Chloro-2-isopropylbenzene		$C_9H_{11}Cl$	2077-13-6	154.636	liq	-74.4	191.1	1.0341^{20}	1.5168^{20}	vs ace, bz, eth, EtOH
2087	1-Chloro-4-isopropylbenzene		$C_9H_{11}Cl$	2621-46-7	154.636	liq	-12.3	198.3	1.0208^{20}	1.5117^{20}	i H_2O; msc EtOH, eth, ace, ctc; vs bz
2088	1-Chloro-4-isothiocyanatobenzene		C_7H_4ClNS	2131-55-7	169.632	nd (al)	46	249.5			i H_2O; s EtOH
2089	Chloromethane	Methyl chloride	CH_3Cl	74-87-3	50.488	col gas	-97.7	-24.09	0.911^{25} (p>1 atm)	1.3389^{20}	sl H_2O; s EtOH; msc eth, ace, bz, chl
2090	4-Chloro-2-methoxyaniline	4-Chloro-2-anisidine	C_7H_8ClNO	93-50-5	157.598	nd or pr (dil al)	52	260			s EtOH, eth, bz, chl
2091	5-Chloro-2-methoxyaniline		C_7H_8ClNO	95-03-4	157.598	nd (dil al)	84				s EtOH; sl lig
2092	(Chloromethoxy)ethane	Chloromethyl ethyl ether	C_3H_7ClO	3188-13-4	94.540			83	1.0188^{15}	1.4040^{20}	
2093	1-Chloro-2-methoxyethane		C_3H_7ClO	627-42-9	94.540			92.5	1.0345^{20}	1.4111^{20}	vs H_2O, eth
2094	[(Chloromethoxy)methyl]benzene		C_8H_9ClO	3587-60-8	156.609			103^{13}	1.1350^{20}	1.5192^{20}	
2095	1-(Chloromethoxy)propane		C_4H_9ClO	3587-57-3	108.566			109	0.9884^{20}	1.4125^{20}	vs eth, EtOH
2096	Chloromethyl acetate		$C_3H_5ClO_2$	625-56-9	108.524			116	1.194^{20}	1.409^{20}	vs eth, EtOH
2097	5-Chloro-2-(methylamino) benzophenone	*N*-Methyl-2-amino-5-chlorobenzophenone	$C_{14}H_{12}ClNO$	1022-13-5	245.704		92				
2098	4-Chloro-*N*-methylaniline		C_7H_8ClN	932-96-7	141.599			240	1.169^{11}	1.5835^{20}	s EtOH, ace, bz
2099	2-Chloro-4-methylaniline		C_7H_8ClN	615-65-6	141.599		7	220	1.151^{20}	1.5748^{22}	sl EtOH, bz
2100	2-Chloro-6-methylaniline		C_7H_8ClN	87-63-8	141.599			215; 97^{10}			
2101	3-Chloro-2-methylaniline		C_7H_8ClN	87-60-5	141.599		1	245		1.5880^{20}	s H_2O, EtOH; i eth, bz
2102	3-Chloro-4-methylaniline		C_7H_8ClN	95-74-9	141.599		26	243			s EtOH; sl ctc
2103	4-Chloro-2-methylaniline	*p*-Chloro-*o*-toluidine	C_7H_8ClN	95-69-2	141.599	lf (al)	30.3	244			s EtOH; sl ctc
2104	5-Chloro-2-methylaniline		C_7H_8ClN	95-79-4	141.599		26	239; 140^{38}			vs EtOH
2105	1-Chloro-2-methyl-9,10-anthracenedione		$C_{15}H_9ClO_2$	129-35-1	256.684		170.5				i EtOH, eth; sl py
2106	(Chloromethyl)benzene	Benzyl chloride	C_7H_7Cl	100-44-7	126.584	liq	-45	179	1.1004^{20}	1.5391^{20}	i H_2O; msc EtOH, eth, chl; sl ctc
2107	3-Chloro-*N*-methylbenzenemethanamine		$C_8H_{10}ClN$	39191-07-6	155.625			88^{4}		1.5350^{25}	s chl
2108	α-(Chloromethyl) benzenemethanol		C_8H_9ClO	1674-30-2	156.609			128^{17}, 121^{11}	1.1926^{20}	1.5523^{20}	s EtOH; vs eth
2109	4-Chloro-α-methylbenzenemethanol		C_8H_9ClO	3391-10-4	156.609			121^{15}		1.5505^{20}	s ctc
2110	5-(Chloromethyl)-1,3-benzodioxole		$C_8H_7ClO_2$	20850-43-5	170.594		20.5	134^{14}	1.312^{25}	1.5660^{20}	
2111	1-Chloro-3-methylbutane	Isopentyl chloride	$C_5H_{11}Cl$	107-84-6	106.594	liq	-104.4	98.9	0.8750^{20}	1.4084^{20}	sl H_2O; msc EtOH, eth; vs chl
2112	2-Chloro-2-methylbutane		$C_5H_{11}Cl$	594-36-5	106.594	liq	-73.5	85.6	0.8653^{20}	1.4055^{20}	sl H_2O; s EtOH, eth, ctc
2113	2-Chloro-3-methylbutane		$C_5H_{11}Cl$	631-65-2	106.594			91.5	0.878^{20}		
2114	1-Chloro-3-methyl-2-butene		C_5H_9Cl	503-60-6	104.578			109	0.9273^{20}	1.4485^{20}	vs ace, eth, EtOH, chl
2115	3-Chloro-3-methyl-1-butyne		C_5H_7Cl	1111-97-3	102.563	liq	-61	76	0.9061^{20}		
2116	(Chloromethyl)cyclopropane		C_4H_7Cl	5911-08-0	90.552	liq	-90.9	88	0.98^{25}	1.4350^{20}	
2117	1-(Chloromethyl)-2,4-dimethylbenzene		$C_9H_{11}Cl$	824-55-5	154.636			215.5; 110^{20}	1.0580^{19}		vs bz, eth, EtOH
2118	(Chloromethyl) dimethylphenylsilane		$C_9H_{13}ClSi$	1833-51-8	184.738			225	1.0240^{25}		s ctc, CS_2
2119	Chloromethyldiphenylsilane		$C_{13}H_{13}ClSi$	144-79-6	232.781			295	1.1277^{20}	1.5742^{20}	
2120	1-Chloro-3-(1-methylethoxy)-2-propanol		$C_6H_{13}ClO_2$	4288-84-0	152.619			182; 87^{20}	1.0910^{20}	1.4370^{25}	s EtOH, eth
2121	1-(Chloromethyl)-4-ethylbenzene		$C_9H_{11}Cl$	1467-05-6	154.636			95^{15}		1.5290^{25}	vs bz, EtOH, chl
2122	(1-Chloro-1-methylethyl)benzene		$C_9H_{11}Cl$	934-53-2	154.636			98^{1}	1.192^{25}	1.5290^{25}	
2123	1-(Chloromethyl)-2-fluorobenzene		C_7H_6ClF	345-35-7	144.574			172; 86^{40}	1.216^{25}	1.5150^{20}	
2124	1-(Chloromethyl)-4-fluorobenzene		C_7H_6ClF	352-11-4	144.574			82^{26}, 76^{20}	1.2143^{20}	1.5130	
2125	2-(Chloromethyl)furan		C_5H_5ClO	617-88-9	116.546			49^{26}	1.1783^{20}	1.4941^{20}	vs bz, eth, EtOH
2126	3-(Chloromethyl)heptane		$C_8H_{17}Cl$	123-04-6	148.674			172	0.8769^{20}	1.4319^{20}	i H_2O; s EtOH, eth, ace, bz; sl ctc

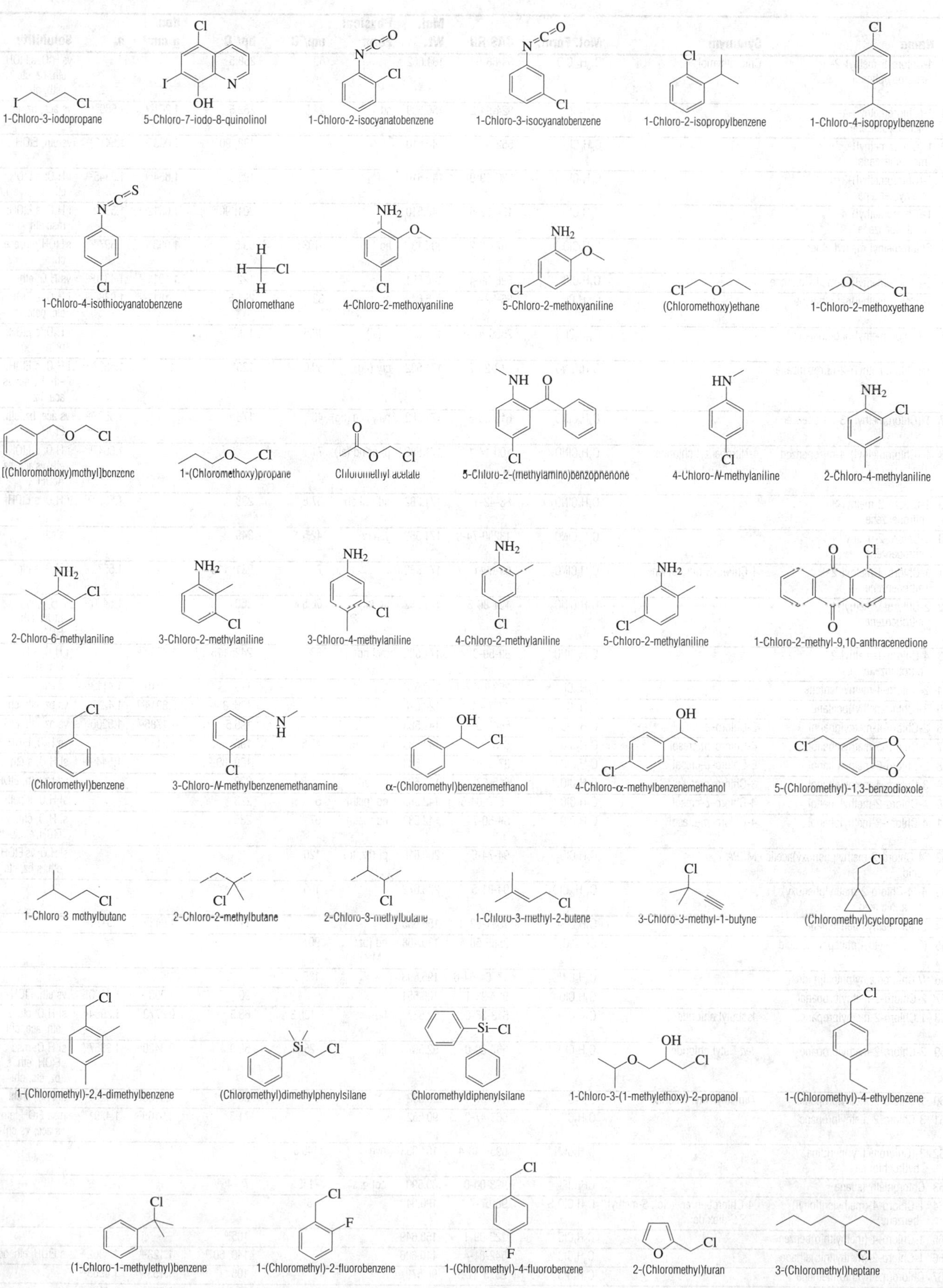

1-Chloro-3-iodopropane
5-Chloro-7-iodo-8-quinolinol
1-Chloro-2-isocyanatobenzene
1-Chloro-3-isocyanatobenzene
1-Chloro-2-isopropylbenzene
1-Chloro-4-isopropylbenzene
1-Chloro-4-isothiocyanatobenzene
Chloromethane
4-Chloro-2-methoxyaniline
5-Chloro-2-methoxyaniline
(Chloromethoxy)ethane
1-Chloro-2-methoxyethane
[(Chloromethoxy)methyl]benzene
1-(Chloromethoxy)propane
Chloromethyl acetate
5-Chloro-2-(methylamino)benzophenone
4-Chloro-*N*-methylaniline
2-Chloro-4-methylaniline
2-Chloro-6-methylaniline
3-Chloro-2-methylaniline
3-Chloro-4-methylaniline
4-Chloro-2-methylaniline
5-Chloro-2-methylaniline
1-Chloro-2-methyl-9,10-anthracenedione
(Chloromethyl)benzene
3-Chloro-*N*-methylbenzenemethanamine
α-(Chloromethyl)benzenemethanol
4-Chloro-α-methylbenzenemethanol
5-(Chloromethyl)-1,3-benzodioxole
1-Chloro-3-methylbutane
2-Chloro-2-methylbutane
2-Chloro-3-methylbutane
1-Chloro-3-methyl-2-butene
3-Chloro-3-methyl-1-butyne
(Chloromethyl)cyclopropane
1-(Chloromethyl)-2,4-dimethylbenzene
(Chloromethyl)dimethylphenylsilane
Chloromethyldiphenylsilane
1-Chloro-3-(1-methylethoxy)-2-propanol
1-(Chloromethyl)-4-ethylbenzene
(1-Chloro-1-methylethyl)benzene
1-(Chloromethyl)-2-fluorobenzene
1-(Chloromethyl)-4-fluorobenzene
2-(Chloromethyl)furan
3-(Chloromethyl)heptane

No.	Name	Synonym	Mol. Form.	CAS RN	Mol. Wt.	Physical Form	mp/°C	bp/°C	den/ g cm^{-3}	n_D	Solubility
2127	4-Chloro-5-methyl-2-isopropylphenol	Chlorothymol	$C_{10}H_{13}ClO$	89-68-9	184.662		63	258.5			vs H_2O; s EtOH, eth, bz, ctc, peth, alk
2128	1-(Chloromethyl)-4-methoxybenzene		C_8H_9ClO	824-94-2	156.609	nd	24.5	262.5	1.261^{20}	1.580^{20}	vs ace, bz, eth
2129	1-(Chloromethyl)-2-methylbenzene		C_8H_9Cl	552-45-4	140.610			198; 90^{20}	1.063^{25}	1.5410^{25}	vs eth, EtOH
2130	1-(Chloromethyl)-3-methylbenzene		C_8H_9Cl	620-19-9	140.610			195.5	1.064^{20}	1.5345^{20}	i H_2O; s EtOH, eth
2131	1-(Chloromethyl)-4-methylbenzene		C_8H_9Cl	104-82-5	140.610			201; 90^{20}	1.0512^{20}	1.5380	i H_2O; s EtOH; msc eth
2132	Chloromethyl methyl ether		C_2H_5ClO	107-30-2	80.513	liq	-103.5	59.5	1.063^{10}	1.397^{20}	s EtOH, eth, ace, chl
2133	2-(Chloromethyl)-2-methyloxirane		C_4H_7ClO	598-09-4	106.551			122	1.1011^{20}	1.4310^{20}	vs H_2O, eth
2134	1-(Chloromethyl)naphthalene		$C_{11}H_9Cl$	86-52-2	176.642	pr	32	291.5	1.1813^{20}	1.6380^{20}	i H_2O; s EtOH, ctc, peth
2135	2-(Chloromethyl)naphthalene		$C_{11}H_9Cl$	2506-41-4	176.642	lf (al)	48.5	169^{20}			i H_2O; s EtOH, peth
2136	1-(Chloromethyl)-2-nitrobenzene		$C_7H_6ClNO_2$	612-23-7	171.582	cry (lig)	50.0	125^{4}		1.5557^{62}	i H_2O; s EtOH, eth, HOAc; vs ace, bz
2137	1-(Chloromethyl)-3-nitrobenzene		$C_7H_6ClNO_2$	619-23-8	171.582	pa ye nd (lig)	46	173^{34}		1.5577^{62}	vs ace, bz, eth, EtOH
2138	1-(Chloromethyl)-4-nitrobenzene	4-Nitrobenzyl chloride	$C_7H_6ClNO_2$	100-14-1	171.582	pl or nd (al)	71			1.5647^{62}	i H_2O; s EtOH, eth; vs ace, bz, AcOEt
2139	1-Chloro-2-methyl-3-nitrobenzene		$C_7H_6ClNO_2$	83-42-1	171.582	nd (dil al)	37.8	238		1.5377^{69}	i H_2O; s EtOH
2140	1-Chloro-2-methyl-4-nitrobenzene		$C_7H_6ClNO_2$	13290-74-9	171.582	ye cry	42.5	249			vs eth
2141	1-Chloro-4-methyl-2-nitrobenzene	4-Chloro-3-nitrotoluene	$C_7H_6ClNO_2$	89-60-1	171.582		7	261; 118^{11}		1.5572^{20}	i H_2O; s ctc
2142	2-Chloro-1-methyl-4-nitrobenzene		$C_7H_6ClNO_2$	121-86-8	171.582	nd (al)	66.5	260		1.5470^{69}	sl H_2O, chl; s EtOH, eth, HOAc
2143	4-Chloro-1-methyl-2-nitrobenzene		$C_7H_6ClNO_2$	89-59-8	171.582	mcl nd	38	242; 115.5^{11}	1.2559^{80}		i H_2O; s EtOH, eth; sl chl
2144	2-Chloro-4-methylpentane		$C_6H_{13}Cl$	25346-32-1	120.620			113	0.8610^{20}	1.4113^{20}	vs eth
2145	3-(Chloromethyl)pentane		$C_6H_{13}Cl$	4737-41-1	120.620			126; 83^{202}	0.8914^{20}	1.4222^{20}	vs bz, eth, chl
2146	2-Chloro-4-methylphenol	2-Chloro-*p*-cresol	C_7H_7ClO	6640-27-3	142.583			195.5	1.1785^{27}	1.5200^{27}	vs bz, eth, EtOH
2147	2-Chloro-5-methylphenol	6-Chloro-*m*-cresol	C_7H_7ClO	615-74-7	142.583	pr (peth)	45.5	196	1.215^{15}		vs H_2O, EtOH
2148	2-Chloro-6-methylphenol	6-Chloro-*o*-cresol	C_7H_7ClO	87-64-9	142.583			189; 80^{20}		1.5449^{20}	sl H_2O; s eth
2149	3-Chloro-4-methylphenol	3-Chloro-*p*-cresol	C_7H_7ClO	615-62-3	142.583	nd (al)	55.5	228			vs bz, eth, EtOH
2150	4-Chloro-2-methylphenol	4-Chloro-*o*-cresol	C_7H_7ClO	1570-64-5	142.583	nd (peth)	51	223			sl H_2O; s peth
2151	4-Chloro-3-methylphenol	4-Chloro-*m*-cresol	C_7H_7ClO	59-50-7	142.583	nd (peth)	67	235			sl H_2O, chl; s EtOH, eth, peth
2152	(4-Chloro-2-methylphenoxy)acetic acid	MCPA	$C_9H_9ClO_3$	94-74-6	200.618	pl (bz, to)	120				sl H_2O; vs EtOH, eth; s bz, ctc
2153	4-(4-Chloro-2-methylphenoxy) butanoic acid		$C_{11}H_{13}ClO_3$	94-81-5	228.672		100				
2154	Chloromethylphenylsilane		C_7H_9ClSi	1631-82-9	156.685			113^{100}	1.043^{20}	1.5171^{20}	
2155	(Chloromethyl)phosphonic acid		CH_4ClO_3P	2565-58-4	130.468	nd (bz/ $MeNO_2$)	90				
2156	*N*-Chloromethylphthalimide		$C_9H_6ClNO_2$	17564-64-6	195.603		135.5				
2157	2-Chloro-2-methylpropanal		C_4H_7ClO	917-93-1	106.551			90	1.053^{15}	1.4160^{16}	vs eth, EtOH
2158	1-Chloro-2-methylpropane	Isobutyl chloride	C_4H_9Cl	513-36-0	92.567	liq	-130.3	68.5	0.8773^{20}	1.3984^{20}	sl H_2O, ctc; s eth, ace, chl
2159	2-Chloro-2-methylpropane	*tert*-Butyl chloride	C_4H_9Cl	507-20-0	92.567	liq	-25.60	50.9	0.8420^{20}	1.3857^{20}	sl H_2O; msc EtOH, eth; s bz, ctc, chl
2160	1-Chloro-2-methylpropene	Dimethylvinyl chloride	C_4H_7Cl	513-37-1	90.552			68	0.9186^{20}	1.4221^{20}	sl H_2O; s chl
2161	3-Chloro-2-methylpropene		C_4H_7Cl	563-47-3	90.552			71.5	0.9165^{20}	1.4291^{20}	msc EtOH, eth; s ace; vs chl
2162	3-(Chloromethyl)pyridine, hydrochloride		$C_6H_7Cl_2N$	6959-48-4	164.033	hyg	143.8				
2163	Chloromethylsilane		CH_5ClSi	993-00-0	80.590	col gas	-135	7; -45^{63}			
2164	1-Chloro-4-(methylsulfonyl) benzene	4-Chlorobenzenethiol, *S*-methyl, *S*,*S*-dioxide	$C_7H_7ClO_2S$	98-57-7	190.648		98				
2165	1-Chloro-4-(methylthio)benzene		C_7H_7ClS	123-09-1	158.649			105^{10}			
2166	1-Chloro-2-(methylthio)ethane		C_3H_7ClS	542-81-4	110.606			140; 60^{30}	1.123^{20}	1.4902^{20}	s EtOH, eth, ace
2167	Chloro(methylthio)methane		C_2H_5ClS	2373-51-5	96.579			105	1.153^{25}	1.4963^{20}	
2168	(Chloromethyl)trimethylsilane		$C_4H_{11}ClSi$	2344-80-1	122.669			98.5	0.879^{25}	1.4175^{20}	

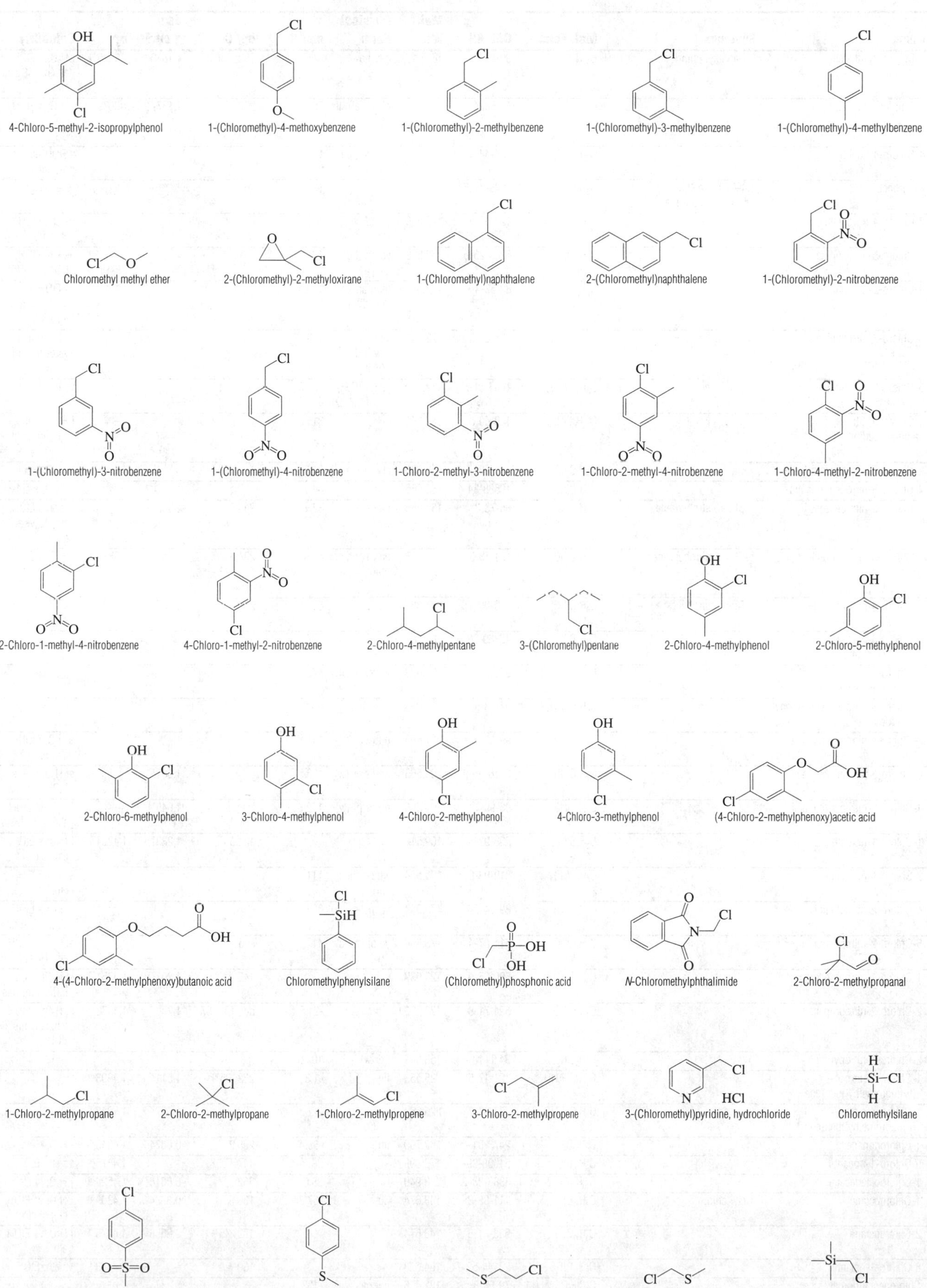
4-Chloro-5-methyl-2-isopropylphenol
1-(Chloromethyl)-4-methoxybenzene
1-(Chloromethyl)-2-methylbenzene
1-(Chloromethyl)-3-methylbenzene
1-(Chloromethyl)-4-methylbenzene
Chloromethyl methyl ether
2-(Chloromethyl)-2-methyloxirane
1-(Chloromethyl)naphthalene
2-(Chloromethyl)naphthalene
1-(Chloromethyl)-2-nitrobenzene
1-(Chloromethyl)-3-nitrobenzene
1-(Chloromethyl)-4-nitrobenzene
1-Chloro-2-methyl-3-nitrobenzene
1-Chloro-2-methyl-4-nitrobenzene
1-Chloro-4-methyl-2-nitrobenzene
2-Chloro-1-methyl-4-nitrobenzene
4-Chloro-1-methyl-2-nitrobenzene
2-Chloro-4-methylpentane
3-(Chloromethyl)pentane
2-Chloro-4-methylphenol
2-Chloro-5-methylphenol
2-Chloro-6-methylphenol
3-Chloro-4-methylphenol
4-Chloro-2-methylphenol
4-Chloro-3-methylphenol
(4-Chloro-2-methylphenoxy)acetic acid
4-(4-Chloro-2-methylphenoxy)butanoic acid
Chloromethylphenylsilane
(Chloromethyl)phosphonic acid
N-Chloromethylphthalimide
2-Chloro-2-methylpropanal
1-Chloro-2-methylpropane
2-Chloro-2-methylpropane
1-Chloro-2-methylpropene
3-Chloro-2-methylpropene
3-(Chloromethyl)pyridine, hydrochloride
Chloromethylsilane
1-Chloro-4-(methylsulfonyl)benzene
1-Chloro-4-(methylthio)benzene
1-Chloro-2-(methylthio)ethane
Chloro(methylthio)methane
(Chloromethyl)trimethylsilane

No.	Name	Synonym	Mol. Form.	CAS RN	Mol. Wt.	Physical Form	mp/°C	bp/°C	den/ g cm^{-3}	n_D	Solubility
2169	1-Chloronaphthalene	1-Naphthyl chloride	$C_{10}H_7Cl$	90-13-1	162.616	oily liq	-2.5	259; 106.5^5	1.1880^{25}	1.6326^{20}	i H_2O; s EtOH, eth, bz, CS_2; sl ctc
2170	2-Chloronaphthalene		$C_{10}H_7Cl$	91-58-7	162.616	pl (dil al), lf	58.0	256	1.1377^{71}	1.6079^{13}	i H_2O; s EtOH, eth, bz, chl, CS_2
2171	4-Chloro-1-naphthol		$C_{10}H_7ClO$	604-44-4	178.615	nd (chl, aq al)	120.5				s EtOH, eth, ace, bz, chl
2172	Chloroneb	1,4-Dichloro-2,5-dimethoxybenzene	$C_8H_8Cl_2O_2$	2675-77-6	207.055		134	268			
2173	2-Chloro-4-nitroaniline		$C_6H_5ClN_2O_2$	121-87-9	172.569	ye nd (w)	108				vs eth, EtOH, HOAc
2174	2-Chloro-5-nitroaniline		$C_6H_5ClN_2O_2$	6283-25-6	172.569	ye nd (lig)	121				vs eth, EtOH, HOAc
2175	4-Chloro-2-nitroaniline		$C_6H_5ClN_2O_2$	89-63-4	172.569	dk oran-ye pr (dil al)	116.5				vs EtOH, eth, HOAc; sl ace, lig
2176	4-Chloro-3-nitroaniline		$C_6H_5ClN_2O_2$	635-22-3	172.569	ye nd or pr (w) nd (peth)	103				s H_2O, eth, chl; vs EtOH; sl lig
2177	5-Chloro-2-nitroaniline		$C_6H_5ClN_2O_2$	1635-61-6	172.569	ye nd (CS_2) ye lf (al, bz)	127.8	sub			vs eth, EtOH
2178	1-Chloro-5-nitro-9,10-anthracenedione		$C_{14}H_6ClNO_4$	129-40-8	287.656		315.3				i H_2O, EtOH, eth, lig; sl bz; s py
2179	2-Chloro-5-nitrobenzaldehyde		$C_7H_4ClNO_3$	6361-21-3	185.565	cry (al)	81.3				vs EtOH, chl
2180	4-Chloro-3-nitrobenzaldehyde		$C_7H_4ClNO_3$	16588-34-4	185.565		64.5				sl H_2O; s chl
2181	1-Chloro-2-nitrobenzene	*o*-Chloronitrobenzene	$C_6H_4ClNO_2$	88-73-3	157.555	mcl nd	32.1	245.5	1.368^{242}		i H_2O; s EtOH, eth, bz; vs ace, tol, py
2182	1-Chloro-3-nitrobenzene	*m*-Chloronitrobenzene	$C_6H_4ClNO_2$	121-73-3	157.555	pa ye orth pr (al)	44.4	235.5	1.343^{50}	1.5374^{80}	i H_2O; s EtOH, eth, bz, chl, CS_2
2183	1-Chloro-4-nitrobenzene	*p*-Chloronitrobenzene	$C_6H_4ClNO_2$	100-00-5	157.555	mcl pr	82	242	1.2979^{90}	1.5376^{100}	i H_2O; sl EtOH; s eth, chl, CS_2
2184	5-Chloro-3-nitro-1,2-benzenediamine		$C_6H_6ClN_3O_2$	42389-30-0	187.584		167				
2185	4-Chloro-3-nitrobenzenesulfonamide		$C_6H_5ClN_2O_4S$	97-09-6	236.633	ye cry (EtOH)	175				
2186	4-Chloro-3-nitrobenzenesulfonyl chloride		$C_6H_3Cl_2NO_4S$	97-08-5	256.064		60.8				
2187	2-Chloro-4-nitrobenzoic acid		$C_7H_4ClNO_4$	99-60-5	201.565	nd (w)	141.8				s H_2O, EtOH, eth, bz
2188	2-Chloro-5-nitrobenzoic acid		$C_7H_4ClNO_4$	2516-96-3	201.565	nd or pr (w)	166.5		1.608^{18}		sl H_2O, ace; s EtOH, eth, bz
2189	4-Chloro-3-nitrobenzoic acid		$C_7H_4ClNO_4$	96-99-1	201.565	nd or pl (w)	182.8		1.645^{18}		i H_2O; sl EtOH, ace
2190	1-Chloro-1-nitroethane		$C_2H_4ClNO_2$	598-92-5	109.512			124.5	1.2837^{20}	1.4224^{20}	i H_2O; s EtOH, ctc, alk
2191	2-Chloro-4-nitrophenol		$C_6H_4ClNO_3$	619-08-9	173.554	wh nd (50% al)	111				s H_2O, EtOH, eth, chl; sl bz
2192	4-Chloro-2-nitrophenol		$C_6H_4ClNO_3$	89-64-5	173.554	ye mcl pr (al)	88.5				i H_2O; s EtOH, eth, chl; sl ace
2193	5-Chloro-2-nitrophenol		$C_6H_4ClNO_3$	611-07-4	173.554	ye pr or nd (w)	41	sub			sl H_2O; s EtOH, eth, HOAc
2194	1-Chloro-1-nitropropane		$C_3H_6ClNO_2$	600-25-9	123.539			142	1.207^{20}	1.4251^{20}	sl H_2O, chl; s EtOH, eth, oils
2195	2-Chloro-2-nitropropane		$C_3H_6ClNO_2$	594-71-8	123.539		-21.5	dec 134; 57^{50}	1.2^{20}	1.4378^{19}	sl H_2O; s EtOH, eth, ctc, oils; i KOH
2196	2-Chloro-3-nitropyridine		$C_5H_3ClN_2O_2$	5470-18-8	158.543	nd (w)	104.0				
2197	1-Chloro-2-nitro-4-(trifluoromethyl)benzene		$C_7H_3ClF_3NO_2$	121-17-5	225.553	liq	-1.3	222; 95^{10}	1.511^{25}	1.4893^{20}	
2198	1-Chloro-4-nitro-2-(trifluoromethyl)benzene		$C_7H_3ClF_3NO_2$	777-37-7	225.553		22	232	1.527^{25}	1.5083^{26}	
2199	1-Chlorononane		$C_9H_{19}Cl$	2473-01-0	162.700	liq	-39.4	205.2	0.8674^{25}	1.4343^{20}	i H_2O; s eth, chl
2200	9-Chloro-1-nonanol		$C_9H_{19}ClO$	51308-99-7	178.699		28	147^{14}		1.4575^{20}	vs eth, EtOH
2201	1-Chlorooctadecane		$C_{18}H_{37}Cl$	3386-33-2	288.940		28.6	352	0.8616^{20}	1.4524^{20}	i H_2O; sl ctc
2202	1-Chlorooctane	Octyl chloride	$C_8H_{17}Cl$	111-85-3	148.674	liq	-57.8	183.5	0.8734^{20}	1.4309^{20}	i H_2O; vs EtOH, eth; sl ctc
2203	2-Chlorooctane		$C_8H_{17}Cl$	628-61-5	148.674			172; 75^{28}	0.8658^{17}	1.4273^{21}	i H_2O; vs EtOH, eth
2204	8-Chloro-1-octanol		$C_8H_{17}ClO$	23144-52-7	164.673			139^{19}		1.4563^{25}	vs eth, EtOH
2205	Chloropentafluoroacetone		C_3ClF_5O	79-53-8	182.476	col gas	-133	8			
2206	Chloropentafluorobenzene		C_6ClF_5	344-07-0	202.509			117.96	1.568^{25}	1.4256^{20}	

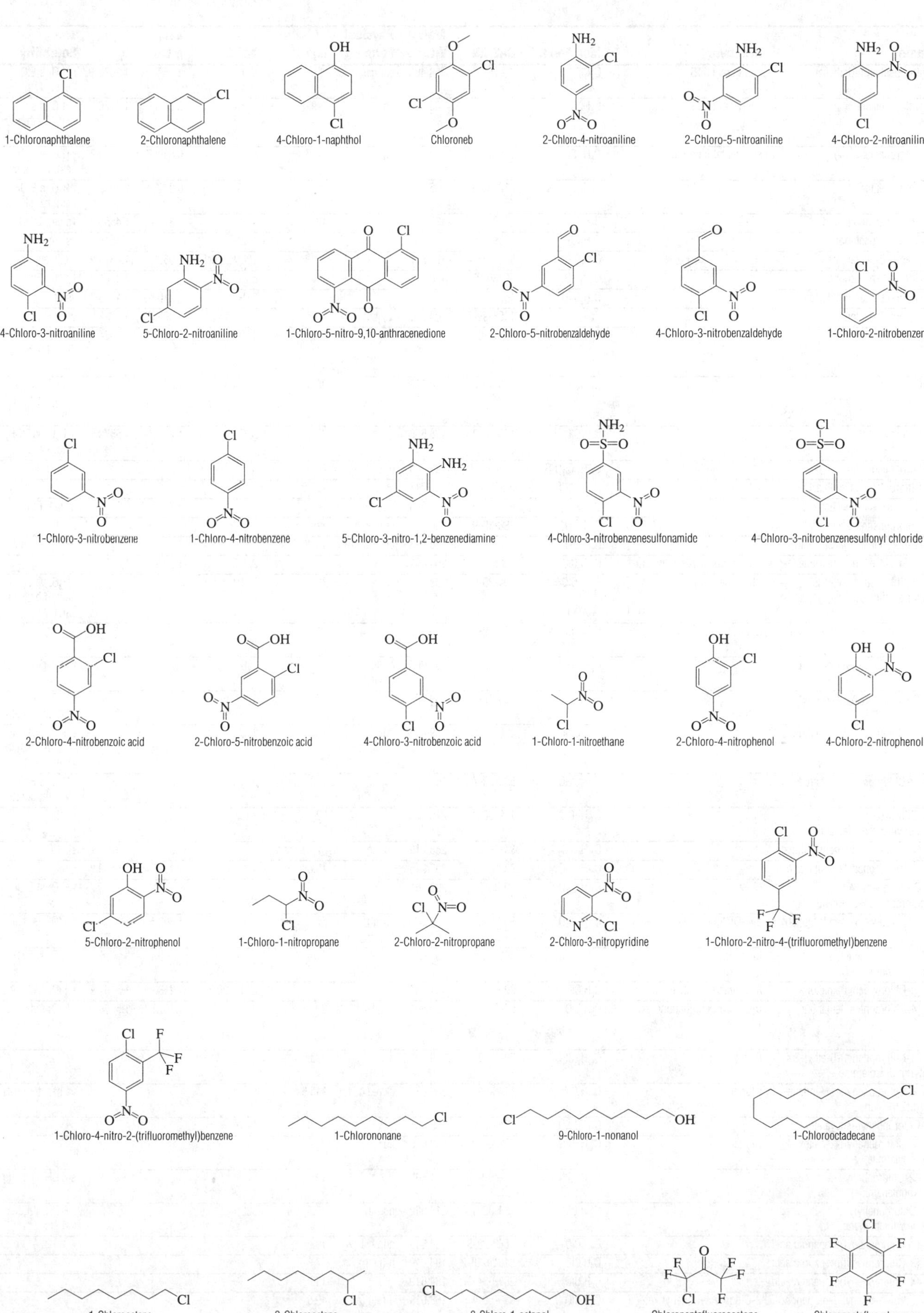
1-Chloronaphthalene
2-Chloronaphthalene
4-Chloro-1-naphthol
Chloroneb
2-Chloro-4-nitroaniline
2-Chloro-5-nitroaniline
4-Chloro-2-nitroaniline
4-Chloro-3-nitroaniline
5-Chloro-2-nitroaniline
1-Chloro-5-nitro-9,10-anthracenedione
2-Chloro-5-nitrobenzaldehyde
4-Chloro-3-nitrobenzaldehyde
1-Chloro-2-nitrobenzene
1-Chloro-3-nitrobenzene
1-Chloro-4-nitrobenzene
5-Chloro-3-nitro-1,2-benzenediamine
4-Chloro-3-nitrobenzenesulfonamide
4-Chloro-3-nitrobenzenesulfonyl chloride
2-Chloro-4-nitrobenzoic acid
2-Chloro-5-nitrobenzoic acid
4-Chloro-3-nitrobenzoic acid
1-Chloro-1-nitroethane
2-Chloro-4-nitrophenol
4-Chloro-2-nitrophenol
5-Chloro-2-nitrophenol
1-Chloro-1-nitropropane
2-Chloro-2-nitropropane
2-Chloro-3-nitropyridine
1-Chloro-2-nitro-4-(trifluoromethyl)benzene
1-Chloro-4-nitro-2-(trifluoromethyl)benzene
1-Chlorononane
9-Chloro-1-nonanol
1-Chlorooctadecane
1-Chlorooctane
2-Chlorooctane
8-Chloro-1-octanol
Chloropentafluoroacetone
Chloropentafluorobenzene

No.	Name	Synonym	Mol. Form.	CAS RN	Mol. Wt.	Physical Form	mp/°C	bp/°C	den/ g cm^{-3}	n_D	Solubility
2207	Chloropentafluoroethane	Refrigerant 115	C_2ClF_5	76-15-3	154.466	col gas	-99.4	-39.1	1.5678^{-42}	1.2678^{-42}	i H_2O; s EtOH, eth
2208	1-Chloropentane	Pentyl chloride	$C_5H_{11}Cl$	543-59-9	106.594	liq	-99.0	108.4	0.8820^{20}	1.4126^{20}	i H_2O; msc EtOH, eth; s bz, ctc; vs chl
2209	2-Chloropentane, (+)	*sec*-Pentyl chloride	$C_5H_{11}Cl$	29882-57-3	106.594	liq	-137	97.0	0.8698^{20}	1.4069^{20}	i H_2O; s EtOH, eth, bz; vs chl
2210	3-Chloropentane		$C_5H_{11}Cl$	616-20-6	106.594	liq	-105	97.5	0.8731^{20}	1.4082^{20}	i H_2O; s EtOH, eth, bz; sl ace
2211	5-Chloropentanoic acid		$C_5H_9ClO_2$	1119-46-6	136.577		18	230	1.3416^{25}	1.4555^{20}	vs eth, EtOH
2212	5-Chloro-1-pentanol		$C_5H_{11}ClO$	5259-98-3	122.593			112^{12}		1.4518^{20}	vs eth, EtOH
2213	5-Chloro-2-pentanone		C_5H_9ClO	5891-21-4	120.577			106^{110}, 76^{34}	1.0523^{20}	1.4375^{20}	s eth, ace; sl ctc
2214	1-Chloro-3-pentanone		C_5H_9ClO	32830-97-0	120.577			68^{20}		1.4361^{20}	vs eth, EtOH
2215	5-Chloropentanoyl chloride		$C_5H_8Cl_2O$	1575-61-7	155.022			83^{12}	1.210^{18}	1.4639^{20}	vs eth
2216	4-Chloro-2-pentene		C_5H_9Cl	1458-99-7	104.578			103; 47^{25}	0.8988^{20}	1.4322^{20}	vs ace, eth, chl
2217	2-Chlorophenol		C_6H_5ClO	95-57-8	128.556		9.4	174.9	1.2634^{20}	1.5524^{20}	sl H_2O, chl; s EtOH, eth; vs bz
2218	3-Chlorophenol		C_6H_5ClO	108-43-0	128.556		32.6	214	1.245^{45}	1.5565^{40}	sl H_2O, chl; s EtOH, eth; vs bz
2219	4-Chlorophenol		C_6H_5ClO	106-48-9	128.556		42.8	220	1.2651^{40}	1.5579^{40}	sl H_2O; vs EtOH, eth, bz; s alk
2220	Chlorophenol Red		$C_{19}H_{12}Cl_2O_5S$	4430-20-0	423.266	grn-br cry	261				sl H_2O; s EtOH
2221	2-Chloro-10*H*-phenothiazine		$C_{12}H_8ClNS$	92-39-7	233.717		198.5				
2222	2-Chlorophenoxyacetic acid		$C_8H_7ClO_3$	614-61-9	186.593	nd (w, al)	148.5				s H_2O, EtOH
2223	3-Chlorophenoxyacetic acid		$C_8H_7ClO_3$	588-32-9	186.593	cry (w)	110				i H_2O
2224	(4-Chlorophenoxy)acetic acid		$C_8H_7ClO_3$	122-88-3	186.593	pr or nd (w)	156.5				vs H_2O; sl chl
2225	1-Chloro-4-phenoxybenzene	4-Chlorophenyl phenyl ether	$C_{12}H_9ClO$	7005-72-3	204.651			284.5	1.2026^{15}	1.599	
2226	3-(4-Chlorophenoxy)-1,2-propanediol	Chlorphenesin	$C_9H_{11}ClO_3$	104-29-0	202.634	cry	78	214^{19}			i H_2O; vs EtOH, eth; s bz, con sulf
2227	2-(3-Chlorophenoxy)propanoic acid	Cloprop	$C_9H_9ClO_3$	101-10-0	200.618	cry	113	$100^{1.5}$			
2228	2-Chloro-*N*-phenylacetamide		C_8H_8ClNO	587-65-5	169.609	nd (dil HOAc)		sub			vs bz, eth, EtOH
2229	*N*-(2-Chlorophenyl)acetamide		C_8H_8ClNO	533-17-5	169.609		88.3				i H_2O; s EtOH, bz, chl; vs eth
2230	*N*-(3-Chlorophenyl)acetamide		C_8H_8ClNO	588-07-8	169.609	nd	79	333			sl H_2O; vs EtOH, eth, bz, CS_2; s chl
2231	*N*-(4-Chlorophenyl)acetamide		C_8H_8ClNO	539-03-7	169.609		179	333	1.385^{22}		i H_2O; s EtOH; vs eth; sl ctc
2232	4-Chloro-α-phenylbenzenemethanol		$C_{13}H_{11}ClO$	119-56-2	218.678		59				sl chl
2233	4-Chlorophenyl benzenesulfonate		$C_{12}H_9ClO_3S$	80-38-6	268.715	col cry	62		1.33		sl H_2O
2234	4-Chloro-1-phenyl-1-butanone		$C_{10}H_{11}ClO$	939-52-6	182.646		19.5	131^{4}	1.137^{25}	1.5459^{20}	
2235	4-Chlorophenyl 4-chlorobenzenesulfonate	Ovex	$C_{12}H_8Cl_2O_3S$	80-33-1	303.161		86.5				i H_2O; sl EtOH; s ace
2236	(2-Chlorophenyl)(4-chlorophenyl) methanone	2,4'-Dichlorodiphenyl ketone	$C_{13}H_8Cl_2O$	85-29-0	251.108	pr (al)	67	214^{22}	1.393^{14}		s EtOH; sl chl
2237	*N'*-(4-Chlorophenyl)-*N,N*-dimethylurea	Monuron	$C_9H_{11}ClN_2O$	150-68-5	198.648	wh pl (MeOH)	170.5				i H_2O; sl EtOH, ace
2238	1-(3-Chlorophenyl)ethanone	*m*-Chloroacetophenone	C_8H_7ClO	99-02-5	154.594			244; 129^{30}	1.2130^{40}	1.5494^{20}	s EtOH, eth, ace
2239	1-(4-Chlorophenyl)ethanone	*p*-Chloroacetophenone	C_8H_7ClO	99-91-2	154.594		20	232	1.1922^{20}	1.5550^{20}	i H_2O; msc EtOH, eth; s chl
2240	5-(4-Chlorophenyl)-6-ethyl-2,4-pyrimidinediamine	Pyrimethamine	$C_{12}H_{13}ClN_4$	58-14-0	248.711		233.5				
2241	2-(4-Chlorophenyl)-1*H*-indene-1,3(2*H*)-dione	Clorindione	$C_{15}H_9ClO_2$	1146-99-2	256.684	dk red nd (al)	145.5				vs bz, eth, EtOH
2242	4-Chlorophenyl isocyanate		C_7H_4ClNO	104-12-1	153.566		31.3	116^{45}			
2243	1-(2-Chlorophenyl)-2-methyl-2-propylamine	Clortermine	$C_{10}H_{14}ClN$	10389-73-8	183.678	liq		117^{16}			
2244	*N*-(2-Chlorophenyl)-3-oxobutanamide		$C_{10}H_{10}ClNO_2$	93-70-9	211.645		106.5				s EtOH; i eth, lig
2245	(4-Chlorophenyl) phenylmethanone		$C_{13}H_9ClO$	134-85-0	216.662	nd (al)	77.5	332			s EtOH, eth, ace; sl ctc
2246	3-(2-Chlorophenyl)propanoic acid		$C_9H_9ClO_2$	1643-28-3	184.619	nd or lf (w)	102				
2247	3-(3-Chlorophenyl)propanoic acid		$C_9H_9ClO_2$	21640-48-2	184.619	lf (peth)	77				
2248	3-(4-Chlorophenyl)propanoic acid		$C_9H_9ClO_2$	2019-34-3	184.619		126				
2249	3-Chloro-1-phenyl-1-propanone	2-Chloroethyl phenyl ketone	C_9H_9ClO	936-59-4	168.619	lf (eth), cry (al, peth)	49.5	113^{4}			

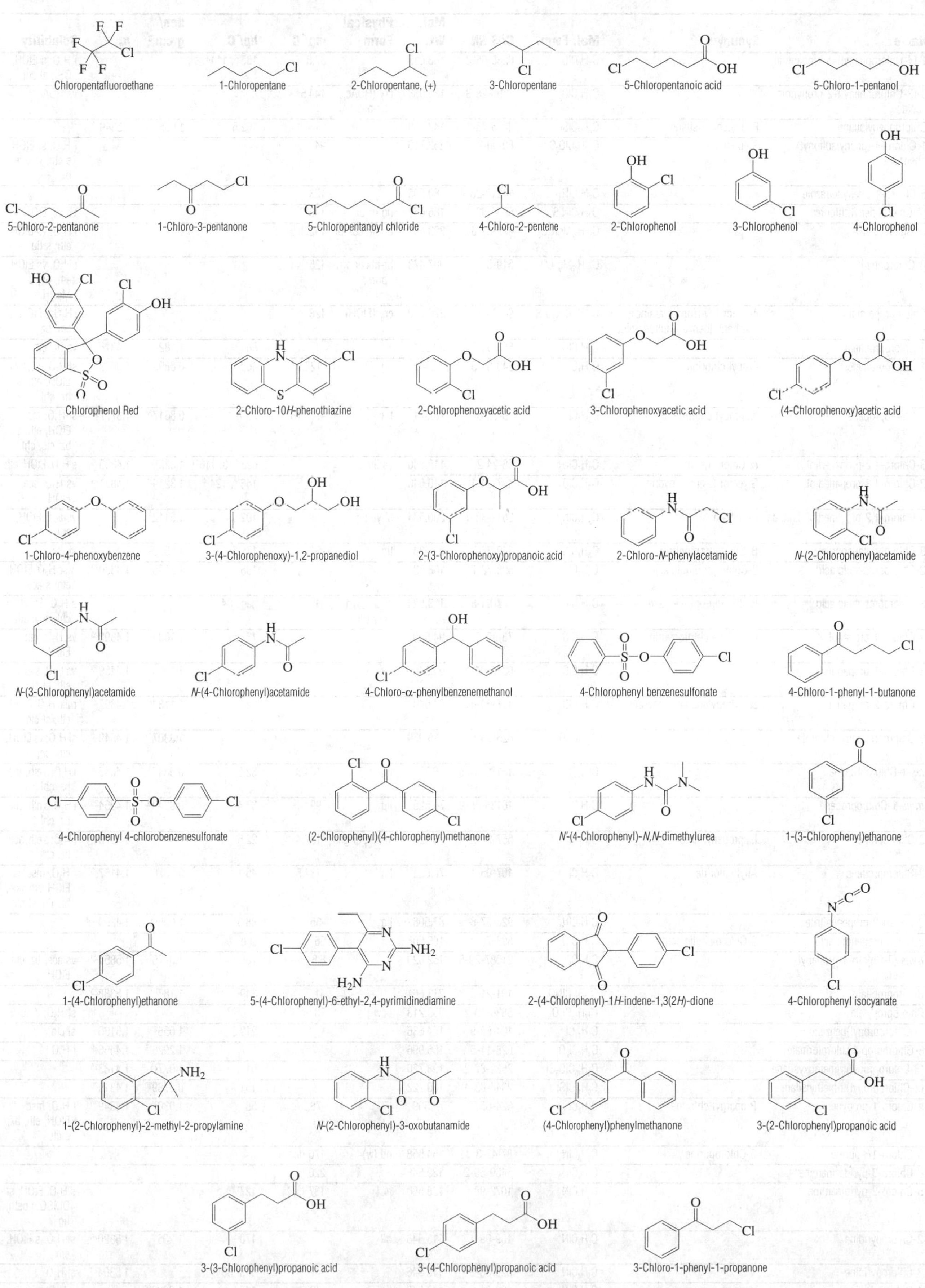
Chloropentafluoroethane
1-Chloropentane
2-Chloropentane, (+)
3-Chloropentane
5-Chloropentanoic acid
5-Chloro-1-pentanol
5-Chloro-2-pentanone
1-Chloro-3-pentanone
5-Chloropentanoyl chloride
4-Chloro-2-pentene
2-Chlorophenol
3-Chlorophenol
4-Chlorophenol
Chlorophenol Red
2-Chloro-10H-phenothiazine
2-Chlorophenoxyacetic acid
3-Chlorophenoxyacetic acid
(4-Chlorophenoxy)acetic acid
1-Chloro-4-phenoxybenzene
3-(4-Chlorophenoxy)-1,2-propanediol
2-(3-Chlorophenoxy)propanoic acid
2-Chloro-N-phenylacetamide
N-(2-Chlorophenyl)acetamide
N-(3-Chlorophenyl)acetamide
N-(4-Chlorophenyl)acetamide
4-Chloro-α-phenylbenzenemethanol
4-Chlorophenyl benzenesulfonate
4-Chloro-1-phenyl-1-butanone
4-Chlorophenyl 4-chlorobenzenesulfonate
(2-Chlorophenyl)(4-chlorophenyl)methanone
N'-(4-Chlorophenyl)-N,N-dimethylurea
1-(3-Chlorophenyl)ethanone
1-(4-Chlorophenyl)ethanone
5-(4-Chlorophenyl)-6-ethyl-2,4-pyrimidinediamine
2-(4-Chlorophenyl)-1H-indene-1,3(2H)-dione
4-Chlorophenyl isocyanate
1-(2-Chlorophenyl)-2-methyl-2-propylamine
N-(2-Chlorophenyl)-3-oxobutanamide
(4-Chlorophenyl)phenylmethanone
3-(2-Chlorophenyl)propanoic acid
3-(3-Chlorophenyl)propanoic acid
3-(4-Chlorophenyl)propanoic acid
3-Chloro-1-phenyl-1-propanone

No.	Name	Synonym	Mol. Form.	CAS RN	Mol. Wt.	Physical Form	mp/°C	bp/°C	den/ g cm^{-3}	n_D	Solubility
2250	1-(4-Chlorophenyl)-1-propanone		C_9H_9ClO	6285-05-8	168.619		37.3	135^{31}, 114^{2}			i H_2O; s EtOH, CS_2; sl chl
2251	3-(3-Chlorophenyl)-2-propynoic acid		$C_9H_5ClO_2$	7396-28-3	180.588	cry (HOAc, bz-peth)	144.5				vs HOAc
2252	Chlorophenylsilane	Phenylchlorosilane	C_6H_7ClSi	4206-75-1	142.659			162.5	1.0683^{20}	1.5340^{20}	
2253	1-Chloro-4-(phenylsulfonyl) benzene	Sulphenone	$C_{12}H_9ClO_2S$	80-00-2	252.716		94				i H_2O; sl EtOH; s eth; vs ace, bz
2254	5-Chloro-1-phenyltetrazole		$C_7H_5ClN_4$	14210-25-4	180.595		123				
2255	(2-Chlorophenyl)thiourea		$C_7H_7ClN_2S$	5344-82-1	186.662	nd or pl	146				vs bz, EtOH
2256	α-Chlorophyll		$C_{55}H_{72}MgN_4O_5$	479-61-8	893.490	bl blk hex pl	152.3				i H_2O; vs EtOH, eth; s lig
2257	β-Chlorophyll		$C_{55}H_{70}MgN_4O_6$	519-62-0	907.473	bl-blk or grn pow	125				i H_2O; vs EtOH, eth, py; s MeOH
2258	Chloropropamide	4-Chloro-*N*-[(propylamino) carbonyl]benzenesulfonamide	$C_{10}H_{13}ClN_2O_3S$	94-20-2	276.739	cry (EtOH)	128				i H_2O; s EtOH; sl eth, bz
2259	2-Chloropropanal		C_3H_5ClO	683-50-1	92.524			86	1.182^{15}	1.431^{17}	vs bz, eth
2260	1-Chloropropane	Propyl chloride	C_3H_7Cl	540-54-5	78.541	liq	-122.9	46.5	0.8899^{20}	1.3879^{20}	sl H_2O, ctc; msc EtOH, eth; s bz, chl
2261	2-Chloropropane	Isopropyl chloride	C_3H_7Cl	75-29-6	78.541	liq	-117.18	35.7	0.8617^{20}	1.3777^{20}	sl H_2O; msc EtOH, eth; s bz, ctc, chl
2262	3-Chloro-1,2-propanediol	α-Chlorohydrin	$C_3H_7ClO_2$	96-24-2	110.540	ye liq		dec 213; 116^{11}	1.325^{18}	1.4809^{20}	s H_2O, EtOH, eth
2263	2-Chloro-1,3-propanediol	Glycerol β-chlorohydrin	$C_3H_7ClO_2$	497-04-1	110.540			146^{18}, 124^{14}	1.3219^{20}	1.4831^{20}	vs H_2O, ace, EtOH
2264	3-Chloro-1,2-propanediol dinitrate	Clonitrate	$C_3H_5ClN_2O_6$	2612-33-1	200.534	sl ye liq		192.5	1.5112^{9}		vs ace, EtOH, chl
2265	3-Chloropropanenitrile	β-Chloropropionitrile	C_3H_4ClN	542-76-7	89.524	liq	-51	175.5	1.1573^{20}	1.4360^{20}	sl ctc
2266	2-Chloropropanoic acid	2-Chloropropionic acid	$C_3H_5ClO_2$	598-78-7	108.524			185	1.2585^{20}	1.4380^{20}	msc H_2O, EtOH, eth; s ace
2267	3-Chloropropanoic acid	β-Chloropropionic acid	$C_3H_5ClO_2$	107-94-8	108.524	lf (w), hyg cry (lig)	41	dec 204			s H_2O, EtOH, chl; msc eth
2268	2-Chloro-1-propanol	Propylene chlorohydrin	C_3H_7ClO	78-89-7	94.540			133.5	1.103^{20}	1.4390^{20}	vs H_2O, eth, EtOH
2269	3-Chloro-1-propanol		C_3H_7ClO	627-30-5	94.540			165	1.1309^{20}	1.4459^{20}	vs H_2O; s EtOH, eth; sl ctc
2270	1-Chloro-2-propanol	*sec*-Propylene chlorohydrin	C_3H_7ClO	127-00-4	94.540			127	1.113^{20}	1.4392^{20}	msc H_2O, EtOH, eth; sl ctc
2271	3-Chloropropanoyl chloride		$C_3H_4Cl_2O$	625-36-5	126.969			144	1.3307^{13}	1.4549^{20}	sl H_2O; vs EtOH, eth, chl
2272	*cis*-1-Chloropropene		C_3H_5Cl	16136-84-8	76.525	liq	-134.8	32.8	0.9347^{20}	1.4055^{20}	i H_2O; s eth, ace, bz, chl
2273	*trans*-1-Chloropropene		C_3H_5Cl	16136-85-9	76.525	liq	-99	37.4	0.9349^{20}	1.4054^{20}	i H_2O; s eth, ace, bz, chl
2274	2-Chloropropene	Isopropenyl chloride	C_3H_5Cl	557-98-2	76.525	vol liq or gas	-137.4	22.6	0.9017^{20}	1.3973^{20}	i H_2O; s eth, ace, bz, chl
2275	3-Chloropropene	Allyl chloride	C_3H_5Cl	107-05-1	76.525	liq	-134.5	45.1	0.9376^{20}	1.4157^{20}	i H_2O; msc EtOH, eth, ace, bz, lig; sl ctc
2276	2-Chloro-2-propenenitrile		C_3H_2ClN	920-37-6	87.508	liq	-65	88.5	1.096^{25}	1.4290^{20}	
2277	2-Chloropropenoic acid	2-Chloroacrylic acid	$C_3H_3ClO_2$	598-79-8	106.508		66	sub			
2278	*trans*-(3-Chloro-1-propenyl) benzene		C_9H_9Cl	21087-29-6	152.620		8.5	106^{13}	1.0926^{20}	1.5851^{20}	vs ace, bz, eth, EtOH
2279	Chloropropham		$C_{10}H_{12}ClNO_2$	101-21-3	213.661		41	149^{2}	1.18^{30}	1.5388^{20}	
2280	Chloropropylate		$C_{17}H_{16}Cl_2O_3$	5836-10-2	339.213	pow	73				sl H_2O; s os
2281	(3-Chloropropyl)benzene		$C_9H_{11}Cl$	104-52-9	154.636			219.5	1.056^{21}	1.5160^{25}	sl ctc
2282	3-Chloropropyl chloroformate		$C_4H_6Cl_2O_2$	628-11-5	156.996			177	1.2926^{25}	1.4456^{20}	i H_2O
2283	(3-Chloropropyl)trimethoxysilane		$C_6H_{15}ClO_3Si$	2530-87-2	198.720			91	1.077^{25}	1.4183^{25}	
2284	(3-Chloropropyl)trimethylsilane		$C_6H_{15}ClSi$	2344-83-4	150.722			151	0.8789^{20}	1.4319^{20}	
2285	3-Chloro-1-propyne	Propargyl chloride	C_3H_3Cl	624-65-7	74.509		-78	58	1.030^{25}	1.4349^{20}	i H_2O; msc EtOH, eth, bz; s ctc
2286	6-Chloro-1*H*-purine	6-Chloropurine	$C_5H_3ClN_4$	87-42-3	154.558	nd (w)	176 dec				
2287	6-Chloro-3-pyridazinamine		$C_4H_4ClN_3$	5469-69-2	129.548		220				
2288	5-Chloro-2-pyridinamine		$C_5H_5ClN_2$	1072-98-6	128.560	pl	137	127^{11}			s H_2O, EtOH; sl DMSO; i peth, lig
2289	2-Chloropyridine		C_5H_4ClN	109-09-1	113.546	oil		170	1.205^{15}	1.5320^{20}	sl H_2O; s EtOH, eth
2290	3-Chloropyridine		C_5H_4ClN	626-60-8	113.546			148; 86^{100}		1.5304^{20}	sl H_2O
2291	4-Chloropyridine		C_5H_4ClN	626-61-9	113.546	liq	-43.5	147.5	1.2000^{25}		s H_2O; msc EtOH

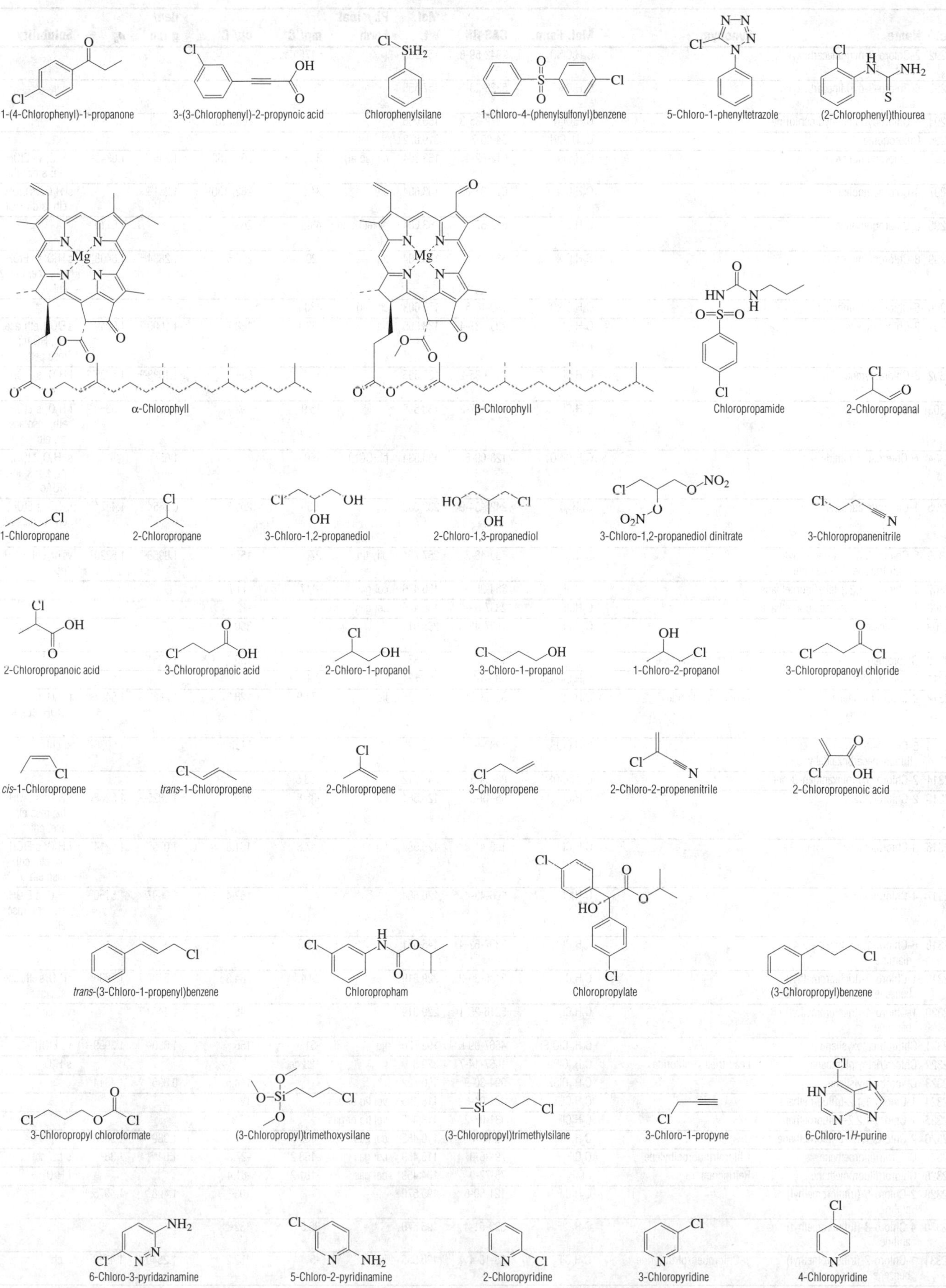

1-(4-Chlorophenyl)-1-propanone
3-(3-Chlorophenyl)-2-propynoic acid
Chlorophenylsilane
1-Chloro-4-(phenylsulfonyl)benzene
5-Chloro-1-phenyltetrazole
(2-Chlorophenyl)thiourea
α-Chlorophyll
β-Chlorophyll
Chloropropamide
2-Chloropropanal
1-Chloropropane
2-Chloropropane
3-Chloro-1,2-propanediol
2-Chloro-1,3-propanediol
3-Chloro-1,2-propanediol dinitrate
3-Chloropropanenitrile
2-Chloropropanoic acid
3-Chloropropanoic acid
2-Chloro-1-propanol
3-Chloro-1-propanol
1-Chloro-2-propanol
3-Chloropropanoyl chloride
cis-1-Chloropropene
trans-1-Chloropropene
2-Chloropropene
3-Chloropropene
2-Chloro-2-propenenitrile
2-Chloropropenoic acid
trans-(3-Chloro-1-propenyl)benzene
Chloropropham
Chloropropylate
(3-Chloropropyl)benzene
3-Chloropropyl chloroformate
(3-Chloropropyl)trimethoxysilane
(3-Chloropropyl)trimethylsilane
3-Chloro-1-propyne
6-Chloro-1H-purine
6-Chloro-3-pyridazinamine
5-Chloro-2-pyridinamine
2-Chloropyridine
3-Chloropyridine
4-Chloropyridine

No.	Name	Synonym	Mol. Form.	CAS RN	Mol. Wt.	Physical Form	mp/°C	bp/°C	den/ g cm-3	n_D	Solubility
2292	2-Chloro-3-pyridinecarboxylic acid		$C_6H_4ClNO_2$	2942-59-8	157.555		>175 dec				
2293	6-Chloro-3-pyridinecarboxylic acid		$C_6H_4ClNO_2$	5326-23-8	157.555		198 dec				
2294	4-Chloropyridine, hydrochloride		$C_5H_5Cl_2N$	7379-35-3	150.006			sub 210			
2295	Chloroquine		$C_{18}H_{26}ClN_3$	54-05-7	319.872		90				
2296	2-Chloroquinoline		C_9H_6ClN	612-62-4	163.604	nd (aq al)	38	266; 153^{22}	1.2464^{25}	1.6342^{25}	i H_2O; vs EtOH, eth; s bz, chl
2297	4-Chloroquinoline		C_9H_6ClN	611-35-8	163.604	cry	34.5	262; 130^{15}	1.251^{25}		sl H_2O; vs EtOH, eth; s dil HCl
2298	6-Chloroquinoline		C_9H_6ClN	612-57-7	163.604	pr (eth), nd (al)	43.8	263		1.6110^{56}	
2299	8-Chloroquinoline		C_9H_6ClN	611-33-6	163.604	liq	-20	288.5	1.2834^{14}	1.6408^{14}	s H_2O; vs EtOH, eth, ace, bz, chl
2300	5-Chloro-8-quinolinol	Cloxyquin	C_9H_6ClNO	130-16-5	179.603	cry (al)	130				
2301	2-Chlorostyrene		C_8H_7Cl	2039-87-4	138.595	liq	-63.1	188.7	1.1000^{20}	1.5649^{20}	s EtOH, eth, ace, ctc, HOAc; msc peth
2302	3-Chlorostyrene		C_8H_7Cl	2039-85-2	138.595			63^{6}	1.1033^{20}	1.5625^{20}	i H_2O; s EtOH, eth
2303	4-Chlorostyrene		C_8H_7Cl	1073-67-2	138.595		15.9	192	1.0868^{20}	1.5660^{20}	i H_2O; s EtOH, eth; msc ace, bz, ctc
2304	*N*-Chlorosuccinimide		$C_4H_4ClNO_2$	128-09-6	133.534	pl (CCl_4)	150		1.65^{25}		sl H_2O, EtOH, bz, lig; s ace, HOAc
2305	1-Chlorotetradecane		$C_{14}H_{29}Cl$	2425-54-9	232.833		4.9	296.8	0.8654^{20}	1.4474^{20}	i H_2O; s EtOH, chl; vs ace, bz; sl ctc
2306	6-Chloro-*N,N,N',N'*-tetraethyl-1,3,5-triazine-2,4-diamine		$C_{11}H_{20}ClN_5$	580-48-3	257.764	oily liq	27	155^{9}	1.0956^{20}	1.5320^{20}	vs bz, chl, EtOH, lig
2307	1-Chloro-1,1,2,2-tetrafluoroethane		C_2HClF_4	354-25-6	136.476	col gas	-117	-11.7			
2308	1-Chloro-1,2,2,2-tetrafluoroethane		C_2HClF_4	2837-89-0	136.476	col gas		-12			
2309	Chlorothalonil		$C_8Cl_4N_2$	1897-45-6	265.911		250	350	1.7^{25}		i H_2O; sl ace, cyhex
2310	Chlorothen	Chloromethapyrilene	$C_{14}H_{18}ClN_3S$	148-65-2	295.831			155^{10}, 192^{5}	1.1751^{25}		
2311	Chlorothiazide		$C_7H_6ClN_3O_4S_2$	58-94-6	295.724		350 dec				
2312	2-Chlorothiophene	2-Thienyl chloride	C_4H_3ClS	96-43-5	118.585	liq	-71.9	128.3	1.2863^{20}	1.5487^{20}	i H_2O; msc EtOH, eth; sl chl
2313	5-Chloro-2-thiophenecarboxaldehyde		C_5H_3ClOS	7283-96-7	146.595			77.5^{5}		1.6036^{25}	sl chl
2314	2-Chloro-9*H*-thioxanthen-9-one		$C_{13}H_7ClOS$	86-39-5	246.712		153.5				
2315	2-Chlorotoluene		C_7H_7Cl	95-49-8	126.584	liq	-35.8	159.0	1.0825^{20}	1.5268^{20}	i H_2O; s EtOH, bz; msc eth, ace, chl
2316	3-Chlorotoluene		C_7H_7Cl	108-41-8	126.584	liq	-47.8	161.8	1.075^{20}	1.5214^{19}	i H_2O; s EtOH, bz, ctc, chl; msc eth
2317	4-Chlorotoluene		C_7H_7Cl	106-43-4	126.584		7.5	162.4	1.0697^{20}	1.5150^{20}	i H_2O; s EtOH, ctc, chl; msc eth
2318	6-Chloro-1,3,5-triazine-2,4-diamine		$C_3H_4ClN_5$	3397-62-4	145.551		>330				
2319	1-Chloro-2-(trichloromethyl) benzene		$C_7H_4Cl_4$	2136-89-2	229.919		29.4	264.3	1.5187^{20}	1.5836^{20}	i H_2O; s eth, ace; sl ctc
2320	1-Chloro-4-(trichloromethyl) benzene		$C_7H_4Cl_4$	5216-25-1	229.919			245	1.4463^{20}		vs ace, eth
2321	Chlorotriethoxysilane		$C_6H_{15}ClO_3Si$	4667-99-6	198.720	liq	-51	156	1.030^{20}	1.3999^{20}	vs EtOH
2322	Chlorotriethylplumbane	Lead triethyl chloride	$C_6H_{15}ClPb$	1067-14-7	329.8		123 dec				s H_2O
2323	Chlorotriethylsilane		$C_6H_{15}ClSi$	994-30-9	150.722			144.5	0.8967^{20}	1.4314^{20}	
2324	1-Chloro-1,1,2-trifluoroethane		$C_2H_2ClF_3$	421-04-5	118.485	vol liq or gas		12			
2325	1-Chloro-1,2,2-trifluoroethane		$C_2H_2ClF_3$	431-07-2	118.485	vol liq or gas		17.3			
2326	2-Chloro-1,1,1-trifluoroethane		$C_2H_2ClF_3$	75-88-7	118.485	col gas	-105.5	6.1	1.389^{0}	1.3090^{0}	
2327	Chlorotrifluoroethene	Chlorotrifluoroethylene	C_2ClF_3	79-38-9	116.469	col gas	-158.2	-27.8	1.54^{-60}	1.38^{0}	s bz, chl
2328	Chlorotrifluoromethane	Refrigerant 13	$CClF_3$	75-72-9	104.459	col gas	-181.2	-81.4			i H_2O
2329	2-Chloro-5-(trifluoromethyl) aniline		$C_7H_5ClF_3N$	121-50-6	195.570			103^{25}	1.428^{25}	1.4975^{20}	
2330	4-Chloro-3-(trifluoromethyl) aniline		$C_7H_5ClF_3N$	320-51-4	195.570		36.5	132^{27}			
2331	1-Chloro-2-(trifluoromethyl) benzene	*o*-Chlorobenzotrifluoride	$C_7H_4ClF_3$	88-16-4	180.555	liq	-6	152.2	1.2540^{30}	1.4513^{25}	s chl
2332	1-Chloro-3-(trifluoromethyl) benzene	*m*-Chlorobenzotrifluoride	$C_7H_4ClF_3$	98-15-7	180.555	liq	-56	137.5	1.3311^{25}	1.4438^{25}	

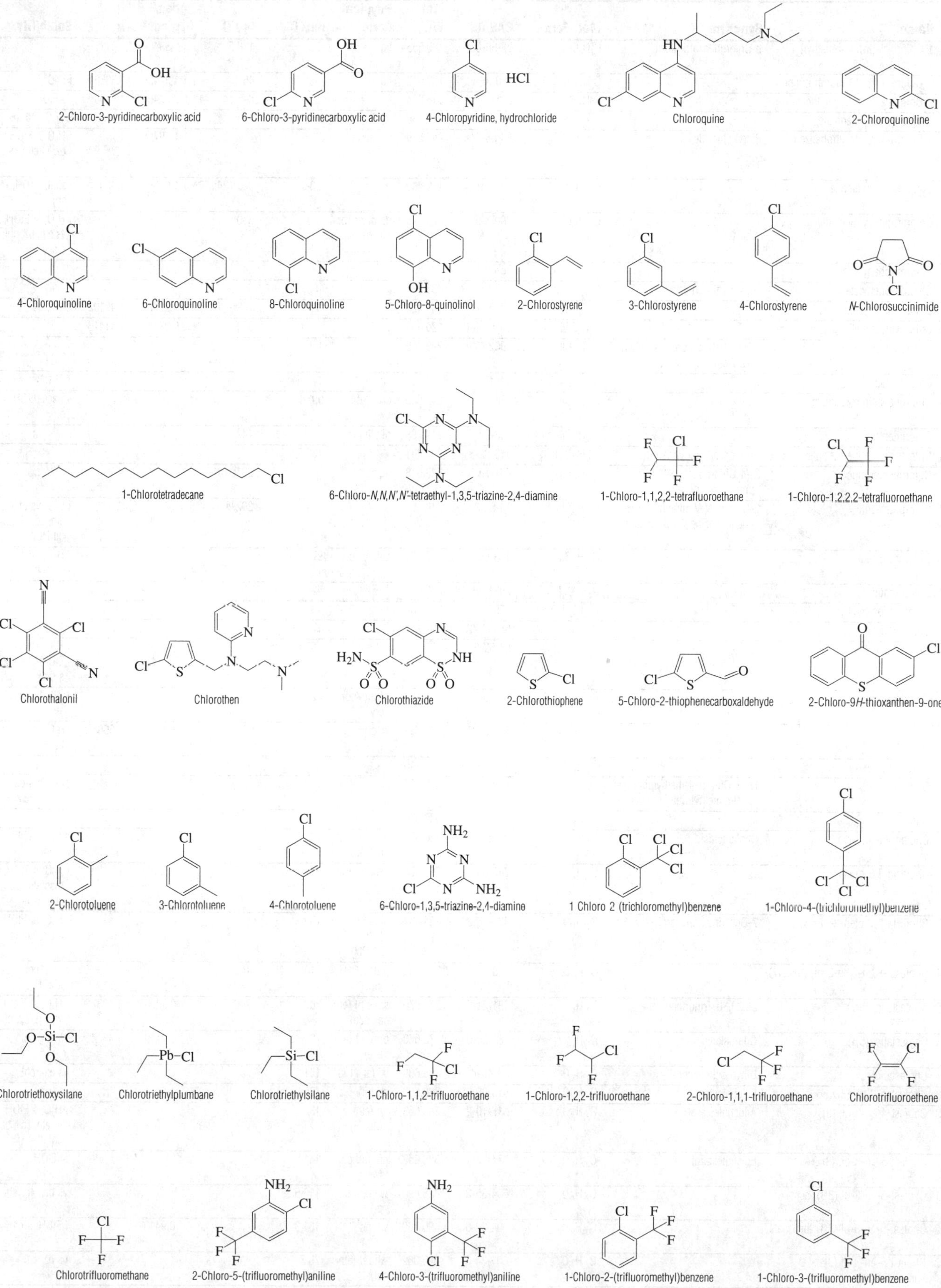

2-Chloro-3-pyridinecarboxylic acid
6-Chloro-3-pyridinecarboxylic acid
4-Chloropyridine, hydrochloride
Chloroquine
2-Chloroquinoline
4-Chloroquinoline
6-Chloroquinoline
8-Chloroquinoline
5-Chloro-8-quinolinol
2-Chlorostyrene
3-Chlorostyrene
4-Chlorostyrene
N-Chlorosuccinimide
1-Chlorotetradecane
6-Chloro-N,N,N',N'-tetraethyl-1,3,5-triazine-2,4-diamine
1-Chloro-1,1,2,2-tetrafluoroethane
1-Chloro-1,2,2,2-tetrafluoroethane
Chlorothalonil
Chlorothen
Chlorothiazide
2-Chlorothiophene
5-Chloro-2-thiophenecarboxaldehyde
2-Chloro-9H-thioxanthen-9-one
2-Chlorotoluene
3-Chlorotoluene
4-Chlorotoluene
6-Chloro-1,3,5-triazine-2,4-diamine
1-Chloro-2-(trichloromethyl)benzene
1-Chloro-4-(trichloromethyl)benzene
Chlorotriethoxysilane
Chlorotriethylplumbane
Chlorotriethylsilane
1-Chloro-1,1,2-trifluoroethane
1-Chloro-1,2,2-trifluoroethane
2-Chloro-1,1,1-trifluoroethane
Chlorotrifluoroethene
Chlorotrifluoromethane
2-Chloro-5-(trifluoromethyl)aniline
4-Chloro-3-(trifluoromethyl)aniline
1-Chloro-2-(trifluoromethyl)benzene
1-Chloro-3-(trifluoromethyl)benzene

No.	Name	Synonym	Mol. Form.	CAS RN	Mol. Wt.	Physical Form	mp/°C	bp/°C	den/ g cm^{-3}	n_D	Solubility
2333	1-Chloro-4-(trifluoromethyl) benzene	*p*-Chlorobenzotrifluoride	$C_7H_4ClF_3$	98-56-6	180.555	liq	-33	138.5	1.3340^{25}	1.4431^{30}	
2334	3-Chloro-1,1,1-trifluoropropane		$C_3H_4ClF_3$	460-35-5	132.512	liq	-106.5	45.1	1.3253^{20}	1.3350^{20}	i H_2O
2335	2-Chloro-2,4,4-trimethylpentane		$C_8H_{17}Cl$	6111-88-2	148.674		-26	dec 147; 44^{16}	0.8746^{20}	1.4308^{20}	vs EtOH
2336	Chlorotrimethylstannane		C_3H_9ClSn	1066-45-1	199.266		38.5	148			s H_2O, chl, os
2337	2-Chloro-1,3,5-trinitrobenzene	Picryl chloride	$C_6H_2ClN_3O_6$	88-88-0	247.549	wh nd or pl (chl, al-lig)	83		1.797^{20}		i H_2O; s EtOH, bz; sl eth; vs ace, tol
2338	Chlorotrinitromethane		$CClN_3O_6$	1943-16-4	185.480		2.3	dec 134; 56^{40}	1.6769^{20}	1.4500^{20}	vs eth, EtOH, chl
2339	Chlorotriphenylmethane		$C_{19}H_{15}Cl$	76-83-5	278.775	nd or pr (bz-peth)	113.5	310			i H_2O; sl EtOH; vs eth, bz, chl; s ace
2340	Chlorotriphenylsilane		$C_{18}H_{15}ClSi$	76-86-8	294.851			241^{35}			
2341	Chlorotriphenylstannane	Triphenyltin chloride	$C_{18}H_{15}ClSn$	639-58-7	385.475		103.5				s chl
2342	Chlorotripropylstannane		$C_9H_{21}ClSn$	2279-76-7	283.426		-23.5	123^{13}	1.2678^{28}	1.49102^{28}	s ctc, os
2343	Chlorovinyldimethylsilane		C_4H_9ClSi	1719-58-0	120.653			83.5	0.8744^{20}	1.4141^{20}	
2344	Chloroxuron	*N'*-[4-(4-Chlorophenoxy)phenyl] -*N,N*-dimethylurea	$C_{15}H_{15}ClN_2O_2$	1982-47-4	290.745		151				
2345	Chlorozotocin		$C_9H_{16}ClN_3O_7$	54749-90-5	313.692	cry	147 dec				s H_2O
2346	Chlorphenesin carbamate		$C_{10}H_{12}ClNO_4$	886-74-8	245.660	cry (bz)	90				vs ace, EtOH, diox
2347	Chlorpheniramine		$C_{16}H_{19}ClN_2$	132-22-9	274.788	oily liq		142^1			
2348	Chlorpheniramine maleate	Chloroprophenpyridamine	$C_{20}H_{23}ClN_2O_4$	113-92-8	390.861		132.5				
2349	Chlorphentermine	2-(4-Chlorobenzyl)-2-propylamine	$C_{10}H_{14}ClN$	461-78-9	183.678	liq		231; 101^2			
2350	Chlorpromazine	2-Chloro-*N,N*-dimethyl-10*H*-phenothiazine-10-propanamine	$C_{17}H_{19}ClN_2S$	50-53-3	318.864			$202^{0.8}$			i H_2O; vs EtOH, eth, bz, chl; s dil HCl
2351	Chlorprothixene		$C_{18}H_{18}ClNS$	113-59-7	315.861	pale ye cry	97				i H_2O, EtOH, eth, chl
2352	Chlorpyrifos		$C_9H_{11}Cl_3NO_3PS$	2921-88-2	350.586		42				
2353	Chlorpyrifos-methyl		$C_7H_7Cl_3NO_3PS$	5598-13-0	322.534		43				
2354	Chlorsulfuron		$C_{12}H_{12}ClN_5O_4S$	64902-72-3	357.773		176				
2355	Chlortetracycline		$C_{22}H_{23}ClN_2O_8$	57-62-5	478.879	gold-ye	168.5				i H_2O, eth; sl EtOH, ace, bz; s diox
2356	Chlorthalidone		$C_{14}H_{11}ClN_2O_4S$	77-36-1	338.765	wh pow or cry	225 dec				s alk, EtOH; sl eth
2357	Chlorthion		$C_8H_9ClNO_5PS$	500-28-7	297.653	ye cry	21	$125^{0.1}$	1.437^{20}	1.5661^{20}	i H_2O; vs bz, eth, EtOH
2358	Chlorthiophos		$C_{11}H_{15}Cl_2O_3PS_2$	21923-23-9	361.245			$150^{0.001}$			
2359	Chlortoluron	*N'*-(3-Chloro-4-methylphenyl)-*N,N*-dimethylurea	$C_{10}H_{13}ClN_2O$	15545-48-9	212.675	cry	147				sl H_2O; s os
2360	Cholane		$C_{24}H_{42}$	548-98-1	330.590	pr (al)	90	$190^{0.001}$			
2361	Cholan-24-oic acid	Cholanic acid	$C_{24}H_{40}O_2$	25312-65-6	360.574	nd (al), cry (HOAc)	163.5				s EtOH, chl, HOAc
2362	Cholesta-3,5-diene		$C_{27}H_{44}$	747-90-0	368.638	wh nd (al)	80	260^{13}	0.925^{100}		i H_2O; s EtOH; msc eth, bz, chl; vs lig
2363	Cholesta-5,7-dien-3-ol, (3β)	7-Dehydrocholesterol	$C_{27}H_{44}O$	434-16-2	384.637	pl (+1w), (eth-MeOH)	150.5				i H_2O; sl EtOH; s eth, ace
2364	Cholesta-8,24-dien-3-ol, (3β,5α)		$C_{27}H_{44}O$	128-33-6	384.637	pl (MeOH) ,nd	110	$160^{0.001}$			s ace, chl, MeOH
2365	Cholestane, (5α)	28,29,30-Trinorlanostane	$C_{27}H_{48}$	481-21-0	372.670	sc or pl (eth-al, ace)	80	250^1	0.9090^{88}	1.4887^{88}	i H_2O; sl EtOH; vs eth, bz, chl
2366	Cholestane, (5β)	Coprostane	$C_{27}H_{48}$	481-20-9	372.670	orth nd (al, ace)	72		0.9119^{87}	1.4884^{88}	vs eth, chl
2367	Cholestanol	Dihydrocholesterol	$C_{27}H_{48}O$	80-97-7	388.669	sc (al,+1w)	141.5				vs eth, chl
2368	Cholestan-3-ol, (3α,5α)	Epicholestanol	$C_{27}H_{48}O$	516-95-0	388.669	nd (al)	185.5				s chl
2369	Cholest-4-en-3-ol, (3β)	Allocholesterol	$C_{27}H_{46}O$	517-10-2	386.653	nd (eth-MeOH)	132				i H_2O; s EtOH; vs eth, ace, bz, chl
2370	Cholest-5-en-3-ol, (3α)	Epicholesterol	$C_{27}H_{46}O$	474-77-1	386.653	cry (al, chl-MeOH)	141.5				sl EtOH
2371	Cholest-5-en-3-ol (3β), acetate		$C_{29}H_{48}O_2$	604-35-3	428.690	wh nd (ace, al)	115.5				vs bz, eth, chl
2372	Cholest-5-en-3-ol (3β), benzoate		$C_{34}H_{50}O_2$	604-32-0	490.760	wh nd	151.3		0.9413^{200}		i EtOH; s eth, chl
2373	Cholest-5-en-3-ol (3β)-, hexadecanoate		$C_{43}H_{76}O_2$	601-34-3	625.062	wh nd (eth al)	79.3				vs bz, chl
2374	Cholest-5-en-3-ol (3β)-, *cis*-9-octadecenoate		$C_{45}H_{78}O_2$	303-43-5	651.100		46.3				s chl

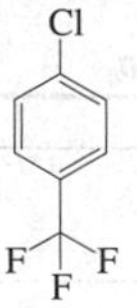

Cl
F F
F
1-Chloro-4-(trifluoromethyl)benzene
3-Chloro-1,1,1-trifluoropropane
2-Chloro-2,4,4-trimethylpentane
Chlorotrimethylstannane 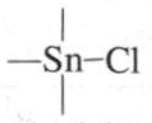
2-Chloro-1,3,5-trinitrobenzene 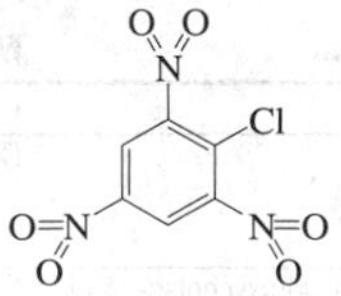
Chlorotrinitromethane 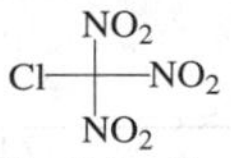
Chlorotriphenylmethane 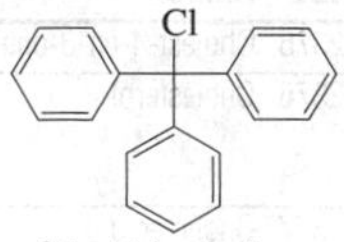
Chlorotriphenylsilane 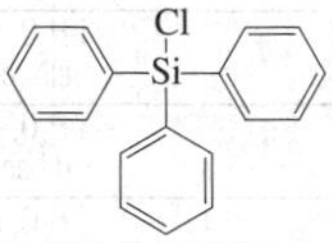
Chlorotriphenylstannane 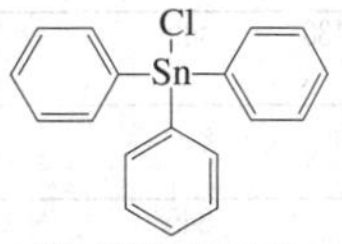
Chlorotripropylstannane 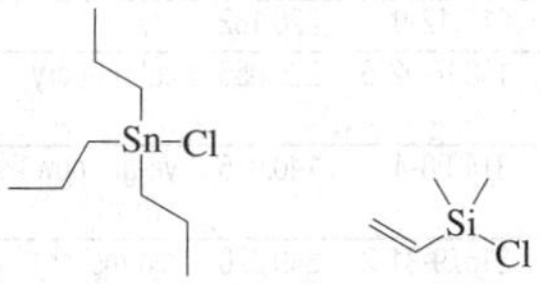
Chlorovinyldimethylsilane
Chloroxuron 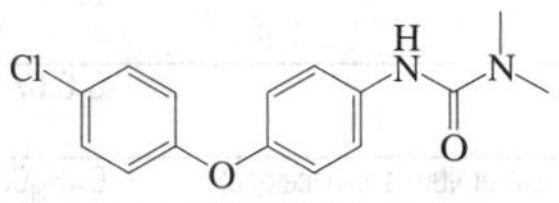
Chlorozotocin 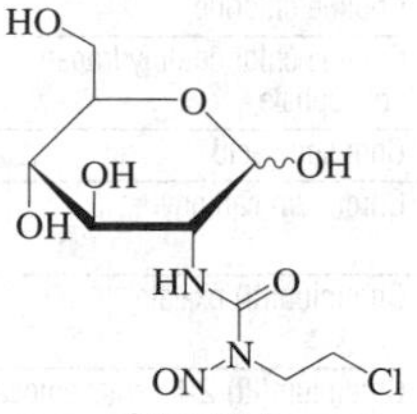
Chlorphenesin carbamate 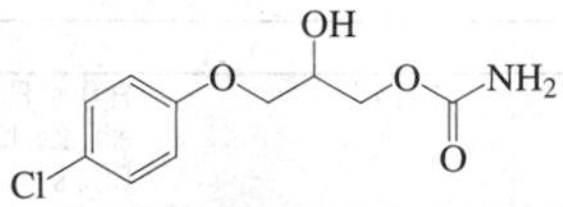
Chlorpheniramine 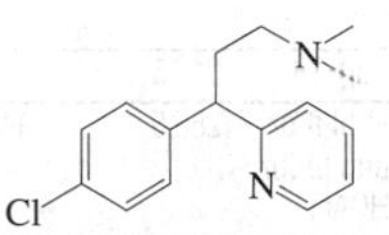
Chlorpheniramine maleate 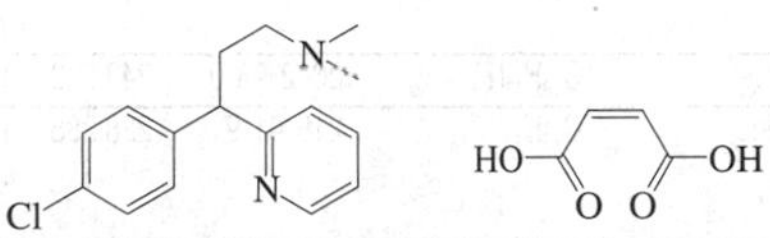
Chlorphentermine 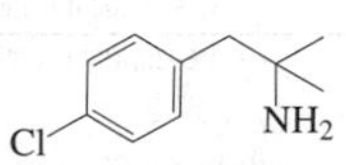
Chlorpromazine 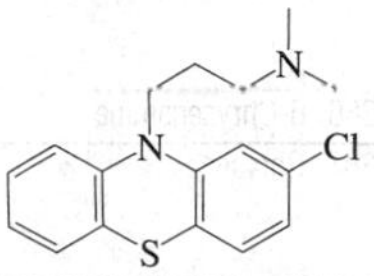
Chlorprothixene 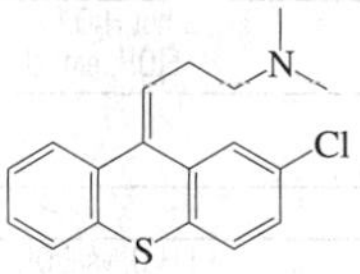
Chlorpyrifos 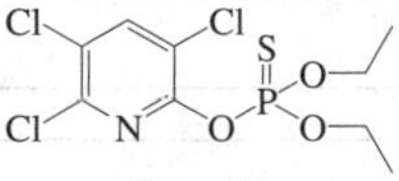
Chlorpyrifos-methyl 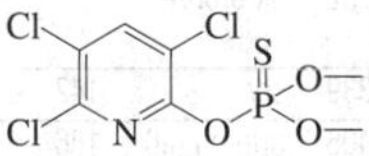
Chlorsulfuron 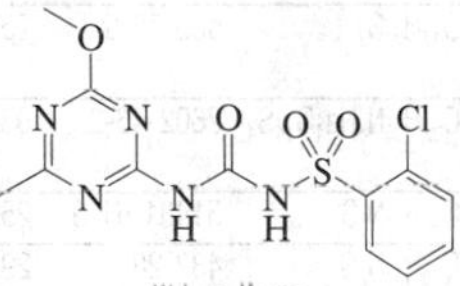
Chlortetracycline 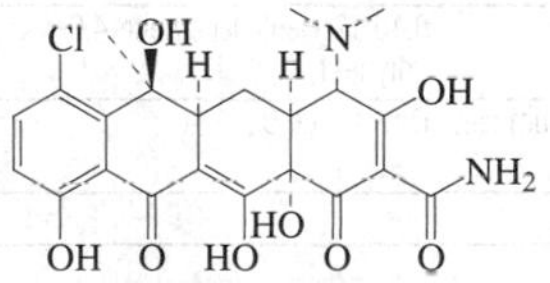
Chlorthalidone 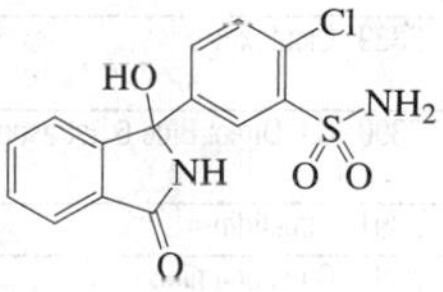
Chlorthion 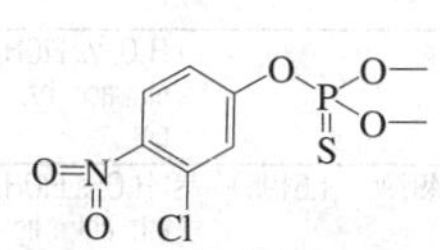
Chlorthiophos 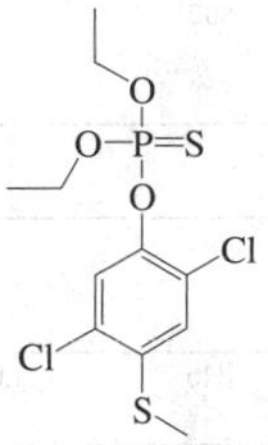
Chlortoluron 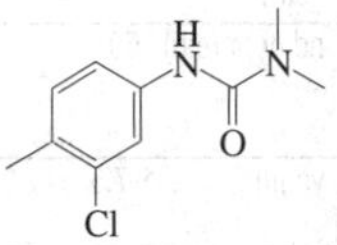
Cholane 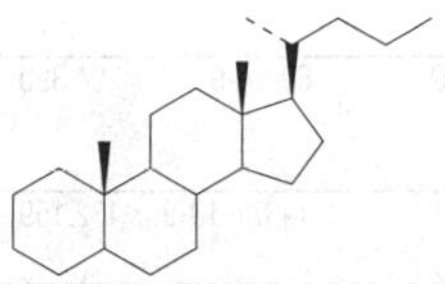
Cholan-24-oic acid 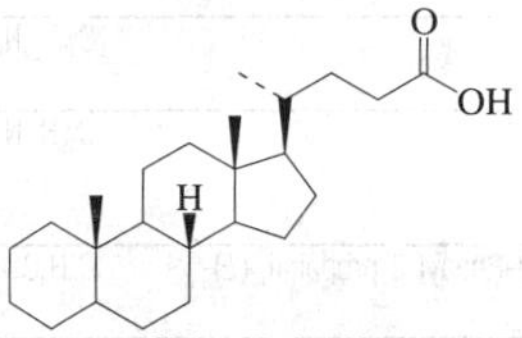
Cholesta-3,5-diene 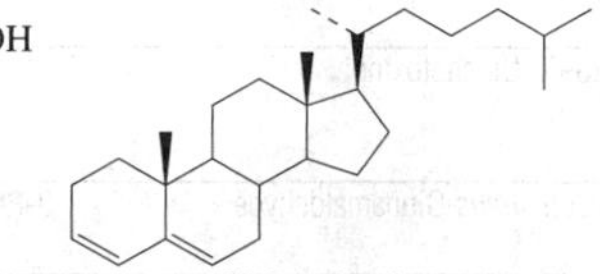
Cholesta-5,7-dien-3-ol, (3β) 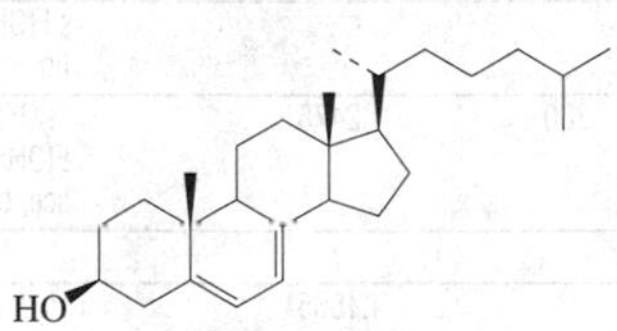
Cholesta-8,24-dien-3-ol, (3β,5α) 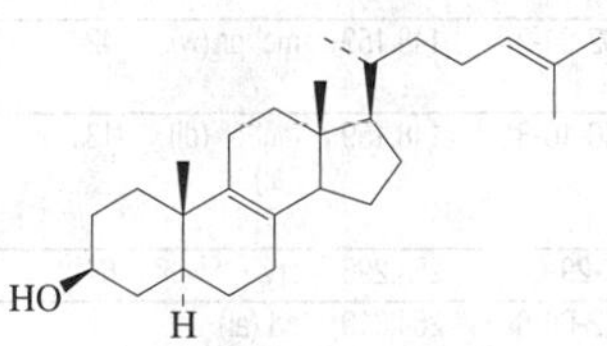
Cholestane, (5α) 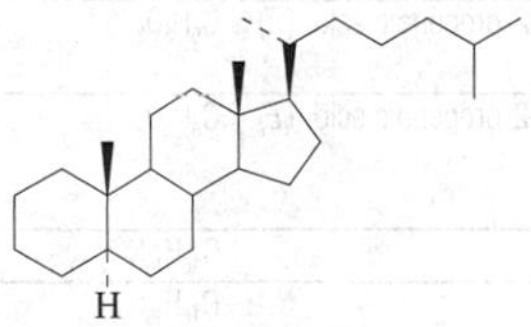
Cholestane, (5β) 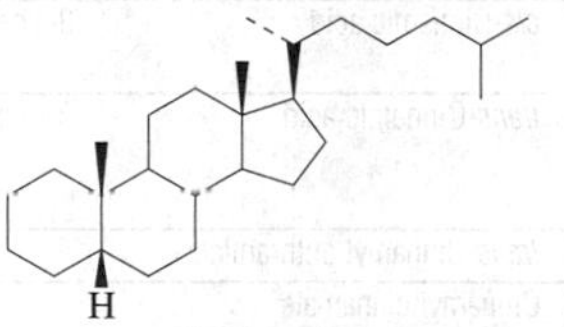
Cholestanol 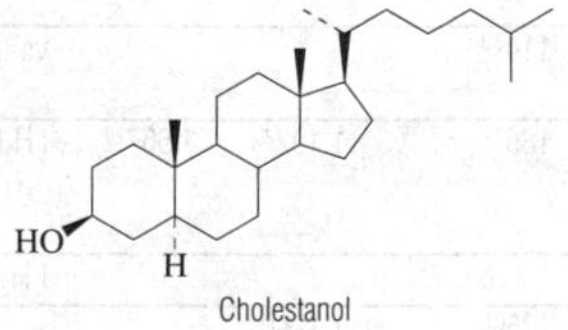
Cholestan-3-ol, (3α,5α) 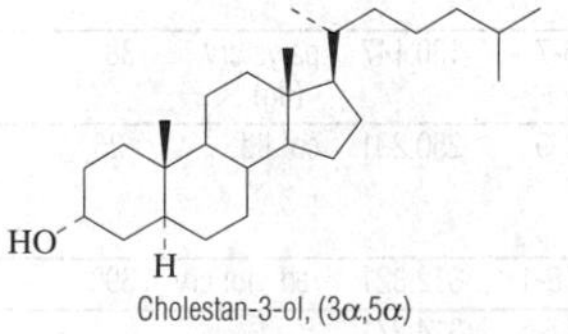
Cholest-4-en-3-ol, (3β) 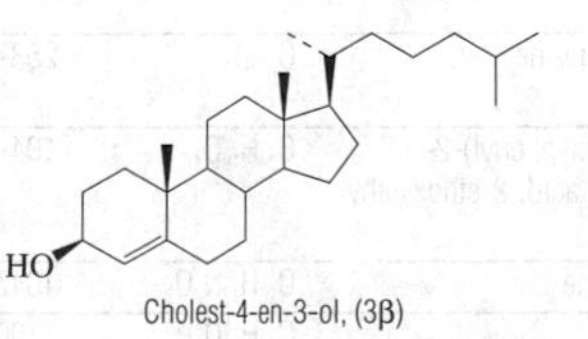
Cholest-5-en-3-ol, (3α) 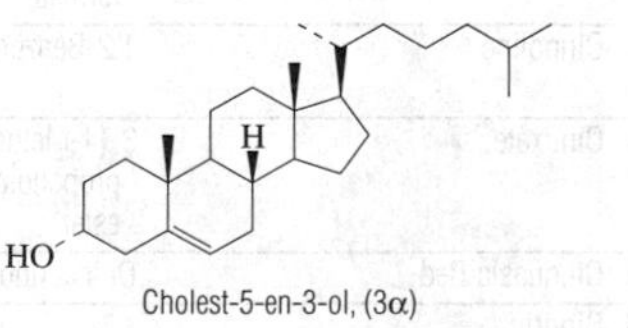
Cholest-5-en-3-ol (3β), acetate 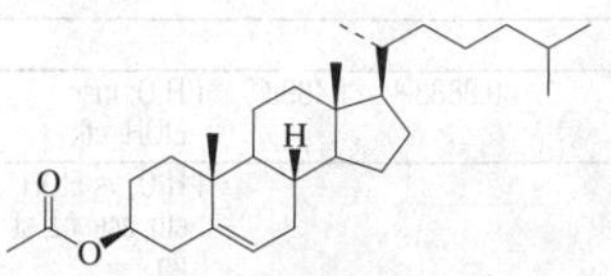
Cholest-5-en-3-ol (3β), benzoate 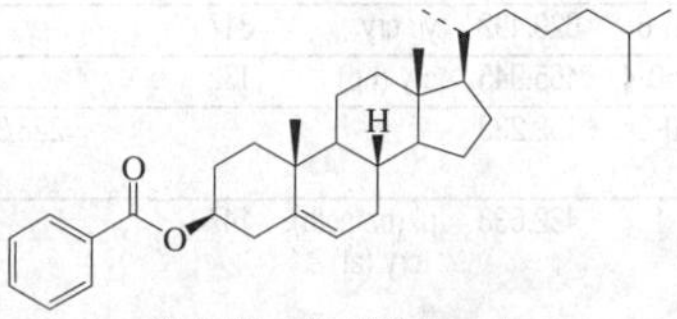
Cholest-5-en-3-ol (3β)-, hexadecanoate 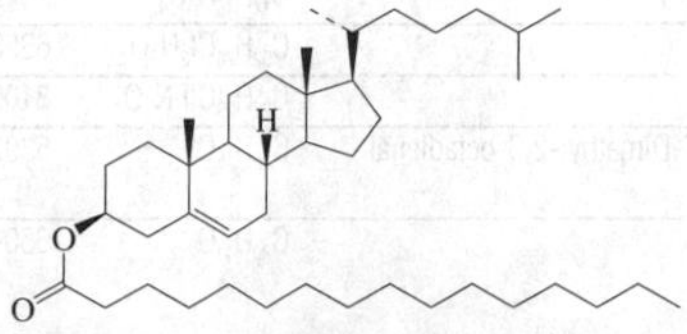
Cholest-5-en-3-ol (3β)-, *cis*-9-octadecenoate

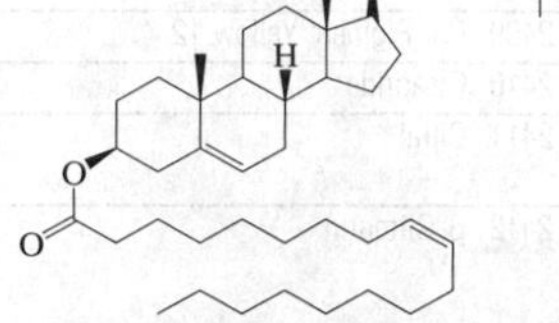

No.	Name	Synonym	Mol. Form.	CAS RN	Mol. Wt.	Physical Form	mp/°C	bp/°C	den/ g cm^{-3}	n_D	Solubility
2375	Cholest-4-en-3-one		$C_{27}H_{44}O$	601-57-0	384.637	nd or pl (al)	81.5	$245^{0.03}$			
2376	Cholesterol		$C_{27}H_{46}O$	57-88-5	386.653	orth or tcl lf (al) nd (eth)	148.5	dec 360; $233^{0.5}$	1.067^{20}		i H_2O; sl EtOH, ace; s bz, HOAc; vs diox
2377	Cholic acid	3,7,12-Trihydroxycholan-24-oic acid, (3α,5β,7α,12α)	$C_{24}H_{40}O_5$	81-25-4	408.572		198				sl H_2O; s EtOH, ace, alk; vs eth, chl
2378	Choline chloride		$C_5H_{14}ClNO$	67-48-1	139.624	hyg cry	305 dec				vs H_2O, EtOH
2379	Choline chloride dihydrogen phosphate	Phosphorylcholine	$C_5H_{15}ClNO_4P$	107-73-3	219.605	visc liq					
2380	Chorismic acid		$C_{10}H_{10}O_6$	617-12-9	226.182	cry	148				s H_2O
2381	Chromium carbonyl		C_6CrO_6	13007-92-6	220.056	col orth cry	dec 130	sub	1.77		i H_2O, EtOH; s eth, chl
2382	Chromium(II) oxalate		C_2CrO_4	814-90-4	140.015	ye-grn pow (hyd)					i H_2O, EtOH; s dil acid
2383	Chromium(III) 2,4-pentanedioate	Chromium acetylacetonate	$C_{15}H_{21}CrO_6$	21679-31-2	349.320	red mcl cry	208	345	1.34		i H_2O; s bz
2384	Chromotrope 2B		$C_{16}H_9N_3Na_2O_{10}S_2$	548-80-1	513.366	red-br pow	300				s H_2O; i EtOH
2385	Chrysamminic acid	1,8-Dihydroxy-2,4,5,7-tetranitro-9,10-anthracenedione	$C_{14}H_4N_4O_{12}$	517-92-0	420.202	ye pl or lf	exp	dec			vs eth, EtOH
2386	6-Chrysenamine	6-Aminochrysene	$C_{18}H_{13}N$	2642-98-0	243.303	lf (al)	210.5				
2387	Chrysene	Benzo[a]phenanthrene	$C_{18}H_{12}$	218-01-9	228.288	red bl fl or orth pl (bz, HOAc)	255.5	448	1.274^{20}		i H_2O; sl EtOH, eth, ace, bz, CS_2; s tol
2388	Ciafos		$C_9H_{10}NO_3PS$	2636-26-2	243.219	ye to red-ye liq	15	$120^{0.09}$ dec		1.5404^{32}	sl H_2O; vs chl, EtOH, ace, MeOH
2389	Cicutoxin	8,10,12-Heptadecatriene-4,6-diyne-1,14-diol	$C_{17}H_{22}O_2$	505-75-9	258.356	pr (eth/peth)	54				s hot H_2O, EtOH, eth, chl
2390	C.I. Direct Blue 6, tetrasodium salt	Direct Blue 6	$C_{32}H_{20}N_6Na_4O_{14}S_4$	2602-46-2	932.752	dk bronze pow					
2391	Cimetidine		$C_{10}H_{16}N_6S$	51481-61-9	252.339	cry	142				
2392	Cinchonamine		$C_{19}H_{24}N_2O$	482-28-0	296.406	orth nd (al) orth pr (MeOH)	186				i H_2O; vs EtOH, eth; s bz, chl
2393	Cinchonidine		$C_{19}H_{22}N_2O$	485-71-2	294.390	or pl or pr (al)	210.5	sub			i H_2O, bz; s EtOH, chl, py; sl eth
2394	Cinchonine		$C_{19}H_{22}N_2O$	118-10-5	294.390	pr nd (al, eth)	265				
2395	Cinchotoxine		$C_{19}H_{22}N_2O$	69-24-9	294.390	nd or pr (eth)	59				i H_2O; vs EtOH, eth, ace, bz, chl
2396	*trans*-Cinnamaldehyde	3-Phenyl-2-propenal, (*E*)-	C_9H_8O	14371-10-9	132.159	ye liq	-7.5	246	1.0497^{20}	1.6195^{20}	sl H_2O; s EtOH, eth, chl; i lig
2397	Cinnamedrine	α-[1-[Methyl(3-phenylallyl) amino]ethyl]benzenemethanol	$C_{19}H_{23}NO$	90-86-8	281.392		75				
2398	*cis*-Cinnamic acid	3-Phenyl-2-propenoic acid, (*Z*)	$C_9H_8O_2$	102-94-3	148.159	mcl pr (w)	42				vs EtOH, HOAc, lig
2399	*trans*-Cinnamic acid	3-Phenyl-2-propenoic acid, (*E*)	$C_9H_8O_2$	140-10-3	148.159	mcl pr (dil al)	133	300	1.2475^4		i H_2O, lig; vs EtOH; s eth, ace, bz
2400	*trans*-Cinnamyl anthranilate		$C_{16}H_{15}NO_2$	87-29-6	253.296	cry	64				
2401	Cinnamyl cinnamate		$C_{18}H_{16}O_2$	122-69-0	264.319	nd (al)	44		1.1565^4		i H_2O; s EtOH, chl; vs eth
2402	Cinnamyl formate	3-Phenyl-2-propen-1-ol, formate	$C_{10}H_{10}O_2$	104-65-4	162.185		0	252	1.086^{25}		
2403	Cinnoline	1,2-Benzodiazine	$C_8H_6N_2$	253-66-7	130.147	pa ye cry (lig)	38	$114^{0.3}$			vs eth, EtOH
2404	Cinoxate	3-(4-Methoxyphenyl)-2-propenoic acid, 2-ethoxyethyl ester	$C_{14}H_{18}O_4$	104-28-9	250.291	col liq	-25	185^2	1.102^{25}	1.567^{20}	i H_2O; msc EtOH
2405	Cinquasia Red	Quinacridone	$C_{20}H_{12}N_2O_2$	1047-16-1	312.321	red-viol cry	390				i H_2O, os
2406	Ciodrin		$C_{14}H_{19}O_6P$	7700-17-6	314.271			$135^{0.03}$	1.19^{25}		
2407	C.I. Pigment Red 170		$C_{26}H_{22}N_4O_4$	2786-76-7	454.478	red solid					
2408	C.I. Pigment Yellow 1		$C_{17}H_{16}N_4O_4$	2512-29-0	340.334	ye cry	256				
2409	C.I. Pigment Yellow 12		$C_{32}H_{26}Cl_2N_6O_4$	6358-85-6	629.492	ye cry	317				
2410	Cisapride		$C_{23}H_{29}ClFN_3O_4$	81098-60-4	465.945	cry (hp)	132				
2411	Citral	3,7-Dimethyl-2,6-octadienal	$C_{10}H_{16}O$	5392-40-5	152.233			228.3	0.8888^{20}	1.4898^{20}	i H_2O; msc EtOH, eth
2412	β-Citraurin		$C_{30}H_{40}O_2$	650-69-1	432.638	pl (bz-peth), cry (al)	147				i H_2O; vs EtOH, eth, ace, bz; sl lig
2413	Citrazinic acid	1,2-Dihydro-6-hydroxy-2-oxo-4-pyridinecarboxylic acid	$C_6H_5NO_4$	99-11-6	155.109	ye pow	>300 dec				s H_2O, alk; sl HCl

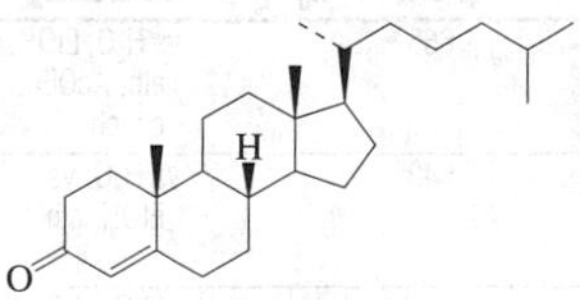

Cholest-4-en-3-one

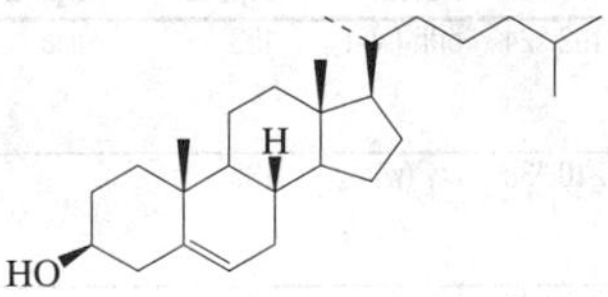

Cholesterol

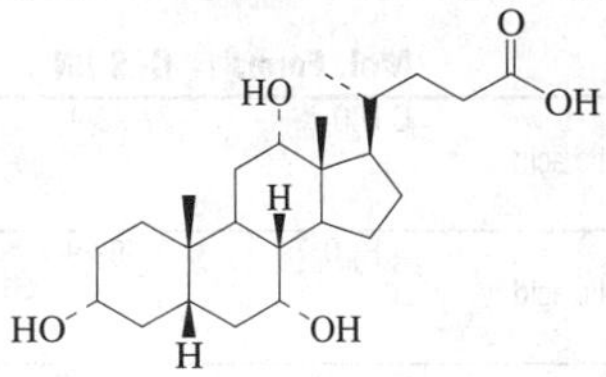

Cholic acid

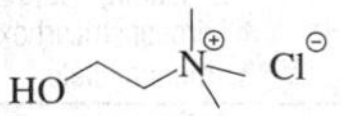

Choline chloride

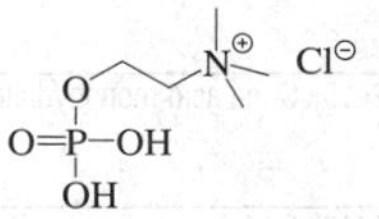

Choline chloride dihydrogen phosphate

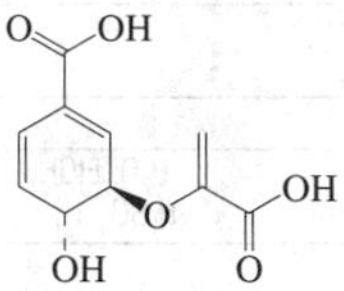

Chorismic acid

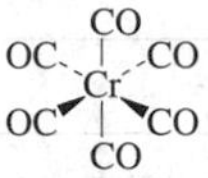

Chromium carbonyl

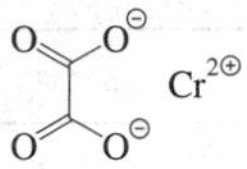

Chromium(II) oxalate

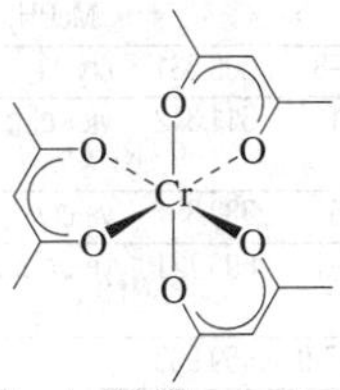

Chromium(III) 2,4-pentanedioate

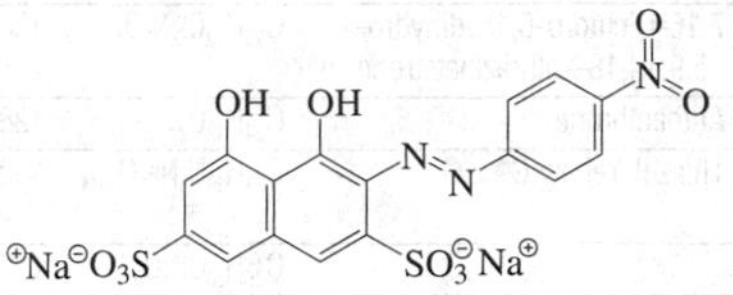

Chromotrope 2B

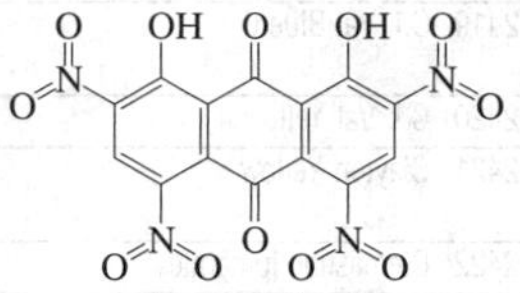

Chrysamminic acid

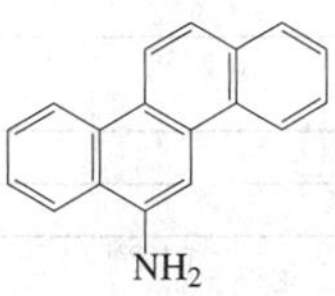

6-Chrysenamine

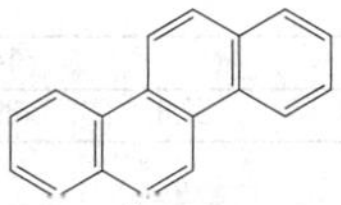

Chrysene

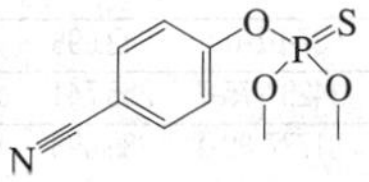

Ciafos

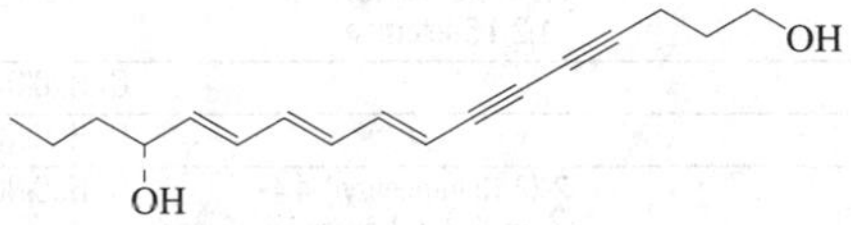

Cicutoxin

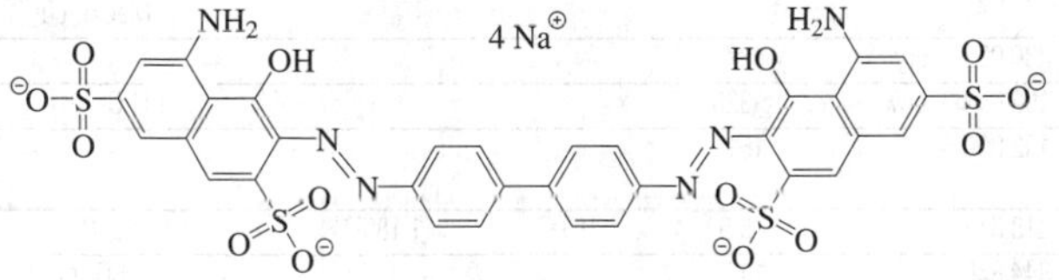

C.I. Direct Blue 6, tetrasodium salt

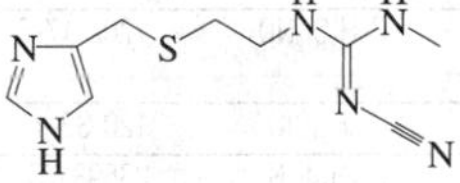

Cimetidine

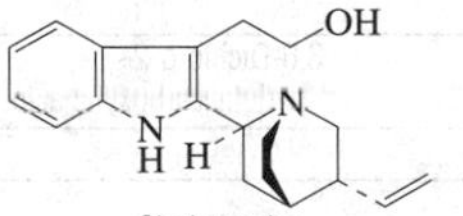

Cinchonamine

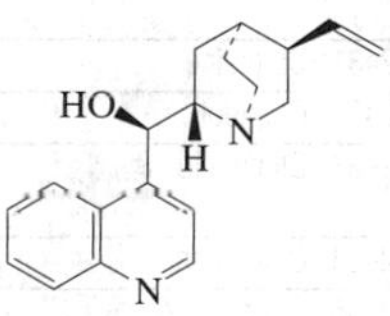

Cinchonidine

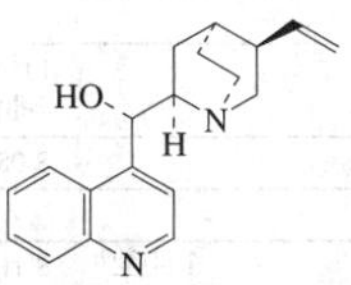

Cinchonine

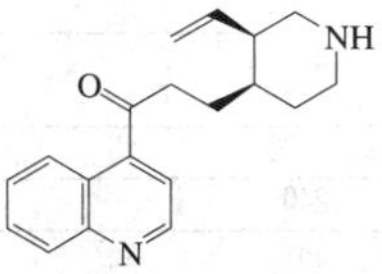

Cinchotoxine

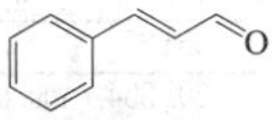

trans-Cinnamaldehyde

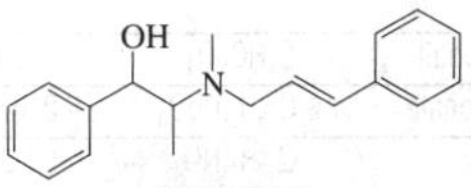

Cinnamedrine

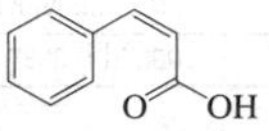

cis-Cinnamic acid

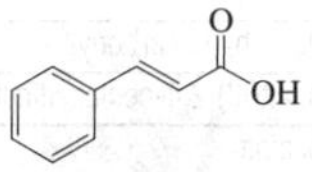

trans-Cinnamic acid

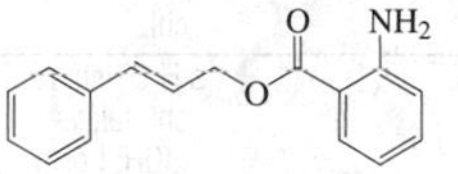

trans-Cinnamyl anthranilate

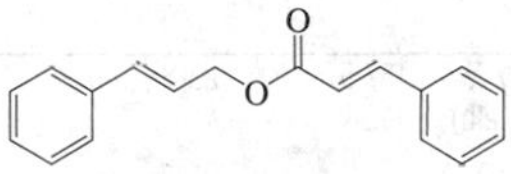

Cinnamyl cinnamate

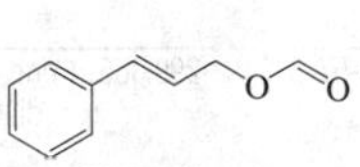

Cinnamyl formate

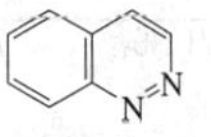

Cinnoline

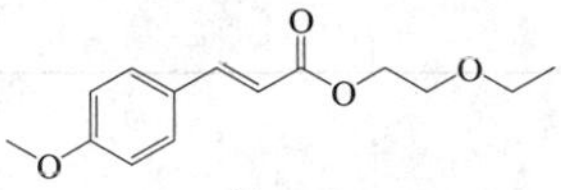

Cinoxate

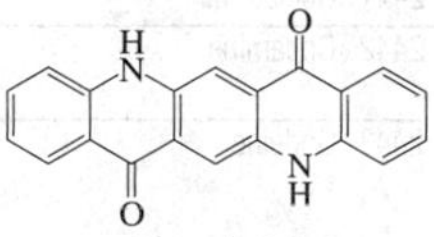

Cinquasia Red

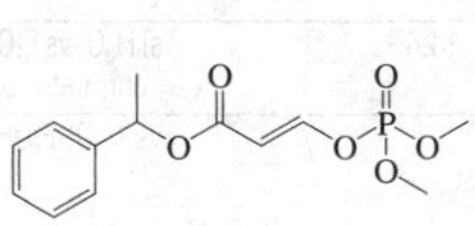

Ciodrin

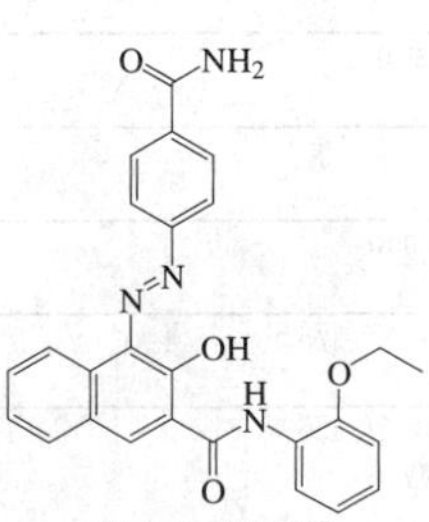

C.I. Pigment Red 170

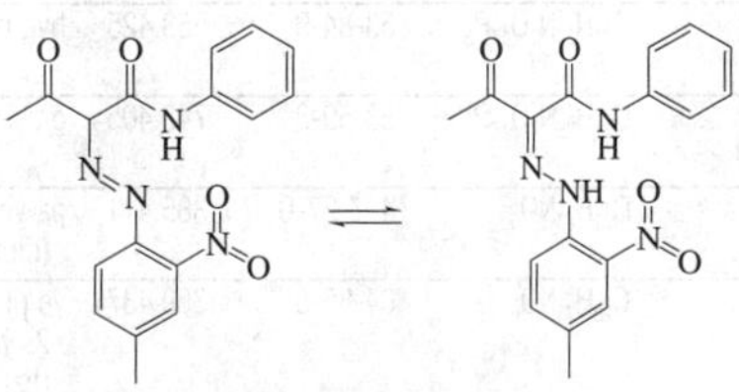

C.I. Pigment Yellow 1

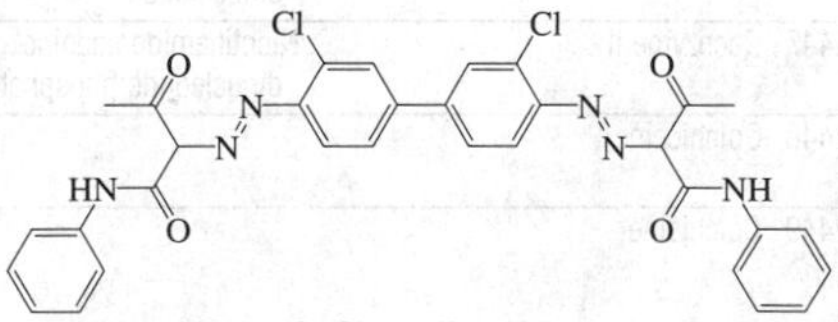

C.I. Pigment Yellow 12

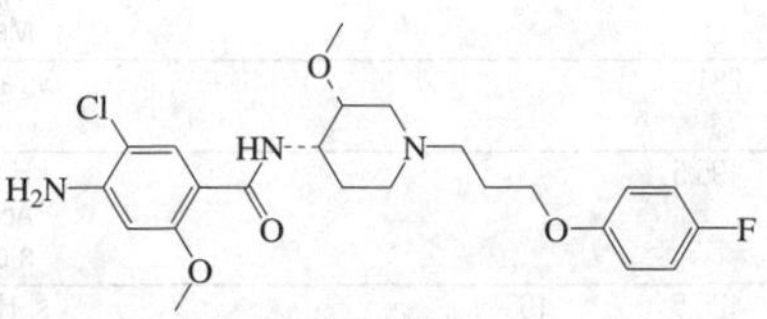

Cisapride

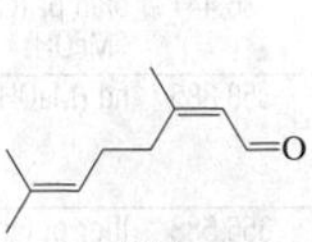

Citral

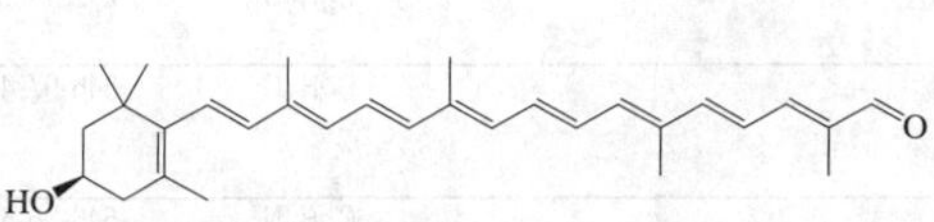

β-Citraurin

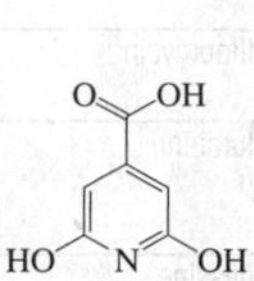

Citrazinic acid

No.	Name	Synonym	Mol. Form.	CAS RN	Mol. Wt.	Physical Form	mp/°C	bp/°C	den/ g cm^{-3}	n_D	Solubility
2414	Citric acid	2-Hydroxy-1,2,3-propanetricarboxylic acid	$C_6H_8O_7$	77-92-9	192.124	orth (w+1)	153	dec	1.665^{20}		vs H_2O, EtOH; s eth, AcOEt; i bz, chl
2415	Citric acid monohydrate	2-Hydroxy-1,2,3-propanetricarboxylic acid, monohydrate	$C_6H_{10}O_8$	5949-29-1	210.138	cry (w)	135		1.542		vs H_2O; vs EtOH, eth
2416	Citrinin	Antimycin	$C_{13}H_{14}O_5$	518-75-2	250.247	ye nd (MeOH)	178 dec				i H_2O; sl EtOH, eth; s ace, bz
2417	Citrulline	*N*5-(Aminocarbonyl)-*L*-ornithine	$C_6H_{13}N_3O_3$	372-75-8	175.185	pr (aq MeOH)	222				s H_2O; i EtOH, MeOH
2418	Citrus Red 2		$C_{18}H_{16}N_2O_3$	6358-53-8	308.331	cry	156				sl H_2O; s EtOH
2419	C.I. Vat Blue 6	7,16-Dichloro-6,15-dihydro-5,9,14,18-anthrazinetetrone	$C_{28}H_{12}Cl_2N_2O_4$	130-20-1	511.312	viol-bl pow					
2420	C.I. Vat Yellow 4	Anthanthrone	$C_{24}H_{12}O_2$	128-66-5	332.351	ye cry					
2421	Clayton Yellow	Thiazol Yellow G	$C_{28}H_{19}N_5Na_2O_6S_4$	1829-00-1	695.721	ye-br pow					s H_2O, EtOH, H_2SO_4
2422	Clemastine fumarate		$C_{25}H_{30}ClNO_5$	14976-57-9	459.963		181				
2423	Clindamycin		$C_{18}H_{33}ClN_2O_5S$	18323-44-9	424.983	ye amorp solid					
2424	Cloconazole		$C_{18}H_{15}ClN_2O$	77175-51-0	310.777		73				s EtOAc
2425	Clofentezine	3,6-Bis(2-chlorophenyl)-1,2,4,5-tetrazine	$C_{14}H_8Cl_2N_4$	74115-24-5	303.147		182				
2426	Clofibrate		$C_{12}H_{15}ClO_3$	637-07-0	242.698			149^{20}			
2427	Cloforex		$C_{13}H_{18}ClNO_2$	14261-75-7	255.741	cry	52.8	$89^{0.005}$			
2428	Clomazone	2-(2-Chlorobenzyl)-4,4-dimethyl-1,2-oxazolidin-3-one	$C_{12}H_{14}ClNO_2$	81777-89-1	239.698				1.192^{20}		
2429	Clomiphene		$C_{26}H_{28}ClNO$	911-45-5	405.959		117				
2430	Clonazepam		$C_{15}H_{10}ClN_3O_3$	1622-61-3	315.711	wh cry	237.5				i H_2O, bz; sl ace, MeOH, chl
2431	Clonidine		$C_9H_9Cl_2N_3$	4205-90-7	230.093	cry	137				
2432	Clopidol		$C_7H_7Cl_2NO$	2971-90-6	192.043	pow	>320				i H_2O
2433	Clopyralid	3,6-Dichloro-2-pyridinecarboxylic acid	$C_6H_3Cl_2NO_2$	1702-17-6	192.000		151				
2434	Clorophene		$C_{13}H_{11}ClO$	120-32-1	218.678		48.5	$161^{3.5}$	1.185^{58}		s ctc, CS_2
2435	Clotrimazole		$C_{22}H_{17}ClN_2$	23593-75-1	344.836	cry	148				sl H_2O, bz; s ace, chl, AcOEt, DMF
2436	Clozapine	Clozaril	$C_{18}H_{19}ClN_4$	5786-21-0	326.824	ye cry	183.5				
2437	Cobalt carbonyl	Dicobalt octacarbonyl	$C_8Co_2O_8$	10210-68-1	341.947	oran cry	51 dec		1.78		i H_2O; s EtOH, eth, CS_2
2438	Cobalt hydrocarbonyl	Tetracarbonylhydrocobalt	C_4HCoO_4	16842-03-8	171.982	ye liq or gas	≈-30	10			s os
2439	Cobalt(III) 2,4-pentanedioate	Cobalt(III) acetylacetonate	$C_{15}H_{21}CoO_6$	21679-46-9	356.257		240				
2440	Cocaine		$C_{17}H_{21}NO_4$	50-36-2	303.354	mcl pr (al)	98	$187^{0.1}$		1.5022^{98}	sl H_2O; vs EtOH, eth, bz, py; s CS_2
2441	Coclaurine		$C_{17}H_{19}NO_3$	486-39-5	285.338	pl (al)	220.5				
2442	Codamine		$C_{20}H_{25}NO_4$	21040-59-5	343.418	pr (bz, eth)	127				vs eth, EtOH, chl
2443	Codeine		$C_{18}H_{21}NO_3$	76-57-3	299.365	orth cry (w, dil al, eth)	157.5	250^{22}, $140^{1.5}$	1.32^{25}		s H_2O, eth, bz, chl, tol; vs EtOH; i peth
2444	Codeine phosphate		$C_{18}H_{24}NO_7P$	52-28-8	397.361	lf or pr (dil al)	227 dec				vs EtOH, chl
2445	Coenzyme A		$C_{21}H_{36}N_7O_{16}P_3S$	85-61-0	767.535	pow; unstab in air					s H_2O
2446	Coenzyme I	Nicotinamide adenine dinucleotide	$C_{21}H_{27}N_7O_{14}P_2$	53-84-9	663.425	hyg pow					s H_2O
2447	Coenzyme II	Nicotinamide adenine dinucleotide phosphate	$C_{21}H_{28}N_7O_{17}P_3$	53-59-8	743.405	gray-wh pow					s H_2O
2448	Colchiceine		$C_{21}H_{23}NO_6$	477-27-0	385.411	pa ye nd (diox)	178.5		1.24^{25}		sl H_2O; vs EtOH, chl; i eth, bz
2449	Colchicine		$C_{22}H_{25}NO_6$	64-86-8	399.437	ye pl (w + 1/2) ye cry (bz)	156				vs H_2O, EtOH
2450	Colistin A		$C_{53}H_{100}N_{16}O_{13}$	7722-44-3	1169.47	amorp pow					sl H_2O, EtOH, hx; s acids, MeOH
2451	Collinomycin		$C_{27}H_{20}O_{12}$	27267-69-2	536.441	oran pr (chl-MeOH)	281				vs ace, diox, chl
2452	Columbin		$C_{20}H_{22}O_6$	546-97-4	358.385	nd (MeOH)	195.5				i H_2O; sl ace, AcOEt, MeOH; s chl
2453	Conessine		$C_{24}H_{40}N_2$	546-06-5	356.588	lf or pl (ace)	125.5	$166^{0.1}$			sl H_2O; s chl, HOAc

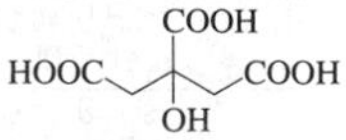

Citric acid

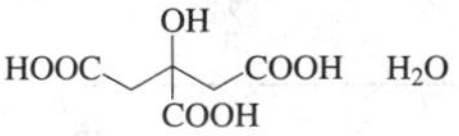

Citric acid monohydrate

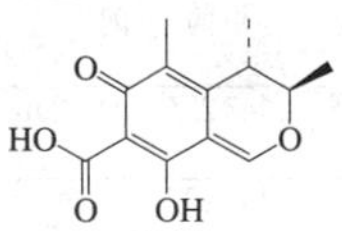

Citrinin

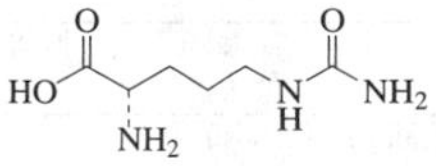

Citrulline

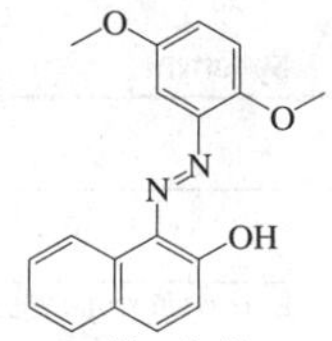

Citrus Red 2

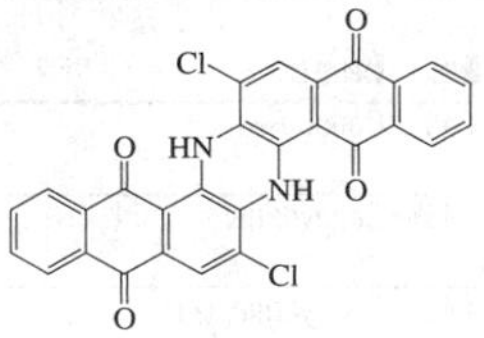

C.I. Vat Blue 6

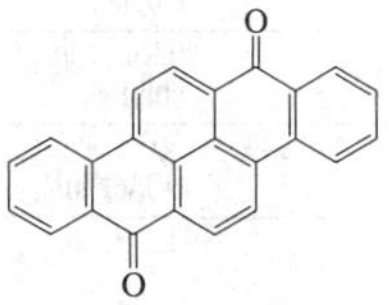

C.I. Vat Yellow 4

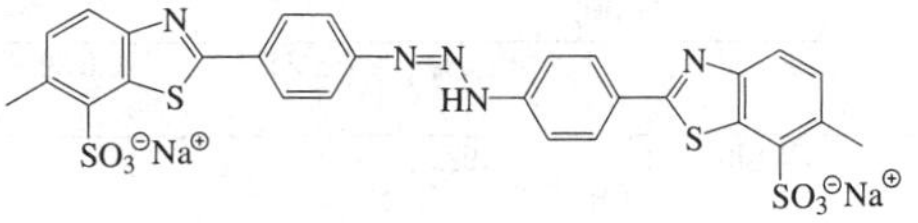

Clayton Yellow

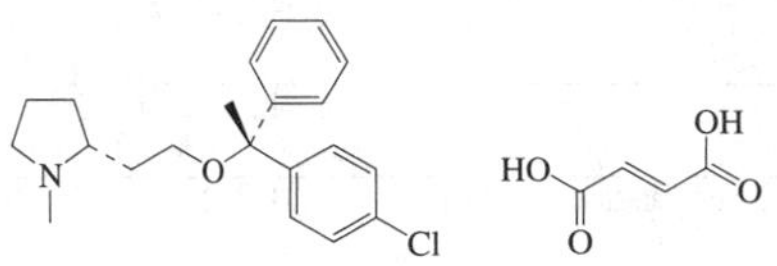

Clemastine fumarate

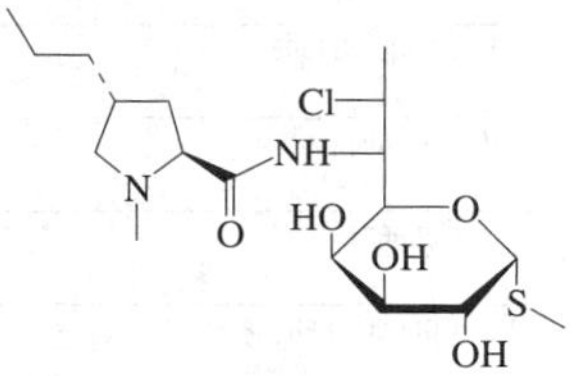

Clindamycin

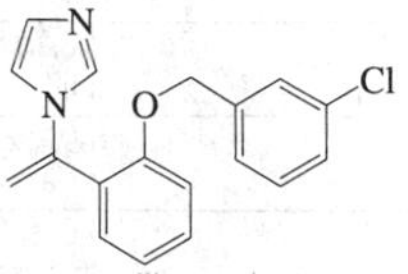

Cloconazole

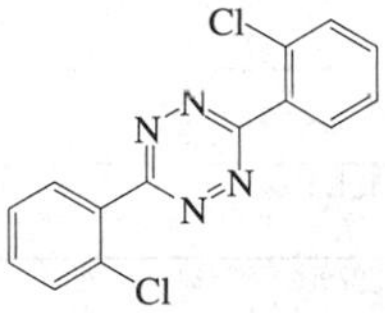

Clofentezine

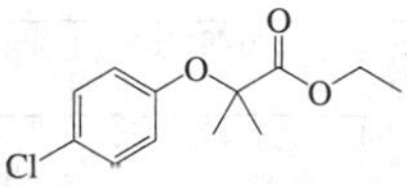

Clofibrate

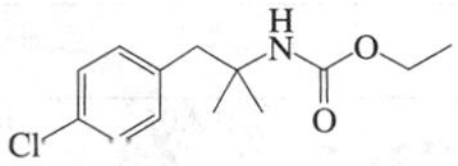

Cloforex

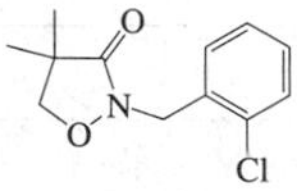

Clomazone

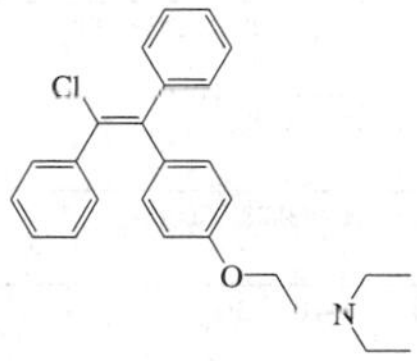

Clomiphene

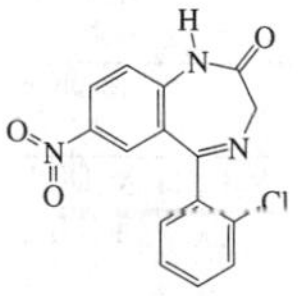

Clonazepam

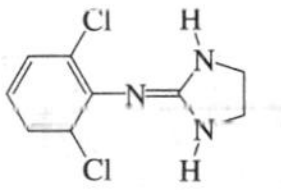

Clonidine

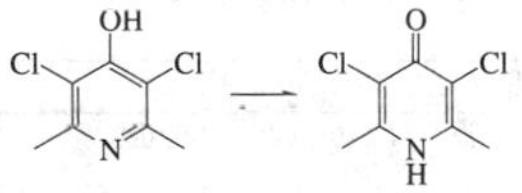

Clopidol

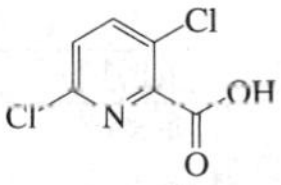

Clopyralid

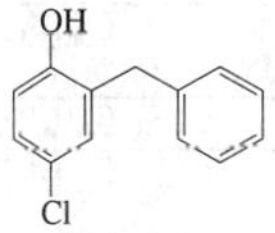

Clorophene

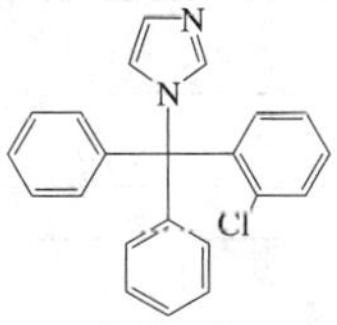

Clotrimazole

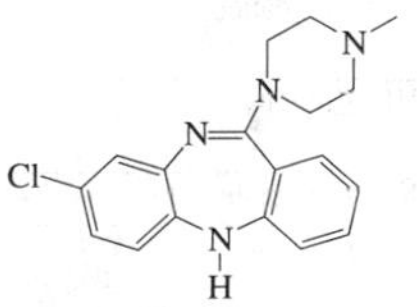

Clozapine

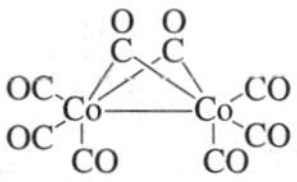

Cobalt carbonyl

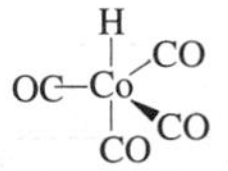

Cobalt hydrocarbonyl

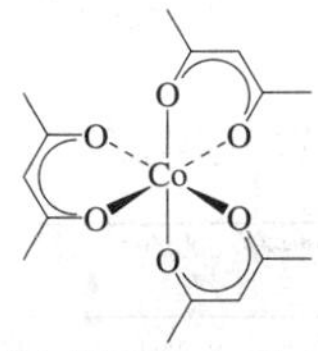

Cobalt(III) 2,4-pentanedioate

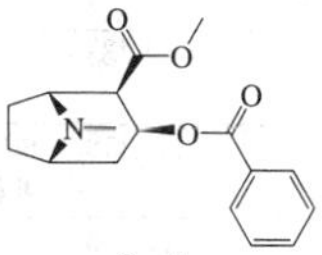

Cocaine

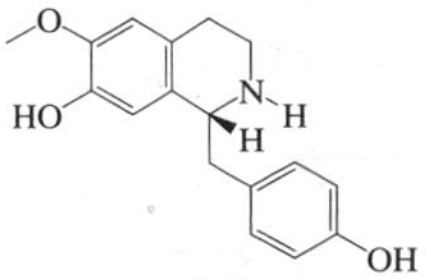

Coclaurine

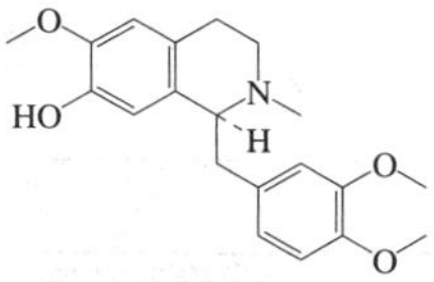

Codamine

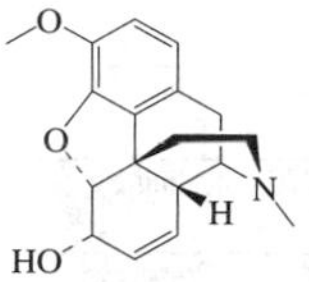

Codeine

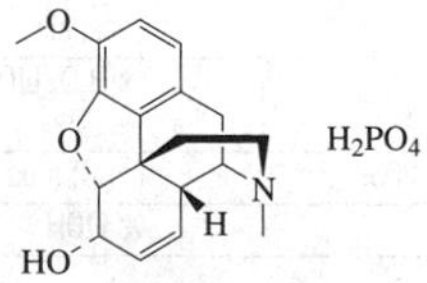

Codeine phosphate

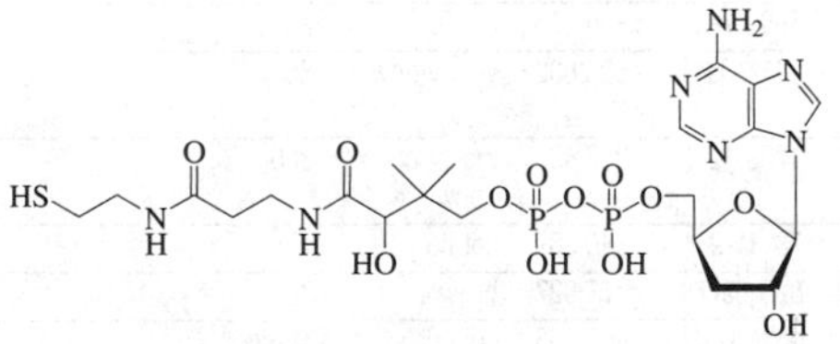

Coenzyme A

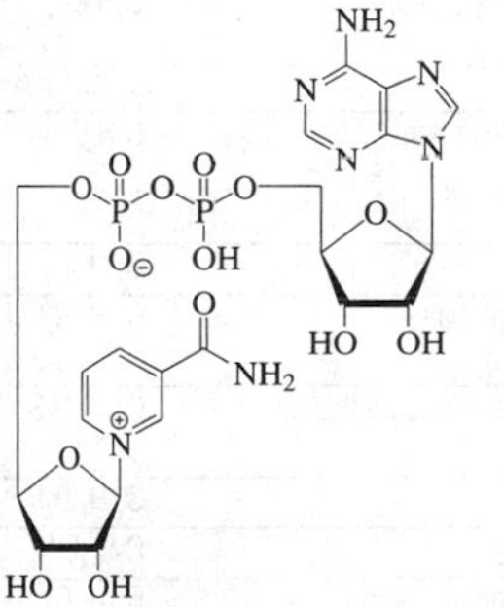

Coenzyme I

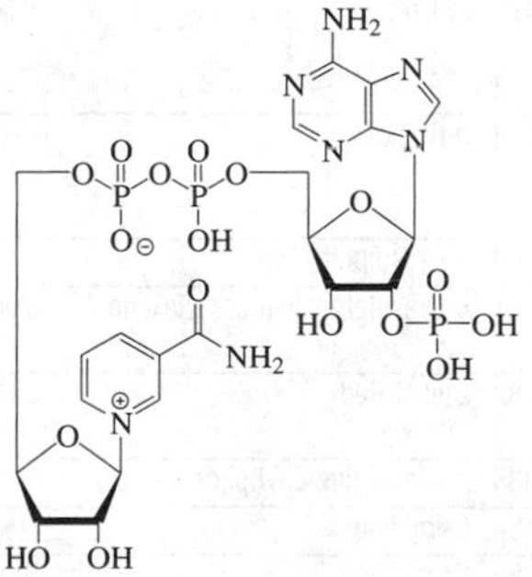

Coenzyme II

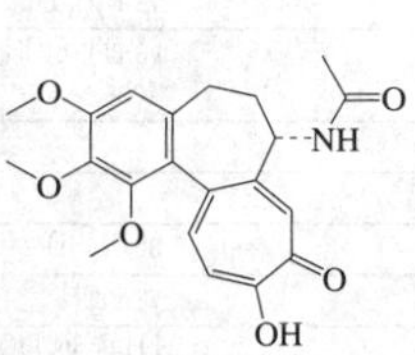

Colchiceine

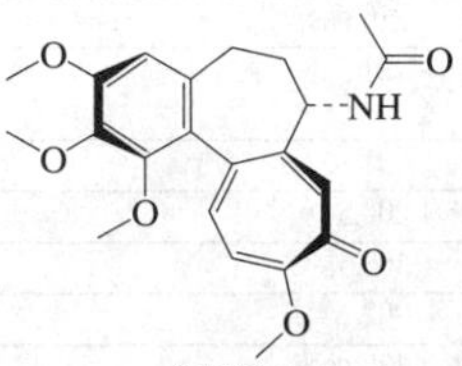

Colchicine

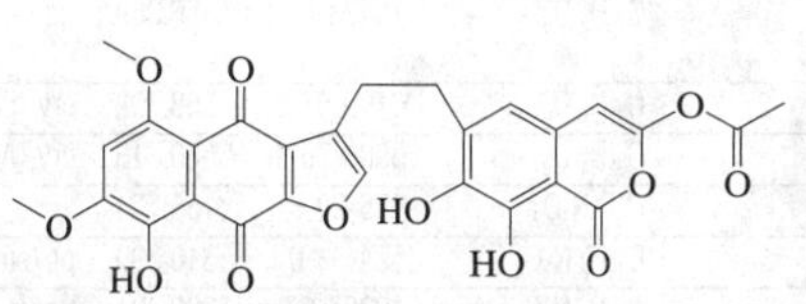

Collinomycin

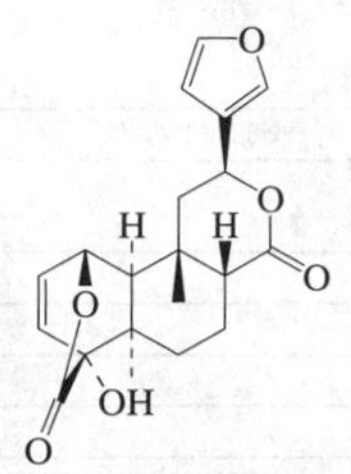

Columbin

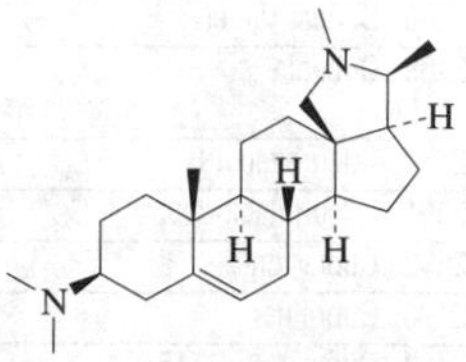

Conessine

No.	Name	Synonym	Mol. Form.	CAS RN	Mol. Wt.	Physical Form	mp/°C	bp/°C	den/ g cm⁻³	n_D	Solubility
2454	Congo Red		$C_{32}H_{22}N_6Na_2O_6S_2$	573-58-0	696.663	pow	>360				sl H_2O; s EtOH; i eth
2455	Conhydrine		$C_8H_{17}NO$	3238-62-8	143.227	nd (peth)	121	226			sl H_2O; vs bz, eth, EtOH
2456	Conhydrine, (+)	2-(α-Hydroxypropyl)piperidine	$C_8H_{17}NO$	495-20-5	143.227	lf (eth)	121	226			sl H_2O; vs eth, EtOH, chl
2457	Coniferin		$C_{16}H_{22}O_8$	531-29-3	342.341	nd (w+2)	186				s H_2O, py; sl EtOH; i eth
2458	Conquinamine		$C_{19}H_{24}N_2O_2$	464-86-8	312.406	ye tetr	123				sl H_2O; s EtOH, eth, chl
2459	Convallatoxin		$C_{29}H_{42}O_{10}$	508-75-8	550.637	pr (eth/ MeOH)	238				s EtOH, ace; sl chl; i eth
2460	Copaene		$C_{15}H_{24}$	3856-25-5	204.352			248.5	0.8996^{20}	1.4894^{20}	i H_2O; s eth, ace, HOAc, lig
2461	Copper(II) ethylacetoacetate	Bis(ethylacetoacetato)copper	$C_{12}H_{18}CuO_6$	14284-06-1	321.813	grn cry (EtOH)	192				s EtOH, chl
2462	Copper(II) gluconate	Cupric gluconate	$C_{12}H_{22}CuO_{14}$	527-09-3	453.841	bl-grn cry	156				sl EtOH; i os
2463	Copper(II) 2,4-pentanedioate	Copper(II) acetylacetonate	$C_{10}H_{14}CuO_4$	13395-16-9	261.762	bl pow	284 dec	sub			sl H_2O; s chl
2464	Copper phthalocyanine	Pigment Blue 15	$C_{32}H_{16}CuN_8$	147-14-8	576.069	bl-purp cry					i H_2O, EtOH; s conc H2SO4
2465	Coronene		$C_{24}H_{12}$	191-07-1	300.352	ye nd (bz)	437.4	525	1.371^{25}		i H_2O, con sulf; sl bz
2466	Corticosterone		$C_{21}H_{30}O_4$	50-22-6	346.461	nd (al, pl) (ace)	181				i H_2O; s EtOH, eth, ace
2467	Corybulbine		$C_{21}H_{25}NO_4$	518-77-4	355.429	nd (al)	237.5				i H_2O; sl EtOH, eth; s ace, bz, HCl
2468	Corycavamine		$C_{21}H_{21}NO_5$	521-85-7	367.396	pr (eth, al)	149				vs EtOH, chl
2469	Corydaline		$C_{22}H_{27}NO_4$	518-69-4	369.454	pr (al)	136				vs bz, eth, EtOH, chl
2470	Corydine		$C_{20}H_{23}NO_4$	476-69-7	341.402	tetr pr (eth)	149				vs eth, EtOH, chl
2471	Corynantheine		$C_{22}H_{26}N_2O_3$	18904-54-6	366.452		165.5				vs EtOH
2472	Cotarnine		$C_{12}H_{15}NO_4$	82-54-2	237.252	nd (bz), cry (eth)	132 dec				sl H_2O; s EtOH, eth, bz, chl, NH_4OH
2473	Coumaphos		$C_{14}H_{16}ClO_5PS$	56-72-4	362.766		93		1.474		
2474	Coumestrol	3,9-Dihydroxy-6*H*-benzofuro[3,2-c][1]benzopyran-6-one	$C_{15}H_8O_5$	479-13-0	268.222	cry rods	385 dec				i H_2O; sl EtOH, ace; i eth
2475	Creatine		$C_4H_9N_3O_2$	57-00-1	131.133	mcl pr (w+1)	303 dec		1.33^{25}		s H_2O; sl EtOH; i eth
2476	Creatinine		$C_4H_7N_3O$	60-27-5	113.118	orth pr (w+2) lf (w)	300 dec				s H_2O; sl EtOH; i eth, ace, chl
2477	*o*-Cresol	2-Methylphenol	C_7H_8O	95-48-7	108.138		31.03	191.04	1.0327^{35}	1.5386^{35}	s H_2O; vs EtOH, eth; msc ace, bz, ctc
2478	*m*-Cresol	3-Methylphenol	C_7H_8O	108-39-4	108.138		12.24	202.27	1.0339^{20}	1.5401^{20}	sl H_2O; msc EtOH, eth, ace, bz, ctc
2479	*p*-Cresol	4-Methylphenol	C_7H_8O	106-44-5	108.138	pr	34.77	201.98	1.0185^{40}	1.5312^{20}	sl H_2O; msc EtOH, eth, ace, bz, ctc
2480	*o*-Cresolphthalein		$C_{22}H_{18}O_4$	596-27-0	346.376	cry (al)	223				vs EtOH
2481	*o*-Cresolphthalein complexone	Metalphthalein	$C_{32}H_{32}N_2O_{12}$	2411-89-4	636.602	ye cry pow	186				i H_2O; s EtOH, ace, alk
2482	Cresol Red	*o*-Cresolsulfonphthalein	$C_{21}H_{18}O_5S$	1733-12-6	382.430	red-br cry pow	>300				vs H_2O, EtOH
2483	*p*-Cresyl diphenyl phosphate		$C_{19}H_{17}O_4P$	78-31-9	340.309	col liq	-40		1.208^{25}		i H_2O; s os
2484	Crimidine		$C_7H_{10}ClN_3$	535-89-7	171.627	br wax	87	143^{4}			vs EtOH
2485	Cromolyn	Cromoglicic acid	$C_{23}H_{16}O_{11}$	16110-51-3	468.366	col cry	241 dec				
2486	Crufomate		$C_{12}H_{19}ClNO_3P$	299-86-5	291.711		60	$118^{0.01}$			
2487	Cryptopine	Cryptocavine	$C_{21}H_{23}NO_5$	482-74-6	369.412	pr or pl (bz) nd (chl-MeOH)	223		1.315^{20}		i H_2O; sl EtOH, eth, bz; s chl, HOAc
2488	Crystal Violet	Gentian violet	$C_{25}H_{30}ClN_3$	548-62-9	407.979	grn pow	215 dec				vs H_2O, chl
2489	Cubebin		$C_{20}H_{20}O_6$	18423-69-3	356.369	nd (al, bz)	131.5				vs eth, EtOH, chl
2490	Cucurbitacin B		$C_{32}H_{46}O_8$	6199-67-3	558.702	cry (EtOH)	181				
2491	Cucurbitacin C		$C_{32}H_{48}O_8$	5988-76-1	560.718	cry (AcOEt)	207.5				
2492	Cupferron		$C_6H_9N_3O_2$	135-20-6	155.154		163.5				sl DMSO
2493	Cupreine		$C_{19}H_{22}N_2O_2$	524-63-0	310.390	pr (eth)	202				vs EtOH
2494	Curan-17-ol, (16α)	Geissoschizoline	$C_{19}H_{26}N_2O$	18397-07-4	298.421	pa ye amor pow	135 dec				i H_2O; vs EtOH, eth, chl

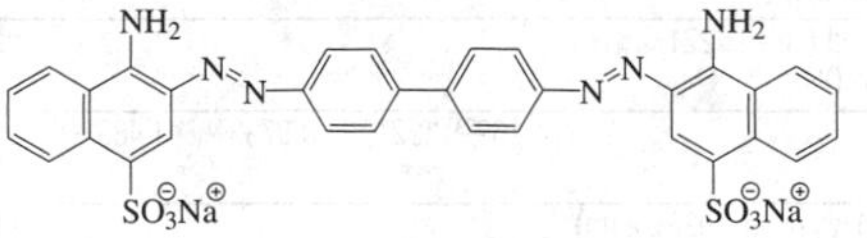

Congo Red

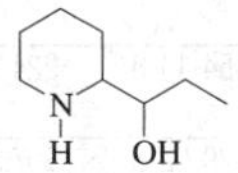

Conhydrine

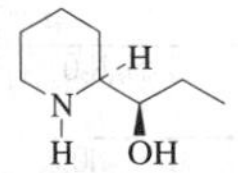

Conhydrine, (+)

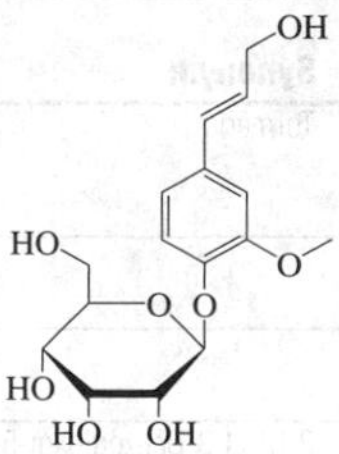

Coniferin

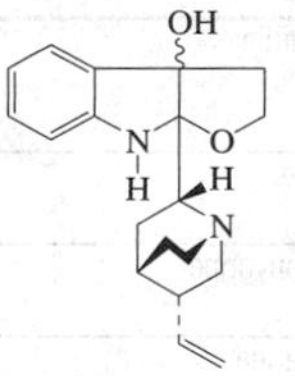

Conquinamine

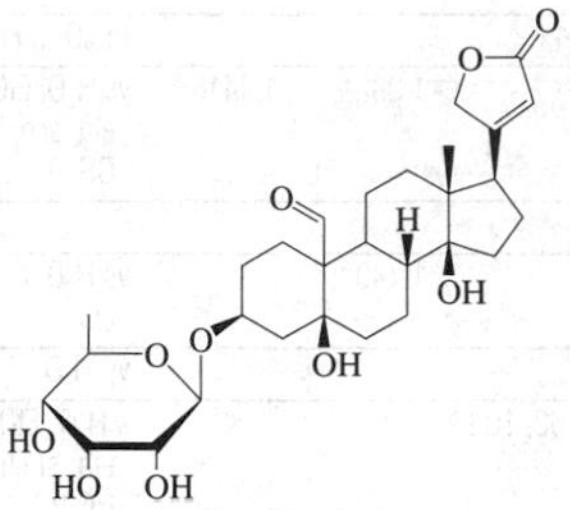

Convallatoxin

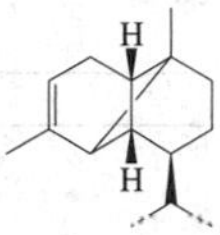

Copaene

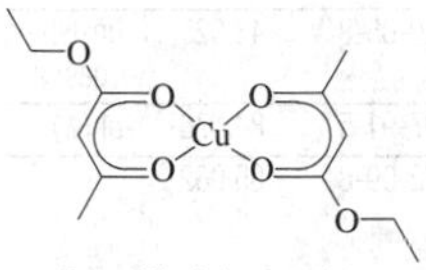

Copper(II) ethylacetoacetate

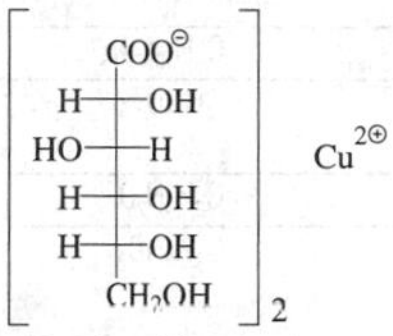

Copper(II) gluconate

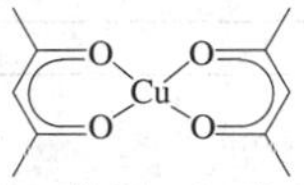

Copper(II) 2,4-pentanedioate

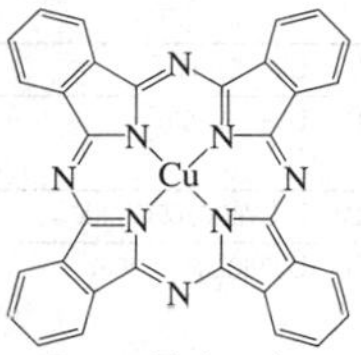

Copper phthalocyanine

Coronene

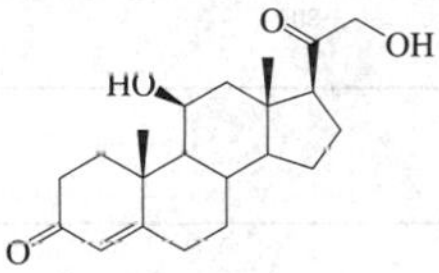

Corticosterone

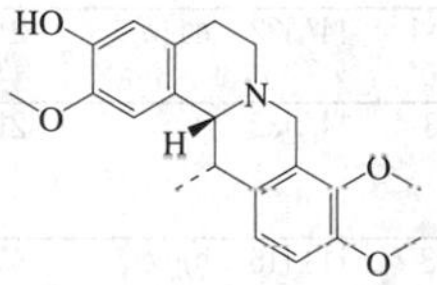

Corybulbine

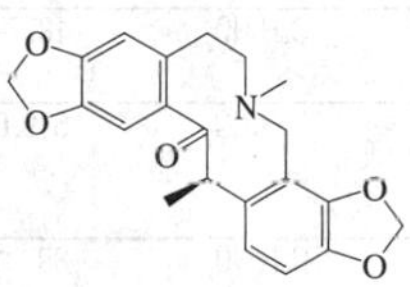

Corycavamine

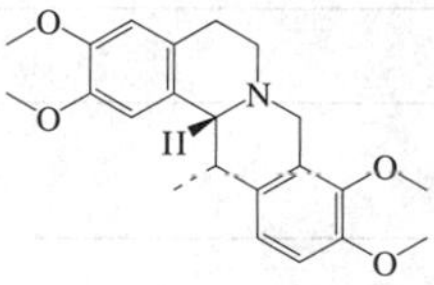

Corydaline

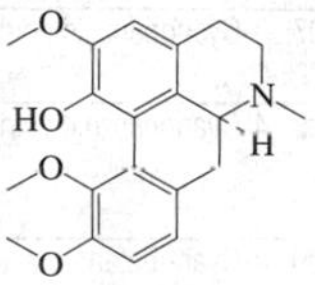

Corydine

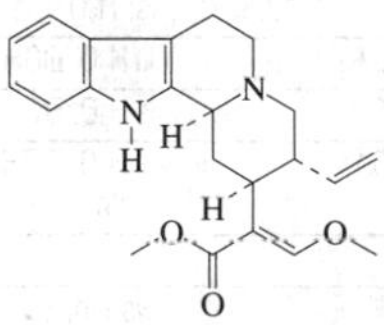

Corynantheine

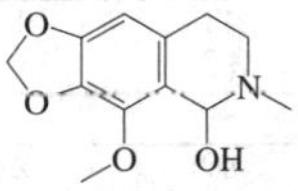

Cotarnine

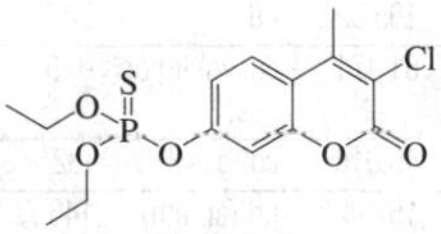

Coumaphos

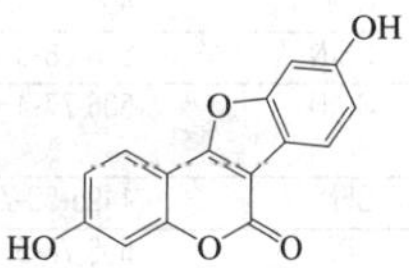

Coumestrol

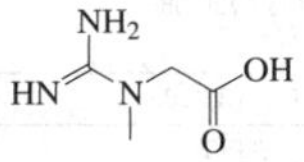

Creatine

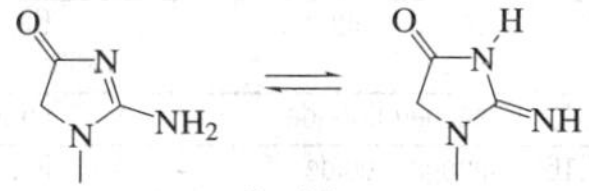

Creatinine

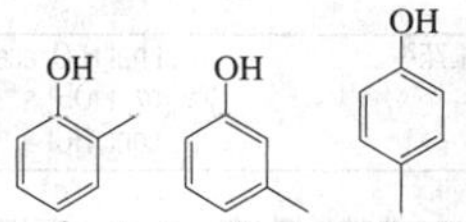

o-Cresol

m-Cresol

p-Cresol

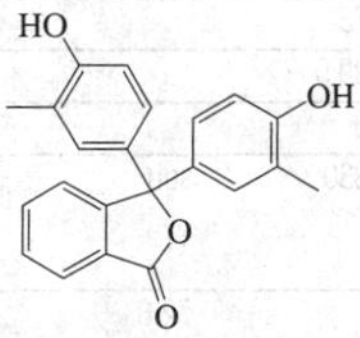

o-Cresolphthalein

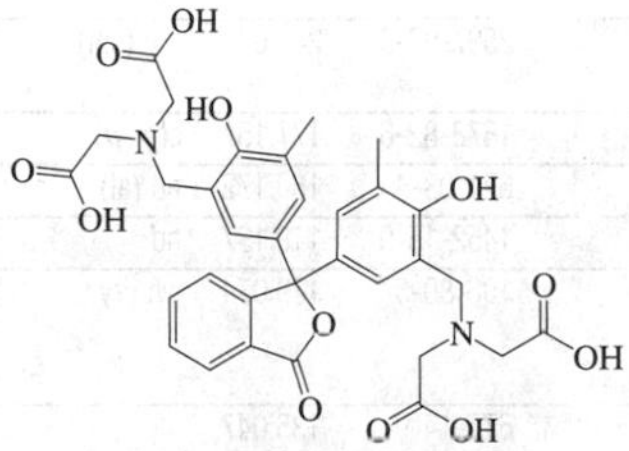

o-Cresolphthalein complexone

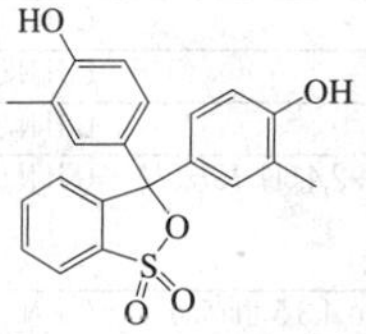

Cresol Red

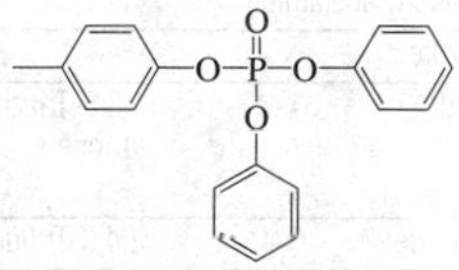

p-Cresyl diphenyl phosphate

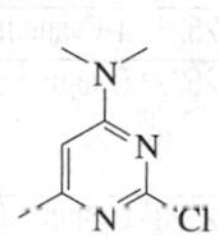

Crimidine

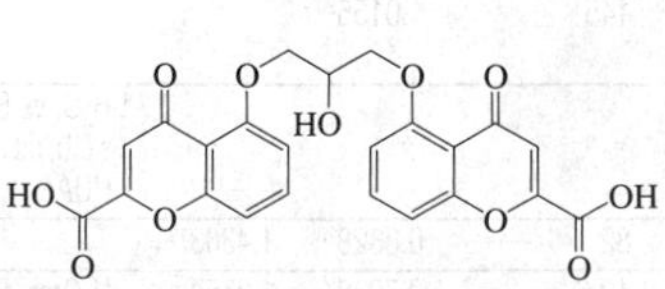

Cromolyn

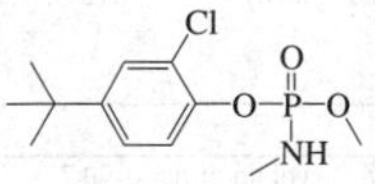

Crufomate

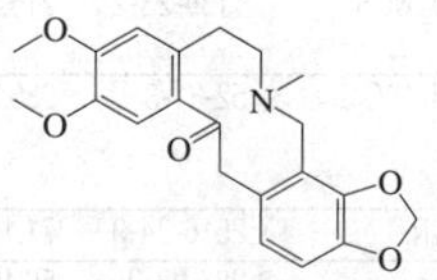

Cryptopine

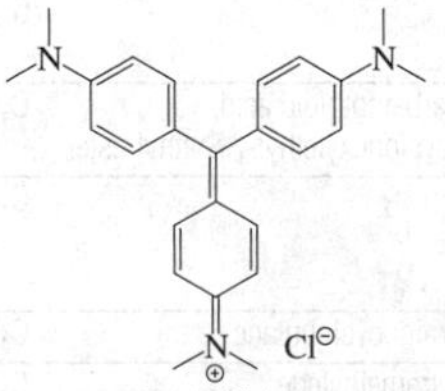

Crystal Violet

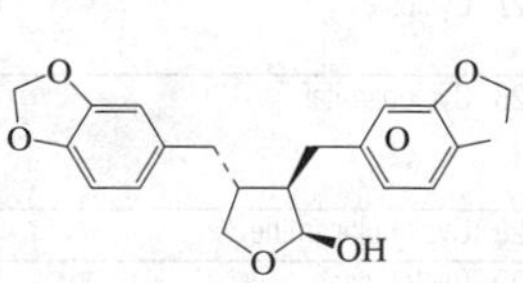

Cubebin

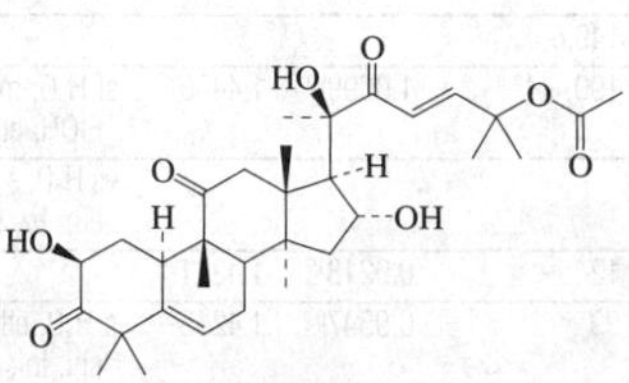

Cucurbitacin B

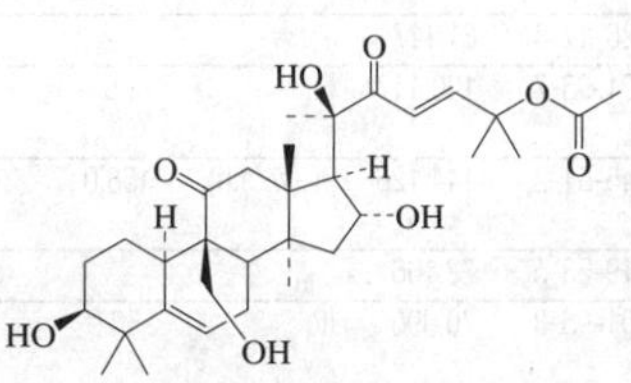

Cucurbitacin C

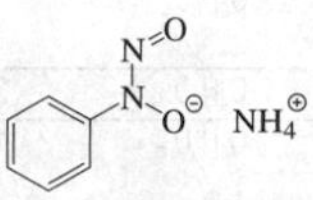

Cupferron

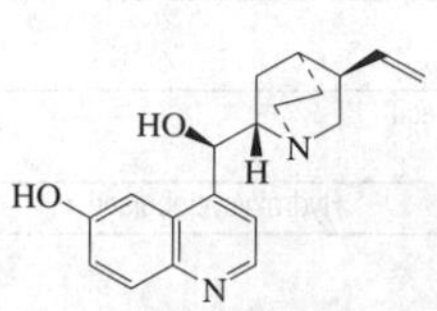

Cupreine

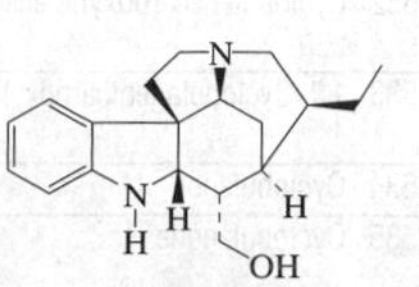

Curan-17-ol, (16α)

No.	Name	Synonym	Mol. Form.	CAS RN	Mol. Wt.	Physical Form	mp/°C	bp/°C	den/g cm^{-3}	n_D	Solubility
2495	Curcumin	Turmeric	$C_{21}H_{20}O_6$	458-37-7	368.380	oran ye pr, orth pr (MeOH)	183				vs EtOH, HOAc
2496	Curine		$C_{36}H_{38}N_2O_6$	436-05-5	594.696	pr, nd (chl-MeOH)	221				vs ace, bz, py
2497	Cuscohygrine		$C_{13}H_{24}N_2O$	454-14-8	224.342	oil		169^{23}, 122^{2}	0.9733^{20}	1.4832^{20}	vs H_2O, bz, eth, EtOH
2498	Cusparine	2-[2-(1,3-Benzodioxol-5-yl) ethyl]-4-methoxyquinoline	$C_{19}H_{17}NO_3$	529-92-0	307.343	(α) wh or ye nd (peth); (β) amber pr	92(α form); 111(β form)				i H_2O; vs ace, bz, eth, EtOH
2499	Cyamemazine		$C_{19}H_{21}N_3S$	3546-03-0	323.455	ye pow	92	$212^{0.25}$			i H_2O; s EtOH
2500	Cyanamide	Cyanogenamide	CH_2N_2	420-04-2	42.040	nd	45.56	140^{19}	1.282^{20}	1.4418^{48}	vs H_2O, EtOH; s eth, ace, bz; sl CS_2
2501	Cyanazine		$C_9H_{13}ClN_6$	21725-46-2	240.692		168				
2502	Cyanic acid	Hydrogen cyanate	CHNO	420-05-3	43.025	unstab liq or gas	-86	23	1.140^{20}		vs H_2O, bz, eth, chl
2503	2-Cyanoacetamide		$C_3H_4N_2O$	107-91-5	84.076	pl (w)	121.5				vs H_2O
2504	Cyanoacetic acid		$C_3H_3NO_2$	372-09-8	85.062		66	dec 160; 108^{15}			s H_2O, EtOH, eth; sl chl, HOAc
2505	Cyanoacetohydrazide	Cyacetacide	$C_3H_5N_3O$	140-87-4	99.091	pr (al)	114.5				vs H_2O, EtOH
2506	Cyanoacetylene		C_3HN	1070-71-9	51.047		5	42.5	0.8167^{17}	1.3868^{25}	sl H_2O; s EtOH
2507	3-Cyanobenzoic acid		$C_8H_5NO_2$	1877-72-1	147.132	nd (w)	219	sub			sl H_2O; s EtOH, eth
2508	4-Cyanobenzoic acid		$C_8H_5NO_2$	619-65-8	147.132		219				s H_2O, EtOH, eth, HOAc; sl tfa
2509	4-Cyanobutanoic acid		$C_5H_7NO_2$	39201-33-7	113.116	hyg cry	45				s H_2O, EtOH, eth, bz
2510	2-Cyanoethyl acrylate		$C_6H_7NO_2$	106-71-8	125.126			108^{12}	1.062^{20}		
2511	Cyanofenphos		$C_{15}H_{14}NO_2PS$	13067-93-1	303.317		83			1.5839^{25}	sl H_2O
2512	Cyanogen	Ethane dinitrile	C_2N_2	460-19-5	52.034	col gas	-27.83	-21.1	0.9537^{-21}		s H_2O, EtOH, eth
2513	Cyanogen bromide	Bromine cyanide	CBrN	506-68-3	105.922	nd	52	61.5	2.015^{20}		s H_2O, EtOH, eth
2514	Cyanogen chloride	Chlorine cyanide	CClN	506-77-4	61.471	col vol liq or gas	-6.5	13	1.186^{20}		s H_2O, EtOH; vs eth
2515	Cyanogen fluoride	Fluorine cyanide	CFN	1495-50-7	45.016	col gas	-82	-46			
2516	Cyanogen iodide	Iodine cyanide	CIN	506-78-5	152.922	nd (al, eth)	146.7	sub	2.84^{18}		vs eth, EtOH
2517	Cyanoguanidine	Dicyanodiamide	$C_2H_4N_4$	461-58-5	84.080		211		1.404^{14}		s H_2O, EtOH, ace; i eth, bz, chl
2518	Cyanomethylmercury	Methylmercurynitrile	C_2H_3HgN	2597-97-9	241.64	cry (chl)	92	subl			vs H_2O, EtOH, bz; s eth
2519	(4-Cyanophenoxy)acetic acid		$C_9H_7NO_3$	1878-82-6	177.157	cry (w)	178				
2520	2-Cyano-*N*-phenylacetamide		$C_9H_8N_2O$	621-03-4	160.172	nd (al)	199.5				
2521	4-Cyanothiazole		$C_4H_2N_2S$	1452-15-9	110.137	nd	58				
2522	Cyanuric acid	1,3,5-Triazine-2,4,6(1*H*,3*H*,5*H*)-trione	$C_3H_3N_3O_3$	108-80-5	129.074	wh cry	>330	sub	1.75^{25}		sl hot H_2O, ace, bz, EtOH; s conc HCl
2523	Cyanuric fluoride	2,4,6-Trifluoro-1,3,5-triazine	$C_3F_3N_3$	675-14-9	135.047			72.8			
2524	Cycasin		$C_8H_{16}N_2O_7$	14901-08-7	252.222	nd (ace aq)	154 dec				
2525	Cyclandelate		$C_{17}H_{24}O_3$	456-59-7	276.371		52	193^{14}			i H_2O
2526	Cyclizine		$C_{18}H_{22}N_2$	82-92-8	266.381	cry (peth)	106				i H_2O; s chl; sl EtOH
2527	Cycloate	Carbamothioic acid, cyclohexylethyl-, *S*-ethyl ester	$C_{11}H_{21}NOS$	1134-23-2	215.356		11.5	145^{10}	1.0156^{30}		
2528	Cyclobarbital		$C_{12}H_{16}N_2O_3$	52-31-3	236.266	lf (w)	173				i H_2O; vs EtOH; s eth, dil alk; sl HOAc
2529	Cyclobutanamine	Aminocyclobutane	C_4H_9N	2516-34-9	71.121			82	0.8328^{20}	1.4363^{19}	
2530	Cyclobutane	Tetramethylene	C_4H_8	287-23-0	56.107	vol liq or gas	-90.7	12.6	0.7038^{0}	1.375^{20}	i H_2O; vs EtOH, ace; msc eth; s bz
2531	Cyclobutanecarbonitrile	Cyanocyclobutane	C_5H_7N	4426-11-3	81.117			149.6			
2532	Cyclobutanecarboxylic acid		$C_5H_8O_2$	3721-95-7	100.117	liq	-1.0	190; 74^{2}	1.0599^{20}	1.4400^{20}	sl H_2O; msc EtOH, eth
2533	1,1-Cyclobutanedicarboxylic acid		$C_6H_8O_4$	5445-51-2	144.126	pr (w, eth)	158.0				vs H_2O; s EtOH, eth, bz; sl lig
2534	Cyclobutanol	Hydroxycyclobutane	C_4H_8O	2919-23-5	72.106			124	0.9218^{15}	1.4371^{20}	
2535	Cyclobutanone		C_4H_6O	1191-95-3	70.090	liq	-50.9	99	0.9547^{0}	1.4215^{20}	s H_2O, eth, bz, chl, tol; vs EtOH; i peth

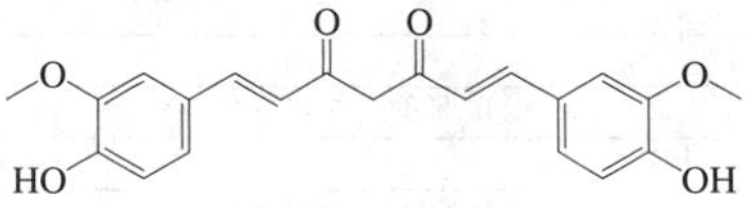

Curcumin

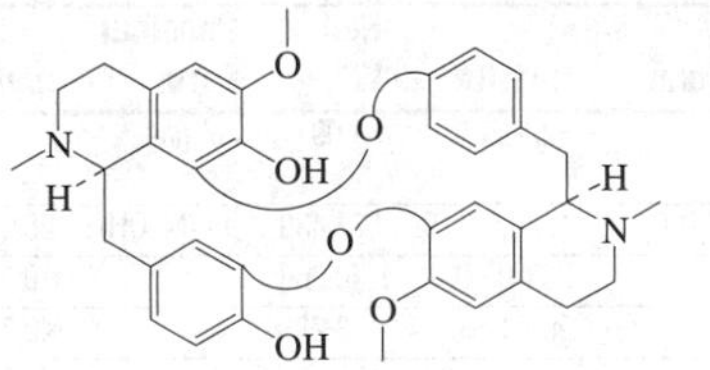

Curine

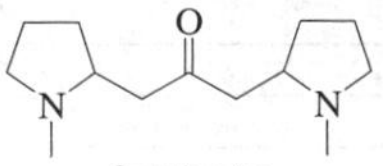

Cuscohygrine

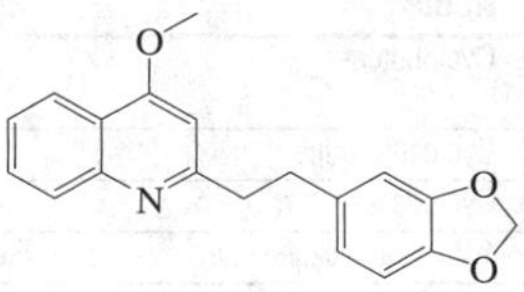

Cusparine

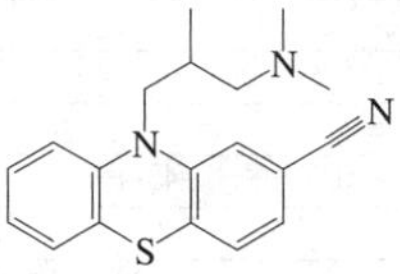

Cyamemazine

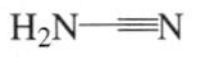

Cyanamide

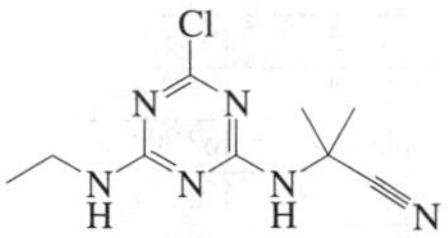

Cyanazine

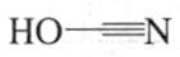

Cyanic acid

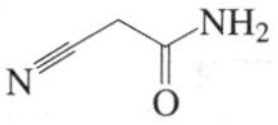

2-Cyanoacetamide

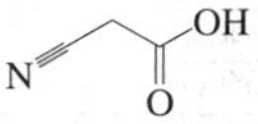

Cyanoacetic acid

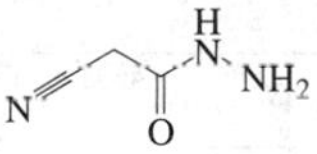

Cyanoacetohydrazide

Cyanoacetylene

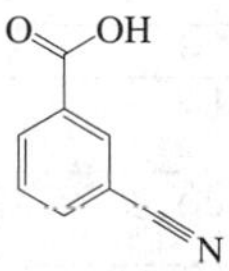

3-Cyanobenzoic acid

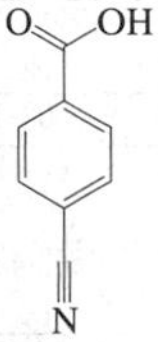

4-Cyanobenzoic acid

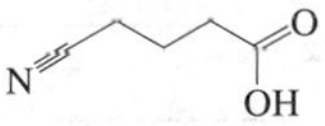

4-Cyanobutanoic acid

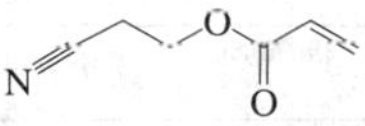

2-Cyanoethyl acrylate

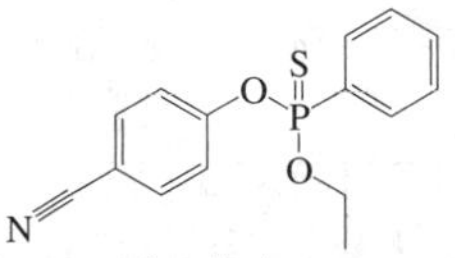

Cyanofenphos

Cyanogen

Cyanogen bromide

Cyanogen chloride

Cyanogen fluoride

Cyanogen iodide

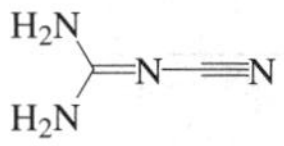

Cyanoguanidine

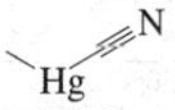

Cyanomethylmercury

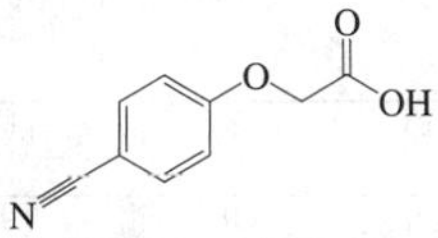

(4-Cyanophenoxy)acetic acid

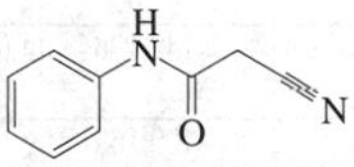

2-Cyano-*N*-phenylacetamide

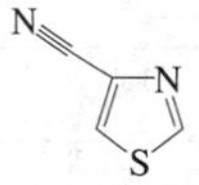

4-Cyanothiazole

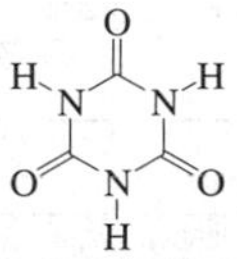

Cyanuric acid

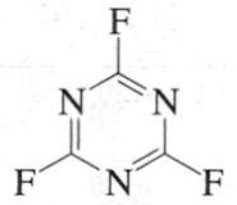

Cyanuric fluoride

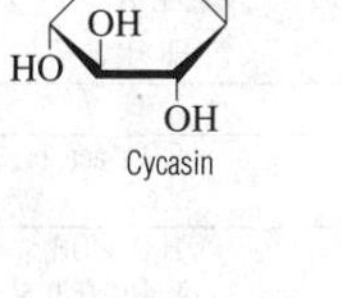

Cycasin

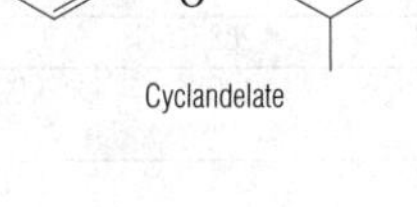

Cyclandelate

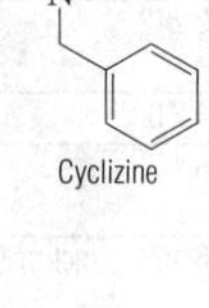

Cyclizine

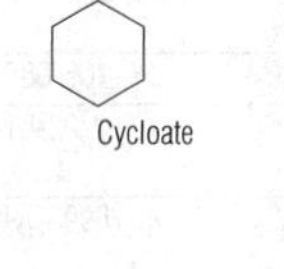

Cycloate

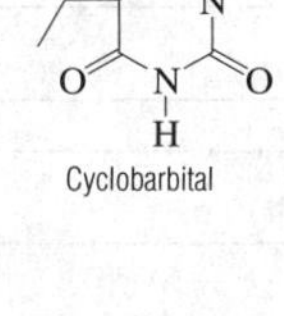

Cyclobarbital

Cyclobutanamine

Cyclobutane

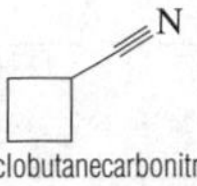

Cyclobutanecarbonitrile

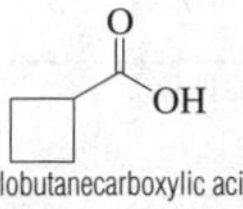

Cyclobutanecarboxylic acid

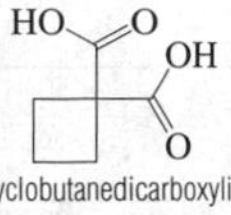

1,1-Cyclobutanedicarboxylic acid

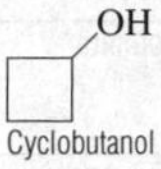

Cyclobutanol

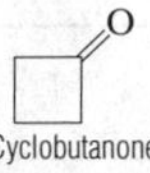

Cyclobutanone

No.	Name	Synonym	Mol. Form.	CAS RN	Mol. Wt.	Physical Form	mp/°C	bp/°C	den/ g cm^{-3}	n_D	Solubility
2536	Cyclobutene		C_4H_6	822-35-5	54.091	col gas		2	0.733^{0}		vs ace; s bz, peth
2537	Cyclochlorotine		$C_{24}H_{31}Cl_2N_5O_7$	12663-46-6	572.439	nd (MeOH)	255 dec				
2538	Cyclodecane		$C_{10}H_{20}$	293-96-9	140.266		10	202	0.8538^{25}	1.4716^{20}	
2539	1,2-Cyclodecanedione	Sebacil	$C_{10}H_{16}O_2$	96-01-5	168.233		40.5	104^{10}			
2540	Cyclodecanol		$C_{10}H_{20}O$	1502-05-2	156.265		40.5	125^{12}	0.9606^{20}	1.4926^{20}	s EtOH
2541	Cyclodecanone		$C_{10}H_{18}O$	1502-06-3	154.249	amor pow	28	106^{13}	0.9654^{20}	1.4806^{20}	vs bz, eth, chl
2542	α-Cyclodextrin	Cyclomaltohexaose	$C_{36}H_{60}O_{30}$	10016-20-3	972.843	hx pl or nd					vs cold H_2O; i hot H_2O
2543	β-Cyclodextrin	Cyclomaltoheptaose	$C_{42}H_{70}O_{35}$	7585-39-9	1134.984	mcl cry (w)	260 dec				
2544	γ-Cyclodextrin	Cyclomaltooctaose	$C_{48}H_{80}O_{40}$	17465-86-0	1297.125	sq pl or rods					
2545	Cyclododecane		$C_{12}H_{24}$	294-62-2	168.319	nd (al)	60.4	247	0.82^{80}		
2546	Cyclododecanol		$C_{12}H_{24}O$	1724-39-6	184.318			286			
2547	Cyclododecanone		$C_{12}H_{22}O$	830-13-7	182.302		59	127^{12}	0.9059^{66}	1.4571^{60}	
2548	1,5,9-Cyclododecatriene	CDT	$C_{12}H_{18}$	4904-61-4	162.271	liq	-17	240	0.84^{100}		
2549	*cis*-Cyclododecene		$C_{12}H_{22}$	1129-89-1	166.303			133^{35}, 71^{2}		1.4840^{20}	vs bz, chl
2550	*trans*-Cyclododecene		$C_{12}H_{22}$	1486-75-5	166.303			113^{17}		1.4850^{20}	vs bz, chl
2551	*cis*-9-Cycloheptadecen-1-one	Civetone	$C_{17}H_{30}O$	542-46-1	250.419		32.5	343; 159^{2}			
2552	1,3-Cycloheptadiene		C_7H_{10}	4054-38-0	94.154	liq	-110.4	120.5	0.868^{25}	1.4978^{20}	
2553	Cycloheptanamine		$C_7H_{15}N$	5452-35-7	113.201			54^{11}		1.4724^{20}	
2554	Cycloheptane		C_7H_{14}	291-64-5	98.186	liq	-8.46	118.4	0.8098^{20}	1.4436^{20}	i H_2O; vs EtOH, eth; s bz, chl
2555	1,2-Cycloheptanedione		$C_7H_{10}O_2$	3008-39-7	126.153		-40	108^{17}	1.0583^{22}	1.4689^{22}	s EtOH
2556	Cycloheptanol		$C_7H_{14}O$	502-41-0	114.185		7.2	185	0.9554^{20}	1.40705^{20}	sl H_2O; vs EtOH, eth
2557	Cycloheptanone	Suberone	$C_7H_{12}O$	502-42-1	112.169			178.5	0.9508^{20}	1.4608^{20}	i H_2O; vs EtOH, eth
2558	1,3,5-Cycloheptatriene	Tropilidene	C_7H_8	544-25-2	92.139	liq; cub cry (-80°C)	-79.5	117; 60.5^{122}	0.8875^{19}	1.5343^{20}	i H_2O; s EtOH, eth; vs bz, chl
2559	2,4,6-Cycloheptatrien-1-one		C_7H_6O	539-80-0	106.122		-7	113^{15}, 84^{6}	1.095^{22}	1.6172^{22}	vs bz, chl
2560	Cycloheptene		C_7H_{12}	628-92-2	96.170	liq	-56	115	0.8228^{20}	1.4552^{20}	i H_2O; s EtOH, eth, bz, chl; sl ctc
2561	1,3-Cyclohexadiene		C_6H_8	592-57-4	80.128	liq	-89	80.5	0.8405^{20}	1.4755^{20}	i H_2O; s EtOH, bz, chl, peth; vs eth
2562	1,4-Cyclohexadiene	1,4-Dihydrobenzene	C_6H_8	628-41-1	80.128	liq	-49.2	85.5	0.8471^{20}	1.4725^{20}	i H_2O; msc EtOH, eth; s bz, chl, peth
2563	3,5-Cyclohexadiene-1,2-dione		$C_6H_4O_2$	583-63-1	108.095	red pl or pr	≈65 dec				s eth, ace, bz; i peth
2564	2,5-Cyclohexadiene-1,4-dione, dioxime		$C_6H_6N_2O_2$	105-11-3	138.124	pa ye nd (w)	240 dec				s H_2O
2565	Cyclohexane	Hexahydrobenzene	C_6H_{12}	110-82-7	84.159		6.59	80.73	0.7739^{25}	1.4235^{25}	i H_2O; msc EtOH, eth, ace, bz, lig, ctc
2566	Cyclohexaneacetic acid		$C_8H_{14}O_2$	5292-21-7	142.196	nd (HCO_2H)	33	245	1.0423^{18}	1.4775^{20}	sl H_2O; s eth, ace
2567	Cyclohexanecarbonitrile	Cyclohexyl cyanide	$C_7H_{11}N$	766-05-2	109.169	liq	11	184; 76^{16}	0.919	1.4505^{20}	
2568	Cyclohexanecarbonyl chloride		$C_7H_{11}ClO$	2719-27-9	146.614			180	1.0962^{15}	1.4711^{29}	
2569	Cyclohexanecarboxaldehyde		$C_7H_{12}O$	2043-61-0	112.169			159.3	0.9035^{20}	1.4496^{20}	s H_2O, eth
2570	Cyclohexanecarboxylic acid	Hexahydrobenzoic acid	$C_7H_{12}O_2$	98-89-5	128.169	mcl pr	31.5	232.5	1.0334^{22}	1.4530^{20}	sl H_2O, ctc; vs EtOH, bz, chl
2571	*cis*-1,2-Cyclohexanediamine	*cis*-1,2-Diaminocyclohexane	$C_6H_{14}N_2$	1436-59-5	114.188	liq		40^{2}	0.952^{20}	1.4951^{20}	
2572	*trans*-1,2-Cyclohexanediamine	*trans*-1,2-Diaminocyclohexane	$C_6H_{14}N_2$	1121-22-8	114.188		14.8	80^{15}, 41^{2}	0.951^{20}		
2573	*trans*-1,4-Cyclohexanedicarboxylic acid		$C_8H_{12}O_4$	619-82-9	172.179	pr (w)	312.5	sub 300			sl H_2O, eth; vs EtOH; s ace; i chl
2574	1,3-Cyclohexanedimethanamine		$C_8H_{18}N_2$	2579-20-6	142.242		<-70	220	0.945^{20}		vs H_2O, eth, EtOH
2575	1,4-Cyclohexanedimethanol		$C_8H_{16}O_2$	105-08-8	144.212		43	283; 167^{10}			
2576	*cis*-1,2-Cyclohexanediol		$C_6H_{12}O_2$	1792-81-0	116.158		100	120^{15}	1.0297^{101}		s EtOH, ace, bz; sl chl
2577	*trans*-1,4-Cyclohexanediol		$C_6H_{12}O_2$	6995-79-5	116.158	mcl pr (ace)	143		1.18^{20}		s H_2O, EtOH, MeOH; i eth; sl ace
2578	1,2-Cyclohexanedione	1,2-Dioxocyclohexane	$C_6H_8O_2$	765-87-7	112.127	cry (peth)	40	194	1.1187^{21}	1.4995^{20}	s H_2O, EtOH, eth, bz
2579	1,3-Cyclohexanedione	Dihydroresorcinol	$C_6H_8O_2$	504-02-9	112.127	pr (bz)	105.5		1.0861^{91}	1.4576^{102}	s H_2O, EtOH, ace, chl; sl eth, bz
2580	1,4-Cyclohexanedione	Tetrahydroquinone	$C_6H_8O_2$	637-88-7	112.127	mcl pl (w), nd (peth)	78	132^{20}	1.0861^{91}		s H_2O, EtOH, eth, ace, bz, chl

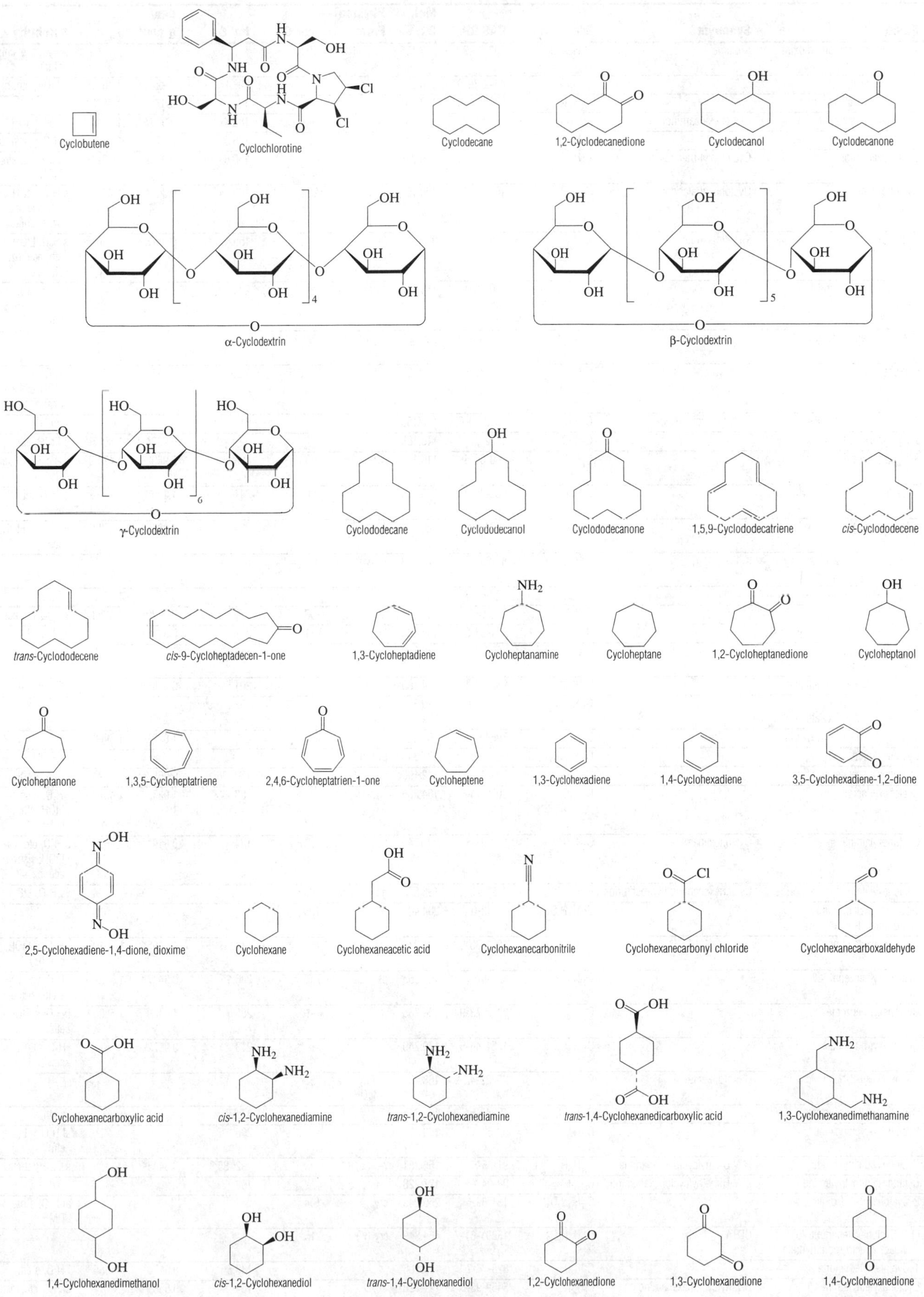

Cyclobutene
Cyclochlorotine
Cyclodecane
1,2-Cyclodecanedione
Cyclodecanol
Cyclodecanone
α-Cyclodextrin
β-Cyclodextrin
γ-Cyclodextrin
Cyclododecane
Cyclododecanol
Cyclododecanone
1,5,9-Cyclododecatriene
cis-Cyclododecene
trans-Cyclododecene
cis-9-Cycloheptadecen-1-one
1,3-Cycloheptadiene
Cycloheptanamine
Cycloheptane
1,2-Cycloheptanedione
Cycloheptanol
Cycloheptanone
1,3,5-Cycloheptatriene
2,4,6-Cycloheptatrien-1-one
Cycloheptene
1,3-Cyclohexadiene
1,4-Cyclohexadiene
3,5-Cyclohexadiene-1,2-dione
2,5-Cyclohexadiene-1,4-dione, dioxime
Cyclohexane
Cyclohexaneacetic acid
Cyclohexanecarbonitrile
Cyclohexanecarbonyl chloride
Cyclohexanecarboxaldehyde
Cyclohexanecarboxylic acid
cis-1,2-Cyclohexanediamine
trans-1,2-Cyclohexanediamine
trans-1,4-Cyclohexanedicarboxylic acid
1,3-Cyclohexanedimethanamine
1,4-Cyclohexanedimethanol
cis-1,2-Cyclohexanediol
trans-1,4-Cyclohexanediol
1,2-Cyclohexanedione
1,3-Cyclohexanedione
1,4-Cyclohexanedione

No.	Name	Synonym	Mol. Form.	CAS RN	Mol. Wt.	Physical Form	mp/°C	bp/°C	den/ g cm^{-3}	n_D	Solubility
2581	1,2-Cyclohexanedione dioxime	Nioxime	$C_6H_{10}N_2O_2$	492-99-9	142.155	nd (w, HOAc)	192				s H_2O, ace, chl; sl tfa
2582	Cyclohexaneethanol		$C_8H_{16}O$	4442-79-9	128.212			208	0.9229^{20}	1.4641^{20}	s EtOH, eth, bz
2583	Cyclohexanemethanamine		$C_7H_{15}N$	3218-02-8	113.201			160	0.87^{25}	1.4630^{20}	
2584	Cyclohexanemethanol	Cyclohexylcarbinol	$C_7H_{14}O$	100-49-2	114.185	liq	-43	183	0.9297^{20}	1.4644^{20}	vs eth, EtOH
2585	Cyclohexanepropanoic acid		$C_9H_{16}O_2$	701-97-3	156.222		16	276.5	0.912^{25}	1.4638^{20}	s H_2O, eth; sl ctc
2586	Cyclohexanethiol	Cyclohexyl mercaptan	$C_6H_{12}S$	1569-69-3	116.224			158.8	0.9782^{20}	1.4921^{20}	vs ace, bz, eth, EtOH
2587	Cyclohexanol	Cyclohexyl alcohol	$C_6H_{12}O$	108-93-0	100.158	hyg nd	25.93	160.84	0.9624^{20}	1.4641^{20}	s H_2O, EtOH, eth, ace; msc bz; sl chl
2588	Cyclohexanone	Pimelic ketone	$C_6H_{10}O$	108-94-1	98.142	liq	-27.9	155.43	0.9478^{20}	1.4507^{20}	s H_2O, EtOH, eth, ace, bz, chl, ctc
2589	Cyclohexanone oxime		$C_6H_{11}NO$	100-64-1	113.157	hex pr (lig)	90	206			s H_2O, EtOH, eth, MeOH; sl chl
2590	Cyclohexanone peroxide		$C_{12}H_{22}O_5$	78-18-2	246.300	cry or long nd	79				
2591	Cyclohexene	Tetrahydrobenzene	C_6H_{10}	110-83-8	82.143	liq	-103.5	82.98	0.8110^{20}	1.4465^{20}	i H_2O; msc EtOH, eth, ace, bz, lig, ctc
2592	1-Cyclohexenecarbonitrile	1-Cyanocyclohexene	C_7H_9N	1855-63-6	107.153			81^{12}			
2593	1-Cyclohexene-1-carboxaldehyde		$C_7H_{10}O$	1192-88-7	110.153			69^{18}	0.9694^{20}	1.5005^{20}	s EtOH, eth
2594	3-Cyclohexene-1-carboxaldehyde		$C_7H_{10}O$	100-50-5	110.153		1.0	105	0.9692^{20}	1.4745^{20}	s ace, MeOH; sl ctc
2595	1-Cyclohexene-1-carboxylic acid		$C_7H_{10}O_2$	636-82-8	126.153		38	241	1.109^{20}	1.4902^{20}	sl H_2O; s EtOH, ace
2596	3-Cyclohexene-1-carboxylic acid		$C_7H_{10}O_2$	4771-80-6	126.153		17	234.5	1.0820^{20}	1.4814^{20}	vs H_2O; s EtOH, ace
2597	4-Cyclohexene-1,2-dicarboxylic acid		$C_8H_{10}O_4$	88-98-2	170.163	pr (w)	173.0				
2598	2-Cyclohexen-1-ol		$C_6H_{10}O$	822-67-3	98.142			164	0.9923^{15}	1.4790^{25}	s EtOH, ace
2599	2-Cyclohexen-1-one		C_6H_8O	930-68-7	96.127	liq	-53	170	0.9620^{25}	1.4883^{20}	vs EtOH; s ace
2600	1-Cyclohexen-1-ylbenzene		$C_{12}H_{14}$	771-98-2	158.239	liq	-11	252	0.9939^{20}	1.5718^{20}	vs MeOH
2601	2-(1-Cyclohexen-1-yl) cyclohexanone		$C_{12}H_{18}O$	1502-22-3	178.270			116^{3}		1.5070^{20}	
2602	1-(1-Cyclohexen-1-yl)ethanone		$C_8H_{12}O$	932-66-1	124.180		73	201.5	0.9655^{20}	1.4881^{20}	s EtOH, eth
2603	3-Cyclohexenylmethyl 3-cyclohexenecarboxylate		$C_{14}H_{20}O_2$	2611-00-9	220.308	liq		153^{7}, $109^{0.6}$			
2604	4-(3-Cyclohexen-1-yl)pyridine		$C_{11}H_{13}N$	70644-46-1	159.228		22.1	226	1.0222^{25}	1.5466^{25}	
2605	Cycloheximide		$C_{15}H_{23}NO_4$	66-81-9	281.349	pl (al)	119				vs EtOH
2606	Cyclohexyl acetate		$C_8H_{14}O_2$	622-45-7	142.196			173; 96^{75}	0.968^{20}	1.442^{20}	vs eth, EtOH
2607	Cyclohexyl acrylate		$C_9H_{14}O_2$	3066-71-5	154.206			183; 88^{20}	1.0275^{20}	1.4673^{20}	i H_2O; msc EtOH, eth; s chl
2608	Cyclohexylamine	Cyclohexanamine	$C_6H_{13}N$	108-91-8	99.174	liq	-17.8	134	0.8191^{20}	1.4625^{15}	s H_2O, ctc; vs EtOH; msc eth, ace, bz
2609	Cyclohexylamine hydrochloride	Cyclohexanamine hydrochloride	$C_6H_{14}ClN$	4998-76-9	135.635	nd (w, al-eth)	206.5				vs H_2O, EtOH
2610	2-(Cyclohexylaminothio) benzothiazole		$C_{13}H_{16}N_2S_2$	95-33-0	264.409		103				
2611	*N*-Cyclohexylaniline		$C_{12}H_{17}N$	1821-36-9	175.270	mcl pr	16	279; 192^{73}	1.0155^{20}	1.5610^{20}	i H_2O; s EtOH, eth, bz
2612	Cyclohexylbenzene		$C_{12}H_{16}$	827-52-1	160.255	pl	7.07	240.1	0.9427^{20}	1.5329^{20}	i H_2O; vs EtOH; s eth; sl ctc
2613	Cyclohexyl benzoate		$C_{13}H_{16}O_2$	2412-73-9	204.265		<-10	285	1.0429^{20}	1.5200^{20}	i H_2O; s EtOH, eth
2614	Cyclohexyl butanoate		$C_{10}H_{18}O_2$	1551-44-6	170.249			213	0.9572^{0}		i H_2O; s EtOH; sl ctc
2615	3-Cyclohexyl-2-butenoic acid	Cicrotoic acid	$C_{10}H_{16}O_2$	25229-42-9	168.233	pr (aq-MeOH)	85.5				
2616	Cyclohexyl chloroformate		$C_7H_{11}ClO_2$	13248-54-9	162.614			87.5^{27}			vs eth
2617	Cyclohexylcyclohexane		$C_{12}H_{22}$	92-51-3	166.303		4	238			sl H_2O; s EtOH, eth
2618	Cyclohexyldiethylamine	*N,N*-Diethylcyclohexanamine	$C_{10}H_{21}N$	91-65-6	155.281			192; 85^{20}	0.8443^{25}		s EtOH; sl ctc
2619	Cyclohexyldimethylamine	*N,N*-Dimethylcyclohexanamine	$C_8H_{17}N$	98-94-2	127.228			162			
2620	2-Cyclohexyl-4,6-dinitrophenol		$C_{12}H_{14}N_2O_5$	131-89-5	266.249	cry	104				sl H_2O; s bz, DMF
2621	(1,2-Cyclohexylenedinitrilo) tetraacetic acid monohydrate	CDTA	$C_{14}H_{24}N_2O_9$	13291-61-7	364.349	cry (w)	215				
2622	1-Cyclohexylethanone		$C_8H_{14}O$	823-76-7	126.196			180.5	0.9176^{20}	1.4565^{16}	i H_2O; s eth
2623	Cyclohexylethylamine	*N*-Ethylcyclohexanamine	$C_8H_{17}N$	5459-93-8	127.228			164	0.868^{0}		sl H_2O, ctc; msc EtOH, eth

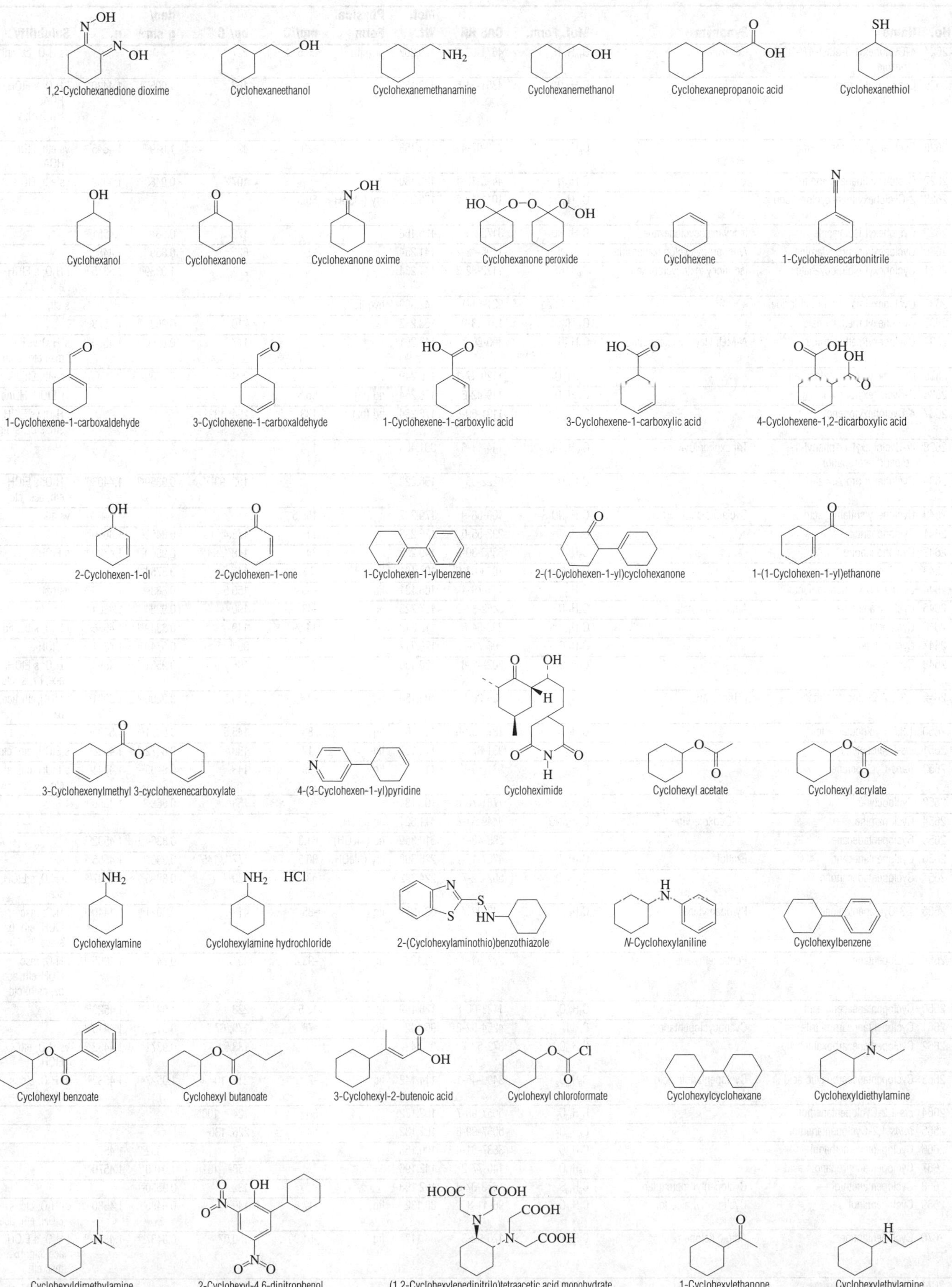

1,2-Cyclohexanedione dioxime
Cyclohexaneethanol
Cyclohexanemethanamine
Cyclohexanemethanol
Cyclohexanepropanoic acid
Cyclohexanethiol
Cyclohexanol
Cyclohexanone
Cyclohexanone oxime
Cyclohexanone peroxide
Cyclohexene
1-Cyclohexenecarbonitrile
1-Cyclohexene-1-carboxaldehyde
3-Cyclohexene-1-carboxaldehyde
1-Cyclohexene-1-carboxylic acid
3-Cyclohexene-1-carboxylic acid
4-Cyclohexene-1,2-dicarboxylic acid
2-Cyclohexen-1-ol
2-Cyclohexen-1-one
1-Cyclohexen-1-ylbenzene
2-(1-Cyclohexen-1-yl)cyclohexanone
1-(1-Cyclohexen-1-yl)ethanone
3-Cyclohexenylmethyl 3-cyclohexenecarboxylate
4-(3-Cyclohexen-1-yl)pyridine
Cycloheximide
Cyclohexyl acetate
Cyclohexyl acrylate
Cyclohexylamine
Cyclohexylamine hydrochloride
2-(Cyclohexylaminothio)benzothiazole
N-Cyclohexylaniline
Cyclohexylbenzene
Cyclohexyl benzoate
Cyclohexyl butanoate
3-Cyclohexyl-2-butenoic acid
Cyclohexyl chloroformate
Cyclohexylcyclohexane
Cyclohexyldiethylamine
Cyclohexyldimethylamine
2-Cyclohexyl-4,6-dinitrophenol
(1,2-Cyclohexylenedinitrilo)tetraacetic acid monohydrate
1-Cyclohexylethanone
Cyclohexylethylamine

No.	Name	Synonym	Mol. Form.	CAS RN	Mol. Wt.	Physical Form	mp/°C	bp/°C	den/ g cm^{-3}	n_D	Solubility
2624	4-Cyclohexyl-3-ethyl-4*H*-1,2,4-triazole	Hexazole	$C_{10}H_{17}N_3$	4671-03-8	179.262	pr (eth)	89.5	227^{10}			vs H_2O, bz, chl
2625	Cyclohexyl formate		$C_7H_{12}O_2$	4351-54-6	128.169			162	1.0057^{0}	1.4430^{20}	i H_2O; s EtOH, HOAc, HCOOH; vs eth
2626	Cyclohexyl hydroperoxide		$C_6H_{12}O_2$	766-07-4	116.158		-20	$42^{0.1}$	1.019^{20}	1.4645^{25}	vs eth, EtOH, HOAc
2627	Cyclohexylideneacetonitrile		$C_8H_{11}N$	4435-18-1	121.180			107^{22}	0.9483^{15}	1.4382^{25}	vs eth, EtOH
2628	2-Cyclohexylidenecyclohexanone		$C_{12}H_{18}O$	1011-12-7	178.270	cry (MeOH aq)	56.5				
2629	Cyclohexyl isocyanate	Isocyanatocyclohexane	$C_7H_{11}NO$	3173-53-3	125.168			172	0.98^{25}	1.4551^{20}	
2630	Cyclohexylisopropylamine	*N*-Isopropylcyclohexanamine	$C_9H_{19}N$	1195-42-2	141.254			62^{12}	0.859^{25}	1.4480^{20}	
2631	Cyclohexyl isothiocyanate	Isothiocyanatocyclohexane	$C_7H_{11}NS$	1122-82-3	141.234			221	1.0339^{20}	1.5375^{20}	i H_2O; s EtOH, eth; sl ctc
2632	Cyclohexylmagnesium chloride		$C_6H_{11}ClMg$	931-51-1	142.909	hyg liq					s eth
2633	Cyclohexyl methacrylate		$C_{10}H_{16}O_2$	101-43-9	168.233			210	0.9626^{20}	1.4578^{20}	
2634	Cyclohexylmethylamine	*N*-Methylcyclohexanamine	$C_7H_{15}N$	100-60-7	113.201			147	0.8660^{23}	1.4560^{20}	sl H_2O; vs EtOH; msc eth; s chl
2635	Cyclohexyl 2-methylpropanoate		$C_{10}H_{18}O_2$	1129-47-1	170.249			204	0.9489^{0}		vs eth, EtOH
2636	2-Cyclohexylphenol		$C_{12}H_{16}O$	119-42-6	176.254	nd (lig)	56.5				vs EtOH, HOAc
2637	4-Cyclohexylphenol		$C_{12}H_{16}O$	1131-60-8	176.254	nd (bz)	133	294; 133^{4}			i H_2O; vs EtOH, eth; s bz; sl lig
2638	α-Cyclohexyl-α-phenyl-1-piperidinepropanol	Trihexphenidyl	$C_{20}H_{31}NO$	144-11-6	301.466		114				
2639	Cyclohexyl propanoate		$C_9H_{16}O_2$	6222-35-1	156.222			193; 93^{35}	0.9359^{20}	1.4403^{20}	i H_2O; s EtOH, eth, ace, ctc
2640	Cyclohexylsulfamic acid	Cyclamic acid	$C_6H_{13}NO_3S$	100-88-9	179.237		169.5				vs alk
2641	Cyclononane		C_9H_{18}	293-55-0	126.239		11	178.4	0.8463^{25}	1.4666^{20}	
2642	Cyclononanone		$C_9H_{16}O$	3350-30-9	140.222		34	148^{24}, 94^{12}	0.9560^{20}	1.4729^{20}	s EtOH
2643	1,4-Cyclooctadiene		C_8H_{12}	1073-07-0	108.181	liq	-53	145	0.8754^{20}		
2644	*cis,cis*-1,5-Cyclooctadiene		C_8H_{12}	111-78-4	108.181	liq	-56.4	150.5	0.883^{20}	1.4905^{25}	vs bz
2645	Cyclooctanamine	Aminocyclooctane	$C_8H_{17}N$	5452-37-9	127.228	liq	-48	190	0.928^{25}	1.4804^{20}	
2646	Cyclooctane		C_8H_{16}	292-64-8	112.213		14.59	149	0.8349^{20}	1.4586^{20}	i H_2O; s bz, lig
2647	Cyclooctanol		$C_8H_{16}O$	696-71-9	128.212		25.1	99^{16}	0.9740^{20}	1.4871^{20}	s EtOH
2648	Cyclooctanone		$C_8H_{14}O$	502-49-8	126.196		29	196	0.9581^{20}	1.4694^{20}	i H_2O; s EtOH, ace, bz; sl ctc
2649	1,3,5,7-Cyclooctatetraene	[8]Annulene	C_8H_8	629-20-9	104.150	liq	-2.4	140.5	0.9206^{20}	1.5381^{20}	s EtOH, eth, ace, bz
2650	1,3,5-Cyclooctatriene		C_8H_{10}	1871-52-9	106.165	liq	-83	145.5	0.8971^{25}	1.5035^{25}	
2651	*cis*-Cyclooctene		C_8H_{14}	931-87-3	110.197	liq	-12	138	0.8472^{20}	1.4698^{20}	s EtOH, eth, ctc
2652	*trans*-Cyclooctene		C_8H_{14}	931-89-5	110.197	liq	-59	143	0.8483^{20}	1.4741^{25}	s EtOH, chl; sl ctc
2653	Cyclooctyne		C_8H_{12}	1781-78-8	108.181			158	0.868^{20}	1.4850^{20}	
2654	Cyclopamine	11-Deoxojervine	$C_{27}H_{41}NO_2$	4449-51-8	411.621	nd (EtOH)	237				
2655	Cyclopentadecane		$C_{15}H_{30}$	295-48-7	210.399	nd (MeOH)	61.3		0.8364^{61}	1.4592^{61}	
2656	Cyclopentadecanol	Exaltol	$C_{15}H_{30}O$	4727-17-7	226.398	cry (MeOH)	80.5	177^{11}, $145^{0.3}$	0.930^{20}	1.4555^{98}	
2657	Cyclopentadecanone		$C_{15}H_{28}O$	502-72-7	224.382		63	$120^{0.3}$	0.8895^{25}	1.4637^{60}	sl H_2O; s EtOH, ace
2658	1,3-Cyclopentadiene	Pyropentylene	C_5H_6	542-92-7	66.102	liq	-85	41	0.8021^{20}	1.4440^{20}	i H_2O; msc EtOH, eth, bz; s ace
2659	Cyclopentane	Pentamethylene	C_5H_{10}	287-92-3	70.133	liq	-93.4	49.3	0.7457^{20}	1.4065^{20}	i H_2O; msc EtOH, eth, ace, bz, peth, ctc
2660	Cyclopentaneacetic acid		$C_7H_{12}O_2$	1123-00-8	128.169	pl	13.5	228	1.0216^{18}	1.4523^{18}	
2661	Cyclopentanecarbonitrile	Cyanocyclopentane	C_6H_9N	4254-02-8	95.142	liq	-76	170; 67^{10}	0.912	1.4410^{20}	
2662	Cyclopentanecarboxaldehyde		$C_6H_{10}O$	872-53-7	98.142			133.5	0.9371^{20}	1.4432^{20}	vs H_2O, eth, EtOH
2663	Cyclopentanecarboxylic acid	Cyclopentanoic acid	$C_6H_{10}O_2$	3400-45-1	114.142	liq	-7	212; 104^{11}	1.0527^{20}	1.4532^{20}	sl H_2O, ctc; s MeOH
2664	*cis*-1,2-Cyclopentanediol		$C_5H_{10}O_2$	5057-98-7	102.132		30	124^{29}, 100^{10}			
2665	*trans*-1,2-Cyclopentanediol		$C_5H_{10}O_2$	5057-99-8	102.132		54.7	226; 136^{21}			
2666	Cyclopentanemethanol		$C_6H_{12}O$	3637-61-4	100.158			163	0.9332^{20}	1.4579^{20}	
2667	Cyclopentanepropanoic acid		$C_8H_{14}O_2$	140-77-2	142.196			158^{26}, 131^{12}	1.0100^{17}	1.4570^{20}	
2668	Cyclopentanethiol	Cyclopentyl mercaptan	$C_5H_{10}S$	1679-07-8	102.198			132.1	0.9550^{20}		
2669	Cyclopentanol	Cyclopentyl alcohol	$C_5H_{10}O$	96-41-3	86.132	liq	-17.5	140.42	0.9488^{20}	1.4530^{20}	sl H_2O, ctc; s EtOH, eth, ace
2670	Cyclopentanone	Adipic ketone	C_5H_8O	120-92-3	84.117	liq	-51.90	130.57	0.9487^{20}	1.4366^{20}	i H_2O; s EtOH, ace, ctc, hx; msc eth
2671	Cyclopentanone oxime		C_5H_9NO	1192-28-5	99.131		57.8	196			vs H_2O, bz

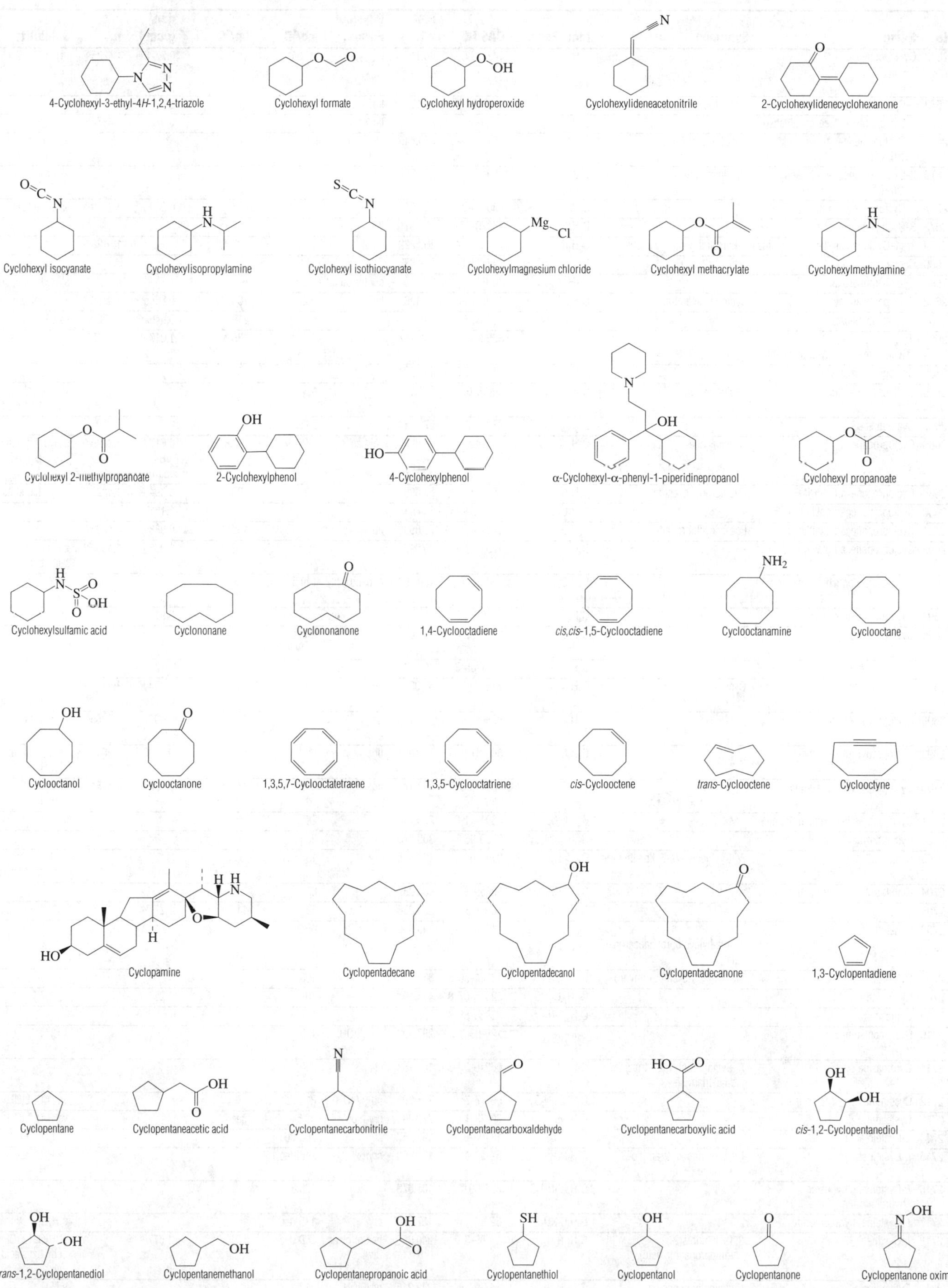

4-Cyclohexyl-3-ethyl-4*H*-1,2,4-triazole
Cyclohexyl formate
Cyclohexyl hydroperoxide
Cyclohexylideneacetonitrile
2-Cyclohexylidenecyclohexanone
Cyclohexyl isocyanate
Cyclohexylisopropylamine
Cyclohexyl isothiocyanate
Cyclohexylmagnesium chloride
Cyclohexyl methacrylate
Cyclohexylmethylamine
Cyclohexyl 2-methylpropanoate
2-Cyclohexylphenol
4-Cyclohexylphenol
α-Cyclohexyl-α-phenyl-1-piperidinepropanol
Cyclohexyl propanoate
Cyclohexylsulfamic acid
Cyclononane
Cyclononanone
1,4-Cyclooctadiene
cis,cis-1,5-Cyclooctadiene
Cyclooctanamine
Cyclooctane
Cyclooctanol
Cyclooctanone
1,3,5,7-Cyclooctatetraene
1,3,5-Cyclooctatriene
cis-Cyclooctene
trans-Cyclooctene
Cyclooctyne
Cyclopamine
Cyclopentadecane
Cyclopentadecanol
Cyclopentadecanone
1,3-Cyclopentadiene
Cyclopentane
Cyclopentaneacetic acid
Cyclopentanecarbonitrile
Cyclopentanecarboxaldehyde
Cyclopentanecarboxylic acid
cis-1,2-Cyclopentanediol
trans-1,2-Cyclopentanediol
Cyclopentanemethanol
Cyclopentanepropanoic acid
Cyclopentanethiol
Cyclopentanol
Cyclopentanone
Cyclopentanone oxime

No.	Name	Synonym	Mol. Form.	CAS RN	Mol. Wt.	Physical Form	mp/°C	bp/°C	den/ g cm^{-3}	n_D	Solubility
2672	Cyclopentene		C_5H_8	142-29-0	68.118	liq	-135.0	44.2	0.7720^{20}	1.4225^{20}	i H_2O; s EtOH, eth, bz, ctc, peth
2673	1-Cyclopentenecarbonitrile	1-Cyanocyclopentene	C_6H_7N	3047-38-9	93.127	liq		81^{30}			
2674	1-Cyclopentene-1-carboxaldehyde		C_6H_8O	6140-65-4	96.127	liq	-32	146	0.970^{21}	1.4872^{17}	
2675	2-Cyclopentene-1-tridecanoic acid, (*S*)	Chaulmoogric acid	$C_{18}H_{32}O_2$	29106-32-9	280.446	pl or lf (al, HOAc)	68.5	247^{20}			vs eth, chl
2676	2-Cyclopentene-1-undecanoic acid, (*R*)	Hydnocarpic acid	$C_{16}H_{28}O_2$	459-67-6	252.392		60.5				vs EtOH, chl, peth
2677	2-Cyclopenten-1-one		C_5H_6O	930-30-3	82.101			136; 40^{12}	0.989^{15}	1.4629^{15}	vs eth, EtOH
2678	3-Cyclopenten-1-one		C_5H_6O	14320-37-7	82.101	liq		28^{17}			
2679	*N*-(1-Cyclopenten-1-yl)pyrrolidine	1-Pyrrolidinylcyclopentene	$C_9H_{15}N$	7148-07-4	137.222			105^{15}		1.5128^{20}	
2680	Cyclopenthiazide		$C_{13}H_{18}ClN_3O_4S_2$	742-20-1	379.883		238				
2681	Cyclopentobarbital		$C_{12}H_{14}N_2O_3$	76-68-6	234.250	cry (w, dil al)	139.5				sl H_2O; vs EtOH
2682	Cyclopentylamine	Cyclopentanamine	$C_5H_{11}N$	1003-03-8	85.148	liq	-82.7	108	0.8689^{20}	1.4728^{25}	s ace, bz, chl
2683	Cyclopentylbenzene		$C_{11}H_{14}$	700-88-9	146.229			219	0.9462^{20}	1.5280^{20}	vs eth
2684	2-Cyclopentylidenecyclopentanone		$C_{10}H_{14}O$	825-25-2	150.217			135^{25}	1.0179^{18}	1.5215^{18}	
2685	Cyclopentyl methyl sulfide		$C_6H_{12}S$	7133-36-0	116.224			156.2			
2686	Cyclophosphamide	Cyclophosphane	$C_7H_{15}Cl_2N_2O_2P$	50-18-0	261.086		43				vs H_2O; sl bz, chl, diox, EtOH
2687	Cycloposine		$C_{33}H_{51}NO_7$	23185-94-6	573.761		268				
2688	Cyclopropane	Trimethylene	C_3H_6	75-19-4	42.080	col gas	-127.58	-32.81	0.617^{25} (p>1 atm)	1.3799^{-42}	s H_2O, bz, peth; vs EtOH, eth
2689	Cyclopropanecarbonitrile	Cyclopropyl cyanide	C_4H_5N	5500-21-0	67.090			135.1	0.8946^{20}	1.4229^{20}	s eth, hx; sl ctc
2690	Cyclopropanecarbonyl chloride		C_4H_5ClO	4023-34-1	104.535			119	1.1516^{20}		
2691	Cyclopropanecarboxaldehyde	Formylcyclopropane	C_4H_6O	1489-69-6	70.090	liq		100	0.938	1.4298^{20}	
2692	Cyclopropanecarboxylic acid		$C_4H_6O_2$	1759-53-1	86.090		18.5	183	1.0885^{20}	1.4390^{20}	s H_2O, EtOH, eth; sl ctc
2693	1,1-Cyclopropanedicarboxylic acid		$C_5H_6O_4$	598-10-7	130.100	pr or nd (chl) pr (w +1)	140.5				vs H_2O, eth
2694	Cyclopropanemethanol		C_4H_8O	2516-33-8	72.106			124	0.911^{25}		sl ctc
2695	Cyclopropanone		C_3H_4O	5009-27-8	56.063		stable only at low temp.				
2696	Cyclopropene		C_3H_4	2781-85-3	40.064	gas		dec -36			
2697	Cyclopropylamine	Cyclopropanamine	C_3H_7N	765-30-0	57.095	liq	-35.39	50.5	0.8240^{20}	1.4210^{20}	msc H_2O; s EtOH, eth, chl
2698	Cyclopropylbenzene		C_9H_{10}	873-49-4	118.175	liq	-31	173.6; 80^{37}	0.9317^{20}	1.5285^{20}	i H_2O; s eth, ace, chl
2699	Cyclopropyl methyl ether		C_4H_8O	540-47-6	72.106	liq	-119	44.7	0.8100^{20}	1.3802^{20}	vs H_2O, bz, eth, EtOH
2700	Cyclopropyl methyl ketone		C_5H_8O	765-43-5	84.117	liq	-68.3	111.3	0.8984^{20}	1.4251^{20}	vs H_2O, eth, EtOH
2701	Cyclotetramethylenetetranitramine	HMX	$C_4H_8N_8O_8$	2691-41-0	296.156	cry	286	exp			
2702	Cyclothiazide		$C_{14}H_{16}ClN_3O_4S_2$	2259-96-3	389.878		234				
2703	Cycluron	*N'*-Cyclooctyl-*N,N*-dimethylurea	$C_{11}H_{22}N_2O$	2163-69-1	198.305	cry	138				sl H_2O; s bz, ace; vs MeOH
2704	Cyfluthrin		$C_{22}H_{18}Cl_2FNO_3$	68359-37-5	434.287		60				
2705	Cygon		$C_5H_{12}NO_3PS_2$	60-51-5	229.258		52	117$^{0.1}$	1.277^{65}		
2706	Cyhalothrin	2,2-Dimethylcyclopropanecarboxylate	$C_{23}H_{19}ClF_3NO_3$	91465-08-6	449.850		49.2				
2707	Cyhexatin	Stannane, tricyclohexylhydroxy-	$C_{18}H_{34}OSn$	13121-70-5	385.172		196				
2708	Cypermethrin		$C_{22}H_{19}Cl_2NO_3$	52315-07-8	416.297		70		1.25^{20}		
2709	Cyprazine		$C_9H_{14}ClN_5$	22936-86-3	227.694		167				
2710	Cyproheptadine		$C_{21}H_{21}N$	129-03-3	287.399	cry (EtOH aq)	113				
2711	Cyromazine	*N*-Cyclopropyl-1,3,5-triazine-2,4,6-triamine	$C_6H_{10}N_6$	66215-27-8	166.183	cry	220				
2712	Cystamine dihydrochloride		$C_4H_{14}Cl_2N_2S_2$	56-17-7	225.203	nd (MeOH)	218 dec				vs H_2O, EtOH
2713	Cysteamine		C_2H_7NS	60-23-1	77.149	cry (sub)	99.5	dec			vs H_2O, EtOH
2714	*L*-Cysteic acid		$C_3H_7NO_5S$	13100-82-8	169.157	cry	260 dec				s H_2O; i EtOH
2715	*L*-Cysteine	Propanoic acid, 2-amino-3-mercapto-, (*R*)-	$C_3H_7NO_2S$	52-90-4	121.159	cry (w)	240 dec				vs H_2O, ace, EtOH
2716	*L*-Cysteine, ethyl ester, hydrochloride		$C_5H_{12}ClNO_2S$	868-59-7	185.673		125.8				vs H_2O
2717	*L*-Cysteine, hydrochloride		$C_3H_8ClNO_2S$	52-89-1	157.620	cry	175 dec				s H_2O
2718	*L*-Cystine	3,3'-Dithiobis(2-aminopropanoic acid)	$C_6H_{12}N_2O_4S_2$	56-89-3	240.300	hex pl or pr (w)	260 dec		1.677^{25}		sl H_2O; i EtOH, eth, bz; s acid, alk
2719	Cytarabine	Cytosine arabinoside	$C_9H_{13}N_3O_5$	147-94-4	243.216	pr (EtOH aq)	212				s H_2O

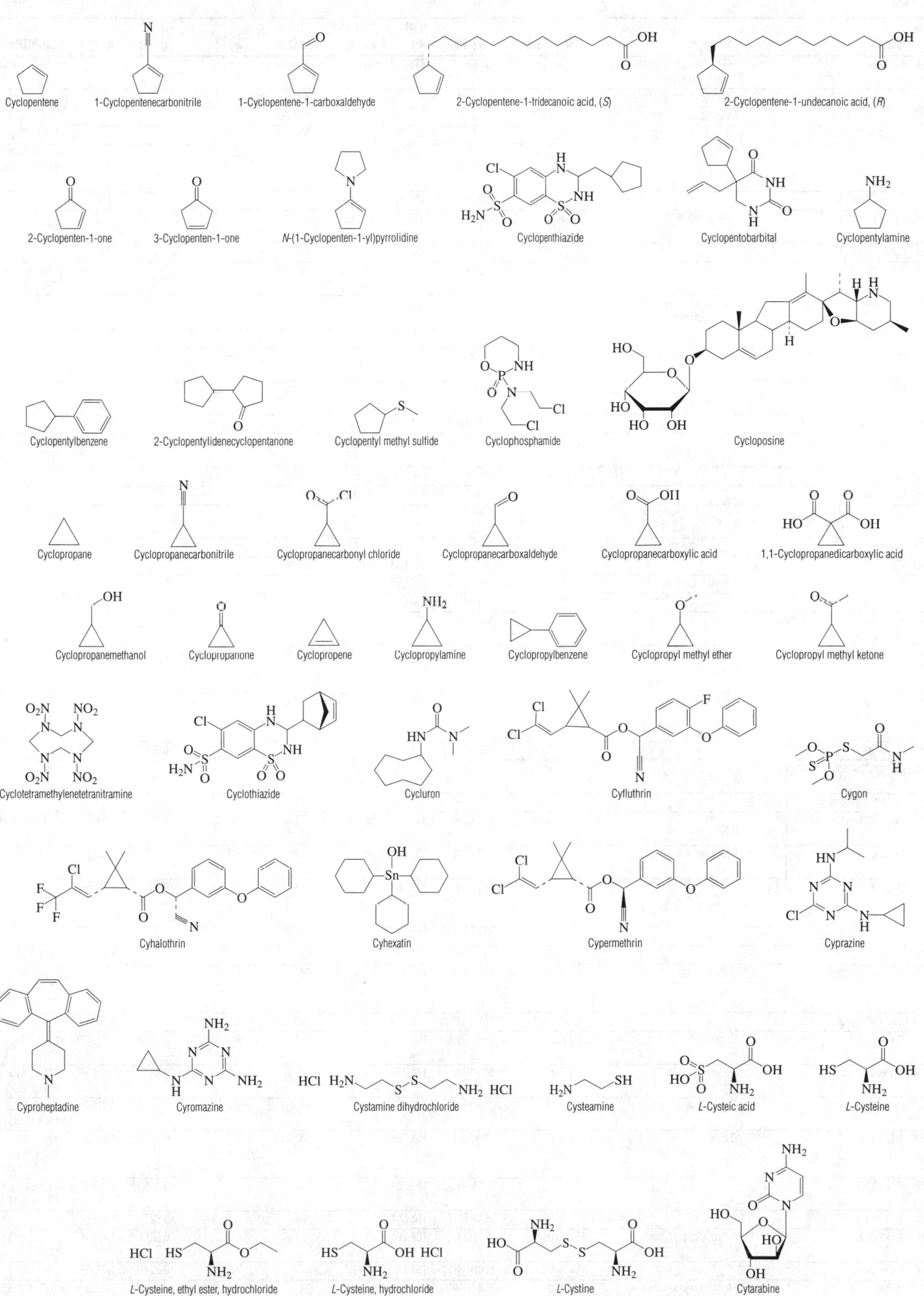
Cyclopentene
1-Cyclopentenecarbonitrile
1-Cyclopentene-1-carboxaldehyde
2-Cyclopentene-1-tridecanoic acid, (S)
2-Cyclopentene-1-undecanoic acid, (R)
2-Cyclopenten-1-one
3-Cyclopenten-1-one
N-(1-Cyclopenten-1-yl)pyrrolidine
Cyclopenthiazide
Cyclopentobarbital
Cyclopentylamine
Cyclopentylbenzene
2-Cyclopentylidenecyclopentanone
Cyclopentyl methyl sulfide
Cyclophosphamide
Cycloposine
Cyclopropane
Cyclopropanecarbonitrile
Cyclopropanecarbonyl chloride
Cyclopropanecarboxaldehyde
Cyclopropanecarboxylic acid
1,1-Cyclopropanedicarboxylic acid
Cyclopropanemethanol
Cyclopropanone
Cyclopropene
Cyclopropylamine
Cyclopropylbenzene
Cyclopropyl methyl ether
Cyclopropyl methyl ketone
Cyclotetramethylenetetranitramine
Cyclothiazide
Cycluron
Cyfluthrin
Cygon
Cyhalothrin
Cyhexatin
Cypermethrin
Cyprazine
Cyproheptadine
Cyromazine
Cystamine dihydrochloride
Cysteamine
L-Cysteic acid
L-Cysteine
L-Cysteine, ethyl ester, hydrochloride
L-Cysteine, hydrochloride
L-Cystine
Cytarabine

No.	Name	Synonym	Mol. Form.	CAS RN	Mol. Wt.	Physical Form	mp/°C	bp/°C	den/ g cm^{-3}	n_D	Solubility
2720	Cytidine	4-Amino-1-β-*D*-ribofuranosyl-2(1*H*)-pyrimidinone	$C_9H_{13}N_3O_5$	65-46-3	243.216	nd (dil al)	230 dec				vs H_2O; sl EtOH
2721	2'-Cytidylic acid	Cytidine 2'-monophosphate	$C_9H_{14}N_3O_8P$	85-94-9	323.196		239 dec				
2722	3'-Cytidylic acid	Cytidine 3'-monophosphate	$C_9H_{14}N_3O_8P$	84-52-6	323.196		233 dec				s H_2O, EtOH
2723	5'-Cytidylic acid	Cytidine 5'-monophosphate	$C_9H_{14}N_3O_8P$	63-37-6	323.196	orth nd	233 dec				vs H_2O, EtOH
2724	Cytisine	Sophorine	$C_{11}H_{14}N_2O$	485-35-8	190.241	pr	153	218^{2}			vs H_2O, EtOH, MeOH; s bz, ace
2725	Cytochalasin B		$C_{29}H_{37}NO_5$	14930-96-2	479.608	nd (ace)	219				
2726	Cytochalasin D	Zygosporin A	$C_{30}H_{37}NO_6$	22144-77-0	507.618	nd (ace/ peth)	270				
2727	Cytochalasin E		$C_{28}H_{33}NO_7$	36011-19-5	495.565		207				
2728	Cytosine		$C_4H_5N_3O$	71-30-7	111.102	mcl or tcl pl (w+1)	322 dec				s H_2O; sl EtOH, chl; i eth
2729	Dacarbazine	5-(3,3-Dimethyl-1-triazenyl)-1*H*-imidazole-4-carboxamide	$C_6H_{10}N_6O$	4342-03-4	182.182	cry	205				
2730	Dactinomycin		$C_{62}H_{86}N_{12}O_{16}$	50-76-0	1255.416		245 dec				
2731	Daidzein	7-Hydroxy-3-(4-hydroxyphenyl)-4*H*-1-benzopyran-4-one	$C_{15}H_{10}O_4$	486-66-8	254.238	pa ye pr (50% al)	323 dec	sub			s EtOH, eth
2732	Daminozide	Butanedioic acid, mono(2,2-dimethylhydrazide)	$C_6H_{12}N_2O_3$	1596-84-5	160.170		154.5				
2733	Dantrolene		$C_{14}H_{10}N_4O_5$	7261-97-4	314.253	cry (DMF aq)	280				
2734	Datiscetin		$C_{15}H_{10}O_6$	480-15-9	286.236	pa ye nd (al, aq HOAc)	277.5				vs ace, eth, EtOH
2735	Daucol		$C_{15}H_{26}O_2$	887-08-1	238.366	cry	114	128^{2}			
2736	Daunorubicin		$C_{27}H_{29}NO_{10}$	20830-81-3	527.520	red nd	208				
2737	Dazomet		$C_5H_{10}N_2S_2$	533-74-4	162.276	nd (bz)	106				reac H_2O; s EtOH
2738	Decabromobiphenyl ether	Bis(pentabromophenyl) ether	$C_{12}Br_{10}O$	1163-19-5	959.167	ye pr (tol)	305				i H_2O
2739	Decachlorobiphenyl		$C_{12}Cl_{10}$	2051-24-3	498.658	cry (bz)	309				i H_2O
2740	1,3-Decadiene	1-Hexyl-1,3-butadiene	$C_{10}H_{18}$	2051-25-4	138.250			169	0.752^{30}		vs bz
2741	1,9-Decadiene		$C_{10}H_{18}$	1647-16-1	138.250			167	0.75^{25}	1.4325^{20}	
2742	2,2',3,3',4,4',5,5',6,6'-Decafluoro-1,1'-biphenyl		$C_{12}F_{10}$	434-90-2	334.112		67.5	206	1.785^{20}		
2743	*cis*-Decahydronaphthalene	*cis*-Decalin	$C_{10}H_{18}$	493-01-6	138.250	liq	-42.9	195.8	0.8965^{20}	1.4810^{20}	i H_2O; msc EtOH; vs eth, ace, chl
2744	*trans*-Decahydronaphthalene	*trans*-Decalin	$C_{10}H_{18}$	493-02-7	138.250	liq	-30.4	187.3	0.8659^{25}	1.4695^{20}	i H_2O; vs EtOH, eth, ace; msc bz; sl MeOH
2745	Decahydro-2-naphthol	Decahydro-β-naphthol	$C_{10}H_{18}O$	825-51-4	154.249			109^{14}	0.996^{25}	1.4992^{20}	
2746	Decamethonium dibromide		$C_{16}H_{38}Br_2N_2$	541-22-0	418.294	cry (MeOH/ ace)	269 dec				i eth
2747	Decamethylcyclopentasiloxane		$C_{10}H_{30}O_5Si_5$	541-02-6	370.770	liq	-38	210	0.9593^{20}	1.3982^{20}	i H_2O
2748	Decamethyltetrasiloxane		$C_{10}H_{30}O_3Si_4$	141-62-8	310.685	liq	-76	194	0.8536^{25}	1.3895^{20}	i H_2O; sl EtOH; s bz, peth
2749	Decanal	Capraldehyde	$C_{10}H_{20}O$	112-31-2	156.265	liq	-4.0	208.5	0.830^{15}	1.4287^{20}	i H_2O; s EtOH, eth, ace; sl ctc
2750	Decane		$C_{10}H_{22}$	124-18-5	142.282	liq	-29.6	174.15	0.7266^{25}	1.4090^{25}	i H_2O; msc EtOH; s eth; sl ctc
2751	1,10-Decanediamine		$C_{10}H_{24}N_2$	646-25-3	172.311		59.73	140^{12}			
2752	Decanedinitrile		$C_{10}H_{16}N_2$	1871-96-1	164.247		7.6	204^{16}	0.913^{20}	1.4474^{20}	i H_2O; s chl
2753	1,10-Decanediol	Decamethylene glycol	$C_{10}H_{22}O_2$	112-47-0	174.281	nd (w, dil al)	74	192^{20}			sl H_2O, eth; vs EtOH; s DMSO; i lig
2754	Decanedioyl dichloride		$C_{10}H_{16}Cl_2O_2$	111-19-3	239.139		-1.3	220^{75}, 165^{11}	1.1212^{20}	1.4684^{18}	
2755	Decanenitrile	Caprinitrile	$C_{10}H_{19}N$	1975-78-6	153.265	liq	-17.9	243; 106^{10}	0.8199^{20}	1.4296^{20}	vs ace, eth, EtOH, chl
2756	1-Decanethiol	Decyl mercaptan	$C_{10}H_{22}S$	143-10-2	174.347	liq	-26	240.6	0.8443^{20}	1.4509^{20}	i H_2O; s EtOH, eth
2757	Decanoic acid	Capric acid	$C_{10}H_{20}O_2$	334-48-5	172.265	nd	31.4	268.7	0.8858^{40}	1.4288^{40}	i H_2O; vs ace, bz, eth, EtOH
2758	1-Decanol	Capric alcohol	$C_{10}H_{22}O$	112-30-1	158.281		6.9	231.1	0.8297^{20}	1.4372^{20}	i H_2O; msc EtOH, eth, ace, bz, chl; s ctc
2759	2-Decanol		$C_{10}H_{22}O$	74742-10-2	158.281	liq	-1.2	211	0.8250^{20}	1.4326^{25}	s EtOH, bz; msc eth, ace; sl ctc
2760	3-Decanol		$C_{10}H_{22}O$	1565-81-7	158.281	liq	-7.5	213; 101^{12}	0.827^{20}	1.434^{20}	
2761	4-Decanol	1-Propylheptyl alcohol	$C_{10}H_{22}O$	2051-31-2	158.281	liq	-11	210.5	0.8261^{20}	1.4320^{20}	i H_2O; s EtOH, ctc
2762	5-Decanol		$C_{10}H_{22}O$	5205-34-5	158.281	liq	8.7	201	0.824^{20}	1.4333^{20}	
2763	2-Decanone	Methyl octyl ketone	$C_{10}H_{20}O$	693-54-9	156.265	nd	14	210; 96^{12}	0.8248^{20}	1.4255^{20}	i H_2O; s EtOH, eth; sl ctc

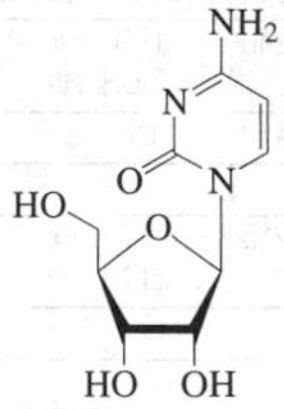

Cytidine
2'-Cytidylic acid 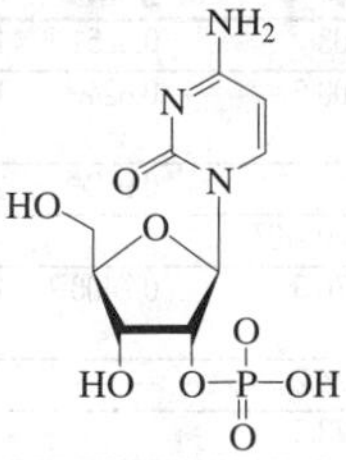
3'-Cytidylic acid 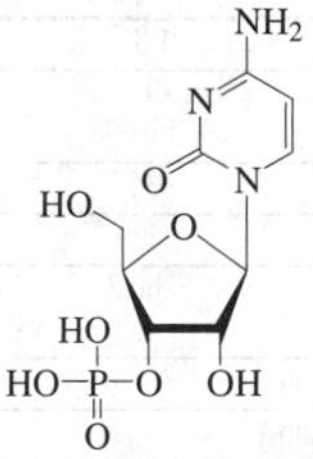
5'-Cytidylic acid 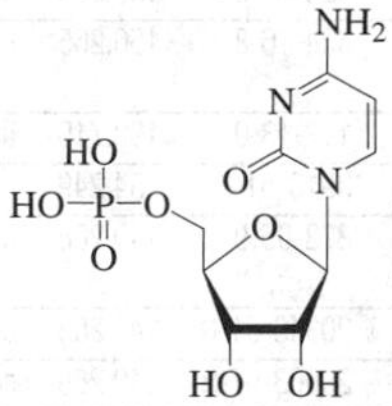
Cytisine 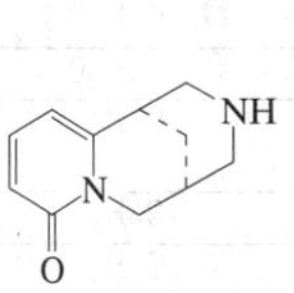
Cytochalasin B 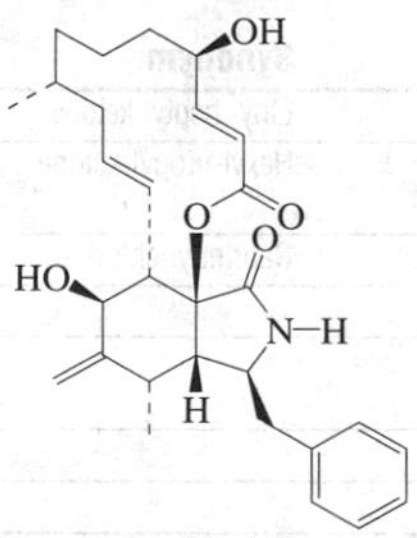
Cytochalasin D 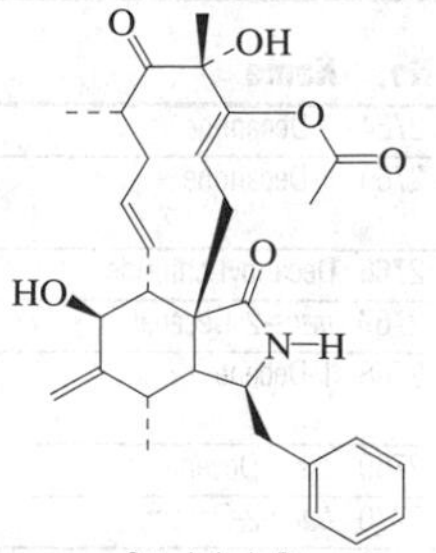
Cytochalasin E 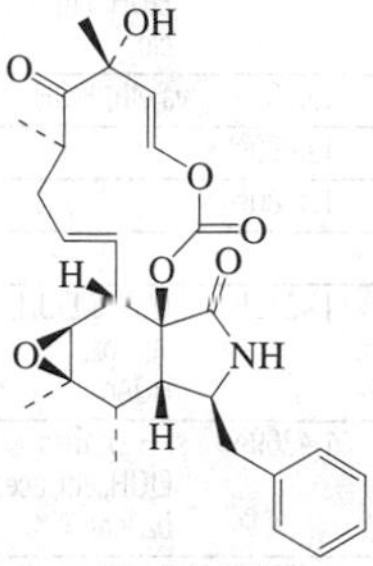
Cytosine 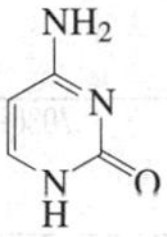
Dacarbazine 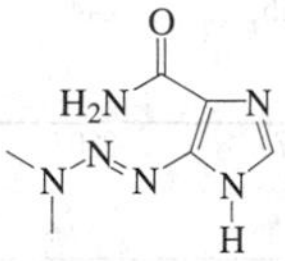
Dactinomycin 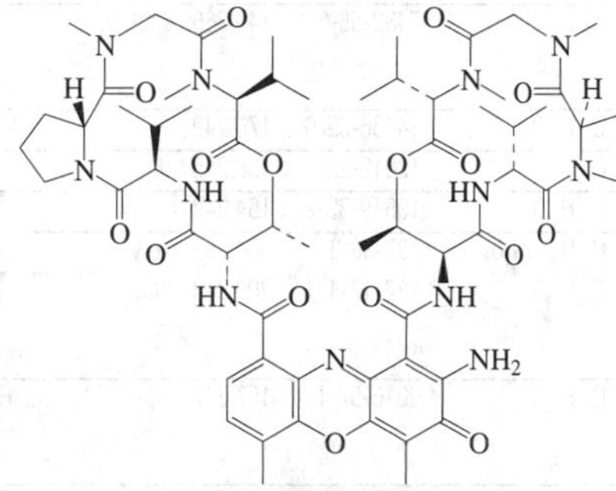
Daidzein 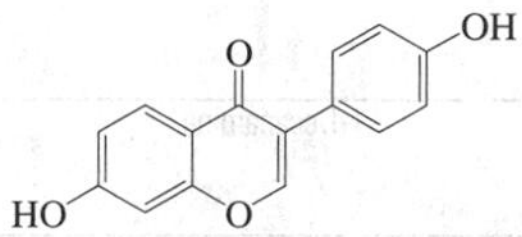
Daminozide 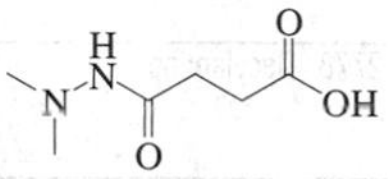
Dantrolene 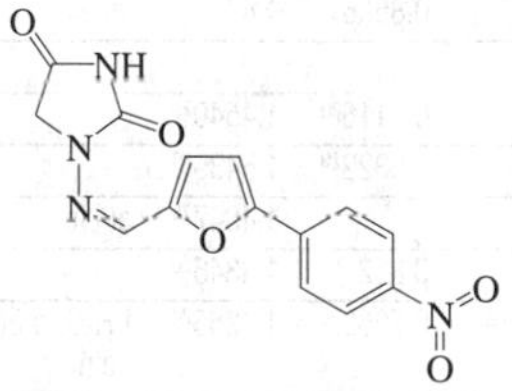
Datiscetin 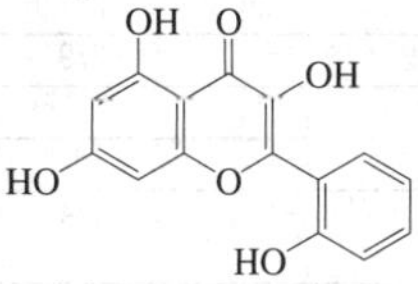
Daucol 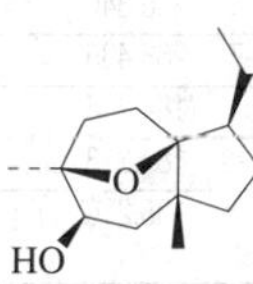
Daunorubicin 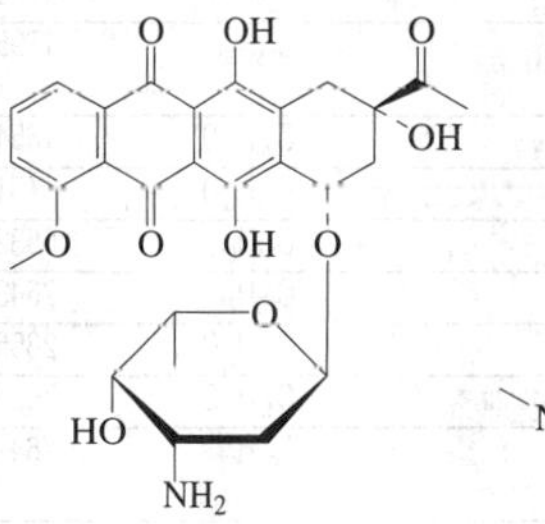
Dazomet 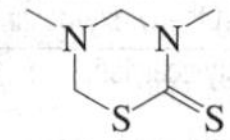
Decabromobiphenyl ether 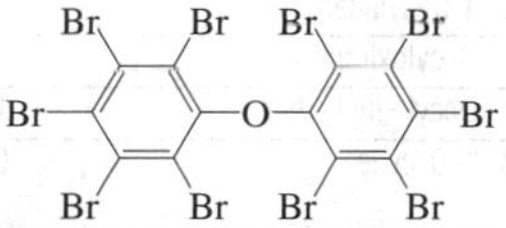
Decachlorobiphenyl 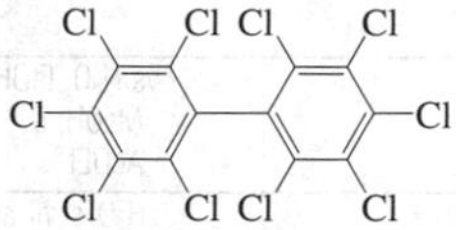
1,3-Decadiene 
1,9-Decadiene 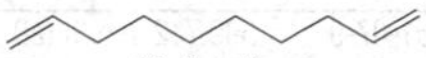
2,2',3,3',4,4',5,5',6,6'-Decafluoro-1,1'-biphenyl 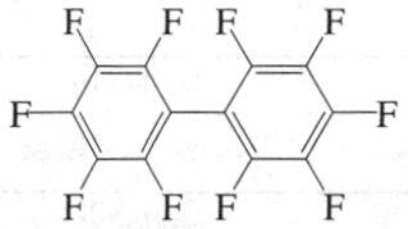
cis-Decahydronaphthalene 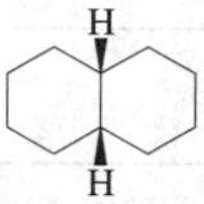
trans-Decahydronaphthalene 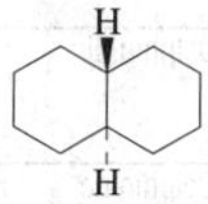
Decahydro-2-naphthol 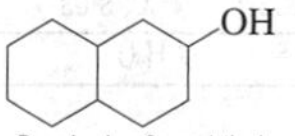
Decamethonium dibromide 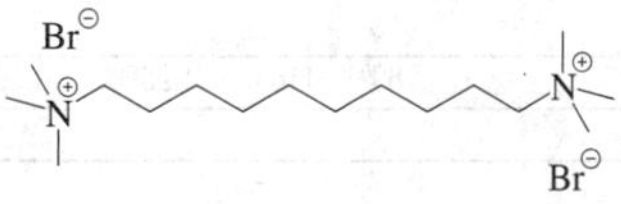
Decamethylcyclopentasiloxane 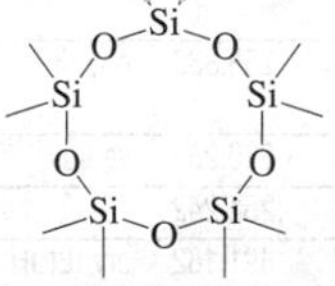
Decamethyltetrasiloxane 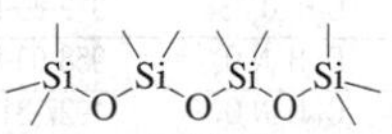
Decanal 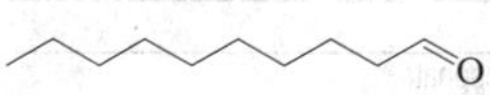
Decane 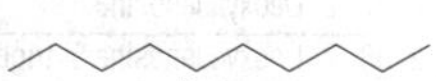
1,10-Decanediamine 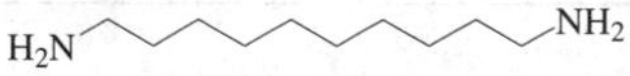
Decanedinitrile 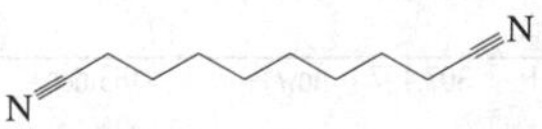
1,10-Decanediol 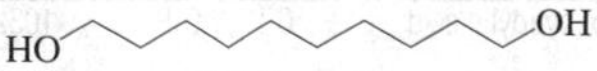
Decanedioyl dichloride 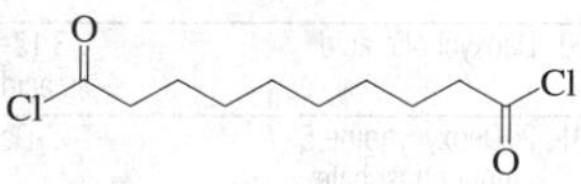
Decanenitrile 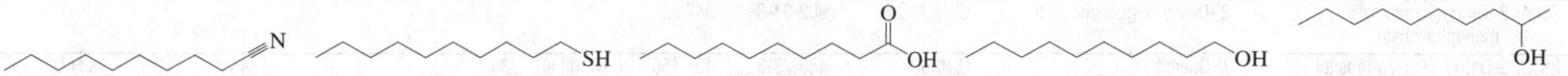
1-Decanethiol
Decanoic acid
1-Decanol
2-Decanol
3-Decanol 
4-Decanol
5-Decanol
2-Decanone

No.	Name	Synonym	Mol. Form.	CAS RN	Mol. Wt.	Physical Form	mp/°C	bp/°C	den/ g cm^{-3}	n_D	Solubility
2764	3-Decanone	Ethyl heptyl ketone	$C_{10}H_{20}O$	928-80-3	156.265	liq	1.3	203	0.8251^{20}	1.4252^{20}	s EtOH, eth, ctc
2765	4-Decanone	Hexyl propyl ketone	$C_{10}H_{20}O$	624-16-8	156.265	liq	-9	206.5	0.824^{20}	1.4240^{21}	i H_2O; msc EtOH, eth
2766	Decanoyl chloride	Caprinoyl chloride	$C_{10}H_{19}ClO$	112-13-0	190.710	liq	-34.5	95	0.919^{25}	1.4410^{20}	s eth, ctc
2767	*trans*-2-Decenal		$C_{10}H_{18}O$	3913-81-3	154.249			230; 107^{11}			
2768	1-Decene		$C_{10}H_{20}$	872-05-9	140.266	liq	-66.3	170.5	0.7408^{20}	1.4215^{20}	i H_2O; msc EtOH, eth
2769	*cis*-2-Decene		$C_{10}H_{20}$	20348-51-0	140.266	col liq		174.2			
2770	*trans*-2-Decene		$C_{10}H_{20}$	20063-97-2	140.266	col liq		173.3			
2771	*cis*-5-Decene		$C_{10}H_{20}$	7433-78-5	140.266	col liq	-112	171; 73^{20}	0.7445^{20}	1.4258^{20}	i H_2O; msc EtOH, eth; sl ctc
2772	*trans*-5-Decene		$C_{10}H_{20}$	7433-56-9	140.266	col liq	-73	171	0.7401^{20}	1.4243^{20}	i H_2O; msc EtOH, eth; sl ctc
2773	9-Decenoic acid	Caproleic acid	$C_{10}H_{18}O_2$	14436-32-9	170.249		26.5	158^{21}, 142^{4}	0.9238^{15}	1.4507^{15}	vs eth, EtOH
2774	9-Decen-1-ol	Decylenic alcohol	$C_{10}H_{20}O$	13019-22-2	156.265			236	0.876^{25}	1.4480^{20}	
2775	3-Decen-2-one	Heptylidene acetone	$C_{10}H_{18}O$	10519-33-2	154.249			$102^{15.3}$	0.8473^{20}	1.4480^{20}	
2776	Declomycin	Demeclocycline	$C_{21}H_{21}ClN_2O_8$	127-33-3	464.853	cry	176 dec				
2777	Decyl acetate		$C_{12}H_{24}O_2$	112-17-4	200.318	liq	-15	244	0.8671^{20}	1.4273^{20}	i H_2O; s EtOH, eth, bz, ctc, HOAc
2778	Decylamine	1-Decanamine	$C_{10}H_{23}N$	2016-57-1	157.297		17	220.5	0.7936^{20}	1.4369^{20}	sl H_2O; msc EtOH, eth, ace, bz, chl
2779	Decylbenzene		$C_{16}H_{26}$	104-72-3	218.377	liq	-14.4	293	0.8555^{20}	1.4832^{20}	vs ace, bz, eth, EtOH
2780	Decylcyclohexane		$C_{16}H_{32}$	1795-16-0	224.425	liq	-0.9	299	0.8186^{20}	1.4534^{20}	
2781	Decylcyclopentane		$C_{15}H_{30}$	1795-21-7	210.399	liq	-22	279	0.8110^{20}	1.4486^{20}	vs ace, bz, eth, EtOH
2782	Decyl decanoate		$C_{20}H_{40}O_2$	1654-86-0	312.531		9.7	219^{15}	0.8586^{20}	1.4423^{20}	vs eth
2783	Decyl formate		$C_{11}H_{22}O_2$	5451-52-5	186.292	liq		243			
2784	11-Decylheneicosane		$C_{31}H_{64}$	55320-06-4	436.840		10.0	282.0^{10}	0.8116^{20}	1.4540^{20}	
2785	1-Decylnaphthalene		$C_{20}H_{28}$	26438-27-7	268.436		15	379	0.9322^{20}	1.5435^{20}	
2786	Decyloxirane		$C_{12}H_{24}O$	2855-19-8	184.318					1.4347^{25}	sl ctc
2787	Decyl vinyl ether	1-(Ethenyloxy)decane	$C_{12}H_{24}O$	765-05-9	184.318		-41	101^{10}	0.812^{20}	1.4346^{20}	
2788	1-Decyne	Octylacetylene	$C_{10}H_{18}$	764-93-2	138.250	liq	-44	174	0.7655^{20}	1.4265^{20}	i H_2O; s EtOH, eth
2789	5-Decyne	Dibutylacetylene	$C_{10}H_{18}$	1942-46-7	138.250	liq	-73	177; 78.8^{25}	0.7690^{20}	1.4331^{20}	i H_2O; s EtOH, eth
2790	Dehydroabietic acid	8,11,13-Abietatrien-18-oic acid	$C_{20}H_{28}O_2$	1740-19-8	300.435	cry (EtOH aq)	172				
2791	Delphinidin		$C_{15}H_{11}ClO_7$	528-53-0	338.697		>350				vs H_2O, EtOH, MeOH; s AcOEt
2792	Delphinine		$C_{33}H_{45}NO_9$	561-07-9	599.712	orth (al)	199				i H_2O; s chl, ace, eth; vs EtOH
2793	Deltamethrin		$C_{22}H_{19}Br_2NO_3$	52918-63-5	505.199		99				
2794	Demecarium bromide		$C_{32}H_{52}Br_2N_4O_4$	56-94-0	716.588	hyg pow	165 dec				vs H_2O; sl ace; i ace, eth
2795	Demeton	Systox	$C_8H_{19}O_3PS_2$	8065-48-3	258.339	oily liq		134^{2}			i H_2O; s EtOH, tol
2796	Demeton-*S*-methyl		$C_6H_{15}O_3PS_2$	919-86-8	230.285	ye liq		$89^{0.15}$, 118^{1}	1.20^{20}	1.5063^{20}	i H_2O; s os
2797	2'-Deoxyadenosine		$C_{10}H_{13}N_5O_3$	958-09-8	251.242						sl H_2O
2798	2'-Deoxyadenosine 5'-triphosphate		$C_{10}H_{16}N_5O_{12}P_3$	1927-31-7	491.182	cry (EtOH aq)					
2799	6-Deoxy-*L*-ascorbic acid		$C_6H_8O_5$	528-81-4	160.125	pr (AcOEt)	168	sub 160			vs H_2O, ace, EtOH
2800	Deoxycholic acid	3,12-Dihydroxycholan-24-oic acid, (3α,5β,12α)	$C_{24}H_{40}O_4$	83-44-3	392.573	cry (al)	177				
2801	2'-Deoxycytidine 5'-monophosphate	2'-Deoxy-5'-cytidylic acid	$C_9H_{14}N_3O_7P$	1032-65-1	307.197	pow	183 dec				
2802	2'-Deoxy-5-fluorouridine	Floxuridine	$C_9H_{11}FN_2O_5$	50-91-9	246.191	cry	150				
2803	2-Deoxy-*D*-glucose		$C_6H_{12}O_5$	154-17-6	164.156		146.5				
2804	2'-Deoxyguanosine 5'-monophosphate	2'-Deoxy-5'-guanylic acid	$C_{10}H_{14}N_5O_7P$	902-04-5	347.222						s H_2O
2805	2-Deoxy-*D*-*chiro*-inositol	*D*-Quercitol	$C_6H_{12}O_5$	488-73-3	164.156	pr (w, dil al)	236		1.5845^{13}		vs H_2O
2806	1-Deoxy-1-(methylamino)-*D*-glucitol	*N*-Methylglucamine	$C_7H_{17}NO_5$	6284-40-8	195.214	cry (MeOH)	128.5				s H_2O
2807	6-Deoxy-3-*O*-methylgalactose	Digitalose	$C_7H_{14}O_5$	4481-08-7	178.183	nd (AcOEt)	119				vs H_2O
2808	*D*-2-Deoxyribose		$C_5H_{10}O_4$	533-67-5	134.131		90				
2809	Deserpidine		$C_{32}H_{38}N_2O_8$	131-01-1	578.652	nd or pr	230.5				i H_2O; s EtOH, chl

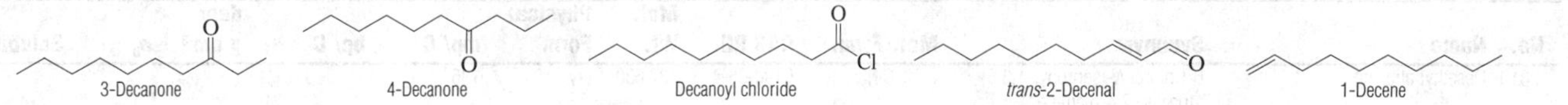

3-Decanone

4-Decanone

Decanoyl chloride

trans-2-Decenal

1-Decene

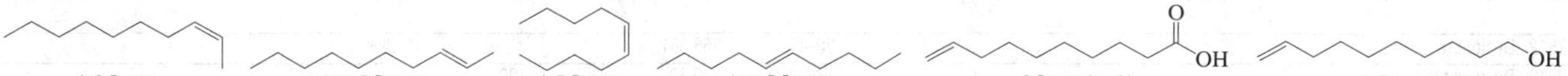

cis-2-Decene

trans-2-Decene

cis-5-Decene

trans-5-Decene

9-Decenoic acid

9-Decen-1-ol

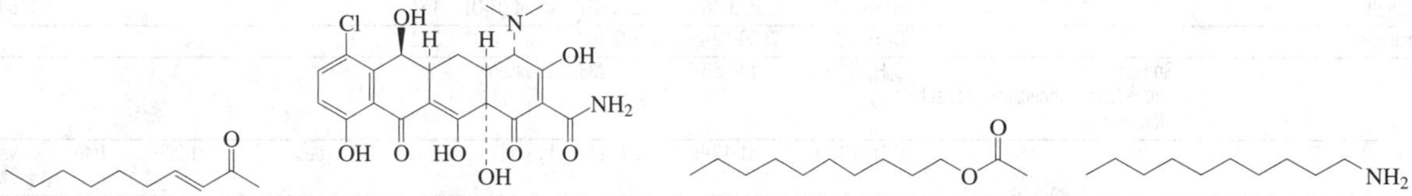

3-Decen-2-one

Declomycin

Decyl acetate

Decylamine

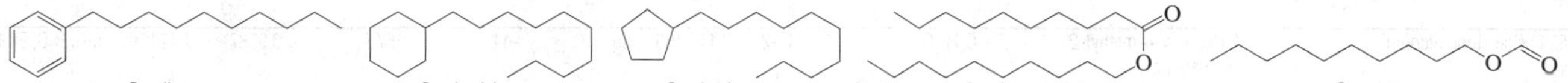

Decylbenzene

Decylcyclohexane

Decylcyclopentane

Decyl decanoate

Decyl formate

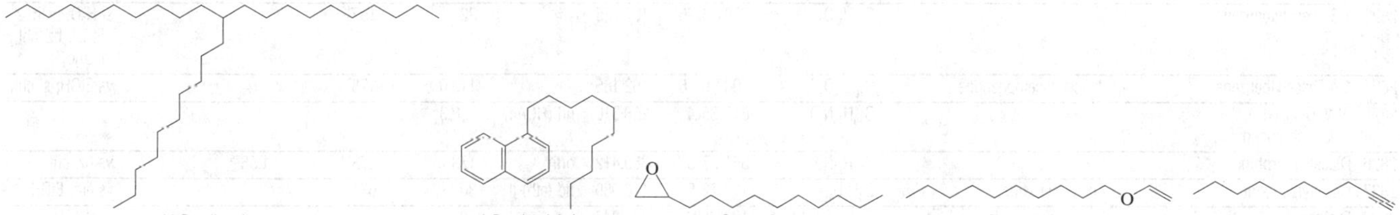

11-Decylheneicosane

1-Decylnaphthalene

Decyloxirane

Decyl vinyl ether

1-Decyne

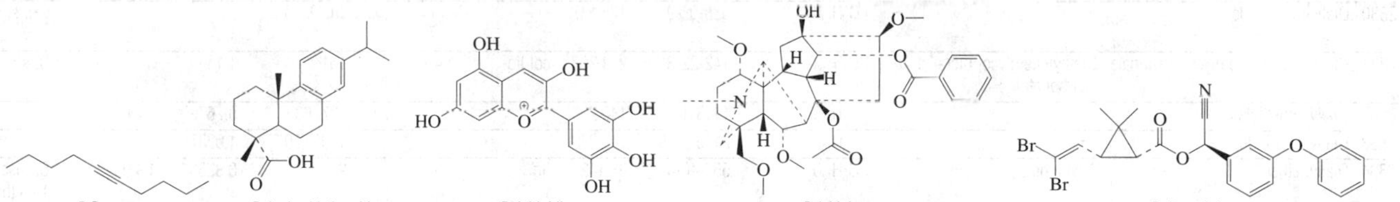

5-Decyne

Dehydroabietic acid

Delphinidin

Delphinine

Deltamethrin

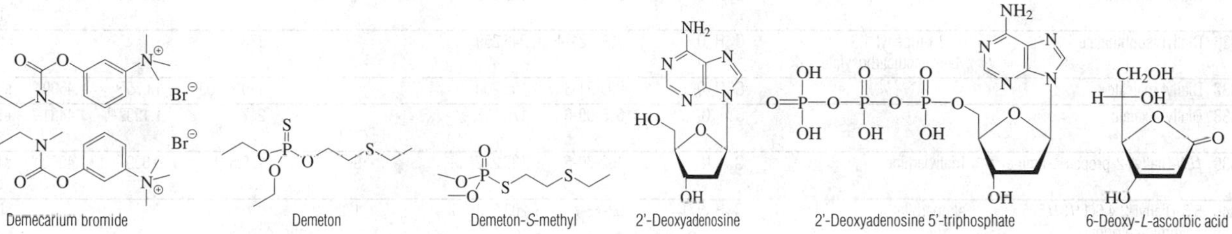

Demecarium bromide

Demeton

Demeton-*S*-methyl

2′-Deoxyadenosine

2′-Deoxyadenosine 5′-triphosphate

6-Deoxy-*L*-ascorbic acid

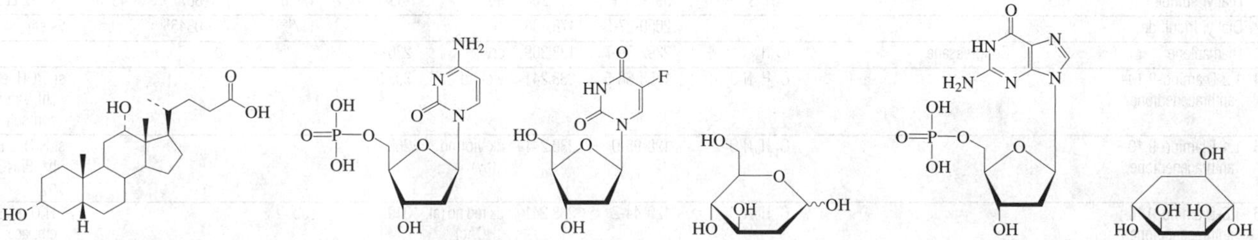

Deoxycholic acid

2′-Deoxycytidine 5′-monophosphate

2′-Deoxy-5-fluorouridine

2-Deoxy-*D*-glucose

2′-Deoxyguanosine 5′-monophosphate

2-Deoxy-*D*-*chiro*-inositol

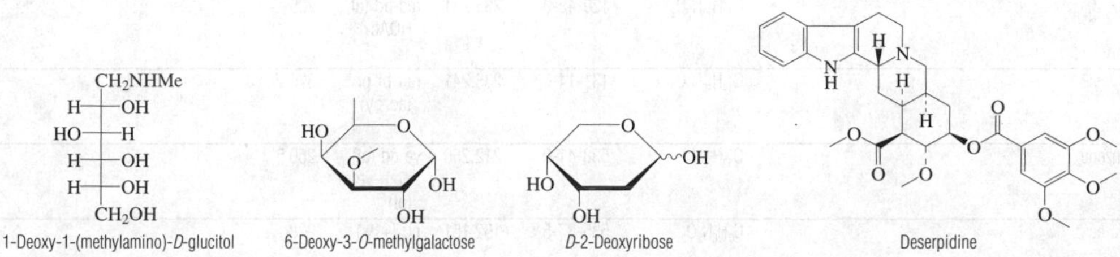

1-Deoxy-1-(methylamino)-*D*-glucitol

6-Deoxy-3-*O*-methylgalactose

D-2-Deoxyribose

Deserpidine

No.	Name	Synonym	Mol. Form.	CAS RN	Mol. Wt.	Physical Form	mp/°C	bp/°C	den/ g cm^{-3}	n_D	Solubility
2810	Desethyl atrazine	6-Chloro-*N*-isopropyl-1,3,5-triazine-2,4-diamine	$C_6H_{10}ClN_5$	6190-65-4	187.630	cry	136				
2811	Desferrioxamine	Deferoxamine	$C_{25}H_{48}N_6O_8$	70-51-9	560.684	cry (EtOH aq)	139				
2812	Desipramine		$C_{18}H_{22}N_2$	50-47-5	266.381			$173^{0.02}$			
2813	Desmedipham		$C_{16}H_{16}N_2O_4$	13684-56-5	300.309		120				
2814	Desmetryne		$C_8H_{15}N_5S$	1014-69-3	213.304	cry	85				
2815	Desthiobiotin		$C_{10}H_{18}N_2O_3$	533-48-2	214.261	lo nd (H_2O)	157				s H_2O
2816	Dexamethasone		$C_{22}H_{29}FO_5$	50-02-2	392.460		262				
2817	Dexon	Sodium dimethylaminobenzenediazosulfonate	$C_8H_{10}N_3NaO_3S$	140-56-7	251.238	ye-br pow					sl H_2O; s DMF
2818	Dexpanthenol		$C_9H_{19}NO_4$	81-13-0	205.252	hyg oil		dec	1.20^{20}	1.497^{20}	vs H_2O, EtOH, MeOH; sl eth
2819	Dextroamphetamine sulfate		$C_{18}H_{28}N_2O_4S$	51-63-8	368.491		>300		1.15^{25}		vs H_2O
2820	Dextromethorphan hydrobromide		$C_{18}H_{26}BrNO$	125-69-9	352.309	wh cry pow	123				s EtOH, chl; i eth
2821	Diacetone alcohol	4-Hydroxy-4-methyl-2-pentanone	$C_6H_{12}O_2$	123-42-2	116.158	liq	-44	167.9	0.9387^{20}	1.4213^{20}	msc H_2O, EtOH, eth; s chl
2822	3,3-Diacetoxy-1-propene		$C_7H_{10}O_4$	869-29-4	158.152	liq	-37.6	180	1.0760^{20}	1.4193^{20}	vs ace, bz, eth, EtOH
2823	1,3-Diacetylbenzene		$C_{10}H_{10}O_2$	6781-42-6	162.185		32	152^{15}			sl H_2O, peth; s EtOH, bz, chl, HOAc
2824	1,4-Diacetylbenzene	4-Acetylacetophenone	$C_{10}H_{10}O_2$	1009-61-6	162.185		113.0	128^{3}			vs EtOH; sl chl
2825	*N,N'*-Diacetyl-4,4'-diaminobiphenyl		$C_{16}H_{16}N_2O_2$	613-35-4	268.310	nd (HOAc)	328.3				
2826	Diacetylmorphine		$C_{21}H_{23}NO_5$	561-27-3	369.412	orth	173	273^{12}	1.56^{25}		vs bz, chl
2827	Diacetylperoxide	Acetyl peroxide	$C_4H_6O_4$	110-22-5	118.089	nd (eth) lf	30	63^{21}			vs eth, EtOH
2828	Dialifor		$C_{14}H_{17}ClNO_4PS_2$	10311-84-9	393.846		68				
2829	Diallate		$C_{10}H_{17}Cl_2NOS$	2303-16-4	270.219			150^{9}			
2830	Diallylcyanamide		$C_7H_{10}N_2$	538-08-9	122.167			142^{90}, 95^{9}			s EtOH; sl eth, ctc
2831	Diallyl diethylene glycol carbonate	Diethylene glycol bis(allyl carbonate)	$C_{12}H_{18}O_7$	142-22-3	274.267	col liq	-4	161^{2}	1.14^{20}		i H_2O; s os
2832	Diallyldimethylsilane		$C_8H_{16}Si$	1113-12-8	140.299			137; 68^{50}	0.7679^{20}	1.4420^{20}	
2833	Diallyl disulfide		$C_6H_{10}S_2$	2179-57-9	146.273			100^{48}, 79^{16}	1.0237^{15}		
2834	Diallyl ether	Allyl ether	$C_6H_{10}O$	557-40-4	98.142	liq	-6	94	0.8260^{20}	1.4163^{20}	i H_2O; msc EtOH, eth; vs ace; s chl
2835	Diallyl fumarate		$C_{10}H_{12}O_4$	2807-54-7	196.200			140^{3}	1.0768^{20}	1.4670^{25}	vs ace, bz, eth, EtOH
2836	Diallyl isophthalate	Di-2-propenyl 1,3-benzenedicarboxylate	$C_{14}H_{14}O_4$	1087-21-4	246.259			176^{5}			
2837	Diallyl maleate		$C_{10}H_{12}O_4$	999-21-3	196.200			129^{10}, 109^{3}	1.075^{20}	1.4699^{20}	s chl
2838	Diallyl oxalate		$C_8H_{10}O_4$	615-99-6	170.163			217	1.1582^{20}	1.4481^{20}	i H_2O; s EtOH, ace, bz; sl chl
2839	*N,N*-Diallyl-2-propen-1-amine	Triallylamine	$C_9H_{15}N$	102-70-5	137.222		94	155.5	0.809^{20}	1.4502^{20}	s EtOH, eth, ace, bz, acid
2840	5,5-Diallyl-2,4,6(1*H*,3*H*,5*H*)-pyrimidinetrione	Allobarbital	$C_{10}H_{12}N_2O_3$	52-43-7	208.213	lf	172				sl H_2O, DMSO; s EtOH, eth, bz
2841	Diallyl sulfide		$C_6H_{10}S$	592-88-1	114.208	liq	-85	138.6	0.8877^{27}	1.4870^{25}	vs eth, EtOH
2842	Diallyl trisulfide		$C_6H_{10}S_3$	2050-87-5	178.338			117^{16}	1.0845^{15}		vs eth
2843	Diamantane	Congressane	$C_{14}H_{20}$	2292-79-7	188.309	cry	236				
2844	1,2-Diamino-9,10-anthracenedione		$C_{14}H_{10}N_2O_2$	1758-68-5	238.241	viol nd	303.5				sl EtOH, eth, chl, xyl; s py, con sulf
2845	1,4-Diamino-9,10-anthracenedione		$C_{14}H_{10}N_2O_2$	128-95-0	238.241	dk viol nd (py)	268				sl H_2O; s EtOH, bz, $PhNO_2$; vs py
2846	1,5-Diamino-9,10-anthracenedione		$C_{14}H_{10}N_2O_2$	129-44-2	238.241	dk red nd (al, HOAc)	319	sub			i H_2O; sl EtOH, eth, ace, bz; s $PhNO_2$
2847	1,8-Diamino-9,10-anthracenedione		$C_{14}H_{10}N_2O_2$	129-42-0	238.241	red nd (al, HOAc)	265				i H_2O; s EtOH, py; sl eth, HOAc
2848	2,6-Diamino-9,10-anthracenedione		$C_{14}H_{10}N_2O_2$	131-14-6	238.241	red-br pr (aq-py)	320 dec				sl H_2O; s EtOH, chl, con sulf, xyl, py
2849	4,4'-Diaminoazobenzene		$C_{12}H_{12}N_4$	538-41-0	212.250	ye nd (al), oran-ye pr (al)	250.5				sl H_2O, lig; s EtOH; vs bz, chl
2850	3,5-Diaminobenzoic acid		$C_7H_8N_2O_2$	535-87-5	152.151	nd (+1w)	228				sl H_2O, tfa; s EtOH; vs eth

Desethyl atrazine

Desferrioxamine

Desipramine

Desmedipham

Desmetryne

Desthiobiotin

Dexamethasone

Dexon

Dexpanthenol

Dextroamphetamine sulfate

Dextromethorphan hydrobromide

Diacetone alcohol

3,3-Diacetoxy-1-propene

1,3-Diacetylbenzene

1,4-Diacetylbenzene

N,N'-Diacetyl-4,4'-diaminobiphenyl

Diacetylmorphine

Diacetylperoxide

Dialifor

Diallate

Diallylcyanamide

Diallyl diethylene glycol carbonate

Diallyldimethylsilane

Diallyl disulfide

Diallyl ether

Diallyl fumarate

Diallyl isophthalate

Diallyl maleate

Diallyl oxalate

N,N-Diallyl-2-propen-1-amine

5,5-Diallyl-2,4,6(1*H*,3*H*,5*H*)-pyrimidinetrione

Diallyl sulfide

Diallyl trisulfide

Diamantane

1,2-Diamino-9,10-anthracenedione

1,4-Diamino-9,10-anthracenedione

1,5-Diamino-9,10-anthracenedione

1,8-Diamino-9,10-anthracenedione

2,6-Diamino-9,10-anthracenedione

4,4'-Diaminoazobenzene

3,5-Diaminobenzoic acid

No.	Name	Synonym	Mol. Form.	CAS RN	Mol. Wt.	Physical Form	mp/°C	bp/°C	den/ g cm^{-3}	n_D	Solubility
2851	2,4-Diaminobutanoic acid		$C_4H_{10}N_2O_2$	305-62-4	118.134	hyg cry					s H_2O; sl EtOH, MeOH
2852	*cis*-2,3-Diamino-2-butenedinitrile		$C_4H_4N_4$	1187-42-4	108.102		178.5		1.41^{20}		
2853	1,8-Diamino-4,5-dihydroxy-9,10-anthracenedione		$C_{14}H_{10}N_2O_4$	128-94-9	270.240	bl nd (xyl)					i H_2O; s bz, xyl, EtOH
2854	4,4'-Diaminodiphenyl ether	4,4-Oxydianiline	$C_{12}H_{12}N_2O$	101-80-4	200.235		189 dec	>300			
2855	4,4'-Diaminodiphenylmethane	4,4'-Methylenedianiline	$C_{13}H_{14}N_2$	101-77-9	198.263	pl or nd (w) pl (bz)	92.5	398; 257^{18}			sl H_2O; vs EtOH, eth, bz
2856	4,4'-Diaminodiphenyl sulfide	4,4'-Thiodianiline	$C_{12}H_{12}N_2S$	139-65-1	216.301	nd (w)	108.5				sl H_2O; vs EtOH, eth, bz; s tfa
2857	3,3'-Diaminodiphenyl sulfone	3,3'-Sulfonyldianiline	$C_{12}H_{12}N_2O_2S$	599-61-1	248.300		168.5				vs H_2O, EtOH
2858	*meso*-2,6-Diaminoheptanedioic acid	2,6-Diaminopimelic acid	$C_7H_{14}N_2O_4$	922-54-3	190.197	nd (w)	314 dec				s H_2O
2859	1,4-Diamino-2-methoxy-9,10-anthracenedione		$C_{15}H_{12}N_2O_3$	2872-48-2	268.267		235				
2860	1,4-Diamino-5-nitro-9,10-anthracenedione		$C_{14}H_9N_3O_4$	82-33-7	283.239		278				
2861	2,4-Diaminophenol		$C_6H_8N_2O$	95-86-3	124.140	lf	79 dec				vs H_2O, ace, EtOH
2862	2,4-Diaminophenol, dihydrochloride		$C_6H_{10}Cl_2N_2O$	137-09-7	197.061	nd	235 dec				vs H_2O
2863	3,7-Diaminophenothiazin-5-ium chloride	Thionine	$C_{12}H_{10}ClN_3S$	581-64-6	263.745						sl H_2O, EtOH, eth; s bz, chl, acid
2864	4-[(2,4-Diaminophenyl)azo] benzenesulfonamide	Prontosil	$C_{12}H_{14}ClN_5O_2S$	103-12-8	327.790		249.5				sl H_2O; s EtOH, ace, oils, fats
2865	1,3-Diamino-2-propanol		$C_3H_{10}N_2O$	616-29-5	90.123	cry	42.8				i eth, bz
2866	4,4'-Diamino-2,2'-stilbenedisulfonic acid	Amsonic acid	$C_{14}H_{14}N_2O_6S_2$	81-11-8	370.400	ye nd	300				sl H_2O
2867	4,6-Diamino-1,3,5-triazin-2(1*H*)-one		$C_3H_5N_5O$	645-92-1	127.105	nd (aq Na_2CO_3)	dec				i H_2O, EtOH, eth, bz, HOAc; s acid, alk
2868	8,8'-Diapo-ψ,ψ-carotenedioic acid	Crocetin	$C_{20}H_{24}O_4$	27876-94-4	328.403	brick red orth	286				sl H_2O, EtOH; i eth, bz; s py; vs NaOH
2869	Diatrizoic acid	*N,N'*-Diacetyl-3,5-diamino-2,4,6-triiodobenzoic acid	$C_{11}H_9I_3N_2O_4$	117-96-4	613.913	cry (EtOH aq)	300				
2870	Diazenedicarboxamide	Azodicarbonamide	$C_2H_4N_4O_2$	123-77-3	116.079		212 dec				
2871	Diazinon		$C_{12}H_{21}N_2O_3PS$	333-41-5	304.345			$87^{0.05}$	1.1088^{20}	1.4922^{20}	
2872	Diazomethane		CH_2N_2	334-88-3	42.040	ye gas	-145	-23			vs eth, diox
2873	Dibenz[a,h]acridine		$C_{21}H_{13}N$	226-36-8	279.335	ye cry	228				
2874	Dibenz[a,j]acridine	7-Azadibenz[a,j]anthracene	$C_{21}H_{13}N$	224-42-0	279.335		216				i H_2O
2875	Dibenz[c,h]acridine		$C_{21}H_{13}N$	224-53-3	279.335	ye cry (EtOH)	189				
2876	Dibenz[a,h]anthracene	1,2:5,6-Dibenzanthracene	$C_{22}H_{14}$	53-70-3	278.346	pl (dil ace)	269.5				i H_2O; sl EtOH; s ace, bz, CS_2
2877	Dibenz[a,j]anthracene		$C_{22}H_{14}$	224-41-9	278.346	oran lf or nd (bz)	197.5				i H_2O, HOAc; sl EtOH, eth, bz; s peth
2878	5*H*-Dibenz[b,f]azepine-5-carboxamide	Carbamazepine	$C_{15}H_{12}N_2O$	298-46-4	236.268		190.2				
2879	Dibenzepin		$C_{18}H_{21}N_3O$	4498-32-2	295.379		117	$185^{0.01}$			
2880	7*H*-Dibenzo[c,g]carbazole		$C_{20}H_{13}N$	194-59-2	267.324	cry (EtOH)	158				
2881	13*H*-Dibenzo[a,i]carbazole		$C_{20}H_{13}N$	239-64-5	267.324		221.3				i H_2O
2882	Dibenzo[b,k]chrysene		$C_{26}H_{16}$	217-54-9	328.405		400				
2883	Dibenzo[b,e][1,4]dioxin	Diphenylene dioxide	$C_{12}H_8O_2$	262-12-4	184.191	nd (MeOH)	120.5				
2884	Dibenzofuran	2,2'-Biphenylene oxide	$C_{12}H_8O$	132-64-9	168.191	lf or nd (al)	86.5	287	1.0886^{99}	1.6079^{99}	i H_2O; s EtOH, ace, bz; vs eth, HOAc
2885	Dibenzo[a,e]pyrene	Naphtho[1,2,3,4-def]chrysene	$C_{24}H_{14}$	192-65-4	302.368	pa ye nd(xyl)	233.5				sl EtOH, ace, bz, HOAc; s tol, con sulf
2886	Dibenzo[a,h]pyrene	Dibenzo[b,def]chrysene	$C_{24}H_{14}$	189-64-0	302.368	oran pl	315				
2887	Dibenzo[a,i]pyrene	Benzo[rst]pentaphene	$C_{24}H_{14}$	189-55-9	302.368		281.5	$275^{0.05}$			
2888	Dibenzo[a,l]pyrene	Dibenzo[def,p]chrysene	$C_{24}H_{14}$	191-30-0	302.368	ye pl (bz/ EtOH)	164.5				
2889	Dibenzothiophene		$C_{12}H_8S$	132-65-0	184.257	nd (dil al, lig)	98.2	332.5			i H_2O; s chl, MeOH; vs EtOH, bz
2890	Dibenz[c,e]oxepin-5,7-dione		$C_{14}H_8O_3$	6050-13-1	224.212	nd (HOAc or bz)	217	sub			i H_2O; sl eth
2891	Dibenzoyl disulfide	Benzoyl disulfide	$C_{14}H_{10}O_2S_2$	644-32-6	274.358	pr(al), sc(chl-peth)	134.5	dec			i H_2O; sl EtOH, eth; s CS_2

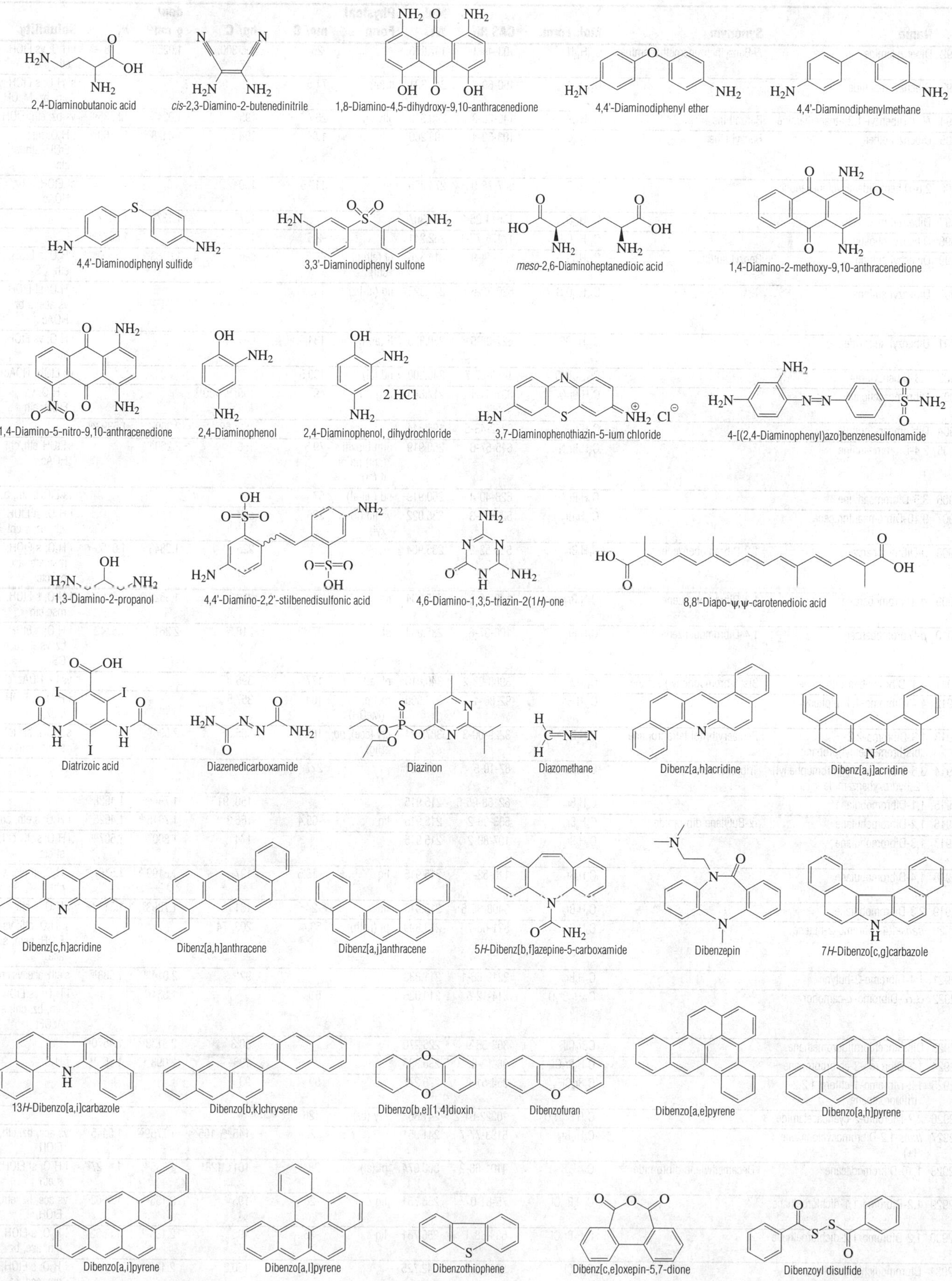

2,4-Diaminobutanoic acid
cis-2,3-Diamino-2-butenedinitrile
1,8-Diamino-4,5-dihydroxy-9,10-anthracenedione
4,4′-Diaminodiphenyl ether
4,4′-Diaminodiphenylmethane
4,4′-Diaminodiphenyl sulfide
3,3′-Diaminodiphenyl sulfone
meso-2,6-Diaminoheptanedioic acid
1,4-Diamino-2-methoxy-9,10-anthracenedione
1,4-Diamino-5-nitro-9,10-anthracenedione
2,4-Diaminophenol
2,4-Diaminophenol, dihydrochloride
3,7-Diaminophenothiazin-5-ium chloride
4-[(2,4-Diaminophenyl)azo]benzenesulfonamide
1,3-Diamino-2-propanol
4,4′-Diamino-2,2′-stilbenedisulfonic acid
4,6-Diamino-1,3,5-triazin-2(1H)-one
8,8′-Diapo-ψ,ψ-carotenedioic acid
Diatrizoic acid
Diazenedicarboxamide
Diazinon
Diazomethane
Dibenz[a,h]acridine
Dibenz[a,j]acridine
Dibenz[c,h]acridine
Dibenz[a,h]anthracene
Dibenz[a,j]anthracene
5H-Dibenz[b,f]azepine-5-carboxamide
Dibenzepin
7H-Dibenzo[c,g]carbazole
13H-Dibenzo[a,i]carbazole
Dibenzo[b,k]chrysene
Dibenzo[b,e][1,4]dioxin
Dibenzofuran
Dibenzo[a,e]pyrene
Dibenzo[a,h]pyrene
Dibenzo[a,i]pyrene
Dibenzo[a,l]pyrene
Dibenzothiophene
Dibenz[c,e]oxepin-5,7-dione
Dibenzoyl disulfide

No.	Name	Synonym	Mol. Form.	CAS RN	Mol. Wt.	Physical Form	mp/°C	bp/°C	den/ g cm^{-3}	n_D	Solubility
2892	Dibenzylamine	*N*-Benzylbenzenemethanamine	$C_{14}H_{15}N$	103-49-1	197.276		-26	dec 300; 270^{250}	1.0256^{22}	1.5781^{20}	i H_2O; vs EtOH, eth; s ctc
2893	Dibenzyl disulfide		$C_{14}H_{14}S_2$	150-60-7	246.391	lf (al)	71.5				sl H_2O; s EtOH, eth, bz, MeOH
2894	*N,N'*-Dibenzyl-1,2-ethanediamine	Benzathine	$C_{16}H_{20}N_2$	140-28-3	240.343	oily liq	26	195^{4}	1.024^{20}	1.5635^{20}	vs bz, eth, EtOH
2895	Dibenzyl ether	Benzyl ether	$C_{14}H_{14}O$	103-50-4	198.260		1.8	298	1.0428^{20}	1.5168^{20}	i H_2O; msc EtOH, eth; s ctc
2896	2,6-Dibenzylidenecyclohexanone		$C_{20}H_{18}O$	897-78-9	274.356		117.5	190^{20}			sl EtOH; s bz, HOAc
2897	Dibenzyl malonate		$C_{17}H_{16}O_4$	15014-25-2	284.307			187^{2}	1.137^{25}	1.5447^{20}	
2898	Dibenzyl phosphite		$C_{14}H_{15}O_3P$	17176-77-1	262.241		-2.5	$162^{0.1}$		1.5521^{18}	
2899	Dibenzyl sulfide	Benzyl sulfide	$C_{14}H_{14}S$	538-74-9	214.326	pl (eth or chl)	49.5	dec	1.0583^{50}		i H_2O; s EtOH, eth, CS_2
2900	Dibenzyl sulfone		$C_{14}H_{14}O_2S$	620-32-6	246.325	nd (al-bz)	152	dec 290			i H_2O; sl EtOH; vs ace; s bz, HOAc
2901	Dibenzyl sulfoxide		$C_{14}H_{14}OS$	621-08-9	230.325	lf (al, w)	134	dec 210			i H_2O; vs EtOH, eth
2902	1,3-Dibenzylurea		$C_{15}H_{16}N_2O$	1466-67-7	240.300	nd (al)	169.5				vs EtOH, HOAc
2903	Dibromoacetic acid		$C_2H_2Br_2O_2$	631-64-1	217.844	hyg cry	49	195^{250}, 130^{16}			vs H_2O; vs EtOH, eth
2904	Dibromoacetonitrile		C_2HBr_2N	3252-43-5	198.844			169; 68^{24}	2.369^{20}	1.5393^{20}	
2905	2,4-Dibromoaniline		$C_6H_5Br_2N$	615-57-6	250.919	orth bipym (chl) nd or lf (al)	79.5	156^{74}	2.260^{20}		s EtOH, eth, chl, HOAc
2906	3,5-Dibromoaniline		$C_6H_5Br_2N$	626-40-4	250.919	nd (dil al)	57				vs EtOH, eth, bz
2907	9,10-Dibromoanthracene		$C_{14}H_8Br_2$	523-27-3	336.022	ye nd (to or xyl)	226	sub			i H_2O; sl EtOH, eth, bz; s chl
2908	*o*-Dibromobenzene	1,2-Dibromobenzene	$C_6H_4Br_2$	583-53-9	235.904		7.1	225	1.9843^{20}	1.6155^{20}	i H_2O; s EtOH; msc eth, ace, bz, ctc
2909	*m*-Dibromobenzene	1,3-Dibromobenzene	$C_6H_4Br_2$	108-36-1	235.904	liq	-7	218	1.9523^{20}	1.6083^{17}	i H_2O; s EtOH; msc eth
2910	*p*-Dibromobenzene	1,4-Dibromobenzene	$C_6H_4Br_2$	106-37-6	235.904	pl	87.43	218.5	2.261^{17}	1.5742	i H_2O; s EtOH, bz; vs eth, ace, CS_2
2911	4,4'-Dibromobenzophenone	Bis(4-bromophenyl) ketone	$C_{13}H_8Br_2O$	3988-03-2	340.010	pl (al)	177	395			vs bz, HOAc, chl
2912	4,4'-Dibromo-1,1'-biphenyl		$C_{12}H_8Br_2$	92-86-4	312.000	mcl pr (MeOH)	164	357.5			i H_2O; sl EtOH; s bz
2913	1,3-Dibromo-2,2-bis(bromomethyl)propane	Pentaerythritol tetrabromide	$C_5H_8Br_4$	3229-00-3	387.734	cry (ace), nd (lig)	163	305.5	2.596^{15}		s EtOH, bz, tol; sl eth, chl
2914	3,5-Dibromo-*N*-(4-bromophenyl)-2-hydroxybenzamide	Tribromsalan	$C_{13}H_8Br_3NO_2$	87-10-5	449.921		227				
2915	1,1-Dibromobutane		$C_4H_8Br_2$	62168-25-6	215.915			158; 91^{101}	1.784^{25}	1.4988^{25}	
2916	1,2-Dibromobutane	α-Butylene dibromide	$C_4H_8Br_2$	533-98-2	215.915	liq	-65.4	166.3	1.7915^{20}	1.4025^{20}	i H_2O; s eth, chl
2917	1,3-Dibromobutane		$C_4H_8Br_2$	107-80-2	215.915			174	1.800^{20}	1.507^{20}	i H_2O; s eth, chl; sl ctc
2918	1,4-Dibromobutane		$C_4H_8Br_2$	110-52-1	215.915	liq	-16.5	197	1.8199^{25}	1.5167^{25}	i H_2O; sl ctc; s chl
2919	2,3-Dibromobutane		$C_4H_8Br_2$	5408-86-6	215.915	liq	-24	161	1.7893^{22}	1.5133^{22}	i H_2O; s eth
2920	*trans*-1,4-Dibromo-2-butene		$C_4H_6Br_2$	821-06-7	213.899	pl (peth)	53.4	203; 74^{14}			sl H_2O, chl; vs EtOH, peth; s ace
2921	1,4-Dibromo-2-butyne		$C_4H_4Br_2$	2219-66-1	211.883			92^{15}	2.014^{18}	1.588^{18}	s eth, ace; vs chl
2922	α,α'-Dibromo-*d*-camphor		$C_{10}H_{14}Br_2O$	514-12-5	310.025		61		1.854^{21}		i H_2O; vs EtOH, eth, bz, chl; s AcOEt
2923	Dibromochlorofluoromethane		CBr_2ClF	353-55-9	226.270			80.3	2.3173^{22}	1.4570^{20}	
2924	1,2-Dibromo-3-chloropropane		$C_3H_5Br_2Cl$	96-12-8	236.333			196	2.093^{14}	1.553^{14}	i H_2O
2925	1,2-Dibromo-1-chloro-1,2,2-trifluoroethane		$C_2Br_2ClF_3$	354-51-8	276.277		50	93			
2926	2,2-Dibromo-2-cyanoacetamide		$C_3H_2Br_2N_2O$	10222-01-2	241.868	cry (bz)	126				
2927	*trans*-1,2-Dibromocyclohexane, (±)		$C_6H_{10}Br_2$	5183-77-7	241.951		-2.0	145^{100}, 105^{20}	1.7759^{20}	1.5445^{19}	vs ace, bz, eth, EtOH
2928	1,10-Dibromodecane	Decamethylene dibromide	$C_{10}H_{20}Br_2$	4101-68-2	300.074	pl (al)	28	161^{9}, 128^{4}	1.335^{30}	1.4927^{25}	i H_2O; sl EtOH; s eth
2929	1,2-Dibromo-1,1-dichloroethane		$C_2H_2Br_2Cl_2$	75-81-0	256.751	liq	-26	195	2.135^{20}	1.5662^{20}	vs ace, bz, eth, EtOH
2930	1,2-Dibromo-1,2-dichloroethane		$C_2H_2Br_2Cl_2$	683-68-1	256.751	liq	-26	195	2.135^{20}	1.5662^{20}	i H_2O; s EtOH, eth, ace, bz
2931	Dibromodichloromethane		CBr_2Cl_2	594-18-3	242.725		38	150.2	2.42^{25}		i H_2O; s EtOH, eth, ace, bz
2932	1,2-Dibromo-1,1-difluoroethane	Genetron 132b-B2	$C_2H_2Br_2F_2$	75-82-1	223.842	liq	-61.3	92.5	2.2238^{20}	1.4456^{20}	

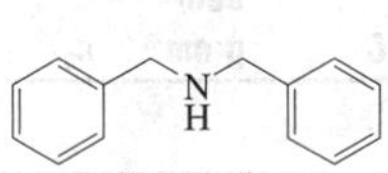

Dibenzylamine

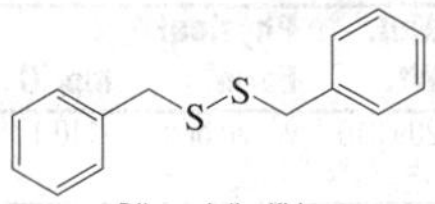

Dibenzyl disulfide

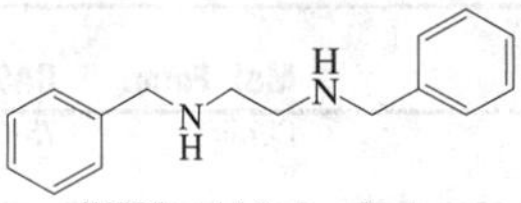

N,N'-Dibenzyl-1,2-ethanediamine

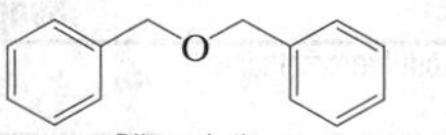

Dibenzyl ether

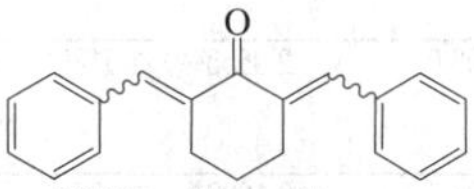

2,6-Dibenzylidenecyclohexanone

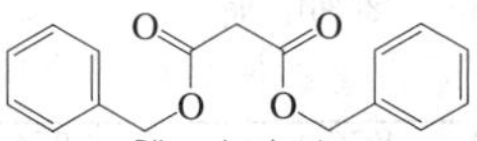

Dibenzyl malonate

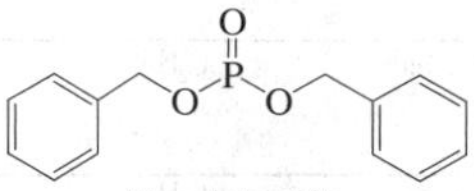

Dibenzyl phosphite

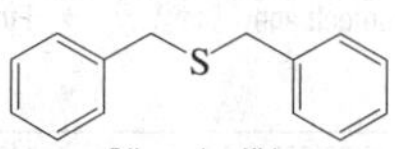

Dibenzyl sulfide

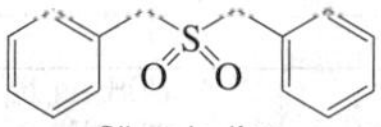

Dibenzyl sulfone

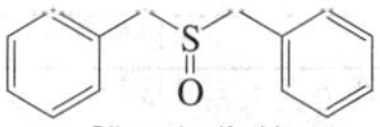

Dibenzyl sulfoxide

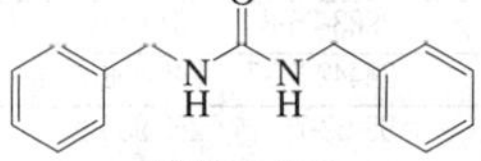

1,3-Dibenzylurea

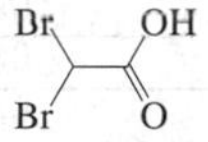

Dibromoacetic acid

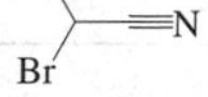

Dibromoacetonitrile

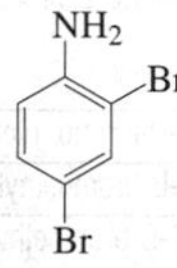

2,4-Dibromoaniline

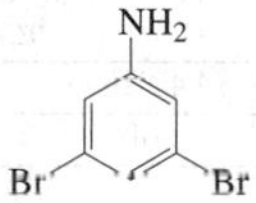

3,5-Dibromoaniline

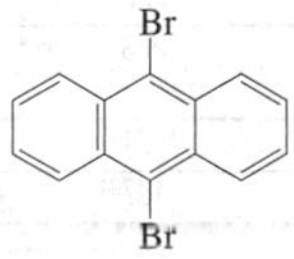

9,10-Dibromoanthracene

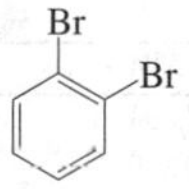

o-Dibromobenzene

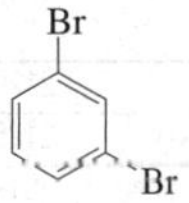

m-Dibromobenzene

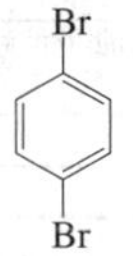

p-Dibromobenzene

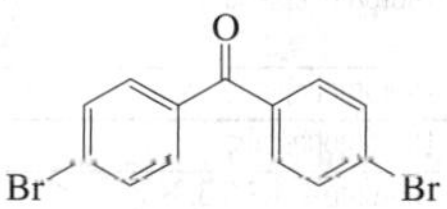

4,4'-Dibromobenzophenone

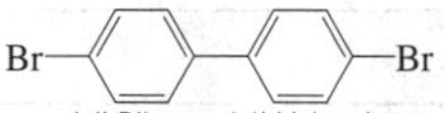

4,4'-Dibromo-1,1'-biphenyl

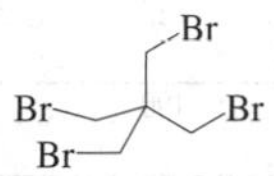

1,3-Dibromo-2,2-bis(bromomethyl)propane

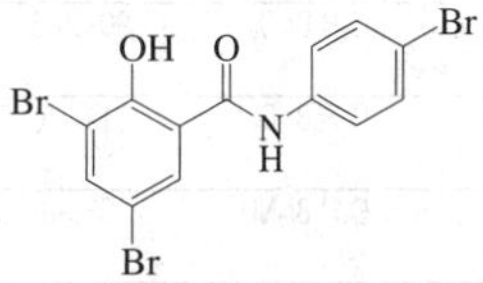

3,5-Dibromo-*N*-(4-bromophenyl)-2-hydroxybenzamide

1,1-Dibromobutane

1,2-Dibromobutane

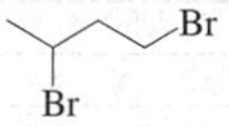

1,3-Dibromobutane

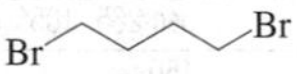

1,4-Dibromobutane

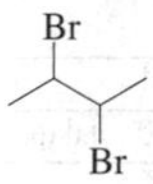

2,3-Dibromobutane

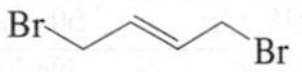

trans-1,4-Dibromo-2-butene

1,4-Dibromo-2-butyne

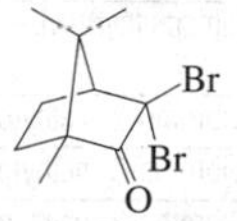

α,α'-Dibromo-*d*-camphor

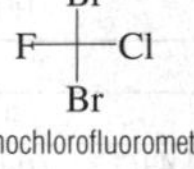

Dibromochlorofluoromethane

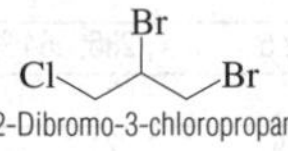

1,2-Dibromo-3-chloropropane

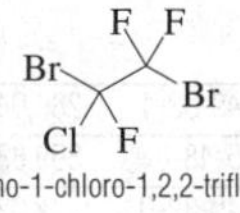

1,2-Dibromo-1-chloro-1,2,2-trifluoroethane

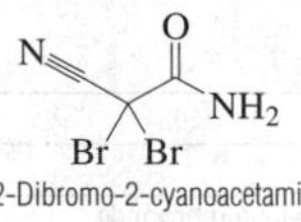

2,2-Dibromo-2-cyanoacetamide

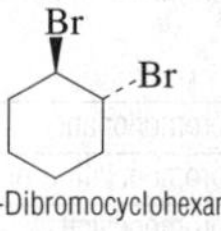

trans-1,2-Dibromocyclohexane, (±)

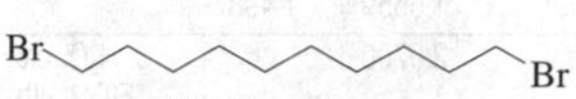

1,10-Dibromodecane

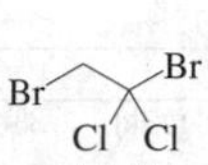

1,2-Dibromo-1,1-dichloroethane

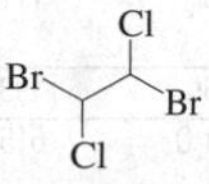

1,2-Dibromo-1,2-dichloroethane

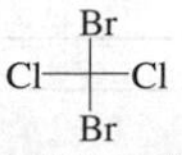

Dibromodichloromethane

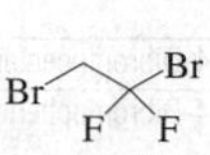

1,2-Dibromo-1,1-difluoroethane

No.	Name	Synonym	Mol. Form.	CAS RN	Mol. Wt.	Physical Form	mp/°C	bp/°C	den/ g cm^{-3}	n_D	Solubility
2933	Dibromodifluoromethane		CBr_2F_2	75-61-6	209.816	vol liq or gas	-110.1	22.76			s H_2O, eth, ace, bz
2934	1,3-Dibromo-5,5-dimethyl-2,4-imidazolidinedione	Dibromantine	$C_5H_6Br_2N_2O_2$	77-48-5	285.922		198 dec				
2935	1,3-Dibromo-2,2-dimethylpropane		$C_5H_{10}Br_2$	5434-27-5	229.941			184; 80^{26}	1.6775^{20}	1.5090	
2936	1,12-Dibromododecane		$C_{12}H_{24}Br_2$	3344-70-5	328.127	nd (al,HOAc)	41	215^{15}			i H_2O; vs EtOH, chl; s eth, HOAc
2937	1,1-Dibromoethane	Ethylidene dibromide	$C_2H_4Br_2$	557-91-5	187.861	liq	-63	108.0	2.0555^{20}	1.5128^{20}	i H_2O; s EtOH, ace, bz; sl chl; vs eth
2938	1,2-Dibromoethane	Ethylene dibromide	$C_2H_4Br_2$	106-93-4	187.861		9.84	131.6	2.1683^{25}	1.5356^{25}	vs ace, bz, eth, EtOH
2939	*cis*-1,2-Dibromoethene	*cis*-1,2-Dibromoethylene	$C_2H_2Br_2$	590-11-4	185.845	liq	-53	112.5	2.2464^{20}	1.5428^{20}	i H_2O; vs EtOH, eth; s ace, bz, chl
2940	*trans*-1,2-Dibromoethene	*trans*-1,2-Dibromoethylene	$C_2H_2Br_2$	590-12-5	185.845	liq	-6.5	108	2.2308^{20}	1.5505^{18}	i H_2O; vs EtOH, eth; s ace, bz, chl
2941	1,2-Dibromo-1-ethoxyethane		$C_4H_8Br_2O$	2983-26-8	231.914			80^{20}	1.7320^{20}	1.5044^{20}	vs EtOH, chl
2942	1,2-Dibromoethyl acetate		$C_4H_6Br_2O_2$	24442-57-7	245.898	liq		89.5^{16}	1.91^{20}		
2943	(1,2-Dibromoethyl)benzene		$C_8H_8Br_2$	93-52-7	263.958		75	133^{19}			s EtOH, eth, bz, chl, HOAc, MeOH, lig
2944	Dibromofluoromethane		$CHBr_2F$	1868-53-7	191.825	liq	-78	64.9	2.421^{20}	1.4685^{20}	i H_2O; s EtOH, eth, ace, bz, chl
2945	1,2-Dibromoheptane		$C_7H_{14}Br_2$	42474-21-5	257.994			228	1.5086^{20}	1.4986^{20}	
2946	1,7-Dibromoheptane	Heptamethylene dibromide	$C_7H_{14}Br_2$	4549-31-9	257.994		41.7	263	1.5306^{20}	1.5034^{20}	i H_2O; s eth, ace, bz, ctc, chl
2947	2,3-Dibromoheptane		$C_7H_{14}Br_2$	21266-88-6	257.994			101^{17}	1.5139^{20}	1.4992^{20}	
2948	3,4-Dibromoheptane		$C_7H_{14}Br_2$	21266-90-0	257.994			107^{24}	1.5182^{20}	1.5010^{20}	
2949	1,2-Dibromo-1,1,2,3,3,3-hexafluoropropane		$C_3Br_2F_6$	661-95-0	309.830			72.8	2.1630^{20}		i H_2O
2950	1,2-Dibromohexane		$C_6H_{12}Br_2$	624-20-4	243.967			103^{36}	1.5774^{20}	1.5024^{20}	vs bz, eth, chl
2951	1,6-Dibromohexane		$C_6H_{12}Br_2$	629-03-8	243.967	liq	-1.2	245.5	1.6025^{25}	1.5054^{25}	i H_2O; s eth, ace, chl; sl ctc
2952	3,4-Dibromohexane		$C_6H_{12}Br_2$	89583-12-0	243.967			80^{13}	1.6027^{20}	1.5043^{20}	
2953	3,5-Dibromo-2-hydroxybenzaldehyde	3,5-Dibromosalicylaldehyde	$C_7H_4Br_2O_2$	90-59-5	279.914	pa ye pr	86	sub			vs bz, eth, chl
2954	3,5-Dibromo-2-hydroxybenzoic acid	3,5-Dibromosalicylic acid	$C_7H_4Br_2O_3$	3147-55-5	295.913	nd	228				s ace
2955	3,5-Dibromo-4-hydroxybenzonitrile	Bromoxynil	$C_7H_3Br_2NO$	1689-84-5	276.913		190				
2956	Dibromomethane	Methylene bromide	CH_2Br_2	74-95-3	173.835	liq	-52.5	97	2.4969^{20}	1.5420^{20}	sl H_2O; msc EtOH, eth, ace; s ctc
2957	1,4-Dibromo-2-methylbenzene	2,5-Dibromotoluene	$C_7H_6Br_2$	615-59-8	249.931		5.6	236	1.8127^{17}	1.5982^{18}	i H_2O
2958	2,4-Dibromo-1-methylbenzene		$C_7H_6Br_2$	31543-75-6	249.931		-9.7	103^{11}	1.8176^{25}	1.5964^{25}	
2959	(Dibromomethyl)benzene		$C_7H_6Br_2$	618-31-5	249.931		1.0	156^{23}	1.8365^{28}	1.6147^{20}	i H_2O; msc EtOH, eth
2960	2,3-Dibromo-2-methylbutane		$C_5H_{10}Br_2$	594-51-4	229.941		7	62^{17}	1.6717^{20}	1.5729^{25}	
2961	2,4-Dibromo-6-methylphenol		$C_7H_6Br_2O$	609-22-3	265.930	nd (peth)	58	dec 265; 105^{4}			s chl
2962	1,2-Dibromo-2-methylpropane		$C_4H_8Br_2$	594-34-3	215.915		10.5	150	1.7827^{20}	1.5119^{20}	s EtOH, eth, chl
2963	1,4-Dibromonaphthalene		$C_{10}H_6Br_2$	83-53-4	285.963		83	310			i H_2O; s EtOH, eth; sl HOAc
2964	2,6-Dibromo-4-nitroaniline		$C_6H_4Br_2N_2O_2$	827-94-1	295.916	ye nd (al, HOAc)	207				sl H_2O; s HOAc
2965	2,6-Dibromo-4-nitrophenol		$C_6H_3Br_2NO_3$	99-28-5	296.901	pa ye pr or lf (al)	145 dec				i H_2O; vs EtOH, eth; sl ace, bz, HOAc
2966	1,9-Dibromononane		$C_9H_{18}Br_2$	4549-33-1	286.047	liq	-22.5	285; 154^{10}	1.4229^{20}		
2967	1,4-Dibromooctafluorobutane		$C_4Br_2F_8$	335-48-8	359.838			97			
2968	1,8-Dibromooctane	Octamethylene dibromide	$C_8H_{16}Br_2$	4549-32-0	272.021		15.5	271	1.4594^{25}	1.4971^{25}	i H_2O; s eth, ctc, chl
2969	1,2-Dibromopentane		$C_5H_{10}Br_2$	3234-49-9	229.941			184	1.668^{18}		
2970	1,4-Dibromopentane		$C_5H_{10}Br_2$	626-87-9	229.941		-34.4	146^{150}, 99^{14}	1.6222^{20}	1.5086^{20}	
2971	1,5-Dibromopentane		$C_5H_{10}Br_2$	111-24-0	229.941	liq	-39.5	222.3	1.6928^{25}	1.5102^{25}	i H_2O; s bz, chl; sl ctc
2972	2,4-Dibromopentane		$C_5H_{10}Br_2$	19398-53-9	229.941			75^{21}, 60^{12}	1.6659^{20}	1.4987^{20}	
2973	2,4-Dibromophenol		$C_6H_4Br_2O$	615-58-7	251.903	nd (peth)	38	238.5	2.0700^{20}		sl H_2O, ctc; vs EtOH, eth, bz
2974	2,6-Dibromophenol		$C_6H_4Br_2O$	608-33-3	251.903	nd (w)	56.5	255; 162^{21}			s H_2O; vs EtOH, eth

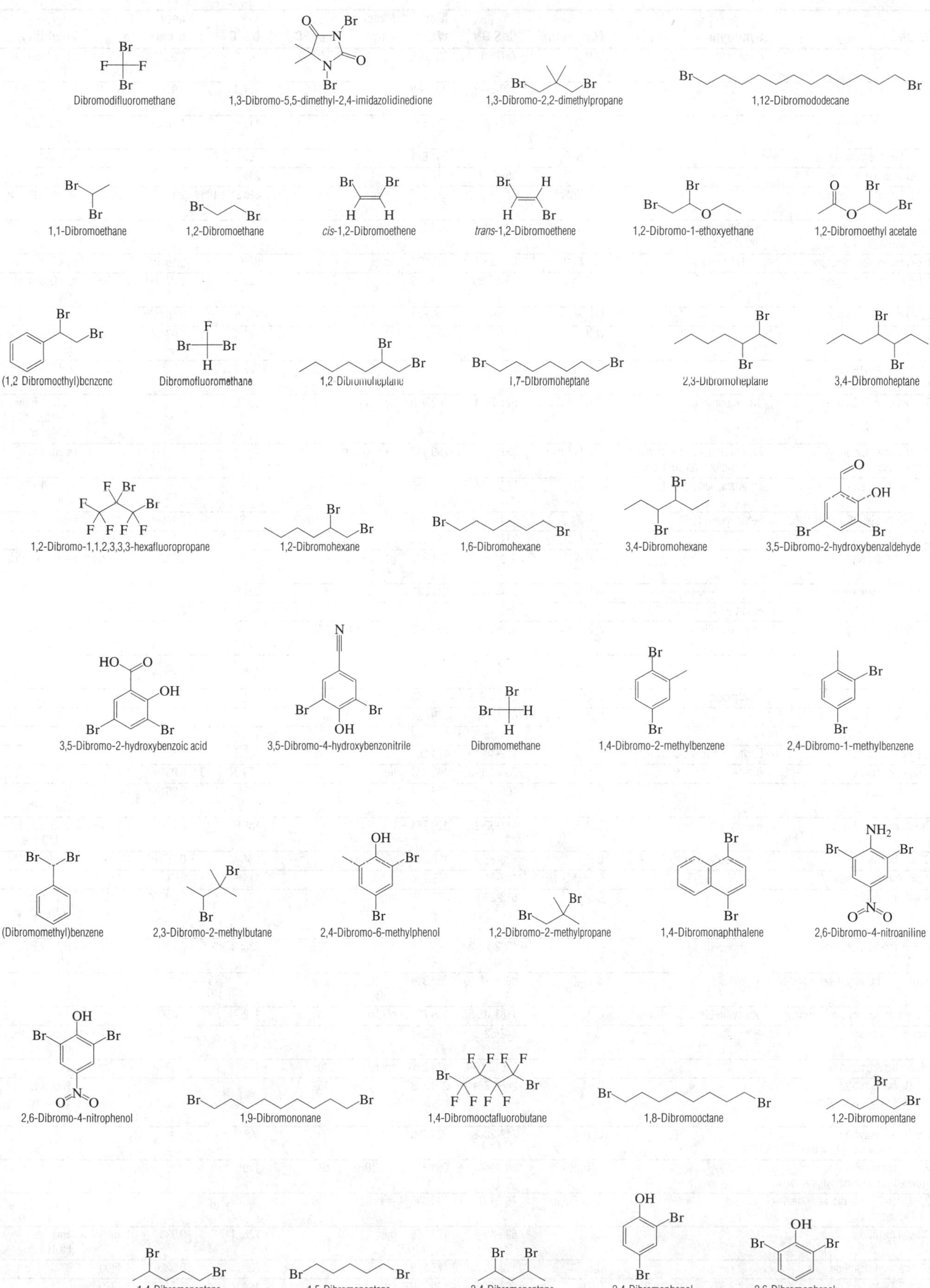

Dibromodifluoromethane
1,3-Dibromo-5,5-dimethyl-2,4-imidazolidinedione
1,3-Dibromo-2,2-dimethylpropane
1,12-Dibromododecane
1,1-Dibromoethane
1,2-Dibromoethane
cis-1,2-Dibromoethene
trans-1,2-Dibromoethene
1,2-Dibromo-1-ethoxyethane
1,2-Dibromoethyl acetate
(1,2-Dibromoethyl)benzene
Dibromofluoromethane
1,2-Dibromoheptane
1,7-Dibromoheptane
2,3-Dibromoheptane
3,4-Dibromoheptane
1,2-Dibromo-1,1,2,3,3,3-hexafluoropropane
1,2-Dibromohexane
1,6-Dibromohexane
3,4-Dibromohexane
3,5-Dibromo-2-hydroxybenzaldehyde
3,5-Dibromo-2-hydroxybenzoic acid
3,5-Dibromo-4-hydroxybenzonitrile
Dibromomethane
1,4-Dibromo-2-methylbenzene
2,4-Dibromo-1-methylbenzene
(Dibromomethyl)benzene
2,3-Dibromo-2-methylbutane
2,4-Dibromo-6-methylphenol
1,2-Dibromo-2-methylpropane
1,4-Dibromonaphthalene
2,6-Dibromo-4-nitroaniline
2,6-Dibromo-4-nitrophenol
1,9-Dibromononane
1,4-Dibromooctafluorobutane
1,8-Dibromooctane
1,2-Dibromopentane
1,4-Dibromopentane
1,5-Dibromopentane
2,4-Dibromopentane
2,4-Dibromophenol
2,6-Dibromophenol

No.	Name	Synonym	Mol. Form.	CAS RN	Mol. Wt.	Physical Form	mp/°C	bp/°C	den/ g cm^{-3}	n_D	Solubility
2975	1,2-Dibromopropane	Propylene dibromide	$C_3H_6Br_2$	78-75-1	201.888	liq	-55.49	141.9	1.9324^{20}	1.5201^{20}	s EtOH, eth, chl; sl ctc
2976	1,3-Dibromopropane		$C_3H_6Br_2$	109-64-8	201.888	liq	-34.5	167.3	1.9701^{25}	1.5204^{25}	i H_2O; s EtOH, eth, chl; sl ctc
2977	2,2-Dibromopropane		$C_3H_6Br_2$	594-16-1	201.888			113	1.880^{20}		vs eth, EtOH, chl
2978	2,3-Dibromopropanoic acid		$C_3H_4Br_2O_2$	600-05-5	231.871		66.5	160^{20}, 138^{12}			vs bz, eth, EtOH
2979	2,3-Dibromo-1-propanol		$C_3H_6Br_2O$	96-13-9	217.887			219	2.120^{20}		
2980	1,3-Dibromo-2-propanol		$C_3H_6Br_2O$	96-21-9	217.887	ye liq		dec 219; 105^{16}	2.1364^{20}	1.5495^{25}	vs ace, eth, EtOH
2981	2,3-Dibromo-1-propanol, phosphate (3:1)	Tris(2,3-dibromopropyl) phosphate	$C_9H_{15}Br_6O_4P$	126-72-7	697.610						s chl
2982	1,3-Dibromo-2-propanone	1,3-Dibromoacetone	$C_3H_4Br_2O$	816-39-7	215.871	nd	26	97^{22}	2.1670^{18}		vs eth, CS_2
2983	1,1-Dibromo-1-propene		$C_3H_4Br_2$	13195-80-7	199.872			125	1.9767^{20}	1.5260^{20}	sl H_2O; s bz, ctc, chl
2984	1,2-Dibromo-1-propene		$C_3H_4Br_2$	26391-16-2	199.872			131.5	2.0076^{20}		
2985	2,3-Dibromo-1-propene		$C_3H_4Br_2$	513-31-5	199.872			141; 37.7^{11}	2.0345^{25}	1.5416^{25}	i H_2O; s eth, ace, chl
2986	3,5-Dibromopyridine		$C_5H_3Br_2N$	625-92-3	236.893	nd (al)	112	222			sl H_2O; s EtOH, eth
2987	5,7-Dibromo-8-quinolinol	Broxyquinoline	$C_9H_5Br_2NO$	521-74-4	302.950	nd (al)	196	sub			i H_2O; s EtOH, ace, bz, chl, HOAc; sl eth
2988	2,6-Dibromoquinone-4-chlorimide	2,6-Dibromo-4-(chloroimino)-2,5-cyclohexadien-1-one	$C_6H_2Br_2ClNO$	537-45-1	299.347	ye pr (al or HOAc)	83				vs EtOH
2989	1,14-Dibromotetradecane	Tetradecamethylene dibromide	$C_{14}H_{28}Br_2$	37688-96-3	356.180	lf (al-eth) cry (al)	50.4	190^{8}			vs eth, EtOH, chl
2990	1,2-Dibromotetrafluoroethane	Refrigerant 114B2	$C_2Br_2F_4$	124-73-2	259.823	liq	-110.32	47.35	2.149^{25}	1.361^{25}	i H_2O
2991	2,3-Dibromothiophene		$C_4H_2Br_2S$	3140-93-0	241.932	liq	-17.5	218.5; 89^{13}		1.6304^{22}	
2992	2,5-Dibromothiophene		$C_4H_2Br_2S$	3141-27-3	241.932	liq	-6	210.3	2.142^{23}	1.6288^{20}	i H_2O; vs EtOH, eth; s ctc
2993	3,4-Dibromothiophene		$C_4H_2Br_2S$	3141-26-2	241.932		4.5	221.5			
2994	1,2-Dibromo-1,1,2-trifluoroethane	Halon 2302	$C_2HBr_2F_3$	354-04-1	241.832			76	2.274^{27}	1.4191^{24}	
2995	2,6-Dibromo-3,4,5-trihydroxybenzoic acid	Dibromogallic acid	$C_7H_4Br_2O_5$	602-92-6	327.912	nd, pr or lf (w+1)	150				vs H_2O, eth, EtOH
2996	3,5-Dibromo-*L*-tyrosine		$C_9H_9Br_2NO_3$	300-38-9	338.980	nd or pl	245				sl H_2O, EtOH; i eth; s alk, acid
2997	Dibucaine	Cinchocaine	$C_{20}H_{29}N_3O_2$	85-79-0	343.463	hyg cry	64				
2998	Dibucaine hydrochloride		$C_{20}H_{30}ClN_3O_2$	61-12-1	379.924		94 dec				s chl
2999	1,4-Dibutoxybenzene		$C_{14}H_{22}O_2$	104-36-9	222.324		45.5	158^{15}			s ctc
3000	1,2-Dibutoxyethane	Ethylene glycol dibutyl ether	$C_{10}H_{22}O_2$	112-48-1	174.281	liq	-69.1	203.3	0.8319^{25}	1.4112^{25}	
3001	Dibutoxymethane	Butylal	$C_9H_{20}O_2$	2568-90-3	160.254	liq	-58.1	179.2	0.8339^{20}	1.4072^{17}	
3002	Dibutyl adipate		$C_{14}H_{26}O_4$	105-99-7	258.354		-32.4	165^{10}	0.9613^{20}	1.4369^{20}	i H_2O; msc EtOH, eth
3003	Dibutylamine	*N*-Butylbutanamine	$C_8H_{19}N$	111-92-2	129.244	liq	-62	159.6	0.7670^{20}	1.4177^{20}	s H_2O, ace, bz; vs EtOH, eth
3004	Di-*sec*-butylamine	*N-sec*-Butyl-2-butanamine	$C_8H_{19}N$	626-23-3	129.244			134	0.7534^{20}	1.4162^{20}	vs H_2O; s EtOH
3005	2-Dibutylaminoethanol		$C_{10}H_{23}NO$	102-81-8	173.296			114^{18}			
3006	*N,N*-Dibutylaniline		$C_{14}H_{23}N$	613-29-6	205.340	liq	-32.2	274.8	0.9037^{20}	1.5186^{20}	i H_2O; msc EtOH, eth; vs ace, bz; s ctc
3007	1,4-Di-*tert*-butylbenzene		$C_{14}H_{22}$	1012-72-2	190.325	nd (MeOH)	79.5	238; 109^{15}	0.9850^{20}		i H_2O; s EtOH, eth
3008	2,5-Di-*tert*-butyl-1,4-benzenediol		$C_{14}H_{22}O_2$	88-58-4	222.324	cry (aq HOAc)	213.5				
3009	Dibutylbis(dodecylthio)stannane	Dibutyltin bis(dodecyl sulfide)	$C_{32}H_{68}S_2Sn$	1185-81-5	635.722	col liq		$122^{0.3}$	1.05^{20}		s tol, hp
3010	Dibutyl carbonate		$C_9H_{18}O_3$	542-52-9	174.237			207	0.9251^{20}	1.4117^{20}	i H_2O; s EtOH, eth
3011	Di-*tert*-butyl carbonate		$C_9H_{18}O_3$	34619-03-9	174.237	cry (al)	40	158			vs EtOH
3012	2,5-Di-*tert*-butyl-2,5-cyclohexadiene-1,4-dione		$C_{14}H_{20}O_2$	2460-77-7	220.308	ye cry (al)	152.5				i H_2O; s EtOH, eth, bz, chl, HOAc
3013	2,6-Di-*tert*-butyl-2,5-cyclohexadiene-1,4-dione		$C_{14}H_{20}O_2$	719-22-2	220.308		69	$60^{0.01}$			
3014	2,6-Di-*tert*-butyl-4-(dimethylaminomethyl)phenol		$C_{17}H_{29}NO$	88-27-7	263.418	pl (EtOH)	94	179^{40}			
3015	2,2-Dibutyl-1,3,2-dioxastannepin-4,7-dione		$C_{12}H_{20}O_4Sn$	78-04-6	346.995	ye solid	110				
3016	Dibutyl disulfide		$C_8H_{18}S_2$	629-45-8	178.359	oil		226; 117^{20}	0.938^{20}	1.4923^{20}	i H_2O; msc EtOH, eth
3017	Di-*tert*-butyl disulfide		$C_8H_{18}S_2$	110-06-5	178.359		-2.5	88^{21}	0.9226^{20}	1.4899^{20}	
3018	*cis*-1,2-Di-*tert*-butylethene	*cis*-2,2,5,5-Tetramethyl-3-hexene	$C_{10}H_{20}$	692-47-7	140.266	liq		144	0.744^{20}	1.4270^{20}	

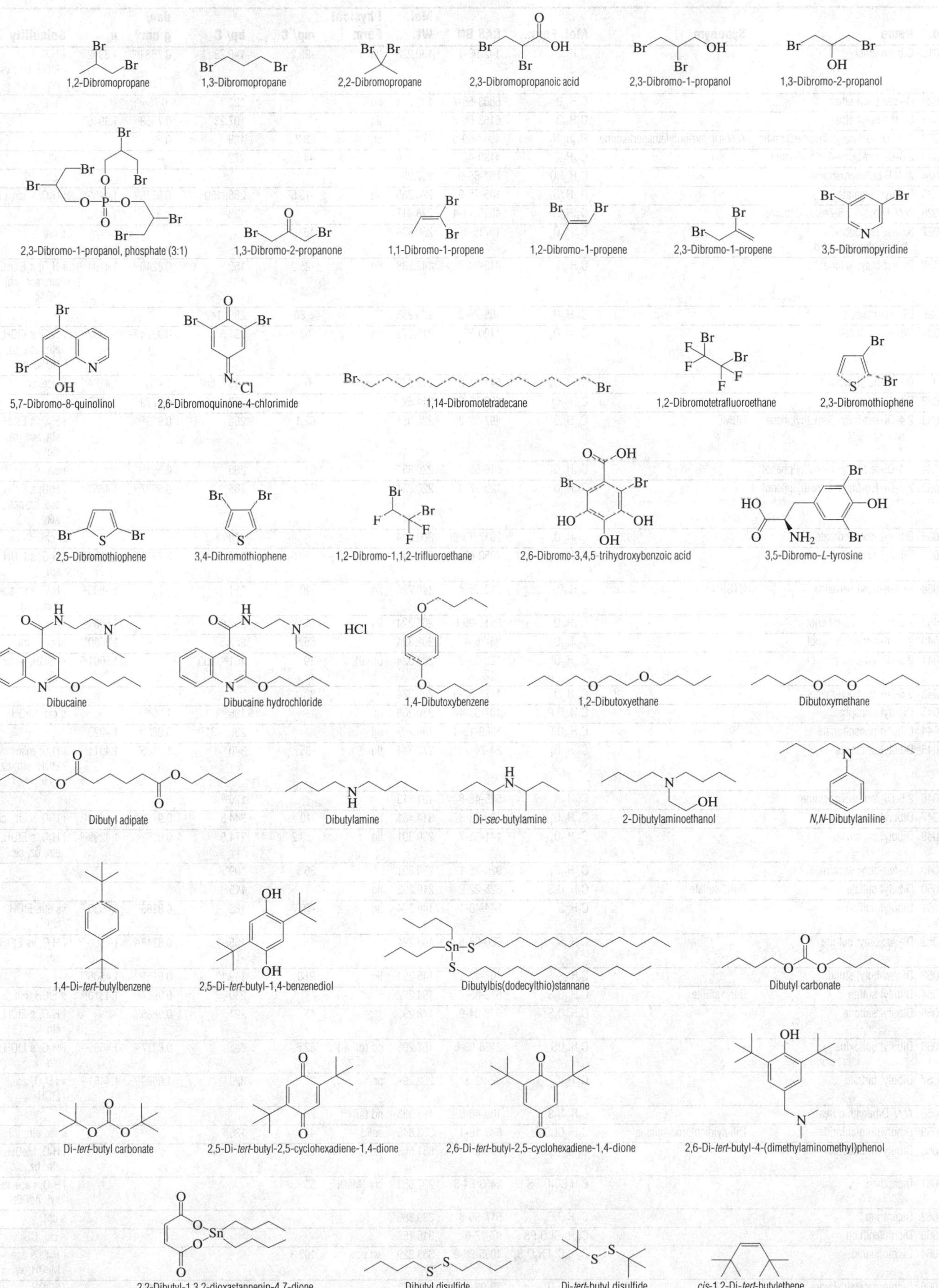

1,2-Dibromopropane
1,3-Dibromopropane
2,2-Dibromopropane
2,3-Dibromopropanoic acid
2,3-Dibromo-1-propanol
1,3-Dibromo-2-propanol
2,3-Dibromo-1-propanol, phosphate (3:1)
1,3-Dibromo-2-propanone
1,1-Dibromo-1-propene
1,2-Dibromo-1-propene
2,3-Dibromo-1-propene
3,5-Dibromopyridine
5,7-Dibromo-8-quinolinol
2,6-Dibromoquinone-4-chlorimide
1,14-Dibromotetradecane
1,2-Dibromotetrafluoroethane
2,3-Dibromothiophene
2,5-Dibromothiophene
3,4-Dibromothiophene
1,2-Dibromo-1,1,2-trifluoroethane
2,6-Dibromo-3,4,5-trihydroxybenzoic acid
3,5-Dibromo-L-tyrosine
Dibucaine
HCl
Dibucaine hydrochloride
1,4-Dibutoxybenzene
1,2-Dibutoxyethane
Dibutoxymethane
Dibutyl adipate
Dibutylamine
Di-sec-butylamine
2-Dibutylaminoethanol
N,N-Dibutylaniline
1,4-Di-tert-butylbenzene
2,5-Di-tert-butyl-1,4-benzenediol
Dibutylbis(dodecylthio)stannane
Dibutyl carbonate
Di-tert-butyl carbonate
2,5-Di-tert-butyl-2,5-cyclohexadiene-1,4-dione
2,6-Di-tert-butyl-2,5-cyclohexadiene-1,4-dione
2,6-Di-tert-butyl-4-(dimethylaminomethyl)phenol
2,2-Dibutyl-1,3,2-dioxastannepin-4,7-dione
Dibutyl disulfide
Di-tert-butyl disulfide
cis-1,2-Di-tert-butylethene

No.	Name	Synonym	Mol. Form.	CAS RN	Mol. Wt.	Physical Form	mp/°C	bp/°C	den/ g cm^{-3}	n_D	Solubility
3019	Dibutyl ether		$C_8H_{18}O$	142-96-1	130.228	liq	-95.2	140.28	0.7684^{20}	1.3992^{20}	i H_2O; msc EtOH, eth; vs ace; sl ctc
3020	Di-*sec*-butyl ether		$C_8H_{18}O$	6863-58-7	130.228	liq		121.1	0.756^{25}		
3021	Di-*tert*-butyl ether		$C_8H_{18}O$	6163-66-2	130.228	liq		107.23	0.7658^{20}	1.3949^{20}	
3022	*N,N'*-Di-*tert*-butylethylenediamine	*N,N'*-Di-*tert*-butylethanediamine	$C_{10}H_{24}N_2$	4062-60-6	172.311	cry	53.3	189	0.69		
3023	2,6-Di-*tert*-butyl-4-ethylphenol		$C_{16}H_{26}O$	4130-42-1	234.376		44	272			i alk
3024	*N,N*-Dibutylformamide		$C_9H_{19}NO$	761-65-9	157.253						s ctc, CS_2
3025	Dibutyl fumarate		$C_{12}H_{20}O_4$	105-75-9	228.285	liq	-13.5	285; 150^4	0.9775^{20}	1.4469^{20}	i H_2O; s ace, chl
3026	*N,N'*-Dibutyl-1,6-hexanediamine		$C_{14}H_{32}N_2$	4835-11-4	228.417			$138^{3.5}$		1.4470^{25}	
3027	3,5-Di-*tert*-butyl-2-hydroxybenzoic acid		$C_{15}H_{22}O_3$	19715-19-6	250.334		163.3				s chl
3028	Di-*tert*-butyl ketone		$C_9H_{18}O$	815-24-7	142.238	liq	-25.2	152	0.8240^{18}	1.4194^{20}	i H_2O; s EtOH, eth, ace, chl, HOAc
3029	Dibutyl maleate		$C_{12}H_{20}O_4$	105-76-0	228.285		<-80	280; 142^{10}			
3030	Dibutyl malonate		$C_{11}H_{20}O_4$	1190-39-2	216.275	liq	-83	251.5	0.9824^{20}	1.4262^{20}	i H_2O; s EtOH, eth, ace, bz, HOAc, ctc
3031	Di-*tert*-butyl malonate		$C_{11}H_{20}O_4$	541-16-2	216.275		-6	113^{31}, 66^2	1.4184^{20}	1.4184^{29}	s ace, chl
3032	Dibutylmercury		$C_8H_{18}Hg$	629-35-6	314.82			223; 105^{10}	1.7779^{20}	1.5057^{20}	
3033	2,4-Di-*tert*-butyl-5-methylphenol	DBMC	$C_{15}H_{24}O$	497-39-2	220.351		62.1	282	0.912^{80}		i H_2O; s EtOH, eth, ace, bz, ctc
3034	2,4-Di-*tert*-butyl-6-methylphenol		$C_{15}H_{24}O$	616-55-7	220.351		51	269	0.891^{80}		i alk
3035	2,6-Di-*tert*-butyl-4-methylphenol		$C_{15}H_{24}O$	128-37-0	220.351		71	265	0.8937^{75}	1.4859^{75}	i H_2O; s EtOH, ace, bz, peth; i alk
3036	Dibutyl nonanedioate		$C_{17}H_{32}O_4$	2917-73-9	300.434			170^2			sl chl
3037	Dibutyl oxalate		$C_{10}H_{18}O_4$	2050-60-4	202.248	liq	-30.5	241; 96^2	0.9873^{20}	1.4234^{20}	i H_2O; s EtOH, eth
3038	Di-*tert*-butyl peroxide	DTBP	$C_8H_{18}O_2$	110-05-4	146.228	liq	-40	111	0.704^{20}	1.3890^{20}	i H_2O; msc ace; s ctc, lig
3039	2,6-Di-*sec*-butylphenol		$C_{14}H_{22}O$	5510-99-6	206.324	liq	-42	257.5		1.5080^{20}	
3040	2,4-Di-*tert*-butylphenol		$C_{14}H_{22}O$	96-76-4	206.324		56.5	263.5		1.5080^{20}	sl ctc; i alk
3041	2,6-Di-*tert*-butylphenol		$C_{14}H_{22}O$	128-39-2	206.324	pr (al)	39	161^{50}, 133^{20}		1.5001^{20}	sl EtOH; s ctc; i alk
3042	3,5-Di-*tert*-butylphenol		$C_{14}H_{22}O$	1138-52-9	206.324		88				
3043	Dibutyl phosphate		$C_8H_{19}O_4P$	107-66-4	210.208	oil		$136^{0.05}$	1.06^{20}		s ctc, BuOH
3044	Dibutyl phosphonate		$C_8H_{19}O_3P$	1809-19-4	194.209	oil		230; 131^{19}	0.985^{25}	1.4220^{20}	
3045	Dibutyl phthalate		$C_{16}H_{22}O_4$	84-74-2	278.344	liq	-35	340	1.0465^{20}	1.4911^{20}	i H_2O; msc EtOH, eth, bz; s ctc
3046	2,6-Di-*tert*-butylpyridine		$C_{13}H_{21}N$	585-48-8	191.313			120^{20}			
3047	Dibutyl sebacate		$C_{18}H_{34}O_4$	109-43-3	314.461	liq	-10	344.5	0.9405^{15}	1.4433^{15}	i H_2O; s eth, ctc
3048	Dibutyl succinate		$C_{12}H_{22}O_4$	141-03-7	230.301	liq	-29.2	274.5	0.9752^{20}	1.4299^{20}	i H_2O; s EtOH, eth, bz, ctc
3049	Di-*tert*-butyl succinate		$C_{12}H_{22}O_4$	926-26-1	230.301		36.5	109^9			
3050	Dibutyl sulfate	Butyl sulfate	$C_8H_{18}O_4S$	625-22-9	210.292	liq		115^6			
3051	Dibutyl sulfide		$C_8H_{18}S$	544-40-1	146.294	liq	-79.7	185	0.8386^{20}	1.4530^{20}	vs eth, EtOH, chl
3052	Di-*sec*-butyl sulfide		$C_8H_{18}S$	626-26-6	146.294			165	0.8348^{20}	1.4506^{20}	i H_2O; vs EtOH, eth
3053	Di-*tert*-butyl sulfide		$C_8H_{18}S$	107-47-1	146.294	liq	-9.0	149.1	0.815^{25}	1.4506^{20}	
3054	Dibutyl sulfite	Butyl sulfite	$C_8H_{18}O_3S$	626-85-7	194.292			230	0.9957^{20}	1.4310^{20}	s EtOH, eth
3055	Dibutyl sulfone		$C_8H_{18}O_2S$	598-04-9	178.293		45	291	0.9885^{47}		i H_2O; s EtOH, eth
3056	Dibutyl sulfoxide		$C_8H_{18}OS$	2168-93-6	162.293	nd (dil al)	32.6	dec	0.8317^{23}	1.4669^{20}	i H_2O; s EtOH, eth
3057	Dibutyl tartrate		$C_{12}H_{22}O_6$	87-92-3	262.299	pr	22	320	1.0909^{20}	1.4451^{20}	vs H_2O, ace, EtOH
3058	*N,N'*-Dibutylthiourea		$C_9H_{20}N_2S$	109-46-6	188.333	nd (al)	78				
3059	Dibutyltin dichloride	Dibutyldichlorostannane	$C_8H_{18}Cl_2Sn$	683-18-1	303.845	solid	43	135^{10}			s hx, eth, thf
3060	Dibutyltin dilaurate		$C_{32}H_{64}O_4Sn$	77-58-7	631.558	ye liq or cry	23				i H_2O, MeOH; s eth, bz, ctc
3061	Dicapthon		$C_8H_9ClNO_5PS$	2463-84-5	297.653	cry (MeOH)	53				i H_2O; s ace, tol, xyl, AcOEt
3062	Dicentrine		$C_{20}H_{21}NO_4$	517-66-8	339.386						s chl
3063	Dichlofenthion		$C_{10}H_{13}Cl_2O_3PS$	97-17-6	315.153						s ctc, CS_2
3064	Dichlofluanid		$C_9H_{11}Cl_2FN_2O_2S_2$	1085-98-9	333.229	wh pow	105.3				i H_2O; s ace, MeOH, xyl
3065	Dichloroacetaldehyde		$C_2H_2Cl_2O$	79-02-7	112.942			90.5	1.436^{25}		sl EtOH

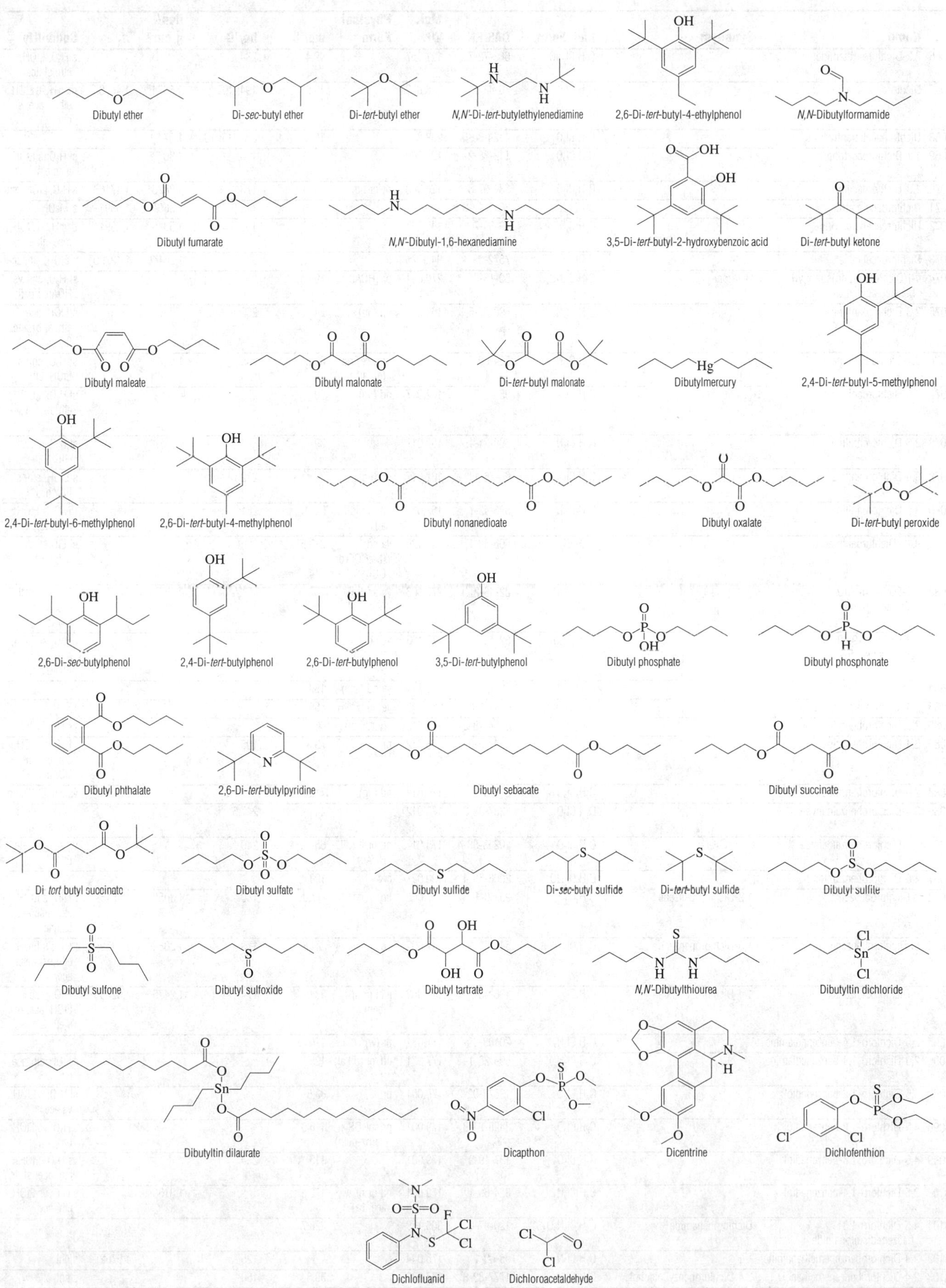
Dibutyl ether
Di-sec-butyl ether
Di-tert-butyl ether
N,N′-Di-tert-butylethylenediamine
2,6-Di-tert-butyl-4-ethylphenol
N,N-Dibutylformamide
Dibutyl fumarate
N,N′-Dibutyl-1,6-hexanediamine
3,5-Di-tert-butyl-2-hydroxybenzoic acid
Di-tert-butyl ketone
Dibutyl maleate
Dibutyl malonate
Di-tert-butyl malonate
Dibutylmercury
2,4-Di-tert-butyl-5-methylphenol
2,4-Di-tert-butyl-6-methylphenol
2,6-Di-tert-butyl-4-methylphenol
Dibutyl nonanedioate
Dibutyl oxalate
Di-tert-butyl peroxide
2,6-Di-sec-butylphenol
2,4-Di-tert-butylphenol
2,6-Di-tert-butylphenol
3,5-Di-tert-butylphenol
Dibutyl phosphate
Dibutyl phosphonate
Dibutyl phthalate
2,6-Di-tert-butylpyridine
Dibutyl sebacate
Dibutyl succinate
Di-tert-butyl succinate
Dibutyl sulfate
Dibutyl sulfide
Di-sec-butyl sulfide
Di-tert-butyl sulfide
Dibutyl sulfite
Dibutyl sulfone
Dibutyl sulfoxide
Dibutyl tartrate
N,N′-Dibutylthiourea
Dibutyltin dichloride
Dibutyltin dilaurate
Dicapthon
Dicentrine
Dichlofenthion
Dichlofluanid
Dichloroacetaldehyde

No.	Name	Synonym	Mol. Form.	CAS RN	Mol. Wt.	Physical Form	mp/°C	bp/°C	den/ g cm^{-3}	n_D	Solubility
3066	2,2-Dichloroacetamide		$C_2H_3Cl_2NO$	683-72-7	127.957		99.4	234			s H_2O, EtOH, eth; sl ace
3067	Dichloroacetic acid		$C_2H_2Cl_2O_2$	79-43-6	128.942		13.5	194; 102^{20}	1.5634^{20}	1.4658^{20}	msc H_2O, EtOH, eth; s ace; sl ctc
3068	Dichloroacetic anhydride		$C_4H_2Cl_4O_3$	4124-30-5	239.869		18.0	dec 215; 100^{10}	1.574^{24}		
3069	1,1-Dichloroacetone		$C_3H_4Cl_2O$	513-88-2	126.969			120	1.304^{18}		sl H_2O; s EtOH; msc eth
3070	1,3-Dichloroacetone		$C_3H_4Cl_2O$	534-07-6	126.969	pr or nd	45	173.4	1.3826^{46}	1.4716^{40}	s H_2O, EtOH, eth
3071	Dichloroacetonitrile		C_2HCl_2N	3018-12-0	109.942			112.5	1.369^{20}	1.4391^{25}	s MeOH
3072	Dichloroacetyl chloride		C_2HCl_3O	79-36-7	147.387			108	1.5315^{16}	1.4591^{20}	dec H_2O, EtOH; msc eth
3073	Dichloroacetylene		C_2Cl_2	7572-29-4	94.927	liq	-66	33	1.261^{20}	1.42790^{20}	s EtOH, eth, ace
3074	4-[(Dichloroamino)sulfonyl] benzoic acid	Halazone	$C_7H_5Cl_2NO_4S$	80-13-7	270.091	pr (HOAc)	195 dec				sl H_2O, chl; vs HOAc; i peth
3075	2,3-Dichloroaniline		$C_6H_5Cl_2N$	608-27-5	162.017	nd (lig)	24	252			s EtOH, ace; vs eth; sl bz, ctc, lig
3076	2,4-Dichloroaniline		$C_6H_5Cl_2N$	554-00-7	162.017	pr (ace) nd (dil al) (lig)	63.5	245	1.567^{20}		sl H_2O, chl; s EtOH, eth
3077	2,5-Dichloroaniline		$C_6H_5Cl_2N$	95-82-9	162.017	nd (lig)	50	251			sl H_2O; s EtOH, eth, bz, chl, CS_2
3078	2,6-Dichloroaniline		$C_6H_5Cl_2N$	608-31-1	162.017		39				sl H_2O; s EtOH, eth
3079	3,4-Dichloroaniline		$C_6H_5Cl_2N$	95-76-1	162.017	nd (lig)	72	272			s EtOH, eth; sl bz, chl
3080	3,5-Dichloroaniline		$C_6H_5Cl_2N$	626-43-7	162.017	nd (lig, dil al)	52	261			i H_2O; s EtOH, eth, ctc, lig
3081	9,10-Dichloroanthracene		$C_{14}H_8Cl_2$	605-48-1	247.120	ye nd (MeCOEt or CCl_4)	213.5				sl EtOH, eth, chl; s bz
3082	1,5-Dichloro-9,10-anthracenedione		$C_{14}H_6Cl_2O_2$	82-46-2	277.103	ye nd (to)	252				i H_2O; sl EtOH, ace; s bz, HOAc
3083	1,8-Dichloro-9,10-anthracenedione		$C_{14}H_6Cl_2O_2$	82-43-9	277.103	ye nd (HOAc)	202.5				i H_2O; sl EtOH; s bz, tol, $PhNO_2$
3084	*trans*-4,4'-Dichloroazobenzene		$C_{12}H_8Cl_2N_2$	1602-00-2	251.111	ye nd (ace)	189				
3085	4,4'-Dichloroazoxybenzene		$C_{12}H_8Cl_2N_2O$	614-26-6	267.110	ye nd (EtOH)	158				
3086	2,3-Dichlorobenzaldehyde		$C_7H_4Cl_2O$	6334-18-5	175.012	cry (dil al)	66				vs eth, EtOH
3087	2,4-Dichlorobenzaldehyde		$C_7H_4Cl_2O$	874-42-0	175.012	pr	73.3	105^{15}			i H_2O; s EtOH, eth, bz, chl, HOAc
3088	2,6-Dichlorobenzaldehyde		$C_7H_4Cl_2O$	83-38-5	175.012	nd (lig)	71.8				vs eth, EtOH, lig
3089	3,4-Dichlorobenzaldehyde		$C_7H_4Cl_2O$	6287-38-3	175.012		44	247.5			i H_2O; s EtOH, eth; sl ctc
3090	3,5-Dichlorobenzaldehyde		$C_7H_4Cl_2O$	10203-08-4	175.012	nd or lf (dil HOAc)	65	240			vs ace, bz, eth, EtOH
3091	2,6-Dichlorobenzamide		$C_7H_5Cl_2NO$	2008-58-4	190.027	cry	198				
3092	*o*-Dichlorobenzene	1,2-Dichlorobenzene	$C_6H_4Cl_2$	95-50-1	147.002	liq	-17.0	180	1.3059^{20}	1.5515^{20}	i H_2O; s EtOH, eth; msc ace, bz, ctc
3093	*m*-Dichlorobenzene	1,3-Dichlorobenzene	$C_6H_4Cl_2$	541-73-1	147.002	liq	-24.8	173	1.2884^{20}	1.5459^{20}	i H_2O; s EtOH, eth, bz; msc ace
3094	*p*-Dichlorobenzene	1,4-Dichlorobenzene	$C_6H_4Cl_2$	106-46-7	147.002	mcl pr, lf (ace)	53.09	174	1.2475^{55}	1.5285^{20}	i H_2O; msc EtOH, ace, bz; s eth, ctc
3095	2,5-Dichloro-1,4-benzenediamine		$C_6H_6Cl_2N_2$	20103-09-7	177.031	pr (w)	170				
3096	2,6-Dichloro-1,4-benzenediamine		$C_6H_6Cl_2N_2$	609-20-1	177.031	nd, pr (dil al)	125				s EtOH, eth, ace, bz
3097	3,5-Dichloro-1,2-benzenediol		$C_6H_4Cl_2O_2$	13673-92-2	179.001	pr	83.5				sl H_2O; s EtOH; vs ace
3098	4,5-Dichloro-1,2-benzenediol		$C_6H_4Cl_2O_2$	3428-24-8	179.001	pr(chl-CS_2) nd(bz-peth)	116.5				s H_2O; vs EtOH, bz
3099	4,6-Dichloro-1,3-benzenediol		$C_6H_4Cl_2O_2$	137-19-9	179.001		113	254			vs H_2O, EtOH, eth, ace; sl lig
3100	2,5-Dichloro-1,4-benzenediol		$C_6H_4Cl_2O_2$	824-69-1	179.001	nd or pr w, ace, bz)	172.5		1.8150^{24}		s H_2O; vs EtOH, eth, ace
3101	4,5-Dichloro-1,3-benzenedisulfonamide	Dichlorphenamide	$C_6H_6Cl_2N_2O_4S_2$	120-97-8	305.159		228.7				
3102	2,4-Dichlorobenzenemethanamine		$C_7H_7Cl_2N$	95-00-1	176.044			125^{13}		1.5762^{25}	s chl
3103	2,4-Dichlorobenzenemethanol	2,4-Dichlorobenzyl alcohol	$C_7H_6Cl_2O$	1777-82-8	177.028		59.5	150^{25}			s chl
3104	*N,N*-Dichlorobenzenesulfonamide		$C_6H_5Cl_2NO_2S$	473-29-0	226.081	ye mcl or pl	76				s EtOH; sl ctc

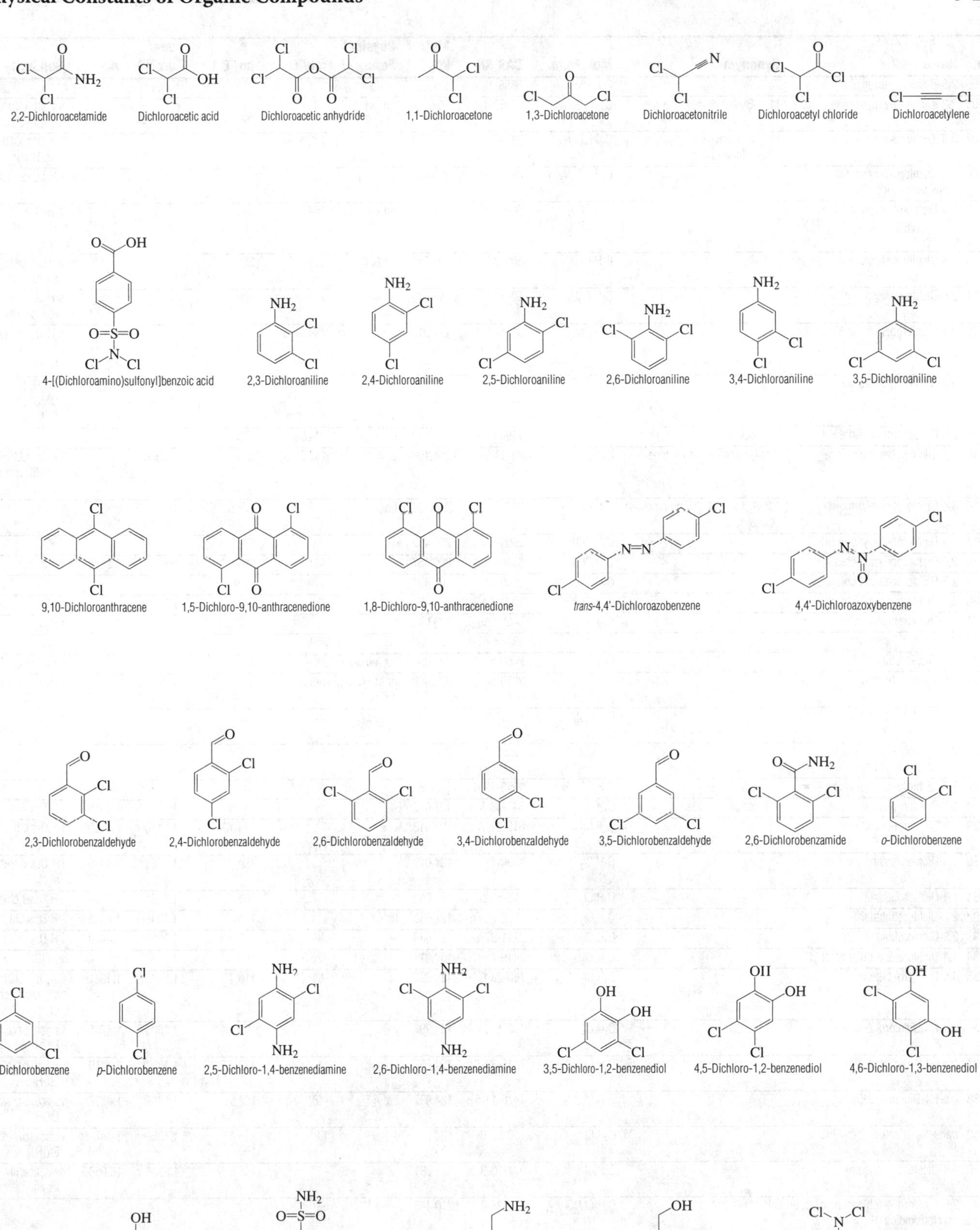

2,2-Dichloroacetamide
Dichloroacetic acid
Dichloroacetic anhydride
1,1-Dichloroacetone
1,3-Dichloroacetone
Dichloroacetonitrile
Dichloroacetyl chloride
Dichloroacetylene
4-[(Dichloroamino)sulfonyl]benzoic acid
2,3-Dichloroaniline
2,4-Dichloroaniline
2,5-Dichloroaniline
2,6-Dichloroaniline
3,4-Dichloroaniline
3,5-Dichloroaniline
9,10-Dichloroanthracene
1,5-Dichloro-9,10-anthracenedione
1,8-Dichloro-9,10-anthracenedione
trans-4,4'-Dichloroazobenzene
4,4'-Dichloroazoxybenzene
2,3-Dichlorobenzaldehyde
2,4-Dichlorobenzaldehyde
2,6-Dichlorobenzaldehyde
3,4-Dichlorobenzaldehyde
3,5-Dichlorobenzaldehyde
2,6-Dichlorobenzamide
o-Dichlorobenzene
m-Dichlorobenzene
p-Dichlorobenzene
2,5-Dichloro-1,4-benzenediamine
2,6-Dichloro-1,4-benzenediamine
3,5-Dichloro-1,2-benzenediol
4,5-Dichloro-1,2-benzenediol
4,6-Dichloro-1,3-benzenediol
2,5-Dichloro-1,4-benzenediol
4,5-Dichloro-1,3-benzenedisulfonamide
2,4-Dichlorobenzenemethanamine
2,4-Dichlorobenzenemethanol
N,N-Dichlorobenzenesulfonamide

No.	Name	Synonym	Mol. Form.	CAS RN	Mol. Wt.	Physical Form	mp/°C	bp/°C	den/g cm^{-3}	n_D	Solubility
3105	2,5-Dichlorobenzenethiol		$C_6H_4Cl_2S$	5858-18-4	179.067			115^{50}			
3106	2,2'-Dichloro-*p*-benzidine	[1,1'-Biphenyl]-4,4'-diamine, 2,2'-dichloro-	$C_{12}H_{10}Cl_2N_2$	84-68-4	253.126	nd (w), pr (al)	165				vs eth, EtOH
3107	3,3'-Dichloro-*p*-benzidine	[1,1'-Biphenyl]-4,4'-diamine, 3,3'-dichloro-	$C_{12}H_{10}Cl_2N_2$	91-94-1	253.126	nd	132.5				i H_2O; s EtOH, bz, HOAc
3108	3,3'-Dichloro-*p*-benzidine dihydrochloride		$C_{12}H_{12}Cl_4N_2$	612-83-9	326.048						i H_2O; vs EtOH
3109	2,4-Dichlorobenzoic acid		$C_7H_4Cl_2O_2$	50-84-0	191.012	nd (w or bz)	164.2	sub			s H_2O, EtOH, eth, bz, chl; sl ace
3110	2,5-Dichlorobenzoic acid		$C_7H_4Cl_2O_2$	50-79-3	191.012	nd (w)	154.4	301			sl H_2O, DMSO; s EtOH, eth
3111	2,6-Dichlorobenzoic acid		$C_7H_4Cl_2O_2$	50-30-6	191.012	nd (al), pr (w)	144	sub			s H_2O, EtOH, eth, bz, chl
3112	3,4-Dichlorobenzoic acid		$C_7H_4Cl_2O_2$	51-44-5	191.012	nd (w, al, bz)	208.5				s H_2O, eth; vs EtOH; sl DMSO
3113	3,5-Dichlorobenzoic acid		$C_7H_4Cl_2O_2$	51-36-5	191.012	nd (al, w)	188	sub			sl H_2O, lig, DMSO; s EtOH, eth
3114	2,6-Dichlorobenzonitrile	Dichlobenil	$C_7H_3Cl_2N$	1194-65-6	172.012	cry (peth)	144.5	270			
3115	4,4'-Dichlorobenzophenone	Bis(4-chlorophenyl) ketone	$C_{13}H_8Cl_2O$	90-98-2	251.108	pl (al)	147.5	353	1.4500^{20}		i H_2O; s EtOH; vs eth, chl; sl ace
3116	3,4-Dichlorobenzotrifluoride	1,2-Dichloro-4-(trifluoromethyl) benzene	$C_7H_3Cl_2F_3$	328-84-7	215.000	liq		173.5; 64^{14}	1.4729^{25}		
3117	2,3-Dichlorobenzoyl chloride		$C_7H_3Cl_3O$	2905-60-4	209.457	liq		140^{14}			
3118	2,4-Dichlorobenzoyl chloride		$C_7H_3Cl_3O$	89-75-8	209.457		16.5	150^{34}, $111^{7.5}$		1.5895^{20}	s ctc
3119	2,5-Dichlorobenzoyl chloride		$C_7H_3Cl_3O$	2905-61-5	209.457	liq		95.4^{1}			
3120	3,4-Dichlorobenzoyl chloride		$C_7H_3Cl_3O$	3024-72-4	209.457		25	242			sl ctc
3121	2,5-Dichlorobiphenyl		$C_{12}H_8Cl_2$	34883-39-1	223.098			182^{30}, 171^{15}			i H_2O
3122	2,6-Dichlorobiphenyl		$C_{12}H_8Cl_2$	33146-45-1	223.098	cry	35.5				i H_2O
3123	3,3'-Dichlorobiphenyl		$C_{12}H_8Cl_2$	2050-67-1	223.098	nd (dil al)	29	320			vs bz, eth, EtOH
3124	4,4'-Dichlorobiphenyl		$C_{12}H_8Cl_2$	2050-68-2	223.098	pr or nd (al, to-peth)	149.3	317	1.4420^{0}		i H_2O; sl EtOH, chl; s bz
3125	1,1-Dichloro-2,2-bis(*p*-chlorophenyl)ethane		$C_{14}H_{10}Cl_4$	72-54-8	320.041		109.5	193^{1}			sl chl
3126	2,2-Dichloro-1,1-bis(4-chlorophenyl)ethene		$C_{14}H_8Cl_4$	72-55-9	318.026		89				
3127	2,3-Dichloro-1,3-butadiene		$C_4H_4Cl_2$	1653-19-6	122.981			98	1.1829^{20}	1.4890^{20}	vs chl
3128	1,1-Dichlorobutane	Butylidene chloride	$C_4H_8Cl_2$	541-33-3	127.013			113.8	1.0863^{20}	1.4355^{20}	i H_2O; s chl
3129	1,2-Dichlorobutane		$C_4H_8Cl_2$	616-21-7	127.013			124.1	1.1116^{25}	1.4450^{20}	i H_2O; s eth, chl; sl ctc
3130	1,3-Dichlorobutane		$C_4H_8Cl_2$	1190-22-3	127.013			134	1.1158^{20}	1.4445^{20}	i H_2O; s eth, chl; sl ctc
3131	1,4-Dichlorobutane		$C_4H_8Cl_2$	110-56-5	127.013	liq	-37.3	161	1.1331^{25}	1.4522^{25}	i H_2O; vs chl
3132	2,2-Dichlorobutane		$C_4H_8Cl_2$	4279-22-5	127.013	liq	-74	104	1.1048^{25}	1.4295	i H_2O; s chl
3133	2,3-Dichlorobutane, (±)		$C_4H_8Cl_2$	2211-67-8	127.013	liq	-80	119; 53^{80}	1.105^{25}	1.4409^{25}	i H_2O
3134	1,4-Dichloro-2,3-butanediol		$C_4H_8Cl_2O_2$	2419-73-0	159.012		126.5	150^{30}			vs EtOH
3135	3,4-Dichloro-1-butene		$C_4H_6Cl_2$	760-23-6	124.997	liq	-61	116	1.1170^{20}	1.4641^{20}	i H_2O; s EtOH, eth, ctc; vs chl, bz
3136	*cis*-1,3-Dichloro-2-butene		$C_4H_6Cl_2$	10075-38-4	124.997			130; 34^{20}	1.1605^{20}	1.4735^{20}	vs ace, bz, eth, EtOH
3137	*trans*-1,3-Dichloro-2-butene		$C_4H_6Cl_2$	7415-31-8	124.997			132; 53^{50}	1.160^{20}	1.4719^{20}	vs ace, bz, eth, EtOH
3138	*cis*-1,4-Dichloro-2-butene		$C_4H_6Cl_2$	1476-11-5	124.997	liq	-48	152.5	1.188^{25}	1.4887^{25}	vs ace, bz, eth, EtOH
3139	*trans*-1,4-Dichloro-2-butene		$C_4H_6Cl_2$	110-57-6	124.997		1.0	155.4	1.183^{25}	1.4871^{25}	vs ace, bz, eth, EtOH
3140	1,4-Dichloro-2-butyne		$C_4H_4Cl_2$	821-10-3	122.981			165.5	1.258^{20}	1.5058^{20}	s eth, ace; sl ctc; vs chl
3141	2,6-Dichloro-4-(chloroimino)-2,5-cyclohexadien-1-one	Gibbs' reagent	$C_6H_2Cl_3NO$	101-38-2	210.445		66				
3142	1,2-Dichloro-4-(chloromethyl) benzene		$C_7H_5Cl_3$	102-47-6	195.474		37.5	241			i H_2O; s EtOH, ctc
3143	2,4-Dichloro-1-(chloromethyl) benzene		$C_7H_5Cl_3$	94-99-5	195.474			120^{13}			
3144	Dichloro(chloromethyl) methylsilane		$C_2H_5Cl_3Si$	1558-33-4	163.506			121.5	1.2858^{20}	1.4500^{20}	
3145	Dichloro(2-chlorovinyl)arsine		$C_2H_2AsCl_3$	541-25-3	207.318	liq	0.1	190	1.888^{20}		
3146	2,5-Dichloro-2,5-cyclohexadiene-1,4-dione		$C_6H_2Cl_2O_2$	615-93-0	176.985	pa ye mcl pr (al)	162.3				i H_2O; sl EtOH; s eth, chl

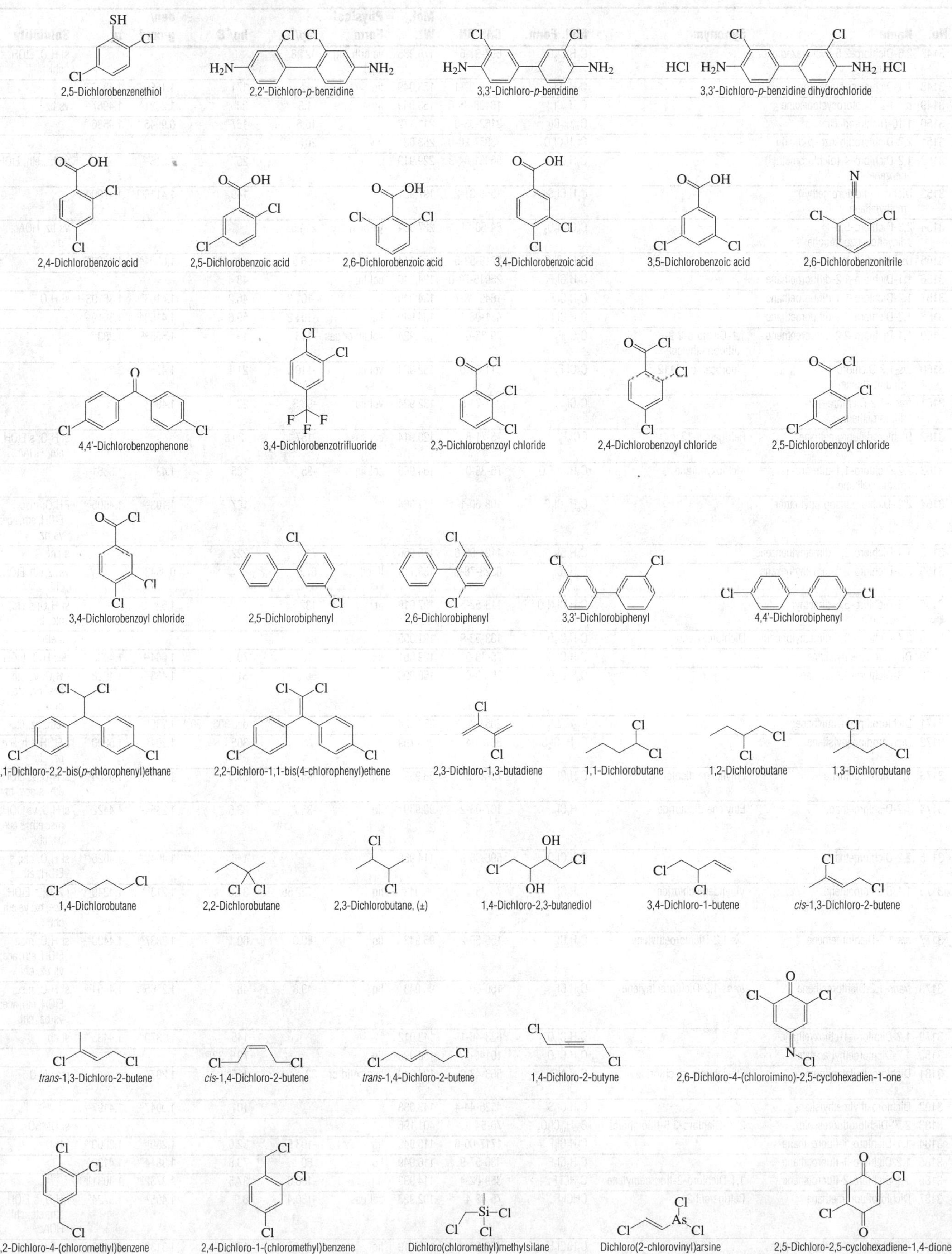
2,5-Dichlorobenzenethiol
2,2'-Dichloro-p-benzidine
3,3'-Dichloro-p-benzidine
3,3'-Dichloro-p-benzidine dihydrochloride
2,4-Dichlorobenzoic acid
2,5-Dichlorobenzoic acid
2,6-Dichlorobenzoic acid
3,4-Dichlorobenzoic acid
3,5-Dichlorobenzoic acid
2,6-Dichlorobenzonitrile
4,4'-Dichlorobenzophenone
3,4-Dichlorobenzotrifluoride
2,3-Dichlorobenzoyl chloride
2,4-Dichlorobenzoyl chloride
2,5-Dichlorobenzoyl chloride
3,4-Dichlorobenzoyl chloride
2,5-Dichlorobiphenyl
2,6-Dichlorobiphenyl
3,3'-Dichlorobiphenyl
4,4'-Dichlorobiphenyl
1,1-Dichloro-2,2-bis(p-chlorophenyl)ethane
2,2-Dichloro-1,1-bis(4-chlorophenyl)ethene
2,3-Dichloro-1,3-butadiene
1,1-Dichlorobutane
1,2-Dichlorobutane
1,3-Dichlorobutane
1,4-Dichlorobutane
2,2-Dichlorobutane
2,3-Dichlorobutane, (±)
1,4-Dichloro-2,3-butanediol
3,4-Dichloro-1-butene
cis-1,3-Dichloro-2-butene
trans-1,3-Dichloro-2-butene
cis-1,4-Dichloro-2-butene
trans-1,4-Dichloro-2-butene
1,4-Dichloro-2-butyne
2,6-Dichloro-4-(chloroimino)-2,5-cyclohexadien-1-one
1,2-Dichloro-4-(chloromethyl)benzene
2,4-Dichloro-1-(chloromethyl)benzene
Dichloro(chloromethyl)methylsilane
Dichloro(2-chlorovinyl)arsine
2,5-Dichloro-2,5-cyclohexadiene-1,4-dione

No.	Name	Synonym	Mol. Form.	CAS RN	Mol. Wt.	Physical Form	mp/°C	bp/°C	den/ g cm^{-3}	n_D	Solubility
3147	2,6-Dichloro-2,5-cyclohexadiene-1,4-dione		$C_6H_2Cl_2O_2$	697-91-6	176.985	ye orth (lig, bz)	121.8				sl H_2O, EtOH; s chl
3148	1,1-Dichlorocyclohexane		$C_6H_{10}Cl_2$	2108-92-1	153.049	liq	-47	171	1.1559^{20}	1.4803^{20}	
3149	*cis*-1,2-Dichlorocyclohexane		$C_6H_{10}Cl_2$	10498-35-8	153.049	liq	-1.5	206.9	1.2021^{20}	1.4967^{20}	vs bz
3150	1,10-Dichlorodecane		$C_{10}H_{20}Cl_2$	2162-98-3	211.172		15.6	167^{28}	0.9945^{25}	1.4586^{25}	
3151	2,7-Dichlorodibenzo-*p*-dioxin		$C_{12}H_6Cl_2O_2$	33857-26-0	253.081	cry	201				
3152	1,2-Dichloro-4-(dichloromethyl) benzene		$C_7H_4Cl_4$	56961-84-3	229.919			257	1.515^{22}		vs bz, eth, EtOH
3153	Dichloro(dichloromethyl) methylsilane		$C_2H_4Cl_4Si$	1558-31-2	197.951			149	1.4116^{20}	1.4700^{20}	
3154	2,3-Dichloro-5,6-dicyanobenzoquinone		$C_8Cl_2N_2O_2$	84-58-2	227.004	ye-oran cry	214.5				vs bz, HOAc, diox
3155	Dichlorodiethylsilane		$C_4H_{10}Cl_2Si$	1719-53-5	157.114		-96.5	dec 129	1.0504^{20}	1.4309^{20}	
3156	1,1-Dichloro-1,2-difluoroethane		$C_2H_2Cl_2F_2$	25915-78-0	134.940	col liq		48.4			
3157	1,2-Dichloro-1,1-difluoroethane		$C_2H_2Cl_2F_2$	1649-08-7	134.940	liq	-101.2	46.2	1.4163^{20}	1.36193^{20}	sl H_2O
3158	1,2-Dichloro-1,2-difluoroethane		$C_2H_2Cl_2F_2$	431-06-1	134.940	liq	-101.2	59.6	1.4163^{20}	1.3619^{20}	
3159	1,1-Dichloro-2,2-difluoroethene	1,1-Dichloro-2,2-difluoroethylene	$C_2Cl_2F_2$	79-35-6	132.924	vol liq or gas	-116	19	1.555^{-20}	1.383^{-20}	
3160	*cis*-1,2-Dichloro-1,2-difluoroethene	Fluorocarbon 1112	$C_2Cl_2F_2$	311-81-9	132.924	vol liq	-119.6	21.1	1.495^{0}		
3161	*trans*-1,2-Dichloro-1,2-difluoroethene		$C_2Cl_2F_2$	381-71-5	132.924	vol liq	-93.3	22	1.494^{0}		
3162	Dichlorodifluoromethane	Refrigerant 12	CCl_2F_2	75-71-8	120.914	col gas	-157.1	-29.8			sl H_2O; s EtOH, eth, HOAc
3163	2,2-Dichloro-1,1-difluoro-1-methoxyethane	Methoxyflurane	$C_3H_4Cl_2F_2O$	76-38-0	164.966	col liq	-35	105	1.43^{20}	1.3861^{20}	
3164	2,2'-Dichlorodiisopropyl ether		$C_6H_{12}Cl_2O$	108-60-1	171.064			187	1.103^{20}	1.4505^{20}	i H_2O; msc EtOH, eth, ace; vs bz
3165	1,4-Dichloro-2,5-dimethylbenzene		$C_8H_8Cl_2$	1124-05-6	175.056		71	222			s chl
3166	2,5-Dichloro-2,5-dimethylhexane		$C_8H_{16}Cl_2$	6223-78-5	183.119	lf, nd	67.5		0.9543^{20}		vs bz, eth, EtOH, chl
3167	1,3-Dichloro-5,5-dimethyl hydantoin		$C_5H_6Cl_2N_2O_2$	118-52-5	197.019	pr	132		1.5^{20}		sl H_2O; s chl, ctc, bz
3168	2,4-Dichloro-3,5-dimethylphenol	Dichloroxylenol	$C_8H_8Cl_2O$	133-53-9	191.055		83				vs eth
3169	Dichlorodimethylsilane		$C_2H_6Cl_2Si$	75-78-5	129.061	liq	-16	70.3	1.064^{25}	1.4038^{20}	dec H_2O, EtOH
3170	2,3-Dichloro-1,4-dioxane		$C_4H_6Cl_2O_2$	95-59-0	156.996		30	81^{10}	1.468^{20}	1.4928^{20}	i H_2O; vs eth, ace, bz, ctc, diox
3171	Dichlorodiphenylmethane		$C_{13}H_{10}Cl_2$	2051-90-3	237.124			dec 305; 190^{21}	1.235^{18}		s eth, bz, ctc
3172	Dichlorodiphenylsilane		$C_{12}H_{10}Cl_2Si$	80-10-4	253.199			305	1.204^{25}	1.5800^{20}	s EtOH, eth, ace, bz, ctc
3173	1,1-Dichloroethane	Ethylidene dichloride	$C_2H_4Cl_2$	75-34-3	98.959	liq	-96.9	57.3	1.1757^{20}	1.4164^{20}	sl H_2O; vs EtOH, eth; s ace, bz
3174	1,2-Dichloroethane	Ethylene dichloride	$C_2H_4Cl_2$	107-06-2	98.959	liq	-35.7	83.5	1.2454^{25}	1.4422^{25}	sl H_2O; vs EtOH; msc eth; s ace, bz, chl
3175	2,2-Dichloroethanol		$C_2H_4Cl_2O$	598-38-9	114.958			146	1.4040^{25}	1.4626^{25}	sl H_2O, ctc; s EtOH, eth
3176	1,1-Dichloroethene	Vinylidene chloride	$C_2H_2Cl_2$	75-35-4	96.943	liq	-122.56	31.6	1.213^{20}	1.4249^{20}	i H_2O; s EtOH, ace, bz; vs eth, chl
3177	*cis*-1,2-Dichloroethene	*cis*-1,2-Dichloroethylene	$C_2H_2Cl_2$	156-59-2	96.943	liq	-80.0	60.1	1.2837^{20}	1.4490^{20}	sl H_2O; msc EtOH, eth, ace; vs bz, chl
3178	*trans*-1,2-Dichloroethene	*trans*-1,2-Dichloroethylene	$C_2H_2Cl_2$	156-60-5	96.943	liq	-49.8	48.7	1.2565^{20}	1.4454^{20}	sl H_2O; msc EtOH, eth, ace; vs bz, chl
3179	1,2-Dichloro-1-ethoxyethane		$C_4H_8Cl_2O$	623-46-1	143.012			145	1.1370^{20}	1.4435^{20}	sl chl
3180	1,2-Dichloroethyl acetate		$C_4H_6Cl_2O_2$	10140-87-1	156.996	liq		79^{33}, 32^{10}			
3181	Dichloroethylaluminum	Ethylaluminum chloride	$C_2H_5AlCl_2$	563-43-9	126.949	hyg solid or liq	32	115^{50}	1.207		reac H_2O
3182	Dichloroethylmethylsilane		$C_3H_8Cl_2Si$	4525-44-4	143.088			101	1.0047^{20}	1.4197^{20}	
3183	2',7'-Dichlorofluorescein	2',7'-Dichloro-3,6-fluorandiol	$C_{20}H_{10}Cl_2O_5$	76-54-0	401.196						sl DMSO
3184	1,1-Dichloro-1-fluoroethane		$C_2H_3Cl_2F$	1717-00-6	116.949	liq	-103.5	32.0	1.250^{10}	1.3600^{10}	i H_2O
3185	1,2-Dichloro-1-fluoroethane		$C_2H_3Cl_2F$	430-57-9	116.949	liq	-60	73.8	1.3814^{20}	1.4132^{20}	
3186	1,1-Dichloro-2-fluoroethene	1,1-Dichloro-2-fluoroethylene	C_2HCl_2F	359-02-4	114.933	liq	-108.8	37.5	1.3732^{16}	1.4031^{16}	
3187	Dichlorofluoromethane	Refrigerant 21	$CHCl_2F$	75-43-4	102.923	col gas	-130.4	8.9	1.405^{9}	1.3724^{9}	i H_2O; s EtOH, eth, ctc, chl, HOAc
3188	(Dichlorofluoromethyl)benzene		$C_7H_5Cl_2F$	498-67-9	179.019	liq	-26.8	179	1.3138^{11}	1.5180^{11}	vs EtOH
3189	1,1-Dichloro-2-fluoropropene		$C_3H_3Cl_2F$	430-95-5	128.960			78	1.3026^{25}	1.4196^{25}	
3190	1,7-Dichloroheptane		$C_7H_{14}Cl_2$	821-76-1	169.092			124^{35}	1.0408^{25}	1.4565^{25}	

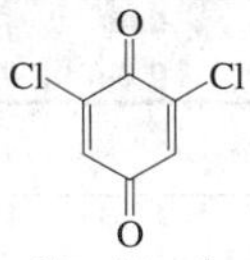

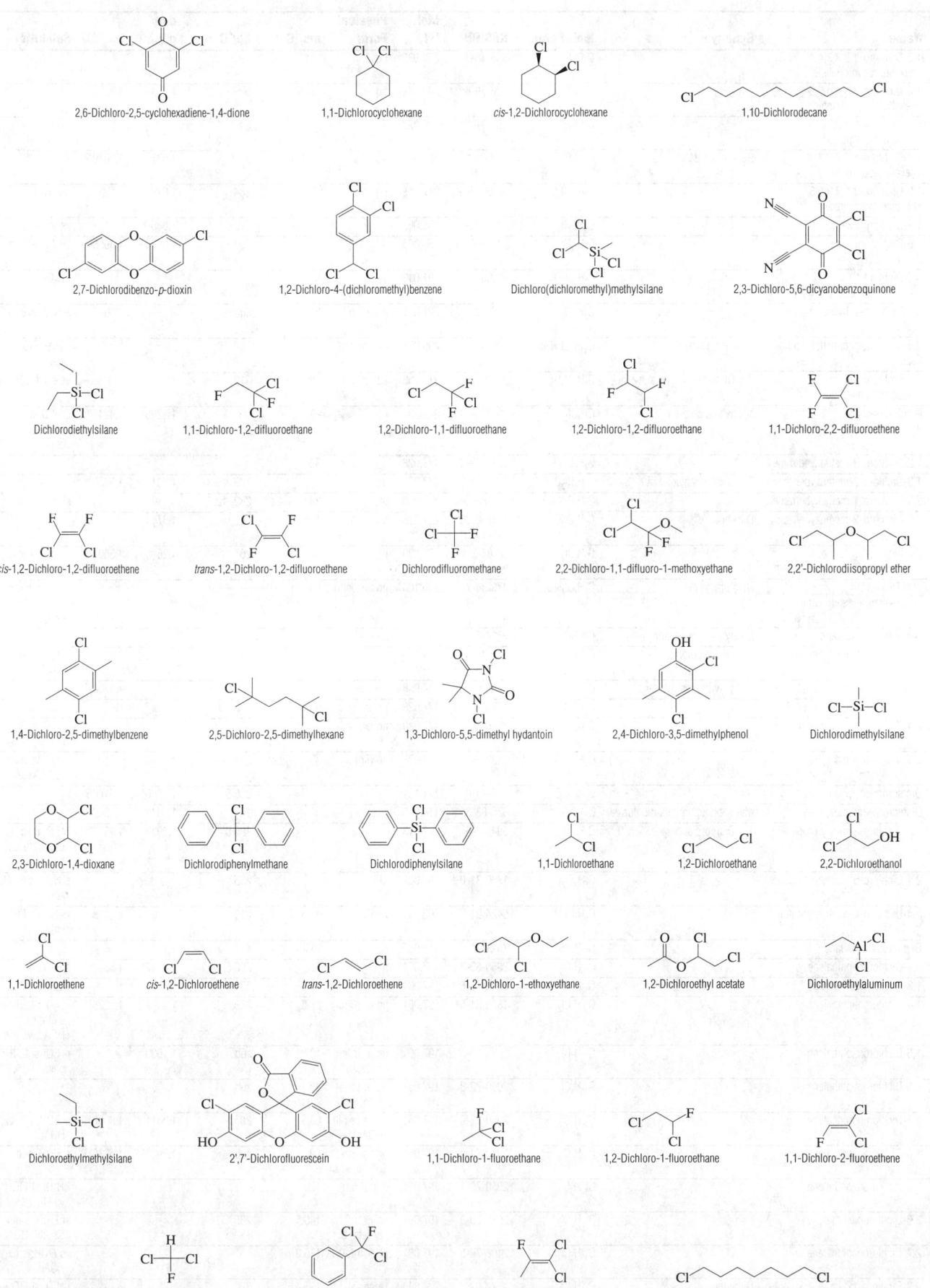

Cl
O
Cl
O
2,6-Dichloro-2,5-cyclohexadiene-1,4-dione 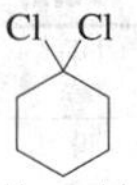
Cl Cl
1,1-Dichlorocyclohexane 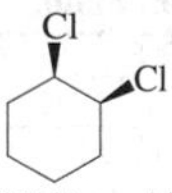
Cl
Cl
cis-1,2-Dichlorocyclohexane 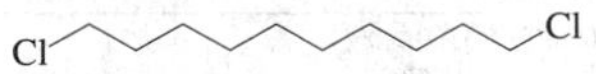
Cl
Cl
1,10-Dichlorodecane 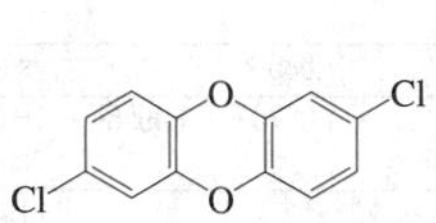
O
Cl
Cl
O
2,7-Dichlorodibenzo-*p*-dioxin 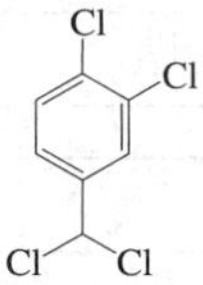
Cl
Cl
Cl Cl
1,2-Dichloro-4-(dichloromethyl)benzene 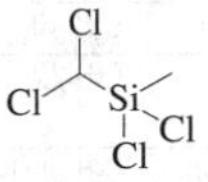
Cl
Cl
Si
Cl
Cl
Dichloro(dichloromethyl)methylsilane 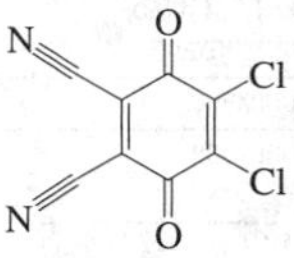
N
O
Cl
Cl
N
O
2,3-Dichloro-5,6-dicyanobenzoquinone 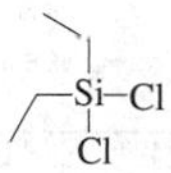
Si–Cl
Cl
Dichlorodiethylsilane 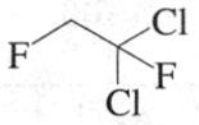
F
Cl
F
Cl
1,1-Dichloro-1,2-difluoroethane 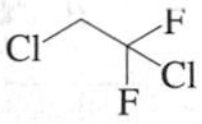
Cl
F
Cl
F
1,2-Dichloro-1,1-difluoroethane 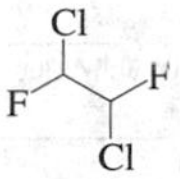
Cl
F
F
Cl
1,2-Dichloro-1,2-difluoroethane 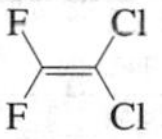
F Cl
F Cl
1,1-Dichloro-2,2-difluoroethene 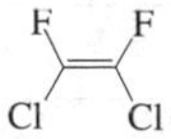
F F
Cl Cl
cis-1,2-Dichloro-1,2-difluoroethene 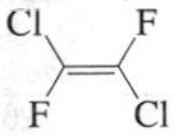
Cl F
F Cl
trans-1,2-Dichloro-1,2-difluoroethene 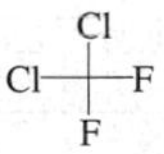
Cl
Cl–F
F
Dichlorodifluoromethane 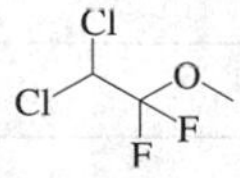
Cl
Cl
O
F
F
2,2-Dichloro-1,1-difluoro-1-methoxyethane 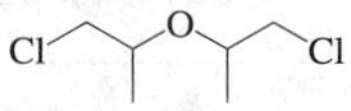
Cl
O
Cl
2,2'-Dichlorodiisopropyl ether 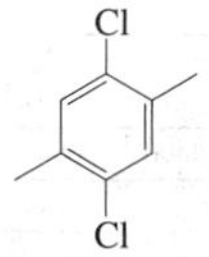
Cl
Cl
1,4-Dichloro-2,5-dimethylbenzene 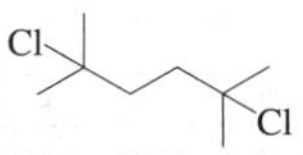
Cl
Cl
2,5-Dichloro-2,5-dimethylhexane 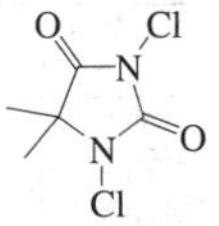
O
N
Cl
N
O
Cl
1,3-Dichloro-5,5-dimethyl hydantoin 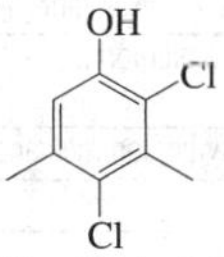
OH
Cl
Cl
2,4-Dichloro-3,5-dimethylphenol 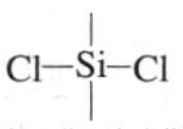
Cl–Si–Cl
Dichlorodimethylsilane 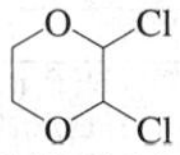
O
Cl
O
Cl
2,3-Dichloro-1,4-dioxane 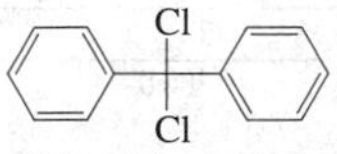
Cl
Cl
Dichlorodiphenylmethane 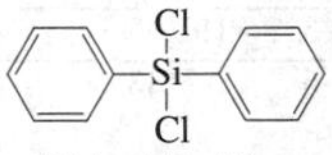
Cl
Si
Cl
Dichlorodiphenylsilane 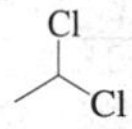
Cl
Cl
1,1-Dichloroethane 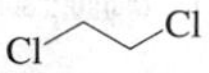
Cl
Cl
1,2-Dichloroethane 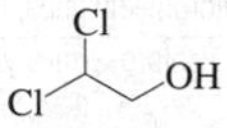
Cl
Cl
OH
2,2-Dichloroethanol 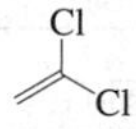
Cl
Cl
1,1-Dichloroethene
Cl Cl
cis-1,2-Dichloroethene
Cl
Cl
trans-1,2-Dichloroethene 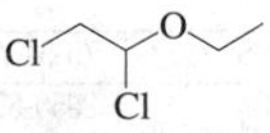
Cl
O
Cl
1,2-Dichloro-1-ethoxyethane 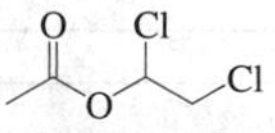
O
Cl
O
Cl
1,2-Dichloroethyl acetate 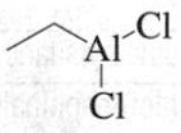
Al
Cl
Cl
Dichloroethylaluminum 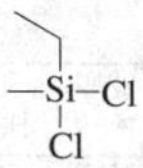
Si–Cl
Cl
Dichloroethylmethylsilane 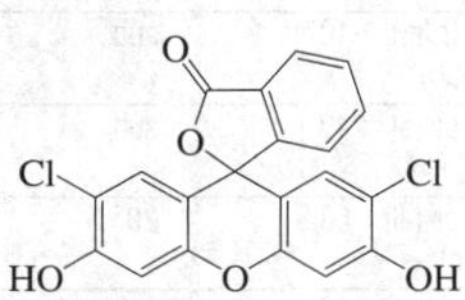
O
O
Cl
Cl
HO
O
OH
2',7'-Dichlorofluorescein 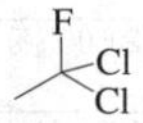
F
Cl
Cl
1,1-Dichloro-1-fluoroethane 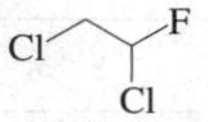
Cl
F
Cl
1,2-Dichloro-1-fluoroethane 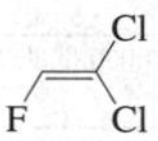
Cl
F
Cl
1,1-Dichloro-2-fluoroethene 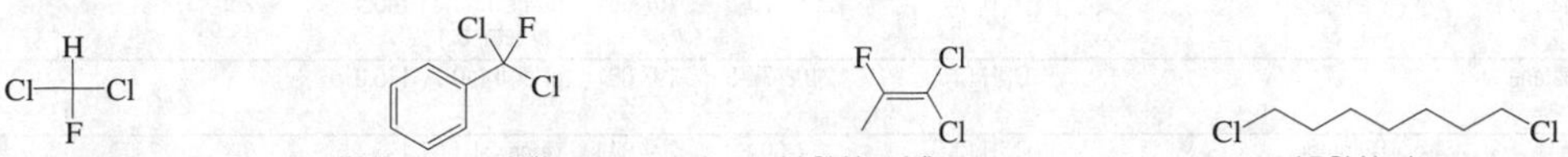
H
Cl–Cl
F
Dichlorofluoromethane
Cl F
Cl
(Dichlorofluoromethyl)benzene
F
Cl
Cl
1,1-Dichloro-2-fluoropropene
Cl
Cl
1,7-Dichloroheptane

No.	Name	Synonym	Mol. Form.	CAS RN	Mol. Wt.	Physical Form	mp/°C	bp/°C	den/ g cm^{-3}	n_D	Solubility
3191	1,2-Dichloro-1,2,3,3,4,4-hexafluorocyclobutane		$C_4Cl_2F_6$	356-18-3	232.939	liq	-24.2	59.5			
3192	1,2-Dichloro-3,3,4,4,5,5-hexafluorocyclopentene		$C_5Cl_2F_6$	706-79-6	244.949	liq	-105.8	90.7	1.6546^{20}	1.3676^{20}	
3193	1,2-Dichloro-1,1,2,3,3,3-hexafluoropropane		$C_3Cl_2F_6$	661-97-2	220.928			34.1			i H_2O
3194	1,3-Dichloro-1,1,2,2,3,3-hexafluoropropane	Refrigerant 216	$C_3Cl_2F_6$	662-01-1	220.928	liq	-125.4	35.7	1.573^{20}	1.3030^{20}	
3195	1,5-Dichloro-1,1,3,3,5,5-hexamethyltrisiloxane		$C_6H_{18}Cl_2O_2Si_3$	3582-71-6	277.369	liq	-53	184	1.018^{20}		dec H_2O
3196	1,2-Dichlorohexane		$C_6H_{12}Cl_2$	2162-92-7	155.065			173	1.085^{15}		vs eth, chl
3197	1,6-Dichlorohexane		$C_6H_{12}Cl_2$	2163-00-0	155.065			204	1.0676^{25}	1.4555^{25}	i H_2O; s eth, ctc, chl
3198	3,5-Dichloro-2-hydroxybenzaldehyde		$C_7H_4Cl_2O_2$	90-60-8	191.012	ye orth (HOAc)	95				i H_2O
3199	3,5-Dichloro-2-hydroxybenzoic acid		$C_7H_4Cl_2O_3$	320-72-9	207.011	nd (dil al) orth pr	220.5	sub			sl H_2O; vs EtOH, eth
3200	2,6-Dichloroindophenol, sodium salt	Tillman's reagent	$C_{12}H_6Cl_2NNaO_2$	620-45-1	290.078	dk grn cry					s H_2O, EtOH, ace
3201	5,6-Dichloro-1,3-isobenzofurandione	4,5-Dichlorophthalic anhydride	$C_8H_2Cl_2O_3$	942-06-3	217.006	tab or pr (to)	188	313			vs eth, EtOH, tol
3202	Dichloromethane	Methylene chloride	CH_2Cl_2	75-09-2	84.933	liq	-97.2	40	1.3266^{20}	1.4242^{20}	sl H_2O; msc EtOH, eth; s ctc
3203	1,2-Dichloro-3-methoxybenzene		$C_7H_6Cl_2O$	1984-59-4	177.028		32				
3204	1,3-Dichloro-2-methoxybenzene	2,6-Dichloroanisole	$C_7H_6Cl_2O$	1984-65-2	177.028	liq	10	105^{20}	1.291	1.5430^{20}	
3205	2,4-Dichloro-1-methoxybenzene		$C_7H_6Cl_2O$	553-82-2	177.028	pr	28.5	232; 125^{10}			sl chl
3206	3,6-Dichloro-2-methoxybenzoic acid	Dicamba	$C_8H_6Cl_2O_3$	1918-00-9	221.038	cry (pent)	115		1.57^{25}		
3207	(Dichloromethyl)benzene	Benzal chloride	$C_7H_6Cl_2$	98-87-3	161.029	liq	-17	205	1.26^{25}	1.5502^{20}	i H_2O; vs eth, EtOH
3208	*N,N*-Dichloro-4-methylbenzenesulfonamide	Dichloramine-T	$C_7H_7Cl_2NO_2S$	473-34-7	240.108	pr(chl-peth)	83				i H_2O; s EtOH, eth, bz, ctc, HOAc
3209	Dichloromethylborane	Methyldichloroborane	CH_3BCl_2	7318-78-7	96.752	col gas		11			
3210	2,3-Dichloro-2-methylbutane	Amylene dichloride	$C_5H_{10}Cl_2$	507-45-9	141.038			129	1.0696^{15}	1.4450^{18}	i H_2O; vs eth, EtOH
3211	1,1-Dichloromethyl methyl ether	Methoxydichloromethane	$C_2H_4Cl_2O$	4885-02-3	114.958			85	1.271^{25}	1.4300^{20}	
3212	2,4-Dichloro-3-methylphenol		$C_7H_6Cl_2O$	17788-00-0	177.028	pr (peth)	58	236; 77^4			vs eth, chl
3213	2,4-Dichloro-6-methylphenol		$C_7H_6Cl_2O$	1570-65-6	177.028	nd (w, peth)	55				sl H_2O; vs EtOH, eth, chl, CS_2
3214	2,6-Dichloro-4-methylphenol		$C_7H_6Cl_2O$	2432-12-4	177.028	nd (lig)	39	231; 138^{28}			i H_2O; vs eth, EtOH, HOAc
3215	Dichloromethylphenylsilane		$C_7H_8Cl_2Si$	149-74-6	191.131			206.5	1.1866^{20}	1.5180^{20}	
3216	Dichloromethylphosphine	Methylphosphonous dichloride	CH_3Cl_2P	676-83-5	116.915			12^{50}	1.304^{20}	1.4940^{20}	
3217	1,2-Dichloro-2-methylpropane	1,2-Dichloroisobutane	$C_4H_8Cl_2$	594-37-6	127.013			106.5	1.093^{20}	1.4370^{20}	i H_2O; msc EtOH, eth, ace, bz, ctc
3218	2,4-Dichloro-5-methylpyrimidine		$C_5H_4Cl_2N_2$	1780-31-0	163.004	pl (al)	26	235			sl H_2O; vs EtOH, eth, bz, chl
3219	2,4-Dichloro-6-methylpyrimidine		$C_5H_4Cl_2N_2$	5424-21-5	163.004	nd (lig)	46.5	219			vs bz, eth, EtOH, chl
3220	Dichloromethylsilane		CH_4Cl_2Si	75-54-7	115.035	liq	-93	41	1.105^{25}		
3221	1,2-Dichloronaphthalene		$C_{10}H_6Cl_2$	2050-69-3	197.061	pl (al)	36	296.5	1.3147^{49}	1.5338^{49}	s EtOH, eth
3222	1,3-Dichloronaphthalene		$C_{10}H_6Cl_2$	2198-75-6	197.061	nd or pr (al)	62.3	291			s EtOH
3223	1,4-Dichloronaphthalene		$C_{10}H_6Cl_2$	1825-31-6	197.061	nd or pr (al, ace)	67.5	288; 147^{12}	1.2997^{76}	1.6228^{76}	i H_2O; sl EtOH; s eth, bz, HOAc; vs ace
3224	1,5-Dichloronaphthalene		$C_{10}H_6Cl_2$	1825-30-5	197.061	nd or lf (al) pr (sub)	107	sub	1.4900^{20}		i H_2O; sl EtOH; s eth
3225	1,6-Dichloronaphthalene		$C_{10}H_6Cl_2$	2050-72-8	197.061	nd or pr (al, peth)	49	sub			
3226	1,7-Dichloronaphthalene		$C_{10}H_6Cl_2$	2050-73-9	197.061	nd or pr (al, HOAc)	63.5	285.5	1.2611^{100}	1.6092^{100}	s EtOH, eth, bz, HOAc
3227	1,8-Dichloronaphthalene		$C_{10}H_6Cl_2$	2050-74-0	197.061	orth pl (hx) nd (al, sub)	89	sub	1.2924^{100}	1.6236^{100}	s EtOH, peth
3228	2,3-Dichloronaphthalene		$C_{10}H_6Cl_2$	2050-75-1	197.061	orth lf (al)	120				i H_2O; sl EtOH; vs eth
3229	2,6-Dichloronaphthalene		$C_{10}H_6Cl_2$	2065-70-5	197.061	nd or lf (al) pl (eth, bz)	140.5	285			sl EtOH; s eth, bz, chl, HOAc
3230	2,7-Dichloronaphthalene		$C_{10}H_6Cl_2$	2198-77-8	197.061	pl or lf (al)	115.0				vs EtOH; s hx, HOAc
3231	2,3-Dichloro-1,4-naphthalenedione	Dichlone	$C_{10}H_4Cl_2O_2$	117-80-6	227.044	ye nd (al)	195				i H_2O; sl EtOH, eth, bz; s chl

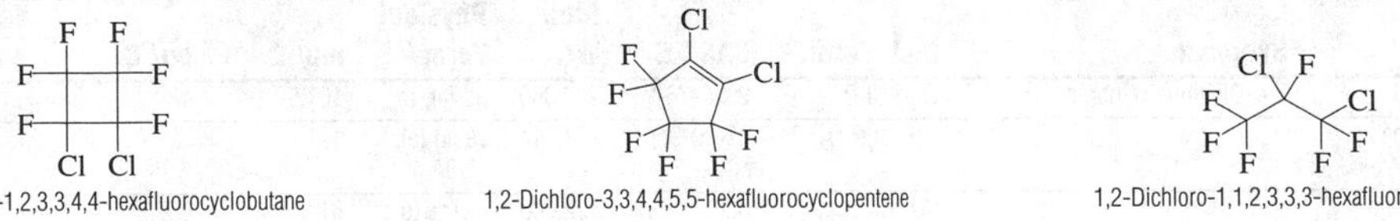

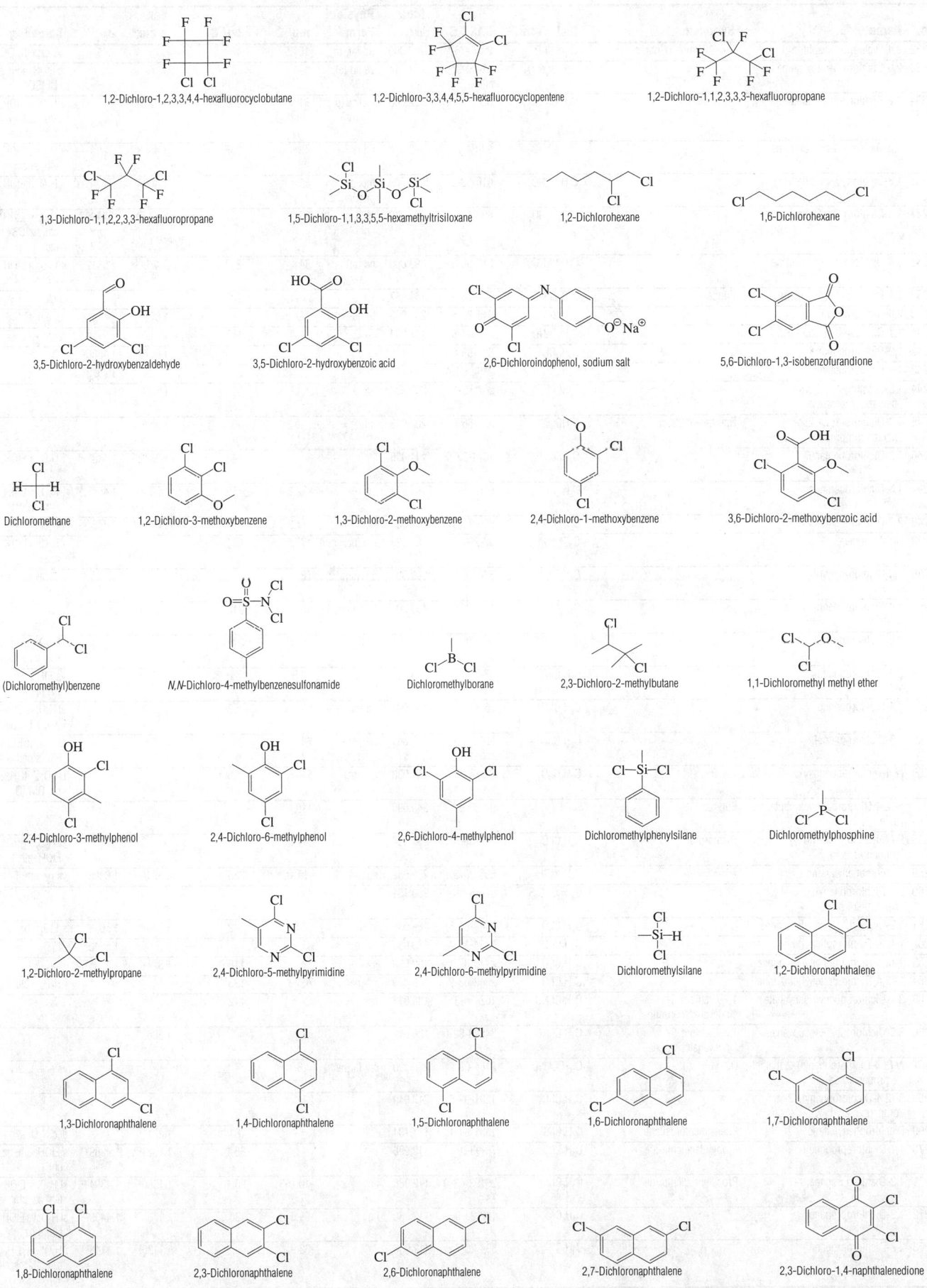

1,2-Dichloro-1,2,3,3,4,4-hexafluorocyclobutane

1,2-Dichloro-3,3,4,4,5,5-hexafluorocyclopentene

1,2-Dichloro-1,1,2,3,3,3-hexafluoropropane

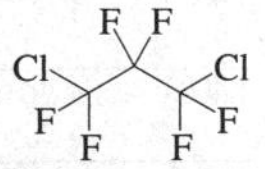

1,3-Dichloro-1,1,2,2,3,3-hexafluoropropane

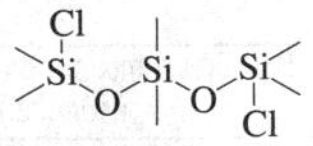

1,5-Dichloro-1,1,3,3,5,5-hexamethyltrisiloxane

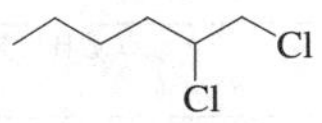

1,2-Dichlorohexane

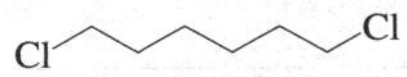

1,6-Dichlorohexane

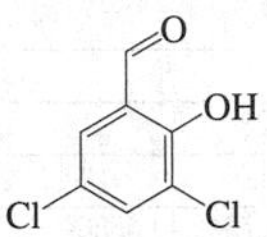

3,5-Dichloro-2-hydroxybenzaldehyde

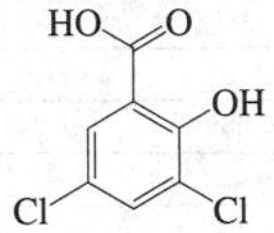

3,5-Dichloro-2-hydroxybenzoic acid

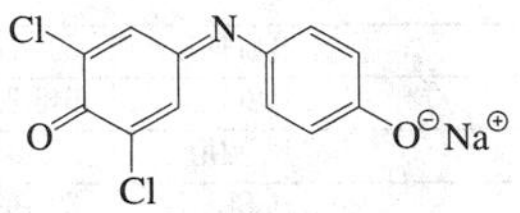

2,6-Dichloroindophenol, sodium salt

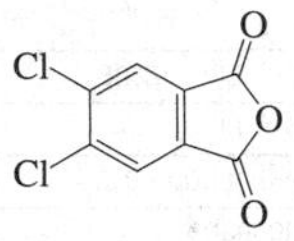

5,6-Dichloro-1,3-isobenzofurandione

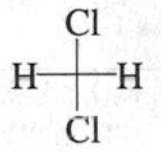

Dichloromethane

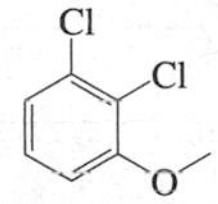

1,2-Dichloro-3-methoxybenzene

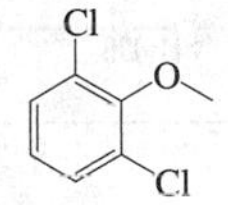

1,3-Dichloro-2-methoxybenzene

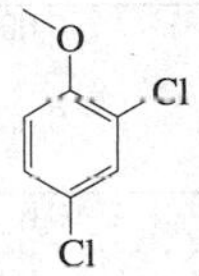

2,4-Dichloro-1-methoxybenzene

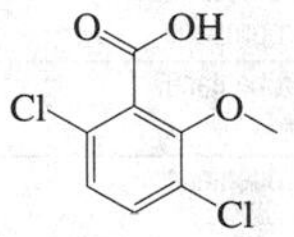

3,6-Dichloro-2-methoxybenzoic acid

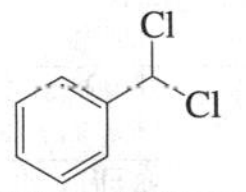

(Dichloromethyl)benzene

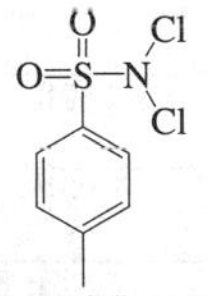

N,N-Dichloro-4-methylbenzenesulfonamide

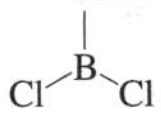

Dichloromethylborane

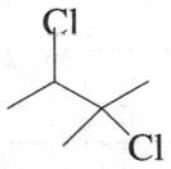

2,3-Dichloro-2-methylbutane

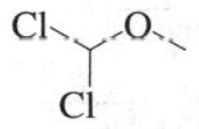

1,1-Dichloromethyl methyl ether

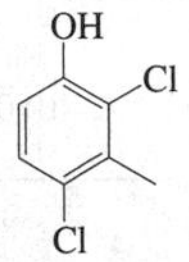

2,4-Dichloro-3-methylphenol

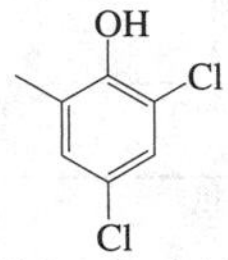

2,4-Dichloro-6-methylphenol

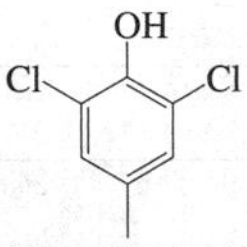

2,6-Dichloro-4-methylphenol

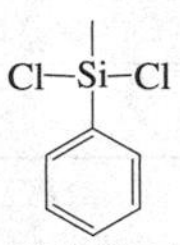

Dichloromethylphenylsilane

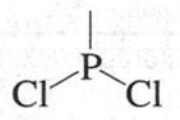

Dichloromethylphosphine

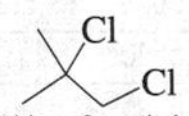

1,2-Dichloro-2-methylpropane

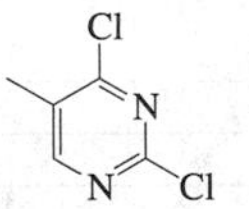

2,4-Dichloro-5-methylpyrimidine

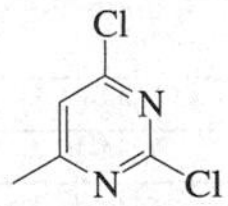

2,4-Dichloro-6-methylpyrimidine

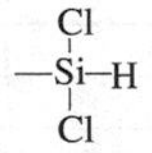

Dichloromethylsilane

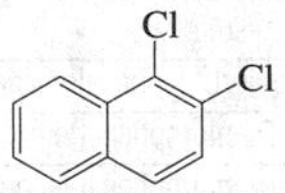

1,2-Dichloronaphthalene

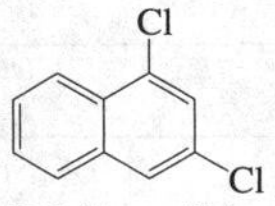

1,3-Dichloronaphthalene

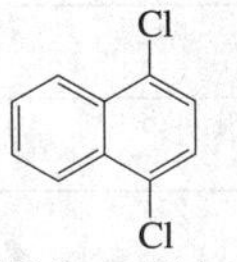

1,4-Dichloronaphthalene

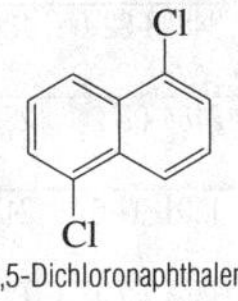

1,5-Dichloronaphthalene

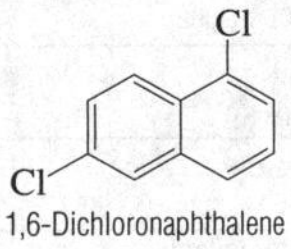

1,6-Dichloronaphthalene

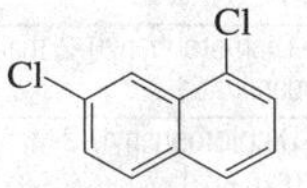

1,7-Dichloronaphthalene

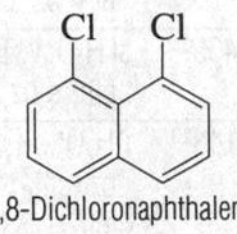

1,8-Dichloronaphthalene

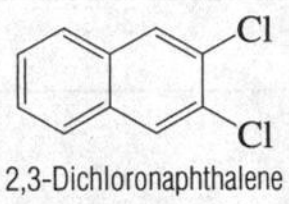

2,3-Dichloronaphthalene

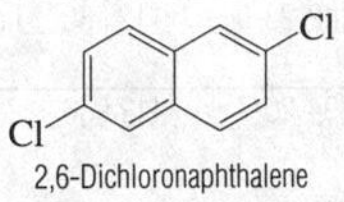

2,6-Dichloronaphthalene

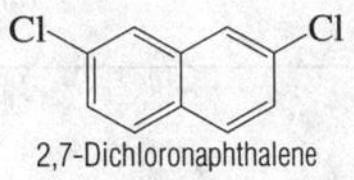

2,7-Dichloronaphthalene

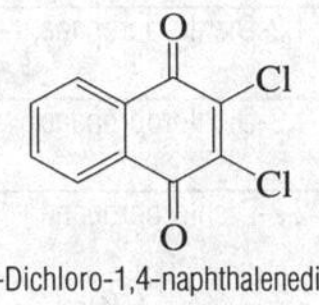

2,3-Dichloro-1,4-naphthalenedione

No.	Name	Synonym	Mol. Form.	CAS RN	Mol. Wt.	Physical Form	mp/°C	bp/°C	den/ g cm^{-3}	n_D	Solubility
3232	2,4-Dichloro-1-naphthol	2,4-Dichloro-α-naphthol	$C_{10}H_6Cl_2O$	2050-76-2	213.060	nd (al, bz)	107.5	180			vs bz, eth, EtOH
3233	2,6-Dichloro-4-nitroaniline		$C_6H_4Cl_2N_2O_2$	99-30-9	207.014	ye nd (al, HOAc)	191				s EtOH, acid; sl DMSO
3234	1,2-Dichloro-3-nitrobenzene		$C_6H_3Cl_2NO_2$	3209-22-1	192.000	mcl nd (peth, HOAc)	61.5	257.5	1.721^{14}		i H_2O; s EtOH, eth, ace, bz, peth; sl chl
3235	1,2-Dichloro-4-nitrobenzene		$C_6H_3Cl_2NO_2$	99-54-7	192.000	nd (al)	43	255.5	1.4558^{75}		i H_2O; s EtOH, eth; sl ctc
3236	1,3-Dichloro-5-nitrobenzene		$C_6H_3Cl_2NO_2$	618-62-2	192.000	mcl pr or lf (HOAc, al)	65.4		1.4000^{100}		i H_2O; s EtOH, eth
3237	1,4-Dichloro-2-nitrobenzene		$C_6H_3Cl_2NO_2$	89-61-2	192.000	pl or pr (al) pl (AcOEt)	56	267	1.439^{75}	1.4390^{75}	i H_2O; s EtOH, eth, bz, CS_2; sl ctc
3238	2,4-Dichloro-1-nitrobenzene		$C_6H_3Cl_2NO_2$	611-06-3	192.000	nd (al)	34	258.5	1.4790^{80}	1.5512^{70}	i H_2O; s EtOH, eth; sl chl
3239	1,1-Dichloro-1-nitroethane	Ethide	$C_2H_3Cl_2NO_2$	594-72-9	143.957			123.5			s ctc
3240	2,6-Dichloro-4-nitrophenol		$C_6H_3Cl_2NO_3$	618-80-4	207.999	br nd (w)	127 exp		1.822^{25}		vs eth, chl
3241	1,1-Dichloro-1-nitropropane		$C_3H_5Cl_2NO_2$	595-44-8	157.984			145	1.312^{20}		s ctc
3242	1,9-Dichlorononane		$C_9H_{18}Cl_2$	821-99-8	197.145			260; 138^{17}	1.0173^{25}	1.4586^{25}	
3243	1,8-Dichlorooctane		$C_8H_{16}Cl_2$	2162-99-4	183.119			241	1.0248^{25}	1.4572^{25}	
3244	1,3-Dichloro-1,1,2,2,3-pentafluoropropane		$C_3HCl_2F_5$	507-55-1	202.938	liq		52	1.55^{25}		
3245	3,3-Dichloro-1,1,1,2,2-pentafluoropropane	Refrigerant 225ca	$C_3HCl_2F_5$	422-56-0	202.938	liq		45.5	1.54^{25}		
3246	1,2-Dichloropentane		$C_5H_{10}Cl_2$	1674-33-5	141.038			148.3	1.0872^{20}	1.4485^{20}	i H_2O; s EtOH; vs chl
3247	1,5-Dichloropentane		$C_5H_{10}Cl_2$	628-76-2	141.038	liq	-72.8	179	1.0956^{25}	1.4545^{25}	i H_2O; s EtOH, eth, bz, ctc
3248	2,3-Dichloropentane		$C_5H_{10}Cl_2$	600-11-3	141.038	liq	-77.3	139	1.0789^{20}	1.4464^{20}	i H_2O
3249	Dichlorophene		$C_{13}H_{10}Cl_2O_2$	97-23-4	269.123	cry (bz, peth)	177.5				i H_2O; s EtOH, ace
3250	2,3-Dichlorophenol		$C_6H_4Cl_2O$	576-24-9	163.001	cry (lig, bz)	58				s EtOH, eth, bz, lig
3251	2,4-Dichlorophenol		$C_6H_4Cl_2O$	120-83-2	163.001	hex nd (bz)	45	210			sl H_2O; s EtOH, eth, bz, chl
3252	2,5-Dichlorophenol		$C_6H_4Cl_2O$	583-78-8	163.001	pr (bz, peth)	59	211			sl H_2O; vs EtOH, eth; s bz, peth
3253	2,6-Dichlorophenol		$C_6H_4Cl_2O$	87-65-0	163.001	nd (peth)	68.5	220; 82^4	1.653^{20}		vs EtOH, eth; s bz, peth
3254	3,4-Dichlorophenol		$C_6H_4Cl_2O$	95-77-2	163.001	nd (bz-peth)	68	253			sl H_2O; vs EtOH, eth; s bz, peth
3255	3,5-Dichlorophenol		$C_6H_4Cl_2O$	591-35-5	163.001	pr (peth)	68	233			sl H_2O; vs EtOH, eth; s peth
3256	(2,4-Dichlorophenoxy)acetic acid	2,4-D	$C_8H_6Cl_2O_3$	94-75-7	221.038	cry (bz)	140.5	$160^{0.4}$			i H_2O; s EtOH; sl bz, DMSO
3257	4-(2,4-Dichlorophenoxy)butanoic acid	Butyrac 118	$C_{10}H_{10}Cl_2O_3$	94-82-6	249.090		118				
3258	2-(2,4-Dichlorophenoxy) propanoic acid	Dichlorprop	$C_9H_8Cl_2O_3$	120-36-5	235.064		117.5				sl H_2O, lig; s EtOH, eth
3259	Dichlorophenylarsine		$C_6H_5AsCl_2$	696-28-6	222.932	liq	-19	255	1.6516^{20}	1.6386^{15}	vs bz, eth, EtOH
3260	2,4-Dichlorophenyl benzenesulfonate	Genite	$C_{12}H_8Cl_2O_3S$	97-16-5	303.161		45.5				s ctc, CS_2
3261	2,2-Dichloro-1-phenylethanone		$C_8H_6Cl_2O$	2648-61-5	189.039	amor	20.5	249	1.340^{16}	1.5686^{20}	s EtOH, bz, ctc
3262	1-(2,4-Dichlorophenyl)ethanone		$C_8H_6Cl_2O$	2234-16-4	189.039		33.5			1.5640^{20}	i H_2O
3263	1-(2,5-Dichlorophenyl)ethanone		$C_8H_6Cl_2O$	2476-37-1	189.039		12	118^{12}	1.321^{30}	1.5595^{30}	
3264	1-(3,4-Dichlorophenyl)ethanone		$C_8H_6Cl_2O$	2642-63-9	189.039	nd (peth)	76	135^{12}			i H_2O; s ctc, lig
3265	3,4-Dichlorophenyl isocyanate	1,2-Dichloro-5-isocyanatobenzene	$C_7H_3Cl_2NO$	102-36-3	188.011	cry	42	112^{12}			
3266	3,5-Dichlorophenyl isocyanate	1,3-Dichloro-5-isocyanatobenzene	$C_7H_3Cl_2NO$	34893-92-0	188.011		33		1.380		
3267	*N*-(3,4-Dichlorophenyl)-2-methyl-2-propenamide	Dicryl	$C_{10}H_9Cl_2NO$	2164-09-2	230.090	cry (al-peth)	128				vs ace, EtOH
3268	3-(2,4-Dichlorophenyl)-2-propenoic acid		$C_9H_6Cl_2O_2$	1201-99-6	217.049		234				s DMSO
3269	Dichlorophenylsilane	Phenyldichlorosilane	$C_6H_6Cl_2Si$	1631-84-1	177.104			181	1.221^{25}		dec H_2O
3270	1,1-Dichloropropane	Propylidene chloride	$C_3H_6Cl_2$	78-99-9	112.986			88.1	1.1321^{20}	1.4289^{20}	s EtOH, eth, bz, chl
3271	1,2-Dichloropropane, (±)	Propylene dichloride	$C_3H_6Cl_2$	26198-63-0	112.986	liq	-100.53	96.4	1.1560^{20}	1.4394^{20}	sl H_2O; s EtOH, eth, bz, chl
3272	1,3-Dichloropropane		$C_3H_6Cl_2$	142-28-9	112.986	liq	-99.5	120.9	1.1785^{25}	1.4455^{25}	sl H_2O; vs EtOH, eth; s bz, chl
3273	2,2-Dichloropropane		$C_3H_6Cl_2$	594-20-7	112.986	liq	-33.9	69.3	1.1136^{20}	1.4148^{20}	i H_2O; s EtOH, bz, chl; msc eth

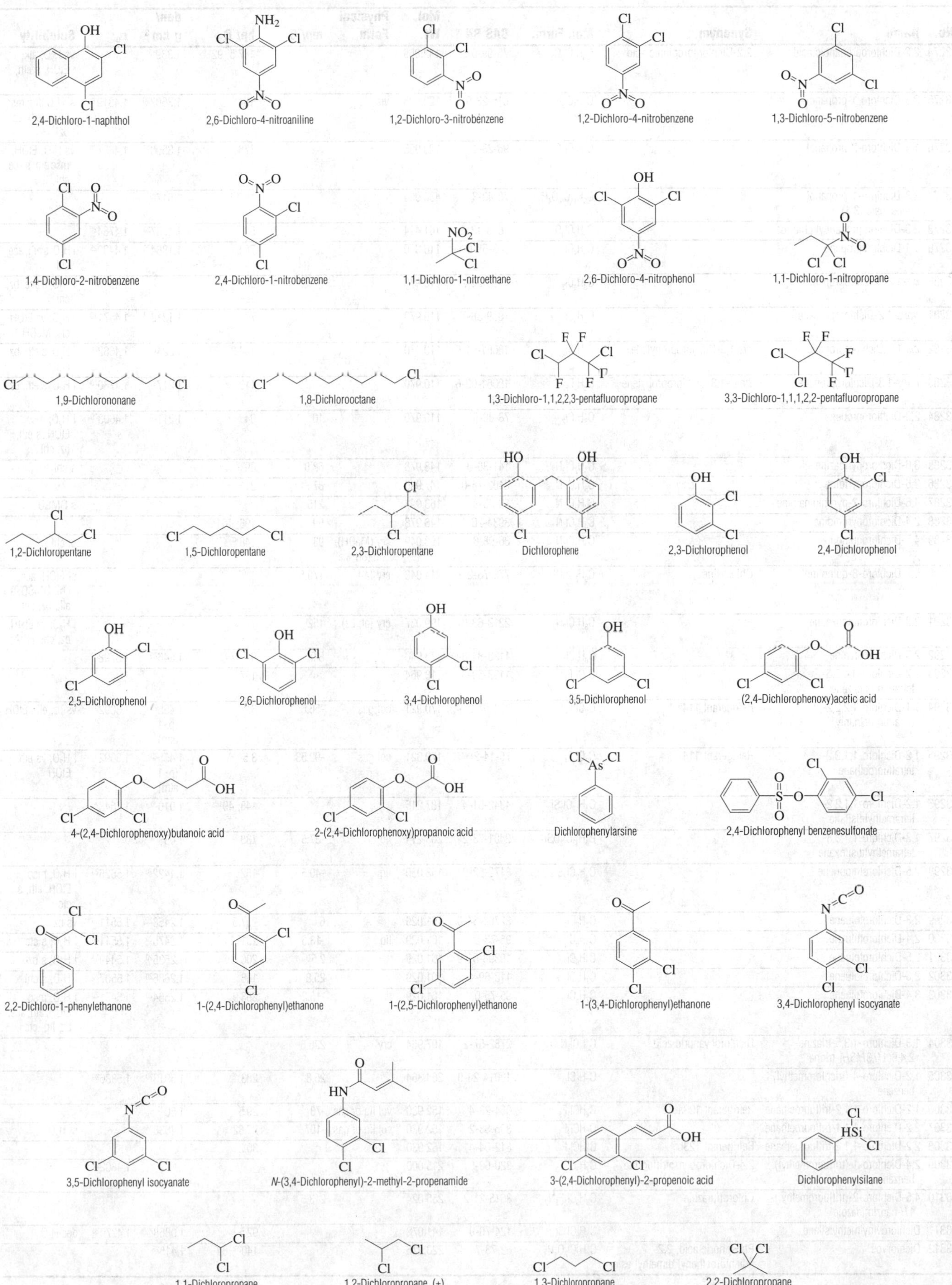
2,4-Dichloro-1-naphthol
2,6-Dichloro-4-nitroaniline
1,2-Dichloro-3-nitrobenzene
1,2-Dichloro-4-nitrobenzene
1,3-Dichloro-5-nitrobenzene
1,4-Dichloro-2-nitrobenzene
2,4-Dichloro-1-nitrobenzene
1,1-Dichloro-1-nitroethane
2,6-Dichloro-4-nitrophenol
1,1-Dichloro-1-nitropropane
1,9-Dichlorononane
1,8-Dichlorooctane
1,3-Dichloro-1,1,2,2,3-pentafluoropropane
3,3-Dichloro-1,1,1,2,2-pentafluoropropane
1,2-Dichloropentane
1,5-Dichloropentane
2,3-Dichloropentane
Dichlorophene
2,3-Dichlorophenol
2,4-Dichlorophenol
2,5-Dichlorophenol
2,6-Dichlorophenol
3,4-Dichlorophenol
3,5-Dichlorophenol
(2,4-Dichlorophenoxy)acetic acid
4-(2,4-Dichlorophenoxy)butanoic acid
2-(2,4-Dichlorophenoxy)propanoic acid
Dichlorophenylarsine
2,4-Dichlorophenyl benzenesulfonate
2,2-Dichloro-1-phenylethanone
1-(2,4-Dichlorophenyl)ethanone
1-(2,5-Dichlorophenyl)ethanone
1-(3,4-Dichlorophenyl)ethanone
3,4-Dichlorophenyl isocyanate
3,5-Dichlorophenyl isocyanate
N-(3,4-Dichlorophenyl)-2-methyl-2-propenamide
3-(2,4-Dichlorophenyl)-2-propenoic acid
Dichlorophenylsilane
1,1-Dichloropropane
1,2-Dichloropropane, (±)
1,3-Dichloropropane
2,2-Dichloropropane

No.	Name	Synonym	Mol. Form.	CAS RN	Mol. Wt.	Physical Form	mp/°C	bp/°C	den/ g cm^{-3}	n_D	Solubility
3274	2,2-Dichloropropanoic acid	2,2-Dichloropropionic acid	$C_3H_4Cl_2O_2$	75-99-0	142.969			187.5; 92^{14}	1.389^{12}		vs H_2O, alk, EtOH; s eth, ctc
3275	2,3-Dichloro-1-propanol		$C_3H_6Cl_2O$	616-23-9	128.985	visc		184	1.3607^{20}	1.4819^{20}	sl H_2O, lig; msc EtOH, eth, ace, bz
3276	1,3-Dichloro-2-propanol		$C_3H_6Cl_2O$	96-23-1	128.985			176	1.3506^{17}	1.4837^{20}	vs H_2O, EtOH; msc eth; s ace, chl
3277	2,3-Dichloro-1-propanol, phosphate (3:1)		$C_9H_{15}Cl_6O_4P$	78-43-3	430.904			$190^{0.1}$	1.517^{22}		
3278	2,3-Dichloropropanoyl chloride		$C_3H_3Cl_3O$	7623-13-4	161.414			53^{17}	1.4757^{20}	1.4764^{20}	
3279	1,1-Dichloropropene		$C_3H_4Cl_2$	563-58-6	110.970			76.5	1.1864^{25}	1.4430^{25}	i H_2O; s eth, ace, chl
3280	*cis*-1,2-Dichloropropene		$C_3H_4Cl_2$	6923-20-2	110.970			93		1.4549^{20}	i H_2O; s ace, bz, chl
3281	*trans*-1,2-Dichloropropene		$C_3H_4Cl_2$	7069-38-7	110.970			77	1.1818^{20}	1.4471^{20}	i H_2O; vs EtOH, ctc, MeOH
3282	*cis*-1,3-Dichloropropene	*cis*-1,3-Dichloropropylene	$C_3H_4Cl_2$	10061-01-5	110.970			104.3	1.224^{20}	1.4682^{20}	i H_2O; s eth, bz, chl
3283	*trans*-1,3-Dichloropropene	*trans*-1,3-Dichloropropylene	$C_3H_4Cl_2$	10061-02-6	110.970			112	1.217^{20}	1.4730^{20}	i H_2O; s eth, bz, chl
3284	2,3-Dichloropropene		$C_3H_4Cl_2$	78-88-6	110.970		10	94	1.211^{20}	1.4603^{20}	i H_2O; msc EtOH; s eth, bz, chl
3285	3,6-Dichloropyridazine		$C_4H_2Cl_2N_2$	141-30-0	148.978		68.8	$89^{0.2}$			s chl
3286	2,6-Dichloropyridine		$C_5H_3Cl_2N$	2402-78-0	147.990		87	211			
3287	4,6-Dichloro-2-pyrimidinamine		$C_4H_3Cl_2N_3$	56-05-3	163.993		215				s DMSO
3288	2,4-Dichloropyrimidine		$C_4H_2Cl_2N_2$	3934-20-1	148.978		59	198; 101^{23}			
3289	4,7-Dichloroquinoline		$C_9H_5Cl_2N$	86-98-6	198.049	cry (MeOH), nd (80% al)	93	148^{10}			sl chl
3290	5,7-Dichloro-8-quinolinol	Chloroxine	$C_9H_5Cl_2NO$	773-76-2	214.048	cry (al)	179.5				sl EtOH, ace, chl, DMSO; s alk, bz, peth
3291	2,3-Dichloroquinoxaline		$C_8H_4Cl_2N_2$	2213-63-0	199.037	cry (al, bz)	152				i H_2O; vs EtOH, bz, chl, HOAc
3292	2,5-Dichlorostyrene		$C_8H_6Cl_2$	1123-84-8	173.040		8.0	93^5, 74^3	1.246^{20}	1.5798^{20}	
3293	1,2-Dichloro-3,4,5,6-tetrafluorobenzene		$C_6Cl_2F_4$	1198-59-0	218.964			157.7			
3294	1,1-Dichloro-1,2,2,2-tetrafluoroethane	Refrigerant 114a	$C_2Cl_2F_4$	374-07-2	170.921	col gas	-56.6	3.4	1.455^{25} (p>1 atm)	1.3092^{0}	vs bz, eth, EtOH
3295	1,2-Dichloro-1,1,2,2-tetrafluoroethane	Refrigerant 114	$C_2Cl_2F_4$	76-14-2	170.921	col gas	-92.53	3.5	1.455^{25} (p>1 atm)	1.3092^{0}	i H_2O; vs eth, EtOH
3296	1,2-Dichloro-1,1,2,2-tetramethyldisilane		$C_4H_{12}Cl_2Si_2$	4342-61-4	187.215			148; 49^{18}	1.010^{20}	1.4548^{20}	
3297	1,3-Dichloro-1,1,3,3-tetramethyldisiloxane		$C_4H_{12}Cl_2OSi_2$	2401-73-2	203.214	liq	-37.5	138	1.038^{20}		
3298	2,5-Dichlorothiophene		$C_4H_2Cl_2S$	3172-52-9	153.030	liq	-40.5	162	1.4422^{20}	1.5626^{20}	i H_2O; msc EtOH, eth; s ctc
3299	2,3-Dichlorotoluene		$C_7H_6Cl_2$	32768-54-0	161.029		6	207.5	1.2458^{20}	1.5511^{20}	vs bz
3300	2,4-Dichlorotoluene		$C_7H_6Cl_2$	95-73-8	161.029	liq	-13.5	201	1.2476^{20}	1.5511^{20}	i H_2O; s ctc
3301	2,5-Dichlorotoluene		$C_7H_6Cl_2$	19398-61-9	161.029		2.5	200	1.2535^{20}	1.5449^{20}	i H_2O; s bz
3302	2,6-Dichlorotoluene		$C_7H_6Cl_2$	118-69-4	161.029		25.8	198	1.2686^{20}	1.5507^{20}	i H_2O; s chl
3303	3,4-Dichlorotoluene		$C_7H_6Cl_2$	95-75-0	161.029	liq	-15.2	208.9	1.2564^{20}	1.5471^{20}	i H_2O; msc EtOH, eth, ace, bz, lig, ctc
3304	1,3-Dichloro-1,3,5-triazine-2,4,6(1*H*,3*H*,5*H*)-trione	Dichlorocyanuric acid	$C_3HCl_2N_3O_3$	2782-57-2	197.964	cry	226.6				
3305	1,2-Dichloro-4-(trichloromethyl) benzene		$C_7H_3Cl_5$	13014-24-9	264.364		25.8	283.1	1.5913^{20}	1.5886^{20}	
3306	1,2-Dichloro-1,1,2-trifluoroethane	Refrigerant 123a	$C_2HCl_2F_3$	354-23-4	152.930	vol liq or gas	-78	29.5	1.50^{25}		
3307	2,2-Dichloro-1,1,1-trifluoroethane		$C_2HCl_2F_3$	306-83-2	152.930	vol liq or gas	-107	27.82	1.4638^{25}		sl H_2O
3308	2,2-Dichloro-1,1,2-trifluoroethane	Refrigerant 123b	$C_2HCl_2F_3$	812-04-4	152.930			30.2			
3309	2,4-Dichloro-1-(trifluoromethyl) benzene	2,4-Dichlorobenzotrifluoride	$C_7H_3Cl_2F_3$	320-60-5	215.000					1.4802^{20}	
3310	4,5-Dichloro-2-(trifluoromethyl)-1*H*-benzimidazole	Chloroflurazole	$C_8H_3Cl_2F_3N_2$	3615-21-2	255.024		213.5				
3311	Dichlorovinylmethylsilane		$C_3H_6Cl_2Si$	124-70-9	141.072			92.5	1.0868^{20}	1.4270^{20}	dec H_2O
3312	Dichlorvos	Phosphoric acid, 2,2-dichloroethenyl dimethyl ester	$C_4H_7Cl_2O_4P$	62-73-7	220.976			140^{20}, 84^{1}	1.415^{25}		

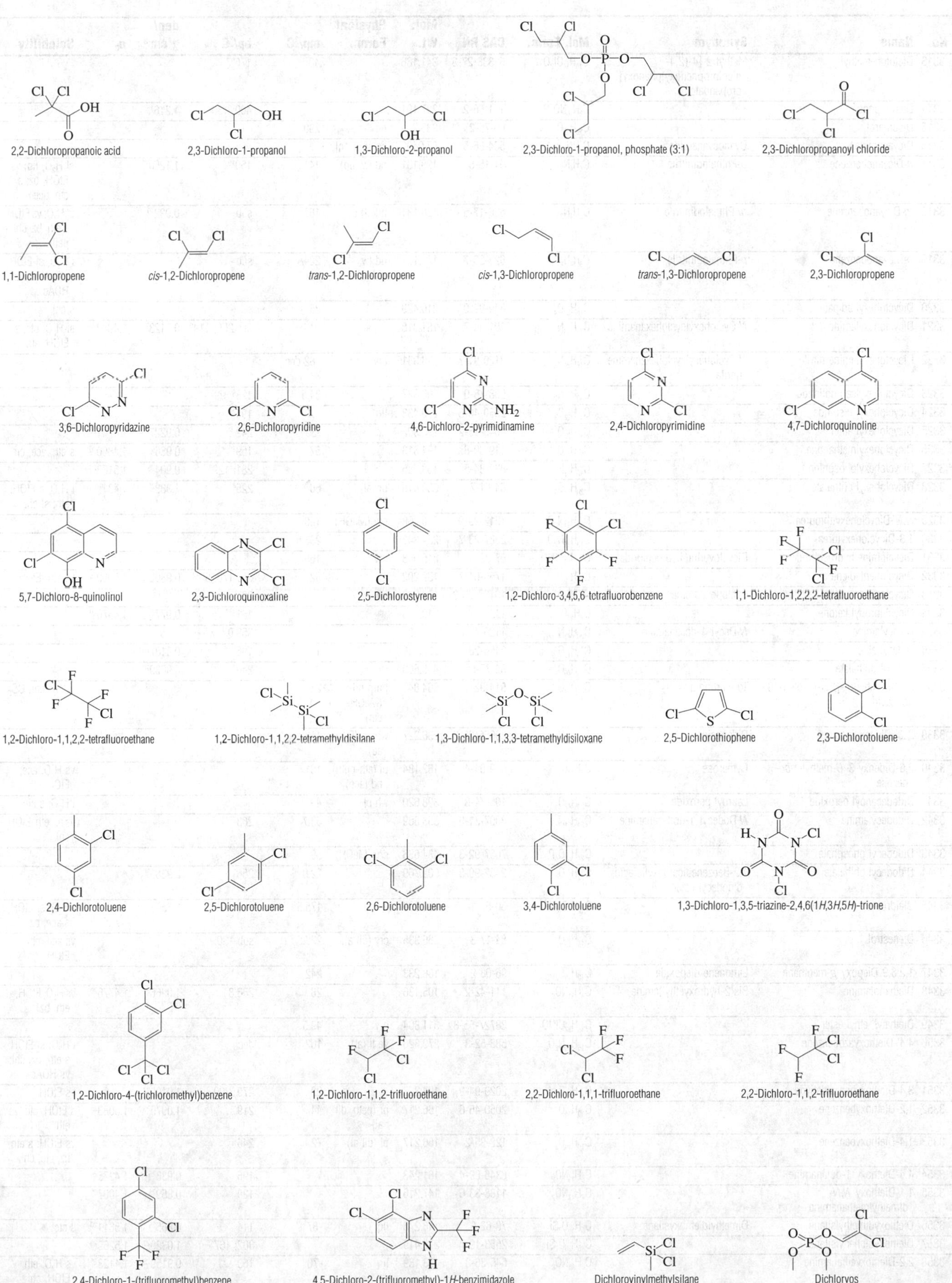
2,2-Dichloropropanoic acid
2,3-Dichloro-1-propanol
1,3-Dichloro-2-propanol
2,3-Dichloro-1-propanol, phosphate (3:1)
2,3-Dichloropropanoyl chloride
1,1-Dichloropropene
cis-1,2-Dichloropropene
trans-1,2-Dichloropropene
cis-1,3-Dichloropropene
trans-1,3-Dichloropropene
2,3-Dichloropropene
3,6-Dichloropyridazine
2,6-Dichloropyridine
4,6-Dichloro-2-pyrimidinamine
2,4-Dichloropyrimidine
4,7-Dichloroquinoline
5,7-Dichloro-8-quinolinol
2,3-Dichloroquinoxaline
2,5-Dichlorostyrene
1,2-Dichloro-3,4,5,6-tetrafluorobenzene
1,1-Dichloro-1,2,2,2-tetrafluoroethane
1,2-Dichloro-1,1,2,2-tetrafluoroethane
1,2-Dichloro-1,1,2,2-tetramethyldisilane
1,3-Dichloro-1,1,3,3-tetramethyldisiloxane
2,5-Dichlorothiophene
2,3-Dichlorotoluene
2,4-Dichlorotoluene
2,5-Dichlorotoluene
2,6-Dichlorotoluene
3,4-Dichlorotoluene
1,3-Dichloro-1,3,5-triazine-2,4,6(1H,3H,5H)-trione
1,2-Dichloro-4-(trichloromethyl)benzene
1,2-Dichloro-1,1,2-trifluoroethane
2,2-Dichloro-1,1,1-trifluoroethane
2,2-Dichloro-1,1,2-trifluoroethane
2,4-Dichloro-1-(trifluoromethyl)benzene
4,5-Dichloro-2-(trifluoromethyl)-1H-benzimidazole
Dichlorovinylmethylsilane
Dichlorvos

No.	Name	Synonym	Mol. Form.	CAS RN	Mol. Wt.	Physical Form	mp/°C	bp/°C	den/ g cm^{-3}	n_D	Solubility
3313	Diclofop-methyl	Methyl 2-[4-(2,4-dichlorophenoxy)phenoxy] propanoate	$C_{16}H_{14}Cl_2O_4$	51338-27-3	341.186		40	$176^{0.1}$			
3314	Dicrotophos		$C_8H_{16}NO_5P$	141-66-2	237.191			400; $130^{0.1}$	1.216^{15}		
3315	Dicumarol		$C_{19}H_{12}O_6$	66-76-2	336.294	nd	290				
3316	Dicyanamide	Cyanocyanamide	C_2HN_3	504-66-5	67.049	aq soln only					
3317	*o*-Dicyanobenzene	*o*-Phthalodinitrile	$C_8H_4N_2$	91-15-6	128.131	nd (w, lig)	141	150^{10}	1.1250^{25}		sl H_2O, lig; vs EtOH, bz; s eth, ace
3318	*m*-Dicyanobenzene	*m*-Phthalodinitrile	$C_8H_4N_2$	626-17-5	128.131	nd(al)	162	sub	0.992^{40}		sl H_2O; vs EtOH; s eth, bz, chl; i peth
3319	*p*-Dicyanobenzene	*p*-Phthalodinitrile	$C_8H_4N_2$	623-26-7	128.131	nd (w, MeOH)	224	sub			i H_2O; sl EtOH, eth; s bz; vs HOAc
3320	Dicyclohexyl adipate		$C_{18}H_{30}O_4$	849-99-0	310.429		35				s chl
3321	Dicyclohexylamine	*N*-Cyclohexylcyclohexanamine	$C_{12}H_{23}N$	101-83-7	181.318		-0.1	dec 256; 114^9	0.9123^{20}	1.4842^{20}	sl H_2O, ctc; s EtOH, eth, bz
3322	Dicyclohexylamine nitrite	*N*-Cyclohexylcyclohexanamine, nitrite	$C_{12}H_{24}N_2O_2$	3129-91-7	228.331	cry	182 dec				
3323	Dicyclohexylcarbodiimide		$C_{13}H_{22}N_2$	538-75-0	206.327		34.5	123^6, $99^{0.5}$			
3324	Dicyclohexyl disulfide		$C_{12}H_{22}S_2$	2550-40-5	230.433	liq		195^{20}			
3325	Dicyclohexyl ether		$C_{12}H_{22}O$	4645-15-2	182.302	liq	-36	242.5	0.9227^{20}	1.4741^{20}	
3326	Dicyclohexylmethanone		$C_{13}H_{22}O$	119-60-8	194.313		57	159^{20}	0.986^0	1.4860^{20}	s eth, ace, ctc
3327	Dicyclohexylphosphine		$C_{12}H_{23}P$	829-84-5	198.285			281; 129^8	0.904^{25}	1.5163^{20}	
3328	Dicyclohexyl phthalate		$C_{20}H_{26}O_4$	84-61-7	330.418	pr (al)	66	225^4	1.383^{20}	1.431^{20}	i H_2O; s EtOH, eth; sl chl
3329	*N,N'*-Dicyclohexylthiourea		$C_{13}H_{24}N_2S$	1212-29-9	240.408	cry (MeOH)	180				
3330	1,3-Dicyclohexylurea		$C_{13}H_{24}N_2O$	2387-23-7	224.342		233.8				
3331	Dicyclomine hydrochloride	Dicycloverine hydrochloride	$C_{19}H_{36}ClNO_2$	67-92-5	345.948	cry	165				
3332	Dicyclopentadiene		$C_{10}H_{12}$	1755-01-7	132.202		32	dec 170; 65^{14}	0.9302^{35}	1.5050^{35}	vs eth, EtOH
3333	Dicyclopentyl ether	Cyclopentyl ether	$C_{10}H_{18}O$	10137-73-2	154.249	liq		80^{13}			
3334	Dicyclopropyl ketone		$C_7H_{10}O$	1121-37-5	110.153			161	0.977^{25}	1.4670^{20}	
3335	Didecylamine	*N*-Decyl-1-decanamine	$C_{20}H_{43}N$	1120-49-6	297.562			359.0			
3336	Didecyl ether		$C_{20}H_{42}O$	2456-28-2	298.546		16	$196^{15.5}$	0.8187^{20}		
3337	Didecyl phthalate		$C_{28}H_{46}O_4$	84-77-5	446.663		2.5	240^3	0.9639^{20}		
3338	3',4'-Didehydro-β,ψ-caroten-16'-oic acid	Torularhodin	$C_{40}H_{52}O_2$	514-92-1	564.840	purp nd (MeOH-eth)	211				vs py, chl, CS_2
3339	2',3'-Dideoxyinosine	Didanosine	$C_{10}H_{12}N_4O_3$	69655-05-6	236.227	wh cry (EtOH aq)	162				
3340	2,6-Dideoxy-3-*O*-methyl-*ribo*-hexose	Cymarose	$C_7H_{14}O_4$	579-04-4	162.184	pr (eth-peth) nd (ace)	101				vs H_2O, ace, EtOH
3341	Didodecanoyl peroxide	Lauroyl peroxide	$C_{24}H_{46}O_4$	105-74-8	398.620	wh pl	49				i H_2O; s chl
3342	Didodecylamine	*N*-Dodecyl-1-dodecanamine	$C_{24}H_{51}N$	3007-31-6	353.669		53.7	263^{27}			vs bz, eth, EtOH, chl
3343	Didodecyl phosphate		$C_{24}H_{51}O_4P$	7057-92-3	434.633	cry (MeOH)	59				
3344	Didodecyl phthalate	1,2-Benzenedicarboxylic acid, didodecyl ester	$C_{32}H_{54}O_4$	2432-90-8	502.769		22.0	256^1	0.9389^{20}		
3345	Dieldrin		$C_{12}H_8Cl_6O$	60-57-1	380.909		175.5		1.75^{25}		i H_2O; sl EtOH; s ace, bz
3346	Dienestrol		$C_{18}H_{18}O_2$	84-17-3	266.335	cry (dil al)	227.5	sub 130			vs ace, eth, EtOH
3347	1,2:8,9-Diepoxy-*p*-menthane	Limonene diepoxide	$C_{10}H_{16}O_2$	96-08-2	168.233		242				
3348	Diethanolamine	Bis(2-hydroxyethyl)amine	$C_4H_{11}NO_2$	111-42-2	105.136		28	268.8	1.0966^{20}	1.4776^{20}	vs H_2O, EtOH; sl eth, bz
3349	Diethatyl, ethyl ester		$C_{16}H_{22}ClNO_3$	38727-55-8	311.804	cry	49.5				
3350	4,4'-Diethoxyazobenzene		$C_{16}H_{18}N_2O_2$	588-52-3	270.326	ye lf (al)	162	dec			i H_2O; sl EtOH; s eth, bz, chl; vs HOAc
3351	3,4-Diethoxybenzaldehyde		$C_{11}H_{14}O_3$	2029-94-9	194.227		22	279; 200^{50}	1.0100^{22}		vs EtOH
3352	1,2-Diethoxybenzene		$C_{10}H_{14}O_2$	2050-46-6	166.217	pr (peth, dil al)	44	219	1.0075^{20}	1.5083^{25}	s EtOH, ctc; vs eth
3353	1,4-Diethoxybenzene		$C_{10}H_{14}O_2$	122-95-2	166.217	pl (dil al)	72	246			vs EtOH; s eth, bz, ctc, chl
3354	4,4-Diethoxy-1-butanamine		$C_8H_{19}NO_2$	6346-09-4	161.243			196	0.933^{25}	1.4275^{20}	
3355	1,1-Diethoxy-*N,N*-dimethylmethanamine		$C_7H_{17}NO_2$	1188-33-6	147.216			129	0.859^{25}	1.4007^{20}	
3356	Diethoxydimethylsilane	Dimethyldiethoxysilane	$C_6H_{16}O_2Si$	78-62-6	148.276	liq	-87	114	0.865^{25}	1.3811^{20}	s ctc
3357	Diethoxydiphenylsilane		$C_{16}H_{20}O_2Si$	2553-19-7	272.415			302; 167^{15}	1.0329^{20}	1.5269^{20}	
3358	2,2-Diethoxyethanamine		$C_6H_{15}NO_2$	645-36-3	133.189	liq	-78	163	0.9159^{25}	1.4123^{25}	vs H_2O, eth, EtOH, chl

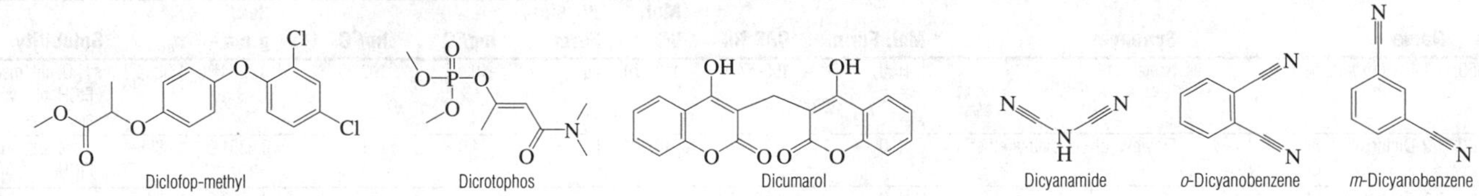

Diclofop-methyl

Dicrotophos

Dicumarol

Dicyanamide

o-Dicyanobenzene

m-Dicyanobenzene

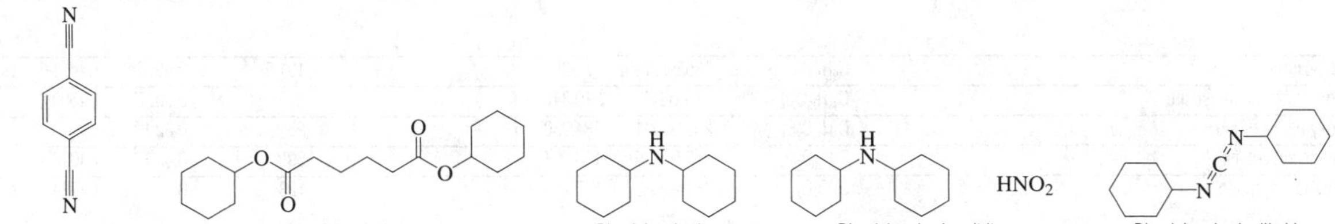

p-Dicyanobenzene

Dicyclohexyl adipate

Dicyclohexylamine

HNO_2

Dicyclohexylamine nitrite

Dicyclohexylcarbodiimide

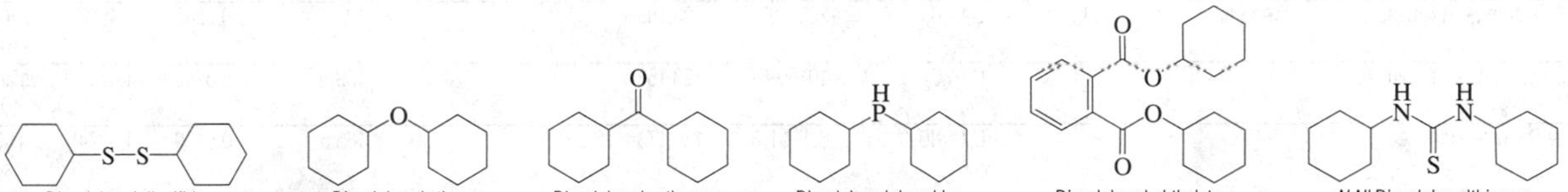

Dicyclohexyl disulfide

Dicyclohexyl ether

Dicyclohexylmethanone

Dicyclohexylphosphine

Dicyclohexyl phthalate

N,N'-Dicyclohexylthiourea

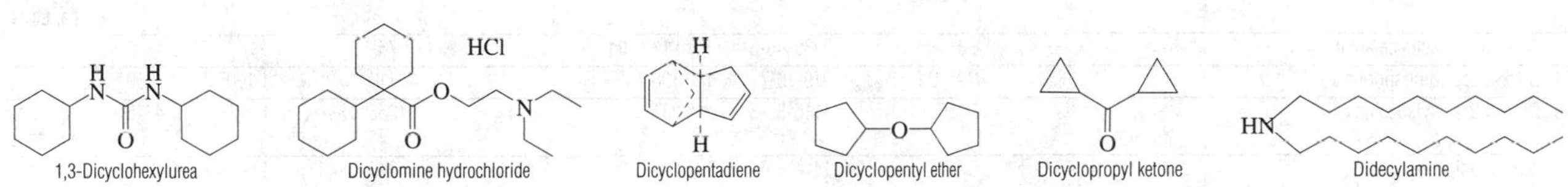

1,3-Dicyclohexylurea

HCl

Dicyclomine hydrochloride

Dicyclopentadiene

Dicyclopentyl ether

Dicyclopropyl ketone

Didecylamine

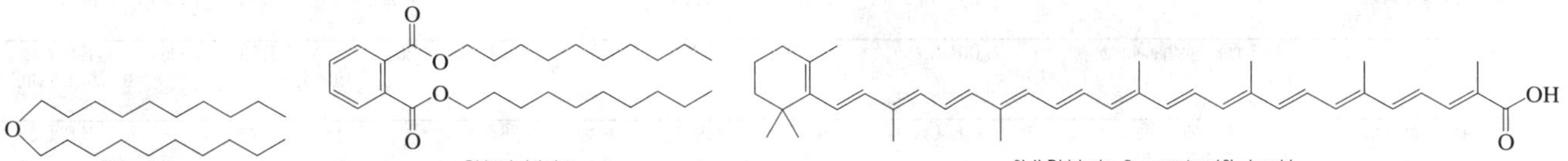

Didecyl ether

Didecyl phthalate

3',4'-Didehydro-β,ψ-caroten-16'-oic acid

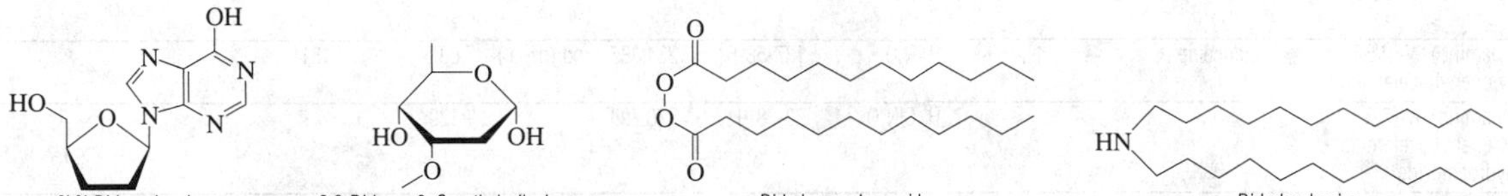

2',3'-Dideoxyinosine

2,6-Dideoxy-3-*O*-methyl-*ribo*-hexose

Didodecanoyl peroxide

Didodecylamine

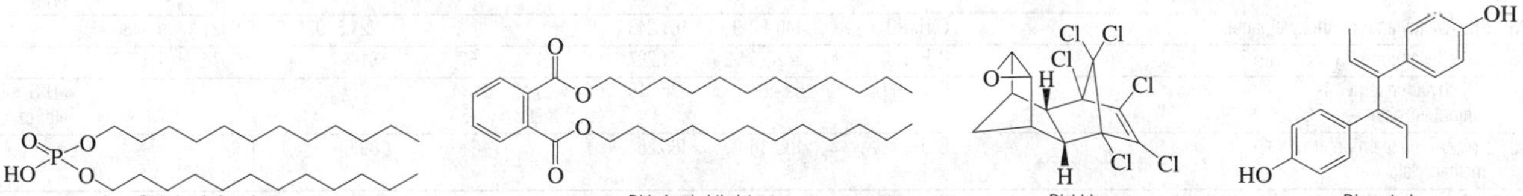

Didodecyl phosphate

Didodecyl phthalate

Dieldrin

Dienestrol

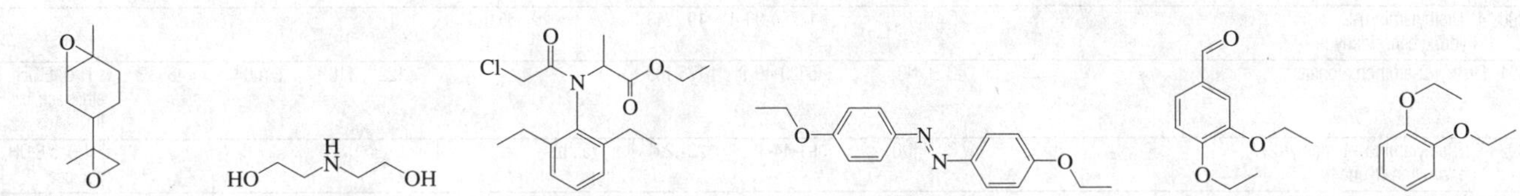

1,2:8,9-Diepoxy-*p*-menthane

Diethanolamine

Diethatyl, ethyl ester

4,4'-Diethoxyazobenzene

3,4-Diethoxybenzaldehyde

1,2-Diethoxybenzene

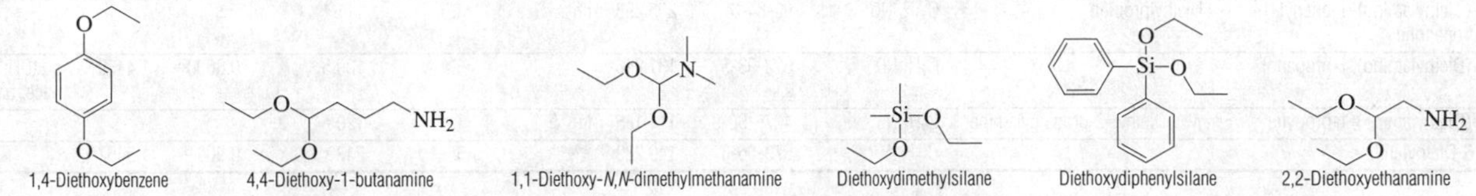

1,4-Diethoxybenzene

4,4-Diethoxy-1-butanamine

1,1-Diethoxy-*N,N*-dimethylmethanamine

Diethoxydimethylsilane

Diethoxydiphenylsilane

2,2-Diethoxyethanamine

No.	Name	Synonym	Mol. Form.	CAS RN	Mol. Wt.	Physical Form	mp/°C	bp/°C	den/ g cm^{-3}	n_D	Solubility
3359	1,1-Diethoxyethane	Acetal	$C_6H_{14}O_2$	105-57-7	118.174	liq	-100	102.25	0.8254^{20}	1.3834^{20}	s H_2O, chl; msc EtOH, eth; vs ace
3360	1,2-Diethoxyethane	Ethylene glycol diethyl ether	$C_6H_{14}O_2$	629-14-1	118.174	liq	-74.0	121.2	0.8351^{25}	1.3898^{25}	vs ace, bz, eth, EtOH
3361	1,1-Diethoxyethene		$C_6H_{12}O_2$	2678-54-8	116.158			68^{100}	0.7932^{20}	1.3643^{21}	
3362	Diethoxymethane		$C_5H_{12}O_2$	462-95-3	104.148	liq	-66.5	88	0.8319^{20}	1.3748^{18}	s H_2O; msc EtOH; vs ace, bz; sl chl
3363	2-(Diethoxymethyl)furan		$C_9H_{14}O_3$	13529-27-6	170.205			191.5	0.9976^{20}	1.4451^{20}	vs EtOH
3364	Diethoxymethylphenylsilane		$C_{11}H_{18}O_2Si$	775-56-4	210.346			218	0.9627^{20}	1.4690^{20}	
3365	Diethoxymethylsilane		$C_5H_{14}O_2Si$	2031-62-1	134.250			98	0.829^{25}		
3366	1,1-Diethoxypentane		$C_9H_{20}O_2$	3658-79-5	160.254			59^{12}	0.829^{22}	1.4029^{22}	
3367	1,1-Diethoxypropane		$C_7H_{16}O_2$	4744-08-5	132.201			123	0.825^{20}	1.3924^{19}	s H_2O, ace, bz; vs EtOH, eth
3368	2,2-Diethoxypropane		$C_7H_{16}O_2$	126-84-1	132.201			114	0.8200^{21}	1.3891^{20}	s EtOH, ace, bz; vs eth; sl ctc
3369	3,3-Diethoxy-1-propene	Acrolein, diethyl acetal	$C_7H_{14}O_2$	3054-95-3	130.185			123.5	0.8543^{15}	1.4000^{20}	sl H_2O; msc EtOH, eth
3370	3,3-Diethoxy-1-propyne		$C_7H_{12}O_2$	10160-87-9	128.169			139	0.8942^{22}	1.4140^{20}	vs ace, eth, EtOH, chl
3371	*N,N*-Diethylacetamide		$C_6H_{13}NO$	685-91-6	115.173			185.5	0.9130^{17}	1.4374^{17}	s H_2O, EtOH; msc eth, ace, bz; sl ctc
3372	Diethyl 2-acetamidomalonate		$C_9H_{15}NO_5$	1068-90-2	217.219	cry (al,bz-peth)	96.3	185^{20}			sl H_2O, eth; s tfa, EtOH
3373	*N,N*-Diethylacetoacetamide		$C_8H_{15}NO_2$	2235-46-3	157.211	liq		76^{13}			
3374	Diethyl acetylphosphonate		$C_6H_{13}O_4P$	919-19-7	180.138			114^{20}	1.1005^{20}	1.4200^{26}	
3375	Diethyl 2-acetylsuccinate		$C_{10}H_{16}O_5$	1115-30-6	216.231			255; 133^{17}	1.081^{20}	1.4346^{20}	i H_2O; s EtOH, eth, bz; sl chl
3376	Diethyl adipate		$C_{10}H_{18}O_4$	141-28-6	202.248	liq	-19.8	245	1.0076^{20}	1.4272^{20}	i H_2O; s EtOH, eth
3377	Diethyl 2-allylmalonate		$C_{10}H_{16}O_4$	2049-80-1	200.232			222.5; 93^6	1.0098^{20}	1.4305^{20}	i H_2O; vs EtOH, eth; s ctc
3378	Diethylamine	*N*-Ethylethanamine	$C_4H_{11}N$	109-89-7	73.137	liq	-49.8	55.5	0.7056^{20}	1.3864^{20}	vs H_2O; msc EtOH; s eth, ctc
3379	Diethylamine hydrochloride	*N*-Ethylethanamine hydrochloride	$C_4H_{12}ClN$	660-68-4	109.598	lf (al-eth)	228.5		1.0477^{22}		vs H_2O, EtOH
3380	(Diethylamino)acetonitrile		$C_6H_{12}N_2$	3010-02-4	112.172			170	0.8660^{20}	1.4260^{20}	s H_2O
3381	4-(Diethylamino)benzaldehyde		$C_{11}H_{15}NO$	120-21-8	177.243	ye nd (w)	41	172^{10}			vs H_2O; s EtOH, eth, bz, ctc
3382	2-(Diethylamino)-*N*-(2,6-dimethylphenyl)acetamide	Lidocaine	$C_{14}H_{22}N_2O$	137-58-6	234.337	nd (bz, al)	68.5	181^4			vs bz, eth, EtOH, chl
3383	2-(Diethylamino)-*N*-(2,6-dimethylphenyl)acetamide, monohydrochloride		$C_{14}H_{23}ClN_2O$	73-78-9	270.798		128				vs H_2O
3384	2-Diethylaminoethanol		$C_6H_{15}NO$	100-37-8	117.189	hyg		163	0.8921^{20}	1.4412^{20}	msc H_2O; s EtOH, eth, ace, bz, peth; sl ctc
3385	2-[2-(Diethylamino)ethoxy]ethanol		$C_8H_{19}NO_2$	140-82-9	161.243			221.5; 92^7	0.9421^{25}	1.4480^{20}	
3386	2-(Diethylamino)ethyl acrylate		$C_9H_{17}NO_2$	2426-54-2	171.237		<-60	81^{10}	0.937^{20}	1.4376^{25}	
3387	2-Diethylaminoethyl 4-aminobenzoate	Procaine	$C_{13}H_{20}N_2O_2$	59-46-1	236.310	nd (w+2) pl (lig or eth)	61				sl H_2O; s EtOH, eth, bz, chl
3388	2-(*N,N*-Diethylamino)ethyl methacrylate		$C_{10}H_{19}NO_2$	105-16-8	185.264			80^{10}	0.92^{30}		
3389	2-(Diethylamino)ethyl 2-phenylbutanoate	Butethamate	$C_{16}H_{25}NO_2$	14007-64-8	263.376			168^{11}		1.4909^{20}	
3390	4-(Diethylamino)-2-hydroxybenzaldehyde		$C_{11}H_{15}NO_2$	17754-90-4	193.243		65.0				
3391	Diethyl 2-aminomalonate		$C_7H_{13}NO_4$	6829-40-9	175.183			122^{16}, 116^{12}	1.100^{16}	1.4353^{16}	vs H_2O, EtOH, eth; s ace, bz; i lig
3392	7-(Diethylamino)-4-methyl-2*H*-1-benzopyran-2-one		$C_{14}H_{17}NO_2$	91-44-1	231.291	cry (al, bz-lig)					sl H_2O; s EtOH, eth, ace
3393	3-(Diethylamino)phenol		$C_{10}H_{15}NO$	91-68-9	165.232	orth bipym (CS_2-lig)	78	276; 170^{15}			s H_2O, EtOH, eth, CS_2; sl lig
3394	2-(Diethylamino)-1-phenyl-1-propanone	Diethylpropion	$C_{13}H_{19}NO$	90-84-6	205.296	liq		111^{14}			
3395	3-(Diethylamino)-1-propanol		$C_7H_{17}NO$	622-93-5	131.216			189.5	0.8600^{20}	1.4439^{20}	s EtOH; s eth, ace, bz; sl chl
3396	3-(Diethylamino)-1-propyne	*N,N*-Diethyl-2-propargylamine	$C_7H_{13}N$	4079-68-9	111.185	liq		120			
3397	2,6-Diethylaniline		$C_{10}H_{15}N$	579-66-8	149.233		1.5	243	0.906^{25}	1.5452^{20}	

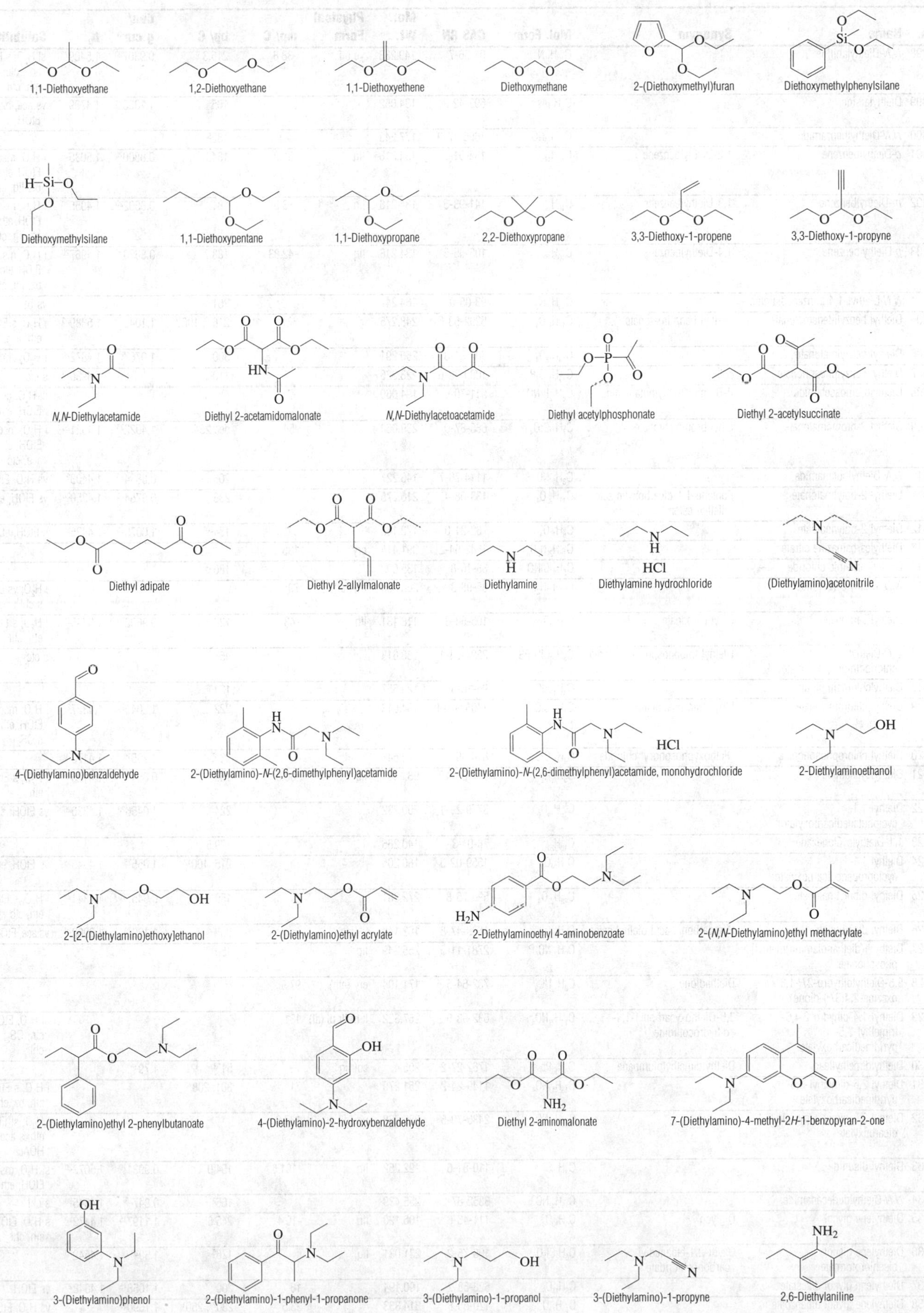
1,1-Diethoxyethane
1,2-Diethoxyethane
1,1-Diethoxyethene
Diethoxymethane
2-(Diethoxymethyl)furan
Diethoxymethylphenylsilane
Diethoxymethylsilane
1,1-Diethoxypentane
1,1-Diethoxypropane
2,2-Diethoxypropane
3,3-Diethoxy-1-propene
3,3-Diethoxy-1-propyne
N,N-Diethylacetamide
Diethyl 2-acetamidomalonate
N,N-Diethylacetoacetamide
Diethyl acetylphosphonate
Diethyl 2-acetylsuccinate
Diethyl adipate
Diethyl 2-allylmalonate
Diethylamine
Diethylamine hydrochloride
(Diethylamino)acetonitrile
4-(Diethylamino)benzaldehyde
2-(Diethylamino)-N-(2,6-dimethylphenyl)acetamide
2-(Diethylamino)-N-(2,6-dimethylphenyl)acetamide, monohydrochloride
2-Diethylaminoethanol
2-[2-(Diethylamino)ethoxy]ethanol
2-(Diethylamino)ethyl acrylate
2-Diethylaminoethyl 4-aminobenzoate
2-(N,N-Diethylamino)ethyl methacrylate
2-(Diethylamino)ethyl 2-phenylbutanoate
4-(Diethylamino)-2-hydroxybenzaldehyde
Diethyl 2-aminomalonate
7-(Diethylamino)-4-methyl-2H-1-benzopyran-2-one
3-(Diethylamino)phenol
2-(Diethylamino)-1-phenyl-1-propanone
3-(Diethylamino)-1-propanol
3-(Diethylamino)-1-propyne
2,6-Diethylaniline

No.	Name	Synonym	Mol. Form.	CAS RN	Mol. Wt.	Physical Form	mp/°C	bp/°C	den/ g cm⁻³	n_D	Solubility
3398	*N,N*-Diethylaniline		$C_{10}H_{15}N$	91-66-7	149.233	ye oil	-38.8	216.3	0.9307^{20}	1.5409^{20}	sl H_2O; s EtOH, ace, ctc; vs eth, chl
3399	Diethylarsine		$C_4H_{11}As$	692-42-2	134.052			105	1.1338^{24}	1.4709	vs ace, bz, eth, EtOH
3400	*N,N*-Diethylbenzamide		$C_{11}H_{15}NO$	1696-17-9	177.243			132^{5}			
3401	*o*-Diethylbenzene	1,2-Diethylbenzene	$C_{10}H_{14}$	135-01-3	134.218	liq	-31.2	184	0.8800^{20}	1.5035^{20}	i H_2O; msc EtOH, eth, ace, bz, lig, ctc
3402	*m*-Diethylbenzene	1,3-Diethylbenzene	$C_{10}H_{14}$	141-93-5	134.218	liq	-83.9	181.1	0.8602^{20}	1.4955^{20}	i H_2O; msc EtOH, eth, ace, bz, lig, ctc
3403	*p*-Diethylbenzene	1,4-Diethylbenzene	$C_{10}H_{14}$	105-05-5	134.218	liq	-42.83	183.7	0.8620^{20}	1.4967^{20}	i H_2O; msc EtOH, eth, ace, bz, lig, ctc
3404	*N,N*-Diethyl-1,4-benzenediamine		$C_{10}H_{16}N_2$	93-05-0	164.247			261			vs bz
3405	Diethyl benzylidenemalonate	Diethyl benzalmalonate	$C_{14}H_{16}O_4$	5292-53-5	248.275		32	216^{30}, 196^{14}	1.1045^{20}	1.5389^{20}	i H_2O; s EtOH, eth, ace, bz
3406	Diethyl benzylmalonate		$C_{14}H_{18}O_4$	607-81-8	250.291			300	1.076^{15}	1.4872^{20}	i H_2O; sl chl
3407	Diethyl benzylphosphonate		$C_{11}H_{17}O_3P$	1080-32-6	228.225			110^{2}		1.4930^{20}	s ctc
3408	Diethylbromoacetamide	2-Bromo-2-ethylbutanamide	$C_6H_{12}BrNO$	511-70-6	194.069		67				sl H_2O, chl; vs EtOH, eth, bz
3409	Diethyl 2-bromomalonate	Ethyl bromomalonate	$C_7H_{11}BrO_4$	685-87-0	239.064		-54	dec 254	1.4022^{25}	1.4521^{20}	i H_2O; msc EtOH, eth; s ace, ctc
3410	*N,N*-Diethylbutanamide		$C_8H_{17}NO$	1114-76-7	143.227			206	0.8884^{20}	1.4403^{25}	vs H_2O, EtOH
3411	Diethyl 2-butylmalonate	Pentane-1,1-dicarboxylic acid, diethyl ester	$C_{11}H_{20}O_4$	133-08-4	216.275			238	0.9764^{20}	1.4250^{20}	vs EtOH, eth
3412	Diethyl 2-butynedioate		$C_8H_{10}O_4$	762-21-0	170.163		0.8	184^{200}	1.0075^{20}	1.4425^{20}	s EtOH, eth, ctc
3413	Diethylcarbamazine citrate		$C_{16}H_{29}N_3O_8$	1642-54-2	391.416	cry	138				
3414	Diethylcarbamic chloride		$C_5H_{10}ClNO$	88-10-8	135.592			186			
3415	*N,N'*-Diethylcarbanilide		$C_{17}H_{20}N_2O$	85-98-3	268.353	cry (al)	79				i H_2O; vs EtOH; s chl
3416	Diethyl carbonate	Ethyl carbonate	$C_5H_{10}O_3$	105-58-8	118.131	liq	-43	126	0.9692^{25}	1.3845^{20}	i H_2O; s EtOH, eth, chl
3417	*O,O*-Diethyl chloridothionophosphate	Diethyl thiophosphoryl chloride	$C_4H_{10}ClO_2PS$	2524-04-1	188.613			45^{3}			s ctc
3418	Diethylchloroaluminum		$C_4H_{10}AlCl$	96-10-6	120.557			134^{70}			
3419	Diethyl chloromalonate	Ethyl chloromalonate	$C_7H_{11}ClO_4$	14064-10-9	194.613			222	1.2040^{20}	1.4327^{20}	i H_2O; msc EtOH, eth, chl; s CS_2
3420	Diethyl chlorophosphonate	Diethoxyphosphoryl chloride	$C_4H_{10}ClO_3P$	814-49-3	172.547			93.5	1.205^{19}	1.4170^{20}	
3421	Diethylcyanamide		$C_5H_{10}N_2$	617-83-4	98.146	liq	-80.6	188	0.854^{20}	1.4126^{25}	i H_2O; s EtOH, eth
3422	Diethyl 1,1-cyclobutanedicarboxylate		$C_{10}H_{16}O_4$	3779-29-1	200.232			224	1.0456^{20}	1.4330^{26}	vs EtOH; sl ctc
3423	1,1-Diethylcyclohexane		$C_{10}H_{20}$	78-01-3	140.266			179.5			
3424	Diethyl 1,1-cyclopropanedicarboxylate		$C_9H_{14}O_4$	1559-02-0	186.205			215; 100^{12}	1.055^{25}	1.4345^{18}	vs EtOH, eth
3425	Diethyl dibutylmalonate		$C_{15}H_{28}O_4$	596-75-8	272.381			150^{12}	0.9457^{20}	1.4341^{20}	i H_2O; s EtOH, eth, ctc
3426	Diethyl dicarbonate	Pyrocarbonic acid diethyl ester	$C_6H_{10}O_5$	1609-47-8	162.140			93^{18}	1.120^{20}	1.3960^{20}	vs ace, EtOH, lig
3427	Diethyl [(diethanolamino)methyl] phosphonate		$C_9H_{22}NO_5P$	2781-11-5	255.249	liq		$150^{0.01}$			
3428	5,5-Diethyldihydro-2*H*-1,3-oxazine-2,4(3*H*)-dione	Diethadione	$C_8H_{13}NO_3$	702-54-5	171.194	cry (eth)	97.5				
3429	Diethyl 1,4-dihydro-2,4,6-trimethyl-3,5-pyridinedicarboxylate	3,5-Diethoxycarbonyl-1,4-dihydrocollidine	$C_{14}H_{21}NO_4$	632-93-9	267.322	lt bl flr pl (al)	131				sl H_2O, EtOH, eth, CS_2; vs chl
3430	Diethyldimethyllead	Diethyldimethylplumbane	$C_6H_{16}Pb$	1762-27-2	295.4	col liq		51^{13}	1.79^{20}		
3431	Diethyl 2,6-dimethyl-3,5-pyridinedicarboxylate		$C_{13}H_{17}NO_4$	1149-24-2	251.279		71	301; 208^{40}			i H_2O; s EtOH, eth, bz, chl, lig
3432	Diethyl 3,5-dimethylpyrrole-2,4-dicarboxylate		$C_{12}H_{17}NO_4$	2436-79-5	239.268	nd (dil al)	137.8				i H_2O; sl EtOH, eth; s ace, bz, HOAc
3433	Diethyl disulfide		$C_4H_{10}S_2$	110-81-6	122.252	liq	-101.5	154.0	0.9931^{20}	1.5073^{20}	sl H_2O; msc EtOH, eth
3434	*N,N*-Diethyldodecanamide		$C_{16}H_{33}NO$	3352-87-2	255.439			166^{2}	0.847^{25}	1.4545^{20}	s chl
3435	Diethylene glycol	Diglycol	$C_4H_{10}O_3$	111-46-6	106.120	liq	-10.4	245.8	1.1197^{15}	1.4472^{20}	s H_2O, EtOH, eth, chl
3436	Diethylene glycol, bischloroformate	Oxydi-2,1-ethanediyl carbonochloridate	$C_6H_8Cl_2O_5$	106-75-2	231.031	liq		126^{5}	1.39^{20}	1.4542^{20}	
3437	Diethylene glycol diacetate		$C_8H_{14}O_5$	628-68-2	190.194		18	200	1.1068^{15}	1.4348^{20}	vs EtOH
3438	Diethylene glycol dibenzoate		$C_{18}H_{18}O_5$	120-55-8	314.333		33.5	280^{24}, 250^{1}	1.1690^{15}		vs H_2O, EtOH

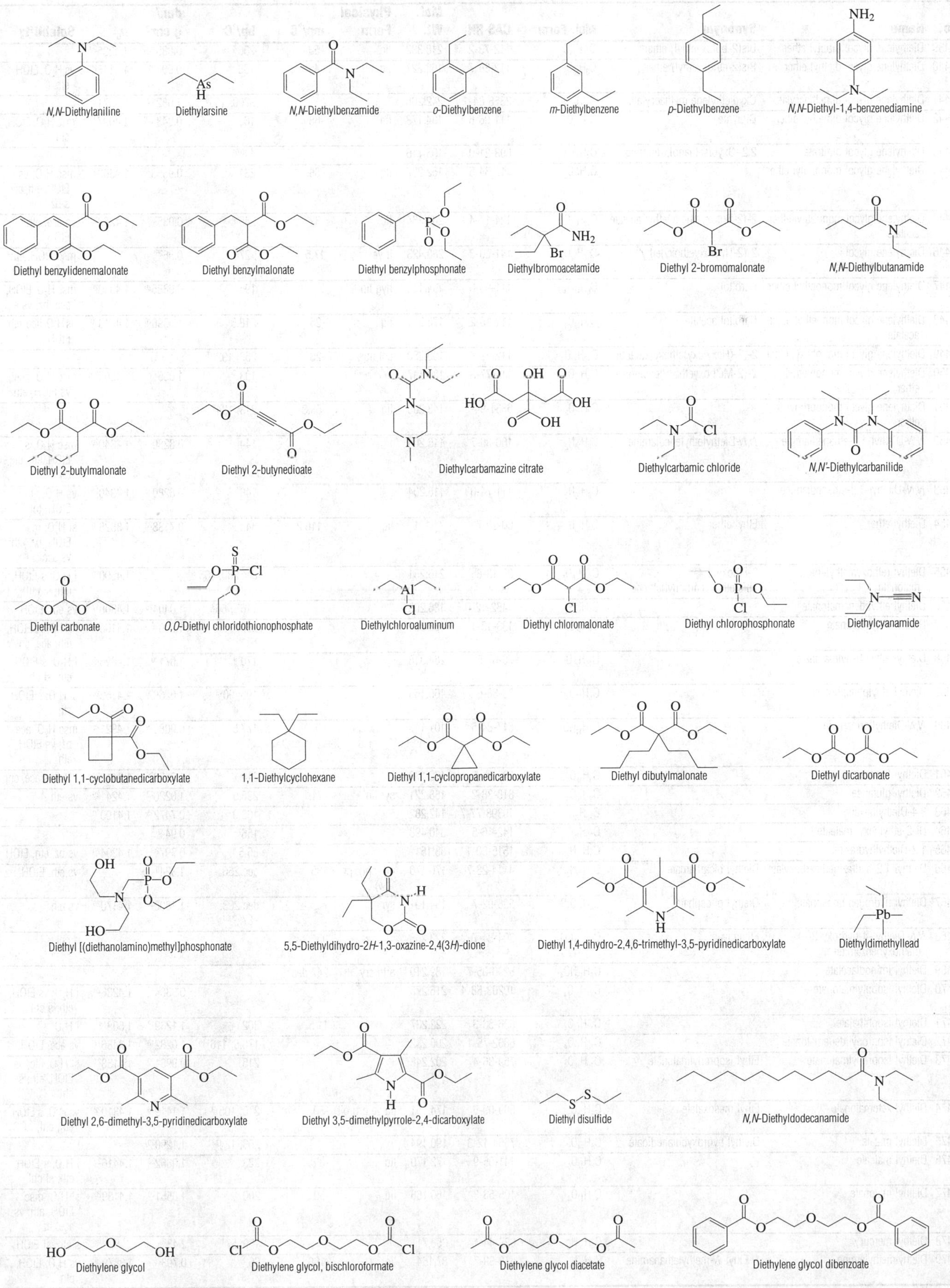
N,N-Diethylaniline
Diethylarsine
N,N-Diethylbenzamide
o-Diethylbenzene
m-Diethylbenzene
p-Diethylbenzene
N,N-Diethyl-1,4-benzenediamine
Diethyl benzylidenemalonate
Diethyl benzylmalonate
Diethyl benzylphosphonate
Diethylbromoacetamide
Diethyl 2-bromomalonate
N,N-Diethylbutanamide
Diethyl 2-butylmalonate
Diethyl 2-butynedioate
Diethylcarbamazine citrate
Diethylcarbamic chloride
N,N'-Diethylcarbanilide
Diethyl carbonate
O,O-Diethyl chloridothionophosphate
Diethylchloroaluminum
Diethyl chloromalonate
Diethyl chlorophosphonate
Diethylcyanamide
Diethyl 1,1-cyclobutanedicarboxylate
1,1-Diethylcyclohexane
Diethyl 1,1-cyclopropanedicarboxylate
Diethyl dibutylmalonate
Diethyl dicarbonate
Diethyl [(diethanolamino)methyl]phosphonate
5,5-Diethyldihydro-2H-1,3-oxazine-2,4(3H)-dione
Diethyl 1,4-dihydro-2,4,6-trimethyl-3,5-pyridinedicarboxylate
Diethyldimethyllead
Diethyl 2,6-dimethyl-3,5-pyridinedicarboxylate
Diethyl 3,5-dimethylpyrrole-2,4-dicarboxylate
Diethyl disulfide
N,N-Diethyldodecanamide
Diethylene glycol
Diethylene glycol, bischloroformate
Diethylene glycol diacetate
Diethylene glycol dibenzoate

No.	Name	Synonym	Mol. Form.	CAS RN	Mol. Wt.	Physical Form	mp/°C	bp/°C	den/ g cm-3	n_D	Solubility
3439	Diethylene glycol dibutyl ether	Bis(2-butoxyethyl) ether	$C_{12}H_{26}O_3$	112-73-2	218.332	liq	-60	256	0.885^{25}	1.4235^{20}	
3440	Diethylene glycol diethyl ether	Bis(2-ethoxyethyl) ether	$C_8H_{18}O_3$	112-36-7	162.227	liq	-45	188	0.9063^{20}	1.4115^{20}	vs H_2O, EtOH; s eth
3441	Diethylene glycol dimethacrylate	Oxydiethylene methacrylate	$C_{12}H_{18}O_5$	2358-84-1	242.268			>200; 150^{8}	1.0821^{20}	1.4571^{25}	
3442	Diethylene glycol dimethyl ether	Diglyme	$C_6H_{14}O_3$	111-96-6	134.173	liq	-68	162	0.9434^{20}	1.4097^{20}	msc H_2O, EtOH, eth
3443	Diethylene glycol dinitrate	2,2'-Oxybisethanol, dinitrate	$C_4H_8N_2O_7$	693-21-0	196.116			$44^{0.01}$			
3444	Diethylene glycol monobutyl ether		$C_8H_{18}O_3$	112-34-5	162.227	liq	-68	231	0.9553^{20}	1.4306^{20}	msc H_2O; vs EtOH, eth, ace; s bz
3445	Diethylene glycol monobutyl ether acetate	2-(2-Butoxyethoxy)ethyl acetate	$C_{10}H_{20}O_4$	124-17-4	204.264	liq	-32	245	0.985^{20}	1.4262^{20}	vs ace, eth, EtOH
3446	Diethylene glycol monododecanoate	2-(2-Hydroxyethoxy)ethyl laurate	$C_{16}H_{32}O_4$	141-20-8	288.423	lt ye	17.5	>270	0.96^{25}		msc EtOH, eth, ace; s bz, tol
3447	Diethylene glycol monoethyl ether	Carbitol	$C_6H_{14}O_3$	111-90-0	134.173	hyg liq		196	0.9885^{20}	1.4300^{20}	msc H_2O, EtOH, ace, bz; vs eth
3448	Diethylene glycol monoethyl ether acetate	Carbitol acetate	$C_8H_{16}O_4$	112-15-2	176.211	liq	-25	218.5	1.0096^{20}	1.4213^{20}	vs H_2O, ace, eth, EtOH
3449	Diethylene glycol monohexyl ether	2-[2-(Hexyloxy)ethoxy]ethanol	$C_{10}H_{22}O_3$	112-59-4	190.280	col liq	-28	258; 192^{100}			
3450	Diethylene glycol monomethyl ether	2-(2-Methoxyethoxy)ethanol	$C_5H_{12}O_3$	111-77-3	120.147			193	1.035^{20}	1.4264^{20}	msc H_2O, ace; vs EtOH, eth
3451	Diethylene glycol monopropyl ether		$C_7H_{16}O_3$	6881-94-3	148.200	liq	-53.3	213; 124^{4}			
3452	*N,N*-Diethyl-1,2-ethanediamine	*N,N*-Diethylethylenediamine	$C_6H_{16}N_2$	100-36-7	116.204			144	0.8280^{20}	1.4340^{20}	msc H_2O; s EtOH, eth, ctc, tol
3453	*N,N'*-Diethyl-1,2-ethanediamine		$C_6H_{16}N_2$	111-74-0	116.204			146	0.8280^{20}	1.4340^{20}	vs H_2O, eth, EtOH, tol
3454	Diethyl ether	Ethyl ether	$C_4H_{10}O$	60-29-7	74.121	liq	-116.2	34.5	0.7138^{20}	1.3526^{20}	sl H_2O; msc EtOH, bz, eth; vs ace
3455	Diethyl (ethoxymethylene) malonate	2-Ethoxy-1,1-bis(ethoxycarbonyl)ethene	$C_{10}H_{16}O_5$	87-13-8	216.231			dec 280; 165^{19}		1.4600^{20}	i H_2O; s EtOH, eth; sl chl
3456	Diethyl ethylidenemalonate		$C_9H_{14}O_4$	1462-12-0	186.205			116^{17}, 86^{3}	1.0404^{20}	1.4308^{17}	vs eth, EtOH
3457	Diethyl ethylmalonate		$C_9H_{16}O_4$	133-13-1	188.221			208; 98^{12}	1.006^{20}	1.4166^{20}	sl H_2O; vs EtOH, eth, ace, chl
3458	Diethyl ethylphenylmalonate		$C_{15}H_{20}O_4$	76-67-5	264.318			170^{19}	1.071^{20}	1.4896^{25}	i H_2O; s EtOH, eth; sl chl
3459	Diethyl ethylphosphonate		$C_6H_{15}O_3P$	78-38-6	166.155			198; 90^{16}	1.0259^{20}	1.4163^{20}	sl H_2O; s EtOH, eth
3460	*N,N*-Diethylformamide		$C_5H_{11}NO$	617-84-5	101.147			177.5	0.9080^{19}	1.4321^{25}	msc H_2O, ace, bz; vs EtOH, eth
3461	Diethyl fumarate		$C_8H_{12}O_4$	623-91-6	172.179		0.8	214	1.0452^{20}	1.4412^{20}	i H_2O; s ace, chl
3462	Diethyl glutarate		$C_9H_{16}O_4$	818-38-2	188.221	syr liq	-24.1	236.5	1.0220^{20}	1.4241^{20}	vs eth
3463	3,4-Diethylhexane		$C_{10}H_{22}$	19398-77-7	142.282			163.9	0.7472^{25}	1.4190^{20}	
3464	Di-2-ethylhexyl maleate		$C_{20}H_{36}O_4$	142-16-5	340.498			156^{7}	0.94^{20}		
3465	1,2-Diethylhydrazine		$C_4H_{12}N_2$	1615-80-1	88.151			85.5	0.797^{26}	1.4204^{20}	vs bz, eth, EtOH
3466	Diethyl 1,2-hydrazinedicarboxylate	Diethyl bicarbamate	$C_6H_{12}N_2O_4$	4114-28-7	176.170	nd (chl), pr (w)	135	dec 250	1.324^{8}		vs eth, EtOH
3467	Diethyl hydrogen phosphate	Diethyl phosphate	$C_4H_{11}O_4P$	598-02-7	154.101	syr		dec 203; $87^{0.0001}$	1.1800^{20}	1.4170^{20}	vs eth
3468	*N,N*-Diethyl-4-hydroxy-3-methoxybenzamide	Ethamivan	$C_{12}H_{17}NO_3$	304-84-7	223.268		95				s chl
3469	Diethyl iminodiacetate		$C_8H_{15}NO_4$	6290-05-7	189.210	orth cry	247 dec				
3470	Diethyl isobutylmalonate		$C_{11}H_{20}O_4$	10203-58-4	216.275				0.9804^{20}	1.4236^{20}	i H_2O; vs EtOH, eth; s chl
3471	Diethyl isophthalate		$C_{12}H_{14}O_4$	636-53-3	222.237		11.5	302	1.1239^{17}	1.508^{18}	i H_2O
3472	Diethyl isopropylidenemalonate		$C_{10}H_{16}O_4$	6802-75-1	200.232			176.5; 116^{14}	1.0282^{18}	1.4486^{17}	vs ace, EtOH
3473	Diethyl isopropylmalonate	Ethyl isopropylmalonate	$C_{10}H_{18}O_4$	759-36-4	202.248			215	0.9961^{20}	1.4188^{21}	sl H_2O, ctc; vs EtOH, eth; s chl
3474	Diethyl ketomalonate	Ethyl mesoxalate	$C_7H_{10}O_5$	609-09-6	174.151	pa ye grn oil	-30	210; 105^{19}	1.1419^{16}	1.4310^{22}	vs H_2O; s EtOH, eth, chl; i CS_2
3475	Diethyl malate	Diethyl hydroxybutanedioate	$C_8H_{14}O_5$	7554-12-3	190.194			253; 124^{13}	1.1290^{20}		
3476	Diethyl maleate		$C_8H_{12}O_4$	141-05-9	172.179	liq	-8.8	223	1.0662^{20}	1.4416^{20}	i H_2O; s EtOH, eth; sl chl
3477	Diethyl malonate		$C_7H_{12}O_4$	105-53-3	160.168	liq	-50	200	1.0551^{20}	1.4139^{20}	sl H_2O; msc EtOH, eth; vs ace, bz
3478	Diethyl mercury		$C_4H_{10}Hg$	627-44-1	258.71			159; 57^{16}	2.43^{20}		s eth; sl EtOH
3479	Diethylmethylamine	*N*-Ethyl-*N*-methylethanamine	$C_5H_{13}N$	616-39-7	87.164	liq	-196	66	0.703^{25}	1.3879^{25}	vs H_2O, EtOH, eth

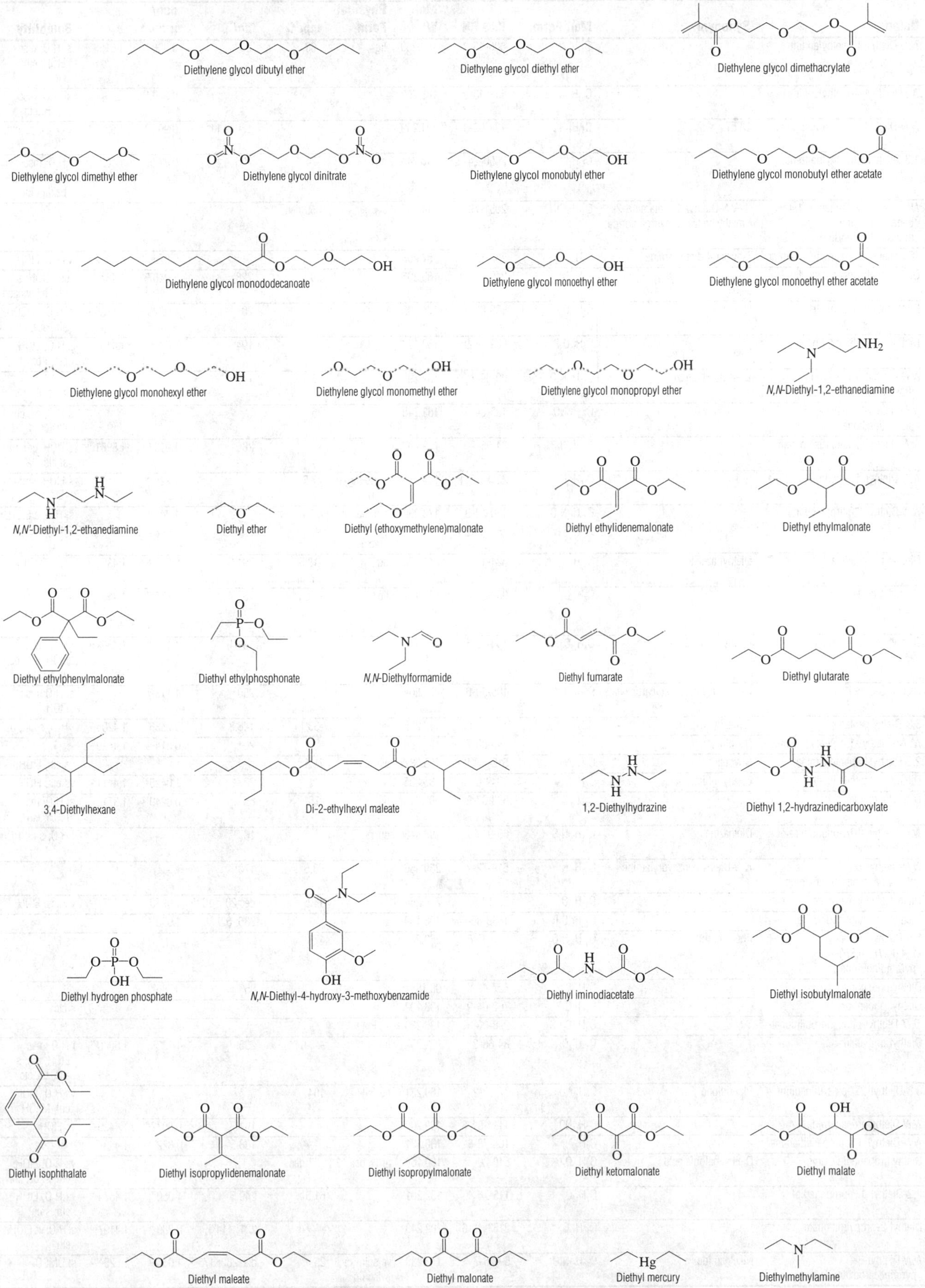

Diethylene glycol dibutyl ether
Diethylene glycol diethyl ether
Diethylene glycol dimethacrylate
Diethylene glycol dimethyl ether
Diethylene glycol dinitrate
Diethylene glycol monobutyl ether
Diethylene glycol monobutyl ether acetate
Diethylene glycol monododecanoate
Diethylene glycol monoethyl ether
Diethylene glycol monoethyl ether acetate
Diethylene glycol monohexyl ether
Diethylene glycol monomethyl ether
Diethylene glycol monopropyl ether
N,N-Diethyl-1,2-ethanediamine
N,N'-Diethyl-1,2-ethanediamine
Diethyl ether
Diethyl (ethoxymethylene)malonate
Diethyl ethylidenemalonate
Diethyl ethylmalonate
Diethyl ethylphenylmalonate
Diethyl ethylphosphonate
N,N-Diethylformamide
Diethyl fumarate
Diethyl glutarate
3,4-Diethylhexane
Di-2-ethylhexyl maleate
1,2-Diethylhydrazine
Diethyl 1,2-hydrazinedicarboxylate
Diethyl hydrogen phosphate
N,N-Diethyl-4-hydroxy-3-methoxybenzamide
Diethyl iminodiacetate
Diethyl isobutylmalonate
Diethyl isophthalate
Diethyl isopropylidenemalonate
Diethyl isopropylmalonate
Diethyl ketomalonate
Diethyl malate
Diethyl maleate
Diethyl malonate
Diethyl mercury
Diethylmethylamine

No.	Name	Synonym	Mol. Form.	CAS RN	Mol. Wt.	Physical Form	mp/°C	bp/°C	den/ g cm^{-3}	n_D	Solubility
3480	*N,N*-Diethyl-2-methylaniline		$C_{11}H_{17}N$	606-46-2	163.260	liq	-60	209	0.9286^{20}	1.5153^{20}	sl H_2O; msc EtOH, eth; s ctc
3481	*N,N*-Diethyl-4-methylaniline		$C_{11}H_{17}N$	613-48-9	163.260			229	0.9242^{16}		sl H_2O; msc EtOH, eth
3482	*N,N*-Diethyl-3-methylbenzamide	DEET	$C_{12}H_{17}NO$	134-62-3	191.269			160^{19}, 111^{1}	0.996^{20}	1.5212^{20}	vs H_2O, bz, eth, EtOH
3483	1,3-Diethyl-5-methylbenzene		$C_{11}H_{16}$	2050-24-0	148.245	liq	-74.1	205	0.8748^{20}	1.5027^{20}	i H_2O; msc EtOH, eth, ace, bz, lig, ctc
3484	*N4,N4*-Diethyl-2-methyl-1,4-benzenediamine, monohydrochloride	4-*N,N*-Diethyl-1,4-diamino-2-methylbenzene, hydrochloride	$C_{11}H_{19}ClN_2$	2051-79-8	214.735	cry	250 dec				
3485	*N,N*-Diethyl-3-methylbutanamide	Isovaleryl diethylamide	$C_9H_{19}NO$	533-32-4	157.253			211	0.8764^{20}	1.4422^{20}	vs eth, EtOH
3486	Diethyl methylenesuccinate		$C_9H_{14}O_4$	2409-52-1	186.205		58.5	228	1.0467^{20}	1.4377^{20}	msc EtOH; s eth, bz; vs ace
3487	Diethyl methylmalonate		$C_8H_{14}O_4$	609-08-5	174.195			201	1.0225^{20}	1.4126^{20}	sl H_2O; vs EtOH, eth, ace, chl
3488	Diethyl methylphosphonate		$C_5H_{13}O_3P$	683-08-9	152.129			194	1.0406^{30}	1.4101^{30}	s H_2O, EtOH, eth; i bz
3489	*N,N*-Diethyl-4-methyl-1-piperazinecarboxamide	Diethylcarbamazine	$C_{10}H_{21}N_3O$	90-89-1	199.293		48	110^{3}			
3490	3,3-Diethyl-5-methyl-2,4-piperidinedione		$C_{10}H_{17}NO_2$	125-64-4	183.248		75.5				s H_2O, bz, chl, EtOH
3491	*N,N*-Diethyl-1-naphthalenamine		$C_{14}H_{17}N$	84-95-7	199.292			285	1.013^{20}	1.5961^{20}	s EtOH, eth, bz; sl ctc
3492	*N,N*-Diethyl-4-nitroaniline		$C_{10}H_{14}N_2O_2$	2216-15-1	194.230	ye nd (lig) pl (al)	77.5		1.225^{25}		s EtOH; sl lig
3493	*N,N*-Diethyl-4-nitrosoaniline		$C_{10}H_{14}N_2O$	120-22-9	178.230	grn mcl pr (eth) grn lf (ace)	87.5		1.24^{15}		sl H_2O; s EtOH, eth, ace, chl
3494	Diethyl nonanedioate	Diethyl azelate	$C_{13}H_{24}O_4$	624-17-9	244.328	liq	-18.5	291.5	0.9729^{20}	1.4351^{20}	i H_2O; s EtOH, eth
3495	Diethyl oxalate		$C_6H_{10}O_4$	95-92-1	146.141	liq	-40.6	185.7	1.0785^{20}	1.4101^{20}	sl H_2O; msc EtOH, eth, ace; s ctc
3496	Diethyl oxobutanedioate	Diethyl oxalacetate	$C_8H_{12}O_5$	108-56-5	188.178			131^{24}	1.131^{20}	1.4561^{17}	i H_2O; msc EtOH, eth, bz; vs ace
3497	Diethyl 3-oxo-1,5-pentanedioate	Diethyl 1,3-acetonedicarboxylate	$C_9H_{14}O_5$	105-50-0	202.204			250	1.113^{20}		sl H_2O; msc EtOH
3498	3,3-Diethylpentane	Tetraethylmethane	C_9H_{20}	1067-20-5	128.255	liq	-33.1	146.3	0.7536^{20}	1.4206^{20}	i H_2O; s eth, bz
3499	N^1,N^1-Diethyl-1,4-pentanediamine	Novoldiamine	$C_9H_{22}N_2$	140-80-7	158.284			201	0.814^{20}	1.4429^{20}	
3500	2,2-Diethyl-4-pentenamide	Novonal	$C_9H_{17}NO$	512-48-1	155.237	wh pow	75.5				vs eth, EtOH
3501	Diethyl 2-pentenedioate	Diethyl glutaconate	$C_9H_{14}O_4$	2049-67-4	186.205			237	1.0496^{20}	1.4411^{20}	vs eth, EtOH
3502	Diethylperoxide		$C_4H_{10}O_2$	628-37-5	90.121	liq	-70	65	0.8240^{19}	1.3715^{17}	sl H_2O; msc EtOH, eth
3503	*N,N*-Diethyl-10*H*-phenothiazine-10-ethanamine	Diethazine	$C_{18}H_{22}N_2S$	60-91-3	298.446	oil		$167^{0.5}$			i H_2O; s dil HCl
3504	*N,N*-Diethyl-α-phenylbenzenemethanamine	*N,N*-Diethylbenzhydrylamine	$C_{17}H_{21}N$	519-72-2	239.356		58.5	170^{17}			
3505	Diethyl phenylmalonate		$C_{13}H_{16}O_4$	83-13-6	236.264		16.5	dec 205; 168^{12}	1.0950^{20}	1.4977^{20}	vs ace, EtOH
3506	Diethyl phenylphosphonite		$C_{10}H_{15}O_2P$	1638-86-4	198.199			235; 62^{1}	1.032^{16}		
3507	5,5-Diethyl-1-phenyl-2,4,6(1*H*,3*H*,5*H*)-pyrimidinetrione	Phenetharbital	$C_{14}H_{16}N_2O_3$	357-67-5	260.288		178				vs EtOH
3508	Diethylphosphine		$C_4H_{11}P$	627-49-6	90.104			85	0.786^{20}		
3509	Diethyl phosphonate		$C_4H_{11}O_3P$	762-04-9	138.102			54^{6}			s ctc
3510	*O,O'*-Diethyl phosphorodithionate		$C_4H_{11}O_2PS_2$	298-06-6	186.233						s H_2O
3511	Diethyl phthalate		$C_{12}H_{14}O_4$	84-66-2	222.237	liq	-40.5	295	1.232^{14}	1.5000^{21}	i H_2O; msc EtOH, eth; s ace, bz, ctc
3512	3,3-Diethyl-2,4-piperidinedione	Piperidione	$C_9H_{15}NO_2$	77-03-2	169.221	nd (w)	104				vs H_2O, EtOH, chl, MeOH
3513	*N,N*-Diethylpropanamide		$C_7H_{15}NO$	1114-51-8	129.200			191	0.8972^{20}	1.4425^{20}	vs EtOH
3514	*N,N*-Diethyl-1,3-propanediamine		$C_7H_{18}N_2$	104-78-9	130.231			168.5	0.822^{20}	1.443^{20}	
3515	Diethylpropanedioic acid	Diethylmalonic acid	$C_7H_{12}O_4$	510-20-3	160.168	pr (w,bz)	127 dec				vs H_2O, EtOH, eth; sl bz, chl
3516	2,2-Diethyl-1,3-propanediol		$C_7H_{16}O_2$	115-76-4	132.201		61.5	240.5	1.050^{20}	1.4574^{25}	vs H_2O, EtOH, eth; s chl
3517	Diethyl 2-propylmalonate		$C_{10}H_{18}O_4$	2163-48-6	202.248			221; 114^{22}	0.989^{20}	1.4197^{20}	sl H_2O; vs EtOH, eth
3518	*N,N*-Diethyl-3-pyridinecarboxamide	Nikethamide	$C_{10}H_{14}N_2O$	59-26-7	178.230	ye solid or visc liq	25	dec 280; 175^{25}	1.060^{25}	1.525^{20}	sl DMSO

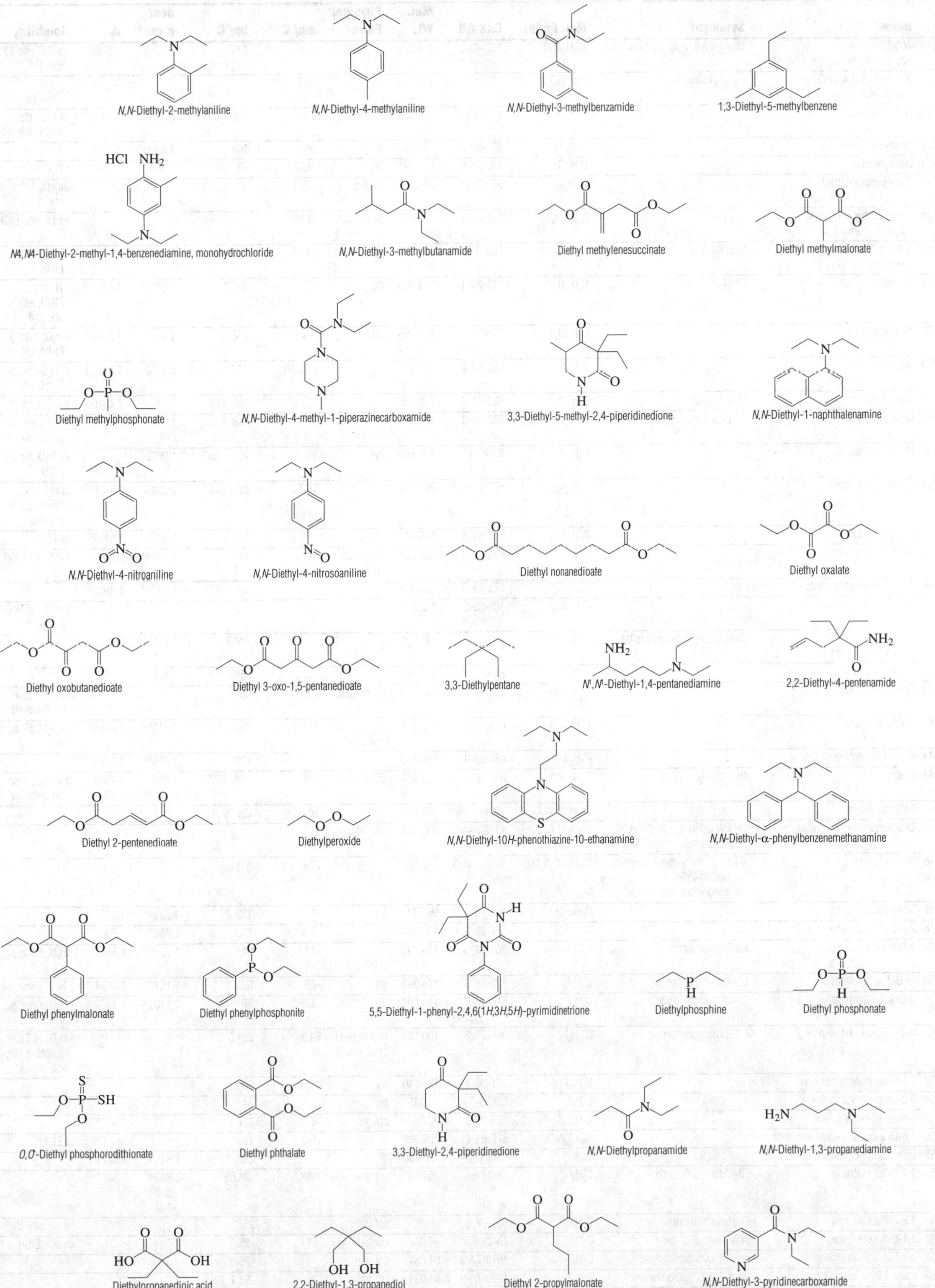
N,N-Diethyl-2-methylaniline
N,N-Diethyl-4-methylaniline
N,N-Diethyl-3-methylbenzamide
1,3-Diethyl-5-methylbenzene
N4,N4-Diethyl-2-methyl-1,4-benzenediamine, monohydrochloride
N,N-Diethyl-3-methylbutanamide
Diethyl methylenesuccinate
Diethyl methylmalonate
Diethyl methylphosphonate
N,N-Diethyl-4-methyl-1-piperazinecarboxamide
3,3-Diethyl-5-methyl-2,4-piperidinedione
N,N-Diethyl-1-naphthalenamine
N,N-Diethyl-4-nitroaniline
N,N-Diethyl-4-nitrosoaniline
Diethyl nonanedioate
Diethyl oxalate
Diethyl oxobutanedioate
Diethyl 3-oxo-1,5-pentanedioate
3,3-Diethylpentane
N',N'-Diethyl-1,4-pentanediamine
2,2-Diethyl-4-pentenamide
Diethyl 2-pentenedioate
Diethylperoxide
N,N-Diethyl-10H-phenothiazine-10-ethanamine
N,N-Diethyl-α-phenylbenzenemethanamine
Diethyl phenylmalonate
Diethyl phenylphosphonite
5,5-Diethyl-1-phenyl-2,4,6(1H,3H,5H)-pyrimidinetrione
Diethylphosphine
Diethyl phosphonate
O,O-Diethyl phosphorodithionate
Diethyl phthalate
3,3-Diethyl-2,4-piperidinedione
N,N-Diethylpropanamide
N,N-Diethyl-1,3-propanediamine
Diethylpropanedioic acid
2,2-Diethyl-1,3-propanediol
Diethyl 2-propylmalonate
N,N-Diethyl-3-pyridinecarboxamide

No.	Name	Synonym	Mol. Form.	CAS RN	Mol. Wt.	Physical Form	mp/°C	bp/°C	den/ g cm-3	n_D	Solubility
3519	*N,N*-Diethyl-4-pyridinecarboxamide	Isonicotinic acid diethylamide	$C_{10}H_{14}N_2O$	530-40-5	178.230			119^{1}		1.525^{20}	vs H_2O, ace, eth, EtOH
3520	3,3-Diethyl-2,4(1*H*,3*H*)-pyridinedione	Pyrithyldione	$C_9H_{13}NO_2$	77-04-3	167.205		90.7				
3521	Diethyl sebacate		$C_{14}H_{26}O_4$	110-40-7	258.354		2.5	305; 188^{19}	0.9646^{20}	1.4306^{20}	sl H_2O, ctc; s EtOH, ace; i bz
3522	Diethyl selenide		$C_4H_{10}Se$	627-53-2	137.08	pa ye	55	108	1.2300^{20}	1.4768^{20}	
3523	Diethylsilane		$C_4H_{12}Si$	542-91-6	88.224	liq	-134.3	57	0.6843^{20}	1.3921^{20}	i H_2O
3524	*trans*-Diethylstilbestrol		$C_{18}H_{20}O_2$	56-53-1	268.351	pl (bz)	170.5				vs eth, EtOH, chl
3525	*trans*-Diethylstilbestrol dipropanoate	Clinestrol	$C_{24}H_{28}O_4$	130-80-3	380.477	pr (MeOH)	104				vs bz, eth, EtOH
3526	*trans*-Diethylstilbestrol monomethyl ether	Mestilbol	$C_{19}H_{22}O_2$	18839-90-2	282.377	nd (bz-peth)	117.5	$190^{0.3}$			vs ace, eth, EtOH
3527	Diethyl succinate	Ethyl succinate	$C_8H_{14}O_4$	123-25-1	174.195	liq	-21	217.7	1.0402^{20}	1.4201^{20}	i H_2O; msc EtOH, eth; s ace, chl
3528	Diethyl sulfate		$C_4H_{10}O_4S$	64-67-5	154.185	oil	-24	208	1.172^{25}	1.3989^{20}	i H_2O; msc EtOH, eth
3529	Diethyl sulfide		$C_4H_{10}S$	352-93-2	90.187	liq	-103.91	92.1	0.8362^{20}	1.4430^{20}	sl H_2O, ctc; s EtOH, eth
3530	Diethyl sulfite	Ethyl sulfite	$C_4H_{10}O_3S$	623-81-4	138.185			158; 51^{13}	1.1^{20}	1.4310^{20}	s EtOH, eth
3531	Diethyl sulfone	Ethyl sulfone	$C_4H_{10}O_2S$	597-35-3	122.186	orth pl	73.5	248	1.357^{20}		s H_2O, eth; vs bz; i peth
3532	Diethyl sulfoxide		$C_4H_{10}OS$	70-29-1	106.186	syr	14	104^{25}, 90^{15}	1.0092^{22}		vs H_2O, eth, EtOH
3533	Diethyl *DL*-tartrate		$C_8H_{14}O_6$	57968-71-5	206.193		18.7	281; 158^{14}	1.2046^{20}	1.4438^{20}	sl H_2O; msc EtOH, eth; s ace, ctc
3534	Diethyl telluride		$C_4H_{10}Te$	627-54-3	185.72	red-ye		137.5	1.599^{15}	1.5182^{15}	vs EtOH
3535	Diethyl terephthalate		$C_{12}H_{14}O_4$	636-09-9	222.237	mcl pr (al, peth)	44	302	1.0989^{45}		i H_2O; vs EtOH, eth
3536	Diethyl thiodipropionate		$C_{10}H_{18}O_4S$	673-79-0	234.313			174^{15}, 121^{2}	1.1034^{20}	1.4655^{20}	
3537	*N,N'*-Diethylthiourea		$C_5H_{12}N_2S$	105-55-5	132.227		78	dec			s H_2O, EtOH; vs eth; sl ctc
3538	*N,N*-Diethyl-1,1,1-trimethylsilanamine	(Diethylamino)trimethylsilane	$C_7H_{19}NSi$	996-50-9	145.319			126.3	0.7627^{20}	1.4112^{20}	
3539	Diethyltrisulfide		$C_4H_{10}S_3$	3600-24-6	154.317		-72.6	85^{26}	1.1082^{20}	1.5689^{13}	
3540	*N,N*-Diethylurea		$C_5H_{12}N_2O$	634-95-7	116.161	pl, nd (eth)	75	$95^{0.02}$			vs H_2O, EtOH, bz, lig; s eth
3541	*N,N'*-Diethylurea		$C_5H_{12}N_2O$	623-76-7	116.161	tab (lig), hyg nd (al)	112.5	263	1.0415^{25}	1.4616^{40}	vs H_2O, EtOH, eth
3542	Diethyl vinylphosphonate		$C_6H_{13}O_3P$	682-30-4	164.139			110^{2}	1.068^{25}	1.4290^{20}	
3543	Diethyl zinc	Zinc diethyl	$C_4H_{10}Zn$	557-20-0	123.531	col liq	-28	118; 80^{200}	1.2065^{20}	1.4936^{20}	dec H_2O; msc eth, peth, bz
3544	Difenoconazole		$C_{19}H_{17}Cl_2N_3O_3$	119446-68-3	406.262		76	$220^{0.03}$			
3545	Difenzoquat methyl sulfate	1*H*-Pyrazolium, 1,2-dimethyl-3,5-diphenyl-, methyl sulfate	$C_{18}H_{20}N_2O_4S$	43222-48-6	360.428		157				
3546	Diflubenzuron	*N*-[[(4-Chlorophenyl)amino] carbonyl]-2,6-difluorobenzamide	$C_{14}H_9ClF_2N_2O_2$	35367-38-5	310.683		239				
3547	Difluoroacetic acid		$C_2H_2F_2O_2$	381-73-7	96.033	liq	-1	133	1.526^{25}	1.3470^{20}	
3548	2,4-Difluoroaniline		$C_6H_5F_2N$	367-25-9	129.108	liq	-7.5	170	1.268^{25}	1.5063^{20}	
3549	*o*-Difluorobenzene	1,2-Difluorobenzene	$C_6H_4F_2$	367-11-3	114.093	liq	-47.1	94	1.1599^{18}	1.4451^{18}	i H_2O; s ace, bz, chl
3550	*m*-Difluorobenzene	1,3-Difluorobenzene	$C_6H_4F_2$	372-18-9	114.093	liq	-69.12	82.6	1.1572^{20}	1.4374^{20}	i H_2O; s ace, bz
3551	*p*-Difluorobenzene	1,4-Difluorobenzene	$C_6H_4F_2$	540-36-3	114.093	liq	-23.55	89	1.1701^{20}	1.4422^{20}	i H_2O; s ace, bz; sl ctc
3552	4,4'-Difluoro-1,1'-biphenyl	4,4'-Difluorodiphenyl	$C_{12}H_8F_2$	398-23-2	190.189	mcl pr (al) lf (w)	94.5	254.5			i H_2O; vs EtOH, bz, chl; s eth, ace
3553	1,1-Difluorocyclohexane		$C_6H_{10}F_2$	371-90-4	120.140	liq		99.5			
3554	3,3-Difluorocyclopropene		$C_3H_2F_2$	56830-75-2	76.045	liq		34			
3555	Difluorodimethylsilane		$C_2H_6F_2Si$	353-66-2	96.152	col gas	-87.5	2.5			
3556	1,5-Difluoro-2,4-dinitrobenzene		$C_6H_2F_2N_2O_4$	327-92-4	204.088		75.5	132^{2}			sl EtOH
3557	Difluorodiphenylsilane		$C_{12}H_{10}F_2Si$	312-40-3	220.290			246; 157^{50}	1.145^{17}	1.5221^{25}	
3558	1,1-Difluoroethane	Ethylidene difluoride	$C_2H_4F_2$	75-37-6	66.050	col gas	-118.6	-24.05	0.896^{25} (p>1 atm)	1.3011^{-72}	
3559	1,2-Difluoroethane	Ethylene difluoride	$C_2H_4F_2$	624-72-6	66.050	vol liq		26			vs bz, eth, chl
3560	1,1-Difluoroethene	Vinylidene fluoride	$C_2H_2F_2$	75-38-7	64.034	col gas	-144	-85.7			vs eth, EtOH
3561	*cis*-1,2-Difluoroethene	*cis*-1,2-Difluoroethylene	$C_2H_2F_2$	1630-77-9	64.034	col gas		-26			
3562	*trans*-1,2-Difluoroethene	*trans*-1,2-Difluoroethylene	$C_2H_2F_2$	1630-78-0	64.034	col gas		-53.1			
3563	Difluoromethane	Methylene fluoride	CH_2F_2	75-10-5	52.024	col gas	-136.8 tp	-51.6	1.2139^{-52}		i H_2O; s EtOH

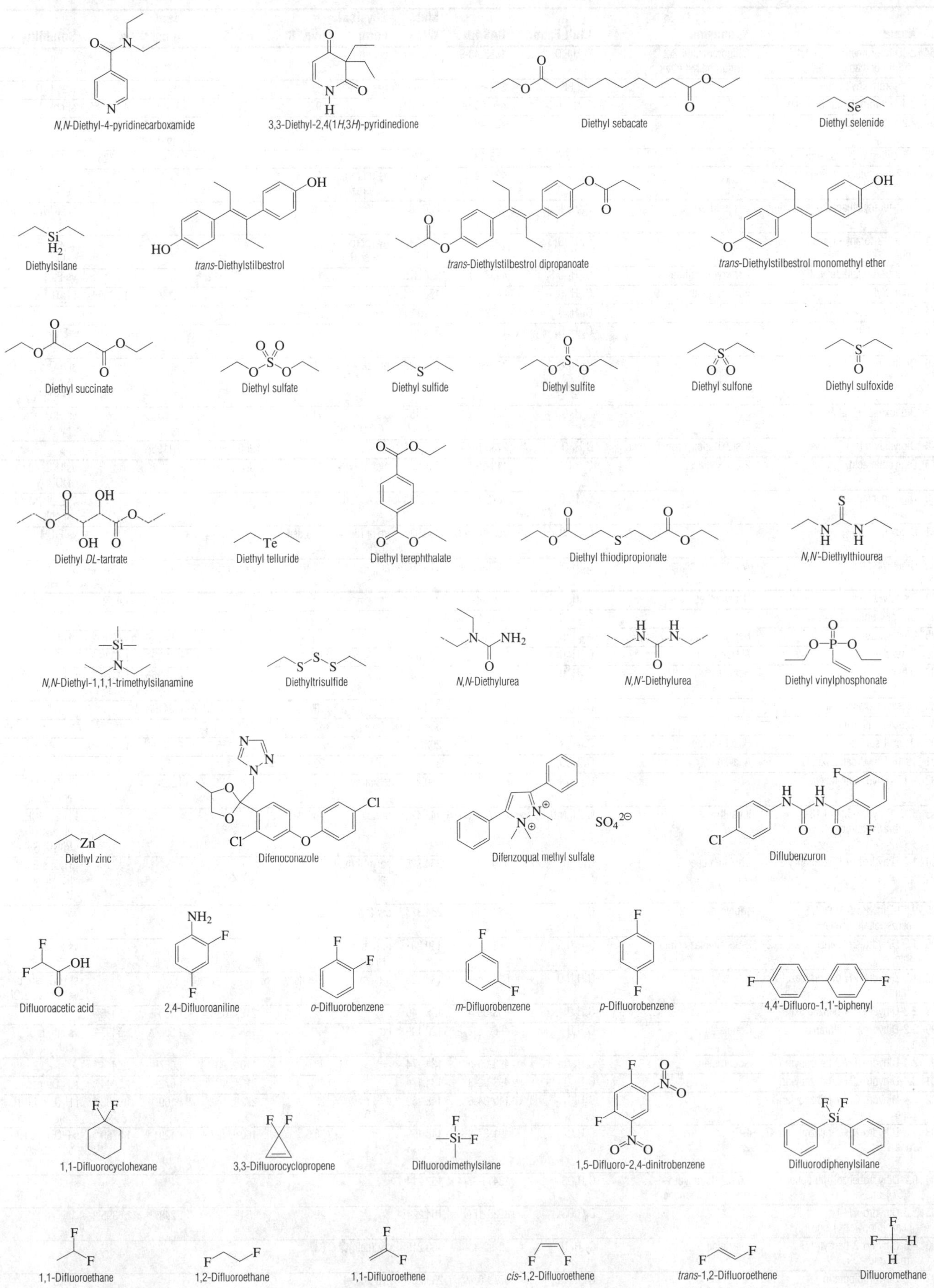
N,N-Diethyl-4-pyridinecarboxamide
3,3-Diethyl-2,4(1H,3H)-pyridinedione
Diethyl sebacate
Diethyl selenide
Diethylsilane
trans-Diethylstilbestrol
trans-Diethylstilbestrol dipropanoate
trans-Diethylstilbestrol monomethyl ether
Diethyl succinate
Diethyl sulfate
Diethyl sulfide
Diethyl sulfite
Diethyl sulfone
Diethyl sulfoxide
Diethyl DL-tartrate
Diethyl telluride
Diethyl terephthalate
Diethyl thiodipropionate
N,N'-Diethylthiourea
N,N-Diethyl-1,1,1-trimethylsilanamine
Diethyltrisulfide
N,N-Diethylurea
N,N'-Diethylurea
Diethyl vinylphosphonate
Diethyl zinc
Difenoconazole
Difenzoquat methyl sulfate
Diflubenzuron
Difluoroacetic acid
2,4-Difluoroaniline
o-Difluorobenzene
m-Difluorobenzene
p-Difluorobenzene
4,4'-Difluoro-1,1'-biphenyl
1,1-Difluorocyclohexane
3,3-Difluorocyclopropene
Difluorodimethylsilane
1,5-Difluoro-2,4-dinitrobenzene
Difluorodiphenylsilane
1,1-Difluoroethane
1,2-Difluoroethane
1,1-Difluoroethene
cis-1,2-Difluoroethene
trans-1,2-Difluoroethene
Difluoromethane

No.	Name	Synonym	Mol. Form.	CAS RN	Mol. Wt.	Physical Form	mp/°C	bp/°C	den/ g cm-3	n_D	Solubility
3564	2-(Difluoromethoxy)-1,1,1-trifluoroethane	Difluoromethyl 2,2,2-trifluoroethyl ether	$C_3H_3F_5O$	1885-48-9	150.047	col liq		29			
3565	Difluoromethylborane		CH_3BF_2	373-64-8	63.843	gas		-78.5^{287}			reac H_2O
3566	2,4-Difluoro-1-nitrobenzene		$C_6H_3F_2NO_2$	446-35-5	159.091		9.8	207	1.4571^{14}	1.5149^{14}	sl chl
3567	2,2-Difluoropropane		$C_3H_6F_2$	420-45-1	80.077	col gas	-104.8	-0.4	0.9205^{20} (p>1 atm)	1.2904^{20}	
3568	1,3-Difluoro-2-propanol		$C_3H_6F_2O$	453-13-4	96.076			127; 55^{34}	1.24^{25}	1.3725^{20}	
3569	Di-2-furanylethanedione		$C_{10}H_6O_4$	492-94-4	190.153	ye nd (al), cry (bz)	166.3				sl H_2O; s EtOH, eth, bz, chl
3570	Di-2-furanylethanedione dioxime	α-Furildioxime	$C_{10}H_8N_2O_4$	522-27-0	220.182		167				sl EtOH, eth, bz, lig
3571	1,5-Di-2-furanyl-1,4-pentadien-3-one		$C_{13}H_{10}O_3$	886-77-1	214.216	hyg pr (peth) ye pr (lig)	60.5	181^{4}			vs eth, EtOH, chl
3572	Difurfuryl disulfide	Furfuryl disulfide	$C_{10}H_{10}O_2S_2$	4437-20-1	226.315		10	167^{13}, $112^{0.5}$			vs EtOH
3573	Difurfuryl ether	Furfuryl ether	$C_{10}H_{10}O_3$	4437-22-3	178.184			101^{2}	1.1405^{20}	1.5088^{20}	i H_2O
3574	Digitonin		$C_{56}H_{92}O_{29}$	11024-24-1	1229.312		237.5				
3575	Digitoxigenin		$C_{23}H_{34}O_4$	143-62-4	374.514		253				s EtOH; vs MeOH
3576	Digitoxin		$C_{41}H_{64}O_{13}$	71-63-6	764.939	pr (dil al)	255.5				sl H_2O; vs EtOH; s eth, chl, MeOH, py
3577	Digitoxose		$C_6H_{12}O_4$	527-52-6	148.157	cry (MeOH+eth)	112				vs H_2O, ace; s py, AcOEt
3578	Diglycidyl ether	Bis(2,3-epoxypropyl) ether	$C_6H_{10}O_3$	2238-07-5	130.141			260	1.1195^{20}		
3579	Diglycolic acid	2,2'-Oxydiacetic acid	$C_4H_6O_5$	110-99-6	134.088	mcl pr (w + 1)	148	dec			vs H_2O, eth, EtOH
3580	Digoxigenin		$C_{23}H_{34}O_5$	1672-46-4	390.513	pr (AcOEt)	222				vs EtOH, MeOH; sl chl
3581	Digoxin		$C_{41}H_{64}O_{14}$	20830-75-5	780.939	trc pl (dil al, py)	249 dec				vs EtOH
3582	Diheptylamine	*N*-Heptyl-1-heptanamine	$C_{14}H_{31}N$	2470-68-0	213.403	nd	31.5	271; 135^{9}	0.7956^{21}		sl H_2O; s EtOH; vs eth
3583	Diheptyl ether	Heptyl ether	$C_{14}H_{30}O$	629-64-1	214.387			258.5	0.8008^{20}	1.4275^{20}	vs eth, EtOH
3584	Diheptyl phthalate		$C_{22}H_{34}O_4$	3648-21-3	362.503			360			
3585	Diheptyl sulfide	Heptyl sulfide	$C_{14}H_{30}S$	629-65-2	230.453		70	298	0.8416^{20}	1.4606^{20}	i H_2O; s eth
3586	Dihexylamine	*N*-Hexyl-1-hexanamine	$C_{12}H_{27}N$	143-16-8	185.349	liq	-13.1	236; 75^{1}	0.7889^{20}	1.4339^{20}	s EtOH, eth
3587	Dihexyl ether	Hexyl ether	$C_{12}H_{26}O$	112-58-3	186.333			226	0.7936^{20}	1.4204^{20}	i H_2O; s eth; sl ctc
3588	Dihexyl hexanedioate		$C_{18}H_{34}O_4$	110-33-8	314.461	liq	-9	348; 182.5^{4}	0.941^{20}		
3589	Dihexyl phthalate		$C_{20}H_{30}O_4$	84-75-3	334.450			210^{5}			
3590	Dihexyl sulfide	Hexyl sulfide	$C_{12}H_{26}S$	6294-31-1	202.399			230; 136^{20}	0.8411^{20}	1.4586^{20}	
3591	15,16-Dihydroaflatoxin G_1	Aflatoxin G_2	$C_{17}H_{14}O_7$	7241-98-7	330.289		239.3				
3592	9,10-Dihydroanthracene		$C_{14}H_{12}$	613-31-0	180.245	tab or pr	111	305	1.215^{20}		i H_2O; s EtOH, eth, bz, chl
3593	6,15-Dihydro-5,9,14,18-anthrazinetetrone	Indanthrene	$C_{28}H_{14}N_2O_4$	81-77-6	442.422	bl nd	485 dec				i H_2O, EtOH, eth, ace, bz; s $PhNO_2$, dil alk
3594	1,2-Dihydrobenz[j]aceanthrylene	Cholanthrene	$C_{20}H_{14}$	479-23-2	254.325	pa ye lf (bz-al)	170.4				i H_2O; s EtOH, bz, HOAc, lig, tol
3595	9,10-Dihydro-9,10[1',2']-benzenoanthracene	Triptycene	$C_{20}H_{14}$	477-75-8	254.325	cry (cyhex)	256				
3596	1,3-Dihydro-2*H*-benzimidazole-2-thione	2-Benzimidazolethiol	$C_7H_6N_2S$	583-39-1	150.201	pl (dil al or NH_3)	298				vs EtOH
3597	1,3-Dihydro-2*H*-benzimidazol-2-one		$C_7H_6N_2O$	615-16-7	134.135	lf (w or al)	318 dec				sl H_2O, eth, bz; s ace; vs EtOH
3598	2,3-Dihydro-1,4-benzodioxin		$C_8H_8O_2$	493-09-4	136.149			212; 103^{6}	1.180^{20}	1.5485^{20}	
3599	2,3-Dihydrobenzofuran	Coumaran	C_8H_8O	496-16-2	120.149	liq	-21.5	188.5	1.058^{25}	1.5497^{20}	vs eth, EtOH, chl
3600	3,4-Dihydro-1*H*-2-benzopyran	Isochroman	$C_9H_{10}O$	493-05-0	134.174		4	110^{25}, 90^{12}	1.067^{25}	1.5444^{20}	
3601	3,4-Dihydro-2*H*-1-benzopyran		$C_9H_{10}O$	493-08-3	134.174		4.8	215; 98^{18}	1.072^{20}	1.5444^{20}	s H_2O; msc os
3602	3,4-Dihydro-2*H*-1-benzopyran-2-one		$C_9H_8O_2$	119-84-6	148.159	lf	25	272	1.169^{18}	1.5563^{20}	i H_2O; sl EtOH, eth, ctc; s chl
3603	2,3-Dihydro-4*H*-1-benzopyran-4-one	4-Chromanone	$C_9H_8O_2$	491-37-2	148.159		36.5	160^{50}, 127^{13}	1.1291^{100}	1.5750	s EtOH; vs eth, ace, bz, chl; sl ctc
3604	6,7-Dihydrobenzo[b]thiophen-4(5*H*)-one	4,5,6,7-Tetrahydro-4-benzothiophenone	C_8H_8OS	13414-95-4	152.214						sl chl
3605	2,3-Dihydro-4*H*-1-benzothiopyran-4-one		C_9H_8OS	3528-17-4	164.224		29	154^{12}	1.2487^{14}	1.6395^{20}	
3606	4,5-Dihydro-2-benzyl-1*H*-imidazole	Tolazoline	$C_{10}H_{12}N_2$	59-98-3	160.215	cry (peth)	67				
3607	7,8-Dihydrobiopterin		$C_9H_{13}N_5O_3$	6779-87-9	239.231	hyg nd (w)					s H_2O

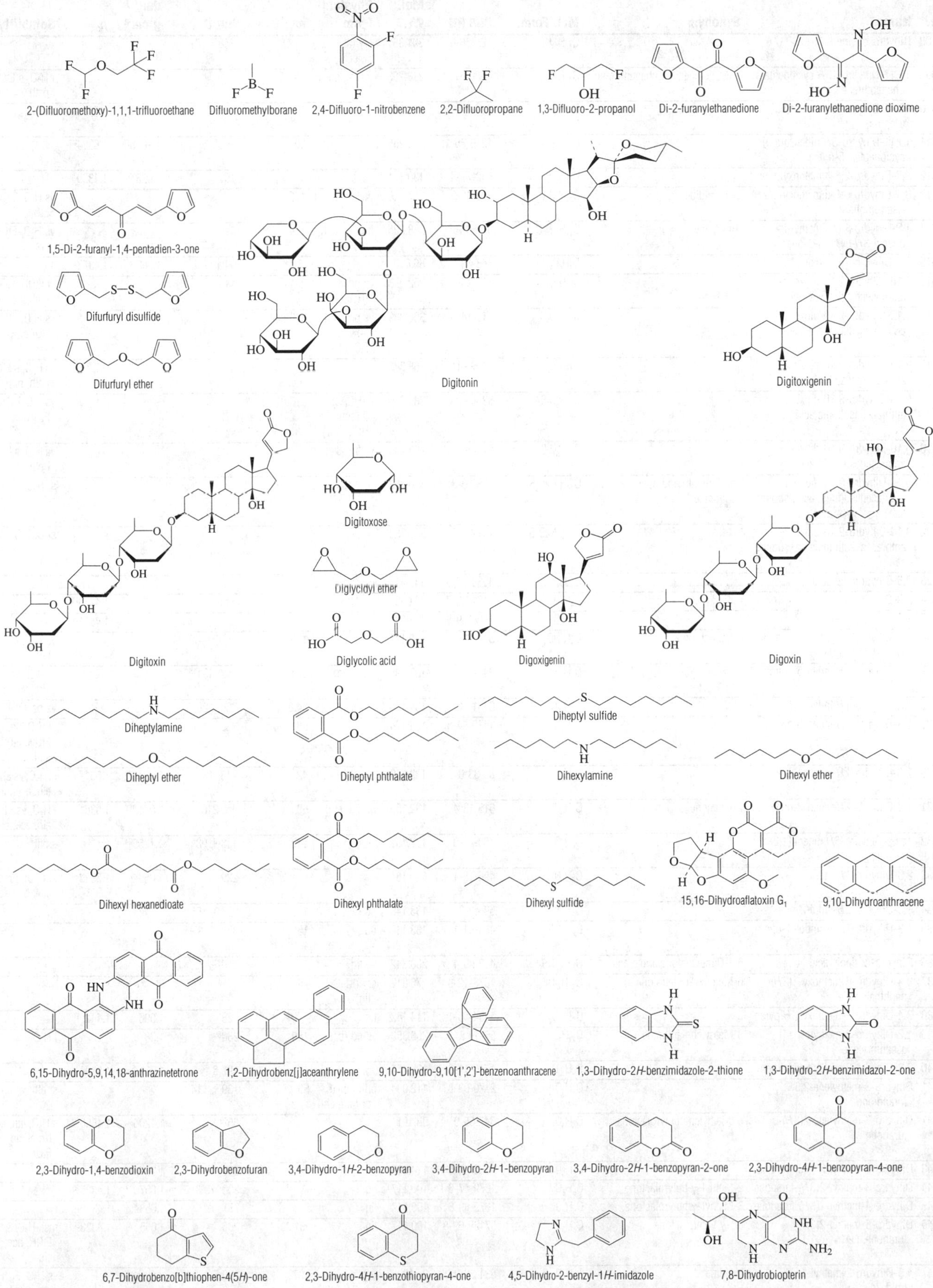

2-(Difluoromethoxy)-1,1,1-trifluoroethane
Difluoromethylborane
2,4-Difluoro-1-nitrobenzene
2,2-Difluoropropane
1,3-Difluoro-2-propanol
Di-2-furanylethanedione
Di-2-furanylethanedione dioxime
1,5-Di-2-furanyl-1,4-pentadien-3-one
Difurfuryl disulfide
Difurfuryl ether
Digitonin
Digitoxigenin
Digitoxin
Digitoxose
Diglycidyl ether
Diglycolic acid
Digoxigenin
Digoxin
Diheptylamine
Diheptyl ether
Diheptyl phthalate
Diheptyl sulfide
Dihexylamine
Dihexyl ether
Dihexyl hexanedioate
Dihexyl phthalate
Dihexyl sulfide
15,16-Dihydroaflatoxin G1
9,10-Dihydroanthracene
6,15-Dihydro-5,9,14,18-anthrazinetetrone
1,2-Dihydrobenz[j]aceanthrylene
9,10-Dihydro-9,10[1',2']-benzenoanthracene
1,3-Dihydro-2H-benzimidazole-2-thione
1,3-Dihydro-2H-benzimidazol-2-one
2,3-Dihydro-1,4-benzodioxin
2,3-Dihydrobenzofuran
3,4-Dihydro-1H-2-benzopyran
3,4-Dihydro-2H-1-benzopyran
3,4-Dihydro-2H-1-benzopyran-2-one
2,3-Dihydro-4H-1-benzopyran-4-one
6,7-Dihydrobenzo[b]thiophen-4(5H)-one
2,3-Dihydro-4H-1-benzothiopyran-4-one
4,5-Dihydro-2-benzyl-1H-imidazole
7,8-Dihydrobiopterin

No.	Name	Synonym	Mol. Form.	CAS RN	Mol. Wt.	Physical Form	mp/°C	bp/°C	den/ g cm^{-3}	n_D	Solubility
3608	Dihydrocodeine		$C_{18}H_{23}NO_3$	125-28-0	301.381	cry (aq, MeOH)	112.5	248^{15}			
3609	16,17-Dihydro-15*H*-cyclopenta[a] phenanthrene	1,2-Cyclopentenophenanthrene	$C_{17}H_{14}$	482-66-6	218.293	nd (al, petr)	135.5				i H_2O; s EtOH, peth
3610	10,11-Dihydro-5*H*-dibenz[b,f] azepine		$C_{14}H_{13}N$	494-19-9	195.260						s chl
3611	10,11-Dihydro-5*H*-dibenzo[a,d] cyclohepten-5-one		$C_{15}H_{12}O$	1210-35-1	208.255		30	203^{7}	1.1635^{20}	1.6324^{20}	
3612	2,5-Dihydro-2,5-dimethoxyfuran		$C_6H_{10}O_3$	332-77-4	130.141			161	1.073^{25}	1.4339^{20}	
3613	3,4-Dihydro-6,7-dimethoxy-1(2*H*) -isoquinolinone	Corydaldine	$C_{11}H_{13}NO_3$	493-49-2	207.226	mcl pr (w, al)	175				vs H_2O, bz, eth, EtOH
3614	1,2-Dihydro-1,5-dimethyl-2-phenyl-3*H*-pyrazol-3-one	Antipyrine	$C_{11}H_{12}N_2O$	60-80-0	188.225	lf or sc (eth, bz)	114	319			vs H_2O, EtOH
3615	2,3-Dihydro-1,4-dioxin		$C_4H_6O_2$	543-75-9	86.090			94.1	1.0836^{20}	1.4372^{20}	s ctc
3616	9,10-Dihydro-9,10-dioxo-2-anthracenecarboxylic acid		$C_{15}H_8O_4$	117-78-2	252.223	ye nd (HOAc)	291	sub			sl EtOH, HOAc; i eth, bz; s ace
3617	9,10-Dihydro-9,10-dioxo-1,5-anthracenedisulfonic acid		$C_{14}H_8O_8S_2$	117-14-6	368.339	ye nd (HCl +4w) pl (dil HOAc)	310 dec				vs H_2O, EtOH, HOAc
3618	9,10-Dihydro-9,10-dioxo-2,6-anthracenedisulfonic acid		$C_{14}H_8O_8S_2$	84-50-4	368.339						vs H_2O; s EtOH; i eth, bz
3619	9,10-Dihydro-9,10-dioxo-1-anthracenesulfonic acid		$C_{14}H_8O_5S$	82-49-5	288.276	lf (HOAc) ye lf (conc HCl, +3w)	216.0				vs H_2O, HOAc; s EtOH
3620	9,10-Dihydro-9,10-dioxo-2-anthracenesulfonic acid		$C_{14}H_8O_5S$	84-48-0	288.276	ye lf (+3w)					vs H_2O; s EtOH; i eth
3621	9,10-Dihydro-9,10-dioxo-1-anthracenesulfonic acid, sodium salt	Sodium anthraquinone-1-sulfonate	$C_{14}H_7NaO_5S$	128-56-3	310.258	ye lf (w)					sl H_2O
3622	9,10-Dihydro-9,10-dioxo-2-anthracenesulfonic acid, sodium salt		$C_{14}H_7NaO_5S$	131-08-8	310.258						sl DMSO
3623	7,8-Dihydrofolic acid		$C_{19}H_{21}N_7O_6$	4033-27-6	443.413	ye cry					
3624	2,3-Dihydrofuran		C_4H_6O	1191-99-7	70.090			54.5	0.927^{25}	1.4239^{20}	
3625	2,5-Dihydrofuran		C_4H_6O	1708-29-8	70.090					1.4311^{20}	
3626	2,3-Dihydro-3-hydroxy-1-methyl-1*H*-indole-5,6-dione	Adrenochrome	$C_9H_9NO_3$	54-06-8	179.172		125 dec				vs H_2O, EtOH; i eth, bz
3627	2,3-Dihydro-1*H*-inden-5-amine		$C_9H_{11}N$	24425-40-9	133.190	nd (peth)	37.5	248; 131^{15}			sl H_2O, chl; s eth, ace, bz
3628	2,3-Dihydro-1*H*-inden-1-ol		$C_9H_{10}O$	6351-10-6	134.174	pl (peth)	54.8	220; 128^{12}			vs bz, EtOH, chl
3629	2,3-Dihydro-1*H*-inden-5-ol		$C_9H_{10}O$	1470-94-6	134.174		58	253			sl H_2O, peth; vs EtOH, eth; s sulf
3630	2,3-Dihydro-1*H*-inden-1-one		C_9H_8O	83-33-0	132.159	ta, nd (w + 3)	42	243; 129^{12}	1.0943^{40}	1.561^{25}	sl H_2O; vs EtOH, eth, ace, chl
3631	1,3-Dihydro-2*H*-inden-2-one	2-Indanone	C_9H_8O	615-13-4	132.159	nd (al, eth)	59	dec 218	1.0712^{69}	1.538^{67}	i H_2O; vs EtOH, eth, ace, chl
3632	1a,6a-Dihydro-6*H*-indeno[1,2-b] oxirene		C_9H_8O	768-22-9	132.159		24.5	113^{20}, 98^{6}	1.1255^{24}		s chl
3633	2,3-Dihydro-1*H*-indole		C_8H_9N	496-15-1	119.164			229	1.069^{20}	1.5923^{20}	sl H_2O; s eth, ace, bz
3634	1,3-Dihydro-2*H*-indol-2-one		C_8H_7NO	59-48-3	133.148	nd (w)	128	227^{23}, 195^{17}			s H_2O, EtOH, eth
3635	2,3-Dihydro-1*H*-isoindol-1-one		C_8H_7NO	480-91-1	133.148	nd (w)	151	338; 103^{18}			vs eth, EtOH, chl
3636	Dihydro-α-lipoic acid	6,8-Dimercaptooctanoic acid	$C_8H_{16}O_2S_2$	462-20-4	208.342	ye liq		$145^{0.2}$			
3637	3,4-Dihydro-6-methoxy-1(2*H*)-naphthalenone	6-Methoxy-α-tetralone	$C_{11}H_{12}O_2$	1078-19-9	176.212	cry (MeOH, lig)	78	171^{11}			
3638	3,4-Dihydro-2-methoxy-2*H*-pyran		$C_6H_{10}O_2$	4454-05-1	114.142	liq		128	1.006	1.4420^{20}	
3639	1,2-Dihydro-3-methylbenz[j] aceanthrylene	3-Methylcholanthrene	$C_{21}H_{16}$	56-49-5	268.352	ye nd (bz)	180	280^{80}	1.28^{20}		i H_2O
3640	2,3-Dihydro-2-methylbenzofuran		$C_9H_{10}O$	1746-11-8	134.174			197.5	1.061^{25}	1.5308	
3641	Dihydro-3-methylene-2,5-furandione		$C_5H_4O_3$	2170-03-8	112.084	orth bipym pr (eth, chl)	69	139^{30}, 114^{18}			sl eth; vs chl
3642	Dihydro-3-methylene-2(3*H*)-furanone	α-Methylene butyrolactone	$C_5H_6O_2$	547-65-9	98.101			85^{10}	1.1206^{20}	1.4650^{20}	s H_2O, eth, ace, bz; sl ctc; vs EtOH
3643	Dihydro-3-methyl-2,5-furandione		$C_5H_6O_3$	4100-80-5	114.100		34	239	1.22^{25}		
3644	Dihydro-3-methyl-2(3*H*)-furanone	2-Methyl-γ-butyrolactone	$C_5H_8O_2$	1679-47-6	100.117	liq		200; 79^{10}	1.0570^{20}	1.4325^{20}	
3645	Dihydro-4-methyl-2(3*H*)-furanone	3-Methyl-γ-butyrolactone	$C_5H_8O_2$	1679-49-8	100.117	liq		76^{11}	1.058^{20}	1.4339^{20}	
3646	Dihydro-5-methyl-2(3*H*)-furanone, (±)	(±)-γ-Valerolactone	$C_5H_8O_2$	57129-69-8	100.117	liq	-31	206	1.0551^{20}	1.4328^{20}	msc H_2O; s EtOH, ace; sl ctc
3647	4,5-Dihydro-2-methyl-1*H*-imidazole	Lysidine	$C_4H_8N_2$	534-26-9	84.120	hyg	107	196.5			vs H_2O, EtOH; i eth; s chl

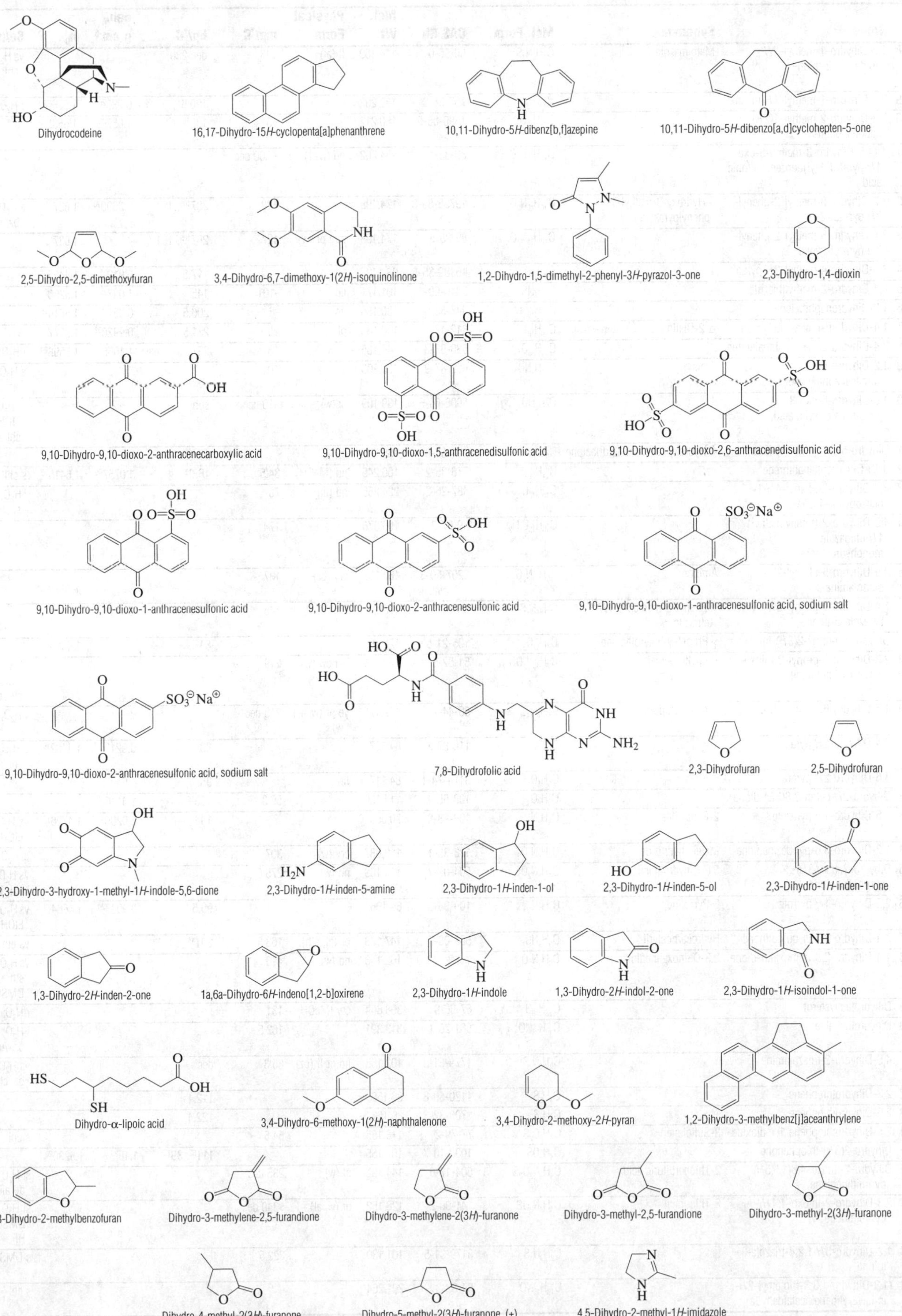
Dihydrocodeine
16,17-Dihydro-15H-cyclopenta[a]phenanthrene
10,11-Dihydro-5H-dibenz[b,f]azepine
10,11-Dihydro-5H-dibenzo[a,d]cyclohepten-5-one
2,5-Dihydro-2,5-dimethoxyfuran
3,4-Dihydro-6,7-dimethoxy-1(2H)-isoquinolinone
1,2-Dihydro-1,5-dimethyl-2-phenyl-3H-pyrazol-3-one
2,3-Dihydro-1,4-dioxin
9,10-Dihydro-9,10-dioxo-2-anthracenecarboxylic acid
9,10-Dihydro-9,10-dioxo-1,5-anthracenedisulfonic acid
9,10-Dihydro-9,10-dioxo-2,6-anthracenedisulfonic acid
9,10-Dihydro-9,10-dioxo-1-anthracenesulfonic acid
9,10-Dihydro-9,10-dioxo-2-anthracenesulfonic acid
9,10-Dihydro-9,10-dioxo-1-anthracenesulfonic acid, sodium salt
9,10-Dihydro-9,10-dioxo-2-anthracenesulfonic acid, sodium salt
7,8-Dihydrofolic acid
2,3-Dihydrofuran
2,5-Dihydrofuran
2,3-Dihydro-3-hydroxy-1-methyl-1H-indole-5,6-dione
2,3-Dihydro-1H-inden-5-amine
2,3-Dihydro-1H-inden-1-ol
2,3-Dihydro-1H-inden-5-ol
2,3-Dihydro-1H-inden-1-one
1,3-Dihydro-2H-inden-2-one
1a,6a-Dihydro-6H-indeno[1,2-b]oxirene
2,3-Dihydro-1H-indole
1,3-Dihydro-2H-indol-2-one
2,3-Dihydro-1H-isoindol-1-one
Dihydro-α-lipoic acid
3,4-Dihydro-6-methoxy-1(2H)-naphthalenone
3,4-Dihydro-2-methoxy-2H-pyran
1,2-Dihydro-3-methylbenz[j]aceanthrylene
2,3-Dihydro-2-methylbenzofuran
Dihydro-3-methylene-2,5-furandione
Dihydro-3-methylene-2(3H)-furanone
Dihydro-3-methyl-2,5-furandione
Dihydro-3-methyl-2(3H)-furanone
Dihydro-4-methyl-2(3H)-furanone
Dihydro-5-methyl-2(3H)-furanone, (±)
4,5-Dihydro-2-methyl-1H-imidazole

No.	Name	Synonym	Mol. Form.	CAS RN	Mol. Wt.	Physical Form	mp/°C	bp/°C	den/ g cm^{-3}	n_D	Solubility
3648	1,3-Dihydro-1-methyl-2*H*-imidazole-2-thione	Methimazole	$C_4H_6N_2S$	60-56-0	114.169	lf (al)	146	dec 280			vs H_2O; s EtOH, chl; sl eth, bz, lig
3649	2,3-Dihydro-1-methyl-1*H*-indene		$C_{10}H_{12}$	767-58-8	132.202			190.6	0.938^{25}	1.5266^{20}	i H_2O
3650	3,4-Dihydro-2-methyl-1(2*H*)-naphthalenone		$C_{11}H_{12}O$	1590-08-5	160.212		15	136^{16}	1.057^{25}	1.5535^{20}	
3651	4-(4,5-Dihydro-3-methyl-5-oxo-1*H*-pyrazol-1-yl)benzenesulfonic acid		$C_{10}H_{10}N_2O_4S$	89-36-1	254.262	nd (w+1)	≈300 dec				
3652	1,2-Dihydro-5-methyl-2-phenyl-3*H*-pyrazol-3-one	5-Hydroxy-3-methyl-1-phenylpyrazole	$C_{10}H_{10}N_2O$	19735-89-8	174.198		128	287^{105}, 191^{17}	1.2600^{20}	1.637	s H_2O, EtOH; sl bz; i peth
3653	2,4-Dihydro-5-methyl-2-phenyl-3*H*-pyrazol-3-one		$C_{10}H_{10}N_2O$	89-25-8	174.198	mcl pr (w)	127	287^{105}, 191^{17}		1.637	
3654	3,6-Dihydro-4-methyl-2*H*-pyran		$C_6H_{10}O$	16302-35-5	98.142			117.5	0.912^{25}	1.4495^{20}	
3655	4,5-Dihydro-2-methylthiazole		C_4H_7NS	2346-00-1	101.171	liq	-101	145	1.067^{25}	1.5200^{20}	
3656	1,2-Dihydronaphthalene		$C_{10}H_{10}$	447-53-0	130.186	liq	-8	206.5	0.9974^{20}	1.5814^{20}	
3657	1,4-Dihydronaphthalene	Δ 2-Dialin	$C_{10}H_{10}$	612-17-9	130.186	pl	25	211.5	0.9928^{33}	1.5577^{20}	
3658	3,4-Dihydro-2(1*H*)-naphthalenone		$C_{10}H_{10}O$	530-93-8	146.185		18	237	1.1055^{27}	1.5598^{20}	i H_2O; s eth, bz
3659	1,2-Dihydro-5-nitroacenaphthylene		$C_{12}H_9NO_2$	602-87-9	199.205		103				s H_2O, EtOH, eth, lig
3660	1,6-Dihydro-6-oxo-3-pyridinecarboxylic acid		$C_6H_5NO_3$	5006-66-6	139.109	nd(w)	310 dec	sub			sl H_2O, tfa; i EtOH, eth, bz, chl
3661	Dihydro-5-pentyl-2(3*H*)-furanone	4-Hydroxynonanoic acid lactone	$C_9H_{16}O_2$	104-61-0	156.222	oil		134^{12}			
3662	9,10-Dihydrophenanthrene		$C_{14}H_{12}$	776-35-2	180.245	nd (MeOH)	34.5	168^{15}	1.0757^{40}	1.6415^{20}	s chl
3663	2,3-Dihydro-2-phenyl-4*H*-1-benzopyran-4-one		$C_{15}H_{12}O_2$	487-26-3	224.255	nd (lig)	76				i H_2O; s ace, bz; sl ctc
3664	4,5-Dihydro-2-(phenylmethyl)-1*H*-imidazole, monohydrochloride		$C_{10}H_{13}ClN_2$	59-97-2	196.676		174				
3665	4,5-Dihydro-5-phenyl-2-oxazolamine	Aminorex	$C_9H_{10}N_2O$	2207-50-3	162.187	cry (bz)	137				
3666	1,4-Dihydro-1-phenyl-5*H*-tetrazole-5-thione	1-Phenyl-5-mercapto-1*H*-tetrazole	$C_7H_6N_4S$	86-93-1	178.215		145				
3667	Dihydro-5-propyl-2(3*H*)-furanone	γ-Propyl-γ-butyrolactone	$C_7H_{12}O_2$	105-21-5	128.169			84^5		1.4385^{25}	
3668	2,3-Dihydro-6-propyl-2-thioxo-4(1*H*)-pyrimidinone	Propylthiouracil	$C_7H_{10}N_2OS$	51-52-5	170.231	w pow (w)	219				sl H_2O, chl, DMSO, EtOH; i eth, bz
3669	1,7-Dihydro-6*H*-purine-6-thione	6-Mercaptopurine	$C_5H_4N_4S$	50-44-2	152.178	ye pr (w, + l w)	313 dec				i H_2O; s alk
3670	3,4-Dihydro-2*H*-pyran		C_5H_8O	110-87-2	84.117			86	0.921^{19}	1.4402^{19}	s H_2O, EtOH; sl chl
3671	3,6-Dihydro-2*H*-pyran		C_5H_8O	3174-74-1	84.117	liq		95	0.94^{19}		
3672	Dihydro-2*H*-pyran-2,6(3*H*)-dione		$C_5H_6O_3$	108-55-4	114.100		56.3	158^{15}	1.4110^{20}		
3673	4,5-Dihydro-1*H*-pyrazole	2-Pyrazoline	$C_3H_6N_2$	109-98-8	70.093			144	1.0200^{17}	1.4796^{17}	vs H_2O, eth, EtOH
3674	1,2-Dihydro-3,6-pyridazinedione	Maleic hydrazide	$C_4H_4N_2O_2$	123-33-1	112.087	cry (w)	307				sl H_2O, EtOH, tfa
3675	Dihydro-2,4(1*H*,3*H*)-pyrimidinedione	5,6-Dihydrouracil	$C_4H_6N_2O_2$	504-07-4	114.103	nd (w)	275.5				vs H_2O; s EtOH, chl, MeOH
3676	2,5-Dihydro-1*H*-pyrrole	3-Pyrroline	C_4H_7N	109-96-6	69.106			90.5	0.9097^{20}	1.4664^{20}	vs H_2O, ace, eth, EtOH
3677	3,4-Dihydro-2(1*H*)-quinolinone	Hydrocarbostyril	C_9H_9NO	553-03-7	147.173	pr (al, eth)	163.5	201^{45}			vs eth, EtOH
3678	1,4-Dihydro-2,3-quinoxalinedione	2,3-Quinoxalinediol	$C_8H_6N_2O_2$	15804-19-0	162.146	nd (w)	410				vs H_2O; sl EtOH, eth; s bz, DMSO, HOAc
3679	Dihydrotachysterol		$C_{28}H_{46}O$	67-96-9	398.664	cry (MeOH)	131				i H_2O; s os
3680	Dihydrothebaine		$C_{19}H_{23}NO_3$	561-25-1	313.391		162.5				i H_2O; s EtOH, bz, AcOEt
3681	4,5-Dihydro-2-thiazolamine		$C_3H_6N_2S$	1779-81-3	102.158	nd or lf (bz)	85.3	dec			vs H_2O, EtOH, bz, chl
3682	2,3-Dihydrothiophene		C_4H_6S	1120-59-8	86.156			112.1			
3683	2,5-Dihydrothiophene		C_4H_6S	1708-32-3	86.156			122.4			
3684	2,5-Dihydrothiophene 1,1-dioxide	3-Sulfolene	$C_4H_6O_2S$	77-79-2	118.155		64.5				s chl
3685	Dihydro-2(3*H*)-thiophenone		C_4H_6OS	1003-10-7	102.155			111^{52}, 39^1	1.18^{25}	1.5230^{20}	
3686	Dihydro-2-thioxo-4,6(1*H*,5*H*)-pyrimidinedione	2-Thiobarbituric acid	$C_4H_4N_2O_2S$	504-17-6	144.152	pl (w)	235 dec				sl H_2O; s EtOH, dil alk, dil HCl
3687	2,3-Dihydro-2-thioxo-4(1*H*)-pyrimidinone	2-Thiouracil	$C_4H_4N_2OS$	141-90-2	128.152	pr (w, al)	>340 dec				sl H_2O, EtOH, DMSO; s anh HF
3688	1,2-Dihydro-3*H*-1,2,4-triazole-3-thione		$C_2H_3N_3S$	3179-31-5	101.130		222.5				s DMSO
3689	(1,3-Dihydro-1,3,3-trimethyl-2*H*-indol-2-ylidene)acetaldehyde		$C_{13}H_{15}NO$	84-83-3	201.264						s chl

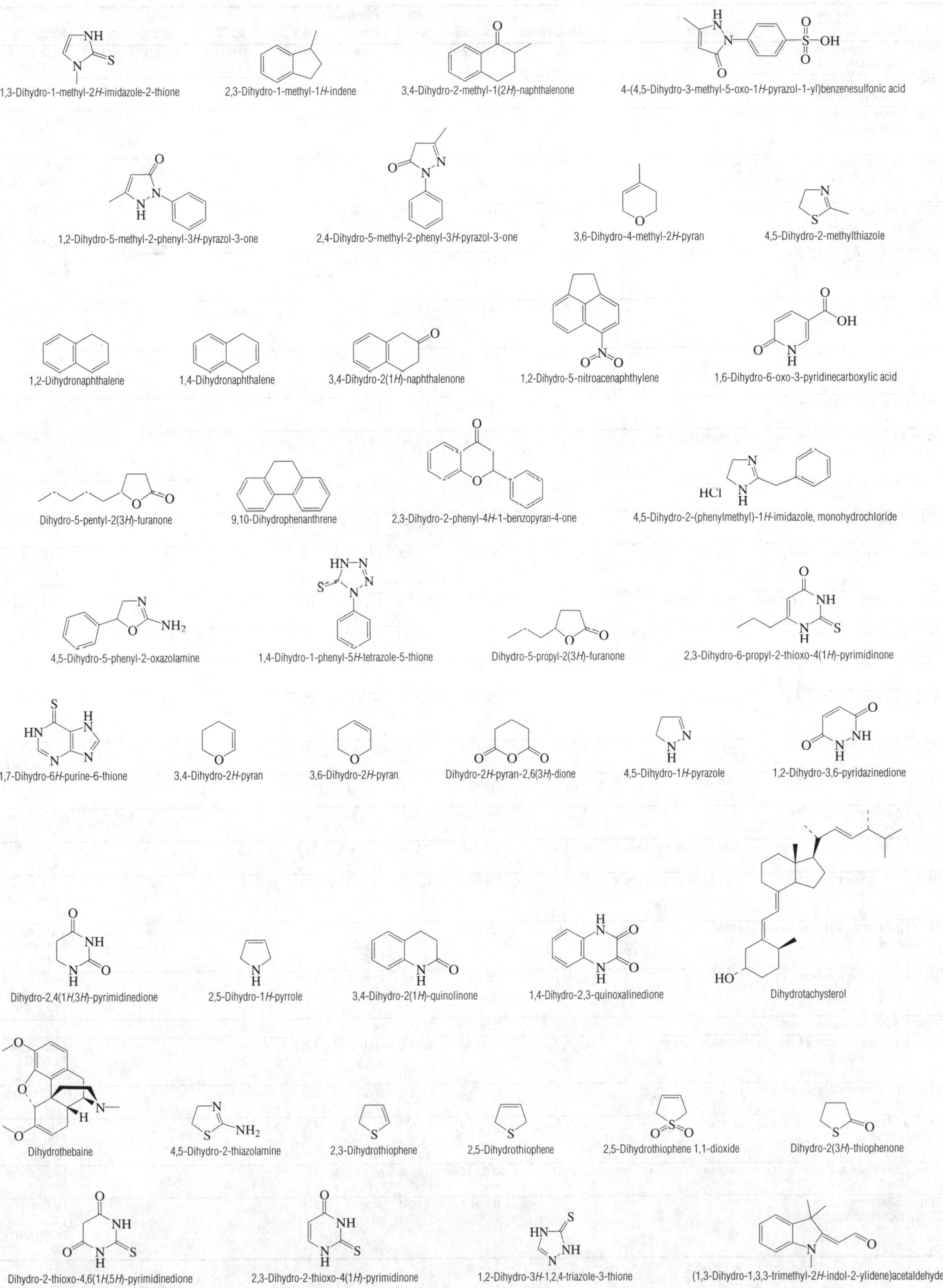
1,3-Dihydro-1-methyl-2H-imidazole-2-thione
2,3-Dihydro-1-methyl-1H-indene
3,4-Dihydro-2-methyl-1(2H)-naphthalenone
4-(4,5-Dihydro-3-methyl-5-oxo-1H-pyrazol-1-yl)benzenesulfonic acid
1,2-Dihydro-5-methyl-2-phenyl-3H-pyrazol-3-one
2,4-Dihydro-5-methyl-2-phenyl-3H-pyrazol-3-one
3,6-Dihydro-4-methyl-2H-pyran
4,5-Dihydro-2-methylthiazole
1,2-Dihydronaphthalene
1,4-Dihydronaphthalene
3,4-Dihydro-2(1H)-naphthalenone
1,2-Dihydro-5-nitroacenaphthylene
1,6-Dihydro-6-oxo-3-pyridinecarboxylic acid
Dihydro-5-pentyl-2(3H)-furanone
9,10-Dihydrophenanthrene
2,3-Dihydro-2-phenyl-4H-1-benzopyran-4-one
4,5-Dihydro-2-(phenylmethyl)-1H-imidazole, monohydrochloride
4,5-Dihydro-5-phenyl-2-oxazolamine
1,4-Dihydro-1-phenyl-5H-tetrazole-5-thione
Dihydro-5-propyl-2(3H)-furanone
2,3-Dihydro-6-propyl-2-thioxo-4(1H)-pyrimidinone
1,7-Dihydro-6H-purine-6-thione
3,4-Dihydro-2H-pyran
3,6-Dihydro-2H-pyran
Dihydro-2H-pyran-2,6(3H)-dione
4,5-Dihydro-1H-pyrazole
1,2-Dihydro-3,6-pyridazinedione
Dihydro-2,4(1H,3H)-pyrimidinedione
2,5-Dihydro-1H-pyrrole
3,4-Dihydro-2(1H)-quinolinone
1,4-Dihydro-2,3-quinoxalinedione
Dihydrotachysterol
Dihydrothebaine
4,5-Dihydro-2-thiazolamine
2,3-Dihydrothiophene
2,5-Dihydrothiophene
2,5-Dihydrothiophene 1,1-dioxide
Dihydro-2(3H)-thiophenone
Dihydro-2-thioxo-4,6(1H,5H)-pyrimidinedione
2,3-Dihydro-2-thioxo-4(1H)-pyrimidinone
1,2-Dihydro-3H-1,2,4-triazole-3-thione
(1,3-Dihydro-1,3,3-trimethyl-2H-indol-2-ylidene)acetaldehyde

No.	Name	Synonym	Mol. Form.	CAS RN	Mol. Wt.	Physical Form	mp/°C	bp/°C	den/ g cm^{-3}	n_D	Solubility
3690	2,3-Dihydro-1,1,3-trimethyl-3-phenyl-1*H*-indene		$C_{18}H_{20}$	3910-35-8	236.352	tcl pr (al)	52.5	308.5	1.0009^{20}	1.5681^{20}	i H_2O; s EtOH, bz, MeOH
3691	1,2-Dihydro-2,2,4-trimethylquinoline		$C_{12}H_{15}N$	147-47-7	173.254		26.5	260; 132^{13}			
3692	1,4-Dihydroxy-9,10-anthracenedione	Quinizarin	$C_{14}H_8O_4$	81-64-1	240.212	ye red lf (eth) dk red nd	200				s H_2O, EtOH, eth, bz, KOH, sulf
3693	1,5-Dihydroxy-9,10-anthracenedione	Anthrarufin	$C_{14}H_8O_4$	117-12-4	240.212	pa ye pl (gl HOAc)	280	sub			i H_2O; sl EtOH, eth, ace, CS_2; s bz
3694	1,8-Dihydroxy-9,10-anthracenedione	Danthron	$C_{14}H_8O_4$	117-10-2	240.212	red or red-ye nd or lf (al)	193	sub			i H_2O; sl EtOH, eth; s ace, HOAc, alk
3695	2,6-Dihydroxy-9,10-anthracenedione		$C_{14}H_8O_4$	84-60-6	240.212	ye nd (al)	360 dec				sl H_2O, EtOH; i eth, bz, chl; s alk
3696	2,7-Dihydroxy-9,10-anthracenedione		$C_{14}H_8O_4$	572-93-0	240.212	ye nd (+1w, dil al) nd (sub)	353.8	sub			i H_2O; s EtOH; sl eth, bz, chl
3697	2,2'-Dihydroxyazobenzene		$C_{12}H_{10}N_2O_2$	2050-14-8	214.219	gold-ye lf (bz), nd (al)	173	$140^{0.001}$			i H_2O; sl EtOH, bz; vs eth; s con alk
3698	2,3-Dihydroxybenzaldehyde		$C_7H_6O_3$	24677-78-9	138.121	ye nd	108	235; 120^{16}			vs ace, EtOH, HOAc
3699	2,4-Dihydroxybenzaldehyde	β-Resorcylaldehyde	$C_7H_6O_3$	95-01-2	138.121	nd (eth-lig)	135	226^{22}			s H_2O, HOAc; vs EtOH, eth, chl; sl bz
3700	2,5-Dihydroxybenzaldehyde		$C_7H_6O_3$	1194-98-5	138.121	ye nd (bz)	100.0				vs H_2O, EtOH, chl
3701	3,4-Dihydroxybenzaldehyde	Protocatechualdehyde	$C_7H_6O_3$	139-85-5	138.121	lf (w, to)	153 dec				s H_2O; vs EtOH, eth
3702	*N*,2-Dihydroxybenzamide	Salicylhydroxamic acid	$C_7H_7NO_3$	89-73-6	153.136	nd (HOAc)	168	sub			sl H_2O, DMSO; vs EtOH, eth; s HOAc
3703	2,5-Dihydroxybenzeneacetic acid	Homogentisic acid	$C_8H_8O_4$	451-13-8	168.148	pr (w+1), lf (al-chl)	153				vs H_2O, EtOH, eth; i bz, chl
3704	2,3-Dihydroxybenzoic acid		$C_7H_6O_4$	303-38-8	154.121	pr or nd (w+1)	205.5		1.542^{20}		s H_2O, EtOH, eth; sl ace
3705	2,4-Dihydroxybenzoic acid	β-Resorcylic acid	$C_7H_6O_4$	89-86-1	154.121	cry (+w)	226 dec				s H_2O, EtOH, eth, bz; i CS_2
3706	2,5-Dihydroxybenzoic acid	Gentisic acid	$C_7H_6O_4$	490-79-9	154.121	nd or pr (w)	199.5				vs H_2O, EtOH, eth; s ace; i bz, chl, CS_2
3707	2,6-Dihydroxybenzoic acid		$C_7H_6O_4$	303-07-1	154.121	nd (+w)	167 dec				s H_2O, EtOH, eth; i chl; sl tfa
3708	3,4-Dihydroxybenzoic acid	Protocatechuic acid	$C_7H_6O_4$	99-50-3	154.121	mcl nd (w+1)	201 dec		1.524^4		sl H_2O; vs EtOH; s eth; i bz
3709	3,5-Dihydroxybenzoic acid		$C_7H_6O_4$	99-10-5	154.121	pr or nd	239				sl H_2O, ace; vs EtOH, eth
3710	2,2'-Dihydroxybenzophenone	Bis(2-hydroxyphenyl) ketone	$C_{13}H_{10}O_3$	835-11-0	214.216		59.5	333			i H_2O; s EtOH, eth, chl
3711	4,4'-Dihydroxybenzophenone	Bis(4-hydroxyphenyl) ketone	$C_{13}H_{10}O_3$	611-99-4	214.216	nd (lig), cry (w)	210		1.133^{131}		sl H_2O; s EtOH, eth, ace; i bz, CS_2
3712	6,7-Dihydroxy-2*H*-1-benzopyran-2-one	Esculetin	$C_9H_6O_4$	305-01-1	178.142	nd (w), pr (HOAc) lf (sub)	276	sub			sl H_2O, eth; s EtOH, ace, chl, AcOEt
3713	7,8-Dihydroxy-2*H*-1-benzopyran-2-one	Daphnetin	$C_9H_6O_4$	486-35-1	178.142	ye nd (dil al)	262	sub			s H_2O, EtOH; sl eth, bz, chl, CS_2
3714	2,4-Dihydroxybutanoic acid		$C_4H_8O_4$	1518-62-3	120.105	liq		96^3			
3715	3,6-Dihydroxycholan-24-oic acid, (3α,5β,6α)	Hyodeoxycholic acid	$C_{24}H_{40}O_4$	83-49-8	392.573	cry (AcOEt)	198.5				sl H_2O, eth, ace, bz; s EtOH, HOAc
3716	3,7-Dihydroxycholan-24-oic acid, (3α,5β,7β)	Ursodiol	$C_{24}H_{40}O_4$	128-13-2	392.573	pl (al)	203				vs EtOH; sl eth
3717	3,7-Dihydroxycholan-24-oic acid, (3α,5β,7α)	Chenodiol	$C_{24}H_{40}O_4$	474-25-9	392.573	nd (EtOAc+hep)	119				i H_2O, bz; vs EtOH, ace; s eth, HOAc
3718	1,25-Dihydroxycholecalciferol	Calcitriol	$C_{27}H_{44}O_3$	32222-06-3	416.636	wh cry pow	115				sl EtOH, MeOH, thf, AcOEt
3719	2,5-Dihydroxy-2,5-cyclohexadiene-1,4-dione		$C_6H_4O_4$	615-94-1	140.094	dk ye nd	211				sl H_2O, ace, DMSO; s EtOH, HOAc; i eth

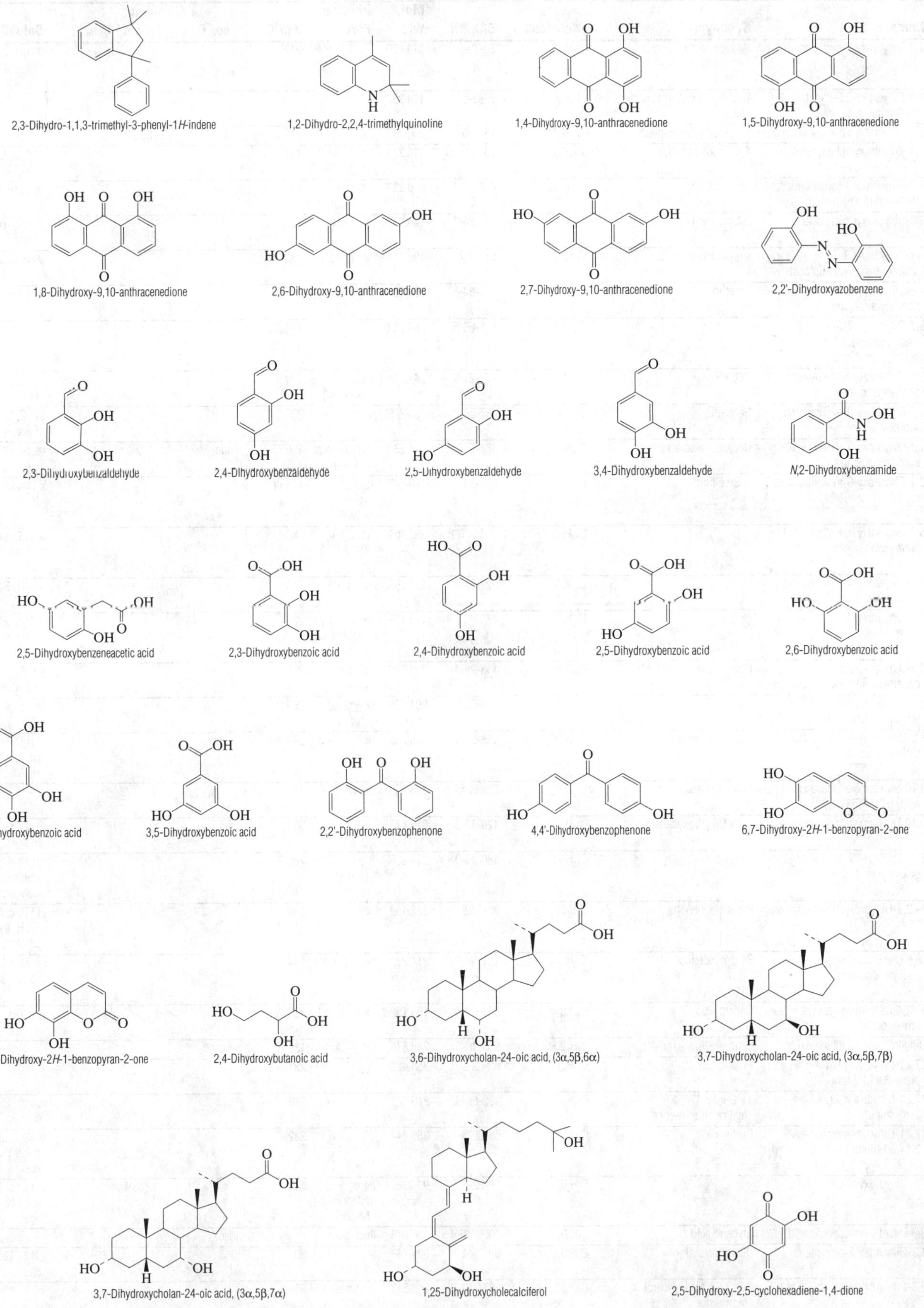
2,3-Dihydro-1,1,3-trimethyl-3-phenyl-1H-indene
1,2-Dihydro-2,2,4-trimethylquinoline
1,4-Dihydroxy-9,10-anthracenedione
1,5-Dihydroxy-9,10-anthracenedione
1,8-Dihydroxy-9,10-anthracenedione
2,6-Dihydroxy-9,10-anthracenedione
2,7-Dihydroxy-9,10-anthracenedione
2,2'-Dihydroxyazobenzene
2,3-Dihydroxybenzaldehyde
2,4-Dihydroxybenzaldehyde
2,5-Dihydroxybenzaldehyde
3,4-Dihydroxybenzaldehyde
N,2-Dihydroxybenzamide
2,5-Dihydroxybenzeneacetic acid
2,3-Dihydroxybenzoic acid
2,4-Dihydroxybenzoic acid
2,5-Dihydroxybenzoic acid
2,6-Dihydroxybenzoic acid
3,4-Dihydroxybenzoic acid
3,5-Dihydroxybenzoic acid
2,2'-Dihydroxybenzophenone
4,4'-Dihydroxybenzophenone
6,7-Dihydroxy-2H-1-benzopyran-2-one
7,8-Dihydroxy-2H-1-benzopyran-2-one
2,4-Dihydroxybutanoic acid
3,6-Dihydroxycholan-24-oic acid, (3α,5β,6α)
3,7-Dihydroxycholan-24-oic acid, (3α,5β,7β)
3,7-Dihydroxycholan-24-oic acid, (3α,5β,7α)
1,25-Dihydroxycholecalciferol
2,5-Dihydroxy-2,5-cyclohexadiene-1,4-dione

No.	Name	Synonym	Mol. Form.	CAS RN	Mol. Wt.	Physical Form	mp/°C	bp/°C	den/ g cm^{-3}	n_D	Solubility
3720	2,3-Dihydroxy-2-cyclopenten-1-one	Reductic acid	$C_5H_6O_3$	80-72-8	114.100		212				s H_2O, EtOH; sl eth, ace, AcOEt; i bz
3721	2,6-Dihydroxy-2,6-dimethyl-4-heptanone	Di(2-hydroxy-2-methylpropyl) ketone	$C_9H_{18}O_3$	3682-91-5	174.237	pale ye cry					
3722	2,2'-Dihydroxydiphenylmethane	2,2'-Methylenebisphenol	$C_{13}H_{12}O_2$	2467-02-9	200.233		118.3	363	1.280^{25}		
3723	4,4'-Dihydroxydiphenyl sulfide	4,4'-Thiobisphenol	$C_{12}H_{10}O_2S$	2664-63-3	218.271	mcl pr or lf (al)	151				sl H_2O, EtOH, eth, CS_2
3724	1,8-Dihydroxy-3-(hydroxymethyl)-9,10-anthracenedione	Aloe-emodol	$C_{15}H_{10}O_5$	481-72-1	270.237	oran ye nd (to, al)	223.5	sub			vs bz, eth, EtOH
3725	2,3-Dihydroxymaleic acid	Dihydroxymaleic acid	$C_4H_4O_6$	526-84-1	148.071	pl (w+2)	155 dec				sl H_2O, eth, MeOH; s EtOH
3726	α,4-Dihydroxy-3-methoxybenzeneacetic acid	Vanilmandelic acid	$C_9H_{10}O_5$	55-10-7	198.172	sc (bz-eth)	132 dec				vs H_2O, ace, eth
3727	7,8-Dihydroxy-6-methoxy-2*H*-1-benzopyran-2-one	Fraxetin	$C_{10}H_8O_5$	574-84-5	208.168	pl (dil al)	231				vs EtOH
3728	5,7-Dihydroxy-3-(4-methoxyphenyl)-4*H*-1-benzopyran-4-one		$C_{16}H_{12}O_5$	491-80-5	284.263		214.8				
3729	(2,6-Dihydroxy-4-methoxyphenyl) phenylmethanone	Cotoin	$C_{14}H_{12}O_4$	479-21-0	244.243	ye pr (chl) lf or nd (w)	130.5				vs ace, bz, eth, EtOH
3730	1,7-Dihydroxy-3-methoxy-9*H*-xanthen-9-one	Gentisin	$C_{14}H_{10}O_5$	437-50-3	258.226	ye orth	266.5				i H_2O; vs EtOH; i ace; s py
3731	1,8-Dihydroxy-3-methyl-9,10-anthracenedione	Chrysophanic acid	$C_{15}H_{10}O_4$	481-74-3	254.238	ye hex or mcl nd (sub)	196	sub	0.92^{25}		vs bz, HOAc
3732	2,4-Dihydroxy-6-methylbenzoic acid	*o*-Orsellinic acid	$C_8H_8O_4$	480-64-8	168.148	nd (dil HOAc, +1w)	176 dec				s EtOH, eth
3733	5,7-Dihydroxy-4-methyl-2*H*-1-benzopyran-2-one		$C_{10}H_8O_4$	2107-76-8	192.169	nd (al), lf (HOAc)	283				sl H_2O, eth, bz, chl; vs EtOH, alk
3734	6,7-Dihydroxy-4-methyl-2*H*-1-benzopyran-2-one		$C_{10}H_8O_4$	529-84-0	192.169	ye nd (dil al)	275				s H_2O, EtOH, HOAc
3735	5,8-Dihydroxy-1,4-naphthalenedione		$C_{10}H_6O_4$	475-38-7	190.153	dk red mcl pr (bz) red-br nd (al)	232	sub			sl H_2O, EtOH, eth; s HOAc
3736	4,5-Dihydroxy-2,7-naphthalenedisulfonic acid	Chromotropic acid	$C_{10}H_8O_8S_2$	148-25-4	320.296	nd or lf (w+2)					s H_2O, alk; i EtOH, eth
3737	5,6-Dihydroxynaphtho[2,3-f] quinoline-7,12-dione	Alizarin Blue	$C_{17}H_9NO_4$	568-02-5	291.258	br-viol nd (bz)	269				vs bz, gl HOAc
3738	1,2-Dihydroxy-3-nitro-9,10-anthracenedione	Alizarin Orange	$C_{14}H_7NO_6$	568-93-4	285.209	oran nd or pl (HOAc)	244 dec	sub			sl H_2O; s EtOH, bz, chl, sulf, HOAc
3739	9,10-Dihydroxyoctadecanedioic acid, (*R**,*R**)-(±)	Phloionic acid	$C_{18}H_{34}O_6$	23843-52-9	346.459	cry (al)	126				
3740	9,10-Dihydroxyoctadecanoic acid	9,10-Dihydroxystearic acid	$C_{18}H_{36}O_4$	120-87-6	316.477		90				i H_2O; sl EtOH, eth
3741	5,7-Dihydroxy-2-phenyl-4*H*-1-benzopyran-4-one	Chrysin	$C_{15}H_{10}O_4$	480-40-0	254.238	lt ye pr (MeOH)	285.5				i H_2O; s EtOH, ace; sl eth, bz, CS_2
3742	1-(2,4-Dihydroxyphenyl)ethanone	Resacetophenone	$C_8H_8O_3$	89-84-9	152.148	nd or lf	146		1.18^{141}		i H_2O, chl; s EtOH, py; sl eth, bz
3743	(2,4-Dihydroxyphenyl) phenylmethanone	Benzoresorcinol	$C_{13}H_{10}O_3$	131-56-6	214.216	nd (w)	144				i H_2O; s EtOH; vs eth; sl bz, chl
3744	3-(3,4-Dihydroxyphenyl)-2-propenoic acid	Caffeic acid	$C_9H_8O_4$	331-39-5	180.158	ye pr, pl (w)	225 dec				vs EtOH
3745	Dihydroxyphenylstibine oxide	Benzenestibonic acid	$C_6H_7O_3Sb$	535-46-6	248.878	nd (HOAc)	139				
3746	17,21-Dihydroxypregna-1,4-diene-3,11,20-trione	Prednisone	$C_{21}H_{26}O_5$	53-03-2	358.428		234 dec				
3747	17,21-Dihydroxypregn-4-ene-3,20-dione	11-Deoxy-17-hydrocorticosterone	$C_{21}H_{30}O_4$	152-58-9	346.461		215				vs ace, EtOH, chl
3748	17,21-Dihydroxypregn-4-ene-3,11,20-trione	Cortisone	$C_{21}H_{28}O_5$	53-06-5	360.444		222				sl H_2O, eth, bz, chl; s EtOH, ace
3749	2,3-Dihydroxypropanal, (±)		$C_3H_6O_3$	56-82-6	90.078	nd or pr (40% MeOH)	145	$145^{0.8}$	1.453^{18}		s H_2O; sl EtOH, eth; i bz, peth, lig
3750	2,3-Dihydroxypropanoic acid, (*R*)	Glyceric acid	$C_3H_6O_4$	6000-40-4	106.078	thick gum		dec			
3751	1,3-Dihydroxy-2-propanone	Dihydroxyacetone	$C_3H_6O_3$	96-26-4	90.078		90				s H_2O, EtOH, eth, ace; i lig
3752	2,3-Dihydroxypropyl decanoate	Decanoic acid glycerol monoester	$C_{13}H_{26}O_4$	2277-23-8	246.343	pr (peth)	53				
3753	2,3-Dihydroxypropyl octanoate	Octanoic acid glycerol monoester	$C_{11}H_{22}O_4$	26402-26-6	218.291	cry (peth)	40				

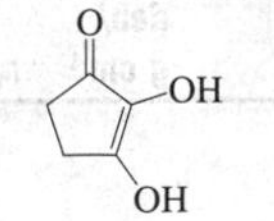

2,3-Dihydroxy-2-cyclopenten-1-one

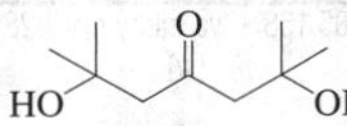

2,6-Dihydroxy-2,6-dimethyl-4-heptanone

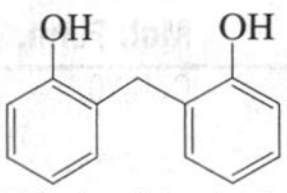

2,2'-Dihydroxydiphenylmethane

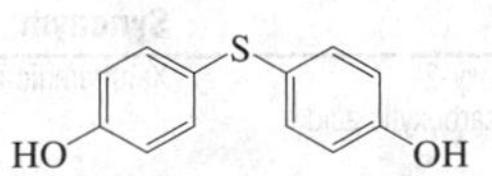

4,4'-Dihydroxydiphenyl sulfide

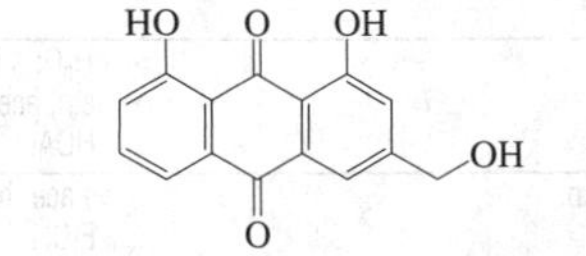

1,8-Dihydroxy-3-(hydroxymethyl)-9,10-anthracenedione

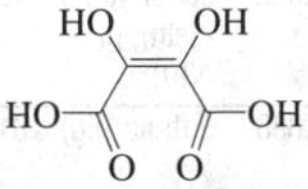

2,3-Dihydroxymaleic acid

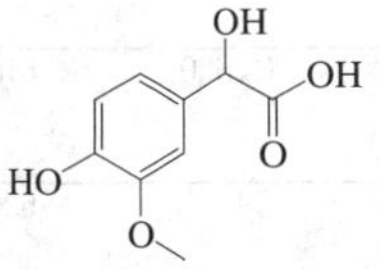

α,4-Dihydroxy-3-methoxybenzeneacetic acid

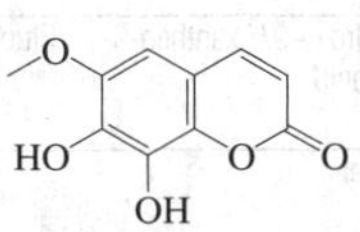

7,8-Dihydroxy-6-methoxy-2*H*-1-benzopyran-2-one

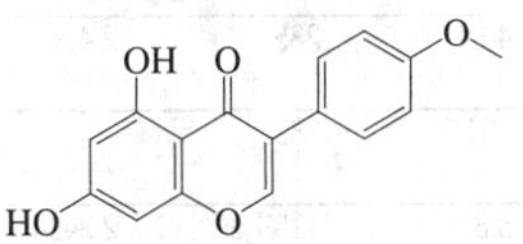

5,7-Dihydroxy-3-(4-methoxyphenyl)-4*H*-1-benzopyran-4-one

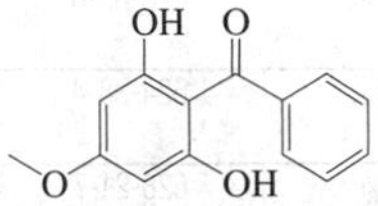

(2,6-Dihydroxy-4-methoxyphenyl)phenylmethanone

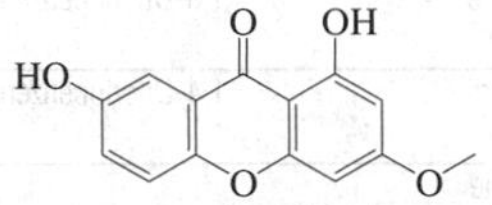

1,7-Dihydroxy-3-methoxy-9*H*-xanthen-9-one

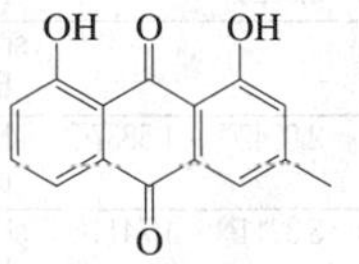

1,8-Dihydroxy-3-methyl-9,10-anthracenedione

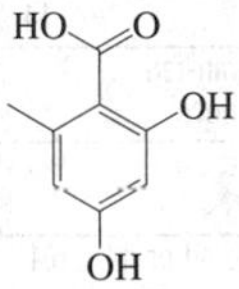

2,4-Dihydroxy-6-methylbenzoic acid

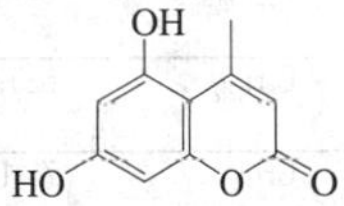

5,7-Dihydroxy-4-methyl-2*H*-1-benzopyran-2-one

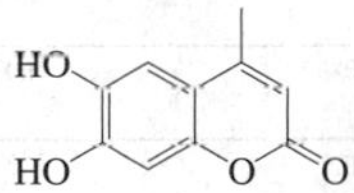

6,7-Dihydroxy-4-methyl-2*H*-1-benzopyran-2-one

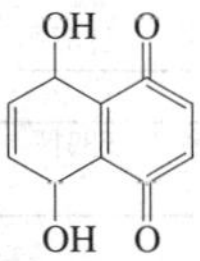

5,8-Dihydroxy-1,4-naphthalenedione

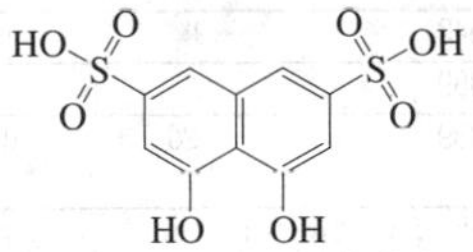

4,5-Dihydroxy-2,7-naphthalenedisulfonic acid

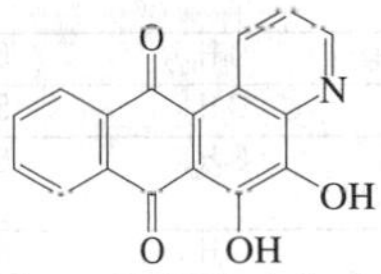

5,6-Dihydroxynaphtho[2,3-f]quinoline-7,12-dione

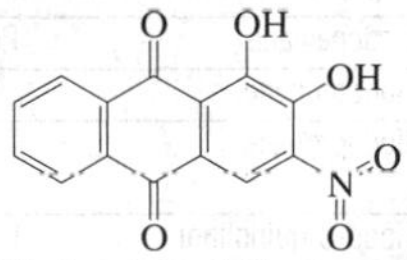

1,2-Dihydroxy-3-nitro-9,10-anthracenedione

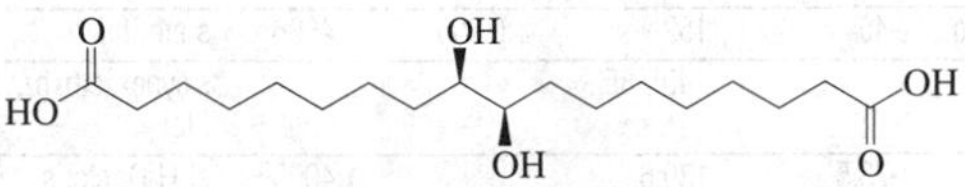

9,10-Dihydroxyoctadecanedioic acid, (*R**,*R**)-(±)

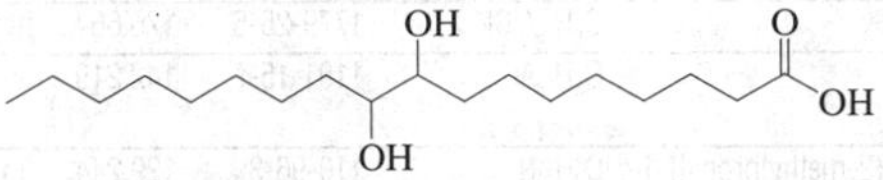

9,10-Dihydroxyoctadecanoic acid

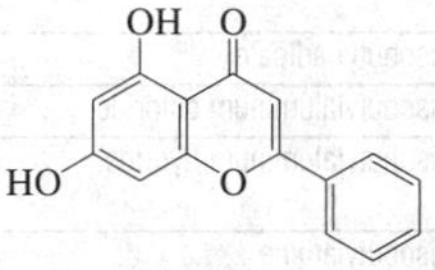

5,7-Dihydroxy-2-phenyl-4*H*-1-benzopyran-4-one

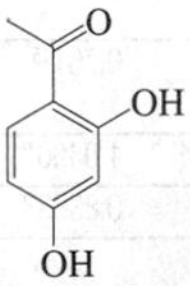

1-(2,4-Dihydroxyphenyl)ethanone

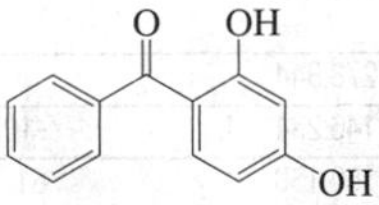

(2,4-Dihydroxyphenyl)phenylmethanone

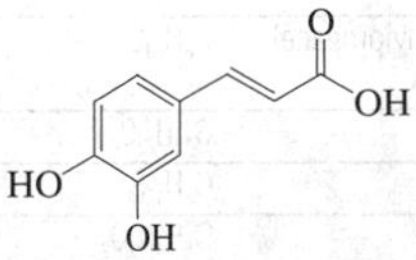

3-(3,4-Dihydroxyphenyl)-2-propenoic acid

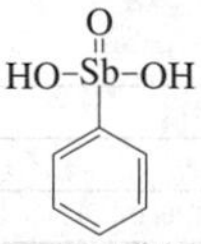

Dihydroxyphenylstibine oxide

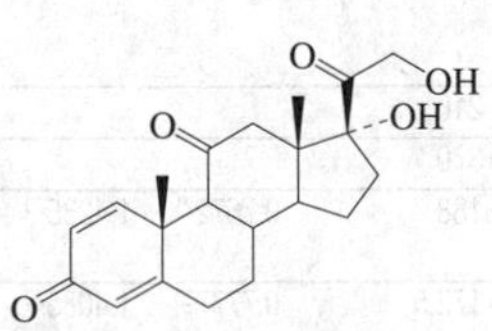

17,21-Dihydroxypregna-1,4-diene-3,11,20-trione

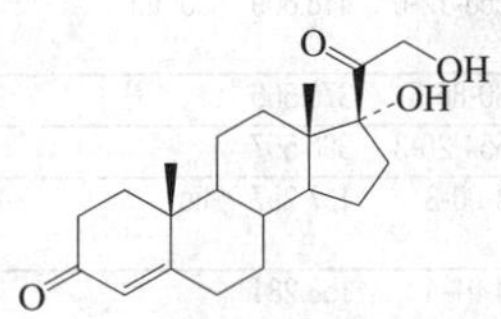

17,21-Dihydroxypregn-4-ene-3,20-dione

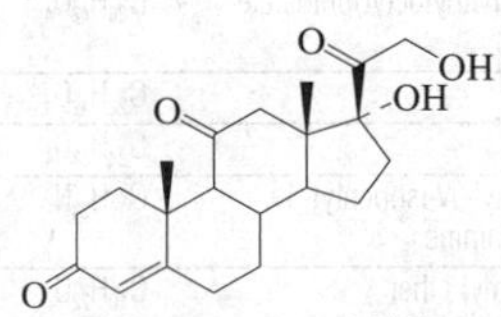

17,21-Dihydroxypregn-4-ene-3,11,20-trione

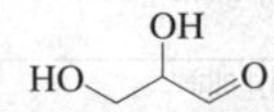

2,3-Dihydroxypropanal, (±)

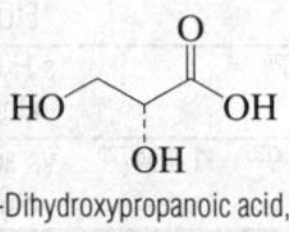

2,3-Dihydroxypropanoic acid, (*R*)

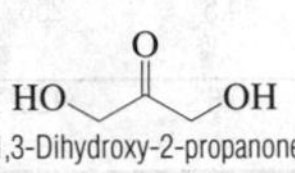

1,3-Dihydroxy-2-propanone

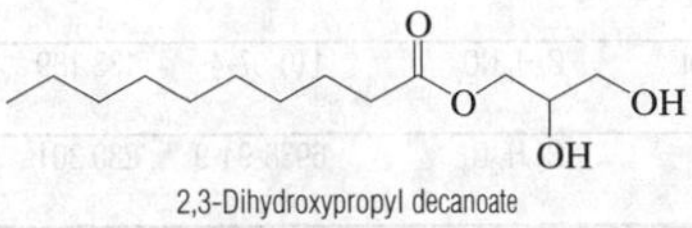

2,3-Dihydroxypropyl decanoate

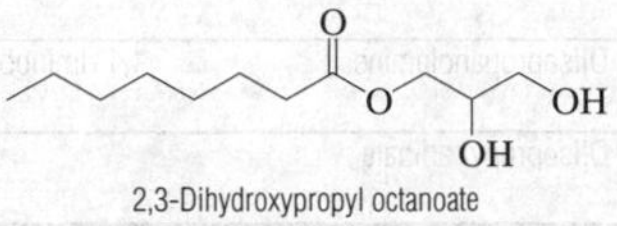

2,3-Dihydroxypropyl octanoate

No.	Name	Synonym	Mol. Form.	CAS RN	Mol. Wt.	Physical Form	mp/°C	bp/°C	den/ g cm^{-3}	n_D	Solubility
3754	4,8-Dihydroxy-2-quinolinecarboxylic acid	Xanthurenic acid	$C_{10}H_7NO_4$	59-00-7	205.168	ye micry cry (w)	289				i H_2O; s EtOH, dil HCl; sl eth, bz
3755	Dihydroxytartaric acid		$C_4H_6O_8$	76-30-2	182.086		114.5				
3756	3,4-Dihydroxy-5-[(3,4,5-trihydroxybenzoyl)oxy]benzoic acid	Digallic acid	$C_{14}H_{10}O_9$	536-08-3	322.224	nf (dil al + 1w)	269 dec				vs ace, EtOH
3757	2-(3,6-Dihydroxy-9*H*-xanthen-9-yl)benzoic acid	Fluorescin	$C_{20}H_{14}O_5$	518-44-5	334.322	col or ye nd (eth), pl (bz)	126				i H_2O; s EtOH, eth, ace, bz, HOAc
3758	Diiodoacetylene		C_2I_2	624-74-8	277.830	orth nd (lig)	81.5	exp			vs ace, bz, eth, EtOH
3759	2,4-Diiodoaniline		$C_6H_5I_2N$	533-70-0	344.920	br nd or orth cry (al)	95.5		2.748^{25}		vs ace, bz, eth, EtOH
3760	*o*-Diiodobenzene	1,2-Diiodobenzene	$C_6H_4I_2$	615-42-9	329.905	pl or pr (lig)	27	287; 100^{3}	2.54^{20}	1.7179^{20}	i H_2O; sl EtOH
3761	*m*-Diiodobenzene	1,3-Diiodobenzene	$C_6H_4I_2$	626-00-6	329.905	orth pl or pr (eth-al)	40.4	285	2.47^{25}		i H_2O; vs eth, EtOH, chl
3762	*p*-Diiodobenzene	1,4-Diiodobenzene	$C_6H_4I_2$	624-38-4	329.905	orth lf (al)	131.5	285			i H_2O; s EtOH; vs eth; sl chl
3763	1,4-Diiodobutane		$C_4H_8I_2$	628-21-7	309.916		5.8	125^{15} dec	2.3494^{25}	1.6184^{25}	i H_2O; sl ctc; s os
3764	1,2-Diiodoethane		$C_2H_4I_2$	624-73-7	281.862	ye mcl pr or orth (eth)	83	200	3.325^{20}	1.871^{20}	sl H_2O; s EtOH, eth, ace, chl
3765	*cis*-1,2-Diiodoethene	*cis*-1,2-Diiodoethylene	$C_2H_2I_2$	590-26-1	279.846		-14	72.5^{16}	3.0625^{20}		i H_2O; s eth, chl
3766	4,4'-Diiodofluorescein		$C_{20}H_{10}I_2O_5$	38577-97-8	584.099	oran-red pow					sl H_2O; s alk, EtOH
3767	1,6-Diiodohexane	Hexamethylene diiodide	$C_6H_{12}I_2$	629-09-4	337.968	nd	9.5	163^{17}, 141^{10}	2.0342^{25}	1.5837^{25}	i H_2O; vs EtOH, eth
3768	Diiodomethane	Methylene iodide	CH_2I_2	75-11-6	267.836	ye nd or lf	6.1	182	3.3211^{20}	1.7411^{20}	sl H_2O, ctc; s EtOH, eth, bz, chl
3769	2,6-Diiodo-4-nitrophenol	Disophenol	$C_6H_3I_2NO_3$	305-85-1	390.902	lt ye cry (gl HOAc)	157				vs EtOH
3770	1,5-Diiodopentane	Pentamethylene diiodide	$C_5H_{10}I_2$	628-77-3	323.942		9	149^{20}, 101^{3}	2.1692^{25}	1.5987^{25}	i H_2O; s eth, chl
3771	1,2-Diiodopropane		$C_3H_6I_2$	598-29-8	295.889				2.490^{18}		vs eth, EtOH
3772	1,3-Diiodopropane	Trimethylene diiodide	$C_3H_6I_2$	627-31-6	295.889		-20	dec 227; 110^{19}	2.5612^{25}	1.6391^{25}	i H_2O; s eth, ctc, chl
3773	5,7-Diiodo-8-quinolinol	Iodoquinol	$C_9H_5I_2NO$	83-73-8	396.951	ye nd (HOAc, xyl)	210				sl H_2O, bz, chl, eth; vs EtOH; s alk
3774	3,5-Diiodo-*L*-tyrosine		$C_9H_9I_2NO_3$	300-39-0	432.981	ye nd (w, 70% al)	213				sl H_2O; i EtOH, eth, bz
3775	Diisobutyl adipate		$C_{14}H_{26}O_4$	141-04-8	258.354			293; 187^{15}	0.9543^{19}	1.4301^{20}	
3776	Diisobutylaluminum chloride		$C_8H_{18}AlCl$	1779-25-5	176.664	hyg col liq	-40	152^{10}	0.905	1.4506^{20}	s eth, hx
3777	Diisobutylaluminum hydride		$C_8H_{19}Al$	1191-15-7	142.219	liq		140^{4}, $85^{0.5}$			s cyhex, eth, bz, tol
3778	Diisobutylamine	2-Methyl-*N*-(2-methylpropyl)-1-propanamine	$C_8H_{19}N$	110-96-3	129.244	liq	-73.5	139.6		1.4090^{20}	sl H_2O, ctc; s EtOH, eth, ace, bz
3779	Diisobutyl carbonate		$C_9H_{18}O_3$	539-92-4	174.237			190	0.9138^{20}	1.4072^{20}	i H_2O; msc EtOH, eth
3780	Diisobutyl ether	1,1'-Oxybis[2-methylpropane]	$C_8H_{18}O$	628-55-7	130.228			122.6	0.761^{15}		i H_2O; msc EtOH, eth
3781	Diisobutyl phthalate		$C_{16}H_{22}O_4$	84-69-5	278.344			296.5; 159^{4}	1.0490^{15}		s ctc
3782	Diisobutyl sulfide		$C_8H_{18}S$	592-65-4	146.294	liq	-105.5	171	0.8363^{10}		
3783	1,3-Diisocyanatobenzene		$C_8H_4N_2O_2$	123-61-5	160.130	cry	51	103^{8}			
3784	1,4-Diisocyanatobenzene		$C_8H_4N_2O_2$	104-49-4	160.130	cry	95	117^{14}			
3785	Diisodecyl phthalate	Bis(8-methylnonyl)phthalate	$C_{28}H_{46}O_4$	26761-40-0	446.663	liq	-50	253^{4}	0.966^{20}		i H_2O; s os
3786	Diisononyl phthalate	Bis(7-methyloctyl)phthalate	$C_{26}H_{42}O_4$	28553-12-0	418.609	col liq					i H_2O; s ace, MeOH; bz, eth
3787	Diisooctyl adipate		$C_{22}H_{42}O_4$	1330-86-5	370.566			210^{4}			
3788	Diisooctyl phthalate		$C_{24}H_{38}O_4$	27554-26-3	390.557			370			
3789	Diisopentylamine	3-Methyl-*N*-isopentyl-1-butanamine	$C_{10}H_{23}N$	544-00-3	157.297	liq	-44	188	0.7672^{21}	1.4235^{20}	i H_2O; s EtOH; msc eth
3790	Diisopentyl ether	Diisoamyl ether	$C_{10}H_{22}O$	544-01-4	158.281			172.5	0.7777^{20}	1.4085^{20}	i H_2O; vs ace, EtOH, chl
3791	Diisopentyl phthalate	Diisoamyl phthalate	$C_{18}H_{26}O_4$	605-50-5	306.397			dec 334	1.0209^{16}	1.4871^{20}	vs EtOH
3792	Diisopentyl sulfide		$C_{10}H_{22}S$	544-02-5	174.347	liq	-74.6	211	0.8323^{20}	1.4520^{20}	i H_2O; msc EtOH; vs eth
3793	Diisopropanolamine	1,1'-Iminobis-2-propanol	$C_6H_{15}NO_2$	110-97-4	133.189	cry	44.5	250; 151^{23}	0.989^{20}		s H_2O, EtOH; sl eth
3794	Diisopropyl adipate		$C_{12}H_{22}O_4$	6938-94-9	230.301		-0.6	$120^{6.5}$	0.9569^{20}	1.4247^{20}	vs ace, eth, EtOH

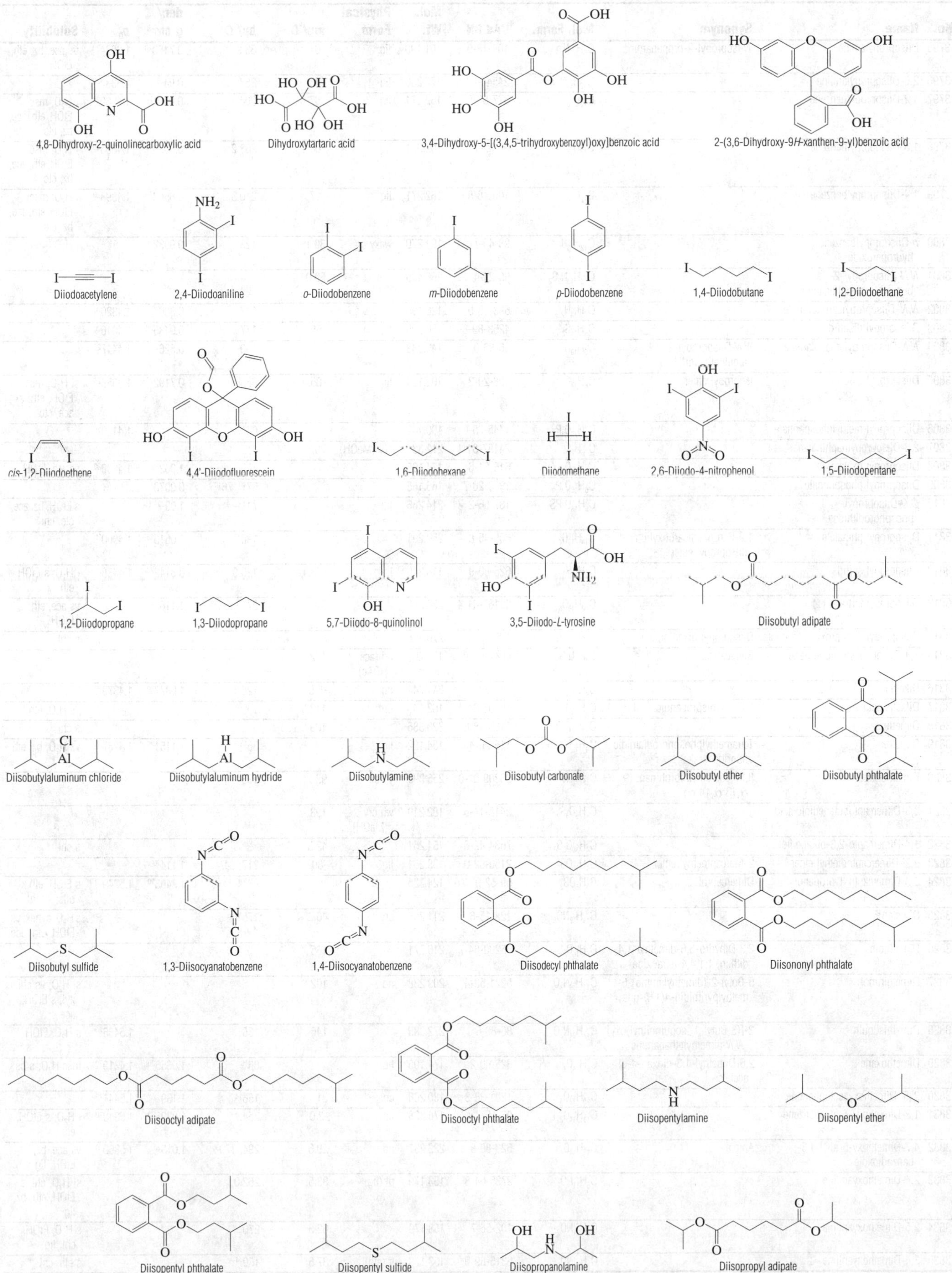

4,8-Dihydroxy-2-quinolinecarboxylic acid
Dihydroxytartaric acid
3,4-Dihydroxy-5-[(3,4,5-trihydroxybenzoyl)oxy]benzoic acid
2-(3,6-Dihydroxy-9H-xanthen-9-yl)benzoic acid
Diiodoacetylene
2,4-Diiodoaniline
o-Diiodobenzene
m-Diiodobenzene
p-Diiodobenzene
1,4-Diiodobutane
1,2-Diiodoethane
cis-1,2-Diiodoethene
4,4'-Diiodofluorescein
1,6-Diiodohexane
Diiodomethane
2,6-Diiodo-4-nitrophenol
1,5-Diiodopentane
1,2-Diiodopropane
1,3-Diiodopropane
5,7-Diiodo-8-quinolinol
3,5-Diiodo-L-tyrosine
Diisobutyl adipate
Diisobutylaluminum chloride
Diisobutylaluminum hydride
Diisobutylamine
Diisobutyl carbonate
Diisobutyl ether
Diisobutyl phthalate
Diisobutyl sulfide
1,3-Diisocyanatobenzene
1,4-Diisocyanatobenzene
Diisodecyl phthalate
Diisononyl phthalate
Diisooctyl adipate
Diisooctyl phthalate
Diisopentylamine
Diisopentyl ether
Diisopentyl phthalate
Diisopentyl sulfide
Diisopropanolamine
Diisopropyl adipate

No.	Name	Synonym	Mol. Form.	CAS RN	Mol. Wt.	Physical Form	mp/°C	bp/°C	den/ g cm^{-3}	n_D	Solubility
3795	Diisopropylamine	*N*-Isopropyl-2-propanamine	$C_6H_{15}N$	108-18-9	101.190	liq	-61	83.9	0.7153^{20}	1.3924^{20}	vs ace, bz, eth, EtOH
3796	2,6-Diisopropylaniline		$C_{12}H_{19}N$	24544-04-5	177.286	liq	-45	257	0.94^{25}	1.5332^{20}	
3797	1,2-Diisopropylbenzene		$C_{12}H_{18}$	577-55-9	162.271	liq	-57	204	0.8701^{20}	1.4960^{20}	i H_2O; msc EtOH, eth, ace, bz, ctc
3798	1,3-Diisopropylbenzene		$C_{12}H_{18}$	99-62-7	162.271	liq	-63.1	203.2	0.8559^{20}	1.4883^{20}	i H_2O; msc EtOH, eth, ace, bz, ctc
3799	1,4-Diisopropylbenzene		$C_{12}H_{18}$	100-18-5	162.271	liq	-17	210.3	0.8568^{20}	1.4898^{20}	i H_2O; msc EtOH, eth, ace, bz, ctc
3800	*p*-Diisopropylbenzene hydroperoxide		$C_{12}H_{18}O_2$	98-49-7	194.270	waxy cry	30.1	123^{1}	0.9932^{20}		i H_2O
3801	*N,N*-Diisopropyl-2-benzothiazolesulfenamide		$C_{13}H_{18}N_2S_2$	95-29-4	266.425		59.0				
3802	*N,N'*-Diisopropylcarbodiimide		$C_7H_{14}N_2$	693-13-0	126.199			147	0.806^{25}	1.4320^{20}	
3803	Diisopropyl disulfide		$C_6H_{14}S_2$	4253-89-8	150.305	liq	-69	177	0.9435^{20}	1.4916^{20}	
3804	*N,N*-Diisopropylethanolamine	*N,N*-Diisopropyl-2-aminoethanol	$C_8H_{19}NO$	96-80-0	145.243			190	0.826^{25}	1.4417^{20}	
3805	Diisopropyl ether	Isopropyl ether	$C_6H_{14}O$	108-20-3	102.174	liq	-85.4	68.4	0.7192^{25}	1.3658^{25}	sl H_2O; msc EtOH, eth; s ace, ctc
3806	Diisopropyl methylphosphonate		$C_7H_{17}O_3P$	1445-75-6	180.182			66^{3}		1.4120^{16}	
3807	2,6-Diisopropylnaphthalene		$C_{16}H_{20}$	24157-81-1	212.330	cry (MeOH)	70				
3808	Diisopropyl oxalate		$C_8H_{14}O_4$	615-81-6	174.195			190	1.002^{20}	1.4100^{20}	vs eth, EtOH
3809	Diisopropyl phosphonate		$C_6H_{15}O_3P$	1809-20-7	166.155			97^{40}, 76^{10}	0.9970^{18}		
3810	*O,O*-Diisopropyl phosphorodithioate		$C_6H_{15}O_2PS_2$	107-56-2	214.286	liq		71^{3}	1.09^{20}		s EtOH, bz, ace, ctc, chl
3811	Diisopropyl phthalate	1,2-Benzenedicarboxylic acid, diisopropyl ester	$C_{14}H_{18}O_4$	605-45-8	250.291			130^{12}	1.0615^{15}	1.4900^{20}	
3812	Diisopropyl sulfide		$C_6H_{14}S$	625-80-9	118.240	liq	-78.1	120.0	0.8142^{20}	1.4438^{20}	i H_2O; s EtOH, eth
3813	Diisopropyl tartrate, (±)		$C_{10}H_{18}O_6$	58167-01-4	234.246		34	275; 154^{12}	1.1166^{20}		vs ace, eth, EtOH
3814	Diisopropyl thioperoxydicarbonate	Diisopropyl dixanthogen	$C_8H_{14}O_2S_4$	105-65-7	270.456		52				s chl
3815	1,4-Diisothiocyanatobenzene	Bitoscanate	$C_8H_4N_2S_2$	4044-65-9	192.261	nd (ace, HOAc)	132				
3816	Diketene		$C_4H_4O_2$	674-82-8	84.074	liq	-6.5	126.1	1.0877^{20}	1.4379^{20}	
3817	Dilactic acid	2,2'-Oxybispropanoic acid	$C_6H_{10}O_5$	19201-34-4	162.140	orth	112.5				vs H_2O, eth
3818	Dimefline		$C_{20}H_{21}NO_3$	1165-48-6	323.386		109.5				s chl
3819	Dimefox	Tetramethylphosphorodiamidic fluoride	$C_4H_{12}FN_2OP$	115-26-4	154.122	liq		86^{15}	1.1151^{20}	1.4267^{20}	vs H_2O, bz, eth
3820	Dimemorfan	3,17-Dimethylmorphinan, (9 α,13 α,14 α)-	$C_{18}H_{25}N$	36309-01-0	255.399	ye oil	92	$133^{0.3}$			
3821	2,3-Dimercaptobutanedioic acid		$C_4H_6O_4S_2$	2418-14-6	182.219	wh cry (MeOH)	193				
3822	1,4-Dimercapto-2,3-butanediol		$C_4H_{10}O_2S_2$	7634-42-6	154.251		42.5				s chl
3823	2,2'-Dimercaptodiethyl ether	2-Mercaptoethyl ether	$C_4H_{10}OS_2$	2150-02-9	138.251	liq	-80	217; 64^{2}	1.114^{20}		
3824	2,3-Dimercapto-1-propanol	Dimercaprol	$C_3H_8OS_2$	59-52-9	124.225			$83^{0.8}$	1.2463^{20}	1.5749^{20}	s EtOH, eth, oils; sl chl
3825	Dimetan®		$C_{11}H_{17}NO_3$	122-15-6	211.258	cry	46	175^{11}			s H_2O, cyhex; vs EtOH, eth, ace
3826	Dimethipin	2,3-Dihydro-5,6-dimethyl-1,4-dithiin, 1,1,4,4-tetraoxide	$C_6H_{10}O_4S_2$	55290-64-7	210.271		165				
3827	Dimethirimol	5-Butyl-2-(dimethylamino)-6-methylpyrimidin-4(1*H*)-one	$C_{11}H_{19}N_3O$	5221-53-4	209.288	nd	102				sl H_2O; vs chl, xyl; s EtOH, ace
3828	Dimethisoquin	2-[(3-Butyl-1-isoquinolinyl)oxy]-*N,N*-dimethylethanamine	$C_{17}H_{24}N_2O$	86-80-6	272.385		146	156^{3}		1.5486^{20}	s H_2O, EtOH
3829	Dimethoxane	2,6-Dimethyl-1,3-dioxan-4-ol acetate	$C_8H_{14}O_4$	828-00-2	174.195	liq		86^{10}	1.0655^{20}	1.4310^{20}	msc H_2O; s os
3830	2',5'-Dimethoxyacetophenone		$C_{10}H_{12}O_3$	1201-38-3	180.200	cry	21	156^{14}	1.139	1.5441^{20}	
3831	1,2-Dimethoxy-4-allylbenzene		$C_{11}H_{14}O_2$	93-15-2	178.228	liq	-2.0	254.7	1.0396^{20}	1.5340^{20}	i H_2O; s EtOH, eth
3832	4,7-Dimethoxy-5-allyl-1,3-benzodioxole	Apiole	$C_{12}H_{14}O_4$	523-80-8	222.237	nd	29.5	294; 179^{35}	1.015^{20}	1.5360^{20}	vs ace, bz, EtOH, lig
3833	2,4-Dimethoxyaniline		$C_8H_{11}NO_2$	2735-04-8	153.179	pl (lig)	33.5	262.0			sl H_2O, chl; s EtOH, eth, bz, lig
3834	2,5-Dimethoxyaniline		$C_8H_{11}NO_2$	102-56-7	153.179		82.5	270			s H_2O, EtOH, chl, lig
3835	3,4-Dimethoxyaniline		$C_8H_{11}NO_2$	6315-89-5	153.179	lf (eth)	87.5	159^{14}			s eth, chl

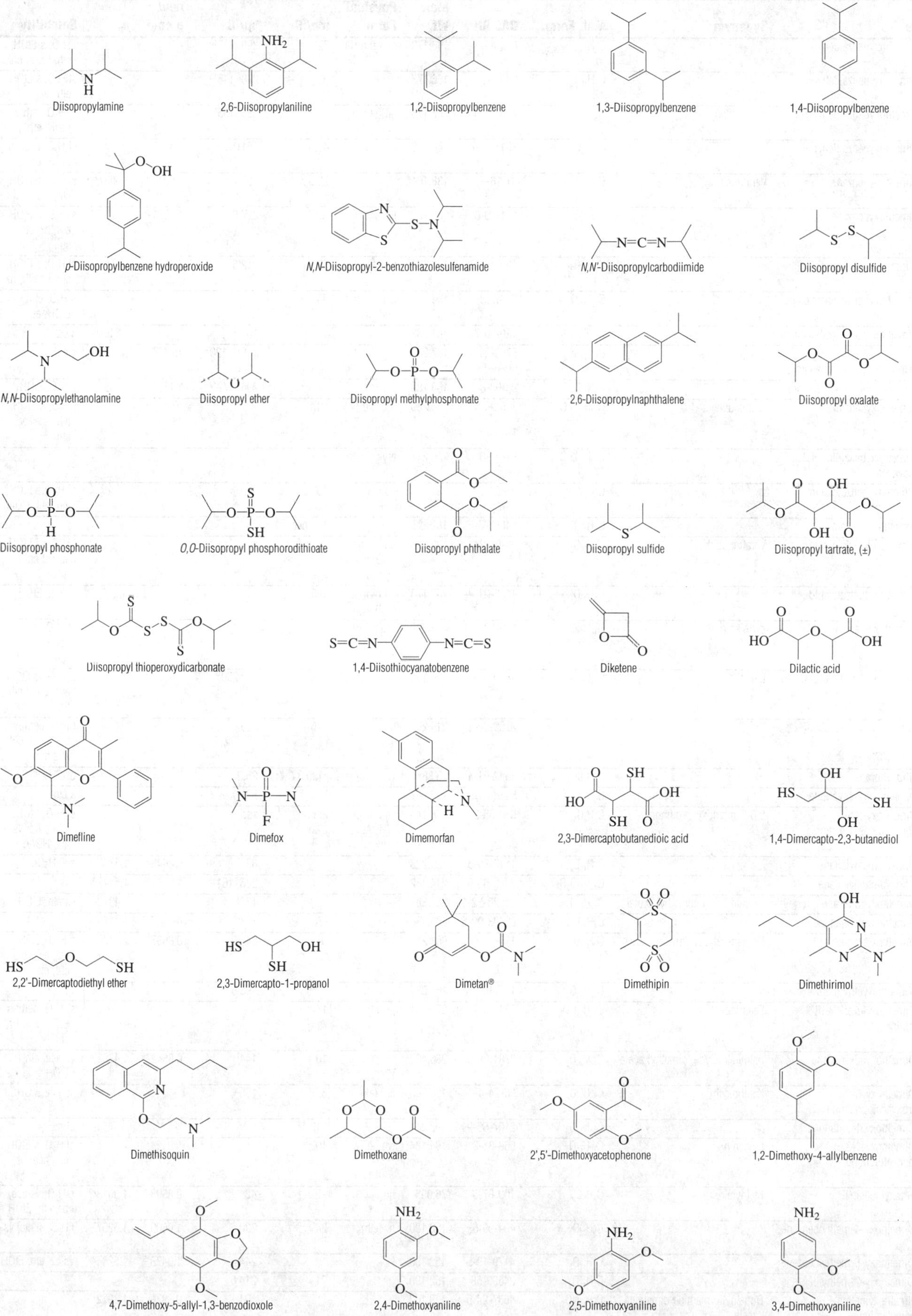

Diisopropylamine
2,6-Diisopropylaniline
1,2-Diisopropylbenzene
1,3-Diisopropylbenzene
1,4-Diisopropylbenzene
p-Diisopropylbenzene hydroperoxide
N,*N*-Diisopropyl-2-benzothiazolesulfenamide
N,*N'*-Diisopropylcarbodiimide
Diisopropyl disulfide
N,*N*-Diisopropylethanolamine
Diisopropyl ether
Diisopropyl methylphosphonate
2,6-Diisopropylnaphthalene
Diisopropyl oxalate
Diisopropyl phosphonate
O,*O*-Diisopropyl phosphorodithioate
Diisopropyl phthalate
Diisopropyl sulfide
Diisopropyl tartrate, (±)
Diisopropyl thioperoxydicarbonate
1,4-Diisothiocyanatobenzene
Diketene
Dilactic acid
Dimefline
Dimefox
Dimemorfan
2,3-Dimercaptobutanedioic acid
1,4-Dimercapto-2,3-butanediol
2,2'-Dimercaptodiethyl ether
2,3-Dimercapto-1-propanol
Dimetan®
Dimethipin
Dimethirimol
Dimethisoquin
Dimethoxane
2',5'-Dimethoxyacetophenone
1,2-Dimethoxy-4-allylbenzene
4,7-Dimethoxy-5-allyl-1,3-benzodioxole
2,4-Dimethoxyaniline
2,5-Dimethoxyaniline
3,4-Dimethoxyaniline

No.	Name	Synonym	Mol. Form.	CAS RN	Mol. Wt.	Physical Form	mp/°C	bp/°C	den/ g cm^{-3}	n_D	Solubility
3836	2,4-Dimethoxybenzaldehyde		$C_9H_{10}O_3$	613-45-6	166.173	nd (al or lig)	72	290; 165^{10}			i H_2O; s EtOH, eth, bz; sl chl
3837	2,5-Dimethoxybenzaldehyde		$C_9H_{10}O_3$	93-02-7	166.173		52	270; 146^{10}			sl H_2O; s EtOH, eth
3838	3,4-Dimethoxybenzaldehyde	Veratraldehyde	$C_9H_{10}O_3$	120-14-9	166.173	nd (eth, lig, to)	43	281; 155^{10}			sl H_2O, chl; vs EtOH, eth
3839	3,5-Dimethoxybenzaldehyde		$C_9H_{10}O_3$	7311-34-4	166.173		46.3	151^{16}			sl H_2O, peth; s EtOH, bz
3840	1,2-Dimethoxybenzene	Veratrole	$C_8H_{10}O_2$	91-16-7	138.164		22.5	206	1.0810^{25}	1.5827^{21}	sl H_2O; s EtOH, eth, ctc
3841	1,3-Dimethoxybenzene		$C_8H_{10}O_2$	151-10-0	138.164	liq	-52	217.5	1.0521^{25}	1.5231^{20}	sl H_2O; s EtOH, eth, bz, ctc, sulf
3842	1,4-Dimethoxybenzene		$C_8H_{10}O_2$	150-78-7	138.164	lf (w)	59	212.6	1.0375^{55}		sl H_2O; s EtOH, chl; vs eth, bz
3843	3,4-Dimethoxybenzeneacetic acid		$C_{10}H_{12}O_4$	93-40-3	196.200	cry (bz-peth) nd (w+1)	98				s H_2O, chl; vs EtOH, eth
3844	3,4-Dimethoxybenzeneethanamine		$C_{10}H_{15}NO_2$	120-20-7	181.232			164^{14}		1.5464^{20}	s ctc
3845	3,4-Dimethoxybenzenemethanamine		$C_9H_{13}NO_2$	5763-61-1	167.205			156^{12}, 120^{3}	1.143^{25}		s chl
3846	3,4-Dimethoxybenzenemethanol		$C_9H_{12}O_3$	93-03-8	168.189	visc oil		298; 172^{12}	1.178^{17}	1.555^{17}	s H_2O, EtOH
3847	3,3'-Dimethoxybenzidine	Dianisidine	$C_{14}H_{16}N_2O_2$	119-90-4	244.289	lf or nd (w)	137				i H_2O; s EtOH, eth, ace, bz, chl
3848	3,3'-Dimethoxybenzidine-4,4'-diisocyanate		$C_{16}H_{12}N_2O_4$	91-93-0	296.277	cry	112				
3849	2,4-Dimethoxybenzoic acid		$C_9H_{10}O_4$	91-52-1	182.173		108.5				sl H_2O; s EtOH, eth, chl, HOAc
3850	2,6-Dimethoxybenzoic acid		$C_9H_{10}O_4$	1466-76-8	182.173		186 dec				
3851	3,4-Dimethoxybenzoic acid	Veratric acid	$C_9H_{10}O_4$	93-07-2	182.173	nd (w or HOAc) orth (sub)	181	sub			i H_2O; vs EtOH, eth; sl chl
3852	3,5-Dimethoxybenzoic acid		$C_9H_{10}O_4$	1132-21-4	182.173	nd (w), pr (al)	185.5	sub			vs eth, EtOH
3853	4,4'-Dimethoxybenzoin	*p*-Anisoin	$C_{16}H_{16}O_4$	119-52-8	272.296	pr (dil al)	114.0				sl H_2O, chl, EtOH, eth; s ace
3854	5,7-Dimethoxy-2*H*-1-benzopyran-2-one	Limettin	$C_{11}H_{10}O_4$	487-06-9	206.195	pr or nd (al)	149	dec 200			sl H_2O; vs EtOH, ace, chl; i eth, lig
3855	4,4'-Dimethoxy-1,1'-biphenyl		$C_{14}H_{14}O_2$	2132-80-1	214.260	lf (bz)	175	sub			i H_2O, peth; vs EtOH, bz, chl; sl eth
3856	Dimethoxyborane		$C_2H_7BO_2$	4542-61-4	73.887	vol liq or gas	-130.6	25.9			dec H_2O
3857	4,4-Dimethoxy-2-butanone		$C_6H_{12}O_3$	5436-21-5	132.157			50^{5}			s ctc
3858	2,6-Dimethoxy-2,5-cyclohexadiene-1,4-dione	2,6-Dimethoxy-*p*-quinone	$C_8H_8O_4$	530-55-2	168.148	ye mcl pr (HOAc)	256	sub			sl H_2O, EtOH, eth; s tfa; vs alk, HOAc
3859	Dimethoxydimethylsilane		$C_4H_{12}O_2Si$	1112-39-6	120.223			82	0.8646^{20}	1.3708^{20}	dec H_2O
3860	Dimethoxydiphenylsilane		$C_{14}H_{16}O_2Si$	6843-66-9	244.362			286; 161^{15}	1.0771^{20}	1.5447^{20}	
3861	1,1-Dimethoxydodecane	Lauraldehyde, dimethyl acetal	$C_{14}H_{30}O_2$	14620-52-1	230.387			133^{5}		1.4310^{25}	vs eth, EtOH
3862	2,2-Dimethoxyethanamine		$C_4H_{11}NO_2$	22483-09-6	105.136		-78	137^{95}	0.966^{25}	1.4170^{20}	
3863	1,2-Dimethoxyethane	Ethylene glycol dimethyl ether	$C_4H_{10}O_2$	110-71-4	90.121	liq	-69.20	84.50	0.8637^{25}	1.3770^{25}	s H_2O, EtOH, eth, ace, bz, chl, ctc
3864	(2,2-Dimethoxyethyl)benzene		$C_{10}H_{14}O_2$	101-48-4	166.217			193.5			
3865	4,8-Dimethoxyfuro[2,3-b] quinoline	Fagarine	$C_{13}H_{11}NO_3$	524-15-2	229.231	pr (al)	142				sl H_2O, peth; s EtOH, eth, bz, chl
3866	1,1-Dimethoxyhexadecane	Palmitaldehyde, dimethyl acetal	$C_{18}H_{38}O_2$	2791-29-9	286.494		10	144^{2}	0.8542^{20}	1.4382^{25}	vs ace, eth, EtOH
3867	2,4-Dimethoxy-6-hydroxyacetophenone	Xanthoxylin	$C_{10}H_{12}O_4$	90-24-4	196.200	cry (al)	82	185^{20}			vs eth, EtOH
3868	5,6-Dimethoxy-1-indanone		$C_{11}H_{12}O_3$	2107-69-9	192.211		119.5				sl ctc
3869	6,7-Dimethoxy-1(3*H*)-isobenzofuranone	Meconin	$C_{10}H_{10}O_4$	569-31-3	194.184	wh nd (w)	102.5				sl H_2O; s EtOH, eth, ace, bz, HOAc, chl
3870	Dimethoxymethane	Methylal	$C_3H_8O_2$	109-87-5	76.095	liq	-105.1	42	0.8593^{20}	1.3513^{20}	s H_2O; vs ace, bz, eth, EtOH
3871	1,2-Dimethoxy-4-methylbenzene		$C_9H_{12}O_2$	494-99-5	152.190	pr (eth)	24	220	1.0509^{25}	1.5257^{25}	i H_2O; sl ctc; vs os
3872	1,3-Dimethoxy-5-methylbenzene		$C_9H_{12}O_2$	4179-19-5	152.190			244	1.0478^{15}	1.5234^{20}	vs bz, eth, EtOH
3873	1,4-Dimethoxy-2-methylbenzene		$C_9H_{12}O_2$	24599-58-4	152.190		21	214.0			
3874	*N*-(Dimethoxymethyl) dimethylamine	Dimethylformamide dimethyl acetal	$C_5H_{13}NO_2$	4637-24-5	119.163			104	0.897^{25}	1.3972^{20}	

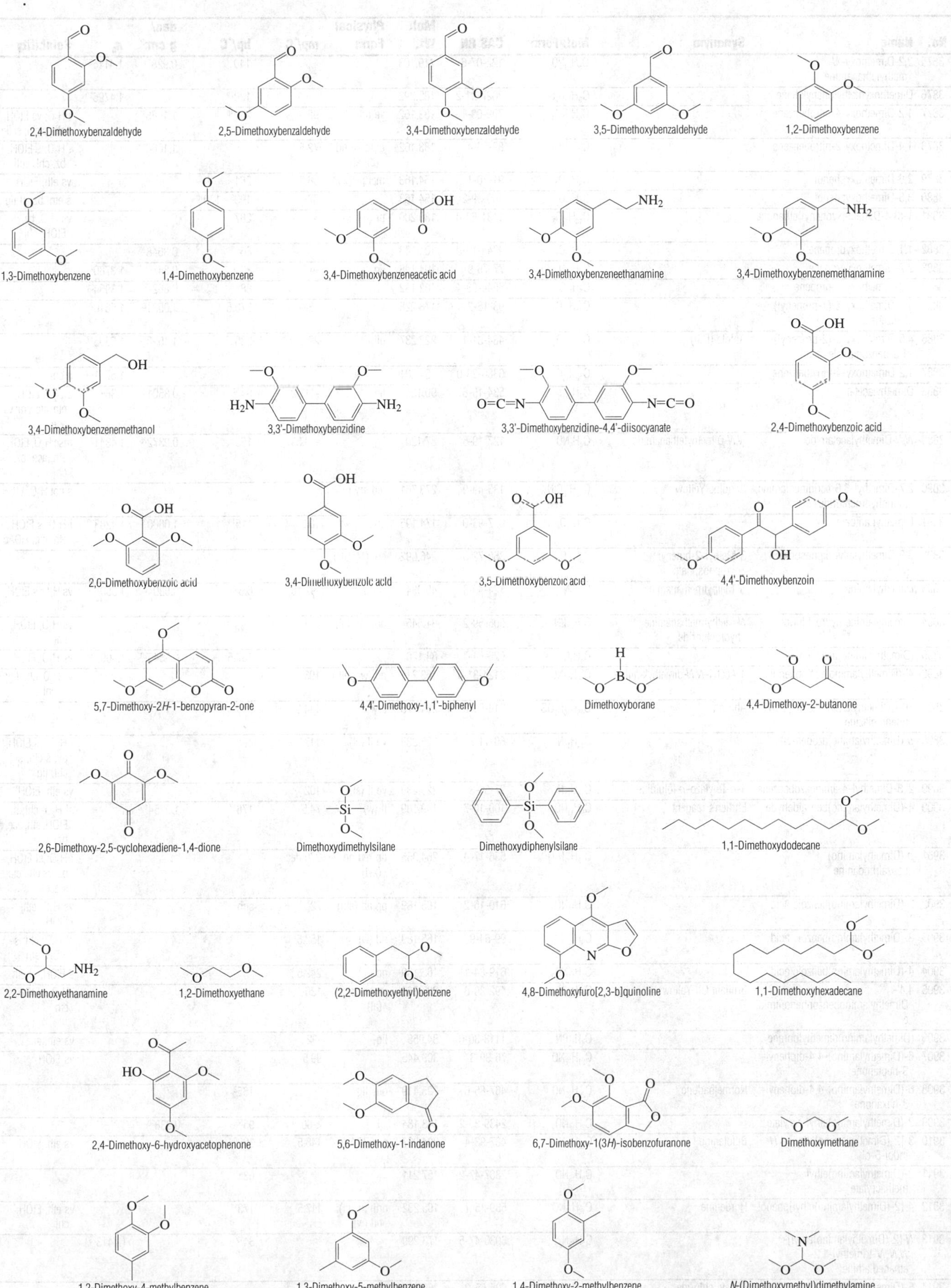

2,4-Dimethoxybenzaldehyde
2,5-Dimethoxybenzaldehyde
3,4-Dimethoxybenzaldehyde
3,5-Dimethoxybenzaldehyde
1,2-Dimethoxybenzene
1,3-Dimethoxybenzene
1,4-Dimethoxybenzene
3,4-Dimethoxybenzeneacetic acid
3,4-Dimethoxybenzeneethanamine
3,4-Dimethoxybenzenemethanamine
3,4-Dimethoxybenzenemethanol
3,3'-Dimethoxybenzidine
3,3'-Dimethoxybenzidine-4,4'-diisocyanate
2,4-Dimethoxybenzoic acid
2,6-Dimethoxybenzoic acid
3,4-Dimethoxybenzoic acid
3,5-Dimethoxybenzoic acid
4,4'-Dimethoxybenzoin
5,7-Dimethoxy-2H-1-benzopyran-2-one
4,4'-Dimethoxy-1,1'-biphenyl
Dimethoxyborane
4,4-Dimethoxy-2-butanone
2,6-Dimethoxy-2,5-cyclohexadiene-1,4-dione
Dimethoxydimethylsilane
Dimethoxydiphenylsilane
1,1-Dimethoxydodecane
2,2-Dimethoxyethanamine
1,2-Dimethoxyethane
(2,2-Dimethoxyethyl)benzene
4,8-Dimethoxyfuro[2,3-b]quinoline
1,1-Dimethoxyhexadecane
2,4-Dimethoxy-6-hydroxyacetophenone
5,6-Dimethoxy-1-indanone
6,7-Dimethoxy-1(3H)-isobenzofuranone
Dimethoxymethane
1,2-Dimethoxy-4-methylbenzene
1,3-Dimethoxy-5-methylbenzene
1,4-Dimethoxy-2-methylbenzene
N-(Dimethoxymethyl)dimethylamine

No.	Name	Synonym	Mol. Form.	CAS RN	Mol. Wt.	Physical Form	mp/°C	bp/°C	den/ g cm⁻³	n_D	Solubility
3875	2,2-Dimethoxy-*N*-methylethanamine		$C_5H_{13}NO_2$	122-07-6	119.163			140	0.928^{25}	1.4115^{20}	
3876	Dimethoxymethylphenylsilane		$C_9H_{14}O_2Si$	3027-21-2	182.292			129^{79}		1.4795^{20}	
3877	1,2-Dimethoxy-4-nitrobenzene		$C_8H_9NO_4$	709-09-1	183.162	ye nd (al-w)	98	230^{15}	1.1888^{133}		i H_2O; vs EtOH, eth; s chl; sl lig
3878	1,4-Dimethoxy-2-nitrobenzene		$C_8H_9NO_4$	89-39-4	183.162	gold-ye nd (dil al)	72.5		1.1666^{132}		i H_2O; s EtOH, bz, chl, sulf
3879	2,6-Dimethoxyphenol		$C_8H_{10}O_3$	91-10-1	154.163	mcl pr (w)	56.5	261			vs eth, EtOH
3880	3,5-Dimethoxyphenol		$C_8H_{10}O_3$	500-99-2	154.163		37	199^{35}, 170^{10}			s eth, bz; sl lig
3881	1-(3,4-Dimethoxyphenyl)ethanone		$C_{10}H_{12}O_3$	1131-62-0	180.200	pr (dil al)	51	287			vs H_2O, bz, EtOH, chl
3882	1,1-Dimethoxypropane		$C_5H_{12}O_2$	4744-10-9	104.148			86	0.8648^{20}		
3883	2,2-Dimethoxypropane		$C_5H_{12}O_2$	77-76-9	104.148	liq	-47	83	0.847^{25}	1.3780^{20}	
3884	3,3-Dimethoxy-1-propene		$C_5H_{10}O_2$	6044-68-4	102.132			88	0.862^{25}	1.3954^{20}	
3885	1,2-Dimethoxy-4-(1-propenyl) benzene		$C_{11}H_{14}O_2$	93-16-3	178.228		18	270.5	1.0521^{20}	1.5616^{20}	
3886	4,5-Dimethoxy-6-(2-propenyl)-1,3-benzodioxole	Apiole (Dill)	$C_{12}H_{14}O_4$	484-31-1	222.237	oil	29.5	285	1.1598^{15}	1.5305^{17}	
3887	1,2-Dimethoxy-4-vinylbenzene		$C_{10}H_{12}O_2$	6380-23-0	164.201					1.5711^{20}	s chl
3888	Dimethylacetal		$C_4H_{10}O_2$	534-15-6	90.121	liq	-113.2	64.5	0.8501^{20}	1.3668^{20}	s H_2O, EtOH, eth, ctc, chl; vs ace
3889	*N,N*-Dimethylacetamide	*N,N*-Dimethylethanamide	C_4H_9NO	127-19-5	87.120	liq	-18.59	165	0.9372^{25}	1.4341^{25}	msc H_2O, EtOH, eth, ace, bz, chl
3890	2,7-Dimethyl-3,6-acridinediamine, monohydrochloride	Acridine Yellow	$C_{15}H_{16}ClN_3$	135-49-9	273.761	red cry pow					s hot H_2O, EtOH
3891	Dimethyl adipate		$C_8H_{14}O_4$	627-93-0	174.195	cry	10.3	115^{13}	1.0600^{20}	1.4283^{20}	i H_2O; s EtOH, eth, ctc, HOAc
3892	3,3-Dimethylallyl diphosphate	3-Methyl-2-butenyl pyrophosphate	$C_5H_{12}O_7P_2$	358-72-5	246.092	cry (MeOH)					
3893	Dimethylamine	*N*-Methylmethanamine	C_2H_7N	124-40-3	45.084	col gas	-92.18	6.88	0.6804^{0}	1.350^{17}	vs H_2O; s EtOH, eth
3894	Dimethylamine hydrochloride	*N*-Methylmethanamine hydrochloride	C_2H_8ClN	506-59-2	81.545	orth nd (al)	171				vs H_2O, EtOH, chl
3895	(Dimethylamino)acetonitrile		$C_4H_8N_2$	926-64-7	84.120			137.5	0.8649^{20}	1.4095^{20}	vs H_2O, EtOH
3896	4'-(Dimethylamino)acetophenone	4-Acetyl-*N,N*-dimethylaniline	$C_{10}H_{13}NO$	2124-31-4	163.216	nd (w, peth)	105.5				vs H_2O, eth, lig; sl chl
3897	10-[(Dimethylamino)acetyl]-10*H*-phenothiazine	Ahistan	$C_{16}H_{16}N_2OS$	518-61-6	284.375	cry	144.5				
3898	*p*-(Dimethylamino)azobenzene		$C_{14}H_{15}N_3$	60-11-7	225.289	ye lf (al)	117	dec			i H_2O; vs EtOH, py; s eth; sl chl, lig
3899	2',3-Dimethyl-4-aminoazobenzene	4-*o*-Tolylazo-*o*-toluidine	$C_{14}H_{15}N_3$	97-56-3	225.289	ye lf (al)	102				vs eth, EtOH
3900	4-(Dimethylamino)benzaldehyde	Ehrlich's reagent	$C_9H_{11}NO$	100-10-7	149.189	lf (w)	74.5	176^{17}	1.0254^{100}		sl H_2O, chl; s EtOH, eth, ace, bz
3901	*p*-(Dimethylamino) benzalrhodanine		$C_{12}H_{12}N_2OS_2$	536-17-4	264.365	dp red nd (xyl)	270 dec				i H_2O; sl EtOH, bz; vs eth, ctc; s ace
3902	2-(Dimethylamino)benzoic acid		$C_9H_{11}NO_2$	610-16-2	165.189	pr, nd (eth)	72	sub			vs H_2O, eth, EtOH
3903	3-(Dimethylamino)benzoic acid		$C_9H_{11}NO_2$	99-64-9	165.189	nd (w)	152.5				sl H_2O, chl; s EtOH, eth
3904	4-(Dimethylamino)benzoic acid		$C_9H_{11}NO_2$	619-84-1	165.189	nd (al)	242.5				s EtOH; sl eth
3905	4,4'-Dimethylaminobenzophenonimide	Brilliant Oil Yellow	$C_{17}H_{21}N_3$	492-80-8	267.369	ye or col pl (al)	136				i H_2O; s EtOH; sl eth
3906	(Dimethylamino)dimethylborane		$C_4H_{12}BN$	1113-30-0	84.956	liq	-92	65			vs eth, ace
3907	6-(Dimethylamino)-4,4-diphenyl-3-heptanone		$C_{21}H_{27}NO$	76-99-3	309.445		99.5				vs EtOH
3908	6-(Dimethylamino)-4,4-diphenyl-3-hexanone	Normethadone	$C_{20}H_{25}NO$	467-85-6	295.419	oily liq		165^{3}			
3909	2-(Dimethylamino)ethyl acrylate		$C_7H_{13}NO_2$	2439-35-2	143.184		<-60	95^{50}	0.938^{20}		
3910	3-[2-(Dimethylamino)ethyl]-1*H*-indol-5-ol	Bufotenine	$C_{12}H_{16}N_2O$	487-93-4	204.267	pr (EtOAc)	146.5	$320^{0.1}$			vs eth, EtOH
3911	2-(Dimethylamino)ethyl methacrylate		$C_8H_{15}NO_2$	2867-47-2	157.211			63^{6}			
3912	4-[2-(Dimethylamino)ethyl]phenol	Hordenine	$C_{10}H_{15}NO$	539-15-1	165.232	orth pr (al), nd (w)	117.5	173^{11}			vs eth, EtOH, chl
3913	*N*-[2-(Dimethylamino)ethyl]-*N,N',N'*-trimethyl-1,2-ethanediamine		$C_9H_{23}N_3$	3030-47-5	173.299			84^{12}		1.4413^{25}	
3914	5-(Dimethylamino)-1-naphthalenesulfonyl chloride	Dansyl chloride	$C_{12}H_{12}ClNO_2S$	605-65-2	269.747		70				

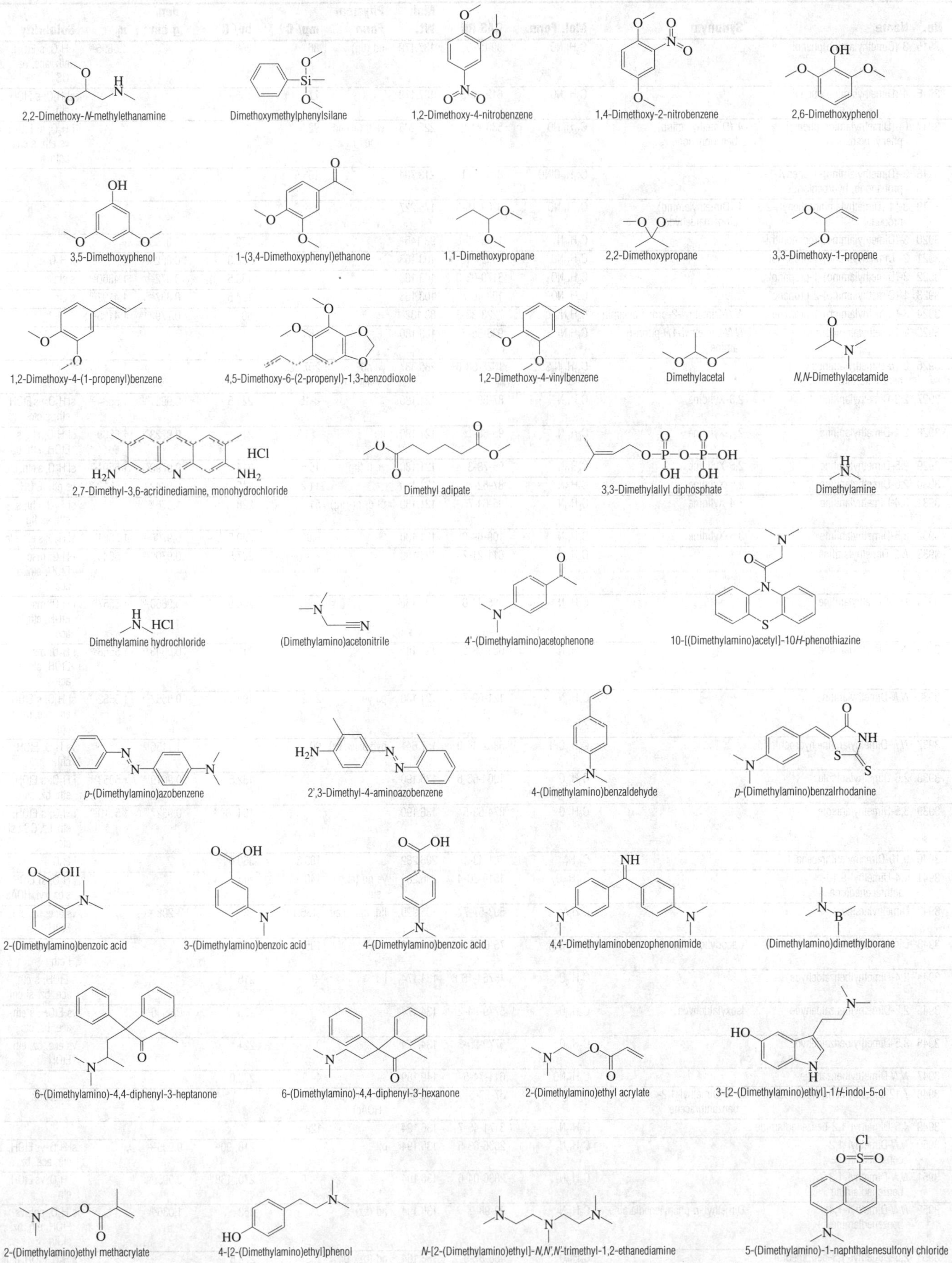

2,2-Dimethoxy-N-methylethanamine
Dimethoxymethylphenylsilane
1,2-Dimethoxy-4-nitrobenzene
1,4-Dimethoxy-2-nitrobenzene
2,6-Dimethoxyphenol
3,5-Dimethoxyphenol
1-(3,4-Dimethoxyphenyl)ethanone
1,1-Dimethoxypropane
2,2-Dimethoxypropane
3,3-Dimethoxy-1-propene
1,2-Dimethoxy-4-(1-propenyl)benzene
4,5-Dimethoxy-6-(2-propenyl)-1,3-benzodioxole
1,2-Dimethoxy-4-vinylbenzene
Dimethylacetal
N,N-Dimethylacetamide
2,7-Dimethyl-3,6-acridinediamine, monohydrochloride
Dimethyl adipate
3,3-Dimethylallyl diphosphate
Dimethylamine
Dimethylamine hydrochloride
(Dimethylamino)acetonitrile
4'-(Dimethylamino)acetophenone
10-[(Dimethylamino)acetyl]-10H-phenothiazine
p-(Dimethylamino)azobenzene
2',3-Dimethyl-4-aminoazobenzene
4-(Dimethylamino)benzaldehyde
p-(Dimethylamino)benzalrhodanine
2-(Dimethylamino)benzoic acid
3-(Dimethylamino)benzoic acid
4-(Dimethylamino)benzoic acid
4,4'-Dimethylaminobenzophenonimide
(Dimethylamino)dimethylborane
6-(Dimethylamino)-4,4-diphenyl-3-heptanone
6-(Dimethylamino)-4,4-diphenyl-3-hexanone
2-(Dimethylamino)ethyl acrylate
3-[2-(Dimethylamino)ethyl]-1H-indol-5-ol
2-(Dimethylamino)ethyl methacrylate
4-[2-(Dimethylamino)ethyl]phenol
N-[2-(Dimethylamino)ethyl]-N,N',N'-trimethyl-1,2-ethanediamine
5-(Dimethylamino)-1-naphthalenesulfonyl chloride

No.	Name	Synonym	Mol. Form.	CAS RN	Mol. Wt.	Physical Form	mp/°C	bp/°C	den/ g cm^{-3}	n_D	Solubility
3915	3-(Dimethylamino)phenol		$C_8H_{11}NO$	99-07-0	137.179	nd (lig)	86	266.5		1.5895^{26}	i H_2O; s EtOH, eth, ace, bz, CS_2
3916	4-(Dimethylamino)phenol		$C_8H_{11}NO$	619-60-3	137.179		77	165^{30}			sl H_2O; s EtOH, eth
3917	[4-(Dimethylamino)phenyl] phenylmethanone	4-(Dimethylamino) benzophenone	$C_{15}H_{15}NO$	530-44-9	225.286	ye lf (al) nd (peth)	92.5				i H_2O; sl EtOH; vs eth; s chl, peth
3918	3-(Dimethylamino)-1-phenyl-1-propanone, hydrochloride		$C_{11}H_{16}ClNO$	879-72-1	213.704		153.5				
3919	3-[4-(Dimethylamino)phenyl]-2-propenal	4-(Dimethylamino) cinnamaldehyde	$C_{11}H_{13}NO$	6203-18-5	175.227		139.5				
3920	3-(Dimethylamino)propanenitrile		$C_5H_{10}N_2$	1738-25-6	98.146			173	0.8705^{20}		
3921	2-(Dimethylamino)-1-propanol		$C_5H_{13}NO$	15521-18-3	103.163			150.3	0.8820^{26}		s H_2O
3922	3-(Dimethylamino)-1-propanol		$C_5H_{13}NO$	3179-63-3	103.163			163.5	0.872^{25}	1.4360^{20}	s ctc
3923	1-(Dimethylamino)-2-propanol		$C_5H_{13}NO$	108-16-7	103.163			124.5	0.837^{25}	1.4193^{20}	s ctc
3924	3-(Dimethylamino)-1-propyne	*N,N*-Dimethyl-2-propargylamine	C_5H_9N	7223-38-3	83.132			80	0.7792^{20}	1.4195^{20}	
3925	2-Dimethylaminopurine	*N,N*-Dimethyl-1*H*-purin-6-amine	$C_7H_9N_5$	938-55-6	163.180		263				
3926	2-(*p*-Dimethylaminostyryl) benzothiazole		$C_{17}H_{16}N_2S$	1628-58-6	280.387	ye nd (MeOH)	207 dec				
3927	2,3-Dimethylaniline	2,3-Xylidine	$C_8H_{11}N$	87-59-2	121.180		<-15	221.5	0.9931^{20}	1.5684^{20}	sl H_2O; vs EtOH, eth; s ctc
3928	2,4-Dimethylaniline	2,4-Xylidine	$C_8H_{11}N$	95-68-1	121.180	liq	-14.3	214	0.9723^{20}	1.5569^{20}	sl H_2O, ctc; s EtOH, eth, bz
3929	2,5-Dimethylaniline	2,5-Xylidine	$C_8H_{11}N$	95-78-3	121.180	ye lf (lig)	15.5	214	0.9790^{21}	1.5591^{21}	sl H_2O; s eth, ctc
3930	2,6-Dimethylaniline	2,6-Xylidine	$C_8H_{11}N$	87-62-7	121.180		11.2	215	0.9842^{20}	1.5610^{20}	vs eth, EtOH
3931	3,4-Dimethylaniline	3,4-Xylidine	$C_8H_{11}N$	95-64-7	121.180	pl or pr (lig)	51	228	1.076^{18}		sl H_2O, chl; s eth; vs lig
3932	3,5-Dimethylaniline	3,5-Xylidine	$C_8H_{11}N$	108-69-0	121.180		9.8	220.5	0.9706^{20}	1.5581^{20}	sl H_2O; s eth, ctc
3933	*N*,2-Dimethylaniline		$C_8H_{11}N$	611-21-2	121.180			207.5	0.9709^{20}	1.5649^{20}	i H_2O; msc EtOH, eth; s ace
3934	*N*,3-Dimethylaniline		$C_8H_{11}N$	696-44-6	121.180			206.5	0.9660^{20}	1.5557^{25}	i H_2O; msc EtOH, eth; s ace
3935	*N*,4-Dimethylaniline		$C_8H_{11}N$	623-08-5	121.180			210	0.9348^{55}	1.5568^{20}	i H_2O; msc EtOH, eth; s ace
3936	*N,N*-Dimethylaniline		$C_8H_{11}N$	121-69-7	121.180	pa ye	2.42	194.15	0.9557^{20}	1.5582^{20}	sl H_2O; s EtOH, eth, ace, bz; vs chl
3937	*N,N*-Dimethylaniline hydrochloride		$C_8H_{12}ClN$	5882-44-0	157.641	hyg pl (w, bz)	90		1.1156^{19}		vs H_2O, EtOH, chl
3938	2,6-Dimethylanisole		$C_9H_{12}O$	1004-66-6	136.190			182.5	0.9619^{14}	1.5053^{14}	i H_2O; s EtOH, eth, bz, ctc
3939	3,5-Dimethylanisole		$C_9H_{12}O$	874-63-5	136.190			194; 89^{15}	0.9627^{15}	1.5110^{20}	i H_2O; s EtOH, eth, bz, CS_2; sl ctc
3940	9,10-Dimethylanthracene		$C_{16}H_{14}$	781-43-1	206.282		183.6	360.0			i H_2O
3941	1,4-Dimethyl-9,10-anthracenedione		$C_{16}H_{12}O_2$	1519-36-4	236.265	ye nd (al, sub)	140.5	sub			i H_2O; sl EtOH; s bz, xyl, HOAc
3942	Dimethylarsine		C_2H_7As	593-57-7	105.999	liq, ign in air	-136.1	36	1.208^{29}		vs ace, bz, eth, EtOH
3943	Dimethylarsinic acid	Cacodylic acid	$C_2H_7AsO_2$	75-60-5	137.998		195	>200			vs H_2O; s EtOH; i eth
3944	2,4-Dimethylbenzaldehyde		$C_9H_{10}O$	15764-16-6	134.174	liq	-9	218			s EtOH; s eth, ace, bz; sl chl
3945	2,5-Dimethylbenzaldehyde	Isoxylaldehyde	$C_9H_{10}O$	5779-94-2	134.174			220	0.9500^{20}		vs EtOH; s eth, ace, bz, ctc
3946	3,5-Dimethylbenzaldehyde		$C_9H_{10}O$	5779-95-3	134.174		9	221			vs ace, bz, eth, EtOH
3947	*N,N*-Dimethylbenzamide		$C_9H_{11}NO$	611-74-5	149.189		44.8	272.0			
3948	7,12-Dimethylbenz[a]anthracene	9,10-Dimethyl-1,2-benzanthracene	$C_{20}H_{16}$	57-97-6	256.341	pa ye pl (al, HOAc)	122.5				vs ace, bz
3949	4,5-Dimethyl-1,2-benzenediamine		$C_8H_{12}N_2$	3171-45-7	136.194		128				
3950	*N,N*-Dimethyl-1,2-benzenediamine		$C_8H_{12}N_2$	2836-03-5	136.194	oil		218; 90^{22}	0.995^{22}		sl H_2O; vs EtOH, eth, ace, bz
3951	*N,N*-Dimethyl-1,3-benzenediamine		$C_8H_{12}N_2$	2836-04-6	136.194		<-20	270; 138^{10}	0.995^{25}		sl H_2O; vs EtOH, eth
3952	*N,N*-Dimethyl-1,4-benzenediamine	Dimethyl-*p*-phenylenediamine	$C_8H_{12}N_2$	99-98-9	136.194	nd (bz)	53	263	1.036^{20}		s H_2O, chl; vs EtOH, eth, bz; sl lig
3953	2,5-Dimethyl-1,3-benzenediol		$C_8H_{10}O_2$	488-87-9	138.164	nd (bz), pr (w)	163	278.5			s H_2O, EtOH, eth

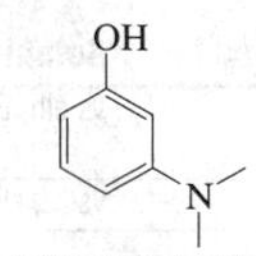

3-(Dimethylamino)phenol

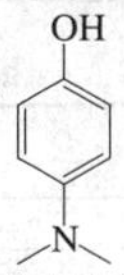

4-(Dimethylamino)phenol

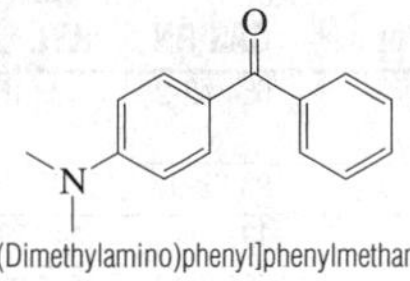

[4-(Dimethylamino)phenyl]phenylmethanone

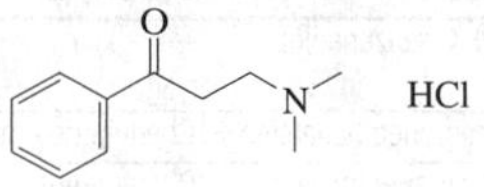

3-(Dimethylamino)-1-phenyl-1-propanone, hydrochloride

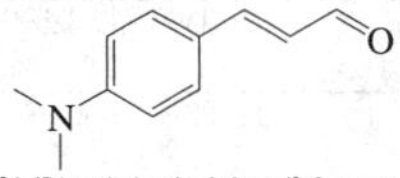

3-[4-(Dimethylamino)phenyl]-2-propenal

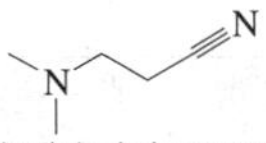

3-(Dimethylamino)propanenitrile

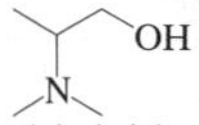

2-(Dimethylamino)-1-propanol

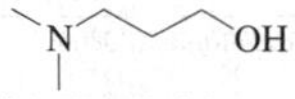

3-(Dimethylamino)-1-propanol

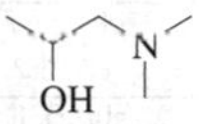

1-(Dimethylamino)-2-propanol

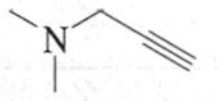

3-(Dimethylamino)-1-propyne

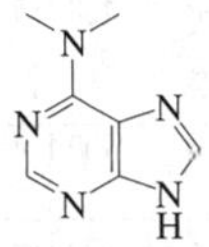

2-Dimethylaminopurine

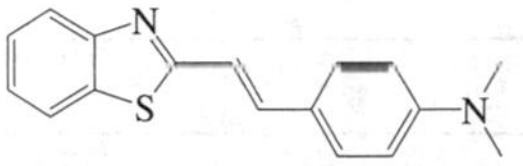

2-(*p*-Dimethylaminostyryl)benzothiazole

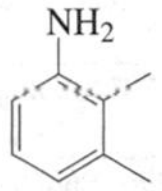

2,3-Dimethylaniline

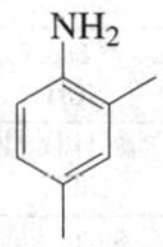

2,4-Dimethylaniline

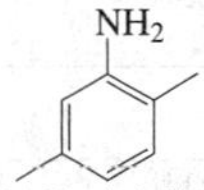

2,5-Dimethylaniline

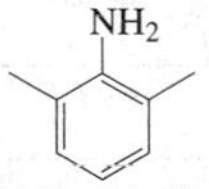

2,6-Dimethylaniline

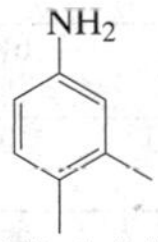

3,4-Dimethylaniline

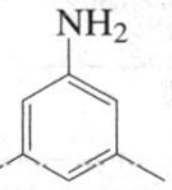

3,5-Dimethylaniline

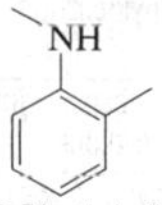

N,2-Dimethylaniline

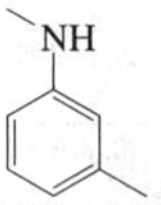

N,3-Dimethylaniline

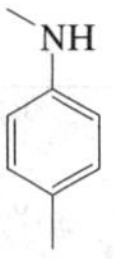

N,4-Dimethylaniline

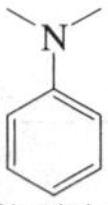

N,*N*-Dimethylaniline

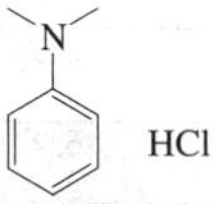

N,*N*-Dimethylaniline hydrochloride

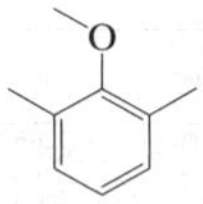

2,6-Dimethylanisole

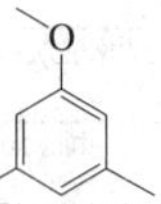

3,5-Dimethylanisole

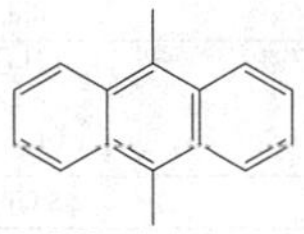

9,10-Dimethylanthracene

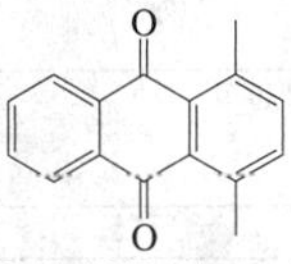

1,4-Dimethyl-9,10-anthracenedione

Dimethylarsine

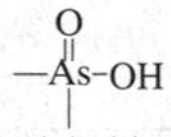

Dimethylarsinic acid

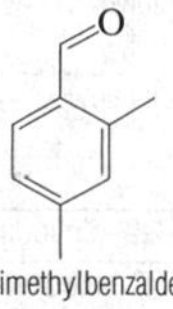

2,4-Dimethylbenzaldehyde

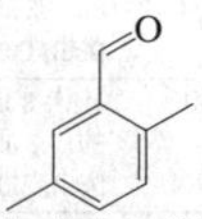

2,5-Dimethylbenzaldehyde

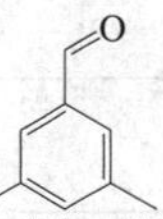

3,5-Dimethylbenzaldehyde

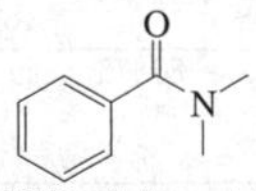

N,*N*-Dimethylbenzamide

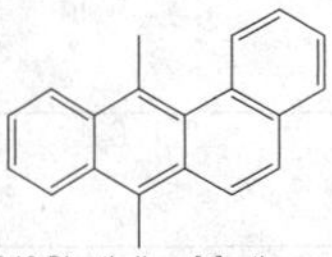

7,12-Dimethylbenz[a]anthracene

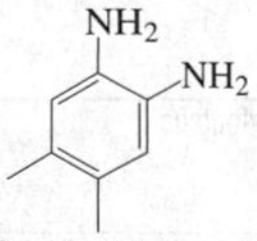

4,5-Dimethyl-1,2-benzenediamine

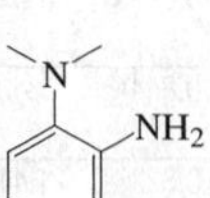

N,*N*-Dimethyl-1,2-benzenediamine

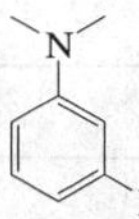

N,*N*-Dimethyl-1,3-benzenediamine

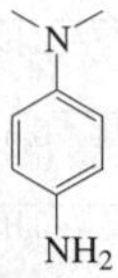

N,*N*-Dimethyl-1,4-benzenediamine

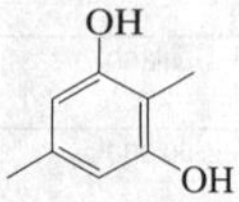

2,5-Dimethyl-1,3-benzenediol

No.	Name	Synonym	Mol. Form.	CAS RN	Mol. Wt.	Physical Form	mp/°C	bp/°C	den/ g cm^{-3}	n_D	Solubility
3954	2,6-Dimethyl-1,4-benzenediol		$C_8H_{10}O_2$	654-42-2	138.164	nd (xyl), cry (w)	152.3				vs eth, EtOH
3955	*N*,β-Dimethylbenzeneethanamine	Phenylpropylmethylamine	$C_{10}H_{15}N$	93-88-9	149.233			207.5	0.915^{25}		vs bz, eth, EtOH
3956	α,α-Dimethylbenzeneethanamine	Phentermine	$C_{10}H_{15}N$	122-09-8	149.233	oily liq		205; 100^{21}			
3957	α,α-Dimethylbenzenemethanamine		$C_9H_{13}N$	585-32-0	135.206			196.5	0.9423^{20}	1.5181^{25}	
3958	α,4-Dimethylbenzenemethanol	1-(4-Methylphenyl)ethanol	$C_9H_{12}O$	536-50-5	136.190			219	0.9668^{25}	1.5246^{20}	i H_2O; vs EtOH, eth
3959	α,α-Dimethylbenzenemethanol	α-Cumyl alcohol	$C_9H_{12}O$	617-94-7	136.190	pr	36	202	0.9735^{20}	1.5325^{20}	i H_2O; s EtOH, eth, bz, HOAc
3960	α,α-Dimethylbenzenepropanol	Benzyl-*tert*-butanol	$C_{11}H_{16}O$	103-05-9	164.244	nd	24.5	121^{13}	0.9626^{21}	1.5077^{21}	i H_2O; vs EtOH, eth, ace, bz
3961	*N*,4-Dimethylbenzenesulfonamide		$C_8H_{11}NO_2S$	640-61-9	185.244	pl (dil al)	78.5		1.340^{25}		vs eth, EtOH
3962	5,6-Dimethyl-1*H*-benzimidazole	Dimedazole	$C_9H_{10}N_2$	582-60-5	146.188	cry (eth)	205.5	sub			s H_2O, EtOH, eth, chl, DMSO
3963	2,4-Dimethylbenzoic acid		$C_9H_{10}O_2$	611-01-8	150.174	mcl or tcl nd (w)	90	268			sl H_2O; s EtOH, ace, bz, chl, HOAc, tol
3964	2,5-Dimethylbenzoic acid		$C_9H_{10}O_2$	610-72-0	150.174	nd (al)	132	sub	1.069^{21}		i H_2O; s EtOH, eth, ace, bz
3965	2,6-Dimethylbenzoic acid		$C_9H_{10}O_2$	632-46-2	150.174	nd (lig)	116	274.5; 155^{17}			sl H_2O, lig; s EtOH, eth
3966	3,4-Dimethylbenzoic acid		$C_9H_{10}O_2$	619-04-5	150.174	pr (al)	167.3				i H_2O; s EtOH, eth, bz
3967	3,5-Dimethylbenzoic acid	Mesitylenic acid	$C_9H_{10}O_2$	499-06-9	150.174	nd (w, al)	171.1	sub			sl H_2O; vs EtOH, eth
3968	4,4'-Dimethylbenzophenone	Bis(4-methylphenyl) ketone	$C_{15}H_{14}O$	611-97-2	210.271	orth (al)	96.5	334			vs ace, bz, eth, EtOH
3969	7,8-Dimethylbenzo[g]pteridine-2,4(1*H*,3*H*)-dione	Lumichrome	$C_{12}H_{10}N_4O_2$	1086-80-2	242.233	ye cry (chl)	300				sl H_2O, EtOH, chl
3970	2,5-Dimethylbenzoxazole		C_9H_9NO	5676-58-4	147.173			218.5	1.0880^{18}	1.5412^{20}	s ctc
3971	*N*,*N*-Dimethylbenzylamine	Dimethylbenzylamine	$C_9H_{13}N$	103-83-3	135.206			181	0.915^{0}	1.5011^{20}	sl H_2O; msc EtOH, eth
3972	*N*,*N*-Dimethyl-*N'*-benzyl-1,2-ethanediamine	*N*-Benzyl-*N'*,*N'*-dimethyl-1,2-ethanediamine	$C_{11}H_{18}N_2$	103-55-9	178.274			145^{30}, 123^{11}	0.9343^{20}	1.5089^{20}	
3973	*N*,*N*-Dimethyl-*N'*-benzyl-*N'*-2-pyridinyl-1,2-ethanediamine	Tripelennamine	$C_{16}H_{21}N_3$	91-81-6	255.358	ye oil		$140^{0.1}$		1.576^{25}	misc H_2O
3974	6,6-Dimethylbicyclo[3.1.1]heptan-2-one, (1*R*)		$C_9H_{14}O$	38651-65-9	138.206	liq	-1	209	0.9807^{20}	1.4787^{20}	vs eth, EtOH
3975	2,3-Dimethylbicyclo[2.2.1]hept-2-ene	2,3-Dimethyl-2-norbornene	C_9H_{14}	529-16-8	122.207			140.5	0.8698^{17}	1.4688^{17}	s eth, ace, bz
3976	6,6-Dimethylbicyclo[3.1.1]hept-2-ene-2-ethanol		$C_{11}H_{18}O$	128-50-7	166.260			235; 110^{10}	0.973^{25}	1.4930^{20}	s chl
3977	2,2'-Dimethylbiphenyl		$C_{14}H_{14}$	605-39-0	182.261	cry (al)	19.5	256	0.9906^{20}	1.5752^{20}	i H_2O; vs EtOH, eth, bz; s ace
3978	3,3'-Dimethylbiphenyl		$C_{14}H_{14}$	612-75-9	182.261		9	280	0.9995^{20}	1.5946^{20}	i H_2O; vs EtOH, eth, bz; s ace
3979	4,4'-Dimethylbiphenyl		$C_{14}H_{14}$	613-33-2	182.261	mcl pr (eth)	125	295	0.917^{121}		i H_2O; sl EtOH; s eth, ace, bz, CS_2
3980	4,4'-Dimethyl-2,2'-bipyridine		$C_{12}H_{12}N_2$	1134-35-6	184.236		171.5				s chl
3981	2,3-Dimethyl-1,3-butadiene	Diisopropenyl	C_6H_{10}	513-81-5	82.143	liq	-76	68.8	0.7222^{25}	1.4394^{20}	s ctc
3982	*N*,*N*-Dimethylbutanamide		$C_6H_{13}NO$	760-79-2	115.173	liq	-40	186; 125^{100}	0.9064^{25}	1.4391^{25}	vs ace, bz, eth, EtOH
3983	3,3-Dimethyl-2-butanamine		$C_6H_{15}N$	3850-30-4	101.190	liq	-20	102	0.7668^{20}	1.4105^{25}	vs H_2O
3984	2,2-Dimethylbutane	Neohexane	C_6H_{14}	75-83-2	86.175	liq	-98.8	49.73	0.6444^{25}	1.3688^{20}	i H_2O; s EtOH, eth; vs ace, bz, peth, ctc
3985	2,3-Dimethylbutane		C_6H_{14}	79-29-8	86.175	liq	-128.10	57.93	0.6616^{20}	1.3750^{20}	i H_2O; s EtOH, eth; vs ace, bz, peth, ctc
3986	2,3-Dimethyl-2,3-butanediol	Pinacol	$C_6H_{14}O_2$	76-09-5	118.174	nd (al,eth)	43.32	174.4			sl H_2O, CS_2; vs EtOH, eth
3987	2,3-Dimethyl-2-butanethiol		$C_6H_{14}S$	1639-01-6	118.240	liq		126.1			
3988	2,2-Dimethylbutanoic acid		$C_6H_{12}O_2$	595-37-9	116.158	liq	-14	186	0.9276^{20}	1.4145^{20}	sl H_2O; s EtOH, eth
3989	2,2-Dimethyl-1-butanol		$C_6H_{14}O$	1185-33-7	102.174		<-15	136.5	0.8283^{20}	1.4208^{20}	sl H_2O; s EtOH, eth
3990	3,3-Dimethyl-1-butanol		$C_6H_{14}O$	624-95-3	102.174	liq	-60	143	0.844^{15}	1.4323^{15}	sl H_2O; s EtOH, eth, ace
3991	2,3-Dimethyl-2-butanol		$C_6H_{14}O$	594-60-5	102.174	liq	-14	118.4	0.8236^{20}	1.4176^{20}	s H_2O; msc EtOH, eth
3992	3,3-Dimethyl-2-butanol, (±)		$C_6H_{14}O$	20281-91-8	102.174		5.6	120.4	0.8122^{25}	1.4148^{20}	sl H_2O; vs EtOH, eth

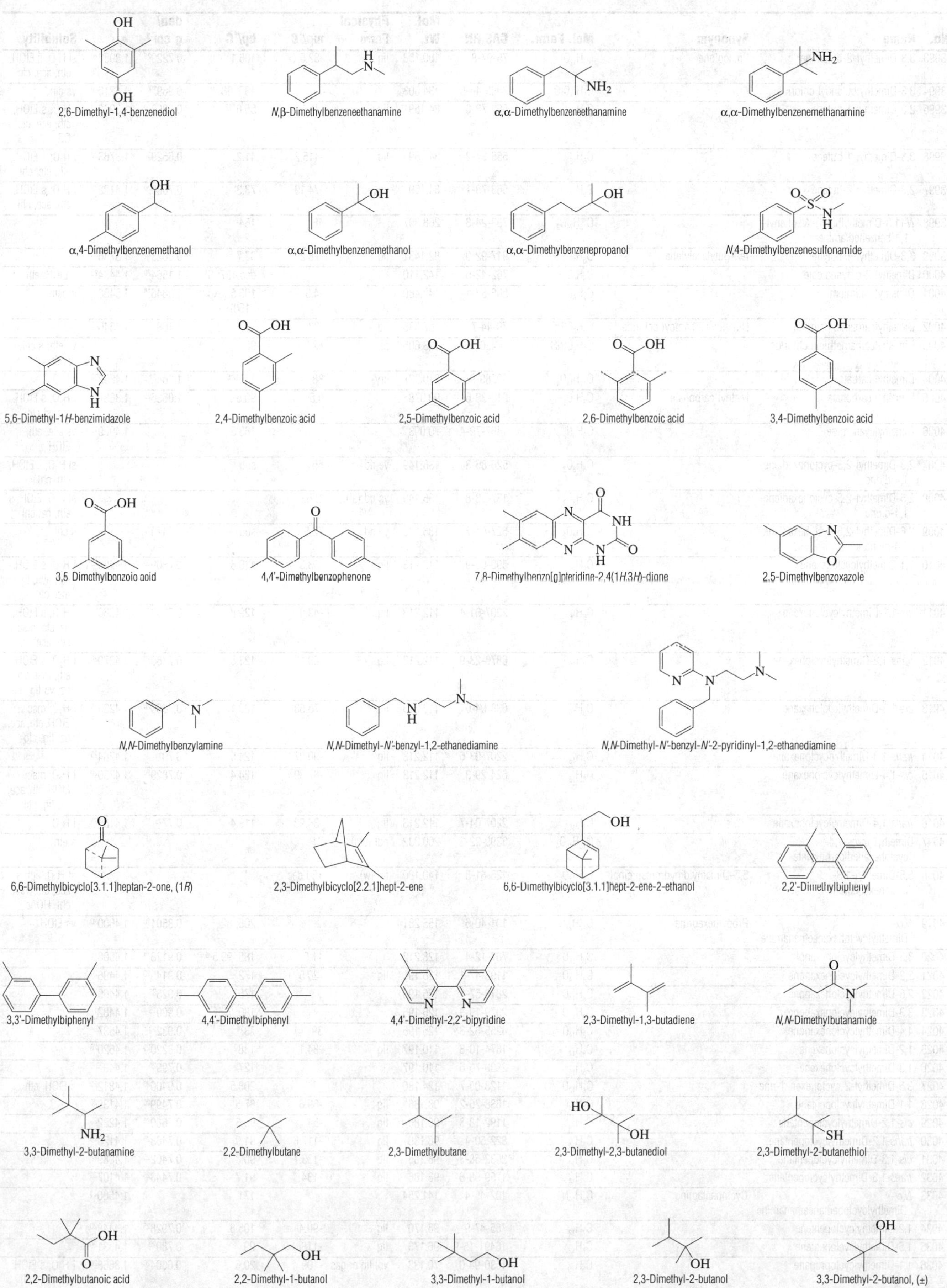

2,6-Dimethyl-1,4-benzenediol
N,β-Dimethylbenzeneethanamine
α,α-Dimethylbenzeneethanamine
α,α-Dimethylbenzenemethanamine
α,4-Dimethylbenzenemethanol
α,α-Dimethylbenzenemethanol
α,α-Dimethylbenzenepropanol
N,4-Dimethylbenzenesulfonamide
5,6-Dimethyl-1H-benzimidazole
2,4-Dimethylbenzoic acid
2,5-Dimethylbenzoic acid
2,6-Dimethylbenzoic acid
3,4-Dimethylbenzoic acid
3,5-Dimethylbenzoic acid
4,4'-Dimethylbenzophenone
7,8-Dimethylbenzo[g]pteridine-2,4(1H,3H)-dione
2,5-Dimethylbenzoxazole
N,N-Dimethylbenzylamine
N,N-Dimethyl-N'-benzyl-1,2-ethanediamine
N,N-Dimethyl-N'-benzyl-N'-2-pyridinyl-1,2-ethanediamine
6,6-Dimethylbicyclo[3.1.1]heptan-2-one, (1R)
2,3-Dimethylbicyclo[2.2.1]hept-2-ene
6,6-Dimethylbicyclo[3.1.1]hept-2-ene-2-ethanol
2,2'-Dimethylbiphenyl
3,3'-Dimethylbiphenyl
4,4'-Dimethylbiphenyl
4,4'-Dimethyl-2,2'-bipyridine
2,3-Dimethyl-1,3-butadiene
N,N-Dimethylbutanamide
3,3-Dimethyl-2-butanamine
2,2-Dimethylbutane
2,3-Dimethylbutane
2,3-Dimethyl-2,3-butanediol
2,3-Dimethyl-2-butanethiol
2,2-Dimethylbutanoic acid
2,2-Dimethyl-1-butanol
3,3-Dimethyl-1-butanol
2,3-Dimethyl-2-butanol
3,3-Dimethyl-2-butanol, (±)

No.	Name	Synonym	Mol. Form.	CAS RN	Mol. Wt.	Physical Form	mp/°C	bp/°C	den/ g cm^{-3}	n_D	Solubility
3993	3,3-Dimethyl-2-butanone	Pinacolone	$C_6H_{12}O$	75-97-8	100.158	liq	-52.5	106.1	0.7229^{25}	1.3952^{20}	sl H_2O; s EtOH, eth, ace, ctc
3994	3,3-Dimethylbutanoyl chloride		$C_6H_{11}ClO$	7065-46-5	134.603			130; 68^{100}	0.969^{20}	1.4210^{20}	vs eth
3995	2,3-Dimethyl-1-butene		C_6H_{12}	563-78-0	84.159	liq	-157.3	55.6	0.6803^{20}	1.3995^{20}	i H_2O; s EtOH, eth, ace, ctc, CS_2
3996	3,3-Dimethyl-1-butene		C_6H_{12}	558-37-2	84.159	liq	-115.2	41.2	0.6529^{20}	1.3763^{20}	i H_2O; s EtOH, eth, ctc, chl
3997	2,3-Dimethyl-2-butene		C_6H_{12}	563-79-1	84.159	liq	-74.19	73.3	0.7080^{20}	1.4122^{20}	i H_2O; s EtOH, eth, ace, chl
3998	*N*-(1,3-Dimethylbutyl)-*N'*-phenyl-1,4-benzenediamine		$C_{18}H_{24}N_2$	793-24-8	268.397		46	164^1			
3999	3,3-Dimethyl-1-butyne	*tert*-Butylacetylene	C_6H_{10}	917-92-0	82.143	liq	-78.2	37.7	0.6623^{25}	1.3736^{20}	
4000	Dimethyl 2-butynedioate		$C_6H_6O_4$	762-42-5	142.110			dec 197; 98^{20}	1.1564^{20}	1.4434^{20}	s EtOH, eth, ctc
4001	Dimethyl cadmium		C_2H_6Cd	506-82-1	142.480		-4.5	105.5 (exp 150)	1.9846^{18}	1.5488	s peth
4002	Dimethylcarbamic chloride	Dimethylcarbamoyl chloride	C_3H_6ClNO	79-44-7	107.539	liq	-33	167	1.168^{25}	1.4540^{20}	
4003	Dimethylcarbamothioic chloride		C_3H_6ClNS	16420-13-6	123.605	pr	42.5	98^{10}			vs eth; s chl, peth
4004	Dimethyl carbate		$C_{11}H_{14}O_4$	39589-98-5	210.227	cry	38	$137^{12.5}$	1.164^{21}	1.4852^{20}	i H_2O
4005	Dimethyl carbonate	Methyl carbonate	$C_3H_6O_3$	616-38-6	90.078		0.5	90.5	1.0636^{25}	1.3687^{20}	i H_2O; s EtOH, eth; sl ctc
4006	Dimethylcyanamide		$C_3H_6N_2$	1467-79-4	70.093			163.5		1.4089^{19}	vs ace, eth, EtOH
4007	2,3-Dimethyl-2,5-cyclohexadiene-1,4-dione		$C_8H_8O_2$	526-86-3	136.149	ye nd	55	sub			sl H_2O; s EtOH, eth, chl
4008	2,5-Dimethyl-2,5-cyclohexadiene-1,4-dione		$C_8H_8O_2$	137-18-8	136.149	ye nd (al)	126.0				sl H_2O, EtOH; s eth, bz, chl
4009	2,6-Dimethyl-2,5-cyclohexadiene-1,4-dione		$C_8H_8O_2$	527-61-7	136.149	ye nd	72.5	sub	1.0479^{28}		s chl
4010	1,1-Dimethylcyclohexane		C_8H_{16}	590-66-9	112.213	liq	-33.3	119.6	0.7809^{20}	1.4290^{20}	i H_2O; s EtOH, eth, ace, bz; msc ctc
4011	*cis*-1,2-Dimethylcyclohexane		C_8H_{16}	2207-01-4	112.213	liq	-49.8	129.8	0.7963^{20}	1.4360^{20}	i H_2O; s EtOH, bz, ctc; msc eth, ace
4012	*trans*-1,2-Dimethylcyclohexane		C_8H_{16}	6876-23-9	112.213	liq	-88.15	123.5	0.7760^{20}	1.4270^{20}	i H_2O; s EtOH, eth; msc ace, bz; vs lig
4013	*cis*-1,3-Dimethylcyclohexane		C_8H_{16}	638-04-0	112.213	liq	-75.53	120.1	0.7660^{20}	1.4229^{20}	i H_2O; msc EtOH, eth, ace, bz, lig, ctc
4014	*trans*-1,3-Dimethylcyclohexane		C_8H_{16}	2207-03-6	112.213	liq	-90.07	124.5	0.79^{15}	1.4284^{25}	
4015	*cis*-1,4-Dimethylcyclohexane		C_8H_{16}	624-29-3	112.213	liq	-87.39	124.4	0.7829^{20}	1.4230^{20}	i H_2O; msc EtOH, eth, ace, bz, lig, ctc
4016	*trans*-1,4-Dimethylcyclohexane		C_8H_{16}	2207-04-7	112.213	liq	-36.93	119.4	0.77^{15}	1.4185^{25}	i H_2O
4017	Dimethyl *trans*-1,4-cyclohexanedicarboxylate		$C_{10}H_{16}O_4$	3399-22-2	200.232	ndl (eth)	71				s eth
4018	5,5-Dimethyl-1,3-cyclohexanedione	5,5-Dimethyldihydroresorcinol	$C_8H_{12}O_2$	126-81-8	140.180	nd (w)	150 dec				sl H_2O, eth; s ace, ctc; vs chl, HOAc
4019	*N*,α-Dimethylcyclohexaneethanamine	Propylhexedrine	$C_{10}H_{21}N$	101-40-6	155.281			205; 82^{10}	0.8501^{20}	1.4600^{20}	vs EtOH
4020	3,3-Dimethylcyclohexanol		$C_8H_{16}O$	767-12-4	128.212		11.5	185; 99.5^{35}	0.9128^{14}	1.4606^{15}	
4021	2,2-Dimethylcyclohexanone		$C_8H_{14}O$	1193-47-1	126.196	liq	-20.5	172	0.9145^{20}	1.4486^{20}	
4022	2,6-Dimethylcyclohexanone		$C_8H_{14}O$	2816-57-1	126.196			175	0.925^{25}	1.4460^{20}	
4023	3,3-Dimethylcyclohexanone		$C_8H_{14}O$	2979-19-3	126.196			180; 72^{25}	0.909^{15}	1.4482^{17}	
4024	4,4-Dimethylcyclohexanone		$C_8H_{14}O$	4255-62-3	126.196		39	73^{14}	0.932^{20}	1.4537^{24}	
4025	1,2-Dimethylcyclohexene		C_8H_{14}	1674-10-8	110.197	liq	-84.1	138	0.8220^{25}	1.4620^{20}	
4026	1,3-Dimethylcyclohexene		C_8H_{14}	2808-76-6	110.197			127	0.799^{25}	1.449^{20}	
4027	3,5-Dimethyl-2-cyclohexen-1-one		$C_8H_{12}O$	1123-09-7	124.180			208.5	0.9400^{20}	1.4812^{20}	s EtOH, eth
4028	1,1-Dimethylcyclopentane		C_7H_{14}	1638-26-2	98.186	liq	-69.8	87.5	0.7499^{25}	1.4136^{20}	
4029	*cis*-1,2-Dimethylcyclopentane		C_7H_{14}	1192-18-3	98.186	liq	-54	99.5	0.7680^{25}	1.4222^{20}	
4030	*trans*-1,2-Dimethylcyclopentane		C_7H_{14}	822-50-4	98.186	liq	-117.6	91.9	0.7468^{25}	1.4120^{20}	
4031	*cis*-1,3-Dimethylcyclopentane		C_7H_{14}	2532-58-3	98.186	liq	-133.7	90.8	0.7402^{25}	1.4089^{20}	
4032	*trans*-1,3-Dimethylcyclopentane		C_7H_{14}	1759-58-6	98.186	liq	-134	91.7	0.7443^{25}	1.4107^{20}	
4033	*N*,α-Dimethylcyclopentaneethanamine	Cyclopentamine	$C_9H_{19}N$	102-45-4	141.254			171		1.4500^{20}	
4034	1,2-Dimethylcyclopentene		C_7H_{12}	765-47-9	96.170	liq	-90.4	105.8	0.7928^{25}	1.4448^{20}	
4035	1,5-Dimethylcyclopentene		C_7H_{12}	16491-15-9	96.170	liq	-118	99	0.780^{20}	1.4331^{20}	
4036	1,1-Dimethylcyclopropane		C_5H_{10}	1630-94-0	70.133	vol liq or gas	-109	20.6	0.6604^{20}	1.3668^{20}	i H_2O; s EtOH; vs eth, sulf

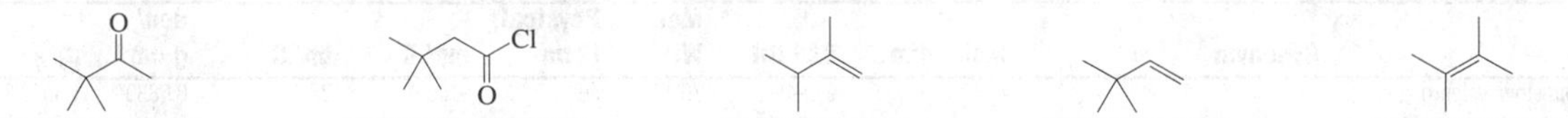

3,3-Dimethyl-2-butanone

3,3-Dimethylbutanoyl chloride

2,3-Dimethyl-1-butene

3,3-Dimethyl-1-butene

2,3-Dimethyl-2-butene

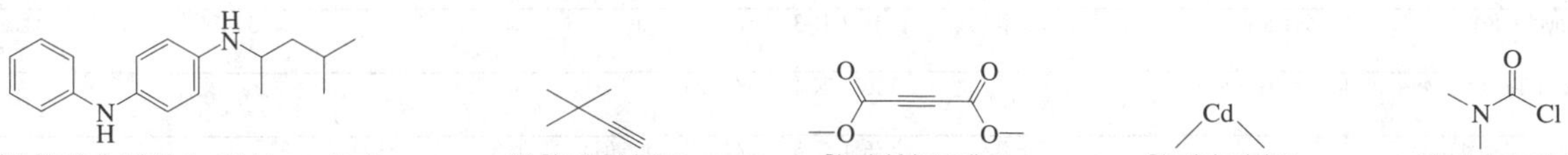

N-(1,3-Dimethylbutyl)-*N*′-phenyl-1,4-benzenediamine

3,3-Dimethyl-1-butyne

Dimethyl 2-butynedioate

Dimethyl cadmium

Dimethylcarbamic chloride

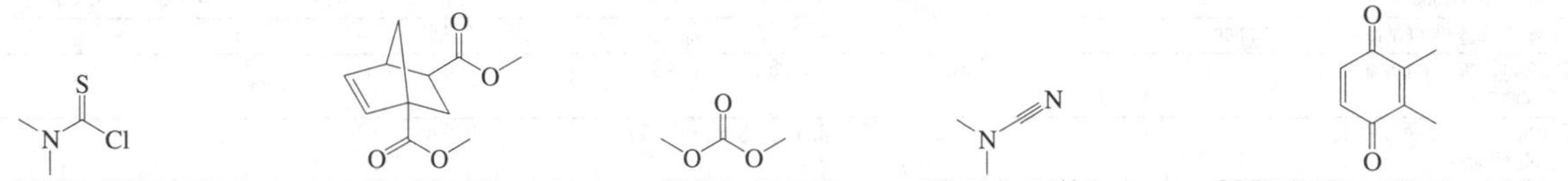

Dimethylcarbamothioic chloride

Dimethyl carbate

Dimethyl carbonate

Dimethylcyanamide

2,3-Dimethyl-2,5-cyclohexadiene-1,4-dione

2,5-Dimethyl-2,5-cyclohexadiene-1,4-dione

2,6-Dimethyl-2,5-cyclohexadiene-1,4-dione

1,1-Dimethylcyclohexane

cis-1,2-Dimethylcyclohexane

trans-1,2-Dimethylcyclohexane

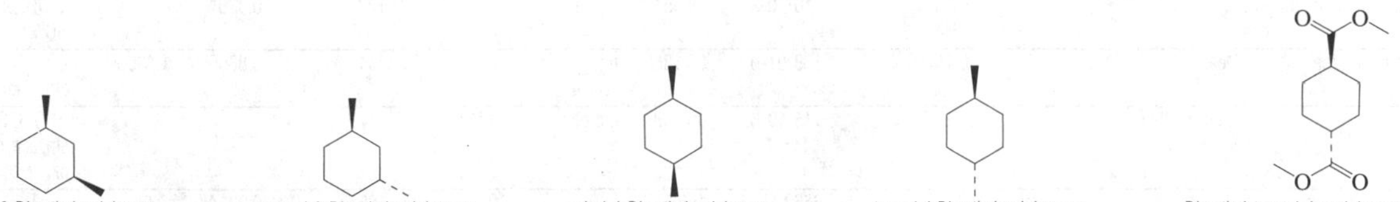

cis-1,3-Dimethylcyclohexane

trans-1,3-Dimethylcyclohexane

cis-1,4-Dimethylcyclohexane

trans-1,4-Dimethylcyclohexane

Dimethyl *trans*-1,4-cyclohexanedicarboxylate

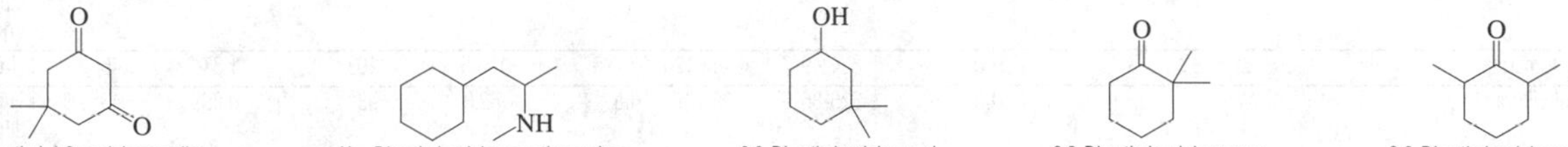

5,5-Dimethyl-1,3-cyclohexanedione

N,α-Dimethylcyclohexaneethanamine

3,3-Dimethylcyclohexanol

2,2-Dimethylcyclohexanone

2,6-Dimethylcyclohexanone

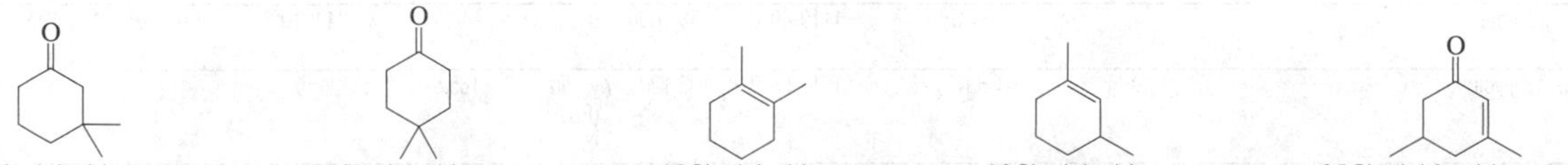

3,3-Dimethylcyclohexanone

4,4-Dimethylcyclohexanone

1,2-Dimethylcyclohexene

1,3-Dimethylcyclohexene

3,5-Dimethyl-2-cyclohexen-1-one

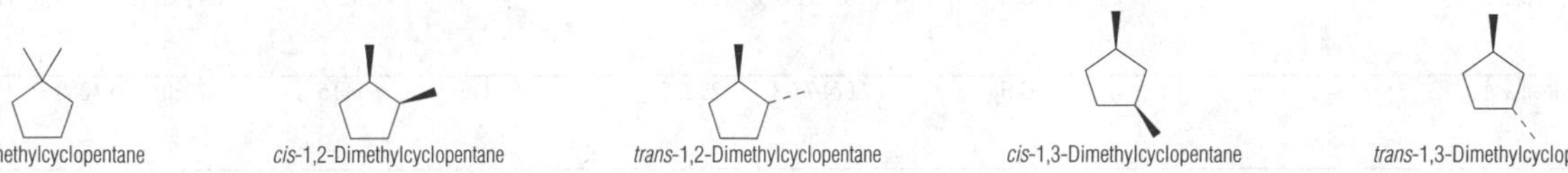

1,1-Dimethylcyclopentane

cis-1,2-Dimethylcyclopentane

trans-1,2-Dimethylcyclopentane

cis-1,3-Dimethylcyclopentane

trans-1,3-Dimethylcyclopentane

N,α-Dimethylcyclopentaneethanamine

1,2-Dimethylcyclopentene

1,5-Dimethylcyclopentene

1,1-Dimethylcyclopropane

No.	Name	Synonym	Mol. Form.	CAS RN	Mol. Wt.	Physical Form	mp/°C	bp/°C	den/ g cm^{-3}	n_D	Solubility
4037	*cis*-1,2-Dimethylcyclopropane		C_5H_{10}	930-18-7	70.133	liq	-140.9	37.0	0.6889^{25}	1.3829^{20}	i H_2O; s EtOH; vs eth; sl ctc
4038	*trans*-1,2-Dimethylcyclopropane		C_5H_{10}	2402-06-4	70.133	vol liq or gas	-149.6	28.2	0.6648^{25}	1.3713^{20}	vs eth, EtOH
4039	Dimethyldecylamine	*N,N*-Dimethyl-1-decanamine	$C_{12}H_{27}N$	1120-24-7	185.349			234.5			
4040	Dimethyldiacetoxysilane	Bis(acetyloxy)dimethylsilane	$C_6H_{12}O_4Si$	2182-66-3	176.243	liq	-12.5	165	1.0540^{20}	1.4030^{20}	
4041	*trans*-Dimethyldiazene	Azomethane	$C_2H_6N_2$	4143-41-3	58.082	gas	-78	1.5	0.743^{0}	1.4199^{19}	vs ace, EtOH, eth; s ctc, hp
4042	2,2-Dimethyl-1,3-dioxane-4,6-dione	Meldrum's acid	$C_6H_8O_4$	2033-24-1	144.126		94				
4043	*cis*-3,6-Dimethyl-1,4-dioxane-2,5-dione		$C_6H_8O_4$	4511-42-6	144.126	orth (eth)	96.8	150^{25}			
4044	2,2-Dimethyl-1,3-dioxolane-4-methanol	Isopropylidene glycerol	$C_6H_{12}O_3$	100-79-8	132.157			82^{10}	1.064^{20}	1.4383^{20}	
4045	Dimethyldiphenoxysilane		$C_{14}H_{16}O_2Si$	3440-02-6	244.362		-23	131^{5}	1.0599^{25}	1.5330^{20}	
4046	2,3-Dimethyl-2,3-diphenylbutane	Dicumene	$C_{18}H_{22}$	1889-67-4	238.368	cry (MeOH)	119.5				
4047	3,3'-Dimethyldiphenylmethane 4,4'-diisocyanate		$C_{17}H_{14}N_2O_2$	139-25-3	278.305						s chl
4048	2,9-Dimethyl-4,7-diphenyl-1,10-phenanthroline		$C_{26}H_{20}N_2$	4733-39-5	360.450		280 dec				
4049	Dimethyldiphenylsilane		$C_{14}H_{16}Si$	778-24-5	212.363			277; 173^{45}	0.9867^{20}	1.5644^{20}	
4050	*N,N'*-Dimethyl-*N,N'*-diphenylurea		$C_{15}H_{16}N_2O$	611-92-7	240.300	pl (al)	122	350			vs H_2O, EtOH, ace; sl eth, bz, CS_2
4051	Dimethyl disulfide	Methyl disulfide	$C_2H_6S_2$	624-92-0	94.199	liq	-84.67	109.74	1.0625^{20}	1.5289^{20}	i H_2O; msc EtOH, eth
4052	*O,O*-Dimethyl dithiophosphate	*O,O*-Dimethyl phosphorodithionate	$C_2H_7O_2PS_2$	756-80-9	158.180	liq		56^{4}	1.29^{20}		
4053	*N,N*-Dimethyldodecylamine oxide		$C_{14}H_{31}NO$	1643-20-5	229.402	hyg nd (tol)	130.5				
4054	1,2-Dimethylenecyclohexane		C_8H_{12}	2819-48-9	108.181			127; 60^{90}	0.8361^{20}	1.4718^{25}	i H_2O; s EtOH, eth, bz, chl; vs ace
4055	*N,N*-Dimethyl-1,2-ethanediamine		$C_4H_{12}N_2$	108-00-9	88.151			104	0.803^{25}	1.4260^{20}	
4056	*N,N'*-Dimethyl-1,2-ethanediamine		$C_4H_{12}N_2$	110-70-3	88.151			120	0.828^{15}		s EtOH, eth, dil HCl
4057	*N,N*-Dimethylethanolamine	Deanol	$C_4H_{11}NO$	108-01-0	89.136	liq	-59	134	0.8866^{20}	1.4300^{20}	msc H_2O, EtOH, eth; s chl
4058	Dimethyl ether	Methyl ether	C_2H_6O	115-10-6	46.068	col gas	-141.5	-24.8			s H_2O, EtOH, eth, ace, chl; sl bz
4059	(1,1-Dimethylethoxy)benzene		$C_{10}H_{14}O$	6669-13-2	150.217	liq	-24	185.5	0.9214^{20}		
4060	[(1,1-Dimethylethoxy)methyl] oxirane		$C_7H_{14}O_2$	7665-72-7	130.185	liq	-70	152	0.898^{20}		
4061	*N,N*-Dimethylformamide	DMF	C_3H_7NO	68-12-2	73.094	liq	-60.48	153	0.9445^{25}	1.4305^{20}	msc H_2O, EtOH, eth, ace, bz; sl lig
4062	Dimethyl fumarate		$C_6H_8O_4$	624-49-7	144.126		103.5	193	1.37^{20}	1.4062^{111}	i H_2O; s ace, chl
4063	2,5-Dimethylfuran		C_6H_8O	625-86-5	96.127	liq	-62.8	93	0.8883^{20}	1.4363^{20}	i H_2O; s EtOH, eth, ace, bz, HOAc, chl
4064	3,4-Dimethyl-2,5-furandione		$C_6H_6O_3$	766-39-2	126.110	pl or lf (dil al)	96	223	1.107^{100}		sl H_2O; vs EtOH, eth, bz, chl
4065	Dimethyl germanium sulfide		C_2H_6GeS	16090-49-6	134.77	col cry	54.5	302			
4066	Dimethyl glutarate		$C_7H_{12}O_4$	1119-40-0	160.168	liq	-42.5	214; 109^{21}	1.0876^{20}	1.4242^{20}	vs EtOH, eth; s chl
4067	*N,N*-Dimethylglycine		$C_4H_9NO_2$	1118-68-9	103.120	hyg nd (PrOH)	185.5				vs H_2O, MeOH; s EtOH, eth, ace
4068	Dimethylglyoxime		$C_4H_8N_2O_2$	95-45-4	116.119	nd (to or dil al)	245.5	sub 234			i H_2O; vs EtOH, eth; sl bz, tol
4069	2,6-Dimethyl-1,5-heptadiene		C_9H_{16}	6709-39-3	124.223	liq	-70	143	0.7648^{25}		
4070	2,2-Dimethylheptane		C_9H_{20}	1071-26-7	128.255	liq	-113	132.7	0.7105^{20}	1.4016^{20}	i H_2O; s eth, ctc; vs ace, chl; msc bz
4071	2,3-Dimethylheptane		C_9H_{20}	3074-71-3	128.255	liq	-116	140.5	0.7260^{20}	1.4088^{20}	i H_2O; msc EtOH, eth, ace, bz, peth, chl
4072	2,4-Dimethylheptane		C_9H_{20}	2213-23-2	128.255			132.9	0.7115^{25}	1.4034^{20}	i H_2O; msc EtOH, eth, ace, bz, chl, peth
4073	2,5-Dimethylheptane		C_9H_{20}	2216-30-0	128.255			136	0.7198^{20}	1.4033^{20}	vs ace, bz, eth, EtOH
4074	2,6-Dimethylheptane		C_9H_{20}	1072-05-5	128.255	liq	-102.9	135.2	0.7089^{20}	1.4011^{20}	sl chl
4075	3,3-Dimethylheptane		C_9H_{20}	4032-86-4	128.255			137.3	0.7254^{20}	1.4087^{20}	i H_2O; msc EtOH; s eth; vs ace, bz

cis-1,2-Dimethylcyclopropane

trans-1,2-Dimethylcyclopropane

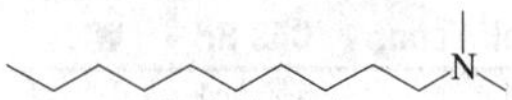

Dimethyldecylamine

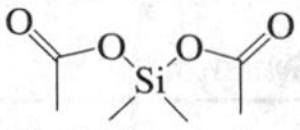

Dimethyldiacetoxysilane

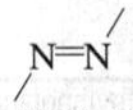

trans-Dimethyldiazene

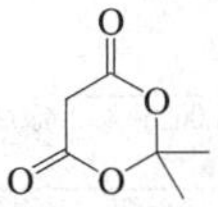

2,2-Dimethyl-1,3-dioxane-4,6-dione

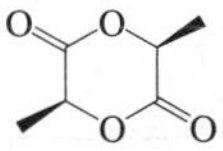

cis-3,6-Dimethyl-1,4-dioxane-2,5-dione

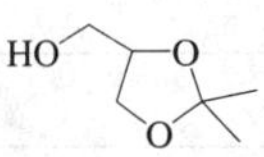

2,2-Dimethyl-1,3-dioxolane-4-methanol

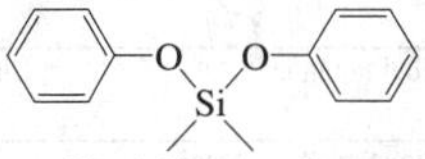

Dimethyldiphenoxysilane

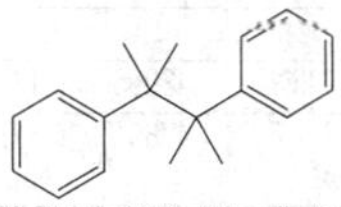

2,3-Dimethyl-2,3-diphenylbutane

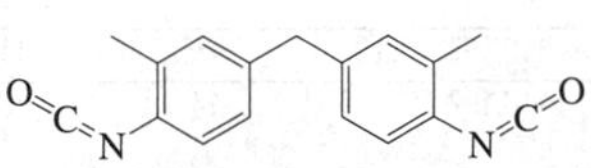

3,3′-Dimethyldiphenylmethane 4,4′-diisocyanate

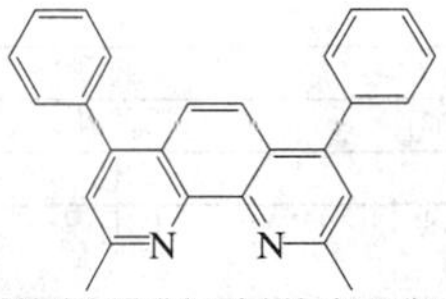

2,9-Dimethyl-4,7-diphenyl-1,10-phenanthroline

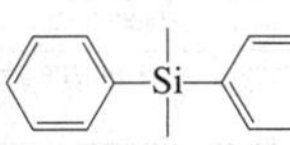

Dimethyldiphenylsilane

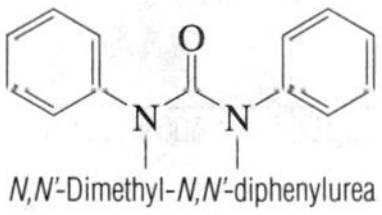

N,N′-Dimethyl-N,N′-diphenylurea

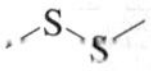

Dimethyl disulfide

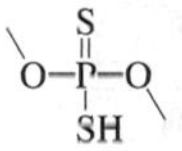

O,O-Dimethyl dithiophosphate

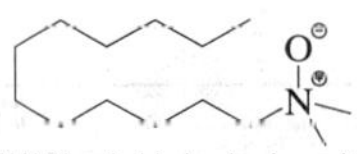

N,N-Dimethyldodecylamine oxide

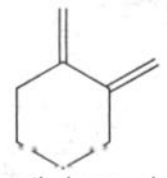

1,2-Dimethylenecyclohexane

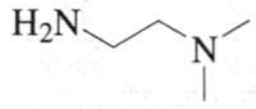

N,N-Dimethyl-1,2-ethanediamine

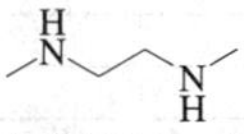

N,N′-Dimethyl-1,2-ethanediamine

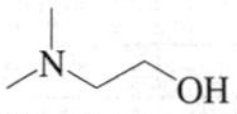

N,N-Dimethylethanolamine

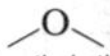

Dimethyl ether

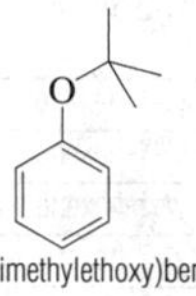

(1,1-Dimethylethoxy)benzene

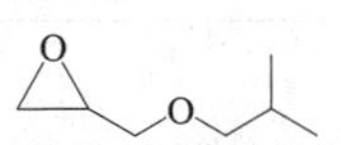

[(1,1-Dimethylethoxy)methyl]oxirane

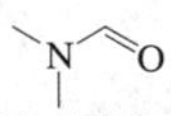

N,N-Dimethylformamide

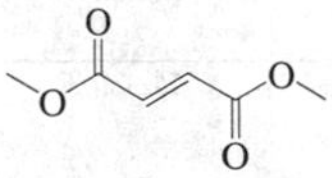

Dimethyl fumarate

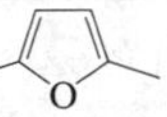

2,5-Dimethylfuran

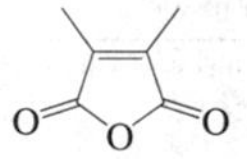

3,4-Dimethyl-2,5-furandione

Dimethyl germanium sulfide

Dimethyl glutarate

N,N-Dimethylglycine

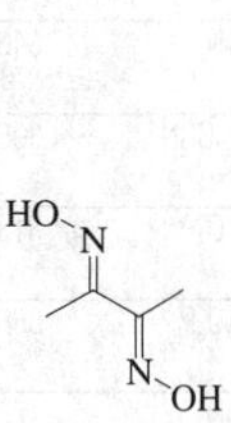

Dimethylglyoxime

2,6-Dimethyl-1,5-heptadiene

2,2-Dimethylheptane

2,3-Dimethylheptane

2,4-Dimethylheptane

2,5-Dimethylheptane

2,6-Dimethylheptane

3,3-Dimethylheptane

No.	Name	Synonym	Mol. Form.	CAS RN	Mol. Wt.	Physical Form	mp/°C	bp/°C	den/ g cm^{-3}	n_D	Solubility
4076	3,4-Dimethylheptane		C_9H_{20}	922-28-1	128.255			140.6	0.7314^{20}	1.4108^{20}	i H_2O; s eth, ctc; vs ace, chl; msc bz
4077	3,5-Dimethylheptane		C_9H_{20}	926-82-9	128.255			136	0.7225^{20}	1.4083^{20}	i H_2O; s eth, ctc; vs ace, chl; msc bz
4078	4,4-Dimethylheptane		C_9H_{20}	1068-19-5	128.255			135.2	0.7221^{20}	1.4076^{20}	i H_2O; s eth, ctc; vs ace, chl; msc bz
4079	Dimethyl heptanedioate	Dimethyl pimelate	$C_9H_{16}O_4$	1732-08-7	188.221		-21	120^{10}, 80^{1}	1.0625^{20}	1.4309^{20}	sl H_2O; s EtOH, eth, bz
4080	2,6-Dimethyl-2-heptanol		$C_9H_{20}O$	13254-34-7	144.254			173	0.8186^{20}	1.4242^{20}	
4081	2,6-Dimethyl-4-heptanol	Diisobutylcarbinol	$C_9H_{20}O$	108-82-7	144.254			174.5	0.8114^{20}	1.4242^{20}	i H_2O; s EtOH, eth; sl ctc
4082	3,5-Dimethyl-4-heptanol		$C_9H_{20}O$	19549-79-2	144.254			186	0.836^{18}	1.4283^{20}	sl H_2O
4083	2,6-Dimethyl-4-heptanone	Diisobutyl ketone	$C_9H_{18}O$	108-83-8	142.238	liq	-41.5	169.4	0.8062^{20}	1.412^{21}	i H_2O; msc EtOH, eth; s ctc
4084	2,6-Dimethyl-5-heptenal		$C_9H_{16}O$	106-72-9	140.222	oil		120^{100}			
4085	*N*,6-Dimethyl-5-hepten-2-amine	Isometheptene	$C_9H_{19}N$	503-01-5	141.254			177			vs eth, EtOH
4086	2,5-Dimethyl-1,5-hexadiene		C_8H_{14}	627-58-7	110.197	liq	-75.6	114.3	0.743^{20}	1.43995^{21}	i H_2O; s ace, chl
4087	2,5-Dimethyl-2,4-hexadiene		C_8H_{14}	764-13-6	110.197		14	134.5	0.7577^{25}	1.4785^{20}	i H_2O; s EtOH, eth, bz, chl
4088	2,2-Dimethylhexane		C_8H_{18}	590-73-8	114.229	liq	-121.1	106.86	0.6953^{20}	1.3935^{20}	vs ace, bz, eth, EtOH
4089	2,3-Dimethylhexane		C_8H_{18}	584-94-1	114.229			115.62	0.6912^{25}	1.4011^{20}	vs ace, bz, EtOH, lig
4090	2,4-Dimethylhexane		C_8H_{18}	589-43-5	114.229			109.5	0.6962^{25}	1.3929^{25}	
4091	2,5-Dimethylhexane		C_8H_{18}	592-13-2	114.229	liq	-91	109.12	0.6901^{25}	1.3925^{20}	i H_2O; msc EtOH, ace, bz; s eth
4092	3,3-Dimethylhexane		C_8H_{18}	563-16-6	114.229	liq	-126.1	111.97	0.7100^{20}	1.4001^{20}	i H_2O; msc EtOH; vs eth, ace, bz
4093	3,4-Dimethylhexane		C_8H_{18}	583-48-2	114.229			117.73	0.7151^{25}	1.4041^{20}	i H_2O; s eth; msc EtOH, ace, bz
4094	2,5-Dimethyl-2,5-hexanediamine		$C_8H_{20}N_2$	23578-35-0	144.258			184; 63^{8}	0.8485^{15}	1.4459^{20}	
4095	2,5-Dimethyl-2,5-hexanediol	1,1,4,4-Tetramethyl-1,4-butanediol	$C_8H_{18}O_2$	110-03-2	146.228	pr (AcOEt) fl (peth)	88.50	214	0.898^{20}		s H_2O; vs EtOH, bz, chl
4096	2,2-Dimethyl-1-hexanol		$C_8H_{18}O$	2370-13-0	130.228			95^{29}			
4097	2,3-Dimethyl-1-hexene		C_8H_{16}	16746-86-4	112.213			110.5	0.7172^{25}	1.4113^{20}	
4098	5,5-Dimethyl-1-hexene		C_8H_{16}	7116-86-1	112.213			104	0.705^{25}	1.4049^{20}	
4099	2,3-Dimethyl-2-hexene		C_8H_{16}	7145-20-2	112.213	liq	-115.1	121.8	0.7366^{25}	1.4268^{20}	
4100	2,5-Dimethyl-2-hexene		C_8H_{16}	3404-78-2	112.213			112.2	0.7182^{20}	1.4140^{20}	
4101	*cis*-2,2-Dimethyl-3-hexene		C_8H_{16}	690-92-6	112.213	liq	-137.4	105.5	0.7086^{25}	1.4099^{20}	
4102	*trans*-2,2-Dimethyl-3-hexene		C_8H_{16}	690-93-7	112.213			100.8	0.6995^{25}	1.4063^{20}	
4103	3,5-Dimethyl-1-hexen-3-ol		$C_8H_{16}O$	3329-48-4	128.212			146.5	0.8382^{20}	1.4342^{20}	
4104	1-(1,5-Dimethyl-4-hexenyl)-4-methylbenzene	α-Curcumene	$C_{15}H_{22}$	644-30-4	202.336			140^{19}	0.8805^{20}	1.4989^{20}	i H_2O; s bz
4105	2,5-Dimethyl-3-hexyne-2,5-diol		$C_8H_{14}O_2$	142-30-3	142.196		95	205	0.947^{20}		s H_2O, chl; vs EtOH, eth, ace, bz
4106	1,1-Dimethylhydrazine		$C_2H_8N_2$	57-14-7	60.098	liq, fumes in air	-57.20	63.9	0.791^{22}	1.4075^{22}	vs H_2O, EtOH, eth, MeOH
4107	1,2-Dimethylhydrazine		$C_2H_8N_2$	540-73-8	60.098	fumes (air)	-8.9	81	0.8274^{20}	1.4209^{20}	msc H_2O, EtOH, eth
4108	1,2-Dimethylhydrazine dihydrochloride		$C_2H_{10}Cl_2N_2$	306-37-6	133.019	pr (w)	170 dec				vs H_2O, EtOH
4109	Dimethyl hydrogen phosphate	Dimethyl phosphate	$C_2H_7O_4P$	813-78-5	126.048			dec 174	1.3225^{20}	1.408^{25}	vs H_2O, ace, EtOH
4110	Dimethyl hydrogen phosphite		$C_2H_7O_3P$	868-85-9	110.049			170.5	1.2002^{20}	1.4036^{20}	s EtOH, py; sl ctc
4111	1,2-Dimethyl-1*H*-imidazole		$C_5H_8N_2$	1739-84-0	96.131			206	1.0051^{11}		vs H_2O, eth, EtOH
4112	2,4-Dimethyl-1*H*-imidazole		$C_5H_8N_2$	930-62-1	96.131		92	267			
4113	5,5-Dimethyl-2,4-imidazolidinedione		$C_5H_8N_2O_2$	77-71-4	128.130	pr (dil al)	178	sub			vs H_2O, EtOH, eth, ace, bz, chl; s DMSO
4114	1,1-Dimethylindan		$C_{11}H_{14}$	4912-92-9	146.229			191	0.919^{20}	1.5135^{25}	
4115	1,3-Dimethyl-1*H*-indole		$C_{10}H_{11}N$	875-30-9	145.201	nd	142	258.5			s eth
4116	2,3-Dimethyl-1*H*-indole		$C_{10}H_{11}N$	91-55-4	145.201		107.5	287			
4117	*N*,*N*-Dimethyl-1*H*-indole-3-ethanamine	*N*,*N*-Dimethyltryptamine	$C_{12}H_{16}N_2$	61-50-7	188.268		46				

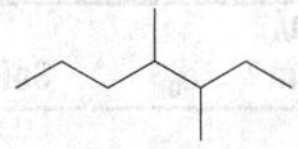

3,4-Dimethylheptane

3,5-Dimethylheptane

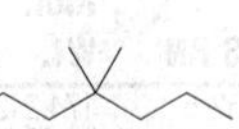

4,4-Dimethylheptane

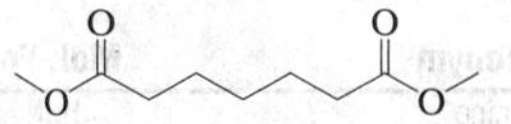

Dimethyl heptanedioate

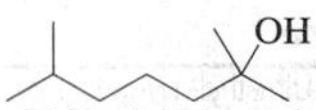

2,6-Dimethyl-2-heptanol

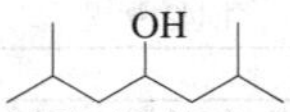

2,6-Dimethyl-4-heptanol

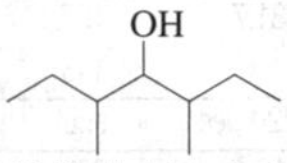

3,5-Dimethyl-4-heptanol

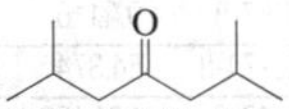

2,6-Dimethyl-4-heptanone

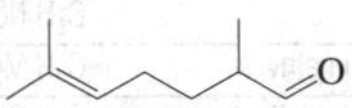

2,6-Dimethyl-5-heptenal

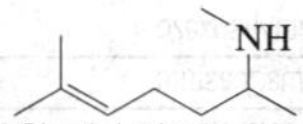

N,6-Dimethyl-5-hepten-2-amine

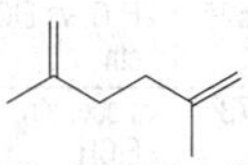

2,5-Dimethyl-1,5-hexadiene

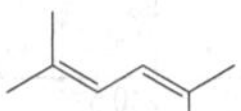

2,5-Dimethyl-2,4-hexadiene

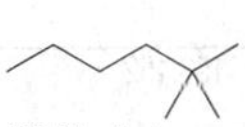

2,2-Dimethylhexane

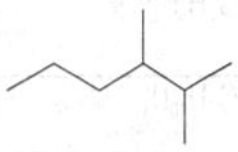

2,3-Dimethylhexane

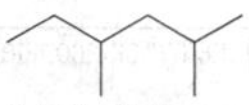

2,4-Dimethylhexane

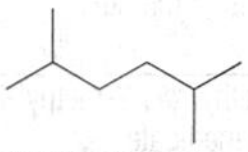

2,5-Dimethylhexane

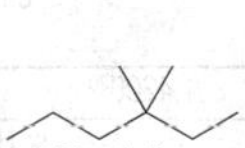

3,3-Dimethylhexane

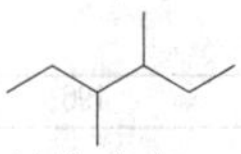

3,4-Dimethylhexane

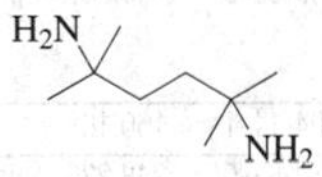

2,5-Dimethyl-2,5-hexanediamine

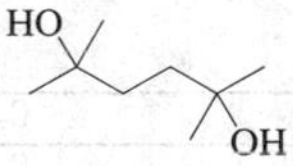

2,5-Dimethyl-2,5-hexanediol

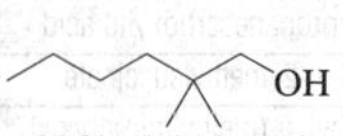

2,2-Dimethyl-1-hexanol

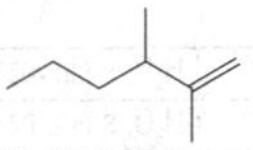

2,3-Dimethyl-1-hexene

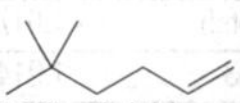

5,5-Dimethyl-1-hexene

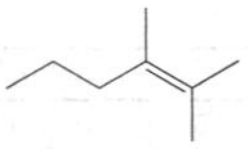

2,3-Dimethyl-2-hexene

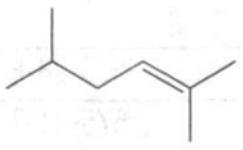

2,5-Dimethyl-2-hexene

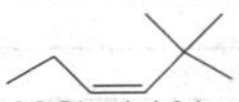

cis-2,2-Dimethyl-3-hexene

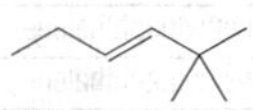

trans-2,2-Dimethyl-3-hexene

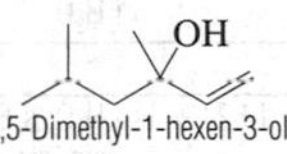

3,5-Dimethyl-1-hexen-3-ol

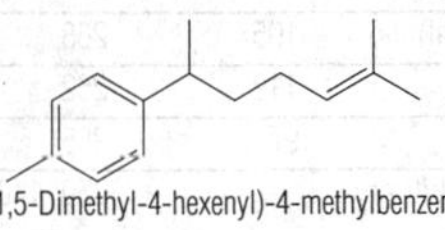

1-(1,5-Dimethyl-4-hexenyl)-4-methylbenzene

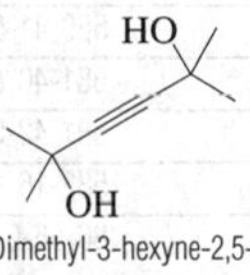

2,5-Dimethyl-3-hexyne-2,5-diol

1,1-Dimethylhydrazine

1,2-Dimethylhydrazine

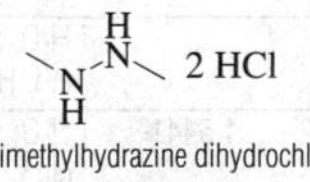

1,2-Dimethylhydrazine dihydrochloride

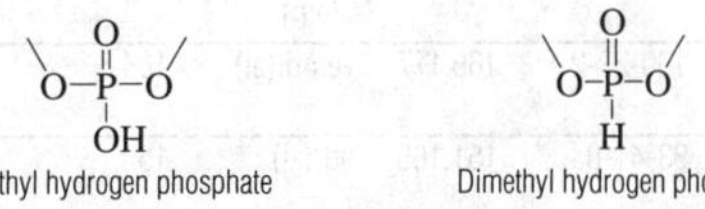

Dimethyl hydrogen phosphate

Dimethyl hydrogen phosphite

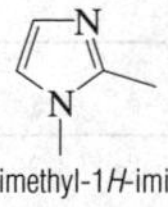

1,2-Dimethyl-1*H*-imidazole

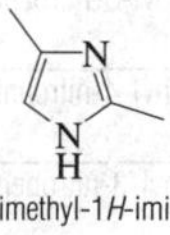

2,4-Dimethyl-1*H*-imidazole

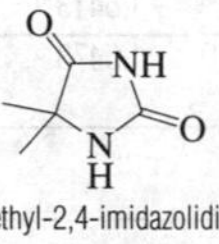

5,5-Dimethyl-2,4-imidazolidinedione

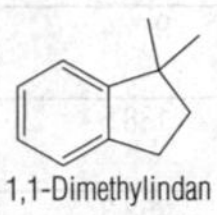

1,1-Dimethylindan

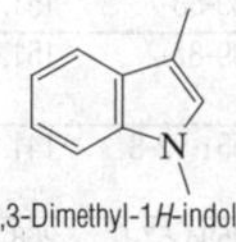

1,3-Dimethyl-1*H*-indole

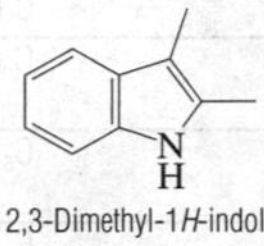

2,3-Dimethyl-1*H*-indole

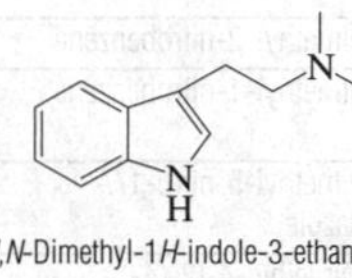

N,*N*-Dimethyl-1*H*-indole-3-ethanamine

No.	Name	Synonym	Mol. Form.	CAS RN	Mol. Wt.	Physical Form	mp/°C	bp/°C	den/ g cm^{-3}	n_D	Solubility
4118	*N,N*-Dimethyl-1*H*-indole-3-methanamine	Gramine	$C_{11}H_{14}N_2$	87-52-5	174.242	nd or pl (ace)	138.5				i H_2O; s EtOH, eth, chl; i peth
4119	Dimethyl isophthalate		$C_{10}H_{10}O_4$	1459-93-4	194.184	nd(dil al)	67.5	282	1.194^{20}	1.5168^{20}	sl H_2O
4120	1,4-Dimethyl-7-isopropylazulene	Guaiazulene	$C_{15}H_{18}$	489-84-9	198.304	bl-viol pl (al)	31.5	167^{12}	0.973^{20}		s EtOH, eth, AcOEt
4121	1,6-Dimethyl-4-isopropylnaphthalene	Cadalene	$C_{15}H_{18}$	483-78-3	198.304			294; 149^{10}	0.9667^{25}	1.5785^{25}	vs oils
4122	2,4-Dimethyl-3-isopropylpentane		$C_{10}H_{22}$	13475-79-1	142.282	liq	-81.7	157.1	0.7545^{25}	1.4246^{20}	
4123	3,5-Dimethylisoxazole		C_5H_7NO	300-87-8	97.116			143	0.99^{25}	1.4421^{20}	
4124	Dimethylmagnesium	Magnesium dimethyl	C_2H_6Mg	2999-74-8	54.374	solid	220 dec	subl			
4125	Dimethyl maleate	Methyl *cis*-butenedioate	$C_6H_8O_4$	624-48-6	144.126	liq	-19	202	1.1606^{20}	1.4416^{20}	sl H_2O, lig; s eth, ctc
4126	Dimethyl malonate	Methyl malonate	$C_5H_8O_4$	108-59-8	132.116	liq	-61.9	181.4	1.528^{20}	1.4135^{20}	sl H_2O; msc EtOH; vs ace, bz; s chl
4127	Dimethylmalonic acid	Dimethylpropanedioc acid	$C_5H_8O_4$	595-46-0	132.116	pr (bz/peth)	192.5	subl			s hot H_2O
4128	Dimethyl mercury		C_2H_6Hg	593-74-8	230.66			93	3.17^{25}	1.5452^{20}	i H_2O; vs EtOH, eth
4129	Dimethyl *cis*-2-methyl-2-butenedioate	Dimethyl citraconate	$C_7H_{10}O_4$	617-54-9	158.152			210.5	1.1153^{20}	1.4473^{20}	vs ace, eth, EtOH
4130	Dimethyl methylenesuccinate		$C_7H_{10}O_4$	617-52-7	158.152	hyg mcl (MeOH)	38	208	1.1241^{18}	1.4457^{20}	s EtOH, eth, MeOH; vs ace
4131	Dimethyl methylmalonate		$C_6H_{10}O_4$	609-02-9	146.141			174	1.0977^{20}	1.4128^{20}	vs ace, eth, EtOH, chl
4132	Dimethyl methylphosphonate		$C_3H_9O_3P$	756-79-6	124.075			181; 79.5^{20}	1.1684^{20}	1.4099^{30}	s H_2O, EtOH, eth
4133	*trans*-2,2-Dimethyl-3-(2-methyl-1-propenyl) cyclopropanecarboxylic acid		$C_{10}H_{16}O_2$	4638-92-0	168.233	pr	20.0	245			vs eth, EtOH, chl
4134	Dimethyl 2-methylsuccinate		$C_7H_{12}O_4$	1604-11-1	160.168			196	1.076^{25}	1.4200^{20}	
4135	Dimethyl *p*-(methylthio)phenyl phosphate		$C_9H_{13}O_4PS$	3254-63-5	248.235	liq			1.273^{21}		sl H_2O; s ace, EtOH, diox, ctc, xyl
4136	2,6-Dimethylmorpholine		$C_6H_{13}NO$	141-91-3	115.173	liq	-88	146.6	0.9329^{20}	1.4460^{20}	msc H_2O, EtOH, bz, lig; s ace; sl chl
4137	Dimethyl morpholinophosphoramidate	Dimethyl 4-morpholinylphosphonate	$C_6H_{14}NO_4P$	597-25-1	195.153	liq		96^{1}			
4138	1,2-Dimethylnaphthalene		$C_{12}H_{12}$	573-98-8	156.223		0.8	266.5	1.0179^{20}	1.6166^{20}	i H_2O; s eth, bz
4139	1,3-Dimethylnaphthalene		$C_{12}H_{12}$	575-41-7	156.223	liq	-6	263	1.0144^{20}	1.6140^{20}	i H_2O; s eth, bz
4140	1,4-Dimethylnaphthalene		$C_{12}H_{12}$	571-58-4	156.223		7.6	268	1.0166^{20}	1.6127^{20}	i H_2O; vs EtOH; msc eth, ace, bz, ctc
4141	1,5-Dimethylnaphthalene		$C_{12}H_{12}$	571-61-9	156.223		82	265			i H_2O; vs bz, eth
4142	1,6-Dimethylnaphthalene		$C_{12}H_{12}$	575-43-9	156.223	liq	-16.9	264	1.0021^{20}	1.6166^{20}	i H_2O; s eth, bz
4143	1,7-Dimethylnaphthalene		$C_{12}H_{12}$	575-37-1	156.223	liq	-13.9	263	1.0115^{20}	1.6083^{20}	i H_2O; s eth, bz
4144	1,8-Dimethylnaphthalene		$C_{12}H_{12}$	569-41-5	156.223		65	270	1.003^{20}		i H_2O; s eth, bz
4145	2,3-Dimethylnaphthalene	Guajen	$C_{12}H_{12}$	581-40-8	156.223	lf (al)	105	268	1.003^{20}	1.5060^{20}	i H_2O; vs bz, eth
4146	2,6-Dimethylnaphthalene		$C_{12}H_{12}$	581-42-0	156.223		112	262	1.003^{20}		i H_2O
4147	2,7-Dimethylnaphthalene		$C_{12}H_{12}$	582-16-1	156.223		97	265	1.003^{20}		
4148	*N,N*-Dimethyl-1-naphthylamine		$C_{12}H_{13}N$	86-56-6	171.238	viol flr cry		250; 140^{13}	1.0423^{20}	1.624^{15}	i H_2O; s EtOH, eth, ctc
4149	*N,N*-Dimethyl-2-naphthylamine		$C_{12}H_{13}N$	2436-85-3	171.238	dk red nd	52.5	305	1.0279^{60}	1.6443^{53}	i H_2O; s EtOH, eth
4150	*N,N*-Dimethyl-2-nitroaniline		$C_8H_{10}N_2O_2$	610-17-3	166.177	ye-oran	-20	146^{20}	1.1794^{20}	1.6102^{20}	s H_2O, eth; vs EtOH, chl
4151	*N,N*-Dimethyl-3-nitroaniline		$C_8H_{10}N_2O_2$	619-31-8	166.177	red mcl pr (eth)	60.5	282.5	1.313^{17}		i H_2O; s EtOH, eth
4152	*N,N*-Dimethyl-4-nitroaniline		$C_8H_{10}N_2O_2$	100-23-2	166.177	ye nd (al)	164.5				i H_2O; s EtOH, eth, HOAc
4153	1,2-Dimethyl-3-nitrobenzene		$C_8H_9NO_2$	83-41-0	151.163	nd (al)	15	240	1.1402^{20}	1.5441^{20}	i H_2O; s EtOH, ctc
4154	1,2-Dimethyl-4-nitrobenzene	4-Nitro-*o*-xylene	$C_8H_9NO_2$	99-51-4	151.163	ye pr (al)	30.5	251; 143^{21}	1.112^{15}	1.5202^{20}	i H_2O; msc EtOH
4155	1,3-Dimethyl-2-nitrobenzene		$C_8H_9NO_2$	81-20-9	151.163		15	226	1.112^{15}	1.5202^{20}	i H_2O; vs EtOH; s ctc
4156	1,3-Dimethyl-5-nitrobenzene		$C_8H_9NO_2$	99-12-7	151.163	nd (al)	75	274			i H_2O; vs EtOH, eth
4157	1,4-Dimethyl-2-nitrobenzene		$C_8H_9NO_2$	89-58-7	151.163	pa ye liq	-25	240.5	1.132^{15}	1.5413^{20}	i H_2O; s EtOH
4158	2,4-Dimethyl-1-nitrobenzene		$C_8H_9NO_2$	89-87-2	151.163		9	247; 122^{18}	1.135^{15}	1.5473^{25}	i H_2O; s eth, ace, bz, chl
4159	1,2-Dimethyl-5-nitro-1*H*-imidazole	Dimetridazole	$C_5H_7N_3O_2$	551-92-8	141.129	nd (w)	138.5				vs eth, EtOH
4160	*N,N*-Dimethyl-4-[2-(4-nitrophenyl)ethenyl]aniline		$C_{16}H_{16}N_2O_2$	4584-57-0	268.310		258.3				

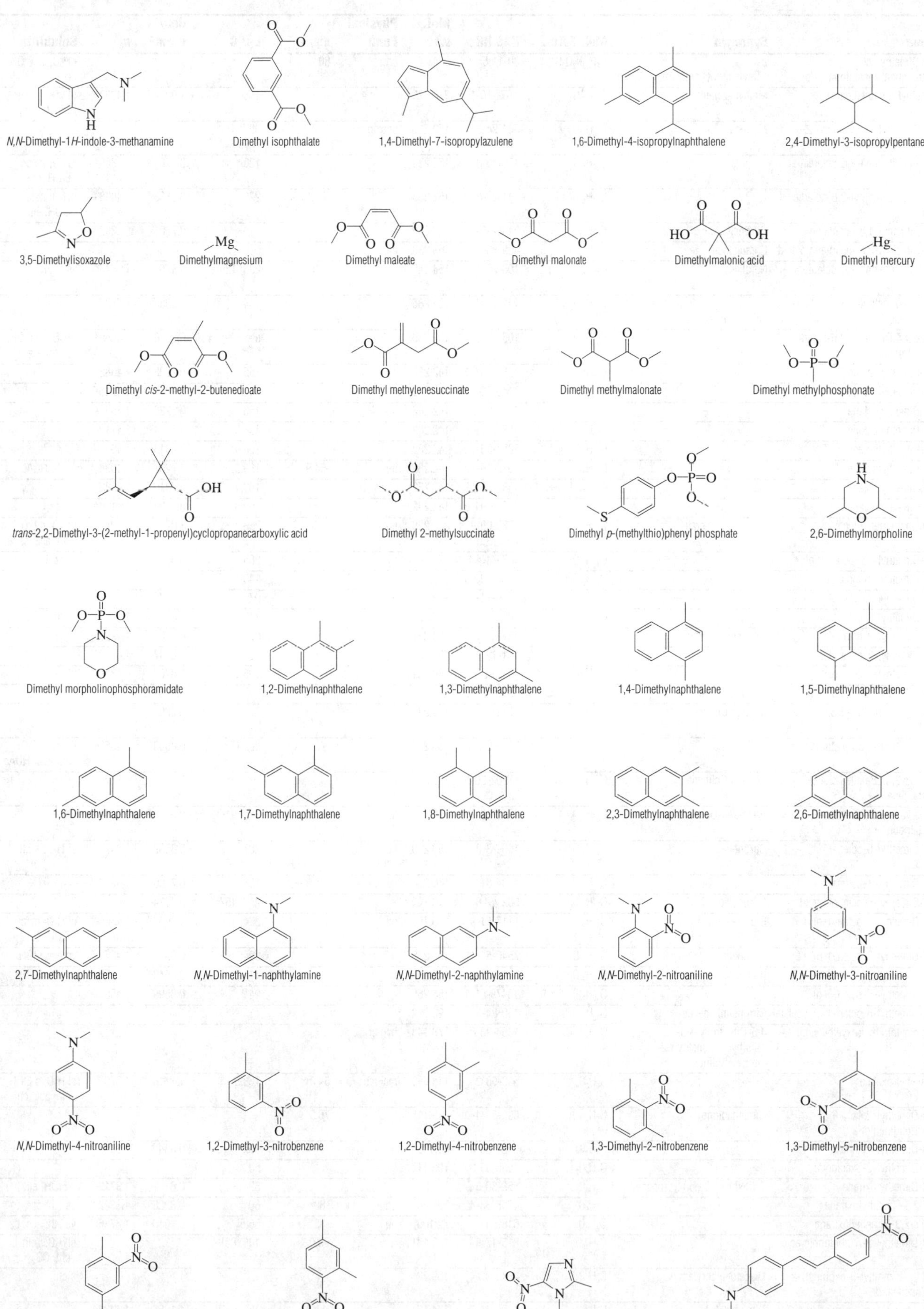
N,N-Dimethyl-1H-indole-3-methanamine
Dimethyl isophthalate
1,4-Dimethyl-7-isopropylazulene
1,6-Dimethyl-4-isopropylnaphthalene
2,4-Dimethyl-3-isopropylpentane
3,5-Dimethylisoxazole
Dimethylmagnesium
Dimethyl maleate
Dimethyl malonate
Dimethylmalonic acid
Dimethyl mercury
Dimethyl cis-2-methyl-2-butenedioate
Dimethyl methylenesuccinate
Dimethyl methylmalonate
Dimethyl methylphosphonate
trans-2,2-Dimethyl-3-(2-methyl-1-propenyl)cyclopropanecarboxylic acid
Dimethyl 2-methylsuccinate
Dimethyl p-(methylthio)phenyl phosphate
2,6-Dimethylmorpholine
Dimethyl morpholinophosphoramidate
1,2-Dimethylnaphthalene
1,3-Dimethylnaphthalene
1,4-Dimethylnaphthalene
1,5-Dimethylnaphthalene
1,6-Dimethylnaphthalene
1,7-Dimethylnaphthalene
1,8-Dimethylnaphthalene
2,3-Dimethylnaphthalene
2,6-Dimethylnaphthalene
2,7-Dimethylnaphthalene
N,N-Dimethyl-1-naphthylamine
N,N-Dimethyl-2-naphthylamine
N,N-Dimethyl-2-nitroaniline
N,N-Dimethyl-3-nitroaniline
N,N-Dimethyl-4-nitroaniline
1,2-Dimethyl-3-nitrobenzene
1,2-Dimethyl-4-nitrobenzene
1,3-Dimethyl-2-nitrobenzene
1,3-Dimethyl-5-nitrobenzene
1,4-Dimethyl-2-nitrobenzene
2,4-Dimethyl-1-nitrobenzene
1,2-Dimethyl-5-nitro-1H-imidazole
N,N-Dimethyl-4-[2-(4-nitrophenyl)ethenyl]aniline

No.	Name	Synonym	Mol. Form.	CAS RN	Mol. Wt.	Physical Form	mp/°C	bp/°C	den/ g cm^{-3}	n_D	Solubility
4161	*N*,4-Dimethyl-*N*-nitrosobenzenesulfonamide	*p*-Tolylsulfonylmethylnitrosamide	$C_8H_{10}N_2O_3S$	80-11-5	214.241	cry	60				i H_2O; vs EtOH, eth
4162	Dimethyl nonanedioate	Methyl azelate	$C_{11}H_{20}O_4$	1732-10-1	216.275		-0.8	156^{20}	1.0082^{20}	1.4367^{20}	i H_2O; s EtOH, ace, bz, ctc
4163	6,6-Dimethyl-2-norpinene-2-carboxaldehyde	Myrtenal	$C_{10}H_{14}O$	564-94-3	150.217	unstab oil		99^{15}			
4164	*cis*-3,7-Dimethyl-2,6-octadienal		$C_{10}H_{16}O$	106-26-3	152.233			120^{20}	0.8869^{20}	1.4869^{20}	i H_2O; msc EtOH, eth
4165	*trans*-3,7-Dimethyl-2,6-octadienal		$C_{10}H_{16}O$	141-27-5	152.233			229	0.8888^{20}	1.4898^{20}	i H_2O; msc EtOH, eth
4166	3,7-Dimethyl-1,6-octadiene	Citronellene	$C_{10}H_{18}$	2436-90-0	138.250				0.7601^{20}	1.4362^{20}	
4167	3,7-Dimethyl-2,6-octadienoic acid	Geranic acid	$C_{10}H_{16}O_2$	459-80-3	168.233	oil					
4168	*cis*-3,7-Dimethyl-2,6-octadien-1-ol	Nerol	$C_{10}H_{18}O$	106-25-2	154.249		<-15	225; 125^{25}	0.8756^{20}	1.4746^{20}	vs EtOH
4169	*cis*-3,7-Dimethyl-2,6-octadien-1-ol acetate		$C_{12}H_{20}O_2$	141-12-8	196.286			134^{25}, 93^{3}	0.905^{15}	1.452^{20}	
4170	*trans*-3,7-Dimethyl-2,6-octadien-1-ol formate		$C_{11}H_{18}O_2$	105-86-2	182.260			dec 229; 113^{25}	0.9086^{25}	1.4659^{20}	i H_2O; vs EtOH; s eth, ace
4171	2,2-Dimethyloctane		$C_{10}H_{22}$	15869-87-1	142.282			155	0.7208^{25}	1.4082^{20}	
4172	2,3-Dimethyloctane		$C_{10}H_{22}$	7146-60-3	142.282			164.3	0.7377^{20}	1.4146^{20}	
4173	2,4-Dimethyloctane		$C_{10}H_{22}$	4032-94-4	142.282			156	0.7226^{25}	1.4091^{20}	
4174	2,5-Dimethyloctane		$C_{10}H_{22}$	15869-89-3	142.282			158.5	0.7264^{25}	1.4112^{20}	
4175	2,6-Dimethyloctane		$C_{10}H_{22}$	2051-30-1	142.282			160.4	0.7313^{20}	1.4097^{20}	
4176	2,7-Dimethyloctane		$C_{10}H_{22}$	1072-16-8	142.282	liq	-54.9	159.9	0.7202^{25}	1.4086^{20}	s eth, HOAc
4177	3,4-Dimethyloctane		$C_{10}H_{22}$	15869-92-8	142.282			163.4	0.7410^{25}	1.4182^{20}	
4178	3,6-Dimethyloctane		$C_{10}H_{22}$	15869-94-0	142.282			160.8	0.7324^{25}	1.4139^{20}	
4179	Dimethyl octanedioate	Dimethyl suberate	$C_{10}H_{18}O_4$	1732-09-8	202.248	liq	-1.6	268	1.0217^{20}	1.4341^{20}	i H_2O; s EtOH, eth, ace; sl ctc
4180	3,7-Dimethyl-1,7-octanediol		$C_{10}H_{22}O_2$	107-74-4	174.281			265	0.937^{20}	1.4599^{20}	sl bz, tol
4181	2,2-Dimethyloctanoic acid		$C_{10}H_{20}O_2$	29662-90-6	172.265			140^{13}			
4182	2,2-Dimethyl-1-octanol		$C_{10}H_{22}O$	2370-14-1	158.281	liq		97	0.84^{20}		
4183	3,7-Dimethyl-1-octanol		$C_{10}H_{22}O$	106-21-8	158.281			212.5	0.832^{25}	1.438^{25}	s eth
4184	2,6-Dimethyl-2-octanol	Tetrahydromyrcenol	$C_{10}H_{22}O$	18479-57-7	158.281			80.5^{10}	0.8023^{25}	1.4220^{25}	
4185	3,6-Dimethyl-3-octanol		$C_{10}H_{22}O$	151-19-9	158.281	liq	-67.5	202.2	0.8347^{22}	1.4370^{20}	
4186	3,7-Dimethyl-3-octanol		$C_{10}H_{22}O$	78-69-3	158.281			205.1	0.826^{25}	1.433^{25}	
4187	*cis*-3,7-Dimethyl-1,3,6-octatriene	*cis*-β-Ocimene	$C_{10}H_{16}$	3338-55-4	136.234				0.799^{20}		
4188	*trans*-3,7-Dimethyl-1,3,6-octatriene	*trans*-β-Ocimene	$C_{10}H_{16}$	3779-61-1	136.234				0.799^{20}		
4189	3,7-Dimethyl-1,3,7-octatriene	α-Ocimene	$C_{10}H_{16}$	502-99-8	136.234			dec 177	0.8000^{20}	1.4862^{20}	i H_2O; s EtOH, eth, chl, HOAc
4190	*cis*, *cis*-2,6-Dimethyl-2,4,6-octatriene	*cis*-*allo*-Ocimene	$C_{10}H_{16}$	17202-20-9	136.234	liq					
4191	*trans*,*trans*-2,6-Dimethyl-2,4,6-octatriene	*trans*-*allo*-Ocimene	$C_{10}H_{16}$	3016-19-1	136.234	liq	-35.4	188; 91^{20}	0.8118^{20}	1.5446^{20}	
4192	3,7-Dimethyl-6-octenal	Citronellal	$C_{10}H_{18}O$	106-23-0	154.249	nd or orth cry		207.5	0.853^{20}	1.4473^{20}	sl H_2O; s EtOH
4193	3,7-Dimethyl-1-octene		$C_{10}H_{20}$	4984-01-4	140.266	col liq		154	0.7396^{20}	1.4212^{20}	
4194	3,7-Dimethyl-6-octenoic acid	Citronellic acid	$C_{10}H_{18}O_2$	502-47-6	170.249			257; 157^{23}	0.9234^{21}		
4195	3,7-Dimethyl-6-octen-1-ol, (*R*)	Citronellol, (+)	$C_{10}H_{20}O$	1117-61-9	156.265	oil		224; 108^{10}	0.8550^{20}	1.4565^{20}	sl H_2O; msc EtOH, eth
4196	3,7-Dimethyl-6-octen-1-ol, (*S*)	Citronellol, (-)	$C_{10}H_{20}O$	7540-51-4	156.265	oil		224; 108^{10}	0.859^{18}	1.4576^{18}	vs eth, EtOH
4197	3,7-Dimethyl-7-octen-1-ol, (*S*)	Rhodinol	$C_{10}H_{20}O$	6812-78-8	156.265			114^{12}	0.8549^{20}	1.4556^{20}	vs eth, EtOH
4198	3,7-Dimethyl-6-octen-3-ol		$C_{10}H_{20}O$	18479-51-1	156.265			94^{14}	0.8695^{15}	1.4569^{15}	
4199	3,7-Dimethyl-6-octen-1-ol, acetate	Citronellol acetate	$C_{12}H_{22}O_2$	150-84-5	198.302			115^{10}			
4200	Dimethyloldihydroxyethyleneurea	4,5-Dihydroxy-1,3-bis(hydroxymethyl)-2-imidazolidinone	$C_5H_{10}N_2O_5$	1854-26-8	178.143	hyg cry					
4201	Dimethyl oxalate		$C_4H_6O_4$	553-90-2	118.089	mcl tab	54.8	163.5	1.1716^{60}	1.379^{82}	sl H_2O; s EtOH, eth, ace, chl
4202	5,5-Dimethyl-2,4-oxazolidinedione	Dimethadione	$C_5H_7NO_3$	695-53-4	129.115		76.5				
4203	3,3-Dimethyloxetane		$C_5H_{10}O$	6921-35-3	86.132			80.6	0.834^{25}	1.3965^{20}	
4204	3,3-Dimethyl-2-oxetanone		$C_5H_8O_2$	1955-45-9	100.117			58^{15}			
4205	2,2-Dimethyloxirane	2-Methyl-1,2-epoxypropane	C_4H_8O	558-30-5	72.106			52	0.8112^{20}	1.3712^{22}	s EtOH, eth
4206	*cis*-2,3-Dimethyloxirane		C_4H_8O	1758-33-4	72.106	liq	-80	60	0.8226^{25}	1.3802^{20}	vs eth, ace, bz
4207	*trans*-2,3-Dimethyloxirane		C_4H_8O	6189-41-9	72.106	liq	-85	56.5	0.8010^{25}	1.3736^{20}	vs eth, ace, bz
4208	3,3-Dimethyl-2-oxobutanoic acid		$C_6H_{10}O_3$	815-17-8	130.141		90.5	189; 80^{15}			sl H_2O; s eth, bz, chl, CS_2
4209	*N*-(1,1-Dimethyl-3-oxobutyl)-2-propenamide	Diacetone acrylamide	$C_9H_{15}NO_2$	2873-97-4	169.221						s chl
4210	Dimethyl 3-oxo-1,5-pentanedioate	Dimethyl 1,3-acetonedicarboxylate	$C_7H_{10}O_5$	1830-54-2	174.151			150^{25}, $77^{0.6}$	1.185^{25}	1.4434^{20}	

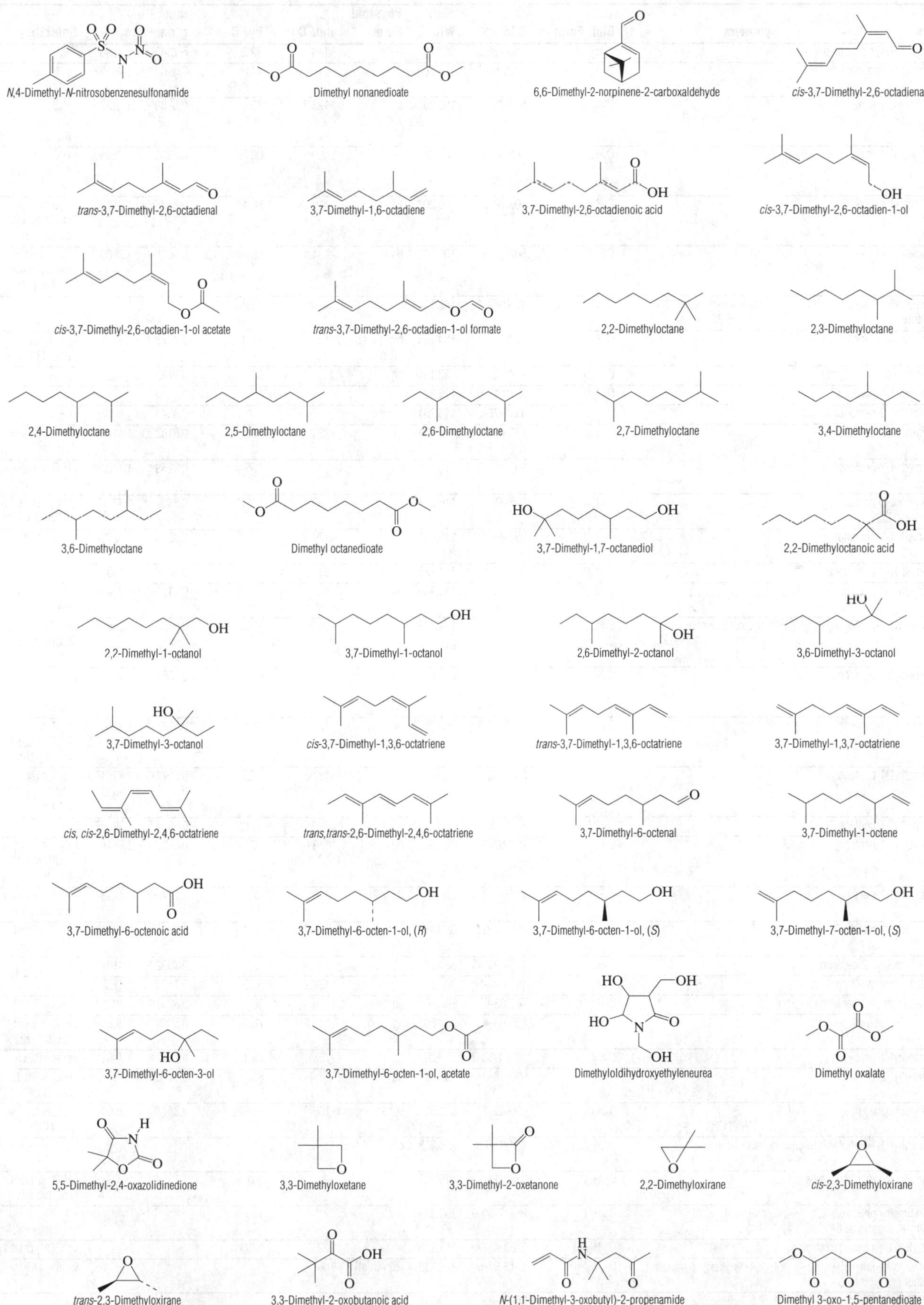

N,4-Dimethyl-N-nitrosobenzenesulfonamide
Dimethyl nonanedioate
6,6-Dimethyl-2-norpinene-2-carboxaldehyde
cis-3,7-Dimethyl-2,6-octadienal
trans-3,7-Dimethyl-2,6-octadienal
3,7-Dimethyl-1,6-octadiene
3,7-Dimethyl-2,6-octadienoic acid
cis-3,7-Dimethyl-2,6-octadien-1-ol
cis-3,7-Dimethyl-2,6-octadien-1-ol acetate
trans-3,7-Dimethyl-2,6-octadien-1-ol formate
2,2-Dimethyloctane
2,3-Dimethyloctane
2,4-Dimethyloctane
2,5-Dimethyloctane
2,6-Dimethyloctane
2,7-Dimethyloctane
3,4-Dimethyloctane
3,6-Dimethyloctane
Dimethyl octanedioate
3,7-Dimethyl-1,7-octanediol
2,2-Dimethyloctanoic acid
2,2-Dimethyl-1-octanol
3,7-Dimethyl-1-octanol
2,6-Dimethyl-2-octanol
3,6-Dimethyl-3-octanol
3,7-Dimethyl-3-octanol
cis-3,7-Dimethyl-1,3,6-octatriene
trans-3,7-Dimethyl-1,3,6-octatriene
3,7-Dimethyl-1,3,7-octatriene
cis, cis-2,6-Dimethyl-2,4,6-octatriene
trans,trans-2,6-Dimethyl-2,4,6-octatriene
3,7-Dimethyl-6-octenal
3,7-Dimethyl-1-octene
3,7-Dimethyl-6-octenoic acid
3,7-Dimethyl-6-octen-1-ol, (*R*)
3,7-Dimethyl-6-octen-1-ol, (*S*)
3,7-Dimethyl-7-octen-1-ol, (*S*)
3,7-Dimethyl-6-octen-3-ol
3,7-Dimethyl-6-octen-1-ol, acetate
Dimethyloldihydroxyethyleneurea
Dimethyl oxalate
5,5-Dimethyl-2,4-oxazolidinedione
3,3-Dimethyloxetane
3,3-Dimethyl-2-oxetanone
2,2-Dimethyloxirane
cis-2,3-Dimethyloxirane
trans-2,3-Dimethyloxirane
3,3-Dimethyl-2-oxobutanoic acid
N-(1,1-Dimethyl-3-oxobutyl)-2-propenamide
Dimethyl 3-oxo-1,5-pentanedioate

No.	Name	Synonym	Mol. Form.	CAS RN	Mol. Wt.	Physical Form	mp/°C	bp/°C	den/ g cm^{-3}	n_D	Solubility
4211	2,4-Dimethyl-1,3-pentadiene		C_7H_{12}	1000-86-8	96.170	liq	-114	93.2	0.7343^{23}	1.4390^{23}	
4212	*N,N*-Dimethylpentanamide		$C_7H_{15}NO$	6225-06-5	129.200		-51	141^{100}	0.8962^{25}	1.4419^{25}	vs H_2O, eth, EtOH
4213	2,2-Dimethylpentane		C_7H_{16}	590-35-2	100.202	liq	-123.7	79.2	0.6739^{20}	1.3822^{20}	i H_2O; s EtOH, eth; msc ace, bz, hp, chl
4214	2,3-Dimethylpentane		C_7H_{16}	565-59-3	100.202			89.78	0.6908^{25}	1.3894^{25}	i H_2O; s EtOH, eth; msc ace, bz, chl
4215	2,4-Dimethylpentane		C_7H_{16}	108-08-7	100.202	liq	-119.2	80.49	0.6727^{20}	1.3815^{20}	i H_2O; s EtOH, eth; msc ace, bz, chl, hp
4216	3,3-Dimethylpentane		C_7H_{16}	562-49-2	100.202	liq	-134.4	86.06	0.6936^{20}	1.3909^{20}	i H_2O; s EtOH, eth; msc ace, bz, hp, chl
4217	3,3-Dimethylpentanedioic acid anhydride	Dihydro-4,4-dimethyl-2*H*-pyran-2,6(3*H*)-dione	$C_7H_{10}O_3$	4160-82-1	142.152		125.8	181^{25}, 156^{20}			
4218	3,3-Dimethylpentanedioic acid		$C_7H_{12}O_4$	4839-46-7	160.168	mcl pl, nd (bz)	103.5	126^{415}, 89^{2}	1.4278^{20}		vs H_2O, EtOH, eth; sl bz; i lig
4219	2,2-Dimethylpentanoic acid		$C_7H_{14}O_2$	1185-39-3	130.185	liq		98^{9}	0.9189^{20}		
4220	2,2-Dimethyl-1-pentanol		$C_7H_{16}O$	2370-12-9	116.201						s chl
4221	2,3-Dimethyl-2-pentanol		$C_7H_{16}O$	4911-70-0	116.201				0.804^{20}		sl H_2O
4222	2,4-Dimethyl-2-pentanol		$C_7H_{16}O$	625-06-9	116.201		<-20	133.1	0.8103^{20}	1.4172^{20}	sl H_2O; s EtOH, eth, ctc
4223	2,2-Dimethyl-3-pentanol		$C_7H_{16}O$	3970-62-5	116.201	liq	-2.5	135	0.8253^{20}	1.4223^{20}	i H_2O; s EtOH, eth
4224	2,3-Dimethyl-3-pentanol		$C_7H_{16}O$	595-41-5	116.201		<-30	139.7	0.833^{20}	1.4287^{20}	sl H_2O, bz; s EtOH, eth
4225	2,4-Dimethyl-3-pentanol		$C_7H_{16}O$	600-36-2	116.201		<-70	138.7	0.8288^{20}	1.4250^{20}	sl H_2O; s EtOH, eth
4226	4,4-Dimethyl-2-pentanone		$C_7H_{14}O$	590-50-1	114.185	liq	-64	126	0.809^{25}	1.4036^{20}	
4227	2,2-Dimethyl-3-pentanone		$C_7H_{14}O$	564-04-5	114.185	liq	-45	125.6	0.8125^{20}	1.4065^{20}	sl H_2O; s EtOH, eth, ace, chl
4228	2,4-Dimethyl-3-pentanone	Diisopropyl ketone	$C_7H_{14}O$	565-80-0	114.185	liq	-69	125.4	0.8108^{20}	1.3999^{20}	sl H_2O; msc EtOH, eth; s bz; sl ctc
4229	2,3-Dimethyl-1-pentene		C_7H_{14}	3404-72-6	98.186	liq	-134.3	84.3	0.7051^{20}	1.4033^{20}	i H_2O; msc EtOH, eth; vs dil sulf
4230	2,4-Dimethyl-1-pentene		C_7H_{14}	2213-32-3	98.186	liq	-124.1	81.6	0.6943^{20}	1.3986^{20}	i H_2O; msc EtOH, eth; s bz, ctc, chl
4231	3,3-Dimethyl-1-pentene		C_7H_{14}	3404-73-7	98.186	liq	-134.3	77.5	0.6974^{20}	1.3984^{20}	i H_2O; msc EtOH, eth; s bz, chl
4232	3,4-Dimethyl-1-pentene		C_7H_{14}	7385-78-6	98.186			80.8	0.6934^{25}	1.3992^{20}	
4233	4,4-Dimethyl-1-pentene		C_7H_{14}	762-62-9	98.186	liq	-136.6	72.5	0.6827^{20}	1.3818^{20}	i H_2O; msc EtOH, eth; s bz, ctc, chl
4234	2,3-Dimethyl-2-pentene		C_7H_{14}	10574-37-5	98.186	liq	-118.3	97.5	0.7277^{20}	1.4208^{20}	i H_2O; s EtOH, eth, bz, chl
4235	2,4-Dimethyl-2-pentene		C_7H_{14}	625-65-0	98.186	liq	-127.7	83.4	0.6954^{20}	1.4040^{20}	i H_2O; s EtOH, eth, bz, ctc
4236	*cis*-3,4-Dimethyl-2-pentene		C_7H_{14}	4914-91-4	98.186	liq	-113.4	89.3	0.7092^{25}	1.4104^{20}	
4237	*trans*-3,4-Dimethyl-2-pentene		C_7H_{14}	4914-92-5	98.186	liq	-124.2	91.5	0.7124^{25}	1.4128^{20}	
4238	*cis*-4,4-Dimethyl-2-pentene		C_7H_{14}	762-63-0	98.186	liq	-135.4	80.4	0.6951^{25}	1.4026^{20}	
4239	*trans*-4,4-Dimethyl-2-pentene		C_7H_{14}	690-08-4	98.186	liq	-115.2	76.7	0.6889^{20}	1.3982^{20}	i H_2O; s EtOH, eth, bz, chl
4240	4,4-Dimethyl-1-pentyne		C_7H_{12}	13361-63-2	96.170	liq	-75.7	76.1	0.7142^{20}	1.3983^{20}	vs bz, eth, chl
4241	4,4-Dimethyl-2-pentyne		C_7H_{12}	999-78-0	96.170	liq	-82.4	83	0.7176^{20}	1.4071^{20}	i H_2O; s eth, bz, chl; sl ctc
4242	Dimethylperoxide		$C_2H_6O_2$	690-02-8	62.068	vol liq or gas	-100	14	0.8677^{0}	1.3503^{0}	sl EtOH, eth; s tol, HOAc
4243	2,9-Dimethyl-1,10-phenanthroline	Neocuproine	$C_{14}H_{12}N_2$	484-11-7	208.258	cry, 1/2w (w, lig)	159.5				
4244	3,4-Dimethylphenol phosphate (3:1)		$C_{24}H_{27}O_4P$	3862-11-1	410.442		72	261^{7}			i H_2O; sl EtOH, chl, hx; s bz
4245	5-(2,5-Dimethylphenoxy)-2,2-dimethylpentanoic acid	Gemfibrozil	$C_{15}H_{22}O_3$	25812-30-0	250.334	cry	62	$159^{0.02}$			
4246	*N*-(2,4-Dimethylphenyl)acetamide		$C_{10}H_{13}NO$	2050-43-3	163.216	nd (al)	129.3	170^{10}			vs EtOH, chl
4247	1-[(2,4-Dimethylphenyl)azo]-2-naphthol	1-(2,4-Xylylazo)-2-naphthol	$C_{18}H_{16}N_2O$	3118-97-6	276.332	red nd (al)	166				vs eth, EtOH
4248	1-[(2,5-Dimethylphenyl)azo]-2-naphthol	1-(2,5-Xylylazo)-2-naphthol	$C_{18}H_{16}N_2O$	85-82-5	276.332	nd (al)	153				
4249	1-(2,4-Dimethylphenyl)ethanone	2,4-Dimethylacetophenone	$C_{10}H_{12}O$	89-74-7	148.201			228	1.0121^{15}	1.5340^{20}	vs eth, EtOH

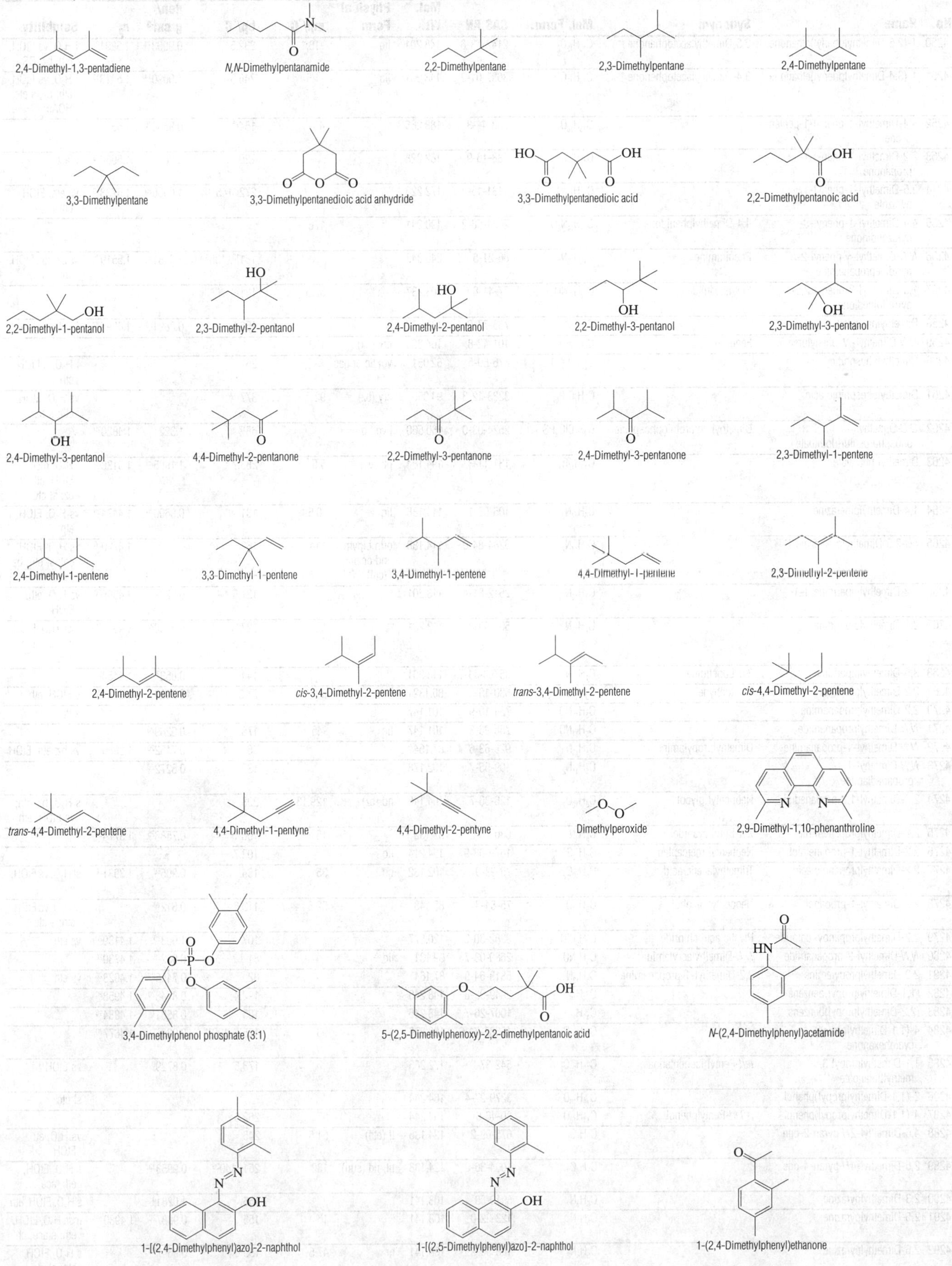

2,4-Dimethyl-1,3-pentadiene
N,N-Dimethylpentanamide
2,2-Dimethylpentane
2,3-Dimethylpentane
2,4-Dimethylpentane
3,3-Dimethylpentane
3,3-Dimethylpentanedioic acid anhydride
3,3-Dimethylpentanedioic acid
2,2-Dimethylpentanoic acid
2,2-Dimethyl-1-pentanol
2,3-Dimethyl-2-pentanol
2,4-Dimethyl-2-pentanol
2,2-Dimethyl-3-pentanol
2,3-Dimethyl-3-pentanol
2,4-Dimethyl-3-pentanol
4,4-Dimethyl-2-pentanone
2,2-Dimethyl-3-pentanone
2,4-Dimethyl-3-pentanone
2,3-Dimethyl-1-pentene
2,4-Dimethyl-1-pentene
3,3-Dimethyl-1-pentene
3,4-Dimethyl-1-pentene
4,4-Dimethyl-1-pentene
2,3-Dimethyl-2-pentene
2,4-Dimethyl-2-pentene
cis-3,4-Dimethyl-2-pentene
trans-3,4-Dimethyl-2-pentene
cis-4,4-Dimethyl-2-pentene
trans-4,4-Dimethyl-2-pentene
4,4-Dimethyl-1-pentyne
4,4-Dimethyl-2-pentyne
Dimethylperoxide
2,9-Dimethyl-1,10-phenanthroline
3,4-Dimethylphenol phosphate (3:1)
5-(2,5-Dimethylphenoxy)-2,2-dimethylpentanoic acid
N-(2,4-Dimethylphenyl)acetamide
1-[(2,4-Dimethylphenyl)azo]-2-naphthol
1-[(2,5-Dimethylphenyl)azo]-2-naphthol
1-(2,4-Dimethylphenyl)ethanone

No.	Name	Synonym	Mol. Form.	CAS RN	Mol. Wt.	Physical Form	mp/°C	bp/°C	den/ g cm⁻³	n_D	Solubility
4250	1-(2,5-Dimethylphenyl)ethanone	2,5-Dimethylacetophenone	$C_{10}H_{12}O$	2142-73-6	148.201	liq	-18.1	232.5	0.9963^{19}	1.5291^{20}	i H_2O; vs EtOH, eth, bz, CS_2
4251	1-(3,4-Dimethylphenyl)ethanone	3,4-Dimethylacetophenone	$C_{10}H_{12}O$	3637-01-2	148.201	liq	-1.5	246.5	1.0090^{14}	1.5413^{15}	i H_2O; vs EtOH, eth, bz; s ctc, HOAc
4252	4,4-Dimethyl-1-phenyl-1-penten-3-one		$C_{13}H_{16}O$	538-44-3	188.265		43	154^{25}	0.9508^{46}	1.5523^{25}	
4253	2,2-Dimethyl-1-phenyl-1-propanone		$C_{11}H_{14}O$	938-16-9	162.228			220	0.963^{26}	1.5086^{19}	s ace
4254	3,5-Dimethyl-1-phenyl-1*H*-pyrazole		$C_{11}H_{12}N_2$	1131-16-4	172.226			272; $145^{12.5}$	1.0566^{20}	1.5738^{19}	vs eth, EtOH, chl
4255	4,4-Dimethyl-1-phenyl-3-pyrazolidinone	4,4-Dimethylphenidone	$C_{11}H_{14}N_2O$	2654-58-2	190.241		176				
4256	*N,N*-Dimethyl-γ-phenyl-2-pyridinepropanamine	Pheniramine	$C_{16}H_{20}N_2$	86-21-5	240.343			181^{13}, $135^{0.5}$	1.0081^{25}	1.5519^{25}	vs bz, eth, EtOH, chl
4257	1,3-Dimethyl-3-phenyl-2,5-pyrrolidinedione	Methsuximide	$C_{12}H_{13}NO_2$	77-41-8	203.237		52.5	$121^{0.1}$			
4258	Dimethylphenylsilane		$C_8H_{12}Si$	766-77-8	136.267			156.5	0.8891^{20}	1.4995^{20}	i H_2O
4259	*N,N*-Dimethyl-*N'*-phenylurea	Fenuron	$C_9H_{12}N_2O$	101-42-8	164.203	cry (hx)	132				
4260	Dimethylphosphine		C_2H_7P	676-59-5	62.051	vol liq or gas		25			i H_2O; s EtOH, eth
4261	Dimethylphosphinic acid		$C_2H_7O_2P$	3283-12-3	94.050	cry (bz)	92	377			vs H_2O, EtOH, eth; s bz
4262	*O,O*-Dimethyl phosphorochloridothioate	Dimethyl chlorothiophosphate	$C_2H_6ClO_2PS$	2524-03-0	160.560	hyg liq		68^{12}	1.322	1.4820^{20}	
4263	Dimethyl phthalate		$C_{10}H_{10}O_4$	131-11-3	194.184	pa ye	5.5	283.7	1.1905^{20}	1.5138^{20}	i H_2O; msc EtOH, eth; s bz; sl ctc
4264	1,4-Dimethylpiperazine		$C_6H_{14}N_2$	106-58-1	114.188	liq	-0.59	131	0.8600^{20}	1.4474^{20}	vs H_2O, EtOH, eth
4265	*cis*-2,5-Dimethylpiperazine		$C_6H_{14}N_2$	6284-84-0	114.188	orth bipym nd or pr (chl)	114	162		1.4720^{20}	vs H_2O, EtOH, chl; sl eth, bz
4266	1,2-Dimethylpiperidine, (±)		$C_7H_{15}N$	2512-81-4	113.201			127.5	0.824^{15}	1.4395^{20}	vs H_2O, eth, EtOH
4267	2,6-Dimethylpiperidine		$C_7H_{15}N$	504-03-0	113.201			127	0.8158^{25}	1.4377^{20}	msc H_2O, EtOH, eth; sl ctc; s acid
4268	3,5-Dimethylpiperidine	3,5-Lupetidine	$C_7H_{15}N$	35794-11-7	113.201			144	0.853^{25}	1.4454^{20}	
4269	2,2-Dimethylpropanal	Pivaldehyde	$C_5H_{10}O$	630-19-3	86.132		6	77.5	0.7923^{17}	1.3791^{20}	s EtOH, eth
4270	2,2-Dimethylpropanamide		$C_5H_{11}NO$	754-10-9	101.147						s tfa
4271	*N,N*-Dimethylpropanamide		$C_5H_{11}NO$	758-96-3	101.147	liq	-45	175	0.9269^{20}		
4272	*N,N*-Dimethyl-1-propanamine	Dimethylpropylamine	$C_5H_{13}N$	926-63-6	87.164			66	0.7152^{20}	1.3860^{20}	vs bz, eth, EtOH
4273	*N,N*-Dimethyl-1,3-propanediamine		$C_5H_{14}N_2$	109-55-7	102.178			132	0.8272^{20}		
4274	2,2-Dimethyl-1,3-propanediol	Neopentyl glycol	$C_5H_{12}O_2$	126-30-7	104.148	nd (bz)	129.13	208			s H_2O, bz, chl; vs EtOH, eth
4275	2,2-Dimethylpropanenitrile	*tert*-Butyl cyanide	C_5H_9N	630-18-2	83.132		15	106.1	0.7586^{25}	1.3774^{20}	
4276	2,2-Dimethyl-1-propanethiol	Neopentyl mercaptan	$C_5H_{12}S$	1679-08-9	104.214	liq		103.7			
4277	2,2-Dimethylpropanoic acid	Trimethylacetic acid	$C_5H_{10}O_2$	75-98-9	102.132	nd	35	164	0.905^{50}	1.3931^{30}	sl H_2O; vs EtOH, eth
4278	2,2-Dimethyl-1-propanol	Neopentyl alcohol	$C_5H_{12}O$	75-84-3	88.148		52.5	113.5	0.812^{20}		sl H_2O; vs EtOH, eth; s ctc
4279	2,2-Dimethylpropanoyl chloride	Pivalic acid chloride	C_5H_9ClO	3282-30-2	120.577			107	1.003^{20}	1.4139^{20}	vs eth
4280	*N,N*-Dimethyl-2-propenamide	*N,N*-Dimethylacrylamide	C_5H_9NO	2680-03-7	99.131	liq		81^{20}	0.962^{25}	1.4730^{20}	
4281	2,2-Dimethylpropylamine	2,2-Dimethyl-1-propanamine	$C_5H_{13}N$	5813-64-9	87.164			82	0.7455^{20}	1.4023^{20}	vs eth
4282	(1,1-Dimethylpropyl)benzene		$C_{11}H_{16}$	2049-95-8	148.245			192.4	0.8748^{20}	1.4958^{20}	
4283	(2,2-Dimethylpropyl)benzene		$C_{11}H_{16}$	1007-26-7	148.245			185	0.8581^{18}	1.4884^{18}	
4284	4-(1,1-Dimethylpropyl) cyclohexanone		$C_{11}H_{20}O$	16587-71-6	168.276		96	125^{16}, 109^{11}	0.920^{25}	1.4677^{20}	
4285	1,1-Dimethylpropyl 3-methylbutanoate	*tert*-Pentyl isopentanoate	$C_{10}H_{20}O_2$	542-37-0	172.265			173.5	0.8729^{0}		vs EtOH
4286	2-(1,1-Dimethylpropyl)phenol		$C_{11}H_{16}O$	3279-27-4	164.244						sl ctc
4287	4-(1,1-Dimethylpropyl)phenol	*p-tert*-Pentylphenol	$C_{11}H_{16}O$	80-46-6	164.244		95	262.5			
4288	4,6-Dimethyl-2*H*-pyran-2-one		$C_7H_8O_2$	675-09-2	124.138	lf (eth)	51.5	245			vs H_2O, eth, EtOH
4289	2,6-Dimethyl-4*H*-pyran-4-one		$C_7H_8O_2$	1004-36-0	124.138	pl, nd (sub)	132	251; 140^{25}	0.9953^{137}		s H_2O, EtOH, eth, ace
4290	2,3-Dimethylpyrazine		$C_6H_8N_2$	5910-89-4	108.141			156	1.0281^{0}		s H_2O, EtOH, eth
4291	2,5-Dimethylpyrazine		$C_6H_8N_2$	123-32-0	108.141		15	155	0.9887^{20}	1.4980^{20}	msc H_2O, EtOH, eth; s ace, chl
4292	2,6-Dimethylpyrazine		$C_6H_8N_2$	108-50-9	108.141	pr	47.5	155.6	0.9647^{50}		s H_2O, EtOH, eth; sl ctc
4293	1,3-Dimethyl-1*H*-pyrazole		$C_5H_8N_2$	694-48-4	96.131			137	0.9561^{17}	1.4734^{15}	vs H_2O

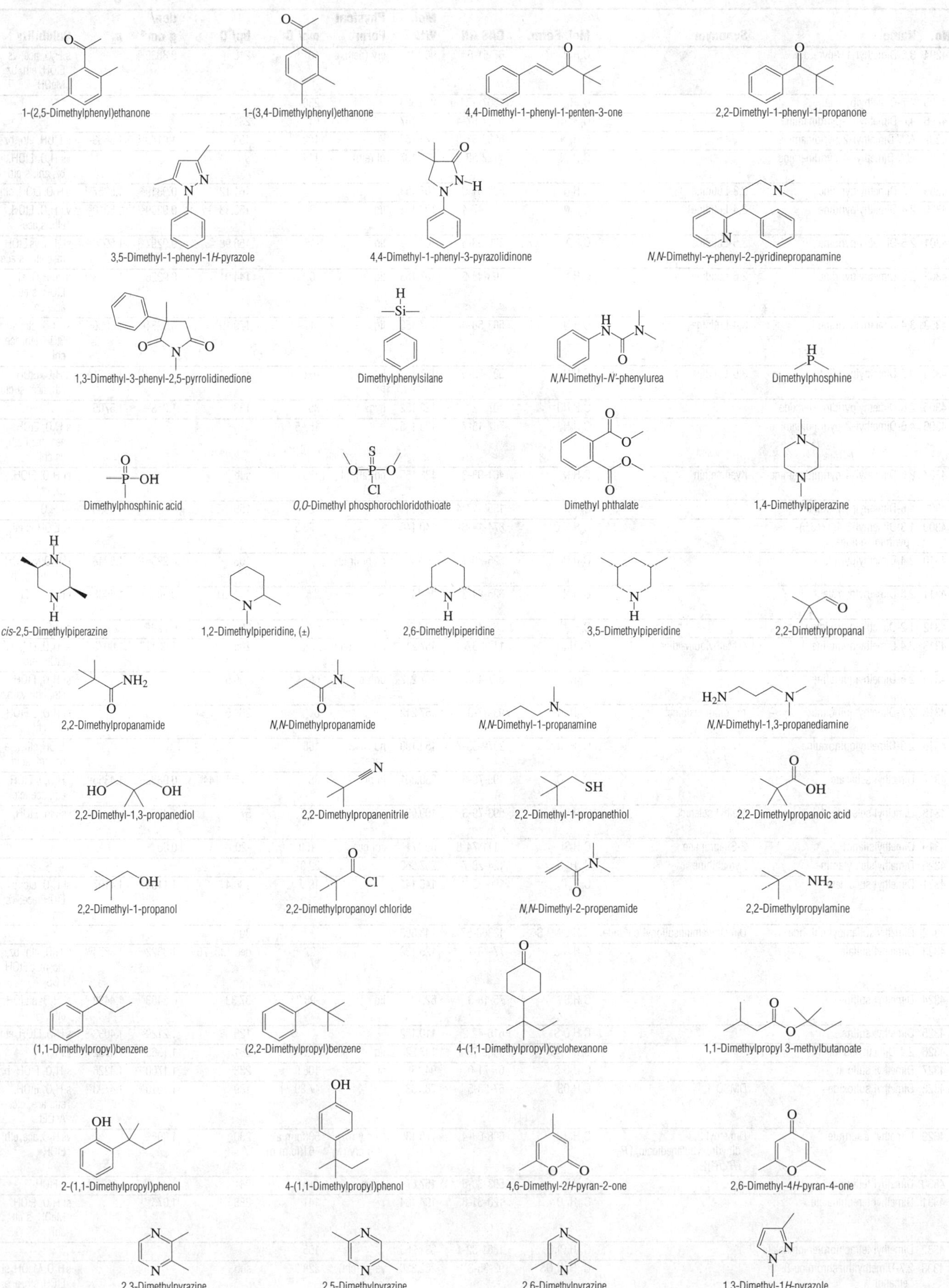

1-(2,5-Dimethylphenyl)ethanone
1-(3,4-Dimethylphenyl)ethanone
4,4-Dimethyl-1-phenyl-1-penten-3-one
2,2-Dimethyl-1-phenyl-1-propanone
3,5-Dimethyl-1-phenyl-1H-pyrazole
4,4-Dimethyl-1-phenyl-3-pyrazolidinone
N,N-Dimethyl-γ-phenyl-2-pyridinepropanamine
1,3-Dimethyl-3-phenyl-2,5-pyrrolidinedione
Dimethylphenylsilane
N,N-Dimethyl-N'-phenylurea
Dimethylphosphine
Dimethylphosphinic acid
O,O-Dimethyl phosphorochloridothioate
Dimethyl phthalate
1,4-Dimethylpiperazine
cis-2,5-Dimethylpiperazine
1,2-Dimethylpiperidine, (±)
2,6-Dimethylpiperidine
3,5-Dimethylpiperidine
2,2-Dimethylpropanal
2,2-Dimethylpropanamide
N,N-Dimethylpropanamide
N,N-Dimethyl-1-propanamine
N,N-Dimethyl-1,3-propanediamine
2,2-Dimethyl-1,3-propanediol
2,2-Dimethylpropanenitrile
2,2-Dimethyl-1-propanethiol
2,2-Dimethylpropanoic acid
2,2-Dimethyl-1-propanol
2,2-Dimethylpropanoyl chloride
N,N-Dimethyl-2-propenamide
2,2-Dimethylpropylamine
(1,1-Dimethylpropyl)benzene
(2,2-Dimethylpropyl)benzene
4-(1,1-Dimethylpropyl)cyclohexanone
1,1-Dimethylpropyl 3-methylbutanoate
2-(1,1-Dimethylpropyl)phenol
4-(1,1-Dimethylpropyl)phenol
4,6-Dimethyl-2H-pyran-2-one
2,6-Dimethyl-4H-pyran-4-one
2,3-Dimethylpyrazine
2,5-Dimethylpyrazine
2,6-Dimethylpyrazine
1,3-Dimethyl-1H-pyrazole

No.	Name	Synonym	Mol. Form.	CAS RN	Mol. Wt.	Physical Form	mp/°C	bp/°C	den/ g cm^{-3}	n_D	Solubility
4294	3,5-Dimethyl-1*H*-pyrazole		$C_5H_8N_2$	67-51-6	96.131	cry (peth, al)	107.5	218	0.8839^{16}		s H_2O, ace; vs EtOH, eth, bz, MeOH
4295	2,7-Dimethylpyrene		$C_{18}H_{14}$	15679-24-0	230.304		230				
4296	4,6-Dimethyl-2-pyridinamine		$C_7H_{10}N_2$	5407-87-4	122.167		61	235			
4297	*N,N*-Dimethyl-2-pyridinamine		$C_7H_{10}N_2$	5683-33-0	122.167		182	196	1.0149^{14}	1.5663^{20}	s EtOH, eth, bz
4298	*N,N*-Dimethyl-4-pyridinamine		$C_7H_{10}N_2$	1122-58-3	122.167	pl (eth)	114				vs H_2O, EtOH, bz, chl; s eth
4299	2,3-Dimethylpyridine	2,3-Lutidine	C_7H_9N	583-61-9	107.153			161.12	0.9319^{25}	1.5057^{20}	s H_2O, EtOH, eth
4300	2,4-Dimethylpyridine	2,4-Lutidine	C_7H_9N	108-47-4	107.153	liq	-64	158.38	0.9309^{20}	1.5010^{20}	vs H_2O, EtOH, eth; s ace
4301	2,5-Dimethylpyridine	2,5-Lutidine	C_7H_9N	589-93-5	107.153	liq	-16	156.98	0.9297^{20}	1.5006^{20}	sl H_2O; vs EtOH; msc eth; s ace
4302	2,6-Dimethylpyridine	2,6-Lutidine	C_7H_9N	108-48-5	107.153	liq	-6.1	144.01	0.9226^{20}	1.4953^{20}	msc H_2O; sl EtOH; s eth, ace, chl
4303	3,4-Dimethylpyridine	3,4-Lutidine	C_7H_9N	583-58-4	107.153	liq	-11	179.10	0.9281^{20}	1.5096^{20}	sl H_2O, ctc; s EtOH, eth, ace, chl
4304	3,5-Dimethylpyridine	3,5-Lutidine	C_7H_9N	591-22-0	107.153	liq	-6.6	171.84	0.9419^{20}	1.5061^{20}	s H_2O, EtOH, eth, ace; sl ctc
4305	2,6-Dimethylpyridine-1-oxide		C_7H_9NO	1073-23-0	123.152	hyg	35	133^{22}	1.073^{25}	1.5706^{20}	
4306	4,6-Dimethyl-2-pyrimidinamine		$C_6H_9N_3$	767-15-7	123.155		153.5				s H_2O, EtOH, ace, bz; i eth; vs chl
4307	2,6-Dimethyl-4-pyrimidinamine	Kyanmethin	$C_6H_9N_3$	461-98-3	123.155	nd (al), pl (bz)	183	sub			sl H_2O, EtOH, bz, chl
4308	4,6-Dimethylpyrimidine		$C_6H_8N_2$	1558-17-4	108.141		25	159		1.4880^{20}	vs H_2O
4309	1,3-Dimethyl-2,4(1*H*,3*H*)-pyrimidinedione		$C_6H_8N_2O_2$	874-14-6	140.140		123.5				sl EtOH; s eth, chl
4310	2,4-Dimethylpyrrole		C_6H_9N	625-82-1	95.142	pa bl flr cry		168	0.9236^{20}	1.5048^{20}	sl H_2O; vs EtOH, eth, bz; s chl
4311	2,5-Dimethylpyrrole		C_6H_9N	625-84-3	95.142		6.5	171; 51^8	0.9353^{20}	1.5036^{20}	i H_2O; vs EtOH, eth
4312	1,2-Dimethylpyrrolidine		$C_6H_{13}N$	765-48-0	99.174	oil		99	0.799^{20}		s H_2O
4313	2,4-Dimethylquinoline	4-Methylquinaldine	$C_{11}H_{11}N$	1198-37-4	157.212	orth pr (eth)		265	1.0611^{15}	1.6075^{20}	sl H_2O, chl; vs EtOH, eth
4314	2,6-Dimethylquinoline		$C_{11}H_{11}N$	877-43-0	157.212	orth pr (eth)	60	266.5			sl H_2O, EtOH, eth, chl; vs bz
4315	2,7-Dimethylquinoline	*m*-Toluquinaldine	$C_{11}H_{11}N$	93-37-8	157.212		61	264.5			sl H_2O; s EtOH, eth, chl
4316	2,3-Dimethylquinoxaline		$C_{10}H_{10}N_2$	2379-55-7	158.199	nd (w+3, ace)	106				s EtOH, eth, ace, bz, chl, acid
4317	Dimethyl sebacate		$C_{12}H_{22}O_4$	106-79-6	230.301	lo pr	38	175^{20}, 144^5	0.9882^{28}	1.4355^{28}	i H_2O; s EtOH, eth, ace, ctc
4318	Dimethyl selenide	Methyl selenide	C_2H_6Se	593-79-3	109.03			57	1.4077^{15}		vs eth, EtOH, chl
4319	Dimethylsilane	2-Silapropane	C_2H_8Si	1111-74-6	60.171	col gas	-150	-20	0.68^{-80}		
4320	Dimethylstearylamine	Dymanthine	$C_{20}H_{43}N$	124-28-7	297.562		22.9				
4321	Dimethyl succinate		$C_6H_{10}O_4$	106-65-0	146.141		19	196.4	1.1198^{20}	1.4197^{20}	sl H_2O, ctc; s EtOH, ace; vs eth
4322	Dimethylsulfamoyl chloride	Dimethylaminosulfonyl chloride	$C_2H_6ClNO_2S$	13360-57-1	143.593			80^{16}			
4323	Dimethyl sulfate		$C_2H_6O_4S$	77-78-1	126.132		-27	dec 188; 76^{15}	1.3322^{20}	1.3874^{20}	s H_2O, eth, bz, ctc; msc EtOH; i CS_2
4324	Dimethyl sulfide		C_2H_6S	75-18-3	62.134	liq	-98.24	37.33	0.8483^{20}	1.4438^{20}	sl H_2O; s EtOH, eth
4325	Dimethyl sulfite		$C_2H_6O_3S$	616-42-2	110.132			126	1.2129^{20}	1.4083^{20}	s H_2O, EtOH, eth
4326	2,4-Dimethylsulfolane		$C_6H_{12}O_2S$	1003-78-7	148.223	liq	-1.5	281	1.1362^{20}	1.4732^{20}	vs lig
4327	Dimethyl sulfone		$C_2H_6O_2S$	67-71-0	94.133	pr	108.9	238	1.1700^{110}	1.4226	s H_2O, EtOH, bz
4328	Dimethyl sulfoxide	DMSO	C_2H_6OS	67-68-5	78.133		17.89	189	1.1010^{25}	1.4793^{20}	s H_2O, EtOH, eth, ace, ctc, AcOEt
4329	Dimethyl *L*-tartrate	Dimethyl 2,3-dihydroxybutanedioate, [*R*-(*R**,*R**)]-	$C_6H_{10}O_6$	608-68-4	178.139	(i) cry (bz) (ii) cry (w)	50(form a); 61(form b)	280	1.306^{45}		vs H_2O, ace, eth, EtOH
4330	Dimethyl telluride		C_2H_6Te	593-80-6	157.67	pa ye		94			vs EtOH
4331	Dimethyl terephthalate		$C_{10}H_{10}O_4$	120-61-6	194.184		141	288	1.075^{141}		sl H_2O, EtOH, MeOH; s eth, chl
4332	Dimethyl tetrachloroterephthalate		$C_{10}H_6Cl_4O_4$	1861-32-1	331.965		155				
4333	2,7-Dimethylthiachromine-8-ethanol		$C_{12}H_{14}N_4OS$	92-35-3	262.330	ye pr (chl)	228.8	sub			s H_2O, MeOH; sl EtOH, eth, ace, chl

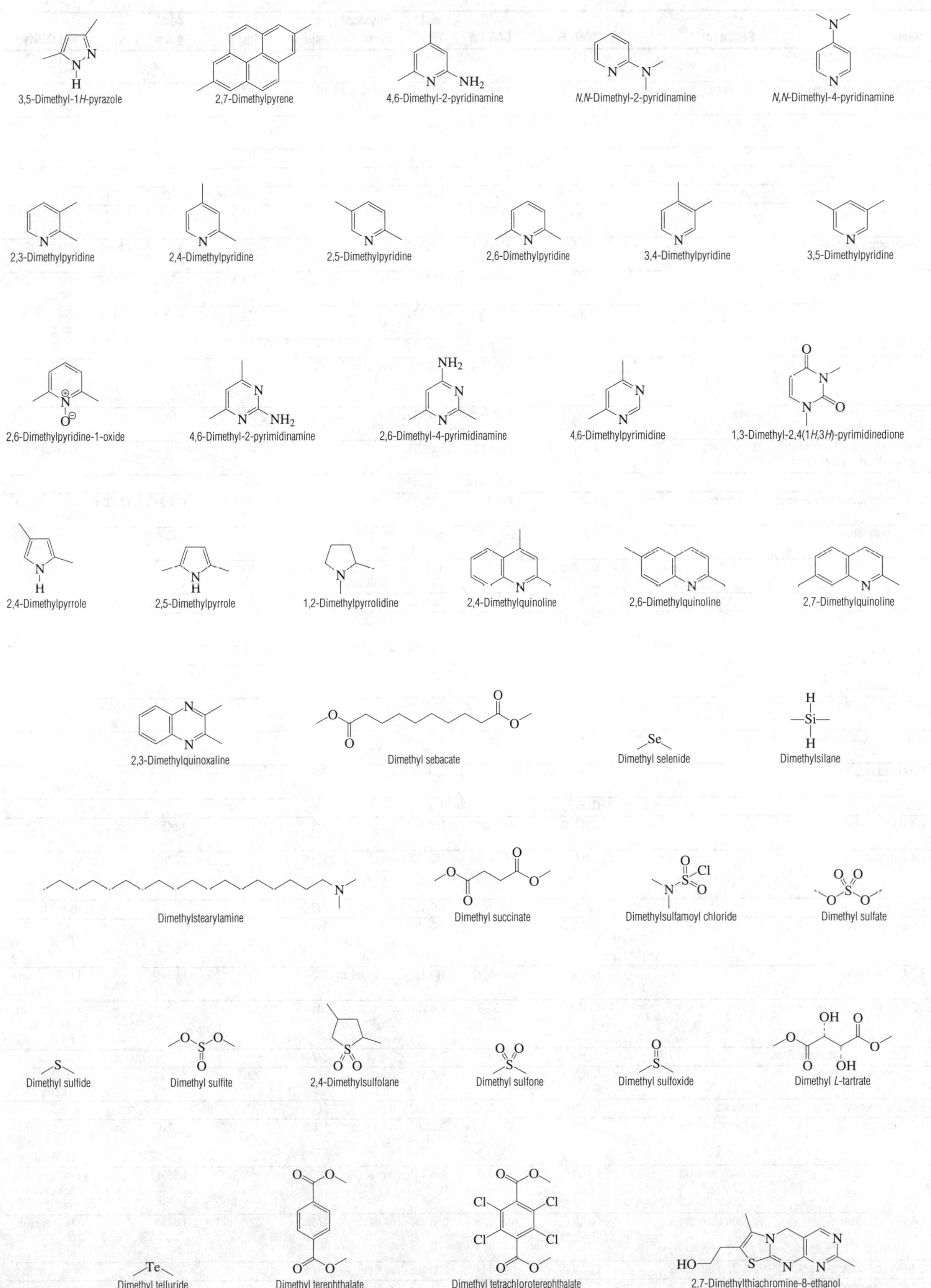

3,5-Dimethyl-1H-pyrazole
2,7-Dimethylpyrene
4,6-Dimethyl-2-pyridinamine
N,N-Dimethyl-2-pyridinamine
N,N-Dimethyl-4-pyridinamine
2,3-Dimethylpyridine
2,4-Dimethylpyridine
2,5-Dimethylpyridine
2,6-Dimethylpyridine
3,4-Dimethylpyridine
3,5-Dimethylpyridine
2,6-Dimethylpyridine-1-oxide
4,6-Dimethyl-2-pyrimidinamine
2,6-Dimethyl-4-pyrimidinamine
4,6-Dimethylpyrimidine
1,3-Dimethyl-2,4(1H,3H)-pyrimidinedione
2,4-Dimethylpyrrole
2,5-Dimethylpyrrole
1,2-Dimethylpyrrolidine
2,4-Dimethylquinoline
2,6-Dimethylquinoline
2,7-Dimethylquinoline
2,3-Dimethylquinoxaline
Dimethyl sebacate
Dimethyl selenide
Dimethylsilane
Dimethylstearylamine
Dimethyl succinate
Dimethylsulfamoyl chloride
Dimethyl sulfate
Dimethyl sulfide
Dimethyl sulfite
2,4-Dimethylsulfolane
Dimethyl sulfone
Dimethyl sulfoxide
Dimethyl L-tartrate
Dimethyl telluride
Dimethyl terephthalate
Dimethyl tetrachloroterephthalate
2,7-Dimethylthiachromine-8-ethanol

No.	Name	Synonym	Mol. Form.	CAS RN	Mol. Wt.	Physical Form	mp/°C	bp/°C	den/ g cm-3	n_D	Solubility
4334	2,5-Dimethyl-1,3,4-thiadiazole		$C_4H_6N_2S$	27464-82-0	114.169		65	202.5			sl H_2O, EtOH, eth
4335	2,7-Dimethylthianthrene	Mesulphen	$C_{14}H_{12}S_2$	135-58-0	244.375	nd (HOAc,al)	123	184^{3}			vs ace, eth, peth, chl
4336	2,4-Dimethylthiazole		C_5H_7NS	541-58-2	113.182			146; 71^{50}	1.0562^{15}	1.5091^{20}	sl H_2O; s EtOH, eth, chl
4337	4,5-Dimethylthiazole		C_5H_7NS	3581-91-7	113.182		83.5	158	1.0699^{20}		vs eth, EtOH
4338	*N,N*-Dimethylthioacetamide		C_4H_9NS	631-67-4	103.186		74.5				
4339	Dimethyl thiodipropionate		$C_8H_{14}O_4S$	4131-74-2	206.260			162^{18}, 148^{18}	1.1559^{20}	1.4740^{20}	
4340	2,3-Dimethylthiophene		C_6H_8S	632-16-6	112.193	liq	-49	141.6	1.0021^{20}	1.5192^{20}	i H_2O; vs EtOH, eth; s bz
4341	2,4-Dimethylthiophene		C_6H_8S	638-00-6	112.193			140.7	0.9938^{20}	1.5104^{20}	i H_2O; s EtOH, eth, bz
4342	2,5-Dimethylthiophene		C_6H_8S	638-02-8	112.193	liq	-62.6	136.5	0.9850^{20}	1.5129^{20}	i H_2O; s EtOH, eth, bz
4343	3,4-Dimethylthiophene		C_6H_8S	632-15-5	112.193			145	0.993^{25}	1.5206^{20}	i H_2O; s EtOH; vs eth
4344	*N,N*-Dimethylthiourea		$C_3H_8N_2S$	6972-05-0	104.174	cry (w)	161.5				
4345	*N,N'*-Dimethylthiourea		$C_3H_8N_2S$	534-13-4	104.174	hyg pl	62				vs H_2O, EtOH, ace; sl eth, bz; i CS_2
4346	2,6-Dimethyl-4-tridecylmorpholine	Tridemorph	$C_{19}H_{39}NO$	24602-86-6	297.519			$141^{1.3}$	0.86		
4347	*N,N*-Dimethyl-*N'*-[3-(trifluoromethyl)phenyl]urea	Fluometuron	$C_{10}H_{11}F_3N_2O$	2164-17-2	232.201		164				vs ace, EtOH
4348	Dimethyl trisulfide		$C_2H_6S_3$	3658-80-8	126.264			41^{6}			
4349	6,10-Dimethyl-3,5,9-undecatrien-2-one	Pseudoionone	$C_{13}H_{20}O$	141-10-6	192.297	pa ye oil		144^{12}	0.8984^{20}	1.5335^{20}	s EtOH, eth, chl, MeOH
4350	*N,N*-Dimethylurea		$C_3H_8N_2O$	598-94-7	88.108	mcl pr (al, chl)	182.1		1.2555^{25}		s H_2O; sl EtOH, tfa; i eth
4351	*N,N'*-Dimethylurea		$C_3H_8N_2O$	96-31-1	88.108	orth bipym (chl-eth)	106.6	269	1.142^{25}		vs H_2O, EtOH; i eth; sl chl
4352	Dimethyl zinc		C_2H_6Zn	544-97-8	95.478	liq, ign in air	-43.0	46	1.386^{10}		s eth; msc peth
4353	Dimetilan		$C_{10}H_{16}N_4O_3$	644-64-4	240.259	col solid	69	205^{13}			s H_2O, chl, EtOH, ace, xyl
4354	Dimorpholamine		$C_{20}H_{38}N_4O_4$	119-48-2	398.541	cry (peth)	41.5	$229^{0.4}$			vs H_2O
4355	*N,N'*-Di-2-naphthyl-1,4-benzenediamine		$C_{26}H_{20}N_2$	93-46-9	360.450		235				i EtOH, eth, bz
4356	Di-2-naphthyl disulfide		$C_{20}H_{14}S_2$	5586-15-2	318.455	nd	139.5		1.144^{145}	1.4555^{20}	i H_2O; vs EtOH, eth; i lig
4357	*N,N'*-Di-1-naphthylurea		$C_{21}H_{16}N_2O$	607-56-7	312.364	nd (py, HOAc)	296	sub			vs py
4358	Diniconazole		$C_{15}H_{17}Cl_2N_3O$	83657-24-3	326.221	cry	149				s H_2O, ace, MeOH, xyl
4359	Dinitramine		$C_{11}H_{13}F_3N_4O_4$	29091-05-2	322.241		98				
4360	2,3-Dinitroaniline		$C_6H_5N_3O_4$	602-03-9	183.122		128		1.646^{50}		i H_2O; s EtOH; sl eth
4361	2,4-Dinitroaniline		$C_6H_5N_3O_4$	97-02-9	183.122	ye nd (ace) grn ye tab (al)	180.0		1.615^{14}		i H_2O; sl EtOH, ace, HCl
4362	2,5-Dinitroaniline		$C_6H_5N_3O_4$	619-18-1	183.122	oran nd (al)	138.0				vs EtOH
4363	2,6-Dinitroaniline		$C_6H_5N_3O_4$	606-22-4	183.122	gold lf (HOAc) ye nd (al)	141.5				i H_2O, lig; sl EtOH; s eth, bz
4364	3,5-Dinitroaniline		$C_6H_5N_3O_4$	618-87-1	183.122	ye nd (dil al)	163		1.601^{50}		i H_2O; s EtOH, eth; sl ace, bz
4365	1,5-Dinitro-9,10-anthracenedione		$C_{14}H_6N_2O_6$	82-35-9	298.207	pa ye nd (xyl)	385	sub			i H_2O; sl EtOH, eth, bz; vs $PhNO_2$
4366	1,8-Dinitro-9,10-anthracenedione		$C_{14}H_6N_2O_6$	129-39-5	298.207		312				
4367	2,4-Dinitrobenzaldehyde		$C_7H_4N_2O_5$	528-75-6	196.117	pa ye pr (al), pl (bz)	72	200^{15}			sl H_2O, chl, lig; s EtOH, eth, bz
4368	3,5-Dinitrobenzamide	Nitromide	$C_7H_5N_3O_5$	121-81-3	211.132	lf (w)	184				vs H_2O
4369	1,2-Dinitrobenzene	*o*-Dinitrobenzene	$C_6H_4N_2O_4$	528-29-0	168.107	nd (bz), pl (al)	116.5	318; 194^{30}	1.3119^{120}	1.565^{17}	i H_2O; s EtOH, bz, chl, AcOEt; sl DMSO
4370	1,3-Dinitrobenzene	*m*-Dinitrobenzene	$C_6H_4N_2O_4$	99-65-0	168.107	orth pl (al)	90.3	291; 167^{14}	1.5751^{18}		sl H_2O; vs EtOH, ace, py; s eth, tol
4371	1,4-Dinitrobenzene	*p*-Dinitrobenzene	$C_6H_4N_2O_4$	100-25-4	168.107	nd (al)	173.5	297; 183^{34}	1.625^{18}		i H_2O; sl EtOH, chl; s ace, bz, tol
4372	2,4-Dinitro-1,3-benzenediol	2,4-Dinitroresorcinol	$C_6H_4N_2O_6$	519-44-8	200.105	ye lf (al)	147.5				sl H_2O, EtOH
4373	2,4-Dinitrobenzenesulfenyl chloride		$C_6H_3ClN_2O_4S$	528-76-7	234.617	ye pr (bz-peth)	99				vs bz, chl, HOAc; sl peth

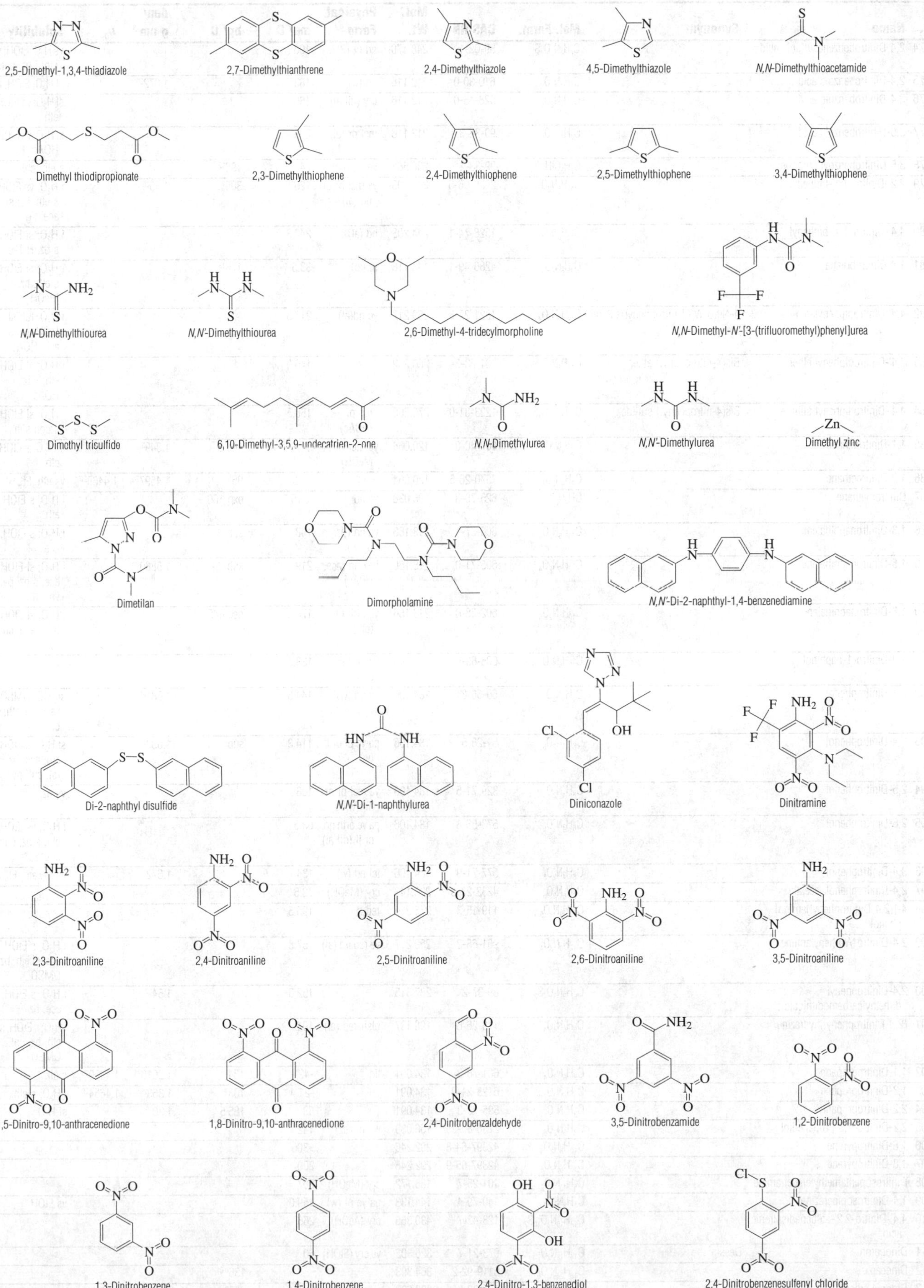
2,5-Dimethyl-1,3,4-thiadiazole
2,7-Dimethylthianthrene
2,4-Dimethylthiazole
4,5-Dimethylthiazole
N,N-Dimethylthioacetamide
Dimethyl thiodipropionate
2,3-Dimethylthiophene
2,4-Dimethylthiophene
2,5-Dimethylthiophene
3,4-Dimethylthiophene
N,N-Dimethylthiourea
N,N'-Dimethylthiourea
2,6-Dimethyl-4-tridecylmorpholine
N,N-Dimethyl-N'-[3-(trifluoromethyl)phenyl]urea
Dimethyl trisulfide
6,10-Dimethyl-3,5,9-undecatrien-2-one
N,N-Dimethylurea
N,N'-Dimethylurea
Dimethyl zinc
Dimetilan
Dimorpholamine
N,N'-Di-2-naphthyl-1,4-benzenediamine
Di-2-naphthyl disulfide
N,N'-Di-1-naphthylurea
Diniconazole
Dinitramine
2,3-Dinitroaniline
2,4-Dinitroaniline
2,5-Dinitroaniline
2,6-Dinitroaniline
3,5-Dinitroaniline
1,5-Dinitro-9,10-anthracenedione
1,8-Dinitro-9,10-anthracenedione
2,4-Dinitrobenzaldehyde
3,5-Dinitrobenzamide
1,2-Dinitrobenzene
1,3-Dinitrobenzene
1,4-Dinitrobenzene
2,4-Dinitro-1,3-benzenediol
2,4-Dinitrobenzenesulfenyl chloride

No.	Name	Synonym	Mol. Form.	CAS RN	Mol. Wt.	Physical Form	mp/°C	bp/°C	den/ g cm⁻³	n_D	Solubility
4374	2,4-Dinitrobenzenesulfonic acid		$C_6H_4N_2O_7S$	89-02-1	248.170	nd (w+3)	108				vs H_2O, EtOH; sl eth; i bz, peth
4375	2,4-Dinitrobenzoic acid		$C_7H_4N_2O_6$	610-30-0	212.116	nd (w)	183		1.672^{20}		sl H_2O, EtOH, bz
4376	3,4-Dinitrobenzoic acid		$C_7H_4N_2O_6$	528-45-0	212.116	cry (dil al)	166				sl H_2O; vs EtOH, eth
4377	3,5-Dinitrobenzoic acid		$C_7H_4N_2O_6$	99-34-3	212.116	mcl pr (al)	205				sl H_2O; vs EtOH, HOAc
4378	3,5-Dinitrobenzoyl chloride		$C_7H_3ClN_2O_5$	99-33-2	230.562	ye nd (bz)	74	196^{12}			s eth, chl
4379	2,2'-Dinitro-1,1'-biphenyl		$C_{12}H_8N_2O_4$	2436-96-6	244.203	ye mcl pr or nd (al)	126	305	1.45^{25}		i H_2O; vs EtOH; s eth, bz; sl ace, lig
4380	4,4'-Dinitro-1,1'-biphenyl		$C_{12}H_8N_2O_4$	1528-74-1	244.203	nd (al)	242.3				i H_2O; sl EtOH; s bz, HOAc
4381	1,4-Dinitrobutane		$C_4H_8N_2O_4$	4286-49-1	148.118	pl (al)	33.5	176^{13}			i H_2O; sl EtOH; s eth, bz, MeOH
4382	4,4'-Dinitrodiphenylamine	4-Nitro-*N*-(4-nitrophenyl)aniline	$C_{12}H_9N_3O_4$	1821-27-8	259.217	ye nd(al)	217.5				i H_2O, tol; sl EtOH, bz; s ace, HOAc
4383	4,4'-Dinitrodiphenyl ether	Bis(4-nitrophenyl) ether	$C_{12}H_8N_2O_5$	101-63-3	260.202		146.0				i H_2O; sl EtOH, eth; s bz, HOAc
4384	4,4'-Dinitrodiphenyl sulfide	Bis(4-nitrophenyl) sulfide	$C_{12}H_8N_2O_4S$	1223-31-0	276.268	oran pl (HOAc)	160.5				i H_2O; sl EtOH; s con sulf
4385	1,1-Dinitroethane		$C_2H_4N_2O_4$	600-40-8	120.064	ye mcl (bz, MeOH)		185.5	1.349^{24}		sl H_2O; s EtOH, eth
4386	1,2-Dinitroethane		$C_2H_4N_2O_4$	7570-26-5	120.064		39.5	95^{5}	1.4597^{20}	1.4468^{20}	vs eth, EtOH
4387	Dinitromethane		$CH_2N_2O_4$	625-76-3	106.038	ye nd	<-15	exp 100			i H_2O; s EtOH, eth
4388	1,3-Dinitronaphthalene		$C_{10}H_6N_2O_4$	606-37-1	218.166	ye nd (bz, py-w)	148	sub			i H_2O; s EtOH, ace
4389	1,5-Dinitronaphthalene		$C_{10}H_6N_2O_4$	605-71-0	218.166	hex nd (ace, HOAc)	219	sub	1.5860^{20}		i H_2O; sl EtOH, ace; s bz, py; vs eth
4390	1,8-Dinitronaphthalene		$C_{10}H_6N_2O_4$	602-38-0	218.166	ye orth pl (chl)	173	dec 445			i H_2O; sl EtOH, bz; s ace, chl, py
4391	2,4-Dinitro-1-naphthol		$C_{10}H_6N_2O_5$	605-69-6	234.165	ye nd (al, chl)	138.8				
4392	2,3-Dinitrophenol		$C_6H_4N_2O_5$	66-56-8	184.106	ye nd (w)	144.5		1.681^{20}		sl H_2O, DMSO; vs EtOH, eth; s bz
4393	2,4-Dinitrophenol		$C_6H_4N_2O_5$	51-28-5	184.106	pa ye pl or lf (w)	114.8	sub	1.683^{24}		sl H_2O; s EtOH, eth, ace, bz, tol, chl, py
4394	2,5-Dinitrophenol		$C_6H_4N_2O_5$	329-71-5	184.106	ye mcl pr or nd (w,lig)	108				vs bz, eth
4395	2,6-Dinitrophenol		$C_6H_4N_2O_5$	573-56-8	184.106	pa ye orth nd or lf (dil al)	63.5				i H_2O; vs EtOH, eth; s bz, chl; sl ctc
4396	3,4-Dinitrophenol		$C_6H_4N_2O_5$	577-71-9	184.106	tcl nd (w)	134		1.672^{25}		vs bz, eth, EtOH
4397	2,4-Dinitrophenol, acetate		$C_8H_6N_2O_6$	4232-27-3	226.143	cry (MeOH)	72.5				
4398	4-[(2,4-Dinitrophenyl)amino] phenol		$C_{12}H_9N_3O_5$	119-15-3	275.216	red lf	195.5				s alk
4399	2,4-Dinitro-*N*-phenylaniline		$C_{12}H_9N_3O_4$	961-68-2	259.217	ye red nd (al)	157.8				i H_2O; s EtOH, ace; sl eth, bz, DMSO
4400	2,4-Dinitrophenyl dimethylcarbamodithioate		$C_9H_9N_3O_4S_2$	89-37-2	287.315		152.5		1.54^{20}		i H_2O; s EtOH, ace, bz
4401	(2,4-Dinitrophenyl)hydrazine		$C_6H_6N_4O_4$	119-26-6	198.137	blsh-red (al)	194				i H_2O; s EtOH; sl eth, bz, chl, DMSO
4402	1,1-Dinitropropane		$C_3H_6N_2O_4$	601-76-3	134.091	liq	-42	184	1.2610^{25}	1.4339^{20}	s alk
4403	1,3-Dinitropropane		$C_3H_6N_2O_4$	6125-21-9	134.091		-21.4	103^{1}	1.353^{26}	1.4654^{20}	i H_2O; s eth
4404	2,2-Dinitropropane		$C_3H_6N_2O_4$	595-49-3	134.091		53	185.5	1.30^{25}		sl H_2O
4405	2,2-Dinitro-1,3-propanediol		$C_3H_6N_2O_6$	2736-80-3	166.089	wh pl (bz)	142				
4406	1,6-Dinitropyrene		$C_{16}H_8N_2O_4$	42397-64-8	292.246		>300				
4407	1,8-Dinitropyrene		$C_{16}H_8N_2O_4$	42397-65-9	292.246		300				
4408	Dinitrosopentamethylenetetramine		$C_5H_{10}N_6O_2$	101-25-7	186.172	cry (MeOH)	207				
4409	1,4-Dinitrosopiperazine		$C_4H_8N_4O_2$	140-79-4	144.133	pa ye pl (w)	159.0				vs EtOH
4410	4,4'-Dinitro-2,2'-stilbenedisulfonic acid		$C_{14}H_{10}N_2O_{10}S_2$	128-42-7	430.366	cry (AcOH)	266				
4411	Dinobuton	Dessin	$C_{14}H_{18}N_2O_7$	973-21-7	326.302	ye cry (EtOH)	60				
4412	Dinocap		$C_{18}H_{24}N_2O_6$	6119-92-2	364.393			$136^{0.01}$			
4413	Dinonyl adipate		$C_{24}H_{46}O_4$	151-32-6	398.620			205^{1}			

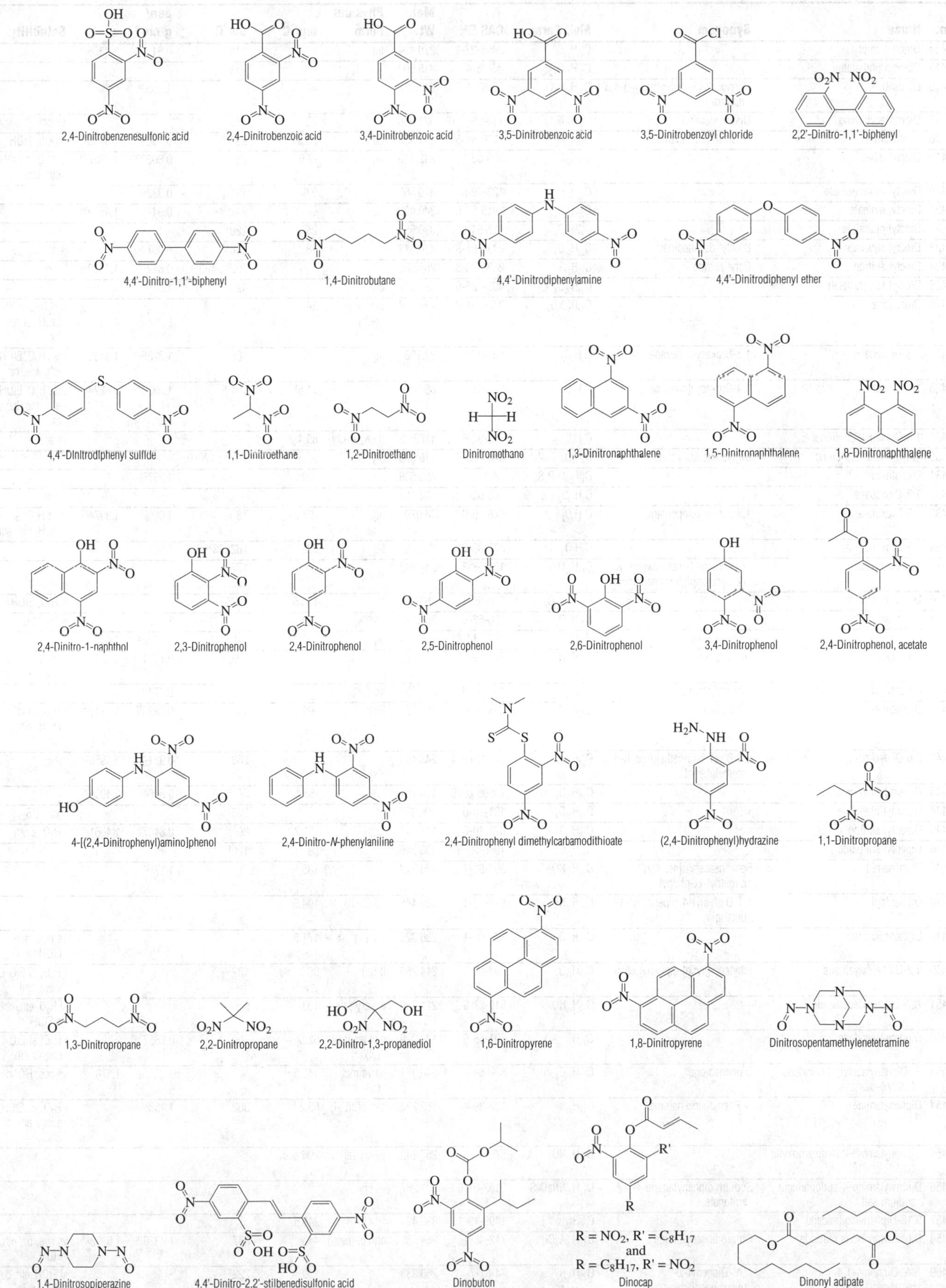

2,4-Dinitrobenzenesulfonic acid
2,4-Dinitrobenzoic acid
3,4-Dinitrobenzoic acid
3,5-Dinitrobenzoic acid
3,5-Dinitrobenzoyl chloride
2,2'-Dinitro-1,1'-biphenyl
4,4'-Dinitro-1,1'-biphenyl
1,4-Dinitrobutane
4,4'-Dinitrodiphenylamine
4,4'-Dinitrodiphenyl ether
4,4'-Dinitrodiphenyl sulfide
1,1-Dinitroethane
1,2-Dinitroethane
Dinitromethane
1,3-Dinitronaphthalene
1,5-Dinitronaphthalene
1,8-Dinitronaphthalene
2,4-Dinitro-1-naphthol
2,3-Dinitrophenol
2,4-Dinitrophenol
2,5-Dinitrophenol
2,6-Dinitrophenol
3,4-Dinitrophenol
2,4-Dinitrophenol, acetate
4-[(2,4-Dinitrophenyl)amino]phenol
2,4-Dinitro-N-phenylaniline
2,4-Dinitrophenyl dimethylcarbamodithioate
(2,4-Dinitrophenyl)hydrazine
1,1-Dinitropropane
1,3-Dinitropropane
2,2-Dinitropropane
2,2-Dinitro-1,3-propanediol
1,6-Dinitropyrene
1,8-Dinitropyrene
Dinitrosopentamethylenetetramine
1,4-Dinitrosopiperazine
4,4'-Dinitro-2,2'-stilbenedisulfonic acid
Dinobuton
R = NO2, R' = C8H17
and
R = C8H17, R' = NO2
Dinocap
Dinonyl adipate

No.	Name	Synonym	Mol. Form.	CAS RN	Mol. Wt.	Physical Form	mp/°C	bp/°C	den/ g cm^{-3}	n_D	Solubility
4414	Dinonyl ether		$C_{18}H_{38}O$	2456-27-1	270.494	liq		318	0.81	1.4356^{20}	
4415	Dinonyl phthalate		$C_{26}H_{42}O_4$	84-76-4	418.609			413			
4416	Dinoseb	Phenol, 2-(1-methylpropyl)-4,6-dinitro-	$C_{10}H_{12}N_2O_5$	88-85-7	240.212		40		1.265^{45}		
4417	Dioctadecylamine	Distearylamine	$C_{36}H_{75}N$	112-99-2	521.988		72.9	268^{2}			vs chl
4418	Dioctylamine	*N*-Octyl-1-octanamine	$C_{16}H_{35}N$	1120-48-5	241.456	nd	35.5	297.5	0.7963^{26}	1.4415^{26}	vs eth, EtOH
4419	Dioctyl ether		$C_{16}H_{34}O$	629-82-3	242.440	liq	-7.6	283	0.8063^{20}	1.4327^{20}	sl H_2O; s EtOH, eth, ctc
4420	Dioctyl hexanedioate		$C_{22}H_{42}O_4$	123-79-5	370.566		9.6	191^{2}	0.922^{25}		
4421	Dioctyl maleate		$C_{20}H_{36}O_4$	2915-53-9	340.498	liq		242$^{0.002}$	0.94^{20}	1.4539^{20}	
4422	Dioctyl phthalate		$C_{24}H_{38}O_4$	117-84-0	390.557		25	220^{4}			
4423	Dioctyl sebacate	Dioctyl decanedioate	$C_{26}H_{50}O_4$	2432-87-3	426.673		18	218$^{0.5}$	0.9074^{25}		s ctc
4424	Dioctyl sulfide	Octyl sulfide	$C_{16}H_{34}S$	2690-08-6	258.506			202^{29}, 180^{10}	0.842^{25}	1.4610^{20}	
4425	Dioctyl terephthalate		$C_{24}H_{38}O_4$	4654-26-6	390.557	cry		425	1.21^{62}		
4426	Dioscorine		$C_{13}H_{19}NO_2$	3329-91-7	221.296	grn-ye pr (eth)	34				s H_2O, ace, chl, EtOH; sl eth, bz
4427	1,3-Dioxane	1,3-Dioxacyclohexane	$C_4H_8O_2$	505-22-6	88.106	liq	-45	106.1	1.0286^{25}	1.4165^{20}	msc H_2O, EtOH, eth, ace, bz
4428	1,4-Dioxane	1,4-Dioxacyclohexane	$C_4H_8O_2$	123-91-1	88.106		11.85	101.5	1.0337^{20}	1.4224^{20}	msc H_2O, EtOH, eth, ace, bz; s ctc
4429	1,4-Dioxane-2,5-dione		$C_4H_4O_4$	502-97-6	116.073	lf (al, al-chl)	85.4				vs ace
4430	1,4-Dioxane-2,6-dione	Diglycollic anhydride	$C_4H_4O_4$	4480-83-5	116.073	cry (bz)	92.5	240.5; 120^{12}			
4431	Dioxathion		$C_{12}H_{26}O_6P_2S_4$	78-34-2	456.538		-20		1.257^{26}		
4432	1,3-Dioxepane		$C_5H_{10}O_2$	505-65-7	102.132						s chl
4433	1,3-Dioxolane	1,3-Dioxacyclopentane	$C_3H_6O_2$	646-06-0	74.079	liq	-97.22	78	1.060^{20}	1.3974^{20}	msc H_2O; s EtOH, eth, ace
4434	1,3-Dioxol-2-one		$C_3H_2O_3$	872-36-6	86.046	liq	22	162; 73^{32}	1.35^{25}		
4435	Dioxybenzone	(2-Hydroxy-4-methoxyphenyl) (2-hydroxyphenyl)methanone	$C_{14}H_{12}O_4$	131-53-3	244.243			172^{1}			
4436	Dioxypyramidon		$C_{13}H_{17}N_3O_3$	519-65-3	263.292	pr	105.5	197^{2}			s H_2O, EtOH
4437	Dipentaerythritol		$C_{10}H_{22}O_7$	126-58-9	254.278	cry (w)	221		1.366^{15}		s hot H_2O
4438	Dipentene	*p*-Menthadiene	$C_{10}H_{16}$	7705-14-8	136.234	liq	-95.5	178	0.8402^{21}	1.4727^{20}	
4439	Dipentylamine	Diamylamine	$C_{10}H_{23}N$	2050-92-2	157.297			202.5	0.7771^{20}	1.4272^{20}	sl H_2O; vs EtOH; msc eth; s ace
4440	Dipentyl *cis*-2-butenedioate	Dipentyl maleate	$C_{14}H_{24}O_4$	10099-71-5	256.339	liq		161^{10}	0.974^{20}		
4441	Dipentyl ether	Amyl ether	$C_{10}H_{22}O$	693-65-2	158.281	liq	-69	190	0.7833^{20}	1.4119^{20}	i H_2O; msc EtOH, eth; s chl
4442	2,6-Di-*tert*-pentyl-4-methylphenol	2,6-Bis(1,1-dimethylpropyl)-4-methylphenol	$C_{17}H_{28}O$	56103-67-4	248.403			283	0.931^{25}	1.4950^{20}	
4443	Di-*tert*-pentyl peroxide		$C_{10}H_{22}O_2$	10508-09-5	174.281		-55	58^{14}, 38^{9}	0.808^{20}	1.4095^{20}	
4444	Dipentyl phthalate		$C_{18}H_{26}O_4$	131-18-0	306.397			205^{11}			s ctc, CS_2
4445	Dipentyl sulfide		$C_{10}H_{22}S$	872-10-6	174.347		-51.3	86$^{3.7}$	0.8407^{20}	1.4561^{20}	i H_2O; s eth
4446	Dipentyl sulfoxide		$C_{10}H_{22}OS$	1986-90-9	190.346		58	120^{1}			
4447	Diphenamid	Benzeneacetamide, *N,N*-dimethyl-α-phenyl-	$C_{16}H_{17}NO$	957-51-7	239.312		135		1.17$^{23.3}$		
4448	Diphenidol	1,1-Diphenyl-4-piperidinyl-1-butanol	$C_{21}H_{27}NO$	972-02-1	309.445	nd (peth)	104.5				
4449	Diphenolic acid		$C_{17}H_{18}O_4$	126-00-1	286.323	cry (w)	171.5				vs H_2O, ace, EtOH
4450	1,2-Diphenoxyethane	Ethylene glycol diphenyl ether	$C_{14}H_{14}O_2$	104-66-5	214.260	lf (al)	98	182^{12}			i H_2O; sl EtOH; s eth, chl
4451	*N,N*-Diphenylacetamide		$C_{14}H_{13}NO$	519-87-9	211.259	wh cry pow	103	sub			sl H_2O, eth, chl; s EtOH
4452	Diphenylacetylene		$C_{14}H_{10}$	501-65-5	178.229	mcl pr or pl (al)	62.5	300	0.9657^{100}		i H_2O; sl EtOH, chl; vs eth
4453	2-(Diphenylacetyl)-1*H*-indene-1,3(2*H*)-dione	Diphenadione	$C_{23}H_{16}O_3$	82-66-6	340.371	pa ye mcl (al)	146.5			1.670	vs ace, HOAc
4454	Diphenylamine	*N*-Phenylbenzenamine	$C_{12}H_{11}N$	122-39-4	169.222	mcl lf(dil al)	53.2	302	1.158^{22}		i H_2O; vs EtOH, ace; s eth; sl chl
4455	Diphenylamine-2,2'-dicarboxylic acid		$C_{14}H_{11}NO_4$	579-92-0	257.242	ye cry (al)	296 dec				
4456	Diphenylamine-4-sulfonic acid, sodium salt	Sodium diphenylamine-4-sulfonate	$C_{12}H_{10}NNaO_3S$	6152-67-6	271.267	ye cry					
4457	9,10-Diphenylanthracene		$C_{26}H_{18}$	1499-10-1	330.421		246.5				
4458	Diphenylarsinous chloride	Chlorodiphenylarsine	$C_{12}H_{10}AsCl$	712-48-1	264.582	orth pl (peth)	44	337	1.4820^{16}	1.6332^{56}	vs ace, bz, eth, EtOH
4459	*N,N'*-Diphenyl-1,4-benzenediamine	*N,N'*-Diphenyl-*p*-phenylenediamine	$C_{18}H_{16}N_2$	74-31-7	260.333		150	222$^{0.5}$			sl EtOH, eth, bz, chl; i acid

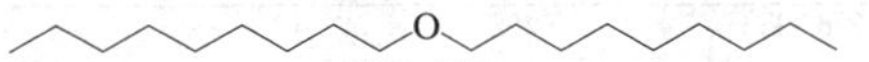

O
OH
O
O
O
O
O
N
O
O
N
O
O
Dinonyl ether
Dinonyl phthalate 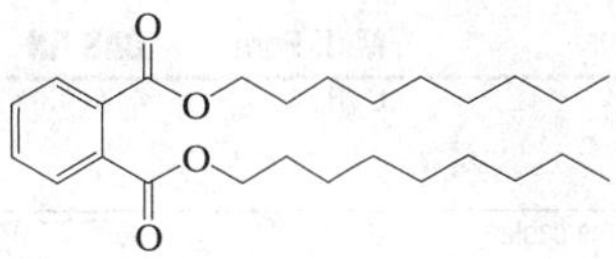
Dinoseb 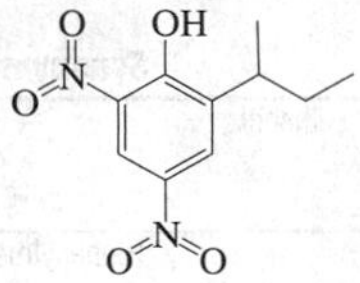 
NH
N
H
O
Dioctadecylamine
Dioctylamine 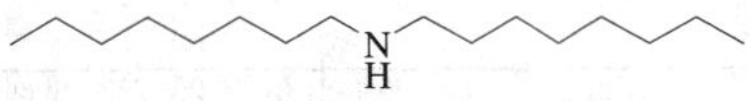
Dioctyl ether 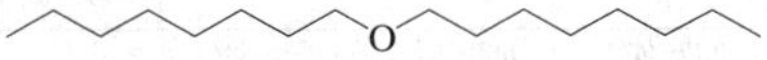
Dioctyl hexanedioate 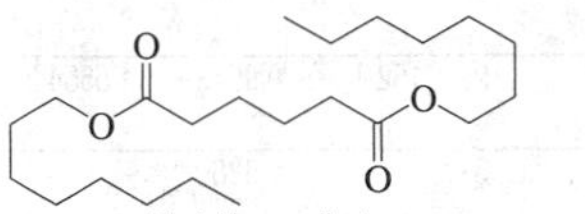
Dioctyl maleate 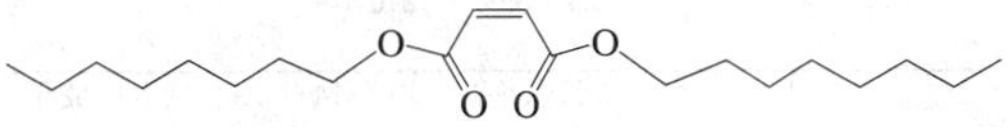
Dioctyl phthalate 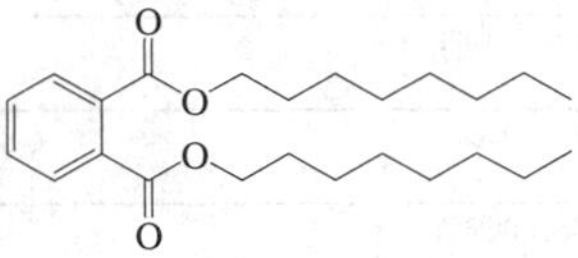 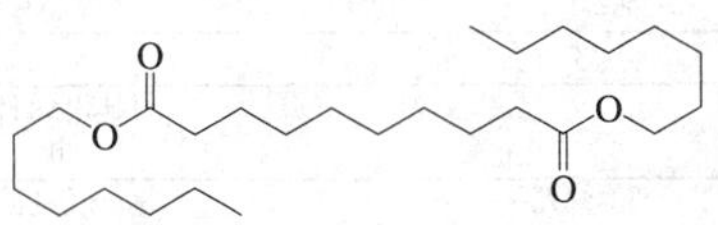
N
Dioctyl sebacate 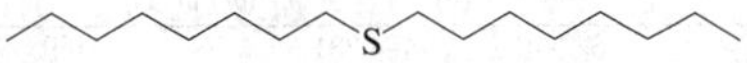
S
Dioctyl sulfide
Dioctyl terephthalate 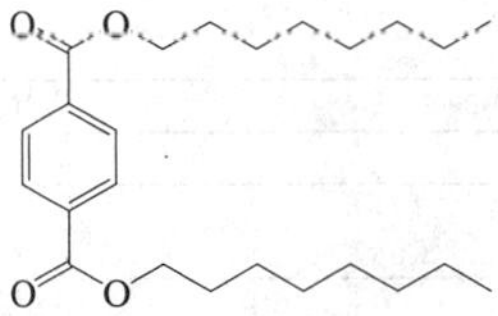
Dioscorine 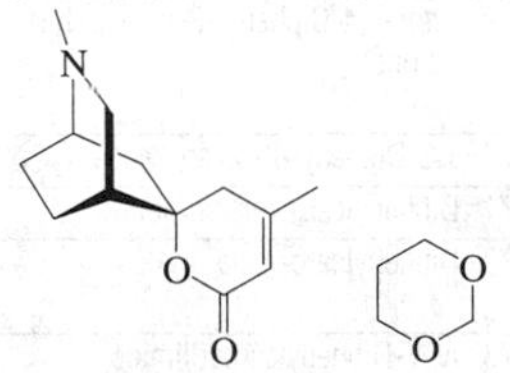
1,3-Dioxane 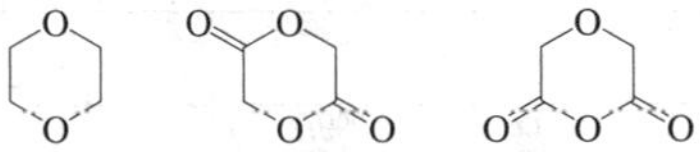
S
P
S
1,4-Dioxane
1,4-Dioxane-2,5-dione
1,4-Dioxane-2,6-dione
Dioxathion 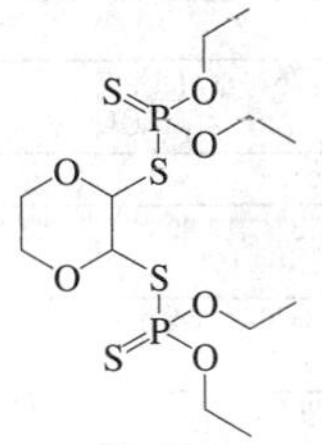
1,3-Dioxepane 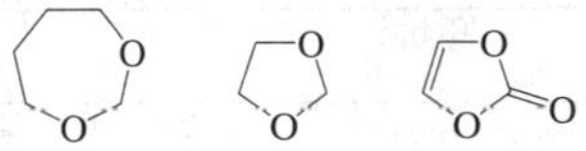
1,3-Dioxolane
1,3-Dioxol-2-one 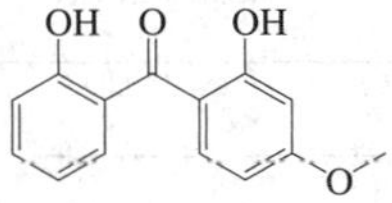
OH O OH
Dioxybenzone
Dioxypyramidon 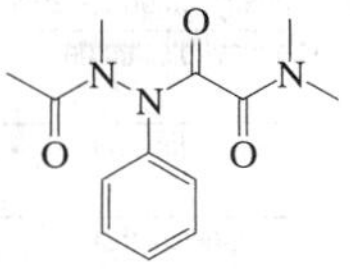 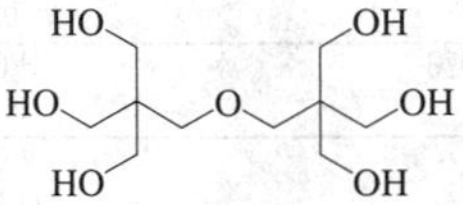
HO
OH
HO
OH
HO
OH
Dipentaerythritol
Dipentene 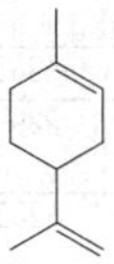
Dipentylamine 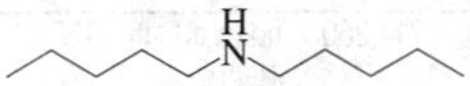
Dipentyl cis-2-butenedioate 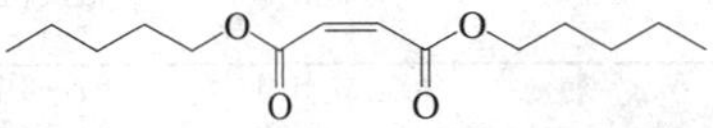
Dipentyl ether 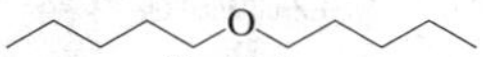 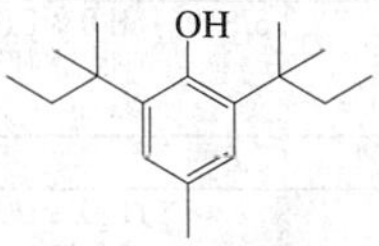
OH
2,6-Di-tert-pentyl-4-methylphenol
Di-tert-pentyl peroxide 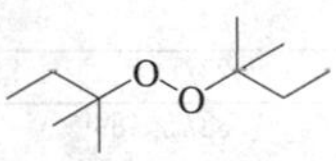
Dipentyl phthalate 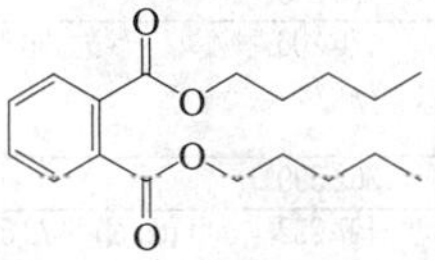
Dipentyl sulfide 
Dipentyl sulfoxide
Diphenamid 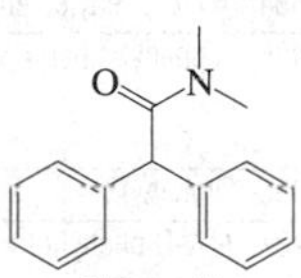 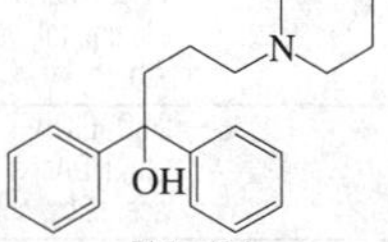
OH
Diphenidol
Diphenolic acid 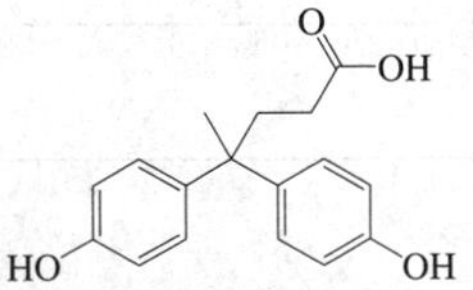
1,2-Diphenoxyethane 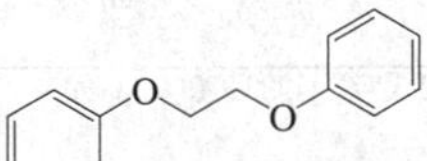
N,N-Diphenylacetamide 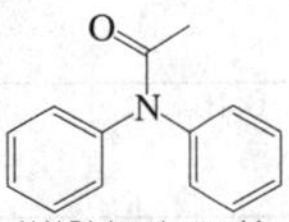
Diphenylacetylene 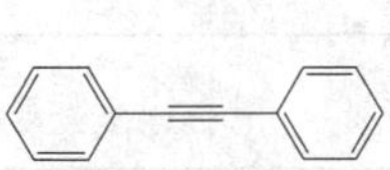
2-(Diphenylacetyl)-1H-indene-1,3(2H)-dione 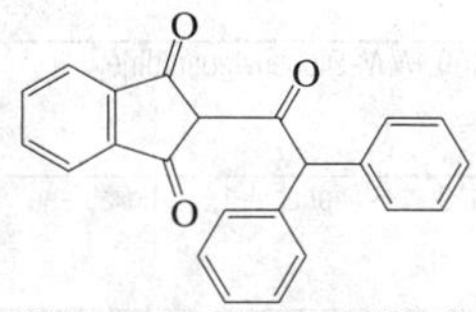
Cl
As
SO₃⁻Na⁺
Diphenylamine
Diphenylamine-2,2′-dicarboxylic acid 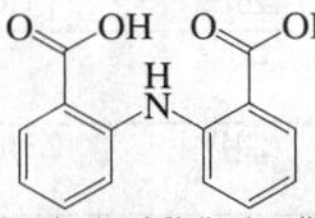
Diphenylamine-4-sulfonic acid, sodium salt 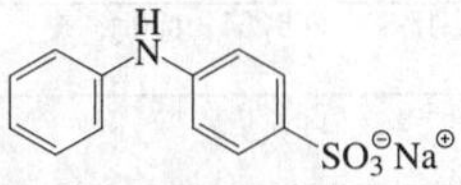
9,10-Diphenylanthracene 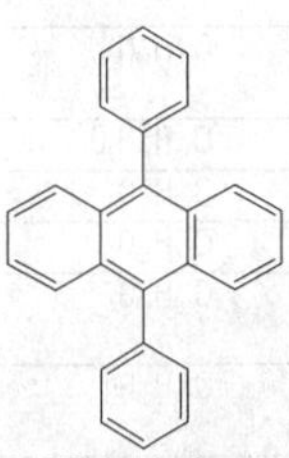
Diphenylarsinous chloride 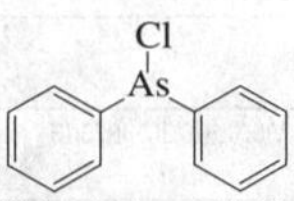
N,N′-Diphenyl-1,4-benzenediamine

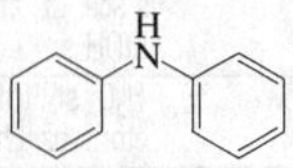

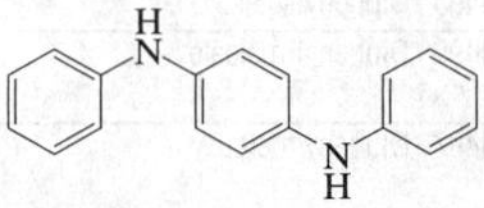

No.	Name	Synonym	Mol. Form.	CAS RN	Mol. Wt.	Physical Form	mp/°C	bp/°C	den/ g cm^{-3}	n_D	Solubility
4460	α,α-Diphenylbenzeneethanol		$C_{20}H_{18}O$	4428-13-1	274.356	nd(bz-lig) pr (peth)	89.5	222^{11}			i H_2O; vs EtOH; sl eth, chl, peth
4461	α,α-Diphenylbenzenemethanethiol	Triphenylmethyl mercaptan	$C_{19}H_{16}S$	3695-77-0	276.395		105.8				
4462	*N,N'*-Diphenyl-[1,1'-biphenyl]-4,4'-diamine	*N,N'*-Diphenylbenzidine	$C_{24}H_{20}N_2$	531-91-9	336.429	lf or pl	247				i H_2O; sl EtOH, eth, bz; vs tol, HOAc
4463	*trans,trans*-1,4-Diphenyl-1,3-butadiene		$C_{16}H_{14}$	538-81-8	206.282	lf (al, HOAc)	154.3	352			vs bz, eth, EtOH, peth
4464	1,4-Diphenyl-1,3-butadiyne	Diphenyldiacetylene	$C_{16}H_{10}$	886-66-8	202.250		86.5				
4465	1,1-Diphenylbutane		$C_{16}H_{18}$	719-79-9	210.314		27	287	0.9928^{20}	1.5664^{20}	i H_2O; s EtOH, eth, bz, chl
4466	1,2-Diphenylbutane		$C_{16}H_{18}$	5223-59-6	210.314			291; 152^{11}	0.9673^{20}	1.5554^{20}	i H_2O; s EtOH, eth, bz, chl
4467	1,4-Diphenylbutane		$C_{16}H_{18}$	1083-56-3	210.314		52.5	317	0.9880^{20}		i H_2O; s EtOH, eth, chl
4468	1,3-Diphenyl-1-butene		$C_{16}H_{16}$	7614-93-9	208.298		47.5	311	0.9996^{20}	1.590^{15}	
4469	*trans*-1,4-Diphenyl-2-butene-1,4-dione		$C_{16}H_{12}O_2$	959-28-4	236.265	ye nd (al, bz)	111				sl EtOH; s bz, HOAc; vs chl; i lig
4470	1,3-Diphenyl-2-buten-1-one	Dypnone	$C_{16}H_{14}O$	495-45-4	222.281			342.5	1.1080^{15}	1.6343^{20}	vs eth, EtOH
4471	Diphenylcarbamic chloride		$C_{13}H_{10}ClNO$	83-01-2	231.677	lf (al)	84.5				
4472	Diphenylcarbazone		$C_{13}H_{12}N_4O$	538-62-5	240.260	oran oran nd (bz) pr (al)	157 dec				i H_2O; vs EtOH, bz, chl
4473	*N,N'*-Diphenylcarbodiimide		$C_{13}H_{10}N_2$	622-16-2	194.231		169	331; 175^{20}			sl H_2O, EtOH, eth; s bz
4474	Diphenyl carbonate	Phenyl carbonate	$C_{13}H_{10}O_3$	102-09-0	214.216	nd (al, bz)	83	306	1.1215^{87}		i H_2O; s EtOH, eth, ctc, HOAc
4475	2,2'-Diphenylcarbonic dihydrazide	*sym*-Diphenylcarbazide	$C_{13}H_{14}N_4O$	140-22-7	242.276	cry (al + 1) cry (HOAc)	170	dec			sl H_2O, eth; s EtOH, ace, bz
4476	Diphenyl chlorophosphonate		$C_{12}H_{10}ClO_3P$	2524-64-3	268.632			314^{272}	1.296^{25}	1.5500^{20}	s tfa
4477	Diphenyl diselenide	Phenyl diselenide	$C_{12}H_{10}Se_2$	1666-13-3	312.13	ye nd	63.5	202^{11}	1.557^{80}	1.743^{20}	s EtOH, eth, xyl, MeOH
4478	Diphenyl disulfide	Phenyl disulfide	$C_{12}H_{10}S_2$	882-33-7	218.337	nd(al) or orth	62	310	1.353^{20}		i H_2O; s EtOH, eth, bz, CS_2
4479	1,1-Diphenylethane		$C_{14}H_{14}$	612-00-0	182.261	liq	-17.9	272.6	0.9997^{20}	1.5756^{20}	i H_2O; msc EtOH, eth; s bz
4480	1,2-Diphenylethane	Dibenzyl	$C_{14}H_{14}$	103-29-7	182.261	mcl pr (MeOH)	52.5	284	0.9780^{25}	1.5476^{60}	i H_2O; s EtOH, eth, CS_2
4481	*N,N'*-Diphenylethanediamide		$C_{14}H_{12}N_2O_2$	620-81-5	240.257	lf (bz)	254	>360			vs bz
4482	*N,N'*-Diphenyl-1,2-ethanediamine	1,2-Dianilinoethane	$C_{14}H_{16}N_2$	150-61-8	212.290	cry (dil al)	74	229^{12}, 178^{2}			i H_2O; s EtOH, eth; sl tfa
4483	1,2-Diphenyl-1,2-ethanediol, (*R**,*R**)-(±)		$C_{14}H_{14}O_2$	655-48-1	214.260	nd (w,al),tab (eth)	122.5	>300			i H_2O, lig; vs EtOH, eth; s ace
4484	1,1-Diphenylethene		$C_{14}H_{12}$	530-48-3	180.245		8.2	277	1.0232^{20}	1.6085^{20}	i H_2O; s eth, chl
4485	Diphenyl ether		$C_{12}H_{10}O$	101-84-8	170.206		26.87	258.0	1.0661^{30}	1.5787^{25}	i H_2O; s EtOH, eth, bz, HOAc; sl chl
4486	Diphenyl 2-ethylhexyl phosphate		$C_{20}H_{27}O_4P$	1241-94-7	362.399			232^{5}	1.090^{25}	1.510^{25}	
4487	*N,N*-Diphenylformamide		$C_{13}H_{11}NO$	607-00-1	197.232	orth (dil al)	73.5	337.5; 189^{13}			i H_2O; s EtOH, eth, bz; sl ctc
4488	2,5-Diphenylfuran		$C_{16}H_{12}O$	955-83-9	220.265	nd or lf (dil al)	91	344			i H_2O; vs EtOH, eth; s ace, bz
4489	*N,N'*-Diphenylguanidine	1,3-Diphenylguanidine	$C_{13}H_{13}N_3$	102-06-7	211.262	mcl nd (al, to)	150	dec 170	1.13^{20}		sl H_2O; s EtOH, ctc chl, tol; vs eth
4490	1,6-Diphenyl-1,3,5-hexatriene		$C_{18}H_{16}$	1720-32-7	232.320	lf (ace)	202.3				i H_2O, EtOH, eth, HOAc; s ace; sl bz, chl
4491	1,1-Diphenylhydrazine		$C_{12}H_{12}N_2$	530-50-7	184.236	tab (lig)	50.5	220^{40}	1.190^{16}		vs bz, eth, EtOH, chl
4492	1,2-Diphenylhydrazine	Hydrazobenzene	$C_{12}H_{12}N_2$	122-66-7	184.236	tab (al-eth)	131		1.158^{16}		vs EtOH; sl bz, DMSO; i HOAc
4493	5,5-Diphenyl-4-imidazolidinone	Doxenitoin	$C_{15}H_{14}N_2O$	3254-93-1	238.284	pl (MeOH)	183				
4494	Diphenyl isophthalate		$C_{20}H_{14}O_4$	744-45-6	318.323		138				s chl
4495	Diphenylketene	Diphenylethenone	$C_{14}H_{10}O$	525-06-4	194.228	red-ye liq		267.5	1.1107^{13}	1.615^{14}	
4496	Diphenyl maleate		$C_{16}H_{12}O_4$	7242-17-3	268.264	pl (lig)	73	226^{15}			vs ace, bz, eth, EtOH
4497	Diphenylmercury	Mercuriodibenzene	$C_{12}H_{10}Hg$	587-85-9	354.80			204^{10}	2.318^{25}		i H_2O; sl EtOH, eth; s bz, chl
4498	Diphenylmethane		$C_{13}H_{12}$	101-81-5	168.234	pr nd	25.4	265.0	1.001^{26}	1.5753^{20}	i H_2O; s EtOH, eth, chl

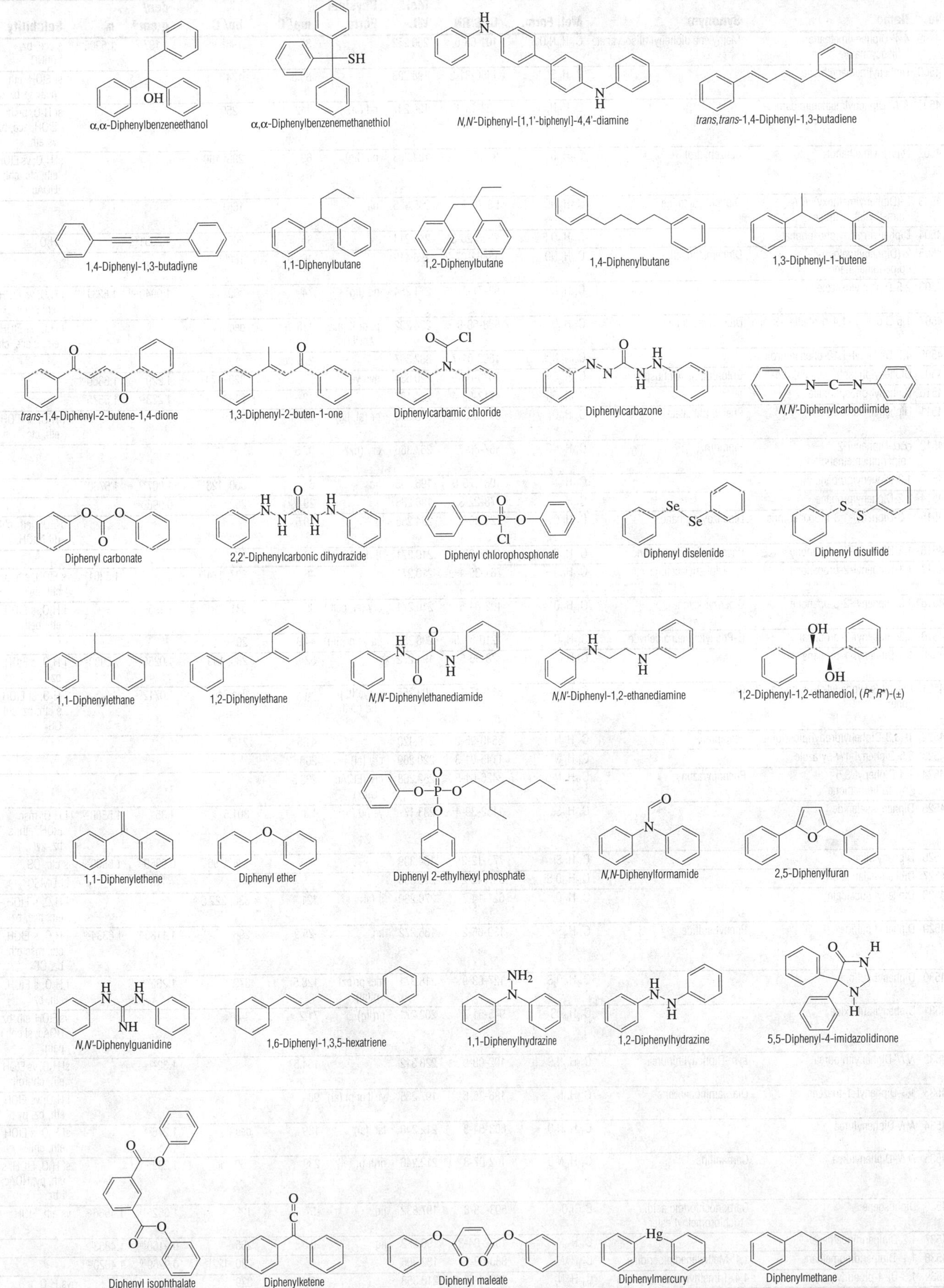
α,α-Diphenylbenzeneethanol
α,α-Diphenylbenzenemethanethiol
N,N'-Diphenyl-[1,1'-biphenyl]-4,4'-diamine
trans,trans-1,4-Diphenyl-1,3-butadiene
1,4-Diphenyl-1,3-butadiyne
1,1-Diphenylbutane
1,2-Diphenylbutane
1,4-Diphenylbutane
1,3-Diphenyl-1-butene
trans-1,4-Diphenyl-2-butene-1,4-dione
1,3-Diphenyl-2-buten-1-one
Diphenylcarbamic chloride
Diphenylcarbazone
N,N'-Diphenylcarbodiimide
Diphenyl carbonate
2,2'-Diphenylcarbonic dihydrazide
Diphenyl chlorophosphonate
Diphenyl diselenide
Diphenyl disulfide
1,1-Diphenylethane
1,2-Diphenylethane
N,N'-Diphenylethanediamide
N,N'-Diphenyl-1,2-ethanediamine
1,2-Diphenyl-1,2-ethanediol, (R*,R*)-(±)
1,1-Diphenylethene
Diphenyl ether
Diphenyl 2-ethylhexyl phosphate
N,N-Diphenylformamide
2,5-Diphenylfuran
N,N'-Diphenylguanidine
1,6-Diphenyl-1,3,5-hexatriene
1,1-Diphenylhydrazine
1,2-Diphenylhydrazine
5,5-Diphenyl-4-imidazolidinone
Diphenyl isophthalate
Diphenylketene
Diphenyl maleate
Diphenylmercury
Diphenylmethane

No.	Name	Synonym	Mol. Form.	CAS RN	Mol. Wt.	Physical Form	mp/°C	bp/°C	den/ g cm^{-3}	n_D	Solubility
4499	4,4'-Diphenylmethane diisocyanate	Methylene diphenyl diisocyanate	$C_{15}H_{10}N_2O_2$	101-68-8	250.252		37	196^{5}	1.197^{70}	1.5906^{50}	s ace, bz, $PhNO_2$
4500	Diphenylmethanethione		$C_{13}H_{10}S$	1450-31-3	198.283		53.5	174^{14}			sl EtOH, eth, peth; vs bz, chl
4501	*N,N'*-Diphenylmethanimidamide		$C_{13}H_{12}N_2$	622-15-1	196.247	nd (al)	142	>250			sl H_2O, peth; s EtOH, ace, bz; vs eth
4502	Diphenylmethanol	Benzohydrol	$C_{13}H_{12}O$	91-01-0	184.233	nd (lig)	69	298; 180^{20}			sl H_2O; vs EtOH, eth, ctc, chl; s HOAc
4503	2-(Diphenylmethoxy)-*N,N*-dimethylethanamine	Diphenhydramine	$C_{17}H_{21}NO$	58-73-1	255.355	oil		165^{3}			
4504	Diphenyl methylphosphonate		$C_{13}H_{13}O_3P$	7526-26-3	248.214		35	205^{13}	1.2051^{20}		i H_2O
4505	2-(Diphenylmethyl)-1-piperidineethanol	Diphemethoxidine	$C_{20}H_{25}NO$	13862-07-2	295.419		106.5	$180^{0.1}$			
4506	2,5-Diphenyloxazole		$C_{15}H_{11}NO$	92-71-7	221.254	nd (lig)	74	360	1.0940^{100}	1.6231^{100}	i H_2O; vs EtOH, eth; sl chl
4507	1,5-Diphenyl-1,4-pentadien-3-one	Dibenzalacetone	$C_{17}H_{14}O$	538-58-9	234.292	pl or lf (ace, AcOEt)	113 dec	dec			i H_2O; sl EtOH, eth; s ace, chl
4508	4,7-Diphenyl-1,10-phenanthroline		$C_{24}H_{16}N_2$	1662-01-7	332.397		220 dec				
4509	Diphenylphosphinous chloride	Chlorodiphenylphosphine	$C_{12}H_{10}ClP$	1079-66-9	220.634	hyg ye liq		320; 174^{5}	1.229	1.6360^{20}	
4510	Diphenyl phosphonate		$C_{12}H_{11}O_3P$	4712-55-4	234.187		12	218^{26}	1.223^{25}	1.5575^{20}	
4511	Diphenyl phthalate	Phenyl phthalate	$C_{20}H_{14}O_4$	84-62-8	318.323	pr (al, lig)	73	253^{14}			i H_2O; sl EtOH, eth, ctc
4512	α,α-Diphenyl-2-piperidinemethanol	Pipradrol	$C_{18}H_{21}NO$	467-60-7	267.366	cry (hx)	97.5				
4513	1,3-Diphenylpropane		$C_{15}H_{16}$	1081-75-0	196.288	liq	6	300; $123^{1.7}$	1.007^{20}	1.5760^{20}	
4514	2,2-Diphenylpropane		$C_{15}H_{16}$	778-22-3	196.288		29	282.5	0.9980^{20}		
4515	1,3-Diphenyl-1,3-propanedione	Dibenzoylmethane	$C_{15}H_{12}O_2$	120-46-7	224.255		70.5				s EtOH, eth, chl, dil NaOH
4516	1,3-Diphenyl-1-propanone	Phenethyl phenyl ketone	$C_{15}H_{14}O$	1083-30-3	210.271	lf (EtOH)	72.5	360			
4517	1,1-Diphenyl-2-propanone	1,1-Diphenylacetone	$C_{15}H_{14}O$	781-35-1	210.271		46	307; 174^{10}		1.5361^{16}	s EtOH, eth, bz, chl, lig
4518	1,3-Diphenyl-2-propanone	Dibenzyl ketone	$C_{15}H_{14}O$	102-04-5	210.271	cry (al, peth)	35	331	1.195^{0}		i H_2O; s EtOH, eth, peth
4519	3,3-Diphenyl-2-propenal	β-Phenylcinnamaldehyde	$C_{15}H_{12}O$	1210-39-5	208.255	pa ye pr (lig)	44.8	205^{14}			
4520	1,1-Diphenyl-1-propene		$C_{15}H_{14}$	778-66-5	194.272		52	280; 149^{11}	1.0250^{20}	1.5880^{20}	i H_2O; s EtOH, bz
4521	*trans*-1,3-Diphenyl-2-propen-1-one	Chalcone	$C_{15}H_{12}O$	614-47-1	208.255	pa ye lf, pr, nd (peth)	59	dec 346	1.0712^{62}		i H_2O; sl EtOH; s eth, bz, chl, CS_2
4522	1-(3,3-Diphenylpropyl)piperidine	Fenpiprane	$C_{20}H_{25}N$	3540-95-2	279.420		41.5	215^{8}			
4523	3,5-Diphenyl-1*H*-pyrazole		$C_{15}H_{12}N_2$	1145-01-3	220.269	cry (al)	200				
4524	1,4-Diphenyl-3,5-pyrazolidinedione	Phenopyrazone	$C_{15}H_{12}N_2O_2$	3426-01-5	252.268	cry (EtOAc, Diox)	233.5				
4525	Diphenyl selenide		$C_{12}H_{10}Se$	1132-39-4	233.17	ye nd (bz)	1.3	301.5	1.351^{20}	1.5500^{20}	i H_2O; msc EtOH, eth; s bz, xyl
4526	Diphenylsilane		$C_{12}H_{12}Si$	775-12-2	184.309			134^{16}, 96^{13}	0.9969^{20}	1.5800^{20}	s ctc, CS_2
4527	Diphenylsilanediol		$C_{12}H_{12}O_2Si$	947-42-2	216.308						sl DMSO
4528	Diphenyl succinate		$C_{16}H_{14}O_4$	621-14-7	270.280	lf (al)	121	330; 222.5^{15}			i H_2O; s EtOH, eth, ace, bz
4529	Diphenyl sulfide	Phenyl sulfide	$C_{12}H_{10}S$	139-66-2	186.272	liq	-25.9	296	1.1136^{20}	1.6334^{20}	i H_2O; s EtOH, ctc; msc eth, bz, CS_2
4530	Diphenyl sulfone		$C_{12}H_{10}O_2S$	127-63-9	218.271	mcl pr(bz) pl(al)	128.5	379	1.252^{20}		i H_2O; s EtOH, eth, bz
4531	Diphenyl sulfoxide		$C_{12}H_{10}OS$	945-51-7	202.271	pr(lig)	71.2	340^{16}			vs EtOH, eth, bz, HOAc; sl chl; i peth
4532	*N,N'*-Diphenylthiourea	*sym*-Diphenylthiourea	$C_{13}H_{12}N_2S$	102-08-9	228.312		154.5		1.32^{25}		sl H_2O; vs EtOH, eth, chl, oils
4533	1,3-Diphenyl-1-triazene	Diazoaminobenzene	$C_{12}H_{11}N_3$	136-35-6	197.235	ye lf or pr (al)	98				i H_2O; vs EtOH, eth, bz, py
4534	*N,N*-Diphenylurea		$C_{13}H_{12}N_2O$	603-54-3	212.246	tab (al)	189	dec	1.276^{25}		sl H_2O; s EtOH, eth, chl
4535	*N,N'*-Diphenylurea	Carbanilide	$C_{13}H_{12}N_2O$	102-07-8	212.246	orth pr (al)	239	260 dec	1.239^{25}		sl H_2O, EtOH; s eth, py, HOAc; i bz
4536	Diphosgene	Carbonochloridic acid, trichloromethyl ester	$C_2Cl_4O_2$	503-38-8	197.832	liq	-57	128	1.6525^{14}	1.4566^{22}	vs eth, EtOH
4537	1,2-Dipiperidinoethane		$C_{12}H_{24}N_2$	1932-04-3	196.332	liq	-0.5	265	0.9160^{25}	1.4853^{25}	
4538	1,1'-Dipiperidinomethane	1,1'-Methylenedipiperidine	$C_{11}H_{22}N_2$	880-09-1	182.306			230; 122^{15}	0.9269^{20}	1.4820^{20}	
4539	1,3-Di-4-piperidylpropane	4,4'-Trimethylenedipiperidine	$C_{13}H_{26}N_2$	16898-52-5	210.358		67.1	329			vs H_2O

4,4′-Diphenylmethane diisocyanate
Diphenylmethanethione
N,N′-Diphenylmethanimidamide
Diphenylmethanol
2-(Diphenylmethoxy)-*N,N*-dimethylethanamine
Diphenyl methylphosphonate
2-(Diphenylmethyl)-1-piperidineethanol
2,5-Diphenyloxazole
1,5-Diphenyl-1,4-pentadien-3-one
4,7-Diphenyl-1,10-phenanthroline
Diphenylphosphinous chloride
Diphenyl phosphonate
Diphenyl phthalate
α,α-Diphenyl-2-piperidinemethanol
1,3-Diphenylpropane
2,2-Diphenylpropane
1,3-Diphenyl-1,3-propanedione
1,3-Diphenyl-1-propanone
1,1-Diphenyl-2-propanone
1,3-Diphenyl-2-propanone
3,3-Diphenyl-2-propenal
1,1-Diphenyl-1-propene
trans-1,3-Diphenyl-2-propen-1-one
1-(3,3-Diphenylpropyl)piperidine
3,5-Diphenyl-1*H*-pyrazole
1,4-Diphenyl-3,5-pyrazolidinedione
Diphenyl selenide
Diphenylsilane
Diphenylsilanediol
Diphenyl succinate
Diphenyl sulfide
Diphenyl sulfone
Diphenyl sulfoxide
N,N′-Diphenylthiourea
1,3-Diphenyl-1-triazene
N,N-Diphenylurea
N,N′-Diphenylurea
Diphosgene
1,2-Dipiperidinoethane
1,1′-Dipiperidinomethane
1,3-Di-4-piperidylpropane

No.	Name	Synonym	Mol. Form.	CAS RN	Mol. Wt.	Physical Form	mp/°C	bp/°C	den/g cm^{-3}	n_D	Solubility
4540	Diploicin		$C_{16}H_{10}Cl_4O_5$	527-93-5	424.059		232				
4541	Di-2-propenoyldiethyleneglycol		$C_{10}H_{14}O_5$	4074-88-8	214.215			200	1.1110^{25}	1.4595^{25}	
4542	Di-2-propenoyl-2,2-dimethyl-1,3-propanediol	2-Propenoic acid, 2,2-dimethyl-1,3-propanediyl ester	$C_{11}H_{16}O_4$	2223-82-7	212.243					1.4542^{25}	
4543	Di-2-propenoyl-1,6-hexanediol	2-Propenoic acid, 1,6-hexanediyl ester	$C_{12}H_{18}O_4$	13048-33-4	226.269				1.010^{25}		
4544	Dipropetryn	6-(Ethylthio)-*N,N'*-diisopropyl-1,3,5-triazine-2,4-diamine	$C_{11}H_{21}N_5S$	4147-51-7	255.384		105				
4545	1,2-Dipropoxyethane		$C_8H_{18}O_2$	18854-56-3	146.228	liq		163.2 dec	0.8312^{25}	1.4013^{25}	
4546	Dipropoxymethane	Formaldehyde, dipropyl acetal	$C_7H_{16}O_2$	505-84-0	132.201	liq	-97.3	140.5	0.8345^{20}	1.3939^{19}	vs ace, bz, eth, EtOH
4547	*N,N*-Dipropylacetamide		$C_8H_{17}NO$	1116-24-1	143.227			209.5	0.8992^{17}	1.4419^{17}	vs EtOH
4548	Dipropyl adipate		$C_{12}H_{22}O_4$	106-19-4	230.301		-15.7	151^{11}	0.9790^{20}	1.4314^{20}	vs eth, EtOH, chl
4549	Dipropylamine	*N*-Propyl-1-propanamine	$C_6H_{15}N$	142-84-7	101.190	liq	-63	109.3	0.7400^{20}	1.4050^{20}	s H_2O, EtOH; msc eth; vs ace, bz
4550	4-[(Dipropylamino)sulfonyl] benzoic acid	Probenecid	$C_{13}H_{19}NO_4S$	57-66-9	285.360		195				
4551	*N,N*-Dipropylaniline		$C_{12}H_{19}N$	2217-07-4	177.286	ye lf		242	0.9104^{20}	1.5271^{20}	i H_2O; s EtOH, eth, ace, bz; sl ctc
4552	Dipropylcarbamothioic acid, *S*-ethyl ester	EPTC	$C_9H_{19}NOS$	759-94-4	189.318			127^{20}	0.9546^{30}		
4553	Dipropyl carbonate		$C_7H_{14}O_3$	623-96-1	146.184			168	0.9435^{20}	1.4008^{20}	sl H_2O; msc EtOH, eth
4554	Dipropyl disulfide		$C_6H_{14}S_2$	629-19-6	150.305	liq	-85.6	195.8	0.9599^{20}	1.4981^{20}	
4555	Dipropylene glycol		$C_6H_{14}O_3$	25265-71-8	134.173			230.5	1.0206^{20}		msc H_2O; s EtOH
4556	Dipropylene glycol dibenzoate		$C_{20}H_{22}O_5$	27138-31-4	342.386			197^{1}			
4557	Dipropylene glycol monomethyl ether	1-(2-Methoxyisopropoxy)-2-propanol	$C_7H_{16}O_3$	34590-94-8	148.200	liq	-80	188.3	0.95	1.4190^{20}	
4558	Dipropyl ether	Propyl ether	$C_6H_{14}O$	111-43-3	102.174	liq	-114.8	90.08	0.7466^{20}	1.3809^{20}	sl H_2O; vs eth, EtOH
4559	Dipropyl fumarate		$C_{10}H_{16}O_4$	14595-35-8	200.232			110^{5}	1.0129^{20}	1.4435^{20}	s EtOH, eth
4560	Dipropyl maleate		$C_{10}H_{16}O_4$	2432-63-5	200.232			126^{12}	1.0245^{20}	1.4434^{20}	i H_2O; s EtOH, eth, ace, bz
4561	Dipropyl oxalate		$C_8H_{14}O_4$	615-98-5	174.195	liq	-44.3	211	1.0188^{20}	1.4158^{20}	sl H_2O; msc EtOH; s eth
4562	5,5-Dipropyl-2,4-oxazolidinedione		$C_9H_{15}NO_3$	512-12-9	185.220		42.5	149^{3}			
4563	Dipropyl succinate		$C_{10}H_{18}O_4$	925-15-5	202.248	liq	-5.9	250.8	1.0020^{20}	1.4250^{20}	vs ace, bz, eth
4564	Dipropyl sulfate		$C_6H_{14}O_4S$	598-05-0	182.238			120^{20}	1.1064^{20}	1.4135^{20}	vs peth
4565	Dipropyl sulfide		$C_6H_{14}S$	111-47-7	118.240	liq	-102.5	142.9	0.814^{17}	1.4487^{20}	i H_2O; s EtOH, eth
4566	Dipropyl sulfone		$C_6H_{14}O_2S$	598-03-8	150.239	cry	29.5		1.0278^{50}	1.4456^{30}	sl H_2O; s EtOH, eth
4567	Dipropyl sulfoxide		$C_6H_{14}OS$	4253-91-2	134.239	nd	22.5	80^{2}	0.9654^{20}	1.4663^{20}	vs eth, EtOH
4568	Dipyridamole		$C_{24}H_{40}N_8O_4$	58-32-2	504.627		163				
4569	Di-2-pyridinyl disulfide, *N,N'*-dioxide	Dipyrithione	$C_{10}H_8N_2O_2S_2$	3696-28-4	252.313	cry (MeOH)	205				
4570	2,2'-Dipyrrolylmethane		$C_9H_{10}N_2$	21211-65-4	146.188	lf or nd (al)	73	164^{12}			vs bz, eth, EtOH
4571	Diquat		$C_{12}H_{12}N_2$	2764-72-9	184.236	Cation					
4572	Diquat dibromide		$C_{12}H_{12}Br_2N_2$	85-00-7	344.044		337		1.24^{20}		
4573	Disodium calcium EDTA	Edetate calcium disodium	$C_{10}H_{12}CaN_2Na_2O_8$	62-33-9	374.268	pow					s H_2O
4574	Disodium hydrogen citrate	Sodium acid citrate	$C_6H_6Na_2O_7$	144-33-2	236.088	wh pow (w)	149 dec				vs H_2O
4575	Disperse Blue No. 1	1,4,5,8-Tetraamino-9,10-anthracenedione	$C_{14}H_{12}N_4O_2$	2475-45-8	268.271	red-br nd	331				
4576	Distearyl thiodipropionate	Dioctadecyl thiobispropanoate	$C_{42}H_{82}O_4S$	693-36-7	683.163	cry	61				
4577	Disulfiram		$C_{10}H_{20}N_2S_4$	97-77-8	296.539		71.5	117^{17}			i H_2O; s EtOH; sl eth; vs chl
4578	Disulfoton		$C_8H_{19}O_2PS_3$	298-04-4	274.405		-25	$108^{0.01}$, 128^{1}	1.144^{20}		
4579	1,2-Dithiane		$C_4H_8S_2$	505-20-4	120.237	nd	32.5	80^{14}, 60^{5}		1.5981^{25}	s eth, bz, chl
4580	1,3-Dithiane		$C_4H_8S_2$	505-23-7	120.237		54	89^{14}		1.5981^{25}	vs bz, eth, chl
4581	1,4-Dithiane		$C_4H_8S_2$	505-29-3	120.237	mcl pr	112.3	199.5			sl H_2O; s EtOH, eth, ctc, CS_2, HOAc
4582	Dithianone		$C_{14}H_4N_2O_2S_2$	3347-22-6	296.324	nd (ace)	220				
4583	Dithiazanine iodide		$C_{23}H_{23}IN_2S_2$	514-73-8	518.476	grn nd (MeOH)	248 dec				i H_2O
4584	2,2'-Dithiobisbenzoic acid	Diphenyl disulfide-2,2'-dicarboxylic acid	$C_{14}H_{10}O_4S_2$	119-80-2	306.357		289.5				i H_2O; s EtOH, eth
4585	3,3'-Dithiobispropanoic acid		$C_6H_{10}O_4S_2$	1119-62-6	210.271		158				
4586	3,3'-Dithiobis-*D*-valine		$C_{10}H_{20}N_2O_4S_2$	20902-45-8	296.407		204.5				

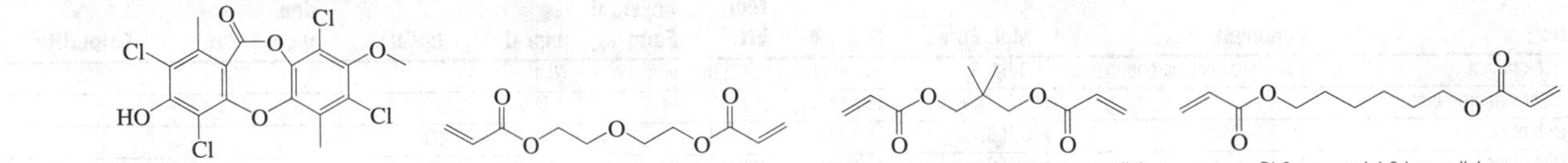

Diploicin

Di-2-propenoyldiethyleneglycol

Di-2-propenoyl-2,2-dimethyl-1,3-propanediol

Di-2-propenoyl-1,6-hexanediol

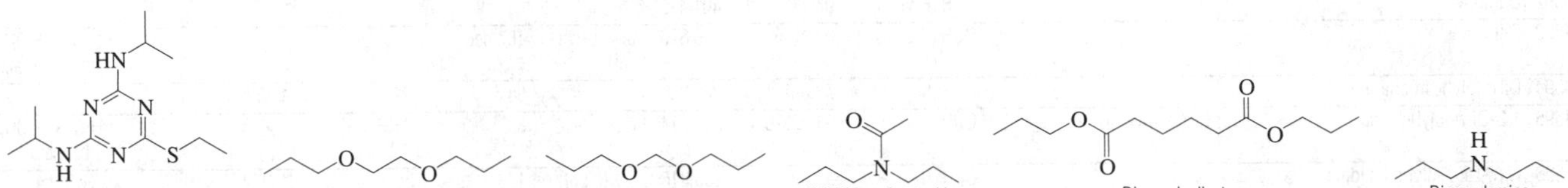

Dipropetryn

1,2-Dipropoxyethane

Dipropoxymethane

N,N-Dipropylacetamide

Dipropyl adipate

Dipropylamine

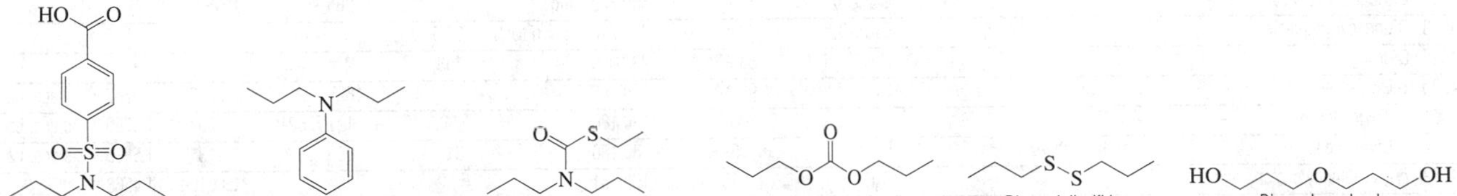

4-[(Dipropylamino)sulfonyl]benzoic acid

N,N-Dipropylaniline

Dipropylcarbamothioic acid, *S*-ethyl ester

Dipropyl carbonate

Dipropyl disulfide

Dipropylene glycol

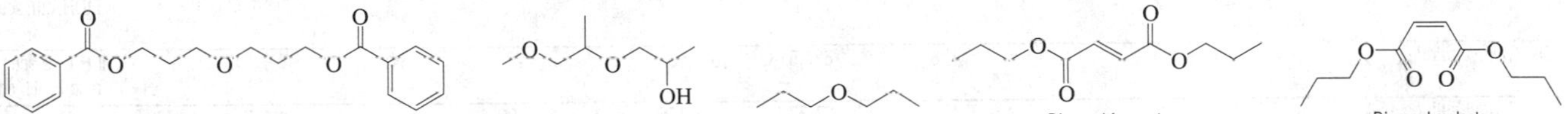

Dipropylene glycol dibenzoate

Dipropylene glycol monomethyl ether

Dipropyl ether

Dipropyl fumarate

Dipropyl maleate

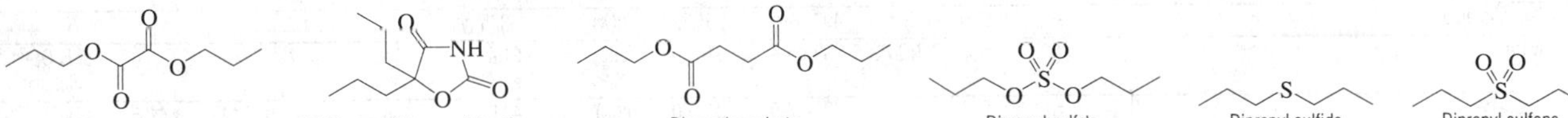

Dipropyl oxalate

5,5-Dipropyl-2,4-oxazolidinedione

Dipropyl succinate

Dipropyl sulfate

Dipropyl sulfide

Dipropyl sulfone

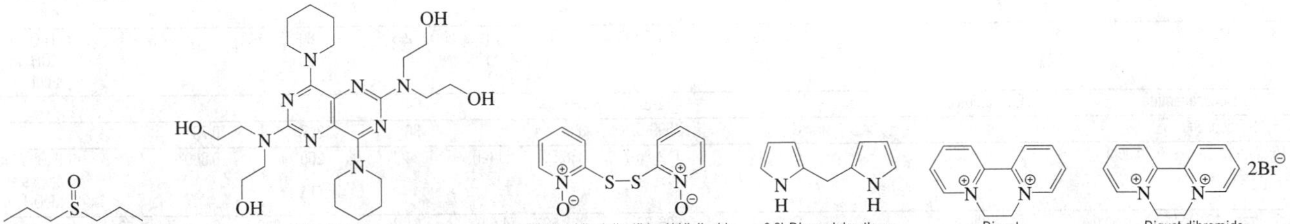

Dipropyl sulfoxide

Dipyridamole

Di-2-pyridinyl disulfide, *N,N'*-dioxide

2,2'-Dipyrrolylmethane

Diquat

Diquat dibromide

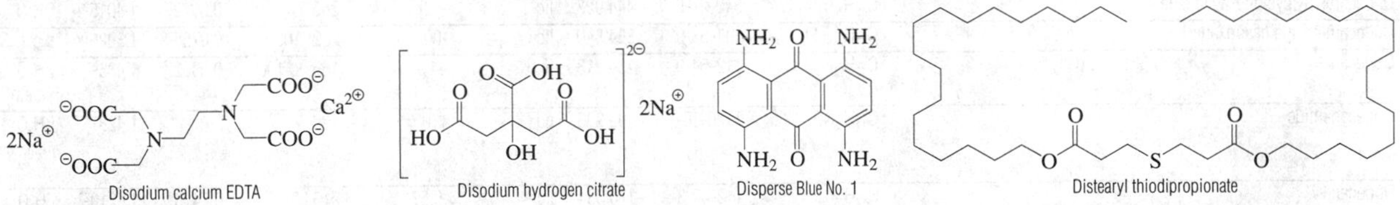

Disodium calcium EDTA

Disodium hydrogen citrate

Disperse Blue No. 1

Distearyl thiodipropionate

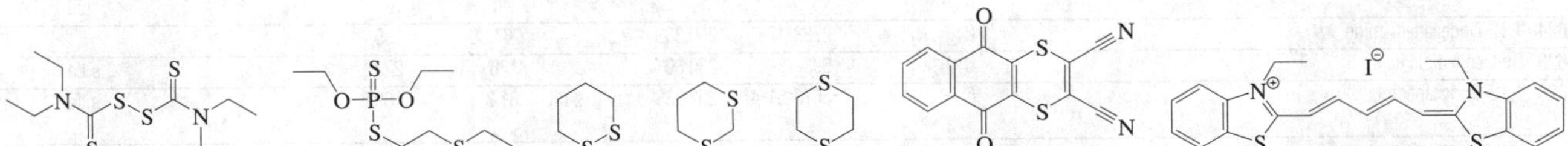

Disulfiram

Disulfoton

1,2-Dithiane

1,3-Dithiane

1,4-Dithiane

Dithianone

Dithiazanine iodide

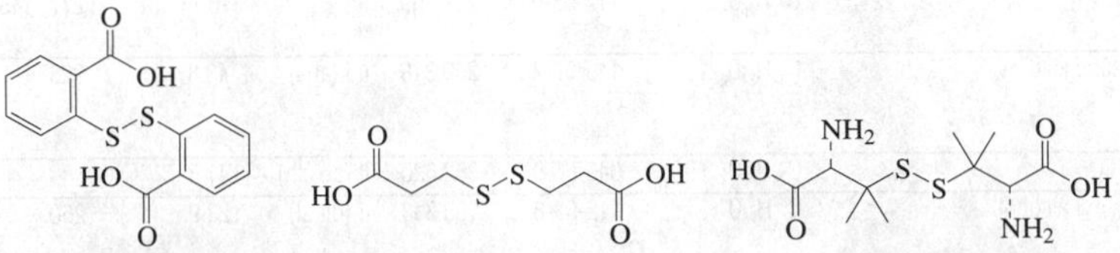

2,2'-Dithiobisbenzoic acid

3,3'-Dithiobispropanoic acid

3,3'-Dithiobis-*D*-valine

No.	Name	Synonym	Mol. Form.	CAS RN	Mol. Wt.	Physical Form	mp/°C	bp/°C	den/ g cm-3	n_D	Solubility
4587	2,5-Dithiobiurea	1,2-Hydrazinedicarbothioamide	$C_2H_6N_4S_2$	142-46-1	150.226	nd (w)	214				
4588	4,4'-Dithiodimorpholine		$C_8H_{16}N_2O_2S_2$	103-34-4	236.355		124.5				s chl
4589	1,2-Dithiolane		$C_3H_6S_2$	557-22-2	106.210		77	90^{27}			
4590	1,3-Dithiolane	1,3-Dithiacyclopentane	$C_3H_6S_2$	4829-04-3	106.210	liq	-50	175	1.259^{17}	1.5975^{15}	s EtOH, eth, xyl
4591	1,3-Dithiolane-2-thione		$C_3H_4S_3$	822-38-8	136.259		35	307			
4592	Dithiopyr		$C_{15}H_{16}F_5NO_2S_2$	97886-45-8	401.416		65				
4593	Dithizone		$C_{13}H_{12}N_4S$	60-10-6	256.326	bl-blk (chl-al)	167 dec				i H_2O; sl EtOH, eth; s chl, alk
4594	Di(*p*-tolyl)carbodiimide		$C_{15}H_{14}N_2$	726-42-1	222.285		58.5	221^{20}	1.1500^{20}		
4595	1,2-Di(*p*-tolyl)ethane	1,2-Bis(*p*-tolyl)ethane	$C_{16}H_{18}$	538-39-6	210.314	lf (al)	85	178^{18}			i H_2O; sl EtOH; s bz, peth
4596	*N,N'*-Di(*o*-tolyl)guanidine		$C_{15}H_{17}N_3$	97-39-2	239.316	cry (dil al)	179		1.10^{20}		sl H_2O, tfa, EtOH; vs eth; s chl
4597	Ditridecyl phthalate		$C_{34}H_{58}O_4$	119-06-2	530.823	liq		$285^{3.5}$	0.952^{25}		
4598	Diundecyl phthalate		$C_{30}H_{50}O_4$	3648-20-2	474.716	cry (EtOH)	35.5				
4599	Diuron		$C_9H_{10}Cl_2N_2O$	330-54-1	233.093		158				
4600	*o*-Divinylbenzene	1,2-Divinylbenzene	$C_{10}H_{10}$	91-14-5	130.186			82^{14}	0.9325^{22}	1.5767^{20}	s ace, bz
4601	*m*-Divinylbenzene	1,3-Divinylbenzene	$C_{10}H_{10}$	108-57-6	130.186		-52.3	121^{76}, 52^{3}	0.9294^{20}	1.5760^{20}	s ace, bz
4602	*p*-Divinylbenzene	1,4-Divinylbenzene	$C_{10}H_{10}$	105-06-6	130.186		31	95^{18}, $34^{0.2}$	0.913^{40}	1.5835^{25}	s ace, bz
4603	*cis*-1,2-Divinylcyclobutane		C_8H_{12}	16177-46-1	108.181			38^{38}	0.8010^{20}	1.4563^{20}	
4604	*trans*-1,2-Divinylcyclobutane		C_8H_{12}	6553-48-6	108.181			112.5	0.7817^{20}	1.4451^{20}	
4605	Divinyl ether		C_4H_6O	109-93-3	70.090	vol liq or gas	-100.6	28.3	0.773^{20}	1.3989^{20}	i H_2O; msc EtOH, eth, ace, chl
4606	Divinyl sulfide	Vinyl sulfide	C_4H_6S	627-51-0	86.156		20	84	0.9174^{15}		sl H_2O; s ace; msc EtOH, eth
4607	Divinyl sulfone	Vinyl sulfone	$C_4H_6O_2S$	77-77-0	118.155	liq	-26	234.5	1.177^{25}	1.4765^{20}	
4608	1,3-Divinyl-1,1,3,3-tetramethyldisiloxane		$C_8H_{18}OSi_2$	2627-95-4	186.399	liq	-99.7	39	0.811^{20}	1.4123^{20}	
4609	Djenkolic acid		$C_7H_{14}N_2O_4S_2$	498-59-9	254.327	nd(w)	≈325 dec				
4610	DMPA		$C_{10}H_{14}Cl_2NO_2PS$	299-85-4	314.169	solid	51.4	150^{2}			sl H_2O; vs bz, ctc, ace
4611	Docosane		$C_{22}H_{46}$	629-97-0	310.600	pl(to), cry (eth)	43.6	368.6	0.7944^{20}	1.4455^{20}	i H_2O; s EtOH, chl; vs eth
4612	Docosanoic acid	Behenic acid	$C_{22}H_{44}O_2$	112-85-6	340.583	nd	81.5	306^{60}	0.8223^{90}	1.4270^{100}	sl H_2O, EtOH, eth
4613	1-Docosanol		$C_{22}H_{46}O$	661-19-8	326.599	cry (ace, chl)	72.5	$180^{0.22}$			sl H_2O, eth; vs EtOH, MeOH; s chl
4614	13-Docosenamide	Erucamide	$C_{22}H_{43}NO$	112-84-5	337.582	cry	94				
4615	1-Docosene		$C_{22}H_{44}$	1599-67-3	308.584		38	367	0.794^{25}		
4616	*cis*-13-Docosenoic acid	Erucic acid	$C_{22}H_{42}O_2$	112-86-7	338.567	nd (al)	34.7	265^{15}	0.860^{55}	1.4758^{20}	i H_2O; s EtOH, ctc; vs eth, MeOH
4617	*trans*-13-Docosenoic acid	Brassidic acid	$C_{22}H_{42}O_2$	506-33-2	338.567	pl (al)	61.9	282^{30}, 256^{10}	0.8585^{57}	1.4347^{100}	
4618	5,7-Dodecadiyne	Dibutylbutadiyne	$C_{12}H_{18}$	1120-29-2	162.271			103^{8}			
4619	Dodecamethylcyclohexasiloxane		$C_{12}H_{36}O_6Si_6$	540-97-6	444.923	liq	-1.5	245	0.9672^{25}	1.4015^{20}	i H_2O
4620	Dodecamethylpentasiloxane		$C_{12}H_{36}O_4Si_5$	141-63-9	384.840	liq	-80	232; 105^{12}	0.8755^{20}	1.3925^{20}	s ctc, CS_2
4621	Dodecanal	Lauraldehyde	$C_{12}H_{24}O$	112-54-9	184.318	lf	44.5	185^{100}, $100^{2.5}$	0.8352^{15}	1.435^{22}	i H_2O; sl EtOH; s eth
4622	Dodecanamide		$C_{12}H_{25}NO$	1120-16-7	199.333	nd	110	199^{12}		1.4287^{110}	i H_2O; s EtOH, ace, ctc; sl eth, bz
4623	Dodecane		$C_{12}H_{26}$	112-40-3	170.334	liq	-9.57	216.32	0.7495^{20}	1.4210^{20}	i H_2O; vs EtOH, eth, ace, ctc, chl
4624	1,12-Dodecanediamine		$C_{12}H_{28}N_2$	2783-17-7	200.363		67.38	135^{3}			
4625	Dodecanedioic acid		$C_{12}H_{22}O_4$	693-23-2	230.301		128	222^{25}	1.15^{25}		s tfa
4626	1,12-Dodecanediol		$C_{12}H_{26}O_2$	5675-51-4	202.333	cry (bz, dil al)	81.3	189^{12}			s tfa
4627	Dodecanenitrile	Lauronitrile	$C_{12}H_{23}N$	2437-25-4	181.318		4	277; 198^{100}	0.8240^{20}	1.4361^{20}	i H_2O; msc EtOH, eth, ace, bz, chl
4628	1-Dodecanethiol		$C_{12}H_{26}S$	112-55-0	202.399	liq	-6.7	277; 143^{15}	0.844^{20}	1.4589^{20}	i H_2O; s EtOH, eth, chl
4629	Dodecanoic acid	Lauric acid	$C_{12}H_{24}O_2$	143-07-7	200.318	nd (al)	43.8	225^{100}	0.8679^{50}	1.4183^{82}	i H_2O; vs EtOH, eth; s ace; msc bz
4630	Dodecanoic anhydride		$C_{24}H_{46}O_3$	645-66-9	382.620	lf (al, eth)	41.8		0.8533^{70}	1.4292^{70}	vs EtOH
4631	1-Dodecanol	Lauryl alcohol	$C_{12}H_{26}O$	112-53-8	186.333	lf (dil al)	23.9	260	0.8309^{24}		i H_2O; s EtOH, eth; sl bz
4632	2-Dodecanol		$C_{12}H_{26}O$	10203-28-8	186.333		19	252	0.8286^{20}	1.4400^{20}	

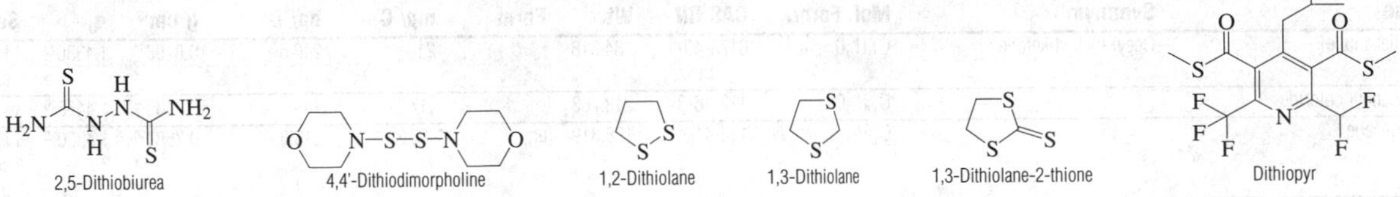

2,5-Dithiobiurea

4,4′-Dithiodimorpholine

1,2-Dithiolane

1,3-Dithiolane

1,3-Dithiolane-2-thione

Dithiopyr

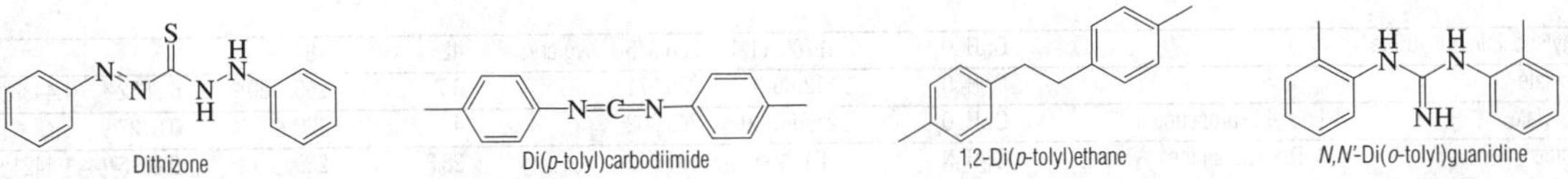

Dithizone

Di(*p*-tolyl)carbodiimide

1,2-Di(*p*-tolyl)ethane

N,N′-Di(*o*-tolyl)guanidine

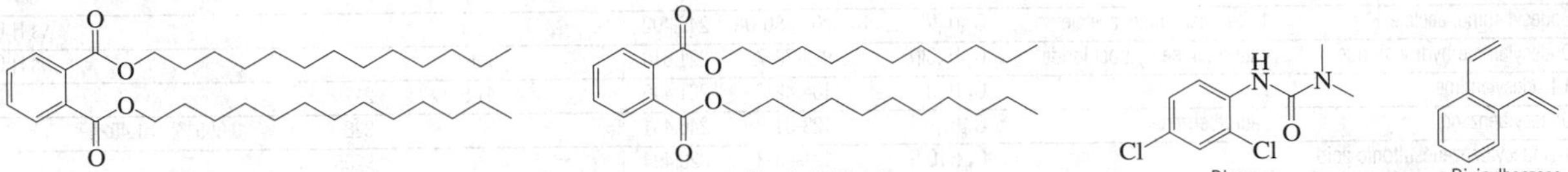

Ditridecyl phthalate

Diundecyl phthalate

Diuron

o-Divinylbenzene

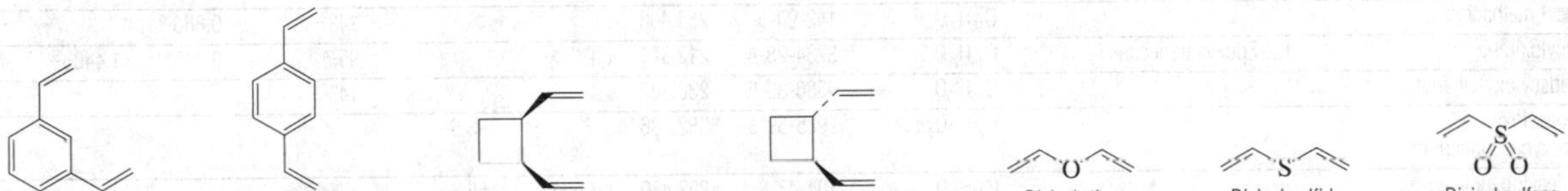

m-Divinylbenzene

p-Divinylbenzene

cis-1,2-Divinylcyclobutane

trans-1,2-Divinylcyclobutane

Divinyl ether

Divinyl sulfide

Divinyl sulfone

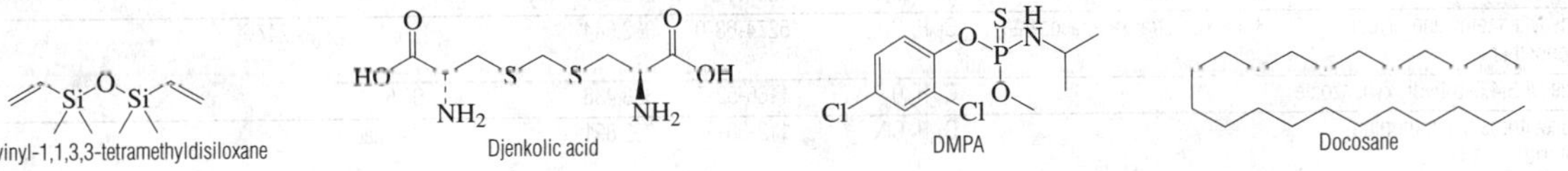

1,3-Divinyl-1,1,3,3-tetramethyldisiloxane

Djenkolic acid

DMPA

Docosane

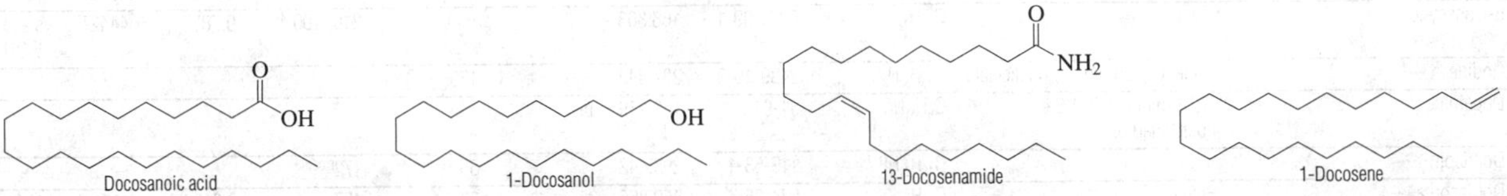

Docosanoic acid

1-Docosanol

13-Docosenamide

1-Docosene

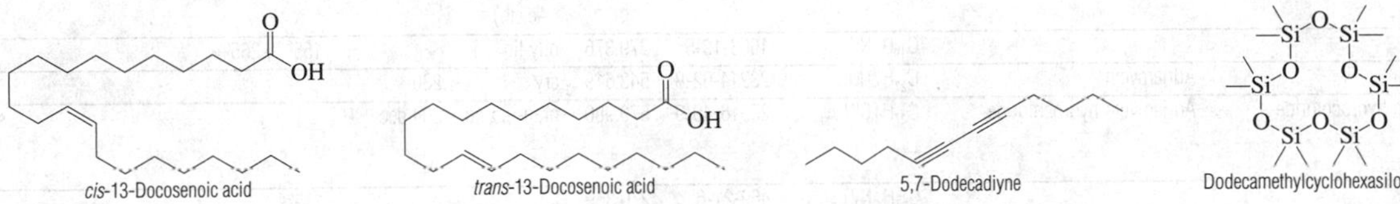

cis-13-Docosenoic acid

trans-13-Docosenoic acid

5,7-Dodecadiyne

Dodecamethylcyclohexasiloxane

Dodecamethylpentasiloxane

Dodecanal

Dodecanamide

Dodecane

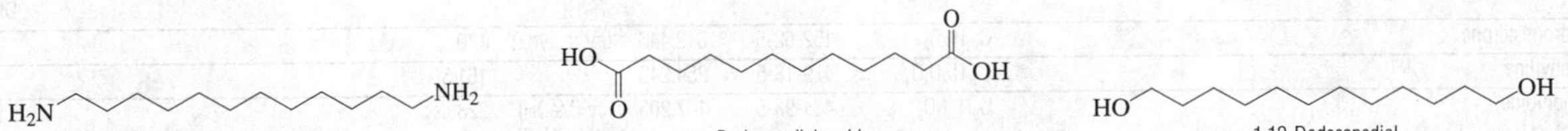

1,12-Dodecanediamine

Dodecanedioic acid

1,12-Dodecanediol

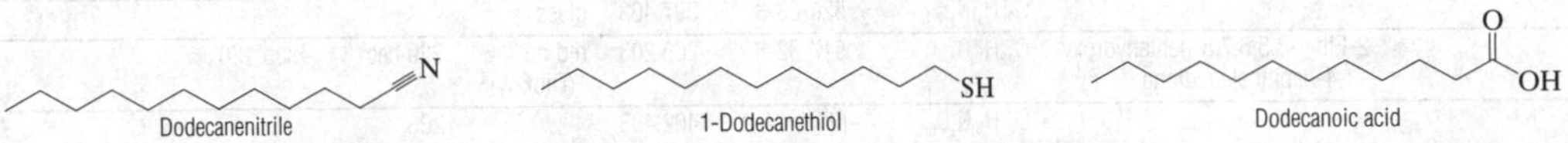

Dodecanenitrile

1-Dodecanethiol

Dodecanoic acid

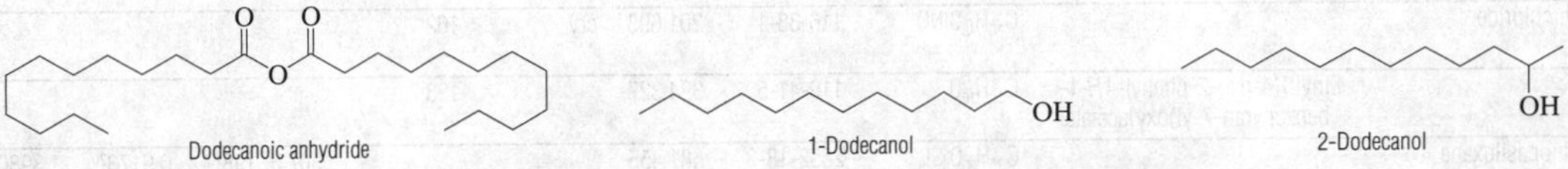

Dodecanoic anhydride

1-Dodecanol

2-Dodecanol

No.	Name	Synonym	Mol. Form.	CAS RN	Mol. Wt.	Physical Form	mp/°C	bp/°C	den/ g cm^{-3}	n_D	Solubility
4633	2-Dodecanone	Decyl methyl ketone	$C_{12}H_{24}O$	6175-49-1	184.318		21	246.5	0.8198^{20}	1.4330^{20}	i H_2O; s EtOH, eth, ace; sl ctc
4634	Dodecanoyl chloride		$C_{12}H_{23}ClO$	112-16-3	218.763		-17	145^{18}	0.9169^{25}	1.4458^{20}	vs eth
4635	1-Dodecene		$C_{12}H_{24}$	112-41-4	168.319	liq	-35.2	213.8	0.7584^{20}	1.4300^{20}	i H_2O; s EtOH, eth, ace, ctc, peth
4636	*trans*-2-Dodecenedioic acid	Traumatic acid	$C_{12}H_{20}O_4$	6402-36-4	228.285	cry (al,ace)	165.5				vs eth, EtOH, chl
4637	2-Dodecenylsuccinic anhydride		$C_{16}H_{26}O_3$	19780-11-1	266.375	hyg cry	42	181^5			
4638	Dodecyl acetate		$C_{14}H_{28}O_2$	112-66-3	228.371		0.7	265; 180^{40}	0.8652^{22}	1.4439^{20}	
4639	Dodecyl acrylate	Lauryl 2-propenoate	$C_{15}H_{28}O_2$	2156-97-0	240.382		4	$120^{0.8}$	0.8727^{20}		
4640	Dodecylamine	1-Dodecanamine	$C_{12}H_{27}N$	124-22-1	185.349		28.3	259	0.8015^{20}	1.4421^{20}	sl H_2O; msc EtOH, eth, bz, chl
4641	Dodecylamine, acetate	1-Dodecanamine, acetate	$C_{14}H_{31}NO_2$	2016-56-0	245.402		69.5				vs H_2O, EtOH
4642	Dodecylamine hydrochloride	Lauryl amine hydrochloride	$C_{12}H_{28}ClN$	929-73-7	221.810		186				vs H_2O, EtOH
4643	4-Dodecylaniline		$C_{18}H_{31}N$	104-42-7	261.446		41.5	211^{10}			
4644	Dodecylbenzene	Laurylbenzene	$C_{18}H_{30}$	123-01-3	246.431		3	328	0.8551^{20}	1.4824^{20}	i H_2O
4645	4-Dodecylbenzenesulfonic acid		$C_{18}H_{30}O_3S$	121-65-3	326.494			>205			
4646	Dodecylcyclohexane		$C_{18}H_{36}$	1795-17-1	252.479		12.5	331	0.8223^{20}	1.4559^{20}	
4647	Dodecyl mercaptoacetate		$C_{14}H_{28}O_2S$	3746-39-2	260.436		1.5	171^3			
4648	Dodecyl methacrylate		$C_{16}H_{30}O_2$	142-90-5	254.408			142^4	0.866^{20}		
4649	Dodecyloxirane	1,2-Epoxytetradecane	$C_{14}H_{28}O$	3234-28-4	212.371	oil		$95^{0.4}$	0.845	1.4408^{20}	
4650	2-(Dodecyloxy)ethanol		$C_{14}H_{30}O_2$	4536-30-5	230.387			$143^{0.8}$			
4651	4-Dodecyloxy-2-hydroxybenzophenone		$C_{25}H_{34}O_3$	2985-59-3	382.536		43.5				
4652	4-Dodecylphenol		$C_{18}H_{30}O$	104-43-8	262.430	nd (bz)	66	175^2			
4653	1-Dodecylpiperidine		$C_{17}H_{35}N$	5917-47-5	253.467	pa ye		161^5, $115^{0.6}$	0.8378^{20}	1.4588^{20}	
4654	Dodecyl sulfate	Lauryl sulfate	$C_{12}H_{26}O_4S$	151-41-7	266.397	cry					s H_2O
4655	Dodecyltetraethylene glycol monoether	3,6,9,12-Tetraoxatetracosan-1-ol	$C_{20}H_{42}O_5$	5274-68-0	362.544			247^{10}			
4656	Dodecyl 3,4,5-trihydroxybenzoate		$C_{19}H_{30}O_5$	1166-52-5	338.438		96.5				s ace
4657	Dodecyltrimethylammonium chloride		$C_{15}H_{34}ClN$	112-00-5	263.891		246 dec				vs H_2O, ace, EtOH, chl
4658	1-Dodecyne	Decylacetylene	$C_{12}H_{22}$	765-03-7	166.303	liq	-19	215	0.7788^{20}	1.4340^{20}	
4659	6-Dodecyne		$C_{12}H_{22}$	6975-99-1	166.303			210; 100^{14}	0.785^{20}	1.4442^{20}	vs ace, eth, EtOH
4660	Dodine	Dodecylguanidine, monoacetate	$C_{15}H_{33}N_3O_2$	2439-10-3	287.442		136				
4661	Dopamine	4(2-Aminoethyl)-1,2-benzenediol	$C_8H_{11}NO_2$	51-61-6	153.179	pr					
4662	Dothiepin		$C_{19}H_{21}NS$	113-53-1	295.442		56	$172^{0.05}$			
4663	Dotriacontane	Bicetyl	$C_{32}H_{66}$	544-85-4	450.866	pl (bz,chl,HOAc,eth)	69.4	467	0.8124^{20}	1.4550^{20}	i H_2O; sl EtOH, chl; s eth, ctc; vs bz
4664	Doxepin		$C_{19}H_{21}NO$	1668-19-5	279.376	oily liq		$155^{0.03}$, $265^{0.2}$			
4665	Doxorubicin	Adriamycin	$C_{27}H_{29}NO_{11}$	23214-92-8	543.519	cry	230				
4666	Doxorubicin hydrochloride	Adriamycin hydrochloride	$C_{27}H_{30}ClNO_{11}$	25316-40-9	579.980	oran-red nd	204 dec				s H_2O, MeOH; i ace, bz, chl, eth, peth
4667	Doxylamine		$C_{17}H_{22}N_2O$	469-21-6	270.369	liq		$139^{0.5}$			
4668	Drimenin		$C_{15}H_{22}O_2$	2326-89-8	234.335	cry	133	$110^{0.1}$			i H_2O
4669	Dromostanolone propanoate	2-Methyl-17-(1-oxopropoxy)androstan-3-one, (2α,5α,17β)	$C_{23}H_{36}O_3$	521-12-0	360.530		128				
4670	Droperidol	Dehydrobenzperidol	$C_{22}H_{22}FN_3O_2$	548-73-2	379.427	cry (w)	146 (hyd)				i H_2O; sl EtOH, eth, bz; s chl, DMF
4671	Dydrogesterone		$C_{21}H_{28}O_2$	152-62-5	312.446	cry (ace/hx)	170				
4672	Dyphylline		$C_{10}H_{14}N_4O_4$	479-18-5	254.243		161.5				
4673	Ecgonidine		$C_9H_{13}NO_2$	484-93-5	167.205	cry (MeOH) (MeOH-eth)	228 dec				vs H_2O
4674	Ecgonine		$C_9H_{15}NO_3$	481-37-8	185.220	mcl pr	205				vs H_2O, EtOH
4675	Echimidine		$C_{20}H_{31}NO_7$	520-68-3	397.463	glass					
4676	Echinochrome A	2-Ethyl-3,5,6,7,8-pentahydroxy-1,4-naphthalenedione	$C_{12}H_{10}O_7$	517-82-8	266.203	red nd (Diox-w)	220 dec	sub 120			sl H_2O; s EtOH, ace; vs eth, bz
4677	Echitamine		$C_{22}H_{30}N_2O_5$	6871-44-9	402.483		206				s H_2O, EtOH, eth, chl, con sulf; i peth
4678	Edrophonium chloride		$C_{10}H_{16}ClNO$	116-38-1	201.693	cry	162				vs H_2O; s EtOH; i eth, chl
4679	Efloxate	Ethyl [(4-oxo-2-phenyl-4*H*-1-benzopyran-7-yl)oxy]acetate	$C_{19}H_{16}O_5$	119-41-5	324.327		123.7				s chl
4680	Eicosamethylnonasiloxane		$C_{20}H_{60}O_8Si_9$	2652-13-3	681.455			307.5; 198^{16}	0.9173^{20}	1.3980^{20}	vs bz

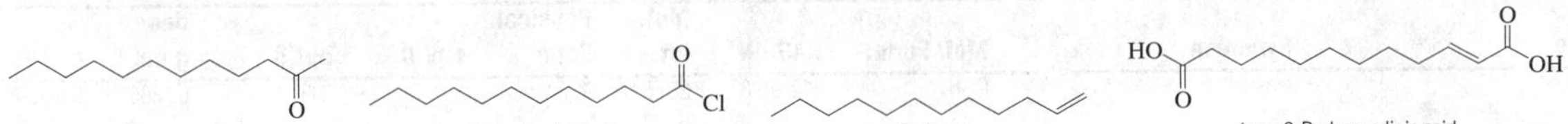

2-Dodecanone

Dodecanoyl chloride

1-Dodecene

trans-2-Dodecenedioic acid

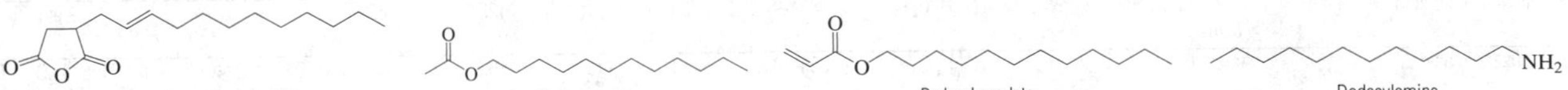

2-Dodecenylsuccinic anhydride

Dodecyl acetate

Dodecyl acrylate

Dodecylamine

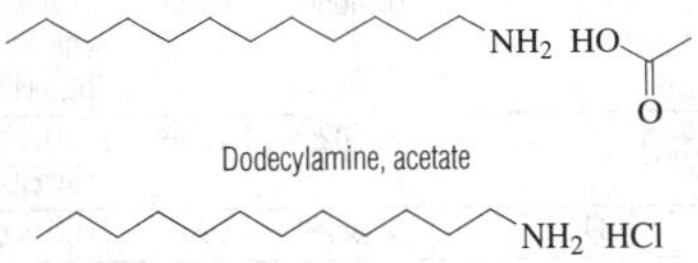

Dodecylamine, acetate

Dodecylamine hydrochloride

4-Dodecylaniline

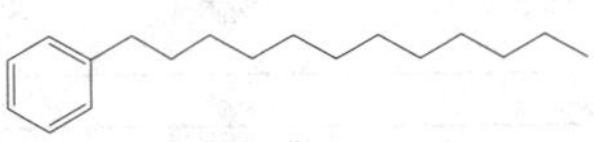

Dodecylbenzene

4-Dodecylbenzenesulfonic acid

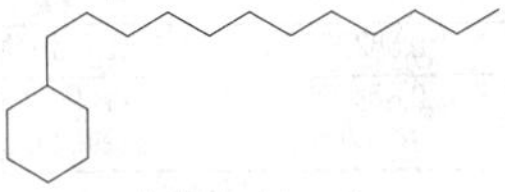

Dodecylcyclohexane

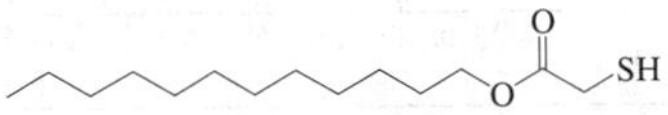

Dodecyl mercaptoacetate

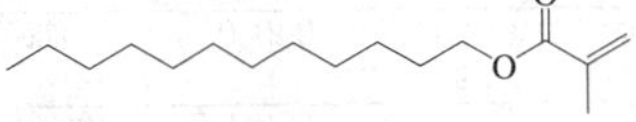

Dodecyl methacrylate

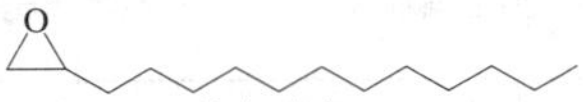

Dodecyloxirane

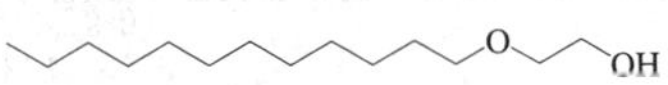

2-(Dodecyloxy)ethanol

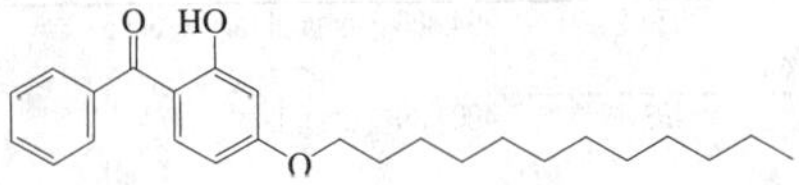

4-Dodecyloxy-2-hydroxybenzophenone

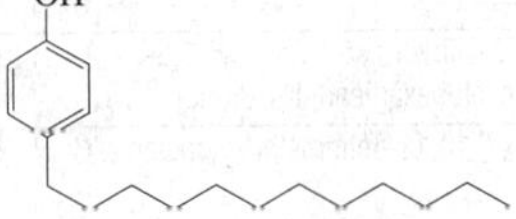

4-Dodecylphenol

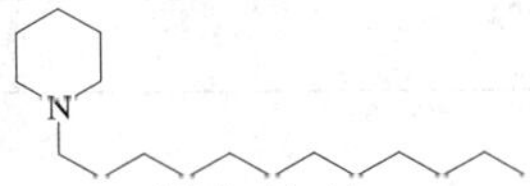

1-Dodecylpiperidine

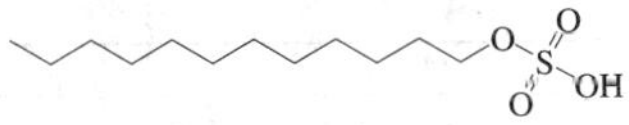

Dodecyl sulfate

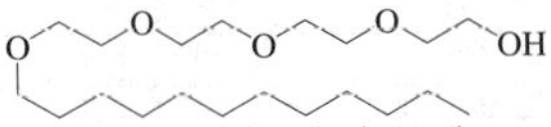

Dodecyltetraethylene glycol monoether

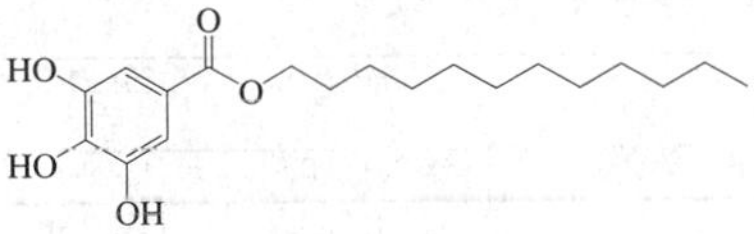

Dodecyl 3,4,5-trihydroxybenzoate

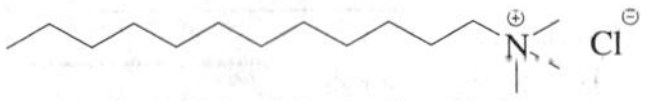

Dodecyltrimethylammonium chloride

1-Dodecyne

6-Dodecyne

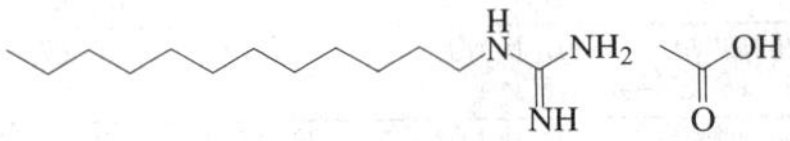

Dodine

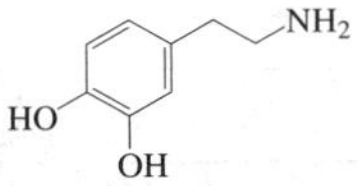

Dopamine

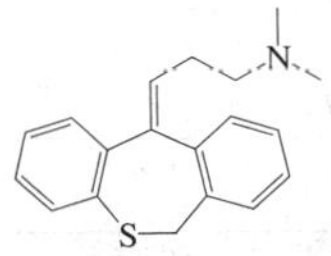

Dothiepin

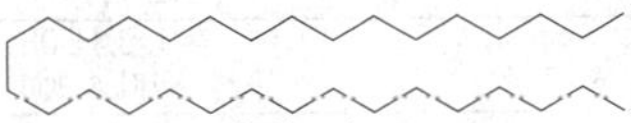

Dotriacontane

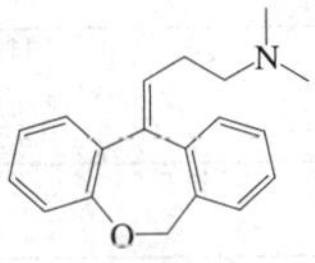

Doxepin

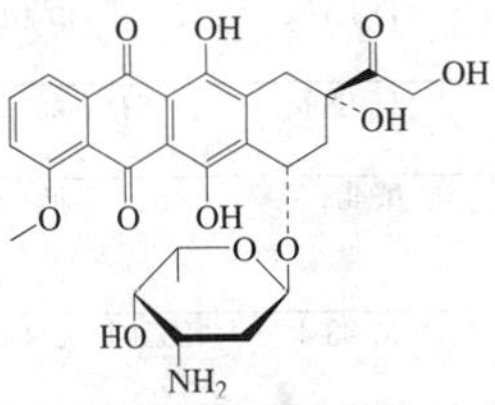

Doxorubicin

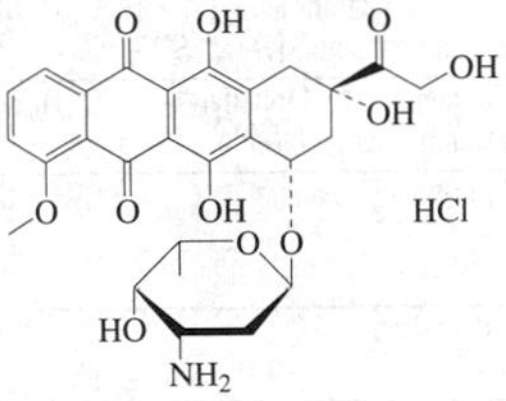

Doxorubicin hydrochloride

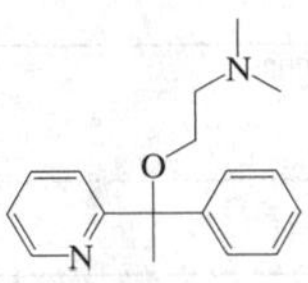

Doxylamine

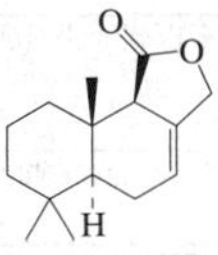

Drimenin

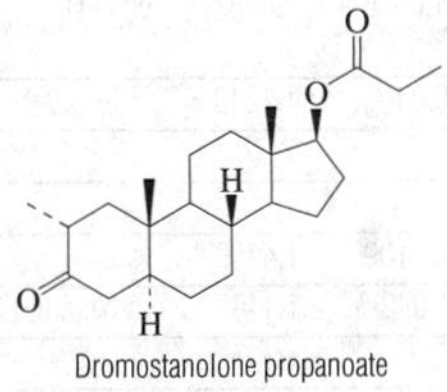

Dromostanolone propanoate

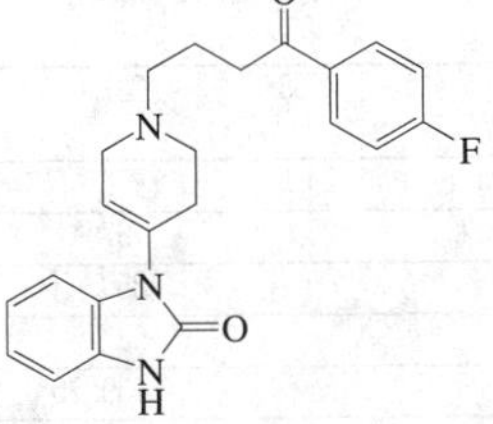

Droperidol

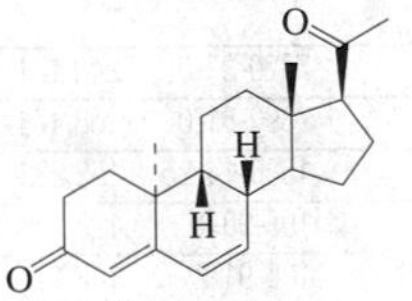

Dydrogesterone

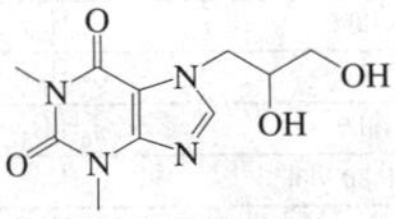

Dyphylline

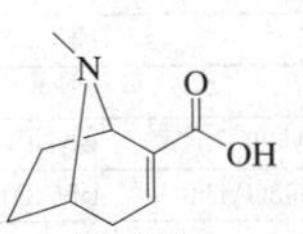

Ecgonidine

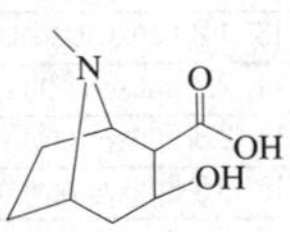

Ecgonine

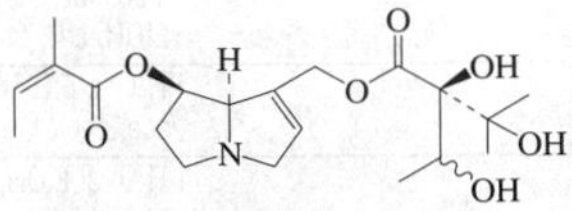

Echimidine

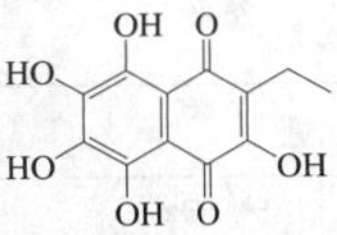

Echinochrome A

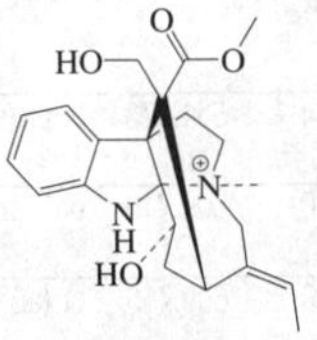

Echitamine

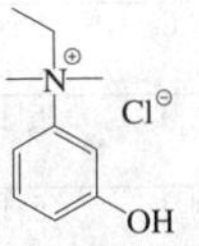

Edrophonium chloride

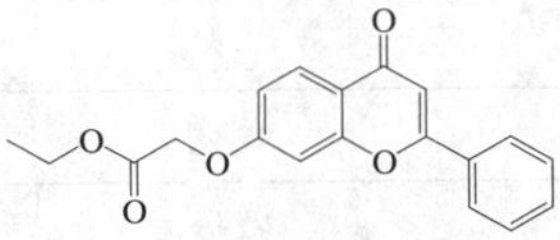

Efloxate

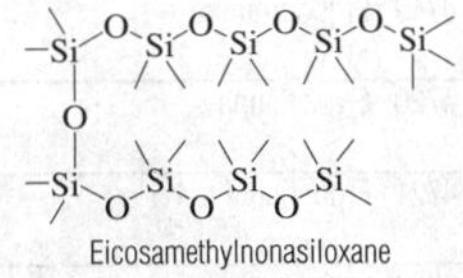

Eicosamethylnonasiloxane

No.	Name	Synonym	Mol. Form.	CAS RN	Mol. Wt.	Physical Form	mp/°C	bp/°C	den/ g cm^{-3}	n_D	Solubility
4681	Eicosane	Icosane	$C_{20}H_{42}$	112-95-8	282.547	lf (al)	36.6	343	0.7886^{20}	1.4425^{20}	i H_2O; s eth, peth, bz; sl chl; vs ace
4682	Eicosanedioic acid	1,18-Octadecanedicarboxylic acid	$C_{20}H_{38}O_4$	2424-92-2	342.514	cry (bz,al)	125.5	233^{2}			s eth
4683	Eicosanoic acid	Arachidic acid	$C_{20}H_{40}O_2$	506-30-9	312.531	pl (al)	76.5	dec 328; 204^{1}	0.8240^{100}	1.425^{100}	i H_2O; sl EtOH; vs eth; s bz, chl
4684	1-Eicosanol	Arachic alcohol	$C_{20}H_{42}O$	629-96-9	298.546	wax (al), cry (chl)	65.4	356; 222^{3}	0.8405^{20}	1.4350^{20}	i H_2O; sl EtOH, chl; vs ace; s bz, peth
4685	5,8,11,14-Eicosatetraenoic acid, (all-*cis*)	Arachidonic acid	$C_{20}H_{32}O_2$	506-32-1	304.467		-49.5	163^{1}	0.9082^{20}	1.4824^{20}	i H_2O; vs ace, eth, EtOH, peth
4686	1-Eicosene		$C_{20}H_{40}$	3452-07-1	280.532		28.5	341; 151^{2}	0.7882^{30}	1.4440^{30}	i H_2O; s bz, peth
4687	*cis*-9-Eicosenoic acid		$C_{20}H_{38}O_2$	29204-02-2	310.515		24.5	220^{6}	0.8882^{25}		
4688	*trans*-9-Eicosenoic acid		$C_{20}H_{38}O_2$	506-31-0	310.515		54				
4689	11-Eicosenoic acid		$C_{20}H_{38}O_2$	2462-94-4	310.515		24	267^{15}	0.8826^{25}		vs EtOH, MeOH
4690	Elaidic acid	*trans*-9-Octadecenoic acid	$C_{18}H_{34}O_2$	112-79-8	282.462	pl (al)	45	288^{100}, 234^{15}	0.8734^{45}	1.4499^{45}	i H_2O; s EtOH, eth, bz, chl
4691	Elaiomycin		$C_{13}H_{26}N_2O_3$	23315-05-1	258.356	ye oil				1.4798^{25}	sl H_2O; s os
4692	1,3-Elemadien-11-ol	Elemol	$C_{15}H_{26}O$	639-99-6	222.366	cry (al)	52.5	142^{12}	0.9345^{18}	1.4980^{18}	
4693	β-Elemene		$C_{15}H_{24}$	33880-83-0	204.352			120^{16}, 104^{11}	0.8749^{20}	1.4935^{20}	
4694	Embelin	2,5-Dihydroxy-3-undecyl-2,5-cyclohexadiene-1,4-dione	$C_{17}H_{26}O_4$	550-24-3	294.386	oran pl (al)	142.5				vs bz, eth, EtOH
4695	Emetine	6',7',10,11-Tetramethoxyemetan	$C_{29}H_{40}N_2O_4$	483-18-1	480.639	amor pow	74				i H_2O; s EtOH, eth, ace; sl bz, chl
4696	Emylcamate	3-Methyl-3-pentanol, carbamate	$C_7H_{15}NO_2$	78-28-4	145.200	nd	57	35^{1}			sl H_2O; vs bz, eth, EtOH
4697	Enallylpropymal		$C_{11}H_{16}N_2O_3$	1861-21-8	224.256	cry (w, dil al)	56.5	177^{12}			vs bz, eth, EtOH, chl
4698	Endosulfan		$C_9H_6Cl_6O_3S$	115-29-7	406.925		106	106$^{0.7}$	1.745^{20}		
4699	Endosulfan sulfate		$C_9H_6Cl_6O_4S$	1031-07-8	422.925	cry (cyhex)	181				
4700	Endothall disodium		$C_8H_8Na_2O_5$	129-67-9	230.13		144		1.431^{20}		
4701	Endrin		$C_{12}H_8Cl_6O$	72-20-8	380.909	cry	dec 245				vs ace, bz, xyl; s ctc, hx
4702	Enflurane		$C_3H_2ClF_5O$	13838-16-9	184.492	liq		56.5	1.5121^{25}	1.3025^{20}	vs os
4703	Ephedrine, (±)	α-[1-(Methylamino)ethyl] benzenemethanol, (*R**,*S**)-(±)-	$C_{10}H_{15}NO$	90-81-3	165.232	nd (eth, peth)	76.5	135^{12}	1.1220^{20}		s H_2O, EtOH, eth, bz, chl
4704	*d*-Ephedrine	α-[1-(Methylamino)ethyl] benzenemethanol, [*S*-(*R**,*S**)]-	$C_{10}H_{15}NO$	321-98-2	165.232	pl (w)	40	225			s H_2O, EtOH, eth, bz, chl
4705	*l*-Ephedrine	α-[1-(Methylamino)ethyl] benzenemethanol, [*R*-(*R**,*S**)]-	$C_{10}H_{15}NO$	299-42-3	165.232	pl (w + 1)	40	225	1.0085^{22}		s H_2O, EtOH, eth, bz, chl
4706	Ephedrine hydrochloride	2-(Methylamino)-1-phenyl-1-propanol, hydrochloride	$C_{10}H_{16}ClNO$	50-98-6	201.693	orth nd	219		1.0208^{20}		
4707	Epichlorohydrin	(Chloromethyl)oxirane	C_3H_5ClO	13403-37-7	92.524	liq	-26	118; 62^{100}	1.1812^{20}	1.4358^{25}	sl H_2O; msc EtOH, eth; s bz, ctc
4708	Epinephrine	*D*-Adrenaline	$C_9H_{13}NO_3$	51-43-4	183.204	br (in air)	211.5				sl H_2O; i EtOH; s HOAc, acid
4709	Epiquinidine		$C_{20}H_{24}N_2O_2$	572-59-8	324.417	cry (AcOEt) lf (eth)	113				vs EtOH; s eth
4710	1,2-Epoxybutane	Ethyloxirane	C_4H_8O	106-88-7	72.106	liq	-150	63.4	0.8297^{20}	1.3851^{20}	vs EtOH, ace; msc eth
4711	1,2-Epoxy-4-(epoxyethyl)cyclohexane	4-Vinyl-1-cyclohexene dioxide	$C_8H_{12}O_2$	106-87-6	140.180		<-55	227	1.0966^{20}	1.4787^{20}	vs H_2O
4712	1,2-Epoxyhexadecane	Tetradecyloxirane	$C_{16}H_{32}O$	7320-37-8	240.424	hyg cry or liq	24.1	178^{12}	0.846	1.2240	
4713	1,2-Epoxyoctadecane	Hexadecyloxirane	$C_{18}H_{36}O$	7390-81-0	268.478	hyg cry	26.1	137$^{0.5}$			
4714	2,3-Epoxy-α-pinane		$C_{10}H_{16}O$	1686-14-2	152.233			85^{24}			
4715	2,3-Epoxypropyl acrylate	Glycidyl acrylate	$C_6H_8O_3$	106-90-1	128.126			53^{10}	1.1109^{20}	1.4490^{20}	vs bz
4716	2,3-Epoxypropyl methacrylate	Glycidol methacrylate	$C_7H_{10}O_3$	106-91-2	142.152			189; 75^{10}	1.042^{20}	1.448^{25}	vs bz, eth, EtOH
4717	Equol		$C_{14}H_{14}O_3$	531-95-3	230.259	cry (aq, al)	189.5				
4718	Ergocornine		$C_{31}H_{39}N_5O_5$	564-36-3	561.673	cry (MeOH)	183 dec				i H_2O; s EtOH, ace, bz, chl, AcOEt
4719	Ergocorninine		$C_{31}H_{39}N_5O_5$	564-37-4	561.673	lo pr (al)	228 dec				vs ace, bz, EtOH, chl
4720	Ergocristine		$C_{35}H_{39}N_5O_5$	511-08-0	609.716	orth (bz)	175 dec				i H_2O; s EtOH, ace, chl
4721	Ergocristinine		$C_{35}H_{39}N_5O_5$	511-07-9	609.716	pr (al)	237 dec				i H_2O; sl EtOH, ace, chl
4722	Ergocryptine		$C_{32}H_{41}N_5O_5$	511-09-1	575.699	pr (al)	213 dec				i H_2O; s EtOH, chl
4723	Ergocryptinine		$C_{32}H_{41}N_5O_5$	511-10-4	575.70	lo pr (al)	245 dec				vs ace, chl

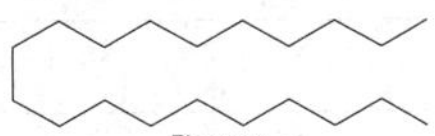

Eicosane

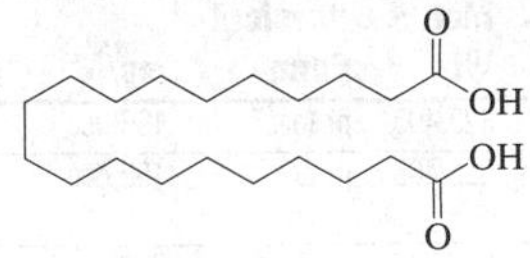

Eicosanedioic acid

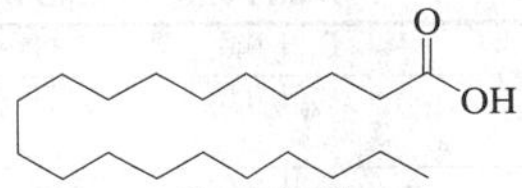

Eicosanoic acid

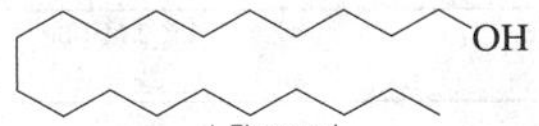

1-Eicosanol

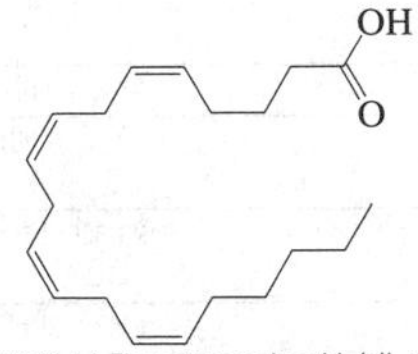

5,8,11,14-Eicosatetraenoic acid, (all-*cis*)

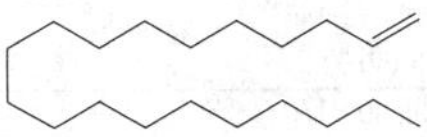

1-Eicosene

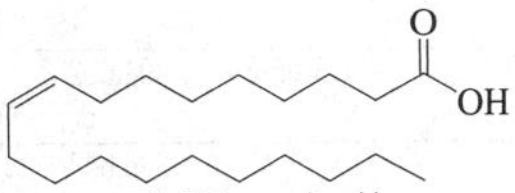

cis-9-Eicosenoic acid

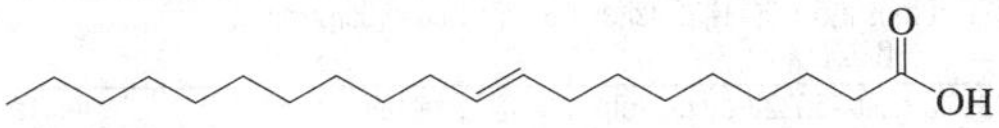

trans-9-Eicosenoic acid

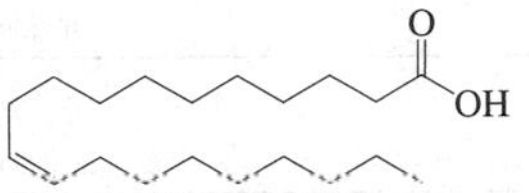

11-Eicosenoic acid

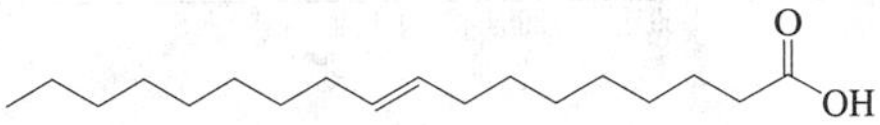

Elaidic acid

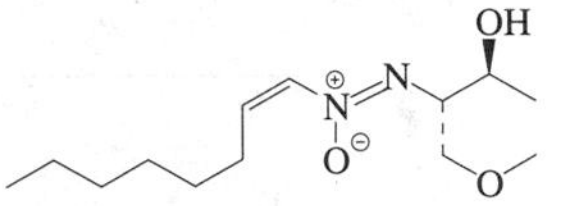

Elaiomycin

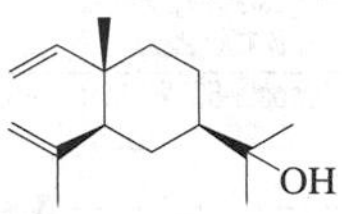

1,3-Elemadien-11-ol

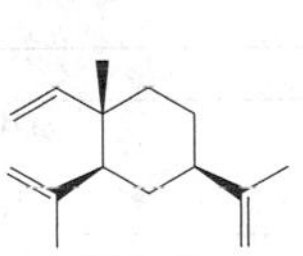

β-Elemene

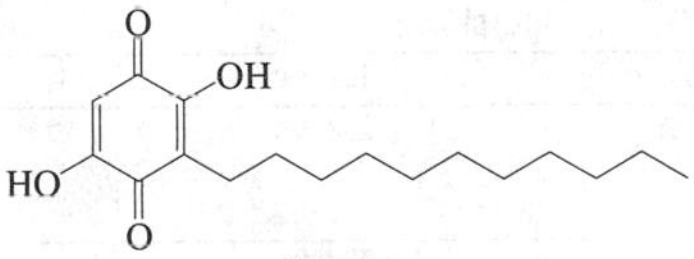

Embelin

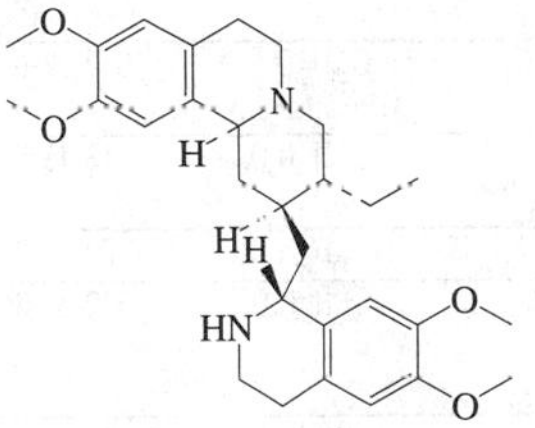

Emetine

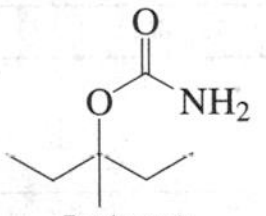

Emylcamate

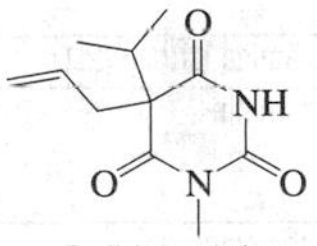

Enallylpropymal

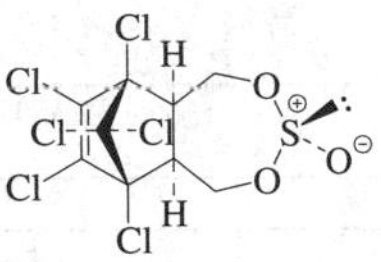

Endosulfan

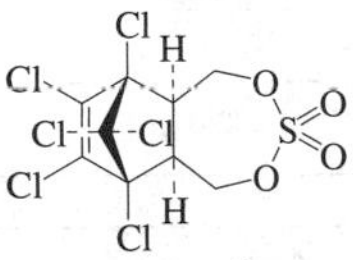

Endosulfan sulfate

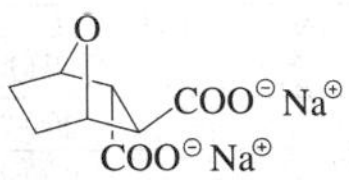

Endothall disodium

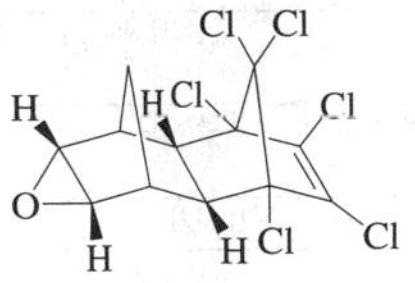

Endrin

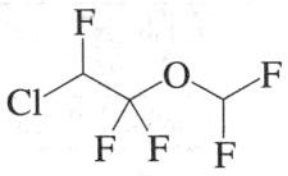

Enflurane

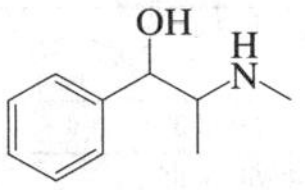

Ephedrine, (±)

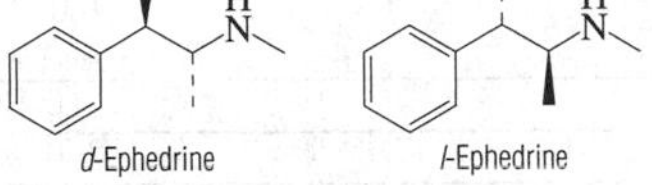

d-Ephedrine

l-Ephedrine

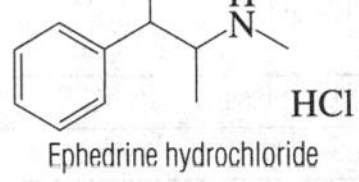

Ephedrine hydrochloride

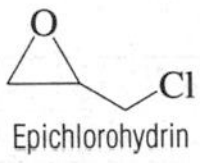

Epichlorohydrin

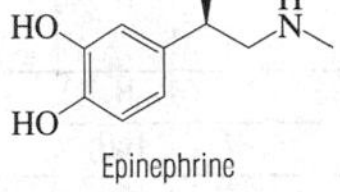

Epinephrine

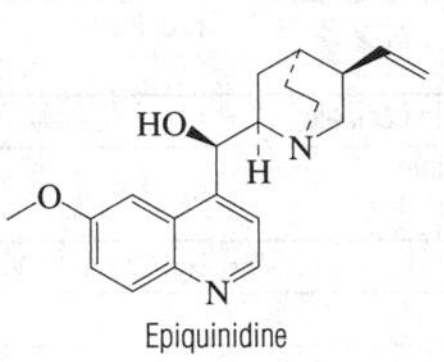

Epiquinidine

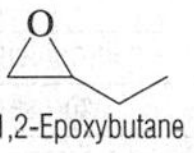

1,2-Epoxybutane

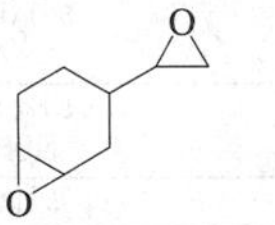

1,2-Epoxy-4-(epoxyethyl)cyclohexane

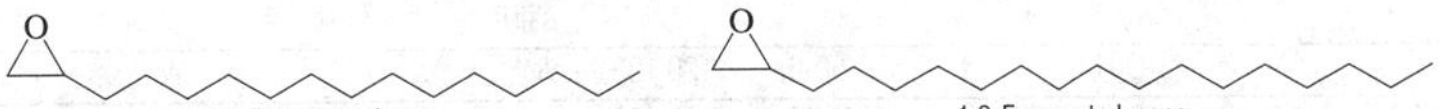

1,2-Epoxyhexadecane

1,2-Epoxyoctadecane

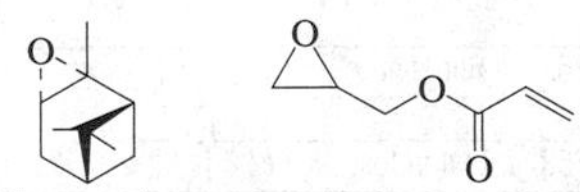

2,3-Epoxy-α-pinane

2,3-Epoxypropyl acrylate

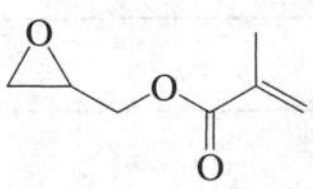

2,3-Epoxypropyl methacrylate

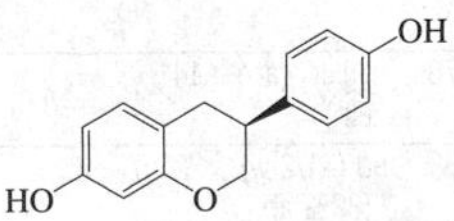

Equol

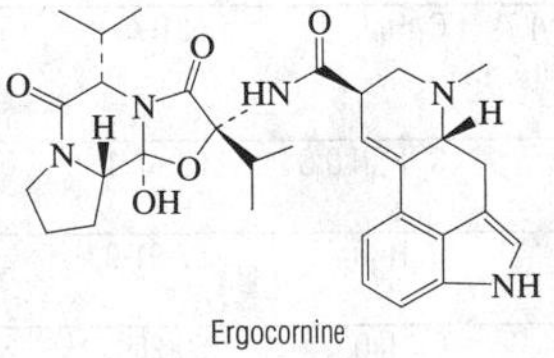

Ergocornine

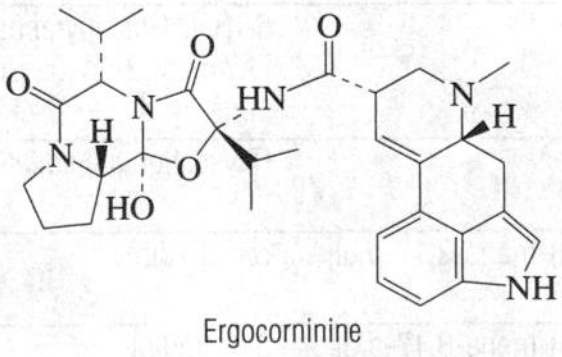

Ergocorninine

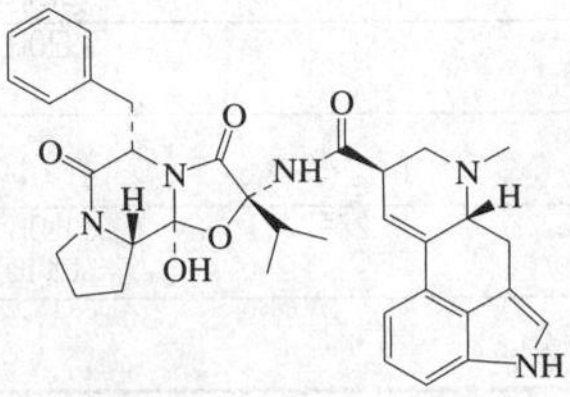

Ergocristine

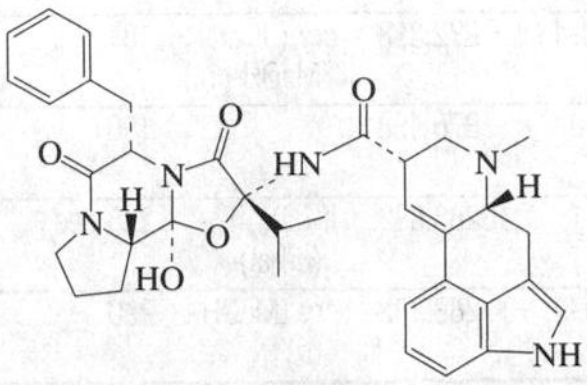

Ergocristinine

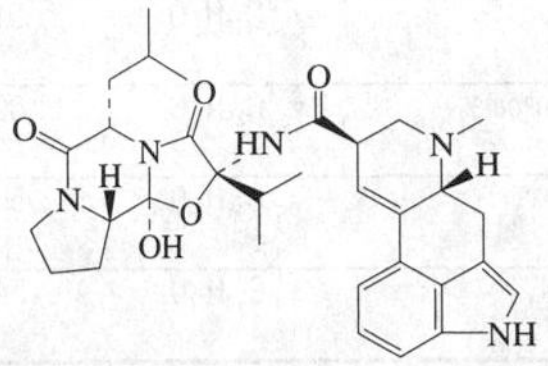

Ergocryptine

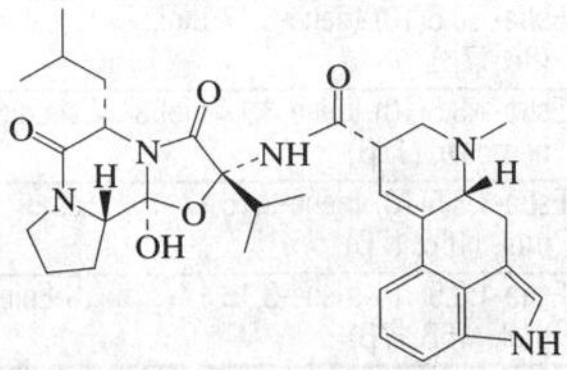

Ergocryptinine

No.	Name	Synonym	Mol. Form.	CAS RN	Mol. Wt.	Physical Form	mp/°C	bp/°C	den/ g cm-3	n_D	Solubility
4724	Ergometrinine		$C_{19}H_{23}N_3O_2$	479-00-5	325.405	pr (ace)	196 dec				vs chl
4725	Ergonovine	Ergometrine	$C_{19}H_{23}N_3O_2$	60-79-7	325.405	pl or nd	162 dec				s H_2O, ace; vs EtOH; sl chl
4726	Ergosine		$C_{30}H_{37}N_5O_5$	561-94-4	547.646	pr (MeOH, AcOEt)	228 dec				s ace, chl; sl MeOH
4727	Ergostane, (5α)		$C_{28}H_{50}$	511-20-6	386.697	lf or pl (ace, eth- MeOH)	85				vs ace, eth, chl
4728	Ergostane, (5β)	Coproergostane	$C_{28}H_{50}$	511-21-7	386.697	nd (ace)	64				vs eth, chl
4729	Ergostan-3-ol, (3β,5α)	Ergostanol	$C_{28}H_{50}O$	6538-02-9	402.696	nd (MeOH-eth)	144.5				i H_2O; s eth, chl
4730	Ergosta-5,7,9(11),22-tetraen-3-ol, (3β,22*E*)	Dehydroergosterol	$C_{28}H_{42}O$	516-85-8	394.632	lf (al) nd (eth),pl (al)	146	$230^{0.5}$			vs ace, bz, eth, EtOH
4731	Ergosta-5,7,22-trien-3-ol, (3β,22*E*)	Ergosterol	$C_{28}H_{44}O$	57-87-4	396.648	pl (+w, al) nd (eth)	170	$250^{0.01}$			i H_2O; sl EtOH, eth, peth; s bz, chl
4732	Ergosta-5,7,22-trien-3-ol, (3β,10α,22*E*)	Pyrocalciferol	$C_{28}H_{44}O$	128-27-8	396.648	nd (MeOH)	94				i H_2O; s EtOH, chl, MeOH
4733	Ergosta-5,7,22-trien-3-ol, (3β,9β,10α,22*E*)	Lumisterol	$C_{28}H_{44}O$	474-69-1	396.648	nd (ace-MeOH)	118				i H_2O; s EtOH, HOAc; vs eth, ace, chl
4734	Ergost-5-en-3-ol, (3β,24*R*)	Campesterol	$C_{28}H_{48}O$	474-62-4	400.680	cry (ace)	157.5				
4735	Ergost-7-en-3-ol, (3β,5α)	γ-Ergostenol	$C_{28}H_{48}O$	516-78-9	400.680	nd (MeOH) cry (PrOH	146				s eth
4736	Ergost-8(14)-en-3-ol, (3β,5α)	α-Ergostenol	$C_{28}H_{48}O$	632-32-6	400.680	lf or nd (MeOH)	131				sl EtOH; s eth, bz, chl
4737	Ergotamine		$C_{33}H_{35}N_5O_5$	113-15-5	581.67	nd (al), pr (bz) pl (ace)	213 dec				vs bz, eth, chl
4738	Ergotamine tartrate (2:1)	Gynergen	$C_{35}H_{38}N_5O_8$	379-79-3	656.706		192 dec				
4739	Ergotaminine		$C_{33}H_{35}N_5O_5$	639-81-6	581.662	orth pl (MeOH) pl (al)	252 dec				i H_2O; sl EtOH, ace, bz; s chl; vs py
4740	Ergothioneine		$C_9H_{15}N_3O_2S$	497-30-3	229.299	nd or lf (dil EtOH)	290 dec				vs H_2O; sl EtOH, ace; i eth, bz, chl
4741	Eriochrome Black T		$C_{20}H_{12}N_3NaO_7S$	1787-61-7	461.380	br-blk pow					s H_2O, EtOH, MeOH
4742	Eriodictyol	3',4',5,7-Tetrahydroxyflavanone, (*S*)	$C_{15}H_{12}O_6$	552-58-9	288.252	pl or nd (EtOH)	267 dec				vs EtOH, HOAc
4743	Erythorbic acid	Isoascorbic acid	$C_6H_8O_6$	89-65-6	176.124	gran cry	168				s H_2O, py; sl ace
4744	β-Erythroidine		$C_{16}H_{19}NO_3$	466-81-9	273.327	cry (al)	99.5				s H_2O, eth, chl; vs EtOH, bz
4745	Erythromycin	Propiocine	$C_{37}H_{67}NO_{13}$	114-07-8	733.927	cry (w)	191				vs ace, eth, EtOH, chl
4746	Erythromycin ethyl succinate		$C_{43}H_{75}NO_{16}$	1264-62-6	862.053	cry (ace aq)	222				
4747	Erythromycin stearate		$C_{55}H_{103}NO_{15}$	643-22-1	1018.405	cry	92				i H_2O; sl EtOH, eth, chl
4748	Erythrophleine	Norcassamidine	$C_{24}H_{39}NO_5$	36150-73-9	421.571	glass	115				s H_2O, EtOH
4749	*D*-Erythrose		$C_4H_8O_4$	583-50-6	120.105	syr					s H_2O; vs EtOH
4750	*L*-Erythrose		$C_4H_8O_4$	533-49-3	120.105	syr					vs H_2O, EtOH
4751	*D*-Erythrose 4-phosphate	2,3-Dihydroxy-4-(phosphonooxy)butanal	$C_4H_9O_7P$	585-18-2	200.084	stab in aq soln only					s H_2O
4752	Erythrosine		$C_{20}H_8I_4O_5$	15905-32-5	835.893	br pow (Na salt)					s H_2; vs eth, EtOH
4753	*L*-Erythrulose		$C_4H_8O_4$	533-50-6	120.105	syr	dec	dec			vs H_2O, EtOH
4754	Esaprazole	*N*-Cyclohexyl-1-piperazineacetamide	$C_{12}H_{23}N_3O$	64204-55-3	225.330		112	$190^{0.5}$			
4755	Esculin	6-(β-*D*-Glucopyranosyloxy)-7-hydroxy-2*H*-1-benzopyran-2-one	$C_{15}H_{16}O_9$	531-75-9	340.283	pr (w+2)	205 (pentahydrate)				sl H_2O, EtOH, eth; s chl, py, HOAc
4756	Eserine sulfate	Physostigmine sulfate	$C_{30}H_{44}N_6O_8S$	64-47-1	648.770	hyg cry (ace-eth)	141				vs ace, EtOH
4757	Estra-1,3,5(10)-triene-3,17-diol, (17α)	α-Estradiol	$C_{18}H_{24}O_2$	57-91-0	272.383	nd (+1/2 w) (80% al)	221.5				i H_2O; s EtOH, ace; sl eth, bz
4758	Estra-1,3,5(10)-triene-3,17-diol (17β)	β-Estradiol	$C_{18}H_{24}O_2$	50-28-2	272.383	pr (80% al),	178.5				vs ace, EtOH, Diox
4759	Estra-1,3,5(10)-triene-3,17-diol, (8α,17β)	Isoestradiol	$C_{18}H_{24}O_2$	517-04-4	272.383	cry (dil MeOH-chl)	181				s EtOH, diox
4760	Estra-1,3,5(10)-triene-3,17-diol 3-benzoate, (17β)	Estradiol benzoate	$C_{25}H_{28}O_3$	50-50-0	376.488		196				
4761	Estra-1,3,5(10)-triene-3,16,17-triol, (16α,17β)	Estriol	$C_{18}H_{24}O_3$	50-27-1	288.382	lf (al), mcl (dil al)	288 dec		1.27^{25}		s EtOH; sl eth, bz, tfa; vs py
4762	Estra-1,3,5(10)-triene-3,16,17-triol, (16β,17β)	16-Epiestriol	$C_{18}H_{24}O_3$	547-81-9	288.382	cry (MeOH-bz)	290				

Ergometrinine

Ergonovine

Ergosine

Ergostane, (5α)

Ergostane, (5β)

Ergostan-3-ol, (3β,5α)

Ergosta-5,7,9(11),22-tetraen-3-ol, (3β,22*E*)

Ergosta-5,7,22-trien-3-ol, (3β,22*E*)

Ergosta-5,7,22-trien-3-ol, (3β,10α,22*E*)

Ergosta-5,7,22-trien-3-ol, (3β,9β,10α,22*E*)

Ergost-5-en-3-ol, (3β,24*R*)

Ergost-7-en-3-ol, (3β,5α)

Ergost-8(14)-en-3-ol, (3β,5α)

Ergotamine

Ergotamine tartrate (2:1)

Ergotaminine

Ergothioneine

Eriochrome Black T

Eriodictyol

Erythorbic acid

β-Erythroidine

Erythromycin

Erythromycin ethyl succinate

Erythromycin stearate

Erythrophleine

D-Erythrose

L-Erythrose

D-Erythrose 4-phosphate

Erythrosine

L-Erythrulose

Esaprazole

Esculin

Eserine sulfate

Estra-1,3,5(10)-triene-3,17-diol, (17α)

Estra-1,3,5(10)-triene-3,17-diol (17β)

Estra-1,3,5(10)-triene-3,17-diol, (8α,17β)

Estra-1,3,5(10)-triene-3,17-diol 3-benzoate, (17β)

Estra-1,3,5(10)-triene-3,16,17-triol, (16α,17β)

Estra-1,3,5(10)-triene-3,16,17-triol, (16β,17β)

No.	Name	Synonym	Mol. Form.	CAS RN	Mol. Wt.	Physical Form	mp/°C	bp/°C	den/ g cm^{-3}	n_D	Solubility
4763	Estrone		$C_{18}H_{22}O_2$	53-16-7	270.367	mcl, orth (al)	260.2		1.236^{25}		i H_2O; sl EtOH, eth, bz; s ace, diox
4764	Ethacrynic acid		$C_{13}H_{12}Cl_2O_4$	58-54-8	303.138		122.5				
4765	Ethalfluralin		$C_{13}H_{14}F_3N_3O_4$	55283-68-6	333.263		57	dec 256			
4766	Ethambutol		$C_{10}H_{24}N_2O_2$	74-55-5	204.310	cry	89				sl H_2O; s bz, chl
4767	Ethane		C_2H_6	74-84-0	30.069	col gas	-182.79	-88.6	0.5446^{-89}		i H_2O; vs bz
4768	Ethanearsonic acid		$C_2H_7AsO_3$	507-32-4	153.997	nd (al), orth nd (w)	99.5	210^{12}			vs H_2O, EtOH
4769	Ethanedial dioxime		$C_2H_4N_2O_2$	557-30-2	88.065	orth pl (w)	178 dec	sub			vs H_2O, EtOH, eth
4770	1,2-Ethanediamine	Ethylenediamine	$C_2H_8N_2$	107-15-3	60.098		11.14	117	0.8979^{20}	1.4565^{20}	vs H_2O; msc EtOH; i eth, bz; s ctc
4771	1,2-Ethanediamine, dihydrochloride	Ethylenediamine dihydrochloride	$C_2H_{10}Cl_2N_2$	333-18-6	133.019					1.633	vs H_2O
4772	1,2-Ethanediol	Ethylene glycol	$C_2H_6O_2$	107-21-1	62.068	liq	-12.69	197.3	1.1135^{20}	1.4318^{20}	msc H_2O, EtOH, ace; s eth, chl; sl bz
4773	1,2-Ethanediol, bis(4-methylbenzenesulfonate)		$C_{16}H_{18}O_6S_2$	6315-52-2	370.440	cry (bz)	128				
4774	1,1-Ethanediol, diacetate	Ethylidene diacetate	$C_6H_{10}O_4$	542-10-9	146.141		18.9	169	1.070^{25}	1.3985^{25}	vs eth, EtOH
4775	1,2-Ethanediol, diacetate	Ethylene glycol diacetate	$C_6H_{10}O_4$	111-55-7	146.141	liq	-31	190	1.1043^{20}	1.4159^{20}	vs H_2O; msc EtOH, eth, ace, bz, CS_2
4776	1,2-Ethanediol, diacrylate	Ethylene glycol diacrylate	$C_8H_{10}O_4$	2274-11-5	170.163	liq		55$^{0.6}$	1.0935^{26}		
4777	1,2-Ethanediol, dibenzoate	Ethylene glycol dibenzoate	$C_{16}H_{14}O_4$	94-49-5	270.280	orth pr (eth)	73.5	dec 360			i H_2O; s eth, chl
4778	1,2-Ethanediol, didodecanoate	Ethylene glycol didodecanoate	$C_{26}H_{50}O_4$	624-04-4	426.673	pl (al)	56.6	188^{20}			vs eth, EtOH
4779	1,2-Ethanediol, diformate	Ethylene glycol diformate	$C_4H_6O_4$	629-15-2	118.089			174	1.193^{0}	1.3580	sl H_2O; s EtOH, eth
4780	1,2-Ethanediol, dihexadecanoate	Ethylene glycol dipalmitate	$C_{34}H_{66}O_4$	624-03-3	538.886	lf or nd (al-chl)	72		0.8594^{78}		i H_2O, EtOH; s eth; vs ace
4781	1,2-Ethanediol, dimethacrylate	Ethylene glycol dimethacrylate	$C_{10}H_{14}O_4$	97-90-5	198.216	liq	-40	260	1.053^{20}	1.4532^{25}	vs bz, EtOH, lig
4782	1,2-Ethanediol, dinitrate	Ethylene glycol dinitrate	$C_2H_4N_2O_6$	628-96-6	152.062	ye liq	-22.3	198.5	1.4918^{20}		vs eth, EtOH
4783	1,2-Ethanediol, distearate	Ethylene glycol distearate	$C_{38}H_{74}O_4$	627-83-8	594.993	lf	79	241^{20}	0.8581^{78}		i H_2O, EtOH; vs eth, ace
4784	1,2-Ethanediol, ditetradecanoate	Ethylene glycol ditetradecanoate	$C_{30}H_{58}O_4$	627-84-9	482.780	cry (eth, ace)	65	208^{20}	0.8600^{80}		i H_2O, EtOH; s eth; vs ace, bz, ctc
4785	1,2-Ethanediol, dithiocyanate	Ethylene glycol dithiocyanate	$C_4H_4N_2S_2$	629-17-4	144.218	orth pl or nd (w)	90	dec	1.4200^{0}		sl H_2O, bz; s EtOH, eth; vs ace
4786	1,2-Ethanediol, monoacetate	Ethylene glycol monoacetate	$C_4H_8O_3$	542-59-6	104.105			188	1.108^{15}		msc H_2O, EtOH, eth
4787	1,2-Ethanediol, monobenzoate	Ethylene glycol monobenzoate	$C_9H_{10}O_3$	94-33-7	166.173		45	150^{10}	1.1101^{30}		vs EtOH
4788	1,2-Ethanediol, monostearate	Ethylene glycol monostearate	$C_{20}H_{40}O_3$	111-60-4	328.530	cry (peth)	60.5	190^{3}	0.8780^{60}	1.4310^{60}	sl EtOH; s eth
4789	1,2-Ethanediol, monosulfite	Ethylene glycol monosulfite	$C_2H_4O_3S$	3741-38-6	108.116	liq	-11	173	1.4402^{20}	1.4463^{20}	vs H_2O, EtOH, eth, ace, bz, AcOEt; sl chl
4790	1,2-Ethanediphosphonic acid	1,2-Diphosphonoethane	$C_2H_8O_6P_2$	6145-31-9	190.029	nd (EtOH/eth)	223				
4791	1,2-Ethanedisulfonic acid	Ethylene disulfonic acid	$C_2H_6O_6S_2$	110-04-3	190.195		173				vs diox
4792	Ethanedithioamide	Rubeanic acid	$C_2H_4N_2S_2$	79-40-3	120.196	red cry	170 dec				sl H_2O, EtOH; s con sulf
4793	1,2-Ethanedithiol	Ethylene dimercaptan	$C_2H_6S_2$	540-63-6	94.199	liq	-41.2	146.1	1.234^{20}	1.5590^{20}	i H_2O; s EtOH, eth, ace, bz; vs alk
4794	1,2-Ethanediyl mercaptoacetate		$C_6H_{10}O_4S_2$	123-81-9	210.271			138$^{1.5}$			
4795	Ethanesulfonic acid	Ethylsulfonic acid	$C_2H_6O_3S$	594-45-6	110.132	hyg	-17	123^{1}	1.3341^{25}	1.4335^{20}	vs H_2O, EtOH
4796	Ethanesulfonyl chloride		$C_2H_5ClO_2S$	594-44-5	128.578	pa ye		174	1.357^{22}	1.4531^{20}	vs eth; s CS_2
4797	Ethanethiol	Ethyl mercaptan	C_2H_6S	75-08-1	62.134	liq	-147.88	35.0	0.8315^{25}	1.4310^{20}	sl H_2O; s EtOH, eth, ace, dil alk
4798	Ethanimidamide		$C_2H_6N_2$	143-37-3	58.082		-35				sl H_2O; s EtOH, acid
4799	Ethanimidamide monohydrochloride	Acetamidine hydrochloride	$C_2H_7ClN_2$	124-42-5	94.543	nd or pr (al) hyg lo pr (al)	177.5				vs H_2O, EtOH
4800	Ethanol	Ethyl alcohol	C_2H_6O	64-17-5	46.068	liq	-114.14	78.29	0.7893^{20}	1.3611^{20}	msc H_2O, EtOH, eth, ace, chl; s bz
4801	Ethanolamine	Glycinol	C_2H_7NO	141-43-5	61.083		10.5	171	1.0180^{20}	1.4541^{20}	msc H_2O, EtOH; sl eth, lig, bz; s chl
4802	Ethanolamine hydrochloride	2-Aminoethanol hydrochloride	C_2H_8ClNO	2002-24-6	97.544	hyg cry (EtOH)	85				

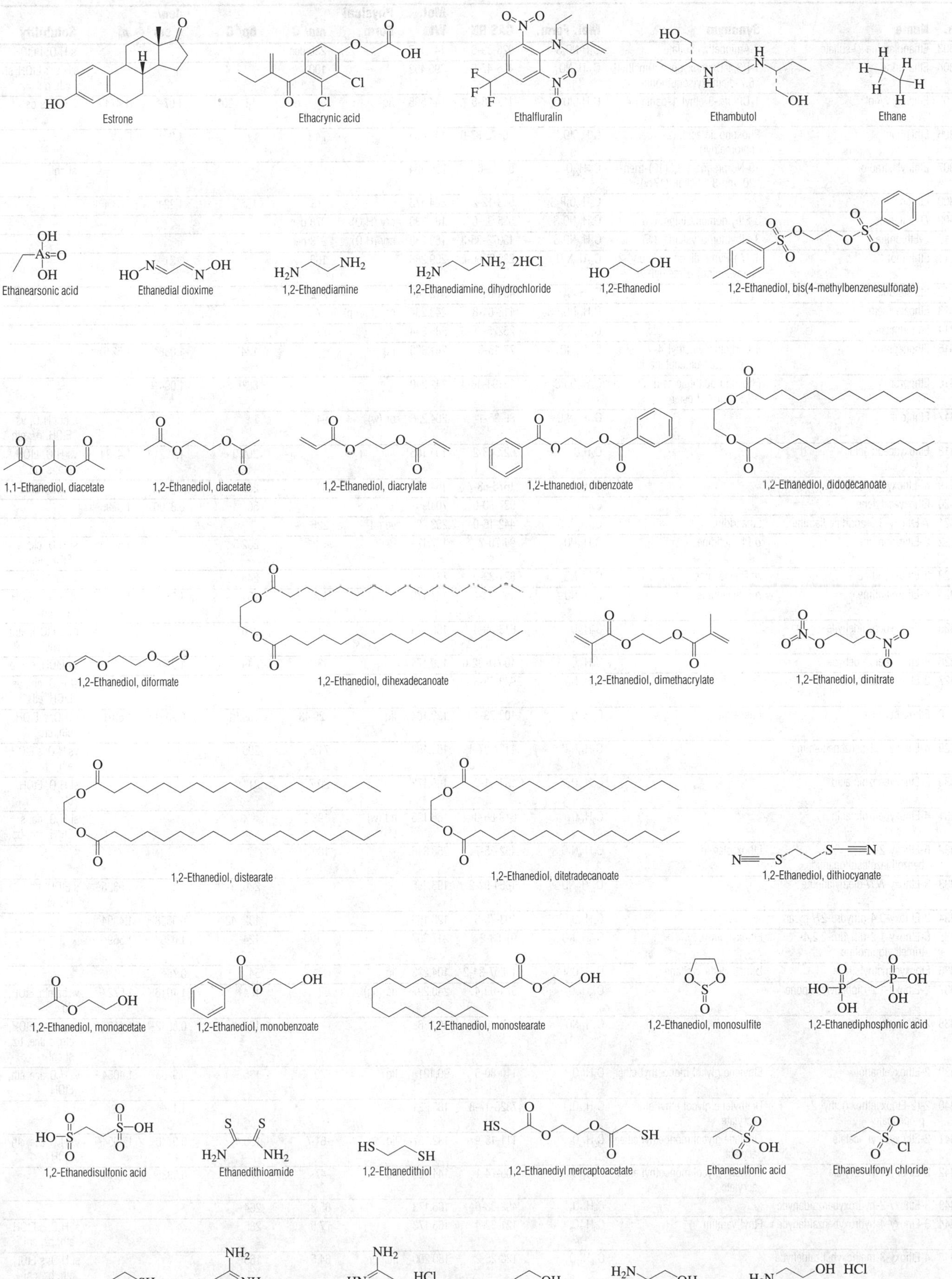

HO
H
Estrone
O
OH
Cl
Cl
Ethacrynic acid
F
N
O
Ethalfluralin
HO
N
H
OH
Ethambutol
H
Ethane
OH
As=O
Ethanearsonic acid
HO
N
OH
Ethanedial dioxime
H2N
NH2
1,2-Ethanediamine
NH2 2HCl
1,2-Ethanediamine, dihydrochloride
1,2-Ethanediol
S
1,2-Ethanediol, bis(4-methylbenzenesulfonate)
1,1-Ethanediol, diacetate
1,2-Ethanediol, diacetate
1,2-Ethanediol, diacrylate
1,2-Ethanediol, dibenzoate
1,2-Ethanediol, didodecanoate
1,2-Ethanediol, diformate
1,2-Ethanediol, dihexadecanoate
1,2-Ethanediol, dimethacrylate
1,2-Ethanediol, dinitrate
1,2-Ethanediol, distearate
1,2-Ethanediol, ditetradecanoate
1,2-Ethanediol, dithiocyanate
1,2-Ethanediol, monoacetate
1,2-Ethanediol, monobenzoate
1,2-Ethanediol, monostearate
1,2-Ethanediol, monosulfite
P
1,2-Ethanediphosphonic acid
1,2-Ethanedisulfonic acid
Ethanedithioamide
SH
1,2-Ethanedithiol
1,2-Ethanediyl mercaptoacetate
Ethanesulfonic acid
Ethanesulfonyl chloride
Ethanethiol
NH
Ethanimidamide
HCl
Ethanimidamide monohydrochloride
Ethanol
Ethanolamine
OH HCl
Ethanolamine hydrochloride

No.	Name	Synonym	Mol. Form.	CAS RN	Mol. Wt.	Physical Form	mp/°C	bp/°C	den/ g cm^{-3}	n_D	Solubility
4803	Ethanolamine *O*-sulfate	2-Aminoethyl sulfate	$C_2H_7NO_4S$	926-39-6	141.147		230 dec				s H_2O; i EtOH
4804	Ethaverine	1-[(3,4-Diethoxyphenyl)methyl]-6,7-diethoxyisoquinoline	$C_{24}H_{29}NO_4$	486-47-5	395.492		100				i H_2O; s EtOH; sl eth, chl
4805	Ethchlorvynol	1-Chloro-3-ethyl-1-penten-4-yn-ol	C_7H_9ClO	113-18-8	144.598	liq		181; $30^{0.1}$	1.07^{25}	1.474^{25}	i H_2O; s os
4806	Ethephon	Phosphonic acid, (2-chloroethyl)-	$C_2H_6ClO_3P$	16672-87-0	144.494		74		1.2		
4807	Ethinylestradiol	19-Norpregna-1,3,5(10)-trien-20-yne-3,17-diol, (17α)-	$C_{20}H_{24}O_2$	57-63-6	296.404						sl chl
4808	Ethion		$C_9H_{22}O_4P_2S_4$	563-12-2	384.476		-13	$165^{0.3}$	1.22^{20}		
4809	*D*-Ethionine	3-Ethylhomocysteine, (*R*)	$C_6H_{13}NO_2S$	535-32-0	163.238	cry (H_2O)	278 dec				
4810	*L*-Ethionine	3-Ethylhomocysteine, (*S*)	$C_6H_{13}NO_2S$	13073-35-3	163.238	cry (H_2O)	273 dec				
4811	Ethirimol	4(1*H*)-Pyrimidinone, 5-butyl-2-(ethylamino)-6-methyl-	$C_{11}H_{19}N_3O$	23947-60-6	209.288		160		1.21^{25}		
4812	Ethisterone		$C_{21}H_{28}O_2$	434-03-7	312.446		272				
4813	Ethoate-methyl		$C_6H_{14}NO_3PS_2$	116-01-8	243.284	cry (tol/hp)	67				
4814	Ethofumesate		$C_{13}H_{18}O_5S$	26225-79-6	286.344		71		1.14		
4815	Ethoheptazine	4-Carbethoxymethyl-4-phenylazacycloheptane	$C_{16}H_{23}NO_2$	77-15-6	261.360	liq		134^{1}	1.038^{26}	1.5210^{26}	
4816	Ethoprop	Phosphorodithioic acid, *O*-ethyl *S,S*-dipropyl ester	$C_8H_{19}O_2PS_2$	13194-48-4	242.340			$88^{0.2}$	1.094^{20}		
4817	Ethotoin		$C_{11}H_{12}N_2O_2$	86-35-1	204.225	pr (w)	94				s hot H_2O; vs EtOH, bz, eth
4818	Ethoxyacetic acid		$C_4H_8O_3$	627-03-2	104.105			206.5	1.1021^{20}	1.4194^{20}	vs H_2O, EtOH, eth; s chl
4819	4'-Ethoxyacetophenone		$C_{10}H_{12}O_2$	1676-63-7	164.201	pl (eth)	39	268			vs eth, EtOH
4820	Ethoxyacetylene		C_4H_6O	927-80-0	70.090			50	0.8000^{20}	1.3796^{20}	
4821	7-Ethoxy-3,9-acridinediamine	Ethacridine	$C_{15}H_{15}N_3O$	442-16-0	253.299	ye nd	226				
4822	2-Ethoxyaniline	*o*-Phenetidine	$C_8H_{11}NO$	94-70-2	137.179		<-21	232.5		1.5560^{20}	sl H_2O, ctc; s EtOH, eth
4823	3-Ethoxyaniline	*m*-Phenetidine	$C_8H_{11}NO$	621-33-0	137.179			248			vs eth, EtOH
4824	4-Ethoxyaniline	*p*-Phenetidine	$C_8H_{11}NO$	156-43-4	137.179		1.2	254	1.0652^{16}	1.5528^{20}	sl H_2O; s EtOH, eth, chl
4825	2-Ethoxybenzaldehyde		$C_9H_{10}O_2$	613-69-4	150.174		21	248			msc EtOH, eth; sl chl
4826	4-Ethoxybenzaldehyde		$C_9H_{10}O_2$	10031-82-0	150.174		13.5	249	1.08^{21}		vs EtOH, eth, bz
4827	2-Ethoxybenzamide	Ethenzamide	$C_9H_{11}NO_2$	938-73-8	165.189	nd (w, al)	133				sl H_2O, chl; vs EtOH, eth
4828	Ethoxybenzene	Phenetole	$C_8H_{10}O$	103-73-1	122.164	liq	-29.43	169.81	0.9651^{20}	1.5076^{20}	i H_2O; s EtOH, eth, ctc
4829	4-Ethoxy-1,2-benzenediamine		$C_8H_{12}N_2O$	1197-37-1	152.193		71.5	295			vs H_2O; s EtOH, eth, chl
4830	2-Ethoxybenzoic acid		$C_9H_{10}O_3$	134-11-2	166.173		20.7	211^{39}			sl H_2O, EtOH, ctc
4831	4-Ethoxybenzoic acid		$C_9H_{10}O_3$	619-86-3	166.173	nd (w)	198.5				sl H_2O, tfa; s EtOH, eth, bz
4832	6-Ethoxy-2-benzothiazolesulfonamide	Ethoxzolamide	$C_9H_{10}N_2O_3S_2$	452-35-7	258.316		189				
4833	3-Ethoxy-*N,N*-diethylaniline		$C_{12}H_{19}NO$	1864-92-2	193.285			286; $97^{0.6}$		1.5325^{25}	s EtOH, bz, HOAc
4834	2-Ethoxy-3,4-dihydro-2*H*-pyran		$C_7H_{12}O_2$	103-75-3	128.169			132; 42^{16}	0.9658^{25}	1.4394^{20}	
4835	6-Ethoxy-1,2-dihydro-2,2,4-trimethylquinoline	Ethoxyquin	$C_{14}H_{19}NO$	91-53-2	217.307			124^{2}	1.026^{25}	1.569^{25}	
4836	Ethoxydimethylsilane	Dimethylethoxysilane	$C_4H_{12}OSi$	14857-34-2	104.223	liq		54	0.76^{20}		
4837	2-Ethoxy-1,2-diphenylethanone		$C_{16}H_{16}O_2$	574-09-4	240.297	nd (lig)	62	194^{20}	1.1016^{17}	1.5727^{17}	vs bz, eth, EtOH, lig
4838	2-Ethoxyethanamine		$C_4H_{11}NO$	110-76-9	89.136			107	0.8512^{20}	1.4101^{20}	msc H_2O, EtOH, eth; s ace, bz; sl chl
4839	2-Ethoxyethanol	Ethylene glycol monoethyl ether	$C_4H_{10}O_2$	110-80-5	90.121	liq	-70	135	0.9253^{25}	1.4054^{25}	vs H_2O, ace, eth, EtOH
4840	2-(2-Ethoxyethoxy)ethyl 2-propenoate	Diethylene glycol ethyl ether acrylate	$C_9H_{16}O_4$	7328-17-8	188.221				1.13^{25}		
4841	2-Ethoxyethyl acetate	Ethylene glycol monoethyl ether acetate	$C_6H_{12}O_3$	111-15-9	132.157	liq	-61.7	156.4	0.9740^{20}	1.4054^{20}	vs H_2O, ace, eth, EtOH
4842	2-Ethoxyethyl acrylate	Ethylene glycol monoethyl ether acrylate	$C_7H_{12}O_3$	106-74-1	144.168	liq	-47	174	0.983^{20}	1.4274^{20}	
4843	3-Ethoxy-2-hydroxybenzaldehyde		$C_9H_{10}O_3$	492-88-6	166.173		65.3	264			
4844	3-Ethoxy-4-hydroxybenzaldehyde	Ethyl vanillin	$C_9H_{10}O_3$	121-32-4	166.173		77.5	285			sl H_2O; s EtOH, eth, bz, chl
4845	4-Ethoxy-3-methoxybenzaldehyde		$C_{10}H_{12}O_3$	120-25-2	180.200	mcl pr	64.5	168^{13}			sl H_2O; s EtOH, eth, bz, chl, HOAc
4846	1-Ethoxy-2-methoxyethane		$C_5H_{12}O_2$	5137-45-1	104.148	liq		103.5	0.8460^{25}	1.3843^{25}	

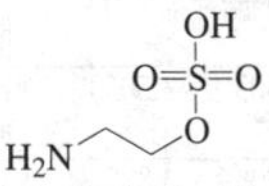

OH
O=S=O
H2N
O
Ethanolamine *O*-sulfate
Ethaverine 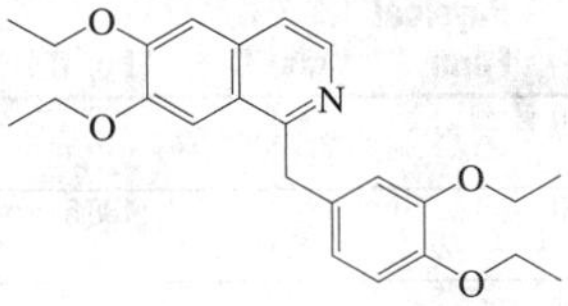 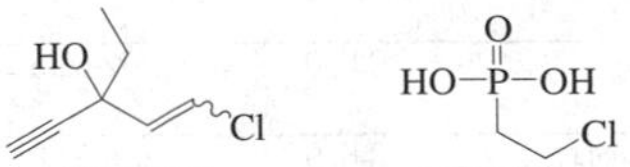
HO
Cl
Ethchlorvynol
O
HO–P–OH
Cl
Ethephon 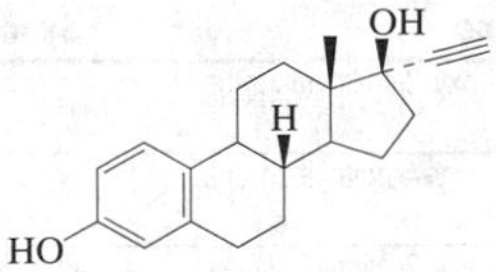
OH
H
HO
Ethinylestradiol 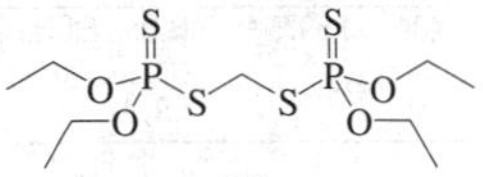
S
S
Ethion 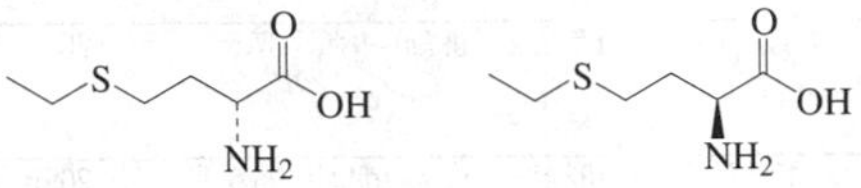
O
OH
NH2
D-Ethionine
O
OH
NH2
L-Ethionine 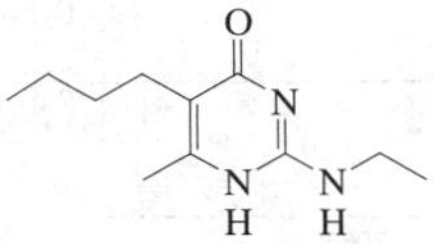
O
N
N
H
H
Ethirimol 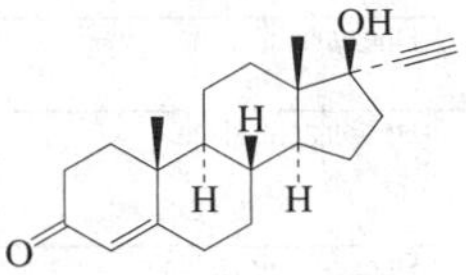
OH
H
H
H
O
Ethisterone
Ethoate-methyl 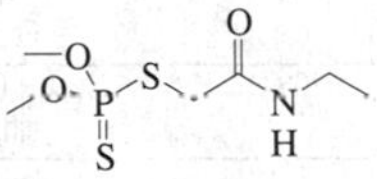
Ethofumesate 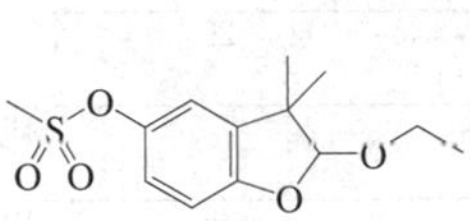
Ethoheptazine 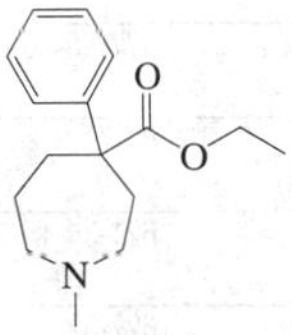
Ethoprop 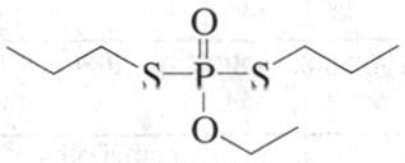
Ethotoin 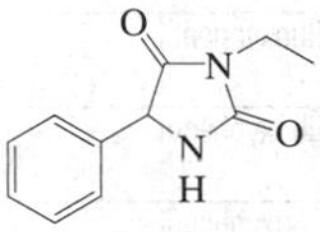
Ethoxyacetic acid 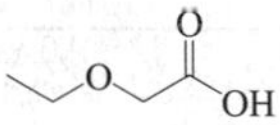
4′-Ethoxyacetophenone 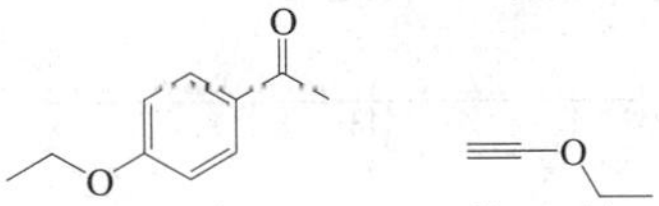
Ethoxyacetylene 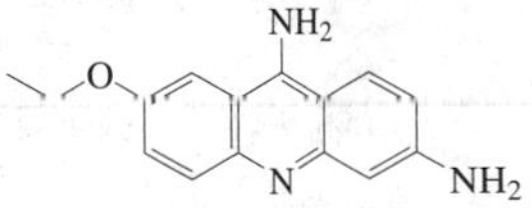
NH2
7-Ethoxy-3,9-acridinediamine 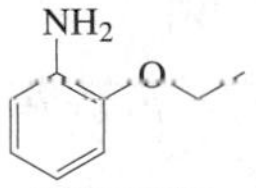
NH2
2-Ethoxyaniline 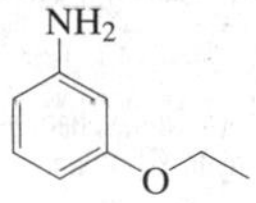
NH2
3-Ethoxyaniline 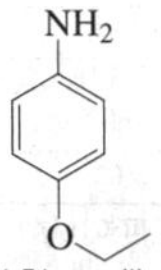
NH2
4-Ethoxyaniline
2-Ethoxybenzaldehyde 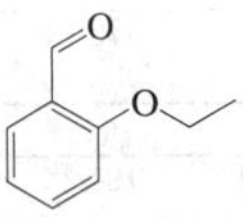
4-Ethoxybenzaldehyde 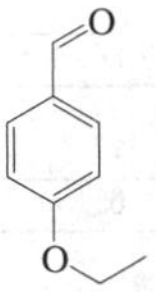
2-Ethoxybenzamide 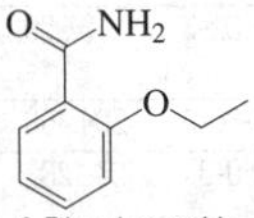
Ethoxybenzene 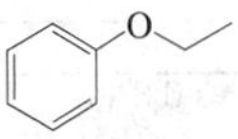 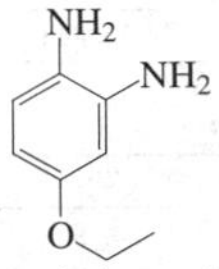
NH2
NH2
4-Ethoxy-1,2-benzenediamine 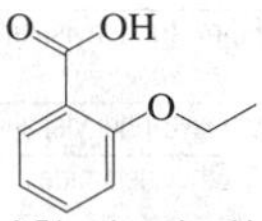
OH
2-Ethoxybenzoic acid 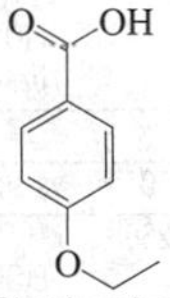
OH
4-Ethoxybenzoic acid 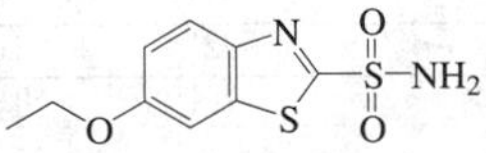
S–NH2
6-Ethoxy-2-benzothiazolesulfonamide
3-Ethoxy-*N,N*-diethylaniline 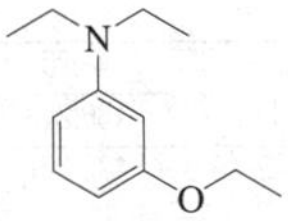
2-Ethoxy-3,4-dihydro-2*H*-pyran 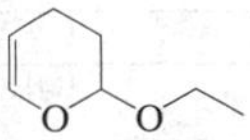
6-Ethoxy-1,2-dihydro-2,2,4-trimethylquinoline 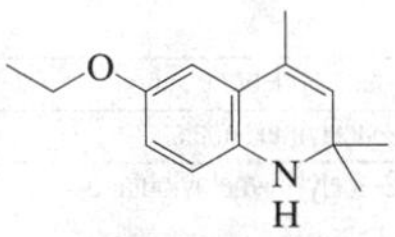 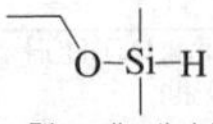
O–Si–H
Ethoxydimethylsilane
2-Ethoxy-1,2-diphenylethanone 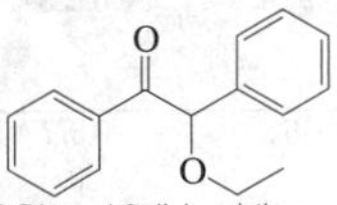
NH2
2-Ethoxyethanamine
OH
2-Ethoxyethanol
2-(2-Ethoxyethoxy)ethyl 2-propenoate 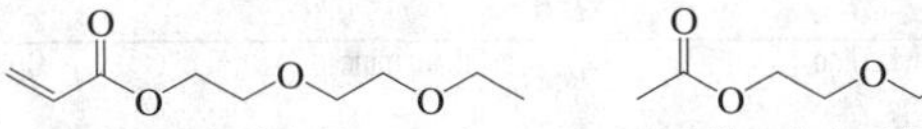
2-Ethoxyethyl acetate
2-Ethoxyethyl acrylate 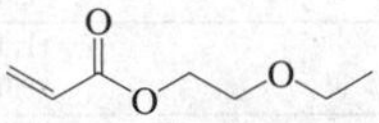 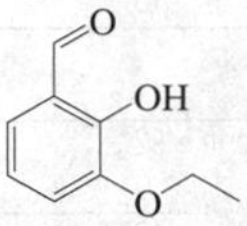
OH
3-Ethoxy-2-hydroxybenzaldehyde 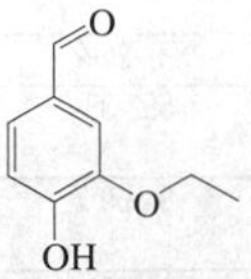
OH
3-Ethoxy-4-hydroxybenzaldehyde
4-Ethoxy-3-methoxybenzaldehyde 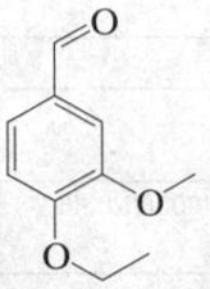
1-Ethoxy-2-methoxyethane

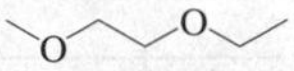

No.	Name	Synonym	Mol. Form.	CAS RN	Mol. Wt.	Physical Form	mp/°C	bp/°C	den/ g cm^{-3}	n_D	Solubility
4847	1-Ethoxy-3-methylbenzene		$C_9H_{12}O$	621-32-9	136.190			192	0.949^{20}	1.513^{20}	i H_2O; s EtOH, eth
4848	1-Ethoxy-4-methylbenzene		$C_9H_{12}O$	622-60-6	136.190			188.5	0.9509^{18}	1.5058^{18}	i H_2O; s EtOH, eth; sl ctc
4849	2-Ethoxy-2-methylbutane	Ethyl *tert*-pentyl ether	$C_7H_{16}O$	919-94-8	116.201			101.4	0.7606^{25}	1.3886^{25}	vs eth, EtOH
4850	(Ethoxymethylene)propanedinitrile		$C_6H_6N_2O$	123-06-8	122.124		66	160^{12}			s EtOH, eth; sl chl
4851	(Ethoxymethyl)oxirane	2,3-Epoxypropyl ethyl ether	$C_5H_{10}O_2$	4016-11-9	102.132			128	0.9700^{20}	1.4320^{20}	s H_2O, EtOH, eth; sl ctc
4852	1-Ethoxynaphthalene		$C_{12}H_{12}O$	5328-01-8	172.222	nd	5.5	280.5	1.060^{20}	1.5953^{25}	i H_2O; vs EtOH, eth
4853	2-Ethoxynaphthalene		$C_{12}H_{12}O$	93-18-5	172.222	pl (al)	37.5	282	1.0640^{20}	1.5975^{36}	i H_2O; s EtOH, eth, tol, lig, CS_2
4854	2-Ethoxy-5-nitroaniline	5-Nitro-*o*-phenetidine	$C_8H_{10}N_2O_3$	136-79-8	182.176	ye nd (dil al)	96.5	205^{14}			vs eth, EtOH
4855	1-Ethoxy-2-nitrobenzene		$C_8H_9NO_3$	610-67-3	167.162	br ye	1.1	267	1.1903^{15}	1.5425^{20}	vs eth, EtOH
4856	1-Ethoxy-4-nitrobenzene		$C_8H_9NO_3$	100-29-8	167.162	pr (dil al, eth)	60	283	1.1176^{100}		sl H_2O, EtOH; vs eth; msc ace, bz; s peth
4857	*N*-(4-Ethoxy-3-nitrophenyl)acetamide		$C_{10}H_{12}N_2O_4$	1777-84-0	224.213	nd (dil al)	124.0				vs ace, bz, EtOH
4858	2-Ethoxyphenol	Catechol monoethyl ether	$C_8H_{10}O_2$	94-71-3	138.164		29	217	1.0903^{25}		sl H_2O, ctc; msc EtOH, eth
4859	3-Ethoxyphenol	Resorcinol monoethyl ether	$C_8H_{10}O_2$	621-34-1	138.164			246; 131^{10}	1.105^{15}		i H_2O; s EtOH, eth, bz; sl chl
4860	4-Ethoxyphenol	Hydroquinone monoethyl ether	$C_8H_{10}O_2$	622-62-8	138.164	pr or lf (w)	66.5	246.5			sl H_2O; vs EtOH, eth; s chl
4861	*N*-(2-Ethoxyphenyl)acetamide		$C_{10}H_{13}NO_2$	581-08-8	179.216	lf(dil al)	79	>240			i H_2O; s EtOH, eth, chl
4862	*N*-(4-Ethoxyphenyl)acetamide	Phenacetin	$C_{10}H_{13}NO_2$	62-44-2	179.216	mcl pr	137.5			1.571	sl H_2O, eth, bz; s EtOH, ace; vs py
4863	*N*-(4-Ethoxyphenyl)-2-hydroxypropanamide	*p*-Lactophenetide	$C_{11}H_{15}NO_3$	539-08-2	209.242		118				s H_2O; vs EtOH; sl eth, bz, chl, peth
4864	(4-Ethoxyphenyl)urea	Dulcin	$C_9H_{12}N_2O_2$	150-69-6	180.203	lf (dil al), pl (w)	173.5	dec			sl H_2O; s EtOH; vs AcOEt
4865	3-Ethoxypropanal		$C_5H_{10}O_2$	2806-85-1	102.132			135.2	0.9165^{20}		
4866	3-Ethoxypropanenitrile		C_5H_9NO	2141-62-0	99.131			171	0.9285^{15}	1.4068^{20}	vs eth, EtOH
4867	8-Ethoxy-5-quinolinesulfonic acid	Actinoquinol	$C_{11}H_{11}NO_4S$	15301-40-3	253.275	br nd (w)	286 dec				s alk
4868	Ethoxytrimethylsilane		$C_5H_{14}OSi$	1825-62-3	118.250			76	0.7573^{20}	1.3741^{20}	i H_2O; s EtOH, eth, ace
4869	Ethoxytriphenylsilane		$C_{20}H_{20}OSi$	1516-80-9	304.458		65	344			s chl
4870	*N*-Ethylacetamide		C_4H_9NO	625-50-3	87.120			205; 104^{18}	0.942^{4}	1.4338^{20}	msc H_2O, EtOH; s chl, HOAc
4871	Ethyl acetate		$C_4H_8O_2$	141-78-6	88.106	liq	-83.8	77.11	0.9003^{20}	1.3723^{20}	s H_2O; msc EtOH, eth; vs ace, bz
4872	Ethyl acetoacetate		$C_6H_{10}O_3$	141-97-9	130.141	liq	-45	180.8	1.0368^{10}	1.4171^{20}	s H_2O; msc EtOH, eth; s bz, chl
4873	4'-Ethylacetophenone		$C_{10}H_{12}O$	937-30-4	148.201			114^{11}			
4874	Ethyl 2-acetylhexanoate		$C_{10}H_{18}O_3$	1540-29-0	186.248			221.5	0.9523^{20}	1.4301^{20}	vs ace, eth
4875	Ethyl 2-acetyl-3-methylbutanoate		$C_9H_{16}O_3$	1522-46-9	172.221			201; 97^{20}	0.9648^{18}	1.4256^{18}	i H_2O; msc EtOH, eth
4876	Ethyl 2-acetylpentanoate		$C_9H_{16}O_3$	1540-28-9	172.221			224; 90^{15}	0.9661^{20}	1.4255^{20}	vs eth, EtOH
4877	Ethyl 2-acetyl-4-pentenoate	Ethyl 2-allylacetoacetate	$C_9H_{14}O_3$	610-89-9	170.205			208	0.9898^{20}	1.4388^{18}	msc EtOH, eth, bz
4878	Ethyl acrylate	Ethyl propenoate	$C_5H_8O_2$	140-88-5	100.117	liq	-71.2	99.4	0.9234^{20}	1.4068^{20}	sl H_2O, DMSO; msc EtOH, eth; s chl
4879	Ethylamine	Ethanamine	C_2H_7N	75-04-7	45.084	vol liq or gas	-80.5	16.5	0.677^{25} (p>1 atm)	1.3663^{20}	msc H_2O, EtOH, eth
4880	Ethylamine hydrochloride	Ethanamine hydrochloride	C_2H_8ClN	557-66-4	81.545	mcl pl (al)	109.5		1.2160^{20}		vs H_2O, EtOH
4881	Ethyl 2-aminoacetate	Glycine, ethyl ester	$C_4H_9NO_2$	459-73-4	103.120			149; 58^{18}	1.0275^{10}	1.4242^{10}	msc H_2O, EtOH, eth, ace, bz; vs lig
4882	Ethyl 2-aminobenzoate		$C_9H_{11}NO_2$	87-25-2	165.189		13	268	1.1174^{20}	1.5646^{20}	vs eth, EtOH
4883	Ethyl 3-aminobenzoate		$C_9H_{11}NO_2$	582-33-2	165.189			294; 160^{5}	1.171^{20}	1.5600^{22}	sl H_2O; vs EtOH, eth; s ctc
4884	Ethyl 4-aminobenzoate	Ethyl aminobenzoate	$C_9H_{11}NO_2$	94-09-7	165.189	nd (w), orth (eth)	92	310			i H_2O; vs EtOH, eth; s chl, acid
4885	Ethyl (aminocarbonyl)carbamate		$C_4H_8N_2O_3$	626-36-8	132.118	nd (w, bz)	196.5	dec			i H_2O, eth; sl EtOH, bz, tfa

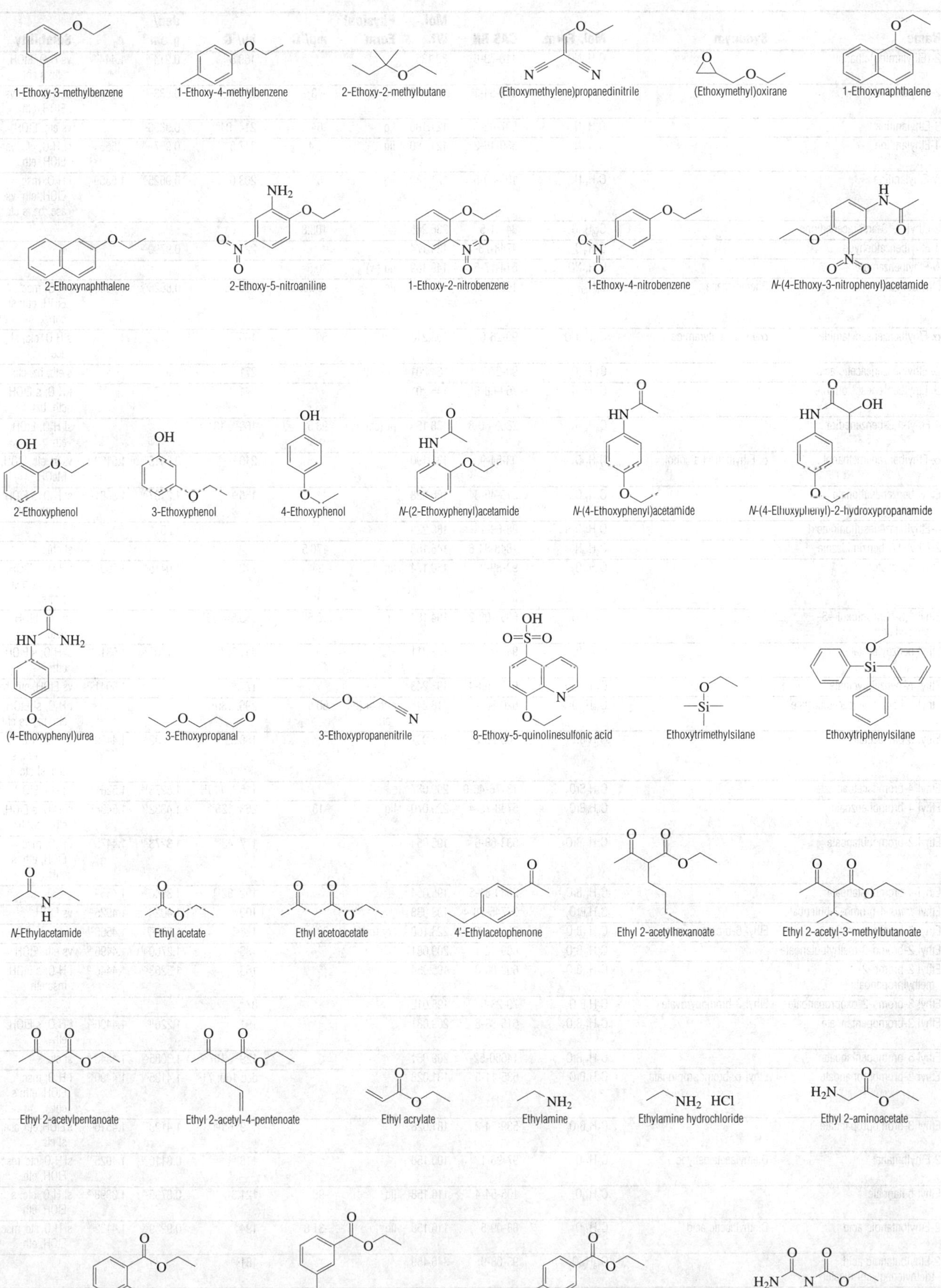
1-Ethoxy-3-methylbenzene
1-Ethoxy-4-methylbenzene
2-Ethoxy-2-methylbutane
(Ethoxymethylene)propanedinitrile
(Ethoxymethyl)oxirane
1-Ethoxynaphthalene
2-Ethoxynaphthalene
2-Ethoxy-5-nitroaniline
1-Ethoxy-2-nitrobenzene
1-Ethoxy-4-nitrobenzene
N-(4-Ethoxy-3-nitrophenyl)acetamide
2-Ethoxyphenol
3-Ethoxyphenol
4-Ethoxyphenol
N-(2-Ethoxyphenyl)acetamide
N-(4-Ethoxyphenyl)acetamide
N-(4-Ethoxyphenyl)-2-hydroxypropanamide
(4-Ethoxyphenyl)urea
3-Ethoxypropanal
3-Ethoxypropanenitrile
8-Ethoxy-5-quinolinesulfonic acid
Ethoxytrimethylsilane
Ethoxytriphenylsilane
N-Ethylacetamide
Ethyl acetate
Ethyl acetoacetate
4'-Ethylacetophenone
Ethyl 2-acetylhexanoate
Ethyl 2-acetyl-3-methylbutanoate
Ethyl 2-acetylpentanoate
Ethyl 2-acetyl-4-pentenoate
Ethyl acrylate
Ethylamine
Ethylamine hydrochloride
Ethyl 2-aminoacetate
Ethyl 2-aminobenzoate
Ethyl 3-aminobenzoate
Ethyl 4-aminobenzoate
Ethyl (aminocarbonyl)carbamate

No.	Name	Synonym	Mol. Form.	CAS RN	Mol. Wt.	Physical Form	mp/°C	bp/°C	den/ g cm⁻³	n_D	Solubility
4886	2-(Ethylamino)ethanol		$C_4H_{11}NO$	110-73-6	89.136			169.5	0.914^{20}	1.444^{20}	vs H_2O, EtOH, eth; s chl
4887	2-Ethylaniline		$C_8H_{11}N$	578-54-1	121.180	liq	-43	209.5	0.983^{22}	1.5584^{22}	sl H_2O, chl; vs EtOH, eth
4888	3-Ethylaniline		$C_8H_{11}N$	587-02-0	121.180	liq	-64	214; 94^{6}	0.9896^{25}		vs eth, EtOH
4889	4-Ethylaniline		$C_8H_{11}N$	589-16-2	121.180	liq	-2.4	217.5	0.9679^{20}	1.5554^{20}	sl H_2O, ctc; vs EtOH, eth
4890	*N*-Ethylaniline		$C_8H_{11}N$	103-69-5	121.180	liq	-63.5	203.0	0.9625^{20}	1.5559^{20}	i H_2O; msc EtOH, eth; vs ace, bz; s ctc
4891	2-Ethyl-9,10-anthracenedione		$C_{16}H_{12}O_2$	84-51-5	236.265		108.8				
4892	4-Ethylbenzaldehyde		$C_9H_{10}O$	4748-78-1	134.174			221	0.9790^{20}		
4893	*N*-Ethylbenzamide		$C_9H_{11}NO$	614-17-5	149.189	nd (w)	70.5				
4894	Ethylbenzene	Phenylethane	C_8H_{10}	100-41-4	106.165	liq	-94.96	136.16	0.8626^{25}	1.4959^{20}	i H_2O; msc EtOH, eth; sl chl
4895	α-Ethylbenzeneacetamide	α-Phenylbutyramide	$C_{10}H_{13}NO$	90-26-6	163.216	cry	86	185^{16}			s H_2O, ctc; sl ace
4896	α-Ethylbenzeneacetic acid		$C_{10}H_{12}O_2$	90-27-7	164.201	pl (eth)	47.5	271			s eth, bz, ctc
4897	α-Ethylbenzeneacetonitrile		$C_{10}H_{11}N$	769-68-6	145.201			241	0.977^{14}		i H_2O; s EtOH, eth, bz
4898	4-Ethyl-1,3-benzenediol		$C_8H_{10}O_2$	2896-60-8	138.164	pr (chl, bz)	98.5	160^{24}, 131^{15}			sl H_2O, EtOH, eth
4899	α-Ethylbenzenemethanol	α-Ethylbenzyl alcohol	$C_9H_{12}O$	93-54-9	136.190			219	0.9915^{25}	1.5169^{23}	vs bz, eth, EtOH, MeOH
4900	Ethyl benzenesulfonate		$C_8H_{10}O_3S$	515-46-8	186.228			156^{15}	1.2167^{20}	1.5081^{20}	sl H_2O; s EtOH; vs eth, chl
4901	4-Ethylbenzenesulfonic acid		$C_8H_{10}O_3S$	98-69-1	186.228				1.23		
4902	2-Ethyl-1*H*-benzimidazole		$C_9H_{10}N_2$	1848-84-6	146.188		176.5				sl chl
4903	Ethyl benzoate		$C_9H_{10}O_2$	93-89-0	150.174	liq	-34	212	1.0415^{25}	1.5007^{20}	i H_2O; s EtOH, ace, bz; msc eth; sl ctc
4904	Ethyl 1,3-benzodioxole-5-carboxylate		$C_{10}H_{10}O_4$	6951-08-2	194.184	pr	18.5	285.5; 135^{6}			vs eth, EtOH, peth
4905	Ethyl benzoylacetate		$C_{11}H_{12}O_3$	94-02-0	192.211		<0	dec 267; 167^{20}	1.1202^{15}	1.5317^{15}	sl H_2O; s EtOH, eth
4906	Ethyl *N*-benzylglycinate		$C_{11}H_{15}NO_2$	6436-90-4	193.243			177^{50}		1.5041^{20}	vs EtOH, eth, bz
4907	Ethyl 2-benzylideneacetoacetate		$C_{13}H_{14}O_3$	620-80-4	218.248	orth pl (dil al)	60.5	296; 180^{17}			i H_2O; sl EtOH, eth, bz; vs chl
4908	Ethyl bromoacetate		$C_4H_7BrO_2$	105-36-2	167.002			168.5	1.5032^{20}	1.4489^{20}	i H_2O; msc EtOH, eth; s ace; sl ctc
4909	Ethyl 4-bromoacetoacetate		$C_6H_9BrO_3$	13176-46-0	209.037			115^{14}, 110^{10}	1.5278^{18}	1.5281^{20}	vs eth, EtOH
4910	Ethyl 4-bromobenzoate		$C_9H_9BrO_2$	5798-75-4	229.070	liq	-18	263; 125^{15}	1.4332^{17}	1.5438^{17}	sl H_2O; s EtOH, eth, ace, bz
4911	Ethyl 2-bromobutanoate		$C_6H_{11}BrO_2$	533-68-6	195.054			177; 43^{5}	1.3273^{20}	1.4475^{20}	i H_2O; msc EtOH, eth; s chl
4912	Ethyl 4-bromobutanoate		$C_6H_{11}BrO_2$	2969-81-5	195.054			192; 82^{10}	1.3540^{20}	1.4559^{20}	
4913	Ethyl *trans*-4-bromo-2-butenoate		$C_6H_9BrO_2$	37746-78-4	193.038			100^{14}	1.402^{16}	1.4925^{20}	vs EtOH
4914	Ethyl 6-bromohexanoate	Ethyl 6-bromocaproate	$C_8H_{15}BrO_2$	25542-62-5	223.108	cry (peth)	33	126^{21}	1.238^{23}	1.4566^{21}	
4915	Ethyl 2-bromo-3-methylbutanoate		$C_7H_{13}BrO_2$	609-12-1	209.081			186	1.2760^{20}	1.4496^{20}	vs eth, EtOH
4916	Ethyl 2-bromo-2-methylpropanoate		$C_6H_{11}BrO_2$	600-00-0	195.054			163	1.3263^{20}	1.4446^{20}	i H_2O; s EtOH; msc eth
4917	Ethyl 3-bromo-2-oxopropanoate	Ethyl 3-bromopyruvate	$C_5H_7BrO_3$	70-23-5	195.012			87^{9}			
4918	Ethyl 2-bromopentanoate		$C_7H_{13}BrO_2$	615-83-8	209.081			191	1.226^{18}	1.4496^{20}	i H_2O; s EtOH, eth
4919	Ethyl 5-bromopentanoate		$C_7H_{13}BrO_2$	14660-52-7	209.081			129^{35}, 107^{20}	1.3085^{20}	1.4543^{20}	sl ctc
4920	Ethyl 2-bromopropanoate	Ethyl α-bromopropionate	$C_5H_9BrO_2$	535-11-5	181.028			dec 160; 71^{26}	1.4135^{20}	1.4490^{20}	i H_2O; msc EtOH, eth; s chl
4921	Ethyl 3-bromopropanoate		$C_5H_9BrO_2$	539-74-2	181.028			179; 65^{15}	1.4123^{18}	1.4516^{20}	s EtOH, eth, ace; sl ctc
4922	2-Ethylbutanal	Diethylacetaldehyde	$C_6H_{12}O$	97-96-1	100.158			118^{160}	0.8110^{20}	1.4025^{20}	sl H_2O, ctc; msc EtOH, eth
4923	Ethyl butanoate		$C_6H_{12}O_2$	105-54-4	116.158	liq	-98	121.3	0.8735^{25}	1.3898^{25}	sl H_2O, ctc; s EtOH, eth
4924	2-Ethylbutanoic acid	Diethylacetic acid	$C_6H_{12}O_2$	88-09-5	116.158	liq	-31.8	194	0.9239^{20}	1.4132^{20}	sl H_2O, ctc; msc EtOH, eth
4925	2-Ethylbutanoic acid, triethyleneglycol diester		$C_{18}H_{34}O_6$	95-08-9	346.459			$181^{3.5}$			
4926	2-Ethyl-1-butanol		$C_6H_{14}O$	97-95-0	102.174		<-15	147	0.8326^{20}	1.4220^{20}	sl H_2O; s EtOH, eth, chl
4927	2-Ethylbutanoyl chloride		$C_6H_{11}ClO$	2736-40-5	134.603			140	0.9825^{20}	1.4234^{20}	vs eth

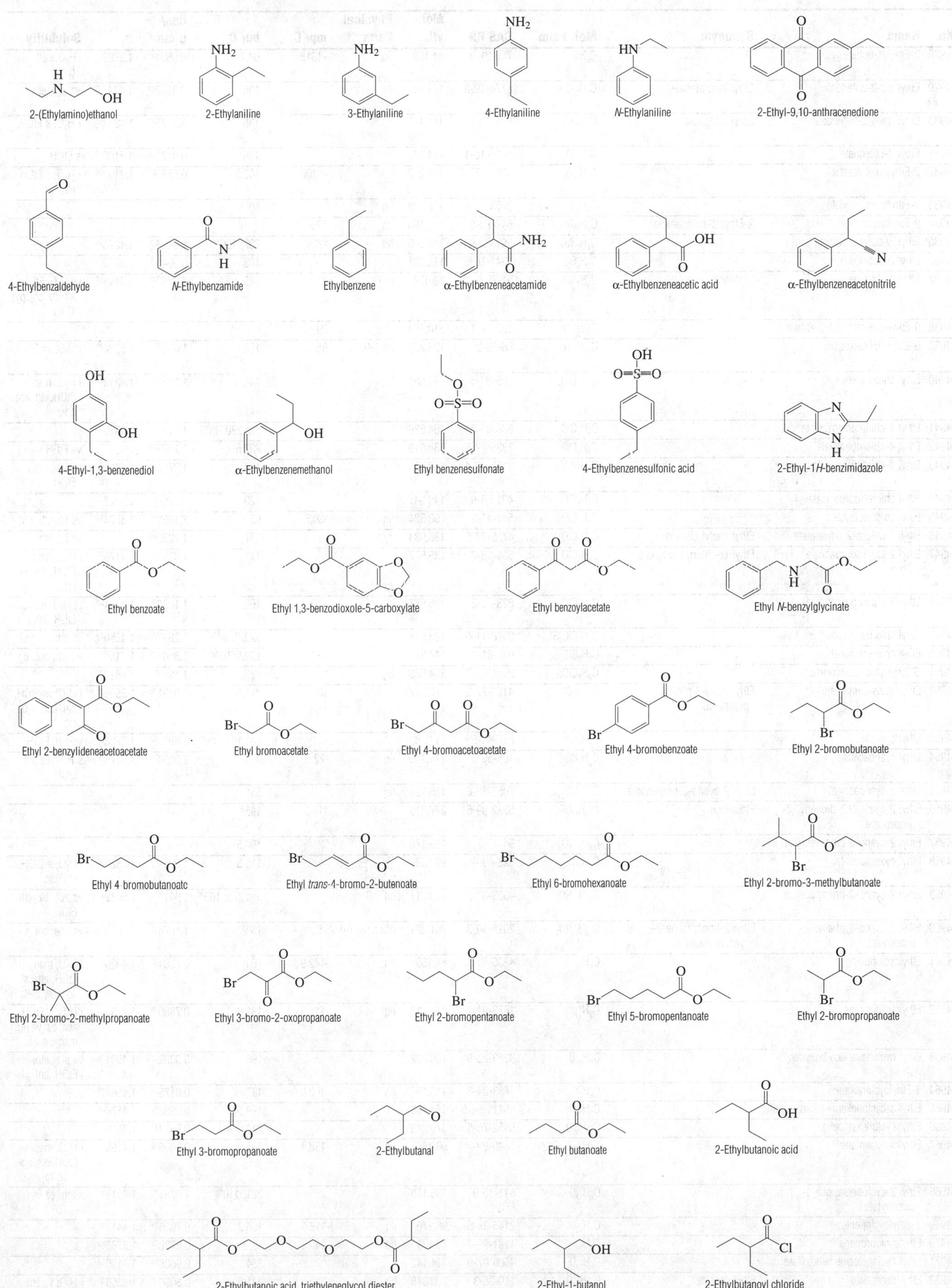

2-(Ethylamino)ethanol
2-Ethylaniline
3-Ethylaniline
4-Ethylaniline
N-Ethylaniline
2-Ethyl-9,10-anthracenedione
4-Ethylbenzaldehyde
N-Ethylbenzamide
Ethylbenzene
α-Ethylbenzeneacetamide
α-Ethylbenzeneacetic acid
α-Ethylbenzeneacetonitrile
4-Ethyl-1,3-benzenediol
α-Ethylbenzenemethanol
Ethyl benzenesulfonate
4-Ethylbenzenesulfonic acid
2-Ethyl-1H-benzimidazole
Ethyl benzoate
Ethyl 1,3-benzodioxole-5-carboxylate
Ethyl benzoylacetate
Ethyl N-benzylglycinate
Ethyl 2-benzylideneacetoacetate
Ethyl bromoacetate
Ethyl 4-bromoacetoacetate
Ethyl 4-bromobenzoate
Ethyl 2-bromobutanoate
Ethyl 4 bromobutanoate
Ethyl trans-4-bromo-2-butenoate
Ethyl 6-bromohexanoate
Ethyl 2-bromo-3-methylbutanoate
Ethyl 2-bromo-2-methylpropanoate
Ethyl 3-bromo-2-oxopropanoate
Ethyl 2-bromopentanoate
Ethyl 5-bromopentanoate
Ethyl 2-bromopropanoate
Ethyl 3-bromopropanoate
2-Ethylbutanal
Ethyl butanoate
2-Ethylbutanoic acid
2-Ethylbutanoic acid, triethyleneglycol diester
2-Ethyl-1-butanol
2-Ethylbutanoyl chloride

No.	Name	Synonym	Mol. Form.	CAS RN	Mol. Wt.	Physical Form	mp/°C	bp/°C	den/ g cm^{-3}	n_D	Solubility
4928	2-Ethyl-1-butene		C_6H_{12}	760-21-4	84.159	liq	-131.5	64.7	0.6894^{20}	1.3969^{20}	i H_2O; s eth, ace, bz, chl
4929	Ethyl *cis*-2-butenoate	Ethyl isocrotonate	$C_6H_{10}O_2$	6776-19-8	114.142			136	0.9182^{20}	1.4242^{20}	vs ace, eth, EtOH
4930	Ethyl *trans*-2-butenoate	Ethyl crotonate	$C_6H_{10}O_2$	623-70-1	114.142			138	0.9175^{20}	1.4243^{20}	i H_2O; s EtOH, eth
4931	Ethyl 3-butenoate		$C_6H_{10}O_2$	1617-18-1	114.142			119	0.9122^{20}	1.4105^{20}	s EtOH
4932	2-Ethylbutyl acetate		$C_8H_{16}O_2$	10031-87-5	144.212		<-100	162.5	0.8790^{20}	1.4109^{20}	i H_2O; s EtOH, eth, ctc
4933	2-Ethylbutyl acrylate		$C_9H_{16}O_2$	3953-10-4	156.222	liq		80^{20}			
4934	2-Ethylbutylamine	2-Ethyl-1-butanamine	$C_6H_{15}N$	617-79-8	101.190	liq	125				
4935	Ethyl *N*-butylcarbamate		$C_7H_{15}NO_2$	591-62-8	145.200	liq	-22	202; 100^{15}	0.9434^{26}	1.4278^{26}	
4936	Ethyl 2-butynoate		$C_6H_8O_2$	4341-76-8	112.127			163	0.9641^{20}	1.4372^{20}	
4937	Ethyl carbamate	Urethane	$C_3H_7NO_2$	51-79-6	89.094	pr (bz, to)	49	185	0.9862^{21}	1.4144^{51}	vs H_2O, EtOH, eth, bz, chl, py; sl lig
4938	9-Ethyl-9*H*-carbazol-3-amine		$C_{14}H_{14}N_2$	132-32-1	210.274		99				
4939	9-Ethyl-9*H*-carbazole		$C_{14}H_{13}N$	86-28-2	195.260	nd (al)	68	190^{10}	1.059^{80}	1.6394^{80}	i H_2O; vs EtOH, eth
4940	Ethyl chloroacetate		$C_4H_7ClO_2$	105-39-5	122.551	liq	-21	144.3	1.1585^{20}	1.4215^{20}	i H_2O; msc EtOH, eth, ace; s bz
4941	Ethyl 4-chloroacetoacetate		$C_6H_9ClO_3$	638-07-3	164.586		-8	dec 220; 115^{14}	1.218^{25}	1.4520^{20}	
4942	Ethyl 4-chlorobenzoate		$C_9H_9ClO_2$	7335-27-5	184.619			237.5	1.1873^{14}		vs EtOH
4943	Ethyl 4-chlorobutanoate		$C_6H_{11}ClO_2$	3153-36-4	150.603			184	1.0756^{20}	1.4311^{20}	vs ace, eth, EtOH
4944	Ethyl chlorofluoroacetate		$C_4H_6ClFO_2$	401-56-9	140.541			129	1.225^{20}	1.3927^{20}	
4945	Ethyl chloroformate		$C_3H_5ClO_2$	541-41-3	108.524	liq	-80.6	95	1.1352^{20}	1.3974^{20}	vs bz, eth, chl
4946	Ethyl 2-chloro-2-oxoacetate	Ethyl oxalyl chloride	$C_4H_5ClO_3$	4755-77-5	136.534	hyg		137	1.2226^{20}		vs bz, eth
4947	Ethyl 2-chloropropanoate	Ethyl α-chloropropionate	$C_5H_9ClO_2$	535-13-7	136.577			147	1.0793^{20}	1.4178^{20}	i H_2O; msc EtOH, eth; sl ctc
4948	Ethyl 3-chloropropanoate		$C_5H_9ClO_2$	623-71-2	136.577			162	1.1086^{20}	1.4254^{20}	sl H_2O; msc EtOH, eth
4949	Ethyl chlorosulfinate		$C_2H_5ClO_2S$	6378-11-6	128.578			52.5^{44}, 32^{16}	1.2837^{20}	1.4550^{25}	vs eth
4950	Ethyl chlorosulfonate		$C_2H_5ClO_3S$	625-01-4	144.577			152.5; 93^{100}	1.3502^{25}	1.416^{20}	vs eth, chl, lig
4951	*S*-Ethyl chlorothioformate		C_3H_5ClOS	2941-64-2	124.589	liq		136	1.195^{20}	1.4820^{20}	
4952	Ethyl *trans*-cinnamate	Ethyl *trans*-3-phenyl-2-propenoate	$C_{11}H_{12}O_2$	4192-77-2	176.212		10	271.5	1.0491^{20}	1.5598^{20}	i H_2O; vs EtOH, eth, ace; s bz, ctc
4953	Ethyl cyanate		C_3H_5NO	627-48-5	71.078			dec 162; 30^{12}	0.89^{20}	1.3788^{25}	vs eth, EtOH
4954	Ethyl cyanoacetate		$C_5H_7NO_2$	105-56-6	113.116	liq	-22.5	205	1.0654^{20}	1.4175^{20}	s H_2O; vs eth, EtOH
4955	Ethyl 2-cyanoacrylate	Ethyl 2-cyano-2-propenoate	$C_6H_7NO_2$	7085-85-0	125.126	liq		55^3			
4956	Ethyl 2-cyano-3,3-diphenyl-2-propenoate	Etocrilene	$C_{18}H_{15}NO_2$	5232-99-5	277.318		110.5	195^3			
4957	Ethyl 2-cyano-3-ethoxyacrylate		$C_8H_{11}NO_3$	94-05-3	169.178		52	190.5			
4958	Ethyl cyanoformate		$C_4H_5NO_2$	623-49-4	99.089			115.5	1.003^{25}	1.3820^{20}	i H_2O; s EtOH, eth, ctc
4959	Ethyl 2-cyano-2-phenylacetate		$C_{11}H_{11}NO_2$	4553-07-5	189.211	oil		dec 275; 165^{20}	1.091^{20}	1.5012^{25}	vs ace, bz, eth, EtOH
4960	Ethyl 2-cyano-3-phenyl-2-propenoate	Ethyl 2-benzylidene-2-cyanoacetate	$C_{12}H_{11}NO_2$	2025-40-3	201.221	(i) nd (al) (ii) oil	51	188^{15}	1.1076^{25}	1.5033	vs ace, chl
4961	Ethylcyclobutane		C_6H_{12}	4806-61-5	84.159	liq	-142.9	70.8	0.7284^{20}	1.4020^{20}	i H_2O; msc EtOH, eth; s ace, bz, peth
4962	Ethylcyclohexane		C_8H_{16}	1678-91-7	112.213	liq	-111.3	131.9	0.7880^{20}	1.4330^{20}	i H_2O; s EtOH, ace, bz; vs lig; msc ctc
4963	Ethyl cyclohexanecarboxylate		$C_9H_{16}O_2$	3289-28-9	156.222			196	0.9362^{20}	1.4501^{15}	vs ace, eth, EtOH, chl
4964	1-Ethylcyclohexene		C_8H_{14}	1453-24-3	110.197	liq	-109.9	137	0.8176^{25}	1.4567^{20}	
4965	Ethyl 3-cyclohexene-1-carboxylate		$C_9H_{14}O_2$	15111-56-5	154.206			194.5	0.9688^{20}	1.4578^{20}	
4966	Ethyl cyclohexylacetate		$C_{10}H_{18}O_2$	5452-75-5	170.249			211	0.9537^{14}	1.451^{14}	
4967	Ethylcyclopentane		C_7H_{14}	1640-89-7	98.186	liq	-138.4	103.5	0.7665^{20}	1.4198^{20}	i H_2O; msc EtOH, eth, ace; s bz, tol
4968	Ethyl 2-cyclopentanone-1-carboxylate		$C_8H_{12}O_3$	611-10-9	156.179			221; 110^{16}	1.0781^{21}	1.4519^{20}	s eth, bz
4969	1-Ethylcyclopentene		C_7H_{12}	2146-38-5	96.170	liq	-118.5	106.3	0.7936^{25}	1.4412^{20}	
4970	Ethylcyclopropane		C_5H_{10}	1191-96-4	70.133	liq	-149.2	35.9	0.6790^{25}	1.3786^{20}	
4971	Ethyl cyclopropanecarboxylate		$C_6H_{10}O_2$	4606-07-9	114.142			134	0.9608^{15}	1.4190^{20}	
4972	Ethyl decanoate	Ethyl caprate	$C_{12}H_{24}O_2$	110-38-3	200.318	liq	-20	241.5	0.8650^{20}	1.4256^{20}	i H_2O; vs eth, EtOH, chl

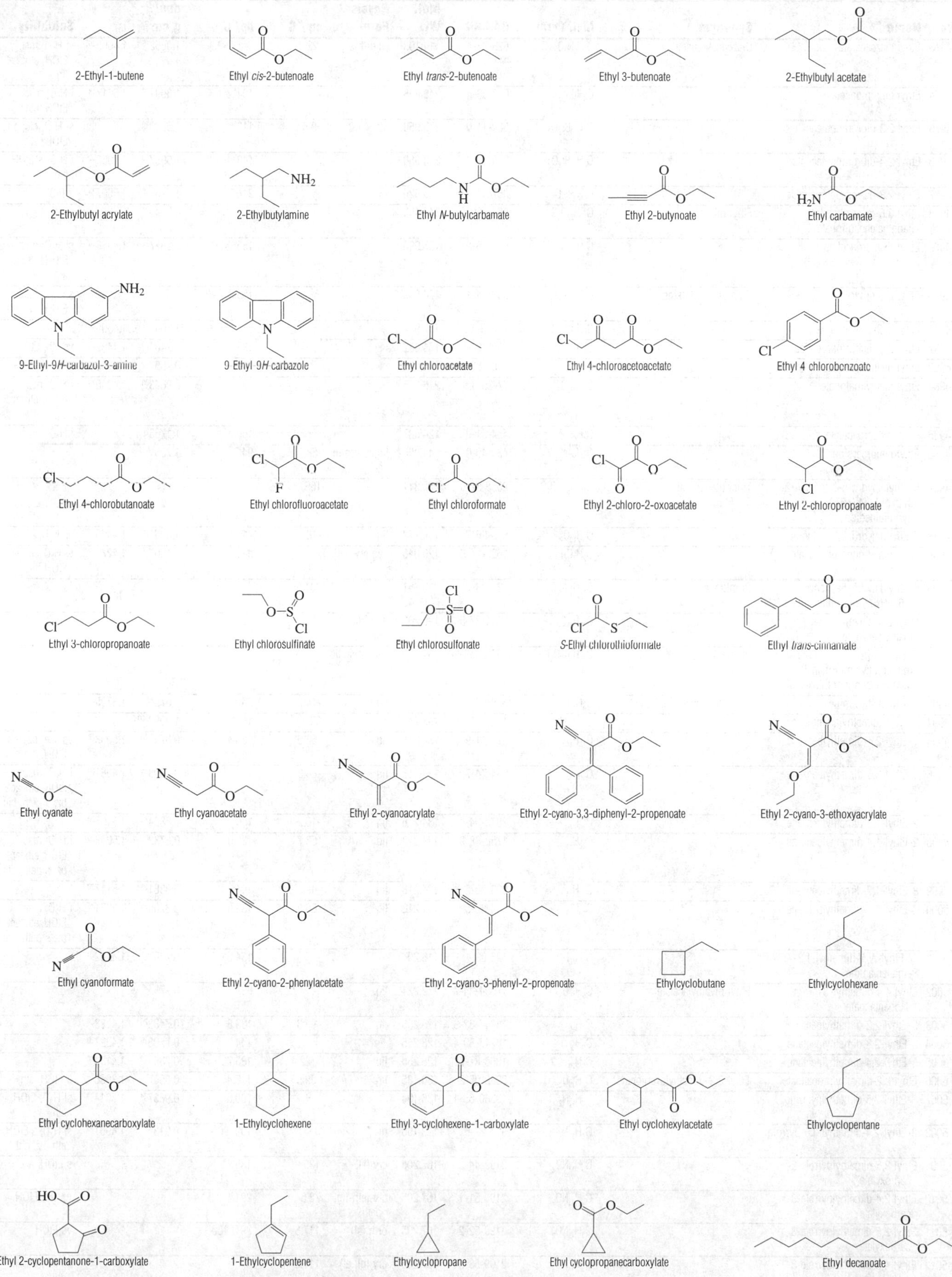

2-Ethyl-1-butene
Ethyl cis-2-butenoate
Ethyl trans-2-butenoate
Ethyl 3-butenoate
2-Ethylbutyl acetate
2-Ethylbutyl acrylate
2-Ethylbutylamine
Ethyl N-butylcarbamate
Ethyl 2-butynoate
Ethyl carbamate
9-Ethyl-9H-carbazol-3-amine
9-Ethyl-9H-carbazole
Ethyl chloroacetate
Ethyl 4-chloroacetoacetate
Ethyl 4-chlorobenzoate
Ethyl 4-chlorobutanoate
Ethyl chlorofluoroacetate
Ethyl chloroformate
Ethyl 2-chloro-2-oxoacetate
Ethyl 2-chloropropanoate
Ethyl 3-chloropropanoate
Ethyl chlorosulfinate
Ethyl chlorosulfonate
S-Ethyl chlorothioformate
Ethyl trans-cinnamate
Ethyl cyanate
Ethyl cyanoacetate
Ethyl 2-cyanoacrylate
Ethyl 2-cyano-3,3-diphenyl-2-propenoate
Ethyl 2-cyano-3-ethoxyacrylate
Ethyl cyanoformate
Ethyl 2-cyano-2-phenylacetate
Ethyl 2-cyano-3-phenyl-2-propenoate
Ethylcyclobutane
Ethylcyclohexane
Ethyl cyclohexanecarboxylate
1-Ethylcyclohexene
Ethyl 3-cyclohexene-1-carboxylate
Ethyl cyclohexylacetate
Ethylcyclopentane
Ethyl 2-cyclopentanone-1-carboxylate
1-Ethylcyclopentene
Ethylcyclopropane
Ethyl cyclopropanecarboxylate
Ethyl decanoate

No.	Name	Synonym	Mol. Form.	CAS RN	Mol. Wt.	Physical Form	mp/°C	bp/°C	den/ g cm^{-3}	n_D	Solubility
4973	Ethyl diazoacetate	Diazoacetic ester	$C_4H_6N_2O_2$	623-73-4	114.103	ye orth cry	-22	dec 140	1.0852^{18}	1.4605^{20}	sl H_2O; msc EtOH, eth, bz, lig
4974	Ethyl dibromoacetate		$C_4H_6Br_2O_2$	617-33-4	245.898			194	1.8991^{20}	1.5017^{13}	i H_2O; msc EtOH, eth
4975	Ethyl 2,3-dibromobutanoate		$C_6H_{10}Br_2O_2$	609-11-0	273.950	nd	58.5	113^{30}	1.6800^{20}		sl H_2O, ctc; s EtOH, eth
4976	Ethyl 2,4-dibromobutanoate		$C_6H_{10}Br_2O_2$	36847-51-5	273.950			149^{52}	1.6987^{20}	1.4960^{20}	i H_2O; s EtOH, eth
4977	Ethyl 2,3-dibromopropanoate		$C_5H_8Br_2O_2$	3674-13-3	259.925			214.5	1.7966^{20}	1.5007^{20}	s EtOH, eth
4978	Ethyl 3,6-di(*tert*-butyl)-1-naphthalenesulfonate	Ethyl dibunate	$C_{20}H_{28}O_3S$	5560-69-0	348.499						s chl
4979	Ethyl dichloroacetate		$C_4H_6Cl_2O_2$	535-15-9	156.996			155; 56^{10}	1.2827^{20}	1.4386^{20}	sl H_2O; msc EtOH, eth; s ace, chl
4980	Ethyldichloroarsine	Dichloroethylarsine	$C_2H_5AsCl_2$	598-14-1	174.889			155.3; 74^{50}	1.66^{20}		s H_2O; misc EtOH, bz
4981	Ethyl dichlorocarbamate		$C_3H_5Cl_2NO_2$	13698-16-3	157.984			66^{18}, 55^{15}	1.304^{30}	1.4595^{20}	
4982	Ethyl 2,3-dichloropropanoate		$C_5H_8Cl_2O_2$	6628-21-3	171.022			183.5	1.2401^{20}	1.4482^{20}	vs eth, EtOH
4983	Ethyl diethoxyacetate		$C_8H_{16}O_4$	6065-82-3	176.211			199	0.985^{25}	1.4100^{20}	
4984	Ethyl diethylmalonate		$C_{11}H_{20}O_4$	77-25-8	216.275			230	0.9643^{30}	1.4240^{20}	i H_2O; msc EtOH, eth; s ctc
4985	Ethyl difluoroacetate		$C_4H_6F_2O_2$	454-31-9	124.087			100	1.1765^{20}		i H_2O
4986	Ethyldifluoroarsine		$C_2H_5AsF_2$	430-40-0	141.980	liq, fumes in air	-38.7	94.3	1.708^{17}		
4987	5-Ethyldihydro-5-*sec*-butyl-2-thioxo-4,6(1*H*,5*H*)-pyrimidinedione	Thiobutabarbital	$C_{10}H_{16}N_2O_2S$	2095-57-0	228.311		169				
4988	5-Ethyldihydro-2(3*H*)-furanone		$C_6H_{10}O_2$	695-06-7	114.142	liq	-18	215.5	1.0261^{20}	1.4495^{20}	vs H_2O, EtOH
4989	Ethyl dihydrogen phosphate		$C_2H_7O_4P$	1623-14-9	126.048	hyg cry		dec	1.430^{25}	1.427	vs H_2O, ace, eth, EtOH
4990	5-Ethyldihydro-5-phenyl-4,6(1*H*,5*H*)-pyrimidinedione	Primidone	$C_{12}H_{14}N_2O_2$	125-33-7	218.251		281.5				
4991	Ethyl 2,4-dihydroxy-6-methylbenzoate		$C_{10}H_{12}O_4$	2524-37-0	196.200	lf (HOAc), pr (al)	132	sub			vs eth, EtOH
4992	*O*-Ethyl *S*-[2-(diisopropylamino)ethyl] methylphosphonothioate	VX Nerve agent	$C_{11}H_{26}NO_2PS$	50782-69-9	267.369	very toxic liq					
4993	Ethyldimethylamine	*N,N*-Dimethylethanamine	$C_4H_{11}N$	598-56-1	73.137	liq	-140	36.5	0.675^{20}	1.3705^{25}	
4994	Ethyl 4-(dimethylamino)benzoate		$C_{11}H_{15}NO_2$	10287-53-3	193.243		66.5	190^{14}	1.0099^{100}		
4995	1-Ethyl-2,4-dimethylbenzene		$C_{10}H_{14}$	874-41-9	134.218	liq	-62.9	188.4	0.8763^{20}	1.5038^{20}	vs ace, bz, eth, EtOH
4996	1-Ethyl-3,5-dimethylbenzene		$C_{10}H_{14}$	934-74-7	134.218	liq	-84.3	183.6	0.8608^{25}	1.4981^{20}	i H_2O; msc EtOH, eth, ace, bz; s peth, ctc
4997	2-Ethyl-1,3-dimethylbenzene		$C_{10}H_{14}$	2870-04-4	134.218	liq	-16.2	190	0.8864^{25}	1.5107^{20}	
4998	2-Ethyl-1,4-dimethylbenzene		$C_{10}H_{14}$	1758-88-9	134.218	liq	-53.7	186.9	0.8732^{25}	1.5043^{20}	i H_2O; msc EtOH, eth, ace, bz; s peth, ctc
4999	3-Ethyl-1,2-dimethylbenzene		$C_{10}H_{14}$	933-98-2	134.218	liq	-49.5	194	0.8881^{25}	1.5117^{20}	
5000	4-Ethyl-1,2-dimethylbenzene		$C_{10}H_{14}$	934-80-5	134.218	liq	-66.9	189.5	0.8706^{25}	1.5031^{20}	i H_2O; msc EtOH, eth, ace, bz; s peth, ctc
5001	*N'*-Ethyl-*N,N*-dimethyl-1,2-ethanediamine		$C_6H_{16}N_2$	123-83-1	116.204			134.5	0.738^{25}	1.4222^{20}	
5002	Ethyl 4,4-dimethyl-3-oxopentanoate	Ethyl pivaloylacetate	$C_9H_{16}O_3$	17094-34-7	172.221	liq		83^{17}	0.97^{18}		
5003	3-Ethyl-2,2-dimethylpentane		C_9H_{20}	16747-32-3	128.255	liq	-99.3	133.8	0.7438^{20}	1.4123^{20}	
5004	3-Ethyl-2,3-dimethylpentane		C_9H_{20}	16747-33-4	128.255			144.7	0.7508^{25}	1.4221^{20}	
5005	3-Ethyl-2,4-dimethylpentane		C_9H_{20}	1068-87-7	128.255	liq	-122.4	136.7	0.7365^{20}	1.4131^{20}	
5006	Ethyl 2,2-dimethylpropanoate	Ethyl 2,2-dimethylpropionate	$C_7H_{14}O_2$	3938-95-2	130.185	liq	-89.5	118.4	0.856^{20}	1.3906^{20}	s EtOH, eth
5007	3-Ethyl-2,5-dimethylpyrazine		$C_8H_{12}N_2$	13360-65-1	136.194			180.5	0.9657^{24}	1.5014^{24}	sl H_2O, EtOH, eth
5008	3-Ethyl-2,4-dimethyl-1*H*-pyrrole		$C_8H_{13}N$	517-22-6	123.196	pr	0	199; 96^{16}	0.913^{20}	1.4961^{20}	sl H_2O; s EtOH, eth, bz, chl
5009	Ethyl 3,5-dimethylpyrrole-2-carboxylate		$C_9H_{13}NO_2$	2199-44-2	167.205	cry (al)	125	$135^{10.5}$			s EtOH, ace
5010	Ethyl 2,4-dimethylpyrrole-3-carboxylate		$C_9H_{13}NO_2$	2199-51-1	167.205	cry (eth-lig, peth)	78.5	291			vs eth, EtOH
5011	Ethyl 2,5-dimethylpyrrole-3-carboxylate		$C_9H_{13}NO_2$	2199-52-2	167.205	orth (al)	117.5	291; 130^{15}			vs EtOH
5012	Ethyl 4,5-dimethylpyrrole-3-carboxylate		$C_9H_{13}NO_2$	2199-53-3	167.205	cry (dil al)	111.3				vs eth, EtOH, chl
5013	Ethyl 2,4-dioxopentanoate		$C_7H_{10}O_4$	615-79-2	158.152		18	214	1.1251^{20}	1.4757^{17}	vs eth, EtOH

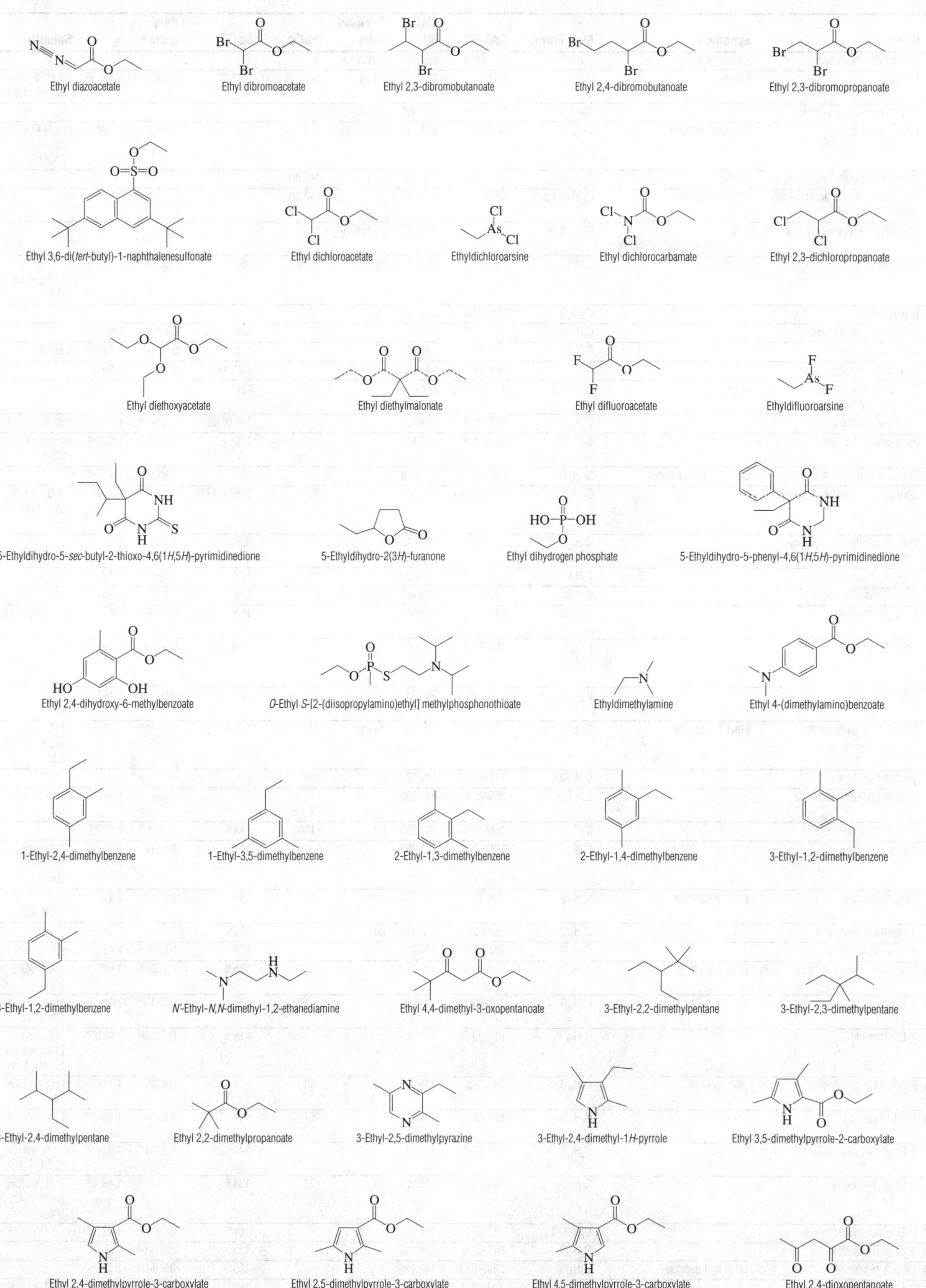

Ethyl diazoacetate
Ethyl dibromoacetate
Ethyl 2,3-dibromobutanoate
Ethyl 2,4-dibromobutanoate
Ethyl 2,3-dibromopropanoate
Ethyl 3,6-di(tert-butyl)-1-naphthalenesulfonate
Ethyl dichloroacetate
Ethyldichloroarsine
Ethyl dichlorocarbamate
Ethyl 2,3-dichloropropanoate
Ethyl diethoxyacetate
Ethyl diethylmalonate
Ethyl difluoroacetate
Ethyldifluoroarsine
5-Ethyldihydro-5-sec-butyl-2-thioxo-4,6(1H,5H)-pyrimidinedione
5-Ethyldihydro-2(3H)-furanone
Ethyl dihydrogen phosphate
5-Ethyldihydro-5-phenyl-4,6(1H,5H)-pyrimidinedione
Ethyl 2,4-dihydroxy-6-methylbenzoate
O-Ethyl S-[2-(diisopropylamino)ethyl] methylphosphonothioate
Ethyldimethylamine
Ethyl 4-(dimethylamino)benzoate
1-Ethyl-2,4-dimethylbenzene
1-Ethyl-3,5-dimethylbenzene
2-Ethyl-1,3-dimethylbenzene
2-Ethyl-1,4-dimethylbenzene
3-Ethyl-1,2-dimethylbenzene
4-Ethyl-1,2-dimethylbenzene
N'-Ethyl-N,N-dimethyl-1,2-ethanediamine
Ethyl 4,4-dimethyl-3-oxopentanoate
3-Ethyl-2,2-dimethylpentane
3-Ethyl-2,3-dimethylpentane
3-Ethyl-2,4-dimethylpentane
Ethyl 2,2-dimethylpropanoate
3-Ethyl-2,5-dimethylpyrazine
3-Ethyl-2,4-dimethyl-1H-pyrrole
Ethyl 3,5-dimethylpyrrole-2-carboxylate
Ethyl 2,4-dimethylpyrrole-3-carboxylate
Ethyl 2,5-dimethylpyrrole-3-carboxylate
Ethyl 4,5-dimethylpyrrole-3-carboxylate
Ethyl 2,4-dioxopentanoate

No.	Name	Synonym	Mol. Form.	CAS RN	Mol. Wt.	Physical Form	mp/°C	bp/°C	den/ g cm^{-3}	n_D	Solubility
5014	*O*-Ethyl dithiocarbonate	Xanthogenic acid	$C_3H_6OS_2$	151-01-9	122.209	unstab liq	-53	25			
5015	Ethylene	Ethene	C_2H_4	74-85-1	28.053	col gas	-169.15	-103.77	0.5678^{-104}	1.363^{-100}	i H_2O; sl EtOH, bz, ace; s eth
5016	Ethylenebisdithiocarbamic acid		$C_4H_8N_2S_4$	111-54-6	212.380	unstab liq					
5017	Ethylene carbonate	Vinylene carbonate	$C_3H_4O_3$	96-49-1	88.062	mcl pl (al)	36.4	248	1.3214^{39}	1.4148^{50}	msc H_2O, EtOH, eth, bz, chl, AcOEt
5018	Ethylenediaminetetraacetic acid	EDTA	$C_{10}H_{16}N_2O_8$	60-00-4	292.242	cry (w)	245 dec				
5019	Ethylenediaminetetraacetic acid, disodium salt, dihydrate	EDTA disodium	$C_{10}H_{18}N_2Na_2O_{10}$	6381-92-6	372.237		242 dec				
5020	*N,N'*-Ethylene distearylamide	*N,N'*-Dioctadecanoylethanediamine	$C_{38}H_{76}N_2O_2$	110-30-5	593.022	cry (EtOH)	149				
5021	Ethyleneimine	Aziridine	C_2H_5N	151-56-4	43.068	liq	-77.9	56	0.832^{25}		msc H_2O; s EtOH; vs eth; sl chl
5022	Ethylestrenol		$C_{20}H_{32}O$	965-90-2	288.467	cry	77				
5023	*N*-Ethyl-1,2-ethanediamine		$C_4H_{12}N_2$	110-72-5	88.151			129	0.837^{25}	1.4385^{20}	
5024	Ethyl ethoxyacetate		$C_6H_{12}O_3$	817-95-8	132.157			158	0.9702^{20}	1.4039^{20}	s EtOH, eth, ace
5025	Ethyl 3-ethoxypropanoate		$C_7H_{14}O_3$	763-69-9	146.184			166; 48^{5}	0.9490^{20}	1.4065^{20}	
5026	Ethyl 2-ethoxy-1(2*H*)-quinolinecarboxylate	EEDQ	$C_{14}H_{17}NO_3$	16357-59-8	247.290		56.5	$126^{0.1}$			s chl
5027	Ethyl 2-ethylacetoacetate		$C_8H_{14}O_3$	607-97-6	158.195			198.0; 80^{10}	0.9847^{16}	1.4214^{25}	msc EtOH, eth
5028	Ethyl ethylcarbamate		$C_5H_{11}NO_2$	623-78-9	117.147			176	0.9813^{20}	1.4215^{20}	vs H_2O, eth, EtOH
5029	Ethyl 2-ethylhexanoate	Ethyl 2-ethylcaproate	$C_{10}H_{20}O_2$	2983-37-1	172.265			90^{28}	0.8586^{25}	1.4123^{25}	
5030	2-Ethyl-*N*-(2-ethylphenyl)aniline		$C_{16}H_{19}N$	64653-59-4	225.329		29	336^{603}, 173^{10}		1.5550^{25}	i H_2O; vs EtOH, eth; sl chl; s acid
5031	*O*-Ethyl ethylthiophosphonyl chloride		$C_4H_{10}ClOPS$	1497-68-3	172.613	liq		$35^{0.7}$	1.15^{20}		
5032	Ethyl fluoroacetate		$C_4H_7FO_2$	459-72-3	106.096			120	1.0912^{20}	1.3755^{20}	vs H_2O
5033	Ethyl 4-fluorobenzoate		$C_9H_9FO_2$	451-46-7	168.164	mcl pr (w)	26	210	1.146^{25}	1.4864^{20}	vs eth, EtOH
5034	*N*-Ethylformamide		C_3H_7NO	627-45-2	73.094			198	0.9552^{20}	1.4320^{20}	msc H_2O, EtOH, eth
5035	Ethyl formate		$C_3H_6O_2$	109-94-4	74.079	liq	-79.6	54.4	0.9208^{20}	1.3609^{20}	s H_2O; msc EtOH, eth; vs ace; sl ctc
5036	2-Ethylfuran		C_6H_8O	3208-16-0	96.127			92.5	0.9018^{20}	1.4403^{20}	s EtOH, eth, ace, bz
5037	Ethyl 2-furancarboxylate	Ethyl 2-furanoate	$C_7H_8O_3$	614-99-3	140.137	lf or pr	34.5	196.8	1.1174^{21}	1.4797^{21}	i H_2O; msc EtOH, eth, ace; s bz
5038	γ-Ethyl *L*-glutamate		$C_7H_{13}NO_4$	1119-33-1	175.183		191				sl H_2O
5039	Ethyl heptafluorobutanoate		$C_6H_5F_7O_2$	356-27-4	242.092			95	1.394^{20}	1.3011^{20}	sl H_2O; s eth, ace
5040	3-Ethylheptane		C_9H_{20}	15869-80-4	128.255	liq	-114.9	143.0	0.7225^{25}	1.4093^{20}	
5041	4-Ethylheptane		C_9H_{20}	2216-32-2	128.255			141.2	0.7241^{25}	1.4096^{20}	i H_2O; s eth; msc EtOH, ace, bz
5042	Ethyl heptanoate	Ethyl oenanthate	$C_9H_{18}O_2$	106-30-9	158.238	liq	-66.1	187	0.8817^{20}	1.4100^{20}	sl H_2O, ctc; s EtOH, eth
5043	2-Ethylheptanoic acid		$C_9H_{18}O_2$	3274-29-1	158.238	liq		153^{31}		1.4255^{27}	
5044	4-Ethyl-4-heptanol		$C_9H_{20}O$	597-90-0	144.254			182	0.8350^{20}	1.4332^{20}	vs eth, EtOH
5045	Ethyl *trans,trans*-2,4-hexadienoate	Ethyl sorbate	$C_8H_{12}O_2$	2396-84-1	140.180			195.5	0.9506^{20}	1.4951^{20}	vs eth, EtOH, chl
5046	2-Ethylhexanal		$C_8H_{16}O$	123-05-7	128.212		<-100	163	0.8540^{20}	1.4142^{20}	i H_2O; s EtOH, eth; sl ctc
5047	3-Ethylhexane		C_8H_{18}	619-99-8	114.229			118.6	0.7136^{20}	1.4018^{20}	i H_2O; msc EtOH, eth, ace, bz, chl; s ctc
5048	2-Ethyl-1,3-hexanediol	Ethohexadiol	$C_8H_{18}O_2$	94-96-2	146.228	liq	-40	244	0.9325^{22}	1.4497^{20}	sl H_2O; s EtOH, eth
5049	Ethyl hexanoate		$C_8H_{16}O_2$	123-66-0	144.212	liq	-67	167	0.873^{20}	1.4073^{20}	sl H_2O; vs eth, EtOH
5050	2-Ethylhexanoic acid		$C_8H_{16}O_2$	149-57-5	144.212			228; 120^{13}	0.9031^{25}	1.4241^{20}	s H_2O, eth, ctc; sl EtOH
5051	2-Ethyl-1-hexanol		$C_8H_{18}O$	104-76-7	130.228	liq	-70	184.6	0.8319^{25}	1.4300^{20}	i H_2O; s EtOH, eth, ace, bz, chl
5052	2-Ethylhexanoyl chloride		$C_8H_{15}ClO$	760-67-8	162.657			101^{40}, 67^{11}	0.939^{25}	1.4335^{20}	
5053	2-Ethyl-2-hexenal		$C_8H_{14}O$	645-62-5	126.196			175	0.8554^{20}		
5054	Ethyl 3-hexenoate	Ethyl hydrosorbate	$C_8H_{14}O_2$	2396-83-0	142.196			166.5	0.8957^{20}	1.4255^{20}	
5055	2-Ethylhexyl acetate		$C_{10}H_{20}O_2$	103-09-3	172.265	liq	-80	199	0.8718^{20}	1.4204^{20}	i H_2O; s EtOH, eth

S SH O
O-Ethyl dithiocarbonate

H H H H
Ethylene

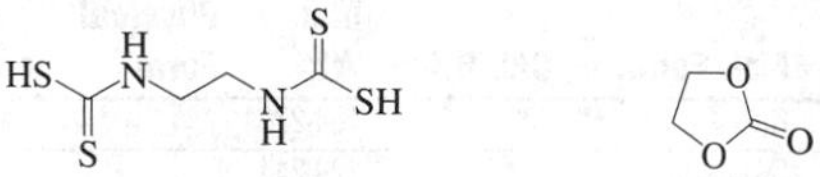

HS H N S N H SH S
Ethylenebisdithiocarbamic acid

O O O
Ethylene carbonate

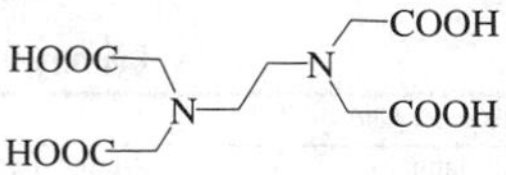

HOOC N N COOH COOH HOOC
Ethylenediaminetetraacetic acid

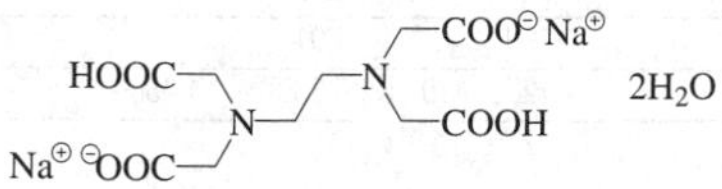

HOOC N N COO⊖ Na⊕ COOH Na⊕ ⊖OOC $2H_2O$
Ethylenediaminetetraacetic acid, disodium salt, dihydrate

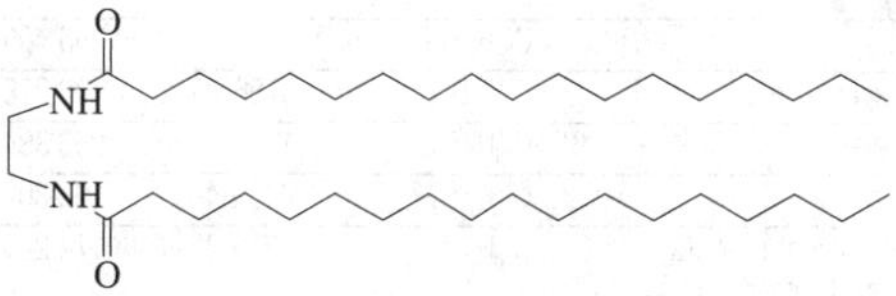

O NH NH O
N,N'-Ethylene distearylamide

N H
Ethyleneimine

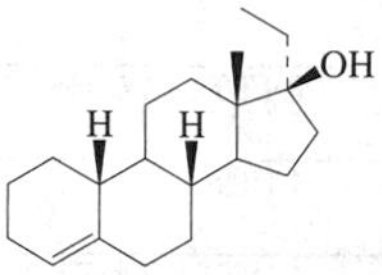

OH H H
Ethylestrenol

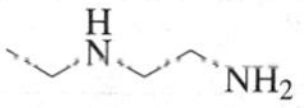

H N NH_2
N-Ethyl-1,2-ethanediamine

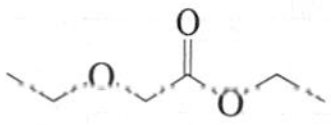

O O O
Ethyl ethoxyacetate

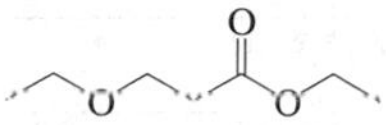

O O O
Ethyl 3-ethoxypropanoate

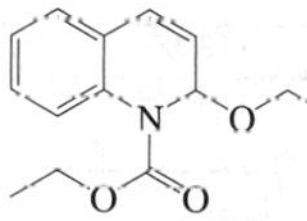

N O O O
Ethyl 2-ethoxy-1(2*H*)-quinolinecarboxylate

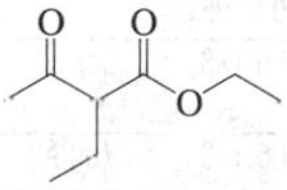

O O O
Ethyl 2-ethylacetoacetate

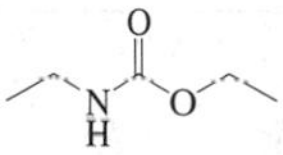

O N H O
Ethyl ethylcarbamate

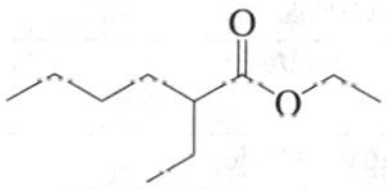

O O
Ethyl 2-ethylhexanoate

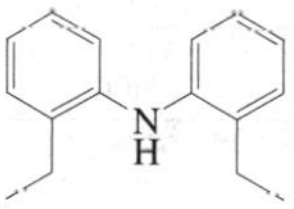

N H
2-Ethyl-*N*-(2-ethylphenyl)aniline

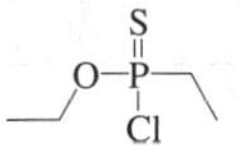

S O P Cl
O-Ethyl ethylthiophosphonyl chloride

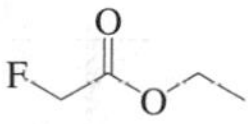

F O O
Ethyl fluoroacetate

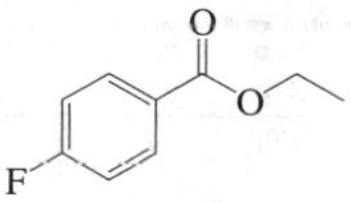

O O F
Ethyl 4-fluorobenzoate

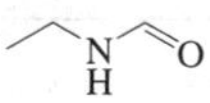

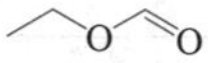

N H O
N-Ethylformamide

O O
Ethyl formate

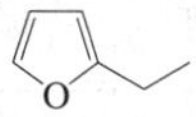

O
2-Ethylfuran

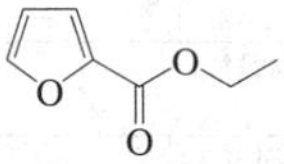

O O O
Ethyl 2-furancarboxylate

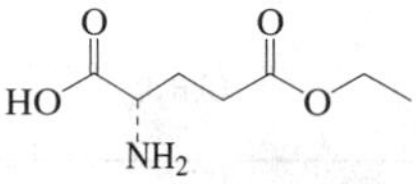

O O HO O NH_2
γ-Ethyl *L*-glutamate

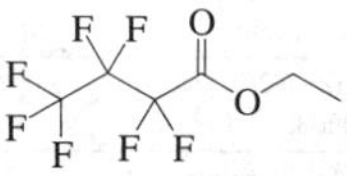

F F O F F F F F O
Ethyl heptafluorobutanoate

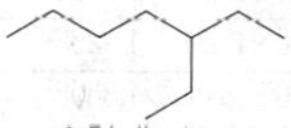

3-Ethylheptane

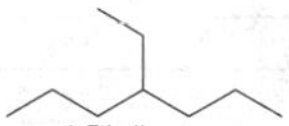

4-Ethylheptane

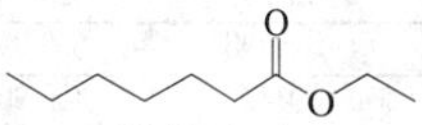

O O
Ethyl heptanoate

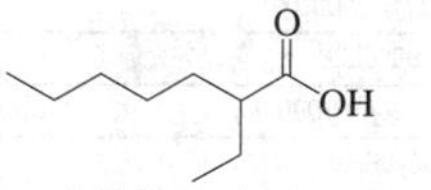

O OH
2-Ethylheptanoic acid

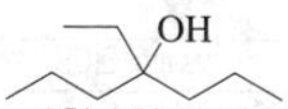

OH
4-Ethyl-4-heptanol

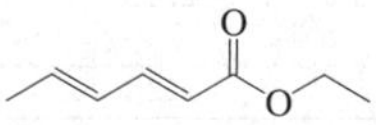

O O
Ethyl *trans,trans*-2,4-hexadienoate

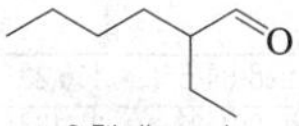

O
2-Ethylhexanal

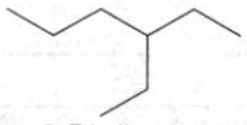

3-Ethylhexane

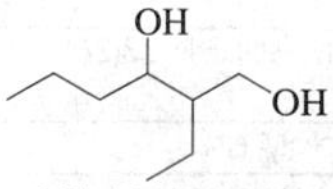

OH OH
2-Ethyl-1,3-hexanediol

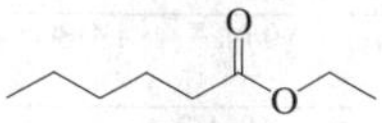

O O
Ethyl hexanoate

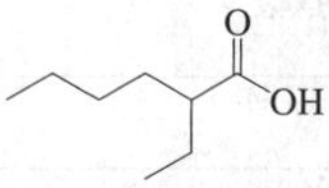

O OH
2-Ethylhexanoic acid

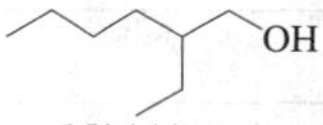

OH
2-Ethyl-1-hexanol

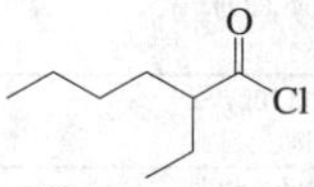

O Cl
2-Ethylhexanoyl chloride

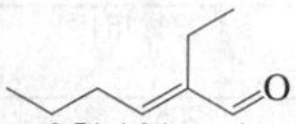

O
2-Ethyl-2-hexenal

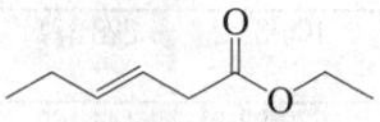

O O
Ethyl 3-hexenoate

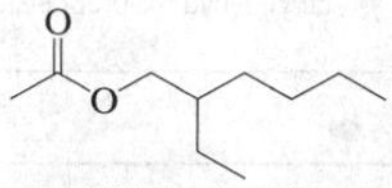

O O
2-Ethylhexyl acetate

No.	Name	Synonym	Mol. Form.	CAS RN	Mol. Wt.	Physical Form	mp/°C	bp/°C	den/ g cm-3	n_D	Solubility
5056	2-Ethylhexyl acrylate		$C_{11}H_{20}O_2$	103-11-7	184.276		-90	125^{60}	0.880^{25}	1.4332^{25}	
5057	2-Ethylhexylamine	2-Ethyl-1-hexanamine	$C_8H_{19}N$	104-75-6	129.244			169.2			sl H_2O
5058	2-Ethylhexyl butyl phthalate	Butyl 2-ethylhexyl phthalate	$C_{20}H_{30}O_4$	85-69-8	334.450	col liq					sl H_2O
5059	2-Ethylhexyl dihydrogen phosphate	Mono(2-ethylhexyl) phosphate	$C_8H_{19}O_4P$	1070-03-7	210.208	liq					s H_2O, bz
5060	2-Ethylhexyl diphenyl phosphite	Forstab	$C_{20}H_{27}O_3P$	15647-08-2	346.400			$152^{0.15}$	1.054^{20}	1.5207^{27}	
5061	Ethyl hexyl ether	1-Ethoxyhexane	$C_8H_{18}O$	5756-43-4	130.228			143	0.7722^{20}	1.4008^{20}	vs eth, EtOH
5062	2-Ethylhexyl 2-hydroxybenzoate	Octisalate	$C_{15}H_{22}O_3$	118-60-5	250.334	liq		190^{21}	1.01		
5063	2-Ethylhexyl methacrylate		$C_{12}H_{22}O_2$	688-84-6	198.302			120^{18}, 110^{14}	0.880^{25}	1.436^{25}	
5064	2-[(2-Ethylhexyl)oxy]ethanol	Ethylene glycol mono(2-ethylhexyl) ether	$C_{10}H_{22}O_2$	1559-35-9	174.281			227.7			
5065	Ethylhydrazine		$C_2H_8N_2$	624-80-6	60.098			101			vs H_2O, ace, eth, EtOH
5066	Ethyl hydrazinecarboxylate	Ethyl carbazate	$C_3H_8N_2O_2$	4114-31-2	104.108	cry	46	dec 198; 93^9			s EtOH, eth; sl chl
5067	Ethyl hydrogen adipate		$C_8H_{14}O_4$	626-86-8	174.195	hyg cry (eth, peth)	29	285	0.9796^{20}	1.4311^{20}	s EtOH, eth, peth
5068	Ethyl hydrogen fumarate		$C_6H_8O_4$	2459-05-4	144.126		70	147^{16}	1.1109^{87}		s EtOH, ace; sl chl
5069	Ethyl hydrogen succinate	Butanedioic acid, monoethyl ester	$C_6H_{10}O_4$	1070-34-4	146.141	pr or nd	8	172^{42}, 119^3	1.1466^{20}	1.4327^{20}	vs H_2O, eth, EtOH
5070	Ethyl hydroperoxide	Ethyl hydrogen peroxide	$C_2H_6O_2$	3031-74-1	62.068	liq	-100	95	0.9332^{20}	1.3800^{20}	vs H_2O, bz, eth, EtOH
5071	Ethyl hydroxyacetate		$C_4H_8O_3$	623-50-7	104.105			160	1.0826^{23}	1.4180^{20}	vs eth, EtOH
5072	Ethyl 3-hydroxybenzoate		$C_9H_{10}O_3$	7781-98-8	166.173	pl (bz)	74		1.0680^{131}		sl H_2O, chl; s EtOH, eth
5073	Ethyl 4-hydroxybenzoate	Ethylparaben	$C_9H_{10}O_3$	120-47-8	166.173	cry (dil al)	117	297.5			sl H_2O, chl, tfa; vs EtOH, eth; i CS_2
5074	Ethyl 3-hydroxybutanoate, (±)		$C_6H_{12}O_3$	35608-64-1	132.157			185; 76^{15}	1.017^{20}	1.4182^{20}	s H_2O, EtOH; sl ctc
5075	Ethyl 2-hydroxy-3-butenoate		$C_6H_{10}O_3$	91890-87-8	130.141			dec 173; 68^{15}	1.0470^{15}	1.436^{13}	vs H_2O, eth, EtOH
5076	α-Ethyl-1-hydroxycyclohexaneacetic acid	Cyclobutyrol	$C_{10}H_{18}O_3$	512-16-3	186.248	cry (eth-peth)	81.5	164^{24}	1.0010^{18}	1.4680^{18}	vs ace, eth, EtOH, chl
5077	*N*-Ethyl-*N*-hydroxyethanamine	*N,N*-Diethylhydroxylamine	$C_4H_{11}NO$	3710-84-7	89.136		10	133	0.8669^{20}	1.4195^{20}	
5078	2-Ethyl-3-hydroxyhexanal		$C_8H_{16}O_2$	496-03-7	144.212			138^{50}, 101^9			
5079	Ethyl 4-hydroxy-3-methoxybenzoate		$C_{10}H_{12}O_4$	617-05-0	196.200	nd (dil al)	44	292			i H_2O; vs EtOH, eth; s chl
5080	Ethyl *cis*-12-hydroxy-9-octadecenoate, (*R*)	Ethyl ricinoleate	$C_{20}H_{38}O_3$	55066-53-0	326.514			258^{13}	0.9180^{20}	1.4618^{22}	
5081	Ethylidenecyclohexane		C_8H_{14}	1003-64-1	110.197			136	0.822^{25}	1.4618^{20}	
5082	5-Ethylidene-2-norbornene	5-Ethylidenebicyclo[2.2.1]hept-2-ene	C_9H_{12}	16219-75-3	120.191	liq		146	0.893	1.4900^{20}	
5083	1-Ethyl-1*H*-imidazole		$C_5H_8N_2$	7098-07-9	96.131			208	0.999^{25}		msc H_2O
5084	Ethyl iodoacetate		$C_4H_7IO_2$	623-48-3	214.002	oil		179	1.8173^{13}	1.5079^{13}	s EtOH, eth
5085	Ethyl isobutylcarbamate	Isobutyl urethane	$C_7H_{15}NO_2$	539-89-9	145.200		<-65	110^{30}	0.9432^{20}	1.4288^{20}	vs eth, EtOH
5086	Ethyl isocyanate		C_3H_5NO	109-90-0	71.078			60	0.9031^{20}	1.3808^{20}	i H_2O; msc EtOH, eth
5087	Ethyl isocyanide		C_3H_5N	624-79-3	55.079		<-66	79	0.7402^{20}	1.3622^{20}	vs H_2O; msc EtOH, eth; s ace
5088	*N*-Ethyl-1*H*-isoindole-1,3(2*H*)-dione		$C_{10}H_9NO_2$	5022-29-7	175.184	nd (al)	79	285.5			s EtOH, eth
5089	Ethyl isopentyl ether		$C_7H_{16}O$	628-04-6	116.201			112.5	0.7688^{21}		vs eth, EtOH
5090	Ethylisopropylamine	*N*-Ethyl-2-propanamine	$C_5H_{13}N$	19961-27-4	87.164			69.6		1.3872^{25}	
5091	1-Ethyl-2-isopropylbenzene		$C_{11}H_{16}$	18970-44-0	148.245			193	0.888^{20}	1.508^{20}	vs ace, bz, eth, EtOH
5092	Ethyl isopropyl ether		$C_5H_{12}O$	625-54-7	88.148			54.1	0.720^{25}	1.3698^{25}	s H_2O, ace, chl; msc EtOH, eth
5093	*N*-Ethyl-*N*-isopropyl-2-propanamine		$C_8H_{19}N$	7087-68-5	129.244			126.5	0.742^{25}	1.4138^{20}	s ctc
5094	Ethyl isopropyl sulfide		$C_5H_{12}S$	5145-99-3	104.214	liq	-122.2	107.5	0.8246^{20}		
5095	Ethyl isothiocyanate		C_3H_5NS	542-85-8	87.144	liq	-5.9	131.5	0.9990^{20}	1.5130^{20}	i H_2O; msc EtOH, eth
5096	Ethyl lactate	Ethyl 2-hydroxypropionate	$C_5H_{10}O_3$	2676-33-7	118.131	liq	-26	154.5	1.0328^{20}	1.4124^{20}	vs H_2O, eth, EtOH
5097	Ethyl laurate		$C_{14}H_{28}O_2$	106-33-2	228.371	liq	-10	271; 154^{15}	0.8618^{20}	1.4311^{20}	i H_2O; vs EtOH; msc eth; sl ctc
5098	Ethyl levulinate		$C_7H_{12}O_3$	539-88-8	144.168			205.8	1.0111^{20}	1.4229^{20}	vs H_2O, EtOH
5099	Ethyl mercaptoacetate		$C_4H_8O_2S$	623-51-8	120.171			157	1.0964^{15}	1.4582^{20}	s EtOH, eth; sl ctc
5100	Ethyl methacrylate		$C_6H_{10}O_2$	97-63-2	114.142			117	0.9135^{20}	1.4147^{20}	sl H_2O, chl; msc EtOH, eth

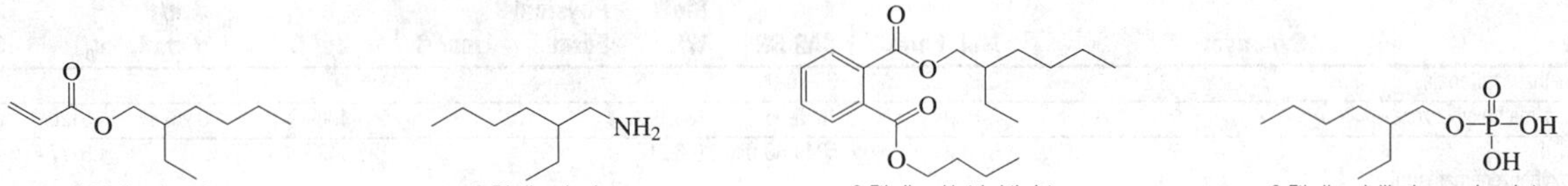

2-Ethylhexyl acrylate

2-Ethylhexylamine

2-Ethylhexyl butyl phthalate

2-Ethylhexyl dihydrogen phosphate

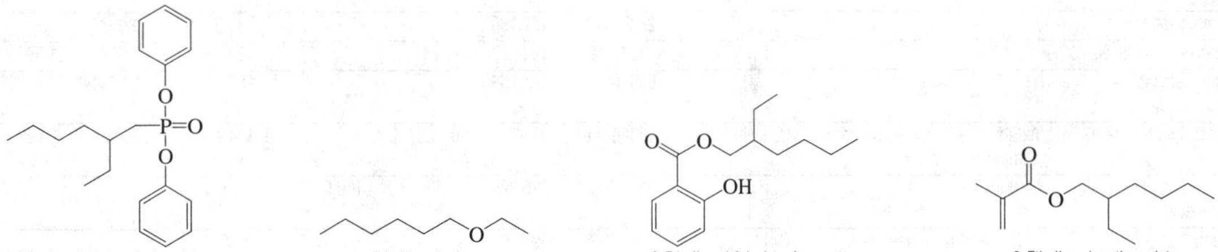

2-Ethylhexyl diphenyl phosphite

Ethyl hexyl ether

2-Ethylhexyl 2-hydroxybenzoate

2-Ethylhexyl methacrylate

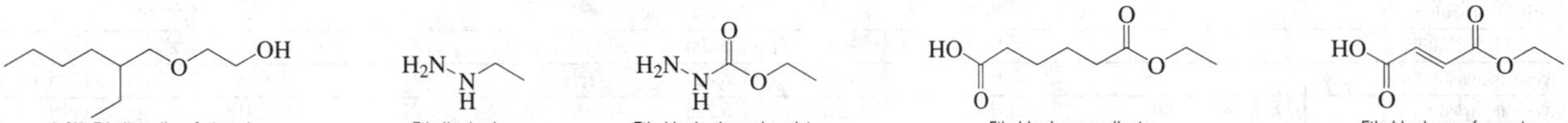

2-[(2-Ethylhexyl)oxy]ethanol

Ethylhydrazine

Ethyl hydrazinecarboxylate

Ethyl hydrogen adipate

Ethyl hydrogen fumarate

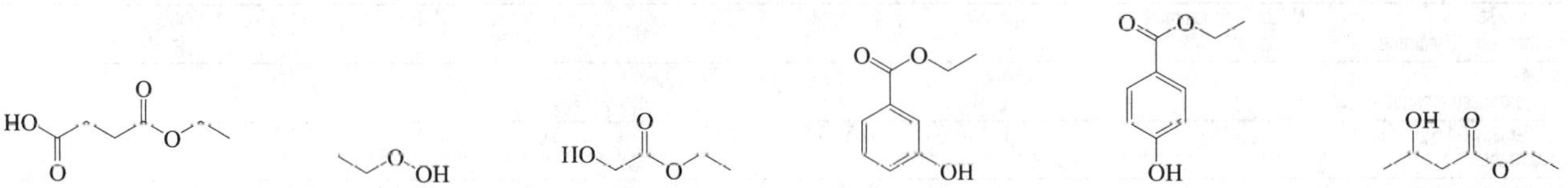

Ethyl hydrogen succinate

Ethyl hydroperoxide

Ethyl hydroxyacetate

Ethyl 3-hydroxybenzoate

Ethyl 4-hydroxybenzoate

Ethyl 3-hydroxybutanoate, (±)

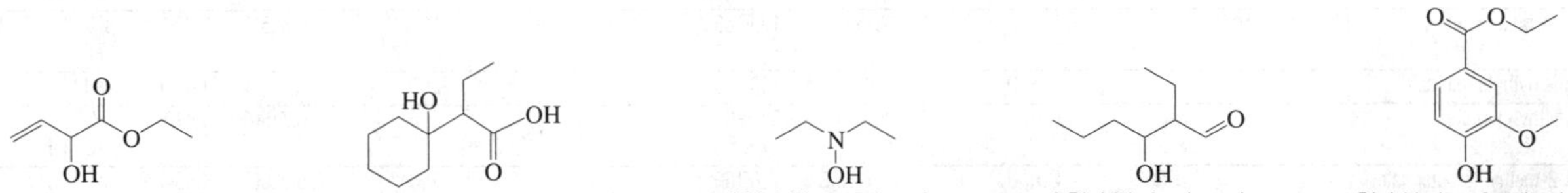

Ethyl 2-hydroxy-3-butenoate

α-Ethyl-1-hydroxycyclohexaneacetic acid

N-Ethyl-*N*-hydroxyethanamine

2-Ethyl-3-hydroxyhexanal

Ethyl 4-hydroxy-3-methoxybenzoate

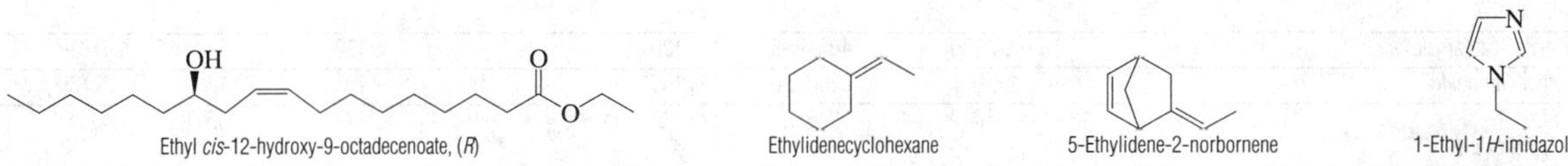

Ethyl *cis*-12-hydroxy-9-octadecenoate, (*R*)

Ethylidenecyclohexane

5-Ethylidene-2-norbornene

1-Ethyl-1*H*-imidazole

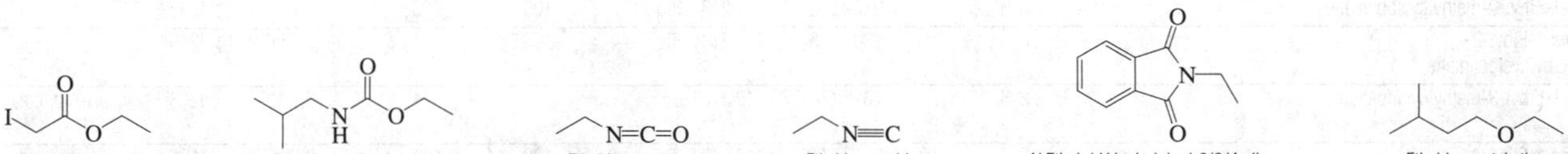

Ethyl iodoacetate

Ethyl isobutylcarbamate

Ethyl isocyanate

Ethyl isocyanide

N-Ethyl-1*H*-isoindole-1,3(2*H*)-dione

Ethyl isopentyl ether

Ethylisopropylamine

1-Ethyl-2-isopropylbenzene

Ethyl isopropyl ether

N-Ethyl-*N*-isopropyl-2-propanamine

Ethyl isopropyl sulfide

Ethyl isothiocyanate

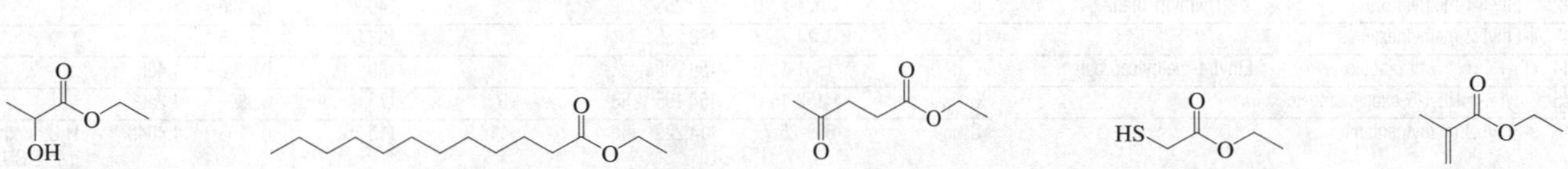

Ethyl lactate

Ethyl laurate

Ethyl levulinate

Ethyl mercaptoacetate

Ethyl methacrylate

No.	Name	Synonym	Mol. Form.	CAS RN	Mol. Wt.	Physical Form	mp/°C	bp/°C	den/ g cm^{-3}	n_D	Solubility
5101	Ethyl methanesulfonate		$C_3H_8O_3S$	62-50-0	124.159			86^{10}			
5102	1-Ethyl-4-methoxybenzene		$C_9H_{12}O$	1515-95-3	136.190			198	0.9624^{15}	1.5120^{20}	vs bz, eth
5103	α-Ethyl-4-methoxybenzenemethanol		$C_{10}H_{14}O_2$	5349-60-0	166.217			143^{20}		1.5277^{20}	s ctc
5104	Ethyl 2-methoxybenzoate		$C_{10}H_{12}O_3$	7335-26-4	180.200			261	1.1124^{20}	1.5224^{20}	vs eth, EtOH
5105	Ethyl 4-methoxybenzoate		$C_{10}H_{12}O_3$	94-30-4	180.200		7.5	269.5	1.1038^{20}	1.5254^{20}	i H_2O; s EtOH, eth
5106	4-Ethyl-2-methoxyphenol		$C_9H_{12}O_2$	2785-89-9	152.190	liq	-7	236.5	1.0931^{18}		
5107	Ethyl (4-methoxyphenyl)acetate		$C_{11}H_{14}O_3$	14062-18-1	194.227			139^{70}	1.097^{25}	1.5075^{20}	
5108	Ethyl 2-methylacetoacetate		$C_7H_{12}O_3$	609-14-3	144.168			187	0.9941^{20}	1.4185^{20}	sl H_2O; s EtOH, eth; vs ace
5109	*N*-Ethyl-2-methylallylamine	*N*-Ethyl-2-methyl-2-propen-1-amine	$C_6H_{13}N$	18328-90-0	99.174	liq		104.7	0.753	1.4221^{20}	msc H_2O
5110	5-Ethyl-5-(2-methylallyl)-2-thiobarbituric acid	Methallatal	$C_{10}H_{14}N_2O_2S$	115-56-0	226.295		160.5				
5111	Ethylmethylamine	*N*-Methylethanamine	C_3H_9N	624-78-2	59.110			36.7			vs H_2O, ace, eth, EtOH
5112	Ethylmethylamine hydrochloride	*N*-Methylethanamine hydrochloride	$C_3H_{10}ClN$	624-60-2	95.571	pl (al-eth)	128		1.0874^{20}		vs H_2O, EtOH; i eth; s chl
5113	2-Ethyl-6-methylaniline		$C_9H_{13}N$	24549-06-2	135.206	liq	-33	231	0.968^{25}	1.5525^{20}	
5114	*N*-Ethyl-2-methylaniline		$C_9H_{13}N$	94-68-8	135.206		<-15	216	0.948^{25}	1.5456^{20}	s EtOH, eth
5115	*N*-Ethyl-3-methylaniline		$C_9H_{13}N$	102-27-2	135.206			221	0.9263^{15}	1.5451^{20}	s EtOH, eth
5116	*N*-Ethyl-4-methylaniline	*N*-Ethyl-4-toluidine	$C_9H_{13}N$	622-57-1	135.206			217	0.9391^{16}		s EtOH, eth
5117	*N*-Ethyl-*N*-methylaniline		$C_9H_{13}N$	613-97-8	135.206			204	0.92^{55}		i H_2O; msc EtOH, eth; s ctc
5118	*N*-Ethyl-α-methylbenzeneethanamine	*N*-Ethylamphetamine	$C_{11}H_{17}N$	457-87-4	163.260			105^{14}		1.4986^{25}	
5119	*N*-Ethyl-4-methylbenzenesulfonamide		$C_9H_{13}NO_2S$	80-39-7	199.270		64				s EtOH
5120	1-Ethyl-2-methyl-1*H*-benzimidazole		$C_{10}H_{12}N_2$	5805-76-5	160.215		51	296	1.073^{25}		
5121	Ethyl 2-methylbenzoate		$C_{10}H_{12}O_2$	87-24-1	164.201		<-10	227; 113^{18}	1.0325^{21}	1.507^{22}	i H_2O; msc EtOH, eth
5122	Ethyl 4-methylbenzoate		$C_{10}H_{12}O_2$	94-08-6	164.201			232	1.0269^{18}	1.5089^{18}	i H_2O; msc EtOH, eth
5123	Ethyl 3-methylbutanoate	Ethyl isovalerate	$C_7H_{14}O_2$	108-64-5	130.185	liq	-99.3	135.0	0.8656^{20}	1.3962^{20}	sl H_2O; vs EtOH, eth
5124	2-Ethyl-2-methylbutanoic acid		$C_7H_{14}O_2$	19889-37-3	130.185		<-20	207		1.4250^{20}	vs EtOH
5125	2-Ethyl-3-methyl-1-butene		C_7H_{14}	7357-93-9	98.186			89	0.7150^{20}	1.410^{20}	i H_2O; s eth, ace, bz, chl
5126	Ethyl *trans*-2-methyl-2-butenoate		$C_7H_{12}O_2$	5837-78-5	128.169			156	0.9200^{20}	1.4340^{20}	
5127	Ethyl 3-methyl-2-butenoate		$C_7H_{12}O_2$	638-10-8	128.169			153.5	0.9199^{21}	1.4345^{20}	
5128	5-Ethyl-5-(1-methylbutyl)-2,4,6(1*H*,3*H*,5*H*)-pyrimidinetrione		$C_{11}H_{18}N_2O_3$	76-74-4	226.272		130				sl H_2O; s EtOH, eth
5129	Ethyl *N*-methylcarbamate		$C_4H_9NO_2$	105-40-8	103.120			170	1.0115^{20}	1.4183^{20}	vs H_2O, EtOH
5130	Ethyl methyl carbonate		$C_4H_8O_3$	623-53-0	104.105	liq	-14	107.5	1.012^{20}	1.3778^{20}	vs eth, EtOH
5131	*trans*-1-Ethyl-4-methylcyclohexane		C_9H_{18}	6236-88-0	126.239	liq	-80.8	149	0.7798^{20}	1.4304^{20}	
5132	1-Ethyl-1-methylcyclopentane		C_8H_{16}	16747-50-5	112.213	liq	-143.8	121.6	0.7767^{25}	1.4272^{20}	vs ace, bz, eth, EtOH
5133	*cis*-1-Ethyl-2-methylcyclopentane		C_8H_{16}	930-89-2	112.213	liq	-106	128	0.7852^{20}	1.4293^{20}	
5134	*trans*-1-Ethyl-2-methylcyclopentane		C_8H_{16}	930-90-5	112.213	liq	-105.9	121.2	0.7649^{25}	1.4219^{20}	
5135	*cis*-1-Ethyl-3-methylcyclopentane		C_8H_{16}	2613-66-3	112.213			121	0.7724^{20}	1.4203^{20}	vs ace, bz, eth, EtOH
5136	*trans*-1-Ethyl-3-methylcyclopentane		C_8H_{16}	2613-65-2	112.213	liq	-108	121	0.7619^{20}	1.4186^{20}	
5137	1-Ethyl-1-methylcyclopropane		C_6H_{12}	53778-43-1	84.159	liq	-130.2	56.8	0.6968^{25}	1.3887^{20}	
5138	2-Ethyl-2-methyl-1,3-dioxolane		$C_6H_{12}O_2$	126-39-6	116.158			118	0.9360^{20}		
5139	Ethyl methyl ether		C_3H_8O	540-67-0	60.095	col gas	-113	7.4	0.7251^{0}	1.3420^{4}	s H_2O, ace, chl; msc EtOH, eth
5140	3-Ethyl-2-methylhexane		C_9H_{20}	16789-46-1	128.255			138	0.7310^{20}	1.4106^{20}	
5141	3-Ethyl-3-methylhexane		C_9H_{20}	3074-76-8	128.255			140.6	0.7371^{25}	1.4140^{20}	
5142	3-Ethyl-4-methylhexane	2,3-Diethylpentane	C_9H_{20}	3074-77-9	128.255			140	0.7420^{20}	1.4134^{20}	
5143	4-Ethyl-2-methylhexane		C_9H_{20}	3074-75-7	128.255			133.8	0.7195^{25}	1.4063^{20}	
5144	Ethyl 4-methylhexanoate	Ethyl 4-methylcaproate	$C_9H_{18}O_2$	1561-10-0	158.238			180	0.8708^{20}	1.4051^{20}	
5145	Ethyl 4-methyl-3-oxopentanoate		$C_8H_{14}O_3$	7152-15-0	158.195	liq	-9	173	0.98^{25}	1.250^{20}	
5146	3-Ethyl-2-methylpentane		C_8H_{18}	609-26-7	114.229	liq	-114.9	115.66	0.7193^{20}	1.4040^{20}	i H_2O; s eth; msc EtOH, ace, bz

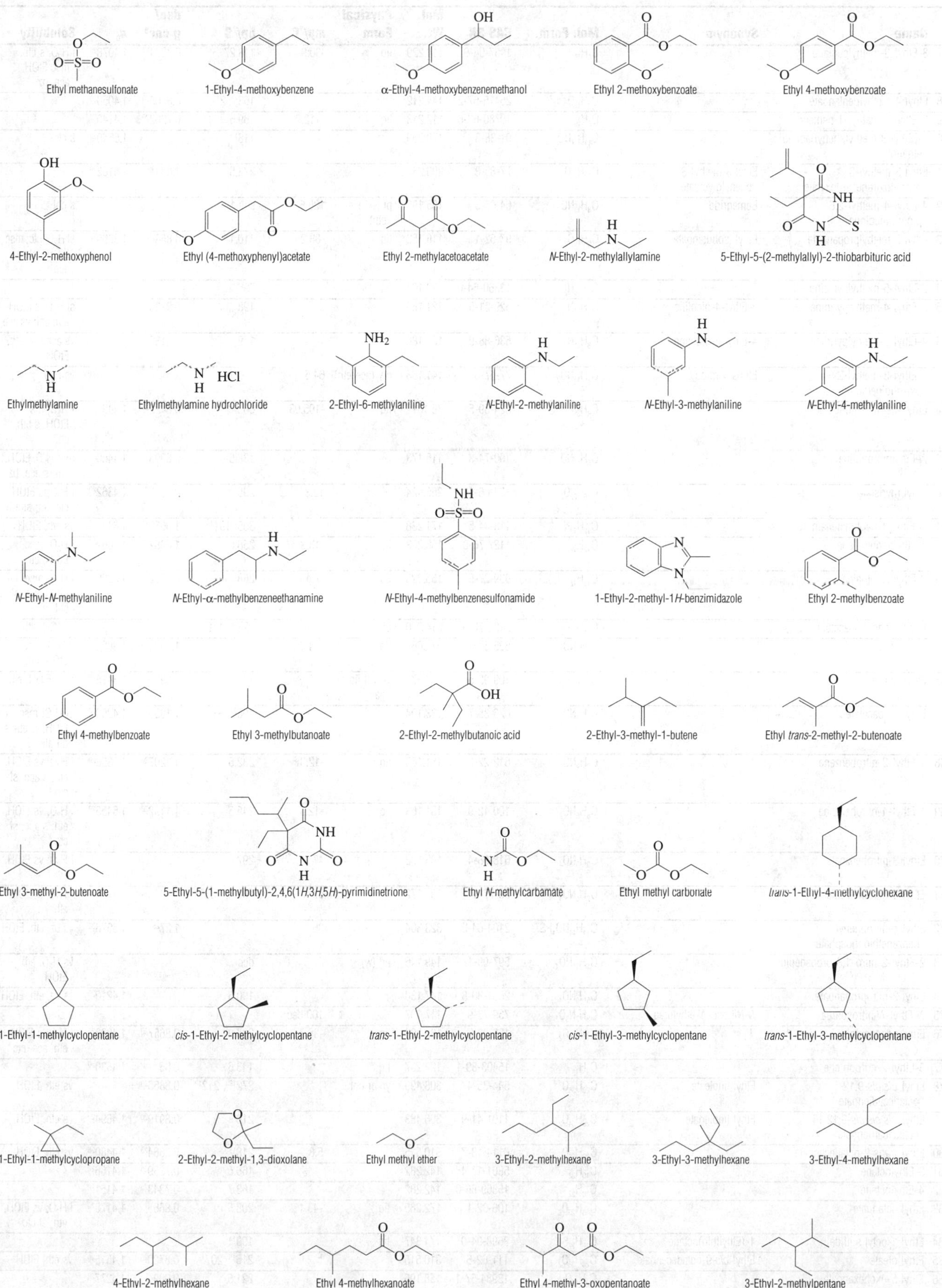
Ethyl methanesulfonate
1-Ethyl-4-methoxybenzene
α-Ethyl-4-methoxybenzenemethanol
Ethyl 2-methoxybenzoate
Ethyl 4-methoxybenzoate
4-Ethyl-2-methoxyphenol
Ethyl (4-methoxyphenyl)acetate
Ethyl 2-methylacetoacetate
N-Ethyl-2-methylallylamine
5-Ethyl-5-(2-methylallyl)-2-thiobarbituric acid
Ethylmethylamine
Ethylmethylamine hydrochloride
2-Ethyl-6-methylaniline
N-Ethyl-2-methylaniline
N-Ethyl-3-methylaniline
N-Ethyl-4-methylaniline
N-Ethyl-N-methylaniline
N-Ethyl-α-methylbenzeneethanamine
N-Ethyl-4-methylbenzenesulfonamide
1-Ethyl-2-methyl-1H-benzimidazole
Ethyl 2-methylbenzoate
Ethyl 4-methylbenzoate
Ethyl 3-methylbutanoate
2-Ethyl-2-methylbutanoic acid
2-Ethyl-3-methyl-1-butene
Ethyl trans-2-methyl-2-butenoate
Ethyl 3-methyl-2-butenoate
5-Ethyl-5-(1-methylbutyl)-2,4,6(1H,3H,5H)-pyrimidinetrione
Ethyl N-methylcarbamate
Ethyl methyl carbonate
trans-1-Ethyl-4-methylcyclohexane
1-Ethyl-1-methylcyclopentane
cis-1-Ethyl-2-methylcyclopentane
trans-1-Ethyl-2-methylcyclopentane
cis-1-Ethyl-3-methylcyclopentane
trans-1-Ethyl-3-methylcyclopentane
1-Ethyl-1-methylcyclopropane
2-Ethyl-2-methyl-1,3-dioxolane
Ethyl methyl ether
3-Ethyl-2-methylhexane
3-Ethyl-3-methylhexane
3-Ethyl-4-methylhexane
4-Ethyl-2-methylhexane
Ethyl 4-methylhexanoate
Ethyl 4-methyl-3-oxopentanoate
3-Ethyl-2-methylpentane

No.	Name	Synonym	Mol. Form.	CAS RN	Mol. Wt.	Physical Form	mp/°C	bp/°C	den/ g cm^{-3}	n_D	Solubility
5147	3-Ethyl-3-methylpentane		C_8H_{18}	1067-08-9	114.229	liq	-90.9	118.27	0.7274^{20}	1.4078^{20}	i H_2O; s eth; msc EtOH, ace, bz
5148	Ethyl 4-methylpentanoate		$C_8H_{16}O_2$	25415-67-2	144.212			163; 52^{10}	0.8705^{20}	1.4050^{20}	
5149	3-Ethyl-2-methyl-1-pentene		C_8H_{16}	19780-66-6	112.213	liq	-112.9	109.5	0.7262^{20}	1.4140^{20}	
5150	2-[Ethyl(3-methylphenyl)amino] ethanol		$C_{11}H_{17}NO$	91-88-3	179.259			$118^{1.5}$		1.5540^{20}	s ctc
5151	Ethyl 3-methyl-3-phenyloxiranecarboxylate	Ethyl 3-methyl-3-phenylglycidate	$C_{12}H_{14}O_3$	77-83-8	206.237			273.5	1.044^{20}	1.5182^{20}	
5152	4-Ethyl-4-methyl-2,6-piperidinedione	Bemegride	$C_8H_{13}NO_2$	64-65-3	155.195	pl (w, ace-eth)	126.5	sub 100			s chl
5153	Ethyl 2-methylpropanoate	Ethyl isobutanoate	$C_6H_{12}O_2$	97-62-1	116.158	liq	-88.2	110.1	0.868^{20}	1.3869^{18}	sl H_2O, ctc; msc EtOH, eth; s ace
5154	2-Ethyl-5-methylpyrazine		$C_7H_{10}N_2$	13360-64-0	122.167	liq		79^{56}			
5155	3-Ethyl-4-methylpyridine	3-Ethyl-4-picoline	$C_8H_{11}N$	529-21-5	121.180			198	0.9286^{17}		sl H_2O; s EtOH, eth, chl; vs ace
5156	4-Ethyl-2-methylpyridine	4-Ethyl-2-picoline	$C_8H_{11}N$	536-88-9	121.180			179	0.9130^{25}		vs ace, bz, eth, EtOH
5157	3-Ethyl-3-methyl-2,5-pyrrolidinedione	Ethosuximide	$C_7H_{11}NO_2$	77-67-8	141.168	cry (ace-eth)	64.5				vs H_2O
5158	Ethyl methyl sulfide		C_3H_8S	624-89-5	76.161	liq	-105.93	66.7	0.8422^{20}	1.4404^{20}	i H_2O; msc EtOH; s eth, chl
5159	*N*-Ethylmorpholine		$C_6H_{13}NO$	100-74-3	115.173			138.5	0.8996^{20}	1.4400^{20}	msc H_2O, EtOH, eth; s ace, bz
5160	Ethyl myristate		$C_{16}H_{32}O_2$	124-06-1	256.424		12.3	295	0.8573^{25}	1.4362^{20}	i H_2O; s EtOH, ctc, lig; sl eth
5161	*N*-Ethyl-1-naphthalenamine		$C_{12}H_{13}N$	118-44-5	171.238			305; 191^{16}	1.0652^{15}	1.6477^{15}	vs eth, EtOH
5162	1-Ethylnaphthalene		$C_{12}H_{12}$	1127-76-0	156.223	liq	-13.9	258.6	1.0082^{20}	1.6062^{20}	i H_2O; msc EtOH, eth
5163	2-Ethylnaphthalene		$C_{12}H_{12}$	939-27-5	156.223	liq	-7.4	258	0.9922^{20}	1.5999^{20}	i H_2O; msc EtOH, eth; sl chl
5164	Ethyl 1-naphthylacetate		$C_{14}H_{14}O_2$	2122-70-5	214.260	oil	88.5	222^{20}, 118^{13}			s EtOH, eth
5165	Ethyl nitrate		$C_2H_5NO_3$	625-58-1	91.066	liq	-94.6	87.2	1.1084^{20}	1.3852^{20}	s H_2O; msc EtOH, eth
5166	Ethyl nitrite		$C_2H_5NO_2$	109-95-5	75.067	ye vol liq or gas		18	0.899^{15}	1.3418^{10}	msc EtOH, eth
5167	Ethyl nitroacetate		$C_4H_7NO_4$	626-35-7	133.104			106^{25}, 83^{6}	1.1953^{20}	1.4250^{20}	sl H_2O; msc EtOH; vs eth; s dil alk
5168	1-Ethyl-2-nitrobenzene		$C_8H_9NO_2$	612-22-6	151.163	liq	-12.3	232.5	1.1207^{20}	1.5356^{20}	i H_2O; vs EtOH, eth; s ace; sl ctc
5169	1-Ethyl-4-nitrobenzene		$C_8H_9NO_2$	100-12-9	151.163	liq	-12.3	245.5	1.1192^{20}	1.5455^{20}	i H_2O; vs EtOH, eth; s ace; sl ctc
5170	Ethyl 3-nitrobenzoate		$C_9H_9NO_4$	618-98-4	195.172		47	297			i H_2O; vs EtOH, eth
5171	Ethyl 4-nitrobenzoate		$C_9H_9NO_4$	99-77-4	195.172		57	186.3; 153^{8}			i H_2O; s EtOH, eth
5172	Ethyl *p*-nitrophenyl benzenethiophosphate		$C_{14}H_{14}NO_4PS$	2104-64-5	323.304		36		1.27^{25}	1.5978^{30}	vs bz, eth, EtOH
5173	2-Ethyl-2-nitro-1,3-propanediol		$C_5H_{11}NO_4$	597-09-1	149.146	nd (w)	57.5	dec			vs H_2O, eth, EtOH
5174	Ethyl 2-nitropropanoate		$C_5H_9NO_4$	2531-80-8	147.130			190.5		1.4210^{20}	vs bz, eth, EtOH
5175	*N*-Ethyl-*N*-nitrosourea	*N*-Nitroso-*N*-ethylurea	$C_3H_7N_3O_2$	759-73-9	117.107		100 dec				s chl
5176	Ethyl nonanoate		$C_{11}H_{22}O_2$	123-29-5	186.292	liq	-36.7	227.0	0.8657^{20}	1.4220^{20}	i H_2O; s EtOH, eth, ace, ctc
5177	5-Ethyl-2-norbornene		C_9H_{14}	15403-89-1	122.207	liq		143.6	0.86	1.4630^{20}	
5178	Ethyl *cis,cis*-9,12-octadecadienoate	Ethyl linoleate	$C_{20}H_{36}O_2$	544-35-4	308.499	ye or col		272^{180}, 212^{12}	0.8865^{20}		vs eth, EtOH
5179	Ethyl *cis,cis,cis*-9,12,15-octadecatrienoate	Ethyl linolenate	$C_{20}H_{34}O_2$	1191-41-9	306.483			218^{15}	0.8919^{20}	1.4694^{20}	vs eth, EtOH
5180	Ethyl *trans*-9-octadecenoate		$C_{20}H_{38}O_2$	6114-18-7	310.515		5.8	218^{15}	0.8664^{25}	1.4480^{25}	vs eth, EtOH
5181	3-Ethyloctane		$C_{10}H_{22}$	5881-17-4	142.282			166.5	0.7359^{25}	1.4156^{20}	
5182	4-Ethyloctane		$C_{10}H_{22}$	15869-86-0	142.282			163.7	0.7343^{25}	1.4151^{20}	
5183	Ethyl octanoate		$C_{10}H_{20}O_2$	106-32-1	172.265	liq	-43.1	208.5	0.866^{18}	1.4178^{20}	i H_2O; vs EtOH, eth; sl ctc
5184	Ethyl 1-octyl sulfide	1-(Ethylthio)octane	$C_{10}H_{22}S$	3698-94-0	174.347	liq		109^{14}			
5185	Ethyl oleate	Ethyl *cis*-9-octadecenoate	$C_{20}H_{38}O_2$	111-62-6	310.515			216^{15}, 207^{13}	0.8720^{20}	1.4515^{20}	vs eth, EtOH
5186	Ethyl 5-oxohexanoate		$C_8H_{14}O_3$	13984-57-1	158.195			221.5	0.989^{25}	1.4277^{20}	
5187	Ethyl 3-oxopentanoate		$C_7H_{12}O_3$	4949-44-4	144.168			191	1.0120^{20}	1.4230^{20}	vs bz, eth, EtOH

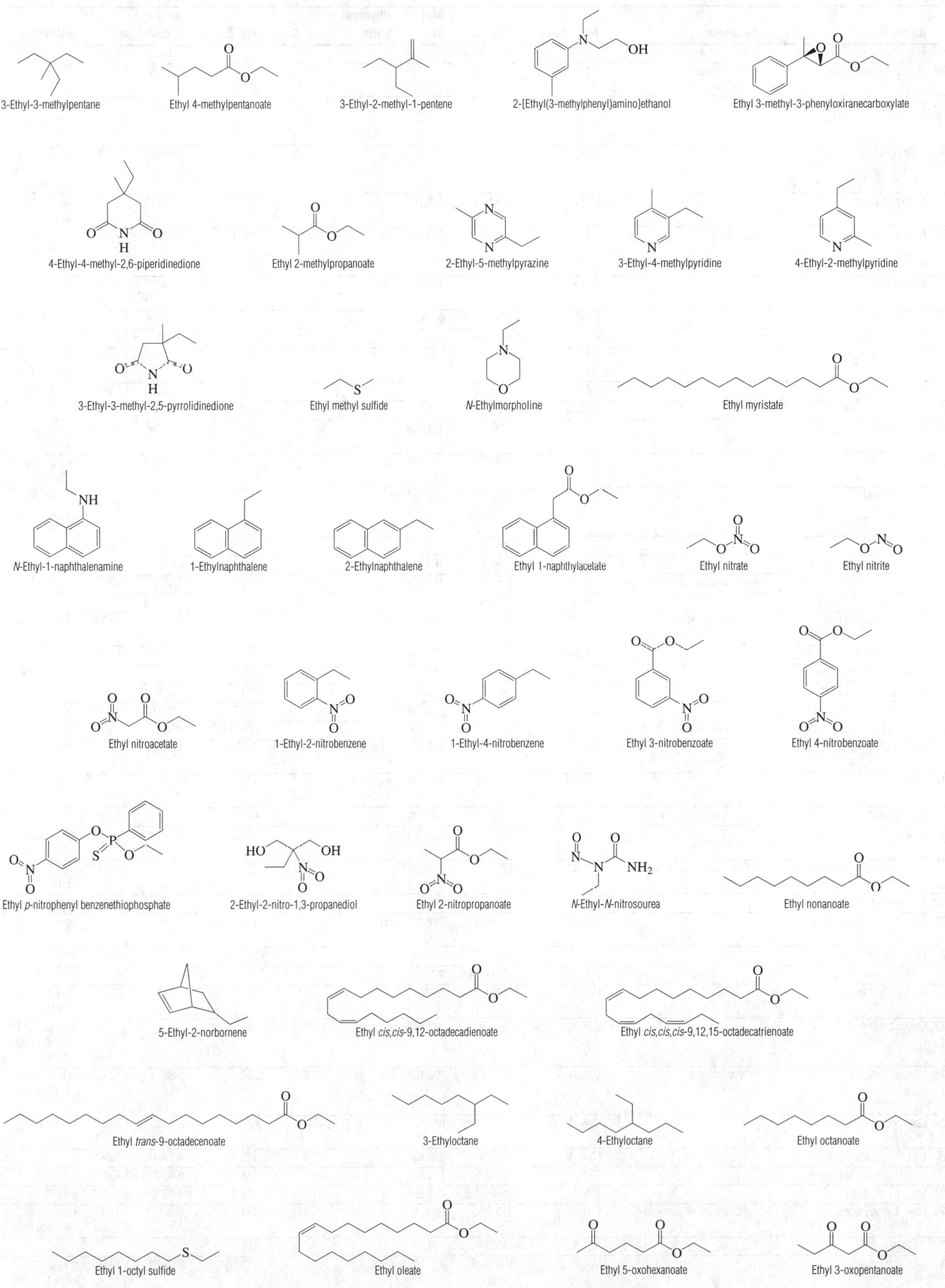

3-Ethyl-3-methylpentane
Ethyl 4-methylpentanoate
3-Ethyl-2-methyl-1-pentene
2-[Ethyl(3-methylphenyl)amino]ethanol
Ethyl 3-methyl-3-phenyloxiranecarboxylate
4-Ethyl-4-methyl-2,6-piperidinedione
Ethyl 2-methylpropanoate
2-Ethyl-5-methylpyrazine
3-Ethyl-4-methylpyridine
4-Ethyl-2-methylpyridine
3-Ethyl-3-methyl-2,5-pyrrolidinedione
Ethyl methyl sulfide
N-Ethylmorpholine
Ethyl myristate
N-Ethyl-1-naphthalenamine
1-Ethylnaphthalene
2-Ethylnaphthalene
Ethyl 1-naphthylacetate
Ethyl nitrate
Ethyl nitrite
Ethyl nitroacetate
1-Ethyl-2-nitrobenzene
1-Ethyl-4-nitrobenzene
Ethyl 3-nitrobenzoate
Ethyl 4-nitrobenzoate
Ethyl p-nitrophenyl benzenethiophosphate
2-Ethyl-2-nitro-1,3-propanediol
Ethyl 2-nitropropanoate
N-Ethyl-N-nitrosourea
Ethyl nonanoate
5-Ethyl-2-norbornene
Ethyl cis,cis-9,12-octadecadienoate
Ethyl cis,cis,cis-9,12,15-octadecatrienoate
Ethyl trans-9-octadecenoate
3-Ethyloctane
4-Ethyloctane
Ethyl octanoate
Ethyl 1-octyl sulfide
Ethyl oleate
Ethyl 5-oxohexanoate
Ethyl 3-oxopentanoate

No.	Name	Synonym	Mol. Form.	CAS RN	Mol. Wt.	Physical Form	mp/°C	bp/°C	den/ g cm-3	n_D	Solubility
5188	Ethyl 2-oxo-2-phenylacetate	Ethyl phenylglyoxylate	$C_{10}H_{10}O_3$	1603-79-8	178.184			256.5	1.1222^{25}	1.5190^{25}	
5189	Ethyl 2-oxopropanoate	Ethyl pyruvate	$C_5H_8O_3$	617-35-6	116.116	liq	-50	155	1.0596^{15}	1.4052^{20}	sl H_2O; s ace; msc EtOH, eth
5190	Ethyl palmitate		$C_{18}H_{36}O_2$	628-97-7	284.478	nd	24	191^{10}	0.8577^{25}	1.4347^{34}	i H_2O; s EtOH, eth, ace, bz, chl
5191	3-Ethylpentane		C_7H_{16}	617-78-7	100.202	liq	-118.55	93.5	0.6982^{20}	1.3934^{20}	i H_2O; s EtOH, eth; msc ace, bz, hp, chl
5192	3-Ethyl-2,4-pentanedione		$C_7H_{12}O_2$	1540-34-7	128.169			178.5	0.9531^{19}	1.4408^{19}	vs eth, EtOH, chl
5193	Ethyl pentanoate	Ethyl valerate	$C_7H_{14}O_2$	539-82-2	130.185	liq	-91.2	146.1	0.8770^{20}	1.4120^{20}	i H_2O; msc EtOH, eth; sl ctc
5194	3-Ethyl-3-pentanol		$C_7H_{16}O$	597-49-9	116.201	liq	-12.5	142	0.8407^{22}	1.4294^{20}	sl H_2O; s EtOH, eth
5195	2-Ethyl-1-pentene		C_7H_{14}	3404-71-5	98.186			94	0.7079^{20}	1.405^{20}	vs bz, eth, EtOH
5196	3-Ethyl-1-pentene		C_7H_{14}	4038-04-4	98.186	liq	-127.5	84.1	0.6917^{25}	1.3982^{20}	
5197	3-Ethyl-2-pentene		C_7H_{14}	816-79-5	98.186			96	0.7204^{20}	1.4148^{20}	i H_2O; s EtOH, eth, bz, chl
5198	Ethyl pentyl ether		$C_7H_{16}O$	17952-11-3	116.201			117.6	0.7622^{20}	1.3927^{20}	vs eth, EtOH
5199	Ethyl 2-pentynoate		$C_7H_{10}O_2$	55314-57-3	126.153			67^{18}	0.962^{25}		
5200	2-Ethylphenol		$C_8H_{10}O$	90-00-6	122.164		18	204.5	1.0146^{25}	1.5367^{20}	vs ace, bz, eth, EtOH
5201	3-Ethylphenol		$C_8H_{10}O$	620-17-7	122.164	liq	-4	218.4	1.0283^{20}		sl H_2O, chl; vs EtOH, eth
5202	4-Ethylphenol		$C_8H_{10}O$	123-07-9	122.164	nd	45.0	217.9		1.5239^{25}	sl H_2O, chl; vs EtOH, eth, bz; s ace
5203	Ethyl phenoxyacetate		$C_{10}H_{12}O_3$	2555-49-9	180.200			247; 110^3	1.0958^{30}	1.5080^{20}	
5204	*N*-Ethyl-*N*-phenylacetamide		$C_{10}H_{13}NO$	529-65-7	163.216		55	260	0.9938^{60}		s H_2O, eth, ctc
5205	Ethyl phenylacetate	Benzeneacetic acid, ethyl ester	$C_{10}H_{12}O_2$	101-97-3	164.201	liq	-29.4	227	1.0333^{20}	1.4980^{20}	vs eth, EtOH
5206	2-(Ethylphenylamino)ethanol		$C_{10}H_{15}NO$	92-50-2	165.232						s chl
5207	Ethyl phenylcarbamate	Phenylurethane	$C_9H_{11}NO_2$	101-99-5	165.189	wh nd (w) pl (dil al)	53	dec 237	1.1064^{30}	1.5376^{30}	i H_2O; vs EtOH, eth; s bz; sl ctc
5208	Ethyl *N*-phenylformimidate		$C_9H_{11}NO$	6780-49-0	149.189			214; 87^{10}	1.0051^{20}	1.5279^{20}	s eth, bz
5209	Ethyl *N*-phenylglycinate		$C_{10}H_{13}NO_2$	2216-92-4	179.216	lf (dil al)	58	273.5			vs eth, EtOH
5210	1-(4-Ethylphenyl)-2-phenylethane		$C_{16}H_{18}$	7439-15-8	210.314	cry		294	1.028^{50}		
5211	Ethyl 3-phenylpropanoate		$C_{11}H_{14}O_2$	2021-28-5	178.228			247.2	1.0147^{20}	1.4954^{20}	vs eth, EtOH
5212	Ethyl 3-phenylpropynoate	Ethyl phenylacetylenecarboxylate	$C_{11}H_{10}O_2$	2216-94-6	174.196			265; $128^{1.6}$	1.055^{25}	1.5520^{20}	s eth
5213	Ethyl phenyl sulfone		$C_8H_{10}O_2S$	599-70-2	170.229	lf (dil al)	42	160^{12}	1.1410^{20}		vs bz, eth, EtOH, chl
5214	Ethylphosphonic acid		$C_2H_7O_3P$	6779-09-5	110.049	hyg pl or nd	61.5	335^8			vs H_2O, eth, EtOH
5215	Ethyl phosphorodichloridate	Ethylphosphoric acid dichloride	$C_2H_5Cl_2O_2P$	1498-51-7	162.940			62^{10}		1.4338^{20}	
5216	5-Ethyl-2-picoline		$C_8H_{11}N$	104-90-5	121.180			178.3	0.9202^{20}	1.4971^{20}	sl H_2O; s EtOH, eth, bz; vs ace
5217	Ethyl 1-piperazinecarboxylate	1-Carbethoxypiperazine	$C_7H_{14}N_2O_2$	120-43-4	158.198			237		1.4760^{25}	vs H_2O, eth, EtOH
5218	1-Ethylpiperidine		$C_7H_{15}N$	766-09-6	113.201			130.8	0.8237^{20}	1.4480^{20}	
5219	Ethyl 4-piperidinecarboxylate		$C_8H_{15}NO_2$	1126-09-6	157.211	col oil		100^{10}		1.4591^{20}	vs H_2O, bz, eth, EtOH
5220	Ethyl 1-piperidinepropanoate		$C_{10}H_{19}NO_2$	19653-33-9	185.264			217; 139^{50}	0.9627^{25}	1.4525^{25}	vs H_2O
5221	1-Ethyl-3-piperidinol		$C_7H_{15}NO$	13444-24-1	129.200			94^{15}		1.4777^{14}	
5222	*N*-Ethyl-1-propanamine		$C_5H_{13}N$	20193-20-8	87.164			81	0.7204^{17}	1.3858^{25}	sl H_2O; vs ace, EtOH
5223	Ethylpropanedioic acid		$C_5H_8O_4$	601-75-2	132.116	pr (w+1)	114	$180^{0.05}$			vs H_2O; s EtOH, eth, bz; i ace; sl tfa
5224	Ethyl propanoate	Ethyl propionate	$C_5H_{10}O_2$	105-37-3	102.132	liq	-73.9	99.1	0.8843^{25}	1.3839^{20}	sl H_2O, ctc; msc EtOH, eth; s ace
5225	Ethyl propyl ether		$C_5H_{12}O$	628-32-0	88.148	liq	-127.5	63.21	0.7386^{20}	1.3695^{20}	vs eth, EtOH, HOAc
5226	2-(1-Ethylpropyl)pyridine		$C_{10}H_{15}N$	7399-50-0	149.233			195.4	0.8981^{20}	1.4850^{25}	
5227	4-(1-Ethylpropyl)pyridine		$C_{10}H_{15}N$	35182-51-5	149.233		125.5	217; 80^{12}	0.9085^{25}	1.40905^{25}	
5228	Ethyl propyl sulfide		$C_5H_{12}S$	4110-50-3	104.214	liq	-117	118.6	0.8370^{20}	1.4462^{20}	s EtOH
5229	Ethyl 2-propynoate	(Ethoxycarbonyl)acetylene	$C_5H_6O_2$	623-47-2	98.101			120	0.9645^{16}	1.4105^{20}	i H_2O; vs EtOH, eth, chl
5230	2-Ethylpyrazine		$C_6H_8N_2$	13925-00-3	108.141			112^{200}			
5231	2-Ethylpyridine		C_7H_9N	100-71-0	107.153	liq	-63.1	148.6	0.9502^{25}	1.4964^{20}	s H_2O; msc EtOH; vs eth, ace; sl ctc

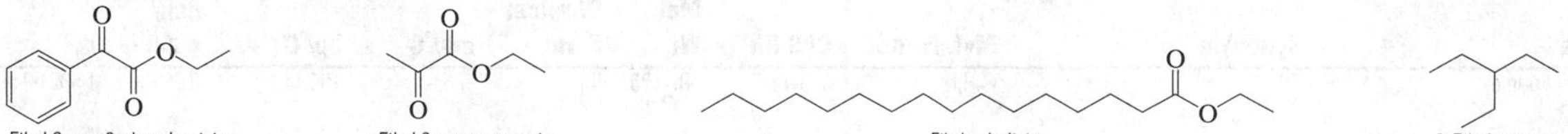

Ethyl 2-oxo-2-phenylacetate
Ethyl 2-oxopropanoate
Ethyl palmitate
3-Ethylpentane
3-Ethyl-2,4-pentanedione 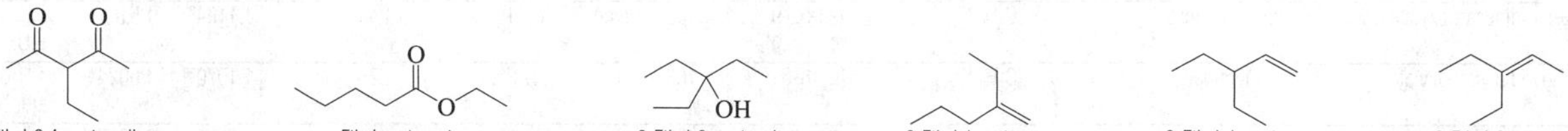
Ethyl pentanoate
3-Ethyl-3-pentanol
2-Ethyl-1-pentene
3-Ethyl-1-pentene
3-Ethyl-2-pentene
Ethyl pentyl ether 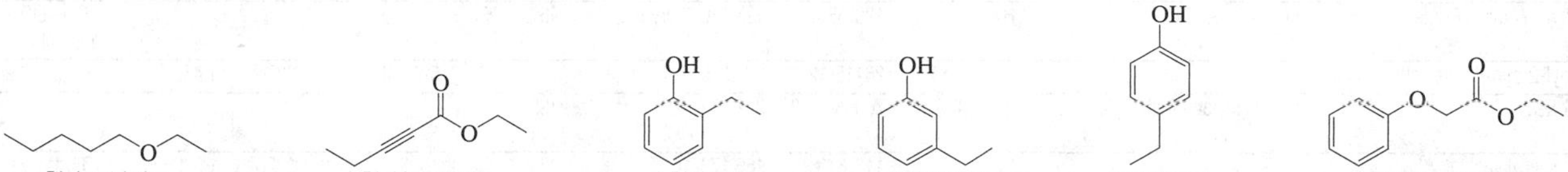
Ethyl 2-pentynoate
2-Ethylphenol
3-Ethylphenol
4-Ethylphenol
Ethyl phenoxyacetate
N-Ethyl-*N*-phenylacetamide 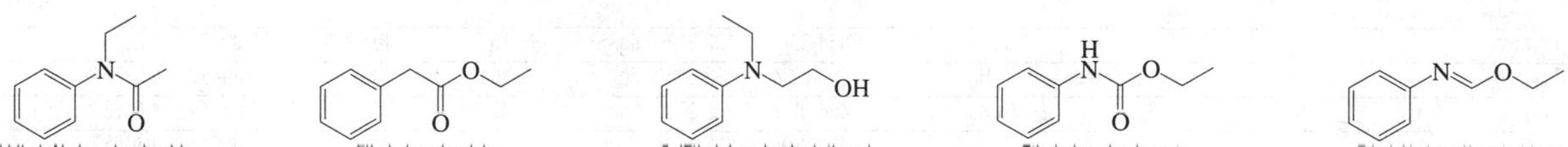
Ethyl phenylacetate
2-(Ethylphenylamino)ethanol
Ethyl phenylcarbamate
Ethyl *N*-phenylformimidate
Ethyl *N*-phenylglycinate 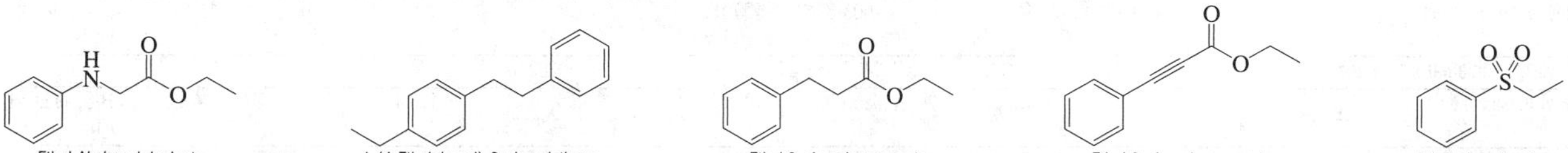
1-(4-Ethylphenyl)-2-phenylethane
Ethyl 3-phenylpropanoate
Ethyl 3-phenylpropynoate
Ethyl phenyl sulfone
Ethylphosphonic acid 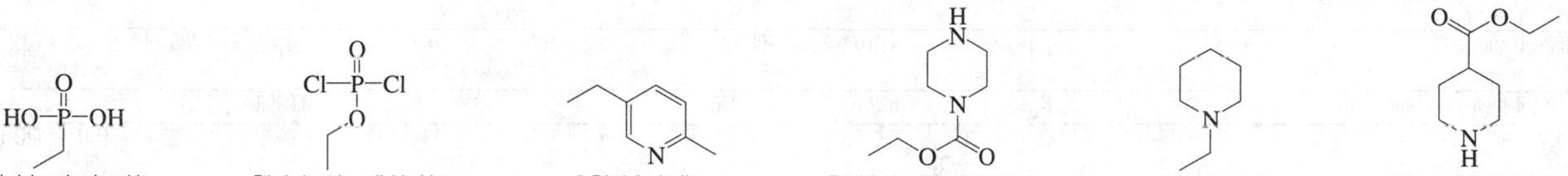
Ethyl phosphorodichloridate
5-Ethyl-2-picoline
Ethyl 1-piperazinecarboxylate
1-Ethylpiperidine
Ethyl 4-piperidinecarboxylate
Ethyl 1-piperidinepropanoate 
1-Ethyl-3-piperidinol
N-Ethyl-1-propanamine
Ethylpropanedioic acid
Ethyl propanoate
Ethyl propyl ether
2-(1-Ethylpropyl)pyridine 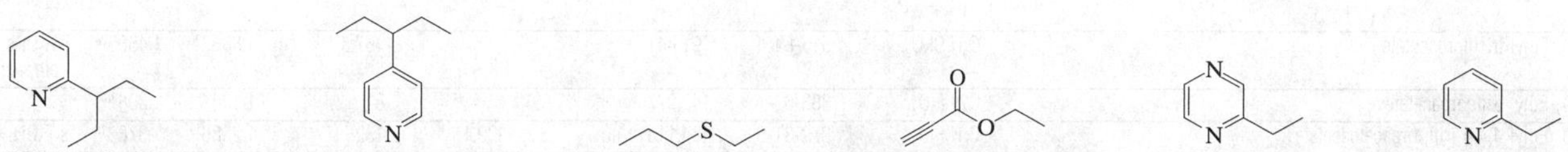
4-(1-Ethylpropyl)pyridine
Ethyl propyl sulfide
Ethyl 2-propynoate
2-Ethylpyrazine
2-Ethylpyridine

No.	Name	Synonym	Mol. Form.	CAS RN	Mol. Wt.	Physical Form	mp/°C	bp/°C	den/ g cm^{-3}	n_D	Solubility
5232	3-Ethylpyridine		C_7H_9N	536-78-7	107.153	liq	-76.9	165	0.9539^{25}	1.5021^{20}	s H_2O, EtOH, eth; vs ace; sl ctc
5233	4-Ethylpyridine		C_7H_9N	536-75-4	107.153	liq	-90.5	168.3	0.9417^{20}	1.5009^{20}	s H_2O, EtOH, eth; vs ace; sl ctc
5234	2-Ethyl-4-pyridinecarbothioamide	Ethionamide	$C_8H_{10}N_2S$	536-33-4	166.243		163				
5235	Ethyl 2-pyridinecarboxylate	Ethyl 2-picolinate	$C_8H_9NO_2$	2524-52-9	151.163	ye cry in air	1	243	1.1194^{20}	1.5104^{20}	vs H_2O, eth, EtOH
5236	Ethyl 3-pyridinecarboxylate	Ethyl nicotinate	$C_8H_9NO_2$	614-18-6	151.163		8.5	224	1.1070^{20}	1.5024^{20}	vs H_2O, EtOH, eth, bz; sl ctc
5237	Ethyl 4-pyridinecarboxylate		$C_8H_9NO_2$	1570-45-2	151.163		23	219.5	1.0091^{15}	1.5017^{20}	sl H_2O; s EtOH, bz; vs eth, chl
5238	*N*-Ethylpyridinium bromide		$C_7H_{10}BrN$	1906-79-2	188.065	cry (al)	111.5				s H_2O, EtOH; i eth
5239	1-Ethyl-1*H*-pyrrole		C_6H_9N	617-92-5	95.142			129.5	0.9009^{20}	1.4841^{20}	vs EtOH
5240	1-Ethyl-1*H*-pyrrole-2,5-dione	*N*-Ethylmaleimide	$C_6H_7NO_2$	128-53-0	125.126	cry (bz)	45.5				sl H_2O; vs EtOH, eth; s chl
5241	1-Ethyl-2-pyrrolidinemethanamine		$C_7H_{16}N_2$	26116-12-1	128.215			59^{16}, 40^{10}	0.887^{25}	1.4665^{20}	
5242	Ethyl Red	2-(4-Diethylaminophenylazo) benzoic acid	$C_{17}H_{19}N_3O_2$	76058-33-8	297.352		135				
5243	Ethyl salicylate		$C_9H_{10}O_3$	118-61-6	166.173		45	150^{10}	1.1326^{20}	1.5296^{20}	i H_2O; msc EtOH; vs eth; s ctc
5244	Ethyl silicate		$C_8H_{20}O_4Si$	78-10-4	208.329	liq	-82.5	168.8	0.9320^{20}	1.3928^{20}	dec H_2O
5245	Ethyl stearate	Ethyl octadecanoate	$C_{20}H_{40}O_2$	111-61-5	312.531		33	199^{10}	1.057^{20}	1.4349^{40}	i H_2O; s EtOH, eth, chl; vs ace
5246	2-Ethylstyrene		$C_{10}H_{12}$	7564-63-8	132.202	liq	-75.5	187.3; 68^{12}	0.9017^{25}	1.5380^{20}	
5247	3-Ethylstyrene		$C_{10}H_{12}$	7525-62-4	132.202	liq	-101	190.0	0.8945^{20}	1.5351^{20}	
5248	4-Ethylstyrene		$C_{10}H_{12}$	3454-07-7	132.202	liq	-49.7	192.3; 86^{20}	0.8884^{25}	1.5376^{20}	
5249	Ethyl sulfate		$C_2H_6O_4S$	540-82-9	126.132			dec 280	1.3657^{20}	1.4105^{20}	vs H_2O
5250	2-(Ethylsulfonyl)ethanol	Ethylsulfonylethyl alcohol	$C_4H_{10}O_3S$	513-12-2	138.185						sl chl
5251	2-Ethyl-5-(3-sulfophenyl) isoxazolium hydroxide, inner salt	Woodward's Reagent K	$C_{11}H_{11}NO_4S$	4156-16-5	253.275		dec 207				
5252	Ethyl tartrate	Ethyl tartrate, acid	$C_6H_{10}O_6$	608-89-9	178.139		90				vs H_2O, EtOH
5253	2-Ethyltetrahydrofuran		$C_6H_{12}O$	1003-30-1	100.158			109	0.8570^{19}	1.4147^{19}	vs ace, bz, eth, EtOH
5254	5-Ethyl-1,3,4-thiadiazol-2-amine		$C_4H_7N_3S$	14068-53-2	129.184		200.8				
5255	*S*-Ethyl thioacetate		C_4H_8OS	625-60-5	104.171			116.4	0.9792^{20}	1.4583^{21}	i H_2O; vs EtOH, eth
5256	(Ethylthio)acetic acid		$C_4H_8O_2S$	627-04-3	120.171		-8.5	164^{83}, 109^{5}	1.1497^{20}		vs H_2O, EtOH, eth
5257	(Ethylthio)benzene	Thiophenetole	$C_8H_{10}S$	622-38-8	138.230			205	1.0211^{20}	1.5670^{20}	s EtOH
5258	Ethyl thiocyanate		C_3H_5NS	542-90-5	87.144	liq	-85.5	146	1.007^{23}	1.4684^{15}	i H_2O; msc EtOH, eth; s chl
5259	2-(Ethylthio)ethanol		$C_4H_{10}OS$	110-77-0	106.186	liq	-100	184	1.0166^{20}	1.4867^{20}	sl H_2O; s EtOH; vs ace
5260	1-(Ethylthio)-4-methylbenzene		$C_9H_{12}S$	622-63-9	152.256			220	0.9996^{20}	1.555^{20}	
5261	2-Ethylthiophene		C_6H_8S	872-55-9	112.193			134	0.9930^{20}	1.5122^{20}	i H_2O; vs EtOH, eth
5262	Ethyl thiophene-2-carboxylate		$C_7H_8O_2S$	2810-04-0	156.203			218	1.1623^{16}	1.5248^{20}	s EtOH, ace; sl ctc
5263	3-Ethyl-2-thioxo-4-thiazolidinone	3-Ethylrhodanine	$C_5H_7NOS_2$	7648-01-3	161.246		35.5				
5264	2-Ethyltoluene		C_9H_{12}	611-14-3	120.191	liq	-79.83	165.2	0.8807^{20}	1.5046^{20}	i H_2O; msc EtOH, eth, ace, bz, peth, ctc
5265	3-Ethyltoluene		C_9H_{12}	620-14-4	120.191	liq	-95.6	161.3	0.8645^{20}	1.4966^{20}	i H_2O; vs EtOH, eth; msc ace, bz
5266	4-Ethyltoluene		C_9H_{12}	622-96-8	120.191	liq	-62.35	162	0.8614^{20}	1.4959^{20}	i H_2O; vs EtOH, eth; msc ace, bz
5267	Ethyl *p*-toluenesulfonate		$C_9H_{12}O_3S$	80-40-0	200.254		34.5	173^{15}	1.166^{48}		i H_2O; s EtOH, eth, AcOEt; sl ctc
5268	Ethyl trichloroacetate		$C_4H_5Cl_3O_2$	515-84-4	191.441			167.5	1.3836^{20}	1.4505^{20}	i H_2O; s EtOH, eth, bz; sl chl
5269	Ethyl trifluoroacetate		$C_4H_5F_3O_2$	383-63-1	142.077			61	1.194^{20}	1.308^{20}	
5270	Ethyl 4,4,4-trifluoroacetoacetate		$C_6H_7F_3O_3$	372-31-6	184.113	liq	-39.1	132	1.2586^{15}	1.3783^{15}	s EtOH, eth
5271	Ethyl trifluoromethanesulfonate		$C_3H_5F_3O_3S$	425-75-2	178.130			115	1.3740^{0}		s eth
5272	Ethyl 3,4,5-trihydroxybenzoate		$C_9H_{10}O_5$	831-61-8	198.172	mcl pr (w+2 1/2) nd (chl)	163.0				sl H_2O, chl; s EtOH, eth, AcOEt

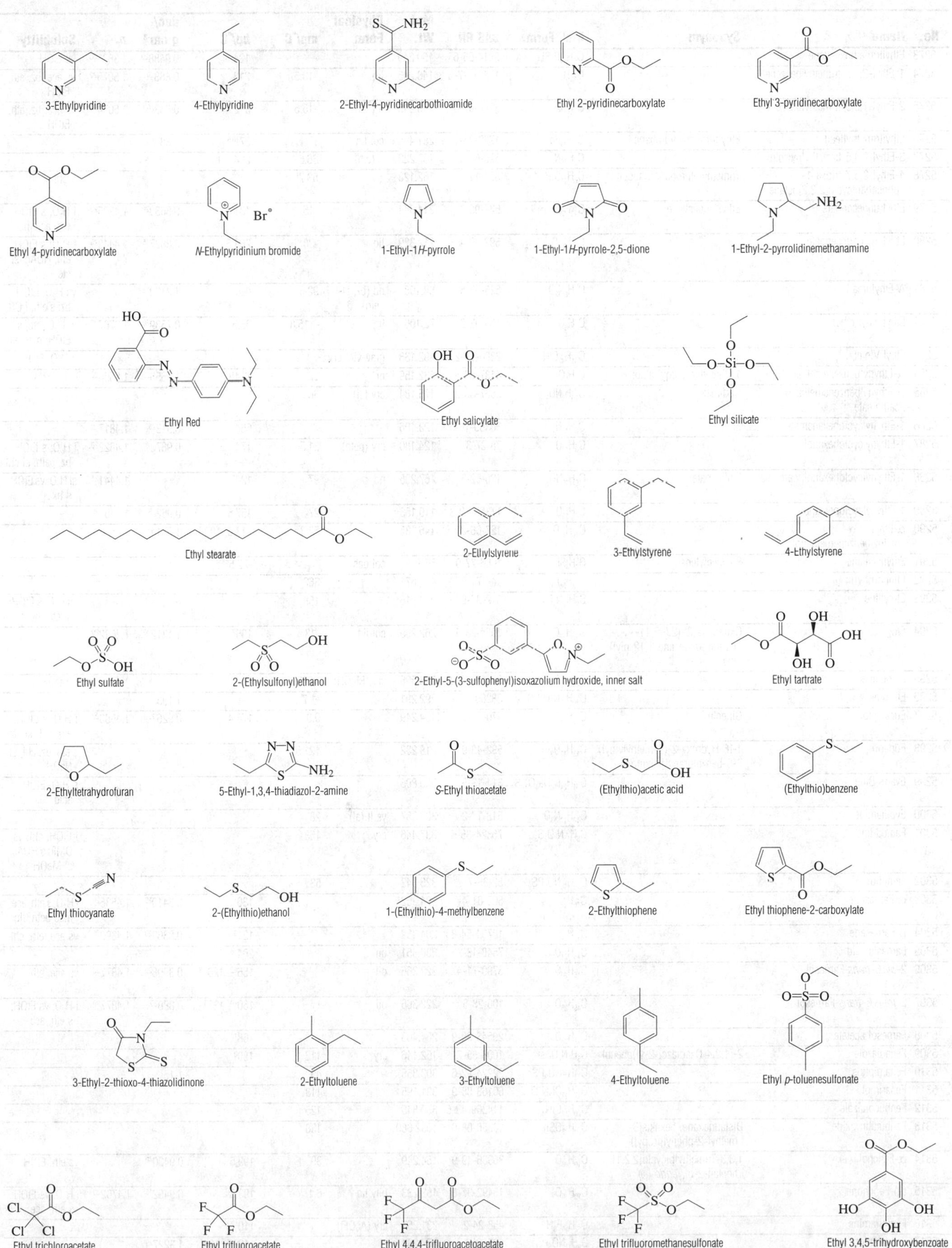
3-Ethylpyridine
4-Ethylpyridine
2-Ethyl-4-pyridinecarbothioamide
Ethyl 2-pyridinecarboxylate
Ethyl 3-pyridinecarboxylate
Ethyl 4-pyridinecarboxylate
N-Ethylpyridinium bromide
1-Ethyl-1H-pyrrole
1-Ethyl-1H-pyrrole-2,5-dione
1-Ethyl-2-pyrrolidinemethanamine
Ethyl Red
Ethyl salicylate
Ethyl silicate
Ethyl stearate
2-Ethylstyrene
3-Ethylstyrene
4-Ethylstyrene
Ethyl sulfate
2-(Ethylsulfonyl)ethanol
2-Ethyl-5-(3-sulfophenyl)isoxazolium hydroxide, inner salt
Ethyl tartrate
2-Ethyltetrahydrofuran
5-Ethyl-1,3,4-thiadiazol-2-amine
S-Ethyl thioacetate
(Ethylthio)acetic acid
(Ethylthio)benzene
Ethyl thiocyanate
2-(Ethylthio)ethanol
1-(Ethylthio)-4-methylbenzene
2-Ethylthiophene
Ethyl thiophene-2-carboxylate
3-Ethyl-2-thioxo-4-thiazolidinone
2-Ethyltoluene
3-Ethyltoluene
4-Ethyltoluene
Ethyl p-toluenesulfonate
Ethyl trichloroacetate
Ethyl trifluoroacetate
Ethyl 4,4,4-trifluoroacetoacetate
Ethyl trifluoromethanesulfonate
Ethyl 3,4,5-trihydroxybenzoate

No.	Name	Synonym	Mol. Form.	CAS RN	Mol. Wt.	Physical Form	mp/°C	bp/°C	den/ g cm^{-3}	n_D	Solubility
5273	Ethyltrimethoxysilane		$C_5H_{14}O_3Si$	5314-55-6	150.249			124.3	0.9488^{20}	1.3838^{20}	vs EtOH
5274	1-Ethyl-2,4,5-trimethylbenzene		$C_{11}H_{16}$	17851-27-3	148.245	liq	-13.5	213	0.883^{20}	1.5075^{20}	vs ace, bz, eth, EtOH
5275	2-Ethyl-1,3,5-trimethylbenzene		$C_{11}H_{16}$	3982-67-0	148.245	liq	-15.5	212.4	0.883^{20}	1.5074^{20}	vs ace, bz, eth, EtOH
5276	Ethyltrimethyllead	Ethyltrimethylplumbane	$C_5H_{14}Pb$	1762-26-1	281.4	col liq		$27^{10.5}$	1.88^{20}		
5277	3-Ethyl-2,4,5-trimethylpyrrole		$C_9H_{15}N$	520-69-4	137.222	lf (eth)	66.5	214; 110^{35}			
5278	4-Ethyl-2,6,7-trioxa-1-phosphabicyclo[2.2.2]octane	Trimethylolpropane phosphite	$C_6H_{11}O_3P$	824-11-3	162.123		53.7				s chl
5279	Ethyl undecanoate	Ethyl undecylate	$C_{13}H_{26}O_2$	627-90-7	214.344		-15	131^{14}	0.8633^{20}	1.4285^{20}	i H_2O; s EtOH, eth, ace, bz
5280	Ethyl 10-undecenoate		$C_{13}H_{24}O_2$	692-86-4	212.329	liq	-38	264.5	0.8827^{15}	1.4449^{25}	i H_2O; s EtOH, eth, HOAc; sl ctc
5281	*N*-Ethylurea		$C_3H_8N_2O$	625-52-5	88.108	nd (bz, al-eth)	92.5	dec	1.2130^{18}		vs H_2O, EtOH, bz; s eth; i CS_2
5282	Ethyl vinyl ether		C_4H_8O	109-92-2	72.106	liq	-115.8	35.5	0.7589^{20}	1.3767^{20}	sl H_2O, ctc; s EtOH; msc eth
5283	Ethyl Violet		$C_{31}H_{42}ClN_3$	2390-59-2	492.138	gray-viol cry					s H_2O, EtOH
5284	α-Ethynylbenzenemethanol	1-Phenylpropargyl alcohol	C_9H_8O	4187-87-5	132.159	pr	22	114^{12}	1.0655^{20}	1.5508^{20}	
5285	α-Ethynylbenzenemethanol carbamate	Carfimate	$C_{10}H_9NO_2$	3567-38-2	175.184	cry (al)	86.5				
5286	1-Ethynylcyclohexanamine		$C_8H_{13}N$	30389-18-5	123.196			65^{20}	0.913^{25}	1.4817^{20}	
5287	1-Ethynylcyclohexanol		$C_8H_{12}O$	78-27-3	124.180	cry (peth)	31.5	174	0.9873^{20}	1.4822^{20}	i H_2O; s EtOH, bz, peth; sl chl
5288	1-Ethynylcyclohexanol, carbamate	Ethinamate	$C_9H_{13}NO_2$	126-52-3	167.205	nd	97	120^3		1.4441^{21}	sl H_2O; vs EtOH; s hx
5289	1-Ethynylcyclopentanol		$C_7H_{10}O$	17356-19-3	110.153		27	157.5	0.962^{25}	1.4751^{20}	
5290	α-Ethynyl-α-methylbenzenemethanol		$C_{10}H_{10}O$	127-66-2	146.185		52.3	217.5; 102^{12}	1.0314^{20}		
5291	Ethynylsilane	Silylacetylene	C_2H_4Si	1066-27-9	56.139	col gas		-22.5			
5292	Etioporphyrin		$C_{32}H_{38}N_4$	448-71-5	478.671		362				
5293	Etofylline		$C_9H_{12}N_4O_3$	519-37-9	224.216		158				vs H_2O; s EtOH; sl eth, bz
5294	Etoglucid	Oxirane, 2,2'-(2,5,8,11-tetraoxadodecane-1,12-diyl) bis-	$C_{12}H_{22}O_6$	1954-28-5	262.299	col liq	-13	196^2	1.1312^{20}	1.4622^{20}	
5295	Etoposide		$C_{29}H_{32}O_{13}$	33419-42-0	588.556	cry (MeOH)	≈243				s MeOH
5296	Etrimfos		$C_{10}H_{17}N_2O_4PS$	38260-54-7	292.291		-1.7		1.195^{20}		
5297	Eucalyptol	Cineole	$C_{10}H_{18}O$	470-82-6	154.249		0.8	176.4	0.9267^{20}	1.4586^{20}	i H_2O; s EtOH, eth, chl; sl ctc
5298	Euparin	1-[6-Hydroxy-2-(1-methylvinyl)-5-benzofuranyl]ethanone	$C_{13}H_{12}O_3$	532-48-9	216.232		121.5				s eth, bz, chl; sl NaOH
5299	Evan's Blue		$C_{34}H_{24}N_6Na_4O_{14}S_4$	314-13-6	960.806						s H_2O, EtOH, acid
5300	Evodiamine		$C_{19}H_{17}N_3O$	518-17-2	303.357	ye lf (al)	28				
5301	Famotidine		$C_8H_{15}N_7O_2S_3$	76824-35-6	337.446	cry	163				i EtOH, chl; vs DMF; s HOAc; sl MeOH
5302	Famphur		$C_{10}H_{16}NO_5PS_2$	52-85-7	325.342		53				
5303	α-Farnesene		$C_{15}H_{24}$	502-61-4	204.352			130^{12}	0.8410^{20}	1.4836^{20}	i H_2O; s eth, ace; msc peth, lig
5304	β-Farnesene		$C_{15}H_{24}$	18794-84-8	204.352			121^9	0.8363^{20}	1.4899^{20}	vs ace, eth, chl
5305	Farnesic acid		$C_{15}H_{24}O_2$	7548-13-2	236.351	oil		204^{16}			
5306	2-*cis*,6-*trans*-Farnesol		$C_{15}H_{26}O$	3790-71-4	222.366	oil		156^{12}, $120^{0.3}$	0.8908^{20}	1.4877^{20}	vs ace, eth, EtOH
5307	2-*trans*,6-*trans*-Farnesol		$C_{15}H_{26}O$	106-28-5	222.366	oil		160^{10}, 137^3	0.888^{20}	1.4877^{20}	i H_2O; vs EtOH; s eth, ace
5308	Farnesol acetate		$C_{17}H_{28}O_2$	29548-30-9	264.403			168^{10}			
5309	Fenadiazole	2-(1,2,4-Oxadiazol-2-yl)phenol	$C_8H_6N_2O_2$	1008-65-7	162.146	cry	112	$180^{0.1}$			
5310	Fenamiphos		$C_{13}H_{22}NO_3PS$	22224-92-6	303.358		49		1.15^{20}		
5311	Fenarimol		$C_{17}H_{12}Cl_2N_2O$	60168-88-9	331.195		118				
5312	Fenbuconazole		$C_{20}H_{19}ClN_4$	114369-43-6	350.845		125				
5313	Fenbutatin oxide	Distannoxane, hexakis(2-methyl-2-phenylpropyl)-	$C_{60}H_{78}OSn_2$	13356-08-6	1052.680		138				
5314	α-Fenchol, (±)	1,3,3-Trimethylbicyclo[2.2.1]heptan-2-ol, *endo*-(±)	$C_{10}H_{18}O$	36386-49-9	154.249		39	199.5	0.9420^{40}		vs eth, EtOH
5315	(±)-Fenchone		$C_{10}H_{16}O$	18492-37-0	152.233	oily liq	6.1	193.5	0.9492^{15}	1.4702^{20}	i H_2O; vs EtOH; s eth, ace
5316	Fenfluramine		$C_{12}H_{16}F_3N$	458-24-2	231.257	cry (AcOEt)		110^{12}			
5317	Fenitrothion		$C_9H_{12}NO_5PS$	122-14-5	277.234			$118^{0.05}$, 164^1	1.3227^{25}		
5318	Fenoxaprop-ethyl		$C_{18}H_{16}ClNO_5$	82110-72-3	361.777		85	$200^{0.001}$			sl H_2O, hx; s eth; vs ace, tol

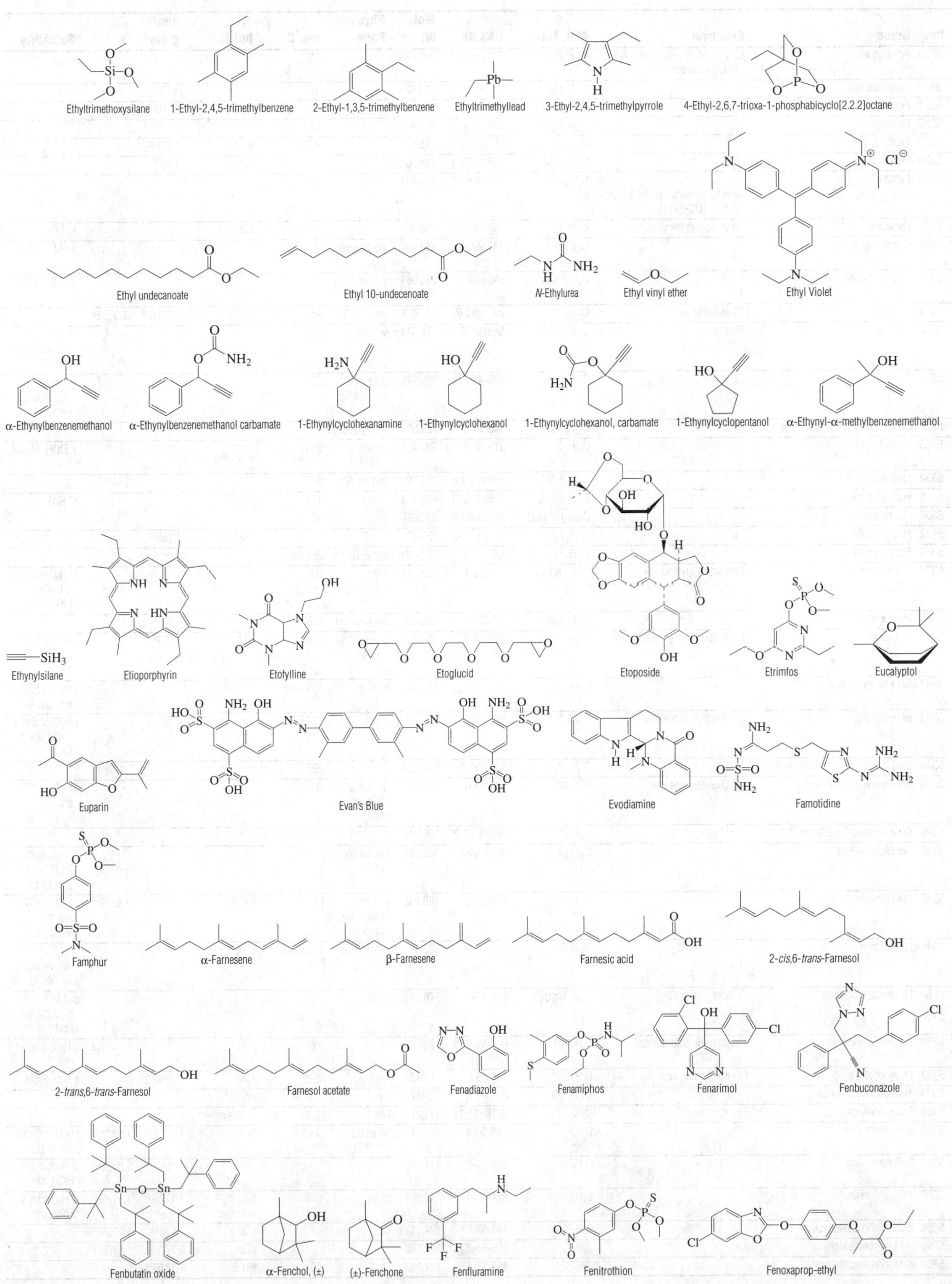

Ethyltrimethoxysilane
1-Ethyl-2,4,5-trimethylbenzene
2-Ethyl-1,3,5-trimethylbenzene
Ethyltrimethyllead
3-Ethyl-2,4,5-trimethylpyrrole
4-Ethyl-2,6,7-trioxa-1-phosphabicyclo[2.2.2]octane
Ethyl undecanoate
Ethyl 10-undecenoate
N-Ethylurea
Ethyl vinyl ether
Ethyl Violet
α-Ethynylbenzenemethanol
α-Ethynylbenzenemethanol carbamate
1-Ethynylcyclohexanamine
1-Ethynylcyclohexanol
1-Ethynylcyclohexanol, carbamate
1-Ethynylcyclopentanol
α-Ethynyl-α-methylbenzenemethanol
Ethynylsilane
Etioporphyrin
Etofylline
Etoglucid
Etoposide
Etrimfos
Eucalyptol
Euparin
Evan's Blue
Evodiamine
Famotidine
Famphur
α-Farnesene
β-Farnesene
Farnesic acid
2-cis,6-trans-Farnesol
2-trans,6-trans-Farnesol
Farnesol acetate
Fenadiazole
Fenamiphos
Fenarimol
Fenbuconazole
Fenbutatin oxide
α-Fenchol, (±)
(±)-Fenchone
Fenfluramine
Fenitrothion
Fenoxaprop-ethyl

No.	Name	Synonym	Mol. Form.	CAS RN	Mol. Wt.	Physical Form	mp/°C	bp/°C	den/ g cm-3	n_D	Solubility
5319	Fenoxycarb	Ethyl 2-(4-phenoxyphenoxy) ethylcarbamate	$C_{17}H_{19}NO_4$	79127-80-3	301.338		53				
5320	Fenpropathrin		$C_{22}H_{23}NO_3$	64257-84-7	349.423		47		1.15^{25}		
5321	Fensulfothion		$C_{11}H_{17}O_4PS_2$	115-90-2	308.354			$140^{0.01}$	1.202^{20}		
5322	Fentanyl		$C_{22}H_{28}N_2O$	437-38-7	336.469		87.5				
5323	Fenthion		$C_{10}H_{15}O_3PS_2$	55-38-9	278.328		7.5	$87^{0.01}$	1.246^{20}		
5324	Fenvalerate		$C_{25}H_{22}ClNO_3$	51630-58-1	419.901			dec	1.15^{25}		
5325	Ferbam	Iron, tris(dimethylcarbamodithioato-*S,S'*)-, (OC-6-11)-	$C_9H_{18}FeN_3S_6$	14484-64-1	416.494		180 dec				
5326	Ferrocene	Dicyclopentadienyl iron	$C_{10}H_{10}Fe$	102-54-5	186.031		172.5	249			i H_2O
5327	Ferrous gluconate		$C_{12}H_{22}FeO_{14}$	299-29-6	446.140	ye-gray pow (w)					s H_2O; i EtOH
5328	Ferrous lactate		$C_6H_{10}FeO_6$	5905-52-2	233.984	grn-wh pow (hyd)					s H_2O; i EtOH
5329	Fichtelite	18-Norabietane	$C_{19}H_{34}$	2221-95-6	262.473	cry	46	236^{43}	0.9380^{22}	1.5052^{20}	
5330	Finasteride	Proscar	$C_{23}H_{36}N_2O_2$	98319-26-7	372.544	wh cry	252				sl H_2O; s chl, EtOH, MeOH, DMSO
5331	Fisetin		$C_{15}H_{10}O_6$	528-48-3	286.236	lt ye nd (dil al, + 1 w)	330				i H_2O; s EtOH, ace; sl eth, bz, peth
5332	Flavine adenine dinucleotide	FAD	$C_{27}H_{33}N_9O_{15}P_2$	146-14-5	785.550	ye cry (w)					
5333	Florantyrone		$C_{20}H_{14}O_3$	519-95-9	302.323	ye cry (HOAc)	208				s EtOH, MeOH
5334	Fluazipop-butyl		$C_{19}H_{20}F_3NO_4$	79241-46-6	383.362	pale ye liq	5				
5335	Flubenzimine		$C_{17}H_{10}F_6N_4S$	37893-02-0	416.343	ye cry	119				sl H_2O
5336	Fluchloralin		$C_{12}H_{13}ClF_3N_3O_4$	33245-39-5	355.697		42				
5337	Flucythrinate	Cythrin	$C_{26}H_{23}F_2NO_4$	70124-77-5	451.463			$108^{0.35}$	1.189^{22}		
5338	Fludrocortisone		$C_{21}H_{29}FO_5$	127-31-1	380.450	cry (EtOH)	261 dec				
5339	Flumethiazide	Trifluoromethylthiazide	$C_8H_6F_3N_3O_4S_2$	148-56-1	329.277	cry	306				sl H_2O; i bz, tol; s MeOH, EtOH, DMF
5340	Fluocinolone acetonide		$C_{24}H_{30}F_2O_6$	67-73-2	452.488	cry (ace/hx)	266 dec				
5341	Fluoranthene	1,2-(1,8-Naphthylene)benzene	$C_{16}H_{10}$	206-44-0	202.250	pa ye nd or pl (al)	110.19	384	1.252^{0}		i H_2O; s EtOH, eth, bz, chl, CS_2
5342	9*H*-Fluoren-2-amine		$C_{13}H_{11}N$	153-78-6	181.233	pl or nd (dil al)	130.3				i H_2O; s EtOH, eth, ctc, CS_2
5343	9*H*-Fluorene	2,2'-Methylenebiphenyl	$C_{13}H_{10}$	86-73-7	166.218	lf (al)	114.77	295	1.203^{0}		i H_2O; sl EtOH; s eth, ace, bz, CS_2
5344	9*H*-Fluorene-9-carboxylic acid		$C_{14}H_{10}O_2$	1989-33-9	210.228		226				
5345	9*H*-Fluorene-2,7-diamine	2,7-Diaminofluorene	$C_{13}H_{12}N_2$	525-64-4	196.247	nd (w), pr (bz), pl (eth)	166				i H_2O; s EtOH, chl
5346	9*H*-Fluorene-9-methanol		$C_{14}H_{12}O$	24324-17-2	196.244		105.0				
5347	9*H*-Fluoren-9-ol		$C_{13}H_{10}O$	1689-64-1	182.217	hex nd (w, peth)	156.0				sl H_2O, peth, EtOH; s eth, ace; vs bz
5348	9*H*-Fluoren-9-one		$C_{13}H_8O$	486-25-9	180.202	ye orth bipym (al, bz-peth)	84	341.5	1.1300^{99}	1.6309^{99}	i H_2O; s EtOH, ace, bz; vs tol; sl ctc
5349	Fluorescein		$C_{20}H_{12}O_5$	2321-07-5	332.306	red orth pr	315 dec				sl H_2O, EtOH, eth; vs ace; s py, MeOH
5350	Fluorescein sodium	CI Acid Yellow 73	$C_{20}H_{10}Na_2O_5$	518-47-8	376.270	ye pow					s H_2O, EtOH, glycerol, dil acid
5351	2-Fluoroacetamide	Fluoroacetic acid amide	C_2H_4FNO	640-19-7	77.057		108	sub			s H_2O, ace; sl chl
5352	Fluoroacetic acid	Fluoroethanoic acid	$C_2H_3FO_2$	144-49-0	78.042	nd	35.2	168	1.3693^{36}		s H_2O, EtOH
5353	Fluoroacetyl chloride		C_2H_2ClFO	359-06-8	96.487	liq		72; 23^{105}			
5354	Fluoroacetylene	Fluoroethyne	C_2HF	2713-09-9	44.027	gas	-196	-105 exp			
5355	2-Fluoroaniline		C_6H_6FN	348-54-9	111.117	pa ye liq	-34.6	175; 55^{12}	1.1513^{21}	1.5421^{20}	i H_2O; s EtOH, eth; sl ctc
5356	3-Fluoroaniline		C_6H_6FN	372-19-0	111.117			188	1.1561^{19}	1.5436^{20}	sl H_2O, chl; s EtOH, eth
5357	4-Fluoroaniline		C_6H_6FN	371-40-4	111.117	pa ye liq	-0.8	182; 85^{19}	1.1725^{20}	1.5195^{20}	sl H_2O, ctc; s EtOH, eth
5358	2-Fluorobenzaldehyde		C_7H_5FO	446-52-6	124.112	liq	-44.5	175	1.178^{25}	1.5234^{20}	
5359	3-Fluorobenzaldehyde		C_7H_5FO	456-48-4	124.112			173	1.17^{25}	1.5206^{20}	
5360	4-Fluorobenzaldehyde		C_7H_5FO	459-57-4	124.112	liq	-10	181.5	1.1810^{19}		

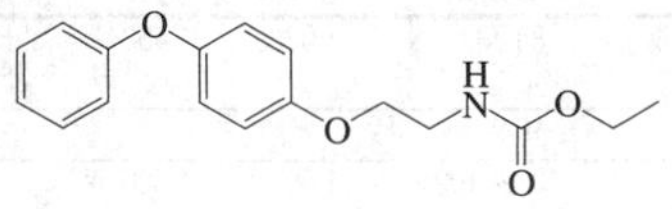

Fenoxycarb

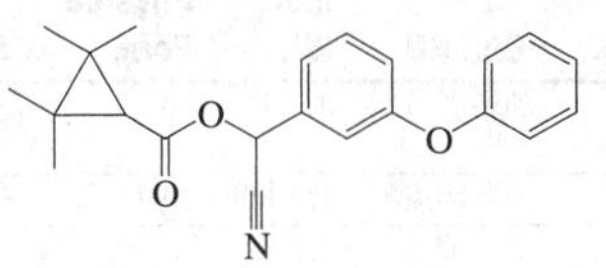

Fenpropathrin

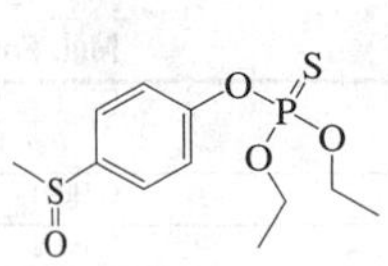

Fensulfothion

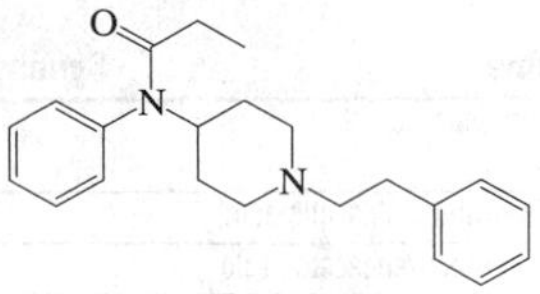

Fentanyl

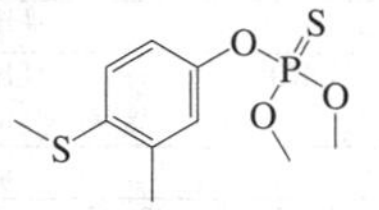

Fenthion

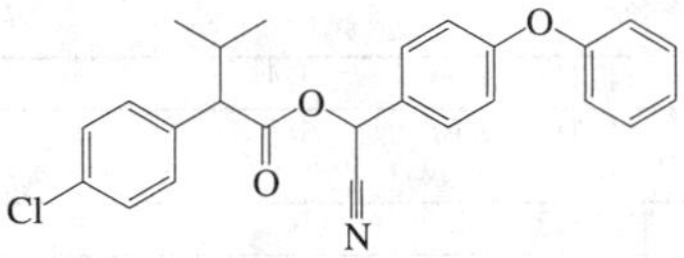

Fenvalerate

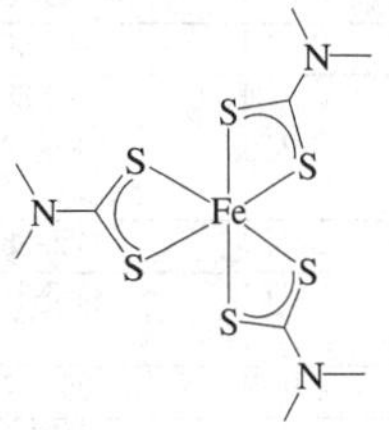

Ferbam

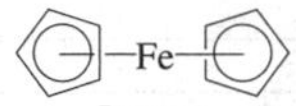

Ferrocene

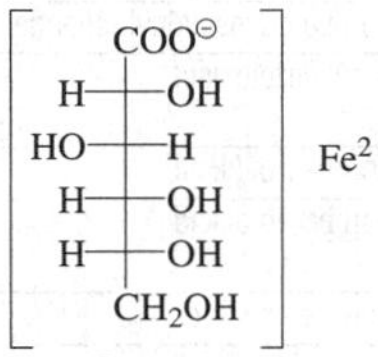

Ferrous gluconate

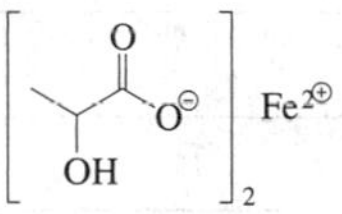

Ferrous lactate

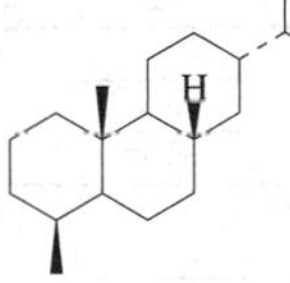

Fichtelite

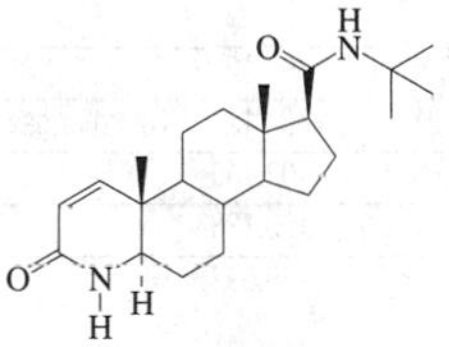

Finasteride

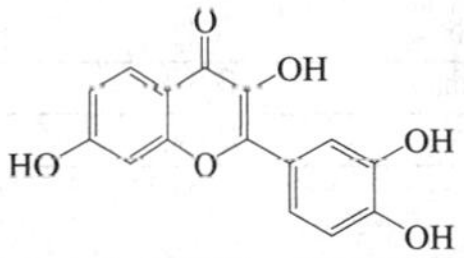

Fisetin

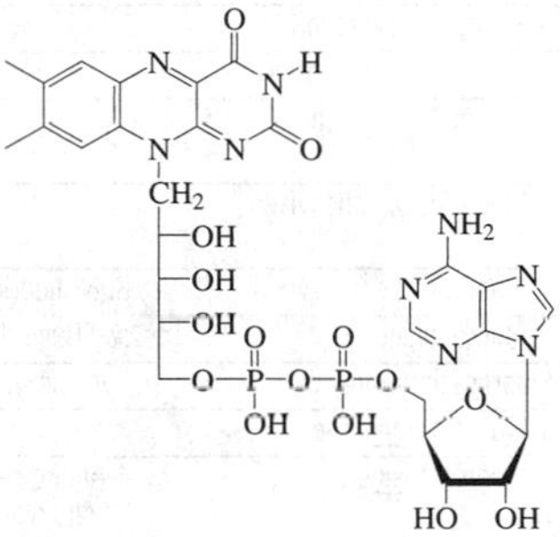

Flavine adenine dinucleotide

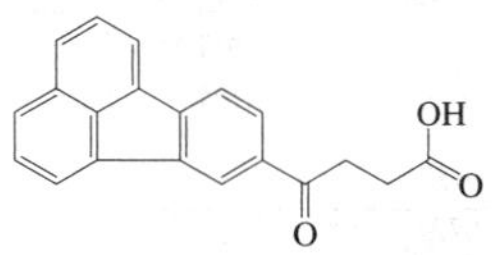

Florantyrone

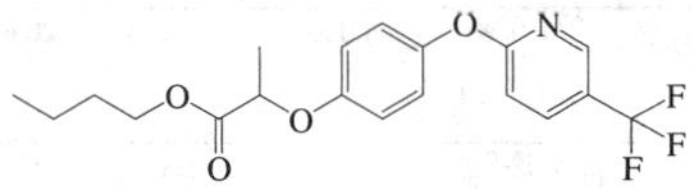

Fluazipop-butyl

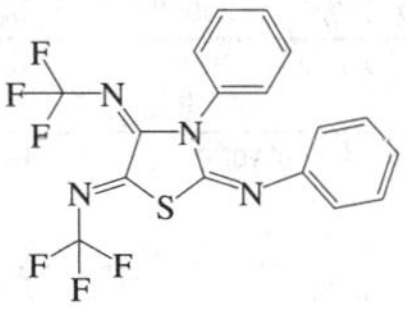

Flubenzimine

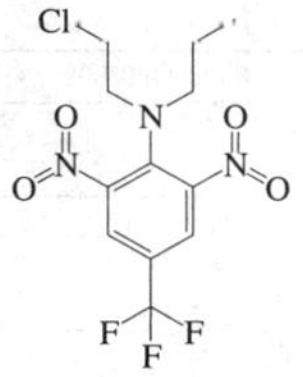

Fluchloralin

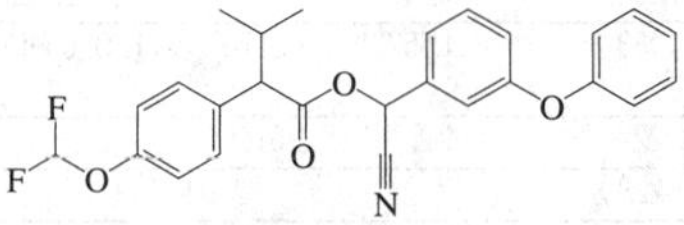

Flucythrinate

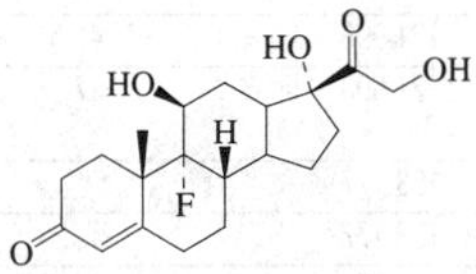

Fludrocortisone

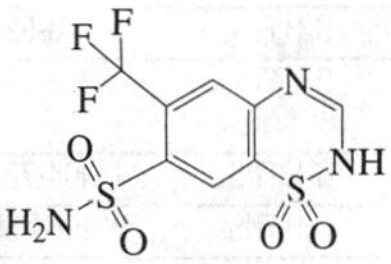

Flumethiazide

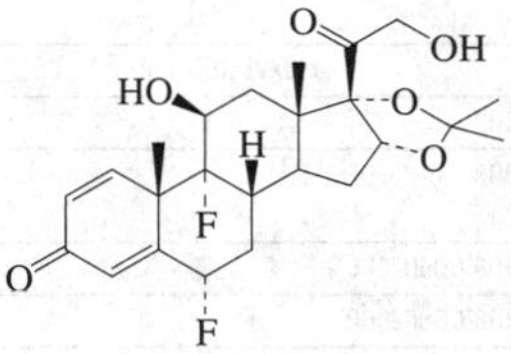

Fluocinolone acetonide

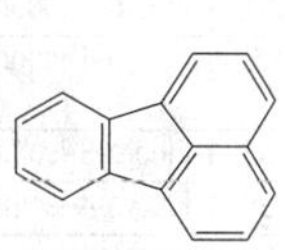

Fluoranthene

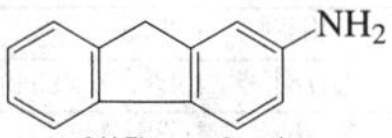

9*H*-Fluoren-2-amine

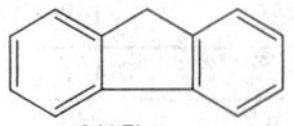

9*H*-Fluorene

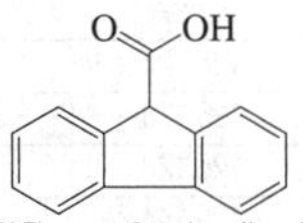

9*H*-Fluorene-9-carboxylic acid

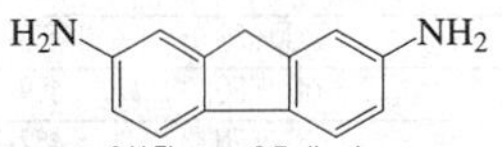

9*H*-Fluorene-2,7-diamine

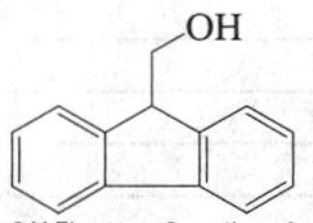

9*H*-Fluorene-9-methanol

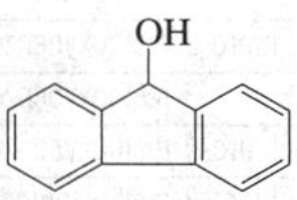

9*H*-Fluoren-9-ol

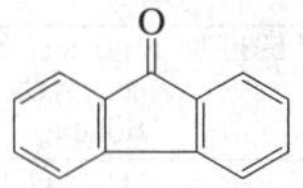

9*H*-Fluoren-9-one

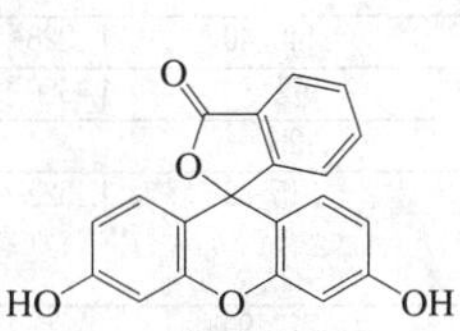

Fluorescein

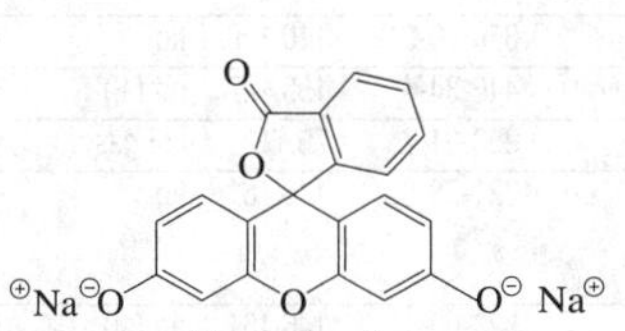

Fluorescein sodium

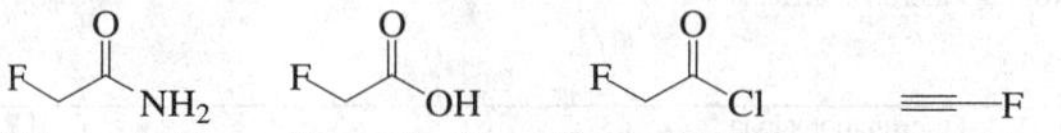

2-Fluoroacetamide

Fluoroacetic acid

Fluoroacetyl chloride

Fluoroacetylene

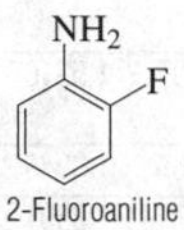

2-Fluoroaniline

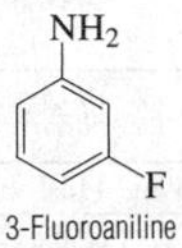

3-Fluoroaniline

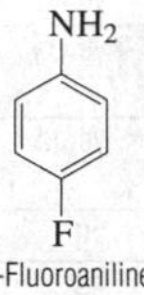

4-Fluoroaniline

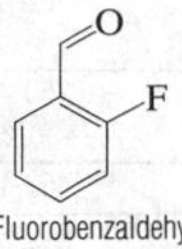

2-Fluorobenzaldehyde

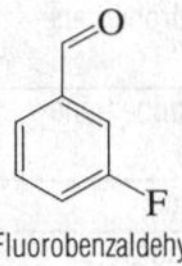

3-Fluorobenzaldehyde

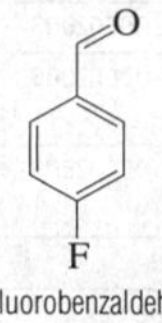

4-Fluorobenzaldehyde

No.	Name	Synonym	Mol. Form.	CAS RN	Mol. Wt.	Physical Form	mp/°C	bp/°C	den/ g cm^{-3}	n_D	Solubility
5361	Fluorobenzene		C_6H_5F	462-06-6	96.102	liq	-42.18	84.73	1.0225^{20}	1.4684^{30}	sl H_2O; vs bz, eth, EtOH, lig
5362	4-Fluorobenzeneacetic acid		$C_8H_7FO_2$	405-50-5	154.139	cry (chl)	86	164^{2}			
5363	2-Fluorobenzeneacetonitrile		C_8H_6FN	326-62-5	135.139			232; 102^{10}	1.059^{25}	1.5009^{20}	
5364	4-Fluorobenzeneacetonitrile		C_8H_6FN	459-22-3	135.139		86.0	228; 119^{18}	1.1390^{20}	1.5002^{20}	
5365	4-Fluorobenzenemethanamine		C_7H_8FN	140-75-0	125.144			183		1.5139^{20}	
5366	4-Fluorobenzenemethanol		C_7H_7FO	459-56-3	126.128		23	210		1.5080^{20}	
5367	4-Fluorobenzenesulfonyl chloride		$C_6H_4ClFO_2S$	349-88-2	194.611	pl or nd	30	106^{9}			vs bz, eth, chl
5368	2-Fluorobenzoic acid		$C_7H_5FO_2$	445-29-4	140.112	nd (a)	126.5		1.460^{25}		sl H_2O; vs EtOH, eth; i bz; s chl
5369	3-Fluorobenzoic acid		$C_7H_5FO_2$	455-38-9	140.112	lf (w)	124		1.474^{25}		sl H_2O; s eth
5370	4-Fluorobenzoic acid		$C_7H_5FO_2$	456-22-4	140.112	pr (w), mcl pr (w)	185		1.479^{25}		sl H_2O, ace; s EtOH, eth
5371	2-Fluorobenzonitrile		C_7H_4FN	394-47-8	121.112			93^{22}			
5372	4-Fluorobenzonitrile		C_7H_4FN	1194-02-1	121.112	nd (peth)	34.8	188.8	1.1070^{55}	1.4925^{55}	sl chl; s peth
5373	2-Fluorobenzoyl chloride		C_7H_4ClFO	393-52-2	158.557		2.0	91^{15}	1.328^{25}	1.5365^{20}	
5374	3-Fluorobenzoyl chloride		C_7H_4ClFO	1711-07-5	158.557	liq	-30	189	1.304^{25}	1.5285^{20}	
5375	4-Fluorobenzoyl chloride		C_7H_4ClFO	403-43-0	158.557		9	82^{20}	1.342^{25}	1.5296^{20}	
5376	2-Fluoro-1,1'-biphenyl		$C_{12}H_9F$	321-60-8	172.197		73.5	248	1.2452^{25}		s EtOH, eth, chl, peth; sl lig
5377	4-Fluoro-1,1'-biphenyl		$C_{12}H_9F$	324-74-3	172.197	pr	74.2	253	1.247^{25}		sl EtOH; s eth, gl HOAc
5378	1-Fluorobutane	Butyl fluoride	C_4H_9F	2366-52-1	76.112	liq	-134	32.5	0.7789^{20}	1.3396^{20}	vs EtOH
5379	2-Fluorobutane	*sec*-Butyl fluoride	C_4H_9F	359-01-3	76.112	vol liq or gas	-121.4	25.1	0.7559^{25}		
5380	Fluorocyclohexane	Cyclohexyl fluoride	$C_6H_{11}F$	372-46-3	102.149		13	101	0.9279^{20}	1.4146^{20}	i H_2O; s py
5381	1-Fluorocyclohexene		C_6H_9F	694-51-9	100.133			96.5		1.4441^{25}	
5382	5-Fluorocytosine	4-Amino-5-fluoro-2-hydroxypyrimidine	$C_4H_4FN_3O$	2022-85-7	129.092	wh cry	296 dec				
5383	1-Fluorodecane	Decyl fluoride	$C_{10}H_{21}F$	334-56-5	160.272	liq	-35	186.2	0.8194^{20}	1.4085	vs eth
5384	Fluorodifen	2-Nitro-1-(4-nitrophenoxy)-4-(trifluoromethyl)benzene	$C_{13}H_7F_3N_2O_5$	15457-05-3	328.200		94				
5385	1-Fluoro-2,4-dinitrobenzene	2,4-Dinitrophenyl fluoride	$C_6H_3FN_2O_4$	70-34-8	186.097		25.8	296	1.4718^{54}	1.5690^{20}	s EtOH; sl chl
5386	Fluoroethane	Ethyl fluoride	C_2H_5F	353-36-6	48.059	col gas	-143.2	-37.7	0.7182^{20} (p>1 atm	1.2656^{20}	sl H_2O; vs EtOH, eth
5387	2-Fluoroethanol	Ethylene fluorohydrin	C_2H_5FO	371-62-0	64.058	liq	-26.4	103.5	1.1040^{20}	1.3647^{18}	msc H_2O, EtOH, eth; vs ace; sl chl
5388	Fluoroethene	Vinyl fluoride	C_2H_3F	75-02-5	46.043	col gas	-160.5	-72			i H_2O; s EtOH, ace
5389	1-Fluoroheptane		$C_7H_{15}F$	661-11-0	118.192	liq	-73	117.9	0.8062^{20}	1.3854^{20}	i H_2O; s eth, ace, bz; vs peth
5390	1-Fluorohexane	Hexyl fluoride	$C_6H_{13}F$	373-14-8	104.165	liq	-103	91.5	0.7995^{20}	1.3738^{20}	s eth, bz
5391	1-Fluoro-2-iodobenzene		C_6H_4FI	348-52-7	221.998	liq	-41.5	188.6		1.5910^{20}	s ace, bz, chl
5392	1-Fluoro-4-iodobenzene		C_6H_4FI	352-34-1	221.998	liq	-27	183	1.9523^{15}	1.5270^{22}	i H_2O; s EtOH, eth, ace
5393	1-Fluoro-3-isothiocyanatobenzene		C_7H_4FNS	404-72-8	153.177			227	1.27^{25}	1.6186^{20}	
5394	1-Fluoro-4-isothiocyanatobenzene		C_7H_4FNS	1544-68-9	153.177		27	228			
5395	Fluoromethane	Methyl fluoride	CH_3F	593-53-3	34.033	col gas	-143.3	-78.4	0.557^{25} (p>1 atm)	1.1674^{25}	sl H_2O, bz, chl; vs EtOH, eth
5396	1-Fluoro-2-methoxybenzene		C_7H_7FO	321-28-8	126.128	liq	-39	154.5	1.5489^{17}	1.4969^{17}	i H_2O; s eth, ctc
5397	1-Fluoro-3-methoxybenzene		C_7H_7FO	456-49-5	126.128	liq	-35	159; 51^{14}	1.104^{25}	1.4876^{20}	
5398	1-Fluoro-4-methoxybenzene		C_7H_7FO	459-60-9	126.128	liq	-45	157	1.1781^{18}	1.4886^{18}	s eth
5399	4-Fluoro-2-methylaniline		C_7H_8FN	452-71-1	125.144		14.2	94^{16}	1.1263^{18}	1.5363^{18}	s eth, ace, bz, ctc
5400	(Fluoromethyl)benzene		C_7H_7F	350-50-5	110.129	liq	-35	140; 40^{14}	1.0228^{25}	1.4892^{25}	s ctc
5401	2-Fluoro-4-methyl-1-nitrobenzene	3-Fluoro-4-nitrotoluene	$C_7H_6FNO_2$	446-34-4	155.127	nd (al)	53.2	97^{3}	1.4380^{25}		
5402	2-Fluoro-2-methylpropane	*tert*-Butyl fluoride	C_4H_9F	353-61-7	76.112	col gas		12.1			
5403	1-Fluoronaphthalene		$C_{10}H_7F$	321-38-0	146.161	liq	-9	215; 80^{11}	1.1322^{20}	1.5939^{20}	i H_2O; s EtOH, eth, bz, chl, HOAc
5404	2-Fluoronaphthalene		$C_{10}H_7F$	323-09-1	146.161	nd (al)	61	212; 90^{16}			i H_2O; s EtOH, eth, bz, chl, HOAc
5405	1-Fluoro-2-nitrobenzene	*o*-Fluoronitrobenzene	$C_6H_4FNO_2$	1493-27-2	141.100	ye liq	-6	dec 215	1.3285^{18}	1.5489^{17}	vs eth, EtOH
5406	1-Fluoro-3-nitrobenzene	*m*-Fluoronitrobenzene	$C_6H_4FNO_2$	402-67-5	141.100	ye cry	41	199; 86^{19}	1.3254^{19}	1.5262^{15}	i H_2O; s EtOH, eth; sl bz
5407	1-Fluoro-4-nitrobenzene	*p*-Fluoronitrobenzene	$C_6H_4FNO_2$	350-46-9	141.100	ye nd	21	205	1.3300^{20}	1.5316^{20}	i H_2O; s EtOH, eth; sl ctc
5408	1-Fluorooctane	Octyl fluoride	$C_8H_{17}F$	463-11-6	132.219	liq	-64	142.3	0.8116^{20}	1.3946^{20}	
5409	1-Fluoropentane	Pentyl fluoride	$C_5H_{11}F$	592-50-7	90.139	liq	-120	62.8	0.7907^{20}	1.3591^{2-}	vs eth, EtOH
5410	2-Fluorophenol		C_6H_5FO	367-12-4	112.101		16.1	151.5	1.120^{25}	1.5144^{20}	s H_2O

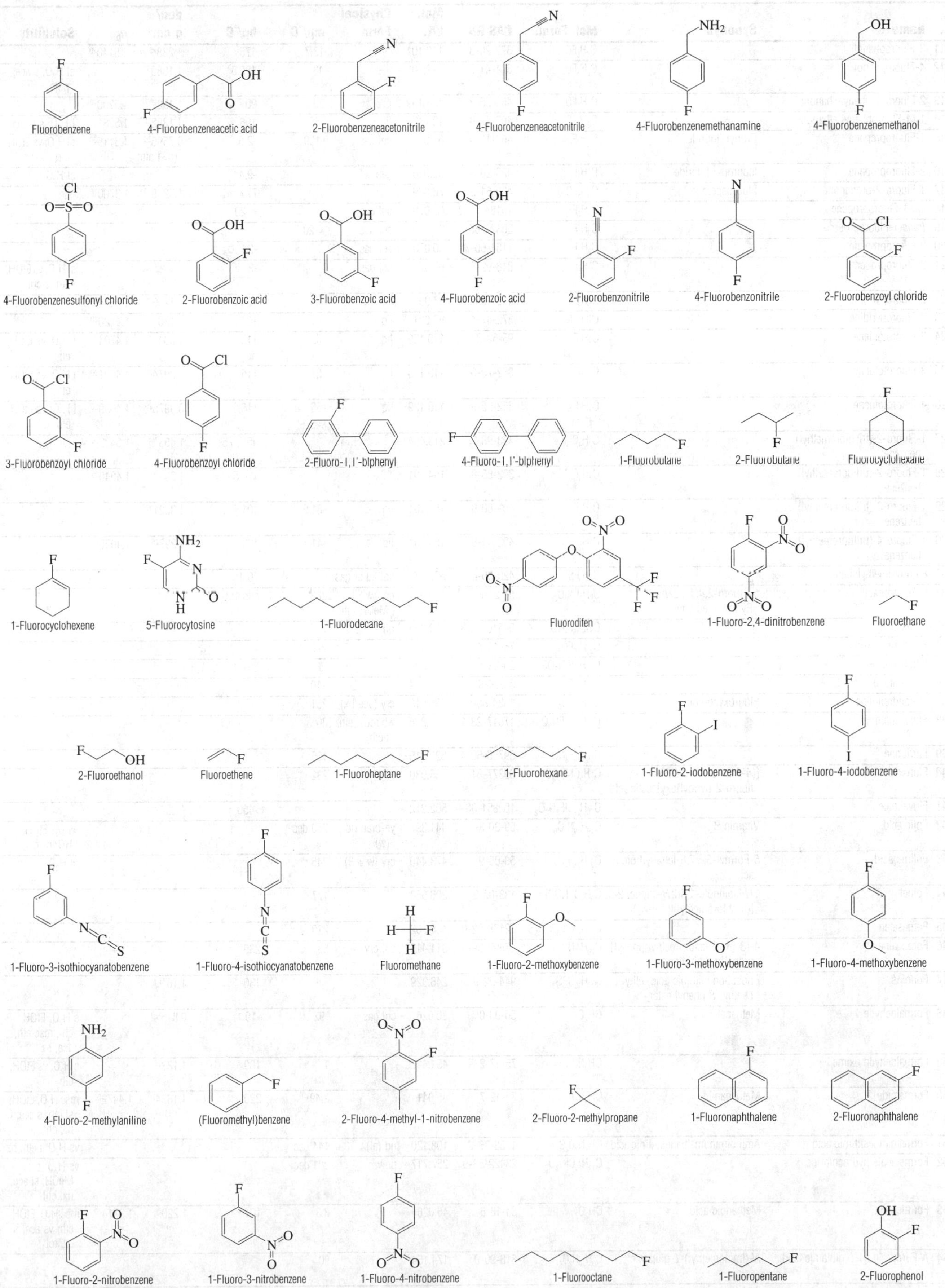
Fluorobenzene
4-Fluorobenzeneacetic acid
2-Fluorobenzeneacetonitrile
4-Fluorobenzeneacetonitrile
4-Fluorobenzenemethanamine
4-Fluorobenzenemethanol
4-Fluorobenzenesulfonyl chloride
2-Fluorobenzoic acid
3-Fluorobenzoic acid
4-Fluorobenzoic acid
2-Fluorobenzonitrile
4-Fluorobenzonitrile
2-Fluorobenzoyl chloride
3-Fluorobenzoyl chloride
4-Fluorobenzoyl chloride
2-Fluoro-1,1'-biphenyl
4-Fluoro-1,1'-biphenyl
1-Fluorobutane
2-Fluorobutane
Fluorocyclohexane
1-Fluorocyclohexene
5-Fluorocytosine
1-Fluorodecane
Fluorodifen
1-Fluoro-2,4-dinitrobenzene
Fluoroethane
2-Fluoroethanol
Fluoroethene
1-Fluoroheptane
1-Fluorohexane
1-Fluoro-2-iodobenzene
1-Fluoro-4-iodobenzene
1-Fluoro-3-isothiocyanatobenzene
1-Fluoro-4-isothiocyanatobenzene
Fluoromethane
1-Fluoro-2-methoxybenzene
1-Fluoro-3-methoxybenzene
1-Fluoro-4-methoxybenzene
4-Fluoro-2-methylaniline
(Fluoromethyl)benzene
2-Fluoro-4-methyl-1-nitrobenzene
2-Fluoro-2-methylpropane
1-Fluoronaphthalene
2-Fluoronaphthalene
1-Fluoro-2-nitrobenzene
1-Fluoro-3-nitrobenzene
1-Fluoro-4-nitrobenzene
1-Fluorooctane
1-Fluoropentane
2-Fluorophenol

No.	Name	Synonym	Mol. Form.	CAS RN	Mol. Wt.	Physical Form	mp/°C	bp/°C	den/ g cm-3	n_D	Solubility
5411	3-Fluorophenol		C_6H_5FO	372-20-3	112.101		13.7	178	1.238^{25}	1.5140^{20}	
5412	4-Fluorophenol		C_6H_5FO	371-41-5	112.101		48	185.5	1.1889^{56}		sl H_2O; s ace, peth
5413	2-Fluoro-1-phenylethanone		C_8H_7FO	450-95-3	138.139	pl	29	90^{12}	1.152^{20}	1.5200^{20}	
5414	1-(4-Fluorophenyl)ethanone		C_8H_7FO	403-42-9	138.139	liq	-45	196	1.1382^{25}	1.5081^{25}	i H_2O; s bz, chl
5415	1-Fluoropropane	Propyl fluoride	C_3H_7F	460-13-9	62.086	col gas	-159	-2.5	0.7596^{20} (p>1 atm	1.3115^{20}	sl H_2O; vs EtOH, eth
5416	2-Fluoropropane	Isopropyl fluoride	C_3H_7F	420-26-8	62.086	gas		-9.4			sl H_2O
5417	1-Fluoro-2-propanone	Fluoroacetone	C_3H_5FO	430-51-3	76.069			77	1.0288^{20}	1.3700^{20}	
5418	*cis*-1-Fluoropropene		C_3H_5F	19184-10-2	60.070	col gas		≈-20			
5419	*trans*-1-Fluoropropene		C_3H_5F	20327-65-5	60.070	col gas	≈-20				
5420	2-Fluoropropene		C_3H_5F	1184-60-7	60.070	col gas		-24			
5421	3-Fluoropropene		C_3H_5F	818-92-8	60.070	col gas		-3			sl H_2O; vs EtOH, eth; s chl
5422	2-Fluoropyridine		C_5H_4FN	372-48-5	97.091			125	1.1280^{20}	1.4574^{20}	
5423	3-Fluoropyridine		C_5H_4FN	372-47-4	97.091	liq		107	1.130	1.4720^{20}	
5424	2-Fluorotoluene		C_7H_7F	95-52-3	110.129	liq	-62	115	1.0041^{13}	1.4704^{20}	i H_2O; vs EtOH, eth
5425	3-Fluorotoluene		C_7H_7F	352-70-5	110.129	liq	-87	115	0.9974^{20}	1.4691^{20}	i H_2O; vs EtOH, eth
5426	4-Fluorotoluene		C_7H_7F	352-32-9	110.129	liq	-56	116.6	0.9975^{20}	1.4699^{20}	i H_2O; vs EtOH, eth
5427	1-Fluoro-2-(trichloromethyl) benzene		$C_7H_4Cl_3F$	488-98-2	213.464			95^{12}, 75^{5}	1.453^{25}	1.5432^{20}	
5428	1-Fluoro-2-(trifluoromethyl) benzene		$C_7H_4F_4$	392-85-8	164.101			114.5	1.293^{25}	1.4040^{25}	
5429	1-Fluoro-3-(trifluoromethyl) benzene		$C_7H_4F_4$	401-80-9	164.101	liq	-81.5	101.5	1.3021^{17}		
5430	1-Fluoro-4-(trifluoromethyl) benzene		$C_7H_4F_4$	402-44-8	164.101	liq	-41.7	103.5	1.293^{25}	1.4025^{20}	
5431	Fluorotrimethylsilane		C_3H_9FSi	420-56-4	92.187	vol liq or gas		16.4			
5432	5-Fluorouracil	5-Fluoro-2,4(1*H*,3*H*)-Pyrimidinedione	$C_4H_3FN_2O_2$	51-21-8	130.077	cry (w, MeOH-eth)	283	sub 190			
5433	Fluoxetine		$C_{17}H_{18}F_3NO$	54910-89-3	309.326	oil					
5434	Fluoxymesterone		$C_{20}H_{29}FO_3$	76-43-7	336.440		270				
5435	Fluphenazine		$C_{22}H_{26}F_3N_3OS$	69-23-8	437.520			$251^{0.3}$			
5436	Fluprednisolone		$C_{21}H_{27}FO_5$	53-34-9	378.434		210				
5437	Flurandrenolide	Fludroxycortide	$C_{24}H_{33}FO_6$	1524-88-5	436.513	cry (ace/hx)	251				
5438	Flurazepam		$C_{21}H_{23}ClFN_3O$	17617-23-1	387.878	wh rods (eth/ peth)	80				
5439	Fluridone		$C_{19}H_{14}F_3NO$	59756-60-4	329.315		155				
5440	Fluroxypyr	[(4-Amino-3,5-dichloro-6-fluoro-2-pyridyl)oxy]acetic acid	$C_7H_5Cl_2FN_2O_3$	69377-81-7	255.030		232				
5441	Fluvalinate		$C_{26}H_{22}ClF_3N_2O_3$	102851-06-9	502.912			>450	1.29^{25}		
5442	Folic acid	Vitamin B_c	$C_{19}H_{19}N_7O_6$	59-30-3	441.397	ye-oran nd (w)	250 dec				vs py, EtOH, HOAc
5443	Folinic acid	5-Formyl-5,6,7,8-tetrahydrofolic acid	$C_{20}H_{23}N_7O_7$	58-05-9	473.440	cry (w + 3)	245 dec				sl H_2O
5444	Folpet	1*H*-Isoindole-1,3(2*H*)-dione, 2-[(trichloromethyl)thio]-	$C_9H_4Cl_3NO_2S$	133-07-3	296.558		177				
5445	Fomesafen		$C_{15}H_{10}ClF_3N_2O_6S$	72178-02-0	438.762		220		1.28^{20}		
5446	Fomocaine	4-[3-[4-(Phenoxymethyl)phenyl] propyl]morpholine	$C_{20}H_{25}NO_2$	17692-39-6	311.419	col cry	53	$239^{1.1}$			
5447	Fonofos	Phosphonodithioic acid, ethyl-, *O*-ethyl *S*-phenyl ester	$C_{10}H_{15}OPS_2$	944-22-9	246.329			$130^{0.1}$	1.16^{25}		
5448	Formaldehyde	Methanal	CH_2O	50-00-0	30.026	col gas	-92	-19.1	0.815^{-20}		s H_2O, EtOH, chl; msc eth, ace, bz
5449	Formaldehyde oxime		CH_3NO	75-17-2	45.041		1.3	109^{15}	1.133^{25}		s H_2O; vs EtOH, eth
5450	Formamide	Methanamide	CH_3NO	75-12-7	45.041		2.49	220	1.1334^{20}	1.4472^{20}	msc H_2O, EtOH; sl eth; s ace; i bz, chl
5451	Formamidinesulfinic acid	Aminoiminomethanesulfinic acid	$CH_4N_2O_2S$	1758-73-2	108.120	nd (al)	144 dec				vs H_2O; i eth, bz
5452	Formetanate hydrochloride		$C_{11}H_{16}ClN_3O_2$	23422-53-9	257.717	pow	201 dec				vs H_2O; s MeOH; sl ace, hx, chl
5453	Formic acid	Methanoic acid	CH_2O_2	64-18-6	46.026		8.3	101	1.220^{20}	1.3714^{20}	msc H_2O, EtOH, eth; vs ace; s bz, tol
5454	*N*-Formimidoyl-*L*-glutamic acid	*N*-(Iminomethyl)-*L*-glutamic acid	$C_6H_{10}N_2O_4$	816-90-0	174.154		90				
5455	Formononetin	7-Hydroxy-3-(4-methoxyphenyl) -4*H*-1-benzopyran-4-one	$C_{16}H_{12}O_4$	485-72-3	268.264		256.5				

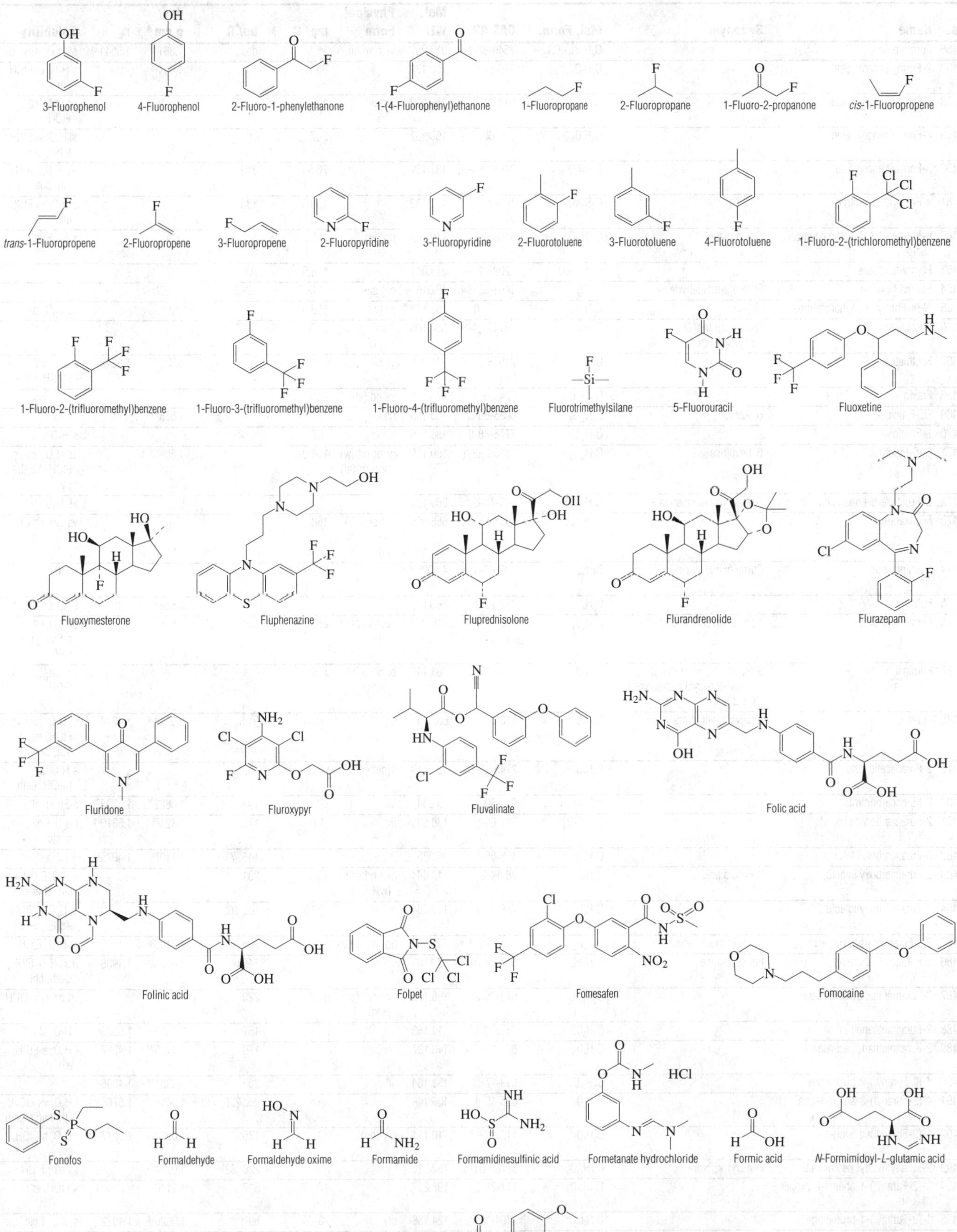
3-Fluorophenol
4-Fluorophenol
2-Fluoro-1-phenylethanone
1-(4-Fluorophenyl)ethanone
1-Fluoropropane
2-Fluoropropane
1-Fluoro-2-propanone
cis-1-Fluoropropene
trans-1-Fluoropropene
2-Fluoropropene
3-Fluoropropene
2-Fluoropyridine
3-Fluoropyridine
2-Fluorotoluene
3-Fluorotoluene
4-Fluorotoluene
1-Fluoro-2-(trichloromethyl)benzene
1-Fluoro-2-(trifluoromethyl)benzene
1-Fluoro-3-(trifluoromethyl)benzene
1-Fluoro-4-(trifluoromethyl)benzene
Fluorotrimethylsilane
5-Fluorouracil
Fluoxetine
Fluoxymesterone
Fluphenazine
Fluprednisolone
Flurandrenolide
Flurazepam
Fluridone
Fluroxypyr
Fluvalinate
Folic acid
Folinic acid
Folpet
Fomesafen
Fomocaine
Fonofos
Formaldehyde
Formaldehyde oxime
Formamide
Formamidinesulfinic acid
Formetanate hydrochloride
Formic acid
N-Formimidoyl-L-glutamic acid
Formononetin

No.	Name	Synonym	Mol. Form.	CAS RN	Mol. Wt.	Physical Form	mp/°C	bp/°C	den/ g cm-3	n_D	Solubility
5456	Formothion		$C_6H_{12}NO_4PS_2$	2540-82-1	257.267	visc ye oil	25.5	dec	1.361^{20}	1.5541^{20}	sl H_2O; misc os
5457	2-Formylbenzoic acid		$C_8H_6O_3$	119-67-5	150.132		98		1.404^{25}		s H_2O; vs EtOH, eth
5458	3-Formylbenzoic acid		$C_8H_6O_3$	619-21-6	150.132	nd (w)	175				vs H_2O, eth, EtOH
5459	4-Formylbenzoic acid		$C_8H_6O_3$	619-66-9	150.132		247				sl H_2O; vs EtOH; s eth, chl
5460	3-Formylbenzonitrile		C_8H_5NO	24964-64-5	131.132		76.5	210			vs H_2O, EtOH, eth, chl
5461	4-Formylbenzonitrile		C_8H_5NO	105-07-7	131.132		100.5	133^{12}			s H_2O; vs EtOH, eth, chl
5462	6-Formyl-2,3-dimethoxybenzoic acid	Opianic acid	$C_{10}H_{10}O_5$	519-05-1	210.183	nd (w)	150				s EtOH, eth
5463	Formylferrocene		$C_{11}H_{10}FeO$	12093-10-6	214.041		118.5	$70^{0.1}$			
5464	Formyl fluoride	Fluoroformaldehyde	CHFO	1493-02-3	48.016	col gas	-142.2	-26.5	1.1950^{-30}		
5465	*N*-(4-Formylphenyl)acetamide		$C_9H_9NO_2$	122-85-0	163.173	pr (w)	158.0				vs H_2O, bz
5466	Fosetyl-Al	Aluminum tris(*O*-ethylphosphonate)	$C_6H_{18}AlO_9P_3$	39148-24-8	354.105		>300				
5467	Fosthietan		$C_6H_{12}NO_3PS_2$	21548-32-3	241.268	ye oil			1.3^{25}	1.5348^{25}	s ace, chl, MeOH, tol
5468	Fraxin		$C_{16}H_{18}O_{10}$	524-30-1	370.308	ye nd (al)	205				
5469	*DL*-Fructose	α-Acrose	$C_6H_{12}O_6$	6035-50-3	180.155	nd	130		1.665^{16}		
5470	*L*-Fructose		$C_6H_{12}O_6$	7776-48-9	180.155	wh cry	102				s H_2O
5471	β-*D*-Fructose	β-Levulose	$C_6H_{12}O_6$	53188-23-1	180.155	pr or nd (w) orth pr (al)	103 dec		1.60^{20}		vs H_2O, ace; s EtOH, MeOH, py
5472	*D*-Fructose 6-phosphate	Hexose monophosphate	$C_6H_{13}O_9P$	643-13-0	260.135						vs H_2O
5473	Fucoxanthin		$C_{42}H_{58}O_6$	3351-86-8	658.905	red pl (eth) hex pl (dil al)	168				vs eth, EtOH
5474	Fulminic acid	Carbyloxime	CHNO	506-85-4	43.025		unstable in pure form				s eth
5475	Fulvene		C_6H_6	497-20-1	78.112			7^{56}	0.8241^{20}	1.4920^{20}	i H_2O; s bz, chl
5476	Fumaric acid	*trans*-2-Butenedioic acid	$C_4H_4O_4$	110-17-8	116.073	nd, mcl pr or lf (w)	287 dec	sub 165	1.635^{20}		sl H_2O, eth, ace; s EtOH, con sulf
5477	Fumigatin	3-Hydroxy-2-methoxy-5-methyl-2,5-cyclohexadiene-1,4-dione	$C_8H_8O_4$	484-89-9	168.148	br nd or pl (peth)	116				vs ace, bz, eth, EtOH
5478	Furan	Oxacyclopentadiene	C_4H_4O	110-00-9	68.074	liq	-85.61	31.5	0.9514^{20}	1.4214^{20}	sl H_2O, chl; vs EtOH, eth; s ace, bz
5479	2-Furanacetic acid		$C_6H_6O_3$	2745-26-8	126.110	lf(peth)	68.5	$102^{0.4}$			s H_2O, bz, MeOH, peth
5480	2-Furancarbonitrile		C_5H_3NO	617-90-3	93.084			147	1.0822^{20}	1.4798^{20}	s EtOH, eth
5481	2-Furancarbonyl chloride		$C_5H_3ClO_2$	527-69-5	130.530	liq	-1.0	173	1.324^{25}	1.5310^{20}	i H_2O; s eth, chl; sl ctc
5482	3-Furancarboxaldehyde		$C_5H_4O_2$	498-60-2	96.085			145; 71^{43}	1.110^{20}	1.4945^{20}	
5483	2-Furancarboxylic acid	2-Furoic acid	$C_5H_4O_3$	88-14-2	112.084	mcl nd or lf (w)	133.5	231			s H_2O, EtOH; vs eth; sl ace
5484	3-Furancarboxylic acid		$C_5H_4O_3$	488-93-7	112.084	nd (w)	122.5	sub 105			sl H_2O; s EtOH, AcOEt; vs eth
5485	2,5-Furandicarboxylic acid	Dehydromucic acid	$C_6H_4O_5$	3238-40-2	156.093	nd (w), lf (al)	342	sub	1.7400^{20}		sl H_2O, EtOH
5486	2-Furanmethanamine	Furfurylamine	C_5H_7NO	617-89-0	97.116			145.5	1.0995^{20}	1.4908^{20}	msc H_2O, EtOH; s eth, chl
5487	2-Furanmethanediol diacetate		$C_9H_{10}O_5$	613-75-2	198.172	nd or pl (eth-peth)	53.3	220			vs bz, eth, EtOH
5488	2-Furanmethanethiol		C_5H_6OS	98-02-2	114.166			157	1.1319^{20}	1.5329^{20}	i H_2O; sl chl
5489	2-Furanmethanol acetate		$C_7H_8O_3$	623-17-6	140.137			179	1.1175^{20}	1.4327^{20}	i H_2O; s EtOH, eth
5490	4-(2-Furanyl)-2-butanone		$C_8H_{10}O_2$	699-17-2	138.164	oil		203	1.0361^{19}	1.4696^{17}	
5491	4-(2-Furanyl)-3-buten-2-one		$C_8H_8O_2$	623-15-4	136.149		39.5	dec 229; 113^{10}	1.0496^{57}	1.5788^{45}	i H_2O; vs EtOH, eth, chl; s peth
5492	1-(2-Furanyl)ethanone		$C_6H_6O_2$	1192-62-7	110.111	cry (lig)	33	175	1.098^{20}	1.5017^{20}	i H_2O; s EtOH, eth
5493	2-Furanylmethyl pentanoate	Furfuryl valerate	$C_{10}H_{14}O_3$	36701-01-6	182.216			228; 82^{1}	1.0284^{20}		vs eth, EtOH
5494	3-(2-Furanyl)-1-phenyl-2-propen-1-one		$C_{13}H_{10}O_2$	717-21-5	198.217		47	317	1.1140^{20}		s EtOH, eth
5495	1-(2-Furanyl)-1-propanone		$C_7H_8O_2$	3194-15-8	124.138	cry	28	88^{14}	1.0626^{28}	1.4922^{25}	s eth; sl ctc
5496	1-(2-Furanyl)-2-propanone	2-Furfuryl methyl ketone	$C_7H_8O_2$	6975-60-6	124.138		29	179.5	1.104^{20}	1.5035^{20}	
5497	3-(2-Furanyl)-2-propenal		$C_7H_6O_2$	623-30-3	122.122		54	135^{14}			i H_2O; msc EtOH; s eth; sl chl
5498	3-(2-Furanyl)-2-propenenitrile	2-Furanacrylonitrile	C_7H_5NO	7187-01-1	119.121		38	96		1.5824^{25}	vs tol

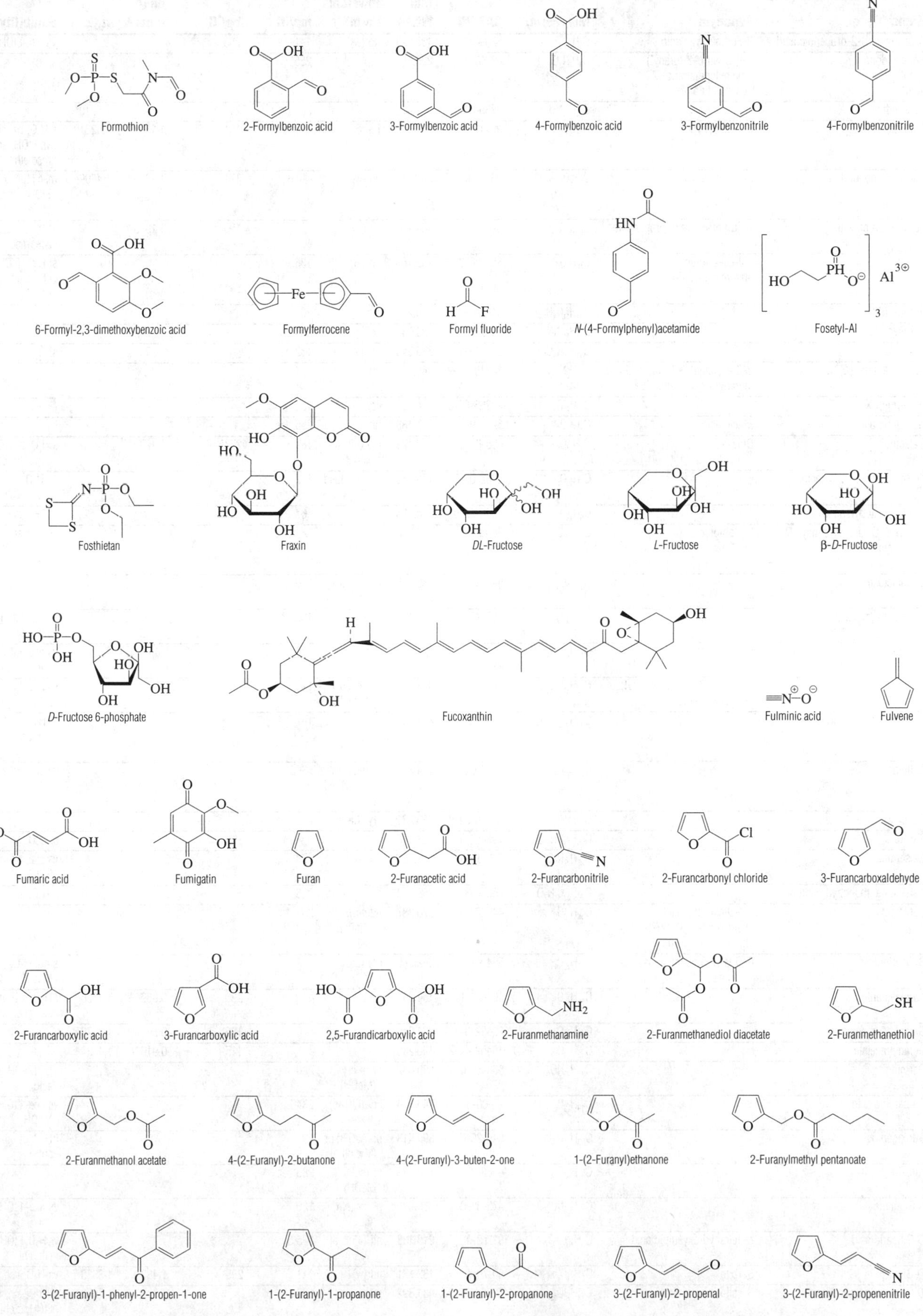

Formothion
2-Formylbenzoic acid
3-Formylbenzoic acid
4-Formylbenzoic acid
3-Formylbenzonitrile
4-Formylbenzonitrile
6-Formyl-2,3-dimethoxybenzoic acid
Formylferrocene
Formyl fluoride
N-(4-Formylphenyl)acetamide
Fosetyl-Al
Fosthietan
Fraxin
DL-Fructose
L-Fructose
β-D-Fructose
D-Fructose 6-phosphate
Fucoxanthin
Fulminic acid
Fulvene
Fumaric acid
Fumigatin
Furan
2-Furanacetic acid
2-Furancarbonitrile
2-Furancarbonyl chloride
3-Furancarboxaldehyde
2-Furancarboxylic acid
3-Furancarboxylic acid
2,5-Furandicarboxylic acid
2-Furanmethanamine
2-Furanmethanediol diacetate
2-Furanmethanethiol
2-Furanmethanol acetate
4-(2-Furanyl)-2-butanone
4-(2-Furanyl)-3-buten-2-one
1-(2-Furanyl)ethanone
2-Furanylmethyl pentanoate
3-(2-Furanyl)-1-phenyl-2-propen-1-one
1-(2-Furanyl)-1-propanone
1-(2-Furanyl)-2-propanone
3-(2-Furanyl)-2-propenal
3-(2-Furanyl)-2-propenenitrile

No.	Name	Synonym	Mol. Form.	CAS RN	Mol. Wt.	Physical Form	mp/°C	bp/°C	den/ g cm⁻³	n_D	Solubility
5499	3-(2-Furanyl)-2-propenoic acid	2-Furanacrylic acid	$C_7H_6O_3$	539-47-9	138.121	nd (w)	141	286			vs eth, EtOH
5500	Furazolidone	3-[[(5-Nitro-2-furanyl) methylene]amino]-2-oxazolidinone	$C_8H_7N_3O_5$	67-45-8	225.159		255				
5501	Furethidine		$C_{21}H_{31}NO_4$	2385-81-1	361.476		28	$210^{0.5}$		1.5219^{20}	
5502	Furfural	2-Furaldehyde	$C_5H_4O_2$	98-01-1	96.085	liq	-38.1	161.7	1.1594^{20}	1.5261^{20}	s H_2O, bz, chl; vs EtOH, ace; msc eth
5503	Furfuryl alcohol	2-Furanmethanol	$C_5H_6O_2$	98-00-0	98.101	col-ye liq	-14.6	171	1.1296^{20}	1.4869^{20}	msc H_2O; vs EtOH, eth; s chl
5504	Furfuryl propanoate	2-Furanmethanol, propanoate	$C_8H_{10}O_3$	623-19-8	154.163			195	1.1085^{20}		sl H_2O; s EtOH, ace; msc eth
5505	Furoin	1,2-Di-2-furanyl-2-hydroxyethanone	$C_{10}H_8O_4$	552-86-3	192.169	nd (al)	138.5				sl H_2O, EtOH, chl; s eth, MeOH
5506	Furonazide		$C_{12}H_{11}N_3O_2$	3460-67-1	229.234		202.3				
5507	Furosemide		$C_{12}H_{11}ClN_2O_5S$	54-31-9	330.743		204 dec				
5508	Fursultiamine		$C_{17}H_{26}N_4O_3S_2$	804-30-8	398.543	col pr	132 dec		1.29		sl H_2O
5509	Furylfuramide, (*E*)	2-(2-Furanyl)-3-(5-nitro-2-furanyl)-2-propenamide	$C_{11}H_8N_2O_5$	18819-45-9	248.192	cry	154				
5510	Fusarenon X		$C_{17}H_{22}O_8$	23255-69-8	354.352	cry	182				
5511	Galactaric acid	Mucic acid	$C_6H_{10}O_8$	526-99-8	210.138	pr (w)	255 dec				
5512	Galactitol	Dulcose	$C_6H_{14}O_6$	608-66-2	182.171	cry (dil MeOH)	189.5	277^1	1.47^{20}		s H_2O; sl EtOH, py; i eth, bz
5513	*D*-Galactonic acid, γ-lactone		$C_6H_{10}O_6$	2782-07-2	178.139	nd (w+1), nd (al)	112				vs H_2O
5514	α-*D*-Galactopyranose		$C_6H_{12}O_6$	3646-73-9	180.155		167				
5515	4-*O*-β-*D*-Galactopyranosyl-*D*-gluconic acid	Lactobionic acid	$C_{12}H_{22}O_{12}$	96-82-2	358.296	syr					vs H_2O; sl EtOH, MeOH, HOAc; i eth
5516	*D*-Galactose		$C_6H_{12}O_6$	59-23-4	180.155	pl or pr (al)pr or nd (w+1)	170				vs H_2O; sl EtOH; i eth, bz; s py
5517	*D*-Galacturonic acid		$C_6H_{10}O_7$	685-73-4	194.139	nd (w)	166 (β)				s H_2O, EtOH; i eth
5518	Galanthamine	Lycoremine	$C_{17}H_{21}NO_3$	357-70-0	287.354	cry (bz)	126.5				vs ace, EtOH, chl
5519	Galipine	2-[2-(3,4-Dimethoxyphenyl) ethyl]-4-methoxyquinoline	$C_{20}H_{21}NO_3$	525-68-8	323.386	pr (al, eth) nd (peth)	115.5				vs ace, bz, eth, EtOH
5520	Gallamine triethiodide		$C_{30}H_{60}I_3N_3O_3$	65-29-2	891.528		147.5				vs H_2O, EtOH; sl eth, ace, bz, chl
5521	Gallein		$C_{20}H_{12}O_7$	2103-64-2	364.31	br-red pow (+1.5w) red (anh)	>300				vs ace, EtOH
5522	Ganciclovir		$C_9H_{13}N_5O_4$	82410-32-0	255.231	cry (MeOH)	250 dec				
5523	Gardol		$C_{15}H_{28}NNaO_3$	137-16-6	293.378						sl H_2O
5524	Gelsemine		$C_{20}H_{22}N_2O_2$	509-15-9	322.401	cry (ace)	178				vs ace, bz, eth, EtOH
5525	Gelsemine, monohydrochloride		$C_{20}H_{23}ClN_2O_2$	35306-33-3	358.862		326				s H_2O; sl EtOH
5526	Genistein	5,7-Dihydroxy-3-(4-hydroxyphenyl)-4*H*-1-benzopyran-4-one	$C_{15}H_{10}O_5$	446-72-0	270.237	nd(eth), pr(dil al)	301 dec				
5527	β-Gentiobiose	6-*O*-β-*D*-Glucopyranosyl-*D*-glucose	$C_{12}H_{22}O_{11}$	554-91-6	342.296	cry (EtOH)	192				s hot H_2O, hot MeOH
5528	*trans*-Geraniol		$C_{10}H_{18}O$	106-24-1	154.249		<-15	230	0.8894^{20}	1.4766^{20}	i H_2O; s EtOH, eth, ace, chl
5529	Geranyl 2-methylpropanoate		$C_{14}H_{24}O_2$	2345-26-8	224.340			136^{13}	0.8997^{15}	1.4576^{20}	
5530	Geranyl acetate		$C_{12}H_{20}O_2$	16409-44-2	196.286			115^{12}	0.9163^{15}	1.4624^{20}	
5531	Germine		$C_{27}H_{43}NO_8$	508-65-6	509.632	pr or cry (MeOH)	220				s bz, MeOH, alk, acid
5532	Gibberellic acid		$C_{19}H_{22}O_6$	77-06-5	346.374	cry (EtOAc)	234				vs ace, EtOH, MeOH
5533	Gitoxigenin		$C_{23}H_{34}O_5$	545-26-6	390.513	pr (AcOEt) pr (+w, dil al)	234				i H_2O; sl eth; s chl
5534	Gitoxin		$C_{41}H_{64}O_{14}$	4562-36-1	780.939	pr (chl-MeOH)	285 dec				
5535	*d*-Glaucine		$C_{21}H_{25}NO_4$	475-81-0	355.429	pl, pr (eth, AcOEt)	120				vs ace, EtOH, chl
5536	*D*-Glucaric acid	*D*-Tetrahydroxyadipic acid	$C_6H_{10}O_8$	87-73-0	210.138	nd (45% al)	125.5				vs H_2O, EtOH; sl eth, chl
5537	*D*-Glucitol	Sorbitol	$C_6H_{14}O_6$	50-70-4	182.171	nd (w)	111	$295^{3.5}$	1.489^{20}	1.3330^{20}	vs H_2O, ace
5538	*D*-Glucitol, hexaacetate	Sorbitol hexaacetate	$C_{18}H_{26}O_{12}$	7208-47-1	434.392	pr (w)	100.8		1.30^{20}		sl H_2O, eth; vs EtOH; s chl, AcOEt

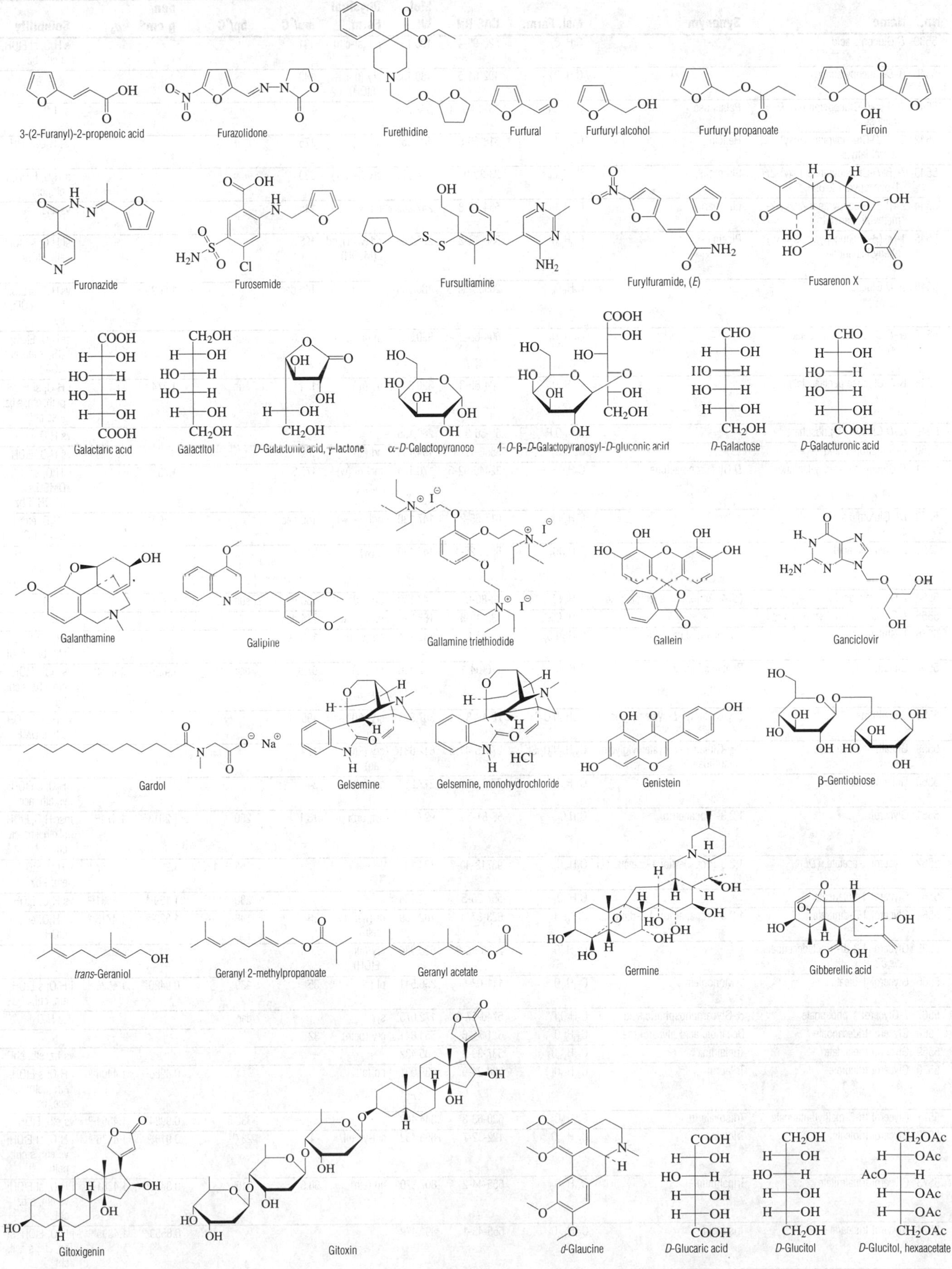

3-(2-Furanyl)-2-propenoic acid
Furazolidone
Furethidine
Furfural
Furfuryl alcohol
Furfuryl propanoate
Furoin
Furonazide
Furosemide
Fursultiamine
Furylfuramide, (E)
Fusarenon X
Galactaric acid
Galactitol
D-Galacturonic acid, γ-lactone
α-D-Galactopyranose
4-O-β-D-Galactopyranosyl-D-gluconic acid
D-Galactose
D-Galacturonic acid
Galanthamine
Galipine
Gallamine triethiodide
Gallein
Ganciclovir
Gardol
Gelsemine
Gelsemine, monohydrochloride
Genistein
β-Gentiobiose
trans-Geraniol
Geranyl 2-methylpropanoate
Geranyl acetate
Germine
Gibberellic acid
Gitoxigenin
Gitoxin
d-Glaucine
D-Glucaric acid
D-Glucitol
D-Glucitol, hexaacetate

No.	Name	Synonym	Mol. Form.	CAS RN	Mol. Wt.	Physical Form	mp/°C	bp/°C	den/ g cm^{-3}	n_D	Solubility
5539	*D*-Gluconic acid		$C_6H_{12}O_7$	526-95-4	196.155	nd (al-eth)	131				s H_2O; sl EtOH; i eth, bz
5540	β-*D*-Glucopyranose		$C_6H_{12}O_6$	492-61-5	180.155	cry (hot EtOH)	149				
5541	6-*O*-α-*D*-Glucopyranosyl-*D*-fructose	Palatinose	$C_{12}H_{22}O_{11}$	13718-94-0	342.296						s H_2O
5542	2-(β-*D*-Glucopyranosyloxy) benzaldehyde	Helicin	$C_{13}H_{16}O_7$	618-65-5	284.262	nd (w)	175				vs H_2O, EtOH
5543	7-(β-*D*-Glucopyranosyloxy)-2*H*-1-benzopyran-2-one	Skimmin	$C_{15}H_{16}O_8$	93-39-0	324.283	cry (w + 1)	220				s H_2O, EtOH; i eth, chl
5544	2-(β-*D*-Glucopyranosyloxy)-2-methylpropanenitrile	Linamarin	$C_{10}H_{17}NO_6$	554-35-8	247.245	nd (w, al)	145				vs ace
5545	1-[4-(β-*D*-Glucopyranosyloxy) phenyl]ethanone	Picein	$C_{14}H_{18}O_7$	530-14-3	298.289	nd (w+1), nd (MeOH)	195.5				sl H_2O; s EtOH, eth, HOAc; i chl
5546	α-*D*-Glucose		$C_6H_{12}O_6$	26655-34-5	180.155		146 dec		1.5620^{18}		vs H_2O; sl EtOH; i ace, AcOEt; s py
5547	α-*D*-Glucose pentaacetate		$C_{16}H_{22}O_{11}$	604-68-2	390.339	pl or nd (al)	113.3	sub			sl H_2O, EtOH, CS_2; s eth, chl, HOAc
5548	β-*D*-Glucose pentaacetate		$C_{16}H_{22}O_{11}$	604-69-3	390.339	nd (al)	134	sub	1.2740^{20}		i H_2O; sl EtOH, peth, eth; s bz; msc chl
5549	α-*D*-Glucose 1-phosphate		$C_6H_{13}O_9P$	59-56-3	260.135						vs H_2O
5550	*D*-Glucuronic acid		$C_6H_{10}O_7$	6556-12-3	194.139	nd (al)	165				vs H_2O, EtOH
5551	*D*-Glucuronic acid γ-lactone	*D*-Glucuronolactone	$C_6H_8O_6$	32449-92-6	176.124	mcl pl (w) cry (al)	177.5		1.76^{20}		s H_2O; sl EtOH, DMSO, MeOH; i bz
5552	*DL*-Glutamic acid		$C_5H_9NO_4$	617-65-2	147.130	orth (al,w)	199 dec		1.4601^{20}		sl H_2O, eth; i EtOH, CS_2, lig
5553	*D*-Glutamic acid		$C_5H_9NO_4$	6893-26-1	147.130	lf (w)	213 dec		1.538^{20}		sl H_2O; i EtOH, eth, ace, bz, HOAc, MeOH
5554	*L*-Glutamic acid	(*S*)-2-Aminopentanedioic acid	$C_5H_9NO_4$	56-86-0	147.130	orth (dil al)	160 dec	sub 175	1.538^{20}		sl H_2O
5555	*L*-Glutamic acid, hydrochloride		$C_5H_{10}ClNO_4$	138-15-8	183.591	orth pl (w)	214 dec				vs H_2O, EtOH
5556	*L*-Glutamine	2-Aminoglutaramic acid	$C_5H_{10}N_2O_3$	56-85-9	146.144	nd (w, dil al)	185 dec				s H_2O; i EtOH, eth, bz, MeOH
5557	Glutaric acid	Pentanedioic acid	$C_5H_8O_4$	110-94-1	132.116	nd (bz)	97.8	dec 303	1.429^{15}	1.4188^{106}	vs H_2O, EtOH, eth; i bz; s chl, lig
5558	Glutathione	*L*-γ-Glutamyl-*L*-cysteinylglycine	$C_{10}H_{17}N_3O_6S$	70-18-8	307.323	cry (50% al)	195				vs H_2O; i EtOH, eth; s DMF
5559	Glutathione disulfide	*L*-γ-Glutamyl-*L*-cysteinylglycine disulfide	$C_{20}H_{32}N_6O_{12}S_2$	27025-41-8	612.631	cry (EtOH aq)	179				
5560	Glutethimide		$C_{13}H_{15}NO_2$	77-21-4	217.264	cry (eth)	84				i H_2O; s EtOH; vs eth, ace
5561	Glycerol	1,2,3-Propanetriol	$C_3H_8O_3$	56-81-5	92.094	syr, orth pl	18.1	290	1.2613^{20}	1.4746^{20}	msc H_2O, EtOH; sl eth; i bz, ctc, chl
5562	Glycerol 1-acetate, (*DL*)	1,2,3-Propanetriol 1-acetate, (±)	$C_5H_{10}O_4$	93713-40-7	134.131			158^{165}, 130^3	1.2060^{20}	1.4157^{20}	s H_2O, EtOH; sl eth; i bz
5563	Glycerol 1-butanoate		$C_7H_{14}O_4$	557-25-5	162.184			280; 117^{10}	1.129^{18}	1.4531^{20}	vs H_2O, EtOH
5564	Glycerol 1,3-dinitrate	1,2,3-Propanetriol, 1,3-dinitrate	$C_3H_6N_2O_7$	623-87-0	182.089	pr (w), cry (eth)	26	148^{15}, $116^{0.6}$	1.523^{20}	1.4715^{20}	vs H_2O, eth, EtOH
5565	Glycerol 1,3-di-9-octadecenoate, *cis,cis*		$C_{39}H_{72}O_5$	2465-32-9	620.986	cry (eth/ EtOH)	50.1				
5566	Glycerol 1-oleate	1-Monoolein	$C_{21}H_{40}O_4$	111-03-5	356.541	pl (al)	35	239^3	0.9420^{20}	1.4626^{20}	i H_2O; s EtOH, eth, chl
5567	*L*-Glycerol 1-phosphate	α-Glycerophosphoric acid	$C_3H_9O_6P$	5746-57-6	172.073	syr		dec			dec H_2O
5568	Glycerol tridecanoate	Decanoic acid glycerol triester	$C_{33}H_{62}O_6$	621-71-6	554.841	cry (peth)	32				
5569	Glycerol trielaidate	Trielaidin	$C_{57}H_{104}O_6$	537-39-3	885.432						vs bz, eth, chl
5570	Glycerol trilaurate	Trilaurin	$C_{39}H_{74}O_6$	538-24-9	639.001	nd (al)			0.8986^{55}	1.4404^{60}	i H_2O; s EtOH, eth, peth; vs ace, bz
5571	Glycerol tri-3-methylbutanoate	Triisovalerin	$C_{18}H_{32}O_6$	620-63-3	344.443			332.5	0.9984^{20}	1.4354^{20}	vs eth, EtOH
5572	Glycerol trioleate	Triolein	$C_{57}H_{104}O_6$	122-32-7	885.432	col-ye oil	-4	237^{18}	0.915^{15}	1.4676^{15}	i H_2O; sl EtOH; vs eth; s chl, peth
5573	Glycerol tripalmitate	Tripalmitin	$C_{51}H_{98}O_6$	555-44-2	807.320	nd (eth)	66.5	315	0.8752^{70}	1.4381^{80}	i H_2O; sl EtOH; vs eth; s bz, chl
5574	Glycerol tristearate	Tristearin	$C_{57}H_{110}O_6$	555-43-1	891.479				0.8559^{90}	1.4395^{80}	i H_2O, EtOH; sl bz, ctc; s ace, chl

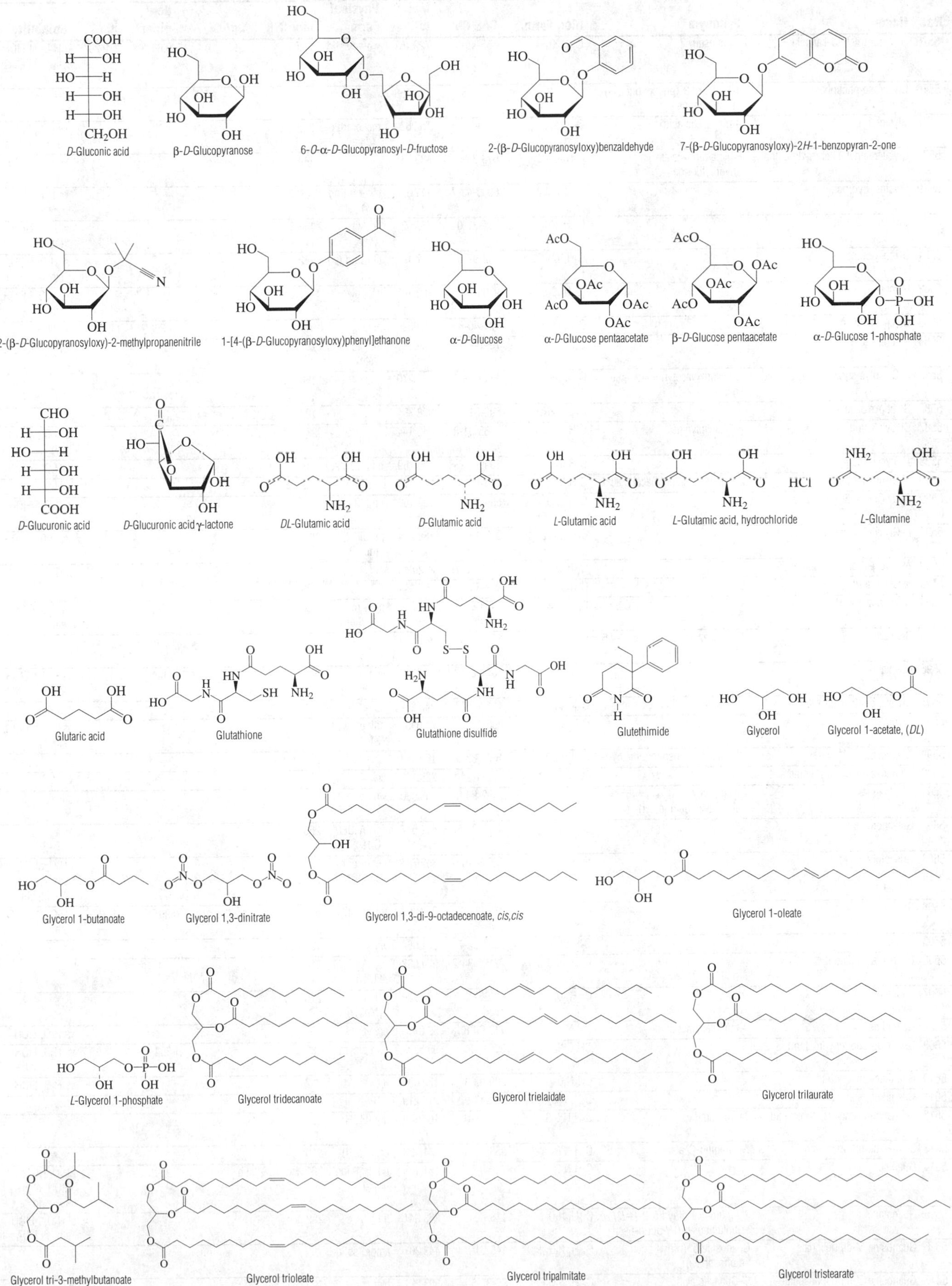
D-Gluconic acid
β-D-Glucopyranose
6-O-α-D-Glucopyranosyl-D-fructose
2-(β-D-Glucopyranosyloxy)benzaldehyde
7-(β-D-Glucopyranosyloxy)-2H-1-benzopyran-2-one
2-(β-D-Glucopyranosyloxy)-2-methylpropanenitrile
1-[4-(β-D-Glucopyranosyloxy)phenyl]ethanone
α-D-Glucose
α-D-Glucose pentaacetate
β-D-Glucose pentaacetate
α-D-Glucose 1-phosphate
D-Glucuronic acid
D-Glucuronic acid γ-lactone
DL-Glutamic acid
D-Glutamic acid
L-Glutamic acid
L-Glutamic acid, hydrochloride
L-Glutamine
Glutaric acid
Glutathione
Glutathione disulfide
Glutethimide
Glycerol
Glycerol 1-acetate, (DL)
Glycerol 1-butanoate
Glycerol 1,3-dinitrate
Glycerol 1,3-di-9-octadecenoate, cis,cis
Glycerol 1-oleate
L-Glycerol 1-phosphate
Glycerol tridecanoate
Glycerol trielaidate
Glycerol trilaurate
Glycerol tri-3-methylbutanoate
Glycerol trioleate
Glycerol tripalmitate
Glycerol tristearate

No.	Name	Synonym	Mol. Form.	CAS RN	Mol. Wt.	Physical Form	mp/°C	bp/°C	den/ g cm⁻³	n_D	Solubility
5575	Glycerol tritetradecanoate	Trimyristin	$C_{45}H_{86}O_6$	555-45-3	723.161	wh-ye solid	58.5	311	0.8848^{60}	1.4428^{60}	i H_2O; sl EtOH, lig; s eth, ace, bz
5576	Glycerone phosphate	1-Hydroxy-3-(phosphonooxy)-2-propanone	$C_3H_7O_6P$	57-04-5	170.058						dec H_2O
5577	Glycine	Aminoacetic acid	$C_2H_5NO_2$	56-40-6	75.067	mcl or trg pr (dil al)	290 dec		1.161^{20}		vs H_2O; i EtOH, eth; sl ace, py
5578	Glycine, ethyl ester, hydrochloride	Ethyl aminoacetate hydrochloride	$C_4H_{10}ClNO_2$	623-33-6	139.581		144				vs H_2O, EtOH
5579	Glycine, hydrochloride		$C_2H_6ClNO_2$	6000-43-7	111.528	hyg orth nd (w)	200.5				vs H_2O
5580	Glycocholic acid		$C_{26}H_{43}NO_6$	475-31-0	465.622	nd (w)	166.5				sl H_2O, eth; vs EtOH
5581	Glycocyamine		$C_3H_7N_3O_2$	352-97-6	117.107	pl or nd (w)	282				sl H_2O, EtOH, eth
5582	Glycogen		$(C_6H_{10}O_5)_x$	9005-79-2	162.140	wh pow					vs H_2O; i EtOH, eth
5583	Glycolaldehyde		$C_2H_4O_2$	141-46-8	60.052	pl	97		1.366^{100}	1.4772^{19}	s chl
5584	Glycolic acid		$C_2H_4O_3$	79-14-1	76.051	orth nd (w) lf (eth)	79.5	100			s H_2O, EtOH, eth
5585	*N*-Glycolylneuraminic acid	*N*-(Hydroxyacetyl)neuraminic acid	$C_{11}H_{19}NO_{10}$	1113-83-3	325.270		186				
5586	Glycopyrrolate		$C_{19}H_{28}BrNO_3$	596-51-0	398.334		192.5				
5587	Glycylalanine	*N*-Alanylglycine	$C_5H_{10}N_2O_3$	1188-01-8	146.144		237 dec				s H_2O; i EtOH, eth
5588	*L*-Glycylasparagine		$C_6H_{11}N_3O_4$	1999-33-3	189.169	nd (EtOH aq)	216				s H_2O; sl EtOH
5589	*N*-Glycylglycine	2-(Aminoacetamido)acetic acid	$C_4H_8N_2O_3$	556-50-3	132.118		263 dec				s H_2O
5590	*N*-(*N*-Glycylglycyl)glycine		$C_6H_{11}N_3O_4$	556-33-2	189.169	nd (dil al)	246 dec				s H_2O; i EtOH, eth
5591	*N*-Glycyl-*L*-leucine		$C_8H_{16}N_2O_3$	869-19-2	188.224	pl (dil al) pl (dil al)	256 dec				vs H_2O; i EtOH
5592	*N*-Glycyl-*L*-phenylalanine		$C_{11}H_{14}N_2O_3$	3321-03-7	222.240		266				s H_2O
5593	*N*-Glycylserine, (*DL*)		$C_5H_{10}N_2O_4$	687-38-7	162.144		198 dec				
5594	Glycyrrhizic acid		$C_{42}H_{62}O_{16}$	1405-86-3	822.931	pl or pr (HOAc)	220 dec				vs H_2O, EtOH; i eth
5595	Glyodin	1*H*-Imidazole, 2-heptadecyl-4,5-dihydro-, monoacetate	$C_{22}H_{44}N_2O_2$	556-22-9	368.596				1.035^{20}		
5596	Glyoxal		$C_2H_2O_2$	107-22-2	58.036	ye pr	15	50.4	1.14^{20}	1.3826^{20}	vs H_2O; s EtOH, eth
5597	Glyoxal bis(2-hydroxyanil)	2,2'-Benzoxazoline	$C_{14}H_{12}N_2O_2$	1149-16-2	240.257		202				s DMSO
5598	Glyoxylic acid		$C_2H_2O_3$	298-12-4	74.035	orth pr (w+1/2)	98				vs H_2O; sl EtOH, eth, bz
5599	Glyphosate	Glycine, *N*-(phosphonomethyl)-	$C_3H_8NO_5P$	1071-83-6	169.074		230 dec				
5600	Glyphosate isopropylamine salt		$C_6H_{17}N_2O_5P$	38641-94-0	228.183	cry					vs H_2O
5601	Glyphosine	Glycine, *N*,*N*-bis(phosphonomethyl)-	$C_4H_{11}NO_8P_2$	2439-99-8	263.080	wh cry					s H_2O
5602	Grayanotoxin I		$C_{22}H_{36}O_7$	4720-09-6	412.517	cry (AcOEt/ C_5H_{12})	268				
5603	Griseofulvin, (+)		$C_{17}H_{17}ClO_6$	126-07-8	352.766	oct or orth cry (bz)	220				i H_2O; sl EtOH, eth, ace, bz, AcOEt, chl
5604	Guaiol		$C_{15}H_{26}O$	489-86-1	222.366	trg pr (al)	91	dec 288; 165^{17}	0.9074^{100}	1.4716^{100}	i H_2O; s EtOH, eth
5605	Guanabenz		$C_8H_8Cl_2N_4$	5051-62-7	231.083	wh solid	228 dec				
5606	Guanadrel sulfate (2:1)		$C_{20}H_{40}N_6O_8S$	22195-34-2	524.632	cry (MeOH/ EtOH)	214				
5607	Guanethidine		$C_{10}H_{22}N_4$	55-65-2	198.309	wh cry (MeOH)	226				
5608	Guanidine	Aminomethanamidine	CH_5N_3	113-00-8	59.071	cry	50				vs H_2O, EtOH
5609	Guanidine monohydrochloride		CH_6ClN_3	50-01-1	95.532	orth bipym (al)	182.3		1.354^{20}		vs H_2O, EtOH
5610	Guanidine mononitrate		$CH_6N_4O_3$	506-93-4	122.084	lf (w)	217	dec			vs H_2O, EtOH
5611	Guanidine, sulfate (2:1)		$C_2H_{12}N_6O_4S$	594-14-9	216.219		292 dec				
5612	2-Guanidinoethanesulfonic acid	Taurocyamine	$C_3H_9N_3O_3S$	543-18-0	167.186	cry (EtOH, ace)	227				
5613	3-Guanidinopropanoic acid	*N*-Amidino-β-alanine	$C_4H_9N_3O_2$	353-09-3	131.133	cry (EtOH)	210				
5614	Guanine		$C_5H_5N_5O$	73-40-5	151.127	nd or pl (aq NH_3)	360 dec	sub			i H_2O, HOAc; sl EtOH, eth; s alk, acid
5615	Guanosine	2-Amino-1,9-dihydro-9-β-*D*-ribofuranosyl-6*H*-purin-6-one	$C_{10}H_{13}N_5O_5$	118-00-3	283.241	nd (w)	239 dec				sl H_2O; i EtOH, eth; vs HOAc
5616	Guanosine 5'-diphosphate	Guanosine 5'-(trihydrogen diphosphate)	$C_{10}H_{15}N_5O_{11}P_2$	146-91-8	443.201	amorp solid					
5617	Guanosine 5'-monophosphate	5'-Guanylic acid	$C_{10}H_{14}N_5O_8P$	85-32-5	363.221	hyg cry	190 dec				sl H_2O

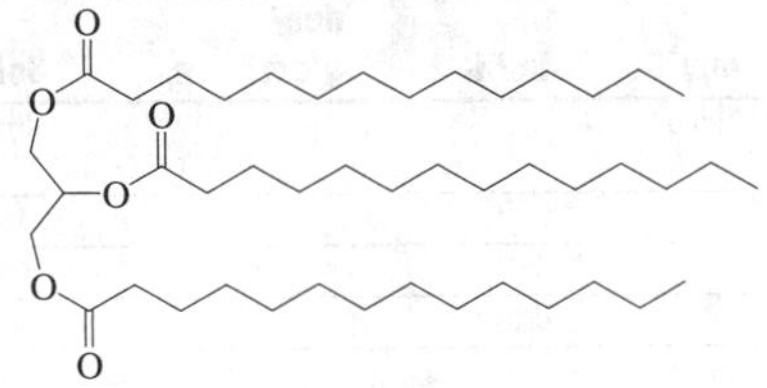

Glycerol tritetradecanoate

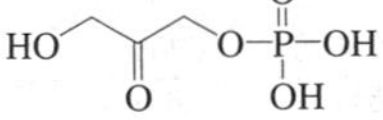

Glycerone phosphate

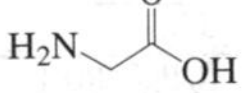

Glycine

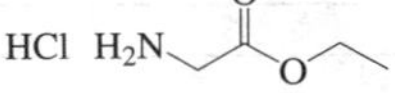

Glycine, ethyl ester, hydrochloride

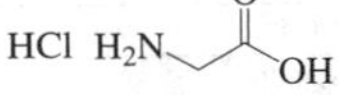

Glycine, hydrochloride

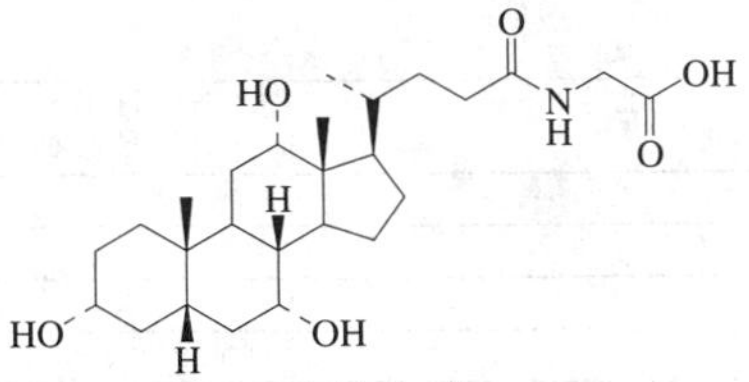

Glycocholic acid

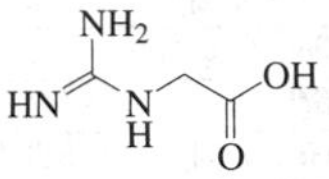

Glycocyamine

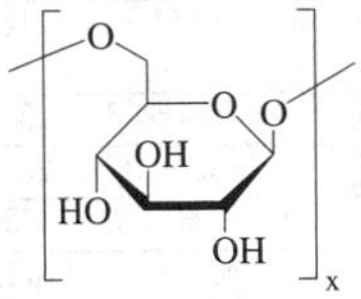

Glycogen

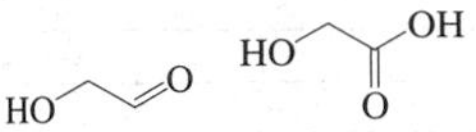

Glycolaldehyde

Glycolic acid

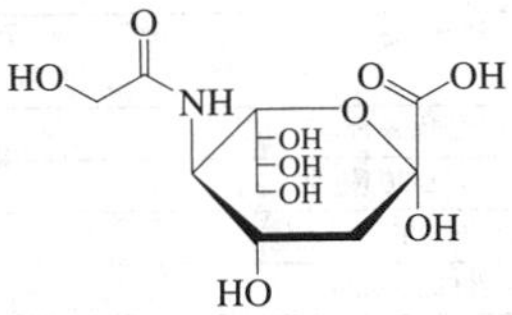

N-Glycolylneuraminic acid

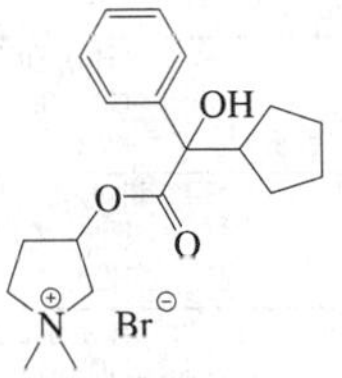

Glycopyrrolate

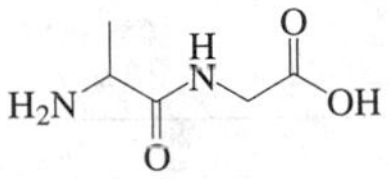

Glycylalanine

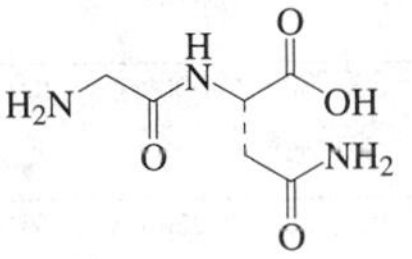

L-Glycylasparagine

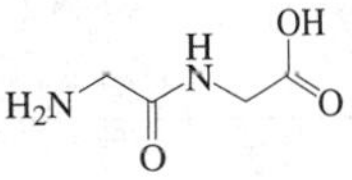

N-Glycylglycine

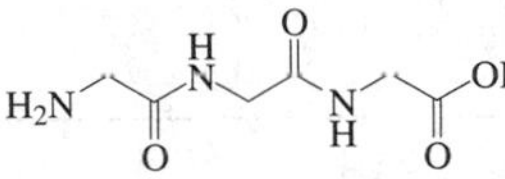

N-(*N*-Glycylglycyl)glycine

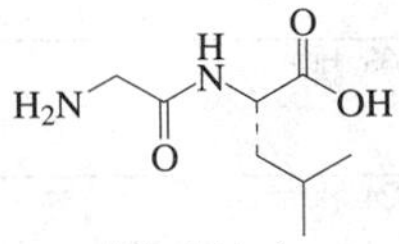

N-Glycyl-*L*-leucine

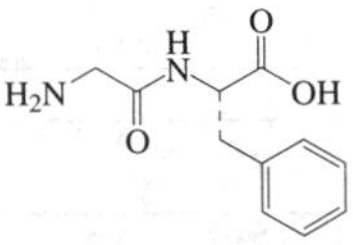

N-Glycyl-*L*-phenylalanine

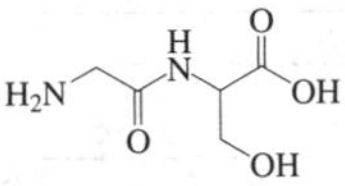

N-Glycylserine, (*DL*)

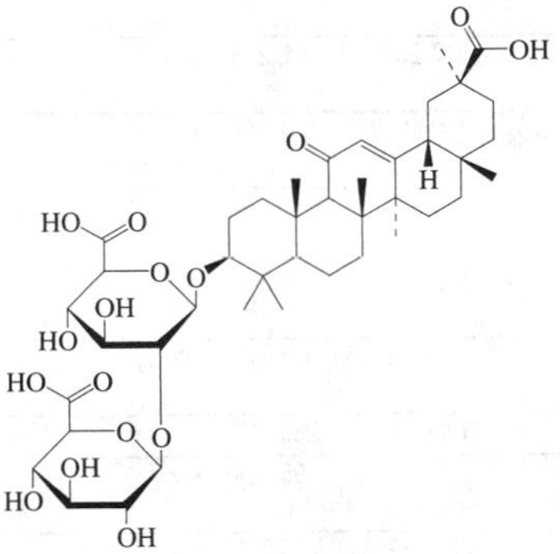

Glycyrrhizic acid

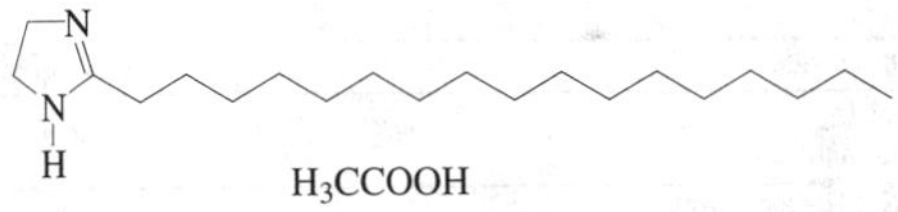

H_3CCOOH

Glyodin

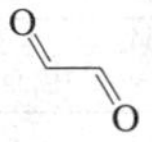

Glyoxal

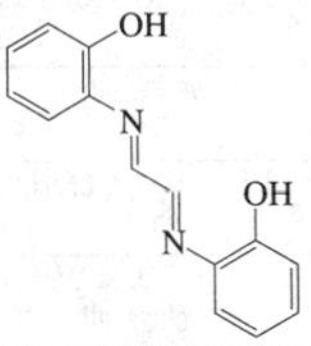

Glyoxal bis(2-hydroxyanil)

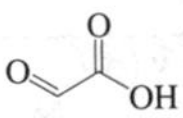

Glyoxylic acid

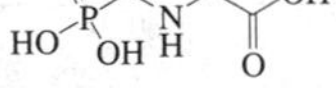

Glyphosate

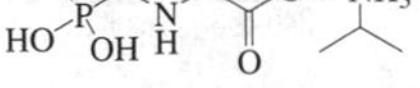

Glyphosate isopropylamine salt

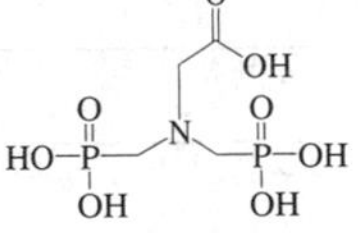

Glyphosine

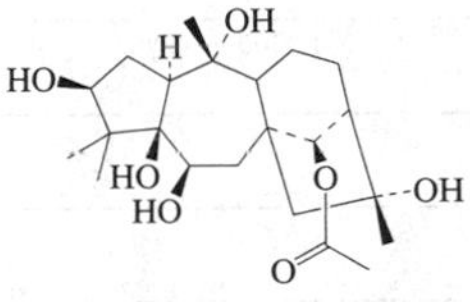

Grayanotoxin I

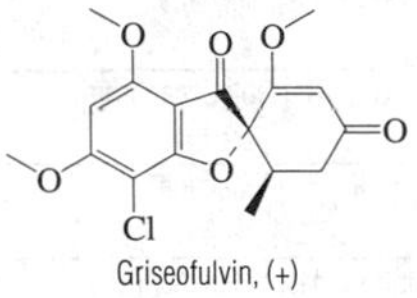

Griseofulvin, (+)

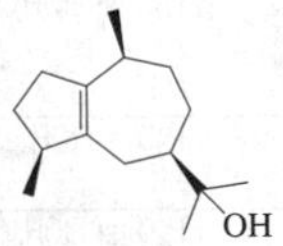

Guaiol

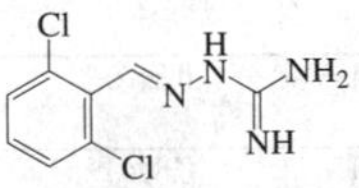

Guanabenz

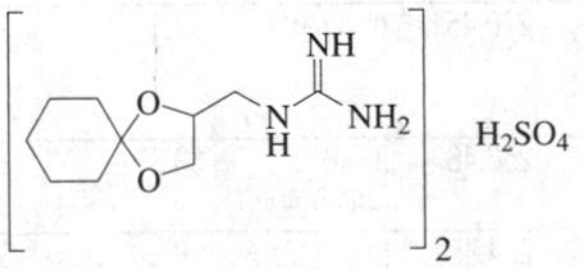

Guanadrel sulfate (2:1)

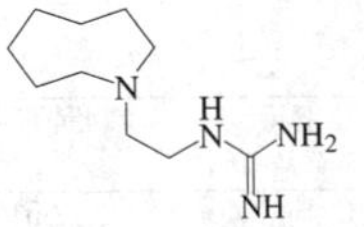

Guanethidine

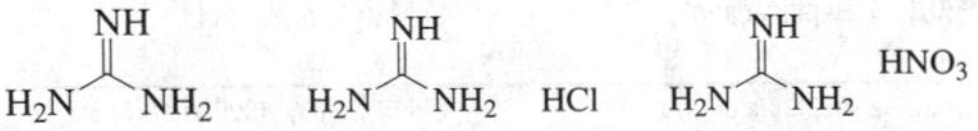

Guanidine

Guanidine monohydrochloride

Guanidine mononitrate

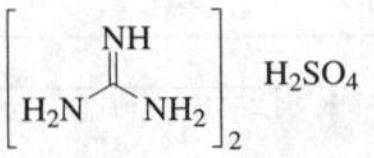

Guanidine, sulfate (2:1)

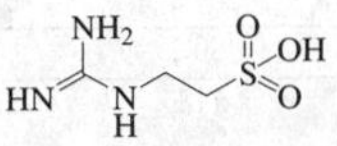

2-Guanidinoethanesulfonic acid

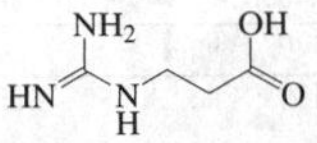

3-Guanidinopropanoic acid

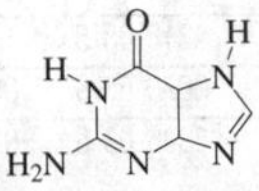

Guanine

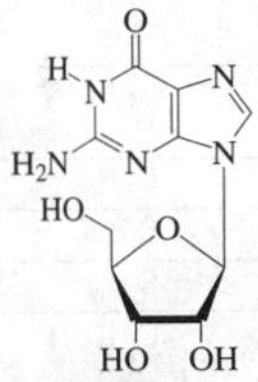

Guanosine

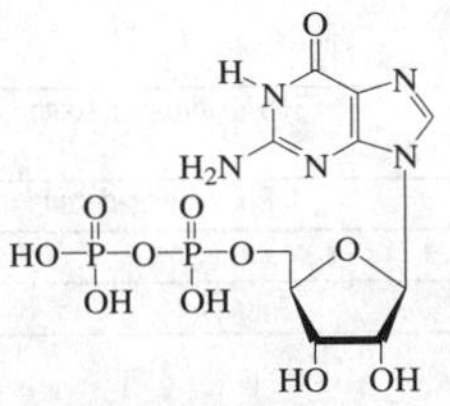

Guanosine 5′-diphosphate

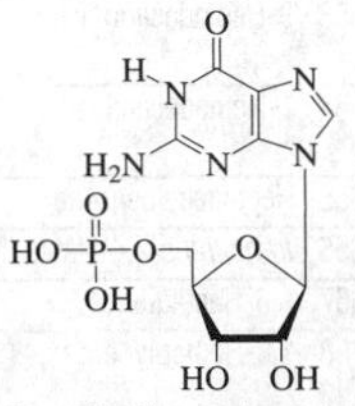

Guanosine 5′-monophosphate

No.	Name	Synonym	Mol. Form.	CAS RN	Mol. Wt.	Physical Form	mp/°C	bp/°C	den/ g cm^{-3}	n_D	Solubility
5618	Guanosine 5'-monophosphate, disodium salt	5'-Guanylic acid, disodium salt	$C_{10}H_{12}N_5Na_2O_8P$	5550-12-9	407.185		195 dec				sl H_2O
5619	Guinea Green B	C.I. Acid Green 3	$C_{37}H_{35}N_2NaO_6S_2$	4680-78-8	690.803	dk grn pow					s H_2O; sl EtOH
5620	*D*-Gulose		$C_6H_{12}O_6$	4205-23-6	180.155	syr		dec			vs H_2O
5621	*L*-Gulose		$C_6H_{12}O_6$	6027-89-0	180.155	syr		dec			vs H_2O
5622	Haloperidol		$C_{21}H_{23}ClFNO_2$	52-86-8	375.865		151.5				
5623	Harmaline	4,9-Dihydro-7-methoxy-1-methyl-3*H*-pyrido[3,4-b]indole	$C_{13}H_{14}N_2O$	304-21-2	214.262	tab (MeOH) orth pr (al)	230				sl H_2O, EtOH, eth; s chl, py
5624	Harman	1-Methyl-9*H*-pyrido[3,4-b] indole	$C_{12}H_{10}N_2$	486-84-0	182.220	bl flr orth cry (hp)	236.5				
5625	Harmine	7-Methoxy-1-methyl-9*H*-pyrido[3,4-b]indole	$C_{13}H_{12}N_2O$	442-51-3	212.246	orth (al), pr (MeOH)	273	sub			sl H_2O, chl, EtOH, eth; s py
5626	HC Blue No. 1		$C_{11}H_{17}N_3O_4$	2784-94-3	255.271	blk cry	100				
5627	HC Blue No. 2		$C_{12}H_{19}N_3O_5$	33229-34-4	285.296	dk bl-blk cry	110				
5628	Hectane		$C_{100}H_{202}$	6703-98-6	1404.67		117				
5629	Hederagenin		$C_{30}H_{48}O_4$	465-99-6	472.700	pr (al)	333				
5630	Helenalin		$C_{15}H_{18}O_4$	6754-13-8	262.302	cry (EtOH)	226				sl H_2O; s EtOH, chl
5631	Helminthosporal		$C_{15}H_{22}O_2$	723-61-5	234.335		58	$117^{0.015}$			
5632	Helvolic acid		$C_{33}H_{44}O_8$	29400-42-8	568.697	nd (dil HOAc)	212 dec				sl H_2O, EtOH; s eth, ace, bz, diox
5633	Hematein		$C_{16}H_{12}O_6$	475-25-2	300.262	red-br cry	250 dec				i H_2O, eth, bz, chl; sl EtOH, HOAc
5634	Hematin		$C_{34}H_{33}FeN_4O_5$	15489-90-4	633.495	br pow (py)	>200				i H_2O, eth; s EtOH, alk; sl py, HOAc
5635	Hematoporphyrin		$C_{34}H_{38}N_4O_6$	14459-29-1	598.689	deep red cry	172.5				i H_2O; s EtOH; sl eth, chl
5636	Hematoxylin		$C_{16}H_{14}O_6$	517-28-2	302.278	ye cry	140				sl H_2O, eth; s alk, EtOH
5637	Hemin		$C_{34}H_{32}ClFeN_4O_4$	16009-13-5	651.941	long blades (gl HOAc)	>300				
5638	Heneicosane		$C_{21}H_{44}$	629-94-7	296.574	cry (w)	40.01	356.5	0.7919^{20}	1.4441^{20}	i H_2O; sl EtOH; s peth
5639	Hentriacontane	Untriacontane	$C_{31}H_{64}$	630-04-6	436.840	lf (AcOEt)	67.9	458	0.781^{68}	1.4278^{90}	sl EtOH, eth, bz, chl; s peth
5640	Heptachlor		$C_{10}H_5Cl_7$	76-44-8	373.318	wh cry	95.5		1.57^9		vs bz, eth, EtOH, lig
5641	Heptachlor epoxide		$C_{10}H_5Cl_7O$	1024-57-3	389.317		160				
5642	2,2',3,3',4,4',6-Heptachlorobiphenyl		$C_{12}H_3Cl_7$	52663-71-5	395.323	cry	117.5				i H_2O
5643	1,1,1,2,3,3,3-Heptachloropropane		C_3HCl_7	3849-33-0	285.211		11	249	1.7921^{34}	1.5427^{21}	vs chl
5644	Heptacontane		$C_{70}H_{142}$	7719-93-9	983.876		107	647			
5645	Heptacosane		$C_{27}H_{56}$	593-49-7	380.734	cry (al, bz) lf (AcOEt)	59.23	442	0.7796^{60}	1.4345^{65}	i H_2O, EtOH; sl eth
5646	Heptadecanal	Margaric aldehyde	$C_{17}H_{34}O$	629-90-3	254.451	nd (peth), cry (al)	36	204^{26}			vs bz, eth
5647	1-Heptadecanamine		$C_{17}H_{37}N$	4200-95-7	255.483		49	336	0.8510^{20}	1.4510^{20}	i H_2O; s EtOH, eth
5648	Heptadecane		$C_{17}H_{36}$	629-78-7	240.468	hex lf	22.0	302.0	0.7780^{20}	1.4369^{20}	i H_2O; sl EtOH, ctc; s eth
5649	Heptadecanenitrile		$C_{17}H_{33}N$	5399-02-0	251.451	cry (al)	34	349	0.8315^{20}	1.4467^{20}	i H_2O; sl EtOH, chl; vs eth
5650	Heptadecanoic acid	Margaric acid	$C_{17}H_{34}O_2$	506-12-7	270.451	pl (peth)	61.3	227^{100}	0.8532^{60}	1.4342^{60}	i H_2O; sl EtOH; s eth, ace, bz, chl
5651	1-Heptadecanol	Margaryl alcohol	$C_{17}H_{36}O$	1454-85-9	256.467	lf (al), cry (ace)	53.9	324	0.8475^{20}		i H_2O; s EtOH, eth
5652	2-Heptadecanone	Pentadecyl methyl ketone	$C_{17}H_{34}O$	2922-51-2	254.451	pl (dil al)	48	320	0.8049^{48}		i H_2O; sl EtOH; s ace, peth; vs bz, eth
5653	9-Heptadecanone		$C_{17}H_{34}O$	540-08-9	254.451	pl (MeOH)	53	251.5; $142^{1.5}$	0.8140^{48}		sl EtOH; s MeOH
5654	1-Heptadecene	Hexahydroaplotaxene	$C_{17}H_{34}$	6765-39-5	238.452		11.5	300	0.7852^{20}	1.4432^{20}	i H_2O; vs eth; s bz; msc lig
5655	Heptadecylbenzene	1-Phenylheptadecane	$C_{23}H_{40}$	14752-75-1	316.564		32	397	0.8546^{20}	1.4810^{20}	
5656	*trans,trans*-2,4-Heptadienal		$C_7H_{10}O$	4313-03-5	110.153			84.5	0.881^{25}	1.5315^{20}	
5657	1,6-Heptadiene		C_7H_{12}	3070-53-9	96.170	liq		90			
5658	1,6-Heptadiyne		C_7H_8	2396-63-6	92.139	liq	-85	112	0.8164^{17}	1.451^{17}	i H_2O; s bz, HOAc
5659	Heptafluorobutanoic acid		$C_4HF_7O_2$	375-22-4	214.039	liq	-17.5	121	1.651^{20}	1.295^{25}	s H_2O, eth, tol; i peth

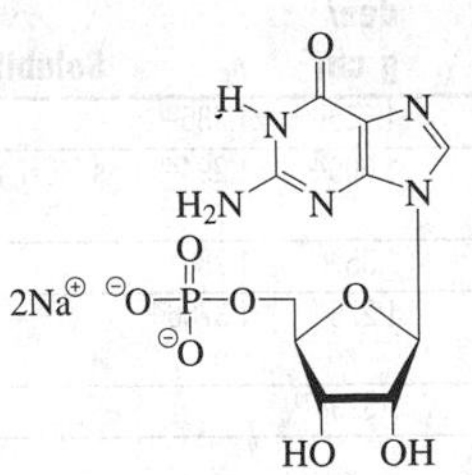

Guanosine 5′-monophosphate, disodium salt

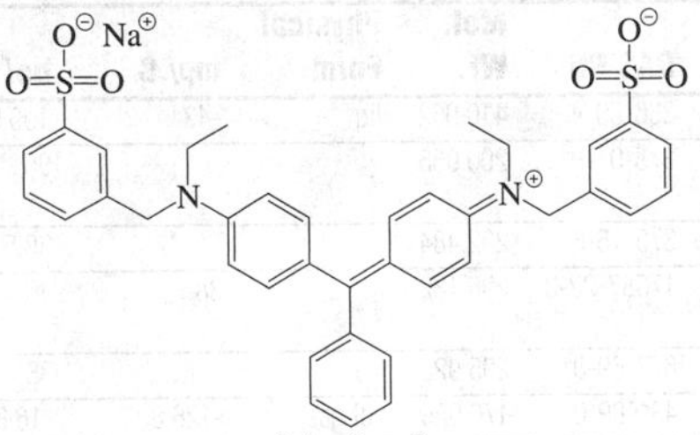

Guinea Green B

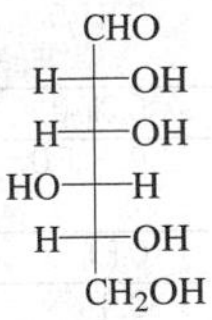

D-Gulose

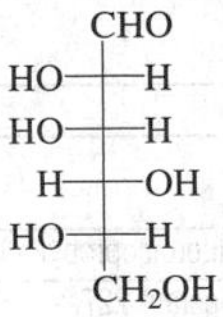

L-Gulose

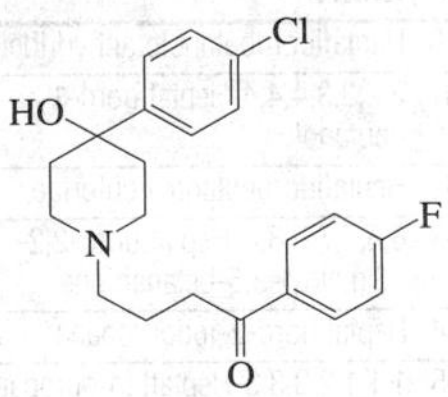

Haloperidol

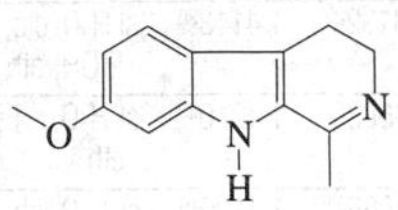

Harmaline

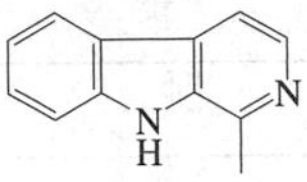

Harman

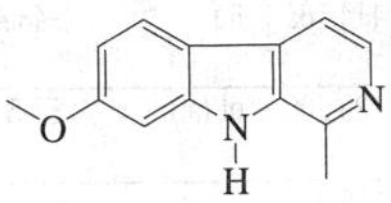

Harmine

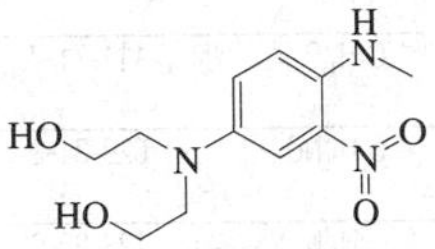

HC Blue No. 1

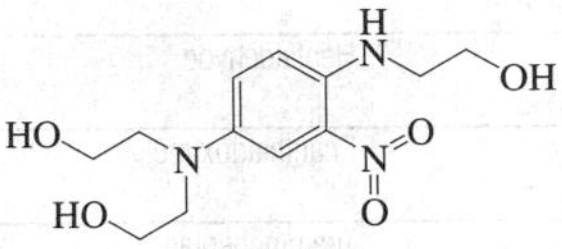

HC Blue No. 2

$H_3C(CH_2)_{98}CH_3$

Hectane

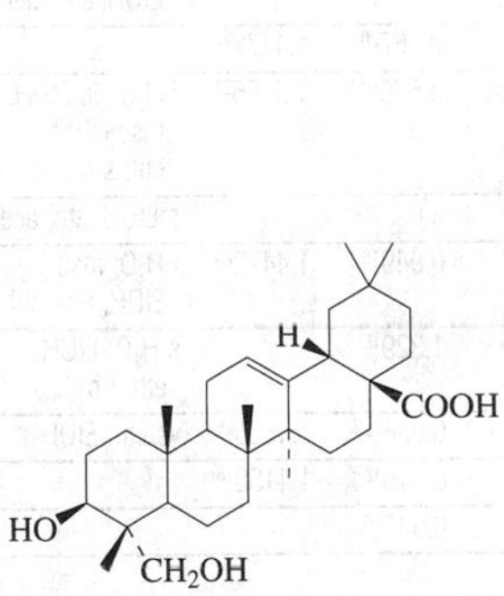

Hederagenin

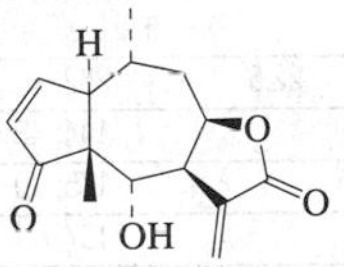

Helenalin

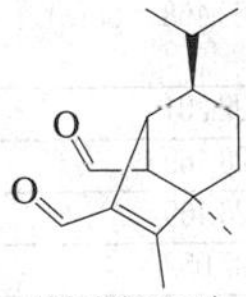

Helminthosporal

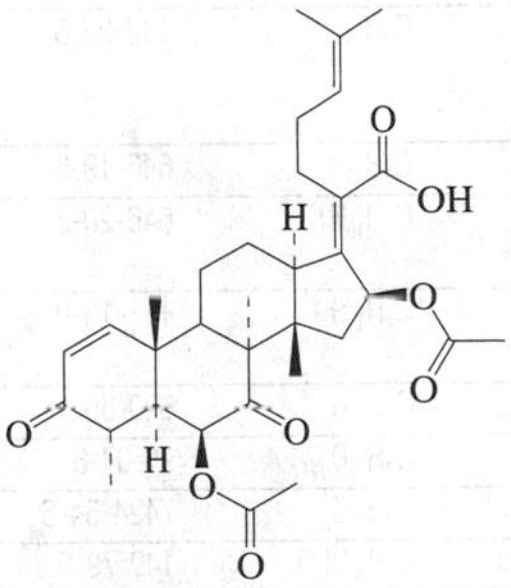

Helvolic acid

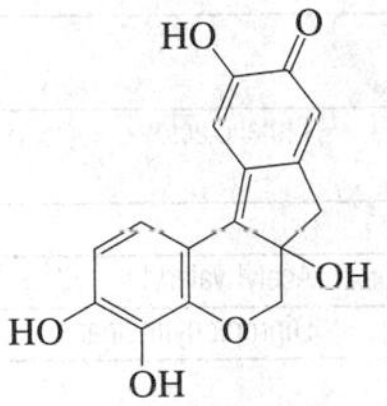

Hematein

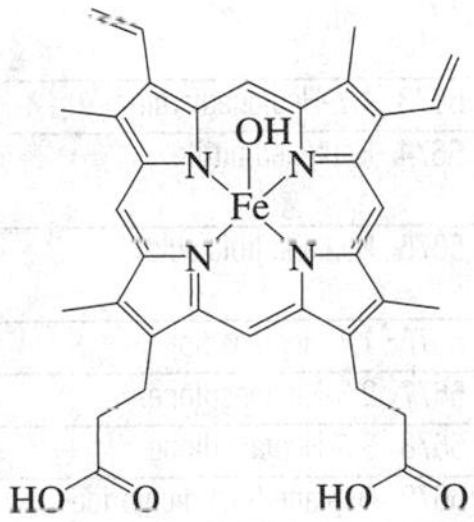

Hematin

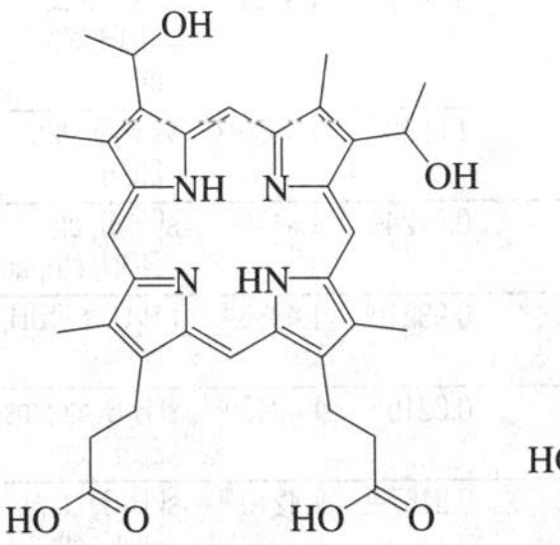

Hematoporphyrin

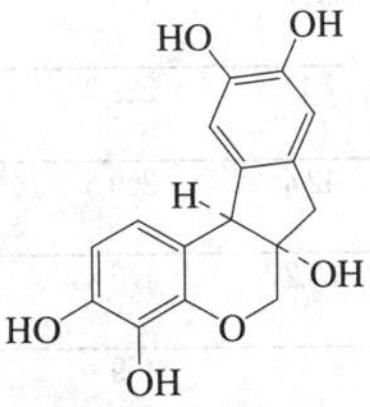

Hematoxylin

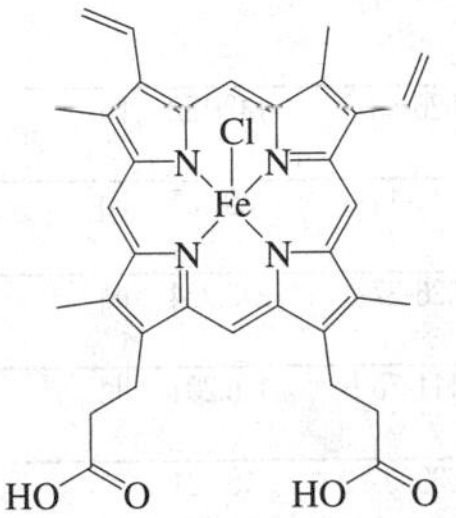

Hemin

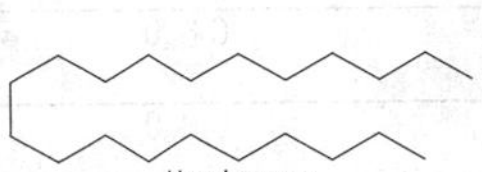

Heneicosane

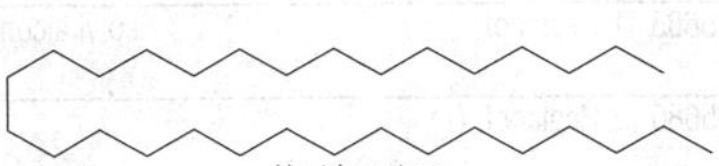

Hentriacontane

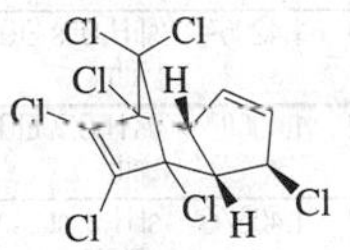

Heptachlor

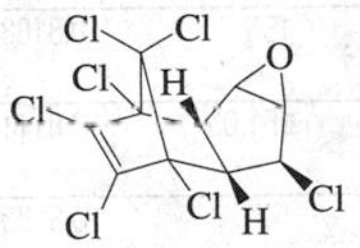

Heptachlor epoxide

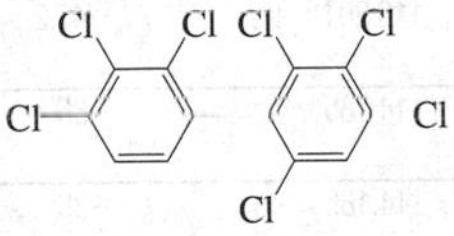

2,2′,3,3′,4,4′,6-Heptachlorobiphenyl

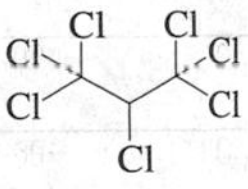

1,1,1,2,3,3,3-Heptachloropropane

$H_3C(CH_2)_{68}CH_3$

Heptacontane

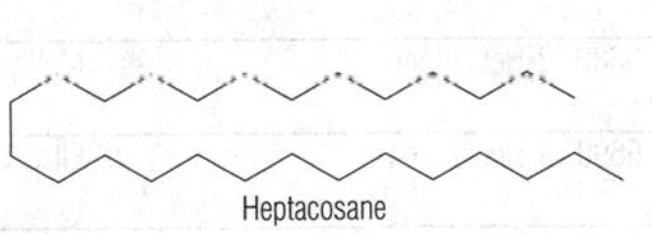

Heptacosane

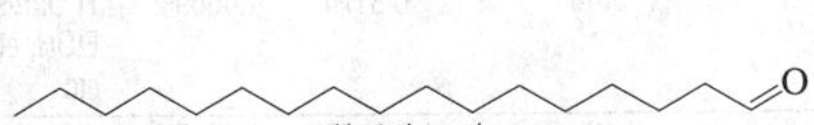

Heptadecanal

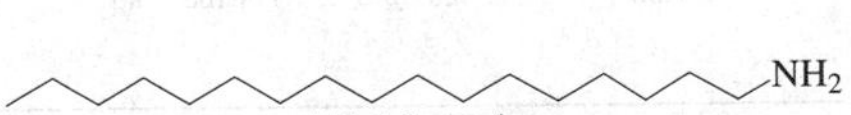

1-Heptadecanamine

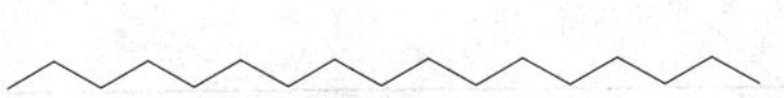

Heptadecane

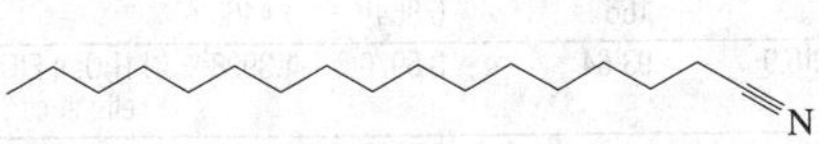

Heptadecanenitrile

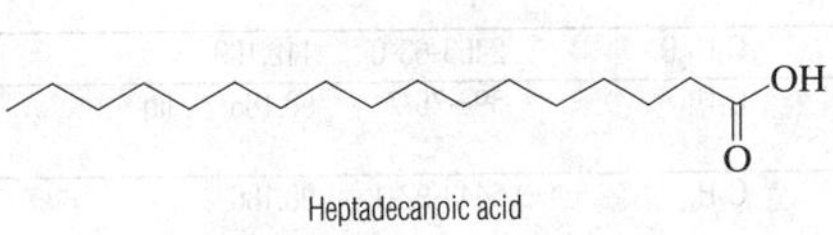

Heptadecanoic acid

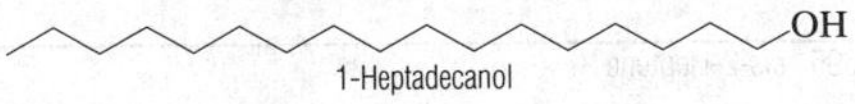

1-Heptadecanol

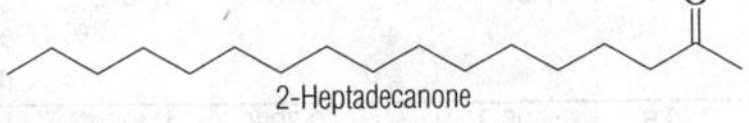

2-Heptadecanone

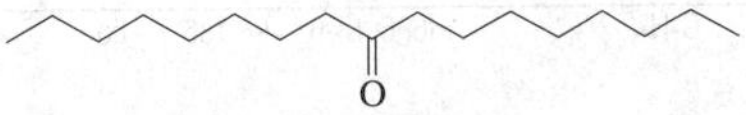

9-Heptadecanone

1-Heptadecene

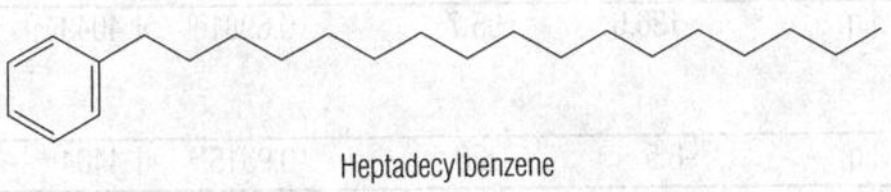

Heptadecylbenzene

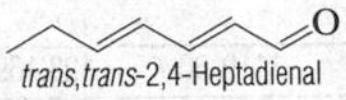

trans,trans-2,4-Heptadienal

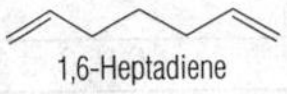

1,6-Heptadiene

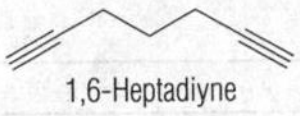

1,6-Heptadiyne

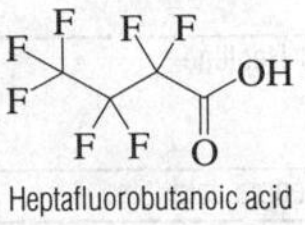

Heptafluorobutanoic acid

No.	Name	Synonym	Mol. Form.	CAS RN	Mol. Wt.	Physical Form	mp/°C	bp/°C	den/ g cm^{-3}	n_D	Solubility
5660	Heptafluorobutanoic anhydride		$C_8F_{14}O_3$	336-59-4	410.062	liq	-43	106.5	1.665^{20}	1.285^{20}	
5661	2,2,3,3,4,4,4-Heptafluoro-1-butanol		$C_4H_3F_7O$	375-01-9	200.055			95	1.600^{20}	1.294^{20}	s EtOH, ace
5662	Heptafluorobutanoyl chloride		C_4ClF_7O	375-16-6	232.484			38.5	1.55^{20}	1.288^{20}	
5663	6,6,7,7,8,8,8-Heptafluoro-2,2-dimethyl-3,5-octanedione		$C_{10}H_{11}F_7O_2$	17587-22-3	296.182		38	46^{5}	1.273^{25}	1.3766^{20}	
5664	Heptafluoro-2-iodopropane	Perfluoroisopropyl iodide	C_3F_7I	677-69-0	295.925			38	1.3298^{20}		
5665	1,1,1,2,3,3,3-Heptafluoropropane	Refrigerant 227ea	C_3HF_7	431-89-0	170.029	col gas	-126.8	-16.34			
5666	2,2,4,4,6,8,8-Heptamethylnonane		$C_{16}H_{34}$	4390-04-9	226.441			246.3			
5667	1,1,1,3,5,5,5-Heptamethyltrisiloxane		$C_7H_{22}O_2Si_3$	1873-88-7	222.506			142	0.8194^{20}	1.3818^{20}	
5668	Heptanal	Heptaldehyde	$C_7H_{14}O$	111-71-7	114.185	liq	-43.4	152.8	0.8132^{25}	1.4113^{20}	sl H_2O, ctc; msc EtOH, eth
5669	Heptanal oxime	Enanthaldoxime	$C_7H_{15}NO$	629-31-2	129.200	pl (al)	57.5	195	0.8583^{55}	1.4210^{20}	sl H_2O; s EtOH, eth
5670	2-Heptanamine	Tuaminoheptane	$C_7H_{17}N$	123-82-0	115.217			142	0.7665^{19}	1.4199^{19}	sl H_2O, chl; s EtOH, eth, peth
5671	4-Heptanamine		$C_7H_{17}N$	16751-59-0	115.217			139.5	0.767^{20}	1.4172^{20}	
5672	Heptane		C_7H_{16}	142-82-5	100.202	liq	-90.55	98.4	0.6795^{25}	1.3855^{25}	i H_2O; vs EtOH; msc eth, bz, chl; s ctc
5673	1,7-Heptanediamine		$C_7H_{18}N_2$	646-19-5	130.231		25.32	224			s EtOH, eth, ace
5674	Heptanedinitrile		$C_7H_{10}N_2$	646-20-8	122.167		-31.4	155^{14}	0.949^{18}	1.4472^{20}	i H_2O; msc EtOH, eth, chl
5675	Heptanedioic acid	Pimelic acid	$C_7H_{12}O_4$	111-16-0	160.168	pr (w)	106	342.0; 272^{100}	1.329^{15}		s H_2O, EtOH, eth; i bz
5676	1,7-Heptanediol		$C_7H_{16}O_2$	629-30-1	132.201		22.5	262	0.9569^{25}	1.4520^{25}	vs eth, EtOH
5677	2,3-Heptanedione	Acetyl valeryl	$C_7H_{12}O_2$	96-04-8	128.169			144; 46^{13}	0.919^{18}	1.4150^{18}	
5678	3,5-Heptanedione	Dipropionylmethane	$C_7H_{12}O_2$	7424-54-6	128.169			175; 79^{30}	0.945^{20}		
5679	Heptanedioyl dichloride		$C_7H_{10}Cl_2O_2$	142-79-0	197.059			137^{15}			
5680	Heptanenitrile		$C_7H_{13}N$	629-08-3	111.185	liq	-64	183; 71^{10}	0.8106^{20}	1.4104^{30}	i H_2O; s eth, ace, bz, HOAc
5681	1-Heptanethiol	Heptyl mercaptan	$C_7H_{16}S$	1639-09-4	132.267	liq	-43	176.9	0.8427^{20}	1.4521^{20}	i H_2O; msc EtOH, eth; s chl
5682	2,4,6-Heptanetrione		$C_7H_{10}O_3$	626-53-9	142.152	lf	49	121^{10}	1.0599^{40}	1.4930^{20}	vs H_2O, eth, EtOH
5683	Heptanoic acid	Enanthic acid	$C_7H_{14}O_2$	111-14-8	130.185	liq	-7.17	222.2	0.9124^{25}	1.4170^{20}	sl H_2O, ctc; s EtOH, eth, ace
5684	Heptanoic anhydride		$C_{14}H_{26}O_3$	626-27-7	242.354	liq	-12.4	269.5	0.9321^{20}	1.4335^{15}	i H_2O; s EtOH, eth
5685	1-Heptanol	Heptyl alcohol	$C_7H_{16}O$	111-70-6	116.201	liq	-33.2	176.45	0.8219^{20}	1.4249^{20}	sl H_2O, ctc; msc EtOH, eth
5686	2-Heptanol, (±)		$C_7H_{16}O$	52390-72-4	116.201			159	0.8167^{20}	1.4210^{20}	sl H_2O, ctc; s EtOH, eth
5687	3-Heptanol, (*S*)		$C_7H_{16}O$	26549-25-7	116.201	liq	-70	157; 66^{18}	0.8227^{20}	1.4201^{20}	sl H_2O, ctc; s EtOH, eth
5688	4-Heptanol	Dipropylcarbinol	$C_7H_{16}O$	589-55-9	116.201	liq	-41.2	156	0.8183^{20}	1.4205^{20}	sl H_2O; s EtOH, eth
5689	2-Heptanone	Methyl pentyl ketone	$C_7H_{14}O$	110-43-0	114.185	liq	-35	151.05	0.8111^{20}	1.4088^{20}	vs H_2O; s EtOH, eth
5690	3-Heptanone	Ethyl butyl ketone	$C_7H_{14}O$	106-35-4	114.185	liq	-39	147	0.8183^{20}	1.4057^{20}	sl H_2O, ctc; msc EtOH, eth
5691	4-Heptanone	Dipropyl ketone	$C_7H_{14}O$	123-19-3	114.185	liq	-33	144	0.8174^{20}	1.4069^{20}	i H_2O; msc EtOH, eth; s ctc
5692	Heptanoyl chloride		$C_7H_{13}ClO$	2528-61-2	148.630	liq	-83.8	125.2	0.9590^{20}	1.4345^{18}	s eth; sl ctc; vs lig
5693	2-Heptenal	Butylacrolein	$C_7H_{12}O$	2463-63-0	112.169			166	0.864^{17}	1.4468^{17}	
5694	1-Heptene		C_7H_{14}	592-76-7	98.186	liq	-118.9	93.64	0.6970^{20}	1.3998^{20}	i H_2O; s EtOH, eth; sl ctc
5695	*cis*-2-Heptene		C_7H_{14}	6443-92-1	98.186			98.4	0.708^{20}	1.406^{20}	i H_2O; s EtOH, eth, ace, bz, chl; sl ctc
5696	*trans*-2-Heptene		C_7H_{14}	14686-13-6	98.186	liq	-109.5	98	0.7012^{20}	1.4045^{20}	i H_2O; s EtOH, eth, ace, bz, peth, chl
5697	*cis*-3-Heptene		C_7H_{14}	7642-10-6	98.186	liq	-136.6	95.8	0.7030^{20}	1.4059^{20}	i H_2O; s EtOH, eth, ace, bz, peth, chl
5698	*trans*-3-Heptene		C_7H_{14}	14686-14-7	98.186	liq	-136.6	95.7	0.6981^{20}	1.4043^{20}	i H_2O; s EtOH, eth, ace, bz, chl; sl ctc
5699	6-Heptenoic acid		$C_7H_{12}O_2$	1119-60-4	128.169	liq	-6.5	226	0.9515^{14}	1.4404^{14}	
5700	1-Hepten-4-ol		$C_7H_{14}O$	3521-91-3	114.185			152.1	0.8384^{22}	1.4347^{20}	

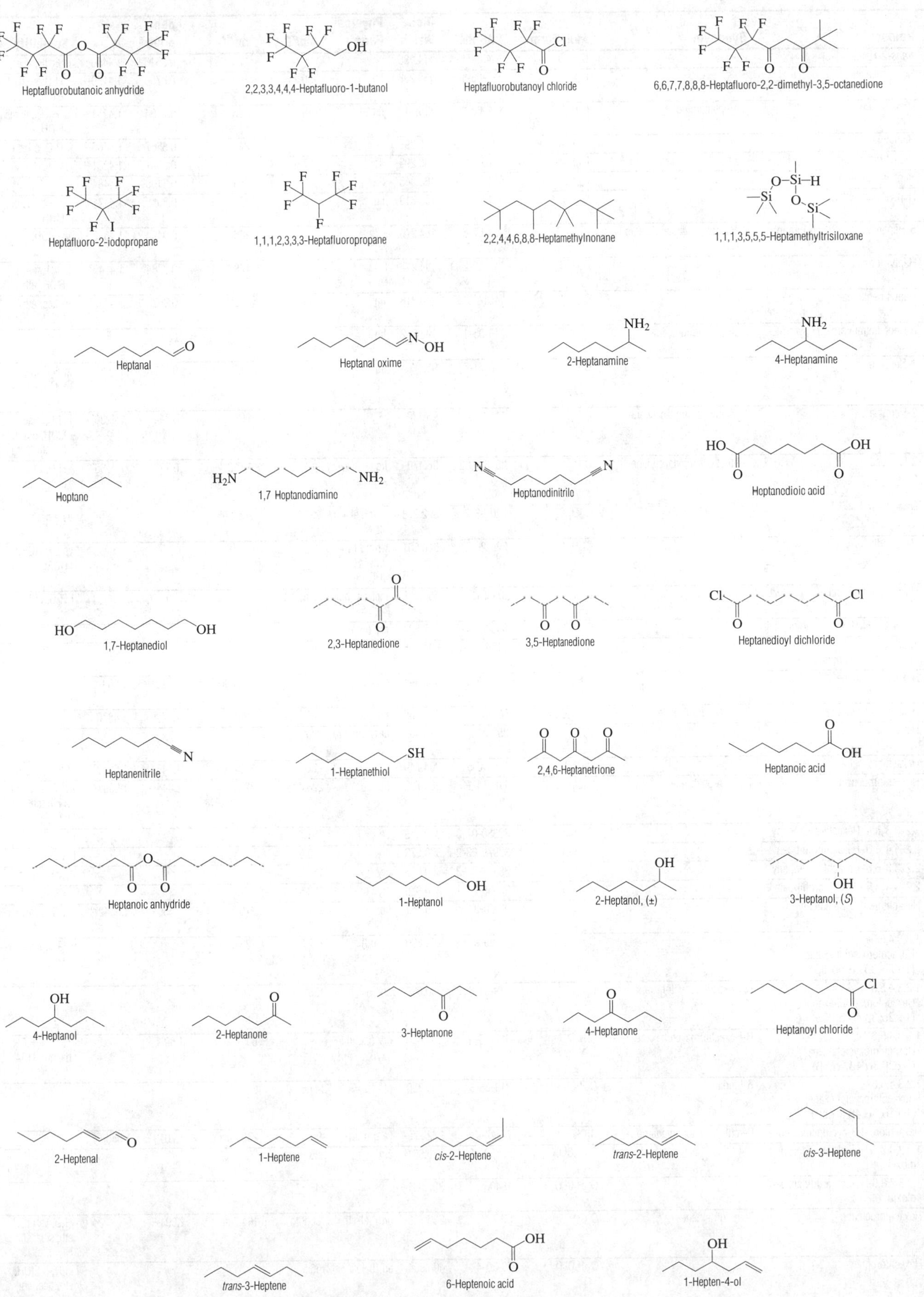
Heptafluorobutanoic anhydride
2,2,3,3,4,4,4-Heptafluoro-1-butanol
Heptafluorobutanoyl chloride
6,6,7,7,8,8,8-Heptafluoro-2,2-dimethyl-3,5-octanedione
Heptafluoro-2-iodopropane
1,1,1,2,3,3,3-Heptafluoropropane
2,2,4,4,6,8,8-Heptamethylnonane
1,1,1,3,5,5,5-Heptamethyltrisiloxane
Heptanal
Heptanal oxime
2-Heptanamine
4-Heptanamine
Heptane
1,7-Heptanediamine
Heptanedinitrile
Heptanedioic acid
1,7-Heptanediol
2,3-Heptanedione
3,5-Heptanedione
Heptanedioyl dichloride
Heptanenitrile
1-Heptanethiol
2,4,6-Heptanetrione
Heptanoic acid
Heptanoic anhydride
1-Heptanol
2-Heptanol, (±)
3-Heptanol, (*S*)
4-Heptanol
2-Heptanone
3-Heptanone
4-Heptanone
Heptanoyl chloride
2-Heptenal
1-Heptene
cis-2-Heptene
trans-2-Heptene
cis-3-Heptene
trans-3-Heptene
6-Heptenoic acid
1-Hepten-4-ol

No.	Name	Synonym	Mol. Form.	CAS RN	Mol. Wt.	Physical Form	mp/°C	bp/°C	den/ g cm^{-3}	n_D	Solubility
5701	*trans*-2-Hepten-1-ol		$C_7H_{14}O$	33467-76-4	114.185			178; 75^{10}	0.8516^{20}	1.4460^{20}	s EtOH, ace
5702	Heptyl acetate		$C_9H_{18}O_2$	112-06-1	158.238	liq	-50.2	193	0.8750^{15}	1.4150^{20}	i H_2O; s EtOH, eth, ctc
5703	Heptylamine	1-Heptanamine	$C_7H_{17}N$	111-68-2	115.217	liq	-18	156	0.7754^{20}	1.4251^{20}	sl H_2O, chl; msc EtOH, eth
5704	Heptylbenzene		$C_{13}H_{20}$	1078-71-3	176.298	liq	-48	240; 109^{10}	0.8567^{20}	1.4865^{20}	i H_2O; s bz, chl
5705	Heptyl butanoate		$C_{11}H_{22}O_2$	5870-93-9	186.292	liq	-57.5	225.8	0.8637^{20}	1.4231^{20}	vs EtOH
5706	Heptylcyclohexane		$C_{13}H_{26}$	5617-41-4	182.345	liq	-30	244	0.8109^{20}	1.4484^{20}	
5707	Heptylcyclopentane		$C_{12}H_{24}$	5617-42-5	168.319	liq	-53	224	0.8010^{20}	1.4421^{20}	vs ace, bz, eth, EtOH
5708	5-Heptyldihydro-2(3*H*)-furanone	4-Hydroxyundecanoic acid lactone	$C_{11}H_{20}O_2$	104-67-6	184.276			286	0.9494^{20}	1.4512^{20}	vs EtOH
5709	Heptyl formate		$C_8H_{16}O_2$	112-23-2	144.212			178.1	0.8784^{20}	1.4140^{20}	i H_2O; msc EtOH, eth
5710	Heptyl pentanoate		$C_{12}H_{24}O_2$	5451-80-9	200.318	liq	-46.4	245.2	0.8623^{20}	1.4254^{15}	vs ace, eth, EtOH
5711	6-Heptyltetrahydro-2*H*-pyran-2-one	5-Dodecanolide	$C_{12}H_{22}O_2$	713-95-1	198.302	liq	-12	$101^{0.03}$			
5712	1-Heptyne		C_7H_{12}	628-71-7	96.170	liq	-81	99.7	0.7328^{20}	1.4087^{20}	sl H_2O; msc EtOH, eth; s bz, chl, peth
5713	2-Heptyne	1-Methyl-2-butylacetylene	C_7H_{12}	1119-65-9	96.170			112	0.744^{25}	1.4230^{20}	i H_2O; msc EtOH, eth; s bz, chl, peth
5714	3-Heptyne	1-Ethyl-2-propylacetylene	C_7H_{12}	2586-89-2	96.170	liq	-130.5	107.2	0.7336^{25}	1.4189^{20}	i H_2O; msc EtOH, eth; s bz, chl, peth
5715	Hesperetin		$C_{16}H_{14}O_6$	520-33-2	302.278	pl (dil al + 1/2 w)	227.5	sub 205			vs eth, EtOH
5716	Hesperidin		$C_{28}H_{34}O_{15}$	520-26-3	610.561	wh nd (dil MeOH, HOAc)	262				vs py, EtOH, HOAc
5717	Hexabromobenzene		C_6Br_6	87-82-1	551.488	mcl nd (bz)	327				i H_2O; sl EtOH, eth; s bz, chl
5718	2,2',4,4',5,5'-Hexabromobiphenyl		$C_{12}H_4Br_6$	59080-40-9	627.584	cry (ctc)	160				
5719	1,2,5,6,9,10-Hexabromocyclododecane		$C_{12}H_{18}Br_6$	3194-55-6	641.695	cry	167				
5720	Hexabromoethane		C_2Br_6	594-73-0	503.445	orth pr (bz)		dec 200	3.823^{20}	1.863	sl EtOH, eth, CS_2
5721	Hexabutyldistannoxane	Bis(tributyltin) oxide	$C_{24}H_{54}OSn_2$	56-35-9	596.105	liq	–45	225^{10}	1.17^{20}	1.4870	
5722	Hexacene		$C_{26}H_{16}$	258-31-1	328.405	dk bl-grn cry (sub)	380	sub			i H_2O, EtOH
5723	Hexachloroacetone		C_3Cl_6O	116-16-5	264.749	liq	-1.0	203	1.7434^{12}	1.5112^{20}	sl H_2O; s ace
5724	Hexachlorobenzene	Perchlorobenzene	C_6Cl_6	118-74-1	284.782	nd (sub)	228.83	325	2.044^{23}	1.5691^{23}	i H_2O; sl EtOH; s eth, chl; vs bz
5725	2,2',3,3',4,4'-Hexachlorobiphenyl		$C_{12}H_4Cl_6$	38380-07-3	360.878	cry	151				i H_2O
5726	2,2',4,4',6,6'-Hexachlorobiphenyl		$C_{12}H_4Cl_6$	33979-03-2	360.878	cry	112.5				i H_2O
5727	2,2',3,3',6,6'-Hexachlorobiphenyl		$C_{12}H_4Cl_6$	38411-22-2	360.878	cry (hx)	114.2				i H_2O
5728	2,2',4,4',5,5'-Hexachlorobiphenyl		$C_{12}H_4Cl_6$	35065-27-1	360.878	cry	103.5				
5729	Hexachloro-1,3-butadiene		C_4Cl_6	87-68-3	260.761	liq	-21	215	1.556^{25}	1.5542^{20}	i H_2O; s EtOH, eth
5730	1,2,3,4,5,6-Hexachlorocyclohexane, (1α,2α,3β,4α,5α,6β)	Lindane	$C_6H_6Cl_6$	58-89-9	290.830	nd (al)	112.5	323.4			vs ace, bz
5731	1,2,3,4,5,6-Hexachlorocyclohexane, (1α,2α,3β,4α,5β,6β)	α-Hexachlorocyclohexane	$C_6H_6Cl_6$	319-84-6	290.830	cry	158				
5732	1,2,3,4,5,6-Hexachlorocyclohexane, (1α,2β,3α,4β,5α,6β)	β-Hexachlorocyclohexane	$C_6H_6Cl_6$	319-85-7	290.830	cry (bz, al, xyl)		$60^{0.50}$	1.89^{19}		i H_2O; sl EtOH, bz, chl, HOAc
5733	1,2,3,4,5,6-Hexachlorocyclohexane, (1α,2α,3α,4β,5α,6β)	δ-Lindane	$C_6H_6Cl_6$	319-86-8	290.830	pl	141.5	$60^{0.36}$			
5734	Hexachloro-1,3-cyclopentadiene	Perchlorocyclopentadiene	C_5Cl_6	77-47-4	272.772	ye grn liq	-9	239; $48^{0.3}$	1.7019^{25}	1.5658^{20}	
5735	1,2,3,6,7,8-Hexachlorodibenzo-*p*-dioxin		$C_{12}H_2Cl_6O_2$	57653-85-7	390.861	cry	285				
5736	1,2,3,7,8,9-Hexachlorodibenzo-*p*-dioxin		$C_{12}H_2Cl_6O_2$	19408-74-3	390.861	cry	243				
5737	Hexachloroethane	Perchloroethane	C_2Cl_6	67-72-1	236.739	orth (al-eth)	186.8 tp	184.7 sp	2.091^{20}		i H_2O; vs EtOH, eth; s bz; sl liq HF
5738	Hexachlorophene		$C_{13}H_6Cl_6O_2$	70-30-4	406.904	nd (bz)	166.5				i H_2O; s EtOH, eth, ace, chl, dil alk

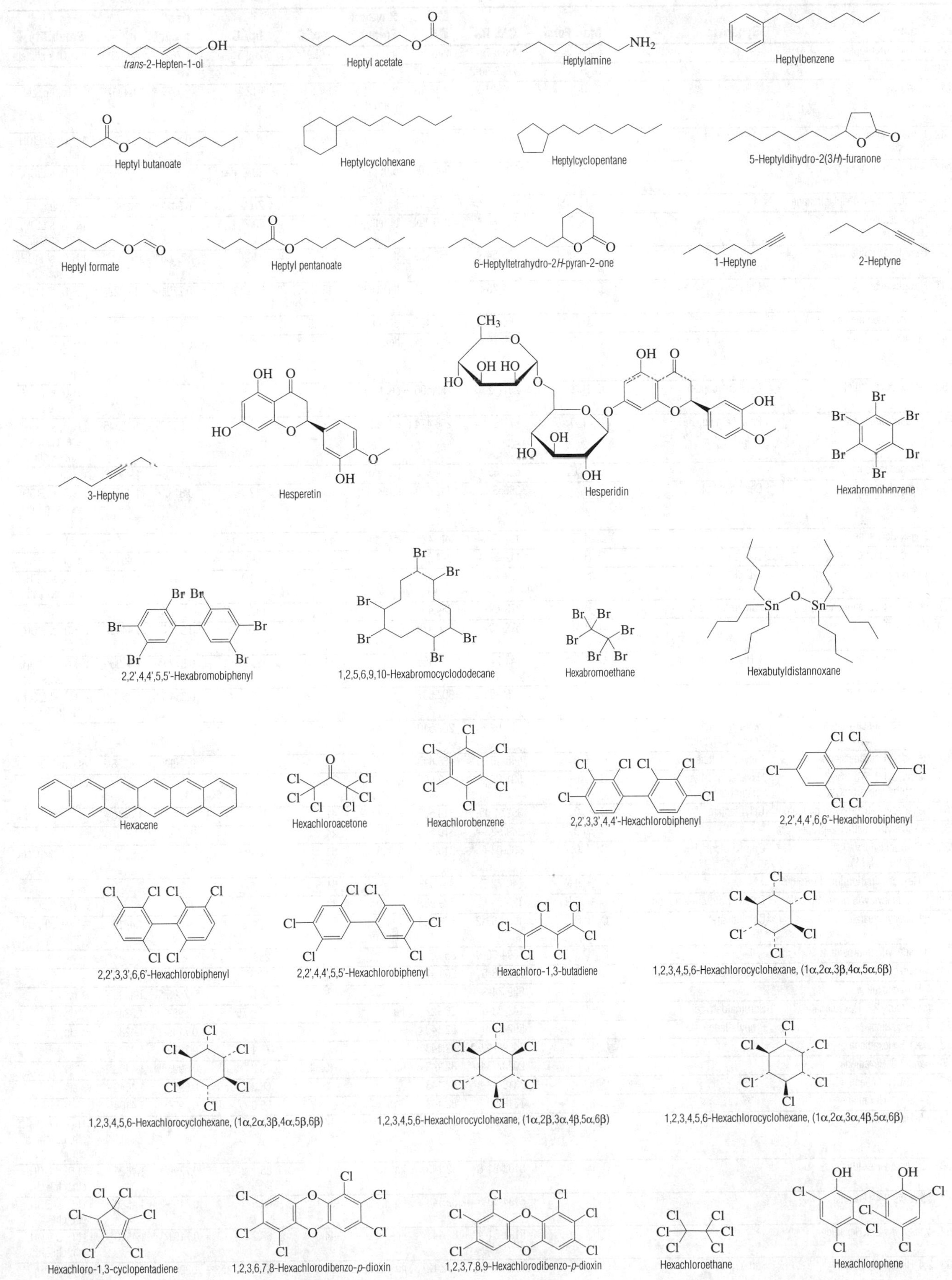
trans-2-Hepten-1-ol
Heptyl acetate
Heptylamine
Heptylbenzene
Heptyl butanoate
Heptylcyclohexane
Heptylcyclopentane
5-Heptyldihydro-2(3H)-furanone
Heptyl formate
Heptyl pentanoate
6-Heptyltetrahydro-2H-pyran-2-one
1-Heptyne
2-Heptyne
3-Heptyne
Hesperetin
Hesperidin
Hexabromobenzene
2,2',4,4',5,5'-Hexabromobiphenyl
1,2,5,6,9,10-Hexabromocyclododecane
Hexabromoethane
Hexabutyldistannoxane
Hexacene
Hexachloroacetone
Hexachlorobenzene
2,2',3,3',4,4'-Hexachlorobiphenyl
2,2',4,4',6,6'-Hexachlorobiphenyl
2,2',3,3',6,6'-Hexachlorobiphenyl
2,2',4,4',5,5'-Hexachlorobiphenyl
Hexachloro-1,3-butadiene
1,2,3,4,5,6-Hexachlorocyclohexane, (1α,2α,3β,4α,5α,6β)
1,2,3,4,5,6-Hexachlorocyclohexane, (1α,2α,3β,4α,5β,6β)
1,2,3,4,5,6-Hexachlorocyclohexane, (1α,2β,3α,4β,5α,6β)
1,2,3,4,5,6-Hexachlorocyclohexane, (1α,2α,3α,4β,5α,6β)
Hexachloro-1,3-cyclopentadiene
1,2,3,6,7,8-Hexachlorodibenzo-p-dioxin
1,2,3,7,8,9-Hexachlorodibenzo-p-dioxin
Hexachloroethane
Hexachlorophene

No.	Name	Synonym	Mol. Form.	CAS RN	Mol. Wt.	Physical Form	mp/°C	bp/°C	den/ g cm^{-3}	n_D	Solubility
5739	Hexachloropropene		C_3Cl_6	1888-71-7	248.750	liq	-72.9	209.5	1.7632^{20}	1.5091^{20}	i H_2O; s ctc, chl
5740	Hexacontane		$C_{60}H_{122}$	7667-80-3	843.611		99.3				
5741	Hexacosane		$C_{26}H_{54}$	630-01-3	366.707	mcl, tcl or orth (bz) cry (eth)	56.1	412.2	0.7783^{60}	1.4357^{60}	vs bz, lig, chl
5742	Hexacosanoic acid	Cerotic acid	$C_{26}H_{52}O_2$	506-46-7	396.690		88.5		0.8198^{100}	1.4301^{100}	i H_2O; vs EtOH, eth
5743	1-Hexacosanol		$C_{26}H_{54}O$	506-52-5	382.706	orth pl (dil al)	80	305^{20} dec			i H_2O; s EtOH, eth
5744	Hexadecamethylheptasiloxane		$C_{16}H_{48}O_6Si_7$	541-01-5	533.147	liq	-78	270	0.9012^{20}	1.3965^{20}	vs bz, lig
5745	Hexadecanal		$C_{16}H_{32}O$	629-80-1	240.424	pl (eth), nd (peth)	35	200^{29}			i H_2O; s EtOH, eth, ace, bz
5746	Hexadecanamide		$C_{16}H_{33}NO$	629-54-9	255.439	lf	107	236^{12}	1.0000^{20}		i H_2O; sl EtOH, bz, ace, eth
5747	Hexadecane	Cetane	$C_{16}H_{34}$	544-76-3	226.441	lf (HOAc)	18.12	286.86	0.7701^{25}	1.4329^{25}	i H_2O; sl EtOH; msc eth; s ctc
5748	Hexadecanedioic acid		$C_{16}H_{30}O_4$	505-54-4	286.407	pl (al)	126.6				vs ace, EtOH
5749	Hexadecanenitrile		$C_{16}H_{31}N$	629-79-8	237.424	hex	31	333	0.8303^{20}	1.4450^{20}	i H_2O; vs EtOH, eth, ace, bz, chl
5750	1-Hexadecanethiol	Cetyl mercaptan	$C_{16}H_{34}S$	2917-26-2	258.506	cry (lig)	19	$125^{0.5}$			i H_2O; sl EtOH, ctc; s eth
5751	Hexadecanoic acid	Palmitic acid	$C_{16}H_{32}O_2$	57-10-3	256.424	nd (al)	62.5	351.5	0.8527^{62}	1.43345^{60}	i H_2O; s EtOH, ace, bz; msc eth; vs chl
5752	Hexadecanoic anhydride		$C_{32}H_{62}O_3$	623-65-4	494.832	lf (peth)	64		0.8388^{83}	1.4364^{68}	vs eth
5753	1-Hexadecanol	Cetyl alcohol	$C_{16}H_{34}O$	36653-82-4	242.440	fl (AcOEt)	49.2	312	0.8187^{50}	1.4283^{79}	i H_2O; sl EtOH; vs eth, bz, chl; s ace
5754	3-Hexadecanone		$C_{16}H_{32}O$	18787-64-9	240.424	lf (peth)	43	184^{17}, 140^{2}			s chl
5755	Hexadecanoyl chloride		$C_{16}H_{31}ClO$	112-67-4	274.869		12	199^{20}	0.9016^{25}	1.4514^{20}	vs eth
5756	1-Hexadecene	1-Cetene	$C_{16}H_{32}$	629-73-2	224.425	lf	2.1	284.9	0.7811^{20}	1.4412^{20}	i H_2O; s EtOH, eth, ctc, peth
5757	*cis*-9-Hexadecenoic acid	Palmitoleic acid	$C_{16}H_{30}O_2$	373-49-9	254.408		0.5	182^{1}			
5758	Hexadecyl acetate		$C_{18}H_{36}O_2$	629-70-9	284.478		-18.5	222^{205}	0.8574^{25}	1.4438^{20}	i H_2O; sl EtOH; s ctc
5759	Hexadecylamine	1-Hexadecanamine	$C_{16}H_{35}N$	143-27-1	241.456	lf	46.8	322.5	0.8129^{20}	1.4496^{20}	i H_2O; vs EtOH, eth, bz; s ace
5760	Hexadecylbenzene		$C_{22}H_{38}$	1459-09-2	302.537		27	385	0.8547^{20}	1.4813^{20}	i H_2O; sl EtOH; vs eth, bz, CS_2
5761	Hexadecyldimethylamine	*N,N*-Dimethyl-1-hexadecanamine	$C_{18}H_{39}N$	112-69-6	269.510			330.0			
5762	Hexadecyl hexadecanoate	Cetyl palmitate	$C_{32}H_{64}O_2$	540-10-3	480.849	mcl lf	54		0.989^{20}	1.4398^{70}	vs eth, EtOH
5763	Hexadecyl 3-hydroxy-2-naphthalenecarboxylate	Hexadecyl 3-hydroxy-2-naphthoate	$C_{27}H_{40}O_3$	531-84-0	412.605	grn-wh fl	72.5				vs bz, HOAc
5764	Hexadecyl 2-hydroxypropanoate	Cetyl lactate	$C_{19}H_{38}O_3$	35274-05-6	314.503	wax	41	219^{10}, 170^{1}		1.4410^{40}	
5765	Hexadecyl 2-methyl-2-propenoate		$C_{20}H_{38}O_2$	2495-27-4	310.515		24	183^{2}	0.87^{20}		
5766	3-(Hexadecyloxy)-1,2-propanediol, (*S*)	Chimyl alcohol	$C_{19}H_{40}O_3$	506-03-6	316.519	lf (hx)	64	$120^{0.005}$			vs ace, peth, chl
5767	1-Hexadecylpyridinium bromide		$C_{21}H_{38}BrN$	140-72-7	384.438		61				
5768	1-Hexadecylpyridinium chloride	Cetylpyridinium chloride	$C_{21}H_{38}ClN$	123-03-5	339.987	wh pow	80				vs H_2O, chl
5769	Hexadecyl stearate	Cetyl stearate	$C_{34}H_{68}O_2$	1190-63-2	508.903	lf or pl (eth, HOAc)	57			1.4410^{70}	vs ace, eth, chl
5770	Hexadecyltrichlorosilane		$C_{16}H_{33}Cl_3Si$	5894-60-0	359.878			269			
5771	Hexadecyl vinyl ether	1-(Ethenyloxy)hexadecane	$C_{18}H_{36}O$	822-28-6	268.478		16	160^{2}	0.821^{27}	1.4444^{25}	
5772	1-Hexadecyne		$C_{16}H_{30}$	629-74-3	222.409		15	284	0.7965^{20}	1.4440^{20}	vs bz
5773	*trans,trans*-2,4-Hexadienal	Sorbinaldehyde	C_6H_8O	142-83-6	96.127	liq	-16.5	174; 76^{30}	0.898^{20}	1.5384^{20}	
5774	1,2-Hexadiene	Propylallene	C_6H_{10}	592-44-9	82.143			76	0.7149^{20}	1.4282^{20}	vs eth, chl
5775	*cis*-1,3-Hexadiene		C_6H_{10}	14596-92-0	82.143			73.1	0.7033^{25}	1.4379^{20}	
5776	*trans*-1,3-Hexadiene		C_6H_{10}	20237-34-7	82.143	liq	-102.4	73.2	0.6995^{25}	1.4406^{20}	
5777	*cis*-1,4-Hexadiene		C_6H_{10}	7318-67-4	82.143			66.3	0.695^{25}	1.4049^{20}	vs eth
5778	*trans*-1,4-Hexadiene		C_6H_{10}	7319-00-8	82.143	liq	-138.7	65.0	0.695^{25}	1.4104^{20}	
5779	1,5-Hexadiene	Biallyl	C_6H_{10}	592-42-7	82.143	liq	-140.7	59.4	0.6878^{25}	1.4042^{20}	i H_2O; s EtOH, eth, bz, chl; sl ctc
5780	*cis,cis*-2,4-Hexadiene		C_6H_{10}	6108-61-8	82.143	liq		85	0.7298^{25}	1.4606^{20}	i H_2O; s EtOH, eth, chl
5781	*trans,cis*-2,4-Hexadiene		C_6H_{10}	5194-50-3	82.143	liq	-96.1	83.5	0.7185^{25}	1.4560^{20}	i H_2O; s EtOH, eth, chl
5782	*trans,trans*-2,4-Hexadiene		C_6H_{10}	5194-51-4	82.143	liq	-44.9	82.2	0.7101^{25}	1.4510^{20}	i H_2O; s EtOH, eth, chl
5783	2,4-Hexadienoic acid	Sorbic acid	$C_6H_8O_2$	110-44-1	112.127	nd (dil al) nd (w)	134.5	dec 228; 153^{50}	1.204^{19}		s H_2O, EtOH, chl; vs eth

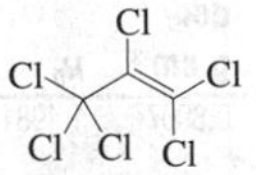

Cl
Cl Cl
Cl Cl Cl
Hexachloropropene

$H_3C(CH_2)_{58}CH_3$
Hexacontane

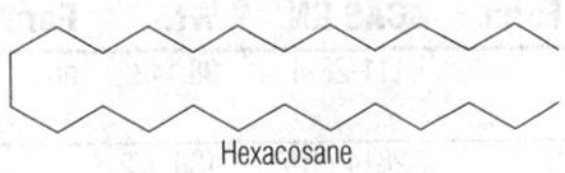

Hexacosane

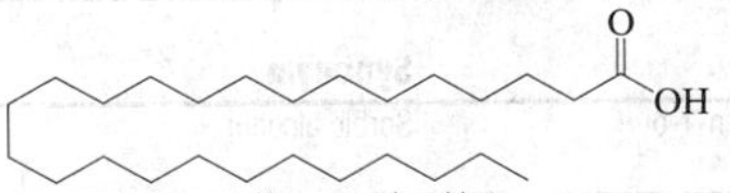

OH
Hexacosanoic acid

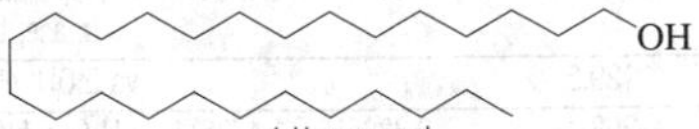

OH
1-Hexacosanol

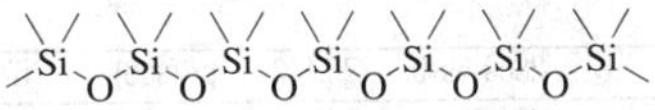

Si O Si O Si O Si O Si O Si O Si
Hexadecamethylheptasiloxane

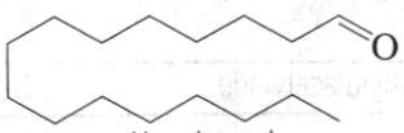

O
Hexadecanal

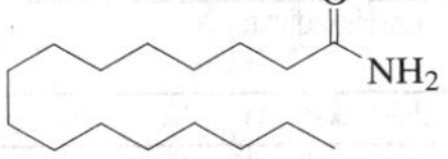

NH_2
Hexadecanamide

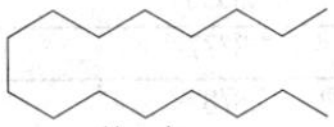

Hexadecane

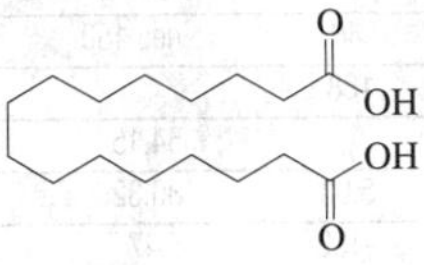

OH OH
Hexadecanedioic acid

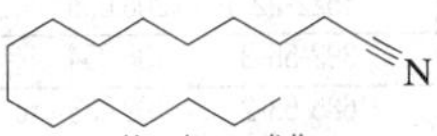

N
Hexadecanenitrile

SH
1-Hexadecanethiol

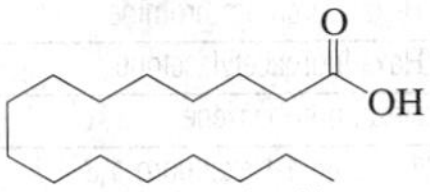

OH
Hexadecanoic acid

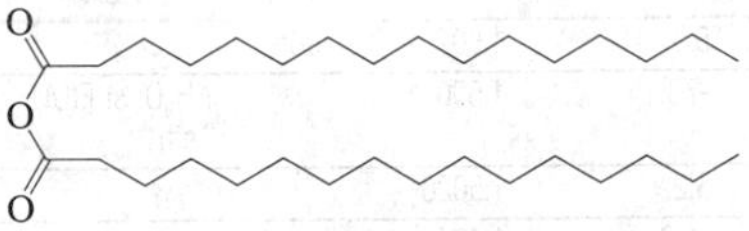

O O O
Hexadecanoic anhydride

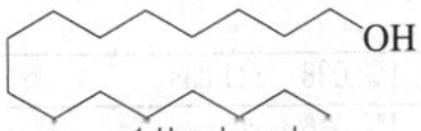

OH
1-Hexadecanol

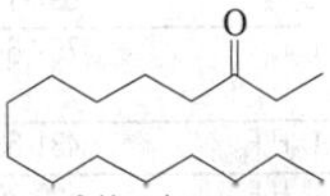

O
3-Hexadecanone

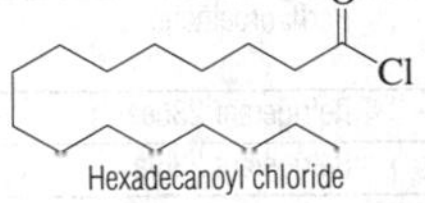

Cl
Hexadecanoyl chloride

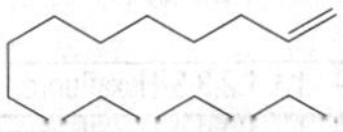

1-Hexadecene

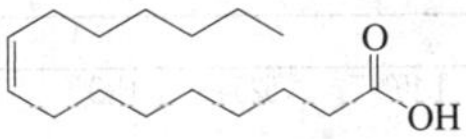

OH
cis-9-Hexadecenoic acid

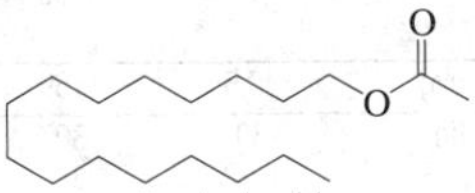

O
Hexadecyl acetate

NH_2
Hexadecylamine

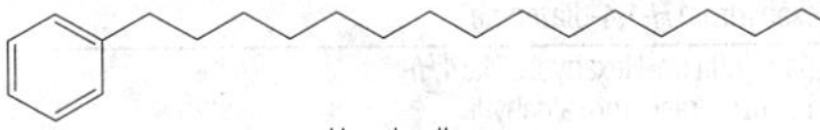

Hexadecylbenzene

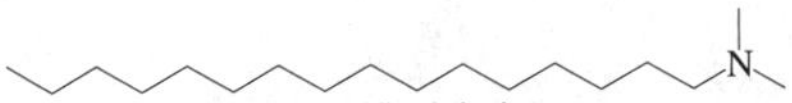

N
Hexadecyldimethylamine

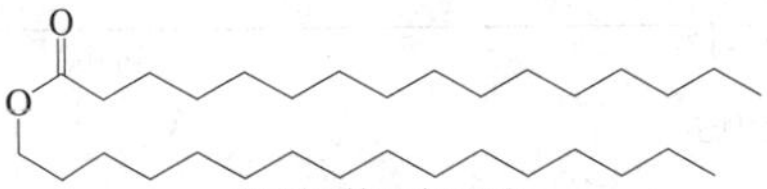

O
Hexadecyl hexadecanoate

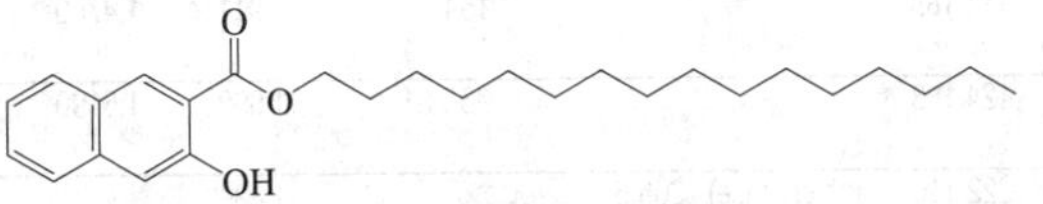

O OH
Hexadecyl 3-hydroxy-2-naphthalenecarboxylate

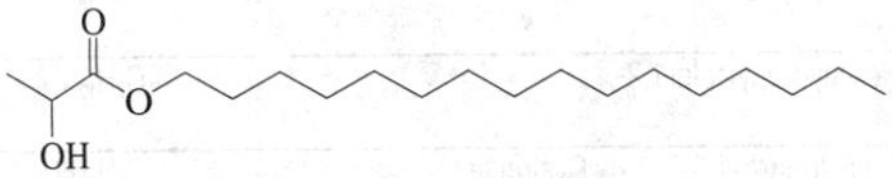

O OH
Hexadecyl 2-hydroxypropanoate

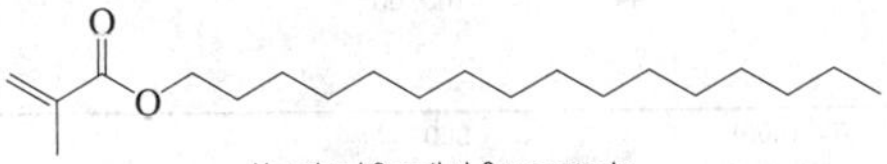

O
Hexadecyl 2-methyl-2-propenoate

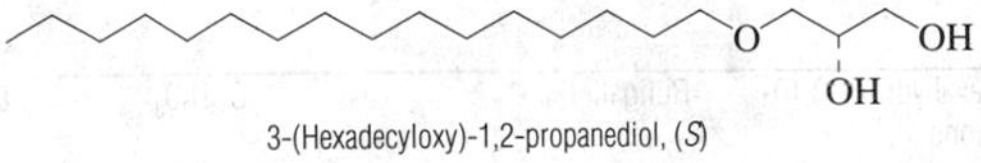

O OH OH
3-(Hexadecyloxy)-1,2-propanediol, (*S*)

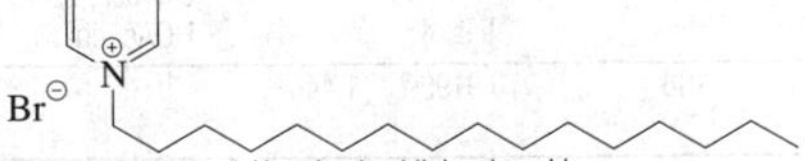

$Br^{\ominus}$ $N^{\oplus}$
1-Hexadecylpyridinium bromide

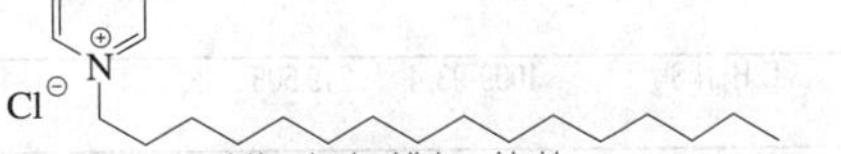

$Cl^{\ominus}$ $N^{\oplus}$
1-Hexadecylpyridinium chloride

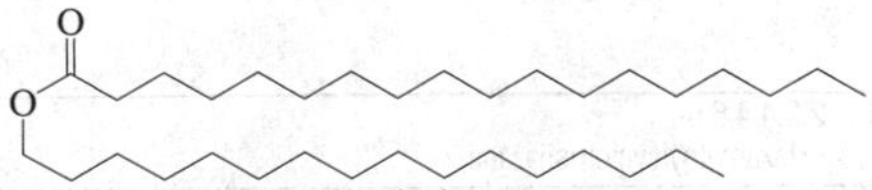

O
Hexadecyl stearate

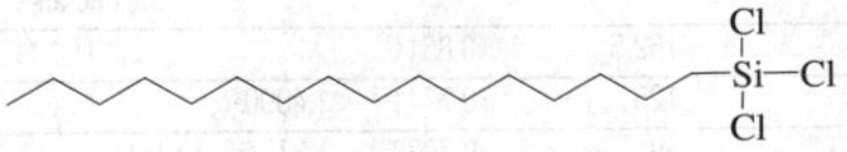

Cl Si Cl Cl
Hexadecyltrichlorosilane

O
Hexadecyl vinyl ether

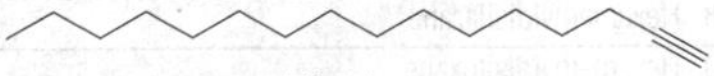

1-Hexadecyne

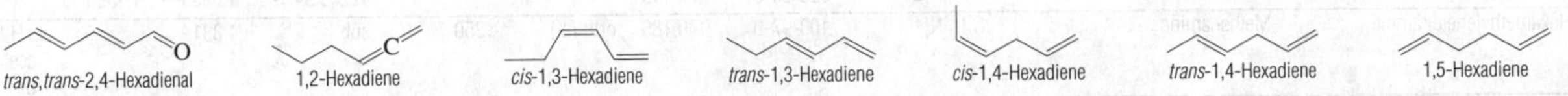

O
trans,trans-2,4-Hexadienal

C
1,2-Hexadiene

cis-1,3-Hexadiene

trans-1,3-Hexadiene

cis-1,4-Hexadiene

trans-1,4-Hexadiene

1,5-Hexadiene

cis,cis-2,4-Hexadiene

trans,cis-2,4-Hexadiene

trans,trans-2,4-Hexadiene

OH
2,4-Hexadienoic acid

No.	Name	Synonym	Mol. Form.	CAS RN	Mol. Wt.	Physical Form	mp/°C	bp/°C	den/ g cm^{-3}	n_D	Solubility
5784	2,4-Hexadien-1-ol	Sorbic alcohol	$C_6H_{10}O$	111-28-4	98.142	nd	30.5	76^{12}	0.8967^{23}	1.4981^{20}	i H_2O; s EtOH, eth
5785	*trans,trans*-2,4-Hexadienoyl chloride		C_6H_7ClO	2614-88-2	130.572			82^{22}	1.0666^{19}	1.5545^{20}	vs ace
5786	1,5-Hexadien-3-yne	Divinylacetylene	C_6H_6	821-08-9	78.112	liq	-88	85	0.7851^{20}	1.5035^{20}	i H_2O; s bz
5787	1,5-Hexadiyne	Bipropargyl	C_6H_6	628-16-0	78.112	liq	-6	86	0.8049^{20}	1.4380^{23}	i H_2O; s EtOH, eth, ace, bz
5788	2,4-Hexadiyne	Dimethyldiacetylene	C_6H_6	2809-69-0	78.112	pr (sub)	67.8	129.5			vs EtOH, eth
5789	Hexaethylbenzene		$C_{18}H_{30}$	604-88-6	246.431	mcl pr (al or bz)	129	298	0.8305^{130}	1.4736^{130}	i H_2O; s EtOH, sulf; vs eth, bz
5790	Hexaethyldisiloxane		$C_{12}H_{30}OSi_2$	994-49-0	246.536			233; 129^{30}	0.8457^{20}	1.4340^{20}	
5791	Hexaethyl tetraphosphate	Ethyl tetraphosphate	$C_{12}H_{30}O_{13}P_4$	757-58-4	506.253	hyg	-40	dec 150	1.2917^{27}	1.4273^{27}	vs ace, bz, EtOH
5792	Hexafluorenium bromide		$C_{36}H_{42}Br_2N_2$	317-52-2	662.539	cry (PrOH)	188				
5793	Hexafluoroacetylacetone		$C_5H_2F_6O_2$	1522-22-1	208.059			54.15	1.485^{20}	1.3333^{20}	
5794	Hexafluorobenzene	Perfluorobenzene	C_6F_6	392-56-3	186.054		5.03	80.32	1.6175^{20}	1.3777^{20}	
5795	1,1,2,3,4,4-Hexafluoro-1,3-butadiene		C_4F_6	685-63-2	162.033	col gas	-132	5.47	1.553^{-20}	1.378^{-20}	
5796	1,1,1,4,4,4-Hexafluoro-2-butyne		C_4F_6	692-50-2	162.033	col gas	-117.4	-24.6			s EtOH, eth, ace, ctc, HOAc
5797	Hexafluorocyclobutene		C_4F_6	697-11-0	162.033	col gas	-60	5.5	1.602^{-20}	1.298^{-20}	
5798	Hexafluoroethane	Perfluoroethane	C_2F_6	76-16-4	138.011	col gas	-100.05	-78.1	1.590^{-78}		i H_2O; sl EtOH, eth
5799	1,1,1,2,3,3-Hexafluoropropane	Refrigerant 236ea	$C_3H_2F_6$	431-63-0	152.038	col gas		6.2	1.5026^{0}		
5800	1,1,1,3,3,3-Hexafluoropropane	Refrigerant 236fa	$C_3H_2F_6$	690-39-1	152.038	col gas	-93.6	-1.0	1.4343^{0}		
5801	1,1,1,3,3,3-Hexafluoro-2-propanol		$C_3H_2F_6O$	920-66-1	168.037	liq	-2.0	59	1.4600^{21}		
5802	Hexahydro-1*H*-azepine	Hexamethylenimine	$C_6H_{13}N$	111-49-9	99.174			138	0.8643^{22}	1.4631^{20}	s H_2O; vs EtOH, eth
5803	Hexahydro-1*H*-1,4-diazepine		$C_5H_{12}N_2$	505-66-8	100.162	hyg	40.5	169			
5804	1,5a,6,9,9a,9b-Hexahydro-4a(4*H*)-dibenzofurancarboxaldehyde		$C_{13}H_{16}O_2$	126-15-8	204.265	liq	-80	307	1.10^{20}	1.5254^{20}	i H_2O
5805	*cis*-1,2,3,5,6,8a-Hexahydro-4,7-dimethyl-1-isopropylnaphthalene, (1*S*)		$C_{15}H_{24}$	483-76-1	204.352			125^{12}	0.9160^{15}	1.5089^{15}	
5806	1,2,4a,5,8,8a-Hexahydro-4,7-dimethyl-1-isopropylnaphthalene, [1*S*-(1α,4aβ,8aα)]		$C_{15}H_{24}$	523-47-7	204.352			274; 136^{11}	0.9230^{20}	1.5059^{20}	vs eth, lig
5807	Hexahydro-1,3-isobenzofurandione	Hexahydrophthalic anhydride	$C_8H_{10}O_3$	85-42-7	154.163		32	145^{18}			
5808	Hexahydro-1-methyl-1*H*-1,4-diazepine		$C_6H_{14}N_2$	4318-37-0	114.188			154	0.9111^{20}	1.4769^{20}	
5809	2,3,4,6,7,8-Hexahydropyrrolo[1,2-a]pyrimidine		$C_7H_{12}N_2$	3001-72-7	124.183			96^{7}, 81^{2}	1.005^{25}	1.5190^{20}	
5810	Hexahydro-1,3,5-trinitro-1,3,5-triazine	Cyclonite	$C_3H_6N_6O_6$	121-82-4	222.116	orth cry (ace)	205.5		1.82^{20}		i H_2O, EtOH, bz; sl eth, MeOH; s ace, HOAc
5811	Hexahydro-1,3,5-triphenyl-1,3,5-triazine		$C_{21}H_{21}N_3$	91-78-1	315.412		144	185; 60^{29}			i H_2O; sl EtOH; s eth, ace, bz, tol
5812	1,2,3,5,6,7-Hexahydroxy-9,10-anthracenedione	Rufigallol	$C_{14}H_8O_8$	82-12-2	304.209	red rhom, red-ye nd (sub)		sub			i H_2O; sl EtOH, eth; s ace, alk
5813	Hexamethylbenzene	Mellitene	$C_{12}H_{18}$	87-85-4	162.271	orth pr or nd (al)	165.5	263.4	1.0630^{25}		i H_2O; s EtOH, eth, ace, bz, HOAc, chl
5814	2,2,4,4,6,6-Hexamethylcyclotrisilazane		$C_6H_{21}N_3Si_3$	1009-93-4	219.508	liq	-10	188	0.9196^{20}	1.448^{20}	
5815	Hexamethylcyclotrisiloxane	Dimethylsiloxane cyclic trimer	$C_6H_{18}O_3Si_3$	541-05-9	222.462		64.5	134	1.1200^{20}		i H_2O
5816	Hexamethyldisilane		$C_6H_{18}Si_2$	1450-14-2	146.378		13.5	113.5	0.7247^{22}	1.4229^{20}	i H_2O; s eth, ace, bz; dec alk
5817	Hexamethyldisilathiane		$C_6H_{18}SSi_2$	3385-94-2	178.443			162.5	0.851^{20}		
5818	Hexamethyldisilazane		$C_6H_{19}NSi_2$	999-97-3	161.393			125	0.7741^{25}	1.4090^{20}	
5819	Hexamethyldisiloxane		$C_6H_{18}OSi_2$	107-46-0	162.377	liq	-66	99	0.7638^{20}	1.3774^{20}	i H_2O
5820	Hexamethylenediamine carbamate	(6-Aminohexyl)carbamic acid	$C_7H_{16}N_2O_2$	143-06-6	160.214	cry	150				
5821	Hexamethylene diisocyanate		$C_8H_{12}N_2O_2$	822-06-0	168.193			122^{10}, 94^{1}	1.0528^{20}	1.4585^{20}	
5822	Hexamethylenetetramine	Methenamine	$C_6H_{12}N_4$	100-97-0	140.186	orth (al)	>250	sub	1.331^{-5}		vs H_2O; s EtOH, ace, chl; sl eth, bz
5823	Hexamethylolmelamine		$C_9H_{18}N_6O_6$	531-18-0	306.275		137				vs H_2O
5824	Hexamethylphosphoric triamide	Tris(dimethylamino)phosphine oxide	$C_6H_{18}N_3OP$	680-31-9	179.200			232.5	1.03^{20}	1.4579^{20}	s EtOH, eth
5825	Hexamethylphosphorous triamide	Tris(dimethylamino)phosphine	$C_6H_{18}N_3P$	1608-26-0	163.201						s chl

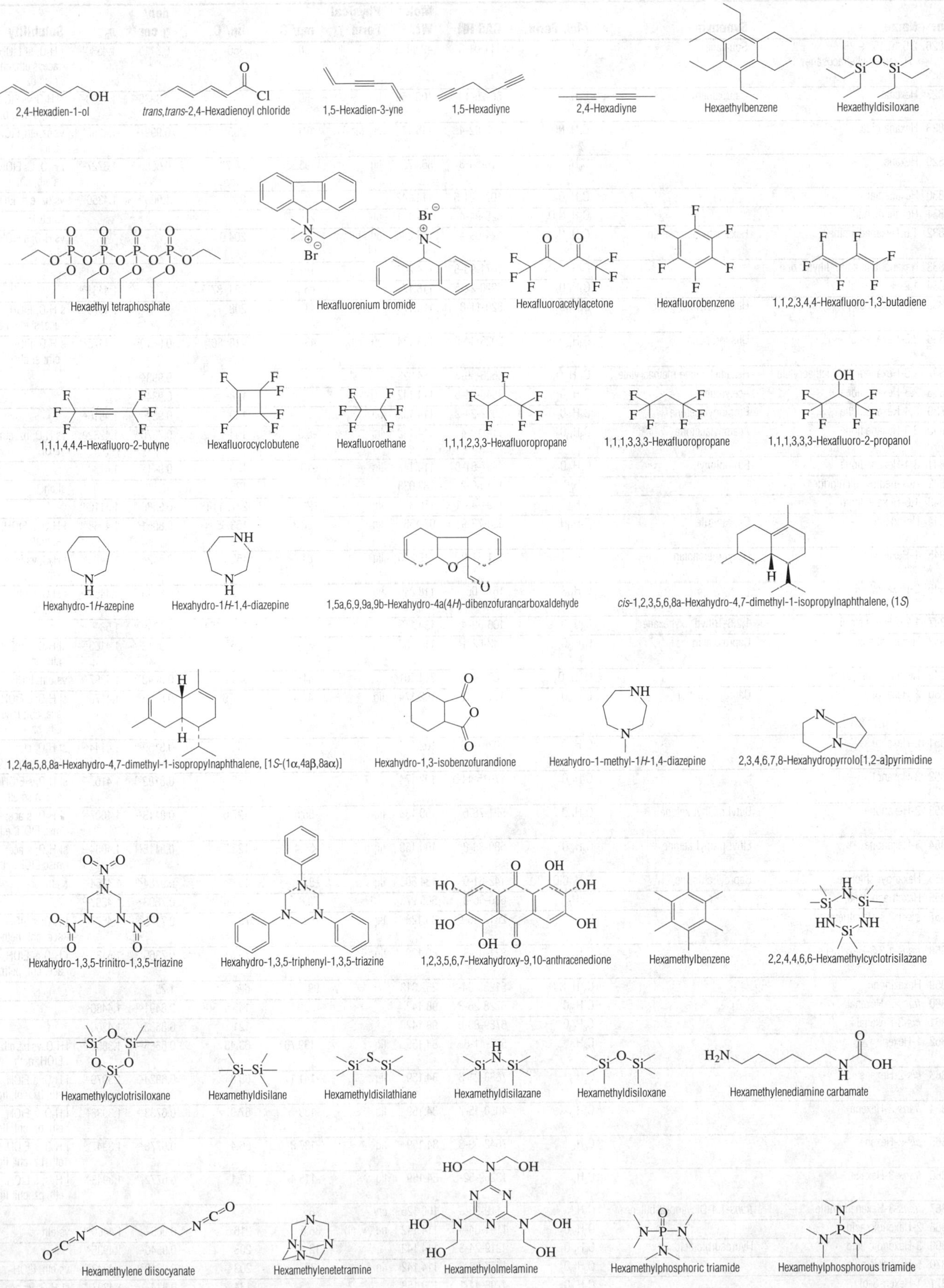

2,4-Hexadien-1-ol
trans,trans-2,4-Hexadienoyl chloride
1,5-Hexadien-3-yne
1,5-Hexadiyne
2,4-Hexadiyne
Hexaethylbenzene
Hexaethyldisiloxane
Hexaethyl tetraphosphate
Hexafluorenium bromide
Hexafluoroacetylacetone
Hexafluorobenzene
1,1,2,3,4,4-Hexafluoro-1,3-butadiene
1,1,1,4,4,4-Hexafluoro-2-butyne
Hexafluorocyclobutene
Hexafluoroethane
1,1,1,2,3,3-Hexafluoropropane
1,1,1,3,3,3-Hexafluoropropane
1,1,1,3,3,3-Hexafluoro-2-propanol
Hexahydro-1H-azepine
Hexahydro-1H-1,4-diazepine
1,5a,6,9,9a,9b-Hexahydro-4a(4H)-dibenzofurancarboxaldehyde
cis-1,2,3,5,6,8a-Hexahydro-4,7-dimethyl-1-isopropylnaphthalene, (1S)
1,2,4a,5,8,8a-Hexahydro-4,7-dimethyl-1-isopropylnaphthalene, [1S-(1α,4aβ,8aα)]
Hexahydro-1,3-isobenzofurandione
Hexahydro-1-methyl-1H-1,4-diazepine
2,3,4,6,7,8-Hexahydropyrrolo[1,2-a]pyrimidine
Hexahydro-1,3,5-trinitro-1,3,5-triazine
Hexahydro-1,3,5-triphenyl-1,3,5-triazine
1,2,3,5,6,7-Hexahydroxy-9,10-anthracenedione
Hexamethylbenzene
2,2,4,4,6,6-Hexamethylcyclotrisilazane
Hexamethylcyclotrisiloxane
Hexamethyldisilane
Hexamethyldisilathiane
Hexamethyldisilazane
Hexamethyldisiloxane
Hexamethylenediamine carbamate
Hexamethylene diisocyanate
Hexamethylenetetramine
Hexamethylolmelamine
Hexamethylphosphoric triamide
Hexamethylphosphorous triamide

No.	Name	Synonym	Mol. Form.	CAS RN	Mol. Wt.	Physical Form	mp/°C	bp/°C	den/ g cm^{-3}	n_D	Solubility
5826	2,6,10,15,19,23-Hexamethyltetracosane	Squalane	$C_{30}H_{62}$	111-01-3	422.813	liq	-38	350	0.8115^{15}	1.4530^{15}	i H_2O; sl EtOH, ace; s eth, chl; msc bz
5827	Hexanal	Caproaldehyde	$C_6H_{12}O$	66-25-1	100.158	liq	-56	131	0.8335^{20}	1.4039^{20}	sl H_2O; vs EtOH, eth; s ace, bz
5828	Hexanamide		$C_6H_{13}NO$	628-02-4	115.173	cry (ace)	101	255	0.999^{20}	1.4200^{110}	vs bz, eth, EtOH, chl
5829	Hexane		C_6H_{14}	110-54-3	86.175	liq	-95.35	68.73	0.6606^{25}	1.3727^{25}	i H_2O; vs EtOH; s eth, chl
5830	Hexanedial		$C_6H_{10}O_2$	1072-21-5	114.142		-8	93^{9}	1.003^{19}	1.4350^{20}	vs bz, eth, EtOH
5831	Hexanediamide		$C_6H_{12}N_2O_2$	628-94-4	144.171	pl	220				vs EtOH
5832	1,6-Hexanediamine	Hexamethylenediamine	$C_6H_{16}N_2$	124-09-4	116.204	orth bipym pl	39.13	204.6			vs H_2O; s EtOH, bz
5833	Hexanedioic acid, dihydrazide		$C_6H_{14}N_4O_2$	1071-93-8	174.201		181.8				
5834	1,2-Hexanediol		$C_6H_{14}O_2$	6920-22-5	118.174		45	224; $87^{1.5}$		1.4431^{20}	
5835	1,6-Hexanediol	Hexamethylene glycol	$C_6H_{14}O_2$	629-11-8	118.174		41.5	208		1.4579^{25}	s H_2O, EtOH, ace; sl eth; i bz
5836	2,5-Hexanediol	Diisopropanol	$C_6H_{14}O_2$	2935-44-6	118.174	cry (eth)	43	218; 86^{1}	0.9610^{20}	1.4475^{20}	s H_2O, EtOH, eth; sl ctc
5837	1,6-Hexanediol dimethacrylate	Hexamethylene methacrylate	$C_{14}H_{22}O_4$	6606-59-3	254.323				0.998^{25}		
5838	2,3-Hexanedione	Acetylbutyryl	$C_6H_{10}O_2$	3848-24-6	114.142			128	0.934^{19}		
5839	2,4-Hexanedione	Propionylacetone	$C_6H_{10}O_2$	3002-24-2	114.142	oil		160	0.959^{20}	1.4516^{20}	
5840	2,5-Hexanedione	Acetonylacetone	$C_6H_{10}O_2$	110-13-4	114.142	liq	-5.5	194	0.7370^{20}	1.4232^{20}	vs H_2O, bz, eth, EtOH
5841	3,4-Hexanedione	Bipropionyl	$C_6H_{10}O_2$	4437-51-8	114.142	liq	-10	130	0.941^{21}	1.4130^{21}	
5842	Hexanedioyl dichloride		$C_6H_8Cl_2O_2$	111-50-2	183.033			126^{12}			sl chl
5843	1,6-Hexanedithiol		$C_6H_{14}S_2$	1191-43-1	150.305	liq	-21	237; 118^{15}	0.9886^{25}	1.5110^{20}	
5844	Hexanenitrile	Capronitrile	$C_6H_{11}N$	628-73-9	97.158	liq	-80.3	163.65	0.8051^{20}	1.4068^{20}	i H_2O; s EtOH, eth; sl chl
5845	1-Hexanethiol	Hexyl mercaptan	$C_6H_{14}S$	111-31-9	118.240	liq	-81	152.7	0.8424^{20}	1.4496^{20}	i H_2O; vs EtOH, eth
5846	2-Hexanethiol		$C_6H_{14}S$	1679-06-7	118.240	liq	-147	142	0.8345^{20}	1.4451^{20}	i H_2O; s EtOH, eth, bz
5847	1,2,6-Hexanetriol	1,2,6-Trihydroxyhexane	$C_6H_{14}O_3$	106-69-4	134.173			170^{3}, 161^{1}	1.1049^{20}	1.58^{20}	
5848	Hexanoic acid	Caproic acid	$C_6H_{12}O_2$	142-62-1	116.158	liq	-3	205.2	0.9212^{25}	1.4163^{20}	sl H_2O; s EtOH, eth, chl
5849	Hexanoic anhydride		$C_{12}H_{22}O_3$	2051-49-2	214.301		-41	dec 255	0.9240^{15}	1.4297^{20}	vs eth, EtOH
5850	1-Hexanol	Caproyl alcohol	$C_6H_{14}O$	111-27-3	102.174	liq	-47.4	157.6	0.8136^{20}	1.4178^{20}	sl H_2O; s EtOH, ace, chl; msc eth, bz
5851	2-Hexanol		$C_6H_{14}O$	20281-86-1	102.174			140	0.8159^{20}	1.4144^{20}	sl H_2O, ctc; s EtOH, eth
5852	3-Hexanol		$C_6H_{14}O$	17015-11-1	102.174			135	0.8182^{20}	1.4167^{20}	sl H_2O; s EtOH, ace; msc eth
5853	2-Hexanone	Butyl methyl ketone	$C_6H_{12}O$	591-78-6	100.158	liq	-55.5	127.6	0.8113^{20}	1.4007^{20}	sl H_2O; s ace; msc EtOH, eth
5854	3-Hexanone	Ethyl propyl ketone	$C_6H_{12}O$	589-38-8	100.158	liq	-55.4	123.5	0.8118^{20}	1.4004^{20}	sl H_2O; s ace; msc EtOH, eth
5855	Hexanoyl chloride	Caproyl chloride	$C_6H_{11}ClO$	142-61-0	134.603	liq	-87	153	0.9784^{20}	1.4264^{20}	s eth, ace
5856	Hexatriacontane		$C_{36}H_{74}$	630-06-8	506.973		75.8	298.4^{3}	0.7803^{80}	1.4397^{80}	
5857	*cis*-1,3,5-Hexatriene		C_6H_8	2612-46-6	80.128	liq	-12	78	0.7175^{20}	1.4577^{20}	i H_2O; s EtOH, ace, chl, peth
5858	*trans*-1,3,5-Hexatriene		C_6H_8	821-07-8	80.128	liq	-12	78.5	0.7369^{15}	1.5135^{20}	i H_2O; s EtOH, ace, chl, peth
5859	Hexazinone		$C_{12}H_{20}N_4O_2$	51235-04-2	252.313		99	dec	1.25		
5860	*trans*-2-Hexenal		$C_6H_{10}O$	6728-26-3	98.142			146.5; 50^{20}	0.8491^{20}	1.4480^{20}	
5861	*cis*-3-Hexenal		$C_6H_{10}O$	6789-80-6	98.142			121	0.8533^{22}	1.4300^{21}	
5862	1-Hexene		C_6H_{12}	592-41-6	84.159	liq	-139.76	63.48	0.6685^{25}	1.3852^{25}	i H_2O; vs bz, eth, EtOH, peth
5863	*cis*-2-Hexene		C_6H_{12}	7688-21-3	84.159	liq	-141.11	68.8	0.6824^{25}	1.3979^{20}	i H_2O; s EtOH, eth, bz, chl, lig
5864	*trans*-2-Hexene		C_6H_{12}	4050-45-7	84.159	liq	-133	67.9	0.6733^{25}	1.3936^{20}	i H_2O; s EtOH, eth, bz, chl, lig
5865	*cis*-3-Hexene		C_6H_{12}	7642-09-3	84.159	liq	-137.8	66.4	0.6778^{20}	1.3947^{20}	i H_2O; s EtOH, eth, bz, chl, lig
5866	*trans*-3-Hexene		C_6H_{12}	13269-52-8	84.159	liq	-115.4	67.1	0.6772^{20}	1.3943^{20}	i H_2O; s EtOH, eth, bz, chl, lig
5867	*trans*-3-Hexenedinitrile	*trans*-1,4-Dicyano-2-butene	$C_6H_6N_2$	1119-85-3	106.125	cry	76				
5868	2-Hexenoic acid		$C_6H_{10}O_2$	1191-04-4	114.142	nd (w, al)	36.5	216.5	0.965^{20}	1.4460^{40}	vs eth
5869	3-Hexenoic acid	Hydrosorbic acid	$C_6H_{10}O_2$	4219-24-3	114.142		12	208	0.9640^{23}	1.4935^{20}	
5870	5-Hexenoic acid		$C_6H_{10}O_2$	1577-22-6	114.142	liq	-37	203	0.9610^{20}	1.4343^{20}	vs eth, EtOH
5871	1-Hexen-3-ol		$C_6H_{12}O$	4798-44-1	100.158			134	0.834^{22}	1.4297^{18}	sl H_2O; vs ace, eth, EtOH

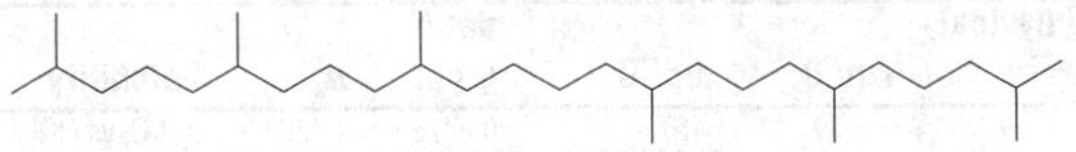

O
NH2
2,6,10,15,19,23-Hexamethyltetracosane
Hexanal 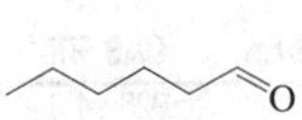
Hexanamide 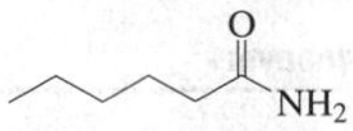
Hexane  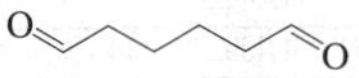
O
O
H2N
NH2
O
H2N
NH2
H2N
N
H
O
H
N
NH2
O
Hexanedial
Hexanediamide 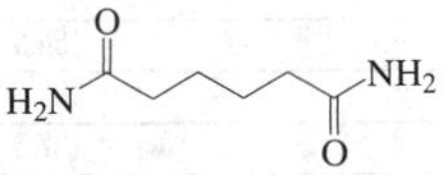
1,6-Hexanediamine 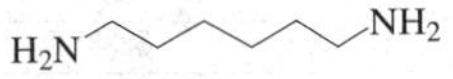
Hexanedioic acid, dihydrazide 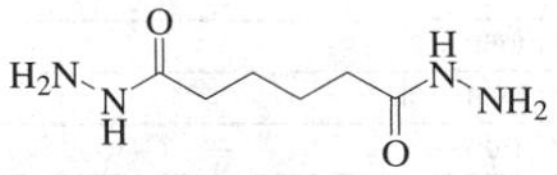 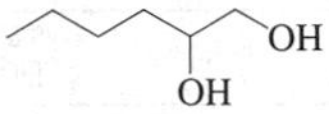
OH
OH
HO
OH
OH
OH
O
O
O
O
1,2-Hexanediol
1,6-Hexanediol 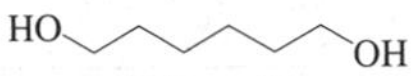
2,5-Hexanediol 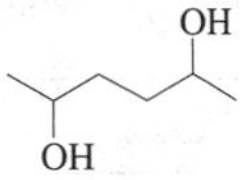
1,6-Hexanediol dimethacrylate 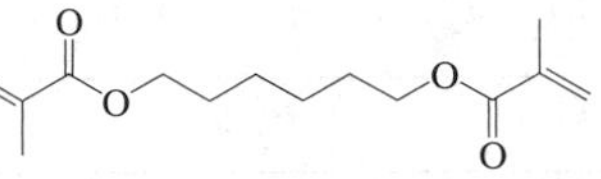
O
O
O
O
O
O
O
Cl
O
Cl
O
2,3-Hexanedione
2,4-Hexanedione
2,5-Hexanedione
3,4-Hexanedione
Hexanedioyl dichloride 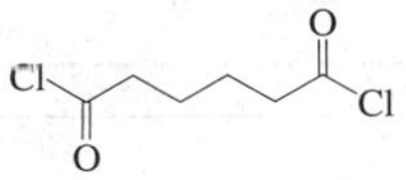 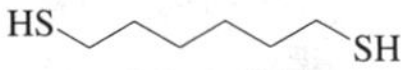
HS
SH
N
SH
SH
HO
OH
OH
1,6-Hexanedithiol
Hexanenitrile 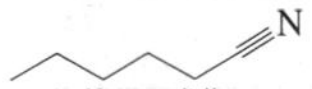
1-Hexanethiol 
2-Hexanethiol 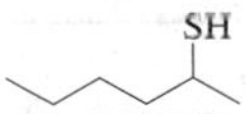
1,2,6-Hexanetriol 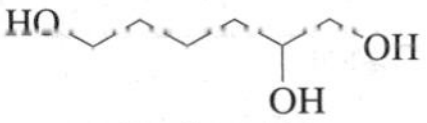 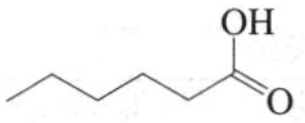
OH
O
O
O
O
OH
OH
OH
Hexanoic acid
Hexanoic anhydride 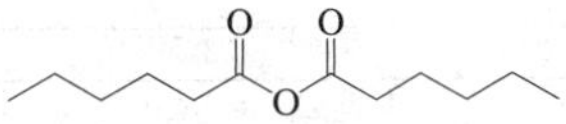
1-Hexanol 
2-Hexanol 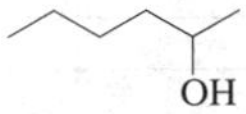
3-Hexanol 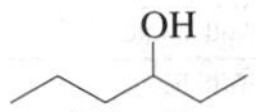 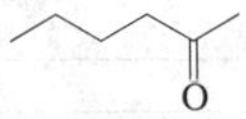
O
O
O
Cl
2-Hexanone
3-Hexanone 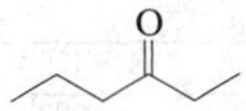
Hexanoyl chloride 
Hexatriacontane 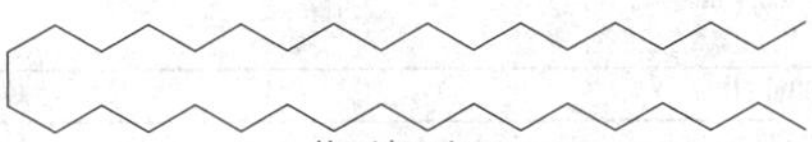 
O
N
N
O
N
N
cis-1,3,5-Hexatriene
trans-1,3,5-Hexatriene 
Hexazinone 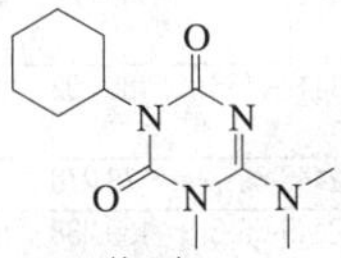 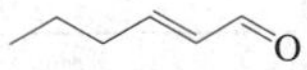
O
trans-2-Hexenal 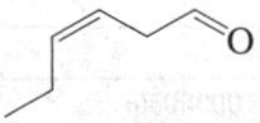
O
cis-3-Hexenal
N
N
1-Hexene
cis-2-Hexene
trans-2-Hexene
cis-3-Hexene
trans-3-Hexene 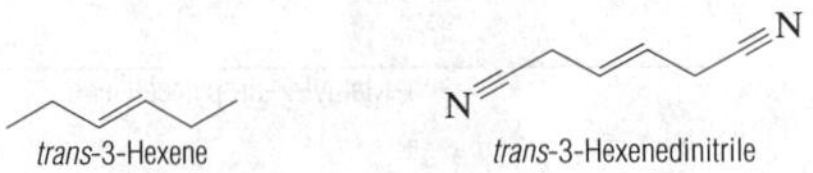
trans-3-Hexenedinitrile
O
OH
O
OH
O
OH
OH
2-Hexenoic acid
3-Hexenoic acid
5-Hexenoic acid 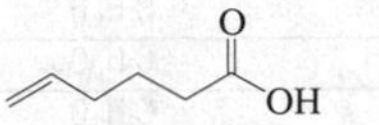
1-Hexen-3-ol

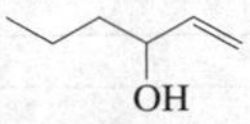

No.	Name	Synonym	Mol. Form.	CAS RN	Mol. Wt.	Physical Form	mp/°C	bp/°C	den/ g cm^{-3}	n_D	Solubility
5872	*cis*-2-Hexen-1-ol		$C_6H_{12}O$	928-94-9	100.158			157	0.8472^{20}	1.4397^{20}	s H_2O; vs EtOH; s eth, ace; sl ctc
5873	*trans*-2-Hexen-1-ol		$C_6H_{12}O$	928-95-0	100.158			157	0.8490^{16}	1.4340^{20}	
5874	*cis*-3-Hexen-1-ol		$C_6H_{12}O$	928-96-1	100.158			156.5	0.8478^{22}	1.4380^{20}	s H_2O; vs EtOH, eth
5875	*trans*-3-Hexen-1-ol		$C_6H_{12}O$	928-97-2	100.158			154.5		1.4374^{20}	
5876	*trans*-4-Hexen-1-ol		$C_6H_{12}O$	928-92-7	100.158			159	0.8513^{20}	1.4402^{20}	
5877	4-Hexen-2-ol		$C_6H_{12}O$	52387-50-5	100.158			137.5	0.8405^{18}	1.4392^{20}	sl H_2O
5878	5-Hexen-2-ol		$C_6H_{12}O$	626-94-8	100.158			139	0.842^{16}		sl H_2O
5879	*cis*-3-Hexen-1-ol, acetate		$C_8H_{14}O_2$	3681-71-8	142.196	liq		66^{12}			
5880	*trans*-2-Hexen-1-ol, acetate		$C_8H_{14}O_2$	2497-18-9	142.196	liq		166; 68^{15}	0.898	1.4270^{20}	
5881	5-Hexen-2-one		$C_6H_{10}O$	109-49-9	98.142			129.5	0.833^{27}	1.4178^{27}	
5882	4-Hexen-3-one		$C_6H_{10}O$	2497-21-4	98.142			138.5	0.8559^{20}	1.4388^{20}	s EtOH, eth; vs ace
5883	Hexestrol		$C_{18}H_{22}O_2$	84-16-2	270.367	nd (bz)	186.5				vs ace, eth, EtOH
5884	Hexobarbital		$C_{12}H_{16}N_2O_3$	56-29-1	236.266		146.5				
5885	Hexocyclium methyl sulfate		$C_{21}H_{36}N_2O_5S$	115-63-9	428.586	cry	205				sl chl; i eth
5886	Hexyl acetate		$C_8H_{16}O_2$	142-92-7	144.212	liq	-80.9	171.5	0.8779^{15}	1.4092^{20}	i H_2O; vs eth, EtOH
5887	*sec*-Hexyl acetate		$C_8H_{16}O_2$	108-84-9	144.212			147.5	0.8805^{25}	1.3980^{20}	sl H_2O; vs eth, EtOH
5888	Hexyl acrylate		$C_9H_{16}O_2$	2499-95-8	156.222		-45	40^{1}	0.878^{20}		
5889	Hexylamine	1-Hexanamine	$C_6H_{15}N$	111-26-2	101.190	liq	-22.9	132.8	0.7660^{20}	1.4180^{20}	sl H_2O; msc EtOH, eth; s chl
5890	Hexylbenzene		$C_{12}H_{18}$	1077-16-3	162.271	liq	-61	226.1	0.8575^{20}	1.4864^{20}	i H_2O; msc eth; s bz, peth
5891	4-Hexyl-1,3-benzenediol	4-Hexylresorcinol	$C_{12}H_{18}O_2$	136-77-6	194.270	nd (bz)	68	334			vs ace, eth, EtOH, chl
5892	Hexyl benzoate		$C_{13}H_{18}O_2$	6789-88-4	206.281			272; 139^{8}	0.9793^{20}		i H_2O; s EtOH, ace
5893	Hexyl butanoate		$C_{10}H_{20}O_2$	2639-63-6	172.265	liq	-78	208	0.8652^{20}	1.4160^{15}	i H_2O; s EtOH; sl chl
5894	Hexylcyclohexane		$C_{12}H_{24}$	4292-75-5	168.319	liq	-43	224	0.8076^{20}	1.4462^{20}	
5895	Hexylcyclopentane		$C_{11}H_{22}$	4457-00-5	154.293	liq	-73	203	0.7965^{20}	1.4392^{20}	vs ace, bz, eth, EtOH
5896	2-Hexyldecanoic acid		$C_{16}H_{32}O_2$	25354-97-6	256.424	visc oil		$145^{0.02}$		1.4432^{24}	
5897	Hexyl formate		$C_7H_{14}O_2$	629-33-4	130.185	liq	-62.6	155.5	0.8813^{20}	1.4071^{20}	i H_2O; msc EtOH, eth
5898	Hexyl hexanoate	Hexyl caproate	$C_{12}H_{24}O_2$	6378-65-0	200.318	liq	-55	246	0.865^{18}	1.4264^{15}	vs ace, bz, eth, EtOH
5899	Hexyl isocyanate		$C_7H_{13}NO$	2525-62-4	127.184			44^{7}			
5900	Hexyl methacrylate		$C_{10}H_{18}O_2$	142-09-6	170.249			162; 86^{17}	0.880^{25}	1.429^{25}	vs ace, bz, eth, EtOH
5901	Hexyl methyl ether		$C_7H_{16}O$	4747-07-3	116.201			126.1			
5902	1-Hexylnaphthalene		$C_{16}H_{20}$	2876-53-1	212.330	liq	-18	322	0.9566^{20}	1.5647^{20}	
5903	Hexyl octanoate		$C_{14}H_{28}O_2$	1117-55-1	228.371	liq	-30.6	277.4	0.8603^{20}	1.4323^{25}	i H_2O; s EtOH, eth, ace
5904	4-(Hexyloxy)benzoic acid		$C_{13}H_{18}O_3$	1142-39-8	222.280	cry	106				
5905	2-(Hexyloxy)ethanol	Ethylene glycol monohexyl ether	$C_8H_{18}O_2$	112-25-4	146.228	liq	-45.1	208	0.8878^{20}	1.4291^{20}	sl H_2O; vs EtOH, eth
5906	Hexyl pentanoate		$C_{11}H_{22}O_2$	1117-59-5	186.292	liq	-63.1	226.3	0.8635^{20}	1.4228^{15}	vs ace, eth, EtOH
5907	4-Hexylphenol		$C_{12}H_{18}O$	2446-69-7	178.270			148^{9}			
5908	1-Hexyl propanoate		$C_9H_{18}O_2$	2445-76-3	158.238	liq	-57.5	190	0.8698^{20}	1.4162^{15}	i H_2O; s EtOH, eth, ace, AcOEt
5909	1-Hexyl-1,2,3,4-tetrahydronaphthalene		$C_{16}H_{24}$	66325-11-9	216.362	liq		305	0.9176^{25}	1.5127^{25}	
5910	1-Hexyne	Butylacetylene	C_6H_{10}	693-02-7	82.143	liq	-131.9	71.3	0.7155^{25}	1.3989^{20}	i H_2O; s EtOH, eth, bz, chl; sl ctc
5911	2-Hexyne	1-Methyl-2-propylacetylene	C_6H_{10}	764-35-2	82.143	liq	-89.6	84.5	0.7315^{20}	1.4138^{20}	i H_2O; msc EtOH, eth; s bz, chl, peth
5912	3-Hexyne	Diethylacetylene	C_6H_{10}	928-49-4	82.143	liq	-103	81	0.7231^{20}	1.4115^{20}	i H_2O; s EtOH, eth, bz, chl, peth
5913	3-Hexyne-2,5-diol		$C_6H_{10}O_2$	3031-66-1	114.142			121^{15}	1.0180^{20}	1.4691^{20}	
5914	3-Hexyn-1-ol	3-Hexynol	$C_6H_{10}O$	1002-28-4	98.142			162; 65^{12}	0.8982^{20}	1.4530^{20}	
5915	1-Hexyn-3-ol		$C_6H_{10}O$	105-31-7	98.142	liq	-80	142	0.8704^{20}	1.4340^{25}	s ctc
5916	5-Hexyn-2-one		C_6H_8O	2550-28-9	96.127			149	0.9065^{20}	1.4366^{20}	

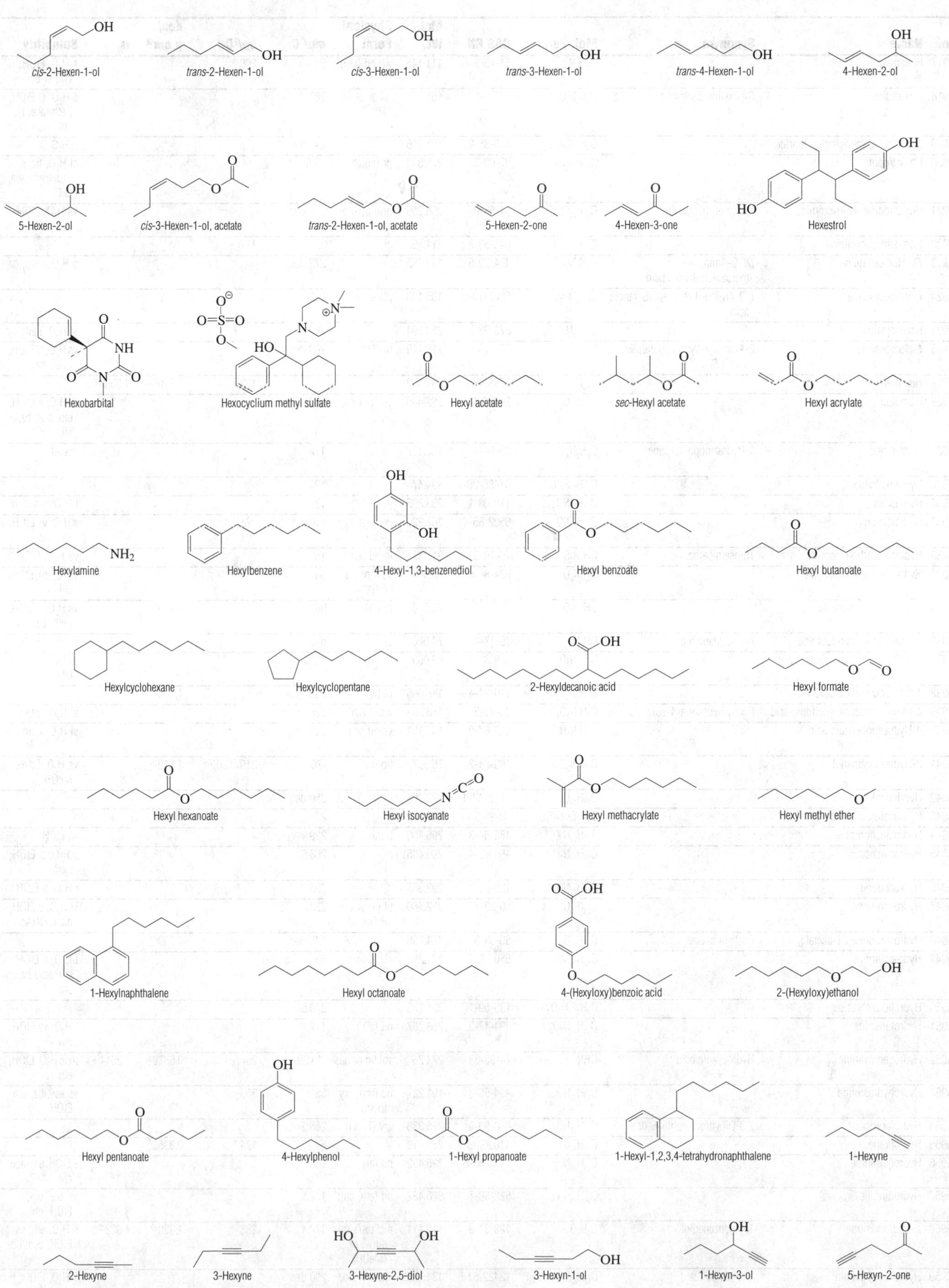
cis-2-Hexen-1-ol
trans-2-Hexen-1-ol
cis-3-Hexen-1-ol
trans-3-Hexen-1-ol
trans-4-Hexen-1-ol
4-Hexen-2-ol
5-Hexen-2-ol
cis-3-Hexen-1-ol, acetate
trans-2-Hexen-1-ol, acetate
5-Hexen-2-one
4-Hexen-3-one
Hexestrol
Hexobarbital
Hexocyclium methyl sulfate
Hexyl acetate
sec-Hexyl acetate
Hexyl acrylate
Hexylamine
Hexylbenzene
4-Hexyl-1,3-benzenediol
Hexyl benzoate
Hexyl butanoate
Hexylcyclohexane
Hexylcyclopentane
2-Hexyldecanoic acid
Hexyl formate
Hexyl hexanoate
Hexyl isocyanate
Hexyl methacrylate
Hexyl methyl ether
1-Hexylnaphthalene
Hexyl octanoate
4-(Hexyloxy)benzoic acid
2-(Hexyloxy)ethanol
Hexyl pentanoate
4-Hexylphenol
1-Hexyl propanoate
1-Hexyl-1,2,3,4-tetrahydronaphthalene
1-Hexyne
2-Hexyne
3-Hexyne
3-Hexyne-2,5-diol
3-Hexyn-1-ol
1-Hexyn-3-ol
5-Hexyn-2-one

No.	Name	Synonym	Mol. Form.	CAS RN	Mol. Wt.	Physical Form	mp/°C	bp/°C	den/ g cm^{-3}	n_D	Solubility
5917	Histamine		$C_5H_9N_3$	51-45-6	111.145	wh nd (chl)	83	209^{18}			s H_2O, EtOH, chl; sl eth
5918	*L*-Histidine	Glyoxaline-5-alanine	$C_6H_9N_3O_2$	71-00-1	155.154	nd or pl (dil al)	287 dec				s H_2O; sl EtOH; i eth, ace, bz, chl
5919	*L*-Histidine, monohydrochloride		$C_6H_{10}ClN_3O_2$	645-35-2	191.615		245 dec				s H_2O
5920	Homatropine		$C_{16}H_{21}NO_3$	87-00-3	275.343	pr (al, eth)	99.5				sl H_2O, bz; s EtOH, eth, ace, chl
5921	Homatropine hydrobromide	Tropanol mandelate	$C_{16}H_{22}BrNO_3$	51-56-9	356.255	orth pym or pl (w)	217 dec				vs H_2O, EtOH
5922	Homochlorocyclizine		$C_{19}H_{23}ClN_2$	848-53-3	314.852	oil		$177^{0.8}$			
5923	*DL*-Homocysteine	*DL*-2-Amino-4-mercaptobutanoic acid	$C_4H_9NO_2S$	454-29-5	135.185		272 dec				s H_2O; i eth, bz
5924	*L*-Homocysteine	*L*-2-Amino-4-mercaptobutanoic acid	$C_4H_9NO_2S$	6027-13-0	135.185	platelets	232				
5925	Homocystine		$C_8H_{16}N_2O_4S_2$	870-93-9	268.354		264				sl H_2O; i eth, bz
5926	*L*-Homoserine	2-Amino-4-hydroxybutanoic acid, (*S*)	$C_4H_9NO_3$	672-15-1	119.119	pr (90% al)	203 dec				vs H_2O; sl EtOH; i eth, bz
5927	Humulene		$C_{15}H_{24}$	6753-98-6	204.352			123^{10}	0.8905^{20}	1.5038^{20}	
5928	Humulon		$C_{21}H_{30}O_5$	26472-41-3	362.460	ye cry (eth)	66.5				sl H_2O; s EtOH, eth, ace, bz, alk
5929	Hydralazine	1-Hydrazinophthalazine	$C_8H_8N_4$	86-54-4	160.177	ye cry (MeOH)	172				s acid
5930	Hydramethylnon		$C_{25}H_{24}F_6N_4$	67485-29-4	494.476		190				
5931	Hydrastine		$C_{21}H_{21}NO_6$	118-08-1	383.395	ye pr (al)	132				i H_2O; s ace, bz
5932	Hydrastinine		$C_{11}H_{13}NO_3$	6592-85-4	207.226	nd (lig), cry (eth)	116.5				s H_2O; vs EtOH, eth, chl
5933	Hydrazinecarbothioamide	Thiosemicarbazide	CH_5N_3S	79-19-6	91.136	lo nd (w)	183				vs H_2O, EtOH
5934	Hydrazinecarboxaldehyde		CH_4N_2O	624-84-0	60.055	ye lf or nd (al)	54				vs bz, eth, EtOH, chl
5935	Hydrazinecarboxamide		CH_5N_3O	57-56-7	75.070	pr (al)	96		1.484^{8}		vs H_2O; s EtOH; i eth, bz, chl
5936	Hydrazinecarboximidamide	Aminoguanidine	CH_6N_4	79-17-4	74.086	cry	dec				vs H_2O, EtOH
5937	1,2-Hydrazinedicarboxaldehyde		$C_2H_4N_2O_2$	628-36-4	88.065	pr (al)	161.0				vs H_2O; sl EtOH, DMSO; i eth
5938	1,2-Hydrazinedicarboxamide		$C_2H_6N_4O_2$	110-21-4	118.095	pl (w)	258		1.604^{17}		
5939	4-Hydrazinobenzenesulfonic acid	Phenylhydrazine-4-sulfonic acid	$C_6H_8N_2O_3S$	98-71-5	188.204	nd, lf (w)	286				sl H_2O, EtOH
5940	4-Hydrazinobenzoic acid		$C_7H_8N_2O_2$	619-67-0	152.151	ye nd or pl (w)	221 dec				sl H_2O; i eth
5941	2-Hydrazinoethanol		$C_2H_8N_2O$	109-84-2	76.097	liq	-70	219; $120^{17.5}$	1.119^{25}		vs H_2O, EtOH, MeOH
5942	Hydrindantin		$C_{18}H_{10}O_6$	5103-42-4	322.268	pr (ace)	250 dec				
5943	Hydrochlorothiazide		$C_7H_8ClN_3O_4S_2$	58-93-5	297.740		274				
5944	Hydrocinchonidine		$C_{19}H_{24}N_2O$	485-64-3	296.406	lf (al)	229				vs EtOH
5945	Hydrocinchonine		$C_{19}H_{24}N_2O$	485-65-4	296.406	pr	268.5				s H_2O; sl EtOH; i eth
5946	Hydrocodone		$C_{18}H_{21}NO_3$	125-29-1	299.365		198				i H_2O; s EtOH
5947	Hydrocortisone		$C_{21}H_{30}O_5$	50-23-7	362.460	pl (al or i-PrOH)	220				sl H_2O; s EtOH, diox, HOAc
5948	Hydrocortisone 21-acetate	Cortisol acetate	$C_{23}H_{32}O_6$	50-03-3	404.496		223 dec		1.289^{20}		
5949	Hydrocotarnine		$C_{12}H_{15}NO_3$	550-10-7	221.252		56				i H_2O; s EtOH, eth, ace, bz, chl
5950	Hydroflumethiazide		$C_8H_8F_3N_3O_4S_2$	135-09-1	331.293		270.5				
5951	Hydrofuramide		$C_{15}H_{12}N_2O_3$	494-47-3	268.267	nd (al)	117				i H_2O; vs EtOH, eth
5952	Hydrogen cyanide	Hydrocyanic acid	CHN	74-90-8	27.026	vol liq or gas	-13.29	26	0.6876^{20}	1.2614^{20}	msc H_2O, EtOH, eth
5953	Hydrohydrastinine		$C_{11}H_{13}NO_2$	494-55-3	191.227	nd (lig), cry (peth)	66	303			vs ace, bz, eth, EtOH
5954	Hydromorphone	7,8-Dihydromorphin-6-one	$C_{17}H_{19}NO_3$	466-99-9	285.338	cry (EtOH)	266.5				
5955	Hydroprene		$C_{17}H_{30}O_2$	41096-46-2	266.419			174^{19}	0.8955^{20}		
5956	Hydroquinidine		$C_{20}H_{26}N_2O_2$	1435-55-8	326.432	nd (al)	168.5				s EtOH, eth, ace, chl
5957	Hydroquinine		$C_{20}H_{26}N_2O_2$	522-66-7	326.432	nd (eth, chl)	172.5				vs ace, eth, EtOH, chl
5958	*p*-Hydroquinone	1,4-Benzenediol	$C_6H_6O_2$	123-31-9	110.111	mcl pr (sub) nd(w) pr (MeOH)	172.4	285	1.330^{20}	1.632^{25}	s H_2O, eth; vs EtOH, ace; i bz
5959	Hydroxocobalamin	Vitamin B-12a	$C_{62}H_{89}CoN_{13}O_{15}P$	13422-51-0	1346.355	red cry (ace aq)	200 dec				s H_2O, EtOH; i ace, eth, bz

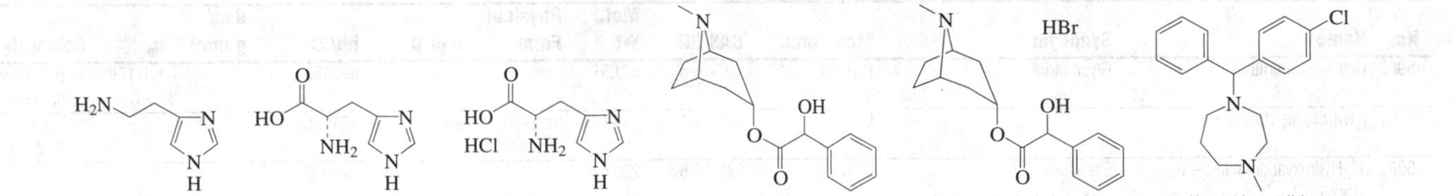

Histamine

L-Histidine

L-Histidine, monohydrochloride

Homatropine

Homatropine hydrobromide

Homochlorocyclizine

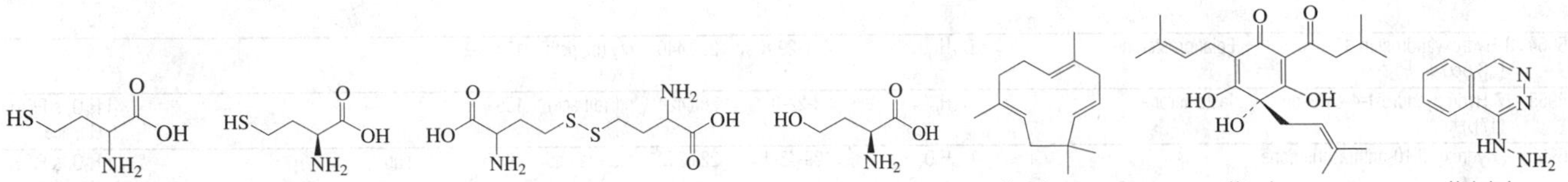

DL-Homocysteine

L-Homocysteine

Homocystine

L-Homoserine

Humulene

Humulon

Hydralazine

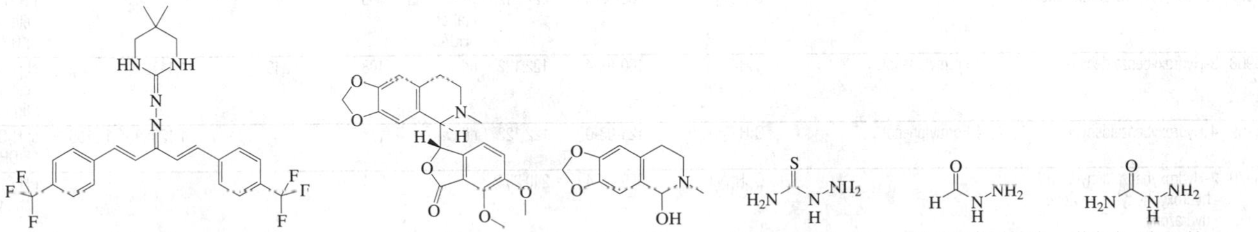

Hydramethylnon

Hydrastine

Hydrastinine

Hydrazinecarbothioamide

Hydrazinecarboxaldehyde

Hydrazinecarboxamide

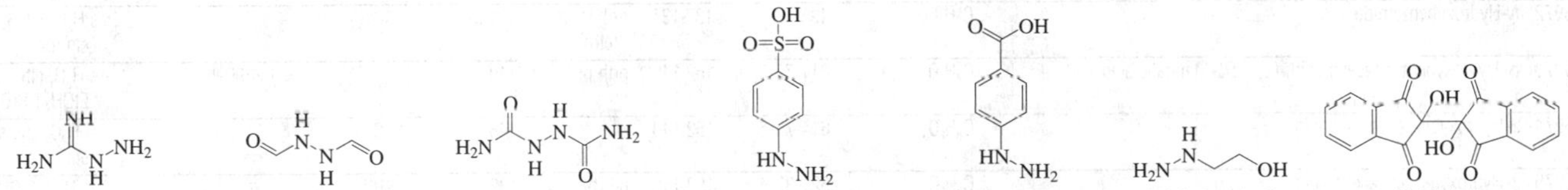

Hydrazinecarboximidamide

1,2-Hydrazinedicarboxaldehyde

1,2-Hydrazinedicarboxamide

4-Hydrazinobenzenesulfonic acid

4-Hydrazinobenzoic acid

2-Hydrazinoethanol

Hydrindantin

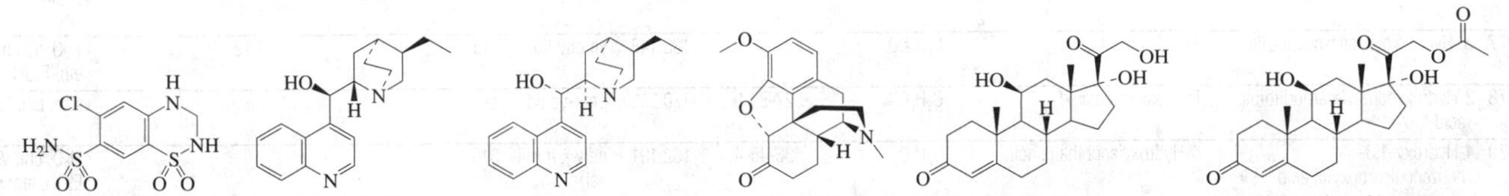

Hydrochlorothiazide

Hydrocinchonidine

Hydrocinchonine

Hydrocodone

Hydrocortisone

Hydrocortisone 21-acetate

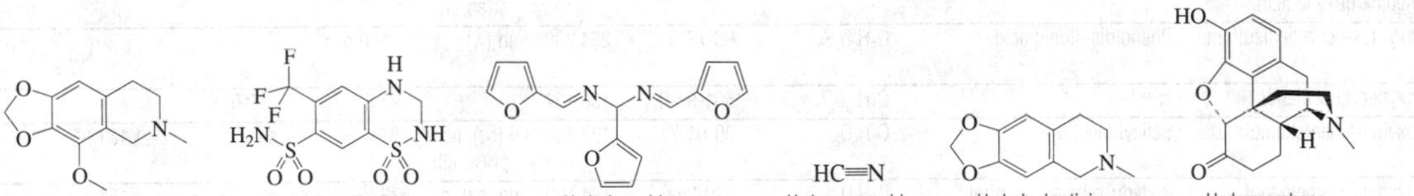

Hydrocotarnine

Hydroflumethiazide

Hydrofuramide

Hydrogen cyanide

Hydrohydrastinine

Hydromorphone

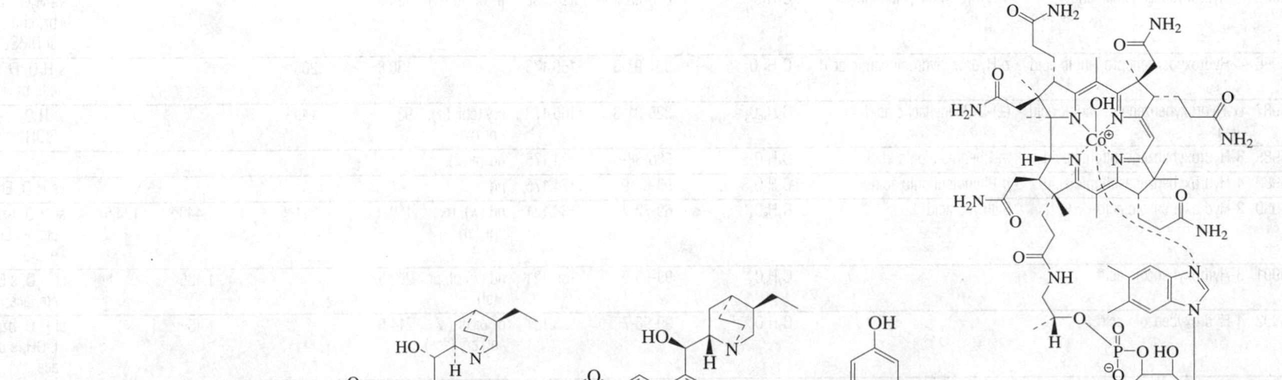

Hydroprene

Hydroquinidine

Hydroquinine

p-Hydroquinone

Hydroxocobalamin

No.	Name	Synonym	Mol. Form.	CAS RN	Mol. Wt.	Physical Form	mp/°C	bp/°C	den/ g cm^{-3}	n_D	Solubility
5960	Hydroxyacetonitrile	Glyconitrile	C_2H_3NO	107-16-4	57.051		<-72	dec 183; 119^{24}		1.4117^{19}	vs H_2O, EtOH, eth; i bz, chl
5961	(Hydroxyacetyl)benzene		$C_8H_8O_2$	582-24-1	136.149	hex pl (al), pl (w or dil al)	90	125^{12}, 56^{1}	1.0963^{99}		s H_2O, EtOH, eth, chl; sl lig
5962	17-Hydroxyandrostan-3-one, (5α,17β)	Stanolone	$C_{19}H_{30}O_2$	521-18-6	290.440		181	sub 135			
5963	3-Hydroxyandrostan-17-one, (3α,5α)	Androsterone	$C_{19}H_{30}O_2$	53-41-8	290.440	lf or nd (al, ace)	185				sl H_2O, chl; s EtOH, eth, ace, bz
5964	3-Hydroxyandrostan-17-one, (3β,5α)	Epiandrosterone	$C_{19}H_{30}O_2$	481-29-8	290.440	cry (bz-peth, ace)	178				
5965	17-Hydroxyandrost-4-en-3-one, (17β)	Testosterone	$C_{19}H_{28}O_2$	58-22-0	288.424	nd (dil ace)	155				i H_2O; s EtOH, eth, ace
5966	1-Hydroxy-9,10-anthracenedione		$C_{14}H_8O_3$	129-43-1	224.212	red-oran nd (al)	193.8	sub			i H_2O; s EtOH, eth, bz; sl liq NH_3
5967	2-Hydroxy-9,10-anthracenedione		$C_{14}H_8O_3$	605-32-3	224.212	ye pl or nd (al or HOAc)	306	sub			i H_2O; s EtOH, eth, aq NH_3, KOH
5968	3-Hydroxybenzaldehyde	3-Formylphenol	$C_7H_6O_2$	100-83-4	122.122	nd (w)	108	240	1.1179^{130}		sl H_2O; s EtOH, eth, ace, bz; i lig
5969	4-Hydroxybenzaldehyde	4-Formylphenol	$C_7H_6O_2$	123-08-0	122.122	nd (w)	117		1.129^{130}	1.5705^{130}	sl H_2O, ace; vs EtOH, eth; s bz
5970	2-Hydroxybenzaldehyde, [(2-hydroxyphenyl)methylene] hydrazone		$C_{14}H_{12}N_2O_2$	959-36-4	240.257		214				i H_2O; s EtOH, chl; vs bz, alk
5971	2-Hydroxybenzamide	Salicylamide	$C_7H_7NO_2$	65-45-2	137.137		142	181.5^{14}	1.175^{140}		sl H_2O, eth, DMSO; s EtOH
5972	*N*-Hydroxybenzamide		$C_7H_7NO_2$	495-18-1	137.137	orth ta, lf (eth)	131 exp				s H_2O, EtOH; sl eth, bz
5973	α-Hydroxybenzeneacetic acid, (±)	*DL*-Mandelic acid	$C_8H_8O_3$	611-72-3	152.148	orth pl	119		1.2890^{20}		s H_2O, eth, EtOH, i-PrOH
5974	2-Hydroxybenzeneacetic acid		$C_8H_8O_3$	614-75-5	152.148		148	240			sl H_2O, chl; s eth
5975	3-Hydroxybenzeneacetic acid		$C_8H_8O_3$	621-37-4	152.148	nd (bz-lig)	132	190^{11}			vs H_2O, EtOH, eth; s bz; sl lig
5976	4-Hydroxybenzeneacetic acid		$C_8H_8O_3$	156-38-7	152.148	nd (w)	152	sub			sl H_2O; vs EtOH, eth
5977	α-Hydroxybenzeneacetonitrile	Mandelonitrile	C_8H_7NO	532-28-5	133.148	ye oily liq	-10		1.12		i H_2O; vs chl, eth, EtOH
5978	2-Hydroxybenzenecarbodithioic acid	Dithiosalicylic acid	$C_7H_6OS_2$	527-89-9	170.252	oran-ye nd	49				vs bz, eth, EtOH
5979	4-Hydroxy-1,3-benzenedicarboxylic acid	4-Hydroxyisophthalic acid	$C_8H_6O_5$	636-46-4	182.131	nd(w), lf (dil al)	310				i H_2O, chl; vs EtOH, eth; s HOAc
5980	5-Hydroxy-1,3-benzenedicarboxylic acid		$C_8H_6O_5$	618-83-7	182.131	nd(w+2) cr(aq-al)		sub			vs bz, eth, EtOH
5981	4-Hydroxy-1,3-benzenedisulfonic acid	Phenoldisulfonic acid	$C_6H_6O_7S_2$	96-77-5	254.238	nd (w)	>100 dec				vs H_2O, EtOH
5982	4-Hydroxybenzeneethanol		$C_8H_{10}O_2$	501-94-0	138.164		91.8	310.0			
5983	2-Hydroxybenzenemethanol	Salicyl alcohol	$C_7H_8O_2$	90-01-7	124.138	lf (bz), nd or pl (w, eth)	87	sub	1.1613^{25}		s H_2O, EtOH, eth, bz; vs chl
5984	3-Hydroxybenzenemethanol	3-Hydroxybenzyl alcohol	$C_7H_8O_2$	620-24-6	124.138	nd (bz), cry (CCl_4)	73	dec 300	1.161^{25}		vs H_2O, EtOH, eth; sl chl
5985	4-Hydroxybenzenemethanol	4-Hydroxybenzyl alcohol	$C_7H_8O_2$	623-05-2	124.138	pr or nd (w)	124.5	252			vs H_2O, EtOH, bz, chl; s eth; sl DMSO
5986	4-Hydroxybenzenepropanoic acid	*p*-Hydroxyhydrocinnamic acid	$C_9H_{10}O_3$	501-97-3	166.173		130.8	209^{14}			s H_2O, EtOH, eth, bz; i CS_2
5987	α-Hydroxybenzenepropanoic acid, (±)	(±)-3-Phenyllactic acid	$C_9H_{10}O_3$	828-01-3	166.173	cry (chl, bz), pr (w)	98	149^{15}			vs H_2O, ace, eth, EtOH
5988	3-Hydroxybenzenesulfonic acid	*m*-Phenolsulfonic acid	$C_6H_6O_4S$	585-38-6	174.175	nd (w+2)					
5989	4-Hydroxybenzenesulfonic acid	*p*-Phenolsulfonic acid	$C_6H_6O_4S$	98-67-9	174.175	nd					vs H_2O, EtOH
5990	2-Hydroxybenzoic acid	Salicylic acid	$C_7H_6O_3$	69-72-7	138.121	nd (w), mcl pr (al)	159.0	211^{20}	1.443^{20}	1.565	sl H_2O, bz, chl, ctc; vs EtOH, eth, ace
5991	3-Hydroxybenzoic acid		$C_7H_6O_3$	99-06-9	138.121	nd (w) pl, pr (al)	202.5		1.485^{25}		sl H_2O; s EtOH, eth, ace; i bz
5992	4-Hydroxybenzoic acid		$C_7H_6O_3$	99-96-7	138.121	pr or pl (w, al) cry (ace)	214.5		1.46^{25}		sl H_2O, bz; vs EtOH; s eth, ace
5993	2-Hydroxybenzoic acid, hydrazide		$C_7H_8N_2O_2$	936-02-7	152.151		148				vs bz, EtOH
5994	2-Hydroxybenzonitrile		C_7H_5NO	611-20-1	119.121		98	149^{14}	1.1052^{100}	1.5372^{100}	sl H_2O; vs EtOH, eth, bz, chl

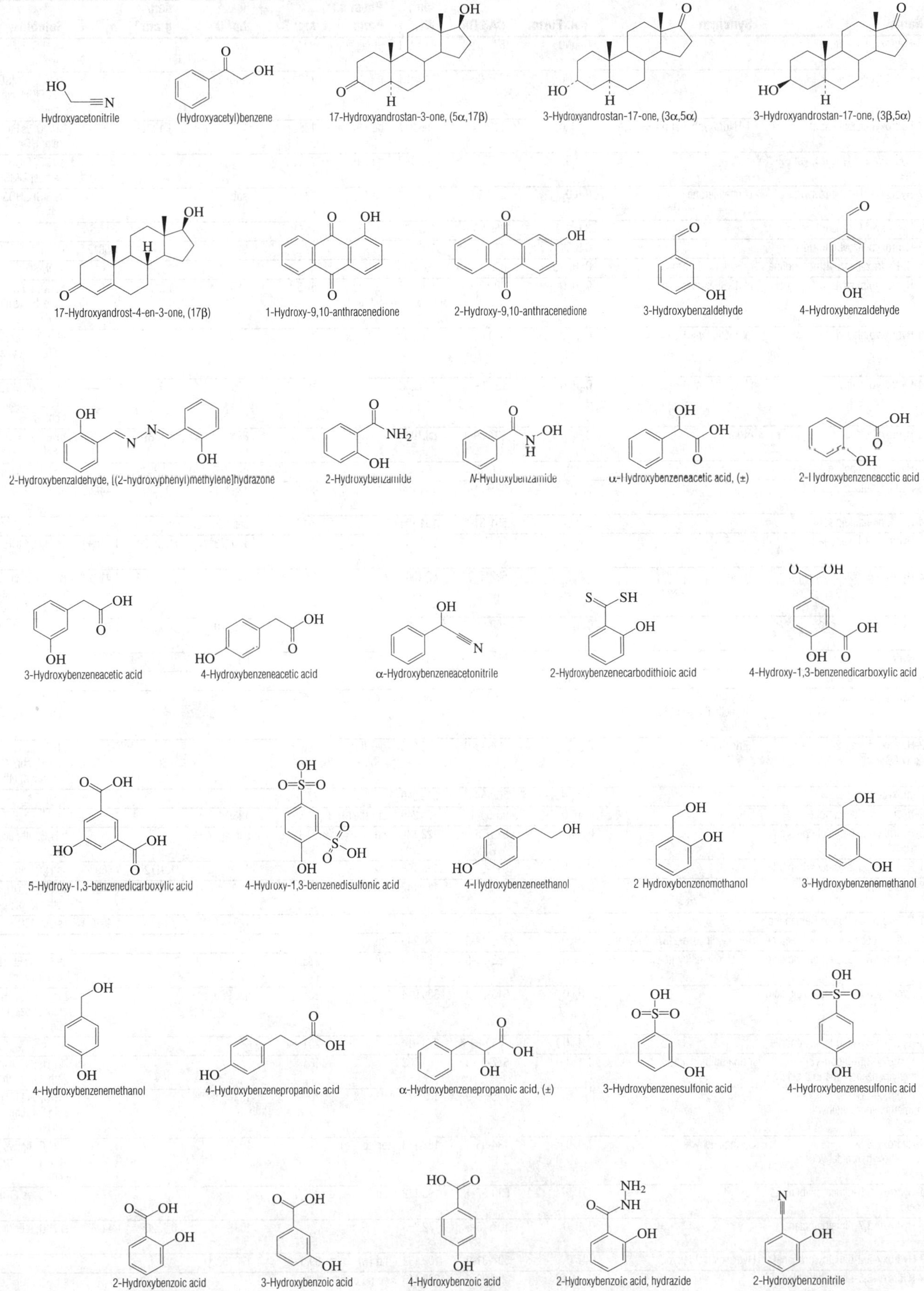

Hydroxyacetonitrile
(Hydroxyacetyl)benzene
17-Hydroxyandrostan-3-one, (5α,17β)
3-Hydroxyandrostan-17-one, (3α,5α)
3-Hydroxyandrostan-17-one, (3β,5α)
17-Hydroxyandrost-4-en-3-one, (17β)
1-Hydroxy-9,10-anthracenedione
2-Hydroxy-9,10-anthracenedione
3-Hydroxybenzaldehyde
4-Hydroxybenzaldehyde
2-Hydroxybenzaldehyde, [(2-hydroxyphenyl)methylene]hydrazone
2-Hydroxybenzamide
N-Hydroxybenzamide
α-Hydroxybenzeneacetic acid, (±)
2-Hydroxybenzeneacetic acid
3-Hydroxybenzeneacetic acid
4-Hydroxybenzeneacetic acid
α-Hydroxybenzeneacetonitrile
2-Hydroxybenzenecarbodithioic acid
4-Hydroxy-1,3-benzenedicarboxylic acid
5-Hydroxy-1,3-benzenedicarboxylic acid
4-Hydroxy-1,3-benzenedisulfonic acid
4-Hydroxybenzeneethanol
2-Hydroxybenzenemethanol
3-Hydroxybenzenemethanol
4-Hydroxybenzenemethanol
4-Hydroxybenzenepropanoic acid
α-Hydroxybenzenepropanoic acid, (±)
3-Hydroxybenzenesulfonic acid
4-Hydroxybenzenesulfonic acid
2-Hydroxybenzoic acid
3-Hydroxybenzoic acid
4-Hydroxybenzoic acid
2-Hydroxybenzoic acid, hydrazide
2-Hydroxybenzonitrile

No.	Name	Synonym	Mol. Form.	CAS RN	Mol. Wt.	Physical Form	mp/˚C	bp/˚C	den/ g cm-3	n_D	Solubility
5995	3-Hydroxybenzonitrile		C_7H_5NO	873-62-1	119.121	pr (al, eth) lf (w)	82.8				vs H_2O, EtOH, eth, bz, chl
5996	4-Hydroxybenzonitrile		C_7H_5NO	767-00-0	119.121	lf (w)	113	148^{1}			sl H_2O, DMSO; vs EtOH, eth, chl
5997	4-Hydroxybenzophenone	4-Hydroxyphenyl phenyl ketone	$C_{13}H_{10}O_2$	1137-42-4	198.217	nd (al), pr (dil al)	135		1.133^{172}		sl H_2O; vs EtOH, eth, HOAc
5998	4-Hydroxy-2*H*-1-benzopyran-2-one		$C_9H_6O_3$	1076-38-6	162.142	nd (w)	213.5				s H_2O, EtOH, eth; sl DMSO
5999	7-Hydroxy-2*H*-1-benzopyran-2-one	Umbelliferone	$C_9H_6O_3$	93-35-6	162.142	nd (w)	230.5	sub			vs EtOH, HOAc, chl
6000	1-Hydroxy-1*H*-benzotriazole		$C_6H_5N_3O$	2592-95-2	135.123		157.8				
6001	2-Hydroxybenzoyl chloride		$C_7H_5ClO_2$	1441-87-8	156.567		19	92^{15}	1.3112^{20}	1.5812^{20}	vs eth
6002	4-(2-Hydroxybenzoyl)morpholine	4-Salicyloylmorpholine	$C_{11}H_{13}NO_3$	3202-84-4	204.202						s DMSO
6003	2-Hydroxybiphenyl	[1,1'-Biphenyl]-2-ol	$C_{12}H_{10}O$	90-43-7	170.206		57.5	286	1.213^{25}		i H_2O; s EtOH, ace, bz; vs eth, py
6004	3-Hydroxybiphenyl	[1,1'-Biphenyl]-3-ol	$C_{12}H_{10}O$	580-51-8	170.206		78	>300			sl H_2O; vs EtOH, eth, bz, py; s chl
6005	4-Hydroxybiphenyl	[1,1'-Biphenyl]-4-ol	$C_{12}H_{10}O$	92-69-3	170.206		166	305			sl H_2O, DMSO; vs EtOH, eth, chl, py
6006	3-Hydroxybutanal	Aldol	$C_4H_8O_2$	107-89-1	88.106			83^{20}	1.103^{20}	1.4238^{20}	msc H_2O, EtOH; s eth; vs ace
6007	2-Hydroxybutanoic acid, (±)		$C_4H_8O_3$	600-15-7	104.105		44.2	dec 260; 140^{14}	1.125^{20}		s H_2O, EtOH, eth
6008	3-Hydroxybutanoic acid, (±)		$C_4H_8O_3$	625-71-8	104.105		49	130^{12}, $94^{0.1}$		1.4424^{20}	vs H_2O, EtOH, eth; i bz
6009	4-Hydroxybutanoic acid		$C_4H_8O_3$	591-81-1	104.105		<-17	dec 180			
6010	1-Hydroxy-2-butanone		$C_4H_8O_2$	5077-67-8	88.106			160; 78^{60}	1.0272^{20}	1.4189^{20}	vs H_2O, EtOH, eth
6011	3-Hydroxy-2-butanone, (±)	Acetoin	$C_4H_8O_2$	52217-02-4	88.106		15	148	1.0044^{20}	1.4171^{20}	msc H_2O; sl EtOH, eth; s ace, chl; i lig
6012	4-Hydroxy-2-butanone		$C_4H_8O_2$	590-90-9	88.106			182; 90^{11}	1.0233^{20}	1.4585^{14}	msc H_2O, EtOH, eth; vs ace
6013	2-Hydroxy-3-butenenitrile		C_4H_5NO	5809-59-6	83.089	liq		94^{17}			
6014	4-Hydroxybutyramide		$C_4H_9NO_2$	927-60-6	103.120		52				
6015	3-Hydroxycamphor	3-Hydroxy-1,7,7-trimethylbicyclo[2.2.1]heptan-2-one	$C_{10}H_{16}O_2$	10373-81-6	168.233	nd (bz-peth)	205.5				vs eth, EtOH, chl
6016	3-Hydroxycholan-24-oic acid, (3α,5β)	Lithocholic acid	$C_{24}H_{40}O_3$	434-13-9	376.573	hex lf (al) pr (dil al)	186				i H_2O, lig; s EtOH, chl, HOAc; sl eth
6017	Hydroxycodeinone		$C_{18}H_{19}NO_4$	508-54-3	313.349		275 dec				
6018	2-Hydroxycyclodecanone	Sebacoin	$C_{10}H_{18}O_2$	96-00-4	170.249	cry (peth)	38.5	136^{14}			
6019	2-Hydroxy-2,4,6-cycloheptatrien-1-one		$C_7H_6O_2$	533-75-5	122.122	nd	50.8	sub 40			s H_2O, eth, ace
6020	1-Hydroxycyclohexanecarbonitrile		$C_7H_{11}NO$	931-97-5	125.168		35	132^{20}	1.0172^{20}	1.4693^{20}	vs H_2O, eth
6021	2-Hydroxycyclohexanone		$C_6H_{10}O_2$	533-60-8	114.142	nd (al)				1.4785^{21}	vs H_2O, EtOH; i eth, bz, peth
6022	1-(1-Hydroxycyclohexyl)ethanone		$C_8H_{14}O_2$	1123-27-9	142.196			125.5; 91^{11}	1.0248^{25}	1.4670^{25}	vs eth, EtOH
6023	4-Hydroxydecanoic acid γ-lactone	5-Hexyldihydro-2(3*H*)-furanone	$C_{10}H_{18}O_2$	706-14-9	170.249	liq		281			
6024	2-Hydroxy-3,5-diiodobenzoic acid	3,5-Diiodosalicylic acid	$C_7H_4I_2O_3$	133-91-5	389.914	nd (al)	235.5				sl H_2O; vs EtOH, eth; i bz, chl
6025	4-Hydroxy-3,5-diiodobenzoic acid		$C_7H_4I_2O_3$	618-76-8	389.914		237	dec 260			i H_2O; vs EtOH, eth; sl bz, chl, lig
6026	4-Hydroxy-3,5-diiodobenzonitrile		$C_7H_3I_2NO$	1689-83-4	370.914		201 dec				
6027	4-Hydroxy-3,5-diiodo-α-phenylbenzenepropanoic acid	Iodoalphionic acid	$C_{15}H_{12}I_2O_3$	577-91-3	494.063		164				i H_2O; s EtOH, eth; sl bz, chl
6028	2-Hydroxy-4,6-dimethoxybenzaldehyde		$C_9H_{10}O_4$	708-76-9	182.173		70	193^{25}, 165^{10}			i H_2O; vs EtOH, eth, bz, chl, HOAc
6029	4-Hydroxy-3,5-dimethoxybenzaldehyde	Syringaldehyde	$C_9H_{10}O_4$	134-96-3	182.173	br nd (lig)	113	192^{14}			sl H_2O, lig; vs EtOH, eth, bz, chl
6030	4-Hydroxy-3,5-dimethoxybenzoic acid		$C_9H_{10}O_5$	530-57-4	198.172	nd (w)	204.5				sl H_2O; vs EtOH
6031	7-Hydroxy-3,7-dimethyloctanal		$C_{10}H_{20}O_2$	107-75-5	172.265			103^{3}	0.9220^{20}	1.4494^{20}	sl H_2O; s EtOH, ace
6032	3-Hydroxy-2,2-dimethylpropanal	Hydroxypivaldehyde	$C_5H_{10}O_2$	597-31-9	102.132	nd (w)	89.5	173; 68^{14}			
6033	2-Hydroxy-3,5-dinitrobenzoic acid		$C_7H_4N_2O_7$	609-99-4	228.116	ye nd or pl (+1w)	182				s H_2O, EtOH, eth, bz

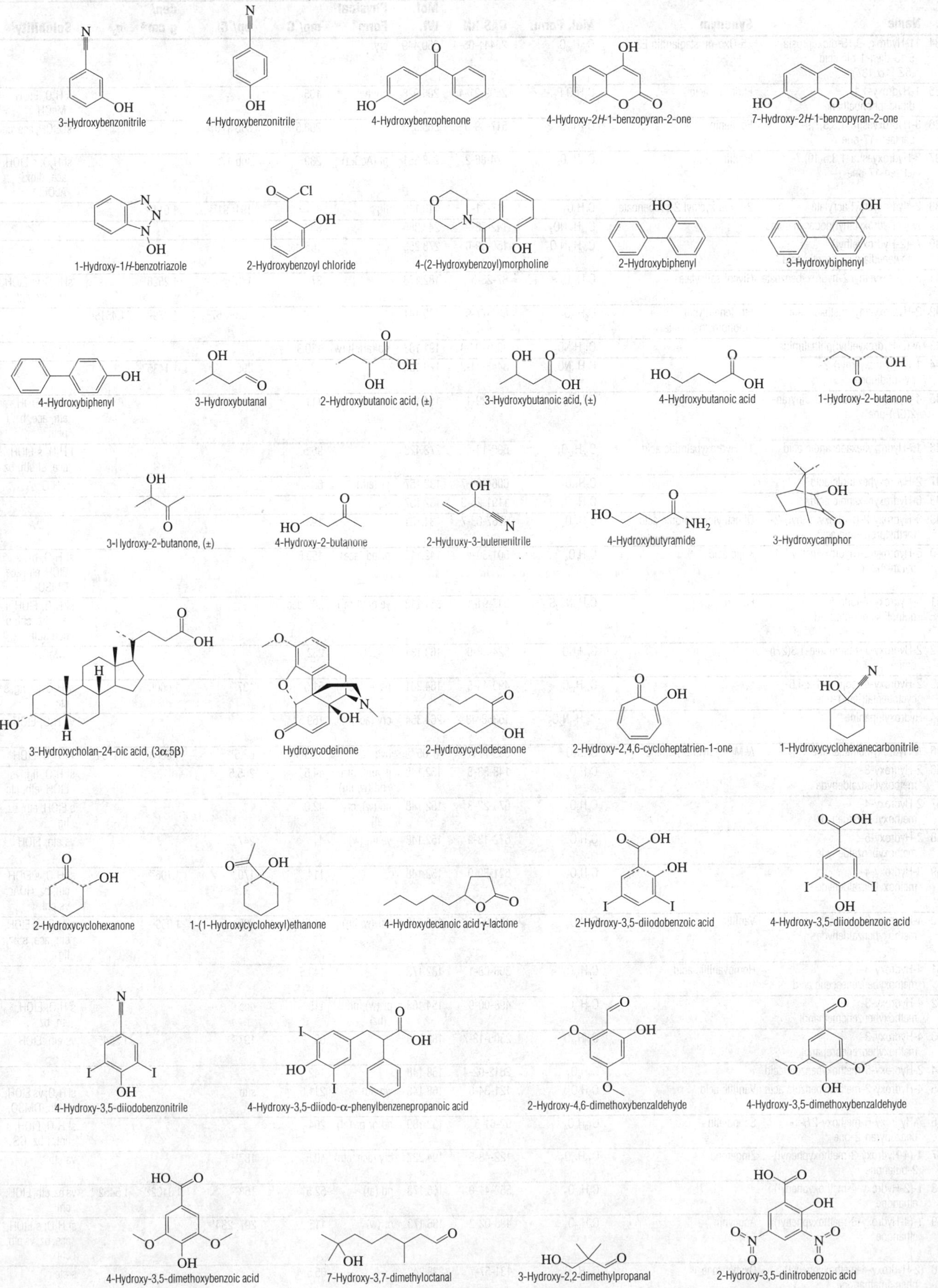
3-Hydroxybenzonitrile
4-Hydroxybenzonitrile
4-Hydroxybenzophenone
4-Hydroxy-2H-1-benzopyran-2-one
7-Hydroxy-2H-1-benzopyran-2-one
1-Hydroxy-1H-benzotriazole
2-Hydroxybenzoyl chloride
4-(2-Hydroxybenzoyl)morpholine
2-Hydroxybiphenyl
3-Hydroxybiphenyl
4-Hydroxybiphenyl
3-Hydroxybutanal
2-Hydroxybutanoic acid, (±)
3-Hydroxybutanoic acid, (±)
4-Hydroxybutanoic acid
1-Hydroxy-2-butanone
3-Hydroxy-2-butanone, (±)
4-Hydroxy-2-butanone
2-Hydroxy-3-butenenitrile
4-Hydroxybutyramide
3-Hydroxycamphor
3-Hydroxycholan-24-oic acid, (3α,5β)
Hydroxycodeinone
2-Hydroxycyclodecanone
2-Hydroxy-2,4,6-cycloheptatrien-1-one
1-Hydroxycyclohexanecarbonitrile
2-Hydroxycyclohexanone
1-(1-Hydroxycyclohexyl)ethanone
4-Hydroxydecanoic acid γ-lactone
2-Hydroxy-3,5-diiodobenzoic acid
4-Hydroxy-3,5-diiodobenzoic acid
4-Hydroxy-3,5-diiodobenzonitrile
4-Hydroxy-3,5-diiodo-α-phenylbenzenepropanoic acid
2-Hydroxy-4,6-dimethoxybenzaldehyde
4-Hydroxy-3,5-dimethoxybenzaldehyde
4-Hydroxy-3,5-dimethoxybenzoic acid
7-Hydroxy-3,7-dimethyloctanal
3-Hydroxy-2,2-dimethylpropanal
2-Hydroxy-3,5-dinitrobenzoic acid

No.	Name	Synonym	Mol. Form.	CAS RN	Mol. Wt.	Physical Form	mp/°C	bp/°C	den/ g cm^{-3}	n_D	Solubility
6034	11-Hydroxy-9,15-dioxoprosta-5,13-dien-1-oic acid, (5*Z*,11α,13*E*)	15-Oxo-prostaglandin E2	$C_{20}H_{30}O_5$	26441-05-4	350.449	cry					
6035	1-Hydroxy-1,1-diphosphonoethane	Etidronic acid	$C_2H_8O_7P_2$	2809-21-4	206.028	cry (w)	105				s H_2O, EtOH, MeOH
6036	3-Hydroxyestra-1,3,5,7,9-pentaen-17-one	Equilenin	$C_{18}H_{18}O_2$	517-09-9	266.335		258.5	sub 170			sl EtOH, ace, chl
6037	3-Hydroxyestra-1,3,5(10),7-tetraen-17-one	Equilin	$C_{18}H_{20}O_2$	474-86-2	268.351	pl (AcOEt)	239	sub 170			sl H_2O; s EtOH, ace, diox, AcOEt
6038	2-Hydroxyethyl acrylate	2-Hydroxyethyl 2-propenoate	$C_5H_8O_3$	818-61-1	116.116	liq		191; 91^{12}	1.011^{23}		
6039	*N*-(2-Hydroxyethyl)dodecanamide		$C_{14}H_{29}NO_2$	142-78-9	243.386		88.5				
6040	*N*-(2-Hydroxyethyl) ethylenediaminetriacetic acid		$C_{10}H_{18}N_2O_7$	150-39-0	278.259	cry	165 dec				
6041	2-Hydroxyethyl 2-hydroxybenzoate	Glycol salicylate	$C_9H_{10}O_4$	87-28-5	182.173		37	173^{15}	1.2526^{15}		sl H_2O; vs EtOH, eth, bz, chl
6042	2-Hydroxyethyl methacrylate	Ethylene glycol monomethacrylate	$C_6H_{10}O_3$	868-77-9	130.141			103^{13}, 67^{3}	1.079^{20}	1.4515^{20}	
6043	*N*-(2-Hydroxyethyl)phthalimide		$C_{10}H_9NO_3$	3891-07-4	191.183	nd (al), lf (w)	130.3				sl H_2O
6044	1-(2-Hydroxyethyl)-2-pyrrolidinone		$C_6H_{11}NO_2$	3445-11-2	129.157		20	295	1.1435^{20}		
6045	4-Hydroxy-4*H*-furo[3,2-c]pyran-2(6*H*)-one	Patulin	$C_7H_6O_4$	149-29-1	154.121	pl or pr (eth, chl)	111				s H_2O, EtOH, eth, ace, bz; i peth
6046	16-Hydroxyhexadecanoic acid	16-Hydroxypalmitic acid	$C_{16}H_{32}O_3$	506-13-8	272.423		96.5				i H_2O; s EtOH, ace; sl eth, bz
6047	2-Hydroxyhexanoic acid		$C_6H_{12}O_3$	6064-63-7	132.157	pr (eth)	60				vs H_2O
6048	6-Hydroxyhexanoic acid		$C_6H_{12}O_3$	1191-25-9	132.157	liq					
6049	3-Hydroxy-2-(hydroxymethyl)-2-methylpropanoic acid	Dimethylolpropionic acid	$C_5H_{10}O_4$	4767-03-7	134.131		190				
6050	5-Hydroxy-2-(hydroxymethyl)-4*H*-pyran-4-one	Kojic acid	$C_6H_6O_4$	501-30-4	142.110	pr nd (ace)	153.5				sl H_2O, bz; s EtOH, eth, ace, DMSO
6051	8-Hydroxy-7-iodo-5-quinolinesulfonic acid	Ferron	$C_9H_6INO_4S$	547-91-1	351.118	ye pr, lf (al)	260 dec				sl H_2O, EtOH; i eth, bz, chl; s con sulf
6052	2-Hydroxy-1*H*-isoindole-1,3(2*H*)-dione		$C_8H_5NO_3$	524-38-9	163.131		232				s DMSO
6053	2-Hydroxy-4-isopropyl-2,4,6-cycloheptatrien-1-one		$C_{10}H_{12}O_2$	499-44-5	164.201	pa ye (peth)	51.5	137^{10}	1.0606^{65}		sl H_2O, bz, lig; s ctc
6054	Hydroxylupanine		$C_{15}H_{24}N_2O_2$	15358-48-2	264.364	cry (ace)	169.5				vs H_2O, EtOH, chl
6055	*N*-Hydroxymethanamine	*N*-Methylhydroxylamine	CH_5NO	593-77-1	47.057	hyg nd	87.5	62.5^{15}	1.0003^{20}	1.4164^{20}	vs H_2O, EtOH
6056	2-Hydroxy-3-methoxybenzaldehyde		$C_8H_8O_3$	148-53-8	152.148	lt ye lf, grn nd (w, lig)	44.5	265.5			sl H_2O, lig; vs EtOH, eth, ctc
6057	2-Hydroxy-4-methoxybenzaldehyde		$C_8H_8O_3$	673-22-3	152.148	nd (w), cry (al)	42.0				s EtOH, eth, bz, lig
6058	2-Hydroxy-5-methoxybenzaldehyde		$C_8H_8O_3$	672-13-9	152.148	ye liq (w)	4	247.5			vs eth, EtOH
6059	3-Hydroxy-4-methoxybenzaldehyde		$C_8H_8O_3$	621-59-0	152.148		114	179^{15}	1.196^{25}		sl H_2O; s EtOH, eth, bz, HOAc; vs chl
6060	4-Hydroxy-3-methoxybenzaldehyde	Vanillin	$C_8H_8O_3$	121-33-5	152.148	tetr (w, lig)	81.5	285	1.056^{25}		sl H_2O; vs EtOH, eth, ace; s bz, lig
6061	4-Hydroxy-3-methoxybenzeneacetic acid	Homovanillic acid	$C_9H_{10}O_4$	306-08-1	182.173		143.5				
6062	4-Hydroxy-3-methoxybenzenemethanol		$C_8H_{10}O_3$	498-00-0	154.163	pr (w), nd (bz)	115	dec			s H_2O, EtOH, eth, bz
6063	4-Hydroxy-3-methoxybenzenepropanol		$C_{10}H_{14}O_3$	2305-13-7	182.216		65	197^{15}		1.5545^{25}	vs eth, EtOH
6064	2-Hydroxy-5-methoxybenzoic acid		$C_8H_8O_4$	2612-02-4	168.148		142				
6065	4-Hydroxy-3-methoxybenzoic acid	Vanillic acid	$C_8H_8O_4$	121-34-6	168.148	wh nd	211.5	sub			sl H_2O; vs EtOH; s eth, DMSO
6066	7-Hydroxy-6-methoxy-2*H*-1-benzopyran-2-one	Scopoletin	$C_{10}H_8O_4$	92-61-5	192.169	nd or pr (al)	204				sl H_2O, EtOH; s chl; i bz, CS_2
6067	4-(4-Hydroxy-3-methoxyphenyl)-2-butanone	Zingerone	$C_{11}H_{14}O_3$	122-48-5	194.227	cry (ace, eth)	40.5	187^{14}			vs eth
6068	1-(2-Hydroxy-4-methoxyphenyl) ethanone		$C_9H_{10}O_3$	552-41-0	166.173	nd (al)	52.5	158^{20}	1.3102^{81}	1.5452^{81}	vs bz, eth, EtOH, chl
6069	1-(4-Hydroxy-3-methoxyphenyl) ethanone	Apocynin	$C_9H_{10}O_3$	498-02-2	166.173	pr (w)	115	297; 234^{15}			sl H_2O; s EtOH, ace, bz; vs eth, chl
6070	(2-Hydroxy-4-methoxyphenyl) phenylmethanone	Oxybenzone	$C_{14}H_{12}O_3$	131-57-7	228.243		65.5				s ctc

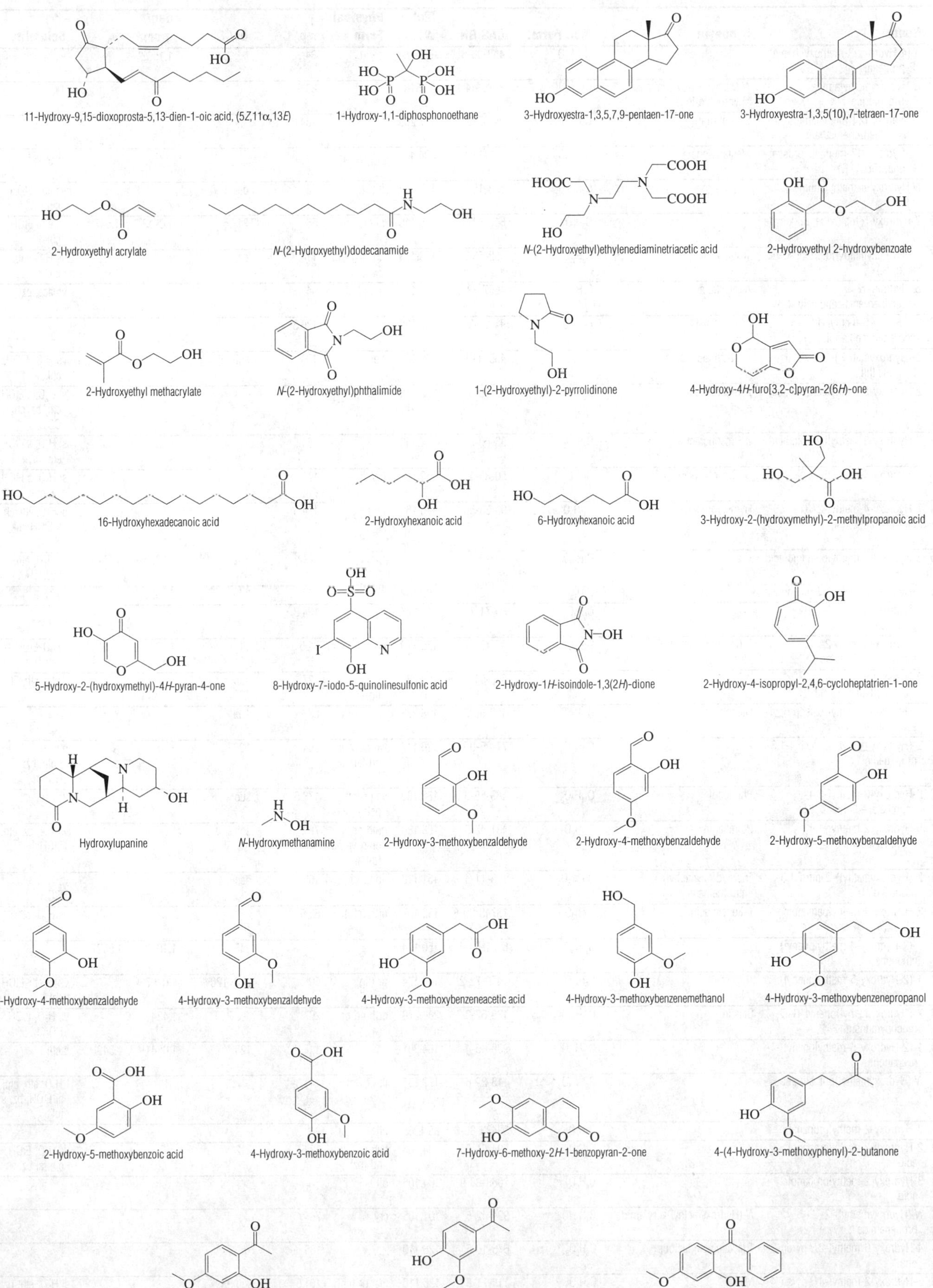
11-Hydroxy-9,15-dioxoprosta-5,13-dien-1-oic acid, (5Z,11α,13E)
1-Hydroxy-1,1-diphosphonoethane
3-Hydroxyestra-1,3,5,7,9-pentaen-17-one
3-Hydroxyestra-1,3,5(10),7-tetraen-17-one
2-Hydroxyethyl acrylate
N-(2-Hydroxyethyl)dodecanamide
N-(2-Hydroxyethyl)ethylenediaminetriacetic acid
2-Hydroxyethyl 2-hydroxybenzoate
2-Hydroxyethyl methacrylate
N-(2-Hydroxyethyl)phthalimide
1-(2-Hydroxyethyl)-2-pyrrolidinone
4-Hydroxy-4H-furo[3,2-c]pyran-2(6H)-one
16-Hydroxyhexadecanoic acid
2-Hydroxyhexanoic acid
6-Hydroxyhexanoic acid
3-Hydroxy-2-(hydroxymethyl)-2-methylpropanoic acid
5-Hydroxy-2-(hydroxymethyl)-4H-pyran-4-one
8-Hydroxy-7-iodo-5-quinolinesulfonic acid
2-Hydroxy-1H-isoindole-1,3(2H)-dione
2-Hydroxy-4-isopropyl-2,4,6-cycloheptatrien-1-one
Hydroxylupanine
N-Hydroxymethanamine
2-Hydroxy-3-methoxybenzaldehyde
2-Hydroxy-4-methoxybenzaldehyde
2-Hydroxy-5-methoxybenzaldehyde
3-Hydroxy-4-methoxybenzaldehyde
4-Hydroxy-3-methoxybenzaldehyde
4-Hydroxy-3-methoxybenzeneacetic acid
4-Hydroxy-3-methoxybenzenemethanol
4-Hydroxy-3-methoxybenzenepropanol
2-Hydroxy-5-methoxybenzoic acid
4-Hydroxy-3-methoxybenzoic acid
7-Hydroxy-6-methoxy-2H-1-benzopyran-2-one
4-(4-Hydroxy-3-methoxyphenyl)-2-butanone
1-(2-Hydroxy-4-methoxyphenyl)ethanone
1-(4-Hydroxy-3-methoxyphenyl)ethanone
(2-Hydroxy-4-methoxyphenyl)phenylmethanone

No.	Name	Synonym	Mol. Form.	CAS RN	Mol. Wt.	Physical Form	mp/°C	bp/°C	den/ g cm-3	n_D	Solubility
6071	3-(4-Hydroxy-3-methoxyphenyl)-2-propenal		$C_{10}H_{10}O_3$	458-36-6	178.184	cry (bz)	84		1.1562^{102}		vs bz, eth, EtOH
6072	*N*-Hydroxymethylamine hydrochloride	*N*-Methylhydroxylamine hydrochloride	CH_6ClNO	4229-44-1	83.518		83.5				
6073	4-Hydroxy-α-[(methylamino) methyl]benzenemethanol	Synephrine	$C_9H_{13}NO_2$	94-07-5	167.205		184.5				
6074	17-Hydroxy-17-methylandrostan-3-one, (5α,17β)	Mestanolone	$C_{20}H_{32}O_2$	521-11-9	304.467		192.5				sl AcOEt
6075	*N*-Hydroxy-4-methylaniline		C_7H_9NO	623-10-9	123.152	lf (bz)	96	dec 117			vs eth, EtOH, chl
6076	2-Hydroxy-5-methylbenzaldehyde		$C_8H_8O_2$	613-84-3	136.149	pl (aq, al)	56	217.5	1.0913^{59}	1.547^{59}	vs eth, EtOH, chl
6077	α-(Hydroxymethyl)benzeneacetic acid, (±)	Tropic acid	$C_9H_{10}O_3$	552-63-6	166.173	nd, pl (al, bz, w)	118	dec			vs H_2O, eth, EtOH
6078	α-Hydroxy-α-methylbenzeneacetic acid, (±)	Atrolactic acid	$C_9H_{10}O_3$	4607-38-9	166.173	nd, pl (lig)	94				vs ace, bz
6079	2-Hydroxy-5-methyl-1,3-benzenedimethanol		$C_9H_{12}O_3$	91-04-3	168.189		130.5				
6080	2-(Hydroxymethyl)-1,4-benzenediol	Gentisyl alcohol	$C_7H_8O_3$	495-08-9	140.137	nd (chl)	100	sub 75			vs H_2O, EtOH, chl
6081	2-Hydroxy-5-methylbenzoic acid	*p*-Cresotic acid	$C_8H_8O_3$	89-56-5	152.148		151				sl H_2O; s EtOH, eth, bz, chl; i CS_2
6082	2-Hydroxy-3-methylbenzoic acid	*o*-Cresotic acid	$C_8H_8O_3$	83-40-9	152.148		165.5				sl H_2O; s EtOH, eth, bz, chl
6083	2-Hydroxy-4-methylbenzoic acid	*m*-Cresotic acid	$C_8H_8O_3$	50-85-1	152.148	cry, lf	177				sl H_2O; s EtOH, bz, chl; vs eth
6084	7-Hydroxy-4-methyl-2*H*-1-benzopyran-2-one	Hymecromone	$C_{10}H_8O_3$	90-33-5	176.169	nd (al)	194.5				sl H_2O, eth, chl; s EtOH, alk, HOAc
6085	3-Hydroxy-3-methylbutanoic acid		$C_5H_{10}O_3$	625-08-1	118.131		<-32	162^{12}	0.9384^{20}	1.5081^{20}	vs H_2O, eth, EtOH
6086	3-Hydroxy-3-methyl-2-butanone		$C_5H_{10}O_2$	115-22-0	102.132			140	0.9526^{20}		s chl
6087	2-Hydroxy-3-methyl-2-cyclopenten-1-one		$C_6H_8O_2$	80-71-7	112.127		104.8				
6088	5-(Hydroxymethyl)-2-furancarboxaldehyde	5-(Hydroxymethyl)-2-furaldehyde	$C_6H_6O_3$	67-47-0	126.110	nd (eth-peth)	31.5	115^{1}	1.2062^{25}	1.5627^{18}	s H_2O, EtOH, bz, chl; sl eth, ctc
6089	2-Hydroxy-6-methyl-3-isopropylbenzoic acid	*o*-Thymotic acid	$C_{11}H_{14}O_3$	548-51-6	194.227	nd (w, bz, lig)	127	sub			vs bz, eth, EtOH
6090	2-Hydroxy-3-methyl-6-isopropyl-2-cyclohexen-1-one	Diosphenol	$C_{10}H_{16}O_2$	490-03-9	168.233		83	109^{10}			
6091	2-(Hydroxymethyl)-2-methyl-1,3-propanediol		$C_5H_{12}O_3$	77-85-0	120.147	wh pow or nd (al)	204	136^{15}			msc H_2O, EtOH; i eth, bz; vs HOAc
6092	2-Hydroxy-3-methyl-1,4-naphthalenedione	Phthiocol	$C_{11}H_8O_3$	483-55-6	188.180	ye pr (eth-peth)	173.5	sub			vs ace, eth
6093	5-Hydroxy-2-methyl-1,4-naphthalenedione	Plumbagin	$C_{11}H_8O_3$	481-42-5	188.180	gold pr or oran-ye nd (dil al)	78.5	sub			vs ace, bz, eth, EtOH
6094	2-(Hydroxymethyl)-2-nitro-1,3-propanediol	Tris(hydroxymethyl) nitromethane	$C_4H_9NO_5$	126-11-4	151.118	nd or pr	165	dec			vs H_2O, eth, EtOH
6095	2-Hydroxy-4-methylpentanoic acid, (*S*)	*L*-Leucic acid	$C_6H_{12}O_3$	13748-90-8	132.157	orth (eth)	81.5				vs H_2O, eth, EtOH
6096	1-(2-Hydroxy-4-methylphenyl) ethanone		$C_9H_{10}O_2$	6921-64-8	150.174		21	245	1.1012^{10}	1.5527^{13}	
6097	1-(2-Hydroxy-5-methylphenyl) ethanone		$C_9H_{10}O_2$	1450-72-2	150.174	pr (lig)	50	210; 120^{20}	1.0797^{53}		vs bz, eth, EtOH, chl
6098	2-(Hydroxymethyl)phenyl-β-*D*-glucopyranoside	Salicin	$C_{13}H_{18}O_7$	138-52-3	286.278	orth nd or lf (w)	207	dec 240	1.434^{20}		vs H_2O, EtOH, HOAc
6099	1-(2-Hydroxy-5-methylphenyl)-1-propanone		$C_{10}H_{12}O_2$	938-45-4	164.201		1.0	$129^{16.5}$	1.0841^{14}	1.549^{13}	s chl
6100	*N*-(Hydroxymethyl)phthalimide		$C_9H_7NO_3$	118-29-6	177.157	lf, pr (to)	141.5				i H_2O, eth, ctc; sl EtOH, bz; s tol
6101	3-Hydroxy-2-methylpropanal		$C_4H_8O_2$	38433-80-6	88.106	oil					
6102	2-Hydroxy-2-methylpropanoic acid		$C_4H_8O_3$	594-61-6	104.105	hyg pr (eth) nd (bz)	82.5	212			vs H_2O, EtOH, eth; sl bz
6103	3-Hydroxy-2-methylpropanoic acid		$C_4H_8O_3$	2068-83-9	104.105	oil					
6104	*N*-(Hydroxymethyl)-2-propenamide	*N*-(Hydroxymethyl)acrylamide	$C_4H_7NO_2$	924-42-5	101.105	cry	76				
6105	4-Hydroxy-6-methyl-2*H*-pyran-2-one	Triacetic acid lactone	$C_6H_6O_3$	675-10-5	126.110		189 dec				
6106	3-Hydroxy-2-methyl-4*H*-pyran-4-one	Maltol	$C_6H_6O_3$	118-71-8	126.110	mcl pr (chl)	161.5	sub 93			sl H_2O, eth, bz; vs chl; s alk; peth

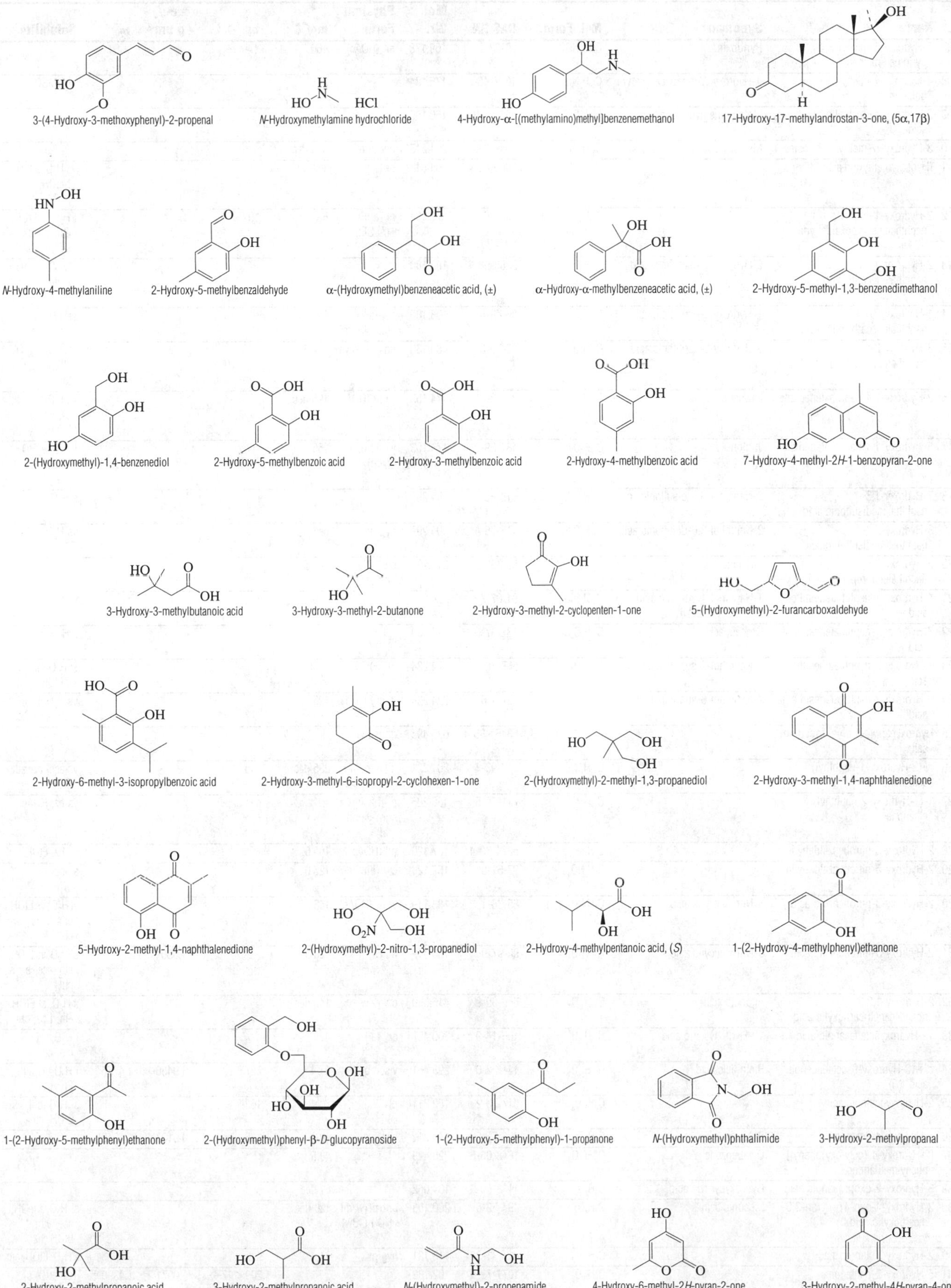

3-(4-Hydroxy-3-methoxyphenyl)-2-propenal
N-Hydroxymethylamine hydrochloride
4-Hydroxy-α-[(methylamino)methyl]benzenemethanol
17-Hydroxy-17-methylandrostan-3-one, (5α,17β)
N-Hydroxy-4-methylaniline
2-Hydroxy-5-methylbenzaldehyde
α-(Hydroxymethyl)benzeneacetic acid, (±)
α-Hydroxy-α-methylbenzeneacetic acid, (±)
2-Hydroxy-5-methyl-1,3-benzenedimethanol
2-(Hydroxymethyl)-1,4-benzenediol
2-Hydroxy-5-methylbenzoic acid
2-Hydroxy-3-methylbenzoic acid
2-Hydroxy-4-methylbenzoic acid
7-Hydroxy-4-methyl-2H-1-benzopyran-2-one
3-Hydroxy-3-methylbutanoic acid
3-Hydroxy-3-methyl-2-butanone
2-Hydroxy-3-methyl-2-cyclopenten-1-one
5-(Hydroxymethyl)-2-furancarboxaldehyde
2-Hydroxy-6-methyl-3-isopropylbenzoic acid
2-Hydroxy-3-methyl-6-isopropyl-2-cyclohexen-1-one
2-(Hydroxymethyl)-2-methyl-1,3-propanediol
2-Hydroxy-3-methyl-1,4-naphthalenedione
5-Hydroxy-2-methyl-1,4-naphthalenedione
2-(Hydroxymethyl)-2-nitro-1,3-propanediol
2-Hydroxy-4-methylpentanoic acid, (S)
1-(2-Hydroxy-4-methylphenyl)ethanone
1-(2-Hydroxy-5-methylphenyl)ethanone
2-(Hydroxymethyl)phenyl-β-D-glucopyranoside
1-(2-Hydroxy-5-methylphenyl)-1-propanone
N-(Hydroxymethyl)phthalimide
3-Hydroxy-2-methylpropanal
2-Hydroxy-2-methylpropanoic acid
3-Hydroxy-2-methylpropanoic acid
N-(Hydroxymethyl)-2-propenamide
4-Hydroxy-6-methyl-2H-pyran-2-one
3-Hydroxy-2-methyl-4H-pyran-4-one

No.	Name	Synonym	Mol. Form.	CAS RN	Mol. Wt.	Physical Form	mp/°C	bp/°C	den/ g cm^{-3}	n_D	Solubility
6107	5-Hydroxy-6-methyl-3,4-pyridinedimethanol	Pyridoxin	$C_8H_{11}NO_3$	65-23-6	169.178	nd (HOAc)	160	$140^{0.0001}$			
6108	4-Hydroxy-1-methyl-2-quinolinone	4-Hydroxy-*N*-methylcarbostyril	$C_{10}H_9NO_2$	1677-46-9	175.184		265				sl DMSO
6109	2-Hydroxy-4-(methylthio)butanoic acid	Methionine hydroxy analog	$C_5H_{10}O_3S$	583-91-5	150.196	oil					
6110	3-Hydroxy-α-methyl-*L*-tyrosine	Methyldopa	$C_{10}H_{13}NO_4$	555-30-6	211.215	cry (MeOH)	300 dec				
6111	(Hydroxymethyl)urea		$C_2H_6N_2O_2$	1000-82-4	90.081	pr (al)	111				vs H_2O; s EtOH, MeOH, HOAc; i eth
6112	2-Hydroxy-1-naphthalenecarboxaldehyde		$C_{11}H_8O_2$	708-06-5	172.181	pr (al), nd (AcOEt)	83	192^{27}			i H_2O; s EtOH, eth, aq alk, sulf, peth
6113	2-Hydroxy-1-naphthalenecarboxylic acid	2-Hydroxy-1-naphthoic acid	$C_{11}H_8O_3$	2283-08-1	188.180		157.3				sl H_2O; vs EtOH; s eth, ace, bz, lig, chl
6114	1-Hydroxy-2-naphthalenecarboxylic acid	1-Hydroxy-2-naphthoic acid	$C_{11}H_8O_3$	86-48-6	188.180	cry (al) nd (al, eth, bz)	195				sl H_2O; vs EtOH, eth; s bz
6115	3-Hydroxy-2-naphthalenecarboxylic acid	3-Hydroxy-2-naphthoic acid	$C_{11}H_8O_3$	92-70-6	188.180	nd (dil al) ye lf (dil al)	222.5				sl H_2O; vs EtOH, eth; s bz, chl, tol
6116	2-Hydroxy-1,4-naphthalenedione	Lawsone	$C_{10}H_6O_3$	83-72-7	174.153	ye pr (HOAc)	195 dec				vs EtOH; i eth, bz, chl; s HOAc
6117	5-Hydroxy-1,4-naphthalenedione	Juglone	$C_{10}H_6O_3$	481-39-0	174.153	ye nd (bz) peth)	155	sub			i H_2O; s EtOH, eth, bz; vs chl; sl lig
6118	7-Hydroxy-1,3-naphthalenedisulfonic acid	2-Naphthol-6,8-disulfonic acid	$C_{10}H_8O_7S_2$	118-32-1	304.297						s H_2O
6119	3-Hydroxy-2,7-naphthalenedisulfonic acid	2-Naphthol-3,6-disulfonic acid	$C_{10}H_8O_7S_2$	148-75-4	304.297	hyg nd	dec				vs H_2O, EtOH
6120	6-Hydroxy-2-naphthalenepropanoic acid	Allenolic acid	$C_{13}H_{12}O_3$	553-39-9	216.232	cry (dil MeOH)	180.5				vs py, EtOH, MeOH
6121	4-Hydroxy-1-naphthalenesulfonic acid	1-Naphthol-4-sulfonic acid	$C_{10}H_8O_4S$	84-87-7	224.234	tab or pl (w)	170 dec				vs H_2O; i eth
6122	7-Hydroxy-1-naphthalenesulfonic acid	Croceic acid	$C_{10}H_8O_4S$	132-57-0	224.234						s H_2O
6123	1-Hydroxy-2-naphthalenesulfonic acid	1-Naphthol-2-sulfonic acid	$C_{10}H_8O_4S$	567-18-0	224.234	pl (w)	>250				sl H_2O, dil HCl; s EtOH; i eth
6124	6-Hydroxy-2-naphthalenesulfonic acid	2-Naphthol-6-sulfonic acid	$C_{10}H_8O_4S$	93-01-6	224.234	lf, cry (w+1)	125				vs H_2O, EtOH; i eth; s HOAc
6125	Hydroxynaphthol blue, trisodium salt		$C_{20}H_{11}N_2Na_3O_{11}S_3$	63451-35-4	623.495	dk red cry					
6126	*N*-(2-Hydroxy-1-naphthyl) acetamide		$C_{12}H_{11}NO_2$	117-93-1	201.221	lf (w, dil al)	235 dec	sub			vs ace, bz, eth, EtOH
6127	1-(1-Hydroxy-2-naphthyl) ethanone		$C_{12}H_{10}O_2$	711-79-5	186.206	pr (bz, lig) grn-ye nd (al)	101	dec 325			vs bz, HOAc
6128	2-Hydroxy-3-nitrobenzaldehyde		$C_7H_5NO_4$	5274-70-4	167.120	nd (HOAc)	109.5				vs bz, EtOH
6129	2-Hydroxy-5-nitrobenzaldehyde		$C_7H_5NO_4$	97-51-8	167.120	cry (dil HOAc)	127.0				s ace
6130	2-Hydroxy-3-nitrobenzoic acid	3-Nitrosalicylic acid	$C_7H_5NO_5$	85-38-1	183.119	ye nd (HOAc, w+1)	148				sl H_2O; vs EtOH, eth; s ace, bz, chl
6131	2-Hydroxy-5-nitrobenzoic acid	5-Nitrosalicylic acid	$C_7H_5NO_5$	96-97-9	183.119	nd (w)	229.5		1.650^{20}		sl H_2O; vs EtOH, eth, ace, bz; s chl
6132	2-Hydroxy-1,2,3-nonadecanetricarboxylic acid	Agaricic acid	$C_{22}H_{40}O_7$	666-99-9	416.549	cry pow	142 dec				s H_2O; sl EtOH, eth; i bz, chl
6133	12-Hydroxyoctadecanoic acid	12-Hydroxysteric acid	$C_{18}H_{36}O_3$	106-14-9	300.477	cry (al)	82				i H_2O; s EtOH, eth, chl
6134	*cis*-12-Hydroxy-9-octadecenoic acid, (*R*)	Ricinoleic acid	$C_{18}H_{34}O_3$	141-22-0	298.461	visc liq	5.5	227^{10}	0.9450^{21}	1.4716^{21}	i H_2O; vs eth, EtOH
6135	2-Hydroxyoctanoic acid		$C_8H_{16}O_3$	617-73-2	160.211	pl	70	162^{10}			sl H_2O, chl; vs EtOH, eth
6136	5-Hydroxy-4-octanone	Butyroin	$C_8H_{16}O_2$	496-77-5	144.212	liq	-10	185	0.9107^{16}	1.4345^{16}	
6137	[2-Hydroxy-4-(octyloxy)phenyl] phenylmethanone	Octabenzone	$C_{21}H_{26}O_3$	1843-05-6	326.429		48.5				
6138	3-Hydroxy-2-oxopropanoic acid	Hydroxypyruvic acid	$C_3H_4O_4$	1113-60-6	104.062		81 dec				
6139	3-Hydroxy-4-oxo-4*H*-pyran-2,6-dicarboxylic acid	Meconic acid	$C_7H_4O_7$	497-59-6	200.103	orth pl (w, dil HCl) (+3w)	120 dec				sl H_2O, MeOH, ace, eth; s EtOH, bz
6140	2-Hydroxypentanoic acid		$C_5H_{10}O_3$	617-31-2	118.131	hyg pl	34	sub			s H_2O, EtOH, eth
6141	5-Hydroxy-2-pentanone		$C_5H_{10}O_2$	1071-73-4	102.132			209; 117^{33}	1.0071^{20}	1.4390^{20}	msc H_2O; s EtOH, eth

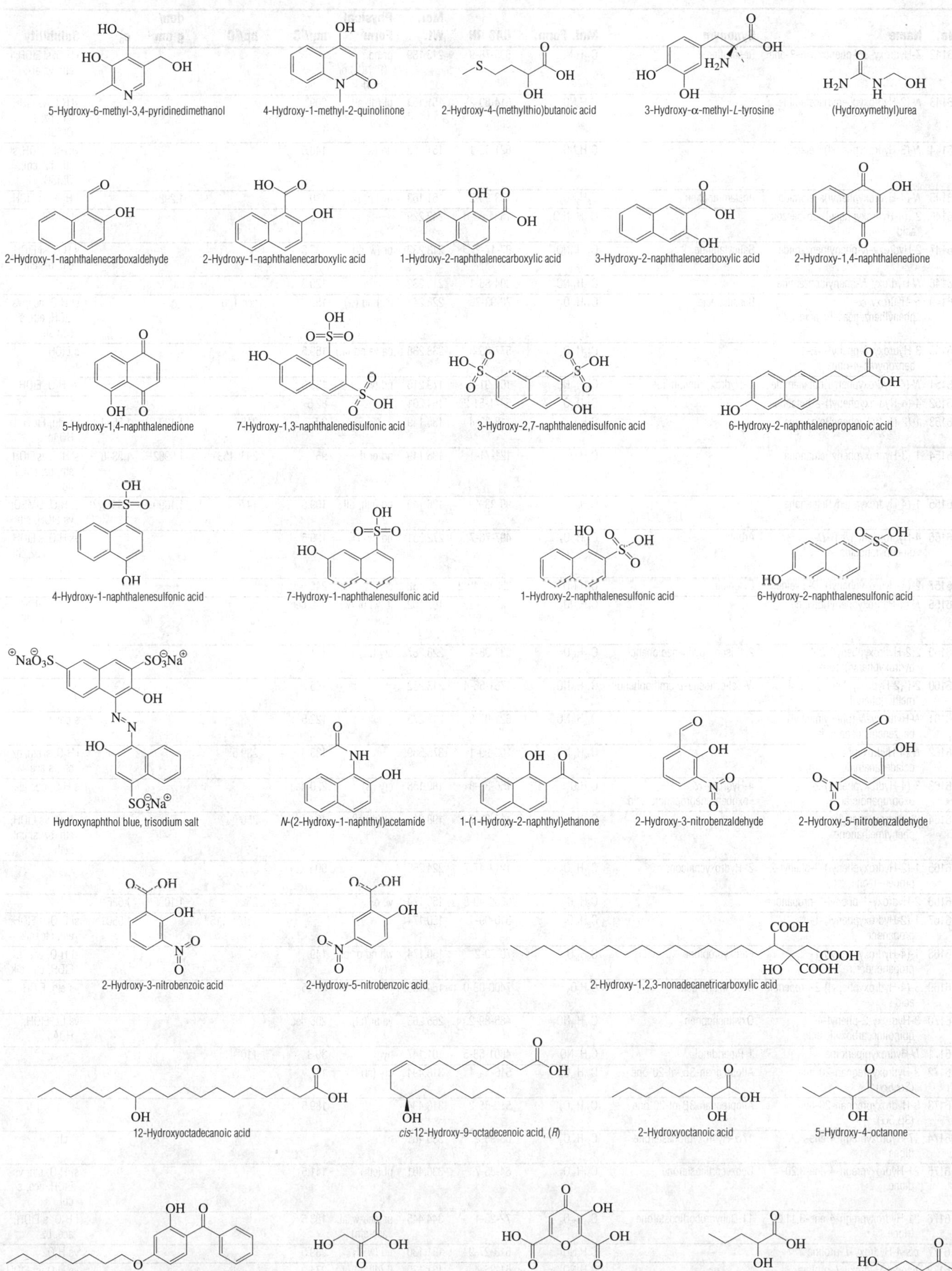

5-Hydroxy-6-methyl-3,4-pyridinedimethanol
4-Hydroxy-1-methyl-2-quinolinone
2-Hydroxy-4-(methylthio)butanoic acid
3-Hydroxy-α-methyl-L-tyrosine
(Hydroxymethyl)urea
2-Hydroxy-1-naphthalenecarboxaldehyde
2-Hydroxy-1-naphthalenecarboxylic acid
1-Hydroxy-2-naphthalenecarboxylic acid
3-Hydroxy-2-naphthalenecarboxylic acid
2-Hydroxy-1,4-naphthalenedione
5-Hydroxy-1,4-naphthalenedione
7-Hydroxy-1,3-naphthalenedisulfonic acid
3-Hydroxy-2,7-naphthalenedisulfonic acid
6-Hydroxy-2-naphthalenepropanoic acid
4-Hydroxy-1-naphthalenesulfonic acid
7-Hydroxy-1-naphthalenesulfonic acid
1-Hydroxy-2-naphthalenesulfonic acid
6-Hydroxy-2-naphthalenesulfonic acid
Hydroxynaphthol blue, trisodium salt
N-(2-Hydroxy-1-naphthyl)acetamide
1-(1-Hydroxy-2-naphthyl)ethanone
2-Hydroxy-3-nitrobenzaldehyde
2-Hydroxy-5-nitrobenzaldehyde
2-Hydroxy-3-nitrobenzoic acid
2-Hydroxy-5-nitrobenzoic acid
2-Hydroxy-1,2,3-nonadecanetricarboxylic acid
12-Hydroxyoctadecanoic acid
cis-12-Hydroxy-9-octadecenoic acid, (R)
2-Hydroxyoctanoic acid
5-Hydroxy-4-octanone
[2-Hydroxy-4-(octyloxy)phenyl]phenylmethanone
3-Hydroxy-2-oxopropanoic acid
3-Hydroxy-4-oxo-4H-pyran-2,6-dicarboxylic acid
2-Hydroxypentanoic acid
5-Hydroxy-2-pentanone

No.	Name	Synonym	Mol. Form.	CAS RN	Mol. Wt.	Physical Form	mp/°C	bp/°C	den/ g cm^{-3}	n_D	Solubility
6142	7-Hydroxy-3*H*-phenoxazin-3-one	Resorufine	$C_{12}H_7NO_3$	635-78-9	213.189	br nd ($PhNO_2$) pr (HCl)					i H_2O; sl EtOH; i eth; vs alk
6143	*N*-(2-Hydroxyphenyl)acetamide		$C_8H_9NO_2$	614-80-2	151.163	pl (dil al)	209				sl H_2O; vs EtOH, eth, bz; s DMSO
6144	*N*-(3-Hydroxyphenyl)acetamide		$C_8H_9NO_2$	621-42-1	151.163	nd (w)	148.5				vs H_2O, EtOH; sl eth, bz, chl, DMSO
6145	*N*-(4-Hydroxyphenyl)acetamide	Acetaminophen	$C_8H_9NO_2$	103-90-2	151.163	mcl pr (w)	170		1.293^{21}		i H_2O; vs EtOH
6146	2-[(4-Hydroxyphenyl)azo]benzoic acid		$C_{13}H_{10}N_2O_3$	1634-82-8	242.229		206				sl DMSO
6147	2-Hydroxy-*N*-phenylbenzamide	Salicylanilide	$C_{13}H_{11}NO_2$	87-17-2	213.232	pr (w, al)	136.5				s H_2O; sl EtOH, eth, bz, chl
6148	*N*-Hydroxy-*N*-phenylbenzamide		$C_{13}H_{11}NO_2$	304-88-1	213.232		120.3				
6149	α-Hydroxy-α-phenylbenzeneacetic acid	Benzilic acid	$C_{14}H_{12}O_3$	76-93-7	228.243	mcl nd (w)	150	dec 180			sl H_2O, ace; vs EtOH, eth; s con sulf
6150	3-Hydroxy-2-phenyl-4*H*-1-benzopyran-4-one		$C_{15}H_{10}O_3$	577-85-5	238.238	pa ye nd (al)	169.5				s EtOH
6151	*N*-(4-Hydroxyphenyl)butanamide	4'-Hydroxybutyranilide	$C_{10}H_{13}NO_2$	101-91-7	179.216	nd (w)	139.5				vs H_2O, EtOH
6152	4-(4-Hydroxyphenyl)-2-butanone		$C_{10}H_{12}O_2$	5471-51-2	164.201		82.5				
6153	1-(2-Hydroxyphenyl)ethanone		$C_8H_8O_2$	118-93-4	136.149		2.5	218	1.1307^{20}	1.5584^{20}	vs eth, EtOH, HOAc
6154	1-(3-Hydroxyphenyl)ethanone		$C_8H_8O_2$	121-71-1	136.149	nd or lf	96	296; 153^5	1.0992^{109}	1.5348^{109}	sl H_2O; vs EtOH, eth, bz, chl; i lig
6155	1-(4-Hydroxyphenyl)ethanone		$C_8H_8O_2$	99-93-4	136.149	nd (eth, dil al)	109.5	147^3	1.1090^{109}	1.5577^{109}	sl H_2O, DMSO; vs EtOH, eth
6156	4-Hydroxyphenyl-β-*D*-glucopyranoside	Arbutin	$C_{12}H_{16}O_7$	497-76-7	272.251	nd (w+1)	199.5				vs H_2O; s EtOH; sl eth; i bz, chl, CS_2
6157	2-(4-Hydroxyphenyl)-*D*-glycine	Oxfenicine	$C_8H_9NO_3$	22818-40-2	167.162	cry	240 dec				
6158	*N*-(4-Hydroxyphenyl)glycine		$C_8H_9NO_3$	122-87-2	167.162	lf (w) pl (w)	246 dec				sl H_2O, EtOH; i eth; s AcOEt, chl
6159	2(2-Hydroxyphenyl)-2(4-hydroxyphenyl)propane	2,4'-Isopropylidenediphenol	$C_{15}H_{16}O_2$	837-08-1	228.287	cry (bz)	111				
6160	2-[[(2-Hydroxyphenyl)imino] methyl]phenol	*N*-Salicylidene-*o*-aminophenol	$C_{13}H_{11}NO_2$	1761-56-4	213.232		185				
6161	*N*-Hydroxy-*N*-(phenylmethyl) benzenemethanamine		$C_{14}H_{15}NO$	621-07-8	213.275		122.5				s chl
6162	*N*-(4-Hydroxyphenyl) octadecanamide		$C_{24}H_{41}NO_2$	103-99-1	375.589		133.8	239.5^{10}			i H_2O; sl eth, bz, chl; s ace
6163	3-(4-Hydroxyphenyl)-2-oxopropanoic acid	4-Hydroxy-α-oxobenzenepropanoic acid	$C_9H_8O_4$	156-39-8	180.158	cry (w)	220 dec				s H_2O; dec alk
6164	(2-Hydroxyphenyl) phenylmethanone		$C_{13}H_{10}O_2$	117-99-7	198.217	pl (dil al)	40	250^{560}			i H_2O; vs EtOH, eth, bz; sl chl, peth
6165	1-(2-Hydroxyphenyl)-3-phenyl-2-propen-1-one	2'-Hydroxychalcone	$C_{15}H_{12}O_2$	1214-47-7	224.255		90				
6166	2-Hydroxy-1-phenyl-1-propanone		$C_9H_{10}O_2$	5650-40-8	150.174	ye oil		251	1.1085^{18}	1.536^{23}	
6167	1-(2-Hydroxyphenyl)-1-propanone		$C_9H_{10}O_2$	610-99-1	150.174			150^{80}, 115^{15}		1.5501^{20}	sl H_2O; s EtOH, eth, ctc, alk
6168	1-(4-Hydroxyphenyl)-1-propanone	Paroxypropione	$C_9H_{10}O_2$	70-70-2	150.174	wh nd or pl (w)	149				sl H_2O, ace; s EtOH, eth, alk
6169	3-(4-Hydroxyphenyl)-2-propenoic acid	*p*-Coumaric acid	$C_9H_8O_3$	7400-08-0	164.158	nd	211.5				vs eth, EtOH
6170	3-Hydroxy-2-phenyl-4-quinolinecarboxylic acid	Oxycinchophen	$C_{16}H_{11}NO_3$	485-89-2	265.263	ye pr (al)	206 dec				vs bz, EtOH, HOAc
6171	*N*-Hydroxypiperidine	1-Piperidinol	$C_5H_{11}NO$	4801-58-5	101.147	hyg	39.3	110^{55}			
6172	3-Hydroxypregnan-20-one, (3α,5α)	Allopregnan-3α-ol-20-one	$C_{21}H_{34}O_2$	516-54-1	318.494	cry (al)	177				
6173	3-Hydroxypregnan-20-one, (3β,5α)	Allopregnan-3β-ol-20-one	$C_{21}H_{34}O_2$	516-55-2	318.494		189.5				
6174	17-Hydroxypregn-4-ene-3,20-dione	17α-Hydroxyprogesterone	$C_{21}H_{30}O_3$	68-96-2	330.461						sl chl
6175	21-Hydroxypregn-4-ene-3,20-dione	Deoxycorticosterone	$C_{21}H_{30}O_3$	64-85-7	330.461	pl (eth)	141.5				sl H_2O, eth; vs EtOH, ace; s chl
6176	21-Hydroxypregn-4-ene-3,11,20-trione	11-Dehydrocorticosterone	$C_{21}H_{28}O_4$	72-23-1	344.445	pr (ace-w, al, ace-eth)	183.5				i H_2O; s EtOH, ace, bz
6177	*cis*-4-Hydroxy-*L*-proline		$C_5H_9NO_3$	618-27-9	131.130	nd (w+1)	239.5				vs H_2O
6178	*trans*-4-Hydroxy-*L*-proline		$C_5H_9NO_3$	51-35-4	131.130	lf (dil al) pr (w)	274				vs H_2O; sl EtOH

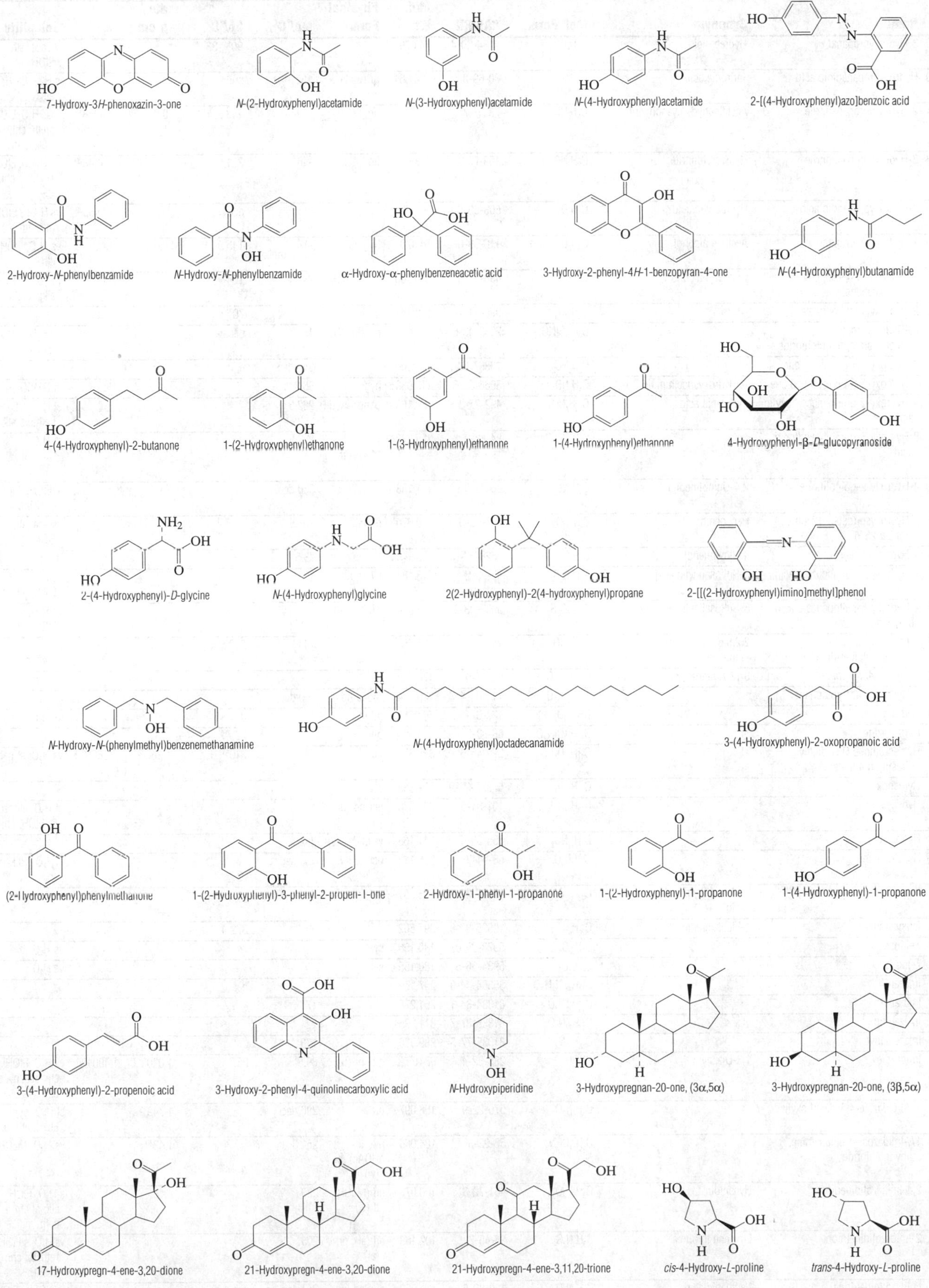
7-Hydroxy-3H-phenoxazin-3-one
N-(2-Hydroxyphenyl)acetamide
N-(3-Hydroxyphenyl)acetamide
N-(4-Hydroxyphenyl)acetamide
2-[(4-Hydroxyphenyl)azo]benzoic acid
2-Hydroxy-N-phenylbenzamide
N-Hydroxy-N-phenylbenzamide
α-Hydroxy-α-phenylbenzeneacetic acid
3-Hydroxy-2-phenyl-4H-1-benzopyran-4-one
N-(4-Hydroxyphenyl)butanamide
4-(4-Hydroxyphenyl)-2-butanone
1-(2-Hydroxyphenyl)ethanone
1-(3-Hydroxyphenyl)ethanone
1-(4-Hydroxyphenyl)ethanone
4-Hydroxyphenyl-β-D-glucopyranoside
2-(4-Hydroxyphenyl)-D-glycine
N-(4-Hydroxyphenyl)glycine
2(2-Hydroxyphenyl)-2(4-hydroxyphenyl)propane
2-[[(2-Hydroxyphenyl)imino]methyl]phenol
N-Hydroxy-N-(phenylmethyl)benzenemethanamine
N-(4-Hydroxyphenyl)octadecanamide
3-(4-Hydroxyphenyl)-2-oxopropanoic acid
(2-Hydroxyphenyl)phenylmethanone
1-(2-Hydroxyphenyl)-3-phenyl-2-propen-1-one
2-Hydroxy-1-phenyl-1-propanone
1-(2-Hydroxyphenyl)-1-propanone
1-(4-Hydroxyphenyl)-1-propanone
3-(4-Hydroxyphenyl)-2-propenoic acid
3-Hydroxy-2-phenyl-4-quinolinecarboxylic acid
N-Hydroxypiperidine
3-Hydroxypregnan-20-one, (3α,5α)
3-Hydroxypregnan-20-one, (3β,5α)
17-Hydroxypregn-4-ene-3,20-dione
21-Hydroxypregn-4-ene-3,20-dione
21-Hydroxypregn-4-ene-3,11,20-trione
cis-4-Hydroxy-L-proline
trans-4-Hydroxy-L-proline

No.	Name	Synonym	Mol. Form.	CAS RN	Mol. Wt.	Physical Form	mp/°C	bp/°C	den/ g cm^{-3}	n_D	Solubility
6179	3-Hydroxypropanal	Hydracrolein	$C_3H_6O_2$	2134-29-4	74.079			90^{18}, $38^{0.2}$			vs ace, eth, EtOH
6180	Hydroxypropanedioic acid	Tartronic acid	$C_3H_4O_5$	80-69-3	120.061	pr (w+1)	157	sub			s H_2O, EtOH; sl eth
6181	2-Hydroxypropanenitrile	Acetaldehyde cyanohydrin	C_3H_5NO	78-97-7	71.078	liq	-40	183	0.9877^{20}	1.4058^{18}	msc H_2O, EtOH; s eth, chl; i CS_2, peth
6182	3-Hydroxypropanenitrile	Hydracrylonitrile	C_3H_5NO	109-78-4	71.078	liq	-46	221	1.0404^{25}	1.4248^{20}	msc H_2O, EtOH; sl eth; s chl; i CS_2
6183	3-Hydroxypropanoic acid	Hydracrylic acid	$C_3H_6O_3$	503-66-2	90.078	syr		dec		1.4489^{20}	vs H_2O; s EtOH; msc eth
6184	1-Hydroxy-2-propanone	Acetone alcohol	$C_3H_6O_2$	116-09-6	74.079	hyg liq	-17	145.5	1.0805^{20}	1.4295^{20}	vs H_2O, EtOH, eth
6185	4-(3-Hydroxy-1-propenyl)-2-methoxyphenol	Coniferyl alcohol	$C_{10}H_{12}O_3$	458-35-5	180.200	pr (eth-lig)	74	164^3			i H_2O; s EtOH, alk; vs eth
6186	2-Hydroxypropyl acrylate		$C_6H_{10}O_3$	999-61-1	130.141	liq		70^2			
6187	(2-Hydroxypropyl) trimethylammonium chloride		$C_6H_{16}ClNO$	2382-43-6	153.650	pr (Bu OH)	165	dec			vs H_2O, EtOH
6188	3-Hydroxy-1*H*-pyridin-2-one		$C_5H_5NO_2$	16867-04-2	111.100		245 dec				
6189	1-Hydroxy-2,5-pyrrolidinedione	*N*-Hydroxysuccinimide	$C_4H_5NO_3$	6066-82-6	115.088	hyg	96.3				sl DMSO
6190	4-Hydroxy-2-quinolinecarboxylic acid	Kynurenic acid	$C_{10}H_7NO_3$	492-27-3	189.168	ye nd (+w, dil al)	282.5				sl H_2O; s EtOH; i eth; vs alk
6191	8-Hydroxy-5-quinolinesulfonic acid		$C_9H_7NO_4S$	84-88-8	225.222	ye lf, nd (+ 1w) (dil HCl)	322.5				sl H_2O
6192	4-Hydroxy-2-quinolinone	2,4-Quinolinediol	$C_9H_7NO_2$	86-95-3	161.158		360 dec				sl EtOH, $PhNO_2$, gl HOAc
6193	3-Hydroxyspirostan-12-one, (3β,5α,25*R*)	Hecogenin	$C_{27}H_{42}O_4$	467-55-0	430.620	pl (eth)	266.5				vs ace, eth, EtOH
6194	4-Hydroxystyrene	4-Vinylphenol	C_8H_8O	2628-17-3	120.149		73.5				
6195	2-Hydroxy-5-sulfobenzoic acid	5-Sulfosalicylic acid	$C_7H_6O_6S$	97-05-2	218.184	hyg nd	120				vs H_2O; vs EtOH, eth
6196	2-Hydroxy-5-sulfobenzoic acid dihydrate	5-Sulfosalicylic acid dihydrate	$C_7H_{10}O_8S$	5965-83-3	254.214	wh cry (w)					vs H_2O; vs EtOH, eth
6197	4-Hydroxy-2,2,6,6-tetramethylpiperidine	2,2,6,6-Tetramethyl-4-piperidinol	$C_9H_{19}NO$	2403-88-5	157.253		130	213.5			
6198	5-Hydroxytryptamine	3-(2-Aminoethyl)indol-5-ol	$C_{10}H_{12}N_2O$	50-67-9	176.214						s H_2O
6199	5-Hydroxy-*DL*-tryptophan		$C_{11}H_{12}N_2O_3$	114-03-4	220.224	rod or nd (al)	300 dec				
6200	Hydroxyurea		$CH_4N_2O_2$	127-07-1	76.055	nd (al)	141	dec			vs H_2O
6201	Hydroxyzine		$C_{21}H_{27}ClN_2O_2$	68-88-2	374.904	oil		$220^{0.5}$			
6202	Hymecromone O,O-diethyl phosphorothioate		$C_{14}H_{17}O_5PS$	299-45-6	328.321	nd	38	$210^{1.0}$ dec	1.260^{38}	1.5685^{37}	vs H_2O; sl peth
6203	Hymenoxone		$C_{15}H_{22}O_5$	57377-32-9	282.333	cry					
6204	Hyoscyamine		$C_{17}H_{23}NO_3$	101-31-5	289.370	tetr nd (dil al)	108.5				sl H_2O, eth, bz; vs EtOH, chl
6205	Hypoglycin A		$C_7H_{11}NO_2$	156-56-9	141.168	ye pl (Me aq)	282				
6206	Hypoxanthine		$C_5H_4N_4O$	68-94-0	136.112	oct nd (w)	150 dec				sl H_2O; s alk, dil acid
6207	Ibuprofen	2-(4-Isobutylphenyl)propanoic acid	$C_{13}H_{18}O_2$	15687-27-1	206.281	col cry	76				sl H_2O; s os
6208	Icosylamine	1-Eicosanamine	$C_{20}H_{43}N$	10525-37-8	297.562			372.4			
6209	*D*-Idose		$C_6H_{12}O_6$	5978-95-0	180.155	syr					vs H_2O
6210	*L*-Idose		$C_6H_{12}O_6$	5934-56-5	180.155	syr					vs H_2O
6211	Imazalil		$C_{14}H_{14}Cl_2N_2O$	35554-44-0	297.179		50	dec	1.243^{23}		
6212	Imazapyr		$C_{13}H_{15}N_3O_3$	81334-34-1	261.276		171				
6213	Imazaquin		$C_{17}H_{17}N_3O_3$	81335-37-7	311.335		221				
6214	Imazethapyr		$C_{15}H_{19}N_3O_3$	81335-77-5	289.330		173				
6215	Imidazole	1,3-Diazole	$C_3H_4N_2$	288-32-4	68.077	mcl pr (bz)	89.5	257	1.0303^{101}	1.4801^{101}	vs H_2O, EtOH; s eth, ace, py; sl bz
6216	1*H*-Imidazole-4,5-dicarboxylic acid		$C_5H_4N_2O_4$	570-22-9	156.097	pr	290 dec		1.749^{25}		sl H_2O, py; i EtOH, eth, bz
6217	1*H*-Imidazole-4-ethanamine, dihydrochloride		$C_5H_{11}Cl_2N_3$	56-92-8	184.066	pl (eth-HOAc), pr (w)	251.3		1.43^{20}		vs H_2O, MeOH
6218	2,4-Imidazolidinedione	Hydantoin	$C_3H_4N_2O_2$	461-72-3	100.076	nd (MeOH), lf (w)	220				s H_2O, EtOH, alk; sl eth; i peth
6219	2-Imidazolidinethione	Ethylene thiourea	$C_3H_6N_2S$	96-45-7	102.158	nd (al), pr (al)	203				vs H_2O; s EtOH; i eth, bz, chl; sl DMSO
6220	Imidazolidinetrione	Parabanic acid	$C_3H_2N_2O_3$	120-89-8	114.059	mcl nd (w)	244 dec	sub 100			s H_2O; vs EtOH

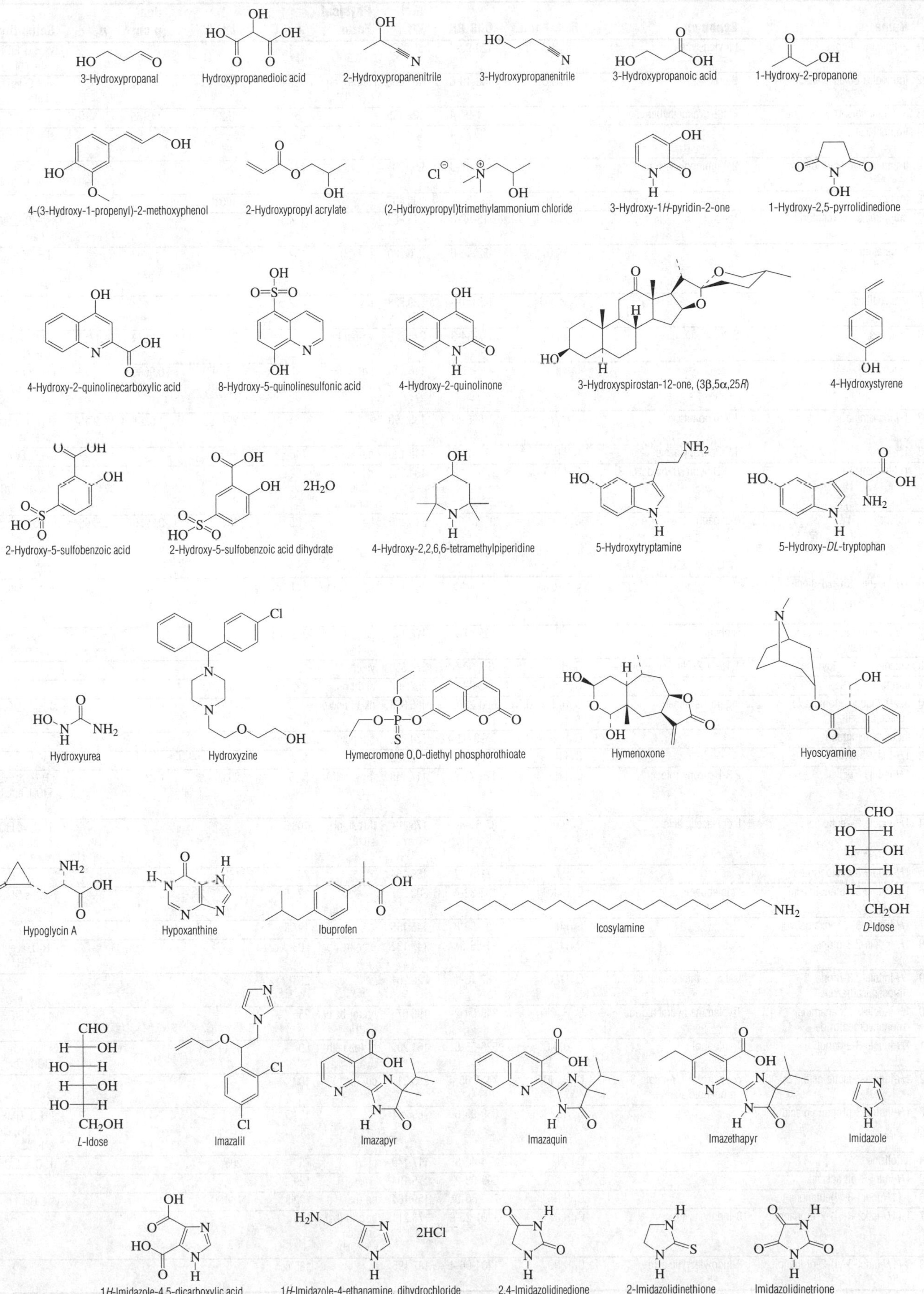

HO
O
3-Hydroxypropanal
OH
HO
OH
O
O
Hydroxypropanedioic acid
OH
N
2-Hydroxypropanenitrile
HO
N
3-Hydroxypropanenitrile
O
HO
OH
3-Hydroxypropanoic acid
O
OH
1-Hydroxy-2-propanone
OH
HO
O
4-(3-Hydroxy-1-propenyl)-2-methoxyphenol
O
O
OH
2-Hydroxypropyl acrylate
Cl
N
OH
(2-Hydroxypropyl)trimethylammonium chloride
OH
N
O
H
3-Hydroxy-1H-pyridin-2-one
O
N
O
OH
1-Hydroxy-2,5-pyrrolidinedione
OH
N
OH
O
4-Hydroxy-2-quinolinecarboxylic acid
OH
O=S=O
N
OH
8-Hydroxy-5-quinolinesulfonic acid
OH
N
O
H
4-Hydroxy-2-quinolinone
O
O
O
H
H
HO
H
3-Hydroxyspirostan-12-one, (3β,5α,25R)
OH
4-Hydroxystyrene
O
OH
OH
O
S
HO
O
2-Hydroxy-5-sulfobenzoic acid
O
OH
OH
2H2O
O
S
HO
O
2-Hydroxy-5-sulfobenzoic acid dihydrate
OH
N
H
4-Hydroxy-2,2,6,6-tetramethylpiperidine
NH2
HO
N
H
5-Hydroxytryptamine
HO
O
OH
NH2
N
H
5-Hydroxy-DL-tryptophan
HO
O
N
NH2
H
Hydroxyurea
Cl
N
N
O
OH
Hydroxyzine
O
O
P
O
O
O
O
S
Hymecromone O,O-diethyl phosphorothioate
H
HO
O
O
O
OH
Hymenoxone
N
OH
O
O
Hyoscyamine
NH2
OH
O
Hypoglycin A
O
H
H
N
N
N
N
Hypoxanthine
OH
O
Ibuprofen
NH2
Icosylamine
CHO
HO
H
H
OH
HO
H
H
OH
CH2OH
D-Idose
CHO
H
OH
HO
H
H
OH
HO
H
CH2OH
L-Idose
N
N
O
Cl
Cl
Imazalil
O
OH
N
N
N
O
H
Imazapyr
O
OH
N
N
N
H
O
Imazaquin
O
OH
N
N
N
H
O
Imazethapyr
N
N
H
Imidazole
OH
O
N
HO
N
O
H
1H-Imidazole-4,5-dicarboxylic acid
H2N
N
2HCl
N
H
1H-Imidazole-4-ethanamine, dihydrochloride
O
H
N
O
N
H
2,4-Imidazolidinedione
H
N
S
N
H
2-Imidazolidinethione
O
H
N
O
O
N
H
Imidazolidinetrione

No.	Name	Synonym	Mol. Form.	CAS RN	Mol. Wt.	Physical Form	mp/°C	bp/°C	den/ g cm^{-3}	n_D	Solubility
6221	2-Imidazolidinone	Ethylene urea	$C_3H_6N_2O$	120-93-4	86.092		131				vs H_2O, EtOH; sl eth, chl
6222	Imidodicarbonic diamide	Biuret	$C_2H_5N_3O_2$	108-19-0	103.080	pl (al), nd (w+1)	190 dec				sl H_2O; vs EtOH; i eth
6223	3,3'-Iminobispropanenitrile	Bis(2-cyanoethyl)amine	$C_6H_9N_3$	111-94-4	123.155		-6	162^{5}	1.0165^{20}		
6224	Iminodiacetic acid		$C_4H_7NO_4$	142-73-4	133.104	orth pr	247.5				sl H_2O; i EtOH, eth
6225	Iminodiacetic acid, dinitrile	2,2'-Iminobisacetonitrile	$C_4H_5N_3$	628-87-5	95.103		78				s H_2O, EtOH; sl eth, bz, chl
6226	Imipramine		$C_{19}H_{24}N_2$	50-49-7	280.407			$160^{0.1}$			
6227	Imipramine hydrochloride	Tofranil	$C_{19}H_{25}ClN_2$	113-52-0	316.868		174.5				vs H_2O; s EtOH; sl ace
6228	Imperatorin		$C_{16}H_{14}O_4$	482-44-0	270.280	cry (al)	102				sl H_2O; s EtOH, eth, bz, peth; vs chl
6229	Indaconitine		$C_{34}H_{47}NO_{10}$	4491-19-4	629.738	cry	202 dec				vs eth, EtOH, chl
6230	Indalone	Butopyronoxyl	$C_{12}H_{18}O_4$	532-34-3	226.269	ye-red liq		263	1.057^{20}	1.475^{25}	i H_2O; vs EtOH, eth, chl
6231	Indan		C_9H_{10}	496-11-7	118.175	liq	-51.38	177.97	0.9639^{20}	1.5378^{20}	i H_2O; msc EtOH, eth; sl chl
6232	1-Indanamine	1-Aminoindane	$C_9H_{11}N$	34698-41-4	133.190			221; 96^{8}	1.038^{15}	1.5613^{20}	sl H_2O; s eth, ace, bz
6233	1*H*-Indazole	1*H*-Benzopyrazole	$C_7H_6N_2$	271-44-3	118.136	nd (al, w)	148	269			s H_2O, EtOH, eth
6234	1*H*-Indazol-3-ol	1,2-Dihydro-3*H*-indazol-3-one	$C_7H_6N_2O$	7364-25-2	134.135	nd or lf (MeOH) pl or nd (al)	252.5				sl H_2O, eth; s MeOH, EtOH
6235	Indene	Indonaphthene	C_9H_8	95-13-6	116.160	liq	-1.5	182	0.9960^{25}	1.5768^{20}	i H_2O; msc EtOH, eth; s ace, bz, py; sl chl
6236	1*H*-Indene-1,3(2*H*)-dione		$C_9H_6O_2$	606-23-5	146.143	nd (eth, lig)	131 dec		1.37^{21}		sl H_2O, ctc; vs EtOH; s eth, bz, alk
6237	1*H*-Indene-1,2,3-trione monohydrate	Ninhydrin	$C_9H_6O_4$	485-47-2	178.142	pa ye pr (w, al)	242 dec				vs H_2O; s EtOH, alk; sl eth
6238	Indeno[1,2,3-cd]pyrene	1,10-(1,2-Phenylene)pyrene	$C_{22}H_{12}$	193-39-5	276.330	ye cry (cy)	162				
6239	Indigo		$C_{16}H_{10}N_2O_2$	482-89-3	262.262	dk bl pow	390 dec	sub 300			
6240	5,5'-Indigodisulfonic acid, disodium salt	Indigo Carmine	$C_{16}H_8N_2Na_2O_8S_2$	860-22-0	466.353	dk-bl pow					sl H_2O, EtOH; i os
6241	Indocyanine green		$C_{43}H_{47}N_2NaO_6S_2$	3599-32-4	774.962	grn pow	244 dec				
6242	1*H*-Indol-5-amine		$C_8H_8N_2$	5192-03-0	132.163		132				
6243	1*H*-Indole	2,3-Benzopyrrole	C_8H_7N	120-72-9	117.149	lf (w, peth) cry (eth)	52.5	253.6	1.22^{25}		s H_2O, bz; vs EtOH, eth, tol; sl ctc
6244	1*H*-Indole-3-acetic acid	Indoleacetic acid	$C_{10}H_9NO_2$	87-51-4	175.184	lf (bz), pl (chl)	168.5				i H_2O; vs EtOH; s eth, ace, bz; sl chl
6245	1*H*-Indole-3-acetonitrile		$C_{10}H_8N_2$	771-51-7	156.184		36	$160^{0.2}$			
6246	1*H*-Indole-3-butanoic acid	Indolebutyric acid	$C_{12}H_{13}NO_2$	133-32-4	203.237		124.5				vs bz; s DMSO; i peth
6247	1*H*-Indole-3-carboxaldehyde		C_9H_7NO	487-89-8	145.158		197.8				
6248	1*H*-Indole-2,3-dione	Isatin	$C_8H_5NO_2$	91-56-5	147.132	oran mcl pr	203 dec				s H_2O, ace, bz; vs EtOH; sl eth
6249	1*H*-Indole-2,3-dione, 3-thiosemicarbazone	Isatin, 3-thiosemicarbazone	$C_9H_8N_4OS$	487-16-1	220.251		283				
6250	1*H*-Indole-3-ethanamine, monohydrochloride	Tryptamine hydrochloride	$C_{10}H_{13}ClN_2$	343-94-2	196.676	nd (al-bz or lig)	255				vs ace, EtOH
6251	1*H*-Indole-3-ethanol	Tryptophol	$C_{10}H_{11}NO$	526-55-6	161.200	pr (bz-peth)	59	174^{2}			vs ace, eth, EtOH, chl
6252	1*H*-Indole-3-lactic acid, (*S*)	α-Hydroxy-1*H*-indole-3-propanoic acid	$C_{11}H_{11}NO_3$	7417-65-4	205.210	cry (peth)	100				
6253	1*H*-Indole-3-propanoic acid		$C_{11}H_{11}NO_2$	830-96-6	189.211		134.5				sl H_2O, DMSO; vs EtOH, eth, ace, bz
6254	Indolizine		C_8H_7N	274-40-8	117.149	pl	75	205			i H_2O; s EtOH
6255	1*H*-Indol-3-ol, acetate		$C_{10}H_9NO_2$	608-08-2	175.184		129				
6256	1-(1*H*-Indol-3-yl)ethanone		$C_{10}H_9NO$	703-80-0	159.184	nd (bz)	192.3	144^{10}			vs EtOH
6257	1-(1*H*-Indol-3-yl)-2-propanone	3-Indolylacetone	$C_{11}H_{11}NO$	1201-26-9	173.211	br orth (bz), nd (aq MeOH)	116				
6258	3-(1*H*-Indol-3-yl)-2-propenoic acid	3-Indolylacrylic acid	$C_{11}H_9NO_2$	1204-06-4	187.195		185 dec				

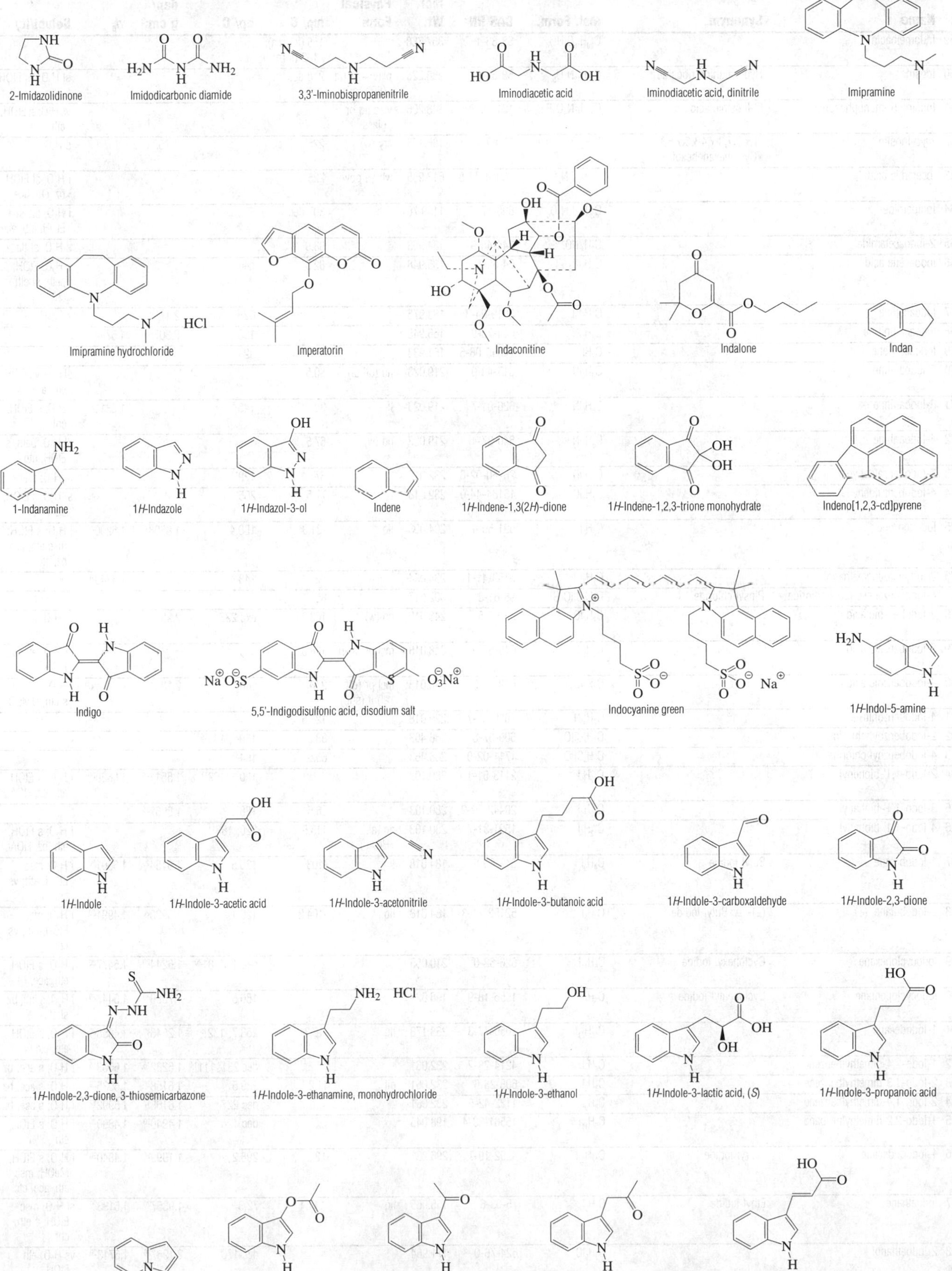
2-Imidazolidinone
Imidodicarbonic diamide
3,3'-Iminobispropanenitrile
Iminodiacetic acid
Iminodiacetic acid, dinitrile
Imipramine
Imipramine hydrochloride
Imperatorin
Indaconitine
Indalone
Indan
1-Indanamine
1H-Indazole
1H-Indazol-3-ol
Indene
1H-Indene-1,3(2H)-dione
1H-Indene-1,2,3-trione monohydrate
Indeno[1,2,3-cd]pyrene
Indigo
5,5'-Indigodisulfonic acid, disodium salt
Indocyanine green
1H-Indol-5-amine
1H-Indole
1H-Indole-3-acetic acid
1H-Indole-3-acetonitrile
1H-Indole-3-butanoic acid
1H-Indole-3-carboxaldehyde
1H-Indole-2,3-dione
1H-Indole-2,3-dione, 3-thiosemicarbazone
1H-Indole-3-ethanamine, monohydrochloride
1H-Indole-3-ethanol
1H-Indole-3-lactic acid, (S)
1H-Indole-3-propanoic acid
Indolizine
1H-Indol-3-ol, acetate
1-(1H-Indol-3-yl)ethanone
1-(1H-Indol-3-yl)-2-propanone
3-(1H-Indol-3-yl)-2-propenoic acid

No.	Name	Synonym	Mol. Form.	CAS RN	Mol. Wt.	Physical Form	mp/°C	bp/°C	den/ g cm^{-3}	n_D	Solubility
6259	Indomethacin		$C_{19}H_{16}ClNO_4$	53-86-1	357.788		155 (form a); 162 (form b				
6260	Inosine	Hypoxanthine riboside	$C_{10}H_{12}N_4O_5$	58-63-9	268.226	pl (w+2), nd (80% al)	218 dec				sl H_2O; vs EtOH
6261	Inosine 5'-monophosphate	5'-Inosinic acid	$C_{10}H_{13}N_4O_8P$	131-99-7	348.206	visc liq or glass					vs H_2O; sl EtOH, eth
6262	*myo*-Inositol	(1α,2α,3α,4β,5α,6β)-Cyclohexanehexol	$C_6H_{12}O_6$	87-89-8	180.155	cry (w)	225		1.752		s H_2O
6263	Iocetamic acid		$C_{12}H_{13}I_3N_2O_3$	16034-77-8	613.955	wh-ye pow	225				i H_2O; sl EtOH, bz, eth, ace
6264	Iodipamide		$C_{20}H_{14}I_6N_2O_6$	606-17-7	1139.761		307 dec				i H_2O, bz; sl EtOH, eth, ace
6265	2-Iodoacetamide		C_2H_4INO	144-48-9	184.963		93.0				s H_2O; sl tfa
6266	Iodoacetic acid		$C_2H_3IO_2$	64-69-7	185.948		82.5	dec			s H_2O, EtOH, peth; sl eth, chl
6267	Iodoacetone		C_3H_5IO	3019-04-3	183.975			62^{12}	2.17^{15}		s EtOH
6268	Iodoacetonitrile		C_2H_2IN	624-75-9	166.948			185	2.307^{25}	1.5744^{20}	
6269	Iodoacetylene		C_2HI	14545-08-5	151.933			32			
6270	2-Iodoaniline		C_6H_6IN	615-43-0	219.023	nd (dil al)	60.5				sl H_2O; vs EtOH, eth, ace
6271	3-Iodoaniline		C_6H_6IN	626-01-7	219.023	lf	33	145^{15}		1.6811^{20}	i H_2O; s EtOH, chl
6272	4-Iodoaniline		C_6H_6IN	540-37-4	219.023	nd (w)	67.5				sl H_2O, peth; s EtOH, eth
6273	2-Iodobenzaldehyde		C_7H_5IO	26260-02-6	232.018		37	129^{14}			sl H_2O; s ace
6274	4-Iodobenzaldehyde		C_7H_5IO	15164-44-0	232.018		77.5	265			sl H_2O; s EtOH, bz
6275	Iodobenzene		C_6H_5I	591-50-4	204.008	liq	-31.3	188.4	1.8308^{20}	1.6200^{20}	i H_2O; s EtOH; msc eth, ace, bz, ctc
6276	2-Iodobenzenemethanol		C_7H_7IO	5159-41-1	234.034		92	148^{32}		1.6349^{20}	
6277	4-Iodobenzenesulfonyl chloride	Pipsyl chloride	$C_6H_4ClIO_2S$	98-61-3	302.517		85				
6278	2-Iodobenzoic acid		$C_7H_5IO_2$	88-67-5	248.018	nd (w)	163	exp 233	2.25^{25}		sl H_2O, ace; vs EtOH, eth
6279	3-Iodobenzoic acid		$C_7H_5IO_2$	618-51-9	248.018	mcl pr (ace)	188.3	sub			sl H_2O, eth; vs EtOH
6280	4-Iodobenzoic acid		$C_7H_5IO_2$	619-58-9	248.018	mcl pr (dil al) lf (sub)	270	sub	2.184^{20}		i H_2O; sl EtOH; s eth, DMSO
6281	4-Iodobenzonitrile		C_7H_4IN	3058-39-7	229.018		127.5				
6282	2-Iodobenzoyl chloride		C_7H_4ClIO	609-67-6	266.463		38.3	159^{27}, 135^{19}			
6283	4-Iodobenzoyl chloride		C_7H_4ClIO	1711-02-0	266.463		65.5	164^{32}			
6284	2-Iodo-1,1'-biphenyl		$C_{12}H_9I$	2113-51-1	280.103			190^{36}, 169^{17}	1.5511^{25}	1.6620^{20}	i H_2O; s EtOH, eth, bz, HOAc
6285	3-Iodo-1,1'-biphenyl		$C_{12}H_9I$	20442-79-9	280.103		26.5	188^{16}	1.5967^{25}		
6286	4-Iodo-1,1'-biphenyl		$C_{12}H_9I$	1591-31-7	280.103	nd (al, HOAc)	113.5	320; 183^{11}			i H_2O; s EtOH, eth, bz, HOAc
6287	1-Iodobutane	Butyl iodide	C_4H_9I	542-69-8	184.018	liq	-103	130.5	1.6154^{20}	1.5001^{20}	i H_2O; msc EtOH, eth; vs chl
6288	2-Iodobutane, (±)	(±)-*sec*-Butyl iodide	C_4H_9I	52152-71-3	184.018	liq	-104.2	120.1	1.5920^{20}	1.4991^{20}	i H_2O; msc EtOH, eth; vs chl
6289	Iodocyclohexane	Cyclohexyl iodide	$C_6H_{11}I$	626-62-0	210.055			dec 180; 81^{20}	1.6244^{20}	1.5477^{20}	i H_2O; s EtOH, eth, ace, bz
6290	Iodocyclopentane	Cyclopentyl iodide	C_5H_9I	1556-18-9	196.029			166.5	1.7096^{20}	1.5447^{20}	i H_2O; s eth, bz; sl ctc
6291	1-Iododecane		$C_{10}H_{21}I$	2050-77-3	268.178	liq	-16.3	263.7; 132^{15}	1.2546^{20}	1.4858^{20}	i H_2O; s EtOH, eth, ctc
6292	1-Iodo-2,4-dimethylbenzene		C_8H_9I	4214-28-2	232.061			dec 231; 111^{14}	1.6282^{16}	1.6008^{16}	i H_2O; s ace, bz
6293	2-Iodo-1,3-dimethylbenzene		C_8H_9I	608-28-6	232.061	oil	11.2	229.5	1.6158^{20}	1.6035^{20}	i H_2O; s ace, bz
6294	2-Iodo-1,4-dimethylbenzene		C_8H_9I	1122-42-5	232.061			dec 227	1.6168^{17}	1.5992^{17}	i H_2O; s ace, bz
6295	1-Iodo-2,2-dimethylpropane		$C_5H_{11}I$	15501-33-4	198.045			dec 128	1.4940^{20}	1.4890^{20}	i H_2O; s EtOH, eth
6296	1-Iodododecane	Lauryl iodide	$C_{12}H_{25}I$	4292-19-7	296.231		0.3	298.2	1.1999^{20}	1.4840^{20}	i H_2O; s EtOH, MeOH; msc eth, ace, ctc
6297	Iodoethane	Ethyl iodide	C_2H_5I	75-03-6	155.965	liq	-111.1	72.3	1.9357^{20}	1.5133^{20}	sl H_2O; msc EtOH; s eth, chl
6298	2-Iodoethanol		C_2H_5IO	624-76-0	171.964			dec 176	2.1967^{20}	1.5713^{20}	vs H_2O, eth, EtOH
6299	Iodoethene	Vinyl iodide	C_2H_3I	593-66-8	153.949			56	2.037^{20}	1.5385^{20}	vs eth, EtOH
6300	(2-Iodoethyl)benzene		C_8H_9I	17376-04-4	232.061	liq		122^{13}	1.603	1.6010^{20}	

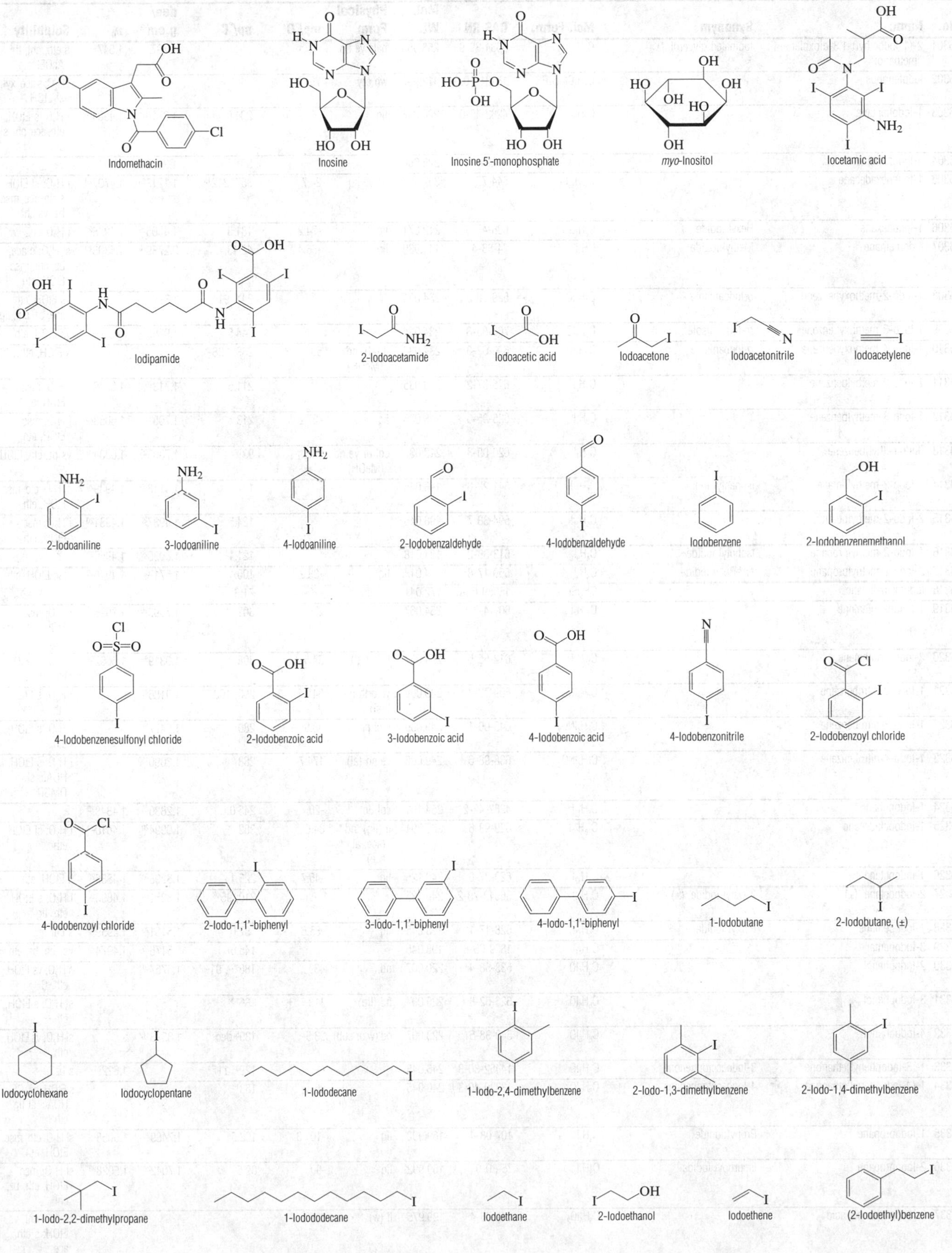
Indomethacin
Inosine
Inosine 5'-monophosphate
myo-Inositol
Iocetamic acid
Iodipamide
2-Iodoacetamide
Iodoacetic acid
Iodoacetone
Iodoacetonitrile
Iodoacetylene
2-Iodoaniline
3-Iodoaniline
4-Iodoaniline
2-Iodobenzaldehyde
4-Iodobenzaldehyde
Iodobenzene
2-Iodobenzenemethanol
4-Iodobenzenesulfonyl chloride
2-Iodobenzoic acid
3-Iodobenzoic acid
4-Iodobenzoic acid
4-Iodobenzonitrile
2-Iodobenzoyl chloride
4-Iodobenzoyl chloride
2-Iodo-1,1'-biphenyl
3-Iodo-1,1'-biphenyl
4-Iodo-1,1'-biphenyl
1-Iodobutane
2-Iodobutane, (±)
Iodocyclohexane
Iodocyclopentane
1-Iododecane
1-Iodo-2,4-dimethylbenzene
2-Iodo-1,3-dimethylbenzene
2-Iodo-1,4-dimethylbenzene
1-Iodo-2,2-dimethylpropane
1-Iodododecane
Iodoethane
2-Iodoethanol
Iodoethene
(2-Iodoethyl)benzene

No.	Name	Synonym	Mol. Form.	CAS RN	Mol. Wt.	Physical Form	mp/°C	bp/°C	den/ g cm-3	n_D	Solubility
6301	2-(1-Iodoethyl)-1,3-dioxolane-4-methanol	Iodinated glycerol	$C_6H_{11}IO_3$	5634-39-9	258.053	pale ye liq			1.797	1.547	s eth, chl, thf, AcOEt
6302	Iodofenphos		$C_8H_8Cl_2IO_3PS$	18181-70-9	412.997	wh cry	76				i H_2O; s ace, xyl; sl EtOH
6303	1-Iodoheptane		$C_7H_{15}I$	4282-40-0	226.098	liq	-48.2	204.0	1.3719^{25}	1.4904^{20}	i H_2O; s EtOH, eth, ace, chl; sl ctc
6304	3-Iodoheptane		$C_7H_{15}I$	31294-92-5	226.098			89^{30}	1.3676^{20}		
6305	1-Iodohexadecane		$C_{16}H_{33}I$	544-77-4	352.337	pa ye liq	24.7	357; 212^{15}	1.1213^{25}	1.4797^{20}	i H_2O; sl EtOH; s eth, ace; msc bz; vs chl
6306	1-Iodohexane	Hexyl iodide	$C_6H_{13}I$	638-45-9	212.071	liq	-74.2	181.3	1.4305^{25}	1.4928^{20}	i H_2O
6307	Iodomethane	Methyl iodide	CH_3I	74-88-4	141.939	liq	-66.4	42.43	2.2789^{20}	1.5308^{20}	sl H_2O; s ace, bz, chl; msc EtOH, eth
6308	1-Iodo-2-methoxybenzene	*o*-Iodoanisole	C_7H_7IO	529-28-2	234.034			241; 91^{2}	1.8^{20}		vs EtOH, eth, ace, bz, chl, lig
6309	1-Iodo-3-methoxybenzene	*m*-Iodoanisole	C_7H_7IO	766-85-8	234.034			244.5	1.9650^{20}		vs EtOH, eth
6310	1-Iodo-4-methoxybenzene	*p*-Iodoanisole	C_7H_7IO	696-62-8	234.034	lf (al), nd (MeOH)	53	238; 138^{25}			s EtOH, eth, chl
6311	1-Iodo-2-methylbenzene		C_7H_7I	615-37-2	218.035			211.5	1.713^{20}	1.6079^{20}	i H_2O; msc EtOH, eth
6312	1-Iodo-3-methylbenzene		C_7H_7I	625-95-6	218.035	liq	-27.2	213	1.705^{20}	1.6053^{20}	i H_2O; msc EtOH, eth
6313	(Iodomethyl)benzene		C_7H_7I	620-05-3	218.035	col or ye nd (MeOH)	24.5	93^{10}	1.7335^{25}	1.6334^{25}	vs bz, eth, EtOH
6314	1-Iodo-3-methylbutane	Isopentyl iodide	$C_5H_{11}I$	541-28-6	198.045			147	1.5118^{20}	1.4939^{20}	sl H_2O, ctc; msc EtOH, eth
6315	2-Iodo-2-methylbutane		$C_5H_{11}I$	594-38-7	198.045			124.5	1.4937^{20}	1.4981^{20}	i H_2O; msc EtOH, eth
6316	1-Iodo-2-methylpropane	Isobutyl iodide	C_4H_9I	513-38-2	184.018			121.1	1.6035^{20}	1.4959^{20}	
6317	2-Iodo-2-methylpropane	*tert*-Butyl iodide	C_4H_9I	558-17-8	184.018	liq	-38.2	100.1	1.571^{25}	1.4918^{20}	msc EtOH, eth
6318	Iodomethylsilane		CH_5ISi	18089-64-0	172.041	col liq	-109.5	71.8			
6319	1-Iodonaphthalene		$C_{10}H_7I$	90-14-2	254.067		2.1	302	1.7399^{20}	1.7026^{20}	i H_2O; msc EtOH, eth, bz, CS_2
6320	2-Iodonaphthalene		$C_{10}H_7I$	612-55-5	254.067	lf (dil al)	54.5	308	1.6319^{99}	1.6662^{99}	i H_2O; vs EtOH, eth, HOAc
6321	1-Iodo-2-nitrobenzene		$C_6H_4INO_2$	609-73-4	249.006	ye orth nd (al)	54	290; 162^{18}	1.9186^{75}		i H_2O; s EtOH, eth
6322	1-Iodo-3-nitrobenzene		$C_6H_4INO_2$	645-00-1	249.006	mcl pr	38.5	280	1.9477^{50}		i H_2O; s EtOH, eth
6323	1-Iodo-4-nitrobenzene		$C_6H_4INO_2$	636-98-6	249.006	ye nd (al)	174.7	288	1.8090^{155}		i H_2O; s EtOH, HOAc; sl DMSO
6324	1-Iodononane		$C_9H_{19}I$	4282-42-2	254.151	col liq	-20	245.0	1.2836^{25}	1.4848^{25}	
6325	1-Iodooctadecane		$C_{18}H_{37}I$	629-93-6	380.391	lf (lig), nd (ace, al-ace)	34.0	383	1.0994^{20}	1.4810^{20}	i H_2O; sl EtOH, eth
6326	1-Iodooctane		$C_8H_{17}I$	629-27-6	240.125	liq	-45.7	225.1	1.3298^{20}	1.4885^{20}	s EtOH, eth
6327	2-Iodooctane, (±)	2-Octyl iodide, (±)	$C_8H_{17}I$	36049-78-2	240.125			210; 95^{16}	1.3251^{20}	1.4896^{20}	i H_2O; s EtOH, eth, lig
6328	1-Iodopentane	Pentyl iodide	$C_5H_{11}I$	628-17-1	198.045	liq	-85.6	157.0	1.5161^{20}	1.4959^{20}	s chl
6329	3-Iodopentane		$C_5H_{11}I$	1809-05-8	198.045			145.5	1.5176^{20}	1.4974^{20}	vs ace, bz, eth
6330	2-Iodophenol		C_6H_5IO	533-58-4	220.007	nd	43	186^{160}, 91^{2}	1.8757^{80}		s H_2O; vs EtOH, eth, CS_2
6331	3-Iodophenol		C_6H_5IO	626-02-8	220.007	nd (lig)	118	186^{100}			sl H_2O; s EtOH, eth
6332	4-Iodophenol		C_6H_5IO	540-38-5	220.007	nd (w or sub)	93.5	139^{5} dec	1.8573^{112}		sl H_2O; vs EtOH, eth
6333	1-(3-Iodophenyl)ethanone	3-Iodoacetophenone	C_8H_7IO	14452-30-3	246.045			129^{8}, 117^{4}		1.622^{20}	s bz
6334	1-(4-Iodophenyl)ethanone	4-Iodoacetophenone	C_8H_7IO	13329-40-3	246.045		86	153^{18}			s EtOH, bz, CS_2, HOAc; sl lig, eth
6335	1-Iodopropane	Propyl iodide	C_3H_7I	107-08-4	169.992	liq	-101.3	102.5	1.7489^{20}	1.5058^{20}	sl H_2O, ctc; msc EtOH, eth
6336	2-Iodopropane	Isopropyl iodide	C_3H_7I	75-30-9	169.992	liq	-90	89.5	1.7042^{20}	1.5028^{20}	sl H_2O; msc EtOH, eth, bz, chl
6337	3-Iodopropanoic acid		$C_3H_5IO_2$	141-76-4	199.975	lf (w)	85				sl H_2O, chl; vs EtOH; s eth, ace
6338	3-Iodo-1-propanol		C_3H_7IO	627-32-7	185.991	visc oil		226; 115^{38}	1.9976^{20}	1.5585^{20}	
6339	3-Iodopropene	Allyl iodide	C_3H_5I	556-56-9	167.976	ye liq	-99.3	103	1.848^{12}	1.5540^{21}	i H_2O; s EtOH, eth, chl

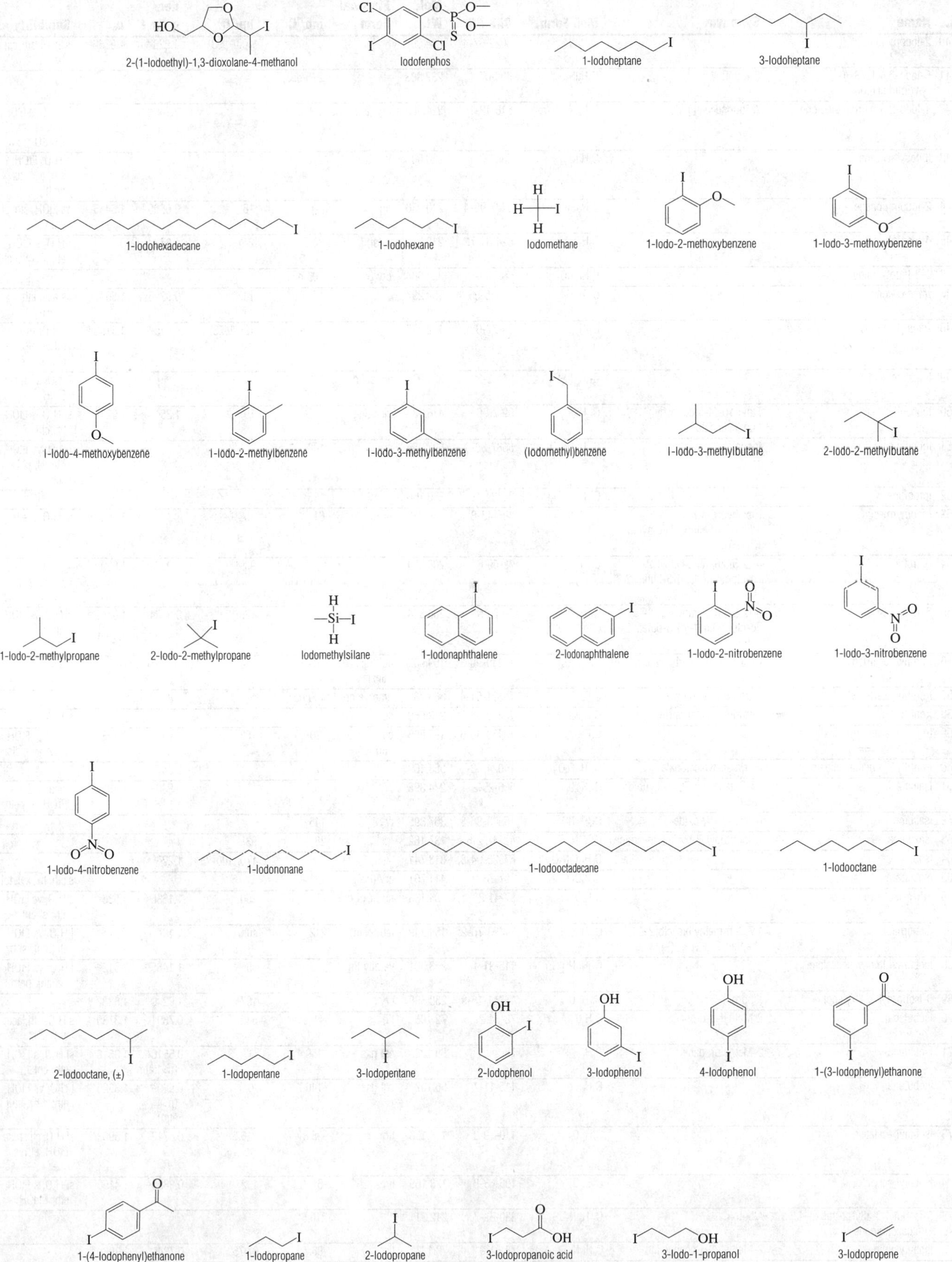

2-(1-Iodoethyl)-1,3-dioxolane-4-methanol
Iodofenphos
1-Iodoheptane
3-Iodoheptane
1-Iodohexadecane
1-Iodohexane
Iodomethane
1-Iodo-2-methoxybenzene
1-Iodo-3-methoxybenzene
1-Iodo-4-methoxybenzene
1-Iodo-2-methylbenzene
1-Iodo-3-methylbenzene
(Iodomethyl)benzene
1-Iodo-3-methylbutane
2-Iodo-2-methylbutane
1-Iodo-2-methylpropane
2-Iodo-2-methylpropane
Iodomethylsilane
1-Iodonaphthalene
2-Iodonaphthalene
1-Iodo-2-nitrobenzene
1-Iodo-3-nitrobenzene
1-Iodo-4-nitrobenzene
1-Iodononane
1-Iodooctadecane
1-Iodooctane
2-Iodooctane, (±)
1-Iodopentane
3-Iodopentane
2-Iodophenol
3-Iodophenol
4-Iodophenol
1-(3-Iodophenyl)ethanone
1-(4-Iodophenyl)ethanone
1-Iodopropane
2-Iodopropane
3-Iodopropanoic acid
3-Iodo-1-propanol
3-Iodopropene

No.	Name	Synonym	Mol. Form.	CAS RN	Mol. Wt.	Physical Form	mp/°C	bp/°C	den/ g cm^{-3}	n_D	Solubility
6340	2-Iodopyridine		C_5H_4IN	5029-67-4	204.997			100^{15}, 93^{13}	1.928^{25}	1.6366^{20}	s EtOH, eth, ace, bz
6341	5-Iodo-2,4(1*H*,3*H*)-pyrimidinedione	5-Iodouracil	$C_4H_3IN_2O_2$	696-07-1	237.983		275 dec				
6342	1-Iodo-2,5-pyrrolidinedione	*N*-Iodosuccinimide	$C_4H_4INO_2$	516-12-1	224.985	cry (ace)	200.5		2.245^{25}		vs H_2O; s EtOH, ace; sl eth, DMSO
6343	Iodosylbenzene		C_6H_5IO	536-80-1	220.007	ye pow	210 exp				s H_2O, EtOH; i eth, ace, bz, peth
6344	2-Iodothiophene		C_4H_3IS	3437-95-4	210.036	liq	-40	181	2.0595^{25}	1.6465^{25}	vs EtOH, eth; sl chl
6345	4-Iodotoluene		C_7H_7I	624-31-7	218.035	lf (al)	36.5	211	1.678^{20}		i H_2O; s EtOH, eth, CS_2; sl chl
6346	*L*-3-Iodotyrosine		$C_9H_{10}INO_3$	70-78-0	307.084	cry (w)	205 dec				
6347	*trans*-α-Ionone, (±)		$C_{13}H_{20}O$	30685-95-1	192.297			146^{28}	0.9298^{21}	1.5041^{20}	vs ace, eth, EtOH
6348	*trans*-β-Ionone		$C_{13}H_{20}O$	79-77-6	192.297			124^{10}, $73^{0.1}$	0.945^{20}	1.5198^{20}	sl H_2O; msc EtOH, eth; s chl
6349	Iopanoic acid		$C_{11}H_{12}I_3NO_2$	96-83-3	570.932	wh solid	156				i H_2O; s dil alk, EtOH
6350	Iophendylate	Ethyl 10-(4-iodophenyl) undecanoate	$C_{19}H_{29}IO_2$	99-79-6	416.336	visc liq		197^{1}	1.25^{20}	1.525^{25}	sl H_2O; s EtOH, bz, chl
6351	Iopodic acid	Ipodate	$C_{12}H_{13}I_3N_2O_2$	5587-89-3	597.956	cry	168				i H_2O; vs EtOH, MeOH, chl, ace
6352	Iprodione		$C_{13}H_{13}Cl_2N_3O_3$	36734-19-7	330.166		136				
6353	Iridomyrmecin	Hexahydro-4,7-dimethylcyclopenta[c]pyran-3(1*H*)-one	$C_{10}H_{16}O_2$	485-43-8	168.233	pr	61	$106^{1.5}$		1.4607^{65}	sl H_2O; s eth
6354	α-Irone	4-(2,5,6,6-Tetramethyl-2-cyclohexen-1-yl)-3-buten-2-one	$C_{14}H_{22}O$	79-69-6	206.324			$90^{0.4}$	0.9362^{20}	1.5002^{20}	
6355	β-Irone	4-(2,5,6,6-Tetramethyl-1-cyclohexen-1-yl)-3-buten-2-one	$C_{14}H_{22}O$	79-70-9	206.324			125^{11}	0.9434^{21}	1.5162^{25}	sl H_2O; vs EtOH, eth, bz, chl
6356	Iron hydrocarbonyl	Hydrogen tetracarbonylferrate(II)	$C_4H_2FeO_4$	17440-90-3	169.902	col liq; unstab	-70	dec			s alk
6357	Iron nonacarbonyl	Diiron nonacarbonyl	$C_9Fe_2O_9$	15321-51-4	363.781	oran-ye cry	100 dec		2.85		
6358	Iron(III) NTA	Nitrilotriacetatoiron(III)	$C_6H_6FeNO_6$	16448-54-7	243.960	solid					s H_2O
6359	Iron pentacarbonyl		C_5FeO_5	13463-40-6	195.896	col to ye oily liq	-20	103	1.5^{20}	1.453^{22}	i H_2O; sl EtOH; s bz, ace, ctc
6360	Iron(III) 2,4-pentanedioate	Ferric acetylacetonate	$C_{15}H_{21}FeO_6$	14024-18-1	353.169		179		5.24		
6361	Isanic acid	17-Octadecene-9,11-diynoic acid	$C_{18}H_{26}O_2$	506-25-2	274.398	cry	39.5		0.9309^{45}	1.49148^{50}	s ace, EtOH, i-PrOH; sl peth
6362	Isatidine	Retrorsine *N*-oxide	$C_{18}H_{25}NO_7$	15503-86-3	367.395	cry	145				
6363	Isaxonine	*N*-Isopropyl-2-pyrimidineamine	$C_7H_{11}N_3$	4214-72-6	137.182		28	93^{12}			
6364	Isazophos		$C_9H_{17}ClN_3O_3PS$	67329-04-8	313.741			170; $100^{0.001}$	1.22^{20}		
6365	Isobenzan		$C_9H_4Cl_8O$	297-78-9	411.751	cry (hp)	121				s eth, bz, xyl, tol
6366	1(3*H*)-Isobenzofuranone		$C_8H_6O_2$	87-41-2	134.133	nd or pl (w)	75	290	1.1636^{99}	1.536^{99}	s H_2O; vs EtOH, eth; sl chl
6367	Isoborneol	1,7,7-Trimethylbicyclo[2.2.1] heptan-2-ol, *exo*-(±)	$C_{10}H_{18}O$	24393-70-2	154.249	tab (peth)	212	sub	1.10^{20}		i H_2O; vs EtOH, eth, chl; sl bz
6368	Isobornyl thiocyanoacetate		$C_{13}H_{19}NO_2S$	115-31-1	253.361	ye oily liq		$95^{0.06}$	1.1465^{25}	1.512^{25}	i H_2O; vs EtOH, bz, chl, peth
6369	6-Isobornyl-3,4-xylenol	Xibornol	$C_{18}H_{26}O$	13741-18-9	258.398	cry	95	167^{3}	1.0240^{20}	1.5382^{20}	
6370	Isobutanal	2-Methyl-1-propanal	C_4H_8O	78-84-2	72.106	liq	-65.9	64.5	0.7891^{20}	1.3730^{20}	s H_2O, eth, ace, chl; sl ctc
6371	Isobutane	2-Methylpropane	C_4H_{10}	75-28-5	58.122	col gas	-159.4	-11.73	0.5510^{25} (p>1 atm	1.3518^{-25}	sl H_2O; s EtOH, eth, chl
6372	Isobutene		C_4H_8	115-11-7	56.107	col gas	-140.7	-6.9	0.589^{25} (p>1 atm)	1.3926^{-25}	i H_2O; vs EtOH, eth; s bz, sulf
6373	Isobutyl acetate		$C_6H_{12}O_2$	110-19-0	116.158	liq	-98.8	116.5	0.8712^{20}	1.3902^{20}	sl H_2O, ctc; msc EtOH, eth; s ace
6374	Isobutyl acrylate		$C_7H_{12}O_2$	106-63-8	128.169	liq	-61	132	0.8896^{20}	1.4150^{20}	sl H_2O; s EtOH, eth, MeOH
6375	5-Isobutyl-3-allyl-2-thioxo-4-imidazolidinone	Albutoin	$C_{10}H_{16}N_2OS$	830-89-7	212.311		210.5				
6376	Isobutylamine	2-Methyl-1-propanamine	$C_4H_{11}N$	78-81-9	73.137	liq	-86.7	67.75	0.724^{25}	1.3988^{19}	
6377	Isobutyl 4-aminobenzoate	Isobutyl *p*-aminobenzoate	$C_{11}H_{15}NO_2$	94-14-4	193.243		64.5				

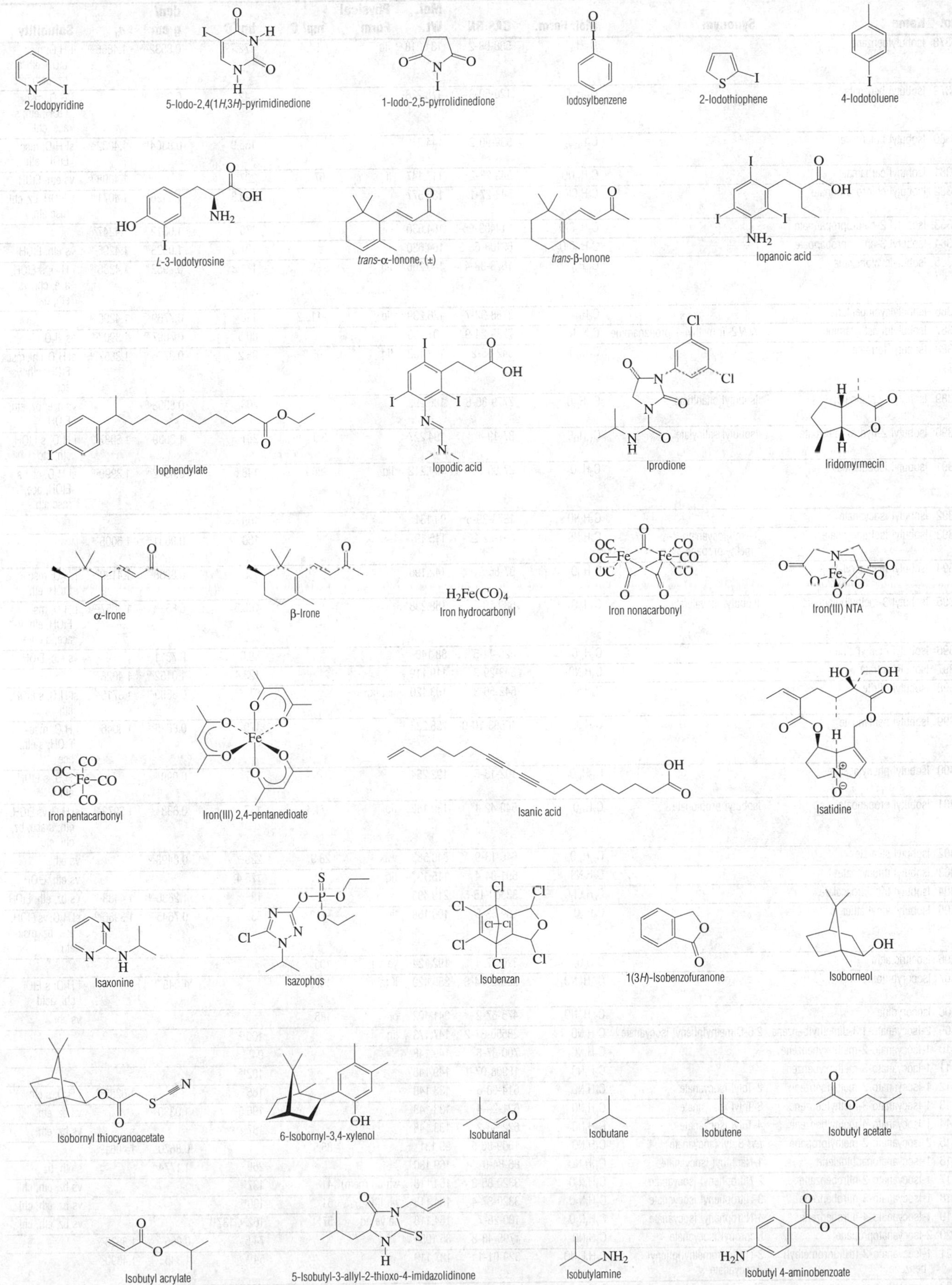
2-Iodopyridine
5-Iodo-2,4(1H,3H)-pyrimidinedione
1-Iodo-2,5-pyrrolidinedione
Iodosylbenzene
2-Iodothiophene
4-Iodotoluene
L-3-Iodotyrosine
trans-α-Ionone, (±)
trans-β-Ionone
Iopanoic acid
Iophendylate
Iopodic acid
Iprodione
Iridomyrmecin
α-Irone
β-Irone
H2Fe(CO)4
Iron hydrocarbonyl
Iron nonacarbonyl
Iron(III) NTA
Iron pentacarbonyl
Iron(III) 2,4-pentanedioate
Isanic acid
Isatidine
Isaxonine
Isazophos
Isobenzan
1(3H)-Isobenzofuranone
Isoborneol
Isobornyl thiocyanoacetate
6-Isobornyl-3,4-xylenol
Isobutanal
Isobutane
Isobutene
Isobutyl acetate
Isobutyl acrylate
5-Isobutyl-3-allyl-2-thioxo-4-imidazolidinone
Isobutylamine
Isobutyl 4-aminobenzoate

No.	Name	Synonym	Mol. Form.	CAS RN	Mol. Wt.	Physical Form	mp/°C	bp/°C	den/ g cm^{-3}	n_D	Solubility
6378	Isobutylbenzene		$C_{10}H_{14}$	538-93-2	134.218	liq	-51.4	172.79	0.8532^{20}	1.4866^{20}	i H_2O; msc EtOH, eth, ace, bz, peth, ctc
6379	Isobutyl benzoate		$C_{11}H_{14}O_2$	120-50-3	178.228			242	0.9990^{20}		i H_2O; msc EtOH, eth; s ace, chl
6380	Isobutyl butanoate		$C_8H_{16}O_2$	539-90-2	144.212			156.9	0.8364^{18}	1.4032^{20}	sl H_2O; msc EtOH, eth
6381	Isobutyl carbamate		$C_5H_{11}NO_2$	543-28-2	117.147	lf	67	207		1.4098^{76}	vs eth, EtOH
6382	Isobutyl chlorocarbonate		$C_5H_9ClO_2$	543-27-1	136.577			128.8	1.0426^{18}	1.4071^{18}	s EtOH, bz, chl; msc eth
6383	Isobutyl 2-chloropropanoate		$C_7H_{13}ClO_2$	114489-96-2	164.630			176	1.0312^{20}	1.4247^{20}	
6384	Isobutyl 3-chloropropanoate		$C_7H_{13}ClO_2$	62108-68-3	164.630			191.3	1.0323^{20}	1.4295^{20}	vs eth, EtOH
6385	Isobutylcyclohexane		$C_{10}H_{20}$	1678-98-4	140.266	liq	-95	171.3	0.7952^{20}	1.4386^{20}	i H_2O; s EtOH, ace, chl; vs eth, bz
6386	Isobutylcyclopentane		C_9H_{18}	3788-32-7	126.239	liq	-115.2	148	0.7769^{25}	1.4298^{20}	
6387	Isobutyldimethylamine	*N,N*,2-Trimethyl-1-propanamine	$C_6H_{15}N$	7239-24-9	101.190			80.5	0.7097^{20}	1.3907^{20}	vs H_2O
6388	Isobutyl formate		$C_5H_{10}O_2$	542-55-2	102.132	liq	-95.8	98.2	0.8776^{20}	1.3857^{20}	sl H_2O, chl; msc EtOH, eth; vs ace
6389	Isobutyl heptanoate	Isobutyl enanthate	$C_{11}H_{22}O_2$	7779-80-8	186.292			208	0.8593^{20}		vs ace, bz, eth, EtOH
6390	Isobutyl 2-hydroxybenzoate	Isobutyl salicylate	$C_{11}H_{14}O_3$	87-19-4	194.227		5.9	261	1.0639^{20}	1.5087^{20}	i H_2O; s EtOH, eth, ctc
6391	Isobutyl isobutanoate		$C_8H_{16}O_2$	97-85-8	144.212	liq	-80.7	148.6	0.8542^{20}	1.3999^{20}	sl H_2O, ctc; s EtOH, ace; msc eth
6392	Isobutyl isocyanate		C_5H_9NO	1873-29-6	99.131			106			
6393	Isobutyl isothiocyanate	1-Isothiocyanato-2-methylpropane	C_5H_9NS	591-82-2	115.197			160	0.9631^{14}	1.5005^{14}	
6394	Isobutyl methacrylate		$C_8H_{14}O_2$	97-86-9	142.196			155	0.8858^{20}	1.4199^{20}	i H_2O; msc EtOH, eth
6395	Isobutyl 3-methylbutanoate	Isobutyl isovalerate	$C_9H_{18}O_2$	589-59-3	158.238			168.5	0.853^{20}	1.4057^{20}	i H_2O; msc EtOH, eth; vs ace; s chl
6396	Isobutyl methyl ether		$C_5H_{12}O$	625-44-5	88.148			58.6	0.7311^{20}		vs eth, EtOH
6397	Isobutyl nitrate		$C_4H_9NO_3$	543-29-3	119.119			123.4	1.0152^{20}	1.4028^{20}	
6398	Isobutyl nitrite		$C_4H_9NO_2$	542-56-3	103.120	col liq		67	0.8699^{22}	1.3715^{22}	sl H_2O; s EtOH, eth
6399	Isobutyl pentanoate		$C_9H_{18}O_2$	10588-10-0	158.238			179	0.8625^{25}	1.4046^{20}	i H_2O; msc EtOH; s eth, ace
6400	Isobutyl phenylacetate		$C_{12}H_{16}O_2$	102-13-6	192.254			247	0.999^{18}		i H_2O; s EtOH, eth
6401	Isobutyl propanoate	Isobutyl propionate	$C_7H_{14}O_2$	540-42-1	130.185	liq	-71.4	137	0.888^{0}	1.3973^{20}	sl H_2O; vs EtOH, eth; s ace, bz, chl, ctc
6402	Isobutyl stearate		$C_{22}H_{44}O_2$	646-13-9	340.583	wax	28.9	223^{15}	0.8498^{20}		vs eth
6403	Isobutyl thiocyanate		C_5H_9NS	591-84-4	115.197	liq	-59	175.4			vs eth, EtOH
6404	Isobutyl trichloroacetate		$C_6H_9Cl_3O_2$	33560-15-5	219.493			188	1.2636^{20}	1.4483^{20}	vs bz, eth, EtOH
6405	Isobutyl vinyl ether		$C_6H_{12}O$	109-53-5	100.158	liq	-112	83	0.7645^{20}	1.3966^{20}	sl H_2O; vs EtOH, ace, bz; msc eth
6406	Isocitric acid		$C_6H_8O_7$	320-77-4	192.124	ye syr	105				
6407	Isocorybulbine		$C_{21}H_{25}NO_4$	22672-74-8	355.429	lf (al)	187.5		1.045^{20}		i H_2O; s EtOH, chl, acid
6408	Isocorydine		$C_{20}H_{23}NO_4$	475-67-2	341.402	pl	185				vs chl
6409	2-Isocyanato-1,3-dimethylbenzene	2,6-Dimethylphenyl isocyanate	C_9H_9NO	28556-81-2	147.173	liq		100^{13}			
6410	1-Isocyanato-2-methoxybenzene		$C_8H_7NO_2$	700-87-8	149.148			94^{17}			
6411	1-Isocyanato-3-methoxybenzene		$C_8H_7NO_2$	18908-07-1	149.148			102^{15}			
6412	1-Isocyanato-2-methylbenzene	2-Tolyl isocyanate	C_8H_7NO	614-68-6	133.148			185		1.5282^{20}	i H_2O; s eth
6413	1-Isocyanato-3-methylbenzene	3-Tolyl isocyanate	C_8H_7NO	621-29-4	133.148			196.5	1.0330^{20}		vs bz, eth
6414	1-Isocyanato-4-methylbenzene	4-Tolyl isocyanate	C_8H_7NO	622-58-2	133.148			187			vs bz, eth
6415	2-Isocyanato-2-methylpropane	*tert*-Butyl isocyanate	C_5H_9NO	1609-86-5	99.131		85.5		0.8670^{7}	1.4061^{20}	
6416	1-Isocyanatonaphthalene	1-Naphthyl isocyanate	$C_{11}H_7NO$	86-84-0	169.180			269	1.1774^{20}		s eth, bz
6417	1-Isocyanato-2-nitrobenzene	2-Nitrophenyl isocyanate	$C_7H_4N_2O_3$	3320-86-3	164.118	wh nd (peth)	41	137^{18}			vs bz, eth, chl
6418	1-Isocyanato-3-nitrobenzene	3-Nitrophenyl isocyanate	$C_7H_4N_2O_3$	3320-87-4	164.118	wh lf (lig)	51	130^{11}			vs bz, eth, chl
6419	1-Isocyanato-4-nitrobenzene	4-Nitrophenyl isocyanate	$C_7H_4N_2O_3$	100-28-7	164.118	pa ye nd	57	162^{20}, 137^{11}			vs bz, eth, chl
6420	2-Isocyanatopropane	Isopropyl isocyanate	C_4H_7NO	1795-48-8	85.105			74.5	0.866^{25}	1.3825^{20}	
6421	1-Isocyanato-3-(trifluoromethyl) benzene	3-(Trifluoromethyl)phenyl isocyanate	$C_8H_4F_3NO$	329-01-1	187.119			54^{11}	1.3455^{20}	1.4690^{20}	
6422	Isocyanobenzene	Phenyl isocyanide	C_7H_5N	931-54-4	103.122	unstab liq		80^{40}	0.98^{15}		

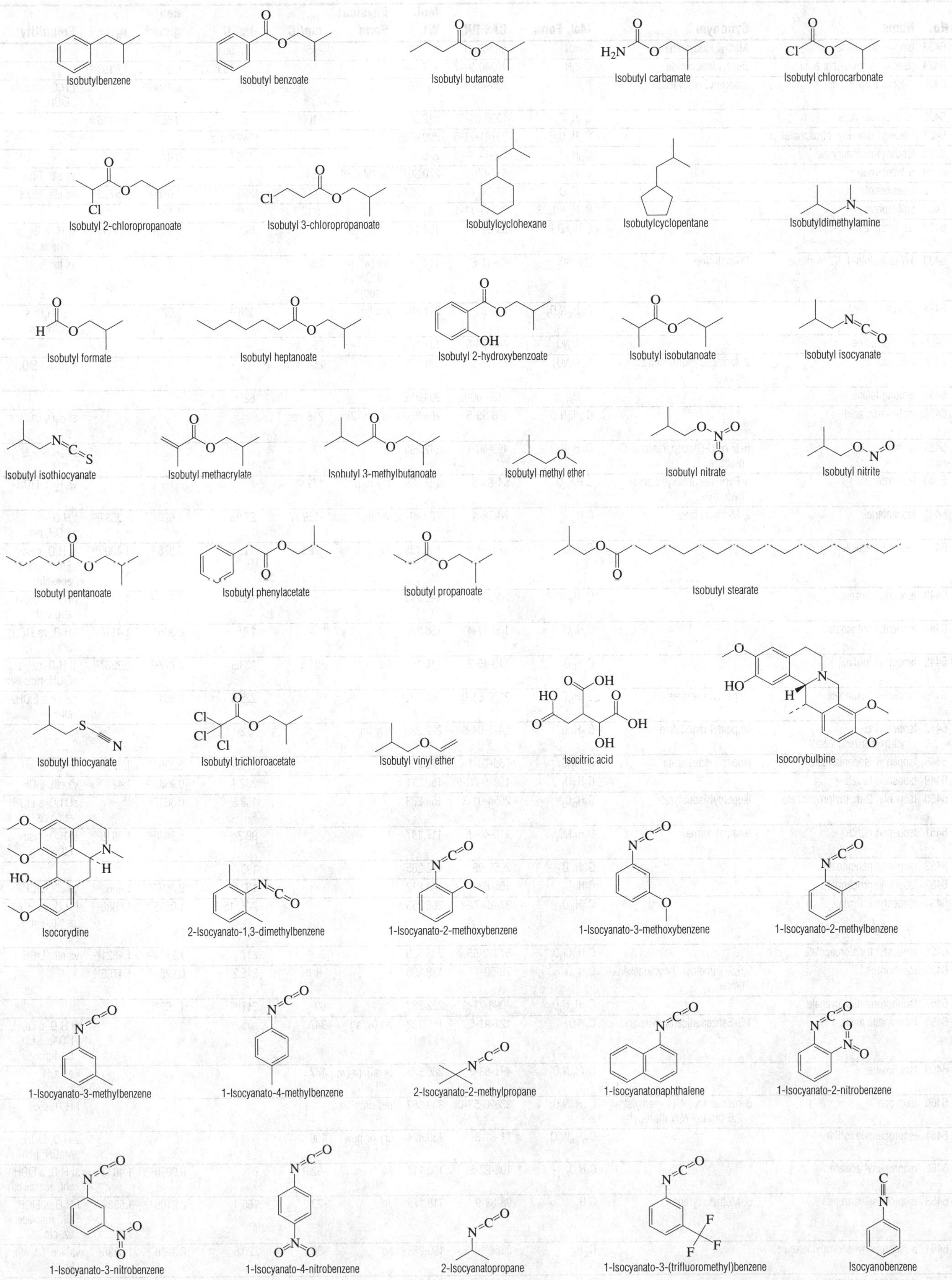
Isobutylbenzene
Isobutyl benzoate
Isobutyl butanoate
Isobutyl carbamate
Isobutyl chlorocarbonate
Isobutyl 2-chloropropanoate
Isobutyl 3-chloropropanoate
Isobutylcyclohexane
Isobutylcyclopentane
Isobutyldimethylamine
Isobutyl formate
Isobutyl heptanoate
Isobutyl 2-hydroxybenzoate
Isobutyl isobutanoate
Isobutyl isocyanate
Isobutyl isothiocyanate
Isobutyl methacrylate
Isobutyl 3-methylbutanoate
Isobutyl methyl ether
Isobutyl nitrate
Isobutyl nitrite
Isobutyl pentanoate
Isobutyl phenylacetate
Isobutyl propanoate
Isobutyl stearate
Isobutyl thiocyanate
Isobutyl trichloroacetate
Isobutyl vinyl ether
Isocitric acid
Isocorybulbine
Isocorydine
2-Isocyanato-1,3-dimethylbenzene
1-Isocyanato-2-methoxybenzene
1-Isocyanato-3-methoxybenzene
1-Isocyanato-2-methylbenzene
1-Isocyanato-3-methylbenzene
1-Isocyanato-4-methylbenzene
2-Isocyanato-2-methylpropane
1-Isocyanatonaphthalene
1-Isocyanato-2-nitrobenzene
1-Isocyanato-3-nitrobenzene
1-Isocyanato-4-nitrobenzene
2-Isocyanatopropane
1-Isocyanato-3-(trifluoromethyl)benzene
Isocyanobenzene

No.	Name	Synonym	Mol. Form.	CAS RN	Mol. Wt.	Physical Form	mp/°C	bp/°C	den/ g cm^{-3}	n_D	Solubility
6423	Isocyanomethane	Methyl isocyanide	C_2H_3N	593-75-9	41.052		-45	exp 59.6	0.756^{4}		
6424	(Isocyanomethyl)benzene	Benzyl isocyanide	C_8H_7N	10340-91-7	117.149			dec 199; 93^{55}	0.972^{15}	1.5193^{20}	
6425	2-Isocyanopropane	Isopropyl isocyanide	C_4H_7N	598-45-8	69.106			87	0.7596^{25}		i H_2O; msc EtOH, eth
6426	Isodecyl acrylate		$C_{13}H_{24}O_2$	1330-61-6	212.329		-100	158^{50}	0.885^{20}	1.4416^{20}	
6427	Isodecyl diphenyl phosphate		$C_{22}H_{31}O_4P$	29761-21-5	390.452			249^{10} dec			
6428	Isodecyl methacrylate		$C_{14}H_{26}O_2$	29964-84-9	226.355			126^{10}	0.876^{20}		
6429	8-Isoestrone		$C_{18}H_{22}O_2$	517-06-6	270.367	pr (MeOH)	254				vs eth, Diox
6430	Isoeugenol		$C_{10}H_{12}O_2$	97-54-1	164.201			266	1.080^{25}	1.5739^{19}	vs eth, EtOH
6431	Isofenphos		$C_{15}H_{24}NO_4PS$	25311-71-1	345.395		<-12	$120^{0.01}$	1.134^{20}		
6432	Isoflurophate		$C_6H_{14}FO_3P$	55-91-4	184.145			62^{9}	1.055^{25}	1.3830^{25}	sl H_2O, lig; s eth; vs oils
6433	1*H*-Isoindole-1,3(2*H*)-dione	Phthalimide	$C_8H_5NO_2$	85-41-6	147.132	nd (w), pr (HOAc) lf (sub)	238				vs bz
6434	Isolan		$C_{10}H_{17}N_3O_2$	119-38-0	211.261	col liq		$118^{2.5}$	1.07^{20}		msc H_2O; s EtOH, xyl
6435	*DL*-Isoleucine		$C_6H_{13}NO_2$	443-79-8	131.173		292 dec				
6436	*L*-Isoleucine	2-Amino-3-methylpentanoic acid	$C_6H_{13}NO_2$	73-32-5	131.173		284 dec				s H_2O; i EtOH
6437	Isolongifolene		$C_{15}H_{24}$	1135-66-6	204.352	liq		$82^{0.4}$			
6438	Isolysergic acid		$C_{16}H_{16}N_2O_2$	478-95-5	268.310	cry (w+2)	218 dec				sl H_2O, EtOH; s py
6439	α-Isomaltose	6-*O*-α-*D*-Glucopyranosyl-*D*-glucose	$C_{12}H_{22}O_{11}$	499-40-1	342.296		120				
6440	Isoniazid	4-Pyridinecarboxylic acid hydrazide	$C_6H_7N_3O$	54-85-3	137.139	cry (al)	171.4				vs H_2O, EtOH
6441	Isopentane	2-Methylbutane	C_5H_{12}	78-78-4	72.149	vol liq or gas	-159.77	27.88	0.6201^{20}	1.3537^{20}	i H_2O; msc EtOH, eth
6442	Isopentyl acetate		$C_7H_{14}O_2$	123-92-2	130.185	liq	-78.5	142.5	0.876^{15}	1.4000^{20}	sl H_2O; msc EtOH, eth; s ace, chl
6443	Isopentylbenzene		$C_{11}H_{16}$	2049-94-7	148.245			195	0.856^{20}	1.4867^{10}	i H_2O; s EtOH, eth; vs bz
6444	Isopentyl butanoate		$C_9H_{18}O_2$	106-27-4	158.238			179	0.865^{19}	1.4110^{20}	i H_2O; vs EtOH, eth
6445	Isopentyl formate		$C_6H_{12}O_2$	110-45-2	116.158	liq	-93.5	123.5	0.877^{20}	1.3967^{20}	sl H_2O, ctc; s EtOH; msc eth
6446	Isopentyl hexanoate	Isopentyl caproate	$C_{11}H_{22}O_2$	2198-61-0	186.292			225.5	0.861^{20}		i H_2O; s EtOH, eth
6447	Isopentyl α-hydroxybenzeneacetate	Isopentyl mandelate	$C_{13}H_{18}O_3$	5421-04-5	222.280	oily liq		172^{11}			
6448	Isopentyl isopentanoate	Isoamyl isovalerate	$C_{10}H_{20}O_2$	659-70-1	172.265			190.4	0.8583^{19}	1.4130^{19}	
6449	Isopentyl lactate		$C_8H_{16}O_3$	19329-89-6	160.211			202.4	0.9589^{25}	1.4240^{25}	vs eth, EtOH
6450	Isopentyl 2-methylpropanoate	Isopentyl isobutyrate	$C_9H_{18}O_2$	2050-01-3	158.238			168.5	0.8627^{20}		sl H_2O; s EtOH, eth, ace
6451	Isopentyl nitrite	Isoamyl nitrite	$C_5H_{11}NO_2$	110-46-3	117.147			99.2	0.8828^{20}	1.3918^{20}	sl H_2O; msc EtOH, eth
6452	Isopentyl pentanoate		$C_{10}H_{20}O_2$	2050-09-1	172.265			193			
6453	Isopentyl propanoate		$C_8H_{16}O_2$	105-68-0	144.212			160.2	0.8697^{20}	1.4069^{20}	vs eth, EtOH
6454	Isopentyl salicylate		$C_{12}H_{16}O_3$	87-20-7	208.253			278; 151^{15}	1.0535^{20}	1.5080^{20}	i H_2O; vs EtOH; s eth, chl; sl ctc
6455	Isopentyl trichloroacetate		$C_7H_{11}Cl_3O_2$	57392-55-9	233.520			217	1.2314^{20}	1.4521^{20}	vs eth, EtOH
6456	Isophorone	3,5,5-Trimethyl-2-cyclohexen-1-one	$C_9H_{14}O$	78-59-1	138.206	liq	-8.1	215.2	0.9255^{20}	1.4766^{18}	
6457	Isophorone diisocyanate		$C_{12}H_{18}N_2O_2$	4098-71-9	222.283		60	217^{100}	1.062^{20}		
6458	Isophthalic acid	1,3-Benzenedicarboxylic acid	$C_8H_6O_4$	121-91-5	166.132	nd (w, al)	347	sub			sl H_2O; s EtOH, HOAc; i eth, bz, lig
6459	Isopilosine		$C_{16}H_{18}N_2O_3$	491-88-3	286.325	pl (al), pr (w, dil al)	187				vs EtOH
6460	Isopropalin	Benzenamine, 4-(1-methylethyl) -2,6-dinitro-*N*,*N*-dipropyl-	$C_{15}H_{23}N_3O_4$	33820-53-0	309.362	red-oran liq					i H_2O; s os
6461	Isopropamide iodide		$C_{23}H_{33}IN_2O$	71-81-8	480.424	cry or pow	190				s H_2O, EtOH, MeOH; i chl
6462	Isopropenyl acetate		$C_5H_8O_2$	108-22-5	100.117	liq	-92.9	94	0.9090^{20}	1.4033^{20}	sl H_2O; s EtOH, chl, ace; vs eth
6463	Isopropenylbenzene	α-Methyl styrene	C_9H_{10}	98-83-9	118.175	liq	-23.2	165.4	0.9106^{20}	1.5386^{20}	i H_2O; s EtOH, eth; msc ace, bz, ctc
6464	*p*-Isopropenylisopropylbenzene		$C_{12}H_{16}$	2388-14-9	160.255	liq	-30.6	220.8	0.8936^{20}	1.5238^{20}	vs ace, bz, eth, EtOH
6465	*p*-Isopropenylstyrene		$C_{11}H_{12}$	16262-48-9	144.213	liq		242	0.93	1.5684^{20}	

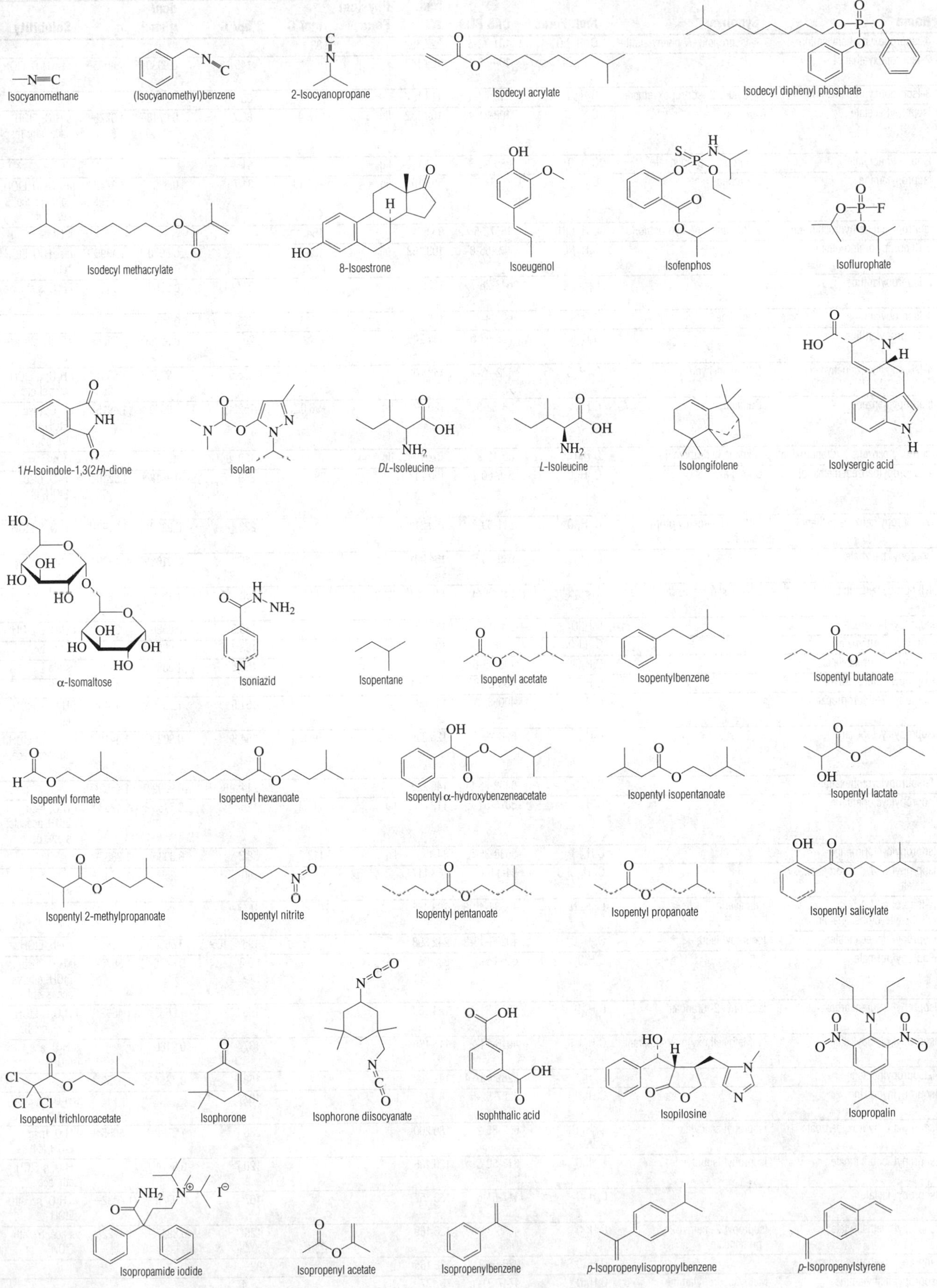

Isocyanomethane
(Isocyanomethyl)benzene
2-Isocyanopropane
Isodecyl acrylate
Isodecyl diphenyl phosphate
Isodecyl methacrylate
8-Isoestrone
Isoeugenol
Isofenphos
Isoflurophate
1H-Isoindole-1,3(2H)-dione
Isolan
DL-Isoleucine
L-Isoleucine
Isolongifolene
Isolysergic acid
α-Isomaltose
Isoniazid
Isopentane
Isopentyl acetate
Isopentylbenzene
Isopentyl butanoate
Isopentyl formate
Isopentyl hexanoate
Isopentyl α-hydroxybenzeneacetate
Isopentyl isopentanoate
Isopentyl lactate
Isopentyl 2-methylpropanoate
Isopentyl nitrite
Isopentyl pentanoate
Isopentyl propanoate
Isopentyl salicylate
Isopentyl trichloroacetate
Isophorone
Isophorone diisocyanate
Isophthalic acid
Isopilosine
Isopropalin
Isopropamide iodide
Isopropenyl acetate
Isopropenylbenzene
p-Isopropenylisopropylbenzene
p-Isopropenylstyrene

No.	Name	Synonym	Mol. Form.	CAS RN	Mol. Wt.	Physical Form	mp/°C	bp/°C	den/ g cm⁻³	n_D	Solubility
6466	4-Isopropoxydiphenylamine	4-Isopropoxy-*N*-phenylaniline	$C_{15}H_{17}NO$	101-73-5	227.302		83				
6467	2-Isopropoxyethanol		$C_5H_{12}O_2$	109-59-1	104.148			145	0.9030^{20}	1.4095^{20}	msc H_2O, EtOH, eth; s ace
6468	3-Isopropoxypropanenitrile	1-Cyano-2-isopropoxyethane	$C_6H_{11}NO$	110-47-4	113.157			65^{10}			s chl
6469	Isopropyl acetate		$C_5H_{10}O_2$	108-21-4	102.132	liq	-73.4	88.7	0.8718^{20}	1.3773^{20}	s H_2O, EtOH, ace, chl; msc eth
6470	Isopropyl acrylate	Isopropyl 2-propenoate	$C_6H_{10}O_2$	689-12-3	114.142	liq		51^{103}			
6471	Isopropylamine	2-Propanamine	C_3H_9N	75-31-0	59.110	liq	-95.13	31.76	0.6891^{20}	1.3742^{20}	msc H_2O, EtOH, eth; vs ace; s bz, chl
6472	Isopropylamine hydrochloride	2-Propanamine hydrochloride	$C_3H_{10}ClN$	15572-56-2	95.571		164				s DMSO
6473	2-(Isopropylamino)ethanol		$C_5H_{13}NO$	109-56-8	103.163		128.5	173	0.8970^{20}	1.4395^{20}	msc H_2O, EtOH, eth
6474	2-Isopropylaniline		$C_9H_{13}N$	643-28-7	135.206			221; 95^{13}	0.9760^{12}		i H_2O; s eth, bz, ctc
6475	4-Isopropylaniline	Cumidine	$C_9H_{13}N$	99-88-7	135.206			225	0.953^{20}		
6476	*N*-Isopropylaniline		$C_9H_{13}N$	768-52-5	135.206			203	0.9526^{25}	1.5380^{20}	s EtOH, eth, ace, bz
6477	4-Isopropylbenzaldehyde	Cuminaldehyde	$C_{10}H_{12}O$	122-03-2	148.201			235.5	0.9755^{20}	1.5301^{20}	i H_2O; s EtOH, eth; sl ctc
6478	Isopropylbenzene	Cumene	C_9H_{12}	98-82-8	120.191	liq	-96.02	152.41	0.8640^{25}	1.4915^{20}	i H_2O; msc EtOH, eth, ace, bz, peth, ctc
6479	Isopropylbenzene hydroperoxide	Cumene hydroperoxide	$C_9H_{12}O_2$	80-15-9	152.190	liq		153; 84^8	1.03^{20}		
6480	4-Isopropylbenzenemethanol	Cumic alcohol	$C_{10}H_{14}O$	536-60-7	150.217		28	249	0.9818^{20}	1.5210^{20}	i H_2O; msc EtOH, eth; vs bz
6481	α-Isopropylbenzenemethanol	1-Phenyl-2-methylpropyl alcohol	$C_{10}H_{14}O$	611-69-8	150.217			223	0.9869^{14}	1.5193^{14}	i H_2O; s EtOH, ace
6482	Isopropyl benzoate		$C_{10}H_{12}O_2$	939-48-0	164.201			216	1.0163^{15}	1.4890^{20}	i H_2O; s EtOH, eth, ace
6483	4-Isopropylbenzoic acid	Cumic acid	$C_{10}H_{12}O_2$	536-66-3	164.201	tcl pl (al)	117.5	sub	1.162^4		sl H_2O; vs EtOH, eth; s peth
6484	Isopropyl butanoate		$C_7H_{14}O_2$	638-11-9	130.185			130.5	0.8588^{20}	1.3936^{20}	i H_2O; s EtOH
6485	Isopropyl carbamate		$C_4H_9NO_2$	1746-77-6	103.120	nd	93	183	0.9951^{66}		
6486	Isopropyl chloroacetate		$C_5H_9ClO_2$	105-48-6	136.577			150.5	1.0888^{20}	1.4382^{20}	vs eth
6487	Isopropyl chloroformate		$C_4H_7ClO_2$	108-23-6	122.551			105		1.4013^{20}	vs eth
6488	Isopropyl 2-chloropropanoate		$C_6H_{11}ClO_2$	40058-87-5	150.603			151.5	1.0315^{20}	1.4149^{20}	i H_2O; s EtOH, eth
6489	Isopropylcyclohexane		C_9H_{18}	696-29-7	126.239	liq	-89.4	154.8	0.8023^{20}	1.4410^{20}	i H_2O; vs EtOH, eth; msc ace, bz
6490	4-Isopropylcyclohexanone		$C_9H_{16}O$	5432-85-9	140.222			214; 139^{100}	0.9099^{30}	1.4552^{25}	
6491	Isopropylcyclopentane		C_8H_{16}	3875-51-2	112.213	liq	-111.4	126.5	0.7765^{20}	1.4258^{20}	i H_2O; msc EtOH, ace, ctc; s eth, bz
6492	Isopropylcyclopropane		C_6H_{12}	3638-35-5	84.159	liq	-112.9	58.3	0.6936^{25}	1.3865^{20}	
6493	Isopropyl (2,4-dichlorophenoxy) acetate		$C_{11}H_{12}Cl_2O_3$	94-11-1	263.117		5	140^1	1.26^{25}	1.5209^{25}	
6494	*N*-Isopropyl-4,4-diphenylcyclohexanamine	Pramiverin	$C_{21}H_{27}N$	14334-40-8	293.446		70	$165^{0.05}$			
6495	Isopropyl dodecanoate	Isopropyl laurate	$C_{15}H_{30}O_2$	10233-13-3	242.398			196^{60}, $105^{0.8}$	0.8536^{20}	1.4280^{25}	vs eth, EtOH
6496	Isopropyl formate		$C_4H_8O_2$	625-55-8	88.106			68.2	0.8728^{20}	1.3678^{20}	sl H_2O; msc EtOH, eth; vs ace; s chl
6497	Isopropyl 2-furancarboxylate	Isopropyl 2-furanoate	$C_8H_{10}O_3$	6270-34-4	154.163			198.5	1.0655^{24}	1.4682^{24}	i H_2O; s EtOH, eth, ace, bz
6498	Isopropyl glycidyl ether	(1-Methylethoxy)methyloxirane	$C_6H_{12}O_2$	4016-14-2	116.158			137	0.9186^{20}		s H_2O, ace, EtOH
6499	4-Isopropylheptane		$C_{10}H_{22}$	52896-87-4	142.282			158.9	0.7354^{25}	1.4153^{20}	
6500	Isopropylhydrazine		$C_3H_{10}N_2$	2257-52-5	74.124	liq		107			s H_2O, bz, EtOH; sl eth
6501	Isopropyl 2-hydroxybenzoate	Isopropyl salicylate	$C_{10}H_{12}O_3$	607-85-2	180.200			238	1.0729^{20}	1.5065^{20}	i H_2O; msc EtOH, eth
6502	Isopropyl isobutanoate	Isopropyl isobutyrate	$C_7H_{14}O_2$	617-50-5	130.185			120.7	0.8471^{21}		i H_2O; s EtOH, eth, ace
6503	Isopropyl lactate		$C_6H_{12}O_3$	617-51-6	132.157			167	0.9980^{20}	1.4082^{25}	vs H_2O, bz, eth, EtOH
6504	Isopropyl methacrylate	Isopropyl 2-methyl-2-propenoate	$C_7H_{12}O_2$	4655-34-9	128.169			125	0.8847^{20}	1.4122^{20}	vs ace, bz, eth, EtOH
6505	Isopropyl methanesulfonate		$C_4H_{10}O_3S$	926-06-7	138.185			82^6			
6506	Isopropylmethylamine	Methylisopropylamine	$C_4H_{11}N$	4747-21-1	73.137			50.4			

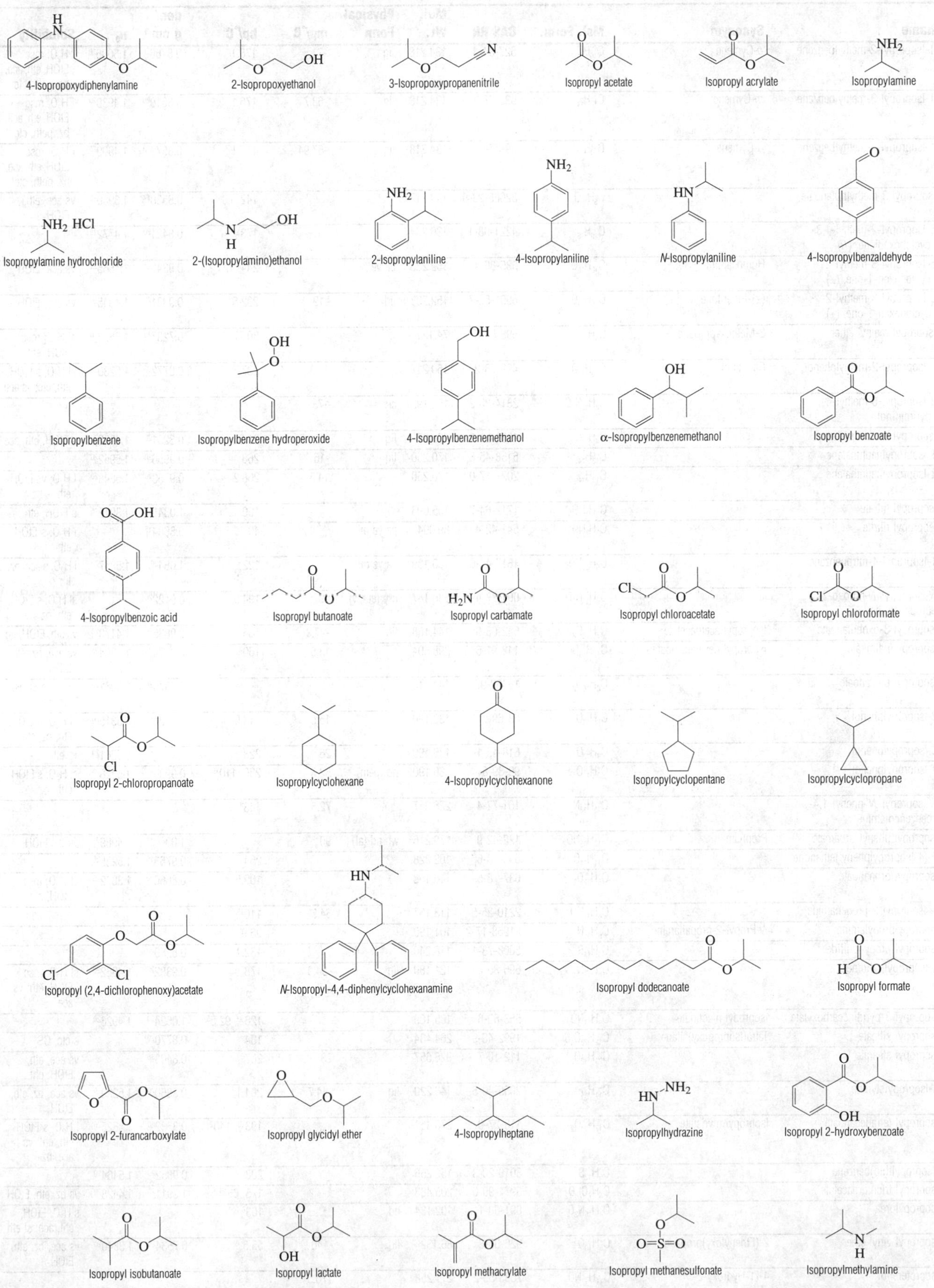
4-Isopropoxydiphenylamine
2-Isopropoxyethanol
3-Isopropoxypropanenitrile
Isopropyl acetate
Isopropyl acrylate
Isopropylamine
Isopropylamine hydrochloride
2-(Isopropylamino)ethanol
2-Isopropylaniline
4-Isopropylaniline
N-Isopropylaniline
4-Isopropylbenzaldehyde
Isopropylbenzene
Isopropylbenzene hydroperoxide
4-Isopropylbenzenemethanol
α-Isopropylbenzenemethanol
Isopropyl benzoate
4-Isopropylbenzoic acid
Isopropyl butanoate
Isopropyl carbamate
Isopropyl chloroacetate
Isopropyl chloroformate
Isopropyl 2-chloropropanoate
Isopropylcyclohexane
4-Isopropylcyclohexanone
Isopropylcyclopentane
Isopropylcyclopropane
Isopropyl (2,4-dichlorophenoxy)acetate
N-Isopropyl-4,4-diphenylcyclohexanamine
Isopropyl dodecanoate
Isopropyl formate
Isopropyl 2-furancarboxylate
Isopropyl glycidyl ether
4-Isopropylheptane
Isopropylhydrazine
Isopropyl 2-hydroxybenzoate
Isopropyl isobutanoate
Isopropyl lactate
Isopropyl methacrylate
Isopropyl methanesulfonate
Isopropylmethylamine

No.	Name	Synonym	Mol. Form.	CAS RN	Mol. Wt.	Physical Form	mp/°C	bp/°C	den/ g cm⁻³	n_D	Solubility
6507	1-Isopropyl-2-methylbenzene	*o*-Cymene	$C_{10}H_{14}$	527-84-4	134.218	liq	-71.5	178.1	0.8766^{20}	1.5006^{20}	i H_2O; msc EtOH, eth, ace, bz, peth, ctc
6508	1-Isopropyl-3-methylbenzene	*m*-Cymene	$C_{10}H_{14}$	535-77-3	134.218	liq	-63.7	175.1	0.8610^{20}	1.4930^{20}	i H_2O; msc EtOH, eth, ace, bz, peth, ctc
6509	1-Isopropyl-4-methylbenzene	*p*-Cymene	$C_{10}H_{14}$	99-87-6	134.218	liq	-67.94	177.1	0.8573^{20}	1.4909^{20}	i H_2O; msc EtOH, eth, ace, bz, peth, ctc
6510	Isopropyl 3-methylbutanoate		$C_8H_{16}O_2$	32665-23-9	144.212			142; 70^{55}	0.8538^{17}	1.3960^{20}	vs ace, eth, EtOH
6511	5-Isopropyl-2-methyl-1,3-cyclohexadiene, (*R*)		$C_{10}H_{16}$	4221-98-1	136.234			173	0.8421^{20}	1.4772^{19}	
6512	5-Isopropyl-3-methyl-2-cyclohexen-1-one, (±)	Homocamfin	$C_{10}H_{16}O$	535-86-4	152.233	pa ye		244; 121^{15}	0.9340^{21}	1.4865^{21}	vs ace, EtOH
6513	6-Isopropyl-3-methyl-2-cyclohexen-1-one, (±)	(±)-Piperitone	$C_{10}H_{16}O$	6091-52-7	152.233	liq	-19	232.5	0.9331^{20}	1.4845^{20}	vs ace, EtOH
6514	Isopropyl methyl ether	2-Methoxypropane	$C_4H_{10}O$	598-53-8	74.121			30.77	0.7237^{15}	1.3576^{20}	sl H_2O; msc EtOH, eth
6515	5-Isopropyl-2-methylphenol	Carvacrol	$C_{10}H_{14}O$	499-75-2	150.217	nd	1	237.7	0.9772^{20}	1.5230^{20}	sl H_2O; s EtOH, eth, ctc; vs ace
6516	2-Isopropyl-6-methyl-4-pyrimidinol		$C_8H_{12}N_2O$	2814-20-2	152.193	cry	173				
6517	Isopropyl methyl sulfide		$C_4H_{10}S$	1551-21-9	90.187	liq	-101.5	84.8	0.8291^{20}	1.4932^{20}	s EtOH, eth, ace
6518	1-Isopropylnaphthalene		$C_{13}H_{14}$	6158-45-8	170.250	liq	-16	268	0.9956^{20}	1.5952^{20}	
6519	2-Isopropylnaphthalene		$C_{13}H_{14}$	2027-17-0	170.250		14.5	268.2	0.9753^{20}	1.5848^{20}	i H_2O; vs EtOH, eth; s bz
6520	Isopropyl nitrate		$C_3H_7NO_3$	1712-64-7	105.093			100	1.034^{19}	1.3912^{16}	s EtOH, eth
6521	Isopropyl nitrite		$C_3H_7NO_2$	541-42-4	89.094	pa ye oil		40	0.8684^{15}		i H_2O; s EtOH, eth
6522	1-Isopropyl-4-nitrobenzene		$C_9H_{11}NO_2$	1817-47-6	165.189	pa ye oil		122^9	1.084^{20}	1.5367^{20}	i H_2O; s ace, bz, lig
6523	*N*-Isopropyl-*N*-nitroso-2-propanamine		$C_6H_{14}N_2O$	601-77-4	130.187	cry (eth,w)	48	194.5	0.9422^{20}		sl H_2O; s EtOH, eth, bz
6524	Isopropyl 3-oxobutanoate	Isopropyl acetoacetate	$C_7H_{12}O_3$	542-08-5	144.168	liq	-27.3	186	0.9835^{20}	1.4173^{20}	vs eth, EtOH, lig
6525	Isopropyl palmitate	Isopropyl hexadecanoate	$C_{19}H_{38}O_2$	142-91-6	298.504		13.5	160^2	0.8404^{38}	1.4364^{25}	vs ace, bz, eth, EtOH
6526	Isopropyl pentanoate		$C_8H_{16}O_2$	18362-97-5	144.212				0.8579^{20}	1.4061^{20}	i H_2O; s EtOH, eth, ace
6527	2-Isopropylphenol		$C_9H_{12}O$	88-69-7	136.190		15.5	213.5	1.012^{20}	1.5315^{20}	sl H_2O; s EtOH, eth, bz, ctc
6528	3-Isopropylphenol		$C_9H_{12}O$	618-45-1	136.190		26	228		1.5261^{20}	vs eth
6529	4-Isopropylphenol		$C_9H_{12}O$	99-89-8	136.190	nd (peth)	62.3	230; 110^{10}	0.990^{20}	1.5228^{20}	sl H_2O; s EtOH, chl
6530	*N*-Isopropyl-*N'*-phenyl-1,4-benzenediamine		$C_{15}H_{18}N_2$	101-72-4	226.317		72.5	148^2			
6531	Isopropyl phenylcarbamate	Propham	$C_{10}H_{13}NO_2$	122-42-9	179.216	wh nd (al)	90		1.09^{20}	1.4989^{91}	vs bz, EtOH
6532	1-(4-Isopropylphenyl)ethanone		$C_{11}H_{14}O$	645-13-6	162.228			254	0.9753^{15}	1.5235^{20}	
6533	Isopropyl propanoate		$C_6H_{12}O_2$	637-78-5	116.158			109.5	0.8660^{20}	1.3872^{20}	sl H_2O; msc EtOH, eth
6534	*N*-Isopropyl-2-propenamide		$C_6H_{11}NO$	2210-25-5	113.157		64.5	110^{15}			
6535	Isopropylpropylamine	*N*-Propyl-2-propanamine	$C_6H_{15}N$	21968-17-2	101.190			96.9			
6536	Isopropyl propyl sulfide		$C_6H_{14}S$	5008-73-1	118.240			132.1	0.8269^{20}		
6537	4-Isopropylpyridine		$C_8H_{11}N$	696-30-0	121.180	liq	-54.9	178	0.9382^{25}	1.4962^{20}	sl H_2O; msc EtOH, eth; vs ace
6538	Isopropyl 3-pyridinecarboxylate	Isopropyl nicotinate	$C_9H_{11}NO_2$	553-60-6	165.189			126^{30}, 92.5^5	1.0624^{20}	1.4926^{20}	
6539	Isopropyl silicate	Tetra(isopropoxy)silane	$C_{12}H_{28}O_4Si$	1992-48-9	264.434			184	0.8770^{20}		s ctc, CS_2
6540	Isopropyl stearate		$C_{21}H_{42}O_2$	112-10-7	326.557		28	207^6	0.8403^{38}		vs ace, eth, EtOH, chl
6541	4-Isopropylstyrene		$C_{11}H_{14}$	2055-40-5	146.229	liq	-44.7	204.1	0.8850^{20}	1.5289^{20}	vs ace, bz, eth, EtOH
6542	Isopropyl tetradecanoate	Isopropyl myristate	$C_{17}H_{34}O_2$	110-27-0	270.451			193^{20}, 140^2	0.8532^{20}	1.4325^{25}	i H_2O; s EtOH, eth, chl; vs ace, bz
6543	(Isopropylthio)benzene		$C_9H_{12}S$	3019-20-3	152.256			208	0.9852^{20}	1.5464^{20}	
6544	Isopropyl trichloroacetate		$C_5H_7Cl_3O_2$	3974-99-0	205.468			175; 66^{15}	1.2911^{25}	1.4428^{20}	vs bz, eth, EtOH
6545	Isopropylurea		$C_4H_{10}N_2O$	691-60-1	102.134	nd		$103^{0.1}$			s H_2O, EtOH, chl, ace; sl eth
6546	Isopropyl vinyl ether	2-(Ethenyloxy)propane	$C_5H_{10}O$	926-65-8	86.132	liq	-140	55.5	0.7534^{20}	1.3840^{20}	vs ace, bz, eth, EtOH
6547	Isoproterenol	4-[1-Hydroxy-2-[isopropylamino]ethyl]-1,2-benzenediol	$C_{11}H_{17}NO_3$	7683-59-2	211.258		170.5				

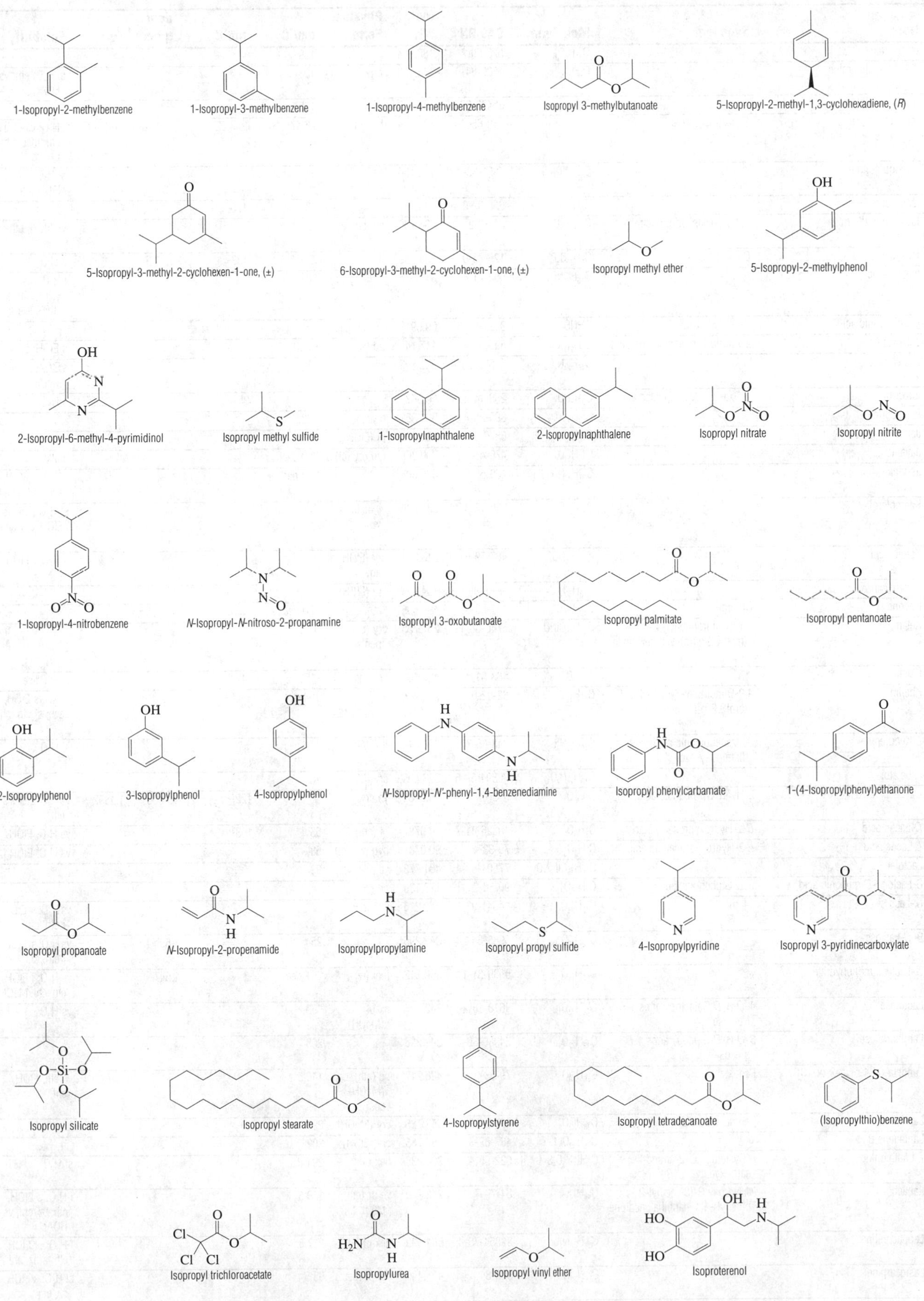
1-Isopropyl-2-methylbenzene
1-Isopropyl-3-methylbenzene
1-Isopropyl-4-methylbenzene
Isopropyl 3-methylbutanoate
5-Isopropyl-2-methyl-1,3-cyclohexadiene, (R)
5-Isopropyl-3-methyl-2-cyclohexen-1-one, (±)
6-Isopropyl-3-methyl-2-cyclohexen-1-one, (±)
Isopropyl methyl ether
5-Isopropyl-2-methylphenol
2-Isopropyl-6-methyl-4-pyrimidinol
Isopropyl methyl sulfide
1-Isopropylnaphthalene
2-Isopropylnaphthalene
Isopropyl nitrate
Isopropyl nitrite
1-Isopropyl-4-nitrobenzene
N-Isopropyl-N-nitroso-2-propanamine
Isopropyl 3-oxobutanoate
Isopropyl palmitate
Isopropyl pentanoate
2-Isopropylphenol
3-Isopropylphenol
4-Isopropylphenol
N-Isopropyl-N'-phenyl-1,4-benzenediamine
Isopropyl phenylcarbamate
1-(4-Isopropylphenyl)ethanone
Isopropyl propanoate
N-Isopropyl-2-propenamide
Isopropylpropylamine
Isopropyl propyl sulfide
4-Isopropylpyridine
Isopropyl 3-pyridinecarboxylate
Isopropyl silicate
Isopropyl stearate
4-Isopropylstyrene
Isopropyl tetradecanoate
(Isopropylthio)benzene
Isopropyl trichloroacetate
Isopropylurea
Isopropyl vinyl ether
Isoproterenol

No.	Name	Synonym	Mol. Form.	CAS RN	Mol. Wt.	Physical Form	mp/°C	bp/°C	den/ g cm^{-3}	n_D	Solubility
6548	Isopsoralen		$C_{11}H_6O_3$	523-50-2	186.164		139				
6549	1-Isoquinolinamine		$C_9H_8N_2$	1532-84-9	144.173	pl(w)	123	164^{8}			sl H_2O, eth; vs EtOH
6550	3-Isoquinolinamine		$C_9H_8N_2$	25475-67-6	144.173		178.5				
6551	Isoquinoline	Benzo[c]pyridine	C_9H_7N	119-65-3	129.159	hyg pl	26.47	243.22	1.0910^{30}	1.6148^{20}	i H_2O; vs EtOH, chl; msc eth, bz
6552	7-Isoquinolinol		C_9H_7NO	7651-83-4	145.158		230				sl H_2O, eth; s EtOH
6553	Isosorbide		$C_6H_{10}O_4$	652-67-5	146.141		63	170^{2}			
6554	Isosorbide dinitrate	1,4:3,6-Dianhydroglucitol	$C_6H_8N_2O_8$	87-33-2	236.136	col cry	52				vs EtOH, eth, ace
6555	Isosystox	Demeton-S	$C_8H_{19}O_3PS_2$	126-75-0	258.339	liq		133^{2}	1.132^{21}		s H_2O
6556	Isothebaine		$C_{19}H_{21}NO_3$	568-21-8	311.375	orth cry (al)	203.5				i H_2O; msc EtOH, chl; sl eth; s MeOH
6557	Isothiocyanic acid		CHNS	3129-90-6	59.091	unstab gas					
6558	*L*-Isovaline	2-Amino-2-methylbutyric acid	$C_5H_{11}NO_2$	595-40-4	117.147	nd (w)	≈300				s EtOH; sl eth
6559	Isoxaben		$C_{18}H_{24}N_2O_4$	82558-50-7	332.395	wh cry	173				s EtOAc, MeCN, MeOH
6560	Isoxazole	1-Oxa-2-azacyclopentadiene	C_3H_3NO	288-14-2	69.062			95	1.078^{20}	1.4298^{17}	s H_2O
6561	Isoxsuprine		$C_{18}H_{23}NO_3$	395-28-8	301.381	cry	103.0				
6562	Jacobine		$C_{18}H_{25}NO_6$	6870-67-3	351.395	pl (EtOH)	228				
6563	Javanicin		$C_{15}H_{14}O_6$	476-45-9	290.268	red cry (al)	208 dec				s alk
6564	Jervine		$C_{27}H_{39}NO_3$	469-59-0	425.604		243 dec				i H_2O; s EtOH, ace, chl; sl eth
6565	Kaempferol		$C_{15}H_{10}O_6$	520-18-3	286.236	ye nd (al, + 1 w)	277				sl H_2O, chl; vs EtOH, eth, ace; i bz
6566	Kainic acid		$C_{10}H_{15}NO_4$	487-79-6	213.231	cry (EtOH aq)	253 dec				s H_2O; i EtOH
6567	Kanamycin A		$C_{18}H_{36}N_4O_{11}$	59-01-8	484.499	cry (EtOH)					
6568	Kepone	Chlordecone	$C_{10}Cl_{10}O$	143-50-0	490.636		350 dec		1.61^{25}		
6569	Ketamine	2-(2-Chlorophenyl)-2-(methylamino)cyclohexanone, (±)	$C_{13}H_{16}ClNO$	6740-88-1	237.725	cry (eth-pentane)	92.5				
6570	Ketene		C_2H_2O	463-51-4	42.036	col gas	-151	-49.8			sl eth, ace
6571	Khellin	4,9-Dimethoxy-7-methyl-5*H*-furo[3,2-g][1]benzopyran-5-one	$C_{14}H_{12}O_5$	82-02-0	260.242	eth, al	154 dec	$190^{0.05}$			i H_2O; s EtOH, ace; sl eth, chl
6572	*L*-Kynurenine	Benzenebutanoic acid, α,2-diamino-γ-oxo-	$C_{10}H_{12}N_2O_3$	343-65-7	208.213	lf (+l/2w)	191 dec				sl H_2O
6573	Labetalol		$C_{19}H_{24}N_2O_3$	36894-69-6	328.405	cry (MeOH)	164				
6574	*DL*-Lactic acid	2-Hydroxypropanoic acid, (±)	$C_3H_6O_3$	598-82-3	90.078	ye cry		122^{15}	1.2060^{21}	1.4392^{20}	vs H_2O, EtOH; sl eth
6575	*D*-Lactic acid	*D*-2-Hydroxypropanoic acid	$C_3H_6O_3$	10326-41-7	90.078	pl (chl)	53	103^{2}			vs H_2O, EtOH
6576	*L*-Lactic acid	*L*-2-Hydroxypropanoic acid	$C_3H_6O_3$	79-33-4	90.078	hyg pr (eth)	53				vs H_2O, EtOH
6577	Lactofen		$C_{19}H_{15}ClF_3NO_7$	77501-63-4	461.773	ye pow (bz)	93				
6578	δ-Lactone-*D*-gluconic acid	δ-*D*-Gluconolactone	$C_6H_{10}O_6$	90-80-2	178.139	nd (al)					
6579	α-Lactose		$C_{12}H_{22}O_{11}$	14641-93-1	342.296	wh pow	222.8				vs H_2O; sl EtOH; i eth, chl
6580	β-*D*-Lactose		$C_{12}H_{22}O_{11}$	5965-66-2	342.296		254		1.59^{20}		vs H_2O; sl EtOH; i eth, chl
6581	α-Lactose monohydrate		$C_{12}H_{24}O_{12}$	5989-81-1	360.312	mcl (w)	201 dec		1.547^{20}		vs H_2O; i EtOH, eth, chl, MeOH
6582	Lactulose	4-*O*-β-*D*-Galactopyranosyl-*D*-fructose	$C_{12}H_{22}O_{11}$	4618-18-2	342.296	hx pl (MeOH)	169				vs H_2O
6583	Laminaribiose	3-*O*-β-*D*-Glucopyranosyl-*D*-glucose	$C_{12}H_{22}O_{11}$	34980-39-7	342.296		205				
6584	Lanosta-8,24-dien-3-ol, (3β)	Lanosterol	$C_{30}H_{50}O$	79-63-0	426.717	nd (eth), cry (MeOH-ace)	140.5				vs eth, EtOH, chl
6585	Lantadene A	Rehmannic acid	$C_{35}H_{52}O_5$	467-81-2	552.785	cry (MeOH)	297				
6586	Lantadene B		$C_{35}H_{52}O_5$	467-82-3	552.785	cry (EtOH)	302				
6587	*L*-Lanthionine	*L*-Cysteine, *S*-(2-amino-2-carboxyethyl)-, (*R*)-	$C_6H_{12}N_2O_4S$	922-55-4	208.235	hex pl	294 dec				sl H_2O
6588	Lapachol	2-Hydroxy-3-(3-methyl-2-butenyl)-1,4-naphthalenedione	$C_{15}H_{14}O_3$	84-79-7	242.270	ye pr (eth, bz) pl (al)	139.5				i H_2O; s EtOH, eth, bz, chl; vs HOAc
6589	Lappaconitine		$C_{32}H_{44}N_2O_8$	32854-75-4	584.699	hex pl (al)	217.5				i H_2O; sl EtOH, eth; s bz, chl
6590	Lasiocarpine		$C_{21}H_{33}NO_7$	303-34-4	411.490	col pl (peth)	95.5				sl H_2O; s EtOH, bz, eth
6591	Laudanidine		$C_{20}H_{25}NO_4$	301-21-3	343.418	hex pr (al)	184.5				vs H_2O, bz

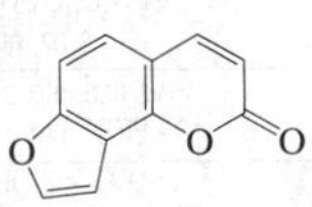

O
NH2
N
HO
H
OH
H
O-N
S
Isopsoralen
1-Isoquinolinamine 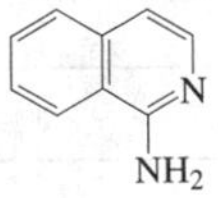
3-Isoquinolinamine 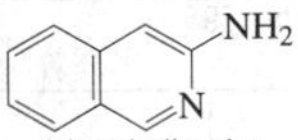
Isoquinoline 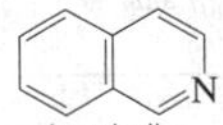
7-Isoquinolinol 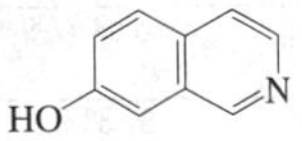
Isosorbide 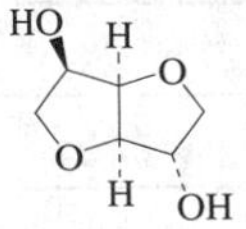
Isosorbide dinitrate 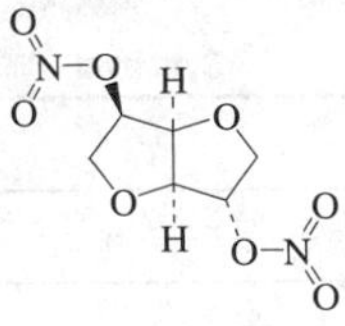
Isosystox 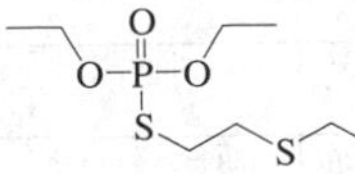
HN=C=S
NH2
OH
N
H
O
Isothebaine
Isothiocyanic acid
L-Isovaline 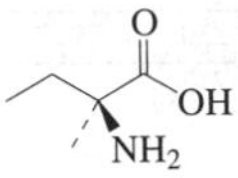
Isoxaben 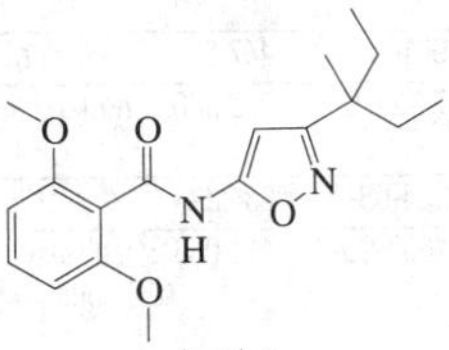
Isoxazole 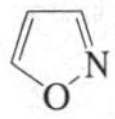
Isoxsuprine 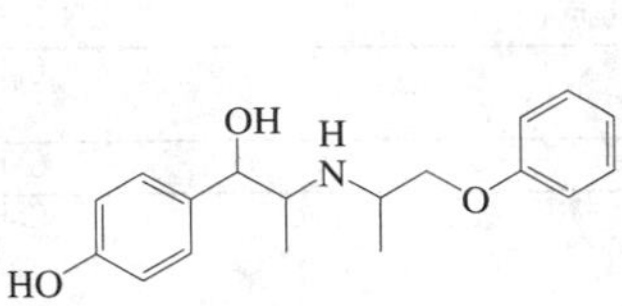
Jacobine 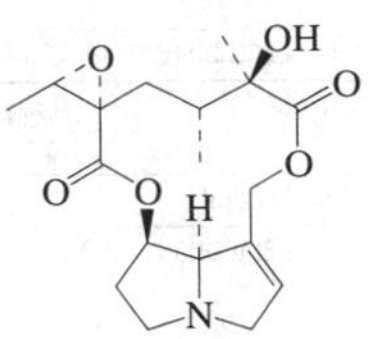 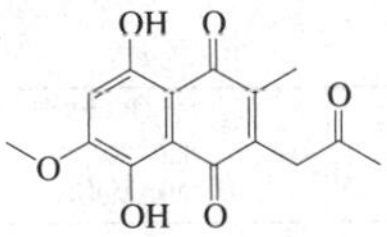
HO
H2N
NH2
OH
Cl
NH
Javanicin
Jervine 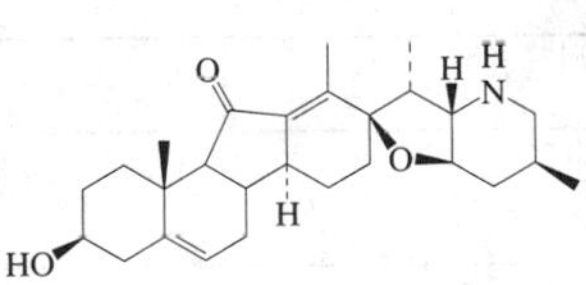
Kaempferol 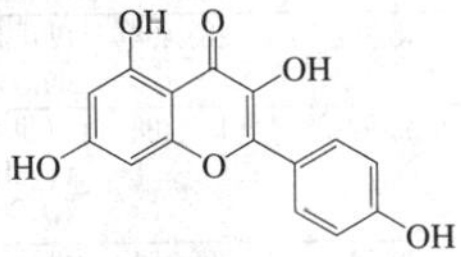
Kainic acid 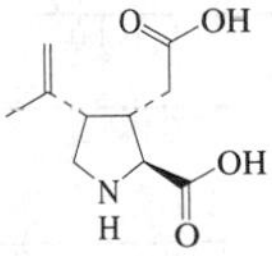
Kanamycin A 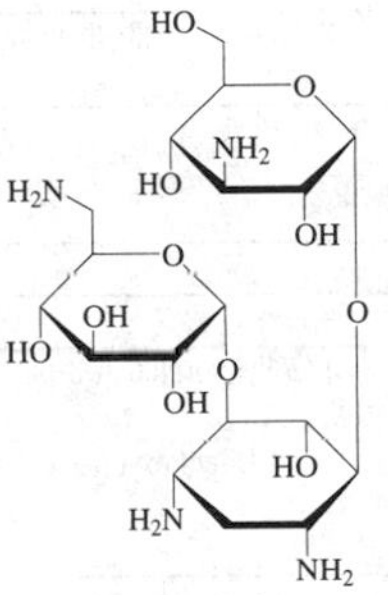
Kepone 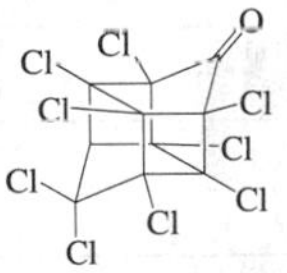
Ketamine 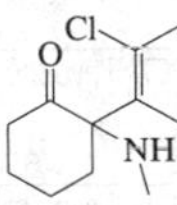
H2C=C=O
NH2
O
Ketene
Khellin 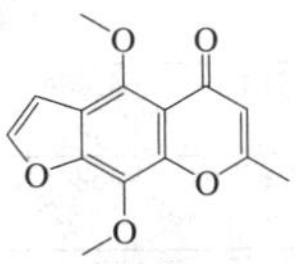
L-Kynurenine 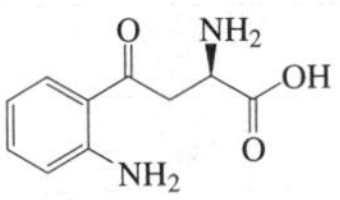
Labetalol 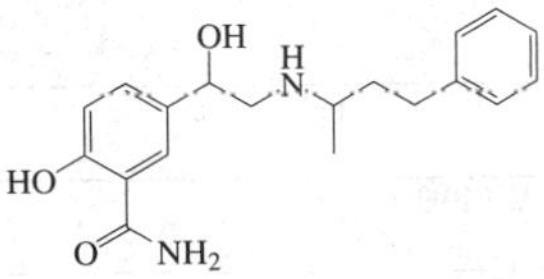
DL-Lactic acid 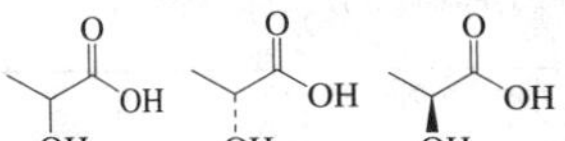
D-Lactic acid
L-Lactic acid
Lactofen 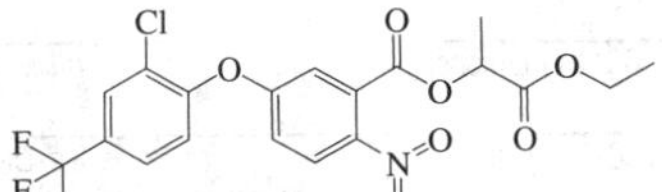 
HO
OH
H2O
δ-Lactone-D-gluconic acid
α-Lactose 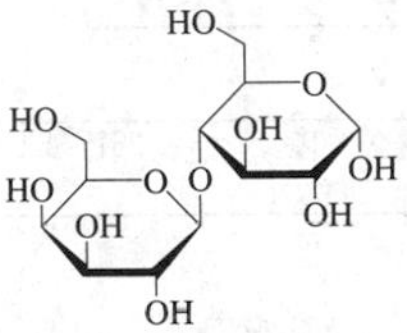
β-D-Lactose 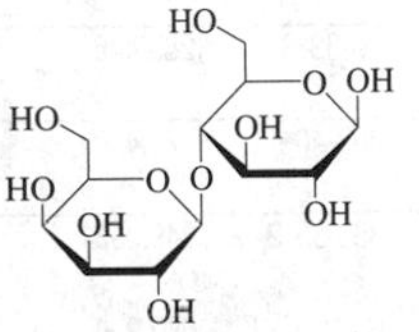
α-Lactose monohydrate 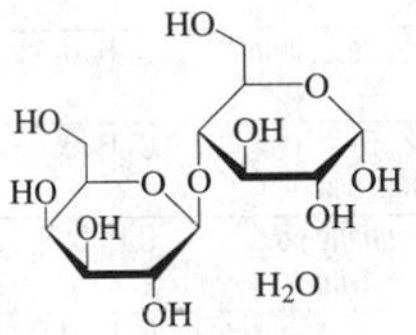
Lactulose 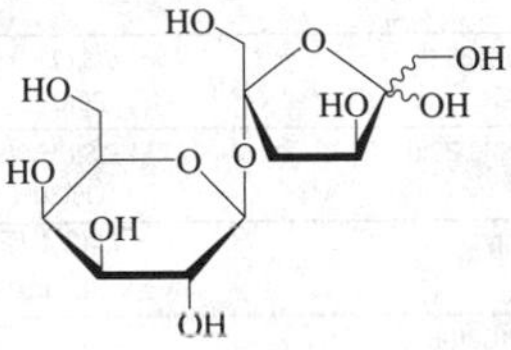
Laminaribiose 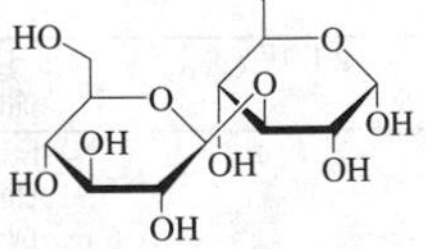
Lanosta-8,24-dien-3-ol, (3β) 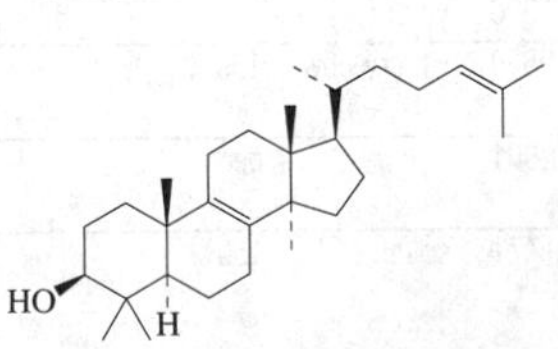
Lantadene A 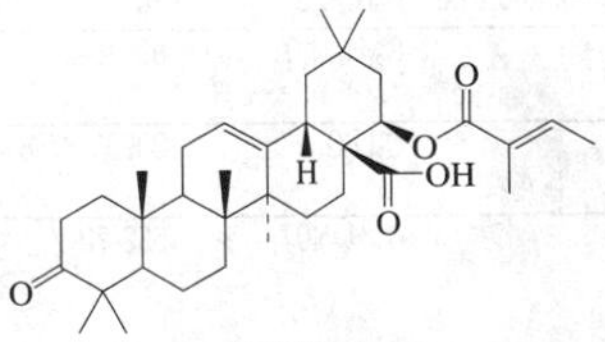
Lantadene B 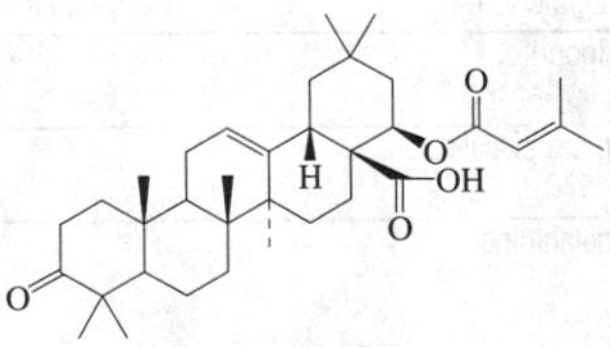
NH2
NH
HO
OH
L-Lanthionine
Lapachol
Lappaconitine 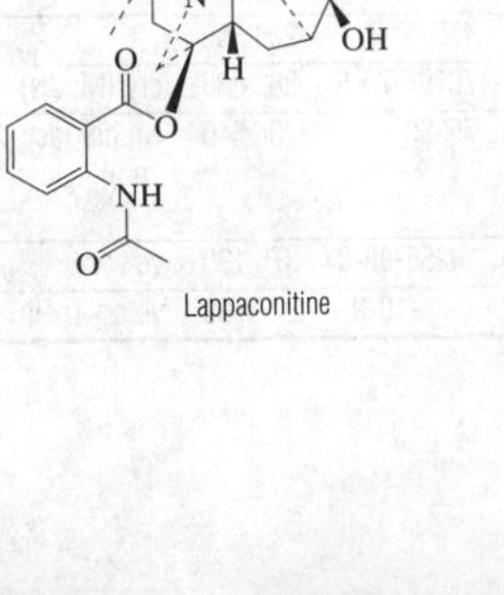
Lasiocarpine
Laudanidine

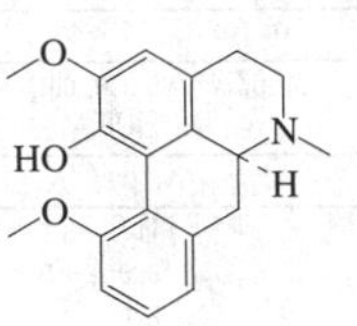

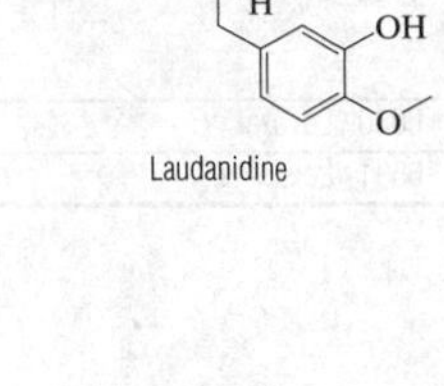

No.	Name	Synonym	Mol. Form.	CAS RN	Mol. Wt.	Physical Form	mp/°C	bp/°C	den/ g cm-3	n_D	Solubility
6592	Laudanine		$C_{20}H_{25}NO_4$	85-64-3	343.418	ye wh pr (dil al, al-chl)	167		1.26^{20}		sl H_2O, EtOH, eth; s bz, chl
6593	Laudanosine		$C_{21}H_{27}NO_4$	2688-77-9	357.444	nd (peth), pr (al)	89				vs ace, eth, EtOH, chl
6594	Laureline		$C_{19}H_{19}NO_3$	81-38-9	309.359	tab (al) cubes (peth)	114				i H_2O; s EtOH, eth, dil acid, con sulf
6595	Laurocapram	1-Dodecylhexahydro-2*H*-azepin-2-one	$C_{18}H_{35}NO$	59227-89-3	281.477	col liq	-7	160^{50}	0.91	1.4701	i H_2O
6596	Lead bis(dimethyldithiocarbamate)		$C_6H_{12}N_2PbS_4$	19010-66-3	447.6	pale ye nd	258				
6597	Ledol		$C_{15}H_{26}O$	577-27-5	222.366	nd (al)	105	292	0.9078^{100}	1.4667^{110}	vs ace, eth, EtOH
6598	Lenacil		$C_{13}H_{18}N_2O_2$	2164-08-1	234.294		290		1.32^{25}		vs py
6599	Leptophos		$C_{13}H_{10}BrCl_2O_2PS$	21609-90-5	412.066	tan waxy solid	71		1.53^{25}		i H_2O; vs bz; s ace, 2-PrOH, xyl
6600	*DL*-Leucine		$C_6H_{13}NO_2$	328-39-2	131.173	lf (w)	293	sub	1.293^{18}		s H_2O; sl EtOH; i eth
6601	*D*-Leucine		$C_6H_{13}NO_2$	328-38-1	131.173	pl (al)	293	sub			sl H_2O
6602	*L*-Leucine	2-Amino-4-methylpentanoic acid	$C_6H_{13}NO_2$	61-90-5	131.173	hex pl (dil al)	293	sub	1.293^{18}		sl H_2O; i EtOH, eth
6603	*N*-Leucylglycine		$C_8H_{16}N_2O_3$	686-50-0	188.224		248 dec				s H_2O; sl EtOH, eth; i ace, bz, chl
6604	Leuprolide		$C_{59}H_{84}N_{16}O_{12}$	53714-56-0	1209.398	fluffy solid					
6605	Leurosine		$C_{46}H_{56}N_4O_9$	23360-92-1	808.959	cry	203				
6606	Levallorphan	17-Allylmorphinan-3-ol	$C_{19}H_{25}NO$	152-02-3	283.408	cry (EtOH aq)	181				
6607	Levodopa	*L*-3,4-Dihydroxyphenylalanine	$C_9H_{11}NO_4$	59-92-7	197.188	pl (dil al) pr or nd ($w+SO_2$)	277 dec				s H_2O; i EtOH, eth, ace, bz; s alk, MeOH
6608	Levopimaric acid		$C_{20}H_{30}O_2$	79-54-9	302.451	orth cry	150				
6609	Levorphanol	17-Methylmorphinan-3-ol	$C_{17}H_{23}NO$	77-07-6	257.371	cry	198				
6610	*d*-Limonene	*p*-Mentha-1,8-diene, (*R*)	$C_{10}H_{16}$	5989-27-5	136.234	oil	-74.0	178	0.8411^{20}	1.4730^{20}	i H_2O; msc EtOH, eth; s ctc
6611	*l*-Limonene	*p*-Mentha-1,8-diene, (*S*)	$C_{10}H_{16}$	5989-54-8	136.234	oil		178; 64.4^{15}	0.843^{20}	1.4746^{20}	i H_2O; vs eth, EtOH
6612	Linalol	3,7-Dimethyl-1,6-octadien-3-ol, (±)-	$C_{10}H_{18}O$	22564-99-4	154.249			198; 86^{13}	0.870^{15}	1.4627	
6613	Linalyl acetate	3,7-Dimethyl-1,6-octadien-3-yl acetate	$C_{12}H_{20}O_2$	115-95-7	196.286	liq		220; $44^{0.2}$	0.895^{20}	1.4460^{20}	i H_2O; misc EtOH, eth
6614	Lincomycin		$C_{18}H_{34}N_2O_6S$	154-21-2	406.537	amor solid					sl H_2O; s EtOH, ace, chl
6615	Linoleic acid	*cis,cis*-9,12-Octadecadienoic acid	$C_{18}H_{32}O_2$	60-33-3	280.446		-7	229^{16}	0.9022^{20}	1.4699^{20}	vs ace, bz, eth, EtOH
6616	Linolenic acid	*cis,cis,cis*-9,12,15-Octadecatrienoic acid	$C_{18}H_{30}O_2$	463-40-1	278.430		-11	231^{17}, $129^{0.05}$	0.9164^{20}	1.4800^{20}	i H_2O; s EtOH, eth; sl bz
6617	Linuron	*N*-(3,4-Dichlorophenyl)-*N'*-methoxy-*N'*-methylurea	$C_9H_{10}Cl_2N_2O_2$	330-55-2	249.093		93				
6618	Liothyronine		$C_{15}H_{12}I_3NO_4$	6893-02-3	650.974	cry	236 dec				i H_2O, EtOH; s dil alk
6619	Lipoamide	1,2-Dithiolane-3-pentanamide	$C_8H_{15}NOS_2$	940-69-2	205.341	cry	128				
6620	α-Lipoic acid	1,2-Dithiolane-3-pentanoic acid	$C_8H_{14}O_2S_2$	1077-28-7	206.326	ye pl (cy)	60	87^{25}			i H_2O
6621	Lisinopril		$C_{21}H_{35}N_3O_7$	83915-83-7	441.519	wh cry pow	159				i EtOH, chl, ace; sl MeOH
6622	Lithium oxalate		$C_2Li_2O_4$	30903-87-8	101.901		dec		2.121^{17}		s H_2O; i EtOH, eth
6623	Lobelanidine		$C_{22}H_{29}NO_2$	552-72-7	339.471	sc (al, eth)	150				i H_2O; s EtOH; sl eth; vs ace, bz, py
6624	Lobelanine		$C_{22}H_{25}NO_2$	579-21-5	335.440	nd (eth, peth)	99				vs ace, bz, EtOH, chl
6625	Lobeline		$C_{22}H_{27}NO_2$	90-69-7	337.455	nd (al, bz)	130.5				sl H_2O; s EtOH, eth, bz, chl; vs ace
6626	Loflucarban		$C_{13}H_9Cl_2FN_2S$	790-69-2	315.192		163.5				
6627	Longifolene	Kuromatsuene	$C_{15}H_{24}$	475-20-7	204.352			258; 126^{15}	0.9319^{18}	1.5040^{20}	i H_2O; s bz
6628	Loratadine	Claritin	$C_{22}H_{23}ClN_2O_2$	79794-75-5	382.883	cry (MeCN)	132				
6629	Lovastatin	Mevacor	$C_{24}H_{36}O_5$	75330-75-5	404.540	wh cry (ace aq)	174				i H_2O; vs chl; s DMF; sl ace, EtOH
6630	Lovozal		$C_{15}H_7Cl_2F_3N_2O_2$	14255-88-0	375.130	ye cry	103				s ace, diox
6631	Loxapine		$C_{18}H_{18}ClN_3O$	1977-10-2	327.808	ye cry (peth)	109.5				

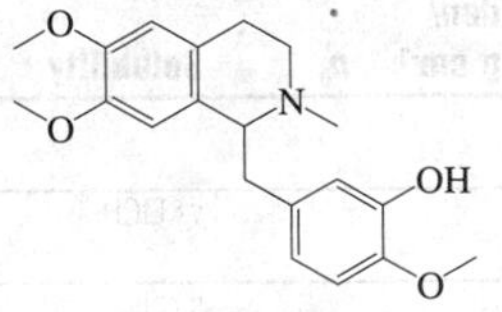

Laudanine

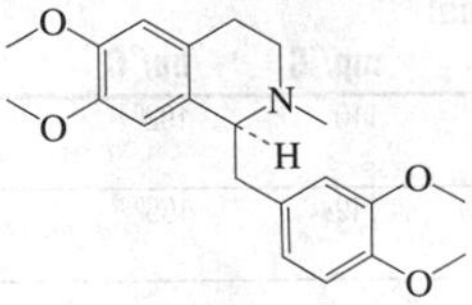

Laudanosine

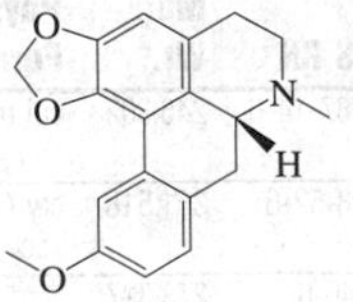

Laureline

Laurocapram

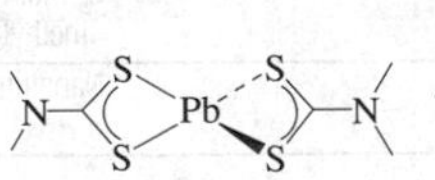

Lead bis(dimethyldithiocarbamate)

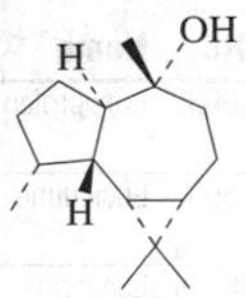

Ledol

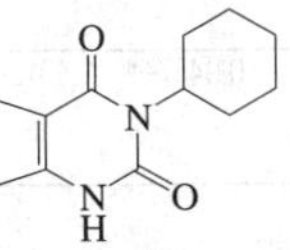

Lenacil

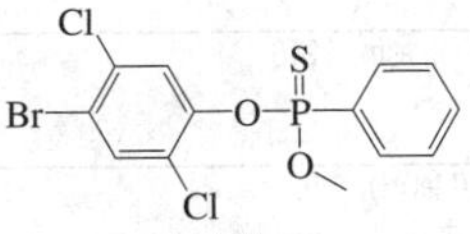

Leptophos

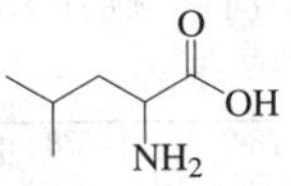

DL-Leucine

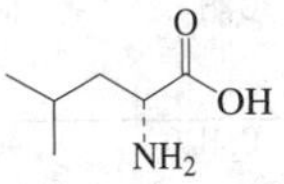

D-Leucine

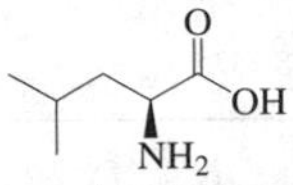

L-Leucine

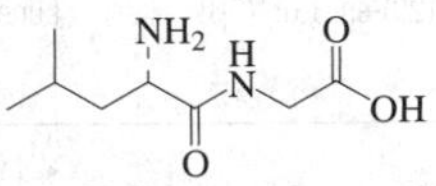

N-Leucylglycine

H-5-oxoPro-His-Trp-Ser-Tyr-D-Leu-Leu-Arg-Pro-NHEt

Leuprolide

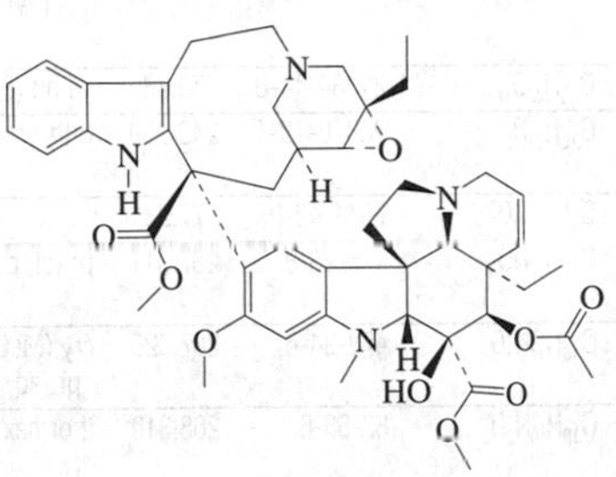

Leurosine

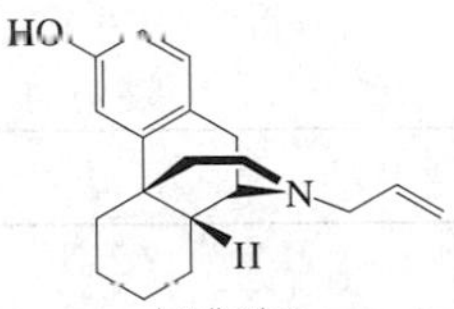

Levallorphan

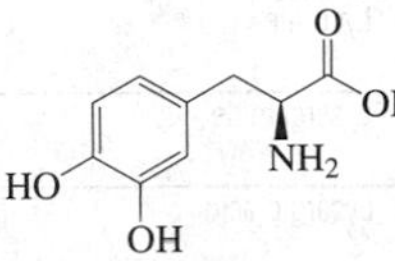

Levodopa

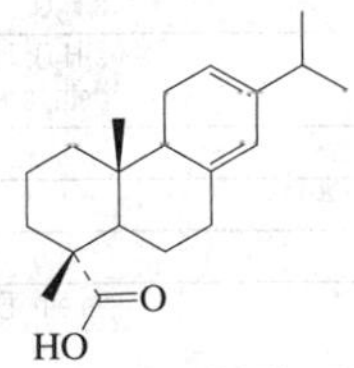

Levopimaric acid

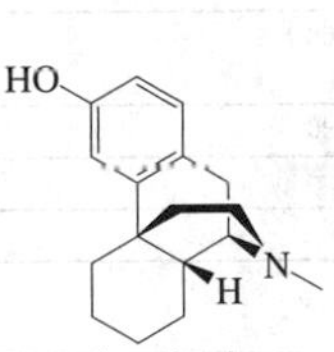

Levorphanol

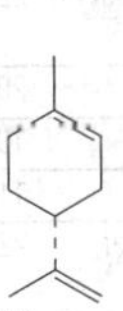

d-Limonene

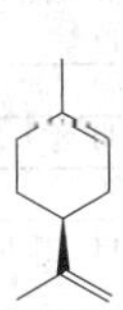

l-Limonene

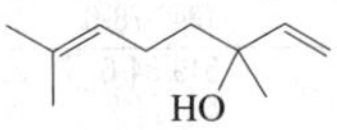

Linalol

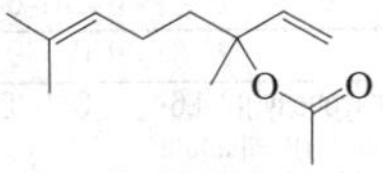

Linalyl acetate

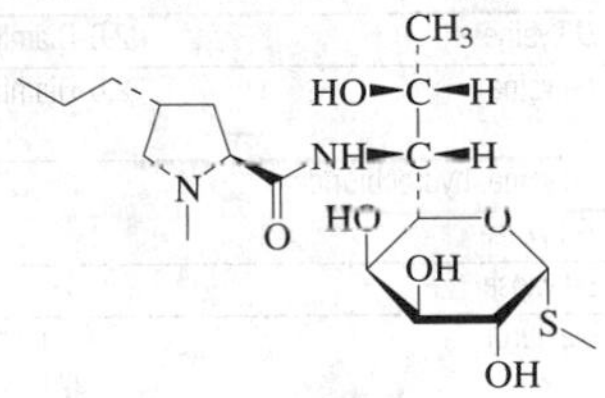

Lincomycin

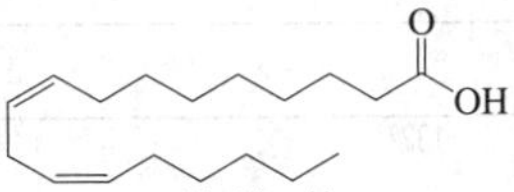

Linoleic acid

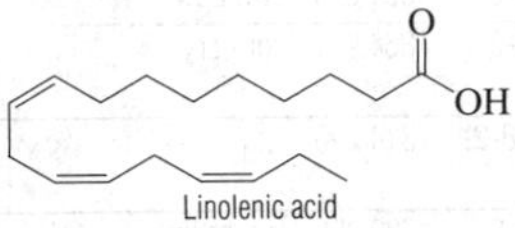

Linolenic acid

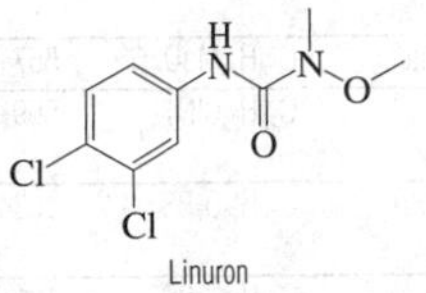

Linuron

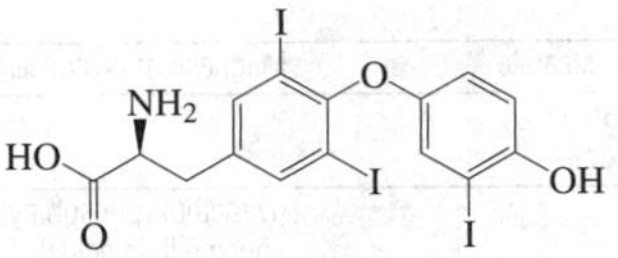

Liothyronine

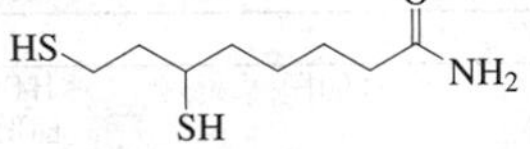

Lipoamide

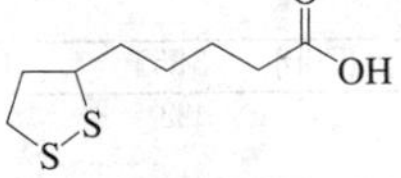

α-Lipoic acid

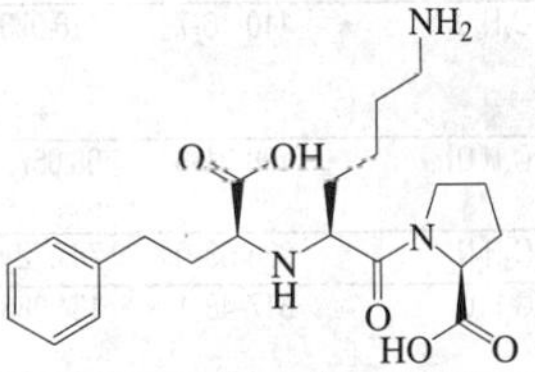

Lisinopril

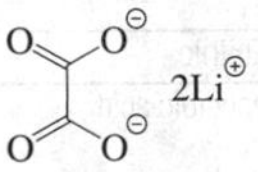

Lithium oxalate

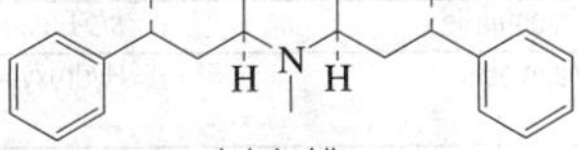

Lobelanidine

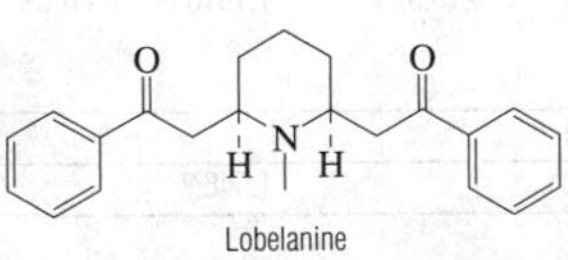

Lobelanine

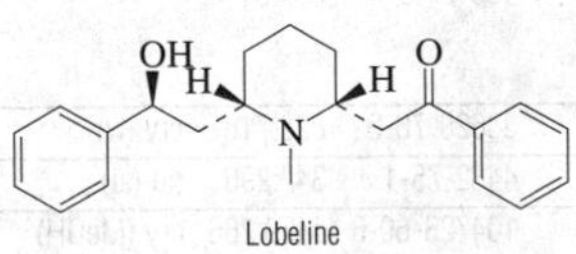

Lobeline

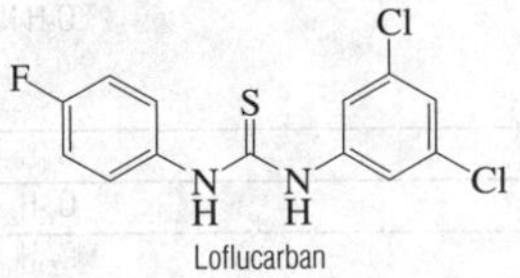

Loflucarban

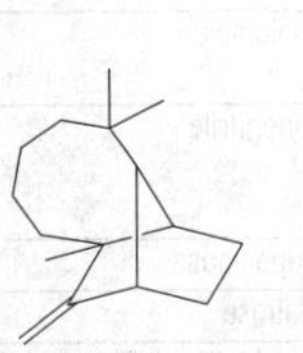

Longifolene

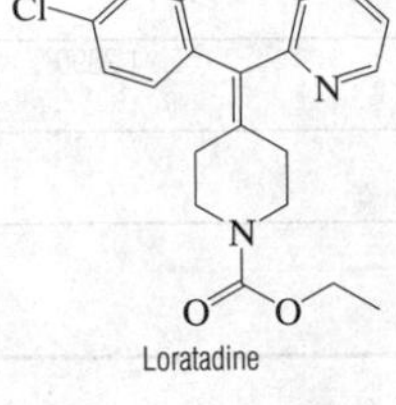

Loratadine

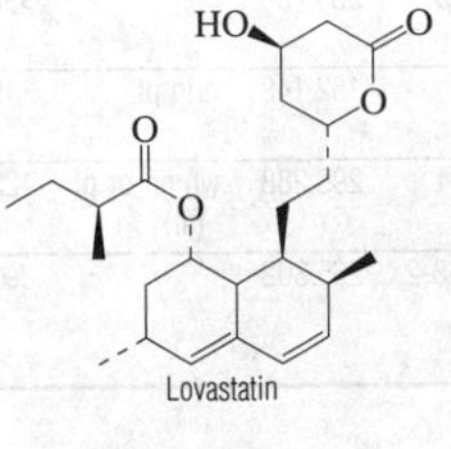

Lovastatin

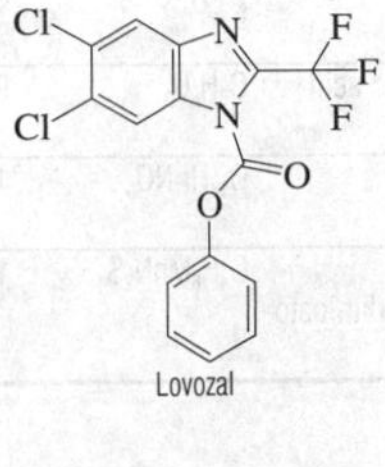

Lovozal

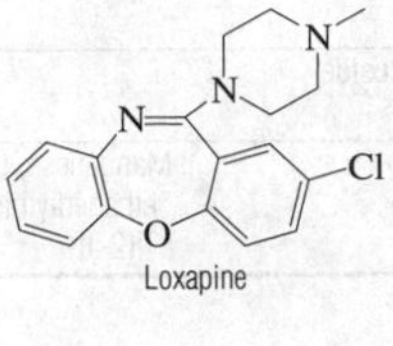

Loxapine

No.	Name	Synonym	Mol. Form.	CAS RN	Mol. Wt.	Physical Form	mp/°C	bp/°C	den/ g cm^{-3}	n_D	Solubility
6632	Loxoprofen	α-Methyl-4-[(2-oxocyclopentyl) methyl]benzeneacetic acid	$C_{15}H_{18}O_3$	68767-14-6	246.302	col oil	110	$192^{0.3}$			
6633	Luciculine	Napelline	$C_{22}H_{33}NO_3$	5008-52-6	361.518	cry (+1w, ace)	149	$165^{0.02}$			vs EtOH
6634	Lunacrine		$C_{16}H_{19}NO_3$	82-40-6	273.327						s chl
6635	Lup-20(29)-ene-3,28-diol, (3β)	Betulin	$C_{30}H_{50}O_2$	473-98-3	442.717	nd (al +1)	250	sub 240			i H_2O; sl EtOH, bz; s eth, AcOEt, lig
6636	Lup-20(29)-en-3-ol, (3β)	Lupeol	$C_{30}H_{50}O$	545-47-1	426.717	nd (al, ace)	216		0.9457^{218}	1.4910^{218}	i H_2O; vs EtOH, eth, ace, bz, chl
6637	Lupulon		$C_{26}H_{38}O_4$	468-28-0	414.578	pr (MeOH)	93				i H_2O; s EtOH, peth, hx
6638	Luteolin		$C_{15}H_{10}O_6$	491-70-3	286.236	ye nd (dil al, + 1 w)	329 dec				sl H_2O; s EtOH, eth, alk, con sulf
6639	Luteoskyrin	8,8'-Dihydroxyrugulosin	$C_{30}H_{22}O_{12}$	21884-44-6	574.489	ye nd (EtOH)	278 dec				
6640	Lycodine		$C_{16}H_{22}N_2$	20316-18-1	242.359	orth pr	99	$190^{1.0}$			s H_2O, chl, eth, EtOH; i peth
6641	Lycomarasmine		$C_9H_{15}N_3O_7$	7611-43-0	277.231		228 dec				
6642	Lycorine		$C_{16}H_{17}NO_4$	476-28-8	287.311	pr (al, py)	280	sub			i H_2O; sl EtOH, eth, chl
6643	Lysergamide		$C_{16}H_{17}N_3O$	478-94-4	267.325	cry (MeOH), pr (aq, ace)	137.5				sl EtOH, ace, os
6644	Lysergic acid		$C_{16}H_{16}N_2O_2$	82-58-6	268.310	lf or hex sc (w)	240 dec				sl H_2O, eth, bz; s EtOH, py
6645	Lysergide		$C_{20}H_{25}N_3O$	50-37-3	323.432		82				
6646	*DL*-Lysine	2,6-Diaminohexanoic acid, (±)	$C_6H_{14}N_2O_2$	70-54-2	146.187		224				sl H_2O
6647	*D*-Lysine	2,6-Diaminohexanoic acid, (*D*)	$C_6H_{14}N_2O_2$	923-27-3	146.187		218 dec				s H_2O
6648	*L*-Lysine	2,6-Diaminohexanoic acid, (*L*)	$C_6H_{14}N_2O_2$	56-87-1	146.187	nd (w, dil al)	224 dec				s H_2O; i EtOH, eth, ace, bz
6649	*L*-Lysine, hydrochloride		$C_6H_{15}ClN_2O_2$	10098-89-2	182.648		263 dec				
6650	*D*-Lyxose		$C_5H_{10}O_5$	1114-34-7	150.130		108		1.545^{20}		
6651	*L*-Lyxose		$C_5H_{10}O_5$	1949-78-6	150.130		110				
6652	Maclurin	(3,4-Dihydroxyphenyl)(2,4,6-trihydroxyphenyl)methanone	$C_{13}H_{10}O_6$	519-34-6	262.214	ye nd (al)	222.5				vs eth, EtOH
6653	Magenta base	Rosaniline	$C_{20}H_{19}N_3$	3248-93-9	301.385	br-red cry	186 dec				
6654	Magenta I	Rosaniline hydrochloride	$C_{20}H_{20}ClN_3$	632-99-5	337.846	grn cry	200 dec				sl H_2O, EtOH; i eth
6655	Magnesium stearate	Magnesium octadecanoate	$C_{36}H_{70}MgO_4$	557-04-0	591.244	wh pow	132				i H_2O; reac acid
6656	Malachite Green		$C_{23}H_{25}ClN_2$	569-64-2	364.911	grn cry					vs H_2O, EtOH, MeOH
6657	Malaoxon	(Dimethoxyphosphinylthio) butanedioic acid	$C_{10}H_{19}O_7PS$	1634-78-2	314.293	liq		$132^{0.1}$			
6658	Malathion		$C_{10}H_{19}O_6PS_2$	121-75-5	330.358	ye-br liq	1.4	$156^{0.7}$ dec	1.2076^{20}	1.4960^{20}	sl H_2O; s EtOH, eth, bz
6659	Maleic acid	*cis*-2-Butenedioic acid	$C_4H_4O_4$	110-16-7	116.073	mcl pr (w)	139		1.590^{20}		vs H_2O, EtOH, ace; s eth; i bz, chl
6660	Maleic anhydride		$C_4H_2O_3$	108-31-6	98.057	nd (chl, eth)	52.56	202	1.314^{60}		s H_2O; s eth, ace, chl; sl lig
6661	Maleonitrile	*cis*-Butenedinitrile	$C_4H_2N_2$	928-53-0	78.072	pr (EtOH)	31.5	111^{20}			
6662	Malic acid	Hydroxybutanedioic acid	$C_4H_6O_5$	617-48-1	134.088		132		1.601^{20}		s H_2O; vs eth, EtOH, MeOH
6663	Malonaldehyde	1,3-Propanedial	$C_3H_4O_2$	542-78-9	72.063	hyg nd	73				
6664	Malonic acid		$C_3H_4O_4$	141-82-2	104.062	tcl (al)	135 dec	sub	1.619^{10}		vs H_2O, py; s EtOH, eth; i bz
6665	Malononitrile		$C_3H_2N_2$	109-77-3	66.061		32	218.5	1.1910^{20}	1.4146^{34}	s H_2O, ace, bz, chl; vs EtOH, eth
6666	Maltopentaose		$C_{30}H_{52}O_{26}$	34620-76-3	828.718	cry (w)	78 (hyd)				
6667	α-Maltose		$C_{12}H_{22}O_{11}$	4482-75-1	342.296	nd (al)	162.5		1.546^{20}		vs H_2O
6668	6-*O*-α-Maltosyl-β-cyclodextrin		$C_{54}H_{90}O_{45}$	104723-60-6	1459.266	cry (MeOH)					
6669	Maltotetraose		$C_{24}H_{42}O_{21}$	34612-38-9	666.577	amorp solid	170 dec				
6670	Malvidin chloride		$C_{17}H_{15}ClO_7$	643-84-5	366.750		>300				sl H_2O; s EtOH, MeOH
6671	Mandelic acid	α-Hydroxybenzeneacetic acid	$C_8H_8O_3$	90-64-2	152.148	orth pl	119		1.2890^{20}		s H_2O, eth, EtOH, i-PrOH
6672	Mandelonitrile glucoside		$C_{14}H_{17}NO_6$	138-53-4	295.288	wh nd or pl (al)	122				vs H_2O, EtOH
6673	Maneb	Manganese, [[1,2-ethanediylbis[carbamodithioato]](2-)]-	$C_4H_6MnN_2S_4$	12427-38-2	265.302		dec 200				

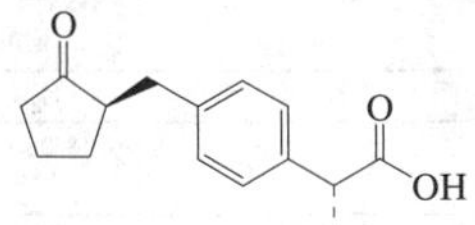

Loxoprofen

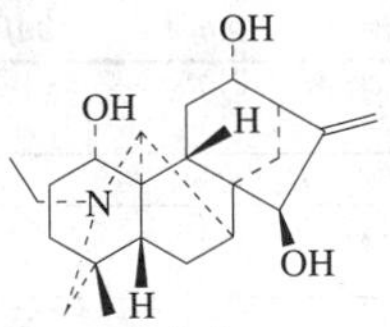

Luciculine

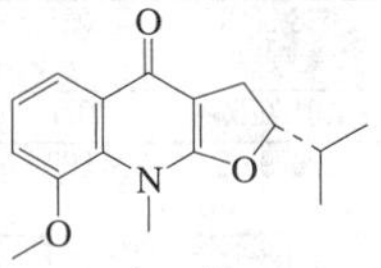

Lunacrine

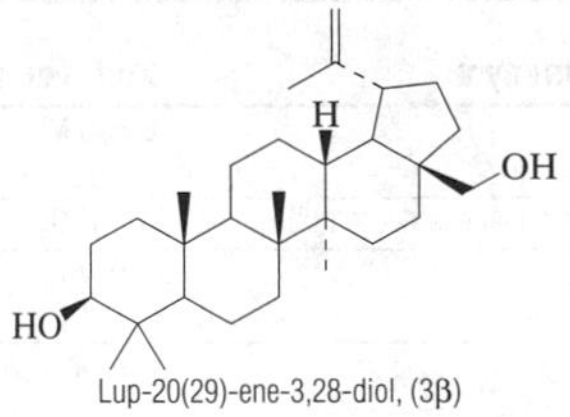

Lup-20(29)-ene-3,28-diol, (3β)

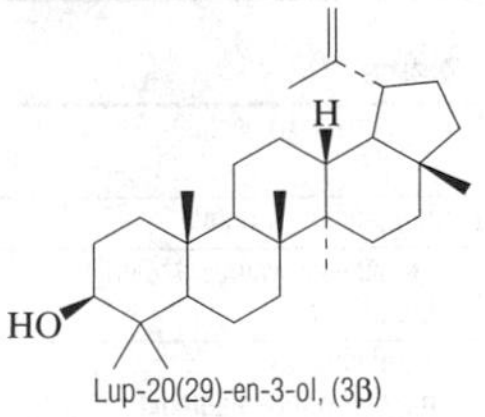

Lup-20(29)-en-3-ol, (3β)

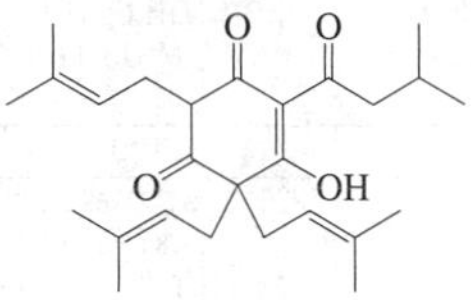

Lupulon

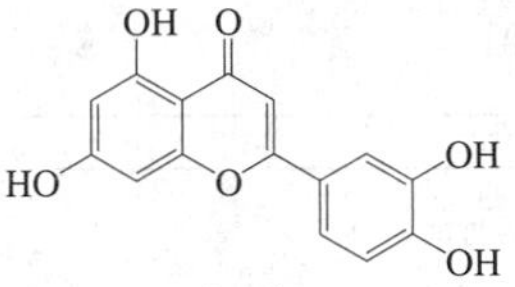

Luteolin

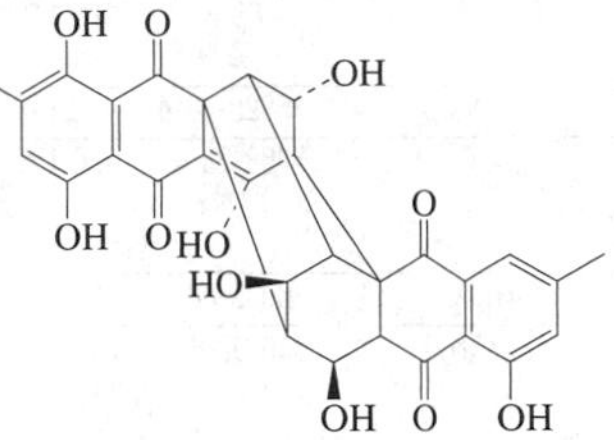

Luteoskyrin

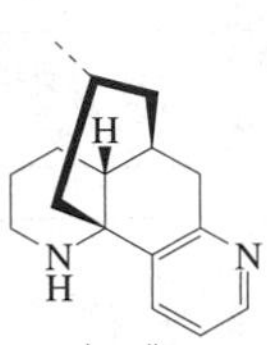

Lycodine

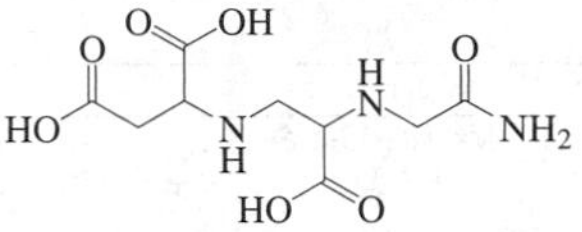

Lycomarasmine

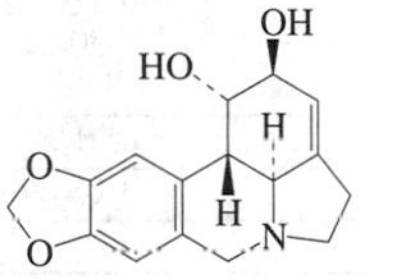

Lycorine

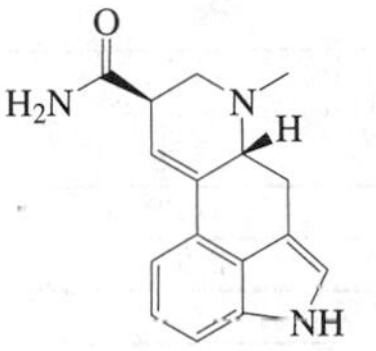

Lysergamide

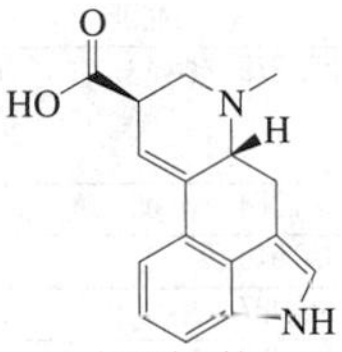

Lysergic acid

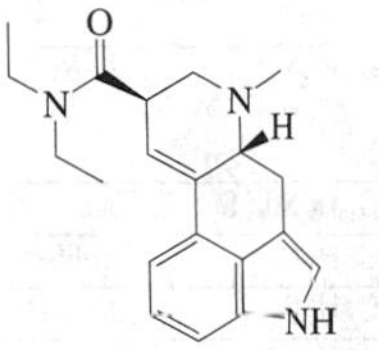

Lysergide

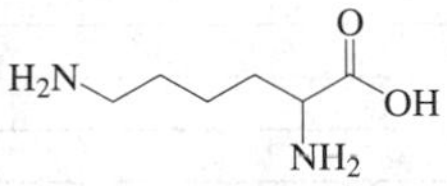

DL-Lysine

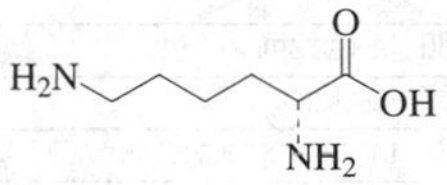

D-Lysine

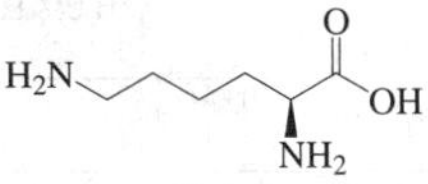

L-Lysine

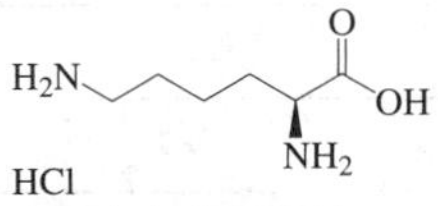

L-Lysine, hydrochloride

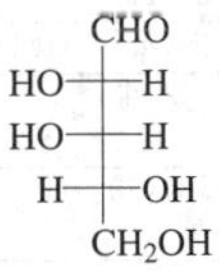

D-Lyxose

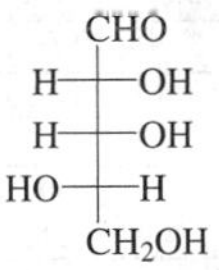

L-Lyxose

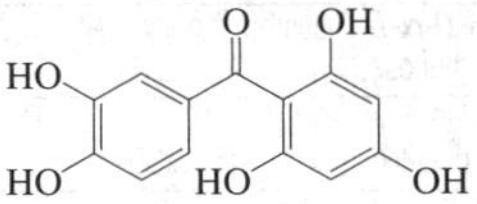

Maclurin

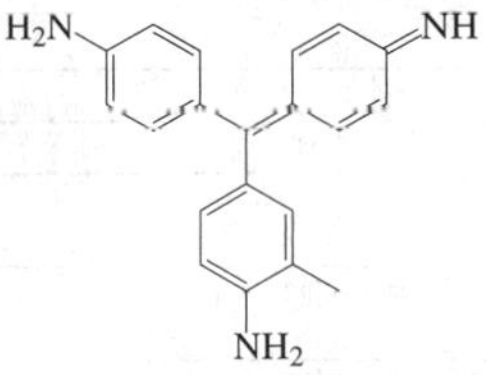

Magenta base

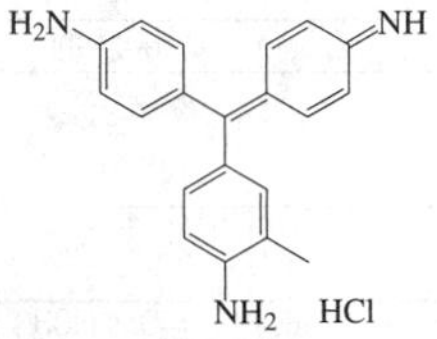

Magenta I

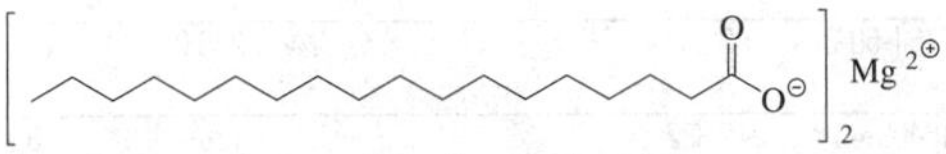

Magnesium stearate

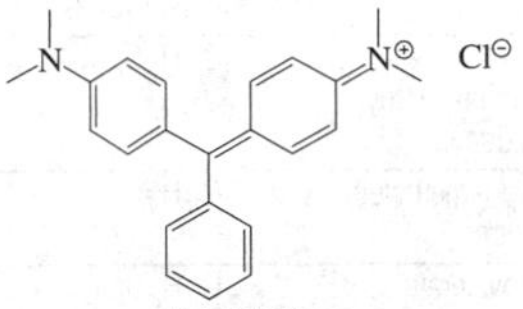

Malachite Green

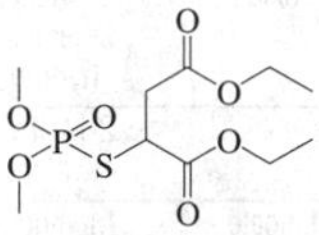

Malaoxon

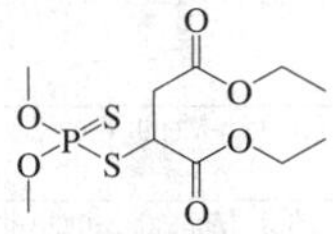

Malathion

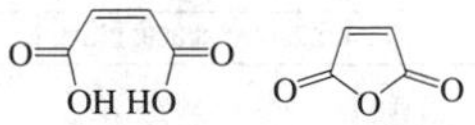

Maleic acid

Maleic anhydride

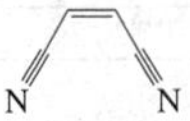

Maleonitrile

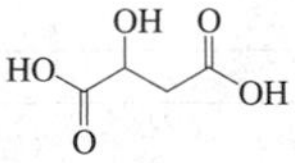

Malic acid

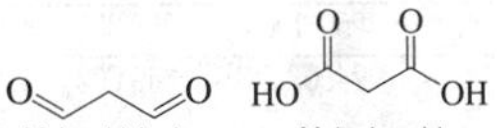

Malonaldehyde

Malonic acid

Malononitrile

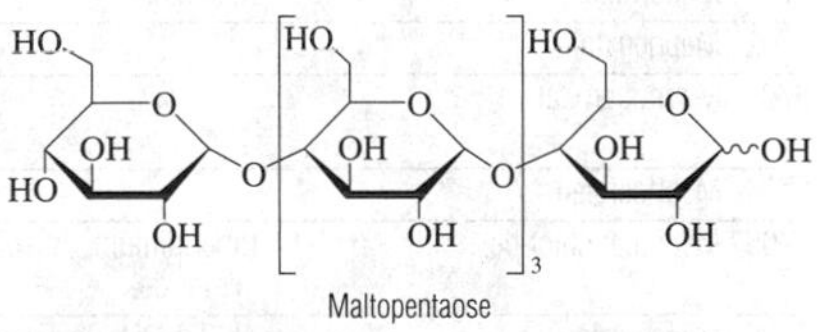

Maltopentaose

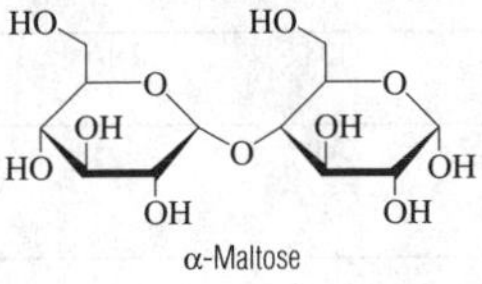

α-Maltose

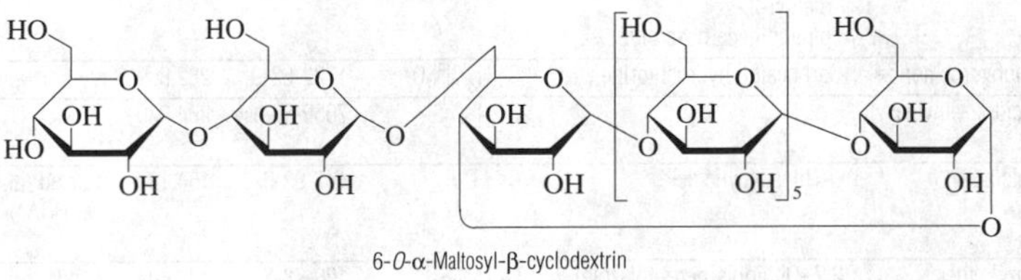

6-*O*-α-Maltosyl-β-cyclodextrin

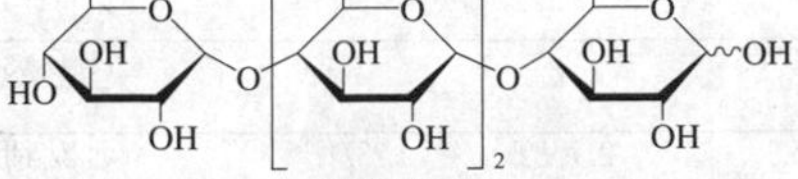

Maltotetraose

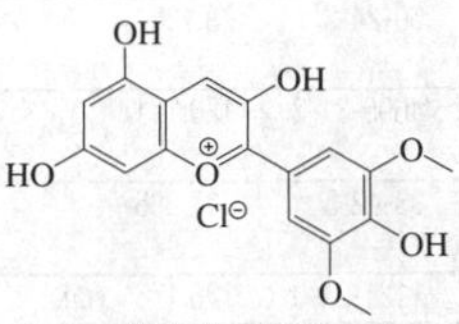

Malvidin chloride

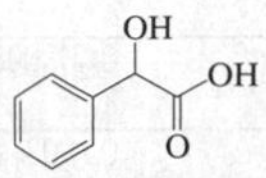

Mandelic acid

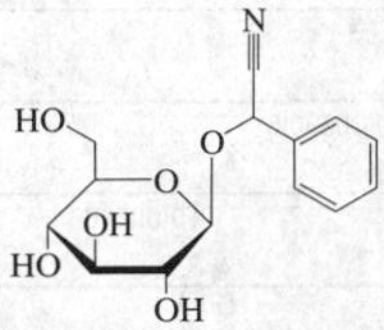

Mandelonitrile glucoside

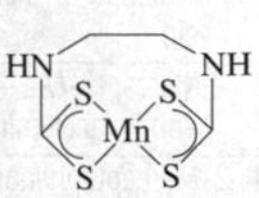

Maneb

No.	Name	Synonym	Mol. Form.	CAS RN	Mol. Wt.	Physical Form	mp/°C	bp/°C	den/ g cm^{-3}	n_D	Solubility
6674	Manganese(II) acetate		$C_4H_6O_4Mn$	638-38-0	173.027	red cry (w)	210				s H_2O, MeOH, HOAc; i ace
6675	Manganese carbonyl	Dimanganese decacarbonyl	$C_{10}Mn_2O_{10}$	10170-69-1	389.977	ye mcl cry	154		1.75		i H_2O; s os
6676	Manganese cyclopentadienyl tricarbonyl		$C_8H_5MnO_3$	12079-65-1	204.062	pale ye cry	77.0	subl			s os
6677	Manganese 2-methylcyclopentadienyl tricarbonyl		$C_9H_7MnO_3$	12108-13-3	218.088	ye liq	1.5	233; 102^{10}	1.388^{20}		i H_2O; misc bz
6678	*D*-Mannitol	Cordycepic acid	$C_6H_{14}O_6$	69-65-8	182.171	orth nd or pr (w)	168	$295^{3.5}$	1.489^{20}	1.3330	vs H_2O; sl EtOH, py; i eth
6679	*D*-Mannitol hexanitrate		$C_6H_8N_6O_{18}$	15825-70-4	452.157	nd (al)	107	exp	1.8^{20}		vs bz, eth, EtOH
6680	*D*-Mannose	Seminose	$C_6H_{12}O_6$	3458-28-4	180.155	nd or orth pr (al)	132 dec		1.539^{20}		vs H_2O; sl EtOH, MeOH; i eth, bz
6681	*L*-Mannose		$C_6H_{12}O_6$	10030-80-5	180.155	cry (al)	132				vs H_2O
6682	Matridin-15-one	Matrine	$C_{15}H_{24}N_2O$	519-02-8	248.364	α-nd or pl; β-orth pr,		223^6		1.5286^{25}	s H_2O, eth, ace; vs EtOH, bz; sl peth
6683	Mazindol		$C_{16}H_{13}ClN_2O$	22232-71-9	284.739	cry (ace/hx)	198				i H_2O; s EtOH
6684	Mebendazole		$C_{16}H_{13}N_3O_3$	31431-39-7	295.292	cry (HOAc/ MeOH)	288.5				i H_2O, EtOH, eth, chl
6685	Mebhydroline		$C_{19}H_{20}N_2$	524-81-2	276.375	cry	95	211^1			i H_2O; sl eth; vs EtOH, ace, MeOH
6686	Mecarbam		$C_{10}H_{20}NO_5PS_2$	2595-54-2	329.374	ye oil		$144^{0.02}$	1.223^{20}		sl H_2O
6687	Meclizine		$C_{25}H_{27}ClN_2$	569-65-3	390.948			230			s CS_2
6688	Medroxyprogesterone		$C_{22}H_{32}O_3$	520-85-4	344.487		214.5				vs chl
6689	Mefenamic acid	2-[(2,3-Dimethylphenyl)amino] benzoic acid	$C_{15}H_{15}NO_2$	61-68-7	241.286	hyg cry	230 dec				s alk; sl eth, chl
6690	Mefloquine		$C_{17}H_{16}F_6N_2O$	53230-10-7	378.311	cry (MeOH aq)	178.2				
6691	Mefluidide		$C_{11}H_{13}F_3N_2O_3S$	53780-34-0	310.292		184				
6692	Melezitose		$C_{18}H_{32}O_{16}$	597-12-6	504.437	cry (w+2)	153		1.5565^{25}		vs H_2O
6693	α-*D*-Melibiose	6-*O*-α-*D*-Galactopyranosyl-*D*-glucose	$C_{12}H_{22}O_{11}$	585-99-9	342.296						vs H_2O; sl EtOH; dec acid
6694	Melinamide	*N*-(1-Phenylethyl)-9,12-octadecadieneamide, (*Z*,*Z*)-	$C_{26}H_{41}NO$	14417-88-0	383.610	oil	<4	$202^{0.07}$		1.5050^{23}	
6695	Melphalan	*L*-Phenylalanine, 4-[bis(2-chloroethyl)amino]-	$C_{13}H_{18}Cl_2N_2O_2$	148-82-3	305.200	nd	183 dec				i H_2O; s EtOH
6696	Menaquinone 7	Vitamin K_2(35)	$C_{46}H_{64}O_2$	2124-57-4	648.999	cry	54				
6697	Menazon		$C_6H_{12}N_5O_2PS_2$	78-57-9	281.296	cry (MeOH)	160				sl H_2O; s thf
6698	*p*-Menthane hydroperoxide	1-Methyl-1-(4-methylcyclohexyl)ethyl hydroperoxide	$C_{10}H_{20}O_2$	80-47-7	172.265			259	0.92		
6699	*p*-Menth-8-en-2-one	2-Methyl-5-(1-methylethenyl) cyclohexanone	$C_{10}H_{16}O$	7764-50-3	152.233			223.0			
6700	Menthol 3-methylbutanoate	Menthol, isovalerate	$C_{15}H_{28}O_2$	16409-46-4	240.382			129^9	0.908^{15}	1.4486^{20}	i H_2O; s EtOH, ace
6701	Meperidine	Pethidine	$C_{15}H_{21}NO_2$	57-42-1	247.334		30	155^5			
6702	Mephenytoin		$C_{12}H_{14}N_2O_2$	50-12-4	218.251		136				
6703	Mephobarbital		$C_{13}H_{14}N_2O_3$	115-38-8	246.261	wh cry (w)	176				sl H_2O, eth, chl; vs EtOH
6704	Mephosfolan		$C_8H_{16}NO_3PS_2$	950-10-7	269.322	ye liq		$120^{0.001}$		1.5354^{26}	s ace, EtOH, bz
6705	Mepiquat chloride	Piperidinium, 1,1-dimethyl-, chloride	$C_7H_{16}ClN$	24307-26-4	149.662		223				
6706	Mepivacaine	*N*-(2,6-Dimethylphenyl)-1-methyl-2-piperidinecarboxamide	$C_{15}H_{22}N_2O$	96-88-8	246.348	cry (eth)	150.5				s CS_2
6707	Mepivacaine monohydrochloride	Carbocaine hydrochloride	$C_{15}H_{23}ClN_2O$	1722-62-9	282.809	cry	263				s H_2O
6708	Mercaptoacetic acid, 2-ethylhexyl ester		$C_{10}H_{20}O_2S$	7659-86-1	204.330			133.5	0.97^{20}		
6709	2-Mercaptobenzoic acid	*o*-Thiosalicylic acid	$C_7H_6O_2S$	147-93-3	154.187	lf or nd (al, w, HOAc)	168.5	sub			s H_2O, EtOH, eth; sl DMSO, lig
6710	Mercaptobenzthiazyl ether	2,2'-Dithiobis[benzothiazole]	$C_{14}H_8N_2S_4$	120-78-5	332.487	ye nd	180		1.50		i H_2O; sl EtOH, bz, ctc, ace
6711	2-Mercaptoethanol		C_2H_6OS	60-24-2	78.133			158; 55^{13}	1.1143^{20}	1.4996^{20}	s H_2O, EtOH, eth, bz
6712	2-Mercapto-2-methylpropanoic acid		$C_4H_8O_2S$	4695-31-2	120.171		47	101^{15}			vs H_2O
6713	2-Mercapto-*N*-2-naphthylacetamide	Thionalide	$C_{12}H_{11}NOS$	93-42-5	217.286		111.5				i H_2O; vs EtOH, os
6714	2-Mercaptophenol		C_6H_6OS	1121-24-0	126.176	oil	5.5	217; 89^8	1.2371^0		vs bz, eth, EtOH

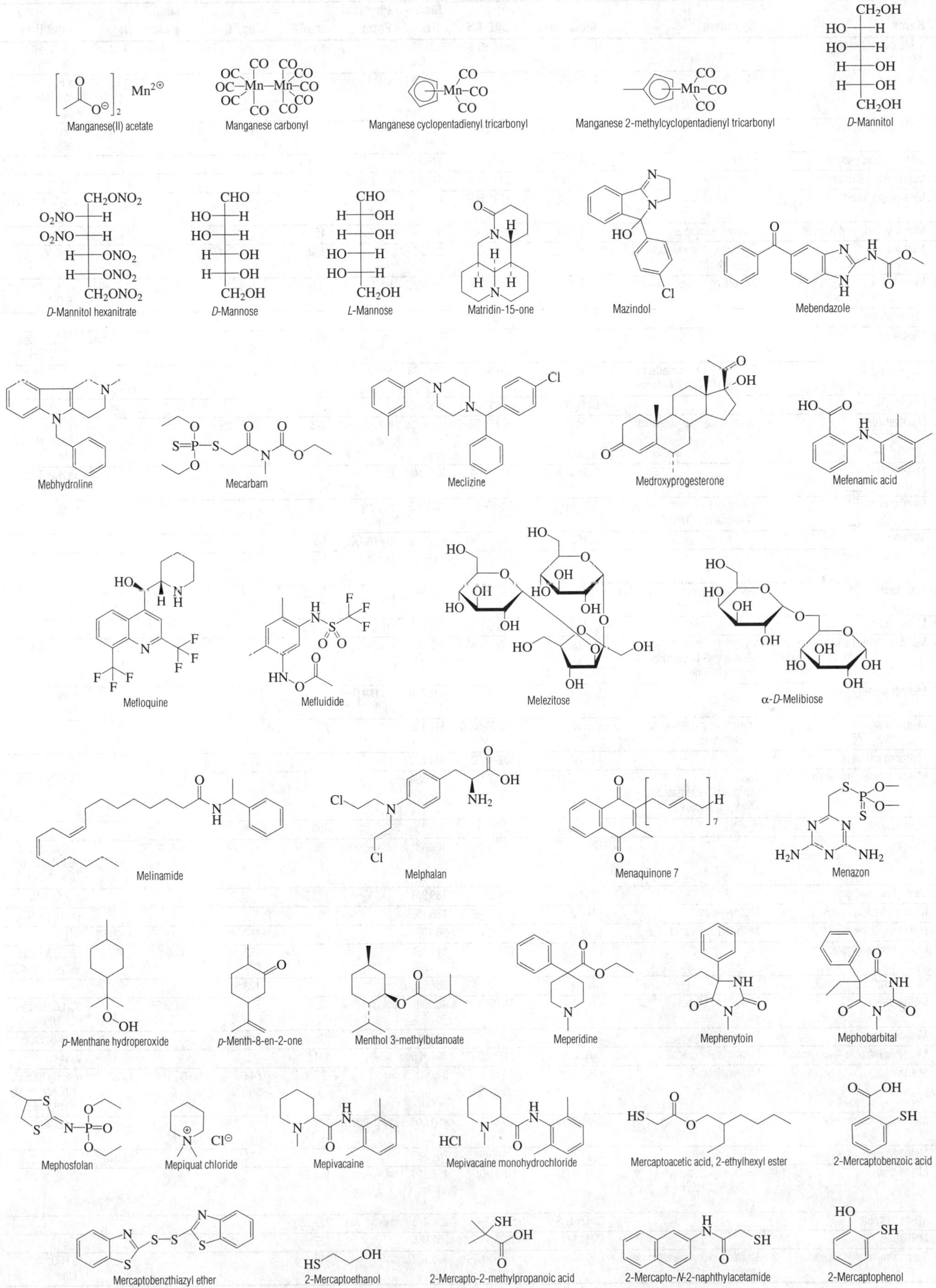

Manganese(II) acetate
Manganese carbonyl
Manganese cyclopentadienyl tricarbonyl
Manganese 2-methylcyclopentadienyl tricarbonyl
D-Mannitol
D-Mannitol hexanitrate
D-Mannose
L-Mannose
Matridin-15-one
Mazindol
Mebendazole
Mebhydroline
Mecarbam
Meclizine
Medroxyprogesterone
Mefenamic acid
Mefloquine
Mefluidide
Melezitose
α-D-Melibiose
Melinamide
Melphalan
Menaquinone 7
Menazon
p-Menthane hydroperoxide
p-Menth-8-en-2-one
Menthol 3-methylbutanoate
Meperidine
Mephenytoin
Mephobarbital
Mephosfolan
Mepiquat chloride
Mepivacaine
Mepivacaine monohydrochloride
Mercaptoacetic acid, 2-ethylhexyl ester
2-Mercaptobenzoic acid
Mercaptobenzthiazyl ether
2-Mercaptoethanol
2-Mercapto-2-methylpropanoic acid
2-Mercapto-N-2-naphthylacetamide
2-Mercaptophenol

No.	Name	Synonym	Mol. Form.	CAS RN	Mol. Wt.	Physical Form	mp/°C	bp/°C	den/ g cm^{-3}	n_D	Solubility
6715	4-Mercaptophenol		C_6H_6OS	637-89-8	126.176	cry	29.5	167^{45}, 135^{11}	1.1285^{25}	1.5101^{25}	s H_2O, EtOH, alk, con sulf
6716	3-Mercapto-1,2-propanediol	Thioglycerol	$C_3H_8O_2S$	96-27-5	108.160	visc		100^{1}	1.2455^{20}	1.5268^{20}	sl H_2O, eth, bz, chl; msc EtOH; vs ace
6717	3-Mercaptopropanoic acid		$C_3H_6O_2S$	107-96-0	106.144	amor	18	111^{15}, 86^{3}	1.218^{21}	1.494^{20}	s H_2O, EtOH, eth, ctc
6718	3-Mercapto-*D*-valine	Penicillamine	$C_5H_{11}NO_2S$	52-67-5	149.212		198.5				
6719	Mercury(II) benzoate	Mercuric benzoate	$C_{14}H_{10}HgO_4$	583-15-3	442.81	cry pow (w)	≈125				i EtOH
6720	Mercury(II) oleate	Mercuric oleate	$C_{36}H_{66}HgO_4$	1191-80-6	763.35	ye-br solid					i H_2O; sl EtOH, eth
6721	Mercury(II) phenyl acetate	Phenylmercuric acetate	$C_8H_8HgO_2$	62-38-4	336.74		153				i H_2O; s chl
6722	Merphos	Phosphorotrithious acid, *S,S,S*-tributyl ester	$C_{12}H_{27}PS_3$	150-50-5	298.511		100	$137^{0.7}$, 176^{15}	1.02^{20}		
6723	Mesityl oxide	Isobutenyl methyl ketone	$C_6H_{10}O$	141-79-7	98.142	liq	-59	130	0.8653^{20}	1.4440^{20}	s H_2O, ace; msc EtOH, eth
6724	Mesoridazine		$C_{21}H_{26}N_2OS_2$	5588-33-0	386.573	oil					
6725	Mestranol		$C_{21}H_{26}O_2$	72-33-3	310.430	cry	151				i H_2O; s diox, eth, EtOH, chl
6726	[2.2]Metacyclophane	Tricyclo[9.3.1.1]hexadeca-1(15),4,6,8(16),11,13-hexaene	$C_{16}H_{16}$	2319-97-3	208.298	orth pr	132.5	290			sl EtOH; s bz, eth
6727	Metalaxyl		$C_{15}H_{21}NO_4$	57837-19-1	279.333		71				
6728	Metaldehyde	Metacetaldehyde (polymer)	$(C_2H_4O)_x$	37273-91-9		tetr nd or pr (al)	246	sub 115			i H_2O, ace; sl EtOH, eth, bz, chl
6729	Metanil Yellow		$C_{18}H_{14}N_3NaO_3S$	587-98-4	375.377	br-ye pow					vs H_2O, EtOH; s bz, eth; sl ace
6730	Metaraminol	2-Amino-1-(3-hydroxyphenyl)-1-propanol, (1*R*,2*S*)	$C_9H_{13}NO_2$	54-49-9	167.205	hyg cry (HCl)					s H_2O
6731	Metaxalone		$C_{12}H_{15}NO_3$	1665-48-1	221.252	cry (AcOEt)	122	$223^{1.5}$			
6732	Methacholine chloride		$C_8H_{18}ClNO_2$	62-51-1	195.688	hyg cry	172				vs H_2O, EtOH, chl
6733	Methacrylic acid	2-Methylpropenoic acid	$C_4H_6O_2$	79-41-4	86.090	pr	16	162.5	1.0153^{20}	1.4314^{20}	s H_2O, chl; msc EtOH, eth
6734	Methacycline		$C_{22}H_{22}N_2O_8$	914-00-1	442.418	cry	205 dec				
6735	Methadone hydrochloride	6-(Dimethylamino)-4,4-diphenyl-3-heptanone hydrochloride	$C_{21}H_{28}ClNO$	1095-90-5	345.906	pl (al-eth)	235				vs H_2O, EtOH
6736	Methallenestril		$C_{18}H_{22}O_3$	517-18-0	286.366	cry (MeOH aq)	139				s eth
6737	Methamidophos	Phosphoramidothioic acid, *O,S*-dimethyl ester	$C_2H_8NO_2PS$	10265-92-6	141.130		46		1.31^{20}		
6738	Methamphetamine		$C_{10}H_{15}N$	537-46-2	149.233			212			
6739	Methamphetamine hydrochloride	*N*,α-Dimethylbenzeneethanamine, hydrochloride, (*S*)-	$C_{10}H_{16}ClN$	51-57-0	185.694		173.8				vs H_2O, EtOH, chl
6740	Methandrostenolone		$C_{20}H_{28}O_2$	72-63-9	300.435		166				
6741	Methane		CH_4	74-82-8	16.043	col gas	-182.47	-161.48	0.4228^{-162}		sl H_2O, ace; s EtOH, eth, bz, tol, MeOH
6742	Methanearsonic acid		CH_5AsO_3	124-58-3	139.971		160.5				s H_2O, EtOH
6743	Methanedisulfonic acid	Methionic acid	$CH_4O_6S_2$	503-40-2	176.169		98				i H_2O; s HNO_3
6744	Methanesulfonic acid	Methylsulfonic acid	CH_4O_3S	75-75-2	96.106		20	167^{10}	1.4812^{18}	1.4317^{18}	s H_2O
6745	Methanesulfonyl chloride		CH_3ClO_2S	124-63-0	114.552			162; 55^{11}	1.4805^{18}	1.4573^{20}	i H_2O; s EtOH, eth
6746	Methanesulfonyl fluoride		CH_3FO_2S	558-25-8	98.097			123.5			
6747	Methanethiol	Methyl mercaptan	CH_4S	74-93-1	48.108	col gas	-123	5.9	0.8665^{20}		sl H_2O, chl; vs EtOH, eth
6748	Methanimidamide	Formamidine	CH_4N_2	463-52-5	44.056	pr	81	dec			vs H_2O, EtOH
6749	Methanimidamide, monoacetate	Formamidine acetate	$C_3H_8N_2O_2$	3473-63-0	104.108		161.5				vs H_2O
6750	Methanol	Methyl alcohol	CH_4O	67-56-1	32.042	liq	-97.53	64.6	0.7914^{20}	1.3288^{20}	msc H_2O, EtOH, eth, ace; vs bz; s chl
6751	Methantheline bromide		$C_{21}H_{26}BrNO_3$	53-46-3	420.340	cry (i-PrOH)	174.5				s H_2O, EtOH, chl; i eth
6752	Methapyrilene		$C_{14}H_{19}N_3S$	91-80-5	261.386			174^{3}		1.5915^{20}	
6753	Metharbital	5,5-Diethyl-1-methyl-2,4,6(1*H*,3*H*,5*H*)-pyrimidinetrione	$C_9H_{14}N_2O_3$	50-11-3	198.218	nd	150.5				s H_2O; sl chl
6754	Methazolamide		$C_5H_8N_4O_3S_2$	554-57-4	236.273	cry (w)	213 dec				
6755	Methazole		$C_9H_6Cl_2N_2O_3$	20354-26-1	261.061		123		1.24^{25}		
6756	Methenamine allyl iodide	Allylhexamethylenetetramine iodide	$C_9H_{17}IN_4$	36895-62-2	308.162	cry	148 dec				vs H_2O; i chl, eth

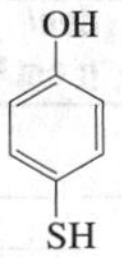

4-Mercaptophenol

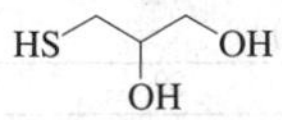

3-Mercapto-1,2-propanediol

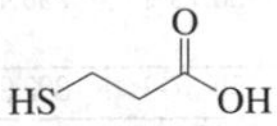

3-Mercaptopropanoic acid

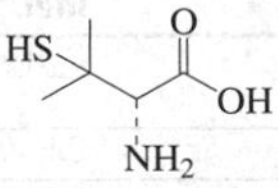

3-Mercapto-*D*-valine

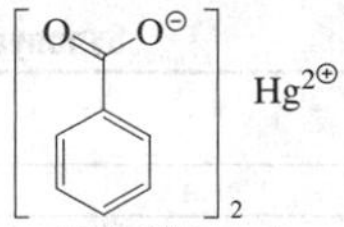

Mercury(II) benzoate

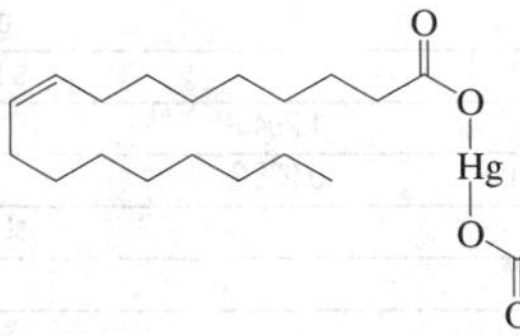

Mercury(II) oleate

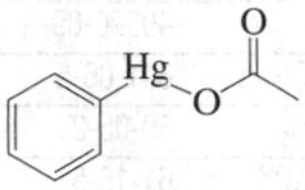

Mercury(II) phenyl acetate

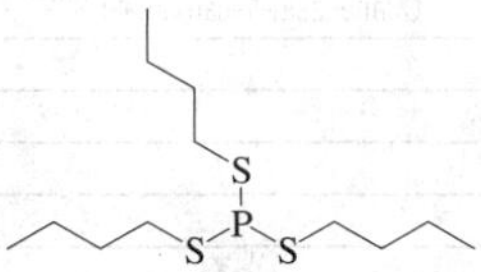

Merphos

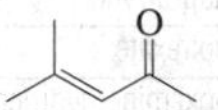

Mesityl oxide

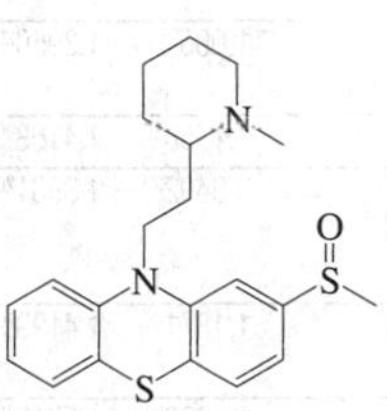

Mesoridazine

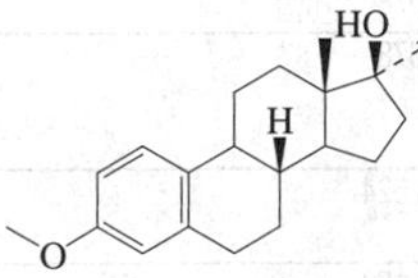

Mestranol

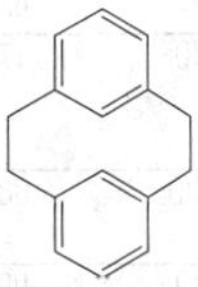

[2.2]Metacyclophane

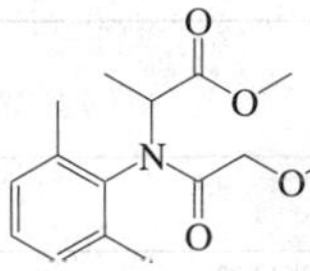

Metalaxyl

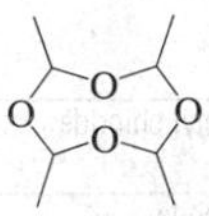

Metaldehyde

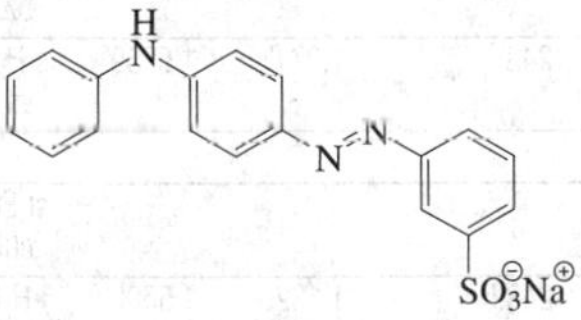

Metanil Yellow

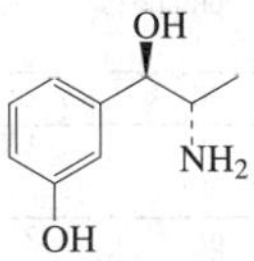

Metaraminol

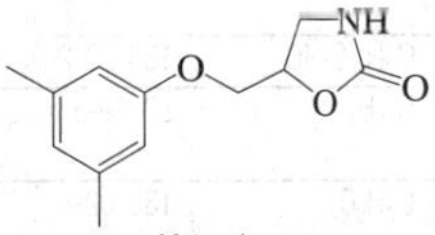

Metaxalone

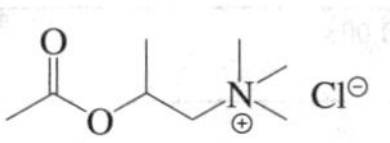

Methacholine chloride

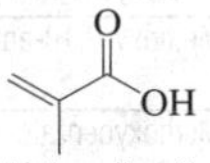

Methacrylic acid

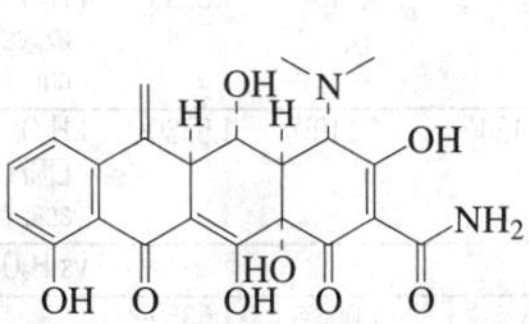

Methacycline

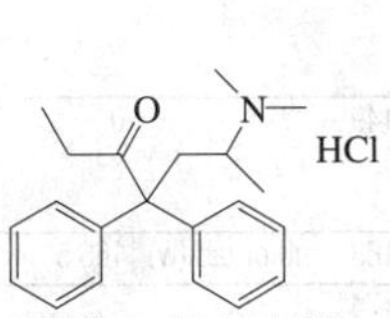

Methadone hydrochloride

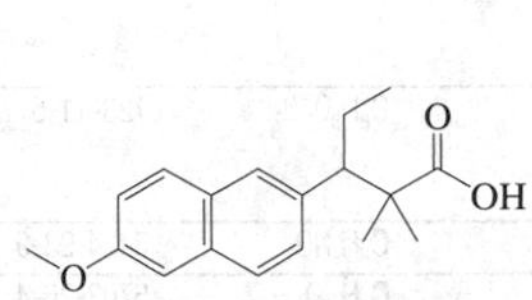

Methallenestril

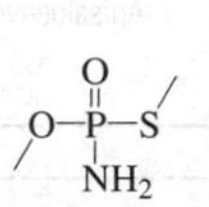

Methamidophos

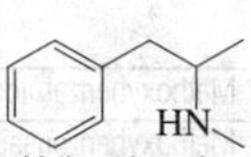

Methamphetamine

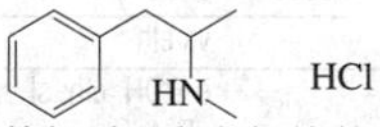

Methamphetamine hydrochloride

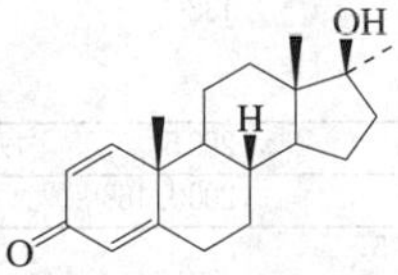

Methandrostenolone

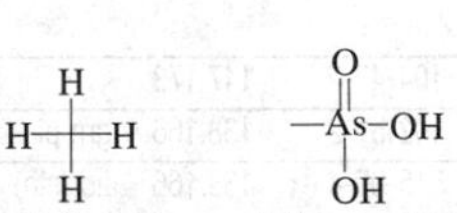

Methane

Methanearsonic acid

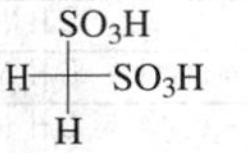

Methanedisulfonic acid

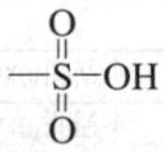

Methanesulfonic acid

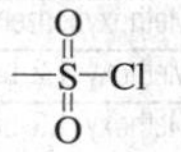

Methanesulfonyl chloride

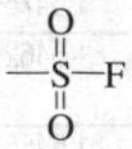

Methanesulfonyl fluoride

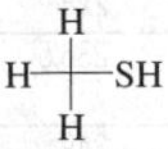

Methanethiol

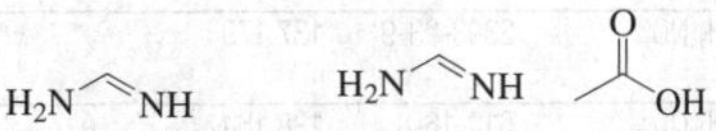

Methanimidamide

Methanimidamide, monoacetate

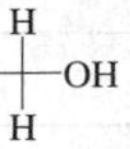

Methanol

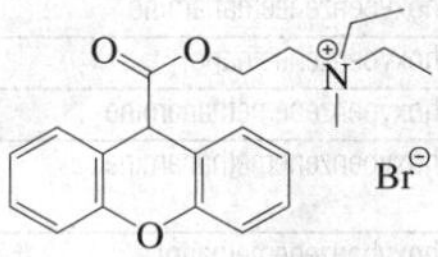

Methantheline bromide

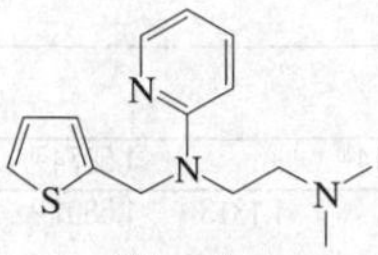

Methapyrilene

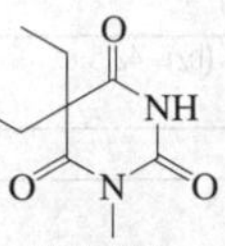

Metharbital

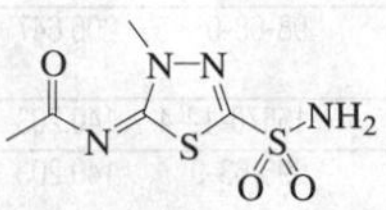

Methazolamide

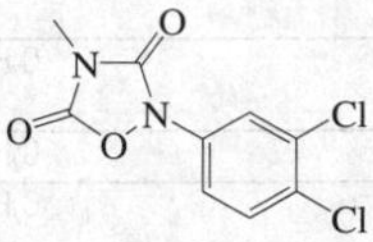

Methazole

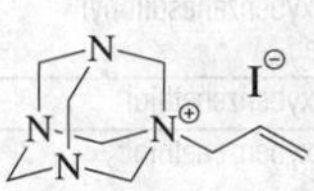

Methenamine allyl iodide

No.	Name	Synonym	Mol. Form.	CAS RN	Mol. Wt.	Physical Form	mp/°C	bp/°C	den/ g cm^{-3}	n_D	Solubility
6757	Methestrol		$C_{20}H_{26}O_2$	130-73-4	298.419	cry (dil HOAc)	145				
6758	Methidathion		$C_6H_{11}N_2O_4PS_3$	950-37-8	302.330		39				
6759	Methiocarb	Phenol, 3,5-dimethyl-4-(methylthio)-, methylcarbamate	$C_{11}H_{15}NO_2S$	2032-65-7	225.308		120				
6760	*L*-Methionine		$C_5H_{11}NO_2S$	63-68-3	149.212	hex pl (dil al)	281 dec				s H_2O; i EtOH, eth, ace, bz, peth; sl HOAc
6761	Methocarbamol	Guaifenesin-1-carbamate	$C_{11}H_{15}NO_5$	532-03-6	241.241	cry (bz)	93				s EtOH
6762	Methomyl		$C_5H_{10}N_2O_2S$	16752-77-5	162.210		78		1.2946^{24}		
6763	Methoprene		$C_{19}H_{34}O_3$	40596-69-8	310.471			$100^{0.05}$	0.926^{20}		
6764	Methoprotryne		$C_{11}H_{21}N_5OS$	841-06-5	271.383	cry	69				sl H_2O; s os
6765	Methotrexate		$C_{20}H_{22}N_8O_5$	59-05-2	454.440	ye cry (w)	190 dec				
6766	Methoxamine hydrochloride		$C_{11}H_{18}ClNO_3$	61-16-5	247.719	cry	214				vs H_2O; i eth, bz, chl
6767	Methoxsalen	9-Methoxy-7*H*-furo[3,2-g][1] benzopyran-7-one	$C_{12}H_8O_4$	298-81-7	216.190	pr (dil al) nd (peth)	148				sl H_2O, eth, ace, peth; vs EtOH
6768	Methoxyacetaldehyde		$C_3H_6O_2$	10312-83-1	74.079			92	1.005^{25}	1.3950^{20}	vs H_2O, ace, eth, EtOH
6769	Methoxyacetic acid		$C_3H_6O_3$	625-45-6	90.078	hyg		203.5	1.1768^{20}	1.4168^{20}	s H_2O, EtOH, eth
6770	Methoxyacetonitrile		C_3H_5NO	1738-36-9	71.078			119	0.9492^{20}	1.3831^{20}	sl H_2O; s EtOH, eth, ace, chl, alk, acid
6771	Methoxyacetyl chloride		$C_3H_5ClO_2$	38870-89-2	108.524			112.5	1.1871^{20}	1.4199^{20}	s eth, ace, ctc; vs chl
6772	2-Methoxyaniline	*o*-Anisidine	C_7H_9NO	90-04-0	123.152		6.2	224	1.0923^{20}	1.5715^{10}	sl H_2O; s EtOH, eth, ace, bz
6773	3-Methoxyaniline	*m*-Anisidine	C_7H_9NO	536-90-3	123.152	liq	-1	251	1.096^{20}	1.5794^{20}	sl H_2O, ctc; s EtOH, eth, ace, bz
6774	4-Methoxyaniline	*p*-Anisidine	C_7H_9NO	104-94-9	123.152	orth pl	57.2	243	1.071^{57}	1.5559^{60}	s H_2O, ace, bz; vs EtOH, eth
6775	2-Methoxyaniline hydrochloride	*o*-Anisidine hydrochloride	$C_7H_{10}ClNO$	134-29-2	159.613	nd	225				
6776	1-Methoxy-9,10-anthracenedione		$C_{15}H_{10}O_3$	82-39-3	238.238		170.3				sl EtOH; vs bz, chl
6777	2-Methoxybenzaldehyde		$C_8H_8O_2$	135-02-4	136.149	pr	37.5	243.5	1.1326^{20}	1.5600^{20}	i H_2O; s EtOH, bz, ctc; vs eth, ace, chl
6778	3-Methoxybenzaldehyde		$C_8H_8O_2$	591-31-1	136.149			231	1.1187^{20}	1.5530^{20}	i H_2O; s EtOH, bz; vs eth, ace, chl
6779	4-Methoxybenzaldehyde	*p*-Anisaldehyde	$C_8H_8O_2$	123-11-5	136.149		0	248; 134^{12}	1.119^{15}	1.5730^{20}	i H_2O; msc EtOH, eth; vs ace, chl; s bz
6780	4-Methoxybenzamide		$C_8H_9NO_2$	3424-93-9	151.163	nd or tab (w)	166.5	295			vs H_2O, EtOH
6781	4-Methoxybenzeneacetaldehyde		$C_9H_{10}O_2$	5703-26-4	150.174			255.5	1.096^{20}	1.5359^{20}	
6782	2-Methoxybenzeneacetic acid		$C_9H_{10}O_3$	93-25-4	166.173	nd (w)	124	100^{2}			s H_2O; vs EtOH, eth, ace, bz, chl
6783	4-Methoxybenzeneacetic acid		$C_9H_{10}O_3$	104-01-8	166.173	pl (w)	87	138^{2}			i H_2O; vs EtOH; s eth, bz; sl chl, lig
6784	4-Methoxybenzeneacetonitrile		C_9H_9NO	104-47-2	147.173			286.5	1.0845^{20}	1.5309^{20}	s EtOH, eth, chl
6785	4-Methoxy-1,2-benzenediamine	4-Methoxy-*o*-phenylenediamine	$C_7H_{10}N_2O$	102-51-2	138.166	grn pl	51	200^{21}, 168^{11}			vs eth
6786	4-Methoxy-1,3-benzenediamine	4-Methoxy-*m*-phenylenediamine	$C_7H_{10}N_2O$	615-05-4	138.166	nd (eth)	67.5				s EtOH, eth; sl DMSO
6787	2-Methoxy-1,4-benzenediamine	2,5-Diaminoanisole	$C_7H_{10}N_2O$	5307-02-8	138.166	cry	107				
6788	3-Methoxy-1,2-benzenediol		$C_7H_8O_3$	934-00-9	140.137	nd	42.8	163^{48}, 129^{10}			s chl
6789	4-Methoxybenzeneethanamine		$C_9H_{13}NO$	55-81-2	151.205			139^{20}		1.5379^{20}	
6790	4-Methoxybenzeneethanol		$C_9H_{12}O_2$	702-23-8	152.190		29	335			
6791	2-Methoxybenzenemethanamine		$C_8H_{11}NO$	6850-57-3	137.179			228	1.051^{25}	1.5475^{20}	
6792	4-Methoxybenzenemethanamine		$C_8H_{11}NO$	2393-23-9	137.179			236.5	1.050^{15}	1.5462^{20}	sl H_2O, EtOH, eth
6793	2-Methoxybenzenemethanol		$C_8H_{10}O_2$	612-16-8	138.164			249	1.0386^{25}	1.5455^{20}	i H_2O; s EtOH; msc eth
6794	3-Methoxybenzenemethanol		$C_8H_{10}O_2$	6971-51-3	138.164		30	252	1.112^{25}	1.5440^{20}	
6795	4-Methoxybenzenemethanol	Anise alcohol	$C_8H_{10}O_2$	105-13-5	138.164	nd	25	259.1	1.109^{26}	1.5420^{25}	s H_2O, ctc; vs EtOH, eth
6796	4-Methoxybenzenesulfonyl chloride		$C_7H_7ClO_3S$	98-68-0	206.647	nd or pr (bz)	42.5	$103^{0.25}$			s EtOH, eth, bz
6797	3-Methoxybenzenethiol		C_7H_8OS	15570-12-4	140.203			224.5; 114^{20}		1.5874^{20}	s chl
6798	4-Methoxybenzenethiol		C_7H_8OS	696-63-9	140.203			228	1.1313^{25}	1.5801^{25}	s EtOH, eth, bz; sl chl

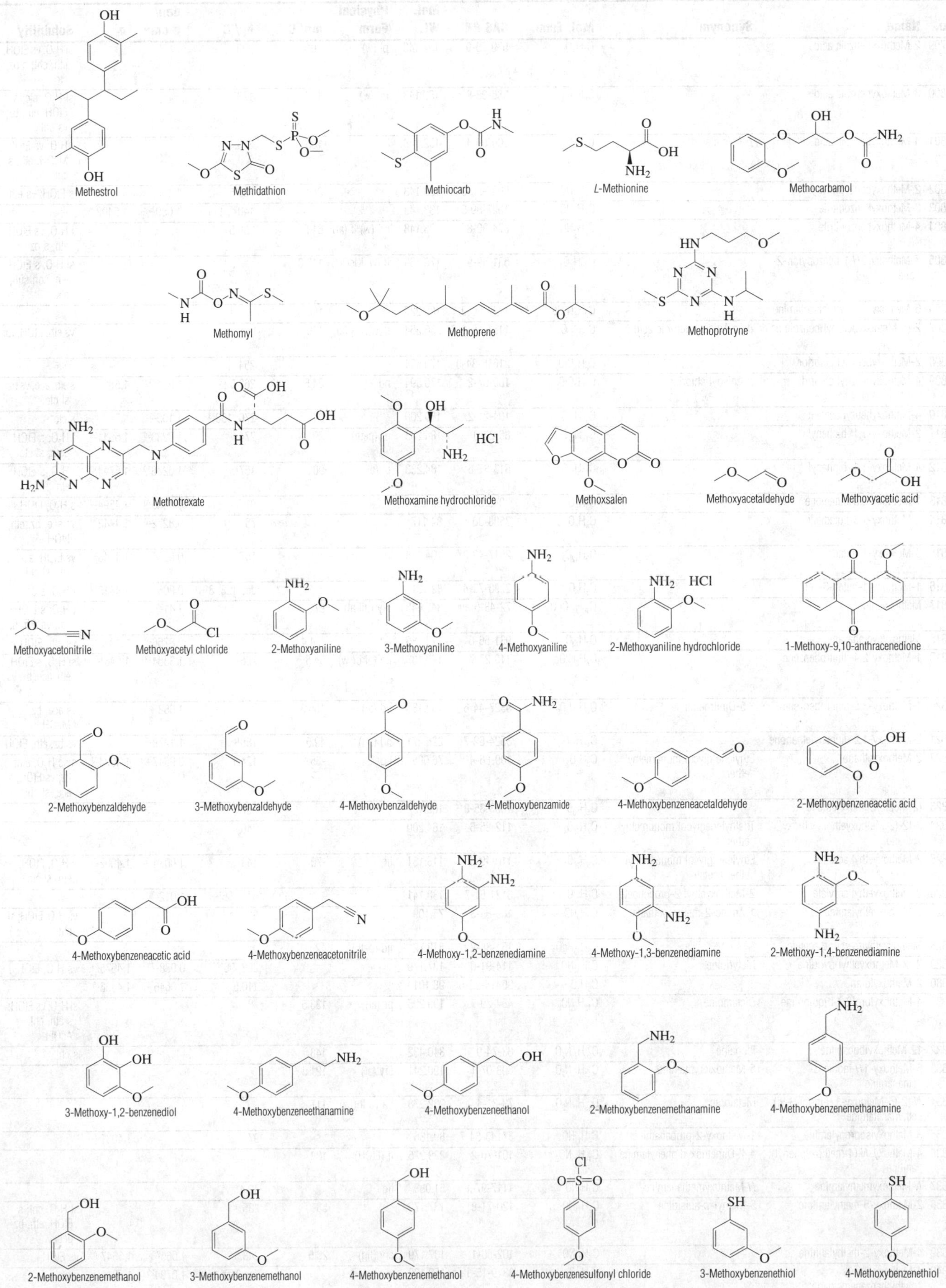

Methestrol
Methidathion
Methiocarb
L-Methionine
Methocarbamol
Methomyl
Methoprene
Methoprotryne
Methotrexate
Methoxamine hydrochloride
Methoxsalen
Methoxyacetaldehyde
Methoxyacetic acid
Methoxyacetonitrile
Methoxyacetyl chloride
2-Methoxyaniline
3-Methoxyaniline
4-Methoxyaniline
2-Methoxyaniline hydrochloride
1-Methoxy-9,10-anthracenedione
2-Methoxybenzaldehyde
3-Methoxybenzaldehyde
4-Methoxybenzaldehyde
4-Methoxybenzamide
4-Methoxybenzeneacetaldehyde
2-Methoxybenzeneacetic acid
4-Methoxybenzeneacetic acid
4-Methoxybenzeneacetonitrile
4-Methoxy-1,2-benzenediamine
4-Methoxy-1,3-benzenediamine
2-Methoxy-1,4-benzenediamine
3-Methoxy-1,2-benzenediol
4-Methoxybenzeneethanamine
4-Methoxybenzeneethanol
2-Methoxybenzenemethanamine
4-Methoxybenzenemethanamine
2-Methoxybenzenemethanol
3-Methoxybenzenemethanol
4-Methoxybenzenemethanol
4-Methoxybenzenesulfonyl chloride
3-Methoxybenzenethiol
4-Methoxybenzenethiol

No.	Name	Synonym	Mol. Form.	CAS RN	Mol. Wt.	Physical Form	mp/°C	bp/°C	den/ g cm^{-3}	n_D	Solubility
6799	2-Methoxybenzoic acid		$C_8H_8O_3$	579-75-9	152.148	pl (w)	101	200			sl H_2O; vs EtOH, eth, chl; s bz, ctc
6800	3-Methoxybenzoic acid		$C_8H_8O_3$	586-38-9	152.148	nd (w)	107	170^{10}			sl H_2O, ctc; s EtOH, eth, bz; vs chl
6801	4-Methoxybenzoic acid	*p*-Anisic acid	$C_8H_8O_3$	100-09-4	152.148		185	276.5			i H_2O; vs EtOH, MeOH, eth; s chl
6802	2-Methoxybenzonitrile		C_8H_7NO	6609-56-9	133.148		24.5	255.5	1.1063^{20}		s EtOH; vs eth
6803	3-Methoxybenzonitrile		C_8H_7NO	1527-89-5	133.148			140^{34}, 111^{13}	1.089^{25}	1.5402^{20}	
6804	4-Methoxybenzonitrile		C_8H_7NO	874-90-8	133.148	nd (w) lf (al)	61.5	256.5			i H_2O; vs EtOH, eth; s bz
6805	7-Methoxy-2*H*-1-benzopyran-2-one		$C_{10}H_8O_3$	531-59-9	176.169	lf (w, MeOH)	118.3				sl H_2O; s EtOH, eth, con sulf, alk
6806	6-Methoxy-2-benzothiazolamine		$C_8H_8N_2OS$	1747-60-0	180.227		166				
6807	2-(4-Methoxybenzoyl)benzoic acid	*o*-(*p*-Anisoyl)benzoic acid	$C_{15}H_{12}O_4$	1151-15-1	256.254	lf (w), cry (al, to)	146				vs eth, EtOH, tol
6808	2-Methoxybenzoyl chloride		$C_8H_7ClO_2$	21615-34-9	170.594			254			
6809	4-Methoxybenzoyl chloride	*p*-Anisoyl chloride	$C_8H_7ClO_2$	100-07-2	170.594	nd	24.5	262.5	1.261^{20}	1.580^{20}	s eth, ace; vs bz; sl ctc
6810	4-Methoxybenzyl acetate		$C_{10}H_{12}O_3$	104-21-2	180.200		84	270; 150^{23}	1.105^{25}		s ctc
6811	2-Methoxy-1,1'-biphenyl		$C_{13}H_{12}O$	86-26-0	184.233	pr (peth)	29	274	1.0233^{99}	1.5641^{99}	i H_2O; s EtOH, peth; sl ctc
6812	4-Methoxy-1,1'-biphenyl		$C_{13}H_{12}O$	613-37-6	184.233	pl (al)	90	157^{10}	1.0278^{100}	1.5744^{100}	i H_2O; s EtOH, eth
6813	1-Methoxy-1,3-butadiene		C_5H_8O	3036-66-6	84.117			91.5	0.8296^{20}	1.4594^{20}	s H_2O, EtOH
6814	2-Methoxy-1,3-butadiene		C_5H_8O	3588-30-5	84.117			75	0.8272^{20}	1.4442^{20}	vs ace, bz, eth, EtOH
6815	3-Methoxy-1-butanol		$C_5H_{12}O_2$	2517-43-3	104.148			157	0.923^{23}	1.4148^{25}	vs EtOH, ace; s eth; sl chl
6816	1-Methoxy-1-buten-3-yne		C_5H_6O	2798-73-4	82.101			dec 123; 39^{23}	0.906^{20}	1.4818^{20}	i H_2O; s chl
6817	Methoxychlor		$C_{16}H_{15}Cl_3O_2$	72-43-5	345.648	cry (dil al)	87		1.41^{25}		i H_2O; s EtOH, ctc; vs eth, bz
6818	Methoxycyclohexane		$C_7H_{14}O$	931-56-6	114.185	liq	-74.4	133	0.8756^{20}	1.4355^{20}	vs eth, EtOH
6819	1-Methoxy-2,4-dinitrobenzene		$C_7H_6N_2O_5$	119-27-7	198.133	nd (al or w)	94.5	206^{12}	1.3364^{131}	1.546^{15}	sl H_2O; s EtOH, eth, ace, bz; vs py
6820	1-Methoxy-3,5-dinitrobenzene	3,5-Dinitroanisole	$C_7H_6N_2O_5$	5327-44-6	198.133	nd (al)	105.3		1.558^{12}		vs ace, bz, MeOH
6821	2-Methoxy-1,2-diphenylethanone		$C_{15}H_{14}O_2$	3524-62-7	226.271	nd (lig)	49.5	188^{15}	1.1278^{14}		vs bz, eth, EtOH
6822	2-Methoxyethanol	Ethylene glycol monomethyl ether	$C_3H_8O_2$	109-86-4	76.095	liq	-85.1	124.1	0.9647^{20}	1.4024^{20}	msc H_2O, eth, bz; vs EtOH; s ace; sl chl
6823	(2-Methoxyethoxy)ethene		$C_5H_{10}O_2$	1663-35-0	102.132			107			
6824	2-[2-(2-Methoxyethoxy)ethoxy] ethanol	Triethyleneglycol monomethyl ether	$C_7H_{16}O_4$	112-35-6	164.200			246			
6825	2-Methoxyethyl acetate	Ethylene glycol monomethyl ether acetate	$C_5H_{10}O_3$	110-49-6	118.131	liq	-70	143	1.0074^{19}	1.4002^{20}	s H_2O, EtOH, eth; sl ctc
6826	2-Methoxyethyl acrylate	2-Methoxyethyl 2-propenoate	$C_6H_{10}O_3$	3121-61-7	130.141			67^{16}, 56^{12}	1.012^{20}		
6827	2-Methoxyethylamine	1-Amino-2-methoxyethane	C_3H_9NO	109-85-3	75.109			95			vs H_2O, EtOH; sl chl
6828	Methoxyethylmercuric acetate		$C_5H_{10}HgO_3$	151-38-2	318.72	nd (peth)	42				
6829	2-(2-Methoxyethyl)pyridine	Metyridine	$C_8H_{11}NO$	114-91-0	137.179			203; 96^{17}	0.988^{20}	1.4975^{20}	vs H_2O, EtOH
6830	2-Methoxyfuran		$C_5H_6O_2$	25414-22-6	98.101			110.5	1.0646^{25}	1.4468^{25}	
6831	4-Methoxyfuro[2,3-b]quinoline	Dictamnine	$C_{12}H_9NO_2$	484-29-7	199.205	pr (al)	133.5				sl H_2O; vs EtOH; s eth, chl, AcOEt
6832	12-Methoxyibogamine	Ibogaine	$C_{20}H_{26}N_2O$	83-74-9	310.432		148				s chl
6833	5-Methoxy-1*H*-indole-3-ethanamine	5-Methoxytryptamine	$C_{11}H_{14}N_2O$	608-07-1	190.241	cry (al)	121.5				
6834	*N*-[2-(5-Methoxy-1*H*-indol-3-yl) ethyl]acetamide	Melatonin	$C_{13}H_{16}N_2O_2$	73-31-4	232.278	pa ye lf (bz)	117				
6835	3-Methoxyisopropylamine	1-Methoxy-2-propanamine	$C_4H_{11}NO$	37143-54-7	89.136			97		1.4031^{25}	
6836	4-Methoxy-*N*-(4-methoxyphenyl) aniline	4,4'-Dimethoxydiphenylamine	$C_{14}H_{15}NO_2$	101-70-2	229.275	lf (EtOH)	103				
6837	*N*-Methoxymethylamine	*N*-Methoxymethanamine	C_2H_7NO	1117-97-1	61.083	liq		42.4			
6838	2-Methoxy-5-methylaniline	5-Methyl-*o*-anisidine	$C_8H_{11}NO$	120-71-8	137.179		53	235			sl H_2O, chl; s EtOH, eth, bz, peth
6839	4-Methoxy-2-methylaniline		$C_8H_{11}NO$	102-50-1	137.179	cry (lig)	29.5	248.5	1.065^{25}	1.5647^{20}	vs EtOH
6840	4-Methoxy-α-methylbenzenemethanol		$C_9H_{12}O_2$	3319-15-1	152.190			dec 310; 140^{17}	1.0794^{20}	1.5310^{25}	s ctc

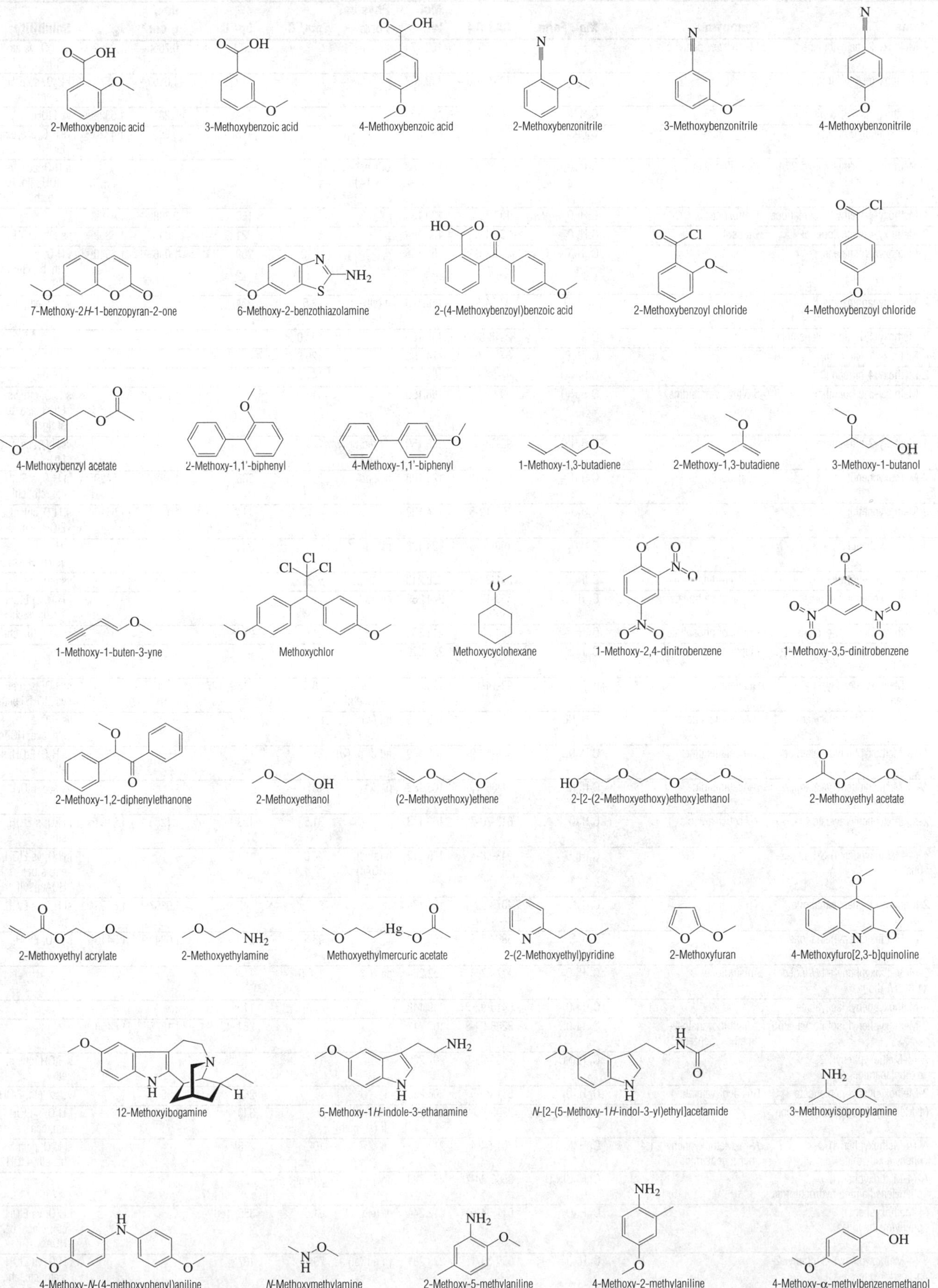

2-Methoxybenzoic acid
3-Methoxybenzoic acid
4-Methoxybenzoic acid
2-Methoxybenzonitrile
3-Methoxybenzonitrile
4-Methoxybenzonitrile
7-Methoxy-2H-1-benzopyran-2-one
6-Methoxy-2-benzothiazolamine
2-(4-Methoxybenzoyl)benzoic acid
2-Methoxybenzoyl chloride
4-Methoxybenzoyl chloride
4-Methoxybenzyl acetate
2-Methoxy-1,1'-biphenyl
4-Methoxy-1,1'-biphenyl
1-Methoxy-1,3-butadiene
2-Methoxy-1,3-butadiene
3-Methoxy-1-butanol
1-Methoxy-1-buten-3-yne
Methoxychlor
Methoxycyclohexane
1-Methoxy-2,4-dinitrobenzene
1-Methoxy-3,5-dinitrobenzene
2-Methoxy-1,2-diphenylethanone
2-Methoxyethanol
(2-Methoxyethoxy)ethene
2-[2-(2-Methoxyethoxy)ethoxy]ethanol
2-Methoxyethyl acetate
2-Methoxyethyl acrylate
2-Methoxyethylamine
Methoxyethylmercuric acetate
2-(2-Methoxyethyl)pyridine
2-Methoxyfuran
4-Methoxyfuro[2,3-b]quinoline
12-Methoxyibogamine
5-Methoxy-1H-indole-3-ethanamine
N-[2-(5-Methoxy-1H-indol-3-yl)ethyl]acetamide
3-Methoxyisopropylamine
4-Methoxy-N-(4-methoxyphenyl)aniline
N-Methoxymethylamine
2-Methoxy-5-methylaniline
4-Methoxy-2-methylaniline
4-Methoxy-α-methylbenzenemethanol

No.	Name	Synonym	Mol. Form.	CAS RN	Mol. Wt.	Physical Form	mp/°C	bp/°C	den/ g cm-3	n_D	Solubility
6841	2-Methoxy-2-methylbutane	Methyl *tert*-pentyl ether	$C_6H_{14}O$	994-05-8	102.174			86.1	0.7660^{25}	1.3862^{25}	sl H_2O; vs eth, EtOH
6842	2-(Methoxymethyl)furan		$C_6H_8O_2$	13679-46-4	112.127			132	1.0163^{20}	1.4570^{20}	i H_2O; s EtOH; vs eth
6843	2-(Methoxymethyl)-5-nitrofuran		$C_6H_7NO_4$	586-84-5	157.125			104^{3}	1.281^{20}	1.5325^{20}	vs EtOH
6844	(Methoxymethyl)oxirane		$C_4H_8O_2$	930-37-0	88.106			113	0.9890^{20}	1.4320^{20}	vs H_2O, ace, eth, EtOH
6845	3-Methoxy-5-methyl-4-oxo-2,5-hexadienoic acid	Penicillic acid	$C_8H_{10}O_4$	90-65-3	170.163	orth or hex pl (+ 1w)	83				s H_2O, ace; vs EtOH, eth, bz; sl peth
6846	4-Methoxy-4-methyl-2-pentanone	Pentoxone	$C_7H_{14}O_2$	107-70-0	130.185			160	0.8980^{25}	1.418^{20}	
6847	2-Methoxy-4-methylphenol	Creosol	$C_8H_{10}O_2$	93-51-6	138.164	pr	5.5	221	1.098^{20}	1.5353^{25}	vs eth, EtOH
6848	1-Methoxynaphthalene		$C_{11}H_{10}O$	2216-69-5	158.196		<-10	269	1.0963^{14}	1.6940^{25}	i H_2O; s EtOH, eth, bz, chl; vs CS_2
6849	2-Methoxynaphthalene		$C_{11}H_{10}O$	93-04-9	158.196	lf (eth), pl (peth)	73.5	274			vs bz, eth, chl
6850	2-Methoxy-1,4-naphthalenedione		$C_{11}H_8O_3$	2348-82-5	188.180		183.0				
6851	4-Methoxy-1-naphthol		$C_{11}H_{10}O_2$	84-85-5	174.196		129.8				
6852	2-Methoxy-4-nitroaniline		$C_7H_8N_2O_3$	97-52-9	168.150		141.0				s DMSO
6853	2-Methoxy-5-nitroaniline	5-Nitro-*o*-anisidine	$C_7H_8N_2O_3$	99-59-2	168.150		118		1.2068^{15}		s H_2O, eth; vs EtOH, ace, bz; sl lig
6854	4-Methoxy-2-nitroaniline		$C_7H_8N_2O_3$	96-96-8	168.150	dk red pr (w or al)	129				vs H_2O, ace, eth, EtOH
6855	2-Methoxyphenol	Guaiacol	$C_7H_8O_2$	90-05-1	124.138	hex pr	32	205	1.1287^{21}	1.5429^{20}	sl H_2O; s EtOH, eth, ctc, chl
6856	3-Methoxyphenol		$C_7H_8O_2$	150-19-6	124.138		<-17	114^{5}	1.131^{25}	1.5510^{20}	sl H_2O, chl; msc EtOH, eth
6857	4-Methoxyphenol		$C_7H_8O_2$	150-76-5	124.138	pl	57	243			s H_2O, bz, ctc; vs EtOH, eth
6858	2-Methoxyphenol benzoate	Guaiacol benzoate	$C_{14}H_{12}O_3$	531-37-3	228.243		57.5				vs eth, chl
6859	2-Methoxyphenol carbonate (2:1)	Guaiacol carbonate	$C_{15}H_{14}O_5$	553-17-3	274.269	cry (al)	89				i H_2O; sl EtOH; s eth; vs chl
6860	2-Methoxyphenol phosphate (3:1)	Guaiacol phosphate	$C_{21}H_{21}O_7P$	563-03-1	416.362		91	277^{3}			vs ace, tol, chl
6861	5-[(2-Methoxyphenoxy)methyl]-2-oxazolidinone	Mephenoxalone	$C_{11}H_{13}NO_4$	70-07-5	223.226		144				
6862	3-(2-Methoxyphenoxy)-1,2-propanediol	Guaifenesin	$C_{10}H_{14}O_4$	93-14-1	198.216	orth pr (eth, eth-peth)	78.5	215^{19}, $127^{0.2}$			s H_2O, bz, chl; vs EtOH; i peth
6863	*N*-(2-Methoxyphenyl)acetamide	*o*-Acetanisidine	$C_9H_{11}NO_2$	93-26-5	165.189	nd (w)	87.5	304			vs H_2O, EtOH; s eth, ace, HOAc
6864	*N*-(3-Methoxyphenyl)acetamide	*m*-Acetanisidine	$C_9H_{11}NO_2$	588-16-9	165.189	nd or pl (w)	81				vs H_2O, EtOH; s eth, ace
6865	*N*-(4-Methoxyphenyl)acetamide	*p*-Acetanisidine	$C_9H_{11}NO_2$	51-66-1	165.189	pl (w)	131				vs ace, EtOH, chl
6866	2-Methoxyphenyl acetate	2-Acetoxyanisole	$C_9H_{10}O_3$	613-70-7	166.173		31.5	123^{13}	1.1285^{25}	1.5101^{25}	i H_2O; s EtOH, eth
6867	4-(4-Methoxyphenyl)-3-buten-2-one		$C_{11}H_{12}O_2$	943-88-4	176.212	lf (al, eth, HOAc)	74.0	187.5^{19}			i H_2O; vs EtOH, eth; s bz, HOAc, sulf
6868	2-Methoxy-1-phenylethanone		$C_9H_{10}O_2$	4079-52-1	150.174	ye liq	8	245; 125^{19}	1.0897^{20}	1.5393^{20}	sl H_2O; s EtOH, ace
6869	1-(3-Methoxyphenyl)ethanone		$C_9H_{10}O_2$	586-37-8	150.174		95.5	240	1.0343^{19}	1.5410^{20}	s H_2O, EtOH, ace, ctc
6870	2-(4-Methoxyphenyl)-1*H*-indene-1,3(2*H*)-dione	Anisindione	$C_{16}H_{12}O_3$	117-37-3	252.264	pa ye cry (HOAc, al)	156.5				
6871	4-Methoxyphenyl isocyanate		$C_8H_7NO_2$	5416-93-3	149.148			110^{10}			
6872	2-Methoxyphenyl isothiocyanate	1-Isothiocyanato-2-methoxybenzene	C_8H_7NOS	3288-04-8	165.213			264; 131^{11}	1.1878^{20}	1.6458^{20}	
6873	*N*-(4-Methoxyphenyl)-3-oxobutanamide		$C_{11}H_{13}NO_3$	5437-98-9	207.226		117.3				s EtOH, chl; sl eth
6874	2-Methoxyphenyl pentanoate	Guaiacol valerate	$C_{12}H_{16}O_3$	531-39-5	208.253			265	1.05^{25}		vs bz, eth, EtOH
6875	(4-Methoxyphenyl)phenyldiazene		$C_{13}H_{12}N_2O$	2396-60-3	212.246	oran-red pl, lf (al, peth)	56	340	1.12^{75}		i H_2O; s EtOH, eth, ace
6876	*N*-(*p*-Methoxyphenyl)-*p*-phenylenediamine	*N*-(4-Methoxyphenyl)-1,4-benzenediamine	$C_{13}H_{14}N_2O$	101-64-4	214.262	nd	102	238^{12}			sl H_2O, peth; vs bz, eth, EtOH
6877	*N*-(4-Methoxyphenyl)-*p*-phenylenediamine hydrochloride		$C_{13}H_{15}ClN_2O$	3566-44-7	250.723	cry	245 dec				
6878	(4-Methoxyphenyl) phenylmethanone		$C_{14}H_{12}O_2$	611-94-9	212.244	pr (eth)	61.5	355; 168^{12}			i H_2O; vs EtOH, eth; s ace, bz, HOAc
6879	3-(4-Methoxyphenyl)-1-phenyl-2-propen-1-one		$C_{16}H_{14}O_2$	959-33-1	238.281	ye nd (al)	79	187^{19}			i H_2O; vs EtOH; s eth, ctc, chl, HOAc

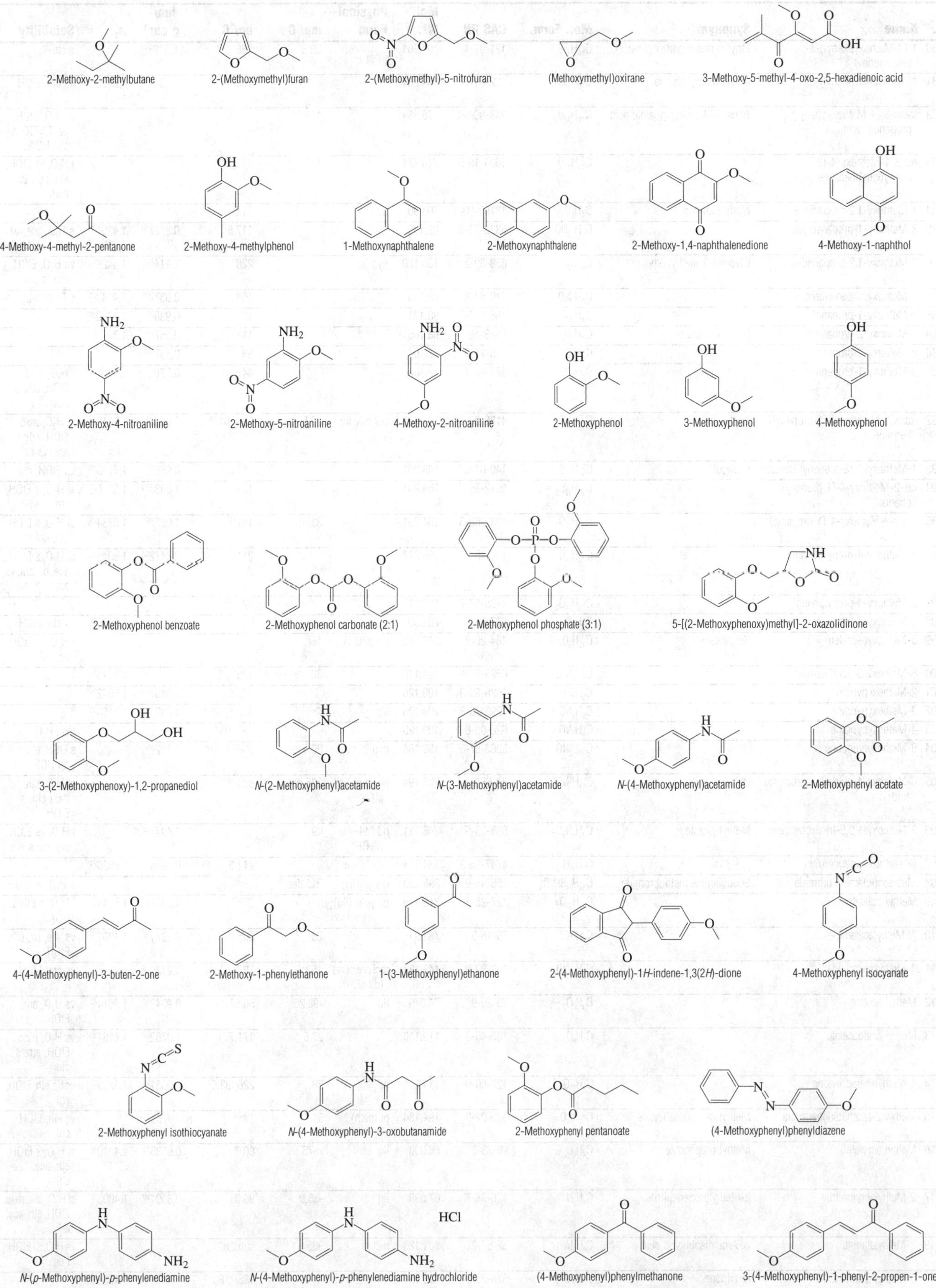
2-Methoxy-2-methylbutane
2-(Methoxymethyl)furan
2-(Methoxymethyl)-5-nitrofuran
(Methoxymethyl)oxirane
3-Methoxy-5-methyl-4-oxo-2,5-hexadienoic acid
4-Methoxy-4-methyl-2-pentanone
2-Methoxy-4-methylphenol
1-Methoxynaphthalene
2-Methoxynaphthalene
2-Methoxy-1,4-naphthalenedione
4-Methoxy-1-naphthol
2-Methoxy-4-nitroaniline
2-Methoxy-5-nitroaniline
4-Methoxy-2-nitroaniline
2-Methoxyphenol
3-Methoxyphenol
4-Methoxyphenol
2-Methoxyphenol benzoate
2-Methoxyphenol carbonate (2:1)
2-Methoxyphenol phosphate (3:1)
5-[(2-Methoxyphenoxy)methyl]-2-oxazolidinone
3-(2-Methoxyphenoxy)-1,2-propanediol
N-(2-Methoxyphenyl)acetamide
N-(3-Methoxyphenyl)acetamide
N-(4-Methoxyphenyl)acetamide
2-Methoxyphenyl acetate
4-(4-Methoxyphenyl)-3-buten-2-one
2-Methoxy-1-phenylethanone
1-(3-Methoxyphenyl)ethanone
2-(4-Methoxyphenyl)-1H-indene-1,3(2H)-dione
4-Methoxyphenyl isocyanate
2-Methoxyphenyl isothiocyanate
N-(4-Methoxyphenyl)-3-oxobutanamide
2-Methoxyphenyl pentanoate
(4-Methoxyphenyl)phenyldiazene
N-(p-Methoxyphenyl)-p-phenylenediamine
HCl
N-(4-Methoxyphenyl)-p-phenylenediamine hydrochloride
(4-Methoxyphenyl)phenylmethanone
3-(4-Methoxyphenyl)-1-phenyl-2-propen-1-one

No.	Name	Synonym	Mol. Form.	CAS RN	Mol. Wt.	Physical Form	mp/°C	bp/°C	den/ g cm^{-3}	n_D	Solubility
6880	1-(4-Methoxyphenyl)-1-propanone	Ethyl 4-methoxyphenyl ketone	$C_{10}H_{12}O_2$	121-97-1	164.201		25.5	266	1.0798^{16}		s ctc
6881	1-(4-Methoxyphenyl)-2-propanone	Anisyl methyl ketone	$C_{10}H_{12}O_2$	122-84-9	164.201		<-15	268	1.0694^{17}	1.5253^{20}	vs eth, EtOH
6882	*trans*-3-(4-Methoxyphenyl)-2-propenoic acid	*trans*-4-Methoxycinnamic acid	$C_{10}H_{10}O_3$	943-89-5	178.184		173.5				sl H_2O, EtOH, bz, DMSO; s ctc, HOAc
6883	*trans*-1-Methoxy-4-(2-phenylvinyl)benzene		$C_{15}H_{14}O$	1694-19-5	210.271		136.5	142.5^{15}			i H_2O; vs EtOH, eth, ace, bz; s peth
6884	1-Methoxy-1,2-propadiene	Methoxyallene	C_4H_6O	13169-00-1	70.090	oil		51.5			
6885	3-Methoxy-1-propanamine		$C_4H_{11}NO$	5332-73-0	89.136			117.5	0.8727^{20}	1.4391^{20}	s H_2O, ace, bz, ctc, chl, MeOH
6886	3-Methoxy-1,2-propanediol	Glycerol 3-methyl ether	$C_4H_{10}O_3$	623-39-2	106.120	hyg liq		220	1.114^{20}	1.442^{25}	vs H_2O, EtOH, ace; s eth
6887	3-Methoxypropanenitrile		C_4H_7NO	110-67-8	85.105			163	0.9379^{20}	1.4043^{20}	s EtOH, eth, chl
6888	2-Methoxy-1-propanol		$C_4H_{10}O_2$	1589-47-5	90.121			130	0.938^{20}	1.4070^{20}	
6889	1-Methoxy-2-propanone	Methoxyacetone	$C_4H_8O_2$	5878-19-3	88.106			116	0.957^{25}	1.3970^{20}	
6890	2-Methoxy-1-propene		C_4H_8O	116-11-0	72.106			38	0.7372^{20}		
6891	3-Methoxy-1-propene		C_4H_8O	627-40-7	72.106			44	0.77^{11}	1.3778^{20}	i H_2O; msc EtOH, eth; s ace
6892	*trans*-1-Methoxy-4-(1-propenyl) benzene	Anethole	$C_{10}H_{12}O$	4180-23-8	148.201	col oily liq	22.5	235; $81^{2.3}$	0.9882^{20}	1.5615^{20}	sl H_2O; msc EtOH, eth; s ace; vs bz
6893	1-Methoxy-4-(2-propenyl)benzene	Estragole	$C_{10}H_{12}O$	140-67-0	148.201			215.5	0.965^{25}	1.5195^{20}	vs EtOH, chl
6894	*cis*-2-Methoxy-4-(1-propenyl) phenol		$C_{10}H_{12}O_2$	5912-86-7	164.201			134^{13}	1.0837^{20}	1.5726^{20}	sl H_2O; s EtOH, eth
6895	*trans*-2-Methoxy-4-(1-propenyl) phenol		$C_{10}H_{12}O_2$	5932-68-3	164.201		33.5	141^{13}	1.0852^{20}	1.5784^{20}	sl H_2O; s EtOH, eth, chl
6896	1-Methoxy-4-propylbenzene		$C_{10}H_{14}O$	104-45-0	150.217			211.5	0.9472^{20}	1.5045^{20}	sl H_2O; s EtOH, ace, bz, chl; vs eth
6897	2-Methoxy-4-propylphenol		$C_{10}H_{14}O_2$	2785-87-7	166.217			121^{10}			
6898	3-Methoxy-1-propyne		C_4H_6O	627-41-8	70.090			63	0.83^{12}	1.5035^{20}	vs eth, EtOH
6899	5-Methoxypsoralen	Bergaptene	$C_{12}H_8O_4$	484-20-8	216.190	nd (EtOH)	188				i H_2O; sl EtOH, bz, chl
6900	6-Methoxy-3-pyridinamine		$C_6H_8N_2O$	6628-77-9	124.140		30	125^{10}, 87^{1}		1.5745^{20}	
6901	2-Methoxypyridine		C_6H_7NO	1628-89-3	109.126			142.5	1.0457^{20}	1.5042^{20}	
6902	3-Methoxypyridine		C_6H_7NO	7295-76-3	109.126	liq		178.5; 65^{15}	1.083	1.5180^{20}	
6903	4-Methoxypyridine		C_6H_7NO	620-08-6	109.126			192; 95^{45}			msc H_2O
6904	6-Methoxyquinoline		$C_{10}H_9NO$	5263-87-6	159.184	hyg lf	26.5	306; 153^{12}	1.152^{20}		s EtOH, eth, chl, dil HCl
6905	6-Methoxy-4-quinolinecarboxylic acid	Quininic acid	$C_{11}H_9NO_3$	86-68-0	203.194	pa ye pr (dil al)	285 dec	sub			sl H_2O, eth, bz, tfa; i chl; s EtOH
6906	2-Methoxy-1,3,5-trinitrobenzene	Methyl picrate	$C_7H_5N_3O_7$	606-35-9	243.131	nd (dil MeOH)	69		1.4947^{80}		i H_2O; vs EtOH, chl, bz; s eth
6907	(2-Methoxyvinyl)benzene		$C_9H_{10}O$	4747-15-3	134.174			211.5	0.9894^{23}	1.5620^{24}	
6908	Methscopolamine bromide	Scopolamine methobromide	$C_{18}H_{24}BrNO_4$	155-41-9	398.293	cry (EtOH)	215 dec				s H_2O; sl EtOH
6909	Methyl abietate		$C_{21}H_{32}O_2$	127-25-3	316.478	pa ye lf (liq)		225^{16}	1.049^{20}	1.5344	i H_2O; s EtOH, HOAc
6910	*N*-Methylacetamide		C_3H_7NO	79-16-3	73.094		28	205	0.9371^{25}	1.4301^{20}	vs ace, bz, eth, EtOH
6911	4-Methylacetanilide		$C_9H_{11}NO$	103-89-9	149.189	mcl cry or nd (dil al)	152	307	1.2120^{15}		vs eth, EtOH
6912	Methyl acetate		$C_3H_6O_2$	79-20-9	74.079	liq	-98.25	56.87	0.9342^{20}	1.3614^{20}	vs H_2O, eth, EtOH
6913	Methyl acetoacetate		$C_5H_8O_3$	105-45-3	116.116		27.5	171.7	1.0762^{20}	1.4184^{20}	vs H_2O; msc EtOH, eth; s ctc
6914	4-Methylacetophenone		$C_9H_{10}O$	122-00-9	134.174	nd	28	226; 93.5^{7}	1.0051^{20}	1.5335^{20}	vs bz, eth, EtOH, chl
6915	Methyl 2-(acetyloxy)benzoate	Methyl *o*-acetylsalicylate	$C_{10}H_{10}O_4$	580-02-9	194.184	pl (peth)	51.5	135^{9}			vs eth, EtOH, chl
6916	Methyl acrylate	Methyl propenoate	$C_4H_6O_2$	96-33-3	86.090	liq	<-75	80.7	0.9535^{20}	1.4040^{20}	sl H_2O; s EtOH, eth, ace, bz, chl
6917	2-Methylacrylonitrile	2-Methylpropenenitrile	C_4H_5N	126-98-7	67.090	liq	-35.8	90.3	0.8001^{20}	1.4003^{20}	sl H_2O, chl; msc EtOH, eth, ace, tol
6918	2-Methylalanine	α-Aminoisobutyric acid	$C_4H_9NO_2$	62-57-7	103.120	mcl pr	335	sub 280			vs H_2O; sl EtOH; i eth
6919	5-Methyl-3-allyl-2,4-oxazolidinedione	Aloxidone	$C_7H_9NO_3$	526-35-2	155.151			138^{35}, $86^{0.5}$		1.4688^{25}	

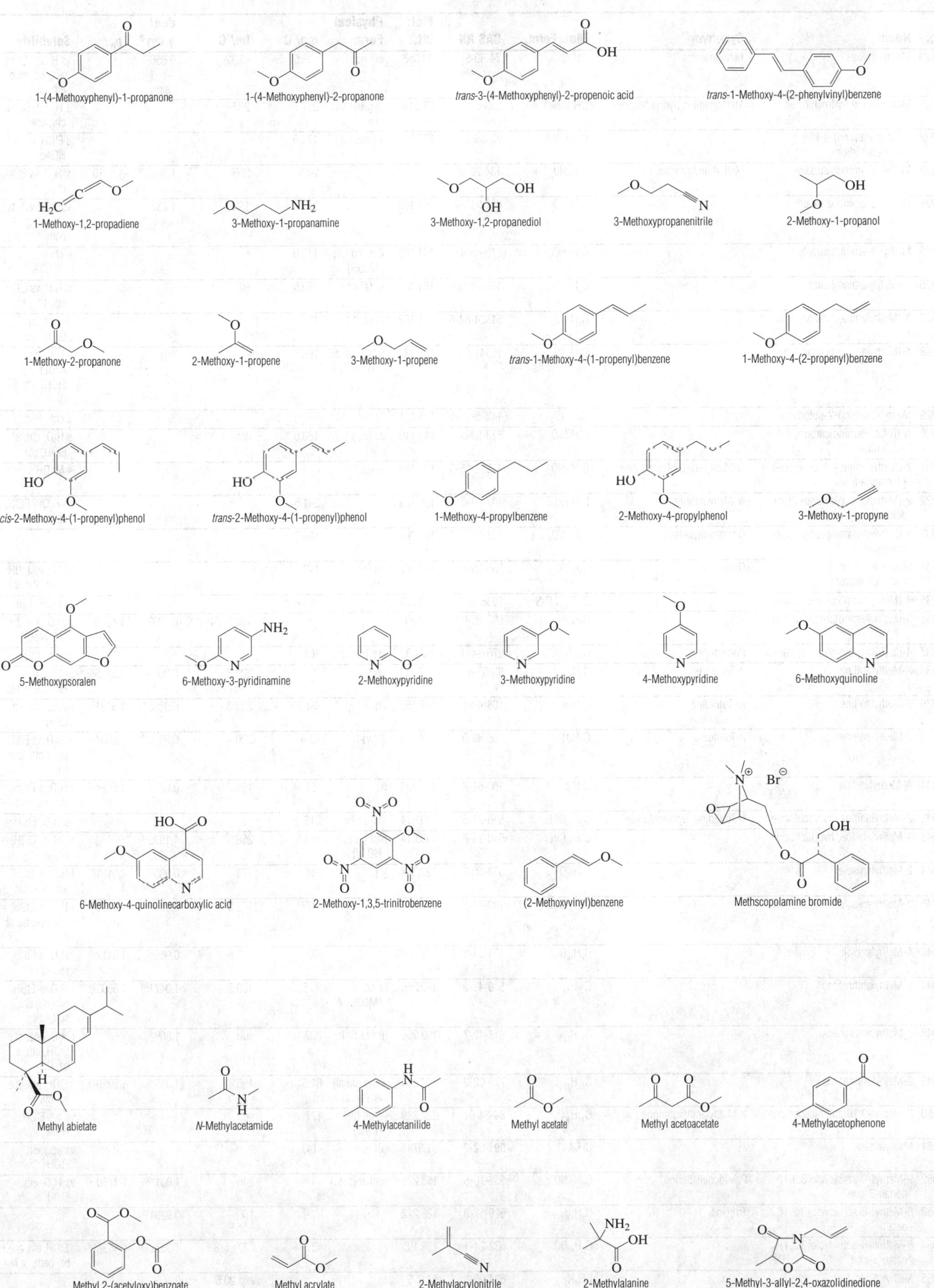
1-(4-Methoxyphenyl)-1-propanone
1-(4-Methoxyphenyl)-2-propanone
trans-3-(4-Methoxyphenyl)-2-propenoic acid
trans-1-Methoxy-4-(2-phenylvinyl)benzene
1-Methoxy-1,2-propadiene
3-Methoxy-1-propanamine
3-Methoxy-1,2-propanediol
3-Methoxypropanenitrile
2-Methoxy-1-propanol
1-Methoxy-2-propanone
2-Methoxy-1-propene
3-Methoxy-1-propene
trans-1-Methoxy-4-(1-propenyl)benzene
1-Methoxy-4-(2-propenyl)benzene
cis-2-Methoxy-4-(1-propenyl)phenol
trans-2-Methoxy-4-(1-propenyl)phenol
1-Methoxy-4-propylbenzene
2-Methoxy-4-propylphenol
3-Methoxy-1-propyne
5-Methoxypsoralen
6-Methoxy-3-pyridinamine
2-Methoxypyridine
3-Methoxypyridine
4-Methoxypyridine
6-Methoxyquinoline
6-Methoxy-4-quinolinecarboxylic acid
2-Methoxy-1,3,5-trinitrobenzene
(2-Methoxyvinyl)benzene
Methscopolamine bromide
Methyl abietate
N-Methylacetamide
4-Methylacetanilide
Methyl acetate
Methyl acetoacetate
4-Methylacetophenone
Methyl 2-(acetyloxy)benzoate
Methyl acrylate
2-Methylacrylonitrile
2-Methylalanine
5-Methyl-3-allyl-2,4-oxazolidinedione

No.	Name	Synonym	Mol. Form.	CAS RN	Mol. Wt.	Physical Form	mp/°C	bp/°C	den/ g cm^{-3}	n_D	Solubility
6920	Methylamine	Methanamine	CH_5N	74-89-5	31.058	col gas	-93.5	-6.32	0.656^{25} (p>1 atm)		vs H_2O; s EtOH, ace, bz; msc eth
6921	Methylamine hydrochloride	Methanamine hydrochloride	CH_6ClN	593-51-1	67.519	hyg tetr tab (al)	227.5	227^{15}			s H_2O, EtOH; i chl, ace
6922	1-(Methylamino)-9,10-anthracenedione		$C_{15}H_{11}NO_2$	82-38-2	237.254	ye-red nd	171.0				s EtOH, bz, chl, HOAc
6923	Methyl 2-aminobenzoate	Methyl anthranilate	$C_8H_9NO_2$	134-20-3	151.163		24.5	256	1.1682^{10}	1.5810	sl H_2O; vs EtOH, eth
6924	Methyl 3-aminobenzoate		$C_8H_9NO_2$	4518-10-9	151.163		39	152^{11}	1.232^{20}		vs EtOH, eth, bz, chl; s lig; sl peth
6925	Methyl 4-aminobenzoate		$C_8H_9NO_2$	619-45-4	151.163	lf or nd (aq MeOH)	113.0				s chl
6926	2-(Methylamino)benzoic acid		$C_8H_9NO_2$	119-68-6	151.163	pl (al or lig)	180.5	80$^{0.01}$			sl H_2O; vs EtOH, eth, bz, chl
6927	3-(Methylamino)benzoic acid		$C_8H_9NO_2$	51524-84-6	151.163	pl (peth)	127				vs ace, bz, EtOH, chl
6928	4-(Methylamino)benzoic acid		$C_8H_9NO_2$	10541-83-0	151.163	nd (bz, w, dil al)	168				s H_2O, bz, AcOEt; vs EtOH, eth; sl tfa
6929	Methyl 3-amino-2-butenoate		$C_5H_9NO_2$	14205-39-1	115.131						s chl
6930	*N*-[(Methylamino)carbonyl] acetamide		$C_4H_8N_2O_2$	623-59-6	116.119	tcl (w, al), pr (w)	180.5	dec			s H_2O, chl; sl EtOH, eth
6931	2-(Methylamino)-2-deoxy-α-*L*-glucopyranose	*N*-Methyl-α-*L*-glucosamine	$C_7H_{15}NO_5$	42852-95-9	193.198	glass					s MeOH
6932	2-(Methylamino)ethanesulfonic acid	*N*-Methyltaurine	$C_3H_9NO_3S$	107-68-6	139.173		241.5				vs H_2O; i EtOH, eth
6933	4-[2-(Methylamino)ethyl]-1,2-benzenediol	Deoxyepinephrine	$C_9H_{13}NO_2$	501-15-5	167.205		188.5				
6934	Methyl 3-amino-4-hydroxybenzoate	Orthocaine	$C_8H_9NO_3$	536-25-4	167.162	nd (bz or HOAc)	143				i H_2O; vs EtOH; s eth, alk; sl bz
6935	4-(Methylamino)phenol sulfate		$C_{14}H_{20}N_2O_6S$	1936-57-8	344.383	cry	260 dec				sl EtOH; i eth
6936	3-(Methylamino)propanenitrile		$C_4H_8N_2$	693-05-0	84.120			102^{49}, 74^{16}	0.8992^{20}	1.4320^{20}	s H_2O, ace, bz, chl, MeOH
6937	4-[2-(Methylamino)propyl]phenol	Pholedrine	$C_{10}H_{15}NO$	370-14-9	165.232	cry (MeOH)	161				vs eth, EtOH
6938	2-Methylaniline	*o*-Toluidine	C_7H_9N	95-53-4	107.153	liq	-14.41	200.3	0.9984^{20}	1.5725^{20}	sl H_2O; msc EtOH, eth, ctc
6939	3-Methylaniline	*m*-Toluidine	C_7H_9N	108-44-1	107.153	liq	-31.3	203.3	0.9889^{20}	1.5681^{20}	vs ace, bz, eth, EtOH
6940	4-Methylaniline	*p*-Toluidine	C_7H_9N	106-49-0	107.153	lf (w+1)	43.6	200.4	0.9619^{20}	1.5534^{45}	sl H_2O; vs EtOH, py; s eth, ace, ctc
6941	*N*-Methylaniline		C_7H_9N	100-61-8	107.153	liq	-57	196.2	0.9891^{20}	1.5684^{20}	i H_2O; s EtOH, eth, ctc, chl
6942	2-Methylaniline, hydrochloride	*o*-Toluidine, hydrochloride	$C_7H_{10}ClN$	636-21-5	143.614	mcl pr (w)	215				vs H_2O, EtOH
6943	4-Methylaniline, hydrochloride		$C_7H_{10}ClN$	540-23-8	143.614	mcl nd (eth-HOAc)	244.5	258	1.1930^{18}		vs H_2O, EtOH, HOAc
6944	2-Methylanisole		$C_8H_{10}O$	578-58-5	122.164	liq	-34.1	171	0.985^{25}	1.5161^{20}	i H_2O; s EtOH, eth, ace, ctc
6945	3-Methylanisole		$C_8H_{10}O$	100-84-5	122.164	liq	-47	175.5	0.969^{25}	1.5130^{20}	i H_2O; s EtOH, eth, ace, bz; sl ctc
6946	4-Methylanisole		$C_8H_{10}O$	104-93-8	122.164	liq	-32	175.5	0.969^{25}	1.5112^{20}	i H_2O; s EtOH, eth, chl
6947	1-Methylanthracene		$C_{15}H_{12}$	610-48-0	192.256	bl nd (MeOH) lf (al)	85.5	199.5	1.0471^{99}	1.6802^{99}	i H_2O; s EtOH, eth, bz, chl, sulf
6948	2-Methylanthracene		$C_{15}H_{12}$	613-12-7	192.256	grn bl flr lf (sub)	209	sub	1.80^{0}		i H_2O, ace; sl EtOH, eth; s bz, CS_2
6949	9-Methylanthracene		$C_{15}H_{12}$	779-02-2	192.256	ye nd (dil al) pr (bz, al)	81.5	196^{12}	1.065^{99}	1.6959^{99}	s EtOH, eth, ace, bz, chl
6950	2-Methyl-9,10-anthracenedione	2-Methylanthraquinone	$C_{15}H_{10}O_2$	84-54-8	222.239	ye nd (al, HOAc)	177	sub			vs bz, EtOH, HOAc
6951	Methylarsine		CH_5As	593-52-2	91.973	col gas	-143	2			vs ace, eth, EtOH
6952	9-Methyl-9-azabicyclo[3.3.1] nonan-3-one	Pseudopelletierine	$C_9H_{15}NO$	552-70-5	153.221	orth pr (peth)	54	246	1.001^{100}	1.4760^{100}	vs H_2O, eth, EtOH
6953	8-Methyl-8-azabicyclo[3.2.1] octane	Tropane	$C_8H_{15}N$	529-17-9	125.212			166	0.9251^{15}		
6954	8-Methyl-8-azabicyclo[3.2.1] octan-3-one		$C_8H_{13}NO$	532-24-1	139.195		43	227; 113^{25}	1.9872^{100}	1.4598^{100}	s EtOH, eth, ace, bz, peth; sl chl
6955	Methyl azide		CH_3N_3	624-90-8	57.055			exp 20.5	0.869^{15}		
6956	Methylazoxymethanol acetate		$C_4H_8N_2O_3$	592-62-1	132.118			191; 49$^{0.45}$			

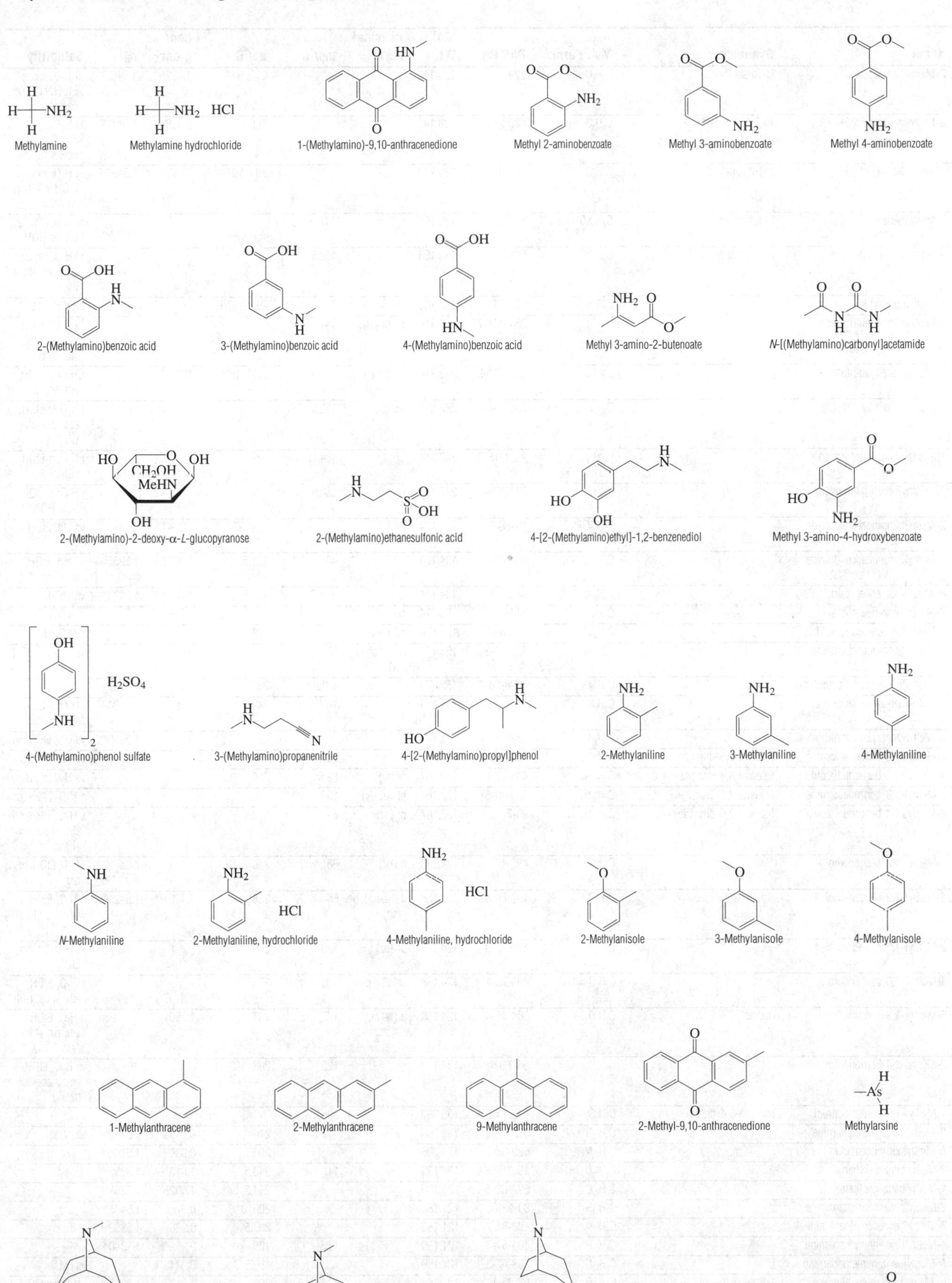

Methylamine
Methylamine hydrochloride
1-(Methylamino)-9,10-anthracenedione
Methyl 2-aminobenzoate
Methyl 3-aminobenzoate
Methyl 4-aminobenzoate
2-(Methylamino)benzoic acid
3-(Methylamino)benzoic acid
4-(Methylamino)benzoic acid
Methyl 3-amino-2-butenoate
N-[(Methylamino)carbonyl]acetamide
2-(Methylamino)-2-deoxy-α-L-glucopyranose
2-(Methylamino)ethanesulfonic acid
4-[2-(Methylamino)ethyl]-1,2-benzenediol
Methyl 3-amino-4-hydroxybenzoate
4-(Methylamino)phenol sulfate
3-(Methylamino)propanenitrile
4-[2-(Methylamino)propyl]phenol
2-Methylaniline
3-Methylaniline
4-Methylaniline
N-Methylaniline
2-Methylaniline, hydrochloride
4-Methylaniline, hydrochloride
2-Methylanisole
3-Methylanisole
4-Methylanisole
1-Methylanthracene
2-Methylanthracene
9-Methylanthracene
2-Methyl-9,10-anthracenedione
Methylarsine
9-Methyl-9-azabicyclo[3.3.1]nonan-3-one
8-Methyl-8-azabicyclo[3.2.1]octane
8-Methyl-8-azabicyclo[3.2.1]octan-3-one
Methyl azide
Methylazoxymethanol acetate

No.	Name	Synonym	Mol. Form.	CAS RN	Mol. Wt.	Physical Form	mp/°C	bp/°C	den/ g cm-3	n_D	Solubility
6957	2-Methylbenzaldehyde	*o*-Tolualdehyde	C_8H_8O	529-20-4	120.149			$200; 94^{10}$	1.0328^{20}	1.5462^{20}	sl H_2O, ctc; s EtOH, eth, bz; vs ace
6958	3-Methylbenzaldehyde	*m*-Tolualdehyde	C_8H_8O	620-23-5	120.149			199	1.0189^{21}	1.5413^{21}	sl H_2O; msc EtOH, eth; vs ace; s bz, chl
6959	4-Methylbenzaldehyde	*p*-Tolualdehyde	C_8H_8O	104-87-0	120.149			$204; 106^{10}$	1.0194^{17}	1.5454^{20}	sl H_2O; msc EtOH, eth, ace; vs chl
6960	2-Methylbenzamide	*o*-Toluamide	C_8H_9NO	527-85-5	135.163		147				sl H_2O, eth, tfa, bz; vs EtOH
6961	4-Methylbenzamide	*p*-Toluamide	C_8H_9NO	619-55-6	135.163		162.5				sl H_2O, bz, chl; vs EtOH, eth; s tfa
6962	*N*-Methylbenzamide		C_8H_9NO	613-93-4	135.163		82	$291; 167^{12}$			s EtOH, ace
6963	7-Methylbenz[a]anthracene		$C_{19}H_{14}$	2541-69-7	242.314	ye pl (al)	141				i H_2O; s EtOH, eth, ace, ctc, HOAc, CS_2
6964	8-Methylbenz[a]anthracene		$C_{19}H_{14}$	2381-31-9	242.314	pl (bz-al), nd (bz-lig)	156.5	$272^3, 160^{0.1}$	1.2310^0		i H_2O; s EtOH, eth, bz, xyl
6965	9-Methylbenz[a]anthracene		$C_{19}H_{14}$	2381-16-0	242.314	nd (al)	152.5				i H_2O; s EtOH, eth, ctc, chl, CS_2, xyl
6966	10-Methylbenz[a]anthracene		$C_{19}H_{14}$	2381-15-9	242.314		184				i H_2O; s EtOH, HOAc
6967	12-Methylbenz[a]anthracene		$C_{19}H_{14}$	2422-79-9	242.314	pl (al)	150.5				i H_2O; s EtOH, CS_2, HOAc
6968	2-Methylbenzeneacetaldehyde		$C_9H_{10}O$	10166-08-2	134.174			$221; 92^{10}$	1.0241^{10}		vs eth, EtOH, chl
6969	4-Methylbenzeneacetaldehyde		$C_9H_{10}O$	104-09-6	134.174		40	221.5	1.0052^{20}	1.5255^{20}	vs eth, EtOH, chl
6970	α-Methylbenzeneacetaldehyde		$C_9H_{10}O$	93-53-8	134.174			203.5	1.0089^{20}	1.5176^{20}	vs EtOH
6971	2-Methylbenzeneacetic acid		$C_9H_{10}O_2$	644-36-0	150.174	nd (w)	89				s H_2O, chl
6972	3-Methylbenzeneacetic acid		$C_9H_{10}O_2$	621-36-3	150.174	nd (w)	62	121^{26}			s H_2O, chl
6973	4-Methylbenzeneacetic acid		$C_9H_{10}O_2$	622-47-9	150.174	nd or pl (al, w)	93	265			vs bz, eth, EtOH
6974	α-Methylbenzeneacetic acid, (±)		$C_9H_{10}O_2$	2328-24-7	150.174		<-20	263	1.1^0	1.5237^{20}	
6975	4-Methylbenzeneacetonitrile		C_9H_9N	2947-61-7	131.174		18	242.5	0.992^{25}	1.5190^{20}	i H_2O; s EtOH, eth, bz, ctc
6976	α-Methylbenzeneacetonitrile		C_9H_9N	1823-91-2	131.174			231	0.9854^{20}	1.5095^{25}	vs eth, EtOH
6977	3-Methyl-1,2-benzenediamine	Toluene-2,3-diamine	$C_7H_{10}N_2$	2687-25-4	122.167		63.5	255			vs ace, bz, EtOH
6978	4-Methyl-1,2-benzenediamine	Toluene-3,4-diamine	$C_7H_{10}N_2$	496-72-0	122.167	pl (lig)	89.5	265			vs H_2O; s lig
6979	2-Methyl-1,3-benzenediamine	Toluene-2,6-diamine	$C_7H_{10}N_2$	823-40-5	122.167	pr (bz, w)	106				s H_2O, EtOH, bz
6980	2-Methyl-1,4-benzenediamine	Toluene-2,5-diamine	$C_7H_{10}N_2$	95-70-5	122.167	pl (bz)	64	273.5			s H_2O, EtOH, eth; sl bz, HOAc
6981	3-Methyl-1,2-benzenediol		$C_7H_8O_2$	488-17-5	124.138	lf (bz)	68	248			s H_2O, EtOH, bz, chl
6982	4-Methyl-1,2-benzenediol		$C_7H_8O_2$	452-86-8	124.138	lf (bz-lig), pr (bz)	65	258	1.1287^{74}	1.5425^{74}	s H_2O, EtOH, eth, ace, chl; sl lig
6983	2-Methyl-1,3-benzenediol		$C_7H_8O_2$	608-25-3	124.138	pr (bz)	120	265			vs H_2O, bz, eth, EtOH
6984	4-Methyl-1,3-benzenediol		$C_7H_8O_2$	496-73-1	124.138	cry (bz-peth)	105	270			s H_2O, EtOH, eth; sl bz, peth
6985	5-Methyl-1,3-benzenediol	Orcinol	$C_7H_8O_2$	504-15-4	124.138	pr(w+1), lf(chl)	107	287	1.290^4		s H_2O, EtOH, eth, bz; sl lig, peth
6986	2-Methyl-1,4-benzenediol		$C_7H_8O_2$	95-71-6	124.138		125	$283; 163^{11}$			vs H_2O, EtOH, eth; s ace; sl bz, lig
6987	4-Methyl-1,2-benzenedithiol	Toluene-3,4-dithiol	$C_7H_8S_2$	496-74-2	156.269		29				s chl
6988	β-Methylbenzeneethanamine		$C_9H_{13}N$	582-22-9	135.206			210	0.9433^4	1.5255^{20}	vs bz, eth, EtOH
6989	*N*-Methylbenzeneethanamine		$C_9H_{13}N$	589-08-2	135.206			206	0.93^{25}	1.5162^{20}	
6990	2-Methylbenzeneethanol		$C_9H_{12}O$	19819-98-8	136.190		1.0	243.5	1.016^{25}	1.5355^{20}	
6991	4-Methylbenzeneethanol		$C_9H_{12}O$	699-02-5	136.190			$244.5; 94^6$	1.0028^{20}	1.5267^{20}	
6992	2-Methylbenzenemethanamine		$C_8H_{11}N$	89-93-0	121.180	liq	-30	$206; 81^{15}$	0.9766^{19}	1.5436^{19}	
6993	3-Methylbenzenemethanamine		$C_8H_{11}N$	100-81-2	121.180			203.5	0.966^{25}	1.5360^{20}	
6994	4-Methylbenzenemethanamine		$C_8H_{11}N$	104-84-7	121.180		12.5	195	0.952^{20}	1.5340^{20}	
6995	*N*-Methylbenzenemethanamine		$C_8H_{11}N$	103-67-3	121.180			180.5	0.9442^{18}		vs H_2O
6996	α-Methylbenzenemethanol	1-Phenylethanol	$C_8H_{10}O$	98-85-1	122.164		20	205	1.013^{25}	1.5265^{20}	i H_2O; vs EtOH, eth
6997	2-Methylbenzenemethanol	*o*-Tolyl alcohol	$C_8H_{10}O$	89-95-2	122.164	nd (peth-eth)	38	$224; 118^{20}$	1.023^{40}		vs eth, EtOH, chl

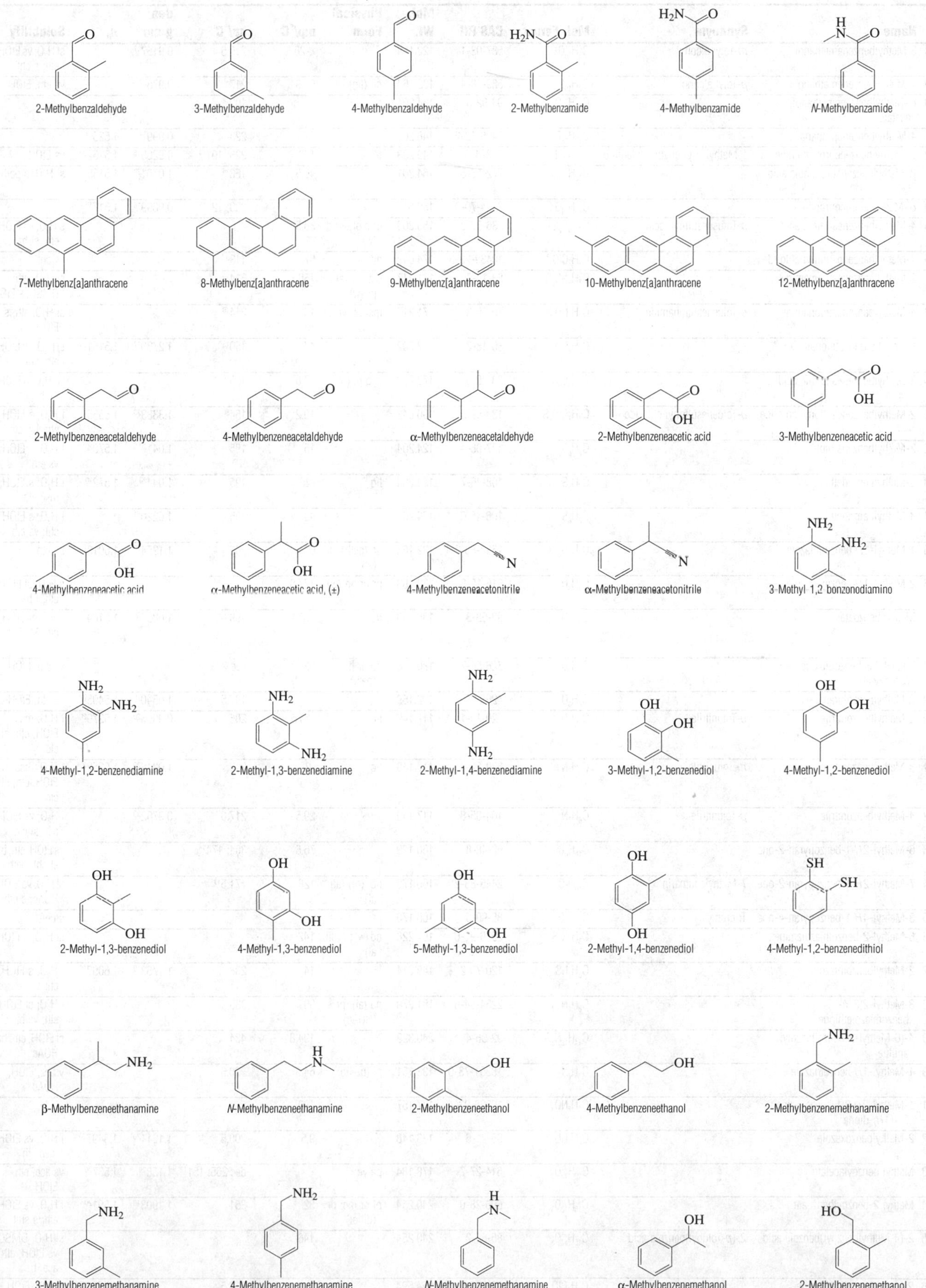
2-Methylbenzaldehyde
3-Methylbenzaldehyde
4-Methylbenzaldehyde
2-Methylbenzamide
4-Methylbenzamide
N-Methylbenzamide
7-Methylbenz[a]anthracene
8-Methylbenz[a]anthracene
9-Methylbenz[a]anthracene
10-Methylbenz[a]anthracene
12-Methylbenz[a]anthracene
2-Methylbenzeneacetaldehyde
4-Methylbenzeneacetaldehyde
α-Methylbenzeneacetaldehyde
2-Methylbenzeneacetic acid
3-Methylbenzeneacetic acid
4-Methylbenzeneacetic acid
α-Methylbenzeneacetic acid, (±)
4-Methylbenzeneacetonitrile
α-Methylbenzeneacetonitrile
3-Methyl-1,2-benzenediamine
4-Methyl-1,2-benzenediamine
2-Methyl-1,3-benzenediamine
2-Methyl-1,4-benzenediamine
3-Methyl-1,2-benzenediol
4-Methyl-1,2-benzenediol
2-Methyl-1,3-benzenediol
4-Methyl-1,3-benzenediol
5-Methyl-1,3-benzenediol
2-Methyl-1,4-benzenediol
4-Methyl-1,2-benzenedithiol
β-Methylbenzeneethanamine
N-Methylbenzeneethanamine
2-Methylbenzeneethanol
4-Methylbenzeneethanol
2-Methylbenzenemethanamine
3-Methylbenzenemethanamine
4-Methylbenzenemethanamine
N-Methylbenzenemethanamine
α-Methylbenzenemethanol
2-Methylbenzenemethanol

No.	Name	Synonym	Mol. Form.	CAS RN	Mol. Wt.	Physical Form	mp/°C	bp/°C	den/ g cm^{-3}	n_D	Solubility
6998	3-Methylbenzenemethanol	*m*-Tolyl alcohol	$C_8H_{10}O$	587-03-1	122.164		<-20	215.5	0.9157^{17}		sl H_2O; vs EtOH, eth; s chl
6999	4-Methylbenzenemethanol	*p*-Tolyl alcohol	$C_8H_{10}O$	589-18-4	122.164	nd (lig)	61.5	217	0.978^{22}		vs eth, EtOH
7000	α-Methylbenzenemethanol, acetate		$C_{10}H_{12}O_2$	93-92-5	164.201	oil		109^{18}			
7001	4-Methylbenzenepropanal		$C_{10}H_{12}O$	5406-12-2	148.201			223	0.999^{14}	1.525^{14}	
7002	α-Methylbenzenepropanamine	1-Methyl-3-phenylpropylamine	$C_{10}H_{15}N$	22374-89-6	149.233		143	223; 101^{14}	0.9289^{15}	1.5152^{20}	vs EtOH
7003	β-Methylbenzenepropanoic acid, (±)		$C_{10}H_{12}O_2$	772-17-8	164.201		46.5	168^{14}	1.0701^{20}	1.5155^{20}	sl H_2O; s peth
7004	α-Methylbenzenepropanol		$C_{10}H_{14}O$	2344-70-9	150.217			239; 123^{15}	0.9899^{16}	1.517^{16}	
7005	4-Methylbenzenesulfinic acid	*p*-Toluenesulfinic acid	$C_7H_8O_2S$	536-57-2	156.203	orth pl or nd (w)	86.5				s H_2O; vs EtOH, eth; sl bz
7006	4-Methylbenzenesulfinyl chloride		C_7H_7ClOS	10439-23-3	174.648	nd	57	$113^{3.5}$			vs chl
7007	2-Methylbenzenesulfonamide		$C_7H_9NO_2S$	88-19-7	171.217	oct cry (al), pr (w)	158.7	214^{10}			sl H_2O, eth, DMSO; s EtOH
7008	4-Methylbenzenesulfonamide	*p*-Toluenesulfonamide	$C_7H_9NO_2S$	70-55-3	171.217	mcl pl (w+2)	138	214^{10}			sl H_2O, eth; s EtOH
7009	Methyl benzenesulfonate		$C_7H_8O_3S$	80-18-2	172.202		4.5	150^{15}	1.2730^{17}	1.5151^{20}	sl H_2O; vs EtOH, eth, chl
7010	2-Methylbenzenesulfonic acid		$C_7H_8O_3S$	88-20-0	172.202	hyg pl (w+2)	67.5	128.8^{25}			vs H_2O; s EtOH; i eth
7011	2-Methylbenzenesulfonyl chloride	*o*-Toluenesulfonyl chloride	$C_7H_7ClO_2S$	133-59-5	190.648		10.2	154^{36}	1.3383^{20}	1.5565^{20}	i H_2O; s EtOH, eth, bz, ctc
7012	2-Methylbenzenethiol		C_7H_8S	137-06-4	124.204		15	195	1.041^{20}	1.570^{20}	i H_2O; s EtOH; vs eth
7013	3-Methylbenzenethiol		C_7H_8S	108-40-7	124.204	liq	-20	195	1.044^{20}	1.572^{20}	i H_2O; s EtOH; msc eth
7014	4-Methylbenzenethiol		C_7H_8S	106-45-6	124.204		43	195	1.0220^{51}		i H_2O; s EtOH, chl; vs eth
7015	1-Methyl-1*H*-benzimidazole		$C_8H_8N_2$	1632-83-3	132.163	nd (peth), pl (al)	66	286	1.1254^{20}	1.6013^{7}	s peth
7016	2-Methyl-1*H*-benzimidazole		$C_8H_8N_2$	615-15-6	132.163	pr or nd (w)	177.8				s H_2O; sl EtOH, eth; i bz
7017	Methyl benzoate		$C_8H_8O_2$	93-58-3	136.149	liq	-12.4	199	1.0837^{25}	1.5164^{20}	i H_2O; s EtOH, ctc, MeOH; msc eth
7018	Methyl 1,3-benzodioxole-5-carboxylate		$C_9H_8O_4$	326-56-7	180.158	nd or lf (peth)	53	dec 273			vs eth, EtOH
7019	2-Methylbenzofuran		C_9H_8O	4265-25-2	132.159			197.5	1.0540^{20}	1.5495^{22}	vs eth, EtOH
7020	2-Methylbenzonitrile	*o*-Tolunitrile	C_8H_7N	529-19-1	117.149	liq	-13.5	205	0.9955^{20}	1.5279^{20}	i H_2O; msc EtOH, eth; sl ctc
7021	3-Methylbenzonitrile	*m*-Tolunitrile	C_8H_7N	620-22-4	117.149	liq	-23	213	1.0316^{20}	1.5252^{20}	i H_2O; msc EtOH, eth; sl ctc
7022	4-Methylbenzonitrile	*p*-Tolunitrile	C_8H_7N	104-85-8	117.149		29.5	217.0	0.9762^{30}		i H_2O; vs EtOH, eth; sl ctc
7023	6-Methyl-2*H*-1-benzopyran-2-one		$C_{10}H_8O_2$	92-48-8	160.170		76.5	304; 174^{14}			vs EtOH, eth, bz; sl chl, peth
7024	7-Methyl-2*H*-1-benzopyran-2-one	7-Methylcoumarin	$C_{10}H_8O_2$	2445-83-2	160.170	nd, (pl) (aq al)	128	171.5^{11}			sl H_2O; vs EtOH, HOAc; s eth
7025	3-Methyl-4*H*-1-benzopyran-4-one	Tricromyl	$C_{10}H_8O_2$	85-90-5	160.170						s chl
7026	6-Methyl-2-benzothiazolamine		$C_8H_8N_2S$	2536-91-6	164.228	nd (w) pr (dil al)	142				sl H_2O; s EtOH
7027	2-Methylbenzothiazole		C_8H_7NS	120-75-2	149.214		14	238	1.1763^{19}	1.6092^{19}	i H_2O; s EtOH, chl
7028	3-Methyl-2(3*H*)-benzothiazolethione		$C_8H_7NS_2$	2254-94-6	181.279	nd (al), pr (HOAc)	90	335			i H_2O; sl EtOH, eth; vs bz, chl
7029	4-(6-Methyl-2-benzothiazolyl) aniline		$C_{14}H_{12}N_2S$	92-36-4	240.323		194.8	434			sl EtOH, eth, bz, HOAc
7030	1-Methyl-1*H*-benzotriazole		$C_7H_7N_3$	13351-73-0	133.151	pl (bz-lig)	64.5	270.5			vs bz, EtOH, HOAc
7031	1-Methyl-2*H*-3,1-benzoxazine-2,4(1*H*)-dione		$C_9H_7NO_3$	10328-92-4	177.157		180				
7032	2-Methylbenzoxazole		C_8H_7NO	95-21-6	133.148		9.5	200.5	1.1211^{20}	1.5497^{20}	i H_2O; vs EtOH; msc eth
7033	Methyl benzoylacetate		$C_{10}H_{10}O_3$	614-27-7	178.184	pa ye		dec 265; 151^{12}	1.158^{29}	1.537^{20}	vs ace, eth, EtOH
7034	Methyl 2-benzoylbenzoate		$C_{15}H_{12}O_3$	606-28-0	240.254	pl or mcl pr (dil al)	52	351	1.1903^{19}	1.591^{20}	i H_2O; vs EtOH, eth; s sulf
7035	2-(4-Methylbenzoyl)benzoic acid	2-(*p*-Toluoyl)benzoic acid	$C_{15}H_{12}O_3$	85-55-2	240.254		146				sl H_2O, DMSO; vs EtOH, eth, ace, bz
7036	2-Methylbenzoyl chloride		C_8H_7ClO	933-88-0	154.594			213.5		1.5549^{20}	vs eth, EtOH
7037	3-Methylbenzoyl chloride		C_8H_7ClO	1711-06-4	154.594	liq	-23	219.5	1.0265^{21}	1.505^{22}	vs eth, EtOH

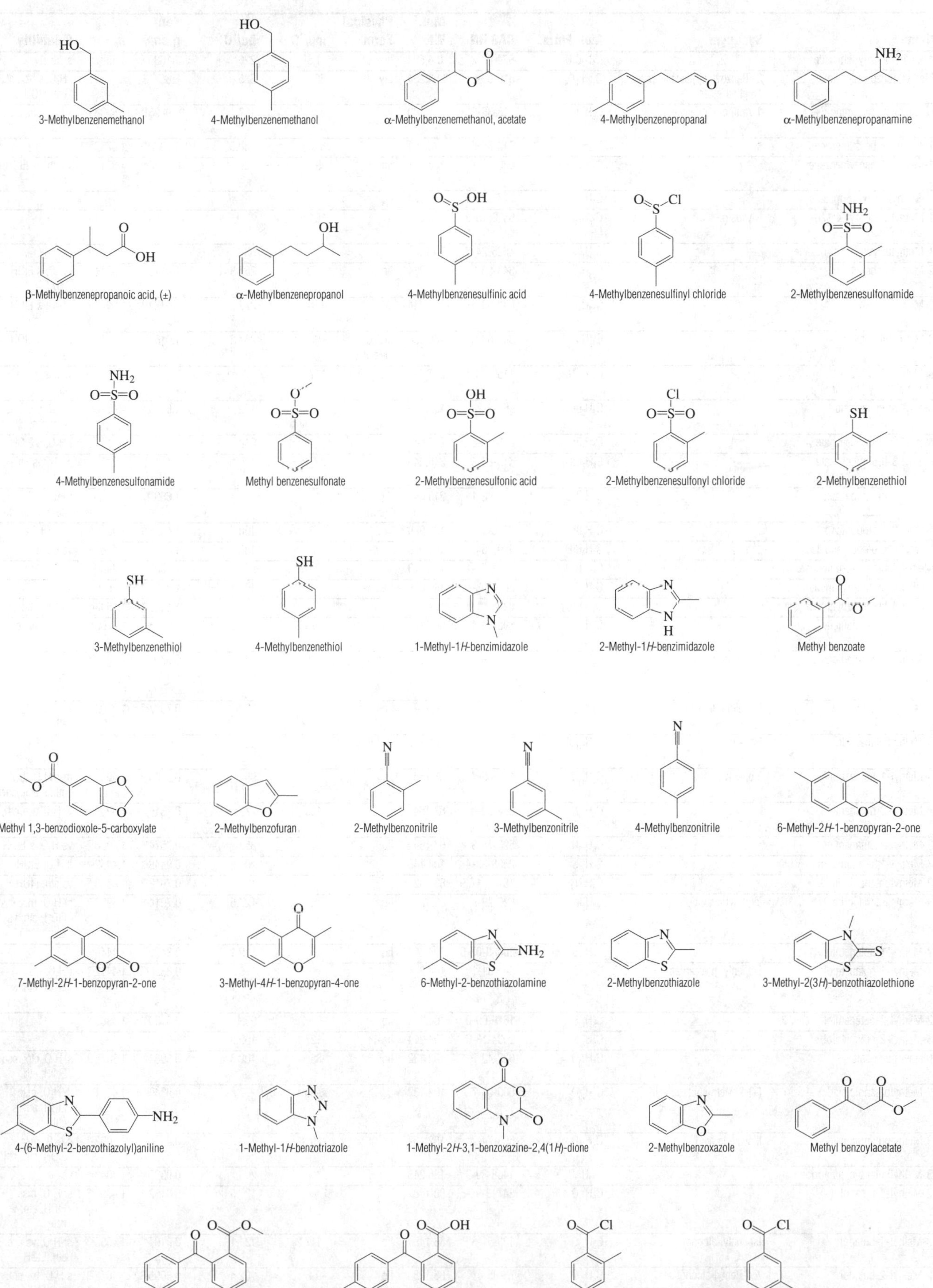
3-Methylbenzenemethanol
4-Methylbenzenemethanol
α-Methylbenzenemethanol, acetate
4-Methylbenzenepropanal
α-Methylbenzenepropanamine
β-Methylbenzenepropanoic acid, (±)
α-Methylbenzenepropanol
4-Methylbenzenesulfinic acid
4-Methylbenzenesulfinyl chloride
2-Methylbenzenesulfonamide
4-Methylbenzenesulfonamide
Methyl benzenesulfonate
2-Methylbenzenesulfonic acid
2-Methylbenzenesulfonyl chloride
2-Methylbenzenethiol
3-Methylbenzenethiol
4-Methylbenzenethiol
1-Methyl-1H-benzimidazole
2-Methyl-1H-benzimidazole
Methyl benzoate
Methyl 1,3-benzodioxole-5-carboxylate
2-Methylbenzofuran
2-Methylbenzonitrile
3-Methylbenzonitrile
4-Methylbenzonitrile
6-Methyl-2H-1-benzopyran-2-one
7-Methyl-2H-1-benzopyran-2-one
3-Methyl-4H-1-benzopyran-4-one
6-Methyl-2-benzothiazolamine
2-Methylbenzothiazole
3-Methyl-2(3H)-benzothiazolethione
4-(6-Methyl-2-benzothiazolyl)aniline
1-Methyl-1H-benzotriazole
1-Methyl-2H-3,1-benzoxazine-2,4(1H)-dione
2-Methylbenzoxazole
Methyl benzoylacetate
Methyl 2-benzoylbenzoate
2-(4-Methylbenzoyl)benzoic acid
2-Methylbenzoyl chloride
3-Methylbenzoyl chloride

No.	Name	Synonym	Mol. Form.	CAS RN	Mol. Wt.	Physical Form	mp/°C	bp/°C	den/ g cm-3	n_D	Solubility
7038	4-Methylbenzoyl chloride		C_8H_7ClO	874-60-2	154.594	liq	-1.5	226	1.1686^{20}	1.5547^{20}	s ctc
7039	Methyl benzoylsalicylate	2-(Benzoyloxy)benzoic acid, methyl ester	$C_{15}H_{12}O_4$	610-60-6	256.254	cry	85	385			i H_2O; s bz, chl, eth, EtOH
7040	α-Methylbenzylamine, (±)	1-Amino-1-phenylethane	$C_8H_{11}N$	618-36-0	121.180		32	187	0.9395^{15}	1.5238^{25}	s H_2O, chl; msc EtOH, eth
7041	1-Methyl-2-benzylbenzene		$C_{14}H_{14}$	713-36-0	182.261		6.6	280.5	1.0020^{20}	1.5763^{20}	
7042	1-Methyl-4-benzylbenzene		$C_{14}H_{14}$	620-83-7	182.261	liq	-30	286	0.9976^{20}	1.5712^{20}	vs eth, bz, EtOH, chl
7043	α-Methylbenzyl formate		$C_9H_{10}O_2$	7775-38-4	150.174	liq					
7044	1-Methyl-2-benzyl-4(1*H*)-quinazolinone	Glycosine	$C_{16}H_{14}N_2O$	6873-15-0	250.294		161.5				
7045	1-Methylbicyclo[3.1.0]hexane		C_7H_{12}	4625-24-5	96.170			93.1			
7046	2-Methylbiphenyl		$C_{13}H_{12}$	643-58-3	168.234	liq	-0.2	255.3	1.0113^{20}	1.5914^{20}	i H_2O; s EtOH, eth
7047	3-Methylbiphenyl		$C_{13}H_{12}$	643-93-6	168.234		2.3	272.7	1.0182^{17}	1.5972^{20}	i H_2O; s EtOH, eth, ctc
7048	4-Methylbiphenyl		$C_{13}H_{12}$	644-08-6	168.234	pl (lig, MeOH)	49.5	267.5	1.015^{27}		i H_2O; s EtOH, eth; sl ctc
7049	4-Methyl-*N,N*-bis(4-methylphenyl)aniline		$C_{21}H_{21}N$	1159-53-1	287.399	cry (HOAc)	117				vs ace, bz, eth, chl
7050	Methyl bromoacetate		$C_3H_5BrO_2$	96-32-2	152.975			132	1.6350^{20}	1.4520^{20}	i H_2O; s EtOH, eth, ace, bz
7051	Methyl 2-bromobenzoate		$C_8H_7BrO_2$	610-94-6	215.045			244			i H_2O; s EtOH
7052	Methyl 3-bromobenzoate		$C_8H_7BrO_2$	618-89-3	215.045	pl	32	125^{15}			sl H_2O; s EtOH, eth
7053	Methyl 4-bromobenzoate		$C_8H_7BrO_2$	619-42-1	215.045	lf (dil al), nd (eth)	81		1.689^{25}		s EtOH, eth, ace, peth; vs bz, chl
7054	Methyl 2-bromobutanoate		$C_5H_9BrO_2$	3196-15-4	181.028			168	1.4528^{20}	1.4029^{25}	vs EtOH
7055	Methyl 4-bromobutanoate		$C_5H_9BrO_2$	4897-84-1	181.028			186.5	1.4^{25}	1.4567^{25}	vs EtOH
7056	Methyl 4-bromo-2-butenoate		$C_5H_7BrO_2$	1117-71-1	179.013			84^{12}	1.490^{19}	1.498^{19}	
7057	Methyl 5-bromopentanoate		$C_6H_{11}BrO_2$	5454-83-1	195.054	liq		101^{14}	1.363	1.4630^{20}	
7058	Methyl 3-bromopropanoate		$C_4H_7BrO_2$	3395-91-3	167.002			105^{60}, 62^{12}	1.4123^{18}	1.4542^{20}	s EtOH, eth, ace
7059	3-Methyl-1,2-butadiene		C_5H_8	598-25-4	68.118	liq	-113.6	40.83	0.6806^{25}	1.4203^{20}	vs ace, bz, eth, EtOH
7060	2-Methyl-1,3-butadiene	Isoprene	C_5H_8	78-79-5	68.118	liq	-145.9	34.0	0.679^{20}	1.4219^{20}	i H_2O; msc EtOH, eth, ace, bz
7061	3-Methylbutanal	Isovaleraldehyde	$C_5H_{10}O$	590-86-3	86.132	liq	-51	92.5	0.7977^{20}	1.3902^{20}	sl H_2O; s EtOH, eth
7062	3-Methylbutanamide	Isovaleramide	$C_5H_{11}NO$	541-46-8	101.147	mcl lf (al)	137	226			s H_2O, EtOH, eth; vs peth
7063	3-Methyl-1-butanamine	Isopentylamine	$C_5H_{13}N$	107-85-7	87.164			96	0.7505^{20}	1.4083^{20}	msc H_2O, EtOH, eth; s ace, chl
7064	2-Methyl-2-butanamine		$C_5H_{13}N$	594-39-8	87.164	liq	-105	77	0.731^{25}	1.3954^{25}	vs H_2O, ace, eth, EtOH
7065	3-Methyl-2-butanamine		$C_5H_{13}N$	598-74-3	87.164	liq	-50	85.5	0.7574^{19}	1.4096^{18}	vs H_2O; s EtOH
7066	3-Methyl-1,3-butanediol		$C_5H_{12}O_2$	2568-33-4	104.148			202.5	0.9448^{20}	1.4452^{20}	s H_2O, EtOH
7067	2-Methylbutanenitrile		C_5H_9N	18936-17-9	83.132			125	0.7913^{15}	1.3933^{20}	vs eth, EtOH
7068	3-Methylbutanenitrile	Isobutyl cyanide	C_5H_9N	625-28-5	83.132	liq	-101	127.5	0.7914^{20}	1.3927^{20}	sl H_2O; msc EtOH, eth; vs ace
7069	2-Methyl-1-butanethiol, (+)		$C_5H_{12}S$	20089-07-0	104.214	liq		119.1	0.8420^{20}	1.4440^{20}	
7070	3-Methyl-1-butanethiol	Isopentyl mercaptan	$C_5H_{12}S$	541-31-1	104.214	liq		116	0.8350^{20}	1.4412^{20}	i H_2O; msc EtOH, eth; s ctc
7071	2-Methyl-2-butanethiol		$C_5H_{12}S$	1679-09-0	104.214	liq		99.1	0.8120^{20}	1.4385^{20}	
7072	3-Methyl-2-butanethiol		$C_5H_{12}S$	2084-18-6	104.214	liq	-127.1	109.8			
7073	Methyl butanoate		$C_5H_{10}O_2$	623-42-7	102.132	liq	-85.8	102.8	0.8984^{20}	1.3878^{20}	sl H_2O, ctc; msc EtOH, eth
7074	2-Methylbutanoic acid	(±)-2-Methylbutyric acid	$C_5H_{10}O_2$	600-07-7	102.132		<-80	177	0.934^{20}	1.4051^{20}	sl H_2O; msc EtOH, eth; s chl
7075	3-Methylbutanoic acid	Isovaleric acid	$C_5H_{10}O_2$	503-74-2	102.132	liq	-29.3	176.5	0.931^{20}	1.4033^{20}	s H_2O; msc EtOH, eth, chl
7076	3-Methylbutanoic anhydride		$C_{10}H_{18}O_3$	1468-39-9	186.248			215	0.9327^{20}	1.4043^{20}	vs eth
7077	2-Methyl-1-butanol, (±)		$C_5H_{12}O$	34713-94-5	88.148			127.5	0.8152^{25}	1.4092^{20}	sl H_2O; msc EtOH, eth; vs ace
7078	3-Methyl-1-butanol	Isopentyl alcohol	$C_5H_{12}O$	123-51-3	88.148	liq	-117.2	131.1	0.8104^{20}	1.4053^{20}	sl H_2O; vs ace, eth, EtOH
7079	2-Methyl-2-butanol	*tert*-Pentyl alcohol	$C_5H_{12}O$	75-85-4	88.148	liq	-9.1	102.4	0.8096^{20}	1.4052^{20}	s H_2O, bz, chl; msc EtOH, eth; vs ace

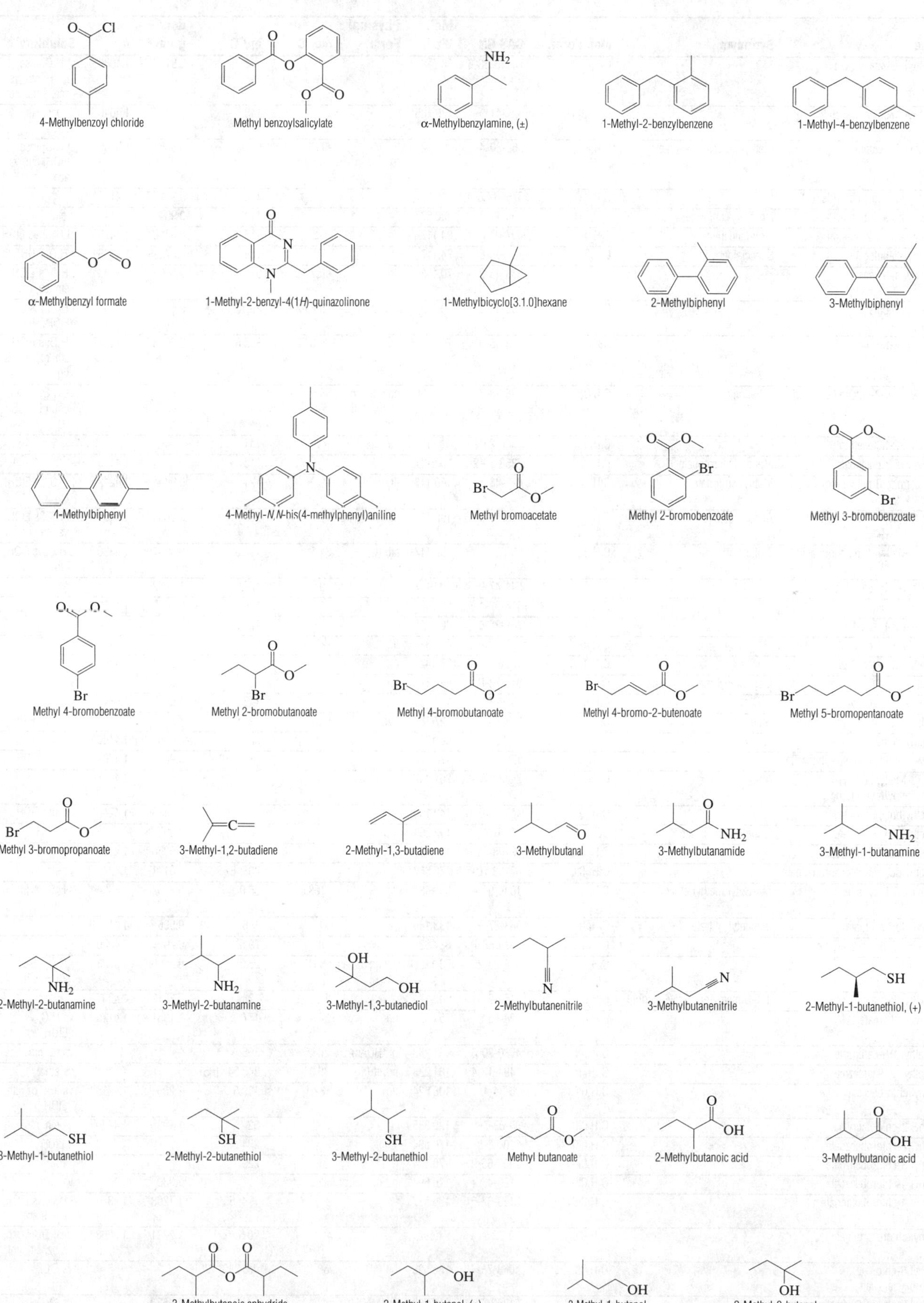

4-Methylbenzoyl chloride
Methyl benzoylsalicylate
α-Methylbenzylamine, (±)
1-Methyl-2-benzylbenzene
1-Methyl-4-benzylbenzene
α-Methylbenzyl formate
1-Methyl-2-benzyl-4(1*H*)-quinazolinone
1-Methylbicyclo[3.1.0]hexane
2-Methylbiphenyl
3-Methylbiphenyl
4-Methylbiphenyl
4-Methyl-*N*,*N*-bis(4-methylphenyl)aniline
Methyl bromoacetate
Methyl 2-bromobenzoate
Methyl 3-bromobenzoate
Methyl 4-bromobenzoate
Methyl 2-bromobutanoate
Methyl 4-bromobutanoate
Methyl 4-bromo-2-butenoate
Methyl 5-bromopentanoate
Methyl 3-bromopropanoate
3-Methyl-1,2-butadiene
2-Methyl-1,3-butadiene
3-Methylbutanal
3-Methylbutanamide
3-Methyl-1-butanamine
2-Methyl-2-butanamine
3-Methyl-2-butanamine
3-Methyl-1,3-butanediol
2-Methylbutanenitrile
3-Methylbutanenitrile
2-Methyl-1-butanethiol, (+)
3-Methyl-1-butanethiol
2-Methyl-2-butanethiol
3-Methyl-2-butanethiol
Methyl butanoate
2-Methylbutanoic acid
3-Methylbutanoic acid
3-Methylbutanoic anhydride
2-Methyl-1-butanol, (±)
3-Methyl-1-butanol
2-Methyl-2-butanol

No.	Name	Synonym	Mol. Form.	CAS RN	Mol. Wt.	Physical Form	mp/°C	bp/°C	den/ g cm^{-3}	n_D	Solubility
7080	3-Methyl-2-butanol, (±)		$C_5H_{12}O$	70116-68-6	88.148			112.9	0.8180^{20}	1.4089^{20}	sl H_2O; msc EtOH, eth; vs ace; s bz, ctc
7081	2-Methyl-1-butanol acetate		$C_7H_{14}O_2$	624-41-9	130.185			140	0.8740^{20}	1.4040^{20}	vs ace, eth, EtOH
7082	3-Methyl-2-butanone	Methyl isopropyl ketone	$C_5H_{10}O$	563-80-4	86.132	liq	-93.1	94.33	0.8051^{20}	1.3880^{20}	sl H_2O; msc EtOH, eth; vs ace; s ctc
7083	2-Methylbutanoyl chloride, (±)		C_5H_9ClO	57526-28-0	120.577			116	0.9917^{20}	1.4170^{20}	
7084	3-Methylbutanoyl chloride	Isovaleryl chloride	C_5H_9ClO	108-12-3	120.577			114	0.9844^{20}	1.4149^{20}	s eth
7085	*trans*-2-Methyl-2-butenal	Tiglic aldehyde	C_5H_8O	497-03-0	84.117	liq		117; 64^{119}	0.8710^{20}	1.4475^{20}	sl H_2O; vs EtOH
7086	3-Methyl-2-butenal	Senecialdehyde	C_5H_8O	107-86-8	84.117			134	0.8722^{20}	1.4528^{20}	s H_2O, EtOH, eth
7087	2-Methyl-1-butene		C_5H_{10}	563-46-2	70.133	liq	-137.53	31.2	0.6504^{20}	1.3778^{20}	i H_2O; s EtOH, eth, bz, ctc
7088	3-Methyl-1-butene		C_5H_{10}	563-45-1	70.133	vol liq or gas	-168.43	20.1	0.6213^{25}	1.3643^{20}	i H_2O; msc EtOH, eth; s bz
7089	2-Methyl-2-butene		C_5H_{10}	513-35-9	70.133	liq	-133.72	38.56	0.6623^{20}	1.3874^{20}	i H_2O; s EtOH, eth, bz, ctc; vs lig
7090	*cis*-2-Methyl-2-butenedioic acid	Citraconic acid	$C_5H_6O_4$	498-23-7	130.100	nd (eth-lig) tcl pr (eth-bz)	93.5		1.617^{25}		vs H_2O; sl eth, chl; i bz, CS_2
7091	3-Methyl-2-butenenitrile		C_5H_7N	4786-24-7	81.117	liq		141			
7092	Methyl *cis*-2-butenoate	Methyl isocrotonate	$C_5H_8O_2$	4358-59-2	100.117			118		1.4175^{20}	
7093	Methyl *trans*-2-butenoate	Methyl crotonate	$C_5H_8O_2$	623-43-8	100.117	liq	-42	121	0.9444^{20}	1.4242^{20}	i H_2O; vs EtOH, eth
7094	*cis*-2-Methyl-2-butenoic acid	Angelic acid	$C_5H_8O_2$	565-63-9	100.117	mcl pr or nd	45.5	185	0.9834^{49}	1.4434^{47}	sl H_2O; s EtOH; vs eth
7095	*trans*-2-Methyl-2-butenoic acid	Tiglic acid	$C_5H_8O_2$	80-59-1	100.117	tab (w)	64.5	198.5	0.9641^{76}	1.4330^{76}	s H_2O; vs EtOH, eth
7096	3-Methyl-2-butenoic acid		$C_5H_8O_2$	541-47-9	100.117		69.5	197	1.0062^{24}		
7097	3-Methyl-2-buten-1-ol		$C_5H_{10}O$	556-82-1	86.132			140	0.848^{25}	1.4412^{20}	
7098	3-Methyl-3-buten-1-ol		$C_5H_{10}O$	763-32-6	86.132			129.9			
7099	2-Methyl-3-buten-2-ol		$C_5H_{10}O$	115-18-4	86.132	liq	-28	97	0.82^{20}		
7100	3-Methyl-3-buten-2-ol		$C_5H_{10}O$	10473-14-0	86.132			114	0.8531^{17}	1.4288^{17}	
7101	3-Methyl-3-buten-2-one	Isopropenyl methyl ketone	C_5H_8O	814-78-8	84.117	liq	-54	98	0.8527^{20}	1.4220^{20}	vs EtOH
7102	3-Methyl-2-butenoyl chloride		C_5H_7ClO	3350-78-5	118.562			146	1.065^{25}	1.4770^{20}	
7103	(3-Methyl-2-butenyl)guanidine	Galegine	$C_6H_{13}N_3$	543-83-9	127.187	hyg	62.5	dec			vs H_2O, EtOH
7104	2-Methyl-1-buten-3-yne	Isopropenylacetylene	C_5H_6	78-80-8	66.102	liq	-113	32	0.6801^{11}	1.4140^{20}	s chl
7105	[(3-Methylbutoxy)methyl]benzene		$C_{12}H_{18}O$	122-73-6	178.270			236; 118^{19}	0.909^{20}	1.4792^{20}	vs eth, EtOH
7106	1-[2-(3-Methylbutoxy)-2-phenylethyl]pyrrolidine	Amixetrine	$C_{17}H_{27}NO$	24622-72-8	261.402			121^2		1.4978^{22}	
7107	2-Methylbutyl acrylate		$C_8H_{14}O_2$	44914-03-6	142.196			160; 45^{10}	0.8936^{20}	1.4240^{20}	vs eth, EtOH
7108	3-Methylbutyl benzoate	Isopentyl benzoate	$C_{12}H_{16}O_2$	94-46-2	192.254			261	0.993^{15}		vs EtOH
7109	3-Methylbutyl 2-chloropropanoate		$C_8H_{15}ClO_2$	62108-69-4	178.657			208	1.0050^{20}	1.4289^{20}	
7110	3-Methylbutyl 3-chloropropanoate		$C_8H_{15}ClO_2$	62108-70-7	178.657			208; 87^{12}	1.0171^{20}	1.4343^{20}	vs eth, EtOH
7111	Methyl *tert*-butyl ether	*tert*-Butyl methyl ether	$C_5H_{12}O$	1634-04-4	88.148	liq	-108.6	55.0	0.7353^{25}	1.3664^{25}	s H_2O; vs EtOH, eth
7112	3-Methylbutyl nitrate	Isopentyl nitrate	$C_5H_{11}NO_3$	543-87-3	133.146			148	0.996^{22}	1.4122^{21}	
7113	2-Methyl-3-butyn-2-amine		C_5H_9N	2978-58-7	83.132		18	79.5	0.79^{25}	1.4235^{20}	
7114	3-Methyl-1-butyne		C_5H_8	598-23-2	68.118	vol liq or gas	-89.7	26.3	0.6660^{20}	1.3723^{20}	i H_2O; msc EtOH, eth
7115	2-Methyl-3-butyn-2-ol	1,1-Dimethylpropargyl alcohol	C_5H_8O	115-19-5	84.117		1.5	104	0.8618^{20}	1.4207^{20}	
7116	Methyl carbamate		$C_2H_5NO_2$	598-55-0	75.067	nd	54	177	1.1361^{56}	1.4125^{56}	vs H_2O; vs EtOH, eth
7117	3-Methyl-9*H*-carbazole		$C_{13}H_{11}N$	4630-20-0	181.233	pl (HOAc)	208.5	365			vs bz, eth
7118	9-Methyl-9*H*-carbazole		$C_{13}H_{11}N$	1484-12-4	181.233	nd, lf (al)	89.34	343.64; 195^{12}			vs eth
7119	Methyl chloroacetate		$C_3H_5ClO_2$	96-34-4	108.524	liq	-32.1	129.5	1.236^{20}	1.4218^{20}	vs ace, bz, eth, EtOH
7120	Methyl 2-chloroacrylate		$C_4H_5ClO_2$	80-63-7	120.535			52^{51}	1.189^{20}	1.4420^{20}	vs eth
7121	Methyl 2-chlorobenzoate		$C_8H_7ClO_2$	610-96-8	170.594			234			s EtOH
7122	Methyl 3-chlorobenzoate		$C_8H_7ClO_2$	2905-65-9	170.594		21	229			
7123	Methyl 4-chlorobenzoate		$C_8H_7ClO_2$	1126-46-1	170.594	nd or mcl pr	43.5		1.382^{20}		vs EtOH
7124	Methyl 4-chlorobutanoate		$C_5H_9ClO_2$	3153-37-5	136.577			174; 55^4	1.1293^{20}	1.4321^{20}	i H_2O; vs EtOH, eth; s ace
7125	Methyl chlorocarbonate		$C_2H_3ClO_2$	79-22-1	94.497			70.5	1.2231^{20}	1.3868^{20}	msc EtOH, eth; s bz, ctc, chl
7126	Methyl 5-chloro-2-hydroxybenzoate		$C_8H_7ClO_3$	4068-78-4	186.593	nd (al)	50	dec 249; 120^{12}			vs EtOH
7127	Methyl 5-chloro-2-nitrobenzoate		$C_8H_6ClNO_4$	51282-49-6	215.592	pl (MeOH)	48.5		1.453^{18}		vs MeOH
7128	Methyl chlorooxoacetate		$C_3H_3ClO_3$	5781-53-3	122.507			119	1.3316^{20}	1.4189^{20}	
7129	Methyl 2-chloropropanoate		$C_4H_7ClO_2$	17639-93-9	122.551			132.5	1.0750^{25}		

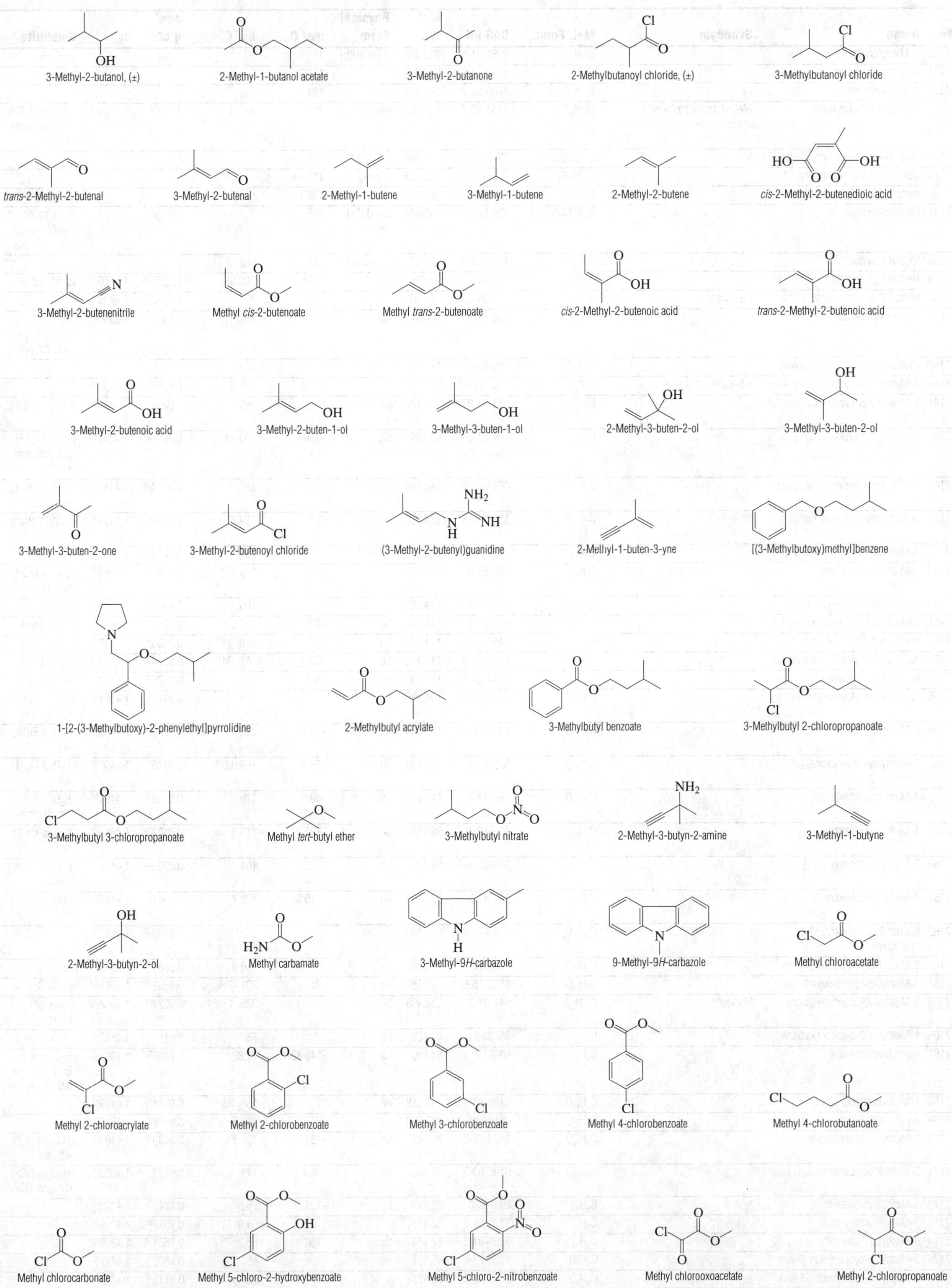
3-Methyl-2-butanol, (±)
2-Methyl-1-butanol acetate
3-Methyl-2-butanone
2-Methylbutanoyl chloride, (±)
3-Methylbutanoyl chloride
trans-2-Methyl-2-butenal
3-Methyl-2-butenal
2-Methyl-1-butene
3-Methyl-1-butene
2-Methyl-2-butene
cis-2-Methyl-2-butenedioic acid
3-Methyl-2-butenenitrile
Methyl *cis*-2-butenoate
Methyl *trans*-2-butenoate
cis-2-Methyl-2-butenoic acid
trans-2-Methyl-2-butenoic acid
3-Methyl-2-butenoic acid
3-Methyl-2-buten-1-ol
3-Methyl-3-buten-1-ol
2-Methyl-3-buten-2-ol
3-Methyl-3-buten-2-ol
3-Methyl-3-buten-2-one
3-Methyl-2-butenoyl chloride
(3-Methyl-2-butenyl)guanidine
2-Methyl-1-buten-3-yne
[(3-Methylbutoxy)methyl]benzene
1-[2-(3-Methylbutoxy)-2-phenylethyl]pyrrolidine
2-Methylbutyl acrylate
3-Methylbutyl benzoate
3-Methylbutyl 2-chloropropanoate
3-Methylbutyl 3-chloropropanoate
Methyl *tert*-butyl ether
3-Methylbutyl nitrate
2-Methyl-3-butyn-2-amine
3-Methyl-1-butyne
2-Methyl-3-butyn-2-ol
Methyl carbamate
3-Methyl-9*H*-carbazole
9-Methyl-9*H*-carbazole
Methyl chloroacetate
Methyl 2-chloroacrylate
Methyl 2-chlorobenzoate
Methyl 3-chlorobenzoate
Methyl 4-chlorobenzoate
Methyl 4-chlorobutanoate
Methyl chlorocarbonate
Methyl 5-chloro-2-hydroxybenzoate
Methyl 5-chloro-2-nitrobenzoate
Methyl chlorooxoacetate
Methyl 2-chloropropanoate

No.	Name	Synonym	Mol. Form.	CAS RN	Mol. Wt.	Physical Form	mp/°C	bp/°C	den/ g cm^{-3}	n_D	Solubility
7130	3-Methylchrysene		$C_{19}H_{14}$	3351-31-3	242.314	lf (bz-peth)	173.3				vs EtOH
7131	5-Methylchrysene		$C_{19}H_{14}$	3697-24-3	242.314		118.3				i H_2O
7132	6-Methylchrysene		$C_{19}H_{14}$	1705-85-7	242.314		161				
7133	Methyl *trans*-cinnamate	Methyl *trans*-3-phenyl-2-propenoate	$C_{10}H_{10}O_2$	1754-62-7	162.185	cry (peth, dil al)	36.5	261.9	1.042^{36}	1.5766^{22}	i H_2O; vs EtOH, eth; s ace, bz; sl chl
7134	*trans-o*-Methylcinnamic acid		$C_{10}H_{10}O_2$	2373-76-4	162.185	cry (EtOH)	175				
7135	*trans-m*-Methylcinnamic acid		$C_{10}H_{10}O_2$	3029-79-6	162.185	cry (w)	115				
7136	*trans-p*-Methylcinnamic acid		$C_{10}H_{10}O_2$	1866-39-3	162.185		198.5				
7137	Methyclothiazide		$C_9H_{11}Cl_2N_3O_4S_2$	135-07-9	360.237	cry (EtOH aq)	225				i H_2O, bz, chl; sl MeOH; vs ace, py
7138	Methyl cyanate		C_2H_3NO	1768-34-9	57.051	unstab gas	-30	exp			
7139	Methyl cyanoacetate		$C_4H_5NO_2$	105-34-0	99.089	liq	-22.5	200.5	1.1225^{25}	1.4176^{20}	vs eth, EtOH
7140	Methyl 2-cyanoacrylate	Mecrylate	$C_5H_5NO_2$	137-05-3	111.100			47^{2}	1.1012^{20}	1.4430	
7141	Methylcyclobutane		C_5H_{10}	598-61-8	70.133	liq	-161.5	36.3	0.6884^{20}	1.3866^{20}	i H_2O; msc EtOH, eth; s ace, bz, peth
7142	Methyl cyclobutanecarboxylate		$C_6H_{10}O_2$	765-85-5	114.142			135.5			
7143	2-Methyl-1,3-cyclohexadiene	4,5-Dihydrotoluene	C_7H_{10}	1489-57-2	94.154			107.5	0.8260^{18}	1.4662^{18}	
7144	2-Methyl-2,5-cyclohexadiene-1,4-dione		$C_7H_6O_2$	553-97-9	122.122	ye pl or nd	69	sub	1.08^{75}		sl H_2O; s EtOH, eth
7145	Methylcyclohexane		C_7H_{14}	108-87-2	98.186	liq	-126.6	100.93	0.7694^{20}	1.4231^{20}	i H_2O; s EtOH, eth; msc ace, bz, lig
7146	Methyl cyclohexanecarboxylate		$C_8H_{14}O_2$	4630-82-4	142.196			183	0.9954^{15}	1.4433^{20}	i H_2O; s EtOH, eth, ace, chl
7147	α-Methylcyclohexanemethanol		$C_8H_{16}O$	1193-81-3	128.212			189	0.928^{25}	1.4656^{20}	vs EtOH, eth; sl ctc
7148	4-Methylcyclohexanemethanol		$C_8H_{16}O$	34885-03-5	128.212			75$^{2.5}$	0.9074^{20}	1.4617^{20}	
7149	1-Methylcyclohexanol		$C_7H_{14}O$	590-67-0	114.185		25	155; 70^{25}	0.9194^{20}	1.4595^{20}	i H_2O; s EtOH, bz, chl
7150	*cis*-2-Methylcyclohexanol		$C_7H_{14}O$	615-38-3	114.185		7	165	0.9360^{20}	1.4640^{20}	vs EtOH
7151	*trans*-2-Methylcyclohexanol, (±)		$C_7H_{14}O$	615-39-4	114.185	liq	-2.0	167.5	0.9247^{20}	1.4616^{20}	vs eth, EtOH
7152	*cis*-3-Methylcyclohexanol, (±)		$C_7H_{14}O$	5454-79-5	114.185	liq	-5.5	168; 94^{12}	0.9155^{20}	1.4752^{20}	vs eth, EtOH
7153	*trans*-3-Methylcyclohexanol, (±)		$C_7H_{14}O$	7443-55-2	114.185	liq	-0.5	167; 84^{13}	0.9214^{30}	1.4580^{20}	vs eth, EtOH
7154	*cis*-4-Methylcyclohexanol		$C_7H_{14}O$	7731-28-4	114.185	liq	-9.2	173	0.9170^{20}	1.4614^{20}	vs eth, EtOH
7155	*trans*-4-Methylcyclohexanol		$C_7H_{14}O$	7731-29-5	114.185			174	0.9118^{21}	1.4561^{20}	sl H_2O; msc EtOH; s eth
7156	2-Methylcyclohexanone, (±)		$C_7H_{12}O$	24965-84-2	112.169	liq	-13.9	165	0.9250^{20}	1.4483^{25}	i H_2O; s EtOH, eth
7157	3-Methylcyclohexanone, (±)		$C_7H_{12}O$	625-96-7	112.169	liq	-73.5	169; 65^{15}	0.9136^{20}	1.4456^{20}	i H_2O; s EtOH, eth
7158	4-Methylcyclohexanone		$C_7H_{12}O$	589-92-4	112.169	liq	-40.6	170	0.9138^{20}	1.4451^{20}	i H_2O; s EtOH, eth; sl ctc
7159	1-Methylcyclohexene		C_7H_{12}	591-49-1	96.170	liq	-120.4	110.3	0.8102^{20}	1.4503^{20}	i H_2O; s eth, bz, ctc
7160	3-Methylcyclohexene, (±)		C_7H_{12}	56688-75-6	96.170	liq	-115.5	104	0.7990^{20}	1.4414^{20}	vs bz, eth, chl, peth
7161	4-Methylcyclohexene		C_7H_{12}	591-47-9	96.170	liq	-115.5	102.7	0.7991^{20}	1.4414^{20}	i H_2O; s EtOH, eth
7162	Methyl 3-cyclohexene-1-carboxylate		$C_8H_{12}O_2$	6493-77-2	140.180			182; 80^{20}	1.0130^{20}	1.4610^{20}	
7163	2-Methyl-2-cyclohexen-1-one		$C_7H_{10}O$	1121-18-2	110.153			178.5	0.966^{20}	1.4833^{20}	s bz
7164	3-Methyl-2-cyclohexen-1-one		$C_7H_{10}O$	1193-18-6	110.153	liq	-21	201	0.9693^{20}	1.49475^{20}	msc H_2O; s bz
7165	3-Methylcyclopentadecanone	Muscone	$C_{16}H_{30}O$	541-91-3	238.408	oily liq		329; 130$^{0.5}$	0.9221^{17}	1.4802^{17}	vs ace, eth, EtOH
7166	1-Methyl-1,3-cyclopentadiene		C_6H_8	96-39-9	80.128	liq		73	0.81^{20}	1.4512^{20}	
7167	Methylcyclopentane		C_6H_{12}	96-37-7	84.159	liq	-142.42	71.8	0.7486^{20}	1.4097^{20}	i H_2O; msc EtOH, eth, ace, bz, lig, ctc
7168	1-Methylcyclopentanol		$C_6H_{12}O$	1462-03-9	100.158	nd	36	136; 53^{30}	0.9044^{23}	1.4429^{23}	
7169	*cis*-2-Methylcyclopentanol		$C_6H_{12}O$	25144-05-2	100.158			148.5	0.9379^{16}	1.4504^{16}	
7170	2-Methylcyclopentanone		$C_6H_{10}O$	1120-72-5	98.142	liq	-75	139.5	0.9139^{20}	1.4364^{20}	s H_2O; vs EtOH, eth, ace
7171	3-Methylcyclopentanone, (±)		$C_6H_{10}O$	6195-92-2	98.142	liq	-58.4	144	0.913^{22}	1.4329^{20}	s H_2O; vs EtOH, eth, ace, HOAc
7172	1-Methylcyclopentene		C_6H_{10}	693-89-0	82.143	liq	-126.5	75.5	0.7748^{25}	1.4322^{20}	
7173	3-Methylcyclopentene		C_6H_{10}	1120-62-3	82.143			64.9	0.7572^{25}	1.4216^{20}	
7174	4-Methylcyclopentene		C_6H_{10}	1759-81-5	82.143	liq	-160.8	65.7	0.7634^{25}	1.4209^{20}	
7175	2-Methyl-2-cyclopenten-1-one		C_6H_8O	1120-73-6	96.127			157	0.9808^{16}	1.4762^{15}	
7176	3-Methyl-2-cyclopenten-1-one		C_6H_8O	2758-18-1	96.127			157.5	0.9712^{20}	1.4714^{20}	

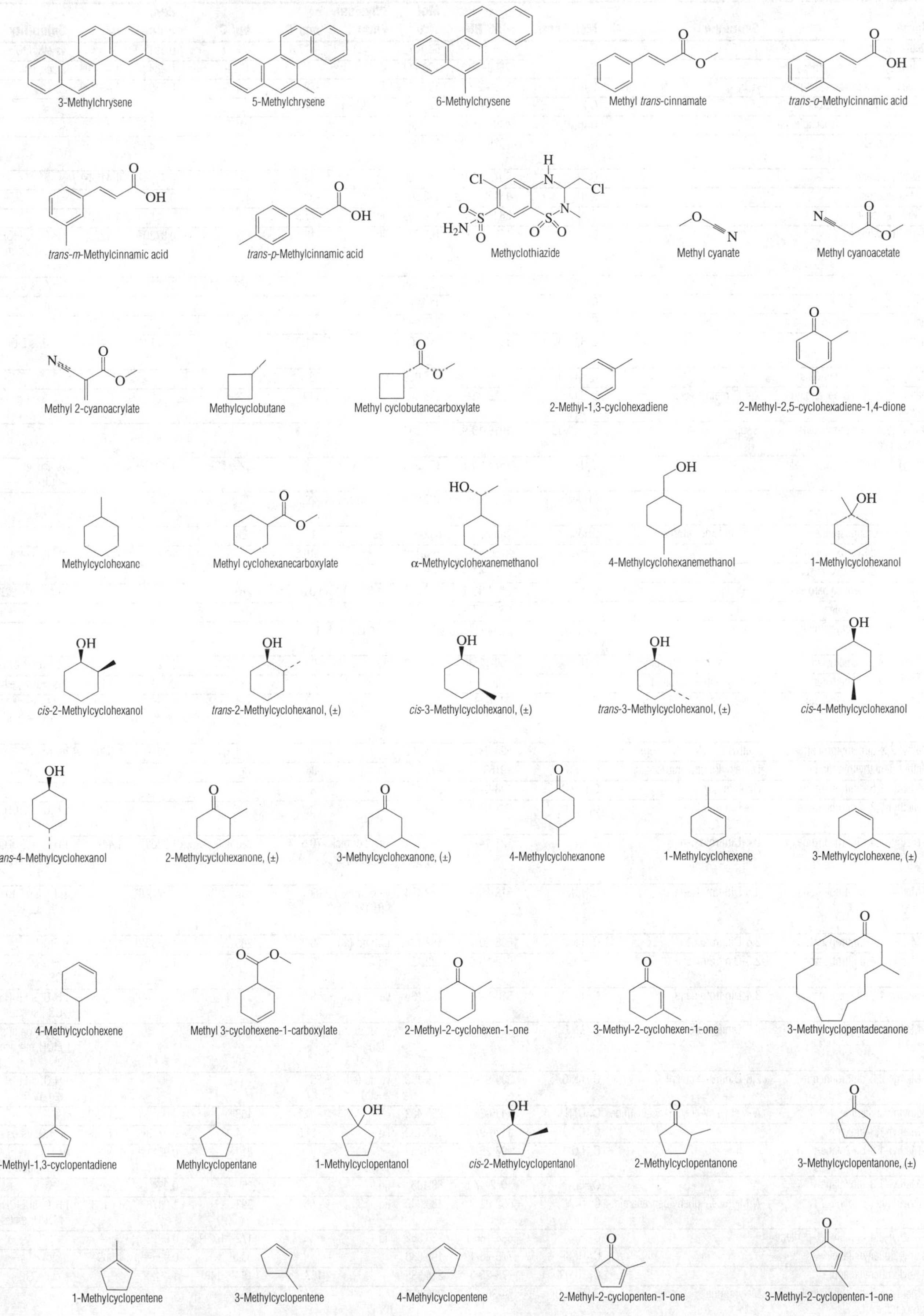
3-Methylchrysene
5-Methylchrysene
6-Methylchrysene
Methyl *trans*-cinnamate
trans-o-Methylcinnamic acid
trans-m-Methylcinnamic acid
trans-p-Methylcinnamic acid
Methyclothiazide
Methyl cyanate
Methyl cyanoacetate
Methyl 2-cyanoacrylate
Methylcyclobutane
Methyl cyclobutanecarboxylate
2-Methyl-1,3-cyclohexadiene
2-Methyl-2,5-cyclohexadiene-1,4-dione
Methylcyclohexane
Methyl cyclohexanecarboxylate
α-Methylcyclohexanemethanol
4-Methylcyclohexanemethanol
1-Methylcyclohexanol
cis-2-Methylcyclohexanol
trans-2-Methylcyclohexanol, (±)
cis-3-Methylcyclohexanol, (±)
trans-3-Methylcyclohexanol, (±)
cis-4-Methylcyclohexanol
trans-4-Methylcyclohexanol
2-Methylcyclohexanone, (±)
3-Methylcyclohexanone, (±)
4-Methylcyclohexanone
1-Methylcyclohexene
3-Methylcyclohexene, (±)
4-Methylcyclohexene
Methyl 3-cyclohexene-1-carboxylate
2-Methyl-2-cyclohexen-1-one
3-Methyl-2-cyclohexen-1-one
3-Methylcyclopentadecanone
1-Methyl-1,3-cyclopentadiene
Methylcyclopentane
1-Methylcyclopentanol
cis-2-Methylcyclopentanol
2-Methylcyclopentanone
3-Methylcyclopentanone, (±)
1-Methylcyclopentene
3-Methylcyclopentene
4-Methylcyclopentene
2-Methyl-2-cyclopenten-1-one
3-Methyl-2-cyclopenten-1-one

No.	Name	Synonym	Mol. Form.	CAS RN	Mol. Wt.	Physical Form	mp/°C	bp/°C	den/ g cm^{-3}	n_D	Solubility
7177	Methylcyclopropane		C_4H_8	594-11-6	56.107	col gas	-177.6	0.7	0.6912^{-20}		vs eth, EtOH
7178	Methyl cyclopropanecarboxylate		$C_5H_8O_2$	2868-37-3	100.117			114.9	0.9848^{20}	1.4144^{19}	s ace, chl
7179	α-Methylcyclopropanemethanol		$C_5H_{10}O$	765-42-4	86.132	liq	-32.1	123.5	0.8805^{20}	1.4316^{20}	
7180	Methyl *L*-cysteine hydrochloride		$C_4H_{10}ClNO_2S$	18598-63-5	171.646	cry (MeOH)	140.5				
7181	Methyl *trans*-2,*cis*-4-decadienoate		$C_{11}H_{18}O_2$	4493-42-9	182.260			$71^{0.15}$	0.9128^{22}	1.4874^{22}	
7182	Methyl *trans*-2,*trans*-4-decadienoate		$C_{11}H_{18}O_2$	7328-33-8	182.260			87^{13}, $70^{0.2}$	0.9082^{22}	1.4918^{22}	
7183	2-Methyldecane		$C_{11}H_{24}$	6975-98-0	156.309	liq	-48.9	189.3	0.7368^{20}	1.4154^{20}	
7184	3-Methyldecane		$C_{11}H_{24}$	13151-34-3	156.309	liq	-92.9	188.1	0.7422^{20}	1.4177^{20}	
7185	4-Methyldecane		$C_{11}H_{24}$	2847-72-5	156.309	liq	-77.5	187		1.4352^{20}	
7186	Methyl decanoate		$C_{11}H_{22}O_2$	110-42-9	186.292	liq	-18	224	0.8730^{20}	1.4259^{20}	i H_2O; vs EtOH, eth; sl ctc; msc chl
7187	Methyl demeton		$C_6H_{15}O_3PS_2$	8022-00-2	230.285	ye liq		$89^{0.15}$, 118^1	1.20^{20}	1.5063^{20}	i H_2O; s os
7188	Methyldiborane(6)		CH_8B_2	23777-55-1	41.697	unstab gas					s eth
7189	Methyl 2,3-dibromopropanoate		$C_4H_6Br_2O_2$	1729-67-5	245.898			206	1.9333^{20}	1.5127^{20}	s EtOH
7190	Methyl dichloroacetate		$C_3H_4Cl_2O_2$	116-54-1	142.969	liq	-51.9	142.9	1.3774^{20}	1.4429^{20}	i H_2O; s EtOH, ctc
7191	Methyl 2,5-dichlorobenzoate		$C_8H_6Cl_2O_2$	2905-69-3	205.039	cry	38				
7192	Methyl (2,4-dichlorophenoxy) acetate	2,4-D methyl ester	$C_9H_8Cl_2O_3$	1928-38-7	235.064		119	141^{18}			
7193	Methyl (3,4-dichlorophenyl) carbamate	Swep	$C_8H_7Cl_2NO_2$	1918-18-9	220.054	nd	114				
7194	Methyl 2,3-dichloropropanoate		$C_4H_6Cl_2O_2$	3674-09-7	156.996			92^{50}, 63^{10}	1.3282^{20}		vs ace, eth, EtOH
7195	Methyldifluoroarsine		CH_3AsF_2	420-24-6	127.954	liq, fumes in air	-29.7	76.5	1.924^{18}		
7196	Methyldifluorophosphine	(Difluoro)methylphosphine	CH_3F_2P	753-59-3	84.006	gas	-110	-28			
7197	Methyl 2,4-dihydroxybenzoate		$C_8H_8O_4$	2150-47-2	168.148		116.5				sl EtOH, ace
7198	Methyl 3,5-dihydroxybenzoate		$C_8H_8O_4$	2150-44-9	168.148		165				
7199	Methyl 3,4-dimethoxybenzoate		$C_{10}H_{12}O_4$	2150-38-1	196.200	nd (dil al)	60.8	283			vs bz, eth, EtOH
7200	Methyldimethoxysilane		$C_3H_{10}O_2Si$	16881-77-9	106.196			61			
7201	3-Methyl-4'-(dimethylamino) azobenzene		$C_{15}H_{17}N_3$	55-80-1	239.316	oran cry	122				
7202	2-Methyl-*N,N*-dimethylaniline	*N,N*-Dimethyl-*o*-toluidine	$C_9H_{13}N$	609-72-3	135.206	liq	-60	194.1	0.9286^{20}	1.5152^{20}	vs eth, EtOH
7203	3-Methyl-*N,N*-dimethylaniline	*N,N*-Dimethyl-*m*-toluidine	$C_9H_{13}N$	121-72-2	135.206			212	0.9410^{20}	1.5492^{20}	msc EtOH, eth
7204	4-Methyl-*N,N*-dimethylaniline	*N,N*-Dimethyl-*p*-toluidine	$C_9H_{13}N$	99-97-8	135.206			211	0.9366^{20}	1.5366^{20}	i H_2O; msc EtOH, eth; s ctc
7205	Methyl 2,2-dimethylpropanoate	Methyl 2,2-dimethylpropionate	$C_6H_{12}O_2$	598-98-1	116.158			101.1	0.891^0	1.3905^{20}	vs eth, EtOH
7206	Methyl dimethylthioborane	Dimethyl(methylthio)borane	C_3H_9BS	19163-05-4	87.979	liq	-84	71			vs ace, eth
7207	2-Methyl-3,5-dinitrobenzamide	Dinitolmide	$C_8H_7N_3O_5$	148-01-6	225.159	cry	181				
7208	1-Methyl-2,3-dinitrobenzene	2,3-Dinitrotoluene	$C_7H_6N_2O_4$	602-01-7	182.134		63				i H_2O; s EtOH, eth; sl chl
7209	1-Methyl-2,4-dinitrobenzene	2,4-Dinitrotoluene	$C_7H_6N_2O_4$	121-14-2	182.134	ye nd or mcl pr (CS_2)	70.5	dec 300	1.3208^{71}	1.442	i H_2O; s EtOH, eth, chl, bz; vs ace, py
7210	1-Methyl-3,5-dinitrobenzene	3,5-Dinitrotoluene	$C_7H_6N_2O_4$	618-85-9	182.134	ye orth nd (HOAc)	93	sub	1.2772^{111}		sl H_2O; s EtOH, eth, bz, chl, CS_2
7211	2-Methyl-1,3-dinitrobenzene	2,6-Dinitrotoluene	$C_7H_6N_2O_4$	606-20-2	182.134	orth nd (al)	66.0	285	1.2833^{111}	1.479	s EtOH, chl
7212	2-Methyl-1,4-dinitrobenzene	2,5-Dinitrotoluene	$C_7H_6N_2O_4$	619-15-8	182.134	nd (al)	52.5		1.282^{111}		s EtOH, bz; vs CS_2
7213	4-Methyl-1,2-dinitrobenzene	3,4-Dinitrotoluene	$C_7H_6N_2O_4$	610-39-9	182.134	ye nd (CS_2)	59.0		1.2594^{111}		i H_2O; s EtOH, CS_2; sl chl
7214	2-Methyl-4,6-dinitrophenol	4,6-Dinitro-*o*-cresol	$C_7H_6N_2O_5$	534-52-1	198.133	ye pr or nd (al)	86.5				sl H_2O, peth; s EtOH, eth, ace, chl
7215	4-Methyl-2,6-dinitrophenol	2,6-Dinitro-*p*-cresol	$C_7H_6N_2O_5$	609-93-8	198.133	ye nd (eth, peth)	85				i H_2O; s EtOH, eth, bz
7216	Methyldioctylamine	*N*-Methyl-*N*-octyl-1-octanamine	$C_{17}H_{37}N$	4455-26-9	255.483		-30.1	158^{10}		1.4424^{20}	
7217	4-Methyl-1,3-dioxane		$C_5H_{10}O_2$	1120-97-4	102.132	liq	-44.5	114	0.9758^{20}	1.4159^{20}	sl H_2O; vs os
7218	2-Methyl-1,3-dioxolane		$C_4H_8O_2$	497-26-7	88.106			81.5	0.9811^{20}	1.4035^{17}	vs H_2O; msc EtOH, eth
7219	4-Methyl-1,3-dioxolane		$C_4H_8O_2$	1072-47-5	88.106	liq		85	0.99^{20}	1.3980^{20}	
7220	Methyldiphenylamine	*N*-Methyl-*N*-phenylbenzenamine	$C_{13}H_{13}N$	552-82-9	183.249	liq	-7.5	293.5	1.0476^{20}	1.6193^{20}	i H_2O; sl EtOH, MeOH; s ctc
7221	4-Methyl-2,4-diphenyl-1-pentene		$C_{18}H_{20}$	6362-80-7	236.352	liq		172^8, $102^{0.2}$	0.99^{25}		
7222	Methyldiphenylsilane		$C_{13}H_{14}Si$	776-76-1	198.336			93.5^1	0.996^{20}	1.5694^{20}	s ctc
7223	Methyldiphenylsilanol		$C_{13}H_{14}OSi$	778-25-6	214.335		167	184^{24}, 148^3	1.0840^{25}		s ctc, CS_2
7224	2-Methyl-1,2-di-3-pyridinyl-1-propanone	Metyrapone	$C_{14}H_{14}N_2O$	54-36-4	226.273		50.5				
7225	Methyl docosanoate	Methyl behenate	$C_{23}H_{46}O_2$	929-77-1	354.610	nd (ace)	54			1.4339^{60}	vs eth, EtOH

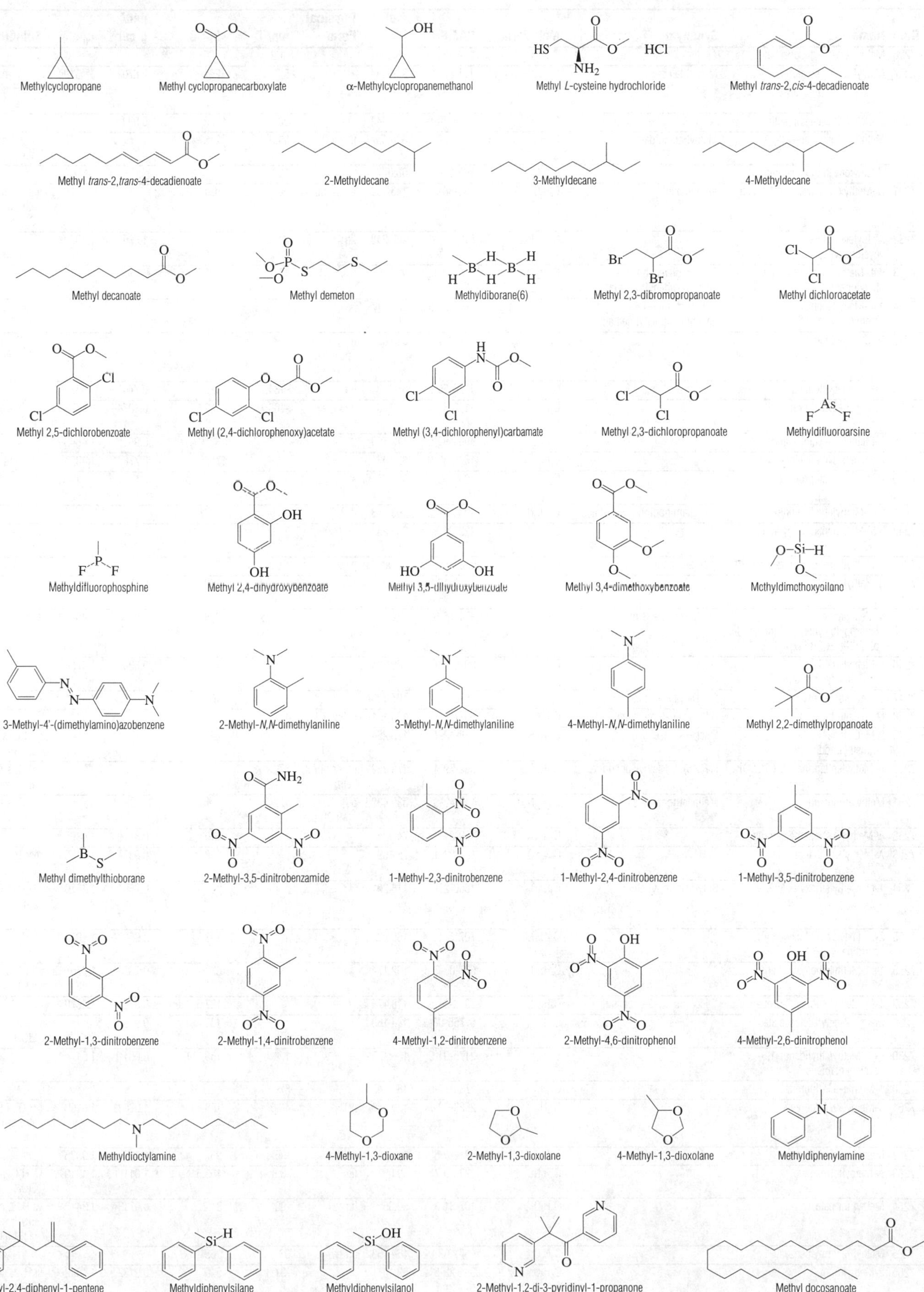
Methylcyclopropane
Methyl cyclopropanecarboxylate
α-Methylcyclopropanemethanol
Methyl L-cysteine hydrochloride
Methyl trans-2,cis-4-decadienoate
Methyl trans-2,trans-4-decadienoate
2-Methyldecane
3-Methyldecane
4-Methyldecane
Methyl decanoate
Methyl demeton
Methyldiborane(6)
Methyl 2,3-dibromopropanoate
Methyl dichloroacetate
Methyl 2,5-dichlorobenzoate
Methyl (2,4-dichlorophenoxy)acetate
Methyl (3,4-dichlorophenyl)carbamate
Methyl 2,3-dichloropropanoate
Methyldifluoroarsine
Methyldifluorophosphine
Methyl 2,4-dihydroxybenzoate
Methyl 3,5-dihydroxybenzoate
Methyl 3,4-dimethoxybenzoate
Methyldimethoxysilane
3-Methyl-4'-(dimethylamino)azobenzene
2-Methyl-N,N-dimethylaniline
3-Methyl-N,N-dimethylaniline
4-Methyl-N,N-dimethylaniline
Methyl 2,2-dimethylpropanoate
Methyl dimethylthioborane
2-Methyl-3,5-dinitrobenzamide
1-Methyl-2,3-dinitrobenzene
1-Methyl-2,4-dinitrobenzene
1-Methyl-3,5-dinitrobenzene
2-Methyl-1,3-dinitrobenzene
2-Methyl-1,4-dinitrobenzene
4-Methyl-1,2-dinitrobenzene
2-Methyl-4,6-dinitrophenol
4-Methyl-2,6-dinitrophenol
Methyldioctylamine
4-Methyl-1,3-dioxane
2-Methyl-1,3-dioxolane
4-Methyl-1,3-dioxolane
Methyldiphenylamine
4-Methyl-2,4-diphenyl-1-pentene
Methyldiphenylsilane
Methyldiphenylsilanol
2-Methyl-1,2-di-3-pyridinyl-1-propanone
Methyl docosanoate

No.	Name	Synonym	Mol. Form.	CAS RN	Mol. Wt.	Physical Form	mp/°C	bp/°C	den/ g cm-3	n_D	Solubility
7226	Methyl *cis*-13-docosenoate		$C_{23}H_{44}O_2$	1120-34-9	352.594		-1.2	220^{5}			
7227	Methyl dodecanoate	Methyl laurate	$C_{13}H_{26}O_2$	111-82-0	214.344		5.2	267	0.8702^{20}	1.4319^{20}	i H_2O; msc EtOH, eth, ace, bz; s chl, ctc
7228	2-Methyldodecanoic acid		$C_{13}H_{26}O_2$	2874-74-0	214.344	pl	22	153^{1}	0.890^{18}		
7229	Methyl eicosanoate	Methyl arachidate	$C_{21}H_{42}O_2$	1120-28-1	326.557	lf (MeOH)	54.5	215^{10}		1.4317^{60}	vs bz, eth, EtOH, chl
7230	(Methyleneamino)acetonitrile		$C_3H_4N_2$	109-82-0	68.077		129				
7231	α-Methylenebenzeneacetic acid	Atropic acid	$C_9H_8O_2$	492-38-6	148.159	lf (al), nd (w)	106.5	dec 267			sl H_2O; s EtOH, eth, bz, chl, CS_2
7232	Methylenebis(4-cyclohexylisocyanate)		$C_{15}H_{22}N_2O_2$	5124-30-1	262.348	liq			1.066	1.4970^{20}	
7233	4,4'-Methylenebis[2,6-di-*tert*-butylphenol]	Bis(3,5-di-*tert*-butyl-4-hydroxyphenyl)methane	$C_{29}H_{44}O_2$	118-82-1	424.658		154	289^{40}, 250^{10}			
7234	4,4'-Methylenebis(*N*-methylaniline)	*N,N'*-Dimethyl-4,4'-diaminodiphenylmethane	$C_{15}H_{18}N_2$	1807-55-2	226.317						s ctc, CS_2
7235	Methylene blue		$C_{16}H_{18}ClN_3S$	61-73-4	319.852	dk grn cry or pow (chl-eth)					s H_2O, EtOH, chl; i eth; sl py
7236	Methylenecyclobutane		C_5H_8	1120-56-5	68.118	liq	-134.7	42.2	0.7401^{20}	1.4210^{20}	
7237	Methylenecyclohexane		C_7H_{12}	1192-37-6	96.170	liq	-106.7	102.5	0.8074^{20}	1.4523^{20}	i H_2O; s eth, bz, chl
7238	2-Methylenecyclohexanol		$C_7H_{12}O$	4065-80-9	112.169			83^{13}	0.955^{20}	1.4843^{20}	
7239	Methylenecyclopentane		C_6H_{10}	1528-30-9	82.143			75.5	0.7787^{20}	1.4355^{20}	s bz, chl
7240	Methylenecyclopropene		C_4H_4	4095-06-1	52.075	solid stab at -196					
7241	2,4'-Methylenedianiline	2,4'-Diaminodiphenylmethane	$C_{13}H_{14}N_2$	1208-52-2	198.263	lf (bz)	88.5	222^{9}			
7242	5,5'-Methylenedisalicylic acid		$C_{15}H_{12}O_6$	122-25-8	288.252	nd (bz)	243.5				vs ace, eth, EtOH
7243	5-Methylene-2(5*H*)-furanone	Protoanemonin	$C_5H_4O_2$	108-28-1	96.085	pa ye oil		73^{11}			sl H_2O; s chl
7244	3-Methyleneheptane		C_8H_{16}	1632-16-2	112.213			120	0.7270^{20}	1.4157^{20}	i H_2O; vs eth, bz, peth
7245	4-Methylene-1-isopropylbicyclo[3.1.0]hexan-3-ol, [1*S*-(1α,3β,5α)]	4(10)-Thujene-3-ol	$C_{10}H_{16}O$	471-16-9	152.233			208	0.9488^{19}	1.4871^{25}	s eth
7246	4-Methylene-1-isopropylcyclohexene		$C_{10}H_{16}$	99-84-3	136.234			173.5	0.838^{22}	1.4754^{22}	
7247	2-Methylenepentanedinitrile	2,4-Dicyano-1-butene	$C_6H_6N_2$	1572-52-7	106.125			103^{5}		1.4561^{20}	s chl
7248	Methylene thiocyanate	Dithiocyanatomethane	$C_3H_2N_2S_2$	6317-18-6	130.191	solid	102				
7249	2-Methylene-1,3,3-trimethylindoline	Fischer's base	$C_{12}H_{15}N$	118-12-7	173.254			244			sl H_2O; s EtOH, eth, bz, chl
7250	*N*-Methylephedrine, [*R*-(*R**,*S**)]	(1*R*,2*S*)-*N*-Methylephedrine	$C_{11}H_{17}NO$	552-79-4	179.259	nd or pl (al, eth)	87.5				i H_2O; s EtOH, eth, MeOH
7251	Methylergonovine	Methylergometrine	$C_{20}H_{25}N_3O_2$	113-42-8	339.432	pr (MeOH,ace)	172				i H_2O; s EtOH, ace
7252	*N*-Methyl-1,2-ethanediamine		$C_3H_{10}N_2$	109-81-9	74.124			115	0.841^{25}	1.4395^{20}	
7253	*N*-Methyl-2-ethanolamine		C_3H_9NO	109-83-1	75.109			158	0.937^{20}	1.4385^{20}	msc H_2O, EtOH, eth
7254	1-(1-Methylethoxy)butane	Butyl isopropyl ether	$C_7H_{16}O$	1860-27-1	116.201			108	0.7594^{15}	1.3870^{15}	i H_2O; s EtOH, eth, ace, con sulf
7255	2-[2-(1-Methylethoxy)ethyl] pyridine		$C_{10}H_{15}NO$	70715-19-4	165.232			133^{50}	0.9502^{25}	1.4820^{25}	vs H_2O
7256	1-(1-Methylethoxy)propane		$C_6H_{14}O$	627-08-7	102.174			83	0.7370^{20}	1.376^{21}	sl H_2O; vs EtOH; s eth, ace
7257	1-(1-Methylethoxy)-2-propanol	1-Isopropoxy-2-propanol	$C_6H_{14}O_2$	3944-36-3	118.174			137.5	0.879^{20}	1.4070^{20}	
7258	Methyl 2-ethylacetoacetate		$C_7H_{12}O_3$	51756-08-2	144.168			182	0.995^{14}		vs ace, eth, EtOH
7259	5-(1-Methylethylidene)-1,3-cyclopentadiene		C_8H_{10}	2175-91-9	106.165		1.4	155; 49^{11}	0.881^{20}	1.5474^{20}	
7260	1-Methyl-9*H*-fluorene		$C_{14}H_{12}$	1730-37-6	180.245		87				
7261	9-Methyl-9*H*-fluorene		$C_{14}H_{12}$	2523-37-7	180.245	pr	46.5	155^{15}	1.0263^{66}	1.610^{66}	i H_2O; s EtOH, eth, ace, bz, chl
7262	Methyl fluorosulfonate		CH_3FO_3S	421-20-5	114.096	col liq	-95	93	1.412	1.3326^{20}	
7263	*N*-Methylformamide		C_2H_5NO	123-39-7	59.067	liq	-3.8	199.51	1.011^{19}	1.4319^{20}	vs H_2O, ace, EtOH
7264	Methyl formate		$C_2H_4O_2$	107-31-3	60.052	liq	-99	31.7	0.9713^{20}	1.3419^{20}	vs H_2O; msc EtOH; s eth, chl, MeOH
7265	Methyl 4-formylbenzoate		$C_9H_8O_3$	1571-08-0	164.158	nd (w)	63	265			
7266	2-Methylfuran		C_5H_6O	534-22-5	82.101	liq	-91.3	64.7	0.9132^{20}	1.4342^{20}	sl H_2O, ctc; s EtOH, eth

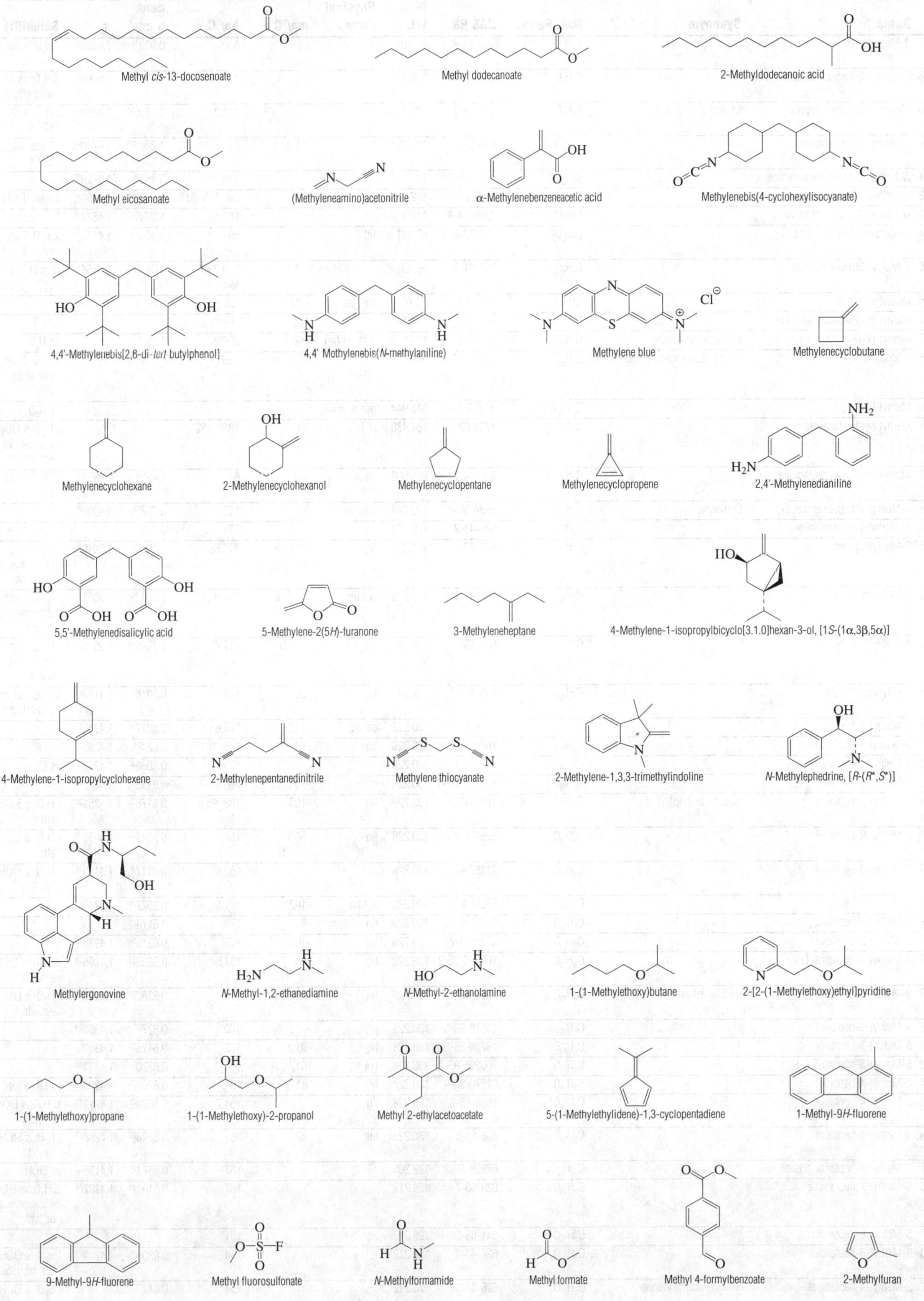
Methyl cis-13-docosenoate
Methyl dodecanoate
2-Methyldodecanoic acid
Methyl eicosanoate
(Methyleneamino)acetonitrile
α-Methylenebenzeneacetic acid
Methylenebis(4-cyclohexylisocyanate)
4,4'-Methylenebis[2,6-di-tert butylphenol]
4,4' Methylenebis(N-methylaniline)
Methylene blue
Methylenecyclobutane
Methylenecyclohexane
2-Methylenecyclohexanol
Methylenecyclopentane
Methylenecyclopropene
2,4'-Methylenedianiline
5,5'-Methylenedisalicylic acid
5-Methylene-2(5H)-furanone
3-Methyleneheptane
4-Methylene-1-isopropylbicyclo[3.1.0]hexan-3-ol, [1S-(1α,3β,5α)]
4-Methylene-1-isopropylcyclohexene
2-Methylenepentanedinitrile
Methylene thiocyanate
2-Methylene-1,3,3-trimethylindoline
N-Methylephedrine, [R-(R*,S*)]
Methylergonovine
N-Methyl-1,2-ethanediamine
N-Methyl-2-ethanolamine
1-(1-Methylethoxy)butane
2-[2-(1-Methylethoxy)ethyl]pyridine
1-(1-Methylethoxy)propane
1-(1-Methylethoxy)-2-propanol
Methyl 2-ethylacetoacetate
5-(1-Methylethylidene)-1,3-cyclopentadiene
1-Methyl-9H-fluorene
9-Methyl-9H-fluorene
Methyl fluorosulfonate
N-Methylformamide
Methyl formate
Methyl 4-formylbenzoate
2-Methylfuran

No.	Name	Synonym	Mol. Form.	CAS RN	Mol. Wt.	Physical Form	mp/°C	bp/°C	den/ g cm^{-3}	n_D	Solubility
7267	3-Methylfuran		C_5H_6O	930-27-8	82.101			65.5	0.923^{18}	1.4330^{19}	i H_2O; s EtOH, eth
7268	5-Methyl-2-furancarboxaldehyde		$C_6H_6O_2$	620-02-0	110.111			187; 89^{26}	1.1072^{18}	1.5264^{20}	s H_2O; vs EtOH; msc eth; sl ctc
7269	Methyl 2-furancarboxylate	Methyl 2-furanoate	$C_6H_6O_3$	611-13-2	126.110			181.3	1.1786^{21}	1.4860^{20}	i H_2O; s EtOH, eth, bz, chl
7270	3-Methyl-2,5-furandione		$C_5H_4O_3$	616-02-4	112.084		7.5	213.5	1.2469^{16}	1.4710^{21}	vs ace, eth, EtOH
7271	*N*-Methyl-2-furanmethanamine		C_6H_9NO	4753-75-7	111.141			149	0.989^{25}	1.4729^{20}	
7272	5-Methyl-2-furanmethanol		$C_6H_8O_2$	3857-25-8	112.127			dec 195; 81^{23}	1.0769^{20}	1.4853^{20}	vs eth, EtOH
7273	α-Methyl-2-furanmethanol		$C_6H_8O_2$	4208-64-4	112.127			162.5	1.0739^{25}	1.4827^{15}	
7274	5-Methyl-2(3*H*)-furanone		$C_5H_6O_2$	591-12-8	98.101	nd	18	56^{12}	1.084^{20}	1.4476^{20}	s H_2O, EtOH, eth, CS_2; sl ctc
7275	5-Methyl-2(5*H*)-furanone		$C_5H_6O_2$	591-11-7	98.101		<-17	209; 98^{15}	1.0810^{20}	1.4454^{20}	msc H_2O; s EtOH, eth
7276	Methylgermane		CH_6Ge	1449-65-6	90.70	col gas	-158	-23			
7277	Methyl β-*D*-glucopyranoside		$C_7H_{14}O_6$	709-50-2	194.182		109				s H_2O
7278	Methyl α-*D*-glucopyranoside	α-Methylglucoside	$C_7H_{14}O_6$	97-30-3	194.182	orth nd (al)	168	200$^{0.2}$	1.46^{30}		vs H_2O
7279	3-Methylglutaric acid	3-Methylpentanedioic acid	$C_6H_{10}O_4$	626-51-7	146.141		87	166$^{0.5}$			s H_2O, EtOH, eth; sl bz, chl; i lig
7280	Methyl Green		$C_{27}H_{35}BrClN_3$	14855-76-6	516.944	grn pow (al)					vs H_2O
7281	Methyl heptadecanoate		$C_{18}H_{36}O_2$	1731-92-6	284.478	pl (al)	30	185^{9}, 152$^{0.05}$			i H_2O; s EtOH, ace, ctc; vs eth, bz
7282	Methyl heptafluorobutanoate		$C_5H_3F_7O_2$	356-24-1	228.066	liq	-86	80	1.483^{20}	1.295^{20}	sl H_2O; s eth, ace
7283	6-Methyl-2-heptanamine, (±)	Octodrine	$C_8H_{19}N$	5984-58-7	129.244	visc liq		155	0.767^{25}	1.4209^{20}	
7284	*N*-Methyl-2-heptanamine		$C_8H_{19}N$	540-43-2	129.244			155			
7285	2-Methylheptane		C_8H_{18}	592-27-8	114.229	liq	-109.02	117.66	0.6980^{20}	1.3949^{20}	i H_2O; msc EtOH, ace, bz; s eth, ctc
7286	3-Methylheptane		C_8H_{18}	589-81-1	114.229	col liq	-120.48	118.9	0.7017^{25}	1.3961^{25}	i H_2O; s EtOH, eth; msc ace, bz, chl
7287	4-Methylheptane		C_8H_{18}	589-53-7	114.229	liq	-121.0	117.72	0.7046^{20}	1.3979^{20}	i H_2O; s eth; msc EtOH, ace, bz
7288	Methyl heptanoate		$C_8H_{16}O_2$	106-73-0	144.212	liq	-56	174	0.8815^{20}	1.4152^{20}	sl H_2O, ctc, ace; s EtOH, eth
7289	2-Methyl-1-heptanol, (±)		$C_8H_{18}O$	111675-77-5	130.228	col liq	-112	175.6	0.8022^{20}	1.424^{20}	
7290	3-Methyl-1-heptanol		$C_8H_{18}O$	1070-32-2	130.228	liq	-90	186; 101^{20}	0.824^{24}	1.4295^{25}	
7291	4-Methyl-1-heptanol		$C_8H_{18}O$	817-91-4	130.228			188	0.8065^{25}	1.4253^{25}	vs EtOH
7292	5-Methyl-1-heptanol, (±)		$C_8H_{18}O$	111767-95-4	130.228	col liq	-104	186.6	0.8153^{25}	1.4272^{25}	
7293	6-Methyl-1-heptanol	Isooctyl alcohol	$C_8H_{18}O$	1653-40-3	130.228	liq	-106	188; 95.8^{20}	0.8176^{25}	1.4251^{25}	i H_2O; s EtOH, eth
7294	2-Methyl-2-heptanol		$C_8H_{18}O$	625-25-2	130.228	liq	-50.4	156	0.8142^{20}	1.4250^{20}	i H_2O; s EtOH, eth
7295	3-Methyl-2-heptanol		$C_8H_{18}O$	31367-46-1	130.228			166.1	0.8177^{25}	1.4199^{25}	i H_2O; s EtOH, eth, ctc
7296	4-Methyl-2-heptanol		$C_8H_{18}O$	56298-90-9	130.228	col liq	-102	171.6	0.8027^{20}	1.424^{20}	
7297	5-Methyl-2-heptanol		$C_8H_{18}O$	54630-50-1	130.228	liq	-61	170	0.8174^{21}		
7298	6-Methyl-2-heptanol		$C_8H_{18}O$	4730-22-7	130.228	liq	-105	174	0.8218^{20}	1.4238^{10}	
7299	2-Methyl-3-heptanol, (±)		$C_8H_{18}O$	100296-26-2	130.228	liq	-85	167.5	0.8235^{20}	1.4265^{20}	sl H_2O; s EtOH, eth, ctc
7300	3-Methyl-3-heptanol	2-Ethyl-2-hexanol	$C_8H_{18}O$	5582-82-1	130.228	liq	-83	163	0.8282^{20}	1.4279^{20}	i H_2O; s EtOH, eth, ctc
7301	4-Methyl-3-heptanol		$C_8H_{18}O$	14979-39-6	130.228	liq	-123	170	0.827^{25}	1.4300^{20}	
7302	5-Methyl-3-heptanol		$C_8H_{18}O$	18720-65-5	130.228	liq	-91.2	172	0.8425^{25}	1.433^{24}	
7303	6-Methyl-3-heptanol, (±)		$C_8H_{18}O$	100295-85-0	130.228	col liq	-61	169	0.8220^{20}	1.4254^{20}	
7304	2-Methyl-4-heptanol		$C_8H_{18}O$	21570-35-4	130.228	liq	-81	164	0.8207^{20}	1.4203	vs eth, EtOH
7305	3-Methyl-4-heptanol		$C_8H_{18}O$	1838-73-9	130.228	liq		164.7	0.8329^{25}	1.4211^{25}	sl H_2O; s EtOH, eth, ctc
7306	4-Methyl-4-heptanol		$C_8H_{18}O$	598-01-6	130.228	liq	-82	161	0.8248^{20}	1.4258^{20}	i H_2O; s EtOH, eth, ctc
7307	6-Methyl-2-heptanol acetate		$C_{10}H_{20}O_2$	67952-57-2	172.265			187	0.8474^{20}	1.413^{20}	vs EtOH
7308	6-Methyl-2-heptanone		$C_8H_{16}O$	928-68-7	128.212			167	0.8151^{20}	1.4162^{20}	sl H_2O; vs EtOH, eth; msc ace, bz, chl
7309	5-Methyl-3-heptanone		$C_8H_{16}O$	541-85-5	128.212	liq		161			
7310	6-Methyl-3-heptanone		$C_8H_{16}O$	624-42-0	128.212			164	0.8304^{20}	1.4209^{20}	i H_2O; s EtOH, eth, bz, ctc
7311	2-Methyl-4-heptanone	Isobutyl propyl ketone	$C_8H_{16}O$	626-33-5	128.212			154	0.813^{22}		i H_2O; s EtOH, eth

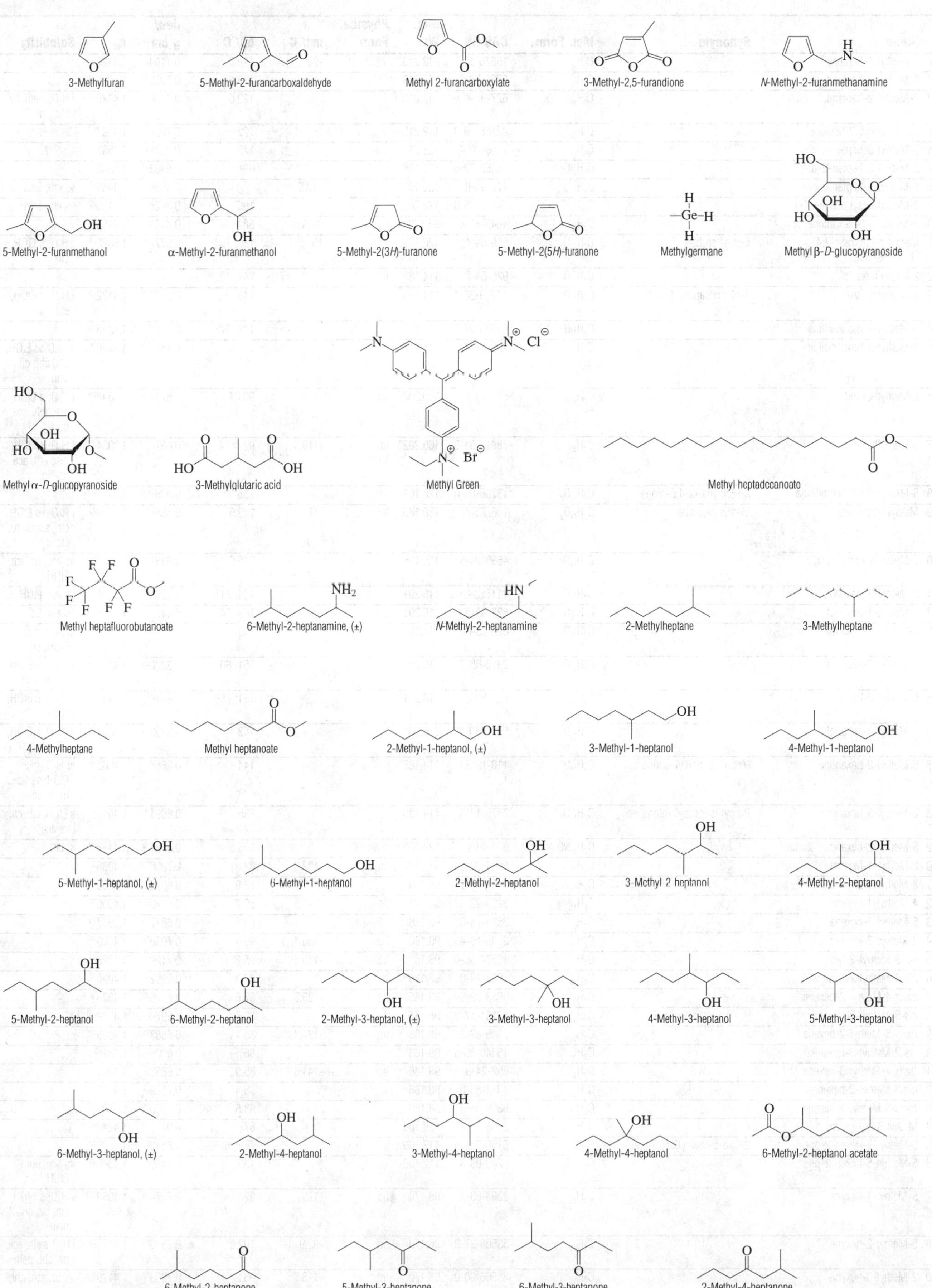
3-Methylfuran
5-Methyl-2-furancarboxaldehyde
Methyl 2-furancarboxylate
3-Methyl-2,5-furandione
N-Methyl-2-furanmethanamine
5-Methyl-2-furanmethanol
α-Methyl-2-furanmethanol
5-Methyl-2(3H)-furanone
5-Methyl-2(5H)-furanone
Methylgermane
Methyl β-D-glucopyranoside
Methyl α-D-glucopyranoside
3-Methylglutaric acid
Methyl Green
Methyl heptadecanoate
Methyl heptafluorobutanoate
6-Methyl-2-heptanamine, (±)
N-Methyl-2-heptanamine
2-Methylheptane
3-Methylheptane
4-Methylheptane
Methyl heptanoate
2-Methyl-1-heptanol, (±)
3-Methyl-1-heptanol
4-Methyl-1-heptanol
5-Methyl-1-heptanol, (±)
6-Methyl-1-heptanol
2-Methyl-2-heptanol
3-Methyl-2-heptanol
4-Methyl-2-heptanol
5-Methyl-2-heptanol
6-Methyl-2-heptanol
2-Methyl-3-heptanol, (±)
3-Methyl-3-heptanol
4-Methyl-3-heptanol
5-Methyl-3-heptanol
6-Methyl-3-heptanol, (±)
2-Methyl-4-heptanol
3-Methyl-4-heptanol
4-Methyl-4-heptanol
6-Methyl-2-heptanol acetate
6-Methyl-2-heptanone
5-Methyl-3-heptanone
6-Methyl-3-heptanone
2-Methyl-4-heptanone

No.	Name	Synonym	Mol. Form.	CAS RN	Mol. Wt.	Physical Form	mp/°C	bp/°C	den/ g cm^{-3}	n_D	Solubility
7312	2-Methyl-1-heptene		C_8H_{16}	15870-10-7	112.213	liq	-90	119.3	0.7104^{25}	1.4123^{20}	
7313	6-Methyl-1-heptene		C_8H_{16}	5026-76-6	112.213			113.2	0.7079^{25}	1.4070^{20}	
7314	2-Methyl-2-heptene		C_8H_{16}	627-97-4	112.213			122.6	0.7200^{25}	1.4170^{20}	i H_2O; s eth, bz, ctc, chl
7315	*cis*-3-Methyl-2-heptene		C_8H_{16}	22768-19-0	112.213			122	0.725^{25}	1.419^{20}	
7316	6-Methyl-5-hepten-2-ol		$C_8H_{16}O$	1569-60-4	128.212			175	0.8545^{20}	1.4505^{20}	
7317	3-Methyl-5-hepten-2-one		$C_8H_{14}O$	38552-72-6	126.196			63^{20}	0.8463^{18}	1.4345^{18}	
7318	6-Methyl-5-hepten-2-one		$C_8H_{14}O$	110-93-0	126.196		173.5		0.8546^{16}	1.4445^{20}	vs eth, EtOH
7319	2-Methylheptyl acetate, (±)		$C_{10}H_{20}O_2$	74112-36-0	172.265			195	0.8626^{14}	1.4146^{20}	vs eth, EtOH
7320	2-Methyl-1,5-hexadiene		C_7H_{12}	4049-81-4	96.170	liq	-128.8	88.1	0.7153^{25}	1.4183^{20}	
7321	Methyl *trans,trans*-2,4-hexadienoate	Methyl sorbate	$C_7H_{10}O_2$	689-89-4	126.153	lf	15	180; 70^{20}	0.9777^{20}	1.5025^{22}	i H_2O; s EtOH, eth
7322	2-Methylhexanal		$C_7H_{14}O$	925-54-2	114.185	liq		141; 132^{60}			
7323	3-Methylhexanal	3-Methylcaproaldehyde	$C_7H_{14}O$	19269-28-4	114.185			143	0.8203^{20}	1.4122^{20}	i H_2O; s EtOH, eth
7324	3-Methyl-1-hexanamine		$C_7H_{17}N$	65530-93-0	115.217			149; 67^{45}	0.772^{26}	1.4249^{25}	
7325	4-Methyl-2-hexanamine		$C_7H_{17}N$	105-41-9	115.217			132.5	0.7655^{20}	1.4150^{25}	sl H_2O; vs EtOH, eth, chl, dil acid
7326	2-Methylhexane		C_7H_{16}	591-76-4	100.202	liq	-118.2	90.04	0.6787^{20}	1.3848^{20}	i H_2O; s EtOH; msc eth, ace, bz, lig, chl
7327	3-Methylhexane		C_7H_{16}	78918-91-9	100.202	liq	-119.4	92	0.687^{21}	1.3854^{25}	i H_2O; s EtOH; msc eth, ace, bz, lig, chl
7328	5-Methyl-2,3-hexanedione	2-Methylhexa-4,5-dione	$C_7H_{12}O_2$	13706-86-0	128.169			138	0.908^{22}	1.4119^{20}	
7329	Methyl hexanoate	Methyl caproate	$C_7H_{14}O_2$	106-70-7	130.185	liq	-71	149.5	0.8846^{20}	1.4049^{20}	i H_2O; vs EtOH, eth; s ace, bz, ctc
7330	2-Methylhexanoic acid		$C_7H_{14}O_2$	4536-23-6	130.185			215.5	0.918^{20}	1.4193^{20}	vs ace, bz, eth, EtOH
7331	2-Methyl-1-hexanol, (±)		$C_7H_{16}O$	111768-04-8	116.201			164; 71^{15}	0.826^{20}	1.4226^{20}	vs eth, EtOH
7332	5-Methyl-1-hexanol		$C_7H_{16}O$	627-98-5	116.201			169; 54^{15}	0.8192^{24}	1.4175^{20}	vs eth, EtOH
7333	2-Methyl-2-hexanol		$C_7H_{16}O$	625-23-0	116.201			143	0.8119^{20}	1.4175^{20}	sl H_2O; msc EtOH, eth
7334	3-Methyl-2-hexanol		$C_7H_{16}O$	2313-65-7	116.201			151; 80^{52}	0.8220^{25}	1.4198^{18}	i H_2O; vs EtOH, eth; s ace
7335	5-Methyl-2-hexanol		$C_7H_{16}O$	627-59-8	116.201			151; 78^{28}	0.814^{20}	1.4180^{20}	sl H_2O; s EtOH, eth
7336	3-Methyl-3-hexanol		$C_7H_{16}O$	597-96-6	116.201			143	0.8233^{20}	1.4231^{20}	sl H_2O; s EtOH, eth, ctc
7337	5-Methyl-2-hexanone	Methyl isopentyl ketone	$C_7H_{14}O$	110-12-3	114.185			144	0.888^{20}	1.4062^{20}	sl H_2O; msc EtOH; vs ace, bz; s ctc
7338	2-Methyl-3-hexanone	Propyl isopropyl ketone	$C_7H_{14}O$	7379-12-6	114.185			135	0.8091^{20}	1.4042^{20}	s EtOH, eth, chl; vs ace
7339	5-Methyl-2-hexanone oxime		$C_7H_{15}NO$	624-44-2	129.200			195.5	0.8881^{20}	1.4448^{20}	sl chl
7340	2-Methyl-1-hexene		C_7H_{14}	6094-02-6	98.186	liq	-102.8	92	0.7000^{20}	1.4035^{20}	
7341	3-Methyl-1-hexene		C_7H_{14}	3404-61-3	98.186			83.9	0.6871^{25}	1.3965^{20}	
7342	4-Methyl-1-hexene		C_7H_{14}	3769-23-1	98.186	liq	-141.5	86.7	0.6942^{25}	1.4000^{20}	
7343	5-Methyl-1-hexene		C_7H_{14}	3524-73-0	98.186			85.3	0.6877^{25}	1.3967^{20}	
7344	2-Methyl-2-hexene		C_7H_{14}	2738-19-4	98.186	liq	-130.4	95.4	0.7038^{25}	1.4106^{20}	
7345	*cis*-3-Methyl-2-hexene		C_7H_{14}	10574-36-4	98.186	liq	-118.5	95.6	0.712^{20}	1.4126^{20}	
7346	*cis*-4-Methyl-2-hexene		C_7H_{14}	3683-19-0	98.186			86.3	0.6952^{25}	1.4026^{20}	
7347	*trans*-4-Methyl-2-hexene		C_7H_{14}	3683-22-5	98.186	liq	-125.7	87.6	0.6925^{25}	1.4025^{20}	
7348	*cis*-5-Methyl-2-hexene		C_7H_{14}	13151-17-2	98.186			89.5	0.697^{25}	1.404^{20}	
7349	*trans*-5-Methyl-2-hexene		C_7H_{14}	7385-82-2	98.186	liq	-124.3	88.1	0.6883^{25}	1.4006^{20}	
7350	*cis*-2-Methyl-3-hexene		C_7H_{14}	15840-60-5	98.186			86	0.690^{25}	1.401^{20}	
7351	*trans*-2-Methyl-3-hexene		C_7H_{14}	692-24-0	98.186	liq	-141.6	85.9	0.6853^{25}	1.4001^{20}	
7352	*cis*-3-Methyl-3-hexene		C_7H_{14}	4914-89-0	98.186			95.4	0.7079^{25}	1.4126^{20}	
7353	*trans*-3-Methyl-3-hexene		C_7H_{14}	3899-36-3	98.186			93.5	0.7050^{25}	1.4109^{20}	
7354	Methyl 3-hexenoate		$C_7H_{12}O_2$	2396-78-3	128.169			67^{34}	0.9132^{25}	1.4240^{23}	
7355	5-Methyl-3-hexen-2-one	2-Oxo-5-methylhex-3-ene	$C_7H_{12}O$	5166-53-0	112.169			77^{50}, 65^{13}	0.8549^{28}	1.4395^{22}	
7356	5-Methyl-5-hexen-2-one		$C_7H_{12}O$	3240-09-3	112.169			150	0.8460^{20}	1.4348^{20}	vs ace, eth, EtOH
7357	5-Methyl-1-hexyne		C_7H_{12}	2203-80-7	96.170	liq	-125	92	0.7274^{20}	1.4059^{-20}	i H_2O; s EtOH, eth, bz, chl, peth
7358	5-Methyl-2-hexyne		C_7H_{12}	53566-37-3	96.170	liq	-92.9	102.5	0.7378^{20}	1.4176^{20}	i H_2O; s eth, ace, bz, chl, peth
7359	2-Methyl-3-hexyne		C_7H_{12}	36566-80-0	96.170	liq	-116.7	95.2	0.7263^{20}	1.4120^{20}	vs bz, eth, chl, peth
7360	Methyl 2-hexynoate		$C_7H_{10}O_2$	18937-79-6	126.153			80^{23}	0.9648^{25}		

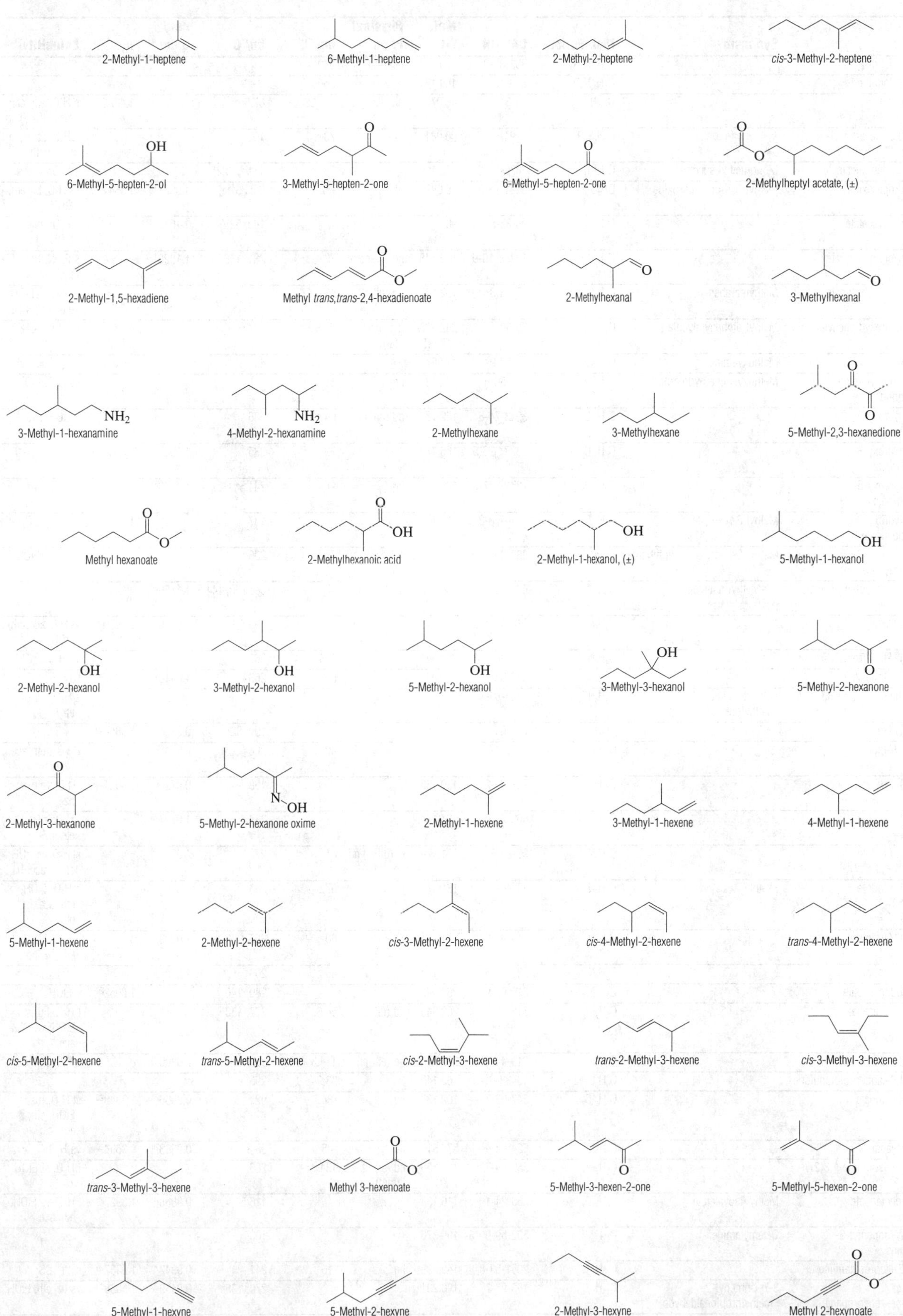
2-Methyl-1-heptene
6-Methyl-1-heptene
2-Methyl-2-heptene
cis-3-Methyl-2-heptene
6-Methyl-5-hepten-2-ol
3-Methyl-5-hepten-2-one
6-Methyl-5-hepten-2-one
2-Methylheptyl acetate, (±)
2-Methyl-1,5-hexadiene
Methyl *trans,trans*-2,4-hexadienoate
2-Methylhexanal
3-Methylhexanal
3-Methyl-1-hexanamine
4-Methyl-2-hexanamine
2-Methylhexane
3-Methylhexane
5-Methyl-2,3-hexanedione
Methyl hexanoate
2-Methylhexanoic acid
2-Methyl-1-hexanol, (±)
5-Methyl-1-hexanol
2-Methyl-2-hexanol
3-Methyl-2-hexanol
5-Methyl-2-hexanol
3-Methyl-3-hexanol
5-Methyl-2-hexanone
2-Methyl-3-hexanone
5-Methyl-2-hexanone oxime
2-Methyl-1-hexene
3-Methyl-1-hexene
4-Methyl-1-hexene
5-Methyl-1-hexene
2-Methyl-2-hexene
cis-3-Methyl-2-hexene
cis-4-Methyl-2-hexene
trans-4-Methyl-2-hexene
cis-5-Methyl-2-hexene
trans-5-Methyl-2-hexene
cis-2-Methyl-3-hexene
trans-2-Methyl-3-hexene
cis-3-Methyl-3-hexene
trans-3-Methyl-3-hexene
Methyl 3-hexenoate
5-Methyl-3-hexen-2-one
5-Methyl-5-hexen-2-one
5-Methyl-1-hexyne
5-Methyl-2-hexyne
2-Methyl-3-hexyne
Methyl 2-hexynoate

No.	Name	Synonym	Mol. Form.	CAS RN	Mol. Wt.	Physical Form	mp/°C	bp/°C	den/ g cm^{-3}	n_D	Solubility
7361	*L*-1-Methylhistidine		$C_7H_{11}N_3O_2$	332-80-9	169.181	pl (DMF aq)	249				
7362	*L*-3-Methylhistidine		$C_7H_{11}N_3O_2$	368-16-1	169.181		250				
7363	Methylhydrazine		CH_6N_2	60-34-4	46.072	liq	-52.36	87.5		1.4325^{20}	s H_2O, eth, ctc; msc EtOH; i lig
7364	Methyl hydrazinecarboxylate	Methyl carbazate	$C_2H_6N_2O_2$	6294-89-9	90.081		73	108^{12}			s H_2O, EtOH; sl bz; i peth
7365	Methyl hydrogen succinate	Monomethyl succinate	$C_5H_8O_4$	3878-55-5	132.116		58	151^{20}, 122^{4}			s H_2O
7366	Methyl hydroperoxide	Methyl hydrogen peroxide	CH_4O_2	3031-73-0	48.042	liq	-72	86; 39^{65}	1.9967^{15}	1.3641^{15}	vs H_2O, bz, eth, EtOH
7367	Methyl hydroxyacetate		$C_3H_6O_3$	96-35-5	90.078			149; 52^{17}	1.1677^{18}		s H_2O; msc EtOH, eth
7368	Methyl 3-hydroxybenzoate		$C_8H_8O_3$	19438-10-9	152.148	nd (bz-peth)	73	281; 178^{17}	1.1528^{100}		s EtOH, bz, peth; sl chl
7369	Methyl 4-hydroxybenzoate	Methylparaben	$C_8H_8O_3$	99-76-3	152.148	nd (dil al)	131	dec 275			sl H_2O; vs EtOH, eth, ace; s tfa
7370	Methyl α-hydroxydiphenylacetate	Methyl diphenylglycolate	$C_{15}H_{14}O_3$	76-89-1	242.270	mcl or tcl cry (al)	75.8	187^{13}			vs eth, EtOH
7371	*O*-Methylhydroxylamine	Methoxyamine	CH_5NO	67-62-9	47.057			49			
7372	*O*-Methylhydroxylamine hydrochloride	Methoxyamine hydrochloride	CH_6ClNO	593-56-6	83.518	pr	150.0				vs H_2O, EtOH
7373	Methyl 4-hydroxy-3-methoxybenzoate		$C_9H_{10}O_4$	3943-74-6	182.173	nd (dil al)	64	286			s EtOH, peth; sl chl
7374	Methyl 2-hydroxy-3-methylbenzoate		$C_9H_{10}O_3$	23287-26-5	166.173		29	235	1.1683^{25}	1.5354^{16}	
7375	Methyl 2-hydroxy-5-methylbenzoate		$C_9H_{10}O_3$	22717-57-3	166.173	liq	-1	244.5	1.1673^{25}	1.5351^{15}	
7376	Methyl 2-hydroxy-2-methylpropanoate	Methyl 2-methyllactate	$C_5H_{10}O_3$	2110-78-3	118.131			137		1.4056^{20}	vs H_2O, EtOH
7377	Methyl 3-hydroxy-2-naphthalenecarboxylate	Methyl 3-hydroxy-2-naphthoate	$C_{12}H_{10}O_3$	883-99-8	202.205	pa ye orth nd (dil MeOH)	75.5	206			i H_2O; s EtOH
7378	Methyl α-hydroxyphenylacetate, (±)	(±)-Methyl mandelate	$C_9H_{10}O_3$	4358-87-6	166.173	pl (bz-lig)	58	dec 250; 144^{20}	1.1756^{20}		vs EtOH, chl
7379	1-Methylimidazol		$C_4H_6N_2$	616-47-7	82.104	liq	-6	195.5	1.0325^{20}	1.4970^{20}	vs H_2O, ace, eth, EtOH
7380	2-Methyl-1*H*-imidazole		$C_4H_6N_2$	693-98-1	82.104		144	267			vs H_2O, EtOH
7381	4-Methyl-1*H*-imidazole		$C_4H_6N_2$	822-36-6	82.104		56	263	1.0416^{14}	1.5037^{14}	vs H_2O, EtOH
7382	*N*-Methyliminodiacetic acid	*N*-(Carboxymethyl)-*N*-methylglycine	$C_5H_9NO_4$	4408-64-4	147.130	cry (w)	226				s H_2O; i EtOH, eth
7383	1-Methyl-1*H*-indene		$C_{10}H_{10}$	767-59-9	130.186			199; 82^{15}	0.970^{25}	1.5616^{20}	
7384	2-Methyl-1*H*-indene		$C_{10}H_{10}$	2177-47-1	130.186		80	208	0.974^{25}	1.5652^{20}	i H_2O; s eth, ace, bz
7385	3-Methyl-1*H*-indene		$C_{10}H_{10}$	767-60-2	130.186			198	0.972^{25}	1.5621^{20}	i H_2O; s eth, ace, bz
7386	1-Methyl-1*H*-indole		C_9H_9N	603-76-9	131.174			237	1.0707^{25}		i H_2O; s EtOH, eth, bz
7387	2-Methyl-1*H*-indole		C_9H_9N	95-20-5	131.174	pl (dil al) nd or lf (w)	61	272	1.07^{20}		sl H_2O; vs EtOH, eth; s ace, bz
7388	3-Methyl-1*H*-indole	Skatole	C_9H_9N	83-34-1	131.174	lf (lig)	97.5	266			s H_2O, EtOH, eth, ace, bz, chl
7389	5-Methyl-1*H*-indole		C_9H_9N	614-96-0	131.174		60	267	1.0202^{78}		s H_2O, EtOH, eth, bz, lig
7390	7-Methyl-1*H*-indole		C_9H_9N	933-67-5	131.174		85	266	1.0202^{100}		
7391	Methyl 2-iodobenzoate		$C_8H_7IO_2$	610-97-9	262.045			280; 146^{16}		1.6052^{20}	s EtOH
7392	Methyl 3-iodobenzoate		$C_8H_7IO_2$	618-91-7	262.045	nd (dil al)	54.5	277; 150^{18}			i H_2O, lig; s EtOH; vs eth, ace
7393	Methyl 4-iodobenzoate		$C_8H_7IO_2$	619-44-3	262.045	nd (eth-al)	114.8	sub	2.0200^{10}		s EtOH, eth
7394	5-Methyl-1,3-isobenzofurandione		$C_9H_6O_3$	19438-61-0	162.142		93.0	295			
7395	Methyl isobutanoate		$C_5H_{10}O_2$	547-63-7	102.132	liq	-84.7	92.5	0.8906^{20}	1.3840^{20}	sl H_2O; msc EtOH, eth; s ace, ctc
7396	Methyl isocyanate		C_2H_3NO	624-83-9	57.051	liq	-45	38.3	0.9588^{20}	1.3694^{20}	vs H_2O
7397	2-Methyl-1*H*-isoindole-1,3(2*H*)-dione		$C_9H_7NO_2$	550-44-7	161.158	nd (al), lf (sub)	134	286			i H_2O; sl EtOH
7398	Methyl isopentanoate	Methyl isovalerate	$C_6H_{12}O_2$	556-24-1	116.158			116.5	0.8808^{20}	1.3927^{20}	i H_2O; vs EtOH, eth, ace
7399	6-Methyl-*N*-isopentyl-2-heptanamine	Octamylamine	$C_{13}H_{29}N$	502-59-0	199.376			100^{7}			
7400	2-Methyl-5-isopropylaniline		$C_{10}H_{15}N$	2051-53-8	149.233	liq	-16	241	0.9942^{20}	1.5387^{20}	s ctc, CS_2
7401	α-Methyl-4-isopropylbenzenepropanal	3-*p*-Cumenyl-2-methylpropionaldehyde	$C_{13}H_{18}O$	103-95-7	190.281			270; 135^{99}	0.9459^{20}	1.5068^{20}	vs bz, eth, EtOH
7402	2-Methyl-5-isopropylbicyclo[3.1.0]hex-2-ene		$C_{10}H_{16}$	2867-05-2	136.234			151	0.8301^{20}	1.4515^{20}	

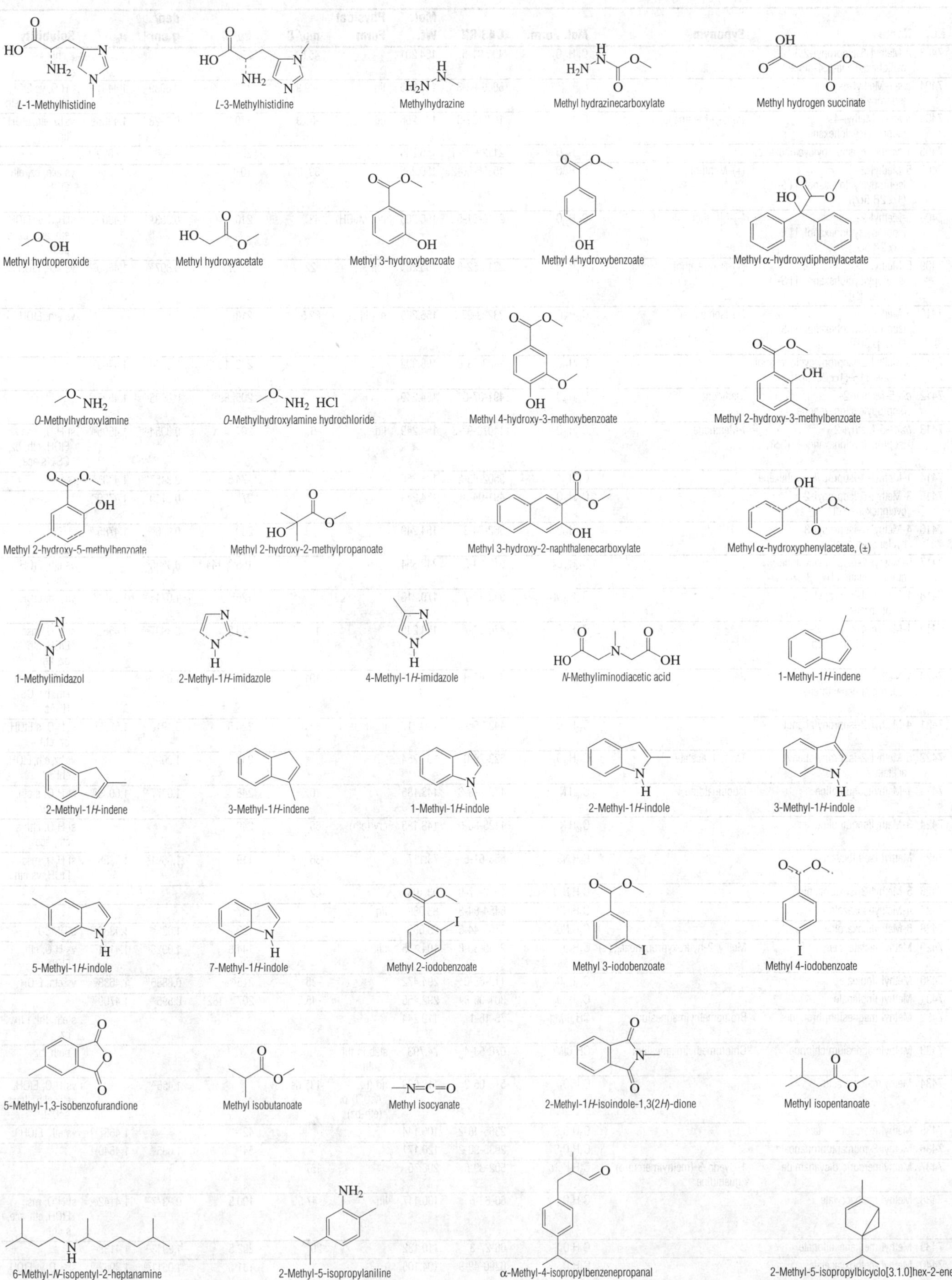

L-1-Methylhistidine
L-3-Methylhistidine
Methylhydrazine
Methyl hydrazinecarboxylate
Methyl hydrogen succinate
Methyl hydroperoxide
Methyl hydroxyacetate
Methyl 3-hydroxybenzoate
Methyl 4-hydroxybenzoate
Methyl α-hydroxydiphenylacetate
O-Methylhydroxylamine
O-Methylhydroxylamine hydrochloride
Methyl 4-hydroxy-3-methoxybenzoate
Methyl 2-hydroxy-3-methylbenzoate
Methyl 2-hydroxy-5-methylbenzoate
Methyl 2-hydroxy-2-methylpropanoate
Methyl 3-hydroxy-2-naphthalenecarboxylate
Methyl α-hydroxyphenylacetate, (±)
1-Methylimidazol
2-Methyl-1H-imidazole
4-Methyl-1H-imidazole
N-Methyliminodiacetic acid
1-Methyl-1H-indene
2-Methyl-1H-indene
3-Methyl-1H-indene
1-Methyl-1H-indole
2-Methyl-1H-indole
3-Methyl-1H-indole
5-Methyl-1H-indole
7-Methyl-1H-indole
Methyl 2-iodobenzoate
Methyl 3-iodobenzoate
Methyl 4-iodobenzoate
5-Methyl-1,3-isobenzofurandione
Methyl isobutanoate
Methyl isocyanate
2-Methyl-1H-isoindole-1,3(2H)-dione
Methyl isopentanoate
6-Methyl-N-isopentyl-2-heptanamine
2-Methyl-5-isopropylaniline
α-Methyl-4-isopropylbenzenepropanal
2-Methyl-5-isopropylbicyclo[3.1.0]hex-2-ene

No.	Name	Synonym	Mol. Form.	CAS RN	Mol. Wt.	Physical Form	mp/°C	bp/°C	den/ g cm^{-3}	n_D	Solubility
7403	2-Methyl-5-isopropyl-2,5-cyclohexadiene-1,4-dione		$C_{10}H_{12}O_2$	490-91-5	164.201		45.5	232			s chl
7404	*cis*-1-Methyl-4-isopropylcyclohexane		$C_{10}H_{20}$	6069-98-3	140.266	liq	-89.9	172	0.8039^{20}	1.4431^{20}	i H_2O; vs EtOH, eth; s bz, peth
7405	*trans*-1-Methyl-4-isopropylcyclohexane	*trans-p*-Menthane	$C_{10}H_{20}$	1678-82-6	140.266	oil	-86.3	170.6	0.7928^{20}	1.4366^{20}	vs bz, eth, EtOH, lig
7406	1-Methyl-4-isopropylcyclohexanol		$C_{10}H_{20}O$	21129-27-1	156.265			208.5	0.90^{20}	1.4619^{20}	
7407	5-Methyl-2-isopropylcyclohexanol, [1*S*-(1α,2β,5α)]-	(+)-Menthol	$C_{10}H_{20}O$	15356-60-2	156.265		39	103^{9}			vs ace, bz, eth, EtOH
7408	5-Methyl-2-isopropylcyclohexanol, [1*R*-(1α,2β,5α)]-	(-)-Menthol	$C_{10}H_{20}O$	2216-51-5	156.265	nd (MeOH)	43	216	0.903^{15}	1.460^{22}	sl H_2O; vs EtOH, eth, ace, bz; s peth
7409	5-Methyl-2-isopropylcyclohexanol, [1*S*-(1α,2α,5β)]-	(+)-Neomenthol	$C_{10}H_{20}O$	2216-52-6	156.265	oil	-22	211.7	0.897^{22}	1.4600^{20}	vs ace, EtOH
7410	5-Methyl-2-isopropylcyclohexanol, [1*S*-(1α,2β,5β)]-	(+)-Isomenthol	$C_{10}H_{20}O$	23283-97-8	156.265	nd(dil al)	82.5	218			vs eth, EtOH
7411	5-Methyl-2-isopropylcyclohexanol acetate, [1*R*-(1α,2α,5β)]		$C_{12}H_{22}O_2$	2623-23-6	198.302			222; 109^{10}	0.9244^{20}	1.4469^{20}	
7412	*cis*-5-Methyl-2-isopropylcyclohexanone	Menthone	$C_{10}H_{18}O$	491-07-6	154.249			205; 89^{15}	0.8995^{20}	1.4527^{20}	
7413	*trans*-5-Methyl-2-isopropylcyclohexanone, (2*S*)	*l*-Menthone	$C_{10}H_{18}O$	14073-97-3	154.249	liq	-6	207	0.8954^{20}	1.4505^{20}	sl H_2O; msc EtOH, eth, bz, CS_2; s ace
7414	1-Methyl-4-isopropylcyclohexene		$C_{10}H_{18}$	5502-88-5	138.250			174.5	0.8457^{15}	1.4735^{20}	
7415	3-Methyl-6-isopropyl-2-cyclohexen-1-ol		$C_{10}H_{18}O$	491-04-3	154.249			$97^{15.5}$	0.9119^{25}	1.4729^{25}	
7416	4-Methyl-1-isopropyl-3-cyclohexen-1-ol		$C_{10}H_{18}O$	562-74-3	154.249			209	0.926^{20}	1.4785^{19}	
7417	5-Methyl-2-isopropylcyclohexyl ethoxyacetate, (1α,2β,5α)		$C_{14}H_{26}O_3$	579-94-2	242.354			155^{20}, 144^{14}	0.9545^{20}		vs eth, EtOH, chl
7418	1-Methyl-4-isopropyl-2-nitrobenzene		$C_{10}H_{13}NO_2$	943-15-7	179.216			126^{10}	1.0744^{20}	1.5301^{20}	vs eth, EtOH
7419	1-Methyl-4-isopropyl-7-oxabicyclo[2.2.1]heptane		$C_{10}H_{18}O$	470-67-7	154.249		1	173.5	0.8997^{20}	1.4562^{20}	sl H_2O; msc EtOH, eth; s bz, lig
7420	1-Methyl-7-isopropylphenanthrene	Retene	$C_{18}H_{18}$	483-65-8	234.336		101	390	1.035^{25}		i H_2O; s EtOH, eth, bz, CS_2, HOAc
7421	4-Methyl-2-isopropylphenol		$C_{10}H_{14}O$	4427-56-9	150.217	nd (HOAc)	36.5	228.5	0.9910^{20}	1.5275^{20}	sl H_2O; s EtOH, bz, chl
7422	5-Methyl-2-isopropylphenyl acetate	Thymol, acetate	$C_{12}H_{16}O_2$	528-79-0	192.254			245	1.009^{9}		vs bz, eth, EtOH, chl
7423	1-Methylisoquinoline	Isoquinaldine	$C_{10}H_9N$	1721-93-3	143.185		10	248	1.0777^{20}	1.6095^{20}	sl H_2O; s eth, ace, bz
7424	3-Methylisoquinoline		$C_{10}H_9N$	1125-80-0	143.185	cry (eth)	68	249			sl H_2O, chl; s eth, ace
7425	Methyl isothiocyanate		C_2H_3NS	556-61-6	73.117		36	119	1.0691^{37}	1.5258	sl H_2O; msc EtOH; vs eth
7426	5-Methyl-3-isoxazolamine		$C_4H_6N_2O$	1072-67-9	98.103		62				
7427	4-Methylisoxazole		C_4H_5NO	6454-84-8	83.089	liq		127			
7428	5-Methylisoxazole		C_4H_5NO	5765-44-6	83.089			122	1.023^{20}	1.4386^{20}	s DMSO
7429	Methyl lactate, (±)	Methyl 2-hydroxypropanoate, (±)	$C_4H_8O_3$	2155-30-8	104.105	oil		144.8	1.0928^{20}	1.4141^{20}	vs H_2O, eth, EtOH
7430	Methyl linoleate		$C_{19}H_{34}O_2$	112-63-0	294.472		-35	215^{20}	0.8886^{10}	1.4638^{20}	vs eth, EtOH
7431	Methyl linolenate		$C_{19}H_{32}O_2$	301-00-8	292.456		-45.5	207^{14}, 182^{3}	0.895^{25}	1.4709^{20}	
7432	Methyl magnesium bromide	Bromomethylmagnesium	CH_3BrMg	75-16-1	119.244						s eth, thf; i hx, bz
7433	Methylmagnesium chloride	Chloromethylmagnesium	CH_3ClMg	676-58-4	74.793	stab in thf soln					i peth, bz
7434	Methylmalonic acid		$C_4H_6O_4$	516-05-2	118.089	nd (bz-AcOEt) pr (eth-bz)	135 dec		1.455^{20}		vs H_2O, EtOH, eth; sl bz, tfa; s AcOEt
7435	Methyl mercaptoacetate		$C_3H_6O_2S$	2365-48-2	106.144			42^{10}		1.4657^{20}	vs eth, EtOH
7436	Methyl 3-mercaptopropanoate		$C_4H_8O_2S$	2935-90-2	120.171			54^{14}	1.085^{25}	1.4640^{20}	
7437	Methylmercuric dicyanamide	1-Cyano-3-(methylmercurio) guanidine	$C_3H_6HgN_4$	502-39-6	298.70		157				
7438	Methyl methacrylate		$C_5H_8O_2$	80-62-6	100.117	liq	-47.55	100.5	0.9377^{25}	1.4142^{20}	sl H_2O; msc EtOH, eth, ace; s chl
7439	Methyl methanesulfonate		$C_2H_6O_3S$	66-27-3	110.132		20	202.5	1.2943^{20}	1.4138^{20}	
7440	Methyl methoxyacetate		$C_4H_8O_3$	6290-49-9	104.105			131	1.0511^{20}	1.3962^{20}	sl H_2O; vs EtOH, eth, ace

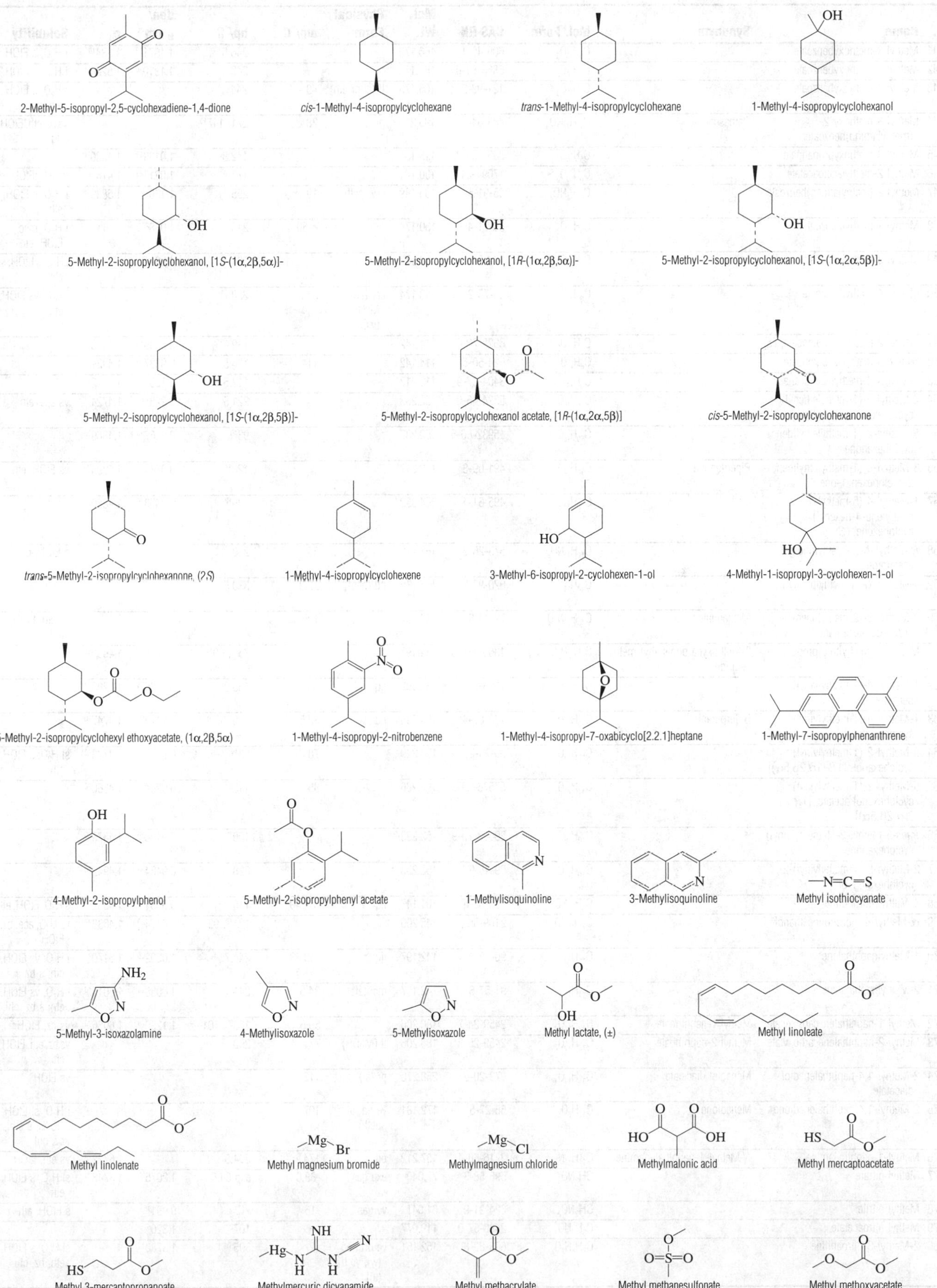

2-Methyl-5-isopropyl-2,5-cyclohexadiene-1,4-dione
cis-1-Methyl-4-isopropylcyclohexane
trans-1-Methyl-4-isopropylcyclohexane
1-Methyl-4-isopropylcyclohexanol
5-Methyl-2-isopropylcyclohexanol, [1S-(1α,2β,5α)]-
5-Methyl-2-isopropylcyclohexanol, [1R-(1α,2β,5α)]-
5-Methyl-2-isopropylcyclohexanol, [1S-(1α,2α,5β)]-
5-Methyl-2-isopropylcyclohexanol, [1S-(1α,2β,5β)]-
5-Methyl-2-isopropylcyclohexanol acetate, [1R-(1α,2α,5β)]
cis-5-Methyl-2-isopropylcyclohexanone
trans-5-Methyl-2-isopropylcyclohexanone, (2S)
1-Methyl-4-isopropylcyclohexene
3-Methyl-6-isopropyl-2-cyclohexen-1-ol
4-Methyl-1-isopropyl-3-cyclohexen-1-ol
5-Methyl-2-isopropylcyclohexyl ethoxyacetate, (1α,2β,5α)
1-Methyl-4-isopropyl-2-nitrobenzene
1-Methyl-4-isopropyl-7-oxabicyclo[2.2.1]heptane
1-Methyl-7-isopropylphenanthrene
4-Methyl-2-isopropylphenol
5-Methyl-2-isopropylphenyl acetate
1-Methylisoquinoline
3-Methylisoquinoline
Methyl isothiocyanate
5-Methyl-3-isoxazolamine
4-Methylisoxazole
5-Methylisoxazole
Methyl lactate, (±)
Methyl linoleate
Methyl linolenate
Methyl magnesium bromide
Methylmagnesium chloride
Methylmalonic acid
Methyl mercaptoacetate
Methyl 3-mercaptopropanoate
Methylmercuric dicyanamide
Methyl methacrylate
Methyl methanesulfonate
Methyl methoxyacetate

No.	Name	Synonym	Mol. Form.	CAS RN	Mol. Wt.	Physical Form	mp/°C	bp/°C	den/ g cm^{-3}	n_D	Solubility
7441	Methyl 2-methoxybenzoate		$C_9H_{10}O_3$	606-45-1	166.173			246.5	1.1571^{19}	1.534^{19}	i H_2O; s EtOH
7442	Methyl 3-methoxybenzoate		$C_9H_{10}O_3$	5368-81-0	166.173			248	1.1310^{20}	1.5224^{20}	i H_2O; s EtOH
7443	Methyl 4-methoxybenzoate		$C_9H_{10}O_3$	121-98-2	166.173	lf (al or eth)	49	244			i H_2O; s EtOH, eth, chl
7444	Methyl 3-methoxy-2-(methylamino)benzoate	Damascenine	$C_{10}H_{13}NO_3$	483-64-7	195.215	pr (al)	28	271; 147^{10}			vs bz, eth, EtOH, lig
7445	Methyl 3-methoxypropanoate		$C_5H_{10}O_3$	3852-09-3	118.131			142.8	1.0139^{15}	1.4030^{20}	
7446	Methyl 2-methylacetoacetate		$C_6H_{10}O_3$	17094-21-2	130.141			177.4	1.0217^{25}	1.416^{24}	vs eth, EtOH
7447	Methyl 2-(methylamino)benzoate		$C_9H_{11}NO_2$	85-91-6	165.189	cry (peth)	19	255	1.120^{15}	1.5839^{15}	i H_2O; s EtOH, eth
7448	Methyl 2-methylbenzoate		$C_9H_{10}O_2$	89-71-4	150.174		<-50	215	1.068^{20}		i H_2O; msc EtOH, eth
7449	Methyl 3-methylbenzoate		$C_9H_{10}O_2$	99-36-5	150.174			221	1.061^{20}		i H_2O; s EtOH; sl ctc
7450	Methyl 4-methylbenzoate		$C_9H_{10}O_2$	99-75-2	150.174	cry (aq MeOH, peth)	33.2	220			i H_2O; vs EtOH, eth
7451	Methyl 2-methyl-2-butenoate, (*E*)		$C_6H_{10}O_2$	6622-76-0	114.142			139	0.9349^{12}	1.4370^{20}	
7452	Methyl 3-methyl-2-butenoate		$C_6H_{10}O_2$	924-50-5	114.142		114	136.5	0.9337^{20}	1.432^{20}	
7453	3-Methyl-4-methylenehexane		C_8H_{16}	3404-67-9	112.213			112.5	0.725^{25}	1.4142^{20}	
7454	2-Methyl-5-(1-methylethenyl) cyclohexanone, (2*R*-*trans*)		$C_{10}H_{16}O$	5524-05-0	152.233			221.5	0.928^{19}	1.4724	vs ace, eth
7455	5-Methyl-2-(1-methylethylidene) cyclohexanone		$C_{10}H_{16}O$	15932-80-6	152.233			93^{10}	0.9367^{20}	1.4869^{20}	
7456	3-Methyl-6-(1-methylethylidene)-2-cyclohexen-1-one	Piperitenone	$C_{10}H_{14}O$	491-09-8	150.217			120^{14}	0.9774^{20}	1.5294^{20}	vs EtOH, eth
7457	1-Methyl-4-(5-methyl-1-methylene-4-hexenyl) cyclohexene, (*S*)		$C_{15}H_{24}$	495-61-4	204.352			129^{10}	0.8673^{20}	1.4880^{20}	
7458	*N*-Methyl-*N*-(2-methylphenyl) acetamide		$C_{10}H_{13}NO$	573-26-2	163.216		55.5	260			s EtOH, chl
7459	4-Methyl-*N*-(4-methylphenyl) aniline		$C_{14}H_{15}N$	620-93-9	197.276	nd (peth)	79.8	330.5			vs eth, peth
7460	2-Methyl-3-(2-methylphenyl)-4(3*H*)-quinazolinone	Methaqualone	$C_{16}H_{14}N_2O$	72-44-6	250.294		120				vs eth, EtOH, chl
7461	Methyl 3-(methylthio)propanoate	2-Methoxycarbonylethyl methyl sulfide	$C_5H_{10}O_2S$	13532-18-8	134.197			75^{13}, 69^{11}	1.077^{25}	1.4650^{20}	
7462	1-Methyl-4-(1-methylvinyl) benzene		$C_{10}H_{12}$	1195-32-0	132.202	liq	-20	185.3	0.8936^{23}	1.5283^{23}	
7463	1-Methyl-4-(1-methylvinyl) cyclohexanol	β-Terpineol	$C_{10}H_{18}O$	138-87-4	154.249	nd	32.5	210; 90^{10}	0.917^{20}	1.4747^{20}	
7464	5-Methyl-2-(1-methylvinyl) cyclohexanol, [1*R*-(1α,2β,5α)]		$C_{10}H_{18}O$	89-79-2	154.249		78	93^{14}	0.911^{20}	1.4723^{20}	sl H_2O; s EtOH, eth
7465	5-Methyl-2-(1-methylvinyl) cyclohexanol acetate, [1*R*-(1α,2β,5α)]		$C_{12}H_{20}O_2$	57576-09-7	196.286		85	113^{8}	0.925^{25}	1.4566^{20}	
7466	*trans*-5-Methyl-2-(1-methylvinyl) cyclohexanone		$C_{10}H_{16}O$	29606-79-9	152.233			100^{18}	0.9198^{20}	1.4675^{20}	
7467	2-Methyl-5-(1-methylvinyl)-2-cyclohexen-1-ol		$C_{10}H_{16}O$	99-48-9	152.233			228	0.9484^{25}	1.4942^{25}	
7468	4-Methylmorpholine		$C_5H_{11}NO$	109-02-4	101.147	liq	-64.40	116	0.9051^{20}	1.4332^{20}	s H_2O, EtOH, eth
7469	α-Methyl-4-morpholineethanol		$C_7H_{15}NO_2$	2109-66-2	145.200			121^{18}, 93^{13}	1.0174^{20}	1.4638^{20}	vs H_2O, ace, bz, EtOH
7470	1-Methylnaphthalene		$C_{11}H_{10}$	90-12-0	142.197	liq	-30.43	244.7	1.0202^{20}	1.6170^{20}	i H_2O; vs EtOH, eth; s bz
7471	2-Methylnaphthalene		$C_{11}H_{10}$	91-57-6	142.197	mcl (al)	34.6	241.1	1.0058^{20}	1.6015^{40}	i H_2O; vs EtOH, eth; s bz, chl
7472	Methyl 1-naphthalenecarboxylate	Methyl 1-naphthoate	$C_{12}H_{10}O_2$	2459-24-7	186.206		59.5	168^{20}, 101$^{0.04}$	1.1290^{20}	1.6086^{20}	vs bz, EtOH
7473	Methyl 2-naphthalenecarboxylate	Methyl 2-naphthoate	$C_{12}H_{10}O_2$	2459-25-8	186.206	lf (MeOH)	77	290			vs bz, eth, EtOH, chl
7474	2-Methyl-1,4-naphthalenediol diacetate	Menadiol diacetate	$C_{15}H_{14}O_4$	573-20-6	258.270	pr (al)	113				vs EtOH
7475	2-Methyl-1,4-naphthalenedione	Menadione	$C_{11}H_8O_2$	58-27-5	172.181	ye nd (al, peth)	107				i H_2O; sl EtOH, HOAc; s eth, bz, chl
7476	Methyl-1-naphthylamine	*N*-Methyl-1-naphthalenamine	$C_{11}H_{11}N$	2216-68-4	157.212	oil	174	294.5		1.6722^{20}	vs eth, EtOH
7477	Methyl nitrate		CH_3NO_3	598-58-3	77.040	exp gas	-83.0	exp 64.6	1.2075^{20}	1.3748^{20}	sl H_2O; s EtOH, eth
7478	Methyl nitrite		CH_3NO_2	624-91-9	61.041	ye gas	-16	-12	0.991^{15}		s EtOH, eth
7479	Methyl nitroacetate		$C_3H_5NO_4$	2483-57-0	119.077			107^{28}	1.320^{0}		
7480	2-Methyl-3-nitroaniline		$C_7H_8N_2O_2$	603-83-8	152.151	ye orth nd (w), ye lf (al)	92	305	1.3780^{15}		sl H_2O; s EtOH, eth, bz, chl

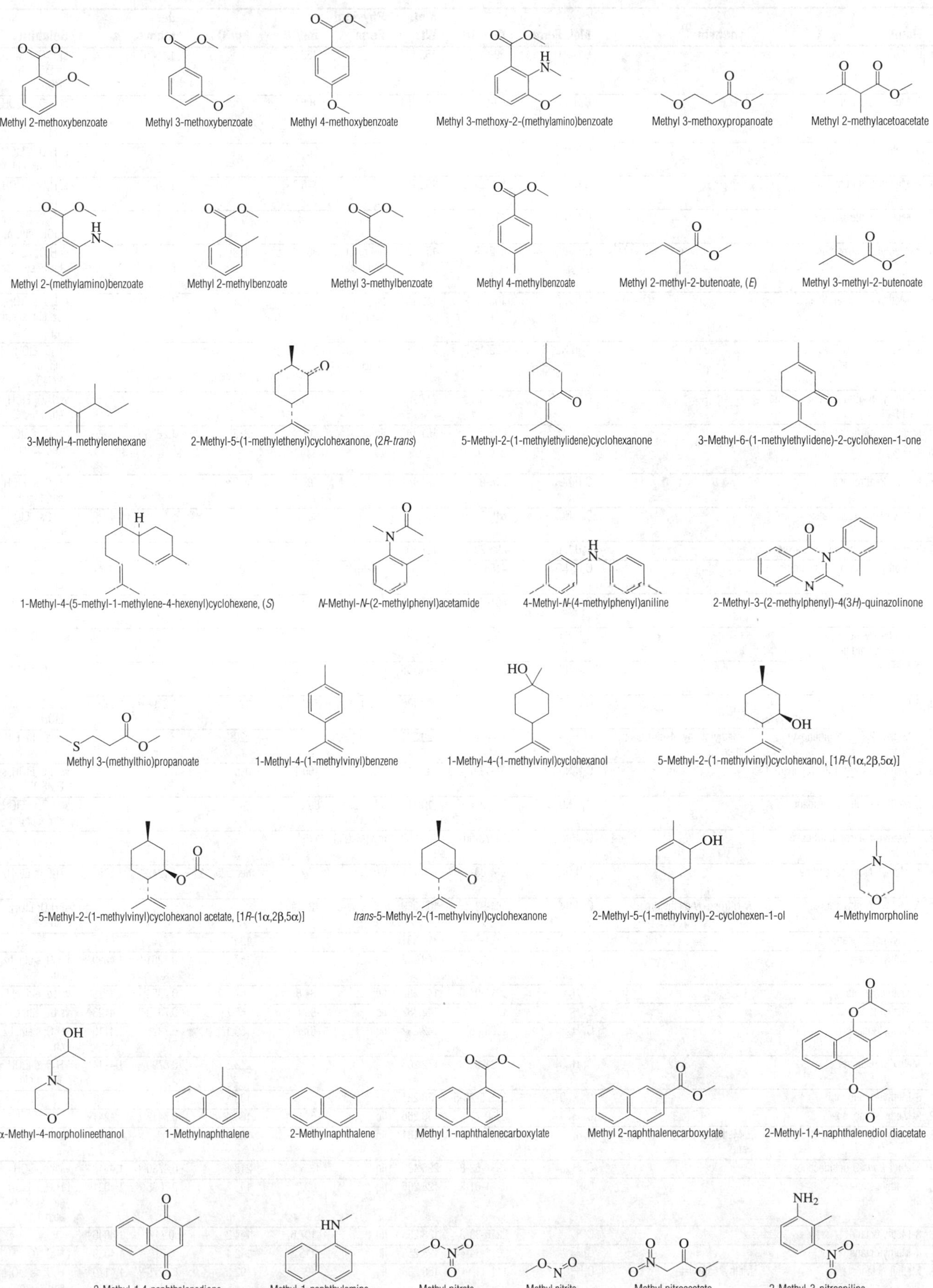

Methyl 2-methoxybenzoate
Methyl 3-methoxybenzoate
Methyl 4-methoxybenzoate
Methyl 3-methoxy-2-(methylamino)benzoate
Methyl 3-methoxypropanoate
Methyl 2-methylacetoacetate
Methyl 2-(methylamino)benzoate
Methyl 2-methylbenzoate
Methyl 3-methylbenzoate
Methyl 4-methylbenzoate
Methyl 2-methyl-2-butenoate, (E)
Methyl 3-methyl-2-butenoate
3-Methyl-4-methylenehexane
2-Methyl-5-(1-methylethenyl)cyclohexanone, (2R-trans)
5-Methyl-2-(1-methylethylidene)cyclohexanone
3-Methyl-6-(1-methylethylidene)-2-cyclohexen-1-one
1-Methyl-4-(5-methyl-1-methylene-4-hexenyl)cyclohexene, (S)
N-Methyl-N-(2-methylphenyl)acetamide
4-Methyl-N-(4-methylphenyl)aniline
2-Methyl-3-(2-methylphenyl)-4(3H)-quinazolinone
Methyl 3-(methylthio)propanoate
1-Methyl-4-(1-methylvinyl)benzene
1-Methyl-4-(1-methylvinyl)cyclohexanol
5-Methyl-2-(1-methylvinyl)cyclohexanol, [1R-(1α,2β,5α)]
5-Methyl-2-(1-methylvinyl)cyclohexanol acetate, [1R-(1α,2β,5α)]
trans-5-Methyl-2-(1-methylvinyl)cyclohexanone
2-Methyl-5-(1-methylvinyl)-2-cyclohexen-1-ol
4-Methylmorpholine
α-Methyl-4-morpholineethanol
1-Methylnaphthalene
2-Methylnaphthalene
Methyl 1-naphthalenecarboxylate
Methyl 2-naphthalenecarboxylate
2-Methyl-1,4-naphthalenediol diacetate
2-Methyl-1,4-naphthalenedione
Methyl-1-naphthylamine
Methyl nitrate
Methyl nitrite
Methyl nitroacetate
2-Methyl-3-nitroaniline

No.	Name	Synonym	Mol. Form.	CAS RN	Mol. Wt.	Physical Form	mp/°C	bp/°C	den/ g cm⁻³	n_D	Solubility
7481	2-Methyl-4-nitroaniline		$C_7H_8N_2O_2$	99-52-5	152.151		133.5		1.1586^{140}		sl H_2O, DMSO; s EtOH, bz, HOAc
7482	2-Methyl-5-nitroaniline		$C_7H_8N_2O_2$	99-55-8	152.151		105.5				sl H_2O; s EtOH, eth, ace, bz, chl
7483	2-Methyl-6-nitroaniline		$C_7H_8N_2O_2$	570-24-1	152.151		96		1.1900^{100}		sl H_2O; s EtOH, eth, bz, chl
7484	4-Methyl-2-nitroaniline		$C_7H_8N_2O_2$	89-62-3	152.151		116.3		1.16^{121}		sl H_2O; s EtOH, chl
7485	4-Methyl-3-nitroaniline		$C_7H_8N_2O_2$	119-32-4	152.151		79.8				sl H_2O, CS_2; s EtOH, eth, bz
7486	*N*-Methyl-2-nitroaniline		$C_7H_8N_2O_2$	612-28-2	152.151	red or oran nd (peth)	38	158^{18}			sl H_2O, lig; s EtOH, eth, ace, bz
7487	*N*-Methyl-4-nitroaniline		$C_7H_8N_2O_2$	100-15-2	152.151	br-ye pr (al) cry (eth)	152	dec	1.201^{155}		i H_2O; s EtOH, bz, chl; sl eth, lig
7488	2-Methyl-1-nitro-9,10-anthracenedione		$C_{15}H_9NO_4$	129-15-7	267.237	pa ye nd (HOAc)	273.0				i H_2O, EtOH; sl eth, bz, chl; s $PhNO_2$
7489	2-Methyl-5-nitrobenzenesulfonic acid		$C_7H_7NO_5S$	121-03-9	217.200		135.8				vs H_2O, EtOH, eth, chl
7490	Methyl 2-nitrobenzoate		$C_8H_7NO_4$	606-27-9	181.147	liq	-13	275	1.2855^{20}		i H_2O; s EtOH, eth, bz, chl; i lig
7491	Methyl 3-nitrobenzoate		$C_8H_7NO_4$	618-95-1	181.147		78	279^{60}			i H_2O; sl EtOH, eth, MeOH
7492	Methyl 4-nitrobenzoate		$C_8H_7NO_4$	619-50-1	181.147		96				i H_2O; s EtOH, eth, chl
7493	2-Methyl-4-nitro-1*H*-imidazole		$C_4H_5N_3O_2$	696-23-1	127.102		253				
7494	*N*-Methyl-*N*-nitromethanamine		$C_2H_6N_2O_2$	4164-28-7	90.081	nd(eth)	58	187	1.1090^{72}	1.4462^{72}	vs H_2O, ace, eth, EtOH
7495	2-Methyl-1-nitronaphthalene		$C_{11}H_9NO_2$	881-03-8	187.195	ye pr or nd (al)	81.5	188^{20}			i H_2O; s EtOH; vs ace
7496	*N*-Methyl-*N'*-nitro-*N*-nitrosoguanidine		$C_2H_5N_5O_3$	70-25-7	147.093						s DMSO
7497	3-Methyl-4-nitrophenol		$C_7H_7NO_3$	2581-34-2	153.136	nd or pr (w)	129				sl H_2O; s EtOH, eth, bz, chl
7498	4-Methyl-2-nitrophenol		$C_7H_7NO_3$	119-33-5	153.136	ye nd (al, w)	36.5	125^{22}	1.2399^{20}	1.5744^{40}	vs ace, bz, eth, EtOH
7499	1-Methyl-2-(4-nitrophenoxy) benzene	2-Methylphenyl 4-nitrophenyl ether	$C_{13}H_{11}NO_3$	2444-29-3	229.231	ye cry (peth)		220^{27}			vs bz, eth, EtOH
7500	2-Methyl-2-nitro-1,3-propanediol		$C_4H_9NO_4$	77-49-6	135.119	mcl	150.1	dec			vs H_2O, EtOH; sl DMSO
7501	2-Methyl-2-nitro-1-propanol		$C_4H_9NO_3$	76-39-1	119.119	nd or pl (MeOH)	89.5	94^{10}			sl H_2O; vs EtOH, eth; s chl
7502	3-Methyl-4-nitroquinoline-*N*-oxide		$C_{10}H_8N_2O_3$	14073-00-8	204.182	cry (MeOH)	179				
7503	*N*-Methyl-*N*-nitrosoaniline		$C_7H_8N_2O$	614-00-6	136.151	ye cry	14.7	dec 225; 121^{13}	1.1240^{20}	1.5769^{20}	i H_2O; s EtOH, eth
7504	*N*-Methyl-*N*-nitrosourea	*N*-Nitroso-*N*-methylurea	$C_2H_5N_3O_2$	684-93-5	103.080	col or ye pl (eth)	123 dec				sl H_2O, EtOH, eth
7505	Methyl nonadecanoate		$C_{20}H_{40}O_2$	1731-94-8	312.531		41.3	190^4			
7506	2-Methylnonane		$C_{10}H_{22}$	871-83-0	142.282	liq	-74.6	167.1	0.7281^{20}	1.4099^{20}	i H_2O; s eth, bz, chl
7507	3-Methylnonane		$C_{10}H_{22}$	5911-04-6	142.282	liq	-84.8	167.9	0.7354^{20}	1.4125^{20}	vs bz, eth, chl
7508	4-Methylnonane		$C_{10}H_{22}$	17301-94-9	142.282	liq	-99	165.7	0.7323^{20}	1.4123^{20}	vs bz, eth, chl
7509	5-Methylnonane		$C_{10}H_{22}$	15869-85-9	142.282	liq	-87.7	165.1	0.7326^{20}	1.4116^{20}	i H_2O; s eth, bz, chl
7510	Methyl nonanoate		$C_{10}H_{20}O_2$	1731-84-6	172.265			213.5	0.8799^{15}	1.4214^{20}	i H_2O; s EtOH, eth; sl ctc
7511	8-Methyl-1-nonanol		$C_{10}H_{22}O$	55505-26-5	158.281			108^{10}			
7512	2-Methyl-1-nonene		$C_{10}H_{20}$	2980-71-4	140.266	liq	-64.2	168.4	0.7412^{25}	1.4241^{20}	
7513	2-Methyl-2-norbornene	2-Methylbicyclo[2.2.1]hept-2-ene	C_8H_{12}	694-92-8	108.181	liq		122			
7514	Methyl *trans*-9-octadecenoate		$C_{19}H_{36}O_2$	1937-62-8	296.488	liq	13.5	218^{24}	0.8730^{20}	1.4513^{20}	vs eth, EtOH
7515	2-Methyloctane		C_9H_{20}	3221-61-2	128.255	liq	-80.3	143.2	0.7095^{25}	1.4031^{20}	i H_2O; s EtOH, eth; sl ctc; vs peth
7516	3-Methyloctane		C_9H_{20}	2216-33-3	128.255	liq	-107.6	144.2	0.717^{25}	1.4040^{25}	
7517	4-Methyloctane		C_9H_{20}	2216-34-4	128.255	liq	-113.3	142.4	0.716^{25}	1.4039^{25}	i H_2O
7518	Methyl octanoate	Methyl caprylate	$C_9H_{18}O_2$	111-11-5	158.238	liq	-40	192.9	0.8775^{20}	1.4170^{20}	i H_2O; vs EtOH, eth; sl ctc
7519	2-Methyloctanoic acid		$C_9H_{18}O_2$	3004-93-1	158.238			138^{14}, 88^4		1.4281^{25}	

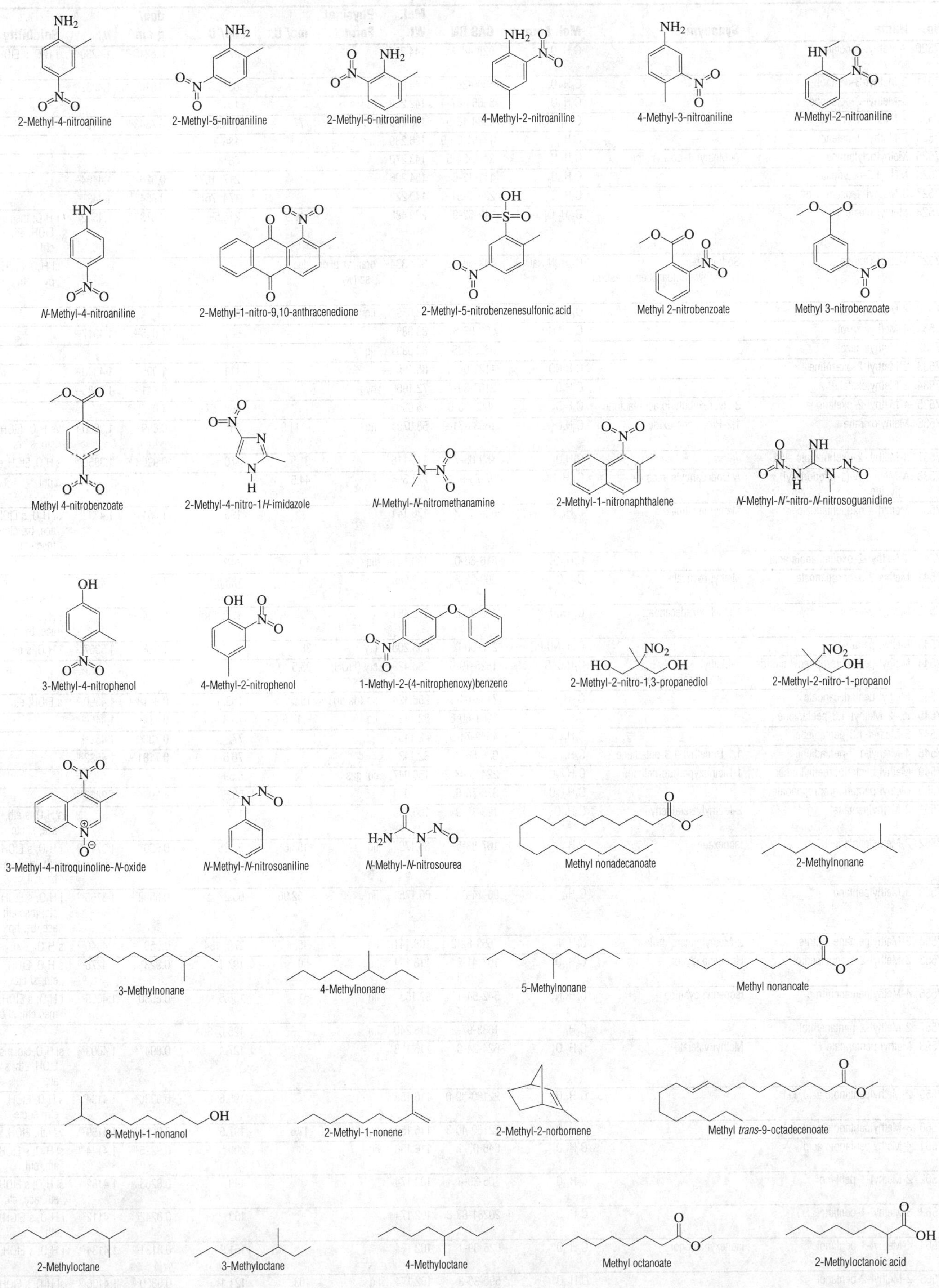
2-Methyl-4-nitroaniline
2-Methyl-5-nitroaniline
2-Methyl-6-nitroaniline
4-Methyl-2-nitroaniline
4-Methyl-3-nitroaniline
N-Methyl-2-nitroaniline
N-Methyl-4-nitroaniline
2-Methyl-1-nitro-9,10-anthracenedione
2-Methyl-5-nitrobenzenesulfonic acid
Methyl 2-nitrobenzoate
Methyl 3-nitrobenzoate
Methyl 4-nitrobenzoate
2-Methyl-4-nitro-1H-imidazole
N-Methyl-N-nitromethanamine
2-Methyl-1-nitronaphthalene
N-Methyl-N′-nitro-N-nitrosoguanidine
3-Methyl-4-nitrophenol
4-Methyl-2-nitrophenol
1-Methyl-2-(4-nitrophenoxy)benzene
2-Methyl-2-nitro-1,3-propanediol
2-Methyl-2-nitro-1-propanol
3-Methyl-4-nitroquinoline-N-oxide
N-Methyl-N-nitrosoaniline
N-Methyl-N-nitrosourea
Methyl nonadecanoate
2-Methylnonane
3-Methylnonane
4-Methylnonane
5-Methylnonane
Methyl nonanoate
8-Methyl-1-nonanol
2-Methyl-1-nonene
2-Methyl-2-norbornene
Methyl trans-9-octadecenoate
2-Methyloctane
3-Methyloctane
4-Methyloctane
Methyl octanoate
2-Methyloctanoic acid

No.	Name	Synonym	Mol. Form.	CAS RN	Mol. Wt.	Physical Form	mp/°C	bp/°C	den/ g cm^{-3}	n_D	Solubility
7520	2-Methyl-2-octanol		$C_9H_{20}O$	628-44-4	144.254			178	0.8210^{20}	1.4280^{20}	i H_2O; s EtOH, eth
7521	3-Methyl-3-octanol		$C_9H_{20}O$	5340-36-3	144.254			83^{18}, 36^{3}	0.8108^{25}	1.4257^{25}	
7522	5-Methyl-2-octanone		$C_9H_{18}O$	58654-67-4	142.238			101^{50}			
7523	2-Methyl-1-octene		C_9H_{18}	4588-18-5	126.239	liq	-77.8	144.8	0.7343^{20}	1.4184^{20}	
7524	7-Methyl-1-octene		C_9H_{18}	13151-06-9	126.239	liq		138.9			
7525	Methyloctylamine	*N*-Methyl-1-octanamine	$C_9H_{21}N$	2439-54-5	143.270			68^{8}			
7526	Methyl 2-octynoate		$C_9H_{14}O_2$	111-12-6	154.206			217; 107^{20}	0.926^{20}	1.4464^{20}	
7527	3-Methyl-1-octyn-3-ol		$C_9H_{16}O$	23580-51-0	140.222			174; 75^{10}	0.8547^{20}	1.443^{10}	
7528	Methyl oleate		$C_{19}H_{36}O_2$	112-62-9	296.488		-19.9	218.5^{20}	0.8739^{20}	1.4522^{20}	i H_2O; msc EtOH, eth; s chl
7529	Methyl Orange	Sodium *p*-dimethylaminoazobenzenesulfonate	$C_{14}H_{14}N_3NaO_3S$	547-58-0	327.334	oran, ye pl or sc (w)	dec				sl H_2O, EtOH, py; i eth
7530	2-Methyloxazole		C_4H_5NO	23012-10-4	83.089	liq		87.5			
7531	4-Methyloxazole		C_4H_5NO	693-93-6	83.089			88	1.015^{25}	1.4317^{20}	
7532	5-Methyloxazole		C_4H_5NO	66333-88-8	83.089	liq		88			
7533	2-Methyl-2-oxazoline		C_4H_7NO	1120-64-5	85.105			111	1.005^{25}	1.4340^{20}	
7534	2-Methyloxetane		C_4H_8O	2167-39-7	72.106	hyg		59	0.841^{25}	1.3885^{20}	
7535	4-Methyl-2-oxetanone	3-Hydroxybutyric acid lactone	$C_4H_6O_2$	3068-88-0	86.090			86^{50}, 57^{9}	1.0555^{20}		
7536	Methyloxirane	1,2-Propylene oxide	C_3H_6O	16033-71-9	58.079	liq	-111.9	35	0.859^{0}	1.3660^{20}	vs H_2O, EtOH, eth; s chl
7537	3-Methyl-2-oxobutanoic acid		$C_5H_8O_3$	759-05-7	116.116		31.5	170.5	0.9968^{20}	1.3850^{16}	s H_2O, EtOH, eth
7538	*N*-Methyl-*N*-(1-oxododecyl) glycine	*N*-Dodecanoylsarcosine	$C_{15}H_{29}NO_3$	97-78-9	271.396		44.5				s chl
7539	Methyl 4-oxopentanoate	Methyl levulinate	$C_6H_{10}O_3$	624-45-3	130.141			196	1.0511^{20}	1.4233^{20}	sl H_2O; s EtOH, ace, bz, ctc; msc eth
7540	4-Methyl-2-oxopentanoic acid		$C_6H_{10}O_3$	816-66-0	130.141	liq	10	84^{15}			
7541	Methyl 2-oxopropanoate	Methyl pyruvate	$C_4H_6O_3$	600-22-6	102.089			135.5	1.154^{0}	1.4046^{25}	sl H_2O; s ace; msc EtOH, eth
7542	Methyl palmitate	Methyl hexadecanoate	$C_{17}H_{34}O_2$	112-39-0	270.451		30	417; 148^{2}	0.8247^{75}		i H_2O; vs EtOH, ace, bz; s eth
7543	Methyl parathion		$C_8H_{10}NO_5PS$	298-00-0	263.208	cry	38		1.358^{20}	1.5367^{25}	i H_2O; s os
7544	Methyl pentachlorophenyl sulfide	*S*-Methyl pentachlorobenzenethiol	$C_7H_3Cl_5S$	1825-19-0	296.429	cry (EtOH)	95.5				
7545	Methyl pentadecanoate		$C_{16}H_{32}O_2$	7132-64-1	256.424	nd (dil al)	18.5	153.5	0.8618^{25}	1.4390^{25}	s EtOH, eth
7546	*cis*-2-Methyl-1,3-pentadiene		C_6H_{10}	1501-60-6	82.143	liq	-117.6	75.8	0.714^{25}	1.446^{20}	
7547	3-Methyl-1,3-pentadiene		C_6H_{10}	4549-74-0	82.143			77	0.730^{25}	1.452^{20}	
7548	4-Methyl-1,3-pentadiene	1,1-Dimethyl-1,3-butadiene	C_6H_{10}	926-56-7	82.143			76.5	0.7181^{20}	1.4532^{20}	
7549	Methyl pentafluoroethyl ether	1-Methoxyperfluoroethane	$C_3H_3F_5O$	22410-44-2	150.047	col gas		5.59			
7550	Methyl pentafluoropropanoate		$C_4H_3F_5O_2$	378-75-6	178.058			59.5	1.390^{25}	1.2869^{25}	
7551	2-Methylpentanal	2-Methylvaleraldehyde	$C_6H_{12}O$	123-15-9	100.158			117			s H_2O; s eth, ace; sl ctc
7552	2-Methylpentane	Isohexane	C_6H_{14}	107-83-5	86.175	liq	-153.6	60.26	0.650^{25}	1.3715^{20}	i H_2O; s EtOH, eth; msc ace, bz, chl
7553	3-Methylpentane		C_6H_{14}	96-14-0	86.175	liq	-162.90	63.27	0.6598^{25}	1.3765^{20}	i H_2O; s EtOH, ctc; msc eth, ace, bz, hp
7554	2-Methylpentanedinitrile	2-Methylglutaronitrile	$C_6H_8N_2$	4553-62-2	108.141	liq	-45	270; 134^{13}	0.950	1.4340^{20}	s H_2O
7555	2-Methyl-2,4-pentanediol	Hexylene glycol	$C_6H_{14}O_2$	107-41-5	118.174	liq	-50	197.1	0.923^{15}	1.4276^{20}	s H_2O, EtOH, eth; sl ctc
7556	4-Methylpentanenitrile	Isopentyl cyanide	$C_6H_{11}N$	542-54-1	97.158	liq	-51	156.5	0.8030^{20}	1.4059^{20}	i H_2O; s EtOH; msc eth; sl ctc
7557	2-Methyl-2-pentanethiol		$C_6H_{14}S$	1633-97-2	118.240	liq		125.0; 36^{30}			
7558	Methyl pentanoate	Methyl valerate	$C_6H_{12}O_2$	624-24-8	116.158			127.4	0.8947^{20}	1.4003^{20}	sl H_2O, ctc; msc EtOH, eth; s ace
7559	2-Methylpentanoic acid, (±)		$C_6H_{12}O_2$	22160-39-0	116.158			195.6	0.9230^{20}	1.413^{20}	s H_2O, EtOH, eth; sl ctc
7560	3-Methylpentanoic acid, (±)		$C_6H_{12}O_2$	22160-40-3	116.158	liq	-41.6	197.5	0.9262^{20}	1.4159^{20}	vs eth, EtOH
7561	4-Methylpentanoic acid		$C_6H_{12}O_2$	646-07-1	116.158	liq	-33	200.5	0.9225^{20}	1.4144^{20}	sl H_2O; s EtOH, eth, chl
7562	2-Methyl-1-pentanol		$C_6H_{14}O$	105-30-6	102.174			149	0.8263^{20}	1.4182^{20}	sl H_2O; s EtOH, eth, ace, ctc
7563	3-Methyl-1-pentanol, (±)		$C_6H_{14}O$	20281-83-8	102.174			153	0.8242^{20}	1.4112^{23}	i H_2O; s EtOH, eth
7564	4-Methyl-1-pentanol	Isohexyl alcohol	$C_6H_{14}O$	626-89-1	102.174			151.9	0.8131^{20}	1.4134^{25}	i H_2O; s EtOH, eth
7565	2-Methyl-2-pentanol		$C_6H_{14}O$	590-36-3	102.174	liq	-103	121.1	0.8350^{16}	1.4100^{20}	sl H_2O; s EtOH, eth

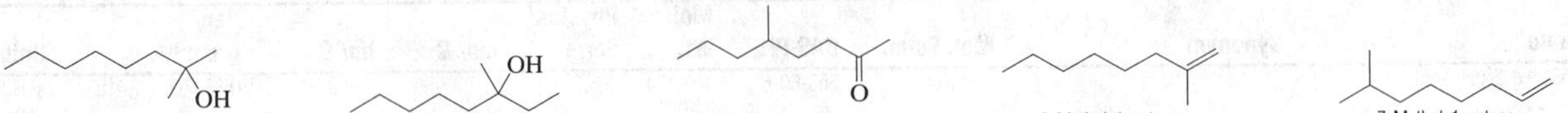

2-Methyl-2-octanol

3-Methyl-3-octanol

5-Methyl-2-octanone

2-Methyl-1-octene

7-Methyl-1-octene

Methyloctylamine

Methyl 2-octynoate

3-Methyl-1-octyn-3-ol

Methyl oleate

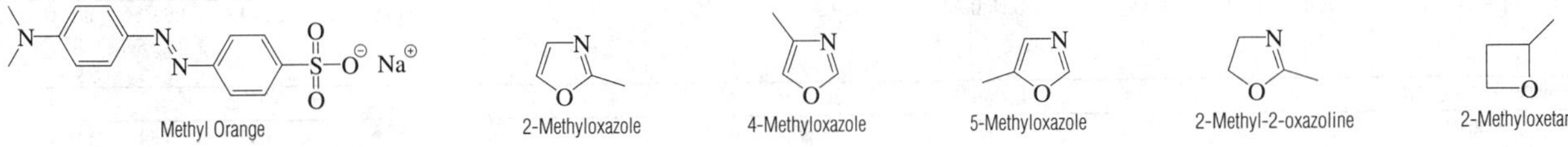

Methyl Orange

2-Methyloxazole

4-Methyloxazole

5-Methyloxazole

2-Methyl-2-oxazoline

2-Methyloxetane

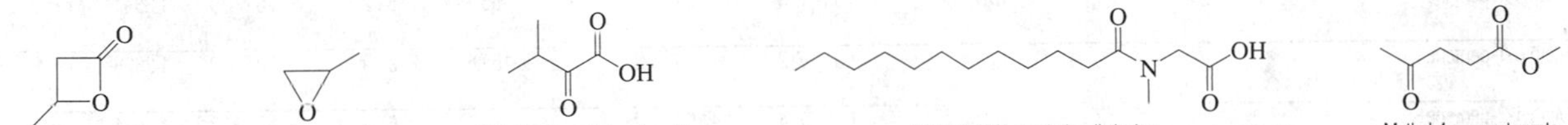

4-Methyl-2-oxetanone

Methyloxirane

3-Methyl-2-oxobutanoic acid

N-Methyl-*N*-(1-oxododecyl)glycine

Methyl 4-oxopentanoate

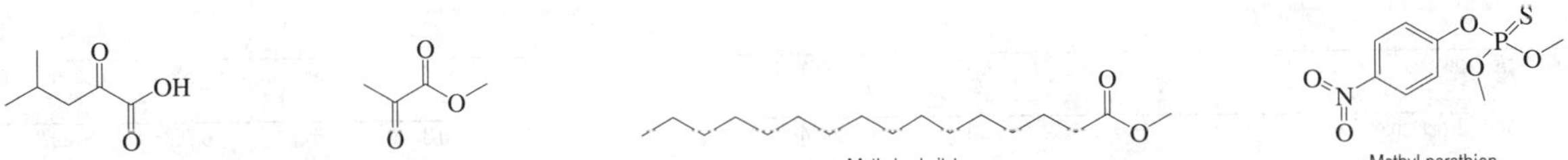

4-Methyl-2-oxopentanoic acid

Methyl 2-oxopropanoate

Methyl palmitate

Methyl parathion

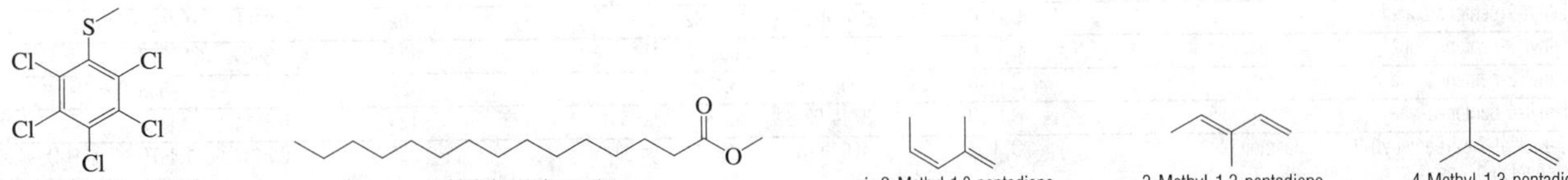

Methyl pentachlorophenyl sulfide

Methyl pentadecanoate

cis-2-Methyl-1,3-pentadiene

3-Methyl-1,3-pentadiene

4-Methyl-1,3-pentadiene

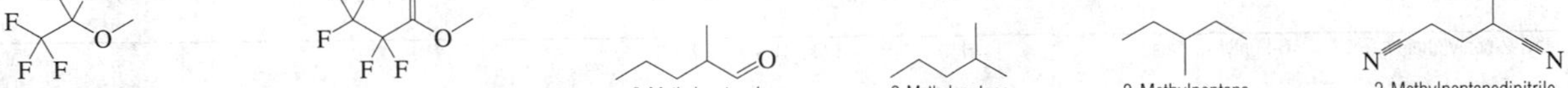

Methyl pentafluoroethyl ether

Methyl pentafluoropropanoate

2-Methylpentanal

2-Methylpentane

3-Methylpentane

2-Methylpentanedinitrile

2-Methyl-2,4-pentanediol

4-Methylpentanenitrile

2-Methyl-2-pentanethiol

Methyl pentanoate

2-Methylpentanoic acid, (±)

3-Methylpentanoic acid, (±)

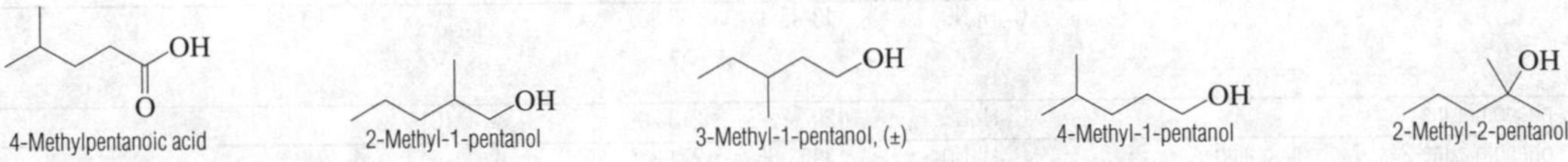

4-Methylpentanoic acid

2-Methyl-1-pentanol

3-Methyl-1-pentanol, (±)

4-Methyl-1-pentanol

2-Methyl-2-pentanol

No.	Name	Synonym	Mol. Form.	CAS RN	Mol. Wt.	Physical Form	mp/°C	bp/°C	den/ g cm^{-3}	n_D	Solubility
7566	3-Methyl-2-pentanol		$C_6H_{14}O$	565-60-6	102.174			134.3	0.8307^{20}	1.4182^{20}	sl H_2O; s EtOH, eth
7567	4-Methyl-2-pentanol		$C_6H_{14}O$	108-11-2	102.174	liq	-90	131.6	0.8075^{20}	1.4100^{20}	sl H_2O, ctc; s EtOH, eth
7568	2-Methyl-3-pentanol		$C_6H_{14}O$	565-67-3	102.174			126.5	0.8243^{20}	1.4175^{20}	sl H_2O; msc EtOH, eth
7569	3-Methyl-3-pentanol		$C_6H_{14}O$	77-74-7	102.174	liq	-23.6	122.4	0.8286^{20}	1.4186^{20}	sl H_2O, ctc; msc EtOH, eth
7570	2-Methyl-1-pentanol acetate		$C_8H_{16}O_2$	7789-99-3	144.212			163	0.870^{25}		vs eth, EtOH
7571	3-Methyl-2-pentanone, (±)	(±)-*sec*-Butyl methyl ketone	$C_6H_{12}O$	55156-16-6	100.158			117.5	0.8130^{20}	1.4002^{20}	sl H_2O; msc EtOH, eth; s chl
7572	4-Methyl-2-pentanone	Isobutyl methyl ketone	$C_6H_{12}O$	108-10-1	100.158	liq	-84	116.5	0.7965^{25}	1.3962^{20}	sl H_2O; msc EtOH, eth, ace, bz; s chl
7573	2-Methyl-3-pentanone	Ethyl isopropyl ketone	$C_6H_{12}O$	565-69-5	100.158			113.5	0.814^{18}	1.3975^{20}	sl H_2O; vs EtOH, bz; msc eth, ace; s chl
7574	4-Methylpentanoyl chloride		$C_6H_{11}ClO$	38136-29-7	134.603			143	0.9725^{20}		
7575	2-Methyl-2-pentenal		$C_6H_{10}O$	623-36-9	98.142			136.5	0.8581^{20}	1.4488^{20}	i H_2O; s EtOH, eth, bz, MeOH
7576	2-Methyl-1-pentene		C_6H_{12}	763-29-1	84.159	liq	-135.7	62.1	0.6799^{20}	1.3920^{20}	i H_2O; s EtOH, bz, chl; sl ctc
7577	3-Methyl-1-pentene		C_6H_{12}	760-20-3	84.159	liq	-153	54.2	0.6675^{20}	1.3841^{20}	i H_2O; s EtOH, bz, chl, peth
7578	4-Methyl-1-pentene		C_6H_{12}	691-37-2	84.159	liq	-153.6	53.9	0.6642^{20}	1.3828^{20}	i H_2O; s EtOH, bz, chl, peth
7579	2-Methyl-2-pentene		C_6H_{12}	625-27-4	84.159	liq	-135	67.3	0.6863^{20}	1.4004^{20}	i H_2O; s EtOH, bz, ctc, chl
7580	3-Methyl-*cis*-2-pentene		C_6H_{12}	922-62-3	84.159	liq	-134.8	67.7	0.6886^{25}	1.4016^{20}	i H_2O; s EtOH, bz, chl, peth
7581	3-Methyl-*trans*-2-pentene		C_6H_{12}	616-12-6	84.159	liq	-138.5	70.4	0.6930^{25}	1.4045^{20}	i H_2O; s EtOH, bz, ctc, chl, peth
7582	4-Methyl-*cis*-2-pentene		C_6H_{12}	691-38-3	84.159	liq	-134.8	56.3	0.6690^{20}	1.3800^{20}	i H_2O; s EtOH, bz, chl, peth
7583	4-Methyl-*trans*-2-pentene		C_6H_{12}	674-76-0	84.159	liq	-140.8	58.6	0.6686^{20}	1.3889^{20}	i H_2O; s EtOH, bz, chl; sl ctc
7584	*trans*-2-Methyl-2-pentenoic acid		$C_6H_{10}O_2$	16957-70-3	114.142	pr	24.4	214; 112^{12}	0.9751^{20}	1.4513^{20}	sl H_2O; s eth, chl, CS_2
7585	4-Methyl-2-pentenoic acid	4,4-Dimethyl-2-butenoic acid	$C_6H_{10}O_2$	10321-71-8	114.142		35	217	0.9529^{21}	1.4489^{21}	vs ace, eth, EtOH
7586	2-Methyl-3-pentenoic acid		$C_6H_{10}O_2$	37674-63-8	114.142			199	0.966^{15}	1.4402^{25}	
7587	4-Methyl-3-penten-2-ol		$C_6H_{12}O$	4325-82-0	100.158			134	0.840^{15}	1.9377^{15}	
7588	3-Methyl-2-penten-4-one		$C_6H_{10}O$	565-62-8	98.142			138		1.4508^{20}	
7589	4-Methyl-4-penten-2-one		$C_6H_{10}O$	3744-02-3	98.142	liq	-72.6	124.2	0.8411^{20}		
7590	*cis*-3-Methyl-2-(2-pentenyl)-2-cyclopenten-1-one	Jasmone	$C_{11}H_{16}O$	488-10-8	164.244	ye oil		258; 134^{12}	0.9437^{22}	1.4979^{22}	sl H_2O; s EtOH, eth, ctc, lig
7591	3-(4-Methyl-3-pentenyl)furan		$C_{10}H_{14}O$	539-52-6	150.217			185.5	0.9017^{20}	1.4705^{21}	
7592	3-Methyl-3-penten-1-yne		C_6H_8	1574-33-0	80.128			66.5	0.739^{20}	1.4332^{20}	s eth, bz
7593	3-Methyl-2-pentyl-2-cyclopenten-1-one		$C_{11}H_{18}O$	1128-08-1	166.260			143^{22}, 116^{12}	0.9165^{18}	1.4767^{20}	
7594	Methyl pentyl ether		$C_6H_{14}O$	628-80-8	102.174			99	0.759^{22}	1.3862^{22}	vs ace, eth, EtOH
7595	5-Methyl-2-pentylphenol	6-Pentyl-*m*-cresol	$C_{12}H_{18}O$	1300-94-3	178.270		24	138^{15}			vs ace, eth, EtOH
7596	Methyl pentyl sulfide		$C_6H_{14}S$	1741-83-9	118.240	liq	-94	145.1	0.8431^{20}	1.4506^{20}	s EtOH, eth, ace, bz, chl
7597	Methyl *tert*-pentyl sulfide	2-Methyl-2-(methylthio)butane	$C_6H_{14}S$	13286-92-5	118.240	liq		150	0.84	1.4570^{20}	
7598	4-Methyl-1-pentyne		C_6H_{10}	7154-75-8	82.143	liq	-104.6	61.2	0.7000^{25}	1.3936^{20}	i H_2O; s bz, chl
7599	4-Methyl-2-pentyne		C_6H_{10}	21020-27-9	82.143	liq	-110.3	73.1	0.7112^{25}	1.4057^{20}	vs bz, chl
7600	3-Methyl-1-pentyn-3-ol	Meparfynol	$C_6H_{10}O$	77-75-8	98.142		30.5	120.5	0.8688^{20}	1.4310^{20}	
7601	Methyl perfluorooctanoate		$C_9H_3F_{15}O_2$	376-27-2	428.095			158	1.684^{20}	1.304^{27}	
7602	1-Methylphenanthrene		$C_{15}H_{12}$	832-69-9	192.256	lf, pl (dil al)	123	354			i H_2O; s EtOH
7603	3-Methylphenanthrene		$C_{15}H_{12}$	832-71-3	192.256	pr or nd (al)	65	350; 145^6			i H_2O; s EtOH, ace; sl chl
7604	4-Methylphenanthrene		$C_{15}H_{12}$	832-64-4	192.256	pl (90% al)	53.5	177^{10}			i H_2O; s EtOH, ctc
7605	Methylphenidate		$C_{14}H_{19}NO_2$	113-45-1	233.307			$136^{0.6}$			i H_2O, peth; s chl, EtOH, eth, AcOEt
7606	10-Methyl-10*H*-phenothiazine		$C_{13}H_{11}NS$	1207-72-3	213.298		101				
7607	10-Methyl-10*H*-phenothiazine-2-acetic acid	Metiazinic acid	$C_{15}H_{13}NO_2S$	13993-65-2	271.335		144				s chl

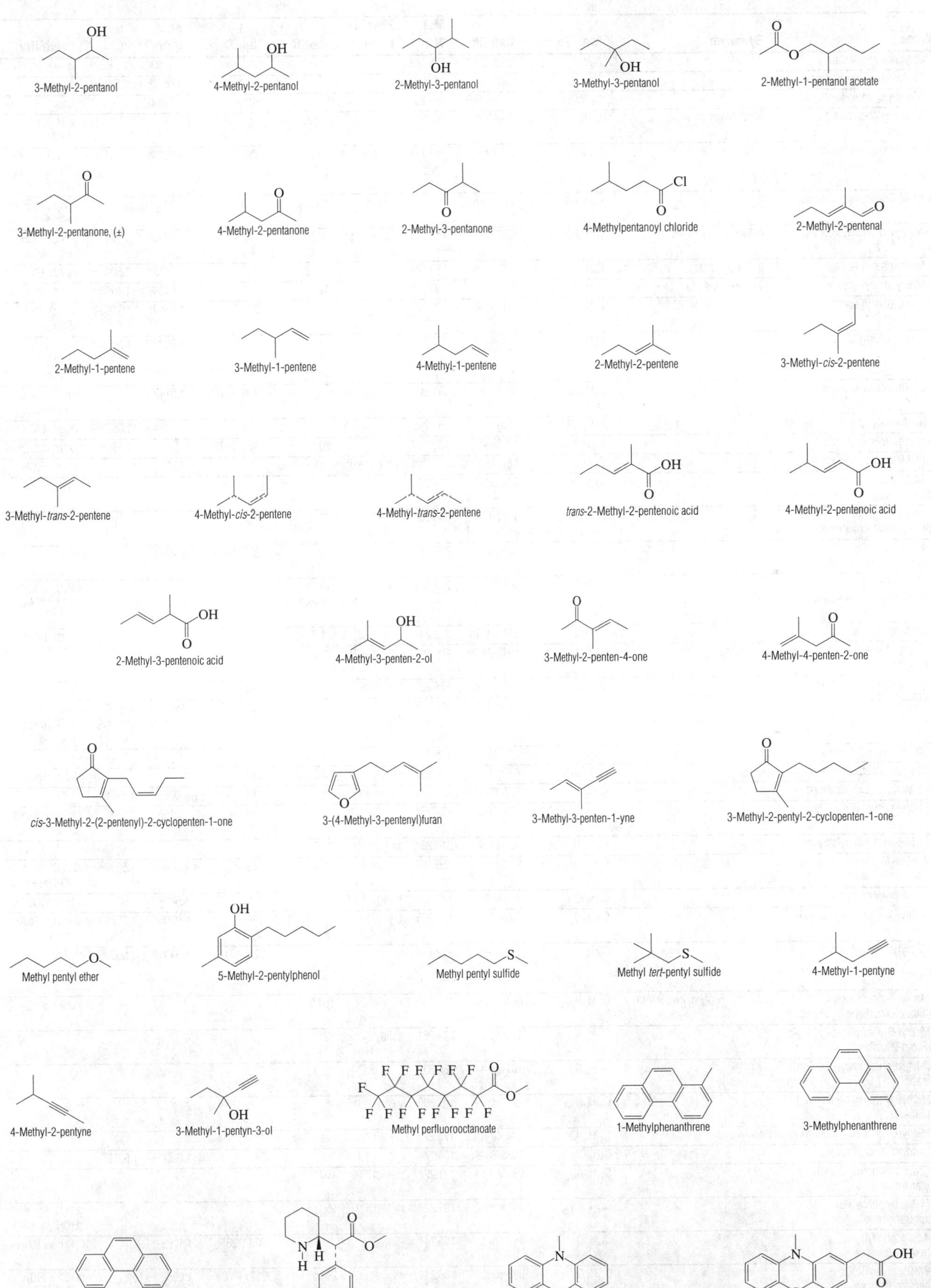

3-Methyl-2-pentanol

4-Methyl-2-pentanol

2-Methyl-3-pentanol

3-Methyl-3-pentanol

2-Methyl-1-pentanol acetate

3-Methyl-2-pentanone, (±)

4-Methyl-2-pentanone

2-Methyl-3-pentanone

4-Methylpentanoyl chloride

2-Methyl-2-pentenal

2-Methyl-1-pentene

3-Methyl-1-pentene

4-Methyl-1-pentene

2-Methyl-2-pentene

3-Methyl-*cis*-2-pentene

3-Methyl-*trans*-2-pentene

4-Methyl-*cis*-2-pentene

4-Methyl-*trans*-2-pentene

trans-2-Methyl-2-pentenoic acid

4-Methyl-2-pentenoic acid

2-Methyl-3-pentenoic acid

4-Methyl-3-penten-2-ol

3-Methyl-2-penten-4-one

4-Methyl-4-penten-2-one

cis-3-Methyl-2-(2-pentenyl)-2-cyclopenten-1-one

3-(4-Methyl-3-pentenyl)furan

3-Methyl-3-penten-1-yne

3-Methyl-2-pentyl-2-cyclopenten-1-one

Methyl pentyl ether

5-Methyl-2-pentylphenol

Methyl pentyl sulfide

Methyl *tert*-pentyl sulfide

4-Methyl-1-pentyne

4-Methyl-2-pentyne

3-Methyl-1-pentyn-3-ol

Methyl perfluorooctanoate

1-Methylphenanthrene

3-Methylphenanthrene

4-Methylphenanthrene

Methylphenidate

10-Methyl-10*H*-phenothiazine

10-Methyl-10*H*-phenothiazine-2-acetic acid

No.	Name	Synonym	Mol. Form.	CAS RN	Mol. Wt.	Physical Form	mp/°C	bp/°C	den/ g cm^{-3}	n_D	Solubility
7608	Methyl phenoxyacetate		$C_9H_{10}O_3$	2065-23-8	166.173			245	1.1493^{20}	1.5155^{20}	vs eth, EtOH
7609	1-Methyl-3-phenoxybenzene		$C_{13}H_{12}O$	3586-14-9	184.233			272	1.051^{25}	1.5727^{20}	
7610	[(2-Methylphenoxy)methyl]oxirane		$C_{10}H_{12}O_2$	2210-79-9	164.201			123^{2}	1.0884^{20}		
7611	3-(2-Methylphenoxy)-1,2-propanediol	Mephenesin	$C_{10}H_{14}O_3$	59-47-2	182.216		70 dec				sl H_2O, eth; s EtOH
7612	*N*-(2-Methylphenyl)acetamide		$C_9H_{11}NO$	120-66-1	149.189	nd (al)	110	296	1.168^{15}		sl H_2O, bz; s EtOH, eth, ace, HOAc
7613	*N*-(3-Methylphenyl)acetamide		$C_9H_{11}NO$	537-92-8	149.189	nd (w)	65.5	303	1.141^{15}		sl H_2O; vs EtOH, eth; s chl
7614	*N*-Methyl-*N*-phenylacetamide	*N*-Methylacetanilide	$C_9H_{11}NO$	579-10-2	149.189	nd (eth), pr (al)	103	256	1.0036^{105}	1.576	s H_2O, EtOH, eth, chl, lig
7615	2-Methylphenyl acetate	*o*-Cresyl acetate	$C_9H_{10}O_2$	533-18-6	150.174			208	1.0533^{15}	1.5002^{20}	vs eth, EtOH
7616	3-Methylphenyl acetate	*m*-Cresyl acetate	$C_9H_{10}O_2$	122-46-3	150.174		12	212	1.043^{20}	1.4978^{20}	vs bz, eth, EtOH
7617	4-Methylphenyl acetate	*p*-Cresyl acetate	$C_9H_{10}O_2$	140-39-6	150.174			212.5	1.0512^{17}	1.5163^{22}	sl H_2O, ctc; s EtOH, eth, chl
7618	Methyl 2-phenylacetate		$C_9H_{10}O_2$	101-41-7	150.174			216.5	1.0622^{16}	1.5075^{20}	i H_2O; msc EtOH, eth; s ace, ctc
7619	2-(Methylphenylamino)ethanol		$C_9H_{13}NO$	93-90-3	151.205			218^{110}, 150^{14}	1.0143^{0}		s H_2O; vs EtOH, eth, ace, bz
7620	2-[(2-Methylphenyl)amino]ethanol		$C_9H_{13}NO$	136-80-1	151.205			285.5	1.0794^{20}	1.5675^{20}	vs eth, EtOH
7621	3-Methyl-*N*-phenylaniline		$C_{13}H_{13}N$	1205-64-7	183.249		30	316; 183^{17}		1.6350^{20}	vs bz, eth, EtOH
7622	*N*-(4-Methylphenyl)benzamide		$C_{14}H_{13}NO$	582-78-5	211.259	orth nd (al)	158		1.202^{15}		vs eth, EtOH
7623	*N*-Methyl-*N*-phenylbenzenemethanamine		$C_{14}H_{15}N$	614-30-2	197.276						s ctc
7624	4-Methyl-α-phenylbenzenemethanol		$C_{14}H_{14}O$	1517-63-1	198.260		52				
7625	α-Methyl-α-phenylbenzenemethanol		$C_{14}H_{14}O$	599-67-7	198.260			285; 190^{12}	1.1059^{15}		
7626	4-Methyl-*N*-phenylbenzenesulfonamide		$C_{13}H_{13}NO_2S$	68-34-8	247.313	(α) tcl, (β) mcl pr (al, bz)	103.5				i H_2O; vs EtOH; s bz, HOAc
7627	4-Methylphenyl benzoate		$C_{14}H_{12}O_2$	614-34-6	212.244	pl (eth-al)	71.5	316			vs eth, EtOH
7628	1-Methyl-*N*-phenyl-*N*-benzyl-4-piperidinamine	Bamipine	$C_{19}H_{24}N_2$	4945-47-5	280.407	cry (MeOH)	115				
7629	Methyl 2-phenylbutanoate		$C_{11}H_{14}O_2$	2294-71-5	178.228	nd (dil al)	77.5	228			vs eth, EtOH
7630	3-Methyl-1-phenyl-1-butanone		$C_{11}H_{14}O$	582-62-7	162.228			236.5	0.9701^{16}	1.5139^{15}	i H_2O; msc EtOH, eth; vs ace
7631	3-Methyl-4-phenyl-3-butenamide	β-Benzalbutyramide	$C_{11}H_{13}NO$	7236-47-7	175.227		133				
7632	Methylphenylcarbamic chloride		C_8H_8ClNO	4285-42-1	169.609	pl (al)	88.5	280			vs eth, EtOH
7633	1-(2-Methylphenyl)ethanone		$C_9H_{10}O$	577-16-2	134.174			214	1.026^{20}	1.5276^{20}	
7634	1-(3-Methylphenyl)ethanone		$C_9H_{10}O$	585-74-0	134.174			220	1.0165^{0}	1.533^{15}	s EtOH, eth, ace; sl ctc
7635	4-(1-Methyl-1-phenylethyl)phenol		$C_{15}H_{16}O$	599-64-4	212.287	pr (peth)	74.5	335			
7636	*N*-Methyl-*N*-phenylformamide		C_8H_9NO	93-61-8	135.163		14.5	243	1.0948^{20}	1.5589^{20}	sl H_2O, ctc; s EtOH, ace
7637	*N*-(2-Methylphenyl)formamide		C_8H_9NO	94-69-9	135.163	lf (al)	62	288	1.086^{55}		s H_2O; vs EtOH
7638	5-Methyl-1-phenyl-1-hexen-3-one		$C_{13}H_{16}O$	2892-18-4	188.265	cry	43	154^{25}	0.9509^{46}	1.5523^{25}	sl H_2O; s EtOH, bz, chl
7639	1-Methyl-1-phenylhydrazine		$C_7H_{10}N_2$	618-40-6	122.167			228; 131^{35}	1.0404^{20}	1.5691^{20}	sl H_2O; msc EtOH, eth, bz, chl
7640	3-Methyl-5-phenyl-2,4-imidazolidinedione	3-Methyl-5-phenylhydantoin	$C_{10}H_{10}N_2O_2$	6846-11-3	190.198		164.5				s chl
7641	1-Methyl-6-phenylimidazo[4,5-b]pyridin-2-amine	PhIP	$C_{13}H_{12}N_4$	105650-23-5	224.261	solid	327				
7642	2-[Methyl(phenylmethyl)amino]ethanol		$C_{10}H_{15}NO$	101-98-4	165.232			134^{14}			
7643	4-Methyl-*N*-(phenylmethylene)aniline		$C_{14}H_{13}N$	2272-45-9	195.260	ye cry	35	318; 178^{11}			vs ace
7644	3-Methyl-2-phenylmorpholine	Phenmetrazine	$C_{11}H_{15}NO$	134-49-6	177.243			139^{12}, 104^{1}			
7645	2-Methyl-2-phenyloxirane		$C_9H_{10}O$	2085-88-3	134.174			84^{17}	1.0228^{20}	1.5232^{20}	
7646	*N*-(2-Methylphenyl)-3-oxo-butanamide		$C_{11}H_{13}NO_2$	93-68-5	191.227	pr (AcOEt)	107.5				vs bz, EtOH
7647	*N*-(4-Methylphenyl)-3-oxobutanamide		$C_{11}H_{13}NO_2$	2415-85-2	191.227	pr (AcOEt)	95				sl H_2O, lig; s EtOH, bz
7648	(2-Methylphenyl)phenylmethanone		$C_{14}H_{12}O$	131-58-8	196.244		<-18	308; 128^{12}	1.1098^{20}		i H_2O; vs EtOH
7649	(3-Methylphenyl)phenylmethanone		$C_{14}H_{12}O$	643-65-2	196.244	oil	2	317; 170^{9}	1.095^{20}		i H_2O; s EtOH, eth, bz, chl, HOAc

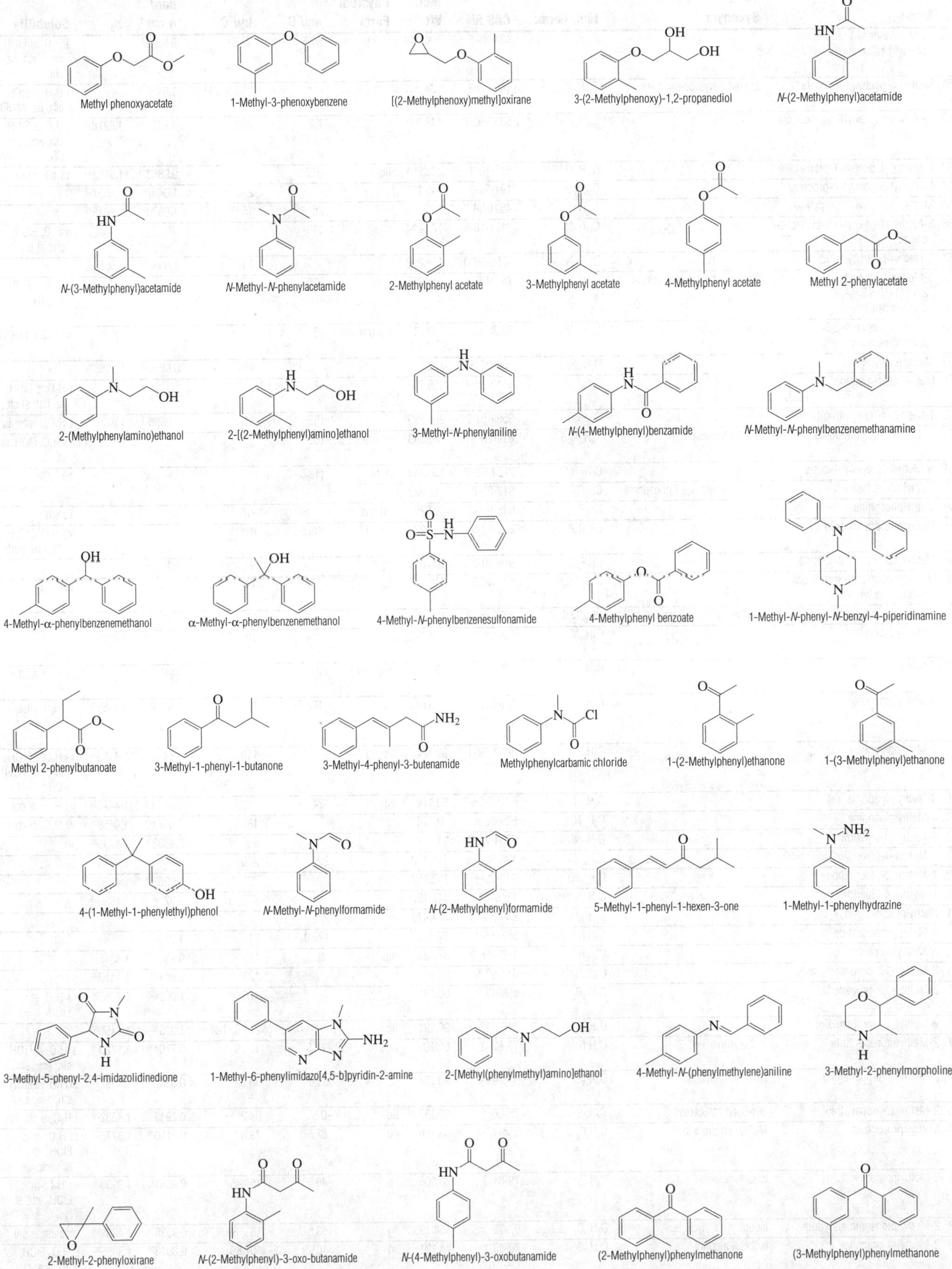

Methyl phenoxyacetate
1-Methyl-3-phenoxybenzene
[(2-Methylphenoxy)methyl]oxirane
3-(2-Methylphenoxy)-1,2-propanediol
N-(2-Methylphenyl)acetamide
N-(3-Methylphenyl)acetamide
N-Methyl-N-phenylacetamide
2-Methylphenyl acetate
3-Methylphenyl acetate
4-Methylphenyl acetate
Methyl 2-phenylacetate
2-(Methylphenylamino)ethanol
2-[(2-Methylphenyl)amino]ethanol
3-Methyl-N-phenylaniline
N-(4-Methylphenyl)benzamide
N-Methyl-N-phenylbenzenemethanamine
4-Methyl-α-phenylbenzenemethanol
α-Methyl-α-phenylbenzenemethanol
4-Methyl-N-phenylbenzenesulfonamide
4-Methylphenyl benzoate
1-Methyl-N-phenyl-N-benzyl-4-piperidinamine
Methyl 2-phenylbutanoate
3-Methyl-1-phenyl-1-butanone
3-Methyl-4-phenyl-3-butenamide
Methylphenylcarbamic chloride
1-(2-Methylphenyl)ethanone
1-(3-Methylphenyl)ethanone
4-(1-Methyl-1-phenylethyl)phenol
N-Methyl-N-phenylformamide
N-(2-Methylphenyl)formamide
5-Methyl-1-phenyl-1-hexen-3-one
1-Methyl-1-phenylhydrazine
3-Methyl-5-phenyl-2,4-imidazolidinedione
1-Methyl-6-phenylimidazo[4,5-b]pyridin-2-amine
2-[Methyl(phenylmethyl)amino]ethanol
4-Methyl-N-(phenylmethylene)aniline
3-Methyl-2-phenylmorpholine
2-Methyl-2-phenyloxirane
N-(2-Methylphenyl)-3-oxo-butanamide
N-(4-Methylphenyl)-3-oxobutanamide
(2-Methylphenyl)phenylmethanone
(3-Methylphenyl)phenylmethanone

No.	Name	Synonym	Mol. Form.	CAS RN	Mol. Wt.	Physical Form	mp/°C	bp/°C	den/ g cm-3	n_D	Solubility
7650	(4-Methylphenyl) phenylmethanone		$C_{14}H_{12}O$	134-84-9	196.244	mcl pr	59.5	228^{70}	0.9926^{0}		i H_2O; sl EtOH, lig; s eth, bz, chl
7651	Methyl 3-phenylpropanoate	Methyl dihydrocinnamate	$C_{10}H_{12}O_2$	103-25-3	164.201			238.5; 91^{4}	1.0455^{25}		i H_2O; s EtOH, eth, bz, AcOEt
7652	1-(4-Methylphenyl)-1-propanone		$C_{10}H_{12}O$	5337-93-9	148.201		7.2	236	0.9926^{20}	1.5278^{20}	i H_2O; s EtOH, eth, ace, bz, CS_2
7653	2-Methyl-1-phenyl-1-propanone		$C_{10}H_{12}O$	611-70-1	148.201	liq	-0.7	220	0.9863^{11}	1.5172^{20}	vs eth, EtOH
7654	2-Methyl-3-phenyl-2-propenal		$C_{10}H_{10}O$	101-39-3	146.185			248; 150^{100}	1.0407^{17}	1.6057^{17}	
7655	Methyl 3-phenyl-2-propynoate		$C_{10}H_8O_2$	4891-38-7	160.170		26	158^{48}, 132^{16}	1.0830^{25}	1.5618^{25}	
7656	3-Methyl-1-phenyl-1*H*-pyrazol-5-amine		$C_{10}H_{11}N_3$	1131-18-6	173.214		116	333			s H_2O, EtOH, chl; sl bz
7657	2-Methyl-5-phenylpyridine		$C_{12}H_{11}N$	3256-88-0	169.222			189^{50}	1.0590^{25}	1.6055^{25}	
7658	5-Methyl-5-phenyl-2,4,6(1*H*,3*H*,5*H*)-pyrimidinetrione	Phenylmethylbarbituric acid	$C_{11}H_{10}N_2O_3$	76-94-8	218.208	cry	220				i H_2O; s EtOH, eth, alk
7659	1-Methyl-3-phenyl-2,5-pyrrolidinedione	Phensuximide	$C_{11}H_{11}NO_2$	86-34-0	189.211	cry (hot al)	72				vs EtOH, MeOH
7660	Methylphenylsilane		$C_7H_{10}Si$	766-08-5	122.240			140	0.8895^{20}	1.5058^{20}	
7661	Methyl phenyl sulfone		$C_7H_8O_2S$	3112-85-4	156.203		88				i H_2O; s EtOH, bz, chl; sl ctc
7662	1-Methyl-4-(phenylthio)benzene		$C_{13}H_{12}S$	3699-01-2	200.299		15.7	317	1.0986^{25}	1.6225^{25}	i H_2O; s ace, bz
7663	(2-Methylphenyl)thiourea	*o*-Tolylthiourea	$C_8H_{10}N_2S$	614-78-8	166.243	nd (dil al, w)	162				vs H_2O, EtOH; sl eth
7664	*N*-Methyl-*N'*-phenylthiourea		$C_8H_{10}N_2S$	2724-69-8	166.243	ta, pl	112.5				vs EtOH
7665	Methyl phosphate	Methyl dihydrogen phosphate	CH_5O_4P	812-00-0	112.022	oil					
7666	Methylphosphine		CH_5P	593-54-4	48.025	col gas		-16			vs eth
7667	Methylphosphonic acid		CH_5O_3P	993-13-5	96.023	hyg pl	108.5	dec			vs H_2O, EtOH, eth; i bz, peth
7668	Methylphosphonic difluoride		CH_3F_2OP	676-99-3	100.005	liq		98; 22^{27}	1.3314^{20}		
7669	Methylphosphonofluoridic acid, isopropyl ester	Sarin	$C_4H_{10}FO_2P$	107-44-8	140.093	liq	-57	147	1.10^{20}		dec H_2O
7670	Methyl phosphorodichloridite	Methyl dichlorophosphite	CH_3Cl_2OP	3279-26-3	132.914	hyg liq	-91	93	1.406	1.4740^{20}	
7671	1-Methylpiperazine		$C_5H_{12}N_2$	109-01-3	100.162			138		1.4378^{20}	vs H_2O, eth, EtOH
7672	2-Methylpiperazine		$C_5H_{12}N_2$	109-07-9	100.162	hyg lf (al)	62	153			vs H_2O; s EtOH, eth, bz, chl
7673	1-Methylpiperidine		$C_6H_{13}N$	626-67-5	99.174	liq	-102.7	107	0.8159^{20}	1.4355^{20}	vs H_2O; msc EtOH, eth; s ctc
7674	2-Methylpiperidine, (±)		$C_6H_{13}N$	3000-79-1	99.174	liq	-2.5	118	0.8436^{24}	1.4459^{20}	vs H_2O; s EtOH, eth; sl chl; i dil KOH
7675	3-Methylpiperidine, (±)		$C_6H_{13}N$	53152-98-0	99.174	liq	-24	125.5	0.8446^{26}	1.4470^{20}	vs H_2O; sl chl
7676	4-Methylpiperidine		$C_6H_{13}N$	626-58-4	99.174			130	0.8674^{25}	1.4458^{20}	vs H_2O; sl chl
7677	1-Methyl-3-piperidinol		$C_6H_{13}NO$	3554-74-3	115.173			93^{26}, 77^{11}	0.9635^{16}	1.4735^{20}	
7678	1-Methyl-4-piperidinol		$C_6H_{13}NO$	106-52-5	115.173		29	200		1.4775^{20}	
7679	1-Methyl-2-piperidinone		$C_6H_{11}NO$	931-20-4	113.157			221; 105^{12}	1.0263^{25}	1.4820^{20}	
7680	1-Methyl-4-piperidinone		$C_6H_{11}NO$	1445-73-4	113.157			85^{45}, 57^{11}	0.971^{25}	1.4580^{25}	
7681	Methylprednisolone		$C_{22}H_{30}O_5$	83-43-2	374.470	cry	232				
7682	2-Methylpropanamide		C_4H_9NO	563-83-7	87.120		129.0	217	1.013^{20}		s chl
7683	*N*-Methylpropanamide		C_4H_9NO	1187-58-2	87.120	liq	-30.9	148	0.9305^{25}	1.4345^{25}	
7684	2-Methyl-1,2-propanediamine		$C_4H_{12}N_2$	811-93-8	88.151			123	0.841^{25}	1.4410^{20}	s ctc
7685	2-Methyl-1,2-propanediol		$C_4H_{10}O_2$	558-43-0	90.121			176	1.0024^{20}	1.4350^{20}	vs H_2O, eth, EtOH
7686	2-Methyl-1,3-propanediol		$C_4H_{10}O_2$	2163-42-0	90.121	liq	-91	211.6; 124^{20}	1.015^{20}	1.4450^{20}	
7687	2-Methylpropanenitrile	Isobutyronitrile	C_4H_7N	78-82-0	69.106	liq	-71.5	103.9	0.7704^{20}	1.3720^{20}	sl H_2O; vs EtOH, eth, ace, chl
7688	2-Methyl-1-propanethiol	Isobutyl mercaptan	$C_4H_{10}S$	513-44-0	90.187		<-70	88.5	0.8357^{20}	1.4387^{20}	sl H_2O; vs EtOH, eth, ace; s ctc
7689	2-Methyl-2-propanethiol	*tert*-Butyl mercaptan	$C_4H_{10}S$	75-66-1	90.187	liq	-0.5	64.2	0.7943^{25}	1.4232^{20}	i H_2O; s ctc, hp
7690	Methyl propanoate	Methyl propionate	$C_4H_8O_2$	554-12-1	88.106	liq	-87.5	79.8	0.9150^{20}	1.3775^{20}	sl H_2O; msc EtOH, eth; s ace, ctc
7691	2-Methylpropanoic acid	Isobutyric acid	$C_4H_8O_2$	79-31-2	88.106	liq	-46	154.45	0.9681^{20}	1.3930^{20}	vs H_2O; msc EtOH, eth; sl ctc
7692	2-Methylpropanoic anhydride	Isobutyric anhydride	$C_8H_{14}O_3$	97-72-3	158.195	liq	-53.5	183; 89^{32}	0.9535^{20}	1.4061^{19}	msc eth; s chl
7693	2-Methyl-1-propanol	Isobutyl alcohol	$C_4H_{10}O$	78-83-1	74.121	liq	-101.9	107.89	0.8018^{20}	1.3955^{20}	s H_2O, EtOH, eth, ace, ctc
7694	2-Methyl-2-propanol	*tert*-Butyl alcohol	$C_4H_{10}O$	75-65-0	74.121		25.69	82.4	0.7887^{20}	1.3878^{20}	msc H_2O, EtOH, eth; s chl

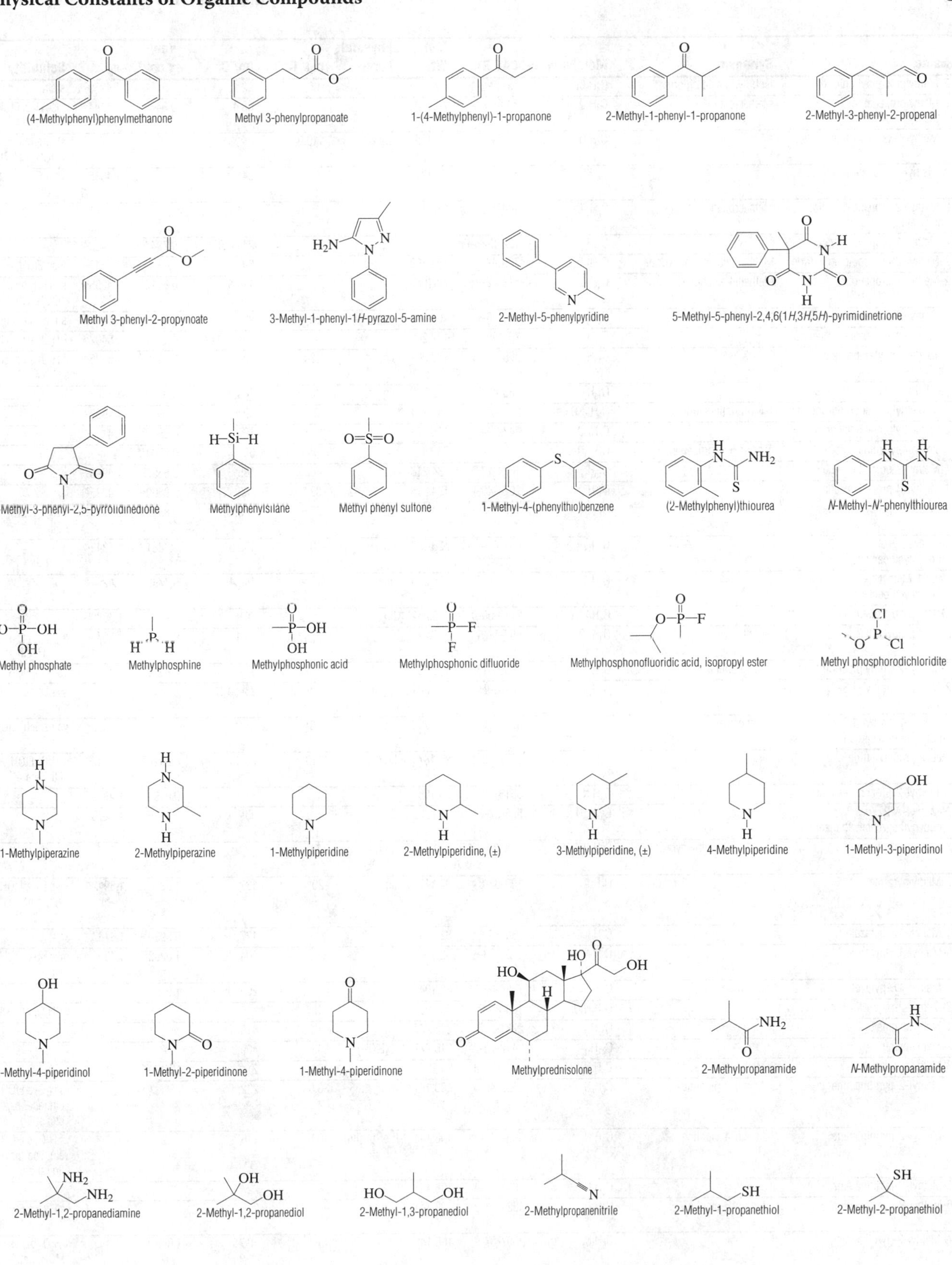
(4-Methylphenyl)phenylmethanone
Methyl 3-phenylpropanoate
1-(4-Methylphenyl)-1-propanone
2-Methyl-1-phenyl-1-propanone
2-Methyl-3-phenyl-2-propenal
Methyl 3-phenyl-2-propynoate
3-Methyl-1-phenyl-1H-pyrazol-5-amine
2-Methyl-5-phenylpyridine
5-Methyl-5-phenyl-2,4,6(1H,3H,5H)-pyrimidinetrione
1-Methyl-3-phenyl-2,5-pyrrolidinedione
Methylphenylsilane
Methyl phenyl sulfone
1-Methyl-4-(phenylthio)benzene
(2-Methylphenyl)thiourea
N-Methyl-N'-phenylthiourea
Methyl phosphate
Methylphosphine
Methylphosphonic acid
Methylphosphonic difluoride
Methylphosphonofluoridic acid, isopropyl ester
Methyl phosphorodichloridite
1-Methylpiperazine
2-Methylpiperazine
1-Methylpiperidine
2-Methylpiperidine, (±)
3-Methylpiperidine, (±)
4-Methylpiperidine
1-Methyl-3-piperidinol
1-Methyl-4-piperidinol
1-Methyl-2-piperidinone
1-Methyl-4-piperidinone
Methylprednisolone
2-Methylpropanamide
N-Methylpropanamide
2-Methyl-1,2-propanediamine
2-Methyl-1,2-propanediol
2-Methyl-1,3-propanediol
2-Methylpropanenitrile
2-Methyl-1-propanethiol
2-Methyl-2-propanethiol
Methyl propanoate
2-Methylpropanoic acid
2-Methylpropanoic anhydride
2-Methyl-1-propanol
2-Methyl-2-propanol

No.	Name	Synonym	Mol. Form.	CAS RN	Mol. Wt.	Physical Form	mp/°C	bp/°C	den/ g cm^{-3}	n_D	Solubility
7695	2-Methylpropanoyl chloride	Isobutyric acid chloride	C_4H_7ClO	79-30-1	106.551	liq	-90	92		1.4079^{20}	s eth
7696	2-Methylpropenal	Methacrolein	C_4H_6O	78-85-3	70.090			68.4	0.840^{25}	1.4144^{20}	msc H_2O, EtOH, eth
7697	2-Methyl-2-propenamide		C_4H_7NO	79-39-0	85.105	cry (bz)	110.5				sl eth, chl; s EtOH, CH_2Cl_2
7698	*N*-Methyl-2-propen-1-amine		C_4H_9N	627-37-2	71.121			64		1.4065^{20}	vs H_2O, ace, eth, EtOH
7699	2-Methyl-2-propene-1,1-diol diacetate	Methacrolein diacetate	$C_8H_{12}O_4$	10476-95-6	172.179			191		1.4241^{20}	
7700	2-Methyl-1-propene, tetramer		$C_{16}H_{32}$	15220-85-6	224.425	liq	-98	244; 109^{15}	0.7944^{20}	1.4482^{20}	
7701	2-Methyl-2-propenoic anhydride	Methacrylic acid anhydride	$C_8H_{10}O_3$	760-93-0	154.163			89^{5}		1.4540^{20}	msc EtOH, eth
7702	2-Methyl-2-propenol	Methallyl alcohol	C_4H_8O	513-42-8	72.106			114.5	0.8515^{20}	1.4255^{20}	vs H_2O; msc EtOH, eth
7703	2-Methyl-2-propenoyl chloride	Methacrylic acid chloride	C_4H_5ClO	920-46-7	104.535	liq	-60	96	1.0871^{20}	1.4435^{20}	s eth, ace, chl
7704	*cis*-(1-Methyl-1-propenyl)benzene		$C_{10}H_{12}$	767-99-7	132.202			194.7; 177^{500}	0.9191^{25}	1.5402^{25}	i H_2O; s bz, chl
7705	*trans*-(1-Methyl-1-propenyl) benzene		$C_{10}H_{12}$	768-00-3	132.202	liq	-23.5	194.7	0.9138^{25}	1.5425^{20}	i H_2O; s bz, chl
7706	(2-Methyl-1-propenyl)benzene		$C_{10}H_{12}$	768-49-0	132.202	liq	-48.0	183; 99^{43}	0.900^{20}	1.5388^{20}	
7707	4-(2-Methylpropenyl)morpholine	1-Morpholinoisobutene	$C_8H_{15}NO$	2403-55-6	141.211		120	89^{20}		1.4663^{20}	
7708	2-(2-Methylpropoxy)ethanol		$C_6H_{14}O_2$	4439-24-1	118.174			160	0.8900^{20}	1.4143^{20}	
7709	Methylpropylamine	*N*-Methyl-1-propanamine	$C_4H_{11}N$	627-35-0	73.137			63	0.7204^{17}		
7710	1-Methyl-2-propylbenzene		$C_{10}H_{14}$	1074-17-5	134.218	liq	-60.3	185	0.8697^{25}	1.4996^{20}	
7711	1-Methyl-3-propylbenzene		$C_{10}H_{14}$	1074-43-7	134.218	liq	-82.5	182	0.8569^{25}	1.4935^{20}	
7712	1-Methyl-4-propylbenzene		$C_{10}H_{14}$	1074-55-1	134.218	liq	-63.6	183.4	0.8544^{25}	1.4922^{20}	i H_2O; s EtOH, eth
7713	*cis*-1-Methyl-2-propylcyclopentane		C_9H_{18}	932-43-4	126.239	liq	-104	152.6	0.7881^{25}	1.4343^{20}	
7714	*trans*-1-Methyl-2-propylcyclopentane		C_9H_{18}	932-44-5	126.239	liq	-123	146.4	0.7735^{25}	1.4274^{20}	
7715	Methyl propyl disulfide		$C_4H_{10}S_2$	2179-60-4	122.252	liq		70^{43}	0.980	1.5080^{20}	
7716	Methyl propyl ether	1-Methoxypropane	$C_4H_{10}O$	557-17-5	74.121			39.1	0.7356^{13}	1.3579^{25}	s H_2O, ace; msc EtOH, eth
7717	1-Methyl-2-propylpiperidine, (*S*)	Methylconiine	$C_9H_{19}N$	35305-13-6	141.254			174	0.8326^{22}	1.4538^{12}	vs ace, EtOH
7718	2-Methyl-2-propyl-1,3-propanediol		$C_7H_{16}O_2$	78-26-2	132.201	cry (hx)	62.5	234; 121^{10}			s H_2O, hx; sl chl
7719	2-Methyl-2-propyl-1,3-propanediol dicarbamate	Meprobamate	$C_9H_{18}N_2O_4$	57-53-4	218.250	cry (w)	105				vs bz, eth, EtOH
7720	Methyl propyl sulfide		$C_4H_{10}S$	3877-15-4	90.187	liq	-113	95.6	0.8424^{20}	1.4442^{20}	s H_2O, EtOH, eth, ace
7721	*N*-Methyl-2-propyn-1-amine		C_4H_7N	35161-71-8	69.106			83	0.819^{25}	1.4332^{20}	
7722	*N*-Methyl-*N*-2-propynylbenzenemethanamine	Pargyline	$C_{11}H_{13}N$	555-57-7	159.228			96^{11}	0.944^{25}	1.5213^{20}	
7723	3-Methylpyrazinamine	2-Amino-3-methylpyrazine	$C_5H_7N_3$	19838-08-5	109.130	nd (hx/AcOEt)	174				
7724	2-Methylpyrazine		$C_5H_6N_2$	109-08-0	94.115	liq	-29	137	1.03^{20}	1.5042^{20}	msc H_2O, EtOH, eth; s ace; sl ctc
7725	1-Methyl-1*H*-pyrazole		$C_4H_6N_2$	930-36-9	82.104			127	0.9929^{13}	1.4787^{13}	
7726	3-Methyl-1*H*-pyrazole		$C_4H_6N_2$	1453-58-3	82.104		36.5	204; 108^{25}	1.0203^{16}	1.4915^{20}	msc H_2O, EtOH, eth
7727	4-Methyl-1*H*-pyrazole	Fomepizole	$C_4H_6N_2$	7554-65-6	82.104			206; 95^{13}	1.015^{20}		
7728	3-Methyl-2-pyrazolin-5-one		$C_4H_6N_2O$	108-26-9	98.103		215				vs H_2O; sl EtOH
7729	1-Methylpyrene		$C_{17}H_{12}$	2381-21-7	216.277		71.3	410			
7730	2-Methylpyrene		$C_{17}H_{12}$	3442-78-2	216.277	fl (EtOH)	143	409.8			
7731	3-Methylpyridazine	3-Methyl-1,2-diazine	$C_5H_6N_2$	1632-76-4	94.115		184	214	1.0450^{26}	1.5145^{20}	
7732	3-Methyl-2-pyridinamine	2-Amino-3-picoline	$C_6H_8N_2$	1603-40-3	108.141	hyg	33.5	222; 95^{8}			vs H_2O; s EtOH, eth, ace, bz, ctc; sl lig
7733	4-Methyl-2-pyridinamine	2-Amino-4-picoline	$C_6H_8N_2$	695-34-1	108.141	lf or pl (lig)	100	116^{11}			vs H_2O; s EtOH, eth, ace, bz; i lig; sl chl
7734	5-Methyl-2-pyridinamine		$C_6H_8N_2$	1603-41-4	108.141		76.5	227			
7735	6-Methyl-2-pyridinamine	2-Amino-6-picoline	$C_6H_8N_2$	1824-81-3	108.141	hyg (lig)	41	208.5			vs H_2O; s EtOH, eth, ace, bz, lig
7736	*N*-Methyl-2-pyridinamine		$C_6H_8N_2$	4597-87-9	108.141		15	200.5	1.048^{29}		s H_2O, bz; vs EtOH, eth, HOAc
7737	*N*-Methyl-4-pyridinamine		$C_6H_8N_2$	1121-58-0	108.141	pl (eth)	118.8				vs H_2O, ace, eth, EtOH
7738	2-Methylpyridine	2-Picoline	C_6H_7N	109-06-8	93.127	liq	-66.68	129.38	0.9443^{20}	1.4957^{20}	vs H_2O, ace; msc EtOH, eth; s ctc

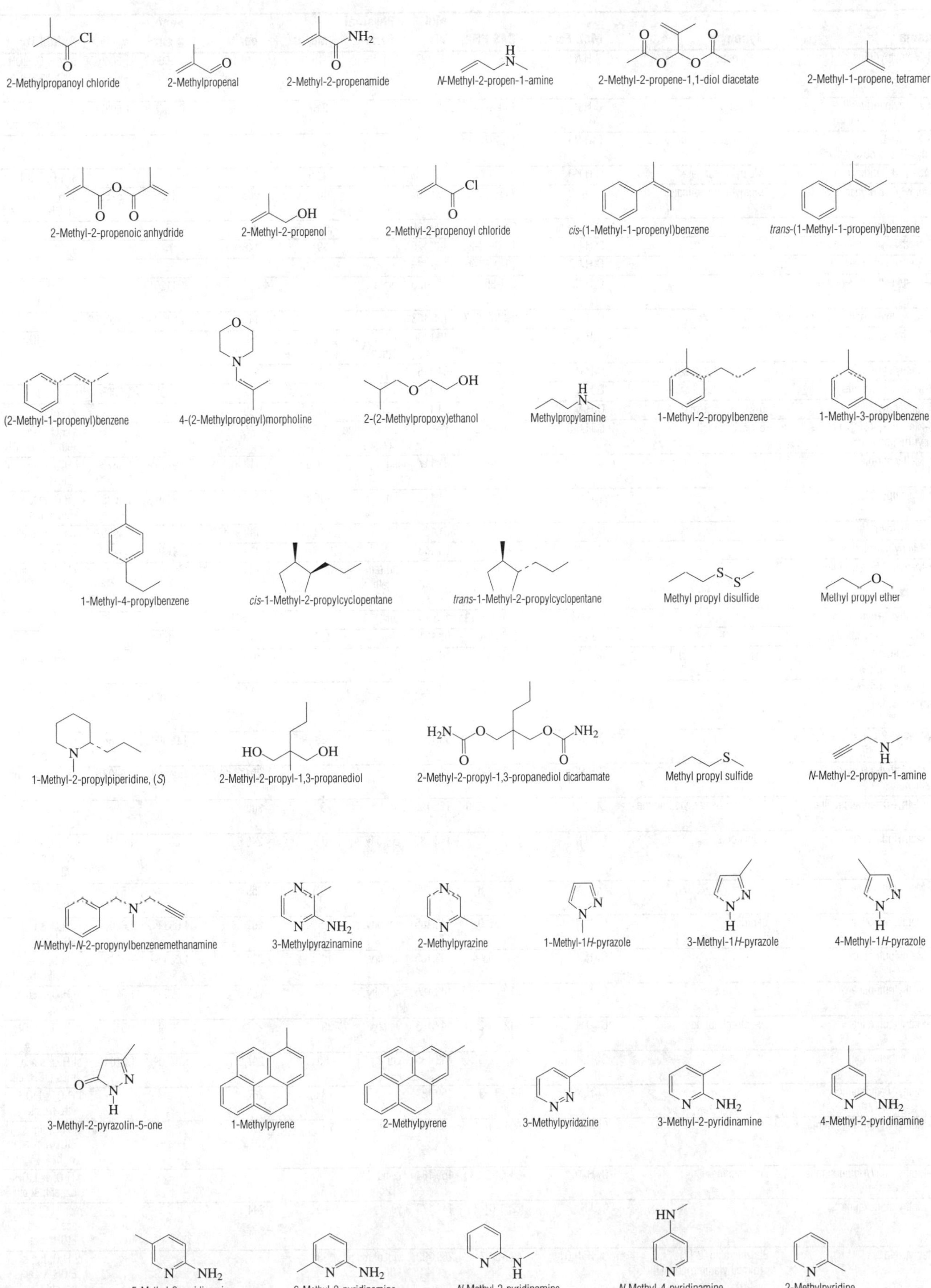
Cl
O
2-Methylpropanoyl chloride
O
2-Methylpropenal
NH2
O
2-Methyl-2-propenamide
H
N
N-Methyl-2-propen-1-amine
O O
O O
2-Methyl-2-propene-1,1-diol diacetate
2-Methyl-1-propene, tetramer
O
O O
2-Methyl-2-propenoic anhydride
OH
2-Methyl-2-propenol
Cl
O
2-Methyl-2-propenoyl chloride
cis-(1-Methyl-1-propenyl)benzene
trans-(1-Methyl-1-propenyl)benzene
(2-Methyl-1-propenyl)benzene
O
N
4-(2-Methylpropenyl)morpholine
O OH
2-(2-Methylpropoxy)ethanol
H
N
Methylpropylamine
1-Methyl-2-propylbenzene
1-Methyl-3-propylbenzene
1-Methyl-4-propylbenzene
cis-1-Methyl-2-propylcyclopentane
trans-1-Methyl-2-propylcyclopentane
S S
Methyl propyl disulfide
O
Methyl propyl ether
N
1-Methyl-2-propylpiperidine, (S)
HO OH
2-Methyl-2-propyl-1,3-propanediol
H2N O O NH2
O O
2-Methyl-2-propyl-1,3-propanediol dicarbamate
S
Methyl propyl sulfide
N
H
N-Methyl-2-propyn-1-amine
N
N-Methyl-N-2-propynylbenzenemethanamine
N
N NH2
3-Methylpyrazinamine
N
N
2-Methylpyrazine
N
N
1-Methyl-1H-pyrazole
N
N
H
3-Methyl-1H-pyrazole
N
N
H
4-Methyl-1H-pyrazole
O N
N
H
3-Methyl-2-pyrazolin-5-one
1-Methylpyrene
2-Methylpyrene
N N
3-Methylpyridazine
N NH2
3-Methyl-2-pyridinamine
N NH2
4-Methyl-2-pyridinamine
N NH2
5-Methyl-2-pyridinamine
N NH2
6-Methyl-2-pyridinamine
N N
H
N-Methyl-2-pyridinamine
HN
N
N-Methyl-4-pyridinamine
N
2-Methylpyridine

No.	Name	Synonym	Mol. Form.	CAS RN	Mol. Wt.	Physical Form	mp/°C	bp/°C	den/ g cm^{-3}	n_D	Solubility
7739	3-Methylpyridine	3-Picoline	C_6H_7N	108-99-6	93.127	liq	-18.14	144.14	0.9566^{20}	1.5040^{20}	msc H_2O, EtOH, eth; vs ace; s ctc
7740	4-Methylpyridine	4-Picoline	C_6H_7N	108-89-4	93.127		3.67	145.36	0.9548^{20}	1.5037^{20}	msc H_2O, EtOH, eth; s ace, ctc
7741	6-Methyl-2-pyridinecarboxaldehyde		C_7H_7NO	1122-72-1	121.137		32	77^{12}			
7742	Methyl 3-pyridinecarboxylate	Methyl nicotinate	$C_7H_7NO_2$	93-60-7	137.137	cry	42.5	204			s H_2O, EtOH, bz
7743	Methyl 4-pyridinecarboxylate	Methyl isonicotinate	$C_7H_7NO_2$	2459-09-8	137.137		16.1	208	1.1599^{20}	1.5135^{20}	sl H_2O, ctc; s EtOH, eth, bz
7744	2-Methylpyridine-1-oxide		C_6H_7NO	931-19-1	109.126		49	260			
7745	3-Methylpyridine-1-oxide		C_6H_7NO	1003-73-2	109.126		39	148^{15}			s chl
7746	4-Methylpyridine-1-oxide		C_6H_7NO	1003-67-4	109.126		185.8				
7747	1-Methyl-2(1*H*)-pyridinone		C_6H_7NO	694-85-9	109.126	nd	31	250	1.1120^{20}		msc H_2O; sl peth, lig
7748	1-(6-Methyl-3-pyridinyl)ethanone		C_8H_9NO	36357-38-7	135.163		17.6	144^{50}	1.0168^{25}	1.5302^{25}	vs H_2O
7749	4-Methyl-2-pyrimidinamine		$C_5H_7N_3$	108-52-1	109.130	pl (w), nd (sub)	160.3	sub			s H_2O, EtOH; sl chl
7750	2-Methylpyrimidine	2-Methyl-1,3-diazine	$C_5H_6N_2$	5053-43-0	94.115	liq	-4	138			msc H_2O
7751	4-Methylpyrimidine	4-Methyl-1,3-diazine	$C_5H_6N_2$	3438-46-8	94.115		32	142	1.030^{16}	1.500^{20}	msc H_2O
7752	5-Methylpyrimidine	5-Methyl-1,3-diazine	$C_5H_6N_2$	2036-41-1	94.115		30.5	153			vs H_2O
7753	6-Methyl-2,4(1*H*,3*H*)-pyrimidinedione	6-Methyluracil	$C_5H_6N_2O_2$	626-48-2	126.114	oct pr or nd (w, al)	275 dec				s H_2O, EtOH; sl eth, tfa; vs NH_3
7754	1-Methylpyrrole		C_5H_7N	96-54-8	81.117	liq	-56.32	112.81	0.9145^{15}	1.4875^{20}	i H_2O; msc EtOH, eth
7755	2-Methylpyrrole		C_5H_7N	636-41-9	81.117	liq	-35.6	147.6	0.9446^{15}	1.5035^{16}	i H_2O; msc EtOH, eth
7756	3-Methylpyrrole		C_5H_7N	616-43-3	81.117	liq	-48.4	142.9; 45^{11}		1.4970^{20}	msc EtOH, eth
7757	*N*-Methylpyrrolidine		$C_5H_{11}N$	120-94-5	85.148			81	0.8188^{20}	1.4247^{20}	vs H_2O, eth
7758	1-Methyl-2,5-pyrrolidinedione		$C_5H_7NO_2$	1121-07-9	113.116	nd (eth-peth, al, ace)	71	234			s H_2O, EtOH; vs eth
7759	*N*-Methyl-2-pyrrolidinethione		C_5H_9NS	10441-57-3	115.197	oil		$100^{0.08}$			
7760	5-Methyl-2-pyrrolidinone		C_5H_9NO	108-27-0	99.131		43	248	1.0458^{20}		
7761	1-(1-Methyl-2-pyrrolidinyl)-2-propanone, (*R*)	Hygrine	$C_8H_{15}NO$	496-49-1	141.211			76.5^{11}		1.4555^{20}	vs EtOH, chl
7762	3-(1-Methyl-2-pyrrolidinyl) pyridine, (±)		$C_{10}H_{14}N_2$	22083-74-5	162.231			244	1.0082^{20}	1.5289^{20}	msc H_2O; vs EtOH, eth, chl; s lig
7763	*N*-Methyl-2-pyrrolidinone		C_5H_9NO	872-50-4	99.131	liq	-23.09	202	1.0230^{25}	1.4684^{20}	vs H_2O; s eth, ace, chl
7764	1-(1-Methyl-1*H*-pyrrol-2-yl) ethanone		C_7H_9NO	932-16-1	123.152			201^{252}, 93^{22}	1.0445^{15}	1.5403^{15}	s EtOH, bz, chl
7765	6-Methyl-8-quinolinamine	8-Amino-6-methylquinoline	$C_{10}H_{10}N_2$	68420-93-9	158.199	nd	73	sub			vs ace, bz, eth, EtOH
7766	2-Methylquinoline	Quinaldine	$C_{10}H_9N$	91-63-4	143.185	col oily liq	-0.8	246.5	1.06^{25}	1.6116^{20}	sl H_2O; s EtOH, eth, ace, ctc, chl
7767	3-Methylquinoline		$C_{10}H_9N$	612-58-8	143.185	pr	16.5	259.8	1.0673^{20}	1.6171^{20}	vs ace, eth, EtOH
7768	4-Methylquinoline	Lepidine	$C_{10}H_9N$	491-35-0	143.185	col oily liq	9.5	262	1.083^{20}	1.6200^{20}	sl H_2O; s EtOH, eth, ace; i alk
7769	5-Methylquinoline		$C_{10}H_9N$	7661-55-4	143.185	col cry	19	262.7	1.0832^{20}	1.6219^{20}	sl H_2O; s ace; msc EtOH, eth
7770	6-Methylquinoline		$C_{10}H_9N$	91-62-3	143.185	col oily liq	-22	258.6	1.0654^{20}	1.6157^{20}	sl H_2O; s EtOH, eth, ace
7771	7-Methylquinoline	*m*-Toluquinoline	$C_{10}H_9N$	612-60-2	143.185	ye cry	39	257.6	1.0609^{20}	1.6150^{20}	sl H_2O; s EtOH, eth, ace
7772	8-Methylquinoline		$C_{10}H_9N$	611-32-5	143.185	col liq	-80	247.5	1.0719^{20}	1.6164^{20}	sl H_2O; s ace; msc EtOH, eth
7773	2-Methyl-8-quinolinol		$C_{10}H_9NO$	826-81-3	159.184		73.8	267			i H_2O; s EtOH, eth, bz, ctc
7774	1-Methyl-2(1*H*)-quinolinone		$C_{10}H_9NO$	606-43-9	159.184	nd (lig)	74	325			sl H_2O, lig; s EtOH, eth, ace; vs bz
7775	1-Methyl-4(1*H*)-quinolinone	Echinopsine	$C_{10}H_9NO$	83-54-5	159.184	α-nd (bz); β-cry (al)					s H_2O; vs EtOH, bz, chl; sl eth
7776	2-Methylquinoxaline		$C_9H_8N_2$	7251-61-8	144.173	ye cry	180.5	244			msc H_2O, eth, ace, bz; vs EtOH; s ctc
7777	Methyl Red	Benzoic acid, 2-[[4-(dimethylamino)phenyl]azo]-	$C_{15}H_{15}N_3O_2$	493-52-7	269.299	viol or red pr (to, bz)	183				sl H_2O, lig; s EtOH; vs ace, bz, chl
7778	Methyl β-*D*-ribofuranoside		$C_6H_{12}O_5$	7473-45-2	164.156		80				

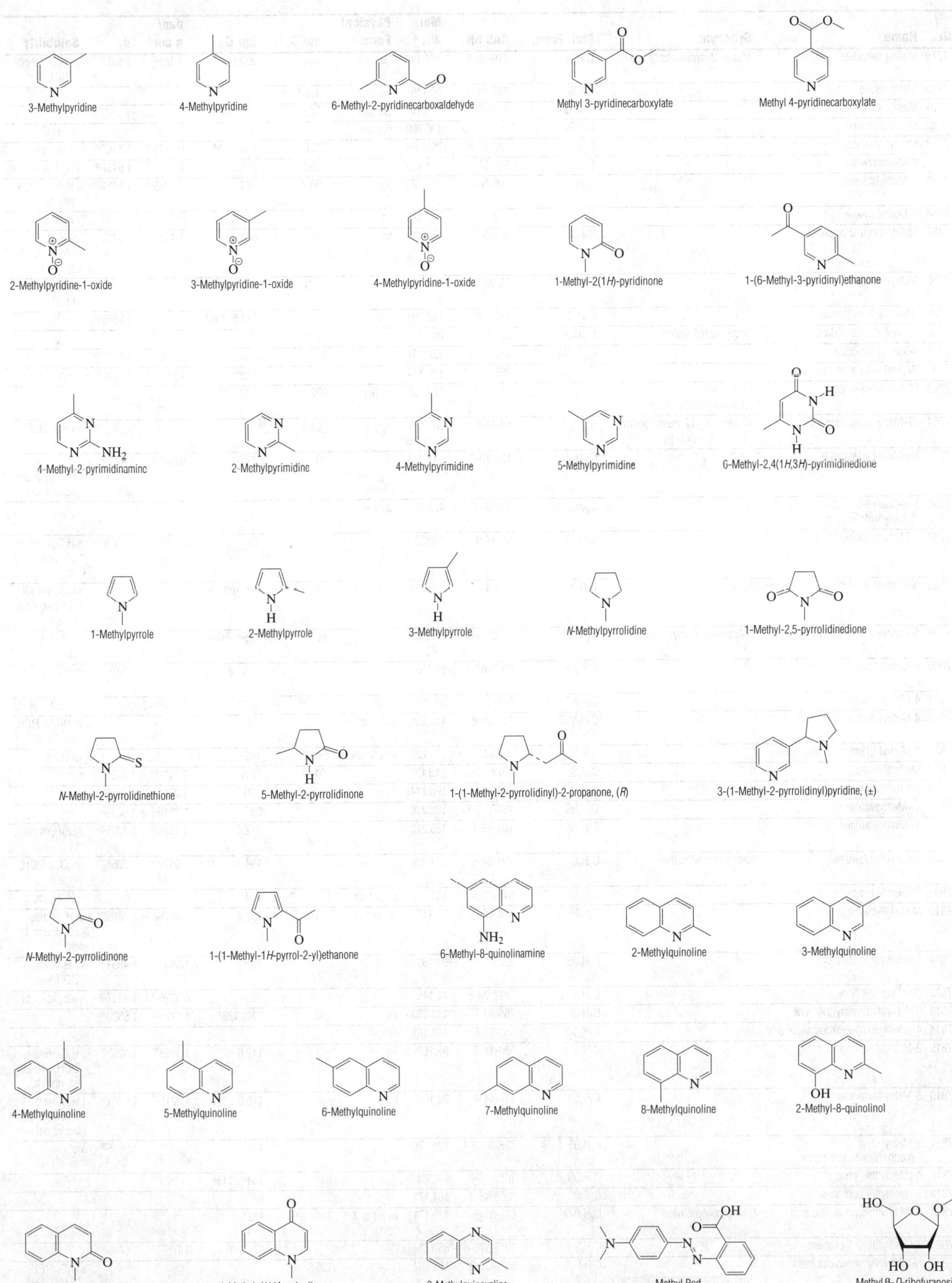
3-Methylpyridine
4-Methylpyridine
6-Methyl-2-pyridinecarboxaldehyde
Methyl 3-pyridinecarboxylate
Methyl 4-pyridinecarboxylate
2-Methylpyridine-1-oxide
3-Methylpyridine-1-oxide
4-Methylpyridine-1-oxide
1-Methyl-2(1H)-pyridinone
1-(6-Methyl-3-pyridinyl)ethanone
4-Methyl-2-pyrimidinamine
2-Methylpyrimidine
4-Methylpyrimidine
5-Methylpyrimidine
6-Methyl-2,4(1H,3H)-pyrimidinedione
1-Methylpyrrole
2-Methylpyrrole
3-Methylpyrrole
N-Methylpyrrolidine
1-Methyl-2,5-pyrrolidinedione
N-Methyl-2-pyrrolidinethione
5-Methyl-2-pyrrolidinone
1-(1-Methyl-2-pyrrolidinyl)-2-propanone, (R)
3-(1-Methyl-2-pyrrolidinyl)pyridine, (±)
N-Methyl-2-pyrrolidinone
1-(1-Methyl-1H-pyrrol-2-yl)ethanone
6-Methyl-8-quinolinamine
2-Methylquinoline
3-Methylquinoline
4-Methylquinoline
5-Methylquinoline
6-Methylquinoline
7-Methylquinoline
8-Methylquinoline
2-Methyl-8-quinolinol
1-Methyl-2(1H)-quinolinone
1-Methyl-4(1H)-quinolinone
2-Methylquinoxaline
Methyl Red
Methyl β-D-ribofuranoside

No.	Name	Synonym	Mol. Form.	CAS RN	Mol. Wt.	Physical Form	mp/°C	bp/°C	den/ g cm^{-3}	n_D	Solubility
7779	Methyl salicylate	Methyl 2-hydroxybenzoate	$C_8H_8O_3$	119-36-8	152.148	liq	-8	222.9	1.181^{25}	1.535^{20}	sl H_2O; vs eth, EtOH, chl
7780	Methylsilane		CH_6Si	992-94-9	46.145	col gas	-156.5	-57.5			
7781	Methyl silyl ether		CH_6OSi	2171-96-2	62.144	col gas	-98.5	-21; -87^{10}			
7782	Methylstannane		CH_6Sn	1631-78-3	136.769	col gas		0			dec H_2O
7783	Methyl stearate		$C_{19}H_{38}O_2$	112-61-8	298.504		39.1	443; 215^{15}	0.8498^{40}	1.4367^{40}	vs eth, chl
7784	2-Methylstyrene		C_9H_{10}	611-15-4	118.175	liq	-68.5	169.8	0.9077^{25}	1.5437^{20}	i H_2O; s bz, chl
7785	3-Methylstyrene		C_9H_{10}	100-80-1	118.175	liq	-86.3	164	0.9076^{25}	1.5411^{20}	i H_2O; s EtOH, eth, bz
7786	4-Methylstyrene		C_9H_{10}	622-97-9	118.175	liq	-34.1	172.8	0.9173^{25}	1.5420	i H_2O; s bz
7787	Methylsuccinic acid		$C_5H_8O_4$	636-60-2	132.116	pr	115	dec	1.4200^{0}	1.4303	vs H_2O, EtOH, MeOH; s eth; sl chl
7788	Methyl sulfate		CH_4O_4S	75-93-4	112.106		<-30	dec 135			vs H_2O, eth, EtOH
7789	(Methylsulfinyl)benzene		C_7H_8OS	1193-82-4	140.203		32.0	263.5; 140^{13}		1.5885^{20}	
7790	1-(Methylsulfinyl)decane	Decyl methyl sulfoxide	$C_{11}H_{24}OS$	3079-28-5	204.373	cry	52.5				
7791	3-Methyl sulfolane		$C_5H_{10}O_2S$	872-93-5	134.197		1	276	1.188^{25}	1.4772^{20}	
7792	(Methylsulfonyl)ethene		$C_3H_6O_2S$	3680-02-2	106.144			122^{24}	1.2117^{20}	1.4636^{20}	s eth, ace
7793	Methyl terephthalate	Methyl 1,4-benzenedicarboxylate	$C_9H_8O_4$	1679-64-7	180.158	nd (w)	222	subl ≈ 230			
7794	17-Methyltestosterone	17-Hydroxy-17-methylandrost-4-en-3-one, (17β)	$C_{20}H_{30}O_2$	58-18-4	302.451		163.5				vs eth, EtOH
7795	Methyl tetradecanoate		$C_{15}H_{30}O_2$	124-10-7	242.398		19	295; 155^{7}	0.8671^{20}	1.425^{45}	i H_2O; msc EtOH, eth, ace, bz, chl, ctc
7796	5-*N*-Methyl-5,6,7,8-tetrahydrofolic acid		$C_{20}H_{25}N_7O_6$	134-35-0	459.456	cry (w)					
7797	2-Methyltetrahydrofuran		$C_5H_{10}O$	96-47-9	86.132			78	0.8552^{20}	1.4059^{21}	s H_2O; vs EtOH, eth, ace, bz; sl ctc
7798	*N*-Methyl-*N*,2,4,6-tetranitroaniline	Tetryl	$C_7H_5N_5O_8$	479-45-8	287.144	ye pr (al)	131.5	exp 180	1.57^{10}		i H_2O; sl EtOH, eth, chl; s ace, bz, py
7799	4-Methyl-2-thiazolamine	2-Amino-4-methylthiazole	$C_4H_6N_2S$	1603-91-4	114.169		45.5	125^{20}, $70^{0.4}$			vs H_2O, EtOH, eth
7800	2-Methylthiazole		C_4H_5NS	3581-87-1	99.155			128		1.510	msc H_2O; s EtOH, ace
7801	4-Methylthiazole		C_4H_5NS	693-95-8	99.155			133.3	1.112^{25}		s H_2O, EtOH, eth
7802	4-Methyl-5-thiazoleethanol		C_6H_9NOS	137-00-8	143.206	col to pa ye		135^{7}	1.196^{24}		vs H_2O; s EtOH, eth, bz, chl
7803	4-Methyl-2(3*H*)-thiazolethione		$C_4H_5NS_2$	5685-06-3	131.220	ye cry (dil al)	89.3	188^{3}			vs EtOH
7804	Methylthiirane		C_3H_6S	1072-43-1	74.145	liq	-91	72.5	0.941^{20}	1.472^{20}	s chl
7805	(Methylthio)acetic acid		$C_3H_6O_2S$	2444-37-3	106.144		13.0	130^{27}	1.221^{20}	1.495^{20}	
7806	2-(Methylthio)aniline		C_7H_9NS	2987-53-3	139.218			234	1.111^{25}	1.6239^{20}	
7807	4-(Methylthio)aniline		C_7H_9NS	104-96-1	139.218			272.5	1.1379^{20}	1.6395^{20}	s EtOH, eth, ace, bz
7808	(Methylthio)benzene	Methyl phenyl sulfide	C_7H_8S	100-68-5	124.204			194.3	1.0579^{20}	1.5868^{20}	i H_2O; s EtOH; vs ace
7809	2-(Methylthio)benzothiazole		$C_8H_7NS_2$	615-22-5	181.279	pr (dil al)	52	174^{22}			s EtOH, chl
7810	Methyl thiocyanate		C_2H_3NS	556-64-9	73.117	col liq	-2.5	132.9	1.0678^{25}	1.4669^{25}	sl H_2O; msc EtOH, eth; s ctc
7811	2-(Methylthio)ethanol		C_3H_8OS	5271-38-5	92.160			70^{20}	1.063^{20}	1.4861^{30}	vs H_2O, eth, EtOH
7812	(Methylthio)ethene		C_3H_6S	1822-74-8	74.145			69.5	0.9026^{20}	1.4837^{20}	s eth, ace, chl
7813	[(Methylthio)methyl]benzene		$C_8H_{10}S$	766-92-7	138.230	liq	-30	210; 120^{48}	1.0274^{20}	1.5620^{20}	
7814	4-(Methylthio)-2-oxobutanoic acid		$C_5H_8O_3S$	583-92-6	148.181	oil					
7815	2-Methylthiophene		C_5H_6S	554-14-3	98.167	liq	-63.4	112.6	1.0193^{20}	1.5203^{20}	i H_2O; msc EtOH, eth, ace, bz, hp, ctc
7816	3-Methylthiophene		C_5H_6S	616-44-4	98.167	liq	-69	115.5	1.0218^{20}	1.5204^{20}	i H_2O; msc EtOH, eth, ace, bz; vs chl
7817	5-Methyl-2-thiophenecarboxaldehyde		C_6H_6OS	13679-70-4	126.176			114^{25}		1.5825^{20}	s chl
7818	4-(Methylthio)phenol		C_7H_8OS	1073-72-9	140.203		84	154^{20}, 113^{6}			
7819	3-(Methylthio)propanal		C_4H_8OS	3268-49-3	104.171			62^{11}			
7820	3-(Methylthio)propanoic acid	*S*-Methylpropiothetin	$C_4H_8O_2S$	646-01-5	120.171	ye oil or fl (hx)	21	132^{13}			
7821	3-(Methylthio)-1-propene		C_4H_8S	10152-76-8	88.172			92	0.8767^{20}	1.4714^{20}	
7822	*N*-Methylthiosemicarbazide	*N*-Methylhydrazinecarbothioamide	$C_2H_7N_3S$	6610-29-3	105.162		136.5				s H_2O, EtOH, DMSO; i eth, bz, lig

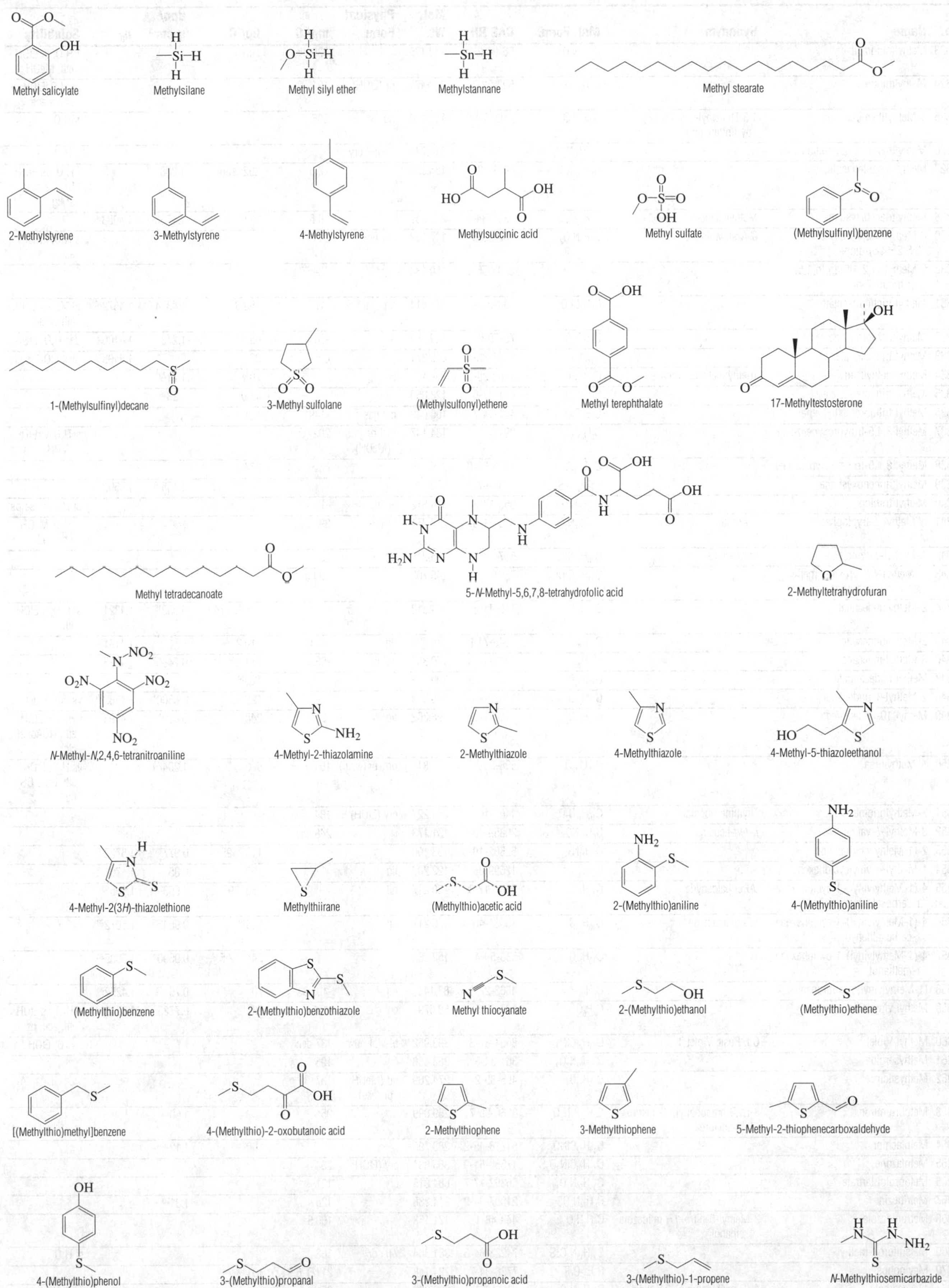
Methyl salicylate
Methylsilane
Methyl silyl ether
Methylstannane
Methyl stearate
2-Methylstyrene
3-Methylstyrene
4-Methylstyrene
Methylsuccinic acid
Methyl sulfate
(Methylsulfinyl)benzene
1-(Methylsulfinyl)decane
3-Methyl sulfolane
(Methylsulfonyl)ethene
Methyl terephthalate
17-Methyltestosterone
Methyl tetradecanoate
5-N-Methyl-5,6,7,8-tetrahydrofolic acid
2-Methyltetrahydrofuran
N-Methyl-N,2,4,6-tetranitroaniline
4-Methyl-2-thiazolamine
2-Methylthiazole
4-Methylthiazole
4-Methyl-5-thiazoleethanol
4-Methyl-2(3H)-thiazolethione
Methylthiirane
(Methylthio)acetic acid
2-(Methylthio)aniline
4-(Methylthio)aniline
(Methylthio)benzene
2-(Methylthio)benzothiazole
Methyl thiocyanate
2-(Methylthio)ethanol
(Methylthio)ethene
[(Methylthio)methyl]benzene
4-(Methylthio)-2-oxobutanoic acid
2-Methylthiophene
3-Methylthiophene
5-Methyl-2-thiophenecarboxaldehyde
4-(Methylthio)phenol
3-(Methylthio)propanal
3-(Methylthio)propanoic acid
3-(Methylthio)-1-propene
N-Methylthiosemicarbazide

No.	Name	Synonym	Mol. Form.	CAS RN	Mol. Wt.	Physical Form	mp/°C	bp/°C	den/ g cm-3	n_D	Solubility
7823	Methylthiouracil		$C_5H_6N_2OS$	56-04-2	142.179		330 dec	sub			i H_2O; sl EtOH, eth, MeOH, bz
7824	Methylthiourea		$C_2H_6N_2S$	598-52-7	90.147	pr (EtOH)	121				vs H_2O, EtOH; sl eth; s ace
7825	1-Methylthymine	1,5-Dimethyl-2,4(1*H*,3*H*)-pyrimidinedione	$C_6H_8N_2O_2$	4160-72-9	140.140	nd (w)	295				s H_2O
7826	Methylthymol blue, sodium salt		$C_{37}H_{40}N_2O_{13}Na_4S$	1945-77-3	844.743	bl-viol cry					s H_2O
7827	Methyl 4-toluenesulfonate		$C_8H_{10}O_3S$	80-48-8	186.228		28.5	292; 186^{22}	1.2087^{40}		i H_2O; vs EtOH, bz; s eth, ctc; sl lig
7828	Methyltriacetoxysilane	Methylsilanetriol, triacetate	$C_7H_{12}O_6Si$	4253-34-3	220.252		40.5	111^{17}	1.1750^{20}	1.4083^{20}	
7829	6-Methyl-1,2,4-triazine-3,5(2*H*,4*H*)-dione	6-Azathymine	$C_4H_5N_3O_2$	932-53-6	127.102	cry (w)	211				s H_2O, EtOH, ace
7830	5-Methyl-[1,2,4]triazolo[1,5-a]pyrimidin-7-ol		$C_6H_6N_4O$	2503-56-2	150.138		>245				
7831	Methyl trichloroacetate		$C_3H_3Cl_3O_2$	598-99-2	177.414	liq	-17.5	153.8	1.4874^{20}	1.4572^{20}	i H_2O; vs EtOH, eth; s ctc
7832	Methyltrichlorosilane		CH_3Cl_3Si	75-79-6	149.480	liq	-90	65.6	1.273^{20}	1.4106^{20}	dec H_2O, EtOH
7833	Methyl tridecanoate		$C_{14}H_{28}O_2$	1731-88-0	228.371		6.5	92^{1}		1.4405^{20}	msc EtOH; s ctc
7834	Methyltriethyllead	Triethylmethylplumbane	$C_7H_{18}Pb$	1762-28-3	309.4	col liq		70^{16}	1.71^{20}		
7835	Methyl trifluoroacetate		$C_3H_3F_3O_2$	431-47-0	128.050			43.0	1.28^{20}		
7836	Methyl trifluoromethyl ether		$C_2H_3F_3O$	421-14-7	100.039	col gas	-149	-23.66			
7837	Methyl 3,4,5-trihydroxybenzoate		$C_8H_8O_5$	99-24-1	184.147	mcl pr (MeOH)	202				sl H_2O; vs EtOH, MeOH
7838	Methyl 3,4,5-trimethoxybenzoate		$C_{11}H_{14}O_5$	1916-07-0	226.226		83	274.5			
7839	Methyltriphenoxysilane		$C_{19}H_{18}O_3Si$	3439-97-2	322.430			269^{100}, 179^{2}	1.135^{20}	1.5599^{20}	
7840	Methyl trithion		$C_9H_{12}ClO_2PS_3$	953-17-3	314.812	ye liq	-18				sl H_2O; misc os
7841	*N*-Methyl-*L*-tryptophan	*L*-Abrine	$C_{12}H_{14}N_2O_2$	526-31-8	218.251	pr (w)	295 dec				sl H_2O, MeOH; i eth; s alk
7842	*N*-Methyl-*L*-tyrosine	Surinamine	$C_{10}H_{13}NO_3$	537-49-5	195.215	nd	293				
7843	α-Methyl-*DL*-tyrosine, methyl ester, hydrochloride		$C_{11}H_{16}ClNO_3$	7361-31-1	245.703		190 dec				s H_2O
7844	2-Methylundecanal		$C_{12}H_{24}O$	110-41-8	184.318			119^{16}, 114^{10}	0.832^{15}	1.4321^{20}	sl H_2O; s EtOH, eth
7845	2-Methylundecane		$C_{12}H_{26}$	7045-71-8	170.334	liq	-45.6	210.2		1.4191^{20}	
7846	3-Methylundecane		$C_{12}H_{26}$	1002-43-3	170.334	col liq	-58.0	211.2	0.7485^{25}	1.4208^{25}	
7847	Methyl undecanoate		$C_{12}H_{24}O_2$	1731-86-8	200.318			123^{10}			
7848	2-Methyl-1-undecanol		$C_{12}H_{26}O$	10522-26-6	186.333			129^{12}	0.8300^{15}	1.4382^{20}	vs eth, EtOH
7849	Methyl 10-undecenoate		$C_{12}H_{22}O_2$	111-81-9	198.302	liq	-27.5	248	0.889^{15}	1.4393^{20}	i H_2O; s EtOH, eth, HOAc; sl ctc
7850	*N*-Methylurea		$C_2H_6N_2O$	598-50-5	74.081	orth pr (w, al)	104.9	dec	1.2040^{0}		vs H_2O, EtOH; i eth, bz; s CS_2, lig
7851	5-Methyluridine	Thymine riboside	$C_{10}H_{14}N_2O_6$	1463-10-1	258.227	cry (EtOH)	184				
7852	3-Methyl-*L*-valine	*L*-*tert*-Leucine	$C_6H_{13}NO_2$	20859-02-3	131.173		248 dec				
7853	2-(1-Methylvinyl)aniline		$C_9H_{11}N$	52562-19-3	133.190			115^{20}, 95^{13}	0.977^{25}	1.5722^{20}	
7854	1-Methyl-4-vinylcyclohexene		C_9H_{14}	17699-86-4	122.207	liq		152	0.85	1.4701^{20}	
7855	4-(1-Methylvinyl)-1-cyclohexene-1-carboxaldehyde, (*R*)	*d*-Perillaldehyde	$C_{10}H_{14}O$	5503-12-8	150.217	oil		238; 99^{7}	0.953^{20}	1.5058^{20}	s ctc
7856	4-(1-Methylvinyl)-1-cyclohexene-1-carboxaldehyde, (*S*)	*l*-Perillaldehyde	$C_{10}H_{14}O$	18031-40-8	150.217	oil		104^{10}	0.9645^{20}	1.5072^{20}	
7857	4-(1-Methylvinyl)-1-cyclohexene-1-methanol		$C_{10}H_{16}O$	536-59-4	152.233			244; 12.5^{12}	0.9690^{20}	1.5005^{20}	
7858	(1-Methylvinyl)cyclopropane		C_6H_{10}	4663-22-3	82.143	liq	-102.3	70	0.751^{20}	1.4252^{20}	
7859	Methyl vinyl ether		C_3H_6O	107-25-5	58.079	col gas	-122	5.5	0.7725^{0}	1.3730^{0}	sl H_2O; vs EtOH, eth, ace, bz
7860	Methyl Violet	C.I. Basic Violet 1	$C_{24}H_{28}ClN_3$	8004-87-3	393.952	bl-viol pow	137 dec				s H_2O, EtOH
7861	Methysergide		$C_{21}H_{27}N_3O_2$	361-37-5	353.458	cry	195				
7862	Methysticin		$C_{15}H_{14}O_5$	495-85-2	274.269	nd (MeOH), pr (ace)	137				
7863	Metobromuron	3-(*p*-Bromophenyl)-1-methoxy-1-methylurea	$C_9H_{11}BrN_2O_2$	3060-89-7	259.099		95		1.60^{20}		
7864	Metolachlor		$C_{15}H_{22}ClNO_2$	51218-45-2	283.795			$100^{0.001}$	1.12^{20}		
7865	Metolazone		$C_{16}H_{16}ClN_3O_3S$	17560-51-9	365.834	cry (EtOH)	254				
7866	Metoprolol tartrate		$C_{34}H_{56}N_2O_{12}$	56392-17-7	684.815	cry	121				
7867	Metribuzin		$C_8H_{14}N_4OS$	21087-64-9	214.288		126		1.31^{20}		
7868	Metronidazole	2-Methyl-5-nitro-1*H*-imidazole-1-ethanol	$C_6H_9N_3O_3$	443-48-1	171.153		160.5				
7869	Metsulfuron-methyl		$C_{14}H_{15}N_5O_6S$	74223-64-6	381.364	wh cry	163				sl H_2O
7870	Mevinphos		$C_7H_{13}O_6P$	7786-34-7	224.148		21 (*E*), 6.9 (*Z*)	$101^{0.3}$			
7871	Mexacarbate	4-(Dimethylamino)-3,5-xylyl methylcarbamate	$C_{12}H_{18}N_2O_2$	315-18-4	222.283	cry	85				vs EtOH, bz, ace

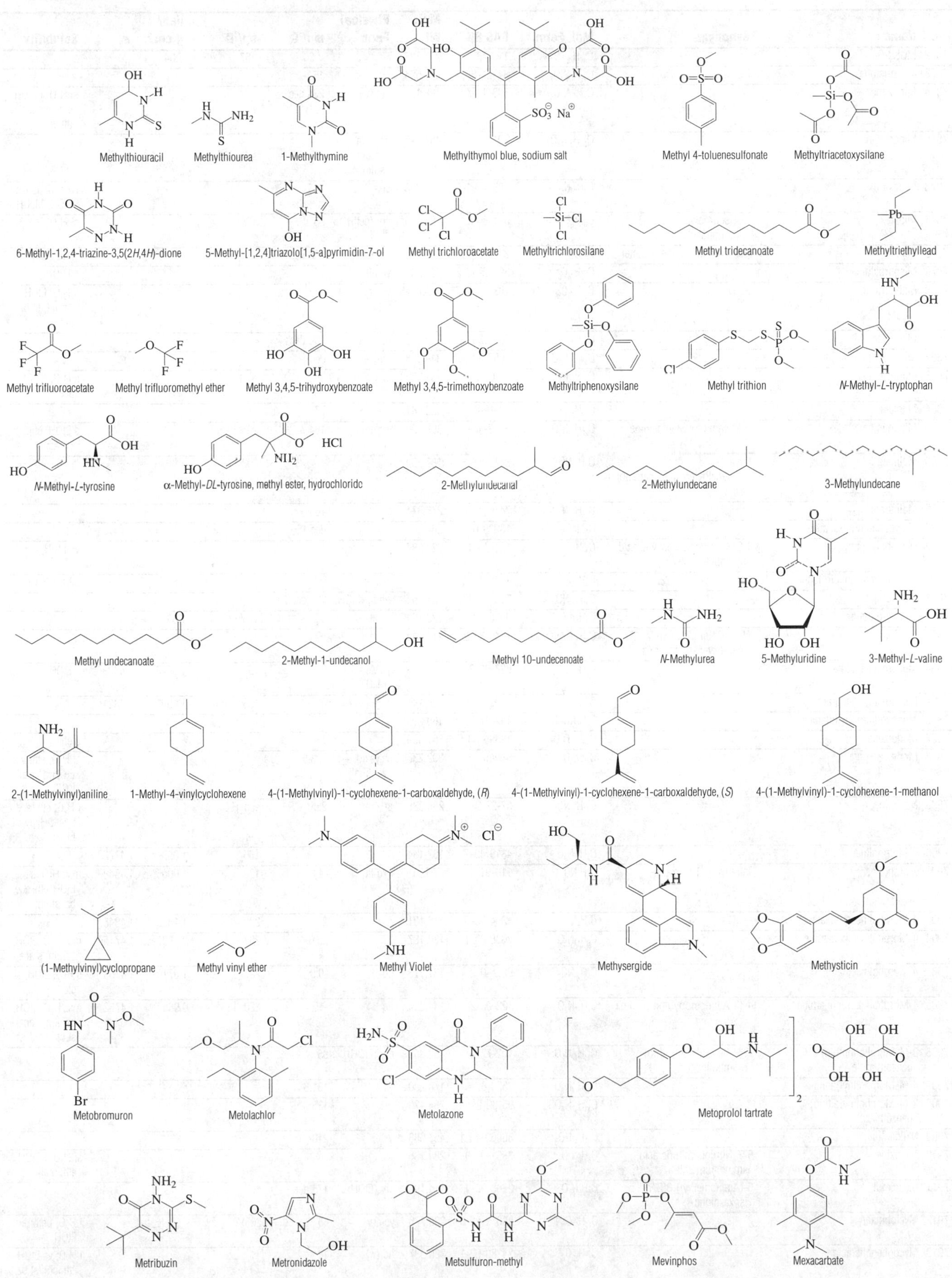

Methylthiouracil
Methylthiourea
1-Methylthymine
Methylthymol blue, sodium salt
Methyl 4-toluenesulfonate
Methyltriacetoxysilane
6-Methyl-1,2,4-triazine-3,5(2H,4H)-dione
5-Methyl-[1,2,4]triazolo[1,5-a]pyrimidin-7-ol
Methyl trichloroacetate
Methyltrichlorosilane
Methyl tridecanoate
Methyltriethyllead
Methyl trifluoroacetate
Methyl trifluoromethyl ether
Methyl 3,4,5-trihydroxybenzoate
Methyl 3,4,5-trimethoxybenzoate
Methyltriphenoxysilane
Methyl trithion
N-Methyl-L-tryptophan
N-Methyl-L-tyrosine
α-Methyl-DL-tyrosine, methyl ester, hydrochloride
2-Methylundecanal
2-Methylundecane
3-Methylundecane
Methyl undecanoate
2-Methyl-1-undecanol
Methyl 10-undecenoate
N-Methylurea
5-Methyluridine
3-Methyl-L-valine
2-(1-Methylvinyl)aniline
1-Methyl-4-vinylcyclohexene
4-(1-Methylvinyl)-1-cyclohexene-1-carboxaldehyde, (R)
4-(1-Methylvinyl)-1-cyclohexene-1-carboxaldehyde, (S)
4-(1-Methylvinyl)-1-cyclohexene-1-methanol
(1-Methylvinyl)cyclopropane
Methyl vinyl ether
Methyl Violet
Methysergide
Methysticin
Metobromuron
Metolachlor
Metolazone
Metoprolol tartrate
Metribuzin
Metronidazole
Metsulfuron-methyl
Mevinphos
Mexacarbate

No.	Name	Synonym	Mol. Form.	CAS RN	Mol. Wt.	Physical Form	mp/°C	bp/°C	den/ g cm^{-3}	n_D	Solubility
7872	MGK 264		$C_{17}H_{25}NO_2$	113-48-4	275.387		<-20	157	1.04		
7873	Mifepristone	RU-486	$C_{29}H_{35}NO_2$	84371-65-3	429.594	cry	150				
7874	Mimosine		$C_8H_{10}N_2O_4$	500-44-7	198.176	tab (w)	228 dec				sl H_2O; i EtOH, eth, ace, bz; s dil alk
7875	Minocycline		$C_{23}H_{27}N_3O_7$	10118-90-8	457.476	ye-oran amorp solid					
7876	Minoxidil		$C_9H_{15}N_5O$	38304-91-5	209.248	cry	248				i ace, bz, chl, sl; EtOH, MeOH
7877	Mipafox	Bis(isopropylamido) fluorophosphate	$C_6H_{16}FN_2OP$	371-86-8	182.175	cry (peth)	65	125^{2}			sl H_2O
7878	Mirex	Hexachloropentadiene dimer	$C_{10}Cl_{12}$	2385-85-5	545.543	cry (bz)	485 dec				vs bz, diox
7879	Misoprostol		$C_{22}H_{38}O_5$	59122-46-2	382.534	ye oil					s H_2O
7880	Mithramycin	Plicamycin	$C_{52}H_{76}O_{24}$	18378-89-7	1085.145	ye cry (ace)	182				s H_2O, EtOH, AcOEt; sl bz, eth
7881	Mitomycin A		$C_{16}H_{19}N_3O_6$	4055-39-4	349.338	purp nd	160 dec				
7882	Mitomycin B		$C_{16}H_{19}N_3O_6$	4055-40-7	349.338	purp-bl nd	dec				
7883	Mitomycin C		$C_{15}H_{18}N_4O_5$	50-07-7	334.328	bl-viol cry	360				s H_2O, MeOH, ace
7884	Mitotane		$C_{14}H_{10}Cl_4$	53-19-0	320.041		77				
7885	Mitragynine	9-Methoxycorynantheidine	$C_{23}H_{30}N_2O_4$	4098-40-2	398.495	wh amor pow	104	235^{5}			s EtOH, chl, HOAc
7886	Molinate	Ethyl 1-hexamethyleneiminecarbothiolate	$C_9H_{17}NOS$	2212-67-1	187.302			202^{10}	1.063^{20}		
7887	Molindone		$C_{16}H_{24}N_2O_2$	7416-34-4	276.374	cry	180				
7888	Molybdenum hexacarbonyl		C_6MoO_6	13939-06-5	264.00		dec 150				s os
7889	Monobutyl phthalate	1,2-Benzenedicarboxylic acid, monobutyl ester	$C_{12}H_{14}O_4$	131-70-4	222.237	pl (ace, al)	73.5				vs EtOH, chl
7890	Monobutyltin trichloride		$C_4H_9Cl_3Sn$	1118-46-3	282.183	hyg liq	-63	93^{10}	0.85^{20}		s bz, CH_2Cl_2
7891	Monocrotaline		$C_{16}H_{23}NO_6$	315-22-0	325.357	wh pr (EtOH)	198 dec				
7892	Monocrotophos		$C_7H_{14}NO_5P$	6923-22-4	223.164		55	$125^{0.0005}$	1.33^{20}		
7893	Monolinuron	*N'*-(4-Chlorophenyl)-*N*-methoxy-*N*-methylurea	$C_9H_{11}ClN_2O_2$	1746-81-2	214.648	solid	77				
7894	Monomethyl adipate		$C_7H_{12}O_4$	627-91-8	160.168	lf (Me_3N-MeOH)	9	158^{10}	1.0623^{20}	1.4283^{20}	s EtOH
7895	Monomethyl glutarate		$C_6H_{10}O_4$	1501-27-5	146.141			158^{27}, 150^{10}	1.169^{25}	1.4381^{20}	
7896	Monosodium *L*-glutamate		$C_5H_8NNaO_4$	142-47-2	169.113						s H_2O
7897	Moquizone		$C_{20}H_{21}N_3O_3$	19395-58-5	351.399		136				s chl
7898	Morin		$C_{15}H_{10}O_7$	480-16-0	302.236	pa ye nd (+1 w, dil al)	303.5				sl H_2O, eth; vs EtOH; s bz, alk; i CS_2
7899	Morphine		$C_{17}H_{19}NO_3$	57-27-2	285.338	pr	255	sub 190			i H_2O, eth, ace; s MeOH, py; sl EtOH
7900	4-Morpholinamine		$C_4H_{10}N_2O$	4319-49-7	102.134			166	1.059^{25}	1.4772^{20}	
7901	Morpholine	Tetrahydro-1,4-oxazine	C_4H_9NO	110-91-8	87.120	hyg liq	-4.8	128	1.0005^{20}	1.4548^{20}	msc H_2O; s EtOH, eth, ace, bz; sl chl
7902	4-Morpholinecarboxaldehyde		$C_5H_9NO_2$	4394-85-8	115.131		21	239	1.1520^{20}	1.4845^{20}	
7903	4-Morpholineethanamine		$C_6H_{14}N_2O$	2038-03-1	130.187		25.6	205	0.9897^{20}	1.4715^{20}	msc H_2O, EtOH, bz, lig; s ace
7904	4-Morpholineethanol		$C_6H_{13}NO_2$	622-40-2	131.173	liq	-0.8	227	1.0710^{20}	1.4763^{20}	s H_2O, EtOH; sl ctc
7905	4-Morpholinepropanamine	4-(3-Aminopropyl)morpholine	$C_7H_{16}N_2O$	123-00-2	144.214	liq	-15	220; 134^{50}	0.9854^{20}	1.4762^{20}	msc H_2O, EtOH, bz, lig; s ace; sl ctc
7906	2-(4-Morpholinothio) benzothiazole	4-(2-Benzothiazolylthio) morpholine	$C_{11}H_{12}N_2OS_2$	102-77-2	252.355	cry (EtOH)	85				
7907	4-(4-Morpholinyl)aniline		$C_{10}H_{14}N_2O$	2524-67-6	178.230		131.6				
7908	2-(4-Morpholinyldithio) benzothiazole		$C_{11}H_{12}N_2OS_3$	95-32-9	284.420		135				
7909	Muldamine		$C_{29}H_{47}NO_3$	36069-45-1	457.688		210				
7910	Murexide	5,5'-Nitrilobarbituric acid, ammonium salt	$C_8H_8N_6O_6$	3051-09-0	284.186						sl H_2O; i EtOH, eth; s alk
7911	Muscimol	5-(Aminomethyl)-3(2*H*)-isoxazolone	$C_4H_6N_2O_2$	2763-96-4	114.103	cry (EtOH)	175 dec				
7912	Myclobutanil		$C_{15}H_{17}ClN_4$	88671-89-0	288.776	ye cry	65	$205^{1.0}$			i H_2O, peth; s EtOH
7913	Mycophenolic acid		$C_{17}H_{20}O_6$	24280-93-1	320.337	nd (w)	141				i H_2O; vs EtOH, eth, chl; sl bz, tol

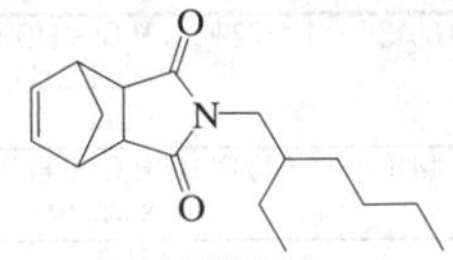

MGK 264

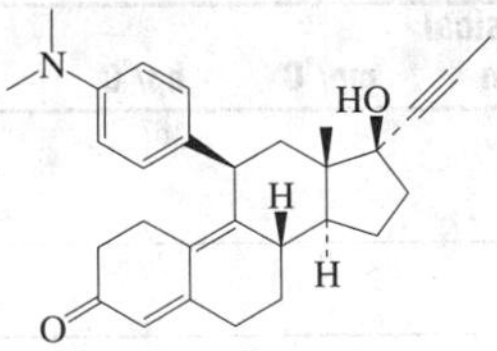

Mifepristone

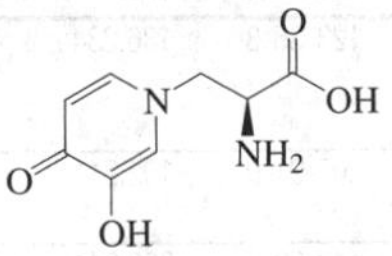

Mimosine

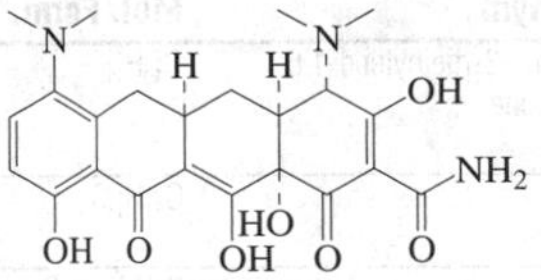

Minocycline

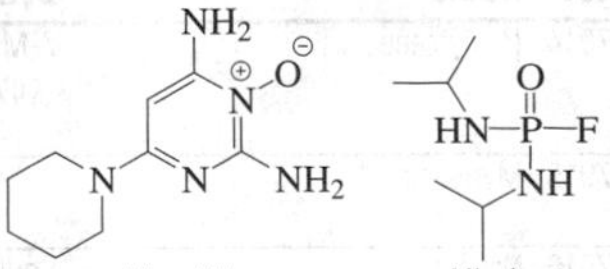

Minoxidil

Mipafox

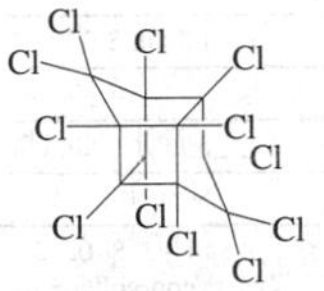

Mithramycin

Mirex

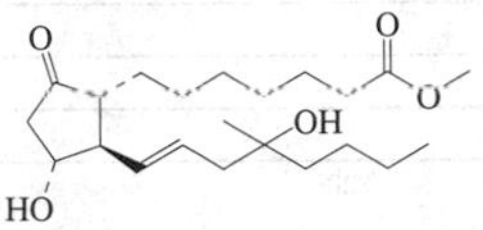

Misoprostol

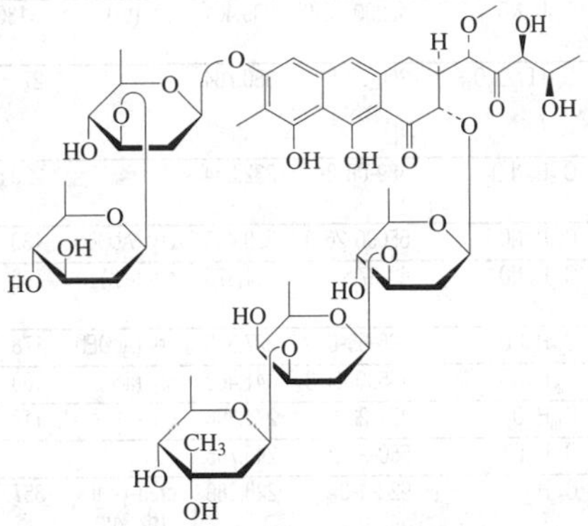

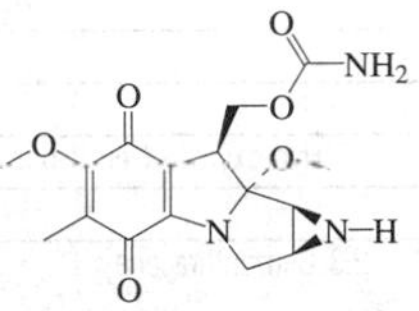

Mitomycin A

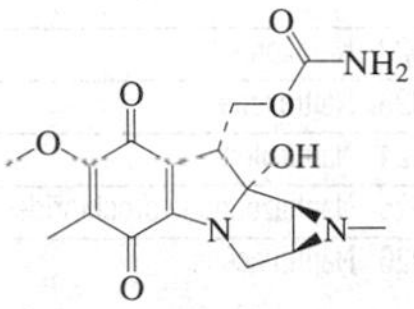

Mitomycin B

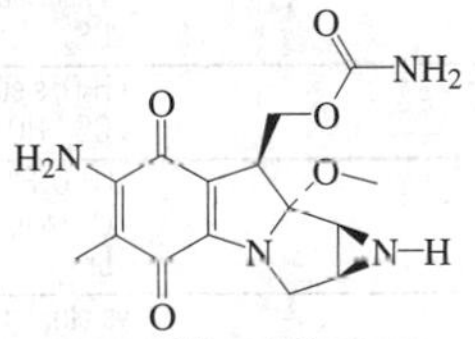

Mitomycin C

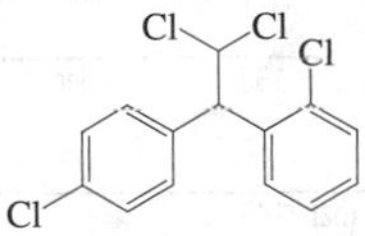

Mitotane

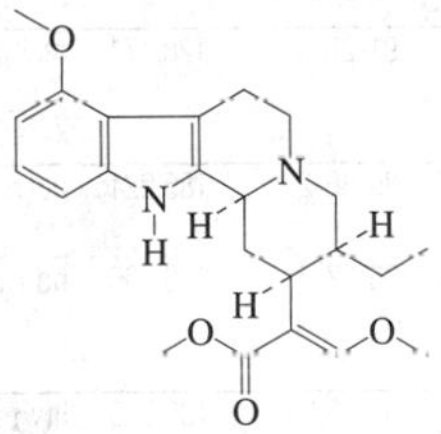

Mitragynine

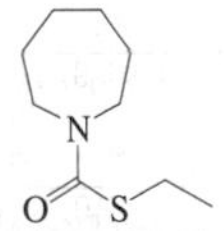

Molinate

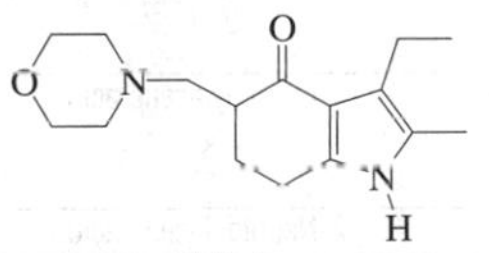

Molindone

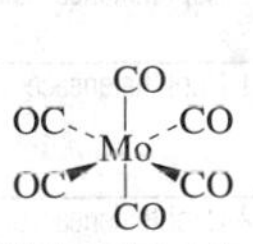

Molybdenum hexacarbonyl

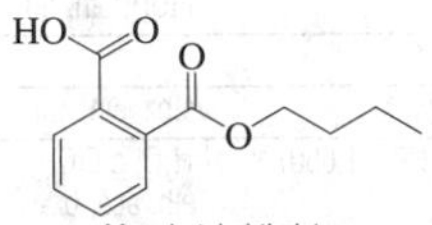

Monobutyl phthalate

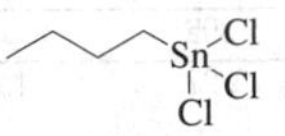

Monobutyltin trichloride

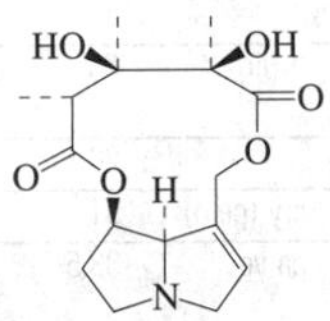

Monocrotaline

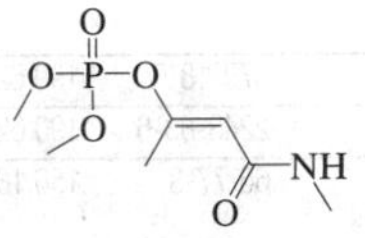

Monocrotophos

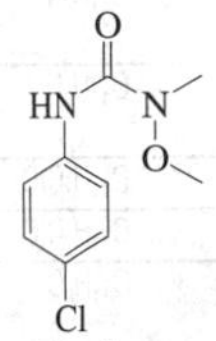

Monolinuron

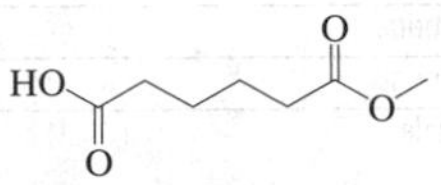

Monomethyl adipate

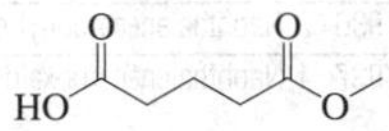

Monomethyl glutarate

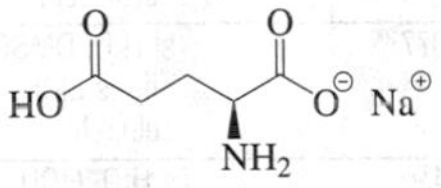

Monosodium *L*-glutamate

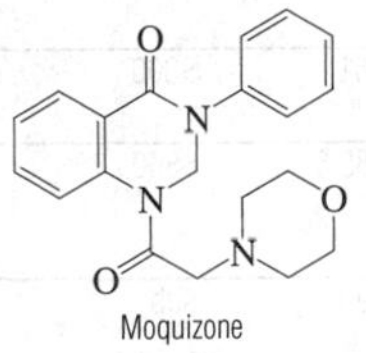

Moquizone

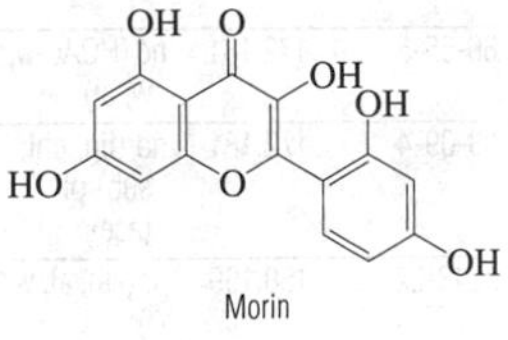

Morin

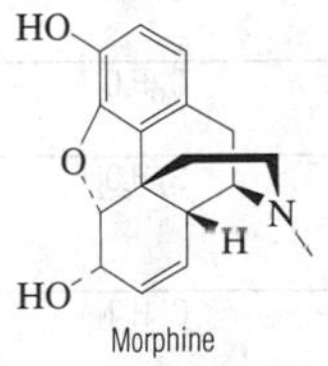

Morphine

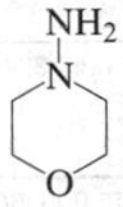

4-Morpholinamine

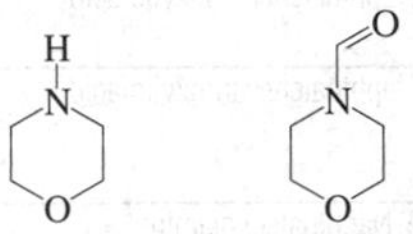

Morpholine

4-Morpholinecarboxaldehyde

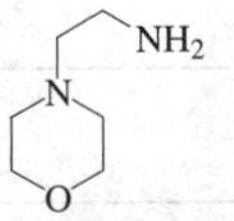

4-Morpholineethanamine

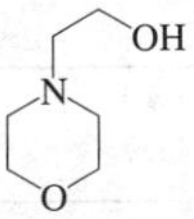

4-Morpholineethanol

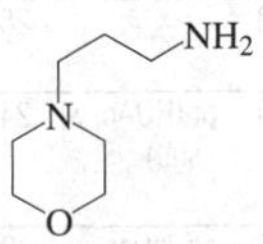

4-Morpholinepropanamine

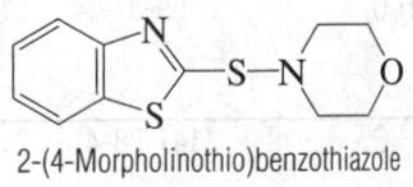

2-(4-Morpholinothio)benzothiazole

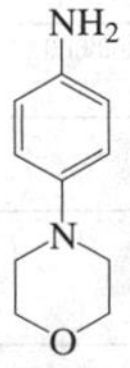

4-(4-Morpholinyl)aniline

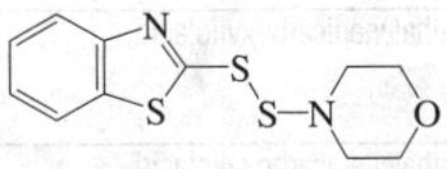

2-(4-Morpholinyldithio)benzothiazole

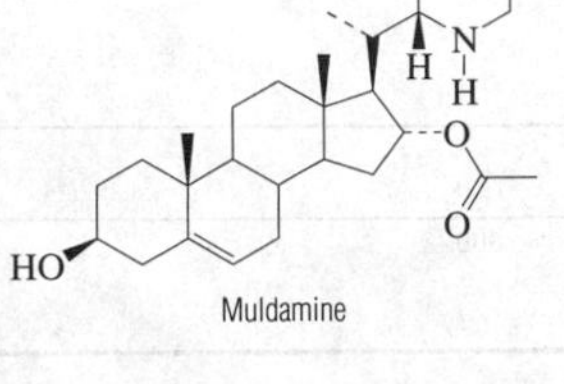

Muldamine

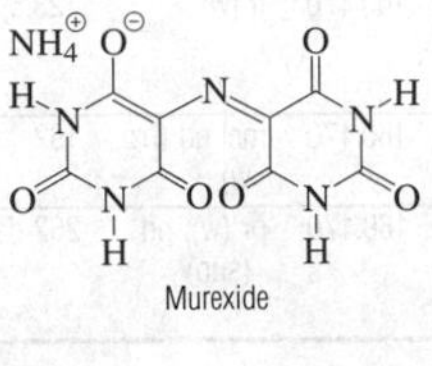

Murexide

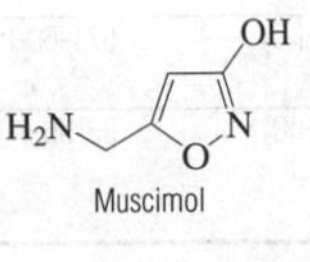

Muscimol

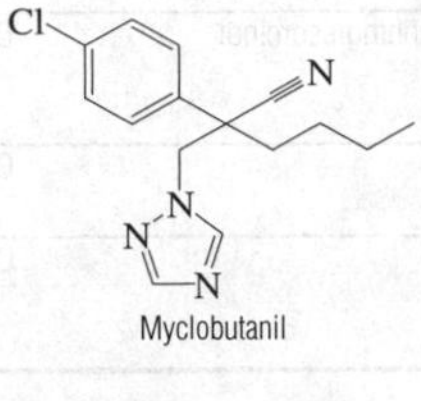

Myclobutanil

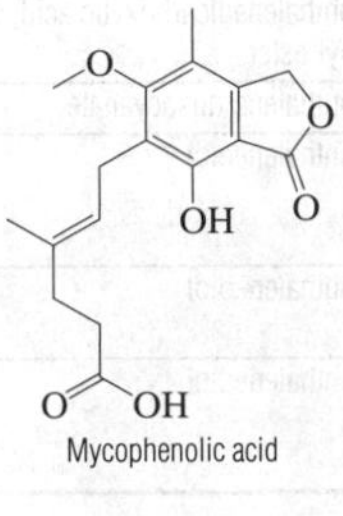

Mycophenolic acid

No.	Name	Synonym	Mol. Form.	CAS RN	Mol. Wt.	Physical Form	mp/°C	bp/°C	den/ g cm^{-3}	n_D	Solubility
7914	β-Myrcene	7-Methyl-3-methylene-1,6-octadiene	$C_{10}H_{16}$	123-35-3	136.234			167	0.8013^{15}	1.4722^{20}	i H_2O; s EtOH, eth, bz, chl, HOAc
7915	Myristicin		$C_{11}H_{12}O_3$	607-91-0	192.211		<-20	276.5	1.1416^{20}	1.5403^{20}	i H_2O; sl EtOH; s eth, bz
7916	Nabam	Sodium ethylenebisdithiocarbamic acid	$C_4H_6N_2Na_2S_4$	142-59-6	256.344	cry (w)					s H_2O
7917	Nadolol		$C_{17}H_{27}NO_4$	42200-33-9	309.401	cry (bz)	≈130				s EtOH; sl chl; i ace, eth, hx
7918	Naled	1,2-Dibromo-2,2-dichloroethylphosphoric acid, dimethyl ester	$C_4H_7Br_2Cl_2O_4P$	300-76-5	380.784		27	$110^{0.5}$	1.96^{20}		
7919	Nalidixic acid		$C_{12}H_{12}N_2O_3$	389-08-2	232.234		229.5				sl EtOH, eth; s chl
7920	Nalmefene		$C_{21}H_{25}NO_3$	55096-26-9	339.429	cry (AcOEt)	189				
7921	Nalorphine	Acetorphin	$C_{19}H_{21}NO_3$	62-67-9	311.375	cry (eth)	208				sl H_2O; s alk, ace, EtOH
7922	Naloxone		$C_{19}H_{21}NO_4$	465-65-6	327.375	cry (AcOEt)	178				i peth; s chl
7923	Naltrexone		$C_{20}H_{23}NO_4$	16590-41-3	341.402	cry (ace)	169				
7924	Nandrolone	17-Hydroxyestr-4-en-3-one	$C_{18}H_{26}O_2$	434-22-0	274.398	cry	112				s EtOH, eth, chl
7925	Naphazoline hydrochloride		$C_{14}H_{15}ClN_2$	550-99-2	246.735						sl H_2O
7926	Naphthacene	2,3-Benzanthracene	$C_{18}H_{12}$	92-24-0	228.288	oran-ye lf (bz, xyl)	357	sub			i H_2O; sl bz; s con sulf
7927	5,12-Naphthacenedione		$C_{18}H_{10}O_2$	1090-13-7	258.271		285 dec				sl ace, bz, gl HOAc
7928	Naphthalene		$C_{10}H_8$	91-20-3	128.171	mcl pl (al)	80.26	217.9	1.0253^{20}	1.5898^{25}	i H_2O; s EtOH; vs eth, ace, bz, CS_2
7929	1-Naphthaleneacetamide		$C_{12}H_{11}NO$	86-86-2	185.221	nd(w, al)		sub 180			i H_2O; s eth, bz, CS_2, HOAc
7930	1-Naphthaleneacetic acid	1-Naphthylacetic acid	$C_{12}H_{10}O_2$	86-87-3	186.206	nd (w)	133	dec			sl H_2O, EtOH; vs eth, ace, chl; s bz
7931	2-Naphthaleneacetic acid	2-Naphthylacetic acid	$C_{12}H_{10}O_2$	581-96-4	186.206	lf(w) cry (bz)	143				vs eth, lig, chl
7932	1-Naphthaleneacetonitrile		$C_{12}H_9N$	132-75-2	167.206		32.5	192^{18}, 163^{12}		1.6192^{20}	s EtOH
7933	1-Naphthalenecarbonitrile		$C_{11}H_7N$	86-53-3	153.181	nd (lig)	37.5	299	1.1080^{25}	1.6298^{18}	i H_2O; vs EtOH, eth; s lig
7934	2-Naphthalenecarbonitrile		$C_{11}H_7N$	613-46-7	153.181	lf (lig)	66	306.5	1.0755^{60}		sl H_2O, chl; s EtOH, eth, lig
7935	1-Naphthalenecarbonyl chloride		$C_{11}H_7ClO$	879-18-5	190.626		20	297.5			
7936	2-Naphthalenecarbonyl chloride		$C_{11}H_7ClO$	2243-83-6	190.626	cry (peth)	51	305			vs bz, eth, chl
7937	1-Naphthalenecarboxaldehyde		$C_{11}H_8O$	66-77-3	156.181	pa ye	33.5	292	1.1503^{20}	1.6507^{20}	i H_2O; s EtOH, eth, ace, bz, sulf
7938	2-Naphthalenecarboxaldehyde		$C_{11}H_8O$	66-99-9	156.181	lf (w)	62	160^{19}	1.0775^{99}	1.6211^{99}	sl H_2O; vs EtOH, eth; s ace
7939	1-Naphthalenecarboxylic acid	1-Naphthoic acid	$C_{11}H_8O_2$	86-55-5	172.181	nd (HOAc-w, w, al)	161	>300	1.398^{25}	1.46	i H_2O; vs eth, EtOH, chl
7940	2-Naphthalenecarboxylic acid	2-Naphthoic acid	$C_{11}H_8O_2$	93-09-4	172.181	nd (lig, chl, sub) pl (ace)	185.5	>300	1.077^{100}		sl H_2O, DMSO, lig; s EtOH, eth, chl
7941	1,5-Naphthalenediamine	1,5-Diaminonaphthalene	$C_{10}H_{10}N_2$	2243-62-1	158.199	pr (eth, al, w)	190	sub	1.4^{25}		s H_2O, EtOH, eth; vs chl
7942	1,8-Naphthalenediamine	1,8-Diaminonaphthalene	$C_{10}H_{10}N_2$	479-27-6	158.199		66.5	205^{12}	1.1265^{90}	1.6828^{99}	vs eth, EtOH
7943	2,3-Naphthalenediamine	2,3-Diaminonaphthalene	$C_{10}H_{10}N_2$	771-97-1	158.199	lf (eth, w)	199		1.0968^{26}	1.6392^{26}	sl H_2O, DMSO; vs EtOH; s eth
7944	1,8-Naphthalenedicarboxylic acid	Naphthalic acid	$C_{12}H_8O_4$	518-05-8	216.190		260				i H_2O; sl EtOH, eth
7945	2,3-Naphthalenedicarboxylic acid		$C_{12}H_8O_4$	2169-87-1	216.190	pr (HOAc, w, sub)	244.5				i H_2O, bz, chl; sl EtOH, eth, DMSO
7946	2,6-Naphthalenedicarboxylic acid		$C_{12}H_8O_4$	1141-38-4	216.190	nd (al or sub)	>300 dec				vs EtOH
7947	2,6-Naphthalenedicarboxylic acid, dimethyl ester		$C_{14}H_{12}O_4$	840-65-3	244.243		190.0				
7948	1,5-Naphthalene diisocyanate	1,5-Diisocyanatonaphthalene	$C_{12}H_6N_2O_2$	3173-72-6	210.188	cry	127	183^{10}			
7949	1,3-Naphthalenediol	Naphthoresorcinol	$C_{10}H_8O_2$	132-86-5	160.170	lf (w)	123.5				s H_2O, EtOH, eth; sl ace, bz, lig
7950	1,4-Naphthalenediol		$C_{10}H_8O_2$	571-60-8	160.170	mcl nd (bz, w)	192				s H_2O, EtOH, eth; sl ace; i bz
7951	1,5-Naphthalenediol		$C_{10}H_8O_2$	83-56-7	160.170	pr (w), nd (sub)	262 dec	sub			sl H_2O, EtOH; vs eth, ace; i bz; s HOAc

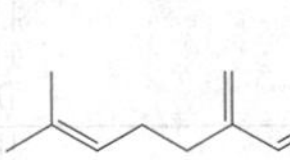

β-Myrcene

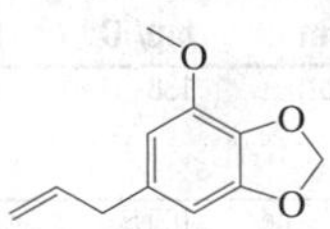

Myristicin

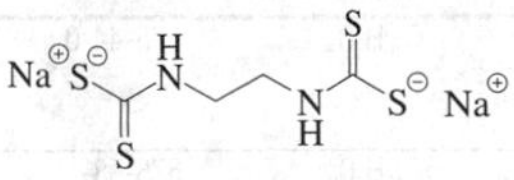

Nabam

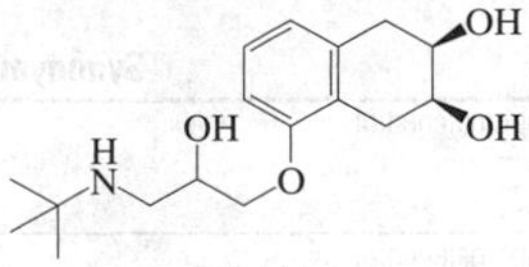

Nadolol

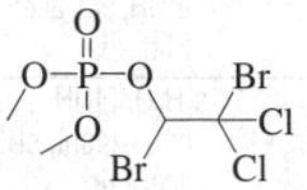

Naled

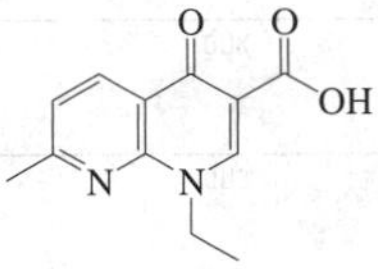

Nalidixic acid

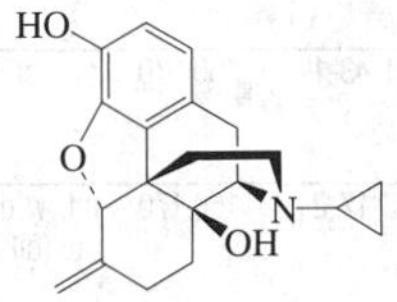

Nalmefene

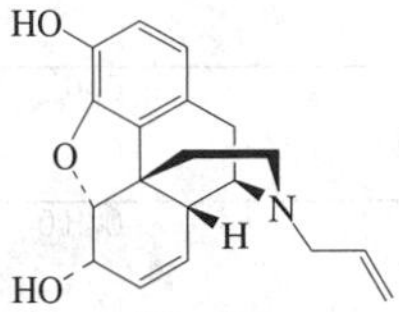

Nalorphine

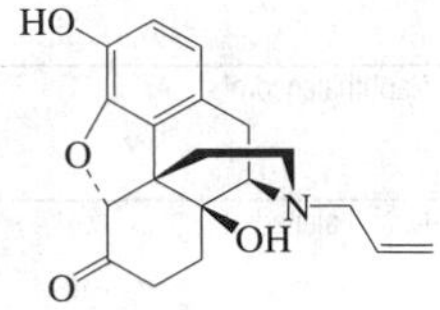

Naloxone

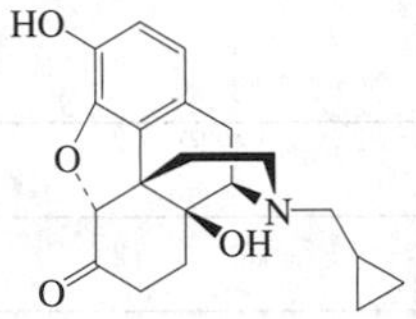

Naltrexone

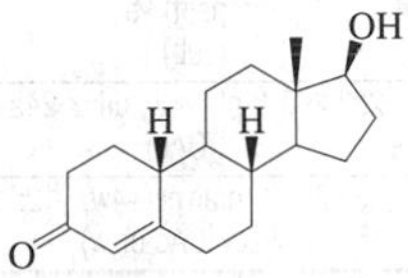

Nandrolone

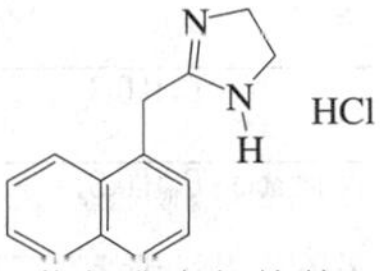

Naphazoline hydrochloride

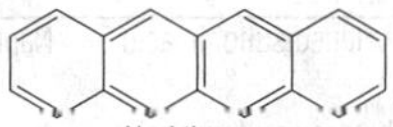

Naphthacene

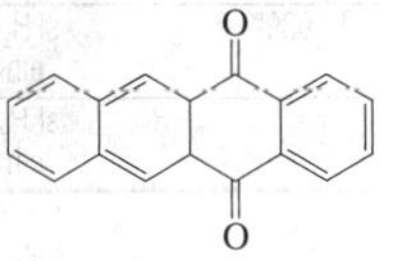

5,12-Naphthacenedione

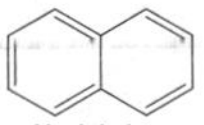

Naphthalene

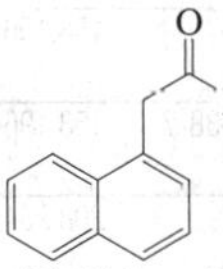

1-Naphthaleneacetamide

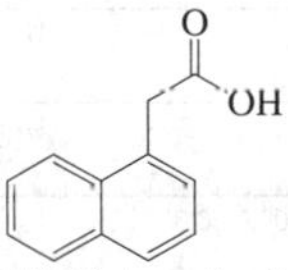

1-Naphthaleneacetic acid

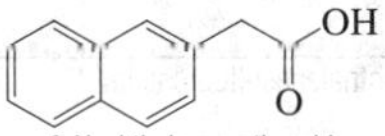

2-Naphthaleneacetic acid

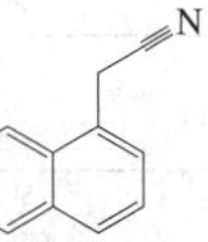

1-Naphthaleneacetonitrile

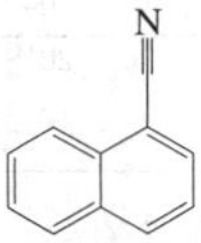

1-Naphthalenecarbonitrile

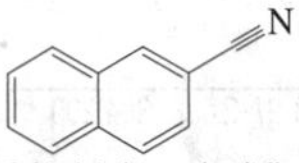

2-Naphthalenecarbonitrile

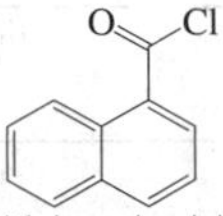

1-Naphthalenecarbonyl chloride

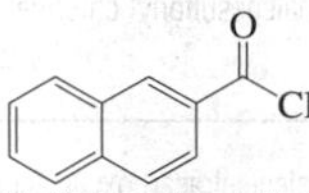

2-Naphthalenecarbonyl chloride

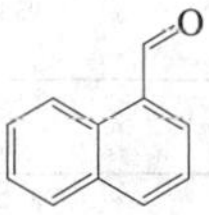

1-Naphthalenecarboxaldehyde

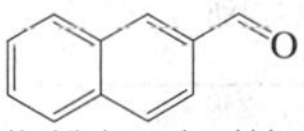

2-Naphthalenecarboxaldehyde

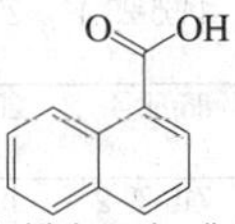

1-Naphthalenecarboxylic acid

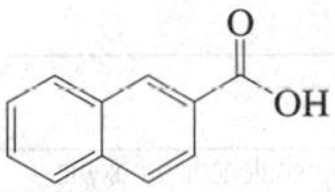

2-Naphthalenecarboxylic acid

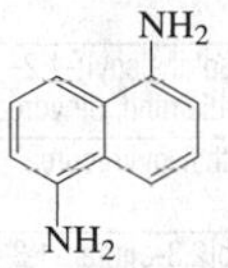

1,5-Naphthalenediamine

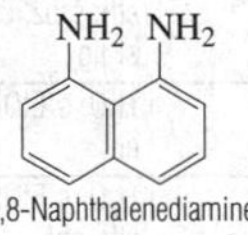

1,8-Naphthalenediamine

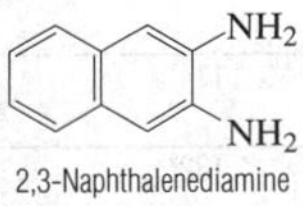

2,3-Naphthalenediamine

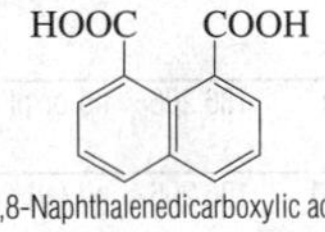

1,8-Naphthalenedicarboxylic acid

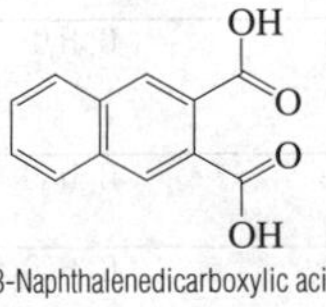

2,3-Naphthalenedicarboxylic acid

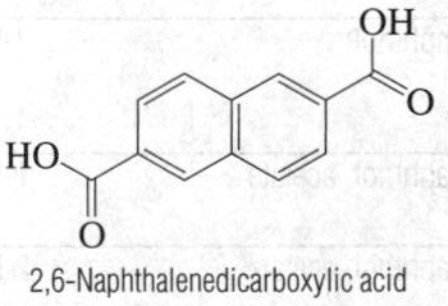

2,6-Naphthalenedicarboxylic acid

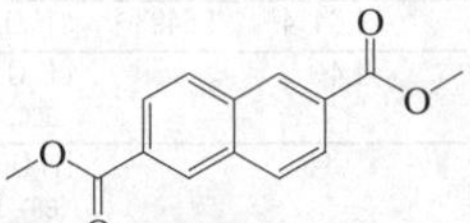

2,6-Naphthalenedicarboxylic acid, dimethyl ester

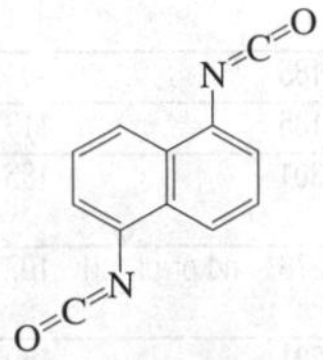

1,5-Naphthalene diisocyanate

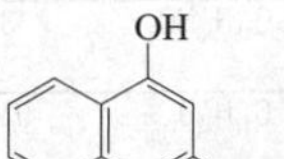

1,3-Naphthalenediol

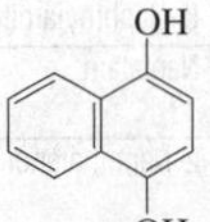

1,4-Naphthalenediol

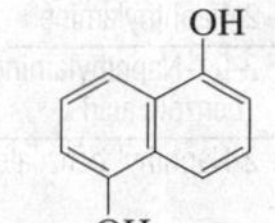

1,5-Naphthalenediol

No.	Name	Synonym	Mol. Form.	CAS RN	Mol. Wt.	Physical Form	mp/°C	bp/°C	den/ g cm^{-3}	n_D	Solubility
7952	1,6-Naphthalenediol		$C_{10}H_8O_2$	575-44-0	160.170	pr (bz)	138	sub			sl H_2O, EtOH; s eth, ace, bz, DMSO
7953	1,7-Naphthalenediol		$C_{10}H_8O_2$	575-38-2	160.170	nd (bz or sub)	180.5	sub			sl H_2O; vs EtOH, eth; s bz, HOAc
7954	2,3-Naphthalenediol		$C_{10}H_8O_2$	92-44-4	160.170	lf (w)	163.5				s H_2O, EtOH, eth, ace, bz, lig, HOAc
7955	2,6-Naphthalenediol		$C_{10}H_8O_2$	581-43-1	160.170	orth pl (w)	220	sub			sl H_2O, bz; s EtOH, eth, ace; i lig
7956	2,7-Naphthalenediol		$C_{10}H_8O_2$	582-17-2	160.170	nd, (w, dil al), pl (dil al)	193	sub			s H_2O, EtOH, eth, bz, chl; sl ace; i lig
7957	1,2-Naphthalenedione	1,2-Naphthoquinone	$C_{10}H_6O_2$	524-42-5	158.154	ye-red nd (eth) oran lf (bz)	146		1.450^{25}		s H_2O, EtOH, eth, sulf; sl lig
7958	1,4-Naphthalenedione	1,4-Naphthoquinone	$C_{10}H_6O_2$	130-15-4	158.154	bt ye nd (al, peth) ye (sub)	128.5	sub			sl H_2O; vs EtOH; s eth, bz, chl, CS_2
7959	1,5-Naphthalenedisulfonic acid	Armstrong's acid	$C_{10}H_8O_6S_2$	81-04-9	288.297	pl (+4w, dil HOAc)	242 dec		1.493^{25}		vs H_2O; s EtOH; i eth
7960	1,6-Naphthalenedisulfonic acid	Naphthalene-1,6-disulfonic acid	$C_{10}H_8O_6S_2$	525-37-1	288.297	oran pr (+4w, HOAc or w)	125 dec				vs H_2O; s EtOH; i eth
7961	2,7-Naphthalenedisulfonic acid	Naphthalene-2,7-disulfonic acid	$C_{10}H_8O_6S_2$	92-41-1	288.297	hyg nd (conc HCl)	199				s H_2O; sl con HCl
7962	1-Naphthalenemethanamine		$C_{11}H_{11}N$	118-31-0	157.212			292	1.0958^{20}		s EtOH, eth, sulf, CS_2
7963	1-Naphthalenemethanol		$C_{11}H_{10}O$	4780-79-4	158.196	nd (w, al), cry (bz-lig)	64	304; 163^{12}	1.1039^{80}		sl H_2O; vs EtOH, eth
7964	2-Naphthalenemethanol		$C_{11}H_{10}O$	1592-38-7	158.196	lf	81.3	178^{12}			sl H_2O; s EtOH, eth
7965	1-Naphthalenesulfonic acid	α-Naphthylsulfonic acid	$C_{10}H_8O_3S$	85-47-2	208.234	pr (+2 w, dil HCl)	140				s H_2O, EtOH; sl eth
7966	2-Naphthalenesulfonic acid	β-Naphthylsulfonic acid	$C_{10}H_8O_3S$	120-18-3	208.234	hyg pl (+1w), cry (+3w, HCl)	91	dec	1.441^{25}		vs H_2O, EtOH; s eth; sl bz
7967	1-Naphthalenesulfonyl chloride		$C_{10}H_7ClO_2S$	85-46-1	226.680	lf (eth)	68	209^{20}, $147^{0.9}$			vs bz, eth, EtOH
7968	2-Naphthalenesulfonyl chloride		$C_{10}H_7ClO_2S$	93-11-8	226.680	pow or lf (bz-peth)	81	201^{13}, $148^{0.5}$			i H_2O; s EtOH, bz, chl; sl peth; vs eth
7969	1,4,5,8-Naphthalenetetracarboxylic acid		$C_{14}H_8O_8$	128-97-2	304.209	lf or nd (w, dil HCl)	320				sl H_2O, bz, chl, EtOH; vs ace
7970	1-Naphthalenethiol	1-Naphthyl mercaptan	$C_{10}H_8S$	529-36-2	160.236			dec 285; 161^{20}	1.1607^{20}	1.6802^{20}	sl H_2O, dil alk; vs EtOH, eth
7971	2-Naphthalenethiol	2-Naphthyl mercaptan	$C_{10}H_8S$	91-60-1	160.236	pl (al)	81	288	1.550^{25}		sl H_2O; vs EtOH, eth, lig
7972	*N*-(1-Naphthalenyl)-1,2-ethanediamine, dihydrochloride		$C_{12}H_{16}Cl_2N_2$	1465-25-4	259.174	hex pr	189				vs H_2O, EtOH
7973	1-Naphthalenylthiourea	ANTU	$C_{11}H_{10}N_2S$	86-88-4	202.275	pr (al)	198				i H_2O; sl EtOH, eth, ace
7974	Naphtho[2,3-c]furan-1,3-dione	2,3-Naphthalenedicarboxylic acid anhydride	$C_{12}H_6O_3$	716-39-2	198.174		246				sl EtOH, chl; s eth, bz
7975	1-Naphthol	1-Naphthalenol	$C_{10}H_8O$	90-15-3	144.170	ye nd (w)	95.0	288; 184^{40}	1.0989^{99}	1.6224^{99}	i H_2O; vs EtOH, eth; s ace, bz; sl ctc
7976	2-Naphthol	2-Naphthalenol	$C_{10}H_8O$	135-19-3	144.170	mcl lf (w)	121.5	285	1.28^{20}		i H_2O; vs EtOH, eth; s bz, chl; sl lig
7977	1-Naphthol, acetate	1-Naphthyl acetate	$C_{12}H_{10}O_2$	830-81-9	186.206	nd or pl (al)	49	114^{1}			i H_2O; s EtOH, eth
7978	2-Naphthol, acetate	2-Naphthyl acetate	$C_{12}H_{10}O_2$	1523-11-1	186.206	nd (al)	71.0	132^{2}			i H_2O; s EtOH, eth, chl
7979	*p*-Naphtholbenzein		$C_{27}H_{18}O_2$	145-50-6	374.431		123				
7980	1*H*,3*H*-Naphtho[1,8-cd]pyran-1,3-dione		$C_{12}H_6O_3$	81-84-5	198.174		275.0				i H_2O, eth, bz; sl EtOH; s HOAc
7981	1-Naphthylamine	α-Naphthylamine	$C_{10}H_9N$	134-32-7	143.185		49.2	300.7	1.0228^{20}	1.6140^{20}	s chl
7982	2-Naphthylamine	β-Naphthylamine	$C_{10}H_9N$	91-59-8	143.185		113	306.2	1.6414^{98}	1.6493^{98}	s H_2O, EtOH, eth
7983	2-[(1-Naphthylamino)carbonyl] benzoic acid	Naptalam	$C_{18}H_{13}NO_3$	132-66-1	291.301		185		1.4^{20}		i H_2O; sl EtOH, ace, bz, tfa
7984	2-Naphthyl benzoate	2-Naphthalenol benzoate	$C_{17}H_{12}O_2$	93-44-7	248.276	nd or pr (al)	107				i H_2O; s EtOH; sl eth, HOAc
7985	*N*-1-Naphthalenylacetamide		$C_{12}H_{11}NO$	575-36-0	185.221		160				s H_2O, EtOH; sl eth
7986	*N*-1-Naphthyl-1,2-ethanediamine	*N*-(1-Naphthyl)ethylenediamine	$C_{12}H_{14}N_2$	551-09-7	186.252	visc lig		204^{9}	1.114^{25}	1.6648^{25}	

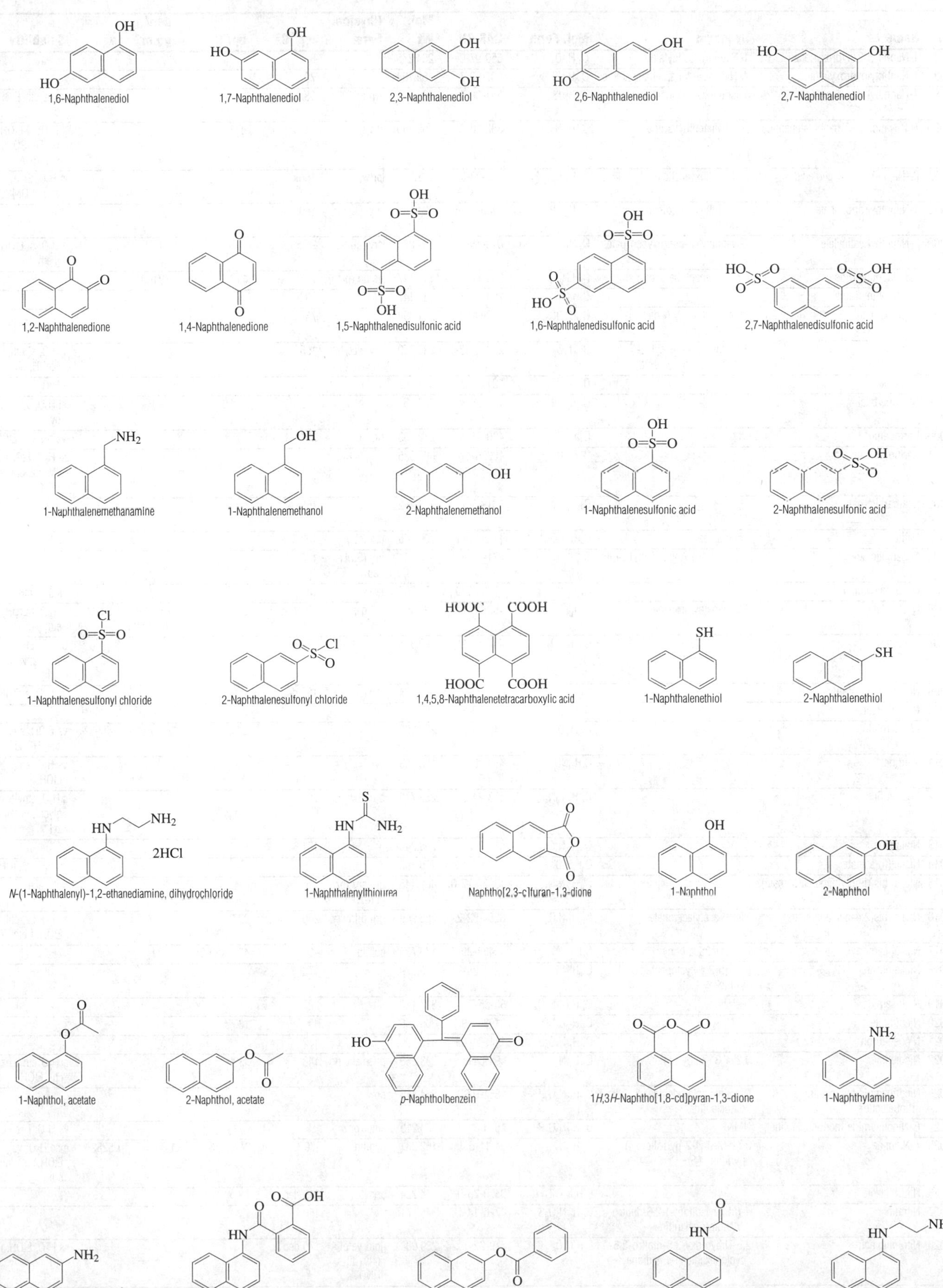
1,6-Naphthalenediol
1,7-Naphthalenediol
2,3-Naphthalenediol
2,6-Naphthalenediol
2,7-Naphthalenediol
1,2-Naphthalenedione
1,4-Naphthalenedione
1,5-Naphthalenedisulfonic acid
1,6-Naphthalenedisulfonic acid
2,7-Naphthalenedisulfonic acid
1-Naphthalenemethanamine
1-Naphthalenemethanol
2-Naphthalenemethanol
1-Naphthalenesulfonic acid
2-Naphthalenesulfonic acid
1-Naphthalenesulfonyl chloride
2-Naphthalenesulfonyl chloride
1,4,5,8-Naphthalenetetracarboxylic acid
1-Naphthalenethiol
2-Naphthalenethiol
2HCl
N-(1-Naphthalenyl)-1,2-ethanediamine, dihydrochloride
1-Naphthalenylthiourea
Naphtho[2,3-c]furan-1,3-dione
1-Naphthol
2-Naphthol
1-Naphthol, acetate
2-Naphthol, acetate
p-Naphtholbenzein
1H,3H-Naphtho[1,8-cd]pyran-1,3-dione
1-Naphthylamine
2-Naphthylamine
2-[(1-Naphthylamino)carbonyl]benzoic acid
2-Naphthyl benzoate
N-1-Naphthalenylacetamide
N-1-Naphthyl-1,2-ethanediamine

No.	Name	Synonym	Mol. Form.	CAS RN	Mol. Wt.	Physical Form	mp/°C	bp/°C	den/ g cm^{-3}	n_D	Solubility
7987	1-Naphthyl 2-hydroxybenzoate	1-Naphthyl salicylate	$C_{17}H_{12}O_3$	550-97-0	264.275		83				vs eth
7988	1-Naphthylhydroxylamine	*N*-Hydroxyl-1-naphthalenamine	$C_{10}H_9NO$	607-30-7	159.184		79				
7989	1-Naphthyl isothiocyanate	1-Isothiocyanatonaphthalene	$C_{11}H_7NS$	551-06-4	185.246	wh nd (al)	58				vs bz, eth, EtOH, chl
7990	*N*-2-Naphthyl-2-naphthalenamine	β,β'-Dinaphthylamine	$C_{20}H_{15}N$	532-18-3	269.340	lf(bz)	172.2	471			i H_2O; sl EtOH, bz, DMSO; s eth, HOAc
7991	(2-Naphthyloxy)acetic acid	2-Naphthoxyacetic acid	$C_{12}H_{10}O_3$	120-23-0	202.205	pr(w)	156				s H_2O, EtOH, eth; sl DMSO
7992	1-Naphthyl phosphate	1-Naphthalenol, dihydrogen phosphate	$C_{10}H_9O_4P$	1136-89-6	224.149	cry	160				
7993	2-Naphthyl salicylate	2-Naphthyl 2-hydroxybenzoate	$C_{17}H_{12}O_3$	613-78-5	264.275	cry (al)	95.5		1.11^{116}		i H_2O; sl EtOH; s eth, bz
7994	1,5-Naphthyridine	1,5-Diazanaphthalene	$C_8H_6N_2$	254-79-5	130.147	ye nd (peth)	75	112^{12}	1.2100^{20}		
7995	1,6-Naphthyridine		$C_8H_6N_2$	253-72-5	130.147		29.5				
7996	Napropamide	Propanamide, *N,N*-diethyl-2-(1-naphthalenyloxy)-	$C_{17}H_{21}NO_2$	15299-99-7	271.355		75				
7997	Naproxen	6-Methoxy-α-methyl-2-naphthaleneacetic acid	$C_{14}H_{14}O_3$	22204-53-1	230.259	cry (ace/hx)	155				i H_2O; sl eth; s MeOH, chl
7998	Narceine		$C_{23}H_{27}NO_8$	131-28-2	445.462		138				i H_2O
7999	Narcobarbital		$C_{11}H_{15}BrN_2O_3$	125-55-3	303.152		115				sl H_2O; s EtOH, py
8000	Naringenin		$C_{15}H_{12}O_5$	480-41-1	272.253	nd (dil al)	251				vs bz, eth, EtOH
8001	Naringin		$C_{27}H_{32}O_{14}$	10236-47-2	580.535	nd (w+8)					sl H_2O, EtOH; i eth, bz, chl; s HOAc
8002	Nealbarbital		$C_{12}H_{18}N_2O_3$	561-83-1	238.282		156				vs ace, eth, EtOH
8003	Nellite	Diamidafos	$C_8H_{13}N_2O_2P$	1754-58-1	200.175	cry (ctc)	103.5				sl AcOEt, bz
8004	Neoabietic acid	8(14),13(15)-Abietadien-18-oic acid	$C_{20}H_{30}O_2$	471-77-2	302.451	cry (EtOH aq)	173				
8005	Neobornylamine		$C_{10}H_{19}N$	2223-67-8	153.265	pow	184				vs ace, eth
8006	Neopentane	2,2-Dimethylpropane	C_5H_{12}	463-82-1	72.149	col gas	-16.4	9.48	0.5852^{25} (p>1 atm	1.3476^{6}	i H_2O; s EtOH, eth, ctc
8007	Neopine		$C_{18}H_{21}NO_3$	467-14-1	299.365	nd (peth)	127.5				s H_2O, EtOH, eth, bz; vs chl; sl lig
8008	Neostigmine bromide		$C_{12}H_{19}BrN_2O_2$	114-80-7	303.195	cry (al-eth)	167 dec				vs H_2O; s EtOH
8009	Nepetalactone		$C_{10}H_{14}O_2$	490-10-8	166.217			$71^{0.05}$	1.0663^{25}	1.4859^{25}	
8010	*cis*-Nerolidol		$C_{15}H_{26}O$	142-50-7	222.366			276; $70^{0.1}$	0.8778^{20}	1.4898^{20}	vs EtOH; s eth, ace, HOAc
8011	Neurine		$C_5H_{13}NO$	463-88-7	103.163	syr					vs H_2O, eth, EtOH
8012	Neutral Red		$C_{15}H_{17}ClN_4$	553-24-2	288.776	grn pow					s H_2O, ethylene glycol, EtOH; i xyl
8013	Nialamide		$C_{16}H_{18}N_4O_2$	51-12-7	298.340		151.6				
8014	Nickel(II) acetate		$C_4H_6NiO_4$	373-02-4	176.782						vs H_2O; s EtOH
8015	Nickel bis(dibutyldithiocarbamate)		$C_{18}H_{36}N_2NiS_4$	13927-77-0	467.445	grn cry (bz/EtOH)	91				s bz, ace
8016	Nickel bis(2,4-pentanedioate)	Nickel acetylacetonate	$C_{10}H_{14}NiO_4$	3264-82-2	256.909	grn orth cry	230	227^{11}			s H_2O, bz, chl, EtOH; i eth
8017	Nickel carbonyl	Nickel tetracarbonyl	C_4NiO_4	13463-39-3	170.734	col liq	-19.3	43 (exp 60)	1.31^{25}		
8018	Nickelocene	Bis(η5-2,4-cyclopentadien-1-yl)nickel	$C_{10}H_{10}Ni$	1271-28-9	188.879		172				
8019	Niclosamide		$C_{13}H_8Cl_2N_2O_4$	50-65-7	327.120		227				
8020	Nicofibrate		$C_{16}H_{16}ClNO_3$	31980-29-7	305.756		49	$180^{0.4}$			
8021	Nicosulfuron		$C_{15}H_{18}N_6O_6S$	111991-09-4	410.405		172				
8022	Nicotelline	3,2':4',3"-Terpyridine	$C_{15}H_{11}N_3$	494-04-2	233.268	prismatic nd	148	>300			sl H_2O, eth; s bz, chl, EtOH
8023	Nicotinamide hypoxanthine dinucleotide	Nicotinic acid adenine dinucleotide	$C_{21}H_{26}N_6O_{15}P_2$	1851-07-6	664.410	pow					
8024	β-Nicotinamide mononucleotide	NMN	$C_{11}H_{15}N_2O_8P$	1094-61-7	334.219	amor pow					vs H_2O; i ace
8025	*L*-Nicotine	3-(1-Methyl-2-pyrrolidinyl) pyridine, (*S*)-	$C_{10}H_{14}N_2$	54-11-5	162.231	hyg liq	-79	247; 125^{18}	1.0097^{20}	1.5282^{20}	msc H_2O; vs EtOH, eth, chl; s lig
8026	Nifurthiazole		$C_8H_6N_4O_4S$	3570-75-0	254.224	cry	215 dec				
8027	Nitralin	4-(Methylsulfonyl)-2,6-dinitro-*N,N*-dipropylaniline	$C_{13}H_{19}N_3O_6S$	4726-14-1	345.371		150				
8028	Nitranilic acid	2,5-Dihydroxy-3,6-dinitro-2,5-cyclohexadiene-1,4-dione	$C_6H_2N_2O_8$	479-22-1	230.088	gold-ye pl (+w, dil HNO_3	170 dec				vs H_2O, EtOH; i eth
8029	Nitrapyrin	Pyridine, 2-chloro-6-(trichloromethyl)-	$C_6H_3Cl_4N$	1929-82-4	230.907		63	136^{11}			

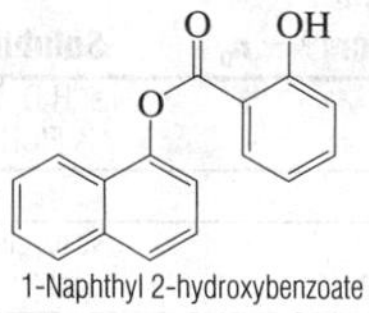

1-Naphthyl 2-hydroxybenzoate
1-Naphthylhydroxylamine 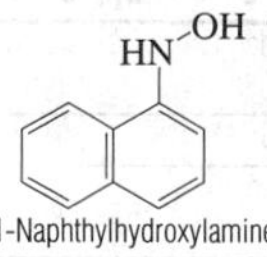
1-Naphthyl isothiocyanate 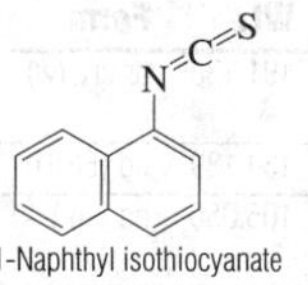
N-2-Naphthyl-2-naphthalenamine 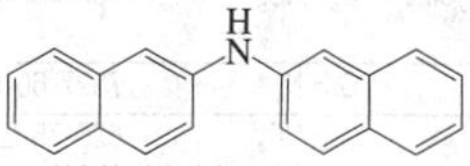
(2-Naphthyloxy)acetic acid 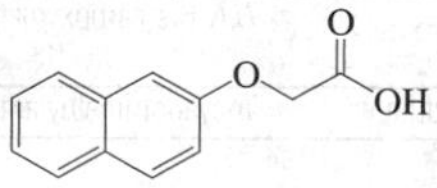
1-Naphthyl phosphate 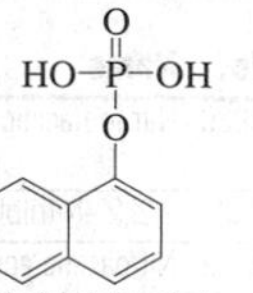

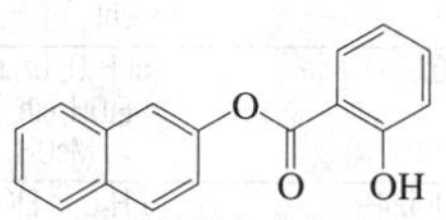

2-Naphthyl salicylate
1,5-Naphthyridine 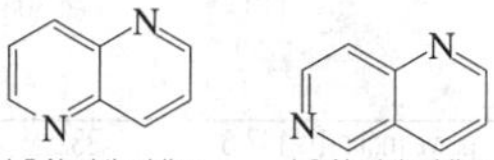
1,6-Naphthyridine
Napropamide 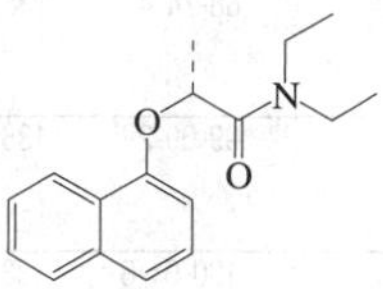
Naproxen 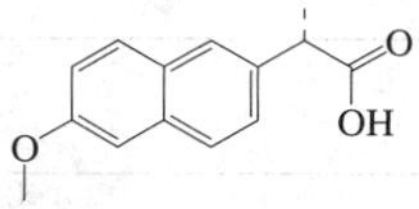
Narceine 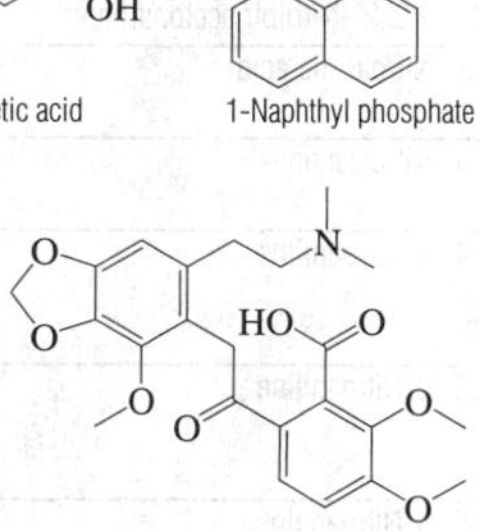

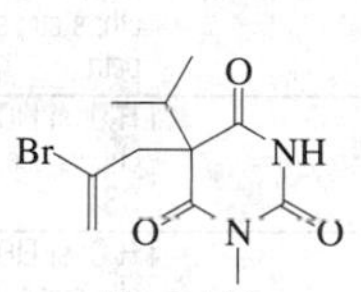

Narcobarbital
Naringenin 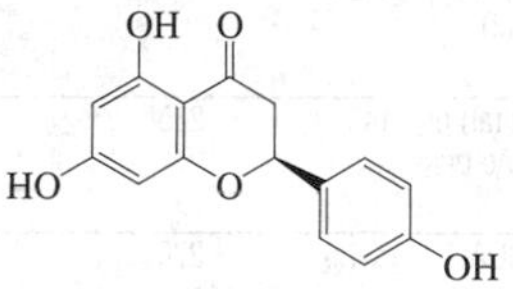
Naringin 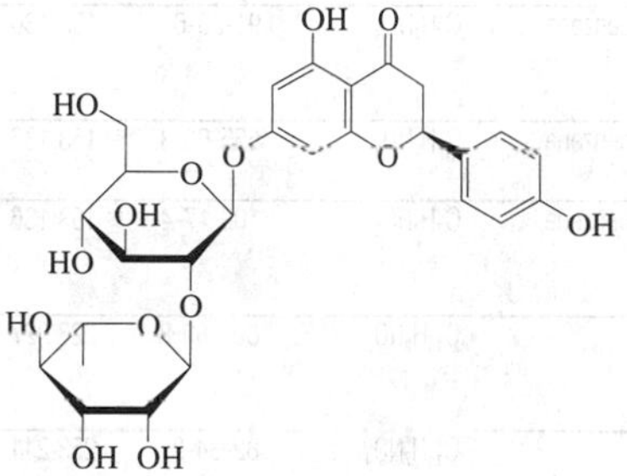
Nealbarbital 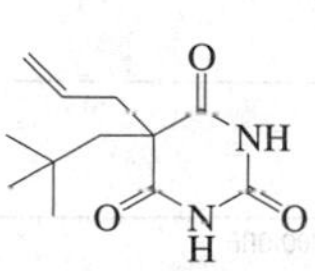
Nellite 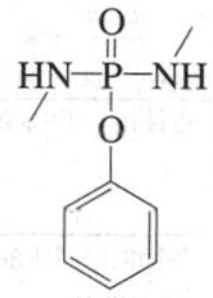

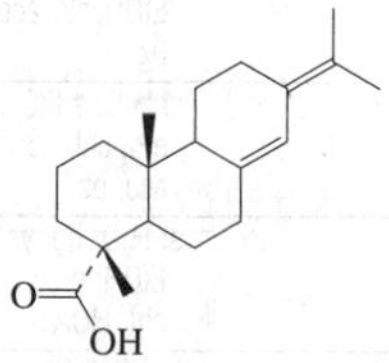

Neoabietic acid
Neobornylamine 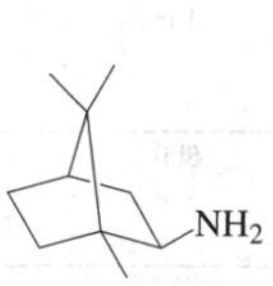
Neopentane 
Neopine 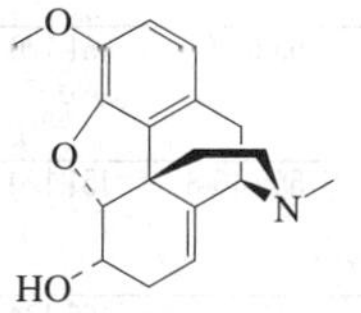
Neostigmine bromide 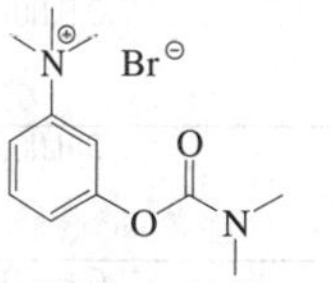
Nepetalactone 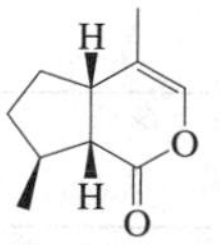
cis-Nerolidol 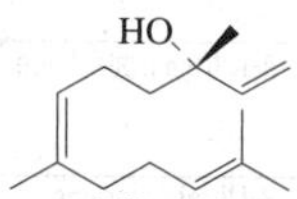

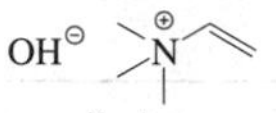

Neurine
Neutral Red 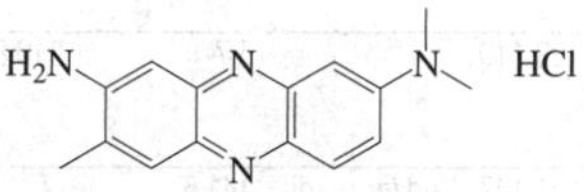
Nialamide 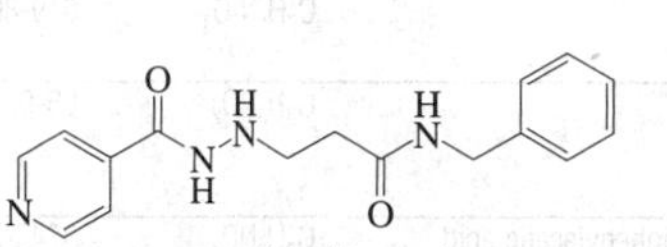
Nickel(II) acetate 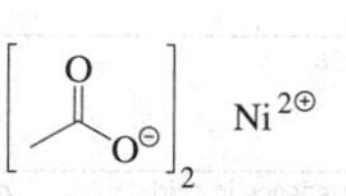

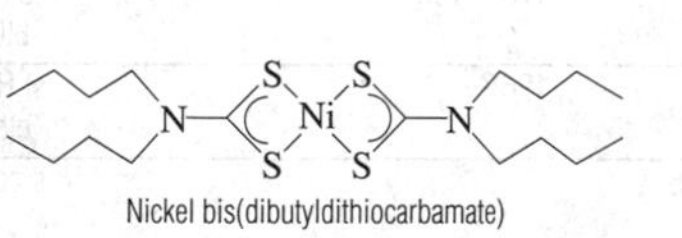

Nickel bis(dibutyldithiocarbamate)
Nickel bis(2,4-pentanedioate) 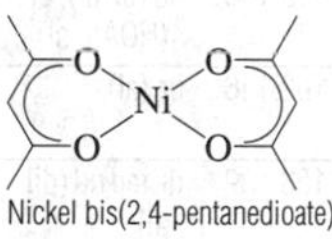
Nickel carbonyl 
Nickelocene
Niclosamide 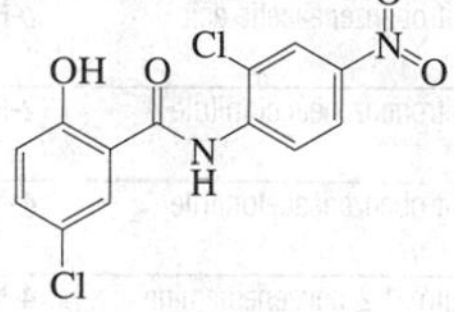

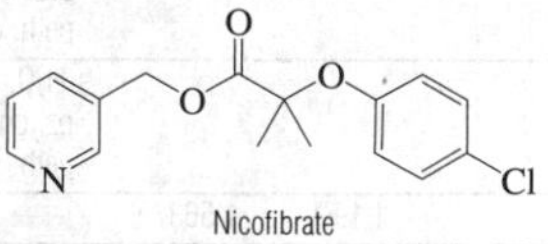

Nicofibrate
Nicosulfuron 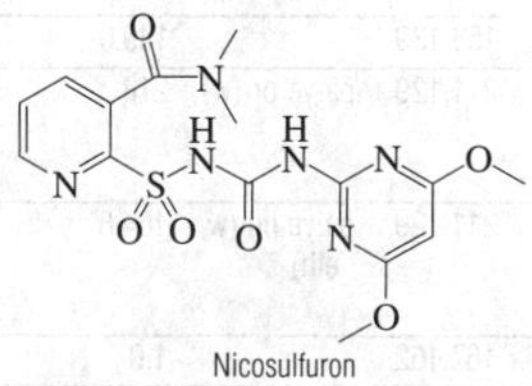
Nicotelline 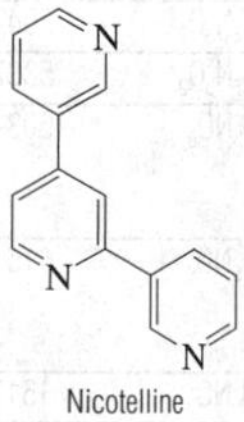
Nicotinamide hypoxanthine dinucleotide 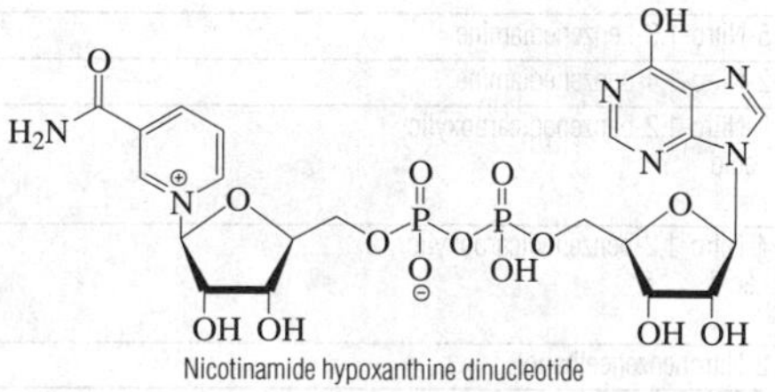

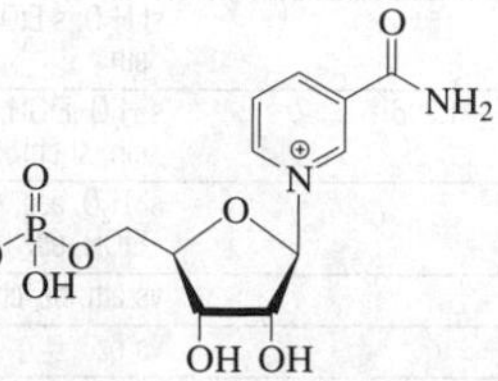

β-Nicotinamide mononucleotide
L-Nicotine 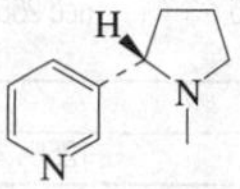
Nifurthiazole 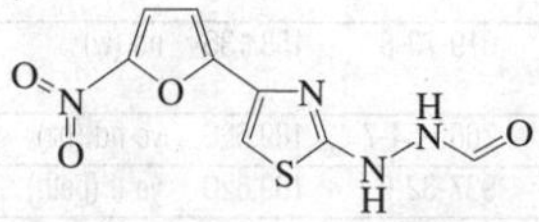
Nitralin 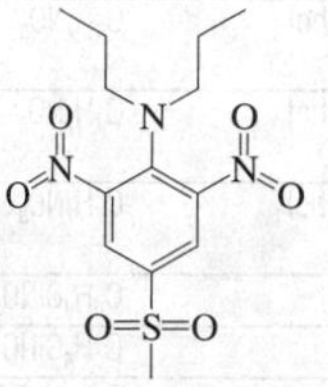
Nitranilic acid 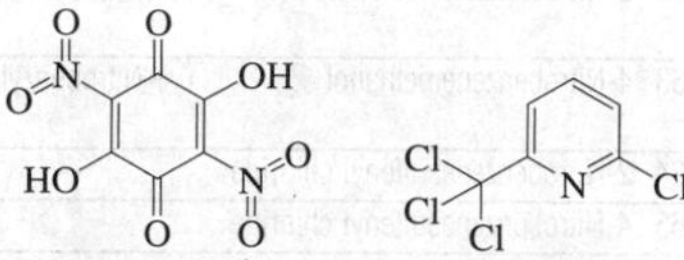
Nitrapyrin

No.	Name	Synonym	Mol. Form.	CAS RN	Mol. Wt.	Physical Form	mp/°C	bp/°C	den/ g cm-3	n_D	Solubility
8030	Nitrilotriacetic acid	*N,N*-Bis(carboxymethyl)glycine	$C_6H_9NO_6$	139-13-9	191.138	pr cry (w)	242 dec				sl H_2O, DMSO; s EtOH
8031	2,2',2''-Nitrilotriacetonitrile	Tricyanotrimethylamine	$C_6H_6N_4$	7327-60-8	134.139	nd (EtOH)	125.5				
8032	Nitroacetic acid		$C_2H_3NO_4$	625-75-2	105.050	nd (chl)	92 dec				vs bz, eth, EtOH, chl
8033	Nitroacetone		$C_3H_5NO_3$	10230-68-9	103.077	pl, nd (eth, bz)	50.3	103^{24}			vs bz, eth, EtOH
8034	2-Nitroaniline		$C_6H_6N_2O_2$	88-74-4	138.124		71.0	284	0.9015^{25}		sl H_2O; s EtOH; vs eth, ace, bz, chl
8035	3-Nitroaniline		$C_6H_6N_2O_2$	99-09-2	138.124		113.4	dec 306	0.9011^{25}		sl H_2O, bz; s EtOH, eth, ace; vs MeOH
8036	4-Nitroaniline		$C_6H_6N_2O_2$	100-01-6	138.124	pa ye mcl nd (w)	147.5	332	1.424^{20}		i H_2O; s EtOH, eth, ace; sl bz, DMSO
8037	2-Nitroanisole	1-Methoxy-2-nitrobenzene	$C_7H_7NO_3$	91-23-6	153.136		10.5	277; 144^{4}	1.2540^{20}	1.5161^{20}	i H_2O; msc EtOH, eth; s ctc
8038	3-Nitroanisole	1-Methoxy-3-nitrobenzene	$C_7H_7NO_3$	555-03-3	153.136	nd (al), pl (bz-lig)	38.5	258	1.373^{18}		i H_2O; s EtOH; vs eth
8039	4-Nitroanisole	1-Methoxy-4-nitrobenzene	$C_7H_7NO_3$	100-17-4	153.136	pr (al), nd (dil al)	54	274	1.2192^{60}	1.5070^{60}	i H_2O; vs EtOH, eth; s ctc; sl peth
8040	9-Nitroanthracene		$C_{14}H_9NO_2$	602-60-8	223.227	ye nd (al) pr (HOAc or xyl)	146	275^{17}			i H_2O; sl EtOH, chl; vs ace, CS_2
8041	1-Nitro-9,10-anthracenedione		$C_{14}H_7NO_4$	82-34-8	253.211	nd (HOAc) ye pr (ace)	231.5	270^{7}			i H_2O; sl EtOH, eth; s ace, bz
8042	2-Nitrobenzaldehyde		$C_7H_5NO_3$	552-89-6	151.120	ye nd (w)	43.5	153^{23}	1.2844^{20}		sl H_2O, chl; vs EtOH, eth, ace, bz
8043	3-Nitrobenzaldehyde		$C_7H_5NO_3$	99-61-6	151.120	lt ye nd (w)	58.5	164^{23}	1.2792^{20}		sl H_2O; s EtOH, eth, chl; vs ace, bz
8044	4-Nitrobenzaldehyde		$C_7H_5NO_3$	555-16-8	151.120	lf, pr (w)	107	sub	1.496^{25}		sl H_2O, lig; vs EtOH; s bz, chl, HOAc
8045	3-Nitrobenzamide		$C_7H_6N_2O_3$	645-09-0	166.134		142.7	312.5			s H_2O, EtOH, eth
8046	4-Nitrobenzamide		$C_7H_6N_2O_3$	619-80-7	166.134	nd (w)	200.7				i H_2O; s EtOH, eth
8047	Nitrobenzene		$C_6H_5NO_2$	98-95-3	123.110		5.7	210.8	1.2037^{20}	1.5562^{20}	sl H_2O, ctc; vs EtOH, eth, ace, bz
8048	2-Nitrobenzeneacetic acid	*o*-Nitrophenylacetic acid	$C_8H_7NO_4$	3740-52-1	181.147	nd (w, pl (dil al)	141.5				s H_2O, EtOH
8049	3-Nitrobenzeneacetic acid	*m*-Nitrophenylacetic acid	$C_8H_7NO_4$	1877-73-2	181.147	nd (w)	122				vs EtOH
8050	4-Nitrobenzeneacetic acid	*p*-Nitrophenylacetic acid	$C_8H_7NO_4$	104-03-0	181.147	pa ye nd (w)	154				sl H_2O; s EtOH, eth, bz
8051	2-Nitrobenzeneacetonitrile	2-Nitrobenzyl cyanide	$C_8H_6N_2O_2$	610-66-2	162.146	nd (dil al), pr (HOAc, al)	84	178^{12}, 138^{1}			vs ace, bz, eth, EtOH
8052	4-Nitrobenzeneacetonitrile	4-Nitrobenzyl cyanide	$C_8H_6N_2O_2$	555-21-5	162.146	pr (al)	117	196^{12}			sl H_2O; s EtOH, eth, bz, chl
8053	4-Nitro-1,2-benzenediamine	4-Nitro-*o*-phenylenediamine	$C_6H_7N_3O_2$	99-56-9	153.139	dk red nd (dil al)	199.5				s acid
8054	4-Nitro-1,3-benzenediamine		$C_6H_7N_3O_2$	5131-58-8	153.139	oran pr (w)	161				
8055	5-Nitro-1,3-benzenediamine		$C_6H_7N_3O_2$	5042-55-7	153.139	red cry (w)	143				
8056	2-Nitro-1,4-benzenediamine		$C_6H_7N_3O_2$	5307-14-2	153.139		140.0				
8057	3-Nitro-1,2-benzenedicarboxylic acid		$C_8H_5NO_6$	603-11-2	211.129	pa ye pr (w)	218				sl H_2O, ace; s EtOH; i bz, peth, chl
8058	4-Nitro-1,2-benzenedicarboxylic acid		$C_8H_5NO_6$	610-27-5	211.129	pa ye nd (w, eth)	164.8				s H_2O, EtOH; i bz, chl, CS_2, peth
8059	2-Nitrobenzeneethanol		$C_8H_9NO_3$	15121-84-3	167.162		1.0	267	1.19^{25}	1.5637^{20}	
8060	4-Nitrobenzeneethanol		$C_8H_9NO_3$	100-27-6	167.162		63	148^{2}			
8061	2-Nitrobenzenemethanol	2-Nitrobenzyl alcohol	$C_7H_7NO_3$	612-25-9	153.136	nd (w)	74	270; 168^{20}			sl H_2O; s EtOH, eth
8062	3-Nitrobenzenemethanol	3-Nitrobenzyl alcohol	$C_7H_7NO_3$	619-25-0	153.136	orth nd (w)	30.5	177^{3}	1.296^{19}		s H_2O, EtOH, eth; sl chl
8063	4-Nitrobenzenemethanol	4-Nitrobenzyl alcohol	$C_7H_7NO_3$	619-73-8	153.136	nd (w)	96.5	dec 255; 185^{12}			sl H_2O, ace; s EtOH, eth
8064	2-Nitrobenzenesulfenyl chloride		$C_6H_4ClNO_2S$	7669-54-7	189.620	ye nd (bz)	75				vs eth, bz, chl
8065	4-Nitrobenzenesulfenyl chloride		$C_6H_4ClNO_2S$	937-32-6	189.620	ye lf (peth)	52	$125^{0.1}$			vs bz
8066	4-Nitrobenzenesulfonamide		$C_6H_6N_2O_4S$	6325-93-5	202.188		180 dec				

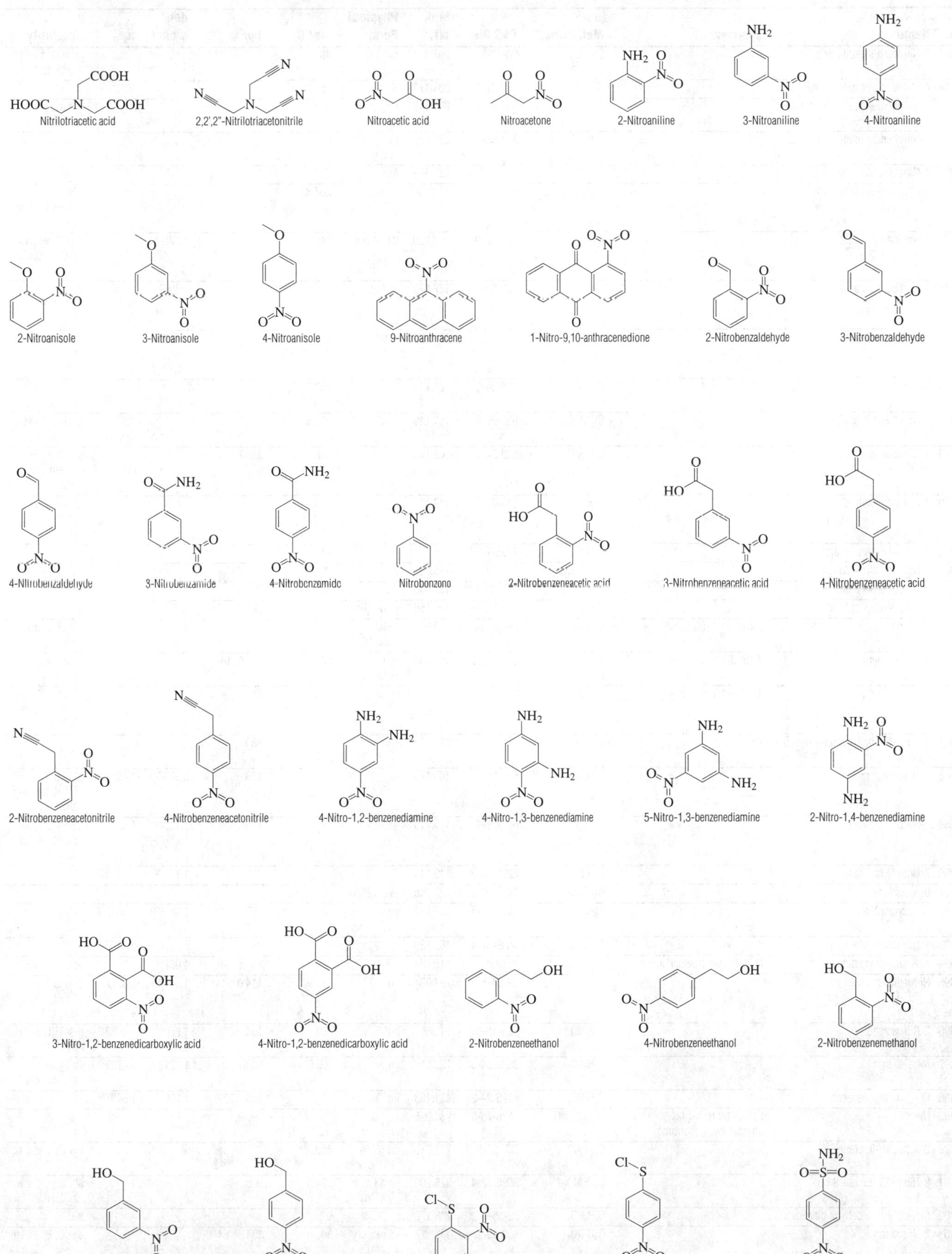
Nitrilotriacetic acid
2,2',2''-Nitrilotriacetonitrile
Nitroacetic acid
Nitroacetone
2-Nitroaniline
3-Nitroaniline
4-Nitroaniline
2-Nitroanisole
3-Nitroanisole
4-Nitroanisole
9-Nitroanthracene
1-Nitro-9,10-anthracenedione
2-Nitrobenzaldehyde
3-Nitrobenzaldehyde
4-Nitrobenzaldehyde
3-Nitrobenzamide
4-Nitrobenzamide
Nitrobenzene
2-Nitrobenzeneacetic acid
3-Nitrobenzeneacetic acid
4-Nitrobenzeneacetic acid
2-Nitrobenzeneacetonitrile
4-Nitrobenzeneacetonitrile
4-Nitro-1,2-benzenediamine
4-Nitro-1,3-benzenediamine
5-Nitro-1,3-benzenediamine
2-Nitro-1,4-benzenediamine
3-Nitro-1,2-benzenedicarboxylic acid
4-Nitro-1,2-benzenedicarboxylic acid
2-Nitrobenzeneethanol
4-Nitrobenzeneethanol
2-Nitrobenzenemethanol
3-Nitrobenzenemethanol
4-Nitrobenzenemethanol
2-Nitrobenzenesulfenyl chloride
4-Nitrobenzenesulfenyl chloride
4-Nitrobenzenesulfonamide

No.	Name	Synonym	Mol. Form.	CAS RN	Mol. Wt.	Physical Form	mp/°C	bp/°C	den/ g cm^{-3}	n_D	Solubility
8067	3-Nitrobenzenesulfonic acid		$C_6H_5NO_5S$	98-47-5	203.173	pl	48				vs H_2O; s EtOH; i eth, bz
8068	4-Nitrobenzenesulfonic acid		$C_6H_5NO_5S$	138-42-1	203.173		95				vs H_2O
8069	2-Nitrobenzenesulfonyl chloride		$C_6H_4ClNO_4S$	1694-92-4	221.619	pr (lig, eth-peth)	68.5				s eth; sl peth
8070	3-Nitrobenzenesulfonyl chloride		$C_6H_4ClNO_4S$	121-51-7	221.619	mcl pr (eth) nd (lig)	64				i H_2O; s EtOH
8071	4-Nitrobenzenesulfonyl chloride		$C_6H_4ClNO_4S$	98-74-8	221.619	mcl pr (peth)	79.5	$143^{1.5}$			s peth
8072	5-Nitro-1*H*-benzimidazole		$C_7H_5N_3O_2$	94-52-0	163.134	nd (w)	207.8				i H_2O, eth, bz, chl; s acid; vs EtOH
8073	2-Nitrobenzoic acid		$C_7H_5NO_4$	552-16-9	167.120	tcl nd (w)	147.5		1.575^{20}		s H_2O, eth; vs EtOH, ace; sl bz, lig
8074	3-Nitrobenzoic acid		$C_7H_5NO_4$	121-92-6	167.120	mcl pr (w)	141.1		1.494^{20}		sl H_2O, bz; vs EtOH, eth, ace; s chl
8075	4-Nitrobenzoic acid		$C_7H_5NO_4$	62-23-7	167.120	mcl lf (w)	242	sub	1.610^{20}		vs ace, eth, EtOH, chl, MeOH
8076	3-Nitrobenzoic acid, hydrazide		$C_7H_7N_3O_3$	618-94-0	181.149		153.5				sl H_2O, EtOH; i eth, bz, chl
8077	4-Nitrobenzoic acid, hydrazide		$C_7H_7N_3O_3$	636-97-5	181.149		215.5				sl H_2O, EtOH; i eth, bz, chl
8078	3-Nitrobenzonitrile		$C_7H_4N_2O_2$	619-24-9	148.119		118	165^{16}			s H_2O, EtOH, bz; vs eth, ace; i peth
8079	4-Nitrobenzonitrile		$C_7H_4N_2O_2$	619-72-7	148.119		150.0				sl H_2O, EtOH, eth; s chl, HOAc
8080	5-Nitro-1*H*-benzotriazole		$C_6H_4N_4O_2$	2338-12-7	164.122		217				
8081	2-Nitrobenzoyl chloride		$C_7H_4ClNO_3$	610-14-0	185.565		20				vs eth; sl ctc
8082	3-Nitrobenzoyl chloride		$C_7H_4ClNO_3$	121-90-4	185.565		36	276.5			vs eth
8083	4-Nitrobenzoyl chloride		$C_7H_4ClNO_3$	122-04-3	185.565	ye nd (lig)	75	203^{105}, 151^{15}			s eth
8084	2-Nitrobiphenyl	2-Nitro-1,1'-biphenyl	$C_{12}H_9NO_2$	86-00-0	199.205	pl (al, MeOH)	37.2	320	1.44^{25}		i H_2O; s EtOH, eth, chl
8085	3-Nitrobiphenyl	3-Nitro-1,1'-biphenyl	$C_{12}H_9NO_2$	2113-58-8	199.205	ye pl or nd (dil al)	62	227^{35}, 143^{9}			i H_2O; s EtOH, eth, HOAc, lig
8086	4-Nitrobiphenyl	4-Nitro-1,1'-biphenyl	$C_{12}H_9NO_2$	92-93-3	199.205	ye nd (al)	114	340			i H_2O; sl EtOH; s eth, bz, chl, HOAc
8087	2-Nitro-1,1-bis(*p*-chlorophenyl) propane		$C_{15}H_{13}Cl_2NO_2$	117-27-1	310.176	cry	81	$180^{0.16}$			
8088	1-Nitrobutane		$C_4H_9NO_2$	627-05-4	103.120			153	0.970^{25}	1.4303^{20}	sl H_2O; msc EtOH, eth; s alk
8089	2-Nitro-1-butanol		$C_4H_9NO_3$	609-31-4	119.119		-47	105^{10}	1.1332^{25}	1.4390^{20}	s H_2O, ace; msc EtOH, eth; sl ctc
8090	3-Nitro-2-butanol		$C_4H_9NO_3$	6270-16-2	119.119			91^{9}, $55^{0.5}$	1.1260^{20}	1.4414^{20}	
8091	6-Nitrochrysene		$C_{18}H_{11}NO_2$	7496-02-8	273.286	ye nd (bz)	≈215 dec				
8092	Nitrocyclohexane		$C_6H_{11}NO_2$	1122-60-7	129.157	liq	-34	205; 95^{22}	1.0610^{20}	1.4612^{19}	i H_2O; s EtOH, lig
8093	1-Nitrodecane		$C_{10}H_{21}NO_2$	4609-87-4	187.280			86^{1}		1.4337^{20}	
8094	*N*-Nitrodiethylamine	*N*-Ethyl-*N*-nitroethanamine	$C_4H_{10}N_2O_2$	7119-92-8	118.134			206.5	1.057^{15}		vs eth, EtOH
8095	Nitroethane		$C_2H_5NO_2$	79-24-3	75.067	liq	-89.5	114.0	1.0448^{25}	1.3917^{20}	sl H_2O; msc EtOH, eth; s ace, chl
8096	2-Nitroethanol		$C_2H_5NO_3$	625-48-9	91.066	liq	-80	194; 102^{10}	1.270^{15}	1.4438^{19}	msc H_2O, EtOH, eth; i bz
8097	Nitroethene		$C_2H_3NO_2$	3638-64-0	73.051	liq	-55.5	98.5	1.2212^{14}	1.4282^{20}	vs EtOH, eth, ace, bz, chl
8098	(2-Nitroethyl)benzene		$C_8H_9NO_2$	6125-24-2	151.163	liq	-23	250; 137^{16}	1.126^{24}	1.5407^{19}	
8099	Nitrofen	2,4-Dichloro-1-(4-nitrophenoxy)benzene	$C_{12}H_7Cl_2NO_3$	1836-75-5	284.095		70				
8100	2-Nitro-9*H*-fluorene		$C_{13}H_9NO_2$	607-57-8	211.216	nd (50% HOAc ace)	159.3				i H_2O; s ace, bz
8101	2-Nitro-9*H*-fluoren-9-one		$C_{13}H_7NO_3$	3096-52-4	225.200	ye nd or lf (HOAc)	224.3	sub			sl EtOH; s ace, sulf, HOAc
8102	5-Nitro-2-furaldehyde diacetate		$C_9H_9NO_7$	92-55-7	243.170		92.0				s chl
8103	2-Nitrofuran		$C_4H_3NO_3$	609-39-2	113.072	ye mcl cry (peth)	30	134^{123}, 84^{13}			s H_2O, EtOH, eth
8104	5-Nitro-2-furancarboxaldehyde		$C_5H_3NO_4$	698-63-5	141.083	pa ye (peth)	35.5	130^{10}			sl H_2O; s peth

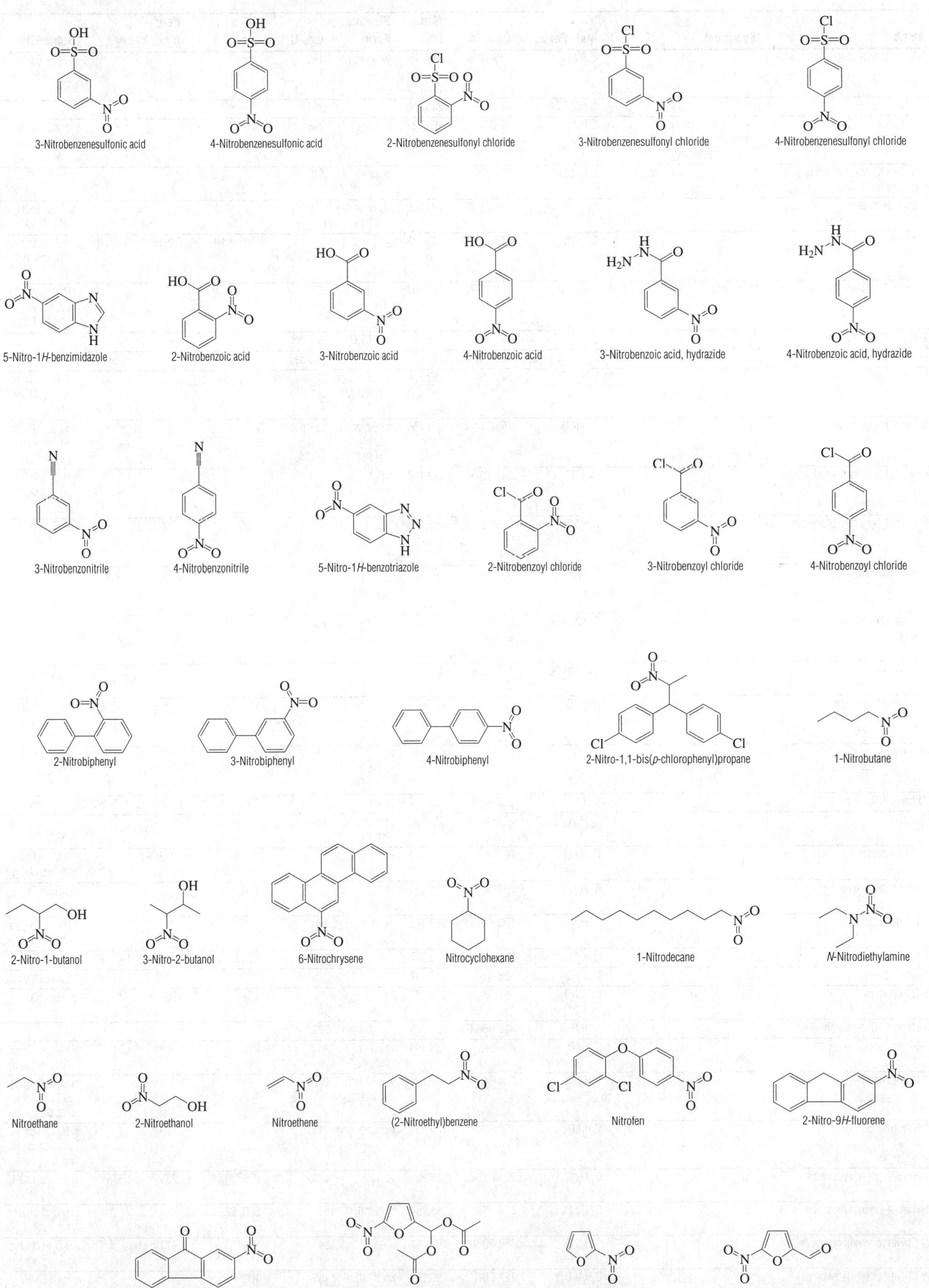

OH
O=S=O
O
N
O
3-Nitrobenzenesulfonic acid
4-Nitrobenzenesulfonic acid
Cl
2-Nitrobenzenesulfonyl chloride
3-Nitrobenzenesulfonyl chloride
4-Nitrobenzenesulfonyl chloride
H
5-Nitro-1H-benzimidazole
HO
2-Nitrobenzoic acid
3-Nitrobenzoic acid
4-Nitrobenzoic acid
H₂N
3-Nitrobenzoic acid, hydrazide
4-Nitrobenzoic acid, hydrazide
3-Nitrobenzonitrile
4-Nitrobenzonitrile
5-Nitro-1H-benzotriazole
2-Nitrobenzoyl chloride
3-Nitrobenzoyl chloride
4-Nitrobenzoyl chloride
2-Nitrobiphenyl
3-Nitrobiphenyl
4-Nitrobiphenyl
2-Nitro-1,1-bis(p-chlorophenyl)propane
1-Nitrobutane
2-Nitro-1-butanol
3-Nitro-2-butanol
6-Nitrochrysene
Nitrocyclohexane
1-Nitrodecane
N-Nitrodiethylamine
Nitroethane
2-Nitroethanol
Nitroethene
(2-Nitroethyl)benzene
Nitrofen
2-Nitro-9H-fluorene
2-Nitro-9H-fluoren-9-one
5-Nitro-2-furaldehyde diacetate
2-Nitrofuran
5-Nitro-2-furancarboxaldehyde

No.	Name	Synonym	Mol. Form.	CAS RN	Mol. Wt.	Physical Form	mp/°C	bp/°C	den/ g cm^{-3}	n_D	Solubility
8105	5-Nitro-2-furancarboxylic acid		$C_5H_3NO_5$	645-12-5	157.082	pa ye pl (w)	186	sub			s H_2O, EtOH, eth; sl ace, bz; i chl
8106	Nitrofurantoin		$C_8H_6N_4O_5$	67-20-9	238.158		263				
8107	Nitrofurazone	2-[(5-Nitro-2-furanyl) methylene] hydrazinecarboxamide	$C_6H_6N_4O_4$	59-87-0	198.137	pa ye nd	238 dec				i H_2O, eth; sl EtOH, DMSO; s alk
8108	Nitrogen mustard *N*-oxide hydrochloride	Mechlorethamine oxide hydrochloride	$C_5H_{12}Cl_3NO$	302-70-5	208.514	pr (ace)	110				s H_2O
8109	Nitroguanidine		$CH_4N_4O_2$	556-88-7	104.069	nd or pr (w)	239 dec				sl H_2O, EtOH; i eth; vs alk
8110	1-Nitrohexane		$C_6H_{13}NO_2$	646-14-0	131.173			193; 84^{21}	0.9396^{20}	1.4270^{20}	i H_2O; s EtOH, eth, ace, bz, alk
8111	3-Nitro-4-hydroxyphenylarsonic acid	Roxarsone	$C_6H_6AsNO_6$	121-19-7	263.037	ye nd or pl (w)	300				sl hot H_2O; i eth, EtOAc; vs MeOH, EtOH
8112	2-Nitro-1*H*-imidazole	Azomycin	$C_3H_3N_3O_2$	527-73-1	113.075	cry (MeOH)	287 dec				
8113	4-Nitro-1*H*-imidazole		$C_3H_3N_3O_2$	3034-38-6	113.075		303 dec				
8114	5-Nitro-1*H*-indazole		$C_7H_5N_3O_2$	5401-94-5	163.134	ye nd or col nd (al)	208				s EtOH, eth, bz; vs ace, HOAc; i lig
8115	6-Nitro-1*H*-indazole		$C_7H_5N_3O_2$	7597-18-4	163.134	nd (w, al, ace)	181 dec				s H_2O, EtOH, eth, bz; vs ace; i lig
8116	4-Nitro-1,3-isobenzofurandione		$C_8H_3NO_5$	641-70-3	193.114	nd (ace, al)	164				i H_2O; s EtOH, ace, HOAc; sl bz
8117	5-Nitro-1,3-isobenzofurandione		$C_8H_3NO_5$	5466-84-2	193.114		120.3	196^{8}			i H_2O, peth; s EtOH, ace; sl eth
8118	2-Nitroisobutane		$C_4H_9NO_2$	594-70-7	103.120		26.23	127.16	0.9501^{28}	1.4015^{20}	msc EtOH, eth, ace, bz; vs chl; i alk
8119	5-Nitro-1*H*-isoindole-1,3(2*H*)-dione		$C_8H_4N_2O_4$	89-40-7	192.129	col nd (w), ye lf (al-ace)	202				vs ace
8120	Nitromersol		$C_7H_5HgNO_3$	133-58-4	351.71						i H_2O; sl ace, EtOH; s alk
8121	*N*-Nitromethanamine		$CH_4N_2O_2$	598-57-2	76.055		38	82^{10}	1.2433^{49}	1.4616^{49}	vs H_2O, EtOH, bz, chl; s eth; sl peth
8122	Nitromethane		CH_3NO_2	75-52-5	61.041	liq	-28.38	101.19	1.1371^{20}	1.3817^{20}	s H_2O, EtOH, eth, ace, ctc, alk
8123	(Nitromethyl)benzene		$C_7H_7NO_2$	622-42-4	137.137	ye liq		226; 135^{25}	1.1596^{20}	1.5323^{20}	vs ace, eth
8124	Nitron		$C_{20}H_{16}N_4$	2218-94-2	312.368	ye lf (al), nd (chl)	189 dec				vs ace, bz, EtOH, chl
8125	1-Nitronaphthalene		$C_{10}H_7NO_2$	86-57-7	173.169	ye nd (al)	61	180^{14}	1.332^{20}		i H_2O; vs EtOH, eth, bz, chl, py
8126	2-Nitronaphthalene		$C_{10}H_7NO_2$	581-89-5	173.169	ye orth nd or pl (al)	79	314; 165^{15}			i H_2O; vs EtOH, eth
8127	1-Nitro-2-naphthol		$C_{10}H_7NO_3$	550-60-7	189.168	ye nd, lf or pr (al)	104	$115^{0.05}$			s H_2O, EtOH; vs eth; sl chl
8128	1-Nitrooctane		$C_8H_{17}NO_2$	629-37-8	159.227		15	208.5	0.9346^{20}	1.4322^{20}	
8129	1-Nitropentane		$C_5H_{11}NO_2$	628-05-7	117.147			172.5	0.9525^{20}	1.4175^{20}	s EtOH, eth, bz
8130	3-Nitropentane		$C_5H_{11}NO_2$	551-88-2	117.147			154	0.957^{0}		vs ace, eth, EtOH
8131	5-Nitro-1,10-phenanthroline		$C_{12}H_7N_3O_2$	4199-88-6	225.203		202.3				
8132	2-Nitrophenol		$C_6H_5NO_3$	88-75-5	139.109	ye nd or pr (eth, al)	44.8	216	1.2942^{40}	1.5723^{50}	sl H_2O; vs EtOH, eth, ace, bz, py
8133	3-Nitrophenol		$C_6H_5NO_3$	554-84-7	139.109	ye mcl (eth, aq Hcl)	96.8	194^{70}	1.2797^{100}		sl H_2O, DMSO; vs EtOH, eth, ace, bz
8134	4-Nitrophenol		$C_6H_5NO_3$	100-02-7	139.109	ye mcl pr (to)	113.6		1.479^{20}		sl H_2O; vs EtOH, eth, ace; s tol, py
8135	1-Nitro-2-phenoxybenzene		$C_{12}H_9NO_3$	2216-12-8	215.204	ye liq	<-20	235^{60}, 184^{8}	1.2539^{22}	1.575^{20}	vs bz, eth, EtOH, chl
8136	1-Nitro-4-phenoxybenzene		$C_{12}H_9NO_3$	620-88-2	215.204	pl (peth), MeOH)	61	320; 225^{30}			i H_2O; sl EtOH, ctc; s eth, bz
8137	*N*-(2-Nitrophenyl)acetamide		$C_8H_8N_2O_3$	552-32-9	180.161		94	$100^{0.1}$	1.419^{15}		s H_2O, EtOH, bz, chl, lig; vs eth
8138	*N*-(3-Nitrophenyl)acetamide		$C_8H_8N_2O_3$	122-28-1	180.161	wh lf (al)	155	$100^{0.0074}$			s H_2O, EtOH, chl; i eth; sl tfa

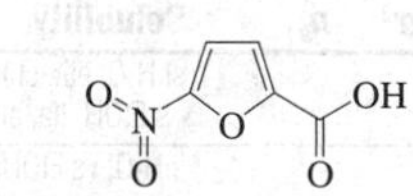

5-Nitro-2-furancarboxylic acid

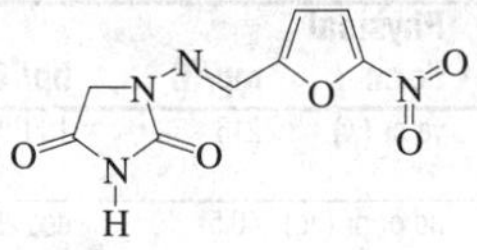

Nitrofurantoin

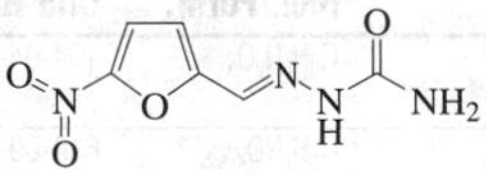

Nitrofurazone

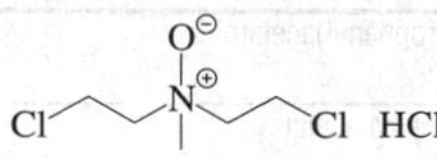

Nitrogen mustard *N*-oxide hydrochloride

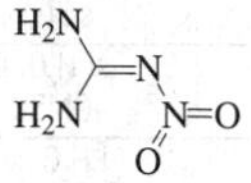

Nitroguanidine

1-Nitrohexane

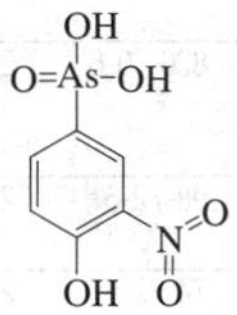

3-Nitro-4-hydroxyphenylarsonic acid

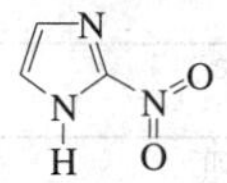

2-Nitro-1*H*-imidazole

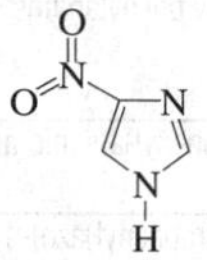

4-Nitro-1*H*-imidazole

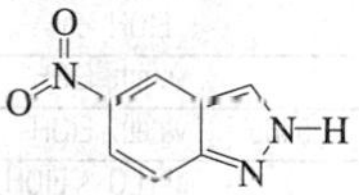

5-Nitro-1*H*-indazole

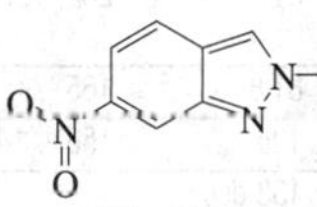

6-Nitro-1*H*-indazole

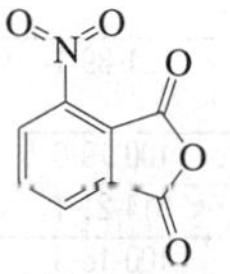

4-Nitro-1,3-isobenzofurandione

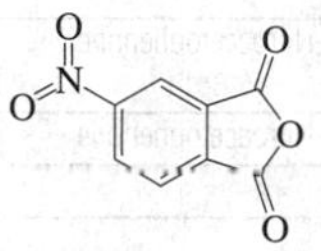

5-Nitro-1,3-isobenzofurandione

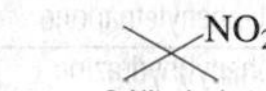

2-Nitroisobutane

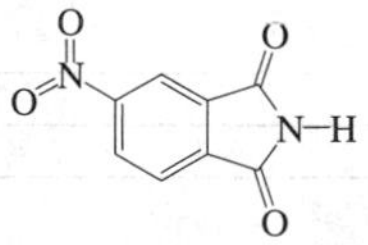

5-Nitro-1*H*-isoindole-1,3(2*H*)-dione

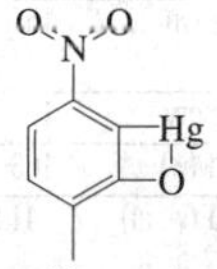

Nitromersol

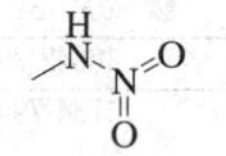

N-Nitromethanamine

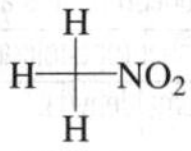

Nitromethane

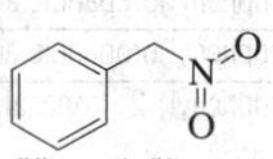

(Nitromethyl)benzene

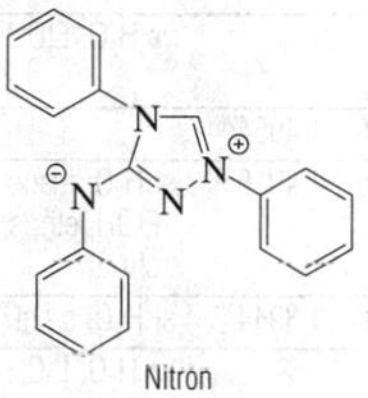

Nitron

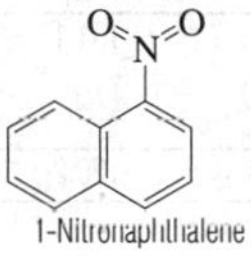

1-Nitronaphthalene

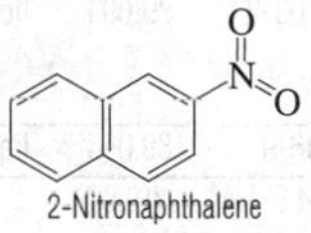

2-Nitronaphthalene

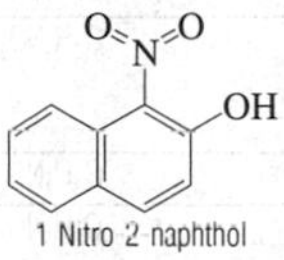

1-Nitro-2-naphthol

1-Nitrooctane

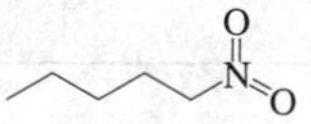

1-Nitropentane

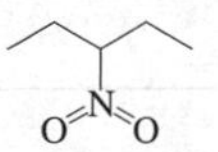

3-Nitropentane

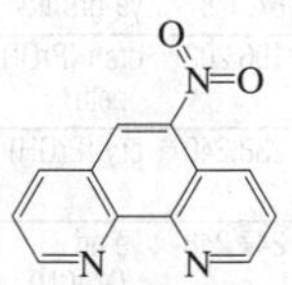

5-Nitro-1,10-phenanthroline

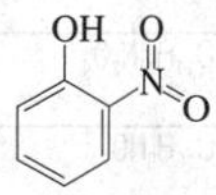

2-Nitrophenol

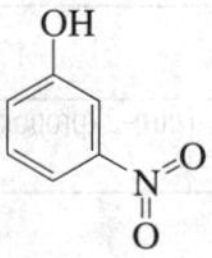

3-Nitrophenol

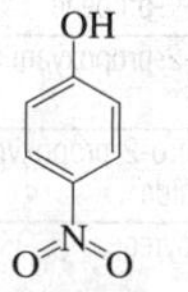

4-Nitrophenol

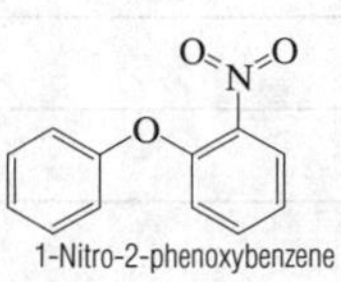

1-Nitro-2-phenoxybenzene

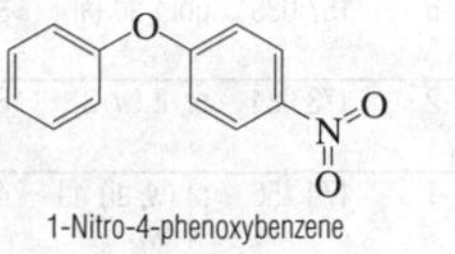

1-Nitro-4-phenoxybenzene

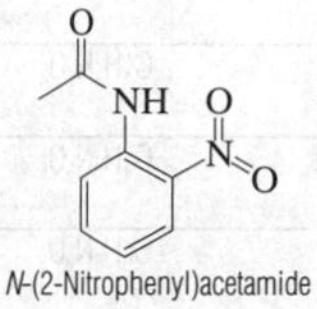

N-(2-Nitrophenyl)acetamide

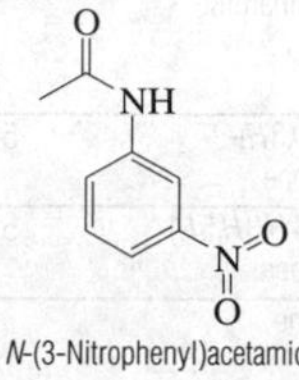

N-(3-Nitrophenyl)acetamide

No.	Name	Synonym	Mol. Form.	CAS RN	Mol. Wt.	Physical Form	mp/°C	bp/°C	den/ g cm^{-3}	n_D	Solubility
8139	*N*-(4-Nitrophenyl)acetamide		$C_8H_8N_2O_3$	104-04-1	180.161	ye pr (w)	216	$100^{0.008}$			sl H_2O, eth, chl; s EtOH, tfa, alk
8140	2-Nitrophenyl acetate		$C_8H_7NO_4$	610-69-5	181.147	nd or pr (lig)	40.5	dec 253; 141^{11}			s H_2O; vs EtOH, eth, ace, bz; sl lig
8141	4-Nitrophenyl acetate		$C_8H_7NO_4$	830-03-5	181.147	lf (dil al)	82.3				vs H_2O, bz; s EtOH, chl, lig
8142	2-Nitro-*N*-phenylaniline		$C_{12}H_{10}N_2O_2$	119-75-5	214.219		75.5	215^{15}	1.3660^{20}		i H_2O; s EtOH; sl ctc
8143	4-Nitro-*N*-phenylaniline		$C_{12}H_{10}N_2O_2$	836-30-6	214.219		135.3	211^{30}			i H_2O; vs EtOH; sl ace; s con sulf
8144	(4-Nitrophenyl)arsonic acid	Nitarsone	$C_6H_6AsNO_5$	98-72-6	247.038	lf or nd (w)	>310 dec				sl H_2O, EtOH, DMSO
8145	4-[(4-Nitrophenyl)azo]-1,3-benzenediol	Magneson	$C_{12}H_9N_3O_4$	74-39-5	259.217	red pow (al or MeOH)	200				i H_2O; sl EtOH, bz, HOAc, tol
8146	1-[(4-Nitrophenyl)azo]-2-naphthol		$C_{16}H_{11}N_3O_3$	6410-10-2	293.276	br-oran pl (to or bz)	257				vs bz, EtOH
8147	(3-Nitrophenyl)boronic acid		$C_6H_6BNO_4$	13331-27-6	166.928		274.5				
8148	1-(2-Nitrophenyl)ethanone	2-Nitroacetophenone	$C_8H_7NO_3$	577-59-3	165.147		28.5	178^{32}, 158^{16}	1.2370^{25}	1.5468^{20}	i H_2O; vs EtOH, eth, chl
8149	1-(3-Nitrophenyl)ethanone	3-Nitroacetophenone	$C_8H_7NO_3$	121-89-1	165.147	nd (al)	81	202; 167^{18}			vs H_2O, eth; sl EtOH, chl
8150	1-(4-Nitrophenyl)ethanone	4-Nitroacetophenone	$C_8H_7NO_3$	100-19-6	165.147	ye pr (al)	81.8	165^{5}			vs eth, EtOH
8151	2-Nitro-1-phenylethanone		$C_8H_7NO_3$	614-21-1	165.147		106	158^{16}, 142^{10}		1.5468^{30}	vs eth, EtOH
8152	(4-Nitrophenyl)hydrazine		$C_6H_7N_3O_2$	100-16-3	153.139	oran-red lf or nd (al)	158 dec				sl H_2O; s EtOH, eth, bz, chl, AcOEt
8153	(4-Nitrophenyl)phenylmethanone		$C_{13}H_9NO_3$	1144-74-7	227.215	nd or lf (al)	138		1.406^{9}		vs bz
8154	3-(4-Nitrophenyl)-1-phenyl-2-propen-1-one	Nitrochalcone	$C_{15}H_{11}NO_3$	1222-98-6	253.253	pa ye nd (al) pl (bz)	164				s EtOH, chl; i eth, lig
8155	4-Nitrophenyl phosphate	4-Nitrophenyl dihydrogen phosphate	$C_6H_6NO_6P$	330-13-2	219.089	ye-wh nd	155				i cold H_2O; s EtOH, chl, bz
8156	3-(2-Nitrophenyl)propanoic acid	2-Nitrobeazenepropanoic acid	$C_9H_9NO_4$	2001-32-3	195.172	ye cry	115				
8157	3-(4-Nitrophenyl)propanoic acid	4-Nitrobenzenepropanoic acid	$C_9H_9NO_4$	16642-79-8	195.172	nd (w)	163				
8158	3-(4-Nitrophenyl)-2-propenal	4-Nitrocinnamaldehyde	$C_9H_7NO_3$	1734-79-8	177.157	nd (w, al)	141.5				s H_2O, eth, ace, bz; vs EtOH
8159	3-(2-Nitrophenyl)-2-propynoic acid	*o*-Nitrophenylpropiolic acid	$C_9H_5NO_4$	530-85-8	191.141		≈157 dec; may explode				sl H_2O; vs EtOH, eth; i CS_2O
8160	1-Nitro-4-(phenylthio)benzene		$C_{12}H_9NO_2S$	952-97-6	231.270	pa ye mcl pr (lig)	56	288^{100}, 240^{25}			vs eth, EtOH
8161	(4-Nitrophenyl)urea	*p*-Nitrophenylurea	$C_7H_7N_3O_3$	556-10-5	181.149	pr (al), nd (dil al)	238				vs H_2O, EtOH
8162	*N*-Nitropiperidine		$C_5H_{10}N_2O_2$	7119-94-0	130.145	liq	-5.5	245; 121^{20}	1.1519^{26}	1.4954^{26}	
8163	1-Nitropropane		$C_3H_7NO_2$	108-03-2	89.094	liq	-108	131.1	0.9961^{25}	1.4018^{20}	sl H_2O; msc EtOH, eth; s chl
8164	2-Nitropropane		$C_3H_7NO_2$	79-46-9	89.094	liq	-91.3	120.2	0.9821^{25}	1.3944^{20}	sl H_2O; s chl
8165	3-Nitropropanoic acid		$C_3H_5NO_4$	504-88-1	119.077		62		1.59^{20}		vs H_2O, EtOH, eth; s chl; i lig
8166	2-Nitro-1-propanol		$C_3H_7NO_3$	2902-96-7	105.093			120^{32}, 100^{12}	1.1841^{25}	1.4379^{20}	s H_2O, EtOH, eth; sl chl
8167	1-Nitro-1-propene		$C_3H_5NO_2$	3156-70-5	87.078			60^{34}, 37^{10}	1.0661^{20}	1.4527^{20}	s eth, ace, chl
8168	2-Nitro-1-propene		$C_3H_5NO_2$	4749-28-4	87.078	ye-grn liq		52^{80}, 32^{30}	1.0559^{25}	1.4358^{20}	s eth, ace, chl
8169	5-Nitro-2-propoxyaniline		$C_9H_{12}N_2O_3$	553-79-7	196.202	oran (PrOH-peth)	49				vs EtOH
8170	*N*-(5-Nitro-2-propoxyphenyl)acetamide	5'-Nitro-2'-propoxyacetanilide	$C_{11}H_{14}N_2O_4$	553-20-8	238.240	cry (PrOH)	102.5				
8171	1-Nitropyrene		$C_{16}H_9NO_2$	5522-43-0	247.248	ye nd (MeCN)	152				
8172	5-Nitro-2-pyridinamine		$C_5H_5N_3O_2$	4214-76-0	139.113	ye lf (dil al)	188				sl H_2O, eth, bz, lig; s EtOH
8173	4-Nitropyridine		$C_5H_4N_2O_2$	1122-61-8	124.098	pl (aq al)	50				
8174	4-Nitropyridine 1-oxide		$C_5H_4N_2O_3$	1124-33-0	140.097		160.5				
8175	5-Nitropyrimidinamine		$C_4H_4N_4O_2$	3073-77-6	140.101	nd (al)	236.5				sl H_2O, DMSO; s EtOH, ace; i eth, bz
8176	5-Nitro-2,4(1*H*,3*H*)-pyrimidinedione	5-Nitrouracil	$C_4H_3N_3O_4$	611-08-5	157.085	gold nd (al)	>300 exp				sl H_2O; s EtOH
8177	5-Nitro-2,4,6(1*H*,3*H*,5*H*)-pyrimidinetrione	5-Nitrobarbituric acid	$C_4H_3N_3O_5$	480-68-2	173.084	pr, lf (w+3)	180.5				s H_2O, EtOH; i eth
8178	5-Nitroquinoline		$C_9H_6N_2O_2$	607-34-1	174.156	pl (w, al) nd (+w)	74	sub			sl H_2O, chl; s EtOH, bz

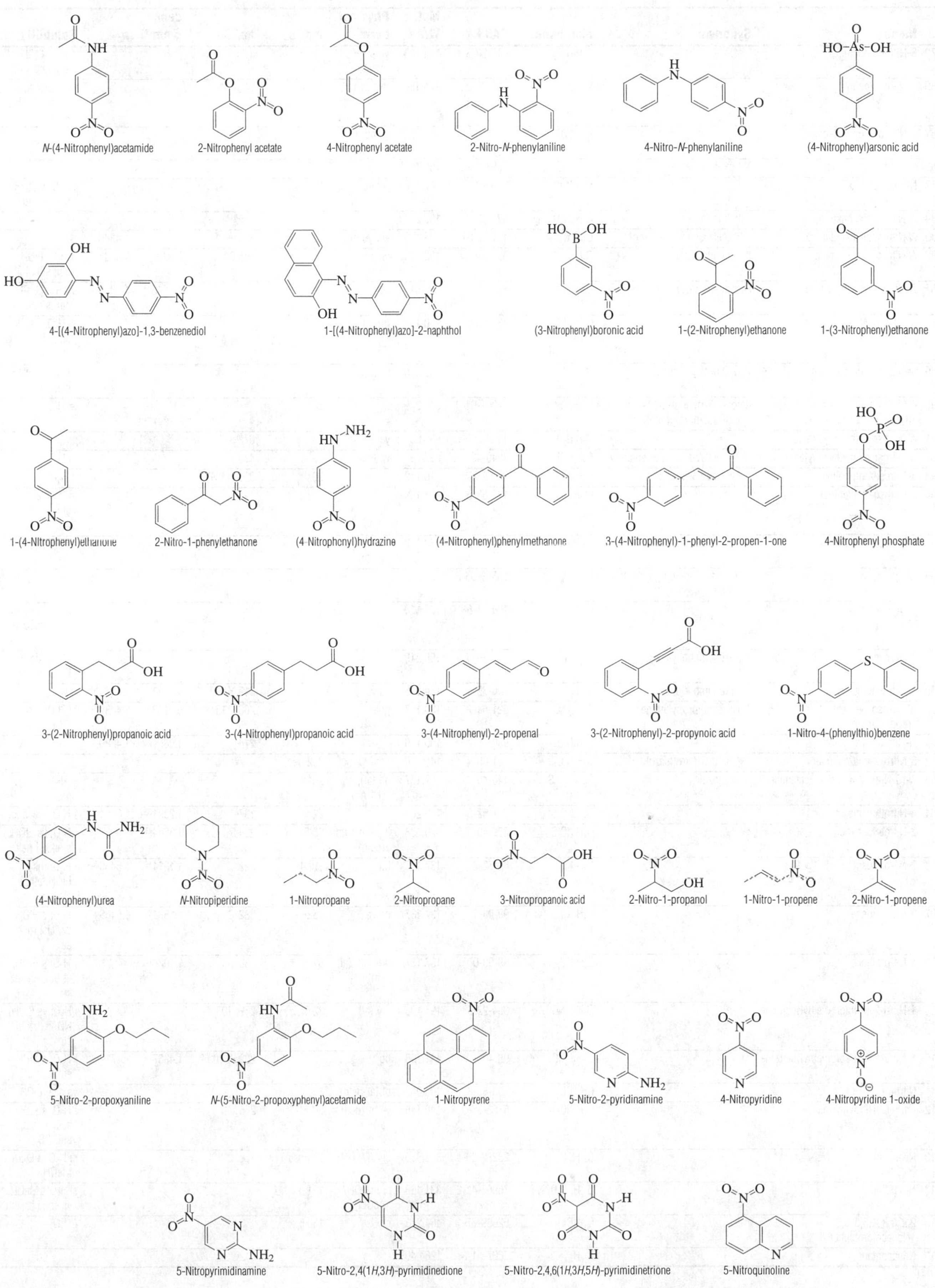

N-(4-Nitrophenyl)acetamide
2-Nitrophenyl acetate
4-Nitrophenyl acetate
2-Nitro-N-phenylaniline
4-Nitro-N-phenylaniline
(4-Nitrophenyl)arsonic acid
4-[(4-Nitrophenyl)azo]-1,3-benzenediol
1-[(4-Nitrophenyl)azo]-2-naphthol
(3-Nitrophenyl)boronic acid
1-(2-Nitrophenyl)ethanone
1-(3-Nitrophenyl)ethanone
1-(4-Nitrophenyl)ethanone
2-Nitro-1-phenylethanone
(4-Nitrophenyl)hydrazine
(4-Nitrophenyl)phenylmethanone
3-(4-Nitrophenyl)-1-phenyl-2-propen-1-one
4-Nitrophenyl phosphate
3-(2-Nitrophenyl)propanoic acid
3-(4-Nitrophenyl)propanoic acid
3-(4-Nitrophenyl)-2-propenal
3-(2-Nitrophenyl)-2-propynoic acid
1-Nitro-4-(phenylthio)benzene
(4-Nitrophenyl)urea
N-Nitropiperidine
1-Nitropropane
2-Nitropropane
3-Nitropropanoic acid
2-Nitro-1-propanol
1-Nitro-1-propene
2-Nitro-1-propene
5-Nitro-2-propoxyaniline
N-(5-Nitro-2-propoxyphenyl)acetamide
1-Nitropyrene
5-Nitro-2-pyridinamine
4-Nitropyridine
4-Nitropyridine 1-oxide
5-Nitropyrimidinamine
5-Nitro-2,4(1H,3H)-pyrimidinedione
5-Nitro-2,4,6(1H,3H,5H)-pyrimidinetrione
5-Nitroquinoline

No.	Name	Synonym	Mol. Form.	CAS RN	Mol. Wt.	Physical Form	mp/°C	bp/°C	den/ g cm^{-3}	n_D	Solubility
8179	6-Nitroquinoline		$C_9H_6N_2O_2$	613-50-3	174.156	ye pl (HCl-HOAc)	153.5	$170^{0.2}$			s H_2O, EtOH; sl eth, chl; vs bz
8180	8-Nitroquinoline		$C_9H_6N_2O_2$	607-35-2	174.156	mcl pr (al)	91.5				sl H_2O, chl; s EtOH, eth, bz, acid
8181	4-Nitroquinoline 1-oxide		$C_9H_6N_2O_3$	56-57-5	190.155	ye nd, pl (ace)	154				
8182	5-Nitro-8-quinolinol	Nitroxoline	$C_9H_6N_2O_3$	4008-48-4	190.155		180				
8183	Nitrosobenzene		C_6H_5NO	586-96-9	107.110	orth or mcl (al-eth)	67	58^{18}			i H_2O; s EtOH, eth, bz, lig
8184	*N*-Nitrosodibutylamine	Dibutylnitrosamine	$C_8H_{18}N_2O$	924-16-3	158.241			105^{8}			
8185	*N*-Nitrosodiethanolamine	2,2'-(Nitrosoimino)ethanol	$C_4H_{10}N_2O_3$	1116-54-7	134.133	wh-ye oil		$125^{0.01}$		1.4849^{20}	
8186	*N*-Nitrosodiethylamine	Diethylnitrosamine	$C_4H_{10}N_2O$	55-18-5	102.134	ye oil		176.9	0.9422^{20}	1.4386^{20}	s H_2O, EtOH, eth; sl chl
8187	*N*-Nitrosodimethylamine	Dimethylnitrosamine	$C_2H_6N_2O$	62-75-9	74.081	ye liq		152	1.0048^{20}	1.4368^{20}	vs H_2O, EtOH, eth; s chl
8188	*p*-Nitroso-*N,N*-dimethylaniline		$C_8H_{10}N_2O$	138-89-6	150.177	grn pl (eth)	92.5		1.145^{20}		sl H_2O; s EtOH, eth, chl, $HCONH_2$
8189	*N*-Nitrosodiphenylamine	*N,N*-Diphenylnitrosamine	$C_{12}H_{10}N_2O$	86-30-6	198.219	ye pl(lig)	66.5				i H_2O; sl EtOH, chl; s bz
8190	4-(*N*-Nitrosomethylamino)-1-(3-pyridyl)-1-butanone	Ketone, 3-pyridyl-3-(*N*-methyl-*N*-nitrosamino)propyl	$C_{10}H_{13}N_3O_2$	64091-91-4	207.229		63				sl H_2O
8191	*N*-Nitrosomethylethylamine		$C_3H_8N_2O$	10595-95-6	88.108	ye liq		67^{40}			
8192	*N*-Nitroso-*N*-methylvinylamine	*N*-Methyl-*N*-nitrosoethenamine	$C_3H_6N_2O$	4549-40-0	86.092	ye liq		47			sl H_2O
8193	4-Nitrosomorpholine	*N*-Nitrosomorpholine	$C_4H_8N_2O_2$	59-89-2	116.119		29	225; 140^{25}			s H_2O
8194	2-Nitroso-1-naphthol		$C_{10}H_7NO_2$	132-53-6	173.169		157 dec				sl H_2O, eth, bz, chl; s EtOH, ace, HOAc
8195	1-Nitroso-2-naphthol	1-Nitroso-β-naphthol	$C_{10}H_7NO_2$	131-91-9	173.169	ye-br nd (peth)	109.5				vs bz, eth
8196	*N*-Nitrosonornicotine	*N'*-Nitroso-3-(2-pyrrolidinyl) pyridine	$C_9H_{11}N_3O$	16543-55-8	177.202			$155^{0.2}$			
8197	4-Nitrosophenol		$C_6H_5NO_2$	104-91-6	123.110	pa ye orth nd (ace, bz)	144 dec				sl H_2O; s EtOH, eth, ace, bz, dil alk
8198	4-Nitroso-*N*-phenylaniline	*p*-Nitrosodiphenylamine	$C_{12}H_{10}N_2O$	156-10-5	198.219		143				sl H_2O, lig; vs EtOH, eth, bz
8199	*N*-Nitrosopiperidine	1-Nitrosopiperidine	$C_5H_{10}N_2O$	100-75-4	114.145	pa ye		219; 109^{20}	1.0631^{18}	1.4933^{18}	s H_2O, HCl
8200	*N*-Nitroso-*N*-propyl-1-propanamine	*N*-Nitrosodipropylamine	$C_6H_{14}N_2O$	621-64-7	130.187	gold		206; 113^{40}	0.9163^{20}	1.4437^{20}	sl H_2O; msc EtOH, eth
8201	*N*-Nitrosopyrrolidine		$C_4H_8N_2O$	930-55-2	100.119			214	1.085^{25}	1.4880^{25}	
8202	5-Nitro-2-thiazolamine	2-Amino-5-nitrothiazole	$C_3H_3N_3O_2S$	121-66-4	145.140	oran-ye pow	202 dec				
8203	*N*-(5-Nitro-2-thiazolyl)acetamide	Aminitrozole	$C_5H_5N_3O_3S$	140-40-9	187.177	nd (al), pl (HOAc)	264.5				s alk
8204	4-Nitrothioanisole		$C_7H_7NO_2S$	701-57-5	169.202		72	137^{2}	1.2391^{80}	1.6401^{20}	i H_2O; s ace, bz
8205	2-Nitrothiophene		$C_4H_3NO_2S$	609-40-5	129.138	lt ye mcl nd (peth)	46.5	224.5	1.3644^{43}		i H_2O; vs EtOH; s alk; sl peth
8206	2-Nitrotoluene		$C_7H_7NO_2$	88-72-2	137.137	liq	-10.4	222	1.1611^{19}	1.5450^{20}	i H_2O; msc EtOH, eth; s ctc
8207	3-Nitrotoluene		$C_7H_7NO_2$	99-08-1	137.137	pa ye	15.5	232	1.1581^{20}	1.5466^{20}	i H_2O; s EtOH, bz, ctc; msc eth
8208	4-Nitrotoluene		$C_7H_7NO_2$	99-99-0	137.137	orth cry (al, eth)	51.63	238.3	1.1038^{75}		i H_2O; s EtOH; vs eth, ace, bz, chl
8209	1-Nitro-2-(trifluoromethyl)benzene		$C_7H_4F_3NO_2$	384-22-5	191.108	cry (al)	32.5	217; 105^{20}			i H_2O; vs EtOH, HOAc, bz; sl ctc
8210	1-Nitro-3-(trifluoromethyl)benzene		$C_7H_4F_3NO_2$	98-46-4	191.108	liq	-2.4	202.8; 103^{40}	1.4357^{15}	1.4719^{20}	i H_2O; s EtOH, eth; sl ctc
8211	Nitrourea		$CH_3N_3O_3$	556-89-8	105.053	pl (al-peth)	158 dec				vs ace, EtOH
8212	*trans*-(2-Nitrovinyl)benzene		$C_8H_7NO_2$	5153-67-3	149.148	ye pr (peth, al)	60	255			i H_2O; s EtOH, ace; vs eth, chl, CS_2
8213	Nivalenol		$C_{15}H_{20}O_7$	23282-20-4	312.316	cry (MeOH)	224 dec				sl H_2O; s EtOH, MeOH
8214	Nizatidine		$C_{12}H_{21}N_5O_2S_2$	76963-41-2	331.458	cry (EtOH/AcOEt)	131				sl H_2O; s MeOH; vs chl; i bz, eth
8215	2,2',3,3',4,5,5',6,6'-Nonachlorobiphenyl		$C_{12}HCl_9$	52663-77-1	464.213	cry	180.5				i H_2O
8216	Nonacontane		$C_{90}H_{182}$	7667-51-8	1264.408			612^{200}			

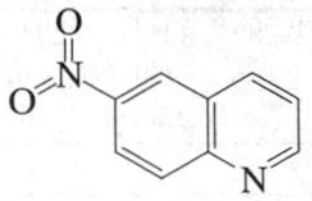

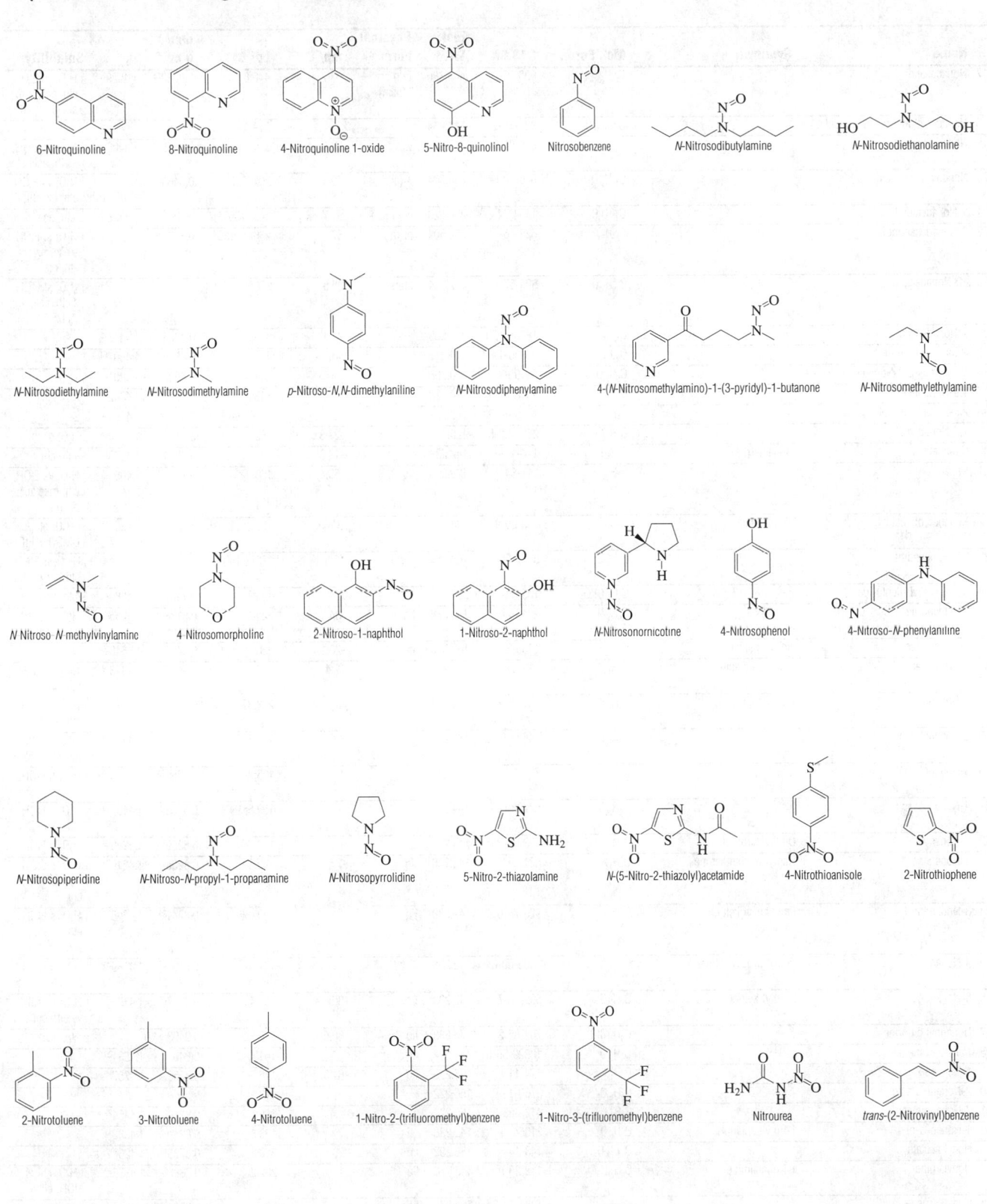

6-Nitroquinoline

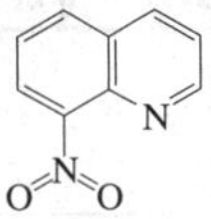

8-Nitroquinoline

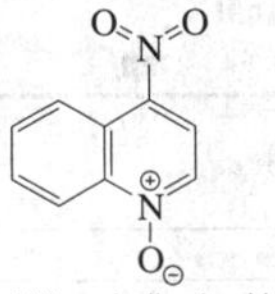

4-Nitroquinoline 1-oxide

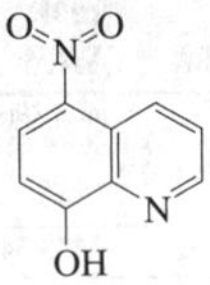

5-Nitro-8-quinolinol

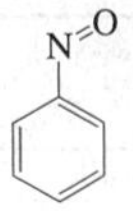

Nitrosobenzene

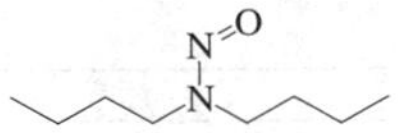

N-Nitrosodibutylamine

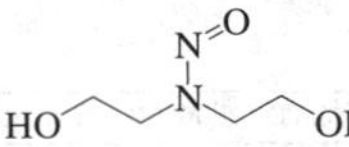

N-Nitrosodiethanolamine

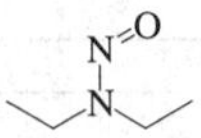

N-Nitrosodiethylamine

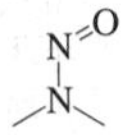

N-Nitrosodimethylamine

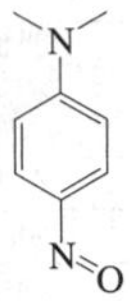

p-Nitroso-*N*,*N*-dimethylaniline

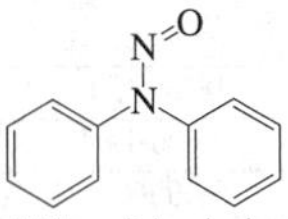

N-Nitrosodiphenylamine

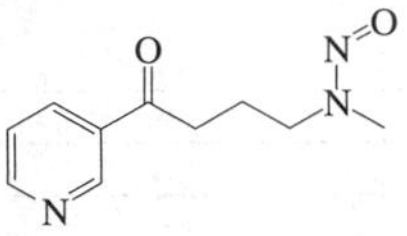

4-(*N*-Nitrosomethylamino)-1-(3-pyridyl)-1-butanone

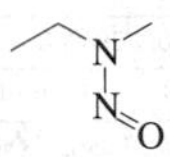

N-Nitrosomethylethylamine

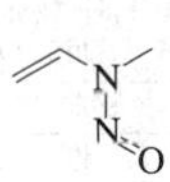

N-Nitroso-*N*-methylvinylamine

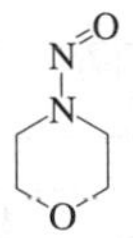

4-Nitrosomorpholine

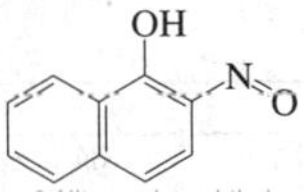

2-Nitroso-1-naphthol

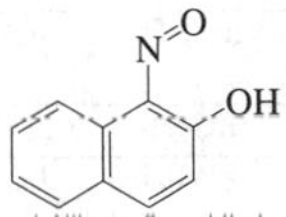

1-Nitroso-2-naphthol

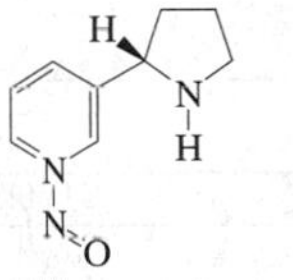

N-Nitrosonornicotine

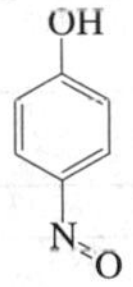

4-Nitrosophenol

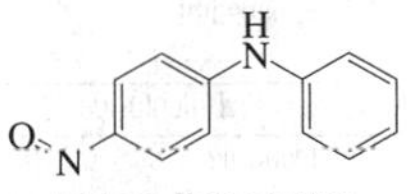

4-Nitroso-*N*-phenylaniline

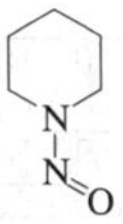

N-Nitrosopiperidine

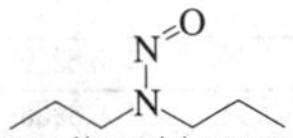

N-Nitroso-*N*-propyl-1-propanamine

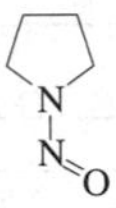

N-Nitrosopyrrolidine

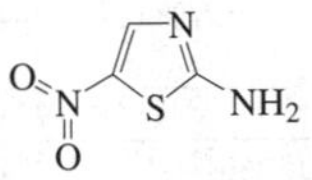

5-Nitro-2-thiazolamine

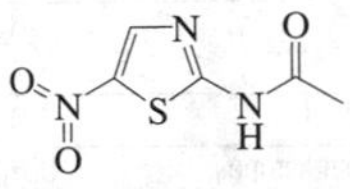

N-(5-Nitro-2-thiazolyl)acetamide

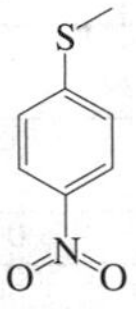

4-Nitrothioanisole

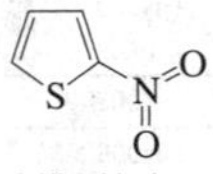

2-Nitrothiophene

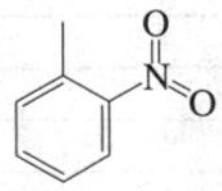

2-Nitrotoluene

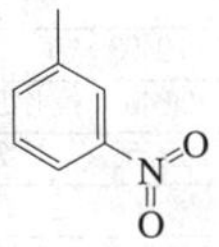

3-Nitrotoluene

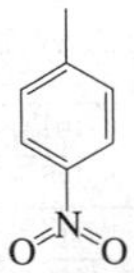

4-Nitrotoluene

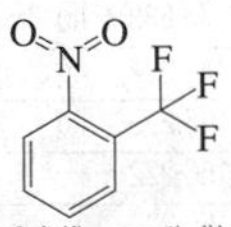

1-Nitro-2-(trifluoromethyl)benzene

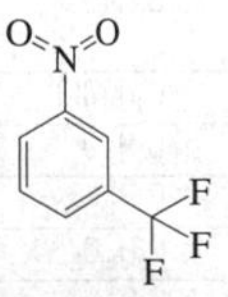

1-Nitro-3-(trifluoromethyl)benzene

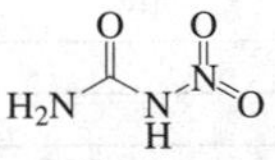

Nitrourea

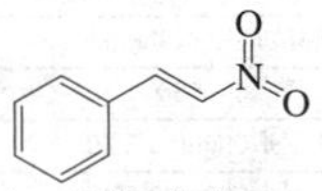

trans-(2-Nitrovinyl)benzene

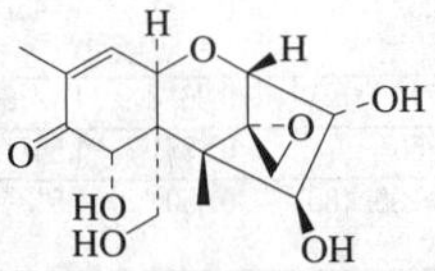

Nivalenol

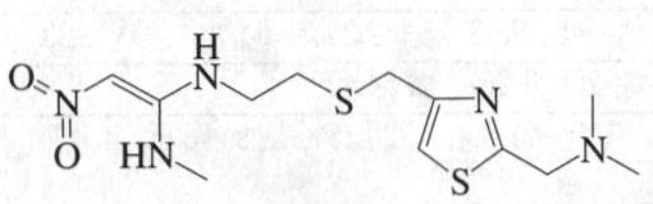

Nizatidine

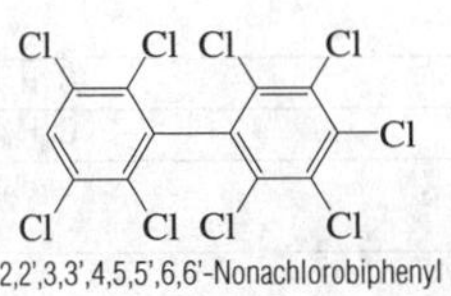

2,2',3,3',4,5,5',6,6'-Nonachlorobiphenyl

$H_3C(CH_2)_{88}CH_3$

Nonacontane

No.	Name	Synonym	Mol. Form.	CAS RN	Mol. Wt.	Physical Form	mp/°C	bp/°C	den/ g cm^{-3}	n_D	Solubility
8217	Nonacosane		$C_{29}H_{60}$	630-03-5	408.786	orth cry (peth)	63.7	440.8	0.8083^{20}	1.4529^{20}	i H_2O; vs EtOH, eth, ace; s bz; sl chl
8218	Nonadecafluorodecanoic acid		$C_{10}HF_{19}O_2$	335-76-2	514.084			219.4			
8219	Nonadecane		$C_{19}H_{40}$	629-92-5	268.521	wax	32.0	329.9	0.7855^{20}	1.4409^{20}	i H_2O; sl EtOH; s eth, ace, ctc
8220	Nonadecanoic acid		$C_{19}H_{38}O_2$	646-30-0	298.504	lf (al)	69.4	297^{100}, 228^{10}	0.8468^{70}		i H_2O; vs EtOH, eth, bz, chl, lig
8221	1-Nonadecanol		$C_{19}H_{40}O$	1454-84-8	284.520	cry (ace)	61.7	345; $166^{0.3}$		1.4328^{75}	s eth, ace
8222	2-Nonadecanone		$C_{19}H_{38}O$	629-66-3	282.504	pr (al)	57	266^{110}, 165^{2}	0.8108^{56}		i H_2O; sl EtOH; s ace, bz; vs eth, ctc
8223	10-Nonadecanone		$C_{19}H_{38}O$	504-57-4	282.504	lf(al)	65.5	>350; $156^{1.1}$			i H_2O; sl EtOH; s eth, ace, lig; vs bz
8224	1-Nonadecene		$C_{19}H_{38}$	18435-45-5	266.505		23.4	329.0	0.7886^{25}	1.4445^{25}	
8225	Nonadecylbenzene		$C_{25}H_{44}$	29136-19-4	344.617		40	419	0.8545^{20}	1.4807^{20}	
8226	*trans,trans*-2,4-Nonadienal		$C_9H_{14}O$	5910-87-2	138.206			98^{10}	0.862^{25}	1.5207^{20}	
8227	1,8-Nonadiene		C_9H_{16}	4900-30-5	124.223			142.5	0.7511^{20}	1.4302^{20}	
8228	2,6-Nonadien-1-ol		$C_9H_{16}O$	7786-44-9	140.222			108^{24}, 98^{11}	0.8604^{25}	1.4598^{25}	
8229	1,8-Nonadiyne		C_9H_{12}	2396-65-8	120.191	liq	-27.3	162	0.8158^{20}	1.4490^{20}	i H_2O; s eth, ace
8230	Nonanal	Nonaldehyde	$C_9H_{18}O$	124-19-6	142.238		-19.3	191	0.8264^{22}	1.4273^{20}	s eth, chl
8231	Nonane		C_9H_{20}	111-84-2	128.255	liq	-53.46	150.82	0.7192^{20}	1.4058^{20}	i H_2O; vs EtOH, eth; msc ace, bz, hp
8232	Nonanedioic acid	Azelaic acid	$C_9H_{16}O_4$	123-99-9	188.221	lf or nd	106.5	357.1; 287^{100}	1.225^{25}	1.4303^{111}	sl H_2O, eth, bz, DMSO; s EtOH
8233	1,9-Nonanediol		$C_9H_{20}O_2$	3937-56-2	160.254	cry (bz)	45.8	173^{20}, 150^{3}			sl H_2O; vs EtOH, eth; s bz; i lig
8234	Nonanedioyl dichloride		$C_9H_{14}Cl_2O_2$	123-98-8	225.112			166^{18}	1.143	1.4680^{20}	s eth; vs bz
8235	Nonanenitrile		$C_9H_{17}N$	2243-27-8	139.238	liq	-34.2	224.4	0.8178^{20}	1.4255^{20}	i H_2O; s EtOH, eth; sl ctc
8236	1-Nonanethiol	Nonyl mercaptan	$C_9H_{20}S$	1455-21-6	160.320	liq	-20.1	220	0.842^{25}	1.4548^{20}	
8237	Nonanoic acid	Pelargonic acid	$C_9H_{18}O_2$	112-05-0	158.238		12.4	254.5	0.9052^{20}	1.4343^{19}	i H_2O; s EtOH, eth, chl
8238	1-Nonanol	Nonyl alcohol	$C_9H_{20}O$	143-08-8	144.254	liq	-5	213.37	0.8280^{20}	1.4333^{20}	i H_2O; s EtOH, eth; sl ctc
8239	2-Nonanol, (±)		$C_9H_{20}O$	74683-66-2	144.254	liq	-35	193.5	0.8471^{20}	1.4353^{20}	i H_2O; vs eth, EtOH
8240	3-Nonanol, (±)		$C_9H_{20}O$	74742-08-8	144.254		22	195; 93^{18}	0.8250^{20}	1.4289^{20}	i H_2O; s EtOH, eth
8241	4-Nonanol		$C_9H_{20}O$	52708-03-9	144.254			192.5; 94^{18}	0.8282^{20}	1.4197^{20}	i H_2O; s EtOH, eth
8242	5-Nonanol	Dibutylcarbinol	$C_9H_{20}O$	623-93-8	144.254		5.6	193; 97^{20}	0.8220^{20}	1.4289^{20}	i H_2O; s EtOH
8243	2-Nonanone	Heptyl methyl ketone	$C_9H_{18}O$	821-55-6	142.238	liq	-7.5	195.3	0.8208^{20}	1.4210^{20}	i H_2O; s EtOH, eth, bz; vs ace, chl
8244	3-Nonanone	Ethyl hexyl ketone	$C_9H_{18}O$	925-78-0	142.238	liq	-8	190; 86^{20}	0.8241^{20}	1.4208^{20}	i H_2O; s EtOH, eth, bz, chl; vs ace
8245	4-Nonanone	Pentyl propyl ketone	$C_9H_{18}O$	4485-09-0	142.238			187.5	0.8190^{25}	1.4189^{20}	i H_2O; s EtOH, eth, chl; vs ace
8246	5-Nonanone	Dibutyl ketone	$C_9H_{18}O$	502-56-7	142.238	liq	-3.8	188.45	0.8217^{20}	1.4195^{20}	i H_2O; s EtOH; vs eth, chl
8247	Nonanoyl chloride		$C_9H_{17}ClO$	764-85-2	176.683	liq	-60.5	215.3	0.9463^{15}		s eth, ace
8248	*trans*-2-Nonenal		$C_9H_{16}O$	18829-56-6	140.222	liq		101^{16}, 89^{12}	0.846	1.4531^{20}	
8249	1-Nonene		C_9H_{18}	124-11-8	126.239	liq	-81.3	146.9	0.7253^{25}	1.4257^{20}	
8250	2-Nonenoic acid		$C_9H_{16}O_2$	3760-11-0	156.222			173^{20}, 136^{5}			
8251	3-Nonenoic acid		$C_9H_{16}O_2$	4124-88-3	156.222		-4.4	156^{18}, 106^{1}	0.9254^{20}	1.4454^{25}	
8252	1-Nonen-3-ol	1-Vinylheptanol	$C_9H_{18}O$	21964-44-3	142.238			193.5	0.824^{21}	1.4382^{15}	
8253	Nonyl acetate		$C_{11}H_{22}O_2$	143-13-5	186.292	liq	-26	210	0.8785^{15}	1.426^{20}	
8254	Nonylamine	1-Nonanamine	$C_9H_{21}N$	112-20-9	143.270	liq	-1	202.2	0.7886^{20}	1.4336^{20}	sl H_2O, chl; s EtOH, eth
8255	Nonylbenzene		$C_{15}H_{24}$	1081-77-2	204.352	liq	-24	280.5	0.8584^{20}	1.4816^{20}	
8256	Nonylcyclohexane		$C_{15}H_{30}$	2883-02-5	210.399	liq	-10	282	0.8163^{20}	1.4519^{20}	
8257	Nonylcyclopentane		$C_{14}H_{28}$	2882-98-6	196.372	liq	-29	262	0.8081^{20}	1.4467^{20}	vs ace, bz, eth, EtOH
8258	Nonyl formate		$C_{10}H_{20}O_2$	5451-92-3	172.265	liq	-33	214	0.86	1.4216^{20}	
8259	1-Nonylnaphthalene		$C_{19}H_{26}$	26438-26-6	254.409		8	366	0.9371^{20}	1.5477^{20}	
8260	4-Nonylphenol		$C_{15}H_{24}O$	104-40-5	220.351	visc ye liq	42	≈295; 180^{10}	0.950^{20}	1.513^{20}	i H_2O; s bz, ctc, hp
8261	1-Nonyne	Heptylacetylene	C_9H_{16}	3452-09-3	124.223	liq	-50	150.8	0.7658^{20}	1.4217^{20}	i H_2O; s eth, bz, ctc

Nonacosane

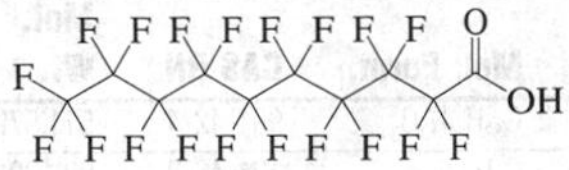

Nonadecafluorodecanoic acid

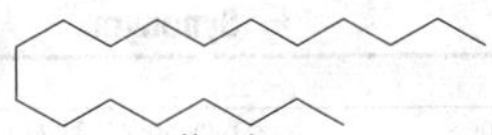

Nonadecane

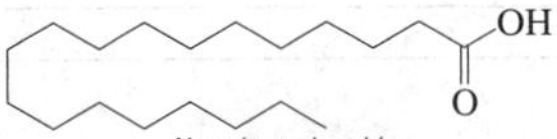

Nonadecanoic acid

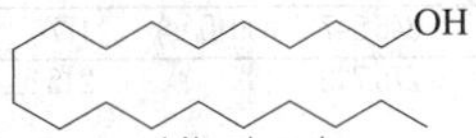

1-Nonadecanol

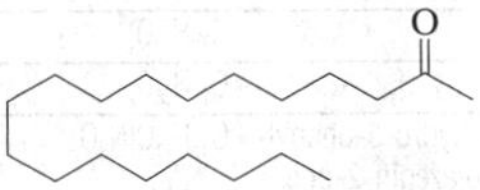

2-Nonadecanone

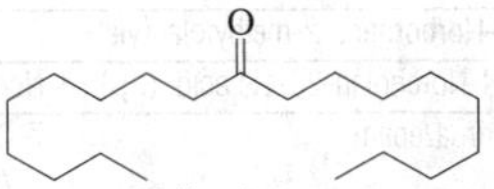

10-Nonadecanone

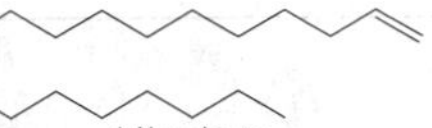

1-Nonadecene

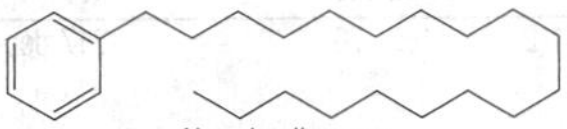

Nonadecylbenzene

trans,trans-2,4-Nonadienal

1,8-Nonadiene

2,6-Nonadien-1-ol

1,8-Nonadiyne

Nonanal

Nonane

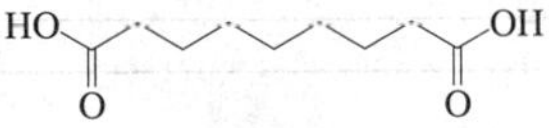

Nonanedioic acid

1,9-Nonanediol

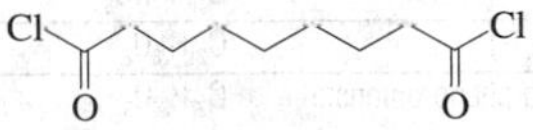

Nonanedioyl dichloride

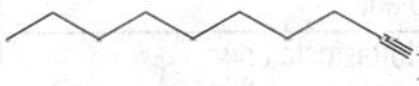

Nonanenitrile

1-Nonanethiol

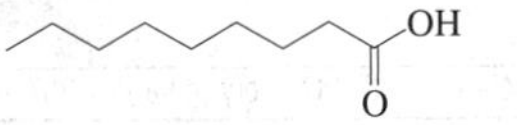

Nonanoic acid

1-Nonanol

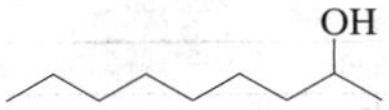

2-Nonanol, (±)

3-Nonanol, (±)

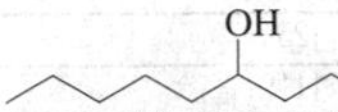

4-Nonanol

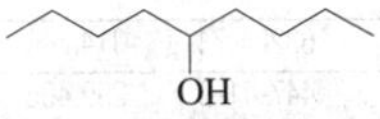

5-Nonanol

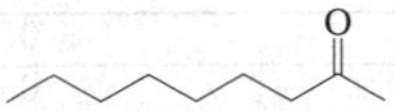

2-Nonanone

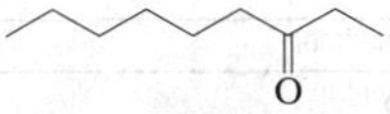

3-Nonanone

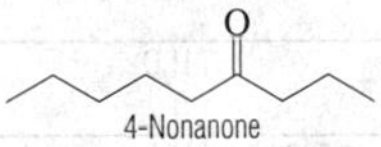

4-Nonanone

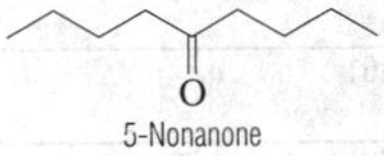

5-Nonanone

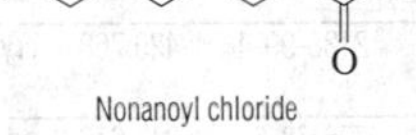

Nonanoyl chloride

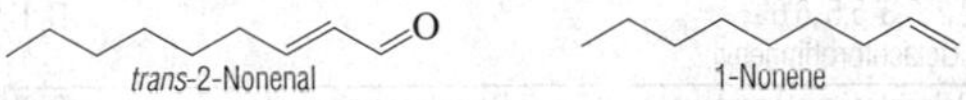

trans-2-Nonenal

1-Nonene

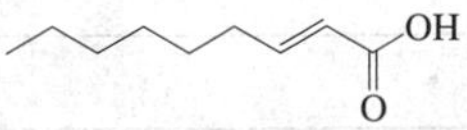

2-Nonenoic acid

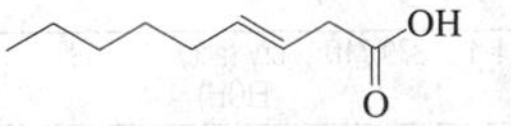

3-Nonenoic acid

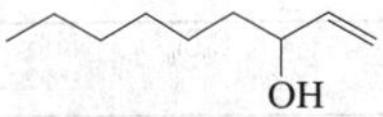

1-Nonen-3-ol

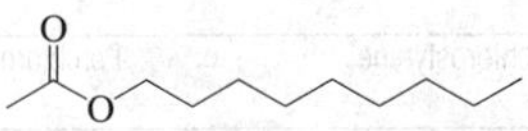

Nonyl acetate

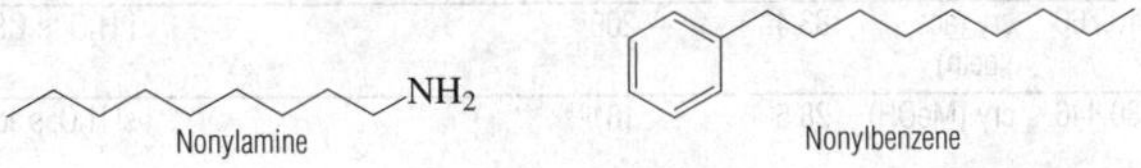

Nonylamine

Nonylbenzene

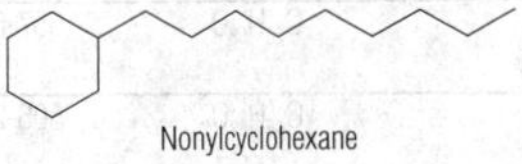

Nonylcyclohexane

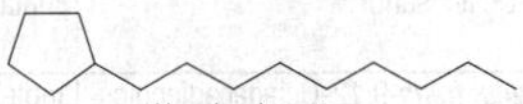

Nonylcyclopentane

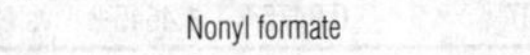

Nonyl formate

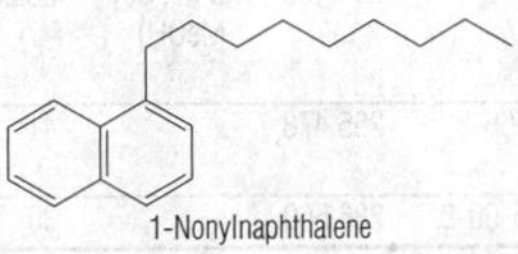

1-Nonylnaphthalene

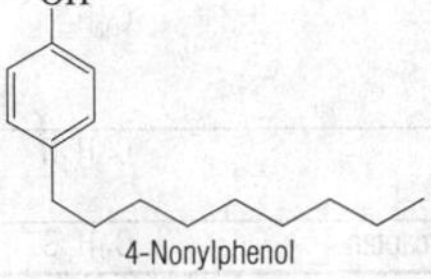

4-Nonylphenol

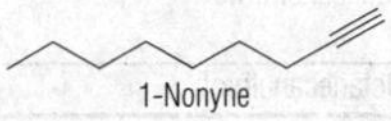

1-Nonyne

No.	Name	Synonym	Mol. Form.	CAS RN	Mol. Wt.	Physical Form	mp/°C	bp/°C	den/ g cm⁻³	n_D	Solubility
8262	Norbormide		$C_{33}H_{25}N_3O_3$	991-42-4	511.570	cry (eth)	194				
8263	2,5-Norbornadiene	Bicyclo[2.2.1]hepta-2,5-diene	C_7H_8	121-46-0	92.139	liq	-19.1	89.5	0.9064^{20}	1.4702^{20}	i H_2O; s EtOH, eth, ace, bz; msc tol
8264	5-Norbornene-2,3-dicarboxylic acid anhydride		$C_9H_8O_3$	826-62-0	164.158		166				
8265	5-Norbornene-2-methylolacrylate		$C_{11}H_{14}O_2$	95-39-6	178.228	col liq		104	1.029^{25}		s os
8266	24-Norcholan-23-oic acid, (5β)	Norcholanic acid	$C_{23}H_{38}O_2$	511-18-2	346.547	nd(HOAc)	177				
8267	Nordazepam	7-Chloro-1,3-dihydro-5-phenyl-2*H*-1,4-benzodiazepin-2-one	$C_{15}H_{11}ClN_2O$	1088-11-5	270.713		216.5				
8268	Nordihydroguaiaretic acid		$C_{18}H_{22}O_4$	500-38-9	302.366	nd(w, al, HOAc)	185.5				sl H_2O; s EtOH, eth, ace, alk; i bz
8269	Norea		$C_{13}H_{22}N_2O$	18530-56-8	222.326		177				
8270	Norepinephrine	Noradrenaline	$C_8H_{11}NO_3$	51-41-2	169.178		217 dec				sl H_2O, EtOH, eth; vs alk, dil HCl
8271	Norethisterone	19-Norpregn-4-en-20-yn-3-one, 17-hydroxy-, (17 α)-	$C_{20}H_{26}O_2$	68-22-4	298.419	cry	204				
8272	Norethynodrel		$C_{20}H_{26}O_2$	68-23-5	298.419	cry (MeOH)	170				
8273	Norflurazon		$C_{12}H_9ClF_3N_3O$	27314-13-2	303.666		184				
8274	Norhyoscyamine		$C_{16}H_{21}NO_3$	537-29-1	275.343	nd	140.5				vs EtOH, chl
8275	*DL*-Norleucine	2-Aminohexanoic acid, (*DL*)	$C_6H_{13}NO_2$	616-06-8	131.173	lf(w)	327 dec		1.172^{25}		s H_2O; sl EtOH; i eth
8276	*L*-Norleucine	2-Aminohexanoic acid, (*L*)	$C_6H_{13}NO_2$	327-57-1	131.173		301 dec				sl H_2O
8277	Normorphine		$C_{16}H_{17}NO_3$	466-97-7	271.311		273				
8278	Norplant	Norgestrel, (-)	$C_{21}H_{28}O_2$	797-63-7	312.446	cry (MeOH)	206				
8279	19-Nortestosterone phenylpropionate	Nandrolone phenpropionate	$C_{27}H_{34}O_3$	62-90-8	406.557	cry	95				
8280	Nortriptyline hydrochloride		$C_{19}H_{22}ClN$	894-71-3	299.838	cry (eth)	214				s H_2O, EtOH; i bz, eth, ace
8281	*DL*-Norvaline	2-Aminopentanoic acid, (±)	$C_5H_{11}NO_2$	760-78-1	117.147	lf(al, w)	303	sub			s H_2O; i EtOH, eth, chl, AcOEt, lig
8282	*L*-Norvaline	2-Aminopentanoic acid, (*S*)	$C_5H_{11}NO_2$	6600-40-4	117.147	cry (dil al)	307				s H_2O
8283	Noscapine		$C_{22}H_{23}NO_7$	128-62-1	413.421	pr or nd (al)	176				i H_2O; s EtOH, bz, chl; sl eth; vs ace
8284	Novobiocin	Streptonivicin	$C_{31}H_{36}N_2O_{11}$	303-81-1	612.624	wh-ye orth cry	154		1.3448		i H_2O; s EtOH, EtOAc, ace, py
8285	Nuarimol		$C_{17}H_{12}ClFN_2O$	63284-71-9	314.740		126				
8286	Nylidrin	Buphenine	$C_{19}H_{25}NO_2$	447-41-6	299.408	cry (MeOH)	111				
8287	Ochratoxin A		$C_{20}H_{18}ClNO_6$	303-47-9	403.813	cry (xyl)	169				
8288	Ochratoxin B		$C_{20}H_{19}NO_6$	4825-86-9	369.368	cry (MeOH)	221				
8289	Ochratoxin C		$C_{22}H_{22}ClNO_6$	4865-85-4	431.866	amorp solid					
8290	Octacaine	3-(Diethylamino)-*N*-phenylbutanamide	$C_{14}H_{22}N_2O$	13912-77-1	234.337	cry	47	200^{1}			vs EtOH, bz, eth
8291	2,2',3,3',5,5',6,6'-Octachlorobiphenyl		$C_{12}H_2Cl_8$	2136-99-4	429.768	cry	161				i H_2O
8292	Octachlorocyclopentene	Perchlorocyclopentene	C_5Cl_8	706-78-5	343.678	nd	40	283	1.8200^{50}	1.5660^{50}	i H_2O; vs EtOH
8293	Octachlorodibenzo-*p*-dioxin		$C_{12}Cl_8O_2$	3268-87-9	459.751	nd	331				
8294	Octachloronaphthalene	Perchloronaphthalene	$C_{10}Cl_8$	2234-13-1	403.731	nd (bz-CCl_4)	197.5	441^{7}, $248^{0.5}$			sl EtOH; vs bz, chl, lig
8295	Octachlorostyrene	Perchlorostyrene	C_8Cl_8	29082-74-4	379.710	cry (ace/ EtOH)	99				
8296	Octacontane		$C_{80}H_{162}$	7667-88-1	1124.142		112	672			
8297	Octacosane		$C_{28}H_{58}$	630-02-4	394.761	mcl or orth (bz-al)	61.1	431.6	0.8067^{20}	1.4330^{70}	i H_2O; msc ace; s bz, chl
8298	Octacosanoic acid	Montanic acid	$C_{28}H_{56}O_2$	506-48-9	424.744		90.9		0.8191^{100}	1.4313^{100}	vs bz, chl
8299	1-Octacosanol	Montanyl alcohol	$C_{28}H_{58}O$	557-61-9	410.760	cry (ace, peth)	83.4	200^{1}			i H_2O; s CS_2
8300	*trans, trans*-9,12-Octadecadienoic acid	Linolelaidic acid	$C_{18}H_{32}O_2$	506-21-8	280.446	cry (MeOH)	28.5	$181^{0.8}$			sl H_2O; s ace, hx
8301	Octadecahydrochrysene		$C_{18}H_{30}$	2090-14-4	246.431		115	353			vs EtOH
8302	Octadecamethyloctasiloxane		$C_{18}H_{54}O_7Si_8$	556-69-4	607.302		-63	186^{20}, 153^{5}	0.913^{25}	1.3970^{20}	vs bz, peth, lig
8303	Octadecanamide		$C_{18}H_{37}NO$	124-26-5	283.493	lf (al)	109	250^{12}			vs eth, chl
8304	Octadecane		$C_{18}H_{38}$	593-45-3	254.495	nd (al, eth-MeOH)	28.2	316.3	0.7768^{28}	1.4390^{20}	i H_2O; sl EtOH; s eth, ace, chl, lig
8305	Octadecanenitrile		$C_{18}H_{35}N$	638-65-3	265.478		41	362	0.8325^{20}	1.4389^{45}	i H_2O; s EtOH; vs eth, ace, chl
8306	1-Octadecanethiol	Stearyl mercaptan	$C_{18}H_{38}S$	2885-00-9	286.560		30	207^{11}	0.8475^{20}	1.4645^{20}	vs eth

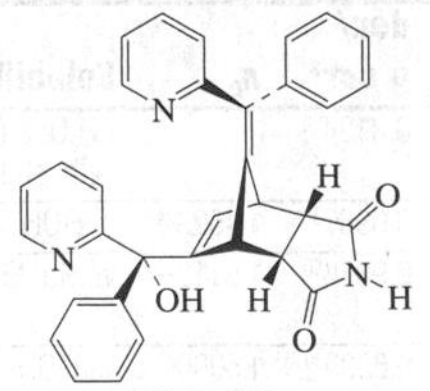

Norbormide

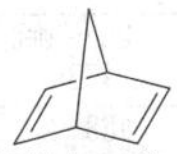

2,5-Norbornadiene

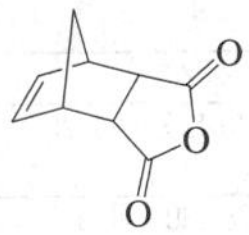

5-Norbornene-2,3-dicarboxylic acid anhydride

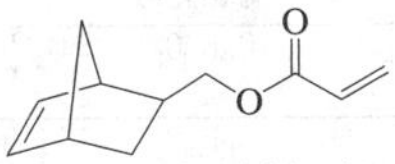

5-Norbornene-2-methylolacrylate

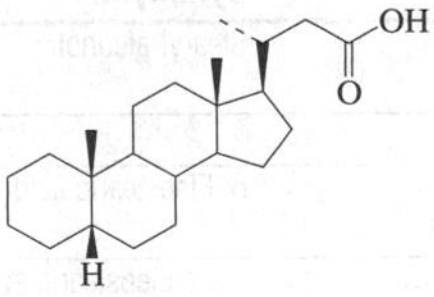

24-Norcholan-23-oic acid, (5β)

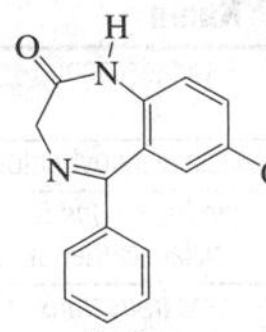

Nordazepam

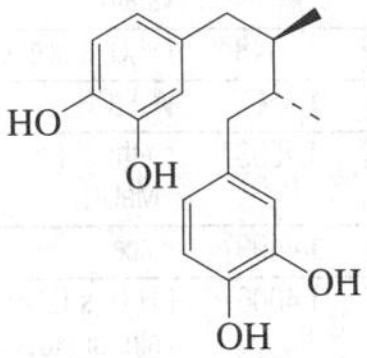

Nordihydroguaiaretic acid

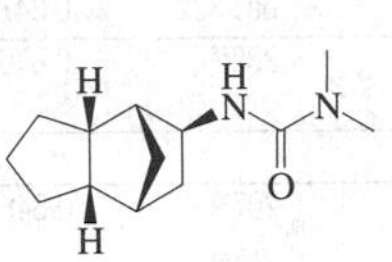

Norea

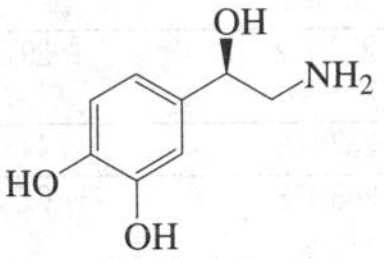

Norepinephrine

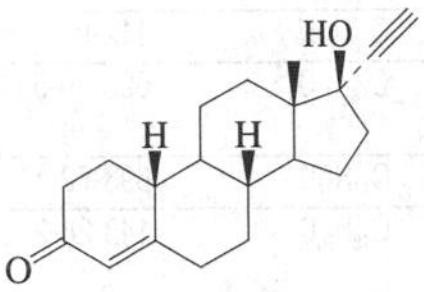

Norethisterone

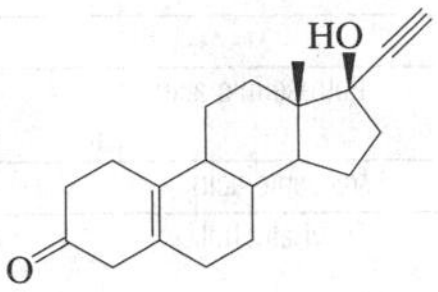

Norethynodrel

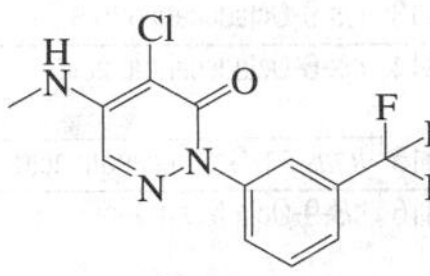

Norflurazon

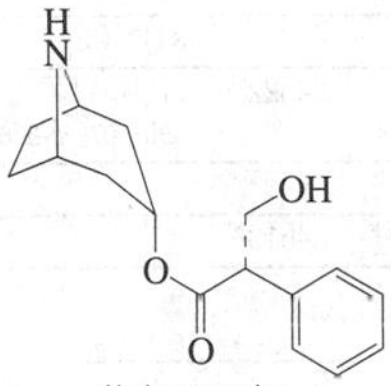

Norhyoscyamine

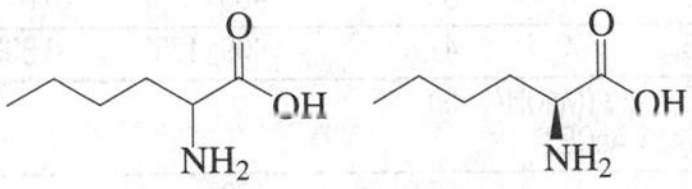

DL-Norleucine

L-Norleucine

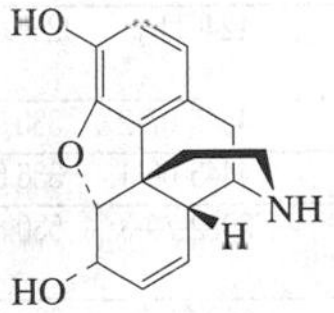

Normorphine

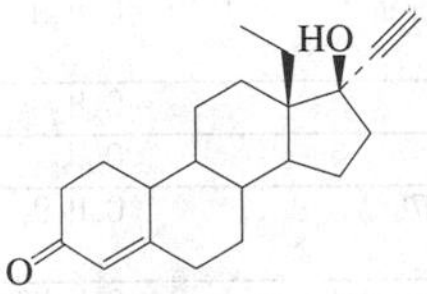

Norplant

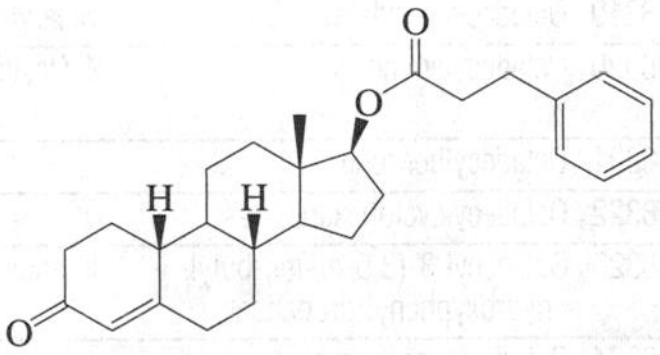

19-Nortestosterone phenylpropionate

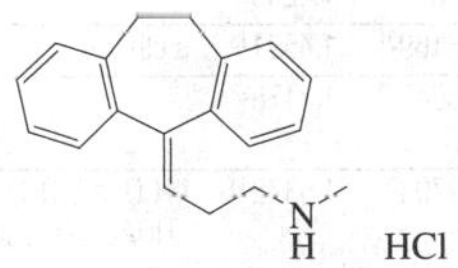

Nortriptyline hydrochloride

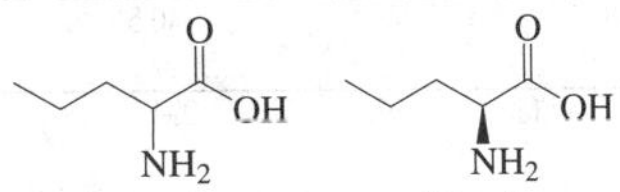

DL-Norvaline

L-Norvaline

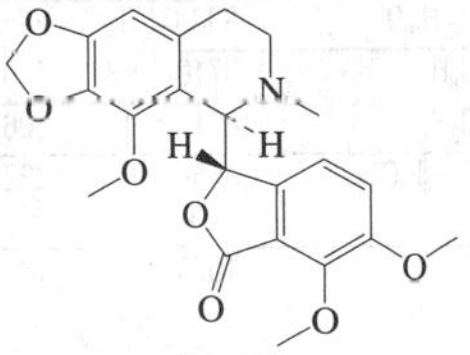

Noscapine

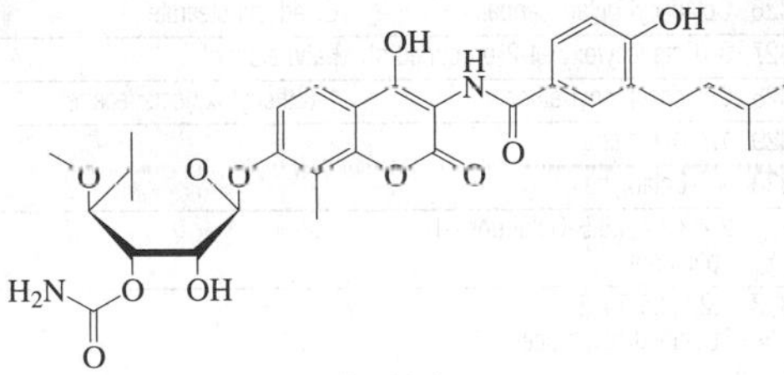

Novobiocin

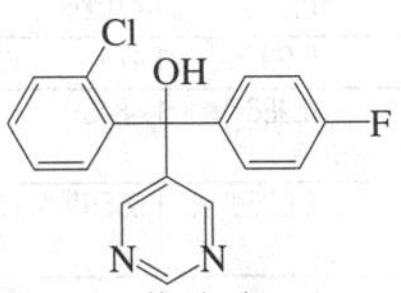

Nuarimol

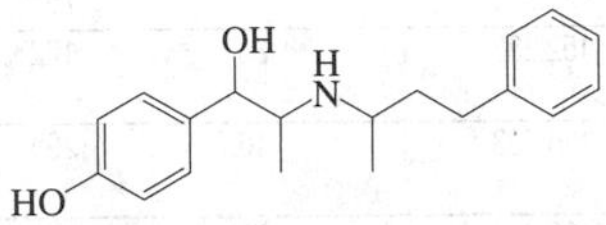

Nylidrin

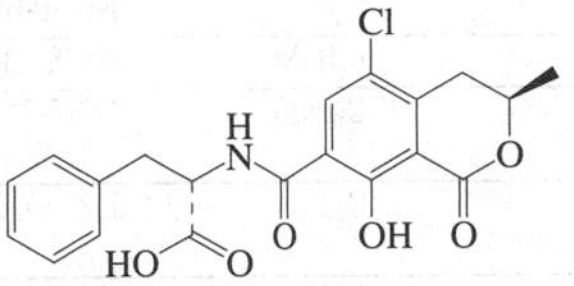

Ochratoxin A

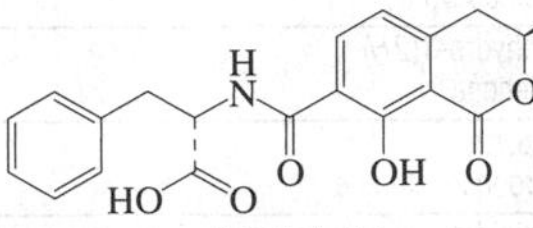

Ochratoxin B

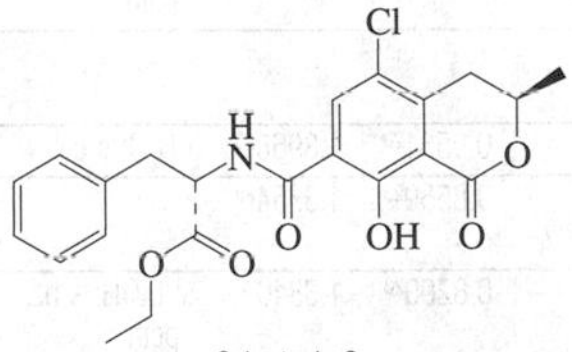

Ochratoxin C

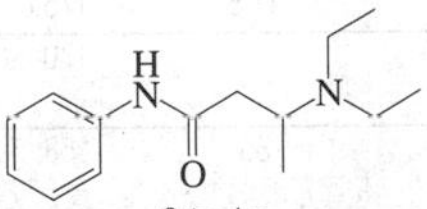

Octacaine

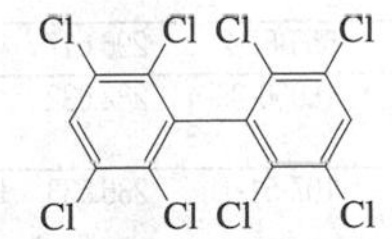

2,2',3,3',5,5',6,6'-Octachlorobiphenyl

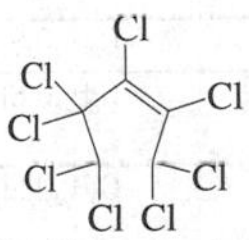

Octachlorocyclopentene

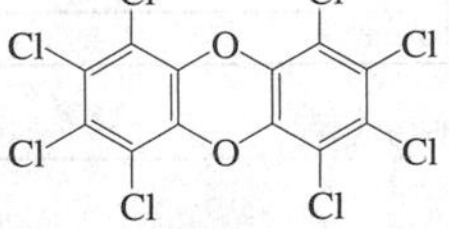

Octachlorodibenzo-*p*-dioxin

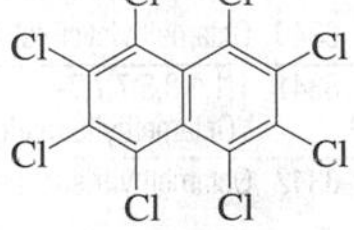

Octachloronaphthalene

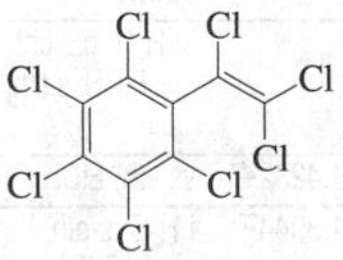

Octachlorostyrene

$H_3C(CH_2)_{78}CH_3$

Octacontane

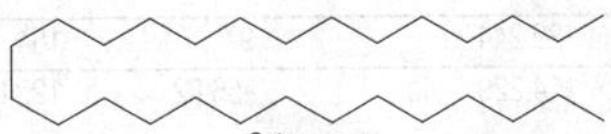

Octacosane

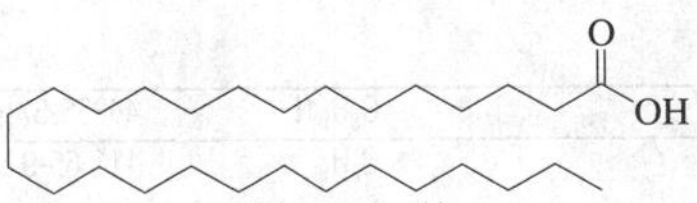

Octacosanoic acid

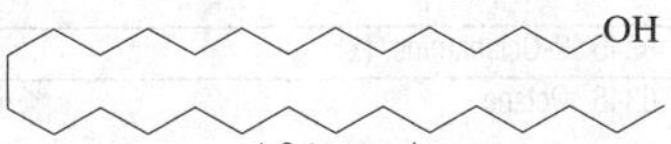

1-Octacosanol

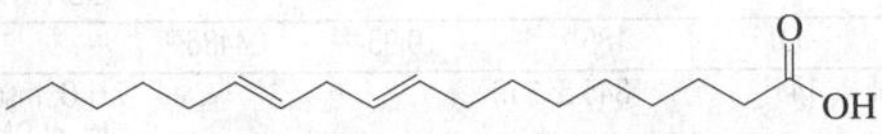

trans, trans-9,12-Octadecadienoic acid

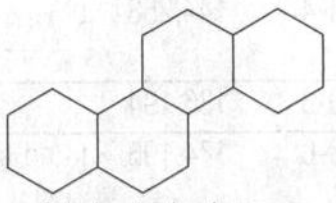

Octadecahydrochrysene

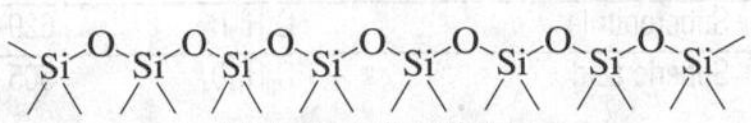

Octadecamethyloctasiloxane

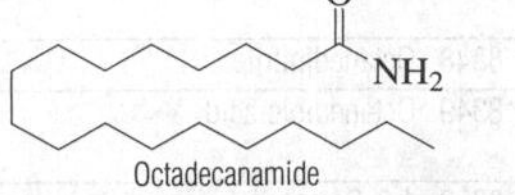

Octadecanamide

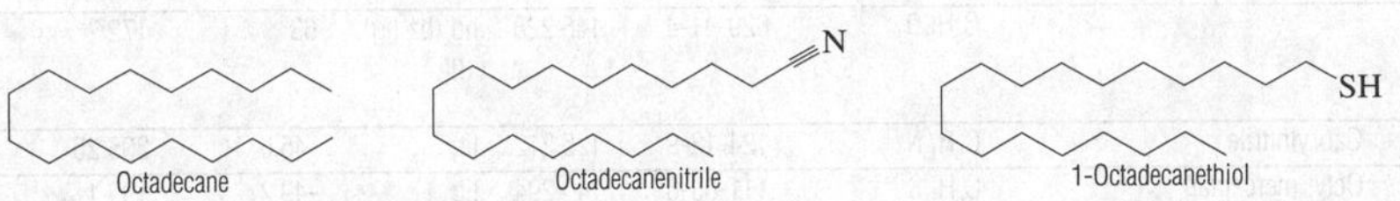

Octadecane

Octadecanenitrile

1-Octadecanethiol

No.	Name	Synonym	Mol. Form.	CAS RN	Mol. Wt.	Physical Form	mp/°C	bp/°C	den/ g cm^{-3}	n_D	Solubility
8307	1-Octadecanol	Stearyl alcohol	$C_{18}H_{38}O$	112-92-5	270.494	lf (al)	57.9	335; 210.5^{15}	0.8124^{59}		i H_2O; s EtOH, eth; sl ace, bz
8308	Octadecanoyl chloride		$C_{18}H_{35}ClO$	112-76-5	302.923		23	215^{15}	0.8969^{0}	1.4523^{24}	sl EtOH
8309	*cis,trans,trans*-9,11,13-Octadecatrienoic acid	α-Eleostearic acid	$C_{18}H_{30}O_2$	506-23-0	278.430	nd (al)	49	235^{12} dec, 170^{1}	0.9028^{50}	1.5112^{50}	vs eth, EtOH
8310	*trans,trans,trans*-9,11,13-Octadecatrienoic acid	*trans*-Eleostearic acid	$C_{18}H_{30}O_2$	544-73-0	278.430	lf (al)	71.5	188^{1}	0.8839^{80}	1.5000^{80}	vs EtOH
8311	*cis*-9-Octadecenamide		$C_{18}H_{35}NO$	301-02-0	281.477		76				vs eth
8312	1-Octadecene		$C_{18}H_{36}$	112-88-9	252.479		17.5	179^{15}, 145^{8}	0.7891^{20}	1.4448^{20}	i H_2O; s ace, ctc
8313	*cis*-9-Octadecenenitrile		$C_{18}H_{33}N$	112-91-4	263.462		-1	dec 332	0.847^{17}	1.4566^{20}	vs EtOH
8314	*cis*-6-Octadecenoic acid	Petroselinic acid	$C_{18}H_{34}O_2$	593-39-5	282.462	lf	29.8	238^{18}	0.8700^{40}	1.4533^{40}	s eth; sl hp, MeOH
8315	*trans*-11-Octadecenoic acid	Vaccenic acid	$C_{18}H_{34}O_2$	693-72-1	282.462		44			1.4499^{60}	s ace
8316	*cis*-9-Octadecen-1-ol	Oleyl alcohol	$C_{18}H_{36}O$	143-28-2	268.478		6.5	207^{15}	0.8489^{20}	1.4606^{20}	i H_2O; s EtOH, eth; sl ctc
8317	*cis*-9-Octadecenylamine	Oleylamine	$C_{18}H_{37}N$	112-90-3	267.494	oil	25	147^{2}			
8318	Octadecyl acetate		$C_{20}H_{40}O_2$	822-23-1	312.531		34.5	208^{9}	0.8510^{30}		vs EtOH
8319	Octadecyl acrylate	Stearyl 2-propenoate	$C_{21}H_{40}O_2$	4813-57-4	324.542						s ctc, CS_2
8320	Octadecylamine	1-Octadecanamine	$C_{18}H_{39}N$	124-30-1	269.510	cry (w)	52.9	346.8	0.8618^{20}	1.4522^{20}	i H_2O; s EtOH, eth, bz; sl ace
8321	Octadecylbenzene		$C_{24}H_{42}$	4445-07-2	330.590		36	400	0.85^{36}	1.479^{36}	
8322	Octadecylcyclohexane		$C_{24}H_{48}$	4445-06-1	336.638		41.6	409; 175^{1}	0.8300^{20}	1.4610^{20}	
8323	Octadecyl 3-(3,5-di-*tert*-butyl-4-hydroxyphenyl)propanoate	Irganox 1076	$C_{35}H_{62}O_3$	2082-79-3	530.865	cry (MeOH/AcOEt)	50				
8324	Octadecyl isocyanate	1-Isocyanatooctadecane	$C_{19}H_{37}NO$	112-96-9	295.503		15.5	172^{5}			
8325	Octadecyl methacrylate	Stearyl methacrylate	$C_{22}H_{42}O_2$	32360-05-7	338.567			195^{6}	0.880^{25}	1.429^{25}	
8326	Octadecyl octadecanoate	Octadecyl stearate	$C_{36}H_{72}O_2$	2778-96-3	536.956	cry (EtOH)	60				
8327	3-(Octadecyloxy)-1,2-propanediol	Batyl alcohol	$C_{21}H_{44}O_3$	544-62-7	344.572		70.5	217^{2}			vs eth
8328	Octadecyl vinyl ether	1-(Ethenyloxy)octadecane	$C_{20}H_{40}O$	930-02-9	296.531		30	182^{3}	0.8138^{40}		sl chl
8329	1,7-Octadiene		C_8H_{14}	3710-30-3	110.197			115.5	0.734^{20}	1.4245^{20}	
8330	1,7-Octadiyne		C_8H_{10}	871-84-1	106.165			135.5; 59^{35}	0.8169^{21}	1.4521^{18}	s eth
8331	2,2,3,3,4,4,5,5-Octafluoro-1-pentanol		$C_5H_4F_8O$	355-80-6	232.072			140.5	1.6647^{20}	1.3178^{20}	
8332	1,2,3,4,5,6,7,8-Octahydroanthracene		$C_{14}H_{18}$	1079-71-6	186.293	pl (al)	78	294	0.9703^{80}	1.5372^{80}	i H_2O; s EtOH, HOAc; vs bz; sl ctc
8333	Octahydroazocine		$C_7H_{15}N$	1121-92-2	113.201		29	52^{15}	0.896^{25}	1.4720^{20}	
8334	Octahydroindene		C_9H_{16}	496-10-6	124.223	liq	-53	167	0.876^{25}	1.4702^{20}	
8335	Octahydroindolizine		$C_8H_{15}N$	13618-93-4	125.212			75^{43}	0.9074^{10}	1.4748	vs eth, EtOH
8336	*trans*-Octahydro-1(2*H*)-naphthalenone		$C_{10}H_{16}O$	21370-71-8	152.233		33	122^{20}	0.986^{20}	1.4849^{21}	
8337	1,2,3,4,5,6,7,8-Octahydrophenanthrene		$C_{14}H_{18}$	5325-97-3	186.293		16.7	295	1.026^{20}	1.5569^{17}	i H_2O; s ace, bz, CS_2, HOAc
8338	*trans*-Octahydro-2*H*-quinolizine-1-methanol, (1*R*)	Lupinine	$C_{10}H_{19}NO$	486-70-4	169.264	orth (peth)	70	270			s H_2O, EtOH, eth, bz, chl; sl peth
8339	2,2,4,4,6,6,8,8-Octamethylcyclotetrasilazane		$C_8H_{28}N_4Si_4$	1020-84-4	292.677		97				
8340	Octamethylcyclotetrasiloxane		$C_8H_{24}O_4Si_4$	556-67-2	296.617		17.5	175.8	0.9561^{20}	1.3968^{20}	i H_2O; s ctc
8341	1,1,1,3,5,7,7,7-Octamethyltetrasiloxane		$C_8H_{26}O_3Si_4$	16066-09-4	282.632			170	0.8559^{20}	1.3854^{20}	
8342	Octamethyltrisiloxane		$C_8H_{24}O_2Si_3$	107-51-7	236.533	liq	-80	153; 51^{17}	0.8200^{20}	1.3840^{20}	sl EtOH; s bz, peth
8343	Octanal	Caprylic aldehyde	$C_8H_{16}O$	124-13-0	128.212			171	0.8211^{20}	1.4217^{20}	vs ace, bz, eth, EtOH
8344	Octanamide		$C_8H_{17}NO$	629-01-6	143.227	lf, pl	108	239	0.8450^{110}		sl H_2O, bz, chl; vs EtOH; s eth, ace
8345	2-Octanamine, (±)		$C_8H_{19}N$	44855-57-4	129.244		97	164	0.7744^{20}	1.4232^{25}	vs eth, EtOH
8346	Octane		C_8H_{18}	111-65-9	114.229	liq	-56.82	125.67	0.6986^{25}	1.3944^{25}	i H_2O; s eth; msc EtOH, ace, bz
8347	1,8-Octanediamine		$C_8H_{20}N_2$	373-44-4	144.258	pl	51.64	225.6			vs H_2O, eth, EtOH
8348	Octanedinitrile	Suberonitrile	$C_8H_{12}N_2$	629-40-3	136.194		-1.8	185^{15}	0.954^{25}	1.4436^{20}	
8349	Octanedioic acid	Suberic acid	$C_8H_{14}O_4$	505-48-6	174.195	lo nd or pl (w)	144	345.5; 219^{20}			i H_2O; msc eth, bz; sl DMSO
8350	1,2-Octanediol		$C_8H_{18}O_2$	1117-86-8	146.228		30	131^{10}, 104$^{0.2}$			
8351	1,8-Octanediol		$C_8H_{18}O_2$	629-41-4	146.228	nd (bz-lig), pr	63	172^{20}			sl H_2O, eth, chl, lig; vs EtOH; s bz
8352	Octanenitrile	Caprylnitrile	$C_8H_{15}N$	124-12-9	125.212	liq	-45.6	205.25	0.8136^{20}	1.4203^{20}	vs eth
8353	1-Octanethiol	Octyl mercaptan	$C_8H_{18}S$	111-88-6	146.294	liq	-49.2	199.1	0.8433^{20}	1.4540^{20}	s EtOH; sl ctc

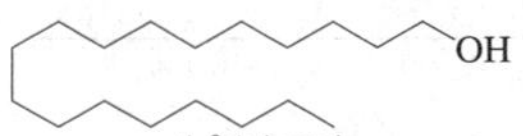

1-Octadecanol

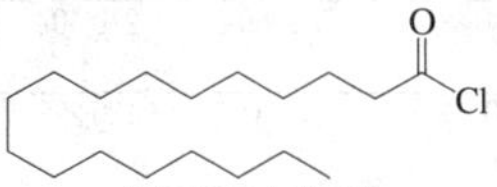

Octadecanoyl chloride

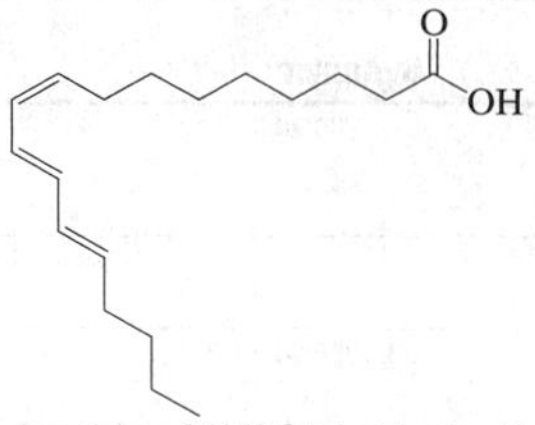

trans,cis,trans-9,11,13-Octadecatrienoic acid

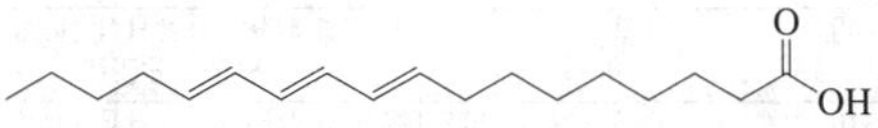

trans,trans,trans-9,11,13-Octadecatrienoic acid

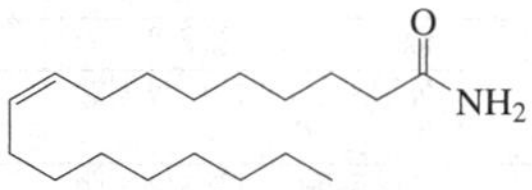

cis-9-Octadecenamide

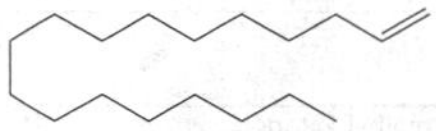

1-Octadecene

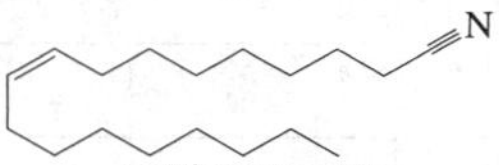

cis-9-Octadecenenitrile

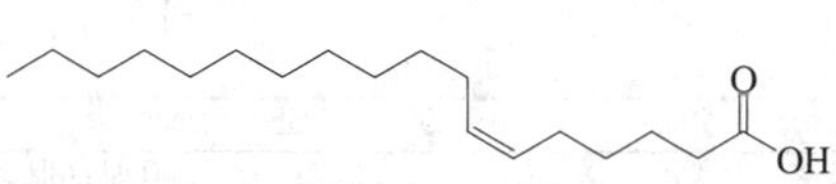

cis-6-Octadecenoic acid

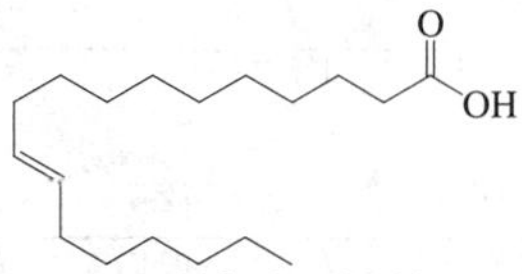

trans-11-Octadecenoic acid

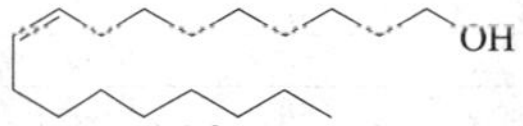

cis-9-Octadecen-1-ol

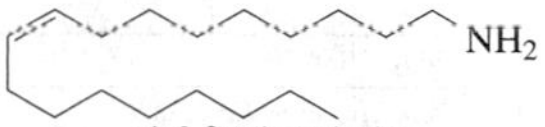

cis-9-Octadecenylamine

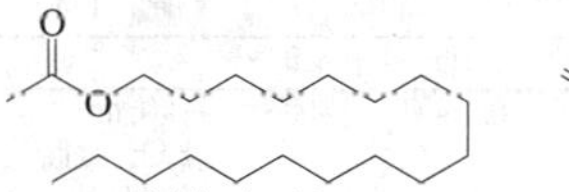

Octadecyl acetate

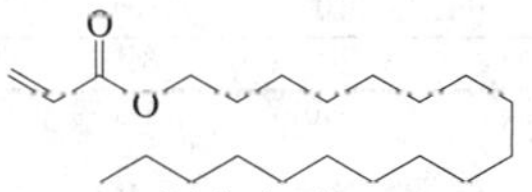

Octadecyl acrylate

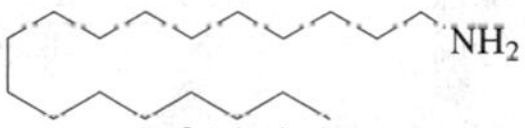

Octadecylamine

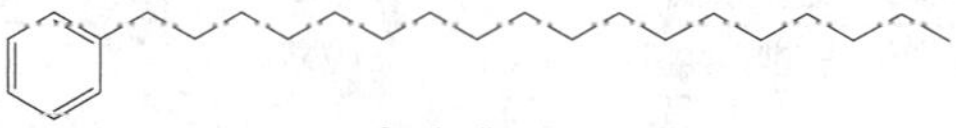

Octadecylbenzene

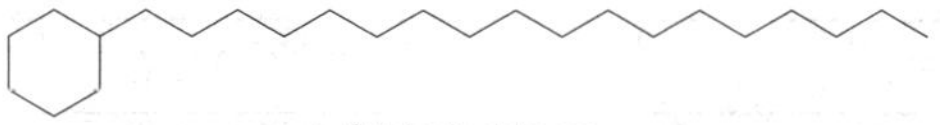

Octadecylcyclohexane

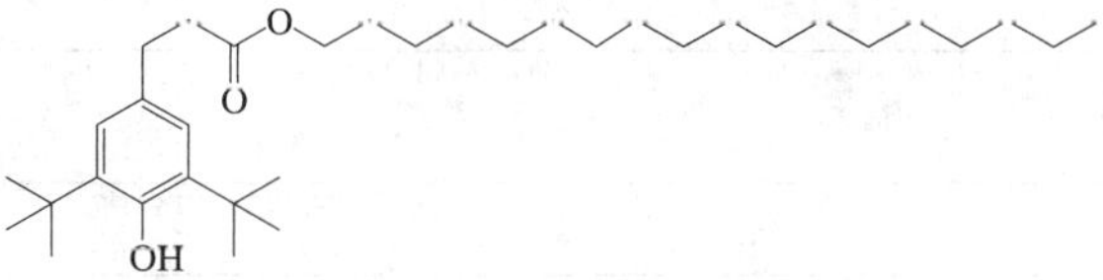

Octadecyl 3-(3,5-di-*tert*-butyl-4-hydroxyphenyl)propanoate

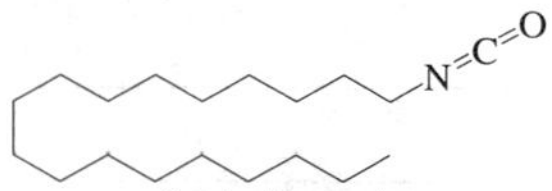

Octadecyl isocyanate

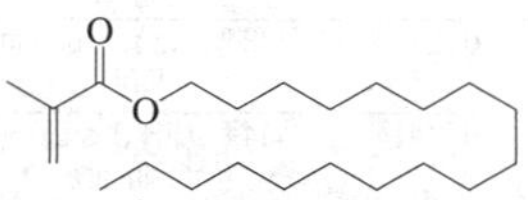

Octadecyl methacrylate

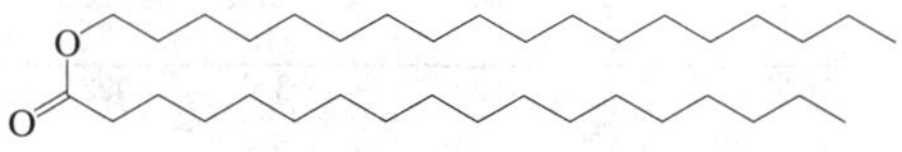

Octadecyl octadecanoate

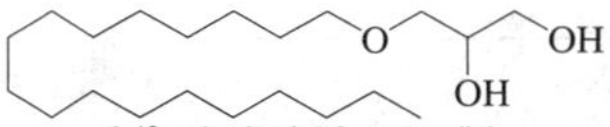

3-(Octadecyloxy)-1,2-propanediol

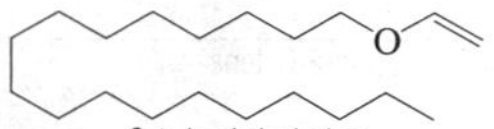

Octadecyl vinyl ether

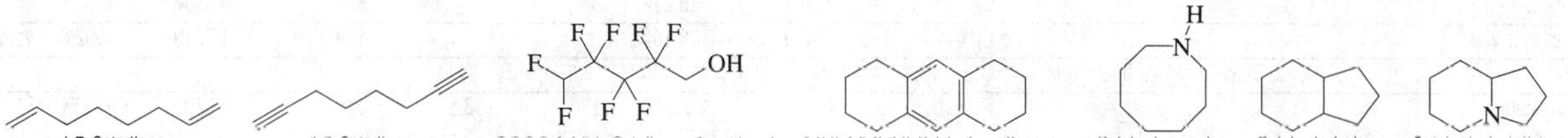

1,7-Octadiene

1,7-Octadiyne

2,2,3,3,4,4,5,5-Octafluoro-1-pentanol

1,2,3,4,5,6,7,8-Octahydroanthracene

Octahydroazocine

Octahydroindene

Octahydroindolizine

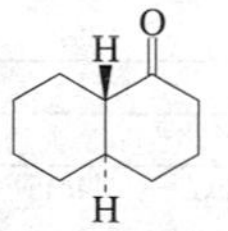

trans-Octahydro-1(2*H*)-naphthalenone

1,2,3,4,5,6,7,8-Octahydrophenanthrene

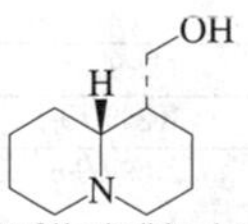

trans-Octahydro-2*H*-quinolizine-1-methanol, (1*R*)

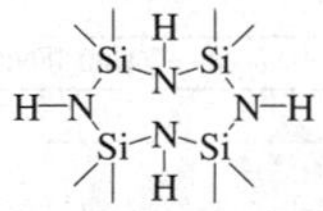

2,2,4,4,6,6,8,8-Octamethylcyclotetrasilazane

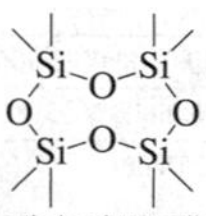

Octamethylcyclotetrasiloxane

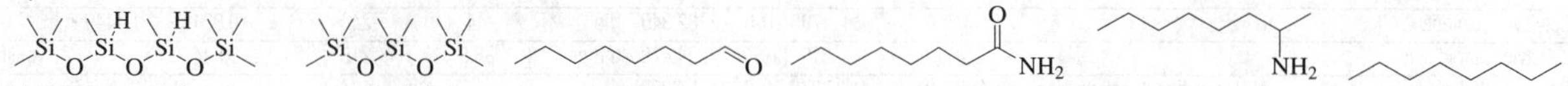

1,1,1,3,5,7,7,7-Octamethyltetrasiloxane

Octamethyltrisiloxane

Octanal

Octanamide

2-Octanamine, (±)

Octane

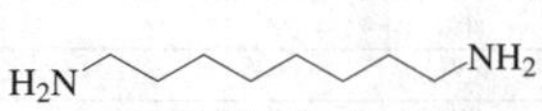

1,8-Octanediamine

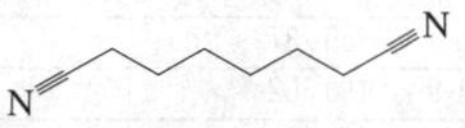

Octanedinitrile

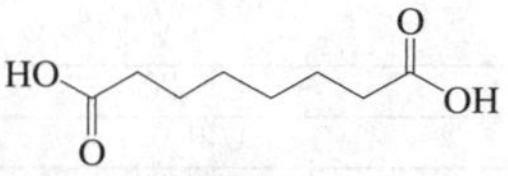

Octanedioic acid

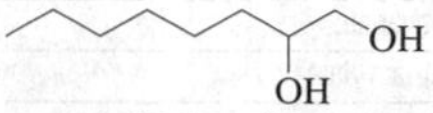

1,2-Octanediol

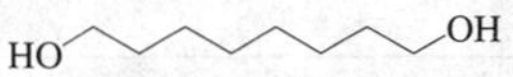

1,8-Octanediol

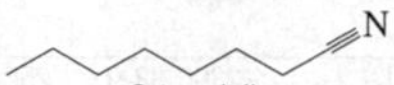

Octanenitrile

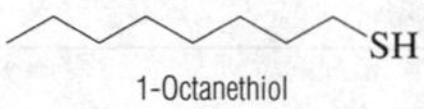

1-Octanethiol

No.	Name	Synonym	Mol. Form.	CAS RN	Mol. Wt.	Physical Form	mp/°C	bp/°C	den/ g cm^{-3}	n_D	Solubility
8354	Octanoic acid	Caprylic acid	$C_8H_{16}O_2$	124-07-2	144.212		16.5	239	0.9073^{25}	1.4285^{20}	sl H_2O; msc EtOH, chl, CH_3CN
8355	Octanoic anhydride		$C_{16}H_{30}O_3$	623-66-5	270.407	liq	-1	282.5	0.9065^{18}	1.4358^{18}	vs ace, eth, EtOH
8356	1-Octanol	Capryl alcohol	$C_8H_{18}O$	111-87-5	130.228	liq	-14.8	195.16	0.8262^{25}	1.4295^{20}	i H_2O; msc EtOH, eth; s ctc
8357	2-Octanol	(±)-*sec*-Caprylic alcohol	$C_8H_{18}O$	4128-31-8	130.228	liq	-31.6	179.3	0.8193^{20}	1.4203^{20}	sl H_2O; s EtOH, eth, ace
8358	3-Octanol		$C_8H_{18}O$	589-98-0	130.228	liq	-45	171	0.8258^{20}		
8359	4-Octanol		$C_8H_{18}O$	74778-22-6	130.228	liq	-40.7	176.3	0.8186^{20}	1.4248^{20}	sl H_2O, ctc; s EtOH
8360	2-Octanone	Hexyl methyl ketone	$C_8H_{16}O$	111-13-7	128.212	liq	-16	172.5	0.820^{20}	1.4151^{20}	sl H_2O; msc EtOH, eth
8361	3-Octanone	Ethyl pentyl ketone	$C_8H_{16}O$	106-68-3	128.212			167.5	0.822^{25}	1.4150^{20}	i H_2O; msc EtOH, eth
8362	4-Octanone	Butyl propyl ketone	$C_8H_{16}O$	589-63-9	128.212			163	0.8146^{25}	1.4173^{14}	i H_2O; msc EtOH, eth; s ctc
8363	Octanoyl chloride		$C_8H_{15}ClO$	111-64-8	162.657	liq	-63	195.6	0.9535^{15}	1.4335^{20}	s eth
8364	Octaphenylcyclotetrasiloxane		$C_{48}H_{40}O_4Si_4$	546-56-5	793.172	nd (bz-al, HOAc)	200.5	330^{1}			i H_2O; sl EtOH; s bz, chl, HOAc
8365	1,3,5,7-Octatetraene		C_8H_{10}	1482-91-3	106.165	cry (bz)	50	sub			s peth, HOAc
8366	*trans*-2-Octenal		$C_8H_{14}O$	2548-87-0	126.196	liq		85^{19}	0.846	1.4500^{20}	
8367	1-Octene	Caprylene	C_8H_{16}	111-66-0	112.213	liq	-101.7	121.29	0.7149^{20}	1.4087^{20}	i H_2O; msc EtOH; s eth, ace; sl ctc
8368	*cis*-2-Octene		C_8H_{16}	7642-04-8	112.213	liq	-100.2	125.6	0.7243^{20}	1.4150^{20}	i H_2O; s EtOH, eth, ace, bz, chl
8369	*trans*-2-Octene		C_8H_{16}	13389-42-9	112.213	liq	-87.7	125	0.7199^{20}	1.4132^{20}	i H_2O; s EtOH, eth, ace, bz; vs chl
8370	*cis*-3-Octene		C_8H_{16}	14850-22-7	112.213	liq	-126	122.9	0.7159^{20}	1.4135^{20}	vs ace, bz, eth, EtOH
8371	*trans*-3-Octene		C_8H_{16}	14919-01-8	112.213	liq	-110	123.3	0.7152^{20}	1.4126^{20}	i H_2O; s EtOH, eth, ace, bz, lig, ctc
8372	*cis*-4-Octene		C_8H_{16}	7642-15-1	112.213	liq	-118.7	122.5	0.7212^{20}	1.4148^{20}	vs ace, bz, eth, EtOH
8373	*trans*-4-Octene		C_8H_{16}	14850-23-8	112.213	liq	-93.8	122.3	0.7141^{20}	1.4114^{20}	i H_2O; s EtOH, eth, ace, bz, lig; sl ctc
8374	1-Octen-3-ol		$C_8H_{16}O$	3391-86-4	128.212			174; 69^{12}	0.8395^{13}	1.4391^{12}	
8375	2-Octen-1-ol		$C_8H_{16}O$	22104-78-5	128.212			88^{11}	0.850^{20}	1.4470^{20}	
8376	1-Octen-3-yne		C_8H_{12}	17679-92-4	108.181			134; 62^{60}	0.7749^{20}	1.4592^{20}	vs eth
8377	Octhilinone	2-Octyl-3(2*H*)-isothiazolone	$C_{11}H_{19}NOS$	26530-20-1	213.340			$120^{0.01}$			
8378	Octyl acetate		$C_{10}H_{20}O_2$	112-14-1	172.265	liq	-38.5	210	0.8705^{20}	1.4150^{20}	i H_2O; s EtOH, eth; sl ctc
8379	Octyl acrylate	Octyl 2-propenoate	$C_{11}H_{20}O_2$	2499-59-4	184.276			229; $57^{0.05}$	0.8810^{20}		
8380	Octylamine	1-Octanamine	$C_8H_{19}N$	111-86-4	129.244		0	179.6	0.7826^{20}	1.4292^{20}	sl H_2O; vs EtOH, eth; s ctc
8381	Octylamine hydrochloride	1-Octanamine hydrochloride	$C_8H_{20}ClN$	142-95-0	165.705		196.5				s H_2O
8382	4-Octylaniline		$C_{14}H_{23}N$	16245-79-7	205.340		20	310; 138^{5}	0.9128^{20}		vs eth
8383	Octylbenzene		$C_{14}H_{22}$	2189-60-8	190.325	liq	-36	264	0.8562^{20}	1.4845^{20}	i H_2O; msc eth, bz
8384	Octyl butanoate		$C_{12}H_{24}O_2$	110-39-4	200.318	liq	-55.6	244.1	0.8629^{20}	1.4267^{15}	vs EtOH
8385	Octylcyclohexane		$C_{14}H_{28}$	1795-15-9	196.372	liq	-20	264	0.8138^{20}	1.4503^{20}	
8386	Octylcyclopentane		$C_{13}H_{26}$	1795-20-6	182.345	liq	-44	243	0.8048^{20}	1.4446^{20}	
8387	2-Octyldecanoic acid		$C_{18}H_{36}O_2$	619-39-6	284.478	nd or lf (al)	38.5	215^{13}	0.8447^{70}		vs eth, EtOH
8388	Octyldimethylamine	*N,N*-Dimethyl-1-octanamine	$C_{10}H_{23}N$	7378-99-6	157.297			194			
8389	Octyl diphenyl phosphate		$C_{20}H_{27}O_4P$	115-88-8	362.399				1.09^{25}		
8390	Octyl formate		$C_9H_{18}O_2$	112-32-3	158.238	liq	-39.1	198.8	0.8744^{20}	1.4208^{15}	i H_2O; s EtOH; msc eth; sl ctc
8391	Octyl isocyanate		$C_9H_{17}NO$	3158-26-7	155.237			78^{6}			
8392	Octyl methacrylate		$C_{12}H_{22}O_2$	2157-01-9	198.302			239.5			
8393	Octyl nitrate		$C_8H_{17}NO_3$	629-39-0	175.226			110^{20}	0.975^{0}		sl H_2O; s EtOH, eth
8394	Octyl nitrite		$C_8H_{17}NO_2$	629-46-9	159.227			174.5	0.862^{17}	1.4127^{20}	sl H_2O; vs EtOH, eth
8395	Octyl octanoate		$C_{16}H_{32}O_2$	2306-88-9	256.424	liq	-18.1	306.8	0.8554^{20}	1.4352^{20}	vs ace, eth, EtOH

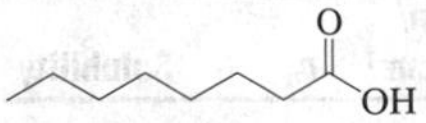

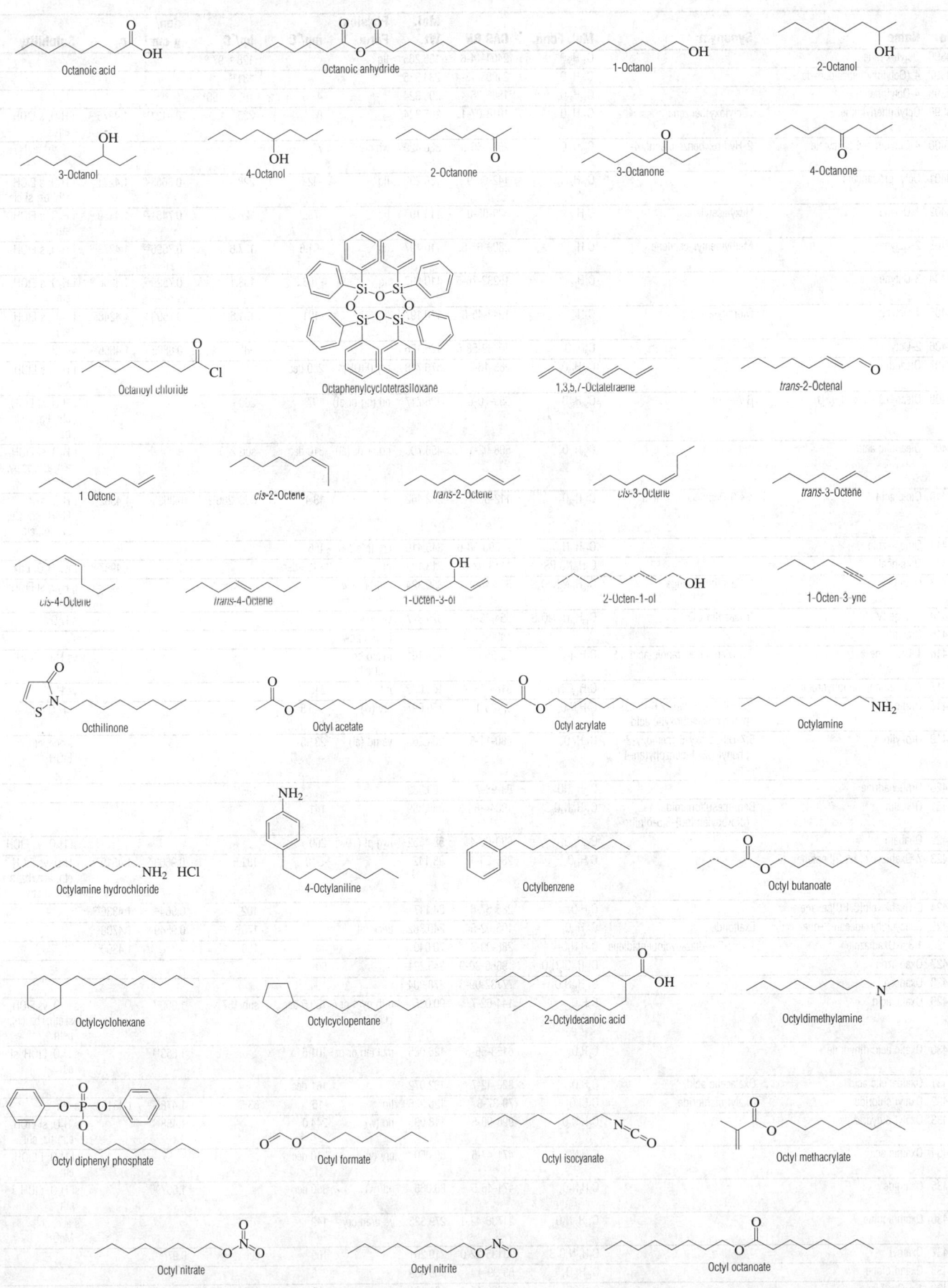

Octanoic acid

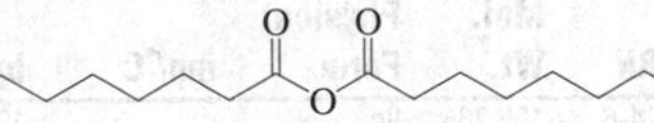

Octanoic anhydride

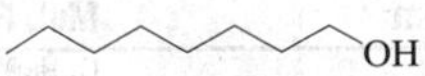

1-Octanol

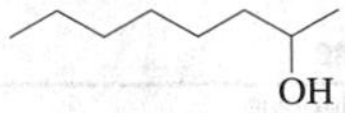

2-Octanol

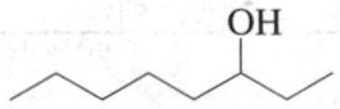

3-Octanol

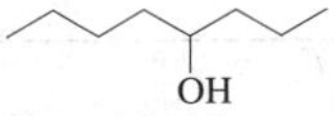

4-Octanol

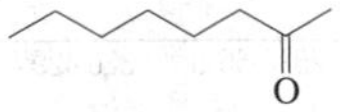

2-Octanone

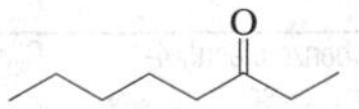

3-Octanone

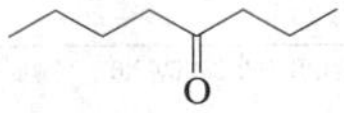

4-Octanone

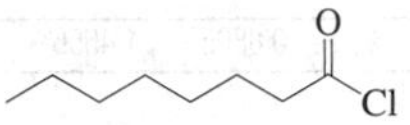

Octanoyl chloride

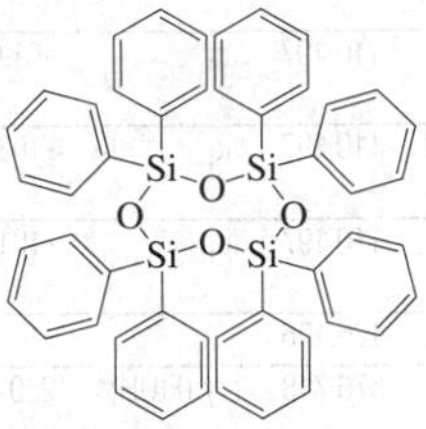

Octaphenylcyclotetrasiloxane

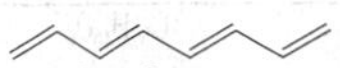

1,3,5,7-Octatetraene

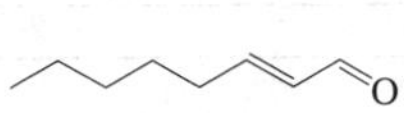

trans-2-Octenal

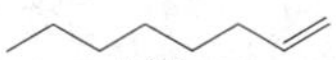

1-Octene

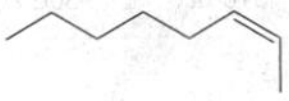

cis-2-Octene

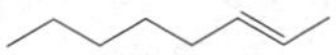

trans-2-Octene

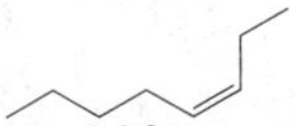

cis-3-Octene

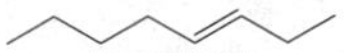

trans-3-Octene

cis-4-Octene

trans-4-Octene

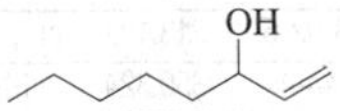

1-Octen-3-ol

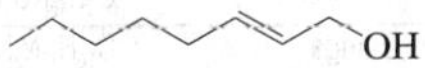

2-Octen-1-ol

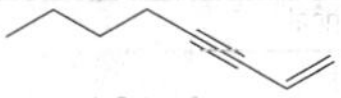

1-Octen-3-yne

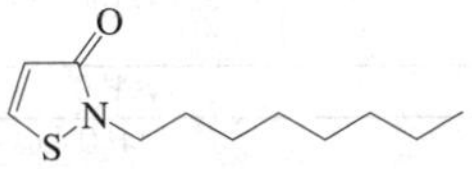

Octhilinone

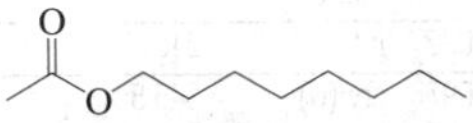

Octyl acetate

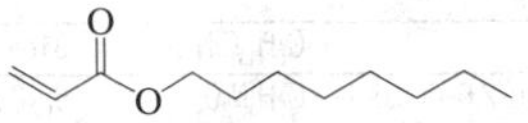

Octyl acrylate

Octylamine

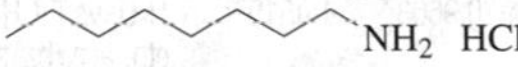

Octylamine hydrochloride

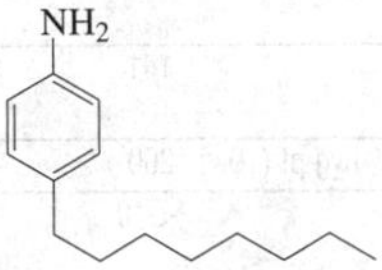

4-Octylaniline

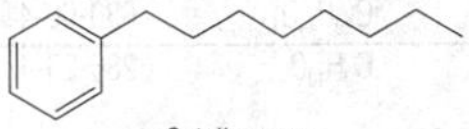

Octylbenzene

Octyl butanoate

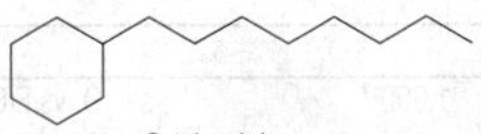

Octylcyclohexane

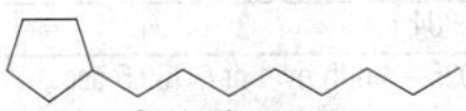

Octylcyclopentane

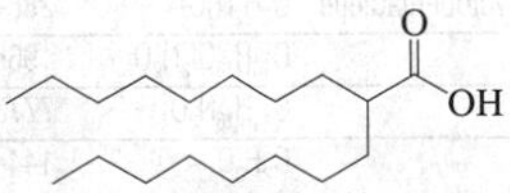

2-Octyldecanoic acid

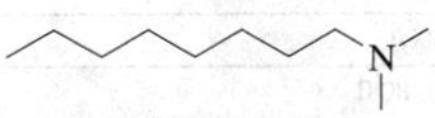

Octyldimethylamine

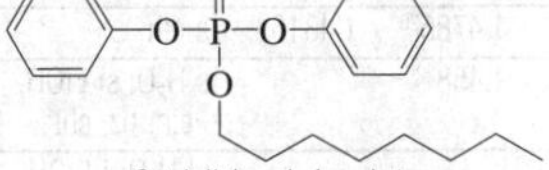

Octyl diphenyl phosphate

Octyl formate

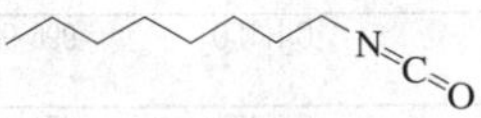

Octyl isocyanate

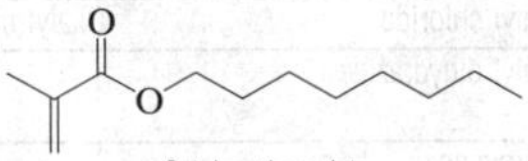

Octyl methacrylate

Octyl nitrate

Octyl nitrite

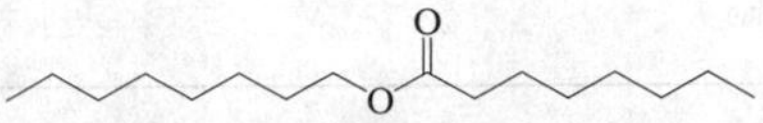

Octyl octanoate

No.	Name	Synonym	Mol. Form.	CAS RN	Mol. Wt.	Physical Form	mp/°C	bp/°C	den/ g cm^{-3}	n_D	Solubility
8396	Octyloxirane		$C_{10}H_{20}O$	2404-44-6	156.265	liq		128^{95}, 97^{30}			
8397	4-(Octyloxy)benzaldehyde		$C_{15}H_{22}O_2$	24083-13-4	234.335			$131^{0.5}$			
8398	4-Octylphenol		$C_{14}H_{22}O$	1806-26-4	206.324		43.0	169^{10}, 150^{4}			
8399	Octyl phenyl ether	(Octyloxy)benzene	$C_{14}H_{22}O$	1818-07-1	206.324		8	285	0.9131^{15}	1.4875^{20}	i H_2O; s EtOH, eth
8400	4-Octylphenyl salicylate	2-Hydroxybenzoic acid, 4-octylphenyl ester	$C_{21}H_{26}O_3$	2512-56-3	326.429	wh cry	73				
8401	Octyl propanoate		$C_{11}H_{22}O_2$	142-60-9	186.292	liq	-42.6	228	0.8663^{20}	1.4221^{15}	i H_2O; s EtOH, eth, bz; sl ctc
8402	1-Octyne	Hexylacetylene	C_8H_{14}	629-05-0	110.197	liq	-79.3	126.3	0.7461^{20}	1.4159^{20}	i H_2O; s EtOH, eth
8403	2-Octyne	Methylpentylacetylene	C_8H_{14}	2809-67-8	110.197	liq	-61.6	137.6	0.7596^{20}	1.4278^{20}	i H_2O; s EtOH, eth
8404	3-Octyne		C_8H_{14}	15232-76-5	110.197	liq	-103.9	133.1	0.7529^{20}	1.4250^{20}	i H_2O; s EtOH, eth
8405	4-Octyne	Dipropylacetylene	C_8H_{14}	1942-45-6	110.197	liq	-101	131.6	0.7509^{20}	1.4248^{20}	i H_2O; s EtOH, eth
8406	2-Octyn-1-ol	2-Octynol	$C_8H_{14}O$	20739-58-6	126.196		-18	98^{15}	0.8805^{20}	1.4556^{20}	vs eth
8407	Oleandrin		$C_{32}H_{48}O_9$	465-16-7	576.718	cry (EtOH)	250 dec				i H_2O; s EtOH, chl
8408	Olean-12-en-3-ol, (3β)	β-Amyrin	$C_{30}H_{50}O$	559-70-6	426.717	nd (lig or al)	197	$260^{0.5}$			i H_2O; sl EtOH, chl, lig; s eth, bz
8409	Oleanolic acid		$C_{30}H_{48}O_3$	508-02-1	456.700	nd or pr (al)	310 dec	sub 280			i H_2O; sl EtOH, eth, ace; vs py, HOAc
8410	Oleic acid	*cis*-9-Octadecenoic acid	$C_{18}H_{34}O_2$	112-80-1	282.462		13.4	360; 286^{100}	0.8935^{20}	1.4582^{20}	i H_2O; msc EtOH, eth, ace, bz, chl, ctc
8411	Omeprazole		$C_{17}H_{19}N_3O_3S$	73590-58-6	345.416	cry (MeCN)	156				
8412	Omethoate		$C_5H_{12}NO_4PS$	1113-02-6	213.192	oil	≈135 dec		1.32^{20}	1.4987^{20}	msc H_2O; i hx
8413	Orange I	1-Naphthol Orange	$C_{16}H_{11}N_2NaO_4S$	523-44-4	350.324	red-br pow					s H_2O; sl EtOH; i bz
8414	Orange IV	Tropaeolin OO	$C_{18}H_{14}N_3NaO_3S$	554-73-4	375.377	ye pow					s H_2O
8415	Orcein			1400-62-0		br-red pow					
8416	*L*-Ornithine	2,5-Diaminopentanoic acid, (*S*)	$C_5H_{12}N_2O_2$	70-26-8	132.161	micro cry (al-eth)	140				vs H_2O, EtOH
8417	*L*-Ornithine, monohydrochloride		$C_5H_{13}ClN_2O_2$	3184-13-2	168.622	nd	215				vs H_2O
8418	Orotic acid	1,2,3,6-Tetrahydro-2,6-dioxo-4-pyrimidinecarboxylic acid	$C_5H_4N_2O_4$	65-86-1	156.097	cry (w)	345.5				sl H_2O; i os
8419	Oroxylin A	5,7-Dihydroxy-6-methoxy-2-phenyl-4*H*-1-benzopyran-4-one	$C_{16}H_{12}O_5$	480-11-5	284.263	ye nd (al)	231.5				vs ace, eth, EtOH
8420	Orphenadrine		$C_{18}H_{23}NO$	83-98-7	269.382			195^{12}			
8421	Oryzalin	Benzenesulfonamide, 4-(dipropylamino)-3,5-dinitro-	$C_{12}H_{18}N_4O_6S$	19044-88-3	346.359		141				
8422	Ouabain		$C_{29}H_{44}O_{12}$	630-60-4	584.652	hyg pl (+9w)	200				sl H_2O; vs EtOH
8423	7-Oxabicyclo[4.1.0]heptane		$C_6H_{10}O$	286-20-4	98.142		<-10	131.5	0.9663^{20}	1.4519^{20}	i H_2O; vs EtOH, eth, ace, bz; s chl; sl ctc
8424	6-Oxabicyclo[3.1.0]hexane		C_5H_8O	285-67-6	84.117			102	0.964^{25}	1.4336^{20}	
8425	Oxacyclohexadecan-2-one	Exaltolide	$C_{15}H_{28}O_2$	106-02-5	240.382	thick oil		176^{15}	0.9549^{20}	1.4708^{20}	
8426	1,3,4-Oxadiazole	1-Oxa-3,4-diazacyclopentadiene	$C_2H_2N_2O$	288-99-3	70.049			150		1.4300^{25}	
8427	Oxadiazon		$C_{15}H_{18}Cl_2N_2O_3$	19666-30-9	345.221		90				
8428	Oxadixyl		$C_{14}H_{18}N_2O_4$	77732-09-3	278.304		104				
8429	Oxalic acid		$C_2H_2O_4$	144-62-7	90.035	orth pym or oct	189.5 dec	sub 157	1.900^{17}		s H_2O; vs EtOH; sl eth; i bz, chl, peth
8430	Oxalic acid dihydrate		$C_2H_6O_6$	6153-56-6	126.065	mcl tab or pr	101.5		1.653^{18}		s H_2O, EtOH; sl eth
8431	Oxaloacetic acid	Oxalacetic acid	$C_4H_4O_5$	328-42-7	132.072		161 dec				
8432	Oxalyl chloride	Oxalyl dichloride	$C_2Cl_2O_2$	79-37-8	126.926	liq	-16	63.5	1.4785^{20}	1.4316^{20}	s eth
8433	Oxalyl dihydrazide		$C_2H_6N_4O_2$	996-98-5	118.095	nd (w)	244.0		1.458^{22}		s H_2O; sl EtOH, eth, bz, chl
8434	Oxamic acid		$C_2H_3NO_3$	471-47-6	89.050	cry (w)	210 dec				sl H_2O; i EtOH, eth
8435	Oxamide		$C_2H_4N_2O_2$	471-46-5	88.065	nd (w)	350 dec		1.667^{20}		sl H_2O, EtOH; i eth
8436	Oxamniquine		$C_{14}H_{21}N_3O_3$	21738-42-1	279.335	ye-oran cry	149				s ace, chl, MeOH
8437	Oxamyl		$C_7H_{13}N_3O_3S$	23135-22-0	219.261		109	dec	0.97^{25}		
8438	Oxandrolone		$C_{19}H_{30}O_3$	53-39-4	306.439		236				
8439	1,4-Oxathiane		C_4H_8OS	15980-15-1	104.171	liq	-17	147	1.1174^{20}		sl H_2O

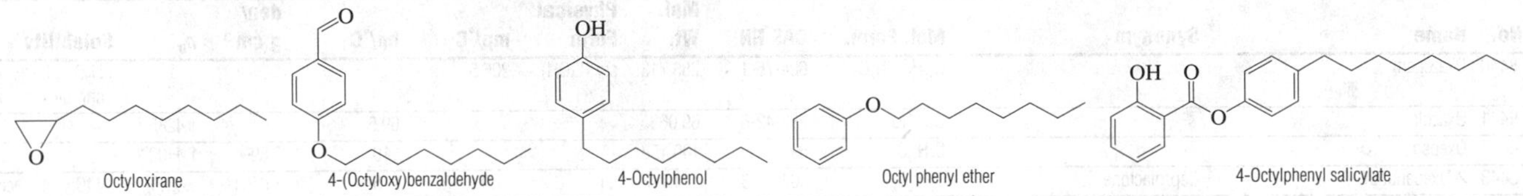

Octyloxirane

4-(Octyloxy)benzaldehyde

4-Octylphenol

Octyl phenyl ether

4-Octylphenyl salicylate

Octyl propanoate

1-Octyne

2-Octyne

3-Octyne

4-Octyne

2-Octyn-1-ol

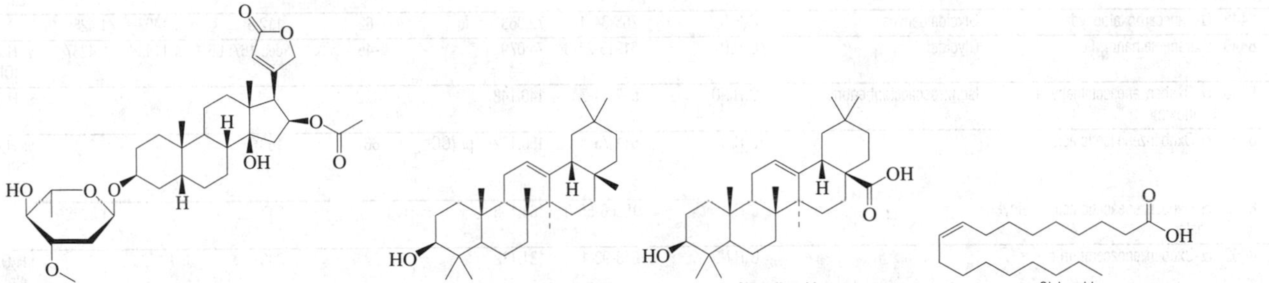

Oleandrin

Olean-12-en-3-ol, (3β)

Oleanolic acid

Oleic acid

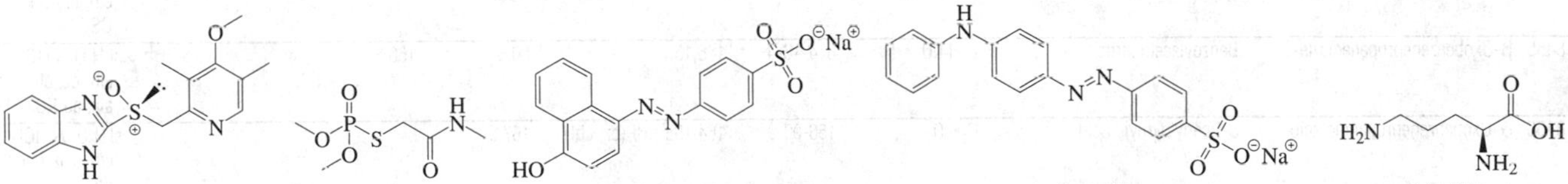

Omeprazole

Omethoate

Orange I

Orange IV

L-Ornithine

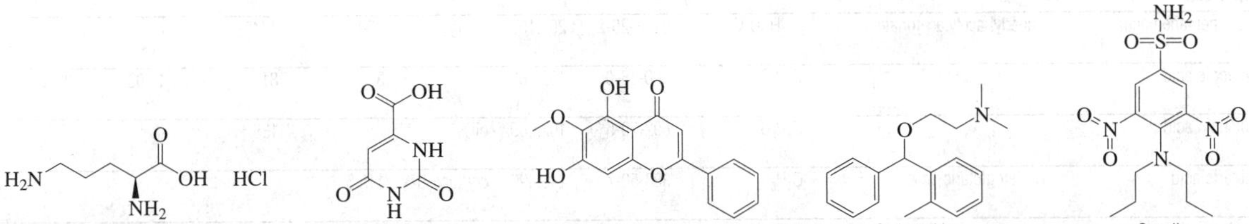

L-Ornithine, monohydrochloride

Orotic acid

Oroxylin A

Orphenadrine

Oryzalin

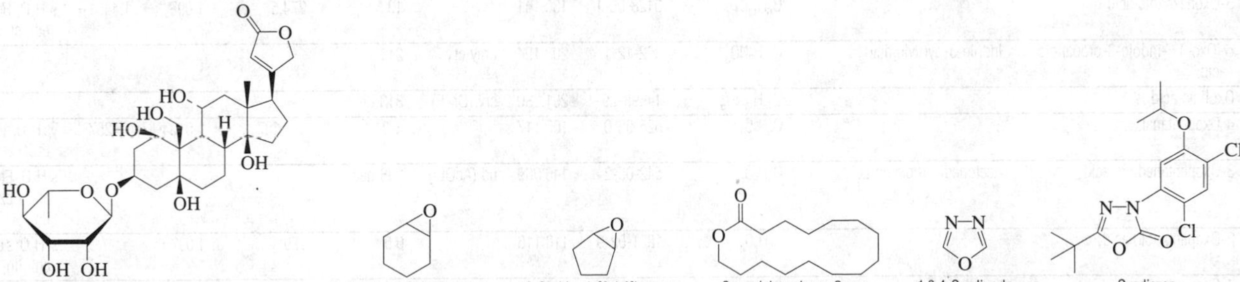

Ouabain

7-Oxabicyclo[4.1.0]heptane

6-Oxabicyclo[3.1.0]hexane

Oxacyclohexadecan-2-one

1,3,4-Oxadiazole

Oxadiazon

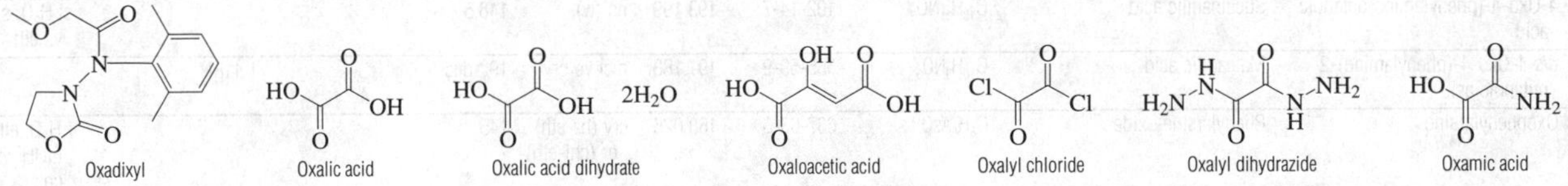

Oxadixyl

Oxalic acid

Oxalic acid dihydrate

Oxaloacetic acid

Oxalyl chloride

Oxalyl dihydrazide

Oxamic acid

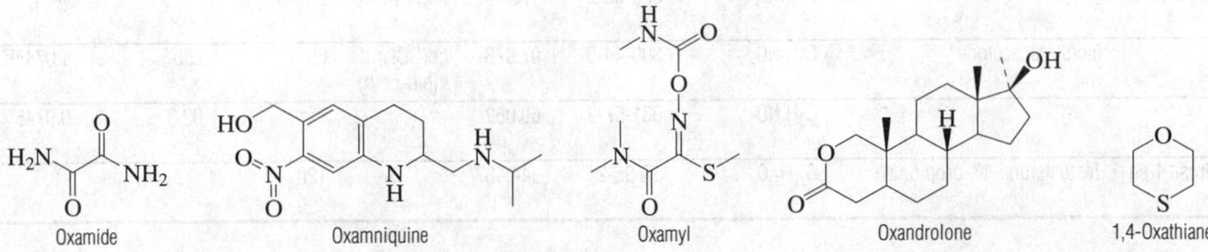

Oxamide

Oxamniquine

Oxamyl

Oxandrolone

1,4-Oxathiane

No.	Name	Synonym	Mol. Form.	CAS RN	Mol. Wt.	Physical Form	mp/°C	bp/°C	den/ g cm-3	n_D	Solubility
8440	Oxazepam		$C_{15}H_{11}ClN_2O_2$	604-75-1	286.713	cry (EtOH)	205.5				i H_2O; s EtOH, chl, diox
8441	Oxazole		C_3H_3NO	288-42-6	69.062			69.5		1.4285^{17}	
8442	Oxepane		$C_6H_{12}O$	592-90-5	100.158			119	0.89^{25}	1.4400^{20}	
8443	2-Oxepanone	Caprolactone	$C_6H_{10}O_2$	502-44-3	114.142	liq	-1.0	215	1.0761^{20}	1.4611^{20}	s EtOH, eth, ace
8444	Oxetane	Trimethylene oxide	C_3H_6O	503-30-0	58.079	liq	-97	47.6	0.8930^{25}	1.3961^{20}	msc H_2O, EtOH; s eth; vs ace
8445	2-Oxetanone	β-Propiolactone	$C_3H_4O_2$	57-57-8	72.063	liq	-33.4	162	1.1460^{20}	1.4105^{20}	msc eth; s chl
8446	3-Oxetanone		$C_3H_4O_2$	6704-31-0	72.063	unstab liq		106	1.137		
8447	Oxirane	Ethylene oxide	C_2H_4O	75-21-8	44.052	vol liq or gas	-112.5	10.6	0.8821^{10}	1.3597^{7}	s H_2O, EtOH, eth, ace, bz
8448	Oxiranecarboxaldehyde	Glycidaldehyde	$C_3H_4O_2$	765-34-4	72.063	liq	-62	112.5	1.1403^{20}	1.4265^{20}	
8449	Oxiranemethanol, (±)	Glycidol	$C_3H_6O_2$	61915-27-3	74.079		-45	dec 167; $66^{2.5}$	1.1143^{25}	1.4287^{20}	vs H_2O, ace, eth, EtOH; s bz, chl
8450	α-Oxobenzeneacetaldehyde aldoxime	Isonitrosoacetophenone	$C_8H_7NO_2$	532-54-7	149.148		129				sl H_2O; s chl
8451	α-Oxobenzeneacetic acid		$C_8H_6O_3$	611-73-4	150.132	pr (CCl_4)	66	163^{15}			vs H_2O; s EtOH, eth; sl ctc; i CS_2
8452	α-Oxobenzeneacetic acid, methyl ester		$C_9H_8O_3$	15206-55-0	164.158			247		1.5268^{20}	
8453	α-Oxobenzeneacetonitrile		C_8H_5NO	613-90-1	131.132		32.5	206			i H_2O; vs EtOH, eth; sl chl
8454	γ-Oxobenzenebutanoic acid		$C_{10}H_{10}O_3$	2051-95-8	178.184	lf (dil al)	116.5				s H_2O, EtOH, eth, bz, chl, CS_2
8455	β-Oxobenzenepropanenitrile	Benzoylacetonitrile	C_9H_7NO	614-16-4	145.158		80.5	160^{10}			sl H_2O; s EtOH, eth, bz, chl, alk, aq KCN
8456	α-Oxobenzenepropanoic acid	3-Phenylpyruvic acid	$C_9H_8O_3$	156-06-9	164.158	lf (bz, chl)	157.5				sl H_2O; vs EtOH, eth; s bz, chl; i lig
8457	2-Oxo-2*H*-1-benzopyran-3-carboxylic acid	Coumarin-3-carboxylic acid	$C_{10}H_6O_4$	531-81-7	190.15	nd (w, bz)	190 dec				vs EtOH
8458	Oxobis(2,4-pentanedione) vanadium	Vanadyl acetylacetonate	$C_{10}H_{14}O_5V$	3153-26-2	265.157	bl cry	258	$174^{0.2}$			i H_2O; s EtOH, MeOH, bz, chl
8459	2-Oxobutanoic acid		$C_4H_6O_3$	600-18-0	102.089		33	81^{16}	1.200^{17}	1.3972^{20}	vs H_2O, EtOH; sl eth
8460	4-Oxobutanoic acid		$C_4H_6O_3$	692-29-5	102.089	oil		135^{14}			s H_2O, EtOH, eth, bz
8461	2-Oxoglutaric acid	α-Ketoglutaric acid	$C_5H_6O_5$	328-50-7	146.099	cry (ace-bz)	115.5				vs H_2O, EtOH, eth; s ace
8462	6-Oxoheptanoic acid		$C_7H_{12}O_3$	3128-07-2	144.168		40.2	251^{280}, 135^{1}		1.4306^{25}	vs H_2O, ace, eth, EtOH
8463	5-Oxohexanoic acid		$C_6H_{10}O_3$	3128-06-1	130.141		13.5	274.5	1.09^{25}	1.4451^{20}	s H_2O, EtOH, eth; sl ctc
8464	α-Oxo-1*H*-indole-3-propanoic acid	Indole-3-pyruvic acid	$C_{11}H_9NO_3$	392-12-1	203.194	gray cry	211				
8465	Oxolinic acid		$C_{13}H_{11}NO_5$	14698-29-4	261.230	cry (DMF)	313 dec				
8466	4-Oxopentanal		$C_5H_8O_2$	626-96-0	100.117		<-21	dec 187	1.0134^{21}	1.4257^{22}	vs H_2O, ace, eth, EtOH
8467	3-Oxopentanedioic acid	Acetonedicarboxylic acid	$C_5H_6O_5$	542-05-2	146.099	nd (AcOEt)	138 dec				s H_2O, EtOH; sl eth; i bz, chl, lig
8468	2-Oxopentanoic acid		$C_5H_8O_3$	1821-02-9	116.116		6.5	179	1.0970^{14}		sl H_2O; s eth, bz, chl, lig, CS_2
8469	4-Oxopentanoic acid	Levulinic acid	$C_5H_8O_3$	123-76-2	116.116	lf or pl	33	dec 245	1.1335^{20}	1.4396^{20}	vs H_2O, EtOH, eth; s chl
8470	4-Oxo-4-(phenylamino)butanoic acid	Succinanilic acid	$C_{10}H_{11}NO_3$	102-14-7	193.199	nd (w)	148.5				sl H_2O; s EtOH; vs eth
8471	*cis*-4-Oxo-4-(phenylamino)-2-butenoic acid	Maleanilic acid	$C_{10}H_9NO_3$	555-59-9	191.183	mcl ye cry	192 dec		1.418^{30}		
8472	Oxophenylarsine	Phenylarsine oxide	C_6H_5AsO	637-03-6	168.025	cry (bz-eth) or (chl-eth)	145				i H_2O, eth; sl EtOH; vs bz, chl
8473	4-Oxo-4-phenyl-2-butenoic acid		$C_{10}H_8O_3$	583-06-2	176.169	nd or pr (tol)	99				sl H_2O, chl, lig; s EtOH, eth, tol
8474	2-Oxopropanal oxime	Isonitrosoacetone	$C_3H_5NO_2$	306-44-5	87.078	nd(CCl_4) lf (eth-peth)	69	sub	1.0744^{67}		s H_2O, eth; sl bz, ctc, chl
8475	2-Oxopropanenitrile		C_3H_3NO	631-57-2	69.062			92.3	0.9745^{20}	1.3764^{20}	s eth, ace, CH_3CN
8476	17-(1-Oxopropoxy)-androst-4-en-3-one, (17β)	Testosterone-17-propionate	$C_{22}H_{32}O_3$	57-85-2	344.487		120				vs eth, py, EtOH

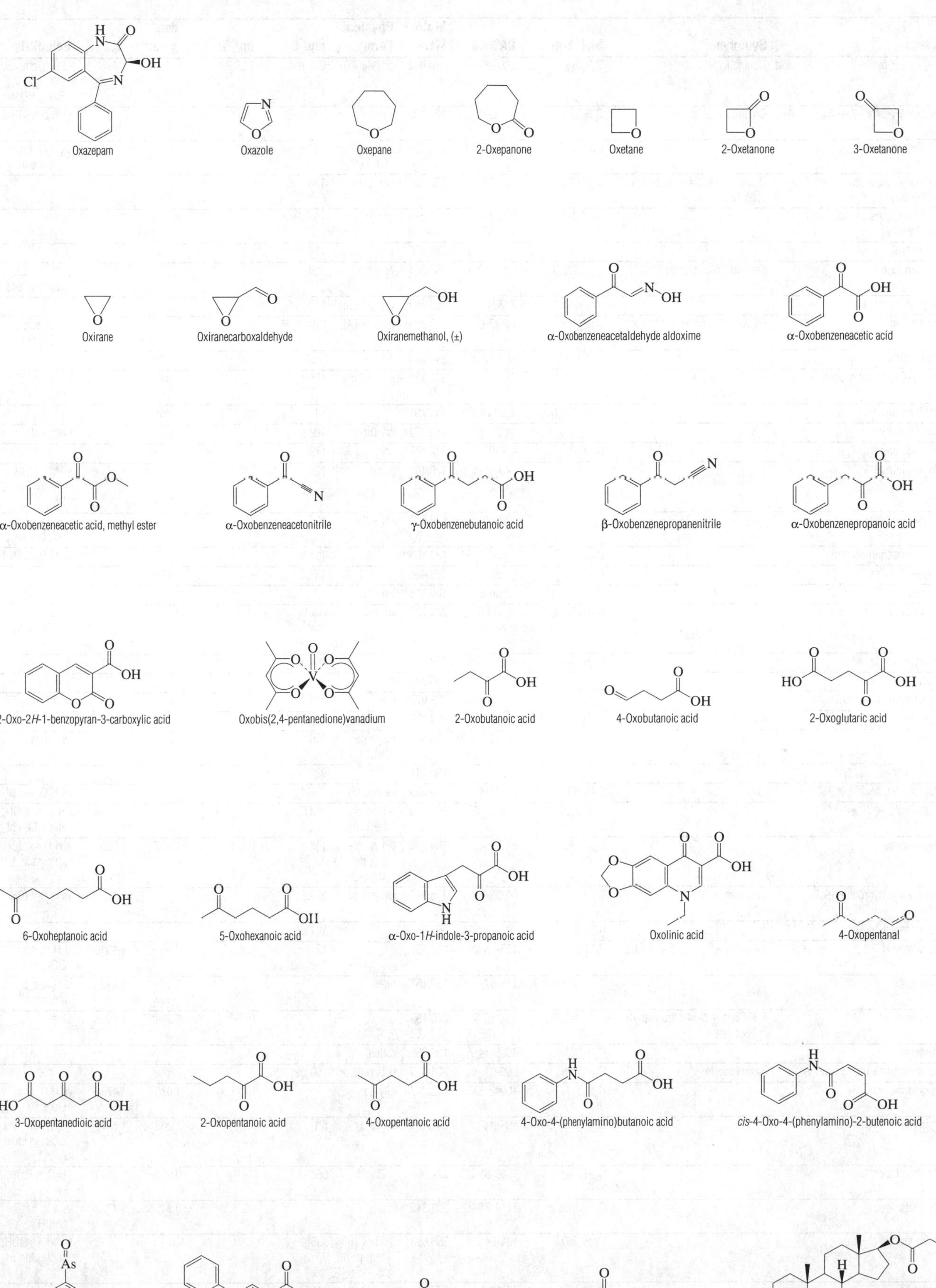

Oxazepam
Oxazole
Oxepane
2-Oxepanone
Oxetane
2-Oxetanone
3-Oxetanone
Oxirane
Oxiranecarboxaldehyde
Oxiranemethanol, (±)
α-Oxobenzeneacetaldehyde aldoxime
α-Oxobenzeneacetic acid
α-Oxobenzeneacetic acid, methyl ester
α-Oxobenzeneacetonitrile
γ-Oxobenzenebutanoic acid
β-Oxobenzenepropanenitrile
α-Oxobenzenepropanoic acid
2-Oxo-2H-1-benzopyran-3-carboxylic acid
Oxobis(2,4-pentanedione)vanadium
2-Oxobutanoic acid
4-Oxobutanoic acid
2-Oxoglutaric acid
6-Oxoheptanoic acid
5-Oxohexanoic acid
α-Oxo-1H-indole-3-propanoic acid
Oxolinic acid
4-Oxopentanal
3-Oxopentanedioic acid
2-Oxopentanoic acid
4-Oxopentanoic acid
4-Oxo-4-(phenylamino)butanoic acid
cis-4-Oxo-4-(phenylamino)-2-butenoic acid
Oxophenylarsine
4-Oxo-4-phenyl-2-butenoic acid
2-Oxopropanal oxime
2-Oxopropanenitrile
17-(1-Oxopropoxy)-androst-4-en-3-one, (17β)

No.	Name	Synonym	Mol. Form.	CAS RN	Mol. Wt.	Physical Form	mp/°C	bp/°C	den/g cm^{-3}	n_D	Solubility
8477	2-Oxo-2*H*-pyran-5-carboxylic acid	Coumalic acid	$C_6H_4O_4$	500-05-0	140.094	pr (MeOH)	207 dec	218^{120}			sl H_2O, eth, ace; i bz, chl; s EtOH, HOAc
8478	4-Oxo-4*H*-pyran-2,6-dicarboxylic acid	Chelidonic acid	$C_7H_4O_6$	99-32-1	184.103	rose mcl nd (al-w,+1w)	262				sl H_2O, EtOH
8479	17-Oxosparteine		$C_{15}H_{24}N_2O$	489-72-5	248.364	ye to col hyg nd (peth)	84	209^{12}			vs H_2O, EtOH, eth; s chl
8480	4,4'-Oxybis(benzenesulfonyl chloride)	Diphenyl ether 4,4'-disulfonyl chloride	$C_{12}H_8Cl_2O_5S_2$	121-63-1	367.225	cry (peth)	128				
8481	4,4'-Oxybis(benzenesulfonyl hydrazide)		$C_{12}H_{14}N_4O_5S_2$	80-51-3	358.393	cry (H_2O)	164 dec				
8482	Oxybutynin		$C_{22}H_{31}NO_3$	5633-20-5	357.486	cry	114				
8483	Oxycarboxin	Carboxin *S,S*-dioxide	$C_{12}H_{13}NO_4S$	5259-88-1	267.301	pr (EtOH)	129				sl H_2O; s bz, EtOH; vs ace
8484	Oxychlordane		$C_{10}H_4Cl_8O$	27304-13-8	423.762	cry (pentane)	100				
8485	Oxycodone	Dihydro-14-hydroxycodeinone	$C_{18}H_{21}NO_4$	76-42-6	315.365	rods (EtOH)	219				i H_2O, eth; s EtOH, chl
8486	Oxydemeton-methyl		$C_6H_{15}O_4PS_2$	301-12-2	246.284		<-20	106$^{0.01}$	1.289^{20}		
8487	10,10'-Oxydiphenoxarsine	10,10'-Oxybis[10*H*-phenoxarsine]	$C_{24}H_{16}As_2O_3$	58-36-6	502.225	col mcl cry	185		1.41		i H_2O; s EtOH, chl; i CH_2Cl_2
8488	Oxyfluorfen		$C_{15}H_{11}ClF_3NO_4$	42874-03-3	361.701		84	dec 358	1.35^{73}		
8489	Oxymetazoline		$C_{16}H_{24}N_2O$	1491-59-4	260.374	cry (bz)	182				i eth, chl
8490	Oxymetholone		$C_{21}H_{32}O_3$	434-07-1	332.477	cry	179				
8491	Oxymethurea		$C_3H_8N_2O_3$	140-95-4	120.107	pr(al)	126	149^{25}			s H_2O, EtOH, MeOH; i eth; sl DMSO
8492	Oxyphenbutazone		$C_{19}H_{20}N_2O_3$	129-20-4	324.373	cry (eth/peth)	124				s EtOH, MeOH, chl, bz, eth
8493	Oxyphenonium bromide		$C_{21}H_{34}BrNO_3$	50-10-2	428.404		191.5				vs H_2O; sl EtOH
8494	Oxytetracycline		$C_{22}H_{24}N_2O_9$	79-57-2	460.434		184.5		1.634^{20}		
8495	Oxytocin		$C_{43}H_{66}N_{12}O_{12}S_2$	50-56-6	1007.187	wh pow					s H_2O, BuOH
8496	Paclobutrazol		$C_{15}H_{20}ClN_3O$	76738-62-0	293.792	wh cry	166		1.22		i H_2O; vs ace, MeOH; s xyl, hx
8497	Palustric acid		$C_{20}H_{30}O_2$	1945-53-5	302.451	cry (MeOH)	164.5				
8498	Pamoic acid		$C_{23}H_{16}O_6$	130-85-8	388.369		315				
8499	Pancuronium dibromide		$C_{35}H_{60}Br_2N_2O_4$	15500-66-0	732.670	cry	215				sl chl
8500	Panose	4-α-Isomaltosylglucose	$C_{18}H_{32}O_{16}$	33401-87-5	504.437		223 dec				
8501	Panthesin		$C_{18}H_{32}N_2O_5S$	135-44-4	388.522	pa ye pow (al)	158				vs H_2O, EtOH
8502	Pantolactone		$C_6H_{10}O_3$	599-04-2	130.141		92				
8503	Pantothenic acid		$C_9H_{17}NO_5$	79-83-4	219.235	ye visc oil					vs H_2O, bz, eth
8504	Papaveraldine		$C_{20}H_{19}NO_5$	522-57-6	353.369	nd (al),cry (bz, peth)	210.5				i H_2O; sl EtOH, eth; s bz, chl
8505	Papaverine		$C_{20}H_{21}NO_4$	58-74-2	339.386	wh pr (al-eth), nd (chl-peth)	147.5	sub 135	1.337^{20}	1.625	sl H_2O; vs EtOH, chl; s ace, bz, py
8506	Papaverine hydrochloride	Cerespan	$C_{20}H_{22}ClNO_4$	61-25-6	375.847	wh mcl pr (w)	224.5				vs H_2O, EtOH
8507	Paraformaldehyde		$(CH_2O)_x$	30525-89-4	30.026		164 dec				
8508	Paraldehyde	2,4,6-Trimethyl-1,3,5-trioxane	$C_6H_{12}O_3$	123-63-7	132.157		12.6	124.3	0.9943^{20}	1.4049^{20}	sl H_2O; msc EtOH, eth, chl
8509	Paramethadione		$C_7H_{11}NO_3$	115-67-3	157.167	liq			1.121^{25}	1.449^{25}	sl H_2O; s EtOH, chl, bz, eth
8510	Paraoxon	*O,O*-Diethyl *O*-(4-nitrophenyl) phosphate	$C_{10}H_{14}NO_6P$	311-45-5	275.195	oily liq		161$^{0.5}$	1.2683^{25}	1.5096	s eth
8511	Paraquat		$C_{12}H_{14}N_2$	4685-14-7	186.252	cation					
8512	Pararosaniline hydrochloride	Basic fuchsin	$C_{19}H_{18}ClN_3$	569-61-9	323.819	pale viol pow	269 dec				
8513	Parasorbic acid		$C_6H_8O_2$	10048-32-5	112.127	oily lig		100^{15}	1.079^{18}	1.4736^{20}	vs H_2O, eth, EtOH
8514	Parathion		$C_{10}H_{14}NO_5PS$	56-38-2	291.261	ye liq	6.1	375	1.2681^{20}	1.5370^{25}	i H_2O; s eth, ace; sl ctc; vs EtOH, AcOEt
8515	Patchouli alcohol		$C_{15}H_{26}O$	5986-55-0	222.366		56		0.9906^{65}	1.5029^{65}	i H_2O; s EtOH, eth
8516	Pebulate		$C_{10}H_{21}NOS$	1114-71-2	203.345			142^{20}	0.9458^{20}	1.4752^{20}	vs ace, bz, MeOH
8517	Pelargonidin chloride		$C_{15}H_{11}ClO_5$	134-04-3	306.698	red br hyg (anh) pr or pl	>350				s H_2O; vs EtOH; sl chl, MeOH
8518	Pellotine		$C_{13}H_{19}NO_3$	83-14-7	237.295	pl (al, peth)	111.5				vs ace, eth, EtOH, peth
8519	Pemoline	2-Amino-5-phenyl-4(5*H*)-oxazolone	$C_9H_8N_2O_2$	2152-34-3	176.172	cry	256 dec				i H_2O, eth, ace; sl hot EtOH

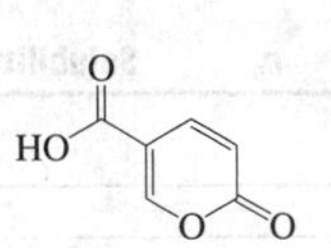

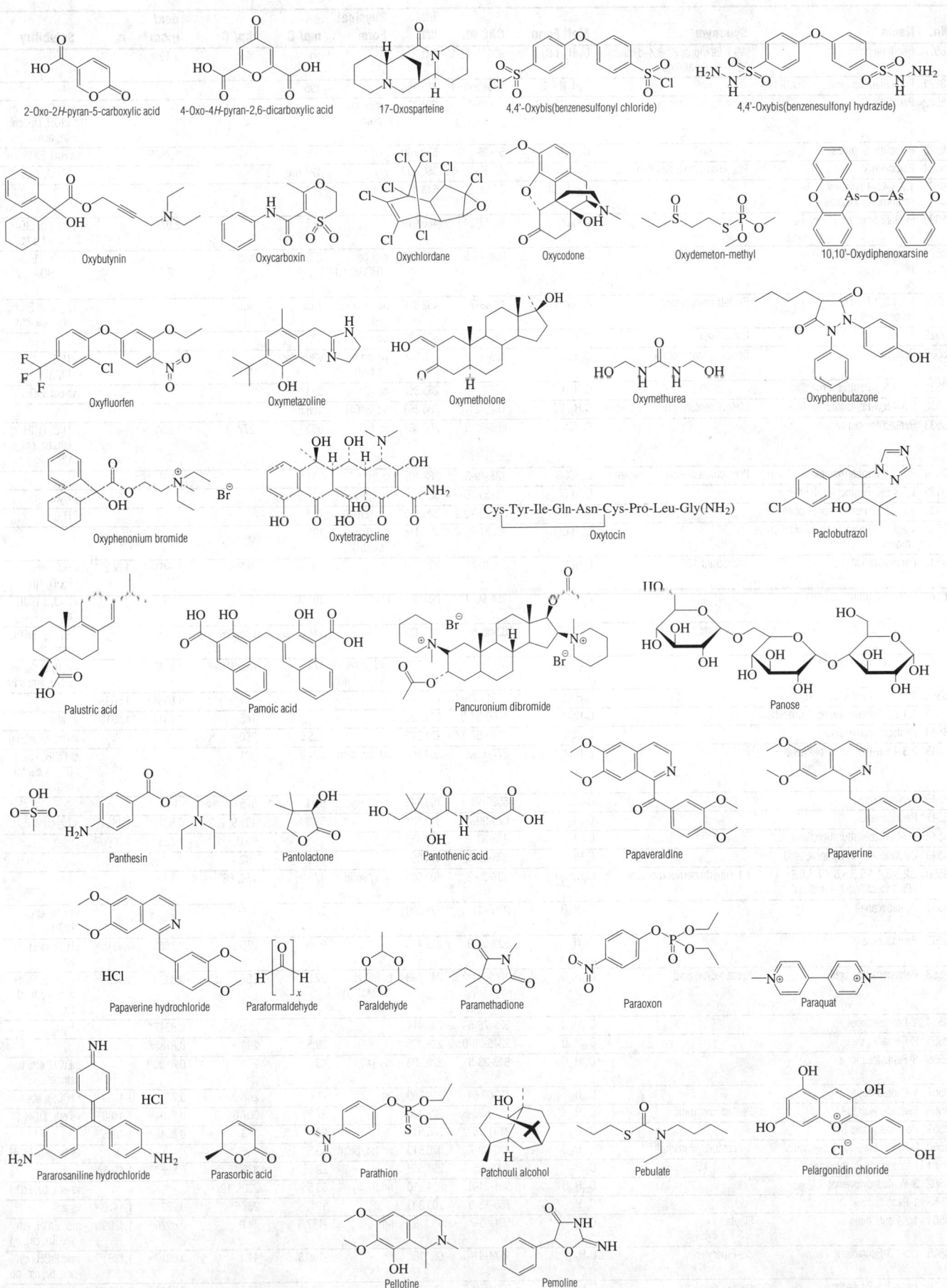

2-Oxo-2*H*-pyran-5-carboxylic acid

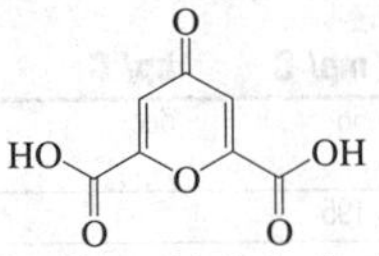

4-Oxo-4*H*-pyran-2,6-dicarboxylic acid

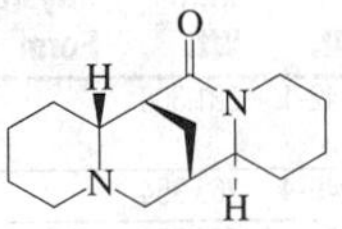

17-Oxosparteine

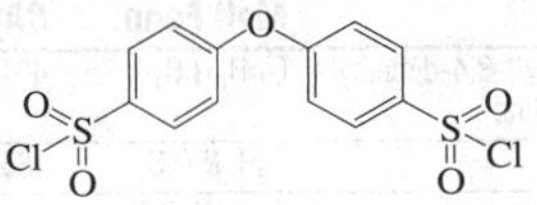

4,4'-Oxybis(benzenesulfonyl chloride)

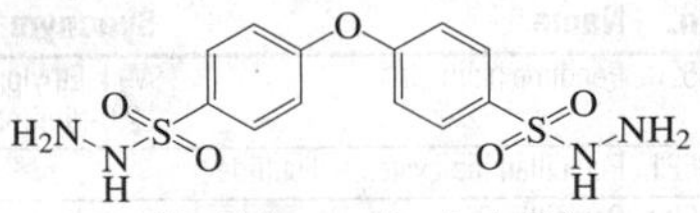

4,4'-Oxybis(benzenesulfonyl hydrazide)

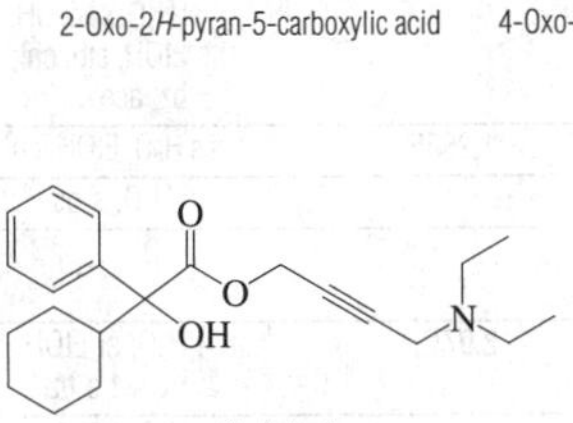

Oxybutynin

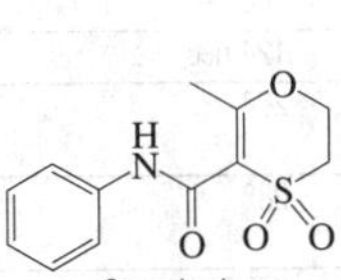

Oxycarboxin

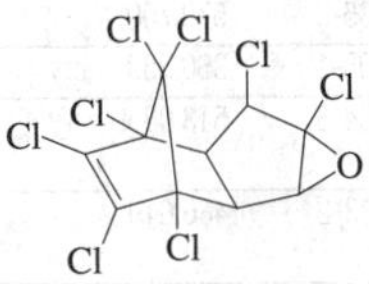

Oxychlordane

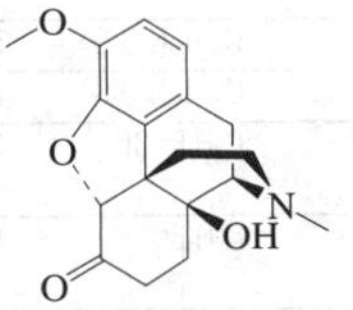

Oxycodone

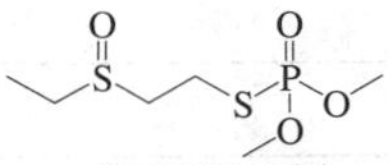

Oxydemeton-methyl

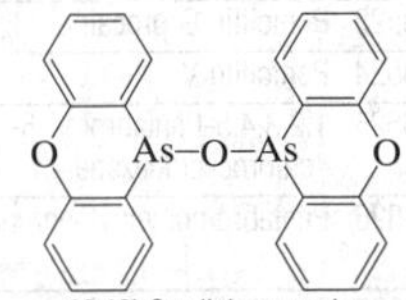

10,10'-Oxydiphenoxarsine

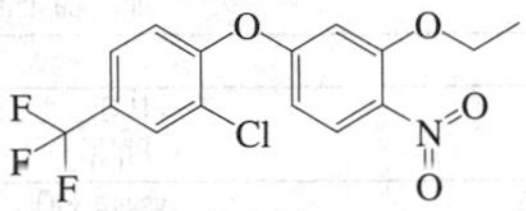

Oxyfluorfen

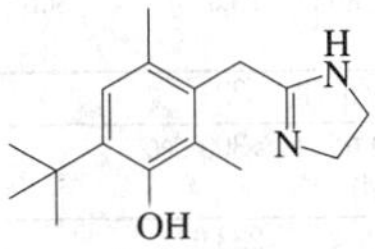

Oxymetazoline

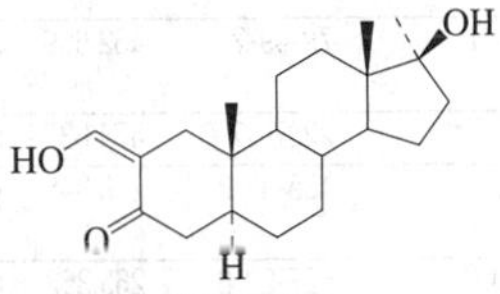

Oxymetholone

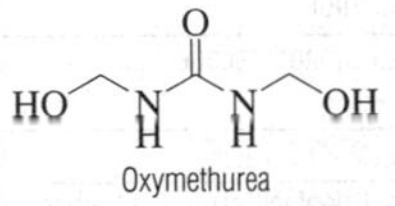

Oxymethurea

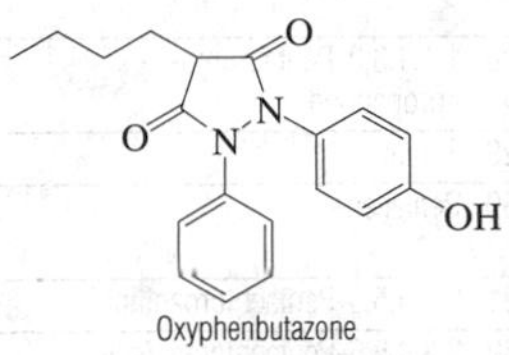

Oxyphenbutazone

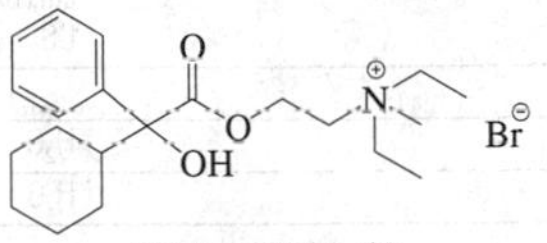

Oxyphenonium bromide

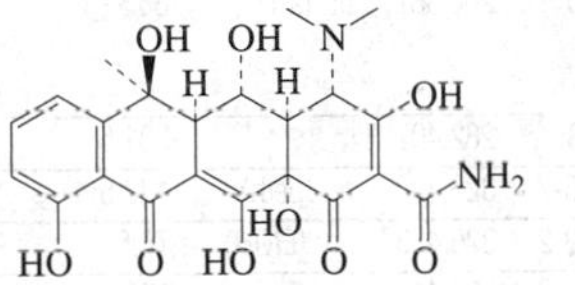

Oxytetracycline

Cys-Tyr-Ile-Gln-Asn-Cys-Pro-Leu-Gly(NH2)

Oxytocin

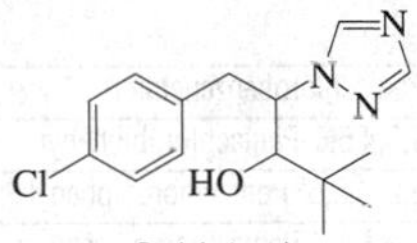

Paclobutrazol

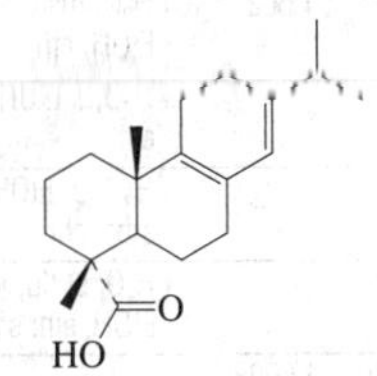

Palustric acid

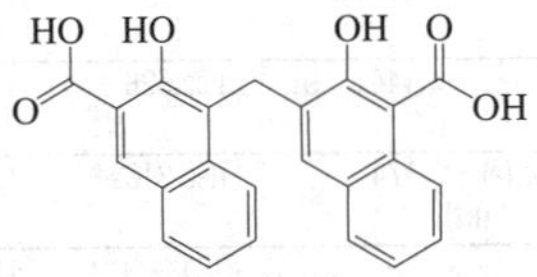

Pamoic acid

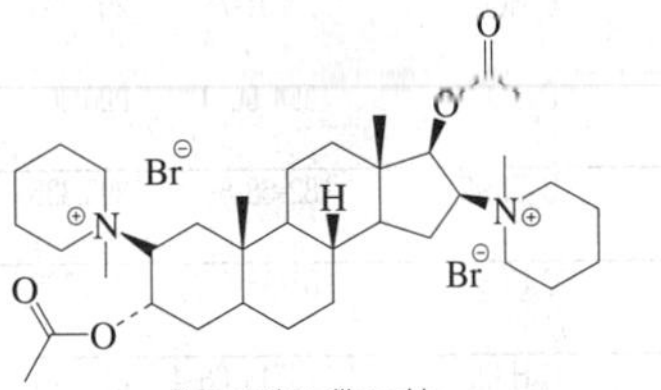

Pancuronium dibromide

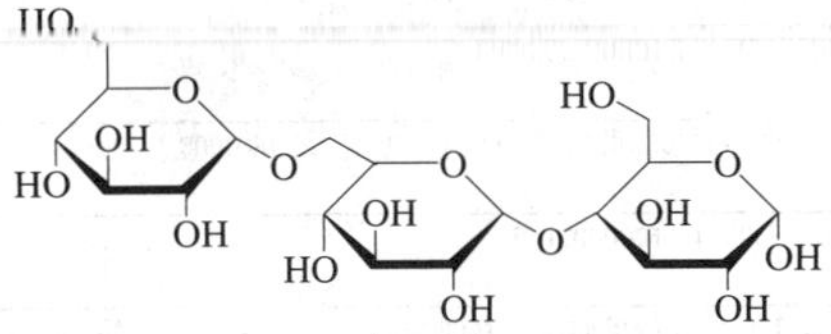

Panose

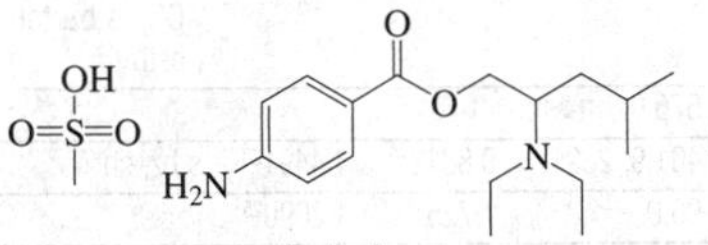

Panthesin

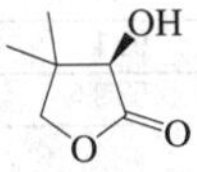

Pantolactone

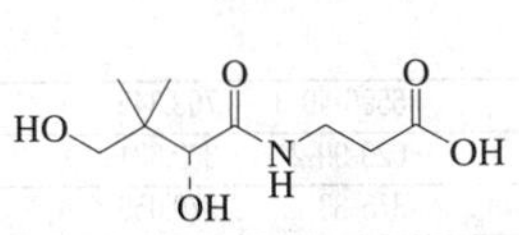

Pantothenic acid

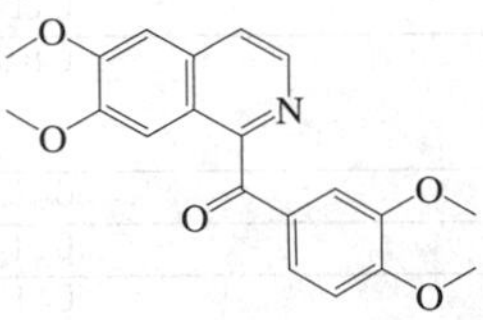

Papaveraldine

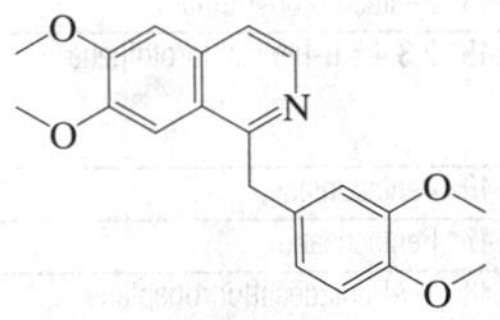

Papaverine

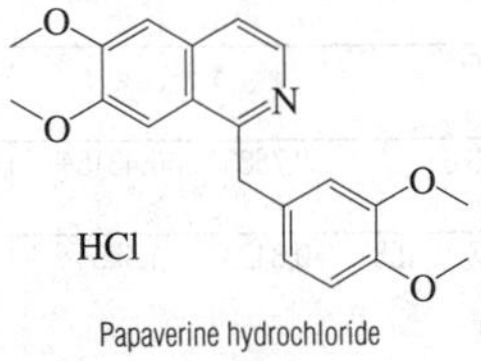

Papaverine hydrochloride

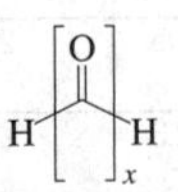

Paraformaldehyde

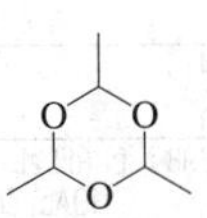

Paraldehyde

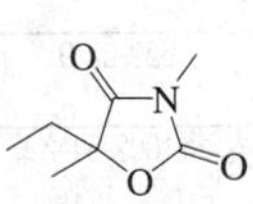

Paramethadione

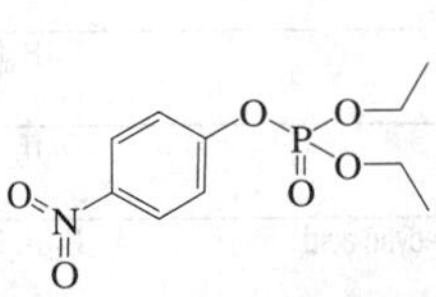

Paraoxon

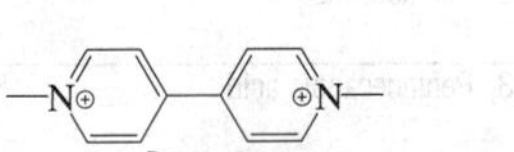

Paraquat

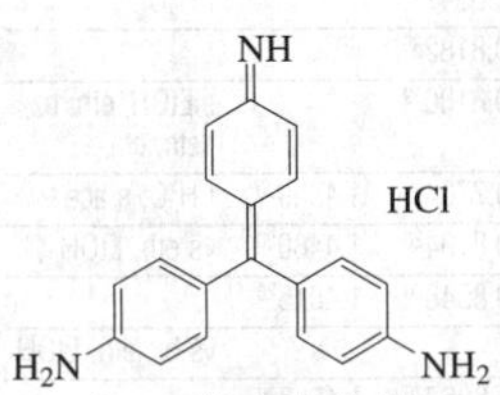

Pararosaniline hydrochloride

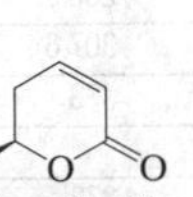

Parasorbic acid

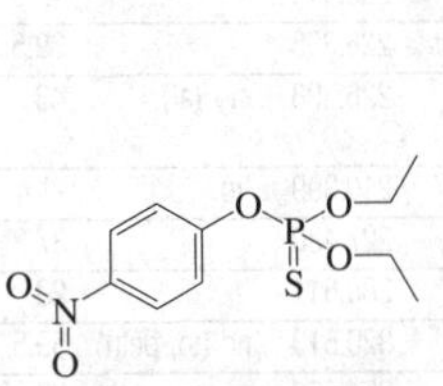

Parathion

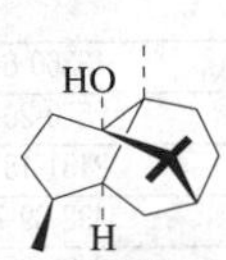

Patchouli alcohol

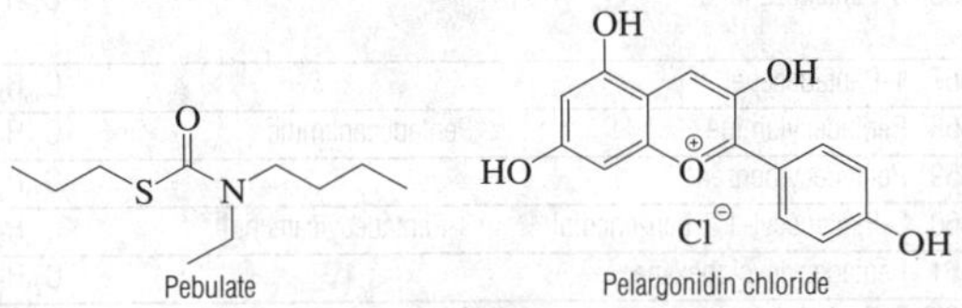

Pebulate

Pelargonidin chloride

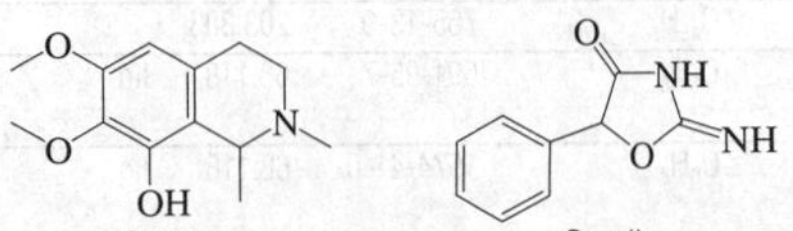

Pellotine

Pemoline

No.	Name	Synonym	Mol. Form.	CAS RN	Mol. Wt.	Physical Form	mp/°C	bp/°C	den/g cm^{-3}	n_D	Solubility
8520	Pendimethalin	*N*-(1-Ethylpropyl)-3,4-dimethyl-2,6-dinitroaniline	$C_{13}H_{19}N_3O_4$	40487-42-1	281.308		56	dec	1.19^{25}		
8521	Penicillamine cysteine disulfide		$C_8H_{16}N_2O_4S_2$	18840-45-4	268.354		195				
8522	Penicillin G	Benzylpenicillinic acid	$C_{16}H_{18}N_2O_4S$	61-33-6	334.390	amor wh pow					sl H_2O; s MeOH, EtOH, eth, chl, bz, ace
8523	Penicillin G procaine		$C_{29}H_{38}N_4O_6S$	54-35-3	570.700		108 dec		1.2555^{25}		s H_2O, EtOH, chl
8524	Penicillin V	Phenoxymethylpenicillin	$C_{16}H_{18}N_2O_5S$	87-08-1	350.389	cry	124 dec				sl H_2O; s os
8525	1,2,3,4,5-Pentabromo-6-chlorocyclohexane		$C_6H_6Br_5Cl$	87-84-3	513.085	cry	204				
8526	Pentabromomethylbenzene		$C_7H_3Br_5$	87-83-2	486.619		288		2.97^{17}		i H_2O; sl EtOH, HOAc; s bz
8527	Pentabromophenol		C_6HBr_5O	608-71-9	488.591	mcl pr (HOAc) nd (al)	229.5	sub			i H_2O; s EtOH, bz, HOAc; sl eth
8528	1,1,1,3,3-Pentabromo-2-propanone	Pentabromoacetone	C_3HBr_5O	79-49-2	452.559	nd (w, al) pr (eth)	79.5	sub			i H_2O; vs EtOH, eth, ace, chl
8529	Pentac	Dienochlor	$C_{10}Cl_{10}$	2227-17-0	474.637	tan cry (peth)	122				
8530	Pentacene	Benzo[b]naphthacene	$C_{22}H_{14}$	135-48-8	278.346	ye grn nd or lf (xyl)	>300 dec				i H_2O; sl bz; s $PhNO_2$
8531	2,3,4,5,6-Pentachloroaniline		$C_6H_2Cl_5N$	527-20-8	265.352	nd (al)	233.0				vs eth, EtOH, lig
8532	2,3,4,5,6-Pentachloroanisole	Methyl pentachlorophenyl ether	$C_7H_3Cl_5O$	1825-21-4	280.363	nd MeOH	108.5				
8533	Pentachlorobenzene		C_6HCl_5	608-93-5	250.337	nd (al)	86	277	1.8342^{16}		i H_2O, EtOH; sl eth, bz, chl, CS_2
8534	Pentachlorobenzenethiol	Pentachlorophenyl mercaptan	C_6HCl_5S	133-49-3	282.402		231.5				
8535	2,3,4,5,6-Pentachlorobiphenyl		$C_{12}H_5Cl_5$	18259-05-7	326.433	nd (peth)	123.5				i H_2O
8536	2,2',4,5,5'-Pentachlorobiphenyl		$C_{12}H_5Cl_5$	37680-73-2	326.433	cry (EtOH)	78.5				i H_2O
8537	1,2,3,4,7-Pentachlorodibenzo-*p*-dioxin		$C_{12}H_3Cl_5O_2$	39227-61-7	356.416	cry (bz/MeOH)	195				
8538	Pentachloroethane	Refrigerant 120	C_2HCl_5	76-01-7	202.294	liq	-28.78	162.0	1.6796^{20}	1.5025^{20}	i H_2O; msc EtOH, eth
8539	Pentachlorofluoroethane		C_2Cl_5F	354-56-3	220.284	col liq	101.3	138	1.74^{25}		i H_2O; s EtOH, eth
8540	Pentachloronitrobenzene	Quintozene	$C_6Cl_5NO_2$	82-68-8	295.335	cry (al)	144	dec 328	1.718^{25}		i H_2O; sl EtOH; s bz, chl
8541	Pentachlorophenol		C_6HCl_5O	87-86-5	266.336	mcl pr (al + 1w) nd (bz)	174	dec 310	1.978^{22}		i H_2O; sl lig; vs EtOH, eth; s bz
8542	1,1,2,2,3-Pentachloropropane		$C_3H_3Cl_5$	16714-68-4	216.321			181^{500}	1.633^{25}	1.5098^{25}	
8543	1,1,2,3,3-Pentachloro-1-propene		C_3HCl_5	1600-37-9	214.305			185	1.6317^{34}	1.5313^{20}	vs eth
8544	Pentachloropyridine		C_5Cl_5N	2176-62-7	251.326		125.5	280			vs bz, EtOH, lig
8545	2,3,4,5,6-Pentachlorotoluene		$C_7H_3Cl_5$	877-11-2	264.364	nd (bz, peth)	224.8	301			sl EtOH, eth, CS_2; s bz, tol, peth
8546	Pentacontane		$C_{50}H_{102}$	6596-40-3	703.345		92.1	575.0			
8547	Pentacosane		$C_{25}H_{52}$	629-99-2	352.681		53.93	401.9; 282^{40}	0.8012^{20}	1.4491^{20}	s bz, chl
8548	1*H*-Pentadecafluoroheptane		C_7HF_{15}	375-83-7	370.059			96.0	1.725^{25}	1.2690^{25}	
8549	Pentadecafluorooctanoic acid		$C_8HF_{15}O_2$	335-67-1	414.069		54.3	192.4			
8550	2,2,3,3,4,4,5,5,6,6,7,7,8,8,8-Pentadecafluoro-1-octanol	1,1-Dihydroperfluorooctanol	$C_8H_3F_{15}O$	307-30-2	400.085	waxy solid	47	164; 68^{8}			
8551	Pentadecanal		$C_{15}H_{30}O$	2765-11-9	226.398	nd	24.5	185^{25}			vs ace, eth, EtOH
8552	Pentadecane		$C_{15}H_{32}$	629-62-9	212.415		9.95	270.6	0.7685^{20}	1.4315^{20}	i H_2O; vs EtOH, eth
8553	Pentadecanoic acid	Pentadecylic acid	$C_{15}H_{30}O_2$	1002-84-2	242.398	pl (dil al, HOAc) cry (peth)	52.3	257^{100}, 158^{1}	0.8423^{80}	1.4254^{80}	i H_2O; vs EtOH, ace; s eth; sl tfa
8554	1-Pentadecanol		$C_{15}H_{32}O$	629-76-5	228.414		43.9	300	0.8347^{25}		i H_2O
8555	2-Pentadecanone		$C_{15}H_{30}O$	2345-28-0	226.398		39.5	294	0.8182^{39}		
8556	8-Pentadecanone		$C_{15}H_{30}O$	818-23-5	226.398	cry (al)	43	291	0.8180^{39}		s EtOH, eth, bz, ctc, chl
8557	1-Pentadecene		$C_{15}H_{30}$	13360-61-7	210.399	liq	-1.4	268.2	0.7764^{20}	1.4389^{20}	i H_2O; s ace
8558	Pentadecylamine	Pentadecanamine	$C_{15}H_{33}N$	2570-26-5	227.430		37.3	307.6	0.8104^{20}	1.4480^{20}	vs eth, EtOH
8559	Pentadecylbenzene		$C_{21}H_{36}$	2131-18-2	288.511		22	373	0.8548^{20}	1.4815^{20}	
8560	3-Pentadecyl-1,2-benzenediol	3-Pentadecylcatechol	$C_{21}H_{36}O_2$	492-89-7	320.510	nd (to, peth)	59.5				vs bz, eth, EtOH
8561	Pentadecylcyclohexane		$C_{21}H_{42}$	6006-95-7	294.558		29	373	0.8267^{20}	1.4588^{20}	
8562	3-Pentadecylphenol		$C_{21}H_{36}O$	501-24-6	304.510	nd (peth)	53.5	230^{8}, $197^{1.5}$			vs ace, bz, EtOH
8563	1-Pentadecyne		$C_{15}H_{28}$	765-13-9	208.383		10	268	0.7928^{20}	1.4419^{20}	vs ace
8564	1,2-Pentadiene	Ethylallene	C_5H_8	591-95-7	68.118	liq	-137.3	44.9	0.6926^{20}	1.4209^{20}	msc EtOH, eth, ace, bz, ctc, hp
8565	*cis*-1,3-Pentadiene	*cis*-Piperylene	C_5H_8	1574-41-0	68.118	liq	-140.8	44.1	0.6910^{20}	1.4363^{20}	msc EtOH, eth, ace, bz, ctc, hp
8566	*trans*-1,3-Pentadiene	*trans*-Piperylene	C_5H_8	2004-70-8	68.118	liq	-87.4	42	0.6710^{25}	1.4301^{20}	

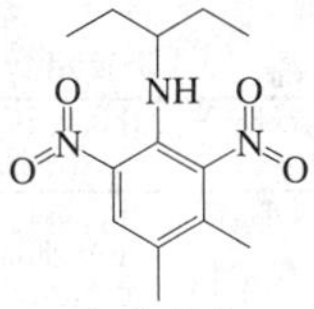

Pendimethalin

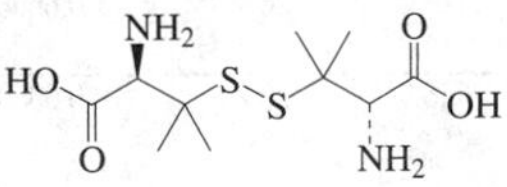

Penicillamine cysteine disulfide

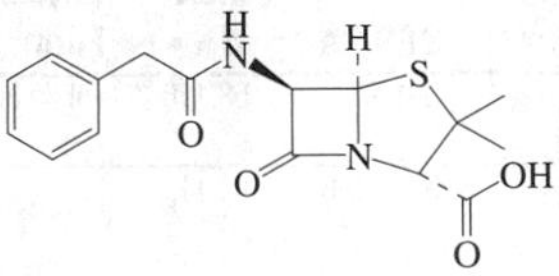

Penicillin G

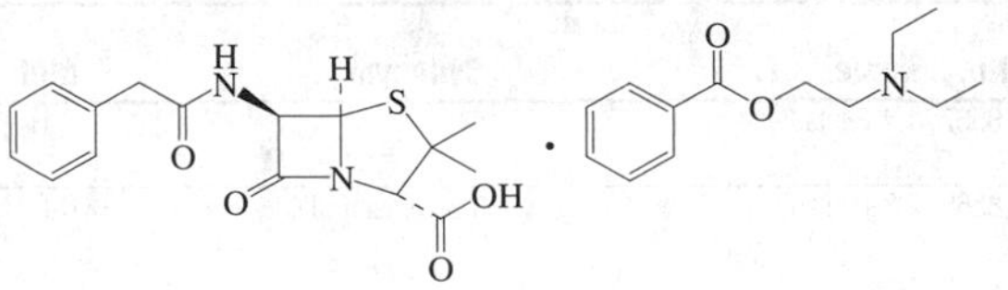

Penicillin G procaine

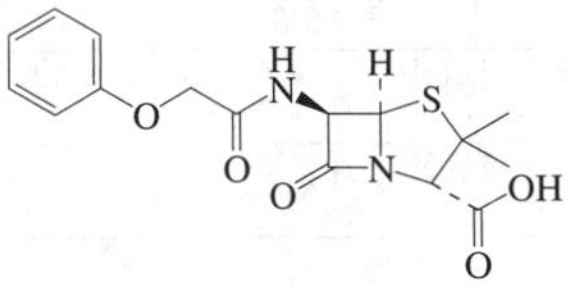

Penicillin V

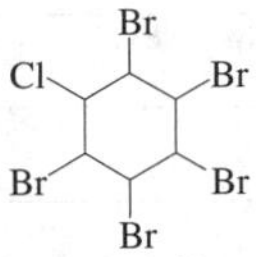

1,2,3,4,5-Pentabromo-6-chlorocyclohexane

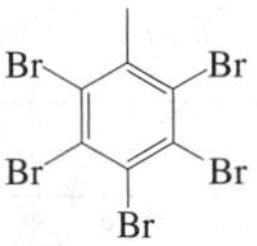

Pentabromomethylbenzene

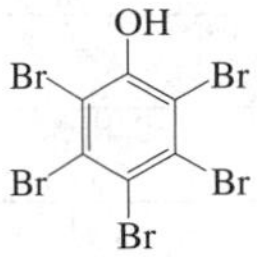

Pentabromophenol

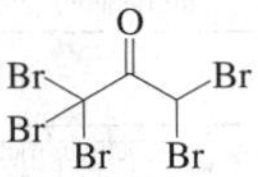

1,1,1,3,3-Pentabromo-2-propanone

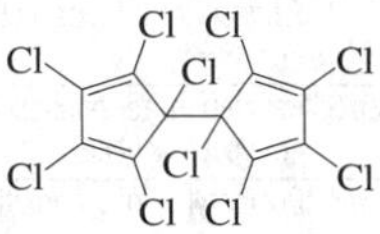

Pentac

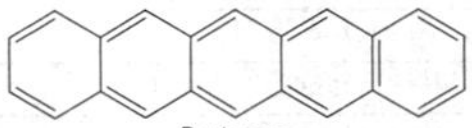

Pentacene

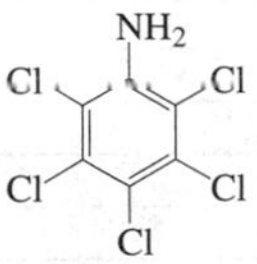

2,3,4,5,6-Pentachloroaniline

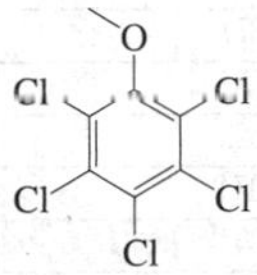

2,3,4,5,6-Pentachloroanisole

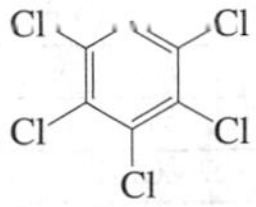

Pentachlorobenzene

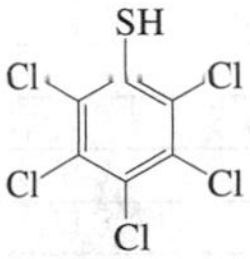

Pentachlorobenzenethiol

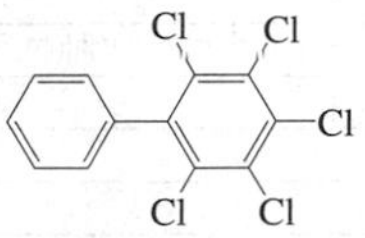

2,3,4,5,6-Pentachlorobiphenyl

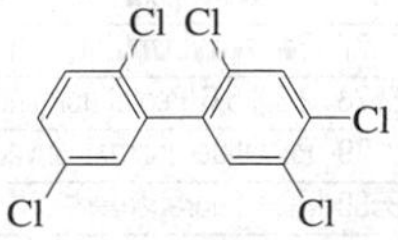

2,2',4,5,5'-Pentachlorobiphenyl

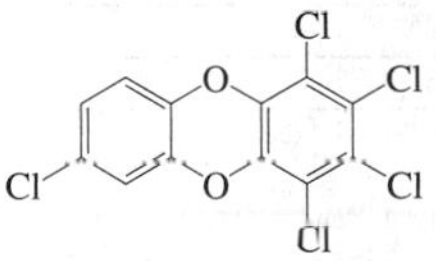

1,2,3,4,7-Pentachlorodibenzo-*p*-dioxin

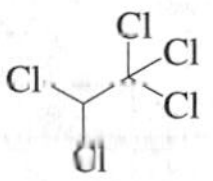

Pentachloroethane

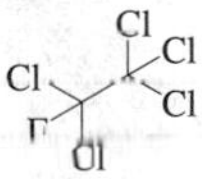

Pentachlorofluoroethane

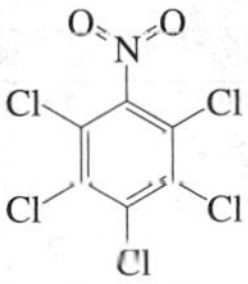

Pentachloronitrobenzene

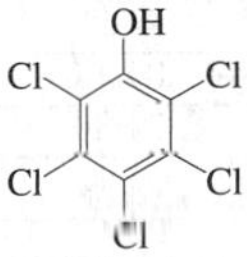

Pentachlorophenol

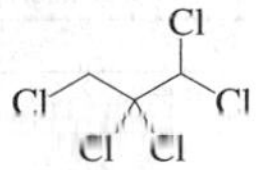

1,1,2,2,3-Pentachloropropane

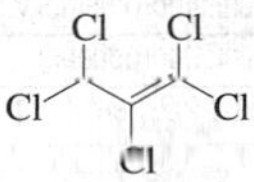

1,1,2,3,3-Pentachloro-1-propene

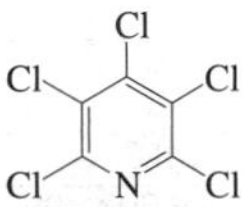

Pentachloropyridine

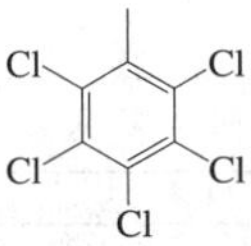

2,3,4,5,6-Pentachlorotoluene

$H_3C(CH_2)_{48}CH_3$

Pentacontane

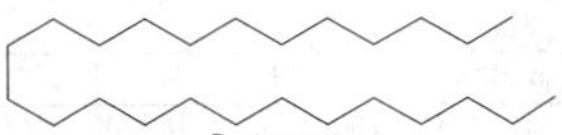

Pentacosane

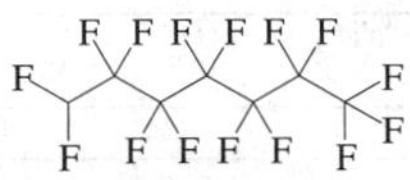

1*H*-Pentadecafluoroheptane

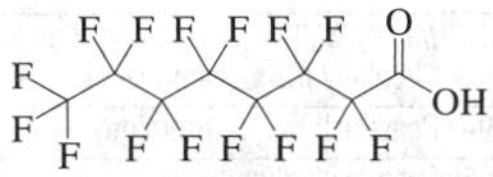

Pentadecafluorooctanoic acid

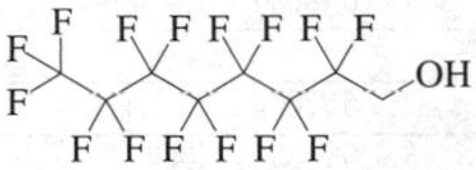

2,2,3,3,4,4,5,5,6,6,7,7,8,8,8-Pentadecafluoro-1-octanol

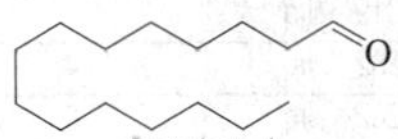

Pentadecanal

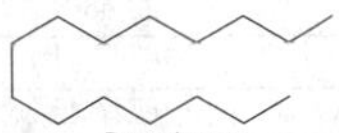

Pentadecane

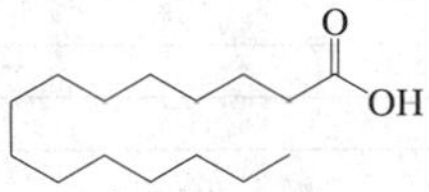

Pentadecanoic acid

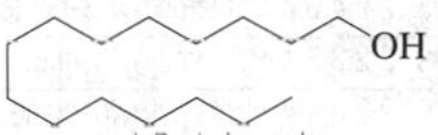

1-Pentadecanol

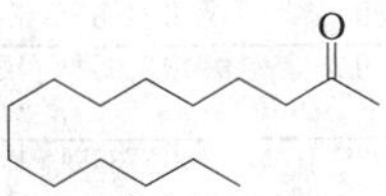

2-Pentadecanone

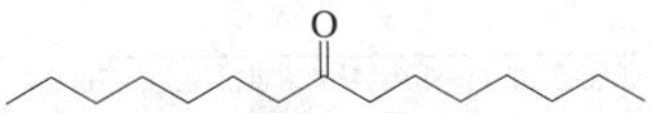

8-Pentadecanone

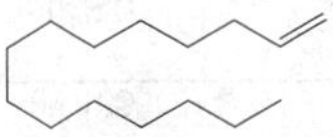

1-Pentadecene

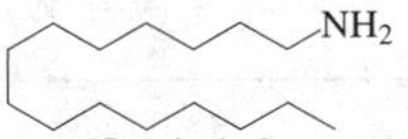

Pentadecylamine

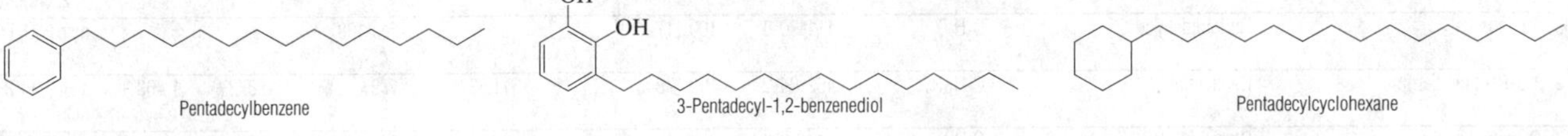

Pentadecylbenzene

3-Pentadecyl-1,2-benzenediol

Pentadecylcyclohexane

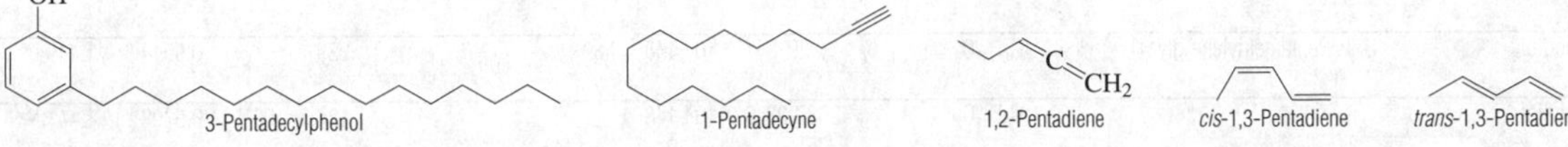

3-Pentadecylphenol

1-Pentadecyne

1,2-Pentadiene

cis-1,3-Pentadiene

trans-1,3-Pentadiene

No.	Name	Synonym	Mol. Form.	CAS RN	Mol. Wt.	Physical Form	mp/°C	bp/°C	den/ g cm^{-3}	n_D	Solubility
8567	1,4-Pentadiene		C_5H_8	591-93-5	68.118	vol liq or gas	-148.2	26	0.6608^{20}	1.3888^{20}	i H_2O; vs EtOH, eth, ace, bz
8568	2,3-Pentadiene	1,3-Dimethylallene	C_5H_8	591-96-8	68.118	liq	-125.6	48.2	0.6950^{20}	1.4284^{20}	i H_2O; msc EtOH, eth, ace, bz, hp, ctc
8569	1,4-Pentadien-3-ol		C_5H_8O	922-65-6	84.117			115.5	0.860^{23}	1.4400^{17}	
8570	1,3-Pentadiyne	Methyldiacetylene	C_5H_4	4911-55-1	64.086	liq	-38.5	55	0.7909^{20}	1.4431^{21}	i H_2O; s eth, bz, chl
8571	Pentaerythritol		$C_5H_{12}O_4$	115-77-5	136.147	cry (dil HCl)	258	sub		1.548	s H_2O; i eth, bz
8572	Pentaerythritol tetraacetate	2,2-Bis[(acetyloxy)methyl]-1,3-propanediol diacetate	$C_{13}H_{20}O_8$	597-71-7	304.293	tetr nd (w, bz)	83.5		1.273^{18}		s H_2O; vs EtOH, eth
8573	Pentaerythritol tetrakis(2-mercaptoacetate)		$C_{13}H_{20}O_8S_4$	10193-99-4	432.553	liq		250^{1}	1.385^{25}	1.5470^{20}	
8574	Pentaerythritol tetramethacrylate	Tetramethylolmethane tetramethacrylate	$C_{21}H_{28}O_8$	3253-41-6	408.442		53.5				
8575	Pentaerythritol tetranitrate		$C_5H_8N_4O_{12}$	78-11-5	316.138	tetr (ace) pr (ace-al)	140.5		1.773^{20}		sl H_2O, EtOH, eth; vs ace; s bz, py
8576	Pentaethylbenzene		$C_{16}H_{26}$	605-01-6	218.377		<-20	277	0.8971^{19}	1.5127^{20}	
8577	Pentaethyl tantalate	Ethanol, tantalum(5+) salt	$C_{10}H_{25}O_5Ta$	6074-84-6	406.251			151^{1}			
8578	2,3,4,5,6-Pentafluoroaniline		$C_6H_2F_5N$	771-60-8	183.079		34	153.5			
8579	Pentafluorobenzaldehyde		C_7HF_5O	653-37-2	196.074		20	167		1.4506^{20}	
8580	Pentafluorobenzene		C_6HF_5	363-72-4	168.064	liq	-47.4	85.74	1.514^{25}	1.3905^{20}	
8581	Pentafluorobenzenethiol		C_6HF_5S	771-62-0	200.129	liq	-24	143	1.501^{25}	1.4645^{20}	
8582	Pentafluorobenzoic acid		$C_7HF_5O_2$	602-94-8	212.074		101	220			
8583	Pentafluorobenzonitrile		C_7F_5N	773-82-0	193.074		1.2	162	1.563^{20}	1.4402^{25}	
8584	Pentafluoroethane		C_2HF_5	354-33-6	120.021	col gas	-100.6	-48.1			
8585	Pentafluoroiodobenzene		C_6F_5I	827-15-6	293.960	liq	-29	166	2.212^{20}	1.4950^{25}	
8586	Pentafluoromethoxybenzene	Methyl pentafluorophenyl ether	$C_7H_3F_5O$	389-40-2	198.090	liq	-37	138.5	1.493^{20}	1.4087^{20}	
8587	Pentafluorophenol		C_6HF_5O	771-61-9	184.063		37.5	145.6		1.4263^{20}	
8588	1,1,1,2,2-Pentafluoropropane	Refrigerant 245cb	$C_3H_3F_5$	1814-88-6	134.048	col gas		-17.4			
8589	2,2,3,3,3-Pentafluoro-1-propanol		$C_3H_3F_5O$	422-05-9	150.047			26^{50}			
8590	2,3,4,5,6-Pentafluorotoluene		$C_7H_3F_5$	771-56-2	182.091	liq	-29.78	117.5	1.440^{20}	1.4016^{25}	
8591	1,1,2,4,4-Pentafluoro-3-(trifluoromethyl)-1,3-butadiene		C_5F_8	384-04-3	212.041			39	1.527^{0}	1.3000^{0}	vs ace, bz, eth
8592	Pentagastrin		$C_{37}H_{49}N_7O_9S$	5534-95-2	767.892	col nd	230 dec				i H_2O, bz, EtOH, eth
8593	*trans*-3,3',4',5,7-Pentahydroxyflavanone, (±)	Taxifolin	$C_{15}H_{12}O_7$	480-18-2	304.252		227 dec				s chl
8594	Pentamethonium bromide		$C_{11}H_{28}Br_2N_2$	541-20-8	348.161		301				sl H_2O
8595	Pentamethylbenzene		$C_{11}H_{16}$	700-12-9	148.245	pr (al)	54.5	232	0.917^{20}	1.527^{20}	i H_2O; vs EtOH, bz; s chl
8596	2,4,6,8,10-Pentamethylcyclopentasiloxane		$C_5H_{20}O_5Si_5$	6166-86-5	300.638	liq	-108	169	0.9985^{20}	1.3912^{20}	
8597	2,2,4,6,6-Pentamethylheptane		$C_{12}H_{26}$	13475-82-6	170.334	liq	-67	177.8	0.7463^{20}	1.4440^{20}	
8598	2,2,4,6,6-Pentamethyl-3-heptene		$C_{12}H_{24}$	123-48-8	168.319	liq		180.5			
8599	2,2,3,3,4-Pentamethylpentane		$C_{10}H_{22}$	16747-44-7	142.282	liq	-36.4	166.1	0.7767^{25}	1.4361^{20}	
8600	2,2,3,4,4-Pentamethylpentane		$C_{10}H_{22}$	16747-45-8	142.282	liq	-38.7	159.3	0.7636^{25}	1.4307^{20}	
8601	Pentamethylphenol		$C_{11}H_{16}O$	2819-86-5	164.244	nd (al, peth, ace)	128	267			i H_2O; s EtOH
8602	1,2,2,6,6-Pentamethylpiperidine	Pempidine	$C_{10}H_{21}N$	79-55-0	155.281			147	0.8580^{0}	1.4550^{21}	
8603	Pentamethylsilanamine		$C_5H_{15}NSi$	2083-91-2	117.266			86	0.7400^{20}	1.4379^{24}	
8604	Pentanal	Valeraldehyde	$C_5H_{10}O$	110-62-3	86.132	liq	-91.5	103	0.8095^{20}	1.3944^{20}	sl H_2O; s EtOH, eth
8605	Pentanamide		$C_5H_{11}NO$	626-97-1	101.147	mcl pl (peth, al)	106	225	0.8735^{110}	1.4183^{110}	vs H_2O, EtOH, eth; sl chl
8606	3-Pentanamine		$C_5H_{13}N$	616-24-0	87.164			89	0.7487^{20}	1.4063^{20}	s EtOH; sl chl
8607	Pentane		C_5H_{12}	109-66-0	72.149	liq	-129.67	36.06	0.6262^{20}	1.3575^{20}	sl H_2O; msc EtOH, eth, ace, bz, chl; s ctc
8608	Pentanedial	Glutaraldehyde	$C_5H_8O_2$	111-30-8	100.117			dec 188			msc H_2O, EtOH; s bz
8609	1,5-Pentanediamine	Cadaverine	$C_5H_{14}N_2$	462-94-2	102.178		11.83	179	0.873^{25}	1.463^{20}	s H_2O, EtOH; sl eth
8610	Pentanedinitrile	Glutaronitrile	$C_5H_6N_2$	544-13-8	94.115	liq	-29	286	0.9911^{15}	1.4295^{20}	vs EtOH, chl
8611	1,2-Pentanediol, (±)		$C_5H_{12}O_2$	91049-43-3	104.148			209	0.9723^{20}	1.4397^{19}	
8612	1,4-Pentanediol		$C_5H_{12}O_2$	626-95-9	104.148			202; 125^{10}	0.9883^{20}	1.4452^{23}	vs H_2O, EtOH, chl
8613	1,5-Pentanediol	Pentamethylene glycol	$C_5H_{12}O_2$	111-29-5	104.148	liq	-18	239	0.9914^{20}	1.4494^{20}	s H_2O, EtOH; sl eth, bz
8614	2,3-Pentanediol		$C_5H_{12}O_2$	42027-23-6	104.148			187.5; 100^{17}	0.9798^{19}	1.4412^{25}	s H_2O, EtOH; sl eth
8615	2,4-Pentanediol	2,4-Amylene glycol	$C_5H_{12}O_2$	625-69-4	104.148			199; 97^{13}	0.9635^{20}	1.4349^{20}	vs H_2O, EtOH

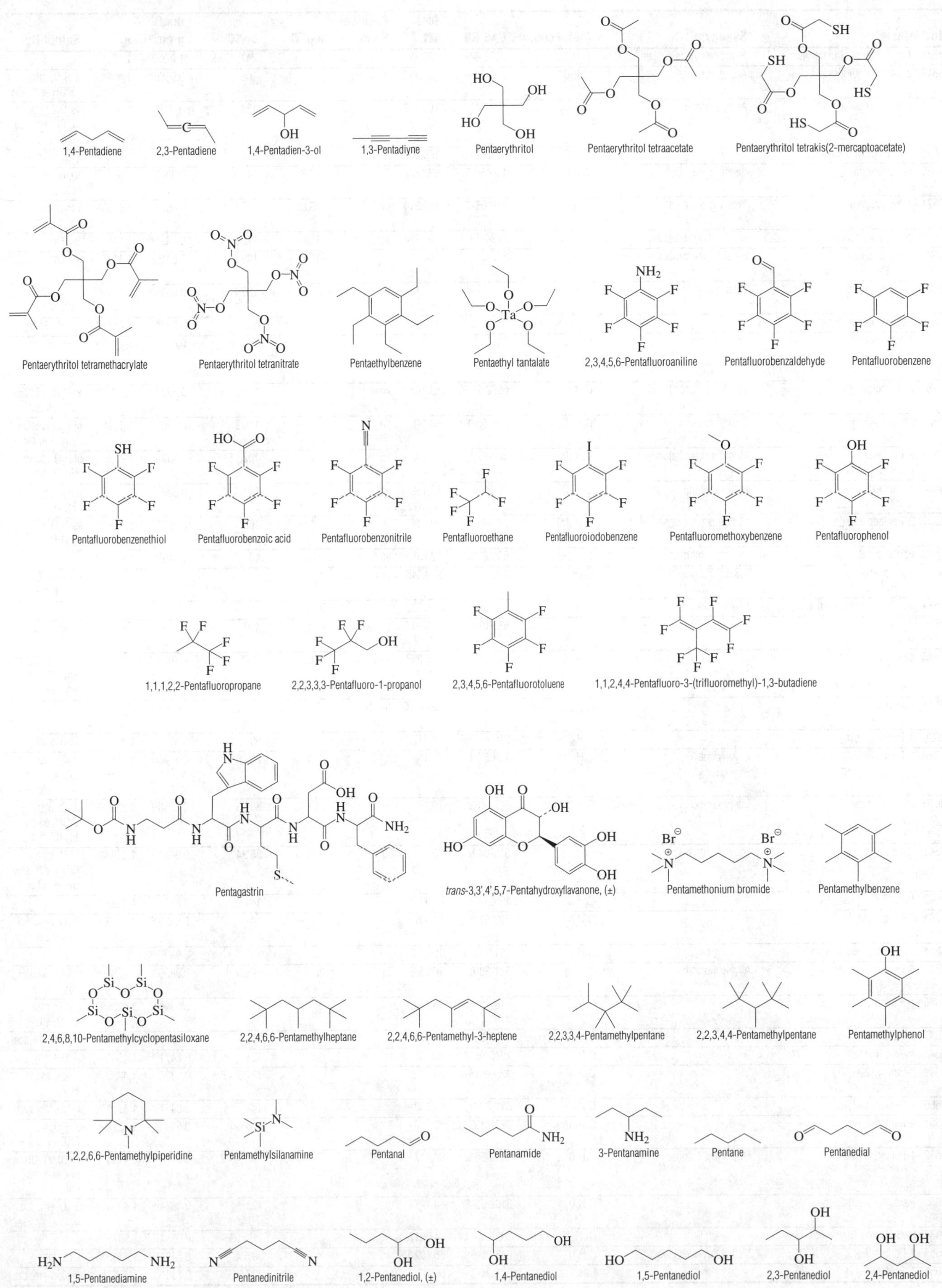

1,4-Pentadiene
2,3-Pentadiene
1,4-Pentadien-3-ol
1,3-Pentadiyne
Pentaerythritol
Pentaerythritol tetraacetate
Pentaerythritol tetrakis(2-mercaptoacetate)
Pentaerythritol tetramethacrylate
Pentaerythritol tetranitrate
Pentaethylbenzene
Pentaethyl tantalate
2,3,4,5,6-Pentafluoroaniline
Pentafluorobenzaldehyde
Pentafluorobenzene
Pentafluorobenzenethiol
Pentafluorobenzoic acid
Pentafluorobenzonitrile
Pentafluoroethane
Pentafluoroiodobenzene
Pentafluoromethoxybenzene
Pentafluorophenol
1,1,1,2,2-Pentafluoropropane
2,2,3,3,3-Pentafluoro-1-propanol
2,3,4,5,6-Pentafluorotoluene
1,1,2,4,4-Pentafluoro-3-(trifluoromethyl)-1,3-butadiene
Pentagastrin
trans-3,3',4',5,7-Pentahydroxyflavanone, (±)
Pentamethonium bromide
Pentamethylbenzene
2,4,6,8,10-Pentamethylcyclopentasiloxane
2,2,4,6,6-Pentamethylheptane
2,2,4,6,6-Pentamethyl-3-heptene
2,2,3,3,4-Pentamethylpentane
2,2,3,4,4-Pentamethylpentane
Pentamethylphenol
1,2,2,6,6-Pentamethylpiperidine
Pentamethylsilanamine
Pentanal
Pentanamide
3-Pentanamine
Pentane
Pentanedial
1,5-Pentanediamine
Pentanedinitrile
1,2-Pentanediol, (±)
1,4-Pentanediol
1,5-Pentanediol
2,3-Pentanediol
2,4-Pentanediol

No.	Name	Synonym	Mol. Form.	CAS RN	Mol. Wt.	Physical Form	mp/°C	bp/°C	den/ g cm⁻³	n_D	Solubility
8616	1,5-Pentanediol diacetate	Pentamethylene acetate	$C_9H_{16}O_4$	6963-44-6	188.221		2	241; 123^{3}	1.0296^{20}	1.4261^{19}	
8617	2,3-Pentanedione	Acetylpropionyl	$C_5H_8O_2$	600-14-6	100.117	dk ye liq		108	0.9565^{19}	1.4014^{19}	s H_2O; msc EtOH, eth, ace
8618	2,4-Pentanedione	Acetylacetone	$C_5H_8O_2$	123-54-6	100.117	liq	-23	138	0.9721^{25}	1.4494^{20}	vs H_2O; msc EtOH, eth, ace, chl
8619	Pentanedioyl dichloride		$C_5H_6Cl_2O_2$	2873-74-7	169.006			217	1.324^{20}	1.4728^{20}	s eth; sl chl
8620	Pentanenitrile	Valeronitrile	C_5H_9N	110-59-8	83.132	liq	-96.2	141.3	0.8008^{20}	1.3971^{20}	s eth, ace, bz; sl ctc
8621	1-Pentanethiol	Pentyl mercaptan	$C_5H_{12}S$	110-66-7	104.214	liq	-75.65	126.6	0.850^{20}	1.4469^{20}	i H_2O; msc EtOH, eth
8622	2-Pentanethiol	*sec*-Pentyl mercaptan	$C_5H_{12}S$	2084-19-7	104.214	liq	-169	112.9	0.8327^{20}	1.4412^{20}	s EtOH, lig
8623	3-Pentanethiol	3-Pentyl mercaptan	$C_5H_{12}S$	616-31-9	104.214	liq	-110.8	105	0.8410^{20}	1.4447^{20}	s EtOH; sl DMSO
8624	Pentanoic acid	Valeric acid	$C_5H_{10}O_2$	109-52-4	102.132	liq	-33.6	186.1	0.9339^{25}	1.4085^{20}	s H_2O, EtOH, eth; sl ctc
8625	Pentanoic anhydride		$C_{10}H_{18}O_3$	2082-59-9	186.248	liq	-56.1	227	0.924^{20}	1.4171^{26}	vs eth, EtOH
8626	1-Pentanol	Amyl alcohol	$C_5H_{12}O$	71-41-0	88.148	liq	-77.6	137.98	0.8144^{20}	1.4101^{20}	sl H_2O; msc EtOH, eth; s ace, chl
8627	2-Pentanol	*sec*-Amyl alcohol	$C_5H_{12}O$	6032-29-7	88.148	liq	-73	119.3	0.8094^{20}	1.4053^{20}	sl H_2O; s EtOH, eth, ctc, chl
8628	3-Pentanol	Diethyl carbinol	$C_5H_{12}O$	584-02-1	88.148	liq	-69	116.25	0.8203^{20}	1.4104^{20}	sl H_2O; s EtOH, eth, ace, ctc
8629	2-Pentanone	Methyl propyl ketone	$C_5H_{10}O$	107-87-9	86.132	liq	-76.8	102.26	0.809^{20}	1.3895^{20}	sl H_2O, ctc; msc EtOH, eth
8630	3-Pentanone	Diethyl ketone	$C_5H_{10}O$	96-22-0	86.132	liq	-39	101.7	0.8098^{25}	1.3905^{25}	s H_2O, ctc; msc EtOH, eth
8631	2-Pentanone oxime	Methyl propyl ketone oxime	$C_5H_{11}NO$	623-40-5	101.147			168	0.9095^{20}	1.4450^{20}	vs H_2O, eth, EtOH
8632	Pentanoyl chloride	Valeroyl chloride	C_5H_9ClO	638-29-9	120.577	liq	-110	109	1.0155^{15}	1.4200^{20}	
8633	Pentaphene	2,3:6,7-Dibenzphenanthrene	$C_{22}H_{14}$	222-93-5	278.346	ye grn lf(xyl)	257				i H_2O; sl EtOH, xyl, eth; s bz
8634	1,2,3,5,6-Pentathiepane	Lenthionine	$C_2H_4S_5$	292-46-6	188.378		60.5				
8635	Pentatriacontane		$C_{35}H_{72}$	630-07-9	492.947	cry (al)	74.6	490	0.8157^{20}	1.4568^{20}	i H_2O; sl eth; s ace
8636	18-Pentatriacontanone		$C_{35}H_{70}O$	504-53-0	506.930	lf (lig)	89.0	$270^{0.1}$	0.793^{95}		i H_2O; sl EtOH, eth, ace, bz, lig, chl
8637	Pentazocine		$C_{19}H_{27}NO$	359-83-1	285.423	cry (MeOH aq)	147				
8638	4-Pentenal		C_5H_8O	2100-17-6	84.117			99	0.852^{20}	1.4191^{20}	i H_2O; s eth, ace
8639	1-Pentene	α-Amylene	C_5H_{10}	109-67-1	70.133	vol liq or gas	-165.12	29.96	0.6405^{20}	1.3715^{20}	i H_2O; msc EtOH, eth; s bz; sl ctc
8640	*cis*-2-Pentene	*cis*-β-Amylene	C_5H_{10}	627-20-3	70.133	liq	-151.36	36.93	0.6556^{20}	1.3830^{20}	i H_2O; msc EtOH, eth; s bz, dil sulf
8641	*trans*-2-Pentene	*trans*-β-Amylene	C_5H_{10}	646-04-8	70.133	liq	-140.21	36.34	0.6431^{25}	1.3793^{20}	i H_2O; msc EtOH, eth; s bz; vs dil sulf
8642	*trans*-3-Pentenenitrile		C_5H_7N	16529-66-1	81.117	liq		144	0.837	1.4220^{20}	
8643	4-Pentenenitrile		C_5H_7N	592-51-8	81.117			140	0.8239^{24}	1.4213^{14}	i H_2O; msc EtOH, eth
8644	*trans*-3-Pentenoic acid		$C_5H_8O_2$	1617-32-9	100.117			193.2	0.989^{19}		
8645	4-Pentenoic acid	Allylacetic acid	$C_5H_8O_2$	591-80-0	100.117	liq	-22.5	188.5	0.9809^{20}	1.4281^{20}	sl H_2O; vs EtOH, eth
8646	1-Penten-3-ol		$C_5H_{10}O$	616-25-1	86.132			115	0.839^{20}	1.4239^{20}	sl H_2O; msc EtOH, eth
8647	*cis*-2-Penten-1-ol		$C_5H_{10}O$	1576-95-0	86.132			138	0.8529^{20}	1.4354^{20}	s EtOH, eth, ace
8648	*trans*-2-Penten-1-ol		$C_5H_{10}O$	1576-96-1	86.132			138	0.8471^{20}	1.4341^{20}	s EtOH, eth, ace
8649	3-Penten-2-ol, (±)		$C_5H_{10}O$	42569-16-4	86.132			121.6; 65^{70}	0.8328^{25}	1.4280^{20}	vs ace, eth, EtOH
8650	4-Penten-1-ol		$C_5H_{10}O$	821-09-0	86.132			141	0.8457^{20}	1.4309^{20}	sl H_2O, ctc; s eth
8651	4-Penten-2-ol		$C_5H_{10}O$	625-31-0	86.132			116	0.8367^{20}	1.4225^{20}	vs H_2O; msc EtOH, eth
8652	1-Penten-3-one	Ethyl vinyl ketone	C_5H_8O	1629-58-9	84.117			103; 44^{90}	0.8468^{20}	1.4195^{20}	i H_2O; s EtOH, eth, ace, bz, chl
8653	*trans*-3-Penten-2-one		C_5H_8O	3102-33-8	84.117			122	0.8624^{20}	1.4350^{20}	s H_2O, eth, ace, ctc
8654	2-(3-Pentenyl)pyridine		$C_{10}H_{13}N$	2057-43-4	147.217			216; 93^{12}	0.9234^{25}	1.5076^{25}	
8655	1-Penten-3-yne	Methylvinylacetylene	C_5H_6	646-05-9	66.102			59.5	0.7401^{20}	1.4496^{20}	vs bz, eth
8656	1-Penten-4-yne		C_5H_6	871-28-3	66.102			42.5	0.738^{16}	1.4125^{16}	i H_2O; s eth, bz

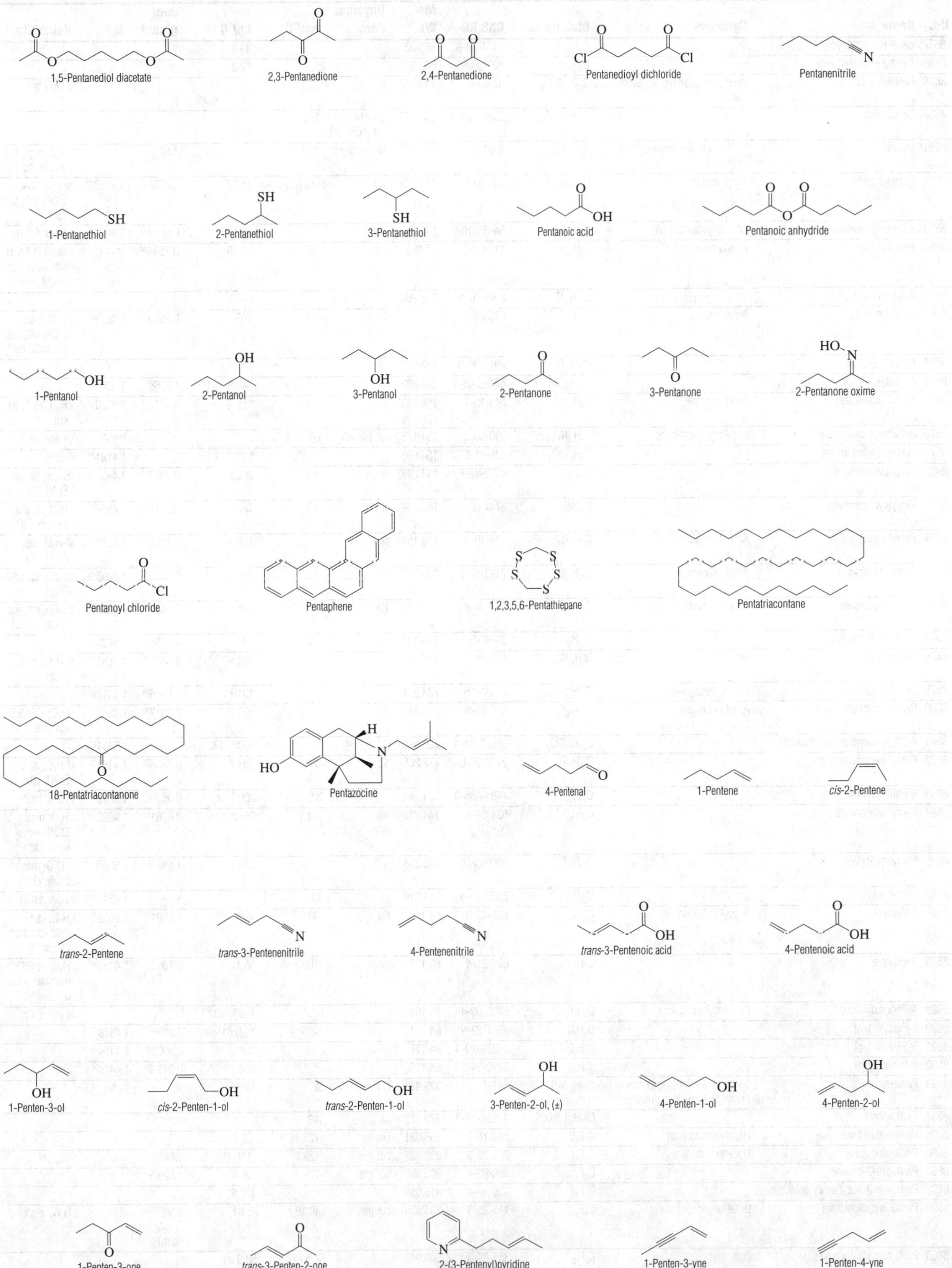
1,5-Pentanediol diacetate
2,3-Pentanedione
2,4-Pentanedione
Pentanedioyl dichloride
Pentanenitrile
1-Pentanethiol
2-Pentanethiol
3-Pentanethiol
Pentanoic acid
Pentanoic anhydride
1-Pentanol
2-Pentanol
3-Pentanol
2-Pentanone
3-Pentanone
2-Pentanone oxime
Pentanoyl chloride
Pentaphene
1,2,3,5,6-Pentathiepane
Pentatriacontane
18-Pentatriacontanone
Pentazocine
4-Pentenal
1-Pentene
cis-2-Pentene
trans-2-Pentene
trans-3-Pentenenitrile
4-Pentenenitrile
trans-3-Pentenoic acid
4-Pentenoic acid
1-Penten-3-ol
cis-2-Penten-1-ol
trans-2-Penten-1-ol
3-Penten-2-ol, (±)
4-Penten-1-ol
4-Penten-2-ol
1-Penten-3-one
trans-3-Penten-2-one
2-(3-Pentenyl)pyridine
1-Penten-3-yne
1-Penten-4-yne

No.	Name	Synonym	Mol. Form.	CAS RN	Mol. Wt.	Physical Form	mp/°C	bp/°C	den/ g cm^{-3}	n_D	Solubility
8657	*cis*-3-Penten-1-yne		C_5H_6	1574-40-9	66.102			44.6			
8658	*trans*-3-Penten-1-yne		C_5H_6	2004-69-5	66.102			52.2			
8659	Pentetic acid	Diethylenetriaminepentaacetic acid	$C_{14}H_{23}N_3O_{10}$	67-43-6	393.347	cry (w)	219				s H_2O, alk
8660	Pentostatin		$C_{11}H_{16}N_4O_4$	53910-25-1	268.270	wh cry (MeOH aq)	222				
8661	Pentryl	2-(*N*,2,4,6-Tetranitroanilino) ethanol	$C_8H_6N_6O_{11}$	4481-55-4	362.167	wh-ye cry	129		1.82		i H_2O, ctc; s chl; vs eth, bz
8662	Pentyl acetate	Amyl acetate	$C_7H_{14}O_2$	628-63-7	130.185	liq	-70.8	149.2	0.8756^{20}	1.4023^{20}	sl H_2O; msc EtOH, eth; s ctc
8663	*sec*-Pentyl acetate (*R*)	*sec*-Amyl acetate (*R*)	$C_7H_{14}O_2$	54638-10-7	130.185			142	0.8803^{18}	1.4012^{20}	vs eth, EtOH
8664	Pentylamine	Amylamine	$C_5H_{13}N$	110-58-7	87.164	liq	-55	104.3	0.7544^{20}	1.448^{20}	msc H_2O, EtOH, eth; vs ace, bz; sl chl
8665	4-*tert*-Pentylaniline		$C_{11}H_{17}N$	2049-92-5	163.260			260.5			
8666	Pentylbenzene	Amylbenzene	$C_{11}H_{16}$	538-68-1	148.245	liq	-75	205.4	0.8585^{20}	1.4878^{20}	i H_2O; msc EtOH, eth, ace, bz, peth, ctc
8667	Pentyl benzoate		$C_{12}H_{16}O_2$	2049-96-9	192.254			137^{15}			
8668	4-Pentylbenzoyl chloride		$C_{12}H_{15}ClO$	49763-65-7	210.699			144^{10}, 121^{8}	1.036^{25}	1.5300^{20}	
8669	Pentyl butanoate	Amyl butyrate	$C_9H_{18}O_2$	540-18-1	158.238	liq	-73.2	186.4	0.8713^{15}	1.4123^{20}	i H_2O; vs EtOH, eth
8670	*tert*-Pentyl carbamate	*tert*-Amyl carbamate	$C_6H_{13}NO_2$	590-60-3	131.173	nd (dil al)	86				vs ace, bz
8671	Pentyl chloroformate		$C_6H_{11}ClO_2$	638-41-5	150.603			61^{15}		1.4181^{18}	s eth
8672	Pentylcyclohexane		$C_{11}H_{22}$	4292-92-6	154.293	liq	-57.5	203.7	0.8037^{20}	1.4437^{20}	vs ace, bz, eth, EtOH
8673	Pentylcyclopentane		$C_{10}H_{20}$	3741-00-2	140.266	liq	-83	180	0.7912^{20}	1.4356^{20}	i H_2O; vs ace, bz, eth, EtOH
8674	Pentyl formate	Amyl formate	$C_6H_{12}O_2$	638-49-3	116.158	liq	-73.5	130.4	0.8853^{20}	1.3992^{20}	sl H_2O; msc EtOH, eth
8675	Pentyl heptanoate	Amyl enanthate	$C_{12}H_{24}O_2$	7493-82-5	200.318	liq	-50	245.4	0.8623^{20}	1.4263^{15}	vs ace, bz, eth, EtOH
8676	Pentyl hexanoate	Amyl caproate	$C_{11}H_{22}O_2$	540-07-8	186.292	liq	-47	226	0.8612^{25}	1.4202^{25}	s EtOH, eth, ace; sl ctc
8677	1-Pentylnaphthalene		$C_{15}H_{18}$	86-89-5	198.304	liq	-22	307	0.9656^{20}	1.5725^{20}	
8678	Pentyl nitrite	Amyl nitrite	$C_5H_{11}NO_2$	463-04-7	117.147			104.5	0.8817^{20}	1.3851^{20}	sl H_2O; msc EtOH, eth
8679	Pentyl nonanoate	Pentyl pelargonate	$C_{14}H_{28}O_2$	61531-45-1	228.371		-27	131^{20}	0.8506^{25}	1.4318^{20}	
8680	Pentyl octanoate	Amyl octanoate	$C_{13}H_{26}O_2$	638-25-5	214.344	liq	-34.8	260.2	0.8613^{20}	1.4262^{25}	i H_2O; s EtOH, eth, ace
8681	4-(Pentyloxy)benzoyl chloride		$C_{12}H_{15}ClO_2$	36823-84-4	226.699			198^{30}, 182^{25}	1.087^{25}	1.5434^{20}	
8682	Pentyl pentanoate		$C_{10}H_{20}O_2$	2173-56-0	172.265	liq	-78.8	203.7	0.8638^{20}	1.4164^{20}	sl H_2O; msc EtOH, eth
8683	4-Pentylphenol		$C_{11}H_{16}O$	14938-35-3	164.244		23	250.5	0.960^{20}	1.5272^{25}	vs eth, EtOH
8684	Pentyl propanoate		$C_8H_{16}O_2$	624-54-4	144.212	liq	-73.1	168.6	0.8761^{25}	1.4096^{15}	i H_2O; msc EtOH, eth; s bz; sl ctc
8685	Pentyl salicylate		$C_{12}H_{16}O_3$	2050-08-0	208.253			270	1.064^{15}	1.506^{20}	sl H_2O; msc EtOH, eth
8686	Pentyl stearate		$C_{23}H_{46}O_2$	6382-13-4	354.610	pl	30			1.4342^{50}	vs eth, EtOH
8687	1-Pentyne	Propylacetylene	C_5H_8	627-19-0	68.118	liq	-90	40.1	0.6901^{20}	1.3852^{20}	i H_2O; vs EtOH; msc eth; s bz, chl; sl ctc
8688	2-Pentyne		C_5H_8	627-21-4	68.118	liq	-109.3	56.1	0.7058^{25}	1.4039^{20}	i H_2O; vs EtOH; msc eth; s bz, chl
8689	4-Pentynoic acid	Propargylacetic acid	$C_5H_6O_2$	6089-09-4	98.101		57.7	110^{30}, 102^{17}			vs eth, EtOH
8690	2-Pentyn-1-ol		C_5H_8O	6261-22-9	84.117	liq	-49.7	154; 61^{15}	0.909^{20}	1.4518^{17}	
8691	3-Pentyn-1-ol		C_5H_8O	10229-10-4	84.117			154	0.9002^{20}	1.4454^{20}	
8692	4-Pentyn-1-ol		C_5H_8O	5390-04-5	84.117			154	0.913^{20}	1.4414^{20}	
8693	Perazine	10-[3-(4-Methyl-1-piperazinyl) propyl]-10*H*-phenothiazine	$C_{20}H_{25}N_3S$	84-97-9	339.498	cry	52	$165^{0.001}$			
8694	Perfluidone		$C_{14}H_{12}F_3NO_4S_2$	37924-13-3	379.375		143				
8695	Perfluoroacetone	Hexafluoroacetone	C_3F_6O	684-16-2	166.021	col gas	-125.45	-27.4			
8696	Perfluorobutane	Decafluorobutane	C_4F_{10}	355-25-9	238.027	col gas	-129.1	-1.9	1.6484^{25}		s bz, chl
8697	Perfluoro-2-butene		C_4F_8	360-89-4	200.030	col gas	-129	1.5	1.5297^{25}		
8698	Perfluoro-2-butyltetrahydrofuran		$C_8F_{16}O$	335-36-4	416.059			102.6			
8699	Perfluorocyclobutane	Octafluorocyclobutane	C_4F_8	115-25-3	200.030	col gas	-40.19	-5.91	1.500^{25} (p>1 atm)		i H_2O; s eth
8700	Perfluorocyclohexane		C_6F_{12}	355-68-0	300.045		62.5 (triple point)	52.8 sp			
8701	Perfluorocyclohexene		C_6F_{10}	355-75-9	262.048			52.0	1.6650^{25}	1.293^{20}	

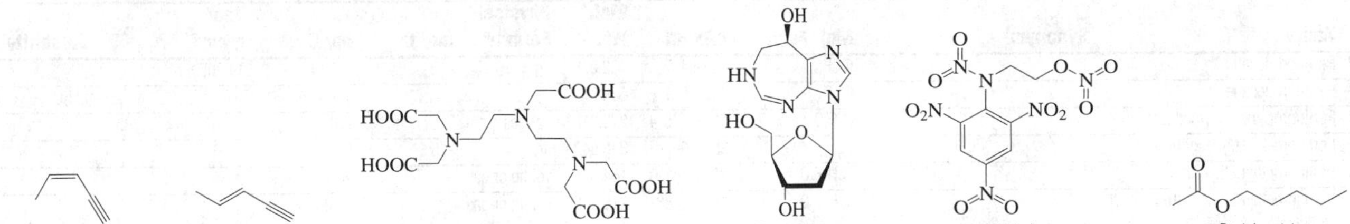

cis-3-Penten-1-yne

trans-3-Penten-1-yne

HOOC, HOOC, N, N, N, COOH, COOH, COOH

Pentetic acid

OH, HN, N, N, N, HO, O, OH

Pentostatin

O, N, N, O, N, O, O, O_2N, NO_2, O, N, O

Pentryl

O, O

Pentyl acetate

O, O

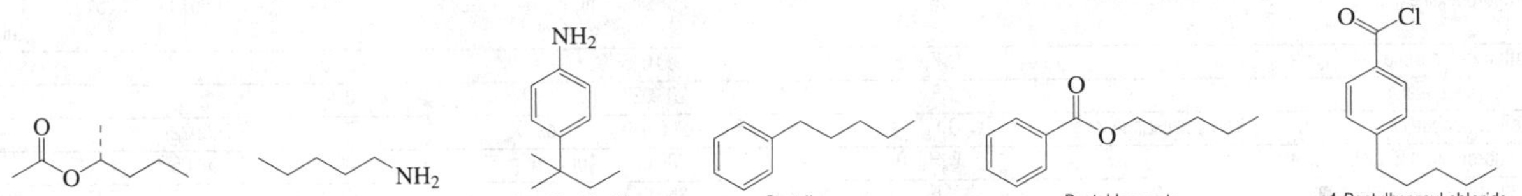

sec-Pentyl acetate (*R*)

NH_2

Pentylamine

NH_2

4-*tert*-Pentylaniline

Pentylbenzene

O, O

Pentyl benzoate

O, Cl

4-Pentylbenzoyl chloride

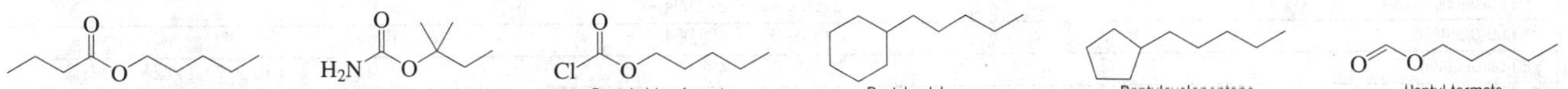

O, O

Pentyl butanoate

O, H_2N, O

tert-Pentyl carbamate

O, Cl, O

Pentyl chloroformate

Pentylcyclohexane

Pentylcyclopentane

O, O

Pentyl formate

O, O

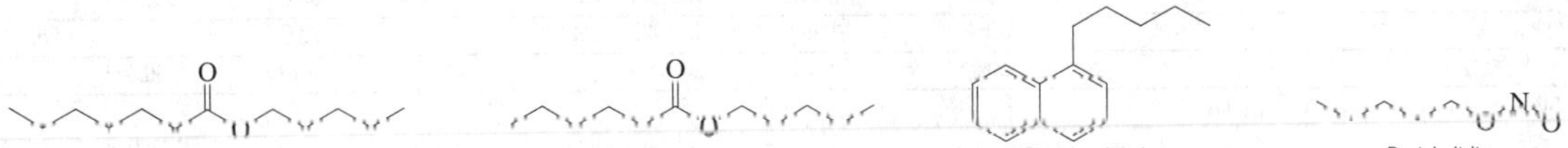

Pentyl heptanoate

O, O

Pentyl hexanoate

1-Pentylnaphthalene

O, N, O

Pentyl nitrite

O, O

Pentyl nonanoate

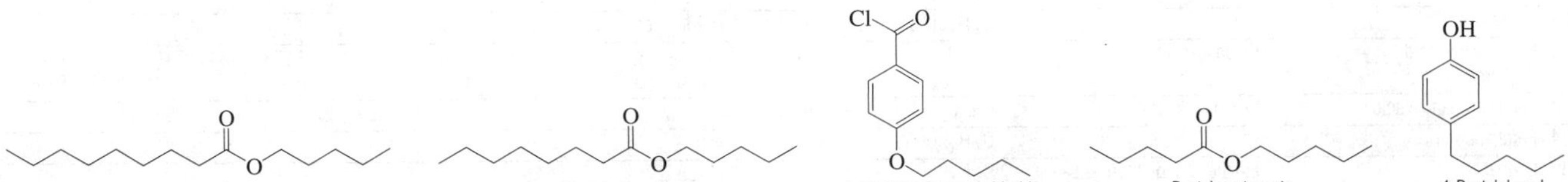

O, O

Pentyl octanoate

Cl, O, O

4-(Pentyloxy)benzoyl chloride

O, O

Pentyl pentanoate

OH

4-Pentylphenol

O, O

Pentyl propanoate

O, O, OH

Pentyl salicylate

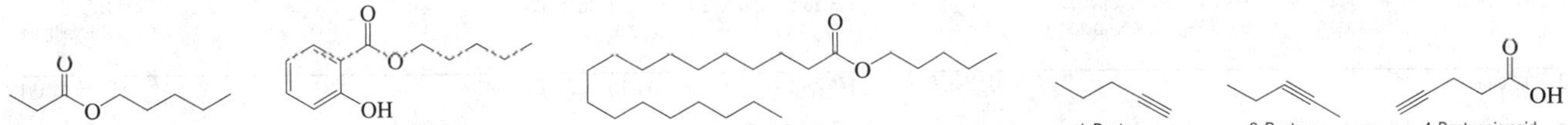

O, O

Pentyl stearate

1-Pentyne

2-Pentyne

O, OH

4-Pentynoic acid

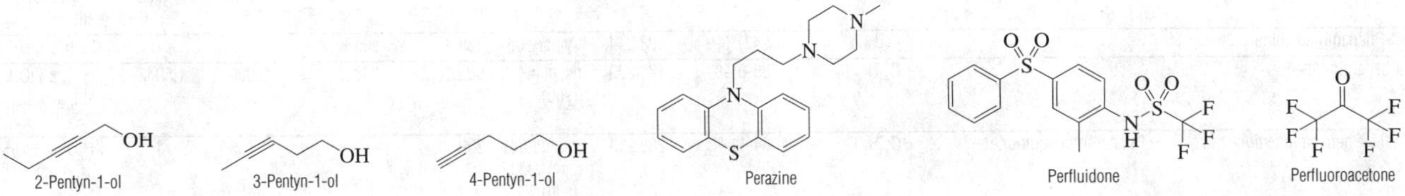

OH

2-Pentyn-1-ol

OH

3-Pentyn-1-ol

OH

4-Pentyn-1-ol

N, N, N, S

Perazine

O, O, S, N, H, O, O, S, F, F, F

Perfluidone

O, F, F, F, F, F, F

Perfluoroacetone

F F F F F F F F F F

Perfluorobutane

F F F F F F F F

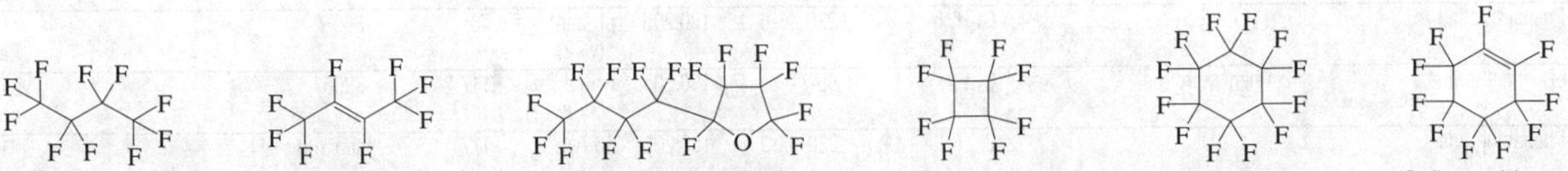

Perfluoro-2-butene

F F F F F F F F F F F F F F F F O F

Perfluoro-2-butyltetrahydrofuran

F F F F F F F F

Perfluorocyclobutane

F F F F F F F F F F F F

Perfluorocyclohexane

F F F F F F F F F F

Perfluorocyclohexene

No.	Name	Synonym	Mol. Form.	CAS RN	Mol. Wt.	Physical Form	mp/°C	bp/°C	den/ g cm^{-3}	n_D	Solubility
8702	Perfluorodecalin		$C_{10}F_{18}$	306-94-5	462.078	liq	-10	142.02	1.9305^{25}		
8703	Perfluorodecane		$C_{10}F_{22}$	307-45-9	538.072			144.2			i H_2O
8704	Perfluorodimethoxymethane		$C_3F_8O_2$	53772-78-4	220.018	col gas	-161	-10			
8705	Perfluoro-2,3-dimethylbutane		C_6F_{14}	354-96-1	338.042	liq	-15	59.8			
8706	Perfluoroethyl ethyl ether		$C_4H_5F_5O$	22052-81-9	164.074	vol liq or gas		28.11			
8707	Perfluoroethyl 2,2,2-trifluoroethyl ether		$C_4H_2F_8O$	156053-88-2	218.045	vol liq or gas		27.89			
8708	Perfluoroheptane		C_7F_{16}	335-57-9	388.049	liq	-51.2	82.5	1.7333^{20}	1.2618^{20}	i H_2O; vs ace, eth, EtOH, chl
8709	Perfluoro-1-heptene		C_7F_{14}	355-63-5	350.053			81.0			
8710	Perfluorohexane		C_6F_{14}	355-42-0	338.042	liq	-88.2	57.14	1.6910^{20}	1.2515^{20}	i H_2O; s eth, bz, chl
8711	Perfluoro-1-hexene		C_6F_{12}	755-25-9	300.045			57.0			vs chl
8712	Perfluoroisobutane		C_4F_{10}	354-92-7	238.027	col gas		0			
8713	Perfluoroisobutene	Perfluoroisobutylene	C_4F_8	382-21-8	200.030	col gas	-130	7	1.5922^{0}		
8714	Perfluoroisopropyl methyl ether		$C_4H_3F_7O$	22052-84-2	200.055	vol liq or gas		29.34	1.4205^{20}		
8715	Perfluoromethylcyclohexane		C_7F_{14}	355-02-2	350.053	liq	-44.7	76.3	1.7878^{25}	1.285^{17}	s ace, bz, ctc, tol, AcOEt
8716	Perfluoro-2-methylpentane		C_6F_{14}	355-04-4	338.042			57.6	1.7326^{20}	1.2564^{22}	i H_2O; s bz
8717	Perfluoro-3-methylpentane		C_6F_{14}	865-71-4	338.042	liq	-115	58.4			s bz
8718	Perfluoronaphthalene		$C_{10}F_8$	313-72-4	272.094		87.5	209			
8719	Perfluorononane		C_9F_{20}	375-96-2	488.064			117.61	1.8001^{20}		
8720	Perfluorooctane		C_8F_{18}	307-34-6	438.057			105.9	1.73^{20}	1.282^{20}	i H_2O
8721	Perfluorooctylsulfonyl fluoride		$C_8F_{18}O_2S$	307-35-7	502.121	liq		154			
8722	Perfluorooxetane		C_3F_6O	425-82-1	166.021	col gas	-117	-28.4			
8723	Perfluoropentane		C_5F_{12}	678-26-2	288.035	vol liq or gas	-10	29.2			i H_2O
8724	Perfluoropropane		C_3F_8	76-19-7	188.019	col gas	-147.70	-36.6			i H_2O
8725	Perfluoropropene		C_3F_6	116-15-4	150.022	col gas	-156.5	-29.6		1.583^{-40}	i H_2O
8726	Perfluoropropyl methyl ether		$C_4H_3F_7O$	375-03-1	200.055			34.23	1.4092^{20}		
8727	Perfluoropyridine	Pentafluoropyridine	C_5F_5N	700-16-3	169.053			83.7			
8728	Perfluorotoluene		C_7F_8	434-64-0	236.062	liq	-65.49	103.55	1.6616^{25}	1.3670^{20}	
8729	Perfluorotripropylamine		$C_9F_{21}N$	338-83-0	521.069			130	1.822^{4}	1.279^{25}	
8730	1*H*-Perimidine		$C_{11}H_8N_2$	204-02-4	168.195	grn cry (dil al)	223.0				i H_2O; s EtOH, eth, ace, bz; sl DMSO
8731	Permethrin		$C_{21}H_{20}Cl_2O_3$	52645-53-1	391.288	cry or ye liq	34	$200^{0.01}$	1.23^{20}		i H_2O; s os
8732	Peroxyacetic acid	Ethaneperoxoic acid	$C_2H_4O_3$	79-21-0	76.051	liq	-0.2	110	1.226^{15}	1.3974^{20}	vs H_2O, eth, sulf; s EtOH
8733	Peroxypropanoic acid	Propaneperoxoic acid	$C_3H_6O_3$	4212-43-5	90.078			exp 119.7		1.4148^{15}	
8734	Perphenazine		$C_{21}H_{26}ClN_3OS$	58-39-9	403.968		97				
8735	Perthane	Ethane, 1,1-dichloro-2,2-bis(*p*-ethylphenyl)-	$C_{18}H_{20}Cl_2$	72-56-0	307.258		56				
8736	Perylene	Dibenz[de,kl]anthracene	$C_{20}H_{12}$	198-55-0	252.309	gold-br, ye pl (bz, HOAc)	277.76		1.35^{25}		i H_2O; sl EtOH, eth; vs ace, chl; s bz
8737	Peucedanin	3-Methoxy-2-isopropyl-7*H*-furo[3,2-g][1]benzopyran-7-one	$C_{15}H_{14}O_4$	133-26-6	258.270	pr or pl (bz-peth)	85	278^{17}			sl H_2O, bz; s EtOH, eth; vs chl, CS_2
8738	Phalloidin		$C_{35}H_{48}N_8O_{11}S$	17466-45-4	788.868	nd (w)	281 (hyd)				s EtOH, MeOH, py
8739	Phalloin		$C_{35}H_{48}N_8O_{10}S$	28227-92-1	772.869	cry (w)	250 dec				
8740	α-Phellandrene	2-Methyl-5-(1-methylethyl)-1,3-cyclohexadiene	$C_{10}H_{16}$	99-83-2	136.234		238	174.9	0.8410^{20}	1.471^{25}	i H_2O; s eth
8741	β-Phellandrene	*p*-Mentha-1(7),2-diene	$C_{10}H_{16}$	555-10-2	136.234			171.5	0.8520^{20}	1.4788^{20}	i H_2O, EtOH; s eth
8742	9-Phenanthrenamine		$C_{14}H_{11}N$	947-73-9	193.244	lt ye cry (al)	138.3	sub			sl eth, bz, chl
8743	Phenanthrene		$C_{14}H_{10}$	85-01-8	178.229	mcl pl (al), lf (sub)	99.24	340	0.9800^{4}	1.5943	i H_2O; s EtOH, eth, ace, bz, CS_2
8744	9,10-Phenanthrenedione	Phenanthrenequinone	$C_{14}H_8O_2$	84-11-7	208.213	oran nd (to) oran-red pl (sub)	209		1.405^{22}		i H_2O; sl EtOH, bz; s eth
8745	Phenanthridine		$C_{13}H_9N$	229-87-8	179.217	nd (dil al)	107.4	348.9			sl H_2O; vs EtOH, eth, bz, CS_2; s ace
8746	1,7-Phenanthroline		$C_{12}H_8N_2$	230-46-6	180.205	pl (anh), nd (w+2)	78	360			s H_2O; vs EtOH; i eth, bz, lig
8747	1,10-Phenanthroline	*o*-Phenanthroline	$C_{12}H_8N_2$	66-71-7	180.205	wh nd (bz) cry (w+1)	117	>300			vs H_2O; s EtOH, ace, bz; i peth
8748	4,7-Phenanthroline		$C_{12}H_8N_2$	230-07-9	180.205	nd (w)	177	sub 100			s H_2O, lig; vs EtOH; sl eth, bz, CS_2

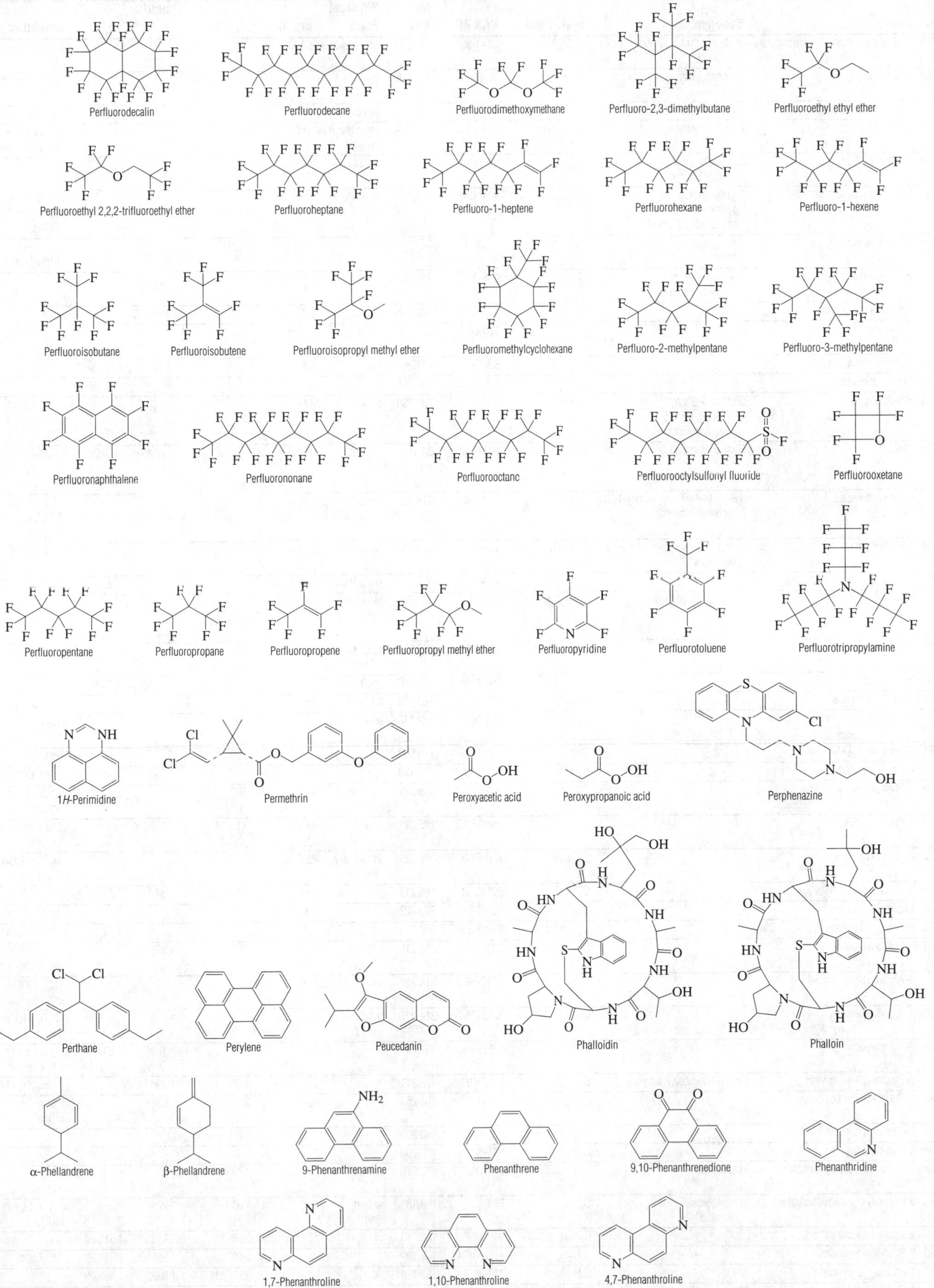
Perfluorodecalin
Perfluorodecane
Perfluorodimethoxymethane
Perfluoro-2,3-dimethylbutane
Perfluoroethyl ethyl ether
Perfluoroethyl 2,2,2-trifluoroethyl ether
Perfluoroheptane
Perfluoro-1-heptene
Perfluorohexane
Perfluoro-1-hexene
Perfluoroisobutane
Perfluoroisobutene
Perfluoroisopropyl methyl ether
Perfluoromethylcyclohexane
Perfluoro-2-methylpentane
Perfluoro-3-methylpentane
Perfluoronaphthalene
Perfluorononane
Perfluorooctane
Perfluorooctylsulfonyl fluoride
Perfluorooxetane
Perfluoropentane
Perfluoropropane
Perfluoropropene
Perfluoropropyl methyl ether
Perfluoropyridine
Perfluorotoluene
Perfluorotripropylamine
1H-Perimidine
Permethrin
Peroxyacetic acid
Peroxypropanoic acid
Perphenazine
Perthane
Perylene
Peucedanin
Phalloidin
Phalloin
α-Phellandrene
β-Phellandrene
9-Phenanthrenamine
Phenanthrene
9,10-Phenanthrenedione
Phenanthridine
1,7-Phenanthroline
1,10-Phenanthroline
4,7-Phenanthroline

No.	Name	Synonym	Mol. Form.	CAS RN	Mol. Wt.	Physical Form	mp/°C	bp/°C	den/ g cm^{-3}	n_D	Solubility
8749	1,10-Phenanthroline monohydrate	*o*-Phenanthroline monohydrate	$C_{12}H_{10}N_2O$	5144-89-8	198.219	wh cry pow	93				s EtOH, ace; sl bz
8750	Phenazine	Dibenzopyrazine	$C_{12}H_8N_2$	92-82-0	180.205	ye-red nd (HOAc)	176.5				sl H_2O, eth; s bz, EtOH
8751	2,3-Phenazinediamine	2,3-Diaminophenazine	$C_{12}H_{10}N_4$	655-86-7	210.234	ye nd	264	sub			vs bz, EtOH
8752	1-Phenazinol	Hemipyocyanine	$C_{12}H_8N_2O$	528-71-2	196.204	ye nd (bz, dil MeOH)	158	sub			sl H_2O, EtOH; s bz, py, dil alk
8753	Phenazopyridine	2,6-Diamino-3-phenylazopyridine	$C_{11}H_{11}N_5$	94-78-0	213.239	red cry	139				
8754	Phenazopyridine hydrochloride	3-(Phenylazo)-2,6-pyridinediamine, monohydrochloride	$C_{11}H_{12}ClN_5$	136-40-3	249.700	ye-red cry					sl H_2O, EtOH; i bz, ace; s HOAc
8755	Phencarbamide		$C_{19}H_{24}N_2OS$	3735-90-8	328.471		48.5	$121^{0.01}$			vs eth, chl, MeOH, peth
8756	Phendimetrazine	3,4-Dimethyl-2-phenylmorpholine	$C_{12}H_{17}NO$	634-03-7	191.269			134^{12}, $78^{0.35}$			
8757	Phenethicillin potassium		$C_{17}H_{19}KN_2O_5S$	132-93-4	402.506	cry (ace)	235				s H_2O
8758	Phenicin		$C_{14}H_{10}O_6$	128-68-7	274.225	ye-br (al)	230.5				sl H_2O; vs EtOH, chl, HOAc
8759	Phenindamine		$C_{19}H_{19}N$	82-88-2	261.361	cry	91		1.17		
8760	Phenmedipham		$C_{16}H_{16}N_2O_4$	13684-63-4	300.309		143				
8761	Phenobarbital	5-Ethyl-5-phenyl-2,4,6(1*H*,3*H*,5*H*)-pyrimidinetrione	$C_{12}H_{12}N_2O_3$	50-06-6	232.234	pl (w)	174				i H_2O, bz; s EtOH, eth; sl DMSO
8762	Phenol	Hydroxybenzene	C_6H_6O	108-95-2	94.111		40.89	181.87	1.0545^{45}	1.5408^{41}	s H_2O, EtOH; vs eth; msc ace, bz
8763	Phenolphthalein	3,3-Bis(4-hydroxyphenyl)-1(3*H*)-isobenzofuranone	$C_{20}H_{14}O_4$	77-09-8	318.323	wh orth nd	262.5		1.277^{32}		i H_2O, bz; vs EtOH, ace; s eth, chl
8764	Phenolphthalin	2-[Bis(4-hydroxyphenyl)methyl] benzoic acid	$C_{20}H_{16}O_4$	81-90-3	320.339	nd (w)	230.5				vs eth, EtOH
8765	Phenolphthalol		$C_{20}H_{18}O_3$	81-92-5	306.355	cry (dil al)	201.5				
8766	Phenol Red	Phenolsulfonphthalein	$C_{19}H_{14}O_5S$	143-74-8	354.376	dk red nd or pl	>300				sl H_2O, EtOH, ace, bz; i eth, chl
8767	10*H*-Phenothiazine	Thiodiphenylamine	$C_{12}H_9NS$	92-84-2	199.271	ye pr (al) ye lf or pl (tol)	187.5	371			vs ace, bz, eth, EtOH
8768	Phenothrin		$C_{23}H_{26}O_3$	26002-80-2	350.450	col liq			1.061^{25}	1.5483^{25}	i H_2O; s ace, xyl
8769	10*H*-Phenoxazine		$C_{12}H_9NO$	135-67-1	183.205	lf (dil al, bz)	156	dec			vs bz, eth, EtOH
8770	Phenoxyacetic acid		$C_8H_8O_3$	122-59-8	152.148	nd or pl (w)	98.5	dec 285			s H_2O; vs EtOH, eth, bz, CS_2
8771	Phenoxyacetyl chloride		$C_8H_7ClO_2$	701-99-5	170.594			225.5			s eth
8772	Phenoxyacetylene		C_8H_6O	4279-76-9	118.133		-36	61^{25}	1.0614^{20}	1.5125^{20}	vs eth, EtOH
8773	2-Phenoxyaniline		$C_{12}H_{11}NO$	2688-84-8	185.221	cry (lig)	45.8	308; 172^{14}			s EtOH; s eth, ace, bz
8774	3-Phenoxyaniline		$C_{12}H_{11}NO$	3586-12-7	185.221	pr (lig)	37	315; 180^{10}	1.1583^{25}		s EtOH, eth, ace, bz; sl lig
8775	4-Phenoxyaniline		$C_{12}H_{11}NO$	139-59-3	185.221	nd (w), cry (dil al)	85.5				s H_2O; vs EtOH, eth; sl lig
8776	3-Phenoxybenzaldehyde		$C_{13}H_{10}O_2$	39515-51-0	198.217		14.0	169^{11}, $140^{0.1}$	1.147^{25}	1.5954^{20}	
8777	Phenoxybenzamine		$C_{18}H_{22}ClNO$	59-96-1	303.827		39				s bz
8778	Phenoxybenzamine hydrochloride		$C_{18}H_{23}Cl_2NO$	63-92-3	340.288		139				sl H_2O; s EtOH
8779	2-Phenoxybenzoic acid		$C_{13}H_{10}O_3$	2243-42-7	214.216	lf (dil al)	113	355	1.1553^{50}		i H_2O; vs EtOH, eth; s chl
8780	3-Phenoxybenzoic acid		$C_{13}H_{10}O_3$	3739-38-6	214.216	nd (aq al)	145.8				i H_2O; s EtOH, eth
8781	4-Phenoxybenzoic acid		$C_{13}H_{10}O_3$	2215-77-2	214.216	pr (chl)	161				sl H_2O; s EtOH, eth, chl
8782	2-Phenoxyethanol		$C_8H_{10}O_2$	122-99-6	138.164	oil	14	245	1.102^{22}	1.534^{20}	i H_2O; s EtOH, eth, chl, alk
8783	2-Phenoxyethyl acrylate	Phenyl Cellosolve acrylate	$C_{11}H_{12}O_3$	48145-04-6	192.211			110^2	1.090^{25}		vs ace, eth, chl
8784	2-Phenoxyethyl butanoate		$C_{12}H_{16}O_3$	23511-70-8	208.253			251; 88^2	1.0388^{21}		vs ace, eth, EtOH
8785	3-Phenoxyphenol		$C_{12}H_{10}O_2$	713-68-8	186.206			175^7			
8786	4-Phenoxyphenol		$C_{12}H_{10}O_2$	831-82-3	186.206		84.0				
8787	2-(3-Phenoxyphenyl)propanoic acid, (±)	Fenoprofen	$C_{15}H_{14}O_3$	31879-05-7	242.270	visc oil		$170^{0.11}$		1.5742^{25}	
8788	3-Phenoxy-1,2-propanediol	Phenylglyceryl ether	$C_9H_{12}O_3$	538-43-2	168.189	nd (eth, peth)	67.5	200^{22}	1.225^{20}		vs H_2O, bz, eth, EtOH
8789	2-Phenoxypropanoic acid		$C_9H_{10}O_3$	940-31-8	166.173	nd (w)	115.5	266; 105^5	1.1865^{20}	1.5184^{20}	
8790	2-Phenoxy-1-propanol		$C_9H_{12}O_2$	4169-04-4	152.190			244	0.9801^{25}	1.4760^{25}	s EtOH, eth
8791	1-Phenoxy-2-propanol		$C_9H_{12}O_2$	770-35-4	152.190			233; 134^{20}	1.0622^{20}	1.5232^{20}	
8792	1-Phenoxy-2-propanone	Phenoxyacetone	$C_9H_{10}O_2$	621-87-4	150.174			229.5	1.0903^{20}	1.5228^{20}	s eth, ace

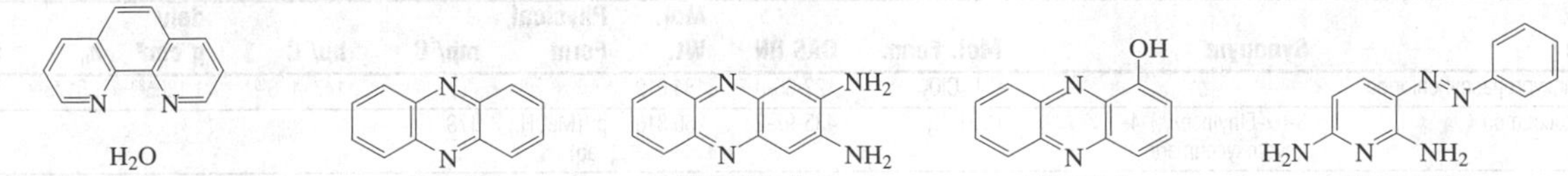

H_2O
1,10-Phenanthroline monohydrate

Phenazine

2,3-Phenazinediamine

1-Phenazinol

Phenazopyridine

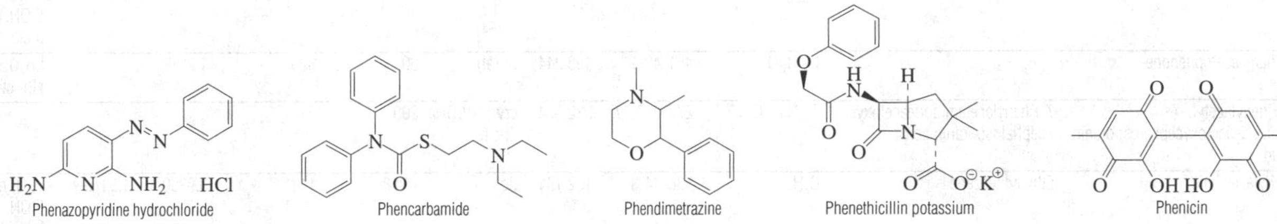

HCl
Phenazopyridine hydrochloride

Phencarbamide

Phendimetrazine

Phenethicillin potassium

Phenicin

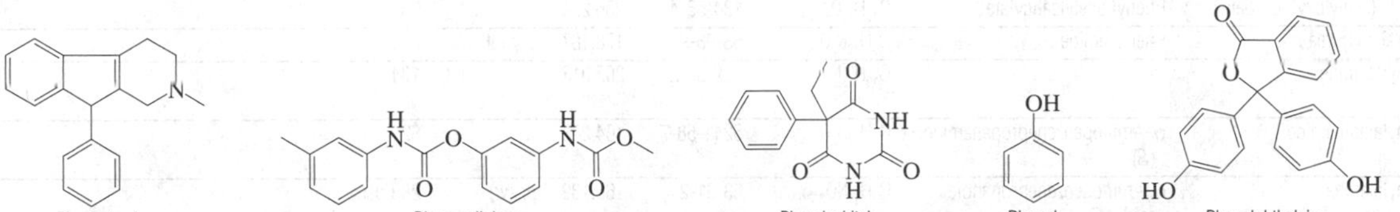

Phenindamine

Phenmedipham

Phenobarbital

Phenol

Phenolphthalein

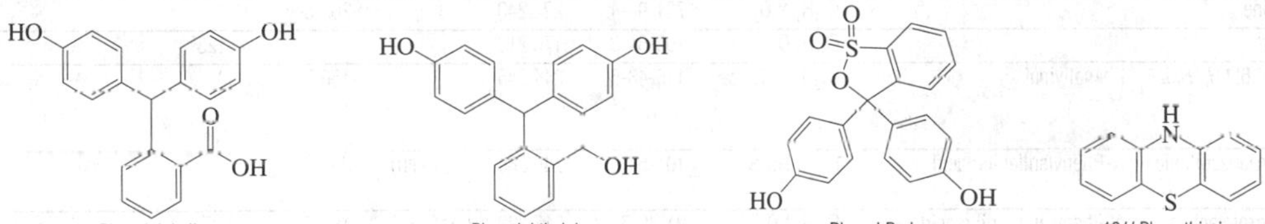

Phenolphthalin

Phenolphthalol

Phenol Red

10*H*-Phenothiazine

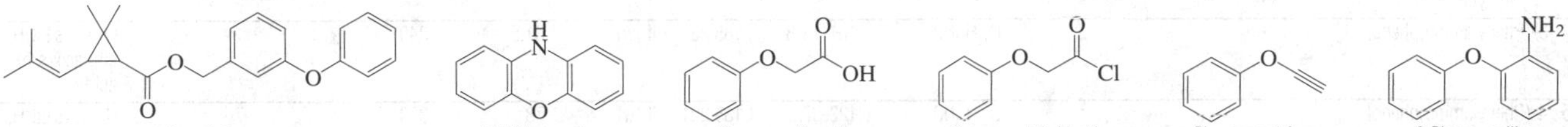

Phenothrin

10*H*-Phenoxazine

Phenoxyacetic acid

Phenoxyacetyl chloride

Phenoxyacetylene

2-Phenoxyaniline

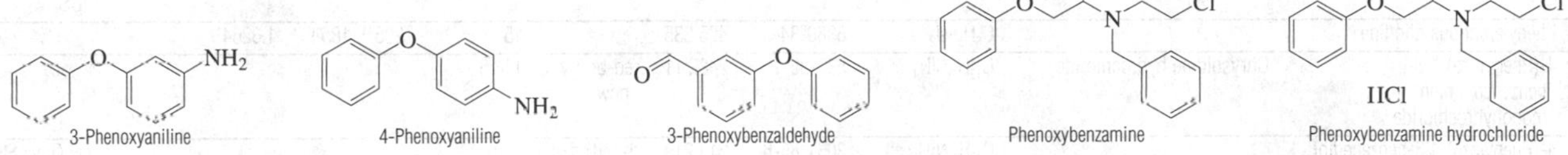

3-Phenoxyaniline

4-Phenoxyaniline

3-Phenoxybenzaldehyde

Phenoxybenzamine

HCl
Phenoxybenzamine hydrochloride

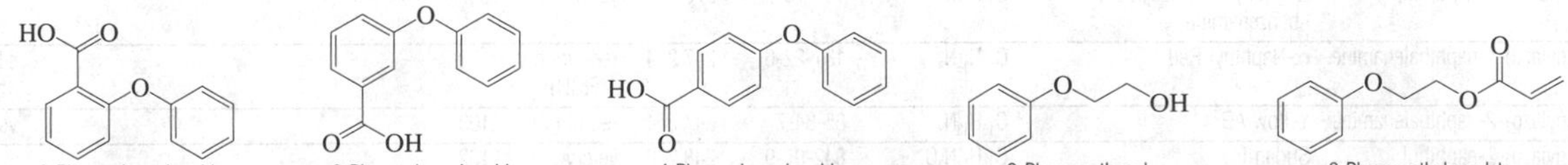

2-Phenoxybenzoic acid

3-Phenoxybenzoic acid

4-Phenoxybenzoic acid

2-Phenoxyethanol

2-Phenoxyethyl acrylate

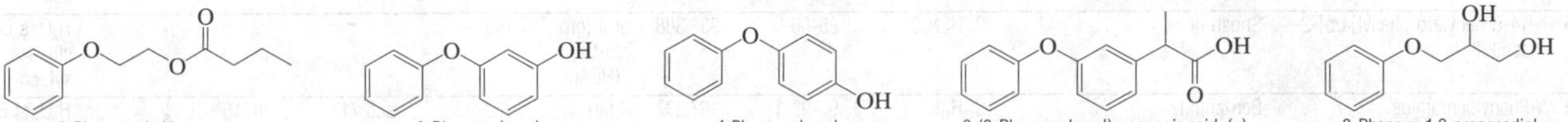

2-Phenoxyethyl butanoate

3-Phenoxyphenol

4-Phenoxyphenol

2-(3-Phenoxyphenyl)propanoic acid, (±)

3-Phenoxy-1,2-propanediol

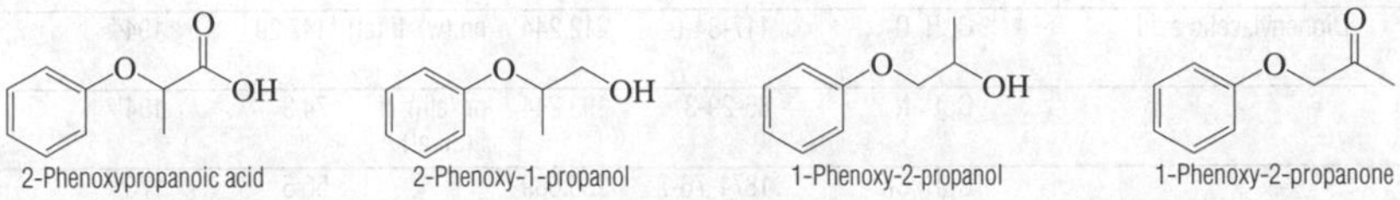

2-Phenoxypropanoic acid

2-Phenoxy-1-propanol

1-Phenoxy-2-propanol

1-Phenoxy-2-propanone

No.	Name	Synonym	Mol. Form.	CAS RN	Mol. Wt.	Physical Form	mp/°C	bp/°C	den/ g cm-3	n_D	Solubility
8793	2-Phenoxypropanoyl chloride		$C_9H_9ClO_2$	122-35-0	184.619			147; 116[10]	1.1865[20]	1.5178[20]	s eth
8794	Phenprocoumon	3-(α-Ethylbenzyl)-4-hydroxycoumarin	$C_{18}H_{16}O_3$	435-97-2	280.318	pr (MeOH aq)	179				
8795	Phenthoate		$C_{12}H_{17}O_4PS_2$	2597-03-7	320.364	ye oil		123[0.01]			sl H_2O; s hx
8796	Phentolamine		$C_{17}H_{19}N_3O$	50-60-2	281.352		175				
8797	Phenyl acetate		$C_8H_8O_2$	122-79-2	136.149			196; 75[8]	1.0780[20]	1.5035[20]	sl H_2O; msc EtOH, eth, chl; s ctc
8798	2-Phenylacetophenone		$C_{14}H_{12}O$	451-40-1	196.244	pl (al)	60	320	1.201[0]		sl H_2O; s EtOH, eth, ctc, chl
8799	*N*-(Phenylacetyl)-7-aminodeacetoxycephalosporanic acid	7-Phenylacetamidodeacetoxy-cephalosporanic acid	$C_{16}H_{16}N_2O_4S$	27255-72-7	332.374	cry (2-PrOH/peth)	200				
8800	Phenylacetylene	Ethynylbenzene	C_8H_6	536-74-3	102.134	liq	-44.8	143	0.9300[20]	1.5470[20]	i H_2O; msc EtOH, eth; s ace; sl chl
8801	(*N*-Phenylacetyl)glycine	Phenaceturic acid	$C_{10}H_{11}NO_3$	500-98-1	193.199	lf (EtOH)	143				
8802	Phenyl 2-(acetyloxy)benzoate	Phenyl acetylsalicylate	$C_{15}H_{12}O_4$	134-55-4	256.254		96				
8803	(Phenylacetyl)urea	Phenacemide	$C_9H_{10}N_2O_2$	63-98-9	178.187	cry (al)	215				vs bz, eth, EtOH
8804	9-Phenylacridine		$C_{19}H_{13}N$	602-56-2	255.313	ye nd, lf (al)	184	404			i H_2O; sl EtOH; s eth; vs bz
8805	*L*-Phenylalaninamide	α-Aminobenzenepropanamide, (*S*)-	$C_9H_{12}N_2O$	5241-58-7	164.203		82				
8806	*L*-Phenylalanine	α-Aminobenzenepropanoic acid, (*S*)	$C_9H_{11}NO_2$	63-91-2	165.189	pr (w)	283 dec				sl H_2O; i EtOH, eth, bz, acid
8807	*L*-Phenylalanine, ethyl ester	Ethyl 2-amino-3-phenylpropionate	$C_{11}H_{15}NO_2$	3081-24-1	193.243		136	148[13]	1.065[15]		sl H_2O
8808	*L*-Phenylalanylglycine		$C_{11}H_{14}N_2O_3$	721-90-4	222.240		262 dec				s H_2O
8809	3-Phenylallyl acetate		$C_{11}H_{12}O_2$	103-54-8	176.212			123[5]			
8810	5-Phenyl-5-allyl-2,4,6(1*H*,3*H*,5*H*)-pyrimidinetrione	Phenallymal	$C_{13}H_{12}N_2O_3$	115-43-5	244.245		156.5				sl H_2O, bz, DMSO; vs EtOH, eth; i lig
8811	4-(Phenylamino)benzenesulfonic acid	*N*-Phenylsulfanilic acid	$C_{12}H_{11}NO_3S$	101-57-5	249.285	pl (al-eth)	206				vs H_2O, EtOH
8812	2-(Phenylamino)benzoic acid	*N*-Phenylanthranilic acid	$C_{13}H_{11}NO_2$	91-40-7	213.232	lf (al)	183.5				i H_2O; vs EtOH; sl eth, bz
8813	Phenyl 4-amino-3-hydroxybenzoate	Phenyl *p*-aminosalicylate	$C_{13}H_{11}NO_3$	133-11-9	229.231		153				
8814	3-(Phenylamino)phenol		$C_{12}H_{11}NO$	101-18-8	185.221	lf (w)	81.5	340			sl H_2O; vs EtOH, eth, ace; s bz, acid
8815	4-(Phenylamino)phenol		$C_{12}H_{11}NO$	122-37-2	185.221	lf (w)	73	330			sl H_2O; vs EtOH, eth, bz, chl; s acid
8816	9-Phenylanthracene		$C_{20}H_{14}$	602-55-1	254.325	bl lf (al) (HOAc)	156	417			i H_2O; s EtOH, eth, bz, chl, CS_2
8817	Phenylarsonous diiodide		$C_6H_5AsI_2$	6380-34-3	405.835		15	205[14], 185[10]	1.6264[15]		
8818	4-(Phenylazo)-1,3-benzenediamine monohydrochloride	Chrysoidine hydrochloride	$C_{12}H_{13}ClN_4$	532-82-1	248.711	red-br cry pow	118.5				vs ace
8819	4-(Phenylazo)-1,3-benzenediol		$C_{12}H_{10}N_2O_2$	2051-85-6	214.219	dk red nd (dil al)	170				i H_2O; vs EtOH, eth, bz, HOAc
8820	4-Phenylazodiphenylamine	*N*-Phenyl-4-(phenylazo) benzenamine	$C_{18}H_{15}N_3$	101-75-7	273.332	ye pl or pr	84.0				i H_2O; vs EtOH, eth, lig
8821	4-(Phenylazo)-1-naphthalenamine	α-Naphthyl Red	$C_{16}H_{13}N_3$	131-22-6	247.294	red-viol cry (EtOH)	123				s EtOH, dil HCl, bz
8822	1-(Phenylazo)-2-naphthalenamine	Yellow AB	$C_{16}H_{13}N_3$	85-84-7	247.294	red pl (al)	103				vs EtOH, HOAc
8823	1-(Phenylazo)-2-naphthol	Sudan I	$C_{16}H_{12}N_2O$	842-07-9	248.278	ye cry	132				
8824	4-(Phenylazo)phenol		$C_{12}H_{10}N_2O$	1689-82-3	198.219	ye lf (bz) oran pr (al)	155	225[20] dec			i H_2O; vs EtOH, eth; s bz, con sulf
8825	1-[[4-(Phenylazo)phenyl]azo]-2-naphthol	Sudan III	$C_{22}H_{16}N_4O$	85-86-9	352.388	br lf (grn lustre) (HOAc)	195				i H_2O; s EtOH, eth, ace, bz, xyl, chl
8826	*N*-Phenylbenzamide	Benzanilide	$C_{13}H_{11}NO$	93-98-1	197.232	lf (al)	163	sub 117	1.315[25]		i H_2O; sl EtOH, eth, HOAc
8827	α-Phenylbenzeneacetaldehyde		$C_{14}H_{12}O$	947-91-1	196.244			dec 315; 157[7]	1.1061[21]	1.5920[21]	i H_2O; vs EtOH, eth, bz
8828	α-Phenylbenzeneacetic acid	Diphenylacetic acid	$C_{14}H_{12}O_2$	117-34-0	212.244	nd (w), lf (al)	147.29	194[25]	1.257[15]		sl H_2O; vs EtOH; s eth, chl
8829	α-Phenylbenzeneacetonitrile		$C_{14}H_{11}N$	86-29-3	193.244	pr (eth), lf (dil al)	74.3	184[16]			s EtOH, chl; vs eth; sl lig
8830	α-Phenylbenzeneacetyl chloride		$C_{14}H_{11}ClO$	1871-76-7	230.689		56.5	170[16]			s lig

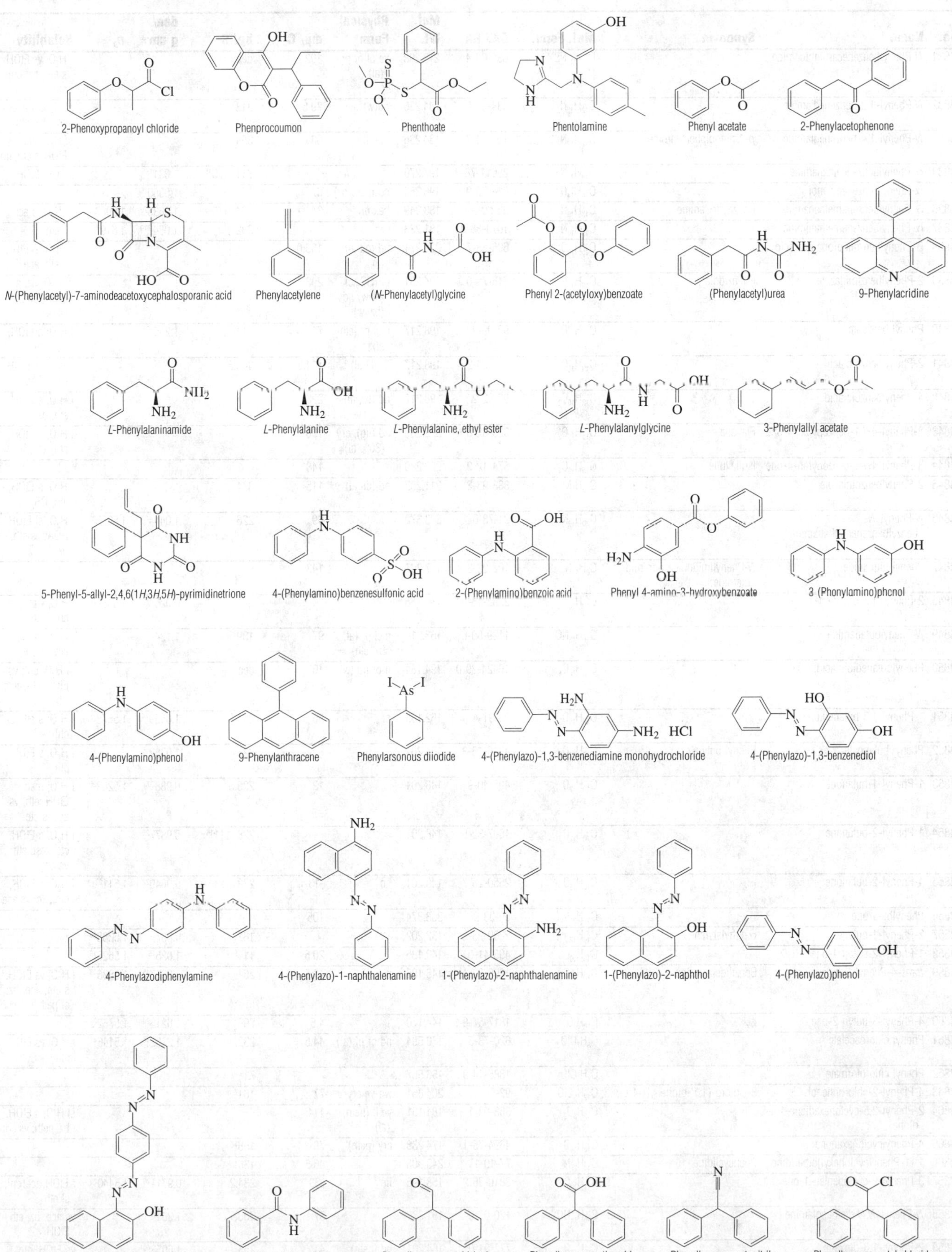
2-Phenoxypropanoyl chloride
Phenprocoumon
Phenthoate
Phentolamine
Phenyl acetate
2-Phenylacetophenone
N-(Phenylacetyl)-7-aminodeacetoxycephalosporanic acid
Phenylacetylene
(N-Phenylacetyl)glycine
Phenyl 2-(acetyloxy)benzoate
(Phenylacetyl)urea
9-Phenylacridine
L-Phenylalaninamide
L-Phenylalanine
L-Phenylalanine, ethyl ester
L-Phenylalanylglycine
3-Phenylallyl acetate
5-Phenyl-5-allyl-2,4,6(1H,3H,5H)-pyrimidinetrione
4-(Phenylamino)benzenesulfonic acid
2-(Phenylamino)benzoic acid
Phenyl 4-amino-3-hydroxybenzoate
3-(Phenylamino)phenol
4-(Phenylamino)phenol
9-Phenylanthracene
Phenylarsonous diiodide
4-(Phenylazo)-1,3-benzenediamine monohydrochloride
4-(Phenylazo)-1,3-benzenediol
4-Phenylazodiphenylamine
4-(Phenylazo)-1-naphthalenamine
1-(Phenylazo)-2-naphthalenamine
1-(Phenylazo)-2-naphthol
4-(Phenylazo)phenol
1-[[4-(Phenylazo)phenyl]azo]-2-naphthol
N-Phenylbenzamide
α-Phenylbenzeneacetaldehyde
α-Phenylbenzeneacetic acid
α-Phenylbenzeneacetonitrile
α-Phenylbenzeneacetyl chloride

No.	Name	Synonym	Mol. Form.	CAS RN	Mol. Wt.	Physical Form	mp/°C	bp/°C	den/ g cm-3	n_D	Solubility
8831	*N*-Phenylbenzenecarbothioamide		$C_{13}H_{11}NS$	636-04-4	213.298	ye pl or pr (al)	102	dec			i H_2O; vs EtOH; s eth, bz, chl; sl lig
8832	*N*-Phenyl-1,2-benzenediamine		$C_{12}H_{12}N_2$	534-85-0	184.236	nd(w)	79.5	313			sl H_2O, lig; s ace, bz, chl
8833	*N*-Phenyl-1,4-benzenediamine	*p*-Aminodiphenylamine	$C_{12}H_{12}N_2$	101-54-2	184.236	nd(al)	66	354			sl H_2O, chl; vs EtOH; s eth, lig
8834	α-Phenylbenzeneethanamine		$C_{14}H_{15}N$	25611-78-3	197.276			311; 175^{15}	1.031^{15}		vs eth, EtOH
8835	α-Phenylbenzeneethanol		$C_{14}H_{14}O$	614-29-9	198.260	nd (peth-bz)	67	177^{15}	1.0360^{70}		
8836	α-Phenylbenzenemethanamine	Benzhydrylamine	$C_{13}H_{13}N$	91-00-9	183.249	hex pl	34	304; 176^{23}	1.0633^{20}	1.5963	sl H_2O; s ace
8837	α-Phenylbenzenemethanimine		$C_{13}H_{11}N$	1013-88-3	181.233			282	1.0847^{19}	1.6191^{19}	vs eth
8838	β-Phenylbenzenepropanoic acid		$C_{15}H_{14}O_2$	606-83-7	226.271	nd (dil al)	156.0				sl H_2O; vs EtOH; s eth, ace
8839	2-Phenylbenzimidazole	Phenzidole	$C_{13}H_{10}N_2$	716-79-0	194.231	pl (HOAc) (al-w) nd (bz, w)	293				sl H_2O, bz; s EtOH, chl, HOAc
8840	Phenyl benzoate		$C_{13}H_{10}O_2$	93-99-2	198.217	mcl pr (eth-al)	71	314	1.235^{20}		i H_2O; s EtOH, eth, chl
8841	2-Phenylbenzoic acid		$C_{13}H_{10}O_2$	947-84-2	198.217	lf (dil al)	114.3	343.5			i H_2O; vs EtOH, bz, HOAc
8842	4-Phenylbenzoic acid		$C_{13}H_{10}O_2$	92-92-2	198.217	nd (bz, al)	228	sub			i H_2O; s EtOH, eth, bz
8843	2-Phenyl-4*H*-1-benzopyran-4-one	Flavone	$C_{15}H_{10}O_2$	525-82-6	222.239	nd (lig), cry (30% al)	100				i H_2O; s EtOH, eth, ace, bz
8844	3-Phenyl-4*H*-1-benzopyran-4-one	Isoflavone	$C_{15}H_{10}O_2$	574-12-9	222.239		148				
8845	2-Phenylbenzothiazole		$C_{13}H_9NS$	883-93-2	211.282	nd (dil al)	115	371			i H_2O; s EtOH, eth, CS_2
8846	*N*-Phenyl-*N*-benzylbenzenemethanamine		$C_{20}H_{19}N$	91-73-6	273.372		69	226^{10}	1.0444^{80}	1.6065^{80}	i H_2O; sl EtOH, HOAc; s eth, bz
8847	Phenyl biguanide	*N*-Phenylimidodicarbonimidic diamide	$C_8H_{11}N_5$	102-02-3	177.207		143				
8848	2-Phenyl-1,3-butadiene		$C_{10}H_{10}$	2288-18-8	130.186			60^{17}	0.925^{20}	1.5489^{20}	i H_2O; s eth, bz, chl
8849	*N*-Phenylbutanamide		$C_{10}H_{13}NO$	1129-50-6	163.216	mcl pr (al, bz, eth)	97	189^{15}	1.134^{25}		i H_2O; vs EtOH, eth; sl chl
8850	Phenylbutanedioic acid, (±)		$C_{10}H_{10}O_4$	10424-29-0	194.184	lf or nd (w)	168	dec			sl H_2O, chl; vs EtOH, eth, ace; i bz
8851	1-Phenyl-1,3-butanedione		$C_{10}H_{10}O_2$	93-91-4	162.185	pr	56	261.5	1.0599^{74}	1.5678^{78}	i H_2O; s eth; sl chl
8852	Phenyl butanoate	Phenyl butyrate	$C_{10}H_{12}O_2$	4346-18-3	164.201			225	1.0382^{15}		i H_2O; s EtOH, eth
8853	1-Phenyl-1-butanone		$C_{10}H_{12}O$	495-40-9	148.201		12	228.5	0.988^{20}	1.5203^{20}	i H_2O; msc EtOH, eth; vs ace; s ctc
8854	1-Phenyl-2-butanone		$C_{10}H_{12}O$	1007-32-5	148.201			228; 111^{16}	0.9877^{20}		i H_2O; s EtOH, ctc; msc eth; vs ace
8855	4-Phenyl-2-butanone		$C_{10}H_{12}O$	2550-26-7	148.201	liq	-13	233.5	0.9849^{22}	1.511^{22}	i H_2O; s EtOH, eth, ctc; vs ace
8856	Phenylbutazone		$C_{19}H_{20}N_2O_2$	50-33-9	308.374		105				
8857	2-Phenyl-1-butene	α-Ethylstyrene	$C_{10}H_{12}$	2039-93-2	132.202			182	0.887^{25}	1.5288^{20}	
8858	1-Phenyl-2-buten-1-one		$C_{10}H_{10}O$	495-41-0	146.185		20.5	111^{9}	1.025^{15}	1.5626^{18}	
8859	*trans*-4-Phenyl-3-buten-2-one	Benzilideneacetone	$C_{10}H_{10}O$	1896-62-4	146.185	pl	41.5	261	1.0097^{45}	1.5836^{45}	i H_2O; vs EtOH; s eth, ace, bz; sl peth
8860	4-Phenyl-3-butyn-2-one		$C_{10}H_8O$	1817-57-8	144.170		4.5	79^{2}	1.0215^{20}	1.5762^{20}	
8861	Phenyl chloroacetate		$C_8H_7ClO_2$	620-73-5	170.594	nd or pl (al)	44.5	232.5	1.2202^{44}	1.5146^{44}	i H_2O; vs EtOH, eth
8862	Phenyl chloroformate		$C_7H_5ClO_2$	1885-14-9	156.567			71^{9}			
8863	4-Phenyl-2-chlorophenol	3-Chloro-(1,1'-biphenyl)-4-ol	$C_{12}H_9ClO$	92-04-6	204.651	wh-ye cry	77	161^{7}			
8864	2-Phenyl-2,5-cyclohexadiene-1,4-dione		$C_{12}H_8O_2$	363-03-1	184.191	ye lf (peth, al)	114				sl H_2O; s EtOH, bz, peth; vs chl
8865	4-Phenylcyclohexanone		$C_{12}H_{14}O$	4894-75-1	174.238	cry (peth)	79	158^{12}			
8866	1-(1-Phenylcyclohexyl)piperidine	Phencyclidine	$C_{17}H_{25}N$	77-10-1	243.388		46.5	$136^{1.0}$			
8867	3-Phenyl-2-cyclopenten-1-one		$C_{11}H_{10}O$	3810-26-2	158.196	liq	-23	234.2	0.9711^{20}	1.5440^{20}	s EtOH, ace, chl; sl eth
8868	*N*-Phenyl-*N,N*-diethanolamine		$C_{10}H_{15}NO_2$	120-07-0	181.232		57	200^{10}	1.201^{60}		vs ace, bz, eth, EtOH
8869	2-Phenyl-1,3-dioxane		$C_{10}H_{12}O_2$	772-01-0	164.201	nd (peth)	41	253	1.6053^{60}		vs EtOH, eth
8870	4-Phenyl-1,3-dioxane		$C_{10}H_{12}O_2$	772-00-9	164.201			247	1.1038^{20}	1.5306^{18}	i H_2O; s os
8871	1-Phenyl-1-dodecanone		$C_{18}H_{28}O$	1674-38-0	260.414		47	201^{9}, 181^{5}	0.8794^{18}	1.4700^{18}	i H_2O; s ace; sl ctc

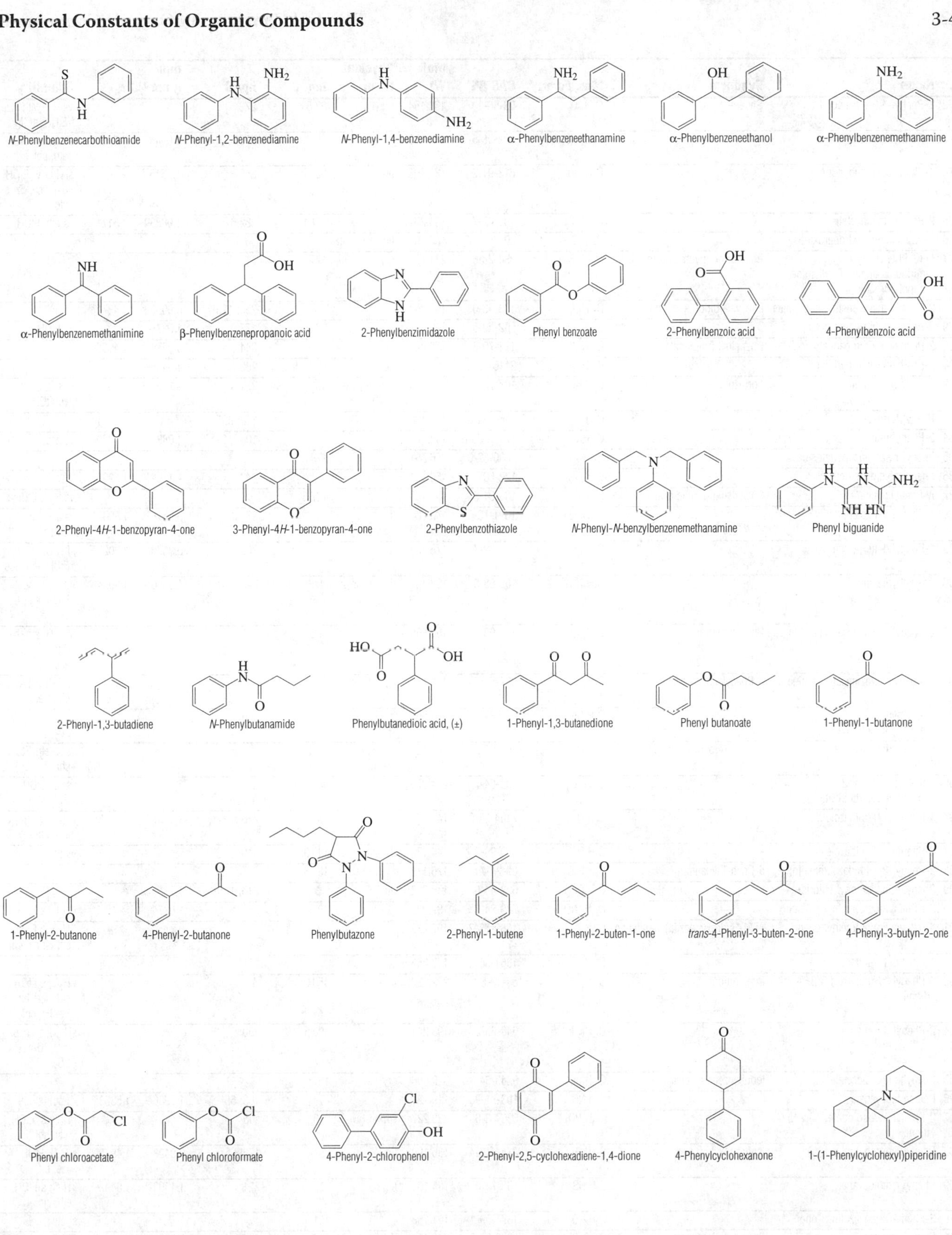

N-Phenylbenzenecarbothioamide
N-Phenyl-1,2-benzenediamine
N-Phenyl-1,4-benzenediamine
α-Phenylbenzeneethanamine
α-Phenylbenzeneethanol
α-Phenylbenzenemethanamine
α-Phenylbenzenemethanimine
β-Phenylbenzenepropanoic acid
2-Phenylbenzimidazole
Phenyl benzoate
2-Phenylbenzoic acid
4-Phenylbenzoic acid
2-Phenyl-4*H*-1-benzopyran-4-one
3-Phenyl-4*H*-1-benzopyran-4-one
2-Phenylbenzothiazole
N-Phenyl-*N*-benzylbenzenemethanamine
Phenyl biguanide
2-Phenyl-1,3-butadiene
N-Phenylbutanamide
Phenylbutanedioic acid, (±)
1-Phenyl-1,3-butanedione
Phenyl butanoate
1-Phenyl-1-butanone
1-Phenyl-2-butanone
4-Phenyl-2-butanone
Phenylbutazone
2-Phenyl-1-butene
1-Phenyl-2-buten-1-one
trans-4-Phenyl-3-buten-2-one
4-Phenyl-3-butyn-2-one
Phenyl chloroacetate
Phenyl chloroformate
4-Phenyl-2-chlorophenol
2-Phenyl-2,5-cyclohexadiene-1,4-dione
4-Phenylcyclohexanone
1-(1-Phenylcyclohexyl)piperidine
3-Phenyl-2-cyclopenten-1-one
N-Phenyl-*N*,*N*-diethanolamine
2-Phenyl-1,3-dioxane
4-Phenyl-1,3-dioxane
1-Phenyl-1-dodecanone

No.	Name	Synonym	Mol. Form.	CAS RN	Mol. Wt.	Physical Form	mp/°C	bp/°C	den/ g cm^{-3}	n_D	Solubility
8872	1-Phenyl-1,2-ethanediol	Styrene glycol	$C_8H_{10}O_2$	93-56-1	138.164	nd (lig)	67.5	273			vs H_2O, eth, bz, EtOH; sl lig
8873	*N*-Phenylethanolamine		$C_8H_{11}NO$	122-98-5	137.179			279.5; 150^{10}	1.0945^{20}	1.5760^{20}	sl H_2O; vs EtOH, eth, chl
8874	1-Phenylethanone oxime		C_8H_9NO	613-91-2	135.163	nd (w)	60	245	1.0515^{78}		sl H_2O; vs EtOH, eth, ace, bz; s ctc
8875	2-Phenylethyl acetate		$C_{10}H_{12}O_2$	103-45-7	164.201	liq	-31.1	232.6	1.0883^{20}	1.5171^{20}	vs eth, EtOH
8876	1-Phenylethyl hydroperoxide		$C_8H_{10}O_2$	3071-32-7	138.164	liq		$50^{0.01}$			
8877	*N*-(2-Phenylethyl) imidodicarbonimidic diamide, monohydrochloride	Phenformin hydrochloride	$C_{10}H_{16}ClN_5$	834-28-6	241.721	cry	177.3				s H_2O
8878	2-Phenylethyl 2-methylpropanoate	Benzylcarbinol isobutyrate	$C_{12}H_{16}O_2$	103-48-0	192.254			250; 123^{15}	0.9950^{15}	1.4871^{20}	
8879	2-Phenylethyl phenylacetate		$C_{16}H_{16}O_2$	102-20-5	240.297		26.5	$177^{4.5}$	1.077^{25}		vs EtOH
8880	2-Phenylethyl propanoate	Phenethyl propionate	$C_{11}H_{14}O_2$	122-70-3	178.228	liq		244	1.02^{25}	1.4950^{20}	
8881	2-(2-Phenylethyl)pyridine		$C_{13}H_{13}N$	2116-62-3	183.249	liq	-1.5	289	1.0465^{0}		
8882	*N*-Phenylformamide	Formanilide	C_7H_7NO	103-70-8	121.137	mcl pr (lig-xyl)	46	271	1.1186^{50}		s H_2O, eth, bz; vs EtOH
8883	Phenyl formate		$C_7H_6O_2$	1864-94-4	122.122	liq		178; 82^{15}			
8884	2-Phenylfuran		$C_{10}H_8O$	17113-33-6	144.170			108^{18}, 82^{5}	1.083^{20}	1.5920^{20}	vs ace, bz
8885	Phenyl α-*D*-glucopyranoside		$C_{12}H_{16}O_6$	4630-62-0	256.251		174				
8886	Phenyl glycidyl ether		$C_9H_{10}O_2$	122-60-1	150.174			243	1.1109^{21}	1.5307^{21}	
8887	*N*-Phenylglycine	Phenylaminoacetic acid	$C_8H_9NO_2$	103-01-5	151.163		127.5				vs H_2O, EtOH
8888	1-Phenyl-1-heptanone		$C_{13}H_{18}O$	1671-75-6	190.281	lf	16.4	283.3	0.9516^{20}	1.5060^{20}	vs ace, eth, EtOH
8889	1-Phenyl-1-hexanone		$C_{12}H_{16}O$	942-92-7	176.254	fl	27	265	0.9576^{20}	1.5027^{25}	sl H_2O, ctc; s EtOH, eth, ace
8890	Phenylhydrazine		$C_6H_8N_2$	100-63-0	108.141	mcl pr or pl	20.6	243.5	1.0986^{20}	1.6084^{10}	s H_2O; msc EtOH, eth, bz; vs ace
8891	2-Phenylhydrazinecarboxamide	Phenicarbazide	$C_7H_9N_3O$	103-03-7	151.165		172				sl H_2O, eth, bz, lig; s EtOH, ace
8892	*N*-Phenylhydrazinecarboxamide	4-Phenylsemicarbazide	$C_7H_9N_3O$	537-47-3	151.165	nd (bz), pl (w)	128				sl H_2O; vs EtOH, chl; i eth
8893	Phenylhydrazine monohydrochloride		$C_6H_9ClN_2$	59-88-1	144.601	lf (al)	244 dec	sub			vs H_2O, EtOH
8894	Phenylhydroxylamine	*N*-Hydroxybenzenamine	C_6H_7NO	100-65-2	109.126	nd (w, bz, peth)	83.5				vs bz, eth, EtOH, chl
8895	Phenyl 1-hydroxy-2-naphthalenecarboxylate		$C_{17}H_{12}O_3$	132-54-7	264.275		96				vs bz, EtOH
8896	1-Phenyl-1*H*-imidazole		$C_9H_8N_2$	7164-98-9	144.173		13	276	1.1397^{15}	1.6025^{25}	i H_2O; vs eth, ace, chl
8897	2-Phenyl-1*H*-imidazole		$C_9H_8N_2$	670-96-2	144.173	lf (bz)	149.3	340			vs EtOH
8898	5-Phenyl-2,4-imidazolidinedione	5-Phenylhydantoin	$C_9H_8N_2O_2$	89-24-7	176.172		184.5				
8899	Phenylimidocarbonyl chloride		$C_7H_5Cl_2N$	622-44-6	174.028	liq		210; 105^{30}	1.28^{15}		
8900	2-[(Phenylimino)methyl]phenol		$C_{13}H_{11}NO$	779-84-0	197.232		49.5		1.087^{25}		i H_2O; s EtOH
8901	4-[(Phenylimino)methyl]phenol	*N*-(4-Hydroxybenzilidene)aniline	$C_{13}H_{11}NO$	1689-73-2	197.232		196.0				i H_2O; s EtOH, eth; sl bz, chl
8902	1-Phenyl-1*H*-indene		$C_{15}H_{12}$	1961-96-2	192.256	oil		158^{7}			
8903	2-Phenyl-1*H*-indene-1,3(2*H*)-dione	Phenindione	$C_{15}H_{10}O_2$	83-12-5	222.239	lf (al, bz)	150				i H_2O; s EtOH, eth, ace, bz, MeOH, chl
8904	2-Phenyl-1*H*-indole		$C_{14}H_{11}N$	948-65-2	193.244		190.5	250^{10}			sl H_2O; s eth, bz, chl, HOAc, CS_2
8905	Phenyliodine diacetate	Iodobenzene diacetate	$C_{10}H_{11}IO_4$	3240-34-4	322.096	cry	161				
8906	Phenyl isocyanate		C_7H_5NO	103-71-9	119.121			163; 55^{13}	1.0956^{20}	1.5368^{20}	vs eth; sl chl
8907	2-Phenyl-1*H*-isoindole-1,3(2*H*)-dione		$C_{14}H_9NO_2$	520-03-6	223.227	wh nd (al)	210	sub			i H_2O; sl EtOH; msc chl
8908	Phenyl isopropyl ether	Isopropoxybenzene	$C_9H_{12}O$	2741-16-4	136.190	liq	-33	176.8	0.9408^{25}	1.4975^{20}	s H_2O, EtOH, ace, bz
8909	Phenyl isothiocyanate		C_7H_5NS	103-72-0	135.187	liq	-21	221	1.1303^{20}	1.6492^{23}	i H_2O; s EtOH, eth, ctc
8910	3-Phenyl-2-isoxazolin-5-one		$C_9H_7NO_2$	1076-59-1	161.158		151				sl chl
8911	Phenyl laurate	Phenyl dodecanoate	$C_{18}H_{28}O_2$	4228-00-6	276.414	lf (al)	24.5	210^{15}	0.9354^{30}		vs ace, eth, EtOH
8912	Phenylmagnesium chloride	Chlorophenylmagnesium	C_6H_5ClMg	100-59-4	136.862	cry					reac H_2O; s thf, eth
8913	Phenylmercuric chloride	Chlorophenylmercury	C_6H_5ClHg	100-56-1	313.15	pl (bz)	251				i H_2O; sl EtOH, bz
8914	Phenylmercuric nitrate		$C_6H_5HgNO_3$	55-68-5	339.70		≈181				
8915	4-(Phenylmethoxy)benzaldehyde		$C_{14}H_{12}O_2$	4397-53-9	212.244		73	217^{13}			

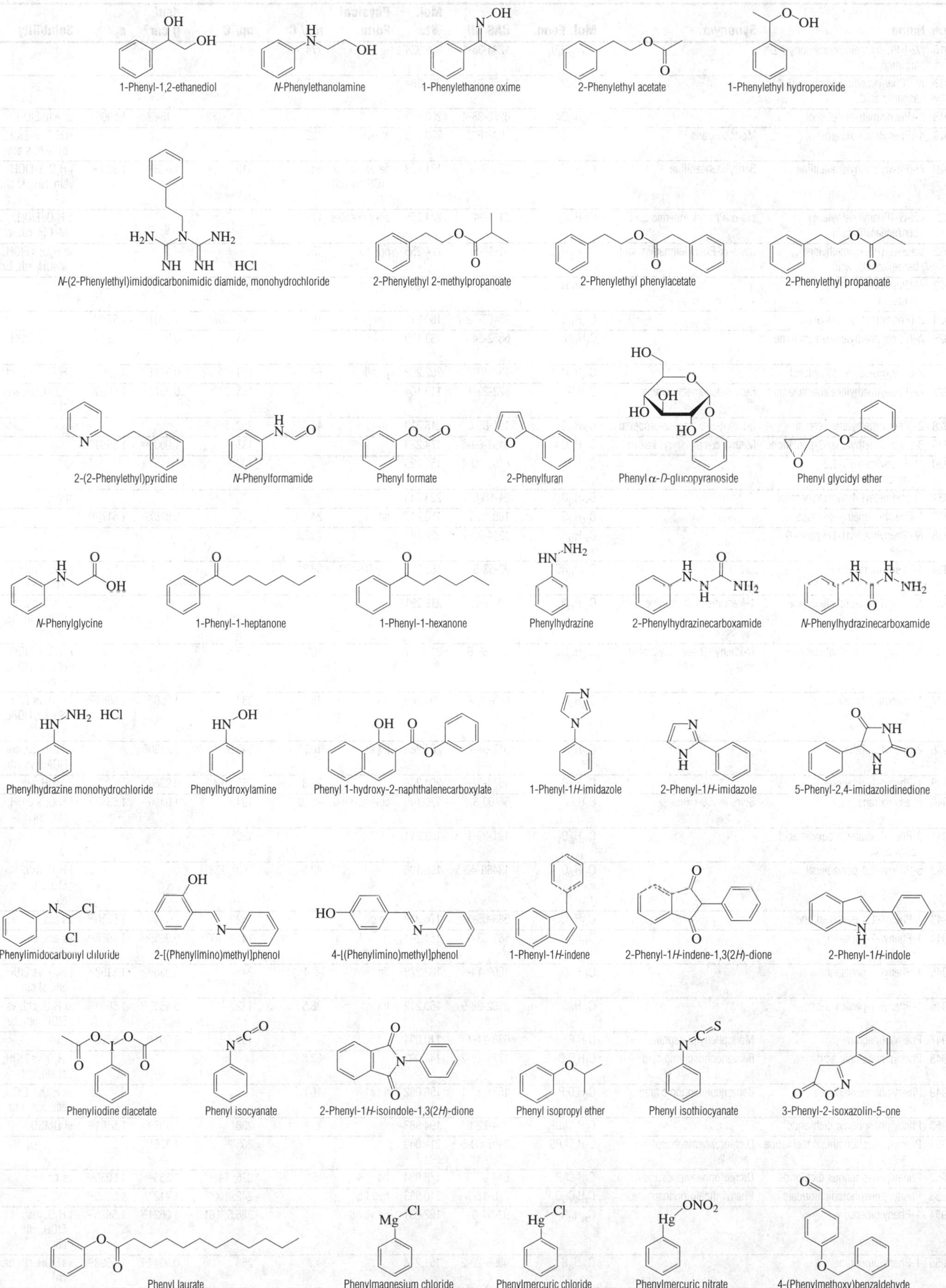
1-Phenyl-1,2-ethanediol
N-Phenylethanolamine
1-Phenylethanone oxime
2-Phenylethyl acetate
1-Phenylethyl hydroperoxide
N-(2-Phenylethyl)imidodicarbonimidic diamide, monohydrochloride
2-Phenylethyl 2-methylpropanoate
2-Phenylethyl phenylacetate
2-Phenylethyl propanoate
2-(2-Phenylethyl)pyridine
N-Phenylformamide
Phenyl formate
2-Phenylfuran
Phenyl α-D-glucopyranoside
Phenyl glycidyl ether
N-Phenylglycine
1-Phenyl-1-heptanone
1-Phenyl-1-hexanone
Phenylhydrazine
2-Phenylhydrazinecarboxamide
N-Phenylhydrazinecarboxamide
Phenylhydrazine monohydrochloride
Phenylhydroxylamine
Phenyl 1-hydroxy-2-naphthalenecarboxylate
1-Phenyl-1H-imidazole
2-Phenyl-1H-imidazole
5-Phenyl-2,4-imidazolidinedione
Phenylimidocarbonyl chloride
2-[(Phenylimino)methyl]phenol
4-[(Phenylimino)methyl]phenol
1-Phenyl-1H-indene
2-Phenyl-1H-indene-1,3(2H)-dione
2-Phenyl-1H-indole
Phenyliodine diacetate
Phenyl isocyanate
2-Phenyl-1H-isoindole-1,3(2H)-dione
Phenyl isopropyl ether
Phenyl isothiocyanate
3-Phenyl-2-isoxazolin-5-one
Phenyl laurate
Phenylmagnesium chloride
Phenylmercuric chloride
Phenylmercuric nitrate
4-(Phenylmethoxy)benzaldehyde

No.	Name	Synonym	Mol. Form.	CAS RN	Mol. Wt.	Physical Form	mp/°C	bp/°C	den/ g cm^{-3}	n_D	Solubility
8916	*N*2-[(Phenylmethoxy)carbonyl]-*L*-arginine		$C_{14}H_{20}N_4O_4$	1234-35-1	308.334		174				
8917	*N*-[(Phenylmethoxy)carbonyl]-*L*-aspartic acid		$C_{12}H_{13}NO_6$	1152-61-0	267.234		117.0				
8918	2-(Phenylmethoxy)phenol		$C_{13}H_{12}O_2$	6272-38-4	200.233			205^{20}, 173^{13}	1.154^{22}	1.5906^{18}	vs eth, EtOH
8919	4-(Phenylmethoxy)phenol	Monobenzone	$C_{13}H_{12}O_2$	103-16-2	200.233	pl (w)	122				sl H_2O; vs EtOH, bz, eth; s ace
8920	*N*-(Phenylmethylene)aniline	Benzylideneaniline	$C_{13}H_{11}N$	538-51-2	181.233	pa ye nd (CS_2) pl (dil al)	54	310	1.038^{55}	1.600^{100}	i H_2O; s EtOH, eth, NH_3; sl chl
8921	*cis*-α-(Phenylmethylene) benzeneacetic acid	*cis*-α-Phenylcinnamic acid	$C_{15}H_{12}O_2$	91-47-4	224.255	silky needles	174				s H_2O, EtOH, MeOH, eth, bz
8922	*trans*-α-(Phenylmethylene) benzeneacetic acid	*trans*-α-Phenylcinnamic acid	$C_{15}H_{12}O_2$	91-48-5	224.255	prisms	138				vs H_2O; s EtOH, MeOH, eth, bz
8923	*N*-(Phenylmethylene) benzenemethanamine		$C_{14}H_{13}N$	780-25-6	195.260			205^{20}			
8924	2-(Phenylmethylene)butanal		$C_{11}H_{12}O$	28467-92-7	160.212		18	243; 157^5	1.0201^{22}	1.578^{20}	
8925	*N*-(Phenylmethylene)ethanamine		$C_9H_{11}N$	6852-54-6	133.190			195	0.937^{20}	1.5378^{15}	i H_2O; s EtOH, eth
8926	2-(Phenylmethylene)heptanal		$C_{14}H_{18}O$	122-40-7	202.292	ye oil	80	174^{20}	0.9711^{20}	1.5381^{20}	i H_2O; s ace, ctc
8927	*N*-(Phenylmethylene)methanamine	Benzylidenemethylamine	C_8H_9N	622-29-7	119.164			185; 92^{34}	0.9671^{14}	1.5526^{20}	s EtOH, eth, ace, chl
8928	2-(Phenylmethylene)octanal	2-Hexyl-3-phenyl-2-propenal	$C_{15}H_{20}O$	101-86-0	216.319	liq	4	252; 169^{20}			
8929	3-(Phenylmethylene)-2-pentanone	Methyl α-ethylstyryl ketone	$C_{12}H_{14}O$	3437-89-6	174.238			137^{12}	1.0005^{22}	1.5650^{22}	
8930	*N*-(Phenylmethyl)-1,2-ethanediamine		$C_9H_{14}N_2$	4152-09-4	150.220			130^{11}			
8931	Phenylmethyl 4-hydroxybenzoate		$C_{14}H_{12}O_3$	94-18-8	228.243						sl chl
8932	1-Phenyl-2-methyl-2-propanol		$C_{10}H_{14}O$	100-86-7	150.217	nd	24	215	0.9787^{16}	1.5173^{16}	
8933	*N*-(Phenylmethyl)-1*H*-purin-6-amine		$C_{12}H_{11}N_5$	1214-39-7	225.249		232.8				
8934	4-Phenylmorpholine		$C_{10}H_{13}NO$	92-53-5	163.216	cry (al-eth)	58.3				i H_2O, EtOH; vs eth
8935	*N*-Phenyl-1-naphthalenamine	1-Naphthylphenylamine	$C_{16}H_{13}N$	90-30-2	219.281		61				sl H_2O, ctc; s EtOH, eth, bz, HOAc
8936	*N*-Phenyl-2-naphthalenamine	*N*-Phenyl-β-naphthylamine	$C_{16}H_{13}N$	135-88-6	219.281		108	395.5			i H_2O; s EtOH, eth, bz, HOAc; sl chl
8937	1-Phenylnaphthalene		$C_{16}H_{12}$	605-02-7	204.266	cry	45	334	1.096^{20}	1.6664^{20}	i H_2O; vs EtOH, eth, bz, HOAc; s ctc
8938	2-Phenylnaphthalene		$C_{16}H_{12}$	612-94-2	204.266	lf (al)	103.5	345.5	1.2180^{20}		s EtOH, bz, chl, HOAc; vs eth
8939	1-Phenyl-1-octanone		$C_{14}H_{20}O$	1674-37-9	204.308		22.8	285; 164^{15}	0.9360^{30}		s EtOH, eth
8940	Phenyloxirane	Styrene-7,8-oxide	C_8H_8O	96-09-3	120.149	colorless liq	-35.6	194.1	1.0490^{25}	1.5342^{20}	i H_2O; s EtOH, eth, chl
8941	3-Phenyloxiranecarboxylic acid, ethyl ester		$C_{11}H_{12}O_3$	121-39-1	192.211			136^5			
8942	5-Phenyl-2,4-pentadienal		$C_{11}H_{10}O$	13466-40-5	158.196		42.5	160^3, $133^{1.0}$			i H_2O; msc EtOH, bz; vs eth
8943	1-Phenyl-1,4-pentanedione		$C_{11}H_{12}O_2$	583-05-1	176.212	ye oil		162^{12}		1.5250^{30}	vs ace
8944	1-Phenyl-1-pentanol		$C_{11}H_{16}O$	583-03-9	164.244			141^{25}, 102^3	0.9655^{20}	1.4086^{25}	vs ace, eth, EtOH
8945	1-Phenyl-1-pentanone		$C_{11}H_{14}O$	1009-14-9	162.228	liq	-9.4	245	0.986^{20}	1.5158^{20}	i H_2O; vs EtOH, eth; sl ctc
8946	1-Phenyl-1-penten-3-one		$C_{11}H_{12}O$	3152-68-9	160.212	lf (lig)	38.5	142^{12}	0.8697^{20}	1.5684^{20}	sl H_2O, chl; vs EtOH, eth, bz
8947	Phenylphosphine	Monophenylphosphine	C_6H_7P	638-21-1	110.094			160.5	1.001^{15}	1.5796^{20}	
8948	Phenylphosphinic acid	Benzenephosphinic acid	$C_6H_7O_2P$	1779-48-2	142.093		83.8				s H_2O; vs EtOH; sl eth, chl
8949	Phenylphosphonic acid	Benzenephosphonic acid	$C_6H_7O_3P$	1571-33-1	158.092	lf (w)	160				vs H_2O; s EtOH, eth, ace; i bz
8950	Phenylphosphonic dichloride		$C_6H_5Cl_2OP$	824-72-6	194.983		1	258	1.197^{25}	1.5581^{25}	sl DMSO
8951	Phenylphosphonothioic dichloride	Dichlorophenylphosphine sulfide	$C_6H_5Cl_2PS$	3497-00-5	211.049			205^{130}	1.376^{13}		
8952	Phenylphosphonous dichloride	Dichlorophenylphosphine	$C_6H_5Cl_2P$	644-97-3	178.984	liq	-51	225; 142^{57}	1.356^{20}	1.6030^{20}	vs bz
8953	Phenyl phosphorodichloridate	Phenyl dichlorophosphate	$C_6H_5Cl_2O_2P$	770-12-7	210.983	hyg liq		242; 100^5	1.412^{20}	1.5230^{20}	
8954	1-Phenylpiperazine		$C_{10}H_{14}N_2$	92-54-6	162.231	pa ye oil		286.5; 161^{15}	1.0621^{20}	1.5875^{20}	i H_2O; msc EtOH, eth; s chl
8955	1-Phenylpiperidine		$C_{11}H_{15}N$	4096-20-2	161.244		4.7	258	0.9944^{25}	1.5598^{25}	vs EtOH, eth, bz, chl
8956	4-Phenylpiperidine		$C_{11}H_{15}N$	771-99-3	161.244		60.5	257	0.9996^{16}		s chl

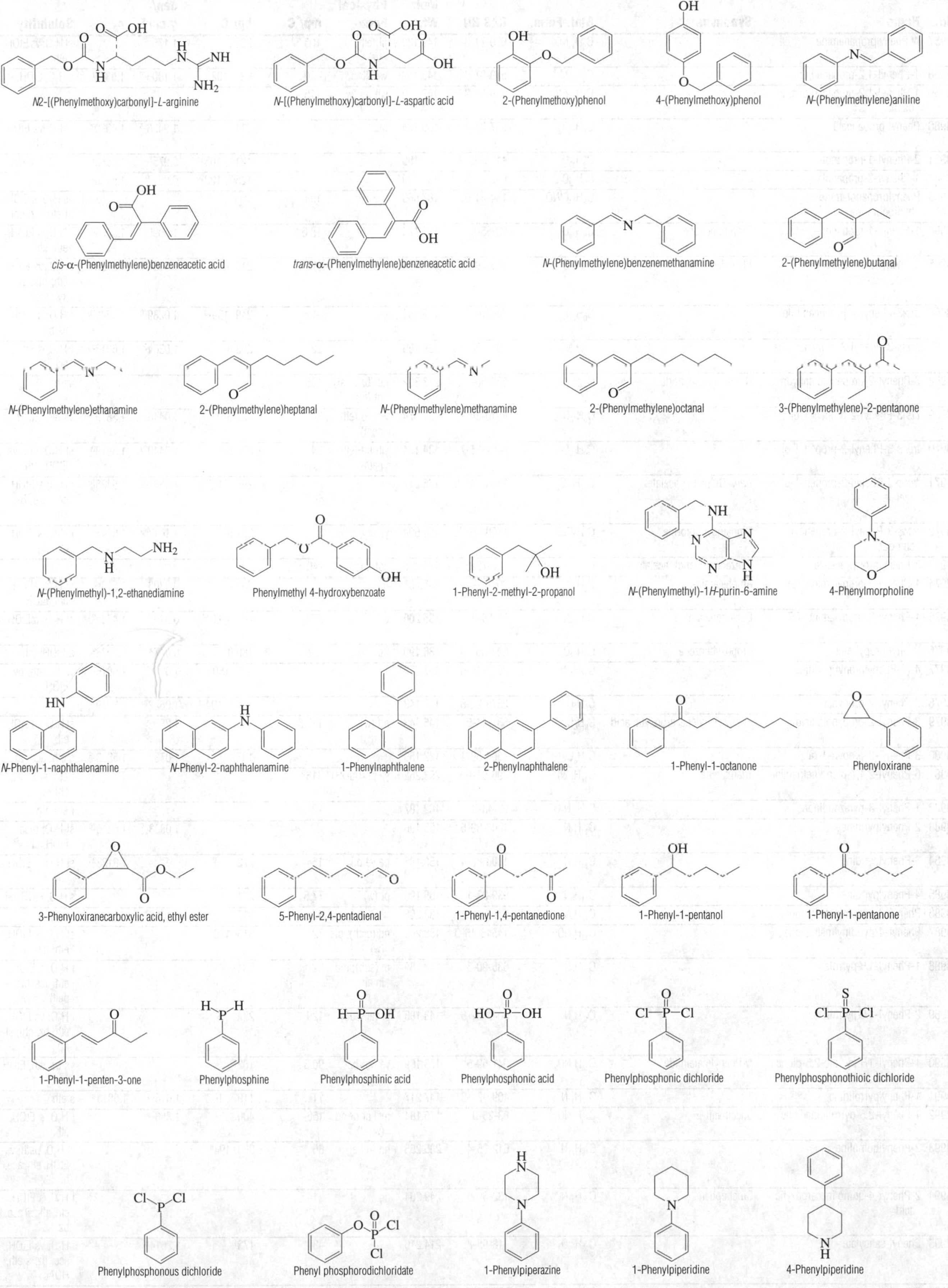
N2-[(Phenylmethoxy)carbonyl]-L-arginine
N-[(Phenylmethoxy)carbonyl]-L-aspartic acid
2-(Phenylmethoxy)phenol
4-(Phenylmethoxy)phenol
N-(Phenylmethylene)aniline
cis-α-(Phenylmethylene)benzeneacetic acid
trans-α-(Phenylmethylene)benzeneacetic acid
N-(Phenylmethylene)benzenemethanamine
2-(Phenylmethylene)butanal
N-(Phenylmethylene)ethanamine
2-(Phenylmethylene)heptanal
N-(Phenylmethylene)methanamine
2-(Phenylmethylene)octanal
3-(Phenylmethylene)-2-pentanone
N-(Phenylmethyl)-1,2-ethanediamine
Phenylmethyl 4-hydroxybenzoate
1-Phenyl-2-methyl-2-propanol
N-(Phenylmethyl)-1H-purin-6-amine
4-Phenylmorpholine
N-Phenyl-1-naphthalenamine
N-Phenyl-2-naphthalenamine
1-Phenylnaphthalene
2-Phenylnaphthalene
1-Phenyl-1-octanone
Phenyloxirane
3-Phenyloxiranecarboxylic acid, ethyl ester
5-Phenyl-2,4-pentadienal
1-Phenyl-1,4-pentanedione
1-Phenyl-1-pentanol
1-Phenyl-1-pentanone
1-Phenyl-1-penten-3-one
Phenylphosphine
Phenylphosphinic acid
Phenylphosphonic acid
Phenylphosphonic dichloride
Phenylphosphonothioic dichloride
Phenylphosphonous dichloride
Phenyl phosphorodichloridate
1-Phenylpiperazine
1-Phenylpiperidine
4-Phenylpiperidine

No.	Name	Synonym	Mol. Form.	CAS RN	Mol. Wt.	Physical Form	mp/°C	bp/°C	den/ g cm^{-3}	n_D	Solubility
8957	*N*-Phenylpropanamide		$C_9H_{11}NO$	620-71-3	149.189	pl (eth, al, bz)	105.5	222.2	1.175^{25}		sl H_2O; vs EtOH, eth
8958	1-Phenyl-1,2-propanedione		$C_9H_8O_2$	579-07-7	148.159	ye oil	<20	222; 102^{12}	1.1006^{20}	1.537^{10}	s H_2O, EtOH, eth
8959	1-Phenyl-1,2-propanedione, 2-oxime		$C_9H_9NO_2$	119-51-7	163.173	wh nd (w)	115				
8960	Phenyl propanoate		$C_9H_{10}O_2$	637-27-4	150.174	pr	20	211	1.0436^{25}	1.4980^{20}	i H_2O; vs EtOH, eth; s bz
8961	2-Phenyl-1-propanol		$C_9H_{12}O$	1123-85-9	136.190			121^{26}, 105^{11}	0.975^{25}	1.5582^{2}	i H_2O; s EtOH
8962	1-Phenyl-2-propanol		$C_9H_{12}O$	698-87-3	136.190			125^{25}, 120^{20}	0.991^{20}	1.5190^{20}	
8963	Phenylpropanolamine hydrochloride		$C_9H_{14}ClNO$	154-41-6	187.666		194				vs H_2O; s EtOH; i eth, bz, chl
8964	1-Phenyl-1-propanone	Propiophenone	$C_9H_{10}O$	93-55-0	134.174		18.6	217.5	1.0096^{20}	1.5269^{20}	i H_2O; s EtOH, eth, chl
8965	1-Phenyl-2-propanone	Phenylacetone	$C_9H_{10}O$	103-79-7	134.174	liq	-15	216.5	1.0157^{20}	1.5168^{20}	i H_2O; vs EtOH, eth; msc bz, xyl; s chl
8966	*cis*-3-Phenyl-2-propenenitrile		C_9H_7N	24840-05-9	129.159	liq	-4.4	249; 139^{30}	1.0289^{20}	1.5843^{20}	i H_2O; s EtOH; vs bz
8967	*trans*-3-Phenyl-2-propenenitrile		C_9H_7N	1885-38-7	129.159		22	263.8	1.0304^{20}	1.6013^{20}	i H_2O; s EtOH, ace, ctc
8968	3-Phenyl-2-propenoic anhydride	Cinnamic anhydride	$C_{18}H_{14}O_3$	538-56-7	278.302	nd (bz or al) pt (al)	136				vs bz
8969	*cis*-3-Phenyl-2-propen-1-ol		$C_9H_{10}O$	4510-34-3	134.174	wh nd (eth-peth)	34	257.5	1.0440^{20}	1.5819^{20}	vs eth, EtOH
8970	*trans*-3-Phenyl-2-propen-1-ol		$C_9H_{10}O$	4407-36-7	134.174	wh nd (eth-peth)	34	257.5	1.0440^{20}	1.5819^{20}	sl H_2O, chl; vs EtOH, eth
8971	*trans*-3-Phenyl-2-propen-1-ol acetate	*trans*-Cinnamyl acetate	$C_{11}H_{12}O_2$	21040-45-9	176.212			265; 145^{15}	1.0567^{20}	1.5425^{20}	i H_2O; s EtOH, eth, ace, bz, chl
8972	*trans*-3-Phenyl-2-propenoyl chloride	Cinnamoyl chloride	C_9H_7ClO	17082-09-6	166.604	ye cry	37.5	257.5	1.1617^{45}	1.614^{42}	i H_2O; s EtOH, ctc, lig
8973	3-Phenylpropyl acetate	Benzenepropanol, acetate	$C_{11}H_{14}O_2$	122-72-5	178.228	col liq	-40	69^{1}			
8974	1-Phenyl-2-propylamine, (±)	Amphetamine	$C_9H_{13}N$	300-62-9	135.206	oil		203	0.9306^{25}	1.518^{26}	sl H_2O, eth; s chl, EtOH
8975	1-Phenyl-2-propylamine, (*S*)	Dexamphetamine	$C_9H_{13}N$	51-64-9	135.206	oil	27.5	203.5; 80^{12}	0.949^{15}	1.4704^{20}	sl H_2O; s EtOH, eth
8976	Phenyl propyl ether	Propoxybenzene	$C_9H_{12}O$	622-85-5	136.190	liq	-27	189.9	0.9474^{20}	1.5014^{20}	s EtOH, eth
8977	4-(3-Phenylpropyl)pyridine		$C_{14}H_{15}N$	2057-49-0	197.276			322; 150^{5}	1.024^{25}	1.5616^{25}	vs bz, eth, py, EtOH
8978	3-Phenyl-2-propynal		C_9H_6O	2579-22-8	130.143			127^{28}, 104^{11}	1.0622^{20}	1.6079^{12}	
8979	3-Phenyl-2-propynoic acid	Phenylacetylenecarboxylic acid	$C_9H_6O_2$	637-44-5	146.143	nd (w)	137.5		1.28^{20}		sl H_2O; vs EtOH, eth
8980	3-Phenyl-2-propyn-1-ol		C_9H_8O	1504-58-1	132.159			137^{15}	1.078^{20}	1.5873^{28}	s eth, ace, bz
8981	6-Phenyl-2,4,7-pteridinetriamine	Triamterene	$C_{12}H_{11}N_7$	396-01-0	253.262	ye pl (BuOH)	316				i eth; sl EtOH, chl
8982	1-Phenyl-3-pyrazolidinone		$C_9H_{10}N_2O$	92-43-3	162.187		126				i eth, lig
8983	2-Phenylpyridine		$C_{11}H_9N$	1008-89-5	155.196			271	1.0833^{25}	1.6210^{20}	sl H_2O; msc EtOH, eth
8984	3-Phenylpyridine		$C_{11}H_9N$	1008-88-4	155.196	pa ye oil	164	272		1.6123^{25}	sl H_2O; s EtOH, eth
8985	4-Phenylpyridine		$C_{11}H_9N$	939-23-1	155.196	pl (w)	77.5	281			s H_2O, EtOH, eth
8986	Phenyl-2-pyridinylmethanone		$C_{12}H_9NO$	91-02-1	183.205		42	317	1.1556^{20}		s chl
8987	Phenyl-4-pyridinylmethanone		$C_{12}H_9NO$	14548-46-0	183.205	nd (peth), pl (w)	72	315; 170^{10}			sl H_2O; s EtOH, eth, bz
8988	1-Phenyl-1*H*-pyrrole		$C_{10}H_9N$	635-90-5	143.185	pl (sub), red in air	62	234			i H_2O; s EtOH, eth, ace, bz; vs peth
8989	2-Phenyl-1*H*-pyrrole		$C_{10}H_9N$	3042-22-6	143.185	pl (al, sub)	129	272			i H_2O; vs EtOH, eth, bz, chl; sl lig
8990	1-Phenyl-1*H*-pyrrole-2,5-dione	*N*-Phenylmaleimide	$C_{10}H_7NO_2$	941-69-5	173.169	ye nd (bz-lig)	90.5	162^{12}			vs bz, eth, EtOH
8991	1-Phenylpyrrolidine		$C_{10}H_{13}N$	4096-21-3	147.217		11	119^{12}, 102^{5}	1.018^{20}	1.5813^{20}	s eth
8992	1-Phenyl-2,5-pyrrolidinedione	Succinanil	$C_{10}H_9NO_2$	83-25-0	175.184	mcl pr or nd (w, al)	156	400	1.356^{25}		i H_2O; s EtOH, eth
8993	2-Phenylquinoline		$C_{15}H_{11}N$	612-96-4	205.255	nd (dil al)	86	363; 194^{6}			sl H_2O, peth; vs EtOH, eth, ace, bz
8994	2-Phenyl-4-quinolinecarboxylic acid	Cinchophen	$C_{16}H_{11}NO_2$	132-60-5	249.264	nd	214.5				i H_2O; s EtOH, eth, alk; sl ace, bz
8995	Phenyl salicylate		$C_{13}H_{10}O_3$	118-55-8	214.216		43	173^{12}	1.2614^{30}		i H_2O; vs EtOH, ace, bz; s eth, HOAc
8996	Phenylsilane		C_6H_8Si	694-53-1	108.214			119	0.8681^{20}	1.5125^{20}	i H_2O

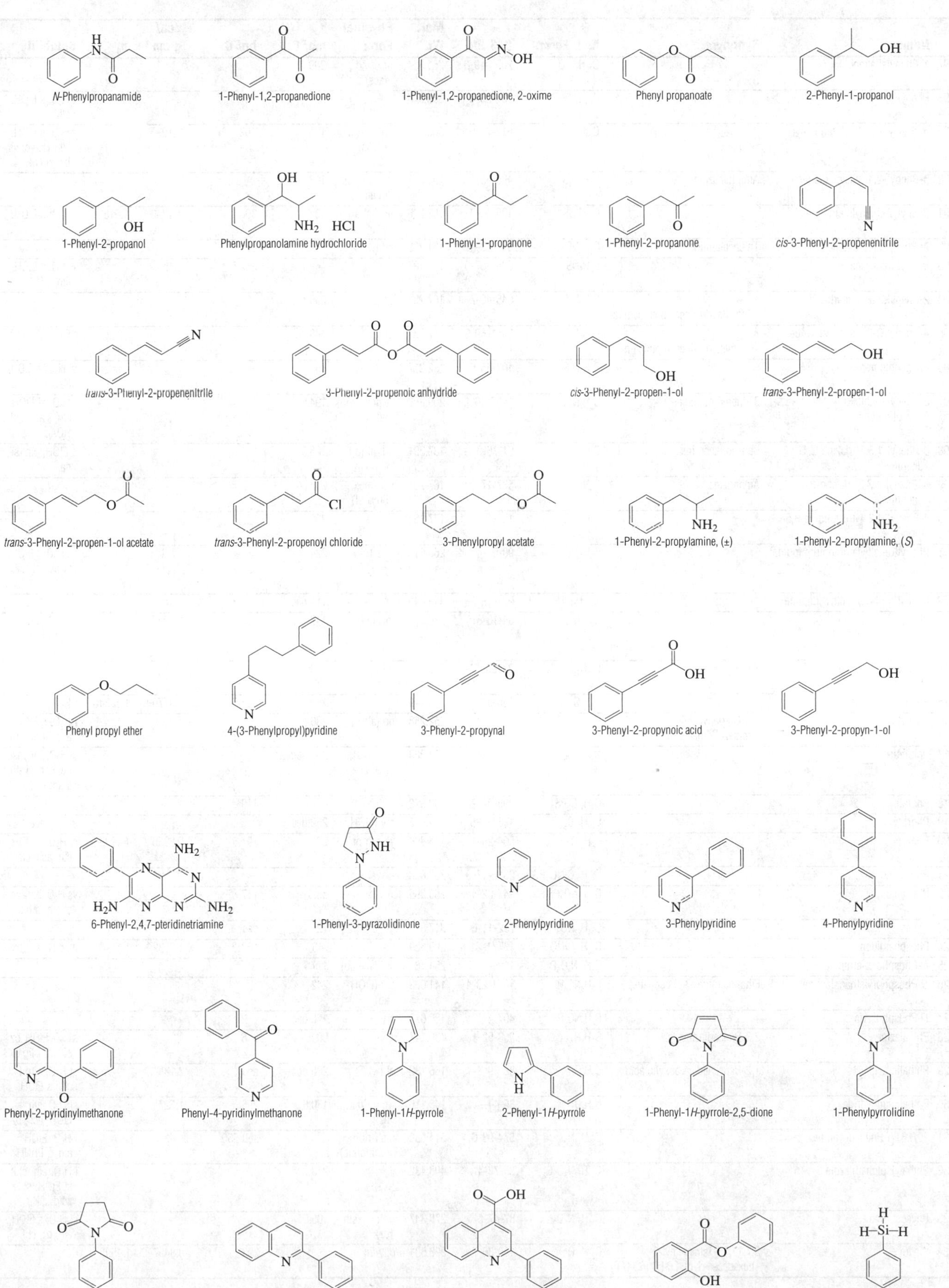

N-Phenylpropanamide
1-Phenyl-1,2-propanedione
1-Phenyl-1,2-propanedione, 2-oxime
Phenyl propanoate
2-Phenyl-1-propanol
1-Phenyl-2-propanol
Phenylpropanolamine hydrochloride
1-Phenyl-1-propanone
1-Phenyl-2-propanone
cis-3-Phenyl-2-propenenitrile
trans-3-Phenyl-2-propenenitrile
3-Phenyl-2-propenoic anhydride
cis-3-Phenyl-2-propen-1-ol
trans-3-Phenyl-2-propen-1-ol
trans-3-Phenyl-2-propen-1-ol acetate
trans-3-Phenyl-2-propenoyl chloride
3-Phenylpropyl acetate
1-Phenyl-2-propylamine, (±)
1-Phenyl-2-propylamine, (S)
Phenyl propyl ether
4-(3-Phenylpropyl)pyridine
3-Phenyl-2-propynal
3-Phenyl-2-propynoic acid
3-Phenyl-2-propyn-1-ol
6-Phenyl-2,4,7-pteridinetriamine
1-Phenyl-3-pyrazolidinone
2-Phenylpyridine
3-Phenylpyridine
4-Phenylpyridine
Phenyl-2-pyridinylmethanone
Phenyl-4-pyridinylmethanone
1-Phenyl-1H-pyrrole
2-Phenyl-1H-pyrrole
1-Phenyl-1H-pyrrole-2,5-dione
1-Phenylpyrrolidine
1-Phenyl-2,5-pyrrolidinedione
2-Phenylquinoline
2-Phenyl-4-quinolinecarboxylic acid
Phenyl salicylate
Phenylsilane

No.	Name	Synonym	Mol. Form.	CAS RN	Mol. Wt.	Physical Form	mp/°C	bp/°C	den/ g cm-3	n_D	Solubility
8997	1-Phenylsilatrane		$C_{12}H_{17}NO_3Si$	2097-19-0	251.354	pr or nd (ace)	209				
8998	Phenyl stearate		$C_{24}H_{40}O_2$	637-55-8	360.574		52	267^{15}			i H_2O; s EtOH, eth
8999	5'-Phenyl-1,1':3',1''-terphenyl		$C_{24}H_{18}$	612-71-5	306.400	orth nd (al or HOAc)	176	462	1.199^{30}		i H_2O; s EtOH, eth, HOAc; vs bz; sl chl
9000	5-Phenyl-2,4-thiazolediamine	Amiphenazole	$C_9H_9N_3S$	490-55-1	191.252	fl (dil al) br in air	163 dec				
9001	Phenyl-2-thienylmethanone		$C_{11}H_8OS$	135-00-2	188.246	nd (dil al)	56.5	300	1.1890^{54}	1.6181^{54}	i H_2O; s EtOH, eth
9002	*N*-Phenylthioacetamide	Thioacetanilide	C_8H_9NS	637-53-6	151.229	nd (w)	75.5	dec			
9003	Phenyl thiocyanate		C_7H_5NS	5285-87-0	135.187			232.5	1.153^{18}		i H_2O; s EtOH, eth
9004	2-Phenylthiosemicarbazide	2-Phenylhydrazinecarbothioamide	$C_7H_9N_3S$	645-48-7	167.231	pr (al)	200 dec				
9005	4-Phenyl-3-thiosemicarbazide	*N*-Phenylhydrazinecarbothioamide	$C_7H_9N_3S$	5351-69-9	167.231	pl (al)	140 dec				i EtOH, lig; sl bz
9006	Phenylthiourea		$C_7H_8N_2S$	103-85-5	152.217	nd (w), pr (al)	154				sl H_2O; s EtOH, NaOH
9007	3-Phenyl-2-thioxo-4-thiazolidinone	3-Phenylrhodanine	$C_9H_7NOS_2$	1457-46-1	209.288	ye pr (HOAc) nd or pr (al)	194.5				i H_2O; sl EtOH, eth; s ace, chl, HOAc
9008	6-Phenyl-1,3,5-triazine-2,4-diamine	Benzoguanamine	$C_9H_9N_5$	91-76-9	187.201	nd, pl (al)	226.5				s EtOH, eth; sl tfa
9009	*N*-Phenyl-1,3,5-triazine-2,4-diamine	Amanozine	$C_9H_9N_5$	537-17-7	187.201	cry (diox, 50% al)	235.5				
9010	4-Phenyl-1,2,4-triazolidine-3,5-dione		$C_8H_7N_3O_2$	15988-11-1	177.161		205.5				
9011	Phenyltrimethylammonium iodide		$C_9H_{14}IN$	98-04-4	263.118	lf (al)	224				vs H_2O; s EtOH, HOAc; sl ace; i chl
9012	Phenyl(triphenylmethyl)diazene		$C_{25}H_{20}N_2$	981-18-0	348.440		111 dec				
9013	Phenylurea		$C_7H_8N_2O$	64-10-8	136.151	mcl pr (w, al)	147	238	1.302^{25}		sl H_2O, eth, DMSO; s EtOH, AcOEt
9014	*trans*-5-(2-Phenylvinyl)-1,3-benzenediol	Pinosylvin	$C_{14}H_{12}O_2$	22139-77-1	212.244	nd (HOAc)	156				vs ace, bz, chl, HOAc
9015	Phenyl vinyl ether		C_8H_8O	766-94-9	120.149			155.5	0.9770^{20}	1.5224^{20}	i H_2O; vs eth
9016	Phenytoin	5,5-Diphenyl-2,4-imidazolidinedione	$C_{15}H_{12}N_2O_2$	57-41-0	252.268	nd (al)	286				i H_2O; s EtOH, ace; sl eth, bz
9017	Phloretin		$C_{15}H_{14}O_5$	60-82-2	274.269	nd (dil al), cry (ace)	263 dec				sl H_2O, chl; msc EtOH, bz; i eth; s ace
9018	Phorate		$C_7H_{17}O_2PS_3$	298-02-2	260.378		<-15	$119^{0.8}$	1.16^{25}		
9019	Phorbol		$C_{20}H_{28}O_6$	17673-25-5	364.432	cry (EtOH)	250 dec				s H_2O, ace
9020	Phorone		$C_9H_{14}O$	504-20-1	138.206	ye-grn pr	28	197.5	0.8850^{20}	1.4998^{20}	sl H_2O; s EtOH, eth, ace, ctc
9021	Phosalone		$C_{12}H_{15}ClNO_4PS_2$	2310-17-0	367.808		46				
9022	Phosfolan		$C_7H_{14}NO_3PS_2$	947-02-4	255.295		36.5	$117^{0.001}$			vs H_2O, bz, ace; sl eth; s hx
9023	Phosmet		$C_{11}H_{12}NO_4PS_2$	732-11-6	317.321		72	dec			
9024	Phosphamidon		$C_{10}H_{19}ClNO_5P$	13171-21-6	299.689	oil	-45	$162^{1.5}$	1.2132^{25}	1.4718^{25}	msc H_2O; s hx
9025	*N*-Phospho-*L*-arginine		$C_6H_{15}N_4O_5P$	1189-11-3	254.181	cry (ace aq)	177				
9026	*O*-Phosphorylethanolamine	Ethanolamine *O*-phosphate	$C_2H_8NO_4P$	1071-23-4	141.063	cry (EtOH aq)	242				
9027	*O*-Phosphoserine		$C_3H_8NO_6P$	407-41-0	185.073	cry	166 dec				
9028	Phthalazine	2,3-Benzodiazine	$C_8H_6N_2$	253-52-1	130.147		90.5	316			s H_2O, EtOH, bz; sl eth; i lig
9029	Phthalic acid	1,2-Benzenedicarboxylic acid	$C_8H_6O_4$	88-99-3	166.132	pl (w)	230 dec	dec	2.18^{191}		sl H_2O, eth; i chl; s EtOH
9030	Phthalic anhydride		$C_8H_4O_3$	85-44-9	148.116	wh nd (al, bz)	130.8	295	1.527^4		sl H_2O, eth; s EtOH, ace, bz
9031	29*H*,31*H*-Phthalocyanine		$C_{32}H_{18}N_8$	574-93-6	514.539	grsh-bl mcl (quinoline)		sub 550			i H_2O, EtOH, eth; s $PhNH_2$
9032	Phthalylsulphathiazole		$C_{17}H_{13}N_3O_5S_2$	85-73-4	403.432		273				i H_2O, eth, chl; sl EtOH; s acid, alk
9033	Physostigmine		$C_{15}H_{21}N_3O_2$	57-47-6	275.347	orth pr (eth, bz)	105.5				sl H_2O; s EtOH, eth, bz, chl
9034	Phytol	3,7,11,15-Tetramethyl-2-hexadecen-1-ol, [*R*-[*R**,*R**-(*E*)]]	$C_{20}H_{40}O$	150-86-7	296.531	oily liq		203^{10}	0.8497^{25}	1.4595^{25}	

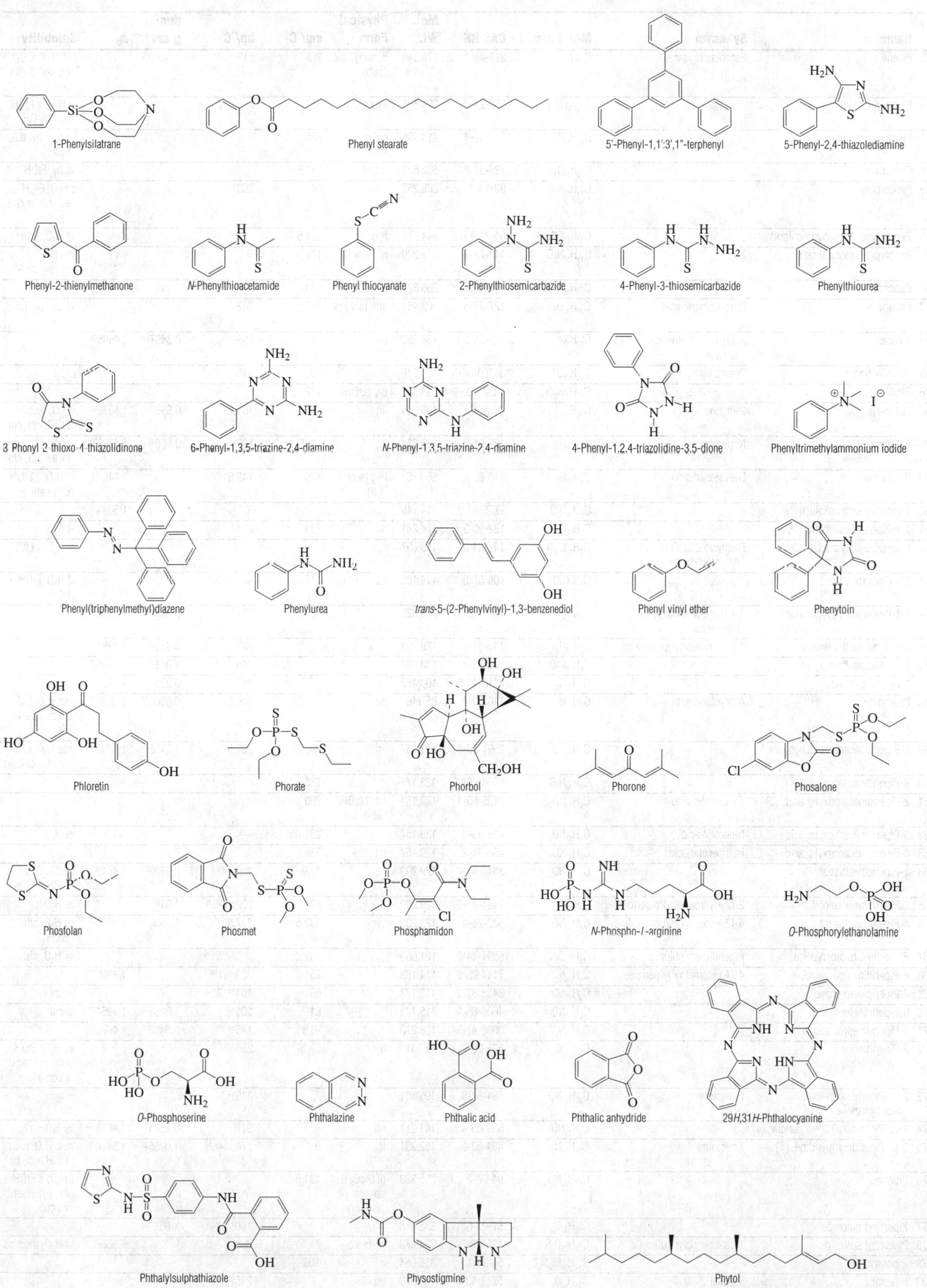

1-Phenylsilatrane
Phenyl stearate
5′-Phenyl-1,1′:3′,1″-terphenyl
5-Phenyl-2,4-thiazolediamine
Phenyl-2-thienylmethanone
N-Phenylthioacetamide
Phenyl thiocyanate
2-Phenylthiosemicarbazide
4-Phenyl-3-thiosemicarbazide
Phenylthiourea
3-Phenyl-2-thioxo-4-thiazolidinone
6-Phenyl-1,3,5-triazine-2,4-diamine
N-Phenyl-1,3,5-triazine-2,4-diamine
4-Phenyl-1,2,4-triazolidine-3,5-dione
Phenyltrimethylammonium iodide
Phenyl(triphenylmethyl)diazene
Phenylurea
trans-5-(2-Phenylvinyl)-1,3-benzenediol
Phenyl vinyl ether
Phenytoin
Phloretin
Phorate
Phorbol
Phorone
Phosalone
Phosfolan
Phosmet
Phosphamidon
N-Phospho-l-arginine
O-Phosphorylethanolamine
O-Phosphoserine
Phthalazine
Phthalic acid
Phthalic anhydride
29H,31H-Phthalocyanine
Phthalylsulphathiazole
Physostigmine
Phytol

No.	Name	Synonym	Mol. Form.	CAS RN	Mol. Wt.	Physical Form	mp/°C	bp/°C	den/ g cm^{-3}	n_D	Solubility
9035	Picene	Benzo[a]chrysene	$C_{22}H_{14}$	213-46-7	278.346	lf, pl (xyl, py, sub)	368	519			i H_2O; sl EtOH, bz, chl; s con sulf
9036	Picrolonic acid		$C_{10}H_8N_4O_5$	550-74-3	264.195	ye nd (al)	116	dec			sl H_2O; s EtOH, eth, MeOH
9037	Picropodophyllin		$C_{22}H_{22}O_8$	477-47-4	414.405	col nd (al, bz)	228				vs ace, bz, eth, EtOH
9038	Picrotoxin		$C_{30}H_{34}O_{13}$	124-87-8	602.583	orth lf	203.5				vs py, EtOH
9039	Pilocarpine		$C_{11}H_{16}N_2O_2$	92-13-7	208.257	nd	34	260^{5}			s H_2O, EtOH; sl eth, bz; vs chl; i peth
9040	Pilocarpine, monohydrochloride		$C_{11}H_{17}ClN_2O_2$	54-71-7	244.718	hyg cry	204.5				vs H_2O, EtOH
9041	Pilocarpine, mononitrate		$C_{11}H_{17}N_3O_5$	148-72-1	271.270	wh pow or cry (al)	178				vs H_2O
9042	Pilosine		$C_{16}H_{18}N_2O_3$	13640-28-3	286.325	nd (al)	179				
9043	Pimaric acid	Dextropimaric acid	$C_{20}H_{30}O_2$	127-27-5	302.451	orth (ace) pr (al)	218.5	282^{18}			vs eth, py, EtOH
9044	Pinane	2,6,6-Trimethylbicyclo[3.1.1] heptane	$C_{10}H_{18}$	473-55-2	138.250	oil	-53	169	0.8467^{21}	1.4605^{21}	
9045	*trans*-2-Pinanol	Pinene hydrate	$C_{10}H_{18}O$	35408-04-9	154.249		60	81^{10}			
9046	Pindolol		$C_{14}H_{20}N_2O_2$	13523-86-9	248.321	cry (EtOH)	172				
9047	α-Pinene	2-Pinene	$C_{10}H_{16}$	80-56-8	136.234	liq	-64	156.2	0.8539^{25}	1.4632^{25}	i H_2O; msc EtOH, eth, chl
9048	β-Pinene	Nopinene	$C_{10}H_{16}$	127-91-3	136.234	liq	-61.5	166	0.860^{25}	1.4768^{25}	i H_2O; s bz, EtOH, eth, chl
9049	Piperazine	Diethylenediamine	$C_4H_{10}N_2$	110-85-0	86.135	hyg pl or lf (al)	106	148.6		1.446^{113}	vs H_2O; s EtOH, chl; i eth
9050	1-Piperazinecarboxaldehyde		$C_5H_{10}N_2O$	7755-92-2	114.145			$95^{0.5}$		1.5094^{20}	
9051	1,4-Piperazinediethanol		$C_8H_{18}N_2O_2$	122-96-3	174.241		135	217^{30}			
9052	Piperazine dihydrochloride	Diethylenediamine dihydrochloride	$C_4H_{12}Cl_2N_2$	142-64-3	159.057						sl H_2O; i EtOH
9053	2,5-Piperazinedione		$C_4H_6N_2O_2$	106-57-0	114.103	tab or pl (w)	312 dec	sub 260			sl H_2O, EtOH; s HCl
9054	1,4-Piperazinedipropanamine	1,4-Bis(3-aminopropyl) piperazine	$C_{10}H_{24}N_4$	7209-38-3	200.325		15	151^{2}	0.973^{25}	1.5015^{20}	
9055	1-Piperazineethanamine	1-(2-Aminoethyl)piperazine	$C_6H_{15}N_3$	140-31-8	129.203			220	0.985^{25}	1.4983^{20}	
9056	1-Piperazineethanol		$C_6H_{14}N_2O$	103-76-4	130.187			246	1.061^{25}	1.5065^{20}	
9057	1-Piperidinamine		$C_5H_{12}N_2$	2213-43-6	100.162			147	0.928^{25}	1.4750^{20}	
9058	Piperidine	Azacyclohexane	$C_5H_{11}N$	110-89-4	85.148	liq	-11.02	106.22	0.8606^{20}	1.4530^{20}	msc H_2O, EtOH; s eth, ace, bz, chl
9059	1-Piperidinecarboxaldehyde		$C_6H_{11}NO$	2591-86-8	113.157	liq	-30.8	222.5	1.0158^{25}	1.4805^{25}	msc H_2O, EtOH, eth, bz, chl, lig
9060	4-Piperidinecarboxamide		$C_6H_{12}N_2O$	39546-32-2	128.171		138.5				
9061	2-Piperidinecarboxylic acid, (*S*)	*L*-Pipecolic acid	$C_6H_{11}NO_2$	3105-95-1	129.157	nd (MeOH/ eth)	260				
9062	3-Piperidinecarboxylic acid	Nipecotic acid	$C_6H_{11}NO_2$	498-95-3	129.157		261 dec				vs H_2O
9063	4-Piperidinecarboxylic acid	Isonipecotic acid	$C_6H_{11}NO_2$	498-94-2	129.157	nd	336				
9064	1-Piperidineethanol		$C_7H_{15}NO$	3040-44-6	129.200		17.9	202; 90^{12}	0.9703^{25}	1.4749^{20}	msc H_2O; vs EtOH
9065	2-Piperidineethanol	2-(2-Hydroxyethyl)piperidine	$C_7H_{15}NO$	1484-84-0	129.200		69	202; 145^{36}	1.01^{27}		vs H_2O
9066	4-Piperidineethanol	4-(2-Hydroxyethyl)piperidine	$C_7H_{15}NO$	622-26-4	129.200	syr	132.5	227.5	1.0059^{15}	1.4907^{20}	vs H_2O, eth, EtOH
9067	Piperidine, hydrochloride	Piperidinium chloride	$C_5H_{12}ClN$	6091-44-7	121.609		142 dec				vs H_2O, chl
9068	4-Piperidinemethanamine	4-(Aminomethyl)piperidine	$C_6H_{14}N_2$	7144-05-0	114.188		25	200; 31^{10}		1.4900^{20}	
9069	2-Piperidinemethanol		$C_6H_{13}NO$	3433-37-2	115.173		69	104^{10}, 80^{1}			sl chl
9070	3-Piperidinemethanol		$C_6H_{13}NO$	4606-65-9	115.173		61	$106^{3.5}$	1.0263^{20}	1.4964^{20}	sl chl
9071	1-Piperidinepropanenitrile		$C_8H_{14}N_2$	3088-41-3	138.210		-6.8	145^{50}	0.9403^{25}	1.4676^{25}	
9072	2-Piperidinone		C_5H_9NO	675-20-7	99.131	hyg	39.5	256			vs H_2O, EtOH, eth; s dil acid; i con alk
9073	2-(1-Piperidinylmethyl) cyclohexanone	Pimeclone	$C_{12}H_{21}NO$	534-84-9	195.301			119^{14}			
9074	1-(2-Piperidinyl)-2-propanone, (±)		$C_8H_{15}NO$	539-00-4	141.211	oil		91^{14}	0.9624^{20}	1.4683^{20}	vs EtOH, chl
9075	3-(2-Piperidinyl)pyridine, (*S*)	Anabasine	$C_{10}H_{14}N_2$	494-52-0	162.231	liq	9	276; 146^{14}	1.0455^{20}	1.5430^{20}	msc H_2O; s EtOH, eth, bz
9076	Piperine		$C_{17}H_{19}NO_3$	94-62-2	285.338	pr (AcOEt) pl or mcl pr (al), cry	131.5				i H_2O; s EtOH, bz, py; sl eth; vs chl
9077	Piperonyl butoxide		$C_{19}H_{30}O_5$	51-03-6	338.438			180^{1}	1.05^{25}		
9078	Piperonyl sulfoxide	Isosafrole octyl sulfoxide	$C_{18}H_{28}O_3S$	120-62-7	324.478	ye-br liq		dec		1.530^{25}	sl H_2O; misc os
9079	Pipobroman		$C_{10}H_{16}Br_2N_2O_2$	54-91-1	356.054		106				
9080	Piprotal	Tropital	$C_{24}H_{40}O_8$	5281-13-0	456.570	liq		$215^{0.04}$			

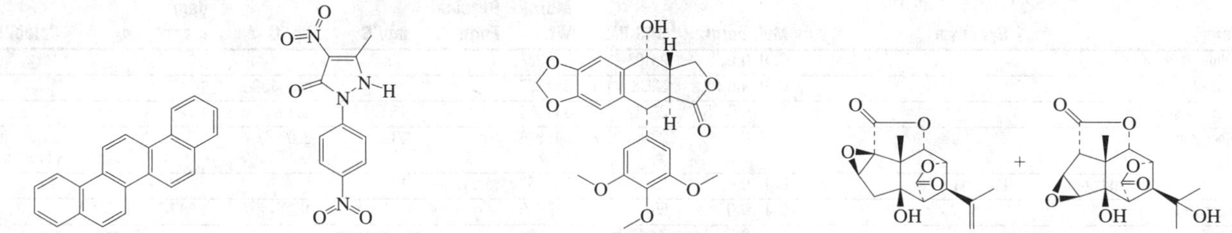

Picene

Picrolonic acid

Picropodophyllin

Picrotoxin

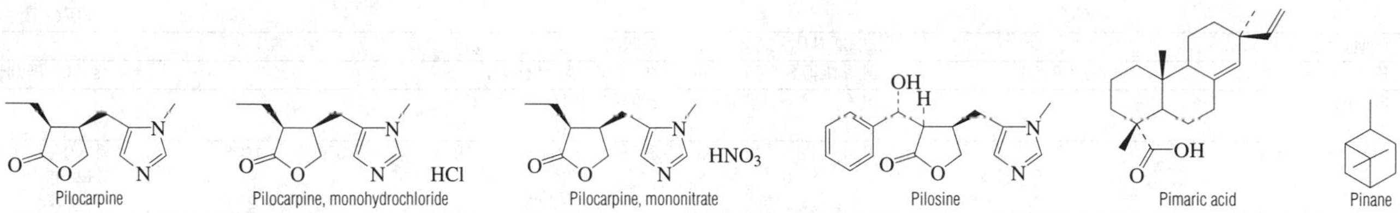

Pilocarpine

Pilocarpine, monohydrochloride

Pilocarpine, mononitrate

Pilosine

Pimaric acid

Pinane

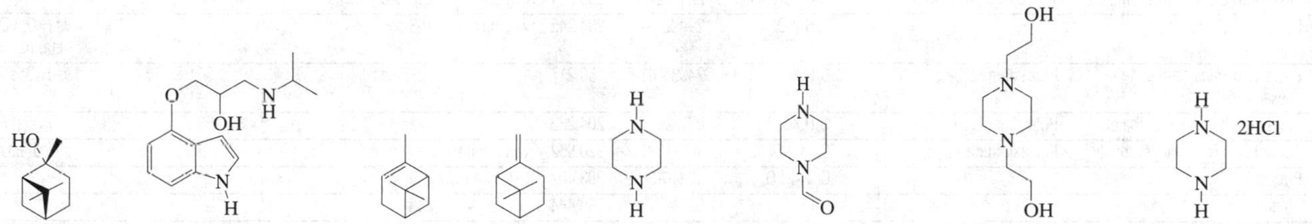

trans-2-Pinanol

Pindolol

α-Pinene

β-Pinene

Piperazine

1-Piperazinecarboxaldehyde

1,4-Piperazinediethanol

Piperazine dihydrochloride

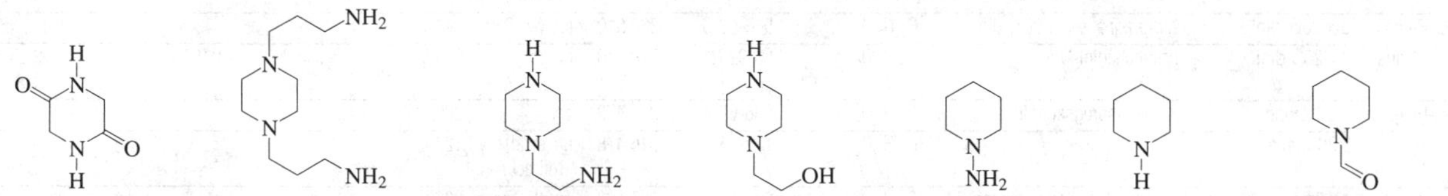

2,5-Piperazinedione

1,4-Piperazinedipropanamine

1-Piperazineethanamine

1-Piperazineethanol

1-Piperidinamine

Piperidine

1-Piperidinecarboxaldehyde

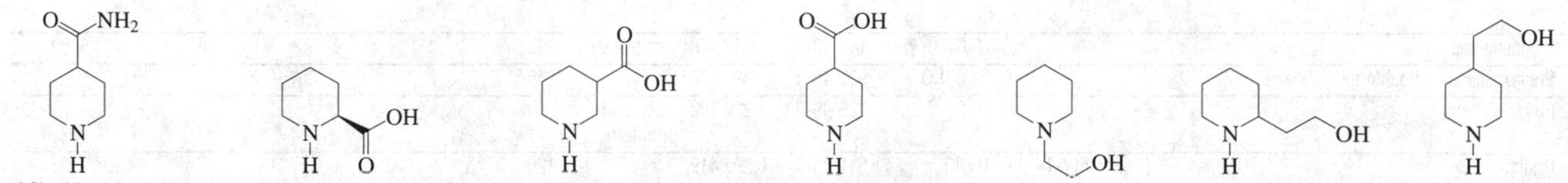

4-Piperidinecarboxamide

2-Piperidinecarboxylic acid, (*S*)

3-Piperidinecarboxylic acid

4-Piperidinecarboxylic acid

1-Piperidineethanol

2-Piperidineethanol

4-Piperidineethanol

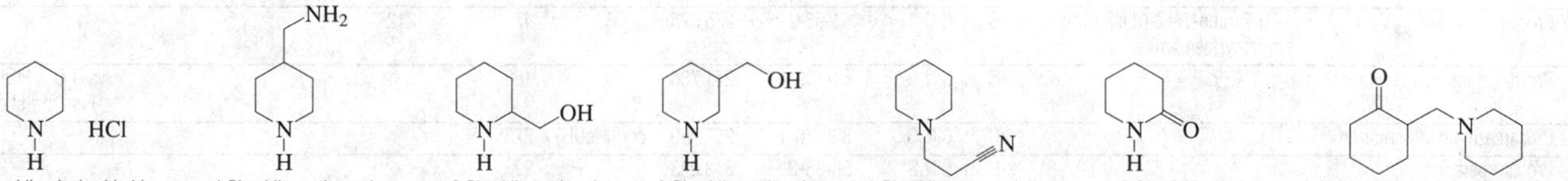

Piperidine, hydrochloride

4-Piperidinemethanamine

2-Piperidinemethanol

3-Piperidinemethanol

1-Piperidinepropanenitrile

2-Piperidinone

2-(1-Piperidinylmethyl)cyclohexanone

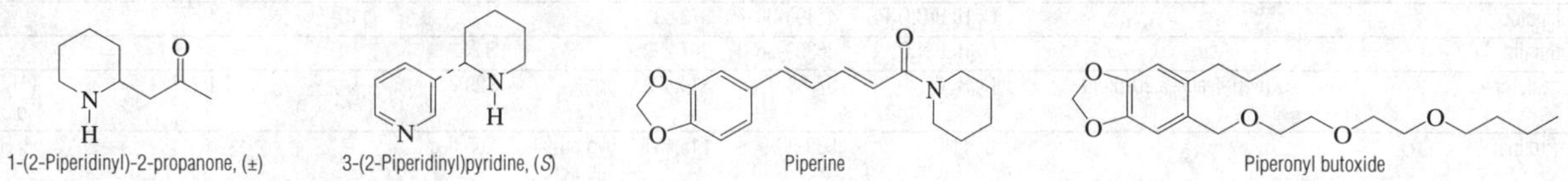

1-(2-Piperidinyl)-2-propanone, (±)

3-(2-Piperidinyl)pyridine, (*S*)

Piperine

Piperonyl butoxide

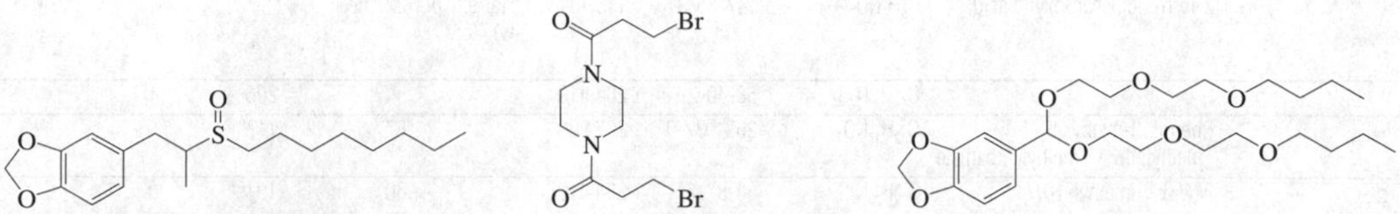

Piperonyl sulfoxide

Pipobroman

Piprotal

No.	Name	Synonym	Mol. Form.	CAS RN	Mol. Wt.	Physical Form	mp/°C	bp/°C	den/ g cm^{-3}	n_D	Solubility
9081	Pirimicarb		$C_{11}H_{18}N_4O_2$	23103-98-2	238.287		90.5				
9082	Pirimiphos-ethyl		$C_{13}H_{24}N_3O_3PS$	23505-41-1	333.387			dec >130	1.14^{20}		
9083	Pirimiphos-methyl		$C_{11}H_{20}N_3O_3PS$	29232-93-7	305.334		15	dec	1.17^{20}		
9084	Pithecolobine		$C_{22}H_{46}N_4O_2$	22368-82-7	398.626	cry	68	$230^{0.007}$			s H_2O, chl, eth, EtOH, peth
9085	2-Pivaloyl-1,3-indandione	Pindone	$C_{14}H_{14}O_3$	83-26-1	230.259	ye cry	109				
9086	Plasmocid		$C_{17}H_{25}N_3O$	551-01-9	287.400			$182^{1.0}$	1.0569^{24}	1.5855^{24}	
9087	Plumericin		$C_{15}H_{14}O_6$	77-16-7	290.268						s chl
9088	Podophyllotoxin		$C_{22}H_{22}O_8$	518-28-5	414.405		183				sl H_2O; vs EtOH; i eth; s ace, bz, HOAc
9089	Polythiazide		$C_{11}H_{13}ClF_3N_3O_4S_3$	346-18-9	439.882		214				
9090	Ponceau 3R	C.I. Food Red 6	$C_{19}H_{16}N_2Na_2O_7S_2$	3564-09-8	494.449	dk red pow					s H_2O; sl EtOH
9091	Populin		$C_{20}H_{22}O_8$	99-17-2	390.384	nd (w+2), pr (al)	180				
9092	21*H*,23*H*-Porphine		$C_{20}H_{14}N_4$	101-60-0	310.352	red or oran lf (chl-MeOH)	360	sub 300	1.336^{25}		i H_2O, eth, ace, bz; sl EtOH; s diox
9093	Potassium benzoate		$C_7H_5KO_2$	582-25-2	160.212	hyg cry					
9094	Potassium dichloroisocyanurate	Troclosene potassium	$C_3Cl_2KN_3O_3$	2244-21-5	236.054	hyg cry	250 dec				
9095	Potassium *D*-gluconate		$C_6H_{11}KO_7$	299-27-4	234.245	ye-wh cry	183 dec				vs H_2O; i EtOH, eth, bz, chl
9096	Potassium *trans,trans*-2,4-hexadienoate	Potassium sorbate	$C_6H_7KO_2$	24634-61-5	150.217		>270 dec		1.361^{25}		vs H_2O; s EtOH
9097	Potassium hydrogen phthalate	Potassium biphthalate	$C_8H_5KO_4$	877-24-7	204.222				1.636^{25}		s H_2O; sl EtOH
9098	Potassium *cis*-9-octadecenoate	Potassium oleate	$C_{18}H_{33}KO_2$	143-18-0	320.552	ye-br solid					s H_2O, EtOH
9099	Prazosin		$C_{19}H_{21}N_5O_4$	19216-56-9	383.402	cry	279				
9100	Prednisolone		$C_{21}H_{28}O_5$	50-24-8	360.444		235				
9101	5α-Pregnane	Allopregnane	$C_{21}H_{36}$	641-85-0	288.511		84.5				
9102	5β-Pregnane	17β-Ethyletiocholane	$C_{21}H_{36}$	481-26-5	288.511	mcl sc or pl (MeOH)	83.5		1.032^{15}		i H_2O; s chl, MeOH
9103	5α-Pregnane-3α,20α-diol	Allopregnane-3α,20α-diol	$C_{21}H_{36}O_2$	566-58-5	320.510	cry (MeOH)	244				
9104	5β-Pregnane-3α,20S-diol	Pregnanediol	$C_{21}H_{36}O_2$	80-92-2	320.510	pl (ace)	243.5		1.15^{25}		sl EtOH, eth; s ace
9105	5α-Pregnane-3,20-dione	3,20-Allopregnanedione	$C_{21}H_{32}O_2$	566-65-4	316.478	cry	200				
9106	5β-Pregnane-3,20-dione		$C_{21}H_{32}O_2$	128-23-4	316.478	nd (dil al) cry (dil ace)	123				i H_2O; vs EtOH; s eth, ace
9107	5-Pregnane-3,17,21-triol-20-one	3,17,21-Trihydroxypregnan-20-one, (3α,5β)	$C_{21}H_{34}O_4$	68-60-0	350.493	cry (EtOAc)	226				
9108	Pregnan-3α-ol-20-one		$C_{21}H_{34}O_2$	128-20-1	318.494	nd (bz), cry (dil al)	149.5				vs EtOH
9109	Pregnenolone		$C_{21}H_{32}O_2$	145-13-1	316.478	nd (dil al)	192				
9110	Prenoxdiazine hydrochloride		$C_{23}H_{28}ClN_3O$	982-43-4	397.940		186.5				
9111	Prephenic acid		$C_{10}H_{10}O_6$	126-49-8	226.182	free acid unstab					
9112	Pridinol	1,1-Diphenyl-3-(1-piperidinyl)-1-propanol	$C_{20}H_{25}NO$	511-45-5	295.419	cry	120				s ace
9113	Prilocaine	*N*-(2-Methylphenyl)-2-(propylamino)propanamide	$C_{13}H_{20}N_2O$	721-50-6	220.310	nd	38	160^{1}		1.5299^{20}	
9114	Procainamide	4-Amino-*N*-[2-(diethylamino) ethyl]benzamide	$C_{13}H_{21}N_3O$	51-06-9	235.325		47	212^{2}			
9115	Procainamide hydrochloride		$C_{13}H_{22}ClN_3O$	614-39-1	271.786		166				vs H_2O; s EtOH; i eth, bz; sl chl
9116	Procarbazine hydrochloride		$C_{12}H_{20}ClN_3O$	366-70-1	257.759	cry (MeOH)	225				
9117	Prochlorperazine		$C_{20}H_{24}ClN_3S$	58-38-8	373.943		228				
9118	Procymidone		$C_{13}H_{11}Cl_2NO_2$	32809-16-8	284.138		166		1.452^{25}		
9119	Prodiamine		$C_{13}H_{17}F_3N_4O_4$	29091-21-2	350.294		124		1.47^{25}		
9120	Profenofos		$C_{11}H_{15}BrClO_3PS$	41198-08-7	373.631			$110^{0.001}$	1.455^{20}		
9121	Profluralin		$C_{14}H_{16}F_3N_3O_4$	26399-36-0	347.290		34				
9122	Progesterone	Pregn-4-ene-3,20-dione	$C_{21}H_{30}O_2$	57-83-0	314.462	pr	129		1.166^{23}		i H_2O; s EtOH, diox, ace
9123	*DL*-Proline		$C_5H_9NO_2$	609-36-9	115.131	hyg nd (al-eth) cry (+w)	205 dec				vs H_2O, EtOH
9124	*L*-Proline	2-Pyrrolidinecarboxylic acid	$C_5H_9NO_2$	147-85-3	115.131	nd (al-eth) pr (w)	221 dec				vs H_2O; sl EtOH, ace, bz; i eth, PrOH
9125	Promazine		$C_{17}H_{20}N_2S$	58-40-2	284.419			$206^{0.3}$			
9126	Promecarb	Phenol, 3-methyl-5-(1-methylethyl)-, methylcarbamate	$C_{12}H_{17}NO_2$	2631-37-0	207.269		87	$117^{0.01}$			
9127	Promethazine	*N,N*,α-Trimethyl-10*H*-phenothiazine-10-ethanamine	$C_{17}H_{20}N_2S$	60-87-7	284.419		60	$191^{0.5}$			i H_2O; vs dil HCl

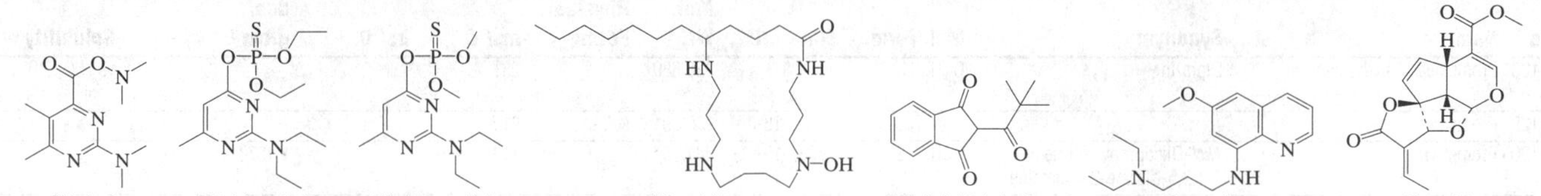

Pirimicarb

Pirimiphos-ethyl

Pirimiphos-methyl

Pithecolobine

2-Pivaloyl-1,3-indandione

Plasmocid

Plumericin

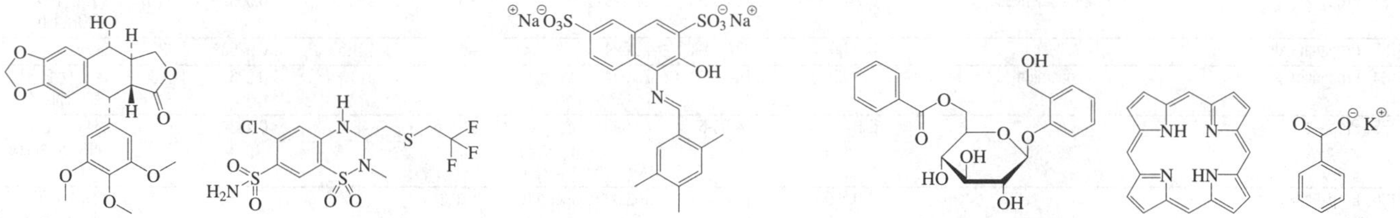

Podophyllotoxin

Polythiazide

Ponceau 3R

Populin

21*H*,23*H*-Porphine

Potassium benzoate

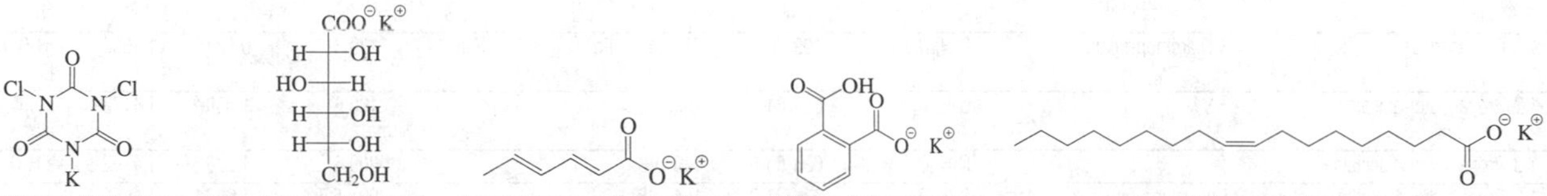

Potassium dichloroisocyanurate

Potassium *D*-gluconate

Potassium *trans,trans*-2,4-hexadienoate

Potassium hydrogen phthalate

Potassium *cis*-9-octadecenoate

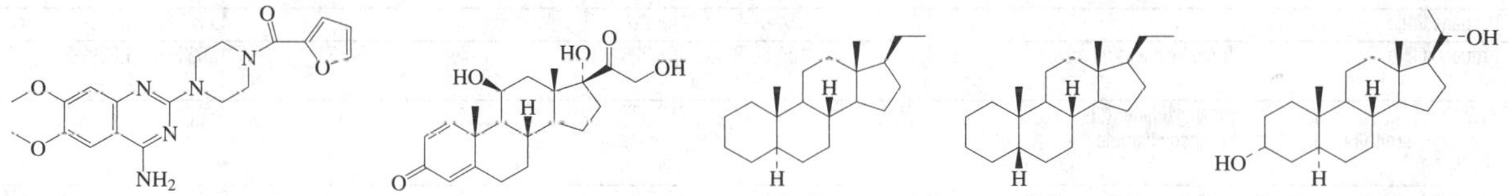

Prazosin

Prednisolone

5α-Pregnane

5β-Pregnane

5α-Pregnane-3α,20α-diol

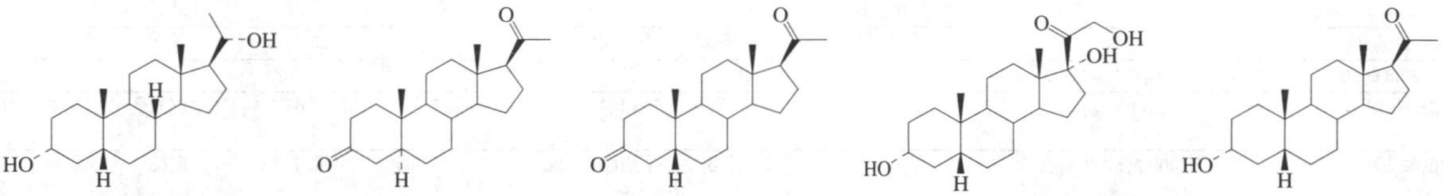

5β-Pregnane-3α,20S-diol

5α-Pregnane-3,20-dione

5β-Pregnane-3,20-dione

5-Pregnane-3,17,21-triol-20-one

Pregnan-3α-ol-20-one

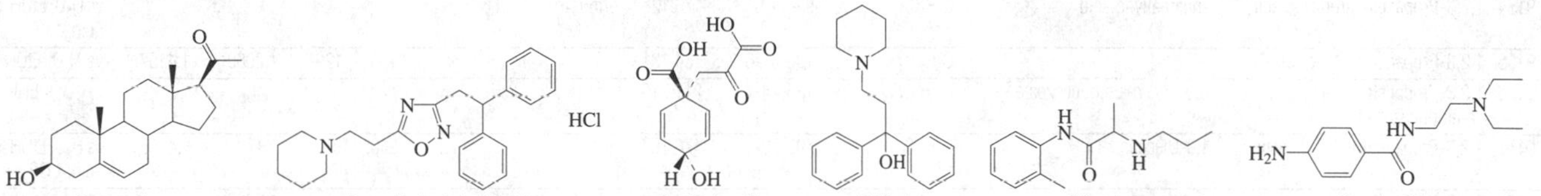

Pregnenolone

Prenoxdiazine hydrochloride

Prephenic acid

Pridinol

Prilocaine

Procainamide

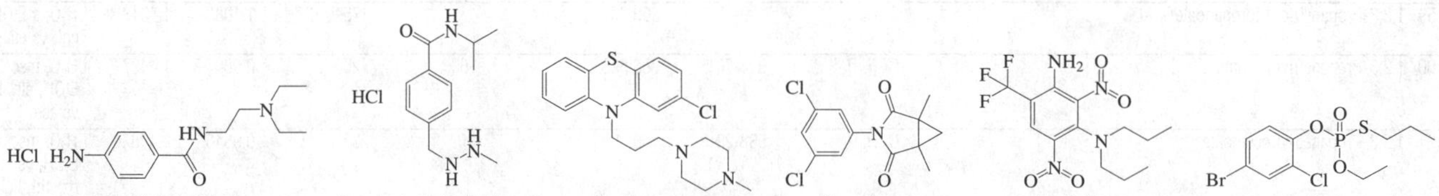

Procainamide hydrochloride

Procarbazine hydrochloride

Prochlorperazine

Procymidone

Prodiamine

Profenofos

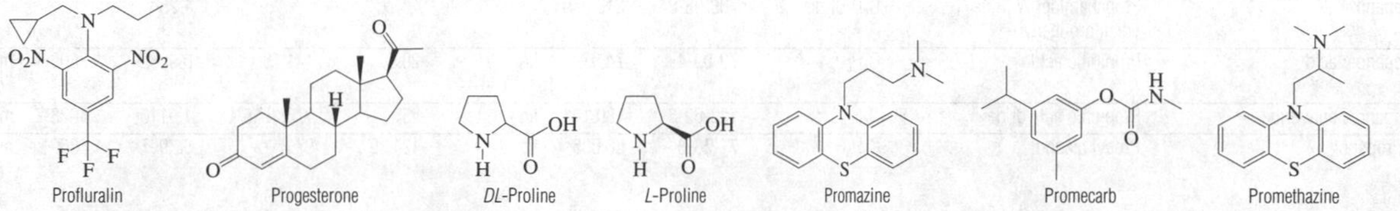

Profluralin

Progesterone

DL-Proline

L-Proline

Promazine

Promecarb

Promethazine

No.	Name	Synonym	Mol. Form.	CAS RN	Mol. Wt.	Physical Form	mp/˚C	bp/˚C	den/ g cm-3	n_D	Solubility
9128	Promethazine hydrochloride	Diprazin	$C_{17}H_{21}ClN_2S$	58-33-3	320.880		231				vs H_2O, EtOH, chl
9129	Prometone		$C_{10}H_{19}N_5O$	1610-18-0	225.291	solid	91.5				
9130	Prometryn	*N,N'*-Diisopropyl-6-(methylthio) -1,3,5-triazine-2,4-diamine	$C_{10}H_{19}N_5S$	7287-19-6	241.357		119		1.157^{20}		
9131	Propachlor	Acetamide, 2-chloro-*N*-(1-methylethyl)-*N*-phenyl-	$C_{11}H_{14}ClNO$	1918-16-7	211.688		77	$110^{0.03}$	1.242^{25}		
9132	Propanal	Propionaldehyde	C_3H_6O	123-38-6	58.079	liq	-80	48	0.8657^{25}	1.3636^{20}	s H_2O; msc EtOH, eth
9133	Propanal oxime		C_3H_7NO	627-39-4	73.094		40	131.5	0.9258^{20}	1.4287^{20}	
9134	Propanamide	Propionamide	C_3H_7NO	79-05-0	73.094	rhom, pl (bz)	81.3	213	0.9262^{110}	1.4180^{110}	vs H_2O, EtOH, eth, chl
9135	Propane		C_3H_8	74-98-6	44.096	col gas	-187.63	-42.1	0.493^{25} (p>1 atm)		s H_2O, EtOH; vs eth, bz; sl ace
9136	Propanediamide		$C_3H_6N_2O_2$	108-13-4	102.092	mcl pr(w)	170.8				s H_2O; i EtOH, eth, bz; sl DMSO
9137	1,2-Propanediamine, (±)	Propylenediamine	$C_3H_{10}N_2$	10424-38-1	74.124	hyg		119.5	0.878^{15}	1.4460^{20}	vs H_2O; i eth; vs chl
9138	1,3-Propanediamine	1,3-Diaminopropane	$C_3H_{10}N_2$	109-76-2	74.124	liq	-10.8	139.8	0.884^{25}	1.4600^{20}	s H_2O; msc EtOH, eth
9139	1,2-Propanediol diacetate		$C_7H_{12}O_4$	623-84-7	160.168			190.5	1.059^{20}	1.4173^{20}	vs H_2O; s EtOH, eth
9140	1,3-Propanediol diacetate		$C_7H_{12}O_4$	628-66-0	160.168			209.5	1.070^{14}	1.4192	vs H_2O; s EtOH
9141	1,2-Propanediol 1-methacrylate	2-Hydroxypropyl methacrylate	$C_7H_{12}O_3$	923-26-2	144.168			90^{9}, $57^{0.5}$	1.066^{25}	1.4458^{20}	
9142	1,2-Propanedione	Pyruvaldehyde	$C_3H_4O_2$	78-98-8	72.063	ye hyg liq		72	1.0455^{20}	1.4002^{18}	s EtOH, eth, bz
9143	Propanedioyl dichloride		$C_3H_2Cl_2O_2$	1663-67-8	140.953			57^{28}	1.4509^{20}	1.4639^{20}	s eth, AcOEt
9144	1,2-Propanedithiol		$C_3H_8S_2$	814-67-5	108.226			152	1.08^{20}	1.532^{20}	s chl
9145	1,3-Propanedithiol	Trimethylene dimercaptan	$C_3H_8S_2$	109-80-8	108.226	liq	-79	172.9	1.0772^{20}	1.5392^{20}	sl H_2O, ctc; msc EtOH, eth, bz
9146	2,2'-[1,3-Propanediylbis(nitrilomethylidyne)]bisphenol	Disalicylidene-1,3-propanediamine	$C_{17}H_{18}N_2O_2$	120-70-7	282.337		54.3				
9147	Propanenitrile	Ethyl cyanide	C_3H_5N	107-12-0	55.079	liq	-92.78	97.14	0.7818^{20}	1.3655^{20}	vs H_2O; s EtOH, eth, ace, bz, ctc
9148	1-Propanesulfonic acid		$C_3H_8O_3S$	5284-66-2	124.159		8	136^{1}	1.2516^{25}		
9149	1-Propanesulfonyl chloride		$C_3H_7ClO_2S$	10147-36-1	142.605			dec 180; 77^{12}	1.267^{20}	1.452^{20}	
9150	1,3-Propane sultone	1,2-Oxathiolane, 2,2-dioxide	$C_3H_6O_3S$	1120-71-4	122.143						s chl
9151	1-Propanethiol	Propyl mercaptan	C_3H_8S	107-03-9	76.161	liq	-113.13	67.8	0.8411^{20}	1.4380^{20}	sl H_2O; s EtOH, eth, ace, bz
9152	2-Propanethiol	Isopropyl mercaptan	C_3H_8S	75-33-2	76.161	liq	-130.5	52.6	0.8143^{20}	1.4255^{20}	sl H_2O; msc EtOH, eth; vs ace; s chl
9153	1,2,3-Propanetriamine	1,2,3-Triaminopropane	$C_3H_{11}N_3$	21291-99-6	89.139	visc oil		190; 92^{9}			s H_2O
9154	1,2,3-Propanetricarboxylic acid	Tricarballylic acid	$C_6H_8O_6$	99-14-9	176.124	orth (eth)	166				vs H_2O, EtOH; sl eth
9155	1,2,3-Propanetriol-1-acetate		$C_5H_{10}O_4$	106-61-6	134.131			158^{165}, 129^{3}	1.2060^{20}	1.4157^{20}	vs H_2O, EtOH
9156	1,2,3-Propanetriol 1-(4-aminobenzoate)	Glyceryl *p*-aminobenzoate	$C_{10}H_{13}NO_4$	136-44-7	211.215						i H_2O; s EtOH
9157	1,2,3-Propanetriol-1,3-diacetate	1,3-Diacetin	$C_7H_{12}O_5$	105-70-4	176.167	hyg liq		260; 149^{12}	1.179^{15}	1.4395^{20}	vs H_2O, EtOH; sl eth; i CS_2
9158	1,2,3-Propanetriol tribenzoate		$C_{24}H_{20}O_6$	614-33-5	404.412	nd (MeOH)	76		1.228^{12}		i H_2O; s EtOH; vs eth, ace, bz, chl
9159	1,2,3-Propanetriol tripropanoate		$C_{12}H_{20}O_6$	139-45-7	260.283			175^{20}, 157^{13}	1.108^{15}	1.4318^{19}	i H_2O; s EtOH, chl; vs eth
9160	1,2,3-Propanetriyl hexanoate		$C_{21}H_{38}O_6$	621-70-5	386.523		-60	>200	0.9867^{20}	1.4427^{20}	i H_2O; msc EtOH, eth, bz; vs ace
9161	1,2,3-Propanetriyl octanoate		$C_{27}H_{50}O_6$	538-23-8	470.682		10	233	0.9540^{20}	1.4482^{20}	i H_2O; msc EtOH; vs eth, bz, chl, lig
9162	Propanidid		$C_{18}H_{27}NO_5$	1421-14-3	337.411			$211^{0.7}$			i H_2O; s EtOH, chl
9163	Propanil	Propanamide, *N*-(3,4-dichlorophenyl)-	$C_9H_9Cl_2NO$	709-98-8	218.079		92		1.25^{25}		
9164	Propanoic acid	Propionic acid	$C_3H_6O_2$	79-09-4	74.079	liq	-20.5	141.15	0.9882^{25}	1.3809^{20}	msc H_2O, EtOH; s eth; sl chl
9165	Propanoic anhydride	Propionic anhydride	$C_6H_{10}O_3$	123-62-6	130.141	liq	-45	170; 67.5^{18}	1.0110^{20}	1.4038^{20}	msc eth; sl ctc
9166	1-Propanol	Propyl alcohol	C_3H_8O	71-23-8	60.095	liq	-124.39	97.2	0.7997^{25}	1.3850^{20}	msc H_2O, EtOH, eth; s ace, chl; vs bz

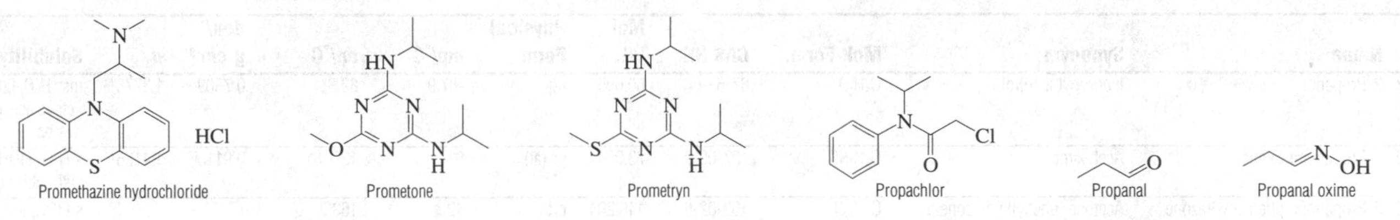

Promethazine hydrochloride | Prometone | Prometryn | Propachlor | Propanal | Propanal oxime

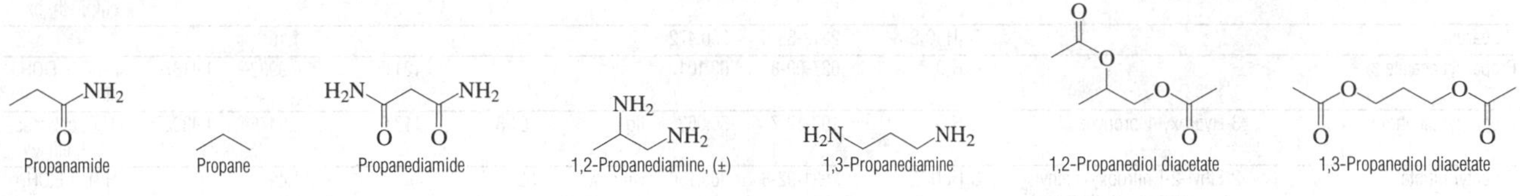

Propanamide | Propane | Propanediamide | 1,2-Propanediamine, (±) | 1,3-Propanediamine | 1,2-Propanediol diacetate | 1,3-Propanediol diacetate

1,2-Propanediol 1-methacrylate | 1,2-Propanedione | Propanedioyl dichloride | 1,2-Propanedithiol | 1,3-Propanedithiol | 2,2'-[1,3-Propanediylbis(nitrilomethylidyne)]bisphenol

Propanenitrile | 1-Propanesulfonic acid | 1-Propanesulfonyl chloride | 1,3-Propane sultone | 1-Propanethiol | 2-Propanethiol | 1,2,3-Propanetriamine

1,2,3-Propanetricarboxylic acid | 1,2,3-Propanetriol-1-acetate | 1,2,3-Propanetriol 1-(4-aminobenzoate) | 1,2,3-Propanetriol-1,3-diacetate

1,2,3-Propanetriol tribenzoate | 1,2,3-Propanetriol tripropanoate | 1,2,3-Propanetriyl hexanoate | 1,2,3-Propanetriyl octanoate

Propanidid | Propanil | Propanoic acid | Propanoic anhydride | 1-Propanol

No.	Name	Synonym	Mol. Form.	CAS RN	Mol. Wt.	Physical Form	mp/°C	bp/°C	den/ g cm^{-3}	n_D	Solubility
9167	2-Propanol	Isopropyl alcohol	C_3H_8O	67-63-0	60.095	liq	-87.9	82.3	0.7809^{25}	1.3776^{20}	msc H_2O, EtOH, eth; s ace, chl; vs bz
9168	2-Propanone oxime	Acetoxime	C_3H_7NO	127-06-0	73.094	pr (al)	61	136; 61^{20}	0.9113^{62}	1.4156^{20}	s H_2O, EtOH, eth, chl, lig
9169	2-Propanone phenylhydrazone	Acetone, phenylhydrazone	$C_9H_{12}N_2$	103-02-6	148.204	orth	42	163^{50}			s EtOH, eth, dil acid
9170	Propanoyl chloride	Propionyl chloride	C_3H_5ClO	79-03-8	92.524	liq	-94	80	1.0646^{20}	1.4032^{20}	s eth
9171	Propanoyl fluoride	Propionyl fluoride	C_3H_5FO	430-71-7	76.069			44	0.972^{15}	1.329^{13}	
9172	Propantheline bromide		$C_{23}H_{30}BrNO_3$	50-34-0	448.393	cry	160				vs H_2O, EtOH, chl; i eth, bz
9173	Propargite		$C_{19}H_{26}O_4S$	2312-35-8	350.472				1.10^{25}		
9174	Propargyl acetate		$C_5H_6O_2$	627-09-8	98.101			121.5	0.9982^{20}	1.4187^{20}	sl H_2O; s EtOH, eth
9175	Propargyl alcohol	3-Hydroxy-1-propyne	C_3H_4O	107-19-7	56.063	liq	-51.8	113.6	0.9478^{20}	1.4322^{20}	s H_2O, chl; msc EtOH, eth
9176	Propatyl nitrate	2-Ethyl-2-[(nitrooxy)methyl]-1,3-propanediol, dinitrate	$C_6H_{11}N_3O_9$	2921-92-8	269.166	wh pow	52		1.49		i H_2O; s EtOH, ace
9177	Propazine	6-Chloro-*N,N'*-diisopropyl-1,3,5-triazine-2,4-diamine	$C_9H_{16}ClN_5$	139-40-2	229.710		213		1.162^{20}		
9178	Propene	Propylene	C_3H_6	115-07-1	42.080	col gas	-185.24	-47.69	0.505^{25} (p>1 atm)	1.3567^{-70}	sl H_2O; vs EtOH, HOAc
9179	*trans*-1-Propene-1,2-dicarboxylic acid	Mesaconic acid	$C_5H_6O_4$	498-24-8	130.100	orth nd or mcl pr (eth)	204.5	sub	1.466^{20}		sl H_2O, bz, CS_2; vs EtOH; s eth, tfa
9180	1-Propene-2,3-dicarboxylic acid	Itaconic acid	$C_5H_6O_4$	97-65-4	130.100	rhom (bz)	175	dec	1.632^{25}		s H_2O, EtOH, ace; sl eth, bz, peth
9181	2-Propene-1-thiol		C_3H_6S	870-23-5	74.145			65	0.925^{23}	1.4832^{20}	i H_2O; msc EtOH, eth; s chl
9182	*cis*-1-Propene-1,2,3-tricarboxylic acid	*cis*-Aconitic acid	$C_6H_6O_6$	585-84-2	174.108	nd (w)	125				s H_2O; sl eth
9183	*trans*-1-Propene-1,2,3-tricarboxylic acid	*trans*-Aconitic acid	$C_6H_6O_6$	4023-65-8	174.108	lf (w) nd (w, eth)	196 dec				vs H_2O, EtOH
9184	1-Propen-1-one	Methylketene	C_3H_4O	6004-44-0	56.063	col gas	-80	-23			vs eth
9185	2-Propenoyl chloride	Acrylic acid chloride	C_3H_3ClO	814-68-6	90.508			75.5	1.1136^{20}	1.4343^{20}	vs chl
9186	*cis*-1-Propenylbenzene		C_9H_{10}	766-90-5	118.175	liq	-61.6	167.5	0.9088^{20}	1.5420^{20}	i H_2O; msc EtOH, eth, ace, bz, peth, ctc
9187	*trans*-1-Propenylbenzene		C_9H_{10}	873-66-5	118.175	liq	-29.3	178.3	0.9023^{25}	1.5506^{20}	i H_2O; msc EtOH, eth, ace, bz
9188	*trans*-5-(1-Propenyl)-1,3-benzodioxole		$C_{10}H_{10}O_2$	4043-71-4	162.185		6.8	253	1.1224^{20}	1.5782^{20}	i H_2O; msc EtOH, eth; vs ace; s chl
9189	4-(1-Propenyl)phenol	*p*-Anol	$C_9H_{10}O$	539-12-8	134.174	lf	94	dec 250			sl H_2O; vs DMF
9190	2-(1-Propenyl)piperidine	β-Coniceine	$C_8H_{15}N$	538-90-9	125.212		8	168	0.8716^{15}		
9191	Propetamphos		$C_{10}H_{20}NO_4PS$	31218-83-4	281.309			$88^{0.005}$	1.1294^{20}		
9192	Propiconazole		$C_{15}H_{17}Cl_2N_3O_2$	60207-90-1	342.221			$180^{0.1}$	1.27^{20}		
9193	Propiomazine		$C_{20}H_{24}N_2OS$	362-29-8	340.482			$240^{0.5}$			
9194	Propionyl-*L*-carnitine	Carnitine, *O*-propanoyl	$C_{10}H_{19}NO_4$	20064-19-1	217.263	hyg pr (2-PrOH)	147 dec				
9195	Propofol		$C_{12}H_{18}O$	2078-54-8	178.270		19	256; 136^{30}	0.955^{20}	1.5140^{20}	
9196	Propoxur	Phenol, 2-(1-methylethoxy)-, methylcarbamate	$C_{11}H_{15}NO_3$	114-26-1	209.242		87	dec	1.12^{20}		
9197	2-Propoxyethanol	Ethylene glycol monopropyl ether	$C_5H_{12}O_2$	2807-30-9	104.148			149.8	0.9112^{20}	1.4133^{20}	s H_2O; vs EtOH, eth
9198	*D*-Propoxyphene	Dextropropoxyphene	$C_{22}H_{29}NO_2$	469-62-5	339.471	cry (peth)	75.5				
9199	*L*-Propoxyphene	Levopropoxyphene	$C_{22}H_{29}NO_2$	2338-37-6	339.471	cry (peth)	75.5				
9200	1-Propoxy-2-propanol	1,2-Propylene glycol 1-propyl ether	$C_6H_{14}O_2$	1569-01-3	118.174			150	0.8886^{20}	1.4130^{20}	
9201	3-Propoxy-1-propene		$C_6H_{12}O$	1471-03-0	100.158			91	0.7764^{20}	1.3919^{20}	vs ace, eth, EtOH
9202	Propranolol		$C_{16}H_{21}NO_2$	525-66-6	259.344	cry (cyhex)	96				
9203	Propyl acetate		$C_5H_{10}O_2$	109-60-4	102.132	liq	-93	101.3	0.8820^{25}	1.3828^{25}	sl H_2O; msc EtOH, eth; s ctc
9204	Propyl acrylate	2-Propenoic acid, propyl ester	$C_6H_{10}O_2$	925-60-0	114.142			122; 63^{100}			
9205	Propylamine	1-Propanamine	C_3H_9N	107-10-8	59.110	liq	-84.75	47.22	0.7173^{20}	1.3870^{20}	msc H_2O; vs EtOH, ace; s bz, chl; sl ctc
9206	Propylamine hydrochloride	1-Propanamine hydrochloride	$C_3H_{10}ClN$	556-53-6	95.571		163.5				s DMSO

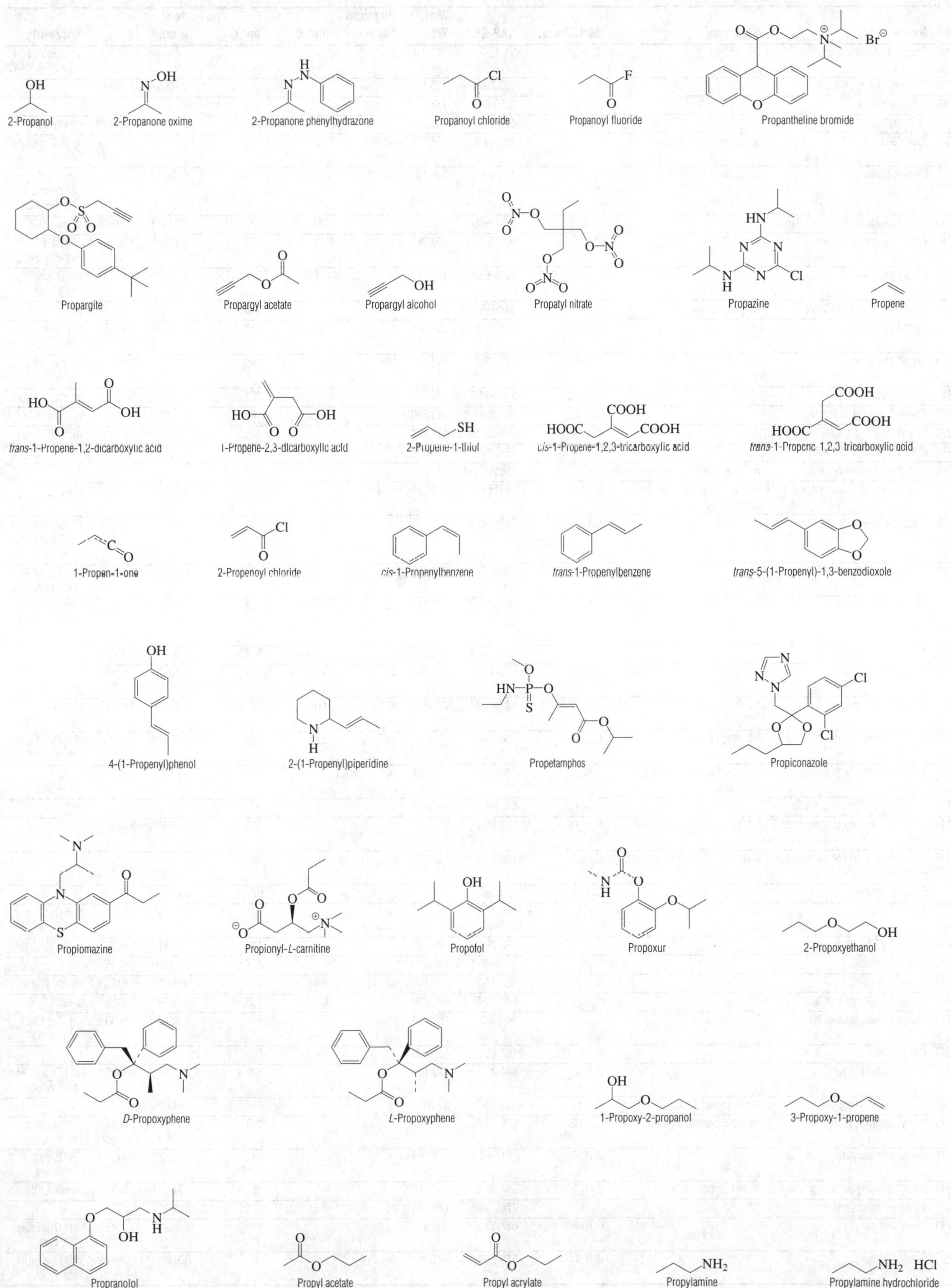

2-Propanol
2-Propanone oxime
2-Propanone phenylhydrazone
Propanoyl chloride
Propanoyl fluoride
Propantheline bromide
Propargite
Propargyl acetate
Propargyl alcohol
Propatyl nitrate
Propazine
Propene
trans-1-Propene-1,2-dicarboxylic acid
1-Propene-2,3-dicarboxylic acid
2-Propene-1-thiol
cis-1-Propene-1,2,3-tricarboxylic acid
trans-1-Propene-1,2,3-tricarboxylic acid
1-Propen-1-one
2-Propenoyl chloride
cis-1-Propenylbenzene
trans-1-Propenylbenzene
trans-5-(1-Propenyl)-1,3-benzodioxole
4-(1-Propenyl)phenol
2-(1-Propenyl)piperidine
Propetamphos
Propiconazole
Propiomazine
Propionyl-*L*-carnitine
Propofol
Propoxur
2-Propoxyethanol
D-Propoxyphene
L-Propoxyphene
1-Propoxy-2-propanol
3-Propoxy-1-propene
Propranolol
Propyl acetate
Propyl acrylate
Propylamine
Propylamine hydrochloride

No.	Name	Synonym	Mol. Form.	CAS RN	Mol. Wt.	Physical Form	mp/°C	bp/°C	den/ g cm^{-3}	n_D	Solubility
9207	Propyl 4-aminobenzoate	Risocaine	$C_{10}H_{13}NO_2$	94-12-2	179.216	pr	75				vs bz, eth, EtOH, chl
9208	2-(Propylamino)ethanol		$C_5H_{13}NO$	16369-21-4	103.163			182	0.9005^{20}	1.4428^{20}	
9209	4-Propylaniline		$C_9H_{13}N$	2696-84-6	135.206			227			
9210	*N*-Propylaniline		$C_9H_{13}N$	622-80-0	135.206			222; 98^{11}	0.9443^{20}	1.5428^{20}	vs eth, EtOH
9211	Propylarsonic acid	1-Propanearsonic acid	$C_3H_9AsO_3$	107-34-6	168.023	nd (al), pl (w)	134.5				vs H_2O, EtOH; i eth
9212	Propylbenzene	Isocumene	C_9H_{12}	103-65-1	120.191	liq	-99.6	159.24	0.8593^{25}	1.4895^{25}	i H_2O; msc EtOH, eth, ace, bz, peth, ctc
9213	α-Propylbenzenemethanol, (*R*)		$C_{10}H_{14}O$	22144-60-1	150.217		16	232	0.9740^{20}	1.5139^{20}	vs eth, EtOH
9214	Propyl benzenesulfonate		$C_9H_{12}O_3S$	80-42-2	200.254			162^{15}	1.1804^{17}	1.5035^{25}	sl H_2O; s EtOH; vs eth, chl
9215	Propyl benzoate		$C_{10}H_{12}O_2$	2315-68-6	164.201	liq	-51.6	211	1.0230^{20}	1.5000^{20}	i H_2O; msc EtOH, eth
9216	5-Propyl-1,3-benzodioxole	Dihydrosafrole	$C_{10}H_{12}O_2$	94-58-6	164.201			228			s ctc
9217	Propyl butanoate		$C_7H_{14}O_2$	105-66-8	130.185	liq	-95.2	143.0	0.8730^{20}	1.4001^{20}	sl H_2O; msc EtOH, eth
9218	Propyl carbamate		$C_4H_9NO_2$	627-12-3	103.120	pr	60	196			vs ace, eth, EtOH
9219	Propyl chloroacetate		$C_5H_9ClO_2$	5396-24-7	136.577			161	1.104^{20}	1.4261^{20}	vs eth
9220	Propyl 2-chlorobutanoate		$C_7H_{13}ClO_2$	62108-71-8	164.630			183	1.0252^{20}		
9221	Propyl chlorocarbonate		$C_4H_7ClO_2$	109-61-5	122.551			115.2	1.0901^{20}	1.4035^{20}	msc EtOH, eth
9222	Propyl 3-chloropropanoate		$C_6H_{11}ClO_2$	62108-66-1	150.603			180	1.0656^{20}	1.4290^{20}	vs eth, EtOH
9223	*S*-Propyl chlorothioformate	*S*-Propyl carbonochloridothioate	C_4H_7ClOS	13889-92-4	138.616	liq		59^{26}			
9224	Propyl *trans*-cinnamate	Propyl *trans*-3-phenyl-2-propenoate	$C_{12}H_{14}O_2$	74513-58-9	190.238			285	1.0433^{0}		i H_2O
9225	Propylcyclohexane		C_9H_{18}	1678-92-8	126.239	liq	-94.9	156	0.7936^{20}	1.4370^{20}	i H_2O; msc EtOH, ace, ctc; s eth, bz
9226	2-Propylcyclohexanone		$C_9H_{16}O$	94-65-5	140.222			197	0.927^{20}	1.4538^{20}	i H_2O; s EtOH, ace; vs eth, bz
9227	Propylcyclopentane		C_8H_{16}	2040-96-2	112.213	liq	-117.3	131	0.7763^{20}	1.4266^{20}	i H_2O; msc EtOH, eth, ace; s bz; vs ctc
9228	1-Propylcyclopentanol		$C_8H_{16}O$	1604-02-0	128.212	liq	-37.5	173.5	0.9040^{25}	1.4502^{25}	
9229	Propylene carbonate	4-Methyl-1,3-dioxolan-2-one	$C_4H_6O_3$	108-32-7	102.089	liq	-48.8	242	1.2047^{20}	1.4189^{20}	vs H_2O, EtOH, eth, ace, bz
9230	1,2-Propylene glycol	1,2-Propanediol	$C_3H_8O_2$	57-55-6	76.095	liq	-60	187.6	1.0361^{20}	1.4324^{20}	msc H_2O, EtOH; s eth, bz, chl
9231	1,3-Propylene glycol	Trimethylene glycol	$C_3H_8O_2$	504-63-2	76.095	liq	-27.7	214.4	1.0538^{20}	1.4398^{20}	msc H_2O, EtOH; vs eth; sl bz
9232	1,2-Propylene glycol 2-*tert*-butyl ether	2-(1,1-Dimethylethoxy)-1-propanol	$C_7H_{16}O_2$	94023-15-1	132.201	liq		152	0.87		
9233	1,2-Propylene glycol dinitrate		$C_3H_6N_2O_6$	6423-43-4	166.089	liq	exp	92^{10}			
9234	1,2-Propylene glycol monomethyl ether	1-Methoxy-2-propanol	$C_4H_{10}O_2$	107-98-2	90.121			119	0.9620^{20}	1.4034^{20}	
9235	1,2-Propylene glycol monomethyl ether acetate	2-Acetoxy-1-methoxypropane	$C_6H_{12}O_3$	108-65-6	132.157	liq		147			
9236	Propyleneimine	2-Methylaziridine	C_3H_7N	75-55-8	57.095			67	0.812^{16}		
9237	Propyl formate		$C_4H_8O_2$	110-74-7	88.106	liq	-92.9	80.9	0.9073^{20}	1.377^{20}	sl H_2O, ctc; msc EtOH, eth
9238	Propyl 3-(2-furyl)acrylate		$C_{10}H_{12}O_3$	623-22-3	180.200			113^{16}, 92^{3}	1.0744^{20}	1.5392^{24}	vs bz, eth, EtOH
9239	4-Propylheptane		$C_{10}H_{22}$	3178-29-8	142.282			157.5	0.7321^{25}	1.4135^{20}	
9240	Propyl hexanoate		$C_9H_{18}O_2$	626-77-7	158.238	liq	-68.7	187	0.8672^{20}	1.4170^{20}	vs eth, EtOH
9241	Propyl 2-hydroxybenzoate		$C_{10}H_{12}O_3$	607-90-9	180.200		97	239	1.0979^{20}	1.5161^{20}	s ctc, CS_2
9242	Propyl 4-hydroxybenzoate	Propylparaben	$C_{10}H_{12}O_3$	94-13-3	180.200	pr (eth)	97		1.0630^{102}	1.5050^{102}	i H_2O; s EtOH, eth; sl chl
9243	Propyliodone		$C_{10}H_{11}I_2NO_3$	587-61-1	447.008		186				
9244	Propyl isobutanoate		$C_7H_{14}O_2$	644-49-5	130.185			135.4	0.8843^{0}	1.3955^{20}	sl H_2O; s EtOH, ace; vs eth
9245	Propyl isocyanate	1-Isocyanatopropane	C_4H_7NO	110-78-1	85.105			83.5	0.908^{25}	1.3970^{20}	
9246	Propyl isothiocyanate	1-Isothiocyanatopropane	C_4H_7NS	628-30-8	101.171			153	0.9781^{16}	1.5085^{16}	sl H_2O; msc EtOH, eth
9247	Propyl methacrylate		$C_7H_{12}O_2$	2210-28-8	128.169			141	0.9022^{20}	1.4190^{20}	i H_2O; msc EtOH, eth
9248	Propyl 3-methylbutanoate	Propyl isopentanoate	$C_8H_{16}O_2$	557-00-6	144.212			155.9	0.8617^{20}	1.4031^{20}	vs eth, EtOH
9249	1-Propylnaphthalene		$C_{13}H_{14}$	2765-18-6	170.250	liq	-8.6	274.5	0.9897^{20}	1.5923^{20}	
9250	Propyl nitrate		$C_3H_7NO_3$	627-13-4	105.093			110	1.0538^{20}	1.3973^{20}	sl H_2O; s EtOH, eth, ctc
9251	Propyl nitrite		$C_3H_7NO_2$	543-67-9	89.094	liq		48	0.886^{20}	1.3604^{20}	sl H_2O; s EtOH, eth

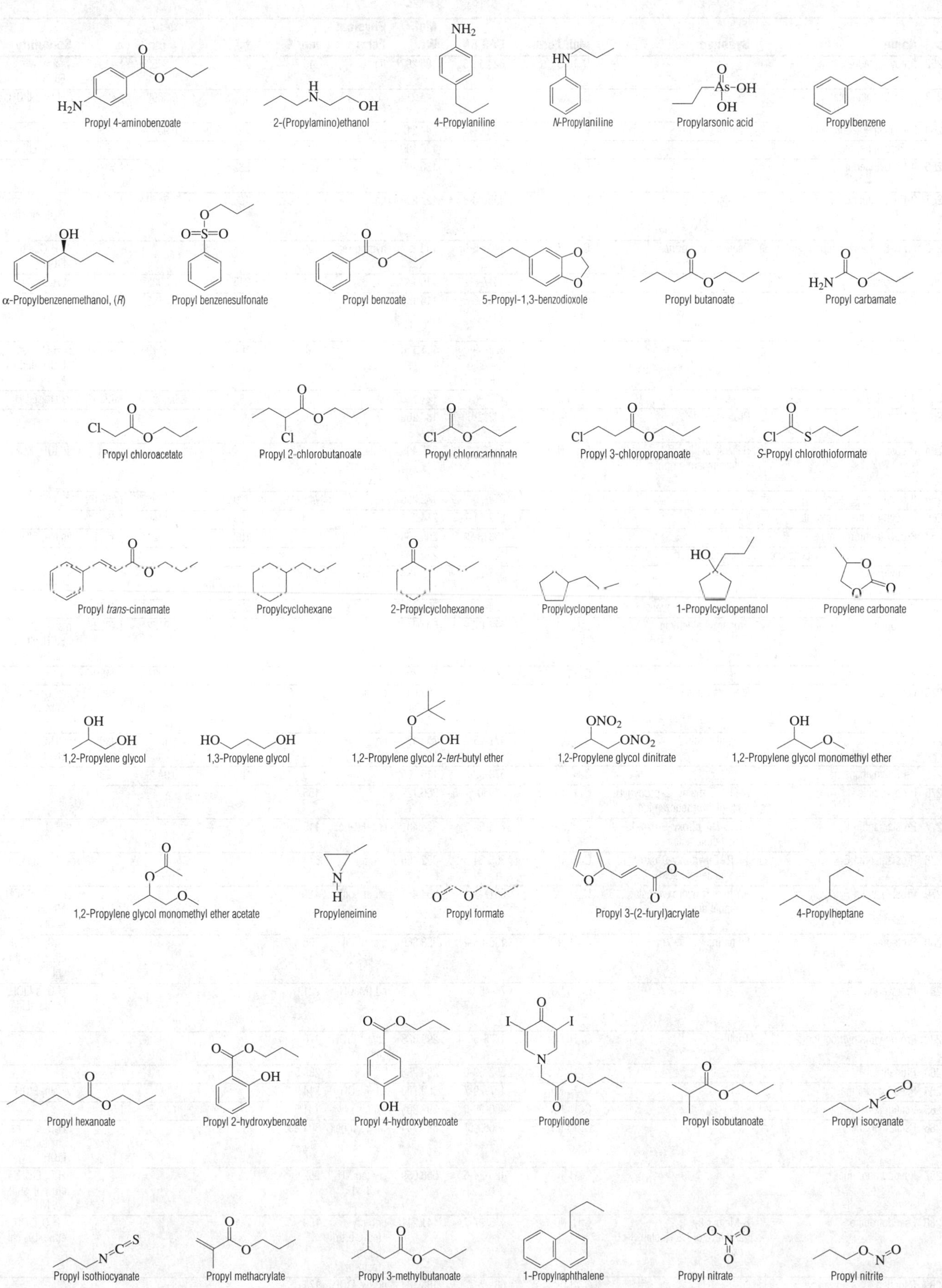
Propyl 4-aminobenzoate
2-(Propylamino)ethanol
4-Propylaniline
N-Propylaniline
Propylarsonic acid
Propylbenzene
α-Propylbenzenemethanol, (R)
Propyl benzenesulfonate
Propyl benzoate
5-Propyl-1,3-benzodioxole
Propyl butanoate
Propyl carbamate
Propyl chloroacetate
Propyl 2-chlorobutanoate
Propyl chlorocarbonate
Propyl 3-chloropropanoate
S-Propyl chlorothioformate
Propyl trans-cinnamate
Propylcyclohexane
2-Propylcyclohexanone
Propylcyclopentane
1-Propylcyclopentanol
Propylene carbonate
1,2-Propylene glycol
1,3-Propylene glycol
1,2-Propylene glycol 2-tert-butyl ether
1,2-Propylene glycol dinitrate
1,2-Propylene glycol monomethyl ether
1,2-Propylene glycol monomethyl ether acetate
Propyleneimine
Propyl formate
Propyl 3-(2-furyl)acrylate
4-Propylheptane
Propyl hexanoate
Propyl 2-hydroxybenzoate
Propyl 4-hydroxybenzoate
Propyliodone
Propyl isobutanoate
Propyl isocyanate
Propyl isothiocyanate
Propyl methacrylate
Propyl 3-methylbutanoate
1-Propylnaphthalene
Propyl nitrate
Propyl nitrite

No.	Name	Synonym	Mol. Form.	CAS RN	Mol. Wt.	Physical Form	mp/°C	bp/°C	den/ g cm^{-3}	n_D	Solubility
9252	Propyl octanoate		$C_{11}H_{22}O_2$	624-13-5	186.292	liq	-46.2	226.4	0.8659^{20}	1.4191^{25}	vs ace, eth, EtOH
9253	Propyl pentanoate		$C_8H_{16}O_2$	141-06-0	144.212	liq	-70.7	167.5	0.8699^{20}	1.4065^{20}	i H_2O; s EtOH, eth, chl
9254	2-Propylpentanoic acid	Valproic acid	$C_8H_{16}O_2$	99-66-1	144.212	col liq		221; 120^{14}	0.904^{25}	1.425^{25}	sl H_2O
9255	2-Propylphenol		$C_9H_{12}O$	644-35-9	136.190		7	220	1.015^{20}		vs eth, EtOH
9256	4-Propylphenol		$C_9H_{12}O$	645-56-7	136.190		22	232.6	1.009^{20}	1.5379^{25}	sl H_2O, ctc; s EtOH
9257	2-Propylpiperidine, (*S*)	Coniine	$C_8H_{17}N$	458-88-8	127.228	liq	-1.0	166.5	0.8440^{20}	1.4512^{22}	sl H_2O, chl; msc EtOH; vs eth; s bz
9258	*trans*-6-Propyl-3-piperidinol, (3*S*)	Pseudoconhydrine	$C_8H_{17}NO$	140-55-6	143.227	hyg nd (eth)	106	236			vs H_2O, eth, EtOH
9259	*N*-Propylpropanamide		$C_6H_{13}NO$	3217-86-5	115.173		154	215; 108^9	0.8985^{25}		sl H_2O, eth
9260	Propyl propanoate	Propyl propionate	$C_6H_{12}O_2$	106-36-5	116.158	liq	-75.9	122.5	0.8755^{25}	1.3909^{25}	sl H_2O, ctc; msc EtOH, eth; s ace
9261	2-Propylpyridine		$C_8H_{11}N$	622-39-9	121.180		1.0	167	0.9119^{20}	1.4925^{20}	sl H_2O; msc EtOH, eth; vs ace
9262	4-Propylpyridine		$C_8H_{11}N$	1122-81-2	121.180			185	0.9381^{15}	1.4966^{20}	vs eth, EtOH
9263	2-Propyl-4-pyridinecarbothioamide	Protionamide	$C_9H_{12}N_2S$	14222-60-7	180.269		136.7				
9264	Propyl Red	Benzoic acid, 2-[[4-(dipropylamino)phenyl]azo]-	$C_{19}H_{23}N_3O_2$	2641-01-2	325.405	viol-bl or purp red cry (al)					s EtOH, KOH
9265	(Propylthio)benzene		$C_9H_{12}S$	874-79-3	152.256	liq	-45	220	0.9995^{20}	1.5571^{20}	
9266	Propyl 4-toluenesulfonate		$C_{10}H_{14}O_3S$	599-91-7	214.281		<-20	189^9	1.144^{20}	1.4998^{20}	
9267	Propyl trichloroacetate		$C_5H_7Cl_3O_2$	13313-91-2	205.468			187	1.3221^{20}	1.4501^{20}	vs eth, EtOH
9268	Propyl 3,4,5-trihydroxybenzoate	Propyl gallate	$C_{10}H_{12}O_5$	121-79-9	212.199	nd (w)	130				sl H_2O
9269	Propylurea		$C_4H_{10}N_2O$	627-06-5	102.134	pr (al)	108.5				sl H_2O, DMSO; s EtOH
9270	Propyl vinyl ether	1-(Ethenyloxy)propane	$C_5H_{10}O$	764-47-6	86.132			65	0.7674^{20}	1.3908^{20}	
9271	2-Propynal	Propargyl aldehyde	C_3H_2O	624-67-9	54.047			60	0.9152^{20}	1.4033^{25}	msc H_2O; s EtOH, eth, ace, bz, tol
9272	2-Propyn-1-amine		C_3H_5N	2450-71-7	55.079			83	0.803^{25}	1.4480^{20}	
9273	Propyne	Methylacetylene	C_3H_4	74-99-7	40.064	col gas	-102.7	-23.2	0.607^{25} (p>1 atm)	1.3863^{-40}	sl H_2O; vs EtOH; s bz, chl
9274	2-Propynoic acid	Propiolic acid	$C_3H_2O_2$	471-25-0	70.047	cry (CS_2)	9	dec 144; 72^{50}	1.1380^{20}	1.4306^{20}	vs H_2O, eth, EtOH, chl
9275	1-Propynylbenzene		C_9H_8	673-32-5	116.160			183	0.942^{15}	1.563^{15}	
9276	Propyzamide	*N*-(1,1-Dimethyl-2-propynyl)-3,5-dichlorobenzamide	$C_{12}H_{11}Cl_2NO$	23950-58-5	256.127		155				
9277	Prostaglandin E_1	11,15-Dihydroxy-9-oxo-13-prostenoic acid	$C_{20}H_{34}O_5$	745-65-3	354.481	cry (EtOAc)	115				s H_2O
9278	Prostaglandin E_2	11,15-Dihydroxy-9-oxo-5,13-prostadienoic acid	$C_{20}H_{32}O_5$	363-24-6	352.465	col cry	67				s H_2O, thf
9279	Prostaglandin $F_{2\alpha}$	9,11,15-Trihydroxyprosta-5,13-dienoic acid	$C_{20}H_{34}O_5$	551-11-1	354.481	oil or solid	≈30				sl H_2O; s EtOH, MeOH, chl, AcOEt
9280	Protopine	Fumarine	$C_{20}H_{19}NO_5$	130-86-9	353.369	mcl pr (al-chl)	208				i H_2O; sl EtOH, eth, bz, peth; s chl
9281	Protoverine		$C_{27}H_{43}NO_9$	76-45-9	525.632	nd (MeOH)	221				i H_2O; s EtOH, bz, aq acid, MeOH
9282	Protriptyline hydrochloride	Triptil	$C_{19}H_{22}ClN$	1225-55-4	299.838	cry (2-PrOH/eth)	170				
9283	Prunetin		$C_{16}H_{12}O_5$	552-59-0	284.263		239.5				
9284	Pseudoaconitine		$C_{36}H_{51}NO_{12}$	127-29-7	689.790	tcl (MeOH)	214				vs eth, EtOH
9285	Pseudocodeine		$C_{18}H_{21}NO_3$	466-96-6	299.365	wh nd	181.5		1.290^{80}	1.574	
9286	Pseudojervine		$C_{33}H_{49}NO_8$	36069-05-3	587.744	wh nd or hex cry	304 dec				i H_2O, eth, bz, chl, tol, peth; s EtOH
9287	Pseudomorphine		$C_{34}H_{36}N_2O_6$	125-24-6	568.659	cry (aq NH_3, + 3 w)	282.5				i H_2O, EtOH, eth, chl, sulf; s py, NH_3
9288	Pseudotropine	8-Methyl-8-azabicyclo[3.2.1]octan-3-ol, *exo*	$C_8H_{15}NO$	135-97-7	141.211	orth cry (eth), orth bipym (peth-bz)	109	241			vs H_2O, EtOH; sl eth; s bz, chl
9289	Psoralen		$C_{11}H_6O_3$	66-97-7	186.164	nd (w, EtOH)	171				

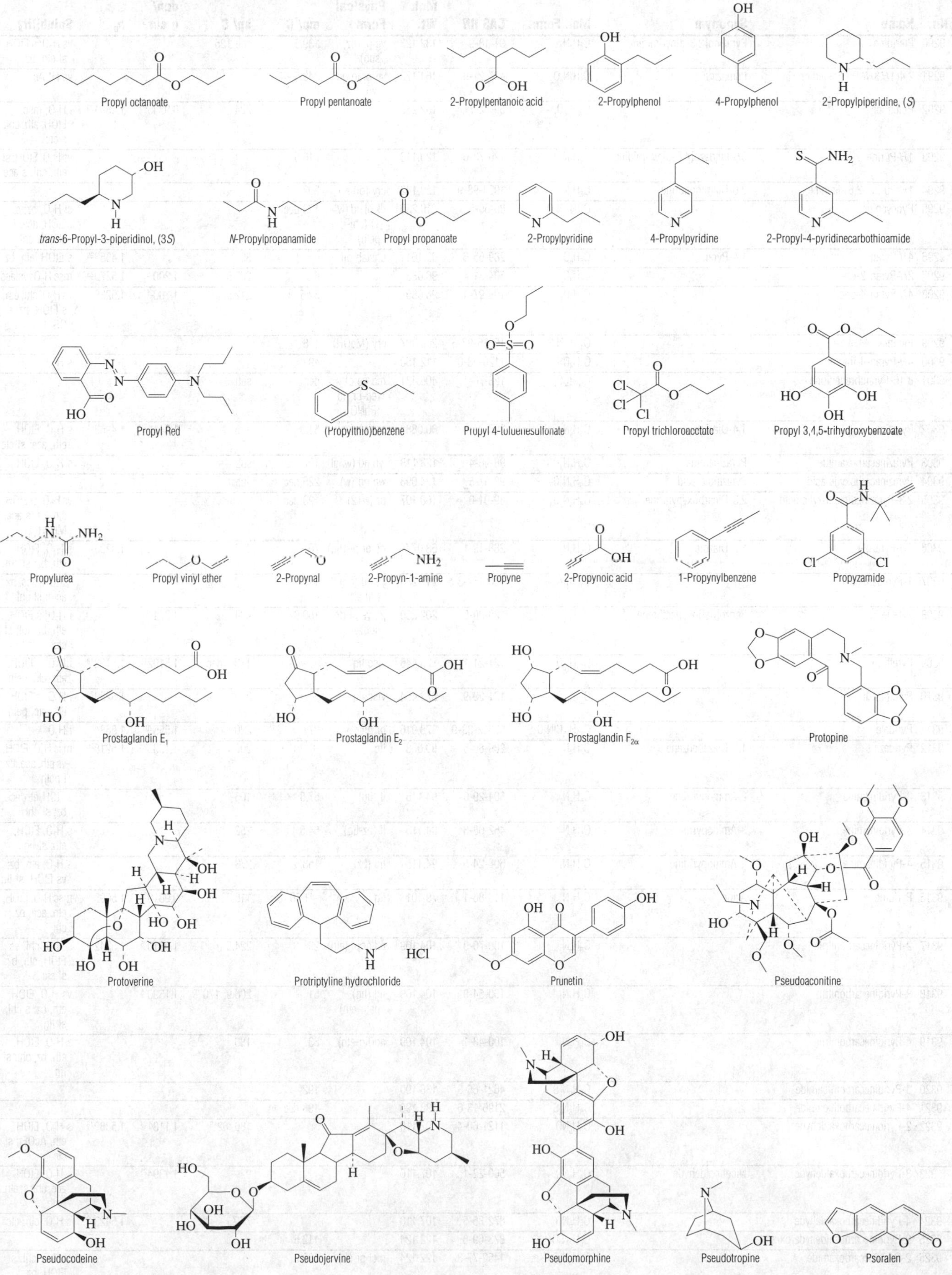

Propyl octanoate
Propyl pentanoate
2-Propylpentanoic acid
2-Propylphenol
4-Propylphenol
2-Propylpiperidine, (S)
trans-6-Propyl-3-piperidinol, (3S)
N-Propylpropanamide
Propyl propanoate
2-Propylpyridine
4-Propylpyridine
2-Propyl-4-pyridinecarbothioamide
Propyl Red
(Propylthio)benzene
Propyl 4-toluenesulfonate
Propyl trichloroacetate
Propyl 3,4,5-trihydroxybenzoate
Propylurea
Propyl vinyl ether
2-Propynal
2-Propyn-1-amine
Propyne
2-Propynoic acid
1-Propynylbenzene
Propyzamide
Prostaglandin E1
Prostaglandin E2
Prostaglandin F2α
Protopine
Protoverine
Protriptyline hydrochloride
Prunetin
Pseudoaconitine
Pseudocodeine
Pseudojervine
Pseudomorphine
Pseudotropine
Psoralen

No.	Name	Synonym	Mol. Form.	CAS RN	Mol. Wt.	Physical Form	mp/°C	bp/°C	den/ g cm^{-3}	n_D	Solubility
9290	Pteridine	Pyrazino[2,3-d]pyrimidine	$C_6H_4N_4$	91-18-9	132.123	ye pl (bz, sub)	139.5	sub 125			vs H_2O; s EtOH; sl eth, bz
9291	2,4(1*H*,3*H*)-Pteridinedione	Lumazine	$C_6H_4N_4O_2$	487-21-8	164.122	ye-oran nd (w)	348.5				vs HOAc
9292	Pulegone		$C_{10}H_{16}O$	89-82-7	152.233			224	0.9346^{45}	1.4894^{20}	i H_2O; msc EtOH, eth, chl; s ctc
9293	1*H*-Purine	6*H*-Imidazo[4,5-d]pyrimidine	$C_5H_4N_4$	120-73-0	120.113		216.5				vs H_2O, EtOH; sl eth, chl; s ace
9294	1*H*-Purine-2,6-diamine	2,6-Diaminopurine	$C_5H_6N_6$	1904-98-9	150.142	cry (dil al)	302				
9295	Pyocyanine		$C_{13}H_{10}N_2O$	85-66-5	210.230	dk bl nd (w + 1) (chl-peth)	133 dec				sl H_2O, bz; s EtOH, ace; i eth; vs chl
9296	4*H*-Pyran	1,4-Pyran	C_5H_6O	289-65-6	82.101	unstab oil		80		1.4559^{20}	s EtOH, eth, bz
9297	2*H*-Pyran-2-one		$C_5H_4O_2$	504-31-4	96.085		8.5	207.5	1.200^{20}	1.5270^{25}	msc H_2O; vs ace
9298	4*H*-Pyran-4-one		$C_5H_4O_2$	108-97-4	96.085		32.5	212.5	1.190^{25}	1.5238	vs H_2O, chl, eth; s EtOH, bz; sl CS_2
9299	Pyrantel		$C_{11}H_{14}N_2S$	15686-83-6	206.307	cry (MeOH)	178				
9300	4*H*-Pyran-4-thione		C_5H_4OS	1120-93-0	112.150		49				s H_2O
9301	8,16-Pyranthrenedione		$C_{30}H_{14}O_2$	128-70-1	406.431	red-ye or red-br nd ($PhNO_2$)	dec	sub			
9302	Pyrazine	1,4-Diazine	$C_4H_4N_2$	290-37-9	80.088	pr (w)	51.0	115	1.0311^{61}	1.4953^{61}	s H_2O, EtOH, eth, ace; sl ctc
9303	Pyrazinecarboxamide	Pyrazinamide	$C_5H_5N_3O$	98-96-4	123.113	wh nd (w, al)	192	sub			s H_2O, EtOH
9304	Pyrazinecarboxylic acid	Pyrazinoic acid	$C_5H_4N_2O_2$	98-97-5	124.098	wh nd (w)	225 dec	sub			
9305	2,3-Pyrazinedicarboxylic acid	2,3-Dicarboxypyrazine	$C_6H_4N_2O_4$	89-01-0	168.107	pr (w+2)	193 dec				vs H_2O; sl EtOH, eth, bz; s ace, MeOH
9306	1*H*-Pyrazole	1,2-Diazole	$C_3H_4N_2$	288-13-1	68.077	nd or pr (lig)	70.7	187		1.4203	s H_2O, EtOH, eth, bz; sl chl
9307	1-Pyrenamine		$C_{16}H_{11}N$	1606-67-3	217.265	ye nd (hx) lf (dil al)	117.5				s EtOH, ace, hx, acid; sl chl
9308	Pyrene	Benzo[def]phenanthrene	$C_{16}H_{10}$	129-00-0	202.250	pa ye pl (to, sub)	150.62	404	1.271^{23}		i H_2O; s EtOH, eth, bz, tol; sl ctc
9309	Pyrethrin I		$C_{21}H_{28}O_3$	121-21-1	328.445	visc liq		$170^{0.1}$ dec	1.5192^{18}	1.5192^{18}	i H_2O; s EtOH, eth, ctc, peth
9310	Pyrethrin II		$C_{22}H_{28}O_5$	121-29-9	372.454	visc liq		$200^{0.1}$ dec		1.5258^{20}	i H_2O; s EtOH, eth, ctc, peth
9311	Pyridate		$C_{19}H_{23}ClN_2O_2S$	55512-33-9	378.916	br oil	27	$220^{0.1}$	1.555^{20}	1.568^{20}	i H_2O
9312	Pyridazine	1,2-Diazabenzene	$C_4H_4N_2$	289-80-5	80.088	liq	-8	208	1.1035^{23}	1.5218^{20}	msc H_2O, EtOH; vs eth, ace, bz; i peth
9313	2-Pyridinamine	2-Aminopyridine	$C_5H_6N_2$	504-29-0	94.115	lf (lig)	57.5	105^{20}			s EtOH, eth, ace, bz; sl chl
9314	3-Pyridinamine	3-Aminopyridine	$C_5H_6N_2$	462-08-8	94.115	lf (bz-lig)	64.5	252			s H_2O, EtOH, eth; sl lig
9315	4-Pyridinamine	4-Aminopyridine	$C_5H_6N_2$	504-24-5	94.115	nd (bz)	158.5	273			s H_2O, eth, bz; vs EtOH; sl lig
9316	Pyridine	Azine	C_5H_5N	110-86-1	79.101	liq	-41.70	115.23	0.9819^{20}	1.5095^{20}	msc H_2O, EtOH, eth, ace, bz, chl
9317	2-Pyridinecarbonitrile		$C_6H_4N_2$	100-70-9	104.109	nd or pr (eth)	29	224.5	1.0810^{25}	1.5242^{25}	s H_2O, chl; vs EtOH, eth, bz; sl ctc
9318	3-Pyridinecarbonitrile		$C_6H_4N_2$	100-54-9	104.109	nd (lig), peth-eth)	51	206.9; 170^{300}	1.1590^{25}		vs H_2O, EtOH, eth, bz; s chl; sl lig
9319	4-Pyridinecarbonitrile		$C_6H_4N_2$	100-48-1	104.109	nd(lig-eth)	83	186			s H_2O, EtOH, eth, bz, chl; sl lig
9320	3-Pyridinecarbothioamide		$C_6H_6N_2S$	4621-66-3	138.190		192				
9321	4-Pyridinecarbothioamide		$C_6H_6N_2S$	2196-13-6	138.190		198 dec				
9322	2-Pyridinecarboxaldehyde		C_6H_5NO	1121-60-4	107.110			180; 62^{13}	1.1181^{25}	1.5389^{18}	s H_2O, EtOH, eth, AcOEt; sl ctc
9323	3-Pyridinecarboxaldehyde	Nicotinaldehyde	C_6H_5NO	500-22-1	107.110			92^{23}	1.1394^{25}		s H_2O, EtOH, ace, chl; sl eth, peth
9324	4-Pyridinecarboxaldehyde		C_6H_5NO	872-85-5	107.110			77^{12}		1.5423^{20}	s H_2O, eth, ctc
9325	2-Pyridinecarboxaldehyde oxime		$C_6H_6N_2O$	873-69-8	122.124		112.5				
9326	2-Pyridinecarboxamide		$C_6H_6N_2O$	1452-77-3	122.124	mcl pr (w)	108.3				sl H_2O, chl; s EtOH, bz

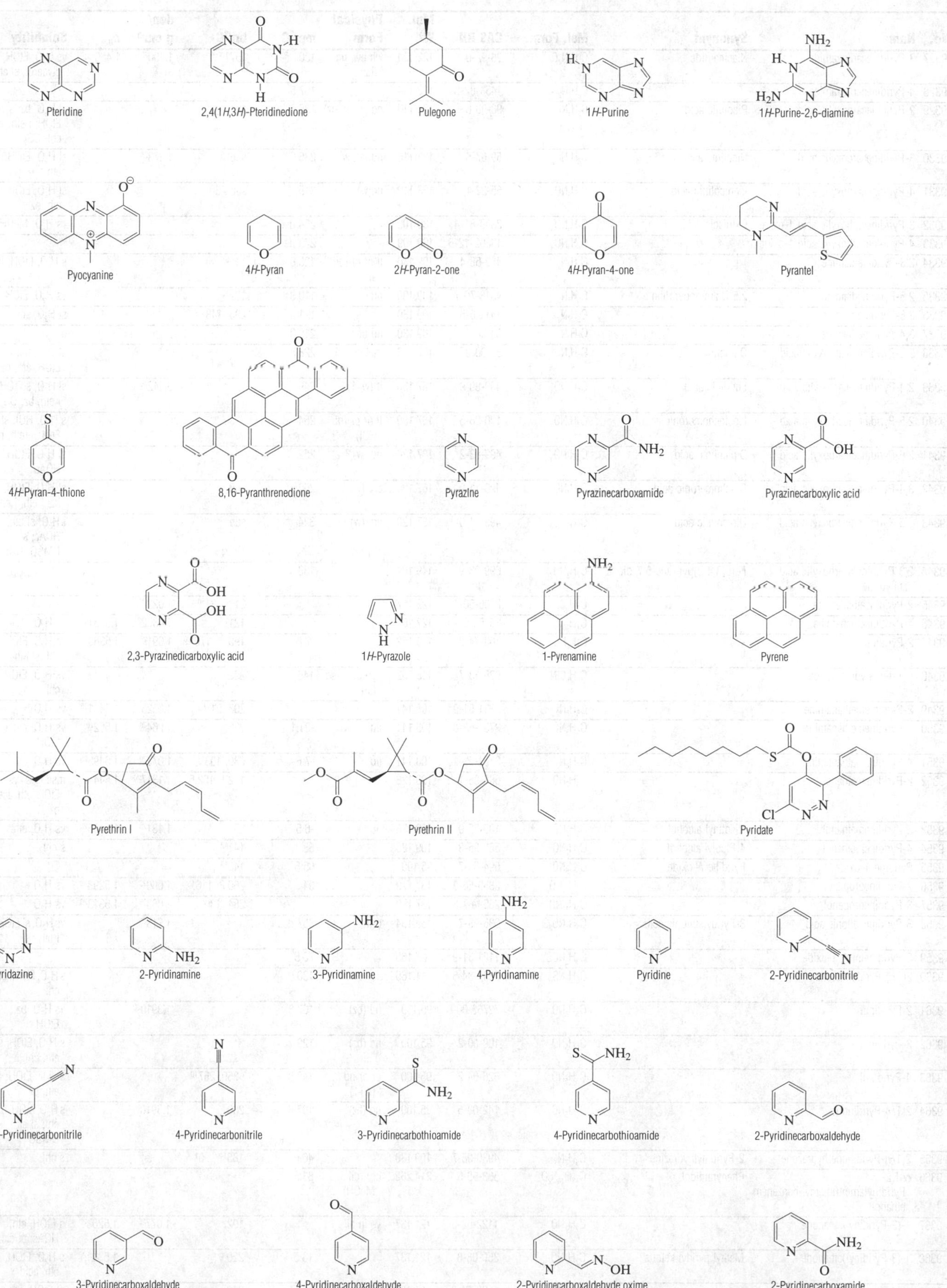
Pteridine
2,4(1H,3H)-Pteridinedione
Pulegone
1H-Purine
1H-Purine-2,6-diamine
Pyocyanine
4H-Pyran
2H-Pyran-2-one
4H-Pyran-4-one
Pyrantel
4H-Pyran-4-thione
8,16-Pyranthrenedione
Pyrazine
Pyrazinecarboxamide
Pyrazinecarboxylic acid
2,3-Pyrazinedicarboxylic acid
1H-Pyrazole
1-Pyrenamine
Pyrene
Pyrethrin I
Pyrethrin II
Pyridate
Pyridazine
2-Pyridinamine
3-Pyridinamine
4-Pyridinamine
Pyridine
2-Pyridinecarbonitrile
3-Pyridinecarbonitrile
4-Pyridinecarbonitrile
3-Pyridinecarbothioamide
4-Pyridinecarbothioamide
2-Pyridinecarboxaldehyde
3-Pyridinecarboxaldehyde
4-Pyridinecarboxaldehyde
2-Pyridinecarboxaldehyde oxime
2-Pyridinecarboxamide

No.	Name	Synonym	Mol. Form.	CAS RN	Mol. Wt.	Physical Form	mp/°C	bp/°C	den/ g cm-3	n_D	Solubility
9327	3-Pyridinecarboxamide	Niacinamide	$C_6H_6N_2O$	98-92-0	122.124	wh pw, nd (bz)	130	$157^{0.0005}$	1.400^{25}	1.466	vs H_2O, EtOH, glycerol; sl chl
9328	4-Pyridinecarboxamide		$C_6H_6N_2O$	1453-82-3	122.124		157.5				
9329	2-Pyridinecarboxylic acid	Picolinic acid	$C_6H_5NO_2$	98-98-6	123.110	nd (w, al, bz)	136.5	sub			sl H_2O, bz; s EtOH; i eth, chl, CS_2
9330	3-Pyridinecarboxylic acid	Nicotinic acid	$C_6H_5NO_2$	59-67-6	123.110	nd (al, w)	236.6	sub	1.473^{25}		sl H_2O, EtOH, eth
9331	4-Pyridinecarboxylic acid	Isonicotinic acid	$C_6H_5NO_2$	55-22-1	123.110	nd(w)	315	sub 260			sl H_2O, EtOH, eth, bz
9332	3-Pyridinecarboxylic acid 1-oxide	Oxiniacic acid	$C_6H_5NO_3$	2398-81-4	139.109	nd	254 dec				vs H_2O, MeOH
9333	4-Pyridinecarboxylic acid 1-oxide		$C_6H_5NO_3$	13602-12-5	139.109		273 dec				
9334	2,3-Pyridinediamine		$C_5H_7N_3$	452-58-4	109.130	lf or pl (dil al)	120.8	149^{5}			s H_2O, EtOH, bz
9335	2,5-Pyridinediamine	2,5-Diaminopyridine	$C_5H_7N_3$	4318-76-7	109.130	nd	110.3	182^{12}			vs H_2O, EtOH
9336	2,6-Pyridinediamine		$C_5H_7N_3$	141-86-6	109.130		121.5	285; 148^{5}			sl H_2O, ace
9337	3,4-Pyridinediamine		$C_5H_7N_3$	54-96-6	109.130	nd or lf	219.3				
9338	2,3-Pyridinedicarboxylic acid	Quinolinic acid	$C_7H_5NO_4$	89-00-9	167.120	mcl pr (w)	228.5				sl H_2O, tfa; i EtOH, eth, bz
9339	2,4-Pyridinedicarboxylic acid	Lutidinic acid	$C_7H_5NO_4$	499-80-9	167.120	lf (w+1)	249		0.942^{25}		sl H_2O; s EtOH; i eth, bz, CS_2
9340	2,5-Pyridinedicarboxylic acid	Isocinchomeronic acid	$C_7H_5NO_4$	100-26-5	167.120	lf or pr (dil HCl)	254				s H_2O, HCl; sl EtOH; i eth, bz
9341	2,6-Pyridinedicarboxylic acid	Dipicolinic acid	$C_7H_5NO_4$	499-83-2	167.120	nd (w+3/2)	252				sl H_2O, EtOH, HOAc
9342	3,4-Pyridinedicarboxylic acid	Cinchomeronic acid	$C_7H_5NO_4$	490-11-9	167.120	cry (w)	256	sub			sl H_2O, EtOH, bz; i eth, i chl
9343	3,5-Pyridinedicarboxylic acid	Dinicotinic acid	$C_7H_5NO_4$	499-81-0	167.120	cry (w)	324	sub			i H_2O; sl eth, HOAc; s DMSO, HCl
9344	2,3-Pyridinedicarboxylic acid anhydride	Furo[3,4-b]pyridine-5,7-dione	$C_7H_3NO_3$	699-98-9	149.104		138				
9345	2-Pyridineethanamine		$C_7H_{10}N_2$	2706-56-1	122.167			213; 131^{50}	1.0220^{25}	1.5335^{25}	
9346	4-Pyridineethanamine		$C_7H_{10}N_2$	13258-63-4	122.167			121^{10}	1.0302^{25}	1.5381^{25}	vs H_2O
9347	2-Pyridineethanol		C_7H_9NO	103-74-2	123.152		-7.8	190^{200}, 170^{100}	1.091^{25}	1.5366^{20}	vs H_2O, EtOH, chl; sl eth
9348	Pyridine hydrochloride		C_5H_6ClN	628-13-7	115.562	hyg pl or sc (al)	146	222			vs H_2O, EtOH, chl
9349	2-Pyridinemethanamine		$C_6H_8N_2$	3731-51-9	108.141			203; 91^{17}	1.0525^{25}	1.5431^{25}	vs H_2O
9350	3-Pyridinemethanamine		$C_6H_8N_2$	3731-52-0	108.141	liq	-21.1	226	1.064^{20}	1.552^{20}	vs H_2O, eth, EtOH
9351	4-Pyridinemethanamine		$C_6H_8N_2$	3731-53-1	108.141	liq	-7.6	230; 103^{11}	1.072^{20}	1.5495^{25}	vs H_2O
9352	2-Pyridinemethanol		C_6H_7NO	586-98-1	109.126			112^{16}, 102.5^{8}	1.1317^{20}	1.5444^{20}	msc H_2O; vs EtOH, eth, ace, bz
9353	3-Pyridinemethanol	Nicotinyl alcohol	C_6H_7NO	100-55-0	109.126	liq	-6.5	266	1.131^{20}	1.5455^{20}	vs H_2O, eth
9354	4-Pyridinemethanol	4-Picolyl alcohol	C_6H_7NO	586-95-8	109.126		53	141^{12}			s chl
9355	Pyridine-1-oxide	Pyridine *N*-oxide	C_5H_5NO	694-59-7	95.100		65.5	146^{13}			
9356	2-Pyridinepropanol		$C_8H_{11}NO$	2859-68-9	137.179		34	260.2; 116^{4}	1.060^{25}	1.5298^{20}	vs H_2O
9357	3-Pyridinepropanol		$C_8H_{11}NO$	2859-67-8	137.179			284; 130^{3}	1.063^{25}	1.5313^{20}	vs H_2O
9358	3-Pyridinesulfonic acid	3-Pyridylsulfonic acid	$C_5H_5NO_3S$	636-73-7	159.164	orth	357 dec		1.713^{25}		vs H_2O; sl EtOH; i eth
9359	2-Pyridinethiol, 1-oxide		C_5H_5NOS	1121-31-9	127.165		70.5				
9360	2(1*H*)-Pyridinethione		C_5H_5NS	2637-34-5	111.166		130.0				s H_2O, EtOH, bz, chl
9361	2-Pyridinol		C_5H_5NO	72762-00-6	95.100	nd (bz)	107.8		1.3910^{20}		vs H_2O, bz, EtOH
9362	3-Pyridinol		C_5H_5NO	109-00-2	95.100	nd (bz)	129				s H_2O, EtOH; sl eth, chl
9363	4-Pyridinol		C_5H_5NO	626-64-2	95.100	pr or nd (w+1)	149.8	>350; 257^{10}			s H_2O, EtOH; i eth, bz
9364	2(1*H*)-Pyridinone		C_5H_5NO	142-08-5	95.100	nd (bz)	107.8	280	1.3910^{20}		s H_2O, EtOH, bz, chl; sl eth, DMSO
9365	2(1*H*)-Pyridinone hydrazone	2-Pyridinylhydrazine	$C_5H_7N_3$	4930-98-7	109.130		46.6	185^{140}, 90^{1}			s chl
9366	α-[(2-Pyridinylamino)methyl]benzenemethanol	Phenyramidol	$C_{13}H_{14}N_2O$	553-69-5	214.262	cry (dil MeOH)	83.5				
9367	1-(2-Pyridinyl)ethanone		C_7H_7NO	1122-62-9	121.137	ye in air		192	1.077^{25}	1.5203^{20}	s EtOH, eth, HOAc; sl ctc
9368	1-(3-Pyridinyl)ethanone	Methyl pyridyl ketone	C_7H_7NO	350-03-8	121.137		13.5	220		1.5341^{20}	s H_2O, EtOH, eth, acid
9369	1-(4-Pyridinyl)ethanone		C_7H_7NO	1122-54-9	121.137		16	212	1.097^{25}	1.5282^{25}	sl EtOH, eth, acid

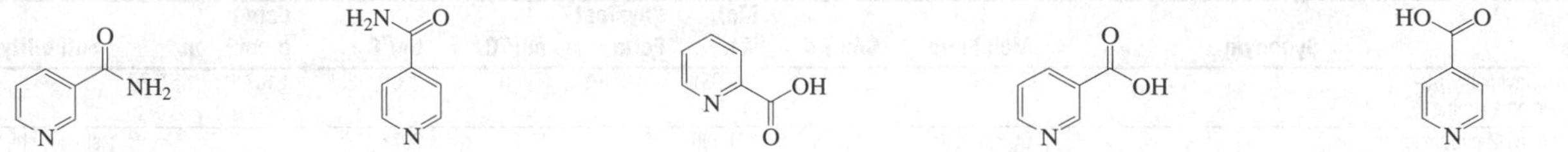

3-Pyridinecarboxamide

4-Pyridinecarboxamide

2-Pyridinecarboxylic acid

3-Pyridinecarboxylic acid

4-Pyridinecarboxylic acid

3-Pyridinecarboxylic acid 1-oxide

4-Pyridinecarboxylic acid 1-oxide

2,3-Pyridinediamine

2,5-Pyridinediamine

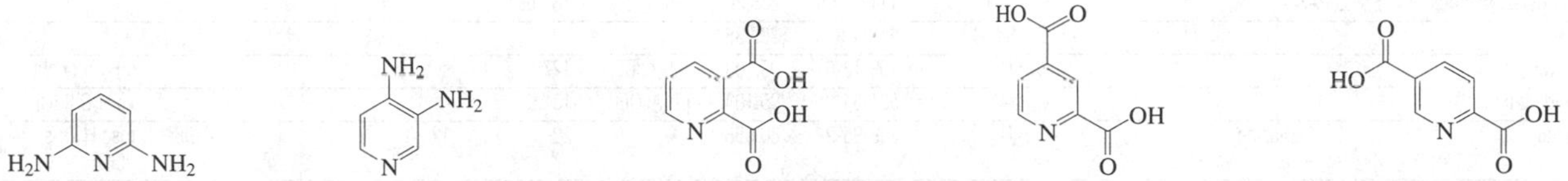

2,6-Pyridinediamine

3,4-Pyridinediamine

2,3-Pyridinedicarboxylic acid

2,4-Pyridinedicarboxylic acid

2,5-Pyridinedicarboxylic acid

2,6-Pyridinedicarboxylic acid

3,4-Pyridinedicarboxylic acid

3,5-Pyridinedicarboxylic acid

2,3-Pyridinedicarboxylic acid anhydride

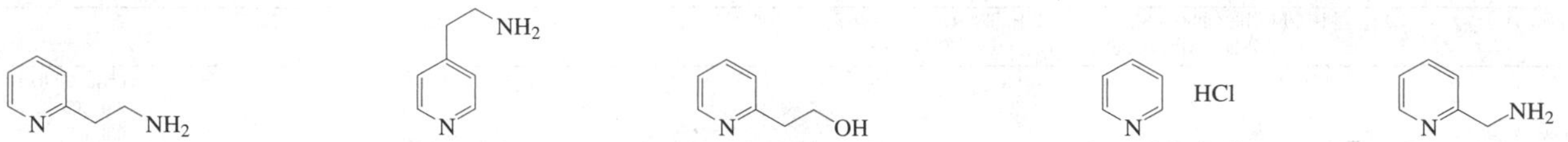

2-Pyridineethanamine

4-Pyridineethanamine

2-Pyridineethanol

Pyridine hydrochloride

2-Pyridinemethanamine

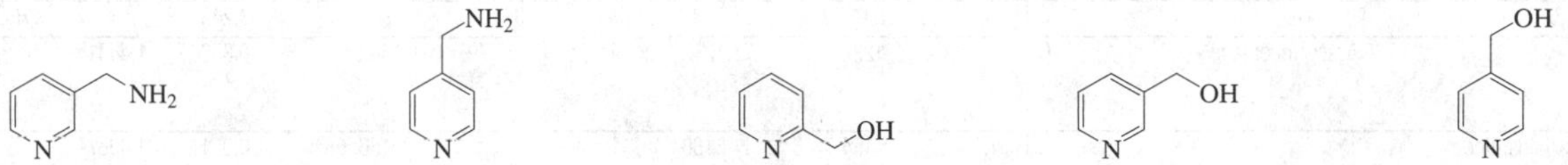

3-Pyridinemethanamine

4-Pyridinemethanamine

2-Pyridinemethanol

3-Pyridinemethanol

4-Pyridinemethanol

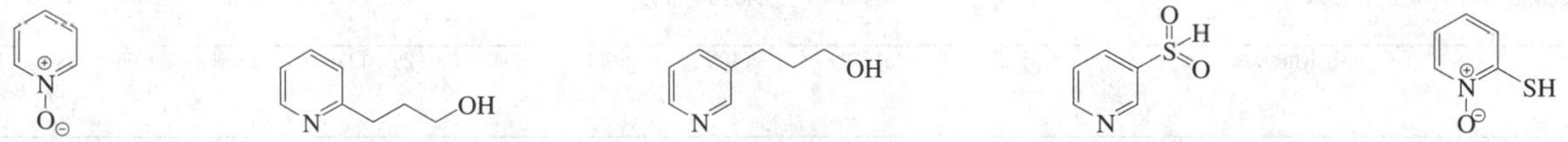

Pyridine-1-oxide

2-Pyridinepropanol

3-Pyridinepropanol

3-Pyridinesulfonic acid

2-Pyridinethiol, 1-oxide

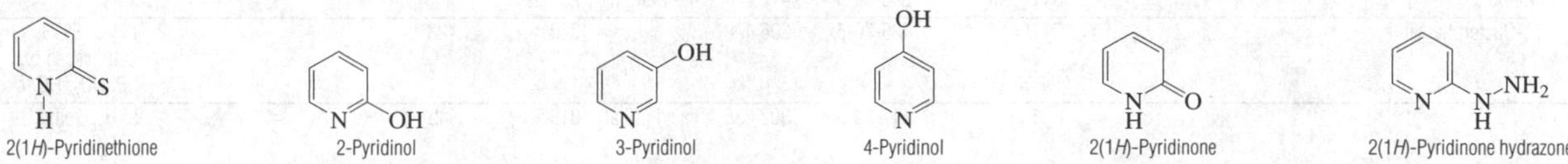

2(1*H*)-Pyridinethione

2-Pyridinol

3-Pyridinol

4-Pyridinol

2(1*H*)-Pyridinone

2(1*H*)-Pyridinone hydrazone

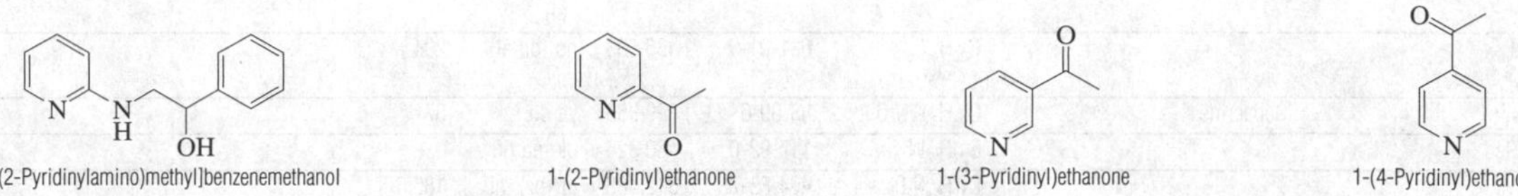

α-[(2-Pyridinylamino)methyl]benzenemethanol

1-(2-Pyridinyl)ethanone

1-(3-Pyridinyl)ethanone

1-(4-Pyridinyl)ethanone

No.	Name	Synonym	Mol. Form.	CAS RN	Mol. Wt.	Physical Form	mp/°C	bp/°C	den/ g cm^{-3}	n_D	Solubility
9370	*N*-(2-Pyridinylmethyl)-2-pyridinemethanamine		$C_{12}H_{13}N_3$	1539-42-0	199.251			200^{10}, 139^{1}	1.1074^{25}	1.5757^{25}	
9371	*N*-2-Pyridinyl-2-pyridinamine		$C_{10}H_9N_3$	1202-34-2	171.198		90.5	307.5			sl H_2O, chl; vs EtOH, eth, ace, bz
9372	Pyridoxal hydrochloride	Vitamin B6	$C_8H_{10}ClNO_3$	65-22-5	203.623	orth	165 dec				vs H_2O; sl EtOH
9373	Pyridoxal 5-phosphate	Pyridoxal 5-(dihydrogen phosphate)	$C_8H_{10}NO_6P$	54-47-7	247.142	wh-ye pow or cry	141				
9374	Pyridoxamine	4-(Aminomethyl)-5-hydroxy-6-methyl-3-pyridinemethanol	$C_8H_{12}N_2O_2$	85-87-0	168.193	cry	198				s EtOH, acid
9375	Pyridoxamine dihydrochloride		$C_8H_{14}Cl_2N_2O_2$	524-36-7	241.115	pl (al)	226 dec				vs H_2O; sl EtOH
9376	Pyridoxine hydrochloride	5-Hydroxy-6-methyl-3,4-pyridinedimethanol hydrochloride	$C_8H_{12}ClNO_3$	58-56-0	205.639	pl (al, ace)	207	sub			vs H_2O
9377	1-(2-Pyridylazo)-2-naphthol	PAN	$C_{15}H_{11}N_3O$	85-85-8	249.267	red-br cry	130				i H_2O; s EtOH, eth, chl
9378	4-(2'-Pyridylazo)resorcinol	PAR	$C_{11}H_9N_3O_2$	1141-59-9	215.208	red-br cry	187 dec				
9379	Pyrilamine		$C_{17}H_{23}N_3O$	91-84-9	285.384			201^{5}			
9380	2-Pyrimidinamine		$C_4H_5N_3$	109-12-6	95.103	nd (AcOEt)	127.5	sub			s H_2O; sl chl
9381	4-Pyrimidinamine		$C_4H_5N_3$	591-54-8	95.103	pl (AcOEt)	151.5				vs H_2O, EtOH
9382	Pyrimidine	1,3-Diazine	$C_4H_4N_2$	289-95-2	80.088		22	123.8		1.4998^{20}	msc H_2O; s EtOH
9383	2,4,5,6(1*H*,3*H*)-Pyrimidinetetrone	Alloxan	$C_4H_2N_2O_4$	50-71-5	142.070		256 dec	sub			vs H_2O; s EtOH, ace, bz, HOAc
9384	2,4,5,6(1*H*,3*H*)-Pyrimidinetetrone 5-oxime	Violuric acid	$C_4H_3N_3O_4$	87-39-8	157.085	pa ye orth	203 dec				sl H_2O; s EtOH
9385	2,4,6-Pyrimidinetriamine		$C_4H_7N_5$	1004-38-2	125.133		248 dec				
9386	Pyriminil		$C_{13}H_{12}N_4O_3$	53558-25-1	272.259	solid	224 dec				
9387	Pyrithione zinc		$C_{10}H_8N_2O_2S_2Zn$	13463-41-7	317.722	wh solid	262				s chl, DMSO, DMF
9388	Pyrocatechol	1,2-Benzenediol	$C_6H_6O_2$	120-80-9	110.111	cry	104.6	245	1.344^{20}	1.604^{25}	vs H_2O, bz, eth, EtOH
9389	*L*-Pyroglutamic acid	5-Oxo-*L*-proline	$C_5H_7NO_3$	98-79-3	129.115		162				s DMSO
9390	Pyrolan		$C_{13}H_{15}N_3O_2$	87-47-8	245.277		50	$161^{0.2}$			s ctc, CS_2
9391	Pyrrobutamine	1-[4-(4-Chlorophenyl)-3-phenyl-2-butenyl]pyrrolidine	$C_{20}H_{22}ClN$	91-82-7	311.849	cry	49	$192^{0.3}$			
9392	Pyrrole	Imidole	C_4H_5N	109-97-7	67.090	liq	-23.39	129.79	0.9698^{20}	1.5085^{20}	sl H_2O; s EtOH, eth, ace, bz, chl
9393	1*H*-Pyrrole-2-carboxaldehyde		C_5H_5NO	1003-29-8	95.100	orth pr (peth)	46.5	218		1.5939^{16}	sl chl, lig
9394	1*H*-Pyrrole-2-carboxylic acid		$C_5H_5NO_2$	634-97-9	111.100	lf (w)	208 dec				s H_2O, EtOH, eth
9395	1*H*-Pyrrole-3-carboxylic acid	3-Pyrrolecarboxylic acid	$C_5H_5NO_2$	931-03-3	111.100	nd (lig)	161.5				
9396	1*H*-Pyrrole-2,5-dione		$C_4H_3NO_2$	541-59-3	97.073	pl (bz)	94	sub	1.2493^{106}		s H_2O, EtOH, eth
9397	Pyrrolidine	Azacyclopentane	C_4H_9N	123-75-1	71.121	col liq	-57.79	86.56	0.8586^{20}	1.4431^{20}	msc H_2O; s EtOH, eth; sl bz, chl
9398	1-Pyrrolidineethanamine		$C_6H_{14}N_2$	7154-73-6	114.188			166; 68^{23}	0.901^{25}	1.4687^{20}	
9399	1-Pyrrolidineethanol		$C_6H_{13}NO$	2955-88-6	115.173			187; 80^{13}	0.9750^{20}	1.4713^{20}	
9400	1-[4-(1-Pyrrolidinyl)-2-butynyl]-2-pyrrolidinone	Oxotremorine	$C_{12}H_{18}N_2O$	70-22-4	206.283	pa ye liq		$124^{0.1}$	0.991^{25}	1.5160^{20}	
9401	3-(2-Pyrrolidinyl)pyridine, (*S*)	Nornicotine	$C_9H_{12}N_2$	494-97-3	148.204	hyg		270	1.0737^{19}	1.5378^{18}	vs H_2O, ace, eth, EtOH
9402	2-Pyrrolidone	γ-Butyrolactam	C_4H_7NO	616-45-5	85.105	cry (peth)	25	251; 133^{12}	1.120^{20}	1.4806^{30}	vs H_2O, EtOH, eth, bz, chl, CS_2
9403	1-(1*H*-Pyrrol-2-yl)ethanone		C_6H_7NO	1072-83-9	109.126	mcl nd (w)	90	220			s H_2O, EtOH, eth
9404	Pyruvic acid		$C_3H_4O_3$	127-17-3	88.062		13.8	dec 165; 54^{10}	1.2272^{20}	1.4280^{20}	msc H_2O, EtOH, eth; s ace
9405	Pyrvinium chloride		$C_{26}H_{28}ClN_3$	548-84-5	417.973	red pow (w)	250 dec				
9406	1,1':4',1":4",1"'-Quaterphenyl		$C_{24}H_{18}$	135-70-6	306.400		320	428^{18}			i H_2O, EtOH, eth, chl; s bz, $PhNO_2$, HOAc
9407	Quercetin		$C_{15}H_{10}O_7$	117-39-5	302.236	ye nd (dil al, + 2 w)	316.5	sub			sl H_2O, eth, MeOH; s EtOH, ace, py
9408	Quercitrin	Quercetin-3-*L*-rhamnoside	$C_{21}H_{20}O_{11}$	522-12-3	448.377	pa ye nd or pl (+2w, dil al)	170				i H_2O, eth; s EtOH, HOAc, MeOH, alk
9409	Quillaic acid		$C_{30}H_{46}O_5$	631-01-6	486.683	nd (dil al)	294				vs ace, eth, py, EtOH
9410	Quinacrine	Mepacrine	$C_{23}H_{30}ClN_3O$	83-89-6	399.956	ye oil	87				
9411	Quinaldine Red		$C_{21}H_{23}IN_2$	117-92-0	430.325	dk red pow					s H_2O; vs EtOH
9412	Quinamine		$C_{19}H_{24}N_2O_2$	464-85-7	312.406	pr (bz), nd (80% al)	185.5				i H_2O; vs EtOH, bz; s eth, ace

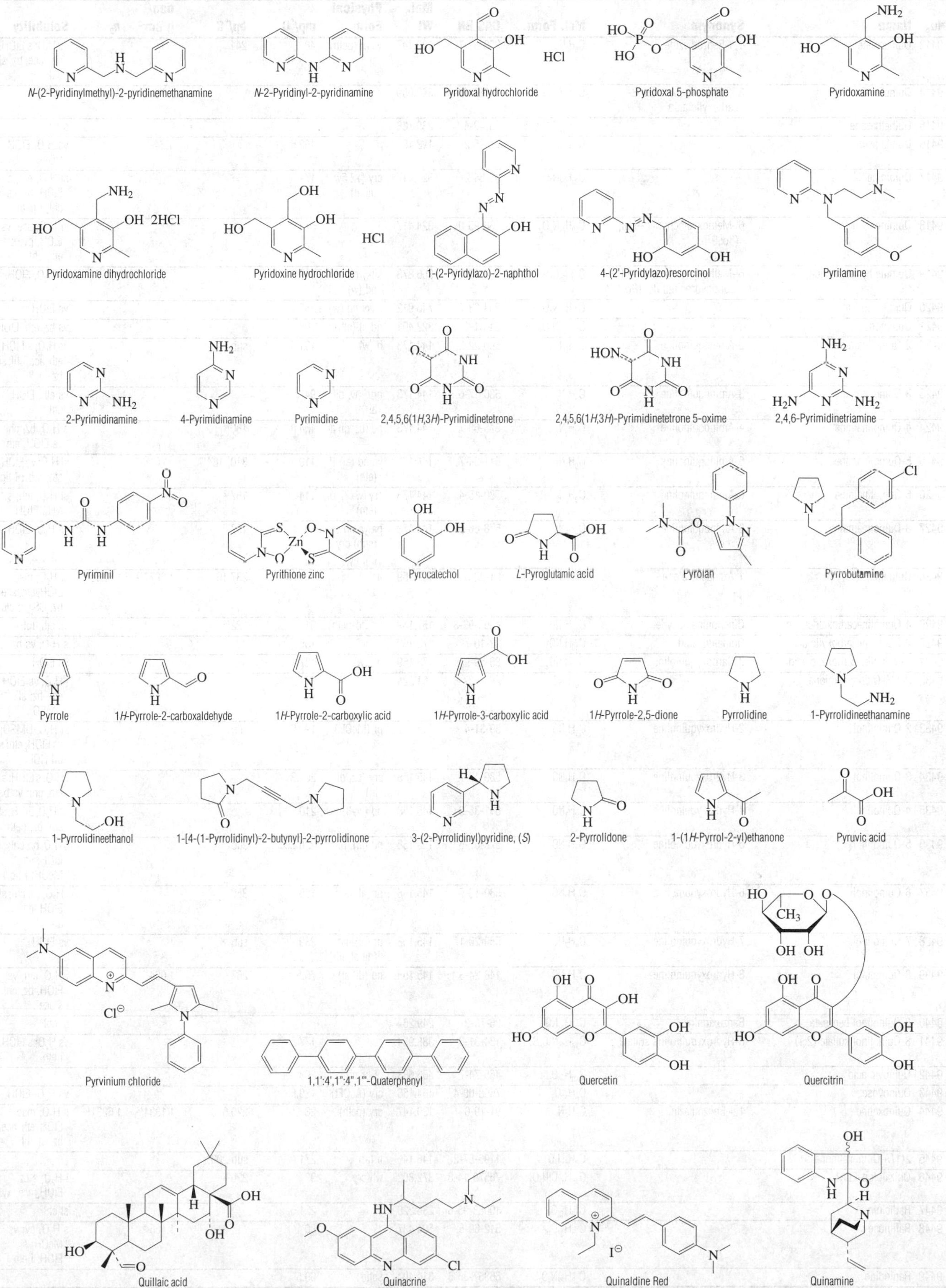

N-(2-Pyridinylmethyl)-2-pyridinemethanamine
N-2-Pyridinyl-2-pyridinamine
Pyridoxal hydrochloride
Pyridoxal 5-phosphate
Pyridoxamine
Pyridoxamine dihydrochloride
Pyridoxine hydrochloride
1-(2-Pyridylazo)-2-naphthol
4-(2'-Pyridylazo)resorcinol
Pyrilamine
2-Pyrimidinamine
4-Pyrimidinamine
Pyrimidine
2,4,5,6(1H,3H)-Pyrimidinetetrone
2,4,5,6(1H,3H)-Pyrimidinetetrone 5-oxime
2,4,6-Pyrimidinetriamine
Pyriminil
Pyrithione zinc
Pyrocatechol
L-Pyroglutamic acid
Pyrolan
Pyrrobutamine
Pyrrole
1H-Pyrrole-2-carboxaldehyde
1H-Pyrrole-2-carboxylic acid
1H-Pyrrole-3-carboxylic acid
1H-Pyrrole-2,5-dione
Pyrrolidine
1-Pyrrolidineethanamine
1-Pyrrolidineethanol
1-[4-(1-Pyrrolidinyl)-2-butynyl]-2-pyrrolidinone
3-(2-Pyrrolidinyl)pyridine, (S)
2-Pyrrolidone
1-(1H-Pyrrol-2-yl)ethanone
Pyruvic acid
Pyrvinium chloride
1,1':4',1'':4'',1'''-Quaterphenyl
Quercetin
Quercitrin
Quillaic acid
Quinacrine
Quinaldine Red
Quinamine

No.	Name	Synonym	Mol. Form.	CAS RN	Mol. Wt.	Physical Form	mp/°C	bp/°C	den/ g cm⁻³	n_D	Solubility
9413	Quinazoline	1,3-Benzodiazine	$C_8H_6N_2$	253-82-7	130.147	ye pl (peth)	48	241			vs H_2O; s EtOH, eth, ace, bz; sl chl
9414	Quinclorac	3,7-Dichloroquinoline-8-carboxylic acid	$C_{10}H_5Cl_2NO_2$	84087-01-4	242.059		274		1.75		
9415	Quinethazone		$C_{10}H_{12}ClN_3O_3S$	73-49-4	289.738						s tfa
9416	Quinic acid		$C_7H_{12}O_6$	77-95-2	192.166		162.5		1.64^{25}		vs H_2O, EtOH, HOAc
9417	Quinidine		$C_{20}H_{24}N_2O_2$	56-54-2	324.417	cry (+2.5w, dil al)	174				sl H_2O, eth; s EtOH, bz; vs chl; i peth
9418	Quinine	6'-Methoxycinchonan-9-ol, (8α,9*R*)	$C_{20}H_{24}N_2O_2$	130-95-0	324.417		57			1.625^{15}	sl H_2O, ace; vs EtOH, py; s eth, chl
9419	Quinine hydrochloride	6'-Methoxycinchonan-9-ol monohydrochloride, (8α,9*R*)	$C_{20}H_{25}ClN_2O_2$	130-89-2	360.878	silky efflor nd (w)	159				vs H_2O, EtOH, chl
9420	Quinine sulfate		$C_{40}H_{50}N_4O_8S$	804-63-7	746.912	silky nd (w)	235.2				vs EtOH
9421	Quininone		$C_{20}H_{22}N_2O_2$	84-31-1	322.401	nd, lf (eth)	108				vs bz, eth, EtOH
9422	2-Quinolinamine	2-Aminoquinoline	$C_9H_8N_2$	580-22-3	144.173	lf (w)	131.5	sub			vs H_2O; s EtOH, eth, ace, chl; sl bz
9423	3-Quinolinamine	3-Aminoquinoline	$C_9H_8N_2$	580-17-6	144.173	orth (w, dil al)	94				vs eth, EtOH, chl
9424	4-Quinolinamine	4-Aminoquinoline	$C_9H_8N_2$	578-68-7	144.173	nd (bz, dil al)	154.8	180^{12}			s H_2O, bz, chl; vs EtOH, eth
9425	5-Quinolinamine	5-Aminoquinoline	$C_9H_8N_2$	611-34-7	144.173	ye nd (al) lf (eth)	110	310; 184^{10}			sl H_2O; vs EtOH, eth; s bz; i lig
9426	6-Quinolinamine	6-Aminoquinoline	$C_9H_8N_2$	580-15-4	144.173	cry (w+2), pr (eth)	114	187^{12}			sl H_2O, eth; s NH_3, EtOH
9427	8-Quinolinamine	8-Aminoquinoline	$C_9H_8N_2$	578-66-5	144.173	pa ye nd (sub) cry (al, lig)	70	157^{19}			vs H_2O, EtOH
9428	Quinoline	1-Azanaphthalene	C_9H_7N	91-22-5	129.159	liq	-14.78	237.16	1.0977^{15}	1.6268^{20}	sl H_2O; msc EtOH, eth, ace, bz, CS_2; s ctc
9429	4-Quinolinecarboxaldehyde	Cinchoninaldehyde	$C_{10}H_7NO$	4363-93-3	157.169	nd (to-peth)	51	122^{4}			vs eth, tol
9430	2-Quinolinecarboxylic acid	Quinaldic acid	$C_{10}H_7NO_2$	93-10-7	173.169		156				s H_2O; vs bz
9431	8-Quinolinecarboxylic acid	8-Carboxyquinoline	$C_{10}H_7NO_2$	86-59-9	173.169	nd (w)	187	sub			vs EtOH
9432	2(1*H*)-Quinolinethione		C_9H_7NS	2637-37-8	161.224		187				i H_2O; vs EtOH, eth, bz; sl DMSO
9433	2-Quinolinol	2-Hydroxyquinoline	C_9H_7NO	59-31-4	145.158	pr (MeOH)	199.5	sub			sl H_2O, DMSO; vs EtOH, eth; s dil HCl
9434	3-Quinolinol	3-Hydroxyquinoline	C_9H_7NO	580-18-7	145.158	cry (bz, dil al)	201.3				i H_2O; s EtOH; sl eth, chl; vs bz
9435	4-Quinolinol	4-Hydroxyquinoline	C_9H_7NO	611-36-9	145.158	nd (w+3)	210				vs H_2O, EtOH; sl eth, bz, peth
9436	5-Quinolinol	5-Hydroxyquinoline	C_9H_7NO	578-67-6	145.158	nd (al), pl	226 dec	sub			s H_2O, bz, chl; sl EtOH; vs MeOH; i lig
9437	6-Quinolinol	6-Hydroxyquinoline	C_9H_7NO	580-16-5	145.158	pr (al, eth)	195	360			i H_2O, bz, chl; sl EtOH, eth; s alk
9438	7-Quinolinol	7-Hydroxyquinoline	C_9H_7NO	580-20-1	145.158	pr (al), nd (dil al-eth)	239	sub			vs EtOH
9439	8-Quinolinol	8-Hydroxyquinoline	C_9H_7NO	148-24-3	145.158	nd (dil al)	75.5	267	1.034^{20}		i H_2O, eth; vs EtOH, bz, chl; s ace
9440	8-Quinolinol benzoate	Benzoxiquine	$C_{16}H_{11}NO_2$	86-75-9	249.264						sl chl
9441	8-Quinolinol sulfate (2:1)	8-Hydroxyquinoline sulfate	$C_{18}H_{16}N_2O_6S$	134-31-6	388.934		177.5				vs H_2O; s EtOH; i eth
9442	Quinovic acid		$C_{30}H_{46}O_5$	465-74-7	486.683	pl or nd	298 dec				
9443	Quinovose		$C_6H_{12}O_5$	7658-08-4	164.156	cry (AcOEt)	139.5				vs H_2O, EtOH
9444	Quinoxaline	1,4-Benzodiazine	$C_8H_6N_2$	91-19-0	130.147	cry (peth)	28	229.5	1.1334^{48}	1.6231^{48}	s H_2O; msc EtOH, eth, ace, bz; sl chl
9445	2(1*H*)-Quinoxalinone		$C_8H_6N_2O$	1196-57-2	146.146	lf (al)	271	sub 200			
9446	Quizalofop-Ethyl		$C_{19}H_{17}ClN_2O_4$	76578-14-8	372.802	wh cry	93	$220^{0.2}$			i H_2O; s bz, EtOH, ace, xyl
9447	Radicinin		$C_{12}H_{12}O_5$	10088-95-6	236.220		221.5				sl chl
9448	Raffinose		$C_{18}H_{32}O_{16}$	512-69-6	504.437		80		1.465^{25}		s H_2O, py; vs MeOH; sl EtOH; i eth
9449	Ranitidine		$C_{13}H_{22}N_4O_3S$	66357-35-5	314.404	solid	69.5				

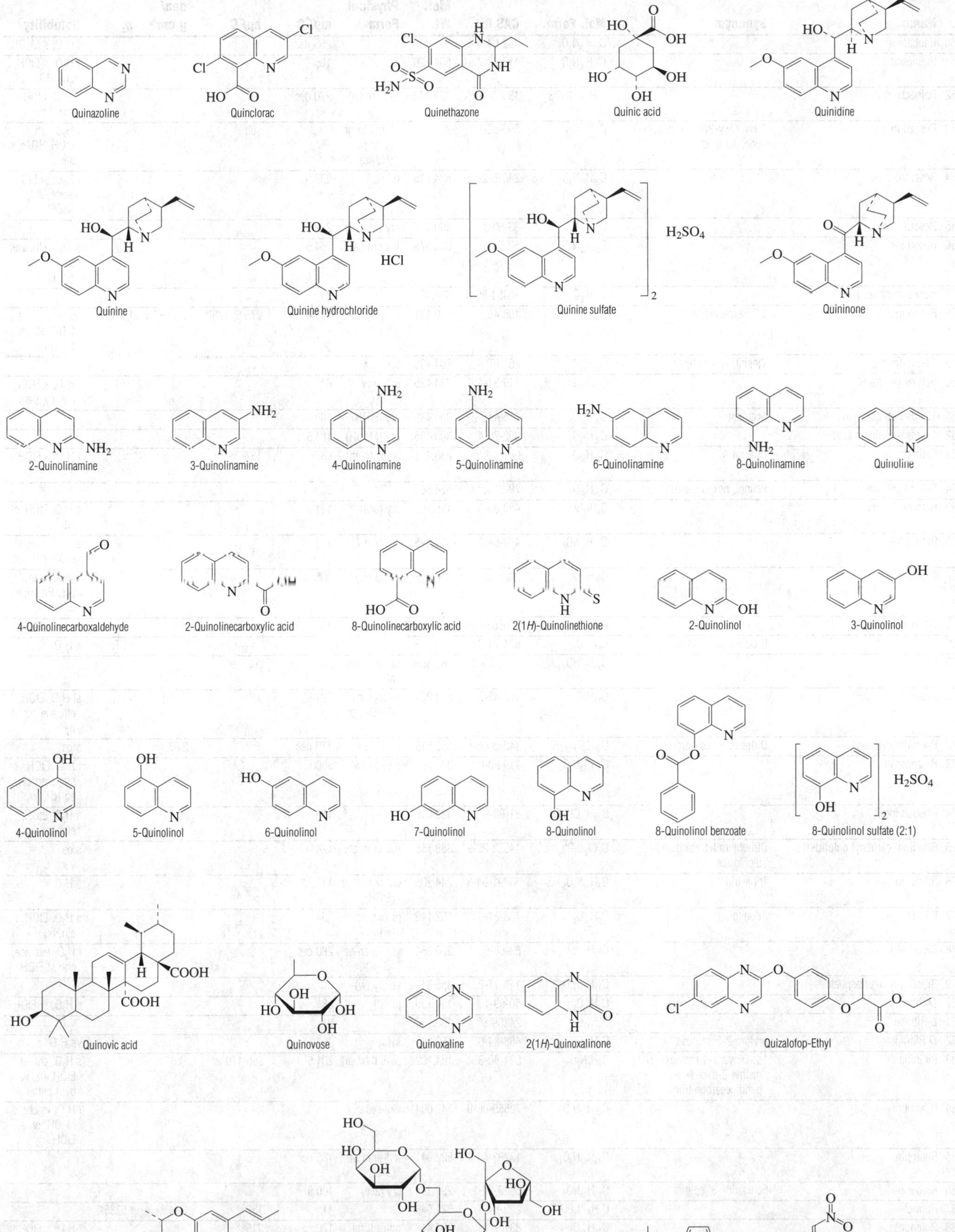
Quinazoline
Quinclorac
Quinethazone
Quinic acid
Quinidine
Quinine
Quinine hydrochloride
Quinine sulfate
Quininone
2-Quinolinamine
3-Quinolinamine
4-Quinolinamine
5-Quinolinamine
6-Quinolinamine
8-Quinolinamine
Quinoline
4-Quinolinecarboxaldehyde
2-Quinolinecarboxylic acid
8-Quinolinecarboxylic acid
2(1H)-Quinolinethione
2-Quinolinol
3-Quinolinol
4-Quinolinol
5-Quinolinol
6-Quinolinol
7-Quinolinol
8-Quinolinol
8-Quinolinol benzoate
8-Quinolinol sulfate (2:1)
Quinovic acid
Quinovose
Quinoxaline
2(1H)-Quinoxalinone
Quizalofop-Ethyl
Radicinin
Raffinose
Ranitidine

No.	Name	Synonym	Mol. Form.	CAS RN	Mol. Wt.	Physical Form	mp/°C	bp/°C	den/ g cm^{-3}	n_D	Solubility
9450	Raubasine		$C_{21}H_{24}N_2O_3$	483-04-5	352.427		258 dec				i H_2O; s MeOH
9451	Raunescine		$C_{31}H_{36}N_2O_8$	117-73-7	564.626		165				i H_2O; s EtOH, chl, HOAc
9452	Reinecke salt		$C_4H_{12}CrN_7OS_4$	13573-16-5	354.440	red cry (w)	270 dec				s H_2O, EtOH, ace; i bz
9453	Resazurin	7-Hydroxy-3*H*-phenoxazin-3-one, 10-oxide	$C_{12}H_7NO_4$	550-82-3	229.189	dk red to gr pr or pl (HOAc)		sub			i H_2O, eth; sl EtOH, HOAc; s alk
9454	Rescinnamine		$C_{35}H_{42}N_2O_9$	24815-24-5	634.716	nd (bz)	238.5				i H_2O; sl EtOH; s ace, chl, AcOEt
9455	Reserpic acid		$C_{22}H_{28}N_2O_5$	83-60-3	400.467	cry (MeOH)	242				
9456	Reserpine		$C_{33}H_{40}N_2O_9$	50-55-5	608.679	lo pr (dil ace)	264.5				sl H_2O, eth, ace; s EtOH, bz, AcOEt
9457	*cis*-Resmethrin, (-)		$C_{22}H_{26}O_3$	10453-86-8	338.439		75				
9458	Resorcinol	1,3-Benzenediol	$C_6H_6O_2$	108-46-3	110.111	nd (bz), pl (w)	109.4	276.5; 178^{16}	1.278^{20}	1.578^{25}	vs H_2O, ctc; s EtOH, eth; sl bz, chl
9459	11-*cis*-Retinal	Vitamin A_1 aldehyde	$C_{20}H_{28}O$	564-87-4	284.435	cry					
9460	Retinal (all *trans*)		$C_{20}H_{28}O$	116-31-4	284.435	oran cry	64				i H_2O; s EtOH, chl, cy, peth
9461	13-*cis*-Retinoic acid	Accutane	$C_{20}H_{28}O_2$	4759-48-2	300.435	cry (EtOH)	189				
9462	13-*trans*-Retinoic acid		$C_{20}H_{28}O_2$	302-79-4	300.435	cry (MeOH)	181.5				
9463	Retinol	Vitamin A	$C_{20}H_{30}O$	68-26-8	286.451	ye pr (peth)	63.5	$137^{0.000001}$			i H_2O; s EtOH, eth, ace, bz
9464	Retinyl palmitate	Retinol, hexadecanoate	$C_{36}H_{60}O_2$	79-81-2	524.860		28				
9465	Retronecine, (+)		$C_8H_{13}NO_2$	480-85-3	155.195	cry (ace)	121				s H_2O, EtOH; sl eth
9466	Retrorsine		$C_{18}H_{25}NO_6$	480-54-6	351.395	cry (AcOEt)	212				sl H_2O, ace; s EtOH, chl; i eth
9467	Rhamnetin		$C_{16}H_{12}O_7$	90-19-7	316.262	ye nd (al)	295				sl H_2O; s EtOH, ace, PhOH; vs dil alk
9468	*DL*-α-Rhamnose		$C_6H_{12}O_5$	116908-82-8	164.156	cry (w)	151				vs H_2O, EtOH
9469	*D*-Rhamnose	6-Deoxy-*D*-mannose	$C_6H_{12}O_5$	634-74-2	164.156						s H_2O
9470	Rheadine		$C_{21}H_{21}NO_6$	2718-25-4	383.395	nd (chl, eth, al)	257	sub			
9471	Rhein		$C_{15}H_8O_6$	478-43-3	284.221	ye or oran nd (MeOH, py)	321	sub			sl H_2O, EtOH, eth, ace, bz; vs py
9472	Rhenium carbonyl	Dirhenium decacarbonyl	$C_{10}O_{10}Re_2$	14285-68-8	652.515	ye-wh cry	170 dec		2.87		s os
9473	Rhizopterin		$C_{15}H_{12}N_6O_4$	119-20-0	340.294	lt ye pl (w)	>300				i H_2O, EtOH, eth; s aq alk, aq NH_3, py
9474	Rhodamine B		$C_{28}H_{32}ClN_2O_3$	81-88-9	480.018		165				s H_2O, EtOH, eth, bz, xyl
9475	Rhodium carbonyl chloride	Dirhodium tetracarbonyl dichloride	$C_4Cl_2O_4Rh_2$	14523-22-9	388.758	red-oran cry	124				s os
9476	Ribavirin	Tribavirin	$C_8H_{12}N_4O_5$	36791-04-5	244.205	col cry (EtOH)	175				s H_2O
9477	Ribitol	Adonitol	$C_5H_{12}O_5$	488-81-3	152.146	pr (w), nd (al)	104				s H_2O, EtOH; i eth, lig
9478	Riboflavin		$C_{17}H_{20}N_4O_6$	83-88-5	376.364	ye or oran-ye nd (w)	280 dec				i H_2O, eth, ace, chl; sl EtOH
9479	Riboflavin-5'-phosphate		$C_{17}H_{21}N_4O_9P$	146-17-8	456.34	ye cry (w)					
9480	*D*-Ribose		$C_5H_{10}O_5$	50-69-1	150.130	pl (al)	88				s H_2O; sl EtOH
9481	*L*-Ribose		$C_5H_{10}O_5$	24259-59-4	150.130		81				
9482	*D*-Ribulose	*erythro*-2-Pentulose	$C_5H_{10}O_5$	488-84-6	150.130	syrup					vs H_2O
9483	Ricinine	1,2-Dihydro-4-methoxy-1-methyl-2-oxo-3-pyridinecarbonitrile	$C_8H_8N_2O_2$	524-40-3	164.162	pr or lf (w, al)	201.5	sub 170			s H_2O, chl; sl EtOH, bz; vs py; i peth
9484	Rifabutin		$C_{46}H_{62}N_4O_{11}$	72559-06-9	847.004	viol-red cry					i H_2O; vs chl; s MeOH; sl EtOH
9485	Rifampin		$C_{43}H_{58}N_4O_{12}$	13292-46-1	822.941	red-oran pl (ace)	185 dec				
9486	Rinderine	Echinatine-3'-epimer	$C_{15}H_{25}NO_5$	6029-84-1	299.364	cry (ace)	100.5				
9487	Ronnel		$C_8H_8Cl_3O_3PS$	299-84-3	321.546		41	$152^{0.4}$	1.44^{32}	1.5335^{35}	
9488	Rotenone		$C_{23}H_{22}O_6$	83-79-4	394.417	nd or lf (al, aq-ace)	176	$215^{0.5}$			i H_2O; s EtOH, ace, bz; sl eth; vs chl
9489	Rubijervine		$C_{27}H_{43}NO_2$	79-58-3	413.636	nd (+1w, dil al)	242				vs bz, EtOH, chl
9490	Rubratoxin B		$C_{26}H_{30}O_{11}$	21794-01-4	518.509	cry (MeCN)	169 dec				

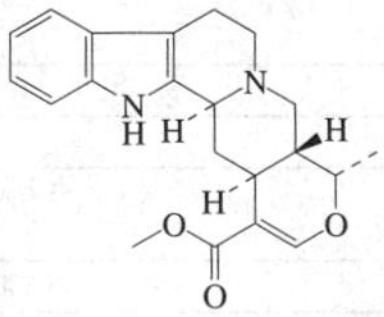

Raubasine

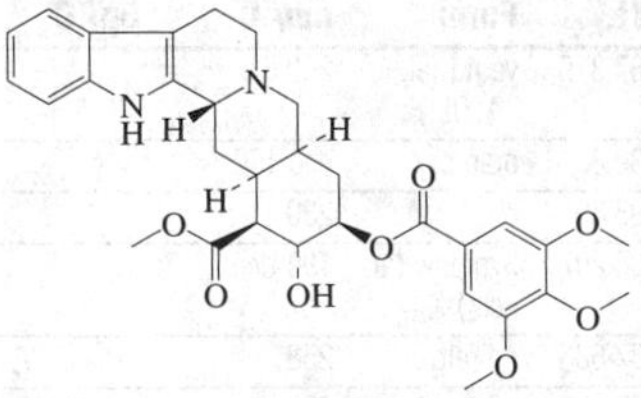

Raunescine

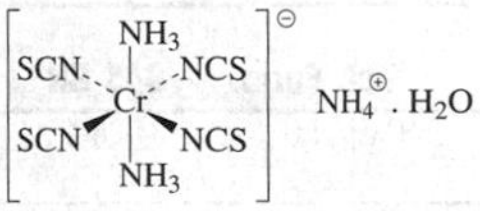

Reinecke salt

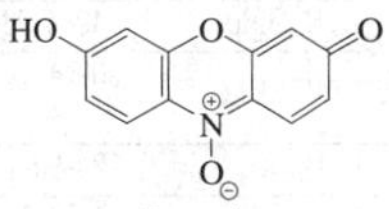

Resazurin

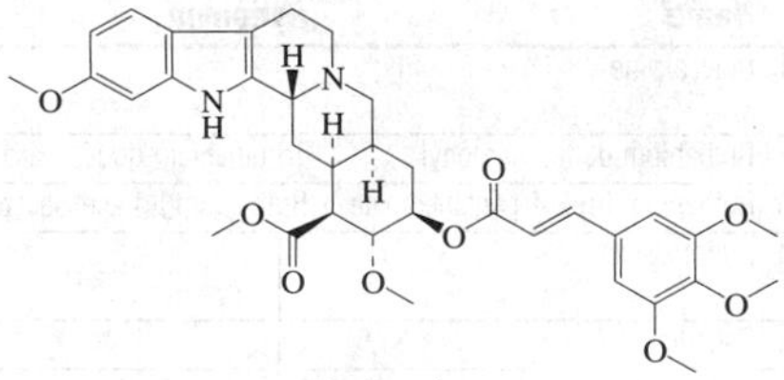

Rescinnamine

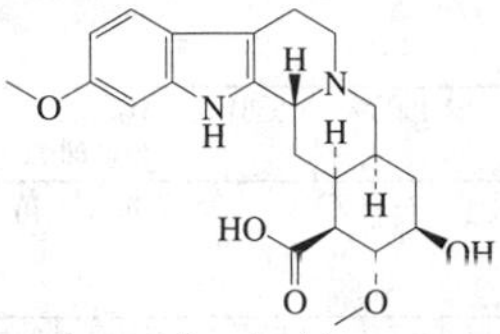

Reserpic acid

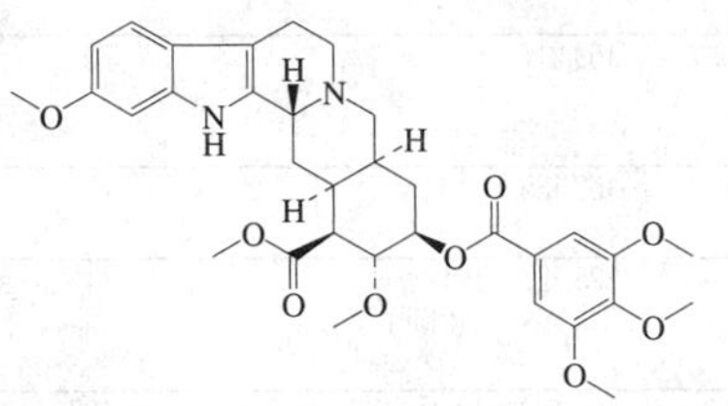

Reserpine

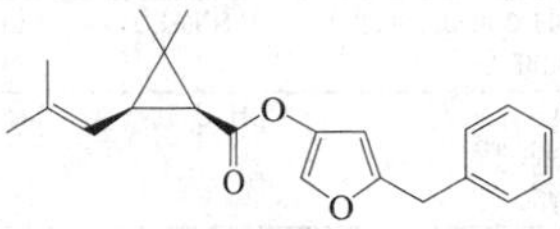

cis-Resmethrin, (-)

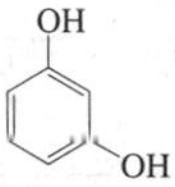

Resorcinol

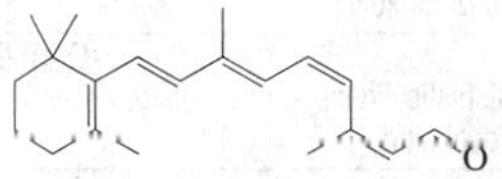

11-*cis*-Retinal

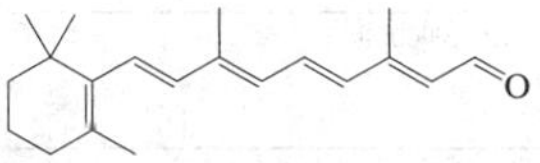

Retinal (all *trans*)

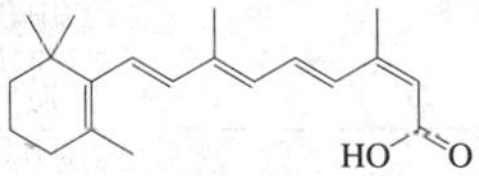

13-*cis*-Retinoic acid

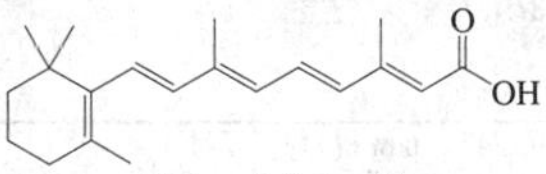

13-*trans*-Retinoic acid

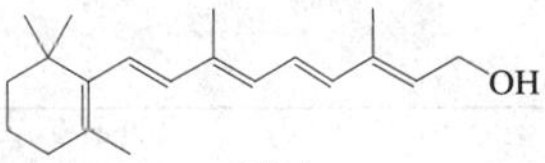

Retinol

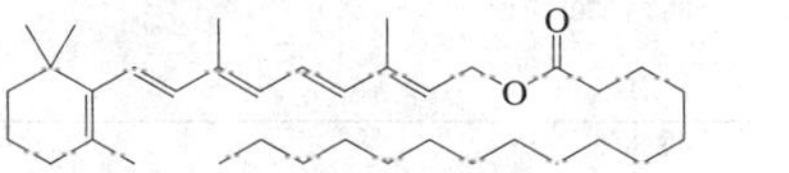

Retinyl palmitate

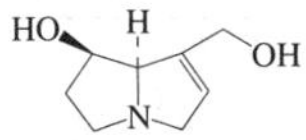

Retronecine, (+)

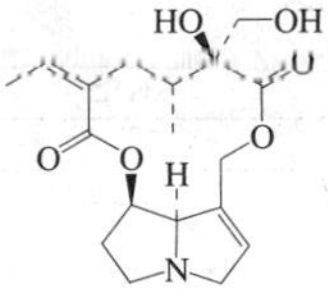

Retrorsine

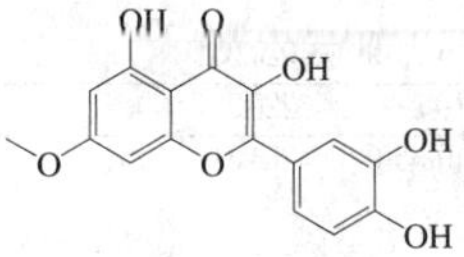

Rhamnetin

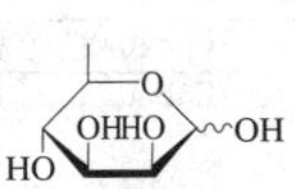

DL-α-Rhamnose

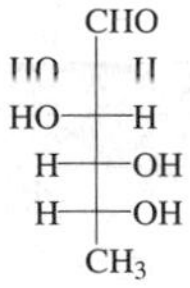

D-Rhamnose

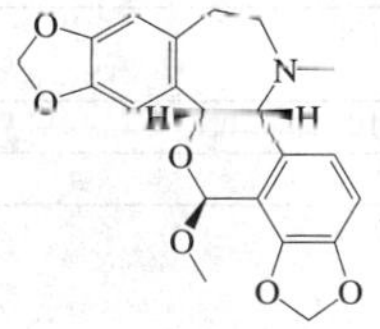

Rheadine

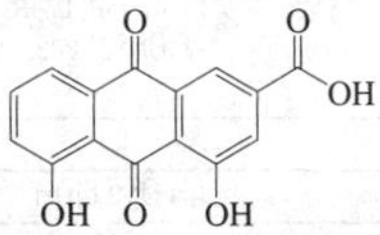

Rhein

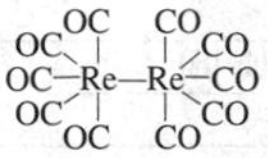

Rhenium carbonyl

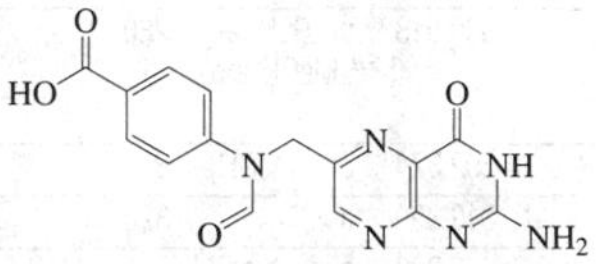

Rhizopterin

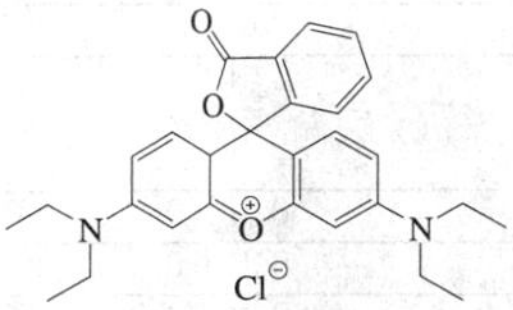

Rhodamine B

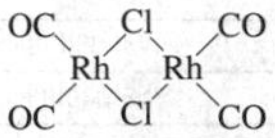

Rhodium carbonyl chloride

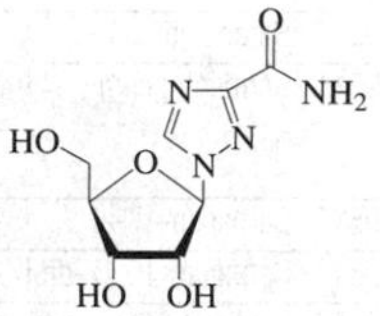

Ribavirin

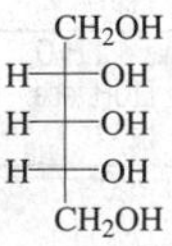

Ribitol

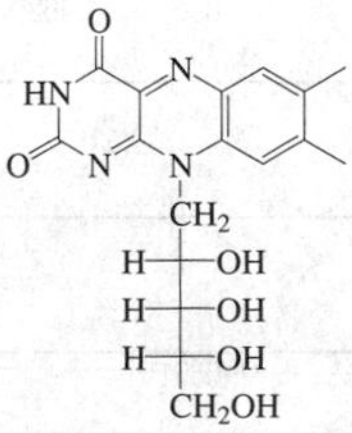

Riboflavin

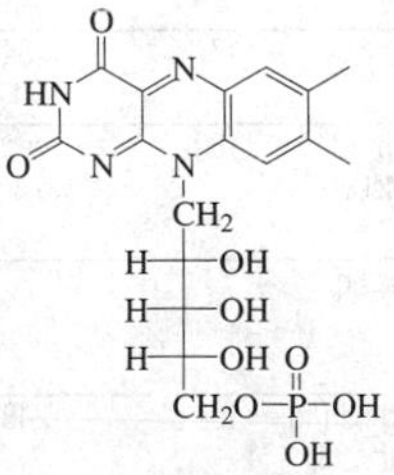

Riboflavin-5′-phosphate

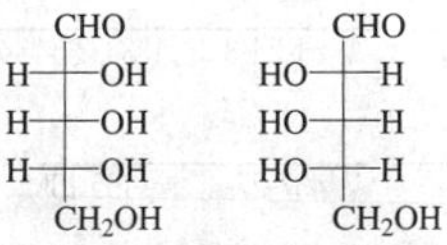

D-Ribose

L-Ribose

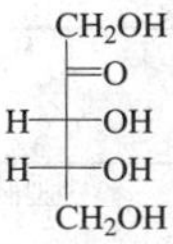

D-Ribulose

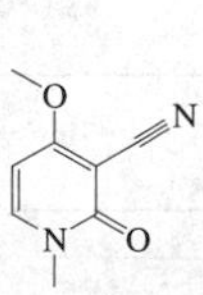

Ricinine

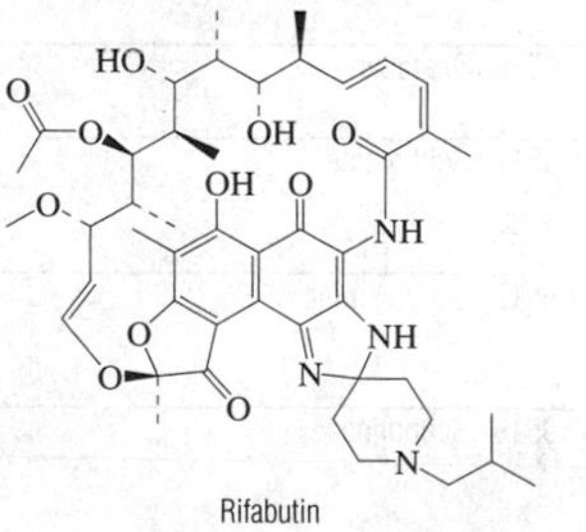

Rifabutin

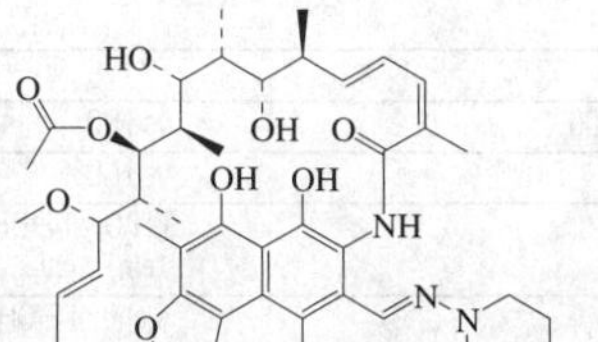

Rifampin

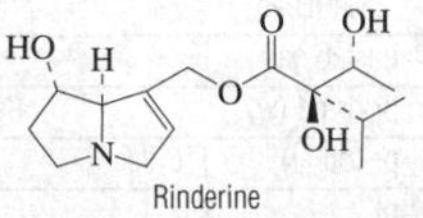

Rinderine

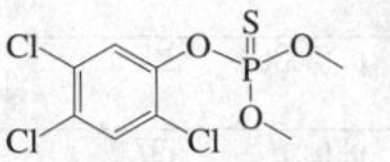

Ronnel

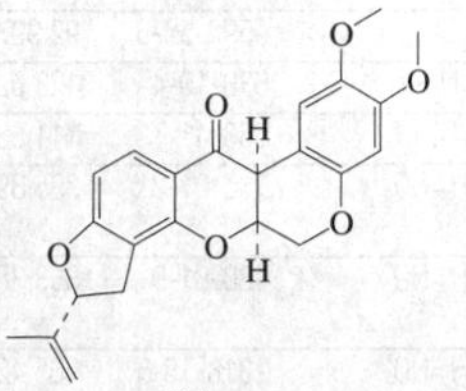

Rotenone

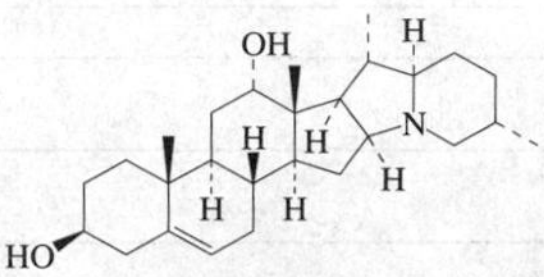

Rubijervine

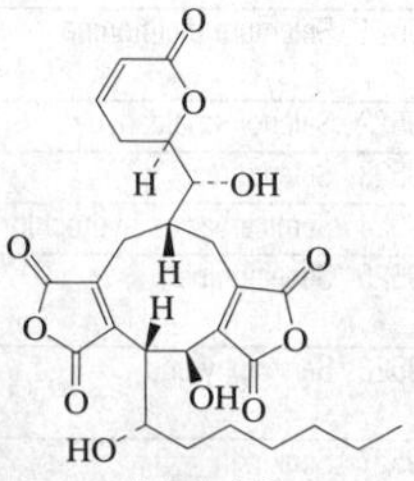

Rubratoxin B

No.	Name	Synonym	Mol. Form.	CAS RN	Mol. Wt.	Physical Form	mp/°C	bp/°C	den/ g cm^{-3}	n_D	Solubility
9491	Rutecarpine		$C_{18}H_{13}N_3O$	84-26-4	287.315	ye nd (al, AcOEt)	259.5				sl EtOH, ace, bz
9492	Ruthenium dodecacarbonyl	Triruthenium dodecacarbonyl	$C_{12}O_{12}Ru_3$	15243-33-1	639.33	oran cry	dec 150				
9493	Ruthenium(III) 2,4-pentanedioate	Ruthenium(III) acetylacetonate	$C_{15}H_{21}O_6Ru$	14284-93-6	398.39		230				
9494	Rutinose		$C_{12}H_{22}O_{10}$	90-74-4	326.297	hyg pow (al, eth)	190 dec				vs H_2O, EtOH
9495	Sabadine		$C_{29}H_{47}NO_8$	124-80-1	537.685	nd (eth)	258				vs ace, EtOH
9496	Saccharin		$C_7H_5NO_3S$	81-07-2	183.185	nd (ace) pr (al), lf (w)	228 dec	sub	0.828^{25}		sl H_2O, bz, eth, chl; s ace, EtOH
9497	Saccharin sodium	1,2-Benzisothiazolin-3-one, 1,1-dioxide, sodium salt	$C_7H_4NNaO_3S$	128-44-9	205.168	wh cry	229				s H_2O
9498	Safranal	2,6,6-Trimethyl-1,3-cyclohexadiene-1-carboxaldehyde	$C_{10}H_{14}O$	116-26-7	150.217			70^{1}	0.9734^{19}	1.5281^{19}	vs EtOH, peth
9499	Safrole	5-(2-Propenyl)-1,3-benzodioxole	$C_{10}H_{10}O_2$	94-59-7	162.185	mcl	11.2	234.5	1.1000^{20}	1.5381^{20}	i H_2O; vs EtOH; msc eth, chl
9500	Salcomine	*N,N'*-Bis(salicylidene)ethylenediaminocobalt(II)	$C_{16}H_{14}CoN_2O_2$	14167-18-1	325.227	red cry (DMF)					s bz, chl, py
9501	Salicylaldehyde	2-Hydroxybenzaldehyde	$C_7H_6O_2$	90-02-8	122.122	liq	-7	197	1.1674^{20}	1.5740^{20}	sl H_2O, chl; msc EtOH; vs ace, bz
9502	Salicylaldoxime		$C_7H_7NO_2$	94-67-7	137.137		57				sl H_2O; vs EtOH, eth, bz; s chl; i lig
9503	Salsoline		$C_{11}H_{15}NO_2$	89-31-6	193.243	pow or cry (al)	221.5				sl H_2O, EtOH; i eth, peth; s chl, alk
9504	Salvarsan dihydrochloride	Arsphenamine	$C_{12}H_{14}As_2Cl_2N_2O_2$	139-93-5	439.001	ye hyg pow	190 dec				vs H_2O
9505	Sanguinarine		$C_{20}H_{15}NO_5$	2447-54-3	349.337	cry (eth, al)	266				vs ace, bz, eth, EtOH
9506	α-Santalol		$C_{15}H_{24}O$	115-71-9	220.351			301.5	0.9679^{20}	1.5023^{20}	i H_2O; s EtOH
9507	β-Santalol		$C_{15}H_{24}O$	77-42-9	220.351			167^{10}	0.9750^{20}	1.5115^{20}	
9508	Santonic acid		$C_{15}H_{20}O_4$	510-35-0	264.318	cry	171	285^{15}			sl H_2O; s chl, eth, HOAc, EtOH
9509	α-Santonin		$C_{15}H_{18}O_3$	481-06-1	246.302	orth (w, eth)	175		1.590^{25}		sl H_2O, EtOH, eth; s bz, chl; i peth
9510	Sarcosine	*N*-Methylglycine	$C_3H_7NO_2$	107-97-1	89.094	cry (al)	212 dec				s H_2O
9511	Sarmentogenin		$C_{23}H_{34}O_5$	76-28-8	390.513	pr (95% al, MeOH-eth)	280				i H_2O, eth, bz; s EtOH; sl ace, chl
9512	Sarpagan-17-al	Vellosimine	$C_{19}H_{20}N_2O$	6874-98-2	292.374	cry (MeOH)	305.5	sub 180			
9513	Sarpagan-10,17-diol	Sarpagine	$C_{19}H_{22}N_2O_2$	482-68-8	310.390	nd	320				i H_2O; s EtOH
9514	Saxitoxin dihydrochloride		$C_{10}H_{19}Cl_2N_7O_4$	35554-08-6	372.209	hyg wh solid					vs H_2O, MeOH, EtOH
9515	Scarlet red		$C_{24}H_{20}N_4O$	85-83-6	380.442	dk br pow or nd	185; dec 260				i H_2O; sl ace, bz; vs chl, peth
9516	Schradan		$C_8H_{24}N_4O_3P_2$	152-16-9	286.250		17	$154^{2.0}$	1.09^{25}	1.462^{25}	vs H_2O, EtOH, chl
9517	Scilliroside		$C_{32}H_{44}O_{12}$	507-60-8	620.684	lo pr (dil MeOH)	169	dec			sl H_2O, ace, chl; vs EtOH, diox; i eth
9518	Scopolamine		$C_{17}H_{21}NO_4$	51-34-3	303.354	visc liq					vs hot H_2O, EtOH, ace; sl bz
9519	Scopoline		$C_8H_{13}NO_2$	487-27-4	155.195	hyg nd (lig, eth, chl, peth)	108.5	248	1.0891^{134}		s H_2O
9520	Sebacic acid	Decanedioic acid	$C_{10}H_{18}O_4$	111-20-6	202.248	lf	130.9	353.9; 295^{100}	1.2705^{20}	1.422^{133}	sl H_2O; s EtOH, eth; i bz
9521	Selenium methionine	Selenomethionine	$C_5H_{11}NO_2Se$	1464-42-2	196.11	hex pl (MeOH aq)	265 dec				
9522	Selenoformaldehyde		CH_2Se	6596-50-5	92.99	unstab gas					
9523	Selenourea	Carbamimidoselenoic acid	CH_4N_2Se	630-10-4	123.02	pr or nd (w)		dec 200			vs H_2O
9524	Semicarbazide hydrochloride		CH_6ClN_3O	563-41-7	111.531	pr (dil al)	176 dec				vs H_2O
9525	Senecionine		$C_{18}H_{25}NO_5$	130-01-8	335.396	pl	232				i H_2O; sl EtOH, eth; s chl
9526	Seneciphylline		$C_{18}H_{23}NO_5$	480-81-9	333.380	pl (AcOEt)	217 dec				s chl; sl EtOH, ace; i eth
9527	Senkirkin		$C_{19}H_{27}NO_6$	2318-18-5	365.420	pl (ace)	197				
9528	*L*-Sepiapterin	6-Lactoyl-7,8-dihydropterin	$C_9H_{11}N_5O_3$	17094-01-8	237.215	ye pow or cry					

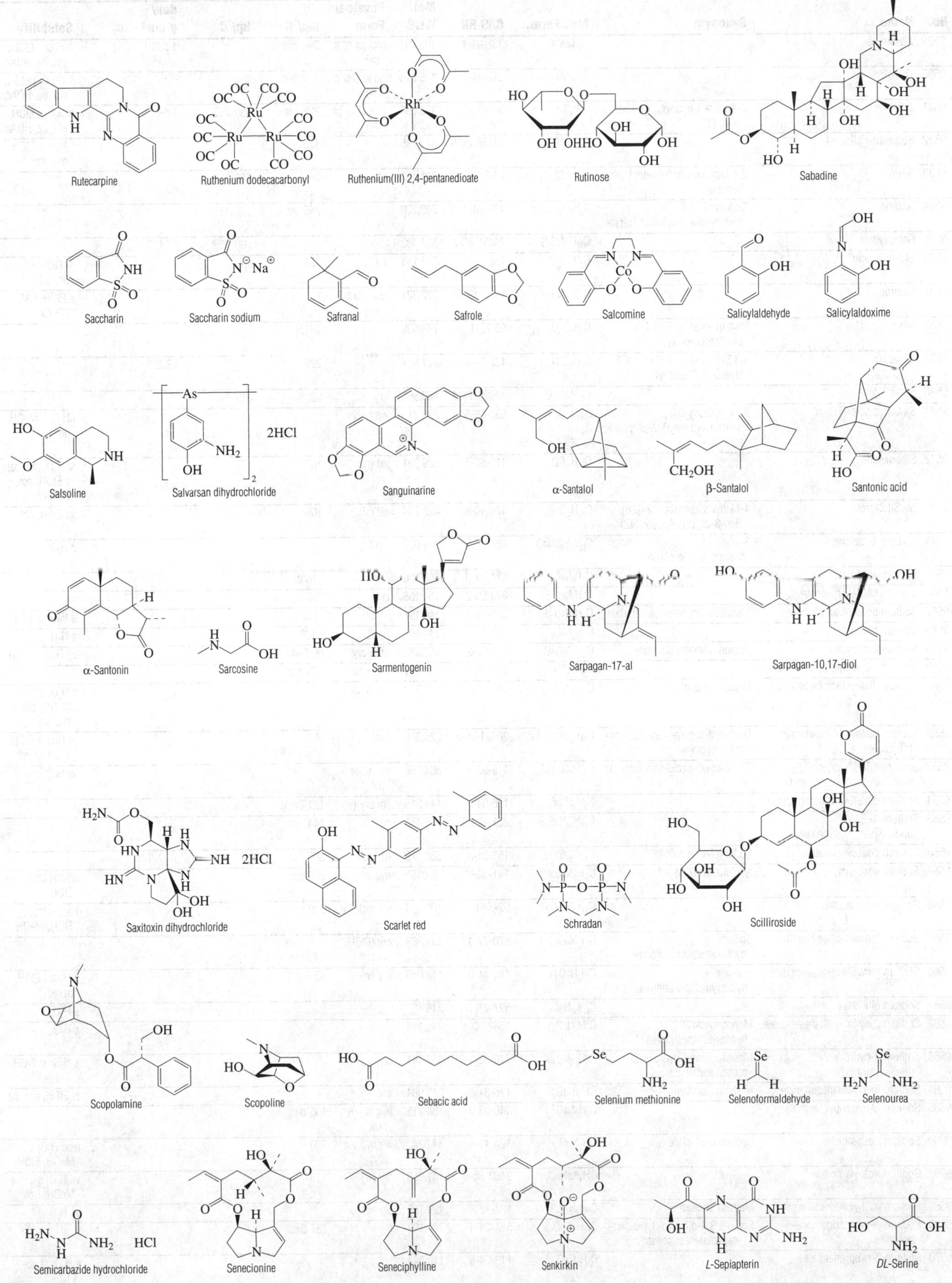
Rutecarpine
Ruthenium dodecacarbonyl
Ruthenium(III) 2,4-pentanedioate
Rutinose
Sabadine
Saccharin
Saccharin sodium
Safranal
Safrole
Salcomine
Salicylaldehyde
Salicylaldoxime
Salsoline
Salvarsan dihydrochloride
Sanguinarine
α-Santalol
β-Santalol
Santonic acid
α-Santonin
Sarcosine
Sarmentogenin
Sarpagan-17-al
Sarpagan-10,17-diol
Saxitoxin dihydrochloride
Scarlet red
Schradan
Scilliroside
Scopolamine
Scopoline
Sebacic acid
Selenium methionine
Selenoformaldehyde
Selenourea
Semicarbazide hydrochloride
Senecionine
Seneciphylline
Senkirkin
L-Sepiapterin
DL-Serine

No.	Name	Synonym	Mol. Form.	CAS RN	Mol. Wt.	Physical Form	mp/°C	bp/°C	den/ g cm⁻³	n_D	Solubility
9529	*DL*-Serine		$C_3H_7NO_3$	302-84-1	105.093	mcl pr or lf (w)	246 dec		1.603^{22}		s H_2O; i EtOH, eth, bz, HOAc
9530	*D*-Serine		$C_3H_7NO_3$	312-84-5	105.093	nd or hex pr (w)	229 dec	dec			vs H_2O; i EtOH, eth, bz, HOAc
9531	*L*-Serine	2-Amino-3-hydroxypropanoic acid, (*S*)	$C_3H_7NO_3$	56-45-1	105.093	hex pl or pr (w)	228 dec	sub 150	1.6^{22}		s H_2O; i EtOH, eth, bz, HOAc
9532	Serpentine alkaloid		$C_{21}H_{20}N_2O_3$	18786-24-8	348.395		175				i H_2O; s EtOH, eth, ace
9533	Sesin	2,4-Dichlorophenoxyethyl benzoate	$C_{15}H_{12}Cl_2O_3$	94-83-7	311.160	cry	66	$185^{1.5}$			
9534	Sesone	Sodium 2-(2,4-dichlorophenoxy)ethyl sulfate	$C_8H_7Cl_2NaO_5S$	136-78-7	309.100		245 dec				
9535	Sethoxydim		$C_{17}H_{29}NO_3S$	74051-80-2	327.482			$>90^{0.00003}$	1.043^{25}		
9536	Shikimic acid		$C_7H_{10}O_5$	138-59-0	174.151	nd	184	subl			sl EtOH; i eth, bz, chl
9537	Siduron		$C_{14}H_{20}N_2O$	1982-49-6	232.321	cry solid	135				s EtOH, DMF, CH_2Cl_2
9538	Silvex	Propanoic acid, 2-(2,4,5-trichlorophenoxy)-	$C_9H_7Cl_3O_3$	93-72-1	269.509		181.6				
9539	Simazine	1,3,5-Triazine-2,4-diamine, 6-chloro-*N,N'*-diethyl-	$C_7H_{12}ClN_5$	122-34-9	201.657		226		1.302^{20}		
9540	Simfibrate		$C_{23}H_{26}Cl_2O_6$	14929-11-4	469.354	cry	52	$225^{0.15}$			
9541	Sinapinic acid	3-(4-Hydroxy-3,5-dimethoxyphenyl)-2-propenoic acid	$C_{11}H_{12}O_5$	530-59-6	224.210	wh pow					i H_2O; s MeOH, ace
9542	Sinomenine		$C_{19}H_{23}NO_4$	115-53-7	329.391	nd (bz)	162				sl H_2O, eth, bz; s EtOH, ace, dil alk
9543	α_1-Sitosterol	4-Methylstigmasta-7,24(28)-dien-3-ol, (3β,4α,5α,24Z)	$C_{30}H_{50}O$	474-40-8	426.717	nd (al)	166				vs EtOH, chl
9544	Sodium arsanilate	Sodium (4-aminophenyl)arsonate	$C_6H_7AsNNaO_3$	127-85-5	239.037	wh cry					s H_2O
9545	Sodium ascorbate		$C_6H_7NaO_6$	134-03-2	198.106	cry	218 dec				
9546	Sodium benzenesulfinate		$C_6H_5NaO_2S$	873-55-2	164.158	cry	300				
9547	Sodium benzenesulfonate	Monosodium benzenesulfonate	$C_6H_5NaO_3S$	515-42-4	180.157		>300				s H_2O; sl EtOH
9548	Sodium benzoate		$C_7H_5NaO_2$	532-32-1	144.104		>300				s H_2O
9549	Sodium cacodylate	Sodium dimethylarsonate	$C_2H_6AsNaO_2$	124-65-2	159.980	gran cry	60 (hyd)				vs H_2O; s EtOH
9550	Sodium 2,2-dichloropropanoate		$C_3H_3Cl_2NaO_2$	127-20-8	164.951	hyg pow	166 dec				
9551	Sodium diethyldithiocarbamate	Dithiocarb sodium	$C_5H_{10}NNaS_2$	148-18-5	171.260	cry (EtOH)	95				s H_2O, EtOH, MeOH, ace; i eth, bz
9552	Sodium diethyldithiocarbamate trihydrate	Diethyldithiocarbamate sodium salt trihydrate	$C_5H_{16}NNaO_3S_2$	20624-25-3	225.306	orth cry (ace)	95				vs H_2O; s EtOH, ace; i bz, eth
9553	Sodium 4,5-dihydroxy-2,7-naphthalenedisulfonic acid	Chromotropic acid disodium salt	$C_{10}H_6Na_2O_8S_2$	129-96-4	364.260	wh nd or lf (w)					vs H_2O
9554	Sodium dimethyldithiocarbamate		$C_3H_7NNaS_2$	128-04-1	144.215	col cry (w)	121 (hyd)				
9555	Sodium 4-dodecylbenzenesulfonate		$C_{18}H_{29}NaO_3S$	2211-98-5	348.476	cry	144				
9556	Sodium dodecyl sulfate	Sodium lauryl sulfate	$C_{12}H_{25}NaO_4S$	151-21-3	288.379	wh pow	205				
9557	Sodium ethanolate	Sodium ethoxide	C_2H_5NaO	141-52-6	68.050	hyg wh pow	260 dec				reac H_2O; s EtOH
9558	Sodium fluoroacetate		$C_2H_2FNaO_2$	62-74-8	100.024	wh mcl cry	200				i ace, chl; sl EtOH, MeOH
9559	Sodium formaldehyde bisulfite	Sodium hydroxymethanesulfonate	CH_3NaO_4S	870-72-4	134.088	cry (EtOH aq)					
9560	Sodium formaldehydesulfoxylate	Sodium hydroxymethanesulfinate	CH_3NaO_3S	149-44-0	118.088	cry (w)	63 (hyd)				s H_2O; i EtOH, bz, eth
9561	Sodium gluconate		$C_6H_{11}NaO_7$	527-07-1	218.137						s H_2O
9562	Sodium 2-hydroxyethanesulfonate	Monosodium 2-hydroxyethanesulfonate	$C_2H_5NaO_4S$	1562-00-1	148.114						s H_2O
9563	Sodium 2-hydroxy-2-propanesulfonate	Monosodium 2-hydroxy-2-propanesulfonate	$C_3H_7NaO_4S$	540-92-1	162.141	cry					s H_2O; sl EtOH
9564	Sodium iodomethanesulfonate	Methiodal sodium	CH_2INaO_3S	126-31-8	243.984	cry					sl EtOH, ace, bz
9565	Sodium *O*-isopropyl xanthate		$C_4H_7NaOS_2$	140-93-2	158.218	hyg wh-ye pow	150 dec				
9566	Sodium methanolate	Sodium methoxide	CH_3NaO	124-41-4	54.024	wh hyg tetr cry	300				reac H_2O; s MeOH, EtOH
9567	Sodium methylarsonate		CH_4AsNaO_3	2163-80-6	161.953	cry (w)	115				vs H_2O; s MeOH; i os
9568	Sodium methyldithiocarbamate	Metham sodium	$C_2H_4NNaS_2$	137-42-8	129.180	cry (w)					vs H_2O
9569	Sodium β-naphthoquinone-4-sulfonate	Sodium 3,4-dihydro-3,4-dioxo-1-naphthalenesulfonate	$C_{10}H_5NaO_5S$	521-24-4	260.199		287 dec				
9570	Sodium 2-oxopropanoate		$C_3H_3NaO_3$	113-24-6	110.044						s H_2O; sl abs EtOH

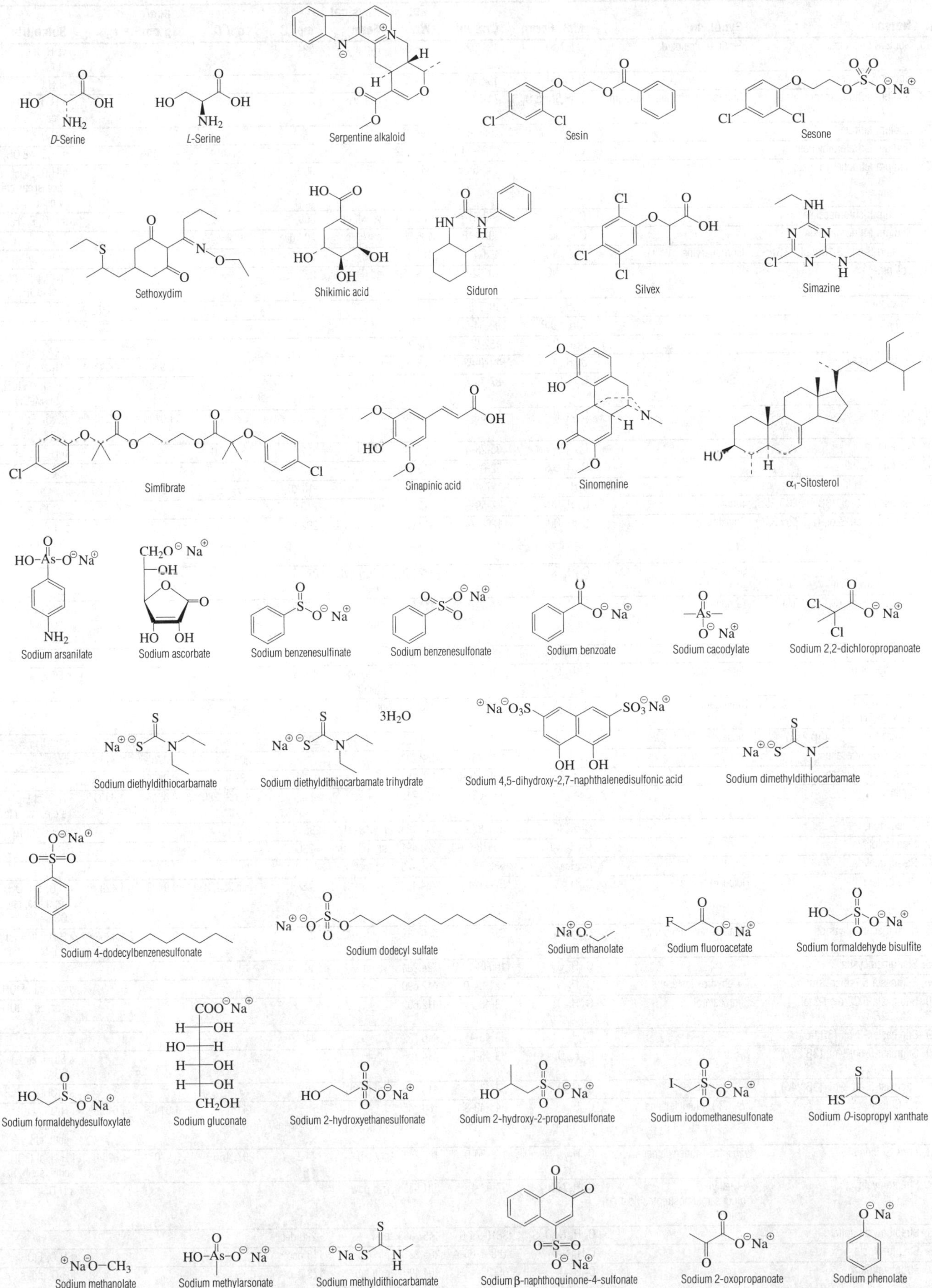
D-Serine
L-Serine
Serpentine alkaloid
Sesin
Sesone
Sethoxydim
Shikimic acid
Siduron
Silvex
Simazine
Simfibrate
Sinapinic acid
Sinomenine
α1-Sitosterol
Sodium arsanilate
Sodium ascorbate
Sodium benzenesulfinate
Sodium benzenesulfonate
Sodium benzoate
Sodium cacodylate
Sodium 2,2-dichloropropanoate
Sodium diethyldithiocarbamate
Sodium diethyldithiocarbamate trihydrate
Sodium 4,5-dihydroxy-2,7-naphthalenedisulfonic acid
Sodium dimethyldithiocarbamate
Sodium 4-dodecylbenzenesulfonate
Sodium dodecyl sulfate
Sodium ethanolate
Sodium fluoroacetate
Sodium formaldehyde bisulfite
Sodium formaldehydesulfoxylate
Sodium gluconate
Sodium 2-hydroxyethanesulfonate
Sodium 2-hydroxy-2-propanesulfonate
Sodium iodomethanesulfonate
Sodium O-isopropyl xanthate
Sodium methanolate
Sodium methylarsonate
Sodium methyldithiocarbamate
Sodium β-naphthoquinone-4-sulfonate
Sodium 2-oxopropanoate
Sodium phenolate

No.	Name	Synonym	Mol. Form.	CAS RN	Mol. Wt.	Physical Form	mp/°C	bp/°C	den/ g cm-3	n_D	Solubility
9571	Sodium phenolate	Sodium phenoxide	C_6H_5NaO	139-02-6	116.093	hyg cry	384				vs H_2O; s EtOH, thf
9572	Sodium propanoate		$C_3H_5NaO_2$	137-40-6	96.061						sl H_2O
9573	Sodium sulfobromophthalein	Sulfobromophthalein sodium	$C_{20}H_8Br_4Na_2O_{10}S_2$	71-67-0	837.998	hyg cry					s H_2O; i EtOH, ace
9574	Sodium tartrate		$C_4H_4Na_2O_6$	868-18-8	194.051						s H_2O
9575	Sodium tartrate dihydrate		$C_4H_8Na_2O_8$	6106-24-7	230.082				1.545^{25}		s H_2O; i EtOH
9576	Sodium tetraphenylborate		$C_{24}H_{20}BNa$	143-66-8	342.217	nd	300				s H_2O, EtOH, ace; sl eth, chl; i peth
9577	Sodium trichloroacetate		$C_2Cl_3NaO_2$	650-51-1	185.369	ye-wh pow	300				s H_2O, EtOH
9578	Sodium trifluoroacetate		$C_2F_3NaO_2$	2923-18-4	136.005	cry	207 dec				
9579	Solanid-5-ene-3,18-diol, (3β)	Isorubijervine	$C_{27}H_{43}NO_2$	468-45-1	413.636	pr(al)	242.5				vs bz, chl
9580	Solanine		$C_{45}H_{73}NO_{15}$	20562-02-1	868.060	nd (EtOH aq)	286 dec				i H_2O, eth, chl; s hot EtOH
9581	Solanone		$C_{13}H_{22}O$	1937-54-8	194.313			60^{1}	0.870^{20}	1.4755^{20}	
9582	Soman		$C_7H_{16}FO_2P$	96-64-0	182.173	liq					
9583	Sophoricoside		$C_{21}H_{20}O_{10}$	152-95-4	432.378		274				
9584	Sorbitan oleate		$C_{24}H_{44}O_6$	1338-43-8	428.602	ye oil			0.986	1.4800^{20}	i H_2O; s EtOH
9585	*L*-Sorbose	*L*-Sorbinose	$C_6H_{12}O_6$	87-79-6	180.155	orth (al)	165		1.612^{17}		s H_2O; sl EtOH, eth, MeOH
9586	Sparteine		$C_{15}H_{26}N_2$	90-39-1	234.380		30.5	325; 173^{8}	1.0196^{20}	1.5312^{20}	vs eth, EtOH, chl
9587	Spinulosin		$C_8H_8O_5$	85-23-4	184.147	red-bl	202.5	sub 120			sl H_2O; s alk
9588	Spironolactone		$C_{24}H_{32}O_4S$	52-01-7	416.574			134			
9589	Spiro[2.2]pentane		C_5H_8	157-40-4	68.118	liq	-107.0	39	0.7266^{20}	1.4120^{20}	
9590	Spirosolan-3-ol, (3β,5α,22β,25*S*)	Tomatidine	$C_{27}H_{45}NO_2$	77-59-8	415.652	pl	210.5				s EtOH, eth
9591	Spirosol-5-en-3-ol, (3β,22α,25*R*)	Solasodine	$C_{27}H_{43}NO_2$	126-17-0	413.636	hex pl (sub)	202				s EtOH, ace, bz, diox, py; sl eth; vs chl
9592	Spirostan-2,3-diol, (2α,3β,5α,25*R*)	Gitogenin	$C_{27}H_{44}O_4$	511-96-6	432.636	lf (bz), nd (eth)	271.5				i H_2O; s EtOH, chl; sl eth
9593	Spirostan-3-ol, (3β,5α,25*R*)	Tigogenin	$C_{27}H_{44}O_3$	77-60-1	416.636	lf (al +1w) pr (ace)	205.5				s EtOH, eth, ace, ctc, MeOH, peth
9594	Spirostan-3-ol, (3β,5β,25*R*)	Smilagenin	$C_{27}H_{44}O_3$	126-18-1	416.636	nd (ace)	185				vs ace, bz, EtOH
9595	Spirostan-3-ol, (3β,5β,25*S*)	Sarsasapogenin	$C_{27}H_{44}O_3$	126-19-2	416.636	lo pr, nd (ace)	200.5				s EtOH, ace, bz, chl
9596	Spirostan-2,3,15-triol, (2α,3β,5α,15β,25*R*)	Digitogenin	$C_{27}H_{44}O_5$	511-34-2	448.635	nd (al)	281.5				vs chl
9597	Spirost-5-en-3-ol, (3β,25*R*)	Diosgenin	$C_{27}H_{42}O_3$	512-04-9	414.620	cry (ace)	205.5				vs EtOH
9598	Spiro[5.5]undecane		$C_{11}H_{20}$	180-43-8	152.277			208	0.8783^{20}	1.4731	
9599	S-Propyl thioacetate		$C_5H_{10}OS$	2307-10-0	118.197			137.9	0.9535^{25}		
9600	Squalene		$C_{30}H_{50}$	111-02-4	410.718	oil	-4.8	421.3; 280^{17}	0.8584^{20}	1.4990^{20}	i H_2O; sl EtOH; s eth, ace, ctc
9601	Stachydrine		$C_7H_{13}NO_2$	471-87-4	143.184	cry (w+1)	235				vs H_2O, EtOH
9602	Stanozolol		$C_{21}H_{32}N_2O$	10418-03-8	328.491	cry (EtOH)	≈236				
9603	Stearaldehyde		$C_{18}H_{36}O$	638-66-4	268.478	nd (peth)		261			
9604	Stearic acid	Octadecanoic acid	$C_{18}H_{36}O_2$	57-11-4	284.478	mcl lf (al)	69.3	dec 350; 232^{15}	0.9408^{20}	1.4299^{80}	i H_2O; sl EtOH, bz; s ace, chl, CS_2
9605	Stearic acid anhydride	Octadecanoic anhydride	$C_{36}H_{70}O_3$	638-08-4	550.939		72		0.8365^{82}	1.4362^{80}	i H_2O, EtOH; sl eth, bz
9606	Sterigmatocystin		$C_{18}H_{12}O_6$	10048-13-2	324.284	ye nd	246 dec				
9607	Stigmasta-5,7-dien-3-ol, (3β)	7-Dehydrositosterol	$C_{29}H_{48}O$	521-04-0	412.690		144.5				vs bz, eth, EtOH
9608	Stigmasta-5,22-dien-3-ol, (3β,22*E*)	Stigmasterol	$C_{29}H_{48}O$	83-48-7	412.690		170				vs bz, eth, EtOH
9609	Stigmastan-3-ol, (3β,5α)		$C_{29}H_{52}O$	83-45-4	416.722		144				
9610	Stigmast-5-en-3-ol, (3β,24*R*)	β-Sitosterol	$C_{29}H_{50}O$	83-46-5	414.706	pl (al)	137				s EtOH, eth, HOAc
9611	Stigmast-5-en-3-ol, (3β,24*S*)	γ-Sitosterol	$C_{29}H_{50}O$	83-47-6	414.706	cry (EtOH)	148				s EtOH
9612	*cis*-Stilbene	*cis*-1,2-Diphenylethene	$C_{14}H_{12}$	645-49-8	180.245		-5	141^{12}	1.0143^{20}	1.6130^{20}	i H_2O; s EtOH, eth, ace, bz, peth, chl
9613	*trans*-Stilbene	*trans*-1,2-Diphenylethene	$C_{14}H_{12}$	103-30-0	180.245	cry (al)	124.2	307; 166^{12}	0.9707^{20}	1.6264^{17}	i H_2O; sl EtOH, chl; vs eth, bz
9614	Streptomycin	*N*-Methyl-*L*-glucosamidinostreptosidostreptidine	$C_{21}H_{39}N_7O_{12}$	57-92-1	581.575	hyg pow					s H_2O
9615	Streptomycin sulfate		$C_{42}H_{84}N_{14}O_{36}S_3$	3810-74-0	1457.383	pow	≈230 dec				
9616	Streptozotocin		$C_8H_{15}N_3O_7$	18883-66-4	265.221	pl	115 dec				s H_2O, EtOH
9617	Strophanthidin		$C_{23}H_{32}O_6$	66-28-4	404.496	orth tab (MeOH-w) lf (w+2)	173 dec				i H_2O, eth; s EtOH, ace, bz, HOAc, chl

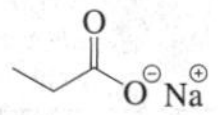

Sodium propanoate

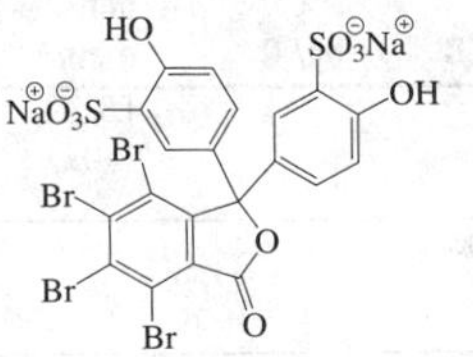

Sodium sulfobromophthalein

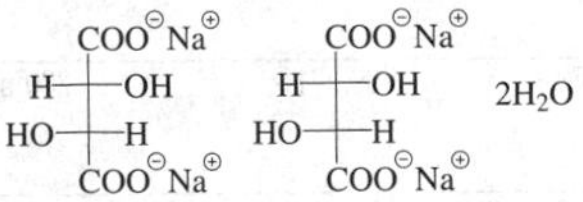

Sodium tartrate

Sodium tartrate dihydrate

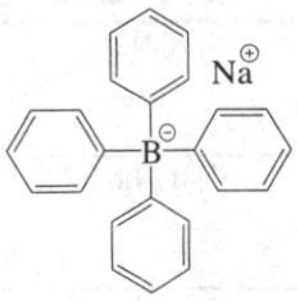

Sodium tetraphenylborate

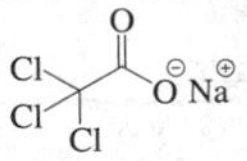

Sodium trichloroacetate

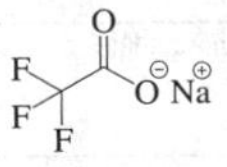

Sodium trifluoroacetate

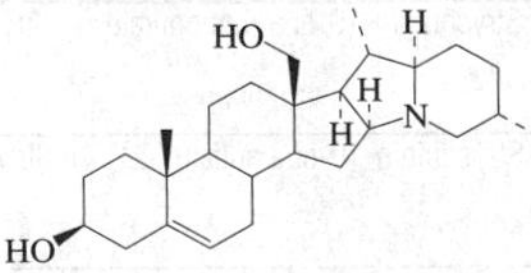

Solanid-5-ene-3,18-diol, (3β)

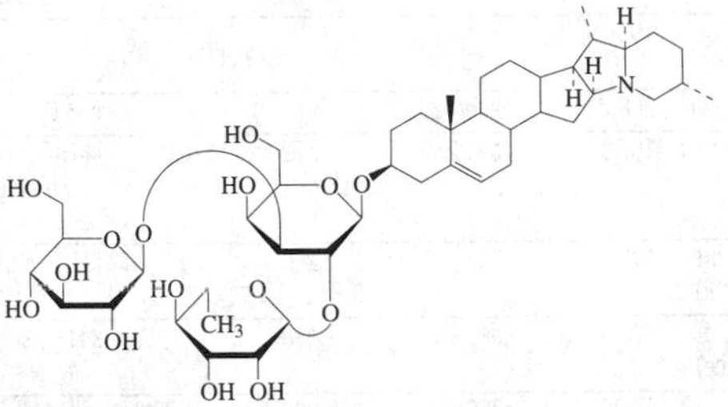

Solanine

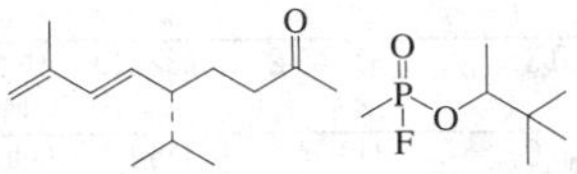

Solanone

Soman

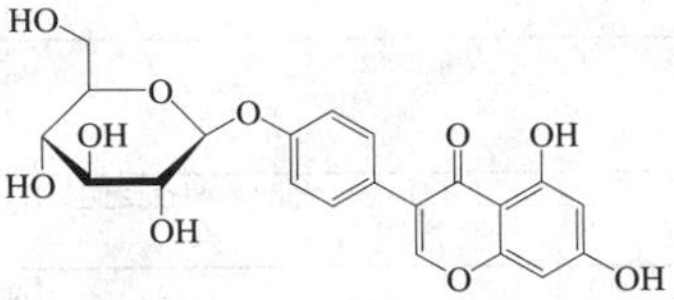

Sophoricoside

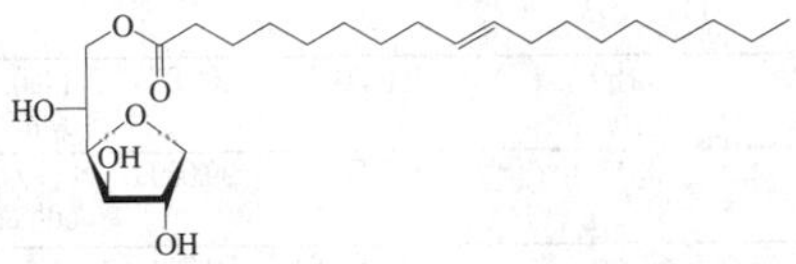

Sorbitan oleate

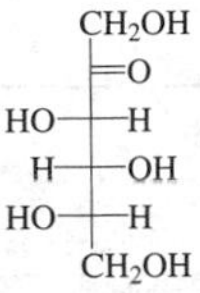

L-Sorbose

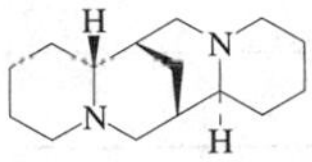

Sparteine

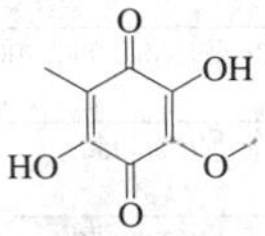

Spinulosin

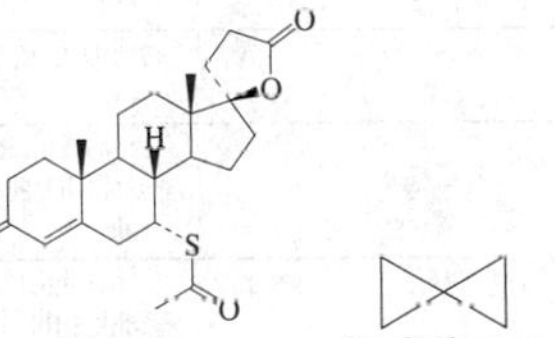

Spironolactone

Spiro[2.2]pentane

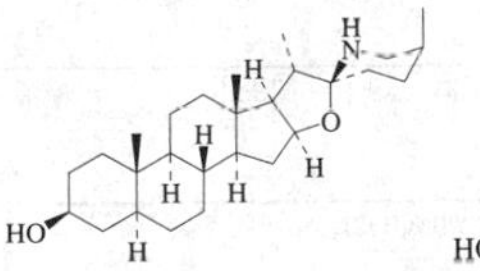

Spirosolan-3-ol, (3β,5α,22β,25*S*)

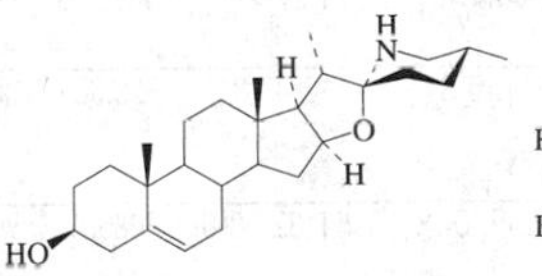

Spirosol-5-en-3-ol, (3β,22α,25*R*)

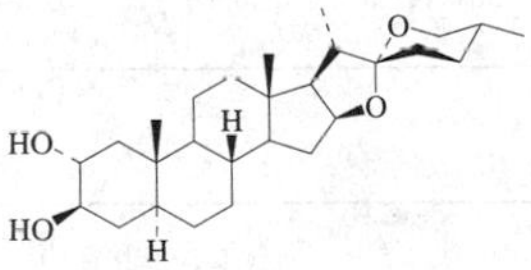

Spirostan-2,3-diol, (2α,3β,5α,25*R*)

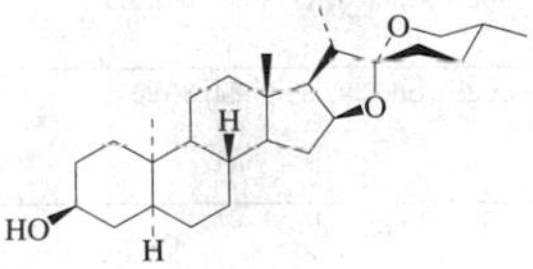

Spirostan-3-ol, (3β,5α,25*R*)

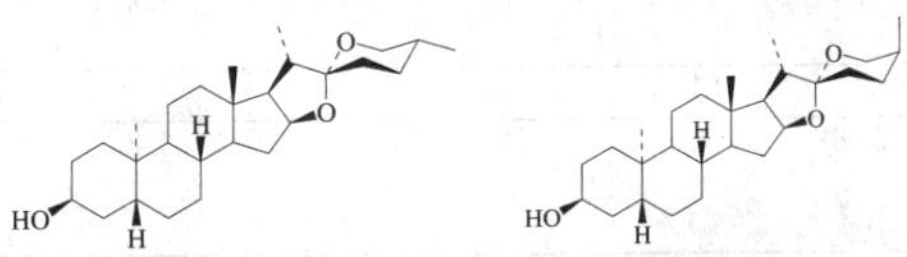

Spirostan-3-ol, (3β,5β,25*R*)

Spirostan-3-ol, (3β,5β,25*S*)

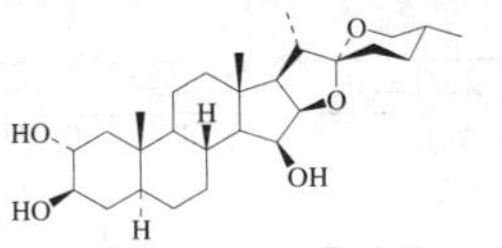

Spirostan-2,3,15-triol, (2α,3β,5α,15β,25*R*)

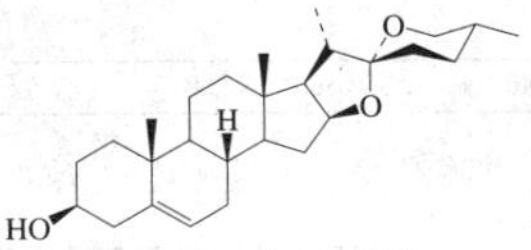

Spirost-5-en-3-ol, (3β,25*R*)

Spiro[5.5]undecane

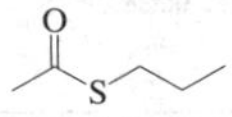

S-Propyl thioacetate

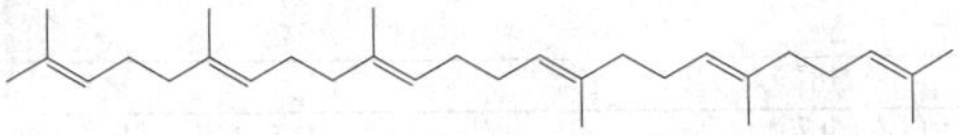

Squalene

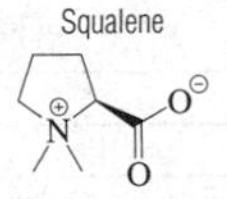

Stachydrine

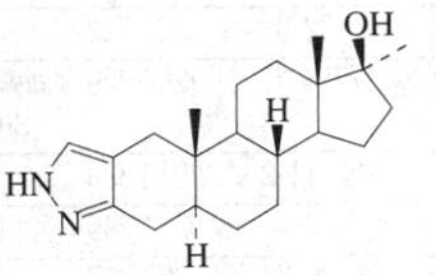

Stanozolol

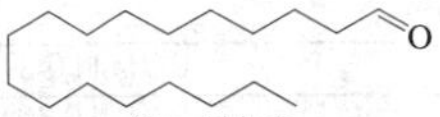

Stearaldehyde

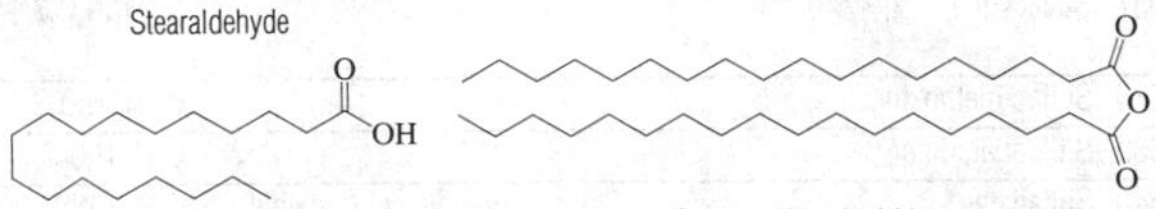

Stearic acid

Stearic acid anhydride

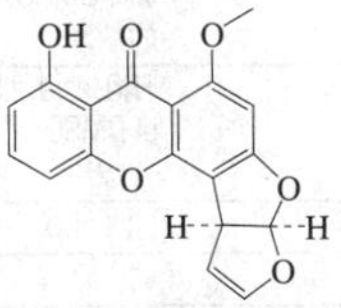

Sterigmatocystin

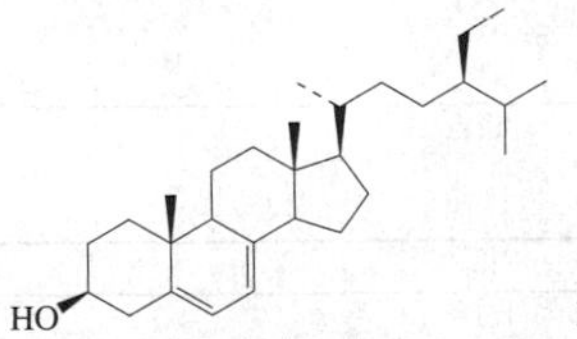

Stigmasta-5,7-dien-3-ol, (3β)

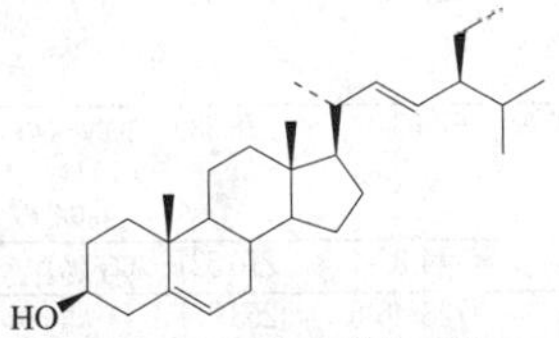

Stigmasta-5,22-dien-3-ol, (3β,22*E*)

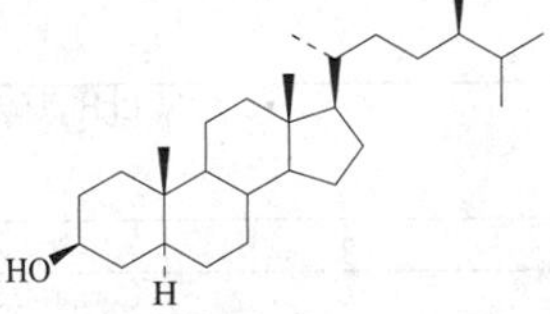

Stigmastan-3-ol, (3β,5α)

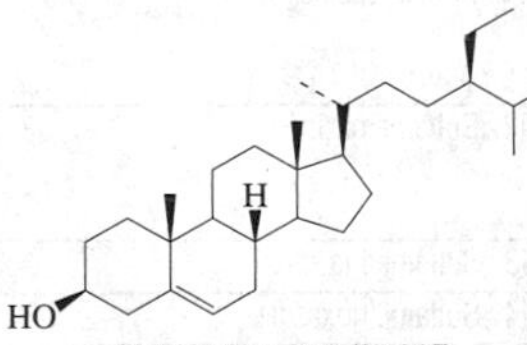

Stigmast-5-en-3-ol, (3β,24*R*)

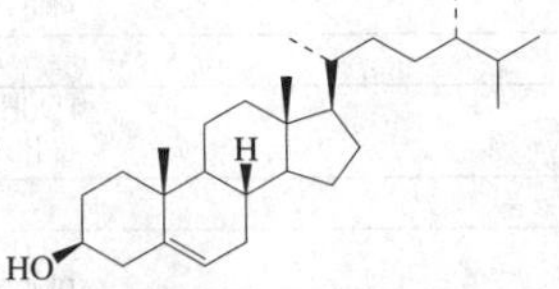

Stigmast-5-en-3-ol, (3β,24*S*)

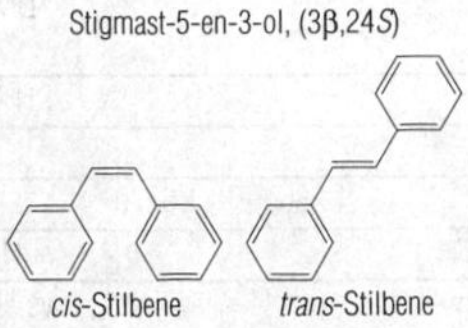

cis-Stilbene

trans-Stilbene

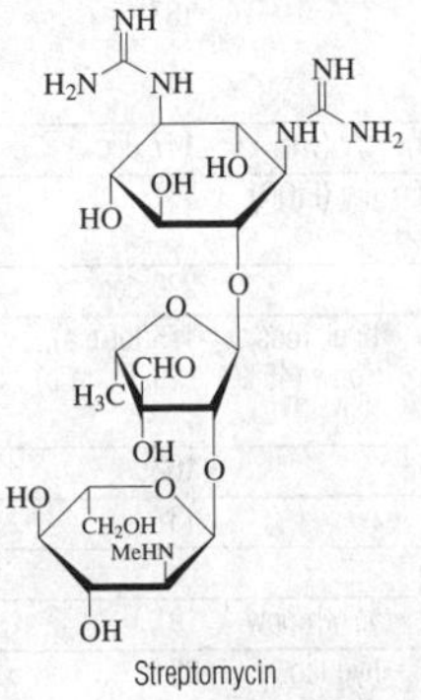

Streptomycin

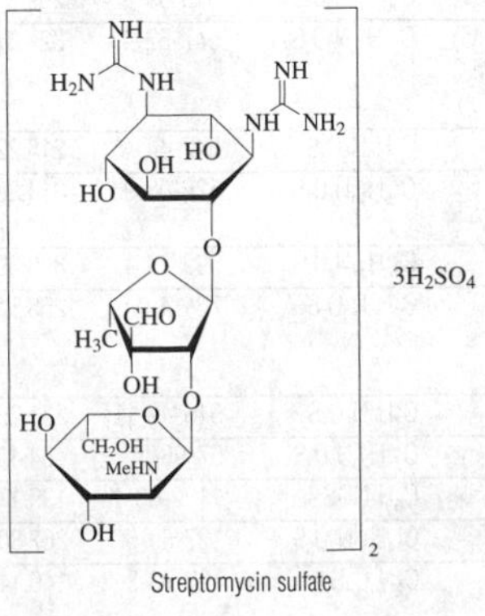

Streptomycin sulfate

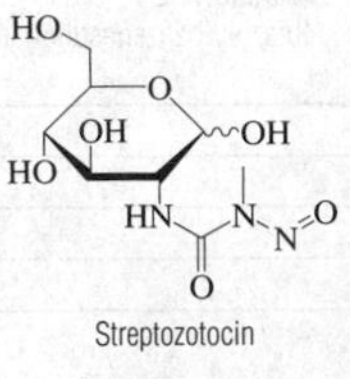

Streptozotocin

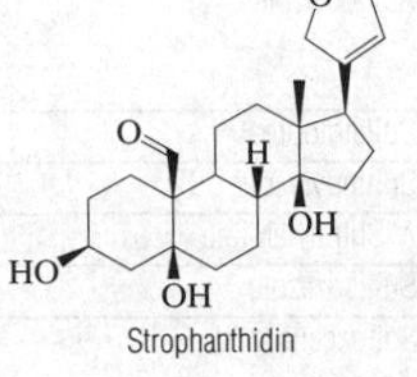

Strophanthidin

No.	Name	Synonym	Mol. Form.	CAS RN	Mol. Wt.	Physical Form	mp/°C	bp/°C	den/ g cm^{-3}	n_D	Solubility
9618	Strychnidin-10-one mononitrate	Strychnine nitrate	$C_{21}H_{23}N_3O_5$	66-32-0	397.425	nd (w)	295		1.627^{25}		vs H_2O, MeOH; sl bz; s chl, EtOH
9619	Strychnidin-10-one sulfate (2:1)	Strychnine sulfate	$C_{42}H_{46}N_4O_8S$	60-41-3	766.901		200 dec				s H_2O, EtOH, MeOH; i eth; sl chl
9620	Strychnine		$C_{21}H_{22}N_2O_2$	57-24-9	334.412	orth pr (al)	287	270^{5}	1.36^{20}		sl H_2O, EtOH, ace, bz; i eth; s chl
9621	Styrene	Vinylbenzene	C_8H_8	100-42-5	104.150	liq	-30.65	145	0.9016^{25}	1.5440^{25}	i H_2O; s EtOH, eth, ace; msc bz; sl ctc
9622	Succimer	2,3-Dimercaptobutanedioic acid, (*R**,*S**)	$C_4H_6O_4S_2$	304-55-2	182.219	wh cry (MeOH)	193				
9623	Succinamide		$C_4H_8N_2O_2$	110-14-5	116.119	orth nd (w)	268 dec	sub 125			s H_2O
9624	Succinic acid		$C_4H_6O_4$	110-15-6	118.089	tcl or mcl pr	187.9	dec 235	1.572^{25}	1.450	sl H_2O, DMSO; s EtOH, eth, ace; i bz
9625	Succinic anhydride		$C_4H_4O_3$	108-30-5	100.073	nd (al), orth pym (chl)	119	261	1.2^{20}		i H_2O; s EtOH, chl; sl eth
9626	Succinimide		$C_4H_5NO_2$	123-56-8	99.089	pl (+1w, al) orth (ace)	126.5	dec 287	1.418^{25}		s H_2O; sl EtOH, eth, ace
9627	Succinonitrile	Butanedinitrile	$C_4H_4N_2$	110-61-2	80.088		57.98	266	0.9867^{60}	1.4173^{60}	vs H_2O; s EtOH, ace, bz, chl; sl eth
9628	Succinylcholine chloride	Suxamethonium chloride	$C_{14}H_{30}Cl_2N_2O_4$	71-27-2	361.305	cry (w)	190				sl EtOH, bz, chl; i eth
9629	Succinylsulphathiazole		$C_{13}H_{13}N_3O_5S_2$	116-43-8	355.389	cry	193.5				i H_2O, eth, chl; sl EtOH, ace; s alk
9630	Sucralfate		$C_{12}H_{54}Al_{16}O_{75}S_8$	54182-58-0	2086.737	wh amorp pow					i H_2O, EtOH, chl; s dil HCl, alk
9631	Sucrose		$C_{12}H_{22}O_{11}$	57-50-1	342.296	mcl	185.5		1.5805^{17}	1.5376	s H_2O, py; sl EtOH; i eth
9632	Sucrose monohexadecanoate	Sucrose palmitate	$C_{28}H_{52}O_{12}$	26446-38-8	580.706	cry	61				s H_2O
9633	Sucrose octaacetate		$C_{28}H_{38}O_{19}$	126-14-7	678.591	nd (al)	86.5	250^{1}	1.27^{16}	1.4660	sl H_2O; s EtOH, eth, ace, bz, chl
9634	Sufentanil		$C_{22}H_{30}N_2O_2S$	56030-54-7	386.550	cry (peth)	96.6				
9635	Sulfabenzamide	*N*-[(4-Aminophenyl)sulfonyl] benzamide	$C_{13}H_{12}N_2O_3S$	127-71-9	276.310	hex pr (60% al)	181.5				
9636	Sulfachlorpyridazine		$C_{10}H_9ClN_4O_2S$	80-32-0	284.722		187				
9637	Sulfacytine		$C_{12}H_{14}N_4O_3S$	17784-12-2	294.329	cry (MeOH/ BuOH)	167				i H_2O; s alk
9638	Sulfadimethoxine		$C_{12}H_{14}N_4O_4S$	122-11-2	310.329		203.5				
9639	Sulfaguanidine		$C_7H_{10}N_4O_2S$	57-67-0	214.245	nd (w)	191.5				
9640	Sulfallate	Carbamodithioic acid, diethyl-, 2-chloro-2-propenyl ester	$C_8H_{14}ClNS_2$	95-06-7	223.787			129^{1}	1.088		
9641	Sulfamerazine		$C_{11}H_{12}N_4O_2S$	127-79-7	264.304	cry	236				sl H_2O, EtOH, ace, DMSO; i eth, chl
9642	Sulfamethazine		$C_{12}H_{14}N_4O_2S$	57-68-1	278.330	pa ye (w+1/ 2) cry (diox-w)	198.5				s H_2O, acid, alk; sl DMSO
9643	Sulfamethiazole		$C_9H_{10}N_4O_2S_2$	144-82-1	270.331	cry (w)	210				sl hot H_2O
9644	Sulfamethoxazole		$C_{10}H_{11}N_3O_3S$	723-46-6	253.277	ye-wh pow	171				i eth
9645	Sulfamethoxypyridazine		$C_{11}H_{12}N_4O_3S$	80-35-3	280.303		182.5				
9646	Sulfamethylthiazole		$C_{10}H_{11}N_3O_2S_2$	515-59-3	269.343		237				vs EtOH
9647	N^4-Sulfanilylsulfanilamide	4-Amino-*N*-[4-(aminosulfonyl) phenyl]benzenesulfonamide	$C_{12}H_{13}N_3O_4S_2$	547-52-4	327.379		137				sl H_2O; s EtOH, eth, ace; i chl, peth
9648	Sulfanilylurea		$C_7H_9N_3O_3S$	547-44-4	215.229	cry (w)	147 dec				
9649	Sulfaphenazole		$C_{15}H_{14}N_4O_2S$	526-08-9	314.363	cry (EtOH)	181				sl EtOH, MeOH, gl HOAc
9650	Sulfasalazine		$C_{18}H_{14}N_4O_5S$	599-79-1	398.393		220 dec				
9651	Sulfathiazole	4-Amino-*N*-2-thiazolylbenzenesulfonamide	$C_9H_9N_3O_2S_2$	72-14-0	255.316	br pl, rods or pow (45% al)	175(form a); 202(form b)				sl H_2O, EtOH, DMSO
9652	Sulfathiourea		$C_7H_9N_3O_2S_2$	515-49-1	231.295		182				i H_2O; sl EtOH
9653	Sulfinpyrazone		$C_{23}H_{20}N_2O_3S$	57-96-5	404.481		137				
9654	*N*-Sulfinylaniline		C_6H_5NOS	1122-83-4	139.175			200	1.236^{25}	1.6270^{20}	
9655	Sulfisoxazole		$C_{11}H_{13}N_3O_3S$	127-69-5	267.304	ye-wh pow	191				sl H_2O
9656	Sulfoacetic acid		$C_2H_4O_5S$	123-43-3	140.115	hyg tab (w+1)	85	dec 245			vs H_2O, ace, EtOH

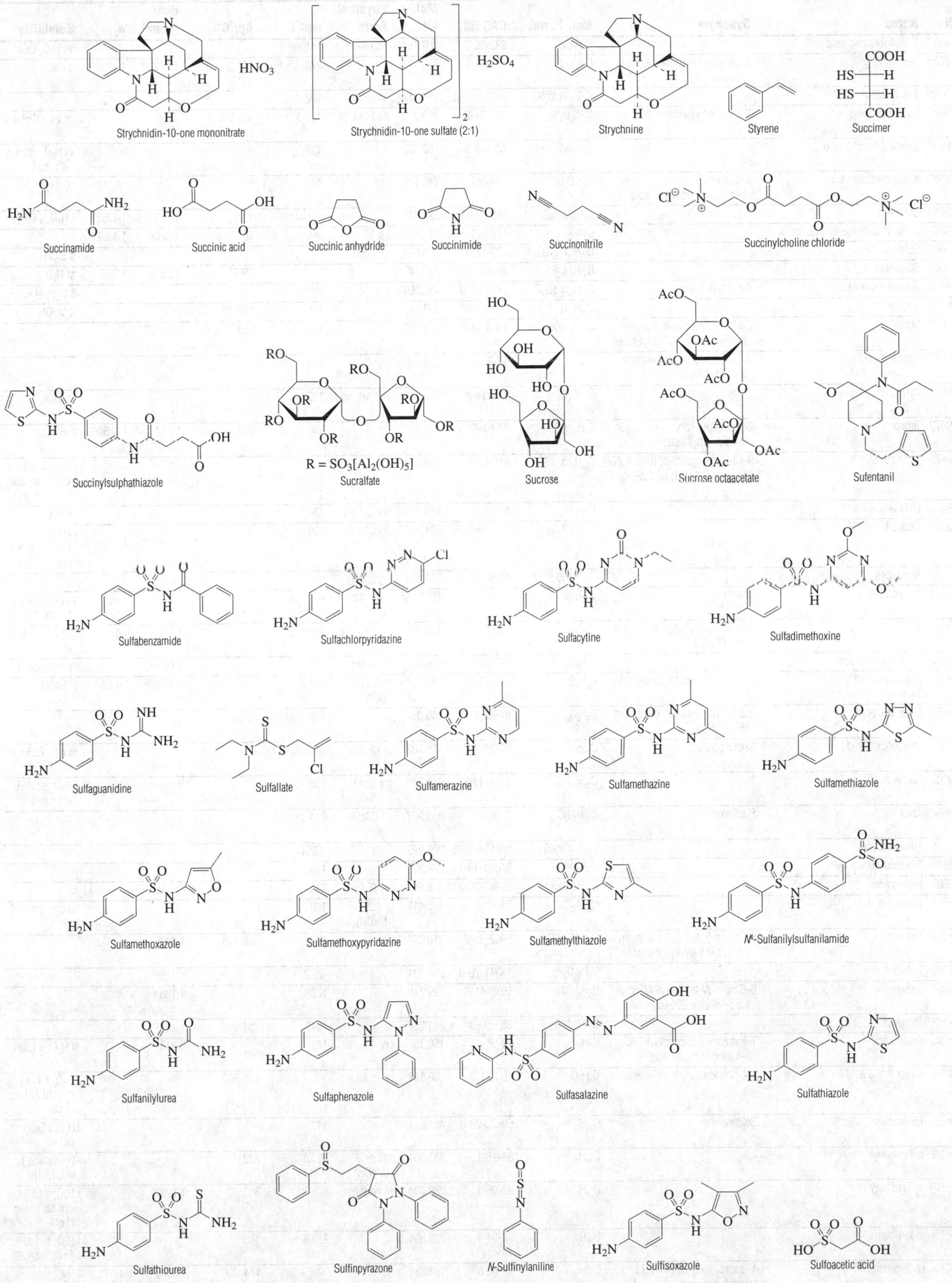

Strychnidin-10-one mononitrate

Strychnidin-10-one sulfate (2:1)

Strychnine

Styrene

Succimer

Succinamide

Succinic acid

Succinic anhydride

Succinimide

Succinonitrile

Succinylcholine chloride

Succinylsulphathiazole

Sucralfate

Sucrose

Sucrose octaacetate

Sufentanil

Sulfabenzamide

Sulfachlorpyridazine

Sulfacytine

Sulfadimethoxine

Sulfaguanidine

Sulfallate

Sulfamerazine

Sulfamethazine

Sulfamethiazole

Sulfamethoxazole

Sulfamethoxypyridazine

Sulfamethylthiazole

N[4]-Sulfanilylsulfanilamide

Sulfanilylurea

Sulfaphenazole

Sulfasalazine

Sulfathiazole

Sulfathiourea

Sulfinpyrazone

N-Sulfinylaniline

Sulfisoxazole

Sulfoacetic acid

No.	Name	Synonym	Mol. Form.	CAS RN	Mol. Wt.	Physical Form	mp/°C	bp/°C	den/g cm^{-3}	n_D	Solubility
9657	2-Sulfobenzoic acid		$C_7H_6O_5S$	632-25-7	202.185	nd (w+3)	141				vs H_2O, EtOH
9658	Sulfolane	Tetrahydrothiophene, 1-1-dioxide	$C_4H_8O_2S$	126-33-0	120.171		27.6	287.3	1.2723^{18}	1.4833^{18}	s chl
9659	Sulfometuron methyl		$C_{15}H_{16}N_4O_5S$	74222-97-2	364.377	wh solid	202				
9660	Sulfonmethane	2,2-Bis(ethylsulfonyl)propane	$C_7H_{16}O_4S_2$	115-24-2	228.330	mcl (w), pr (al)	125.8	dec 300			vs bz, EtOH, chl
9661	Sulfonyldiacetic acid		$C_4H_6O_6S$	123-45-5	182.152		187				vs H_2O, EtOH; s eth, sulf
9662	4-Sulfophthalic acid	4-Sulfo-1,2-benzenedicarboxylic acid	$C_8H_6O_7S$	89-08-7	246.195	cry	139				
9663	Sulfotep		$C_8H_{20}O_5P_2S_2$	3689-24-5	322.320			137^{2}	1.196^{25}	1.4753^{25}	i H_2O; s EtOH
9664	Sulfuryl chloride isocyanate		$CClNO_3S$	1189-71-5	141.534	liq	-44	107	1.626^{25}	1.4467^{20}	
9665	Sulphan Blue		$C_{27}H_{31}N_2NaO_6S_2$	129-17-9	566.664	viol pow					s EtOH
9666	Sulprofos		$C_{12}H_{19}O_2PS_3$	35400-43-2	322.447			$156^{0.1}$	1.20^{20}	1.5859	sl H_2O
9667	Sunset Yellow FCF	C.I. Food Yellow 3	$C_{16}H_{10}N_2Na_2O_7S_2$	2783-94-0	452.369	cry	>300				s H_2O; sl EtOH
9668	Suprasterol II		$C_{28}H_{44}O$	562-71-0	396.648	pr	110	$190^{0.005}$			s MeOH
9669	Sutan	Carbamothioic acid, bis(2-methylpropyl)-, *S*-ethyl ester	$C_{11}H_{23}NOS$	2008-41-5	217.372			138^{21}	0.9402^{25}		
9670	Symclosene	1,3,5-Trichloro-1,3,5-triazine-2,4,6(1*H*,3*H*,5*H*)-trione	$C_3Cl_3N_3O_3$	87-90-1	232.409		246.7 dec				
9671	Syringin		$C_{17}H_{24}O_9$	118-34-3	372.368	cry (w), nd (al)	192				vs EtOH
9672	Tabun	Dimethylphosphoroamidocyanidic acid, ethyl ester	$C_5H_{11}N_2O_2P$	77-81-6	162.127	liq	-50	240	1.077	1.4250^{20}	msc H_2O
9673	Tachysterol	9,10-Secoergosta-5(10),6,8,22-tetraen-3-ol, (3β,6*E*,22*E*)-	$C_{28}H_{44}O$	115-61-7	396.648						i H_2O, MeOH; s EtOH, eth, ace, bz
9674	*D*-Tagatose		$C_6H_{12}O_6$	87-81-0	180.155	cry (dil al)	134.5				vs H_2O
9675	Talbutal		$C_{11}H_{16}N_2O_3$	115-44-6	224.256	cry	109				i H_2O, peth; s EtOH, ace, eth, chl
9676	Tamoxifen		$C_{26}H_{29}NO$	10540-29-1	371.514	cry (peth)	97				
9677	Tannic acid	Tannin	$C_{76}H_{52}O_{46}$	1401-55-4	1701.198	ye-br amorp pow	≈210 dec				vs EtOH, ace; i bz, chl, eth, ctc
9678	*DL*-Tartaric acid	2,3-Dihydroxybutanedioic acid, (*R**, *R**)-(±)-	$C_4H_6O_6$	133-37-9	150.087	mcl pr (w, al +1w)	206		1.788^{25}		s H_2O, EtOH; sl eth; i bz
9679	*meso*-Tartaric acid		$C_4H_6O_6$	147-73-9	150.087	tcl pl (w)	147		1.666^{20}		vs H_2O, EtOH
9680	*D*-Tartaric acid	2,3-Dihydroxybutanedioic acid, [*S*-(*R**,*R**)]-	$C_4H_6O_6$	147-71-7	150.087	mcl, orth pr (w+1)	172.5		1.7598^{20}	1.4955^{20}	sl DMSO
9681	*L*-Tartaric acid	2,3-Dihydroxybutanedioic acid, [*R*-(*R**,*R**)]-	$C_4H_6O_6$	87-69-4	150.087		169				
9682	Taurocholic acid	Cholaic acid	$C_{26}H_{45}NO_7S$	81-24-3	515.703	pr (al-eth)	125 dec				vs H_2O, EtOH; sl eth, AcOEt
9683	Taxine A		$C_{35}H_{47}NO_{10}$	1361-49-5	641.749	cry (ace)	205				i H_2O; s EtOH, eth, chl
9684	Taxol	Paclitaxel	$C_{47}H_{51}NO_{14}$	33069-62-4	853.907	nd (MeOH aq)	214 dec				
9685	Tebuconazole		$C_{16}H_{23}ClN_3O$	107534-96-3	308.826		102.4				
9686	Tebuthiuron		$C_9H_{16}N_4OS$	34014-18-1	228.314		163 dec				
9687	Teniposide		$C_{32}H_{32}O_{13}S$	29767-20-2	656.653	cry (EtOH)	244				
9688	Tephrosin		$C_{23}H_{22}O_7$	76-80-2	410.417	pr (chl-MeOH)	198				vs ace, eth, chl
9689	Terbacil	5-Chloro-3-*tert*-butyl-6-methyl-2,4(1*H*,3*H*)-pyrimidinedione	$C_9H_{13}ClN_2O_2$	5902-51-2	216.664		176	sub 175	1.34^{25}		
9690	Terbufos		$C_9H_{21}O_2PS_3$	13071-79-9	288.431		-29.2	$69^{0.01}$	1.105^{24}		
9691	Terbuthylazine	6-Chloro-*N*-*tert*-butyl-*N'*-ethyl-1,3,5-triazine-2,4-diamine	$C_9H_{16}ClN_5$	5915-41-3	229.710		178		1.188^{20}		
9692	Terbutryn		$C_{10}H_{19}N_5S$	886-50-0	241.357		104	$157^{0.06}$	1.115^{20}		
9693	Terebic acid	Tetrahydro-2,2-dimethyl-5-oxo-3-furancarboxylic acid	$C_7H_{10}O_4$	79-91-4	158.152	cry	175		0.815		sl H_2O; s EtOH
9694	Terephthalic acid	1,4-Benzenedicarboxylic acid	$C_8H_6O_4$	100-21-0	166.132	nd (sub)		sub 300			i H_2O, EtOH, eth, chl, HOAc; sl ctc
9695	Terfenadine	Seldane	$C_{32}H_{41}NO_2$	50679-08-8	471.674		147				i H_2O; s EtOH; sl hx
9696	*o*-Terphenyl		$C_{18}H_{14}$	84-15-1	230.304	mcl pr (MeOH)	56.20	332			i H_2O; s ace, bz, chl, MeOH
9697	*m*-Terphenyl		$C_{18}H_{14}$	92-06-8	230.304	ye nd (al)	87	363	1.199^{20}		i H_2O; s EtOH, eth, bz, HOAc; sl chl
9698	*p*-Terphenyl		$C_{18}H_{14}$	92-94-4	230.304		213.9	376			i H_2O; sl EtOH; s eth, bz, CS_2
9699	α-Terpinene	4-Isopropyl-1-methyl-1,3-cyclohexadiene	$C_{10}H_{16}$	99-86-5	136.234			174	0.8375^{19}	1.477^{19}	i H_2O; msc EtOH, eth

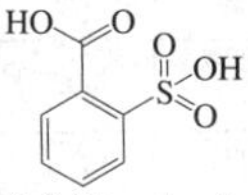

2-Sulfobenzoic acid

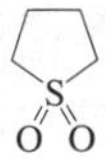

Sulfolane

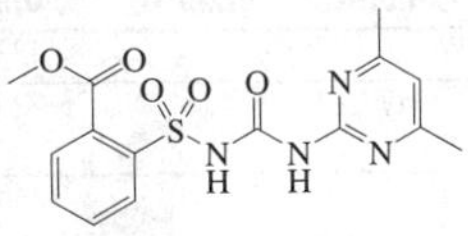

Sulfometuron methyl

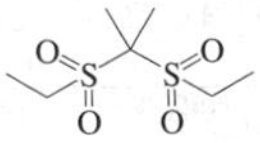

Sulfonmethane

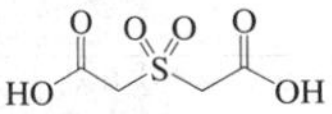

Sulfonyldiacetic acid

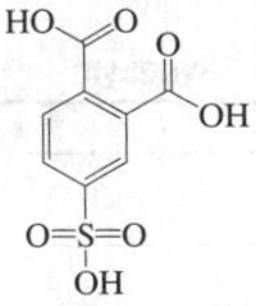

4-Sulfophthalic acid

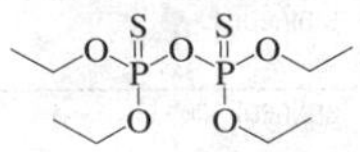

Sulfotep

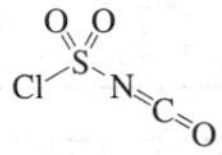

Sulfuryl chloride isocyanate

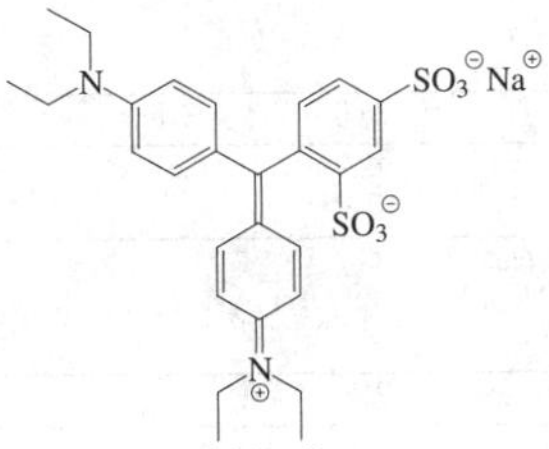

Sulphan Blue

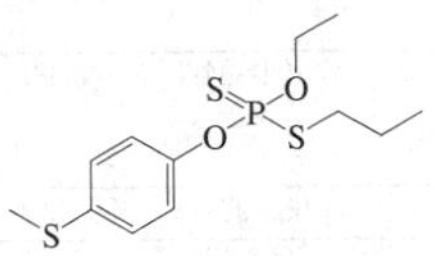

Sulprofos

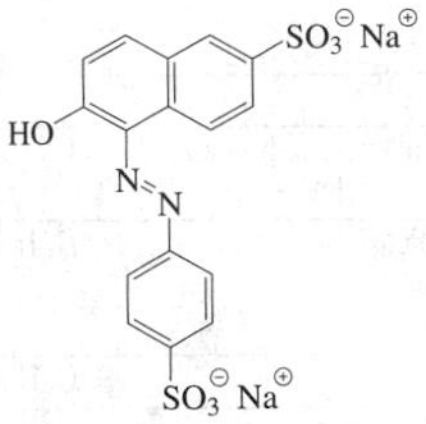

Sunset Yellow FCF

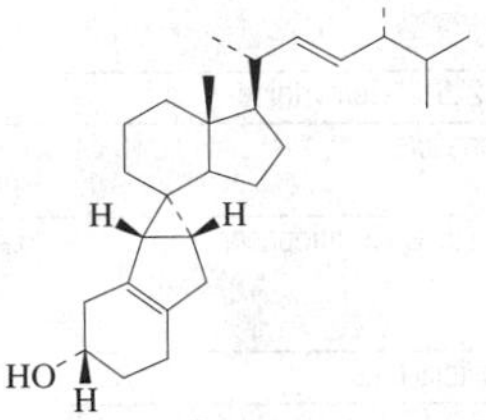

Suprasterol II

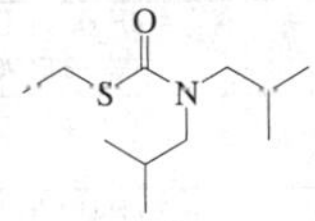

Sutan

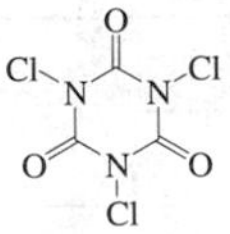

Symclosene

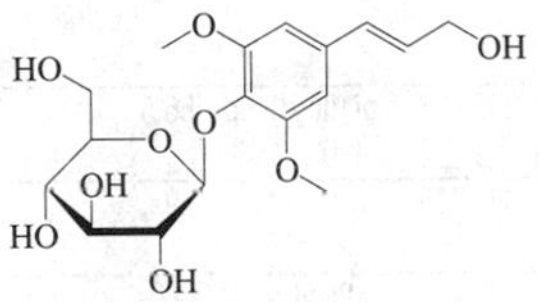

Syringin

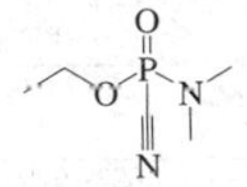

Tabun

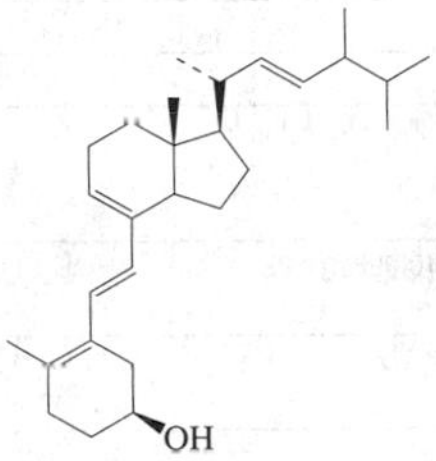

Tachysterol

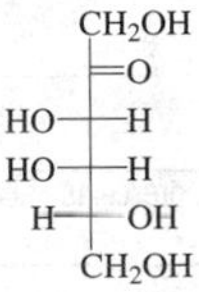

D-Tagatose

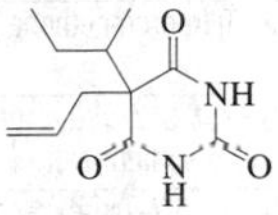

Talbutal

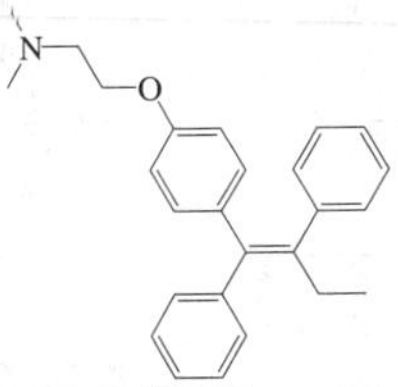

Tamoxifen

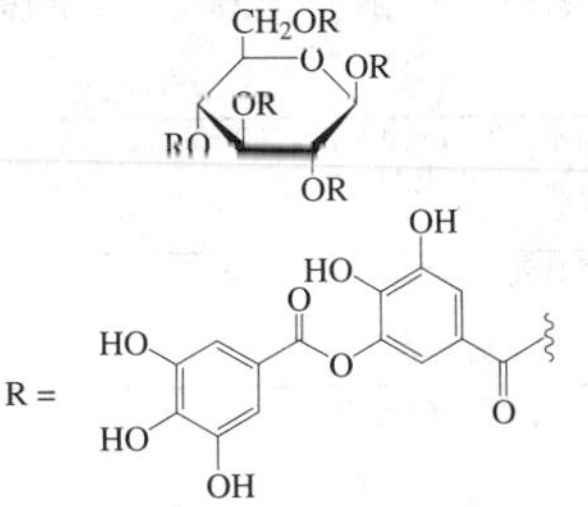

Tannic acid

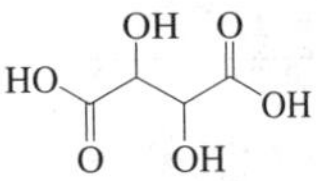

DL-Tartaric acid

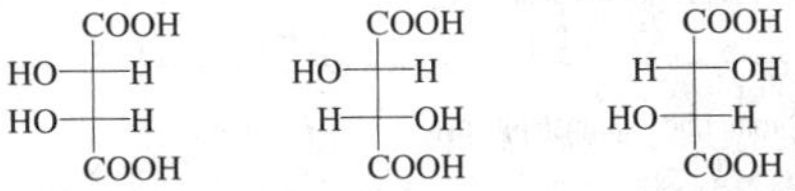

meso-Tartaric acid

D-Tartaric acid

L-Tartaric acid

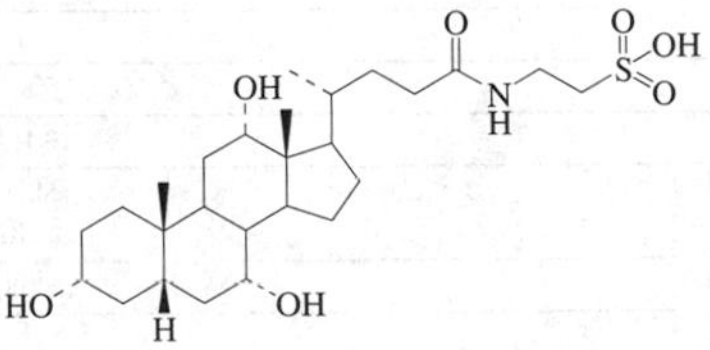

Taurocholic acid

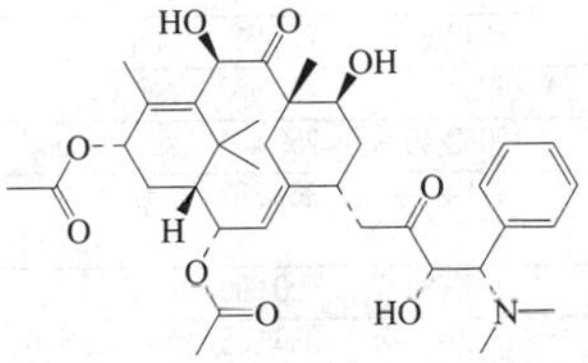

Taxine A

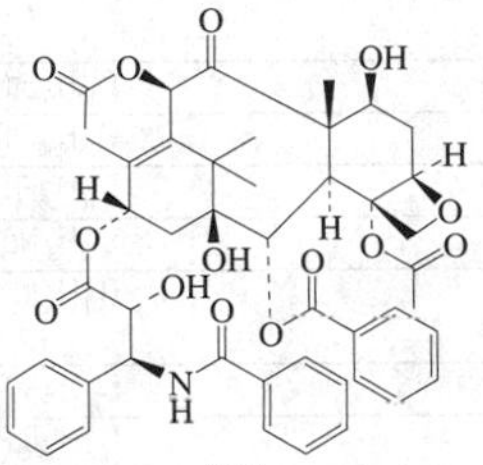

Taxol

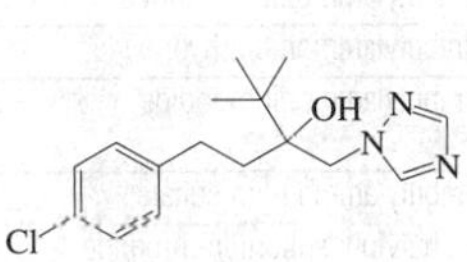

Tebuconazole

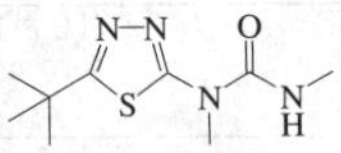

Tebuthiuron

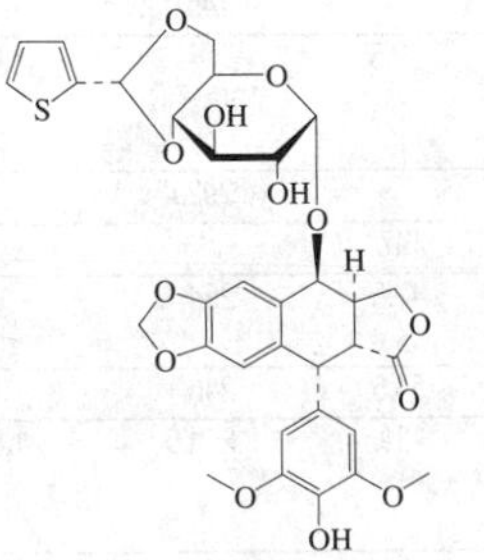

Teniposide

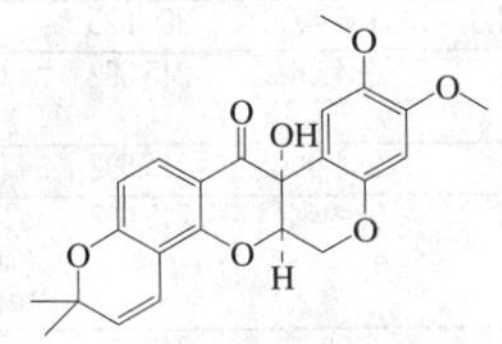

Tephrosin

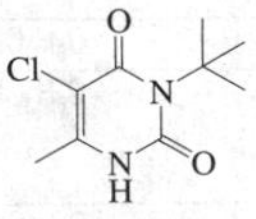

Terbacil

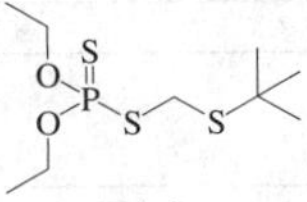

Terbufos

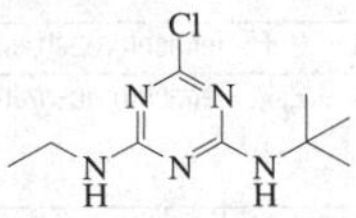

Terbuthylazine

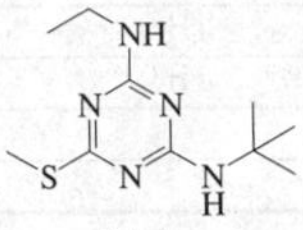

Terbutryn

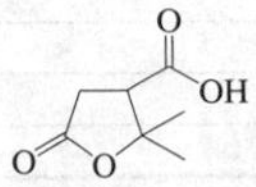

Terebic acid

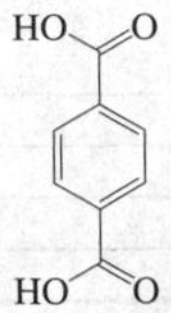

Terephthalic acid

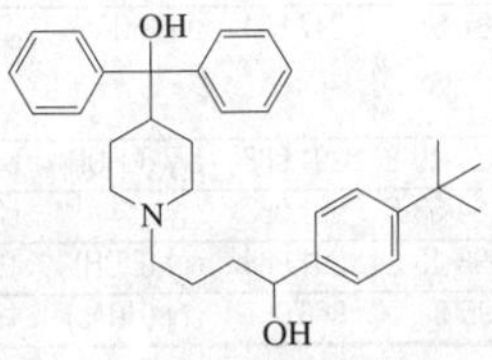

Terfenadine

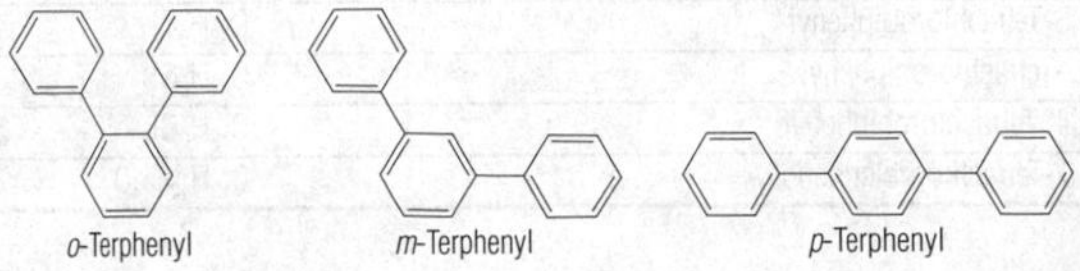

o-Terphenyl

m-Terphenyl

p-Terphenyl

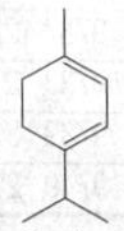

α-Terpinene

No.	Name	Synonym	Mol. Form.	CAS RN	Mol. Wt.	Physical Form	mp/°C	bp/°C	den/ g cm⁻³	n_D	Solubility
9700	γ-Terpinene		$C_{10}H_{16}$	99-85-4	136.234			183	0.849^{20}	1.4765^{14}	
9701	α-Terpineol		$C_{10}H_{18}O$	2438-12-2	154.249	cry (peth)	40.5	220	0.9337^{20}	1.4831^{20}	sl H_2O; vs ace, bz, eth, EtOH
9702	α-Terpineol acetate		$C_{12}H_{20}O_2$	80-26-2	196.286			140^{40}, 105^{11}	0.9659^{21}	1.4689^{21}	i H_2O; s EtOH, eth, bz
9703	Terpinolene	*p*-Mentha-1,4(8)-diene	$C_{10}H_{16}$	586-62-9	136.234			186	0.8632^{15}	1.4883^{20}	i H_2O; msc EtOH, eth; s bz, ctc
9704	2,2':6',2"-Terpyridine		$C_{15}H_{11}N_3$	1148-79-4	233.268		88.0	370			
9705	Terrazole	1,2,4-Thiadiazole, 5-ethoxy-3-(trichloromethyl)-	$C_5H_5Cl_3N_2OS$	2593-15-9	247.530		19.9	95^{1}	1.503^{25}		
9706	2,2':5',2"-Terthiophene	α-Terthienyl	$C_{12}H_8S_3$	1081-34-1	248.387	ye-oran pl (MeOH)	93				i H_2O; sl EtOH; s bz, eth, ace, peth
9707	Testolactone		$C_{19}H_{24}O_3$	968-93-4	300.392	cry (ace)	218				
9708	3,6,9,12-Tetraazatetradecane-1,14-diamine	Pentaethylenehexamine	$C_{10}H_{28}N_6$	4067-16-7	232.369	liq			0.950	1.5096^{20}	
9709	Tetrabenazine		$C_{19}H_{27}NO_3$	58-46-8	317.422		128				s chl
9710	1,2,4,5-Tetrabromobenzene		$C_6H_2Br_4$	636-28-2	393.696	mcl pr (CS_2)	182	3.1^{20}	3.072^{20}		i H_2O; vs eth
9711	1,1,2,2-Tetrabromoethane	Acetylene tetrabromide	$C_2H_2Br_4$	79-27-6	345.653	ye visc liq	0	243.5; 151^{54}	2.9655^{20}	1.6353^{20}	i H_2O; msc EtOH, eth; s ace, bz; sl ctc
9712	Tetrabromoethene	Tetrabromoethylene	C_2Br_4	79-28-7	343.637	pl (dil al), nd (al)	56.5	226			i H_2O; s EtOH, eth, ace; vs chl
9713	2',4',5',7'-Tetrabromofluorescein, disodium salt	Eosine YS	$C_{20}H_6Br_4Na_2O_5$	17372-87-1	691.855	ye-red cry	295.5				vs EtOH
9714	4,5,6,7-Tetrabromo-1,3-isobenzofurandione		$C_8Br_4O_3$	632-79-1	463.700	nd (xyl, HOAc)	280				i H_2O, EtOH; sl bz; s $PhNO_2$
9715	Tetrabromomethane	Carbon tetrabromide	CBr_4	558-13-4	331.627	mcl tab (dil al)	92.3	189.5	2.9608^{100}	1.5942^{100}	i H_2O; s EtOH, eth, chl; vs CS_2
9716	2,3,4,5-Tetrabromo-6-methylphenol	3,4,5,6-Tetrabromo-*o*-cresol	$C_7H_4Br_4O$	576-55-6	423.722	ye nd (chl, HOAc)	208	dec			i H_2O; s EtOH, eth, bz, chl; sl lig, HOAc
9717	3',3",5',5"-Tetrabromophenolphthalein		$C_{20}H_{10}Br_4O_4$	76-62-0	633.907	nd (al, eth)	296				i H_2O; sl EtOH; vs eth; s alk, HOAc
9718	3',3",5',5"-Tetrabromophenolphthalein ethyl ester		$C_{22}H_{14}Br_4O_4$	1176-74-5	661.960	ye cry (bz)	210				
9719	3',3",5',5"-Tetrabromophenolphthalein ethyl ester, potassium salt		$C_{22}H_{13}Br_4KO_4$	62637-91-6	700.050		210				
9720	Tetrabutylammonium bromide	TMAB	$C_{16}H_{36}BrN$	1643-19-2	322.368		99				s chl
9721	Tetrabutylammonium chloride		$C_{16}H_{36}ClN$	1112-67-0	277.917	cry	74				
9722	Tetrabutylammonium fluoride		$C_{16}H_{36}FN$	429-41-4	261.462	cry (w)	37				
9723	Tetrabutylammonium hydroxide		$C_{16}H_{37}NO$	2052-49-5	259.471	stab in soln					s H_2O, MeOH
9724	Tetrabutylammonium iodide		$C_{16}H_{36}IN$	311-28-4	369.368	lf (w, bz)	148				sl H_2O, chl; vs EtOH
9725	Tetrabutylammonium sulfate		$C_{32}H_{72}N_2O_4S$	2472-88-0	580.990		170				sl chl
9726	Tetrabutylphosphonium bromide		$C_{16}H_{36}BrP$	3115-68-2	339.335	cry (ace/eth)	102				
9727	Tetrabutyl silicate	Silicic acid, tetrabutyl ester	$C_{16}H_{36}O_4Si$	4766-57-8	320.541			256; 120^{3}	0.8990^{20}	1.4128^{20}	
9728	Tetrabutylstannane		$C_{16}H_{36}Sn$	1461-25-2	347.167	liq	-97	145^{10}, $95^{0.28}$	1.06^{20}		
9729	*N,N,N',N'*-Tetrabutylthioperoxydicarbonic diamide	Bis(dibutylthiocarbamoyl) disulfide	$C_{18}H_{36}N_2S_4$	1634-02-2	408.752		39.5		1.03^{20}		i H_2O; sl EtOH; s eth
9730	Tetrabutyl titanate	Titanium(IV) butoxide	$C_{16}H_{36}O_4Ti$	5593-70-4	340.322			292.4			
9731	Tetracaine hydrochloride		$C_{15}H_{25}ClN_2O_2$	136-47-0	300.825		147				
9732	1,2,3,4-Tetrachlorobenzene		$C_6H_2Cl_4$	634-66-2	215.892	nd (al)	47.5	254			i H_2O; sl EtOH; vs eth, CS_2
9733	1,2,3,5-Tetrachlorobenzene		$C_6H_2Cl_4$	634-90-2	215.892	nd (al)	54.5	246			i H_2O
9734	1,2,4,5-Tetrachlorobenzene		$C_6H_2Cl_4$	95-94-3	215.892	nd, mcl pr (eth, al or bz)	139.5	244.5	1.858^{22}		i H_2O; sl EtOH; s eth, bz, chl, CS_2
9735	3,4,5,6-Tetrachloro-1,2-benzenediol		$C_6H_2Cl_4O_2$	1198-55-6	247.891	cry (dil al, bz)	194				sl H_2O
9736	2,3,5,6-Tetrachloro-1,4-benzenediol		$C_6H_2Cl_4O_2$	87-87-6	247.891	nd (HOAc)		sub			i H_2O, bz, ctc; vs EtOH, eth; sl HOAc
9737	2,2',4',5-Tetrachlorobiphenyl		$C_{12}H_6Cl_4$	41464-40-8	291.988	cry (MeOH)	66.5				i H_2O
9738	2,3,4,5-Tetrachlorobiphenyl		$C_{12}H_6Cl_4$	33284-53-6	291.988	cry	92.2				i H_2O
9739	3,3',4',4'-Tetrachlorobiphenyl		$C_{12}H_6Cl_4$	32598-13-3	291.988	cry (EtOH)	180				
9740	2,2',6,6'-Tetrachlorobisphenol A		$C_{15}H_{12}Cl_4O_2$	79-95-8	366.067	cry (HOAc)	136				

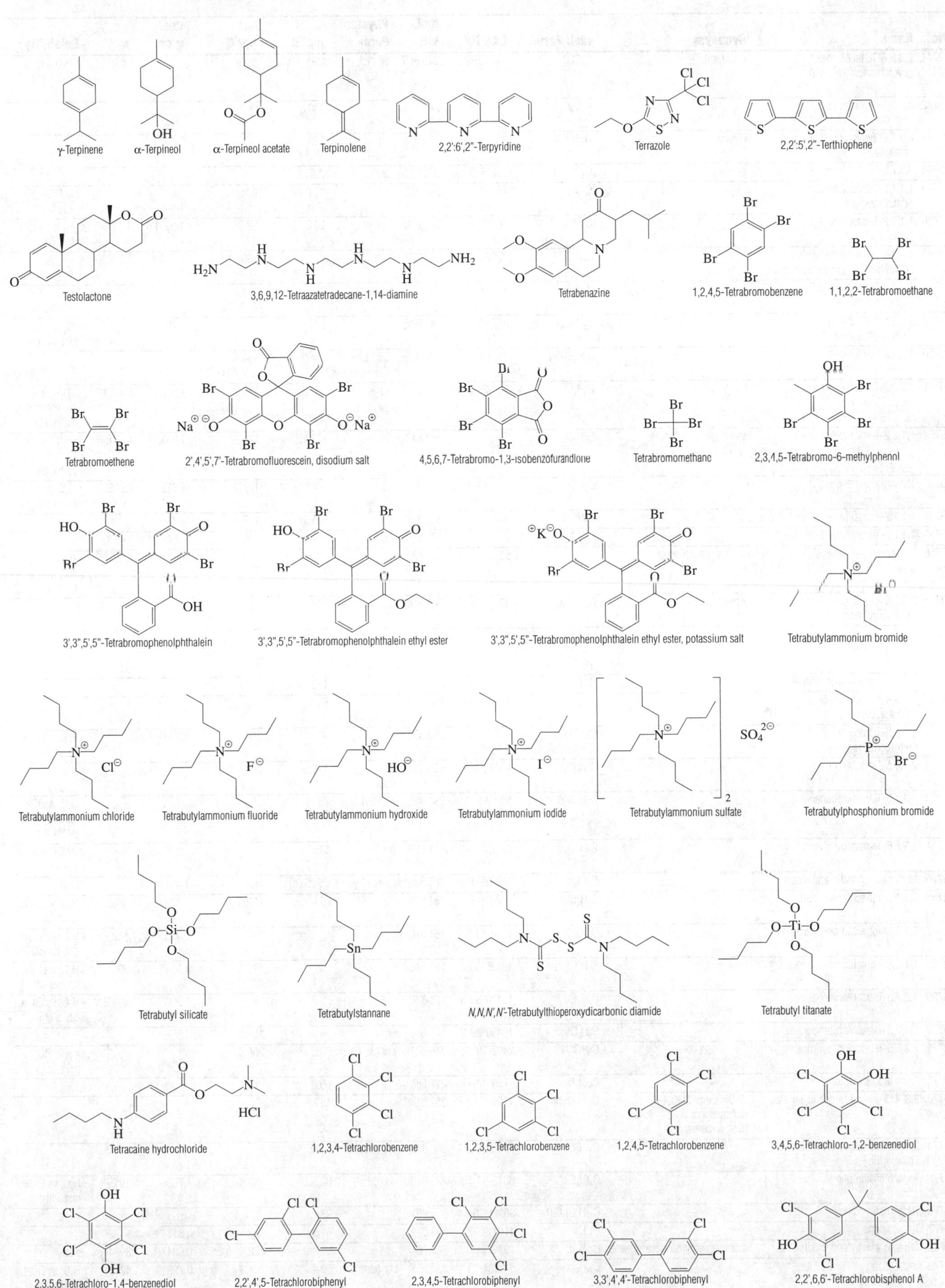
γ-Terpinene
α-Terpineol
α-Terpineol acetate
Terpinolene
2,2':6',2"-Terpyridine
Terrazole
2,2':5',2"-Terthiophene
Testolactone
3,6,9,12-Tetraazatetradecane-1,14-diamine
Tetrabenazine
1,2,4,5-Tetrabromobenzene
1,1,2,2-Tetrabromoethane
Tetrabromoethene
2',4',5',7'-Tetrabromofluorescein, disodium salt
4,5,6,7-Tetrabromo-1,3-isobenzofurandione
Tetrabromomethane
2,3,4,5-Tetrabromo-6-methylphenol
3',3",5',5"-Tetrabromophenolphthalein
3',3",5',5"-Tetrabromophenolphthalein ethyl ester
3',3",5',5"-Tetrabromophenolphthalein ethyl ester, potassium salt
Tetrabutylammonium bromide
Tetrabutylammonium chloride
Tetrabutylammonium fluoride
Tetrabutylammonium hydroxide
Tetrabutylammonium iodide
Tetrabutylammonium sulfate
Tetrabutylphosphonium bromide
Tetrabutyl silicate
Tetrabutylstannane
N,N,N',N'-Tetrabutylthioperoxydicarbonic diamide
Tetrabutyl titanate
Tetracaine hydrochloride
1,2,3,4-Tetrachlorobenzene
1,2,3,5-Tetrachlorobenzene
1,2,4,5-Tetrachlorobenzene
3,4,5,6-Tetrachloro-1,2-benzenediol
2,3,5,6-Tetrachloro-1,4-benzenediol
2,2',4',5-Tetrachlorobiphenyl
2,3,4,5-Tetrachlorobiphenyl
3,3',4',4'-Tetrachlorobiphenyl
2,2',6,6'-Tetrachlorobisphenol A

No.	Name	Synonym	Mol. Form.	CAS RN	Mol. Wt.	Physical Form	mp/°C	bp/°C	den/ g cm^{-3}	n_D	Solubility
9741	2,3,5,6-Tetrachloro-2,5-cyclohexadiene-1,4-dione	Chloranil	$C_6Cl_4O_2$	118-75-2	245.875	ye mcl, pr (bz) ye lf (HOAc)	290	sub			i H_2O, liq; sl EtOH, chl; s eth
9742	3,4,5,6-Tetrachloro-3,5-cyclohexadiene-1,2-dione		$C_6Cl_4O_2$	2435-53-2	245.875		130.5				
9743	2,3,7,8-Tetrachlorodibenzo-*p*-dioxin	Dioxin	$C_{12}H_4Cl_4O_2$	1746-01-6	321.971	nd	295				
9744	2,3,7,8-Tetrachlorodibenzofuran		$C_{12}H_4Cl_4O$	51207-31-9	305.971	cry	227				
9745	1,1,1,2-Tetrachloro-2,2-difluoroethane		$C_2Cl_4F_2$	76-11-9	203.830		41.0	92.8	1.649^{25}		i H_2O; s EtOH, eth, chl
9746	1,1,2,2-Tetrachloro-1,2-difluoroethane		$C_2Cl_4F_2$	76-12-0	203.830		24.8	92.8	1.5951^{50}	1.4130^{25}	i H_2O; s EtOH, eth, chl
9747	1,2,3,4-Tetrachloro-5,5-dimethoxy-1,3-cyclopentadiene		$C_7H_6Cl_4O_2$	2207-27-4	263.934			109	1.501^{25}	1.5282^{20}	
9748	1,2,3,4-Tetrachloro-5,6-dimethylbenzene		$C_8H_6Cl_4$	877-08-7	243.946		228				i H_2O; s EtOH, eth, bz
9749	1,2,3,5-Tetrachloro-4,6-dimethylbenzene		$C_8H_6Cl_4$	877-09-8	243.946		223		1.703^{25}		i H_2O, EtOH, eth, bz, chl
9750	1,1,2,2-Tetrachloro-1,2-dimethyldisilane		$C_2H_6Cl_4Si_2$	4518-98-3	228.052			154			
9751	1,1,1,2-Tetrachloroethane		$C_2H_2Cl_4$	630-20-6	167.849	liq	-70.2	130.2	1.5406^{20}	1.4821^{20}	sl H_2O; s ace, bz, chl; msc EtOH, eth
9752	1,1,2,2-Tetrachloroethane	Acetylene tetrachloride	$C_2H_2Cl_4$	79-34-5	167.849	liq	-42.4	145.2	1.5953^{20}	1.4940^{20}	sl H_2O; s ace, bz, chl; msc EtOH, eth
9753	Tetrachloroethene	Perchloroethylene	C_2Cl_4	127-18-4	165.833	liq	-22.3	121.3	1.6230^{20}	1.5059^{20}	i H_2O; msc EtOH, eth, bz
9754	1,1,1,2-Tetrachloro-2-fluoroethane		C_2HCl_4F	354-11-0	185.839	liq	-95.3	117.1			
9755	1,1,2,2-Tetrachloro-1-fluoroethane		C_2HCl_4F	354-14-3	185.839	liq	-82.6	116.7	1.5497^{17}	1.4390^{20}	
9756	Tetrachloromethane	Carbon tetrachloride	CCl_4	56-23-5	153.823	liq	-22.62	76.8	1.5940^{20}	1.4601^{20}	i H_2O; s EtOH, ace; msc eth, bz, chl
9757	2,3,5,6-Tetrachloro-4-methoxyphenol	Drosophilin A	$C_7H_4Cl_4O_2$	484-67-3	261.918		116				
9758	2,3,4,6-Tetrachloro-5-methylphenol		$C_7H_4Cl_4O$	10460-33-0	245.918	nd (peth)	189.5				i H_2O; s EtOH, eth, ace, bz, KOH
9759	1,2,3,4-Tetrachloronaphthalene		$C_{10}H_4Cl_4$	20020-02-4	265.951		199				
9760	1,2,3,4-Tetrachloro-5-nitrobenzene		$C_6HCl_4NO_2$	879-39-0	260.890		66				
9761	1,2,4,5-Tetrachloro-3-nitrobenzene		$C_6HCl_4NO_2$	117-18-0	260.890		99.5	304	1.744^{25}		i H_2O; s EtOH, bz, chl
9762	2,3,4,5-Tetrachlorophenol		$C_6H_2Cl_4O$	4901-51-3	231.891	nd (peth, sub)	116.5	sub			vs EtOH
9763	2,3,4,6-Tetrachlorophenol		$C_6H_2Cl_4O$	58-90-2	231.891	nd (lig)	70	150^{15}			i H_2O; s EtOH, bz, chl, HOAc; vs NaOH
9764	2,3,5,6-Tetrachlorophenol		$C_6H_2Cl_4O$	935-95-5	231.891	lf (lig)	115				sl H_2O; vs bz; s lig
9765	Tetrachlorophthalic anhydride		$C_8Cl_4O_3$	117-08-8	285.896		254.5	sub	1.49^{275}		sl eth
9766	1,1,1,2-Tetrachloropropane		$C_3H_4Cl_4$	812-03-3	181.876	liq	-64	152.5	1.473^{20}	1.4867^{20}	i H_2O; vs EtOH; s eth, chl
9767	1,1,1,3-Tetrachloropropane		$C_3H_4Cl_4$	1070-78-6	181.876			157	1.4509^{20}	1.4825^{20}	i H_2O; vs EtOH, eth, bz, chl
9768	1,1,2,3-Tetrachloropropane		$C_3H_4Cl_4$	18495-30-2	181.876			179.5	1.513^{17}	1.5037^{17}	i H_2O; s EtOH, chl; vs eth
9769	1,2,2,3-Tetrachloropropane		$C_3H_4Cl_4$	13116-53-5	181.876			165	1.500^{18}	1.4940^{18}	i H_2O; vs EtOH, eth; s chl
9770	1,1,2,3-Tetrachloropropene		$C_3H_2Cl_4$	10436-39-2	179.860	liq		167.2; 59^{17}	1.55^{20}		
9771	2,3,5,6-Tetrachloropyridine		C_5HCl_4N	2402-79-1	216.881	cry (aq al)	90.5	250.5			vs eth, EtOH, peth
9772	Tetrachloropyrimidine		$C_4Cl_4N_2$	1780-40-1	217.868		69.0				
9773	3,3',4',5-Tetrachlorosalicylanilide	3,5-Dichloro-*N*-(3,4-dichlorophenyl)-2-hydroxybenzamide	$C_{13}H_7Cl_4NO_2$	1154-59-2	351.013		161				
9774	2,3,5,6-Tetrachloroterphthaloyl dichloride		$C_8Cl_6O_2$	719-32-4	340.803	cry (ctc)	146.5				
9775	Tetrachlorothiophene		C_4Cl_4S	6012-97-1	221.920	nd (dil al)	30.5	233.4	1.7036^{30}	1.5915^{30}	i H_2O; vs EtOH; msc eth
9776	Tetrachlorovinphos		$C_{10}H_9Cl_4O_4P$	961-11-5	365.961		97				
9777	Tetracontane		$C_{40}H_{82}$	4181-95-7	563.079		81.5	522; 400^{50}	0.8171^{25}	1.4572^{25}	
9778	Tetracosamethylundecasiloxane	Tetracosamethylhendecasiloxane	$C_{24}H_{72}O_{10}Si_{11}$	107-53-9	829.764			322.8; 202^{47}	0.9247^{25}	1.3994^{20}	vs bz
9779	Tetracosane		$C_{24}H_{50}$	646-31-1	338.654	cry (eth)	50.4	391.3	0.7991^{20}	1.4283^{70}	i H_2O; sl EtOH; vs eth

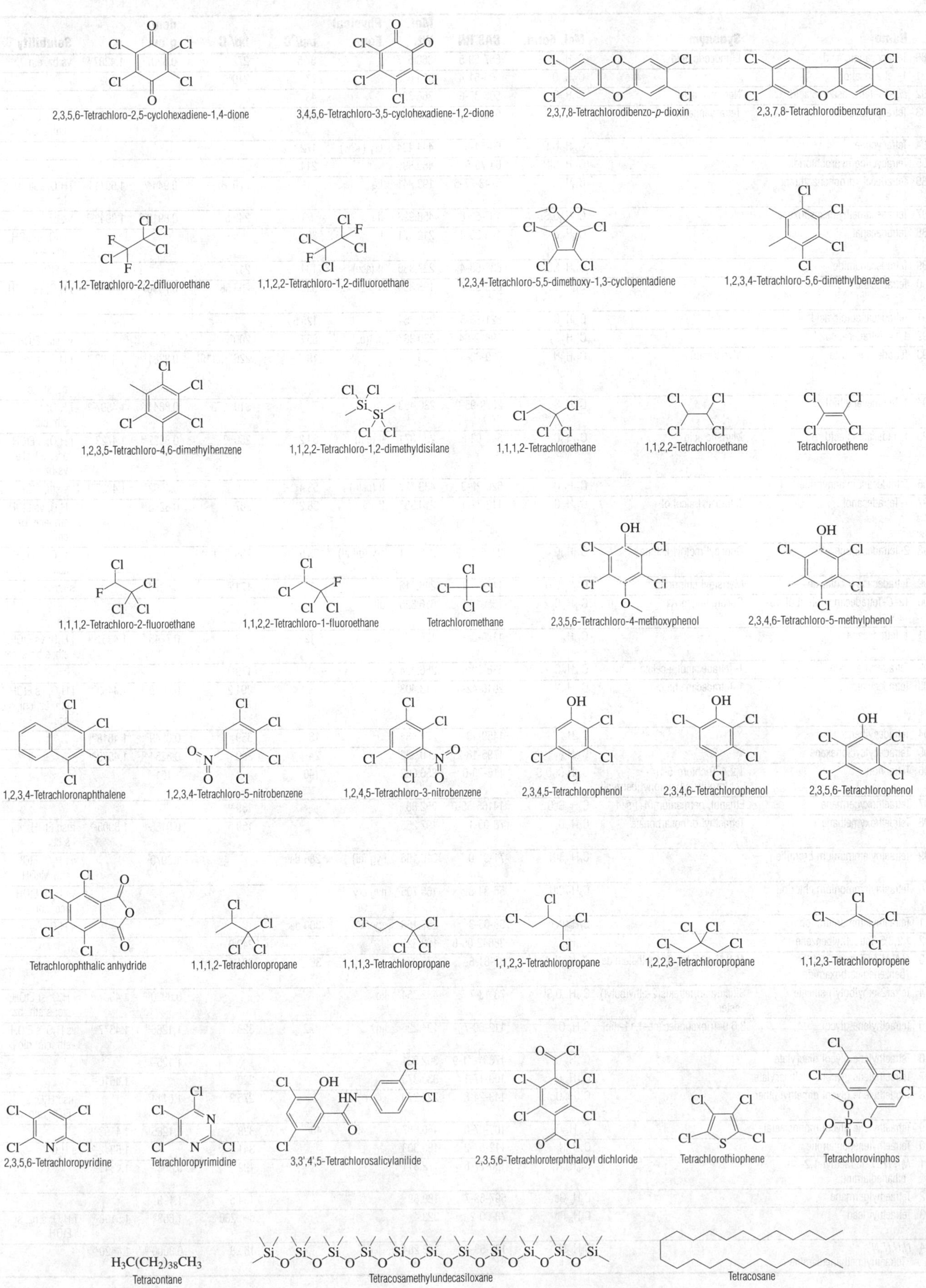
2,3,5,6-Tetrachloro-2,5-cyclohexadiene-1,4-dione
3,4,5,6-Tetrachloro-3,5-cyclohexadiene-1,2-dione
2,3,7,8-Tetrachlorodibenzo-p-dioxin
2,3,7,8-Tetrachlorodibenzofuran
1,1,1,2-Tetrachloro-2,2-difluoroethane
1,1,2,2-Tetrachloro-1,2-difluoroethane
1,2,3,4-Tetrachloro-5,5-dimethoxy-1,3-cyclopentadiene
1,2,3,4-Tetrachloro-5,6-dimethylbenzene
1,2,3,5-Tetrachloro-4,6-dimethylbenzene
1,1,2,2-Tetrachloro-1,2-dimethyldisilane
1,1,1,2-Tetrachloroethane
1,1,2,2-Tetrachloroethane
Tetrachloroethene
1,1,1,2-Tetrachloro-2-fluoroethane
1,1,2,2-Tetrachloro-1-fluoroethane
Tetrachloromethane
2,3,5,6-Tetrachloro-4-methoxyphenol
2,3,4,6-Tetrachloro-5-methylphenol
1,2,3,4-Tetrachloronaphthalene
1,2,3,4-Tetrachloro-5-nitrobenzene
1,2,4,5-Tetrachloro-3-nitrobenzene
2,3,4,5-Tetrachlorophenol
2,3,4,6-Tetrachlorophenol
2,3,5,6-Tetrachlorophenol
Tetrachlorophthalic anhydride
1,1,1,2-Tetrachloropropane
1,1,1,3-Tetrachloropropane
1,1,2,3-Tetrachloropropane
1,2,2,3-Tetrachloropropane
1,1,2,3-Tetrachloropropene
2,3,5,6-Tetrachloropyridine
Tetrachloropyrimidine
3,3',4',5-Tetrachlorosalicylanilide
2,3,5,6-Tetrachloroterphthaloyl dichloride
Tetrachlorothiophene
Tetrachlorovinphos
H3C(CH2)38CH3
Tetracontane
Tetracosamethylundecasiloxane
Tetracosane

No.	Name	Synonym	Mol. Form.	CAS RN	Mol. Wt.	Physical Form	mp/°C	bp/°C	den/ g cm^{-3}	n_D	Solubility
9780	Tetracosanoic acid	Lignoceric acid	$C_{24}H_{48}O_2$	557-59-5	368.637		87.5	272^{10}	0.8207^{100}	1.4287^{100}	vs bz, eth
9781	1-Tetracosanol		$C_{24}H_{50}O$	506-51-4	354.653		77	$210^{0.4}$			
9782	*cis*-15-Tetracosenoic acid	Nervonic acid	$C_{24}H_{46}O_2$	506-37-6	366.621		43				
9783	Tetracyanoethene	Tetracyanoethylene	C_6N_4	670-54-2	128.091		199	223	1.348^{25}	1.560^{25}	sl eth, bz, ctc, chl; s ace
9784	Tetracycline		$C_{22}H_{24}N_2O_8$	60-54-8	444.434	cry (+3w)	172 dec				
9785	Tetracycline hydrochloride		$C_{22}H_{25}ClN_2O_8$	64-75-5	480.895		214				
9786	Tetradecahydrophenanthrene		$C_{14}H_{24}$	5743-97-5	192.341	liq	-3	270; 87^{2}	0.944^{20}	1.5011^{20}	i H_2O; s eth, ace, bz
9787	Tetradecamethylhexasiloxane		$C_{14}H_{42}O_5Si_6$	107-52-8	458.993	liq	-59	245.5	0.8910^{20}	1.3948^{20}	vs bz
9788	Tetradecanal		$C_{14}H_{28}O$	124-25-4	212.371	lf	30				i H_2O; s EtOH, eth, ace
9789	Tetradecanamide		$C_{14}H_{29}NO$	638-58-4	227.386	lf (ace)	104	217^{12}			vs EtOH
9790	Tetradecane		$C_{14}H_{30}$	629-59-4	198.388		5.82	253.58	0.7596^{20}	1.4290^{20}	i H_2O; vs EtOH, eth; s ctc
9791	Tetradecanedioic acid		$C_{14}H_{26}O_4$	821-38-5	258.354		125.5				
9792	1,14-Tetradecanediol		$C_{14}H_{30}O_2$	19812-64-7	230.387	nd (bz)	85.8	200^{9}			vs eth, EtOH
9793	Tetradecanenitrile	Myristonitrile	$C_{14}H_{27}N$	629-63-0	209.371		19	226^{100}, 119^{1}	0.8281^{19}	1.4392^{23}	i H_2O; msc EtOH, eth, ace, bz; sl ctc
9794	1-Tetradecanethiol		$C_{14}H_{30}S$	2079-95-0	230.453		7	310; 178^{22}	0.8641^{20}	1.4597^{20}	i H_2O; s EtOH, eth, ctc
9795	Tetradecanoic acid	Myristic acid	$C_{14}H_{28}O_2$	544-63-8	228.371	lf (eth)	54.2	250^{100}	0.8622^{54}	1.4723^{70}	i H_2O; s EtOH, ace, chl; sl eth; vs bz
9796	Tetradecanoic anhydride		$C_{28}H_{54}O_3$	626-29-9	438.727	lf (peth)	53.4		0.8502^{70}	1.4335^{70}	vs eth, EtOH
9797	1-Tetradecanol	Tetradecyl alcohol	$C_{14}H_{30}O$	112-72-1	214.387	lf	38.2	287	0.8236^{38}		i H_2O; vs EtOH, eth, ace, bz, chl
9798	2-Tetradecanone	Dodecyl methyl ketone	$C_{14}H_{28}O$	2345-27-9	212.371	cry (dil al)	33.5	205^{100}, 134^{13}			i H_2O; s EtOH, ace
9799	Tetradecanoyl chloride	Myristoyl chloride	$C_{14}H_{27}ClO$	112-64-1	246.816		-1	171^{16}	0.9078^{25}		s eth
9800	12-*O*-Tetradecanoylphorbol-13-acetate	Cocarcinogen A1	$C_{36}H_{56}O_8$	16561-29-8	616.825	oil					
9801	1-Tetradecene		$C_{14}H_{28}$	1120-36-1	196.372	liq	-12	233	0.7745^{25}	1.4351^{20}	i H_2O; vs EtOH, eth; s bz; sl ctc
9802	Tetradecyl acetate	1-Tetradecanol, acetate	$C_{16}H_{32}O_2$	638-59-5	256.424			173^{10}			
9803	Tetradecylamine	1-Tetradecanamine	$C_{14}H_{31}N$	2016-42-4	213.403		83.1	291.2	0.8079^{20}	1.4463^{20}	i H_2O; vs EtOH, eth, bz, chl; s ace
9804	Tetradecylbenzene		$C_{20}H_{34}$	1459-10-5	274.484		16	359	0.8549^{20}	1.4818^{20}	
9805	Tetradecylcyclohexane		$C_{20}H_{40}$	1795-18-2	280.532		24	360	0.8254^{20}	1.4579^{20}	
9806	Tetradifon	1,2,4-Trichloro-5-[(4-chlorophenyl)sulfonyl]benzene	$C_{12}H_6Cl_4O_2S$	116-29-0	356.052		146		1.151^{20}		
9807	Tetraethoxygermane	Ethanol, germanium(4+) salt	$C_8H_{20}GeO_4$	14165-55-0	252.88			139^{200}			
9808	Tetraethoxymethane	Tetraethyl orthocarbonate	$C_9H_{20}O_4$	78-09-1	192.253			159.5	0.9186^{20}	1.3905^{25}	msc EtOH, eth; s ctc
9809	Tetraethylammonium bromide		$C_8H_{20}BrN$	71-91-0	210.156	hyg (al)	286 dec		1.3970^{20}		vs H_2O, EtOH, chl, MeOH
9810	Tetraethylammonium chloride		$C_8H_{20}ClN$	56-34-8	165.705	hyg cry					vs H_2O, EtOH, ace, chl
9811	Tetraethylammonium iodide		$C_8H_{20}IN$	68-05-3	257.156	cry (w)	300 dec				s H_2O
9812	1,2,3,5-Tetraethylbenzene		$C_{14}H_{22}$	38842-05-6	190.325			249.0			
9813	*N,N,N',N'*-Tetraethyl-1,2-benzenedicarboxamide	*N,N,N',N'*-Tetraethylphthalamide	$C_{16}H_{24}N_2O_2$	83-81-8	276.374		36	204^{16}			
9814	Tetra(2-ethylbutyl) silicate	Silicic acid, tetrakis(2-ethylbutyl) ester	$C_{24}H_{52}O_4Si$	78-13-7	432.754	liq			0.8920^{20}	1.4307^{20}	i H_2O; sl EtOH, ctc; s eth, bz
9815	Tetraethylene glycol	3,6,9-Trioxaundecane-1,11-diol	$C_8H_{18}O_5$	112-60-7	194.226	liq	-6.2	328	1.1285^{15}	1.4577^{20}	vs H_2O; s EtOH, eth, ctc, diox
9816	Tetraethylene glycol diacrylate		$C_{14}H_{22}O_7$	17831-71-9	302.321				1.125^{25}		
9817	Tetraethylene glycol dimethacrylate		$C_{16}H_{26}O_7$	109-17-1	330.373			220^{1}		1.4610^{25}	
9818	Tetraethylene glycol dimethyl ether		$C_{10}H_{22}O_5$	143-24-8	222.279			275.3	1.0114^{20}		msc H_2O; s EtOH, eth, ctc
9819	Tetraethylene glycol monostearate		$C_{26}H_{52}O_6$	106-07-0	460.687		40	328	1.1285^{15}	1.4593^{20}	
9820	Tetraethylenepentamine		$C_8H_{23}N_5$	112-57-2	189.303			341.5		1.5042^{20}	s H_2O
9821	*N,N,N',N'*-Tetraethyl-1,2-ethanediamine		$C_{10}H_{24}N_2$	150-77-6	172.311			192	0.808^{25}	1.4343^{20}	
9822	Tetraethylgermane		$C_8H_{20}Ge$	597-63-7	188.89			164.5	1.199		
9823	Tetraethyl lead		$C_8H_{20}Pb$	78-00-2	323.4			dec 200	1.653^{20}	1.5198^{20}	i H_2O; s bz; sl EtOH
9824	*N,N,N',N'*-Tetraethylmethanediamine		$C_9H_{22}N_2$	102-53-4	158.284			165.8	0.8000^{20}	1.4420^{25}	

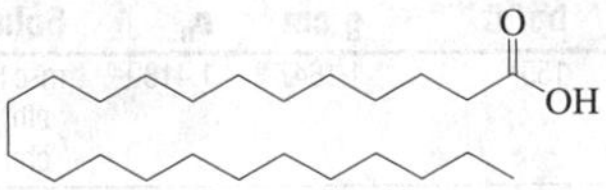

Tetracosanoic acid

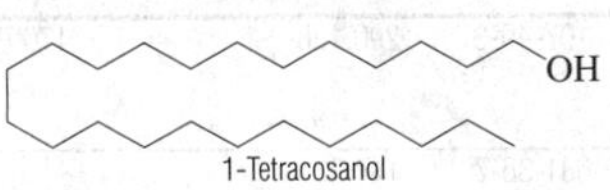

1-Tetracosanol

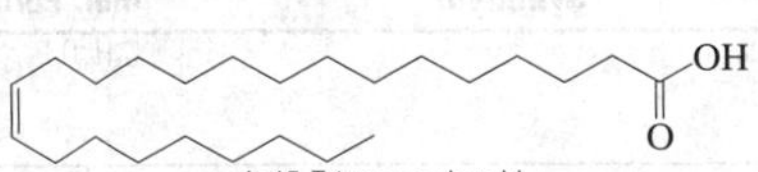

cis-15-Tetracosenoic acid

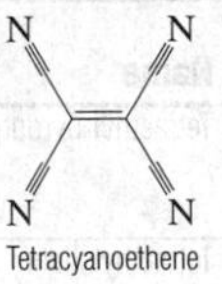

Tetracyanoethene

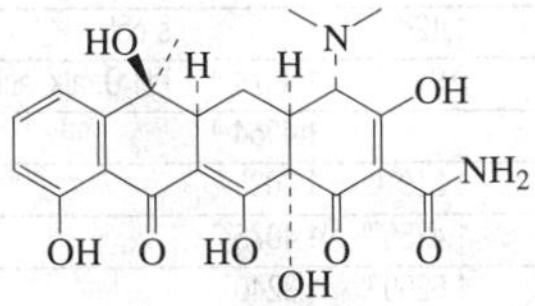

Tetracycline

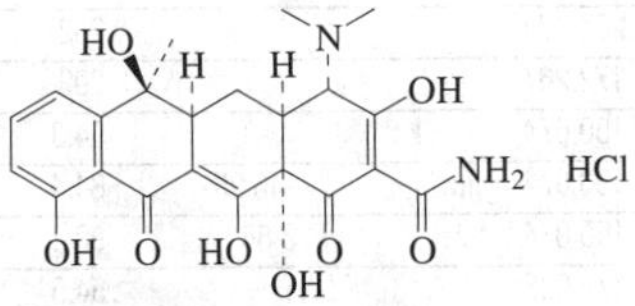

Tetracycline hydrochloride

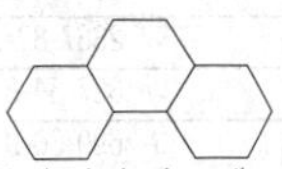

Tetradecahydrophenanthrene

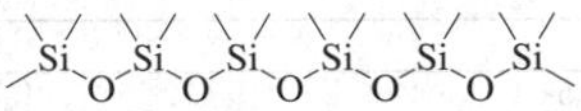

Tetradecamethylhexasiloxane

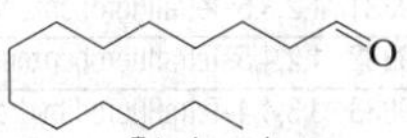

Tetradecanal

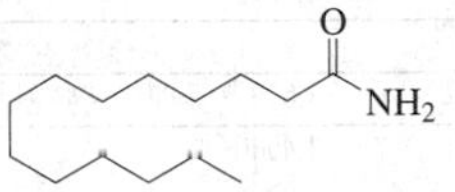

Tetradecanamide

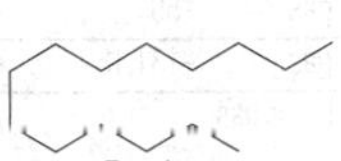

Tetradecane

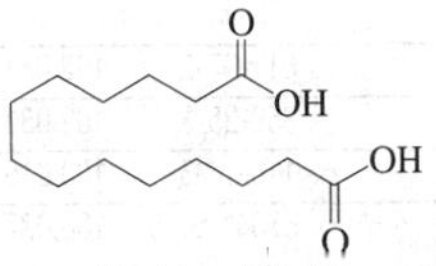

Tetradecanedioic acid

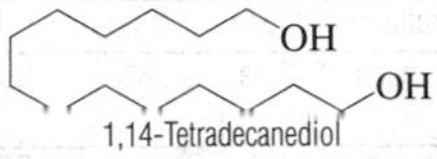

1,14-Tetradecanediol

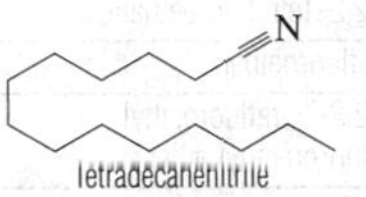

Tetradecanenitrile

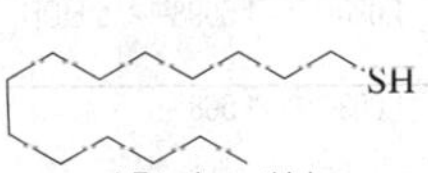

1-Tetradecanethiol

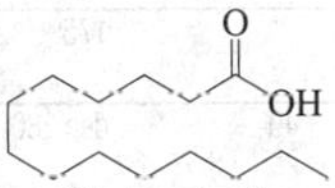

Tetradecanoic acid

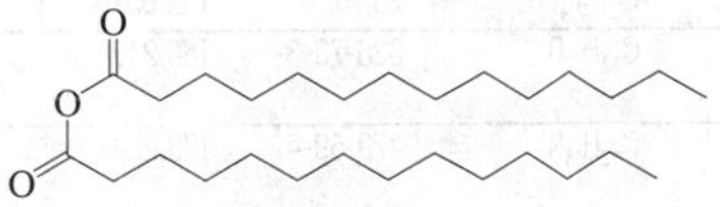

Tetradecanoic anhydride

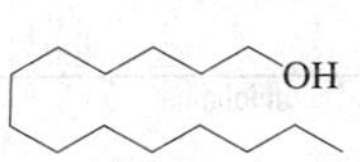

1-Tetradecanol

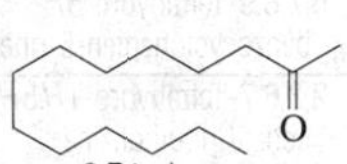

2-Tetradecanone

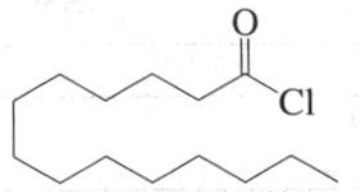

Tetradecanoyl chloride

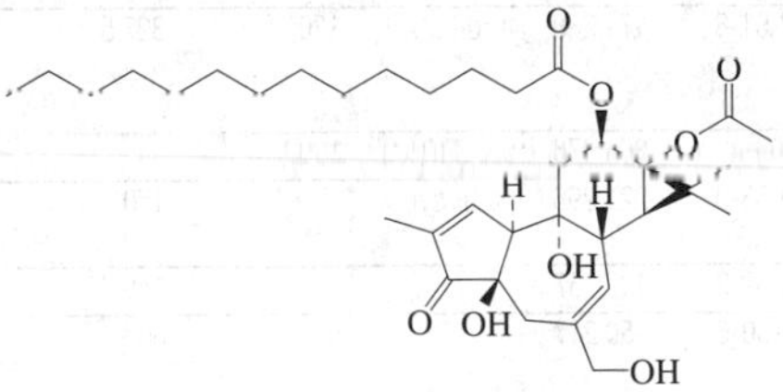

12-*O*-Tetradecanoylphorbol-13-acetate

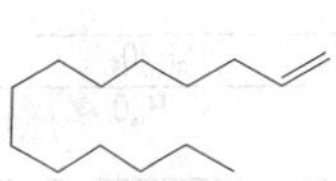

1-Tetradecene

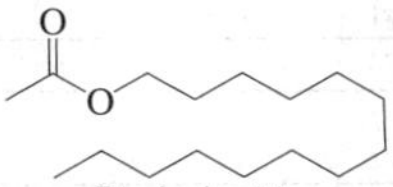

Tetradecyl acetate

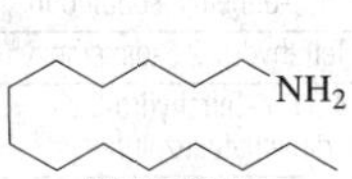

Tetradecylamine

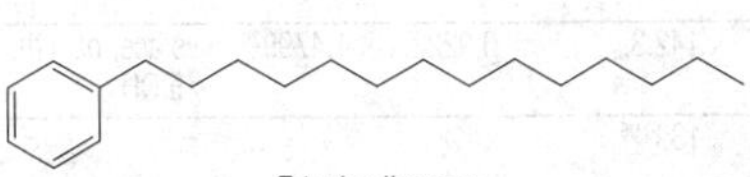

Tetradecylbenzene

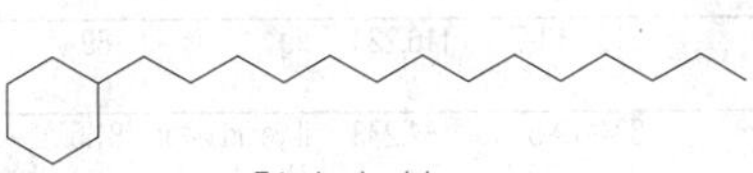

Tetradecylcyclohexane

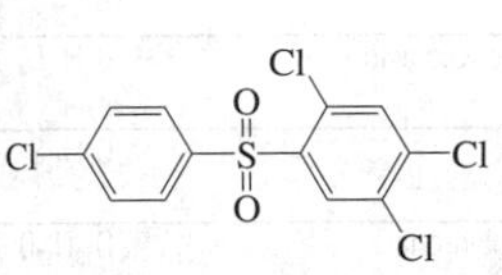

Tetradifon

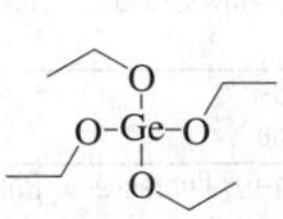

Tetraethoxygermane

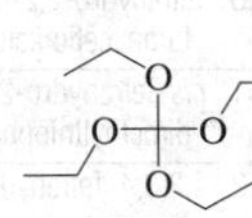

Tetraethoxymethane

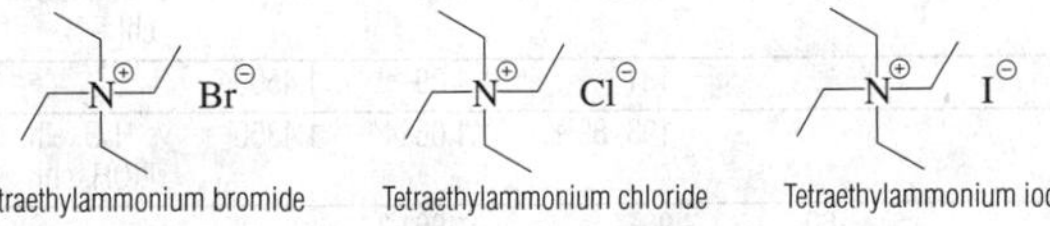

Tetraethylammonium bromide

Tetraethylammonium chloride

Tetraethylammonium iodide

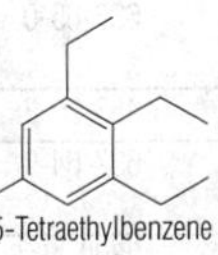

1,2,3,5-Tetraethylbenzene

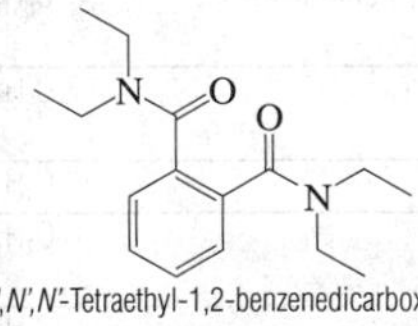

N,N,N',N'-Tetraethyl-1,2-benzenedicarboxamide

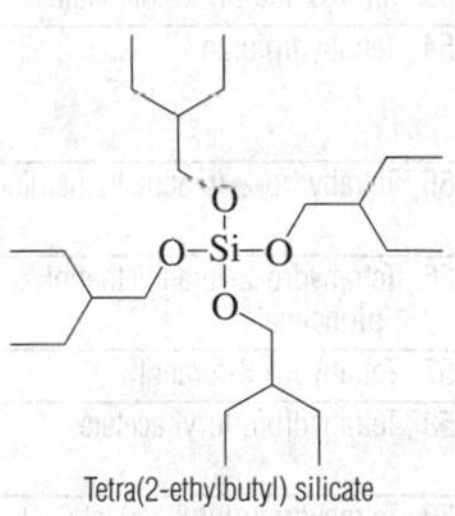

Tetra(2-ethylbutyl) silicate

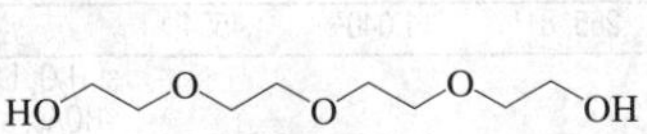

Tetraethylene glycol

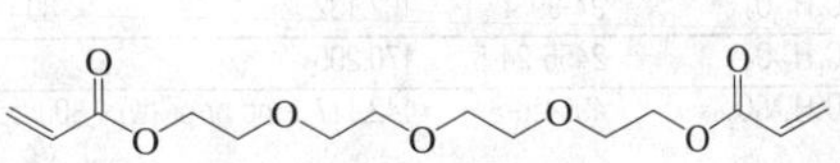

Tetraethylene glycol diacrylate

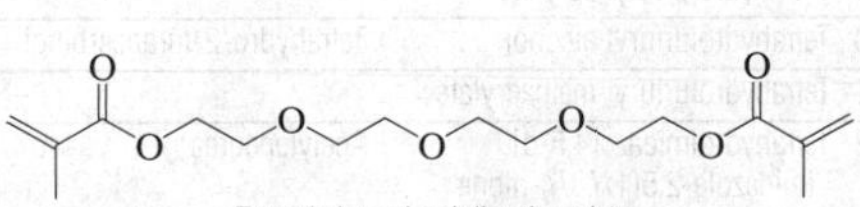

Tetraethylene glycol dimethacrylate

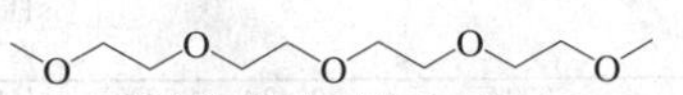

Tetraethylene glycol dimethyl ether

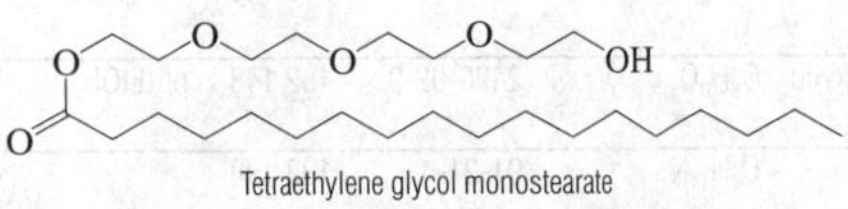

Tetraethylene glycol monostearate

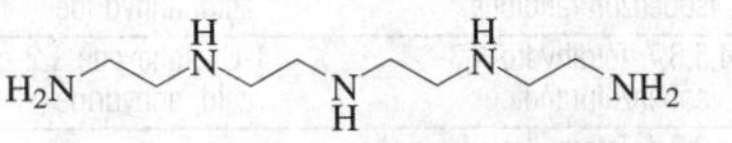

Tetraethylenepentamine

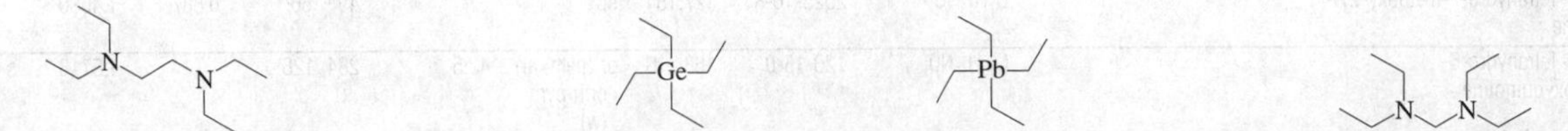

N,N,N',N'-Tetraethyl-1,2-ethanediamine

Tetraethylgermane

Tetraethyl lead

N,N,N',N'-Tetraethylmethanediamine

No.	Name	Synonym	Mol. Form.	CAS RN	Mol. Wt.	Physical Form	mp/°C	bp/°C	den/ g cm^{-3}	n_D	Solubility
9825	Tetraethyl pyrophosphate		$C_8H_{20}O_7P_2$	107-49-3	290.188		170 dec	155^{3}	1.1847^{20}	1.4180^{20}	msc H_2O, EtOH, eth, ace, xyl, chl; sl ctc
9826	Tetraethylsilane		$C_8H_{20}Si$	631-36-7	144.331			154.7	0.7658^{20}	1.4268^{20}	i H_2O
9827	Tetraethylstannane	Tin tetraethyl	$C_8H_{20}Sn$	597-64-8	234.955	liq	-112	181; 64^{12}	1.187^{25}	1.4730^{20}	
9828	Tetraethylthiodicarbonic diamide	Sulfiram	$C_{10}H_{20}N_2S_3$	95-05-6	264.474			232^{3}	1.12^{20}		s chl
9829	Tetraethylurea		$C_9H_{20}N_2O$	1187-03-7	172.267			209	0.919^{20}	1.4474^{20}	i H_2O, alk, acid
9830	1,2,3,4-Tetrafluorobenzene		$C_6H_2F_4$	551-62-2	150.074			94.3		1.4054^{20}	
9831	1,2,3,5-Tetrafluorobenzene		$C_6H_2F_4$	2367-82-0	150.074	liq	-46.25	84.4	1.319^{25}	1.4035^{20}	
9832	1,2,4,5-Tetrafluorobenzene		$C_6H_2F_4$	327-54-8	150.074		3.88	90.2	1.4255^{20}	1.4075^{20}	
9833	3,3,4,4-Tetrafluorodihydro-2,5-furandione		$C_4F_4O_3$	699-30-9	172.035			54.5	1.6209^{20}	1.3240^{20}	
9834	1,1,2,2-Tetrafluoro-1,2-dinitroethane		$C_2F_4N_2O_4$	356-16-1	192.026	liq	-41.5	58.5	1.6024^{25}	1.3265^{25}	i H_2O; s ace
9835	1,1,1,2-Tetrafluoroethane		$C_2H_2F_4$	811-97-2	102.031	col gas	-103.3	-26.5	1.2072^{25}		i H_2O; s eth
9836	1,1,2,2-Tetrafluoroethane		$C_2H_2F_4$	359-35-3	102.031	col gas	-89	-19.9			
9837	Tetrafluoroethene	Tetrafluoroethylene	C_2F_4	116-14-3	100.015	col gas	-131.15	-75.9	1.519^{-76}		i H_2O
9838	1,2,2,2-Tetrafluoroethyl difluoromethyl ether	Refrigerant 236me	$C_3H_2F_6O$	57041-67-5	168.037	vol liq or gas		23.35	1.4540^{23}		
9839	Tetrafluoromethane	Carbon tetrafluoride	CF_4	75-73-0	88.005	col gas	-183.60	-128.0	3.034^{25}		i H_2O; s bz, chl
9840	2,2,3,3-Tetrafluoro-1-propanol		$C_3H_4F_4O$	76-37-9	132.057	liq	-15	109.5	1.4853^{20}	1.3197^{20}	s EtOH, ace, chl
9841	6,7,8,9-Tetrahydro-5*H*-benzocyclohepten-5-one		$C_{11}H_{12}O$	826-73-3	160.212			175^{40}, 124^{7}	1.080^{20}	1.5698^{20}	s EtOH
9842	2,3,6,7-Tetrahydro-1*H*,5*H*-benzo[ij]quinolizine	Julolidine	$C_{12}H_{15}N$	479-59-4	173.254		40	dec 280; 155^{17}	1.003^{20}	1.568^{25}	
9843	1,2,3,6-Tetrahydro-2,3'-bipyridine, (*S*)	Anatabine	$C_{10}H_{12}N_2$	581-49-7	160.215			145^{10}	1.091^{19}	1.5676^{20}	msc H_2O; s EtOH, eth, bz
9844	2,3,4,9-Tetrahydro-1*H*-carbazole		$C_{12}H_{13}N$	942-01-8	171.238	lf (dil al)	120	327.5			i H_2O; s EtOH; vs eth, bz, MeOH
9845	Tetrahydrocortisone		$C_{21}H_{32}O_5$	53-05-4	364.476	cry (EtOAc)	190				
9846	1,2,3,4-Tetrahydro-6,7-dimethoxy-1,2-dimethylisoquinoline, (±)	Carnegine	$C_{13}H_{19}NO_2$	490-53-9	221.296	pa br syr		170^{1}			vs H_2O, eth, EtOH
9847	Tetrahydro-2,5-dimethoxyfuran		$C_6H_{12}O_3$	696-59-3	132.157			145.7	1.02^{25}	1.4180^{20}	
9848	4,5,6,7-Tetrahydro-3,6-dimethylbenzofuran		$C_{10}H_{14}O$	494-90-6	150.217		86	80^{18}	0.972^{15}		
9849	1,2,3,4-Tetrahydro-1,5-dimethylnaphthalene		$C_{12}H_{16}$	21564-91-0	160.255			239	0.941^{20}	1.526^{20}	
9850	Tetrahydro-2,2-dimethyl-5-oxo-3-furanacetic acid	Terpenylic acid	$C_8H_{12}O_4$	26754-48-3	172.179	lf or pr (w+1)	90				vs H_2O
9851	*cis*-Tetrahydro-2,5-dimethylthiophene		$C_6H_{12}S$	5161-13-7	116.224	liq	-89	142.3	0.9222^{20}	1.4799^{20}	vs ace, bz, eth, EtOH
9852	1,2,3,4-Tetrahydro-9*H*-fluoren-9-one	Phentydrone	$C_{13}H_{12}O$	634-19-5	184.233	lt ye nd or pr (pentane)	81.5	$139^{0.05}$			
9853	5,6,7,8-Tetrahydrofolic acid		$C_{19}H_{23}N_7O_6$	135-16-0	445.429	pow					s H_2O
9854	Tetrahydrofuran	Tetramethylene oxide	C_4H_8O	109-99-9	72.106	liq	-108.44	65	0.8833^{25}	1.4050^{25}	s H_2O, chl; vs EtOH, eth, ace, bz
9855	Tetrahydro-2-furanmethanamine	Tetrahydrofurfurylamine	$C_5H_{11}NO$	4795-29-3	101.147			153	0.9752^{20}	1.4551^{20}	vs H_2O, eth, EtOH
9856	Tetrahydro-2-furanmethanol propanoate		$C_8H_{14}O_3$	637-65-0	158.195			205.5	1.044^{20}		vs eth, EtOH, chl
9857	Tetrahydro-3-furanol		$C_4H_8O_2$	453-20-3	88.106			181	1.09^{25}	1.4500^{20}	
9858	Tetrahydrofurfuryl acetate		$C_7H_{12}O_3$	637-64-9	144.168			193; 89^{18}	1.0624^{20}	1.4350^{25}	vs H_2O, eth, EtOH, chl
9859	Tetrahydrofurfuryl acrylate		$C_8H_{12}O_3$	2399-48-6	156.179		<-60	96^{6}	1.061^{20}		
9860	Tetrahydrofurfuryl alcohol	Tetrahydro-2-furancarbinol	$C_5H_{10}O_2$	97-99-4	102.132		<-80	178	1.0524^{20}	1.4520^{20}	vs ace, eth
9861	Tetrahydrofurfuryl methacrylate		$C_9H_{14}O_3$	2455-24-5	170.205			265; 81^{4}	1.040^{25}	1.4554^{25}	
9862	Tetrahydroimidazo[4,5-d] imidazole-2,5(1*H*,3*H*)-dione	Acetyleneurea	$C_4H_6N_4O_2$	496-46-8	142.117	nd or pr (w)	300 dec				sl H_2O; i EtOH, HOAc; s eth, HCl, alk
9863	*cis*-3a,4,7,7a-Tetrahydro-1,3-isobenzofurandione	4-Cyclohexene-1,2-dicarboxylic acid, anhydride	$C_8H_8O_3$	935-79-5	152.148	cry (peth)	103.5				s EtOH, ace, chl, bz; sl peth
9864	4,5,6,7-Tetrahydro-1,3-isobenzofurandione	1-Cyclohexene-1,2-dicarboxylic acid, anhydride	$C_8H_8O_3$	2426-02-0	152.148	pl (EtOH)	74		1.2^{105}		s EtOH, ace, chl; vs eth
9865	1,2,3,4-Tetrahydroisoquinoline		$C_9H_{11}N$	91-21-4	133.190		<-15	232.5	1.0642^{24}	1.5668^{20}	i H_2O; s EtOH, bz, acid, xyl
9866	3,4,5,6-Tetrahydro-7-methoxy-2*H*-azepine		$C_7H_{13}NO$	2525-16-8	127.184	liq		49^{16}, 66^{24}	0.887	1.4630^{20}	
9867	1,2,3,4-Tetrahydro-6-methoxyquinoline		$C_{10}H_{13}NO$	120-15-0	163.216	pr (peth, al) orth pym (w)	42.5	284; 128^{1}		1.5718^{20}	s chl
9868	1,2,3,4-Tetrahydro-1-methylnaphthalene		$C_{11}H_{14}$	1559-81-5	146.229			220.6	0.9583^{20}	1.5353^{20}	

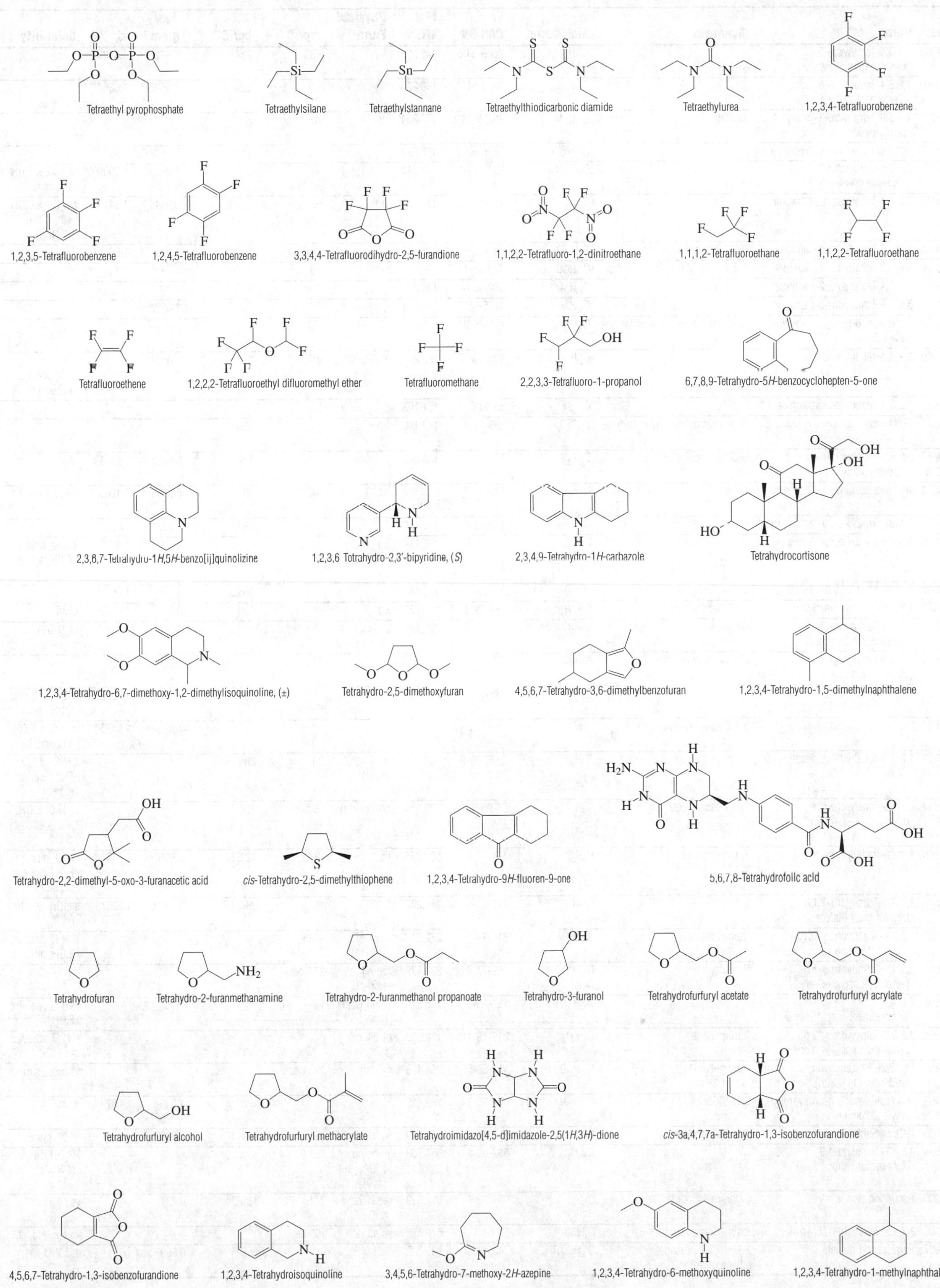

Tetraethyl pyrophosphate
Tetraethylsilane
Tetraethylstannane
Tetraethylthiodicarbonic diamide
Tetraethylurea
1,2,3,4-Tetrafluorobenzene
1,2,3,5-Tetrafluorobenzene
1,2,4,5-Tetrafluorobenzene
3,3,4,4-Tetrafluorodihydro-2,5-furandione
1,1,2,2-Tetrafluoro-1,2-dinitroethane
1,1,1,2-Tetrafluoroethane
1,1,2,2-Tetrafluoroethane
Tetrafluoroethene
1,2,2,2-Tetrafluoroethyl difluoromethyl ether
Tetrafluoromethane
2,2,3,3-Tetrafluoro-1-propanol
6,7,8,9-Tetrahydro-5H-benzocyclohepten-5-one
2,3,6,7-Tetrahydro-1H,5H-benzo[ij]quinolizine
1,2,3,6-Tetrahydro-2,3'-bipyridine, (S)
2,3,4,9-Tetrahydro-1H-carbazole
Tetrahydrocortisone
1,2,3,4-Tetrahydro-6,7-dimethoxy-1,2-dimethylisoquinoline, (±)
Tetrahydro-2,5-dimethoxyfuran
4,5,6,7-Tetrahydro-3,6-dimethylbenzofuran
1,2,3,4-Tetrahydro-1,5-dimethylnaphthalene
Tetrahydro-2,2-dimethyl-5-oxo-3-furanacetic acid
cis-Tetrahydro-2,5-dimethylthiophene
1,2,3,4-Tetrahydro-9H-fluoren-9-one
5,6,7,8-Tetrahydrofolic acid
Tetrahydrofuran
Tetrahydro-2-furanmethanamine
Tetrahydro-2-furanmethanol propanoate
Tetrahydro-3-furanol
Tetrahydrofurfuryl acetate
Tetrahydrofurfuryl acrylate
Tetrahydrofurfuryl alcohol
Tetrahydrofurfuryl methacrylate
Tetrahydroimidazo[4,5-d]imidazole-2,5(1H,3H)-dione
cis-3a,4,7,7a-Tetrahydro-1,3-isobenzofurandione
4,5,6,7-Tetrahydro-1,3-isobenzofurandione
1,2,3,4-Tetrahydroisoquinoline
3,4,5,6-Tetrahydro-7-methoxy-2H-azepine
1,2,3,4-Tetrahydro-6-methoxyquinoline
1,2,3,4-Tetrahydro-1-methylnaphthalene

No.	Name	Synonym	Mol. Form.	CAS RN	Mol. Wt.	Physical Form	mp/°C	bp/°C	den/ g cm^{-3}	n_D	Solubility
9869	1,2,3,4-Tetrahydro-5-methylnaphthalene		$C_{11}H_{14}$	2809-64-5	146.229	liq	-23	234	0.9720^{20}	1.5439^{20}	
9870	1,2,3,4-Tetrahydro-6-methylnaphthalene		$C_{11}H_{14}$	1680-51-9	146.229	liq	-40	229	0.9537^{20}	1.5357^{20}	
9871	1,2,3,6-Tetrahydro-1-methyl-4-phenylpyridine	MPTP	$C_{12}H_{15}N$	28289-54-5	173.254	cry	41	$87^{0.8}$			
9872	Tetrahydro-3-methyl-2*H*-thiopyran		$C_6H_{12}S$	5258-50-4	116.224	liq	-60	158	0.9473^{20}	1.4922^{20}	
9873	5,6,7,8-Tetrahydro-1-naphthalenamine		$C_{10}H_{13}N$	2217-41-6	147.217		38	279	1.0625^{16}	1.5900^{20}	sl H_2O; s EtOH, eth, acid
9874	1,2,3,4-Tetrahydronaphthalene	Tetralin	$C_{10}H_{12}$	119-64-2	132.202	liq	-35.7	207.6	0.9645^{25}	1.5413^{20}	i H_2O; vs EtOH, eth; s chl, $PhNH_2$
9875	1,2,3,4-Tetrahydro-1-naphthol	1,2,3,4-Tetrahydro-α-naphthol	$C_{10}H_{12}O$	529-33-9	148.201		34.5	255; 103^{2}	1.0996^{20}	1.5638^{20}	
9876	5,6,7,8-Tetrahydro-1-naphthol	5,6,7,8-Tetrahydro-α-naphthol	$C_{10}H_{12}O$	529-35-1	148.201		70	266; 143^{11}	1.0556^{75}		
9877	1,2,3,4-Tetrahydro-2-naphthol	Tetralol	$C_{10}H_{12}O$	530-91-6	148.201		15.5	140^{12}			
9878	5,6,7,8-Tetrahydro-2-naphthol	5,6,7,8-Tetrahydro-β-naphthol	$C_{10}H_{12}O$	1125-78-6	148.201		57	275.5	1.0552^{65}		
9879	Tetrahydro-6-pentyl-2*H*-pyran-2-one	5-Hydroxydecanoic acid lactone	$C_{10}H_{18}O_2$	705-86-2	170.249	liq	-27	121^{3}			
9880	1,2,3,4-Tetrahydrophenanthrene		$C_{14}H_{14}$	1013-08-7	182.261	lf (MeOH)	33.5	173^{11}	1.0601^{40}		i H_2O; s EtOH, eth, ace, bz, HOAc, chl, lig
9881	1,2,3,6-Tetrahydrophthalimide		$C_8H_9NO_2$	85-40-5	151.163	cry (EtOH)	137				
9882	Tetrahydro-6-propyl-2*H*-pyran-2-one	5-Hydroxyoctanoic acid lactone	$C_8H_{14}O_2$	698-76-0	142.196	liq	-13	126^{15}			
9883	2,3,4,5-Tetrahydro-6-propylpyridine	γ-Coniceine	$C_8H_{15}N$	1604-01-9	125.212			174	0.8753^{15}	1.4661^{16}	
9884	Tetrahydropyran	Oxane	$C_5H_{10}O$	142-68-7	86.132	liq	-49.1	88	0.8814^{20}	1.4200^{20}	s EtOH, eth, bz, ctc
9885	Tetrahydro-2*H*-pyran-2-methanol		$C_6H_{12}O_2$	100-72-1	116.158			185	1.027^{25}	1.458^{20}	
9886	Tetrahydro-2*H*-pyran-2-one		$C_5H_8O_2$	542-28-9	100.117	liq	-12.5	219	1.1082^{20}	1.4503^{20}	s H_2O; msc EtOH, eth; sl ctc
9887	Tetrahydro-4*H*-pyran-4-one		$C_5H_8O_2$	29943-42-8	100.117			166.5	1.084^{25}	1.4520^{20}	
9888	1,2,5,6-Tetrahydropyridine	Δ^3-Piperidine	C_5H_9N	694-05-3	83.132	liq	-48	108	0.911^{25}	1.4800^{20}	s chl
9889	1,2,5,6-Tetrahydro-3-pyridinecarboxylic acid	Guvacine	$C_6H_9NO_2$	498-96-4	127.141	pr (w), rods (+1w dil al)	295 dec				vs H_2O
9890	3,4,5,6-Tetrahydro-2(1*H*)-pyrimidinethione	Hexahydropyrimidine-2-thione	$C_4H_8N_2S$	2055-46-1	116.185		211		1.33^{20}		
9891	1,2,3,4-Tetrahydroquinoline		$C_9H_{11}N$	635-46-1	133.190	nd	20	251	1.0588^{20}	1.6062^{19}	s H_2O, chl; msc EtOH, eth
9892	5,6,7,8-Tetrahydroquinoline	2,3-Cyclohexenopyridine	$C_9H_{11}N$	10500-57-9	133.190			222	1.0304^{13}	1.5435^{20}	sl H_2O; s EtOH, eth, ace, bz
9893	1,2,3,4-Tetrahydroquinoxaline		$C_8H_{10}N_2$	3476-89-9	134.178	lf (w, eth, peth)	99	289			s H_2O, chl; vs EtOH, eth, bz; sl peth
9894	6,7,8,9-Tetrahydro-5*H*-tetrazolo[1,5-a]azepine	Pentylenetetrazole	$C_6H_{10}N_4$	54-95-5	138.170	cry (bz-lig)	59.5	194^{12}			vs H_2O, EtOH, ace; s eth, bz; sl chl
9895	Tetrahydrothiophene	Thiacyclopentane	C_4H_8S	110-01-0	88.172	liq	-96.2	121.1	0.9987^{20}	1.4871^{18}	i H_2O; msc EtOH, eth, ace, bz; s chl
9896	1,2,3,4-Tetrahydro-1,1,6-trimethylnaphthalene		$C_{13}H_{18}$	475-03-6	174.282			240; 90^{4}	0.9303^{20}	1.5257^{20}	s EtOH, eth, bz, chl
9897	1,2,5,8-Tetrahydroxy-9,10-anthracenedione	Quinalizarin	$C_{14}H_8O_6$	81-61-8	272.210	oran nd	>275				sl H_2O, ace, bz, EtOH, eth
9898	2,3,4,6-Tetrahydroxy-5*H*-benzocyclohepten-5-one	Purpurogallin	$C_{11}H_8O_5$	569-77-7	220.179	red nd (gl HOAc)	274 dec				
9899	2,2',4,4'-Tetrahydroxybenzophenone		$C_{13}H_{10}O_5$	131-55-5	246.215	ye nd (w+1)	197				vs H_2O, ace, eth, EtOH
9900	2,3,5,6-Tetrahydroxy-2,5-cyclohexadiene-1,4-dione	Tetroquinone	$C_6H_4O_6$	319-89-1	172.092	bl-blk cry					sl H_2O, eth, ctc; vs EtOH
9901	11,17,20,21-Tetrahydroxypregn-4-en-3-one, (11β,20*R*)	4-Pregnene-11β,17α,20β,21-tetrol-3-on	$C_{21}H_{32}O_5$	116-58-5	364.476	cry (aq ace)	125 dec				vs ace, EtOH
9902	*N,N,N',N'*-Tetra(2-hydroxypropyl) ethylenediamine	ENTPROL	$C_{14}H_{32}N_2O_4$	102-60-3	292.415				1.030^{25}	1.478^{25}	sl chl
9903	Tetraiodoethene	Tetraiodoethylene	C_2I_4	513-92-8	531.639	ye lf, pr (eth)	187	sub	2.983^{20}		vs bz, chl
9904	4,5,6,7-Tetraiodo-1,3-isobenzofurandione		$C_8I_4O_3$	632-80-4	651.702	ye pr, nd (HOAc) nd (sub)	327.5	sub			i H_2O, EtOH, bz; sl HOAc
9905	Tetraiodomethane	Carbon tetraiodide	CI_4	507-25-5	519.629	red lf (bz, chl)	171	$135^{1.5}$	4.23^{20}		vs py, chl
9906	2,3,4,5-Tetraiodo-1*H*-pyrrole	Iodopyrrole	C_4HI_4N	87-58-1	570.676	ye nd (al)	150 dec				vs ace, eth, chl
9907	Tetraisobutyl titanate	2-Methyl-1-propanol, titanium(4+) salt	$C_{16}H_{36}O_4Ti$	7425-80-1	340.322			256^{500}	0.960^{50}		dec H_2O

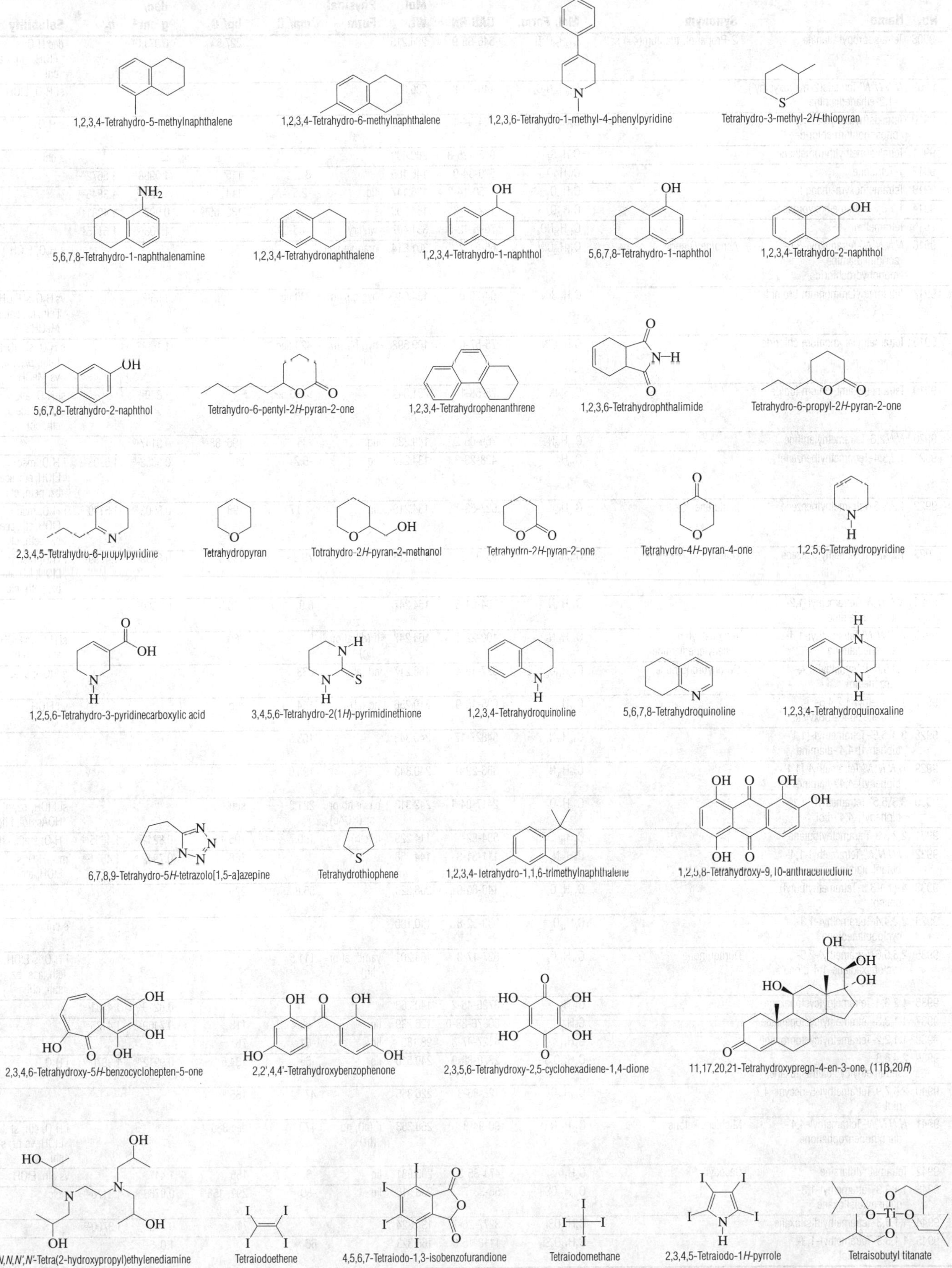
1,2,3,4-Tetrahydro-5-methylnaphthalene
1,2,3,4-Tetrahydro-6-methylnaphthalene
1,2,3,6-Tetrahydro-1-methyl-4-phenylpyridine
Tetrahydro-3-methyl-2H-thiopyran
5,6,7,8-Tetrahydro-1-naphthalenamine
1,2,3,4-Tetrahydronaphthalene
1,2,3,4-Tetrahydro-1-naphthol
5,6,7,8-Tetrahydro-1-naphthol
1,2,3,4-Tetrahydro-2-naphthol
5,6,7,8-Tetrahydro-2-naphthol
Tetrahydro-6-pentyl-2H-pyran-2-one
1,2,3,4-Tetrahydrophenanthrene
1,2,3,6-Tetrahydrophthalimide
Tetrahydro-6-propyl-2H-pyran-2-one
2,3,4,5-Tetrahydro-6-propylpyridine
Tetrahydropyran
Tetrahydro-2H-pyran-2-methanol
Tetrahydro-2H-pyran-2-one
Tetrahydro-4H-pyran-4-one
1,2,5,6-Tetrahydropyridine
1,2,5,6-Tetrahydro-3-pyridinecarboxylic acid
3,4,5,6-Tetrahydro-2(1H)-pyrimidinethione
1,2,3,4-Tetrahydroquinoline
5,6,7,8-Tetrahydroquinoline
1,2,3,4-Tetrahydroquinoxaline
6,7,8,9-Tetrahydro-5H-tetrazolo[1,5-a]azepine
Tetrahydrothiophene
1,2,3,4-Tetrahydro-1,1,6-trimethylnaphthalene
1,2,5,8-Tetrahydroxy-9,10-anthracenedione
2,3,4,6-Tetrahydroxy-5H-benzocyclohepten-5-one
2,2′,4,4′-Tetrahydroxybenzophenone
2,3,5,6-Tetrahydroxy-2,5-cyclohexadiene-1,4-dione
11,17,20,21-Tetrahydroxypregn-4-en-3-one, (11β,20R)
N,N,N′,N′-Tetra(2-hydroxypropyl)ethylenediamine
Tetraiodoethene
4,5,6,7-Tetraiodo-1,3-isobenzofurandione
Tetraiodomethane
2,3,4,5-Tetraiodo-1H-pyrrole
Tetraisobutyl titanate

No.	Name	Synonym	Mol. Form.	CAS RN	Mol. Wt.	Physical Form	mp/°C	bp/°C	den/ g cm^{-3}	n_D	Solubility
9908	Tetraisopropyl titanate	2-Propanol, titanium(4+) salt	$C_{12}H_{28}O_4Ti$	546-68-9	284.215			227.5	0.9711^{20}		dec H_2O; s EtOH, eth, bz, chl
9909	*N,N,N',N'*-Tetrakis(2-hydroxyethyl) -1,2-ethanediamine		$C_{10}H_{24}N_2O_4$	140-07-8	236.309						sl H_2O, EtOH
9910	Tetrakis(hydroxymethyl) phosphonium chloride		$C_4H_{12}ClO_4P$	124-64-1	190.562		152.5				s H_2O
9911	Tetrakis(methylthio)methane		$C_5H_{12}S_4$	6156-25-8	200.409						s chl
9912	1-Tetralone		$C_{10}H_{10}O$	529-34-0	146.185		8	115^{6}	1.0988^{16}	1.5672^{20}	
9913	Tetramethoxymethane		$C_5H_{12}O_4$	1850-14-2	136.147	liq	-2.5	114	1.023^{25}	1.3845^{20}	
9914	1,1,3,3-Tetramethoxypropane		$C_7H_{16}O_4$	102-52-3	164.200			183; 66^{12}	0.997^{25}	1.4081^{20}	
9915	Tetramethrin		$C_{19}H_{25}NO_4$	7696-12-0	331.407	wh cry	≈65-80		1.108^{20}	1.5175^{21}	
9916	*N,N,N',N'*-Tetramethyl-3,6-acridinediamine, monohydrochloride	Acridine Orange	$C_{17}H_{20}ClN_3$	65-61-2	301.814	oran-ye soln					s H_2O, EtOH
9917	Tetramethylammonium bromide		$C_4H_{12}BrN$	64-20-0	154.049	hyg bipym	230 dec		1.56^{25}		vs H_2O; sl EtOH; i eth, bz, chl; s MeOH
9918	Tetramethylammonium chloride		$C_4H_{12}ClN$	75-57-0	109.598	hyg bipym (dil al)	420 dec		1.169^{20}		s H_2O; sl EtOH; i eth, bz, chl; vs MeOH
9919	Tetramethylammonium iodide		$C_4H_{12}IN$	75-58-1	201.049		>230 dec		1.829^{25}		sl H_2O, alk, EtOH, ace; i eth, chl
9920	*N,N*,2,6-Tetramethylaniline		$C_{10}H_{15}N$	769-06-2	149.233	liq	-36	196; 88^{20}	0.9147^{20}		
9921	1,2,3,4-Tetramethylbenzene		$C_{10}H_{14}$	488-23-3	134.218	liq	-6.2	205	0.9052^{20}	1.5203^{20}	i H_2O; msc EtOH, eth, ace, bz, peth, ctc
9922	1,2,3,5-Tetramethylbenzene	Isodurene	$C_{10}H_{14}$	527-53-7	134.218	liq	-23.7	198	0.8903^{20}	1.5130^{20}	i H_2O; msc EtOH, eth, ace, bz, peth, ctc
9923	1,2,4,5-Tetramethylbenzene	Durene	$C_{10}H_{14}$	95-93-2	134.218		79.3	196.8	0.8380^{81}	1.4790^{81}	i H_2O; msc EtOH, eth, ace, bz, peth, ctc
9924	*N,N,N',N'*-Tetramethyl-1,2-benzenediamine		$C_{10}H_{16}N_2$	704-01-8	164.247		8.9	215.5	0.9560^{20}		
9925	*N,N,N',N'*-Tetramethyl-1,4-benzenediamine	Tetramethyl-*p*-phenylenediamine	$C_{10}H_{16}N_2$	100-22-1	164.247	lf (dil al or lig)	51	260			sl H_2O; vs EtOH, eth, bz, chl
9926	2,3,5,6-Tetramethyl-1,4-benzenediol	Durohydroquinone	$C_{10}H_{14}O_2$	527-18-4	166.217	nd (al)	233				s EtOH; sl eth
9927	Tetramethyl 1,2,4,5-benzenetetracarboxylate		$C_{14}H_{14}O_8$	635-10-9	310.256	nd (al)	144	sub			vs EtOH
9928	3,3',5,5'-Tetramethyl-[1,1'-biphenyl]-4,4'-diamine		$C_{16}H_{20}N_2$	54827-17-7	240.343		168.5				
9929	*N,N,N',N'*-Tetramethyl-[1,1'-biphenyl]-4,4'-diamine		$C_{16}H_{20}N_2$	366-29-0	240.343		196.0				
9930	3,3',5,5'-Tetramethyl-[1,1'-biphenyl]-4,4'-diol		$C_{16}H_{18}O_2$	2417-04-1	242.313	pa ye nd or pr (HOAc)	221.8	sub			sl EtOH, bz, gl HOAc, tol; i lig
9931	2,2,3,3-Tetramethylbutane		C_8H_{18}	594-82-1	114.229	lf (eth)	100.7	106.45	0.8242^{20}	1.4695^{20}	i H_2O; s eth, chl
9932	*N,N,N',N'*-Tetramethyl-1,4-butanediamine		$C_8H_{20}N_2$	111-51-3	144.258			168	0.7942^{15}	1.4621^{25}	msc H_2O; s EtOH, eth
9933	4-(1,1,3,3-Tetramethylbutyl) phenol		$C_{14}H_{22}O$	140-66-9	206.324		85.8	279			
9934	2,2,4,4-Tetramethyl-1,3-cyclobutanedione		$C_8H_{12}O_2$	933-52-8	140.180						s chl
9935	2,3,5,6-Tetramethyl-2,5-cyclohexadiene-1,4-dione	Duroquinone	$C_{10}H_{12}O_2$	527-17-3	164.201	ye nd (al or lig)	111.5				i H_2O; s EtOH, eth, ace, bz, sulf, chl
9936	1,2,3,4-Tetramethylcyclohexane		$C_{10}H_{20}$	3726-45-2	140.266				0.8219^{20}	1.4531^{20}	
9937	1,1,3,3-Tetramethylcyclopentane		C_9H_{18}	50876-33-0	126.239	liq	-88.4	118	0.7469^{25}	1.4125^{20}	
9938	1,1,2,2-Tetramethylcyclopropane		C_7H_{14}	4127-47-3	98.186	liq	-81	76			
9939	2,4,6,8-Tetramethylcyclotetrasiloxane		$C_4H_{16}O_4Si_4$	2370-88-9	240.510	liq	-65	134.5	0.9912^{20}	1.3870^{20}	i H_2O
9940	2,4,7,9-Tetramethyl-5-decyne-4,7-diol		$C_{14}H_{26}O_2$	126-86-3	226.355		47	165^{40}			
9941	*N,N,N',N'*-Tetramethyl-4,4'-diaminobenzophenone	Michler's ketone	$C_{17}H_{20}N_2O$	90-94-8	268.353	lf (al), nd (bz)	179	dec 360			i H_2O, eth; sl EtOH; vs bz; s chl
9942	Tetramethyldiarsine	Cacodyl	$C_4H_{12}As_2$	471-35-2	209.981	liq	-6	165	1.447^{15}		vs eth, EtOH
9943	1,1,3,3-Tetramethyl-1,3-diphenyldisiloxane		$C_{16}H_{22}OSi_2$	56-33-7	286.516	liq	-80	292; 156^{13}	0.9763^{20}	1.5176^{20}	s ctc
9944	1,1,3,3-Tetramethyldisiloxane		$C_4H_{14}OSi_2$	3277-26-7	134.324			71	0.756^{20}	1.3700^{20}	
9945	1,1,3,3-Tetramethyl-1,3-disiloxanediol		$C_4H_{14}O_3Si_2$	1118-15-6	166.323		66		1.095^{25}		

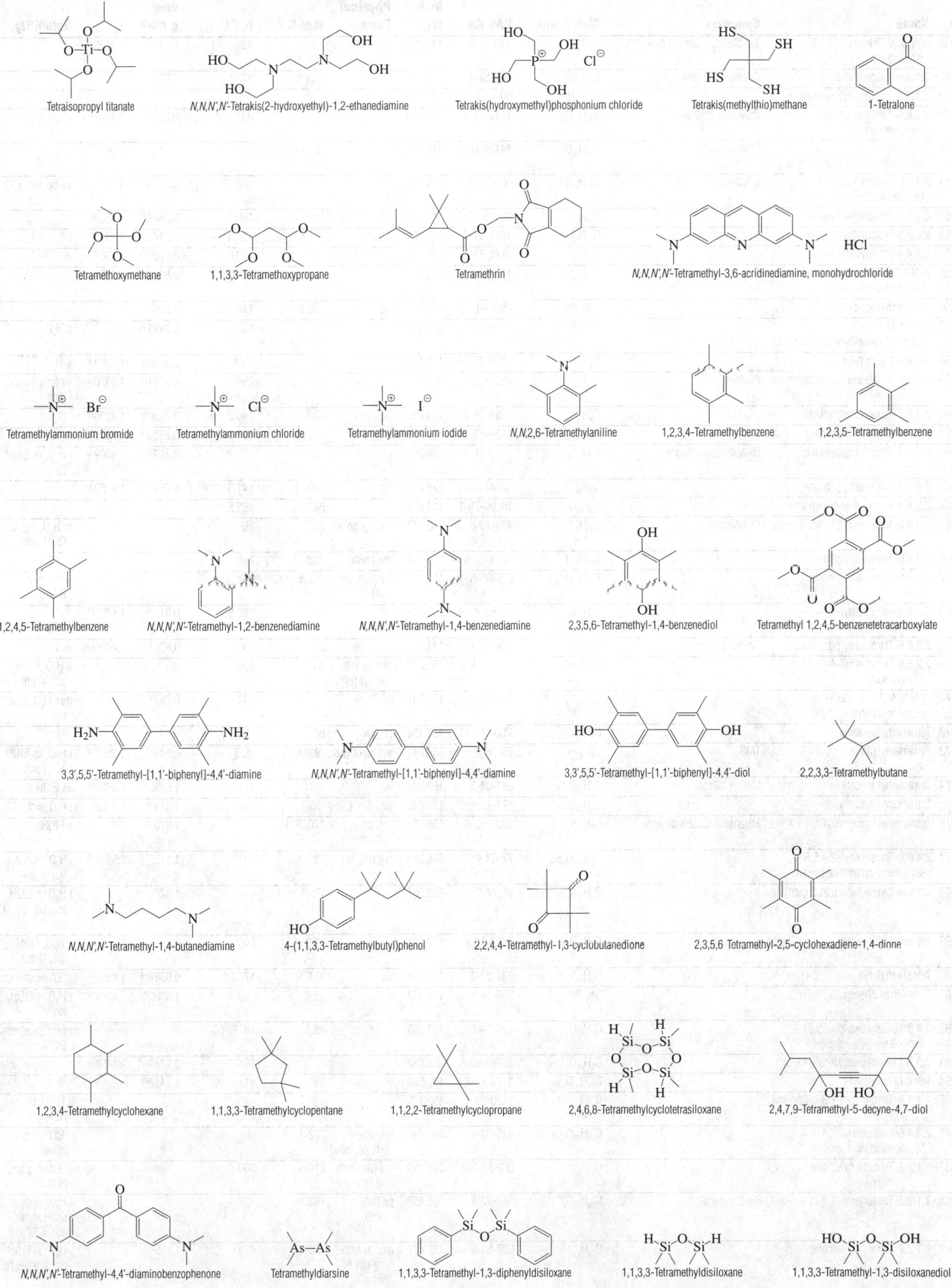

Tetraisopropyl titanate
N,N,N',N'-Tetrakis(2-hydroxyethyl)-1,2-ethanediamine
Tetrakis(hydroxymethyl)phosphonium chloride
Tetrakis(methylthio)methane
1-Tetralone
Tetramethoxymethane
1,1,3,3-Tetramethoxypropane
Tetramethrin
N,N,N',N'-Tetramethyl-3,6-acridinediamine, monohydrochloride
Tetramethylammonium bromide
Tetramethylammonium chloride
Tetramethylammonium iodide
N,N,2,6-Tetramethylaniline
1,2,3,4-Tetramethylbenzene
1,2,3,5-Tetramethylbenzene
1,2,4,5-Tetramethylbenzene
N,N,N',N'-Tetramethyl-1,2-benzenediamine
N,N,N',N'-Tetramethyl-1,4-benzenediamine
2,3,5,6-Tetramethyl-1,4-benzenediol
Tetramethyl 1,2,4,5-benzenetetracarboxylate
3,3',5,5'-Tetramethyl-[1,1'-biphenyl]-4,4'-diamine
N,N,N',N'-Tetramethyl-[1,1'-biphenyl]-4,4'-diamine
3,3',5,5'-Tetramethyl-[1,1'-biphenyl]-4,4'-diol
2,2,3,3-Tetramethylbutane
N,N,N',N'-Tetramethyl-1,4-butanediamine
4-(1,1,3,3-Tetramethylbutyl)phenol
2,2,4,4-Tetramethyl-1,3-cyclobutanedione
2,3,5,6 Tetramethyl-2,5-cyclohexadiene-1,4-dione
1,2,3,4-Tetramethylcyclohexane
1,1,3,3-Tetramethylcyclopentane
1,1,2,2-Tetramethylcyclopropane
2,4,6,8-Tetramethylcyclotetrasiloxane
2,4,7,9-Tetramethyl-5-decyne-4,7-diol
N,N,N',N'-Tetramethyl-4,4'-diaminobenzophenone
Tetramethyldiarsine
1,1,3,3-Tetramethyl-1,3-diphenyldisiloxane
1,1,3,3-Tetramethyldisiloxane
1,1,3,3-Tetramethyl-1,3-disiloxanediol

No.	Name	Synonym	Mol. Form.	CAS RN	Mol. Wt.	Physical Form	mp/°C	bp/°C	den/ g cm⁻³	n_D	Solubility
9946	*N,N,N',N'*-Tetramethyl-1,2-ethanediamine	1,2-Dimethylaminoethane	$C_6H_{16}N_2$	110-18-9	116.204	liq	-55	121	0.77^{25}	1.4179^{20}	
9947	Tetramethylgermane	Germanium tetramethyl	$C_4H_{12}Ge$	865-52-1	132.78			32^{500}	1.006		
9948	1,1,3,3-Tetramethylguanidine		$C_5H_{13}N_3$	80-70-6	115.177						s ctc
9949	2,2,6,6-Tetramethyl-3,5-heptanedione	Dipivaloylmethane	$C_{11}H_{20}O_2$	1118-71-4	184.276			93^{35}, 72^{6}	0.883^{25}	1.4589^{20}	sl ctc
9950	3,7,11,15-Tetramethylhexadecanoic acid	Phytanic acid	$C_{20}H_{40}O_2$	14721-66-5	312.531		-65				
9951	3,7,11,15-Tetramethyl-1-hexadecen-3-ol	Isophytol	$C_{20}H_{40}O$	505-32-8	296.531	oil		$108^{0.01}$	0.8519^{20}	1.4571^{20}	vs bz, eth, EtOH
9952	2,2,3,3-Tetramethylhexane		$C_{10}H_{22}$	13475-81-5	142.282	liq	-54	160.3	0.7609^{25}	1.4282^{20}	
9953	2,2,5,5-Tetramethylhexane		$C_{10}H_{22}$	1071-81-4	142.282	liq	-12.6	137.4	0.7148^{25}	1.4055^{20}	
9954	3,3,4,4-Tetramethylhexane		$C_{10}H_{22}$	5171-84-6	142.282			170.0	0.7789^{25}	1.4368^{20}	
9955	*N,N,N',N'*-Tetramethyl-1,6-hexanediamine		$C_{10}H_{24}N_2$	111-18-2	172.311			209.5	0.806^{25}	1.4359^{20}	
9956	Tetramethyl lead		$C_4H_{12}Pb$	75-74-1	267.3	liq	-30.2	110	1.995^{20}		
9957	*N,N,N',N'*-Tetramethylmethanediamine		$C_5H_{14}N_2$	51-80-9	102.178			83	0.7491^{18}		s H_2O
9958	Tetramethyloxirane		$C_6H_{12}O$	5076-20-0	100.158			90.4	0.8156^{16}	1.3984^{16}	s H_2O
9959	2,6,10,14-Tetramethylpentadecane	Pristane	$C_{19}H_{40}$	1921-70-6	268.521			296	0.7791^{25}	1.4370^{25}	vs bz, eth, chl, peth
9960	2,2,3,3-Tetramethylpentane		C_9H_{20}	7154-79-2	128.255	liq	-9.75	140.2	0.7530^{25}	1.4236^{20}	
9961	2,2,3,4-Tetramethylpentane		C_9H_{20}	1186-53-4	128.255	liq	-121.0	133.0	0.7389^{20}	1.4147^{20}	
9962	2,2,4,4-Tetramethylpentane	Di-*tert*-butylmethane	C_9H_{20}	1070-87-7	128.255	liq	-66.54	122.29	0.7195^{20}	1.4069^{20}	i H_2O; vs EtOH, bz
9963	2,3,3,4-Tetramethylpentane		C_9H_{20}	16747-38-9	128.255	liq	-102.1	141.5	0.7547^{20}	1.4222^{20}	
9964	2,2,4,4-Tetramethyl-3-pentanol		$C_9H_{20}O$	14609-79-1	144.254		52	165.5			
9965	2,3,4,5-Tetramethylphenol	Prehnitenol	$C_{10}H_{14}O$	488-70-0	150.217	nd (lig, aq al)	85.3	266			sl H_2O, lig; vs EtOH, eth
9966	2,3,4,6-Tetramethylphenol		$C_{10}H_{14}O$	3238-38-8	150.217	cry (peth)	80.5	240			s EtOH
9967	2,3,5,6-Tetramethylphenol		$C_{10}H_{14}O$	527-35-5	150.217	nd (lig), pr (al)	118.5	247			s chl, peth, HOAc
9968	2,2,6,6-Tetramethyl-4-piperidinamine		$C_9H_{20}N_2$	36768-62-4	156.268		17	188.5	0.912^{25}	1.4706^{20}	
9969	2,2,6,6-Tetramethylpiperidine	Norpempidine	$C_9H_{19}N$	768-66-1	141.254		28	156	0.8367^{16}	1.4455^{20}	vs eth
9970	2,2,6,6-Tetramethyl-4-piperidinone		$C_9H_{17}NO$	826-36-8	155.237	orth pl (eth-w) nd (eth)	36	205			s H_2O, EtOH, eth; sl chl
9971	*N,N,N',N'*-Tetramethyl-1,3-propanediamine		$C_7H_{18}N_2$	110-95-2	130.231			144	0.7837^{18}		msc H_2O, EtOH, eth
9972	Tetramethylpyrazine		$C_8H_{12}N_2$	1124-11-4	136.194	cry (w)	86	190			
9973	Tetramethylsilane	TMS	$C_4H_{12}Si$	75-76-3	88.224	vol liq or gas	-99.06	26.6	0.648^{19}	1.3587^{20}	i H_2O; vs EtOH, eth; i sulf
9974	Tetramethyl silicate	Methyl silicate	$C_4H_{12}O_4Si$	681-84-5	152.222	liq	-1.0	121	1.0232^{20}	1.3683^{20}	vs EtOH
9975	Tetramethylstannane		$C_4H_{12}Sn$	594-27-4	178.848	liq	-55.1	78	1.314^{25}	1.4386	i H_2O; s ctc, CS_2
9976	Tetramethylsuccinonitrile	Tetramethylbutanedinitrile	$C_8H_{12}N_2$	3333-52-6	136.194	mcl pl, lf, pr (dil al)	170.5		1.070^{25}		s EtOH
9977	2,4,6,8-Tetramethyl-2,4,6,8-tetraphenylcyclotetrasiloxane		$C_{28}H_{32}O_4Si_4$	77-63-4	544.894	cry (HOAc)	99	237^{115}	1.1183^{20}	1.5461^{20}	i H_2O; msc ace, hp
9978	Tetramethylthiodicarbonic diamide		$C_6H_{12}N_2S_3$	97-74-5	208.367		109.5		1.37^{25}		i H_2O; s EtOH, ace, bz, chl; sl eth
9979	Tetramethylthiourea		$C_5H_{12}N_2S$	2782-91-4	132.227		79.3	245			s H_2O, EtOH, chl; sl eth
9980	Tetramethylurea		$C_5H_{12}N_2O$	632-22-4	116.161	liq	-0.6	176.5	0.9687^{20}	1.4496^{23}	sl EtOH, eth, ctc
9981	Tetranitromethane		CN_4O_8	509-14-8	196.033		13.8	126.1	1.6380^{20}	1.4384^{20}	i H_2O; s EtOH, eth
9982	2,4,8,10-Tetraoxaspiro[5.5] undecane		$C_7H_{12}O_4$	126-54-5	160.168		48.3	147^{53}, 68^{1}			vs H_2O, ace, eth, EtOH
9983	2,5,8,11-Tetraoxatridecan-13-ol		$C_9H_{20}O_5$	23783-42-8	208.252			164^{11}	0.987^{25}	1.4453^{20}	
9984	Tetraphenoxysilane		$C_{24}H_{20}O_4Si$	1174-72-7	400.500		49	417; 236^{1}	1.1412^{60}		
9985	1,1,4,4-Tetraphenyl-1,3-butadiene		$C_{28}H_{22}$	1450-63-1	358.475		203.5				s EtOH, bz, chl, HOAc
9986	2,3,4,5-Tetraphenyl-2,4-cyclopentadien-1-one		$C_{29}H_{20}O$	479-33-4	384.468	blk-viol lf (HOAc, xyl)	222.3				s EtOH, bz, xyl, HOAc
9987	1,1,2,2-Tetraphenylethane		$C_{26}H_{22}$	632-50-8	334.453	cry (bz), orth nd (chl)	214.5	360			sl EtOH; s bz, HOAc
9988	1,1,2,2-Tetraphenyl-1,2-ethanediol	Benzopinacol	$C_{26}H_{22}O_2$	464-72-2	366.452	pr (bz), cry (ace)	182				i H_2O, peth; sl EtOH; s eth, ace, CS_2
9989	1,1,2,2-Tetraphenylethene		$C_{26}H_{20}$	632-51-9	332.437	mcl or orth (bz-eth or chl-al)	225	420	1.155^{0}		i H_2O; sl EtOH, chl, eth; vs bz
9990	Tetraphenylgermane	Germanium tetraphenyl	$C_{24}H_{20}Ge$	1048-05-1	381.06		229.0				

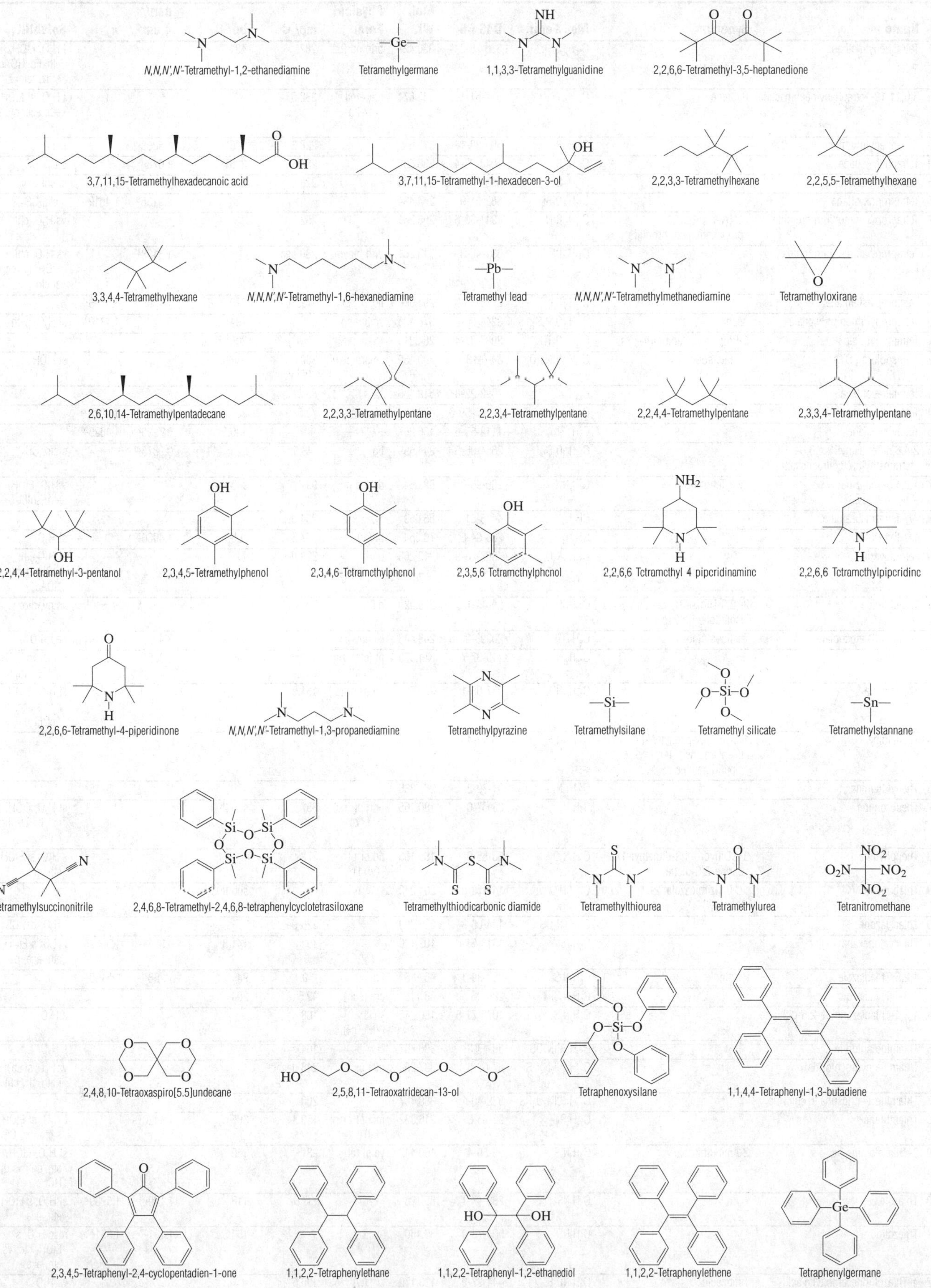

N,N,N',N'-Tetramethyl-1,2-ethanediamine
Tetramethylgermane
1,1,3,3-Tetramethylguanidine
2,2,6,6-Tetramethyl-3,5-heptanedione
3,7,11,15-Tetramethylhexadecanoic acid
3,7,11,15-Tetramethyl-1-hexadecen-3-ol
2,2,3,3-Tetramethylhexane
2,2,5,5-Tetramethylhexane
3,3,4,4-Tetramethylhexane
N,N,N',N'-Tetramethyl-1,6-hexanediamine
Tetramethyl lead
N,N,N',N'-Tetramethylmethanediamine
Tetramethyloxirane
2,6,10,14-Tetramethylpentadecane
2,2,3,3-Tetramethylpentane
2,2,3,4-Tetramethylpentane
2,2,4,4-Tetramethylpentane
2,3,3,4-Tetramethylpentane
2,2,4,4-Tetramethyl-3-pentanol
2,3,4,5-Tetramethylphenol
2,3,4,6-Tetramethylphenol
2,3,5,6-Tetramethylphenol
2,2,6,6-Tetramethyl-4-piperidinamine
2,2,6,6-Tetramethylpiperidine
2,2,6,6-Tetramethyl-4-piperidinone
N,N,N',N'-Tetramethyl-1,3-propanediamine
Tetramethylpyrazine
Tetramethylsilane
Tetramethyl silicate
Tetramethylstannane
Tetramethylsuccinonitrile
2,4,6,8-Tetramethyl-2,4,6,8-tetraphenylcyclotetrasiloxane
Tetramethylthiodicarbonic diamide
Tetramethylthiourea
Tetramethylurea
Tetranitromethane
2,4,8,10-Tetraoxaspiro[5.5]undecane
2,5,8,11-Tetraoxatridecan-13-ol
Tetraphenoxysilane
1,1,4,4-Tetraphenyl-1,3-butadiene
2,3,4,5-Tetraphenyl-2,4-cyclopentadien-1-one
1,1,2,2-Tetraphenylethane
1,1,2,2-Tetraphenyl-1,2-ethanediol
1,1,2,2-Tetraphenylethene
Tetraphenylgermane

No.	Name	Synonym	Mol. Form.	CAS RN	Mol. Wt.	Physical Form	mp/°C	bp/°C	den/ g cm^{-3}	n_D	Solubility
9991	Tetraphenylmethane		$C_{25}H_{20}$	630-76-2	320.427	orth nd (bz, sub)	282	431			i H_2O, EtOH, eth, lig, HOAc; s bz, tol
9992	5,6,11,12-Tetraphenylnaphthacene	Rubrene	$C_{42}H_{28}$	517-51-1	532.671	oran-red (bz-lig)	332.5				i H_2O; sl EtOH, eth, ace, py; s bz
9993	Tetraphenylplumbane		$C_{24}H_{20}Pb$	595-89-1	515.6		228.3	126^{13}	1.5298^{20}		s chl
9994	Tetraphenylsilane		$C_{24}H_{20}Si$	1048-08-4	336.502		236.5	228^{3}	1.078^{20}		s ctc, CS_2
9995	Tetraphenylstannane		$C_{24}H_{20}Sn$	595-90-4	427.126		228	420			sl chl
9996	Tetrapropoxysilane		$C_{12}H_{28}O_4Si$	682-01-9	264.434			226	0.9158^{20}	1.4012^{20}	s ctc, CS_2
9997	Tetrapropylammonium bromide	*N,N,N*-Tripropyl-1-propanaminium bromide	$C_{12}H_{28}BrN$	1941-30-6	266.261		252				vs H_2O, chl
9998	Tetrapropylammonium iodide		$C_{12}H_{28}IN$	631-40-3	313.261	orth bipym	280 dec		1.3138^{25}		vs H_2O, chl; s EtOH, HOAc; sl eth
9999	Tetrapropylstannane		$C_{12}H_{28}Sn$	2176-98-9	291.060	liq	-109.1	228	1.1065^{20}	1.4745^{20}	
10000	Tetrapropyl thiodiphosphate	Aspon	$C_{12}H_{28}O_5P_2S_2$	3244-90-4	378.425	amber liq		$104^{0.1}$	1.12^{25}	1.4710^{21}	sl H_2O, peth
10001	Tetrapropyl titanate	1-Propanol, titanium(4+) salt	$C_{12}H_{28}O_4Ti$	3087-37-4	284.215			206^{100}			
10002	Tetrasodium EDTA	Edetate sodium	$C_{10}H_{12}N_2Na_4O_8$	64-02-8	380.169	amorp pow	300 (dihydrate)				sl EtOH
10003	Tetratetracontane		$C_{44}H_{90}$	7098-22-8	619.186		85.6				
10004	Tetratriacontane		$C_{34}H_{70}$	14167-59-0	478.920	pl (eth)	72.5	285.4^{3}	0.7728^{90}	1.4296^{90}	
10005	Tetravinylsilane		$C_8H_{12}Si$	1112-55-6	136.267			130.2	0.7999^{20}	1.4625^{20}	
10006	2,4,6,8-Tetravinyl-2,4,6,8-tetramethylcyclotetrasiloxane		$C_{12}H_{24}O_4Si_4$	2554-06-5	344.659	liq	-43.5	224; 111^{12}	0.9875^{20}		s ctc, CS_2
10007	1,2,4,5-Tetrazine	*sym*-Tetrazine	$C_2H_2N_4$	290-96-0	82.064	dk red pr	99	sub			s H_2O, EtOH, eth, sulf
10008	1*H*-Tetrazol-5-amine		CH_3N_5	4418-61-5	85.069		204 dec				
10009	1*H*-Tetrazole		CH_2N_4	288-94-8	70.054	pl (al)	157.3	sub	1.4060^{20}		sl H_2O
10010	Tetrodotoxin		$C_{11}H_{17}N_3O_8$	4368-28-9	319.268	cry	225 dec				sl H_2O,eth, EtOH; s dil HOAc
10011	Thalidomide	2-(2,6-Dioxo-3-piperidinyl)-1*H*-isoindole-1,3(2*H*)-dione	$C_{13}H_{10}N_2O_4$	50-35-1	258.229	nd	270				vs py, diox
10012	Thallium(I) ethanolate	Thallous ethoxide	C_2H_5OTl	20398-06-5	249.443	cloudy liq	-3	dec 130	3.49		dec H_2O
10013	Thebaine		$C_{19}H_{21}NO_3$	115-37-7	311.375	pl (eth), pr (dil al)	193	sub 91	1.305^{20}		i H_2O; vs EtOH, chl; sl eth; s bz
10014	Thebainone		$C_{18}H_{21}NO_3$	467-98-1	299.365	nd or pr (al)	151.5				sl H_2O, EtOH, eth; s ace, bz, AcOEt
10015	Thenaldine	1-Methyl-*N*-phenyl-*N*-(2-thienylmethyl)-4-piperidinamine	$C_{17}H_{22}N_2S$	86-12-4	286.435		96	$159^{0.02}$			
10016	Thenyldiamine		$C_{14}H_{19}N_3S$	91-79-2	261.386			170^{1}		1.5915^{20}	
10017	Theobromine		$C_7H_8N_4O_2$	83-67-0	180.165	orth or mcl nd (w)	357	sub 290			sl H_2O, EtOH; i eth, bz, ctc, lig, chl
10018	Theophylline	3,7-Dihydro-1,3-dimethyl-1*H*-purine-2,6-dione	$C_7H_8N_4O_2$	58-55-9	180.165	nd or pl (w+1)	273				s H_2O; sl EtOH, eth, chl
10019	Thiabendazole	1*H*-Benzimidazole, 2-(4-thiazolyl)-	$C_{10}H_7N_3S$	148-79-8	201.248			sub 305			
10020	Thiacetazone		$C_{10}H_{12}N_4OS$	104-06-3	236.293		225 dec				i H_2O, os, CS_2
10021	Thiacyclohexane		$C_5H_{10}S$	1613-51-0	102.198		19	141.8	0.9861^{20}	1.5067^{20}	i H_2O; s EtOH, eth, ace, bz
10022	1,2,5-Thiadiazole	Piazthiole	$C_2H_2N_2S$	288-39-1	86.115	liq	-50.1	94	1.268^{25}	1.5150^{25}	
10023	1,3,4-Thiadiazole		$C_2H_2N_2S$	289-06-5	86.115	cry (sub)	42.5	204			
10024	1,3,4-Thiadiazolidine-2,5-dithione		$C_2H_2N_2S_3$	1072-71-5	150.245	ye cry (MeOH)	168				s H_2O
10025	Thiamine chloride		$C_{12}H_{17}ClN_4OS$	59-43-8	300.807	cry	164				s H_2O
10026	Thiamine hydrochloride		$C_{12}H_{18}Cl_2N_4OS$	67-03-8	337.268	mcl pl	248 dec				vs H_2O; sl EtOH; i eth, bz, chl
10027	Thiamine *O*-phosphate, chloride		$C_{12}H_{18}ClN_4O_4PS$	532-40-1	380.787		200				
10028	Thianthrene		$C_{12}H_8S_2$	92-85-3	216.322	mcl pr or pl (al)	159.3	365	1.4420^{20}		i H_2O; sl EtOH; s eth, bz, CS_2
10029	2-Thiazolamine	2-Aminothiazole	$C_3H_4N_2S$	96-50-4	100.142	ye pl (al)	93	140^{11}			sl H_2O, EtOH, eth, chl; vs dil HCl
10030	Thiazole		C_3H_3NS	288-47-1	85.128		-33.62	118	1.1998^{17}	1.5969^{20}	sl H_2O; s EtOH, eth, ace
10031	Thiazolidine		C_3H_7NS	504-78-9	89.160			164.5	1.131^{25}	1.551^{20}	msc H_2O; s EtOH, ctc; vs eth, ace
10032	4-Thiazolidinecarboxylic acid	Timonacic	$C_4H_7NO_2S$	444-27-9	133.170	cry (w)	196.5				vs H_2O
10033	2,4-Thiazolidinedione		$C_3H_3NO_2S$	2295-31-0	117.127	pl (w), pr (al)	128	179^{19}			vs eth

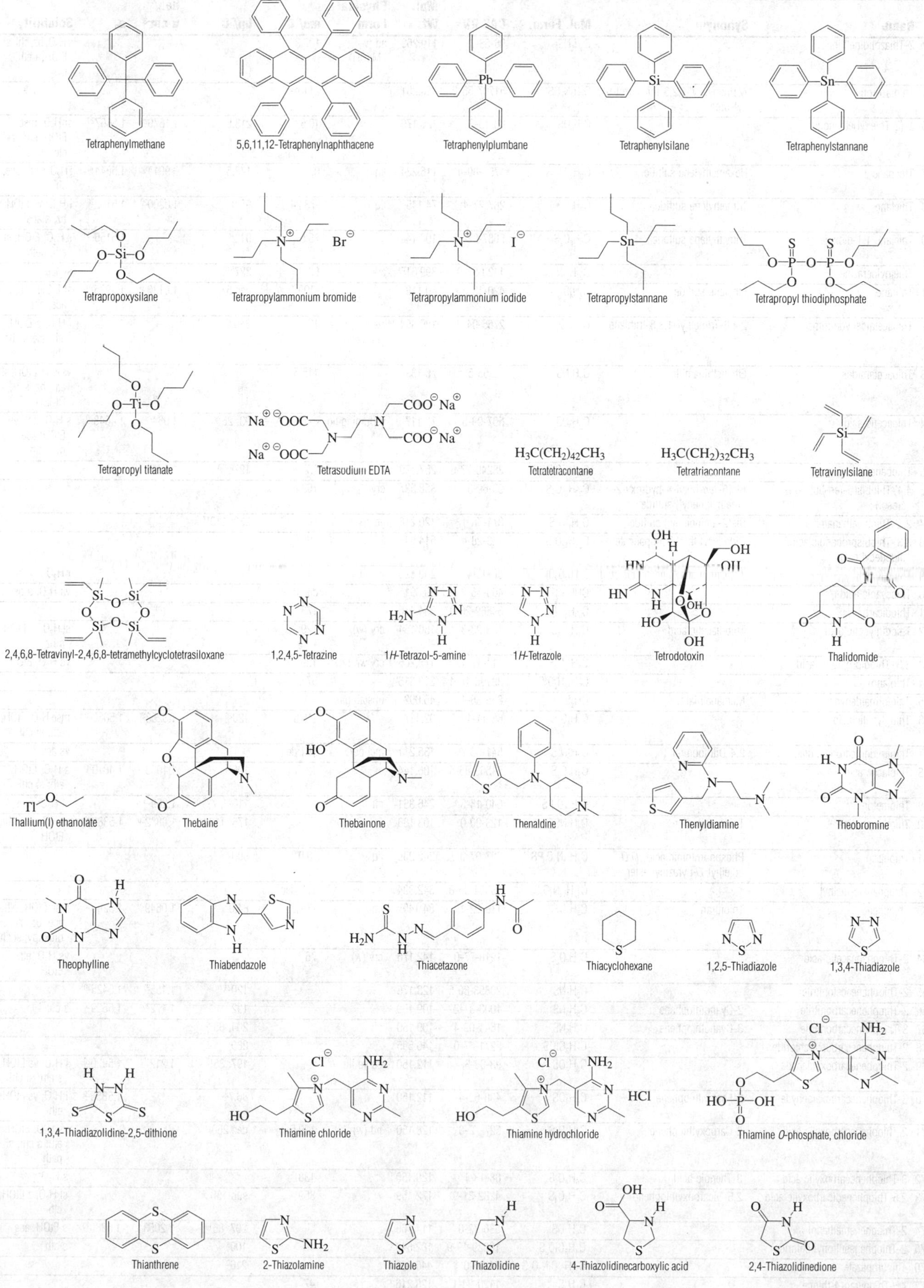

Tetraphenylmethane
5,6,11,12-Tetraphenylnaphthacene
Tetraphenylplumbane
Tetraphenylsilane
Tetraphenylstannane
Tetrapropoxysilane
Tetrapropylammonium bromide
Tetrapropylammonium iodide
Tetrapropylstannane
Tetrapropyl thiodiphosphate
Tetrapropyl titanate
Tetrasodium EDTA
Tetratetracontane
Tetratriacontane
Tetravinylsilane
2,4,6,8-Tetravinyl-2,4,6,8-tetramethylcyclotetrasiloxane
1,2,4,5-Tetrazine
1*H*-Tetrazol-5-amine
1*H*-Tetrazole
Tetrodotoxin
Thalidomide
Thallium(I) ethanolate
Thebaine
Thebainone
Thenaldine
Thenyldiamine
Theobromine
Theophylline
Thiabendazole
Thiacetazone
Thiacyclohexane
1,2,5-Thiadiazole
1,3,4-Thiadiazole
1,3,4-Thiadiazolidine-2,5-dithione
Thiamine chloride
Thiamine hydrochloride
Thiamine *O*-phosphate, chloride
Thianthrene
2-Thiazolamine
Thiazole
Thiazolidine
4-Thiazolidinecarboxylic acid
2,4-Thiazolidinedione

No.	Name	Synonym	Mol. Form.	CAS RN	Mol. Wt.	Physical Form	mp/°C	bp/°C	den/ g cm^{-3}	n_D	Solubility
10034	2-Thiazolidinethione		$C_3H_5NS_2$	96-53-7	119.209	nd (w, MeOH)	107.3				s H_2O, bz, chl; sl EtOH; i eth, CS_2
10035	Thidiazuron	*N*-Phenyl-*N'*-1,2,3-thiadiazol-5-yl-urea	$C_9H_8N_4OS$	51707-55-2	220.251		211 dec				
10036	1-(2-Thienyl)ethanone		C_6H_6OS	88-15-3	126.176		10.5	213.5	1.1679^{20}	1.5667^{20}	sl H_2O; msc EtOH, eth; s ctc
10037	Thiepane	Hexamethylene sulfide	$C_6H_{12}S$	4753-80-4	116.224	liq	0.5	173.5	0.991^{20}	1.5044^{18}	i H_2O; s eth, ace, chl
10038	Thietane	Trimethylene sulfide	C_3H_6S	287-27-4	74.145	liq	-73.24	95.0	1.0200^{20}	1.5102^{20}	i H_2O; vs EtOH, bz; s ace
10039	Thietane 1,1-dioxide	Trimethylene sulfone	$C_3H_6O_2S$	5687-92-3	106.144		75.5	91.2^{14}		1.5156^{20}	s H_2O, EtOH; sl eth, peth
10040	Thiethylperazine		$C_{22}H_{29}N_3S_2$	1420-55-9	399.615	cry	63	$227^{0.01}$			sl ace
10041	Thiirane	Ethylene sulfide	C_2H_4S	420-12-2	60.118		-109	dec 57	1.0130^{20}	1.4935^{20}	sl EtOH, eth; s ace, chl
10042	Thioacetaldehyde trimer	2,4,6-Trimethyl-1,3,5-trithiane	$C_6H_{12}S_3$	2765-04-0	180.354	α-mcl pl; β-nd (ace)	101	246.5			i H_2O; s EtOH, eth, ace; vs bz, chl
10043	Thioacetamide	Ethanethioamide	C_2H_5NS	62-55-5	75.133		115.5				vs H_2O, EtOH; sl eth, bz; s DMSO
10044	Thioacetic acid		C_2H_4OS	507-09-5	76.117	ye fuming liq	<-17	93; 26^{35}	1.064^{20}	1.4648^{20}	s H_2O, chl; vs EtOH, ace; msc eth
10045	Thiobencarb		$C_{12}H_{16}ClNOS$	28249-77-6	257.779		1.7	$127^{0.008}$	1.16^{20}		
10046	4,4'-Thiobis(6-*tert*-butyl-*m*-cresol)	Bis(5-*tert*-butyl-4-hydroxy-2-methylphenyl) sulfide	$C_{22}H_{30}O_2S$	96-69-5	358.537	cry	163				
10047	2,2'-Thiobisethanamine	Bis(2-aminoethyl) sulfide	$C_4H_{12}N_2S$	871-76-1	120.216	ye cry		232; 119^{17}			
10048	3,3'-Thiobispropanoic acid, didodecyl ester	Didodecyl thiobispropanoate	$C_{30}H_{58}O_4S$	123-28-4	514.845		39				
10049	Thioctic acid	1,2-Dithiolane-3-pentanoic acid	$C_8H_{14}O_2S_2$	62-46-4	206.326	ye nd	61	162			i H_2O
10050	Thiocyanic acid		CHNS	463-56-9	59.091		dec 0				vs H_2O; s os
10051	Thiodicarb		$C_{10}H_{18}N_4O_4S_3$	59669-26-0	354.470		173		1.4^{20}		
10052	Thiodiglycolic acid	Thiodiacetic acid	$C_4H_6O_4S$	123-93-3	150.154	cry (w)	129				sl H_2O; vs EtOH; s bz
10053	3,3'-Thiodipropionic acid		$C_6H_{10}O_4S$	111-17-1	178.206	cry wh pow	129				vs H_2O, EtOH
10054	Thiofanox		$C_9H_{18}N_2O_2S$	39196-18-4	218.316		57				
10055	Thioformaldehyde	Methanethial	CH_2S	865-36-1	46.092	unstab gas					
10056	Thioglycolic acid		$C_2H_4O_2S$	68-11-1	92.117		-16.5	120^{20}	1.3253^{20}	1.5080^{20}	msc H_2O, EtOH, eth; sl chl
10057	Thioimidodicarbonic diamide	2,4-Dithiobiuret	$C_2H_5N_3S_2$	541-53-7	135.211	mcl cry	181 dec				vs ace
10058	Thiolactic acid		$C_3H_6O_2S$	71563-86-5	106.144		12	106^{15}	1.1938^{20}	1.4810^{20}	s H_2O, EtOH, eth; sl chl
10059	Thiometon		$C_6H_{15}O_2PS_3$	640-15-3	246.351	oil		$110^{0.1}$, $77^{0.01}$	1.209^{20}		sl H_2O; s os
10060	Thiomorpholine	Thiamorpholine	C_4H_9NS	123-90-0	103.186			175; 110^{100}	1.0882^{20}	1.5386^{20}	vs H_2O, ace, eth, EtOH
10061	Thionazin	Phosphorothioic acid, *O,O*-diethyl *O*-pyrazinyl ester	$C_8H_{13}N_2O_3PS$	297-97-2	248.239	liq	-0.9	80			
10062	Thiophanate-methyl		$C_{12}H_{14}N_4O_4S_2$	23564-05-8	342.394		172 dec				
10063	Thiophene	Thiofuran	C_4H_4S	110-02-1	84.140	liq	-38.21	84.0	1.0649^{20}	1.5289^{20}	msc EtOH, eth, ace, bz, ctc, diox, py; sl chl
10064	2-Thiopheneacetic acid		$C_6H_6O_2S$	1918-77-0	142.176	cry (w)	76				vs H_2O, eth, EtOH
10065	2-Thiopheneacetonitrile		C_6H_5NS	20893-30-5	123.176			120^{23}	1.155^{25}	1.5425^{20}	
10066	2-Thiophenecarbonitrile	2-Cyanothiophene	C_5H_3NS	1003-31-2	109.150			192	1.172^{25}	1.5629^{20}	s chl
10067	3-Thiophenecarbonitrile	3-Cyanothiophene	C_5H_3NS	1641-09-4	109.150	oil		204; 85^{15}			
10068	2-Thiophenecarbonyl chloride		C_5H_3ClOS	5271-67-0	146.595			280			
10069	2-Thiophenecarboxaldehyde		C_5H_4OS	98-03-3	112.150	pa ye liq		197; 85^{16}	1.2127^{21}	1.5920^{20}	i H_2O; vs EtOH; s eth; sl chl
10070	3-Thiophenecarboxaldehyde	3-Formylthiophene	C_5H_4OS	498-62-4	112.150			86.7^{20}		1.5855^{20}	i H_2O; vs EtOH, eth
10071	2-Thiophenecarboxylic acid	2-Carboxythiophene	$C_5H_4O_2S$	527-72-0	128.150	nd (w)	129.5	dec 260			vs H_2O, EtOH, eth; s chl; sl peth
10072	3-Thiophenecarboxylic acid	3-Thenoic acid	$C_5H_4O_2S$	88-13-1	128.150		138				s H_2O
10073	2,5-Thiophenedicarboxylic acid	2,5-Dicarboxythiophene	$C_6H_4O_4S$	4282-31-9	172.159		359	sub 150			sl H_2O; s EtOH, eth
10074	2-Thiophenemethanol		C_5H_6OS	636-72-6	114.166			207; 86^{10}	1.2053^{16}	1.5280^{20}	s EtOH, ace
10075	2-Thiophenesulfonyl chloride		$C_4H_3ClO_2S_2$	16629-19-9	182.649		28	100^{6}			s eth
10076	Thiopropazate		$C_{23}H_{28}ClN_3O_2S$	84-06-0	446.005			$216^{0.1}$			
10077	4*H*-Thiopyran-4-thione		$C_5H_4S_2$	1120-94-1	128.216		47				

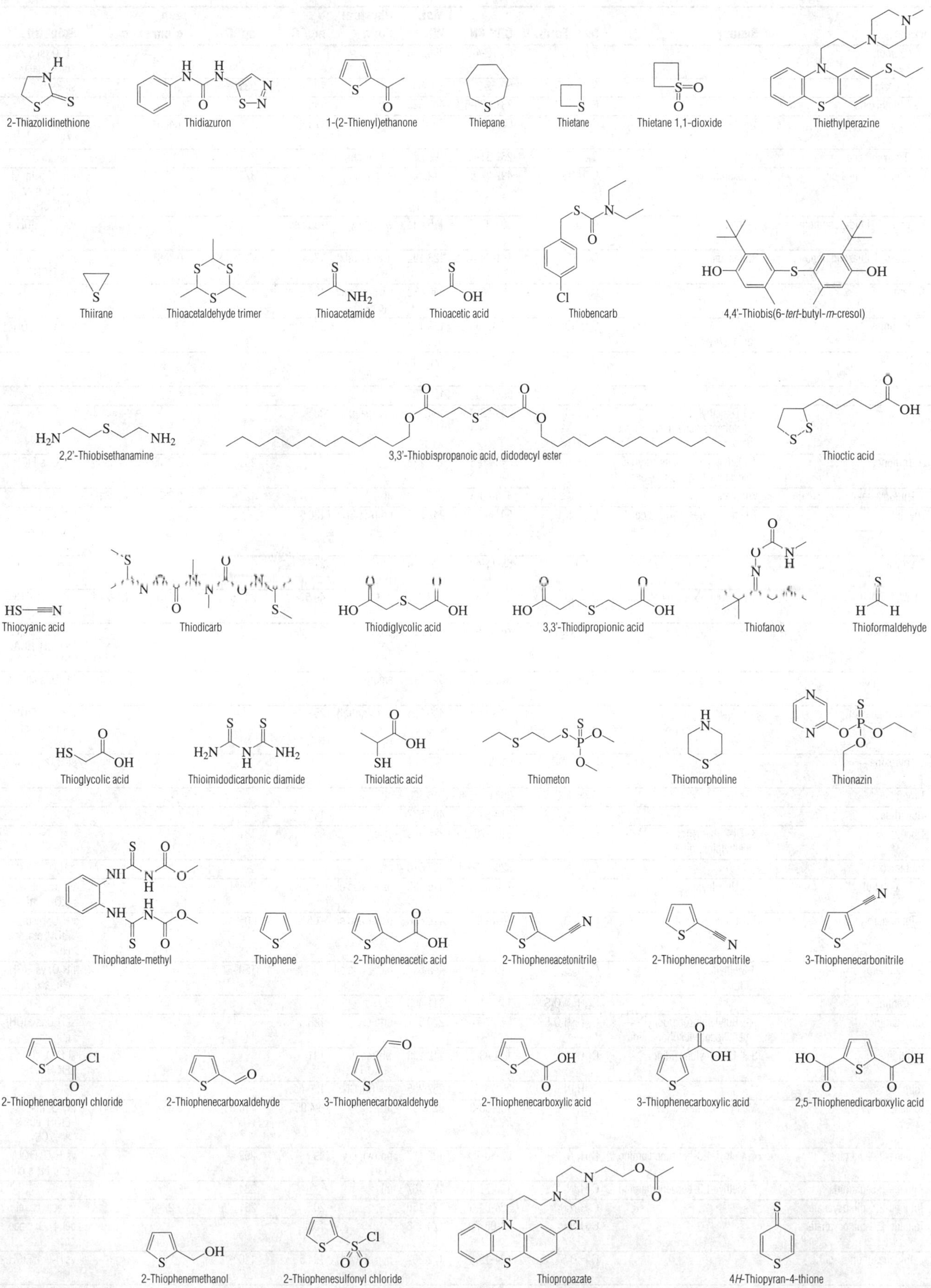
2-Thiazolidinethione
Thidiazuron
1-(2-Thienyl)ethanone
Thiepane
Thietane
Thietane 1,1-dioxide
Thiethylperazine
Thiirane
Thioacetaldehyde trimer
Thioacetamide
Thioacetic acid
Thiobencarb
4,4′-Thiobis(6-*tert*-butyl-*m*-cresol)
2,2′-Thiobisethanamine
3,3′-Thiobispropanoic acid, didodecyl ester
Thioctic acid
Thiocyanic acid
Thiodicarb
Thiodiglycolic acid
3,3′-Thiodipropionic acid
Thiofanox
Thioformaldehyde
Thioglycolic acid
Thioimidodicarbonic diamide
Thiolactic acid
Thiometon
Thiomorpholine
Thionazin
Thiophanate-methyl
Thiophene
2-Thiopheneacetic acid
2-Thiopheneacetonitrile
2-Thiophenecarbonitrile
3-Thiophenecarbonitrile
2-Thiophenecarbonyl chloride
2-Thiophenecarboxaldehyde
3-Thiophenecarboxaldehyde
2-Thiophenecarboxylic acid
3-Thiophenecarboxylic acid
2,5-Thiophenedicarboxylic acid
2-Thiophenemethanol
2-Thiophenesulfonyl chloride
Thiopropazate
4*H*-Thiopyran-4-thione

No.	Name	Synonym	Mol. Form.	CAS RN	Mol. Wt.	Physical Form	mp/°C	bp/°C	den/ g cm-3	n_D	Solubility
10078	Thioquinox		$C_9H_4N_2S_3$	93-75-4	236.336	br-ye pow	180				i H_2O; sl ace, EtOH, peth
10079	Thioridazine		$C_{21}H_{26}N_2S_2$	50-52-2	370.58	cry	73	$230^{0.02}$			sl ace
10080	*cis*-Thiothixene		$C_{23}H_{29}N_3O_2S_2$	3313-26-6	443.625	cry	148				
10081	Thiourea	Thiocarbamide	CH_4N_2S	62-56-6	76.121	orth (al)	178		1.405^{25}		s H_2O, EtOH; i eth
10082	9*H*-Thioxanthene	Dibenzothiapyran	$C_{13}H_{10}S$	261-31-4	198.283	nd (al-chl)	128.5	341			s chl
10083	9*H*-Thioxanthen-9-one	Thioxanthone	$C_{13}H_8OS$	492-22-8	212.267	ye nd (chl)	209	373			i H_2O, peth; sl EtOH; s bz, chl, CS_2
10084	2-Thioxo-4-imidazolidinone	2-Thiohydantoin	$C_3H_4N_2OS$	503-87-7	116.141	wh nd (w)	230 dec				vs H_2O, EtOH; s eth, alk
10085	2-Thioxo-4-thiazolidinone	Rhodanine	$C_3H_3NOS_2$	141-84-4	133.192	lt ye pr (al, w)	170		0.868^{25}		sl H_2O, DMSO; vs EtOH, eth
10086	Thiram		$C_6H_{12}N_2S_4$	137-26-8	240.432	wh or ye mcl (chl-al)	155.6	129^{20}			vs chl
10087	*L*-Threonine	2-Amino-3-hydroxybutanoic acid, [*R*-(*R**,*S**)]	$C_4H_9NO_3$	72-19-5	119.119		256 dec				s H_2O; i EtOH, eth, chl
10088	*D*-Threose		$C_4H_8O_4$	95-43-2	120.105	hyg-syr or nd (w)	129				
10089	*L*-Threose		$C_4H_8O_4$	95-44-3	120.105						vs H_2O
10090	Thujic acid	5,5-Dimethyl-1,3,6-cycloheptatriene-1-carboxylic acid	$C_{10}H_{12}O_2$	499-89-8	164.201	cry (peth)	88.5				
10091	α-Thujone	4-Methyl-1-(1-methylethyl)-bicyclo[3.1.0]hexan-3-one, (*l*)	$C_{10}H_{16}O$	546-80-5	152.233			201.2	0.9109^{25}	1.4490^{15}	i H_2O; s EtOH
10092	3-Thujopsene	Widdrene	$C_{15}H_{24}$	470-40-6	204.352	liq		122^{12}	0.932^{24}	1.5031^{25}	
10093	Thymidine	Thymine 2-desoxyriboside	$C_{10}H_{14}N_2O_5$	50-89-5	242.228	nd (AcOEt)	186.5				s H_2O, EtOH, ace, py, HOAc; sl chl
10094	Thymine		$C_5H_6N_2O_2$	65-71-4	126.114		316				sl H_2O, EtOH, eth, DMSO
10095	Thymol	2-Isopropyl-5-methylphenol	$C_{10}H_{14}O$	89-83-8	150.217		49.5	232.5	0.970^{25}	1.5227^{20}	i H_2O; vs EtOH, eth, chl, AcOEt
10096	Thymol Blue		$C_{27}H_{30}O_5S$	76-61-9	466.589	grn-red (al, eth)	222 dec				sl H_2O, ace, bz; s EtOH, HOAc, $PhNH_2$
10097	Thymol iodide		$C_{20}H_{24}I_2O_2$	552-22-7	550.213	amorp					i H_2O; s eth; vs EtOH
10098	Thymolphthalein		$C_{28}H_{30}O_4$	125-20-2	430.536	pr or nd (al)	253				i H_2O; s EtOH, eth, ace; sl DMSO
10099	*L*-Thyroxine		$C_{15}H_{11}I_4NO_4$	51-48-9	776.871	nd	235				sl H_2O; i EtOH, bz
10100	Timolol		$C_{13}H_{24}N_4O_3S$	26839-75-8	316.420	oil					
10101	Tiocarlide		$C_{23}H_{32}N_2O_2S$	910-86-1	400.577		146				
10102	Tipepidine	3-(Di-2-thienylmethylene)-1-methylpiperidine	$C_{15}H_{17}NS_2$	5169-78-8	275.433	ye cry	65	$181^{4.5}$			
10103	Tobramycin		$C_{18}H_{37}N_5O_9$	32986-56-4	467.516	cry					s H_2O
10104	β-Tocopherol	5,8-Dimethyltocol	$C_{28}H_{48}O_2$	148-03-8	416.680	pa ye visc oil		$205^{0.1}$			vs ace, eth, EtOH, chl
10105	γ-Tocopherol	7,8-Dimethyltocol	$C_{28}H_{48}O_2$	7616-22-0	416.680	pa ye visc oil	-1.5	$205^{0.1}$			i H_2O; msc EtOH, eth, ace, chl
10106	δ-Tocopherol	8-Methyltocol	$C_{27}H_{46}O_2$	119-13-1	402.653	pa ye visc oil		$150^{0.001}$			i H_2O; vs EtOH, eth, ace, chl
10107	Tolazamide		$C_{14}H_{21}N_3O_3S$	1156-19-0	311.400	cry	172				
10108	Tolbutamide	*N*-[(Butylamino)carbonyl]-4-methylbenzenesulfonamide	$C_{12}H_{18}N_2O_3S$	64-77-7	270.347	orth cry	128.5		1.245^{25}		sl H_2O; s EtOH, eth, chl
10109	*o*-Tolidine	3,3'-Dimethylbenzidine	$C_{14}H_{16}N_2$	119-93-7	212.290	wh-red lf (EtOH aq)	131				sl H_2O, chl; vs EtOH, eth
10110	Tolmetin		$C_{15}H_{15}NO_3$	26171-23-3	257.285	cry (MeCN)	156 dec				
10111	Toluene	Methylbenzene	C_7H_8	108-88-3	92.139	liq	-94.95	110.63	0.8623^{25}	1.4941^{25}	i H_2O; msc EtOH, eth; s ace, CS_2
10112	Toluene-2,4-diamine	4-Methyl-1,3-benzenediamine	$C_7H_{10}N_2$	95-80-7	122.167	nd (w), cry (al)	99	292			vs H_2O, EtOH, eth, bz; s chl
10113	Toluene-3,5-diamine	5-Methyl-1,3-benzenediamine	$C_7H_{10}N_2$	108-71-4	122.167	oil		284			
10114	Toluene-2,4-diisocyanate		$C_9H_6N_2O_2$	584-84-9	174.156		20.5	251	1.2244^{20}		vs ace, bz, eth
10115	Toluene-2,6-diisocyanate		$C_9H_6N_2O_2$	91-08-7	174.156		18.3				dec H_2O; s ace, bz
10116	*p*-Toluenesulfonic acid		$C_7H_8O_3S$	104-15-4	172.202	hyg pl (w+1) mcl lf or pl	104.5	140^{20}			vs H_2O; s EtOH, eth
10117	*p*-Toluenesulfonic acid monohydrate	4-Methylbenzenesulfonic acid, monohydrate	$C_7H_{10}O_4S$	6192-52-5	190.217		105.3				s H_2O

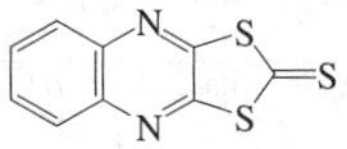

Thioquinox

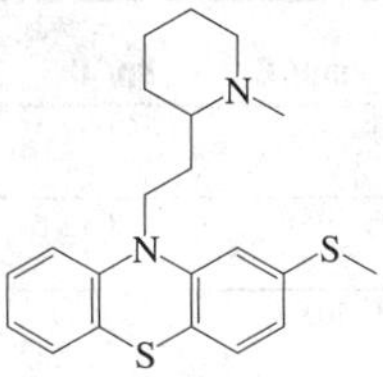

Thioridazine

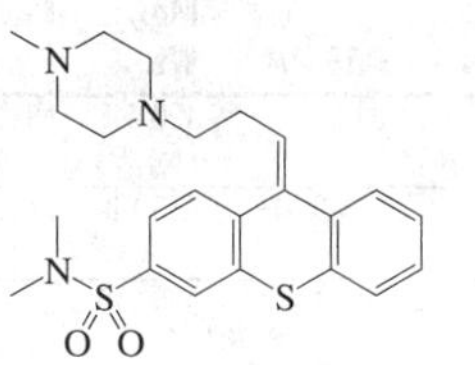

cis-Thiothixene

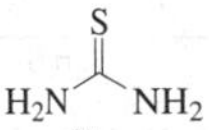

Thiourea

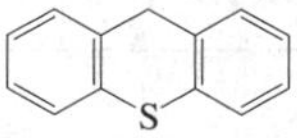

9*H*-Thioxanthene

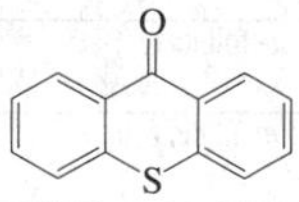

9*H*-Thioxanthen-9-one

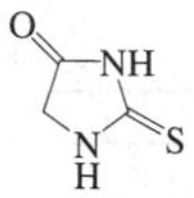

2-Thioxo-4-imidazolidinone

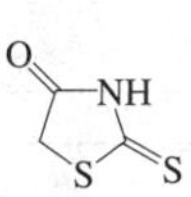

2-Thioxo-4-thiazolidinone

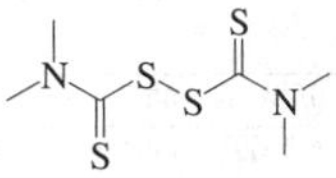

Thiram

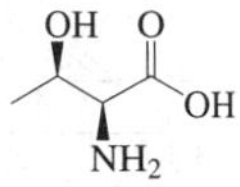

L-Threonine

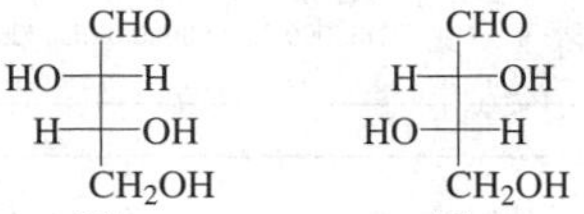

D-Threose

L-Threose

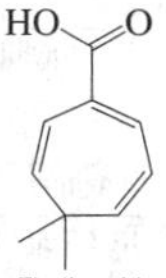

Thujic acid

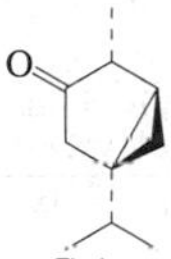

α-Thujone

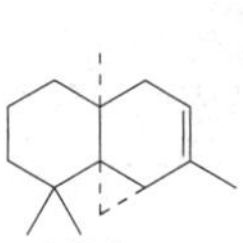

3-Thujopsene

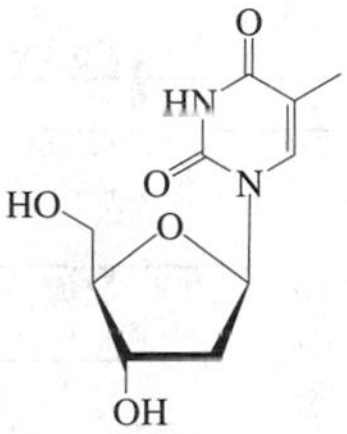

Thymidine

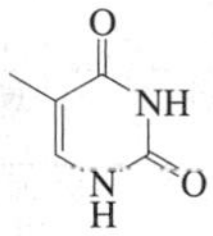

Thymine

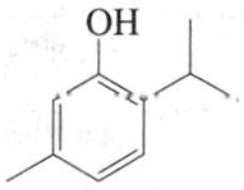

Thymol

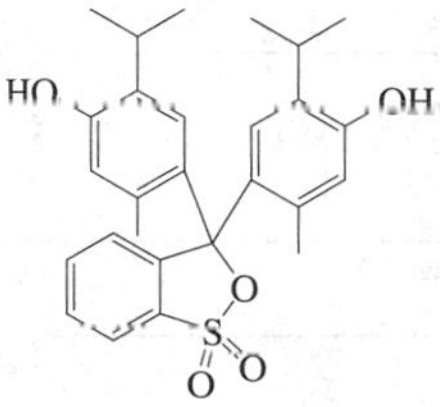

Thymol Blue

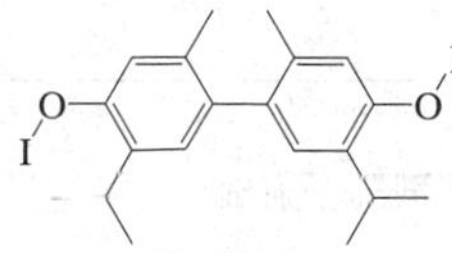

Thymol iodide

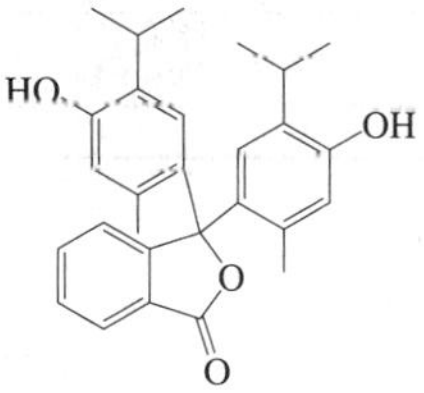

Thymolphthalein

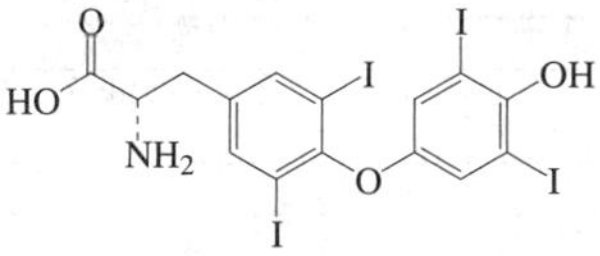

L-Thyroxine

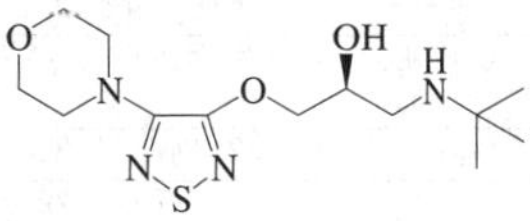

Timolol

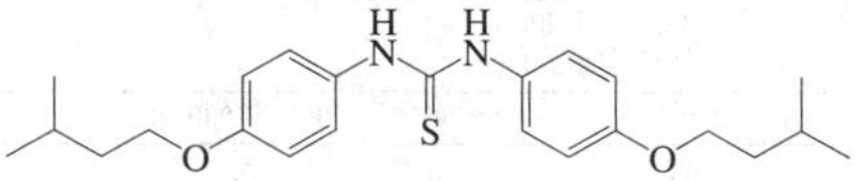

Tiocarlide

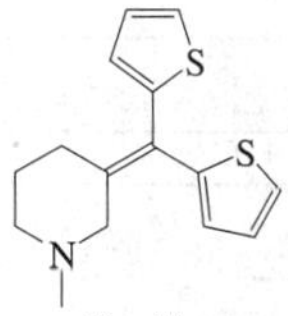

Tipepidine

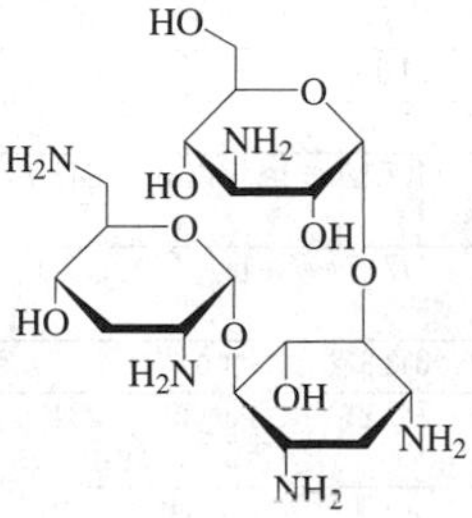

Tobramycin

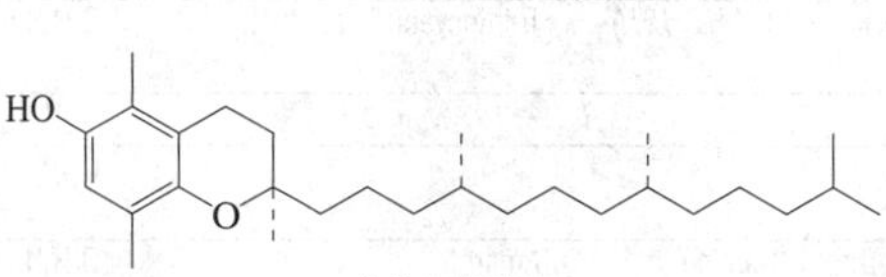

β-Tocopherol

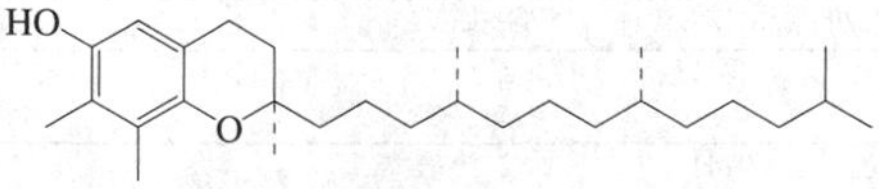

γ-Tocopherol

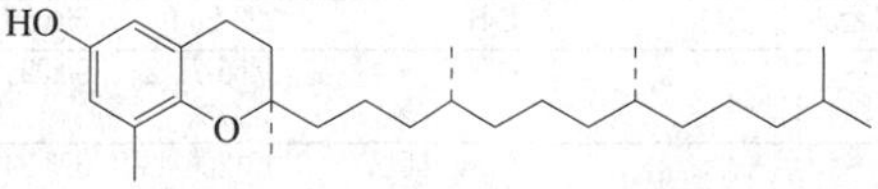

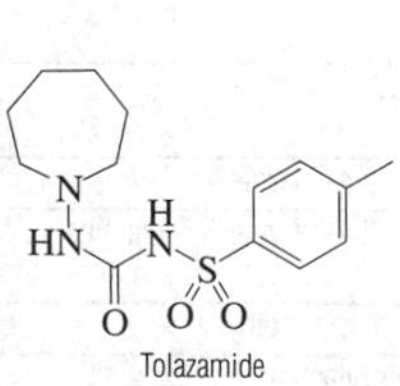

δ-Tocopherol

Tolazamide

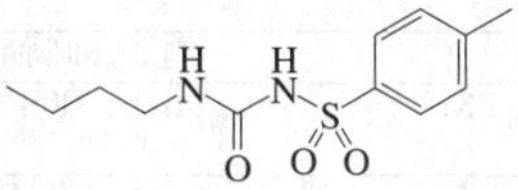

Tolbutamide

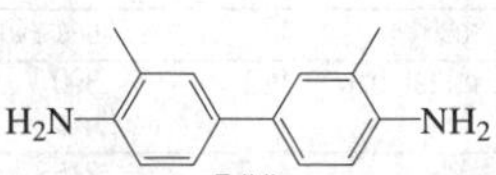

o-Tolidine

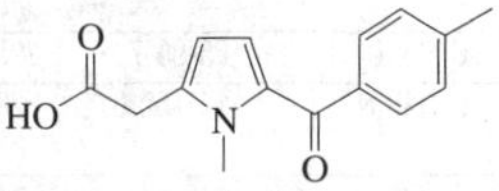

Tolmetin

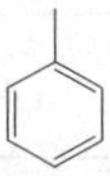

Toluene

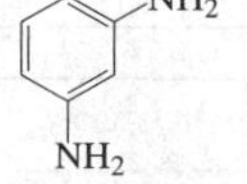

Toluene-2,4-diamine

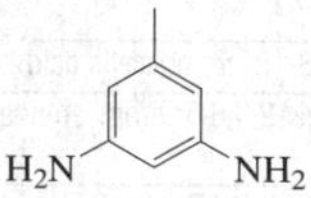

Toluene-3,5-diamine

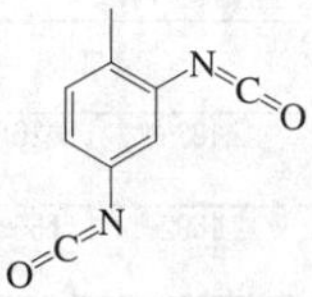

Toluene-2,4-diisocyanate

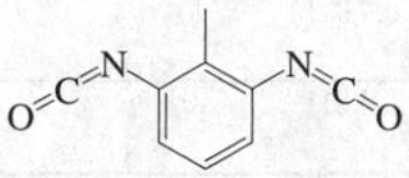

Toluene-2,6-diisocyanate

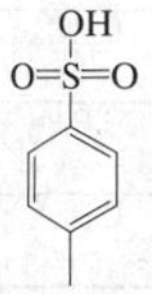

p-Toluenesulfonic acid

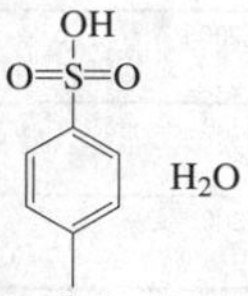

p-Toluenesulfonic acid monohydrate

No.	Name	Synonym	Mol. Form.	CAS RN	Mol. Wt.	Physical Form	mp/°C	bp/°C	den/ g cm^{-3}	n_D	Solubility
10118	*p*-Toluenesulfonyl chloride		$C_7H_7ClO_2S$	98-59-9	190.648	tcl (eth, peth)	71	145^{15}			i H_2O; s EtOH, eth, chl; vs bz
10119	*o*-Toluic acid		$C_8H_8O_2$	118-90-1	136.149	pr or nd (w)	103.5	259	1.062^{115}	1.512^{115}	i H_2O; vs EtOH, eth; s chl
10120	*m*-Toluic acid		$C_8H_8O_2$	99-04-7	136.149		109.9		1.054^{112}	1.509	sl H_2O, chl; vs EtOH, eth
10121	*p*-Toluic acid		$C_8H_8O_2$	99-94-5	136.149		179.6				i H_2O; vs EtOH, eth, MeOH; sl tfa
10122	*N*-*o*-Tolylbiguanide	*N*-(2-Methylphenyl) imidodicarbonimidic diamide	$C_9H_{13}N_5$	93-69-6	191.233	nd or pl (w+1)	145.0				sl H_2O; vs EtOH, ace; i bz, chl, eth
10123	Tomatine		$C_{50}H_{83}NO_{21}$	17406-45-0	1034.188	nd (MeOH)	270				vs EtOH, diox
10124	Tralomethrin		$C_{22}H_{19}Br_4NO_3$	66841-25-6	665.007	oran-ye solid					
10125	Tranylcypromine	2-Phenylcyclopropylamine	$C_9H_{11}N$	155-09-9	133.190	cry	44	127^{32}			
10126	Trehalose		$C_{12}H_{22}O_{11}$	99-20-7	342.296	orth cry	203		1.58^{24}		vs H_2O; s EtOH; i eth, bz
10127	Triacetamide		$C_6H_9NO_3$	641-06-5	143.140	nd (eth)	79				vs eth
10128	Triacetin	Glycerol triacetate	$C_9H_{14}O_6$	102-76-1	218.203	col oily liq	-78	259	1.1583^{20}	1.4301^{20}	sl H_2O; msc EtOH, eth, bz; vs ace
10129	Triacontane		$C_{30}H_{62}$	638-68-6	422.813	orth (eth, bz)	65.1	452.0	0.8097^{20}	1.4352^{70}	i H_2O; sl EtOH; s eth; vs bz
10130	Triacontanoic acid		$C_{30}H_{60}O_2$	506-50-3	452.796	sc, nd (al, ace)	93.6			1.4323^{100}	vs bz, CS_2, chl
10131	1-Triacontanol	Myricyl alcohol	$C_{30}H_{62}O$	593-50-0	438.812	nd (eth),pl (bz)	88		0.777^{95}		vs bz, eth, EtOH
10132	Triadimenol	Mercury, chloro(2-methoxyethyl)-	C_3H_7ClHgO	123-88-6	295.13	cry	115				i H_2O; s EtOH, ace
10133	Triallate		$C_{10}H_{16}Cl_3NOS$	2303-17-5	304.664		29	$117^{0.0003}$	1.273^{25}		
10134	Triallyl phosphate		$C_9H_{15}O_4P$	1623-19-4	218.186		-50	108^{7}	1.0815^{20}		sl chl
10135	1,3,5-Triallyl-1,3,5-triazine-2,4,6(1*H*,3*H*,5*H*)-trione		$C_{12}H_{15}N_3O_3$	1025-15-6	249.265		20.5	149^{4}, $105^{0.5}$	1.1590^{20}		
10136	Triamcinolone	Fluoxiprednisolone	$C_{21}H_{27}FO_6$	124-94-7	394.433	cry	270				
10137	Triamiphos		$C_{12}H_{19}N_6OP$	1031-47-6	294.292	cry (EtOH aq)	167				sl H_2O; s os
10138	Triasulfuron		$C_{14}H_{16}ClN_5O_5S$	82097-50-5	401.826		186				
10139	1,2,4-Triazine		$C_3H_3N_3$	290-38-0	81.076	pa ye oil	16.5	157		1.5149^{25}	
10140	1,3,5-Triazine		$C_3H_3N_3$	290-87-9	81.076		80.3	114	1.38^{25}		s EtOH, eth
10141	1,2,4-Triazine-3,5(2*H*,4*H*)-dione		$C_3H_3N_3O_2$	461-89-2	113.075		276.8				
10142	1,3,5-Triazine-2,4,6-triamine	Melamine	$C_3H_6N_6$	108-78-1	126.120	mcl pr (w)	345 dec	sub	1.573^{16}	1.872^{20}	sl H_2O, EtOH; i eth
10143	1,3,5-Triazine-2,4,6(1*H*,3*H*,5*H*)-trithione	Trithiocyanuric acid	$C_3H_3N_3S_3$	638-16-4	177.271	ye pr	>300	100^{22}			
10144	Triazofos		$C_{12}H_{16}N_3O_3PS$	24017-47-8	313.312	ye-br oil	5		1.2514^{20}		i H2O; s os
10145	Triazolam		$C_{17}H_{12}Cl_2N_4$	28911-01-5	343.210	tan cry (2-PrOH)	234				
10146	1*H*-1,2,4-Triazol-3-amine	Amitrole	$C_2H_4N_4$	61-82-5	84.080	cry (w, al)	159				vs H_2O, EtOH; i eth, ace; s chl; sl AcOEt
10147	1*H*-1,2,3-Triazole		$C_2H_3N_3$	288-36-8	69.065	hyg cry	23	204	1.1861^{25}	1.4854^{25}	s H_2O; s eth, ace; i lig
10148	1*H*-1,2,4-Triazole	Pyrrodiazole	$C_2H_3N_3$	288-88-0	69.065	nd (bz/EtOH)	120.5	260 dec			s H_2O, EtOH
10149	1*H*-1,2,4-Triazole-3,5-diamine		$C_2H_5N_5$	1455-77-2	99.095		211.5				s H_2O, EtOH; i eth, bz
10150	Tribenuron-methyl		$C_{15}H_{17}N_5O_6S$	101200-48-0	395.391	solid	141				
10151	Tribenzylamine	*N*,*N*-Bis(phenylmethyl) benzenemethanamine	$C_{21}H_{21}N$	620-40-6	287.399	pl (eth), mcl (al)	91.5	385	0.9912^{95}		sl H_2O, EtOH; s eth, ctc
10152	Tribromoacetaldehyde	Bromal	C_2HBr_3O	115-17-3	280.740			174	2.6649^{25}	1.5939^{20}	vs ace, eth, EtOH
10153	Tribromoacetic acid		$C_2HBr_3O_2$	75-96-7	296.740	mcl	132	dec 245			s H_2O, EtOH, eth
10154	2,4,6-Tribromoaniline		$C_6H_4Br_3N$	147-82-0	329.815	nd (al, bz)	122	300	2.35^{20}		i H_2O; sl EtOH; s eth, chl
10155	1,2,4-Tribromobenzene		$C_6H_3Br_3$	615-54-3	314.800		44.5	275			i H_2O; s EtOH; vs eth, ace; sl bz
10156	1,3,5-Tribromobenzene		$C_6H_3Br_3$	626-39-1	314.800	nd or pr (al)	122.8	271			i H_2O; sl EtOH; s eth, bz, chl
10157	1,1,2-Tribromobutane		$C_4H_7Br_3$	3675-68-1	294.811			216.2	2.1835^{20}	1.5626^{17}	vs eth, EtOH, chl
10158	1,2,2-Tribromobutane		$C_4H_7Br_3$	3675-69-2	294.811			213.8	2.1692^{20}	1.568^{20}	vs eth, EtOH, chl

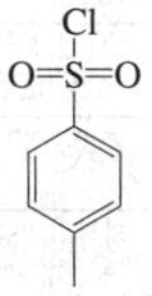

p-Toluenesulfonyl chloride

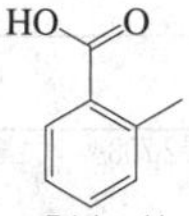

o-Toluic acid

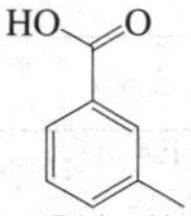

m-Toluic acid

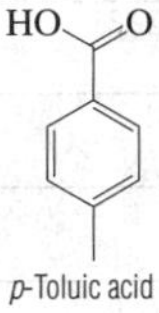

p-Toluic acid

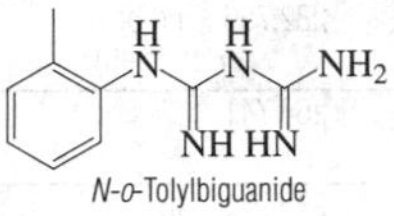

N-*o*-Tolylbiguanide

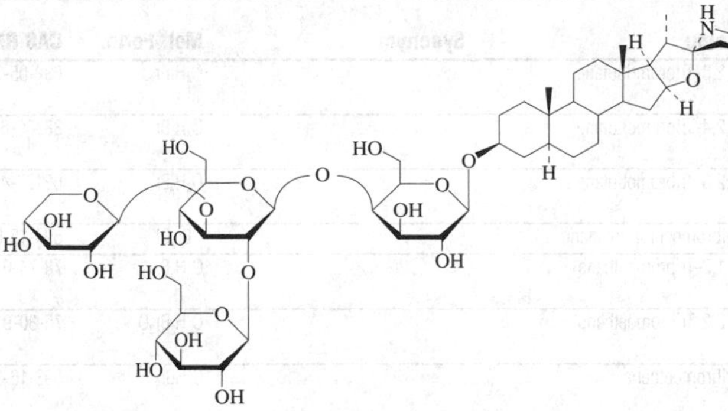

Tomatine

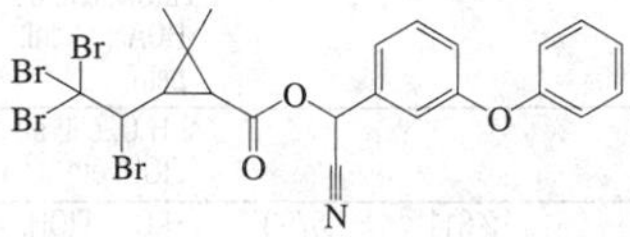

Tralomethrin

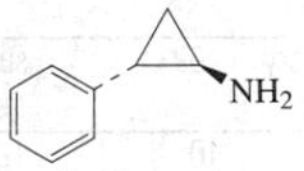

Tranylcypromine

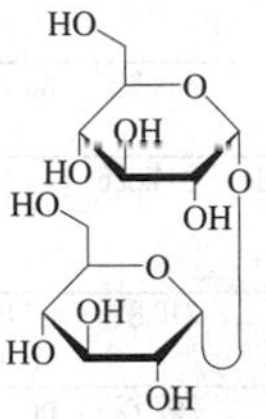

Trehalose

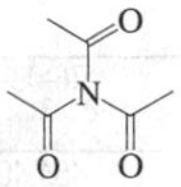

Triacetamide

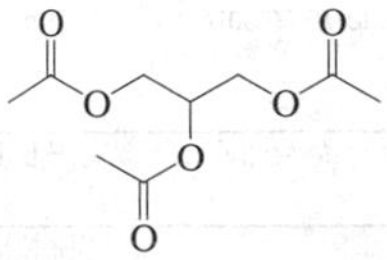

Triacetin

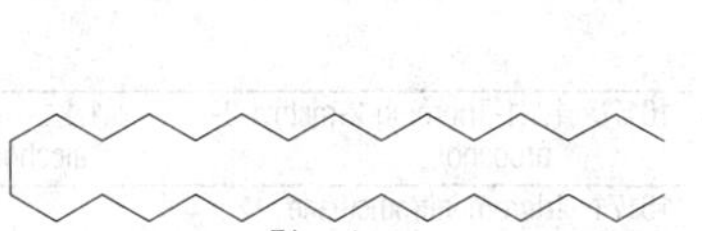

Triacontane

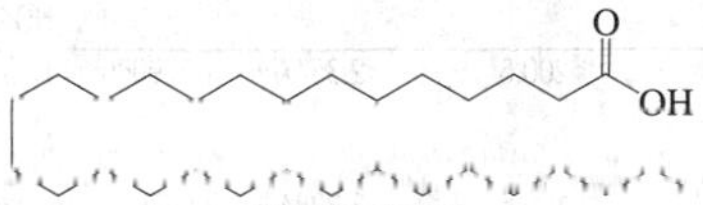

Triacontanoic acid

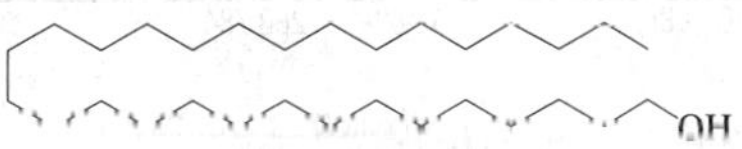

1-Triacontanol

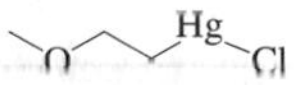

Triadimenol

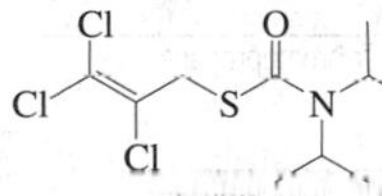

Triallate

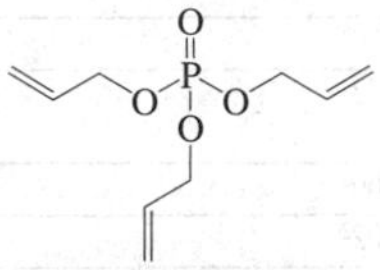

Triallyl phosphate

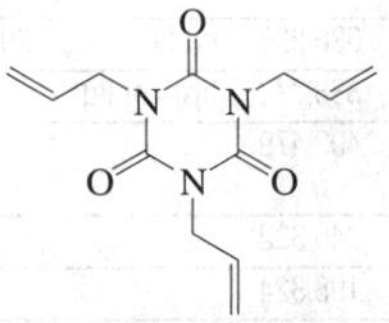

1,3,5-Triallyl-1,3,5-triazine-2,4,6(1*H*,3*H*,5*H*)-trione

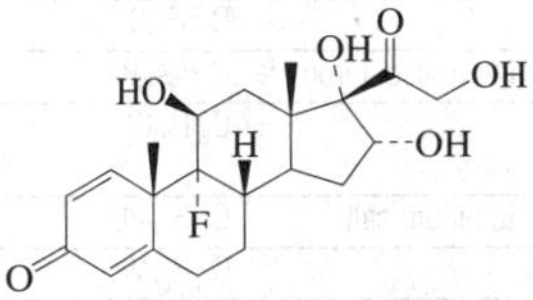

Triamcinolone

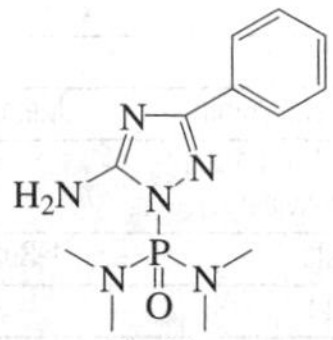

Triamiphos

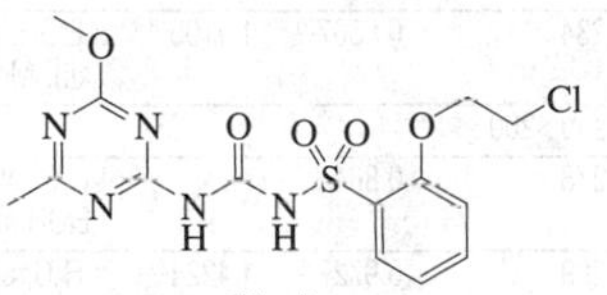

Triasulfuron

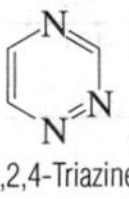

1,2,4-Triazine

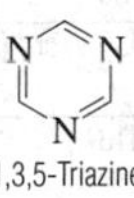

1,3,5-Triazine

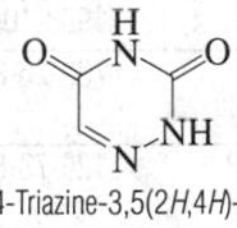

1,2,4-Triazine-3,5(2*H*,4*H*)-dione

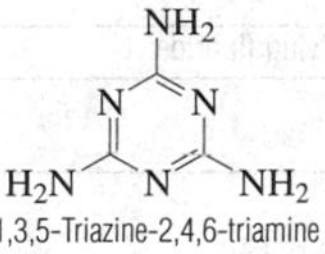

1,3,5-Triazine-2,4,6-triamine

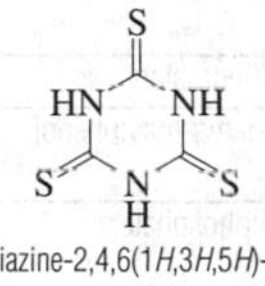

1,3,5-Triazine-2,4,6(1*H*,3*H*,5*H*)-trithione

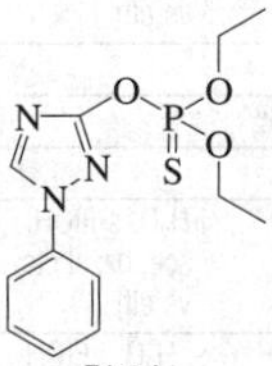

Triazofos

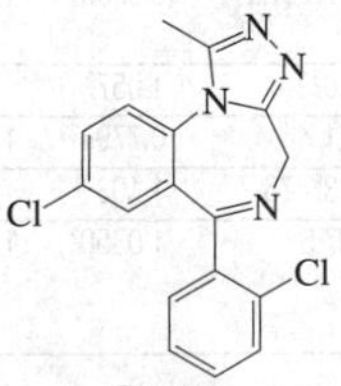

Triazolam

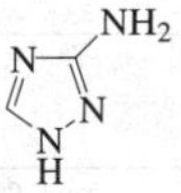

1*H*-1,2,4-Triazol-3-amine

1*H*-1,2,3-Triazole

1*H*-1,2,4-Triazole

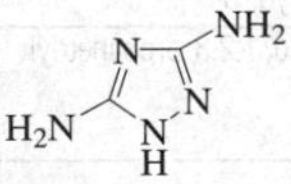

1*H*-1,2,4-Triazole-3,5-diamine

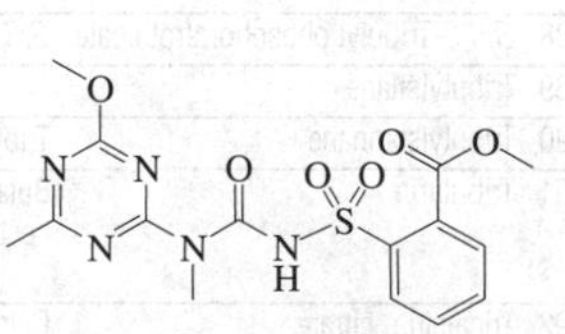

Tribenuron-methyl

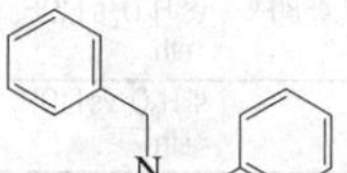

Tribenzylamine

Tribromoacetaldehyde

Tribromoacetic acid

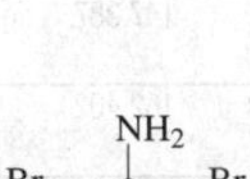

2,4,6-Tribromoaniline

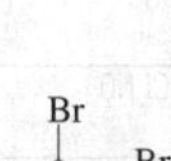

1,2,4-Tribromobenzene

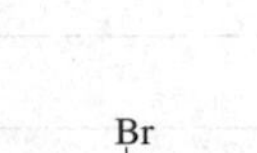

1,3,5-Tribromobenzene

1,1,2-Tribromobutane

1,2,2-Tribromobutane

No.	Name	Synonym	Mol. Form.	CAS RN	Mol. Wt.	Physical Form	mp/°C	bp/°C	den/ g cm^{-3}	n_D	Solubility
10159	1,2,3-Tribromobutane		$C_4H_7Br_3$	632-05-3	294.811	liq	-19	220	2.1907^{20}	1.5680^{20}	vs eth, EtOH, chl
10160	1,2,4-Tribromobutane		$C_4H_7Br_3$	38300-67-3	294.811	liq	-18	215	2.170^{20}	1.5608^{20}	vs eth, EtOH, chl
10161	2,2,3-Tribromobutane		$C_4H_7Br_3$	62127-47-3	294.811		0.9	206	2.1723^{20}	1.5602^{20}	i H_2O; s EtOH, eth, chl; sl ctc
10162	Tribromochloromethane		CBr_3Cl	594-15-0	287.176	lf (eth)	55	158.5	2.71^{15}		vs eth
10163	1,1,2-Tribromoethane		$C_2H_3Br_3$	78-74-0	266.757	liq	-29.3	188.93	2.6210^{20}	1.5933^{20}	i H_2O; s EtOH, eth, bz, ctc
10164	2,2,2-Tribromoethanol		$C_2H_3Br_3O$	75-80-9	282.756	nd or pr (peth)	81	92^{10}			vs bz, eth, EtOH
10165	Tribromoethene		C_2HBr_3	598-16-3	264.741			164	2.708^{20}	1.6045^{16}	sl H_2O; vs EtOH; s eth, ace, chl
10166	Tribromofluoromethane		CBr_3F	353-54-8	270.721	liq	-73.6	108			i H_2O; s EtOH
10167	Tribromomethane	Bromoform	$CHBr_3$	75-25-2	252.731		8.69	149.1	2.8788^{25}	1.5948^{25}	sl H_2O; msc EtOH, eth; s bz, lig, chl
10168	1,3,5-Tribromo-2-methoxybenzene		$C_7H_5Br_3O$	607-99-8	344.826	nd (al)	88	298	2.491^{25}		sl H_2O, EtOH; vs ace, bz; s ctc
10169	2,4,6-Tribromo-3-methylphenol	2,4,6-Tribromo-*m*-cresol	$C_7H_5Br_3O$	4619-74-3	344.826		84				s EtOH, eth, bz, HOAc; sl chl, peth
10170	1,1,1-Tribromo-2-methyl-2-propanol	1,1,1-Tribromo-*tert*-butyl alcohol	$C_4H_7Br_3O$	76-08-4	310.810	nd (lig) cry (dil al)	169	sub			sl H_2O, chl; s EtOH, eth
10171	Tribromonitromethane		CBr_3NO_2	464-10-8	297.729	pr	10	127^{18}	2.811^{12}	1.5790^{20}	i H_2O; s EtOH, eth; vs ace, bz
10172	2,4,6-Tribromophenol		$C_6H_3Br_3O$	118-79-6	330.799	nd (al), pr (bz)	95.5	286	2.55^{20}		i H_2O; vs EtOH; s eth, bz, HOAc, chl
10173	1,1,2-Tribromopropane		$C_3H_5Br_3$	14602-62-1	280.784			200.5	2.3547^{20}	1.5790^{20}	i H_2O; s EtOH, chl, HOAc; vs eth
10174	1,2,2-Tribromopropane		$C_3H_5Br_3$	14476-30-3	280.784			190.5	2.2984^{20}	1.5670^{20}	vs eth, EtOH, chl
10175	1,2,3-Tribromopropane		$C_3H_5Br_3$	96-11-7	280.784		16.9	222.1	2.4208^{20}	1.5862^{20}	i H_2O; vs EtOH, eth; sl ctc
10176	2,3,5-Tribromothiophene		C_4HBr_3S	3141-24-0	320.828	nd (al)	29	260			s chl
10177	Tribromotrimethyldialuminum	Methyl aluminum sesquibromide	$C_3H_9Al_2Br_3$	12263-85-3	338.778	hyg col liq		110^{50}			
10178	Tributyl 2-(acetyloxy)-1,2,3-propanetricarboxylate		$C_{20}H_{34}O_8$	77-90-7	402.479			173^{1}			sl chl
10179	Tributyl aluminate	1-Butanol, aluminum salt	$C_{12}H_{27}AlO_3$	3085-30-1	246.322			260^{5}			
10180	Tributylaluminum		$C_{12}H_{27}Al$	1116-70-7	198.324			102^{2}			
10181	Tributylamine	*N,N*-Dibutyl-1-butanamine	$C_{12}H_{27}N$	102-82-9	185.349	liq	-70	216.5	0.7770^{20}	1.4299^{20}	sl H_2O, ctc; vs EtOH, eth; s ace, bz
10182	Tributyl borate		$C_{12}H_{27}BO_3$	688-74-4	230.151	oil	<-70	234	0.8567^{20}	1.4106^{18}	s EtOH, bz; vs eth, MeOH
10183	Tributylfluorostannane	Tributyltin fluoride	$C_{12}H_{27}FSn$	1983-10-4	309.050	nd	≈260	sub >200			
10184	2,4,6-Tri-*tert*-butylphenol		$C_{18}H_{30}O$	732-26-3	262.430	cry (al, peth)	131	278	0.864^{27}		i H_2O, alk; s EtOH, ace, ctc
10185	Tributyl phosphate		$C_{12}H_{27}O_4P$	126-73-8	266.313			289	0.9727^{25}	1.4224^{25}	s H_2O, eth, bz, CS_2; msc EtOH
10186	Tributylphosphine		$C_{12}H_{27}P$	998-40-3	202.316			240; 150^{50}	0.812^{25}	1.4619^{20}	
10187	Tributyl phosphite	Tributoxyphosphine	$C_{12}H_{27}O_3P$	102-85-2	250.314			137^{26}, 122^{12}	0.9259^{20}	1.4321^{19}	s EtOH; sl ctc; vs eth
10188	*S,S,S*-Tributyl phosphorotrithioate	*S,S,S*-Tributyl trithiophosphate	$C_{12}H_{27}OPS_3$	78-48-8	314.510		<-25	$150^{0.3}$	1.057^{20}		
10189	Tributylsilane		$C_{12}H_{28}Si$	998-41-4	200.436			221	0.7794^{20}	1.4380^{20}	
10190	Tributylstannane	Tributyltin hydride	$C_{12}H_{28}Sn$	688-73-3	291.060	liq		113^{8}, $76^{0.7}$	1.103^{20}		
10191	Tributyrin	Butanoic acid, 1,2,3-propanetriyl ester	$C_{15}H_{26}O_6$	60-01-5	302.363	liq	-75	307.5	1.0350^{20}	1.4359^{20}	i H_2O; s EtOH, ace, bz; sl ctc; vs eth
10192	Tricalcium citrate	Calcium citrate	$C_{12}H_{10}Ca_3O_{14}$	813-94-5	498.433	cry (w)	≈100 dec (hyd)				sl H_2O; i EtOH
10193	Trichlorfon		$C_4H_8Cl_3O_4P$	52-68-6	257.437		77	$100^{0.1}$	1.73^{20}		
10194	Trichloroacetaldehyde	Chloral	C_2HCl_3O	75-87-6	147.387	liq	-57.5	97.8	1.512^{20}	1.4580^{20}	vs H_2O; s EtOH, eth
10195	2,2,2-Trichloroacetamide		$C_2H_2Cl_3NO$	594-65-0	162.402		142	240			sl H_2O; vs EtOH, eth
10196	Trichloroacetic acid		$C_2HCl_3O_2$	76-03-9	163.387	hyg cry	59.2	196.5	1.6126^{64}	1.4603^{61}	vs H_2O; s EtOH, eth; sl ctc
10197	Trichloroacetic anhydride		$C_4Cl_6O_3$	4124-31-6	308.759			dec 223; 139^{60}	1.6908^{20}		vs eth, HOAc
10198	Trichloroacetonitrile		C_2Cl_3N	545-06-2	144.387	liq	-42	85.7	1.4403^{25}	1.4409^{20}	i H_2O
10199	Trichloroacetyl chloride		C_2Cl_4O	76-02-8	181.832			117.9	1.6202^{20}	1.4695^{20}	msc eth
10200	2,3,4-Trichloroaniline		$C_6H_4Cl_3N$	634-67-3	196.462	nd (lig)	73	292			vs EtOH

1,2,3-Tribromobutane

1,2,4-Tribromobutane

2,2,3-Tribromobutane

Tribromochloromethane

1,1,2-Tribromoethane

2,2,2-Tribromoethanol

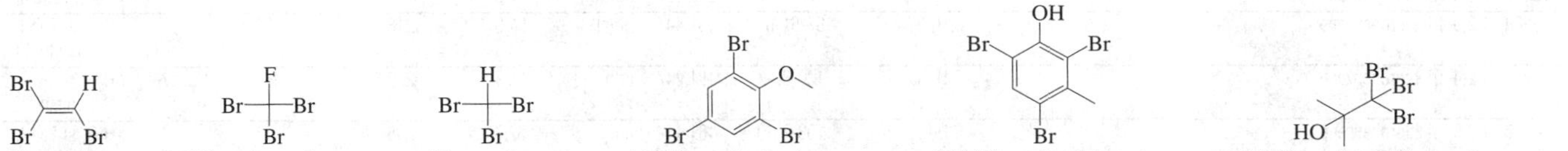

Tribromoethene

Tribromofluoromethane

Tribromomethane

1,3,5-Tribromo-2-methoxybenzene

2,4,6-Tribromo-3-methylphenol

1,1,1-Tribromo-2-methyl-2-propanol

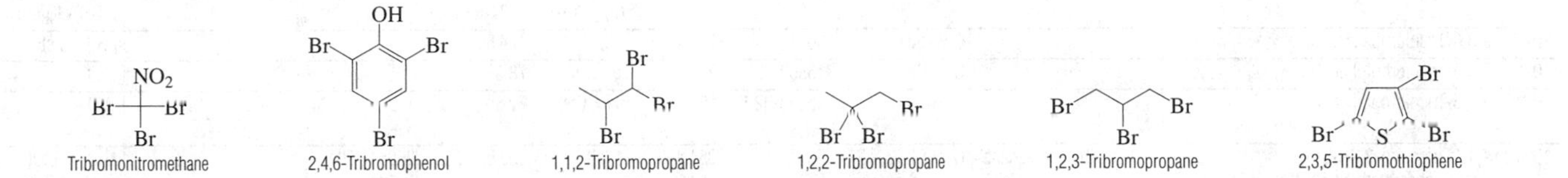

Tribromonitromethane

2,4,6-Tribromophenol

1,1,2-Tribromopropane

1,2,2-Tribromopropane

1,2,3-Tribromopropane

2,3,5-Tribromothiophene

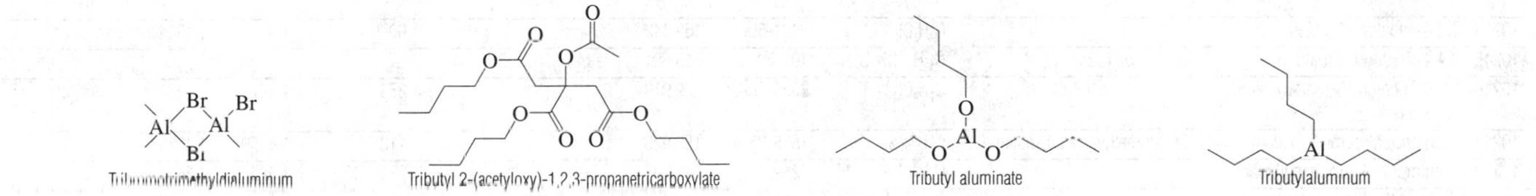

Tribromotrimethyldialuminum

Tributyl 2-(acetyloxy)-1,2,3-propanetricarboxylate

Tributyl aluminate

Tributylaluminum

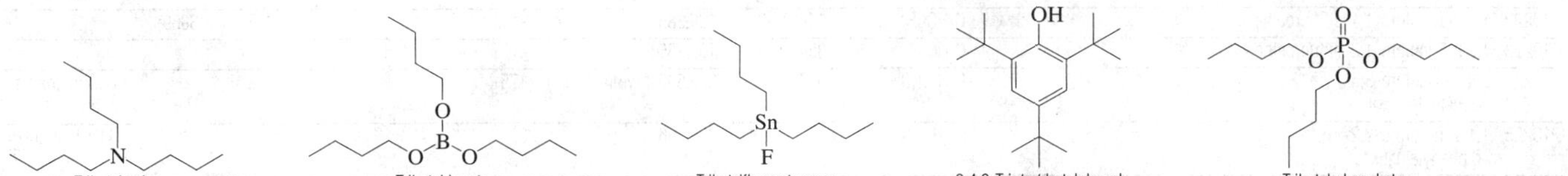

Tributylamine

Tributyl borate

Tributylfluorostannane

2,4,6-Tri-*tert*-butylphenol

Tributyl phosphate

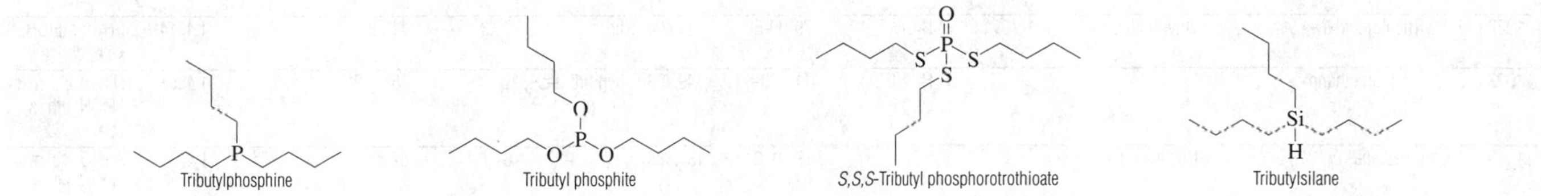

Tributylphosphine

Tributyl phosphite

S,S,S-Tributyl phosphorotrithioate

Tributylsilane

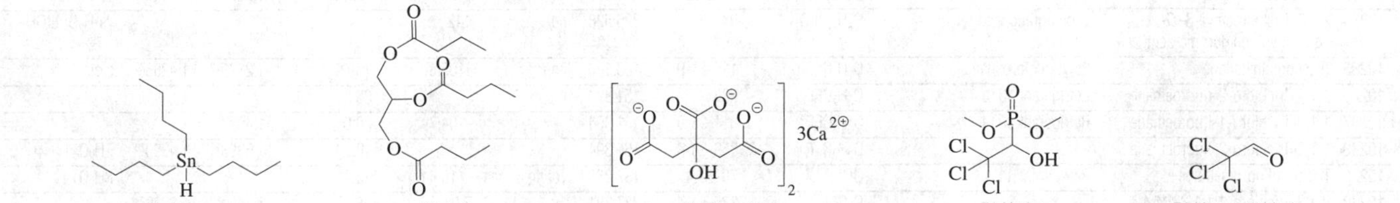

Tributylstannane

Tributyrin

Tricalcium citrate

Trichlorfon

Trichloroacetaldehyde

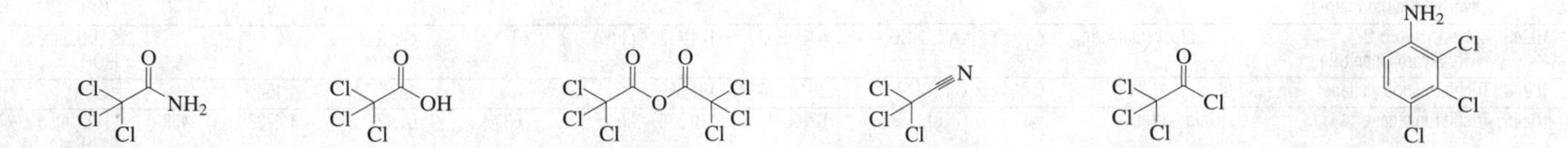

2,2,2-Trichloroacetamide

Trichloroacetic acid

Trichloroacetic anhydride

Trichloroacetonitrile

Trichloroacetyl chloride

2,3,4-Trichloroaniline

No.	Name	Synonym	Mol. Form.	CAS RN	Mol. Wt.	Physical Form	mp/°C	bp/°C	den/ g cm^{-3}	n_D	Solubility
10201	2,4,5-Trichloroaniline		$C_6H_4Cl_3N$	636-30-6	196.462	nd (lig)	96.5	270			s EtOH, eth; vs CS_2; sl lig
10202	2,4,6-Trichloroaniline		$C_6H_4Cl_3N$	634-93-5	196.462	cry (al), nd (lig or peth)	78.5	262			i H_2O; s EtOH, eth, chl; vs CS_2
10203	2,3,6-Trichlorobenzaldehyde		$C_7H_3Cl_3O$	4659-47-6	209.457	nd (lig)	87.3				vs ace, bz, eth
10204	1,2,3-Trichlorobenzene		$C_6H_3Cl_3$	87-61-6	181.447	pl (al)	51.3	218.5	1.4533^{25}		i H_2O; sl EtOH, chl; vs eth, bz
10205	1,2,4-Trichlorobenzene		$C_6H_3Cl_3$	120-82-1	181.447	orth	16.92	213.5	1.459^{25}	1.5717^{20}	i H_2O; sl EtOH, chl; vs eth
10206	1,3,5-Trichlorobenzene		$C_6H_3Cl_3$	108-70-3	181.447	nd	62.8	208			i H_2O; sl EtOH; vs eth, bz; s chl
10207	2,3,6-Trichlorobenzeneacetic acid	Chlorfenac	$C_8H_5Cl_3O_2$	85-34-7	239.484		161				
10208	3,4,5-Trichloro-1,2-benzenediol		$C_6H_3Cl_3O_2$	56961-20-7	213.446	(i) pr (HOAc) (ii) pr (bz)	115(form a); 134(form b)				sl H_2O; vs eth, EtOH, HOAc
10209	2,3,6-Trichlorobenzoic acid		$C_7H_3Cl_3O_2$	50-31-7	225.457		124.5				sl H_2O; s eth
10210	2,4,5-Trichlorobiphenyl		$C_{12}H_7Cl_3$	15862-07-4	257.543	cry	78.5				i H_2O
10211	2,4,6-Trichlorobiphenyl		$C_{12}H_7Cl_3$	35693-92-6	257.543	cry (EtOH aq)	62.5	172^{15}			i H_2O
10212	1,1,1-Trichloro-2,2-bis(4-chlorophenyl)ethane	Dichlorodiphenyltrichloroethane (DDT)	$C_{14}H_9Cl_5$	50-29-3	354.486	nd (al)	108.5	260; $186^{0.05}$			i H_2O; sl EtOH; vs eth, ace, bz, py
10213	2,2,3-Trichlorobutanal	2,2,3-Trichlorobutyraldehyde	$C_4H_5Cl_3O$	76-36-8	175.441			164	1.3956^{20}	1.4755^{20}	vs H_2O, eth, EtOH
10214	2,3,4-Trichloro-1-butene		$C_4H_5Cl_3$	2431-50-7	159.442			60^{20}, 40^{10}	1.3430^{20}	1.4944^{20}	vs ace, chl
10215	3,4,4'-Trichlorocarbanilide	Triclocarban	$C_{13}H_9Cl_3N_2O$	101-20-2	315.581	fine pl	256				
10216	1,2,4-Trichloro-5-(chloromethyl) benzene		$C_7H_4Cl_4$	3955-26-8	229.919			273	1.547^{20}		vs ace, eth, EtOH
10217	Trichloro(chloromethyl)silane	(Chloromethyl)trichlorosilane	CH_2Cl_4Si	1558-25-4	183.925			118	1.4650^{20}	1.4555^{20}	
10218	Trichloro(4-chlorophenyl)silane		$C_6H_4Cl_4Si$	825-94-5	245.994			233; 116^{20}	1.4062^{20}	1.5418^{20}	
10219	Trichloro(3-chloropropyl)silane		$C_3H_6Cl_4Si$	2550-06-3	211.978			181.5	1.3590^{20}	1.4668^{20}	
10220	Trichloro(dichloromethyl)silane	(Dichloromethyl)trichlorosilane	$CHCl_5Si$	1558-24-3	218.370			145	1.5518^{20}	1.4714^{20}	
10221	1,1,1-Trichloro-2,2-difluoroethane		$C_2HCl_3F_2$	354-12-1	169.385			73			
10222	1,2,2-Trichloro-1,1-difluoroethane		$C_2HCl_3F_2$	354-21-2	169.385		-140	71.9	1.5447^{20}	1.3889^{20}	
10223	1,2,2-Trichloro-1,2-difluoroethane		$C_2HCl_3F_2$	354-15-4	169.385		-174	72.5			
10224	2,4,6-Trichloro-3,5-dimethylphenol		$C_8H_7Cl_3O$	6972-47-0	225.500	ye nd (peth)	175				i H_2O; s chl; vs peth
10225	1,1,1-Trichloro-2,2-diphenylethane		$C_{14}H_{11}Cl_3$	2971-22-4	285.596		65				s EtOH; sl chl
10226	Trichlorododecylsilane	Dodecyltrichlorosilane	$C_{12}H_{25}Cl_3Si$	4484-72-4	303.772			155^{10}		1.4581^{20}	
10227	1,1,1-Trichloro-3,4-epoxybutane	(2,2,2-Trichloroethyl)oxirane	$C_4H_5Cl_3O$	3083-25-8	175.441	liq		110^{100}			
10228	1,1,1-Trichloroethane	Methyl chloroform	$C_2H_3Cl_3$	71-55-6	133.404	liq	-30.01	74.09	1.3390^{20}	1.4379^{20}	sl H_2O; s EtOH, chl; msc eth
10229	1,1,2-Trichloroethane	Vinyl trichloride	$C_2H_3Cl_3$	79-00-5	133.404	liq	-36.3	113.8	1.4397^{20}	1.4714^{20}	i H_2O; s EtOH, eth, chl
10230	2,2,2-Trichloroethanol		$C_2H_3Cl_3O$	115-20-8	149.403	hyg orth tab or pl	19	152; 52^{11}		1.4861^{20}	sl H_2O, ctc; msc EtOH, eth; s alk
10231	Trichloroethene	Trichloroethylene	C_2HCl_3	79-01-6	131.388	liq	-84.7	87.21	1.4642^{20}	1.4773^{20}	sl H_2O, ctc; msc EtOH, eth; s ace
10232	2,2,2-Trichloro-1-ethoxyethanol	Chloral alcoholate	$C_4H_7Cl_3O_2$	515-83-3	193.457		56.5	115.5	1.143^{40}		s H_2O, EtOH, eth
10233	Trichloroethoxysilane		$C_2H_5Cl_3OSi$	1825-82-7	179.505	liq	-135	101.9	1.2274^{20}	1.4045^{20}	vs EtOH
10234	2,2,2-Trichloroethyl-β-*D*-glucopyranosiduronic acid	Urochloralic acid	$C_8H_{11}Cl_3O_7$	97-25-6	325.528	nd	142				vs H_2O, EtOH
10235	Trichloroethylsilane	Ethyltrichlorosilane	$C_2H_5Cl_3Si$	115-21-9	163.506	liq	-105.6	100.5	1.2373^{20}	1.4256^{20}	s ctc
10236	1,1,1-Trichloro-2-fluoroethane	Refrigerant 131b	$C_2H_2Cl_3F$	2366-36-1	151.394	liq		86.5			
10237	1,1,2-Trichloro-1-fluoroethane	Refrigerant 131a	$C_2H_2Cl_3F$	811-95-0	151.394		-104.7	88.0	1.492^{20}		
10238	1,1,2-Trichloro-2-fluoroethane		$C_2H_2Cl_3F$	359-28-4	151.394			102.4	1.5393^{20}	1.4390^{20}	i H_2O
10239	Trichlorofluoromethane	Refrigerant 11	CCl_3F	75-69-4	137.368	vol liq or gas	-110.44	23.7			i H_2O
10240	2,2,3-Trichloro-1,1,1,3,4,4,4-heptafluorobutane		$C_4Cl_3F_7$	335-44-4	287.391		2.0	98	1.7484^{20}	1.3530^{20}	
10241	Trichlorohexylsilane	Hexyltrichlorosilane	$C_6H_{13}Cl_3Si$	928-65-4	219.612			190	1.1100^{20}		dec H_2O
10242	*N*-(2,2,2-Trichloro-1-hydroxyethyl)formamide	Chloral formamide	$C_3H_4Cl_3NO_2$	515-82-2	192.429	cry	120				vs ace, eth, EtOH
10243	3,3,3-Trichloro-2-hydroxypropanenitrile	Chlorocyanohydrin	$C_3H_2Cl_3NO$	513-96-2	174.413	pl (w)	61	dec 217			vs H_2O, eth, EtOH
10244	Trichloroisobutylsilane		$C_4H_9Cl_3Si$	18169-57-8	191.559			143.3	1.154^{20}		dec H_2O
10245	Trichloromethane	Chloroform	$CHCl_3$	67-66-3	119.378	liq	-63.41	61.17	1.4788^{25}	1.4459^{20}	sl H_2O; msc EtOH, eth, bz; s ace, ctc
10246	Trichloromethanesulfenyl chloride	Perchloromethyl mercaptan	CCl_4S	594-42-3	185.888	ye oil		149	1.6947^{20}	1.5484^{20}	s eth

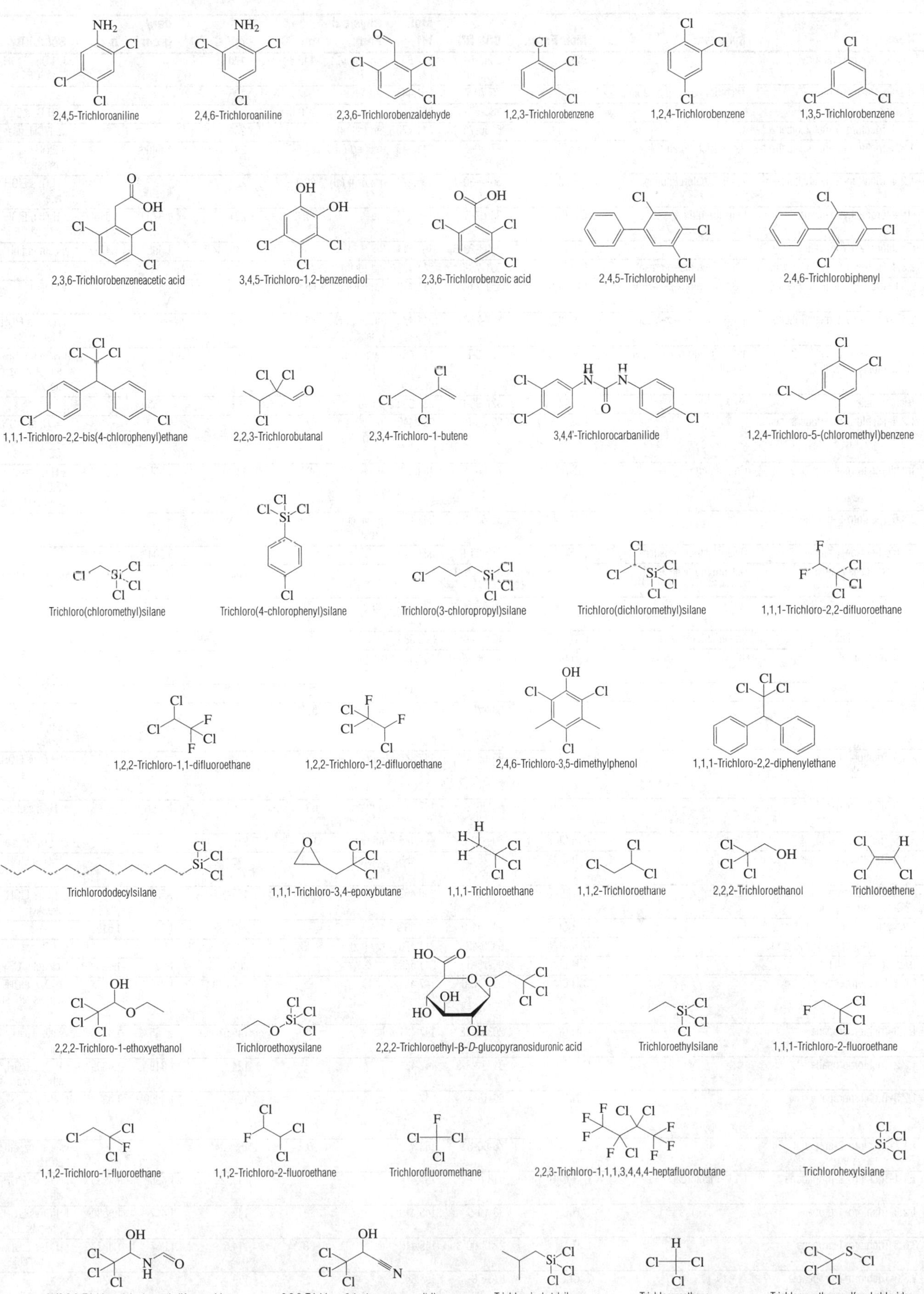
2,4,5-Trichloroaniline
2,4,6-Trichloroaniline
2,3,6-Trichlorobenzaldehyde
1,2,3-Trichlorobenzene
1,2,4-Trichlorobenzene
1,3,5-Trichlorobenzene
2,3,6-Trichlorobenzeneacetic acid
3,4,5-Trichloro-1,2-benzenediol
2,3,6-Trichlorobenzoic acid
2,4,5-Trichlorobiphenyl
2,4,6-Trichlorobiphenyl
1,1,1-Trichloro-2,2-bis(4-chlorophenyl)ethane
2,2,3-Trichlorobutanal
2,3,4-Trichloro-1-butene
3,4,4'-Trichlorocarbanilide
1,2,4-Trichloro-5-(chloromethyl)benzene
Trichloro(chloromethyl)silane
Trichloro(4-chlorophenyl)silane
Trichloro(3-chloropropyl)silane
Trichloro(dichloromethyl)silane
1,1,1-Trichloro-2,2-difluoroethane
1,2,2-Trichloro-1,1-difluoroethane
1,2,2-Trichloro-1,2-difluoroethane
2,4,6-Trichloro-3,5-dimethylphenol
1,1,1-Trichloro-2,2-diphenylethane
Trichlorododecylsilane
1,1,1-Trichloro-3,4-epoxybutane
1,1,1-Trichloroethane
1,1,2-Trichloroethane
2,2,2-Trichloroethanol
Trichloroethene
2,2,2-Trichloro-1-ethoxyethanol
Trichloroethoxysilane
2,2,2-Trichloroethyl-β-D-glucopyranosiduronic acid
Trichloroethylsilane
1,1,1-Trichloro-2-fluoroethane
1,1,2-Trichloro-1-fluoroethane
1,1,2-Trichloro-2-fluoroethane
Trichlorofluoromethane
2,2,3-Trichloro-1,1,1,3,4,4,4-heptafluorobutane
Trichlorohexylsilane
N-(2,2,2-Trichloro-1-hydroxyethyl)formamide
3,3,3-Trichloro-2-hydroxypropanenitrile
Trichloroisobutylsilane
Trichloromethane
Trichloromethanesulfenyl chloride

No.	Name	Synonym	Mol. Form.	CAS RN	Mol. Wt.	Physical Form	mp/°C	bp/°C	den/ g cm^{-3}	n_D	Solubility
10247	Trichloromethanesulfonyl chloride		CCl_4O_2S	2547-61-7	217.887	cry (al-w)	140.5	170			i H_2O; s EtOH, eth, CS_2
10248	Trichloromethanethiol	Trichloromethyl mercaptan	$CHCl_3S$	75-70-7	151.443	oran oil		125^{15}			
10249	Trichloromethiazide		$C_8H_8Cl_3N_3O_4S_2$	133-67-5	380.657		270 dec				sl H_2O; s EtOH
10250	1,2,4-Trichloro-5-methoxybenzene		$C_7H_5Cl_3O$	6130-75-2	211.473	nd (dil al)	77.5	254			vs EtOH, ace
10251	1,3,5-Trichloro-2-methoxybenzene	2,4,6-Trichloroanisole	$C_7H_5Cl_3O$	87-40-1	211.473	mcl nd (al)	61.5	241	1.640^{25}		s EtOH, bz, chl; vs ace
10252	1,2,4-Trichloro-5-methylbenzene	2,4,5-Trichlorotoluene	$C_7H_5Cl_3$	6639-30-1	195.474	nd or lf (al)	82.4	231			i H_2O; s EtOH, ace
10253	(Trichloromethyl)benzene	Benzotrichloride	$C_7H_5Cl_3$	98-07-7	195.474	liq	-4.42	221	1.3723^{20}	1.5580^{20}	i H_2O; s EtOH, eth, bz
10254	(Trichloromethyl)oxirane		$C_3H_3Cl_3O$	3083-23-6	161.414			149; 44^{13}	1.495^{20}	1.4737^{25}	vs eth; s chl
10255	2,3,4-Trichloro-6-methylphenol	4,5,6-Trichloro-*o*-cresol	$C_7H_5Cl_3O$	551-78-0	211.473	nd (peth)	77				
10256	2,3,6-Trichloro-4-methylphenol	2,3,6-Trichloro-*p*-cresol	$C_7H_5Cl_3O$	551-77-9	211.473	nd (HOAc, peth)	66.5				vs EtOH
10257	2,4,6-Trichloro-3-methylphenol	2,4,6-Trichloro-*m*-cresol	$C_7H_5Cl_3O$	551-76-8	211.473	nd or pl (w, peth)	46	265			i H_2O; vs EtOH, MeOH, chl
10258	1,1,1-Trichloro-2-methyl-2-propanol	1,1,1-Trichloro-*tert*-butyl alcohol	$C_4H_7Cl_3O$	57-15-8	177.457	hyg nd (w + 1)	97	167			i H_2O; s EtOH, eth, ace, bz, lig, chl
10259	Trichloronate		$C_{10}H_{12}Cl_3O_2PS$	327-98-0	333.599	ye liq		$108^{0.01}$	1.365^{20}		
10260	1,2,4-Trichloro-5-nitrobenzene		$C_6H_2Cl_3NO_2$	89-69-0	226.445	pr (al), nd (al)	57.5	288	1.790^{23}		i H_2O; sl EtOH; s eth, bz, chl, CS_2
10261	Trichloronitromethane	Chloropicrin	CCl_3NO_2	76-06-2	164.376	liq	-64	112	1.6558^{20}	1.4611^{20}	s H_2O; msc EtOH, ace, bz, MeOH, HOAc
10262	3,4,6-Trichloro-2-nitrophenol		$C_6H_2Cl_3NO_3$	82-62-2	242.444	pa ye cry (peth)	92.5				
10263	Trichlorooctadecylsilane	Octadecyltrichlorosilane	$C_{18}H_{37}Cl_3Si$	112-04-9	387.932			223^{10}	0.984^{25}	1.4602^{20}	
10264	Trichlorooctylsilane	Octyltrichlorosilane	$C_8H_{17}Cl_3Si$	5283-66-9	247.666			232		1.4480^{20}	dec H_2O, EtOH; s ctc
10265	1,2,3-Trichloro-1,1,2,3,3-pentafluoropropane		$C_3Cl_3F_5$	76-17-5	237.383	liq	-72	73.7	1.6631^{20}	1.3512^{20}	
10266	Trichloropentylsilane	Amyltrichlorosilane	$C_5H_{11}Cl_3Si$	107-72-2	205.586			172; 60.5^{15}	1.1330^{20}	1.4503^{20}	
10267	2,3,4-Trichlorophenol		$C_6H_3Cl_3O$	15950-66-0	197.446	nd (bz, lig, sub)	83.5	sub			s EtOH, eth, bz, alk, HOAc
10268	2,3,5-Trichlorophenol		$C_6H_3Cl_3O$	933-78-8	197.446	nd (al)	62	248^{250}			vs eth, EtOH
10269	2,3,6-Trichlorophenol		$C_6H_3Cl_3O$	933-75-5	197.446	nd (dil al, lig)	58				sl H_2O; vs EtOH, eth, bz; s HOAc
10270	2,4,5-Trichlorophenol		$C_6H_3Cl_3O$	95-95-4	197.446	nd (al, peth)	69	247			sl H_2O; vs EtOH, eth, bz; s HOAc
10271	2,4,6-Trichlorophenol		$C_6H_3Cl_3O$	88-06-2	197.446	orth nd (HOAc)	69	246	1.4901^{75}		sl H_2O; s EtOH, eth, HOAc
10272	3,4,5-Trichlorophenol		$C_6H_3Cl_3O$	609-19-8	197.446	nd (lig)	101	275			sl H_2O, lig; s eth
10273	2,4,5-Trichlorophenoxyacetic acid	2,4,5-T	$C_8H_5Cl_3O_3$	93-76-5	255.483	cry (bz)	153	dec			i H_2O; s EtOH; vs bz
10274	2-(2,4,5-Trichlorophenoxy)ethyl 2,2-dichloropropanoate	Pentanate	$C_{11}H_9Cl_5O_3$	136-25-4	366.452		49	$162^{0.5}$	1.55^{50}		i H_2O; s EtOH, ace, xyl
10275	Trichloro(2-phenylethyl)silane		$C_8H_9Cl_3Si$	940-41-0	239.602			242; 98^{5}	1.2397^{20}	1.5185^{20}	
10276	(2,4,6-Trichlorophenyl)hydrazine		$C_6H_5Cl_3N_2$	5329-12-4	211.476	cry (bz)	143				s H_2O, bz
10277	Trichlorophenylsilane		$C_6H_5Cl_3Si$	98-13-5	211.549			201	1.321^{20}	1.5230^{20}	s ctc, chl, CS_2
10278	1,1,2-Trichloropropane		$C_3H_5Cl_3$	598-77-6	147.431			132.0; 117^{500}	1.372^{15}		i H_2O; s EtOH, chl; vs eth; sl ctc
10279	1,1,3-Trichloropropane		$C_3H_5Cl_3$	20395-25-9	147.431	liq	-59	145.5	1.3557^{20}	1.4718^{20}	vs eth, EtOH, chl
10280	1,2,2-Trichloropropane		$C_3H_5Cl_3$	3175-23-3	147.431			124	1.318^{25}	1.4609^{20}	i H_2O; s EtOH, eth; vs chl
10281	1,2,3-Trichloropropane		$C_3H_5Cl_3$	96-18-4	147.431	liq	-14.7	157	1.3889^{20}	1.4852^{20}	sl H_2O, ctc; s EtOH, eth; vs chl
10282	1,1,1-Trichloro-2-propanol		$C_3H_5Cl_3O$	76-00-6	163.430		50.5	163; 54^{12}			vs ace, bz, eth, EtOH
10283	1,1,1-Trichloro-2-propanone	1,1,1-Trichloroacetone	$C_3H_3Cl_3O$	918-00-3	161.414			149; 28^{10}	1.435^{20}	1.4635^{17}	i H_2O; vs EtOH, eth
10284	1,2,3-Trichloro-1-propene		$C_3H_3Cl_3$	96-19-5	145.415			142	1.412^{20}	1.5030^{20}	i H_2O; vs EtOH, eth; s bz, chl
10285	3,3,3-Trichloro-1-propene		$C_3H_3Cl_3$	2233-00-3	145.415	liq	-30	114.5	1.367^{20}	1.4827^{20}	i H_2O; s EtOH, eth, bz, chl
10286	2,3,3-Trichloro-2-propenoyl chloride		C_3Cl_4O	815-58-7	193.843			158		1.5271^{18}	vs bz
10287	Trichloropropylsilane	Propyltrichlorosilane	$C_3H_7Cl_3Si$	141-57-1	177.533			123.5	1.195^{20}	1.4310^{20}	

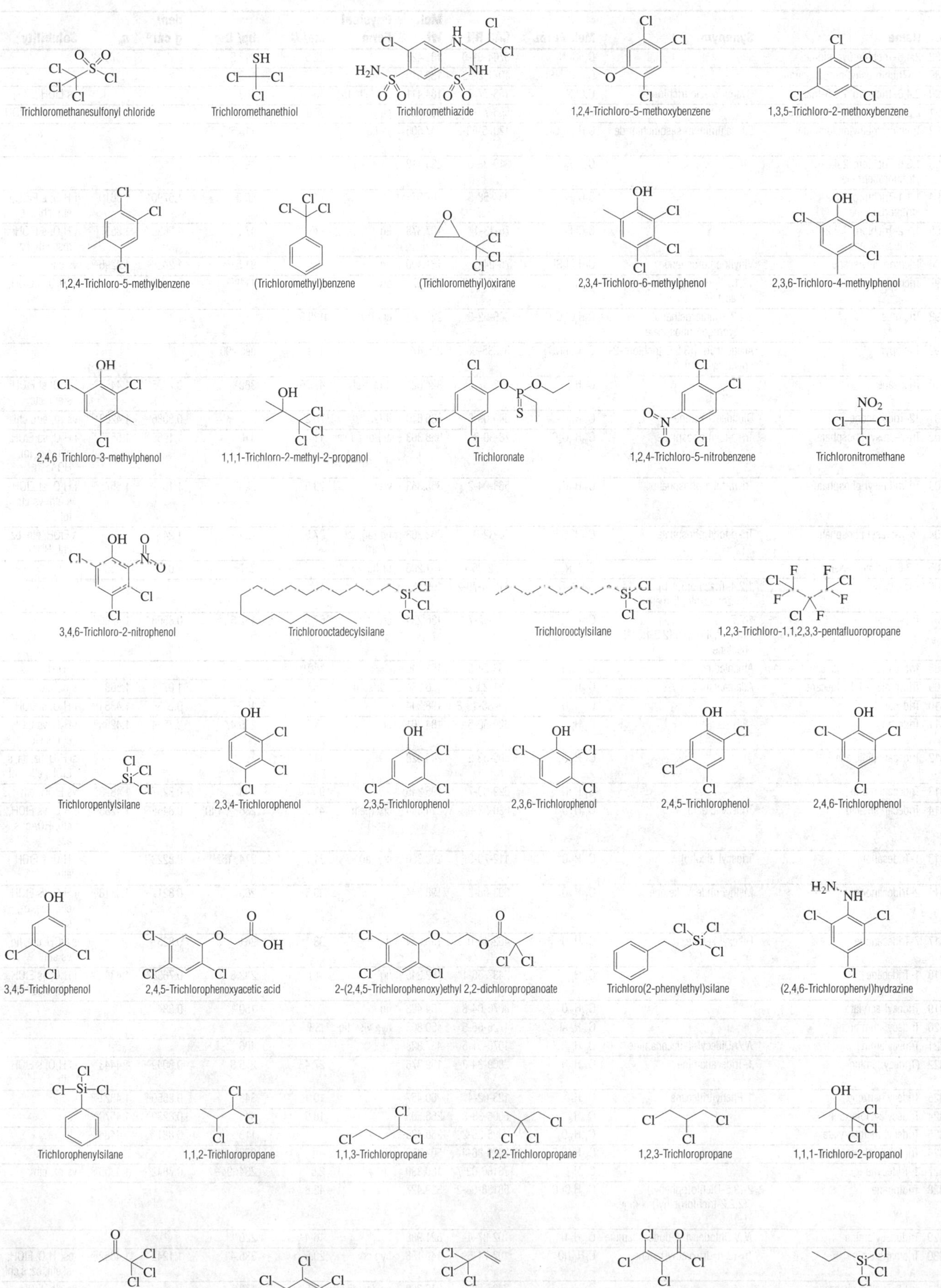
Trichloromethanesulfonyl chloride
Trichloromethanethiol
Trichloromethiazide
1,2,4-Trichloro-5-methoxybenzene
1,3,5-Trichloro-2-methoxybenzene
1,2,4-Trichloro-5-methylbenzene
(Trichloromethyl)benzene
(Trichloromethyl)oxirane
2,3,4-Trichloro-6-methylphenol
2,3,6-Trichloro-4-methylphenol
2,4,6 Trichloro-3-methylphenol
1,1,1-Trichloro-2-methyl-2-propanol
Trichloronate
1,2,4-Trichloro-5-nitrobenzene
Trichloronitromethane
3,4,6-Trichloro-2-nitrophenol
Trichlorooctadecylsilane
Trichlorooctylsilane
1,2,3-Trichloro-1,1,2,3,3-pentafluoropropane
Trichloropentylsilane
2,3,4-Trichlorophenol
2,3,5-Trichlorophenol
2,3,6-Trichlorophenol
2,4,5-Trichlorophenol
2,4,6-Trichlorophenol
3,4,5-Trichlorophenol
2,4,5-Trichlorophenoxyacetic acid
2-(2,4,5-Trichlorophenoxy)ethyl 2,2-dichloropropanoate
Trichloro(2-phenylethyl)silane
(2,4,6-Trichlorophenyl)hydrazine
Trichlorophenylsilane
1,1,2-Trichloropropane
1,1,3-Trichloropropane
1,2,2-Trichloropropane
1,2,3-Trichloropropane
1,1,1-Trichloro-2-propanol
1,1,1-Trichloro-2-propanone
1,2,3-Trichloro-1-propene
3,3,3-Trichloro-1-propene
2,3,3-Trichloro-2-propenoyl chloride
Trichloropropylsilane

No.	Name	Synonym	Mol. Form.	CAS RN	Mol. Wt.	Physical Form	mp/°C	bp/°C	den/ g cm^{-3}	n_D	Solubility
10288	2,4,6-Trichloropyrimidine		$C_4HCl_3N_2$	3764-01-0	183.423		22.5	212.5		1.5700^{20}	
10289	3-(Trichlorosilyl)propanenitrile		$C_3H_4Cl_3NSi$	1071-22-3	188.516			109^{30}			
10290	2,4,6-Trichloro-1,3,5-triazine	Cyanuric acic trichloride	$C_3Cl_3N_3$	108-77-0	184.411	cry (eth, bz)	154	192			vs EtOH
10291	2,2',2"-Trichlorotriethylamine		$C_6H_{12}Cl_3N$	555-77-1	204.525	pa ye	-2.0	143^{15}			vs bz, eth, EtOH
10292	Trichlorotriethyldialuminum	Ethylaluminum sesquichloride	$C_6H_{15}Al_2Cl_3$	12075-68-2	247.505	ye liq		115.5^{50}, $36.2^{0.2}$			
10293	1,3,5-Trichloro-2,4,6-trifluorobenzene		$C_6Cl_3F_3$	319-88-0	235.418			198.4			
10294	1,1,1-Trichloro-2,2,2-trifluoroethane		$C_2Cl_3F_3$	354-58-5	187.375		14.37	45.5	1.5790^{20}	1.3610^{35}	i H_2O; s EtOH, eth, chl
10295	1,1,2-Trichloro-1,2,2-trifluoroethane		$C_2Cl_3F_3$	76-13-1	187.375	liq	-36.22	47.7	1.5635^{25}	1.3557^{25}	i H_2O; s EtOH; msc eth, bz
10296	Trichlorovinylsilane	Vinyltrichlorosilane	$C_2H_3Cl_3Si$	75-94-5	161.490	liq	-95	91.5	1.2426^{20}	1.4295^{20}	vs chl
10297	Trichodermin	12,13-Epoxytrichothec-9-en-4-ol acetate	$C_{17}H_{24}O_4$	4682-50-2	292.371	cry	59	$111^{0.05}$			sl H_2O; s EtOH, chl
10298	Triclofos	2,2,2-Trichloroethanol dihydrogen phosphate	$C_2H_4Cl_3O_4P$	306-52-5	229.383	cry (bz)	120.5				
10299	Triclopyr	Acetic acid, [(3,5,6-trichloro-2-pyridinyl)oxy]-	$C_7H_4Cl_3NO_3$	55335-06-3	256.471		149	dec 290			
10300	Tricosane		$C_{23}H_{48}$	638-67-5	324.627	lf (eth-al)	47.76	380	0.7785^{48}	1.4468^{20}	i H_2O; sl EtOH; s eth, ctc
10301	12-Tricosanone	Diundecyl ketone	$C_{23}H_{46}O$	540-09-0	338.610	lf (al)	70.2		0.8086^{69}	1.4283^{80}	vs bz, eth, chl
10302	Tri-*o*-cresyl phosphate	Tri-*o*-tolyl phosphate	$C_{21}H_{21}O_4P$	78-30-8	368.363	col or pa ye	11	410	1.1955^{20}	1.5575^{20}	i H_2O; vs EtOH, eth, ctc, tol; s HOAc
10303	Tri-*m*-cresyl phosphate	Tri-*m*-tolyl phosphate	$C_{21}H_{21}O_4P$	563-04-2	368.363	wax	25.5	260^{15}	1.150^{25}	1.5575^{20}	i H_2O; sl EtOH; s eth; vs ctc, tol
10304	Tri-*p*-cresyl phosphate	Tri-*p*-tolyl phosphate	$C_{21}H_{21}O_4P$	78-32-0	368.363	nd (al), tab (eth)	77.5	224^{35}	1.247^{25}		s EtOH, eth, bz, chl, HOAc
10305	1,3,6-Tricyanohexane		$C_9H_{11}N_3$	1772-25-4	161.203	br liq		257^2	1.040	1.4660^{20}	
10306	Tricyclazole	1,2,4-Triazolo[3,4-b] benzothiazole, 5-methyl-	$C_9H_7N_3S$	41814-78-2	189.237		187				
10307	Tricyclene	1,7,7-Trimethyltricyclo[2.2.1.0^{2,6}] heptane	$C_{10}H_{16}$	508-32-7	136.234	cry (al)	67.5	152.5	0.8668^{80}	1.4296^{80}	
10308	Tricyclo[3.3.1.1^{3,7}]decan-1-amine	Amantadine	$C_{10}H_{17}N$	768-94-5	151.249		180				sl H_2O
10309	Tricyclo[3.3.1.1^{3,7}]decane	Adamantane	$C_{10}H_{16}$	281-23-2	136.234	nd (sub)	268	sub	1.07^{25}	1.568	s bz, ctc
10310	Tridecanal		$C_{13}H_{26}O$	10486-19-8	198.344		14	156^{13}	0.8356^{18}	1.4384^{18}	i H_2O; s EtOH
10311	Tridecane		$C_{13}H_{28}$	629-50-5	184.361	liq	-5.4	235.47	0.7564^{20}	1.4256^{20}	i H_2O; vs EtOH, eth; s ctc
10312	Tridecanedioic acid		$C_{13}H_{24}O_4$	505-52-2	244.328		114				sl H_2O, bz, tfa; s EtOH, eth, chl
10313	Tridecanenitrile		$C_{13}H_{25}N$	629-60-7	195.345		9.7	293	0.8257^{20}	1.4378^{20}	vs EtOH, eth
10314	Tridecanoic acid	Tridecylic acid	$C_{13}H_{26}O_2$	638-53-9	214.344	cry (peth ace)	41.5	236^{100}, 140^1	0.8458^{80}	1.4286^{60}	i H_2O; vs EtOH, eth, HOAc; s ace
10315	1-Tridecanol	Tridecyl alcohol	$C_{13}H_{28}O$	112-70-9	200.360	cry (al)	31.7	274; 152^{14}	0.8223^{31}		i H_2O; s EtOH, eth
10316	2-Tridecanone	Methyl undecyl ketone	$C_{13}H_{26}O$	593-08-8	198.344		30.5	263	0.8217^{30}	1.4318^{20}	i H_2O; vs EtOH, eth, ace, bz, chl
10317	7-Tridecanone	Dihexyl ketone	$C_{13}H_{26}O$	462-18-0	198.344	lf (al)	33	261	0.825^{30}		s EtOH, chl, lig; vs eth
10318	1-Tridecene		$C_{13}H_{26}$	2437-56-1	182.345	liq	-13	232.8	0.7658^{20}	1.4340^{20}	i H_2O; vs EtOH, eth; s bz
10319	Tridecyl acrylate		$C_{16}H_{30}O_2$	3076-04-8	254.408	liq		150^{10}	0.88^{20}		
10320	Tridecylaluminum		$C_{30}H_{63}Al$	1726-66-5	450.803	hyg visc liq	-38				
10321	Tridecylamine	*N,N*-Didecyl-1-decanamine	$C_{30}H_{63}N$	1070-01-5	437.828			406			
10322	(Tridecyl)amine	1-Tridecanamine	$C_{13}H_{29}N$	2869-34-3	199.376		27.4	275.8	0.8049^{20}	1.4443^{20}	sl H_2O; s EtOH, eth
10323	Tridecylbenzene	1-Phenyltridecane	$C_{19}H_{32}$	123-02-4	260.457		10	346	0.8550^{20}	1.4821^{20}	
10324	Tridecylcyclohexane		$C_{19}H_{38}$	6006-33-3	266.505		18.5	346	0.8239^{20}	1.4570^{20}	
10325	Tridecyl methacrylate		$C_{17}H_{32}O_2$	2495-25-2	268.435			118^1	0.881^{20}	1.448^{25}	
10326	Tri(decyl) phosphite		$C_{30}H_{63}O_3P$	2929-86-4	502.793	liq		255^3, $180^{0.1}$			
10327	1-Tridecyne		$C_{13}H_{24}$	26186-02-7	180.330		2.5	234; 94^{25}	0.7842^{20}	1.4309^{20}	vs bz, eth
10328	Tridiphane	2-(3,5-Dichlorophenyl)-2-(2,2,2-trichloroethyl)oxirane, (±)	$C_{10}H_7Cl_5O$	58138-08-2	320.427		42.8				
10329	Tridodecylamine	*N,N*-Didodecyl-1-dodecanamine	$C_{36}H_{75}N$	102-87-4	521.988		16.4	$220^{0.03}$			
10330	Triethanolamine	Tris(2-hydroxyethyl)amine	$C_6H_{15}NO_3$	102-71-6	149.188	hyg cry	20.5	335.4	1.1242^{20}	1.4852^{20}	msc H_2O, EtOH; sl eth, bz; s chl
10331	1,3,5-Triethoxybenzene		$C_{12}H_{18}O_3$	2437-88-9	210.269	cry (al, dil al)	43.5	170^{24}			vs eth, EtOH
10332	Triethoxy(3-chloropropyl)silane	(3-Chloropropyl)triethoxysilane	$C_9H_{21}ClO_3Si$	5089-70-3	240.800	col gas		-149			

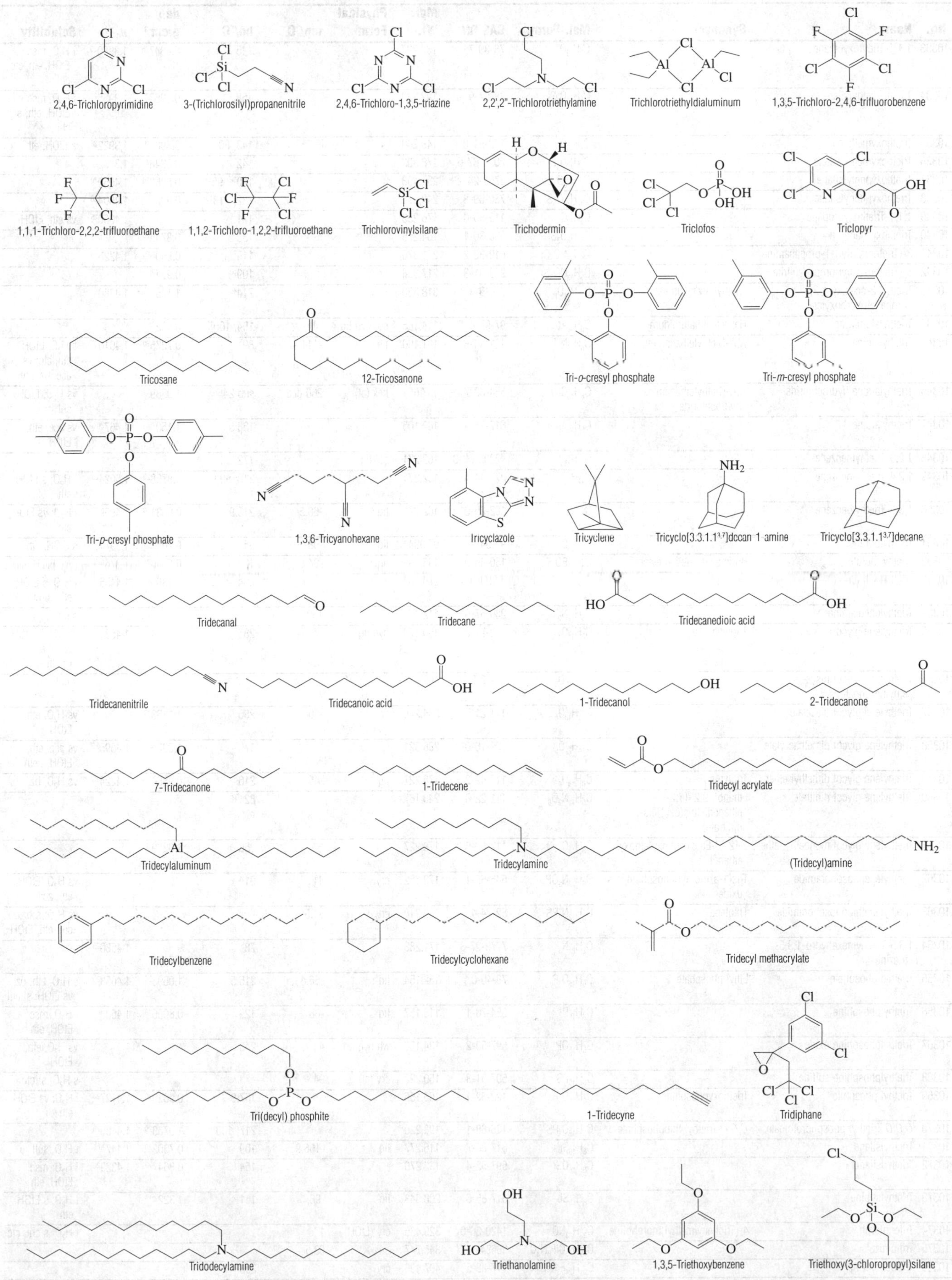
2,4,6-Trichloropyrimidine
3-(Trichlorosilyl)propanenitrile
2,4,6-Trichloro-1,3,5-triazine
2,2',2"-Trichlorotriethylamine
Trichlorotriethyldialuminum
1,3,5-Trichloro-2,4,6-trifluorobenzene
1,1,1-Trichloro-2,2,2-trifluoroethane
1,1,2-Trichloro-1,2,2-trifluoroethane
Trichlorovinylsilane
Trichodermin
Triclofos
Triclopyr
Tricosane
12-Tricosanone
Tri-o-cresyl phosphate
Tri-m-cresyl phosphate
Tri-p-cresyl phosphate
1,3,6-Tricyanohexane
Tricyclazole
Tricyclene
Tricyclo[3.3.1.1^3,7]decan-1-amine
Tricyclo[3.3.1.1^3,7]decane
Tridecanal
Tridecane
Tridecanedioic acid
Tridecanenitrile
Tridecanoic acid
1-Tridecanol
2-Tridecanone
7-Tridecanone
1-Tridecene
Tridecyl acrylate
Tridecylaluminum
Tridecylamine
(Tridecyl)amine
Tridecylbenzene
Tridecylcyclohexane
Tridecyl methacrylate
Tri(decyl) phosphite
1-Tridecyne
Tridiphane
Tridodecylamine
Triethanolamine
1,3,5-Triethoxybenzene
Triethoxy(3-chloropropyl)silane

No.	Name	Synonym	Mol. Form.	CAS RN	Mol. Wt.	Physical Form	mp/°C	bp/°C	den/ g cm^{-3}	n_D	Solubility
10333	1,1,1-Triethoxyethane		$C_8H_{18}O_3$	78-39-7	162.227			145	0.8847^{25}	1.3980^{20}	i H_2O; msc EtOH, eth, ctc, chl
10334	Triethoxyethylsilane		$C_8H_{20}O_3Si$	78-07-9	192.329			158.5	0.8963^{20}	1.3955^{20}	i H_2O; msc EtOH, eth; s chl
10335	Triethoxymethane		$C_7H_{16}O_3$	122-51-0	148.200			143; 60^{20}	0.8909^{20}	1.3922^{20}	s EtOH, eth
10336	Triethoxymethylsilane		$C_7H_{18}O_3Si$	2031-67-6	178.302			142	0.8948^{25}	1.3832^{20}	
10337	Triethoxypentylsilane		$C_{11}H_{26}O_3Si$	2761-24-2	234.408			100^{30}, 95^{13}	0.8862^{20}	1.4059^{20}	
10338	Triethoxyphenylsilane		$C_{12}H_{20}O_3Si$	780-69-8	240.371			232; 113^{10}	0.996^{25}	1.4604^{20}	
10339	1,1,1-Triethoxypropane		$C_9H_{20}O_3$	115-80-0	176.253			171		1.4000^{25}	vs eth, EtOH
10340	Triethoxysilane		$C_6H_{16}O_3Si$	998-30-1	164.275			133.5	0.8745^{20}		
10341	3-(Triethoxysilyl)-1-propanamine		$C_9H_{23}NO_3Si$	919-30-2	221.370			119^{29}	0.9506^{20}	1.4225^{20}	
10342	3-(Triethoxysilyl)propanenitrile		$C_9H_{19}NO_3Si$	919-31-3	217.338	liq		109^{10}	0.974^{20}		
10343	Triethyl 2-acetoxy-1,2,3-propanetricarboxylate	Triethyl acetylcitrate	$C_{14}H_{22}O_8$	77-89-4	318.320			214^{40}	1.135^{25}	1.4380	
10344	Triethylaluminum	Hexaethyldialuminum	$C_6H_{15}Al$	97-93-8	114.165	col hyg liq	-46	194; 100^{13}	0.832^{25}		
10345	Triethylamine	*N,N*-Diethylethanamine	$C_6H_{15}N$	121-44-8	101.190	liq	-114.7	89	0.7275^{20}	1.4010^{20}	s H_2O, EtOH, eth, ctc; vs ace, bz, chl
10346	Triethylamine hydrochloride	*N,N*-Diethylethanamine hydrochloride	$C_6H_{16}ClN$	554-68-7	137.651	hex (al)	260 dec	sub 245	1.0689^{21}		vs H_2O, EtOH, chl
10347	Triethylarsine		$C_6H_{15}As$	617-75-4	162.105			138.5	1.150^{20}	1.467^{20}	vs ace, eth, EtOH
10348	1,2,3-Triethylbenzene		$C_{12}H_{18}$	42205-08-3	162.271	col liq	-26	172			
10349	1,2,4-Triethylbenzene		$C_{12}H_{18}$	877-44-1	162.271			218; 99^{15}	0.8738^{20}	1.5024^{20}	i H_2O; s EtOH, eth
10350	1,3,5-Triethylbenzene		$C_{12}H_{18}$	102-25-0	162.271	liq	-66.5	215.9	0.8631^{20}	1.4969^{20}	i H_2O; vs EtOH, eth
10351	Triethylborane		$C_6H_{15}B$	97-94-9	97.994	liq	-93	95	0.70^{23}	1.3971	s EtOH, eth
10352	Triethyl borate	Boric acid, triethyl ester	$C_6H_{15}BO_3$	150-46-9	145.992	liq	-84.8	120	0.8546^{20}	1.3749^{20}	msc EtOH, eth
10353	Triethyl citrate		$C_{12}H_{20}O_7$	77-93-0	276.283			294	1.1369^{20}	1.4455^{20}	i H_2O; s EtOH, eth; sl ctc
10354	Triethylenediamine		$C_6H_{12}N_2$	280-57-9	112.172		159				s chl
10355	Triethylene glycol	Triglycol	$C_6H_{14}O_4$	112-27-6	150.173	hyg liq	-7	285	1.1274^{15}	1.4531^{20}	msc H_2O, EtOH, bz; sl eth, chl; i peth
10356	Triethylene glycol bis(2-ethylhexanoate)		$C_{22}H_{42}O_6$	94-28-0	402.564						s chl
10357	Triethylene glycol diacetate		$C_{10}H_{18}O_6$	111-21-7	234.246	liq	-50	286	1.1153^{20}		vs H_2O, eth, EtOH
10358	Triethylene glycol dimethacrylate		$C_{14}H_{22}O_6$	109-16-0	286.321			170^{5}	1.092^{20}	1.4595^{25}	vs ace, eth, EtOH, peth
10359	Triethylene glycol dimethyl ether	Triglyme	$C_8H_{18}O_4$	112-49-2	178.227	liq	-45	216	0.986^{20}	1.4224^{20}	vs H_2O, bz
10360	Triethylene glycol dinitrate	Ethanol, 2,2'-[1,2-ethanediylbis(oxy)]bis-, dinitrate	$C_6H_{12}N_2O_8$	111-22-8	240.167			82$^{0.03}$			
10361	Triethylene glycol monoethyl ether	2-[2-(2-Ethoxyethoxy)ethoxy] ethanol	$C_8H_{18}O_4$	112-50-5	178.227			256	1.0209^{20}		
10362	Triethylenephosphoramide	Tris(1-aziridinyl)phosphine, oxide	$C_6H_{12}N_3OP$	545-55-1	173.152	cry	41	91^{23}			vs H_2O, EtOH, eth, ace
10363	Triethylenethiophosphoramide	Thiotepa	$C_6H_{12}N_3PS$	52-24-4	189.218	cry	51.5				vs H_2O; s bz, chl, eth, EtOH
10364	1,3,5-Triethylhexahydro-1,3,5-triazine		$C_9H_{21}N_3$	7779-27-3	171.283			78^{6}		1.4580^{25}	
10365	Triethyl phosphate	Ethyl phosphate	$C_6H_{15}O_4P$	78-40-0	182.154	liq	-56.4	215.5	1.0695^{20}	1.4053^{20}	s H_2O, eth, bz; vs EtOH; sl chl
10366	Triethylphosphine		$C_6H_{15}P$	554-70-1	118.157	liq	-88	129	0.8006^{19}	1.458^{15}	i H_2O; msc EtOH, eth
10367	Triethylphosphine oxide		$C_6H_{15}OP$	597-50-2	134.156	wh hyg nd	48	243			vs H_2O, eth, EtOH
10368	Triethylphosphine sulfide		$C_6H_{15}PS$	597-51-3	150.222	cry (al)	94				s H_2O; sl ctc
10369	Triethyl phosphite	Triethoxyphosphine	$C_6H_{15}O_3P$	122-52-1	166.155			157.9	0.9629^{20}	1.4127^{20}	i H_2O; vs EtOH, eth
10370	*O,O,O*-Triethyl phosphorothioate	*O,O,O*-Triethyl thiophosphate	$C_6H_{15}O_3PS$	126-68-1	198.220			217; 100^{16}	1.0768^{20}	1.4480^{20}	
10371	Triethylsilane		$C_6H_{16}Si$	617-86-7	116.277	liq	-156.9	109	0.7302^{20}	1.447^{20}	i H_2O, sulf
10372	Triethylsilanol		$C_6H_{16}OSi$	597-52-4	132.276			154	0.8647^{20}	1.4329^{20}	i H_2O; msc EtOH, eth
10373	Triethylstibine		$C_6H_{15}Sb$	617-85-6	208.943	liq	-98	161.4	1.3224^{15}		i H_2O; s EtOH, eth
10374	Trifenmorph	4-(Triphenylmethyl)morpholine	$C_{23}H_{23}NO$	1420-06-0	329.435	cry (EtOH)	176				i H_2O; s chl, ctc
10375	Triflumizole		$C_{15}H_{15}ClF_3N_3O$	68694-11-1	345.747		63.5				
10376	Trifluoperazine		$C_{21}H_{24}F_3N_3S$	117-89-5	407.496	cry		206$^{0.7}$			
10377	Trifluoperazine dihydrochloride	Stelazine	$C_{21}H_{26}Cl_2F_3N_3S$	440-17-5	480.417		241.5				

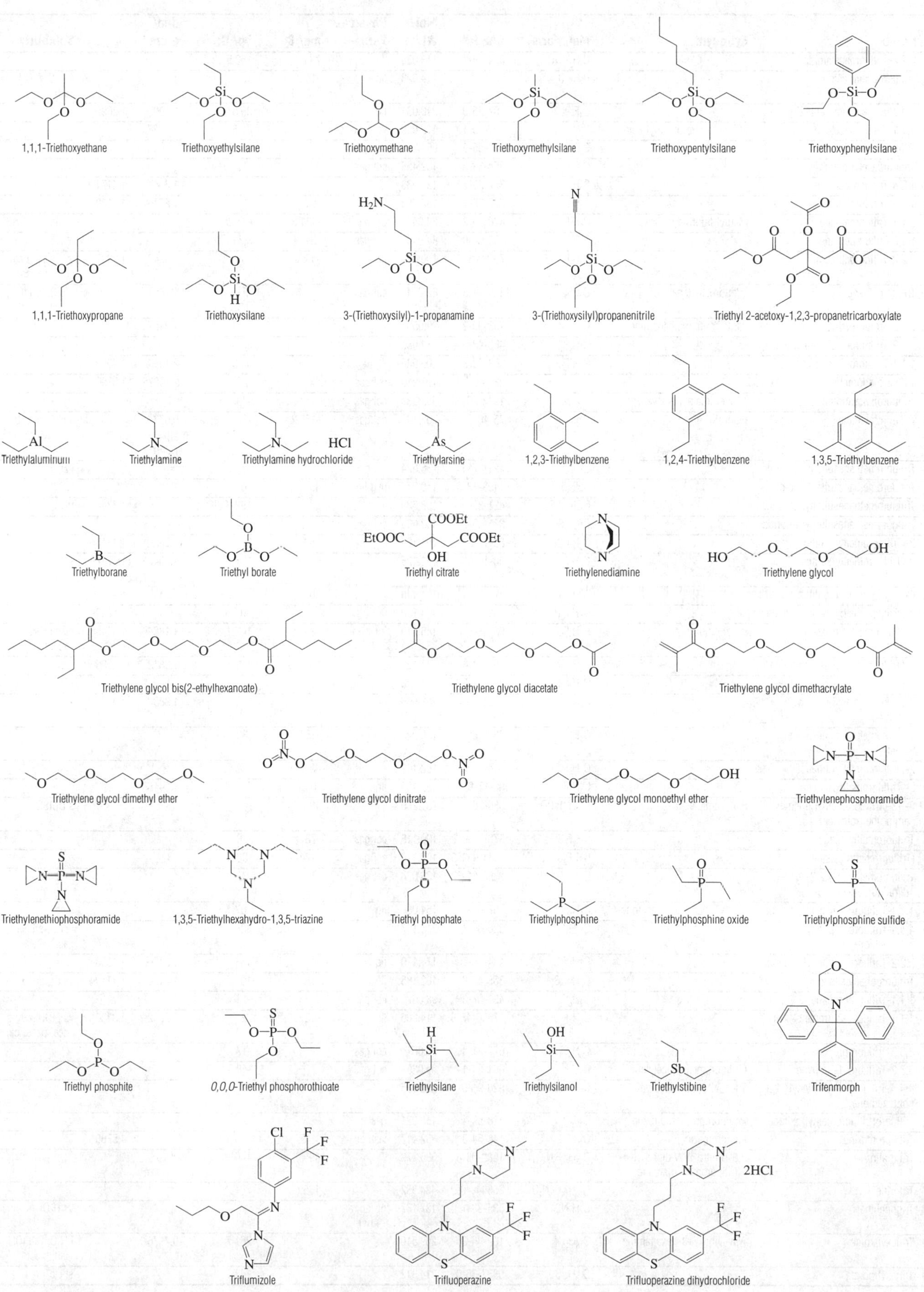

1,1,1-Triethoxyethane
Triethoxyethylsilane
Triethoxymethane
Triethoxymethylsilane
Triethoxypentylsilane
Triethoxyphenylsilane
1,1,1-Triethoxypropane
Triethoxysilane
3-(Triethoxysilyl)-1-propanamine
3-(Triethoxysilyl)propanenitrile
Triethyl 2-acetoxy-1,2,3-propanetricarboxylate
Triethylaluminum
Triethylamine
Triethylamine hydrochloride
Triethylarsine
1,2,3-Triethylbenzene
1,2,4-Triethylbenzene
1,3,5-Triethylbenzene
Triethylborane
Triethyl borate
Triethyl citrate
Triethylenediamine
Triethylene glycol
Triethylene glycol bis(2-ethylhexanoate)
Triethylene glycol diacetate
Triethylene glycol dimethacrylate
Triethylene glycol dimethyl ether
Triethylene glycol dinitrate
Triethylene glycol monoethyl ether
Triethylenephosphoramide
Triethylenethiophosphoramide
1,3,5-Triethylhexahydro-1,3,5-triazine
Triethyl phosphate
Triethylphosphine
Triethylphosphine oxide
Triethylphosphine sulfide
Triethyl phosphite
O,O,O-Triethyl phosphorothioate
Triethylsilane
Triethylsilanol
Triethylstibine
Trifenmorph
Triflumizole
Trifluoperazine
Trifluoperazine dihydrochloride

No.	Name	Synonym	Mol. Form.	CAS RN	Mol. Wt.	Physical Form	mp/°C	bp/°C	den/ g cm^{-3}	n_D	Solubility
10378	2,2,2-Trifluoroacetamide		$C_2H_2F_3NO$	354-38-1	113.038		73.8	162.5			
10379	Trifluoroacetic acid		$C_2HF_3O_2$	76-05-1	114.023	liq	-15.2	73	1.5351^{25}		s H_2O, EtOH, eth, ace
10380	Trifluoroacetic acid anhydride		$C_4F_6O_3$	407-25-0	210.031	liq	-65	39.5	1.490^{25}	1.269^{25}	
10381	1,1,1-Trifluoroacetone	Methyl trifluoromethyl ketone	$C_3H_3F_3O$	421-50-1	112.050	vol liq or gas		21.5	1.252^{25}		
10382	Trifluoroacetonitrile		C_2F_3N	353-85-5	95.023	col gas		-68.8			
10383	Trifluoroacetyl chloride		C_2ClF_3O	354-32-5	132.468	col gas	-146	-18			
10384	1,2,4-Trifluorobenzene		$C_6H_3F_3$	367-23-7	132.083			90	1.264^{25}	1.4171^{20}	
10385	1,3,5-Trifluorobenzene		$C_6H_3F_3$	372-38-3	132.083	liq	-5.5	75.5	1.277^{25}	1.4140^{20}	
10386	1,1,1-Trifluoroethane	Methyl fluoroform	$C_2H_3F_3$	420-46-2	84.040	col gas	-111.3	-47.25			s eth, chl
10387	1,1,2-Trifluoroethane		$C_2H_3F_3$	430-66-0	84.040	col gas	-84	3.7			
10388	2,2,2-Trifluoroethanol		$C_2H_3F_3O$	75-89-8	100.039	liq	-43.5	74	1.3842^{20}	1.2907^{22}	vs EtOH; s eth, ace, bz, chl
10389	Trifluoroethene	Trifluoroethylene	C_2HF_3	359-11-5	82.024	col gas		-51	1.26^{-70}		i H_2O; sl EtOH; s eth
10390	2,2,2-Trifluoroethylamine	2,2,2-Trifluoroethanamine	$C_2H_4F_3N$	753-90-2	99.055			36	1.245^{25}		
10391	2,2,2-Trifluoroethyl methyl ether		$C_3H_5F_3O$	460-43-5	114.066			31.62			
10392	1,1,1-Trifluoro-2-iodoethane		$C_2H_2F_3I$	353-83-3	209.936			54.5	2.13^{25}	1.4009^{20}	
10393	Trifluoroiodomethane		CF_3I	2314-97-8	195.910	col gas		-22.5	2.3607^{-32}	1.3790^{-32}	
10394	Trifluoroisocyanomethane	Trifluoromethyl isocyanide	C_2F_3N	19480-01-4	95.023	col gas		-80			
10395	Trifluoromethane	Fluoroform	CHF_3	75-46-7	70.014	col gas	-155.2	-82.1	0.673^{25} (p>1 atm)		s H_2O, ace, bz; vs EtOH; sl chl
10396	Trifluoromethanesulfenyl chloride		$CClF_3S$	421-17-0	136.524	col gas		-0.7			i H_2O
10397	Trifluoromethanesulfonic acid		CHF_3O_3S	1493-13-6	150.077	hyg liq	45	162			vs eth
10398	Trifluoromethanesulfonyl chloride		$CClF_3O_2S$	421-83-0	168.523			162; 62^{18}		1.3344^{20}	i H_2O
10399	Trifluoromethanesulfonyl fluoride		CF_4O_2S	335-05-7	152.069	col gas		-21.7			
10400	2-(Trifluoromethyl)aniline		$C_7H_6F_3N$	88-17-5	161.125		35.5	68^{15}	1.282^{25}	1.4810^{20}	
10401	3-(Trifluoromethyl)aniline		$C_7H_6F_3N$	98-16-8	161.125		5.5	187; 74^{10}	1.3047^{12}	1.4787^{20}	sl H_2O; s EtOH, eth
10402	4-(Trifluoromethyl)aniline		$C_7H_6F_3N$	455-14-1	161.125		38	117.5^{60}	1.283^{27}	1.4815^{25}	
10403	4-(Trifluoromethyl)benzaldehyde		$C_8H_5F_3O$	455-19-6	174.120			80^{25}		1.4630^{20}	
10404	(Trifluoromethyl)benzene	Benzotrifluoride	$C_7H_5F_3$	98-08-8	146.110	liq	-28.95	102.1	1.1884^{20}	1.4146^{20}	msc EtOH, eth, ace, bz, ctc
10405	3-(Trifluoromethyl)benzonitrile		$C_8H_4F_3N$	368-77-4	171.120		14.5	189	1.2813^{20}	1.4508^{20}	
10406	4-(Trifluoromethyl)benzonitrile		$C_8H_4F_3N$	455-18-5	171.120		37.5				
10407	3-(Trifluoromethyl)benzoyl chloride		$C_8H_4ClF_3O$	2251-65-2	208.565	oil		186; 80^{16}	1.383	1.4770^{20}	
10408	Trifluoromethyl difluoromethyl ether		C_2HF_5O	3822-68-2	136.020	col gas	-157	-38			
10409	2-(Trifluoromethyl)phenol		$C_7H_5F_3O$	444-30-4	162.109		45.5	147.5			
10410	3-(Trifluoromethyl)phenol		$C_7H_5F_3O$	98-17-9	162.109	liq	-0.9	178	1.3418^{25}		
10411	2-[[3-(Trifluoromethyl)phenyl]amino]benzoic acid	Flufenamic acid	$C_{14}H_{10}F_3NO_2$	530-78-9	281.230		133.5				s DMSO
10412	Trifluoromethylsilane		CH_3F_3Si	373-74-0	100.116	col gas	-73	-30			
10413	(Trifluoromethyl)silane		CH_3F_3Si	10112-11-5	100.116	col gas	-124	-38.3			
10414	Trifluoromethyl 1,1,2,2-tetrafluoroethyl ether		C_3HF_7O	2356-61-8	186.028	col gas	-141	-3			
10415	1,1,1-Trifluoro-2,4-pentanedione	1,1,1-Trifluoroacetylacetone	$C_5H_5F_3O_2$	367-57-7	154.088	liq		107			s os
10416	4,4,4-Trifluoro-1-phenyl-1,3-butanedione		$C_{10}H_7F_3O_2$	326-06-7	216.157	cry	39	224			i H_2O; s EtOH, ace
10417	2,2,2-Trifluoro-1-phenylethanone		$C_8H_5F_3O$	434-45-7	174.120	liq	-40	153	1.279^{20}	1.4583^{20}	
10418	Trifluorophenylsilane		$C_6H_5F_3Si$	368-47-8	162.185	liq	-18	101.5	1.2169^{20}	1.4110^{20}	vs bz, EtOH
10419	1,1,1-Trifluoropropane		$C_3H_5F_3$	421-07-8	98.067	col gas		-13			
10420	1,1,1-Trifluoro-2-propanol, (±)		$C_3H_5F_3O$	17556-48-8	114.066	liq	-52	78	1.2632^{25}	1.3130^{25}	vs EtOH, eth; s ace, bz; sl ctc
10421	3,3,3-Trifluoropropene		$C_3H_3F_3$	677-21-4	96.051	col gas		-17			
10422	3,3,3-Trifluoro-1-propyne	(Trifluoromethyl)acetylene	C_3HF_3	661-54-1	94.035	col gas		-48.3			
10423	4,4,4-Trifluoro-1-(2-thienyl)-1,3-butanedione	Thenoyltrifluoroacetone	$C_8H_5F_3O_2S$	326-91-0	222.185		42.8	97^{8}			
10424	Trifluoro(trifluoromethyl)oxirane	Perfluoropropylene oxide	C_3F_6O	428-59-1	166.021	gas		-27.4			
10425	Triflupromazine	Fluopromazine	$C_{18}H_{19}F_3N_2S$	146-54-3	352.417	visc oil		$176^{0.7}$		1.5780^{23}	
10426	Trifluralin	2,6-Dinitro-*N*,*N*-dipropyl-4-(trifluoromethyl)aniline	$C_{13}H_{16}F_3N_3O_4$	1582-09-8	335.279		49	$140^{4.2}$			
10427	Triforine		$C_{10}H_{14}Cl_6N_4O_2$	26644-46-2	434.962		155 dec				
10428	Trigonelline		$C_7H_7NO_2$	535-83-1	137.137	pr (aq, al, +1w)					vs H_2O
10429	Trihexylamine	*N*,*N*-Dihexyl-1-hexanamine	$C_{18}H_{39}N$	102-86-3	269.510			261.7	0.7976^{21}		i H_2O; vs EtOH, eth; s acid
10430	Trihexyl borate		$C_{18}H_{39}BO_3$	5337-36-0	314.312			143^{2}			sl ctc

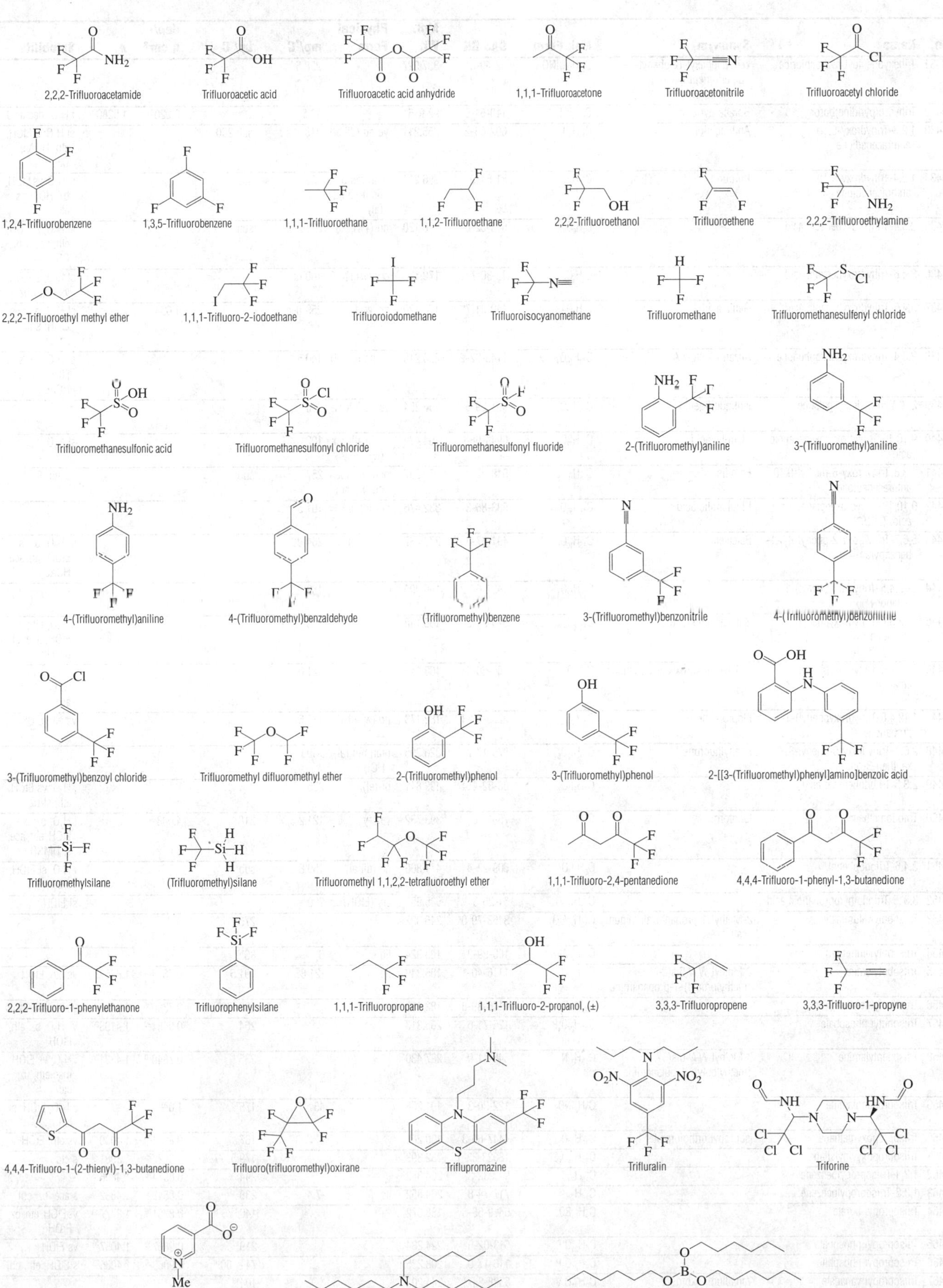
2,2,2-Trifluoroacetamide
Trifluoroacetic acid
Trifluoroacetic acid anhydride
1,1,1-Trifluoroacetone
Trifluoroacetonitrile
Trifluoroacetyl chloride
1,2,4-Trifluorobenzene
1,3,5-Trifluorobenzene
1,1,1-Trifluoroethane
1,1,2-Trifluoroethane
2,2,2-Trifluoroethanol
Trifluoroethene
2,2,2-Trifluoroethylamine
2,2,2-Trifluoroethyl methyl ether
1,1,1-Trifluoro-2-iodoethane
Trifluoroiodomethane
Trifluoroisocyanomethane
Trifluoromethane
Trifluoromethanesulfenyl chloride
Trifluoromethanesulfonic acid
Trifluoromethanesulfonyl chloride
Trifluoromethanesulfonyl fluoride
2-(Trifluoromethyl)aniline
3-(Trifluoromethyl)aniline
4-(Trifluoromethyl)aniline
4-(Trifluoromethyl)benzaldehyde
(Trifluoromethyl)benzene
3-(Trifluoromethyl)benzonitrile
4-(Trifluoromethyl)benzonitrile
3-(Trifluoromethyl)benzoyl chloride
Trifluoromethyl difluoromethyl ether
2-(Trifluoromethyl)phenol
3-(Trifluoromethyl)phenol
2-[[3-(Trifluoromethyl)phenyl]amino]benzoic acid
Trifluoromethylsilane
(Trifluoromethyl)silane
Trifluoromethyl 1,1,2,2-tetrafluoroethyl ether
1,1,1-Trifluoro-2,4-pentanedione
4,4,4-Trifluoro-1-phenyl-1,3-butanedione
2,2,2-Trifluoro-1-phenylethanone
Trifluorophenylsilane
1,1,1-Trifluoropropane
1,1,1-Trifluoro-2-propanol, (±)
3,3,3-Trifluoropropene
3,3,3-Trifluoro-1-propyne
4,4,4-Trifluoro-1-(2-thienyl)-1,3-butanedione
Trifluoro(trifluoromethyl)oxirane
Triflupromazine
Trifluralin
Triforine
Trigonelline
Trihexylamine
Trihexyl borate

No.	Name	Synonym	Mol. Form.	CAS RN	Mol. Wt.	Physical Form	mp/°C	bp/°C	den/ g cm^{-3}	n_D	Solubility
10431	Trihexyphenidyl hydrochloride	α-Cyclohexyl-α-phenyl-1-piperidinepropanol hydrochloride	$C_{20}H_{32}ClNO$	52-49-3	337.927		258.5				
10432	Trihydro(pyridine)boron	Borane pyridine	C_5H_8BN	110-51-0	92.936		10.5		0.920^{20}	1.5280^{25}	i H_2O; dec acid
10433	1,2,3-Trihydroxy-9,10-anthracenedione	Anthragallol	$C_{14}H_8O_5$	602-64-2	256.211	ye nd (dil al)	313	sub 290			sl H_2O; s EtOH, eth, HOAc, CS_2
10434	1,2,4-Trihydroxy-9,10-anthracenedione	Purpurin	$C_{14}H_8O_5$	81-54-9	256.211	oran red or oran-ye nd (al)	259	sub			sl H_2O; vs EtOH, bz, HOAc; s eth
10435	2,3,4-Trihydroxybenzoic acid		$C_7H_6O_5$	610-02-6	170.120	nd (+w)	221	sub			sl H_2O; s EtOH, eth, ace; i bz, CS_2
10436	2,4,6-Trihydroxybenzoic acid		$C_7H_6O_5$	83-30-7	170.120	cry (w+1)	100 dec				sl H_2O; s EtOH; vs eth; i bz
10437	3,4,5-Trihydroxybenzoic acid	Gallic acid	$C_7H_6O_5$	149-91-7	170.120	pr (w+1)	253 dec		1.694^{6}		sl H_2O, eth; vs EtOH; s ace; i bz, chl
10438	2,3,4-Trihydroxybenzophenone	Alizarin Yellow A	$C_{13}H_{10}O_4$	1143-72-2	230.216	ye nd (dil al)	140.5				sl H_2O, bz; s EtOH, eth, ace, HOAc
10439	2',4,4'-Trihydroxychalcone	Isoliquiritigenin	$C_{15}H_{12}O_4$	961-29-5	256.254	ye nd (EtOH-w)	200				
10440	9,10,16-Trihydroxyhexadecanoic acid	Aleuritic acid	$C_{16}H_{32}O_5$	6949-98-0	304.422	lf (dil al), nd (w)	102				sl H_2O
10441	1,3,8-Trihydroxy-6-methyl-9,10-anthracenedione	Emodin	$C_{15}H_{10}O_5$	518-82-1	270.237	oran-red mcl nd (HOAc)	257	sub			vs eth, EtOH
10442	9,10,18-Trihydroxyoctadecanoic acid, (*R**,*R**)	Phloionolic acid	$C_{18}H_{36}O_5$	583-86-8	332.476	cry (dil al)	101.5				
10443	5,6,7-Trihydroxy-2-phenyl-4*H*-1-benzopyran-4-one	Baicalein	$C_{15}H_{10}O_5$	491-67-8	270.237	ye pr (al)	264 dec				sl H_2O, bz; s EtOH, eth, ace, HOAc
10444	1-(2,4,5-Trihydroxyphenyl)-1-butanone		$C_{10}H_{12}O_4$	1421-63-2	196.200		153.8				
10445	1-(2,3,4-Trihydroxyphenyl) ethanone	Gallacetophenone	$C_8H_8O_4$	528-21-2	168.148		173				s H_2O, eth; vs EtOH, ace; sl bz, chl
10446	1-(2,4,6-Trihydroxyphenyl) ethanone	2',4',6'-Trihydroxyacetophenone	$C_8H_8O_4$	480-66-0	168.148		221.0				sl H_2O, chl, bz; vs EtOH, eth, ace
10447	1-(2,4,6-Trihydroxyphenyl)-1-propanone	Flopropione	$C_9H_{10}O_4$	2295-58-1	182.173	nd (w, +1w)	175.5				vs eth, EtOH
10448	2,6,7-Trihydroxy-9-phenyl-3*H*-xanthen-3-one	Phenylfluorone	$C_{19}H_{12}O_5$	975-17-7	320.295	oran red (al-HCl)	>300				
10449	2,3,5-Triiodobenzoic acid		$C_7H_3I_3O_2$	88-82-4	499.811	pr (al)	225				i H_2O; vs EtOH, eth; sl bz
10450	Triiodomethane	Iodoform	CHI_3	75-47-8	393.732	ye cry	121.2	218	4.008^{25}		i H_2O, bz; s EtOH, eth, ace; sl DMSO
10451	2,4,6-Triiodophenol		$C_6H_3I_3O$	609-23-4	471.800	nd (dil al)	159.8	sub			i H_2O; sl EtOH; s eth, ace
10452	3,3',5-Triiodothyropropanoic acid		$C_{15}H_{11}I_3O_4$	51-26-3	635.959	cry (EtOH)	200				sl EtOH
10453	Triisobutyl aluminate	2-Methyl-1-propanol, aluminum salt	$C_{12}H_{27}AlO_3$	3453-79-0	246.322			275^{50}			
10454	Triisobutylaluminum		$C_{12}H_{27}Al$	100-99-2	198.324	liq	6	86^{10}			
10455	Triisobutylamine	2-Methyl-*N*,*N*-bis(2-methylpropyl)-1-propanamine	$C_{12}H_{27}N$	1116-40-1	185.349	liq	-21.8	191.5	0.7684^{20}	1.4252^{17}	vs eth, EtOH
10456	Triisobutylborane		$C_{12}H_{27}B$	1116-39-8	182.153			188; 86^{20}	0.7380^{25}	1.4188^{23}	vs bz, eth, EtOH
10457	Triisobutyl phosphate		$C_{12}H_{27}O_4P$	126-71-6	266.313			264	0.9681^{20}	1.4193^{20}	vs H_2O, bz, eth, EtOH
10458	Triisopentylamine	3-Methyl-*N*,*N*-bis(3-methylbutyl)-1-butanamine	$C_{15}H_{33}N$	645-41-0	227.430			235	0.7848^{20}	1.4331^{20}	i H_2O; vs EtOH; msc eth, bz, ctc
10459	Triisopropanolamine		$C_9H_{21}NO_3$	122-20-3	191.268		45	175^{10}	1.0^{20}		s H_2O, EtOH; sl chl
10460	Triisopropoxymethane	Isopropyl orthoformate	$C_{10}H_{22}O_3$	4447-60-3	190.280			167	0.8621^{20}	1.4000^{20}	vs eth, EtOH
10461	Triisopropoxyvinylsilane		$C_{11}H_{24}O_3Si$	18023-33-1	232.393			179.5; 77^{20}	0.8627^{25}	1.3981^{20}	s ctc
10462	1,2,4-Triisopropylbenzene		$C_{15}H_{24}$	948-32-3	204.352			244	0.8574^{25}	1.4896^{25}	
10463	1,3,5-Triisopropylbenzene		$C_{15}H_{24}$	717-74-8	204.352	liq	-7.4	238	0.8545^{20}	1.4882^{20}	s ace, bz, chl
10464	Triisopropyl borate		$C_9H_{21}BO_3$	5419-55-6	188.072			140; 75^{76}	0.8251^{20}	1.3777^{20}	vs EtOH, eth, bz, PrOH
10465	Triisopropyl phosphate		$C_9H_{21}O_4P$	513-02-0	224.234			219	0.9867^{20}	1.4057^{20}	vs EtOH
10466	Triisopropyl phosphite		$C_9H_{21}O_3P$	116-17-6	208.235			74^{20}, 60^{10}	0.9063^{20}	1.4085^{25}	s EtOH, eth, chl
10467	Triisopropyl vanadate	Vanadium, oxotris(2-propanolato)-, (T-4)-	$C_9H_{21}O_4V$	5588-84-1	244.203			104^{10}			

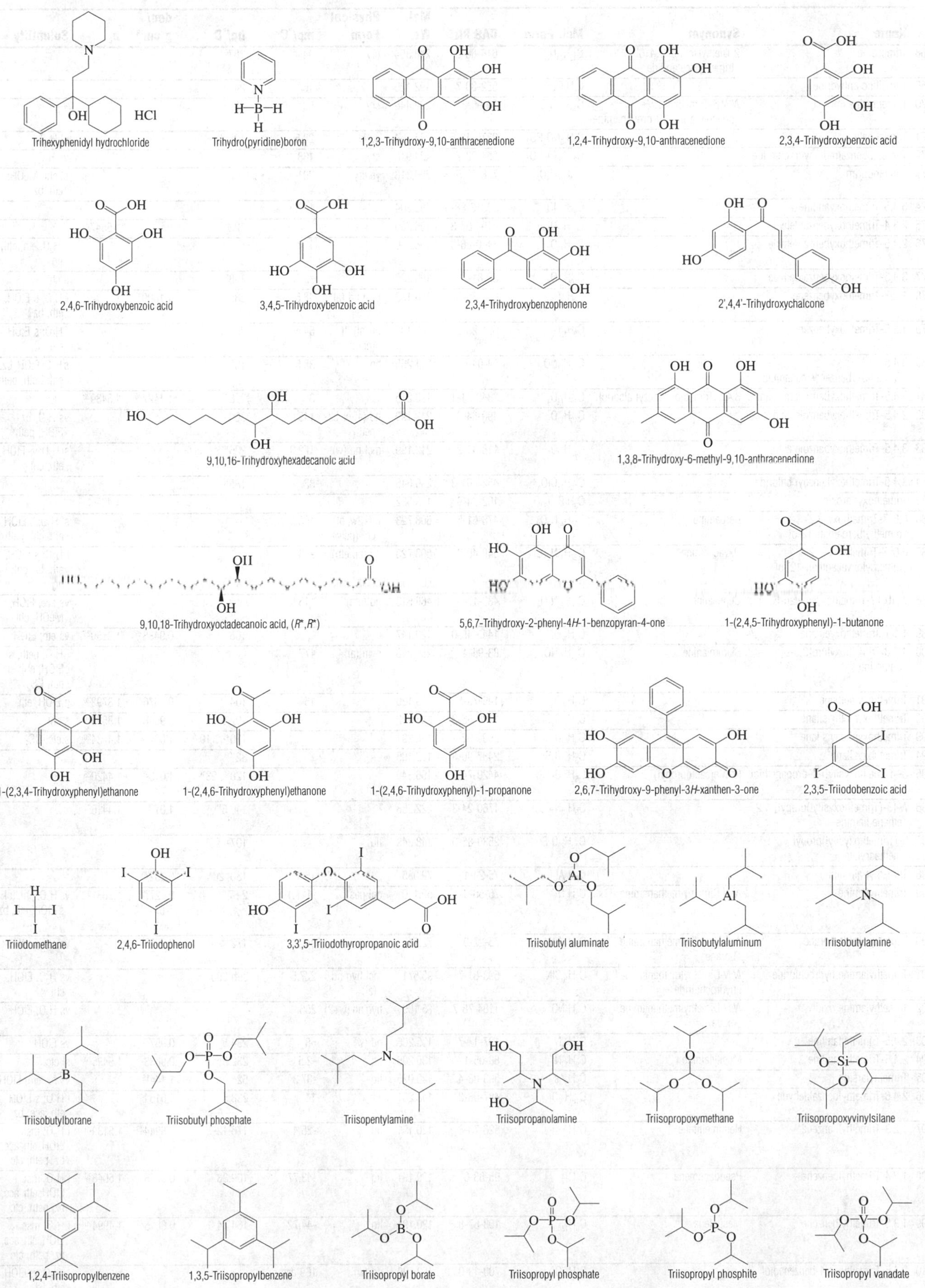

Trihexyphenidyl hydrochloride
Trihydro(pyridine)boron
1,2,3-Trihydroxy-9,10-anthracenedione
1,2,4-Trihydroxy-9,10-anthracenedione
2,3,4-Trihydroxybenzoic acid
2,4,6-Trihydroxybenzoic acid
3,4,5-Trihydroxybenzoic acid
2,3,4-Trihydroxybenzophenone
2',4,4'-Trihydroxychalcone
9,10,16-Trihydroxyhexadecanoic acid
1,3,8-Trihydroxy-6-methyl-9,10-anthracenedione
9,10,18-Trihydroxyoctadecanoic acid, (R^*,R^*)
5,6,7-Trihydroxy-2-phenyl-4*H*-1-benzopyran-4-one
1-(2,4,5-Trihydroxyphenyl)-1-butanone
1-(2,3,4-Trihydroxyphenyl)ethanone
1-(2,4,6-Trihydroxyphenyl)ethanone
1-(2,4,6-Trihydroxyphenyl)-1-propanone
2,6,7-Trihydroxy-9-phenyl-3*H*-xanthen-3-one
2,3,5-Triiodobenzoic acid
Triiodomethane
2,4,6-Triiodophenol
3,3',5-Triiodothyropropanoic acid
Triisobutyl aluminate
Triisobutylaluminum
Triisobutylamine
Triisobutylborane
Triisobutyl phosphate
Triisopentylamine
Triisopropanolamine
Triisopropoxymethane
Triisopropoxyvinylsilane
1,2,4-Triisopropylbenzene
1,3,5-Triisopropylbenzene
Triisopropyl borate
Triisopropyl phosphate
Triisopropyl phosphite
Triisopropyl vanadate

No.	Name	Synonym	Mol. Form.	CAS RN	Mol. Wt.	Physical Form	mp/°C	bp/°C	den/ g cm-3	n_D	Solubility
10468	Trimecaine	2-Diethylamino-2',4',6'-trimethylacetanilide	$C_{15}H_{24}N_2O$	616-68-2	248.364	cry	44	187^{6}			
10469	Trimellitic anhydride		$C_9H_4O_5$	552-30-7	192.125		162	241^{14}			
10470	Trimeprazine	*N,N*,β-Trimethyl-10*H*-phenothiazine-10-propanamine	$C_{18}H_{22}N_2S$	84-96-8	298.446	cry	68	$162^{0.3}$			
10471	Trimethoate		$C_9H_{20}NO_3PS_2$	2275-18-5	285.364	solid	28.5	$135^{0.1}$			sl H_2O
10472	Trimethobenzamide hydrochloride		$C_{21}H_{29}ClN_2O_5$	554-92-7	424.918	cry	188				vs H_2O
10473	Trimethoprim		$C_{14}H_{18}N_4O_3$	738-70-5	290.318	ye cry	201				sl chl, MeOH; i eth, bz
10474	3,4,5-Trimethoxyaniline		$C_9H_{13}NO_3$	24313-88-0	183.204		112.8				
10475	2,3,4-Trimethoxybenzaldehyde		$C_{10}H_{12}O_4$	2103-57-3	196.200			$122^{0.5}$		1.5547^{20}	
10476	2,4,5-Trimethoxybenzaldehyde		$C_{10}H_{12}O_4$	4460-86-0	196.200		114				s H_2O, eth, chl, lig
10477	3,4,5-Trimethoxybenzaldehyde		$C_{10}H_{12}O_4$	86-81-7	196.200		72.5	148^{5}			s chl
10478	1,2,3-Trimethoxybenzene		$C_9H_{12}O_3$	634-36-6	168.189	orth nd (al)	48.5	235	1.1009^{45}		i H_2O; s EtOH, eth, bz
10479	1,3,5-Trimethoxybenzene		$C_9H_{12}O_3$	621-23-8	168.189	pr (al), lf (peth)	54.5	255.5			i H_2O; s EtOH, eth, bz
10480	3,4,5-Trimethoxybenzeneethanamine	Mescaline	$C_{11}H_{17}NO_3$	54-04-6	211.258	cry	35.5	180^{12}			s H_2O, EtOH, bz, chl; i eth, peth
10481	3,4,5-Trimethoxybenzenemethanol	3,4,5-Trimethoxybenzyl alcohol	$C_{10}H_{14}O_4$	3840-31-1	198.216		3	228^{25}	1.1427^{20}	1.5439^{20}	
10482	2,4,5-Trimethoxybenzoic acid		$C_{10}H_{12}O_5$	490-64-2	212.199	nd (al or bz-peth)	145	300			vs H_2O, bz, EtOH, peth
10483	3,4,5-Trimethoxybenzoic acid		$C_{10}H_{12}O_5$	118-41-2	212.199	mcl nd (w)	172.3	226^{10}			sl H_2O; vs EtOH, eth, chl
10484	3,4,5-Trimethoxybenzoyl chloride		$C_{10}H_{11}ClO_4$	4521-61-3	230.645		82	185^{18}			
10485	Trimethoxyboroxin		$C_3H_9B_3O_6$	102-24-9	173.532					1.40^{25}	
10486	6,6',7-Trimethoxy-2,2'-dimethylberbaman-12-ol	Berbamine	$C_{37}H_{40}N_2O_6$	478-61-5	608.723	lf (+2w, al) cry (peth)	198.5				sl H_2O; s EtOH, eth, chl, peth
10487	6,6',7-Trimethoxy-2,2'-dimethyloxyacanthan-12'-ol	Oxyacanthine	$C_{37}H_{40}N_2O_6$	548-40-3	608.723	nd (al, eth)	216.5				i H_2O; s EtOH, eth, bz, chl; i lig
10488	7',10,11-Trimethoxyemetan-6'-ol	Cephaeline	$C_{28}H_{38}N_2O_4$	483-17-0	466.613	nd (eth)	115.5				vs ace, EtOH, MeOH, chl
10489	1,1,1-Trimethoxyethane		$C_5H_{12}O_3$	1445-45-0	120.147			108	0.9438^{25}	1.3859^{25}	vs eth, EtOH
10490	4,7,8-Trimethoxyfuro[2,3-b] quinoline	Skimmianine	$C_{14}H_{13}NO_4$	83-95-4	259.258	pym (al)	177				i H_2O, peth; s EtOH, chl; sl eth, CS_2
10491	Trimethoxymethane		$C_4H_{10}O_3$	149-73-5	106.120		15	104	0.9676^{20}	1.3793^{20}	s EtOH, eth
10492	Trimethoxymethylsilane		$C_4H_{12}O_3Si$	1185-55-3	136.222			102.5	0.9548^{20}	1.3696^{20}	s chl
10493	Trimethoxyphenylsilane		$C_9H_{14}O_3Si$	2996-92-1	198.291			130^{45}, 110^{20}	1.064^{20}	1.4734^{20}	s ctc, CS_2
10494	Trimethoxysilane		$C_3H_{10}O_3Si$	2487-90-3	122.195			32^{100}			
10495	3-(Trimethoxysilyl)-1-propanethiol	(3-Mercaptopropyl) trimethoxysilane	$C_6H_{16}O_3SSi$	4420-74-0	196.340			128^{50}, 93^{10}	1.015^{25}	1.4420^{25}	
10496	*N*-[3-(Trimethoxysilyl)propyl]-1,2-ethanediamine		$C_8H_{22}N_2O_3Si$	1760-24-3	222.358			140.5^{15}	1.01^{25}	1.4416^{25}	
10497	3-(Trimethoxysilyl)propyl methacrylate		$C_{10}H_{20}O_5Si$	2530-85-0	248.349	liq		107^{5}, 95^{1}			
10498	Trimethyl aluminum		C_3H_9Al	75-24-1	72.085		15.4	130; 20^{8}	0.752^{20}		
10499	Trimethylamine	*N,N*-Dimethylmethanamine	C_3H_9N	75-50-3	59.110	col gas	-117.1	2.87	0.627^{25} (p>1 atm)	1.3631^{0}	vs H_2O, chl, tol; s EtOH, eth, bz
10500	Trimethylamine borane	*N,N*-Dimethylmethanamine borane	$C_3H_{12}BN$	75-22-9	72.945		94	172	0.792^{25}		vs eth, EtOH
10501	Trimethylamine hydrochloride	*N,N*-Dimethylmethanamine hydrochloride	$C_3H_{10}ClN$	593-81-7	95.571	mcl hyg nd (al)	277.5	sub 200			vs H_2O, EtOH, chl
10502	Trimethylamine oxide	*N,N*-Dimethylmethanamine oxide	C_3H_9NO	1184-78-7	75.109	hyg nd (w+2)	256				vs H_2O, EtOH
10503	2,4,5-Trimethylaniline		$C_9H_{13}N$	137-17-7	135.206	nd (w)	68	234.5	0.957^{25}		vs EtOH
10504	2,4,6-Trimethylaniline	Mesitylamine	$C_9H_{13}N$	88-05-1	135.206	liq	-2.5	232.5	0.9633^{25}	1.5495^{20}	sl ctc
10505	Trimethylarsine		C_3H_9As	593-88-4	120.025	liq	-87.3	52	1.144^{15}		vs bz, eth, EtOH
10506	2,4,6-Trimethylbenzaldehyde		$C_{10}H_{12}O$	487-68-3	148.201		14	238.5	1.0154^{25}		i H_2O; s EtOH, eth, ace, bz
10507	1,2,3-Trimethylbenzene	Hemimellitene	C_9H_{12}	526-73-8	120.191	liq	-25.4	176.12	0.8944^{20}	1.5139^{20}	i H_2O; msc EtOH, eth, ace, bz, peth, ctc
10508	1,2,4-Trimethylbenzene	Pseudocumene	C_9H_{12}	95-63-6	120.191	liq	-43.77	169.38	0.8758^{20}	1.5048^{20}	i H_2O; msc EtOH, eth, ace, bz, peth, ctc
10509	1,3,5-Trimethylbenzene	Mesitylene	C_9H_{12}	108-67-8	120.191	liq	-44.72	164.74	0.8615^{25}	1.4994^{20}	i H_2O; msc EtOH, eth, ace, bz, peth, ctc
10510	2,3,5-Trimethyl-1,4-benzenediol		$C_9H_{12}O_2$	700-13-0	152.190	nd (w)	169 dec				sl H_2O; vs EtOH, eth, bz

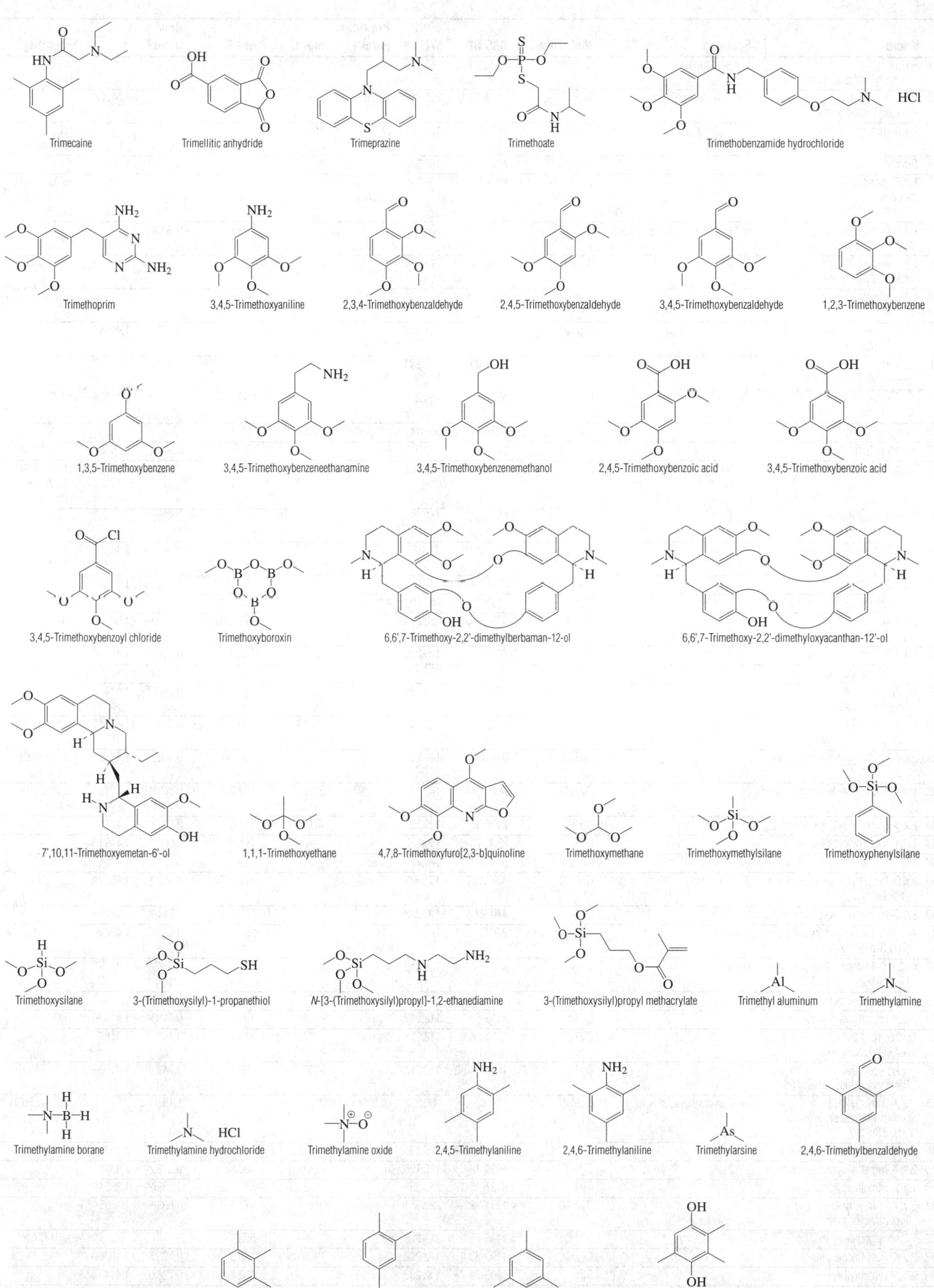
Trimecaine
Trimellitic anhydride
Trimeprazine
Trimethoate
Trimethobenzamide hydrochloride
Trimethoprim
3,4,5-Trimethoxyaniline
2,3,4-Trimethoxybenzaldehyde
2,4,5-Trimethoxybenzaldehyde
3,4,5-Trimethoxybenzaldehyde
1,2,3-Trimethoxybenzene
1,3,5-Trimethoxybenzene
3,4,5-Trimethoxybenzeneethanamine
3,4,5-Trimethoxybenzenemethanol
2,4,5-Trimethoxybenzoic acid
3,4,5-Trimethoxybenzoic acid
3,4,5-Trimethoxybenzoyl chloride
Trimethoxyboroxin
6,6',7-Trimethoxy-2,2'-dimethylberbaman-12-ol
6,6',7-Trimethoxy-2,2'-dimethyloxyacanthan-12'-ol
7',10,11-Trimethoxyemetan-6'-ol
1,1,1-Trimethoxyethane
4,7,8-Trimethoxyfuro[2,3-b]quinoline
Trimethoxymethane
Trimethoxymethylsilane
Trimethoxyphenylsilane
Trimethoxysilane
3-(Trimethoxysilyl)-1-propanethiol
N-[3-(Trimethoxysilyl)propyl]-1,2-ethanediamine
3-(Trimethoxysilyl)propyl methacrylate
Trimethyl aluminum
Trimethylamine
Trimethylamine borane
Trimethylamine hydrochloride
Trimethylamine oxide
2,4,5-Trimethylaniline
2,4,6-Trimethylaniline
Trimethylarsine
2,4,6-Trimethylbenzaldehyde
1,2,3-Trimethylbenzene
1,2,4-Trimethylbenzene
1,3,5-Trimethylbenzene
2,3,5-Trimethyl-1,4-benzenediol

No.	Name	Synonym	Mol. Form.	CAS RN	Mol. Wt.	Physical Form	mp/°C	bp/°C	den/ g cm^{-3}	n_D	Solubility
10511	*N*,α,α-Trimethylbenzeneethanamine	Mephentermine	$C_{11}H_{17}N$	100-92-5	163.260	liq		95^{9}			i H_2O; s eth; vs EtOH
10512	Trimethyl 1,2,4-benzenetricarboxylate	Trimethyl trimellitate	$C_{12}H_{12}O_6$	2459-10-1	252.219	visc oil	-13	194^{12}	1.261	1.5230^{20}	
10513	2,4,6-Trimethylbenzoic acid		$C_{10}H_{12}O_2$	480-63-7	164.201	pr (lig)	156.5				sl H_2O; s EtOH, eth, ace, chl
10514	Trimethylbenzylsilane		$C_{10}H_{16}Si$	770-09-2	164.320			190.5	0.8933^{20}	1.4941^{20}	
10515	1,7,7-Trimethylbicyclo[2.2.1] heptane		$C_{10}H_{18}$	464-15-3	138.250	hex pl(al), pr(MeOH)		161			i H_2O; s EtOH, eth, AcOEt, MeOH
10516	1,3,3-Trimethylbicyclo[2.2.1] heptan-2-ol, (1*S*-*endo*)	α-Fenchyl alcohol, (*l*)	$C_{10}H_{18}O$	512-13-0	154.249	pr	48	94^{20}	0.9034^{84}		
10517	1,7,7-Trimethylbicyclo[2.2.1] heptan-2-ol acetate, *endo*	Bornyl acetate	$C_{12}H_{20}O_2$	76-49-3	196.286		29	221			
10518	1,7,7-Trimethylbicyclo[2.2.1]hept-2-ene		$C_{10}H_{16}$	464-17-5	136.234	cry (al)	113	146			vs bz, eth, EtOH
10519	4,6,6-Trimethylbicyclo[3.1.1]hept-3-en-2-ol, (1α,2α,5α)		$C_{10}H_{16}O$	1820-09-3	152.233		24	92^{10}	0.9657^{25}	1.4908^{25}	
10520	4,6,6-Trimethylbicyclo[3.1.1]hept-3-en-2-ol, (1α,2β,5α)		$C_{10}H_{16}O$	1845-30-3	152.233		15.5	90^{10}	0.9684^{25}	1.4912^{25}	
10521	2,7,7-Trimethylbicyclo[3.1.1]hept-2-en-6-one	Chrysanthenone	$C_{10}H_{14}O$	473-06-3	150.217			88^{12}		1.4720^{22}	vs EtOH
10522	Trimethylborane		C_3H_9B	593-90-8	55.914	col gas	-161.5	-20.2			
10523	Trimethyl borate		$C_3H_9BO_3$	121-43-7	103.912	liq	-29.3	67.5	0.915^{25}	1.3568^{20}	vs eth, MeOH
10524	2,2,3-Trimethylbutane	Triptane	C_7H_{16}	464-06-2	100.202	liq	-24.6	80.86	0.6901^{20}	1.3864^{20}	i H_2O; s EtOH, eth; vs ace, bz, peth, ctc
10525	2,3,3-Trimethyl-2-butanol		$C_7H_{16}O$	594-83-2	116.201	cry (dil al +1/2w)	17	131	0.8380^{25}	1.4233^{22}	sl H_2O; vs ace, eth, EtOH
10526	2,3,3-Trimethyl-1-butene		C_7H_{14}	594-56-9	98.186	liq	-109.9	77.9	0.7050^{20}	1.4025^{20}	i H_2O; s eth, bz, chl, MeOH
10527	Trimethylchlorosilane		C_3H_9ClSi	75-77-4	108.642	liq	-40	60	0.856^{25}	1.3870^{20}	
10528	Trimethyl citrate		$C_9H_{14}O_7$	1587-20-8	234.203	tcl	79.3	285; 176^{16}			vs eth, EtOH
10529	2,6,6-Trimethyl-2,4-cycloheptadien-1-one	Eucarvone	$C_{10}H_{14}O$	503-93-5	150.217			210; 105^{22}	0.9490^{20}	1.5087^{20}	s eth, ace
10530	1,1,2-Trimethylcyclohexane		C_9H_{18}	7094-26-0	126.239	liq	-29	145.2	0.7963^{25}	1.4382^{20}	
10531	1,1,3-Trimethylcyclohexane		C_9H_{18}	3073-66-3	126.239	liq	-65.7	136.6	0.7749^{25}	1.4295^{20}	i H_2O
10532	1α,2β,4β-1,2,4-Trimethylcyclohexane		C_9H_{18}	7667-60-9	126.239	liq	-83.5	142.9	0.7870^{25}	1.4341^{20}	
10533	1α,3α,5β-1,3,5-Trimethylcyclohexane	*trans*-1,3,5-Trimethylcyclohexane	C_9H_{18}	1795-26-2	126.239	liq	-107.4	140.5	0.7794^{20}	1.4307^{20}	vs bz, eth, lig
10534	*cis*-3,3,5-Trimethylcyclohexanol		$C_9H_{18}O$	933-48-2	142.238		37.3	202; 92^{12}	0.9006^{16}	1.4550^{16}	i H_2O; s EtOH, eth, chl
10535	*trans*-3,3,5-Trimethylcyclohexanol		$C_9H_{18}O$	767-54-4	142.238	cry (eth)	55.8	189.2	0.8631^{60}		i H_2O; s EtOH, eth, chl
10536	2,2,6-Trimethylcyclohexanone		$C_9H_{16}O$	2408-37-9	140.222	liq	-31.8	178.5	0.9043^{18}	1.4470^{20}	
10537	2,4,4-Trimethylcyclohexanone		$C_9H_{16}O$	2230-70-8	140.222			191	0.902^{20}	1.4493^{20}	
10538	3,3,5-Trimethylcyclohexanone	Dihydroisophorone	$C_9H_{16}O$	873-94-9	140.222	ye oil		189	0.8919^{19}	1.4454^{15}	
10539	2,6,6-Trimethyl-1-cyclohexene-1-carboxaldehyde	β-Cyclocitral	$C_{10}H_{16}O$	432-25-7	152.233			112^{29}, 97^{15}	0.959^{15}	1.4971^{15}	
10540	3,5,5-Trimethyl-2-cyclohexen-1-ol	Isophorol	$C_9H_{16}O$	470-99-5	140.222			69^{5}	0.914^{20}	1.4717^{20}	
10541	4-(2,6,6-Trimethyl-1-cyclohexen-1-yl)-3-buten-2-ol	β-Ionol	$C_{13}H_{22}O$	22029-76-1	194.313			130^{14}	0.9243^{20}	1.4969^{20}	s EtOH, eth, ace
10542	4-(2,6,6-Trimethyl-2-cyclohexen-1-yl)-3-buten-2-ol	α-Ionol	$C_{13}H_{22}O$	25312-34-9	194.313	oil		127^{14}	0.9189^{20}	1.4735^{20}	
10543	1,1,2-Trimethylcyclopentane		C_8H_{16}	4259-00-1	112.213	liq	-21.6	114; 53^{100}	0.7660^{20}	1.4199^{20}	
10544	1,1,3-Trimethylcyclopentane		C_8H_{16}	4516-69-2	112.213	liq	-142.4	104.9	0.7439^{25}	1.4112^{20}	i H_2O
10545	1α,2α,4β-1,2,4-Trimethylcyclopentane		C_8H_{16}	4850-28-6	112.213	liq	-132.6	116.7	0.7592^{25}	1.4186^{20}	
10546	1α,2β,4α-1,2,4-Trimethylcyclopentane		C_8H_{16}	16883-48-0	112.213	liq	-130.8	109.3	0.7430^{25}	1.4106^{20}	
10547	1,2,2-Trimethyl-1,3-cyclopentanedicarboxylic acid, (1*R*, 3*S*)	(+)-Camphoric acid	$C_{10}H_{16}O_4$	124-83-4	200.232	pr, lf (w)	187		1.186^{20}		sl H_2O; vs EtOH, eth; s ace; i bz, chl
10548	2,2,4-Trimethylcyclopentanone		$C_8H_{14}O$	28056-54-4	126.196	liq	-40.6	158	0.877^{25}	1.4300^{20}	
10549	2,4,4-Trimethylcyclopentanone		$C_8H_{14}O$	4694-12-6	126.196	liq	-25.6	160.5	0.8785^{18}	1.433^{18}	
10550	1,1,2-Trimethylcyclopropane		C_6H_{12}	4127-45-1	84.159	liq	-138.2	54	0.6897^{25}	1.3864^{20}	
10551	3,7,11-Trimethyl-2,6,10-dodecatrienal		$C_{15}H_{24}O$	19317-11-4	220.351			172^{14}	0.893^{18}	1.4995	
10552	Trimethylgallium		C_3H_9Ga	1445-79-0	114.826			55.7			dec H_2O (exp)
10553	2,2,6-Trimethylheptane		$C_{10}H_{22}$	1190-83-6	142.282	liq	-105	148.9	0.7200^{25}	1.4078^{20}	
10554	2,5,5-Trimethylheptane		$C_{10}H_{22}$	1189-99-7	142.282			152.8	0.7362^{25}	1.4149^{20}	
10555	3,3,5-Trimethylheptane		$C_{10}H_{22}$	7154-80-5	142.282			155.7	0.7248^{20}	1.4170^{20}	i H_2O; s bz, ctc, chl

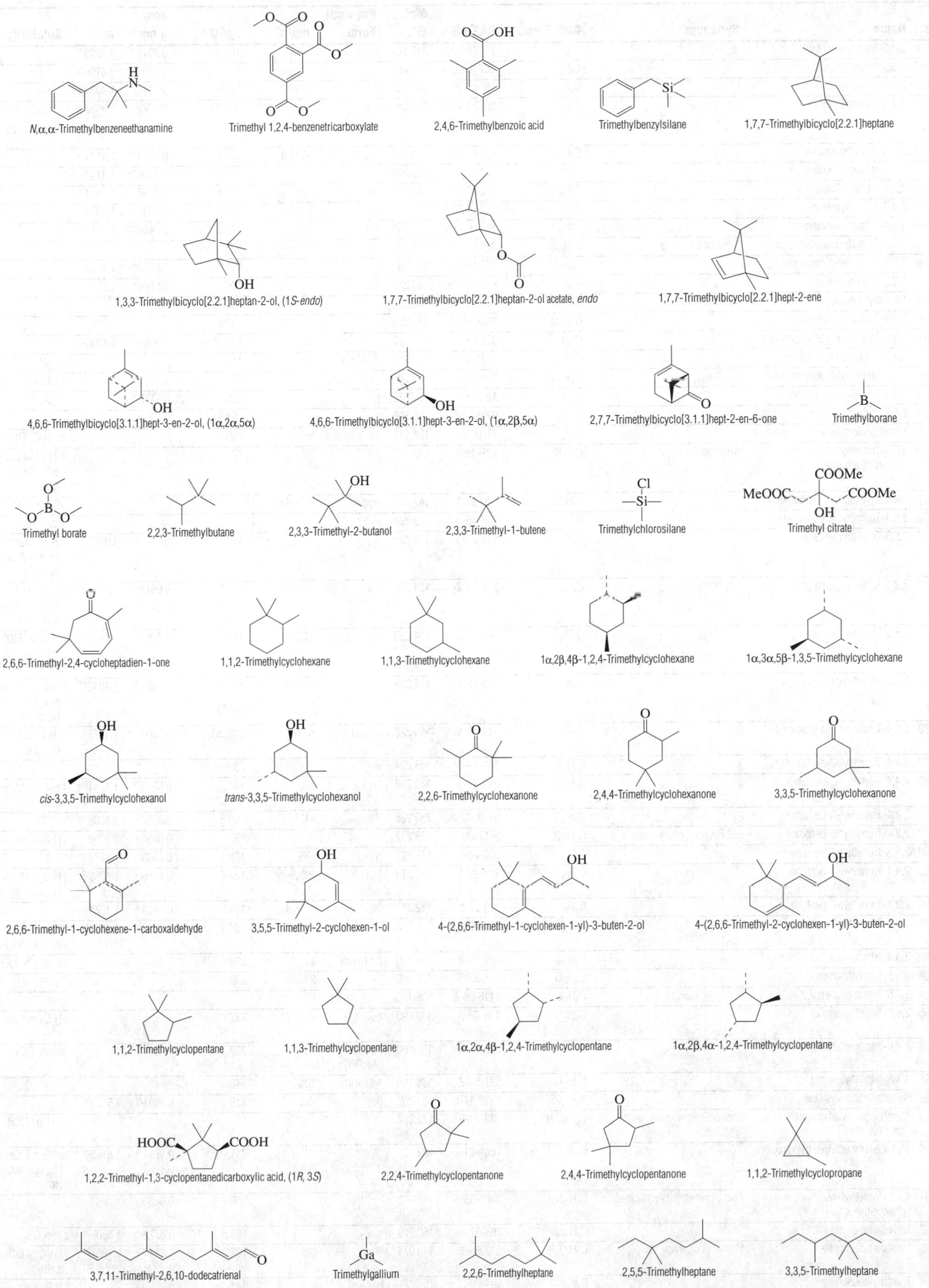

N,α,α-Trimethylbenzeneethanamine
Trimethyl 1,2,4-benzenetricarboxylate
2,4,6-Trimethylbenzoic acid
Trimethylbenzylsilane
1,7,7-Trimethylbicyclo[2.2.1]heptane
1,3,3-Trimethylbicyclo[2.2.1]heptan-2-ol, (1S-endo)
1,7,7-Trimethylbicyclo[2.2.1]heptan-2-ol acetate, endo
1,7,7-Trimethylbicyclo[2.2.1]hept-2-ene
4,6,6-Trimethylbicyclo[3.1.1]hept-3-en-2-ol, (1α,2α,5α)
4,6,6-Trimethylbicyclo[3.1.1]hept-3-en-2-ol, (1α,2β,5α)
2,7,7-Trimethylbicyclo[3.1.1]hept-2-en-6-one
Trimethylborane
Trimethyl borate
2,2,3-Trimethylbutane
2,3,3-Trimethyl-2-butanol
2,3,3-Trimethyl-1-butene
Trimethylchlorosilane
Trimethyl citrate
2,6,6-Trimethyl-2,4-cycloheptadien-1-one
1,1,2-Trimethylcyclohexane
1,1,3-Trimethylcyclohexane
1α,2β,4β-1,2,4-Trimethylcyclohexane
1α,3α,5β-1,3,5-Trimethylcyclohexane
cis-3,3,5-Trimethylcyclohexanol
trans-3,3,5-Trimethylcyclohexanol
2,2,6-Trimethylcyclohexanone
2,4,4-Trimethylcyclohexanone
3,3,5-Trimethylcyclohexanone
2,6,6-Trimethyl-1-cyclohexene-1-carboxaldehyde
3,5,5-Trimethyl-2-cyclohexen-1-ol
4-(2,6,6-Trimethyl-1-cyclohexen-1-yl)-3-buten-2-ol
4-(2,6,6-Trimethyl-2-cyclohexen-1-yl)-3-buten-2-ol
1,1,2-Trimethylcyclopentane
1,1,3-Trimethylcyclopentane
1α,2α,4β-1,2,4-Trimethylcyclopentane
1α,2β,4α-1,2,4-Trimethylcyclopentane
1,2,2-Trimethyl-1,3-cyclopentanedicarboxylic acid, (1R, 3S)
2,2,4-Trimethylcyclopentanone
2,4,4-Trimethylcyclopentanone
1,1,2-Trimethylcyclopropane
3,7,11-Trimethyl-2,6,10-dodecatrienal
Trimethylgallium
2,2,6-Trimethylheptane
2,5,5-Trimethylheptane
3,3,5-Trimethylheptane

No.	Name	Synonym	Mol. Form.	CAS RN	Mol. Wt.	Physical Form	mp/°C	bp/°C	den/g cm^{-3}	n_D	Solubility
10556	3,4,5-Trimethylheptane		$C_{10}H_{22}$	20278-89-1	142.282			162.5	0.7519^{25}	1.4229^{20}	
10557	2,2,3-Trimethylhexane		C_9H_{20}	16747-25-4	128.255			133.6	0.7257^{25}	1.4106^{20}	
10558	2,2,4-Trimethylhexane		C_9H_{20}	16747-26-5	128.255	liq	-120	126.5	0.711^{20}	1.4033^{20}	
10559	2,2,5-Trimethylhexane		C_9H_{20}	3522-94-9	128.255	liq	-105.7	124.09	0.7072^{20}	1.3997^{20}	i H_2O; vs EtOH, eth, ace, bz; s ctc
10560	2,3,3-Trimethylhexane		C_9H_{20}	16747-28-7	128.255	liq	-116.8	137.7	0.7345^{25}	1.4141^{20}	
10561	2,3,4-Trimethylhexane		C_9H_{20}	921-47-1	128.255			139.1	0.7354^{25}	1.4144^{20}	
10562	2,3,5-Trimethylhexane		C_9H_{20}	1069-53-0	128.255	liq	-127.9	131.4	0.7218^{20}	1.4051^{20}	
10563	2,4,4-Trimethylhexane		C_9H_{20}	16747-30-1	128.255	liq	-113.4	130.7	0.7201^{25}	1.4074^{20}	
10564	3,3,4-Trimethylhexane		C_9H_{20}	16747-31-2	128.255	liq	-101.2	140.5	0.7414^{25}	1.4178^{20}	
10565	3,5,5-Trimethylhexanoic acid	Isononanoic acid	$C_9H_{18}O_2$	3302-10-1	158.238	liq		121^{10}, 85^{4}			
10566	3,5,5-Trimethyl-1-hexanol		$C_9H_{20}O$	3452-97-9	144.254			194	0.8236^{25}	1.4300^{25}	
10567	1,2,3-Trimethylindene		$C_{12}H_{14}$	4773-83-5	158.239	liq		100.5^{10}	0.9714^{20}	1.5521^{20}	
10568	Trimethylindium	Indium trimethyl	C_3H_9In	3385-78-2	159.921			135.7	1.568^{19}		
10569	2,3,3-Trimethyl-3*H*-indole		$C_{11}H_{13}N$	1640-39-7	159.228			107^{11}			
10570	Trimethyl(4-methylphenyl)silane		$C_{10}H_{16}Si$	3728-43-6	164.320		38	192; 73^{10}	0.8666^{20}	1.4900^{20}	
10571	1,4,5-Trimethylnaphthalene		$C_{13}H_{14}$	2131-41-1	170.250	lf (MeOH)	63	145^{12}			i H_2O
10572	1,3,5-Trimethyl-2-nitrobenzene		$C_9H_{11}NO_2$	603-71-4	165.189	orth pr (al)	44	255	1.51^{25}		vs EtOH
10573	2,6,8-Trimethyl-4-nonanol		$C_{12}H_{26}O$	123-17-1	186.333			225.4	0.8178^{20}		sl ctc
10574	2,4,7-Trimethyloctane		$C_{11}H_{24}$	62016-38-0	156.309			168.1			
10575	Trimethylolpropane		$C_6H_{14}O_3$	77-99-6	134.173	wh pow or pl	58	160^{5}			vs H_2O, EtOH
10576	3,5,5-Trimethyl-2,4-oxazolidinedione	Trimethadione	$C_6H_9NO_3$	127-48-0	143.140		46	79^{5}			s H_2O; vs EtOH, eth, ace, bz; i peth
10577	Trimethyloxonium fluoborate		$C_3H_9BF_4O$	420-37-1	147.907	hyg nd	148 dec				vs ace, chl
10578	2,4,4-Trimethyl-2-pentanamine		$C_8H_{19}N$	107-45-9	129.244						s chl
10579	2,2,3-Trimethylpentane		C_8H_{18}	564-02-3	114.229	liq	-112.2	110	0.7161^{20}	1.4030^{20}	i H_2O; msc EtOH, eth, ace, hp; s bz
10580	2,2,4-Trimethylpentane	Isooctane	C_8H_{18}	540-84-1	114.229	liq	-107.3	99.22	0.6878^{25}	1.3884^{25}	i H_2O; msc EtOH, ace, hp; s eth, ctc
10581	2,3,3-Trimethylpentane		C_8H_{18}	560-21-4	114.229	liq	-100.9	114.8	0.7262^{20}	1.4075^{20}	i H_2O; vs EtOH; msc eth, ace, bz, hp
10582	2,3,4-Trimethylpentane		C_8H_{18}	565-75-3	114.229	liq	-109.2	113.5	0.7191^{20}	1.4042^{20}	i H_2O; vs EtOH; msc eth, ace, bz; sl ctc
10583	2,2,4-Trimethyl-1,3-pentanediol		$C_8H_{18}O_2$	144-19-4	146.228	pl (bz)	51.5	235; 81^{1}	0.936^{15}	1.4513^{15}	sl H_2O; vs EtOH, eth; s bz, chl
10584	2,4,4-Trimethyl-2-pentanethiol		$C_8H_{18}S$	141-59-3	146.294	liq		76^{50}			
10585	2,4,4-Trimethyl-2-pentanol		$C_8H_{18}O$	690-37-9	130.228	liq	-20	147.5	0.8225^{20}	1.4284^{20}	i H_2O; sl EtOH; s eth
10586	2,2,4-Trimethyl-3-pentanol		$C_8H_{18}O$	5162-48-1	130.228	liq	-13	150.5	0.8297^{20}	1.4288^{20}	
10587	2,2,4-Trimethyl-3-pentanone	*tert*-Butyl isopropyl ketone	$C_8H_{16}O$	5857-36-3	128.212			135.1	0.8065^{20}	1.4060	i H_2O; s eth, ace
10588	2,3,3-Trimethyl-1-pentene		C_8H_{16}	560-23-6	112.213	liq	-69	108.3	0.7308^{25}	1.4174^{20}	
10589	2,4,4-Trimethyl-1-pentene		C_8H_{16}	107-39-1	112.213	liq	-93.5	101.4	0.7150^{20}	1.4086^{20}	i H_2O; s eth, bz, ctc, chl, lig
10590	2,3,4-Trimethyl-2-pentene		C_8H_{16}	565-77-5	112.213	liq	-113.4	116.5	0.7434^{20}	1.4274^{20}	
10591	2,4,4-Trimethyl-2-pentene		C_8H_{16}	107-40-4	112.213	liq	-106.3	104.9	0.7218^{20}	1.4160^{20}	i H_2O; s eth, bz, ctc, chl; vs lig
10592	2,3,4-Trimethylphenol		$C_9H_{12}O$	526-85-2	136.190	nd (peth)	81	236			vs bz, eth, EtOH
10593	2,3,5-Trimethylphenol		$C_9H_{12}O$	697-82-5	136.190		94.5	233			
10594	2,3,6-Trimethylphenol		$C_9H_{12}O$	2416-94-6	136.190		63				
10595	2,4,5-Trimethylphenol		$C_9H_{12}O$	496-78-6	136.190	nd (lig)	72	232			i H_2O; vs EtOH, eth
10596	2,4,6-Trimethylphenol		$C_9H_{12}O$	527-60-6	136.190	nd (peth, MeOH)	73	220			vs eth, EtOH
10597	3,4,5-Trimethylphenol		$C_9H_{12}O$	527-54-8	136.190	nd (peth)	108	248.5			
10598	Trimethylphenoxysilane		$C_9H_{14}OSi$	1529-17-5	166.292	liq	-55	119	0.8681^{20}	1.5125^{20}	
10599	Trimethylphenylammonium chloride	Phenyltrimethylammonium chloride	$C_9H_{14}ClN$	138-24-9	171.667						vs H_2O, EtOH
10600	1-(2,4,6-Trimethylphenyl)ethanone		$C_{11}H_{14}O$	1667-01-2	162.228			241; 120^{12}	0.9754^{20}	1.5175^{20}	i H_2O; s EtOH, eth, ace, bz, ctc
10601	1,1,1-Trimethyl-*N*-phenylsilanamine	Phenyl(trimethylsilyl)amine	$C_9H_{15}NSi$	3768-55-6	165.308			206	0.940^{20}		
10602	Trimethylphenylsilane		$C_9H_{14}Si$	768-32-1	150.293			169.5	0.8722^{20}	1.4907^{20}	s ctc, CS_2
10603	Trimethyl phosphate	Methyl phosphate	$C_3H_9O_4P$	512-56-1	140.074	liq	-46	197.2	1.2144^{20}	1.3967^{20}	vs H_2O; sl EtOH; s eth
10604	Trimethylphosphine		C_3H_9P	594-09-2	76.077	liq	-85	37.5			i H_2O; s eth

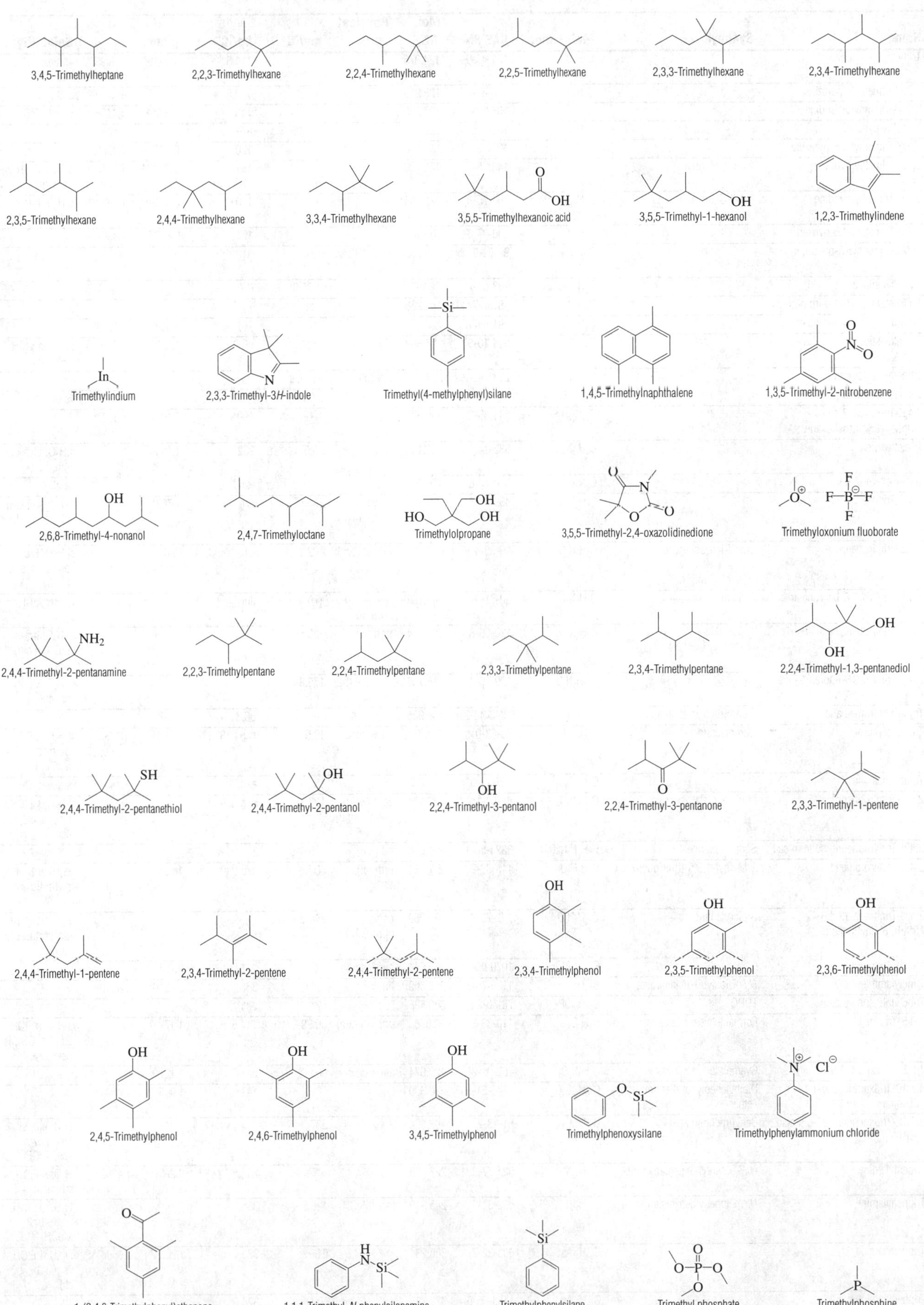

3,4,5-Trimethylheptane
2,2,3-Trimethylhexane
2,2,4-Trimethylhexane
2,2,5-Trimethylhexane
2,3,3-Trimethylhexane
2,3,4-Trimethylhexane
2,3,5-Trimethylhexane
2,4,4-Trimethylhexane
3,3,4-Trimethylhexane
O
OH
3,5,5-Trimethylhexanoic acid
OH
3,5,5-Trimethyl-1-hexanol
1,2,3-Trimethylindene
In
Trimethylindium
N
2,3,3-Trimethyl-3H-indole
Si
Trimethyl(4-methylphenyl)silane
1,4,5-Trimethylnaphthalene
O
N
O
1,3,5-Trimethyl-2-nitrobenzene
OH
2,6,8-Trimethyl-4-nonanol
2,4,7-Trimethyloctane
OH
HO
OH
Trimethylolpropane
O
N
O
O
3,5,5-Trimethyl-2,4-oxazolidinedione
O
F
F—B—F
F
Trimethyloxonium fluoborate
NH2
2,4,4-Trimethyl-2-pentanamine
2,2,3-Trimethylpentane
2,2,4-Trimethylpentane
2,3,3-Trimethylpentane
2,3,4-Trimethylpentane
OH
OH
2,2,4-Trimethyl-1,3-pentanediol
SH
2,4,4-Trimethyl-2-pentanethiol
OH
2,4,4-Trimethyl-2-pentanol
OH
2,2,4-Trimethyl-3-pentanol
O
2,2,4-Trimethyl-3-pentanone
2,3,3-Trimethyl-1-pentene
2,4,4-Trimethyl-1-pentene
2,3,4-Trimethyl-2-pentene
2,4,4-Trimethyl-2-pentene
OH
2,3,4-Trimethylphenol
OH
2,3,5-Trimethylphenol
OH
2,3,6-Trimethylphenol
OH
2,4,5-Trimethylphenol
OH
2,4,6-Trimethylphenol
OH
3,4,5-Trimethylphenol
O
Si
Trimethylphenoxysilane
N
Cl
Trimethylphenylammonium chloride
O
1-(2,4,6-Trimethylphenyl)ethanone
H
N
Si
1,1,1-Trimethyl-N-phenylsilanamine
Si
Trimethylphenylsilane
O
O—P—O
O
Trimethyl phosphate
P
Trimethylphosphine

No.	Name	Synonym	Mol. Form.	CAS RN	Mol. Wt.	Physical Form	mp/°C	bp/°C	den/g cm^{-3}	n_D	Solubility
10605	Trimethyl phosphite		$C_3H_9O_3P$	121-45-9	124.075			111.5	1.0518^{20}	1.4095^{20}	vs EtOH, eth; sl ctc
10606	1,2,4-Trimethylpiperazine		$C_7H_{16}N_2$	120-85-4	128.215			149.5		1.4433^{20}	s ctc
10607	2,2,4-Trimethylpiperidine		$C_8H_{17}N$	101257-71-0	127.228			148	0.832^{15}	1.4458^{20}	vs eth, EtOH
10608	Trimethylpyrazine		$C_7H_{10}N_2$	14667-55-1	122.167			87^{35}			
10609	1,3,5-Trimethyl-1*H*-pyrazole		$C_6H_{10}N_2$	1072-91-9	110.156		37	170	0.9269^{40}	1.4589^{57}	
10610	2,3,6-Trimethylpyridine	2,3,6-Collidine	$C_8H_{11}N$	1462-84-6	121.180			171.6	0.9220^{25}	1.5053^{20}	s H_2O, EtOH, eth, ace, bz
10611	2,4,6-Trimethylpyridine	2,4,6-Collidine	$C_8H_{11}N$	108-75-8	121.180	liq	-46	170.6	0.9166^{22}	1.4959^{25}	s H_2O, EtOH, eth, ace, ctc
10612	1,2,5-Trimethyl-1*H*-pyrrole		$C_7H_{11}N$	930-87-0	109.169			171	0.807^{25}	1.4969^{20}	
10613	*N,N*,2-Trimethyl-6-quinolinamine		$C_{12}H_{14}N_2$	92-99-9	186.252	ye pr (HOAc, AcOEt)	101	319			s ctc, CS_2
10614	Trimethylsilane		$C_3H_{10}Si$	993-07-7	74.197	col gas	-135.9	6.7			
10615	1-(Trimethylsilyl)-1*H*-imidazole		$C_6H_{12}N_2Si$	18156-74-6	140.258						s chl
10616	3-(Trimethylsilyl)-1-propanol		$C_6H_{16}OSi$	2917-47-7	132.276			141; 82^{24}	0.822^{25}	1.4298^{20}	
10617	Trimethylstibine		C_3H_9Sb	594-10-5	166.863	liq	-62	80.6	1.523^{15}	1.42^{15}	i H_2O; s EtOH, eth, CS_2
10618	Trimethylsulfonium iodide		C_3H_9IS	2181-42-2	204.072	cry (eth)	211 dec				
10619	Trimethylthiourea		$C_4H_{10}N_2S$	2489-77-2	118.200	pr (bz-lig)	87.5				vs bz, EtOH, chl
10620	2,4,6-Trimethyl-2,4,6-triphenylcyclotrisiloxane		$C_{21}H_{24}O_3Si_3$	546-45-2	408.671		100	$190^{1.5}$	1.1062^{20}	1.5397^{20}	
10621	Trimethylurea		$C_4H_{10}N_2O$	632-14-4	102.134	pr (eth)	75.5	232.5	1.1900^{20}		s H_2O, EtOH; sl eth, bz
10622	Trinitroacetonitrile		$C_2N_4O_6$	630-72-8	176.044	wax	41.5	exp 220			vs eth
10623	2,4,6-Trinitroaniline		$C_6H_4N_4O_6$	489-98-5	228.119	dk ye pr (HOAc)	193.5	exp	1.762^{10}		i H_2O; sl EtOH, eth; s ace, bz, AcOEt
10624	1,3,5-Trinitrobenzene	*sym*-Trinitrobenzene	$C_6H_3N_3O_6$	99-35-4	213.104	orth pl (bz) lf (w)	122.9	315	1.4775^{152}		sl H_2O, EtOH, eth; vs ace; s bz, py
10625	2,4,6-Trinitro-1,3-benzenediol	Styphnic acid	$C_6H_3N_3O_8$	82-71-3	245.103	hex ye cry (dil al)	175.5	sub			vs eth, EtOH
10626	2,4,6-Trinitrobenzoic acid		$C_7H_3N_3O_8$	129-66-8	257.114	orth (w)	228 dec				sl H_2O, bz; vs EtOH; s eth, ace
10627	2,4,7-Trinitro-9*H*-fluoren-9-one		$C_{13}H_5N_3O_7$	129-79-3	315.195	pa ye nd (bz, HOAc)	175.8				sl H_2O; vs ace, bz, chl
10628	Trinitrofluoromethane	Fluorotrinitromethane	CFN_3O_6	1840-42-2	169.025			86.3	1.59^{20}		
10629	Trinitroglycerol	Nitroglycerin	$C_3H_5N_3O_9$	55-63-0	227.087	pa ye tcl or orth	13.5	exp 218; $93^{0.31}$	1.5931^{20}	1.4786^{12}	sl H_2O; s EtOH, bz; msc eth; vs ace, chl
10630	Trinitromethane		CHN_3O_6	517-25-9	151.035		15	exp	1.479^{20}	1.4451^{24}	vs ace, EtOH
10631	2,4,6-Trinitrophenol	Picric acid	$C_6H_3N_3O_7$	88-89-1	229.104	ye lf (w), pr (eth) pl (al)	122.5	exp 300		1.763	sl H_2O; s EtOH, eth, bz, chl; vs ace
10632	2,4,6-Trinitrophenol, sodium salt	Sodium picrate	$C_6H_2N_3NaO_7$	3324-58-1	251.086	nd (w)	270.4				
10633	2,4,6-Trinitrotoluene	2-Methyl-1,3,5-trinitrobenzene	$C_7H_5N_3O_6$	118-96-7	227.131	orth (al)	80.5	exp 240	1.654^{25}		i H_2O; sl EtOH; s eth; vs ace, bz
10634	2,4,6-Trinitro-*N*-(2,4,6-trinitrophenyl)aniline	Dipicrylamine	$C_{12}H_5N_7O_{12}$	131-73-7	439.208	pa ye pr(HOAc)	244 dec				i H_2O, EtOH, bz, ctc; sl eth, ace; vs py
10635	Trioctylaluminum		$C_{24}H_{51}Al$	1070-00-4	366.644	hyg visc liq	-62		0.701		
10636	Trioctylamine	*N,N*-Dioctyl-1-octanamine	$C_{24}H_{51}N$	1116-76-3	353.669	liq	-34.6	366	0.8110^{20}	1.4510^{19}	
10637	Trioctylphosphine oxide	TOPO	$C_{24}H_{51}OP$	78-50-2	386.635		52	201^{2}			
10638	1,3,5-Trioxane	Formaldehyde, trimer	$C_3H_6O_3$	110-88-3	90.078	orth nd (eth)	60.29	114.5	1.17^{65}		vs H_2O; s EtOH, eth, bz, CS_2; i peth
10639	1,3,5-Trioxane-2,4,6-triimine	Cyamelide	$C_3H_3N_3O_3$	462-02-2	129.074	amor pow	dec	dec	1.127^{15}		vs eth, EtOH
10640	4,7,10-Trioxatridecane-1,13-diamine	Diethyleneglycol diaminopropyl ether	$C_{10}H_{24}N_2O_3$	4246-51-9	220.309	liq		147^{4}	1.005	1.4640^{20}	
10641	3,7,12-Trioxocholan-24-oic acid, (5β)	Dehydrocholic acid	$C_{24}H_{34}O_5$	81-23-2	402.524		237				i H_2O, eth; sl EtOH, bz; s ace, AcOEt
10642	Tripentylamine	*N,N*-Dipentyl-1-pentanamine	$C_{15}H_{33}N$	621-77-2	227.430			242.5	0.7907^{20}	1.4366^{20}	i H_2O; s EtOH, eth, acid
10643	Triphenylamine	*N,N*-Diphenylbenzenamine	$C_{18}H_{15}N$	603-34-9	245.319	mcl (MeOH, bz)	126.5	365	1.18^{25}		i H_2O; sl EtOH; s eth, bz, MeOH
10644	Triphenylarsine		$C_{18}H_{15}As$	603-32-7	306.234		61	360	1.2634^{18}	1.6888^{21}	i H_2O; sl EtOH; vs eth, bz; s chl
10645	Triphenylarsine oxide		$C_{18}H_{15}AsO$	1153-05-5	322.233		192	324.0			

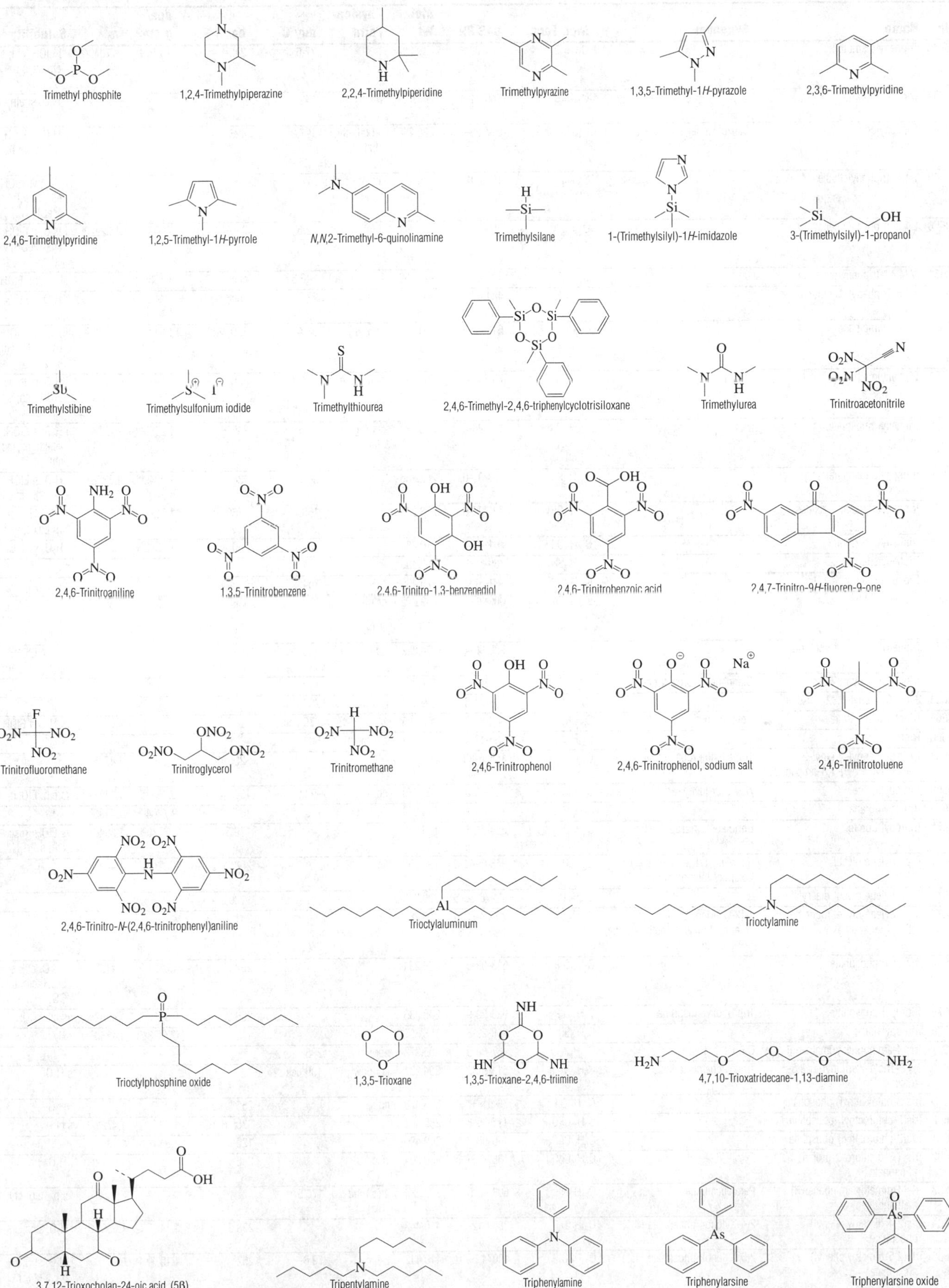
Trimethyl phosphite
1,2,4-Trimethylpiperazine
2,2,4-Trimethylpiperidine
Trimethylpyrazine
1,3,5-Trimethyl-1H-pyrazole
2,3,6-Trimethylpyridine
2,4,6-Trimethylpyridine
1,2,5-Trimethyl-1H-pyrrole
N,N,2-Trimethyl-6-quinolinamine
Trimethylsilane
1-(Trimethylsilyl)-1H-imidazole
3-(Trimethylsilyl)-1-propanol
Trimethylstibine
Trimethylsulfonium iodide
Trimethylthiourea
2,4,6-Trimethyl-2,4,6-triphenylcyclotrisiloxane
Trimethylurea
Trinitroacetonitrile
2,4,6-Trinitroaniline
1,3,5-Trinitrobenzene
2,4,6-Trinitro-1,3-benzenediol
2,4,6-Trinitrobenzoic acid
2,4,7-Trinitro-9H-fluoren-9-one
Trinitrofluoromethane
Trinitroglycerol
Trinitromethane
2,4,6-Trinitrophenol
2,4,6-Trinitrophenol, sodium salt
2,4,6-Trinitrotoluene
2,4,6-Trinitro-N-(2,4,6-trinitrophenyl)aniline
Trioctylaluminum
Trioctylamine
Trioctylphosphine oxide
1,3,5-Trioxane
1,3,5-Trioxane-2,4,6-triimine
4,7,10-Trioxatridecane-1,13-diamine
3,7,12-Trioxocholan-24-oic acid, (5β)
Tripentylamine
Triphenylamine
Triphenylarsine
Triphenylarsine oxide

No.	Name	Synonym	Mol. Form.	CAS RN	Mol. Wt.	Physical Form	mp/°C	bp/°C	den/ g cm^{-3}	n_D	Solubility
10646	Triphenylbismuthine		$C_{18}H_{15}Bi$	603-33-8	440.292		77.6	242^{14}	1.715^{75}	1.7040^{75}	sl EtOH, chl; s eth, ace, bz, CS_2
10647	Triphenylborane		$C_{18}H_{15}B$	960-71-4	242.123	wh cry	142				i H_2O; sl eth; s bz, lig
10648	Triphenylene	Benzo[1]phenanthrene	$C_{18}H_{12}$	217-59-4	228.288	nd (al, chl, bz)	197.8	425			i H_2O; s EtOH, HOAc; vs bz, chl
10649	1,1,2-Triphenylethane		$C_{20}H_{18}$	1520-42-9	258.357	mcl lf (dil al), nd (al)	57				i H_2O; vs EtOH, eth, bz; sl MeOH
10650	1,1,2-Triphenylethene		$C_{20}H_{16}$	58-72-0	256.341	lf (al)	72.5	220^{14}	1.0373^{78}	1.6292^{78}	i H_2O; s EtOH, chl, MeOH; vs eth
10651	*N,N',N''*-Triphenylguanidine		$C_{19}H_{17}N_3$	101-01-9	287.358	nd or pr (al)	146.5	dec	1.163^{20}		sl H_2O; s EtOH
10652	2,4,5-Triphenyl-1*H*-imidazole		$C_{21}H_{16}N_2$	484-47-9	296.365	nd (al)	275	sub			i H_2O; s EtOH, eth
10653	Triphenylmethane		$C_{19}H_{16}$	519-73-3	244.330	orth (al)	93.4	359; 200^{10}	1.014^{99}	1.5839^{99}	i H_2O; sl EtOH; vs eth, py, chl; s bz
10654	Triphenylmethanol		$C_{19}H_{16}O$	76-84-6	260.329	pl (al), trg (bz)	164.2	380	1.199^{0}		i H_2O, peth; vs EtOH, eth; s ace, bz
10655	Triphenyl phosphate		$C_{18}H_{15}O_4P$	115-86-6	326.283	cry (lig), pr (al) nd (eth)	50.5	245^{11}	1.2055^{50}		i H_2O; s EtOH; vs eth, bz, ctc, chl
10656	Triphenylphosphine		$C_{18}H_{15}P$	603-35-0	262.286		80	188^{1}	1.0749^{80}	1.6358^{80}	i H_2O; s EtOH, bz, chl; vs eth
10657	Triphenylphosphine oxide		$C_{18}H_{15}OP$	791-28-6	278.285	pr	156.5	>360	1.2124^{23}		sl H_2O, eth, chl; vs EtOH, bz
10658	Triphenyl phosphite		$C_{18}H_{15}O_3P$	101-02-0	310.284		25	360	1.1842^{20}	1.5900^{20}	i H_2O; vs EtOH
10659	Triphenylsilane		$C_{18}H_{16}Si$	789-25-3	260.406						s ctc, CS_2
10660	Triphenylsilanol		$C_{18}H_{16}OSi$	791-31-1	276.405		154.8		1.1777^{20}		s ctc, CS_2
10661	Triphenylstibine		$C_{18}H_{15}Sb$	603-36-1	353.072	pr (peth)	53.5	>360	1.4343^{25}	1.6948^{42}	i H_2O; s EtOH; vs eth, ace, bz, chl
10662	Triphenyltetrazolium chloride		$C_{19}H_{15}ClN_4$	298-96-4	334.802	nd (al,chl)	243 dec				s H_2O, EtOH, ace, chl; i eth
10663	Triphenyltin hydroxide	Stannane, hydroxytriphenyl-	$C_{18}H_{16}OSn$	76-87-9	367.029		119		1.54^{20}		
10664	2,4,6-Triphenyl-1,3,5-triazine		$C_{21}H_{15}N_3$	493-77-6	309.364		257				
10665	Tripotassium citrate	Potassium citrate	$C_6H_5K_3O_7$	866-84-2	306.395	wh cry (w)	275 dec				vs H_2O; i EtOH
10666	Triprolidine		$C_{19}H_{22}N_2$	486-12-4	278.391	cry (peth)	60				
10667	Tri-2-propenoyl-2-ethyl-2-(hydroxymethyl)-1,3-propanediol	Trimethylolpropane triacrylate	$C_{15}H_{20}O_6$	15625-89-5	296.316			$>200^{1}$		1.4735^{20}	
10668	Tripropylamine	*N,N*-Dipropyl-1-propanamine	$C_9H_{21}N$	102-69-2	143.270	liq	-93.5	156	0.7558^{20}	1.4181^{20}	vs eth, EtOH
10669	Tripropylborane		$C_9H_{21}B$	1116-61-6	140.074	liq	-56	159	0.7204^{25}	1.4135^{22}	
10670	Tripropyl borate	Boric acid, tripropyl ester	$C_9H_{21}BO_3$	688-71-1	188.072			179.5	0.8576^{20}	1.3948^{20}	vs EtOH; msc eth; s PrOH
10671	Tripropylene glycol	[(1-Methyl-1,2-ethanediyl) bis(oxy)]bispropanol	$C_9H_{20}O_4$	24800-44-0	192.253	liq		268; 115^{2}	1.02^{20}	1.4440^{20}	
10672	Tripropylene glycol diacrylate		$C_{15}H_{24}O_6$	42978-66-5	300.348			$>120^{1}$			
10673	Tripropylene glycol monomethyl ether	1-[2-(2-Methoxy-1-methylethoxy)-1-methylethoxy]-2-propanol	$C_{10}H_{22}O_4$	20324-33-8	206.280			241.3			
10674	Tripropyl phosphate		$C_9H_{21}O_4P$	513-08-6	224.234			252	1.0121^{20}	1.4165^{20}	sl H_2O, chl; s EtOH, eth, tol, CS_2
10675	Tripropyl phosphite	Tripropoxyphosphine	$C_9H_{21}O_3P$	923-99-9	208.235			206.5	0.9417^{20}	1.4282^{20}	vs eth, EtOH
10676	Tripropylsilane		$C_9H_{22}Si$	998-29-8	158.357			172	0.7723^{0}	1.4280^{20}	i H_2O
10677	Tris(4-aminophenyl)methanol	C.I. Basic Red 9	$C_{19}H_{19}N_3O$	467-62-9	305.373	purp cry	205				
10678	2,4,6-Tris(1-aziridinyl)-1,3,5-triazine	Triethylenemelamine	$C_9H_{12}N_6$	51-18-3	204.231	cry pow	139 dec				s H_2O
10679	Tris(2-butoxyethyl) phosphate		$C_{18}H_{39}O_7P$	78-51-3	398.473	liq		255^{10}	1.02^{25}		i H_2O
10680	Tris(2-chloroethyl) phosphate		$C_6H_{12}Cl_3O_4P$	115-96-8	285.489			330; 194^{10}	1.39^{25}	1.4721^{20}	s ctc
10681	Tris(2-chloroethyl) phosphite		$C_6H_{12}Cl_3O_3P$	140-08-9	269.490			120^{3}	1.3443^{26}	1.4868^{20}	
10682	Tris(1,3-dichloro-2-propyl) phosphate	Fyrol FR-2	$C_9H_{15}Cl_6O_4P$	13674-87-8	430.904	visc liq		236^{5}		1.5022^{20}	i H_2O
10683	Tris(4-dimethylaminophenyl) methane	Paraleucaniline	$C_{25}H_{31}N_3$	603-48-5	373.534	lf (al), nd (bz)	176.5				vs bz, eth, chl
10684	Tris(2,4-dimethylphenyl) phosphate	2,4-Xylenol, phosphate (3:1)	$C_{24}H_{27}O_4P$	3862-12-2	410.442			233.5	1.142^{38}	1.5550^{20}	i H_2O; s bz, chl, hx
10685	Tris(2,5-dimethylphenyl) phosphate	2,5-Xylenol, phosphate (3:1)	$C_{24}H_{27}O_4P$	19074-59-0	410.442		79.8	262^{8}	1.197^{25}		i H_2O; sl EtOH, hx; s eth, bz, ctc

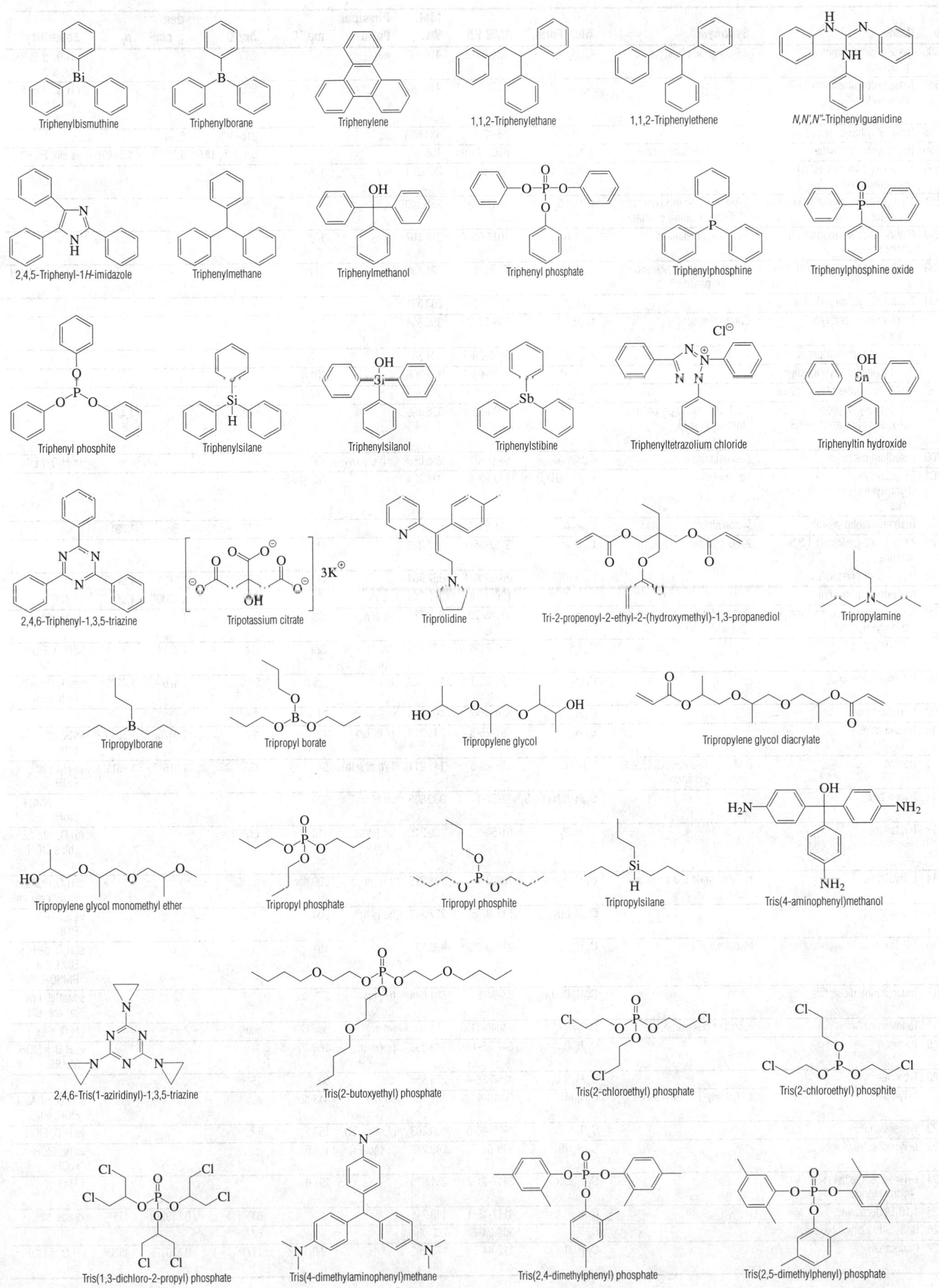

Triphenylbismuthine
Triphenylborane
Triphenylene
1,1,2-Triphenylethane
1,1,2-Triphenylethene
N,N',N''-Triphenylguanidine
2,4,5-Triphenyl-1H-imidazole
Triphenylmethane
Triphenylmethanol
Triphenyl phosphate
Triphenylphosphine
Triphenylphosphine oxide
Triphenyl phosphite
Triphenylsilane
Triphenylsilanol
Triphenylstibine
Triphenyltetrazolium chloride
Triphenyltin hydroxide
2,4,6-Triphenyl-1,3,5-triazine
Tripotassium citrate
Triprolidine
Tri-2-propenoyl-2-ethyl-2-(hydroxymethyl)-1,3-propanediol
Tripropylamine
Tripropylborane
Tripropyl borate
Tripropylene glycol
Tripropylene glycol diacrylate
Tripropylene glycol monomethyl ether
Tripropyl phosphate
Tripropyl phosphite
Tripropylsilane
Tris(4-aminophenyl)methanol
2,4,6-Tris(1-aziridinyl)-1,3,5-triazine
Tris(2-butoxyethyl) phosphate
Tris(2-chloroethyl) phosphate
Tris(2-chloroethyl) phosphite
Tris(1,3-dichloro-2-propyl) phosphate
Tris(4-dimethylaminophenyl)methane
Tris(2,4-dimethylphenyl) phosphate
Tris(2,5-dimethylphenyl) phosphate

No.	Name	Synonym	Mol. Form.	CAS RN	Mol. Wt.	Physical Form	mp/°C	bp/°C	den/ g cm^{-3}	n_D	Solubility
10686	Tris(2,6-dimethylphenyl) phosphate	2,6-Xylenol, phosphate (3:1)	$C_{24}H_{27}O_4P$	121-06-2	410.442	wax	137.8	263^{6}			i H_2O; sl EtOH, hx; s bz
10687	Tris(3,5-dimethylphenyl) phosphate		$C_{24}H_{27}O_4P$	25653-16-1	410.442		46.2	290^{10}			i H_2O; sl EtOH, chl, hx; s HOAc
10688	Tris(2-ethylhexyl) phosphate		$C_{24}H_{51}O_4P$	78-42-2	434.633	liq		215^{5}	0.99^{20}		
10689	Tris(ethylthio)methane	Triethyl orthothioformate	$C_7H_{16}S_3$	6267-24-9	196.397			dec 235; 127^{12}	1.053^{20}	1.5410^{15}	vs eth, EtOH
10690	1,3,5-Tris(2-hydroxyethyl) isocyanuric acid		$C_9H_{15}N_3O_6$	839-90-7	261.231	cry	136				
10691	1,1,1-Tris(hydroxymethyl)ethane trinitrate	2-Methyl-2-[(nitrooxy)methyl]-1,3-propanediol, dinitrate	$C_5H_9N_3O_9$	3032-55-1	255.140			$83^{0.05}$			
10692	*N,N',N''*-Tris(hydroxymethyl) melamine	Trimethylolmelamine	$C_6H_{12}N_6O_3$	1017-56-7	216.197	cry	148				
10693	Tris(hydroxymethyl)methylamine	2-Amino-2-(hydroxymethyl)-1,3-propanediol	$C_4H_{11}NO_3$	77-86-1	121.135		171.5	219^{10}			vs H_2O; s MeOH
10694	Tris(methoxyethoxy)vinylsilane		$C_{11}H_{24}O_6Si$	1067-53-4	280.391						s ctc
10695	Tris(4-methoxyphenyl) chloroethene	Chlorotrianisene	$C_{23}H_{21}ClO_3$	569-57-3	380.864		115				
10696	Tris(2-methylphenyl)phosphine		$C_{21}H_{21}P$	6163-58-2	304.366		127.0				
10697	Tris(3-methylphenyl)phosphine		$C_{21}H_{21}P$	6224-63-1	304.366		101.0				
10698	Tris(4-methylphenyl)phosphine		$C_{21}H_{21}P$	1038-95-5	304.366		147.0				
10699	Tris(2-methyl-2-propenoyl)-2-ethyl-2-hydroxymethyl-1,3-propanediol	1,1,1-Trimethylolpropane trimethacrylate	$C_{18}H_{26}O_6$	3290-92-4	338.395			$>200^{1}$		1.470^{25}	
10700	Trisodium citrate	Sodium citrate	$C_6H_5Na_3O_7$	68-04-2	258.069	wh cry (w)	300				vs H_2O; i EtOH
10701	Trisodium *N*-hydroxyethylethylenediaminetriacetate	Versen-Ol	$C_{10}H_{15}N_2Na_3O_7$	139-89-9	344.204		288 (hyd)				
10702	Tris(perfluorobutyl)amine	Trinonafluorobutylamine	$C_{12}F_{27}N$	311-89-7	671.092			178	1.884^{25}	1.291^{25}	s ace
10703	2,4,6-Tris(2-pyridinyl)-1,3,5-triazine	2,4,6-Tripyridyl-*s*-triazine	$C_{18}H_{12}N_6$	3682-35-7	312.328		210				
10704	Tris(*o*-tolyl) phosphite		$C_{21}H_{21}O_3P$	2622-08-4	352.364		11	238^{11}, 197^{2}	1.1423^{20}	1.5740^{28}	s eth; sl chl
10705	Tris(*p*-tolyl) phosphite		$C_{21}H_{21}O_3P$	620-42-8	352.364	pa ye	52	252^{10}	1.1280^{25}	1.5703^{28}	vs eth
10706	Tris(triphenylphosphine) rhodium carbonyl hydride	Carbonylhydrotris(triphenylphosphine)rhodium	$C_{55}H_{46}OP_3Rh$	17185-29-4	918.781	ye cry	121		1.33		sl bz, chl
10707	1,3,5-Trithiane		$C_3H_6S_3$	291-21-4	138.275	hex (bz), pr (w) nd (al)	220	sub	1.6374^{24}		sl H_2O, EtOH, eth; s bz
10708	Trithiocarbonic acid		CH_2S_3	594-08-1	110.222	red oil	-26.9	57.8	1.476^{25}	1.8225^{20}	dec H_2O, EtOH; vs tol, chl
10709	Tritriacontane		$C_{33}H_{68}$	630-05-7	464.893		71.2				
10710	Tropacocaine		$C_{15}H_{19}NO_2$	537-26-8	245.318	pl or tab	49	dec	1.0426^{100}	1.5080^{100}	vs bz, eth, EtOH, peth
10711	Tropine	8-Methyl-8-azabicyclo[3.2.1] octan-3-ol, *endo*	$C_8H_{15}NO$	120-29-6	141.211	hyg pl (eth)	64	233	1.016^{100}	1.4811^{100}	vs H_2O, eth, EtOH
10712	Trypan blue		$C_{34}H_{24}N_6Na_4O_{14}S_4$	72-57-1	960.805	dk bl cry	300				s H_2O, acid; i EtOH
10713	Tryptamine		$C_{10}H_{12}N_2$	61-54-1	160.215	nd (al-bz, lig)	118	$137^{0.15}$			i H_2O, eth, bz, chl; s EtOH, ace
10714	*L*-Tryptophan	α-Aminoindole-3-propionic acid, (*l*)	$C_{11}H_{12}N_2O_2$	73-22-3	204.225	lf or pl (dil al)	289 dec				sl H_2O, HOAc; s EtOH; i eth, chl
10715	Tsuduranine		$C_{18}H_{19}NO_3$	517-97-5	297.349	nd (eth)	204				vs ace, eth, EtOH
10716	T-2 Toxin	Mycotoxin T2	$C_{24}H_{34}O_9$	21259-20-1	466.522	nd	151				sl H_2O, peth; s EtOH, chl, DMSO
10717	Tubocurarine dichloride		$C_{37}H_{42}Cl_2N_2O_6$	57-94-3	681.644	hyg cry	275 dec				s MeOH; i py, bz, ace, eth
10718	Tungsten carbonyl	Tungsten hexacarbonyl	C_6O_6W	14040-11-0	351.90	wh cry	dec 170	sub	2.65		i H_2O; s os
10719	Turanose		$C_{12}H_{22}O_{11}$	547-25-1	342.296	pr (w-al, MeOH)	168				vs H_2O; s EtOH, MeOH
10720	Tybamate		$C_{13}H_{26}N_2O_4$	4268-36-4	274.356	cry	50	$151^{0.06}$			
10721	*L*-Tyrosine	4-Hydroxy-*L*-phenylalanine	$C_9H_{11}NO_3$	60-18-4	181.188	nd (w)	343 dec	sub			sl H_2O, HOAc; i EtOH, eth
10722	Tyrosineamide		$C_9H_{12}N_2O_2$	4985-46-0	180.203	pl or pl (al)	153.5				vs H_2O, EtOH
10723	*L*-Tyrosine, ethyl ester		$C_{11}H_{15}NO_3$	949-67-7	209.242	pr (AcOEt)	108.5				vs bz, EtOH, AcOEt
10724	*L*-Tyrosine, methyl ester, hydrochloride		$C_{10}H_{14}ClNO_3$	3417-91-2	231.676		191.0				s H_2O
10725	1,10-Undecadiyne		$C_{11}H_{16}$	4117-15-1	148.245		-17	83^{12}	0.8182^{21}	1.453^{21}	vs ace, bz
10726	Undecafluorocyclohexane		C_6HF_{11}	308-24-7	282.054			62.0			
10727	Undecanal		$C_{11}H_{22}O$	112-44-7	170.292		-2.0	117^{18}	0.8251^{23}	1.4520^{20}	i H_2O; s EtOH, eth

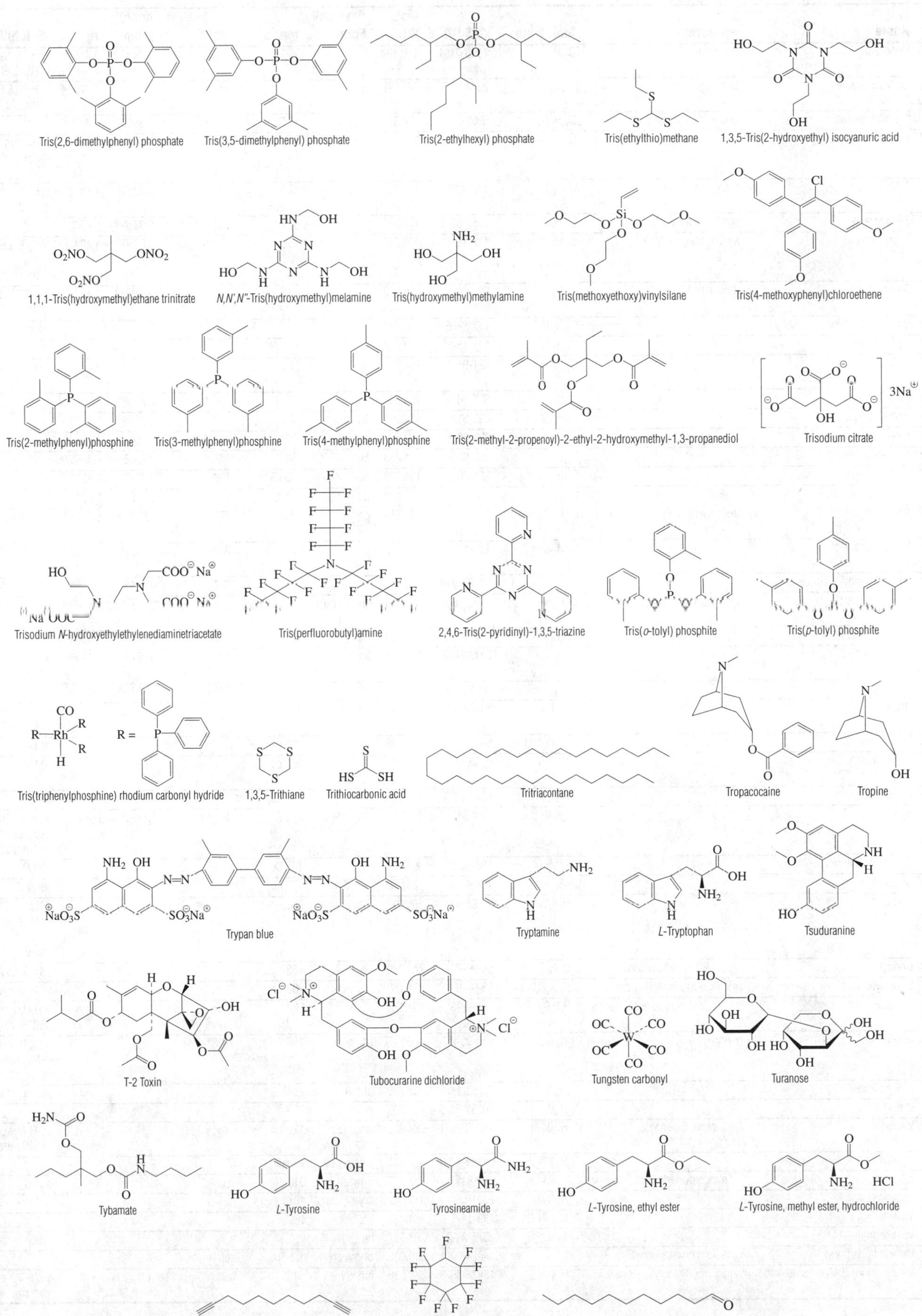
Tris(2,6-dimethylphenyl) phosphate
Tris(3,5-dimethylphenyl) phosphate
Tris(2-ethylhexyl) phosphate
Tris(ethylthio)methane
1,3,5-Tris(2-hydroxyethyl) isocyanuric acid
1,1,1-Tris(hydroxymethyl)ethane trinitrate
N,N',N''-Tris(hydroxymethyl)melamine
Tris(hydroxymethyl)methylamine
Tris(methoxyethoxy)vinylsilane
Tris(4-methoxyphenyl)chloroethene
Tris(2-methylphenyl)phosphine
Tris(3-methylphenyl)phosphine
Tris(4-methylphenyl)phosphine
Tris(2-methyl-2-propenoyl)-2-ethyl-2-hydroxymethyl-1,3-propanediol
Trisodium citrate
Trisodium N-hydroxyethylethylenediaminetriacetate
Tris(perfluorobutyl)amine
2,4,6-Tris(2-pyridinyl)-1,3,5-triazine
Tris(o-tolyl) phosphite
Tris(p-tolyl) phosphite
Tris(triphenylphosphine) rhodium carbonyl hydride
1,3,5-Trithiane
Trithiocarbonic acid
Tritriacontane
Tropacocaine
Tropine
Trypan blue
Tryptamine
L-Tryptophan
Tsuduranine
T-2 Toxin
Tubocurarine dichloride
Tungsten carbonyl
Turanose
Tybamate
L-Tyrosine
Tyrosineamide
L-Tyrosine, ethyl ester
L-Tyrosine, methyl ester, hydrochloride
1,10-Undecadiyne
Undecafluorocyclohexane
Undecanal

No.	Name	Synonym	Mol. Form.	CAS RN	Mol. Wt.	Physical Form	mp/°C	bp/°C	den/ g cm^{-3}	n_D	Solubility
10728	Undecane	Hendecane	$C_{11}H_{24}$	1120-21-4	156.309	liq	-25.5	195.9	0.7402^{20}	1.4164^{20}	i H_2O; msc EtOH, eth
10729	Undecanenitrile	Decyl cyanide	$C_{11}H_{21}N$	2244-07-7	167.292			253	0.8254^{30}	1.4293^{30}	i H_2O; s EtOH, eth, ctc
10730	1-Undecanethiol	Undecyl mercaptan	$C_{11}H_{24}S$	5332-52-5	188.374	liq	-1.5	257.4	0.8448^{20}	1.4585^{20}	
10731	Undecanoic acid		$C_{11}H_{22}O_2$	112-37-8	186.292	cry (ace)	28.6	280	0.8907^{20}	1.4294^{45}	i H_2O; vs EtOH, ace; s eth; msc bz
10732	1-Undecanol	Undecyl alcohol	$C_{11}H_{24}O$	112-42-5	172.308		15.9	245	0.8298^{20}	1.4392^{20}	i H_2O; s EtOH; vs eth
10733	2-Undecanol	*sec*-Undecyl alcohol	$C_{11}H_{24}O$	1653-30-1	172.308	col liq	0	229.7	0.8234^{25}	1.4352^{25}	
10734	2-Undecanone	Methyl nonyl ketone	$C_{11}H_{22}O$	112-12-9	170.292		15	231.5	0.8250^{20}	1.4291^{20}	i H_2O; s EtOH, eth, ace, bz, chl
10735	6-Undecanone	Butyl hexyl ketone	$C_{11}H_{22}O$	927-49-1	170.292		14.5	228	0.8308^{20}	1.4270^{20}	i H_2O; vs EtOH, eth
10736	Undecanoyl chloride		$C_{11}H_{21}ClO$	17746-05-3	204.737						sl ctc
10737	10-Undecenal		$C_{11}H_{20}O$	112-45-8	168.276						sl ctc
10738	1-Undecene		$C_{11}H_{22}$	821-95-4	154.293	liq	-49.2	192.7	0.7503^{20}	1.4261^{20}	i H_2O; s eth, chl, lig
10739	*cis*-2-Undecene		$C_{11}H_{22}$	821-96-5	154.293	liq	-66.5	196.1	0.7576^{20}		
10740	*trans*-2-Undecene		$C_{11}H_{22}$	693-61-8	154.293	liq	-48.3	192.5	0.7528^{20}	1.4292^{20}	
10741	*cis*-4-Undecene		$C_{11}H_{22}$	821-98-7	154.293	liq	-97	192.6	0.7541^{20}	1.4302^{20}	
10742	*trans*-4-Undecene		$C_{11}H_{22}$	693-62-9	154.293	liq	-63.7	193	0.7508^{20}	1.4285^{20}	
10743	*cis*-5-Undecene		$C_{11}H_{22}$	764-96-5	154.293	liq	-106.5	192.3	0.7537^{20}	1.4302^{20}	
10744	*trans*-5-Undecene		$C_{11}H_{22}$	764-97-6	154.293	liq	-61.1	192	0.7497^{20}	1.4285^{20}	vs eth, chl, lig
10745	10-Undecenoic acid	Undecylenic acid	$C_{11}H_{20}O_2$	112-38-9	184.276	cry	24.5	275	0.9072^{24}	1.4486^{24}	i H_2O; s EtOH, eth; sl ctc
10746	10-Undecen-1-ol		$C_{11}H_{22}O$	112-43-6	170.292	liq	-1.0	250	0.8495^{15}	1.4500^{20}	i H_2O; s EtOH, eth; sl ctc
10747	10-Undecenoyl chloride		$C_{11}H_{19}ClO$	38460-95-6	202.721			127^{13}	0.944^{20}	1.454^{20}	
10748	Undecylamine	1-Undecanamine	$C_{11}H_{25}N$	7307-55-3	171.324	cry (eth, al)	17	242	0.7979^{20}	1.4398^{20}	s H_2O, EtOH; i eth; sl ctc
10749	Undecylbenzene		$C_{17}H_{28}$	6742-54-7	232.404	liq	-5	316	0.8553^{20}	1.4828^{20}	
10750	1-Undecyne		$C_{11}H_{20}$	2243-98-3	152.277	liq	-25	196	0.7728^{20}	1.4306^{20}	vs ace, bz, eth, EtOH
10751	2-Undecyne		$C_{11}H_{20}$	60212-29-5	152.277	liq	-30.1	204.2	0.7827^{20}	1.4391^{20}	
10752	Uracil		$C_4H_4N_2O_2$	66-22-8	112.087	nd (w)	338				sl H_2O; vs EtOH, eth; s dil NH_3
10753	Uracil mustard		$C_8H_{11}Cl_2N_3O_2$	66-75-1	252.098		206 dec				sl H_2O
10754	Uranyl acetate dihydrate		$C_4H_{10}O_8U$	6159-44-0	424.146	ye cry (HOAc)	80 dec		2.89		sl EtOH
10755	Urazole		$C_2H_3N_3O_2$	3232-84-6	101.064	lf (w)	249 dec				
10756	Urea	Carbamide	CH_4N_2O	57-13-6	60.055	tetr pr (al)	133.3	dec	1.3230^{20}	1.484	vs H_2O, EtOH; i eth, bz; s HOAc, py
10757	Urea hydrochloride		CH_5ClN_2O	506-89-8	96.516		145 dec				s H_2O
10758	Urea nitrate		$CH_5N_3O_4$	124-47-0	123.069	mcl lf (w)	152 dec		1.690^{20}		vs EtOH
10759	Uric acid		$C_5H_4N_4O_3$	69-93-2	168.111	orth pr or pl	dec	dec	1.89^{25}		i H_2O, EtOH, eth; s alk, glycerol; sl acid
10760	Uridine	1-β-*D*-Ribofuranosyluracil	$C_9H_{12}N_2O_6$	58-96-8	244.200	nd (aq al)	165				s H_2O, EtOH, py
10761	5′-Uridylic acid	Uridine 5′-phosphoric acid	$C_9H_{13}N_2O_9P$	58-97-9	324.180	pr (MeOH)	202 dec				vs H_2O; s MeOH
10762	Urocanic acid	Imidazole-4-acrylic acid	$C_6H_6N_2O_2$	104-98-3	138.124		227				s H_2O, ace; i EtOH, eth
10763	Urs-12-en-3-ol, (3β)	α-Amyrin	$C_{30}H_{50}O$	638-95-9	426.717	nd (al)	186	243$^{0.5}$			s EtOH, eth, bz, chl, HOAc; sl peth
10764	Ursolic acid		$C_{30}H_{48}O_3$	77-52-1	456.700	pl (al)	284				vs ace, eth, chl
10765	Uzarin		$C_{35}H_{54}O_{14}$	20231-81-6	698.796	pr	269				
10766	Vacciniin	*D*-Glucose, 6-benzoate	$C_{13}H_{16}O_7$	14200-76-1	284.262	amor (aq ace, +1w)	122				vs H_2O, ace, EtOH, eth
10767	Validamycin A		$C_{20}H_{35}NO_{13}$	37248-47-8	497.491	amorp pow	95 dec				
10768	*L*-Valine	2-Aminoisovaleric acid	$C_5H_{11}NO_2$	72-18-4	117.147	lf (w-al)	315	sub	1.23^{25}		s H_2O
10769	Valinomycin		$C_{54}H_{90}N_6O_{18}$	2001-95-8	1111.322	cry	187				
10770	Valium		$C_{16}H_{13}ClN_2O$	439-14-5	284.739		132				
10771	Vamidothion		$C_8H_{18}NO_4PS_2$	2275-23-2	287.337	oil					i peth; s os
10772	Vanadium carbonyl	Vanadium hexacarbonyl	C_6O_6V	14024-00-1	219.002	bl-grn cry	dec 60	sub			
10773	Vanadium(III) 2,4-pentanedioate	Vanadium(III) acetylacetonate	$C_{15}H_{21}O_6V$	13476-99-8	348.266	brn cry	≈185	sub	≈1.0		s MeOH, ace, bz, chl

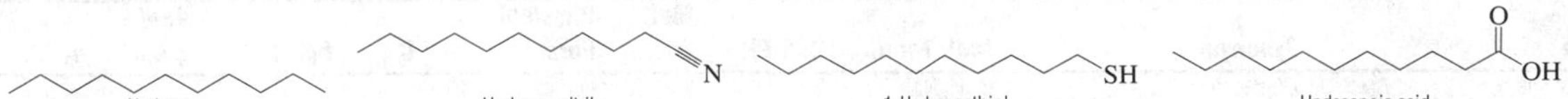

Undecane

Undecanenitrile

1-Undecanethiol

Undecanoic acid

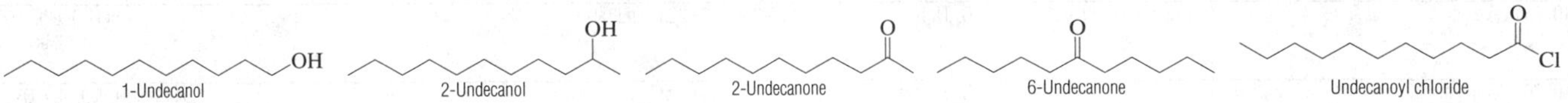

1-Undecanol

2-Undecanol

2-Undecanone

6-Undecanone

Undecanoyl chloride

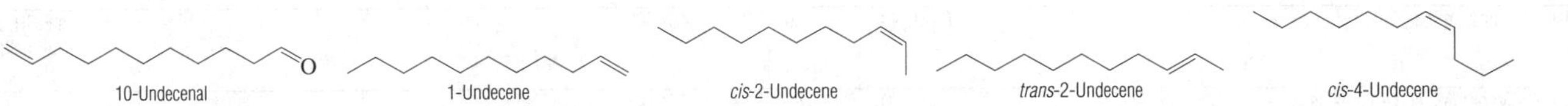

10-Undecenal

1-Undecene

cis-2-Undecene

trans-2-Undecene

cis-4-Undecene

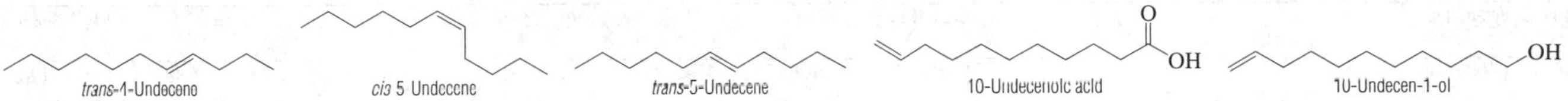

trans-4-Undecene

cis-5-Undecene

trans-5-Undecene

10-Undecenoic acid

10-Undecen-1-ol

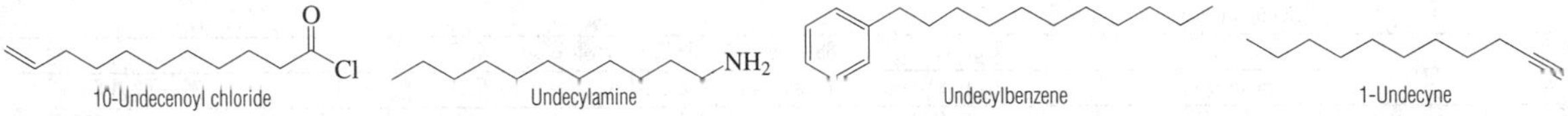

10-Undecenoyl chloride

Undecylamine

Undecylbenzene

1-Undecyne

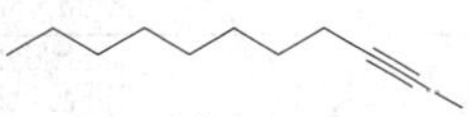

2-Undecyne

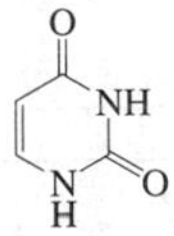

Uracil

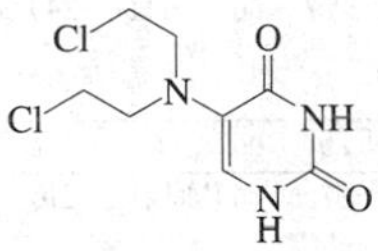

Uracil mustard

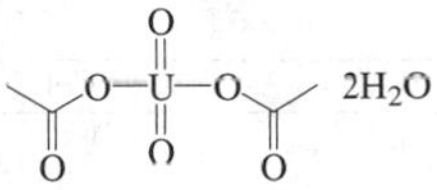

Uranyl acetate dihydrate

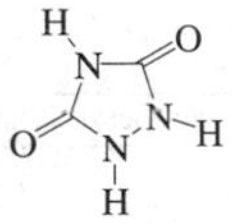

Urazole

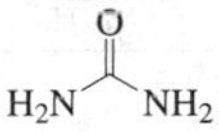

Urea

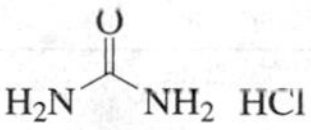

Urea hydrochloride

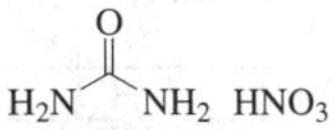

Urea nitrate

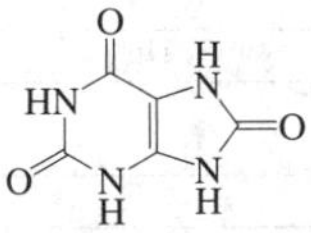

Uric acid

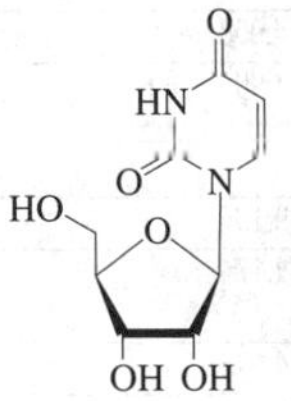

Uridine

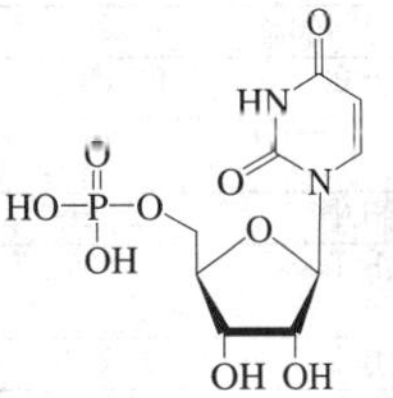

5'-Uridylic acid

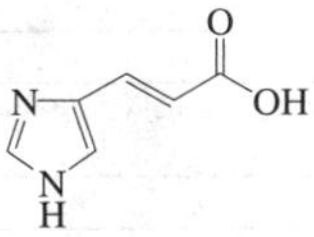

Urocanic acid

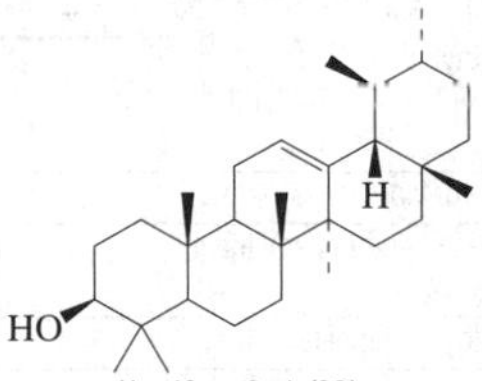

Urs-12-en-3-ol, (3β)

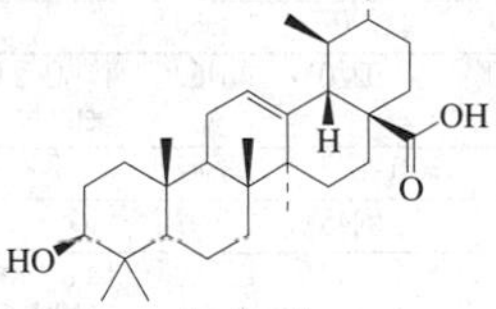

Ursolic acid

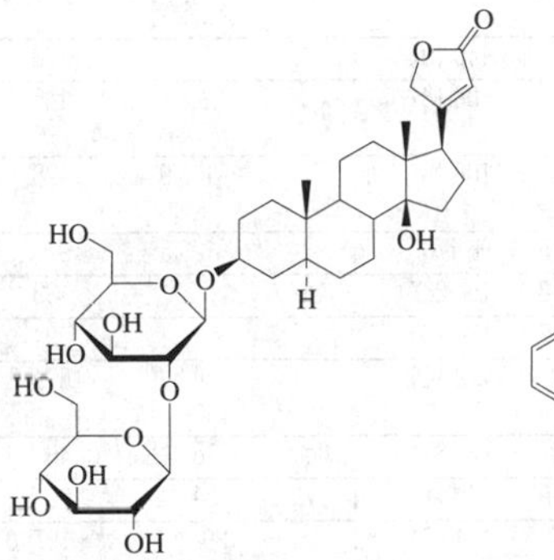

Uzarin

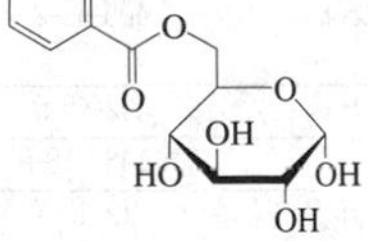

Vacciniin

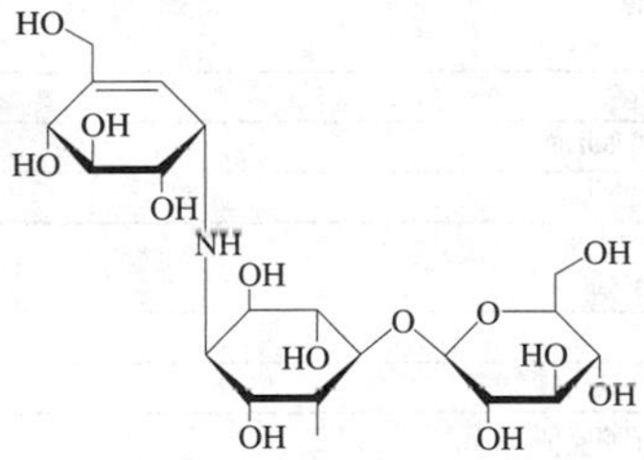

Validamycin A

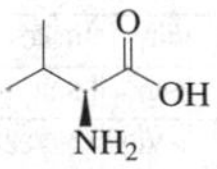

L-Valine

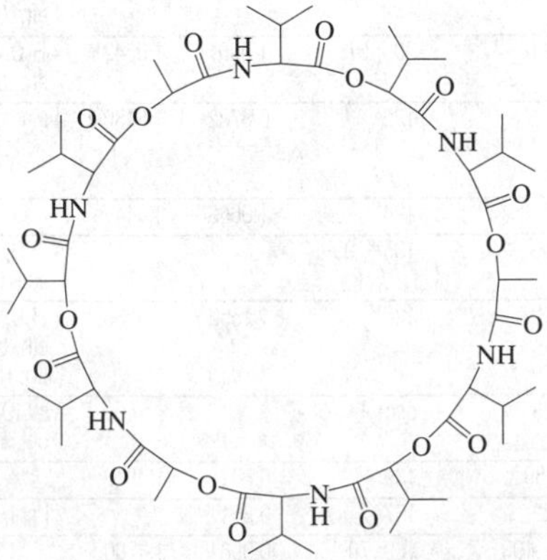

Valinomycin

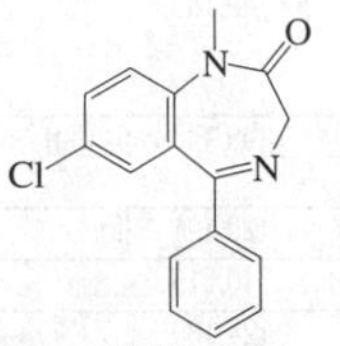

Valium

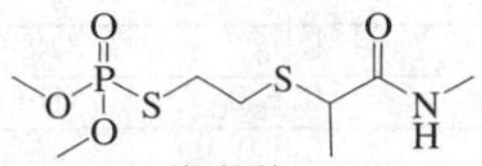

Vamidothion

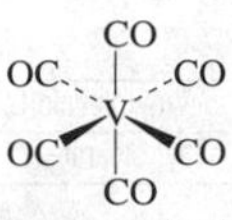

Vanadium carbonyl

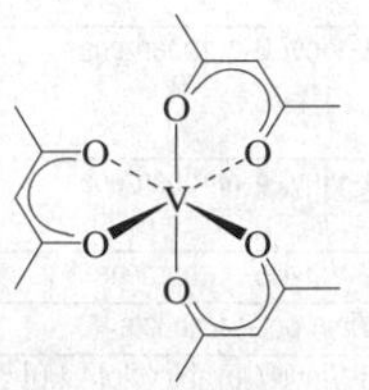

Vanadium(III) 2,4-pentanedioate

No.	Name	Synonym	Mol. Form.	CAS RN	Mol. Wt.	Physical Form	mp/°C	bp/°C	den/ g cm^{-3}	n_D	Solubility
10774	*DL*-Vasicine	*DL*-Peganine	$C_{11}H_{12}N_2O$	6159-56-4	188.225	nd (al)	210.8				sl H_2O, eth, bz; s EtOH, ace, chl
10775	*L*-Vasicine	*L*-Peganine	$C_{11}H_{12}N_2O$	6159-55-3	188.225	nd (al)	211.5				sl H_2O, eth, bz; s EtOH, ace, chl
10776	Verapamil		$C_{27}H_{38}N_2O_4$	52-53-9	454.602	ye oil		$245^{0.01}$		1.5448^{25}	i H_2O; vs EtOH, ace; sl bz, hx
10777	Veratramine		$C_{27}H_{39}NO_2$	60-70-8	409.605	nd	206				s EtOH, bz, chl, dil acid; i dil alk
10778	Veratramine, 3-glucoside		$C_{33}H_{49}NO_7$	475-00-3	571.745	nd (aq. MeOH)	242 dec				
10779	Veratridine		$C_{36}H_{51}NO_{11}$	71-62-5	673.790	ye amorp pow	180				i H_2O; sl eth
10780	*d*-Verbenone		$C_{10}H_{14}O$	18309-32-5	150.217		9.8	227.5	0.9978^{20}	1.4993^{18}	s H_2O, EtOH, ace, bz
10781	Vernolate	Carbamothioic acid, dipropyl-, *S*-propyl ester	$C_{10}H_{21}NOS$	1929-77-7	203.345			150^{30}	0.952^{20}		
10782	Versalide		$C_{18}H_{26}O$	88-29-9	258.398	cry	46.5	130^{2}			s EtOH
10783	α-Vetivone	Isonootkatone	$C_{15}H_{22}O$	15764-04-2	218.335	cry (peth)	51.5	144^{2}	1.0035^{20}	1.5370^{20}	vs ace
10784	β-Vetivone		$C_{15}H_{22}O$	18444-79-6	218.335	cry (peth)	44.5	141^{2}	1.0001^{20}	1.5309^{20}	s ace
10785	Vicine	2,6-Diamino-5-(β-*D*-glucopyranosyloxy)-4(1*H*)-pyrimidinone	$C_{10}H_{16}N_4O_7$	152-93-2	304.257	nd (w, dil al, +1 w)	240 dec				sl H_2O, EtOH; vs acid, alk
10786	Vidarabine	β-*D*-9-Arabinofuranosyladenine	$C_{10}H_{15}N_5O_5$	5536-17-4	285.257	nd (w)	257				
10787	Vinblastine		$C_{46}H_{58}N_4O_9$	865-21-4	810.975	nd (MeOH)	216				i H_2O; s EtOH, ace, chl, AcOEt
10788	Vincamine		$C_{21}H_{26}N_2O_3$	1617-90-9	354.442		231.5				
10789	Vinclozolin		$C_{12}H_9Cl_2NO_3$	50471-44-8	286.110		108	$131^{0.05}$	1.51		
10790	Vincristine		$C_{46}H_{56}N_4O_{10}$	57-22-7	824.958		219				
10791	Vinyl acetate		$C_4H_6O_2$	108-05-4	86.090	liq	-93.2	72.8	0.9256^{25}	1.3926^{25}	sl H_2O; msc EtOH; s eth, ace, bz, chl
10792	4-Vinylaniline		C_8H_9N	1520-21-4	119.164		23.5	116^{9}	1.010^{20}	1.6250^{22}	s ace, bz
10793	α-Vinylbenzenemethanol	1-Phenylallyl alcohol	$C_9H_{10}O$	4393-06-0	134.174				1.0249^{21}	1.5406^{20}	sl H_2O; s EtOH, eth, bz, chl
10794	Vinyl butanoate		$C_6H_{10}O_2$	123-20-6	114.142			116.7; 64^{130}	0.9006^{20}		
10795	Vinyl *trans*-2-butenoate	Vinyl crotonate	$C_6H_8O_2$	3234-54-6	112.127						s ctc
10796	9-Vinyl-9*H*-carbazole		$C_{14}H_{11}N$	1484-13-5	193.244	cry (al)	66				i H_2O; sl EtOH; vs eth
10797	Vinylcyclohexane		C_8H_{14}	695-12-5	110.197			128	0.8166^{19}	1.455^{19}	
10798	1-Vinylcyclohexene		C_8H_{12}	2622-21-1	108.181			145	0.8623^{15}	1.4915^{20}	i H_2O; s eth, bz; vs MeOH
10799	4-Vinylcyclohexene		C_8H_{12}	100-40-3	108.181	liq	-108.9	128	0.8299^{20}	1.4639^{20}	i H_2O; s eth, bz, peth
10800	Vinylcyclopentane		C_7H_{12}	3742-34-5	96.170	liq	-126.5	97	0.7834^{20}	1.4360^{20}	
10801	Vinyldiethoxymethylsilane		$C_7H_{16}O_2Si$	5507-44-8	160.287			133	0.8620^{20}	1.4001^{20}	
10802	Vinylethoxydimethylsilane		$C_6H_{14}OSi$	5356-83-2	130.260			99	0.790^{20}	1.3983^{20}	
10803	1-Vinyl-4-fluorobenzene		C_8H_7F	405-99-2	122.140		-34.5	67.4^{50}, 30^{4}	1.0220^{20}	1.5150^{20}	i H_2O; s EtOH, eth, bz
10804	Vinyl formate		$C_3H_4O_2$	692-45-5	72.063	visc liq	-78	46	0.965^{20}	1.3842^{20}	
10805	2-Vinylfuran		C_6H_6O	1487-18-9	94.111	liq	-94	99.5	0.9445^{19}	1.4992^{19}	
10806	1-Vinyl-2-methoxybenzene		$C_9H_{10}O$	612-15-7	134.174	nd	29	197; 83^{12}	1.0049^{17}	1.5388^{20}	vs ace, bz, eth, EtOH
10807	1-Vinyl-3-methoxybenzene		$C_9H_{10}O$	626-20-0	134.174			91^{15}, 70^{5}	0.9919^{20}	1.5586^{23}	i H_2O; s EtOH, eth, bz
10808	1-Vinyl-4-methoxybenzene		$C_9H_{10}O$	637-69-4	134.174		2.0	205; 91^{13}	1.0001^{13}	1.5642^{13}	i H_2O; s EtOH, eth, bz; sl ctc
10809	6-Vinyl-6-methyl-1-isopropyl-3-(1-methylethylidene) cyclohexene, (*S*)		$C_{15}H_{24}$	5951-67-7	204.352			125^{8}	0.8782^{20}	1.5130^{26}	vs ace, bz
10810	1-Vinylnaphthalene		$C_{12}H_{10}$	826-74-4	154.207			124^{15}	1.0656^{20}	1.644^{20}	
10811	2-Vinylnaphthalene		$C_{12}H_{10}$	827-54-3	154.207		66	135^{18}, 95^{2}			i H_2O; s EtOH, ace, bz
10812	1-Vinyl-3-nitrobenzene		$C_8H_7NO_2$	586-39-0	149.148		-10	120^{11}	1.1552^{32}	1.5836^{20}	i H_2O; s EtOH, eth, bz, chl, lig, HOAc
10813	1-Vinyl-4-nitrobenzene		$C_8H_7NO_2$	100-13-0	149.148	pr (lig)	29	dec			vs EtOH, eth; s chl, HOAc, lig
10814	2-Vinyl-5-norbornene	5-Vinylbicyclo[2.2.1]hept-2-ene	C_9H_{12}	3048-64-4	120.191	liq	-80	139	0.841	1.4810^{20}	
10815	Vinyl octadecanoate	Vinyl stearate	$C_{20}H_{38}O_2$	111-63-7	310.515		29	167^{2}	0.8517^{20}		sl chl
10816	3-Vinyl-7-oxabicyclo[4.1.0] heptane		$C_8H_{12}O$	106-86-5	124.180		<-100	169; 70^{20}	0.9581^{20}	1.4700^{20}	

DL-Vasicine

L-Vasicine

Verapamil

Veratramine

Veratramine, 3-glucoside

Veratridine

d-Verbenone

Vernolate

Versalide

α-Vetivone

β-Vetivone

Vicine

Vidarabine

Vinblastine

Vincamine

Vinclozolin

Vincristine

Vinyl acetate

4-Vinylaniline

α-Vinylbenzenemethanol

Vinyl butanoate

Vinyl trans-2-butenoate

9-Vinyl-9H-carbazole

Vinylcyclohexane

1-Vinylcyclohexene

4-Vinylcyclohexene

Vinylcyclopentane

Vinyldiethoxymethylsilane

Vinylethoxydimethylsilane

1-Vinyl-4-fluorobenzene

Vinyl formate

2-Vinylfuran

1-Vinyl-2-methoxybenzene

1-Vinyl-3-methoxybenzene

1-Vinyl-4-methoxybenzene

6-Vinyl-6-methyl-1-isopropyl-3-(1-methylethylidene)cyclohexene, (S)

1-Vinylnaphthalene

2-Vinylnaphthalene

1-Vinyl-3-nitrobenzene

1-Vinyl-4-nitrobenzene

2-Vinyl-5-norbornene

Vinyl octadecanoate

3-Vinyl-7-oxabicyclo[4.1.0]heptane

No.	Name	Synonym	Mol. Form.	CAS RN	Mol. Wt.	Physical Form	mp/°C	bp/°C	den/ g cm^{-3}	n_D	Solubility
10817	Vinyloxirane		C_4H_6O	930-22-3	70.090			68	0.9006^{25}	1.4168^{20}	s EtOH, eth, bz
10818	2-(Vinyloxy)ethanol	Ethylene glycol monovinyl ether	$C_4H_8O_2$	764-48-7	88.106			141.6	0.9821^{20}	1.4564^{17}	s H_2O, EtOH, eth, bz; i lig
10819	Vinyl propanoate	Vinyl propionate	$C_5H_8O_2$	105-38-4	100.117			91.2			
10820	2-Vinylpyridine		C_7H_7N	100-69-6	105.138			159.5	0.9983^{20}	1.5495^{20}	sl H_2O; vs EtOH, eth, ace, chl
10821	3-Vinylpyridine		C_7H_7N	1121-55-7	105.138			162	0.9879^{20}	1.5530^{20}	sl H_2O; s EtOH, eth
10822	4-Vinylpyridine		C_7H_7N	100-43-6	105.138	red to dk-br		121^{150}, 79^{33}	0.9879^{20}	1.5449^{20}	s H_2O, EtOH, chl; sl eth
10823	1-Vinyl-2-pyrrolidinone		C_6H_9NO	88-12-0	111.141		13.5	193^{400}, 93^{11}	1.04^{20}		
10824	Vinylsilane		C_2H_6Si	7291-09-0	58.155	col gas	-171.6	-22.8			
10825	Vinyl sulfoxide		C_4H_6OS	1115-15-7	102.155	liq		86^{18}			
10826	Vinyltriacetoxysilane	Vinylsilanetriol, triacetate	$C_8H_{12}O_6Si$	4130-08-9	232.263			115^{10}	1.169^{20}	1.4226^{20}	
10827	Vinyltriethoxysilane		$C_8H_{18}O_3Si$	78-08-0	190.313			160; 62^{20}	0.901^{20}	1.3960^{25}	s chl
10828	Vinyltrimethylsilane		$C_5H_{12}Si$	754-05-2	100.235			55	0.65^{20}	1.3914^{20}	i H_2O
10829	Violaxanthin		$C_{40}H_{56}O_4$	126-29-4	600.871	red pr (MeOH, al-eth)	208				s EtOH, eth, CS_2; i peth
10830	Viquidil		$C_{20}H_{24}N_2O_2$	84-55-9	324.417	red ye amor	60				vs eth, EtOH, chl
10831	Visnadine		$C_{21}H_{24}O_7$	477-32-7	388.412	nd	85.5				i H_2O; s EtOH, eth
10832	Visnagin	4-Methoxy-7-methyl-5*H*-furo[3,2-g][1]benzopyran-5-one	$C_{13}H_{10}O_4$	82-57-5	230.216	nd (w, MeOH)	144.5				sl H_2O, EtOH; vs chl
10833	Vitamin B12	Cyanocobalamin	$C_{63}H_{88}CoN_{14}O_{14}P$	68-19-9	1355.365		>300				
10834	Vitamin D2		$C_{28}H_{44}O$	50-14-6	396.648	pr (ace)	116.5	sub			i H_2O; s EtOH, eth, ace, chl
10835	Vitamin D3	9,10-Secocholesta-5,7,10(19)-trien-3-ol, (3β,5*Z*,7*E*)-	$C_{27}H_{44}O$	67-97-0	384.637		84.5				i H_2O; s os
10836	Vitamin E	α-Tocopherol	$C_{29}H_{50}O_2$	59-02-9	430.706	pale ye oil	3.0	$210^{0.1}$	0.950^{25}	1.5045^{25}	i H_2O; s EtOH, eth, ace, chl
10837	Vitamin E acetate		$C_{31}H_{52}O_3$	58-95-7	472.743		-27.5	$184^{0.01}$	0.9533^{21}	1.497^{20}	i H_2O; sl EtOH; s eth, ace, chl
10838	Vitamin K1		$C_{31}H_{46}O_2$	84-80-0	450.696		-20	$142^{0.001}$	0.964^{25}	1.5250^{25}	i H_2O; s EtOH, eth, ace, bz, peth, chl
10839	Vomicine	4-Hydroxy-19-methyl-16,19-secostrychnidine-10,16-dione	$C_{22}H_{24}N_2O_4$	125-15-5	380.437	nd (80% al) pr (ace)	282				sl EtOH, eth, ace; vs chl; s AcOEt
10840	Warfarin	Coumadin	$C_{19}H_{16}O_4$	81-81-2	308.328	cry (al)	161				i H_2O; s EtOH, ace, diox
10841	9*H*-Xanthene	10*H*-9-Oxaanthracene	$C_{13}H_{10}O$	92-83-1	182.217	ye lf (al)	100.5	311			i H_2O; sl EtOH, ctc; s eth, bz, chl
10842	9*H*-Xanthen-9-ol		$C_{13}H_{10}O_2$	90-46-0	198.217	nd (aq al)	125				sl H_2O; s EtOH, eth, chl
10843	Xanthine		$C_5H_4N_4O_2$	69-89-6	152.112	ye pl (w)	dec	sub			i H_2O
10844	Xanthone		$C_{13}H_8O_2$	90-47-1	196.202	nd (al)	174	351; 146^3			i H_2O; s EtOH, eth, bz, chl; sl peth
10845	Xanthopterin		$C_6H_5N_5O_2$	119-44-8	179.137	hyg ye amor or oran pow (HOAc)	>410 dec	99^{18}	1.559^{25}		i H_2O; sl EtOH, eth; vs acid, alk
10846	Xanthosine		$C_{10}H_{12}N_4O_6$	146-80-5	284.225	pr cry (w)					sl cold H_2O; vs hot H_2O; dec acid
10847	Xanthoxyletin		$C_{15}H_{14}O_4$	84-99-1	258.270	pr (MeOH, peth)	133				i H_2O; s EtOH, ace; sl eth; vs bz, alk
10848	Xanthyletin	8,8-Dimethyl-2*H*,8*H*-benzo[1,2-b:5,4-b']dipyran-2-one	$C_{14}H_{12}O_3$	553-19-5	228.243	pr (MeOH)	131.5	$142^{0.1}$			s EtOH, peth
10849	*p*-Xenylcarbimide	4-Isocyanato-1,1'-biphenyl	$C_{13}H_9NO$	92-95-5	195.216	nd	56	dec 283			vs eth
10850	Xibenolol		$C_{15}H_{25}NO_2$	81584-06-7	251.366	cry	57	$135^{0.7}$			s EtOH
10851	*o*-Xylene	1,2-Dimethylbenzene	C_8H_{10}	95-47-6	106.165	liq	-25.2	144.5	0.8755^{25}	1.5018^{25}	i H_2O; msc EtOH, eth, ace, bz, peth, ctc
10852	*m*-Xylene	1,3-Dimethylbenzene	C_8H_{10}	108-38-3	106.165	liq	-47.8	139.07	0.8598^{25}	1.4944^{25}	i H_2O; msc EtOH, eth, ace, bz; s chl
10853	*p*-Xylene	1,4-Dimethylbenzene	C_8H_{10}	106-42-3	106.165	mcl pr (al)	13.25	138.23	0.8565^{25}	1.4929^{25}	i H_2O; msc EtOH, eth, ace, bz; s chl

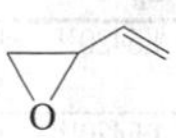

Vinyloxirane
2-(Vinyloxy)ethanol 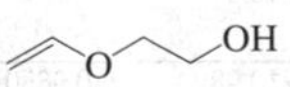
Vinyl propanoate 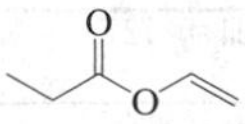
2-Vinylpyridine 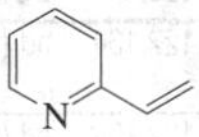
3-Vinylpyridine 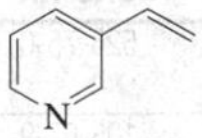
4-Vinylpyridine 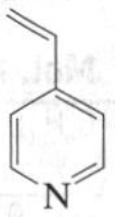
1-Vinyl-2-pyrrolidinone 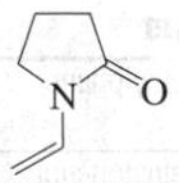
Vinylsilane 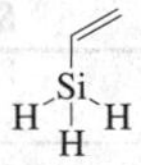
Vinyl sulfoxide 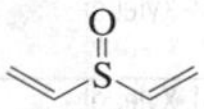

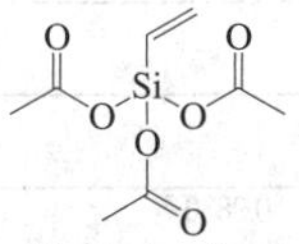

Vinyltriacetoxysilane
Vinyltriethoxysilane 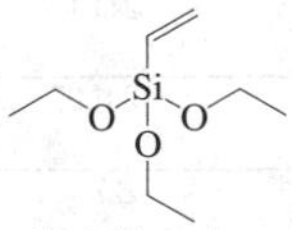
Vinyltrimethylsilane 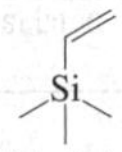
Violaxanthin 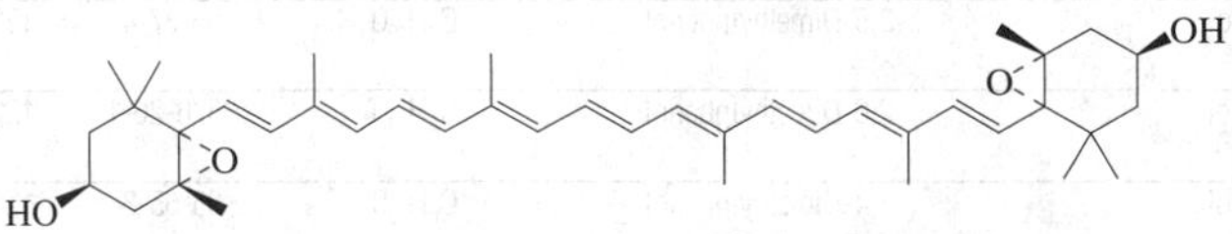

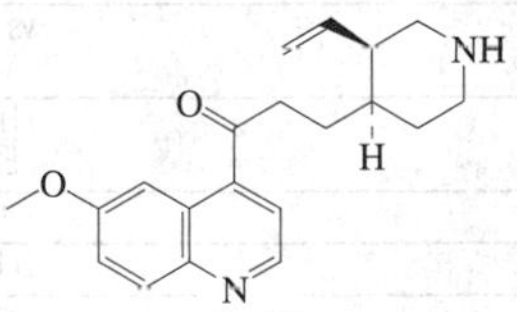

R = CN

Viquidil
Visnadine 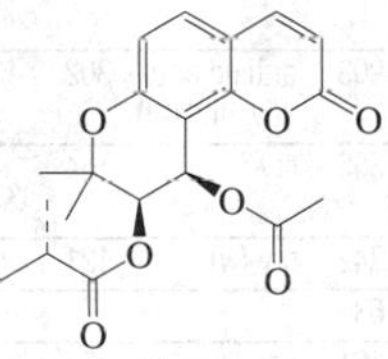
Visnagin 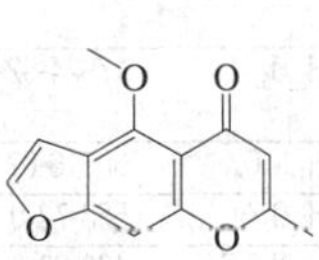 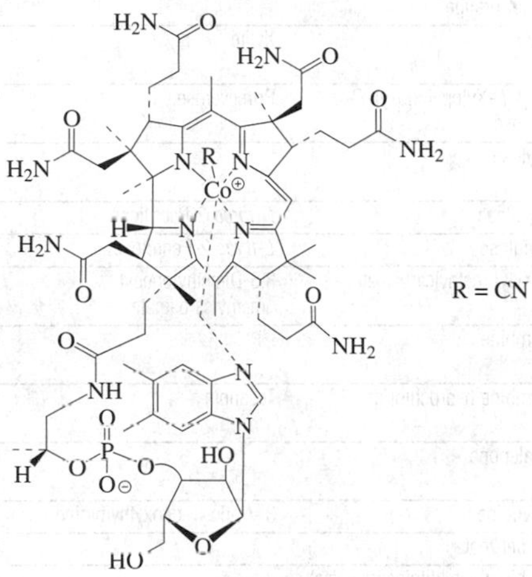
Vitamin B12

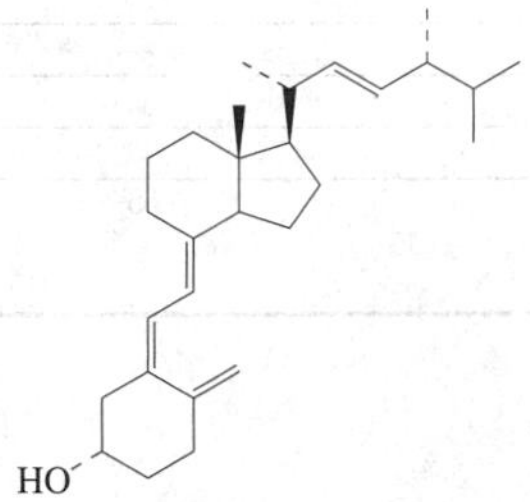

Vitamin D2
Vitamin D3 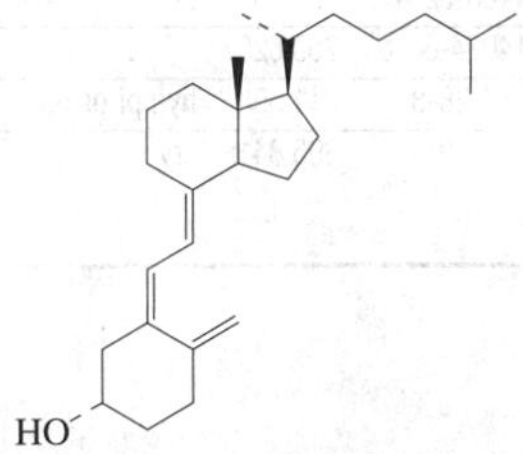
Vitamin E 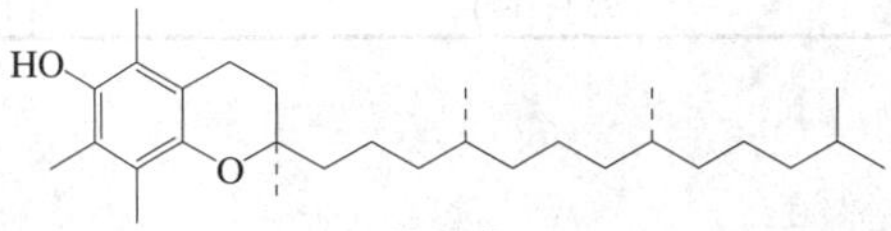

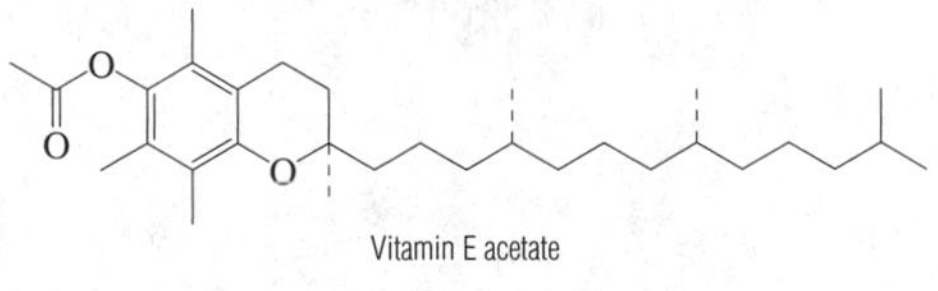

Vitamin E acetate
Vitamin K1 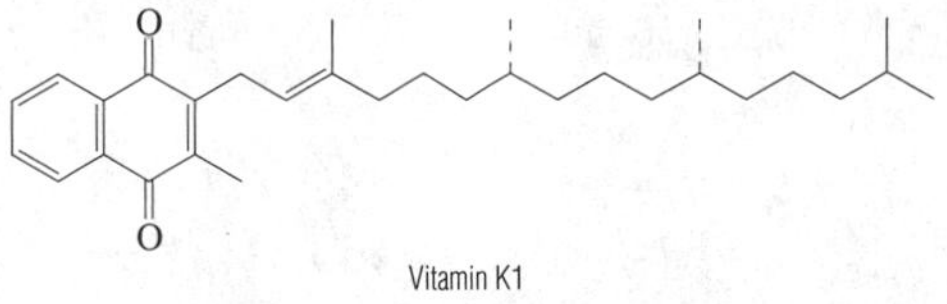
Vomicine 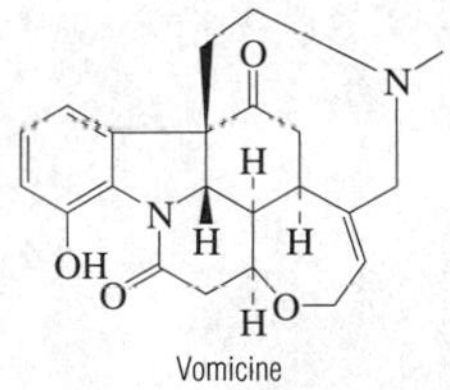

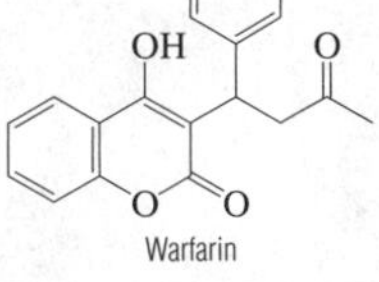

Warfarin
9*H*-Xanthene 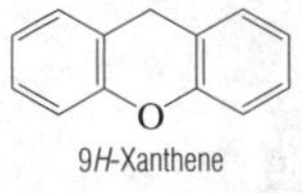
9*H*-Xanthen-9-ol 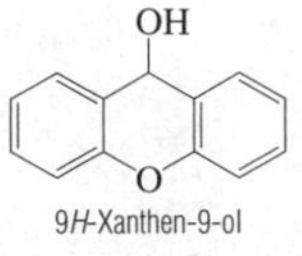
Xanthine 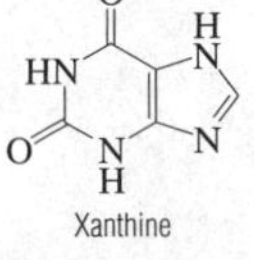
Xanthone 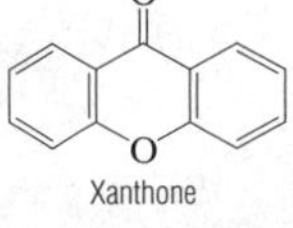
Xanthopterin 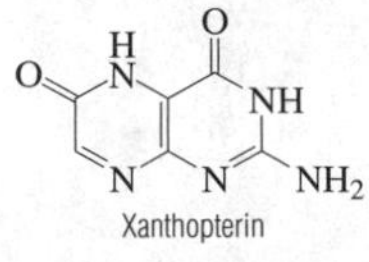
Xanthosine 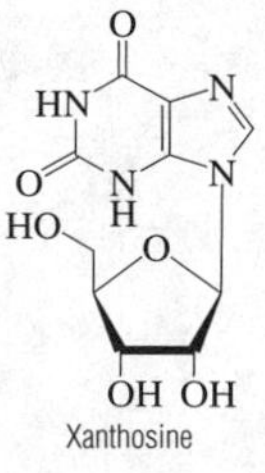

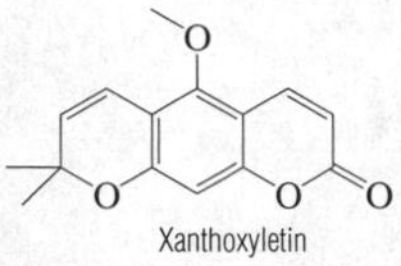

Xanthoxyletin
Xanthyletin 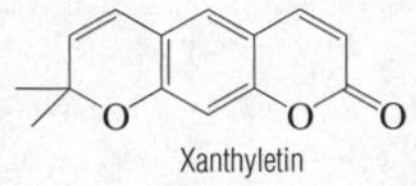
p-Xenylcarbimide 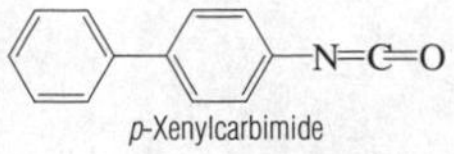
Xibenolol 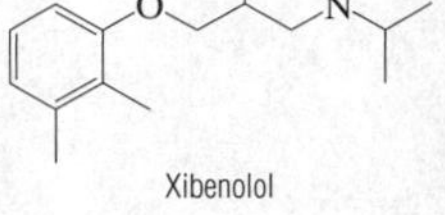
o-Xylene 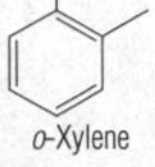
m-Xylene 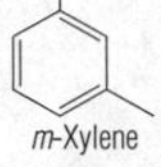
p-Xylene

No.	Name	Synonym	Mol. Form.	CAS RN	Mol. Wt.	Physical Form	mp/°C	bp/°C	den/ g cm-3	n_D	Solubility
10854	2,3-Xylenol	2,3-Dimethylphenol	$C_8H_{10}O$	526-75-0	122.164	nd (w, dil al)	72.5	216.9		1.5420[20]	sl H_2O; s EtOH, eth
10855	2,4-Xylenol	2,4-Dimethylphenol	$C_8H_{10}O$	105-67-9	122.164	nd (w)	24.5	210.98	0.9650[20]	1.5420[14]	sl H_2O; msc EtOH, eth; s ctc
10856	2,5-Xylenol	2,5-Dimethylphenol	$C_8H_{10}O$	95-87-4	122.164	nd (w), pr (al-eth)	74.8	211.1			s H_2O, EtOH; vs eth; sl chl
10857	2,6-Xylenol	2,6-Dimethylphenol	$C_8H_{10}O$	576-26-1	122.164	lf or nd (al)	45.8	201.07			s H_2O, EtOH, eth, ctc
10858	3,4-Xylenol	3,4-Dimethylphenol	$C_8H_{10}O$	95-65-8	122.164		65.1	227	0.9830[20]		sl H_2O; s EtOH, ctc; msc eth
10859	3,5-Xylenol	3,5-Dimethylphenol	$C_8H_{10}O$	108-68-9	122.164	nd (w, peth)	63.4	221.74	0.9680[20]		s H_2O, EtOH, ctc
10860	Xylenol orange		$C_{31}H_{32}N_2O_{13}S$	1611-35-4	672.656	dk red cry	286 dec				s H_2O
10861	Xylitol	Xylite	$C_5H_{12}O_5$	87-99-0	152.146	mcl (al)	93.5	216			vs H_2O, py, EtOH
10862	6-*O*-β-*D*-Xylopyranosyl-*D*-glucose	Primeverose	$C_{11}H_{20}O_{10}$	26531-85-1	312.271	cry (MeOH)	210				vs H_2O, MeOH
10863	*D*-Xylose		$C_5H_{10}O_5$	58-86-6	150.130	mcl nd	90.5		1.525[20]		vs H_2O; s EtOH; sl eth
10864	*D*-Xylulose	*D-threo*-2-Pentulose	$C_5H_{10}O_5$	551-84-8	150.130	visc liq					s H_2O
10865	*L*-Xylulose	*L-threo*-2-Pentulose	$C_5H_{10}O_5$	527-50-4	150.130	syrup					vs H_2O
10866	3,5-Xylyl methylcarbamate	3,5-Dimethylphenyl methylcarbamate	$C_{10}H_{13}NO_2$	2655-14-3	179.216	cry	99				sl H_2O; s os
10867	Yohimbine		$C_{21}H_{26}N_2O_3$	146-48-5	354.442	nd (dil al)	241	sub 160			sl H_2O, bz; s EtOH, eth, chl
10868	Yohimbine hydrochloride	Tosanpin	$C_{21}H_{27}ClN_2O_3$	65-19-0	390.903	orth nd or pl (w, dil HCl)	302				vs H_2O
10869	Zearalenone		$C_{18}H_{22}O_5$	17924-92-4	318.365	cry	164				i H_2O; s alk, bz, EtOH, eth
10870	Zidovudine	3'-Azido-3'-deoxythymidine	$C_{10}H_{13}N_5O_4$	30516-87-1	267.242	cry (w)	121				
10871	Zinc benzoate		$C_{14}H_{10}O_4Zn$	553-72-0	307.636						sl H_2O
10872	Zinc bis(dibutyldithiocarbamate)		$C_{18}H_{36}N_2S_4Zn$	136-23-2	474.161	cry	138				
10873	Zinc *N,N'*-ethylenebisdithiocarbamate	Zineb	$C_4H_6N_2S_4Zn$	12122-67-7	275.773		157 dec				
10874	Zinc gluconate		$C_{12}H_{22}O_{14}Zn$	4468-02-4	455.704	pow					
10875	Zinc 2,4-pentanedioate	Zinc acetylacetonate	$C_{10}H_{14}O_4Zn$	14024-63-6	263.625						s DMSO
10876	Zinc propanoate		$C_6H_{10}O_4Zn$	557-28-8	211.550	hyg pl or nd					sl EtOH
10877	Ziram	Zinc, bis(dimethylcarbamodithioato-*S,S'*)-, (T-4)-	$C_6H_{12}N_2S_4Zn$	137-30-4	305.841	cry	250		1.66[25]		i H_2O; sl bz; s chl

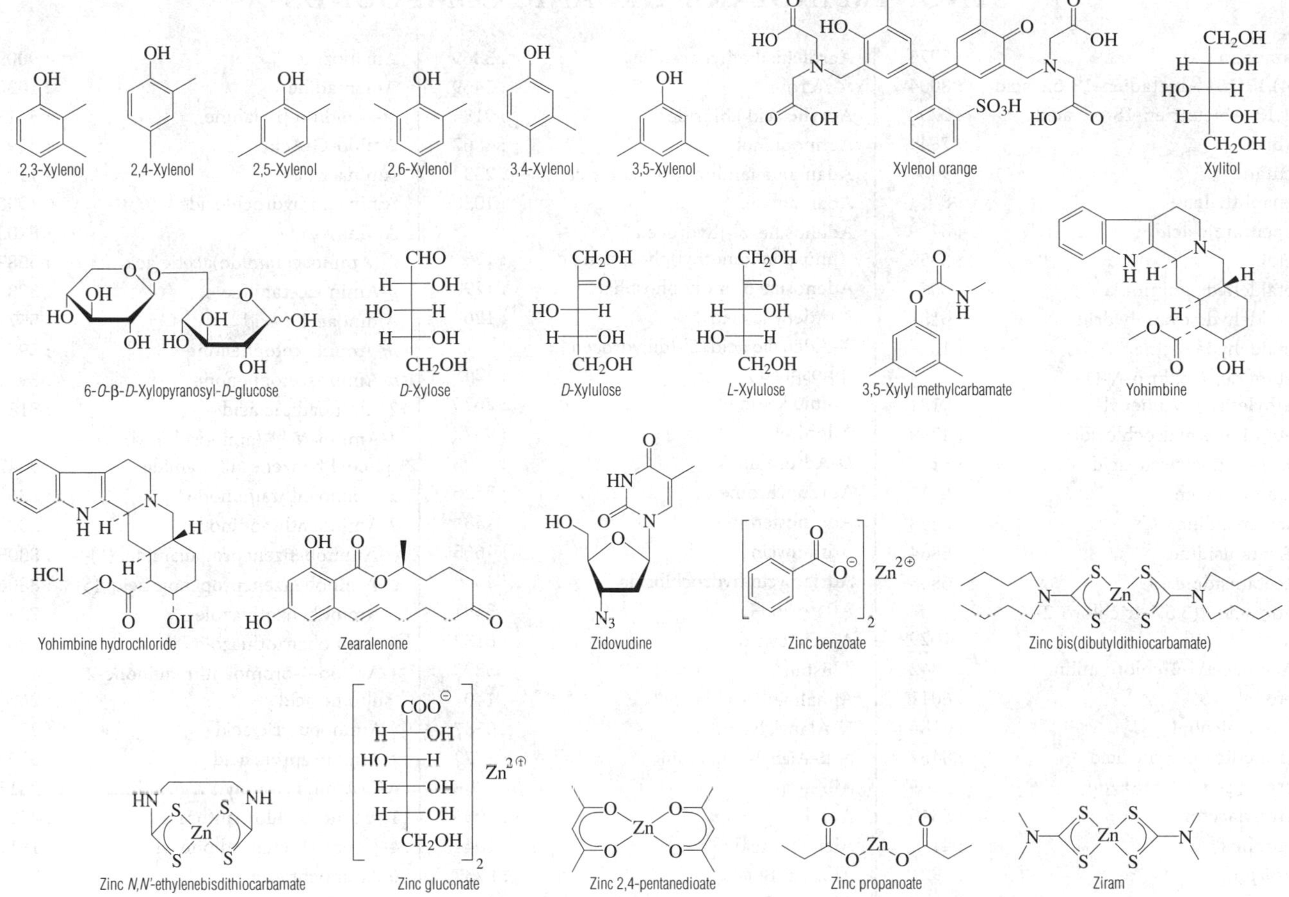
2,3-Xylenol
2,4-Xylenol
2,5-Xylenol
2,6-Xylenol
3,4-Xylenol
3,5-Xylenol
Xylenol orange
Xylitol
6-O-β-D-Xylopyranosyl-D-glucose
D-Xylose
D-Xylulose
L-Xylulose
3,5-Xylyl methylcarbamate
Yohimbine
Yohimbine hydrochloride
Zearalenone
Zidovudine
Zinc benzoate
Zinc bis(dibutyldithiocarbamate)
Zinc N,N'-ethylenebisdithiocarbamate
Zinc gluconate
Zinc 2,4-pentanedioate
Zinc propanoate
Ziram

SYNONYM INDEX OF ORGANIC COMPOUNDS

Abamectin : 537
8(14),13(15)-Abietadien-18-oic acid : 8004
8,11,13-Abietatrien-18-oic acid : 2790
L-Abrine : 7841
Accutane : 9461
Acenaphthalene : 8
Aceneuramic acid : 84
Acetal : 3359
Acetaldehyde ammonia : 306
Acetaldehyde cyanohydrin : 6181
Acetaldehyde oxime : 16
Acetamide, 2-chloro-*N*-(1-methylethyl)-*N*-phenyl- : 9131
Acetamidine hydrochloride : 4799
ε-Acetamidocaproic acid : 51
Acetaminophen : 6145
o-Acetanisidine : 6863
m-Acetanisidine : 6864
p-Acetanisidine : 6865
Acetic acid, [(3,5,6-trichloro-2-pyridinyl)oxy]- : 10299
N-Acetoacetyl-4-chloroaniline : 1842
Acetoin : 6011
Acetone alcohol : 6184
Acetonedicarboxylic acid : 8467
Acetone, phenylhydrazone : 9169
Acetonylacetone : 5840
Acetophos : 41
Acetorphin : 7921
Acetoxime : 9168
Acetoxyacetone : 91
2-Acetoxyanisole : 6866
2-Acetoxy-1-methoxypropane : 9235
Acetozone : 56
Aceturic acid : 71
Acetylacetone : 8618
4-Acetylacetophenone : 2824
α-Acetylbutyrolactone : 65
Acetylbutyryl : 5838
Acetylcysteine : 64
4-Acetyl-*N,N*-dimethylaniline : 3896
Acetylene tetrabromide : 9711
Acetylene tetrachloride : 9752
Acetyleneurea : 9862
N-Acetyl-*D*-glucosamine : 47
N-Acetyl-*D*-mannosamine : 48
3-Acetyl NAD : 101
Acetyl peroxide : 2827
Acetylpropionyl : 8617
Acetylsalicylic acid : 86
Acetylsulfanilamide : 428
Acetylsulfanilyl chloride : 44
Acetylthiourea : 431
Acetyl valeryl : 5677
cis-Aconitic acid : 9182
trans-Aconitic acid : 9183
Acridine Orange : 9916
Acridine Yellow : 3890
Acrolein, diethyl acetal : 3369
α-Acrose : 5469
Acrylic acid chloride : 9185
Actinoquinol : 4867
Adamantanamine hydrochloride : 233
Adamantane : 10309
Adenosine, 5'-[hydrogen (phosphonomethyl)phosphonate] : 125
Adenosine 5'-monophosphate : 129
3'-Adenylic acid : 126
3'-Adenylic acid, 5'-(dihydrogen phosphate) : 124
Adipic ketone : 2670
Adonitol : 9477
D-Adrenaline : 4708
Adrenochrome : 3626
Adrenosterone : 458
Adriamycin : 4665
Adriamycin hydrochloride : 4666
Aflatoxin G_2 : 3591
Agaricic acid : 6132
Ahistan : 3897
Ajmaline : 140
N-Alanylglycine : 5587
N-β-Alanyl-*L*-histidine : 1773
Albutoin : 6375
Aldol : 6006
Aleuritic acid : 10440
Alizarin Blue : 3737
Alizarin Orange : 3738
Alizarin Yellow A : 10438
Allenolic acid : 6120
Allidochlor : 1970
Alliin : 207
Allobarbital : 2840
Allocholesterol : 2369
Allopregnane : 9101
Allopregnane-3α,20α-diol : 9103
3,20-Allopregnanedione : 9105
Allopregnan-3α-ol-20-one : 6172
Allopregnan-3β-ol-20-one : 6173
Alloxan : 9383
Allylacetic acid : 8645
Allyl bromide : 1339
Allyl chloride : 2275
Allyl cyanide : 1450
Allyl ether : 2834
Allyl 2-furanoate : 189
Allylhexamethylenetetramine iodide : 6756
Allyl iodide : 6339
17-Allylmorphinan-3-ol : 6606
Allylphenylamine : 172
Allyl *trans*-3-phenyl-2-propenoate : 180
Aloe-emodol : 3724
Aloxidone : 6919
Aluminon : 536
Aluminum ethoxide : 221
Aluminum tris(*O*-ethylphosphonate) : 5466
Amanozine : 9009
Amantadine : 10308
N-Amidino-β-alanine : 5613
Amido-G-Acid : 359
Aminacrine : 112
Aminazin hydrochloride : 1990
Aminitrozole : 8203
2-(Aminoacetamido)acetic acid : 5589
p-Aminoacetanilide : 393
Aminoacetic acid : 5577
m-Aminoacetophenone : 397
p-Aminoacetophenone : 398
2-Aminoadipic acid : 318
4-Amino-*N*-[4-(aminosulfonyl) phenyl]benzenesulfonamide : 9647
1-Aminoanthraquinone : 236
2-Aminoanthraquinone : 237
α-Aminobenzenepropanamide, (*S*)- : 8805
α-Aminobenzenepropanoic acid, (*S*) : 8806
2-Aminobenzothiazole : 735
6-Aminobenzothiazole : 736
1-Amino-4-bromoanthraquinone-2-sulfonic acid : 267
γ-Aminobutyric acid : 271
ε-Aminocaproic acid : 320
*N*5-(Aminocarbonyl)-*L*-ornithine : 2417
1-Amino-5-chloroanthraquinone : 282
4-Amino-4'-chlorodiphenyl : 1919
6-Aminochrysene : 2386
3-Amino-4-cyanopyrazole : 416
Aminocyclobutane : 2529
Aminocyclooctane : 2645
[(4-Amino-3,5-dichloro-6-fluoro-2-pyridyl)oxy]acetic acid : 5440
4-Amino-*N*-[2-(diethylamino)ethyl] benzamide : 9114
2-Amino-1,9-dihydro-9-β-*D*-ribofuranosyl-6*H*-purin-6-one : 5615
p-Aminodiphenylamine : 8833
2-Aminoethanol hydrochloride : 4802
4(2-Aminoethyl)-1,2-benzenediol : 4661
3-(2-Aminoethyl)indol-5-ol : 6198
1-(2-Aminoethyl)piperazine : 9055
2-Aminoethyl sulfate : 4803
4-Amino-5-fluoro-2-hydroxypyrimidine : 5382
2-Aminoglutaramic acid : 5556
Aminoguanidine : 5936
2-Aminohexanoic acid, (*DL*) : 8275
2-Aminohexanoic acid, (*L*) : 8276
(6-Aminohexyl)carbamic acid : 5820
p-Aminohippuric acid : 262
2-Amino-3-hydroxybutanoic acid, [*R*-(*R**,*S**)] : 10087
2-Amino-4-hydroxybutanoic acid, (*S*) : 5926
4-Amino-3-hydroxybutanoic acid trimethylbetaine : 1772
2-Amino-2-(hydroxymethyl)-1,3-propanediol : 10693

1-Amino-6-hydroxynaphthalene : 371
2-Amino-1-(3-hydroxyphenyl)-1-propanol, (1*R*,2*S*) : 6730
2-Amino-3-hydroxypropanoic acid, (*S*) : 9531
Aminoiminomethanesulfinic acid : 5451
1-Aminoindane : 6232
α-Aminoindole-3-propionic acid, (*l*) : 10714
2-Aminoisobutanol : 354
α-Aminoisobutyric acid : 6918
5-Amino-1*H*-isoindole-1,3(2*H*)-dione : 403
2-Aminoisovaleric acid : 10768
5-Aminolevulinic acid : 384
DL-2-Amino-4-mercaptobutanoic acid : 5923
L-2-Amino-4-mercaptobutanoic acid : 5924
Aminomethanamidine : 5608
1-Amino-2-methoxyethane : 6827
1-Amino-2-methylanthraquinone : 342
2-Amino-2-methylbutyric acid : 6558
4-(Aminomethyl)-5-hydroxy-6-methyl-3-pyridinemethanol : 9374
5-(Aminomethyl)-3(2*H*)-isoxazolone : 7911
2-Amino-3-methylpentanoic acid : 6436
2-Amino-4-methylpentanoic acid : 6602
4-(Aminomethyl)piperidine : 9068
2-Amino-3-methylpyrazine : 7723
8-Amino-6-methylquinoline : 7765
2-Amino-4-methylthiazole : 7799
Aminometradine : 309
3-Amino-2-naphthoic acid : 357
8-Amino-β-naphthol : 373
1-Amino-2-naphthol-4-sulfonic acid : 334
6-Aminonicotinic acid : 418
2-Amino-5-nitrothiazole : 8202
(*S*)-2-Aminopentanedioic acid : 5554
2-Aminopentanoic acid, (±) : 8281
2-Aminopentanoic acid, (*S*) : 8282
p-Aminophenylacetic acid : 245
4-Amino-3-phenylbutyric acid : 344
1-Amino-1-phenylethane : 7040
1-Amino-2-phenylethane : 640
2-Amino-5-phenyl-4(5*H*)-oxazolone : 8519
N-[(4-Aminophenyl)sulfonyl]benzamide : 9635
2-Amino-3-picoline : 7732
2-Amino-4-picoline : 7733
2-Amino-6-picoline : 7735
DL-2-Aminopropanoic acid : 142
2-Aminopropanoic acid, (*R*) : 143
2-Aminopropanoic acid, (*S*) : 144
3-Aminopropanoic acid : 145
3-Aminopropionitrile : 405
p-Aminopropiophenone : 400
α-(α-Aminopropyl)benzyl alcohol : 409
3-(Aminopropyl)diethanolamine : 1004
N-(3-Aminopropyl)ethylenediamine : 316
4-(3-Aminopropyl)morpholine : 7905
2-Aminopyridine : 9313
3-Aminopyridine : 9314
4-Aminopyridine : 9315
2-Aminoquinoline : 9422
3-Aminoquinoline : 9423
4-Aminoquinoline : 9424
5-Aminoquinoline : 9425
6-Aminoquinoline : 9426
8-Aminoquinoline : 9427
Aminorex : 3665
3-Aminorhodanine : 432
4-Amino-1-β-*D*-ribofuranosyl-2(1*H*)-pyrimidinone : 2720
4-Amino-1-β-*D*-ribofuranosyl-1,3,5-triazine-2(1*H*)-one : 541
p-Aminosalicylic acid : 326
p-Aminosalicylic acid hydrazide : 324
α-Aminosuccinamic acid : 520
L-Aminosuccinic acid : 525
N-[5-(Aminosulfonyl)-1,3,4-thiadiazol-2-yl]acetamide : 19
N-[5-(Aminosulfonyl)-1,3,4-thiadiazol-2-yl]butanamide : 1439
2-Aminothiazole : 10029
4-Amino-*N*-2-thiazolylbenzenesulfonamide : 9651
5-Aminouracil : 420
Amiphenazole : 9000
Amitrole : 10146
Amixetrine : 7106
Amphetamine : 8974
Amsonic acid : 2866
Amyl acetate : 8662
sec-Amyl acetate (*R*) : 8663
Amyl alcohol : 8626
sec-Amyl alcohol : 8627
Amylamine : 8664
Amylbenzene : 8666
Amyl butyrate : 8669
Amyl caproate : 8676
tert-Amyl carbamate : 8670
Amyl enanthate : 8675
α-Amylene : 8639
cis-β-Amylene : 8640
trans-β-Amylene : 8641
Amylene dichloride : 3210
2,4-Amylene glycol : 8615
Amyl ether : 4441
Amyl formate : 8674
Amyl nitrite : 8678
Amyl octanoate : 8680
Amyltrichlorosilane : 10266
α-Amyrin : 10763
β-Amyrin : 8408
Anabasine : 9075
Anatabine : 9843
4-Androstene-3,17-dione : 457
Androsterone : 5963
Anethole : 6892
Angelic acid : 7094
Aniline mustard : 926
p-Anisaldehyde : 6779
Anise alcohol : 6795
p-Anisic acid : 6801
o-Anisidine : 6772
m-Anisidine : 6773
p-Anisidine : 6774
o-Anisidine hydrochloride : 6775
Anisindione : 6870
p-Anisoin : 3853
o-(*p*-Anisoyl)benzoic acid : 6807
p-Anisoyl chloride : 6809
Anisyl methyl ketone : 6881
[6]Annulene : 595
[8]Annulene : 2649
p-Anol : 9189
Anthanthrone : 2420
Anthorine : 530
Anthragallol : 10433
Anthralin : 491
o-Anthranilic acid : 467
m-Anthranilic acid : 468
p-Anthranilic acid : 469
Anthranol : 493
Anthraquinone : 487
Anthrarobin : 490
Anthrarufin : 3693
1-Anthroic acid : 482
2-Anthroic acid : 483
9-Anthroic acid : 484
Anthrone : 494
Antimycin : 2416
Antipyrine : 3614
ANTU : 7973
Apiole : 3832
Apiole (Dill) : 3886
Apocynin : 6069
β-*D*-9-Arabinofuranosyladenine : 10786
Arachic alcohol : 4684
Arachidic acid : 4683
Arachidonic acid : 4685
Arbutin : 6156
Armstrong's acid : 7959
Arsanilic acid : 394
Arsphenamine : 9504
L-α-Aspartyl-*L*-phenylalanine, 2-methyl ester : 523
Aspidospermine : 79
Aspon : 10000
ATP : 127
Atrolactic acid : 6078
Atropic acid : 7231
12-Azabenz[a]anthracene : 584
Azacyclohexane : 9058
Azacyclopentane : 9397
7-Azadibenz[a,j]anthracene : 2874
1-Azanaphthalene : 9428
6-Azathymine : 7829
Azelaic acid : 8232
3'-Azido-3'-deoxythymidine : 10870
Azine : 9316
Aziridine : 5021
2,2'-Azobis[2-methylpropionitrile] : 562
Azodicarbonamide : 2870
Azomethane : 4041
Azomycin : 8112
Badische acid : 367
Baicalein : 10443

Bamipine : 7628
Basic fuchsin : 8512
Batyl alcohol : 8327
Behenic acid : 4612
Bemegride : 5152
Bendazol : 787
Benz[e]acephenanthrylene : 700
β-Benzalbutyramide : 7631
Benzal chloride : 3207
Benzanilide : 8826
1,2-Benzanthracene : 592
2,3-Benzanthracene : 7926
Benzathine : 2894
Benzazide : 755
Benzenamine : 466
Benzenamine hydrochloride : 471
Benzenamine, 4-(1-methylethyl)-2,6-dinitro-*N*,*N*-dipropyl- : 6460
Benzeneacetamide, *N*,*N*-dimethyl-α-phenyl- : 4447
Benzeneacetic acid, ethyl ester : 5205
Benzenebutanoic acid, α,2-diamino-γ-oxo- : 6572
Benzenecarboxaldehyde : 585
Benzenecarboxylic acid : 712
1,2-Benzenedicarboxylic acid : 9029
1,3-Benzenedicarboxylic acid : 6458
1,4-Benzenedicarboxylic acid : 9694
1,2-Benzenedicarboxylic acid, didodecyl ester : 3344
1,2-Benzenedicarboxylic acid, diisopropyl ester : 3811
1,2-Benzenedicarboxylic acid, monobutyl ester : 7889
1,2-Benzenediol : 9388
1,3-Benzenediol : 9458
1,4-Benzenediol : 5958
Benzenemethanamine : 781
Benzenemethanol : 780
Benzenephosphinic acid : 8948
Benzenephosphonic acid : 8949
Benzenepropanol, acetate : 8973
Benzenestibonic acid : 3745
Benzenesulfonamide, 4-(dipropylamino)-3,5-dinitro- : 8421
Benzhydrylamine : 8836
Benzilic acid : 6149
Benzilideneacetone : 8859
1*H*-Benzimidazole, 2-(4-thiazolyl)- : 10019
2-Benzimidazolethiol : 3596
1,2-Benzisothiazolin-3-one, 1,1-dioxide, sodium salt : 9497
Benzo[a]chrysene : 9035
1,2-Benzodiazine : 2403
1,3-Benzodiazine : 9413
1,4-Benzodiazine : 9444
2,3-Benzodiazine : 9028
1,3-Benzodioxol-4-ol, 2,2-dimethyl-, methylcarbamate : 578
2-[2-(1,3-Benzodioxol-5-yl)ethyl]-4-methoxyquinoline : 2498
Benzoguanamine : 9008
Benzohydrol : 4502
Benzoic acid amide : 591
Benzoic acid, bromide : 760
Benzoic acid, chloride : 761
Benzoic acid, 2-[[4-(dimethylamino)phenyl]azo]- : 7777
Benzoic acid, 2-[[4-(dipropylamino)phenyl]azo]- : 9264
Benzoic acid, fluoride : 765
Benzoic acid, hydrazide : 711
Benzoic acid, iodide : 767
Benzo[b]naphthacene : 8530
Benzo[rst]pentaphene : 2887
Benzo[a]phenanthrene : 2387
Benzo[l]phenanthrene : 10648
Benzo[def]phenanthrene : 9308
Benzopinacol : 9988
1*H*-Benzopyrazole : 6233
2,3-Benzopyrene : 729
Benzo[c]pyridine : 6551
2,3-Benzopyrrole : 6243
Benzoresorcinol : 3743
Benzosulfonazole : 737
(2-Benzothiazolylthio)methyl thiocyanate : 1390
4-(2-Benzothiazolylthio)morpholine : 7906
Benzotrichloride : 10253
Benzotrifluoride : 10404
Benzo[a]triphenylene : 687
2,2'-Benzoxazoline : 5597
Benzoxiquine : 9440
Benzoylacetonitrile : 8455
Benzoyl *tert*-butyl peroxide : 1638
2-Benzoyl-4-chloroaniline : 287
Benzoyl disulfide : 2891
2-(Benzoyloxy)benzoic acid, methyl ester : 7039
Benzoylpas : 754
1,12-Benzperylene : 716
1,2-Benzpyrene : 730
N-Benzylbenzenemethanamine : 2892
Benzyl bromide : 1247
Benzyl-*tert*-butanol : 3960
Benzyl carbazate : 1737
Benzylcarbinol isobutyrate : 8878
Benzyl chloride : 2106
Benzyl cyanide : 601
N-Benzyl-*N*',*N*'-dimethyl-1,2-ethanediamine : 3972
Benzyl ether : 2895
Benzylideneaniline : 8920
Benzylidene hydrazine : 586
Benzylidenemethylamine : 8927
Benzyl isobutyrate : 810
Benzyl isocyanide : 6424
Benzyl laurate : 795
Benzyl nicotinate : 828
3-(Benzyloxy)-*L*-alanine : 832
N-(Benzyloxycarbonyl)-*L*-proline : 830
Benzylpenicillinic acid : 8522
o-Benzylphenol : 819
p-Benzylphenol : 820
Benzyl *trans*-3-phenyl-2-propenoate : 794
*N*4-Benzylsulfanilamide : 782
Benzyl sulfide : 2899
p-(Benzylsulfonamido)benzoic acid : 834
Berbamine : 10486
Bergaptene : 6899
Beryllium acetylacetonate : 844
Besylic acid : 662
Betulin : 6635
Biallyl : 5779
Bicetyl : 4663
Bicine : 1001
Bicyclo[5.3.0]decapentaene : 568
Bicyclo[2.2.1]hepta-2,5-diene : 8263
1,1'-Binaphthyl : 871
2,3,1',8'-Binaphthylene : 702
p-Biphenylamine : 265
[1,1'-Biphenyl]-4,4'-diamine : 680
[1,1'-Biphenyl]-4,4'-diamine, 2,2'-dichloro- : 3106
[1,1'-Biphenyl]-4,4'-diamine, 3,3'-dichloro- : 3107
2,2'-Biphenylene oxide : 2884
[1,1'-Biphenyl]-2-ol : 6003
[1,1'-Biphenyl]-3-ol : 6004
[1,1'-Biphenyl]-4-ol : 6005
Bipropargyl : 5787
Bipropionyl : 5841
2,3'-Bipyridyl : 894
2,4'-Bipyridyl : 895
3,3'-Bipyridyl : 896
N,*N*'-Bis(acetoacetyl)-3,3'-dimethylbenzidine : 899
Bis(acetyloxy)dimethylsilane : 4040
2,2-Bis[(acetyloxy)methyl]-1,3-propanediol diacetate : 8572
Bis[(aminocarbonyl)amino]acetic acid : 156
Bis(2-aminoethyl) sulfide : 10047
Bis(3-aminopropyl)amine : 413
1,4-Bis(3-aminopropyl)piperazine : 9054
Bis(4-bromophenyl) ketone : 2911
Bis(2-butoxyethyl) ether : 3439
Bis(2-butoxyethyl) phthalate : 625
Bis(5-*tert*-butyl-4-hydroxy-2-methylphenyl) sulfide : 10046
N,*N*-Bis(carboxymethyl)glycine : 8030
Bis(4-chlorophenyl) ketone : 3115
3,6-Bis(2-chlorophenyl)-1,2,4,5-tetrazine : 2425
Bis(2-cyanoethyl)amine : 6223
Bis(η5-2,4-cyclopentadien-1-yl)nickel : 8018
Bis(3,5-di-*tert*-butyl-4-hydroxyphenyl)methane : 7233
Bis(dibutylthiocarbamoyl) disulfide : 9729
4,4'-Bis(dimethylamino)benzhydrol : 966
4,4'-Bis(dimethylamino)thiobenzophenone : 965
2,6-Bis(1,1-dimethylpropyl)-4-methylphenol : 4442
1,4-Bis(2,3-epoxypropoxy)butane : 1411
Bis(2,3-epoxypropyl) ether : 3578
Bis(2-ethoxyethyl) ether : 3440
Bis(ethylacetoacetato)copper : 2461
Bis(2-ethylhexyl) phosphite : 984

2,2-Bis(ethylsulfonyl)propane : 9660
Bis(2-hydroxyethyl)amine : 3348
Bis(2-hydroxyethyl) 1,4-benzenedicarboxylate : 1006
Bis(hydroxymethyl)acetylene : 1680
3,3-Bis(4-hydroxyphenyl)-1(3*H*)-isobenzofuranone : 8763
Bis(2-hydroxyphenyl) ketone : 3710
Bis(4-hydroxyphenyl) ketone : 3711
2-[Bis(4-hydroxyphenyl)methyl] benzoic acid : 8764
Bis(isopropylamido)fluorophosphate : 7877
Bis(2-methoxyethyl) phthalate : 626
Bis(8-methylnonyl)phthalate : 3785
Bis(7-methyloctyl)phthalate : 3786
Bis(4-methylphenyl) ketone : 3968
Bis(4-nitrophenyl) ether : 4383
Bis(4-nitrophenyl) sulfide : 4384
Bis(pentabromophenyl) ether : 2738
Bisphenol A : 1015
Bisphenol AD : 1014
Bisphenol A dimethacrylate : 1016
Bisphenol B : 1013
Bisphenol C : 1010
Bisphenol S : 1017
N,N-Bis(phenylmethyl) benzenemethanamine : 10151
N,N'-Bis(salicylidene) ethylenediaminocobalt(II) : 9500
1,2-Bis(*p*-tolyl)ethane : 4595
Bis(tributyltin) oxide : 5721
Bitoscanate : 3815
Biuret : 6222
Borane pyridine : 10432
Boric acid, triethyl ester : 10352
Boric acid, tripropyl ester : 10670
Bornyl acetate : 10517
d-Bornyl isovalerate : 1068
Brassidic acid : 4617
Brilliant Oil Yellow : 3905
Bromal : 10152
Bromcresol Green : 1156
Bromcresol Purple : 1157
Bromex : 913
Bromindione : 1322
Bromine cyanide : 2513
Bromisovalum : 279
p-Bromoacetanilide : 1318
ω-Bromoacetophenone : 1081
p-Bromoacetophenone : 1320
5-Bromoacetylsalicylic acid : 88
5-Bromoanthranilic acid : 266
p-Bromobenzyl bromide : 1117
p-Bromobenzyl chloride : 1148
α-Bromobenzyl cyanide : 1099
5-Bromo-3-*sec*-butyl-6-methyluracil : 1074
DL-α-Bromobutyric acid : 1125
α-Bromo-γ-butyrolactone : 1174
N'-(4-Bromo-3-chlorophenyl)-*N*-methoxy-*N*-methylurea : 1821
2-Bromoethanamine hydrobromide : 1200
2-Bromo-2-ethylbutanamide : 3408
2-Bromoethyl ethyl ether : 1198
Bromoform : 10167
α-Bromoisobutyric acid : 1281
β-Bromoisovaleric acid : 1254
p-Bromomandelic acid : 1231
2-(Bromomethyl)-1*H*-isoindole-1,3(2*H*)-dione : 1278
Bromomethylmagnesium : 7432
1-Bromo-β-naphthol : 1290
p-Bromophenacyl bromide : 1119
2-Bromo-4'-phenylacetophenone : 1082
(*p*-Bromophenyl)hydrazine : 1321
p-Bromophenyl isocyanate : 1239
3-(*p*-Bromophenyl)-1-methoxy-1-methylurea : 7863
4-Bromophenyl phenyl ether : 1316
Bromophosgene : 1755
3-Bromo-1-propanamine hydrobromide : 1342
β-Bromopropionic acid : 1331
5-Bromosalicylhydroxamic acid : 1175
Bromosaligenin : 1232
Bromosuccinic acid : 1123
5-Bromouracil : 1349
Bromoxynil : 2955
Bromphenol Blue : 1315
Bromthymol Blue : 1358
Bronner acid : 369
Bronopol : 1296
p-Brosyl chloride : 1101
Broxyquinoline : 2987
Buclosamide : 1535
Bufexamac : 1480
Bufotenine : 3910
Buphenine : 8286
Busulfan : 1414
Butallylonal : 1085
Butamben : 1498
1-Butanamine : 1494
2-Butanamine, (±)- : 1495
1-Butanamine hydrochloride : 1497
Butanedinitrile : 9627
Butanedioic acid, mono(2,2-dimethylhydrazide) : 2732
Butanedioic acid, monoethyl ester : 5069
1,4-Butanediol bis(3-aminopropyl) ether : 910
Butanoic acid, 1,2,3-propanetriyl ester : 10191
1-Butanol, aluminum salt : 10179
2-Butanol, aluminum salt : 219
cis-Butenedinitrile : 6661
cis-2-Butenedioic acid : 6659
trans-2-Butenedioic acid : 5476
Butethal : 1572
Butethamate : 3389
Butopyronoxyl : 6230
Butoxybenzene : 1651
2-(2-Butoxyethoxy)ethyl acetate : 3445
Butylacetylene : 5910
tert-Butylacetylene : 3999
Butylacrolein : 5693
N-tert-Butylacrylamide : 1656
Butylal : 3001
Butyl alcohol : 1431
sec-Butyl alcohol : 1432
tert-Butyl alcohol : 7694
N-[(Butylamino)carbonyl]-4-methylbenzenesulfonamide : 10108
2-(Butylamino)-*N*-(2-chloro-6-methylphenyl)acetamide : 1428
Butylated hydroxyanisole : 1585
N-tert-Butyl-2-benzothiazolesulfenamide : 1502
Butyl benzyl phthalate : 791
Butylbis(2-hydroxyethyl)amine : 996
Butyl bromide : 1121
(±)-*sec*-Butyl bromide : 1122
tert-Butyl bromide : 1280
N-Butylbutanamine : 3003
N-sec-Butyl-2-butanamine : 3004
4-*tert*-Butyl-*N*-*sec*-butyl-2,6-dinitroaniline : 1485
Butyl caproate : 1581
tert-Butyl carbonazidate : 1510
Butyl chloride : 1925
(±)-*sec*-Butyl chloride : 1926
tert-Butyl chloride : 2159
tert-Butyl cyanide : 4275
N-Butylcyclohexanamine : 1548
5-Butyl-1-cyclohexyl-2,4,6(1*H*,3*H*,5*H*)-pyrimidinetrione : 1386
5-Butyl-2-(dimethylamino)-6-methylpyrimidin-4(1*H*)-one : 3827
Butyl enanthate : 1580
1-Butylene : 1441
α-Butylene chlorohydrin : 1932
α-Butylene dibromide : 2916
1,3-Butylene glycol : 1406
N-Butyl-*N*-ethyl-2,6-dinitro-4-(trifluoromethyl)aniline : 569
Butyl 2-ethylhexyl phthalate : 5058
N-sec-Butyl-*N'*-ethyl-6-methoxy-1,3,5-triazine-2,4-diamine : 1389
Butyl fluoride : 5378
sec-Butyl fluoride : 5379
tert-Butyl fluoride : 5402
Butyl hexadecanoate : 1634
Butyl hexyl ketone : 10735
N-Butyl-*N*-(4-hydroxybutyl) nitrosamine : 1626
Butylidene chloride : 3128
Butyl iodide : 6287
(±)-*sec*-Butyl iodide : 6288
tert-Butyl iodide : 6317
tert-Butyl isocyanate : 6415
Butyl isopropyl ether : 7254
tert-Butyl isopropyl ketone : 10587
2-[(3-Butyl-1-isoquinolinyl)oxy]-*N,N*-dimethylethanamine : 3828
Butyl levulinate : 1633
Butylmalonic acid : 1653
Butyl mercaptan : 1425
sec-Butyl mercaptan : 1426
tert-Butyl mercaptan : 7689
tert-Butyl methyl ether : 7111

Butyl methyl ketone : 5853
(±)-*sec*-Butyl methyl ketone : 7571
Butyl *cis*-9-octadecenoate : 1631
Butylparaben : 1587
Butyl pelargonate : 1628
Butyl propionate : 1654
Butyl propyl ketone : 8362
Butyl ricinoleate : 1588
p-*tert*-Butylstyrene : 1672
Butyl sulfate : 3050
Butyl sulfite : 3054
Butyl 2-thiophenecarboxylate : 1663
Butyl *o*-toluate : 1609
Butyl *p*-toluate : 1610
2-*tert*-Butyltoluene : 1606
3-*tert*-Butyltoluene : 1607
4-*tert*-Butyltoluene : 1608
Butyrac 118 : 3257
Butyraldehyde : 1398
Butyramide : 1400
Butyric acid : 1429
Butyric anhydride : 1430
Butyroin : 6136
γ-Butyrolactam : 9402
n-Butyryl chloride : 1437
Cacodyl : 9942
Cacodylic acid : 3943
Cadalene : 4121
Cadaverine : 8609
Caffeic acid : 3744
Calcitriol : 3718
Calcium acetylacetonate : 1701
Calcium carbimide : 1696
Calcium citrate : 10192
cAMP : 123
Campesterol : 4734
(+)-Camphoric acid : 10547
Canavanine : 337
Capraldehyde : 2749
Capric acid : 2757
Capric alcohol : 2758
Caprinitrile : 2755
Caprinoyl chloride : 2766
Caproaldehyde : 5827
Caproic acid : 5848
Caprolactone : 8443
Caproleic acid : 2773
Capronitrile : 5844
Caproyl alcohol : 5850
Caproyl chloride : 5855
Capryl alcohol : 8356
Caprylene : 8367
Caprylic acid : 8354
(±)-*sec*-Caprylic alcohol : 8357
Caprylic aldehyde : 8343
Caprylnitrile : 8352
Carbamazepine : 2878
Carbamic acid, 1*H*-benzimidazol-2-yl-, methyl ester : 1732
Carbamide : 10756
Carbamimidoselenoic acid : 9523
Carbamodithioic acid, diethyl-, 2-chloro-2-propenyl ester : 9640
Carbamothioic acid, bis(2-methylpropyl)-, *S*-ethyl ester : 9669
Carbamothioic acid, cyclohexylethyl-, *S*-ethyl ester : 2527
Carbamothioic acid, dipropyl-, *S*-propyl ester : 10781
Carbamyl chloride : 1726
Carbanilide : 4535
Carbarsone : 277
4-Carbethoxymethyl-4-phenylazacycloheptane : 4815
1-Carbethoxypiperazine : 5217
Carbitol : 3447
Carbitol acetate : 3448
Carbobenzoxy chloride : 793
Carbocaine hydrochloride : 6707
Carbocysteine : 1767
Carbohydrazide : 1745
Carbon bisulfide : 1743
Carbonic anhydride : 1741
Carbonic chloride fluoride : 1757
Carbonochloridic acid, trichloromethyl ester : 4536
Carbon oxide : 1746
Carbon selenide : 1742
Carbon tetrabromide : 9715
Carbon tetrachloride : 9756
Carbon tetrafluoride : 9839
Carbon tetraiodide : 9905
4,4'-Carbonyldiphthalic anhydride : 721
Carbonylhydrotris(triphenylphosphine)rhodium : 10706
Carbonyl selenide : 1752
Carbonyl sulfide : 1753
Carboxin *S,S*-dioxide : 8483
N-(Carboxymethyl)-*N*-methylglycine : 7382
8-Carboxyquinoline : 9431
2-Carboxythiophene : 10071
1-Carboxy-*N,N,N*-trimethylmethanaminium, inner salt : 845
Carbromal : 278
Carbutamide : 274
Carbyloxime : 5474
Carfimate : 5285
Carmustine : 931
Carnegine : 9846
Carnitine, *O*-propanoyl : 9194
γ-Carotene : 1776
Carvacrol : 6515
Carzenide : 427
Catechol monoethyl ether : 4858
CDT : 2548
CDTA : 2621
Cellocidin : 1677
Cephaeline : 10488
Cerespan : 8506
Cerotic acid : 5742
Cetane : 5747
1-Cetene : 5756
Cetyl alcohol : 5753
Cetyl lactate : 5764
Cetyl mercaptan : 5750
Cetyl palmitate : 5762
Cetylpyridinium chloride : 5768
Cetyl stearate : 5769
Chalcone : 4521
Chaulmoogric acid : 2675
Chavicol : 202
Chelidonic acid : 8478
Chenodiol : 3717
Chimyl alcohol : 5766
Chloral : 10194
Chloral alcoholate : 10232
Chloral ammonia : 433
Chloral formamide : 10242
Chloramben : 295
Chloraminophenamide : 283
Chloranil : 9741
Chlordecone : 6568
Chlorfenac : 10207
Chlorindanol : 1984
Chlorine cyanide : 2514
ω-Chloroacetophenone : 1845
m-Chloroacetophenone : 2238
p-Chloroacetophenone : 2239
2-Chloroacrylic acid : 2277
4-Chloro-2-anisidine : 2090
N-Chlorobenzenesulfonamide sodium : 1814
p-Chlorobenzenesulfonic acid : 1889
4-Chlorobenzenethiol, *S*-methyl, *S,S*-dioxide : 2164
o-Chlorobenzotrifluoride : 2331
m-Chlorobenzotrifluoride : 2332
p-Chlorobenzotrifluoride : 2333
2-Chlorobenzyl chloride : 1950
3-Chlorobenzyl chloride : 1951
4-Chlorobenzyl chloride : 1952
2-(2-Chlorobenzyl)-4,4-dimethyl-1,2-oxazolidin-3-one : 2428
2-(4-Chlorobenzyl)-2-propylamine : 2349
3-Chloro-(1,1'-biphenyl)-4-ol : 8863
6-Chloro-*N*-*tert*-butyl-*N'*-ethyl-1,3,5-triazine-2,4-diamine : 9691
5-Chloro-3-*tert*-butyl-6-methyl-2,4(1*H*,3*H*)-pyrimidinedione : 9689
1-Chloro-4-[[(4-chlorophenyl)methyl]thio]benzene : 1819
4-Chloro-*o*-cresol : 2150
6-Chloro-*o*-cresol : 2148
4-Chloro-*m*-cresol : 2151
6-Chloro-*m*-cresol : 2147
2-Chloro-*p*-cresol : 2146
3-Chloro-*p*-cresol : 2149
Chlorocyanohydrin : 10243
1-Chloro-2,2-difluoroethylene : 1982
7-Chloro-1,3-dihydro-5-phenyl-2*H*-1,4-benzodiazepin-2-one : 8267
6-Chloro-*N,N'*-diisopropyl-1,3,5-triazine-2,4-diamine : 9177
2-Chloro-*N,N*-dimethyl-10*H*-phenothiazine-10-propanamine : 2350
Chlorodiphenylarsine : 4458

Chlorodiphenylphosphine : 4509
2-Chloroethanamine hydrochloride : 2029
β-Chloroethyl acetate : 2027
Chloroethylene carbonate : 2014
1-Chloro-3-ethyl-1-penten-4-yn-ol : 4805
2-Chloroethyl phenyl ketone : 2249
3-Chloro-4-fluoronitrobenzene : 2056
Chloroflurazole : 3310
Chloroform : 10245
4-(Chloroformyl)phthalic anhydride : 671
α-Chlorohydrin : 2262
5-Chloroisatoic anhydride : 1907
6-Chloro-*N*-isopropyl-1,3,5-triazine-2,4-diamine : 2810
6-Chlorometanilic acid : 284
Chloromethapyrilene : 2310
1-Chloro-2-methoxybenzene : 1855
1-Chloro-3-methoxybenzene : 1856
1-Chloro-4-methoxybenzene : 1857
N-Chloro-4-methylbenzenesulfonamide sodium : 1815
Chloromethyl cyanide : 1844
Chloromethyl *O,O*-diethyl dithiophosphate : 1834
Chloromethyl ethyl ether : 2092
Chloromethylmagnesium : 7433
(Chloromethyl)oxirane : 4707
N'-(3-Chloro-4-methylphenyl)-*N,N*-dimethylurea : 2359
(Chloromethyl)trichlorosilane : 10217
o-Chloronitrobenzene : 2181
m-Chloronitrobenzene : 2182
p-Chloronitrobenzene : 2183
4-Chloro-3-nitrotoluene : 2141
3-Chlorophenacyl bromide : 1137
p-Chlorophenacyl bromide : 1149
N'-[4-(4-Chlorophenoxy)phenyl]-*N,N*-dimethylurea : 2344
N-[[(4-Chlorophenyl)amino]carbonyl]-2,6-difluorobenzamide : 3546
4-Chloro-*o*-phenylenediamine : 1874
2-Chloro-*p*-phenylenediamine : 1876
Chlorophenylmagnesium : 8912
Chlorophenylmercury : 8913
N'-(4-Chlorophenyl)-*N*-methoxy-*N*-methylurea : 7893
2-(2-Chlorophenyl)-2-(methylamino) cyclohexanone, (±) : 6569
1-[4-(4-Chlorophenyl)-3-phenyl-2-butenyl]pyrrolidine : 9391
4-Chlorophenyl phenyl ether : 2225
2-Chlorophenyl phenyl ketone : 1902
Chloropicrin : 10261
Chloroprene : 1923
Chloroprophenpyridamine : 2348
2-Chloropropionic acid : 2266
β-Chloropropionic acid : 2267
β-Chloropropionitrile : 2265
4-Chloro-*N*-[(propylamino)carbonyl] benzenesulfonamide : 2258
3-Chloropropyl ether : 950
(3-Chloropropyl)triethoxysilane : 10332
6-Chloropurine : 2286
Chlorothymol : 2127
p-Chloro-*o*-toluidine : 2103
2-Chloro-*p*-toluidine-5-sulfonic acid : 288
Chlorotrianisene : 10695
Chlorotrifluoroethylene : 2327
5-[2-Chloro-4-(trifluoromethyl) phenoxy]-2-nitrobenzoic acid : 109
2-Chloro-1,7,7-trimethylbicyclo[2.2.1]heptane, *endo* : 1067
Chloroxine : 3290
Chloroxylenol : 2000
Chlorphenesin : 2226
Chlorzoxazone : 1910
Cholaic acid : 9682
Cholanic acid : 2361
Cholanthrene : 3594
4-Chromanone : 3603
1,2-Chromene : 724
Chromium acetylacetonate : 2383
Chromotropic acid : 3736
Chromotropic acid disodium salt : 9553
Chrysanthenone : 10521
Chrysin : 3741
Chrysoidine hydrochloride : 8818
Chrysophanic acid : 3731
C.I. Acid Green 3 : 5619
C.I. Acid Yellow 73 : 5350
C.I. Basic Red 9 : 10677
C.I. Basic Violet 1 : 7860
Cicrotoic acid : 2615
C.I. Direct Red 2, disodium salt : 723
C.I. Food Red 6 : 9090
C.I. Food Yellow 3 : 9667
Cinchocaine : 2997
Cinchomeronic acid : 9342
Cinchoninaldehyde : 9429
Cinchophen : 8994
Cineole : 5297
Cinnamic anhydride : 8968
Cinnamoyl chloride : 8972
trans-Cinnamyl acetate : 8971
Citraconic acid : 7090
Citronellal : 4192
Citronellene : 4166
Citronellic acid : 4194
Citronellol, (+) : 4195
Citronellol, (-) : 4196
Citronellol acetate : 4199
Civetone : 2551
Claritin : 6628
1,7-Cleve's acid : 370
Clinestrol : 3525
Clomethiazole : 2043
Clonitrate : 2264
Cloprop : 2227
Clorindione : 2241
Clortermine : 2243
Clostebol : 2070
Cloxyquin : 2300
Clozaril : 2436
Cobalt(III) acetylacetonate : 2439
Cobalt(II) bis(acetylacetonate) : 1043
Cocaethylene : 769
Cocarcinogen A1 : 9800
Coenzyme R : 874
2,3,6-Collidine : 10610
2,4,6-Collidine : 10611
Congressane : 2843
β-Coniceine : 9190
γ-Coniceine : 9883
Coniferyl alcohol : 6185
Coniine : 9257
Copper(II) acetylacetonate : 2463
Coproergostane : 4728
Coprostane : 2366
Cordycepic acid : 6678
Cortisol acetate : 5948
Cortisone : 3748
Corydaldine : 3613
Cotoin : 3729
Coumadin : 10840
Coumalic acid : 8477
Coumaran : 3599
p-Coumaric acid : 6169
Coumarilic acid : 706
Coumarin : 727
Coumarin-3-carboxylic acid : 8457
Coumarone : 705
Creosol : 6847
o-Cresolsulfonphthalein : 2182
p-Cresotic acid : 6081
o-Cresotic acid : 6082
m-Cresotic acid : 6083
o-Cresyl acetate : 7615
m-Cresyl acetate : 7616
p-Cresyl acetate : 7617
Croceic acid : 6122
Crocetin : 2868
Cromoglicic acid : 2485
trans-Crotonaldehyde : 1440
Crotonic acid : 1452
Crotonic acid anhydride : 1454
Crotononitrile : 1449
cis-Crotyl alcohol : 1455
trans-Crotyl alcohol : 1456
Cryptocavine : 2487
Cryptoxanthin : 1780
Cumene : 6478
Cumene hydroperoxide : 6479
3-*p*-Cumenyl-2-methylpropionaldehyde : 7401
Cumic acid : 6483
Cumic alcohol : 6480
Cumidine : 6475
Cuminaldehyde : 6477
α-Cumyl alcohol : 3959
Cupric gluconate : 2462
Cupriethylenediamine dichloride : 978
α-Curcumene : 4104
Cyacetacide : 2505
Cyamelide : 10639
Cyanocobalamin : 10833
Cyanocyanamide : 3316
Cyanocyclobutane : 2531

1-Cyanocyclohexene : 2592
Cyanocyclopentane : 2661
1-Cyanocyclopentene : 2673
Cyanogenamide : 2500
1-Cyano-2-isopropoxyethane : 6468
1-Cyano-3-(methylmercurio) guanidine : 7437
5-Cyano-1-pentylamine : 319
2-Cyanothiophene : 10066
3-Cyanothiophene : 10067
Cyanuric acic trichloride : 10290
Cyclamic acid : 2640
Cyclobutyrol : 5076
β-Cyclocitral : 10539
Cycloheptyl bromide : 1158
2,5-Cyclohexadiene-1,4-dione : 733
Cyclohexanamine : 2608
Cyclohexanamine hydrochloride : 2609
(1α,2α,3α,4β,5α,6β)-Cyclohexanehexol : 6262
1-Cyclohexene-1,2-dicarboxylic acid, anhydride : 9864
4-Cyclohexene-1,2-dicarboxylic acid, anhydride : 9863
2,3-Cyclohexenopyridine : 9892
Cyclohexyl alcohol : 2587
Cyclohexyl bromide : 1159
Cyclohexylcarbinol : 2584
Cyclohexyl chloride : 1961
Cyclohexyl cyanide : 2567
N-Cyclohexylcyclohexanamine : 3321
N-Cyclohexylcyclohexanamine, nitrite : 3322
2-Cyclohexylcyclohexanone : 860
Cyclohexyl fluoride : 5380
Cyclohexyl iodide : 6289
Cyclohexyl mercaptan : 2586
Cyclohexyl phenyl ketone : 762
α-Cyclohexyl-α-phenyl-1-piperidinepropanol hydrochloride : 10431
N-Cyclohexyl-1-piperazineacetamide : 4754
Cycloleucine : 290
Cyclomaltoheptaose : 2543
Cyclomaltohexaose : 2542
Cyclomaltooctaose : 2544
Cyclonite : 5810
N'-Cyclooctyl-*N,N*-dimethylurea : 2703
Cyclopentamine : 4033
Cyclopentanamine : 2682
Cyclopentanoic acid : 2663
1,2-Cyclopentenophenanthrene : 3609
Cyclopentyl alcohol : 2669
Cyclopentyl bromide : 1163
Cyclopentyl chloride : 1964
Cyclopentyl ether : 3333
Cyclopentyl iodide : 6290
Cyclopentyl mercaptan : 2668
Cyclophosphane : 2686
Cyclopropanamine : 2697
Cyclopropyl cyanide : 2689
N-Cyclopropyl-1,3,5-triazine-2,4,6-triamine : 2711
Cycloserine : 341
Cymarose : 3340
o-Cymene : 6507
m-Cymene : 6508
p-Cymene : 6509
L-Cysteine, *S*-(2-amino-2-carboxyethyl)-, (*R*)- : 6587
Cythrin : 5337
Cytidine 2'-monophosphate : 2721
Cytidine 3'-monophosphate : 2722
Cytidine 5'-monophosphate : 2723
Cytosine arabinoside : 2719
2,4-D : 3256
2,4-D 2-Butoxyethyl ester : 1478
2,4-D Butyl ester : 1552
2,4-D Methyl ester : 7192
Damascenine : 7444
Dansyl chloride : 3914
Danthron : 3694
Daphnetin : 3713
Dapsone : 908
DBMC : 3033
Deanol : 4057
Decafluorobutane : 8696
Decahydro-β-naphthol : 2745
cis-Decalin : 2743
trans-Decalin : 2744
Decamethylene dibromide : 2928
Decamethylene glycol : 2753
1-Decanamine : 2778
Decanoic acid glycerol monoester : 3752
Decanoic acid glycerol triester : 5568
Decylacetylene : 4658
Decyl cyanide : 10729
N-Decyl-1-decanamine : 3335
Decylenic alcohol : 2774
Decyl fluoride : 5383
Decyl mercaptan : 2756
Decyl methyl ketone : 4633
Decyl methyl sulfoxide : 7790
DEET : 3482
Deferoxamine : 2811
Dehydroacetic acid : 82
Dehydrobenzperidol : 4670
Dehydrobilirubin : 869
7-Dehydrocholesterol : 2363
Dehydrocholic acid : 10641
11-Dehydrocorticosterone : 6176
Dehydroergosterol : 4730
Dehydromucic acid : 5485
7-Dehydrositosterol : 9607
Dehydrotestosterone : 1060
Demeclocycline : 2776
Demeton-S : 6555
Denmert : 1468
11-Deoxojervine : 2654
Deoxycorticosterone : 6175
2'-Deoxy-5'-cytidylic acid : 2801
Deoxyepinephrine : 6933
2'-Deoxy-5'-guanylic acid : 2804
11-Deoxy-17-hydrocorticosterone : 3747
6-Deoxy-*D*-mannose : 9469
3-Deoxypseudaconitine : 867
Dessin : 4411
Dexamphetamine : 8975
Dextropimaric acid : 9043
Dextropropoxyphene : 9198
1,3-Diacetin : 9157
Diacetonamine : 347
Diacetone acrylamide : 4209
1,4-Diacetoxy-2-butyne : 1681
Diacetyl : 1415
N,N'-Diacetyl-3,5-diamino-2,4,6-triiodobenzoic acid : 2869
Diacetylene : 1396
Δ 2-Dialin : 3657
Diallylamine : 205
Diallyl phthalate : 627
Diamidafos : 8003
2,5-Diaminoanisole : 6787
cis-1,2-Diaminocyclohexane : 2571
trans-1,2-Diaminocyclohexane : 2572
4,4'-Diaminodiphenylamine : 395
2,4'-Diaminodiphenylmethane : 7241
2,7-Diaminofluorene : 5345
2,6-Diamino-5-(β-*D*-glucopyranosyloxy)-4(1*H*)-pyrimidinone : 10785
2,6-Diaminohexanoic acid, (±) : 6646
2,6-Diaminohexanoic acid, (*D*) : 6647
2,6-Diaminohexanoic acid, (*L*) : 6648
1,5-Diaminonaphthalene : 7941
1,8-Diaminonaphthalene : 7942
2,3-Diaminonaphthalene : 7943
2,5-Diaminopentanoic acid, (*S*) : 8416
2,3-Diaminophenazine : 8751
2,6-Diamino-3-phenylazopyridine : 8753
2,6-Diaminopimelic acid : 2858
1,3-Diaminopropane : 9138
2,3-Diaminopropionic acid : 235
2,6-Diaminopurine : 9294
2,5-Diaminopyridine : 9335
1,3-Diamino-2-thiourea : 1751
Diamylamine : 4439
1,4:3,6-Dianhydroglucitol : 6554
1,2-Dianilinoethane : 4482
Dianisidine : 3847
1,2-Diazabenzene : 9312
1,5-Diazanaphthalene : 7994
1,3-Diazine : 9382
1,4-Diazine : 9302
Diazoacetic ester : 4973
Diazoaminobenzene : 4533
1,2-Diazole : 9306
1,3-Diazole : 6215
Dibenamine : 2037
Dibenamine hydrochloride : 2038
Dibenzalacetone : 4507
1,2:5,6-Dibenzanthracene : 2876
Dibenz[de,kl]anthracene : 8736
Dibenzo[b,def]chrysene : 2886
Dibenzo[def,p]chrysene : 2888
Dibenzo[a,jk]fluorene : 701
Dibenzopyrazine : 8750
Dibenzo[b,e]pyridine : 113

Dibenzopyrolle : 1730
Dibenzothiapyran : 10082
Dibenzoylmethane : 4515
2,3:6,7-Dibenzphenanthrene : 8633
Dibenzyl : 4480
Dibenzyl ketone : 4518
Dibromantine : 2934
1,3-Dibromoacetone : 2982
1,2-Dibromobenzene : 2908
1,3-Dibromobenzene : 2909
1,4-Dibromobenzene : 2910
4,4'-Dibromobenzilic acid isopropyl ester : 1343
2,6-Dibromo-4-(chloroimino)-2,5-cyclohexadien-1-one : 2988
1,2-Dibromo-2,2-dichloroethylphosphoric acid, dimethyl ester : 7918
1,2-Dibromo-2,4-dicyanobutane : 1118
cis-1,2-Dibromoethylene : 2939
trans-1,2-Dibromoethylene : 2940
Dibromogallic acid : 2995
3,5-Dibromosalicylaldehyde : 2953
3,5-Dibromosalicylic acid : 2954
2,5-Dibromotoluene : 2957
Dibutylacetylene : 2789
Dibutylbutadiyne : 4618
N,*N* Dibutyl 1 butanamine : 10181
Dibutylcarbinol : 8242
Dibutyldichlorostannane : 3059
N,*N*'-Di-*tert*-butylethanediamine : 3022
Dibutyl ketone : 8246
Di-*tert*-butylmethane : 9962
Dibutylnitrosamine : 8184
Dibutyltin bis(dodecyl sulfide) : 3009
Dicamba : 3206
3,3'-Dicarboxybenzidine : 679
2,3-Dicarboxypyrazine : 9305
2,5-Dicarboxythiophene : 10073
Dichlobenil : 3114
Dichlone : 3231
Dichloramine-T : 3208
2,6-Dichloroanisole : 3204
1,2-Dichlorobenzene : 3092
1,3-Dichlorobenzene : 3093
1,4-Dichlorobenzene : 3094
2,4-Dichlorobenzotrifluoride : 3309
2,4-Dichlorobenzyl alcohol : 3103
3,3'-Dichloro-[1,1'-biphenyl]-4,4'-diamine : 3108
2,4-Dichloro-6-(*o*-chloroanilino)-*s*-triazine : 464
Dichlorocyanuric acid : 3304
3,5-Dichloro-*N*-(3,4-dichlorophenyl)-2-hydroxybenzamide : 9773
1,1-Dichloro-2,2-difluoroethylene : 3159
7,16-Dichloro-6,15-dihydro-5,9,14,18-anthrazinetetrone : 2419
2,5-Dichloro-3,6-dihydroxy-2,5-cyclohexadiene-1,4-dione : 1818
1,4-Dichloro-2,5-dimethoxybenzene : 2172
2,4'-Dichlorodiphenyl ketone : 2236
Dichlorodiphenyltrichloroethane (DDT) : 10212
Dichloroethylarsine : 4980
cis-1,2-Dichloroethylene : 3177
trans-1,2-Dichloroethylene : 3178
Dichloroethyl ether : 929
2',7'-Dichloro-3,6-fluorandiol : 3183
1,1-Dichloro-2-fluoroethylene : 3186
2,2'-Dichlorohydrazobenzene : 945
1,2-Dichloroisobutane : 3217
1,2-Dichloro-5-isocyanatobenzene : 3265
1,3-Dichloro-5-isocyanatobenzene : 3266
(Dichloromethyl) dimethylchlorosilane : 1973
(Dichloromethyl)trichlorosilane : 10220
2,4-Dichloro-α-naphthol : 3232
2,4-Dichloro-1-(4-nitrophenoxy) benzene : 8099
2,4-Dichlorophenoxyethyl benzoate : 9533
Di(4-chlorophenoxy)methane : 941
N'-(3,4-Dichlorophenyl)-*N*-methoxy-*N*-methylurea : 6617
Dichlorophenylphosphine : 8952
Dichlorophenylphosphine sulfide : 8951
Di(*p*-chlorophenyl)thiourea : 948
2-(3,5-Dichlorophenyl)-2-(2,2,2-trichloroethyl)oxirane, (±) : 10328
4,5-Dichlorophthalic anhydride : 3201
2,2-Dichloropropionic acid : 3274
cis-1,3-Dichloropropylene : 3282
trans-1,3-Dichloropropylene : 3283
3,6-Dichloro-2-pyridinecarboxylic acid : 2433
3,7-Dichloroquinoline-8-carboxylic acid : 9414
1,2-Dichloro-4-(trifluoromethyl) benzene : 3116
Dichloroxylenol : 3168
Dichlorphenamide : 3101
Dichlorprop : 3258
Dicobalt octacarbonyl : 2437
Dicryl : 3267
Dictamnine : 6831
Dicumene : 4046
Dicumyl peroxide : 1027
trans-1,4-Dicyano-2-butene : 5867
2,4-Dicyano-1-butene : 7247
Dicyanodiamide : 2517
Dicyclopentadienyl iron : 5326
Dicycloverine hydrochloride : 3331
Didanosine : 3339
N,*N*-Didecyl-1-decanamine : 10321
N,*N*-Didodecyl-1-dodecanamine : 10329
Didodecyl thiobispropanoate : 10048
Dienochlor : 8529
Diepoxybutane : 875
Diethadione : 3428
Diethanol-*m*-toluidine : 1003
Diethazine : 3503
3,5-Diethoxycarbonyl-1,4-dihydrocollidine : 3429
1-[(3,4-Diethoxyphenyl)methyl]-6,7-diethoxyisoquinoline : 4804
Diethoxyphosphoryl chloride : 3420
Diethylacetaldehyde : 4922
Diethylacetic acid : 4924
Diethyl 1,3-acetonedicarboxylate : 3497
Diethylacetylene : 5912
2-(Diethylamino)ethyl benzilate : 575
2-Diethylaminoethyl benzilate hydrochloride : 576
3-[2-(Diethylamino)ethyl]-3-phenyl-2(3*H*)-benzofuranone : 443
2-(4-Diethylaminophenylazo)benzoic acid : 5242
3-(Diethylamino)-*N*-phenylbutanamide : 8290
2-Diethylamino-2',4',6'-trimethylacetanilide : 10468
(Diethylamino)trimethylsilane : 3538
Diethyl azelate : 3494
5,5-Diethylbarbituric acid : 571
Diethyl benzalmalonate : 3405
1,2-Diethylbenzene : 3401
1,3-Diethylbenzene : 3402
1,4-Diethylbenzene : 3403
N,*N*-Diethylbenzhydrylamine : 3504
Diethyl bicarbamate : 3466
Diethylcarbamazine : 3489
Diethyl carbinol : 8628
N,*N*-Diethylcyclohexanamine : 2618
4-*N*,*N*-Diethyl-1,4-diamino-2-methylbenzene, hydrochloride : 3484
Diethyldiazine 1-oxide : 567
2,2'-Diethyldihexyl ether : 982
Diethyldimethylplumbane : 3430
Diethyldithiocarbamate sodium salt trihydrate : 9552
Diethylenediamine : 9049
Diethylenediamine dihydrochloride : 9052
Diethylene glycol bis(allyl carbonate) : 2831
Diethylene glycol diaminopropyl ether : 10640
Diethylene glycol divinyl ether : 1056
Diethylene glycol ethyl ether acrylate : 4840
Diethylenetriamine : 903
Diethylenetriaminepentaacetic acid : 8659
N,*N*-Diethylethanamine : 10345
N,*N*-Diethylethanamine hydrochloride : 10346
N,*N*-Diethylethylenediamine : 3452
Diethyl glutaconate : 3501
Diethyl hydroxybutanedioate : 3475
N,*N*-Diethylhydroxylamine : 5077
Diethyl ketone : 8630
Diethylmalonic acid : 3515
5,5-Diethyl-1-methyl-2,4,6(1*H*,3*H*,5*H*)-pyrimidinetrione : 6753
O,*O*-Diethyl *O*-(4-nitrophenyl) phosphate : 8510
Diethylnitrosamine : 8186
Diethyl oxalacetate : 3496
2,3-Diethylpentane : 5142
Diethyl phosphate : 3467
N,*N*-Diethyl-2-propargylamine : 3396
N,*N*-Diethyl-2-propen-1-amine : 185
Diethylpropion : 3394

Diethyl thiophosphoryl chloride : 3417
1,2-Difluorobenzene : 3549
1,3-Difluorobenzene : 3550
1,4-Difluorobenzene : 3551
4,4'-Difluorodiphenyl : 3552
cis-1,2-Difluoroethylene : 3561
trans-1,2-Difluoroethylene : 3562
Difluoromethyl ether : 962
(Difluoro)methylphosphine : 7196
Difluoromethyl 2,2,2-trifluoroethyl ether : 3564
1,2-Di-2-furanyl-2-hydroxyethanone : 5505
Digallic acid : 3756
Digitalose : 2807
Digitogenin : 9596
Diglycidyl resorcinol ether : 974
Diglycol : 3435
Diglycolamine : 307
Diglycollic anhydride : 4430
Diglyme : 3442
N,N-Dihexyl-1-hexanamine : 10429
Dihexyl ketone : 10317
1,2-Dihydroacenaphthylene : 7
1,4-Dihydrobenzene : 2562
Dihydrocholesterol : 2367
2,3-Dihydro-5,6-dimethyl-1,4-dithiin, 1,1,4,4-tetraoxide : 3826
3,7-Dihydro-1,3-dimethyl-1*H*-purine-2,6-dione : 10018
Dihydro-4,4-dimethyl-2*H*-pyran-2,6(3*H*)-dione : 4217
Dihydro-14-hydroxycodeinone : 8485
1,2-Dihydro-6-hydroxy-2-oxo-4-pyridinecarboxylic acid : 2413
1,2-Dihydro-3*H*-indazol-3-one : 6234
Dihydroisophorone : 10538
1,2-Dihydro-4-methoxy-1-methyl-2-oxo-3-pyridinecarbonitrile : 9483
4,9-Dihydro-7-methoxy-1-methyl-3*H*-pyrido[3,4-b]indole : 5623
7,8-Dihydromorphin-6-one : 5954
1,1-Dihydroperfluorooctanol : 8550
1,5-Dihydro-4*H*-pyrazolo[3,4-d]pyrimidin-4-one : 163
Dihydroresorcinol : 2579
Dihydrosafrole : 9216
4,5-Dihydrotoluene : 7143
5,6-Dihydrouracil : 3675
Dihydroxyacetone : 3751
2,5-Dihydroxyacetophenone : 66
1,2-Dihydroxy-9,10-anthracenedione : 151
3,9-Dihydroxy-6*H*-benzofuro[3,2-c][1]benzopyran-6-one : 2474
4,5-Dihydroxy-1,3-bis(hydroxymethyl)-2-imidazolidinone : 4200
2,3-Dihydroxybutanedioic acid, (*R**,*R**)-(±)- : 9678
2,3-Dihydroxybutanedioic acid, (*R**,*R**)- : 9679
2,3-Dihydroxybutanedioic acid, [*S*-(*R**,*R**)]- : 9680
2,3-Dihydroxybutanedioic acid, [*R*-(*R**,*R**)]- : 9681
3,3'-Dihydroxy-β,κ-caroten-6'-one, (3*R*,3'*S*,5'*R*) : 1721
3,12-Dihydroxycholan-24-oic acid, (3α,5β,12α) : 2800
2,2'-Dihydroxy-4,4'-dimethoxybenzophenone : 1008
2,5-Dihydroxy-3,6-dinitro-2,5-cyclohexadiene-1,4-dione : 8028
5,7-Dihydroxy-2-(4-hydroxyphenyl)-4*H*-1-benzopyran-4-one : 498
5,7-Dihydroxy-3-(4-hydroxyphenyl)-4*H*-1-benzopyran-4-one : 5526
Dihydroxymaleic acid : 3725
5,7-Dihydroxy-2-(4-methoxyphenyl)-4*H*-1-benzopyran-4-one : 4
5,7-Dihydroxy-6-methoxy-2-phenyl-4*H*-1-benzopyran-4-one : 8419
Di(2-hydroxy-2-methylpropyl) ketone : 3721
11,15-Dihydroxy-9-oxo-5,13-prostadienoic acid : 9278
11,15-Dihydroxy-9-oxo-13-prostenoic acid : 9277
L-3,4-Dihydroxyphenylalanine : 6607
(3,4-Dihydroxyphenyl)(2,4,6-trihydroxyphenyl)methanone : 6652
2,3-Dihydroxy-4-(phosphonooxy)butanal : 4751
8,8'-Dihydroxyrugulosin : 6639
9,10-Dihydroxystearic acid : 3740
1,8-Dihydroxy-2,4,5,7-tetranitro-9,10-anthracenedione : 2385
2,5-Dihydroxy-3-undecyl-2,5-cyclohexadiene-1,4-dione : 4694
1,2-Diiodobenzene : 3760
1,3-Diiodobenzene : 3761
1,4-Diiodobenzene : 3762
cis-1,2-Diiodoethylene : 3765
3,5-Diiodosalicylic acid : 6024
Diiron nonacarbonyl : 6357
Diisoamyl ether : 3790
Diisoamyl phthalate : 3791
Diisobutylcarbinol : 4081
Diisobutyl ketone : 4083
1,5-Diisocyanatonaphthalene : 7948
Diisopropanol : 5836
Diisopropenyl : 3981
1,3-Diisopropenylbenzene : 1024
N,N-Diisopropyl-2-aminoethanol : 3804
Diisopropyl dixanthogen : 3814
Diisopropyl ketone : 4228
N,N'-Diisopropyl-6-(methylthio)-1,3,5-triazine-2,4-diamine : 9130
Dimanganese decacarbonyl : 6675
Dimedazole : 3962
Dimercaprol : 3824
2,3-Dimercaptobutanedioic acid, (*R**,*S**) : 9622
2,2'-Dimercaptodiethyl sulfide : 1018
6,8-Dimercaptooctanoic acid : 3636
Dimethadione : 4202
4,4'-Dimethoxydiphenylamine : 6836
4,9-Dimethoxy-7-methyl-5*H*-furo[3,2-g][1]benzopyran-5-one : 6571
2-[2-(3,4-Dimethoxyphenyl)ethyl]-4-methoxyquinoline : 5519
(Dimethoxyphosphinylthio) butanedioic acid : 6657
2,6-Dimethoxy-*p*-quinone : 3858
2,3-Dimethoxystrychnidin-10-one, monohydrochloride : 1384
2,3-Dimethoxystrychnidin-10-one, sulfate, heptahydrate : 1385
Dimethyl 1,3-acetonedicarboxylate : 4210
2,4-Dimethylacetophenone : 4249
2,5-Dimethylacetophenone : 4250
3,4-Dimethylacetophenone : 4251
Dimethylacetylene : 1676
N,N-Dimethylacrylamide : 4280
1,3-Dimethylallene : 8568
Dimethylaluminum chloride : 1989
4-(Dimethylamino)benzophenone : 3917
4-(Dimethylamino)cinnamaldehyde : 3919
6-(Dimethylamino)-4,4-diphenyl-3-heptanone hydrochloride : 6735
1,2-Dimethylaminoethane : 9946
Dimethylaminosulfonyl chloride : 4322
4-(Dimethylamino)-3,5-xylyl methylcarbamate : 7871
9,10-Dimethyl-1,2-benzanthracene : 3948
1,2-Dimethylbenzene : 10851
1,3-Dimethylbenzene : 10852
1,4-Dimethylbenzene : 10853
N,α-Dimethylbenzeneethanamine, hydrochloride, (*S*)- : 6739
3,3'-Dimethylbenzidine : 10109
8,8-Dimethyl-2*H*,8*H*-benzo[1,2-b:5,4-b']dipyran-2-one : 10848
Dimethylbenzylamine : 3971
1,1-Dimethyl-1,3-butadiene : 7548
N,N-Dimethyl-1-butanamine : 1554
4,4-Dimethyl-2-butenoic acid : 7585
Dimethylcarbamoyl chloride : 4002
Dimethyl chlorothiophosphate : 4262
Dimethyl citraconate : 4129
5,5-Dimethyl-1,3,6-cycloheptatriene-1-carboxylic acid : 10090
N,N-Dimethylcyclohexanamine : 2619
2,2-Dimethylcyclopropanecarboxylate : 2706
N,N-Dimethyl-1-decanamine : 4039
Dimethyldiacetylene : 5788
N,N'-Dimethyl-4,4'-diaminodiphenylmethane : 7234
Dimethyldiethoxysilane : 3356
5,5-Dimethyldihydroresorcinol : 4018
Dimethyl 2,3-dihydroxybutanedioate, [*R*-(*R**,*R**)]- : 4329
2,6-Dimethyl-1,3-dioxan-4-ol acetate : 3829
N,N-Dimethylethanamide : 3889
N,N-Dimethylethanamine : 4993
2-(1,1-Dimethylethoxy)-1-propanol : 9232
Dimethylethoxysilane : 4836
1,1-Dimethylethyl formate : 1578
Dimethylformamide dimethyl acetal : 3874
N,N-Dimethyl-1-hexadecanamine : 5761

Dimethyl ketazine : 36
N,N-Dimethylmethanamine : 10499
N,N-Dimethylmethanamine borane : 10500
N,N-Dimethylmethanamine hydrochloride : 10501
N,N-Dimethylmethanamine oxide : 10502
2,2-Dimethyl-3-methylenebicyclo[2.2.1]heptane, (1*R*)- : 1706
2,2-Dimethyl-3-methylenebicyclo[2.2.1]heptane, (1*S*)- : 1707
Dimethyl(methylthio)borane : 7206
3,17-Dimethylmorphinan, (9 α,13 α,14 α)- : 3820
Dimethyl 4-morpholinylphosphonate : 4137
Dimethylnitrosamine : 8187
2,3-Dimethyl-2-norbornene : 3975
3,7-Dimethyl-1,6-octadien-3-ol, (±)- : 6612
3,7-Dimethyl-1,6-octadien-3-yl acetate : 6613
N,N-Dimethyl-1-octanamine : 8388
1,3-Dimethylolethyleneurea : 1009
Dimethylolpropionic acid : 6049
4,4-Dimethylphenidone : 4255
2,3-Dimethylphenol : 10854
2,4-Dimethylphenol : 10855
2,5-Dimethylphenol : 10856
2,6-Dimethylphenol : 10857
3,4-Dimethylphenol : 10858
3,5-Dimethylphenol : 10859
2-[(2,3-Dimethylphenyl)amino] benzoic acid : 6689
Dimethyl-*p*-phenylenediamine : 3952
2,6-Dimethylphenyl isocyanate : 6409
3,5-Dimethylphenyl methylcarbamate : 10866
N-(2,6-Dimethylphenyl)-1-methyl-2-piperidinecarboxamide : 6706
3,4-Dimethyl-2-phenylmorpholine : 8756
Dimethyl phosphate : 4109
Dimethylphosphoroamidocyanidic acid, ethyl ester : 9672
O,O-Dimethyl phosphorodithionate : 4052
Dimethyl pimelate : 4079
2,2-Dimethyl-1-propanamine : 4281
2,2-Dimethylpropane : 8006
Dimethylpropanedioc acid : 4127
1,1-Dimethylpropargyl alcohol : 7115
N,N-Dimethyl-2-propargylamine : 3924
N,N-Dimethyl-2-propen-1-amine : 186
Dimethylpropylamine : 4272
N-(1,1-Dimethyl-2-propynyl)-3,5-dichlorobenzamide : 9276
N,N-Dimethyl-1*H*-purin-6-amine : 3925
1,5-Dimethyl-2,4(1*H*,3*H*)-pyrimidinedione : 7825
Dimethylsiloxane cyclic trimer : 5815
Dimethyl suberate : 4179
5,8-Dimethyltocol : 10104
7,8-Dimethyltocol : 10105
N,N-Dimethyl-*o*-toluidine : 7202
N,N-Dimethyl-*m*-toluidine : 7203
N,N-Dimethyl-*p*-toluidine : 7204
5-(3,3-Dimethyl-1-triazenyl)-1*H*-imidazole-4-carboxamide : 2729
N,N-Dimethyltryptamine : 4117
Dimethylvinyl chloride : 2160
Dimetridazole : 4159
β,β'-Dinaphthylamine : 7990
Dinicotinic acid : 9343
Dinitolmide : 7207
3,5-Dinitroanisole : 6820
o-Dinitrobenzene : 4369
m-Dinitrobenzene : 4370
p-Dinitrobenzene : 4371
4,4'-Dinitrobibenzyl : 1041
4,4'-Dinitrocarbanilide : 1042
4,6-Dinitro-*o*-cresol : 7214
2,6-Dinitro-*p*-cresol : 7215
2,6-Dinitro-*N,N*-dipropyl-4-(trifluoromethyl)aniline : 10426
2,4-Dinitrophenyl fluoride : 5385
2,4-Dinitroresorcinol : 4372
2,3-Dinitrotoluene : 7208
2,4-Dinitrotoluene : 7209
2,5-Dinitrotoluene : 7212
2,6-Dinitrotoluene : 7211
3,4-Dinitrotoluene : 7213
3,5-Dinitrotoluene : 7210
N,N'-Dioctadecanoylethanediamine : 5020
Dioctadecyl thiobispropanoate : 4576
Dioctyl decanedioate : 4423
N,N-Dioctyl-1-octanamine : 10636
Di-*sec*-octyl phthalate : 986
Diosgenin : 9597
Diosphenol : 6090
1,3-Dioxacyclohexane : 4427
1,4-Dioxacyclohexane : 4428
1,3-Dioxacyclopentane : 4433
trans-1,7-Dioxadispiro[4.0.4.2]dodeca-3,9-diene-2,8-dione : 459
Dioxin : 9743
1,2-Dioxocyclohexane : 2578
2-(2,6-Dioxo-3-piperidinyl)-1*H*-isoindole-1,3(2*H*)-dione : 10011
2,5-Di-*tert*-pentylhydroquinone : 971
Dipentyl maleate : 4440
N,N-Dipentyl-1-pentanamine : 10642
Diphemethoxidine : 4505
Diphenadione : 4453
Diphenhydramine : 4503
o,o'-Diphenic acid : 883
Diphenyl : 876
Diphenylacetic acid : 8828
1,1-Diphenylacetone : 4517
N,N-Diphenylbenzenamine : 10643
N,N'-Diphenylbenzidine : 4462
sym-Diphenylcarbazide : 4475
Diphenyldiacetylene : 4464
cis-Diphenyldiazene : 559
trans-Diphenyldiazene : 558
Diphenyldiazene 1-oxide, (*E*) : 565
Diphenyldiazene 1-oxide, (*Z*) : 566
Diphenyl disulfide-2,2'-dicarboxylic acid : 4584
Diphenylene dioxide : 2883
Diphenylethanedione : 681
cis-1,2-Diphenylethene : 9612
trans-1,2-Diphenylethene : 9613
Diphenylethenone : 4495
Diphenyl ether 4,4'-disulfonyl chloride : 8480
1,3-Diphenylguanidine : 4489
5,5-Diphenyl-2,4-imidazolidinedione : 9016
Diphenyl ketone : 718
Diphenyl ketoxime : 720
N,N-Diphenylnitrosamine : 8189
N,N'-Diphenyl-*p*-phenylenediamine : 4459
1,1-Diphenyl-4-piperidinyl-1-butanol : 4448
1,1-Diphenyl-3-(1-piperidinyl)-1-propanol : 9112
sym-Diphenylthiourea : 4532
Diphos : 973
1,2-Diphosphonoethane : 4790
Dipicolinic acid : 9341
Dipicrylamine : 10634
Dipicryl sulfide : 1055
Dipivaloylmethane : 9949
Diprazin : 9128
Di-2-propenyl 1,3-benzenedicarboxylate : 2836
Dipropionylmethane : 5678
Dipropylacetylene : 8405
Dipropylcarbinol : 5688
Dipropyl ketone : 5691
Dipropyl phthalate : 628
N,N-Dipropyl-1-propanamine : 10668
α,α'-Dipyridyl : 893
γ,γ'-Dipyridyl : 897
Dipyrithione : 4569
Direct Blue 6 : 2390
Diresorcinol : 889
Dirhenium decacarbonyl : 9472
Dirhodium tetracarbonyl dichloride : 9475
Disalicylidene-1,2-ethanediamine : 992
Disalicylidene-1,3-propanediamine : 9146
Disophenol : 3769
Distannoxane, hexakis(2-methyl-2-phenylpropyl)- : 5313
Distearylamine : 4417
1,3-Dithiacyclopentane : 4590
3-(Di-2-thienylmethylene)-1-methylpiperidine : 10102
3,3'-Dithiobis(2-aminopropanoic acid) : 2718
2,2'-Dithiobis[benzothiazole] : 6710
2,4-Dithiobiuret : 10057
Dithiocarb sodium : 9551
Dithiocyanatomethane : 7248
1,2-Dithiolane-3-pentanamide : 6619
1,2-Dithiolane-3-pentanoic acid : 10049
Dithiosalicylic acid : 5978
Di-*p*-Tolyl disulfide : 1025
Di-*p*-tolylmercury : 1028
Di-*p*-tolyl sulfide : 1030
Di-*p*-tolyl sulfone : 1031
Ditridecyl thiodipropionate : 1049
Diundecyl ketone : 10301

Divinyl : 1393
Divinylacetylene : 5786
1,2-Divinylbenzene : 4600
1,3-Divinylbenzene : 4601
1,4-Divinylbenzene : 4602
DMF : 4061
DMSO : 4328
Docusate sodium : 988
1-Dodecanamine : 4640
1-Dodecanamine, acetate : 4641
5-Dodecanolide : 5711
N-Dodecanoylsarcosine : 7538
N-Dodecyl-1-dodecanamine : 3342
Dodecylguanidine, monoacetate : 4660
1-Dodecylhexahydro-2*H*-azepin-2-one : 6595
Dodecyl methyl ketone : 9798
Dodecyltrichlorosilane : 10226
Dopamine hydrochloride : 311
Dowicide 9 : 1967
Doxenitoin : 4493
Drosophilin A : 9757
DTBP : 3038
Dulcin : 4864
Dulcose : 5512
Durene : 9923
Durindone Red : 852
Durohydroquinone : 9926
Duroquinone : 9935
Dymanthine : 4320
Dypnone : 4470
Echinatine-3′-epimer : 9486
Echinopsine : 7775
Edetate calcium disodium : 4573
Edetate sodium : 10002
EDTA : 5018
EDTA disodium : 5019
EEDQ : 5026
Ehrlich's reagent : 3900
1-Eicosanamine : 6208
Elemol : 4692
cis-Eleostearic acid : 8309
trans-Eleostearic acid : 8310
Emodin : 10441
Enanthaldoxime : 5669
Enanthic acid : 5683
ENTPROL : 9902
Eosine YS : 9713
Epiandrostanediol : 454
Epiandrosterone : 5964
Epicholestanol : 2368
Epicholesterol : 2370
16-Epiestriol : 4762
2,3-Epoxy-4-oxo-7,10-dodecadienamide, (2*R*,3*S*)- : 1805
2,3-Epoxypropyl ethyl ether : 4851
1,2-Epoxytetradecane : 4649
12,13-Epoxytrichothec-9-en-4-ol acetate : 10297
EPTC : 4552
Equilenin : 6036
Equilin : 6037
Ergometrine : 4725
Ergostanol : 4729
α-Ergostenol : 4736
γ-Ergostenol : 4735
Ergosterol : 4731
Erucamide : 4614
Erucic acid : 4616
Erythritol : 1423
Erythrityl tetranitrate : 1424
Esculetin : 3712
α-Estradiol : 4757
β-Estradiol : 4758
Estradiol benzoate : 4760
Estragole : 6893
Estriol : 4761
Ethacridine : 4821
Ethamivan : 3468
Ethanal : 14
Ethanamide : 17
Ethanamine : 4879
Ethanamine hydrochloride : 4880
Ethane, 1,1-dichloro-2,2-bis(*p*-ethylphenyl)- : 8735
Ethaneperoxoic acid : 8732
Ethanethioamide : 10043
Ethanoic acid : 21
Ethanolamine *O*-phosphate : 9026
Ethanol, 2,2′-[1,2-ethanediylbis(oxy)] bis-, dinitrate : 10360
Ethanol, germanium(4+) salt : 9807
Ethanol, tantalum(5+) salt : 8577
Ethanoyl bromide : 57
Ethanoyl chloride : 58
Ethanoyl fluoride : 69
Ethanoyl iodide : 74
Ethene : 5015
1-(Ethenyloxy)butane : 1673
1-(Ethenyloxy)decane : 2787
1-(Ethenyloxy)hexadecane : 5771
2-(Ethenyloxy)-2-methylpropane : 1674
1-(Ethenyloxy)octadecane : 8328
1-(Ethenyloxy)propane : 9270
2-(Ethenyloxy)propane : 6546
3-(Ethenyloxy)-1-propene : 213
Ethenzamide : 4827
Ethide : 3239
Ethinamate : 5288
Ethionamide : 5234
Ethohexadiol : 5048
Ethosuximide : 5157
2-Ethoxy-1,1-bis(ethoxycarbonyl) ethene : 3455
(Ethoxycarbonyl)acetylene : 5229
N-(Ethoxycarbonyl)phthalimide : 1734
2-[2-(2-Ethoxyethoxy)ethoxy]ethanol : 10361
1-Ethoxyhexane : 5061
Ethoxyquin : 4835
Ethoxzolamide : 4832
Ethylacetylene : 1675
Ethyl alcohol : 4800
Ethylallene : 8564
Ethyl 2-allylacetoacetate : 4877
Ethylaluminum chloride : 3181
Ethylaluminum sesquichloride : 10292
Ethyl aminoacetate hydrochloride : 5578
Ethyl aminobenzoate : 4884
Ethyl 2-amino-3-phenylpropionate : 8807
N-Ethylamphetamine : 5118
N-Ethylbenzenemethanamine : 796
α-Ethylbenzyl alcohol : 4899
Ethylbenzylaniline : 797
3-(α-Ethylbenzyl)-4-hydroxycoumarin : 8794
Ethyl 2-benzylidene-2-cyanoacetate : 4960
N-Ethyl-bis(3-phenylpropyl)amine : 223
Ethyl bromide : 1192
Ethyl 6-bromocaproate : 4914
Ethyl bromomalonate : 3409
Ethyl α-bromopropionate : 4920
Ethyl 3-bromopyruvate : 4917
N-Ethyl-1-butanamine : 1565
2-Ethyl-1-butanamine : 4934
Ethyl *tert*-butyl ether : 1569
Ethyl butyl ketone : 5690
Ethyl caprate : 4972
Ethyl carbazate : 5066
Ethyl carbonate : 3416
Ethyl chloride : 2018
Ethyl chloromalonate : 3419
Ethyl α-chloropropionate : 4947
Ethyl crotonate : 4930
Ethyl cyanide : 9147
Ethyl 2-cyano-2-propenoate : 4955
N-Ethylcyclohexanamine : 2623
Ethyl dibunate : 4978
N-Ethyldiethanolamine : 999
Ethyl 2,2-dimethylpropionate : 5006
Ethylene bromohydrin : 1193
Ethylene chlorohydrin : 2020
Ethylenediamine : 4770
Ethylenediamine dihydrochloride : 4771
Ethylene dibromide : 2938
Ethylene dichloride : 3174
Ethylene difluoride : 3559
Ethylene dimercaptan : 4793
Ethylene disulfonic acid : 4791
Ethylene fluorohydrin : 5387
Ethylene glycol : 4772
Ethylene glycol diacetate : 4775
Ethylene glycol diacrylate : 4776
Ethylene glycol dibenzoate : 4777
Ethylene glycol dibutyl ether : 3000
Ethylene glycol didodecanoate : 4778
Ethylene glycol diethyl ether : 3360
Ethylene glycol diformate : 4779
Ethylene glycol dimethacrylate : 4781
Ethylene glycol dimethyl ether : 3863
Ethylene glycol dinitrate : 4782
Ethylene glycol dipalmitate : 4780
Ethylene glycol diphenyl ether : 4450
Ethylene glycol distearate : 4783
Ethylene glycol ditetradecanoate : 4784
Ethylene glycol dithiocyanate : 4785
Ethylene glycol monoacetate : 4786

Ethylene glycol monoallyl ether : 200
Ethylene glycol monobenzoate : 4787
Ethylene glycol monobenzyl ether : 817
Ethylene glycol monobutyl ether : 1473
Ethylene glycol monobutyl ether acetate : 1477
Ethylene glycol monoethyl ether : 4839
Ethylene glycol monoethyl ether acetate : 4841
Ethylene glycol monoethyl ether acrylate : 4842
Ethylene glycol mono(2-ethylhexyl) ether : 5064
Ethylene glycol monohexyl ether : 5905
Ethylene glycol monomethacrylate : 6042
Ethylene glycol monomethyl ether : 6822
Ethylene glycol monomethyl ether acetate : 6825
Ethylene glycol monopropyl ether : 9197
Ethylene glycol monostearate : 4788
Ethylene glycol monosulfite : 4789
Ethylene glycol monovinyl ether : 10818
Ethylene oxide : 8447
Ethylene sulfide : 10041
Ethylene thiourea : 6219
Ethylene urea : 6221
N-Ethylethanamine : 3378
N-Ethylethanamine hydrochloride : 3379
Ethyl ether : 3454
Ethyl 2-ethylcaproate : 5029
17β-Ethyletiocholane : 9102
Ethyl fluoride : 5386
Ethyl 2-furanoate : 5037
Ethyl heptyl ketone : 2764
Ethyl 1-hexamethyleneiminecarbothiolate : 7886
2-Ethyl-1-hexanamine : 5057
2-Ethyl-2-hexanol : 7300
Ethyl hexyl ketone : 8244
3-Ethylhomocysteine, (*R*) : 4809
3-Ethylhomocysteine, (*S*) : 4810
Ethyl hydrogen peroxide : 5070
Ethyl hydrosorbate : 5054
Ethyl 2-hydroxypropionate : 5096
5-Ethylidenebicyclo[2.2.1]hept-2-ene : 5082
Ethylidene diacetate : 4774
Ethylidene dibromide : 2937
Ethylidene dichloride : 3173
Ethylidene difluoride : 3558
Ethyl iodide : 6297
Ethyl 10-(4-iodophenyl)undecanoate : 6350
Ethyl isobutanoate : 5153
Ethyl isocrotonate : 4929
5-Ethyl-5-isopentyl-2,4,6(1*H*,3*H*,5*H*)-pyrimidinetrione : 442
Ethyl isopropyl ketone : 7573
Ethyl isopropylmalonate : 3473
Ethyl isovalerate : 5123
Ethyl linoleate : 5178
Ethyl linolenate : 5179
N-Ethylmaleimide : 5240
Ethyl mercaptan : 4797
Ethyl mesoxalate : 3474
Ethyl 4-methoxyphenyl ketone : 6880
Ethyl 4-methylcaproate : 5144
N-Ethyl-*N*-methylethanamine : 3479
Ethyl 3-methyl-3-phenylglycidate : 5151
N-Ethyl-2-methyl-2-propen-1-amine : 5109
Ethyl nicotinate : 5236
N-Ethyl-*N*-nitroethanamine : 8094
2-Ethyl-2-[(nitrooxy)methyl]-1,3-propanediol, dinitrate : 9176
Ethyl octadecanoate : 5245
Ethyl *cis*-9-octadecenoate : 5185
Ethyl oenanthate : 5042
Ethyl oxalyl chloride : 4946
Ethyloxirane : 4710
Ethyl [(4-oxo-2-phenyl-4*H*-1-benzopyran-7-yl)oxy]acetate : 4679
Ethylparaben : 5073
2-Ethyl-3,5,6,7,8-pentahydroxy-1,4-naphthalenedione : 4676
Ethyl *tert*-pentyl ether : 4849
Ethyl pentyl ketone : 8361
Ethyl 2-(4-phenoxyphenoxy) ethylcarbamate : 5319
Ethyl phenylacetylenecarboxylate : 5212
Ethyl phenylglyoxylate : 5188
Ethyl *trans*-3-phenyl-2-propenoate : 4952
5-Ethyl-5-phenyl-2,4,6(1*H*,3*H*,5*H*) pyrimidinetrione : 8761
Ethyl phosphate : 10365
Ethylphosphoric acid dichloride : 5215
Ethyl 2-picolinate : 5235
3-Ethyl-4-picoline : 5155
4-Ethyl-2-picoline : 5156
Ethyl pivaloylacetate : 5002
N-Ethyl-2-propanamine : 5090
Ethyl propenoate : 4878
Ethyl propionate : 5224
1-Ethyl-2-propylacetylene : 5714
N-(1-Ethylpropyl)-3,4-dimethyl-2,6-dinitroaniline : 8520
Ethyl propyl ketone : 5854
Ethyl pyruvate : 5189
3-Ethylrhodanine : 5263
Ethyl ricinoleate : 5080
Ethyl sorbate : 5045
α-Ethylstyrene : 8857
Ethyl succinate : 3527
Ethyl sulfite : 3530
Ethyl sulfone : 3531
Ethylsulfonic acid : 4795
Ethylsulfonylethyl alcohol : 5250
Ethyl tartrate, acid : 5252
Ethyl tetraphosphate : 5791
6-(Ethylthio)-*N*,*N'*-diisopropyl-1,3,5-triazine-2,4-diamine : 4544
1-(Ethylthio)octane : 5184
N-Ethyl-4-toluidine : 5116
Ethyltrichlorosilane : 10235
Ethyltrimethylplumbane : 5276
Ethyl undecylate : 5279
Ethyl valerate : 5193
Ethyl vanillin : 4844
Ethyl vinyl ketone : 8652
Ethyne : 67
Ethynylbenzene : 8800
Ethynyl methyl ketone : 1686
Etidronic acid : 6035
Etiocholanic acid : 453
Etocrilene : 4956
Eucarvone : 10529
Eugenol : 196
1,3,4-Eugenol acetate : 197
Exaltol : 2656
Exaltolide : 8425
FAD : 5332
Fagarine : 3865
Felbinac : 877
α-Fenchyl alcohol, (*l*) : 10516
Fenoprofen : 8787
Fenpiprane : 4522
Fenticlor : 994
Fenuron : 4259
Ferric acetylacetonate : 6360
Ferron : 6051
Fischer's base : 7249
Flavone : 8843
Flopropione : 10447
Floxuridine : 2802
Fludroxycortide : 5437
Flufenamic acid : 10411
Fluometuron : 4347
Fluopromazine : 10425
9*H*-Fluorene-9-carboxylic acid, 2-chloro-9-hydroxy- : 1831
Fluorescin : 3757
Fluorine cyanide : 2515
Fluoroacetic acid amide : 5351
Fluoroacetone : 5417
Fluoroethanoic acid : 5352
Fluoroethyne : 5354
Fluoroform : 10395
Fluoroformaldehyde : 5464
o-Fluoronitrobenzene : 5405
m-Fluoronitrobenzene : 5406
p-Fluoronitrobenzene : 5407
3-Fluoro-4-nitrotoluene : 5401
5-Fluoro-2,4(1*H*,3*H*)-Pyrimidinedione : 5432
Fluorotrinitromethane : 10628
Fluoxiprednisolone : 10136
Fomepizole : 7727
Formaldehyde, dipropyl acetal : 4546
Formaldehyde, trimer : 10638
Formamidine : 6748
Formamidine acetate : 6749
Formanilide : 8882
Formylcyclopropane : 2691
3-Formylphenol : 5968
4-Formylphenol : 5969
5-Formyl-5,6,7,8-tetrahydrofolic acid : 5443
3-Formylthiophene : 10070
Forstab : 5060
Fraxetin : 3727

Fuchsin, acid : 108
Fumaric acid dichloride : 1447
Fumarine : 9280
2-Furaldehyde : 5502
2-Furanacrylic acid : 5499
2-Furanacrylonitrile : 5498
2-Furanmethanol : 5503
2-Furanmethanol, propanoate : 5504
2-(2-Furanyl)-3-(5-nitro-2-furanyl)-2-propenamide : 5509
Furfurylamine : 5486
Furfuryl disulfide : 3572
Furfuryl ether : 3573
2-Furfuryl methyl ketone : 5496
Furfuryl valerate : 5493
α-Furildioxime : 3570
2-Furoic acid : 5483
Furo[3,4-b]pyridine-5,7-dione : 9344
Fusaric acid : 1659
Fyrol FR-2 : 10682
4-*O*-β-*D*-Galactopyranosyl-*D*-fructose : 6582
6-*O*-α-*D*-Galactopyranosyl-*D*-glucose : 6693
Galegine : 7103
Gallacetophenone : 10445
Gallic acid : 10437
Geissoschizoline : 2494
Gemfibrozil : 4245
Genetron 132b-B2 : 2932
Genite : 3260
Gentian violet : 2488
Gentisic acid : 3706
Gentisin : 3730
Gentisyl alcohol : 6080
Geranic acid : 4167
trans-Geranyl bromide : 1185
Germanium tetramethyl : 9947
Germanium tetraphenyl : 9990
Gibbs' reagent : 3141
Gitogenin : 9592
Glucamine : 292
δ-*D*-Gluconolactone : 6578
3-*O*-β-*D*-Glucopyranosyl-*D*-glucose : 6583
6-*O*-α-*D*-Glucopyranosyl-*D*-glucose : 6439
6-*O*-β-*D*-Glucopyranosyl-*D*-glucose : 5527
6-(β-*D*-Glucopyranosyloxy)-7-hydroxy-2*H*-1-benzopyran-2-one : 4755
D-Glucosamine : 293
D-Glucose, 6-benzoate : 10766
D-Glucuronolactone : 5551
L-Glutamic acid, 5-[2-[4-(hydroxymethyl)phenyl]hydrazide] : 139
L-γ-Glutamyl-*L*-cysteinylglycine : 5558
L-γ-Glutamyl-*L*-cysteinylglycine disulfide : 5559
Glutaraldehyde : 8608
Glutaronitrile : 8610
Glyceric acid : 3750
Glycerol β-chlorohydrin : 2263
Glycerol 3-methyl ether : 6886
Glycerol triacetate : 10128
α-Glycerophosphoric acid : 5567
Glyceryl *p*-aminobenzoate : 9156
Glycidaldehyde : 8448
Glycidol : 8449
Glycidol methacrylate : 4716
Glycidyl acrylate : 4715
Glycine, *N*,*N*-bis(phosphonomethyl)- : 5601
Glycine, ethyl ester : 4881
Glycine, *N*-(phosphonomethyl)- : 5599
Glycinol : 4801
Glycol salicylate : 6041
Glyconitrile : 5960
Glycosine : 7044
Glyoxaline-5-alanine : 5918
Gramine : 4118
Guaiacol : 6855
Guaiacol benzoate : 6858
Guaiacol carbonate : 6859
Guaiacol phosphate : 6860
Guaiacol valerate : 6874
Guaiazulene : 4120
Guaifenesin : 6862
Guaifenesin-1-carbamate : 6761
Guajen : 4145
Guanosine 5'-(trihydrogen diphosphate) : 5616
5'-Guanylic acid : 5617
5'-Guanylic acid, disodium salt : 5618
Guvacine : 9889
Gynergen : 4738
Halazone : 3074
Halon 1011 : 1147
Halon 1121 : 1168
Halon 1211 : 1142
Halon 2302 : 2994
Halothane : 1155
Hecogenin : 6193
Helicin : 5542
Hemimellitene : 10507
Hemimellitic acid : 668
Hemipyocyanine : 8752
Hendecane : 10728
8,10,12-Heptadecatriene-4,6-diyne-1,14-diol : 2389
Heptaldehyde : 5668
Heptamethylene chlorohydrin : 2064
Heptamethylene dibromide : 2946
1-Heptanamine : 5703
Heptylacetylene : 8261
Heptyl alcohol : 5685
Heptyl bromide : 1218
2-Heptyl bromide : 1219
4-Heptyl bromide : 1220
Heptyl chloride : 2060
Heptyl ether : 3583
N-Heptyl-1-heptanamine : 3582
Heptylidene acetone : 2775
Heptyl mercaptan : 5681
Heptyl methyl ketone : 8243
Heptyl sulfide : 3585
α-Hexachlorocyclohexane : 5731
β-Hexachlorocyclohexane : 5732
1,4,5,6,7,7-Hexachloro-5-norbornene-2,3-dicarboxylic acid : 1828
Hexachloropentadiene dimer : 7878
1-Hexadecanamine : 5759
6-Hexadecanoylascorbic acid : 519
Hexadecyl 3-hydroxy-2-naphthoate : 5763
Hexadecyloxirane : 4713
Hexaethyldialuminum : 10344
Hexafluoroacetone : 8695
Hexahydroaplotaxene : 5654
Hexahydrobenzene : 2565
Hexahydrobenzoic acid : 2570
Hexahydro-4,7-dimethylcyclopenta[c]pyran-3(1*H*)-one : 6353
Hexahydrophthalic anhydride : 5807
Hexahydropyrimidine-2-thione : 9890
Hexamethylenediamine : 5832
Hexamethylene diiodide : 3767
Hexamethylene glycol : 5835
Hexamethylene methacrylate : 5837
Hexamethylene sulfide : 10037
Hexamethylenimine : 5802
1-Hexanamine : 5889
Hexanedinitrile : 133
1,6-Hexanedioic acid : 132
6-Hexanelactam : 1719
5-Hexanoic acid : 5870
Hexazole : 2624
Hexose monophosphate : 5472
Hexylacetylene : 8402
Hexyl bromide : 1223
1-Hexyl-1,3-butadiene : 2740
Hexyl caproate : 5898
Hexyl chloride : 2066
2-Hexyl chloride : 2067
3-Hexyl chloride : 2068
5-Hexyldihydro-2(3*H*)-furanone : 6023
Hexylene glycol : 7555
Hexyl ether : 3587
Hexyl fluoride : 5390
N-Hexyl-1-hexanamine : 3586
Hexyl iodide : 6306
Hexyl mercaptan : 5845
Hexyl methyl ketone : 8360
2-[2-(Hexyloxy)ethoxy]ethanol : 3449
2-Hexyl-3-phenyl-2-propenal : 8928
Hexyl propyl ketone : 2765
4-Hexylresorcinol : 5891
Hexyl sulfide : 3590
Hexyltrichlorosilane : 10241
3-Hexynol : 5914
Hippuric acid : 766
HMX : 2701
HN1 : 1948
Homocamfin : 6512
Homogentisic acid : 3703
Homovanillic acid : 6061
Hordenine : 3912
Hydantoin : 6218
Hydnocarpic acid : 2676
Hydracrolein : 6179
Hydracrylic acid : 6183

Hydracrylonitrile : 6182
1,2-Hydrazinedicarbothioamide : 4587
1-Hydrazinophthalazine : 5929
Hydrazobenzene : 4492
Hydrocarbostyril : 3677
Hydrocinnamic acid : 653
Hydrocinnamic aldehyde : 650
Hydrocinnamonitrile : 651
Hydrocinnamyl alcohol : 654
Hydrocyanic acid : 5952
Hydrogen cyanate : 2502
Hydrogen tetracarbonylferrate(II) : 6356
Hydroquinone monoethyl ether : 4860
Hydrosorbic acid : 5869
N-Hydroxyacetamide : 30
2-Hydroxyacetophenone : 6153
3-Hydroxyacetophenone : 6154
4-Hydroxyacetophenone : 6155
N-(Hydroxyacetyl)neuraminic acid : 5585
Hydroxyaluminum distearate : 220
γ-Hydroxy-β-aminobutyric acid : 328
1-Hydroxy-3-aminomethyl-3,5,5-trimethylcyclohexane : 356
Hydroxyamphetamine : 412
2-Hydroxybenzaldehyde : 9501
N-Hydroxybenzenamine : 8894
Hydroxybenzene : 8762
N-(4-Hydroxybenzilidene)aniline : 8901
2-Hydroxybenzoic acid, 4-octylphenyl ester : 8400
3-Hydroxybenzyl alcohol : 5984
4-Hydroxybenzyl alcohol : 5985
2-Hydroxybicyclopentyl : 862
Hydroxybutanedioic acid : 6662
4'-Hydroxybutyranilide : 6151
3-Hydroxybutyric acid lactone : 7535
2'-Hydroxychalcone : 6165
2-Hydroxy-5-chloroaniline : 289
Hydroxycyclobutane : 2534
5-Hydroxydecanoic acid lactone : 9879
3-(4-Hydroxy-3,5-dimethoxyphenyl)-2-propenoic acid : 9541
2-Hydroxy-1,2-diphenylethanone, (±) : 714
17-Hydroxyestr-4-en-3-one : 7924
2-(2-Hydroxyethoxy)ethyl laurate : 3446
2-(2-Hydroxyethyl)piperidine : 9065
4-(2-Hydroxyethyl)piperidine : 9066
2-Hydroxyethyl 2-propenoate : 6038
p-Hydroxyhydrocinnamic acid : 5986
Hydroxyhydroquinone : 674
7-Hydroxy-3-(4-hydroxyphenyl)-4*H*-1-benzopyran-4-one : 2731
α-Hydroxy-1*H*-indole-3-propanoic acid : 6252
4-Hydroxyisophthalic acid : 5979
4-[1-Hydroxy-2-[isopropylamino] ethyl]-1,2-benzenediol : 6547
N-Hydroxyl-1-naphthalenamine : 7988
3-Hydroxy-2-methoxy-5-methyl-2,5-cyclohexadiene-1,4-dione : 5477
7-Hydroxy-3-(4-methoxyphenyl)-4*H*-1-benzopyran-4-one : 5455
(2-Hydroxy-4-methoxyphenyl)(2-hydroxyphenyl)methanone : 4435
N-(Hydroxymethyl)acrylamide : 6104
17-Hydroxy-17-methylandrost-4-en-3-one, (17β) : 7794
2-Hydroxy-3-(3-methyl-2-butenyl)-1,4-naphthalenedione : 6588
4-Hydroxy-*N*-methylcarbostyril : 6108
5-Hydroxymethylcytosine : 332
5-(Hydroxymethyl)-2-furaldehyde : 6088
4-Hydroxy-4-methyl-2-pentanone : 2821
1-Hydroxy-4-[(4-methylphenyl) amino]-9,10-anthracenedione : 154
5-Hydroxy-3-methyl-1-phenylpyrazole : 3652
5-Hydroxy-6-methyl-3,4-pyridinedimethanol hydrochloride : 9376
4-Hydroxy-19-methyl-16,19-secostrychnidine-10,16-dione : 10839
1-[6-Hydroxy-2-(1-methylvinyl)-5-benzofuranyl]ethanone : 5298
2-Hydroxy-1-naphthoic acid : 6113
1-Hydroxy-2-naphthoic acid : 6114
3-Hydroxy-2-naphthoic acid : 6115
4-Hydroxynonanoic acid lactone : 3661
5-Hydroxyoctanoic acid lactone : 9882
4-Hydroxy-α-oxobenzenepropanoic acid : 6163
4'β-Hydroxy-19-oxogomphoside : 1703
16-Hydroxypalmitic acid : 6046
7-Hydroxy-3*H*-phenoxazin-3-one, 10-oxide : 9453
4-Hydroxy-*L*-phenylalanine : 10721
4-Hydroxyphenyl phenyl ketone : 5997
1-Hydroxy-3-(phosphonooxy)-2-propanone : 5576
Hydroxypivaldehyde : 6032
17α-Hydroxyprogesterone : 6174
2-Hydroxy-1,2,3-propanetricarboxylic acid : 2414
2-Hydroxy-1,2,3-propanetricarboxylic acid, monohydrate : 2415
2-Hydroxypropanoic acid, (±) : 6574
D-2-Hydroxypropanoic acid : 6575
L-2-Hydroxypropanoic acid : 6576
N-(2-Hydroxypropyl)ethylenediamine : 310
2-Hydroxypropyl methacrylate : 9141
2-(α-Hydroxypropyl)piperidine : 2456
3-Hydroxy-1-propyne : 9175
Hydroxypyruvic acid : 6138
2-Hydroxyquinoline : 9433
3-Hydroxyquinoline : 9434
4-Hydroxyquinoline : 9435
5-Hydroxyquinoline : 9436
6-Hydroxyquinoline : 9437
7-Hydroxyquinoline : 9438
8-Hydroxyquinoline : 9439
8-Hydroxyquinoline sulfate : 9441
8-Hydroxysantonin : 516
12-Hydroxysteric acid : 6133
N-Hydroxysuccinimide : 6189
3-Hydroxy-1,7,7-trimethylbicyclo[2.2.1]heptan-2-one : 6015
4-Hydroxyundecanoic acid lactone : 5708
Hygrine : 7761
Hymecromone : 6084
Hyodeoxycholic acid : 3715
Hypoxanthine riboside : 6260
Ibogaine : 6832
Icosane : 4681
Imidazole-4-acrylic acid : 10762
1*H*-Imidazole, 2-heptadecyl-4,5-dihydro-, monoacetate : 5595
6*H*-Imidazo[4,5-d]pyrimidine : 9293
Imidodicarbonimidic diamide : 866
Imidole : 9392
2,2'-Iminobisacetonitrile : 6225
1,1'-Iminobis-2-propanol : 3793
N-(Iminomethyl)-*L*-glutamic acid : 5454
2-Indanone : 3631
Indanthrene : 3593
Indigo Carmine : 6240
Indium trimethyl : 10568
Indoleacetic acid : 6244
Indolebutyric acid : 6246
Indole-3-pyruvic acid : 8464
3-Indolylacetone : 6257
3-Indolylacrylic acid : 6258
Indonaphthene : 6235
5' Inosinic acid : 6261
Iodinated glycerol : 6301
Iodine cyanide : 2516
3-Iodoacetophenone : 6333
4-Iodoacetophenone : 6334
Iodoalphionic acid : 6027
o-Iodoanisole : 6308
m-Iodoanisole : 6309
p-Iodoanisole : 6310
Iodobenzene diacetate : 8905
Iodochlorhydroxyquin : 2083
Iododocosanoic acid, calcium salt : 1699
Iodoform : 10450
Iodopyrrole : 9906
Iodoquinol : 3773
N-Iodosuccinimide : 6342
o-Iodotoluene : 6311
m-Iodotoluene : 6312
5-Iodouracil : 6341
α-Ionol : 10542
β-Ionol : 10541
Ipodate : 6351
Irganox 1076 : 8323
Iron, tris(dimethylcarbamodithioato-*S,S'*)-, (OC-6-11)- : 5325
Isatin : 6248
Isatin, 3-thiosemicarbazone : 6249
Isoascorbic acid : 4743
Isobutenyl methyl ketone : 6723
Isobutryic anhydride : 7692
Isobutyl alcohol : 7693
5-Isobutyl-5-allyl-2,4,6(1*H*,3*H*,5*H*)-pyrimidinetrione : 1397
Isobutyl *p*-aminobenzoate : 6377
2-Isobutylaminoethyl 4-aminobenzoate : 1465

Isobutyl bromide : 1279
Isobutyl chloride : 2158
Isobutyl cyanide : 7068
Isobutyl enanthate : 6389
Isobutyl iodide : 6316
Isobutyl isovalerate : 6395
Isobutyl mercaptan : 7688
Isobutyl methyl ketone : 7572
2-(4-Isobutylphenyl)propanoic acid : 6207
Isobutyl propionate : 6401
Isobutyl propyl ketone : 7311
Isobutyl salicylate : 6390
Isobutyl urethane : 5085
Isobutyric acid : 7691
Isobutyric acid chloride : 7695
Isobutyronitrile : 7687
Isochroman : 3600
Isocinchomeronic acid : 9340
Isocoumarin : 726
Isocrotonic acid : 1451
Isocrotononitrile : 1448
Isocumene : 9212
4-Isocyanato-1,1′-biphenyl : 10849
Isocyanatocyclohexane : 2629
1-Isocyanatooctadecane : 8324
1-Isocyanatopropane : 9245
Isodurene : 9922
Isoestradiol : 4759
Isoflavone : 8844
Isoguanine : 302
Isohexane : 7552
Isohexyl alcohol : 7564
1*H*-Isoindole-1,3(2*H*)-dione, 2-[(trichloromethyl)thio]- : 5444
Isoliquiritigenin : 10439
4-α-Isomaltosylglucose : 8500
(+)-Isomenthol : 7410
Isometheptene : 4085
Isonicotinic acid : 9331
Isonicotinic acid diethylamide : 3519
Isonipecotic acid : 9063
Isonitrosoacetone : 8474
Isonitrosoacetophenone : 8450
Isononanoic acid : 10565
Isonootkatone : 10783
Isooctane : 10580
Isooctyl alcohol : 7293
Isopentyl alcohol : 7078
Isopentylamine : 7063
Isopentyl benzoate : 7108
Isopentyl bromide : 1252
Isopentyl caproate : 6446
Isopentyl chloride : 2111
Isopentyl cyanide : 7556
Isopentyl iodide : 6314
Isopentyl isobutyrate : 6450
Isopentyl isovalerate : 6448
Isopentyl mandelate : 6447
Isopentyl mercaptan : 7070
Isopentyl nitrate : 7112
Isophorol : 10540
Isophytol : 9951
Isoprene : 7060
Isopropanolamine : 408
Isopropenylacetylene : 7104
Isopropenyl chloride : 2274
Isopropenyl methyl ketone : 7101
Isopropoxybenzene : 8908
4-Isopropoxy-*N*-phenylaniline : 6466
1-Isopropoxy-2-propanol : 7257
Isopropyl acetoacetate : 6524
Isopropyl alcohol : 9167
5-Isopropyl-5-allyl-2,4,6(1*H*,3*H*,5*H*)-pyrimidinetrione : 503
N-Isopropylbenzenemethanamine : 804
Isopropyl bromide : 1328
Isopropyl chloride : 2261
N-Isopropylcyclohexanamine : 2630
Isopropyl ether : 3805
Isopropyl fluoride : 5416
Isopropyl 2-furanoate : 6497
Isopropyl hexadecanoate : 6525
2,4′-Isopropylidenediphenol : 6159
Isopropylidene glycerol : 4044
Isopropyl iodide : 6336
Isopropyl isobutyrate : 6502
Isopropyl isocyanate : 6420
Isopropyl isocyanide : 6425
Isopropyl laurate : 6495
Isopropyl mercaptan : 9152
4-Isopropyl-1-methyl-1,3-cyclohexadiene : 9699
2-Isopropyl-5-methylphenol : 10095
Isopropyl 2-methyl-2-propenoate : 6504
Isopropyl myristate : 6542
Isopropyl nicotinate : 6538
Isopropyl orthoformate : 10460
N-Isopropyl-2-propanamine : 3795
Isopropyl 2-propenoate : 6470
N-Isopropyl-2-pyrimidineamine : 6363
Isopropyl salicylate : 6501
Isoquinaldine : 7423
Isorubijervine : 9579
Isosafrole octyl sulfoxide : 9078
1-Isothiocyanatobutane : 1595
2-Isothiocyanatobutane, (±) : 1596
Isothiocyanatocyclohexane : 2631
1-Isothiocyanato-2-methoxybenzene : 6872
(Isothiocyanatomethyl)benzene : 805
1-Isothiocyanato-2-methylpropane : 6393
2-Isothiocyanato-2-methylpropane : 1597
1-Isothiocyanatonaphthalene : 7989
1-Isothiocyanatopropane : 9246
Isovaleraldehyde : 7061
Isovaleramide : 7062
Isovaleric acid : 7075
Isovaleryl chloride : 7084
Isovaleryl diethylamide : 3485
Isoxylaldehyde : 3945
Itaconic acid : 9180
Jasmone : 7590
Juglone : 6117
Julolidine : 9842
Kafocin : 1799
α-Ketoglutaric acid : 8461
Ketone, 3-pyridyl-3-(*N*-methyl-*N*-nitrosamino)propyl : 8190
Kojic acid : 6050
Kuromatsuene : 6627
Kyanmethin : 4307
Kynurenic acid : 6190
Lactobionic acid : 5515
p-Lactophenetide : 4863
6-Lactoyl-7,8-dihydropterin : 9528
Lanosterol : 6584
Lauraldehyde : 4621
Lauraldehyde, dimethyl acetal : 3861
Lauric acid : 4629
Lauronitrile : 4627
Lauroyl peroxide : 3341
Lauryl alcohol : 4631
Lauryl amine hydrochloride : 4642
Laurylbenzene : 4644
Lauryl bromide : 1190
Lauryl chloride : 2017
Lauryl iodide : 6296
Lauryl 2-propenoate : 4639
Lauryl sulfate : 4654
Lawsone : 6116
Lead triethyl chloride : 2322
Lenthionine : 8634
Lepidine : 7768
Lethane 384 : 1475
L-Leucic acid : 6095
L-*tert*-Leucine : 7852
Levopropoxyphene : 9199
Levulinic acid : 8469
β-Levulose : 5471
Lidocaine : 3382
Lignoceric acid : 9780
Limettin : 3854
Limonene diepoxide : 3347
Linamarin : 5544
Lindane : 5730
δ-Lindane : 5733
Linolelaidic acid : 8300
Lithocholic acid : 6016
Lomustine : 2036
Lumazine : 9291
Lumichrome : 3969
Luminol : 300
Lumisterol : 4733
Lupeol : 6636
3,5-Lupetidine : 4268
Lupinine : 8338
2,3-Lutidine : 4299
2,4-Lutidine : 4300
2,5-Lutidine : 4301
2,6-Lutidine : 4302
3,4-Lutidine : 4303
3,5-Lutidine : 4304
Lutidinic acid : 9339
trans-Lycopene : 1777
Lycoremine : 5518
Lycoxanthin : 1782
Lysidine : 3647

Magnesium dimethyl : 4124
Magnesium octadecanoate : 6655
Magneson : 8145
Maleamic acid : 383
Maleanilic acid : 8471
Maleic hydrazide : 3674
Maltol : 6106
DL-Mandelic acid : 5973
Mandelonitrile : 5977
Manganese, [[1,2-ethanediylbis[carbamodithioato]](2-)]- : 6673
Margaric acid : 5650
Margaric aldehyde : 5646
Margaryl alcohol : 5651
Matrine : 6682
MCPA : 2152
Mechlorethamine : 1949
Mechlorethamine oxide hydrochloride : 8108
Meconic acid : 6139
Meconin : 3869
Mecrylate : 7140
Me-IQ : 303
Melamine : 10142
Melatonin : 6834
Meldrum's acid : 4042
Mellitene : 5813
Mellitic acid : 643
Menadiol diacetate : 7474
Menadione : 7475
p-Menthadiene : 4438
p-Mentha-1,8-diene, (*R*) : 6610
p-Mentha-1,8-diene, (*S*) : 6611
p-Mentha-1(7),2-diene : 8741
p-Mentha-1,4(8)-diene : 9703
p-Mentha-1,8-dien-6-one, (*R*) : 1787
p-Mentha-1,8-dien-6-one, (*S*) : 1788
trans-p-Menthane : 7405
(+)-Menthol : 7407
(-)-Menthol : 7408
Menthol, isovalerate : 6700
Menthone : 7412
l-Menthone : 7413
Mepacrine : 9410
Meparfynol : 7600
Mephenesin : 7611
Mephenoxalone : 6861
Mephentermine : 10511
Meprobamate : 7719
2-Mercaptobenzothiazole : 738
2-Mercaptoethyl ether : 3823
1-(3-Mercapto-2-methyl-1-oxypropyl)proline : 1724
(3-Mercaptopropyl)trimethoxysilane : 10495
6-Mercaptopurine : 3669
Mercuric benzoate : 6719
Mercuric oleate : 6720
Mercuriodibenzene : 4497
Mercury, chloro(2-methoxyethyl)- : 10132
Mesaconic acid : 9179
Mesalamine : 327
Mescaline : 10480
Mesitylamine : 10504
Mesitylene : 10509
Mesitylenic acid : 3967
Mestanolone : 6074
Mestilbol : 3526
Mesulphen : 4335
Metacetaldehyde (polymer) : 6728
Metalphthalein : 2481
Metanilic acid : 252
Methacrolein : 7696
Methacrolein diacetate : 7699
Methacrylic acid anhydride : 7701
Methacrylic acid chloride : 7703
2-(Methacryloyloxy)ethyl acetoacetate : 26
Methallatal : 5110
Methallyl alcohol : 7702
Metham sodium : 9568
Methanal : 5448
Methanamide : 5450
Methanamine : 6920
Methanamine hydrochloride : 6921
Methanethial : 10055
Methanoic acid : 5453
Methaqualone : 7460
Methenamine : 5822
N,N'-Methenyl-*o*-phenylenediamine : 683
Methimazole : 3648
Methiodal sodium : 9564
Methionamine : 78
Methionic acid : 6743
Methionine hydroxy analog : 6109
Methoxyacetone : 6889
Methoxyallene : 6884
Methoxyamine : 7371
Methoxyamine hydrochloride : 7372
Methoxybenzene : 474
2-Methoxycarbonylethyl methyl sulfide : 7461
6'-Methoxycinchonan-9-ol, (8α,9*R*) : 9418
6'-Methoxycinchonan-9-ol monohydrochloride, (8α,9*R*) : 9419
trans-4-Methoxycinnamic acid : 6882
9-Methoxycorynantheidine : 7885
Methoxydichloromethane : 3211
2-(2-Methoxyethoxy)ethanol : 3450
2-Methoxyethyl 2-propenoate : 6826
Methoxyflurane : 3163
9-Methoxy-7*H*-furo[3,2-g][1]benzopyran-7-one : 6767
1-(2-Methoxyisopropoxy)-2-propanol : 4557
3-Methoxy-2-isopropyl-7*H*-furo[3,2-g][1]benzopyran-7-one : 8737
N-Methoxymethanamine : 6837
2-Methoxy-*N*-(2-methoxyethyl)ethanamine : 1020
6-Methoxy-9-methyl-1,3-dioxolo[4,5-h]quinolin-8(9*H*)-one : 1790
1-[2-(2-Methoxy-1-methylethoxy)-1-methylethoxy]-2-propanol : 10673
4-Methoxy-7-methyl-5*H*-furo[3,2-g][1]benzopyran-5-one : 10832
6-Methoxy-α-methyl-2-naphthaleneacetic acid : 7997
7-Methoxy-1-methyl-9*H*-pyrido[3,4-b]indole : 5625
1-Methoxy-2-nitrobenzene : 8037
1-Methoxy-3-nitrobenzene : 8038
1-Methoxy-4-nitrobenzene : 8039
1-Methoxyperfluoroethane : 7549
N-(4-Methoxyphenyl)-1,4-benzenediamine : 6876
4-Methoxy-*o*-phenylenediamine : 6785
4-Methoxy-*m*-phenylenediamine : 6786
3-(4-Methoxyphenyl)-2-propenoic acid, 2-ethoxyethyl ester : 2404
1-Methoxy-2-propanamine : 6835
2-Methoxypropane : 6514
1-Methoxypropane : 7716
1-Methoxy-2-propanol : 9234
6-Methoxy-α-tetralone : 3637
5-Methoxytryptamine : 6833
Methsuximide : 4257
N-Methylacetanilide : 7614
Methylacetylene : 9273
Methyl *o*-acetylsalicylate : 6915
Methylal : 3870
Methyl alcohol : 6750
Methylallene : 1392
Methyl aluminum sesquibromide : 10177
N-Methyl-2-amino-5-chlorobenzophenone : 2097
α-[1-(Methylamino)ethyl]benzenemethanol, (*R**,*S**)-(±)- : 4703
α-[1-(Methylamino)ethyl]benzenemethanol, [*S*-(*R**,*S**)]- : 4704
α-[1-(Methylamino)ethyl]benzenemethanol, [*R*-(*R**,*S**)]- : 4705
2-(Methylamino)-1-phenyl-1-propanol, hydrochloride : 4706
Methyl [(4-aminophenyl)sulfonyl]carbamate : 528
5-Methyl-*o*-anisidine : 6838
Methyl anthranilate : 6923
2-Methylanthraquinone : 6950
Methyl arachidate : 7229
8-Methyl-8-azabicyclo[3.2.1]octan-3-ol, *endo* : 10711
8-Methyl-8-azabicyclo[3.2.1]octan-3-ol, *exo* : 9288
Methyl azelate : 4162
2-Methylaziridine : 9236
Methyl behenate : 7225
Methylbenzene : 10111
4-Methyl-1,3-benzenediamine : 10112
5-Methyl-1,3-benzenediamine : 10113
Methyl 1,4-benzenedicarboxylate : 7793
4-Methylbenzenesulfonic acid, monohydrate : 10117
2-Methylbicyclo[2.2.1]hept-2-ene : 7513
3-Methyl-*N,N*-bis(3-methylbutyl)-1-butanamine : 10458
2-Methyl-*N,N*-bis(2-methylpropyl)-1-propanamine : 10455

N-Methylbis(2,4-xylyliminomethyl) amine : 437
Methyl bromide : 1242
N-Methyl-1-butanamine : 1605
2-Methylbutane : 6441
Methyl *cis*-butenedioate : 4125
3-Methyl-2-butenyl pyrophosphate : 3892
1-Methyl-2-butylacetylene : 5713
(±)-2-Methylbutyric acid : 7074
2-Methyl-γ-butyrolactone : 3644
3-Methyl-γ-butyrolactone : 3645
3-Methylcaproaldehyde : 7323
Methyl caproate : 7329
Methyl caprylate : 7518
Methyl carbazate : 7364
Methyl carbonate : 4005
Methyl chloride : 2089
Methyl chloroform : 10228
3-Methylcholanthrene : 3639
Methylconiine : 7717
7-Methylcoumarin : 7024
Methyl crotonate : 7093
Methyl cyanide : 38
N-Methylcyclohexanamine : 2634
3-Methylcyclohexyl bromide : 1258
5-Methylcytosine : 355
Methyldiacetylene : 8570
3-Methyl-1,2-diazine : 7731
2-Methyl-1,3-diazine : 7750
4-Methyl-1,3-diazine : 7751
5-Methyl-1,3-diazine : 7752
Methyldichloroborane : 3209
Methyl 5-(2,4-dichlorophenoxy)-2-nitrobenzoate : 864
Methyl 2-[4-(2,4-dichlorophenoxy) phenoxy]propanoate : 3313
Methyl dichlorophosphite : 7670
Methyldiethanolamine : 1002
Methyl dihydrocinnamate : 7651
Methyl dihydrogen phosphate : 7665
Methyl 2,2-dimethylpropionate : 7205
4-Methyl-1,3-dioxolan-2-one : 9229
Methyl diphenylglycolate : 7370
Methyl disulfide : 4051
Methyldopa : 6110
2,2′-Methylenebiphenyl : 5343
4,4-Methylene bis(2-chloroaniline) : 901
2,2′-Methylenebisphenol : 3722
Methylene bromide : 2956
α-Methylene butyrolactone : 3642
Methylene chloride : 3202
4,4′-Methylenedianiline : 2855
Methylene diphenyl diisocyanate : 4499
1,1′-Methylenedipiperidine : 4538
Methylene fluoride : 3563
Methylene iodide : 3768
(1*R*,2*S*)-*N*-Methylephedrine : 7250
2-Methyl-1,2-epoxypropane : 4205
Methylergometrine : 7251
N-Methylethanamine : 5111
N-Methylethanamine hydrochloride : 5112
[(1-Methyl-1,2-ethanediyl)bis(oxy)] bispropanol : 10671
Methyl ether : 4058
(1-Methylethoxy)methyloxirane : 6498
Methyl ethyl ketone : 1433
Methyl ethyl ketone peroxide : 1436
Methyl α-ethylstyryl ketone : 8929
Methyl fluoride : 5395
Methyl fluoroform : 10386
Methyl 2-furanoate : 7269
N-Methylglucamine : 2806
N-Methyl-*L*-glucosamidinostreptosidostreptidine : 9614
N-Methyl-α-*L*-glucosamine : 6931
α-Methylglucoside : 7278
2-Methylglutaronitrile : 7554
N-Methylglycine : 9510
7-Methylguanine : 299
Methyl hexadecanoate : 7542
2-Methylhexa-4,5-dione : 7328
N-Methylhydrazinecarbothioamide : 7822
Methyl hydrogen peroxide : 7366
Methyl 2-hydroxybenzoate : 7779
N-Methylhydroxylamine : 6055
N-Methylhydroxylamine hydrochloride : 6072
Methyl 3-hydroxy-2-naphthoate : 7377
Methyl 2-hydroxypropanoate, (±) : 7429
Methyl iodide : 6307
Methyl isocrotonate : 7092
Methyl isocyanide : 6423
Methyl isonicotinate : 7743
3-Methyl-*N*-isopentyl-1-butanamine : 3789
Methyl isopentyl ketone : 7337
Methylisopropylamine : 6506
1-Methyl-4-isopropyl-2,3-dioxabicyclo[2.2.2]oct-5-ene : 517
Methyl isopropyl ketone : 7082
Methyl isovalerate : 7398
Methylketene : 9184
Methyl laurate : 7227
Methyl levulinate : 7539
Methyl malonate : 4126
(±)-Methyl mandelate : 7378
Methyl mercaptan : 6747
Methylmercurynitrile : 2518
N-Methylmethanamine : 3893
N-Methylmethanamine hydrochloride : 3894
1-Methyl-1-(4-methylcyclohexyl)ethyl hydroperoxide : 6698
7-Methyl-3-methylene-1,6-octadiene : 7914
2-Methyl-5-(1-methylethenyl) cyclohexanone : 6699
4-Methyl-1-(1-methylethyl)-bicyclo[3.1.0]hexan-3-one, (*l*) : 10091
2-Methyl-5-(1-methylethyl)-1,3-cyclohexadiene : 8740
Methyl 2-methyllactate : 7376
2-Methyl-*N*-(2-methylpropyl)-1-propanamine : 3778
2-Methyl-2-(methylthio)butane : 7597
17-Methylmorphinan-3-ol : 6609
N-Methyl-1-naphthalenamine : 7476
Methyl 1-naphthoate : 7472
Methyl 2-naphthoate : 7473
Methyl nicotinate : 7742
2-Methyl-5-nitro-1*H*-imidazole-1-ethanol : 7868
2-Methyl-2-[(nitrooxy)methyl]-1,3-propanediol, dinitrate : 10691
N-Methyl-*N*-nitrosoethenamine : 8192
Methyl nonyl ketone : 10734
N-Methyl-1-octanamine : 7525
Methyl octyl ketone : 2763
N-Methyl-*N*-octyl-1-octanamine : 7216
α-Methyl-4-[(2-oxocyclopentyl) methyl]benzeneacetic acid : 6632
2-Methyl-17-(1-oxopropoxy) androstan-3-one, (2α,5α,17β) : 4669
Methylparaben : 7369
S-Methyl pentachlorobenzenethiol : 7544
Methyl pentachlorophenyl ether : 8532
Methyl pentafluorophenyl ether : 8586
3-Methylpentanedioic acid : 7279
3-Methyl-3-pentanol, carbamate : 4696
Methylpentylacetylene : 8403
2-Methylpentyl bromide : 1271
Methyl *tert*-pentyl ether : 6841
Methyl pentyl ketone : 5689
2-Methylphenol : 2477
3-Methylphenol : 2478
4-Methylphenol : 2479
α-[1-[Methyl(3-phenylallyl)amino] ethyl]benzenemethanol : 2397
N-Methyl-*N*-phenylbenzenamine : 7220
1-(4-Methylphenyl)ethanol : 3958
3-Methyl-5-phenylhydantoin : 7640
N-(2-Methylphenyl) imidodicarbonimidic diamide : 10122
Methylphenyl ketazine : 40
Methyl phenyl ketone : 39
1-Methyl-2-phenylmethylhydrazine : 809
2-Methylphenyl 4-nitrophenyl ether : 7499
Methyl *trans*-3-phenyl-2-propenoate : 7133
1-Methyl-3-phenylpropylamine : 7002
N-(2-Methylphenyl)-2-(propylamino) propanamide : 9113
Methyl phenyl sulfide : 7808
1-Methyl-*N*-phenyl-*N*-(2-thienylmethyl)-4-piperidinamine : 10015
Methyl phosphate : 10603
Methylphosphonous dichloride : 3216
Methyl picrate : 6906
10-[3-(4-Methyl-1-piperazinyl) propyl]-10*H*-phenothiazine : 8693
2-Methyl-1-propanal : 6370
2-Methyl-1-propanamine : 6376
N-Methyl-1-propanamine : 7709
2-Methyl-2-propanamine : 1496
2-Methylpropane : 6371
2-Methyl-2-propanethiol : 1574
2-Methyl-1-propanol, aluminum salt : 10453
2-Methyl-1-propanol, titanium(4+) salt : 9907
2-Methylpropenenitrile : 6917
Methyl propenoate : 6916

2-Methylpropenoic acid : 6733
Methyl propionate : 7690
S-Methylpropiothetin : 7820
1-Methyl-2-propylacetylene : 5911
N-(2-Methylpropyl)-2,6,8-decatrienamide : 135
Methyl propyl ketone : 8629
Methyl propyl ketone oxime : 8631
4-(1-Methylpropyl)phenol : 1645
(1-Methylpropyl)urea : 1670
1-Methyl-2-propynyl(3-chlorophenyl) carbamate : 1822
1-Methyl-9*H*-pyrido[3,4-b]indole : 5624
Methyl pyridyl ketone : 9368
3-(1-Methyl-2-pyrrolidinyl)pyridine, (*S*)- : 8025
Methyl pyruvate : 7541
4-Methylquinaldine : 4313
Methyl selenide : 4318
Methylsilanetriol, triacetate : 7828
Methyl silicate : 9974
Methyl sorbate : 7321
4-Methylstigmasta-7,24(28)-dien-3-ol, (3β,4α,5α,24Z) : 9543
α-Methyl styrene : 6463
Methylsulfonic acid : 6744
4-(Methylsulfonyl)-2,6-dinitro-*N*,*N*-dipropylaniline : 8027
N-Methyltaurine : 6932
8-Methyltocol : 10106
Methyl trifluoromethyl ketone : 10381
2-Methyl-1,3,5-trinitrobenzene : 10633
Methyl undecyl ketone : 10316
6-Methyluracil : 7753
2-Methylvaleraldehyde : 7551
Methyl valerate : 7558
Methylvinylacetylene : 8655
Methyl vinyl ketone : 1459
Metiazinic acid : 7607
Metyrapone : 7224
Metyridine : 6829
Mevacor : 6629
Michler's Base : 964
Michler's ethyl ketone : 959
Michler's ketone : 9941
Monobenzone : 8919
Monobutyl maleate : 1529
Monobutyl succinate : 1583
Mono(2-ethylhexyl) phosphate : 5059
Monomethyl succinate : 7365
1-Monoolein : 5566
Monophenylphosphine : 8947
Monosodium benzenesulfonate : 9547
Monosodium 2-hydroxyethanesulfonate : 9562
Monosodium 2-hydroxy-2-propanesulfonate : 9563
Montanic acid : 8298
Montanyl alcohol : 8299
Monuron : 2237
1-Morpholinoisobutene : 7707
MPTP : 9871
Mucic acid : 5511
Muscone : 7165
Musk ketone : 1556
Mustard gas : 932
Mycotoxin T2 : 10716
Myricyl alcohol : 10131
Myristic acid : 9795
Myristonitrile : 9793
Myristoyl chloride : 9799
Myrtenal : 4163
Nandrolone phenpropionate : 8279
Napelline : 6633
2,3-Naphthalenedicarboxylic acid anhydride : 7974
Naphthalene-1,6-disulfonic acid : 7960
Naphthalene-2,7-disulfonic acid : 7961
1,4,5,8-Naphthalenetetracarboxylic acid anhydride : 725
1-Naphthalenol : 7975
2-Naphthalenol : 7976
2-Naphthalenol benzoate : 7984
1-Naphthalenol, dihydrogen phosphate : 7992
Naphthalic acid : 7944
Naphtho[1,2,3,4-def]chrysene : 2885
1-Naphthoic acid : 7939
2-Naphthoic acid : 7940
1-Naphthol-8-amino-3,6-disulfonic acid : 333
2-Naphthol-3,6-disulfonic acid : 6119
2-Naphthol-6,8-disulfonic acid : 6118
1-Naphthol Orange : 8413
1-Naphthol-2-sulfonic acid : 6123
1-Naphthol-4-sulfonic acid : 6121
2-Naphthol-6-sulfonic acid : 6124
β-Naphthoquinoline : 731
1,2-Naphthoquinone : 7957
1,4-Naphthoquinone : 7958
Naphthoresorcinol : 7949
2-Naphthoxyacetic acid : 7991
1-Naphthyl acetate : 7977
2-Naphthyl acetate : 7978
1-Naphthylacetic acid : 7930
2-Naphthylacetic acid : 7931
α-Naphthylamine : 7981
β-Naphthylamine : 7982
1-Naphthylamine-4,6-disulfonic acid : 362
1-Naphthylamine-4,7-disulfonic acid : 361
2-Naphthylamine-1,5-disulfonic acid : 360
1-Naphthylamine-4-sulfonic acid : 364
1-Naphthylamine-5-sulfonic acid : 365
1-Naphthylamine-8-sulfonic acid : 368
2-Naphthylamine-1-sulfonic acid : 363
2-Naphthylamine-5-sulfonic acid : 366
1-Naphthyl bromide : 1287
1-Naphthyl chloride : 2169
1,2-(1,8-Naphthylene)benzene : 5341
N-(1-Naphthyl)ethylenediamine : 7986
1-Naphthyl isocyanate : 6416
1-Naphthyl mercaptan : 7970
2-Naphthyl mercaptan : 7971
1-Naphthylphenylamine : 8935
α-Naphthyl Red : 8821
1-Naphthyl salicylate : 7987
α-Naphthylsulfonic acid : 7965
β-Naphthylsulfonic acid : 7966
Naptalam : 7983
Neocupróine : 4243
Neohexane : 3984
(+)-Neomenthol : 7409
Neopentyl alcohol : 4278
Neopentyl glycol : 4274
Neopentyl mercaptan : 4276
Neophyl chloride : 1997
Nerol : 4168
Nervonic acid : 9782
Niacinamide : 9327
Nickel acetylacetonate : 8016
Nickel tetracarbonyl : 8017
Nicotinaldehyde : 9323
Nicotinamide adenine dinucleotide : 2446
Nicotinamide adenine dinucleotide phosphate : 2447
Nicotinic acid : 9330
Nicotinic acid adenine dinucleotide : 8023
Nicotinyl alcohol : 9353
Nicoumalone : 10
Nikethamide : 3518
Ninhydrin : 6237
Nioxime : 2581
Nipecotic acid : 9062
Nitarsone : 8144
5,5'-Nitrilobarbituric acid, ammonium salt : 7910
Nitrilotriacetatoiron(III) : 6358
2-Nitroacetophenone : 8148
3-Nitroacetophenone : 8149
4-Nitroacetophenone : 8150
5-Nitro-*o*-anisidine : 6853
5-Nitrobarbituric acid : 8177
2-Nitrobenzenepropanoic acid : 8156
4-Nitrobenzenepropanoic acid : 8157
2-Nitrobenzyl alcohol : 8061
3-Nitrobenzyl alcohol : 8062
4-Nitrobenzyl alcohol : 8063
4-Nitrobenzyl chloride : 2138
2-Nitrobenzyl cyanide : 8051
4-Nitrobenzyl cyanide : 8052
2-Nitro-1,1'-biphenyl : 8084
3-Nitro-1,1'-biphenyl : 8085
4-Nitro-1,1'-biphenyl : 8086
p-Nitrobromobenzene : 1294
Nitrochalcone : 8154
4-Nitrocinnamaldehyde : 8158
3-[[(5-Nitro-2-furanyl)methylene] amino]-2-oxazolidinone : 5500
2-[(5-Nitro-2-furanyl)methylene] hydrazinecarboxamide : 8107
Nitrogen mustard hydrochloride : 930
Nitroglycerin : 10629
Nitromide : 4368
1-Nitro-2-naphthol : 8127
2-Nitro-1-(4-nitrophenoxy)-4-(trifluoromethyl)benzene : 5384

4-Nitro-*N*-(4-nitrophenyl)aniline : 4382
5-Nitro-*o*-phenetidine : 4854
Nitrophenide : 1039
o-Nitrophenylacetic acid : 8048
m-Nitrophenylacetic acid : 8049
p-Nitrophenylacetic acid : 8050
4-Nitrophenyl dihydrogen phosphate : 8155
4-Nitro-*o*-phenylenediamine : 8053
2-Nitrophenyl isocyanate : 6417
3-Nitrophenyl isocyanate : 6418
4-Nitrophenyl isocyanate : 6419
o-Nitrophenylpropiolic acid : 8159
p-Nitrophenylurea : 8161
5'-Nitro-2'-propoxyacetanilide : 8170
3-Nitrosalicylic acid : 6130
5-Nitrosalicylic acid : 6131
p-Nitrosodiphenylamine : 8198
N-Nitrosodipropylamine : 8200
N-Nitroso-*N*-ethylurea : 5175
2,2'-(Nitrosoimino)ethanol : 8185
N-Nitroso-*N*-methylurea : 7504
N-Nitrosomorpholine : 8193
1-Nitroso-β-naphthol : 8195
1-Nitrosopiperidine : 8199
N'-Nitroso-3-(2-pyrrolidinyl)pyridine : 8196
5-Nitrouracil : 8176
Nitroxoline : 8182
4-Nitro-*o*-xylene : 4154
NMN : 8024
Nonaisoprenol : 850
Nonaldehyde : 8230
1-Nonanamine : 8254
Nonyl alcohol : 8238
Nonyl mercaptan : 8236
Nopinene : 9048
18-Norabietane : 5329
Noradrenaline : 8270
Norcarane : 854
Norcassamidine : 4748
Norcholanic acid : 8266
Norgestrel, (-) : 8278
Normethadone : 3908
Nornicotine : 9401
Norpempidine : 9969
19-Norpregna-1,3,5(10)-trien-20-yne-3,17-diol, (17α)- : 4807
19-Norpregn-4-en-20-yn-3-one, 17-hydroxy-, (17 α)- : 8271
Novoldiamine : 3499
Novonal : 3500
cis-β-Ocimene : 4187
trans-β-Ocimene : 4188
α-Ocimene : 4189
cis-allo-Ocimene : 4190
trans-allo-Ocimene : 4191
Octabenzone : 6137
cis,cis-9,12-Octadecadienoic acid : 6615
1-Octadecanamine : 8320
1,18-Octadecanedicarboxylic acid : 4682
Octadecanoic acid : 9604
Octadecanoic anhydride : 9605
cis,cis,cis-9,12,15-Octadecatrienoic acid : 6616
17-Octadecene-9,11-diynoic acid : 6361
cis-9-Octadecenoic acid : 8410
trans-9-Octadecenoic acid : 4690
Octadecyl stearate : 8326
Octadecyltrichlorosilane : 10263
Octafluorocyclobutane : 8699
Octamethylene dibromide : 2968
Octamylamine : 7399
1-Octanamine : 8380
1-Octanamine hydrochloride : 8381
Octanoic acid glycerol monoester : 3753
Octatropine methylbromide : 475
Octisalate : 5062
Octodrine : 7283
Octopine : 1765
Octylacetylene : 2788
Octyl bromide : 1299
Octyl chloride : 2202
Octyl fluoride : 5408
2-Octyl iodide, (±) : 6327
2-Octyl-3(2*H*)-isothiazolone : 8377
Octyl mercaptan : 8353
N-Octyl-1-octanamine : 4418
(Octyloxy)benzene : 8399
Octyl 2-propenoate : 8379
Octyl sulfide : 4424
Octyltrichlorosilane : 10264
2-Octynol : 8406
Oleyl alcohol : 8316
Oleylamine : 8317
Opianic acid : 5462
Orcinol : 6985
o-Orsellinic acid : 3732
Orthanilic acid : 251
Orthocaine : 6934
Ovex : 2235
10*H*-9-Oxaanthracene : 10841
1-Oxa-2-azacyclopentadiene : 6560
1-Oxa-3-azaindene : 749
Oxaceprol : 72
Oxacyclopentadiene : 5478
1-Oxa-3,4-diazacyclopentadiene : 8426
2-(1,2,4-Oxadiazol-2-yl)phenol : 5309
Oxalacetic acid : 8431
Oxalyl dichloride : 8432
Oxane : 9884
1,2-Oxathiane 2,2-dioxide : 1421
1,2-Oxathiolane, 2,2-dioxide : 9150
Oxfenicine : 6157
Oxiniacic acid : 9332
Oxirane, 2,2'-(2,5,8,11-tetraoxadodecane-1,12-diyl)bis- : 5294
19-Oxogomphoside : 1693
Oxolan-2-one : 1688
2-Oxo-5-methylhex-3-ene : 7355
5-Oxo-*L*-proline : 9389
15-Oxo-prostaglandin E2 : 6034
Oxotremorine : 9400
Oxyacanthine : 10487
Oxybenzone : 6070
2,2'-Oxybis[*N*,*N*-dimethylethanamine] : 963
2,2'-Oxybisethanol, dinitrate : 3443
1,1'-Oxybis[2-methylpropane] : 3780
10,10'-Oxybis[10*H*-phenoxarsine] : 8487
2,2'-Oxybispropanoic acid : 3817
Oxycinchophen : 6170
2,2'-Oxydiacetic acid : 3579
4,4-Oxydianiline : 2854
Oxydi-2,1-ethanediyl carbonochloridate : 3436
Oxydiethylene methacrylate : 3441
Paclitaxel : 9684
Palatinose : 5541
Palmitaldehyde, dimethyl acetal : 3866
Palmitic acid : 5751
Palmitoleic acid : 5757
PAN : 9377
PAR : 9378
Parabanic acid : 6220
Paraleucaniline : 10683
Pargyline : 7722
Paroxypropione : 6168
Patulin : 6045
DL-Peganine : 10774
Pelargonic acid : 8237
Pempidine : 8602
Penicillamine : 6718
Penicillic acid : 6845
Penicin : 386
Pentabromoacetone : 8528
Pentachlorophenyl mercaptan : 8534
Pentadecanamine : 8558
3-Pentadecylcatechol : 8560
Pentadecylic acid : 8553
Pentadecyl methyl ketone : 5652
Pentaerythritol dibromide : 917
Pentaerythritol dichlorohydrin : 939
Pentaerythritol tetraacrylate : 1011
Pentaerythritol tetrabromide : 2913
Pentaerythritol triacrylate : 1012
Pentaethylenehexamine : 9708
Pentafluoropyridine : 8727
Pentamethylene : 2659
Pentamethylene acetate : 8616
Pentamethylene diiodide : 3770
Pentamethylene glycol : 8613
Pentanate : 10274
Pentane-1,1-dicarboxylic acid, diethyl ester : 3411
Pentanedioic acid : 5557
Pent-4-en-1-al : 8638
Pentoxone : 6846
Pentoxyverine : 1733
erythro-2-Pentulose : 9482
D-threo-2-Pentulose : 10864
L-threo-2-Pentulose : 10865
tert-Pentyl alcohol : 7079
Pentyl bromide : 1305
tert-Pentyl bromide : 1253
Pentyl chloride : 2208
sec-Pentyl chloride : 2209

6-Pentyl-*m*-cresol : 7595
Pentylenetetrazole : 9894
Pentyl fluoride : 5409
Pentyl iodide : 6328
tert-Pentyl isopentanoate : 4285
Pentyl mercaptan : 8621
sec-Pentyl mercaptan : 8622
3-Pentyl mercaptan : 8623
Pentyl pelargonate : 8679
p-*tert*-Pentylphenol : 4287
Pentyl propyl ketone : 8245
Perbenzoic acid : 607
Perchlorobenzene : 5724
Perchlorocyclopentadiene : 5734
Perchlorocyclopentene : 8292
Perchloroethane : 5737
Perchloroethylene : 9753
Perchloromethyl mercaptan : 10246
Perchloronaphthalene : 8294
Perchlorostyrene : 8295
Perfluorobenzene : 5794
Perfluoroethane : 5798
Perfluoroisobutylene : 8713
Perfluoroisopropyl iodide : 5664
Perfluoropropylene oxide : 10424
d-Perillaldehyde : 7855
l-Perillaldehyde : 7856
Pethidine : 6701
Petroselinic acid : 8314
Phenacaine hydrochloride : 977
Phenacemide : 8803
Phenacetin : 4862
Phenaceturic acid : 8801
Phenacylamine : 396
Phenallymal : 8810
Phenanthrenequinone : 8744
o-Phenanthroline : 8747
o-Phenanthroline monohydrate : 8749
9-Phenanthryl bromide : 1311
Phenarsazine chloride : 1985
Phencyclidine : 8866
Phenetharbital : 3507
Phenethyl alcohol : 642
Phenethyl phenyl ketone : 4516
Phenethyl propionate : 8880
o-Phenetidine : 4822
m-Phenetidine : 4823
p-Phenetidine : 4824
Phenetole : 4828
Phenformin hydrochloride : 8877
Phenicarbazide : 8891
Phenindione : 8903
Pheniramine : 4256
Phenmetrazine : 7644
Phenol, 3,5-dimethyl-4-(methylthio)-, methylcarbamate : 6759
Phenoldisulfonic acid : 5981
Phenol, 2-(1-methylethoxy)-, methylcarbamate : 9196
Phenol, 3-methyl-5-(1-methylethyl)-, methylcarbamate : 9126
Phenol, 2-(1-methylpropyl)-4,6-dinitro- : 4416
m-Phenolsulfonic acid : 5988
p-Phenolsulfonic acid : 5989
Phenolsulfonphthalein : 8766
Phenopyrazone : 4524
Phenoxyacetone : 8792
3-Phenoxybenzyl bromide : 1276
Phenoxymethylpenicillin : 8524
4-[3-[4-(Phenoxymethyl)phenyl]propyl]morpholine : 5446
Phenprobamate : 655
Phensuximide : 7659
Phentermine : 3956
Phentydrone : 9852
Phenylacetaldehyde : 596
N-Phenylacetamide : 18
α-Phenylacetamide : 597
7-Phenylacetamidodeacetoxycephalosporanic acid : 8799
Phenylacetic acid : 598
Phenylacetone : 8965
Phenylacetyl chloride : 602
Phenylacetylenecarboxylic acid : 8979
Phenyl acetylsalicylate : 8802
L-Phenylalanine, 4-[bis(2-chloroethyl)amino] : 6695
1-Phenylallyl alcohol : 10793
Phenylaminoacetic acid : 8887
Phenyl *p*-aminosalicylate : 8813
N-Phenylanthranilic acid : 8812
Phenylarsine oxide : 8472
3-(Phenylazo)-2,6-pyridinediamine, monohydrochloride : 8754
N-Phenylbenzenamine : 4454
N-Phenylbenzenemethanamine : 785
4-Phenylbenzophenone : 759
Phenyl bromide : 1096
2-Phenylbutane : 1514
α-Phenylbutyramide : 4895
Phenyl butyrate : 8852
Phenyl carbonate : 4474
Phenyl Cellosolve acrylate : 8783
Phenyl chloride : 1865
2-Phenyl-6-chlorophenol : 1920
Phenylchlorosilane : 2252
β-Phenylcinnamaldehyde : 4519
cis-α-Phenylcinnamic acid : 8921
trans-α-Phenylcinnamic acid : 8922
Phenyl cyanide : 715
2-Phenylcyclopropylamine : 10125
Phenyl dichlorophosphate : 8953
Phenyldichlorosilane : 3269
Phenyl diselenide : 4477
Phenyl disulfide : 4478
Phenyl dodecanoate : 8911
2,2′-*p*-Phenylenebis(4-methyl-5-phenyloxazole) : 1029
o-Phenylenediamine : 611
m-Phenylenediamine : 612
p-Phenylenediamine : 613
1,10-(1,2-Phenylene)pyrene : 6238
Phenylethane : 4894
1-Phenylethanol : 6996
Phenylethanolamine : 343
N-(1-Phenylethyl)-9,12-octadecadieneamide, (*Z*,*Z*)- : 6694
Phenylfluorone : 10448
Phenylglyceryl ether : 8788
α-Phenylglycine : 244
1-Phenylheptadecane : 5655
5-Phenylhydantoin : 8898
2-Phenylhydrazinecarbothioamide : 9004
N-Phenylhydrazinecarbothioamide : 9005
Phenylhydrazine-4-sulfonic acid : 5939
N-Phenylimidodicarbonimidic diamide : 8847
Phenyl isocyanide : 6422
(±) 3-Phenyllactic acid : 5987
N-Phenylmaleimide : 8990
Phenyl mercaptan : 666
1-Phenyl-5-mercapto-1*H*-tetrazole : 3666
Phenylmercuric acetate : 6721
Phenylmethylbarbituric acid : 7658
1-Phenyl-2-methylpropyl alcohol : 6481
2-Phenyl-4*H*-naphtho[1,2-b]pyran-4-one : 699
N-Phenyl-β-naphthylamine : 8936
N-Phenyl-4-(phenylazo)benzenamine : 8820
Phenyl phthalate : 4511
1-Phenylpropargyl alcohol : 5284
3-Phenyl-2-propenal, (*E*)- : 2396
3-Phenyl-2-propenoic acid, (*E*) : 2399
3-Phenyl-2-propenoic acid, (*Z*) : 2398
3-Phenyl-2-propen-1-ol, formate : 2402
Phenylpropylmethylamine : 3955
3-Phenylpyruvic acid : 8456
3-Phenylrhodanine : 9007
Phenylseleninic acid : 657
4-Phenylsemicarbazide : 8892
N-Phenylsulfanilic acid : 8811
Phenyl sulfide : 4529
Phenylsulfonyl chloride : 663
Phenylsulfonyl fluoride : 664
N-Phenyl-*N*′-1,2,3-thiadiazol-5-ylurea : 10035
1-Phenyltridecane : 10323
Phenyltrimethylammonium chloride : 10599
Phenyl(trimethylsilyl)amine : 10601
Phenylurethane : 5207
5-Phenylvaleric acid : 648
Phenyramidol : 9366
Phenzidole : 8839
PhIP : 7641
Phloionic acid : 3739
Phloionolic acid : 10442
Phloroglucinol : 675
Pholedrine : 6937
Phosgene : 1756
Phosphonic acid, (2-chloroethyl)- : 4806
Phosphonodithioic acid, ethyl-, *O*-ethyl *S*-phenyl ester : 5447
Phosphoramidothioic acid, acetyl-, *O*,*S*-dimethyl ester : 11

Phosphoramidothioic acid, *O,S*-dimethyl ester : 6737
Phosphoric acid, 2,2-dichloroethenyl dimethyl ester : 3312
Phosphorodithioic acid, *O*-ethyl *S,S*-dipropyl ester : 4816
Phosphorothioic acid, *O,O*-diethyl *O*-pyrazinyl ester : 10061
Phosphorotrithious acid, *S,S,S*-tributyl ester : 6722
Phosphorylcholine : 2379
Phthalamide : 623
Phthalimide : 6433
o-Phthalodinitrile : 3317
m-Phthalodinitrile : 3318
p-Phthalodinitrile : 3319
Phthaloyl chloride : 617
Phthiocol : 6092
Physostigmine sulfate : 4756
Phytanic acid : 9950
Piazthiole : 10022
Picein : 5545
Picloram : 434
2-Picoline : 7738
3-Picoline : 7739
4-Picoline : 7740
Picolinic acid : 9329
4-Picolyl alcohol : 9354
Picramic acid : 304
Picric acid : 10631
Picryl chloride : 2337
Pigment Blue 15 : 2464
Pimeclone : 9073
Pimelic acid : 5675
Pimelic ketone : 2588
Pinacol : 3986
Pinacolone : 3993
Pindone : 9085
2-Pinene : 9047
Pinene hydrate : 9045
Pinosylvin : 9014
L-Pipecolic acid : 9061
Δ^3-Piperidine : 9888
Piperidinium chloride : 9067
Piperidinium, 1,1-dimethyl-, chloride : 6705
1-Piperidinol : 6171
Piperidione : 3512
Piperinic acid : 698
Piperitenone : 7456
(±)-Piperitone : 6513
Piperonal : 692
Piperonylic acid : 693
cis-Piperylene : 8565
trans-Piperylene : 8566
Pipradrol : 4512
Pipsyl chloride : 6277
Pivaldehyde : 4269
Pivalic acid chloride : 4279
Plicamycin : 7880
Plumbagin : 6093
Potassium biphthalate : 9097
Potassium citrate : 10665
Potassium oleate : 9098
Potassium sorbate : 9096
Pramiverin : 6494
Pramoxine : 1483
Prednisone : 3746
Pregnanediol : 9104
5α-Pregnan-20β-ol-3-one : 162
Pregn-4-ene-3,20-dione : 9122
4-Pregnene-11β,17α,20β,21-tetrol-3-on : 9901
Prehnitenol : 9965
Primeverose : 10862
Primidone : 4990
Pristane : 9959
Probenecid : 4550
Procaine : 3387
Proflavine : 114
Prontosil : 2864
1,2-Propadiene-1,3-dione : 1754
Propallylonal : 1086
Propanamide, *N*-(3,4-dichlorophenyl)- : 9163
Propanamide, *N,N*-diethyl-2-(1-naphthalenyloxy)- : 7996
1-Propanamine : 9205
2-Propanamine : 6471
1-Propanamine hydrochloride : 9206
2-Propanamine hydrochloride : 6472
1-Propanearsonic acid : 9211
1,3-Propanedial : 6663
1,2-Propanediol : 9230
Propaneperoxoic acid : 8733
1,2,3-Propanetriol : 5561
1,2,3-Propanetriol, 1,3-dinitrate : 5564
Propanoic acid, 2-amino-3-mercapto-, (*R*)- : 2715
Propanoic acid, 2-(2,4,5-trichlorophenoxy)- : 9538
Propanolamine : 407
1-Propanol, titanium(4+) salt : 10001
2-Propanol, titanium(4+) salt : 9908
2-Propanone : 33
Propargylacetic acid : 8689
Propargyl aldehyde : 9271
Propargyl bromide : 1345
Propargyl chloride : 2285
2-Propenal : 116
2-Propenamide : 117
2-Propen-1-amine : 171
Propenenitrile : 119
2-Propenoic acid : 118
2-Propenoic acid, 2,2-dimethyl-1,3-propanediyl ester : 4542
2-Propenoic acid, 1,6-hexanediyl ester : 4543
2-Propenoic acid, propyl ester : 9204
2-Propen-1-ol : 170
2-Propenylbenzene : 173
5-(2-Propenyl)-1,3-benzodioxole : 9499
1-(2-Propenyl)cyclohexene : 182
2-Propenyl propanoate : 204
Propham : 6531
Propiocine : 4745
β-Propiolactone : 8445
Propiolic acid : 9274
Propionaldehyde : 9132
Propionamide : 9134
Propionic acid : 9164
Propionic anhydride : 9165
Propionylacetone : 5839
Propionyl chloride : 9170
Propionyl fluoride : 9171
Propiophenone : 8964
Propoxybenzene : 8976
Propylacetylene : 8687
Propyl alcohol : 9166
Propylallene : 5774
Propyl bromide : 1327
gamma-Propyl-*gamma*-butyrolactone : 3667
S-Propyl carbonochloridothioate : 9223
Propyl chloride : 2260
Propyl cyanide : 1419
Propylene : 9178
Propylene chlorohydrin : 2268
sec-Propylene chlorohydrin : 2270
Propylenediamine : 9137
Propylene dibromide : 2975
Propylene dichloride : 3271
1,2-Propylene glycol 1-propyl ether : 9200
1,2-Propylene oxide : 7536
Propyl ether : 4558
Propyl fluoride : 5415
Propyl gallate : 9268
1-Propylheptyl alcohol : 2761
Propylhexedrine : 4019
Propylidene chloride : 3270
Propyl iodide : 6335
Propyl isopentanoate : 9248
Propyl isopropyl ketone : 7338
Propyl mercaptan : 9151
Propylparaben : 9242
Propyl *trans*-3-phenyl-2-propenoate : 9224
N-Propyl-1-propanamine : 4549
N-Propyl-2-propanamine : 6535
Propyl propionate : 9260
Propylthiouracil : 3668
Propyltrichlorosilane : 10287
Proscar : 5330
Protionamide : 9263
Protoanemonin : 7243
Protocatechualdehyde : 3701
Protocatechuic acid : 3708
Pseudoconhydrine : 9258
Pseudocumene : 10508
Pseudoionone : 4349
Pseudopelletierine : 6952
1*H*-Purin-6-amine : 121
1*H*-Purine, 6-[(1-methyl-4-nitro-1*H*-imidazol-5-yl)thio]- : 545
Purpurin : 10434
Purpurogallin : 9898
Putrescine : 1403
1,4-Pyran : 9296
Pyrazinamide : 9303
Pyrazinoic acid : 9304

Pyrazino[2,3-d]pyrimidine : 9290
2-Pyrazoline : 3673
1*H*-Pyrazolium, 1,2-dimethyl-3,5-diphenyl-, methyl sulfate : 3545
3(2*H*)-Pyridazinone, 5-amino-4-chloro-2-phenyl- : 1832
4-Pyridinecarboxylic acid hydrazide : 6440
Pyridine, 2-chloro-6-(trichloromethyl)- : 8029
Pyridine *N*-oxide : 9355
2-Pyridinylhydrazine : 9365
5*H*-Pyrido[4,3-b]indole : 1740
Pyridoxal 5-(dihydrogen phosphate) : 9373
Pyridoxin : 6107
3-Pyridylsulfonic acid : 9358
Pyrimethamine : 2240
4(1*H*)-Pyrimidinone, 5-butyl-2-(ethylamino)-6-methyl- : 4811
Pyrithyldione : 3520
Pyrocalciferol : 4732
Pyrocarbonic acid diethyl ester : 3426
Pyrogallol : 673
Pyromellitic acid : 665
Pyropentylene : 2658
Pyrrodiazole : 10148
3-Pyrrolecarboxylic acid : 9395
2-Pyrrolidinecarboxylic acid : 9124
1-Pyrrolidinylcyclopentene : 2679
3-Pyrroline : 3676
Pyruvaldehyde : 9142
Quercetin-3-*L*-rhamnoside : 9408
D-Quercitol : 2805
Quinacridone : 2405
Quinaldic acid : 9430
Quinaldine : 7766
Quinalizarin : 9897
Quininic acid : 6905
Quinizarin : 3692
2,4-Quinolinediol : 6192
Quinolinic acid : 9338
2,3-Quinoxalinediol : 3678
Quintozene : 8540
Quinuclidine : 539
3-Quinuclidinol : 540
Reductic acid : 3720
Refrigerant 11 : 10239
Refrigerant 12 : 3162
Refrigerant 13 : 2328
Refrigerant 21 : 3187
Refrigerant 22 : 1983
Refrigerant 114 : 3295
Refrigerant 114a : 3294
Refrigerant 114B2 : 2990
Refrigerant 115 : 2207
Refrigerant 120 : 8538
Refrigerant 123a : 3306
Refrigerant 123b : 3308
Refrigerant 131a : 10237
Refrigerant 131b : 10236
Refrigerant 142b : 1980
Refrigerant 216 : 3194
Refrigerant 225ca : 3245
Refrigerant 227ea : 5665
Refrigerant 236ea : 5799
Refrigerant 236fa : 5800
Refrigerant 236me : 9838
Refrigerant 245cb : 8588
Refrigerant C317 : 2059
Refrigerant 1112 : 3160
Rehmannic acid : 6585
Resacetophenone : 3742
Resorcinol monoethyl ether : 4859
β-Resorcylaldehyde : 3699
β-Resorcylic acid : 3705
Resorufine : 6142
Retene : 7420
Retinol, hexadecanoate : 9464
Retrorsine *N*-oxide : 6362
Rhodanine : 10085
Rhodinol : 4197
β-*D*-Ribofuranoside, adenine-9 : 122
2-β-*D*-Ribofuranosyl-1,2,4-triazine-3,5(2*H*,4*H*)-dione : 546
1-β-*D*-Ribofuranosyluracil : 10760
Ricinoleic acid : 6134
Risocaine : 9207
Rosaniline : 6653
Rosaniline hydrochloride : 6654
Roxarsone : 8111
RU-486 : 7073
Rubeanic acid : 4792
Rubixanthin : 1781
Rubrene : 9992
Rufigallol : 5812
Ruthenium(III) acetylacetonate : 9493
Salicin : 6098
Salicyl alcohol : 5983
Salicylamide : 5971
Salicylamide *O*-acetic acid : 280
Salicylanilide : 6147
Salicylhydroxamic acid : 3702
Salicylic acid : 5990
N-Salicylidene-*o*-aminophenol : 6160
4-Salicyloylmorpholine : 6002
Salsalate : 1768
Sarin : 7669
Sarpagine : 9513
Sarsasapogenin : 9595
Scopolamine methobromide : 6908
Scopoletin : 6066
Sebacil : 2539
Sebacoin : 6018
9,10-Secocholesta-5,7,10(19)-trien-3-ol, (3β,5*Z*,7*E*)- : 10835
9,10-Secoergosta-5(10),6,8,22-tetraen-3-ol, (3β,6*E*,22*E*)- : 9673
Seldane : 9695
Selenomethionine : 9521
Seminose : 6680
Semioxamazide : 382
Semustine : 2042
Senecialdehyde : 7086
2-Silapropane : 4319
Silicic acid, tetrabutyl ester : 9727
Silicic acid, tetrakis(2-ethylbutyl) ester : 9814
β-Sitosterol : 9610
γ-Sitosterol : 9611
Skatole : 7388
Skimmianine : 10490
Skimmin : 5543
Smilagenin : 9594
Sodium acid citrate : 4574
Sodium alizarinesulfonate : 152
Sodium (4-aminophenyl)arsonate : 9544
Sodium anthraquinone-1-sulfonate : 3621
Sodium citrate : 10700
Sodium 2-(2,4-dichlorophenoxy)ethyl sulfate : 9534
Sodium 3,4-dihydro-3,4-dioxo-1-naphthalenesulfonate : 9569
Sodium *p*-dimethylaminoazobenzenesulfonate : 7529
Sodium dimethylaminobenzenediazosulfonate : 2817
Sodium dimethylarsonate : 9549
Sodium diphenylamine-4-sulfonate : 4456
Sodium ethoxide : 9557
Sodium ethylenebisdithiocarbamic acid : 7916
Sodium hydroxymethanesulfinate : 9560
Sodium hydroxymethanesulfonate : 9559
Sodium lauryl sulfate : 9556
Sodium methoxide : 9566
Sodium phenoxide : 9571
Sodium picrate : 10632
Solasodine : 9591
Sophorine : 2724
Sorbic acid : 5783
Sorbic alcohol : 5784
Sorbinaldehyde : 5773
L-Sorbinose : 9585
Sorbitol : 5537
Sorbitol hexaacetate : 5538
Spermine : 911
Squalane : 5826
Stannane, hydroxytriphenyl- : 10663
Stannane, tricyclohexylhydroxy- : 2707
Stanolone : 5962
Stearyl alcohol : 8307
Stearyl mercaptan : 8306
Stearyl methacrylate : 8325
Stearyl 2-propenoate : 8319
Stelazine : 10377
Stigmasterol : 9608
Streptonivicin : 8284
Strychnine nitrate : 9618
Strychnine sulfate : 9619
Stylophorine : 1810
Styphnic acid : 10625
Styrene glycol : 8872
Styrene-7,8-oxide : 8940
Suberic acid : 8349
Suberone : 2557
Suberonitrile : 8348
Succinanil : 8992
Succinanilic acid : 8470

Succinyl chloride : 1417
Sucrose palmitate : 9632
Sudan I : 8823
Sudan III : 8825
Sulfacetamide : 401
Sulfadiazine : 424
Sulfanilamide : 250
Sulfanilic acid : 253
p-Sulfanilyl fluoride : 254
Sulfapyrazine : 415
Sulfapyridine : 419
Sulfaquinoxaline : 426
4,4'-Sulfinyldianiline : 909
Sulfiram : 9828
4-Sulfo-1,2-benzenedicarboxylic acid : 9662
Sulfobromophthalein sodium : 9573
3-Sulfolene : 3684
Sulfonethylmethane : 990
3,3'-Sulfonyldianiline : 2857
5-Sulfosalicylic acid : 6195
5-Sulfosalicylic acid dihydrate : 6196
Sulphenone : 2253
Surinamine : 7842
Suxamethonium chloride : 9628
Swep : 7193
Synephrine : 6073
Syringaldehyde : 6029
Systox : 2795
2,4,5-T : 10273
2,4,5-T Butoxyethyl ester : 1479
2,4,5-T Butyl ester : 1666
Tannin : 9677
Tartar emetic : 495
Tartronic acid : 6180
Taurine : 305
Taurocyamine : 5612
Taxifolin : 8593
Temephos : 1
Terpenylic acid : 9850
β-Terpineol : 7463
3,2':4',3''-Terpyridine : 8022
α-Terthienyl : 9706
Testosterone : 5965
Testosterone-17-propionate : 8476
1,4,5,8-Tetraamino-9,10-anthracenedione : 4575
Tetrabromobisphenol A : 1077
3,4,5,6-Tetrabromo-*o*-cresol : 9716
Tetrabromoethylene : 9712
Tetracarbonylhydrocobalt : 2438
2,2',4,4'-Tetrachlorodiphenyl ether : 958
Tetracosamethylhendecasiloxane : 9778
Tetracyanoethylene : 9783
Tetradecamethylene dibromide : 2989
1-Tetradecanamine : 9803
1-Tetradecanol, acetate : 9802
Tetradecyl alcohol : 9797
Tetradecyloxirane : 4712
Tetraethylmethane : 3498
Tetraethyl orthocarbonate : 9808
N,N,N',N'-Tetraethylphthalamide : 9813
Tetrafluoroethylene : 9837
Tetrahelicene : 717
Tetrahydrobenzene : 2591
4,5,6,7-Tetrahydro-4-benzothiophenone : 3604
DL-Tetrahydroberberine : 1714
Tetrahydro-2,2-dimethyl-5-oxo-3-furancarboxylic acid : 9693
1,2,3,6-Tetrahydro-2,6-dioxo-4-pyrimidinecarboxylic acid : 8418
Tetrahydro-2-furancarbinol : 9860
Tetrahydro-2,5-furandimethanol : 463
Tetrahydrofurfurylamine : 9855
1,2,5,6-Tetrahydro-1-methyl-3-pyridinecarboxylic acid : 511
Tetrahydromyrcenol : 4184
1,2,3,4-Tetrahydro-α-naphthol : 9875
5,6,7,8-Tetrahydro-α-naphthol : 9876
5,6,7,8-Tetrahydro-β-naphthol : 9878
Tetrahydro-1,4-oxazine : 7901
Tetrahydroquinone : 2580
Tetrahydrothiophene, 1-1-dioxide : 9658
D-Tetrahydroxyadipic acid : 5536
3',4',5,7-Tetrahydroxyflavanone, (*S*) : 4742
Tetraiodoethylene : 9903
Tetra(isopropoxy)silane : 6539
Tetralin : 9874
Tetralol : 9877
6',7',10,11-Tetramethoxyemetan : 4695
Tetramethylbutanedinitrile : 9976
1,1,4,4-Tetramethyl-1,4-butanediol : 4095
4-(2,5,6,6-Tetramethyl-1-cyclohexen-1-yl)-3-buten-2-one : 6355
4-(2,5,6,6-Tetramethyl-2-cyclohexen-1-yl)-3-buten-2-one : 6354
Tetramethylene : 2530
Tetramethylenedithiol : 1418
Tetramethylene glycol : 1407
Tetramethylene oxide : 9854
3,7,11,15-Tetramethyl-2-hexadecen-1-ol, [*R*-[*R**,*R**-(*E*)]] : 9034
cis-2,2,5,5-Tetramethyl-3-hexene : 3018
Tetramethylolmethane tetramethacrylate : 8574
Tetramethyl-*p*-phenylenediamine : 9925
Tetramethylphosphorodiamidic fluoride : 3819
2,2,6,6-Tetramethyl-4-piperidinol : 6197
2-(*N*,2,4,6-Tetranitroanilino)ethanol : 8661
3,6,9,12-Tetraoxatetracosan-1-ol : 4655
sym-Tetrazine : 10007
Tetroquinone : 9900
Tetryl : 7798
Thallous ethoxide : 10012
3-Thenoic acid : 10072
Thenoyltrifluoroacetone : 10423
Thiacyclopentane : 9895
1,2,4-Thiadiazole, 5-ethoxy-3-(trichloromethyl)- : 9705
Thiamorpholine : 10060
Thianaphthene : 743
Thiazolsulfone : 402
Thiazol Yellow G : 2421
2-Thienyl bromide : 1356
2-Thienyl chloride : 2312
Thioacetanilide : 9002
2-Thiobarbituric acid : 3686
Thiobenzyl alcohol : 647
4,4'-Thiobisphenol : 3723
Thiobutabarbital : 4987
Thiocarbamide : 10081
α-Thiocyanatotoluene : 837
1-Thiocyanobutane : 1661
2-Thiocytosine : 422
Thiodiacetic acid : 10052
4,4'-Thiodianiline : 2856
2,2'-Thiodiethanol : 1005
Thiodiphenylamine : 8767
Thiofuran : 10063
Thioglycerol : 6716
Thioguanine : 301
2-Thiohydantoin : 10084
Thionalide : 6713
Thionaphthene-2-carboxylic acid : 744
Thionine : 2863
Thiophenetole : 5257
Thiophosgene : 1750
o-Thiosalicylic acid : 6709
Thiosemicarbazide : 5933
Thiosinamine : 208
Thiotepa : 10363
2-Thiouracil : 3687
Thioxanthone : 10083
4(10)-Thujene-3-ol : 7245
Thymine 2-desoxyriboside : 10093
Thymine riboside : 7851
Thymol, acetate : 7422
o-Thymotic acid : 6089
Tiglic acid : 7095
Tiglic aldehyde : 7085
Tigogenin : 9593
Tillman's reagent : 3200
Timonacic : 10032
Tin tetraethyl : 9827
Titanium(IV) butoxide : 9730
TMAB : 9720
TMS : 9973
α-Tocopherol : 10836
Tofranil : 6227
Tolazoline : 3606
o-Tolualdehyde : 6957
m-Tolualdehyde : 6958
p-Tolualdehyde : 6959
o-Toluamide : 6960
p-Toluamide : 6961
Toluene-2,3-diamine : 6977
Toluene-2,5-diamine : 6980
Toluene-2,6-diamine : 6979
Toluene-3,4-diamine : 6978
Toluene-α,α-diol, diacetate : 801
Toluene-3,4-dithiol : 6987
p-Toluenesulfinic acid : 7005
p-Toluenesulfonamide : 7008
o-Toluenesulfonyl chloride : 7011
o-Toluidine : 6938
m-Toluidine : 6939

p-Toluidine : 6940
o-Toluidine, hydrochloride : 6942
o-Tolunitrile : 7020
m-Tolunitrile : 7021
p-Tolunitrile : 7022
2-(*p*-Toluoyl)benzoic acid : 7035
m-Toluquinaldine : 4315
m-Toluquinoline : 7771
o-Tolyl alcohol : 6997
m-Tolyl alcohol : 6998
p-Tolyl alcohol : 6999
4-*o*-Tolylazo-*o*-toluidine : 3899
p-Tolyl ether : 1026
2 Tolyl isocyanate : 6412
p-Tolylsulfonylmethylnitrosamide : 4161
o-Tolylthiourea : 7663
Tomatidine : 9590
TOPO : 10637
Torularhodin : 3338
Tosanpin : 10868
Tranexamic acid : 346
Traumatic acid : 4636
Triacetic acid lactone : 6105
Triadimefon : 573
Triallylamine : 2839
1,2,3-Triaminopropane : 9153
Triamterene : 8981
1,2,3-Triaza-1*H*-indene : 745
1,3,5-Triazine-2,4-diamine, 6-chloro-*N,N'*-diethyl- : 9539
1,3,5-Triazine-2,4,6(1*H*,3*H*,5*H*)-trione : 2522
1,2,4-Triazolo[3,4-b]benzothiazole, 5-methyl- : 10306
Tribavirin : 9476
1,1,1-Tribromo-*tert*-butyl alcohol : 10170
2,4,6-Tribromo-*m*-cresol : 10169
Tribromsalan : 2914
Tributoxyphosphine : 10187
Tributyltin acetate : 92
Tributyltin fluoride : 10183
Tributyltin hydride : 10190
S,S,S-Tributyl trithiophosphate : 10188
Tricarballylic acid : 9154
1,1,1-Trichloroacetone : 10283
2,4,6-Trichloroanisole : 10251
1,1,1-Trichloro-*tert*-butyl alcohol : 10258
Trichlorobutylsilane : 1667
2,2,3-Trichlorobutyraldehyde : 10213
1,2,4-Trichloro-5-[(4-chlorophenyl) sulfonyl]benzene : 9806
4,5,6-Trichloro-*o*-cresol : 10255
2,4,6-Trichloro-*m*-cresol : 10257
2,3,6-Trichloro-*p*-cresol : 10256
2,2,2-Trichloroethanol dihydrogen phosphate : 10298
Trichloroethylene : 10231
(2,2,2-Trichloroethyl)oxirane : 10227
Trichloromethyl mercaptan : 10248
Trichloro-2-propenylsilane : 209
3',4',5-Trichlorosalicylanilide : 1974
2,4,5-Trichlorotoluene : 10252
1,3,5-Trichloro-1,3,5-triazine-2,4,6(1*H*,3*H*,5*H*)-trione : 9670
Triclocarban : 10215
Tricromyl : 7025
Tricyanotrimethylamine : 8031
Tricyclo[9.3.1.1]hexadeca-1(15),4,6,8(16),11,13-hexaene : 6726
1-Tridecanamine : 10322
Tridecyl alcohol : 10315
Tridecylic acid : 10314
Tridemorph : 4346
Trielaidin : 5569
Triethanolamine hydrochloride : 995
Triethoxyphosphine : 10369
Triethyl acetylcitrate : 10343
Triethyleneglycol monomethyl ether : 6824
Triethylenemelamine : 10678
Triethylenetetramine : 901
Triethylmethylplumbane : 7834
Triethyl orthothioformate : 10689
O,O,O-Triethyl thiophosphate : 10370
1,1,1-Trifluoroacetylacetone : 10415
2,2,2-Trifluoroethanamine : 10390
Trifluoroethylene : 10389
(Trifluoromethyl)acetylene : 10422
Trifluoromethyl isocyanide : 10394
3-(Trifluoromethyl)phenyl isocyanate : 6421
Trifluoromethylthiazide : 5339
2,4,6-Trifluoro-1,3,5-triazine : 2523
Triglycol : 10355
Triglyme : 10359
Trihexphenidyl : 2638
2',4',6'-Trihydroxyacetophenone : 10446
3,7,12-Trihydroxycholan-24-oic acid, (3α,5β,7α,12α) : 2377
1,2,6-Trihydroxyhexane : 5847
3,17,21-Trihydroxypregnan-20-one, (3α,5β) : 9107
9,11,15-Trihydroxyprosta-5,13-dienoic acid : 9279
Triisovalerin : 5571
Trilaurin : 5570
Trimellitic acid : 669
Trimethadione : 10576
3,4,5-Trimethoxybenzyl alcohol : 10481
Trimethylacetic acid : 4277
2,6,6-Trimethylbicyclo[3.1.1]heptane : 9044
1,3,3-Trimethylbicyclo[2.2.1]heptan-2-ol, *endo*-(±) : 5314
1,7,7-Trimethylbicyclo[2.2.1]heptan-2-ol, *exo*-(±) : 6367
1,7,7-Trimethylbicyclo[2.2.1]heptan-2-one, (±) : 1709
1,7,7-Trimethylbicyclo[2.2.1]heptan-2-one, (1*R*) : 1710
1,7,7-Trimethylbicyclo[2.2.1]heptan-2-one, (1*S*) : 1711
2,6,6-Trimethyl-1,3-cyclohexadiene-1-carboxaldehyde : 9498
trans-1,3,5-Trimethylcyclohexane : 10533
3,5,5-Trimethyl-2-cyclohexen-1-one : 6456
1,2,2-Trimethyl-1,3-cyclopentanedicarboxylic acid, (1*RS*,3*SR*) : 1712
Trimethylene : 2688
Trimethylene diiodide : 3772
Trimethylene dimercaptan : 9145
4,4'-Trimethylenedipiperidine : 4539
Trimethylene glycol : 9231
Trimethylene oxide : 8444
Trimethylene sulfide : 10038
Trimethylene sulfone : 10039
Trimethylolmelamine : 10692
Trimethylolpropane phosphite : 5278
Trimethylolpropane triacrylate : 10667
1,1,1-Trimethylolpropane trimethacrylate : 10699
6,6,9-Trimethyl-3-pentyl-6*H*-dibenzo[b,d]pyran-1-ol : 1716
N,N,α-Trimethyl-10*H*-phenothiazine-10-ethanamine : 9127
N,N,β-Trimethyl-10*H*-phenothiazine-10-propanamine : 10470
N,N,2-Trimethyl-1-propanamine : 6387
1,7,7 Trimethyltricyclo[2.2.1.$0^{2,6}$] heptane : 10307
Trimethyl trimellitate : 10512
2,4,6-Trimethyl-1,3,5-trioxane : 8508
2,4,6-Trimethyl-1,3,5-trithiane : 10042
Trimyristin : 5575
sym-Trinitrobenzene : 10624
Trinonafluorobutylamine : 10702
28,29,30-Trinorlanostane : 2365
Triolein : 5572
3,6,9-Trioxaundecane-1,11-diol : 9815
Tripalmitin : 5573
Tripelennamine : 3973
Triphenylmethyl bromide : 1373
Triphenylmethyl mercaptan : 4461
4-(Triphenylmethyl)morpholine : 10374
Triphenyltin acetate : 93
Triphenyltin chloride : 2341
Triphosgene : 1048
Tripropoxyphosphine : 10675
N,N,N-Tripropyl-1-propanaminium bromide : 9997
Triptane : 10524
Triptil : 9282
Triptycene : 3595
2,4,6-Tripyridyl-*s*-triazine : 10703
Triruthenium dodecacarbonyl : 9492
Tris(1-aziridinyl)phosphine, oxide : 10362
Tris(2,3-dibromopropyl) phosphate : 2981
Tris(dimethylamino)phosphine : 5825
Tris(dimethylamino)phosphine oxide : 5824
Tris(2-hydroxyethyl)amine : 10330
Tris(hydroxymethyl)nitromethane : 6094
Tristearin : 5574
Trithiocyanuric acid : 10143
Tri-*o*-tolyl phosphate : 10302
Tri-*m*-tolyl phosphate : 10303
Tri-*p*-tolyl phosphate : 10304
Troclosene potassium : 9094

Tropaeolin OO : 8414
Tropane : 6953
Tropanol mandelate : 5921
Tropic acid : 6077
Tropilidene : 2558
Tropital : 9080
Tryptamine hydrochloride : 6250
Tryptophol : 6251
Tuaminoheptane : 5670
Tungsten hexacarbonyl : 10718
Turmeric : 2495
Tyramine : 315
Umbelliferone : 5999
1-Undecanamine : 10748
Undecyl alcohol : 10732
sec-Undecyl alcohol : 10733
Undecylenic acid : 10745
Undecyl mercaptan : 10730
Untriacontane : 5639
Uramil : 423
Urethane : 4937
Uridine 5′-phosphoric acid : 10761
Urochloralic acid : 10234
Ursodiol : 3716
Vaccenic acid : 8315
Valeraldehyde : 8604
Valeric acid : 8624
(±)-γ-Valerolactone : 3646
Valeronitrile : 8620
Valeroyl chloride : 8632
Valproic acid : 9254
Vanadium(III) acetylacetonate : 10773
Vanadium hexacarbonyl : 10772
Vanadium, oxotris(2-propanolato)-, (T-4)- : 10467
Vanadyl acetylacetonate : 8458
Vanillic acid : 6065
Vanillin : 6060
Vanilmandelic acid : 3726
Vellosimine : 9512
Veratraldehyde : 3838
Veratric acid : 3851
Veratrole : 3840
Versen-Ol : 10701
Vicianose : 506
Vinylacetylene : 1464
Vinylbenzene : 9621
5-Vinylbicyclo[2.2.1]hept-2-ene : 10814
Vinyl bromide : 1194
Vinyl chloride : 2022
Vinyl crotonate : 10795
4-Vinyl-1-cyclohexene dioxide : 4711
Vinylene carbonate : 5017
Vinyl fluoride : 5388
1-Vinylheptanol : 8252
Vinylidene chloride : 3176
Vinylidene fluoride : 3560
Vinyl iodide : 6299
4-Vinylphenol : 6194
Vinyl propionate : 10819
Vinylsilanetriol, triacetate : 10826
Vinyl stearate : 10815
Vinyl sulfide : 4606
Vinyl sulfone : 4607
Vinyl trichloride : 10229
Vinyltrichlorosilane : 10296
Violuric acid : 9384
Vitamin A : 9463
Vitamin A_1 aldehyde : 9459
Vitamin B6 : 9372
Vitamin B-12a : 5959
Vitamin Bc : 5442
Vitamin C : 518
Vitamin K_2(35) : 6696
VX Nerve agent : 4992
Widdrene : 10092
Woodward's Reagent K : 5251
Xanthogenic acid : 5014
Xanthophyll : 1779
Xanthoxylin : 3867
Xanthurenic acid : 3754
Xibornol : 6369
m-Xylene diamine : 629
2,4-Xylenol, phosphate (3:1) : 10684
2,5-Xylenol, phosphate (3:1) : 10685
2,6-Xylenol, phosphate (3:1) : 10686
2,3-Xylidine : 3927
2,4-Xylidine : 3928
2,5-Xylidine : 3929
2,6-Xylidine : 3930
3,4-Xylidine : 3931
3,5-Xylidine : 3932
Xylite : 10861
1-(2,4-Xylylazo)-2-naphthol : 4247
1-(2,5-Xylylazo)-2-naphthol : 4248
Yellow AB : 8822
Zeaxanthin : 1778
Zinc acetylacetonate : 10875
Zinc, bis(dimethylcarbamodithioato-*S,S′*)-, (T-4)- : 10877
Zinc diethyl : 3543
Zineb : 10873
Zingerone : 6067
Zoxazolamine : 1908
Zygosporin A : 2726

DIAMAGNETIC SUSCEPTIBILITY OF SELECTED ORGANIC COMPOUNDS

When a material is placed in a magnetic field H, a magnetization M is induced in the material which is related to H by $M = \kappa H$, where κ is called the volume susceptibility. Since H and M have the same dimensions, κ is dimensionless. A more useful parameter is the molar susceptibility χ_m, defined by

$$\chi_m = \kappa V_m = \kappa M/\rho$$

where V_m is the molar volume of the substance, M the molar mass, and ρ the mass density. When the cgs system is used, the customary unit for χ_m is $cm^3\ mol^{-1}$; the corresponding SI unit is $m^3\ mol^{-1}$. Substances with no unpaired electrons are called diamagnetic; they have negative values of χ_m.

This table gives values of the diamagnetic susceptibility for about 400 common organic compounds. All values refer to room temperature and atmospheric pressure and to the physical form that is stable under these conditions. Substances are arranged by molecular formula in Hill order. A more extensive table may be found in Reference 1.

In keeping with customary practice, the molar susceptibility is given here in units appropriate to the cgs system. These values should be multiplied by 4π to obtain values for use in SI equations (where the magnetic field strength H has units of $A\ m^{-1}$).

References

1. *Landolt-Börnstein, Numerical Data and Functional Relationships in Science and Technology, New Series,* II/16, *Diamagnetic Susceptibility,* Gupta, R. R., Ed., Springer-Verlag, Heidelberg, 1986.
2. Barter, C., Meisenheimer, R. G., and Stevenson, D. P., *J. Phys. Chem.* 64, 1312, 1960.
3. Broersma, S., *J. Chem. Phys.* 17, 873, 1949.

Molecular formula	Compound	$-\chi_m/10^{-6}\ cm^3\ mol^{-1}$
$CBrCl_3$	Bromotrichloromethane	73.2
CBr_4	Tetrabromomethane	93.7
$CClF_3$	Chlorotrifluoromethane	45.3
$CClN$	Cyanogen chloride	32.4
CCl_2F_2	Dichlorodifluoromethane	52.2
CCl_2O	Carbonyl chloride	47.9
CCl_3F	Trichlorofluoromethane	58.7
CCl_3NO_2	Trichloronitromethane	75.3
CCl_4	Tetrachloromethane	66.8
$CHBrCl_2$	Bromodichloromethane	66.3
$CHBr_3$	Tribromomethane	82.6
$CHCl_3$	Trichloromethane	58.9
CHI_3	Triiodomethane	117.1
CH_2BrCl	Bromochloromethane	55.1
CH_2Br_2	Dibromomethane	65.1
CH_2Cl_2	Dichloromethane	46.6
CH_2I_2	Diiodomethane	93.1
CH_2N_2	Cyanamide	24.8
CH_2O	Formaldehyde	18.6
CH_2O_2	Formic acid	19.9
CH_3Br	Bromomethane	42.8
CH_3Cl	Chloromethane	32.0
CH_3F	Fluoromethane	17.8
CH_3I	Iodomethane	57.2
CH_3NO	Formamide	23.0
CH_3NO_2	Nitromethane	21.0
CH_4	Methane	17.4
CH_4N_2O	Urea	33.5
CH_4O	Methanol	21.4
CH_5N	Methylamine	27.0
CI_4	Tetraiodomethane	136
CN_4O_8	Tetranitromethane	43.0
C_2ClF_3	Chlorotrifluoroethylene	49.1
C_2Cl_4	Tetrachloroethylene	81.6
C_2Cl_6	Hexachloroethane	112.8
C_2HCl_3	Trichloroethylene	65.8
C_2HCl_3O	Trichloroacetaldehyde	73.0
C_2HCl_3O	Dichloroacetyl chloride	69.0
$C_2HCl_3O_2$	Trichloroacetic acid	73.0
C_2HCl_5	Pentachloroethane	99.1
$C_2HF_3O_2$	Trifluoroacetic acid	43.3
C_2H_2	Acetylene	20.8
$C_2H_2Br_4$	1,1,2,2-Tetrabromoethane	123.4
$C_2H_2Cl_2$	1,1-Dichloroethylene	49.2
$C_2H_2Cl_2$	*cis*-1,2-Dichloroethylene	51.0
$C_2H_2Cl_2$	*trans*-1,2-Dichloroethylene	48.9
$C_2H_2Cl_4$	1,1,2,2-Tetrachloroethane	89.8
C_2H_3Cl	Chloroethylene	35.9
C_2H_3ClO	Acetyl chloride	39.3
C_2H_3N	Acetonitrile	27.8
C_2H_4	Ethylene	18.8
$C_2H_4Br_2$	1,2-Dibromoethane	78.9
$C_2H_4Cl_2$	1,1-Dichloroethane	57.4
$C_2H_4Cl_2$	1,2-Dichloroethane	59.6
C_2H_4O	Acetaldehyde	22.2
C_2H_4O	Ethylene oxide	30.5
$C_2H_4O_2$	Acetic acid	31.8
$C_2H_4O_2$	Methyl formate	31.1
C_2H_5Br	Bromoethane	78.8
C_2H_5Cl	Chloroethane	69.9
C_2H_5I	Iodoethane	69.1
C_2H_5NO	Acetamide	33.9

Molecular formula	Compound	$-\chi_m/10^{-6}$ cm^3 mol^{-1}
$C_2H_5NO_2$	Nitroethane	35.4
$C_2H_5NO_2$	Glycine	39.6
C_2H_6	Ethane	26.8
C_2H_6O	Ethanol	33.7
C_2H_6O	Dimethyl ether	26.3
$C_2H_6O_2$	Ethylene glycol	38.9
C_2H_6S	Ethanethiol	47.0
C_2H_6S	Dimethyl sulfide	44.9
$C_2H_8N_2$	1,2-Ethanediamine	46.5
C_2N_2	Cyanogen	21.6
C_3H_4	Allene	25.3
$C_3H_4O_2$	Vinyl formate	34.7
C_3H_5Br	3-Bromopropene	58.6
C_3H_5Cl	2-Chloropropene	47.8
C_3H_5Cl	3-Chloropropene	47.8
C_3H_5N	Propanenitrile	38.6
C_3H_6	Propene	30.7
C_3H_6	Cyclopropane	39.2
C_3H_6O	Allyl alcohol	36.7
C_3H_6O	Propanal	34.2
C_3H_6O	Acetone	33.8
C_3H_6O	Methyloxirane	42.5
$C_3H_6O_2$	Propanoic acid	43.2
$C_3H_6O_2$	Ethyl formate	42.4
C_3H_7Br	1-Bromopropane	65.6
C_3H_7Br	2-Bromopropane	65.1
C_3H_7Cl	1-Chloropropane	56.0
C_3H_7I	1-Iodopropane	84.3
C_3H_7N	Allylamine	40.1
$C_3H_7NO_2$	1-Nitropropane	45.0
$C_3H_7NO_2$	2-Nitropropane	45.4
$C_3H_7NO_2$	Ethyl carbamate	57.0
C_3H_8	Propane	38.6
C_3H_8O	1-Propanol	44.8
C_3H_8O	2-Propanol	45.7
$C_3H_8O_2$	1,3-Propylene glycol	50.2
$C_3H_8O_2$	Dimethoxymethane	47.3
$C_3H_8O_3$	Glycerol	57.1
$C_4H_2O_3$	Maleic anhydride	35.8
$C_4H_4N_2$	Pyrazine	37.8
$C_4H_4N_2$	Pyrimidine	43.1
C_4H_4O	Furan	43.1
$C_4H_4O_3$	Succinic anhydride	47.5
$C_4H_4O_4$	Maleic acid	49.6
$C_4H_4O_4$	Fumaric acid	49.1
C_4H_4S	Thiophene	57.3
C_4H_5N	Pyrrole	48.6
C_4H_6	1,2-Butadiene	35.6
C_4H_6	1,3-Butadiene	32.1
$C_4H_6O_2$	Vinyl acetate	46.4
$C_4H_6O_3$	Acetic anhydride	52.8
$C_4H_6O_4$	Succinic acid	58.0
$C_4H_6O_4$	Dimethyl oxalate	55.7
C_4H_7N	Butanenitrile	50.4
C_4H_8	1-Butene	41.0
C_4H_8	*cis*-2-Butene	42.6
C_4H_8	*trans*-2-Butene	43.3
C_4H_8	Isobutene	40.8
C_4H_8	Cyclobutane	40.0
C_4H_8O	Ethyl vinyl ether	47.9
C_4H_8O	1,2-Epoxybutane	54.8
C_4H_8O	Butanal	45.9
C_4H_8O	2-Butanone	45.6
$C_4H_8O_2$	Butanoic acid	55.2
$C_4H_8O_2$	2-Methylpropanoic acid	56.1
$C_4H_8O_2$	Propyl formate	55.0
$C_4H_8O_2$	Ethyl acetate	54.1
$C_4H_8O_2$	Methyl propanoate	54.5
$C_4H_8O_2$	1,4-Dioxane	52.2
C_4H_9Br	1-Bromobutane	77.1
C_4H_9Br	1-Bromo-2-methylpropane	79.9
C_4H_9Cl	1-Chlorobutane	67.1
C_4H_9Cl	2-Chlorobutane	67.4
C_4H_9I	1-Iodobutane	93.6
C_4H_9N	Pyrrolidine	54.8
C_4H_9NO	Morpholine	55.0
C_4H_{10}	Butane	50.3
C_4H_{10}	Isobutane	50.5
$C_4H_{10}O$	1-Butanol	56.4
$C_4H_{10}O$	2-Butanol	57.6
$C_4H_{10}O$	2-Methyl-1-propanol	57.6
$C_4H_{10}O$	2-Methyl-2-propanol	56.6
$C_4H_{10}O$	Diethyl ether	55.5
$C_4H_{10}O_2$	1,3-Butanediol	61.8
$C_4H_{10}O_2$	1,4-Butanediol	61.8
$C_4H_{10}S$	1-Butanethiol	70.2
$C_4H_{11}N$	Butylamine	58.9
$C_4H_{11}N$	Isobutylamine	59.8
$C_4H_{11}N$	Diethylamine	56.8
$C_5H_4O_2$	Furfural	47.2
C_5H_5N	Pyridine	48.7
$C_5H_6O_2$	Furfuryl alcohol	61.0
$C_5H_7NO_2$	Ethyl cyanoacetate	67.3
C_5H_8	2-Methyl-1,3-butadiene	46.0

Molecular formula	Compound	$-\chi_m/10^{-6}$ cm^3 mol^{-1}
C_5H_8O	Cyclopentanone	51.6
$C_5H_8O_2$	Methyl methacrylate	57.3
$C_5H_8O_2$	2,4-Pentanedione	54.9
C_5H_{10}	1-Pentene	54.6
C_5H_{10}	2-Methyl-2-butene	54.7
C_5H_{10}	Cyclopentane	56.2
$C_5H_{10}O$	Cyclopentanol	64.0
$C_5H_{10}O$	Pentanal	57.5
$C_5H_{10}O$	2-Pentanone	57.5
$C_5H_{10}O$	3-Pentanone	57.7
$C_5H_{10}O_2$	Pentanoic acid	66.5
$C_5H_{10}O_2$	3-Methylbutanoic acid	67.7
$C_5H_{10}O_2$	Butyl formate	65.8
$C_5H_{10}O_2$	Isobutyl formate	66.8
$C_5H_{10}O_2$	Propyl acetate	65.9
$C_5H_{10}O_2$	Isopropyl acetate	67.0
$C_5H_{10}O_2$	Ethyl propanoate	66.3
$C_5H_{10}O_2$	Tetrahydrofurfuryl alcohol	69.4
$C_5H_{10}O_3$	Diethyl carbonate	75.4
$C_5H_{11}N$	Piperidine	64.2
C_5H_{12}	Pentane	61.5
C_5H_{12}	Isopentane	63.0
C_5H_{12}	Neopentane	63.0
$C_5H_{12}O$	1-Pentanol	67.0
$C_5H_{12}O$	2-Pentanol	69.1
$C_5H_{12}O_2$	1,5-Pentanediol	73.5
$C_5H_{13}N$	Pentylamine	69.3
C_6Cl_6	Hexachlorobenzene	147.0
$C_6H_4ClNO_2$	1-Chloro-2-nitrobenzene	75.5
$C_6H_4ClNO_2$	1-Chloro-3-nitrobenzene	77.2
$C_6H_4ClNO_2$	1-Chloro-4-nitrobenzene	74.7
$C_6H_4Cl_2$	*o*-Dichlorobenzene	84.4
$C_6H_4Cl_2$	*m*-Dichlorobenzene	84.1
$C_6H_4Cl_2$	*p*-Dichlorobenzene	81.7
$C_6H_4O_2$	*p*-Benzoquinone	36
C_6H_5Br	Bromobenzene	78.4
C_6H_5Cl	Chlorobenzene	69.5
C_6H_5ClO	*o*-Chlorophenol	77.3
C_6H_5ClO	*m*-Chlorophenol	77.6
C_6H_5ClO	*p*-Chlorophenol	77.7
C_6H_5F	Fluorobenzene	58.4
C_6H_5I	Iodobenzene	92.0
$C_6H_5NO_2$	Nitrobenzene	61.9
$C_6H_5NO_3$	*o*-Nitrophenol	68.9
$C_6H_5NO_3$	*m*-Nitrophenol	65.9
$C_6H_5NO_3$	*p*-Nitrophenol	66.9
C_6H_6	Benzene	54.8
C_6H_6ClN	*o*-Chloroaniline	79.5
C_6H_6ClN	*m*-Chloroaniline	76.6
C_6H_6ClN	*p*-Chloroaniline	76.7
$C_6H_6N_2O_2$	*o*-Nitroaniline	67.4
$C_6H_6N_2O_2$	*m*-Nitroaniline	69.7
$C_6H_6N_2O_2$	*p*-Nitroaniline	68.0
C_6H_6O	Phenol	60.6
$C_6H_6O_2$	*p*-Hydroquinone	64.7
$C_6H_6O_2$	Pyrocatechol	68.2
$C_6H_6O_2$	Resorcinol	67.2
C_6H_7N	Aniline	62.4
C_6H_7N	4-Methylpyridine	59.8
C_6H_8	1,4-Cyclohexadiene	48.7
$C_6H_8N_2$	*o*-Phenylenediamine	72.5
$C_6H_8N_2$	*m*-Phenylenediamine	70.4
$C_6H_8N_2$	*p*-Phenylenediamine	70.7
C_6H_{10}	1,5 Hexadiene	55.1
C_6H_{10}	1-Hexyne	64.5
C_6H_{10}	Cyclohexene	58.0
$C_6H_{10}O$	Cyclohexanone	62.0
$C_6H_{10}O_3$	Ethyl acetoacetate	71.7
$C_6H_{10}O_4$	Diethyl oxalate	81.7
C_6H_{12}	1-Hexene	66.4
C_6H_{12}	2,3-Dimethyl-2-butene	65.9
C_6H_{12}	Cyclohexane	68
C_6H_{12}	Methylcyclopentane	70.2
$C_6H_{12}O$	Hexanal	69.4
$C_6H_{12}O$	2-Hexanone	69.2
$C_6H_{12}O$	3-Hexanone	69.0
$C_6H_{12}O$	4-Methyl-2-pentanone	69.7
$C_6H_{12}O$	Cyclohexanol	73.4
$C_6H_{12}O_2$	Hexanoic acid	78.1
$C_6H_{12}O_2$	Isopentyl formate	78.4
$C_6H_{12}O_2$	Isobutyl acetate	78.7
$C_6H_{12}O_2$	Propyl propanoate	77.7
$C_6H_{12}O_3$	Paraldehyde	86.1
C_6H_{14}	Hexane	74.1
C_6H_{14}	2-Methylpentane	75.3
C_6H_{14}	3-Methylpentane	75.5
C_6H_{14}	2,2-Dimethylbutane	76.2
C_6H_{14}	2,3-Dimethylbutane	76.2
$C_6H_{14}O$	1-Hexanol	79.5
$C_6H_{14}O$	4-Methyl-2-pentanol	80.4
$C_6H_{14}O$	Dipropyl ether	79.4
$C_6H_{14}O_2$	1,6-Hexanediol	84.3
$C_6H_{14}O_2$	1,1-Diethoxyethane	81.4
$C_6H_{14}O_6$	*D*-Glucitol	107.8

Molecular formula	Compound	$-\chi_m/10^{-6}$ cm³ mol⁻¹
$C_6H_{15}N$	Triethylamine	83.3
C_7H_5N	Benzonitrile	65.2
C_7H_6O	Benzaldehyde	60.7
$C_7H_6O_2$	Salicylaldehyde	66.8
$C_7H_6O_3$	Salicylic acid	75
C_7H_7Br	*p*-Bromotoluene	88.7
C_7H_7Cl	*o*-Chlorotoluene	82.4
C_7H_7Cl	*m*-Chlorotoluene	79.7
C_7H_7Cl	*p*-Chlorotoluene	80.3
C_7H_7Cl	(Chloromethyl)benzene	81.6
C_7H_7NO	Benzamide	72.0
$C_7H_7NO_2$	*o*-Nitrotoluene	72.2
$C_7H_7NO_2$	*m*-Nitrotoluene	72.7
$C_7H_7NO_2$	*p*-Nitrotoluene	73.3
C_7H_8	Toluene	65.6
C_7H_8O	*o*-Cresol	73.3
C_7H_8O	*m*-Cresol	72.2
C_7H_8O	*p*-Cresol	72.4
C_7H_8O	Benzyl alcohol	71.8
C_7H_8O	Anisole	72.2
C_7H_9N	*o*-Methylaniline	74.9
C_7H_9N	*m*-Methylaniline	74.6
C_7H_9N	*p*-Methylaniline	72.5
C_7H_9N	*N*-Methylaniline	74.1
C_7H_9N	2,4-Dimethylpyridine	71.3
C_7H_9N	2,6-Dimethylpyridine	72.5
C_7H_9NO	*o*-Methoxyaniline [*o*-Anisidine]	79.1
$C_7H_{12}O_4$	Diethyl malonate	92.6
C_7H_{14}	1-Heptene	77.8
C_7H_{14}	Cycloheptane	73.9
C_7H_{14}	Methylcyclohexane	78.9
$C_7H_{14}O$	1-Heptanal	81.0
$C_7H_{14}O$	2-Heptanone	80.5
$C_7H_{14}O$	3-Heptanone	80.7
$C_7H_{14}O$	4-Heptanone	80.5
$C_7H_{14}O$	2,4-Dimethyl-3-pentanone	81.1
$C_7H_{14}O_2$	Heptanoic acid	89.0
$C_7H_{14}O_2$	Pentyl acetate	88.9
$C_7H_{14}O_2$	Isopentyl acetate	89.4
$C_7H_{14}O_2$	Butyl propanoate	89.1
$C_7H_{14}O_2$	Ethyl 3-methylbutanoate	91.1
C_7H_{16}	Heptane	85.2
C_7H_{16}	3-Ethylpentane	86.2
C_7H_{16}	2,2-Dimethylpentane	87.0
C_7H_{16}	2,3-Dimethylpentane	87.5
C_7H_{16}	2,4-Dimethylpentane	87.5
C_7H_{16}	3,3-Dimethylpentane	89.5
$C_7H_{16}O$	1-Heptanol	91.7
$C_7H_{16}O$	4-Heptanol	92.1
$C_8H_4O_3$	Phthalic anhydride	66.7
$C_8H_6O_4$	Phthalic acid	83.6
$C_8H_6O_4$	Isophthalic acid	84.6
$C_8H_6O_4$	Terephthalic acid	83.5
C_8H_7N	Benzeneacetonitrile	76.9
C_8H_7N	Indole	85.0
C_8H_8	Styrene	68.2
C_8H_8O	Acetophenone	72.5
$C_8H_8O_2$	*o*-Toluic acid	84.3
$C_8H_8O_2$	*m*-Toluic acid	83.0
$C_8H_8O_2$	*p*-Toluic acid	82.4
$C_8H_8O_2$	Benzeneacetic acid	82.4
$C_8H_8O_2$	Methyl benzoate	81.6
$C_8H_8O_3$	Methyl salicylate	86.6
C_8H_{10}	Ethylbenzene	77.3
C_8H_{10}	*o*-Xylene	77.7
C_8H_{10}	*m*-Xylene	76.4
C_8H_{10}	*p*-Xylene	77.0
$C_8H_{10}O$	Phenetole	84.5
$C_8H_{11}N$	*N*-Ethylaniline	85.6
$C_8H_{11}N$	*N,N*-Dimethylaniline	85.1
$C_8H_{11}N$	2,4,6-Trimethylpyridine	83.1
$C_8H_{14}O_4$	Ethyl succinate	105.0
C_8H_{16}	1-Octene	88.8
C_8H_{16}	Cyclooctane	85.3
$C_8H_{16}O_2$	Octanoic acid	99.5
$C_8H_{16}O_2$	Hexyl acetate	100.9
$C_8H_{17}Cl$	1-Chlorooctane	114.9
C_8H_{18}	Octane	96.6
C_8H_{18}	4-Methylheptane	97.3
C_8H_{18}	3-Ethylhexane	97.8
C_8H_{18}	3,4-Dimethylhexane	99.1
C_8H_{18}	2,2,4-Trimethylpentane	99.1
C_8H_{18}	2,3,4-Trimethylpentane	99.8
$C_8H_{18}O$	1-Octanol	101.6
$C_8H_{19}N$	Dibutylamine	103.7
C_9H_7N	Quinoline	86.1
C_9H_7N	Isoquinoline	83.9
C_9H_8	Indene	83
C_9H_{10}	Isopropenylbenzene	80.0
$C_9H_{10}O_2$	Ethyl benzoate	93.8
$C_9H_{10}O_2$	Benzyl acetate	93.2
C_9H_{12}	Propylbenzene	89.1

Molecular formula	Compound	$-\chi_m/10^{-6}$ cm³ mol⁻¹
C_9H_{12}	Isopropylbenzene [Cumene]	89.5
C_9H_{12}	1,3,5-Trimethylbenzene [Mesitylene]	92.3
C_9H_{18}	1-Nonene	100.1
$C_9H_{18}O$	2,6-Dimethyl-4-heptanone	104.3
C_9H_{20}	Nonane	108.1
$C_{10}H_7Br$	1-Bromonaphthalene	123.6
$C_{10}H_7Cl$	1-Chloronaphthalene	107.6
$C_{10}H_8$	Naphthalene	91.6
$C_{10}H_8$	Azulene	123.7
$C_{10}H_8O$	1-Naphthol	96.2
$C_{10}H_8O$	2-Naphthol	96.8
$C_{10}H_9N$	1-Naphthalenamine	92.5
$C_{10}H_9N$	2-Naphthalenamine	98.0
$C_{10}H_{10}O_2$	Safrole	97.5
$C_{10}H_{10}O_4$	Dimethyl terephthalate	101.6
$C_{10}H_{14}$	Butylbenzene	100.7
$C_{10}H_{14}$	*tert*-Butylbenzene	101.8
$C_{10}H_{14}$	Isobutylbenzene	101.7
$C_{10}H_{14}$	*p*-Cymene	102.8
$C_{10}H_{14}$	1,2,4,5-Tetramethylbenzene	101.2
$C_{10}H_{14}O$	*p-tert*-Butylphenol	108.0
$C_{10}H_{15}N$	*N,N*-Diethylaniline	107.9
$C_{10}H_{16}$	*d*-Limonene	98.0
$C_{10}H_{16}$	α-Pinene	100.7
$C_{10}H_{16}$	β-Pinene	101.9
$C_{10}H_{16}O$	Camphor, (+)	103.0
$C_{10}H_{18}$	*cis*-Decahydronaphthalene	107.0
$C_{10}H_{18}$	*trans*-Decahydronaphthalene	107.6
$C_{10}H_{22}$	Decane	119.5
$C_{11}H_{10}$	1-Methylnaphthalene	102.9
$C_{11}H_{10}$	2-Methylnaphthalene	102.7
$C_{11}H_{24}$	Undecane	131.8
$C_{12}H_8$	Acenaphthylene	111.6
$C_{12}H_9N$	Carbazole	119.9
$C_{12}H_{10}$	Acenaphthene	109.9
$C_{12}H_{10}$	Biphenyl	103.3
$C_{12}H_{10}N_2$	Azobenzene	106.8
$C_{12}H_{11}N$	Diphenylamine	108.4
$C_{12}H_{14}O_4$	Diethyl phthalate	127.5
$C_{12}H_{18}$	Hexamethylbenzene	122.5
$C_{12}H_{24}O_2$	Dodecanoic acid	113.0
$C_{13}H_9N$	Acridine	118.8
$C_{13}H_{10}O$	Benzophenone	109.6
$C_{13}H_{12}$	Diphenylmethane	116.0
$C_{13}H_{28}$	Tridecane	153.7
$C_{14}H_8O_2$	9,10-Anthracenedione	113.0
$C_{14}H_{10}$	Anthracene	129.8
$C_{14}H_{10}$	Phenanthrene	127.6
$C_{14}H_{10}$	Diphenylacetylene	116
$C_{14}H_{10}O_2$	Benzil	106.8
$C_{14}H_{12}O_2$	Benzyl benzoate	132.2
$C_{14}H_{14}$	1,2-Diphenylethane	127.8
$C_{14}H_{28}O_2$	Tetradecanoic acid [Myristic acid]	176.0
$C_{14}H_{30}$	Tetradecane	166.2
$C_{16}H_{10}$	Pyrene	147
$C_{16}H_{32}O_2$	Hexadecanoic acid [Palmitic acid]	198.6
$C_{16}H_{34}$	Hexadecane	187.6
$C_{16}H_{34}O$	1-Hexadecanol	183.5
$C_{18}H_{12}$	Chrysene	148.0
$C_{18}H_{14}$	*o*-Terphenyl	150.4
$C_{18}H_{14}$	*m*-Terphenyl	155.5
$C_{18}H_{14}$	*p*-Terphenyl	156.0
$C_{18}H_{34}O_2$	*cis*-9-Octadecenoic acid [Oleic acid]	208.5
$C_{18}H_{36}O_2$	Octadecanoic acid [Stearic acid]	220.8
$C_{20}H_{12}$	Perylene	167.5

Section 4
Properties of the Elements and Inorganic Compounds

The Elements 4-1
Physical Constants of Inorganic Compounds 4-43
Formula Index of Inorganic Compounds 4-102
Physical Properties of the Rare Earth Metals 4-115
Melting, Boiling, Triple, and Critical Point Temperatures of the Elements 4-121
Heat Capacity of the Elements at 25 °C 4-123
Vapor Pressure of the Metallic Elements — Equations 4-124
Vapor Pressure of the Metallic Elements — Data 4-126
Density of Molten Elements and Representative Salts 4-127
Magnetic Susceptibility of the Elements and Inorganic Compounds 4-130
Index of Refraction of Inorganic Liquids 4-136
Physical and Optical Properties of Minerals 4-137
Crystallographic Data on Minerals 4-144

THE ELEMENTS

C. R. Hammond

One of the most striking facts about the elements is their unequal distribution and occurrence in nature. Present knowledge of the chemical composition of the universe, obtained from the study of the spectra of stars and nebulae, indicates that hydrogen is by far the most abundant element and may account for more than 90% of the atoms or about 75% of the mass of the universe. Helium atoms make up most of the remainder. All of the other elements together contribute only slightly to the total mass.

The chemical composition of the universe is undergoing continuous change. Hydrogen is being converted into helium, and helium is being changed into heavier elements. As time goes on, the ratio of heavier elements increases relative to hydrogen. Presumably, the process is not reversible.

Burbidge, Burbidge, Fowler, and Hoyle, and more recently, Peebles, Penzias, and others have studied the synthesis of elements in stars. To explain all of the features of the nuclear abundance curve — obtained by studies of the composition of the Earth, meteorites, stars, etc. — it is necessary to postulate that the elements were originally formed by at least eight different processes: (1) hydrogen burning, (2) helium burning, (3) χ process, (4) *e* process, (5) *s* process, (6) *r* process, (7) *p* process, and (8) the X process. The X process is thought to account for the existence of light nuclei such as D, Li, Be, and B. Common metals such as Fe, Cr, Ni, Cu, Ti, Zn, etc. were likely produced early in the history of our galaxy. It is also probable that most of the heavy elements on Earth and elsewhere in the universe were originally formed in supernovae, or in the hot interior of stars.

Studies of the solar spectrum have led to the identification of 67 elements in the sun's atmosphere; however, all elements cannot be identified with the same degree of certainty. Other elements may be present in the sun, although they have not yet been detected spectroscopically. The element helium was discovered on the sun before it was found on Earth. Some elements such as scandium are relatively more plentiful in the sun and stars than here on Earth.

Minerals in lunar rocks brought back from the moon on the Apollo missions consist predominantly of *plagioclase* {$(Ca,Na)(Al,Si)O_4O_8$} and *pyroxene* {$(Ca,Mg,Fe)_2Si_2O_6$} — two minerals common in terrestrial volcanic rock. No new elements have been found on the moon that cannot be accounted for on Earth; however, three minerals, *armalcolite* {$(Fe,Mg)Ti_2O_5$}, *pyroxferroite* {$CaFe_6(SiO_3)_7$}, and *tranquillityite* {$Fe_8(Zr,Y)Ti_3Si_3O_2$}, are new. The oldest known terrestrial rocks are about 4 billion years old. One rock, known as the "Genesis Rock," brought back from the Apollo 15 Mission, is about 4.15 billion years old. This is only about one-half billion years younger than the supposed age of the moon and solar system. Lunar rocks appear to be relatively enriched in refractory elements such as chromium, titanium, zirconium, and the rare earths, and impoverished in volatile elements such as the alkali metals, in chlorine, and in noble metals such as nickel, platinum, and gold.

Even older than the "Genesis Rock" are *carbonaceous chondrites,* a type of meteorite that has fallen to Earth and has been studied. These are some of the most primitive objects of the solar system yet found. The grains making up these objects probably condensed directly out the gaseous nebula from which the sun and planets were born. Most of the condensation of the grains probably was completed within 50,000 years of the time the disk of the nebula was first formed — about 4.6 billion years ago. It is now thought that this type of meteorite may contain a small percentage of presolar dust grains. The relative abundances of the elements of these meteorites are about the same as the abundances found in the solar chromosphere.

The X-ray fluorescent spectrometer sent with the Viking I spacecraft to Mars shows that the Martian soil contains about 12 to 16% iron, 14 to 15% silicon, 3 to 8% calcium, 2 to 7% aluminum, and one-half to 2% titanium. The gas chromatograph — mass spectrometer on Viking II found no trace of organic compounds.

F. W. Clarke and others have carefully studied the composition of rocks making up the crust of the Earth. Oxygen accounts for about 47% of the crust, by weight, while silicon comprises about 28% and aluminum about 8%. These elements, plus iron, calcium, sodium, potassium, and magnesium, account for about 99% of the composition of the crust.

Many elements such as tin, copper, zinc, lead, mercury, silver, platinum, antimony, arsenic, and gold, which are so essential to our needs and civilization, are among some of the rarest elements in the Earth's crust. These are made available to us only by the processes of concentration in ore bodies. Some of the so-called *rare-earth* elements have been found to be much more plentiful than originally thought and are about as abundant as uranium, mercury, lead, or bismuth. The least abundant rare-earth or *lanthanide* element, thulium, is now believed to be more plentiful on Earth than silver, cadmium, gold, or iodine, for example. Rubidium, the 16th most abundant element, is more plentiful than chlorine while its compounds are little known in chemistry and commerce.

It is now thought that at least 24 elements are essential to living matter. The four most abundant in the human body are hydrogen, oxygen, carbon, and nitrogen. The seven next most common, in order of abundance, are calcium, phosphorus, chlorine, potassium, sulfur, sodium, and magnesium. Iron, copper, zinc, silicon, iodine, cobalt, manganese, molybdenum, fluorine, tin, chromium, selenium, and vanadium are needed and play a role in living matter. Boron is also thought essential for some plants, and it is possible that aluminum, nickel, and germanium may turn out to be necessary.

Ninety-one elements occur naturally on Earth. Minute traces of plutonium-244 have been discovered in rocks mined in southern California. This discovery supports the theory that heavy elements were produced during creation of the solar system. While technetium and promethium have not yet been found naturally on Earth, they have been found to be present in stars. Technetium has been identified in the spectra of certain "late" type stars, and promethium lines have been identified in the spectra of a faintly visible star HR465 in Andromeda. Promethium must have been made near the star's surface for no known isotope of this element has a half-life longer than 17.7 years.

It has been suggested that californium is present in certain stellar explosions known as supernovae; however, this has not been proved. At present no elements are found elsewhere in the universe that cannot be accounted for here on Earth.

All atomic mass numbers from 1 to 238 are found naturally on Earth except for masses 5 and 8. About 285 relatively stable and 67 naturally radioactive isotopes occur on Earth totaling 352. In addition, the neutron, technetium, promethium, and the transuranic elements (lying beyond uranium) have now been produced artificially. In June 1999, scientists at the Lawrence Berkeley National Laboratory reported that they had found evidence of an isotope of Element 118 and its immediate decay products of Elements 116, 114, and 112. This sequence of events tended to reinforce the theory that was predicted since the 1970s that an "island of stability" existed for nuclei with approximately 114 protons and 184 neutrons. This "island" refers to nuclei in which the decay lasts for a period of time instead of a decay that occurs instantaneously. However, on July 27, 2001, researchers at LBNL reported that their laboratory and the facilities at the GSI Laboratory in Germany and at Japanese laboratories failed to confirm the results of their earlier experiments where the fusion of a krypton atom with a lead target resulted in Element 118, with chains of decay leading to Elements 116, 114, and 112, and on down to Element 106. Therefore, the discovery was reported to be spurious. However, with the announcement it was said that different experiments at the Livermore Laboratory and Joint Institute for Nuclear Research in Dubna, Russia indicated that Element 116 had since been created directly. (See also under Elements 116 and 118.)

Laboratory processes have now extended the radioactive element mass numbers beyond 238 to about 280. Each element from atomic numbers 1 to 110 is known to have at least one radioactive isotope. As of December 2001, about 3286 isotopes and isomers were thought to be known and recognized. Many stable and radioactive isotopes are now produced and distributed by the Oak Ridge National Laboratory, Oak Ridge, TN, U.S.A., to customers licensed by the U.S. Department of Energy.

The nucleus of an atom is characterized by the number of protons it contains, denoted by Z, and by the number of neutrons, N. Isotopes of an element have the same value of Z, but different values of N. The *mass number A*, is the sum of Z and N. For example, Uranium-238 has a mass number of 238, and contains 92 protons and 146 neutrons.

There is evidence that the definition of chemical elements must be broadened to include the electron. Several compounds known as *electrides* have recently been made of alkaline metal elements and electrons. A relatively stable combination of a positron and electron, known as *positronium*, has also been studied.

The well-known proton, neutron, and electron are now thought to be members of a group that includes other fundamental particles that have been discovered or hypothesized by physicists. These very elemental particles, of which all matter is made, are now thought to belong to one of two families: namely, **quarks** or **leptons**. Each of these two families consists of six particles. Also, there are four different force carriers that lead to interactions between particles. The six members or "flavors" of the quark family are called **up**, **charm**, **top**, **down**, **strange**, and **bottom**. The force carriers for the quarks are the **gluon** and the **photon**. The six members of the lepton family are the **e neutrino**, the **mu neutrino**, the **tau neutrino**, the **electron**, the **muon particle**, and the **tau particle**. The force carriers for these are the **w boson** and the **z boson**. Furthermore, it appears that each of these particles has an anti-particle that has an opposite electrical charge from the above particles.

Quarks are not found individually, but are found with other quarks arranged to form composites known as **hadrons**. There are two basic types of hadrons: **baryons**, composed of three quarks, and **mesons**, composed of a quark and an anti-quark. Examples of baryons are the neutron and the proton. Neutrons are made of two down quarks and one up quark. Protons are made of two up quarks and one down quark. An example of the meson is the **pion**. This particle is made of an up quark and a down anti-quark. Such particles are unstable and tend to decay rapidly. The anti-particle of the proton is the anti-proton. The exception to the rule is the electron, whose anti-particle is the **positron**.

In recent years a search has been made for a hypothetical particle known as the **Higgs particle** or **Higgs boson**, suggested in 1966 by Peter Higgs of the University of Edinburgh, which could possibly explain why the carriers of the "electro-weak" field (w and z bosons) have mass. The Higgs particle is thought to be responsible possibly for the mass of objects throughout the universe.

Many physicists now hold that all matter and energy in the universe are controlled by four fundamental forces: the **electromagnetic force**, **gravity**, a **weak nuclear force**, and a **strong nuclear force**. The **gluon** binds quarks together by carrying the strong nuclear force. Each of these natural forces is passed back and forth among the basic particles of matter by the force carriers mentioned above. The electromagnetic force is carried by the photon, the weak nuclear force by the **intermediate vector boson**, and the gravity by the **graviton**.

For more complete information on these fundamental particles, please consult recent articles and books on nuclear or particle physics.

The available evidence leads to the conclusion that elements 89 (actinium) through 103 (lawrencium) are chemically similar to the rare-earth or lanthanide elements (elements 57 to 71, inclusive). These elements therefore have been named *actinides* after the first member of this series. Those elements beyond uranium that have been produced artificially have the following names and symbols: neptunium, 93 (Np); plutonium, 94 (Pu); americium, 95 (Am); curium, 96 (Cm); berkelium, 97 (Bk); californium, 98 (Cf); einsteinium, 99 (Es); fermium, 100 (Fm); mendelevium, 101 (Md); nobelium, 102 (No); lawrencium, 103 (Lr); rutherfordium, 104 (Rf); dubnium, 105 (Db); seaborgium, 106 (Sg); bohrium, 107 (Bh); hassium, 108 (Hs); meitnerium, 109 (Mt); darmstadtium, 110 (Ds); and roentgenium, 111 (Rg). As of 2005, evidence has been reported for elements 112, 113, 114, 115, 116, and 118, but these elements have not been officially recognized or named. IUPAC recommends that until the existence of a new element is proven to their satisfaction, the elements are to have names and symbols derived according to these precise and simple rules: The name is based on the digits in the element's atomic number. Each digit is replaced with these expressions, with the end using the usual –ium suffix as follows: **0 nil, 1 un, 2 bi, 3 tri, 4 quad, 5 pent, 6 hex, 7 sept, 8 oct, 9 enn**. Double letter i's are not used, as for example Ununbiium, but would be Ununbium. The symbol used would be the first letter of the three main syllables. For example, Element 126 would be Unbihexium, with the symbol Ubh. (See J. Chatt, *Pure Appl. Chem.* 51, 381, 1979; W. H. Koppenol, *Pure Appl. Chem.* 74, 787, 2002.)

There are many claims in the literature of the existence of various allotropic modifications of the elements, some of which are based on doubtful or incomplete evidence. Also, the physical properties of an element may change drastically by the presence of small amounts of impurities. With new methods of purification, which are now able to produce elements with 99.9999% purity, it has been necessary to restudy the properties of the elements. For example, the melting point of thorium changes by several hundred degrees by the presence of a small percentage of ThO_2 as an impurity. Ordinary commercial tungsten is brittle and can be worked only with difficulty. Pure tungsten, however, can be cut with a hacksaw, forged, spun, drawn, or extruded. In general, the value of a physical property given here applies to the pure element, when it is known.

Many of the chemical elements and their compounds are toxic and should be handled with due respect and care. In recent years there has been greatly increased knowledge and awareness of the health hazards associated with chemicals, radioactive materials, and other agents. Anyone working with the elements and certain of their compounds should become thoroughly familiar with the proper safeguards to be taken. Information on specific hazards and recommended exposure limits may also be found in Section 16. Reference should also be made to publications such as the following:

1. *Code of Federal Regulations, Title 29, Labor.* With additions found in issues of the *Federal Register*.
2. *Code of Federal Regulations, Title 10, Energy.* With additions found in issues of the *Federal Register*. (Published by the U.S. Government Printing Office. Supt. of Documents.)
3. *Occupational Safety and Health Reporter* (latest edition with amendments and corrections), Bureau of National Affairs, Washington, D.C.
4. *Atomic Energy Law Reporter,* Commerce Clearing House, Chicago, IL.
5. *Nuclear Regulation Reporter,* Commerce Clearing House, Chicago, IL.
6. *TLVs® Threshold Limit Values for Chemical Substances and Physical Agents* is issued annually by the American Conference of Governmental Industrial Hygienists, Cincinnati, Ohio.
7. *The Sigma Aldrich Library of Regulatory and Safety Data. Vol. 3,* Robert E. Lenga and Kristine L. Volonpal, Sigma Chemical Co. and Aldrich Chemical Co., Inc. 1993.
8. *Hazardous Chemicals Desk Reference,* 4th ed., Richard J. Lewis, Sr., John Wiley & Sons, New York, 1997.
9. *Sittig's Handbook of Toxic and Hazardous Chemicals and Carcinogens,* 3rd ed., Noyes Publications, 2001/2.
10. *Sax's Dangerous Properties of Industrial Materials,* Richard J. Lewis and N. Irving Sax, John Wiley & Sons, New York, 1999.
11. *World Wide Limits for Toxic and Hazardous Chemicals in Air, Water, and Soil,* Marshall Sittig, Noyes Publishers.

The prices of elements as indicated in this article are intended to be only a rough guide. Prices may vary, over time, widely with supplier, quantity, and purity.

The density of gases is given in grams per liter at 0 °C and a pressure of 1 atm.

Actinium — (Gr. *aktis, aktinos,* beam or ray), Ac; at. wt. (227); at. no. 89; m.p. 1050 °C, b.p. 3198 °C; sp. gr. 10.07 (calc.). Discovered by Andre Debierne in 1899 and independently by F. Giesel in 1902. Occurs naturally in association with uranium minerals. Thirty-four isotopes and isomers are now recognized. All are radioactive. Actinium-227, a decay product of uranium-235, is an alpha and beta emitter with a 21.77-year half-life. Its principal decay products are thorium-227 (18.72-day half-life), radium-223 (11.4-day half-life), and a number of short-lived products including radon, bismuth, polonium, and lead isotopes. In equilibrium with its decay products, it is a powerful source of alpha rays. Actinium metal has been prepared by the reduction of actinium fluoride with lithium vapor at about 1100 to 1300 °C. The chemical behavior of actinium is similar to that of the rare earths, particularly lanthanum. Purified actinium comes into equilibrium with its decay products at the end of 185 days, and then decays according to its 21.77-year half-life. It is about 150 times as active as radium, making it of value in the production of neutrons. Actinium-225, with a purity of 99%, is available from the Oak Ridge National Laboratory to holders of a permit for about $500/millicurie, plus packing charges.

Aluminum — (L. *alumen, alum*), Al; at. wt. 26.9815386(8); at. no. 13; m.p. 660.32 °C; b.p. 2519 °C; sp. gr. 2.6989 (20 °C); valence 3. The ancient Greeks and Romans used *alum* in medicine as an astringent, and as a mordant in dyeing. In 1761 de Morveau proposed the name *alumine* for the base in alum, and Lavoisier, in 1787, thought this to be the oxide of a still undiscovered metal. Wohler is generally credited with having isolated the metal in 1827, although an impure form was prepared by Oersted two years earlier. In 1807, Davy proposed the name *alumium* for the metal, undiscovered at that time, and later agreed to change it to *aluminum.* Shortly thereafter, the name *aluminium* was adopted to conform with the "ium" ending of most elements, and this spelling is now in use elsewhere in the world. *Aluminium* was also the accepted spelling in the U.S. until 1925, at which time the American Chemical Society officially decided to use the name *aluminum* thereafter in their publications. The method of obtaining aluminum metal by the electrolysis of alumina dissolved in *cryolite* was discovered in 1886 by Hall in the U.S. and at about the same time by Heroult in France. Cryolite, a natural ore found in Greenland, is no longer widely used in commercial production, but has been replaced by an artificial mixture of sodium, aluminum, and calcium fluorides. *Bauxite,* an impure hydrated oxide ore, is found in large deposits in Jamaica, Australia, Suriname, Guyana, Russia, Arkansas, and elsewhere. The Bayer process is most commonly used today to refine bauxite so it can be accommodated in the Hall–Heroult refining process used to make most aluminum. Aluminum can now be produced from clay, but the process is not economically feasible at present. Aluminum is the most abundant metal to be found in the Earth's crust (8.1%), but is never found free in nature. In addition to the minerals mentioned above, it is found in feldspars, granite, and in many other common minerals. Twenty-two isotopes and isomers are known. Natural aluminum is made of one isotope, ^{27}Al. Pure aluminum, a silvery-white metal, possesses many desirable characteristics. It is light, nontoxic, has a pleasing appearance, can easily be formed, machined, or cast, has a high thermal conductivity, and has excellent corrosion resistance. It is nonmagnetic and nonsparking, stands second among metals in the scale of malleability, and sixth in ductility. It is extensively used for kitchen utensils, outside building decoration, and in thousands of industrial applications where a strong, light, easily constructed material is needed. Although its electrical conductivity is only about 60% that of copper, it is used in electrical transmission lines because of its light weight. Pure aluminum is soft and lacks strength, but it can be alloyed with small amounts of copper, magnesium, silicon, manganese, and other elements to impart a variety of useful properties. These alloys are of vital importance in the construction of modern aircraft and rockets. Aluminum, evaporated in a vacuum, forms a highly reflective coating for both visible light and radiant heat. These coatings soon form a thin layer of the protective oxide and do not deteriorate as do silver coatings. They have found application in coatings for telescope mirrors, in making decorative paper, packages, toys, and in many other uses. The compounds of greatest importance are aluminum oxide, the sulfate, and the soluble sulfate with potassium (alum). The oxide, alumina, occurs naturally as ruby, sapphire, corundum, and emery, and is used in glassmaking and refractories. Synthetic ruby and sapphire have found application in the construction of lasers for producing coherent light. In 1852, the price of aluminum was about \$1200/kg, and just before Hall's discovery in 1886, about \$25/kg. The price rapidly dropped to 60¢ and has been as low as 33¢/kg. The price in December 2001 was about 64¢/lb or \$1.40/kg.

Americium — (the Americas), Am; at. wt. 243; at. no. 95; m.p. 1176 °C; b.p. 2011 °C; sp. gr. 12; valence 2, 3, 4, 5, or 6. Americium was the fourth transuranium element to be discovered; the isotope ^{241}Am was identified by Seaborg, James, Morgan, and Ghiorso late in 1944 at the wartime Metallurgical Laboratory of the University of Chicago as the result of successive neutron capture reactions by plutonium isotopes in a nuclear reactor:

$$^{239}\text{Pu}(n,\gamma) \rightarrow {}^{240}\text{Pu}(n,\gamma) \rightarrow {}^{241}\text{Pu} \xrightarrow{\beta} {}^{241}\text{Am}$$

Since the isotope ^{241}Am can be prepared in relatively pure form by extraction as a decay product over a period of years from strongly neutron-bombarded plutonium, ^{241}Pu, this isotope is used for much of the chemical investigation of this element. Better suited is the isotope ^{243}Am due to its longer half-life (7.37×10^3 years as compared to 432.2 years for ^{241}Am). A mixture of the isotopes ^{241}Am, ^{242}Am, and ^{243}Am can be prepared by intense neutron irradiation of ^{241}Am according to the reactions $^{241}\text{Am}(n, \gamma) \rightarrow {}^{242}\text{Am}(n, \gamma) \rightarrow {}^{243}\text{Am}$. Nearly isotopically pure, ^{243}Am can be prepared by a sequence of neutron bombardments and chemical separations as follows: neutron bombardment of ^{241}Am yields ^{242}Pu by the reactions $^{241}\text{Am}(n, \gamma) \rightarrow {}^{242}\text{Am} \rightarrow {}^{242}\text{Pu}$, after chemical separation the ^{242}Pu can be transformed to ^{243}Am via the reactions $^{242}\text{Pu}(n, \gamma) \rightarrow {}^{243}\text{Pu} \rightarrow {}^{243}\text{Am}$, and the ^{243}Am can be chemically separated. Fairly pure ^{242}Pu can be prepared more simply by very intense neutron irradiation of ^{239}Pu as the result of successive neutron-capture reactions. Seventeen radioactive isotopes and isomers are now recognized. Americium metal has been prepared by reducing the trifluoride with barium vapor at 1000 to 1200 °C or the dioxide by lanthanum metal. The luster of freshly prepared americium metal is white and more silvery than plutonium or neptunium prepared in the same manner. It appears to be more malleable than uranium or neptunium and tarnishes slowly in dry air at room temperature. Americium is thought to exist in two forms: an alpha form which has a double hexagonal close-packed structure and a loose-packed cubic beta form. Americium must be handled with great care to avoid personal contamination. As little as 0.03 μCi of ^{241}Am is the maximum permissible total body burden. The alpha activity from ^{241}Am is about three times that of radium. When gram quantities of ^{241}Am are handled, the intense gamma activity makes exposure a serious problem. Americium dioxide, AmO_2, is the most important oxide. AmF_3, AmF_4, $AmCl_3$, $AmBr_3$, AmI_3, and other compounds have been prepared. The isotope ^{241}Am has been used as a portable source for gamma radiography. It has also been used as a radioactive glass thickness gage for the flat glass industry, and as a source of ionization for smoke detectors. Americium-243 (99%) is available from the Oak Ridge National Laboratory at a cost of about \$750/g plus packing charges.

Antimony — (Gr. *anti* plus *monos* - a metal not found alone), Sb; at. wt. 121.760(1); at. no. 51; m.p. 630.63 °C; b.p. 1587 °C; sp. gr. 6.68 (20 °C); valence 0, −3, +3, or +5. Antimony was recognized in compounds by the ancients and was known as a metal at the beginning of the 17th century and possibly much earlier. It is not abundant, but is found in over 100 mineral species. It is sometimes found native, but more frequently as the sulfide, *stibnite* (Sb_2S_3); it is also found as antimonides of the heavy metals, and as oxides. It is extracted from the sulfide by roasting to the oxide, which is reduced by salt and scrap iron; from its oxides it is also prepared by reduction with carbon. Two allotropic forms of antimony exist: the normal stable, metallic form, and the amorphous gray form. The so-called explosive antimony is an ill-defined material always containing an appreciable amount of halogen; therefore, it no longer warrants consideration as a separate allotrope. The yellow form, obtained by oxidation of *stibine*, SbH_3, is probably impure, and is not a distinct form. Natural antimony is made of two stable isotopes, ^{121}Sb and ^{123}Sb. Forty-five other radioactive isotopes and isomers are now recognized. Metallic antimony is an extremely brittle metal of a flaky, crystalline texture. It is bluish white and has a metallic luster. It is not acted on by air at room temperature, but burns brilliantly when heated with the formation of white fumes of Sb_2O_3. It is a poor conductor of heat and electricity, and has a hardness of 3 to 3.5. Antimony, available commercially with a purity of 99.999 + %, is finding use in semiconductor technology for making infrared detectors, diodes, and Hall-effect devices. Commercial-grade antimony is widely used in alloys with percentages ranging from 1 to 20. It greatly increases the hardness and mechanical strength of lead. Batteries, antifriction alloys, type metal, small arms and tracer bullets, cable sheathing, and minor products use about half the metal produced. Compounds taking up the other half are oxides, sulfides, sodium antimonate, and antimony trichloride. These are used in manufacturing flame-proofing compounds, paints, ceramic enamels, glass, and pottery. Tartar emetic (hydrated potassium antimonyl tartrate) has been used in medicine. Antimony and many of its compounds are toxic. Antimony costs about \$1.30/kg for the commercial metal or about \$12/g (99.999%).

Argon — (Gr. *argos*, inactive), Ar; at. wt. 39.948(1); at. no. 18; m.p. −189.36 °C; b.p. −185.85 °C; t_c −122.28 °C; density 1.7837 g/L. Its presence in air was suspected by Cavendish in 1785, discovered by Lord Rayleigh and Sir William Ramsay in 1894. The gas is prepared by fractionation of liquid air, the atmosphere containing 0.94% argon. The atmosphere of Mars contains 1.6% of ^{40}Ar and 5 p.p.m. of ^{36}Ar. Argon is two and one half times as soluble in water as nitrogen, having about the same solubility as oxygen. It is recognized by the characteristic lines in the red end of the spectrum. It is used in electric light bulbs and in fluorescent tubes at a pressure of about 400 Pa, and in filling photo tubes, glow tubes, etc. Argon is also used as an inert gas shield for arc welding and cutting, as a blanket for the production of titanium and other reactive elements, and as a protective atmosphere for growing silicon and germanium crystals. Argon is colorless and odorless, both as a gas and liquid. It is available in high-purity form. Commercial argon is available at a cost of about 3¢ per cubic foot. Argon is considered to be a very inert gas and is not known to form true chemical compounds, as do krypton, xenon, and radon. However, it does form a hydrate having a dissociation pressure of 105 atm at 0 °C. Ion molecules such as $(ArKr)^+$, $(ArXe)^+$, $(NeAr)^+$ have been observed spectroscopically. Argon also forms a clathrate with β-hydroquinone. This clathrate is stable and can be stored for a considerable time, but a true chemical bond does not exist. Van der Waals' forces act to hold the argon. In August 2000, researchers at the University of Helsinki, Finland reported they made a new argon compound HArF by shining UV light on frozen argon that contained a small amount of HF. Naturally occurring argon is a mixture of three isotopes. Seventeen other radioactive isotopes are now known to exist. Commercial argon is priced at about $70/300 cu. ft. or 8.5 cu. meters.

Arsenic — (L. *arsenicum,* Gr. *arsenikon,* yellow orpiment, identified with *arsenikos,* male, from the belief that metals were different sexes; Arabic, *Az-zernikh,* the orpiment from Persian *zerni-zar,* gold), As; at. wt. 74.92160(2); at. no. 33; valence –3, 0, +3 or +5. Elemental arsenic occurs in two solid modifications: yellow, and gray or metallic, with specific gravities of 1.97, and 5.75, respectively. Gray arsenic, the ordinary stable form, has a triple point of 817 °C and sublimes at 616 °C and has a critical temperature of 1400 °C. Several other allotropic forms of arsenic are reported in the literature. It is believed that Albertus Magnus obtained the element in 1250 A.D. In 1649 Schroeder published two methods of preparing the element. It is found native, in the sulfides *realgar* and *orpiment,* as arsenides and sulfarsenides of heavy metals, as the oxide, and as arsenates. *Mispickel,* arsenopyrite, (FeSAs) is the most common mineral, from which on heating the arsenic sublimes leaving ferrous sulfide. The element is a steel gray, very brittle, crystalline, semimetallic solid; it tarnishes in air, and when heated is rapidly oxidized to arsenous oxide (As_2O_3) with the odor of garlic. Arsenic and its compounds are poisonous. Exposure to arsenic and its compounds should not exceed 0.01 mg/m^3 as elemental As during an 8-h work day. Arsenic is also used in bronzing, pyrotechny, and for hardening and improving the sphericity of shot. The most important compounds are white arsenic (As_2O_3), the sulfide, Paris green $3Cu(AsO_2)_2 \cdot Cu(C_2H_3O_2)_2$, calcium arsenate, and lead arsenate; the last three have been used as agricultural insecticides and poisons. Marsh's test makes use of the formation and ready decomposition of arsine (AsH_3). Arsenic is available in high-purity form. It is finding increasing uses as a doping agent in solid-state devices such as transistors. Gallium arsenide is used as a laser material to convert electricity directly into coherent light. Natural arsenic is made of one isotope ^{75}As. Thirty other radioactive isotopes and isomers are known. Arsenic (99%) costs about $75/50g. Purified arsenic (99.9995%) costs about $50/g.

Astatine — (Gr. *astatos,* unstable), At; at. wt. (210); at. no. 85; m.p. 302 °C; valence probably 1, 3, 5, or 7. Synthesized in 1940 by D. R. Corson, K. R. MacKenzie, and E. Segre at the University of California by bombarding bismuth with alpha particles. The longest-lived isotope, ^{210}At, has a half-life of only 8.1 hours. Thirty-six other isotopes and isomers are now known. Minute quantities of ^{215}At, ^{218}At, and ^{219}At exist in equilibrium in nature with naturally occurring uranium and thorium isotopes, and traces of ^{217}At are in equilibrium with ^{233}U and ^{239}Np resulting from interaction of thorium and uranium with naturally produced neutrons. The total amount of astatine present in the Earth's crust, however, is probably less than 1 oz. Astatine can be produced by bombarding bismuth with energetic alpha particles to obtain the relatively long-lived $^{209-211}$At, which can be distilled from the target by heating it in air. Only about 0.05 μg of astatine has been prepared to date. The "time of flight" mass spectrometer has been used to confirm that this highly radioactive halogen behaves chemically very much like other halogens, particularly iodine. The interhalogen compounds AtI, AtBr, and AtCl are known to form, but it is not yet known if astatine forms diatomic astatine molecules. HAt and CH_3At (methyl astatide) have been detected. Astatine is said to be more metallic that iodine, and, like iodine, it probably accumulates in the thyroid gland.

Barium — (Gr. *barys,* heavy), Ba; at. wt. 137.327(7), at. no. 56; m.p. 727 °C; b.p. 1897 °C; sp. gr. 3.62 (20 °C); valence 2. Baryta was distinguished from lime by Scheele in 1774; the element was discovered by Sir Humphrey Davy in 1808. It is found only in combination with other elements, chiefly in *barite* or *heavy spar* (sulfate) and *witherite* (carbonate) and is prepared by electrolysis of the chloride. Large deposits of barite are found in China, Germany, India, Morocco, and in the U.S. Barium is a metallic element, soft, and when pure is silvery white like lead; it belongs to the alkaline earth group, resembling calcium chemically. The metal oxidizes very easily and should be kept under petroleum or other suitable oxygen-free liquids to exclude air. It is decomposed by water or alcohol. The metal is used as a "getter" in vacuum tubes. The most important compounds are the peroxide (BaO_2), chloride, sulfate, carbonate, nitrate, and chlorate. Lithopone, a pigment containing barium sulfate and zinc sulfide, has good covering power, and does not darken in the presence of sulfides. The sulfate, as permanent white or *blanc fixe,* is also used in paint, in X-ray diagnostic work, and in glassmaking. *Barite* is extensively used as a weighting agent in oilwell drilling fluids, and also in making rubber. The carbonate has been used as a rat poison, while the nitrate and chlorate give green colors in pyrotechny. The impure sulfide phosphoresces after exposure to the light. The compounds and the metal are not expensive. Barium metal (99.2 + % pure) costs about $3/g. All barium compounds that are water or acid soluble are poisonous. Naturally occurring barium is a mixture of seven stable isotopes. Thirty-six other radioactive isotopes and isomers are known to exist.

Berkelium — (*Berkeley,* home of the University of California), Bk; at. wt. (247); at. no. 97; m.p. 996 °C; valence 3 or 4; sp. gr. 14 (est.). Berkelium, the eighth member of the actinide transition series, was discovered in December 1949 by Thompson, Ghiorso, and Seaborg, and was the fifth transuranium element synthesized. It was produced by cyclotron bombardment of milligram amounts of ^{241}Am with helium ions at Berkeley, California. The first isotope produced had a mass number of 243 and decayed with a half-life of 4.5 hours. Thirteen isotopes are now known and have been synthesized. The existence of ^{249}Bk, with a half-life of 320 days, makes it feasible to isolate berkelium in weighable amounts so that its properties can be investigated with macroscopic quantities. One of the first visible amounts of a pure berkelium compound, berkelium chloride, was produced in 1962. It weighed 3 billionth of a gram. Berkelium probably has not yet been prepared in elemental form, but it is expected to be a silvery metal, easily soluble in dilute mineral acids, and readily oxidized by air or oxygen at elevated temperatures to form the oxide. X-ray diffraction methods have been used to identify the following compounds: BkO_2, BkO_3, BkF_3, BkCl, and BkOCl. As with other actinide elements, berkelium tends to accumulate in the skeletal system. The maximum permissible body burden of ^{249}Bk in the human skeleton is about 0.0004 μg. Because of its rarity, berkelium presently has no commercial or technological use. Berkelium most likely resembles terbium with respect to chemical properties. Berkelium-249 is available from O.R.N.L. at a cost of $185/μg plus packing charges.

Beryllium — (Gr. *beryllos, beryl;* also called Glucinium or Glucinum, Gr. *glykys,* sweet), Be; at. wt. 9.012182(3); at no. 4; m.p. 1287 °C; b.p. 2471 °C; sp. gr. 1.848 (20 °C); valence 2. Discovered as the oxide by Vauquelin in beryl and in emeralds in 1798. The metal was isolated in 1828 by Wohler and by Bussy independently by the action of potassium on beryllium chloride. Beryllium is found in some 30 mineral species, the most important of which are *bertrandite, beryl, chrysoberyl,* and *phenacite. Aquamarine* and *emerald* are precious forms of *beryl.* Beryllium minerals are found in the U.S., Brazil, Russia, Kazakhstan, and elsewhere. Colombia is known for its emeralds. *Beryl* ($3BeO \cdot Al_2O_3 \cdot 6SiO_2$) and *bertrandite* ($4BeO \cdot 2SiO_2 \cdot H_2O$) are the most important commercial sources of the element and its compounds. Most of the metal is now prepared by reducing beryllium fluoride with magnesium metal. Beryllium metal did not become readily available to industry until 1957. The metal, steel gray in color, has many desirable properties. It is one of the lightest of all metals, and has one of the highest melting points of the light metals. Its modulus of elasticity is about one third greater than that of steel. It resists attack by concentrated nitric acid, has excellent thermal conductivity, and is nonmagnetic. It has a high permeability to X-rays, and when bombarded by alpha particles, as from radium or polonium, neutrons are produced in the ratio of about 30 neutrons/million alpha particles. At ordinary temperatures beryllium resists oxidation in air, although its ability to scratch glass is probably due to the formation of a thin layer of the oxide. Beryllium is used as an alloying agent in producing beryllium copper, which is extensively used for springs, electrical contacts, spot-welding electrodes, and nonsparking tools. It has found application as a structural material for high-speed aircraft, missiles, spacecraft, and communication satellites. It is being used in the windshield frame, brake discs, support beams, and other structural components of the space shuttle. Because beryllium is relatively transparent to X-rays, ultra-thin Be-foil is finding use in X-ray lithography for reproduction of microminiature integrated circuits. Natural beryllium is made of ^{9}Be and is stable. Eight other radioactive isotopes are known.

Beryllium is used in nuclear reactors as a reflector or moderator for it has a low thermal neutron absorption cross section. It is used in gyroscopes, computer parts, and instruments where lightness, stiffness, and dimensional stability are required. The oxide has a very high melting point and is also used in nuclear work and ceramic applications. Beryllium and its salts are toxic and should be handled with the greatest of care. Beryllium and its compounds should not be tasted to verify the sweetish nature of beryllium (as did early experimenters). The metal, its alloys, and its salts can be handled safely if certain work codes are observed, but no attempt should be made to work with beryllium before becoming familiar with proper safeguards. Beryllium metal is available at a cost of about $5/g (99.5% pure).

Bismuth — (Ger. *Weisse Masse,* white mass; later *Wisuth* and *Bisemutum*), Bi; at. wt. 208.98040(1); at. no. 83; m.p. 271.4 °C; b.p. 1564 °C; sp. gr. 9.79 (20 °C); valence 3 or 5. In early times bismuth was confused with tin and lead. Claude Geoffroy the Younger showed it to be distinct from lead in 1753. It is a white crystalline, brittle metal with a pinkish tinge. It occurs native. The most important ores are *bismuthinite* or bismuth glance (Bi_2S_3) and *bismite* (Bi_2O_3). Peru, Japan, Mexico, Bolivia, and Canada are major bismuth producers. Much of the bismuth produced in the U.S. is obtained as a by-product in refining lead, copper, tin, silver, and gold ores. Bismuth is the most diamagnetic of all metals, and the thermal conductivity is lower than any metal, except mercury. It has a high electrical resistance, and has the highest Hall effect of any metal (i.e., greatest increase in electrical resistance when placed in a magnetic field). "Bismanol" is a permanent magnet of high coercive force, made of MnBi, by the U.S. Naval Surface Weapons Center. Bismuth expands 3.32% on solidification. This property makes bismuth alloys particularly suited to the making of sharp castings of objects subject to damage by high temperatures. With other metals such as tin, cadmium, etc., bismuth forms low-melting alloys that are extensively used for safety devices in fire detection and extinguishing systems. Bismuth is used in producing malleable irons and is finding use as a catalyst for making acrylic fibers. When bismuth is heated in air it burns with a blue flame, forming yellow fumes of the oxide. The metal is also used as a thermocouple material, and has found application as a carrier for U^{235} or U^{233} fuel in atomic reactors. Its soluble salts are characterized by forming insoluble basic salts on the addition of water, a property sometimes used in detection work. Bismuth oxychloride is used extensively in cosmetics. Bismuth subnitrate and subcarbonate are used in medicine. Natural bismuth contains only one isotope ^{209}Bi. Forty-four isotopes and isomers of bismuth are known. Bismuth metal (99.5%) costs about $250/kg.

Bohrium — (Named after Niels Bohr [1885–1962], Danish atomic and nuclear physicist.) Bh; at. wt. [264]. at. no. 107. Bohrium is expected to have chemical properties similar to rhenium. This element was synthesized and unambiguously identified in 1981 using the Universal Linear Accelerator (UNILAC) at the Gesellschaft für Schwerionenforschung (G.S.I.) in Darmstadt, Germany. The discovery team was led by Armbruster and Münzenberg. The reaction producing the element was proposed and applied earlier by a Dubna Group led by Oganessian in 1976. A target of ^{209}Bi was bombarded by a beam of ^{54}Cr ions. In 1983 experiments at Dubna using the 157-inch cyclotron, produced $^{262}107$ by the reaction $^{209}Bi + {}^{54}Cr$. The alpha decay of ^{246}Cf, the sixth member in the decay chain of $^{262}107$, served to establish a 1-neutron reaction channel. The IUPAC adopted the name ***Bohrium*** with the symbol Bh for Element 107 in August 1997. Five isotopes of bohrium are now recognized. One isotope of bohrium appears to have a relatively long life of 15 seconds. Work on this relatively long-lived isotope has been performed with the 88-inch cyclotron at the Lawrence-Berkeley National Laboratory.

Boron — (Ar. *Buraq,* Pers. *Burah*), B; at. wt. 10.811(7); at. no. 5; m.p. 2075 °C; b.p. 4000 °C; sp. gr. of crystals 2.34, of amorphous variety 2.37; valence 3. Boron compounds have been known for thousands of years, but the element was not discovered until 1808 by Sir Humphry Davy and by Gay-Lussac and Thenard. The element is not found free in nature, but occurs as orthoboric acid usually in certain volcanic spring waters and as borates in *borax* and *colemanite. Ulexite,* another boron mineral, is interesting as it is nature's own version of "fiber optics." Important sources of boron are the ores *rasorite (kernite)* and *tincal (borax ore).* Both of these ores are found in the Mojave Desert. *Tincal* is the most important source of boron from the Mojave. Extensive *borax* deposits are also found in Turkey. Boron exists naturally as 19.9% ^{10}B isotope and 80.1% ^{11}B isotope. Ten other isotopes of boron are known. High-purity crystalline boron may be prepared by the vapor phase reduction of boron trichloride or tribromide with hydrogen on electrically heated filaments. The impure, or amorphous, boron, a brownish-black powder, can be obtained by heating the trioxide with magnesium powder. Boron of 99.9999% purity has been produced and is available commercially. Elemental boron has an energy band gap of 1.50 to 1.56 eV, which is higher than that of either silicon or germanium. It has interesting optical characteristics, transmitting portions of the infrared, and is a poor conductor of electricity at room temperature, but a good conductor at high temperature. Amorphous boron is used in pyrotechnic flares to provide a distinctive green color, and in rockets as an igniter. By far the most commercially important boron compound in terms of dollar sales is $Na_2B_4O_7 \cdot 5H_2O$. This pentahydrate is used in very large quantities in the manufacture of insulation fiberglass and sodium perborate bleach. Boric acid is also an important boron compound with major markets in textile fiberglass and in cellulose insulation as a flame retardant. Next in order of importance is borax ($Na_2B_4O_7 \cdot 10H_2O$) which is used principally in laundry products. Use of borax as a mild antiseptic is minor in terms of dollars and tons. Boron compounds are also extensively used in the manufacture of borosilicate glasses. The isotope boron-10 is used as a control for nuclear reactors, as a shield for nuclear radiation, and in instruments used for detecting neutrons. Boron nitride has remarkable properties and can be used to make a material as hard as diamond. The nitride also behaves like an electrical insulator but conducts heat like a metal. It also has lubricating properties similar to graphite. The hydrides are easily oxidized with considerable energy liberation, and have been studied for use as rocket fuels. Demand is increasing for boron filaments, a high-strength, lightweight material chiefly employed for advanced aerospace structures. Boron is similar to carbon in that it has a capacity to form stable covalently bonded molecular networks. Carboranes, metalloboranes, phosphacarboranes, and other families comprise thousands of compounds. Crystalline boron (99.5%) costs about \$6/g. Amorphous boron (94–96%) costs about \$1.50/g. Elemental boron and the borates are not considered to be toxic, and they do not require special care in handling. However, some of the more exotic boron hydrogen compounds are definitely toxic and do require care.

Bromine — (Gr. *bromos,* stench), Br; at. wt. 79.904(1); at. no. 35; m.p. −7.2 °C; b.p. 58.8 °C; t_c 315 °C; density of gas 7.59 g/l, liquid 3.12 (20 °C); valence 1, 3, 5, or 7. Discovered by Balard in 1826, but not prepared in quantity until 1860. A member of the halogen group of elements, it is obtained from natural brines from wells in Michigan and Arkansas. Little bromine is extracted today from seawater, which contains only about 85 ppm. Bromine is the only liquid nonmetallic element. It is a heavy, mobile, reddish-brown liquid, volatilizing readily at room temperature to a red vapor with a strong disagreeable odor, resembling chlorine, and having a very irritating effect on the eyes and throat; it is readily soluble in water or carbon disulfide, forming a red solution, is less active than chlorine but more so than iodine; it unites readily with many elements and has a bleaching action; when spilled on the skin it produces painful sores. It presents a serious health hazard, and maximum safety precautions should be taken when handling it. Much of the bromine output in the U.S. was used in the production of ethylene dibromide, a lead scavenger used in making gasoline antiknock compounds. Lead in gasoline, however, has been drastically reduced, due to environmental considerations. This will greatly affect future production of bromine. Bromine is also used in making fumigants, flameproofing agents, water purification compounds, dyes, medicinals, sanitizers, inorganic bromides for photography, etc. Organic bromides are also important. Natural bromine is made of two isotopes, ^{79}Br and ^{81}Br. Thirty-four isotopes and isomers are known. Bromine (99.8%) costs about \$70/kg.

Cadmium — (L. *cadmia;* Gr. *kadmeia* - ancient name for calamine, zinc carbonate), Cd; at. wt. 112.411(8); at. no. 48; m.p. 321.07 °C; b.p. 767 °C; sp. gr. 8.69 (20 °C); valence 2. Discovered by Stromeyer in 1817 from an impurity in zinc carbonate. Cadmium most often occurs in small quantities associated with zinc ores, such as *sphalerite* (ZnS). *Greenockite* (CdS) is the only mineral of any consequence bearing cadmium. Almost all cadmium is obtained as a by-product in the treatment of zinc, copper, and lead ores. It is a soft, bluish-white metal which is easily cut with a knife. It is similar in many respects to zinc. It is a component of some of the lowest melting alloys; it is used in bearing alloys with low coefficients of friction and great resistance to fatigue; it is used extensively in electroplating, which accounts for about 60% of its use. It is also used in many types of solder, for standard E.M.F. cells, for Ni-Cd batteries, and as a barrier to control atomic fission. The market for Ni-Cd batteries is expected to grow significantly. Cadmium compounds are used in black and white television phosphors and in blue and green phosphors for color TV tubes. It forms a number of salts, of which the sulfate is most common; the sulfide is used as a yellow pigment. Cadmium and solutions of its compounds are toxic. Failure to appreciate the toxic properties of cadmium may cause workers to be unwittingly exposed to dangerous fumes. Some silver solders, for example, contain cadmium and should be handled with care. Serious toxicity problems have been found from long-term exposure and work with cadmium plating baths. Cadmium is present in certain phosphate rocks. This has raised concerns that the long-term use of certain phosphate fertilizers might pose a health hazard from levels of cadmium that might enter the food chain. In 1927 the International Conference on Weights and Measures redefined the meter in terms of the wavelength of the red cadmium spectral line (i.e., 1 m = 1,553,164.13 wavelengths). This definition has been changed (see under Krypton). The current price of cadmium is about 50¢/g (99.5%). It is available in high purity form for about \$550/kg. Natural cadmium is made of eight isotopes. Thirty-four other isotopes and isomers are now known and recognized.

Calcium — (L. *calx,* lime), Ca; at. wt. 40.078(4); at. no. 20; m.p. 842 °C; b.p. 1484 °C; sp. gr. 1.54 (20 °C); valence 2. Though lime was prepared by the Romans in the first century under the name calx, the metal was not discovered until 1808. After learning that Berzelius and Pontin prepared calcium amalgam by electrolyzing lime in mercury, Davy was able to isolate the impure metal. Calcium is a metallic element, fifth in abundance in the Earth's crust, of which it forms more than 3%. It is an essential constituent of leaves, bones, teeth, and shells. Never found in nature uncombined, it occurs abundantly as *limestone* ($CaCO_3$), *gypsum* ($CaSO_4 \cdot 2H_2O$), and *fluorite* (CaF_2); *apatite* is the fluorophosphate or chlorophosphate of calcium. The metal has a silvery color, is rather hard, and is prepared by electrolysis of the fused chloride to which calcium fluoride is added to lower the melting point. Chemically it is one of the alkaline earth elements; it readily forms a white coating of oxide in air, reacts with water, burns with a yellow-red flame, largely forming the oxide. The metal is used as a reducing agent in preparing other metals such as thorium, uranium, zirconium, etc., and is used as a deoxidizer, desulfurizer, and inclusion modifier for various ferrous and nonferrous alloys. It is also used as an alloying agent for aluminum, beryllium, copper, lead, and magnesium alloys, and serves as a "getter" for residual gases in vacuum tubes. Its natural and prepared compounds are widely used. Quicklime (CaO), made by heating limestone and changed into slaked lime by the careful addition of water, is the great cheap base of the chemical industry with countless uses. Mixed with sand it hardens as mortar and plaster by taking up carbon dioxide from the air. Calcium from limestone is an important element in Portland cement. The solubility of the carbonate in water containing carbon dioxide causes the formation of caves with stalactites and stalagmites and is responsible for hardness in water. Other important compounds are the carbide (CaC_2), chloride ($CaCl_2$), cyanamide ($CaCN_2$), hypochlorite ($Ca(OCl)_2$), nitrate ($Ca(NO_3)_2$), and sulfide (CaS). Calcium sulfide is phosphorescent after being exposed to light. Natural calcium contains six isotopes. Sixteen other radioactive isotopes are known. Metallic calcium (99.5%) costs about $200/kg.

Californium — (State and University of California), Cf; at. wt. (251); m.p. 900 °C; sp. gr. 15.1; at. no. 98. Californium, the sixth transuranium element to be discovered, was produced by Thompson, Street, Ghioirso, and Seaborg in 1950 by bombarding microgram quantities of ^{242}Cm with 35 MeV helium ions in the Berkeley 60-inch cyclotron. Californium (III) is the only ion stable in aqueous solutions, all attempts to reduce or oxidize californium (III) having failed. The isotope ^{249}Cf results from the beta decay of ^{249}Bk while the heavier isotopes are produced by intense neutron irradiation by the reactions:

$$^{249}\mathrm{Bk}(n,\gamma)\rightarrow{}^{250}\mathrm{Bk}\xrightarrow{\beta}{}^{250}\mathrm{Cf}\text{ and }{}^{249}\mathrm{Cf}(n,\gamma)\rightarrow{}^{250}\mathrm{Cf}$$

followed by

$$^{250}\mathrm{Cf}(n,\gamma)\rightarrow{}^{251}\mathrm{Cf}(n,\gamma)\rightarrow{}^{252}\mathrm{Cf}$$

The existence of the isotopes ^{249}Cf, ^{250}Cf, ^{251}Cf, and ^{252}Cf makes it feasible to isolate californium in weighable amounts so that its properties can be investigated with macroscopic quantities. Californium-252 is a very strong neutron emitter. One microgram releases 170 million neutrons per minute, which presents biological hazards. Proper safeguards should be used in handling californium. Twenty isotopes of californium are now recognized. ^{249}Cf and ^{252}Cf have half-lives of 351 years and 900 years, respectively. In 1960 a few tenths of a microgram of californium trichloride, $CfCl_3$, californium oxychloride, CfOCl, and californium oxide, Cf_2O_3, were first prepared. Reduction of californium to its metallic state has not yet been accomplished. Because californium is a very efficient source of neutrons, many new uses are expected for it. It has already found use in neutron moisture gages and in well-logging (the determination of water and oil-bearing layers). It is also being used as a portable neutron source for discovery of metals such as gold or silver by on-the-spot activation analysis. ^{252}Cf is now being offered for sale by the Oak Ridge National Laboratory (O.R.N.L.) at a cost of $60/μg and ^{249}Cf at a cost of $185/μg plus packing charges. It has been suggested that californium may be produced in certain stellar explosions, called *supernovae,* for the radioactive decay of ^{254}Cf (55-day half-life) agrees with the characteristics of the light curves of such explosions observed through telescopes. This suggestion, however, is questioned. Californium is expected to have chemical properties similar to dysprosium.

Carbon — (L. *carbo,* charcoal), C; at. wt. 12.0107(8); at. no. 6; sublimes at 3825 °C; triple point (graphite-liquid-gas), 4489 °C; sp. gr. amorphous 1.8 to 2.1, graphite 1.9 to 2.3, diamond 3.15 to 3.53 (depending on variety); gem diamond 3.513 (25 °C); valence 2, 3, or 4. Carbon, an element of prehistoric discovery, is very widely distributed in nature. It is found in abundance in the sun, stars, comets, and atmospheres of most planets. Carbon in the form of microscopic diamonds is found in some meteorites. Natural diamonds are found in *kimberlite* or *lamporite* of ancient formations called "pipes," such as found in South Africa, Arkansas, and elsewhere. Diamonds are now also being recovered from the ocean floor off the Cape of Good Hope. About 30% of all industrial diamonds used in the U.S. are now made synthetically. The energy of the sun and stars can be attributed at least in part to the well-known carbon-nitrogen cycle. Carbon is found free in nature in three allotropic forms: amorphous, graphite, and diamond. Graphite is one of the softest known materials while diamond is one of the hardest. Graphite exists in two forms: alpha and beta. These have identical physical properties, except for their crystal structure. Naturally occurring graphites are reported to contain as much as 30% of the rhombohedral (beta) form, whereas synthetic materials contain only the alpha form. The hexagonal alpha type can be converted to the beta by mechanical treatment, and the beta form reverts to the alpha on heating it above 1000 °C. Of recent interest is the discovery of all-carbon molecules, known as "buckyballs" or fullerenes, which have a number of unusual properties. These interesting molecules, consisting of 60 or 70 carbon atoms linked together, seem capable of withstanding great pressure and trapping foreign atoms inside their network of carbon. They are said to be capable of magnetism and superconductivity and have potential as a nonlinear optical material. Buckyball films are reported to remain superconductive at temperatures as high as 45 K. In combination, carbon is found as carbon dioxide in the atmosphere of the Earth and dissolved in all natural waters. It is a component of great rock masses in the form of carbonates of calcium (limestone), magnesium, and iron. Coal, petroleum, and natural gas are chiefly hydrocarbons. Carbon is unique among the elements in the vast number and variety of compounds it can form. With hydrogen, oxygen, nitrogen, and other elements, it forms a very large number of compounds, carbon atom often being linked to carbon atom. There are close to ten million known carbon compounds, many thousands of which are vital to organic and life processes. Without carbon, the basis for life would be impossible. While it has been thought that silicon might take the place of carbon in forming a host of similar compounds, it is now not possible to form stable compounds with very long chains of silicon atoms. The atmosphere of Mars contains 96.2% CO_2. Some of the most important compounds of carbon are carbon dioxide (CO_2), carbon monoxide (CO), carbon disulfide (CS_2), chloroform ($CHCl_3$), carbon tetrachloride (CCl_4), methane (CH_4), ethylene (C_2H_4), acetylene (C_2H_2), benzene (C_6H_6), ethyl alcohol (C_2H_5OH), acetic acid (CH_3COOH), and their derivatives. Carbon has fifteen isotopes. Natural carbon consists of 98.89% ^{12}C and 1.11% ^{13}C. In 1961 the International Union of Pure and Applied Chemistry adopted the isotope carbon-12 as the basis for atomic weights. Carbon-14, an isotope with a half-life of 5715 years, has been widely used to date such materials as wood, archeological specimens, etc. A new brittle form of carbon, known as "glassy carbon," has been developed. It can be obtained with high purity. It has a high resistance to corrosion, has good thermal stability, and is structurally impermeable to both gases and liquids. It has a randomized structure, making it useful in ultra-high technology applications, such as crystal growing, crucibles for high-temperature use, etc. Glassy carbon is available at a cost of about $35/10g. Fullerene powder is available at a cost of about $55/10mg (99%$C_{10}$). Diamond powder (99.9%) costs about $40/g.

Cerium — (named for the asteroid *Ceres,* which was discovered in 1801 only 2 years before the element), Ce; at. wt. 140.116(1); at. no. 58; m.p. 799 °C; b.p. 3443 °C; sp. gr. 6.770 (25 °C); valence 3 or 4. Discovered in 1803 by Klaproth and by Berzelius and Hisinger; metal prepared by Hillebrand and Norton in 1875. Cerium is the most abundant of the metals of the so-called rare earths. It is found in a number of minerals including *allanite* (also known as *orthite*), *monazite, bastnasite, cerite,* and *samarskite.* Monazite and bastnasite are presently the two most important sources of cerium. Large deposits of monazite found on the beaches of Travancore, India, in river sands in Brazil, and deposits of *allanite* in the western United States, and *bastnasite* in Southern California will supply cerium, thorium, and the other rare-earth metals for many years to come. Metallic cerium is prepared by metallothermic reduction techniques, such as by reducing cerous fluoride with calcium, or by electrolysis of molten cerous chloride or other cerous halides. The metallothermic technique is used to produce high-purity cerium. Cerium is especially interesting because of its variable electronic structure. The energy of the inner 4f level is nearly the same as that of the outer or valence electrons, and only small amounts of energy are required to change the relative occupancy of these electronic levels. This gives rise to dual valency states. For example, a volume change of about 10% occurs when cerium is subjected to high pressures or low temperatures. It appears that the valence changes from about 3 to 4 when it is cooled or compressed. The low temperature behavior of cerium is complex. Four allotropic modifications are thought to exist: cerium at room temperature and at atmospheric pressure is known as γ cerium. Upon cooling to −16 °C, γ cerium changes to β cerium. The remaining γ cerium starts to change to α cerium when cooled to −172 °C, and the transformation is complete at −269 °C. α Cerium has a density of 8.16; δ cerium exists above 726 °C. At atmospheric pressure, liquid cerium is more dense than its solid form at the melting point. Cerium is an iron-gray lustrous metal. It is malleable, and oxidizes very readily at room temperature, especially in moist air. Except for europium, cerium is the most reactive of the "rare-earth" metals. It slowly decomposes in cold water, and rapidly in hot water. Alkali solutions and dilute and concentrated acids attack the metal rapidly. The pure metal is likely to ignite if scratched with a knife. Ceric salts are orange red or yellowish; cerous salts are usually white. Cerium is a component of misch metal, which is extensively used in the manufacture of pyrophoric alloys for cigarette lighters, etc. Natural cerium is stable and contains four isotopes. Thirty-two other radioactive isotopes and isomers are known. While cerium is not radioactive, the impure commercial grade may contain traces of thorium, which is radioactive. The oxide is an important constituent of incandescent gas mantles and it is emerging as a hydrocarbon catalyst in "self-cleaning" ovens. In this application it can be incorporated into oven walls to prevent the collection of cooking residues. As ceric sulfate it finds extensive use as a volumetric oxidizing agent in quantitative analysis. Cerium compounds are used in the manufacture of glass, both as a component and as a decolorizer. The oxide is finding increased use as a glass polishing agent instead of rouge, for it is much faster than rouge in polishing glass surfaces. Cerium compounds are finding use in automobile exhaust catalysts. Cerium is also finding use in making permanent magnets. Cerium, with other rare earths, is used in carbon-arc lighting, especially in the motion picture industry. It is also finding use as an important catalyst in petroleum refining and in metallurgical and nuclear applications. In small lots, cerium costs about $5/g (99.9%).

Cesium — (L. *caesius,* sky blue), Cs; at. wt. 132.9054519(2); at. no. 55; m.p. 28.44 °C; b.p. 671 °C; sp. gr. 1.873 (20 °C); valence 1. Cesium was discovered spectroscopically by Bunsen and Kirchhoff in 1860 in mineral water from Durkheim. Cesium, an alkali metal, occurs in *lepidolite, pollucite* (a hydrated silicate of aluminum and cesium), and in other sources. One of the world's richest sources of cesium is located at Bernic Lake, Manitoba. The deposits are estimated to contain 300,000 tons of pollucite, averaging 20% cesium. It can be isolated by electrolysis of the fused cyanide and by a number of other methods. Very pure, gas-free cesium can be prepared by thermal decomposition of cesium azide. The metal is characterized by a spectrum containing two bright lines in the blue along with several others in the red, yellow, and green. It is silvery [illegible] soft, and ductile. It is the most electropositive and most alkaline element. Cesium, gallium, and mercury are the only three metals that are liquid at room temperature. Cesium reacts explosively with cold water, and reacts with ice at temperatures above −116 °C. Cesium hydroxide, the strongest base known, attacks glass. Because of its great affinity for oxygen the metal is used as a "getter" in electron tubes. It is also used in photoelectric cells, as well as a catalyst in the hydrogenation of certain organic compounds. The metal has recently found application in ion propulsion systems. Cesium is used in atomic clocks, which are accurate to 5 s in 300 years. A second of time is now defined as being the duration of 9,192,631,770 periods of the radiation corresponding to the transition between the two hyper-fine levels of the ground state of the cesium-133 atom. Its chief compounds are the chloride and the nitrate. Cesium has 52 isotopes and isomers with masses ranging from 112 to 148. The present price of cesium is about $50/g (99.98%) sealed in a glass ampoule.

Chlorine — (Gr. *chloros,* greenish yellow), Cl; at. wt. 35.453(2); at. no. 17; m.p. −101.5 °C; b.p. −34.04 °C; t_c 143.8 °C; density 3.214 g/L; sp. gr. 1.56 (−33.6 °C); valence 1, 3, 5, or 7. Discovered in 1774 by Scheele, who thought it contained oxygen; named in 1810 by Davy, who insisted it was an element. In nature it is found in the combined state only, chiefly with sodium as common salt (NaCl), *carnallite* ($KMgCl_3 \cdot 6H_2O$), and *sylvite* (KCl). It is a member of the halogen (salt-forming) group of elements and is obtained from chlorides by the action of oxidizing agents and more often by electrolysis; it is a greenish-yellow gas, combining directly with nearly all elements. At 10 °C one volume of water dissolves 3.10 volumes of chlorine, at 30 °C only 1.77 volumes. Chlorine is widely used in making many everyday products. It is used for producing safe drinking water the world over. Even the smallest water supplies are now usually chlorinated. It is also extensively used in the production of paper products, dyestuffs, textiles, petroleum products, medicines, antiseptics, insecticides, foodstuffs, solvents, paints, plastics, and many other consumer products. Most of the chlorine produced is used in the manufacture of chlorinated compounds for sanitation, pulp bleaching, disinfectants, and textile processing. Further use is in the manufacture of chlorates, chloroform, carbon tetrachloride, and in the extraction of bromine. Organic chemistry demands much from chlorine, both as an oxidizing agent and in substitution, since it often brings desired properties in an organic compound when substituted for hydrogen, as in one form of synthetic rubber. Chlorine is a respiratory irritant. The gas irritates the mucous membranes and the liquid burns the skin. As little as 3.5 ppm can be detected as an odor, and 1000 ppm is likely to be fatal after a few deep breaths. It was used as a war gas in 1915. Natural chlorine contains two isotopes. Twenty other isotopes and isomers are known.

Chromium — (Gr. *chroma,* color), Cr; at. wt. 51.9961(6); at. no. 24; m.p. 1907 °C; b.p. 2671 °C; sp. gr. 7.15 (20 °C); valence chiefly 2, 3, or 6. Discovered in 1797 by Vauquelin, who prepared the metal the next year, chromium is a steel-gray, lustrous, hard metal that takes a high polish. The principal ore is *chromite* ($FeCr_2O_4$), which is found in Zimbabwe, Russia, South Africa, Turkey, Iran, Albania, Finland, Democratic Republic of Madagascar, the Philippines, and elsewhere. The U.S. has no appreciable chromite ore reserves. The metal is usually produced by reducing the oxide with aluminum. Chromium is used to harden steel, to manufacture stainless steel, and to form many useful alloys. Much is used in plating to produce a hard, beautiful surface and to prevent corrosion. Chromium is used to give glass an emerald green color. It finds wide use as a catalyst. All compounds of chromium are colored; the most important are the chromates of sodium and potassium (K_2CrO_4) and the dichromates ($K_2Cr_2O_7$) and the potassium and ammonium chrome alums, as $KCr(SO_4)_2 \cdot 12H_2O$. The dichromates are used as oxidizing agents in quantitative analysis, also in tanning leather. Other compounds are of industrial value; lead chromate is chrome yellow, a valued pigment. Chromium compounds are used in the textile industry as mordants, and by the aircraft and other industries for anodizing aluminum. The refractory industry has found chromite useful for forming bricks and shapes, as it has a high melting point, moderate thermal expansion, and stability of crystalline structure. Chromium is an essential trace element for human health. Many chromium compounds, however, are acutely or chronically toxic, and some are carcinogenic. They should be handled with proper safeguards. Natural chromium contains four isotopes. Twenty other isotopes are known. Chromium metal (99.95%) costs about $1000/kg. Commercial grade chromium (99%) costs about $75/kg.

Cobalt — (*Kobald,* from the German, goblin or evil spirit, *cobalos,* Greek, mine), Co; at. wt. 58.933195(5); at. no. 27; m.p. 1495 °C; b.p. 2927 °C; sp. gr. 8.9 (20 °C); valence 2 or 3. Discovered by Brandt about 1735. Cobalt occurs in the mineral *cobaltite, smaltite,* and *erythrite,* and is often associated with nickel, silver, lead, copper, and iron ores, from which it is most frequently obtained as a by-product. It is also present in meteorites. Important ore deposits are found in Congo-Kinshasa, Australia, Zambia, Russia, Canada, and elsewhere. The U.S. Geological Survey has announced that the bottom of the north central Pacific Ocean may have cobalt-rich deposits at relatively shallow depths in waters close to the Hawaiian Islands and other U.S. Pacific territories. Cobalt is a brittle, hard metal, closely resembling iron and nickel in appearance. It has a magnetic permeability of about two thirds that of iron. Cobalt tends to exist as a mixture of two allotropes over a wide temperature range; the β-form predominates below 400 °C, and the α above that temperature. The transformation is sluggish and accounts in part for the wide variation in reported data on physical properties of cobalt. It is alloyed with iron, nickel and other metals to make Alnico, an alloy of unusual magnetic strength with many important uses. Stellite alloys, containing cobalt, chromium, and tungsten, are used for high-speed, heavy-duty, high-temperature cutting tools, and for dies. Cobalt is also used in other magnet steels and stainless steels, and in alloys used in jet turbines and gas turbine generators. The metal is used in electroplating because of its appearance, hardness, and resistance to oxidation. The salts have been used for centuries for the production of brilliant and permanent blue colors in porcelain, glass, pottery, tiles, and enamels. It is the principal ingredient in Sevre's and Thenard's blue. A solution of the chloride ($CoCl_2 \cdot 6H_2O$) is used as sympathetic ink. The cobalt ammines are of interest; the oxide and the nitrate are important. Cobalt carefully used in the form of the chloride, sulfate, acetate, or nitrate has been found effective in correcting a certain mineral deficiency disease in animals. Soils should contain 0.13 to 0.30 ppm of cobalt for proper animal nutrition. Cobalt is found in Vitamin B-12, which is essential for human nutrition. Cobalt of 99.9+% purity is priced at about $250/kg. Cobalt-60, an artificial isotope, is an important gamma ray source, and is extensively used as a tracer and a radiotherapeutic agent. Single compact sources of Cobalt-60 vary from about $1 to $10/curie, depending on quantity and specific activity. Thirty isotopes and isomers of cobalt are known.

Columbium — See Niobium.

Copper — (L. *cuprum,* from the island of Cyprus), Cu; at. wt. 63.546(3); at. no. 29; f.p. 1084.62 °C; b.p. 2562 °C; sp. gr. 8.96 (20 °C); valence 1 or 2. The discovery of copper dates from prehistoric times. It is said to have been mined for more than 5000 years. It is one of man's most important metals. Copper is reddish colored, takes on a bright metallic luster, and is malleable, ductile, and a good conductor of heat and electricity (second only to silver in electrical conductivity). The electrical industry is one of the greatest users of copper. Copper occasionally occurs native, and is found in many minerals such as *cuprite, malachite, azurite, chalcopyrite,* and *bornite.* Large copper ore deposits are found in the U.S., Chile, Zambia, Zaire, Peru, and Canada. The most important copper ores are the sulfides, oxides, and carbonates. From these, copper is obtained by smelting, leaching, and by electrolysis. Its alloys, brass and bronze, long used, are still very important; all American coins are now copper alloys; monel and gun metals also contain copper. The most important compounds are the oxide and the sulfate, blue vitriol; the latter has wide use as an agricultural poison and as an algicide in water purification. Copper compounds such as Fehling's solution are widely used in analytical chemistry in tests for sugar. High-purity copper (99.999 + %) is readily available commercially. The price of commercial copper has fluctuated widely. The price of copper in December 2001 was about $1.50/kg. Natural copper contains two isotopes. Twenty-six other radioactive isotopes and isomers are known.

Curium — (Pierre and Marie Curie), Cm; at. wt. (247); at. no. 96; m.p. 1345 °C; sp. gr. 13.51 (calc.); valence 3 and 4. Although curium follows americium in the periodic system, it was actually known before americium and was the third transuranium element to be discovered. It was identified by Seaborg, James, and Ghiorso in 1944 at the wartime Metallurgical Laboratory in Chicago as a result of helium-ion bombardment of ^{239}Pu in the Berkeley, California, 60-inch cyclotron. Visible amounts (30 μg) of ^{242}Cm, in the form of the hydroxide, were first isolated by Werner and Perlman of the University of California in 1947. In 1950, Crane, Wallmann, and Cunningham found that the magnetic susceptibility of microgram samples of CmF_3 was of the same magnitude as that of GdF_3. This provided direct experimental evidence for assigning an electronic configuration to Cm^{+3}. In 1951, the same workers prepared curium in its elemental form for the first time. Sixteen isotopes of curium are now known. The most stable, ^{247}Cm, with a half-life of 16 million years, is so short compared to the Earth's age that any primordial curium must have disappeared long ago from the natural scene. Minute amounts of curium probably exist in natural deposits of uranium, as a result of a sequence of neutron captures and β decays sustained by the very low flux of neutrons naturally present in uranium ores. The presence of natural curium, however, has never been detected. ^{242}Cm and ^{244}Cm are available in multigram quantities. ^{248}Cm has been produced only in milligram amounts. Curium is similar in some regards to gadolinium, its rare-earth homolog, but it has a more complex crystal structure. Curium is silver in color, is chemically reactive, and is more electropositive than aluminum. CmO_2, Cm_2O_3, CmF_3, CmF_4, $CmCl_3$, $CmBr_3$, and CmI_3 have been prepared. Most compounds of trivalent curium are faintly yellow in color. ^{242}Cm generates about three watts of thermal energy per gram. This compares to one-half watt per gram of ^{238}Pu. This suggests use for curium as a power source. ^{244}Cm is now offered for sale by the O.R.N.L. at $185/mg plus packing charges. ^{248}Cm is available at a cost of $160/μg, plus packing charges, from the O.R.N.L. Curium absorbed into the body accumulates in the bones, and is therefore very toxic as its radiation destroys the red-cell forming mechanism. The maximum permissible total body burden of ^{244}Cm (soluble) in a human being is 0.3 μCi (microcurie).

Darmstadtium — (Darmstadt, city in Germany), Ds. In 1987 Oganessian et al., at Dubna, claimed discovery of this element. Their experiments indicated the spontaneous fissioning nuclide $^{272}110$ with a half-life of 10 ms. More recently a group led by Armbruster at G.S.I. in Darmstadt, Germany, reported evidence of $^{269}110$, which was produced by bombarding lead for many days with more than 10^{18} nickel atoms. A detector searched each collision for Element 110's distinct decay sequence. On November 9, 1994, evidence of 110 was detected. In 2003 IUPAC approved the name darmstadtium, symbol Ds, for Element 110. Seven isotopes of Element 110 are now recognized.

Deuterium — an isotope of hydrogen — see Hydrogen.

Dubnium — (named after the Joint Institute of Nuclear Research in Dubna, Russia). Db; at. wt. [262]; at. no. 105. In 1967 G. N. Flerov reported that a Soviet team working at the Joint Institute for Nuclear Research at Dubna may have produced a few atoms of $^{260}105$ and $^{261}105$ by bombarding ^{243}Am with ^{22}Ne. Their evidence was based on time-coincidence measurements of alpha energies. More recently, it was reported that early in 1970 Dubna scientists synthesized Element 105 and that by the end of April 1970 "had investigated all the types of decay of the new element and had determined its chemical properties." In late April 1970, it was announced that Ghiorso, Nurmia, Harris, K. A. Y. Eskola, and P. L. Eskola, working at the University of California at Berkeley, had positively identified Element 105. The discovery was made by bombarding a target of ^{249}Cf with a beam of 84 MeV nitrogen nuclei in the Heavy Ion Linear Accelerator (HILAC). When a ^{15}N nucleus is absorbed by a ^{249}Cf nucleus, four neutrons are emitted and a new atom of $^{260}105$ with a half-life of 1.6 s is formed. While the first atoms of Element 105 are said to have been detected conclusively on March 5, 1970, there is evidence that Element 105 had been formed in Berkeley experiments a year earlier by the method described. Ghiorso and his associates have attempted to confirm Soviet findings by more sophisticated methods without success.

In October 1971, it was announced that two new isotopes of Element 105 were synthesized with the heavy ion linear accelerator by A. Ghiorso and co-workers at Berkeley. Element $^{261}105$ was produced both by bombarding ^{250}Cf with ^{15}N and by bombarding ^{249}Bk with ^{16}O. The isotope emits 8.93-MeV α particles and decays to ^{257}Lr with a half-life of about 1.8 s. Element $^{262}105$ was produced by bombarding ^{249}Bk with ^{18}O. It emits 8.45 MeV α particles and decays to ^{258}Lr with a half-life of about 40 s. Nine isotopes of Dubnium are now recognized. Soon after the discovery the names ***Hahnium*** and ***Joliotium,*** named after Otto Hahn and Jean-Frederic Joliot and Mme. Joliot-Curie, were suggested as names for Element 105. The IUPAC in August 1997 finally resolved the issue, naming Element 105 Dubnium with the symbol Db. Dubnium is thought to have properties similar to tantalum.

Dysprosium — (Gr. *dysprositos,* hard to get at), Dy; at. wt. 162.500(1); at. no. 66; m.p. 1412 °C, b.p. 2567 °C; sp. gr. 8.551 (25 °C); valence 3. Dysprosium was discovered in 1886 by Lecoq de Boisbaudran, but not isolated. Neither the oxide nor the metal was available in relatively pure form until the development of ion-exchange separation and metallographic reduction techniques by Spedding and associates about 1950. Dysprosium occurs along with other so-called rare-earth or lanthanide elements in a variety of minerals such as *xenotime, fergusonite, gadolinite, euxenite, polycrase,* and *blomstrandine.* The most important sources, however, are from *monazite* and *bastnasite.* Dysprosium can be prepared by reduction of the trifluoride with calcium. The element has a metallic, bright silver luster. It is relatively stable in air at room temperature, and is readily attacked and dissolved, with the evolution of hydrogen, by dilute and concentrated mineral acids. The metal is soft enough to be cut with a knife and can be machined without sparking if overheating is avoided. Small amounts of impurities can greatly affect its physical properties. While dysprosium has not yet found many applications, its thermal neutron absorption cross-section and high melting point suggest metallurgical uses in nuclear control applications and for alloying with special stainless steels. A dysprosium oxide-nickel cermet has found use in cooling nuclear reactor rods. This cermet absorbs neutrons readily without swelling or contracting under prolonged neutron bombardment. In combination with vanadium and other rare earths, dysprosium has been used in making laser materials. Dysprosium-cadmium chalcogenides, as sources of infrared radiation, have been used for studying chemical reactions. The cost of dysprosium metal has dropped in recent years since the development of ion-exchange and solvent extraction techniques, and the discovery of large ore bodies. Thirty-two isotopes and isomers are now known. The metal costs about $6/g (99.9% purity).

Einsteinium — (Albert Einstein [1879–1955]), Es; at. wt. (252); m.p. 860 °C (est.); at. no. 99. Einsteinium, the seventh transuranic element of the actinide series to be discovered, was identified by Ghiorso and co-workers at Berkeley in December 1952 in debris from the first large thermonuclear explosion, which took place in the Pacific in November 1952. The isotope produced was the 20-day ^{253}Es isotope. In 1961, a sufficient amount of einsteinium was produced to permit separation of a macroscopic amount of ^{253}Es. This sample weighed about 0.01 μg. A special magnetic-type balance was used in making this determination. ^{253}Es so produced was used to produce mendelevium. About 3 μg of einsteinium has been produced at Oak Ridge National Laboratories by irradiating for several years kilogram quantities of ^{239}Pu in a reactor to produce ^{242}Pu. This was then fabricated into pellets of plutonium oxide and aluminum powder, and loaded into target rods for an initial 1-year irradiation at the Savannah River Plant, followed by irradiation in a HFIR (High Flux Isotopic Reactor). After 4 months in the HFIR the targets were removed for chemical separation of the einsteinium from californium. Nineteen isotopes and isomers of einsteinium are now recognized. ^{254}Es has the longest half-life (276 days). Tracer studies using ^{253}Es show that einsteinium has chemical properties typical of a heavy trivalent, actinide element. Einsteinium is extremely radioactive. Great care must be taken when handling it.

Element 112 — In late February 1996, Siguard Hofmann and his collaborators at GSI Darmstadt announced their discovery of Element 112, having 112 protons and 165 neutrons, with an atomic mass of 277. This element was made by bombarding a lead target with high-energy zinc ions. A single nucleus of Element 112 was detected, which decayed after less than 0.001 sec by emitting an α particle, consisting of two protons and two neutrons. This created Element 110_{273}, which in turn decayed by emitting an α particle to form a new isotope of Element 108 and so on. Evidence indicates that nuclei with 162 neutrons are held together more strongly than nuclei with a smaller or larger number of neutrons. This suggests a narrow "peninsula" of relatively stable isotopes around Element 114. GSI scientists are experimenting to bombard targets with ions heavier than zinc to produce Elements 113 and 114. A name has not yet been suggested for Element 112, although the IUPAC suggested the temporary name of ununbium, with the symbol of Uub, when the element was discovered. Element 112 is expected to have properties similar to mercury.

Element 113 — (Ununtrium) See Element 115.

Element 114 — (Ununquadium) Symbol Uuq. Element 114 is the first new element to be discovered since 1996. This element was found by a Russian–American team, including Livermore researchers, by bombarding a sheet of plutonium with a rare form of calcium hoping to make the atoms stick together in a new element. Radiation showed that the new element broke into smaller pieces. Data of radiation collected at the Russian Joint Institute for Nuclear Research in November and December 1998 were analyzed in January 1999. It was found that some of the heavy atoms created when 114 decayed lived up to 30 seconds, which was longer than ever seen before for such a heavy element. This isotope decayed into a previously unknown isotope of Element 112, which itself lasted 15 minutes. That isotope, in turn, decayed to a previously undiscovered isotope of Element 108, which survived 17 minutes. Isotopes of these and those with longer life-times have been predicted for some time by theorists. It appears that these isotopes are on the edge of the "island of stability," and that some of the isotopes in this region might last long enough for studies of their nuclear behavior and for a chemical evaluation to be made. No name has yet been suggested for Element 114; however, the temporary name of ununquadium with symbol Uuq may be used.

Element 115— (Ununpentium) On February 2, 2004, it was reported that Element 115 had been discovered at the Joint Institute for Nuclear Research (JINR) in Dubna, Russia. Four atoms of this element were produced by JINR physicists and collaborators from the Lawrence Livermore (California) Laboratory using a 248-MeV beam of calcium-48 ions striking a target of americium-243 atoms. The nuclei of these atoms are said to have a life of 90 milliseconds. The relatively long lifetime of Element 115 suggests that these experiments might be getting closer to the "island of stability" long sought to exist by some nuclear physicists. These atoms were thought to decay first to Element 113 by the emission of an alpha particle, then decay further to Element 111 by alpha emission again, and then by three more alpha decay processes to Element 105 (dubnium), which after a long delay from the time of the initial interaction, fissioned. This experiment entailed separating four atoms from trillions of other atoms. A gas-filled separator, employing chemistry, was important in this experiment. Names for Elements 115, Element 113, and Element 111 have not yet been chosen.

Element 116 — (Ununhexium) Symbol Uuh. As of January 2004 it is questionable if this element has been discovered.

Element 117 — (Ununseptium) Symbol Uus. As of January 2004, this element remains undiscovered.

Element 118 — (Ununoctium) Symbol Uuo. In June 1999 it was announced that Elements 118 and 116 had been discovered at the Lawrence Berkeley National Laboratory. A lead target was bombarded for more than 10 days with roughly 1 quintillion krypton ions. The team reported that three atoms of Element 118 were made, which quickly decayed into Elements 116, 114, and elements of lower atomic mass. It was said that the isotopes of Element 118 lasted only about 200 milliseconds, while the isotope of Element 116 lasted only 1.2 milliseconds. It was hoped that these elements might be members of "an island of stability," which had long been sought. At that time it was hoped that a target of bismuth might be bombarded with krypton ions to make Element 119, which, in turn, would decay into Elements 117, 115, and 113.

On July 27, 2001 researchers at the Lawrence Berkeley Laboratory announced that their discovery of Element 118 was being retracted because workers at the GSI Laboratory in Germany and at Japanese laboratories failed to confirm their results. However, it was reported that different experiments at the Livermore Laboratory and Joint Institute from Nuclear Research in Dubna, Russia indicated that Element 116 had since been created.

Researchers at the Australian National Laboratory suggest that super-heavy elements may be more difficult to make than previously thought. Their data suggest the best way to encourage fusion in making super-heavy elements is to combine the lightest projectiles possible with the heaviest possible targets. This would minimize a so-called "quasi-fission process" in which a projectile nucleus steals protons and neutrons from a target nucleus. In this process the two nuclei are said to fly apart without ever having actually combined.

Erbium — (*Ytterby*, a town in Sweden), Er; at. wt. 167.259(3); at. no. 68; m.p. 1529 °C; b.p. 2868 °C; sp. gr. 9.066 (25 °C); valence 3, Erbium, one of the so-called rare-earth elements of the lanthanide series, is found in the minerals mentioned under dysprosium above. In 1842 Mosander separated "yttria," found in the mineral *gadolinite,* into three fractions which he called *yttria, erbia,* and *terbia.* The names *erbia* and *terbia* became confused in this early period. After 1860, Mosander's *terbia* was known as *erbia,* and after 1877, the earlier known *erbia* became *terbia.* The *erbia* of this period was later shown to consist of five oxides, now known as *erbia, scandia, holmia, thulia* and *ytterbia.* By 1905 Urbain and James independently succeeded in isolating fairly pure Er_2O_3. Klemm and Bommer first produced reasonably pure erbium metal in 1934 by reducing the anhydrous chloride with potassium vapor. The pure metal is soft and malleable and has a bright, silvery, metallic luster. As with other rare-earth metals, its properties depend to a certain extent on the impurities present. The metal is fairly stable in air and does not oxidize as rapidly as some of the other rare-earth metals. Naturally occurring erbium is a mixture of six isotopes, all of which are stable. Twenty-seven radioactive isotopes of erbium are also recognized. Recent production techniques, using ion-exchange reactions, have resulted in much lower prices of the rare-earth metals and their compounds in recent years. The cost of 99.9% erbium metal is about $21/g. Erbium is finding nuclear and metallurgical uses. Added to vanadium, for example, erbium lowers the hardness and improves workability. Most of the rare-earth oxides have sharp absorption bands in the visible, ultraviolet, and near infrared. This property, associated with the electronic structure, gives beautiful pastel colors to many of the rare-earth salts. Erbium oxide gives a pink color and has been used as a colorant in glasses and porcelain enamel glazes.

Europium — (Europe), Eu; at. wt. 151.964(1); at. no. 63; m.p. 822 °C; b.p. 1596 °C; sp. gr. 5.244 (25 °C); valence 2 or 3. In 1890 Boisbaudran obtained basic fractions from samarium-gadolinium concentrates that had spark spectral lines not accounted for by samarium or gadolinium. These lines subsequently have been shown to belong to europium. The discovery of europium is generally credited to Demarcay, who separated the rare earth in reasonably pure form in 1901. The pure metal was not isolated until recent years. Europium is now prepared by mixing Eu_2O_3 with a 10% excess of lanthanum metal and heating the mixture in a tantalum crucible under high vacuum. The element is collected as a silvery-white metallic deposit on the walls of the crucible. As with other rare-earth metals, except for lanthanum, europium ignites in air at about 150 to 180 °C. Europium is about as hard as lead and is quite ductile. It is the most reactive of the rare-earth metals, quickly oxidizing in air. It resembles calcium in its reaction with water. *Bastnasite* and *monazite* are the principal ores containing europium. Europium has been identified spectroscopically in the sun and certain stars. Europium isotopes are good neutron absorbers and are being studied for use in nuclear control applications. Europium oxide is now widely used as a phosphor activator and europium-activated yttrium vanadate is in commercial use as the red phosphor in color TV tubes. Europium-doped plastic has been used as a laser material. With the development of ion-exchange techniques and special processes, the cost of the metal has been greatly reduced in recent years. Natural europium contains two stable isotopes. Thirty-five other radioactive isotopes and isomers are known. Europium is one of the rarest and most costly of the rare-earth metals. It is priced at about $60/g (99.9% pure).

Fermium — (Enrico Fermi [1901–1954], nuclear physicist), Fm; at. wt. [257]; at. no. 100; m.p. 1527 °C. Fermium, the eighth transuranium element of the actinide series to be discovered, was identified by Ghiorso and co-workers in 1952 in the debris from a thermonuclear explosion in the Pacific in work involving the University of California Radiation Laboratory, the Argonne National Laboratory, and the Los Alamos Scientific Laboratory. The isotope produced was the 20-hour ^{255}Fm. During 1953 and early 1954, while discovery of elements 99 and 100 was withheld from publication for security reasons, a group from the Nobel Institute of Physics in Stockholm bombarded ^{238}U with ^{16}O ions, and isolated a 30-min α-emitter, which they ascribed to $^{250}100$, without claiming discovery of the element. This isotope has since been identified positively, and the 30-min half-life confirmed. The chemical properties of fermium have been studied solely with tracer amounts, and in normal aqueous media only the (III) oxidation state appears to exist. The isotope ^{254}Fm and heavier isotopes can be produced by intense neutron irradiation of lower elements such as plutonium by a process of successive neutron capture interspersed with beta decays until these mass numbers and atomic numbers are reached. Twenty isotopes and isomers of fermium are known to exist. ^{257}Fm, with a half-life of about 100.5 days, is the longest lived. ^{250}Fm, with a half-life of 30 min, has been shown to be a product of decay of Element $^{254}102$. It was by chemical identification of ^{250}Fm that production of Element 102 (nobelium) was confirmed. Fermium would probably have chemical properties resembling erbium.

Fluorine — (L. and F. *fluere*, flow, or flux), F; at. wt. 18.9984032(5); at. no. 9; m.p. −219.67 °C (1 atm); b.p. −188.12 °C (1 atm); t_c −129.02 °C; density 1.696 g/L (0 °C, 1 atm); liq. den. at b.p. 1.50 g/cm^3; valence 1. In 1529, Georgius Agricola described the use of fluorspar as a flux, and as early as 1670 Schwandhard found that glass was etched when exposed to fluorspar treated with acid. Scheele and many later investigators, including Davy, Gay-Lussac, Lavoisier, and Thenard, experimented with hydrofluoric acid, some experiments ending in tragedy. The element was finally isolated in 1886 by Moisson after nearly 74 years of continuous effort. Fluorine occurs chiefly in *fluorspar* (CaF_2) and *cryolite* (Na_2AlF_6), and is in *topaz* and other minerals. It is a member of the halogen family of elements, and is obtained by electrolyzing a solution of potassium hydrogen fluoride in anhydrous hydrogen fluoride in a vessel of metal or transparent fluorspar. Modern commercial production methods are essentially variations on the procedures first used by Moisson. Fluorine is the most electronegative and reactive of all elements. It is a pale yellow, corrosive gas, which reacts with practically all organic and inorganic substances. Finely divided metals, glass, ceramics, carbon, and even water burn in fluorine with a bright flame. Until World War II, there was no commercial production of elemental fluorine. The atom bomb project and nuclear energy applications, however, made it necessary to produce large quantities. Safe handling techniques have now been developed and it is possible at present to transport liquid fluorine by the ton. Fluorine and its compounds are used in producing uranium (from the hexafluoride) and more than 100 commercial fluorochemicals, including many well-known high-temperature plastics. Hydrofluoric acid is extensively used for etching the glass of light bulbs, etc. Fluorochlorohydrocarbons have been extensively used in air conditioning and refrigeration. However, in recent years the U.S. and other countries have been phasing out ozone-depleting substances, such as the fluorochlorohydrocarbons that have been used in these applications. It has been suggested that fluorine might be substituted for hydrogen wherever it occurs in organic compounds, which could lead to an astronomical number of new fluorine compounds. The presence of fluorine as a soluble fluoride in drinking water to the extent of 2 ppm may cause mottled enamel in teeth, when used by children acquiring permanent teeth; in smaller amounts, however, fluorides are said to be beneficial and used in water supplies to prevent dental cavities. Elemental fluorine has been studied as a rocket propellant as it has an exceptionally high specific impulse value. Compounds of fluorine with rare gases have now been confirmed. Fluorides of xenon, radon, and krypton are among those known. Elemental fluorine and the fluoride ion are highly toxic. The free element has a characteristic pungent odor, detectable in concentrations as low as 20 ppb, which is below the safe working level. The recommended maximum allowable concentration for a daily 8-hour time-weighted exposure is 1 ppm. Fluorine is known to have fourteen isotopes.

Francium — (France), Fr; at. no. 87; at. wt. [223]; m.p. 27 °C; valence 1. Discovered in 1939 by Mlle. Marguerite Perey of the Curie Institute, Paris. Francium, the heaviest known member of the alkali metal series, occurs as a result of an alpha disintegration of actinium. It can also be made artificially by bombarding thorium with protons. While it occurs naturally in uranium minerals, there is probably less than an ounce of francium at any time in the total crust of the earth. It has the highest equivalent weight of any element, and is the most unstable of the first 101 elements of the periodic system. Thirty-six isotopes and isomers of francium are recognized. The longest lived ^{223}Fr(Ac, K), a daughter of ^{227}Ac, has a half-life of 21.8 min. This is the only isotope of francium occurring in nature. Because all known isotopes of francium are highly unstable, knowledge of the chemical properties of this element comes from radiochemical techniques. No weighable quantity of the element has been prepared or isolated. The chemical properties of francium most closely resemble cesium. In 1996, researchers Orozco, Sprouse, and co-workers at the State University of New York, Stony Brook, reported that they had produced francium atoms by bombarding ^{18}O atoms at a gold target heated almost to its melting point. Collisions between gold and oxygen nuclei created atoms of francium-210 which had 87 protons and 123 neutrons. This team reported they had generated about 1 million francium-210 ions per second and held 1000 or more atoms at a time for about 20 secs in a magnetic trap they had devised before the atoms decayed or escaped. Enough francium was trapped so that a videocamera could capture the light given off by the atoms as they fluoresced. A cluster of about 10,000 francium atoms appeared as a glowing sphere about 1 mm in diameter. It is thought that the francium atoms could serve as miniature laboratories for probing interactions between electrons and quarks.

Gadolinium — (*gadolinite*, a mineral named for Gadolin, a Finnish chemist), Gd; at. wt. 157.25(3); at. no. 64; m.p. 1313 °C; b.p. 3273 °C; sp. gr. 7.901 (25 °C); valence 3. Gadolinia, the oxide of gadolinium, was separated by Marignac in 1880 and Lecoq de Boisbaudran independently isolated the element from Mosander's "yttria" in 1886. The element was named for the mineral *gadolinite* from which this rare earth was originally obtained. Gadolinium is found in several other minerals, including *monazite* and *bastnasite*, which are of commercial importance. The element has been isolated only in recent years. With the development of ion-exchange and solvent extraction techniques, the availability and price of gadolinium and the other rare-earth metals have greatly improved. Thirty-one isotopes and isomers of gadolinium are now recognized; seven are stable and occur naturally. The metal can be prepared by the reduction of the anhydrous fluoride with metallic calcium. As with other related rare-earth metals, it is silvery white, has a metallic luster, and is malleable and ductile. At room temperature, gadolinium crystallizes in the hexagonal, close-packed α form. Upon heating to 1235 °C, α gadolinium transforms into the β form, which has a body-centered cubic structure. The metal is relatively stable in dry air, but in moist air it tarnishes with the formation of a loosely adhering oxide film which splits off and exposes more surface to oxidation. The metal reacts slowly with water and is soluble in dilute acid. Gadolinium has the highest thermal neutron capture cross-section of any known element (49,000 barns). Natural gadolinium is a mixture of seven isotopes. Two of these, ^{155}Gd and ^{157}Gd, have excellent capture characteristics, but they are present naturally in low concentrations. As a result, gadolinium has a very fast burnout rate and has limited use as a nuclear control rod material. It has been used in making gadolinium yttrium garnets, which have microwave applications. Compounds of gadolinium are used in making phosphors for color TV tubes. The metal has unusual superconductive properties. As little as 1% gadolinium has been found to improve the workability and resistance of iron, chromium, and related alloys to high temperatures and oxidation. Gadolinium ethyl sulfate has extremely low noise characteristics and may find use in duplicating the performance of amplifiers, such as the maser. The metal is ferromagnetic. Gadolinium is unique for its high magnetic moment and for its special Curie temperature (above which ferromagnetism vanishes) lying just at room temperature. This suggests uses as a magnetic component that senses hot and cold. The price of the metal is about \$5/g (99.9% purity).

Gallium — (L. *Gallia*, France), Ga; at. wt. 69.723(1); at. no. 31; m.p. 29.76 °C; b.p. 2204 °C; sp. gr. 5.904 (29.6 °C) solid; sp. gr. 6.095 (29.6 °C) liquid; valence 2 or 3. Predicted and described by Mendeleev as ekaaluminum, and discovered spectroscopically by Lecoq de Boisbaudran in 1875, who in the same year obtained the free metal by electrolysis of a solution of the hydroxide in KOH, Gallium is often found as a trace element in *diaspore, sphalerite, germanite, bauxite,* and *coal.* Some flue dusts from burning coal have been shown to contain as much as 1.5% gallium. It is the only metal, except for mercury, cesium, and rubidium, which can be liquid near room temperatures; this makes possible its use in high-temperature thermometers. It has one of the longest liquid ranges of any metal and has a low vapor pressure even at high temperatures. There is a strong tendency for gallium to supercool below its freezing point. Therefore, seeding may be necessary to initiate solidification. Ultra-pure gallium has a beautiful, silvery appearance, and the solid metal exhibits a conchoidal fracture similar to glass. The metal expands 3.1% on solidifying; therefore, it should not be stored in glass or metal containers, as they may break as the metal solidifies. Gallium wets glass or porcelain, and forms a brilliant mirror when it is painted on glass. It is widely used in doping semiconductors and producing solid-state devices such as transistors. High-purity gallium is attacked slowly only by mineral acids. Magnesium gallate containing divalent impurities such as Mn^{+2} is finding use in commercial ultraviolet activated powder phosphors. Gallium nitride has been used to produce blue light-emitting diodes such as those used in CD and DVD readers. Gallium has found application in the Gallex Detector Experiment located in the Gran Sasso Underground Laboratory in Italy. This underground facility has been built by the Italian Istituto Nazionale di Fisica Nucleare in the middle of a highway tunnel through the Abruzzese mountains, about 150 km east of Rome. In this experiment, 30.3 tons of gallium in the form of 110 tons of $GaCl_3$-HCl solution are being used to detect solar neutrinos. The production of ^{71}Ge from gallium is being measured. Gallium arsenide is capable of converting electricity directly into coherent light. Gallium readily alloys with most metals, and has been used as a component in low melting alloys. Its toxicity appears to be of a low order, but it should be handled with care until more data are forthcoming. Natural gallium contains two stable isotopes. Twenty-six other isotopes, one of which is an isomer, are known. The metal can be supplied in ultrapure form (99.99999+%). The cost is about \$5/g (99.999%).

Germanium — (L. *Germania,* Germany), Ge; at. wt. 72.64(2); at. no. 32; m.p. 938.25 °C; b.p. 2833 °C; sp. gr. 5.323 (25 °C); valence 2 and 4. Predicted by Mendeleev in 1871 as ekasilicon, and discovered by Winkler in 1886. The metal is found in *argyrodite,* a sulfide of germanium and silver; in *germanite,* which contains 8% of the element; in zinc ores; in coal; and in other minerals. The element is frequently obtained commercially from flue dusts of smelters processing zinc ores, and has been recovered from the by-products of combustion of certain coals. Its presence in coal insures a large reserve of the element in the years to come. Germanium can be separated from other metals by fractional distillation of its volatile tetrachloride. The tetrachloride may then be hydrolyzed to give GeO_2; the dioxide can be reduced with hydrogen to give the metal. Recently developed zone-refining techniques permit the production of germanium of ultra-high purity. The element is a gray-white metalloid, and in its pure state is crystalline and brittle, retaining its luster in air at room temperature. It is a very important semiconductor material. Zone-refining techniques have led to production of crystalline germanium for semiconductor use with an impurity of only one part in 10^{10}. Doped with arsenic, gallium, or other elements, it is used as a transistor element in thousands of electronic applications. Its application in fiber optics and infrared optical systems now provides the largest use for germanium. Germanium is also finding many other applications including use as an alloying agent, as a phosphor in fluorescent lamps, and as a catalyst. Germanium and germanium oxide are transparent to the infrared and are used in infrared spectrometers and other optical equipment, including extremely sensitive infrared detectors. Germanium oxide's high index of refraction and dispersion make it useful as a component of glasses used in wide-angle camera lenses and microscope objectives. The field of organogermanium chemistry is becoming increasingly important. Certain germanium compounds have a low mammalian toxicity, but a marked activity against certain bacteria, which makes them of interest as chemotherapeutic agents. The cost of germanium is about \$10/g (99.999% purity). Thirty isotopes and isomers are known, five of which occur naturally.

Gold — (Sanskrit *Jval;* Anglo-Saxon *gold*), Au (L. *aurum,* gold); at. wt. 196.966569(4); at. no. 79; m.p. 1064.18 °C; b.p. 2856 °C; sp. gr. ~19.3 (20 °C); valence 1 or 3. Known and highly valued from earliest times, gold is found in nature as the free metal and in tellurides; it is very widely distributed and is almost always associated with quartz or pyrite. It occurs in veins and alluvial deposits, and is often separated from rocks and other minerals by sluicing and panning operations. About 25% of the world's gold output comes from South Africa, and about two thirds of the total U.S. production now comes from South Dakota and Nevada. The metal is recovered from its ores by cyaniding, amalgamating, and smelting processes. Refining is also frequently done by electrolysis. Gold occurs in sea water to the extent of 0.1 to 2 mg/ton, depending on the location where the sample is taken. As yet, no method has been found for recovering gold from sea water profitably. It is estimated that all the gold in the world, so far refined, could be placed in a single cube 60 ft on a side. Of all the elements, gold in its pure state is undoubtedly the most beautiful. It is metallic, having a yellow color when in a mass, but when finely divided it may be black, ruby, or purple. The Purple of Cassius is a delicate test for auric gold. It is the most malleable and ductile metal; 1 oz. of gold can be beaten out to 300 ft². It is a soft metal and is usually alloyed to give it more strength. It is a good conductor of heat and electricity, and is unaffected by air and most reagents. It is used in coinage and is a standard for monetary systems in many countries. It is also extensively used for jewelry, decoration, dental work, and for plating. It is used for coating certain space satellites, as it is a good reflector of infrared and is inert. Gold, like other precious metals, is measured in troy weight; when alloyed with other metals, the term *carat* is used to express the amount of gold present, 24 carats being pure gold. For many years the value of gold was set by the U.S. at \$20.67/troy ounce; in 1934 this value was fixed by law at \$35.00/troy ounce, 9/10th fine. On March 17, 1968, because of a gold crisis, a two-tiered pricing system was established whereby gold was still used to settle international accounts at the old \$35.00/troy ounce price while the price of gold on the private market would be allowed to fluctuate. Since this time, the price of gold on the free market has fluctuated widely. The price of gold on the free market reached a price of \$620/troy oz. in January 1980. More recently, the U.K. and other nations, including the I.M.F. have sold or threatened to sell a sizeable portion of their gold reserves. This has caused wide fluctuations in the price of gold. Because this has damaged the economy of some countries, a moratorium for a few years has been declared. This has tended to stabilize temporarily the price of gold. The most common gold compounds are auric chloride ($AuCl_3$) and chlorauric acid ($HAuCl_4$), the latter being used in photography for toning the silver image. Gold has forty-eight recognized isotopes and isomers; ^{198}Au, with a half-life of 2.7 days, is used for treating cancer and other diseases. Disodium aurothiomalate is administered intramuscularly as a treatment for arthritis. A mixture of one part nitric acid with three of hydrochloric acid is called *aqua regia* (because it dissolved gold, the King of Metals). Gold is available commercially with a purity of 99.999+%. For many years the temperature assigned to the freezing point of gold has been 1063.0 °C; this has served as a calibration point for the International Temperature Scales (ITS-27 and ITS-48) and the International Practical Temperature Scale (IPTS-48). In 1968, a new International Practical Temperature Scale (IPTS-68) was adopted, which demanded that the freezing point of gold be changed to 1064.43 °C. In 1990 a new International Temperature Scale (ITS-90) was adopted bringing the t.p.

(triple point) of H_2O (t_{90} (°C)) to 0.01 °C and the freezing point of gold to 1064.18 °C. The specific gravity of gold has been found to vary considerably depending on temperature, how the metal is precipitated, and cold-worked. As of December 2001, gold was priced at about $275/troy oz. ($8.50/g).

Hafnium — (*Hafnia,* Latin name for Copenhagen), Hf; at. wt. 178.49(2); at. no. 72; m.p. 2233 °C; b.p. 4603 °C; sp. gr. 13.31 (20 °C); valence 4. Hafnium was thought to be present in various minerals and concentrations many years prior to its discovery, in 1923, credited to D. Coster and G. von Hevesey. On the basis of the Bohr theory, the new element was expected to be associated with zirconium. It was finally identified in *zircon* from Norway, by means of X-ray spectroscopic analysis. It was named in honor of the city in which the discovery was made. Most zirconium minerals contain 1 to 5% hafnium. It was originally separated from zirconium by repeated recrystallization of the double ammonium or potassium fluorides by von Hevesey and Jantzen. Metallic hafnium was first prepared by van Arkel and deBoer by passing the vapor of the tetraiodide over a heated tungsten filament. Almost all hafnium metal now produced is made by reducing the tetrachloride with magnesium or with sodium (Kroll Process). Hafnium is a ductile metal with a brilliant silver luster. Its properties are considerably influenced by the impurities of zirconium present. Of all the elements, zirconium and hafnium are two of the most difficult to separate. Their chemistry is almost identical; however, the density of zirconium is about half that of hafnium. Very pure hafnium has been produced, with zirconium being the major impurity. Natural hafnium contains six isotopes, one of which is slightly radioactive. Hafnium has a total of 41 recognized isotopes and isomers. Because hafnium has a good absorption cross section for thermal neutrons (almost 600 times that of zirconium), has excellent mechanical properties, and is extremely corrosion resistant, it is used for reactor control rods. Such rods are used in nuclear submarines. Hafnium has been successfully alloyed with iron, titanium, niobium, tantalum, and other metals. Hafnium carbide is the most refractory binary composition known, and the nitride is the most refractory of all known metal nitrides (m.p. 3310 °C). Hafnium is used in gas-filled and incandescent lamps, and is an efficient "getter" for scavenging oxygen and nitrogen. Finely divided hafnium is pyrophoric and can ignite spontaneously in air. Care should be taken when machining the metal or when handling hot sponge hafnium. At 700 °C hafnium rapidly absorbs hydrogen to form the composition $HfH_{1.86}$. Hafnium is resistant to concentrated alkalis, but at elevated temperatures reacts with oxygen, nitrogen, carbon, boron, sulfur, and silicon. Halogens react directly to form tetrahalides. The price of the metal is about $2/g. The yearly demand for hafnium in the U.S. is now in excess of 50,000 kg.

Hahnium — A name previously used for Element 105, now named *dubnium.*

Hassium — (named for the German state, Hesse) Hs; at. wt. [277]; at. no. 108. This element was first synthesized and identified in 1964 by the same G.S.I. Darmstadt Group who first identified *Bohrium* and *Meitnerium.* Presumably this element has chemical properties similar to osmium. Isotope $^{265}108$ was produced using a beam of ^{58}Fe projectiles, produced by the Universal Linear Accelerator (UNILAC) to bombard a ^{208}Pb target. Discovery of *Bohrium* and *Meitnerium* was made using detection of isotopes with odd proton and neutron numbers. Elements having even atomic numbers have been thought to be less stable against spontaneous fusion than odd elements. The production of $^{265}108$ in the same reaction as was used at G.S.I. was confirmed at Dubna with detection of the seventh member of the decay chain ^{253}Es. Isotopes of *Hassium* are believed to decay by spontaneous fission, explaining why 109 was produced before 108. Isotope $^{265}108$ and $^{266}108$ are thought to decay to $^{261}106$, which in turn decay to $^{257}104$ and $^{253}102$. The IUPAC adopted the name *Hassium* after the German state of Hesse in September 1997. In June 2001 it was announced that hassium is now the heaviest element to have its chemical properties analyzed. A research team at the UNILAC heavy-ion accelerator in Darmstadt, Germany built an instrument to detect and analyze hassium. Atoms of curium-248 were collided with atoms of magnesium-26, producing about 6 atoms of hassium with a half-life of 9 sec. This was sufficiently long to obtain data showing that hassium atoms react with oxygen to form hassium oxide molecules. These condensed at a temperature consistent with the behavior of Group 8 elements. This experiment appears to confirm hassium's location under osmium in the periodic table.

Helium — (Gr. *helios,* the sun), He; at. wt. 4.002602(2); at. no. 2; b.p. — 268.93 °C; t_c −267.96 °C; density 0.1785 g/L (0 °C, 1 atm); liquid density 0.125 g/mL at. b.p.; valence usually 0. Evidence of the existence of helium was first obtained by Janssen during the solar eclipse of 1868 when he detected a new line in the solar spectrum; Lockyer and Frankland suggested the name *helium* for the new element; in 1895, Ramsay discovered helium in the uranium mineral *cleveite,* and it was independently discovered in cleveite by the Swedish chemists Cleve and Langlet about the same time. Rutherford and Royds in 1907 demonstrated that α particles are helium nuclei. Except for hydrogen, helium is the most abundant element found throughout the universe. Helium is extracted from natural gas; all natural gas contains at least trace quantities of helium. It has been detected spectroscopically in great abundance, especially in the hotter stars, and it is an important component in both the proton–proton reaction and the carbon cycle, which account for the energy of the sun and stars. The fusion of hydrogen into helium provides the energy of the hydrogen bomb. The helium content of the atmosphere is about 1 part in 200,000. It is present in various radioactive minerals as a decay product. Much of the world's supply of helium is obtained from wells in Texas, Colorado, and Kansas. The only other known helium extraction plants, outside the United States, in 1999 were in Poland, Russia, China, Algeria, and India. The cost of helium has fallen from $2500/ft^3 in 1915 to about 2.5¢/cu.ft. (.028 cu meters) in 1999. Helium has the lowest melting point of any element and has found wide use in cryogenic research, as its boiling point is close to absolute zero. Its use in the study of superconductivity is vital. Using liquid helium, Kurti and co-workers, and others, have succeeded in obtaining temperatures of a few microkelvins by the adiabatic demagnetization of copper nuclei, starting from about 0.01 K. Liquid helium (He^4) exists in two forms: He^4I and He^4II, with a sharp transition point at 2.174 K (3.83 cm Hg). He^4I (above this temperature) is a normal liquid, but He^4II (below it) is unlike any other known substance. It expands on cooling; its conductivity for heat is enormous; and neither its heat conduction nor viscosity obeys normal rules. It has other peculiar properties. Helium is the only liquid that cannot be solidified by lowering the temperature. It remains liquid down to absolute zero at ordinary pressures, but it can readily be solidified by increasing the pressure. Solid ^3He and ^4He are unusual in that both can readily be changed in volume by more than 30% by application of pressure. The specific heat of helium gas is unusually high. The density of helium vapor at the normal boiling point is also very high, with the vapor expanding greatly when heated to room temperature. Containers filled with helium gas at 5 to 10 K should be treated as though they contained liquid helium due to the large increase in pressure resulting from warming the gas to room temperature. While helium normally has a 0 valence, it seems to have a weak tendency to combine with certain other elements. Means of preparing helium diflouride have been studied, and species such as HeNe and the molecular ions He^+ and He^{++} have been investigated. Helium is widely used as an inert gas shield for arc welding; as a protective gas in growing silicon and germanium crystals, and in titanium and zirconium production; as a cooling medium for nuclear reactors, and as a gas for supersonic wind tunnels. A mixture of helium and oxygen is used as an artificial atmosphere for divers and others working under pressure. Different ratios of He/O_2 are used for different depths at which the diver is operating. Helium is extensively used for filling balloons as it is a much safer gas than hydrogen. One of the recent largest uses for helium has been for pressurizing liquid fuel rockets. A Saturn booster such as used on the Apollo lunar missions required about 13 million ft^3 of helium for a firing, plus more for checkouts. Liquid helium's use in magnetic resonance imaging (MRI) continues to increase as the medical profession accepts and develops new uses for the equipment. This equipment is providing accurate diagnoses of problems where exploratory surgery has previously been required to determine problems. Another medical application that is being developed uses MRI to determine by blood analysis whether a patient has any form of cancer. Lifting gas applications are increasing. Various companies in addition to Goodyear, are now using "blimps" for advertising. The Navy and the Air Force are investigating the use of airships to provide early warning systems to detect low-flying cruise missiles. The Drug Enforcement Agency has used radar-equipped blimps to detect drug smugglers along the southern border of the U.S. In addition, NASA is currently using helium-filled balloons to sample the atmosphere in Antarctica to determine what is depleting the ozone layer that protects Earth from harmful U.V. radiation. Research on and development of materials which become superconductive at temperatures well above the boiling point of helium could have a major impact on the demand for helium. Less costly refrigerants having boiling points considerably higher could replace the present need to cool such superconductive materials to the boiling point of helium. Natural helium contains two stable isotopes ^3He and ^4He. ^3He is present in very small quantities. Six other isotopes of helium are now recognized.

Holmium — (L. *Holmia*, for Stockholm), Ho; at. wt. 164.93032(2); at. no 67; m.p. 1472 °C; b.p. 2700 °C; sp. gr. 8.795 (25 °C); valence + 3. The spectral absorption bands of holmium were noticed in 1878 by the Swiss chemists Delafontaine and Soret, who announced the existence of an "Element X." Cleve, of Sweden, later independently discovered the element while working on erbia earth. The element is named after Cleve's native city. Pure holmia, the yellow oxide, was prepared by Homberg in 1911. Holmium occurs in *gadolinite, monazite,* and in other rare-earth minerals. It is commercially obtained from monazite, occurring in that mineral to the extent of about 0.05%. It has been isolated by the reduction of its anhydrous chloride or fluoride with calcium metal. Pure holmium has a metallic to bright silver luster. It is relatively soft and malleable, and is stable in dry air at room temperature, but rapidly oxidizes in moist air and at elevated temperatures. The metal has unusual magnetic properties. Few uses have yet been found for the element. The element, as with other rare earths, seems to have a low acute toxic rating. Natural holmium consists of one isotope ^{165}Ho, which is not radioactive. Holmium has 49 other isotopes known, all of which are radioactive. The price of 99.9% holmium metal is about $20/g.

Hydrogen — (Gr. *hydro*, water, and *genes,* forming), H; at. wt. 1.00794(7); at. no. 1; m.p. −259.1 °C; b.p. −252.76 °C; t_c −240.18; density 0.08988 g/L; density (liquid) 0.0708 g/mL (−253 °C); density (solid) 0.0706 g/mL (−262 °C); valence 1. Hydrogen was prepared many years before it was recognized as a distinct substance by Cavendish in 1766. It was named by Lavoisier. Hydrogen is the most abundant of all elements in the universe, and it is thought that the heavier elements were, and still are, being built from hydrogen and helium. It has been estimated that hydrogen makes up more than 90% of all the atoms or three quarters of the mass of the universe. It is found in the sun and most stars, and plays an important part in the proton–proton reaction and carbon–nitrogen cycle, which accounts for the energy of the sun and stars. It is thought that hydrogen is a major component of the planet Jupiter and that at some depth in the planet's interior the pressure is so great that solid molecular hydrogen is converted into solid metallic hydrogen. In 1973, it was reported that a group of Russian experimenters may have produced metallic hydrogen at a pressure of 2.8 Mbar. At the transition the density changed from 1.08 to 1.3 g/cm^3. Earlier, in 1972, a Livermore (California) group also reported on a similar experiment in which they observed a pressure-volume point centered at 2 Mbar. It has been predicted that metallic hydrogen may be metastable; others have predicted it would be a superconductor at room temperature. On Earth, hydrogen occurs chiefly in combination with oxygen in water, but it is also present in organic matter such as living plants, petroleum, coal, etc. It is present as the free element in the atmosphere, but only to the extent of less than 1 ppm by volume. It is the lightest of all gases, and combines with other elements, sometimes explosively, to form compounds. Great quantities of hydrogen are required commercially for the fixation of nitrogen from the air in the Haber ammonia process and for the hydrogenation of fats and oils. It is also used in large quantities in methanol production, in hydrodealkylation, hydrocracking, and hydrodesulfurization. It is also used as a rocket fuel, for welding, for production of hydrochloric acid, for the reduction of metallic ores, and for filling balloons. The lifting power of 1 ft^3 of hydrogen gas is about 0.076 lb at 0 °C, 760 mm pressure. Production of hydrogen in the U.S. alone now amounts to about 3 billion cubic feet per year. It is prepared by the action of steam on heated carbon, by decomposition of certain hydrocarbons with heat, by the electrolysis of water, or by the displacement from acids by certain metals. It is also produced by the action of sodium or potassium hydroxide on aluminum. Liquid hydrogen is important in cryogenics and in the study of superconductivity, as its melting point is only a 20 °C above absolute zero. Hydrogen consists of three isotopes, most of which is ^1H. The ordinary isotope of hydrogen, H, is known as *protium*. In 1932, Urey announced the discovery of a stable isotope, deuterium (^2H or D) with an atomic weight of 2. Deuterium is present in natural hydrogen to the extent of 0.015%. Two years later an unstable isotope, tritium (^3H), with an atomic weight of 3 was discovered. Tritium has a half-life of about 12.32 years. Tritium atoms are also present in natural hydrogen but in a much smaller proportion. Tritium is readily produced in nuclear reactors and is used in the production of the hydrogen bomb. It is also used as a radioactive agent in making luminous paints, and as a tracer. On August 27, 2001 Russian, French, and Japanese physicists working at the Joint Institute for Nuclear Research near Moscow reported they had made "super-heavy hydrogen," which had a nucleus with one proton and four neutrons. Using an accelerator, they used a beam of helium-6 nuclei to strike a hydrogen target, which resulted in the occasional production of a hydrogen-5 nucleus plus a helium-2 nucleus. These unstable particles quickly disintegrated. This resulted in two protons from the He-2, a triton, and two neutrons from the H-5 breakup. Deuterium gas is readily available, without permit, at about $1/l. Heavy water, deuterium oxide (D_2O), which is used as a moderator to slow down neutrons, is available without permit at a cost of 6c to $1/g, depending on quantity and purity. About 1000 tons (4,400,000 kg) of deuterium oxide (heavy water) are now in use at the Sudbury (Ontario) Neutrino Observatory. This observatory is taking data to provide new revolutionary insight into the properties of neutrinos and into the core of the sun. The heavy water is on loan from Atomic Energy of Canada, Ltd. (AECL). The observatory and detectors are located 6800 ft (2072 m) deep in the Creighton mine of the International Nickel Co., near Sudbury. The heavy water is contained in an acrylic vessel, 12 m in diameter. Neutrinos react with the heavy water to produce Cherenkov radiation. This light is then detected with 9600 photomultiplier tubes surrounding the vessel. The detector laboratory is immensely clean to reduce background radiation, which otherwise hides the very weak signals from neutrinos. Quite apart from isotopes, it has been shown that hydrogen gas under ordinary conditions is a mixture of two kinds of molecules, known as *ortho-* and *para-*hydrogen, which differ from one another by the spins of their electrons and nuclei. Normal hydrogen at room temperature contains 25% of the *para* form and 75% of the *ortho* form. The *ortho* form cannot be prepared in the pure state. Since the two forms differ in energy, the physical properties also differ. The melting and boiling points of *para*hydrogen are about 0.1 °C lower than those of normal hydrogen. Consideration is being given to an entire economy based on solar- and nuclear-generated hydrogen. Located in remote regions, power plants would electrolyze sea water; the hydrogen produced would travel to distant cities by pipelines. Pollution-free hydrogen could replace natural gas, gasoline, etc., and could serve as a reducing agent in metallurgy, chemical processing, refining, etc. It could also be used to convert trash into methane and ethylene. Public acceptance, high capital investment, and the high present cost of hydrogen with respect to current fuels are but a few of the problems facing establishment of such an economy. Hydrogen is being investigated as a substitute for deep-sea diving applications below 300 m. Hydrogen is readily available from air product suppliers.

Indium — (from the brilliant indigo line in its spectrum), In; at. wt. 114.818(3); at. no. 49; m.p. 156.60 °C; b.p. 2072 °C; sp. gr. 7.31 (20 °C); valence 1, 2, or 3. Discovered by Reich and Richter, who later isolated the metal. Indium is most frequently associated with zinc materials, and it is from these that most commercial indium is now obtained; however, it is also found in iron, lead, and copper ores. Until 1924, a gram or so constituted the world's supply of this element in isolated form. It is probably about as abundant as silver. About 4 million troy ounces of indium are now produced annually in the Free World. Canada is presently producing more than 1,000,000 troy ounces annually. The present cost of indium is about $2 to $10/g, depending on quantity and purity. It is available in ultrapure form. Indium is a very soft, silvery-white metal with a brilliant luster. The pure metal gives a high-pitched "cry" when bent. It wets glass, as does gallium. It has found application in making low-melting alloys; an alloy of 24% indium–76% gallium is liquid at room temperature. Indium is used in making bearing alloys, germanium transistors, rectifiers, thermistors, liquid crystal displays, high definition television, batteries, and photoconductors. It can be plated onto metal and evaporated onto glass, forming a mirror as good as that made with silver but with more resistance to atmospheric corrosion. There is evidence that indium has a low order of toxicity; however, care should be taken until further information is available. Seventy isotopes and isomers are now recognized (more than any other element). Natural indium contains two isotopes. One is stable. The other, ^{115}In, comprising 95.71% of natural indium is slightly radioactive with a very long half-life.

Iodine — (Gr. *iodes*, violet), I; at. wt. 126.90447(3); at. no. 53; m.p. 113.7 °C; b.p. 184.4 °C; t_c 546 °C; density of the gas 11.27 g/L; sp. gr. solid 4.93 (20 °C); valence 1, 3, 5, or 7. Discovered by Courtois in 1811. Iodine, a halogen, occurs sparingly in the form of iodides in sea water from which it is assimilated by seaweeds, in Chilean saltpeter and nitrate-bearing earth, known as *caliche* in brines from old sea deposits, and in brackish waters from oil and salt wells. Ultrapure iodine can be obtained from the reaction of potassium iodide with copper sulfate. Several other methods of isolating the element are known. Iodine is a bluish-black, lustrous solid, volatilizing at ordinary temperatures into a blue-violet gas with an irritating odor; it forms compounds with many elements, but is less active than the other halogens, which displace it from iodides. Iodine exhibits some metallic-like properties. It dissolves readily in chloroform, carbon tetrachloride, or carbon disulfide to form beautiful purple solutions. It is only slightly soluble in water. Iodine compounds are important in organic chemistry and very useful in medicine. Forty-two isotopes and isomers are recognized. Only one stable isotope, ^{127}I, is found in nature. The artificial radioisotope ^{131}I, with a half-life of 8 days, has been used in treating the thyroid gland. The most common compounds are the iodides of sodium and potassium (KI) and the iodates (KIO_3). Lack of iodine is the cause of goiter. Iodides and thyroxin, which contains iodine, are used internally in medicine, and a solution of KI and iodine in alcohol is used for external wounds. Potassium iodide finds use in photography. The deep blue color with starch solution is characteristic of the free element. Care should be taken in handling and using iodine, as contact with the skin can cause lesions; iodine vapor is intensely irritating to the eyes and mucous membranes. Elemental iodine costs about 25 to 75¢/g depending on purity and quantity.

Iridium — (L. *iris*, rainbow), Ir; at. wt. 192.217(3); at. no. 77; m.p. 2446 °C; b.p. 4428 °C; sp. gr. 22.562 (20 °C); valence 3 or 4. Discovered in 1803 by Tennant in the residue left when crude platinum is dissolved by aqua regia. The name iridium is appropriate, for its salts are highly colored. Iridium, a metal of the platinum family, is white, similar to platinum, but with a slight yellowish cast. It is very hard and brittle, making it very hard to machine, form, or work. It is the most corrosion-resistant metal known, and was used in making the standard meter bar of Paris, which is a 90% platinum–10% iridium alloy. This meter bar was replaced in 1960 as a fundamental unit of length (see under Krypton). Iridium is not attacked by any of the acids nor by aqua regia, but is attacked by molten salts, such as NaCl and NaCN. Iridium occurs uncombined in nature with platinum and other metals of this family in alluvial deposits. It is recovered as a by-product from the nickel mining industry. The largest reserves and production of the platinum group of metals, which includes iridium, is in South Africa, followed by Russia and Canada. The U.S. has only one active mine, located at Nye, MT. The presence of iridium has recently been used in examining the Cretaceous-Tertiary (K-T) boundary. Meteorites contain small amounts of iridium. Because iridium is found widely distributed at the K-T boundary, it has been suggested that a large meteorite or asteroid collided with the Earth, killing the dinosaurs, and creating a large dust cloud and crater. Searches for such a crater point to one in the Yucatan, known as Chicxulub. Iridium has found use in making crucibles and apparatus for use at high temperatures. It is also used for electrical contacts. Its principal use is as a hardening agent for platinum. With osmium, it forms an alloy that is used for tipping pens and compass bearings. The specific gravity of iridium is only very slightly lower than that of osmium, which has been generally credited as being the heaviest known element. Calculations of the densities of iridium and osmium from the space lattices give values of 22.65 and 22.61 g/cm^3, respectively. These values may be more reliable than actual physical measurements. At present, therefore, we know that either iridium or osmium is the densest known element, but the data do not yet allow selection between the two. Natural iridium contains two stable isotopes. Forty-five other isotopes, all radioactive, are now recognized. Iridium (99.9%) costs about $100/g.

Iron — (Anglo-Saxon, *iron*), Fe (L. *ferrum*); at. wt. 55.845(2); at. no. 26; m.p. 1538 °C; b.p. 2861 °C; sp. gr. 7.874 (20 °C); valence 2, 3, 4, or 6. The use of iron is prehistoric. Genesis mentions that Tubal-Cain, seven generations from Adam, was "an instructor of every artificer in brass and iron." A remarkable iron pillar, dating to about A.D. 400, remains standing today in Delhi, India. This solid shaft of wrought iron is about 7¼ m high by 40 cm in diameter. Corrosion to the pillar has been minimal although it has been exposed to the weather since its erection. Iron is a relatively abundant element in the universe. It is found in the sun and many types of stars in considerable quantity. It has been suggested that the iron we have here on Earth may have originated in a supernova. Iron is a very difficult element to produce in ordinary nuclear reactions, such as would take place in the sun. Iron is found native as a principal component of a class of iron–nickel meteorites known as *siderites*, and is a minor constituent of the other two classes of meteorites. The core of the Earth, 2150 miles in radius, is thought to be largely composed of iron with about 10% occluded hydrogen. The metal is the fourth most abundant element, by weight, making up the crust of the Earth. The most common ore is *hematite* (Fe_2O_3). Magnetite (Fe_3O_4) is frequently seen as *black sands* along beaches and banks of streams. *Lodestone* is another form of magnetite. *Taconite* is becoming increasingly important as a commercial ore. Iron is a vital constituent of plant and animal life, and appears in hemoglobin. The pure metal is not often encountered in commerce, but is usually alloyed with carbon or other metals. The pure metal is very reactive chemically, and rapidly corrodes, especially in moist air or at elevated temperatures. It has four allotropic forms, or ferrites, known as α, β, γ, and δ, with transition points at 700, 928, and 1530 °C. The α form is magnetic, but when transformed into the β form, the magnetism disappears although the lattice remains unchanged. The relations of these forms are peculiar. Pig iron is an alloy containing about 3% carbon with varying amounts of S, Si, Mn, and P. It is hard, brittle, fairly fusible, and is used to produce other alloys, including steel. Wrought iron contains only a few tenths of a percent of carbon, is tough, malleable, less fusible, and usually has a "fibrous" structure. Carbon steel is an alloy of iron with carbon, with small amounts of Mn, S, P, and Si. Alloy steels are carbon steels with other additives such as nickel, chromium, vanadium, etc. Iron is the cheapest and most abundant, useful, and important of all metals. Natural iron contains four isotopes. Twenty-six other isotopes and isomers, all radioactive, are now recognized.

Krypton (Gr. *kryptos*, hidden), Kr, at. wt. 83.798(2); at. no. 36; m.p. −157.36 °C; b.p. −153.34 ± 0.10 °C; t_c −63.67 °C; density 3.733 g/L (0 °C); valence usually 0. Discovered in 1898 by Ramsay and Travers in the residue left after liquid air had nearly boiled away, krypton is present in the air to the extent of about 1 ppm. The atmosphere of Mars has been found to contain 0.3 ppm of krypton. It is one of the "noble" gases. It is characterized by its brilliant green and orange spectral lines. Naturally occurring krypton contains six stable isotopes. Thirty other unstable isotopes and isomers are now recognized. The spectral lines of krypton are easily produced and some are very sharp. In 1960 it was internationally agreed that the fundamental unit of length, the meter, should be defined in terms of the orange-red spectral line of ^{86}Kr. This replaced the standard meter of Paris, which was defined in terms of a bar made of a platinum-iridium alloy. In October 1983 the meter was again redefined by the International Bureau of Weights and Measures as being the length of path traveled by light in a vacuum during a time interval of 1/299,792,458 of a second. Solid krypton is a white crystalline substance with a face-centered cubic structure that is common to all the rare gases. While krypton is generally thought of as a noble gas that normally does not combine with other elements, the existence of some krypton compounds has been established. Krypton difluoride has been prepared in gram quantities and can be made by several methods. A higher fluoride of krypton and a salt of an oxyacid of krypton also have been prepared. Molecule-ions of $ArKr^+$ and KrH^+ have been identified and investigated, and evidence is provided for the formation of KrXe or $KrXe^+$. Krypton clathrates have been prepared with hydroquinone and phenol. ^{85}Kr has found recent application in chemical analysis. By imbedding the isotope in various solids, *kryptonates* are formed. The activity of these kryptonates is sensitive to chemical reactions at the surface. Estimates of the concentration of reactants are therefore made possible. Krypton is used in certain photographic flash lamps for high-speed photography. Uses thus far have been limited because of its high cost. Krypton gas presently costs about $690/100 L.

Kurchatovium — See Rutherfordium.

Lanthanum — (Gr. *lanthanein*, to lie hidden), La; at. wt. 138.90547(7); at. no. 57; m.p. 920 °C; b.p. 3464 °C; sp. gr. 6.145 (25 °C); valence 3. Mosander in 1839 extracted a new earth *lanthana*, from impure cerium nitrate, and recognized the new element. Lanthanum is found in rare-earth minerals such as *cerite, monazite, allanite,* and *bastnasite.* Monazite and bastnasite are principal ores in which lanthanum occurs in percentages up to 25 and 38%, respectively. Misch metal, used in making lighter flints, contains about 25% lanthanum. Lanthanum was isolated in relatively pure form in 1923. Ion-exchange and solvent extraction techniques have led to much easier isolation of the so-called "rare-earth" elements. The availability of lanthanum and other rare earths has improved greatly in recent years. The metal can be produced by reducing the anhydrous fluoride with calcium. Lanthanum is silvery white, malleable, ductile, and soft enough to be cut with a knife. It is one of the most reactive of the rare-earth metals. It oxidizes rapidly when exposed to air. Cold water attacks lanthanum slowly, and hot water attacks it much more rapidly. The metal reacts directly with elemental carbon, nitrogen, boron, selenium, silicon, phosphorus, sulfur, and with halogens. At 310 °C, lanthanum changes from a hexagonal to a face-centered cubic structure, and at 865 °C it again transforms into a body-centered cubic structure. Natural lanthanum is a mixture of two isotopes, one of which is stable and one of which is radioactive with a very long half-life. Thirty other radioactive isotopes are recognized. Rare-earth compounds containing lanthanum are extensively used in carbon lighting applications, especially by the motion picture industry for studio lighting and projection. This application consumes about 25% of the rare-earth compounds produced. La_2O_3 improves the alkali resistance of glass, and is used in making special optical glasses. Small amounts of lanthanum, as an additive, can be used to produce nodular cast iron. There is current interest in hydrogen sponge alloys containing lanthanum. These alloys take up to 400 times their own volume of hydrogen gas, and the process is reversible. Heat energy is released every time they do so; therefore these alloys have possibilities in energy conservation systems. Lanthanum and its compounds have a low to moderate acute toxicity rating; therefore, care should be taken in handling them. The metal costs about $2/g (99.9%).

Lawrencium — (Ernest O. Lawrence [1901–1958], inventor of the cyclotron), Lr; at. no. 103; at. mass no. [262]; valence + 3(?). This member of the 5f transition elements (actinide series) was discovered in March 1961 by A. Ghiorso, T. Sikkeland, A. E. Larsh, and R. M. Latimer. A 3-μg californium target, consisting of a mixture of isotopes of mass number 249, 250, 251, and 252, was bombarded with either ^{10}B or ^{11}B. The electrically charged transmutation nuclei recoiled with an atmosphere of helium and were collected on a thin copper conveyor tape which was then moved to place collected atoms in front of a series of solid-state detectors. The isotope of element 103 produced in this way decayed by emitting an 8.6-MeV alpha particle with a half-life of 8 s. In 1967, Flerov and associates of the Dubna Laboratory reported their inability to detect an alpha emitter with a half-life of 8 s which was assigned by the Berkeley group to $^{257}103$. This assignment has been changed to ^{258}Lr or ^{259}Lr. In 1965, the Dubna workers found a longer-lived lawrencium isotope, ^{256}Lr, with a half-life of 35 s. In 1968, Ghiorso and associates at Berkeley were able to use a few atoms of this isotope to study the oxidation behavior of lawrencium. Using solvent extraction techniques and working very rapidly, they extracted lawrencium ions from a buffered aqueous solution into an organic solvent, completing each extraction in about 30 s. It was found that lawrencium behaves differently from dipositive nobelium and more like the tripositive elements earlier in the actinide series. Ten isotopes of lawrencium are now recognized.

Lead — (Anglo-Saxon *lead*), Pb (L. *plumbum*); at. wt. 207.2(1); at. no. 82; m.p. 327.46 °C; b.p. 1749 °C; sp. gr. 11.35 (20 °C); valence 2 or 4. Long known, mentioned in Exodus. The alchemists believed lead to be the oldest metal and associated it with the planet Saturn. Native lead occurs in nature, but it is rare. Lead is obtained chiefly from *galena* (PbS) by a roasting process. *Anglesite* ($PbSO_4$), *cerussite* ($PbCO_3$), and *minim* (Pb_3O_4) are other common lead minerals. Lead is a bluish-white metal of bright luster, is very soft, highly malleable, ductile, and a poor conductor of electricity. It is very resistant to corrosion; lead pipes bearing the insignia of Roman emperors, used as drains from the baths, are still in service. It is used in containers for corrosive liquids (such as sulfuric acid) and may be toughened by the addition of a small percentage of antimony or other metals. Natural lead is a mixture of four stable isotopes: ^{204}Pb (1.4%), ^{206}Pb (24.1%), ^{207}Pb (22.1%), and ^{208}Pb (52.4%). Lead isotopes are the end products of each of the three series of naturally occurring radioactive elements: ^{206}Pb for the uranium series, ^{207}Pb for the actinium series, and ^{208}Pb for the thorium series. Forty-three other isotopes of lead, all of which are radioactive, are recognized. Its alloys include solder, type metal, and various antifriction metals. Great quantities of lead, both as the metal and as the dioxide, are used in storage batteries. Lead is also used for cable covering, plumbing, and ammunition. The metal is very effective as a sound absorber, is used as a radiation shield around X-ray equipment and nuclear reactors, and is used to absorb vibration. Lead, alloyed with tin, is used in making organ pipes. White lead, the basic carbonate, sublimed white lead ($PbSO_4$), chrome yellow ($PbCrO_4$), red lead (Pb_3O_4), and other lead compounds are used extensively in paints, although in recent years the use of lead in paints has been drastically curtailed to eliminate or reduce health hazards. Lead oxide is used in producing fine "crystal glass" and "flint glass" of a high index of refraction for achromatic lenses. The nitrate and the acetate are soluble salts. Lead salts such as lead arsenate have been used as insecticides, but their use in recent years has been practically eliminated in favor of less harmful organic compounds. Care must be used in handling lead as it is a cumulative poison. Environmental concern with lead poisoning led to elimination of lead tetraethyl in gasoline. The U.S. Occupational Safety and Health Administration (OSHA) has recommended that industries limit airborne lead to 50 μg/cu. meter. Lead is priced at about 90¢/kg (99.9%).

Lithium — (Gr. *lithos*, stone), Li; at. wt. 6.941(2); at. no. 3; m.p. 180.5 °C; b.p. 1342 °C; sp. gr. 0.534 (20 °C); valence 1. Discovered by Arfvedson in 1817. Lithium is the lightest of all metals, with a density only about half that of water. It does not occur free in nature; combined it is found in small amounts in nearly all igneous rocks and in the waters of many mineral springs. *Lepidolite, spodumene, petalite,* and *amblygonite* are the more important minerals containing it. Lithium is presently being recovered from brines of Searles Lake, in California, and from Nevada, Chile, and Argentina. Large deposits of spodumene are found in North Carolina. The metal is produced electrolytically from the fused chloride. Lithium is silvery in appearance, much like Na and K, other members of the alkali metal series. It reacts with water, but not as vigorously as sodium. Lithium imparts a beautiful crimson color to a flame, but when the metal burns strongly the flame is a dazzling white. Since World War II, the production of lithium metal and its compounds has increased greatly. Because the metal has the highest specific heat of any solid element, it has found use in heat transfer applications; however, it is corrosive and requires special handling. The metal has been used as an alloying agent, is of interest in synthesis of organic compounds, and has nuclear applications. It ranks as a leading contender as a battery anode material because it has a high electrochemical potential. Lithium is used in special glasses and ceramics. The glass for the 200-inch telescope at Mt. Palomar contains lithium as a minor ingredient. Lithium chloride is one of the most hygroscopic materials known, and it, as well as lithium bromide, is used in air conditioning and industrial drying systems. Lithium stearate is used as an all-purpose and high-temperature lubricant. Other lithium compounds are used in dry cells and storage batteries. Seven isotopes of lithium are recognized. Natural lithium contains two isotopes. The metal is priced at about $1.50/g (99.9%).

Lutetium — (Lutetia, ancient name for Paris, sometimes called *cassiopeium* by the Germans), Lu; at. wt. 174.967(1); at. no. 71; m.p. 1663 °C; b.p. 3402 °C; sp. gr. 9.841 (25 °C); valence 3. In 1907, Urbain described a process by which Marignac's ytterbium (1879) could be separated into the two elements, ytterbium (neoytterbium) and lutetium. These elements were identical with "aldebaranium" and "cassiopeium," independently discovered by von Welsbach about the same time. Charles James of the University of New Hampshire also independently prepared the very pure oxide, *lutecia*, at this time. The spelling of the element was changed from *lutecium* to *lutetium* in 1949. Lutetium occurs in very small amounts in nearly all minerals containing yttrium, and is present in *monazite* to the extent of about 0.003%, which is a commercial source. The pure metal has been isolated only in recent years and is one of the most difficult to prepare. It can be prepared by the reduction of anhydrous $LuCl_3$ or LuF_3 by an alkali or alkaline earth metal. The metal is silvery white and relatively stable in air. While new techniques, including ion-exchange reactions, have been developed to separate the various rare-earth elements, lutetium is still the most costly of all rare earths. It is priced at about $100/g (99.9%). ^{176}Lu occurs naturally (97.41%) with ^{175}Lu (2.59%), which is radioactive with a very long half-life of about 4×10^{10} years. Lutetium has 50 isotopes and isomers that are now recognized. Stable lutetium nuclides, which emit pure beta radiation after thermal neutron activation, can be used as catalysts in cracking, alkylation, hydrogenation, and polymerization. Virtually no other commercial uses have been found yet for lutetium. While lutetium, like other rare-earth metals, is thought to have a low toxicity rating, it should be handled with care until more information is available.

Magnesium — (*Magnesia,* district in Thessaly) Mg; at. wt. 24.3050(6); at. no. 12; m.p. 650 °C; b.p. 1090 °C; sp. gr. 1.738 (20 °C); valence 2. Compounds of magnesium have long been known. Black recognized magnesium as an element in 1755. It was isolated by Davy in 1808, and prepared in coherent form by Bussy in 1831. Magnesium is the eighth most abundant element in the Earth's crust. It does not occur uncombined, but is found in large deposits in the form of *magnesite, dolomite,* and other minerals. The metal is now principally obtained in the U.S. by electrolysis of fused magnesium chloride derived from brines, wells, and sea water. Magnesium is a light, silvery-white, and fairly tough metal. It tarnishes slightly in air, and finely divided magnesium readily ignites upon heating in air and burns with a dazzling white flame. It is used in flashlight photography, flares, and pyrotechnics, including incendiary bombs. It is one third lighter than aluminum, and in alloys is essential for airplane and missile construction. The metal improves the mechanical, fabrication, and welding characteristics of aluminum when used as an alloying agent. Magnesium is used in producing nodular graphite in cast iron, and is used as an additive to conventional propellants. It is also used as a reducing agent in the production of pure uranium and other metals from their salts. The hydroxide (*milk of magnesia*), chloride, sulfate (*Epsom salts*), and citrate are used in medicine. Dead-burned magnesite is employed for refractory purposes such as brick and liners in furnaces and converters. Calcined magnesia is also used for water treatment and in the manufacture of rubber, paper, etc. Organic magnesium compounds (Grignard's reagents) are important. Magnesium is an important element in both plant and animal life. Chlorophylls are magnesium-centered porphyrins. The adult daily requirement of magnesium is about 300 mg/day, but this is affected by various factors. Great care should be taken in handling magnesium metal, especially in the finely divided state, as serious fires can occur. Water should not be used on burning magnesium or on magnesium fires. Natural magnesium contains three isotopes. Twelve other isotopes are recognized. Magnesium metal costs about $100/kg (99.8%).

Manganese — (L. *magnes*, magnet, from magnetic properties of pyrolusite; It. *manganese*, corrupt form of *magnesia*), Mn; at. wt. 54.938045(5); at. no. 25; m.p. 1246 °C; b.p. 2061 °C; sp. gr. 7.21 to 7.44, depending on allotropic form; valence 1, 2, 3, 4, 6, or 7. Recognized by Scheele, Bergman, and others as an element and isolated by Gahn in 1774 by reduction of the dioxide with carbon. Manganese minerals are widely distributed; oxides, silicates, and carbonates are the most common. The discovery of large quantities of manganese nodules on the floor of the oceans holds promise as a source of manganese. These nodules contain about 24% manganese together with many other elements in lesser abundance. Most manganese today is obtained from ores found in Ukraine, Brazil, Australia, Republic of So. Africa, Gabon, China, and India. *Pyrolusite* (MnO_2) and *rhodochrosite* ($MnCO_3$) are among the most common manganese minerals. The metal is obtained by reduction of the oxide with sodium, magnesium, aluminum, or by electrolysis. It is gray-white, resembling iron, but is harder and very brittle. The metal is reactive chemically, and decomposes in cold water slowly. Manganese is used to form many important alloys. In steel, manganese improves the rolling and forging qualities, strength, toughness, stiffness, wear resistance, hardness, and hardenability. With aluminum and antimony, especially with small amounts of copper, it forms highly ferromagnetic alloys. Manganese metal is ferromagnetic only after special treatment. The pure metal exists in four allotropic forms. The alpha form is stable at ordinary temperature; gamma manganese, which changes to alpha at ordinary temperatures, is soft, easily cut, and capable of being bent. The dioxide (pyrolusite) is used as a depolarizer in dry cells, and is used to "decolorize" glass that is colored green by impurities of iron. Manganese by itself colors glass an amethyst color, and is responsible for the color of true amethyst. The dioxide is also used in the preparation of oxygen and chlorine, and in drying black paints. The permanganate is a powerful oxidizing agent and is used in quantitative analysis and in medicine. Manganese is widely distributed throughout the animal kingdom. It is an important trace element and may be essential for utilization of vitamin B_1. Twenty-seven isotopes and isomers are known. Manganese metal (99.95%) is priced at about $800/kg. Metal of 99.6% purity is priced at about $80/kg.

Meitnerium — (Lise Meitner [1878–1968], Austrian–Swedish physicist and mathematician), Mt; at. wt [268]; at. no. 109. In 1982, Element 109 was made and identified by physicists at the Heavy Ion Research Laboratory (G.S.I.), Darmstadt, Germany, by bombarding a target of ^{209}Bi with accelerated nuclei of ^{58}Fe. The production of Element 109 has been extremely small. It took a week of target bombardment (10^{11} nuclear encounters) to produce a single atom of 109. Oganessian and his team at Dubna in 1984 repeated the Darmstadt experiment using a tenfold irradiation dose. One fission event from seven alpha decays of 109 was observed, thus indirectly confirming the existence of isotope $^{266}109$. In August 1997, the IUPAC adopted the name *meitnerium* for this element, honoring L. Meitner. Four isotopes of *meitnerium* are now recognized.

Mendelevium — (Dmitri Mendeleev [1834–1907]), Md; at. wt. (258); at. no. 101; m.p. 827 °C; valence +2, +3. Mendelevium, the ninth transuranium element of the actinide series to be discovered, was first identified by Ghiorso, Harvey, Choppin, Thompson, and Seaborg early in 1955 as a result of the bombardment of the isotope ^{253}Es with helium ions in the Berkeley 60-inch cyclotron. The isotope produced was ^{256}Md, which has a half-life of 78 min. This first identification was notable in that ^{256}Md was synthesized on a one-atom-at-a-time basis. Nineteen isotopes and isomers are now recognized. ^{258}Md has a half-life of 51.5 days. This isotope has been produced by the bombardment of an isotope of einsteinium with ions of helium. It now appears possible that eventually enough ^{258}Md can be made so that some of its physical properties can be determined. ^{256}Md has been used to elucidate some of the chemical properties of mendelevium in aqueous solution. Experiments seem to show that the element possesses a moderately stable dipositive (II) oxidation state in addition to the tripositive (III) oxidation state, which is characteristic of actinide elements.

Mercury — (Planet *Mercury*), Hg (*hydrargyrum*, liquid silver); at. wt. 200.59(2); at. no. 80; t.p. −38.83 °C; b.p. 356.62 °C; t_c 1477 °C; sp. gr. 13.546 (20 °C); valence 1 or 2. Known to ancient Chinese and Hindus; found in Egyptian tombs of 1500 B.C. Mercury is the only common metal liquid at ordinary temperatures. It only rarely occurs free in nature. The chief ore is *cinnabar* (HgS). Spain and China produce about 75% of the world's supply of the metal. The commercial unit for handling mercury is the "flask," which weighs 76 lb (34.46 kg). The metal is obtained by heating cinnabar in a current of air and by condensing the vapor. It is a heavy, silvery-white metal; a rather poor conductor of heat, as compared with other metals, and a fair conductor of electricity. It easily forms alloys with many metals, such as gold, silver, and tin, which are called *amalgams.* Its ease in amalgamating with gold is made use of in the recovery of gold from its ores. The metal is widely used in laboratory work for making thermometers, barometers, diffusion pumps, and many other instruments. It is used in making mercury-vapor lamps and advertising signs, etc. and is used in mercury switches and other electrical apparatus. Other uses are in making pesticides, mercury cells for caustic soda and chlorine production, dental preparations, antifouling paint, batteries, and catalysts. The most important salts are mercuric chloride $HgCl_2$ (corrosive sublimate — a violent poison), mercurous chloride Hg_2Cl_2 (calomel, occasionally still used in medicine), mercury fulminate ($Hg(ONC)_2$), a detonator widely used in explosives, and mercuric sulfide (HgS, vermillion, a high-grade paint pigment). Organic mercury compounds are important. It has been found that an electrical discharge causes mercury vapor to combine with neon, argon, krypton, and xenon. These products, held together with van der Waals' forces, correspond to HgNe, HgAr, HgKr, and HgXe. Mercury is a virulent poison and is readily absorbed through the respiratory tract, the gastrointestinal tract, or through unbroken skin. It acts as a cumulative poison and dangerous levels are readily attained in air. Air saturated with mercury vapor at 20 °C contains a concentration that exceeds the toxic limit many times. The danger increases at higher temperatures. *It is therefore important that mercury be handled with care.* Containers of mercury should be securely covered and spillage should be avoided. If it is necessary to heat mercury or mercury compounds, it should be done in a well-ventilated hood. Methyl mercury is a dangerous pollutant and is now widely found in water and streams. The triple point of mercury, −38.8344 °C, is a fixed point on the International Temperature Scale (ITS-90). Mercury (99.98%) is priced at about $110/kg. Native mercury contains seven isotopes. Thirty-six other isotopes and isomers are known.

Molybdenum — (Gr. *molybdos*, lead), Mo; at. wt. 95.94(2); at. no. 42; m.p. 2623 °C; b.p. 4639 °C; sp. gr. 10.22 (20 °C); valence 2, 3, 4?, 5?, or 6. Before Scheele recognized molybdenite as a distinct ore of a new element in 1778, it was confused with graphite and lead ore. The metal was prepared in an impure form in 1782 by Hjelm. Molybdenum does not occur native, but is obtained principally from *molybdenite* (MoS_2). *Wulfenite* ($PbMoO_4$) and *powellite* ($Ca(MoW)O_4$) are also minor commercial ores. Molybdenum is also recovered as a by-product of copper and tungsten mining operations. The U.S., Canada, Chile, and China produce most of the world's molybdenum ores. The metal is prepared from the powder made by the hydrogen reduction of purified molybdic trioxide or ammonium molybdate. The metal is silvery white, very hard, but is softer and more ductile than tungsten. It has a high elastic modulus, and only tungsten and tantalum, of the more readily available metals, have higher melting points. It is a valuable alloying agent, as it contributes to the hardenability and toughness of quenched and tempered steels. It also improves the strength of steel at high temperatures. It is used in certain nickel-based alloys, such as the Hastelloys® which are heat-resistant and corrosion-resistant to chemical solutions. Molybdenum oxidizes at elevated temperatures. The metal has found recent application as electrodes for electrically heated glass furnaces and forehearths. It is also used in nuclear energy applications and for missile and aircraft parts. Molybdenum is valuable as a catalyst in the refining of petroleum. It has found application as a filament material in electronic and electrical applications. Molybdenum is an essential trace element in plant nutrition. Some lands are barren for lack of this element in the soil. Molybdenum sulfide is useful as a lubricant, especially at high temperatures where oils would decompose. Almost all ultra-high strength steels with minimum yield points up to 300,000 lb/in.^{2} contain molybdenum in amounts from 0.25 to 8%. Natural molybdenum contains seven isotopes. Thirty other isotopes and isomers are known, all of which are radioactive. Molybdenum metal costs about $1/g (99.999% purity). Molybdenum metal (99.9%) costs about $160/kg.

Neodymium — (Gr. *neos*, new, and *didymos*, twin), Nd; at. wt. 144.242(3); at. no. 60; m.p. 1016 °C; b.p. 3074 °C; sp. gr. 7.008 (25 °C); valence 3. In 1841 Mosander extracted from *cerite* a new rose-colored oxide, which he believed contained a new element. He named the element *didymium*, as it was *an inseparable twin brother of lanthanum.* In 1885 von Welsbach separated didymium into two new elemental components, *neodymia* and *praseodymia*, by repeated fractionation of ammonium didymium nitrate. While the free metal is in *misch metal*, long known and used as a pyrophoric alloy for light flints, the element was not isolated in relatively pure form until 1925. Neodymium is present in misch metal to the extent of about 18%. It is present in the minerals *monazite* and *bastnasite*, which are principal sources of rare-earth metals. The element may be obtained by separating neodymium salts from other rare earths by ion-exchange or solvent extraction techniques, and by reducing anhydrous halides such as NdF_3 with calcium metal. Other separation techniques are possible. The metal has a bright silvery metallic luster. Neodymium is one of the more reactive rare-earth metals and quickly tarnishes in air, forming an oxide that splits off and exposes metal to oxidation. The metal, therefore, should be kept under light mineral oil or sealed in a plastic material. Neodymium exists in two allotropic forms, with a transformation from a double hexagonal to a body-centered cubic structure taking place at 863 °C. Natural neodymium is a mixture of seven isotopes, one of which has a very long half-life. Twenty-seven other radioactive isotopes and isomers are recognized. Didymium, of which neodymium is a component, is used for coloring glass to make welder's goggles. By itself, neodymium colors glass delicate shades ranging from pure violet through wine-red and warm gray. Light transmitted through such glass shows unusually sharp absorption bands. The glass has been used in astronomical work to produce sharp bands by which spectral lines may be calibrated. Glass containing neodymium can be used as a laser material to produce coherent light. Neodymium salts are also used as a colorant for enamels. The element is also being used with iron and boron to produce extremely strong magnets. These are the most compact magnets commercially available. The price of the metal is about \$4/g. Neodymium has a low-to-moderate acute toxic rating. As with other rare earths, neodymium should be handled with care.

Neon — (Gr. *neos*, new), Ne; at. wt. 20.1797(6); at. no. 10; t.p. −248.609 °C; b.p. −246.053 °C; t_c −228.7 °C; density of gas 0.89990 g/L (1 atm, 0 °C); density of liquid at b.p. 1.204 g/cm³; valence 0. Discovered by Ramsay and Travers in 1898. Neon is a rare gaseous element present in the atmosphere to the extent of 1 part in 65,000 of air. It is obtained by liquefaction of air and separated from the other gases by fractional distillation. Natural neon is a mixture of three isotopes. Fourteen other unstable isotopes are known. It is very inert element; however, it is said to form a compound with fluorine. It is still questionable if true compounds of neon exist, but evidence is mounting in favor of their existence. The following ions are known from optical and mass spectrometric studies: Ne^+, $(NeAr)^+$, $(NeH)^+$, and $(HeNe^+)$. Neon also forms an unstable hydrate. In a vacuum discharge tube, neon glows reddish orange. Of all the rare gases, the discharge of neon is the most intense at ordinary voltages and currents. Neon is used in making the common neon advertising signs, which accounts for its largest use. It is also used to make high-voltage indicators, lightning arrestors, wave meter tubes, and TV tubes. Neon and helium are used in making gas lasers. Liquid neon is now commercially available and is finding important application as an economical cryogenic refrigerant. It has over 40 times more refrigerating capacity per unit volume than liquid helium and more than three times that of liquid hydrogen. It is compact, inert, and is less expensive than helium when it meets refrigeration requirements. Neon costs about \$800/80 cu. ft. (2265 l).

Neptunium — (Planet *Neptune*), Np; at. wt. (237); at. no. 93; m.p. 644 °C; sp. gr. 20.25 (20 °C); valence 3, 4, 5, and 6. Neptunium was the first synthetic transuranium element of the actinide series discovered; the isotope ^{239}Np was produced by McMillan and Abelson in 1940 at Berkeley, California, as the result of bombarding uranium with cyclotron-produced neutrons. The isotope ^{237}Np (half-life of 2.14 × 10^6 years) is currently obtained in gram quantities as a by-product from nuclear reactors in the production of plutonium. Twenty-three isotopes and isomers of neptunium are now recognized. Trace quantities of the element are actually found in nature due to transmutation reactions in uranium ores produced by the neutrons which are present. Neptunium is prepared by the reduction of NpF_3 with barium or lithium vapor at about 1200 °C. Neptunium metal has a silvery appearance, is chemically reactive, and exists in at least three structural modifications: α-neptunium, orthorhombic, density 20.25 g/cm³, β-neptunium (above 280 °C), tetragonal, density (313 °C) 19.36 g/cm³; γ-neptunium (above 577 °C), cubic, density (600 °C) 18.0 g/cm³. Neptunium has four ionic oxidation states in solution: Np^{+3} (pale purple), analogous to the rare earth ion Pm^{+3}, Np^{+4} (yellow green); NpO^+ (green blue); and NpO^{++} (pale pink). These latter oxygenated species are in contrast to the rare earths that exhibit only simple ions of the (II), (III), and (IV) oxidation states in aqueous solution. The element forms tri- and tetrahalides such as NpF_3, NpF_4, $NpCl_4$, $NpBr_3$, NpI_3, and oxides of various compositions such as are found in the uranium-oxygen system, including Np_3O_8 and NpO_2.

Nickel — (Ger. *Nickel*, Satan or Old Nick's and from *kupfernickel*, Old Nick's copper), Ni; at. wt. 58.6934(2); at. no. 28; m.p. 1455 °C; b.p. 2913 °C; sp. gr. 8.902 (25 °C); valence 0, 1, 2, 3. Discovered by Cronstedt in 1751 in kupfernickel (*niccolite*). Nickel is found as a constituent in most meteorites and often serves as one of the criteria for distinguishing a meteorite from other minerals. Iron meteorites, or *siderites*, may contain iron alloyed with from 5 to nearly 20% nickel. Nickel is obtained commercially from *pentlandite* and *pyrrhotite* of the Sudbury region of Ontario, a district that produces much of the world's nickel. It is now thought that the Sudbury deposit is the result of an ancient meteorite impact. Large deposits of nickel, cobalt, and copper have recently been developed at Voisey's Bay, Labrador. Other deposits of nickel are found in Russia, New Caledonia, Australia, Cuba, Indonesia, and elsewhere. Nickel is silvery white and takes on a high polish. It is hard, malleable, ductile, somewhat ferromagnetic, and a fair conductor of heat and electricity. It belongs to the iron-cobalt group of metals and is chiefly valuable for the alloys it forms. It is extensively used for making stainless steel and other corrosion-resistant alloys such as Invar®, Monel®, Inconel®, and the Hastelloys®. Tubing made of a copper-nickel alloy is extensively used in making desalination plants for converting sea water into fresh water. Nickel is also now used extensively in coinage and in making nickel steel for armor plate and burglar-proof vaults, and is a component in Nichrome®, Permalloy®, and constantan. Nickel added to glass gives a green color. Nickel plating is often used to provide a protective coating for other metals, and finely divided nickel is a catalyst for hydrogenating vegetable oils. It is also used in ceramics, in the manufacture of Alnico magnets, and in batteries. The sulfate and the oxides are important compounds. Natural nickel is a mixture of five stable isotopes; twenty-five other unstable isotopes are known. Nickel sulfide fume and dust, as well as other nickel compounds, are carcinogens. Nickel metal (99.9%) is priced at about \$2/g or less in larger quantities.

Niobium — (*Niobe*, daughter of Tantalus), Nb; or Columbium (*Columbia*, name for America); at. wt. 92.90638(2); at. no. 41; m.p. 2477 °C; b.p. 4744 °C, sp. gr. 8.57 (20 °C); valence 2, 3, 4?, 5. Discovered in 1801 by Hatchett in an ore sent to England more that a century before by John Winthrop the Younger, first governor of Connecticut. The metal was first prepared in 1864 by Blomstrand, who reduced the chloride by heating it in a hydrogen atmosphere. The name *niobium* was adopted by the International Union of Pure and Applied Chemistry in 1950 after 100 years of controversy. Most leading chemical societies and government organizations refer to it by this name. Some metallurgists and commercial producers, however, still refer to the metal as "columbium." The element is found in *niobite* (or *columbite), niobite-tantalite, pyrochlore,* and *euxenite.* Large deposits of niobium have been found associated with *carbonatites* (carbon-silicate rocks), as a constituent of *pyrochlore.* Extensive ore reserves are found in Canada, Brazil, Congo-Kinshasa, Rwanda, and Australia. The metal can be isolated from tantalum, and prepared in several ways. It is a shiny, white, soft, and ductile metal, and takes on a bluish cast when exposed to air at room temperatures for a long time. The metal starts to oxidize in air at 200 °C, and when processed at even moderate temperatures must be placed in a protective atmosphere. It is used in arc-welding rods for stabilized grades of stainless steel. Thousands of pounds of niobium have been used in advanced air frame systems such as were used in the Gemini space program. It has also found use in super-alloys for applications such as jet engine components, rocket subassemblies, and heat-resisting equipment. The element has superconductive properties; superconductive magnets have been made with Nb-Zr wire, which retains its superconductivity in strong magnetic fields. Natural niobium is composed of only one isotope, ^{93}Nb. Forty-seven other isotopes and isomers of niobium are now recognized. Niobium metal (99.9% pure) is priced at about 50¢/g.

Nitrogen — (L. *nitrum*, Gr. *nitron*, native soda; genes, *forming,* N; at. wt. 14.0067(2); at. no. 7; m.p. −210.00 °C; b.p. −195.798 °C; t_c −146.94 °C; density 1.2506 g/L; sp. gr. liquid 0.808 (−195.8 °C), solid 1.026 (−252 °C); valence 3 or 5. Discovered by Daniel Rutherford in 1772, but Scheele, Cavendish, Priestley, and others about the same time studied "burnt or dephlogisticated air," as air without oxygen was then called. Nitrogen makes up 78% of the air, by volume. The atmosphere of Mars, by comparison, is 2.6% nitrogen. The estimated amount of this element in our atmosphere is more than 4000 trillion tons. From this inexhaustible source it can be obtained by liquefaction and fractional distillation. Nitrogen molecules give the orange-red, blue-green, blue-violet, and deep violet shades to the aurora. The element is so inert that Lavoisier named it *azote*, meaning without life, yet its compounds are so active as to be most important in foods, poisons, fertilizers, and explosives. Nitrogen can be also easily prepared by heating a water solution of ammonium nitrite. Nitrogen, as a gas, is colorless, odorless, and a generally inert element. As a liquid it is also colorless and odorless, and is similar in appearance to water. Two allotropic forms of solid nitrogen exist, with the transition from the α to the β form taking place at −237 °C. When nitrogen is heated, it combines directly with magnesium, lithium, or calcium; when mixed with oxygen and subjected to electric sparks, it forms first nitric oxide (NO) and then the dioxide (NO_2); when heated under pressure with a catalyst with hydrogen, ammonia is formed (Haber process). The ammonia thus formed is of the utmost importance as it is used in fertilizers, and it can be oxidized to nitric acid (Ostwald process). The ammonia industry is the largest consumer of nitrogen. Large amounts of gas are also used by the electronics industry, which uses the gas as a blanketing medium during production of such components as transistors, diodes, etc. Large quantities of nitrogen are used in annealing stainless steel and other steel mill products. The drug industry also uses large quantities. Nitrogen is used as a refrigerant both for the immersion freezing of food products and for transportation of foods. Liquid nitrogen is also used in missile work as a purge for components, insulators for space chambers, etc., and by the oil industry to build up great pressures in wells to force crude oil upward. Sodium and potassium nitrates are formed by the decomposition of organic matter with compounds of the metals present. In certain dry areas of the world these saltpeters are found in quantity. Ammonia, nitric acid, the nitrates, the five oxides (N_2O, NO, N_2O_3, NO_2, and N_2O_5), TNT, the cyanides, etc. are but a few of the important compounds. Nitrogen gas prices vary from 2¢ to \$2.75 per 100 ft^3 (2.83 cu. meters), depending on purity, etc. Production of elemental nitrogen in the U.S. is more than 9 million short tons per year. Natural nitrogen contains two isotopes, ^{14}N and ^{15}N. Ten other isotopes are known.

Nobelium — (Alfred Nobel [1833–1896], inventor of dynamite), No; at. wt. [259]; at. no. 102; valence +2, +3. Nobelium was unambiguously discovered and identified in April 1958 at Berkeley by A. Ghiorso, T. Sikkeland, J. R. Walton, and G. T. Seaborg, who used a new double-recoil technique. A heavy-ion linear accelerator (HILAC) was used to bombard a thin target of curium (95% ^{244}Cm and 4.5% ^{246}Cm) with ^{12}C ions to produce 102^{254} according to the ^{246}Cm (^{12}C, 4*n*) reaction. Earlier in 1957 workers of the U.S., Britain, and Sweden announced the discovery of an isotope of Element 102 with a 10-min half-life at 8.5 MeV, as a result of bombarding ^{244}Cm with ^{13}C nuclei. On the basis of this experiment the name *nobelium* was assigned and accepted by the Commission on Atomic Weights of the International Union of Pure and Applied Chemistry. The acceptance of the name was premature, for both Russian and American efforts now completely rule out the possibility of any isotope of Element 102 having a half-life of 10 min in the vicinity of 8.5 MeV. Early work in 1957 on the search for this element, in Russia at the Kurchatov Institute, was marred by the assignment of 8.9 ± 0.4 MeV alpha radiation with a half-life of 2 to 40 sec, which was too indefinite to support claim to discovery. Confirmatory experiments at Berkeley in 1966 have shown the existence of $^{254}102$ with a 55-s half-life, $^{252}102$ with a 2.3-s half-life, and $^{257}102$ with a 25-s half-life. Twelve isotopes are now recognized, one of which — $^{255}102$ — has a half-life of 3.1 min. In view of the discoverer's traditional right to name an element, the Berkeley group, in 1967, suggested that the hastily given name *nobelium,* along with the symbol No, be retained.

Osmium — (Gr. *osme*, a smell), Os; at. wt. 190.23(3); at. no. 76; m.p. 3033 °C; b.p. 5012 °C; sp. gr. 22.587; valence 0 to +8, more usually +3, +4, +6, and +8. Discovered in 1803 by Tennant in the residue left when crude platinum is dissolved by *aqua regia.* Osmium occurs in *iridosmine* and in platinum-bearing river sands of the Urals, North America, and South America. It is also found in the nickel-bearing ores of the Sudbury, Ontario, region along with other platinum metals. While the quantity of platinum metals in these ores is very small, the large tonnages of nickel ores processed make commercial recovery possible. The metal is lustrous, bluish white, extremely hard, and brittle even at high temperatures. It has the highest melting point and the lowest vapor pressure of the platinum group. The metal is very difficult to fabricate, but the powder can be sintered in a hydrogen atmosphere at a temperature of 2000 °C. The solid metal is not affected by air at room temperature, but the powdered or spongy metal slowly gives off osmium tetroxide, which is a powerful oxidizing agent and has a strong smell. The tetroxide is highly toxic, and boils at 130 °C (760 mm). Concentrations in air as low as 10^{-7} g/m^3 can cause lung congestion, skin damage, or eye damage. The tetroxide has been used to detect fingerprints and to stain fatty tissue for microscope slides. The metal is almost entirely used to produce very hard alloys, with other metals of the platinum group, for fountain pen tips, instrument pivots, phonograph needles, and electrical contacts. The price of 99.9% pure osmium powder — the form usually supplied commercially — is about \$100/g, depending on quantity and supplier. Natural osmium contains seven isotopes, one of which, ^{186}Os, is radioactive with a very long half-life. Thirty-four other isotopes and isomers are known, all of which are radioactive. The measured densities of iridium and osmium seem to indicate that osmium is slightly more dense than iridium, so osmium has generally been credited with being the heaviest known element. Calculations of the density from the space lattice, which may be more reliable for these elements than actual measurements, however, give a density of 22.65 for iridium compared to 22.61 for osmium. At present, therefore, we know either iridium or osmium is the heaviest element, but the data do not allow selection between the two.

Oxygen — (Gr. *oxys*, sharp, acid, and *genes*, forming; acid former), O; at. wt. 15.9994(3); at. no. 8; t.p. −218.79 °C; t_c −118.56 °C; valence 2. For many centuries, workers occasionally realized air was composed of more than one component. The behavior of oxygen and nitrogen as components of air led to the advancement of the phlogiston theory of combustion, which captured the minds of chemists for a century. Oxygen was prepared by several workers, including Bayen and Borch, but they did not know how to collect it, did not study its properties, and did not recognize it as an elementary substance. Priestley is generally credited with its discovery, although Scheele also discovered it independently. Oxygen is the third most abundant element found in the sun, and it plays a part in the carbon–nitrogen cycle, one process thought to give the sun and stars their energy. Oxygen under excited conditions is responsible for the bright red and yellow-green colors of the aurora. Oxygen, as a gaseous element, forms 21% of the atmosphere by volume from which it can be obtained by liquefaction and fractional distillation. The atmosphere of Mars contains about 0.15% oxygen. The element and its compounds make up 49.2%, by weight, of the Earth's crust. About two thirds of the human body and nine tenths of water is oxygen. In the laboratory it can be prepared by the electrolysis of water or by heating potassium chlorate with manganese dioxide as a catalyst. The gas is colorless, odorless, and tasteless. The liquid and solid forms are a pale blue color and are strongly paramagnetic. Ozone (O_3), a highly active compound, is formed by the action of an electrical discharge or ultraviolet light on oxygen. Ozone's presence in the atmosphere (amounting to the equivalent of a layer 3 mm thick at ordinary pressures and temperatures) is of vital importance in preventing harmful ultraviolet rays of the sun from reaching the Earth's surface. There has been recent concern that pollutants in the atmosphere may have a detrimental effect on this ozone layer. Ozone is toxic and exposure should not exceed 0.2 mg/m^3 (8-hour time-weighted average — 40-hour work week). Undiluted ozone has a bluish color. Liquid ozone is bluish black, and solid ozone is violet-black. Oxygen is very reactive and capable of combining with most elements. It is a component of hundreds of thousands of organic compounds. It is essential for respiration of all plants and animals and for practically all combustion. In hospitals it is frequently used to aid respiration of patients. Its atomic weight was used as a standard of comparison for each of the other elements until 1961 when the International Union of Pure and Applied Chemistry adopted carbon 12 as the new basis. Oxygen has thirteen recognized isotopes. Natural oxygen is a mixture of three isotopes. Oxygen 18 occurs naturally, is stable, and is available commercially. Water (H_2O with 1.5% ^{18}O) is also available. Commercial oxygen consumption in the U.S. is estimated to be 20 million short tons per year and the demand is expected to increase substantially in the next few years. Oxygen enrichment of steel blast furnaces accounts for the greatest use of the gas. Large quantities are also used in making synthesis gas for ammonia and methanol, ethylene oxide, and for oxy-acetylene welding. Air separation plants produce about 99% of the gas, electrolysis plants about 1%. The gas costs 5¢/ft^3 (\$1.75/cu. meter) in small quantities.

Palladium — (named after the asteroid *Pallas*, discovered about the same time; Gr. *Pallas*, goddess of wisdom), Pd; at. wt. 106.42(1) at. no. 46; m.p. 1554.8 °C; b.p. 2963 °C; sp. gr. 12.02 (20 °C); valence 2, 3, or 4. Discovered in 1803 by Wollaston. Palladium is found along with platinum and other metals of the platinum group in deposits of Russia, South Africa, Canada (Ontario), and elsewhere. Natural palladium contains six stable isotopes. Twenty-nine other isotopes are recognized, all of which are radioactive. It is frequently found associated with the nickel-copper deposits such as those found in Ontario. Its separation from the platinum metals depends upon the type of ore in which it is found. It is a steel-white metal, does not tarnish in air, and is the least dense and lowest melting of the platinum group of metals. When annealed, it is soft and ductile; cold working greatly increases its strength and hardness. Palladium is attacked by nitric and sulfuric acid. At room temperatures the metal has the unusual property of absorbing up to 900 times its own volume of hydrogen, possibly forming Pd_2H. It is not yet clear if this a true compound. Hydrogen readily diffuses through heated palladium and this provides a means of purifying the gas. Finely divided palladium is a good catalyst and is used for hydrogenation and dehydrogenation reactions. It is alloyed and used in jewelry trades. White gold is an alloy of gold decolorized by the addition of palladium. Like gold, palladium can be beaten into leaf as thin as 1/250,000 in. The metal is used in dentistry, watchmaking, and in making surgical instruments and electrical contacts. Palladium recently has been substituted for higher priced platinum in catalytic converters by some automobile companies. This has caused a large increase in the cost of palladium. The prices of the two metals are now, in 2002, about the same. Palladium, however, is less resistant to poisoning by sulfur and lead than platinum, but it may prove useful in controlling emissions from diesel vehicles. The metal sells for about $350/tr. oz. ($11/g).

Phosphorus — (Gr. *phosphoros,* light bearing; ancient name for the planet Venus when appearing before sunrise), P; at. wt. 30.973762(2); at. no. 15; m.p. (white) 44.15 °C; b.p. 280.5 °C; sp. gr. (white) 1.82, (red) 2.16, (black) 2.25 to 2.69; valence 3 or 5. Discovered in 1669 by Brand, who prepared it from urine. Phosphorus exists in four or more allotropic forms: white (or yellow), red, and black (or violet). White phosphorus has two modifications: α and β with a transition temperature at –3.8 °C. Never found free in nature, it is widely distributed in combination with minerals. Twenty-one isotopes of phosphorus are recognized. *Phosphate* rock, which contains the mineral *apatite*, an impure tricalcium phosphate, is an important source of the element. Large deposits are found in the Russia, China, Morocco, and in Florida, Tennessee, Utah, Idaho, and elsewhere. Phosphorus in an essential ingredient of all cell protoplasm, nervous tissue, and bones. Ordinary phosphorus is a waxy white solid; when pure it is colorless and transparent. It is insoluble in water, but soluble in carbon disulfide. It takes fire spontaneously in air, burning to the pentoxide. It is very poisonous, 50 mg constituting an approximate fatal dose. Exposure to white phosphorus should not exceed 0.1 mg/m^3 (8-hour time-weighted average — 40-hour work week). White phosphorus should be kept under water, as it is dangerously reactive in air, and it should be handled with forceps, as contact with the skin may cause severe burns. When exposed to sunlight or when heated in its own vapor to 250 °C, it is converted to the red variety, which does not phosphoresce in air as does the white variety. This form does not ignite spontaneously and it is not as dangerous as white phosphorus. It should, however, be handled with care as it does convert to the white form at some temperatures and it emits highly toxic fumes of the oxides of phosphorus when heated. The red modification is fairly stable, sublimes with a vapor pressure of 1 atm at 417 °C, and is used in the manufacture of safety matches, pyrotechnics, pesticides, incendiary shells, smoke bombs, tracer bullets, etc. White phosphorus may be made by several methods. By one process, tricalcium phosphate, the essential ingredient of phosphate rock, is heated in the presence of carbon and silica in an electric furnace or fuel-fired furnace. Elementary phosphorus is liberated as vapor and may be collected under water. If desired, the phosphorus vapor and carbon monoxide produced by the reaction can be oxidized at once in the presence of moisture to produce phosphoric acid, an important compound in making super-phosphate fertilizers. In recent years, concentrated phosphoric acids, which may contain as much as 70 to 75% P_2O_5 content, have become of great importance to agriculture and farm production. World-wide demand for fertilizers has caused record phosphate production. Phosphates are used in the production of special glasses, such as those used for sodium lamps. Bone-ash, calcium phosphate, is also used to produce fine chinaware and to produce monocalcium phosphate used in baking powder. Phosphorus is also important in the production of steels, phosphor bronze, and many other products. Trisodium phosphate is important as a cleaning agent, as a water softener, and for preventing boiler scale and corrosion of pipes and boiler tubes. Organic compounds of phosphorus are important. Amorphous (red) phosphorus costs about $70/kg (99%).

Platinum — (It. *platina,* silver), Pt; at. wt. 195.084(9); at. no. 78; m.p. 1768.2 °C; b.p. 3825 °C; sp. gr. 21.45 (20 °C); valence 1?, 2, 3, or 4. Discovered in South America by Ulloa in 1735 and by Wood in 1741. The metal was used by pre-Columbian Indians. Platinum occurs native, accompanied by small quantities of iridium, osmium, palladium, ruthenium, and rhodium, all belonging to the same group of metals. These are found in the alluvial deposits of the Ural mountains and in Columbia. *Sperrylite* ($PtAs_2$), occurring with the nickel-bearing deposits of Sudbury, Ontario, is a source of a considerable amount of metal. The large production of nickel offsets there being only one part of the platinum metals in two million parts of ore. The largest supplier of the platinum group of metals is now South Africa, followed by Russia and Canada. Platinum is a beautiful silvery-white metal, when pure, and is malleable and ductile. It has a coefficient of expansion almost equal to that of soda–lime–silica glass, and is therefore used to make sealed electrodes in glass systems. The metal does not oxidize in air at any temperature, but is corroded by halogens, cyanides, sulfur, and caustic alkalis. It is insoluble in hydrochloric and nitric acid, but dissolves when they are mixed as *aqua regia,* forming chloroplatinic acid (H_2PtCl_6), an important compound. Natural platinum contains six isotopes, one of which, ^{190}Pt, is radioactive with a long half-life. Thirty-seven other radioactive isotopes and isomers are recognized. The metal is used extensively in jewelry, wire, and vessels for laboratory use, and in many valuable instruments including thermocouple elements. It is also used for electrical contacts, corrosion-resistant apparatus, and in dentistry. Platinum–cobalt alloys have magnetic properties. One such alloy made of 76.7% Pt and 23.3% Co, by weight, is an extremely powerful magnet that offers a B-H (max) almost twice that of Alnico V. Platinum resistance wires are used for constructing high-temperature electric furnaces. The metal is used for coating missile nose cones, jet engine fuel nozzles, etc., which must perform reliably for long periods of time at high temperatures. The metal, like palladium, absorbs large volumes of hydrogen, retaining it at ordinary temperatures but giving it up at red heat. In the finely divided state platinum is an excellent catalyst, having long been used in the contact process for producing sulfuric acid. It is also used as a catalyst in cracking petroleum products. There is also much current interest in the use of platinum as a catalyst in fuel cells and in its use as antipollution devices for automobiles. Platinum anodes are extensively used in cathodic protection systems for large ships and ocean-going vessels, pipelines, steel piers, etc. Pure platinum wire will glow red hot when placed in the vapor of methyl alcohol. It acts here as a catalyst, converting the alcohol to formaldehyde. This phenomenon has been used commercially to produce cigarette lighters and hand warmers. Hydrogen and oxygen explode in the presence of platinum. The price of platinum has varied widely; more than a century ago it was used to adulterate gold. It was nearly eight times as valuable as gold in 1920. The price in January 2002 was about $430/troy oz. ($15/g), higher than the price of gold.

Plutonium — (planet *Pluto*), Pu; at. wt. (244); at. no. 94; sp. gr. (α modification) 19.84 (25 °C); m.p. 640 °C; b.p. 3228 °C; valence 3, 4, 5, or 6. Plutonium was the second transuranium element of the actinide series to be discovered. The isotope ^{238}Pu was produced in 1940 by Seaborg, McMillan, Kennedy, and Wahl by deuteron bombardment of uranium in the 60-inch cyclotron at Berkeley, California. Plutonium also exists in trace quantities in naturally occurring uranium ores. It is formed in much the same manner as neptunium, by irradiation of natural uranium with the neutrons that are present. By far of greatest importance is the isotope Pu^{239}, with a half-life of 24,100 years, produced in extensive quantities in nuclear reactors from natural uranium:

$$^{238}U(n,\gamma) \rightarrow ^{239}U \xrightarrow{\beta} ^{239}Np \xrightarrow{\beta} ^{239}Pu$$

Nineteen isotopes of plutonium are now known. Plutonium has assumed the position of dominant importance among the transuranium elements because of its successful use as an explosive ingredient in nuclear weapons and the place it holds as a key material in the development of industrial use of nuclear power. One kilogram is equivalent to about 22 million kilowatt hours of heat energy. The complete detonation of a kilogram of plutonium produces an explosion equal to about 20,000 tons of chemical explosive. Its importance depends on the nuclear property of being readily fissionable with neutrons and its availability in quantity. The world's nuclear-power reactors are now producing about 20,000 kg of plutonium/yr. By 1982 it was estimated that about 300,000 kg had accumulated. The various nuclear applications of plutonium are well known. ^{238}Pu has been used in the Apollo lunar missions to power seismic and other equipment on the lunar surface. As with neptunium and uranium, plutonium metal can be prepared by reduction of the trifluoride with alkaline-earth metals. The metal has a silvery appearance and takes on a yellow tarnish when slightly oxidized. It is chemically reactive. A relatively large piece of plutonium is warm to the touch because of the energy given off in alpha decay. Larger pieces will produce enough heat to boil water. The metal readily dissolves in concentrated hydrochloric acid, hydroiodic acid, or perchloric acid with formation of the Pu^{+3} ion. The metal exhibits six allotropic modifications having various crystalline structures. The densities of these vary from 16.00 to 19.86 g/cm^3. Plutonium also exhibits four ionic valence states in aqueous solutions: Pu^{+3}(blue lavender), Pu^{+4} (yellow brown), PuO^+ (pink?), and PuO^{+2} (pink orange). The ion PuO^+ is unstable in aqueous solutions, disproportionating into Pu^{+4} and PuO^{+2}. The Pu^{+4} thus formed, however, oxidizes the PuO^+ into PuO^{+2}, itself being reduced to Pu^{+3}, giving finally Pu^{+3} and PuO^{+2}. Plutonium forms binary compounds with oxygen: PuO, PuO_2, and intermediate oxides of variable composition; with the halides: PuF_3, PuF_4, $PuCl_3$, $PuBr_3$, PuI_3; with carbon, nitrogen, and silicon: PuC, PuN, $PuSi_2$. Oxyhalides are also well known: PuOCl, PuOBr, PuOI. Because of the high rate of emission of alpha particles and the element being specifically absorbed by bone marrow, plutonium, as well as all of the other transuranium elements except neptunium, are radiological poisons and must be handled with very special equipment and precautions. Plutonium is a very dangerous radiological hazard. Precautions must also be taken to prevent the unintentional formation of a critical mass. Plutonium in liquid solution is more likely to become critical than solid plutonium. The shape of the mass must also be considered where criticality is concerned. Plutonium-239 is available to authorized users from the O.R.N.L. at a cost of about \$4.80/mg (99.9%) plus packing costs.

Polonium — (Poland, native country of Mme. Curie [1867–1934]), Po; at. wt. (209); at. no. 84; m.p. 254 °C; b.p. 962 °C; sp. gr. 9.20; valence −2, 0, +2, +3(?), +4, and +6. Polonium was the first element discovered by Mme. Curie in 1898, while seeking the cause of radioactivity of pitchblende from Joachimsthal, Bohemia. The electroscope showed it separating with bismuth. Polonium is also called Radium F. Polonium is a very rare natural element. Uranium ores contain only about 100 μg of the element per ton. Its abundance is only about 0.2% of that of radium. In 1934, it was found that when natural bismuth (^{209}Bi) was bombarded by neutrons, ^{210}Bi, the parent of polonium, was obtained. Milligram amounts of polonium may now be prepared this way, by using the high neutron fluxes of nuclear reactors. Polonium-210 is a low-melting, fairly volatile metal, 50% of which is vaporized in air in 45 hours at 55 °C. It is an alpha emitter with a half-life of 138.39 days. A milligram emits as many alpha particles as 5 g of radium. The energy released by its decay is so large (140 W/g) that a capsule containing about half a gram reaches a temperature above 500 °C. The capsule also presents a contact gamma-ray dose rate of 0.012 Gy/h. A few curies (1 curie = 3.7×10^{10} Bq) of polonium exhibit a blue glow, caused by excitation of the surrounding gas. Because almost all alpha radiation is stopped within the solid source and its container, giving up its energy, polonium has attracted attention for uses as a lightweight heat source for thermoelectric power in space satellites. Thirty-eight isotopes and isomers of polonium are known, with atomic masses ranging from 192 to 218. All are radioactive. Polonium-210 is the most readily available. Isotopes of mass 209 (half-life 102 years) and mass 208 (half-life 2.9 years) can be prepared by alpha, proton, or deuteron bombardment of lead or bismuth in a cyclotron, but these are expensive to produce. Metallic polonium has been prepared from polonium hydroxide and some other polonium compounds in the presence of concentrated aqueous or anhydrous liquid ammonia. Two allotropic modifications are known to exist. Polonium is readily dissolved in dilute acids, but is only slightly soluble in alkalis. Polonium salts of organic acids char rapidly; halide amines are reduced to the metal. Polonium can be mixed or alloyed with beryllium to provide a source of neutrons. It has been used in devices for eliminating static charges in textile mills, etc.; however, beta sources are more commonly used and are less dangerous. It is also used on brushes for removing dust from photographic films. The polonium for these is carefully sealed and controlled, minimizing hazards to the user. Polonium-210 is very dangerous to handle in even milligram or microgram amounts, and special equipment and strict control are necessary. Damage arises from the complete absorption of the energy of the alpha particle into tissue. The maximum permissible body burden for ingested polonium is only 0.03 μCi, which represents a particle weighing only 6.8×10^{-12} g. Weight for weight it is about 2.5×10^{11} times as toxic as hydrocyanic acid. The maximum allowable concentration for soluble polonium compounds in air is about 2×10^{11} μCi/cm^3. Polonium-209 is available on special order from the Oak Ridge National Laboratory at a cost of \$3600/μCi plus packing costs.

Potassium — (English, *potash* — pot ashes; L. *kalium*, Arab. *qali*, alkali), K; at. wt. 39.0983(1); at. no. 19; m.p. 63.5 °C; b.p. 759 °C; sp. gr. 0.89; valence 1. Discovered in 1807 by Davy, who obtained it from caustic potash (KOH); this was the first metal isolated by electrolysis. The metal is the seventh most abundant and makes up about 2.4% by weight of the Earth's crust. Most potassium minerals are insoluble and the metal is obtained from them only with great difficulty. Certain minerals, however, such as *sylvite, carnallite, langbeinite,* and *polyhalite* are found in ancient lake and sea beds and form rather extensive deposits from which potassium and its salts can readily be obtained. Potash is mined in Germany, New Mexico, California, Utah, and elsewhere. Large deposits of potash, found at a depth of some 1000 m in Saskatchewan, promise to be important in coming years. Potassium is also found in the ocean, but is present only in relatively small amounts compared to sodium. The greatest demand for potash has been in its use for fertilizers. Potassium is an essential constituent for plant growth and it is found in most soils. Potassium is never found free in nature, but is obtained by electrolysis of the hydroxide, much in the same manner as prepared by Davy. Thermal methods also are commonly used to produce potassium (such as by reduction of potassium compounds with CaC_2, C, Si, or Na). It is one of the most reactive and electropositive of metals. Except for lithium, it is the lightest known metal. It is soft, easily cut with a knife, and is silvery in appearance immediately after a fresh surface is exposed. It rapidly oxidizes in air and should be preserved in a mineral oil. As with other metals of the alkali group, it decomposes in water with the evolution of hydrogen. It catches fire spontaneously on water. Potassium and its salts impart a violet color to flames. Twenty-one isotopes, one of which is an isomer, of potassium are known. Ordinary potassium is composed of three isotopes, one of which is ^{40}K (0.0117%), a radioactive isotope with a half-life of 1.26×10^9 years. The radioactivity presents no appreciable hazard. An alloy of sodium and potassium (NaK) is used as a heat-transfer medium. Many potassium salts are of utmost importance, including the hydroxide, nitrate, carbonate, chloride, chlorate, bromide, iodide, cyanide, sulfate, chromate, and dichromate. Metallic potassium is available commercially for about $1200/kg (98% purity) or $75/g (99.95% purity).

Praseodymium — (Gr. *prasios,* green, and *didymos,* twin), Pr; at. wt. 140.90765(2); at. no. 59; m.p. 931 °C; b.p. 3520 °C; sp. gr. 6.773; valence 3. In 1841 Mosander extracted the rare earth *didymia* from *lanthana*; in 1879, Lecoq de Boisbaudran isolated a new earth, *samaria,* from didymia obtained from the mineral *samarskite.* Six years later, in 1885, von Welsbach separated didymia into two others, *praseodymia* and *neodymia,* which gave salts of different colors. As with other rare earths, compounds of these elements in solution have distinctive sharp spectral absorption bands or lines, some of which are only a few Angstroms wide. The element occurs along with other rare-earth elements in a variety of minerals. *Monazite* and *bastnasite* are the two principal commercial sources of the rare-earth metals. Ion-exchange and solvent extraction techniques have led to much easier isolation of the rare earths and the cost has dropped greatly. Thirty-seven isotopes and isomers are now recognized. Praseodymium can be prepared by several methods, such as by calcium reduction of the anhydrous chloride or fluoride. Misch metal, used in making cigarette lighters, contains about 5% praseodymium metal. Praseodymium is soft, silvery, malleable, and ductile. It was prepared in relatively pure form in 1931. It is somewhat more resistant to corrosion in air than europium, lanthanum, cerium, or neodymium, but it does develop a green oxide coating that splits off when exposed to air. As with other rare-earth metals it should be kept under a light mineral oil or sealed in plastic. The rare-earth oxides, including Pr_2O_3, are among the most refractory substances known. Along with other rare earths, it is widely used as a core material for carbon arcs used by the motion picture industry for studio lighting and projection. Salts of praseodymium are used to color glasses and enamels; when mixed with certain other materials, praseodymium produces an intense and unusually clean yellow color in glass. Didymium glass, of which praseodymium is a component, is a colorant for welder's goggles. The metal (99.9% pure) is priced at about $4/g.

Promethium — (*Prometheus,* who, according to mythology, stole fire from heaven), Pm; at. no. 61; at. wt. (145); m.p. 1042 °C; b.p. 3000 °C (est.); sp. gr. 7.264 (25 °C); valence 3. In 1902 Branner predicted the existence of an element between neodymium and samarium, and this was confirmed by Moseley in 1914. Unsuccessful searches were made for this predicted element over two decades, and various investigators proposed the names "illinium," "florentium," and "cyclonium" for this element. In 1941, workers at Ohio State University irradiated neodymium and praseodymium with neutrons, deuterons, and alpha particles, resp., and produced several new radioactivities, which most likely were those of Element 61. Wu and Segre, and Bethe, in 1942, confirmed the formation; however, chemical proof of the production of Element 61 was lacking because of the difficulty in separating the rare earths from each other at that time. In 1945, Marinsky, Glendenin, and Coryell made the first chemical identification by using ion-exchange chromatography. Their work was done by fission of uranium and by neutron bombardment of neodymium. These investigators named the newly discovered element. Searches for the element on Earth have been fruitless, and it now appears that promethium is completely missing from the Earth's crust. Promethium, however, has been reported to be in the spectrum of the star HR[465] in Andromeda. It must be formed near the star's surface, for no known isotope of promethium has a half-life longer than 17.7 years. Thirty-five isotopes and isomers of promethium, with atomic masses from 130 to 158 are now known. Promethium-145, with a half-life of 17.7 years, is the most useful. Promethium-145 has a specific activity of 940 Ci/g. It is a soft beta emitter; although no gamma rays are emitted, X-radiation can be generated when beta particles impinge on elements of a high atomic number, and great care must be taken in handling it. Promethium salts luminesce in the dark with a pale blue or greenish glow, due to their high radioactivity. Ion-exchange methods led to the preparation of about 10 g of promethium from atomic reactor fuel processing wastes in early 1963. Little is yet generally known about the properties of metallic promethium. Two allotropic modifications exist. The element has applications as a beta source for thickness gages, and it can be absorbed by a phosphor to produce light. Light produced in this manner can be used for signs or signals that require dependable operation; it can be used as a nuclear-powered battery by capturing light in photocells that convert it into electric current. Such a battery, using ^{147}Pm, would have a useful life of about 5 years. It is being used for fluorescent lighting starters and coatings for self-luminous watch dials. Promethium shows promise as a portable X-ray source, and it may become useful as a heat source to provide auxiliary power for space probes and satellites. More than 30 promethium compounds have been prepared. Most are colored.

Protactinium — (Gr. *protos,* first), Pa; at. wt. 231.03588(2); at. no. 91; m.p. 1572 °C; sp. gr. 15.37 (calc.); valence 4 or 5. The first isotope of Element 91 to be discovered was ^{234}Pa, also known as UX_2, a short-lived member of the naturally occurring ^{238}U decay series. It was identified by K. Fajans and O. H. Gohring in 1913 and they named the new element *brevium.* When the longer-lived isotope ^{231}Pa was identified by Hahn and Meitner in 1918, the name protoactinium was adopted as being more consistent with the characteristics of the most abundant isotope. Soddy, Cranson, and Fleck were also active in this work. The name *protoactinium* was shortened to *protactinium* in 1949. In 1927, Grosse prepared 2 mg of a white powder, which was shown to be Pa_2O_5. Later, in 1934, from 0.1 g of pure Pa_2O_5 he isolated the element by two methods, one of which was by converting the oxide to an iodide and "cracking" it in a high vacuum by an electrically heated filament by the reaction

$$2PaI_5 \rightarrow 2Pa + 5I_2$$

Protactinium has a bright metallic luster that it retains for some time in air. The element occurs in *pitchblende* to the extent of about 1 part ^{231}Pa to 10 million of ore. Ores from Congo-Kinshasa have about 3 ppm. Protactinium has twenty-eight isotopes and isomers, the most common of which is ^{231}Pr with a half-life of 32,500 years. A number of protactinium compounds are known, some of which are colored. The element is superconductive below 1.4 K. The element is a dangerous toxic material and requires precautions similar to those used when handling plutonium. In 1959 and 1961, it was announced that thc Great Britain Atomic Energy Authority extracted by a 12-stage process 125 g of 99.9% protactinium, the world's only stock of the metal for many years to come. The extraction was made from 60 tons of waste material at a cost of about $500,000. Protactinium is one of the rarest and most expensive naturally occurring elements.

Radium — (L. *radius,* ray), Ra; at. wt. (226); at. no. 88; m.p. 696 °C; sp. gr. 5; valence 2. Radium was discovered in 1898 by M. and Mme. Curie in the *pitchblende* or *uraninite* of North Bohemia (Czech Republic), where it occurs. There is about 1 g of radium in 7 tons of pitchblende. The element was isolated in 1911 by Mme. Curie and Debierne by the electrolysis of a solution of pure radium chloride, employing a mercury cathode; on distillation in an atmosphere of hydrogen this amalgam yielded the pure metal. Originally, radium was obtained from the rich pitchblende ore found at Joachimsthal, Bohemia. The *carnotite* sands of Colorado furnish some radium, but richer ores are found in the Republic of Congo-Kinshasa and the Great Bear Lake region of Canada. Radium is present in all uranium minerals, and could be extracted, if desired, from the extensive wastes of uranium processing. Large uranium deposits are located in Ontario, New Mexico, Utah, Australia, and elsewhere. Radium is obtained commercially as the bromide or chloride; it is doubtful if any appreciable stock of the isolated element now exists. The pure metal is brilliant white when freshly prepared, but blackens on exposure to air, probably due to formation of the nitride. It exhibits luminescence, as do its salts; it decomposes in water and is somewhat more volatile than barium. It is a member of the alkaline-earth group of metals. Radium imparts a carmine red color to a flame. Radium emits alpha, beta, and gamma rays and when mixed with beryllium produce neutrons. One gram of ^{226}Ra undergoes 3.7×10^{10} disintegrations per s. The *curie (Ci)* is defined as that amount of radioactivity which has the same disintegration rate as 1 g of ^{226}Ra. Thirty-six isotopes are now known; radium 226, the common isotope, has a half-life of 1599 years. One gram of radium produces about 0.0001 mL (stp) of emanation, or radon gas, per day. This is pumped from the radium and sealed in minute tubes, which are used in the treatment of cancer and other diseases. One gram of radium yields about 4186 kJ per year. Radium is used in producing self-luminous paints, neutron sources, and in medicine for the treatment of cancer. Some of the more recently discovered radioisotopes, such as ^{60}Co, are now being used in place of radium. Some of these sources are much more powerful, and others are safer to use. Radium loses about 1% of its activity in 25 years, being transformed into elements of lower atomic weight. Lead is a final product of disintegration. Stored radium should be ventilated to prevent build-up of radon. Inhalation, injection, or body exposure to radium can cause cancer and other body disorders. The maximum permissible burden in the total body for ^{226}Ra is 7400 becquerel.

Radon — (from *radium*; called *niton* at first, L. *nitens,* shining), Rn; at. wt. (222); at. no. 86; m.p. −71 °C; b.p. −61.7 °C; t_c 104 °C; density of gas 9.73 g/L; sp. gr. liquid 4.4 at −62 °C, solid 4; valence usually 0. The element was discovered in 1900 by Dorn, who called it *radium emanation.* In 1908 Ramsay and Gray, who named it *niton,* isolated the element and determined its density, finding it to be the heaviest known gas. It is essentially inert and occupies the last placc in the zero group of gases in the Periodic Table. Since 1923, it has been called radon. Thirty-seven isotopes and isomers are known. Radon-222, coming from radium, has a half-life of 3.823 days and is an alpha emitter; Radon-220, emanating naturally from thorium and called *thoron,* has a half-life of 55.6 s and is also an alpha emitter. Radon-219 emanates from actinium and is called *actinon.* It has a half-life of 3.9 s and is also an alpha emitter. It is estimated that every square mile of soil to a depth of 6 inches contains about 1 g of radium, which releases radon in tiny amounts to the atmosphere. Radon is present in some spring waters, such as those at Hot Springs, Arkansas. On the average, one part of radon is present to 1×10^{21} part of air. At ordinary temperatures radon is a colorless gas; when cooled below the freezing point, radon exhibits a brilliant phosphorescence which becomes yellow as the temperature is lowered and orange-red at the temperature of liquid air. It has been reported that fluorine reacts with radon, forming radon fluoride. Radon clathrates have also been reported. Radon is still produced for therapeutic use by a few hospitals by pumping it from a radium source and sealing it in minute tubes, called seeds or needles, for application to patients. This practice has now been largely discontinued as hospitals can order the seeds directly from suppliers, who make up the seeds with the desired activity for the day of use. Care must be taken in handling radon, as with other radioactive materials. The main hazard is from inhalation of the element and its solid daughters, which are collected on dust in the air. Good ventilation should be provided where radium, thorium, or actinium is stored to prevent build-up of this element. Radon build-up is a health consideration in uranium mines. Recently radon build-up in homes has been a concern. Many deaths from lung cancer are caused by radon exposure. In the U.S. it is recommended that remedial action be taken if the air from radon in homes exceeds 4 pCi/L.

Rhenium — (L. *Rhenus,* Rhine), Re; at. wt. 186.207(1); at. no. 75; m.p. 3185 °C; b.p. 5596 °C; sp. gr. 20.8 (20 °C); valence −1, +1, 2, 3, 4, 5, 6, 7. Discovery of rhenium is generally attributed to Noddack, Tacke, and Berg, who announced in 1925 they had detected the element in platinum ores and *columbite.* They also found the element in *gadolinite* and *molybdenite*. By working up 660 kg of molybdenite they were able in 1928 to extract 1 g of rhenium. The price in 1928 was $10,000/g. Rhenium does not occur free in nature or as a compound in a distinct mineral species. It is, however, widely spread throughout the Earth's crust to the extent of about 0.001 ppm. Commercial rhenium in the U.S. today is obtained from molybdenite roaster-flue dusts obtained from copper-sulfide ores mined in the vicinity of Miami, Arizona, and elsewhere in Arizona and Utah. Some molybdenites contain from 0.002 to 0.2% rhenium. It is estimated that in 1999 about 16,000 kg of rhenium was being produced. The total estimated world reserves of rhenium is 11,000,000 kg. Natural rhenium is a mixture of two isotopes, one of which has a very long half-life. Thirty-nine other unstable isotopes are recognized. Rhenium metal is prepared by reducing ammonium perrhenate with hydrogen at elevated temperatures. The element is silvery white with a metallic luster; its density is exceeded by that of only platinum, iridium, and osmium, and its melting point is exceeded by that of only tungsten and carbon. It has other useful properties. The usual commercial form of the element is a powder, but it can be consolidated by pressing and resistance-sintering in a vacuum or hydrogen atmosphere. This produces a compact shape in excess of 90% of the density of the metal. Annealed rhenium is very ductile, and can be bent, coiled, or rolled. Rhenium is used as an additive to tungsten and molybdenum-based alloys to impart useful properties. It is widely used for filaments for mass spectrographs and ion gages. Rhenium-molybdenum alloys are superconductive at 10 K. Rhenium is also used as an electrical contact material as it has good wear resistance and withstands arc corrosion. Thermocouples made of Re-W are used for measuring temperatures up to 2200 °C, and rhenium wire has been used in photoflash lamps for photography. Rhenium catalysts are exceptionally resistant to poisoning from nitrogen, sulfur, and phosphorus, and are used for hydrogenation of fine chemicals, hydrocracking, reforming, and disproportionation of olefins. Rhenium has recently become especially important as a catalyst for petroleum refining and in making super-alloys for jet engines. Rhenium costs about $16/g (99.99% pure). Little is known of its toxicity; therefore, it should be handled with care until more data are available.

Rhodium — (Gr. *rhodon*, rose), Rh; at. wt. 102.90550(2); at. no. 45; m.p. 1964 °C; b.p. 3695 °C; sp. gr. 12.41 (20 °C); valence 2, 3, 4, 5, and 6. Wollaston discovered rhodium in 1803-4 in crude platinum ore he presumably obtained from South America. Rhodium occurs native with other platinum metals in river sands of the Urals and in North and South America. It is also found with other platinum metals in the copper-nickel sulfide ores of the Sudbury, Ontario region. Although the quantity occurring here is very small, the large tonnages of nickel processed make the recovery commercially feasible. The annual world production of rhodium in 1999 was only about 9000 kg. The metal is silvery white and at red heat slowly changes in air to the sesquioxide. At higher temperatures it converts back to the element. Rhodium has a higher melting point and lower density than platinum. Its major use is as an alloying agent to harden platinum and palladium. Such alloys are used for furnace windings, thermocouple elements, bushings for glass fiber production, electrodes for aircraft spark plugs, and laboratory crucibles. It is useful as an electrical contact material as it has a low electrical resistance, a low and stable contact resistance, and is highly resistant to corrosion. Plated rhodium, produced by electroplating or evaporation, is exceptionally hard and is used for optical instruments. It has a high reflectance and is hard and durable. Rhodium is also used for jewelry, for decoration, and as a catalyst. Fifty-two isotopes and isomers are now known. Rhodium metal (powder) costs about $180/g (99.9%).

Roentgenium — (Wilhelm Roentgen, discoverer of X-rays), Rg. On December 20, 1994, scientists at GSI Darmstadt, Germany announced they had detected three atoms of a new element with 111 protons and 161 neutrons. This element was made by bombarding ^{83}Bi with ^{28}Ni. Signals of Element 111 appeared for less than 0.002 s, then decayed into lighter elements including Element ^{268}109 and Element ^{264}107. These isotopes had not previously been observed. In 2004 IUPAC approved the name roentgenium for Element 111. Roentgenium is expected to have properties similar to gold.

Rubidium — (L. *rubidus*, deepest red), Rb; at. wt. 85.4678(3); at. no. 37; m.p. 39.30 °C; b.p. 688 °C; sp. gr. (solid) 1.532 (20 °C), (liquid) 1.475 (39 °C); valence 1, 2, 3, 4. Discovered in 1861 by Bunsen and Kirchhoff in the mineral *lepidolite* by use of the spectroscope. The element is much more abundant than was thought several years ago. It is now considered to be the 16th most abundant element in the Earth's crust. Rubidium occurs in *pollucite, carnallite, leucite,* and *zinnwaldite,* which contains traces up to 1%, in the form of the oxide. It is found in lepidolite to the extent of about 1.5%, and is recovered commercially from this source. Potassium minerals, such as those found at Searles Lake, California, and potassium chloride recovered from brines in Michigan also contain the element and are commercial sources. It is also found along with cesium in the extensive deposits of *pollucite* at Bernic Lake, Manitoba. Rubidium can be liquid at room temperature. It is a soft, silvery-white metallic element of the alkali group and is the second most electropositive and alkaline element. It ignites spontaneously in air and reacts violently in water, setting fire to the liberated hydrogen. As with other alkali metals, it forms amalgams with mercury and it alloys with gold, cesium, sodium, and potassium. It colors a flame yellowish violet. Rubidium metal can be prepared by reducing rubidium chloride with calcium, and by a number of other methods. It must be kept under a dry mineral oil or in a vacuum or inert atmosphere. Thirty-five isotopes and isomers of rubidium are known. Naturally occurring rubidium is made of two isotopes, ^{85}Rb and ^{87}Rb. Rubidium-87 is present to the extent of 27.83% in natural rubidium and is a beta emitter with a half-life of 4.9×10^{10} years. Ordinary rubidium is sufficiently radioactive to expose a photographic film in about 30 to 60 days. Rubidium forms four oxides: Rb_2O, Rb_2O_2, Rb_2O_3, Rb_2O_4. Because rubidium can be easily ionized, it has been considered for use in "ion engines" for space vehicles; however, cesium is somewhat more efficient for this purpose. It is also proposed for use as a working fluid for vapor turbines and for use in a thermoelectric generator using the magnetohydrodynamic principle where rubidium ions are formed by heat at high temperature and passed through a magnetic field. These conduct electricity and act like an armature of a generator thereby generating an electric current. Rubidium is used as a getter in vacuum tubes and as a photocell component. It has been used in making special glasses. $RbAg_4I_5$ is important, as it has the highest room-temperature conductivity of any known ionic crystal. At 20 °C its conductivity is about the same as dilute sulfuric acid. This suggests use in thin film batteries and other applications. The present cost in small quantities is about $50/g (99.8% pure).

Ruthenium — (L. *Ruthenia*, Russia), Ru; at. wt. 101.07(2); at. no. 44, m.p. 2334 °C; b.p. 4150 °C; sp. gr. 12.1 (20 °C); valence 0, 1, 2, 3, 4, 5, 6, 7, 8. Berzelius and Osann in 1827 examined the residues left after dissolving crude platinum from the Ural mountains in *aqua regia*. While Berzelius found no unusual metals, Osann thought he found three new metals, one of which he named ruthenium. In 1844 Klaus, generally recognized as the discoverer, showed that Osann's ruthenium oxide was very impure and that it contained a new metal. Klaus obtained 6 g of ruthenium from the portion of crude platinum that is insoluble in *aqua regia*. A member of the platinum group, ruthenium occurs native with other members of the group of ores found in the Ural mountains and in North and South America. It is also found along with other platinum metals in small but commercial quantities in *pentlandite* of the Sudbury, Ontario, nickel-mining region, and in *pyroxinite* deposits of South Africa. Natural ruthenium contains seven isotopes. Twenty-eight other isotopes and isomers are known, all of which are radioactive. The metal is isolated commercially by a complex chemical process, the final stage of which is the hydrogen reduction of ammonium ruthenium chloride, which yields a powder. The powder is consolidated by powder metallurgy techniques or by argon-arc welding. Ruthenium is a hard, white metal and has four crystal modifications. It does not tarnish at room temperatures, but oxidizes in air at about 800 °C. The metal is not attacked by hot or cold acids or *aqua regia*, but when potassium chlorate is added to the solution, it oxidizes explosively. It is attacked by halogens, hydroxides, etc. Ruthenium can be plated by electrodeposition or by thermal decomposition methods. The metal is one of the most effective hardeners for platinum and palladium, and is alloyed with these metals to make electrical contacts for severe wear resistance. A ruthenium–molybdenum alloy is said to be superconductive at 10.6 K. The corrosion resistance of titanium is improved a hundredfold by addition of 0.1% ruthenium. It is a versatile catalyst. Hydrogen sulfide can be split catalytically by light using an aqueous suspension of CdS particles loaded with ruthenium dioxide. It is thought this may have application to removal of H_2S in oil refining and other industrial processes. Compounds in at least eight oxidation states have been found, but of these, the +2. +3. and +4 states are the most common. Ruthenium tetroxide, like osmium tetroxide, is highly toxic. In addition, it may explode. Ruthenium compounds show a marked resemblance to those of osmium. The metal is priced at about $25/g (99.95% pure).

Rutherfordium — (Ernest Rutherford [1871–1937], New Zealand, Canadian, and British physicist); Rf; at. wt. [261]; at. no. 104. In 1964, workers of the Joint Nuclear Research Institute at Dubna (Russia) bombarded plutonium with accelerated 113 to 115 MeV neon ions. By measuring fission tracks in a special glass with a microscope, they detected an isotope that decays by spontaneous fission. They suggested that this isotope, which has a half-life of 0.3 ± 0.1 s, might be ^{260}104, produced by the following reaction:

$$^{242}_{94}\mathrm{Pu} + ^{22}_{10}\mathrm{Ne} \rightarrow ^{260} 104 + 4\mathrm{n}$$

Element 104, the first *transactinide* element, is expected to have chemical properties similar to those of hafnium. It would, for example, form a relatively volatile compound with chlorine (a tetrachloride). The Soviet scientists have performed experiments aimed at chemical identification, and have attempted to show that the 0.3-s activity is more volatile than that of the relatively nonvolatile actinide trichlorides. This experiment does not fulfill the test of chemically separating the new element from all others, but it provides important evidence for evaluation. New data, reportedly issued by Soviet scientists, have reduced the half-life of the isotope they worked with from 0.3 to 0.15 s. The Dubna scientists suggest the name *kurchatovium* and symbol *Ku* for Element 104, in honor of Igor Vasilevich Kurchatov (1903–1960), late Head of Soviet Nuclear Research. The Dubna Group also has proposed the name *dubnium* for Element 104. In 1969, Ghiorso, Nurmia, Harris, K. A. Y. Eskola, and P. I. Eskola of the University of California at Berkeley reported they had positively identified two, and possibly three, isotopes of Element 104. The group also indicated that after repeated attempts so far they have been unable to produce isotope ^{260}104 reported by the Dubna groups in 1964. The discoveries at Berkeley were made by bombarding a target of ^{249}Cf with ^{12}C nuclei of 71 MeV, and ^{13}C nuclei of 69 MeV. The combination of ^{12}C with ^{249}Cf followed by instant emission of four neutrons produced Element ^{257}104. This isotope has a half-life of 4 to 5 s, decaying by emitting an alpha particle into ^{253}No, with a half-life of 105 s. The same reaction, except with the emission of three neutrons, was thought to have produced ^{258}104 with a half-life of about 1/100 s. Element ^{259}104 is formed by the merging of a ^{13}C nuclei with ^{249}Cf, followed by emission of three neutrons. This isotope has a half-life of 3 to 4 s, and decays by emitting an alpha particle into ^{255}No, which has a half-life of 185 s. Thousands of atoms of ^{257}104 and ^{259}104 have been detected. The Berkeley group believes its identification of ^{258}104 was correct. Eleven isotopes of Element 104 have now been identified. The Berkeley group proposed the name *rutherfordium* (symbol Rf) for the new element, in honor of Ernest Rutherford. This name was formally adapted by IUPAC in August 1997.

Samarium — (*Samarskite*, a mineral), Sm; at. wt. 150.36(3); at. no. 62; m.p. 1072 °C; b.p. 1794 °C; sp. gr (α) 7.520 (25 °C); valence 2 or 3. Discovered spectroscopically by its sharp absorption lines in 1879 by Lecoq de Boisbaudran in the mineral *samarskite*, named in honor of a Russian mine official, Col. Samarski. Samarium is found along with other members of the rare-earth-elements in many minerals, including *monazite* and *bastnasite*, which are commercial sources. The largest producer of rare earth minerals is now China, followed by the U.S., India, and Russia. It occurs in monazite to the extent of 2.8%. While *misch metal* containing about 1% of samarium metal has long been used, samarium has not been isolated in relatively pure form until recently. Ion-exchange and solvent extraction techniques have recently simplified separation of the rare earths from one another; more recently, electrochemical deposition, using an electrolytic solution of lithium citrate and a mercury electrode, is said to be a simple, fast, and highly specific way to separate the rare earths. Samarium metal can be produced by reducing the oxide with barium or lanthanum. Samarium has a bright silver luster and is reasonably stable in air. Three crystal modifications of the metal exist, with transformations at 734 and 922 °C. The metal ignites in air at about 150 °C. Thirty-three isotopes and isomers of samarium are now recognized. Natural samarium is a mixture of seven isotopes, three of which are unstable but have long half-lives. Samarium, along with other rare earths, is used for carbon-arc lighting for the motion picture industry. The sulfide has excellent high-temperature stability and good thermoelectric efficiencies up to 1100 °C. $SmCo_5$ has been used in making a new permanent magnet material with the highest resistance to demagnetization of any known material. It is said to have an intrinsic coercive force as high as 2200 kA/m. Samarium oxide has been used in optical glass to absorb the infrared. Samarium is used to dope calcium fluoride crystals for use in optical masers or lasers. Compounds of the metal act as sensitizers for phosphors excited in the infrared; the oxide exhibits catalytic properties in the dehydration and dehydrogenation of ethyl alcohol. It is used in infrared absorbing glass and as a neutron absorber in nuclear reactors. The metal is priced at about \$3.50/g (99.9%). Little is known of the toxicity of samarium; therefore, it should be handled carefully.

Scandium — (L. *Scandia*, Scandinavia), Sc; at. wt. 44.955912(6); at. no. 21; m.p. 1541 °C; b.p. 2836 °C; sp. gr. 2.989 (25 °C); valence 3. On the basis of the Periodic System, Mendeleev predicted the existence of *ekaboron*, which would have an atomic weight between 40 of calcium and 48 of titanium. The element was discovered by Nilson in 1878 in the minerals *euxenite* and *gadolinite*, which had not yet been found anywhere except in Scandinavia. By processing 10 kg of euxenite and other residues of rare-earth minerals, Nilson was able to prepare about 2 g of scandium oxide of high purity. Cleve later pointed out that Nilson's scandium was identical with Mendeleev's ekaboron. Scandium is apparently a much more abundant element in the sun and certain stars than here on Earth. It is about the 23rd most abundant element in the sun, compared to the 50th most abundant on Earth. It is widely distributed on Earth, occurring in very minute quantities in over 800 mineral species. The blue color of beryl (aquamarine variety) is said to be due to scandium. It occurs as a principal component in the rare mineral *thortveitite*, found in Scandinavia and Malagasy. It is also found in the residues remaining after the extraction of tungsten from Zinnwald *wolframite*, and in *wiikite* and *bazzite*. Most scandium is presently being recovered from *thortveitite* or is extracted as a by-product from uranium mill tailings. Metallic scandium was first prepared in 1937 by Fischer, Brunger, and Grieneisen, who electrolyzed a eutectic melt of potassium, lithium, and scandium chlorides at 700 to 800 °C. Tungsten wire and a pool of molten zinc served as the electrodes in a graphite crucible. Pure scandium is now produced by reducing scandium fluoride with calcium metal. The production of the first pound of 99% pure scandium metal was announced in 1960. Scandium is a silver-white metal that develops a slightly yellowish or pinkish cast upon exposure to air. It is relatively soft, and resembles yttrium and the rare-earth metals more than it resembles aluminum or titanium. It is a very light metal and has a much higher melting point than aluminum, making it of interest to designers of spacecraft. Scandium is not attacked by a 1:1 mixture of conc. HNO_3 and 48% HF. Scandium reacts rapidly with many acids. Twenty-three isotopes and isomers of scandium are recognized. The metal is expensive, costing about \$200/g with a purity of about 99.9%. About 20 kg of scandium (as Sc_2O_3) are now being used yearly in the U.S. to produce high-intensity lights, and the radioactive isotope ^{46}Sc is used as a tracing agent in refinery crackers for crude oil, etc. Scandium iodide added to mercury vapor lamps produces a highly efficient light source resembling sunlight, which is important for indoor or night-time color TV. Little is yet known about the toxicity of scandium; therefore, it should be handled with care.

Seaborgium — (Glenn T. Seaborg [1912–1999], American chemist and nuclear physicist). Sg; at. wt. [266]; at no. 106. The discovery of *Seaborgium*, Element 106, took place in 1974 almost simultaneously at the Lawrence-Berkeley Laboratory and at the Joint Institute for Nuclear Research at Dubna, Russia. The Berkeley Group, under direction of Ghiorso, used the Super-Heavy Ion Linear Accelerator (Super HILAC) as a source of heavy ^{18}O ions to bombard a 259-μg target of ^{249}Cf. This resulted in the production and positive identification of $^{263}106$, which decayed with a half-life of 0.9 ± 0.2 s by the emission of alpha particles as follows:

$$^{263}106 \xrightarrow{\alpha} {}^{259}104 \xrightarrow{\alpha} {}^{255}\mathrm{No} \xrightarrow{\alpha}.$$

The Dubna Team, directed by Flerov and Organessian, produced heavy ions of ^{54}Cr with their 310-cm heavy-ion cyclotron to bombard ^{207}Pb and ^{208}Pb and found a product that decayed with a half-life of 7 ms. They assigned $^{259}106$ to this isotope. It is now thought seven isotopes of *Seaborgium* have been identified. Two of the isotopes are believed to have half-lives of about 30 s. *Seaborgium* most likely would have properties resembling tungsten. The IUPAC adopted the name *Seaborgium* in August 1997. Normally the naming of an element is not given until after the death of the person for which the element is named; however, in this case, it was named while Dr. Seaborg was still alive.

Selenium — (Gr. *Selene*, moon), Se; at. wt. 78.96(3); at. no. 34; m.p. (gray) 221 °C; b.p. (gray) 685 °C; sp. gr. (gray) 4.79, (vitreous) 4.28; valence −2, +4, or +6. Discovered by Berzelius in 1817, who found it associated with tellurium, named for the Earth. Selenium is found in a few rare minerals, such as *crooksite* and *clausthalite*. In years past it has been obtained from flue dusts remaining from processing copper sulfide ores, but the anode muds from electrolytic copper refineries now provide the source of most of the world's selenium. Selenium is recovered by roasting the muds with soda or sulfuric acid, or by smelting them with soda and niter. Selenium exists in several allotropic forms. Three are generally recognized, but as many as six have been claimed. Selenium can be prepared with either an amorphous or crystalline structure. The color of amorphous selenium is either red, in powder form, or black, in vitreous form. Crystalline monoclinic selenium is a deep red; crystalline hexagonal selenium, the most stable variety, is a metallic gray. Natural selenium contains six stable isotopes. Twenty-nine other isotopes and isomers have been characterized. The element is a member of the sulfur family and resembles sulfur both in its various forms and in its compounds. Selenium exhibits both photovoltaic action, where light is converted directly into electricity, and photoconductive action, where the electrical resistance decreases with increased illumination. These properties make selenium useful in the production of photocells and exposure meters for photographic use, as well as solar cells. Selenium is also able to convert a.c. electricity to d.c., and is extensively used in rectifiers. Below its melting point, selenium is a p-type semiconductor and is finding many uses in electronic and solid-state applications. It is used in xerography for reproducing and copying documents, letters, etc., but recently its use in this application has been decreasing in favor of certain organic compounds. It is used by the glass industry to decolorize glass and to make ruby-colored glasses and enamels. It is also used as a photographic toner, and as an additive to stainless steel. Elemental selenium has been said to be practically nontoxic and is considered to be an essential trace element; however, hydrogen selenide and other selenium compounds are extremely toxic, and resemble arsenic in their physiological reactions. Hydrogen selenide in a concentration of 1.5 ppm is intolerable to man. Selenium occurs in some soils in amounts sufficient to produce serious effects on animals feeding on plants, such as locoweed, grown in such soils. Selenium (99.5%) is priced at about $250/kg. It is also available in high-purity form at a cost of about $350/kg (99.999%).

Silicon — (L. *silex, silicis,* flint), Si; at. wt. 28.0855(3); at. no. 14; m.p. 1414 °C; b.p. 3265 °C; sp. gr. 2.33 (25 °C); valence 4. Davy in 1800 thought silica to be a compound and not an element; later in 1811, Gay Lussac and Thenard probably prepared impure amorphous silicon by heating potassium with silicon tetrafluoride. Berzelius, generally credited with the discovery, in 1824 succeeded in preparing amorphous silicon by the same general method as used earlier, but he purified the product by removing the fluosilicates by repeated washings. Deville in 1854 first prepared crystalline silicon, the second allotropic form of the element. Silicon is present in the sun and stars and is a principal component of a class of meteorites known as "aerolites." It is also a component of *tektites,* a natural glass of uncertain origin. Natural silicon contains three isotopes. Twenty-four other radioactive isotopes are recognized. Silicon makes up 25.7% of the Earth's crust, by weight, and is the second most abundant element, being exceeded only by oxygen. Silicon is not found free in nature, but occurs chiefly as the oxide and as silicates. *Sand, quartz, rock crystal, amethyst, agate, flint, jasper,* and *opal* are some of the forms in which the oxide appears. *Granite, hornblende, asbestos, feldspar, clay mica,* etc. are but a few of the numerous silicate minerals. Silicon is prepared commercially by heating silica and carbon in an electric furnace, using carbon electrodes. Several other methods can be used for preparing the element. Amorphous silicon can be prepared as a brown powder, which can be easily melted or vaporized. Crystalline silicon has a metallic luster and grayish color. The Czochralski process is commonly used to produce single crystals of silicon used for solid-state or semiconductor devices. Hyperpure silicon can be prepared by the thermal decomposition of ultra-pure trichlorosilane in a hydrogen atmosphere, and by a vacuum float zone process. This product can be doped with boron, gallium, phosphorus, or arsenic to produce silicon for use in transistors, solar cells, rectifiers, and other solid-state devices that are used extensively in the electronics and space-age industries. Hydrogenated amorphous silicon has shown promise in producing economical cells for converting solar energy into electricity. Silicon is a relatively inert element, but it is attacked by halogens and dilute alkali. Most acids, except hydrofluoric, do not affect it. Silicones are important products of silicon. They may be prepared by hydrolyzing a silicon organic chloride, such as dimethyl silicon chloride. Hydrolysis and condensation of various substituted chlorosilanes can be used to produce a very great number of polymeric products, or silicones, ranging from liquids to hard, glasslike solids with many useful properties. Elemental silicon transmits more than 95% of all wavelengths of infrared, from 1.3 to 6.7 μm. Silicon is one of man's most useful elements. In the form of sand and clay it is used to make concrete and brick; it is a useful refractory material for high-temperature work, and in the form of silicates it is used in making enamels, pottery, etc. Silica, as sand, is a principal ingredient of glass, one of the most inexpensive of materials with excellent mechanical, optical, thermal, and electrical properties. Glass can be made in a very great variety of shapes, and is used as containers, window glass, insulators, and thousands of other uses. Silicon tetrachloride can be used to iridize glass. Silicon is important in plant and animal life. Diatoms in both fresh and salt water extract silica from the water to build up their cell walls. Silica is present in ashes of plants and in the human skeleton. Silicon is an important ingredient in steel; silicon carbide is one of the most important abrasives and has been used in lasers to produce coherent light of 4560 Å. A remarkable material, first discovered in 1930, is *Aerogel,* which is now used by NASA in their space missions to collect cometary and interplanet dust. *Aerogel* is a highly insulative material that has the lowest density of any known solid. One form of *Aerogel* is 99.9% air and 0.1% SiO_2 by volume. It is 1000 times less dense than glass. It has been called "blue smoke" or "solid smoke." A block of *Aerogel* as large as a person may weigh less than a pound and yet support the weight of 1000 lbs (455 kg). This material is expected to trap cometary particles traveling at speeds of 32 km/sec. *Aerogel* is said to be non-toxic and non-inflammable. It has high thermal insulating qualities that could be used in home insulation. Its light weight may have aircraft applications. Regular grade silicon (99.5%) costs about $160/kg. Silicon (99.9999%) pure costs about $200/kg; hyperpure silicon is available at a higher cost. Miners, stonecutters, and other engaged in work where siliceous dust is breathed in large quantities often develop a serious lung disease known as *silicosis.*

Silver — (Anglo-Saxon, *Seolfor siolfur*), Ag (L. *argentum*), at. wt. 107.8682(2); at. no. 47; m.p. 961.78 °C; b.p. 2162 °C; sp. gr. 10.50 (20 °C); valence 1, 2. Silver has been known since ancient times. It is mentioned in Genesis. Slag dumps in Asia Minor and on islands in the Aegean Sea indicate that man learned to separate silver from lead as early as 3000 B.C. Silver occurs native and in ores such as *argentite* (Ag_2S) and *horn silver* (AgCl); lead, lead-zinc, copper, gold, and copper-nickel ores are principal sources. Mexico, Canada, Peru, and the U.S. are the principal silver producers in the western hemisphere. Silver is also recovered during electrolytic refining of copper. Commercial fine silver contains at least 99.9% silver. Purities of 99.999+% are available commercially. Pure silver has a brilliant white metallic luster. It is a little harder than gold and is very ductile and malleable, being exceeded only by gold and perhaps palladium. Pure silver has the highest electrical and thermal conductivity of all metals, and possesses the lowest contact resistance. It is stable in pure air and water, but tarnishes when exposed to ozone, hydrogen sulfide, or air containing sulfur. The alloys of silver are important. Sterling silver is used for jewelry, silverware, etc. where appearance is paramount. This alloy contains 92.5% silver, the remainder being copper or some other metal. Silver is of utmost importance in photography, about 30% of the U.S. industrial consumption going into this application. It is used for dental alloys. Silver is used in making solder and brazing alloys, electrical contacts, and high capacity silver–zinc and silver–cadmium batteries. Silver paints are used for making printed circuits. It is used in mirror production and may be deposited on glass or metals by chemical deposition, electrodeposition, or by evaporation. When freshly deposited, it is the best reflector of visible light known, but is rapidly tarnishes and loses much of its reflectance. It is a poor reflector of ultraviolet. Silver fulminate ($Ag_2C_2N_2O_2$), a powerful explosive, is sometimes formed during the silvering process. Silver iodide is used in seeding clouds to produce rain. Silver chloride has interesting optical properties as it can be made transparent; it also is a cement for glass. Silver nitrate, or *lunar caustic,* the most important silver compound, is used extensively in photography. While silver itself is not considered to be toxic, most of its salts are poisonous. Natural silver contains two stable isotopes. Fifty-six other radioactive isotopes and isomers are known. Silver compounds can be absorbed in the circulatory system and reduced silver deposited in the various tissues of the body. A condition, known as *argyria*, results with a greyish pigmentation of the skin and mucous membranes. Silver has germicidal effects and kills many lower organisms effectively without harm to higher animals. Silver for centuries has been used traditionally for coinage by many countries of the world. In recent times, however, consumption of silver has at times greatly exceeded the output. In 1939, the price of silver was fixed by the U.S. Treasury at 71¢/troy oz., and at 90.5¢/troy oz. in 1946. In November 1961 the U.S. Treasury suspended sales of nonmonetized silver, and the price stabilized for a time at about $1.29, the melt-down value of silver U.S. coins. The Coinage Act of 1965 authorized a change in the metallic composition of the three U.S. subsidiary denominations to clad or composite type coins. This was the first change in U.S. coinage since the monetary system was established in 1792. Clad dimes and quarters are made of an outer layer of 75% Cu and 25% Ni bonded to a central core of pure Cu. The composition of the one- and five-cent pieces remains unchanged. One-cent coins are 95% Cu and 5% Zn. Five-cent coins are 75% Cu and 25% Ni. Old silver dollars are 90% Ag and 10% Cu. Earlier subsidiary coins of 90% Ag and 10% Cu officially were to circulate alongside the clad coins; however, in practice they have largely disappeared (Gresham's Law), as the value of the silver is now greater than their exchange value. Silver coins of other countries have largely been replaced with coins made of other metals. On June 24, 1968, the U.S. Government ceased to redeem U.S. Silver Certificates with silver. Since that time, the price of silver has fluctuated widely. As of January 2002, the price of silver was about $4.10/troy oz. (13¢/g); however the price has fluctuated considerably due to market instability. The price of silver in 2001 was only about four times the cost of the metal about 150 years ago. This has largely been caused by Central Banks disposing of some of their silver reserves and the development of more productive mines with better refining methods. Also, silver has been displaced by other metals or processes, such as digital photography.

Sodium — (English, *soda*; Medieval Latin, *sodanum*, headache remedy), Na (L. *natrium*); at. wt. 22.98976928(2); at. no. 11; m.p. 97.80 °C; b.p. 883 °C; sp. gr. 0.971 (20 °C); valence 1. Long recognized in compounds, sodium was first isolated by Davy in 1807 by electrolysis of caustic soda. Sodium is present in fair abundance in the sun and stars. The D lines of sodium are among the most prominent in the solar spectrum. Sodium is the sixth most abundant element on earth, comprising about 2.6% of the Earth's crust; it is the most abundant of the alkali group of metals of which it is a member. The most common compound is sodium chloride, but it occurs in many other minerals, such as *soda niter, cryolite, amphibole, zeolite, sodalite*, etc. It is a very reactive element and is never found free in nature. It is now obtained commercially by the electrolysis of absolutely dry fused sodium chloride. This method is much cheaper than that of electrolyzing sodium hydroxide, as was used several years ago. Sodium is a soft, bright, silvery metal that floats on water, decomposing it with the evolution of hydrogen and the formation of the hydroxide. It may or may not ignite spontaneously on water, depending on the amount of oxide and metal exposed to the water. It normally does not ignite in air at temperatures below 115 °C. Sodium should be handled with respect, as it can be dangerous when improperly handled. Metallic sodium is vital in the manufacture of sodamide and esters, and in the preparation of organic compounds. The metal may be used to improve the structure of certain alloys, to descale metal, to purify molten metals, and as a heat transfer agent. An alloy of sodium with potassium, NaK, is also an important heat transfer agent. Sodium compounds are important to the paper, glass, soap, textile, petroleum, chemical, and metal industries. Soap is generally a sodium salt of certain fatty acids. The importance of common salt to animal nutrition has been recognized since prehistoric times. Among the many compounds that are of the greatest industrial importance are common salt (NaCl), soda ash (Na_2CO_3), baking soda ($NaHCO_3$), caustic soda (NaOH), Chile saltpeter ($NaNO_3$), di- and tri-sodium phosphates, sodium thiosulfate (hypo, $Na_2S_2O_3 \cdot 5H_2O$), and borax ($Na_2B_4O_7 \cdot 10H_2O$). Seventeen isotopes of sodium are recognized. Metallic sodium is priced at about $575/kg (99.95%). On a volume basis, it is the cheapest of all metals. Sodium metal should be handled with great care. It should be kept in an inert atmosphere and contact with water and other substances with which sodium reacts should be avoided.

Strontium — (*Strontian*, town in Scotland), Sr; at. wt. 87.62(1); at. no. 38; m.p. 777 °C; b.p. 1382 °C; sp. gr. 2.64; valence 2. Isolated by Davey by electrolysis in 1808; however, Adair Crawford in 1790 recognized a new mineral (strontianite) as differing from other barium minerals (baryta). Strontium is found chiefly as *celestite* ($SrSO_4$) and *strontianite* ($SrCO_3$). *Celestite* is found in Mexico, Turkey, Iran, Spain, Algeria, and in the U.K. The U.S. has no active *celestite* mines. The metal can be prepared by electrolysis of the fused chloride mixed with potassium chloride, or is made by reducing strontium oxide with aluminum in a vacuum at a temperature at which strontium distills off. Three allotropic forms of the metal exist, with transition points at 235 and 540 °C. Strontium is softer than calcium and decomposes water more vigorously. It does not absorb nitrogen below 380 °C. It should be kept under mineral oil to prevent oxidation. Freshly cut strontium has a silvery appearance, but rapidly turns a yellowish color with the formation of the oxide. The finely divided metal ignites spontaneously in air. Volatile strontium salts impart a beautiful crimson color to flames, and these salts are used in pyrotechnics and in the production of flares. Natural strontium is a mixture of four stable isotopes. Thirty-two other unstable isotopes and isomers are known to exist. Of greatest importance is ^{90}Sr with a half-life of 29 years. It is a product of nuclear fallout and presents a health problem. This isotope is one of the best long-lived high-energy beta emitters known, and is used in SNAP (Systems for Nuclear Auxiliary Power) devices. These devices hold promise for use in space vehicles, remote weather stations, navigational buoys, etc., where a lightweight, long-lived, nuclear-electric power source is needed. The major use for strontium at present is in producing glass for color television picture tubes. All color TV and cathode ray tubes sold in the U.S. are required by law to contain strontium in the face plate glass to block X-ray emission. Strontium also improves the brilliance of the glass and the quality of the picture. It has also found use in producing ferrite magnets and in refining zinc. Strontium titanate is an interesting optical material as it has an extremely high refractive index and an optical dispersion greater than that of diamond. It has been used as a gemstone, but it is very soft. It does not occur naturally. Strontium metal (99% pure) costs about $220/kg.

Sulfur — (Sanskrit, *sulvere*; L. *sulphurium*), S; at. wt. 32.065(5); at. no. 16; m.p. 115.21 °C; b.p. 444.61 °C; t_c 1041 °C; sp. gr. (rhombic) 2.07, (monoclinic) 2.00 (20 °C); valence 2, 4, or 6. Known to the ancients; referred to in Genesis as *brimstone*. Sulfur is found in meteorites. A dark area near the crater Aristarchus on the moon has been studied by R. W. Wood with ultraviolet light. This study suggests strongly that it is a sulfur deposit. Sulfur occurs native in the vicinity of volcanoes and hot springs. It is widely distributed in nature as *iron pyrites, galena, sphalerite, cinnabar, stibnite, gypsum, Epsom salts, celestite, barite,* etc. Sulfur is commercially recovered from wells sunk into the salt domes along the Gulf Coast of the U.S. It is obtained from these wells by the Frasch process, which forces heated water into the wells to melt the sulfur, which is then brought to the surface. Sulfur also occurs in natural gas and petroleum crudes and must be removed from these products. Formerly this was done chemically, which wasted the sulfur. New processes now permit recovery, and these sources promise to be very important. Large amounts of sulfur are being recovered from Alberta gas fields. Sulfur is a pale yellow, odorless, brittle solid that is insoluble in water but soluble in carbon disulfide. In every state, whether gas, liquid or solid, elemental sulfur occurs in more than one allotropic form or modification; these present a confusing multitude of forms whose relations are not yet fully understood. Amorphous or "plastic" sulfur is obtained by fast cooling of the crystalline form. X-ray studies indicate that amorphous sulfur may have a helical structure with eight atoms per spiral. Crystalline sulfur seems to be made of rings, each containing eight sulfur atoms that fit together to give a normal X-ray pattern. Twenty-one isotopes of sulfur are now recognized. Four occur in natural sulfur, none of which is radioactive. A finely divided form of sulfur, known as *flowers of sulfur*, is obtained by sublimation. Sulfur readily forms sulfides with many elements. Sulfur is a component of black gunpowder, and is used in the vulcanization of natural rubber and a fungicide. It is also used extensively is making phosphatic fertilizers. A tremendous tonnage is used to produce sulfuric acid, the most important manufactured chemical. It is used in making sulfite paper and other papers, as a fumigant, and in the bleaching of dried fruits. The element is a good electrical insulator. Organic compounds containing sulfur are very important. Calcium sulfate, ammonium sulfate, carbon disulfide, sulfur dioxide, and hydrogen sulfide are but a few of the many other important compounds of sulfur. Sulfur is essential to life. It is a minor constituent of fats, body fluids, and skeletal minerals. Carbon disulfide, hydrogen sulfide, and sulfur dioxide should be handled carefully. Hydrogen sulfide in small concentrations can be metabolized, but in higher concentrations it can quickly cause death by respiratory paralysis. It is insidious in that it quickly deadens the sense of smell. Sulfur dioxide is a dangerous component in atmospheric pollution. Sulfur (99.999%) costs about $575/kg.

Tantalum — (Gr. *Tantalos*, mythological character, father of *Niobe*), Ta; at. wt. 180.94788(2); at. no. 73; m.p. 3017 °C; b.p. 5458 °C; sp. gr. 16.4; valence 2?, 3, 4?, or 5. Discovered in 1802 by Ekeberg, but many chemists thought niobium and tantalum were identical elements until Rose, in 1844, and Marignac, in 1866, showed that niobic and tantalic acids were two different acids. The early investigators only isolated the impure metal. The first relatively pure ductile tantalum was produced by von Bolton in 1903. Tantalum occurs principally in the mineral *columbite-tantalite* (Fe, Mn)(Nb, Ta)$_2$O$_6$. Tantalum ores are found in Australia, Brazil, Rwanda, Zimbabwe, Congo-Kinshasa, Nigeria, and Canada. Separation of tantalum from niobium requires several complicated steps. Several methods are used to commercially produce the element, including electrolysis of molten potassium fluorotantalate, reduction of potassium fluorotantalate with sodium, or reacting tantalum carbide with tantalum oxide. Thirty-four isotopes and isomers of tantalum are known to exist. Natural tantalum contains two isotopes, one of which is radioactive with a very long half-life. Tantalum is a gray, heavy, and very hard metal. When pure, it is ductile and can be drawn into fine wire, which is used as a filament for evaporating metals such as aluminum. Tantalum is almost completely immune to chemical attack at temperatures below 150 °C, and is attacked only by hydrofluoric acid, acidic solutions containing the fluoride ion, and free sulfur trioxide. Alkalis attack it only slowly. At high temperatures, tantalum becomes much more reactive. The element has a melting point exceeded only by tungsten and rhenium. Tantalum is used to make a variety of alloys with desirable properties such as high melting point, high strength, good ductility, etc. Scientists at Los Alamos have produced a tantalum carbide graphite composite material that is said to be one of the hardest materials ever made. The compound has a melting point of 3738 °C. Tantalum has good "gettering" ability at high temperatures, and tantalum oxide films are stable and have good rectifying and dielectric properties. Tantalum is used to make electrolytic capacitors and vacuum furnace parts, which account for about 60% of its use. The metal is also widely used to fabricate chemical process equipment, nuclear reactors, and aircraft and missile parts. Tantalum is completely immune to body liquids and is a nonirritating metal. It has, therefore, found wide use in making surgical appliances. Tantalum oxide is used to make special glass with a high index of refraction for camera lenses. The metal has many other uses. The price of (99.9%) tantalum is about $2/g.

Technetium — (Gr. *technetos*, artificial), Tc; at. wt. (98); at. no. 43; m.p. 2157 °C; b.p. 4265 °C; sp. gr. 11.50 (calc.); valence 0, +2, +4, +5, +6, and +7. Element 43 was predicted on the basis of the periodic table, and was erroneously reported as having been discovered in 1925, at which time it was named *masurium*. The element was actually discovered by Perrier and Segre in Italy in 1937. It was found in a sample of molybdenum that was bombarded by deuterons in the Berkeley cyclotron, and which E. Lawrence sent to these investigators. Technetium was the first element to be produced artificially. Since its discovery, searches for the element in terrestrial materials have been made without success. If it does exist, the concentration must be very small. Technetium has been found in the spectrum of S-, M-, and N-type stars, and its presence in stellar matter is leading to new theories of the production of heavy elements in the stars. Forty-three isotopes and isomers of technetium, with mass numbers ranging from 86 to 113, are known. ^{97}Tc has a half-life of 2.6 × 10^6 years. ^{98}Tc has a half-life of 4.2 × 10^6 years. The isomeric isotope ^{95m}Tc, with a half-life of 61 days, is useful for tracer work, as it produces energetic gamma rays. Technetium metal has been produced in kilogram quantities. The metal was first prepared by passing hydrogen gas at 1100 °C over Tc_2S_7. It is now conveniently prepared by the reduction of ammonium pertechnetate with hydrogen. Technetium is a silvery-gray metal that tarnishes slowly in moist air. Until 1960, technetium was available only in small amounts and the price was as high as $2800/g, but the price is now of the order of $100/g. The chemistry of technetium is similar to that of rhenium. Technetium dissolves in nitric acid, aqua regia, and concentrated sulfuric acid, but is not soluble in hydrochloric acid of any strength. The element is a remarkable corrosion inhibitor for steel. It is reported that mild carbon steels may be effectively protected by as little as 55 ppm of $KTcO_4$ in aerated distilled water at temperatures up to 250 °C. This corrosion protection is limited to closed systems, since technetium is radioactive and must be confined. ^{99}Tc has a specific activity of 6.2 × 10^8 Bq/g. Activity of this level must not be allowed to spread. ^{99}Tc is a contamination hazard and should be handled in a glove box. The metal is an excellent superconductor at 11K and below.

Tellurium — (L. *tellus,* earth), Te; at. wt. 127.60(3); at. no. 52; m.p. 449.51 °C; b.p. 988 °C; sp. gr. 6.23 (20 °C); valence −2, 4, or 6. Discovered by Muller von Reichenstein in 1782; named by Klaproth, who isolated it in 1798. Tellurium is occasionally found native, but is more often found as the telluride of gold (*calaverite*), and combined with other metals. It is recovered commercially from the anode muds produced during the electrolytic refining of blister copper. The U.S., Canada, Peru, and Japan are the largest producers of the element. Crystalline tellurium has a silvery-white appearance, and when pure exhibits a metallic luster. It is brittle and easily pulverized. Amorphous tellurium is formed by precipitating tellurium from a solution of telluric or tellurous acid. Whether this form is truly amorphous, or made of minute crystals, is open to question. Tellurium is a p-type semiconductor, and shows greater conductivity in certain directions, depending on alignment of the atoms. Its conductivity increases slightly with exposure to light. It can be doped with silver, copper, gold, tin, or other elements. In air, tellurium burns with a greenish-blue flame, forming the dioxide. Molten tellurium corrodes iron, copper, and stainless steel. Tellurium and its compounds are probably toxic and should be handled with care. Workmen exposed to as little as 0.01 mg/m^3 of air, or less, develop "tellurium breath," which has a garlic-like odor. Forty-two isotopes and isomers of tellurium are known, with atomic masses ranging from 106 to 138. Natural tellurium consists of eight isotopes, two of which are radioactive with very long half-lives. Tellurium improves the machinability of copper and stainless steel, and its addition to lead decreases the corrosive action of sulfuric acid on lead and improves its strength and hardness. Tellurium catalysts are used in the oxidation of organic compounds and are used in hydrogenation and halogenation reactions. Tellurium is also used in electronic and semiconductor devices. It is also used as a basic ingredient in blasting caps, and is added to cast iron for chill control. Tellurium is used in ceramics. Bismuth telluride has been used in thermoelectric devices. Tellurium costs about 50¢/g, with a purity of about 99.5%. The metal with a purity of 99.9999% costs about $5/g.

Terbium — (*Ytterby*, village in Sweden), Tb; at. wt. 158.92534(2); at. no. 65; m.p. 1356 °C; b.p. 3230 °C; sp. gr. 8.230; valence 3, 4. Discovered by Mosander in 1843. Terbium is a member of the lanthanide or "rare earth" group of elements. It is found in *cerite, gadolinite,* and other minerals along with other rare earths. It is recovered commercially from *monazite* in which it is present to the extent of 0.03%, from *xenotime,* and from *euxenite,* a complex oxide containing 1% or more of terbia. Terbium has been isolated only in recent years with the development of ion-exchange techniques for separating the rare-earth elements. As with other rare earths, it can be produced by reducing the anhydrous chloride or fluoride with calcium metal in a tantalum crucible. Calcium and tantalum impurities can be removed by vacuum remelting. Other methods of isolation are possible. Terbium is reasonably stable in air. It is a silver-gray metal, and is malleable, ductile, and soft enough to be cut with a knife. Two crystal modifications exist, with a transformation temperature of 1289 °C. Forty-two isotopes and isomers are recognized. The oxide is a chocolate or dark maroon color. Sodium terbium borate is used as a laser material and emits coherent light at 0.546 μm. Terbium is used to dope calcium fluoride, calcium tungstate, and strontium molybdate, used in solid-state devices. The oxide has potential application as an activator for green phosphors used in color TV tubes. It can be used with ZrO_2 as a crystal stabilizer of fuel cells that operate at elevated temperature. Few other uses have been found. The element is priced at about $40/g (99.9%). Little is known of the toxicity of terbium. It should be handled with care as with other lanthanide elements.

Thallium — (Gr. *thallos,* a green shoot or twig), Tl; at. wt. 204.3833(2); at. no. 81; m.p. 304 °C; b.p. 1473 °C; sp. gr. 11.85 (20 °C); valence 1, or 3. Thallium was discovered spectroscopically in 1861 by Crookes. The element was named after the beautiful green spectral line, which identified the element. The metal was isolated both by Crookes and Lamy in 1862 about the same time. Thallium occurs in *crooksite, lorandite,* and *hutchinsonite.* It is also present in *pyrites* and is recovered from the roasting of this ore in connection with the production of sulfuric acid. It is also obtained from the smelting of lead and zinc ores. Extraction is somewhat complex and depends on the source of the thallium. Manganese nodules, found on the ocean floor, contain thallium. When freshly exposed to air, thallium exhibits a metallic luster, but soon develops a bluish-gray tinge, resembling lead in appearance. A heavy oxide builds up on thallium if left in air, and in the presence of water the hydroxide is formed. The metal is very soft and malleable. It can be cut with a knife. Forty-seven isotopes of thallium, with atomic masses ranging from 179 to 210 are recognized. Natural thallium is a mixture of two isotopes. The element and its compounds are toxic and should be handled carefully. Contact of the metal with skin is dangerous, and when melting the metal adequate ventilation should be provided. Thallium is suspected of carcinogenic potential for man. Thallium sulfate has been widely employed as a rodenticide and ant killer. It is odorless and tasteless, giving no warning of its presence. Its use, however, has been prohibited in the U.S. since 1975 as a household insecticide and rodenticide. The electrical conductivity of thallium sulfide changes with exposure to infrared light, and this compound is used in photocells. Thallium bromide-iodide crystals have been used as infrared optical materials. Thallium has been used, with sulfur or selenium and arsenic, to produce low melting glasses which become fluid between 125 and 150 °C. These glasses have properties at room temperatures similar to ordinary glasses and are said to be durable and insoluble in water. Thallium oxide has been used to produce glasses with a high index of refraction. Thallium has been used in treating ringworm and other skin infections; however, its use has been limited because of the narrow margin between toxicity and therapeutic benefits. A mercury–thallium alloy, which forms a eutectic at 8.5% thallium, is reported to freeze at −60 °C, some 20° below the freezing point of mercury. Thallium metal (99.999%) costs about $2/g.

Thorium — (*Thor*, Scandinavian god of war), Th; at. wt. 232.03806(2); at. no. 90; m.p. 1750 °C; b.p. 4788 °C; sp. gr. 11.72; valence +2(?), +3(?), +4. Discovered by Berzelius in 1828. Thorium occurs in *thorite* ($ThSiO_4$) and in *thorianite* (ThO_2 + UO_2). Large deposits of thorium minerals have been reported in New England and elsewhere, but these have not yet been exploited. Thorium is now thought to be about three times as abundant as uranium and about as abundant as lead or molybdenum. The metal is a source of nuclear power. There is probably more energy available for use from thorium in the minerals of the Earth's crust than from both uranium and fossil fuels. Any sizable demand for thorium as a nuclear fuel is still several years in the future. Work has been done in developing thorium cycle converter-reactor systems. Several prototypes, including the HTGR (high-temperature gas-cooled reactor) and MSRE (molten salt converter reactor experiment), have operated. While the HTGR reactors are efficient, they are not expected to become important commercially for many years because of certain operating difficulties. Thorium is recovered commercially from the mineral *monazite*, which contains from 3 to 9% ThO_2 along with rare-earth minerals. Much of the internal heat the Earth produces has been attributed to thorium and uranium. Several methods are available for producing thorium metal: it can be obtained by reducing thorium oxide with calcium, by electrolysis of anhydrous thorium chloride in a fused mixture of sodium and potassium chlorides, by calcium reduction of thorium tetrachloride mixed with anhydrous zinc chloride, and by reduction of thorium tetrachloride with an alkali metal. Thorium was originally assigned a position in Group IV of the periodic table. Because of its atomic weight, valence, etc., it is now considered to be the second member of the *actinide* series of elements. When pure, thorium is a silvery-white metal which is air stable and retains its luster for several months. When contaminated with the oxide, thorium slowly tarnishes in air, becoming gray and finally black. The physical properties of thorium are greatly influenced by the degree of contamination with the oxide. The purest specimens often contain several tenths of a percent of the oxide. High-purity thorium has been made. Pure thorium is soft, very ductile, and can be cold-rolled, swaged, and drawn. Thorium is dimorphic, changing at 1400 °C from a cubic to a body-centered cubic structure. Thorium oxide has a melting point of 3300 °C, which is the highest of all oxides. Only a few elements, such as tungsten, and a few compounds, such as tantalum carbide, have higher melting points. Thorium is slowly attacked by water, but does not dissolve readily in most common acids, except hydrochloric. Powdered thorium metal is often pyrophoric and should be carefully handled. When heated in air, thorium turnings ignite and burn brilliantly with a white light. The principal use of thorium has been in the preparation of the Welsbach mantle, used for portable gas lights. These mantles, consisting of thorium oxide with about 1% cerium oxide and other ingredients, glow with a dazzling light when heated in a gas flame. Thorium is an important alloying element in magnesium, imparting high strength and creep resistance at elevated temperatures. Because thorium has a low work-function and high electron emission, it is used to coat tungsten wire used in electronic equipment. The oxide is also used to control the grain size of tungsten used for electric lamps; it is also used for high-temperature laboratory crucibles. Glasses containing thorium oxide have a high refractive index and low dispersion. Consequently, they find application in high quality lenses for cameras and scientific instruments. Thorium oxide has also found use as a catalyst in the conversion of ammonia to nitric acid, in petroleum cracking, and in producing sulfuric acid. Thorium has not found many uses due to its radioactive nature and its handling and disposal problems. Thirty isotopes of thorium are known with atomic masses ranging from 210 to 237. All are unstable. ^{232}Th occurs naturally and has a half-life of 1.4×10^{10} years. It is an alpha emitter. ^{232}Th goes through six alpha and four beta decay steps before becoming the stable isotope ^{208}Pb. ^{232}Th is sufficiently radioactive to expose a photographic plate in a few hours. Thorium disintegrates with the production of "thoron" (^{220}Rn), which is an alpha emitter and presents a radiation hazard. Good ventilation of areas where thorium is stored or handled is therefore essential. Thorium metal (99.8%) costs about $25/g.

Thulium — (*Thule*, the earliest name for Scandinavia), Tm; at. wt. 168.93421(2); at. no. 69; m.p. 1545 °C; b.p. 1950 °C; sp. gr. 9.321 (25 °C); valence 3. Discovered in 1879 by Cleve. Thulium occurs in small quantities along with other rare earths in a number of minerals. It is obtained commercially from *monazite*, which contains about 0.007% of the element. Thulium is the least abundant of the rare-earth elements, but with new sources recently discovered, it is now considered to be about as rare as silver, gold, or cadmium. Ion-exchange and solvent extraction techniques have recently permitted much easier separation of the rare earths, with much lower costs. Only a few years ago, thulium metal was not obtainable at any cost; in 1996 the oxide cost $20/g. Thulium metal powder now costs $70/g (99.9%). Thulium can be isolated by reduction of the oxide with lanthanum metal or by calcium reduction of the anhydrous fluoride. The pure metal has a bright, silvery luster. It is reasonably stable in air, but the metal should be protected from moisture in a closed container. The element is silver-gray, soft, malleable, and ductile, and can be cut with a knife. Forty-one isotopes and isomers are known, with atomic masses ranging from 146 to 176. Natural thulium, which is 100% ^{169}Tm, is stable. Because of the relatively high price of the metal, thulium has not yet found many practical applications. ^{169}Tm bombarded in a nuclear reactor can be used as a radiation source in portable X-ray equipment. ^{171}Tm is potentially useful as an energy source. Natural thulium also has possible use in *ferrites* (ceramic magnetic materials) used in microwave equipment. As with other lanthanides, thulium has a low-to-moderate acute toxicity rating. It should be handled with care.

Tin — (Anglo-Saxon, *tin*), Sn (L. *stannum*); at. wt. 118.710(7); at. no. 50; m.p. 231.93 °C; b.p. 2602 °C; sp. gr. (gray) 5.77, (white) 7.29; valence 2, 4. Known to the ancients. Tin is found chiefly in *cassiterite* (SnO_2). Most of the world's supply comes from China, Indonesia, Peru, Brazil, and Bolivia. The U.S. produces almost none, although occurrences have been found in Alaska and Colorado. Tin is obtained by reducing the ore with coal in a reverberatory furnace. Ordinary tin is composed of ten stable isotopes; thirty-six unstable isotopes and isomers are also known. Ordinary tin is a silver-white metal, is malleable, somewhat ductile, and has a highly crystalline structure. Due to the breaking of these crystals, a "tin cry" is heard when a bar is bent. The element has two allotropic forms at normal pressure. On warming, gray, or α tin, with a cubic structure, changes at 13.2 °C into white, or β tin, the ordinary form of the metal. White tin has a tetragonal structure. When tin is cooled below 13.2 °C, it changes slowly from white to gray. This change is affected by impurities such as aluminum and zinc, and can be prevented by small additions of antimony or bismuth. This change from the α to β form is called the tin pest. Tin–lead alloys are used to make organ pipes. There are few if any uses for gray tin. Tin takes a high polish and is used to coat other metals to prevent corrosion or other chemical action. Such tin plate over steel is used in the so-called tin can for preserving food. Alloys of tin are very important. Soft solder, type metal, fusible metal, pewter, bronze, bell metal, Babbitt metal, white metal, die casting alloy, and phosphor bronze are some of the important alloys using tin. Tin resists distilled sea and soft tap water, but is attacked by strong acids, alkalis, and acid salts. Oxygen in solution accelerates the attack. When heated in air, tin forms SnO_2, which is feebly acid, forming stannate salts with basic oxides. The most important salt is the chloride ($SnCl_2 \cdot H_2O$), which is used as a reducing agent and as a mordant in calico printing. Tin salts sprayed onto glass are used to produce electrically conductive coatings. These have been used for panel lighting and for frost-free windshields. Most window glass is now made by floating molten glass on molten tin (float glass) to produce a flat surface (Pilkington process). Of recent interest is a crystalline tin–niobium alloy that is superconductive at very low temperatures. This promises to be important in the construction of superconductive magnets that generate enormous field strengths but use practically no power. Such magnets, made of tin–niobium wire, weigh but a few pounds and produce magnetic fields that, when started with a small battery, are comparable to that of a 100 ton electromagnet operated continuously with a large power supply. The small amount of tin found in canned foods is quite harmless. The agreed limit of tin content in U.S. foods is 300 mg/kg. The trialkyl and triaryl tin compounds are used as biocides and must be handled carefully. Over the past 25 years the price of commercial tin has varied from 50¢/lb ($1.10/kg) to about $6/kg. Tin (99.99% pure) costs about $260/kg.

Titanium — (L. *Titans,* the first sons of the Earth, myth.), Ti; at. wt. 47.867(1); at. no. 22; m.p. 1668 °C; b.p. 3287 °C; sp. gr. 4.51; valence 2, 3, or 4. Discovered by Gregor in 1791; named by Klaproth in 1795. Impure titanium was prepared by Nilson and Pettersson in 1887; however, the pure metal (99.9%) was not made until 1910 by Hunter by heating $TiCl_4$ with sodium in a steel bomb. Titanium is present in meteorites and in the sun. Rocks obtained during the Apollo 17 lunar mission showed presence of 12.1% TiO_2. Analyses of rocks obtained during earlier Apollo missions show lower percentages. Titanium oxide bands are prominent in the spectra of M-type stars. The element is the ninth most abundant in the crust of the Earth. Titanium is almost always present in igneous rocks and in the sediments derived from them. It occurs in the minerals *rutile, ilmenite,* and *sphene,* and is present in titanates and in many iron ores. Deposits of ilmenite and rutile are found in Florida, California, Tennessee, and New York. Australia, Norway, Malaysia, India, and China are also large suppliers of titanium minerals. Titanium is present in the ash of coal, in plants, and in the human body. The metal was a laboratory curiosity until Kroll, in 1946, showed that titanium could be produced commercially by reducing titanium tetrachloride with magnesium. This method is largely used for producing the metal today. The metal can be purified by decomposing the iodide. Titanium, when pure, is a lustrous, white metal. It has a low density, good strength, is easily fabricated, and has excellent corrosion resistance. It is ductile only when it is free of oxygen. The metal burns in air and is the only element that burns in nitrogen. Titanium is resistant to dilute sulfuric and hydrochloric acid, most organic acids, moist chlorine gas, and chloride solutions. Natural titanium consists of five isotopes with atomic masses from 46 to 50. All are stable. Eighteen other unstable isotopes are known. The metal is dimorphic. The hexagonal α form changes to the cubic β form very slowly at about 880 °C. The metal combines with oxygen at red heat, and with chlorine at 550 °C. Titanium is important as an alloying agent with aluminum, molybdenum, manganese, iron, and other metals. Alloys of titanium are principally used for aircraft and missiles where lightweight strength and ability to withstand extremes of temperature are important. Titanium is as strong as steel, but 45% lighter. It is 60% heavier than aluminum, but twice as strong. Titanium has potential use in desalination plants for converting sea water into fresh water. The metal has excellent resistance to sea water and is used for propeller shafts, rigging, and other parts of ships exposed to salt water. A titanium anode coated with platinum has been used to provide cathodic protection from corrosion by salt water. Titanium metal is considered to be physiologically inert; however, titanium powder may be a carcinogenic hazard. When pure, titanium dioxide is relatively clear and has an extremely high index of refraction with an optical dispersion higher than diamond. It is produced artificially for use as a gemstone, but it is relatively soft. Star sapphires and rubies exhibit their asterism as a result of the presence of TiO_2. Titanium dioxide is extensively used for both house paint and artist's paint, as it is permanent and has good covering power. Titanium oxide pigment accounts for the largest use of the element. Titanium paint is an excellent reflector of infrared, and is extensively used in solar observatories where heat causes poor seeing conditions. Titanium tetrachloride is used to iridize glass. This compound fumes strongly in air and has been used to produce smoke screens. The price of titanium metal (99.9%) is about $1100/kg.

Tungsten — (Swedish, *tung sten,* heavy stone); also known as *wolfram* (from *wolframite,* said to be named from *wolf rahm* or *spumi lupi,* because the ore interfered with the smelting of tin and was supposed to devour the tin), W; at. wt. 183.84(1); at. no. 74; m.p. 3422 °C; b.p. 5555 °C; sp. gr. 19.3 (20 °C); valence 2, 3, 4, 5, or 6. In 1779 Peter Woulfe examined the mineral now known as *wolframite* and concluded it must contain a new substance. Scheele, in 1781, found that a new acid could be made from *tung sten* (a name first applied about 1758 to a mineral now known as *scheelite*). Scheele and Berman suggested the possibility of obtaining a new metal by reducing this acid. The de Elhuyar brothers found an acid in *wolframite* in 1783 that was identical to the acid of *tungsten* (tungstic acid) of Scheele, and in that year they succeeded in obtaining the element by reduction of this acid with charcoal. Tungsten occurs in *wolframite,* (Fe, Mn)WO_4; *scheelite,* $CaWO_4$; *huebnerite,* $MnWO_4$; and *ferberite,* $FeWO_4$. Important deposits of tungsten occur in California, Colorado, Bolivia, Russia, and Portugal. China is reported to have about 75% of the world's tungsten resources. Natural tungsten contains five stable isotopes. Thirty-two other unstable isotopes and isomers are recognized. The metal is obtained commercially by reducing tungsten oxide with hydrogen or carbon. Pure tungsten is a steel-gray to tin-white metal. Very pure tungsten can be cut with a hacksaw, and can be forged, spun, drawn, and extruded. The impure metal is brittle and can be worked only with difficulty. Tungsten has the highest melting point of all metals, and at temperatures over 1650 °C has the highest tensile strength. The metal oxidizes in air and must be protected at elevated temperatures. It has excellent corrosion resistance and is attacked only slightly by most mineral acids. The thermal expansion is about the same as borosilicate glass, which makes the metal useful for glass-to-metal seals. Tungsten and its alloys are used extensively for filaments for electric lamps, electron and television tubes, and for metal evaporation work; for electrical contact points for automobile distributors; X-ray targets; windings and heating elements for electrical furnaces; and for numerous spacecraft and high-temperature applications. High-speed tool steels, Hastelloy®, Stellite®, and many other alloys contain tungsten. Tungsten carbide is of great importance to the metal-working, mining, and petroleum industries. Calcium and magnesium tungstates are widely used in fluorescent lighting; other salts of tungsten are used in the chemical and tanning industries. Tungsten disulfide is a dry, high-temperature lubricant, stable to 500 °C. Tungsten bronzes and other tungsten compounds are used in paints. Zirconium tungstate has found recent applications (see under Zirconium). Tungsten powder (99.999%) costs about $2900/kg.

Uranium — (Planet *Uranus*), U; at. wt. 238.02891(3); at. no. 92; m.p. 1135 °C; b.p. 4131 °C; sp. gr. 19.1; valence 2, 3, 4, 5, or 6. Yellow-colored glass, containing more than 1% uranium oxide and dating back to 79 A.D., has been found near Naples, Italy. Klaproth recognized an unknown element in *pitchblende* and attempted to isolate the metal in 1789. The metal apparently was first isolated in 1841 by Peligot, who reduced the anhydrous chloride with potassium. Uranium is not as rare as it was once thought. It is now considered to be more plentiful than mercury, antimony, silver, or cadmium, and is about as abundant as molybdenum or arsenic. It occurs in numerous minerals such as *pitchblende, uraninite, carnotite, autunite, uranophane, davidite,* and *tobernite*. It is also found in *phosphate rock, lignite, monazite sands,* and can be recovered commercially from these sources. Large deposits of uranium ore occur in Utah, Colorado, New Mexico, Canada, and elsewhere. Uranium can be made by reducing uranium halides with alkali or alkaline earth metals or by reducing uranium oxides by calcium, aluminum, or carbon at high temperatures. The metal can also be produced by electrolysis of KUF_5 or UF_4, dissolved in a molten mixture of $CaCl_2$ and NaCl. High-purity uranium can be prepared by the thermal decomposition of uranium halides on a hot filament. Uranium exhibits three crystallographic modifications as follows:

$$\alpha \xrightarrow{688^\circ C} \beta \xrightarrow{776^\circ C} \gamma$$

Uranium is a heavy, silvery-white metal that is pyrophoric when finely divided. It is a little softer than steel, and is attacked by cold water in a finely divided state. It is malleable, ductile, and slightly paramagnetic. In air, the metal becomes coated with a layer of oxide. Acids dissolve the metal, but it is unaffected by alkalis. Uranium has twenty-three isotopes, one of which is an isomer and all of which are radioactive. Naturally occurring uranium contains 99.2745% by weight ^{238}U, 0.720% ^{235}U, and 0.0055% ^{234}U. Studies show that the percentage weight of ^{235}U in natural uranium varies by as much as 0.1%, depending on the source. The U.S.D.O.E. has adopted the value of 0.711 as being their "official" percentage of ^{235}U in natural uranium. Natural uranium is sufficiently radioactive to expose a photographic plate in an hour or so. Much of the internal heat of the Earth is thought to be attributable to the presence of uranium and thorium. ^{238}U, with a half-life of 4.46×10^9 years, has been used to estimate the age of igneous rocks. The origin of uranium, the highest member of the naturally occurring elements — except perhaps for traces of neptunium or plutonium — is not clearly understood, although it has been thought that uranium might be a decay product of elements of higher atomic weight, which may have once been present on Earth or elsewhere in the universe. These original elements may have been formed as a result of a primordial "creation," known as "the big bang," in a supernova, or in some other stellar processes. The fact that recent studies show that most trans-uranic elements are extremely rare with very short half-lives indicates that it may be necessary to find some alternative explanation for the very large quantities of radioactive uranium we find on Earth. Studies of meteorites from other parts of the solar system show a relatively low radioactive content, compared to terrestrial rocks. Uranium is of great importance as a nuclear fuel. ^{238}U can be converted into fissionable plutonium by the following reactions:

$$^{238}U(n,\gamma) \rightarrow ^{239}U \xrightarrow{\beta} ^{239}Np \xrightarrow{\beta} ^{239}Pu$$

This nuclear conversion can be brought about in "breeder" reactors where it is possible to produce more new fissionable material than the fissionable material used in maintaining the chain reaction. ^{235}U is of even greater importance, for it is the key to the utilization of uranium. ^{235}U, while occurring in natural uranium to the extent of only 0.72%, is so fissionable with slow neutrons that a self-sustaining fission chain reaction can be made to occur in a reactor constructed from natural uranium and a suitable moderator, such as heavy water or graphite, alone. ^{235}U can be concentrated by gaseous diffusion and other physical processes, if desired, and used directly as a nuclear fuel, instead of natural uranium, or used as an explosive. Natural uranium, slightly enriched with ^{235}U by a small percentage, is used to fuel nuclear power reactors for the generation of electricity. Natural thorium can be irradiated with neutrons as follows to produce the important isotope ^{233}U.

$$^{232}Th(n,\gamma) \rightarrow ^{233}Th \xrightarrow{\beta} ^{233}Pa \xrightarrow{\beta} ^{233}U$$

While thorium itself is not fissionable, ^{233}U is, and in this way may be used as a nuclear fuel. One pound of completely fissioned uranium has the fuel value of over 1500 tons of coal. The uses of nuclear fuels to generate electrical power, to make isotopes for peaceful purposes, and to make explosives are well known. The estimated world-wide production of the 437 nuclear power reactors in operation in 1998 amounted to about 352,000 megawatt hours. In 1998 the U.S. had about 107 commercial reactors with an output of about 100,000 megawatt-hours. Some nuclear-powered electric generating plants have recently been closed because of safety concerns. There are also serious problems with nuclear waste disposal that have not been completely resolved. Uranium in the U.S. is controlled by the U.S. Nuclear Regulatory Commission, under the Department of Energy. Uses are being found for the large quantities of "depleted" uranium now available, where uranium-235 has been lowered to about 0.2%. Depleted uranium has been used for inertial guidance devices, gyrocompasses, counterweights for aircraft control surfaces, ballast for missile reentry vehicles, and as a shielding material for tanks, etc. Concerns, however, have been raised over its low radioactive properties. Uranium metal is used for X-ray targets for production of high-energy X-rays. The nitrate has been used as photographic toner, and the acetate is used in analytical chemistry. Crystals of uranium nitrate are triboluminescent. Uranium salts have also been used for producing yellow "vaseline" glass and glazes. Uranium and its compounds are highly toxic, both from a chemical and radiological standpoint. Finely divided uranium metal, being pyrophoric, presents a fire hazard. The maximum permissible total body burden of natural uranium (based on radiotoxicity) is 0.2 μCi for soluble compounds. Recently, the natural presence of uranium and thorium in many soils has become of concern to homeowners because of the generation of radon and its daughters (see under Radon). Uranium metal is available commercially at a cost of about $6/g (99.7%) in air-tight glass under argon.

Vanadium — (Scandinavian goddess, *Vanadis*), V; at. wt. 50.9415(1); at. no. 23; m.p. 1910 °C; b.p. 3407 °C; sp. gr. 6.0 (18.7 °C); valence 2, 3, 4, or 5. Vanadium was first discovered by del Rio in 1801. Unfortunately, a French chemist incorrectly declared that del Rio's new element was only impure chromium; del Rio thought himself to be mistaken and accepted the French chemist's statement. The element was rediscovered in 1830 by Sefstrom, who named the element in honor of the Scandinavian goddess *Vanadis* because of its beautiful multicolored compounds. It was isolated in nearly pure form by Roscoe, in 1867, who reduced the chloride with hydrogen. Vanadium of 99.3 to 99.8% purity was not produced until 1927. Vanadium is found in about 65 different minerals among which *carnotite, roscoelite, vanadinite,* and *patronite* are important sources of the metal. Vanadium is also found in phosphate rock and certain iron ores, and is present in some crude oils in the form of organic complexes. It is also found in small percentages in meteorites. Commercial production from petroleum ash holds promise as an important source of the element. China, South Africa, and Russia supply much of the world's vanadium ores. High-purity ductile vanadium can be obtained by reduction of vanadium trichloride with magnesium or with magnesium–sodium mixtures. Much of the vanadium metal being produced is now made by calcium reduction of V_2O_5 in a pressure vessel, an adaptation of a process developed by McKechnie and Seybolt. Natural vanadium is a mixture of two isotopes, ^{50}V (0.25%) and ^{51}V (99.75%). ^{50}V is slightly radioactive, having a long half-life. Twenty other unstable isotopes are recognized. Pure vanadium is a bright white metal, and is soft and ductile. It has good corrosion resistance to alkalis, sulfuric and hydrochloric acid, and salt water, but the metal oxidizes readily above 660 °C. The metal has good structural strength and a low-fission neutron cross section, making it useful in nuclear applications. Vanadium is used in producing rust-resistant, spring, and high-speed tool steels. It is an important carbide stabilizer in making steels. About 80% of the vanadium now produced is used as ferrovanadium or as a steel additive. Vanadium foil is used as a bonding agent in cladding titanium to steel. Vanadium pentoxide is used in ceramics and as a catalyst. It is also used in producing a superconductive magnet with a field of 175,000 gauss. Vanadium and its compounds are toxic and should be handled with care. Ductile vanadium is commercially available. Vanadium metal (99.7%) costs about $3/g.

Wolfram — see Tungsten.

Xenon — (Gr. *xenon,* stranger), Xe; at. wt. 131.293(6); at. no. 54; m.p. −111.74 °C; b.p. −108.09 °C; t_c 16.58 °C; density (gas) 5.887 ± 0.009 g/L, sp. gr (liquid) 2.95 (−109 °C); valence usually 0. Discovered by Ramsay and Travers in 1898 in the residue left after evaporating liquid air components. Xenon is a member of the so-called noble or "inert" gases. It is present in the atmosphere to the extent of about one part in twenty million. Xenon is present in the Martian atmosphere to the extent of 0.08 ppm. The element is found in the gases evolved from certain mineral springs, and is commercially obtained by extraction from liquid air. Natural xenon is composed of nine stable isotopes. In addition to these, thirty-five unstable isotopes and isomers have been characterized. Before 1962, it had generally been assumed that xenon and other noble gases were unable to form compounds. However, it is now known that xenon, as well as other members of the zero valence elements, do form compounds. Among the compounds of xenon now reported are xenon hydrate, sodium perxenate, xenon deuterate, difluoride, tetrafluoride, hexafluoride, and $XePtF_6$ and $XeRhF_6$. Xenon trioxide, which is highly explosive, has been prepared. More than 80 xenon compounds have been made with xenon chemically bonded to fluorine and oxygen. Some xenon compounds are colored. Metallic xenon has been produced, using several hundred kilobars of pressure. Xenon in a vacuum tube produces a beautiful blue glow when excited by an electrical discharge. The gas is used in making electron tubes, stroboscopic lamps, bactericidal lamps, and lamps used to excite ruby lasers for generating coherent light. Xenon is used in the atomic energy field in bubble chambers, probes, and other applications where its high molecular weight is of value. The perxenates are used in analytical chemistry as oxidizing agents. ^{133}Xe and ^{135}Xe are produced by neutron irradiation in air-cooled nuclear reactors. ^{133}Xe has useful applications as a radioisotope. The element is available in sealed glass containers for about $20/L of gas at standard pressure. Xenon is not toxic, but its compounds are highly toxic because of their strong oxidizing characteristics.

Ytterbium — (*Ytterby,* village in Sweden), Yb; at. wt. 173.04(3); at. no. 70; m.p. 824 °C; b.p. 1196 °C; sp. gr (α) 6.903 (β) 6.966; valence 2, 3. Marignac in 1878 discovered a new component, which he called *ytterbia,* in the Earth then known as *erbia.* In 1907, Urbain separated ytterbia into two components, which he called *neoytterbia* and *lutecia.* The elements in these earths are now known as *ytterbium* and *lutetium,* respectively. These elements are identical with *aldebaranium* and *cassiopeium,* discovered independently and at about the same time by von Welsbach. Ytterbium occurs along with other rare earths in a number of rare minerals. It is commercially recovered principally from *monazite sand,* which contains about 0.03%. Ion-exchange and solvent extraction techniques developed in recent years have greatly simplified the separation of the rare earths from one another. The element was first prepared by Klemm and Bonner in 1937 by reducing ytterbium trichloride with potassium. Their metal was mixed, however, with KCl. Daane, Dennison, and Spedding prepared a much purer form in 1953 from which the chemical and physical properties of the element could be determined. Ytterbium has a bright silvery luster, is soft, malleable, and quite ductile. While the element is fairly stable, it should be kept in closed containers to protect it from air and moisture. Ytterbium is readily attacked and dissolved by dilute and concentrated mineral acids and reacts slowly with water. Ytterbium has three allotropic forms with transformation points at −13° and 795 °C. The beta form is a room-temperature, face-centered, cubic modification, while the high-temperature gamma form is a body-centered cubic form. Another body-centered cubic phase has recently been found to be stable at high pressures at room temperatures. The beta form ordinarily has metallic-type conductivity, but becomes a semiconductor when the pressure is increased above 16,000 atm. The electrical resistance increases tenfold as the pressure is increased to 39,000 atm and drops to about 80% of its standard temperature-pressure resistivity at a pressure of 40,000 atm. Natural ytterbium is a mixture of seven stable isotopes. Twenty-six other unstable isotopes and isomers are known. Ytterbium metal has possible use in improving the grain refinement, strength, and other mechanical properties of stainless steel. One isotope is reported to have been used as a radiation source as a substitute for a portable X-ray machine where electricity is unavailable. Few other uses have been found. Ytterbium metal is available with a purity of about 99.9% for about $10/g. Ytterbium has a low acute toxicity rating.

Yttrium — (*Ytterby*, village in Sweden near Vauxholm), Y; at. wt. 88.90585(2); at. no. 39; m.p. 1522 °C; b.p. 3345 °C; sp. gr. 4.469 (25 °C); valence 3. *Yttria*, which is an earth containing yttrium, was discovered by Gadolin in 1794. *Ytterby* is the site of a quarry which yielded many unusually minerals containing rare earths and other elements. This small town, near Stockholm, bears the honor of giving names to *erbium, terbium*, and *ytterbium* as well as *yttrium*. In 1843 Mosander showed that yttria could be resolved into the oxides (or earths) of three elements. The name yttria was reserved for the most basic one; the others were named *erbia* and *terbia*. Yttrium occurs in nearly all of the rare-earth minerals. Analysis of lunar rock samples obtained during the Apollo missions show a relatively high yttrium content. It is recovered commercially from *monazite sand*, which contains about 3%, and from *bastnasite*, which contains about 0.2%. Wohler obtained the impure element in 1828 by reduction of the anhydrous chloride with potassium. The metal is now produced commercially by reduction of the fluoride with calcium metal. It can also be prepared by other techniques. Yttrium has a silver-metallic luster and is relatively stable in air. Turnings of the metal, however, ignite in air if their temperature exceeds 400 °C, and finely divided yttrium is very unstable in air. Yttrium oxide is one of the most important compounds of yttrium and accounts for the largest use. It is widely used in making YVO_4 europium, and Y_2O_3 europium phosphors to give the red color in color television tubes. Many hundreds of thousands of pounds are now used in this application. Yttrium oxide also is used to produce yttrium iron garnets, which are very effective microwave filters. Yttrium iron, aluminum, and gadolinium garnets, with formulas such as $Y_3Fe_5O_{12}$ and $Y_3Al_5O_{12}$, have interesting magnetic properties. Yttrium iron garnet is also exceptionally efficient as both a transmitter and transducer of acoustic energy. Yttrium aluminum garnet, with a hardness of 8.5, is also finding use as a gemstone (simulated diamond). Small amounts of yttrium (0.1 to 0.2%) can be used to reduce the grain size in chromium, molybdenum, zirconium, and titanium, and to increase strength of aluminum and magnesium alloys. Alloys with other useful properties can be obtained by using yttrium as an additive. The metal can be used as a deoxidizer for vanadium and other nonferrous metals. The metal has a low cross section for nuclear capture. ^{90}Y, one of the isotopes of yttrium, exists in equilibrium with its parent ^{90}Sr, a product of atomic explosions. Yttrium has been considered for use as a nodulizer for producing nodular cast iron, in which the graphite forms compact nodules instead of the usual flakes. Such iron has increased ductility. Yttrium is also finding application in laser systems and as a catalyst for ethylene polymerization. It also has potential use in ceramic and glass formulas, as the oxide has a high melting point and imparts shock resistance and low expansion characteristics to glass. Natural yttrium contains but one isotope, ^{89}Y. Forty-three other unstable isotopes and isomers have been characterized. Yttrium metal of 99.9% purity is commercially available at a cost of about $5/g.

Zinc — (Ger. *Zink*, of obscure origin), Zn; at. wt. 65.409(4); at. no. 30; m.p. 419.53 °C; b.p. 907 °C; sp. gr. 7.134 (25 °C); valence 2. Centuries before zinc was recognized as a distinct element, zinc ores were used for making brass. Tubal-Cain, seven generations from Adam, is mentioned as being an "instructor in every artificer in brass and iron." An alloy containing 87% zinc has been found in prehistoric ruins in Transylvania. Metallic zinc was produced in the 13th century A.D. in India by reducing calamine with organic substances such as wool. The metal was rediscovered in Europe by Marggraf in 1746, who showed that it could be obtained by reducing *calamine* with charcoal. The principal ores of zinc are *sphalerite* or *blende* (sulfide), *smithsonite* (carbonate), *calamine* (silicate), and *franklinite* (zinc, manganese, iron oxide). Canada, Japan, Belgium, Germany, and the Netherlands are suppliers of zinc ores. Zinc is also mined in Alaska, Tennessee, Missouri, and elsewhere in the U.S. Zinc can be obtained by roasting its ores to form the oxide and by reduction of the oxide with coal or carbon, with subsequent distillation of the metal. Other methods of extraction are possible. Naturally occurring zinc contains five stable isotopes. Twenty-five other unstable isotopes and isomers are recognized. Zinc is a bluish-white, lustrous metal. It is brittle at ordinary temperatures but malleable at 100 to 150 °C. It is a fair conductor of electricity, and burns in air at high red heat with evolution of white clouds of the oxide. The metal is employed to form numerous alloys with other metals. Brass, nickel silver, typewriter metal, commercial bronze, spring brass, German silver, soft solder, and aluminum solder are some of the more important alloys. Large quantities of zinc are used to produce die castings, used extensively by the automotive, electrical, and hardware industries. An alloy called *Prestal*®, consisting of 78% zinc and 22% aluminum, is reported to be almost as strong as steel but as easy to mold as plastic. It is said to be so plastic that it can be molded into form by relatively inexpensive die casts made of ceramics and cement. It exhibits superplasticity. Zinc is also extensively used to galvanize other metals such as iron to prevent corrosion. Neither zinc nor zirconium is ferromagnetic; but $ZrZn_2$ exhibits ferromagnetism at temperatures below 35 K. Zinc oxide is a unique and very useful material to modern civilization. It is widely used in the manufacture of paints, rubber products, cosmetics, pharmaceuticals, floor coverings, plastics, printing inks, soap, storage batteries, textiles, electrical equipment, and other products. It has unusual electrical, thermal, optical, and solid-state properties that have not yet been fully investigated. Lithopone, a mixture of zinc sulfide and barium sulfate, is an important pigment. Zinc sulfide is used in making luminous dials, X-ray and TV screens, and fluorescent lights. The chloride and chromate are also important compounds. Zinc is an essential element in the growth of human beings and animals. Tests show that zinc-deficient animals require 50% more food to gain the same weight as an animal supplied with sufficient zinc. Zinc is not considered to be toxic, but when freshly formed ZnO is inhaled a disorder known as the *oxide shakes* or *zinc chills* sometimes occurs. It is recommended that where zinc oxide is encountered good ventilation be provided. The commercial price of zinc in January 2002 was roughly 40¢/lb ($90 kg). Zinc metal with a purity of 99.9999% is priced at about $5/g.

Zirconium — (Syriac, *zargun*, color of gold), Zr; at. wt. 91.224(2); at. no. 40; m.p. 1855 °C; b.p. 4409 °C; sp. gr. 6.52 (20 °C); valence +2, +3, and +4. The name *zircon* may have originated from the Syriac word *zargono*, which describes the color of certain gemstones now known as *zircon, jargon, hyacinth, jacinth,* or *ligure*. This mineral, or its variations, is mentioned in biblical writings. These minerals were not known to contain this element until Klaproth, in 1789, analyzed a *jargon* from Sri Lanka and found a new earth, which Werner named zircon (*silex circonius*), and Klaproth called *Zirkonerde (zirconia)*. The impure metal was first isolated by Berzelius in 1824 by heating a mixture of potassium and potassium zirconium fluoride in a small iron tube. Pure zirconium was first prepared in 1914. Very pure zirconium was first produced in 1925 by van Arkel and de Boer by an iodide decomposition process they developed. Zirconium is found in abundance in S-type stars, and has been identified in the sun and meteorites. Analyses of lunar rock samples obtained during the various Apollo missions to the moon show a surprisingly high zirconium oxide content, compared with terrestrial rocks. Naturally occurring zirconium contains five isotopes. Thirty-one other radioactive isotopes and isomers are known to exist. *Zircon*, $ZrSiO_4$, the principal ore, is found in deposits in Florida, South Carolina, Australia, South Africa, and elsewhere. *Baddeleyite*, found in Brazil, is an important zirconium mineral. It is principally pure ZrO_2 in crystalline form having a hafnium content of about 1%. Zirconium also occurs in some 30 other recognized mineral species. Zirconium is produced commercially by reduction of the chloride with magnesium (the Kroll Process), and by other methods. It is a grayish-white lustrous metal. When finely divided, the metal may ignite spontaneously in air, especially at elevated temperatures. The solid metal is much more difficult to ignite. The inherent toxicity of zirconium compounds is low. Hafnium is invariably found in zirconium ores, and the separation is difficult. Commercial-grade zirconium contains from 1 to 3% hafnium. Zirconium has a low absorption cross section for neutrons, and is therefore used for nuclear energy applications, such as for cladding fuel elements. Commercial nuclear power generation now takes more than 90% of zirconium metal production. Reactors of the size now being made may use as much as a half-million lineal feet of zirconium alloy tubing. Reactor-grade zirconium is essentially free of hafnium. *Zircaloy*® is an important alloy developed specifically for nuclear applications. Zirconium is exceptionally resistant to corrosion by many common acids and alkalis, by sea water, and by other agents. It is used extensively by the chemical industry where corrosive agents are employed. Zirconium is used as a getter in vacuum tubes, as an alloying agent in steel, in surgical appliances, photoflash bulbs, explosive primers, rayon spinnerets, lamp filaments, etc. It is used in poison ivy lotions in the form of the carbonate as it combines with *urushiol*. With niobium, zirconium is superconductive at low temperatures and is used to make superconductive magnets. Alloyed with zinc, zirconium becomes magnetic at temperatures below 35 K. Zirconium oxide (zircon) has a high index of refraction and is used as a gem material. The impure oxide, zirconia, is used for laboratory crucibles that will withstand heat shock, for linings of metallurgical furnaces, and by the glass and ceramic industries as a refractory material. Its use as a refractory material accounts for a large share of all zirconium consumed. Zirconium tungstate is an unusual material that shrinks, rather than expands, when heated. A few other compounds are known to possess this property, but they tend to shrink in one direction, while they stretch out in others in order to maintain an overall volume. Zirconium tungstate shrinks in all directions over a wide temperature range of from near absolute zero to +777 °C. It is being considered for use in composite materials where thermal expansion may be a problem. Zirconium of about 99.5% purity is available at a cost of about $2000/kg or about $4/g.

PHYSICAL CONSTANTS OF INORGANIC COMPOUNDS

The compounds in this table were selected on the basis of their laboratory and industrial importance, as well as their value in illustrating trends in the variation of physical properties with position in the periodic table. An effort has been made to include the most frequently encountered inorganic substances; a limited number of organometallics are also covered. Many, if not most, of the compounds that are solids at ambient temperature can exist in more than one crystalline modification. In the absence of other information, the data given here can be assumed to apply to the most stable or common crystalline form. In many cases, however, two or more forms are of practical importance, and separate entries will be found in the table.

Compounds are arranged primarily in alphabetical order by the most commonly used name. However, adjustments are made in many instances so as to bring closely related compounds together. For example, hydrides of elements such as boron, silicon, and germanium are grouped together immediately following the entry for the parent element, since they would otherwise be scattered throughout the table. Likewise, the oxoacids of an element are given in one group whenever a strict alphabetical order would separate them (e.g., sulfuric acid and fluorosulfuric acid). The Formula Index following the table provides another means of locating a compound. An index to CAS Registry Numbers is also available in the electronic versions of the *Handbook* or by request via e-mail (william.haynes@taylorandfrancis.com).

The following data fields appear in the table:

- **Name**: Systematic name for the substance. The valence state of a metallic element is indicated by a Roman numeral, e.g., copper in the +1 state is written as copper(I) rather than cuprous, iron in the +3 state is iron(III) rather than ferric.
- **Formula**: The simplest descriptive formula is given, but this does not necessarily specify the actual structure of the compound. For example, aluminum chloride is designated as $AlCl_3$, even though a more accurate representation of the structure in the solid phase (and, under some conditions, in the gas phase) is Al_2Cl_6 A few exceptions are made, such as the use of Hg_2^{+2} for the mercury(I) ion.
- **CAS Registry Number**: Chemical Abstracts Service Registry Number. An asterisk* following the CAS RN for a hydrate indicates that the number refers to the anhydrous compound. In most cases the generic CAS RN for the compound is given rather than the number for a specific crystalline form or mineral.
- **Mol. Weight**: Molecular weight (relative molar mass) as calculated with the 2005 IUPAC Recommended Atomic Weights. The number of decimal places corresponds to the number of places in the atomic weight of the least accurately known element (e.g., one place for lead compounds, two places for compounds of selenium, germanium, etc.); a maximum of three places is given. For compounds of radioactive elements for which IUPAC makes no recommendation, the mass number of the isotope with longest half-life is used.
- **Physical Form**: The crystal system is given, when available, for compounds that are solid at room temperature, together with color and other descriptive features. Abbreviations are listed below.
- **mp**: Normal melting point in °C. The notation tp indicates the temperature where solid, liquid, and gas are in equilibrium at a pressure greater than one atmosphere (i.e., the normal melting point does not exist). When available, the triple point pressure is listed.
- **bp**: Normal boiling point in °C (referred to 101.325 kPa or 760 mmHg pressure). The notation sp following the number indicates the temperature where the pressure of the vapor in equilibrium with the solid reaches 101.325 kPa. See Reference 8, p. 23, for further discussion of sublimation points and triple points. A notation "sublimes" without a temperature being given indicates that there is a perceptible sublimation pressure above the solid at ambient temperatures.
- **Density**: Density values for solids and liquids are always in units of grams per cubic centimeter and can be assumed to refer to temperatures near room temperature unless otherwise stated. Values for gases are the calculated ideal gas densities in grams per liter at 25 °C and 101.325 kPa; the unit is always specified for a gas value.
- **Aqueous Solubility**: Solubility is expressed as the number of grams of the compound (excluding any water of hydration) that will dissolve in 100 grams of water. The temperature in °C is given as a superscript. Solubility at other temperatures can be found for many compounds in the table "Aqueous Solubility of Inorganic Compounds at Various Temperatures" in Section 8.
- **Qualitative Solubility**: Qualitative information on the solubility in other solvents (and in water, if quantitative data are unavailable) is given here. The abbreviations are:

i	insoluble
sl	slightly soluble
s	soluble
vs	very soluble
reac	reacts with the solvent

Data were taken from a wide variety of reliable sources, including monographs, treatises, review articles, evaluated compilations and databases, and in many cases the primary literature. Some of the most useful references for the properties covered here are listed below.

List of Abbreviations

Ac - acetyl
ace - acetone
acid - acid solutions
alk - alkaline solutions
amorp - amorphous
anh - anhydrous
aq - aqueous
blk - black
brn - brown
bz - benzene
chl - chloroform
col - colorless
conc - concentrated
cry - crystals, crystalline
cub - cubic
cyhex - cyclohexane
dec - decomposes
dil - dilute
diox - dioxane
eth - ethyl ether
EtOH - ethanol
exp - explodes, explosive
extrap - extrapolated
flam - flammable
gl - glass, glassy
grn - green
hc - hydrocarbon solvents
hex - hexagonal, hexane

hp - heptane
HT - high temperature
hyd - hydrate
hyg - hygroscopic
i - insoluble in
liq - liquid
LT - low temperature
MeOH - methanol
monocl - monoclinic
octahed - octahedral
oran - orange
orth - orthorhombic
os - organic solvents
peth - petroleum ether
pow - powder
prec - precipitate
pur - purple
py - pyridine
reac - reacts with
refrac - refractory
rhom - rhombohedral
r.t. - room temperature
s - soluble in
silv - silvery
sl - slightly soluble in
soln - solution
sp - sublimation point
stab - stable
subl - sublimes
temp - temperature
tetr - tetragonal
thf - tetrahydrofuran
tol - toluene
tp - triple point
trans - transition, transformation
tricl - triclinic
trig - trigonal
unstab - unstable
viol - violet
visc - viscous
vs - very soluble in
wh - white
xyl - xylene
yel - yellow

References

1. Phillips, S. L., and Perry, D.L., *Handbook of Inorganic Compounds*, CRC Press, Boca Raton, FL, 1995.
2. Trotman-Dickenson, A. F., Executive Editor, *Comprehensive Inorganic Chemistry*, Vol. 1-5, Pergamon Press, Oxford, 1973.
3. Greenwood, N. N., and Earnshaw, A., *Chemistry of the Elements, Second Edition*, Butterworth-Heinemann, Oxford, 1997.
4. Wiberg, N., Wiberg, E., and Holleman, H. F., *Inorganic Chemistry, 34th Edition*, Academic Press, San Diego, 2001.
5. *GMELIN Handbook of Inorganic and Organometallic Chemistry*, Springer-Verlag, Heidelberg.
6. Chase, M.W., Davies, C.A., Downey, J.R., Frurip, D. J., McDonald, R.A., and Syverud, A.N.; *JANAF Thermochemical Tables, Third Edition, J. Phys. Chem. Ref. Data*, Vol. 14, Suppl. 1, 1985; Chase, M. W., *NIST-JANAF Thermochemical Tables, Fourth Edition, J. Phys. Chem. Ref. Data*, Monograph No. 9, 1998.
7. *Landolt-Börnstein, Numerical Data and Functional Relationships in Science and Technology, New Series*, IV/19A, "Thermodynamic Properties of Inorganic Materials compiled by SGTE", Springer-Verlag, Heidelberg; Part 1, 1999; Part 2; 1999; Part 3, 2000; Part 4, 2001.
8. Lide, D. R., and Kehiaian, H.V., *CRC Handbook of Thermophysical and Thermochemical Data*, CRC Press, Boca Raton, FL, 1994.
9. *Kirk-Othmer Concise Encyclopedia of Chemical Technology*, Wiley-Interscience, New York, 1985.
10. *Dictionary of Inorganic Compounds*, Chapman & Hall, New York, 1992.
11. Massalski, T. B., Ed., *Binary Alloy Phase Diagrams, 2nd Edition*, ASM International, Metals Park, Ohio, 1990.
12. Dinsdale, A.T., "SGTE Data for Pure Elements", *CALPHAD*, 15, 317–425, 1991.
13. Madelung, O., *Semiconductors: Group IV Elements and III-IV Compounds*, Springer-Verlag, Heidelberg, 1991.
14. Lidin, R. A., Andreeva, L. L., and Molochko, V. A., *Constants of Inorganic Substances*, Begell House, New York, 1995.
15. Gurvich, L. V., Veyts, I. V., and Alcock, C. B., *Thermodynamic Properties of Individual Substances, Fourth Edition*, Hemisphere Publishing Corp., New York, 1989.
16. *The Combined Chemical Dictionary on CDROM, Version 9:1*, Chapman & Hall/CRC, Boca Raton, FL, 2005.
17. Macdonald, F., Editor, *Chapman & Hall/CRC Combined Chemical Dictionary*, <http://www.chemnetbase.com/scripts/ccdweb.exe>.
18. Sangeeta, G., and LaGraff, J. R., *Inorganic Materials Chemistry, Second Edition*, CRC Press, Boca Raton, FL, 2005.
19. Stern, K. H., *High Temperature Properties and Thermal Decomposition of Inorganic Salts with Oxyanions*, CRC Press, Boca Raton, FL, 2001.
20. Donnay, J.D.H., and Ondik, H.M., *Crystal Data Determinative Tables, Third Edition, Volumes 2 and 4, Inorganic Compounds*, Joint Committee on Powder Diffraction Standards, Swarthmore, PA, 1973.
21. Robie, R., Bethke, P. M., and Beardsley, K. M., *Selected X-ray Crystallographic Data, Molar Volumes, and Densities of Minerals and Related Substances*, U.S. Geological Survey Bulletin 1248, 1967.
22. Carmichael, R. S., *Practical Handbook of Physical Properties of Rocks and Minerals*, CRC Press, Boca Raton, FL, 1989.
23. Deer, W. A., Howie, R.A., and Zussman, J., An *Introduction to the Rock-Forming Minerals, 2nd Edition*, Longman Scientific & Technical, Harlow, Essex, 1992.
24. Linstrom, P. J., and Mallard, W. G., Editors, NIST Chemistry WebBook, NIST Standard Reference Database No. 69, June 2005, National Institute of Standards and Technology, Gaithersburg, MD 20899, <http://webbook.nist.gov>.
25. *Phase Diagrams for Ceramists, Volumes 1–8; ACerS-NIST Phase Equilibrium Diagrams, Volumes 9–13*, American Ceramic Society, Westerville, Ohio, 1964–2001.

No.	Name	Formula	CAS Reg No.	Mol. weight	Physical form	mp/°C	bp/°C	Density g cm^{-3}	Solubility g/100 g H_2O	Qualitative solubility
1	Actinium	Ac	7440-34-8	227	silv metal; cub	1050	3198	10		
2	Actinium bromide	$AcBr_3$	33689-81-5	467	wh hex cry		800 subl	5.85		s H_2O
3	Actinium chloride	$AcCl_3$	22986-54-5	333	wh hex cry		960 subl	4.81		
4	Actinium fluoride	AcF_3	33689-80-4	284	wh hex cry			7.88		i H_2O
5	Actinium iodide	AcI_3	33689-82-6	608	wh cry					s H_2O
6	Actinium oxide	Ac_2O_3	12002-61-8	502	wh hex cry	1977		9.19		i H_2O
7	Aluminum	Al	7429-90-5	26.982	silv-wh metal; cub cry	660.32	2519	2.70		i H_2O; s acid, alk
8	Aluminum acetate	$Al(C_2H_3O_2)_3$	139-12-8	204.113	wh hyg solid	dec				s H_2O; sl ace
9	Aluminum diacetate	$Al(OH)(C_2H_3O_2)_2$	142-03-0	162.078	wh amorp powder					i H_2O
10	Aluminum ammonium sulfate	$AlNH_4(SO_4)_2$	7784-25-0	237.146	wh powder					sl H_2O; i EtOH
11	Aluminum ammonium sulfate dodecahydrate	$AlNH_4(SO_4)_2 \cdot 12H_2O$	7784-26-1	453.329	col cry or powder	94.5	>280 dec	1.65		s H_2O; i EtOH
12	Aluminum antimonide	AlSb	25152-52-7	148.742	brn cub cry	1065		4.26		
13	Aluminum arsenide	AlAs	22831-42-1	101.903	oran cub cry; hyg	1740		3.76		
14	Aluminum borate	$2Al_2O_3 \cdot B_2O_3$	11121-16-7	273.543	needles	≈1050				i H_2O
15	Aluminum borohydride	$Al(BH_4)_3$	16962-07-5	71.510	flam liq	-64.5	44.5			reac H_2O
16	Aluminum bromate nonahydrate	$Al(BrO_3)_3 \cdot 9H_2O$	11126-81-1*	572.826	wh hyg cry	62	>100 dec			s H_2O
17	Aluminum bromide	$AlBr_3$	7727-15-3	266.694	wh-yel monocl cry; hyg	97.5	255	3.2		reac H_2O; s bz, tol

No.	Name	Formula	CAS Reg No.	Mol. weight	Physical form	mp/°C	bp/°C	Density g cm^{-3}	Solubility g/100 g H_2O	Qualitative solubility
18	Aluminum bromide hexahydrate	$AlBr_3 \cdot 6H_2O$	7784-11-4	374.785	col-yel hyg cry	93		2.54		s H_2O, EtOH, CS_2
19	Aluminum carbide	Al_4C_3	1299-86-1	143.958	yel hex cry	2100	>2200 dec	2.36		reac H_2O
20	Aluminum chlorate nonahydrate	$Al(ClO_3)_3 \cdot 9H_2O$	15477-33-5	439.473	hyg cry					vs H_2O; s EtOH
21	Aluminum chloride	$AlCl_3$	7446-70-0	133.341	wh hex cry or powder; hyg	192.6	180 sp	2.48	45.1^{25}	s bz, ctc, chl
22	Aluminum chloride hexahydrate	$AlCl_3 \cdot 6H_2O$	7784-13-6	241.432	col hyg cry	100 dec		2.398	45.1^{25}	s EtOH, eth
23	Dichloromethylaluminum	$AlCl_2CH_3$	917-65-7	112.923	cry	72.7	95^{100}			s bz, eth, hc
24	Chlorodiethylaluminum	$AlCl(C_2H_5)_2$	96-10-6	120.557	col liq	-74		0.96		reac H_2O
25	Chlorodiisobutylaluminum	$AlCl(C_4H_9)_2$	1779-25-5	176.664	hyg col liq	-40		0.95		s eth, hx
26	Aluminum diboride	AlB_2	12041-50-8	48.604	powder	>920 dec		3.19		s dil HCl
27	Aluminum dodecaboride	AlB_{12}	12041-54-2	156.714	yel-brn prisms	2070		2.55		s hot HNO_3; i acid, alk
28	Aluminum ethanolate	$Al(C_2H_5O)_3$	555-75-9	162.163	liq, condenses to wh solid	140				reac H_2O; sl xyl
29	Aluminum fluoride	AlF_3	7784-18-1	83.977	wh hex cry	2250 tp (220 MPa)	1276 sp	3.10	0.50^{25}	
30	Aluminum fluoride monohydrate	$AlF_3 \cdot H_2O$	32287-65-3	101.992	orth cry			2.17	0.50^{25}	
31	Aluminum fluoride trihydrate	$AlF_3 \cdot 3H_2O$	15098-87-0	138.023	wh hyg cry			1.914	0.50^{25}	
32	Aluminum hexafluorosilicate nonahydrate	$Al_2(SiF_6)_3 \cdot 9H_2O$	17099-70-6	642.320	hex prisms	>500 dec				s H_2O
33	Aluminum hydride	AlH_3	7784-21-6	30.006	col hex cry	>150 dec				reac H_2O
34	Aluminum hydroxide	$Al(OH)_3$	21645-51-2	78.004	wh amorp powder			2.42		i H_2O; s alk, acid
35	Aluminum hydroxychloride	$Al_2(OH)_5Cl \cdot 2H_2O$	1327-41-9	210.483	gl solid					s H_2O
36	Aluminum iodide	AlI_3	7784-23-8	407.695	wh leaflets	188.28	382	3.98		reac H_2O
37	Aluminum iodide hexahydrate	$AlI_3 \cdot 6H_2O$	10090-53-6	515.786	yel hyg cry powder					vs H_2O; s EtOH, eth
38	Aluminum lactate	$Al(C_3H_5O_3)_3$	18917-91-4	294.192	powder					vs H_2O
39	Aluminum molybdate	$Al_2(MoO_4)_3$	15123-80-5	533.78	wh pow	≈950				
40	Aluminum nitrate	$Al(NO_3)_3$	13473-90-0	212.997	wh hyg solid	dec			68.9^{25}	vs EtOH; sl ace
41	Aluminum nitrate nonahydrate	$Al(NO_3)_3 \cdot 9H_2O$	7784-27-2	375.134	wh hyg monocl cry	73	135 dec	1.72	68.9^{25}	vs EtOH; i pyr
42	Aluminum nitride	AlN	24304-00-5	40.989	blue-wh hex cry	3000		3.255		reac H_2O
43	Aluminum oleate	$Al(C_{18}H_{33}O_2)_3$	688-37-9	871.342	yel solid					i H_2O; s EtOH, bz
44	Aluminum oxalate monohydrate	$Al_2(C_2O_4)_3 \cdot H_2O$	814-87-9	336.035	wh pow					i H_2O, EtOH; s acid
45	Aluminum oxide (α)	Al_2O_3	1344-28-1	101.961	wh powder; hex	2054	2977	3.99		i H_2O, os; sl alk
46	Aluminum oxide (γ)	Al_2O_3	1344-28-1	101.961	soft wh pow	trans to corundum 1200		3.97		i H_2O; s acid; sl alk
47	Aluminum oxyhydroxide (boehmite)	$AlO(OH)$	1318-23-6	59.989	wh orth cry	trans to diaspore 227		3.07		i H_2O; s hot acid, alk
48	Aluminum oxyhydroxide (diaspore)	$AlO(OH)$	14457-84-2	59.989	orth cry	dec 450		3.38		i H_2O; s acid, alk
49	Aluminum palmitate	$Al(C_{15}H_{31}COO)_3$	555-35-1	793.230	wh-yel powder					i H_2O, EtOH; s peth
50	Aluminum 2,4-pentanedioate	$Al(CH_3COCHCOCH_3)_3$	13963-57-0	324.306	pale yel prisms	194.6	315	1.27		i H20; s bz, EtOH; sl hex
51	Aluminum perchlorate	$Al(ClO_4)_3$	14452-39-2	325.334	wh hyg cry				55^{0}	s H_2O, eth; i ctc
52	Aluminum perchlorate nonahydrate	$Al(ClO_4)_3 \cdot 9H_2O$	14452-39-2	487.471	wh hyg cry	82 dec		2.0	182.4^{0}	
53	Aluminum phosphate	$AlPO_4$	7784-30-7	121.953	wh rhomb plates	>1460		2.56		i H_2O; sl acid
54	Aluminum phosphate dihydrate	$AlPO_4 \cdot 2H_2O$	13477-75-3	157.984	wh rhom cry	dec 1500		2.54		i H_2O
55	Aluminum phosphate trihydroxide	$Al_2(OH)_3PO_4$	12004-29-4	199.957	wh or yel monocl cry			2.7		
56	Aluminum metaphosphate	$Al(PO_3)_3$	32823-06-6	263.898	col powder; tetr	≈1525		2.78		i H_2O
57	Aluminum hypophosphite	$Al(H_2PO_2)_3$	7784-22-7	221.948	cry powder	220 dec				i H_2O; s alk, acid
58	Aluminum phosphide	AlP	20859-73-8	57.956	grn or yel cub cry	2550		2.40		reac H_2O
59	Aluminum selenide	Al_2Se_3	1302-82-5	290.84	yel-brown powder	960		3.437		reac H_2O
60	Aluminum silicate (andalusite)	Al_2SiO_5	12183-80-1	162.046	gray-grn cry			3.145		
61	Aluminum silicate (kyanite)	Al_2SiO_5	1302-76-7	162.046	blue or gray tricl cry	dec 1000		3.68		
62	Aluminum silicate (mullite)	$3Al_2O_3 \cdot 2SiO_2$	1302-93-8	426.052	col orth cry	1750		3.17		i H_2O, acid, HF
63	Aluminum silicate (sillimanite)	Al_2SiO_5	12141-45-6	162.046	wh orth cry	1816		3.25		
64	Aluminum silicate dihydrate	$Al_2O_3 \cdot 2SiO_2 \cdot 2H_2O$	1332-58-7	258.161	wh-yel powder; tricl			2.59		i H_2O, acid, alk
65	Aluminum stearate	$Al(C_{18}H_{35}O_2)_3$	637-12-7	877.390	wh powder	115		1.070		i H_2O, EtOH, eth; s alk
66	Aluminum monostearate	$Al(OH)_2(C_{18}H_{35}O_2)$	7047-84-9	344.467	yel-wh pow	155		1.02		i H_2O
67	Aluminum distearate	$Al(OH)(C_{18}H_{35}O_2)_2$	300-92-5	610.928	wh pow	145				i H_2O
68	Aluminum sulfate	$Al_2(SO_4)_3$	10043-01-3	342.151	wh cry	1040 dec			38.5^{25}	i EtOH
69	Aluminum sulfate octadecahydrate	$Al_2(SO_4)_3 \cdot 18H_2O$	7784-31-8	666.426	col monocl cry	86 dec		1.69	38.5^{25}	
70	Aluminum sulfide	Al_2S_3	1302-81-4	150.158	yel-gray powder	1100		2.02		
71	Aluminum telluride	Al_2Te_3	12043-29-7	436.76	gray-blk hex cry	≈895		4.5		

No.	Name	Formula	CAS Reg No.	Mol. weight	Physical form	mp/°C	bp/°C	Density g cm⁻³	Solubility g/100 g H_2O	Qualitative solubility
72	Aluminum thiocyanate	$Al(SCN)_3$	538-17-0	201.229	yel powder					s H_2O; i EtOH, eth
73	Aluminum titanate	Al_2TiO_5	12004-39-6	181.827	refrac solid	1860				
74	Aluminum zirconium	Al_2Zr	12004-50-1	145.187	metallic solid	1645				
75	Americium	Am	7440-35-9	243	silv metal; hex or cub	1176	2011	12		s acid
76	Americium(III) oxide	Am_2O_3	12254-64-7	534	tan hex cry			11.77		s acid
77	Americium(III) bromide	$AmBr_3$	14933-38-1	483	wh orth cry			6.85		s H_2O
78	Americium(III) chloride	$AmCl_3$	13464-46-5	349	pink hex cry	500		5.87		
79	Americium(III) fluoride	AmF_3	13708-80-0	300	pink hex cry	1393		9.53		
80	Americium(III) iodide	AmI_3	13813-47-3	624	yel ortho cry	≈950		6.9		
81	Americium(IV) fluoride	AmF_4	15947-41-8	319	tan monocl cry			7.23		
82	Americium(IV) oxide	AmO_2	12005-67-3	275	blk cub cry	>1000 dec		11.68		s acid
83	Ammonia	NH_3	7664-41-7	17.031	col gas	-77.73	-33.33	0.696 g/L		vs H_2O; s EtOH, eth
84	Ammonium acetate	$NH_4C_2H_3O_2$	631-61-8	77.083	wh hyg cry	114		1.073	148^{4}	s EtOH; sl ace
85	Ammonium azide	NH_4N_3	12164-94-2	60.059	orth cry; flam	160	exp	1.346	20.2^{30}	
86	Ammonium benzoate	$NH_4C_7H_5O_2$	1863-63-4	139.152	wh cry or powder	198		1.26		s H_2O; sl EtOH
87	Ammonium bromate	NH_4BrO_3	13843-59-9	145.941	col hex cry	exp				vs H_2O
88	Ammonium bromide	NH_4Br	12124-97-9	97.943	wh hyg tetr cry	542 dec	396 sp	2.429	78.3^{25}	s EtOH, ace; sl eth
89	Ammonium caprylate	$NH_4C_8H_{15}O_2$	5972-76-9	161.243	hyg monocl cry	≈75				reac H_2O; s EtOH; i chl, bz
90	Ammonium carbamate	NH_2COONH_4	1111-78-0	78.071	cry powder					vs H_2O; s EtOH
91	Ammonium carbonate	$(NH_4)_2CO_3$	506-87-6	96.086	col cry powder	58 dec			100^{15}	
92	Ammonium chlorate	NH_4ClO_3	10192-29-7	101.490	wh cry	102 exp		1.80	28.7^{0}	
93	Ammonium chloride	NH_4Cl	12125-02-9	53.492	col cub cry	520.1 tp (dec)	338 sp	1.519	39.5^{25}	
94	Ammonium chromate	$(NH_4)_2CrO_4$	7788-98-9	152.071	yel cry	185 dec		1.90	37^{25}	sl ace, MeOH; i EtOH
95	Ammonium chromic sulfate dodecahydrate	$NH_4Cr(SO_4)_2 \cdot 12H_2O$	10022-47-6	478.343	blue-viol cry	94 dec		1.72		s H_2O; sl EtOH
96	Ammonium cobalt(II) phosphate	$CoNH_4PO_4$	14590-13-7	171.943	red-viol powder (hyd)					i H_2O; s acid
97	Ammonium cobalt(II) phosphate monohydrate	$CoNH_4PO_4 \cdot H_2O$	16827-96-6	189.959	red-purp orth plates	dec 450				s acid
98	Ammonium cobalt(II) sulfate hexahydrate	$(NH_4)_2Co(SO_4)_2 \cdot 6H_2O$	13586-38-4	395.227	red monocl prisms			1.90		s H_2O; i EtOH
99	Ammonium copper(II) chloride	$CuCl_2 \cdot 2NH_4Cl$	10060-13-6*	241.435	yel hyg orth cry					s H_2O
100	Ammonium copper(II) chloride dihydrate	$CuCl_2 \cdot 2NH_4Cl \cdot 2H_2O$	10060-13-6	277.465	blue-grn tetr cry	110 dec		1.993		s H_2O, EtOH
101	Ammonium cyanide	NH_4CN	12211-52-8	44.056	col tetr cry	dec		1.10		vs H_2O
102	Ammonium dichromate	$(NH_4)_2Cr_2O_7$	7789-09-5	252.065	oran-red monocl cry; hyg	180 dec		2.155	35.6^{20}	
103	Ammonium dihydrogen arsenate	$NH_4H_2AsO_4$	13462-93-6	158.975	tetr cry	300 dec		2.311	52.7^{25}	
104	Ammonium dihydrogen phosphate	$NH_4H_2PO_4$	7722-76-1	115.026	wh tetr cry	190		1.80	40.4^{25}	sl EtOH; i ace
105	Ammonium *O,O*-diethyldithiophosphate	$(C_2H_5O)_2P(S)SNH_4$	1068-22-0	203.264	cry	165				
106	Ammonium dithiocarbamate	NH_4NH_2CSS	513-74-6	110.202	yel ortho cry	99 dec		1.45		s H_2O
107	Ammonium ferricyanide trihydrate	$(NH_4)_3Fe(CN)_6 \cdot 3H_2O$	14221-48-8*	320.110	red cry					s H_2O; i EtOH
108	Ammonium ferrocyanide trihydrate	$(NH_4)_4Fe(CN)_6 \cdot 3H_2O$	14481-29-9*	338.149	yel cry	dec				s H_2O; i EtOH
109	Ammonium fluoride	NH_4F	12125-01-8	37.037	wh hex cry; hyg	238		1.015	83.5^{25}	sl EtOH
110	Ammonium fluorosulfonate	NH_4SO_3F	13446-08-7	117.100	col needles	245				s H_2O, EtOH, MeOH
111	Ammonium formate	NH_4CHO_2	540-69-2	63.057	hyg cry	116		1.27	143^{20}	s EtOH
112	Ammonium heptafluorotantalate	$(NH_4)_2TaF_7$	12022-02-5	350.014	hyg cry					
113	Ammonium hexabromoosmate(IV)	$(NH_4)_2OsBr_6$	24598-62-7	705.73	small blk cubes					sl H_2O; s glycerol; i EtOH
114	Ammonium hexabromoplatinate(IV)	$(NH_4)_2PtBr_6$	17363-02-9	710.585	powder	145 dec			0.59^{20}	
115	Ammonium hexachloroiridate(III)	$(NH_4)_3IrCl_6$	15752-05-3	459.050	grn pow					
116	Ammonium hexachloroiridate(IV)	$(NH_4)_2IrCl_6$	16940-92-4	441.012	blk cry powder	dec		2.856	1.09^{25}	
117	Ammonium hexachloroosmate(IV)	$(NH_4)_2OsCl_6$	12125-08-5	439.03	red cry or powder		subl	2.93		s H_2O, EtOH
118	Ammonium hexachloropalladate(IV)	$(NH_4)_2PdCl_6$	19168-23-1	355.22	red-brn hyg cry	dec		2.418		
119	Ammonium hexachloroplatinate(IV)	$(NH_4)_2PtCl_6$	16919-58-7	443.879	red-oran cub cry	380 dec		3.065	0.5^{20}	i EtOH
120	Ammonium hexachlororuthenate(IV)	$(NH_4)_2RuCl_6$	18746-63-9	349.87	red cry					
121	Ammonium hexafluoroaluminate	$(NH_4)_3AlF_6$	7784-19-2	195.087	cub cry			1.78		s H_2O
122	Ammonium hexafluorogallate	$(NH_4)_3GaF_6$	14639-94-2	237.828	col cub cry	>200 dec		2.10		
123	Ammonium hexafluorogermanate	$(NH_4)_2GeF_6$	16962-47-3	222.71	wh cry	380	subl	2.564		s H_2O; i EtOH
124	Ammonium hexafluorophosphate	NH_4PF_6	16941-11-0	163.003	wh cub cry	58 dec		2.180		vs H_2O; s ace, EtOH, MeOH
125	Ammonium hexafluorosilicate	$(NH_4)_2SiF_6$	16919-19-0	178.153	wh cub or trig cry	dec		2.011	22.7^{25}	i EtOH, ace
126	Ammonium hexafluorotitanate	$(NH_4)_2TiF$	16962-40-6	197.934	wh solid					s H_2O

No.	Name	Formula	CAS Reg No.	Mol. weight	Physical form	mp/°C	bp/°C	Density g cm^{-3}	Solubility g/100 g H_2O	Qualitative solubility
127	Ammonium hexafluorozirconate(IV)	$(NH_4)_2ZrF_6$	16919-31-6	241.291	wh hex cry			1.154		s H_2O
128	Ammonium hydrogen arsenate	$(NH_4)_2HAsO_4$	7784-44-3	176.004	wh powder			1.99		s H_2O
129	Ammonium hydrogen carbonate	NH_4HCO_3	1066-33-7	79.056	col or wh prisms	107 dec		1.586	24.8^{25}	i EtOH, bz
130	Ammonium hydrogen citrate	$(NH_4)_2HC_6H_5O_7$	3012-65-5	226.184	col cry			1.48		vs H_2O; sl EtOH
131	Ammonium hydrogen fluoride	NH_4HF_2	1341-49-7	57.044	wh orth cry	125	240 dec	1.50	60.2^{20}	
132	Ammonium hydrogen malate	$NH_4C_4H_5O_5$	5972-71-4	151.118	orth cry	160		1.15		s H_2O; sl EtOH
133	Ammonium hydrogen oxalate monohydrate	$NH_4HC_2O_4 \cdot H_2O$	5972-72-5*	125.081	col rhomb cry	dec		1.56		sl H_2O, EtOH
134	Ammonium hydrogen phosphate	$(NH_4)_2HPO_4$	7783-28-0	132.055	wh cry	155 dec		1.619	69.5^{25}	i EtOH, ace
135	Ammonium hydrogen phosphite monohydrate	$(NH_4)_2HPO_3 \cdot H_2O$	51503-61-8	134.071	hyg cry					s H_2O
136	Ammonium hydrogen selenate	NH_4HSeO_4	10294-60-7	162.01	rhom cry	dec		2.162		
137	Ammonium hydrogen sulfate	NH_4HSO_4	7803-63-6	115.110	wh hyg cry	147		1.78	100^{20}	i EtOH, ace, py
138	Ammonium hydrogen sulfide	NH_4HS	12124-99-1	51.112	wh tetr or orth cry	dec		1.17	128^{0}	sl ace; i bz, eth
139	Ammonium hydrogen sulfite	NH_4HSO_3	10192-30-0	99.110	col cry	dec		2.03	71.8^{0}	
140	Ammonium hydrogen tartrate	$NH_4HC_4H_4O_6$	3095-65-6	167.117	wh cry			1.68		sl H_2O; s alk; i EtOH
141	Ammonium hydroxide	NH_4OH	1336-21-6	35.046	exists only in soln					
142	Ammonium hypophosphite	$NH_4H_2PO_2$	7803-65-8	83.028	wh hyg cry	dec				vs H_2O; sl EtOH; i ace
143	Ammonium iodate	NH_4IO_3	13446-09-8	192.941	wh powder	150		3.3	3.84^{25}	
144	Ammonium iodide	NH_4I	12027-06-4	144.943	wh tetr cry; hyg	551 dec	405 sp	2.514	178^{25}	sl EtOH, MeOH
145	Ammonium iron(II) sulfate hexahydrate	$(NH_4)_2Fe(SO_4)_2 \cdot 6H_2O$	7783-85-9	392.139	blue-grn monocl cry	≈100 dec		1.86		s H_2O; i EtOH
146	Ammonium iron(III) chromate	$NH_4Fe(CrO_4)_2$	7789-08-4	305.871	red powder					i H_2O
147	Ammonium iron(III) oxalate trihydrate	$(NH_4)_3Fe(C_2O_4)_3 \cdot 3H_2O$	13268-42-3	428.063	grn monocl cry; hyg	≈160 dec		1.780		vs H_2O; i EtOH
148	Ammonium iron(III) sulfate dodecahydrate	$NH_4Fe(SO_4)_2 \cdot 12H_2O$	7783-83-7	482.192	col to viol cry	≈37		1.71		vs H_2O; i EtOH
149	Ammonium lactate	$NH_4C_3H_5O_3$	52003-58-4	107.108	col cry	92				s H_2O, EtOH; sl MeOH; i ace, eth
150	Ammonium magnesium chloride hexahydrate	$NH_4MgCl_3 \cdot 6H_2O$	39733-35-2	256.794	hyg cry	dec 100		1.46	17^{20}	s H_2O
151	Ammonium mercuric chloride dihydrate	$(NH_4)_2HgCl_4 \cdot 2H_2O$	33445-15-7*	414.51	powder					s H_2O; sl EtOH
152	Ammonium metatungstate hexahydrate	$(NH_4)_6W_7O_{24} \cdot 6H_2O$	12028-48-7	1887.19	wh cry					s H_2O; i EtOH
153	Ammonium metavanadate	NH_4VO_3	7803-55-6	116.979	wh-yel cry	200 dec		2.326	4.8^{20}	
154	Ammonium molybdate(VI) tetrahydrate	$(NH_4)_6Mo_7O_{24} \cdot 4H_2O$	12054-85-2	1235.86	col or grn-yel cry	90 dec		2.498	43	i EtOH
155	Ammonium dimolybdate	$(NH_4)_2Mo_2O_7$	27546-07-2	339.95	cry					s H_2O
156	Ammonium molybdophosphate	$(NH_4)_3PO_4 \cdot 12MoO_3$	12026-66-3	1876.35	grn or yel cry	dec			0.02^{20}	sl H_2O; s alk
157	Ammonium nitrate	NH_4NO_3	6484-52-2	80.043	wh hyg cry; orth	169.7	dec 200-260	1.72	213^{25}	sl MeOH
158	Ammonium nitrite	NH_4NO_2	13446-48-5	64.044	wh-yel cry	60 exp		1.69	221^{25}	i eth
159	Ammonium nitroferricyanide	$(NH_4)_2Fe(CN)_5NO$	14402-70-1	252.016	red-brn cry					s H_2O, EtOH
160	Ammonium oleate	$NH_4C_{18}H_{33}O_2$	544-60-5	299.493	yel-brn paste	21				s H_2O; sl ace
161	Ammonium oxalate	$(NH_4)_2C_2O_4$	1113-38-8	124.096	col sol			1.5	5.20^{25}	
162	Ammonium oxalate monohydrate	$(NH_4)_2C_2O_4 \cdot H_2O$	6009-70-7	142.110	wh orth cry	dec		1.50	5.20^{25}	sl EtOH
163	Ammonium palmitate	$NH_4C_{15}H_{31}CO_2$	593-26-0	273.455	yel-wh powder	22				s H_2O; sl bz, xyl; i ace, EtOH, ctc
164	Ammonium pentaborate tetrahydrate	$NH_4B_5O_8 \cdot 4H_2O$	12007-89-5	272.150	wh cry				7.03^{18}	
165	Ammonium pentachlororhodate(III) monohydrate	$(NH_4)_2RhCl_5 \cdot H_2O$	63771-33-5	334.262	red cry	dec 210				
166	Ammonium pentachlorozincate	$(NH_4)_3ZnCl_5$	14639-98-6	296.789	hyg orth cry			1.81		vs H_2O
167	Ammonium perchlorate	NH_4ClO_4	7790-98-9	117.490	wh orth cry	dec, exp		1.95	24.5^{25}	s MeOH; sl EtOH, ace; i eth
168	Ammonium permanganate	NH_4MnO_4	13446-10-1	136.975	purp rhomb cry	70 dec		2.22	7.9^{15}	
169	Ammonium peroxydisulfate	$(NH_4)_2S_2O_8$	7727-54-0	228.202	monocl cry or wh powder	dec		1.982	83.5^{25}	
170	Ammonium perrhenate	NH_4ReO_4	13598-65-7	268.244	col powder			3.97	6.23^{20}	
171	Ammonium phosphate trihydrate	$(NH_4)_3PO_4 \cdot 3H_2O$	10361-65-6*	203.133	wh prisms				25.0^{25}	i ace
172	Ammonium phosphomolybdate monohydrate	$(NH_4)_3PO_4 \cdot 12MoO_3 \cdot H_2O$	54723-94-3	1894.36	yel cry or powder	dec			0.02	
173	Ammonium phosphotungstate dihydrate	$(NH_4)_3PO_4 \cdot 12WO_3 \cdot 2H_2O$	1311-90-6	2967.18	cry powder					sl H_2O
174	Ammonium picrate	$NH_4C_6H_2N_3O_7$	131-74-8	246.135	yel orth cry	exp		1.72		sl H_2O
175	Ammonium polysulfide	$(NH_4)_2S_x$	9080-17-5		yel unstab soln					reac acids
176	Ammonium salicylate	$NH_4C_7H_5O_3$	528-94-9	155.151	wh cry powder					vs H_2O; s EtOH

No.	Name	Formula	CAS Reg No.	Mol. weight	Physical form	mp/°C	bp/°C	Density g cm^{-3}	Solubility g/100 g H_2O	Qualitative solubility
177	Ammonium selenate	$(NH_4)_2SeO_4$	7783-21-3	179.04	wh monocl cry	dec		2.194	117^{25}	i EtOH, ace
178	Ammonium selenite	$(NH_4)_2SeO_3$	7783-19-9	163.04	wh or red hyg cry	dec			121^{25}	
179	Ammonium stearate	$NH_4C_{18}H_{35}O_2$	1002-89-7	301.509	yel-wh powder	22		0.89		sl H_2O, bz; s EtOH, MeOH; i ace
180	Ammonium sulfamate	$NH_4NH_2SO_3$	7773-06-0	114.124	wh hyg cry	131	160 dec			vs H_2O; sl EtOH
181	Ammonium sulfate	$(NH_4)_2SO_4$	7783-20-2	132.140	wh or brn orth cry	280 dec		1.77	76.4^{25}	i EtOH, ace
182	Ammonium sulfide	$(NH_4)_2S$	12135-76-1	68.142	yel-oran cry	≈0 dec				s H_2O, EtOH, alk
183	Ammonium sulfite	$(NH_4)_2SO_3$	17026-44-7	116.140	wh hyg cry				64.2^{25}	
184	Ammonium sulfite monohydrate	$(NH_4)_2SO_3 \cdot H_2O$	7783-11-1	134.155	col cry	dec		1.41	64.2^{25}	i EtOH, ace
185	Ammonium tartrate	$(NH_4)_2C_4H_4O_6$	3164-29-2	184.147	wh cry	dec		1.601		s H_2O
186	Ammonium tellurate	$(NH_4)_2TeO_4$	13453-06-0	227.68	wh powder	dec		3.024		
187	Ammonium tetraborate tetrahydrate	$(NH_4)_2B_4O_7 \cdot 4H_2O$	12228-87-4	263.377	wh tetr cry	dec 87				vs H_2O; s HNO_3
188	Ammonium tetrachloroaluminate	NH_4AlCl_4	7784-14-7	186.833	wh hyg solid	304				s H_2O, eth
189	Ammonium tetrachloropalladate(II)	$(NH_4)_2PdCl_4$	13820-40-1	284.31	grn cry or red-brn pow					s H_2O
190	Ammonium tetrachloroplatinate(II)	$(NH_4)_2PtCl_4$	13820-41-2	372.973	red cry	dec		2.936		s H_2O; i EtOH
191	Ammonium tetrachlorozincate	$(NH_4)_2ZnCl_4$	14639-97-5	243.298	wh orth plates; hyg	150 dec		1.879		vs H_2O
192	Ammonium tetrafluoroantimonate	NH_4SbF_4	14972-90-8	215.793	col cry					s H_2O
193	Ammonium tetrafluoroborate	NH_4BF_4	13826-83-0	104.844	wh powder; orth	487 dec		1.871	25^{20}	
194	Ammonium tetrathiocyanodiammonochromate(III) monohydrate	$NH_4[Cr(NH_3)_2(SCN)_4] \cdot H_2O$	13573-16-5	354.440	red cry	270 dec				s H_2O, EtOH, ace; i bz
195	Ammonium tetrathiomolybdate	$(NH_4)_2MoS_4$	15060-55-6	260.28	red cry	100 dec				vs H_2O
196	Ammonium tetrathiotungstate	$(NH_4)_2WS_4$	13862-78-7	348.18	oran cry	dec		2.71		s H_2O
197	Ammonium tetrathiovandate	$(NH_4)_3VS_4$	14693-56-2	233.317	dark viol cry					
198	Ammonium thiocyanate	NH_4SCN	1762-95-4	76.121	col hyg cry	≈149	dec	1.30	181^{25}	vs EtOH; s ace; i chl
199	Ammonium thiosulfate	$(NH_4)_2S_2O_3$	7783-18-8	148.205	wh cry	150 dec		1.678		vs H_2O; i EtOH, eth
200	Ammonium titanium oxalate monohydrate	$(NH_4)_2TiO(C_2O_4)_2 \cdot H_2O$	10580-03-7	293.996	hyg cry					vs H_2O
201	Ammonium tungstate(VI)	$(NH_4)_{10}W_{12}O_{41}$	11120-25-5	3042.44	cry powder			2.3		s H_2O; i EtOH
202	Ammonium tungstate(VI) pentahydrate	$(NH_4)_{10}W_{12}O_{41} \cdot 5H_2O$	1311-93-9	3132.52	cry pow or plates			2.3		vs H_2O; i EtOH
203	Ammonium uranate(VI)	$(NH_4)_2U_2O_7$	7783-22-4	624.131	red-yel amorp powder					i H_2O, alk; s acid
204	Ammonium uranium fluoride	$UO_2(NH_4)_3F_5$	18433-40-4	419.135	grn-yel monocl cry					s H_2O; i EtOH
205	Ammonium valerate	$NH_4C_4H_9CO_2$	42739-38-8	119.163	hyg cry	108				vs H_2O, EtOH; s eth
206	Antimony (gray)	Sb	7440-36-0	121.760	silv metal; hex	630.628	1587	6.68		i dil acid
207	Antimony (black)	Sb	7440-36-0	121.760	blk amorp solid	trans gray 0				
208	Stibine	SbH_3	7803-52-3	124.784	col gas; flam	-88	-17	5.100 g/L		sl H_2O; s EtOH
209	Trimethylstibine	$Sb(CH_3)_3$	594-10-5	166.863	col flam liq	-62	81	1.52		
210	Pentamethylstibine	$Sb(CH_3)_5$	15120-50-0	196.933	col hyg liq	-19	127			reac H_2O
211	Tetramethyldistibine	$[Sb(CH_3)_2]_2$	41422-43-9	303.658	yel flam liq or red solid	17				
212	Antimony arsenide	SbAs	12322-34-8	196.682	hex cry	≈680		6.0		
213	Antimony potassium tartrate trihydrate	$K_2(SbC_4H_2O_6)_2 \cdot 3H_2O$	28300-74-5	667.873	col cry			2.6		sl H_2O
214	Antimony(III) acetate	$Sb(C_2H_3O_2)_3$	3643-76-3	298.891	wh pow					
215	Antimony(III) bromide	$SbBr_3$	7789-61-9	361.472	yel orth cry; hyg	97	288	4.35		reac H_2O; s ace, bz, chl
216	Antimony(III) chloride	$SbCl_3$	10025-91-9	228.119	col orth cry; hyg	73.4	220.3	3.14	987^{25}	s acid, EtOH, bz, ace
217	Antimony(III) fluoride	SbF_3	7783-56-4	178.755	wh orth cry; hyg	287	376	4.38	492^{25}	
218	Antimony(III) iodide	SbI_3	7790-44-5	502.473	red rhomb cry	171	400	4.92		reac H_2O; s EtOH, ace; i ctc
219	Antimony(III) iodide sulfide	SbIS	13816-38-1	280.729	dark red prisms or needles	400				
220	Antimony(III) oxide (senarmontite)	Sb_2O_3	1309-64-4	291.518	col cub cry	570 trans	1425	5.58		sl H_2O; i os
221	Antimony(III) oxide (valentinite)	Sb_2O_3	1309-64-4	291.518	wh orth cry	655	1425	5.7		sl H_2O; i os
222	Antimony(III) oxychloride	SbOCl	7791-08-4	173.212	wh momo cry	170 dec				reac H_2O; i EtOH, eth
223	Antimony(III) phosphate	$SbPO_4$	12036-46-3	216.731	cry pow					reac H_2O
224	Antimony(III) potassium oxalate trihydrate	$K_3Sb(C_2O_4)_3 \cdot 3H_2O$	5965-33-3*	557.158	cry pow					s H_2O
225	Antimony(III) selenide	Sb_2Se_3	1315-05-5	480.40	grn orth cry	611		5.81		sl H_2O
226	Antimony(III) sulfate	$Sb_2(SO_4)_3$	7446-32-4	531.708	wh cry powder; hyg	dec		3.62		sl H_2O
227	Antimony(III) sulfide	Sb_2S_3	1345-04-6	339.715	gray-blk orth cry	550		4.562		i H_2O; s conc HCl
228	Antimony(III) telluride	Sb_2Te_3	1327-50-0	626.32	gray cry	620		6.5		

No.	Name	Formula	CAS Reg No.	Mol. weight	Physical form	mp/°C	bp/°C	Density g cm⁻³	Solubility g/100 g H_2O	Qualitative solubility
229	Antimony(III,V) oxide	Sb_2O_4	1332-81-6	307.518	yel orth cry			6.64		
230	Antimony(V) chloride	$SbCl_5$	7647-18-9	299.025	col or yel liq	4	140 dec	2.34		reac H_2O; s chl, ctc
231	Antimony(V) fluoride	SbF_5	7783-70-2	216.752	hyg visc liq	8.3	141	3.10		reac H_2O
232	Antimony(V) dichlorotrifluoride	$SbCl_2F_3$	7791-16-4	249.661	visc liq					reac H_2O
233	Antimony(V) oxide	Sb_2O_5	1314-60-9	323.517	yel powder; cub	dec		3.78	0.3^{20}	
234	Antimony(V) sulfide	Sb_2S_5	1315-04-4	403.845	oran-yel powder	75 dec		4.120		i H_2O; s acid, alk
235	Argon	Ar	7440-37-1	39.948	col gas	-189.36 tp (69 kPa)	-185.847	1.633 g/L		sl H_2O
236	Arsenic (gray)	As	7440-38-2	74.922	gray metal; rhomb	817 tp (3.70 MPa)	616 sp	5.75		i H_2O
237	Arsenic (black)	As	7440-38-2	74.922	blk amorp solid	trans gray As 270		4.9		
238	Arsenic (yellow)	As	7440-38-2	74.922	soft yel cub cry	trans gray As 358		1.97		s CS_2
239	Arsine	AsH_3	7784-42-1	77.946	col gas	-116	-62.5	3.186 g/L		sl H_2O
240	Diarsine	As_2H_4	15942-63-9	153.875	unstab liq		≈100			
241	Arsenic acid	H_3AsO_4	7778-39-4	141.944	exists only in soln					
242	Arsenic acid hemihydrate	$H_3AsO_4 \cdot 0.5H_2O$	7778-39-4*	150.951	wh hyg cry	36.1		2.5		vs H_2O, EtOH
243	Arsenious acid	H_3AsO_3	13464-58-9	125.944	exists only in soln					
244	Arsenic diiodide	As_2I_4	13770-56-4	657.461	red cry	137				reac H_2O; s os
245	Arsenic hemiselenide	As_2Se	1303-35-1	228.80	blk cry					i H_2O, os; dec acid, alk
246	Arsenic sulfide	As_4S_4	12279-90-2	427.946	red monocl cry	320	565	3.5		i H_2O; sl bz; s alk
247	Arsenic(III) bromide	$AsBr_3$	7784-33-0	314.634	col or yel orth cry; hyg	31.1	221	3.40		reac H_2O; s hc, ctc; vs eth, bz
248	Arsenic(III) chloride	$AsCl_3$	7784-34-1	181.281	col liq	-16	130	2.150		reac H_2O; vs chl, ctc, eth
249	Arsenic(III) ethoxide	$As(C_2H_5O)_3$	3141-12-6	210.103	liq		166	1.21		
250	Arsenic(III) fluoride	AsF_3	7784-35-2	131.917	col liq	-5.9	57.13	2.7		reac H_2O; s EtOH, eth, bz
251	Arsenic(III) iodide	AsI_3	7784-45-4	455.635	red hex cry	141	424	4.73		sl H_2O, EtOH, eth; s bz, tol
252	Arsenic(III) oxide (arsenolite)	As_2O_3	1327-53-3	197.841	wh cub cry	274	460	3.86	2.05^{25}	
253	Arsenic(III) oxide (claudetite)	As_2O_3	1327-53-3	197.841	wh monocl cry	314	460	3.74	2.05^{25}	s dil acid, alk; i EtOH
254	Arsenic(III) selenide	As_2Se_3	1303-36-2	386.72	brn-blk solid	260		4.75		i H_2O; s alk
255	Arsenic(III) sulfide	As_2S_3	1303-33-9	246.038	yel-oran monocl cry	312	707	3.46		i H_2O; s alk
256	Arsenic(III) telluride	As_2Te_3	12044-54-1	532.64	blk monocl cry	621		6.50		
257	Arsenic(V) chloride	$AsCl_5$	22441-45-8	252.187	stab at low temp	≈-50 dec				
258	Arsenic(V) fluoride	AsF_5	7784-36-3	169.914	col gas	-79.8	-52.8	6.945 g/L		reac H_2O; s EtOH, bz, eth
259	Arsenic(V) oxide	As_2O_5	1303-28-2	229.840	wh amorp powder	315		4.32	65.8^{20}	vs EtOH
260	Arsenic(V) selenide	As_2Se_5	1303-37-3	544.64	blk solid	dec				i H_2O, EtOH, eth; s alk
261	Arsenic(V) sulfide	As_2S_5	1303-34-0	310.168	brn-yel amorp solid	dec				i H_2O; s alk
262	Astatine	At	7440-68-8	210	cry	302				s HNO_3, os
263	Barium	Ba	7440-39-3	137.327	silv-yel metal; cub	727	1897	3.62		reac H_2O; sl EtOH
264	Barium acetate	$Ba(C_2H_3O_2)_2$	543-80-6	255.416	wh powder			2.47	79.2^{25}	
265	Barium acetate monohydrate	$Ba(C_2H_3O_2)_2 \cdot H_2O$	5908-64-5	273.431	wh cry	110 dec		2.19	79.2^{25}	sl EtOH
266	Barium aluminate	$BaAl_2O_4$	12004-04-5	255.288	hex cry	1827				
267	Barium aluminide	$BaAl_4$	12672-79-6	245.253	metallic solid	1097				
268	Barium azide	$Ba(N_3)_2$	18810-58-7	221.367	monocl cry; exp	≈120 dec		2.936	17.3^{20}	sl EtOH; i eth
269	Barium bismuthate	$BaBiO_3$	12785-50-1	394.305	bronze cry	1040 dec				
270	Barium bromate	$Ba(BrO_3)_2$	13967-90-3	393.131	col monocl cry				0.79^{25}	s ace
271	Barium bromate monohydrate	$Ba(BrO_3)_2 \cdot H_2O$	10326-26-8	411.147	wh monocl cry	260 dec		3.99	0.831^{25}	i EtOH
272	Barium bromide	$BaBr_2$	10553-31-8	297.135	wh orth cry	857	1835	4.781	100^{25}	
273	Barium bromide dihydrate	$BaBr_2 \cdot 2H_2O$	7791-28-8	333.166	wh cry	75 dec		3.7	100^{25}	s MeOH; i EtOH, ace, diox
274	Barium calcium tungstate	Ba_2CaWO_6	15552-14-4	594.57	cub cry	1420				
275	Barium carbide	BaC_2	50813-65-5	161.348	gray tetr cry	dec		3.74		reac H_2O
276	Barium carbonate	$BaCO_3$	513-77-9	197.336	wh orth cry	1380 dec; 1555 (high pres.)		4.308	0.0014^{20}	s acid
277	Barium chlorate	$Ba(ClO_3)_2$	13477-00-4	304.229	wh cry	414			37.9^{25}	sl EtOH, ace
278	Barium chlorate monohydrate	$Ba(ClO_3)_2 \cdot H_2O$	10294-38-9	322.245	wh monocl cry	120 dec		3.179	37.9^{25}	s acid; sl EtOH, ace
279	Barium chloride	$BaCl_2$	10361-37-2	208.233	wh orth cry; hyg	961	1560	3.9	37.0^{25}	
280	Barium chloride dihydrate	$BaCl_2 \cdot 2H_2O$	10326-27-9	244.264	wh monocl cry	≈120 dec		3.097	37.0^{25}	i EtOH
281	Barium chloride fluoride	BaClF	13718-55-3	191.778	wh cry					
282	Barium chromate(V)	$Ba_3(CrO_4)_2$	12345-14-1	643.968	grn-blk hex cry			5.25		s H_2O

No.	Name	Formula	CAS Reg No.	Mol. weight	Physical form	mp/°C	bp/°C	Density g cm^{-3}	Solubility g/100 g H_2O	Qualitative solubility
283	Barium chromate(VI)	$BaCrO_4$	10294-40-3	253.321	yel orth cry	1380		4.50	0.00026^{20}	reac acid
284	Barium citrate monohydrate	$Ba_3(C_6H_5O_7)_2 \cdot H_2O$	512-25-4*	808.195	gray-wh cry					s H_2O, acid
285	Barium copper yttrium oxide	$BaCuY_2O_5$	82642-06-6	458.682	grn cry; not superconductor					
286	Barium copper yttrium oxide	$Ba_2Cu_3YO_7$	109064-29-1	666.194	blk solid; HT superconductor					
287	Barium copper yttrium oxide	$Ba_2Cu_4YO_8$	114104-80-2	745.739	HT superconductor					
288	Barium copper yttrium oxide	$Ba_4Cu_7Y_2O_{15}$	124365-83-9	1411.933	HT superconductor					
289	Barium cyanide	$Ba(CN)_2$	542-62-1	189.361	wh cry powder					vs H_2O; s EtOH
290	Barium dichromate dihydrate	$BaCr_2O_7 \cdot 2H_2O$	10031-16-0	389.346	brn-red needles	dec				reac H_2O
291	Barium disilicate	$BaSi_2O_5$	12650-28-1	273.495	wh orth cry	1420		3.70		
292	Barium dithionate dihydrate	$BaS_2O_6 \cdot 2H_2O$	13845-17-5	333.484	wh cry	140 dec		4.54	22.1^{20}	sl EtOH
293	Barium ferrite	$BaFe_{12}O_{19}$	11138-11-7	1111.456	magnetic solid					
294	Barium ferrocyanide hexahydrate	$Ba_2Fe(CN)_6 \cdot 6H_2O$	13821-06-2*	594.694	yel monocl cry	80 dec				i H_2O, EtOH
295	Barium fluoride	BaF_2	7787-32-8	175.324	wh cub cry	1368	2260	4.893	0.161^{25}	
296	Barium formate	$Ba(CHO_2)_2$	541-43-5	227.362	cry			3.21		s H_2O; i EtOH
297	Barium hexaboride	BaB_6	12046-08-1	202.193	blk cub cry	2070		4.36		i H_2O; s acid; i EtOH
298	Barium hexafluorogermanate	$BaGeF_6$		323.96	wh cry	≈665		4.56		
299	Barium hexafluorosilicate	$BaSiF_6$	17125-80-3	279.403	wh orth needles	300 dec		4.29		i H_2O, EtOH; sl acid
300	Barium hydride	BaH_2	13477-09-3	139.343	gray orth cry	1200		4.16		reac H_2O
301	Barium hydrogen phosphate	$BaHPO_4$	10048-98-3	233.306	wh cry powder	400 dec		4.16	0.015^{20}	s dil acid
302	Barium hydrosulfide	$Ba(HS)_2$	25417-81-6	203.473	yel hyg cry					s H_2O
303	Barium hydrosulfide tetrahydrate	$Ba(HS)_2 \cdot 4H_2O$	12230-74-9	275.534	yel rhomb cry	50 dec				s H_2O
304	Barium hydroxide	$Ba(OH)_2$	17194-00-2	171.342	wh powder	408			4.91^{25}	
305	Barium hydroxide monohydrate	$Ba(OH)_2 \cdot H_2O$	22326-55-2	189.357	wh powder			3.743	4.91^{25}	s acid
306	Barium hydroxide octahydrate	$Ba(OH)_2 \cdot 8H_2O$	12230-71-6	315.464	wh monocl cry	78 dec		2.18	4.91^{25}	
307	Barium hypophosphite monohydrate	$Ba(H_2PO_2)_2 \cdot H_2O$	14871-79-5*	285.320	monocl plates			2.90		s H_2O; i EtOH
308	Barium iodate	$Ba(IO_3)_2$	10567-69-8	487.132	wh cry powder	476 dec		5.23	0.0396^{25}	
309	Barium iodate monohydrate	$Ba(IO_3)_2 \cdot H_2O$	7787-34-0	505.148	cry	130 dec		5.00	0.0396^{25}	s acid; i EtOH
310	Barium iodide	BaI_2	13718-50-8	391.136	wh orth cry	711		5.15	221^{25}	
311	Barium iodide dihydrate	$BaI_2 \cdot 2H_2O$	7787-33-9	427.167	col cry	740 dec		5.0	221^{25}	s EtOH, ace
312	Barium manganate(VI)	$BaMnO_4$	7787-35-1	256.263	grn-gray hyg cry			4.85	0.00041^{20}	
313	Barium metaborate monohydrate	$Ba(BO_2)_2 \cdot H_2O$	26124-86-7	240.962	wh powder	>900		3.3		sl H_2O
314	Barium metaborate dihydrate	$Ba(BO_2)_2 \cdot 2H_2O$	23436-05-7	258.977	wh prec	dec			1.3^{25}	sl H_2O
315	Barium metaphosphate	$Ba(PO_3)_2$	13466-20-1	295.271	wh powder	1560				i H_2O; sl acid
316	Barium metasilicate	$BaSiO_3$	13255-26-0	213.411	col rhomb powder	1605		4.40		i H_2O; s acid
317	Barium molybdate	$BaMoO_4$	7787-37-3	297.27	wh powder	1450		4.975	0.0021^{20}	
318	Barium niobate	$Ba(NbO_3)_2$	12009-14-2	419.136	yel orth cry	1455		5.44		i H_2O
319	Barium nitrate	$Ba(NO_3)_2$	10022-31-8	261.336	wh cub cry	590		3.24	10.3^{25}	sl EtOH, ace
320	Barium nitride	Ba_3N_2	12047-79-9	439.994	yel-brn cry	>500 dec		4.78		reac H_2O
321	Barium nitrite	$Ba(NO_2)_2$	13465-94-6	229.338	col hex cry	267		3.234	79.5^{25}	
322	Barium nitrite monohydrate	$Ba(NO_2)_2 \cdot H_2O$	7787-38-4	247.353	yel-wh hex cry	217 dec		3.18	79.5^{25}	i EtOH
323	Barium orthovanadate	$Ba_3(VO_4)_2$	39416-30-3	641.859	hex cry	707		5.14		
324	Barium oxalate	BaC_2O_4	516-02-9	225.346	wh powder	400 dec		2.658	0.0075	
325	Barium oxalate monohydrate	$BaC_2O_4 \cdot H_2O$	13463-22-4	243.361	wh cry powder			2.66	0.0075^{20}	s acid
326	Barium oxide	BaO	1304-28-5	153.326	wh-yel powder; cub and hex	1973		5.72(cub)	1.5^{20}	s dil acid, EtOH; i ace
327	Barium 2,4-pentanedioate octahydrate	$Ba(CH_3COCHCOCH_3)_2 \cdot 8H_2O$	12084-29-6*	479.665	col hyg cry	320 (anh)				
328	Barium perchlorate	$Ba(ClO_4)_2$	13465-95-7	336.228	col hex cry	505		3.20	312^{25}	vs EtOH
329	Barium perchlorate trihydrate	$Ba(ClO_4)_2 \cdot 3H_2O$	10294-39-0	390.274	col cry			2.74	312^{25}	s MeOH; sl EtOH, ace; i eth
330	Barium permanganate	$Ba(MnO_4)_2$	7787-36-2	375.198	brn-viol cry	200 dec		3.77	62.5^{20}	reac EtOH
331	Barium peroxide	BaO_2	1304-29-6	169.326	gray-wh tetr cry	450 dec		4.96	0.091^{20}	reac dil acid
332	Barium plumbate	$BaPbO_3$	12047-25-5	392.5	orth cry					
333	Barium potassium chromate	$BaK_2(CrO_4)_2$	27133-66-0	447.511	yel hex cry			3.63		vs H_2O
334	Barium pyrophosphate	$Ba_2P_2O_7$	13466-21-2	448.597	wh powder	1430		3.9	0.0088^{20}	s acid
335	Barium selenate	$BaSeO_4$	7787-41-9	280.29	wh rhomb cry	dec		4.75	0.015^{20}	
336	Barium selenide	BaSe	1304-39-8	216.29	cub cry powder	1780		5.02		reac H_2O
337	Barium selenite	$BaSeO_3$	13718-59-7	264.29	solid					i H_2O
338	Barium silicide	$BaSi_2$	1304-40-1	193.498	gray lumps	1180				reac H_2O
339	Barium sodium niobate	$Ba_2Na(NbO_3)_5$	12323-03-4	1002.167	wh orth cry	1437		5.40		i H_2O
340	Barium stannate	$BaSnO_3$	12009-18-6	304.035	cub cry			7.24		sl H_2O
341	Barium stannate trihydrate	$BaSnO_3 \cdot 3H_2O$	12009-18-6*	358.081	wh cry powder					sl H_2O; s acid

No.	Name	Formula	CAS Reg No.	Mol. weight	Physical form	mp/°C	bp/°C	Density g cm^{-3}	Solubility g/100 g H_2O	Qualitative solubility
342	Barium stearate	$Ba(C_{18}H_{35}O_2)_2$	6865-35-6	704.266	wh powder	160		1.145		i H_2O, EtOH
343	Barium strontium niobate	$BaSr(NbO_3)_4$	37185-09-4	788.57	pale yel solid					
344	Barium strontium tungstate	Ba_2SrWO_6	14871-56-8	642.11	hyg pow	1400				
345	Barium sulfate	$BaSO_4$	7727-43-7	233.390	wh orth cry	1580		4.49	0.00031^{20}	i EtOH
346	Barium sulfide	BaS	21109-95-5	169.392	col cub cry or gray powder	2227		4.3	8.94^{25}	
347	Barium sulfite	$BaSO_3$	7787-39-5	217.390	wh monocl cry	dec		4.44	0.0011^{25}	i EtOH
348	Barium tartrate	$BaC_4H_4O_6$	5908-81-6	285.398	wh cry			2.98		s H_2O; i EtOH
349	Barium tetracyanoplatinate(II) tetrahydrate	$BaPt(CN)_4 \cdot 4H_2O$	13755-32-3	508.543	yel powder or cry			2.076		sl H_2O; i EtOH
350	Barium tetraiodomercurate(II)	$BaHgI_4$	10048-99-4	845.54	yel-red hyg cry					vs H_2O, EtOH
351	Barium thiocyanate	$Ba(SCN)_2$	2092-17-3	253.491	hyg cry				167^{25}	s ace, MeOH, EtOH
352	Barium thiocyanate dihydrate	$Ba(SCN)_2 \cdot 2H_2O$	2092-17-3*	289.522	hyg wh cry				167^{25}	s EtOH
353	Barium thiocyanate trihydrate	$Ba(SCN)_2 \cdot 3H_2O$	68016-36-4	307.537	wh needles; hyg			2.286	167^{25}	s EtOH
354	Barium thiosulfate	BaS_2O_3	35112-53-9	249.455	wh cry powder	220 dec			0.2^{20}	i EtOH
355	Barium thiosulfate monohydrate	$BaS_2O_3 \cdot H_2O$	7787-40-8	267.471	wh cry powder	dec		3.5	0.2	i EtOH
356	Barium titanate ($BaTiO_3$)	$BaTiO_3$	12047-27-7	233.192	wh tetr cry	1625		6.02		i H_2O
357	Barium titanate ($BaTi_2O_5$)	$BaTi_2O_5$	12009-27-7	313.058	wh solid					
358	Barium titanate ($BaTi_4O_9$)	$BaTi_4O_9$	12009-31-3	472.790	wh solid					
359	Barium titanium silicate	$BaTi(SiO_3)_3$	15491-35-7	413.446	rhom blue-pur cry					
360	Barium tungstate	$BaWO_4$	7787-42-0	385.17	wh tetr cry	1475	1730	5.04	0.0016^{20}	
361	Barium uranium oxide	BaU_2O_7	10380-31-1	725.381	oran-yel powder					i H_2O; s acid
362	Barium yttrium tungsten oxide	$Ba_3Y_3WO_9$	37265-86-4	1006.53	cub cry	1470				
363	Barium zirconate	$BaZrO_3$	12009-21-1	276.549	gray-wh cub cry	2500		5.52		i H_2O, alk; sl acid
364	Barium zirconium silicate	$BaO \cdot ZrO_2 \cdot SiO_2$		336.634	wh pow					i H2O, alk, sl acid; s HF
365	Berkelium (α form)	Bk	7440-40-6	247	hex cry	trans to · 930		14.78		
366	Berkelium (β form)	Bk	7440-40-6	247	cub cry	986		13.25		
367	Beryllium	Be	7440-41-7	9.012	hex cry	1287	2471	1.85		s acid, alk
368	Beryllium acetate	$Be(C_2H_3O_2)_2$	543-81-7	127.101	wh cry	60 dec				i H_2O, EtOH
369	Beryllium basic acetate	$Be_4O(C_2H_3O_2)_6$	1332-52-1	406.312	wh cry	285	330	1.25		i H_2O; s eth, os
370	Beryllium aluminate	$BeAl_2O_4$	12004-06-7	126.973	orth cry			3.65		
371	Beryllium aluminum metasilicate	$Be_3Al_2(SiO_3)_6$	1302-52-9	537.502	col or grn-yel cry; hex			2.64		
372	Beryllium boride (BeB_2)	BeB_2	12228-40-9	30.634	refrac solid	>1970				
373	Beryllium boride (BeB_6)	BeB_6	12429-94-6	73.878	red solid	2070				
374	Beryllium boride (Be_2B)	Be_2B	12536-51-5	28.835	pink cry	1520				
375	Beryllium boride (Be_4B)	Be_4B	12536-52-6	46.860	refrac solid	1160				
376	Beryllium borohydride	$Be(BH_4)_2$	17440-85-6	36.682	solid	125 dec	subl			reac H_2O
377	Beryllium bromide	$BeBr_2$	7787-46-4	168.820	orth cry; hyg	508	473 sp	3.465		vs H_2O; s EtOH, pyr
378	Beryllium carbide	Be_2C	506-66-1	30.035	red cub cry	2127		1.90		reac H_2O
379	Beryllium carbonate tetrahydrate	$BeCO_3 \cdot 4H_2O$	60883-64-9	93.085	wh solid	100 dec			0.36^{0}	
380	Beryllium basic carbonate	$Be_2(OH)_2(CO_3)_2$	66104-24-3	181.069	wh powder					i H_2O; s acid, alk
381	Beryllium chloride	$BeCl_2$	7787-47-5	79.918	wh-yel orth cry; hyg	415	482	1.90	71.5^{25}	s EtOH, eth, py; i bz, tol
382	Beryllium fluoride	BeF_2	7787-49-7	47.009	tetr cry or gl; hyg	552	1283	2.1		vs H_2O; sl EtOH
383	Beryllium formate	$Be(CHO_2)_2$	1111-71-3	99.047	powder	>250 dec				reac H_2O; i os
384	Beryllium hydride	BeH_2	7787-52-2	11.028	wh amorp solid	250 dec		0.65		reac H_2O; i eth, tol
385	Beryllium hydrogen phosphate	$BeHPO_4$	13598-15-7	104.991	cry					i H_2O
386	Beryllium hydroxide (α)	$Be(OH)_2$	13327-32-7	43.027	wh powder or cry	≈200 dec		1.92		sl H_2O, alk; s acid
387	Beryllium hydroxide (β)	$Be(OH)_2$	13327-32-7	43.027	col tetr cry	dec 138				i H_2O; s, acid, alk
388	Beryllium iodide	BeI_2	7787-53-3	262.821	hyg needles	480	590	4.32		reac H_2O; s EtOH
389	Beryllium nitrate trihydrate	$Be(NO_3)_2 \cdot 3H_2O$	13597-99-4	187.068	yel-wh hyg cry	≈30	dec		107^{20}	s EtOH
390	Beryllium nitride	Be_3N_2	1304-54-7	55.050	gray refrac cry; cub	2200		2.71		reac acid, alk
391	Beryllium oxalate trihydrate	$BeC_2O_4 \cdot 3H_2O$	15771-43-4	151.077	rhom cry	dec 320				vs H_2O
392	Beryllium oxide	BeO	1304-56-9	25.011	wh hex cry	2578		3.01		i H_2O; sl acid, alk
393	Beryllium 2,4-pentanedioate	$Be(CH_3COCHCOCH_3)_2$	10210-64-7	207.228	monocl cry powder	108	270	1.168		i H_2O; vs EtOH, eth
394	Beryllium perchlorate tetrahydrate	$Be(ClO_4)_2 \cdot 4H_2O$	7787-48-6	279.975	hyg cry	250 dec			198^{25}	
395	Beryllium selenate tetrahydrate	$BeSeO_4 \cdot 4H_2O$	10039-31-3	224.03	orth cry	100 dec		2.03		vs H_2O
396	Beryllium sulfate	$BeSO_4$	13510-49-1	105.075	col tetr cry; hyg	1127		2.5	41.3^{25}	
397	Beryllium sulfate dihydrate	$BeSO_4 \cdot 2H_2O$	14215-00-0	141.105	col cry	dec 92				
398	Beryllium sulfate tetrahydrate	$BeSO_4 \cdot 4H_2O$	7787-56-6	177.136	col tetr cry	≈100 dec		1.71	41.3^{25}	i EtOH
399	Beryllium sulfide	BeS	13598-22-6	41.077	col cub cry	dec		2.36		reac hot H_2O
400	Bismuth	Bi	7440-69-9	208.980	gray-wh soft metal	271.406	1564	9.79		s acid
401	Bismuth acetate	$Bi(C_2H_3O_2)_3$	22306-37-2	386.111	col tablets	250				i H_2O
402	Bismuth subacetate	$BiOC_2H_3O_2$	5142-76-7	284.023	thin cry plates					i H_2O; s dil acid

No.	Name	Formula	CAS Reg No.	Mol. weight	Physical form	mp/°C	bp/°C	Density g cm^{-3}	Solubility g/100 g H_2O	Qualitative solubility
403	Bismuth antimonide	BiSb	12323-19-2	330.740	cry	475				
404	Bismuth arsenate	$BiAsO_4$	13702-38-0	347.900	wh monocl cry			7.14		i H_2O; sl conc HNO_3
405	Bismuth basic carbonate	$(BiO)_2CO_3$	5892-10-4	509.969	wh powder			6.86		i H_2O; s acid
406	Bismuth basic dichromate	$Bi_2O_3 \cdot 2CrO_3$		665.948	red-oran amorp pow					i H_2O; s acid, alk
407	Bismuth citrate	$BiC_6H_5O_7$	813-93-4	398.080	wh powder			3.458		i H_2O; sl EtOH
408	Bismuth hydride	BiH_3	18288-22-7	212.004	col gas; unstab	-67	≈17	8.665 g/L		
409	Bismuth hydroxide	$Bi(OH)_3$	10361-43-0	260.002	wh-yel amorp powder			4.962		i H_2O; s acid
410	Bismuth germanium oxide	$2Bi_2O_3 \cdot 3GeO_2$	12233-56-6	1245.84	wh pow	1044				
411	Bismuth hexafluoro-2,4-pentanedioate	$Bi(CF_3COCHCOCF_3)_3$	142617-56-9	830.132	powder	96				
412	Bismuth molybdate	Bi_2MoO_6	13565-96-3	609.90	yel solid			9.32		
413	Bismuth molybdate	$Bi_2(MoO_4)_3$	51898-99-8	897.77	monocl cry			5.95		
414	Bismuth nitrate pentahydrate	$Bi(NO_3)_3 \cdot 5H_2O$	10035-06-0	485.071	col tricl cry; hyg	≈75 dec		2.83		reac H_2O; s ace; i EtOH
415	Bismuth subnitrate	$Bi_5O(OH)_9(NO_3)_4$	1304-85-4	1461.987	hyg cry powder	260 dec		4.928		i H_2O, EtOH; s dil acid
416	Bismuth oleate	$Bi(C_{18}H_{33}O_2)_3$	52951-38-9	1053.340	soft yel-brn solid					i H_2O; s eth; sl bz
417	Bismuth oxalate	$Bi_2(C_2O_4)_3$	6591-55-5	682.018	wh powder					i H_2O, EtOH; s dil acid
418	Bismuth oxide	Bi_2O_3	1304-76-3	465.959	yel monocl cry or powder	825	1890	8.9		i H_2O; s acid
419	Bismuth tetroxide	Bi_2O_4	12048-50-9	481.959	red-oran powder	305		5.6		reac H_2O
420	Bismuth oxybromide	BiOBr	7787-57-7	304.883	col tetr cry	560 dec		8.08		i H_2O, EtOH; s acid
421	Bismuth oxychloride	BiOCl	7787-59-9	260.432	wh tetr cry	575 dec		7.72		i H_2O
422	Bismuth oxyiodide	BiOI	7787-63-5	351.883	red tetr cry	300 dec		7.92		i H_2O, EtOH, chl; s HCl
423	Bismuth oxynitrate	$BiONO_3$	10361-46-3	286.985	wh powder	260 dec		4.93		i H_2O, EtOH; s acid
424	Bismuth phosphate	$BiPO_4$	10049-01-1	303.951	monocl cry			6.32		sl H_2O, dil acid; i EtOH
425	Bismuth potassium iodide	BiK_4I_7	41944-01-8	1253.704	red cry					reac H_2O; s alk iodide soln
426	Bismuth selenide	Bi_2Se_3	12068-69-8	654.84	blk hex cry	710 dec		7.5		i H_2O
427	Bismuth stannate pentahydrate	$Bi_2(SnO_3)_3 \cdot 5H_2O$	12777-45-6	1008.162	wh cry					i H_2O
428	Bismuth sulfate	$Bi_2(SO_4)_3$	7787-68-0	706.149	wh needles or powder	405 dec		5.08		reac H_2O, EtOH
429	Bismuth sulfide	Bi_2S_3	1345-07-9	514.156	blk-brn orth cry	850		6.78		i H_2O; s acid
430	Bismuth telluride	Bi_2Te_3	1304-82-1	800.76	gray hex plates	580		7.74		i H_2O; s EtOH
431	Bismuth tribromide	$BiBr_3$	7787-58-8	448.692	yel cub cry	219	462	5.72		reac H_2O; s dil acid, ace; i EtOH
432	Bismuth trichloride	$BiCl_3$	7787-60-2	315.339	col or yel cub cry; hyg	234	441	4.75		reac H_2O; s acid, EtOH, ace
433	Bismuth trifluoride	BiF_3	7787-61-3	265.975	wh-gray cub cry	727	900	8.3		i H_2O
434	Bismuth pentafluoride	BiF_5	7787-62-4	303.972	wh tetr needles; hyg	151.4	230	5.55		reac H_2O
435	Bismuth triiodide	BiI_3	7787-64-6	589.693	blk-brn hex cry	408.6	542	5.778	0.00078^{20}	s EtOH
436	Bismuth trimethyl	$Bi(CH_3)_3$	593-91-9	254.083	col flam liq	-86	110	2.3		
437	Bismuth titanate	$Bi_4(TiO_4)_3$	12048-51-0	1171.516	wh orth cry			7.85		
438	Bismuth tungstate	$Bi_2(WO_4)_3$	13595-87-4	1161.47	wh powder					
439	Bismuth vanadate	$BiVO_4$	14059-33-7	323.920	orth cry	trans 500		6.25		i H_2O; s acid
440	Bismuth zirconate	$2Bi_2O_3 \cdot 3ZrO_2$	37306-42-6	1301.587	wh pow					
441	Boron	B	7440-42-8	10.811	blk rhomb cry	2075	4000	2.34		i H_2O
442	Diborane	B_2H_6	19287-45-7	27.670	col gas; flam	-164.85	-92.49	1.131 g/L		reac H_2O
443	Tetraborane(10)	B_4H_{10}	18283-93-7	53.323	unstab col gas	-120	18	2.180 g/L		reac H_2O
444	Pentaborane(9)	B_5H_9	19624-22-7	63.126	flam col liq	-46.74	60.10	0.60		reac hot H_2O
445	Pentaborane(11)	B_5H_{11}	18433-84-6	65.142	col liq; unstab	-122	65			reac H_2O
446	Hexaborane(10)	B_6H_{10}	23777-80-2	74.945	col liq	-62.3	108 dec	0.67		reac hot H_2O
447	Hexaborane(12)	B_6H_{12}	12008-19-4	76.961	col liq	-82.3	≈85			reac H_2O
448	Nonaborane(15)	B_9H_{15}	19465-30-6	112.418	col liq	2.7				
449	Decaborane(14)	$B_{10}H_{14}$	17702-41-9	122.221	wh orth cry	98.78	213	0.94		sl H_2O; s EtOH, bz, CS_2, ctc
450	Decaborane(16)	$B_{10}H_{16}$	71595-75-0	124.237	col cry	≈81	dec 170	subl		
451	Dodecaborane(16)	$B_{12}H_{16}$	89711-39-7	145.859	col cry	65				s bz, hx
452	Tridecaborane(19)	$B_{13}H_{19}$	43093-20-5	159.694	yel cry	44				s hx, CH2Cl2
453	Tetradecaborane(18)	$B_{14}H_{18}$	55606-55-8	169.497	visc yel oil		dec 100			s cyhex, CS_2
454	Hexadecaborane(20)	$B_{16}H_{20}$	28265-11-4	193.135	col cry	≈110				s ctc, cyhex, thf
455	Octadecaborane(22)	$B_{18}H_{22}$	11071-61-7	216.773	yel cry	180				s os
456	Tetrabromodiborane	B_2Br_4	14355-29-4	341.238	col liq	≈1	dec 20			

No.	Name	Formula	CAS Reg No.	Mol. weight	Physical form	mp/°C	bp/°C	Density g cm^{-3}	Solubility g/100 g H_2O	Qualitative solubility
457	Tetrachlorodiborane	B_2Cl_4	13701-67-2	163.434	col liq; flam	-92.6	66.5			reac H_2O
458	Tetrafluorodiborane	B_2F_4	13965-73-6	97.616	col gas; flam	-56	-34.0	3.990 g/L		reac H_2O
459	Borane carbonyl	BH_3CO	13205-44-2	41.845	col gas	-137	-64	1.710 g/L		reac H_2O
460	Borazine	$B_3N_3H_6$	6569-51-3	80.501	col liq	-58	53	0.824		reac H_2O
461	Boric acid	H_3BO_3	10043-35-3	61.833	col tricl cry	170.9		1.5	5.80^{25}	sl EtOH
462	Metaboric acid (α form)	HBO_2	13460-50-9	43.818	col orth cry; hyg	176		1.784		s H_2O
463	Metaboric acid (β form)	HBO_2	13460-50-9	43.818	col monocl cry; hyg	201		2.045		s H_2O
464	Metaboric acid (γ form)	HBO_2	13460-50-9	43.818	col cub cry	236		2.487		s H_2O
465	Tetrafluoroboric acid	HBF_4	16872-11-0	87.813	col liq		130 dec	≈1.8		vs H_2O, EtOH
466	Boron arsenide	BAs	12005-69-5	85.733	brn cub cry	1100 dec		5.22		
467	Boron carbide	B_4C	12069-32-8	55.255	hard blk cry	2350	>3500	2.50		i H_2O, acid
468	Boron nitride	BN	10043-11-5	24.818	wh powder; hex or cub cry	2967		2.18		i H_2O, acid
469	Boron oxide	B_2O_3	1303-86-2	69.620	col gl or hex cry; hyg	450		2.55	2.2^{20}	s EtOH
470	Boron phosphide	BP	20205-91-8	41.785	red cub cry or powder	1125 dec				reac H_2O, acid
471	Boron silicide	B_6Si	12008-29-6	92.952	blk cry	1980				
472	Boron sulfide	B_2S_3	12007-33-9	117.817	yel amorp solid	563		≈1.7		
473	Boron tribromide	BBr_3	10294-33-4	250.523	col liq; hyg	-46	91.3	2.6		reac H_2O, EtOH
474	Boron trichloride	BCl_3	10294-34-5	117.170	col liq or gas	-107.3	12.5	4.789 g/L		reac H_2O, EtOH
475	Boron trifluoride	BF_3	7637-07-2	67.806	col gas	-126.8	-99.9	2.772 g/L		s H_2O
476	Boron trifluoride etherate	$BF_3(C_2H_5)_2O$	109-63-7	141.927	liq	-60.4	125.5	1.125^{25}		reac H_2O; vs eth, EtOH
477	Boron triiodide	BI_3	13517-10-7	391.524	wh needles	49.7	209.5	3.35		i H_2O
478	Bromine	Br_2	7726-95-6	159.808	red liq	-7.2	58.8	3.1028		sl H_2O
479	Bromic acid	$HBrO_3$	7789-31-3	128.910	stab only in aq soln					s H_2O
480	Hypobromous acid	HOBr	13517-11-8	96.911	exists aq soln					s H_2O
481	Bromine dioxide	BrO_2	21255-83-4	111.903	unstab yel cry	≈0 dec				
482	Bromine monoxide	Br_2O	21308-80-5	175.807	unstab brn solid	-17.5 dec				
483	Dibromine trioxide	Br_2O_3	53809-75-9	207.806	oran needles (LT)	dec -40				
484	Dibromine pentoxide	Br_2O_5	58572-43-3	239.805	col cry (low temp)	-20 dec				
485	Bromine azide	BrN_3	13973-87-0	121.924	red cry; exp	≈45	exp			
486	Bromine chloride	BrCl	13863-41-7	115.357	dark red liq (<5°C)	-66	5 dec			reac H_2O; s eth, CS_2
487	Bromine fluoride	BrF	13863-59-7	98.902	unstab red-brn gas	≈-33	≈20 dec	4.043 g/L		
488	Bromine trifluoride	BrF_3	7787-71-5	136.899	col hyg liq	8.77	125.8	2.803		reac H_2O
489	Bromine pentafluoride	BrF_5	7789-30-2	174.896	col liq	-60.5	41.3	2.460		reac H_2O (exp)
490	Bromosyl trifluoride	$BrOF_3$	61519-37-7	152.898	col liq	-5	dec >20			reac H_2O
491	Bromyl fluoride	BrO_2F	22585-64-4	130.901	col liq	-9	dec 55			reac H_2O
492	Perbromyl fluoride	BrO_3F	37265-91-1	146.900	col gas	-110	dec 20			reac H_2O
493	Cadmium	Cd	7440-43-9	112.411	silv-wh metal	321.069	767	8.69		i H_2O; reac acid
494	Cadmium acetate	$Cd(C_2H_3O_2)_2$	543-90-8	230.500	col cry	255		2.34		s H_2O, EtOH
495	Cadmium acetate dihydrate	$Cd(C_2H_3O_2)_2 \cdot 2H_2O$	5743-04-4	266.529	wh cry	130 dec		2.01		vs H_2O; s EtOH
496	Cadmium antimonide	CdSb	12014-29-8	234.171	orth cry	456		6.92		
497	Cadmium arsenide	Cd_3As_2	12006-15-4	487.076	gray tetr cry	721		6.25		
498	Cadmium azide	$Cd(N_3)_2$	14215-29-3	196.451	yel-wh orth cry; exp	exp		3.24		
499	Cadmium borotungstate octadecahydrate	$Cd_5(BW_{12}O_{40}) \cdot 18H_2O$	1306-26-9	3743.20	yel cry					vs H_2O
500	Cadmium bromide	$CdBr_2$	7789-42-6	272.219	wh-yel hex cry; hyg	568	863	5.19	115^{25}	sl ace, eth
501	Cadmium bromide tetrahydrate	$CdBr_2 \cdot 4H_2O$	13464-92-1	344.281	wh-yel cry				115^{25}	s ace, EtOH
502	Cadmium carbonate	$CdCO_3$	513-78-0	172.420	wh hex cry	500 dec		5.026		i H_2O; s acid
503	Cadmium chlorate dihydrate	$Cd(ClO_3)_2 \cdot 2H_2O$	22750-54-5*	315.344	col hyg cry	80 dec		2.28	2.64^{0}	
504	Cadmium chloride	$CdCl_2$	10108-64-2	183.317	rhom cry; hyg	568	964	4.08	120^{25}	s ace; sl EtOH; i eth
505	Cadmium chloride monohydrate	$CdCl_2 \cdot H_2O$	34330-64-8	201.332	wh cry				120^{25}	
506	Cadmium chloride hemipentahydrate	$CdCl_2 \cdot 2.5H_2O$	7790-78-5	228.354	wh rhomb leaflets			3.327	120^{25}	s ace
507	Cadmium chromate	$CdCrO_4$	14312-00-6	228.405	yel orth cry			4.5		i H_2O
508	Cadmium cyanide	$Cd(CN)_2$	542-83-6	164.445	wh cub cry			2.23	1.7^{15}	
509	Cadmium dichromate monohydrate	$CdCr_2O_7 \cdot H_2O$	69239-51-6	346.414	oran solid					s H_2O
510	Cadmium 2-ethylhexanoate	$Cd(C_8H_{15}O_2)_2$	2420-98-6	398.818	powder					
511	Cadmium fluoride	CdF_2	7790-79-6	150.408	cub cry	1075	1750	6.33	4.36^{25}	s acid; i EtOH
512	Cadmium hydroxide	$Cd(OH)_2$	21041-95-2	146.426	wh trig or hex cry	130 dec		4.79	0.00015^{20}	s dil acid
513	Cadmium iodate	$Cd(IO_3)_2$	7790-81-0	462.216	wh powder			6.48	0.091^{25}	s HNO_3

No.	Name	Formula	CAS Reg No.	Mol. weight	Physical form	mp/°C	bp/°C	Density g cm^{-3}	Solubility g/100 g H_2O	Qualitative solubility
514	Cadmium iodide	CdI_2	7790-80-9	366.220	col hex flakes	388	744	5.64	86.2^{25}	vs H_2O; s EtOH, eth, ace
515	Cadmium metasilicate	$CdSiO_3$	13477-19-5	188.495	grn monocl cry	1252		5.10		
516	Cadmium molybdate	$CdMoO_4$	13972-68-4	272.35	col tetr cry	≈900 dec		5.4		i H_2O; s acid
517	Cadmium niobate	$Cd_2Nb_2O_7$	12187-14-3	522.631	cub cry	≈1410		6.28		i H_2O
518	Cadmium nitrate	$Cd(NO_3)_2$	10325-94-7	236.420	wh cub cry; hyg	360		3.6	156^{25}	s EtOH
519	Cadmium nitrate tetrahydrate	$Cd(NO_3)_2 \cdot 4H_2O$	10022-68-1	308.482	col orth cry; hyg	59.5		2.45	156^{25}	s EtOH, ace
520	Cadmium oxalate	CdC_2O_4	814-88-0	200.430	wh solid			3.32	0.0060^{25}	
521	Cadmium oxalate trihydrate	$CdC_2O_4 \cdot 3H_2O$	20712-42-9	254.476	wh amorp powder	340 dec			0.0060^{25}	i EtOH; s dil acid
522	Cadmium oxide	CdO	1306-19-0	128.410	brn cub cry		1559 sp	8.15		i H_2O; s dil acid
523	Cadmium 2,4-pentanedioate	$Cd(CH_3COCHCOCH_3)_2$	14689-45-3	310.627	wh solid or red cry	235				
524	Cadmium perchlorate hexahydrate	$Cd(ClO_4)_2 \cdot 6H_2O$	10326-28-0	419.404	wh hex cry			2.37	191.5^{25}	
525	Cadmium phosphate	$Cd_3(PO_4)_2$	13477-17-3	527.176	powder	≈1500				i H_2O
526	Cadmium phosphide	Cd_3P_2	12014-28-7	399.181	grn tetr needles	700		5.96		s dil HCl
527	Cadmium selenate dihydrate	$CdSeO_4 \cdot 2H_2O$	10060-09-0	291.40	orth cry	100 dec		3.62	70.5^{25}	
528	Cadmium selenide	$CdSe$	1306-24-7	191.37	wh cub cry	1240		5.81		i H_2O
529	Cadmium selenite	$CdSeO_3$	13814-59-0	239.37	col prisms					
530	Cadmium stearate	$Cd(C_{18}H_{35}O_2)_2$	2223-93-0	679.350	wh cry pow	134		1.21		
531	Cadmium succinate	$CdC_4H_4O_4$	141-00-4	228.484	wh pow or needles				0.37^{40}	sl H_2O; i EtOH
532	Cadmium sulfate	$CdSO_4$	10124-36-4	208.474	col orth cry	1000		4.69	76.7^{25}	i EtOH
533	Cadmium sulfate monohydrate	$CdSO_4 \cdot H_2O$	7790-84-3	226.489	monocl cry	105		3.79	76.7^{25}	
534	Cadmium sulfate octahydrate	$CdSO_4 \cdot 8H_2O$	15244-35-6	352.596	col monocl cry	40 dec		3.08	76.7^{25}	
535	Cadmium sulfide	CdS	1306-23-6	144.476	yel-oran hex cry	≈1480		4.826		i H_2O; s acid
536	Cadmium sulfite	$CdSO_3$	13477-23-1	192.474	col prisms	dec ≈400			0.05^{20}	sl H_2O
537	Cadmium telluride	$CdTe$	1306-25-8	240.01	brn-blk cub cry	1042		6.2		i H_2O, dil acid
538	Cadmium tellurite	$CdTeO_3$	15851-44-2	288.01	col monocl cry	695	dec 1050			
539	Cadmium tetrafluoroborate	$Cd(BF_4)_2$	14486-19-2	286.020	col hyg liq			1.6		vs H_2O, EtOH
540	Cadmium titanate	$CdTiO_3$	12014-14-1	208.276	orth cry			6.5		
541	Cadmium tungstate	$CdWO_4$	7790-85-4	360.25	wh monocl cry			8.0		i H_2O, acid; s NH_4OH
542	Calcium	Ca	7440-70-2	40.078	silv-wh metal	842	1484	1.54		reac H_2O; i bz
543	Calcium acetate	$Ca(C_2H_3O_2)_2$	62-54-4	158.167	wh hyg cry	160 dec		1.50		s H_2O; sl EtOH
544	Calcium acetate monohydrate	$Ca(C_2H_3O_2)_2 \cdot H_2O$	5743-26-0	176.182	wh needles or powder	≈150 dec				s H_2O; sl EtOH
545	Calcium acetate dihydrate	$Ca(C_2H_3O_2)_2 \cdot 2H_2O$	14977-17-4	194.196	long col needles					s H_2O
546	Calcium aluminate	$CaAl_2O_4$	12042-68-1	158.039	wh monocl cry	1605		2.98		reac H_2O
547	Calcium aluminate (β form)	$Ca_3Al_2O_6$	12042-78-3	270.193	wh cub cry; refr	1535		3.04		i H_2O
548	Calcium arsenate	$Ca_3(AsO_4)_2$	7778-44-1	398.072	wh powder	dec		3.6	0.0036^{20}	s dil acid
549	Calcium arsenite (1:1)	$CaAsO_3$	52740-16-6	162.998	wh powder					sl H_2O; s acid
550	Calcium borate hexahydrate	$CaB_4O_7 \cdot 6H_2O$	13701-64-9*	303.409	wh cry pow	1162 (anh)				
551	Calcium boride	CaB_6	12007-99-7	104.944	refrac solid	2235		2.49		
552	Calcium bromate	$Ca(BrO_3)_2$	10102-75-7	295.882	wh pow	180				
553	Calcium bromate monohydrate	$Ca(BrO_3)_2 \cdot H_2O$	10102-75-7*	313.898	wh monocl cry	dec 180		3.33		vs H_2O
554	Calcium bromide	$CaBr_2$	7789-41-5	199.886	rhom cry; hyg	742	1815	3.38	156^{25}	s EtOH, ace
555	Calcium bromide dihydrate	$CaBr_2 \cdot 2H_2O$	22208-73-7	235.917	wh cry pow					vs H_2O
556	Calcium bromide hexahydrate	$CaBr_2 \cdot 6H_2O$	13477-28-6	307.977	wh hyg powder	38 dec		2.29	156^{25}	
557	Calcium carbide	CaC_2	75-20-7	64.099	gray-blk orth cry	2300		2.22		reac H_2O
558	Calcium carbonate (aragonite)	$CaCO_3$	471-34-1	100.087	wh orth cry or powder	trans calcite 450		2.930	0.00066^{20}	s dil acid
559	Calcium carbonate (calcite)	$CaCO_3$	471-34-1	100.087	wh hex cry or powder	dec 700-900		2.710	0.00066^{20}	s dil acid
560	Calcium carbonate (vaterite)	$CaCO_3$	471-34-1	100.087	col hex cry			2.653	0.0011^{25}	s dil acid
561	Calcium chlorate	$Ca(ClO_3)_2$	10137-74-3	206.980	wh cry	340			197^{25}	
562	Calcium chlorate dihydrate	$Ca(ClO_3)_2 \cdot 2H_2O$	10035-05-9	243.011	wh monocl cry; hyg	100 dec		2.711	197^{25}	s EtOH
563	Calcium chloride	$CaCl_2$	10043-52-4	110.984	wh cub cry or powder; hyg	775	1935	2.15	81.3^{25}	vs EtOH
564	Calcium chloride monohydrate	$CaCl_2 \cdot H_2O$	13477-29-7	128.999	wh hyg cry	260 dec		2.24	81.3^{25}	s EtOH
565	Calcium chloride dihydrate	$CaCl_2 \cdot 2H_2O$	10035-04-8	147.015	hyg flakes or powder	175 dec		1.85	81.3^{25}	vs EtOH
566	Calcium chloride tetrahydrate	$CaCl_2 \cdot 4H_2O$	25094-02-4	183.046	col tricl cry			1.83		
567	Calcium chloride hexahydrate	$CaCl_2 \cdot 6H_2O$	7774-34-7	219.075	wh hex cry; hyg	30 dec		1.71	81.3^{25}	
568	Calcium chlorite	$Ca(ClO_2)_2$	14674-72-7	174.982	wh cub cry			2.71		reac H_2O
569	Calcium chromate	$CaCrO_4$	13765-19-0	156.072	yel cry	1000 dec				sl H_2O; i EtOH, ace
570	Calcium chromate dihydrate	$CaCrO_4 \cdot 2H_2O$	10060-08-9	192.102	yel orth cry	dec 200		2.50	13.2^{20}	s dil acids
571	Calcium citrate tetrahydrate	$Ca_3(C_6H_5O_7)_2 \cdot 4H_2O$	5785-44-4	570.494	wh needles or pow	dec 100			0.096^{23}	i eth
572	Calcium cyanamide	$CaCN_2$	156-62-7	80.102	col hex cry	≈1340	subl	2.29		reac H_2O
573	Calcium cyanide	$Ca(CN)_2$	592-01-8	92.112	wh rhomb cry; hyg					s H_2O, EtOH

No.	Name	Formula	CAS Reg No.	Mol. weight	Physical form	mp/°C	bp/°C	Density g cm^{-3}	Solubility g/100 g H_2O	Qualitative solubility
574	Calcium dichromate trihydrate	$CaCr_2O_7 \cdot 3H_2O$	14307-33-6*	310.112	red-oran cry	100 dec		2.37		vs H_2O; reac EtOH; i eth, ctc
575	Calcium dihydrogen phosphate monohydrate	$Ca(H_2PO_4)_2 \cdot H_2O$	10031-30-8	252.068	col tricl plates	100 dec		2.220		sl H_2O; s dil acid
576	Calcium 2-ethylhexanoate	$Ca(C_8H_{15}O_2)_2$	136-51-6	326.485	powder					
577	Calcium ferrocyanide dodecahydrate	$Ca_2Fe(CN)_6 \cdot 12H_2O$		508.289	yel tricl cry	dec		1.68	87^{25}	vs H_2O; i EtOH
578	Calcium fluoride	CaF_2	7789-75-5	78.075	wh cub cry or powder	1418	2500	3.18	0.0016^{25}	sl acid
579	Calcium fluorophosphate	$Ca_5(PO_4)_3F$	12015-73-5	504.302	col hex cry	1650		3.201		i H_2O
580	Calcium fluorophosphate dihydrate	$CaPO_3F \cdot 2H_2O$	37809-19-1	174.079	col monocl cry				0.42^{27}	i os
581	Calcium formate	$Ca(CHO_2)_2$	544-17-2	130.113	orth cry	300 dec		2.02	16.6^{20}	i EtOH
582	Calcium hexaborate pentahydrate	$2CaO \cdot 3B_2O_3 \cdot 5H_2O$	12291-65-5	411.091	col monocl cry	dec 375 (exp)		2.42	1^{25}	sl acid
583	Calcium hexafluoro-2,4-pentanedioate	$Ca(CF_3COCHCOCF_3)_2$	121012-90-6	454.180	powder	135				
584	Calcium hexafluorosilicate dihydrate	$CaSiF_6 \cdot 2H_2O$	16925-39-6	218.185	col tetr cry			2.25	0.52^{20}	i ace; reac hot H_2O
585	Calcium hydride	CaH_2	7789-78-8	42.094	gray orth cry or powder	1000		1.7		reac H_2O, EtOH
586	Calcium hydrogen phosphate	$CaHPO_4$	7757-93-9	136.057	wh tricl cry	dec		2.92	0.02^{25}	i EtOH
587	Calcium hydrogen phosphate dihydrate	$CaHPO_4 \cdot 2H_2O$	7789-77-7	172.088	monocl cry	≈100 dec		2.31	0.02^{25}	i EtOH; s dil acid
588	Calcium hydrogen sulfite	$CaH_2(SO_3)_2$	13780-03-5	202.220				1.06		s H_2O
589	Calcium hydrosulfide hexahydrate	$Ca(HS)_2 \cdot 6H_2O$		214.315	col cry	dec				s H_2O, EtOH
590	Calcium hydroxide	$Ca(OH)_2$	1305-62-0	74.093	soft hex cry			≈2.2	0.160^{20}	s acid
591	Calcium hydroxide phosphate	$Ca_5(OH)(PO_4)_3$	12167-74-7	502.311	col hex cry	dec >900		3.155		i H_2O
592	Calcium hypochlorite	$Ca(OCl)_2$	7778-54-3	142.983	powder	100		2.350		
593	Calcium hypophosphite	$Ca(H_2PO_2)_2$	7789-79-9	170.055	wh monocl cry	300 dec				s H_2O; i EtOH
594	Calcium iodate	$Ca(IO_3)_2$	7789-80-2	389.883	wh monocl cry			4.52	0.306^{25}	s HNO_3; i EtOH
595	Calcium iodide	CaI_2	10102-68-8	293.887	hyg hex cry	783	1100	3.96	215^{25}	s MeOH, EtOH, ace; i eth
596	Calcium iodide hexahydrate	$CaI_2 \cdot 6H_2O$	71626-98-7	401.978	wh hex needles or powder	42 dec		2.55	215^{25}	vs EtOH
597	Calcium metaborate	$Ca(BO_2)_2$	13701-64-9	125.698	powder				0.13^{20}	
598	Calcium metasilicate	$CaSiO_3$	1344-95-2	116.162	wh monocl cry	1540		2.92		i H_2O
599	Calcium molybdate	$CaMoO_4$	7789-82-4	200.02	wh tetr cry	1520		4.35	0.0011^{20}	i EtOH; s conc acid
600	Calcium nitrate	$Ca(NO_3)_2$	10124-37-5	164.087	wh cub cry; hyg	561		2.5	144^{25}	s EtOH, MeOH, ace
601	Calcium nitrate tetrahydrate	$Ca(NO_3)_2 \cdot 4H_2O$	13477-34-4	236.149	wh cry	≈40 dec		1.82	144^{25}	s EtOH, ace
602	Calcium nitride	Ca_3N_2	12013-82-0	148.247	red-brn cub cry	1195		2.67		s H_2O, acid; i EtOH
603	Calcium nitrite	$Ca(NO_2)_2$	13780-06-8	132.089	wh-yel hex cry; hyg	392		2.23	94.6^{25}	sl EtOH
604	Calcium nitrite monohydrate	$Ca(NO_2)_2 \cdot H_2O$	10031-34-2	150.104	col or yel cry	dec 100				vs H_2O; sl EtOH
605	Calcium oleate	$Ca(C_{18}H_{33}O_2)_2$	142-17-6	602.985	pale yel solid	dec 140			0.04^{25}	sl H_2O; s bz; i EtOH, ace, eth
606	Calcium oxalate	CaC_2O_4	563-72-4	128.097	wh cry powder			2.2	0.00061^{20}	
607	Calcium oxalate monohydrate	$CaC_2O_4 \cdot H_2O$	5794-28-5	146.112	cub cry	200 dec		2.2	0.00061^{20}	s dil acid
608	Calcium oxide	CaO	1305-78-8	56.077	gray-wh cub cry	2613		3.34		reac H_2O; s acid
609	Calcium oxide silicate	Ca_3OSiO_4	12168-85-3	228.317	refrac solid	2150				
610	Calcium palmitate	$Ca(C_{16}H_{31}O_2)_2$	542-42-7	550.910	wh-yel pow	dec 155				i H_2O, EtOH, eth, ace; sl bz
611	Calcium perborate heptahydrate	$Ca(BO_3)_2 \cdot 7H_2O$		283.803	gray-wh pow					s H_2O, acid
612	Calcium 2,4-pentanedioate	$Ca(CH_3COCHCOCH_3)_2$	19372-44-2	238.294	cry	dec 175				
613	Calcium perchlorate	$Ca(ClO_4)_2$	13477-36-6	238.979	wh cry	270 dec		2.65	188^{25}	s EtOH, MeOH
614	Calcium perchlorate tetrahydrate	$Ca(ClO_4)_2 \cdot 4H_2O$	15627-86-8	311.041	wh cry					vs H_2O
615	Calcium permanganate	$Ca(MnO_4)_2$	10118-76-0	277.949	purp hyg cry			2.4	331^{20}	reac EtOH
616	Calcium peroxide	CaO_2	1305-79-9	72.077	wh-yel tetr cry; hyg	≈200 dec		2.9		sl H_2O; s acid
617	Calcium phosphate	$Ca_3(PO_4)_2$	7758-87-4	310.177	wh amorp powder	1670		3.14	0.00012^{20}	i EtOH; s dil acid
618	Calcium phosphide	Ca_3P_2	1305-99-3	182.182	red-brn hyg cry	≈1600		2.51		reac H_2O; i EtOH, eth
619	Calcium phosphonate monohydrate	$CaHPO_3 \cdot H_2O$	25232-60-4	138.073	col monocl cry	dec 150				sl H_2O; i EtOH
620	Calcium plumbate	Ca_2PbO_4	12013-69-3	351.4	oran-brn orth cry	dec		5.71		i H_2O; s acid
621	Calcium propanoate	$Ca(C_3H_5O_2)_2$	4075-81-4	186.219	monocl cry, powder					s H_2O; sl MeOH, EtOH; i ace, bz
622	Calcium pyrophosphate	$Ca_2P_2O_7$	7790-76-3	254.099	wh powder	1353		3.09		i H_2O; s dil acid
623	Calcium selenate dihydrate	$CaSeO_4 \cdot 2H_2O$	7790-74-1	219.07	wh monocl cry			2.75	8.3^{18}	
624	Calcium selenide	$CaSe$	1305-84-6	119.04	wh-brn cub cry	1400 dec		3.8		reac H_2O
625	Calcium silicide (CaSi)	$CaSi$	12013-55-7	68.164	orth cry	1324		2.39		

No.	Name	Formula	CAS Reg No.	Mol. weight	Physical form	mp/°C	bp/°C	Density g cm⁻³	Solubility g/100 g H_2O	Qualitative solubility
626	Calcium silicide ($CaSi_2$)	$CaSi_2$	12013-56-8	96.249	gray hex cry	1040		2.50		i cold H_2O; reac hot H_2O; s acid
627	Calcium stannate trihydrate	$CaSnO_3 \cdot 3H_2O$	12013-46-6*	260.832	wh cry pow	dec ≈350				i H_2O
628	Calcium stearate	$Ca(C_{18}H_{35}O_2)_2$	1592-23-0	607.017	granular powder	180				i H_2O, EtOH
629	Calcium succinate trihydrate	$CaC_4H_4O_4 \cdot 3H_2O$	140-99-8	210.196	needles					sl H_2O; s dil acid; i EtOH
630	Calcium sulfate	$CaSO_4$	7778-18-9	136.141	orth cry	1460		2.96	0.205^{25}	
631	Calcium sulfate hemihydrate	$CaSO_4 \cdot 0.5H_2O$	10034-76-1	145.149	wh powder				0.205^{25}	
632	Calcium sulfate dihydrate	$CaSO_4 \cdot 2H_2O$	10101-41-4	172.171	monocl cry or powder	150 dec		2.32	0.205^{20}	i os
633	Calcium sulfide	CaS	20548-54-3	72.143	wh-yel cub cry; hyg	2524		2.59		sl H_2O; i EtOH
634	Calcium sulfite dihydrate	$CaSO_3 \cdot 2H_2O$	10257-55-3	156.172	wh powder				0.0070^{25}	sl EtOH; s acid
635	Calcium tartrate tetrahydrate	$CaC_4H_4O_6 \cdot 4H_2O$	3164-34-9*	260.210	wh pow				0.04^{10}	s dil acid; sl EtOH
636	Calcium telluride	CaTe	12013-57-9	167.68	wh cub cry	1600 dec		4.87		
637	Calcium tetrahydroaluminate	$Ca(AlH_4)_2$	16941-10-9	102.105	gray powder; flam					reac H_2O; s thf; i eth, bz
638	Calcium thiocyanate tetrahydrate	$Ca(SCN)_2 \cdot 4H_2O$	2092-16-2	228.304	hyg cry	160 dec				vs H_2O; s EtOH, ace
639	Calcium thiosulfate hexahydrate	$CaS_2O_3 \cdot 6H_2O$	10124-41-1	260.298	tricl cry	45 dec		1.87		s H_2O; i EtOH
640	Calcium titanate	$CaTiO_3$	12049-50-2	135.943	cub cry	1980		3.98		
641	Calcium tungstate	$CaWO_4$	7790-75-2	287.92	wh tetr cry	1620		6.06	0.2^{18}	s hot acid
642	Calcium zirconate	$CaZrO_3$	12013-47-7	179.300	powder	2550				
643	Californium	Cf	7440-71-3	251	hex or cub metal	900		15.1		
644	Carbon (diamond)	C	7782-40-3	12.011	col cub cry	4440 (12.4 GPa)		3.513		i H_2O
645	Carbon (graphite)	C	7782-42-5	12.011	soft blk hex cry	4489 tp (10.3 MPa)	3825 sp	2.2		i H_2O
646	Carbon black	C	1333-86-4	12.011	fine blk pow					i H_2O
647	Carbon (fullerene-C_{60})	C_{60}	99685-96-8	720.642	yel needles or plates	>280				s os
648	Carbon (fullerene-C_{70})	C_{70}	115383-22-7	840.749	red-brn solid	>280				s bz, tol
649	Fullerene fluoride	$C_{60}F_{60}$	134929-59-2	1860.546	col plates	287				vs ace; s thf; i chl
650	Carbon monoxide	CO	630-08-0	28.010	col gas	-205.02	-191.5	1.145 g/L		sl H_2O; s chl, EtOH
651	Carbon dioxide	CO_2	124-38-9	44.010	col gas	-56.558 tp	-78.464 sp	1.799 g/L		s H_2O
652	Carbon suboxide	C_3O_2	504-64-3	68.031	col gas	-112.5	6.8	2.781 g/L		reac H_2O
653	Carbon disulfide	CS_2	75-15-0	76.141	col or yel liq	-112.1	46	1.2632^{20}		i H_2O; vs EtOH, bz, os
654	Carbon subsulfide	C_3S_2	627-34-9	100.162	red liq	-1	90 dec	1.27		reac H_2O
655	Carbon diselenide	CSe_2	506-80-9	169.93	yel liq	-43.7	125.5	2.6823^{20}		i H_2O; vs ctc, tol
656	Carbon oxysulfide	COS	463-58-1	60.075	col gas	-138.8	-50	2.456 g/L		s H_2O, EtOH
657	Carbon oxyselenide	COSe	1603-84-5	106.97	col gas; unstab	-124.4	-21.7	4.372 g/L		reac H_2O
658	Carbon sulfide selenide	CSSe	5951-19-9	123.04	yel liq	-85	84.5	1.99		i H_2O
659	Carbon sulfide telluride	CSTe	10340-06-4	171.68	red-yel liq; unstab	-54	20 dec			reac H_2O
660	Carbonyl bromide	$COBr_2$	593-95-3	187.818	col liq		64.5	2.5		reac H_2O
661	Carbonyl chloride	$COCl_2$	75-44-5	98.916	col gas	-127.78	8	4.043 g/L		sl H_2O; s bz, tol
662	Carbonyl fluoride	COF_2	353-50-4	66.007	col gas	-111.26	-84.57	2.698 g/L		reac H_2O
663	Cyanogen	C_2N_2	460-19-5	52.034	col gas	-27.83	-21.1	2.127 g/L		sl H_2O, eth; s EtOH
664	Cyanogen azide	N_3CN	764-05-6	68.038	col oily liq	exp				
665	Cyanogen bromide	BrCN	506-68-3	105.922	wh hyg needles	52	61.5	2.015		s H_2O, EtOH, eth
666	Cyanogen chloride	ClCN	506-77-4	61.471	col vol liq or gas	-6.55	13	2.513 g/L		s H_2O, EtOH, eth
667	Cyanogen fluoride	FCN	1495-50-7	45.016	col gas	-82	-46	1.840 g/L		
668	Cyanogen iodide	ICN	506-78-5	152.922	col needles	146.7		2.84		s H_2O, EtOH, eth
669	Cerium	Ce	7440-45-1	140.116	silv metal; cub or hex	799	3443	6.770		s dil acid
670	Cerium boride	CeB_6	12008-02-5	204.982	blue refrac solid; hex	2550		4.87		i H_2O, HCl
671	Cerium carbide	CeC_2	12012-32-7	164.137	red hex cry	2250		5.47		reac H_2O
672	Cerium carbide	Ce_2C_3	12115-63-8	316.264	yel-brn cub cry	1505		6.9		
673	Cerium nitride	CeN	25764-08-3	154.123	refrac cub cry	2557		7.89		
674	Cerium silicide	$CeSi_2$	12014-85-6	196.287	tetr cry	1420		5.31		i H_2O
675	Cerium(II) hydride	CeH_2	13569-50-1	142.132	cub cry			5.45		reac H_2O
676	Cerium(II) iodide	CeI_2	19139-47-0	393.925	bronze cry	808				
677	Cerium(II) sulfide	CeS	12014-82-3	172.181	yel cub cry	2445		5.9		
678	Cerium(III) acetate sesqihydrate	$Ce(C_2H_3O_2)_3 \cdot 1.5H_2O$	17829-82-2		col cry	dec 115			26^{15}	s H_2O
679	Cerium(III) ammonium nitrate tetrahydrate	$(NH_4)_2Ce(NO_3)_5 \cdot 4H_2O$	13083-04-0	558.279	col monocl cry	74				vs H_2O
680	Cerium(III) ammonium sulfate tetrahydrate	$NH_4Ce(SO_4)_2 \cdot 4H_2O$	21995-38-0*	422.341	monocl cry					s H_2O
681	Cerium(III) bromide	$CeBr_3$	14457-87-5	379.828	wh hex cry; hyg	732	1457			s H_2O

No.	Name	Formula	CAS Reg No.	Mol. weight	Physical form	mp/°C	bp/°C	Density g cm^{-3}	Solubility g/100 g H_2O	Qualitative solubility
682	Cerium(III) bromide heptahydrate	$CeBr_3 \cdot 7H_2O$	7789-56-2	505.935	col hyg needles	732				s H_2O, EtOH
683	Cerium(III) carbonate	$Ce_2(CO_3)_3$	537-01-9	460.259	wh pow	dec 500				i H_2O; s acid
684	Cerium(III) carbonate pentahydrate	$Ce_2(CO_3)_3 \cdot 5H_2O$	72520-94-6	550.335	wh powder					i H_2O; s dil acid
685	Cerium(III) chloride	$CeCl_3$	7790-86-5	246.475	wh hex cry	807		3.97		s H_2O, EtOH
686	Cerium(III) chloride heptahydrate	$CeCl_3 \cdot 7H_2O$	18618-55-8	372.582	yel orth cry; hyg	90 dec				vs H_2O, EtOH
687	Cerium(III) fluoride	CeF_3	7758-88-5	197.111	wh hex cry; hyg	1430	2180	6.157		i H_2O
688	Cerium(III) hydride	CeH_3	13864-02-3	143.140	blk pow or blue-blk cry	dec (flam)				reac H_2O
689	Cerium(III) hydroxide	$Ce(OH)_3$	15785-09-8	191.138	wh solid					i H_2O; s acid
690	Cerium(III) iodide	CeI_3	7790-87-6	520.829	yel orth cry; hyg	760				s H_2O
691	Cerium(III) iodide nonahydrate	$CeI_3 \cdot 9H_2O$	7790-87-6*	682.967	wh-red cry					vs H_2O; s EtOH
692	Cerium(III) nitrate hexahydrate	$Ce(NO_3)_3 \cdot 6H_2O$	10108-73-3*	434.222	col-red cry	150 dec			176^{25}	s ace
693	Cerium(III) oxalate nonahydrate	$Ce(C_2O_4)_3 \cdot 9H_2O$	13266-83-6	706.426	wh pow	dec				i H_2O, EtOH; s acid
694	Cerium(III) oxide	Ce_2O_3	1345-13-7	328.230	yel-grn cub cry	2210	3730	6.2		i H_2O; s acid
695	Cerium(III) 2,4-pentanedioate trihydrate	$Ce(CH_3COCHCOCH_3)_3 \cdot 3H_2O$	15653-01-7	491.486	yel hyg cry	≈150				vs EtOH
696	Cerium(III) perchlorate hexahydrate	$Ce(ClO_4)_3 \cdot 6H_2O$	36907-38-7	546.559	hyg col cry	dec 200				s H_2O, EtOH
697	Cerium(III) selenate	$Ce_2(SeO_4)_3$		709.11	rhom cry			4.46		s H_2O
698	Cerium(III) sulfate	$Ce_2(SO_4)_3$	13454-94-9	568.420	col hyg cry	920 dec				s H_2O
699	Cerium(III) sulfate octahydrate	$Ce_2(SO_4)_3 \cdot 8H_2O$	13454-94-9	712.542	wh orth cry	≈250 dec		2.87		s H_2O
700	Cerium(III) sulfide	Ce_2S_3	12014-93-6	376.427	red cub cry	2450		5.02		i H_2O
701	Cerium(III) tungstate	$Ce_2(WO_4)_3$	13454-74-5	1023.75	yel tetr cry	1089		6.77		i H_2O
702	Cerium(IV) ammonium nitrate	$(NH_4)_2Ce(NO_3)_6$	16774-21-3	548.223	red-oran cry					vs H_2O
703	Cerium(IV) ammonium sulfate dihydrate	$(NH_4)_4Ce(SO_4)_4 \cdot 2H_2O$	10378-47-9	632.551	cry pow	dec 450				
704	Cerium(IV) fluoride	CeF_4	10060-10-3	216.110	wh hyg powder	≈600 dec		4.77		i H_2O
705	Cerium(IV) hydroxide	$Ce(OH)_4$	12014-56-1	208.146	yel-wh pow					i H_2O; s conc acid
706	Cerium(IV) oxide	CeO_2	1306-38-3	172.115	wh-yel powder; cub	2480		7.216		i H_2O, dil acid
707	Cerium(IV) sulfate tetrahydrate	$Ce(SO_4)_2 \cdot 4H_2O$	10294-42-5	404.303	yel-oran orth cry	100 dec		3.91	9.00^{20}	
708	Cesium	Cs	7440-46-2	132.905	silv-wh metal	28.5	671	1.873		reac H_2O
709	Cesium acetate	$CsC_2H_3O_2$	3396-11-0	191.949	hyg lumps	194			10^{11}	
710	Cesium aluminum sulfate dodecahydrate	$CsAl(SO_4)_2 \cdot 12H_2O$	7784-17-0	568.196	col cub cry	117 dec		1.97		s H_2O; i EtOH
711	Cesium amide	$CsNH_2$	22205-57-8	148.928	wh tetr cry			3.70		
712	Cesium azide	CsN_3	22750-57-8	174.925	hyg tetr cry; exp	326		≈3.5	224^{0}	
713	Cesium bromate	$CsBrO_3$	13454-75-6	260.807	col hex cry			4.11	3.83^{25}	
714	Cesium bromide	CsBr	7787-69-1	212.809	wh cub cry; hyg	636	≈1300	4.43	123^{25}	s EtOH; i ace
715	Cesium carbonate	Cs_2CO_3	534-17-8	325.820	wh monocl cry; hyg	793		4.24	261^{15}	s EtOH, eth
716	Cesium chlorate	$CsClO_3$	13763-67-2	216.356	col hex cry	342		3.57	7.78^{25}	sl H_2O
717	Cesium chloride	CsCl	7647-17-8	168.358	wh cub cry; hyg	646	1297	3.988	191^{25}	s EtOH
718	Cesium chromate(IV)	Cs_4CrO_4	56320-90-2	647.616	yel hex cry	982		4.24		vs H_2O
719	Cesium cyanide	CsCN	21159-32-0	158.923	wh cub cry; hyg	350		3.34		vs H_2O
720	Cesium dibromoiodate	$CsIBr_2$	18278-82-5	419.617	dark oran cry	dec				s H_2O
721	Cesium fluoride	CsF	13400-13-0	151.903	wh cub cry; hyg	703		4.64	573^{25}	s MeOH; i diox, py
722	Cesium fluoroborate	$CsBF_4$	18909-69-8	219.710	wh orth cry	555 dec		3.2	1.6^{17}	sl H_2O
723	Cesium formate	$CsCHO_2$	3495-36-1	177.923	wh cry			1.017		vs H_2O
724	Cesium hexafluorogermanate	Cs_2GeF_6		452.44	wh cry	≈675		4.10		sl cold H_2O; s hot H_2O
725	Cesium hydride	CsH	58724-12-2	133.913	wh cub cry; flam	528		3.42		reac H_2O
726	Cesium hydrogen carbonate	$CsHCO_3$	15519-28-5	193.922	rhom cry	175 dec			209^{15}	s EtOH
727	Cesium hydrogen fluoride	$CsHF_2$	12280-52-3	171.910	tetr cry	170		3.86		
728	Cesium hydrogen sulfate	$CsHSO_4$	7789-16-4	229.976	col rhom prisms	dec		3.352		s H_2O
729	Cesium hydroxide	CsOH	21351-79-1	149.912	wh-yel hyg cry	342.3		3.68	300^{30}	s EtOH
730	Cesium iodate	$CsIO_3$	13454-81-4	307.807	wh monocl cry			4.85	2.6^{25}	
731	Cesium iodide	CsI	7789-17-5	259.809	col cub cry; hyg	632	≈1280	4.51	84.8^{25}	s EtOH, MeOH, ace
732	Cesium metaborate	$CsBO_2$	92141-86-1	175.715	cub cry	732		≈3.7		
733	Cesium molybdate	Cs_2MoO_4	13597-64-3	425.75	wh cry	956.3			67^{18}	s H_2O
734	Cesium nitrate	$CsNO_3$	7789-18-6	194.910	wh hex or cub cry	409		3.66	27.9^{25}	s ace; sl EtOH
735	Cesium nitrite	$CsNO_2$	13454-83-6	178.911	yel cry	406				s H_2O
736	Cesium oxide	Cs_2O	20281-00-9	281.810	yel-oran hex cry	495		4.65		vs H_2O
737	Cesium superoxide	CsO_2	12018-61-0	164.904	yel tetr cry	432		3.77		reac H_2O
738	Cesium trioxide	Cs_2O_3	12134-22-4	313.809	brn cry	≈400		4.25		reac H_2O
739	Cesium perchlorate	$CsClO_4$	13454-84-7	232.356	wh orth cry; hyg	≈600 dec		3.327	2.00^{25}	
740	Cesium periodate	$CsIO_4$	13478-04-1	323.807	wh rhom prisms			4.26	2.2^{15}	
741	Cesium sulfate	Cs_2SO_4	10294-54-9	361.874	wh orth cry or hex prisms; hyg	1005		4.24	182^{25}	i EtOH, ace, py

No.	Name	Formula	CAS Reg No.	Mol. weight	Physical form	mp/°C	bp/°C	Density g cm^{-3}	Solubility g/100 g H_2O	Qualitative solubility
742	Cesium sulfide	Cs_2S	12214-16-3	297.876	yel orth hyg cry	520				vs H_2O
743	Cesium trifluoroacetate	$Cs(C_2F_3O_2)$	21907-50-6	245.920	hyg solid	115				vs H_2O
744	Chlorine	Cl_2	7782-50-5	70.906	grn-yel gas	-101.5	-34.04	2.898 g/L		sl H_2O
745	Hypochlorous acid	$HOCl$	7790-92-3	52.460	grn-yel; stable only in aq soln					s H_2O
746	Chloric acid	$HClO_3$	7790-93-4	84.459	exists only in aq soln					vs H_2O
747	Perchloric acid	$HClO_4$	7601-90-3	100.459	col hyg liq	-112	≈90 dec	1.77		s H_2O
748	Chlorine monoxide	Cl_2O	7791-21-1	86.905	yel-brn gas	-120.6	2.2	3.552 g/L		vs H_2O
749	Chlorine dioxide	ClO_2	10049-04-4	67.452	oran-grn gas	-59	11	2.757 g/L		sl H_2O
750	Dichlorine trioxide	Cl_2O_3	17496-59-2	118.904	dark brn solid	exp <25				
751	Dichlorine hexoxide	Cl_2O_6	12442-63-6	166.902	red liq	3.5	≈200			reac H_2O
752	Dichlorine heptoxide	Cl_2O_7	10294-48-1	182.902	col oily liq; exp	-91.5	82	1.9		reac H_2O
753	Chlorine fluoride	ClF	7790-89-8	54.451	col gas	-155.6	-101.1	2.226 g/L		reac H_2O
754	Chlorine trifluoride	ClF_3	7790-91-2	92.448	gas	-76.34	11.75	3.779 g/L		reac H_2O
755	Chlorine pentafluoride	ClF_5	13637-63-3	130.445	col gas	-103	-13.1	5.332 g/L		
756	Chlorosyl trifluoride	$ClOF_3$	30708-80-6	108.447	col liq	-42	27			reac H_2O
757	Chloryl fluoride	ClO_2F	13637-83-7	86.450	col gas	-115	-6	3.534 g/L		reac H_2O
758	Chloryl trifluoride	ClO_2F_3	38680-84-1	124.447	col gas	-81.2	-21.6	5.087 g/L		reac H_2O
759	Perchloryl fluoride	ClO_3F	7616-94-6	102.449	col gas	-147	-46.75	4.187 g/L		
760	Chlorine perchlorate	$ClOClO_3$	27218-16-2	134.904	unstab yel liq	-117	≈45 dec	1.81^{0}		
761	Chromium	Cr	7440-47-3	51.996	blue-wh metal; cub	1907	2671	7.15		reac dil acid
762	Chromic acid	H_2CrO_4	7738-94-5	118.010	aq soln only					s H_2O
763	Chromium antimonide	$CrSb$	12053-12-2	173.756	hex cry	1110		7.11		
764	Chromium arsenide	Cr_2As	12254-85-2	178.914	tetr cry			7.04		
765	Chromium boride (CrB)	CrB	12006-79-0	62.807	refrac orth cry	2100		6.1		
766	Chromium boride (CrB_2)	CrB_2	12007-16-8	73.618	refrac solid; hex	2200		5.22		
767	Chromium boride (Cr_2B)	Cr_2B	12006-80-3	114.803	refrac solid	1875				
768	Chromium boride (Cr_5B_3)	Cr_5B_3	12007-38-4	292.414	tetr cry	1900		6.10		
769	Chromium carbide	Cr_3C_2	12012-35-0	180.009	gray orth cry	1895		6.68		
770	Chromium carbonyl	$Cr(CO)_6$	13007-92-6	220.056	col orth cry	130 dec	subl	1.77		i H_2O, EtOH; s eth, chl
771	Chromium nitride (CrN)	CrN	24094-93-7	66.003	gray cub cry	1080 dec		5.9		
772	Chromium nitride (Cr_2N)	Cr_2N	12053-27-9	117.999	hex cry	1650		6.8		
773	Chromium phosphide	CrP	26342-61-0	82.970	orth cry			5.25		
774	Chromium selenide	$CrSe$	12053-13-3	130.96	hex cry	≈1500		6.1		
775	Chromium silicide ($CrSi_2$)	$CrSi_2$	12018-09-6	108.167	gray hex cry	1490		4.91		
776	Chromium silicide (Cr_3Si)	Cr_3Si	12018-36-9	184.074	cub cry	1770		6.4		
777	Chromium(II) acetate monohydrate	$Cr(C_2H_3O_2)_2 \cdot H_2O$	628-52-4*	188.100	red monocl cry			1.79		sl H_2O
778	Chromium(II) bromide	$CrBr_2$	10049-25-9	211.804	wh monocl cry; aq soln blue	842		4.236		s H_2O, EtOH
779	Chromium(II) chloride	$CrCl_2$	10049-05-5	122.902	wh hyg needles; aq soln blue	824	1120	2.88		s H_2O
780	Chromium(II) chloride tetrahydrate	$Cr(H_2O)_4Cl_2 \cdot 4H_2O$	13931-94-7	267.024	blue hyg cry	51 dec				s H_2O
781	Chromium(II) fluoride	CrF_2	10049-10-2	89.993	blue-grn monocl cry	894		3.79		sl H_2O; i EtOH
782	Chromium(II) formate monohydrate	$Cr(CHOO)_2 \cdot H_2O$	4493-37-2	160.046	red needles					s H_2O
783	Chromium(II) iodide	CrI_2	13478-28-9	305.805	red-brn cry; hyg	867		5.1		s H_2O
784	Chromium(II) oxalate monohydrate	$CrC_2O_4 \cdot H_2O$	814-90-4*	158.030	yel-grn powder			2.468		sl H_2O
785	Chromium(II) sulfate pentahydrate	$CrSO_4 \cdot 5H_2O$	13825-86-0	238.135	blue cry				21^{0}	s dil acid; sl EtOH; i ace
786	Chromium(II,III) oxide	Cr_3O_4	12018-34-7	219.986	cub cry			6.1		
787	Chromium(III) acetate	$Cr(C_2H_3O_2)_3$	1066-30-4	229.127	blue-grn pwd					sl H_2O
788	Chromium(III) acetate monohydrate	$Cr(C_2H_3O_2)_3 \cdot H_2O$	25013-82-5	247.143	gray-grn pow					sl H_2O; i EtOH
789	Chromium(III) acetate hexahydrate	$Cr(C_2H_3O_2)_3 \cdot 6H_2O$	1066-30-4*	337.220	blue needles					s H_2O
790	Chromium(III) acetate hydroxide	$Cr(C_2H_3O_2)_2(OH)$	39430-51-8	187.092	viol cry pow					vs H_2O
791	Chromium(III) bromide	$CrBr_3$	10031-25-1	291.708	dark grn hex cry	812		4.68		s hot H_2O, bz
792	Chromium(III) bromide hexahydrate (α)	$CrBr_3(H_2O)_4 \cdot 2H_2O$	18721-05-6	399.799	grn hyg cry					s H_2O, EtOH
793	Chromium(III) bromide hexahydrate (β)	$Cr(H_2O)_6Br_3$	10031-25-1*	399.799	viol hyg cry					s H_2O; i EtOH, eth
794	Chromium(III) chloride	$CrCl_3$	10025-73-7	158.355	red-viol cry	1152	1300 dec	2.76		sl H_2O
795	Chromium(III) chloride hexahydrate	$[CrCl_2(H_2O)_4]Cl \cdot 2H_2O$	10060-12-5	266.446	grn monocl cry; hyg					s H_2O, EtOH; sl ace; i eth
796	Chromium(III) fluoride	CrF_3	7788-97-8	108.991	grn needles	1425		3.8		i H_2O, EtOH
797	Chromium(III) fluoride trihydrate	$CrF_3 \cdot 3H_2O$	16671-27-5	163.037	grn hex cry			2.2		sl H_2O
798	Chromium(III) hydroxide sulfate	$Cr(OH)SO_4$	12336-95-7	165.066	grn cry					
799	Chromium(III) fluoride nonahydrate	$Cr(H_2O)_6F_3 \cdot 3H_2O$	102430-09-1	271.129	rhom viol cry					sl H_2O

No.	Name	Formula	CAS Reg No.	Mol. weight	Physical form	mp/°C	bp/°C	Density g cm^{-3}	Solubility g/100 g H_2O	Qualitative solubility
800	Chromium(III) hydroxide trihydrate	$Cr(OH)_3 \cdot 3H_2O$	1308-14-1	157.063	blue-grn powder					i H_2O; s acid
801	Chromium(III) iodide	CrI_3	13569-75-0	432.709	dark grn hex cry	500 dec		5.32		sl H_2O
802	Chromium(III) nitrate	$Cr(NO_3)_3$	13548-38-4	238.011	grn hyg powder	>60 dec				vs H_2O
803	Chromium(III) nitrate nonahydrate	$Cr(NO_3)_3 \cdot 9H_2O$	7789-02-8	400.148	grn-blk monocl cry	66.3	>100 dec	1.80		vs H_2O
804	Chromium(III) oxide	Cr_2O_3	1308-38-9	151.990	grn hex cry	2320	≈3000	5.22		i H_2O, EtOH; sl acid, alk
805	Chromium(III) 2,4-pentanedioate	$Cr(CH_3COCHCOCH_3)_3$	21679-31-2	349.320	red monocl cry	208	345	1.34		i H_2O; s bz
806	Chromium(III) perchlorate	$Cr(ClO_4)_3$	27535-70-2	350.348	grn-blue cry				58^{25}	vs H_2O
807	Chromium(III) phosphate	$CrPO_4$	7789-04-0	146.967	blue orth cry	>1800		4.6		i H_2O, acid, aqua regia
808	Chromium(III) phosphate hemiheptahydrate	$CrPO_4 \cdot 3.5H_2O$	84359-31-9	210.021	blue-grn powder			2.15		i H_2O; s acid
809	Chromium(III) phosphate hexahydrate	$CrPO_4 \cdot 6H_2O$	84359-31-9	255.059	viol cry	>500 dec		2.121		i H_2O; s acid, alk
810	Chromium(III) potassium oxalate trihydrate	$K_3Cr(C_2O_4)_3 \cdot 3H_2O$	15275-09-9	487.394	blue-grn monocl cry					s H_2O
811	Chromium(III) potassium sulfate dodecahydrate	$CrK(SO_4)_2 \cdot 12H_2O$	7788-99-0	499.403	viol-blk cub cry	89 dec		1.83		s H_2O; i EtOH
812	Chromium(III) sulfate	$Cr_2(SO_4)_3$	10101-53-8	392.180	red pow	dec >700		3.1	64^{25}	s H_2O; vs acid
813	Chromium(III) sulfate octadecahydrate	$Cr_2(SO_4)_3 \cdot 18H_2O$	10101-53-8*	716.455	viol cry	dec 115		1.7		reac H_2O
814	Chromium(III) sulfide	Cr_2S_3	12018-22-3	200.187	brn-blk hex cry			3.8		
815	Chromium(III) telluride	Cr_2Te_3	12053-39-3	486.79	hex cry	≈1300		7.0		
816	Chromium(IV) chloride	$CrCl_4$	15597-88-3	193.808	gas, stable at HT		>600 dec	7.922 g/L		
817	Chromium(IV) fluoride	CrF_4	10049-11-3	127.990	grn cry	277		2.89		reac H_2O
818	Chromium(IV) oxide	CrO_2	12018-01-8	83.995	brn-blk tetr powder	≈400 dec		4.89		i H_2O; s acid
819	Chromium(V) fluoride	CrF_5	14884-42-5	146.988	red orth cry	34	117			reac H_2O
820	Chromium(V) oxide	Cr_2O_5	12218-36-9	183.989	blk needles	dec 200				
821	Chromium(VI) fluoride	CrF_6	13843-28-2	165.986	yel solid; stable at low temp	-100 dec				
822	Chromium(VI) oxide	CrO_3	1333-82-0	99.994	red orth cry	197	≈250 dec	2.7	169^{25}	
823	Chromium(VI) tetrafluoride oxide	$CrOF_4$	23276-90-6	143.989	dark red solid	55				reac H_2O, ace, dmso
824	Chromium(VI) dichloride dioxide	CrO_2Cl_2	14977-61-8	154.901	red liq	-96.5	117	1.91		reac H_2O; s ctc, chl, bz
825	Chromium(VI) difluoride dioxide	CrO_2F_2	7788-96-7	121.992	red-viol cry	30	subl			reac H_2O
826	Cobalt	Co	7440-48-4	58.933	gray metal; hex or cub	1495	2927	8.86		s dil acid
827	Cobaltocene	$Co(C_5H_5)_2$	1277-43-6	189.119	blk-purp cry	173				
828	Cobalt antimonide	CoSb	12052-42-5	180.693	hex cry	1202		8.8		
829	Cobalt arsenic sulfide	CoAsS	12254-82-9	165.920	silv-wh solid			≈6.1		
830	Cobalt arsenide (CoAs)	CoAs	27016-73-5	133.855	orth cry	1180		8.22		
831	Cobalt arsenide ($CoAs_2$)	$CoAs_2$	12044-42-7	208.776	monocl cry			7.2		
832	Cobalt arsenide ($CoAs_3$)	$CoAs_3$	12256-04-1	283.698	cub cry	942		6.84		
833	Cobalt boride (CoB)	CoB	12006-77-8	69.744	refrac solid	1460		7.25		reac H_2O, HNO_3
834	Cobalt boride (Co_2B)	Co_2B	12045-01-1	128.677	refrac solid	1280		8.1		
835	Cobalt carbonyl	$Co_2(CO)_8$	10210-68-1	341.947	oran cry	51 dec		1.78		i H_2O; s EtOH, eth, CS_2
836	Cobalt disulfide	CoS_2	12013-10-4	123.063	cub cry			4.3		
837	Cobalt dodecacarbonyl	$Co_4(CO)_{12}$	17786-31-1	571.854	blk cry	60 dec		2.09		
838	Cobalt phosphide	Co_2P	12134-02-0	148.840	gray needles	1386		6.4		i H_2O; s HNO_3
839	Cobalt silicide	$CoSi_2$	12017-12-8	115.104	gray cub cry	1326		4.9		s hot HCl
840	Cobalt(II) acetate	$Co(C_2H_3O_2)_2$	71-48-7	177.022	pink cry					vs H_2O; s EtOH
841	Cobalt(II) acetate tetrahydrate	$Co(C_2H_3O_2)_2 \cdot 4H_2O$	6147-53-1	249.082	red monocl cry			1.705		s H_2O, EtOH, dil acid
842	Cobalt(II) aluminate	$CoAl_2O_4$	13820-62-7	176.894	blue cub cry			4.37		i H_2O
843	Cobalt(II) arsenate octahydrate	$Co_3(AsO_4)_2 \cdot 8H_2O$	24719-19-5	598.760	red monocl needles	400 dec	1000 dec	3.0		i H_2O; s dil acid
844	Cobalt(II) bromate hexahydrate	$Co(BrO_3)_2 \cdot 6H_2O$	13476-01-2	422.829	viol cry			≈2.5		vs H_2O
845	Cobalt(II) bromide	$CoBr_2$	7789-43-7	218.741	grn hex cry; hyg	678		4.91	113.2^{20}	s MeOH, EtOH, ace
846	Cobalt(II) bromide hexahydrate	$CoBr_2 \cdot 6H_2O$	13762-12-4	326.832	red hyg cry	47 dec	100 dec	2.46	113.2	
847	Cobalt(II) carbonate	$CoCO_3$	513-79-1	118.942	pink rhomb cry	dec 280		4.2	0.00014^{20}	i EtOH
848	Cobalt(II) basic carbonate	$2CoCO_3 \cdot 3Co(OH)_2 \cdot H_2O$	7542-09-8	534.743	red-viol cry	dec				i H_2O; s acid
849	Cobalt(II) chlorate hexahydrate	$Co(ClO_3)_2 \cdot 6H_2O$		333.927	dark red hyg cry	dec 61				s H_2O
850	Cobalt(II) chloride	$CoCl_2$	7646-79-9	129.839	blue hyg leaflets	737	1049	3.36	56.2^{25}	s EtOH, eth, ace, py
851	Cobalt(II) chloride dihydrate	$CoCl_2 \cdot 2H_2O$	16544-92-6	165.870	viol-blue cry			2.477	56.2^{25}	

No.	Name	Formula	CAS Reg No.	Mol. weight	Physical form	mp/°C	bp/°C	Density g cm⁻³	Solubility g/100 g H_2O	Qualitative solubility
852	Cobalt(II) chloride hexahydrate	$CoCl_2 \cdot 6H_2O$	7791-13-1	237.930	pink-red monocl cry	87 dec		1.924	56.2^{25}	s EtOH, ace, eth
853	Cobalt(II) chromate	$CoCrO_4$	24613-38-5	174.927	yel-brn orth cry			≈4.0		i H_2O; s acid
854	Cobalt(II) chromite	$CoCr_2O_4$	13455-25-9	226.923	blue-grn cub cry			5.14		i H_2O, conc acid
855	Cobalt(II) citrate dihydrate	$Co_3(C_6H_5O_7)_2 \cdot 2H_2O$	18727-04-3	265.170	rose red cry	dec 150			0.8^{15}	
856	Cobalt(II) cyanide	$Co(CN)_2$	542-84-7	110.967	blue hyg cry			1.872		i H_2O
857	Cobalt(II) cyanide dihydrate	$Co(CN)_2 \cdot 2H_2O$	20427-11-6	146.998	pink-brn needles					i H_2O, acid
858	Cobalt(II) diiron tetroxide	$CoFe_2O_4$	12052-28-7	234.621	blk solid					s hot HCl
859	Cobalt(II) ferricyanide	$Co_3[Fe(CN)_6]_2$	14049-81-1	600.698	red needles					i H_2O, HCl; s NH_4OH
860	Cobalt(II) fluoride	CoF_2	10026-17-2	96.930	red tetr cry	1127	≈1400	4.46	1.4^{25}	s acid
861	Cobalt(II) fluoride tetrahydrate	$CoF_2 \cdot 4H_2O$	13817-37-3	168.992	red orth cry	dec		2.22	1.4^{25}	
862	Cobalt(II) formate dihydrate	$Co(CHO_2)_2 \cdot 2H_2O$	6424-20-0	184.998	red cry powder	140 dec		2.13	5.03^{20}	i EtOH
863	Cobalt(II) hexafluoro-2,4-pentanedioate	$Co(CF_3COCHCOCF_3)_2$	19648-83-0	473.035	powder	197				
864	Cobalt(II) hexafluorosilicate hexahydrate	$CoSiF_6 \cdot 6H_2O$	12021-68-0	309.100	pale red cry			2.087	76.8^{22}	
865	Cobalt(II) hydroxide	$Co(OH)_2$	21041-93-0	92.948	blue-grn cry	≈160 dec		3.60		sl H_2O; s acid
866	Cobalt(II) hydroxide monohydrate	$Co(OH)_2 \cdot H_2O$	35340-84-2	110.963	blue solid	136 dec				
867	Cobalt(II) iodate	$Co(IO_3)_2$	13455-28-2	408.738	blk-viol needles	200 dec		5.09	0.46^{20}	
868	Cobalt(II) iodide	CoI_2	15238-00-3	312.742	blk hex cry; hyg	520		5.60	203^{25}	
869	Cobalt(II) iodide dihydrate	$CoI_2 \cdot 2H_2O$	13455-29-3	348.773	hyg grn cry	dec 100				
870	Cobalt(II) iodide hexahydrate	$CoI_2 \cdot 6H_2O$	15238-00-3*	420.833	red hex prisms	130 dec		2.90	203^{25}	s EtOH, eth, ace
871	Cobalt(II) molybdate	$CoMoO_4$	13762-14-6	218.87	blk monocl cry	1040		4.7		
872	Cobalt(II) molybdate monohydrate	$CoMoO_4 \cdot H_2O$	18601-87-1	236.89	blk pow					
873	Cobalt(II) nitrate	$Co(NO_3)_2$	10141-05-6	182.942	pale red powder	100 dec		2.49	103^{25}	
874	Cobalt(II) nitrate hexahydrate	$Co(NO_3)_2 \cdot 6H_2O$	10026-22-9	291.034	red monocl cry; hyg	≈55		1.88	103^{25}	s EtOH
875	Cobalt(II) nitrite	$Co(NO_2)_2$	18488-96-5	150.944					0.49^{25}	
876	Cobalt(II) oleate	$Co(C_{18}H_{33}O_2)_2$	14666-94-5	621.840	brn amorp pow					i H_2O; s EtOH, eth
877	Cobalt(II) orthosilicate	Co_2SiO_4	12017-08-2	209.950	red-viol orth cry	1345		4.63		i H_2O; s dil HCl
878	Cobalt(II) oxalate	CoC_2O_4	814-89-1	146.952	pink powder	250 dec		3.02	0.0037^{20}	s acid, NH_4OH
879	Cobalt(II) oxalate dihydrate	$CoC_2O_4 \cdot 2H_2O$	5965-38-8	182.982	pink needles	dec			0.0037	sl acid; s NH_4OH
880	Cobalt(II) oxide	CoO	1307-96-6	74.932	gray cub cry	1830		6.44		i H_2O; s acid
881	Cobalt(II) 2,4-pentanedioate	$Co(CH_3COCHCOCH_3)_2$	14024-48-7	257.149	bl-viol cry	167				
882	Cobalt(II) perchlorate	$Co(ClO_4)_2$	13455-31-7	257.834	red needles			3.33	113^{25}	i EtOH, ace
883	Cobalt(II) perchlorate hexahydrate	$Co(ClO_4)_2 \cdot 6H_2O$	13478-33-6	365.926	dark red cry	dec 170		3.33		vs H_2O
884	Cobalt(II) phosphate octahydrate	$Co_3(PO_4)_2 \cdot 8H_2O$	10294-50-5	510.865	pink amorp powder			2.77		i H_2O; s acid
885	Cobalt(II) potassium sulfate hexahydrate	$CoK_2(SO_4)_2 \cdot 6H_2O$	10026-20-7	437.347	red monocl cry	75 dec		2.22		vs H_2O
886	Cobalt(II) selenate pentahydrate	$CoSeO_4 \cdot 5H_2O$	14590-19-3	291.97	red tricl cry	dec		2.51	55^{15}	
887	Cobalt(II) selenide	CoSe	1307-99-9	137.89	yel hex cry	1055		7.65		i H_2O, alk; s aqua regia
888	Cobalt(II) selenite dihydrate	$CoSeO_3 \cdot 2H_2O$	19034-13-0	221.92	blue-red powder					i H_2O
889	Cobalt(II) stannate	Co_2SnO_4	12139-93-4	300.574	grn-blue cub cry			6.30		i H_2O; s alk
890	Cobalt(II) stearate	$Co(C_{18}H_{35}O_2)_2$	1002-88-6	625.872	purp solid	74		1.13		
891	Cobalt(II) sulfate	$CoSO_4$	10124-43-3	154.996	red orth cry	>700		3.71	38.3^{25}	
892	Cobalt(II) sulfate monohydrate	$CoSO_4 \cdot H_2O$	13455-34-0	173.011	red monocl cry			3.08	38.3^{25}	
893	Cobalt(II) sulfate heptahydrate	$CoSO_4 \cdot 7H_2O$	10026-24-1	281.102	pink monocl cry	41 dec		2.03	38.3^{25}	sl EtOH, MeOH
894	Cobalt(II) sulfide	CoS	1317-42-6	90.998	blk amorp powder	1117		5.45		i H_2O; s acid
895	Cobalt(II) telluride	CoTe	12017-13-9	186.53	hex cry			≈8.8		
896	Cobalt(II) thiocyanate	$Co(SCN)_2$	3017-60-5	175.097	yel-brn powder				103^{25}	s EtOH, MeOH, ace, eth
897	Cobalt(II) thiocyanate trihydrate	$Co(SCN)_2 \cdot 3H_2O$	97126-35-7	229.143	viol rhomb cry				103^{25}	s EtOH, eth, ace
898	Cobalt(II) titanate	$CoTiO_3$	12017-01-5	154.798	grn rhomb cry			5.0		
899	Cobalt(II) tungstate	$CoWO_4$	12640-47-0	306.77	blue monocl cry			≈7.8		i H_2O; s hot conc acid
900	Cobalt(II,III) oxide	Co_3O_4	1308-06-1	240.798	blk cub cry	900 dec		6.11		i H_2O; s acid, alk
901	Cobalt(III) acetate	$Co(C_2H_3O_2)_3$	917-69-1	236.064	grn hyg cry	100 dec				s H_2O, EtOH
902	Cobalt(III) ammonium tetranitrodiammine	$NH_4[Co(NH_3)_2(NO_2)_4]$	13600-89-0	295.054	red-brn orth cry			1.97		s H_2O
903	Cobalt(III) fluoride	CoF_3	10026-18-3	115.928	brn hex cry	927		3.88		reac H_2O; s EtOH, eth, bz
904	Cobalt(III) fluoride dihydrate	$Co_2F_6 \cdot 2H_2O$	54496-71-8	267.887	red rhomb cry			2.19		s H_2O; i EtOH
905	Cobalt(III) hexammine chloride	$Co(NH_3)_6Cl_3$	10534-89-1	267.475	red monocl cry			1.71		s H_2O; i EtOH
906	Cobalt(III) hydroxide	$Co(OH)_3$	1307-86-4	109.955	brn powder	dec		≈4		i H_2O; s acid
907	Cobalt(III) nitrate	$Co(NO_3)_3$	15520-84-0	244.948	grn cub cry; hyg			≈3.0		s H_2O; reac os
908	Cobalt(III) oxide	Co_2O_3	1308-04-9	165.864	gray-blk powder	895 dec		5.18		i H_2O; s conc acid

No.	Name	Formula	CAS Reg No.	Mol. weight	Physical form	mp/°C	bp/°C	Density g cm⁻³	Solubility g/100 g H_2O	Qualitative solubility
909	Cobalt(III) oxide monohydrate	$Co_2O_3 \cdot H_2O$	12016-80-7	183.880	brn-blk hex cry	150 dec				i H_2O; s acid
910	Cobalt(III) 2,4-pentanedioate	$Co(CH_3COCHCOCH_3)_3$	21679-46-9	356.257	dark grn cry	213				s bz, ace
911	Cobalt(III) potassium nitrite sesquihydrate	$CoK_3(NO_2)_6 \cdot 1.5H_2O$	13782-01-9*	479.284	yel cub cry			2.6		sl H_2O; reac acid; i EtOH
912	Cobalt(III) sulfide	Co_2S_3	1332-71-4	214.061	blk cub cry			4.8		reac acid
913	Cobalt(II) titanate	Co_2TiO_4	12017-38-8	229.731	grn-blk cub cry			5.1		s conc HCl
914	Copper	Cu	7440-50-8	63.546	red metal; cub	1084.62	2562	8.96		sl dil acid
915	Copper arsenide	Cu_3As	12005-75-3	265.560	dark gray solid	827				
916	Copper nitride	Cu_3N	1308-80-1	204.645	cub cry	300 dec		5.84		
917	Copper phosphide	CuP_2	12019-11-3	125.494	monocl cry	≈900		4.20		
918	Copper silicide	Cu_5Si	12159-07-8	345.816	solid	825				
919	Copper(I) acetate	$CuC_2H_3O_2$	598-54-9	122.590	col cry	dec	subl			reac H_2O
920	Copper(I) acetylide	Cu_2C_2	1117-94-8	151.113	red amorp powder; exp					
921	Copper(I) azide	CuN_3	14336-80-2	105.566	tetr cry; exp					
922	Copper(I) bromide	CuBr	7787-70-4	143.450	wh cub cry; hyg	483	1345	4.98	0.0012^{20}	i ace
923	Copper(I) chloride	CuCl	7758-89-6	98.999	wh cub cry	423	1490	4.14	0.0047^{20}	i EtOH, ace
924	Copper(I) cyanide	CuCN	544-92-3	89.564	wh powder or grn orth cry	474	dec	2.9		i H_2O, EtOH; s KCN soln
925	Copper(I) fluoride	CuF	13478-41-6	82.544	cub cry			7.1		
926	Copper(I) hydride	CuH	13517-00-5	64.554	red-brn solid	60 dec				
927	Copper(I) iodide	CuI	7681-65-4	190.450	wh cub cry	591	≈1290	5.67	0.000020^{20}	i dil acid
928	Copper(I) mercury iodide	Cu_2HgI_4	13876-85-2	835.30	red cry powder	trans ≈60 (brn)				i H_2O, EtOH
929	Copper(I) oxide	Cu_2O	1317-39-1	143.091	red-brn cub cry	1244	1800 dec	6.0		i H_2O
930	Copper(I) selenide	Cu_2Se	20405-64-5	206.05	blue-blk tetr cry	1113		6.84		i H_2O; s acid
931	Copper(I) sulfide	Cu_2S	22205-45-4	159.157	blue-blk orth cry	1129		5.6		i H_2O; sl acid
932	Copper(I) sulfite hemihydrate	$Cu_2SO_3 \cdot 0.5H_2O$	13982-53-1*	216.164	wh-yel hex cry					sl H_2O; s acid, alk; i EtOH, eth
933	Copper(I) sulfite monohydrate	$Cu_2SO_3 \cdot H_2O$	35788-00-2	225.171	cry			3.83		sl H_2O; s HCl
934	Copper(I) telluride	Cu_2Te	12019-52-2	254.69	blue hex cry	1127		4.6		
935	Copper(I) thiocyanate	CuSCN	1111-67-7	121.629	wh-yel amorp powder	1084		2.85		i H_2O, dil acid, EtOH, ace; s eth
936	Copper(I,II) sulfite dihydrate	$Cu_2SO_3 \cdot CuSO_3 \cdot 2H_2O$	13814-81-8	386.795	red prisms or powder					i H_2O, EtOH; s HCl
937	Copper(II) acetate	$Cu(C_2H_3O_2)_2$	142-71-2	181.635	blue-grn hyg powder					
938	Copper(II) acetate monohydrate	$Cu(C_2H_3O_2)_2 \cdot H_2O$	6046-93-1	199.650	grn monocl cry	115	240 dec	1.88		s H_2O, EtOH; sl eth
939	Copper(II) acetate metaarsenite	$Cu(C_2H_3O_2)_2 \cdot 3Cu(AsO_2)_2$	12002-03-8	1013.795	grn cry powder					i H_2O; reac acid
940	Copper(II) basic acetate	$Cu(C_2H_3O_2)_2 \cdot CuO \cdot 6H_2O$	52503-64-7	369.271	blue-grn cry or powder					sl H_2O, EtOH; s dil acid, NH_4OH
941	Copper(II) acetylide	CuC_2	12540-13-5	87.567	brn-blk solid; exp	exp 100				
942	Copper(II) arsenate	$Cu_3(AsO_4)_2$	7778-41-8	468.476	blue-grn cry					i H_2O, EtOH; s dil acid
943	Copper(II) arsenite	$CuHAsO_3$	10290-12-7	187.474	yel-grn powder					i H_2O, EtOH; s acid
944	Copper(II) azide	$Cu(N_3)_2$	14215-30-6	147.586	brn orth cry; exp			≈2.6		
945	Copper(II) borate	$Cu(BO_2)_2$	39290-85-2	149.166	blue-grn powder			3.859		i H_2O; s acid
946	Copper(II) bromide	$CuBr_2$	7789-45-9	223.354	blk monocl cry; hyg	498	900	4.710	126^{25}	vs H_2O; s EtOH, ace; i bz, eth
947	Copper(II) butanoate monohydrate	$Cu(C_4H_7O_2)_2 \cdot H_2O$	540-16-9	255.756	grn monocl plates					s H_2O, diox, bz; sl EtOH
948	Copper(II) carbonate	$CuCO_3$	1184-64-1	123.555	cry					i H_2O
949	Copper(II) carbonate hydroxide	$CuCO_3 \cdot Cu(OH)_2$	12069-69-1	221.116	grn monocl cry	200 dec		4.0		i H_2O, EtOH; s dil acid
950	Copper(II) chlorate hexahydrate	$Cu(ClO_3)_2 \cdot 6H_2O$	14721-21-2	338.540	blue-grn hyg cry	65	100 dec		164^{18}	vs EtOH
951	Copper(II) chloride	$CuCl_2$	7447-39-4	134.452	yel-brn monocl cry; hyg	598	993	3.4	75.7^{25}	s EtOH, ace
952	Copper(II) chloride dihydrate	$CuCl_2 \cdot 2H_2O$	10125-13-0	170.483	grn-blue orth cry; hyg	100 dec		2.51	75.7^{20}	vs EtOH, MeOH; s ace; i eth
953	Copper(II) chloride hydroxide	$Cu_2(OH)_3Cl$	1332-65-6	213.567	pale grn cry					i H_2O; s acid
954	Copper(II) chromate	$CuCrO_4$	13548-42-0	179.540	red-brn cry					i H_2O; s EtOH
955	Copper(II) basic chromate	$CuCrO_4 \cdot 2Cu(OH)_2$	12433-14-6	374.661	brn pow	dec 260				i H_2O; s HNO_3
956	Copper(II) chromite	$CuCr_2O_4$	12018-10-9	231.536	gray-blk tetr cry			5.4		i H_2O, dil acid
957	Copper(II) citrate hemipentahydrate	$Cu_2C_6H_4O_7 \cdot 2.5H_2O$	10402-15-0	360.221	blue-grn cry	100 dec				sl H_2O; s dil acid
958	Copper(II) cyanide	$Cu(CN)_2$	14763-77-0	115.580	grn powder					i H_2O; s acid, alk
959	Copper(II) cyclohexanebutanoate	$Cu(C_{10}H_{17}O_2)_2$	2218-80-6	402.028	powder	126 dec				
960	Copper(II) dichromate dihydrate	$CuCr_2O_7 \cdot 2H_2O$	13675-47-3	315.565	red-brn tricl cry			2.286		vs H_2O
961	Copper(II) ethanolate	$Cu(C_2H_5O)_2$	2850-65-9	153.667	blue hyg solid	120 dec				i os
962	Copper(II) ethylacetoacetate	$Cu(C_2H_5CO_2CHCOCH_3)_2$	14284-06-1	321.813	grn cry	192				s EtOH, chl

No.	Name	Formula	CAS Reg No.	Mol. weight	Physical form	mp/°C	bp/°C	Density g cm^{-3}	Solubility g/100 g H_2O	Qualitative solubility
963	Copper(II) 2-ethylhexanoate	$Cu(C_8H_{15}O_2)_2$	149-11-1	349.953	powder	252 dec				
964	Copper(II) ferrate	$CuFe_2O_4$	12018-79-0	239.234	blk cry					
965	Copper(II) ferrocyanide	$Cu_2Fe(CN)_6$	13601-13-3	339.041	red-br cub cry or powder			2.2		i H_2O, acid, os
966	Copper(II) ferrous sulfide	$CuFeS_2$	1308-56-1	183.521	yel tetr cry	950		4.2		i H_2O, HCl; s HNO_3
967	Copper(II) fluoride	CuF_2	7789-19-7	101.543	wh monocl cry	836	1676	4.23	0.075^{25}	
968	Copper(II) fluoride dihydrate	$CuF_2 \cdot 2H_2O$	13454-88-1	137.574	blue monocl cry	130 dec		2.934	0.075^{25}	
969	Copper(II) formate	$Cu(CHO_2)_2$	544-19-4	153.581	blue cry				12.5^{20}	i os
970	Copper(II) formate tetrahydrate	$Cu(CHO_2)_2 \cdot 4H_2O$	5893-61-8	225.641	blue monocl cry				12.5	sl EtOH; i os
971	Copper(II) gluconate	$CuC_{12}H_{22}O_{14}$	527-09-3	453.841	bl-grn cry	156				sl EtOH; i os
972	Copper(II) hexafluoro-2,4-pentanedioate	$Cu(CF_3COCHCOCF_3)_2$	14781-45-4	477.648	cry	98	220 dec			s MeOH, ace, tol
973	Copper(II) hexafluorosilicate tetrahydrate	$CuSiF_6 \cdot 4H_2O$	12062-24-7	277.684	blue monocl cry	dec		2.56	99.7^{17}	sl EtOH
974	Copper(II) hydroxide	$Cu(OH)_2$	20427-59-2	97.561	blue-grn powder			3.37		i H_2O; s acid, conc alk
975	Copper(II) iodate	$Cu(IO_3)_2$	13454-89-2	413.351	grn monocl cry	dec		5.241	0.15^{20}	s dil acid
976	Copper(II) iodate monohydrate	$Cu(IO_3)_2 \cdot H_2O$	13454-90-5	431.367	blue tricl cry	248 dec		4.872	0.15^{20}	s dil H_2SO_4
977	Copper(II) molybdate	$CuMoO_4$	13767-34-5	223.48	grn cry	≈500		3.4	0.038	
978	Copper(II) nitrate	$Cu(NO_3)_2$	3251-23-8	187.555	blue-grn orth cry; hyg	255	subl		145^{25}	s diox; reac eth
979	Copper(II) nitrate trihydrate	$Cu(NO_3)_2 \cdot 3H_2O$	10031-43-3	241.602	blue rhomb cry	114	170 dec	2.32	145^{25}	vs EtOH
980	Copper(II) nitrate hexahydrate	$Cu(NO_3)_2 \cdot 6H_2O$	13478-38-1	295.647	blue rhomb cry; hyg			2.07	145^{25}	s EtOH
981	Copper(II) oleate	$Cu(C_{18}H_{33}O_2)_2$	1120-44-1	626.453	blue-grn solid					i H_2O; sl EtOH; s eth
982	Copper(II) oxalate	CuC_2O_4	814-91-5	151.565	blue-wh powder	310 dec			0.0026^{20}	i EtOH, eth; s NH_4OH
983	Copper(II) oxalate hemihydrate	$CuC_2O_4 \cdot 0.5H_2O$	814-91-5*	144.573	blue-wh cry	200 dec			0.0026^{20}	s NH_4OH
984	Copper(II) oxide	CuO	1317-38-0	79.545	blk powder or monocl cry	1227		6.31		i H_2O, EtOH; s dil acid
985	Copper(II) 2,4-pentanedioate	$Cu(CH_3COCHCOCH_3)_2$	13395-16-9	261.762	blue powder	284 dec	subl			sl H_2O; s chl
986	Copper(II) oxychloride hemiheptahydrate	$CuCl_2 \cdot 3CuO \cdot 3.5H_2O$	1332-40-7		blue-grn pow	dec 140				i H_2O; s acid, NH_4OH
987	Copper(II) perchlorate	$Cu(ClO_4)_2$	13770-18-8	262.447	grn hyg cry	130 dec			146^{30}	s eth, diox; i bz, ctc
988	Copper(II) perchlorate hexahydrate	$Cu(ClO_4)_2 \cdot 6H_2O$	10294-46-9	370.539	blue monocl cry; hyg	82	120 dec	2.22	146^{30}	vs EtOH, HOAc, ace; sl eth
989	Copper(II) phosphate	$Cu_3(PO_4)_2$	7798-23-4	380.581	blue-grn tricl cry					i H_2O; s acid, NH_4OH
990	Copper(II) phosphate trihydrate	$Cu_3(PO_4)_2 \cdot 3H_2O$	10031-48-8	434.627	blue-grn orth cry					i H_2O; s acid, NH_4OH
991	Copper(II) phthalocyanine	$CuC_{32}H_{16}N_8$	147-14-8	576.069	bl-purp cry					i H_2O, EtOH; s conc H_2SO_4
992	Copper(II) selenate pentahydrate	$CuSeO_4 \cdot 5H_2O$	10031-45-5	296.58	blue tricl cry	80 dec		2.56	27.4^{25}	s acid, NH_4OH; sl ace; i EtOH
993	Copper(II) selenide	CuSe	1317-41-5	142.51	blue-blk needles or plates	550 dec		5.99		reac acid
994	Copper(II) selenite dihydrate	$CuSeO_3 \cdot 2H_2O$	15168-20-4	226.54	blue orth cry			3.31		i H_2O; s acid, NH_4OH
995	Copper(II) silicate dihydrate	$CuSiO_3 \cdot 2H_2O$	26318-99-0	175.661	grn-blue orth cry					
996	Copper(II) stannate	$CuSnO_3$	12019-07-7	230.254	blue pow					
997	Copper(II) stearate	$Cu(C_{18}H_{35}O_2)_2$	660-60-6	630.485	blue-grn amorp powder	≈250				i H_2O, EtOH, eth; s py
998	Copper(II) sulfate	$CuSO_4$	7758-98-7	159.609	wh-grn amorp powder or rhomb cry	560 dec		3.60	22.0^{25}	i EtOH
999	Copper(II) sulfate pentahydrate	$CuSO_4 \cdot 5H_2O$	7758-99-8	249.685	blue tricl cry	110 dec		2.286	22.0^{25}	s MeOH; sl EtOH
1000	Copper(II) sulfate, basic	$Cu_3(OH)_4SO_4$	1332-14-5	354.730	grn rhomb cry			3.88		i H_2O
1001	Copper(II) sulfide	CuS	1317-40-4	95.611	blk hex cry	trans 507		4.76		i H_2O, EtOH, dil acid, alk
1002	Copper(II) tartrate trihydrate	$CuC_4H_4O_6 \cdot 3H_2O$	815-82-7	265.663	blue-grn powder					sl H_2O; s acid, alk
1003	Copper(II) telluride	CuTe	12019-23-7	191.15	yel orth cry	trans ≈400		7.09		
1004	Copper(II) tellurite	$CuTeO_3$	13812-58-3	239.14	blk glassy solid					i H_2O
1005	Copper(II) tetrafluoroborate	$Cu(BF_4)_2$	14735-84-3	237.155	solid					s H_2O
1006	Copper(II) titanate	$CuTiO_3$	12019-08-8	159.411	gray pow					
1007	Copper(II) 1,1,1-trifluoro-2,4-pentanedioate	$Cu(CF_3COCHCOCH_3)_2$	14324-82-4	369.705	blue-purp cry	197	dec 260			s EtOH, tol
1008	Copper(II) tungstate	$CuWO_4$	13587-35-4	311.38	yel-brn powder			7.5		
1009	Copper(II) tungstate dihydrate	$CuWO_4 \cdot 2H_2O$	13587-35-4*	347.41	grn powder					i H_2O; sl HOAc; reac conc acid
1010	Copper(II) vanadate	$Cu(VO_3)_2$	12789-09-2	261.425	powder					

No.	Name	Formula	CAS Reg No.	Mol. weight	Physical form	mp/°C	bp/°C	Density g cm⁻³	Solubility g/100 g H_2O	Qualitative solubility
1011	Curium	Cm	7440-51-9	247	silv metal; hex or cub	1345	≈3100	13.51		
1012	Dysprosium	Dy	7429-91-6	162.500	silv metal; hex	1412	2567	8.55		s dil acid
1013	Dysprosium boride	DyB_4	12310-43-9	205.744	tetr cry	2500		6.98		
1014	Dysprosium nitride	DyN	12019-88-4	176.507	cub cry			9.93		
1015	Dysprosium silicide	$DySi_2$	12133-07-2	218.671	orth cry	1550		5.2		
1016	Dysprosium(II) bromide	$DyBr_2$	83229-05-4	322.308	blk solid					
1017	Dysprosium(II) chloride	$DyCl_2$	13767-31-2	233.406	blk cry	721 dec				reac H_2O
1018	Dysprosium(II) iodide	DyI_2	36377-94-3	416.309	purp cry	659				reac H_2O
1019	Dysprosium(III) acetate tetrahydrate	$Dy(C_2H_3O_2)_3 \cdot 4H_2O$	15280-55-4	411.693	yel needles	dec 120				s H_2O; sl EtOH
1020	Dysprosium(III) bromide	$DyBr_3$	14456-48-5	402.212	wh hyg cry	879				s H_2O
1021	Dysprosium(III) carbonate tetrahydrate	$Dy_2(CO_3)_3 \cdot 4H_2O$	38245-35-1	577.088	wh cry pow					i H_2O
1022	Dysprosium(III) chloride	$DyCl_3$	10025-74-8	268.859	wh or yel cry	718	1530	3.67		s H_2O, MeOH
1023	Dysprosium(III) chloride hexahydrate	$DyCl_3 \cdot 6H_2O$	15059-52-6	376.950	bright yel cry	dec 162				
1024	Dysprosium(III) fluoride	DyF_3	13569-80-7	219.495	grn cry	1157				
1025	Dysprosium(III) hydride	DyH_3	13537-09-2	165.524	hex cry			7.1		
1026	Dysprosium(III) hydroxide	$Dy(OH)_3$	1308-85-6	213.522	yel or wh needles	205 dec				i H_2O
1027	Dysprosium(III) iodide	DyI_3	15474-63-2	543.213	grn cry	978				
1028	Dysprosium(III) nitrate pentahydrate	$Dy(NO_3)_3 \cdot 5H_2O$	10143-38-1*	438.591	yel cry	88.6			208.4^{25}	
1029	Dysprosium(III) oxide	Dy_2O_3	1308-87-8	372.998	wh cub cry	2228	3900	7.81		s acid
1030	Dysprosium(III) sulfate octahydrate	$Dy_2(SO_4)_3 \cdot 8H_2O$	10031-50-2	757.310	pale yel cry	110				sl H_2O
1031	Dysprosium(III) sulfide	Dy_2S_3	12133-10-7	421.195	red-brn monocl cry			6.08		
1032	Dysprosium(III) telluride	Dy_2Te_3	12159-43-2	707.80	solid	≈1550				
1033	Einsteinium	Es	7429-92-7	252	metal; cub	860				
1034	Erbium	Er	7440-52-0	167.259	silv metal; hex	1529	2868	9.07		i H_2O; s acid
1035	Erbium boride	ErB_4	12310-44-0	210.503	tetr cry	2450		7.0		
1036	Erbium acetate tetrahydrate	$Er(C_2H_3O_2)_3 \cdot 4H_2O$	15280-57-6	416.452	pink or wh cry			2.11		s H_2O
1037	Erbium bromide	$ErBr_3$	13536-73-7	406.971	viol hyg cry	950	≈1460			s H_2O, thf
1038	Erbium bromide hexahydrate	$ErBr_3 \cdot 6H_2O$	14890-44-9	515.062	pink cry					s H_2O
1039	Erbium chloride	$ErCl_3$	10138-41-7	273.618	viol monocl cry; hyg	776		4.1		s H_2O
1040	Erbium chloride hexahydrate	$ErCl_3 \cdot 6H_2O$	10025-75-9	381.709	pink hyg cry	dec				s H_2O; sl EtOH
1041	Erbium fluoride	ErF_3	13760-83-3	224.254	pink orth cry	1146		7.8		i H_2O
1042	Erbium hydride	ErH_3	13550-53-3	170.283	hex cry			≈7.6		
1043	Erbium hydroxide	$Er(OH)_3$	14646-16-3	218.281	pink solid					i H_2O
1044	Erbium iodide	ErI_3	13813-42-8	547.972	viol hex cry; hyg	1014		≈5.5		s H_2O
1045	Erbium nitrate pentahydrate	$Er(NO_3)_3 \cdot 5H_2O$	10168-80-6*	443.350	red cry	130 dec			240.8^{25}	s EtOH, ace
1046	Erbium nitride	ErN	12020-21-2	181.266	cub cry			10.6		
1047	Erbium oxide	Er_2O_3	12061-16-4	382.516	pink powder	2344	3920	8.64		i H_2O; s acid
1048	Erbium silicide	$ErSi_2$	12020-28-9	223.430	orth cry			7.26		
1049	Erbium sulfate	$Er_2(SO_4)_3$	13478-49-4	622.706	hyg powder	dec		3.68	13^{20}	
1050	Erbium sulfate octahydrate	$Er_2(SO_4)_3 \cdot 8H_2O$	10031-52-4	766.828	pink monocl cry	dec		3.20	13^{20}	
1051	Erbium sulfide	Er_2S_3	12159-66-9	430.713	red-brn monocl cry	1730		6.07		
1052	Erbium telluride	Er_2Te_3	12020-39-2	717.32	orth cry	1213		7.11		
1053	Europium	Eu	7440-53-1	151.964	soft silv metal; cub	822	1529	5.24		reac H_2O
1054	Europium boride	EuB_6	12008-05-8	216.830	cub cry	≈2600		4.91		
1055	Europium nitride	EuN	12020-58-5	165.971	cub cry			8.7		
1056	Europium silicide	$EuSi_2$	12434-24-1	208.135	tetr cry	1500		5.46		
1057	Europium(II) bromide	$EuBr_2$	13780-48-8	311.772	wh cry	683				s H_2O
1058	Europium(II) chloride	$EuCl_2$	13769-20-5	222.870	wh orth cry	731		4.9		s H_2O
1059	Europium(II) fluoride	EuF_2	14077-39-5	189.961	grn-yel cub cry	≈1380		6.5		
1060	Europium(II) iodide	EuI_2	22015-35-6	405.773	grn cry	580				s H_2O
1061	Europium(II) selenide	EuSe	12020-66-5	230.92	brn cub cry			6.45		
1062	Europium(II) sulfate	$EuSO_4$	10031-54-6	248.027	col orth cry			4.99		i H_2O
1063	Europium(II) sulfide	EuS	12020-65-4	184.029	cub cry			5.7		
1064	Europium(II) telluride	EuTe	12020-69-8	279.56	blk cub cry	1526		6.48		
1065	Europium(III) bromide	$EuBr_3$	13759-88-1	391.676	gray cry	dec				s H_2O
1066	Europium(III) chloride	$EuCl_3$	10025-76-0	258.323	grn-yel needles	623		4.89		
1067	Europium(III) chloride hexahydrate	$EuCl_3 \cdot 6H_2O$	13759-92-7	366.414	wh-yel hyg cry	850		4.89		s H_2O
1068	Europium(III) fluoride	EuF_3	13765-25-8	208.959	wh hyg cry	1276				i H_2O
1069	Europium(III) iodide	EuI_3	13759-90-5	532.677	col cry; unstab	≈875				
1070	Europium(III) nitrate hexahydrate	$Eu(NO_3)_3 \cdot 6H_2O$	10031-53-5	446.070	wh-pink hyg cry	85 dec			193^{25}	
1071	Europium(III) oxalate	$Eu_2(C_2O_4)_3$	3269-12-3	567.985	wh solid					i H_2O; s acid

No.	Name	Formula	CAS Reg No.	Mol. weight	Physical form	mp/°C	bp/°C	Density g cm^{-3}	Solubility g/100 g H_2O	Qualitative solubility
1072	Europium(III) oxide	Eu_2O_3	1308-96-9	351.926	pink powder	2291	3790	7.42		i H_2O; s acid
1073	Europium(III) perchlorate hexahydrate	$Eu(ClO_4)_3 \cdot 6H_2O$	36907-40-1	558.407	wh or pink cry					s H_2O, EtOH
1074	Europium(III) sulfate	$Eu_2(SO_4)_3$	13537-15-0	592.116	pale pink cry			4.99	2.1^{20}	
1075	Europium(III) sulfate octahydrate	$Eu_2(SO_4)_3 \cdot 8H_2O$	10031-52-4	736.238	pink cry	375 dec			2.1^{20}	
1076	Fermium	Fm	7440-72-4	257	metal	1527				
1077	Fluorine	F_2	7782-41-4	37.997	pale yel gas	-219.67 tp	-188.12	1.553 g/L		reac H_2O
1078	Fluorine monoxide	F_2O	7783-41-7	53.996	col gas	-223.8	-144.3	2.207 g/L		sl H_2O
1079	Difluorine dioxide	F_2O_2	7783-44-0	69.996	red-oran solid, unstab gas	-163.5	-57 (extrap)	2.861 g/L		
1080	Fluorine tetroxide	F_2O_4	107782-11-6	101.995	red-brn solid	-191	dec -185			
1081	Fluorine nitrate	FNO_3	7789-26-6	81.003	col gas	-175	-46	3.311 g/L		reac H_2O, EtOH, eth; s ace
1082	Fluorine perchlorate	$FOClO_3$	10049-03-3	118.449	col gas; exp	-167.3	-16	4.841 g/L		reac H_2O
1083	Francium	Fr	7440-73-5	223.000	short-lived alkali metal	27				
1084	Gadolinium	Gd	7440-54-2	157.25	silv metal; hex	1313	3273	7.90		s dil acid
1085	Gadolinium boride	GdB_6	12008-06-9	222.12	blk-brn cub cry	2510		5.31		
1086	Gadolinium nitride	GdN	25764-15-2	171.26	cub cry			9.10		
1087	Gadolinium silicide	$GdSi_2$	12134-75-7	213.42	orth cry	1540		5.9		
1088	Gadolinium(II) iodide	GdI_2	13814-72-7	411.06	bronze cry	831				
1089	Gadolinium(II) selenide	GdSe	12024-81-6	236.21	cub cry	2170		8.1		
1090	Gadolinium(III) acetate tetrahydrate	$Gd(C_2H_3O_2)_2 \cdot 4H_2O$	15280-53-2	406.44	wh tricl cry	dec		1.61		s H_2O
1091	Gadolinium(III) bromide	$GdBr_3$	13818-75-2	396.96	wh monocl cry; hyg	770		4.56		
1092	Gadolinium(III) chloride	$GdCl_3$	10138-52-0	263.61	wh monocl cry; hyg	602		4.52		s H_2O
1093	Gadolinium(III) chloride hexahydrate	$GdCl_3 \cdot 6H_2O$	19423-81-5	371.70	col hyg cry			2.424		s H_2O
1094	Gadolinium(III) fluoride	GdF_3	13765-26-9	214.25	wh cry	1232				
1095	Gadolinium(III) iodide	GdI_3	13572-98-0	537.96	yel cry	930				
1096	Gadolinium(III) nitrate pentahydrate	$Gd(NO_3)_3 \cdot 5H_2O$	52788-53-1	433.34	wh cry	92 dec		2.41	190^{25}	
1097	Gadolinium(III) nitrate hexahydrate	$Gd(NO_3)_3 \cdot 6H_2O$	19598-90-4	451.36	hyg tricl cry	91 dec		2.33	190^{25}	s EtOH
1098	Gadolinium(III) oxalate decahydrate	$Gd_2(C_2O_4)_3 \cdot 10H_2O$	22992-15-0	758.71	wh monocl pow	dec 110				i H_2O; sl acid
1099	Gadolinium(III) oxide	Gd_2O_3	12064-62-9	362.50	wh hyg powder	2339	3900	7.41		i H_2O; s acid
1100	Gadolinium(III) sulfate	$Gd_2(SO_4)_3$	13628-54-1	602.69	col cry	500 dec		4.1	2.60^{20}	sl H_2O
1101	Gadolinium(III) sulfate octahydrate	$Gd_2(SO_4)_3 \cdot 8H_2O$	13450-87-8	746.81	col monocl cry	400 dec		4.14	2.3^{20}	sl H_2O
1102	Gadolinium(III) sulfide	Gd_2S_3	12134-77-9	410.70	yel cub cry			6.1		
1103	Gadolinium(III) telluride	Gd_2Te_3	12160-99-5	697.30	orth cry	1255		7.7		
1104	Gallium	Ga	7440-55-3	69.723	silv liq or gray orth cry	29.7666 tp	2204	5.91		reac alk
1105	Gallium antimonide	GaSb	12064-03-8	191.483	brn cub cry	712		5.6137		
1106	Gallium arsenide	GaAs	1303-00-0	144.645	gray cub cry	1238		5.3176		
1107	Gallium nitride	GaN	25617-97-4	83.730	gray hex cry	>2500		6.1		
1108	Gallium phosphide	GaP	12063-98-8	100.697	yel cub cry	1457		4.138		
1109	Gallium suboxide	Ga_2O	12024-20-3	155.445	brn powder	>660	>800 dec	4.77		
1110	Gallium(II) chloride	$GaCl_2$	24597-12-4	140.629	wh orth cry	172.4	535	2.74		
1111	Gallium(II) selenide	GaSe	12024-11-2	148.68	hex cry	960		5.03		
1112	Gallium(II) sulfide	GaS	12024-10-1	101.788	hex cry	965		3.86		
1113	Gallium(II) telluride	GaTe	12024-14-5	197.32	monocl cry	824		5.44		
1114	Gallium(III) bromide	$GaBr_3$	13450-88-9	309.435	wh orth cry	123	279	3.69		
1115	Gallium(III) chloride	$GaCl_3$	13450-90-3	176.082	col needles or gl solid	77.9	201	2.47		
1116	Gallium(III) fluoride	GaF_3	7783-51-9	126.718	wh powder or col needles	>1000		4.47		i H_2O
1117	Gallium(III) fluoride trihydrate	$GaF_3 \cdot 3H_2O$	22886-66-4	180.764	wh cry	>140 dec				sl H_2O
1118	Gallium(III) hydride	GaH_3	13572-93-5	72.747	visc liq	-15	≈0 dec			
1119	Gallium(III) hydroxide	$Ga(OH)_3$	12023-99-3	120.745	unstab prec					
1120	Gallium(III) iodide	GaI_3	13450-91-4	450.436	monocl cry	212	340	4.5		
1121	Gallium(III) nitrate	$Ga(NO_3)_3$	13494-90-1	255.738	wh cry powder					s H_2O, EtOH, eth
1122	Gallium(III) oxide	Ga_2O_3	12024-21-4	187.444	wh cry	1807		≈6.0		s hot acid
1123	Gallium(III) oxide hydroxide	GaOOH	20665-52-5	102.730	orth cry			5.23		
1124	Gallium(III) 2,4-pentanedioate	$Ga(CH_3COCHCOCH_3)_3$	14405-43-7	367.047	wh powder	193	subl	1.42		
1125	Gallium(III) perchlorate hexahydrate	$Ga(ClO_4)_3 \cdot 6H_2O$	17835-81-3	476.166	cry	dec 175				
1126	Gallium(III) selenide	Ga_2Se_3	12024-24-7	376.33	cub cry	937		4.92		
1127	Gallium(III) sulfate	$Ga_2(SO_4)_3$	13494-91-2	427.634	hex cry					
1128	Gallium(III) sulfate octadecahydrate	$Ga_2(SO_4)_3 \cdot 18H_2O$	13780-42-2	751.909	octahed cry					s H_2O, EtOH
1129	Gallium(III) sulfide	Ga_2S_3	12024-22-5	235.641	monocl cry	1090		3.7		

No.	Name	Formula	CAS Reg No.	Mol. weight	Physical form	mp/°C	bp/°C	Density g cm^{-3}	Solubility g/100 g H_2O	Qualitative solubility
1130	Gallium(III) telluride	Ga_2Te_3	12024-27-0	522.25	cub cry	790		5.57		
1131	Germanium	Ge	7440-56-4	72.64	gray-wh cub cry	938.25	2833	5.3234		i H_2O, dil acid, alk
1132	Germane	GeH_4	7782-65-2	76.67	col gas; flam	-165	-88.1	3.133 g/L		i H_2O
1133	Digermane	Ge_2H_6	13818-89-8	151.33	col liq; flam	-109	29	1.98^{-109}		
1134	Trigermane	Ge_3H_8	14691-44-2	225.98	col liq	-105.6	110.5	2.20^{-105}		i H_2O
1135	Tetragermane	Ge_4H_{10}	14691-47-5	300.64	col liq		176.9			i H_2O
1136	Pentagermane	Ge_5H_{12}	15587-39-0	375.30	col liq		234			i H_2O
1137	Bromogermane	GeH_3Br	13569-43-2	155.57	col liq	-32	52	2.34		reac H_2O
1138	Chlorogermane	GeH_3Cl	13637-65-5	111.12	col liq	-52	28	1.75		reac H_2O
1139	Chlorotrifluorogermane	GeF_3Cl	14188-40-0	165.09	gas	-66.2	-20.3	6.747 g/L		
1140	Dibromogermane	GeH_2Br_2	13769-36-3	234.46	col liq	-15	89	2.80		reac H_2O
1141	Dichlorogermane	GeH_2Cl_2	15230-48-5	145.56	col liq	-68	69.5	1.90		reac H_2O
1142	Dichlorodifluorogermane	GeF_2Cl_2	24422-21-7	181.54	col gas	-51.8	-2.8	7.419 g/L		
1143	Dichlorodimethylgermane	$Ge(CH_3)_2Cl_2$	1529-48-2	173.62	liq	-22	124	1.49		
1144	Fluorogermane	GeH_3F	13537-30-9	94.66	col gas			3.868 g/L		reac H_2O
1145	Iodogermane	GeH_3I	13573-02-9	202.57	liq	-15	≈90			reac H_2O
1146	Methylgermane	GeH_3CH_3	1449-65-6	90.70	col gas	-158	-23	3.706 g/L		
1147	Tribromogermane	$GeHBr_3$	14779-70-5	313.36	col liq	-25	dec			reac H_2O
1148	Trichlorogermane	$GeHCl_3$	1184-65-2	180.01	liq	-71	75.3	1.93		reac H_2O
1149	Trichlorofluorogermane	$GeCl_3F$	24422-20-6	198.00	liq	-49.8	37.5			
1150	Germanium(II) bromide	$GeBr_2$	24415-00-7	232.45	yel monocl cry	122	150 dec			reac H_2O
1151	Germanium(II) chloride	$GeCl_2$	10060-11-4	143.55	wh-yel hyg powder	dec				reac H_2O; s eth, bz
1152	Germanium(II) fluoride	GeF_2	13940-63-1	110.64	wh orth cry; hyg	110	130 dec	3.64		reac H_2O
1153	Germanium(II) iodide	GeI_2	13573-08-5	326.45	oran-yel hex cry	428	550 dec	5.4		reac H_2O
1154	Germanium(II) oxide	GeO	20619-16-3	88.64	blk solid	700 dec				
1155	Germanium(II) selenide	GeSe	12065-10-0	151.60	gray orth cry or brn powder	675		5.6		
1156	Germanium(II) sulfide	GeS	12025-32-0	104.71	gray orth cry	658		4.1		
1157	Germanium(II) telluride	GeTe	12025-39-7	200.24	cub cry	724		6.16		i H_2O, s conc HNO_3
1158	Germanium(IV) bromide	$GeBr_4$	13450-92-5	392.26	wh cry	26.1	186.35	3.132		reac H_2O
1159	Germanium(IV) chloride	$GeCl_4$	10038-98-9	214.45	col liq	-51.50	86.55	1.88		reac H_2O; s bz, eth, EtOH, ctc
1160	Germanium(IV) fluoride	GeF_4	7783-58-6	148.63	col gas	-15 tp	-36.5 sp	6.074 g/L		reac H_2O
1161	Germanium(IV) iodide	GeI_4	13450-95-8	580.26	red-oran cub cry	146	348	4.322		reac H_2O
1162	Germanium(IV) nitride	Ge_3N_4	12065-36-0	273.95	orth cry	900 dec				i H_2O, acid, aqua regia
1163	Germanium(IV) oxide	GeO_2	1310-53-8	104.64	wh hex cry	1116		4.25		i H_2O
1164	Germanium(IV) selenide	$GeSe_2$	12065-11-1	230.56	yel-oran orth cry	707 dec		4.56		
1165	Germanium(IV) sulfide	GeS_2	12025-34-2	136.77	blk orth cry	530		3.01		
1166	Gold	Au	7440-57-5	196.967	soft yel metal	1064.18	2856	19.3		s aqua regia
1167	Bromoauric(III) acid pentahydrate	$HAuBr_4 \cdot 5H_2O$	17083-68-0	607.667	red-brn hyg cry	27				s H_2O, EtOH
1168	Chloroauric(III) acid tetrahydrate	$HAuCl_4 \cdot 4H_2O$	16903-35-8	411.848	yel monocl cry; hyg			≈3.9		vs H_2O, EtOH; s eth
1169	Gold(I) bromide	AuBr	10294-27-6	276.871	yel-gray tetr cry	165 dec		8.20		i H_2O
1170	Gold(I) chloride	AuCl	10294-29-8	232.420	yel orth cry	289 dec		7.6	0.000031^{20}	
1171	Gold(I) cyanide	AuCN	506-65-0	222.985	yel hex cry	dec		7.2		i H_2O, EtOH, eth, dil acid
1172	Gold(I) iodide	AuI	10294-31-2	323.871	yel-grn powder; tetr	120 dec		8.25		i H_2O, s CN soln
1173	Gold(I) sulfide	Au_2S	1303-60-2	425.998	brn-blk cub cry; unstab	240 dec		≈11		i H_2O, acid; s aqua regia
1174	Gold(III) bromide	$AuBr_3$	10294-28-7	436.679	red-br monocl cry	≈160 dec				s H_2O, EtOH
1175	Gold(III) chloride	$AuCl_3$	13453-07-1	303.326	red monocl cry	>160 dec		4.7	68^{20}	
1176	Gold(III) cyanide trihydrate	$Au(CN)_3 \cdot 3H_2O$	535-37-5*	329.065	wh hyg cry	50 dec				vs H_2O; sl EtOH
1177	Gold(III) fluoride	AuF_3	14720-21-9	253.962	oran-yel hex cry	>300	subl	6.75		
1178	Gold(III) hydroxide	$Au(OH)_3$	1303-52-2	247.989	brn powder	≈100 dec				i H_2O; s acid
1179	Gold(III) iodide	AuI_3	31032-13-0	577.680	unstab grn powder	20 dec				
1180	Gold(III) oxide	Au_2O_3	1303-58-8	441.931	brn powder	≈150 dec				i H_2O; s acid
1181	Gold(III) selenate	$Au_2(SeO_4)_3$	10294-32-3	822.81	yel cry					i H_2O; s acid
1182	Gold(III) selenide	Au_2Se_3	1303-62-4	630.81	blk amorp solid	dec		4.65		s aqua regia
1183	Gold(III) sulfide	Au_2S_3	1303-61-3	490.128	unstab blk powder	200 dec				
1184	Hafnium	Hf	7440-58-6	178.49	gray metal; hex	2233	4603	13.3		s HF
1185	Hafnium boride	HfB_2	12007-23-7	200.11	gray hex cry	3100		10.5		
1186	Hafnium carbide	HfC	12069-85-1	190.50	refrac cub cry	≈3000		12.2		
1187	Hafnium hydride	HfH_2	12770-26-2	180.51	refrac tetr cry			11.4		
1188	Hafnium nitride	HfN	25817-87-2	192.50	yel-brn cub cry	3310		13.8		
1189	Hafnium phosphide	HfP	12325-59-6	209.46	hex cry			9.78		

No.	Name	Formula	CAS Reg No.	Mol. weight	Physical form	mp/°C	bp/°C	Density g cm^{-3}	Solubility g/100 g H_2O	Qualitative solubility
1190	Hafnium silicide	$HfSi_2$	12401-56-8	234.66	gray orth cry	≈1700		7.6		
1191	Hafnocene dichloride	$Hf(C_5H_5)_2Cl_2$	12116-66-4	379.58	col hyg cry	235				s bz, chl; sl thf, eth; i hex
1192	Hafnium(II) bromide	$HfBr_2$	13782-95-1	338.30	blue-blk cry	dec 400				
1193	Hafnium(II) chloride	$HfCl_2$	13782-92-8	249.40	blk solid	dec 400				
1194	Hafnium(III) bromide	$HfBr_3$	13782-96-2	418.20	blue-blk cry	dec 350				
1195	Hafnium(III) chloride	$HfCl_3$	13782-93-9	284.85	blk solid	dec				
1196	Hafnium(III) iodide	HfI_3	13779-73-2	559.20	blk cry	dec				
1197	Hafnium(IV) bromide	$HfBr_4$	13777-22-5	498.11	wh cub cry	424 tp	323 sp	4.90		reac H_2O
1198	Hafnium(IV) chloride	$HfCl_4$	13499-05-3	320.30	wh monocl cry	432 tp	317 sp			reac H_2O
1199	Hafnium(IV) fluoride	HfF_4	13709-52-9	254.48	wh monocl cry	1025	970 sp	7.1		reac H_2O
1200	Hafnium(IV) iodide	HfI_4	13777-23-6	686.11	yel-oran cub cry	449 tp	394 sp	5.6		reac H_2O
1201	Hafnium(IV) oxide	HfO_2	12055-23-1	210.49	wh cub cry	2800	≈5400	9.68		i H_2O
1202	Hafnium(IV) oxychloride octahydrate	$HfOCl_2 \cdot 8H_2O$	14456-34-9	409.52	wh tetr cry	dec				s H_2O
1203	Hafnium(IV) selenide	$HfSe_2$	12162-21-9	336.41	brn hex cry			7.46		
1204	Hafnium(IV) silicate	$HfSiO_4$	13870-13-8	270.57	tetr cry	2758		7.0		
1205	Hafnium(IV) sulfate	$Hf(SO_4)_2$	15823-43-5	370.62	wh cry	>500 dec				
1206	Hafnium(IV) sulfide	HfS_2	18855-94-2	242.62	purp-brn hex cry			6.03		
1207	Hafnium(IV) titanate	$HfTiO_4$	12055-24-2	290.36	wh pow	1980 dec				
1208	Helium	He	7440-59-7	4.003	col gas		-268.93	0.164 g/L		sl H_2O; i EtOH
1209	Holmium	Ho	7440-60-0	164.930	silv metal; hex	1472	2700	8.80		s dil acid
1210	Holmium acetate	$Ho(C_2H_3O_2)_3$	25519-09-9	342.062	yel cry	dec 327				s H_2O
1211	Holmium bromide	$HoBr_3$	13825-76-8	404.642	yel hyg cry	919	1470			
1212	Holmium chloride	$HoCl_3$	10138-62-2	271.289	yel monocl cry; hyg	720	1500	3.7		s H_2O
1213	Holmium chloride hexahydrate	$HoCl_3 \cdot 6H_2O$	14914-84-2	379.381	hyg yel cry	160 dec				s H_2O
1214	Holmium fluoride	HoF_3	13760-78-6	221.925	pink-yel orth cry; hyg	1143	>2200	7.664		s H_2O
1215	Holmium iodide	HoI_3	13813-41-7	545.643	yel hex cry	994		5.4		
1216	Holmium nitrate pentahydrate	$Ho(NO_3)_3 \cdot 5H_2O$	14483-18-2	441.022	hyg oran cry					s H_2O, EtOH, ace
1217	Holmium nitride	HoN	12029-81-1	178.937	cub cry			10.6		
1218	Holmium oxalate decahydrate	$Ho_2(C_2O_4)_3 \cdot 10H_2O$	28965-57-3	774.070	yel solid	dec 40				
1219	Holmium oxide	Ho_2O_3	12055-62-8	377.859	yel cub cry	2330	3900	8.41		s acid
1220	Holmium silicide	$HoSi_2$	12136-24-2	221.101	hex cry			7.1		
1221	Holmium sulfide	Ho_2S_3	12162-59-3	426.056	yel-oran monocl cry			5.92		
1222	Hydrazine	N_2H_4	302-01-2	32.045	col oily liq	1.54	113.55	1.0036		vs H_2O, EtOH, MeOH
1223	Hydrazine acetate	$N_2H_4 \cdot CH_3COOH$	13255-48-6	92.097	cry	100				
1224	Hydrazine azide	$N_2H_4 \cdot HN_3$	14662-04-5	75.074	hyg wh prism	75 exp				vs H_2O
1225	Hydrazine monohydrate	$N_2H_4 \cdot H_2O$	7803-57-8	50.060	fuming liq	-51.7	119	1.030		vs H_2O, EtOH; i chl, eth
1226	Hydrazine hydrobromide	$N_2H_4 \cdot HBr$	13775-80-9	112.957	wh monocl cry flakes	84	≈190 dec	2.3		s H_2O, EtOH
1227	Hydrazine hydrochloride	$N_2H_4 \cdot HCl$	2644-70-4	68.506	wh orth cry	89	240 dec	1.5		s H_2O; i os
1228	Hydrazine dihydrochloride	$N_2H_4 \cdot 2HCl$	5341-61-7	104.967	wh orth cry	198 dec		1.42		s H_2O; sl EtOH
1229	Hydrazine hydroiodide	$N_2H_4 \cdot HI$	10039-55-1	159.957	hyg cry	125				s H_2O
1230	Hydrazine nitrate	$N_2H_4 \cdot HNO_3$	13464-97-6	95.058	monocl cry; exp	70				vs H_2O
1231	Hydrazine dinitrate	$N_2H_4 \cdot 2HNO_3$	13464-98-7	158.071	needles	104 dec			20^{35}	s H_2O
1232	Hydrazine perchlorate hemihydrate	$N_2H_4 \cdot HClO_4 \cdot 0.5H_2O$	13762-65-7		solid	137	exp	1.94		reac H_2O, s EtOH; i eth, bz
1233	Hydrazine sulfate	$N_2H_4 \cdot H_2SO_4$	10034-93-2	130.124	col orth cry	254		1.378		sl H_2O; i EtOH
1234	Dihydrazine sulfate	$(N_2H_4)_2 \cdot H_2SO_4$	13464-80-7	162.169	hyg wh cry flakes	104	dec >180		200^{25}	vs H_2O; i os
1235	Hydrazoic acid	HN_3	7782-79-8	43.028	col liq; exp	-80	35.7			s H_2O
1236	Hydroxylamine	H_2NOH	7803-49-8	33.030	wh orth flakes or needles	33.1	58	1.21		vs H_2O, MeOH
1237	Hydroxylamine hydrobromide	$H_2NOH \cdot HBr$	41591-55-3	113.942	monocl cry			2.35		s H_2O
1238	Hydroxylamine hydrochloride	$H_2NOH \cdot HCl$	5470-11-1	69.491	col monocl cry	159 dec	exp	1.68	94^{25}	vs H_2O
1239	Hydroxylamine perchlorate	$H_2NOH \cdot HClO_4$		133.489	orth cry	88	dec 120			
1240	Hydroxylamine sulfate	$2(H_2NOH) \cdot H_2SO_4$	10039-54-0	164.138	cry	170				vs H_2O
1241	Hydrogen	H_2	1333-74-0	2.016	col gas; flam	-259.198 tp	-252.762	0.082 g/L		sl H_2O
1242	Hydrogen-*d*$_2$	D_2	7782-39-0	4.028	col gas	-254.42	-249.48	0.164 g/L		
1243	Hydrogen-*t*$_2$	T_2	10028-17-8	6.032	col gas	-252.53	-248.11	0.246 g/L		
1244	Hydrogen-*d*$_1$	HD	13983-20-5	3.022	col gas	-256.55	-251.02	0.123 g/L		
1245	Hydrogen-*t*$_1$	HT	14885-60-0	4.024	col gas	-254.7	-249.6			
1246	Hydrogen-*d*$_1$,*t*$_1$	DT	14885-61-1	5.030	col gas	-253.5	-238.9			
1247	Hydrogen bromide	HBr	10035-10-6	80.912	col gas	-86.80	-66.38	3.307 g/L		vs H_2O; s EtOH
1248	Hydrogen bromide-*d*	DBr	13536-59-9	81.918	col gas	-87.54	-66.9			s H_2O

No.	Name	Formula	CAS Reg No.	Mol. weight	Physical form	mp/°C	bp/°C	Density g cm^{-3}	Solubility g/100 g H_2O	Qualitative solubility
1249	Hydrogen chloride	HCl	7647-01-0	36.461	col gas	-114.17	-85	1.490 g/L		vs H_2O
1250	Hydrogen chloride dihydrate	$HCl \cdot 2H_2O$	13465-05-9	72.492	col liq	-17.7		1.46		
1251	Hydrogen chloride-*d*	DCl	7698-05-7	37.467	col gas	-114.72	-84.4			s H_2O
1252	Hydrogen cyanide	HCN	74-90-8	27.026	col liq or gas	-13.29	26	0.6876^{20}		vs H_2O, EtOH; sl eth
1253	Hydrogen fluoride	HF	7664-39-3	20.006	col gas	-83.36	20	0.818 g/L		vs H_2O, EtOH; sl eth
1254	Hydrogen iodide	HI	10034-85-2	127.912	col or yel gas	-50.76	-35.55	5.228 g/L		vs H_2O; s os
1255	Hydrogen iodide-*d*	DI	14104-45-1	128.918	col gas	-51.93	-36.2			s H_2O
1256	Hydrogen peroxide	H_2O_2	7722-84-1	34.015	col liq	-0.43	150.2	1.44		vs H_2O
1257	Hydrogen selenide	H_2Se	7783-07-5	80.98	col gas; flam	-65.73	-41.25	3.310 g/L		s H_2O
1258	Hydrogen sulfide	H_2S	7783-06-4	34.081	col gas; flam	-85.5	-59.55	1.393 g/L		s H_2O
1259	Hydrogen disulfide	H_2S_2	13465-07-1	66.146	col liq		70.7	1.334		
1260	Hydrogen telluride	H_2Te	7783-09-7	129.62	col gas	-49	-2	5.298 g/L		s H_2O, EtOH, alk
1261	Indium	In	7440-74-6	114.818	soft wh metal	156.60	2072	7.31		s acid
1262	Indium antimonide	$InSb$	1312-41-0	236.578	blk cub cry	524		5.7747		
1263	Indium arsenide	$InAs$	1303-11-3	189.740	gray cub cry	942		5.67		i acid
1264	Indium nitride	InN	25617-98-5	128.825	brn hex cry	1100		6.88		
1265	Indium phosphide	InP	22398-80-7	145.792	blk cub cry	1062		4.81		sl acid
1266	Indium(I) bromide	$InBr$	14280-53-6	194.722	oran-red orth cry	285	656	4.96		reac H_2O
1267	Indium(I) chloride	$InCl$	13465-10-6	150.271	yel cub cry	225	608	4.19		reac H_2O
1268	Indium(I) iodide	InI	13966-94-4	241.722	orth cry	364.4	712	5.32		
1269	Indium(II) bromide	$InBr_2$	21264-43-7	274.626	orth cry			4.22		reac H_2O
1270	Indium(II) chloride	$InCl_2$	13465-11-7	185.724	col orth cry	235		3.64		reac H_2O
1271	Indium(II) sulfide	InS	12030-14-7	146.883	red-brn orth cry	692		5.2		
1272	Indium(III) bromide	$InBr_3$	13465-09-3	354.530	hyg yel-wh monocl cry	420		4.74	414^{20}	
1273	Indium(III) chloride	$InCl_3$	10025-82-8	221.177	yel monocl cry; hyg	583		4.0	195.1^{22}	s EtOH
1274	Indium(III) chloride tetrahydrate	$InCl_3 \cdot 4H_2O$	22519-64-8	293.239	wh cry					s H_2O
1275	Indium(III) fluoride	InF_3	7783-52-0	171.813	wh hex cry; hyg	1172	>1200	4.39		sl H_2O; s dil acid
1276	Indium(III) fluoride trihydrate	$InF_3 \cdot 3H_2O$	14166-78-0	225.859	wh cry	100 dec				s H_2O
1277	Indium(III) hydroxide	$In(OH)_3$	20661-21-6	165.840	cub cry			4.4		
1278	Indium(III) iodide	InI_3	13510-35-5	495.531	yel-red monocl cry; hyg	207		4.69	1308^{22}	
1279	Indium(III) nitrate trihydrate	$In(NO_3)_3 \cdot 3H_2O$	13770-61-1	354.879	col cry	dec 100				
1280	Indium(III) oxide	In_2O_3	1312-43-2	277.634	yel cub cry	1912		7.18		i H_2O; s hot acid
1281	Indium(III) perchlorate octahydrate	$In(ClO_4)_3 \cdot 8H_2O$	13465-15-1	557.292	wh cry	≈80	200 dec			
1282	Indium(III) phosphate	$InPO_4$	14693-82-4	209.789	wh orth cry			4.9		i H_2O
1283	Indium(III) selenide	In_2Se_3	1312-42-1	466.52	blk hex cry	660		5.8		
1284	Indium(III) sulfate	$In_2(SO_4)_3$	13464-82-9	517.824	hyg wh powder			3.44	117^{20}	
1285	Indium(III) sulfide	In_2S_3	12030-24-9	325.831	oran cub cry	1050		4.45		
1286	Indium(III) telluride	In_2Te_3	1312-45-4	612.44	blk cub cry	667		5.75		
1287	Iodine	I_2	7553-56-2	253.809	blue-blk plates	113.7	184.4	4.933	0.03^{20}	s bz, EtOH, eth, ctc, chl
1288	Iodic acid	HIO_3	7782-68-5	175.910	col orth cry	110 dec		4.63	308^{25}	i EtOH, eth
1289	Periodic acid dihydrate	$HIO_4 \cdot 2H_2O$	10450-60-9	227.940	monocl hyg cry	122 dec				s H_2O, EtOH; sl eth
1290	Iodine tetroxide	I_2O_4	12399-08-5	317.807	yel cry	130	dec >85	4.2		sl H_2O
1291	Iodine pentoxide	I_2O_5	12029-98-0	333.806	hyg wh cry	≈300 dec		4.98	253.4^{20}	i EtOH, eth, CS_2
1292	Iodine hexoxide	I_2O_6	65355-99-9	349.805	yel solid	dec 150				reac H_2O
1293	Iodine nonaoxide	I_4O_9	73560-00-6	651.613	hyg yel powder	75 dec				
1294	Iodine bromide	IBr	7789-33-5	206.808	blk orth cry	40	116 dec	4.3		s H_2O, EtOH, eth
1295	Iodine chloride	ICl	7790-99-0	162.357	red cry or oily liq	27.38	94.4 dec	3.24		reac H_2O; s EtOH
1296	Iodine trichloride	ICl_3	865-44-1	233.263	yel tricl cry; hyg	101 tp (16 atm)	64 sp dec	3.2		reac H_2O; s EtOH, bz
1297	Iodine fluoride	IF	13873-84-2	145.902	wh pow (-78°C)	-14 dec				
1298	Iodine trifluoride	IF_3	22520-96-3	183.899	yel solid, stable at low temp	-28 dec				
1299	Iodine pentafluoride	IF_5	7783-66-6	221.896	yel liq	9.43	100.5	3.19		reac H_2O
1300	Iodine heptafluoride	IF_7	16921-96-3	259.893	col gas	6.5 tp	4.8 sp	10.62 g/L		s H_2O
1301	Iodosyl trifluoride	IOF_3	19058-78-7	199.898	hyg col needles	dec >110				reac H_2O
1302	Iodosyl pentafluoride	IOF_5	16056-61-4	237.895	col liq	4.5				
1303	Iodyl trifluoride	IO_2F_3	25402-50-0	215.898	yel solid	41	subl			
1304	Periodyl fluoride	IO_3F	30708-86-2	193.900	col cry	dec >100				
1305	Iridium	Ir	7439-88-5	192.217	silv-wh metal; cub	2446	4428	22.562^{20}		s aqua regia
1306	Iridium carbonyl	$Ir_4(CO)_{12}$	11065-24-0	1104.989	yel cry	210 dec				
1307	Iridium(III) bromide	$IrBr_3$	10049-24-8	431.929	red-brn monocl cry			6.82		i H_2O, acid, alk

No.	Name	Formula	CAS Reg No.	Mol. weight	Physical form	mp/°C	bp/°C	Density g cm^{-3}	Solubility g/100 g H_2O	Qualitative solubility
1308	Iridium(III) bromide tetrahydrate	$IrBr_3 \cdot 4H_2O$	10049-24-8*	503.991	grn-brn cry					s H_2O; i EtOH
1309	Iridium(III) chloride	$IrCl_3$	10025-83-9	298.576	brn monocl cry	763 dec		5.30		i H_2O, acid, alk
1310	Iridium(III) fluoride	IrF_3	23370-59-4	249.212	blk hex cry	250 dec		≈8.0		i H_2O, dil acid
1311	Iridium(III) iodide	IrI_3	7790-41-2	572.930	dark brn monocl cry			≈7.4		i H_2O, acid, bz, chl; s alk
1312	Iridium(III) oxide	Ir_2O_3	1312-46-5	432.432	blue-blk cry	1000 dec				i H_2O; sl hot HCl
1313	Iridium(III) 2,4-pentanedioate	$Ir(CH_3COCHCOCH_3)_3$	15635-87-7	489.541	oran-yel cry	270	subl			sl H_2O; s tol, chl, ace, MeOH
1314	Iridium(III) sulfide	Ir_2S_3	12136-42-4	480.629	orth cry			10.2		
1315	Iridium(IV) chloride	$IrCl_4$	10025-97-5	334.029	brn hyg solid	≈700 dec				s H_2O, EtOH
1316	Iridium(IV) oxide	IrO_2	12030-49-8	224.216	brn tetr cry	1100 dec		11.7		
1317	Iridium(IV) sulfide	IrS_2	12030-51-2	256.347	orth cry			9.3		
1318	Iridium(VI) fluoride	IrF_6	7783-75-7	306.207	yel cub cry; hyg	44	53.6	4.8		reac H_2O
1319	Iron	Fe	7439-89-6	55.845	silv-wh or gray met	1538	2861	7.87		s dil acid
1320	Ferrocene	$Fe(C_5H_5)_2$	102-54-5	186.031	oran needles	172.5	249			i H_2O; s EtOH, eth, bz, dil HNO_3
1321	Tetracarbonyldihydroiron	$Fe(CO)_4H_2$	12002-28-7	169.902	col liq, stab low temp	-70	dec -20			s alk
1322	Iron pentacarbonyl	$Fe(CO)_5$	13463-40-6	195.896	yel oily liq; flam	-20.5	103	1.46		i H_2O; s eth, bz, ace
1323	Iron nonacarbonyl	$Fe_2(CO)_9$	15321-51-4	363.781	oran-yel cry	100 dec		2.85		
1324	Iron dodecacarbonyl	$Fe_3(CO)_{12}$	12088-65-2	503.656	dark grn cry	140		2.00		
1325	Iron arsenide	FeAs	12044-16-5	130.767	gray orth cry	1030		7.85		
1326	Iron boride (FeB)	FeB	12006-84-7	66.656	refrac solid; orth	1658		≈7		
1327	Iron boride (Fe_2B)	Fe_2B	12006-86-9	122.501	refrac solid; tetr	1389		7.3		
1328	Iron carbide	Fe_3C	12011-67-5	179.546	gray cub cry	1227		7.694		
1329	Iron phosphide (FeP)	FeP	26508-33-8	86.819	rhom cry			6.07		
1330	Iron phosphide (Fe_2P)	Fe_2P	1310-43-6	142.664	gray hex needles	1370		6.8		i H_2O, dil acid, alk
1331	Iron phosphide (Fe_3P)	Fe_3P	12023-53-9	198.509	gray solid	1100		6.74		i H_2O
1332	Iron disulfide	FeS_2	1317-66-4	119.975	blk cub cry	>600 dec		5.02		i H_2O
1333	Iron silicide	FeSi	12022-95-6	83.931	gray cub cry	1410		6.1		
1334	Iron disilicide	$FeSi_2$	12022-99-0	112.016	gray tetr cry	1220		4.74		
1335	Iron(II) acetate	$Fe(C_2H_3O_2)_2$	3094-87-9	173.934	wh cry	190 dec				s H_2O
1336	Iron(II) acetate tetrahydrate	$Fe(C_2H_3O_2)_2 \cdot 4H_2O$	3094-87-9*	245.994	grn cry	dec				s H_2O, EtOH
1337	Iron(II) aluminate	$Fe(AlO_2)_2$	12068-49-4	173.806	blk cub cry			4.3		
1338	Iron(II) arsenate	$Fe_3(AsO_4)_2$	10102-50-8	445.373	grn powder					i H_2O
1339	Iron(II) arsenate hexahydrate	$Fe_3(AsO_4)_2 \cdot 6H_2O$	10102-50-8*	553.465	grn amorp powder	dec				i H_2O; s acid
1340	Iron(II) bromide	$FeBr_2$	7789-46-0	215.653	yel-brn hex cry; hyg	691	dec	4.636	120^{25}	vs EtOH
1341	Iron(II) bromide hexahydrate	$FeBr_2 \cdot 6H_2O$	13463-12-2	323.744	grn hyg cry	27 dec		4.64	120^{25}	s EtOH
1342	Iron(II) carbonate	$FeCO_3$	563-71-3	115.854	gray-brn hex cry			3.944	0.000062^{20}	
1343	Iron(II) chloride	$FeCl_2$	7758-94-3	126.751	wh hex cry; hyg	677	1023	3.16	65.0^{25}	vs EtOH, ace; sl bz
1344	Iron(II) chloride dihydrate	$FeCl_2 \cdot 2H_2O$	16399-77-2	162.782	wh-grn monocl cry	120 dec		2.39	65.0^{25}	
1345	Iron(II) chloride tetrahydrate	$FeCl_2 \cdot 4H_2O$	13478-10-9	198.813	grn monocl cry	105 dec		1.93	65.0^{25}	s EtOH
1346	Iron(II) chromite	$FeCr_2O_4$	1308-31-2	223.835	blk cub cry			5.0		
1347	Iron(II) fluoride	FeF_2	7789-28-8	93.842	wh tetr cry	1100		4.09		sl H_2O; s dil HF; i EtOH, eth
1348	Iron(II) fluoride tetrahydrate	$FeF_2 \cdot 4H_2O$	13940-89-1	165.904	col hex cry			2.20		
1349	Iron(II) hydroxide	$Fe(OH)_2$	18624-44-7	89.860	wh-grn hex cry			3.4	0.000052^{20}	
1350	Iron(II) iodide	FeI_2	7783-86-0	309.654	red-viol hex cry; hyg	594		5.3		s H_2O, EtOH, eth
1351	Iron(II) iodide tetrahydrate	$FeI_2 \cdot 4H_2O$	7783-86-0*	381.716	blk hyg leaflets	90 dec		2.87		s H_2O, EtOH
1352	Iron(II) molybdate	$FeMoO_4$	13718-70-2	215.78	brn-yel monocl cry	1115		5.6		i H_2O
1353	Iron(II) nitrate	$Fe(NO_3)_2$	14013-86-6	179.854	grn solid				87.5^{25}	
1354	Iron(II) nitrate hexahydrate	$Fe(NO_3)_2 \cdot 6H_2O$	14013-86-6*	287.946	grn solid	60 dec			87.5^{25}	
1355	Iron(II) orthosilicate	Fe_2SiO_4	10179-73-4	203.774	brn orth cry			4.30		
1356	Iron(II) oxalate dihydrate	$FeC_2O_4 \cdot 2H_2O$	6047-25-2	179.894	yel cry	150 dec		2.28	0.078^{25}	s acid
1357	Iron(II) oxide	FeO	1345-25-1	71.844	blk cub cry	1377		6.0		i H_2O, alk; s acid
1358	Iron(II) 2,4-pentanedioate	$Fe(CH_3COCHCOCH_3)_2$	14024-17-0	254.061	oran-brn cry	170	subl			sl bz, tol
1359	Iron(II) perchlorate	$Fe(ClO_4)_2$	13933-23-8	254.746	grn-wh hyg needles	>100 dec			210^{25}	
1360	Iron(II) phosphate octahydrate	$Fe_3(PO_4)_2 \cdot 8H_2O$	14940-41-1	501.600	gray-blue monocl cry; hyg			2.58		i H_2O; s acid
1361	Iron(II) selenide	FeSe	1310-32-3	134.81	blk hex cry			6.7		i H_2O
1362	Iron(II) sulfate	$FeSO_4$	7720-78-7	151.908	wh orth cry; hyg			3.65	29.5^{25}	
1363	Iron(II) sulfate monohydrate	$FeSO_4 \cdot H_2O$	17375-41-6	169.923	wh-yel monocl cry	300 dec		3.0	29.5^{25}	

No.	Name	Formula	CAS Reg No.	Mol. weight	Physical form	mp/°C	bp/°C	Density g cm^{-3}	Solubility g/100 g H_2O	Qualitative solubility
1364	Iron(II) sulfate heptahydrate	$FeSO_4 \cdot 7H_2O$	7782-63-0	278.014	blue-grn monocl cry	≈60 dec		1.895	29.5^{25}	i EtOH
1365	Iron(II) sulfide	FeS	1317-37-9	87.910	col hex or tetr cry; hyg	1188	dec	4.7		i H_2O; reac acid
1366	Iron(II) tantalate	$Fe(TaO_3)_2$	12140-41-9	513.737	brn tetr cry			7.33		
1367	Iron(II) tartrate	$FeC_4H_4O_6$		203.916	wh cry				0.88	vs acid; s NH_4OH
1368	Iron(II) telluride	$FeTe$	12125-63-2	183.45	tetr cry	914		6.8		
1369	Iron(II) thiocyanate trihydrate	$Fe(SCN)_2 \cdot 3H_2O$	6010-09-9	226.055	grn monocl cry					s H_2O, EtOH, eth
1370	Iron(II) titanate	$FeTiO_3$	12168-52-4	151.710	blk rhomb cry	≈1470		4.72		
1371	Iron(II) tungstate	$FeWO_4$	13870-24-1	303.68	monocl cry			7.51		
1372	Iron(II,III) oxide	Fe_3O_4	1317-61-9	231.533	blk cub cry or amorp powder	1597		5.17		i H_2O; s acid
1373	Iron(III) acetate, basic	$FeOH(C_2H_3O_2)_2$	10450-55-2	190.941	brn-red amorp powder					i H_2O; s EtOH, acid
1374	Iron(III) ammonium citrate	$Fe(NH_4)_3(C_6H_5O_7)_2$	1185-57-5	488.160	red or brn pow; hyg					s H_2O; i EtOH
1375	Iron(III) arsenate dihydrate	$FeAsO_4 \cdot 2H_2O$	10102-49-5	230.795	grn-brn powder	dec		3.18		i H_2O; s dil acid
1376	Iron(III) bromide	$FeBr_3$	10031-26-2	295.557	dark red hex cry; hyg	dec		4.5	455^{25}	s EtOH, eth
1377	Iron(III) chloride	$FeCl_3$	7705-08-0	162.204	grn hex cry; hyg	307.6	≈316	2.90	91.2^{25}	s EtOH, eth, ace
1378	Iron(III) chloride hexahydrate	$FeCl_3 \cdot 6H_2O$	10025-77-1	270.295	yel-oran monocl cry; hyg	37 dec		1.82	91.2^{25}	s EtOH, eth, ace
1379	Iron(III) chromate	$Fe_2(CrO_4)_3$	10294-52-7	459.671	yel powder					i H_2O, EtOH; s acid
1380	Iron(III) citrate pentahydrate	$FeC_6H_5O_7 \cdot 5H_2O$	3522-50-7	335.021	red-brn cry					s H_2O; i EtOH
1381	Iron(III) dichromate	$Fe_2(Cr_2O_7)_3$	10294-53-8	759.654	red-brn solid					s H_2O, acid
1382	Iron(III) ferrocyanide	$Fe_4[Fe(CN)_6]_3$	14038-43-8	859.229	dark blue powder			1.80		i H_2O, dil acid, os
1383	Iron(III) fluoride	FeF_3	7783-50-8	112.840	grn hex cry	>1000		3.87	5.92^{25}	i EtOH, eth, bz
1384	Iron(III) fluoride trihydrate	$FeF_3 \cdot 3H_2O$	15469-38-2	166.886	yel-brn tetr cry			2.3	5.92^{25}	
1385	Iron(III) formate	$Fe(CHO_2)_3$	555-76-0	190.897	red-yel cry powder					s H_2O; sl EtOH
1386	Iron(III) hydroxide	$Fe(OH)_3$	1309-33-7	106.867	yel monocl cry			3.12		
1387	Iron(III) hydroxide oxide	$FeO(OH)$	20344-49-4	88.852	red-brn orth cry			4.26		i H_2O; s acid
1388	Iron(III) metavanadate	$Fe(VO_3)_3$	65842-03-7	352.665	gray-brn powder					i H_2O, EtOH; s acid
1389	Iron(III) nitrate	$Fe(NO_3)_3$	10421-48-4	241.860	cry				82.5^{20}	
1390	Iron(III) nitrate hexahydrate	$Fe(NO_3)_3 \cdot 6H_2O$	13476-08-9	349.951	viol cub cry	35 dec			82.5^{20}	
1391	Iron(III) nitrate nonahydrate	$Fe(NO_3)_3 \cdot 9H_2O$	7782-61-8	403.997	viol-gray hyg cry	47 dec		1.68	82.5^{20}	vs EtOH, ace
1392	Iron(III) oxalate	$Fe_2(C_2O_4)_3$	19469-07-9	375.747	yel amorp powder	100 dec				s H_2O, acid; i alk
1393	Iron(III) oxide	Fe_2O_3	1309-37-1	159.688	red-brn hex cry	1539		5.25		i H_2O; s acid
1394	Iron(III) 2,4-pentanedioate	$Fe(CH_3COCHCOCH_3)_3$	14024-18-1	353.169	red-oran cry	179		5.24		sl H_2O; s os
1395	Iron(III) perchlorate hexahydrate	$Fe(ClO_4)_3 \cdot 6H_2O$	32963-81-8	462.288	viol cry					
1396	Iron(III) phosphate dihydrate	$FePO_4 \cdot 2H_2O$	10045-86-0	186.847	gray-wh orth cry			2.87		i H_2O; s HCl
1397	Iron(III) pyrophosphate nonahydrate	$Fe_4(P_2O_7)_3 \cdot 9H_2O$	10058-44-3	907.348	yel powder					i H_2O; s acid
1398	Iron(III) hypophosphite	$Fe(H_2PO_2)_3$	7783-84-8	250.811	wh-gray powder					i H_2O
1399	Iron(III) sodium pyrophosphate	$FeNaP_2O_7$	10045-87-1	252.778	wh powder			1.5		i H_2O; s HCl
1400	Iron(III) sulfate	$Fe_2(SO_4)_3$	10028-22-5	399.878	gray-wh rhomb cry; hyg			3.10	440^{20}	sl EtOH; i ace
1401	Iron(III) sulfate nonahydrate	$Fe_2(SO_4)_3 \cdot 9H_2O$	13520-56-4	562.015	yel hex cry	400 dec		2.1	440^{20}	
1402	Iron(III) thiocyanate	$Fe(SCN)_3$	4119-52-2	230.092	red-viol hyg cry	dec				s H_2O, EtOH, ace; i tol, chl
1403	Krypton	Kr	7439-90-9	83.798	col gas	-157.38 tp (73.2 kPa)	-153.34	3.425 g/L		sl H_2O
1404	Krypton difluoride	KrF_2	13773-81-4	121.795	col tetr cry	≈25 dec		3.24		reac H_2O
1405	Krypton fluoride hexafluoroantimonate	$KrFSb_2F_{11}$	39578-36-4	555.299	wh solid	dec 45				
1406	Lanthanum	La	7439-91-0	138.905	silv metal; hex	920	3464	6.15		s dil acid
1407	Lanthanum aluminum oxide	$LaAlO_3$	12003-65-5	213.885	wh rhom cry	trans cub 500				
1408	Lanthanum boride	LaB_6	12008-21-8	203.771	blk cub cry; refrac	2715		4.76		
1409	Lanthanum bromate nonahydrate	$LaBrO_3 \cdot 9H_2O$		684.749	hex cry	dec 100		5.06		vs H_2O
1410	Lanthanum bromide	$LaBr_3$	13536-79-3	378.617	wh hex cry; hyg	788		5.1		s H_2O
1411	Lanthanum carbide	LaC_2	12071-15-7	162.926	tetr cry	2360		5.29		
1412	Lanthanum carbonate octahydrate	$La_2(CO_3)_3 \cdot 8H_2O$	6487-39-4	601.960	wh cry powder			2.6		i H_2O; s dil acid
1413	Lanthanum chloride	$LaCl_3$	10099-58-8	245.264	wh hex cry; hyg	858		3.84	95.7^{25}	
1414	Lanthanum chloride heptahydrate	$LaCl_3 \cdot 7H_2O$	20211-76-1	371.371	wh tricl cry; hyg	91 dec			95.7^{25}	s EtOH
1415	Lanthanum fluoride	LaF_3	13709-38-1	195.900	wh hex cry; hyg	1493		5.9		i H_2O, acid
1416	Lanthanum hydride	LaH_3	13864-01-2	141.929	blk cub cry			5.36		
1417	Lanthanum hydroxide	$La(OH)_3$	14507-19-8	189.927	wh amorp solid	dec			0.000020^{20}	
1418	Lanthanum iodate	$La(IO_3)_3$	13870-19-4	663.614	col cry				1.7	
1419	Lanthanum iodide	LaI_3	13813-22-4	519.619	wh orth cry; hyg	778		5.6		s H_2O
1420	Lanthanum nitrate hexahydrate	$La(NO_3)_3 \cdot 6H_2O$	10277-43-7	433.011	wh hyg tricl cry	≈40 dec			200^{25}	vs EtOH; s ace

No.	Name	Formula	CAS Reg No.	Mol. weight	Physical form	mp/°C	bp/°C	Density g cm^{-3}	Solubility g/100 g H_2O	Qualitative solubility
1421	Lanthanum nitride	LaN	25764-10-7	152.912	cub cry			6.73		
1422	Lanthanum oxide	La_2O_3	1312-81-8	325.809	wh amorp powder	2304	3620	6.51		i H_2O; s dil acid
1423	Lanthanum perchlorate hexahydrate	$La(ClO_4)_3 \cdot 6H_2O$	36907-37-6	475.021	hyg col cry	dec 100				vs H_2O; s EtOH
1424	Lanthanum silicide	$LaSi_2$	12056-90-5	195.076	gray tetr cry	1520		5.0		
1425	Lanthanum sulfate	$La_2(SO_4)_3$	10099-60-2	565.999	hyg wh pow	1150				sl H_2O
1426	Lanthanum sulfate octahydrate	$La_2(SO_4)_3 \cdot 8H_2O$	57804-25-8	702.058	col cry	dec		2.82		sl H_2O
1427	Lanthanum sulfate nonahydrate	$La_2(SO_4)_3 \cdot 9H_2O$	10294-62-9	728.136	hex cry			2.82	2.7^{20}	i EtOH
1428	Lanthanum monosulfide	LaS	12031-30-0	170.970	yel cub cry	2300		5.61		
1429	Lanthanum sulfide	La_2S_3	12031-49-1	374.006	red cub cry	2110		4.9		
1430	Lawrencium	Lr	22537-19-5	262	metal	1627				
1431	Lead	Pb	7439-92-1	207.2	soft silv-gray metal; cub	327.462	1749	11.3		s conc acid
1432	Plumbane	PbH_4	15875-18-0	211.2	unstab col gas		-13			
1433	Lead(II) acetate	$Pb(C_2H_3O_2)_2$	301-04-2	325.3	wh cry	280	dec	3.25	44.3^{20}	
1434	Lead(II) acetate trihydrate	$Pb(C_2H_3O_2)_2 \cdot 3H_2O$	6080-56-4	379.3	col cry	75 dec		2.55		vs H_2O; sl EtOH
1435	Lead(II) acetate, basic	$Pb(C_2H_3O_2)_2 \cdot 2Pb(OH)_2$	1335-32-6	807.7	wh powder	dec			6.3^{0}	
1436	Lead(II) antimonate	$Pb_3(SbO_4)_2$	13510-89-9	993.1	oran-yel powder			6.58		i H_2O, dil acid
1437	Lead(II) arsenate	$Pb_3(AsO_4)_2$	3687-31-8	899.4	wh cry	1042 dec		5.8		i H_2O; s HNO_3
1438	Lead(II) arsenite	$Pb(AsO_2)_2$	10031-13-7	421.0	wh powder			5.85		i H_2O; s dil HNO_3
1439	Lead(II) azide	$Pb(N_3)_2$	13424-46-9	291.2	col orth needles; exp	exp ≈350		4.7	0.023^{18}	vs HOAc
1440	Lead(II) borate monohydrate	$Pb(BO_2)_2 \cdot H_2O$	10214-39-8	310.8	wh powder	500 dec		5.6		i H_2O; s dil HNO_3
1441	Lead(II) bromate monohydrate	$Pb(BrO_3)_2 \cdot H_2O$	10031-21-7	481.0	col cry	≈180 dec		5.53	1.33^{20}	
1442	Lead(II) bromide	$PbBr_2$	10031-22-8	367.0	wh orth cry	371	892	6.69	0.975^{25}	i EtOH
1443	Lead(II) butanoate	$Pb(C_4H_7O_2)_2$	819-73-8	381.4	col solid	≈90				i H_2O; s dil HNO_3
1444	Lead(II) carbonate	$PbCO_3$	598-63-0	267.2	col orth cry	≈315 dec		6.582		i H_2O
1445	Lead(II) carbonate, basic	$Pb(OH)_2 \cdot 2PbCO_3$	1319-46-6	775.6	wh hex cry	400 dec		≈6.5		i H_2O, EtOH; s acid
1446	Lead(II) chlorate	$Pb(ClO_3)_2$	10294-47-0	374.1	col hyg cry	230 dec		3.9	144^{18}	vs EtOH
1447	Lead(II) chloride	$PbCl_2$	7758-95-4	278.1	wh orth needles or powder	501	951	5.98	1.08^{25}	s alk
1448	Lead(II) chloride fluoride	PbClF	13847-57-9	261.7	tetr cry			7.05	0.035^{20}	
1449	Lead(II) chlorite	$Pb(ClO_2)_2$	13453-57-1	342.1	yel monocl cry	dec 126			0.2^{25}	sl H_2O; s alk
1450	Lead(II) chromate	$PbCrO_4$	7758-97-6	323.2	yel-oran monocl cry	844		6.12	0.000017^{20}	s alk, dil acid
1451	Lead(II) chromate(VI) oxide	$PbCrO_4 \cdot PbO$	18454-12-1	546.4	red powder					i H_2O
1452	Lead(II) citrate trihydrate	$Pb_3(C_6H_5O_7)_2 \cdot 3H_2O$	512-26-5	1053.8	wh cry powder					s H_2O; sl EtOH
1453	Lead(II) cyanide	$Pb(CN)_2$	592-05-2	259.2	wh-yel powder					sl H_2O; reac acid
1454	Lead(II) 2-ethylhexanoate	$Pb(C_7H_{15}CO_2)_2$	301-08-6	493.6	visc liq			1.56		
1455	Lead(II) fluoride	PbF_2	7783-46-2	245.2	wh orth cry	830	1293	8.44	0.0670^{25}	
1456	Lead(II) fluoroborate	$Pb(BF_4)_2$	13814-96-5	380.8	stab only in aq soln					s H_2O
1457	Lead(II) formate	$Pb(CHO_2)_2$	811-54-1	297.2	wh prisms or needles	190 dec		4.63	1.6^{16}	i EtOH
1458	Lead(II) hexafluoro-2,4-pentanedioate	$Pb(CF_3COCHCOCF_3)_2$	19648-88-5	621.3	cry	155	210			
1459	Lead(II) hydrogen arsenate	$PbHAsO_4$	7784-40-9	347.1	wh monocl cry	280 dec		5.943		i H_2O; s HNO_3, alk
1460	Lead(II) hydrogen phosphate	$PbHPO_4$	15845-52-0	303.2	wh monocl cry	dec		5.66		
1461	Lead(II) hydroxide	$Pb(OH)_2$	19783-14-3	241.2	wh powder	145 dec		5.69	0.00012^{20}	s acid
1462	Lead(II) iodate	$Pb(IO_3)_2$	25659-31-8	557.0	wh orth cry			6.50	0.0025^{25}	
1463	Lead(II) iodide	PbI_2	10101-63-0	461.0	yel hex cry or powder	410	872 dec	6.16	0.076^{25}	i EtOH
1464	Lead(II) lactate	$Pb(C_3H_5O_3)_2$	18917-82-3	385.3	wh cry powder					s H_2O, hot EtOH
1465	Lead(II) molybdate	$PbMoO_4$	10190-55-3	367.1	yel tetr cry	≈1060		6.7		i H_2O; s HNO_3, NaOH
1466	Lead(II) niobate	$Pb(NbO_3)_2$	12034-88-7	489.0	rhom or tetr cry	1343		6.6		i H_2O
1467	Lead(II) nitrate	$Pb(NO_3)_2$	10099-74-8	331.2	col cub cry	470		4.53	59.7^{25}	sl EtOH
1468	Lead(II) oleate	$Pb(C_{18}H_{33}O_2)_2$	1120-46-3	770.1	wax-like solid					i H_2O; s EtOH, bz, eth
1469	Lead(II) oxalate	PbC_2O_4	814-93-7	295.2	wh powder	300 dec		5.28	0.00025^{20}	s dil HNO_3
1470	Lead(II) oxide (litharge)	PbO	1317-36-8	223.2	red tetr cry	trans to massicot 489		9.35		i H_2O, EtOH; s dil HNO_3
1471	Lead(II) oxide (massicot)	PbO	1317-36-8	223.2	yel orth cry	887		9.64		i H_2O, EtOH; s dil HNO_3
1472	Lead(II) oxide hydrate	$3PbO \cdot H_2O$	1311-11-1	687.6	wh powder			7.41		i H_2O; s dil acid
1473	Lead(II) 2,4-pentanedioate	$Pb(CH_3COCHCOCH_3)_2$	15282-88-9	405.4	cry	143				
1474	Lead(II) perchlorate	$Pb(ClO_4)_2$	13453-62-8	406.1	wh cry				441^{25}	
1475	Lead(II) perchlorate trihydrate	$Pb(ClO_4)_2 \cdot 3H_2O$	13637-76-8	460.1	wh cry	100 dec		2.6	441^{25}	s EtOH

No.	Name	Formula	CAS Reg No.	Mol. weight	Physical form	mp/°C	bp/°C	Density g cm⁻³	Solubility g/100 g H_2O	Qualitative solubility
1476	Lead(II) phosphate	$Pb_3(PO_4)_2$	7446-27-7	811.5	wh hex cry	1014		7.01		i H_2O, EtOH
1477	Lead(II) hypophosphite	$Pb(H_2PO_2)_2$	10294-58-3	337.2	hyg cry powder	dec				sl H_2O; i EtOH
1478	Lead(II) metasilicate	$PbSiO_3$	10099-76-0	283.3	wh monocl cry powder	764		6.49		i H_2O, os
1479	Lead(II) orthosilicate	Pb_2SiO_4	13566-17-1	506.5	monocl cry	743		7.60		
1480	Lead(II) hexafluorosilicate dihydrate	$PbSiF_6 \cdot 2H_2O$	1310-03-8	385.3	col cry	dec				vs H_2O
1481	Lead(II) selenate	$PbSeO_4$	7446-15-3	350.2	orth cry			6.37	0.013^{25}	s conc acid
1482	Lead(II) selenide	PbSe	12069-00-0	286.2	gray cub cry	1078		8.1		i H_2O; s HNO_3
1483	Lead(II) selenite	$PbSeO_3$	7488-51-9	334.2	wh monocl cry	≈500		7.0		i H_2O
1484	Lead(II) sodium thiosulfate	$Na_4Pb(S_2O_3)_3$	10101-94-7	635.5	wh cry					sl H_2O
1485	Lead(II) stearate	$Pb(C_{18}H_{35}O_2)_2$	1072-35-1	774.1	wh powder	≈100		1.4		i H_2O; s hot EtOH
1486	Lead(II) sulfate	$PbSO_4$	7446-14-2	303.3	orth cry	1087		6.29	0.0044^{25}	i acid; sl alk
1487	Lead(II) sulfide	PbS	1314-87-0	239.3	blk powder or silv cub cry	1113		7.60		i H_2O; s acid
1488	Lead(II) sulfite	$PbSO_3$	7446-10-8	287.3	wh powder	dec				i H_2O; s HNO_3
1489	Lead(II) tantalate	$Pb(TaO_3)_2$	12065-68-8	665.1	orth cry			7.9		i H_2O
1490	Lead(II) telluride	PbTe	1314-91-6	334.8	gray cub cry	924		8.164		i H_2O, acid
1491	Lead(II) thiocyanate	$Pb(SCN)_2$	592-87-0	323.4	wh-yel powder			3.82	0.05^{20}	
1492	Lead(II) thiosulfate	PbS_2O_3	13478-50-7	319.3	wh cry	dec		5.18		i H_2O; s acid
1493	Lead(II) titanate	$PbTiO_3$	12060-00-3	303.1	yel tetr cry			7.9		i H_2O; reac HCl
1494	Lead(II) tungstate (stolzite)	$PbWO_4$	7759-01-5	455.0	yel tetr cry	1130		8.24	0.03^{20}	s alk
1495	Lead(II) tungstate (raspite)	$PbWO_4$	7759-01-5	455.0	monocl cry	trans 400		8.46	0.03^{20}	s alk
1496	Lead(II) metavanadate	$Pb(VO_3)_2$	10099-79-3	405.1	yel powder					i H_2O; reac HNO_3
1497	Lead(II) zirconate	$PbZrO_3$	12060-01-4	346.4	col orth cry			≈8		i H_2O, alk; s acid
1498	Lead(II,IV) oxide	Pb_2O_3	1314-27-8	462.4	blk monocl cry or red amorp powder	530 dec		10.05		i H_2O; s alk; reac conc HCl
1499	Lead(II,II,IV) oxide	Pb_3O_4	1314-41-6	685.6	red tetr cry	830		8.92		i H_2O, EtOH; s hot HCl
1500	Lead(IV) acetate	$Pb(C_2H_3O_2)_4$	546-67-8	443.4	col monocl cry	≈175		2.23		reac H_2O, EtOH; s bz, chl
1501	Lead(IV) bromide	$PbBr_4$	13701-91-2	526.8	unstab liq					
1502	Lead(IV) chloride	$PbCl_4$	13463-30-4	349.0	yel oily liq	-15	≈50 dec			
1503	Lead(IV) fluoride	PbF_4	7783-59-7	283.2	wh tetr cry; hyg	≈600		6.7		
1504	Lead(IV) oxide	PbO_2	1309-60-0	239.2	red tetr cry or brn powder	290 dec		9.64		
1505	Lithium	Li	7439-93-2	6.941	soft silv-wh metal	180.50	1342	0.534		reac H_2O
1506	Lithium acetate	$LiC_2H_3O_2$	546-89-4	65.985	cry	286			45.0^{25}	vs EtOH
1507	Lithium acetate dihydrate	$LiC_2H_3O_2 \cdot 2H_2O$	6108-17-4	102.016	wh rhomb cry	58 dec		1.3	45.0^{25}	s EtOH
1508	Lithium aluminum hydride	$LiAlH_4$	16853-85-3	37.955	gray-wh monocl cry	>125 dec		0.917		reac H_2O, EtOH; s eth, thf
1509	Lithium aluminum silicate	$LiAlSi_2O_6$	12068-40-5	186.090	wh monocl cry	1430		3.188		
1510	Lithium amide	$LiNH_2$	7782-89-0	22.964	tetr cry	380		1.18		reac H_2O
1511	Lithium arsenate	Li_3AsO_4	13478-14-3	159.743	col orth cry			3.07		sl H_2O; s HOAc
1512	Lithium azide	LiN_3	19597-69-4	48.961	hyg monocl cry; exp			1.83		vs H_2O
1513	Lithium borohydride	$LiBH_4$	16949-15-8	21.784	wh-gray orth cry or powder	268	380 dec	0.66		s alk, eth, thf
1514	Lithium bromate	$LiBrO_3$	13550-28-2	134.843	hyg col orth cry	260			65.4^{25}	vs H_2O
1515	Lithium bromide	LiBr	7550-35-8	86.845	wh cub cry; hyg	550	≈1300	3.464	181^{25}	s EtOH, eth
1516	Lithium bromide monohydrate	$LiBr \cdot H_2O$	23303-71-1	104.860	wh orth cry	trans cub 33		3.46	145^{4}	vs H_2O
1517	Lithium carbide	Li_2C_2	1070-75-3	37.903	wh hyg cry			1.65		reac H_2O; i os
1518	Lithium carbonate	Li_2CO_3	554-13-2	73.891	wh monocl cry	732	1300 dec	2.11	1.30^{25}	s acid; i EtOH
1519	Lithium chlorate	$LiClO_3$	13453-71-9	90.392	col hyg rhom needles	127.6	300 dec	1.119	459^{25}	vs EtOH; sl ace
1520	Lithium chloride	LiCl	7447-41-8	42.394	wh cub cry or powder; hyg	610	1383	2.07	84.5^{25}	s EtOH, ace, py
1521	Lithium chloride monohydrate	$LiCl \cdot H_2O$	16712-20-2	60.409	hyg wh tetr cry	dec 98		1.78	45.9^{25}	vs H_2O
1522	Lithium chromate dihydrate	$Li_2CrO_4 \cdot 2H_2O$	7789-01-7	165.906	yel orth cry; hyg	75 dec		2.15		vs H_2O; s EtOH
1523	Lithium citrate tetrahydrate	$Li_3C_6H_5O_7 \cdot 4H_2O$	6680-58-6	281.983	wh cry	210 (anh)			75^{25}	vs H_2O; sl EtOH
1524	Lithium cobaltite	$LiCoO_2$	12190-79-3	97.873	dark gray pow					i H_2O
1525	Lithium cyanide	LiCN	2408-36-8	32.959	wh orth cry	160				
1526	Lithium hydride-*d*	LiD	13587-16-1	8.955	hyg wh cry	680		0.82		reac H_2O
1527	Lithium dichromate dihydrate	$Li_2Cr_2O_7 \cdot 2H_2O$	10022-48-7	265.901	yel-red hyg cry	130 dec		2.34		vs H_2O
1528	Lithium dihydrogen phosphate	LiH_2PO_4	13453-80-0	103.928	col hyg cry	>100		2.461	126^{0}	
1529	Lithium diisopropylamide	$LiN(C_3H_7)_2$	4111-54-0	107.123	hyg col cry	dec				s eth; i hc
1530	Lithium ferrosilicon	LiFeSi	64082-35-5	90.872	dark brittle cry					reac H_2O

No.	Name	Formula	CAS Reg No.	Mol. weight	Physical form	mp/°C	bp/°C	Density g cm^{-3}	Solubility g/100 g H_2O	Qualitative solubility
1531	Lithium fluoride	LiF	7789-24-4	25.939	wh cub cry or powder	848.2	1673	2.640	0.134^{25}	s acid
1532	Lithium formate monohydrate	$Li(CHO_2) \cdot H_2O$	6108-23-2	69.974	col-wh cry			1.46		s H_2O
1533	Lithium hexafluoroantimonate	$LiSbF_6$	18424-17-4	242.691	hyg pow	dec				
1534	Lithium hexafluoroarsenate	$LiAsF_6$	29935-35-1	310.672	rhom wh cry; hyg					
1535	Lithium hexafluorophosphate	$LiPF_6$	21324-40-3	151.905	wh pow					
1536	Lithium hexafluorosilicate	Li_2SiF_6	17347-95-4	155.958	col hex cry	dec 350				sl ace
1537	Lithium hexafluorostannate	Li_2SnF_6	17029-16-2	246.582	wh pow					
1538	Lithium hydride	LiH	7580-67-8	7.949	gray cub cry or powder; hyg	692		0.78		reac H_2O, EtOH
1539	Lithium hydrogen carbonate	$LiHCO_3$	5006-97-3	67.958	wh pow					sl H_2O
1540	Lithium hydroxide	$LiOH$	1310-65-2	23.948	col tetr cry	473	1626	1.45	12.5^{25}	sl EtOH
1541	Lithium hydroxide monohydrate	$LiOH \cdot H_2O$	1310-66-3	41.964	wh monocl cry or powder			1.51	12.5^{25}	sl EtOH
1542	Lithium hypochlorite	$LiOCl$	13840-33-0	58.393	wh pow					vs H_2O
1543	Lithium iodate	$LiIO_3$	13765-03-2	181.843	wh hyg hex cry	450		4.502	77.9^{25}	i EtOH
1544	Lithium iodide	LiI	10377-51-2	133.845	wh cub cry; hyg	469	1171	4.06	165^{25}	
1545	Lithium iodide trihydrate	$LiI \cdot 3H_2O$	7790-22-9	187.891	wh hyg cry	73		2.38	165^{25}	vs EtOH, ace
1546	Lithium manganate	Li_2MnO_3	12163-00-7	116.818	red-brn monocl cry			3.90		i H_2O
1547	Lithium metaborate	$LiBO_2$	13453-69-5	49.751	wh monocl cry; hyg	844		2.18	2.6^{20}	sl H_2O; s EtOH
1548	Lithium metaborate dihydrate	$LiBO_2 \cdot 2H_2O$	15293-74-0	85.782	wh cry pow			1.8		s H_2O
1549	Lithium metaphosphate	$LiPO_3$	13762-75-9	85.913	wh cry or gl solid			1.8		i H_2O
1550	Lithium metasilicate	Li_2SiO_3	10102-24-6	89.966	wh orth needles	1201		2.52		i cold H_2O; reac dil acid
1551	Lithium molybdate	Li_2MoO_4	13568-40-6	173.82	hyg wh cry	702		2.66	44.8^{25}	s H_2O
1552	Lithium niobate	$LiNbO_3$	12031-63-9	147.845	wh hex cry	1240		4.30		
1553	Lithium nitrate	$LiNO_3$	7790-69-4	68.946	col hex cry; hyg	253		2.38	102^{25}	s EtOH
1554	Lithium nitride	Li_3N	26134-62-3	34.830	red hex cry	813		1.27		reac H_2O
1555	Lithium nitrite	$LiNO_2$	13568-33-7	52.947	wh hyg cry	222				vs H_2O
1556	Lithium nitrite monohydrate	$LiNO_2 \cdot H_2O$	13568-33-7*	70.962	col needles	>100		1.615	139.5^{25}	vs H_2O, EtOH
1557	Lithium orthosilicate	$LiSiO_4$	13453-84-4	99.025	wh rhom cry	1256		2.39		
1558	Lithium oxalate	$Li_2C_2O_4$	30903-87-8	101.901	col cry	dec		2.121^{17}		s H_2O; i EtOH, eth
1559	Lithium phosphate	Li_3PO_4	10377-52-3	115.794	wh orth cry	1205		2.46	0.027^{25}	
1560	Lithium oxide	Li_2O	12057-24-8	29.881	wh cub cry	1438		2.013		
1561	Lithium perchlorate	$LiClO_4$	7791-03-9	106.392	wh orth cry or powder	236	430 dec	2.428	58.7^{25}	s EtOH, ace, eth
1562	Lithium perchlorate trihydrate	$LiClO_4 \cdot 3H_2O$	13453-78-6	160.438	wh hex cry	95 dec		1.84		vs H_2O, EtOH, ace; i eth
1563	Lithium peroxide	Li_2O_2	12031-80-0	45.881	wh hex cry			2.31		s H_2O; i EtOH
1564	Lithium selenate monohydrate	$Li_2SeO_4 \cdot H_2O$	7790-71-8	174.86	monocl cry			2.56		vs H_2O
1565	Lithium selenite monohydrate	$Li_2SeO_3 \cdot H_2O$	15593-51-8	158.86	hyg cry					
1566	Lithium stearate	$LiC_{18}H_{35}O_2$	4485-12-5	290.411	cry	≈220				
1567	Lithium sulfate	Li_2SO_4	10377-48-7	109.945	wh monocl cry; hyg	860		2.21	34.2^{25}	
1568	Lithium sulfate monohydrate	$Li_2SO_4 \cdot H_2O$	10102-25-7	127.960	col cry	130 dec		2.06	34.2^{25}	sl EtOH
1569	Lithium sulfide	Li_2S	12136-58-2	45.947	wh cub cry; hyg	1372		1.64		
1570	Lithium tantalate	$LiTaO_3$	12031-66-2	235.887	wh pow	1650				
1571	Lithium tetraborate	$Li_2B_4O_7$	12007-60-2	169.122	wh tetr cry	917			2.9^{20}	sl H_2O
1572	Lithium tetraborate pentahydrate	$Li_2B_4O_7 \cdot 5H_2O$	1303-94-2	259.198	wh cry pow	dec 200				vs H_2O; i EtOH
1573	Lithium tetracyanoplatinate pentahydrate	$Li_2Pt(CN)_4 \cdot 5H_2O$	14402-73-4	403.112	grn-yel cry					sl H_2O
1574	Lithium tetrafluoroborate	$LiBF_4$	14283-07-9	93.746	hyg wh pow	dec				vs H_2O
1575	Lithium thiocyanate	$LiSCN$	556-65-0	65.024	wh hyg cry				120^{25}	
1576	Lithium titanate	Li_2TiO_3	12031-82-2	109.747	wh pow	1325				i H_2O
1577	Lithium tungstate	Li_2WO_4	13568-45-1	261.72	wh trig pow	740		3.71		s H_2O
1578	Lithium vanadate	$LiVO_3$	15060-59-0	105.881	yel pow		subl 1400			
1579	Lithium zirconate	Li_2ZrO_3	12031-83-3	153.104	wh solid					
1580	Lutetium	Lu	7439-94-3	174.967	silv metal; hex	1663	3402	9.84		s dil acid
1581	Lutetium boride	LuB_4	12688-52-7	218.211	tetr cry	2600		≈7.0		
1582	Lutetium bromide	$LuBr_3$	14456-53-2	414.679	wh hyg cry	1025				vs H_2O
1583	Lutetium chloride	$LuCl_3$	10099-66-8	281.326	wh monocl cry; hyg	925		3.98		s H_2O
1584	Lutetium chloride hexahydrate	$LuCl_3 \cdot 6H_2O$	15230-79-2	389.417	col cry	dec 150				s H_2O, EtOH
1585	Lutetium fluoride	LuF_3	13760-81-1	231.962	orth cry	1182	2200	8.3		i H_2O
1586	Lutetium iodide	LuI_3	13813-45-1	555.680	brn hex cry; hyg	1050		≈5.6		vs H_2O
1587	Lutetium iron oxide	$Lu_3Fe_5O_{12}$	12023-71-1	996.119	cry					
1588	Lutetium nitrate	$Lu(NO_3)_3$	10099-67-9	360.982	hyg col solid					s H_2O, EtOH

No.	Name	Formula	CAS Reg No.	Mol. weight	Physical form	mp/°C	bp/°C	Density g cm^{-3}	Solubility g/100 g H_2O	Qualitative solubility
1589	Lutetium nitride	LuN	12125-25-6	188.974	cub cry			11.6		
1590	Lutetium oxide	Lu_2O_3	12032-20-1	397.932	wh cub cry or powder	2427	3980	9.41		
1591	Lutetium perchlorate hexahydrate	$Lu(ClO_4)_3 \cdot 6H_2O$	14646-29-8	581.410	col cry	dec 350 (anh)				s H_2O, MeOH
1592	Lutetium sulfate	$Lu_2(SO_4)_3$	14986-89-1	638.122	wh pow	dec >850				vs H_2O
1593	Lutetium sulfate octahydrate	$Lu_2(SO_4)_3 \cdot 8H_2O$	13473-77-3	782.244	col cry				42.3^{20}	s H_2O
1594	Lutetium sulfide	Lu_2S_3	12163-20-1	446.129	gray rhomb cry	1750 dec		6.26		
1595	Lutetium telluride	Lu_2Te_3	12163-22-3	732.73	orth cry			7.8		
1596	Magnesium	Mg	7439-95-4	24.305	silv-wh metal	650	1090	1.74		s dil acid
1597	Magnesium acetate	$Mg(C_2H_3O_2)_2$	142-72-3	142.394	wh orth/monocl cry	323 dec		1.50	65.6^{25}	
1598	Magnesium acetate monohydrate	$Mg(C_2H_3O_2)_2 \cdot H_2O$	60582-92-5	160.409	orth cry			1.55		
1599	Magnesium acetate tetrahydrate	$Mg(C_2H_3O_2)_2 \cdot 4H_2O$	16674-78-5	214.454	col monocl cry; hyg	80 dec		1.45	65.6^{25}	vs EtOH
1600	Magnesium aluminate	$Mg(AlO_2)_2$	12068-51-8	142.266	col cub cry	2105		3.55		i H_2O
1601	Magnesium aluminum silicate	$Mg_2Al_3(AlSi_5O_{18})$	1302-88-1	584.953	blue cry			2.6		
1602	Magnesium amide	$Mg(NH_2)_2$	7803-54-5	56.350	wh powder; flam	dec		1.39		reac H_2O
1603	Magnesium ammonium phosphate hexahydrate	$MgNH_4PO_4 \cdot 6H_2O$	13478-16-5	245.407	wh pow	dec		1.71		i H_2O, EtOH; s acid
1604	Magnesium antimonide	Mg_3Sb_2	12057-75-9	316.435	hex cry	1245		3.99		
1605	Magnesium arsenide	Mg_3As_2	12044-49-4	222.758	solid	≈1200		3.15		i H_2O
1606	Magnesium diboride	MgB_2	12007-25-9	45.927	hex cry	800 dec		2.57		
1607	Magnesium hexaboride	MgB_6	12008-22-9	89.171	refrac solid	1100 dec				i H_2O
1608	Magnesium dodecaboride	MgB_{12}	12230-32-9	154.037	refrac solid	1300 dec				
1609	Magnesium bromate hexahydrate	$Mg(BrO_3)_2 \cdot 6H_2O$	7789-36-8	388.201	col cub cry	200 dec		2.29	98^{25}	
1610	Magnesium bromide	$MgBr_2$	7789-48-2	184.113	wh hex cry; hyg	711		3.72	102^{25}	
1611	Magnesium bromide hexahydrate	$MgBr_2 \cdot 6H_2O$	13446-53-2	292.204	col monocl cry	165 dec		2.0	102^{25}	s EtOH
1612	Magnesium carbonate	$MgCO_3$	546-93-0	84.314	wh hex cry	990		3.010	0.18^{20}	i EtOH; s acid
1613	Magnesium carbonate dihydrate	$MgCO_3 \cdot 2H_2O$	5145-48-2	120.345	col tricl cry			2.8		i H_2O, ace, NH_4OH
1614	Magnesium carbonate trihydrate	$MgCO_3 \cdot 3H_2O$	14457-83-1	138.360	col monocl cry	165		1.8	0.18^{16}	
1615	Magnesium carbonate pentahydrate	$MgCO_3 \cdot 5H_2O$	61042-72-6	174.390	wh monocl cry	dec >400		3.04	0.38^{16}	
1616	Magnesium carbonate hydroxide tetrahydrate	$4MgCO_3 \cdot Mg(OH)_2 \cdot 4H_2O$	39409-82-0	467.636	wh monocl cry			2.3		
1617	Magnesium carbonate hydroxide pentahydrate	$4MgCO_3 \cdot Mg(OH)_2 \cdot 5H_2O$	56378-72-4	485.652	wh pow	dec 700				i H_2O; s dil acid; i EtOH
1618	Magnesium carbonate dihydroxide trihydrate	$MgCO_3 \cdot Mg(OH)_2 \cdot 3H_2O$	12143-96-3	196.680	wh monocl cry	dec		2.04		
1619	Magnesium chlorate hexahydrate	$Mg(ClO_3)_2 \cdot 6H_2O$	13446-19-0	299.299	wh hyg cry	≈35 dec		1.80	142^{25}	sl EtOH
1620	Magnesium chloride	$MgCl_2$	7786-30-3	95.211	wh hex leaflets; hyg	714	1412	2.325	56.0^{25}	
1621	Magnesium chloride hexahydrate	$MgCl_2 \cdot 6H_2O$	7791-18-6	203.302	wh hyg cry	≈100 dec		1.56	56.0^{25}	s EtOH
1622	Magnesium chromate heptahydrate	$MgCrO_4 \cdot 7H_2O$	13423-61-5*	266.405	yel rhom cry			1.695	54.8^{25}	
1623	Magnesium chromite	$MgCr_2O_4$	12053-26-8	192.295	deep grn cry	2390		4.4		
1624	Magnesium citrate	$Mg_3(C_6H_5O_7)_2$	3344-18-1	451.114	wh cry					sl H_2O
1625	Magnesium citrate tetradecahydrate	$Mg_3(C_6H_5O_7)_2 \cdot 14H_2O$	3344-18-1*	703.328	wh cry pow					sl H_2O, s acid
1626	Magnesium fluoride	MgF_2	7783-40-6	62.302	wh tetr cry	1263	2227	3.148	0.013^{25}	
1627	Magnesium formate dihydrate	$Mg(CHO_2)_2 \cdot 2H_2O$	6150-82-9	150.370	wh cry	dec				s H_2O; i EtOH
1628	Magnesium germanate	Mg_2GeO_4	12025-13-7	185.25	wh prec					i H_2O
1629	Magnesium germanide	Mg_2Ge	1310-52-7	121.25	cub cry	1117		3.09		
1630	Magnesium hydride	MgH_2	7693-27-8	26.321	wh tetr cry	327		1.45		reac H_2O
1631	Magnesium hydrogen phosphate trihydrate	$MgHPO_4 \cdot 3H_2O$	7757-86-0	174.331	wh powder	550 dec		2.13		sl H_2O; s dil acid
1632	Magnesium hydroxide	$Mg(OH)_2$	1309-42-8	58.320	wh hex cry	350		2.37	0.00069^{20}	s dil acid
1633	Magnesium iodate tetrahydrate	$Mg(IO_3)_2 \cdot 4H_2O$	7790-32-1*	446.172	col monocl cry	210 dec		3.3	11.1^{25}	
1634	Magnesium iodide	MgI_2	10377-58-9	278.114	wh hex cry; hyg	634		4.43	146^{25}	
1635	Magnesium iodide hexahydrate	$MgI_2 \cdot 6H_2O$	66778-21-0	386.205	wh monocl cry			2.35		
1636	Magnesium iodide octahydrate	$MgI_2 \cdot 8H_2O$	7790-31-0	422.236	wh orth cry; hyg	41 dec		2.10	146^{25}	s EtOH
1637	Magnesium metaborate octahydrate	$Mg(BO_2)_2 \cdot 8H_2O$	13703-82-7*	254.047	wh pow	988 (anh)				sl H_2O
1638	Magnesium metasilicate	$MgSiO_3$	13776-74-4	100.389	wh monocl cry	≈1550 dec		3.19		i H_2O; sl HF
1639	Magnesium metatitanate	$MgTiO_3$	12032-30-3	120.170	col hex cry	1565		3.85		
1640	Magnesium molybdate	$MgMoO_4$	12013-21-7	184.24	wh pow	≈1060		2.2	15.9^{25}	s H_2O
1641	Magnesium nitrate	$Mg(NO_3)_2$	10377-60-3	148.314	wh cub cry			≈2.3	71.2^{25}	
1642	Magnesium nitrate dihydrate	$Mg(NO_3)_2 \cdot 2H_2O$	15750-45-5	184.345	wh cry	≈100 dec		1.45	71.2^{25}	s EtOH
1643	Magnesium nitrate hexahydrate	$Mg(NO_3)_2 \cdot 6H_2O$	13446-18-9	256.406	col monocl cry; hyg	≈95 dec		1.46	71.2^{25}	s EtOH
1644	Magnesium nitride	Mg_3N_2	12057-71-5	100.928	yel cub cry	≈1500 dec		2.71		
1645	Magnesium nitrite trihydrate	$Mg(NO_2)_2 \cdot 3H_2O$	15070-34-5	170.362	wh hyg prisms	100 dec			129.9^{25}	s EtOH

No.	Name	Formula	CAS Reg No.	Mol. weight	Physical form	mp/°C	bp/°C	Density g cm^{-3}	Solubility g/100 g H_2O	Qualitative solubility
1646	Magnesium orthosilicate	Mg_2SiO_4	26686-77-1	140.694	wh orth cry	1897		3.21		i H_2O
1647	Magnesium orthotitanate	Mg_2TiO_4	12032-52-9	160.475	wh cub cry	1840		3.53		
1648	Magnesium oxalate	MgC_2O_4	547-66-0	112.324	wh powder				0.038^{25}	
1649	Magnesium oxalate dihydrate	$MgC_2O_4 \cdot 2H_2O$	6150-88-5	148.354	wh powder				0.038^{25}	i EtOH; s dil acid
1650	Magnesium oxide	MgO	1309-48-4	40.304	wh cub cry	2825	3600	3.6		sl H_2O; i EtOH
1651	Magnesium perborate heptahydrate	$Mg(BO_3)_2 \cdot 7H_2O$	14635-87-1	268.030	wh pow					sl H_2O
1652	Magnesium perchlorate	$Mg(ClO_4)_2$	10034-81-8	223.206	wh hyg powder	250 dec		2.2	100^{25}	
1653	Magnesium perchlorate hexahydrate	$Mg(ClO_4)_2 \cdot 6H_2O$	13446-19-0	331.298	wh hyg cry	190 dec		1.98	100^{25}	s EtOH
1654	Magnesium permanganate hexahydrate	$Mg(MnO_4)_2 \cdot 6H_2O$	10377-62-5	370.268	blue-blk cry	dec		2.18		s H_2O
1655	Magnesium peroxide	MgO_2	1335-26-8	56.304	wh cub cry	100 dec		≈3.0		i H_2O; s dil acid
1656	Magnesium phosphate pentahydrate	$Mg_3(PO_4)_2 \cdot 5H_2O$	7757-87-1*	352.934	wh cry	400 dec			0.00009^{20}	s dil acid
1657	Magnesium phosphate octahydrate	$Mg_3(PO_4)_2 \cdot 8H_2O$	13446-23-6	406.980	wh monocl cry			2.17	0.00009^{20}	s acid
1658	Magnesium pyrophosphate	$Mg_2P_2O_7$	13446-24-7	222.553	col monocl plates	1395		2.56		
1659	Magnesium pyrophosphate trihydrate	$Mg_2P_2O_7 \cdot 3H_2O$	10102-34-8	276.600	wh powder	100 dec		2.56		i H_2O; s acid
1660	Magnesium phosphide	Mg_3P_2	12057-74-8	134.863	yel cub cry			2.06		reac H_2O
1661	Magnesium selenate hexahydrate	$MgSeO_4 \cdot 6H_2O$	13446-28-1	275.35	wh monocl cry			1.928	55.5^{25}	
1662	Magnesium selenide	MgSe	1313-04-8	103.27	brn cub cry			4.2		reac H_2O
1663	Magnesium selenite hexahydrate	$MgSeO_3 \cdot 6H_2O$	15593-61-0	259.36	col hex cry			2.09		i H_2O; s dil acid
1664	Magnesium hexafluorosilicate hexahydrate	$MgSiF_6 \cdot 6H_2O$	60950-56-3	274.472	wh cry	120 dec		1.79	39.3^{18}	i EtOH
1665	Magnesium silicide	Mg_2Si	22831-39-6	76.696	gray cub cry	1102		1.99		reac H_2O
1666	Magnesium stannide	Mg_2Sn	1313-08-2	167.320	blue cub cry	771		3.60		s H_2O, dil HCl
1667	Magnesium sulfate	$MgSO_4$	7487-88-9	120.368	col orth cry	1137		2.66	35.7^{25}	
1668	Magnesium sulfate monohydrate	$MgSO_4 \cdot H_2O$	14168-73-1	138.383	col monocl cry	150 dec		2.57	35.7^{25}	
1669	Magnesium sulfate heptahydrate	$MgSO_4 \cdot 7H_2O$	10034-99-8	246.474	col orth cry	150 dec		1.67	35.7^{25}	sl EtOH
1670	Magnesium sulfide	MgS	12032-36-9	56.370	red-brn cub cry	2226		2.68		reac H_2O
1671	Magnesium sulfite trihydrate	$MgSO_3 \cdot 3H_2O$	19086-20-5	158.414	col orth cry			2.12	0.79^{25}	
1672	Magnesium sulfite hexahydrate	$MgSO_3 \cdot 6H_2O$	13446-29-2	212.460	wh hex cry	200 dec		1.72	0.79^{25}	i EtOH
1673	Magnesium tetrahydrogen phosphate dihydrate	$Mg(H_2PO_4)_2 \cdot 2H_2O$	15609-80-0	254.311	wh hyg cry	dec 90				s H_2O; i EtOH
1674	Magnesium thiocyanate tetrahydrate	$Mg(SCN)_2 \cdot 4H_2O$	306-61-6	212.531	wh hyg cry					vs H_2O, EtOH
1675	Magnesium thiosulfate hexahydrate	$MgS_2O_3 \cdot 6H_2O$	13446-30-5	244.525	col cry	170 dec		1.82	93^{25}	i EtOH
1676	Magnesium trisilicate	$Mg_2Si_3O_8$	14987-04-3	260.862	wh powder					i H_2O, EtOH
1677	Magnesium tungstate	$MgWO_4$	13573-11-0	272.14	wh monocl cry			6.89	0.016^{20}	i EtOH
1678	Magnesium vanadate	$Mg_2V_2O_7$	13568-63-3	262.489	tricl cry			3.1		
1679	Magnesium zirconate	$MgZrO_3$	12032-31-4	163.527	col cry	2060		4.23		
1680	Magnesium zirconium silicate	$MgO \cdot ZrO_2 \cdot SiO_2$	52110-05-1	223.612	wh solid					i H_2O, alk; sl acid
1681	Manganese	Mn	7439-96-5	54.938	hard gray metal	1246	2061	7.3		s dil acids
1682	Manganocene	$Mn(C_5H_5)_2$	1271-27-8	185.124	yel-brn cry	173				s py, thf; sl bz
1683	Manganese antimonide (MnSb)	MnSb	12032-82-5	176.698	hex cry	840		6.9		
1684	Manganese antimonide (Mn_2Sb)	Mn_2Sb	12032-97-2	231.636	tetr cry	948		7.0		
1685	Manganese boride (MnB)	MnB	12045-15-7	65.749	orth cry	1890		6.45		
1686	Manganese boride (MnB_2)	MnB_2	12228-50-1	76.560	hex cry	1827		5.3		
1687	Manganese boride (Mn_2B)	Mn_2B	12045-16-8	120.687	red-brn tetr cry	1580		7.20		
1688	Manganese carbide	Mn_3C	12266-65-8	176.825	refrac solid	1520		6.89		
1689	Manganese carbonyl	$Mn_2(CO)_{10}$	10170-69-1	389.977	yel monocl cry	154		1.75		i H_2O; s os
1690	Manganese pentacarbonyl bromide	$Mn(CO)_5Br$	14516-54-2	274.893	oran-yel cry					s os
1691	Manganese phosphide (MnP)	MnP	12032-78-9	85.912	orth cry	1147		5.49		
1692	Manganese phosphide (Mn_2P)	Mn_2P	12333-54-9	140.850	hex cry	1327		6.0		
1693	Manganese silicide	$MnSi_2$	12032-86-9	111.109	gray solid	1152 dec				
1694	Manganese(II) acetate tetrahydrate	$Mn(C_2H_3O_2)_2 \cdot 4H_2O$	6156-78-1	245.087	red monocl cry	80		1.59		s H_2O, EtOH
1695	Manganese(II) bromide	$MnBr_2$	13446-03-2	214.746	pink hex cry	698		4.385	151^{25}	
1696	Manganese(II) bromide tetrahydrate	$MnBr_2 \cdot 4H_2O$	10031-20-6	286.808	red hyg cry	64 dec			151^{25}	
1697	Manganese(II) carbonate	$MnCO_3$	598-62-9	114.947	pink hex cry	>200 dec		3.70	0.00008^{20}	s dil acid
1698	Manganese(II) chloride	$MnCl_2$	7773-01-5	125.844	pink trig cry; hyg	650	1190	2.977	77.3^{25}	s py, EtOH; i eth
1699	Manganese(II) chloride tetrahydrate	$MnCl_2 \cdot 4H_2O$	13446-34-9	197.906	red monocl cry; hyg	87.5		1.913	77.3^{25}	s EtOH; i eth
1700	Manganese(II) dihydrogen phosphate dihydrate	$Mn(H_2PO_4)_2 \cdot 2H_2O$	18718-07-5	284.944	col hyg cry					s H_2O; i EtOH
1701	Manganese(II) fluoride	MnF_2	7782-64-1	92.935	red tetr cry	900		3.98	1.02^{25}	i EtOH
1702	Manganese(II) hydroxide	$Mn(OH)_2$	18933-05-6	88.953	pink hex cry	dec		3.26	0.00034^{20}	
1703	Manganese(II) hypophosphite monohydrate	$Mn(H_2PO_2)_2 \cdot H_2O$	10043-84-2	202.931	pink cry	>250			15^{20}	s H_2O

No.	Name	Formula	CAS Reg No.	Mol. weight	Physical form	mp/°C	bp/°C	Density g cm^{-3}	Solubility g/100 g H_2O	Qualitative solubility
1704	Manganese(II) iodide	MnI_2	7790-33-2	308.747	wh hex cry; hyg	638		5.04		s H_2O, EtOH
1705	Manganese(II) iodide tetrahydrate	$MnI_2 \cdot 4H_2O$	7790-33-2*	380.809	red cry					vs H_2O; s EtOH
1706	Manganese(II) metasilicate	$MnSiO_3$	7759-00-4	131.022	red orth cry	1291		3.48		i H_2O
1707	Manganese(II) molybdate	$MnMoO_4$	14013-15-1	214.88	yel monocl cry			4.05		
1708	Manganese(II) nitrate	$Mn(NO_3)_2$	10377-93-2	178.947	col orth cry; hyg			2.2	161^{25}	s diox, thf
1709	Manganese(II) nitrate tetrahydrate	$Mn(NO_3)_2 \cdot 4H_2O$	20694-39-7	251.009	pink hyg cry	37.1 dec		2.13	161^{25}	s EtOH
1710	Manganese(II) nitrate hexahydrate	$Mn(NO_3)_2 \cdot 6H_2O$	10377-66-9	287.039	rose monocl cry	28 dec		1.8	161^{25}	vs EtOH
1711	Manganese(II) orthosilicate	Mn_2SiO_4	13568-32-6	201.960	orth cry			4.11		i H_2O
1712	Manganese(II) oxalate dihydrate	$MnC_2O_4 \cdot 2H_2O$	6556-16-7	178.987	wh cry powder	150 dec		2.45	0.032^{20}	s acid
1713	Manganese(II) oxide	MnO	1344-43-0	70.937	grn cub cry or powder	1842		5.37		i H_2O; s acid
1714	Manganese(II) perchlorate hexahydrate	$Mn(ClO_4)_2 \cdot 6H_2O$	15364-94-0	361.931	pink hex cry			2.10		
1715	Manganese(II) pyrophosphate	$Mn_2P_2O_7$	53731-35-4	283.819	wh monocl cry	1196		3.71		i H_2O
1716	Manganese(II) selenide	MnSe	1313-22-0	133.90	gray cub cry	1460		5.45		i H_2O
1717	Manganese(II) sulfate	$MnSO_4$	7785-87-7	151.001	wh orth cry	700	850 dec	3.25	63.7^{25}	
1718	Manganese(II) sulfate monohydrate	$MnSO_4 \cdot H_2O$	10034-96-5	169.016	red monocl cry			2.95	63.7^{25}	i EtOH
1719	Manganese(II) sulfate tetrahydrate	$MnSO_4 \cdot 4H_2O$	10101-68-5	223.062	red monocl cry	38 dec		2.20	63.7^{25}	i EtOH
1720	Manganese(II) sulfide (α form)	MnS	18820-29-6	87.003	grn cub cry	1610		4.0		i H_2O; s dil acid
1721	Manganese(II) sulfide (β form)	MnS	18820-29-6	87.003	red cub cry			3.3		i H_2O; s dil acid
1722	Manganese(II) sulfide (γ form)	MnS	18820-29-6	87.003	red hex cry			≈3.3		i H_2O; s dil acid
1723	Manganese(II) telluride	MnTe	12032-88-1	182.54	hex cry	≈1150		6.0		
1724	Manganese(II) tetraborate octahydrate	$MnB_4O_7 \cdot 8H_2O$	12228-91-0	354.300	red solid					i H_2O, EtOH; s dil acid
1725	Manganese(II) titanate	$MnTiO_3$	12032-74-5	150.803	red hex cry	1360		4.55		
1726	Manganese(II) tungstate	$MnWO_4$	13918-22-4	302.78	wh monocl cry			7.2	0.0054^{20}	
1727	Manganese(II,III) oxide	Mn_3O_4	1317-35-7	228.812	brn tetr cry	1567		4.84		i H_2O; s HCl
1728	Manganese(III) acetate dihydrate	$Mn(C_2H_3O_2)_3 \cdot 2H_2O$	19513-05-4	268.100	brn cry					s eth, HOAc
1729	Manganese(III) fluoride	MnF_3	7783-53-1	111.933	red monocl cry; hyg	>600 dec		3.54		reac H_2O
1730	Manganese(III) hydroxide	MnO(OH)	1332-63-4	87.945	blk monocl cry	250 dec		≈4.3		i H_2O
1731	Manganese(III) oxide	Mn_2O_3	1317-34-6	157.874	blk cub cry	1080 dec		≈5.0		i H_2O
1732	Manganese(IV) oxide	MnO_2	1313-13-9	86.937	blk tetr cry	535 dec		5.08		i H_2O, HNO_3
1733	Manganese(VII) oxide	Mn_2O_7	12057-92-0	221.872	grn oil; exp	5.9	95 exp	2.40		vs H_2O
1734	Mendelevium	Md	7440-11-1	258	metal	827				
1735	Mercury	Hg	7439-97-6	200.59	heavy silv liq	-38.8290	356.62	13.5336		i H_2O
1736	Dimethyl mercury	$Hg(CH_3)_2$	593-74-8	230.66	liq		93	3.17		i H_2O; vs EtOH, eth
1737	Mercury(I) acetate	$Hg_2(C_2H_3O_2)_2$	631-60-7	519.27	col scales	dec				sl H_2O; i EtOH, eth
1738	Mercury(I) bromate	$Hg_2(BrO_3)_2$	13465-33-3	656.98	col cry	dec				i H_2O; sl acid
1739	Mercury(I) bromide	Hg_2Br_2	15385-58-7	560.99	wh tetr cry or powder	345 dec		7.307		i H_2O, EtOH, eth
1740	Mercury(I) carbonate	Hg_2CO_3	6824-78-8	461.19	yel-brn cry	130 dec			0.0000045	i EtOH
1741	Mercury(I) chlorate	$Hg_2(ClO_3)_2$	10294-44-7	568.08	wh rhom cry	≈250 dec		6.409		sl H_2O; s EtOH
1742	Mercury(I) chloride	Hg_2Cl_2	10112-91-1	472.09	wh tetr cry	525 tp	383 sp	7.16	0.0004^{25}	i EtOH, eth
1743	Mercury(I) chromate	Hg_2CrO_4	13465-34-4	517.17	brn-red solid					i H_2O EtOH; s conc HNO_3
1744	Mercury(I) fluoride	Hg_2F_2	13967-25-4	439.18	yel cub cry	570 dec	subl	8.73		reac H_2O
1745	Mercury(I) iodate	$Hg_2(IO_3)_2$	13465-35-5	750.99	yel-wh pow	dec 175			0.0032^{20}	
1746	Mercury(I) iodide	Hg_2I_2	15385-57-6	654.99	yel amorp powder	290		7.70		i H_2O, EtOH, eth
1747	Mercury(I) nitrate	$Hg_2(NO_3)_2$	10415-75-5	525.19	cry					sl H_2O
1748	Mercury(I) nitrate dihydrate	$Hg_2(NO_3)_2 \cdot 2H_2O$	14836-60-3	561.22	col cry	70 dec		4.8		sl H_2O
1749	Mercury(I) nitrite	$Hg_2(NO_2)_2$	13492-25-6	493.19	yel cry	100 dec		7.3		reac H_2O
1750	Mercury(I) oxalate	$Hg_2C_2O_4$	2949-11-3	489.20	cry					i H_2O; sl HNO_3
1751	Mercury(I) oxide	Hg_2O	15829-53-5	417.18	prob mixture of HgO+Hg	100 dec		9.8		i H_2O; s HNO_3
1752	Mercury(I) perchlorate tetrahydrate	$Hg_2(ClO_4)_2 \cdot 4H_2O$	65202-12-2	672.14	cry	64			442^{25}	
1753	Mercury(I) sulfate	Hg_2SO_4	7783-36-0	497.24	wh-yel cry powder			7.56	0.051^{25}	s dil HNO_3
1754	Mercury(I) sulfide	Hg_2S	51595-71-2	433.25	unstab blk pow	dec				i H_2O
1755	Mercury(I) thiocyanate	$Hg_2(SCN)_2$	13465-37-7	517.34	col powder	dec			0.03^{25}	s HCl, KCNS
1756	Mercury(I) tungstate	Hg_2WO_4	38705-19-0	649.02	yel amorp solid	dec				i H_2O, EtOH
1757	Mercury(II) acetate	$Hg(C_2H_3O_2)_2$	1600-27-7	318.68	wh-yel cry or powder	179 dec		3.28	25^{10}	s EtOH
1758	Mercury(II) amide chloride	$Hg(NH_2)Cl$	10124-48-8	252.07	wh solid		subl	5.38		i H_2O, EtOH; s warm acid
1759	Mercury(II) benzoate monohydrate	$Hg(C_7H_5O_2)_2 \cdot H_2O$	32839-04-6	460.83	wh cry	165			1.2^{15}	sl EtOH
1760	Mercury(II) bromate	$Hg(BrO_3)_2$	26522-91-8	456.39	cry	130 dec			0.15	s acid
1761	Mercury(II) bromide	$HgBr_2$	7789-47-1	360.40	wh rhomb cry or powder	241	318	6.05	0.61^{25}	sl chl; s EtOH, MeOH
1762	Mercury(II) chlorate	$Hg(ClO_3)_2$	13465-30-0	367.49	wh needles	dec		4.998	25	

No.	Name	Formula	CAS Reg No.	Mol. weight	Physical form	mp/°C	bp/°C	Density g cm^{-3}	Solubility g/100 g H_2O	Qualitative solubility
1763	Mercury(II) chloride	$HgCl_2$	7487-94-7	271.50	wh orth cry	277	304	5.6	7.31^{25}	sl bz; s EtOH, MeOH, ace, eth
1764	Mercury(II) chromate	$HgCrO_4$	13444-75-2	316.58	red monocl cry			6.06		sl H_2O
1765	Mercury(II) cyanide	$Hg(CN)_2$	592-04-1	252.62	col tetr cry	320 dec		4.00	11.4^{25}	s EtOH; sl eth
1766	Mercury(II) dichromate	$HgCr_2O_7$	7789-10-8	416.58	red cry powder					i H_2O; s acid
1767	Mercury(II) fluoride	HgF_2	7783-39-3	238.59	wh cub cry; hyg	645 dec		8.95		reac H_2O
1768	Mercury(II) fulminate	$Hg(CNO)_2$	628-86-4	284.62	gray cry	exp		4.42		sl H_2O; s EtOH, NH_4OH
1769	Mercury(II) hydrogen arsenate	$HgHAsO_4$	7784-37-4	340.52	yel powder					i H_2O; s acid
1770	Mercury(II) iodate	$Hg(IO_3)_2$	7783-32-6	550.40	wh powder	175 dec				i H_2O
1771	Mercury(II) iodide (yellow)	HgI_2	7774-29-0	454.40	yel tetr cry or powder	256	351	6.28	0.0055^{25}	sl EtOH, ace, eth
1772	Mercury(II) iodide (red)	HgI_2	7774-29-0	454.40	red pow	trans to yel 127			0.006^{25}	sl EtOH, ace, eth, chl
1773	Mercury(II) nitrate	$Hg(NO_3)_2$	10045-94-0	324.60	col hyg cry	79		4.3		s H_2O; i EtOH
1774	Mercury(II) nitrate monohydrate	$Hg(NO_3)_2 \cdot H_2O$	7783-34-8	342.62	wh-yel hyg cry			4.3		s H_2O, dil acid
1775	Mercury(II) nitrate dihydrate	$Hg(NO_3)_2 \cdot 2H_2O$	22852-67-1	360.63	monocl cry			4.78		s H_2O
1776	Mercury(II) oxalate	HgC_2O_4	3444-13-1	288.61	powder	165 dec				i H_2O
1777	Mercury(II) oxide	HgO	21908-53-2	216.59	red or yel orth cry	500 dec		11.14		i H_2O, EtOH; s dil acid
1778	Mercury(II) oxide sulfate	$(Hg_3O_2)SO_4$	1312-03-4	729.83	yel powder					i H_2O; s acid
1779	Mercury(II) oxycyanide	$Hg(CN)_2 \cdot HgO$	1335-31-5	469.21	wh orth cry	exp		4.44	11.4^{25}	
1780	Mercury(II) perchlorate trihydrate	$Hg(ClO_4)_2 \cdot 3H_2O$	7616-83-3	453.54	cry					
1781	Mercury(II) phosphate	$Hg_3(PO_4)_2$	7782-66-3	791.71	wh-yel powder					i H_2O, EtOH; s acid
1782	Mercury(II) selenide	HgSe	20601-83-6	279.55	gray cub cry		subl	8.21		i H_2O
1783	Mercury(II) sulfate	$HgSO_4$	7783-35-9	296.65	wh monocl cry			6.47		reac H_2O
1784	Mercury(II) sulfide (black)	HgS	1344-48-5	232.66	blk cub cry or powder	850		7.70		i H_2O; s acid, EtOH
1785	Mercury(II) sulfide (red)	HgS	1344-48-5	232.66	red hex cry	trans to blk HgS 344		8.17		i H_2O, acid; s aqua regia
1786	Mercury(II) telluride	HgTe	12068-90-5	328.19	gray cub cry	673		8.63		
1787	Mercury(II) thiocyanate	$Hg(SCN)_2$	592-85-8	316.75	monocl cry	≈165 dec		3.71	0.070^{25}	s dil HCl
1788	Mercury(II) tungstate	$HgWO_4$	37913-38-5	448.43	yel cry	dec				i H_2O, EtOH
1789	Molybdenum	Mo	7439-98-7	95.94	gray-blk metal; cub	2623	4639	10.2		i H_2O, dil acid, alk
1790	Molybdophosphoric acid	$H_3P(Mo_3O_{10})_4$	51429-74-4	1825.25	bright yel cry					
1791	Molybdenum boride (Mo_2B)	Mo_2B	12006-99-4	202.69	refrac tetr cry	2000		9.2		
1792	Molybdenum boride (Mo_2B_5)	Mo_2B_5	12007-97-5	245.94	refrac hex cry	1600		≈7.2		
1793	Molybdenum carbide (MoC)	MoC	12011-97-1	107.95	refrac solid; cub	2577				
1794	Molybdenum carbide (Mo_2C)	Mo_2C	12069-89-5	203.89	gray orth cry	2687		9.18		
1795	Molybdenum carbonyl	$Mo(CO)_6$	13939-06-5	264.00	wh orth cry	148	155 dec	1.96		i H_2O; s bz; sl eth
1796	Molybdenum nitride (MoN)	MoN	12033-19-1	109.95	hex cry	1750		9.20		
1797	Molybdenum nitride (Mo_2N)	Mo_2N	12033-31-7	205.89	gray cub cry	790 dec		9.46		
1798	Molybdenum phosphide	MoP	12163-69-8	126.91	blk hex cry			7.34		
1799	Molybdenum silicide ($MoSi_2$)	$MoSi_2$	12136-78-6	152.11	gray tetr cry	≈1900		6.2		i H_2O; s HF
1800	Molybdenum(II) bromide	$MoBr_2$	13446-56-5	255.75	yel-red cry	dec 700		4.88		i H_2O, EtOH
1801	Molybdenum(II) chloride	$MoCl_2$	13478-17-6	166.85	yel cry	dec 500		3.71		i H_2O
1802	Molybdenum(II) iodide	MoI_2	14055-74-4	349.75	blk hyg cry	700		5.28		i H_2O
1803	Molybdenum(III) bromide	$MoBr_3$	13446-57-6	335.65	grn hex cry	dec 500		4.89		i H_2O, EtOH
1804	Molybdenum(III) chloride	$MoCl_3$	13478-18-7	202.30	dark red monocl cry	dec 400		3.74		i H_2O, os
1805	Molybdenum(III) fluoride	MoF_3	20193-58-2	152.94	yel-brn hex cry	>600		4.64		i H_2O
1806	Molybdenum(III) iodide	MoI_3	14055-75-5	476.65	blk solid	927				i H_2O
1807	Molybdenum(III) oxide	Mo_2O_3	1313-29-7	239.88	gray-blk powder					i H_2O; sl acid
1808	Molybdenum(IV) bromide	$MoBr_4$	13520-59-7	415.56	blk cry	dec 110				reac H_2O
1809	Molybdenum(IV) chloride	$MoCl_4$	13320-71-3	237.75	blk cry	317				reac H_2O, sl chl; i eth, bz
1810	Molybdenum(IV) fluoride	MoF_4	23412-45-5	171.93	grn cry	dec				reac H_2O
1811	Molybdenum(IV) iodide	MoI_4	14055-76-6	603.56	blk cry	dec 100				i H_2O
1812	Molybdenum(IV) oxide	MoO_2	18868-43-4	127.94	brn-viol tetr cry	≈1800 dec		6.47		i H_2O, acid, alk
1813	Molybdenum(IV) selenide	$MoSe_2$	12058-18-3	253.86	gray hex cry	>1200		6.90		
1814	Molybdenum(IV) sulfide	MoS_2	1317-33-5	160.07	blk powder or hex cry	1750		5.06		i H_2O; s conc acid
1815	Molybdenum(IV) telluride	$MoTe_2$	12058-20-7	351.14	gray hex cry			7.7		
1816	Molybdenum(V) chloride	$MoCl_5$	10241-05-1	273.21	grn-blk monocl cry; hyg	194	268	2.93		reac H_2O; s EtOH, eth
1817	Molybdenum(V) fluoride	MoF_5	13819-84-6	190.93	yel monocl cry	67	213.6	3.5		reac H_2O
1818	Molybdenum(V) oxytrichloride	$MoOCl_3$	13814-74-9	218.30	blk monocl cry	310	subl			reac H_2O
1819	Molybdenum(VI) acid monohydrate	$H_2MoO_4 \cdot H_2O$	7782-91-4	179.97	wh powder			3.1		sl H_2O; s alk

No.	Name	Formula	CAS Reg No.	Mol. weight	Physical form	mp/°C	bp/°C	Density g cm^{-3}	Solubility g/100 g H_2O	Qualitative solubility
1820	Molybdenum(VI) dioxydichloride	MoO_2Cl_2	13637-68-8	198.85	yel-oran solid	176	250	3.31		reac H_2O
1821	Molybdenum(VI) dioxydifluoride	MoO_2F_2	13824-57-2	165.94	pale lilac cry	subl 270		3.5		i MeCN, chl; sl HF
1822	Molybdenum(VI) fluoride	MoF_6	7783-77-9	209.93	wh cub cry or col liq; hyg	17.5	34.0	2.54		reac H_2O; vs hex, ctc
1823	Molybdenum(VI) metaphosphate	$Mo(PO_3)_6$	133863-98-6	569.77	yel powder			3.28		i H_2O, acid
1824	Molybdenum(VI) oxide	MoO_3	1313-27-5	143.94	wh-yel rhomb cry	802	1155	4.70	0.14^{20}	sl H_2O; s alk, acid
1825	Molybdenum(VI) oxytetrachloride	$MoOCl_4$	13814-75-0	253.75	grn hyg powder	105	159			
1826	Molybdenum(VI) oxytetrafluoride	$MoOF_4$	14459-59-7	187.93	volatile solid	97.2	186.0			
1827	Molybdenum(VI) sulfide	MoS_3	12033-29-3	192.14	blk solid	350 dec				i H_2O, os
1828	Neodymium	Nd	7440-00-8	144.242	silv metal; hex	1016	3074	7.01		
1829	Neodymium boride	NdB_6	12008-23-0	209.108	blk cub cry	2610		4.93		
1830	Neodymium nitride	NdN	25764-11-8	158.249	blk cub cry			7.69		
1831	Neodymium(II) acetate	$Nd(C_2H_3O_2)_2$	6192-13-8	321.373	red-purp cry					s H_2O
1832	Neodymium(II) chloride	$NdCl_2$	25469-93-6	215.148	grn hyg solid	841				
1833	Neodymium(III) bromate nonahydrate	$Nd(BrO_3)_3 \cdot 9H_2O$	15162-92-2	690.086	red hex cry	66 dec				
1834	Neodymium(III) bromide	$NdBr_3$	13536-80-6	383.954	viol orth cry; hyg	682	1540	5.3		s H_2O
1835	Neodymium(III) chloride	$NdCl_3$	10024-93-8	250.601	viol hex cry	759	1600	4.13	100^{25}	vs EtOH; i oth, ohl
1836	Neodymium(III) chloride hexahydrate	$NdCl_3 \cdot 6H_2O$	13477-89-9	358.692	purp cry	124 dec		2.3	100^{25}	s EtOH
1837	Neodymium(III) fluoride	NdF_3	13709-42-7	201.237	viol hex cry; hyg	1377	2300	6.51		i H_2O
1838	Neodymium(III) hydroxide	$Nd(OH)_3$	16469-17-3	195.264	blue solid	dec 210				i H_2O
1839	Neodymium(III) iodide	NdI_3	13813-24-6	524.955	grn orth cry; hyg	787		5.85		s H_2O
1840	Neodymium(III) nitrate	$Nd(NO_3)_3$	10045-95-1	330.257	viol hyg cry				152^{25}	s EtOH
1841	Neodymium(III) nitrate hexahydrate	$Nd(NO_3)_3 \cdot 6H_2O$	14517-29-4	438.348	purp hyg cry				152^{25}	s EtOH, ace
1842	Neodymium(III) oxide	Nd_2O_3	1313-97-9	336.482	blue hex cry; hyg	2233	3760	7.24		i H_2O; s dil acid
1843	Neodymium(III) sulfate	$Nd_2(SO_4)_3$	13477-91-3	576.672	pink needles	≈700 dec			7.1^{20}	
1844	Neodymium(III) sulfate octahydrate	$Nd_2(SO_4)_3 \cdot 8H_2O$	13477-91-3	720.794	red cry	350 dec		2.85		sl H_2O
1845	Neodymium(III) sulfide	Nd_2S_3	12035-32-4	384.679	orth cry	2207		5.46		
1846	Neodymium(III) telluride	Nd_2Te_3	12035-35-7	671.28	gray orth cry	1377		7.0		
1847	Neodymium(III) tris(cyclopentadienyl)	$Nd(C_5H_5)_3$	1273-98-9	339.522	red-blue cry	380				s thf
1848	Neon	Ne	7440-01-9	20.180	col gas	-248.609 tp (43 kPa)	-246.053	0.825 g/L		sl H_2O
1849	Neptunium	Np	7439-99-8	237	silv metal	644		20.2		s HCl
1850	Neptunium(IV) oxide	NpO_2	12035-79-9	269	grn cub cry	2547		11.1		
1851	Nickel	Ni	7440-02-0	58.693	wh metal; cub	1455	2913	8.90		i H_2O; sl dil acid
1852	Nickelocene	$Ni(C_5H_5)_2$	1271-28-9	188.879	grn cry	173				
1853	Nickel aluminide (NiAl)	NiAl	12003-78-0	85.675	metallic solid	1638				
1854	Nickel antimonide	NiSb	12035-52-8	180.453	hex cry	1147		8.74		
1855	Nickel arsenide	NiAs	27016-75-7	133.615	hex cry	967		7.77		
1856	Nickel boride (NiB)	NiB	12007-00-0	69.504	grn refrac solid	1035		7.13		
1857	Nickel boride (Ni_2B)	Ni_2B	12007-01-1	128.198	refrac solid	1125		7.90		
1858	Nickel boride (Ni_3B)	Ni_3B	12007-02-2	186.891	refrac solid	1166		8.17		
1859	Nickel carbonyl	$Ni(CO)_4$	13463-39-3	170.734	col liq	-19.3	42.1 (exp ≈60)	1.31		i H_2O; s EtOH, bz, ace, ctc
1860	Nickel phosphide	Ni_2P	12035-64-2	148.361	hex cry	1100		7.33		
1861	Nickel silicide ($NiSi_2$)	$NiSi_2$	12201-89-7	114.864	cub cry	993		4.83		
1862	Nickel silicide (Ni_2Si)	Ni_2Si	12059-14-2	145.473	orth cry	1255		7.40		
1863	Nickel subsulfide	Ni_3S_2	12035-72-2	240.210	yel hex cry	789		5.87		
1864	Nickel(II) acetate tetrahydrate	$Ni(C_2H_3O_2)_2 \cdot 4H_2O$	6018-89-9	248.842	grn monocl cry	250 dec		1.74	16^{20}	s H_2O, EtOH
1865	Nickel(II) ammonium chloride hexahydrate	$NH_4NiCl_3 \cdot 6H_2O$	16122-03-5*	291.182	grn hyg cry			1.65		s H_2O
1866	Nickel(II) ammonium sulfate	$Ni(NH_4)_2(SO_4)_2$	15699-18-0	286.895	blue-grn cry	dec 250				sl H_2O
1867	Nickel(II) ammonium sulfate hexahydrate	$Ni(NH_4)_2(SO_4)_2 \cdot 6H_2O$	7785-20-8	394.987	blue-grn cry	dec 130		1.92	6.5^{20}	s H_2O; i EtOH
1868	Nickel(II) arsenate octahydrate	$Ni_3(AsO_4)_2 \cdot 8H_2O$	7784-48-7	598.040	yel-grn powder	dec		4.98		i H_2O; s acid
1869	Nickel(II) bromide	$NiBr_2$	13462-88-9	218.501	yel hex cry; hyg	963	subl	5.10	131^{20}	
1870	Nickel(II) bromide trihydrate	$NiBr_2 \cdot 3H_2O$	13462-88-9*	272.547	yel-grn hyg cry	200 dec				vs H_2O; s EtOH, eth
1871	Nickel(II) carbonate	$NiCO_3$	3333-67-3	118.702	grn rhomb cry			4.389	0.0043^{20}	s dil acid
1872	Nickel(II) chlorate hexahydrate	$Ni(ClO_3)_2 \cdot 6H_2O$	13477-94-6	333.687	grn cub cry	dec 80		2.07		vs H_2O
1873	Nickel(II) chloride	$NiCl_2$	7718-54-9	129.599	yel hex cry; hyg	1031	985 sp	3.51	67.5^{25}	s EtOH
1874	Nickel(II) chloride hexahydrate	$NiCl_2 \cdot 6H_2O$	7791-20-0	237.690	grn monocl cry				67.5^{25}	s EtOH
1875	Nickel(II) chromate	$NiCrO_4$	14721-18-7	174.687	red solid					sl H_2O
1876	Nickel(II) cyanide tetrahydrate	$Ni(CN)_2 \cdot 4H_2O$	13477-95-7	182.789	grn plates	200 dec				i H_2O; sl dil acid; s NH_4OH
1877	Nickel(II) fluoride	NiF_2	10028-18-9	96.690	yel tetr cry	1380		4.7	2.56^{25}	i EtOH, eth
1878	Nickel(II) fluoride tetrahydrate	$NiF_2 \cdot 4H_2O$	13940-83-5	168.752	grn pow					sl H_2O

No.	Name	Formula	CAS Reg No.	Mol. weight	Physical form	mp/°C	bp/°C	Density g cm^{-3}	Solubility g/100 g H_2O	Qualitative solubility
1879	Nickel(II) hydroxide	$Ni(OH)_2$	12054-48-7	92.708	grn hex cry	230 dec		4.1	0.00015^{20}	
1880	Nickel(II) hydroxide monohydrate	$Ni(OH)_2 \cdot H_2O$	36897-37-7	110.723	grn powder				0.00015^{20}	s dil acid
1881	Nickel(II) iodate	$Ni(IO_3)_2$	13477-98-0	408.498	yel needles			5.07	1.1^{30}	sl H_2O
1882	Nickel(II) iodate tetrahydrate	$Ni(IO_3)_2 \cdot 4H_2O$	13477-99-1	480.560	yel hex cry	dec 100		5.07		sl H_2O
1883	Nickel(II) iodide	NiI_2	13462-90-3	312.502	blk hex cry; hyg	800	subl	5.22	154^{25}	
1884	Nickel(II) iodide hexahydrate	$NiI_2 \cdot 6H_2O$	7790-34-3	420.593	grn monocl cry; hyg				154^{25}	vs EtOH
1885	Nickel(II) nitrate	$Ni(NO_3)_2$	13138-45-9	182.702	grn cry				99.2^{25}	s EtOH
1886	Nickel(II) nitrate hexahydrate	$Ni(NO_3)_2 \cdot 6H_2O$	13478-00-7	290.794	grn monocl cry; hyg	56 dec		2.05	99.2^{25}	s EtOH
1887	Nickel(II) oxalate dihydrate	$NiC_2O_4 \cdot 2H_2O$	6018-94-6	182.742	grn-wh solid	dec 150			0.0012^{25}	i H_2O; s acid, NH_4OH
1888	Nickel(II) oxide	NiO	1313-99-1	74.692	grn cub cry	1957		6.72		i H_2O; s acid
1889	Nickel(II) perchlorate hexahydrate	$Ni(ClO_4)_2 \cdot 6H_2O$	13637-71-3*	365.686	grn hex needles	140			158.8^{25}	s EtOH, ace
1890	Nickel(II) phosphate octahydrate	$Ni_3(PO_4)_2 \cdot 8H_2O$	10381-36-9*	510.145	grn plates					s acid
1891	Nickel(II) selenate hexahydrate	$NiSeO_4 \cdot 6H_2O$	15060-62-5*	309.74	grn tetr cry			2.314	35.5^{20}	
1892	Nickel(II) selenide	NiSe	1314-05-2	137.65	yel-grn hex cry	980		7.2		
1893	Nickel(II) stannate dihydrate	$NiSnO_3 \cdot 2H_2O$	12035-38-0	261.432	grn pow	dec 120				
1894	Nickel(II) sulfate	$NiSO_4$	7786-81-4	154.756	grn-yel orth cry	840 dec		4.01	40.4^{25}	
1895	Nickel(II) sulfate hexahydrate	$NiSO_4 \cdot 6H_2O$	10101-97-0	262.847	blue-grn tetr cry	≈100 dec		2.07	40.4^{25}	sl EtOH
1896	Nickel(II) sulfate heptahydrate	$NiSO_4 \cdot 7H_2O$	10101-98-1	280.862	grn orth cry			1.98	40.4^{25}	s EtOH
1897	Nickel(II) sulfide	NiS	16812-54-7	90.758	yel hex cry	976		5.5		i H_2O
1898	Nickel(II) thiocyanate	$Ni(SCN)_2$	13689-92-4	174.857	grn pwd				55.0^{25}	
1899	Nickel(II) titanate	$NiTiO_3$	12035-39-1	154.558	brn hex cry			5.0		
1900	Nickel(II,III) sulfide	Ni_3S_4	12137-12-1	304.340	cub cry	995		4.77		
1901	Nickel(III) oxide	Ni_2O_3	1314-06-3	165.385	gray-blk cub cry	≈600 dec				i H_2O; s hot acid
1902	Niobium	Nb	7440-03-1	92.906	gray metal; cub	2477	4744	8.57		i acid
1903	Niobocene dichloride	$Nb(C_5H_5)_2Cl_2$	12793-14-5	293.998	hyg blk cry					sl tol
1904	Niobium boride (NbB)	NbB	12045-19-1	103.717	gray orth cry	2270		7.5		
1905	Niobium boride (NbB_2)	NbB_2	12007-29-3	114.528	gray hex cry	3050		6.97		
1906	Niobium carbide (NbC)	NbC	12069-94-2	104.917	gray cub cry	3608	4300	7.82		i H_2O, acid
1907	Niobium carbide (Nb_2C)	Nb_2C	12011-99-3	197.824	refrac hex cry	3080		7.8		i H_2O
1908	Niobium nitride	NbN	24621-21-4	106.913	gray cry; cub	2300		8.47		i HCl, acid
1909	Niobium phosphide	NbP	12034-66-1	123.880	tetr cry			6.5		
1910	Niobium silicide	$NbSi_2$	12034-80-9	149.077	gray hex cry	1950		5.7		
1911	Niobium(II) oxide	NbO	12034-57-0	108.905	gray cub cry	1937		7.30		
1912	Niobium(III) bromide	$NbBr_3$	15752-41-7	332.618	dark brn solid		subl 400			
1913	Niobium(III) chloride	$NbCl_3$	13569-59-0	199.265	blk solid					
1914	Niobium(III) fluoride	NbF_3	15195-53-6	149.901	blue cub cry			4.2		
1915	Niobium(III) iodide	NbI_3	13870-20-7	473.619	blk solid	dec 510				
1916	Niobium(IV) bromide	$NbBr_4$	13842-75-6	412.522	dark brn cry		subl 300	4.72		reac H_2O
1917	Niobium(IV) chloride	$NbCl_4$	13569-70-5	234.718	viol-blk monocl cry	dec 800	275 subl	3.2		reac H_2O
1918	Niobium(IV) fluoride	NbF_4	13842-88-1	168.900	blk tetr cry; hyg	>350 dec		4.01		
1919	Niobium(IV) iodide	NbI_4	13870-21-8	600.524	gray orth cry	503		5.6		reac H_2O
1920	Niobium(IV) oxide	NbO_2	12034-59-2	124.905	wh tetr cry or powder	1901		5.9		
1921	Niobium(IV) selenide	$NbSe_2$	12034-77-4	250.83	gray hex cry	>1300		6.3		
1922	Niobium(IV) sulfide	NbS_2	12136-97-9	157.036	blk rhomb cry			4.4		
1923	Niobium(IV) telluride	$NbTe_2$	12034-83-2	348.11	hex cry			7.6		
1924	Niobium(V) bromide	$NbBr_5$	13478-45-0	492.426	oran orth cry	265.2	361.6	4.36		s H_2O, EtOH
1925	Niobium(V) chloride	$NbCl_5$	10026-12-7	270.171	yel monocl cry; hyg	205.8	247.4	2.78		reac H_2O; s HCl, ctc
1926	Niobium(V) dioxyfluoride	NbO_2F	15195-33-2	143.903	wh cub cry			4.0		
1927	Niobium(V) ethoxide	$Nb(OC_2H_5)_5$	3236-82-6	318.209	col hyg liq	5	203	1.258		reac H_2O; s peth
1928	Niobium(V) fluoride	NbF_5	7783-68-8	187.898	col monocl cry; hyg	80	234	2.70		reac H_2O; sl CS_2, chl
1929	Niobium(V) iodide	NbI_5	13779-92-5	727.428	yel-blk monocl cry	327		5.32		reac H_2O
1930	Niobium(V) oxide	Nb_2O_5	1313-96-8	265.810	wh orth cry	1500		4.47		i H_2O; s HF
1931	Niobium(V) oxybromide	$NbOBr_3$	14459-75-7	348.617	yel-brn cry	≈320 dec	subl			
1932	Niobium(V) oxychloride	$NbOCl_3$	13597-20-1	215.264	wh tetr cry		subl	3.72		
1933	Nitrogen	N_2	7727-37-9	28.013	col gas	-210.0	-195.798	1.145 g/L		sl H_2O; i EtOH
1934	Nitramide	NO_2NH_2	7782-94-7	62.028	unstab wh cry	72 dec				s H_2O, EtOH, ace, eth; i chl
1935	Nitric acid	HNO_3	7697-37-2	63.013	col liq; hyg	-41.6	83	1.5129^{20}		vs H_2O
1936	Nitrous acid	HNO_2	7782-77-6	47.014	stab only in soln					
1937	Nitrous oxide	N_2O	10024-97-2	44.012	col gas	-90.8	-88.48	1.799 g/L		sl H_2O; s EtOH, eth
1938	Nitric oxide	NO	10102-43-9	30.006	col gas	-163.6	-151.74	1.226 g/L		sl H_2O

No.	Name	Formula	CAS Reg No.	Mol. weight	Physical form	mp/°C	bp/°C	Density g cm^{-3}	Solubility g/100 g H_2O	Qualitative solubility
1939	Nitrogen dioxide	NO_2	10102-44-0	46.006	brn gas; equil with N_2O_4		see N_2O_4	1.880 g/L		reac H_2O
1940	Nitrogen trioxide	N_2O_3	10544-73-7	76.011	blue solid or liq (low temp)	-101.1	≈3 dec	1.4^{2}		reac H_2O
1941	Nitrogen tetroxide	N_2O_4	10544-72-6	92.011	col liq; equil with NO_2	-9.3	21.15	1.45^{20}		reac H_2O
1942	Nitrogen pentoxide	N_2O_5	10102-03-1	108.010	col hex cry		33 sp	2.0		s chl; sl ctc
1943	Nitrogen tribromide	NBr_3	15162-90-0	253.719	unstab solid	exp -100				
1944	Nitrogen trichloride	NCl_3	10025-85-1	120.366	yel oily liq; exp	-40	71	1.653		i H_2O; s CS_2, bz, ctc
1945	Nitrogen trifluoride	NF_3	7783-54-2	71.002	col gas	-206.79	-128.75	2.902 g/L		i H_2O
1946	Nitrogen triiodide	NI_3	13444-85-4	394.720	unstab blk cry; exp					
1947	Nitrogen chloride difluoride	$NClF_2$	13637-87-1	87.457	col gas	-195	-67	3.575 g/L		
1948	Chloramine	NH_2Cl	10599-90-3	51.476	yel liq	-66				s H_2O, EtOH, eth; sl bz, ctc
1949	Fluoramine	NH_2F	15861-05-9	35.021	unstab gas	≈-110		1.431 g/L		
1950	Difluoramine	NHF_2	10405-27-3	53.012	col gas	-116	-23	2.167 g/L		
1951	*cis*-Difluorodiazine	N_2F_2	13812-43-6	66.010	col gas	<-195	-105.75	2.698 g/L		
1952	*trans*-Difluorodiazine	N_2F_2	13776-62-0	66.010	col gas	-172	-111.45	2.698 g/L		
1953	Tetrafluorohydrazine	N_2F_4	10036-47-2	104.007	col gas	-164.5	-74	4.251 g/L		
1954	Nitrosyl bromide	NOBr	13444-87-6	109.910	red gas	-56	≈0	4.492 g/L		reac H_2O
1955	Nitrosyl chloride	NOCl	2696-92-6	65.459	yel gas	-59.6	-5.5	2.676 g/L		reac H_2O
1956	Nitrosyl fluoride	NOF	7789-25-5	49.004	col gas	-132.5	-59.9	2.003 g/L		
1957	Trifluoramine oxide	NOF_3	13847-65-9	87.001	col gas	-161	-87.5	3.556 g/L		
1958	Nitryl chloride	NO_2Cl	13444-90-1	81.459	col gas	-145	-15	3.330 g/L		
1959	Nitryl fluoride	NO_2F	10022-50-1	65.004	col gas	-166	-72.4	2.657 g/L		reac H_2O
1960	Nitrogen selenide	N_4Se_4	12033-88-4	371.87	red monocl cry; hyg	exp		4.2		i H_2O, eth, EtOH; sl bz, CS_2
1961	Nobelium	No	10028-14-5	259.000	metal	827				
1962	Osmium	Os	7440-04-2	190.23	blue-wh metal; hex	3033	5012	22.587^{20}		s aqua regia
1963	Osmocene	$Os(C_5H_5)_2$	1273-81-0	320.42	col cry	229				
1964	Osmium carbonyl	$Os_3(CO)_{12}$	15696-40-9	906.81	yel cry	224		3.48		
1965	Osmium pentacarbonyl	$Os(CO)_5$	16406-49-8	330.28	col liq	-15	dec 100			s os
1966	Osmium nonacarbonyl	$Os_2(CO)_9$	28411-13-4	632.55	oran-yel cry	65 dec				s hc
1967	Osmium(II) chloride	$OsCl_2$	13444-92-3	261.14	hyg brn solid	dec >450				s EtOH, eth
1968	Osmium(III) bromide	$OsBr_3$	59201-51-3	429.94	dark gray cry	340 dec				i H_2O, os, acid
1969	Osmium(III) chloride	$OsCl_3$	13444-93-4	296.59	gray cub cry	450 dec				i H_2O, os; s conc acid
1970	Osmium(IV) chloride	$OsCl_4$	10026-01-4	332.04	red-blk orth cry	323 dec		4.38		reac H_2O; i os
1971	Osmium(IV) fluoride	OsF_4	54120-05-7	266.22	yel cry	230				reac H_2O
1972	Osmium(IV) oxide	OsO_2	12036-02-1	222.23	yel-brn tetr cry	dec 500		11.4		i H_2O, acid
1973	Osmium(V) fluoride	OsF_5	31576-40-6	285.22	hyg blue-grn cry	70	233			reac H_2O
1974	Osmium(VI) fluoride	OsF_6	13768-38-2	304.22	yel cub cry	33.4	47.5	4.1		reac H_2O
1975	Osmium(VI) tetrachloride oxide	$OsOCl_4$	36509-15-6	348.04	dark brn hyg cry	32	200			reac H_2O; s hc
1976	Osmium(VIII) oxide	OsO_4	20816-12-0	254.23	yel monocl cry	40.6	131.2	5.1	6.44^{20}	sl H_2O; s ctc, bz, EtOH, eth
1977	Oxygen	O_2	7782-44-7	31.999	col gas	-218.79	-182.953	1.308 g/L		sl H_2O, EtOH, os
1978	Ozone	O_3	10028-15-6	47.998	blue gas	-193	-111.35	1.962 g/L		sl H_2O
1979	Palladium	Pd	7440-05-3	106.42	silv-wh metal; cub	1554.8	2963	12.0		s aqua regia
1980	Palladium(II) acetate	$Pd(C_2H_3O_2)_2$	3375-31-3	224.51	oran-brn cry	205 dec				i H_2O; s MeCN, eth, ace
1981	Palladium(II) bromide	$PdBr_2$	13444-94-5	266.23	red-blk monocl cry; hyg	250 dec		≈5.2		i H_2O
1982	Palladium(II) chloride	$PdCl_2$	7647-10-1	177.33	red rhomb cry; hyg	679		4.0		s H_2O, EtOH, ace
1983	Palladium(II) chloride dihydrate	$PdCl_2 \cdot 2H_2O$	7647-10-1*	213.36	brn cry					s H_2O, EtOH, ace
1984	Palladium(II) cyanide	$Pd(CN)_2$	2035-66-7	158.45	yel solid	dec				
1985	Palladium(II) fluoride	PdF_2	13444-96-7	144.42	viol tetr cry; hyg	952		5.76		reac H_2O
1986	Palladium(II) iodide	PdI_2	7790-38-7	360.23	blk cry	360 dec		6.0		i H_2O, EtOH, eth
1987	Palladium(II) nitrate	$Pd(NO_3)_2$	10102-05-3	230.43	brn hyg cry	dec				sl H_2O; s dil HNO_3
1988	Palladium(II) oxide	PdO	1314-08-5	122.42	grn-blk tetr cry	750 dec		8.3		i H_2O, acid; sl aqua regia
1989	Palladium(II) 2,4-pentanedioate	$Pd(CH_3COCHCOCH_3)_2$	14024-61-4	304.64	oran-yel cry	205 dec				s bz, chl
1990	Palladium(II) sulfate dihydrate	$PdSO_4 \cdot 2H_2O$	13566-03-5	238.51	grn-brn cry	dec				
1991	Palladium(II) sulfide	PdS	12125-22-3	138.49	gray tetr cry			6.7		
1992	*cis*-Dichlorodiamminepalladium(II)	$Pd(NH_3)_2Cl_2$	15684-18-1	211.39	yel pow				0.025^{25}	
1993	*trans*-Dichlorodiamminepalladium (II)	$Pd(NH_3)_2Cl_2$	13782-33-7	211.39	yel solid			2.50		

No.	Name	Formula	CAS Reg No.	Mol. weight	Physical form	mp/°C	bp/°C	Density g cm⁻³	Solubility g/100 g H_2O	Qualitative solubility
1994	Phosphorus (white)	P	7723-14-0	30.974	col waxlike cub cry	44.15	280.5	1.823		i H_2O; sl bz, EtOH, chl; s CS_2
1995	Phosphorus (red)	P	7723-14-0	30.974	red-viol amorp powder	579.2	431 sp	2.16		i H_2O, os
1996	Phosphorus (black)	P	7723-14-0	30.974	blk orth cry or amorp solid	610		2.69		i os
1997	Phosphine	PH_3	7803-51-2	33.998	col gas; flam	-133.8	-87.75	1.390 g/L		i H_2O; sl EtOH, eth
1998	Diphosphine	P_2H_4	13445-50-6	65.980	col liq	-99	63.5 dec			reac H_2O
1999	Diphosphorus tetrachloride	P_2Cl_4	13497-91-1	203.760	col oily liq	-28	≈180 dec			
2000	Diphosphorus tetrafluoride	P_2F_4	13824-74-3	137.942	col gas	-86.5	-6.2	5.638 g/L		
2001	Diphosphorus tetraiodide	P_2I_4	13455-00-0	569.566	red tricl needles	125.5	dec	3.89		
2002	Phosphonium chloride	PH_4Cl	24567-53-1	70.459	gas		-27 sp	2.880 g/L		reac H_2O
2003	Phosphonium iodide	PH_4I	12125-09-6	161.910	col tetr cry	18.5	62.5	2.86		reac H_2O, EtOH
2004	Phosphoric acid	H_3PO_4	7664-38-2	97.995	col visc liq	42.4	407		548^{20}	s EtOH
2005	Phosphotungstic acid	$H_3PW_{12}O_{40}$	12067-99-1	2880.05	wh-yel cry	89				vs H_2O; s EtOH, eth
2006	Phosphonic acid	H_3PO_3	13598-36-2	81.996	wh hyg cry	74.4	200	1.65	309^{0}	vs EtOH
2007	Phosphinic acid	HPH_2O_2	6303-21-5	65.997	hyg cry or col oily liq	26.5	130	1.49		vs H_2O, EtOH, eth
2008	Metaphosphoric acid	HPO_3	37267-86-0	79.980	gl solid; hyg					sl H_2O; s EtOH
2009	Hypophosphoric acid	$H_4P_2O_6$	7803-60-3	161.976	col orth cry	73 dec				vs H_2O
2010	Diphosphoric acid	$H_4P_2O_7$	2466-09-3	177.975	wh cry	71.5			709^{23}	
2011	Difluorophosphoric acid	HPO_2F_2	13779-41-4	101.978	col liq	≈-94	110 dec	1.583		reac H_2O
2012	Hexafluorophosphoric acid	HPF_6	16940-81-1	145.972	col oily liq	25 dec				reac H_2O
2013	Fluorophosphonic acid	H_2PFO_3	13537-32-1	99.986	col visc liq	<-70		1.82		vs H_2O
2014	Phosphorus nitride (P_3N_5)	P_3N_5	12136-91-3	162.955	yel-brn solid	800 dec				i H_2O; s os
2015	Phosphorus sesquisulfide	P_4S_3	1314-85-8	220.090	yel-grn orth cry	173	407	2.03		i H_2O; s bz; vs CS_2
2016	Phosphorus heptasulfide	P_4S_7	12037-82-0	348.350	pale yel monocl cry	308	523	2.19		sl CS_2
2017	Phosphonitrilic chloride trimer	$(PNCl_2)_3$	940-71-6	347.659	wh hyg cry	128.8		1.98		reac H_2O
2018	Phosphorus(III) bromide	PBr_3	7789-60-8	270.686	col liq	-41.5	173.2	2.8		reac H_2O, EtOH; s ace, CS_2
2019	Phosphorus(III) dibromide fluoride	PBr_2F	15597-39-4	209.780	col liq	-115	78.5			
2020	Phosphorus(III) bromide difluoride	$PBrF_2$	15597-40-7	148.875	col gas	-133.8	-16.1	6.085 g/L		
2021	Phosphorus(III) chloride	PCl_3	7719-12-2	137.333	col liq	-93	76	1.574		reac H_2O, EtOH; s bz, chl, eth
2022	Phosphorus(III) dichloride fluoride	PCl_2F	15597-63-4	120.878	col gas	-144	13.85	4.941 g/L		
2023	Phosphorus(III) chloride difluoride	$PClF_2$	14335-40-1	104.424	col gas	-164.8	-47.3	4.268 g/L		
2024	Phosphorus(III) fluoride	PF_3	7783-55-3	87.969	col gas	-151.5	-101.8	3.596 g/L		reac H_2O
2025	Phosphorus(III) iodide	PI_3	13455-01-1	411.687	red-oran hex cry; hyg	61.2	227 dec	4.18		reac H_2O; s EtOH
2026	Phosphorus(III) oxide	P_2O_3	1314-24-5	109.946	col monocl cry or liq	23.8	173	2.13		reac H_2O
2027	Tetraphosphorus(III) hexoxide	P_4O_6	12440-00-5	219.891	soft wh cry	23.8	175.4			
2028	Phosphorus(III) selenide	P_2Se_3	1314-86-9	298.83	oran-red cry	245	≈380	1.31		reac H_2O; s bz, ctc, CS_2, ace
2029	Phosphorus(III) sulfide	P_2S_3	12165-69-4	158.143	yel solid	290	490			reac H_2O; s EtOH, eth, CS_2
2030	Phosphorus(V) bromide	PBr_5	7789-69-7	430.494	yel orth cry, hyg	≈100 dec		3.61		reac H_2O, EtOH; s CS_2, ctc
2031	Phosphorus(V) tetrabromide fluoride	PBr_4F		369.588	pale yel cry	87 dec				
2032	Phosphorus(V) dibromide trifluoride	PBr_2F_3	13445-58-4	247.777	yel-red liq	-20	15 dec			
2033	Phosphorus(V) chloride	PCl_5	10026-13-8	208.239	wh-yel tetr cry; hyg	167 tp	160 sp	2.1		reac H_2O; s CS_2, ctc
2034	Phosphorus(V) tetrachloride fluoride	PCl_4F	13498-11-8	191.784	col liq	-59	30 dec			
2035	Phosphorus(V) trichloride difluoride	PCl_3F_2	13537-23-0	175.330	col liq	-63				
2036	Phosphorus(V) dichloride trifluoride	PCl_2F_3	13454-99-4	158.875	col gas	-125	7.1	6.494 g/L		
2037	Phosphorus(V) chloride tetrafluoride	$PClF_4$	13498-11-8	142.421	col gas	-132	-43.4	5.821 g/L		
2038	Phosphorus(V) fluoride	PF_5	7647-19-0	125.966	col gas	-93.8	-84.6	5.149 g/L		reac H_2O
2039	Phosphorus(V) oxide	P_2O_5	1314-56-3	141.945	wh orth cry; hyg	562	605	2.30		reac H_2O, EtOH
2040	Phosphorus(V) selenide	P_2Se_5	1314-82-5	456.75	blk-purp amorp solid					reac hot H_2O, ctc; i CS_2
2041	Phosphorus(V) sulfide	P_2S_5	1314-80-3	222.273	grn-yel hyg cry	285	515	2.03		reac H_2O; s CS_2
2042	Phosphonic difluoride	POF_2H	14939-34-5	85.978	volatile liq	>-120	≈60 (gas unstab)			

No.	Name	Formula	CAS Reg No.	Mol. weight	Physical form	mp/°C	bp/°C	Density g cm^{-3}	Solubility g/100 g H_2O	Qualitative solubility
2043	Phosphoryl bromide	$POBr_3$	7789-59-5	286.685	faint oran plates	55	191.7	2.822		reac H_2O; s bz, eth, chl
2044	Phosphoryl dibromide chloride	$POBr_2Cl$	13550-31-7	242.234	yel solid	31	165			
2045	Phosphoryl dibromide fluoride	$POBr_2F$	14014-19-8	225.779	col liq	-117.2	110.1			
2046	Phosphoryl bromide dichloride	$POBrCl_2$	13455-03-3	197.783	col liq	11	136.5	2.104^{14}		
2047	Phosphoryl bromide difluoride	$POBrF_2$	14014-18-7	164.874	col liq	-84.8	31.6			
2048	Phosphoryl bromide chloride fluoride	POBrClF	14518-81-1	181.328	col liq		79			
2049	Phosphoryl chloride	$POCl_3$	10025-87-3	153.332	col liq	1.18	105.5	1.645		reac H_2O, EtOH
2050	Phosphoryl dichloride fluoride	$POCl_2F$	13769-76-1	136.877	col liq	-80.1	52.9			
2051	Phosphoryl chloride difluoride	$POClF_2$	13769-75-0	120.423	col gas	-96.4	3.1	4.922 g/L		
2052	Phosphoryl fluoride	POF_3	13478-20-1	103.968	col gas	-39.1 tp	-39.7 sp	4.250 g/L		reac H_2O
2053	Phosphoryl iodide	POI_3	13455-04-4	427.686	viol cry	53				
2054	Phosphorothioc tribromide	$PSBr_3$	3931-89-3	302.751	yel cry	37.8	212 dec	2.85		
2055	Phosphorothioc dibromide fluoride	$PSBr_2F$	13706-10-0	241.845	yel liq	-75.2	125.3			
2056	Phosphorothioc bromide difluoride	$PSBrF_2$	13706-09-7	180.940	yel liq	-136.9	35.5			
2057	Phosphorothioc trichloride	$PSCl_3$	3982-91-0	169.398	fuming liq	-36.2	125	1.635		reac H_2O; s bz, ctc, chl, CS_2
2058	Phosphorothioc dichloride fluoride	$PSCl_2F$	155698-29-6	152.943	col liq	-96.0	64.7			
2059	Phosphorothioc chloride difluoride	$PSClF_2$	2524-02-9	136.489	col gas	-155.2	6.3	5.579 g/L		
2060	Phosphorothioc trifluoride	PSF_3	2404-52-6	120.034	col gas	-148.8	-52.25	4.906 g/L		
2061	Phosphorothioc triiodide	PSI_3	63972-04-3	443.752	yel cry	48	dec			
2062	Platinum	Pt	7440-06-4	195.084	silv-gray metal; cub	1768.2	3825	21.5		i acid; s aqua regia
2063	Hexachloroplatinic acid	H_2PtCl_6	16941-12-1	409.818	hyg yel-brn cry	60				s H_2O, EtOH
2064	Hydrogen hexahydroxyplatinate(IV)	$H_2Pt(OH)_6$	51850-20-5	299.144	yel needles	dec 100				s H_2O, acid, dil alk
2065	Platinum(II) bromide	$PtBr_2$	13455-12-4	354.892	red-brn powder	250 dec		6.65		i H_2O
2066	Platinum(II) chloride	$PtCl_2$	10025-65-7	265.990	grn hex cry	581 dec		6.0		i H_2O, EtOH, eth; s HCl
2067	Platinum(II) cyanide	$Pt(CN)_2$	592-06-3	247.118	pale yel cry					i H_2O, acid, alk
2068	Platinum(II) iodide	PtI_2	7790-39-8	448.893	blk powder	325 dec		6.4		i H_2O
2069	Platinum(II) oxide	PtO	12035-82-4	211.083	blk tetr cry	325 dec		14.1		i H_2O, EtOH; s aqua regia
2070	Platinum(II) sulfide	PtS	12038-20-9	227.149	tetr cry			10.25		
2071	Platinum(III) bromide	$PtBr_3$	25985-07-3	434.796	grn-blk cry	200 dec				
2072	Platinum(III) chloride	$PtCl_3$	25909-39-1	301.443	grn-blk cry	435 dec		5.26		
2073	Platinum(IV) bromide	$PtBr_4$	68938-92-1	514.700	brn-blk cry	180 dec			0.41^{20}	sl EtOH, eth
2074	Platinum(IV) chloride	$PtCl_4$	37773-49-2	336.896	red-brn cub cry	327 dec		4.30	142^{25}	
2075	Platinum(IV) chloride pentahydrate	$PtCl_4 \cdot 5H_2O$	13454-96-1	426.972	red cry			2.43		s H_2O, EtOH
2076	Platinum(IV) fluoride	PtF_4	13455-15-7	271.078	red cry	600				
2077	Platinum(IV) iodide	PtI_4	7790-46-7	702.702	brn-blk powder	130 dec				s H_2O
2078	Platinum(IV) oxide	PtO_2	1314-15-4	227.083	blk hex cry	450		11.8		i H_2O; s conc acid, dil alk
2079	Platinum(IV) sulfide	PtS_2	12038-21-0	259.214	hex cry			7.85		
2080	Platinum(VI) fluoride	PtF_6	13693-05-5	309.074	red cub cry	61.3	69.1	≈4.0		
2081	*cis*-Diamminedichloroplatinum	$Pt(NH_3)_2Cl_2$	15663-27-1	300.051	yel solid	270 dec			0.253^{25}	
2082	*trans*-Diamminedichloroplatinum	$Pt(NH_3)_2Cl_2$	14913-33-8	300.051	pale yel solid	270 dec			0.036^{25}	s DMF, DMSO
2083	Hexachloroplatinic acid hexahydrate	$H_2PtCl_6 \cdot 6H_2O$	16941-12-1	517.909	brn-yel hyg cry	60		2.43	140^{18}	vs EtOH
2084	Platinum silicide	PtSi	12137-83-6	223.170	orth cry	1229		12.4		
2085	Plutonium	Pu	7440-07-5	244	silv-wh metal; monocl	640	3228	19.7		
2086	Plutonium nitride	PuN	12033-54-4	258	gray cub cry	2550		14.4		
2087	Plutonium(II) oxide	PuO	12035-83-5	260	cub cry			14.0		
2088	Plutonium(III) bromide	$PuBr_3$	15752-46-2	484	grn orth cry	681		6.75		s H_2O
2089	Plutonium(III) chloride	$PuCl_3$	13569-62-5	350	grn hex cry	760		5.71		s H_2O
2090	Plutonium(III) fluoride	PuF_3	13842-83-6	301	purp hex cry	1396		9.33		i H_2O; sl acid
2091	Plutonium(III) iodide	PuI_3	13813-46-2	625	grn orth cry; hyg	777		6.92		s H_2O
2092	Plutonium(III) oxide	Pu_2O_3	12036-34-9	536	blk cub cry	2085		10.5		
2093	Plutonium(IV) fluoride	PuF_4	13709-56-3	320	red-brn monocl cry	1037		7.1		
2094	Plutonium(IV) oxide	PuO_2	12059-95-9	276	yel-brn cub cry	2390		11.5		
2095	Plutonium(VI) fluoride	PuF_6	13693-06-6	358	red-brn orth cry	51.6		5.08		
2096	Polonium	Po	7440-08-6	209	silv metal; cub	254	962	9.20		
2097	Polonium(IV) chloride	$PoCl_4$	10026-02-5	351	yel hyg cry	≈300	390			s H_2O, EtOH, ace
2098	Polonium(IV) oxide	PoO_2	7446-06-2	241	yel cub cry	500 dec		8.9		
2099	Potassium	K	7440-09-7	39.098	soft silv-wh metal; cub	63.5	759	0.89		reac H_2O
2100	Potassium acetate	$KC_2H_3O_2$	127-08-2	98.142	wh hyg cry	309		1.57	269^{25}	s EtOH; i eth

No.	Name	Formula	CAS Reg No.	Mol. weight	Physical form	mp/°C	bp/°C	Density g cm^{-3}	Solubility g/100 g H_2O	Qualitative solubility
2101	Potassium aluminate trihydrate	$K_2Al_2O_4 \cdot 3H_2O$	12003-63-3*	250.204	wh orth cry			2.13		vs H_2O; i EtOH
2102	Potassium aluminum silicate	$KAlSi_3O_8$	1327-44-2	278.332	col monocl cry			2.56		i H_2O
2103	Potassium aluminum sulfate	$KAl(SO_4)_2$	10043-67-1	258.205	wh hyg powder				5.9^{20}	
2104	Potassium aluminum sulfate dodecahydrate	$KAl(SO_4)_2 \cdot 12H_2O$	7784-24-9	474.389	col cry	≈100 dec		1.72	5.9^{20}	
2105	Potassium amide	KNH_2	17242-52-3	55.121	wh/yel-grn hyg cry	335				reac H_2O, EtOH
2106	Potassium arsenate	K_3AsO_4	13464-36-3	256.215	col cry			2.8	125^{25}	
2107	Potassium arsenite	$KAsO_2$	13464-35-2	146.019	wh hyg powder					s H_2O; sl EtOH
2108	Potassium azide	KN_3	20762-60-1	81.118	tetr cry; exp			2.04	49.7^{17}	
2109	Potassium borohydride	KBH_4	13762-51-1	53.941	wh cub cry	≈500 dec		1.11		s H_2O
2110	Potassium bromate	$KBrO_3$	7758-01-2	167.000	wh hex cry	434 dec		3.27	8.17^{25}	i EtOH
2111	Potassium bromide	KBr	7758-02-3	119.002	col cub cry; hyg	734	1435	2.74	67.8^{25}	sl EtOH
2112	Potassium carbonate	K_2CO_3	584-08-7	138.206	wh monocl cry; hyg	899	dec	2.29	111^{25}	i EtOH
2113	Potassium carbonate sesquihydrate	$K_2CO_3 \cdot 1.5H_2O$	6381-79-9	165.229	granular cry				111^{20}	
2114	Potassium chlorate	$KClO_3$	3811-04-9	122.549	wh monocl cry	357	dec	2.34	8.61^{25}	
2115	Potassium chloride	KCl	7447-40-7	74.551	wh cub cry	771		1.988	35.5^{25}	i eth, ace
2116	Potassium chlorochromate	$KCrO_3Cl$	16037-50-6	174.545	oran cry			2.5		reac H_2O; s ace, acid
2117	Potassium chromate	K_2CrO_4	7789-00-6	194.191	yel orth cry	974		2.73	65.0^{25}	
2118	Potassium citrate monohydrate	$K_3C_6H_5O_7 \cdot H_2O$	6100-05-6	324.410	col hyg cry	180 dec		1.98	172^{20}	vs H_2O; sl EtOH
2119	Potassium cobalt(II) selenate hexahydrate	$K_2Co(SeO_4)_2 \cdot 6H_2O$	28041-86-3	531.14	red monocl cry			2.51		
2120	Potassium cyanate	KCNO	590-28-3	81.115	wh tetr cry	≈700 dec		2.05	75^{25}	sl EtOH
2121	Potassium cyanide	KCN	151-50-8	65.116	wh cub cry; hyg	622		1.55	69.9^{20}	sl EtOH
2122	Potassium cyanoaurite	$KAu(CN)_2$	13967-50-5	288.099	col cry			3.45	14^{20}	s H_2O; sl EtOH; i eth, ace
2123	Potassium dichromate	$K_2Cr_2O_7$	7778-50-9	294.185	oran-red tricl cry	398	≈500 dec	2.68	15.1^{25}	
2124	Potassium dihydrogen arsenate	KH_2AsO_4	7784-41-0	180.034	col cry	288		2.87	19^{6}	i EtOH
2125	Potassium dihydrogen phosphate	KH_2PO_4	7778-77-0	136.085	wh tetr cry	253 dec		2.34	25.0^{25}	sl EtOH
2126	Potassium dihydrogen phosphonate	KH_2PO_3	13977-65-6	120.086	col monocl hyg cry					
2127	Potassium dithionate	$K_2S_2O_6$	13455-20-4	238.323	col hex cry	dec		2.27		sl H_2O; i EtOH
2128	Potassium ferricyanide	$K_3Fe(CN)_6$	13746-66-2	329.244	red cry	dec		1.89	48.8^{25}	
2129	Potassium ferrocyanide trihydrate	$K_4Fe(CN)_6 \cdot 3H_2O$	14459-95-1	422.388	yel monocl cry	60 dec		1.85	36.0^{25}	i EtOH, eth
2130	Potassium fluoride	KF	7789-23-3	58.096	wh cub cry	858	1502	2.48	102^{25}	
2131	Potassium fluoride dihydrate	$KF \cdot 2H_2O$	13455-21-5	94.127	monocl cry	41 dec		2.5	102^{25}	
2132	Potassium fluoroborate	KBF_4	14075-53-7	125.903	col orth cry	530		2.505	0.55^{25}	sl EtOH
2133	Potassium fluorotantalate	K_2TaF_7	16924-00-8	392.134	col cry	730		5.24	0.5^{0}	
2134	Potassium formate	$KCHO_2$	590-29-4	84.116	col hyg cry	167		1.91	331^{18}	
2135	Potassium hexachloroosmate(IV)	K_2OsCl_6	16871-60-6	481.15	red cub cry					vs H_2O; sl EtOH
2136	Potassium hexachloroplatinate	K_2PtCl_6	16921-30-5	485.999	yel-oran cub cry	250 dec		3.50	0.77^{20}	i EtOH
2137	Potassium hexacyanocobaltate	$K_3Co(CN)_6$	13963-58-1	332.332	yel monocl cry	dec		1.91		vs H_2O; i EtOH
2138	Potassium hexafluoromanganate(IV)	K_2MnF_6	16962-31-5	247.125	yel hex cry					reac H_2O
2139	Potassium hexafluorosilicate	K_2SiF_6	16871-90-2	220.273	wh cry	dec		2.27	0.084^{20}	i EtOH
2140	Potassium hexafluorozirconate(IV)	K_2ZrF_6	16923-95-8	283.411	col monocl cry			3.48	0.78^{2}	
2141	Potassium hydride	KH	7693-26-7	40.106	cub cry	619		1.43		reac H_2O
2142	Potassium hydrogen arsenate	K_2HAsO_4	21093-83-4	218.125	col monocl prisms	300 dec			18.7^{6}	i EtOH
2143	Potassium hydrogen carbonate	$KHCO_3$	298-14-6	100.115	col monocl cry	≈100 dec		2.17	36.2^{25}	i EtOH
2144	Potassium hydrogen fluoride	KHF_2	7789-29-9	78.103	col tetr cry	238.8		2.37	39.2^{20}	i EtOH
2145	Potassium hydrogen iodate	$KH(IO_3)_2$	13455-24-8	389.911	col cry	dec			1.3^{15}	sl H_2O; i EtOH
2146	Potassium hydrogen oxalate hemihydrate	$KHC_2O_4 \cdot 0.5H_2O$	127-95-7		wh cry	dec		2.09	2.5^{20}	sl EtOH
2147	Potassium hydrogen phosphate	K_2HPO_4	7758-11-4	174.176	wh hyg cry	dec			168^{25}	s EtOH
2148	Potassium hydrogen phosphite	K_2HPO_3	13492-26-7	158.177	wh hyg powder	dec			170^{20}	i EtOH
2149	Potassium hydrogen selenite	$KHSeO_3$	7782-70-9	167.06	hyg orth cry	>100 dec				s H_2O; sl EtOH
2150	Potassium hydrogen sulfate	$KHSO_4$	7646-93-7	136.169	wh monocl cry; hyg	≈200		2.32	50.6^{25}	
2151	Potassium hydrogen sulfide	KHS	1310-61-8	72.171	wh hex cry; hyg	≈450		1.69		s H_2O, EtOH
2152	Potassium hydrogen sulfide hemihydrate	$KHS \cdot 0.5H_2O$	1310-61-8*	81.179	wh-yel hyg cry	≈175		1.7		vs H_2O, EtOH
2153	Potassium hydrogen sulfite	$KHSO_3$	7773-03-7	120.169	wh cry powder	190 dec			49^{20}	i EtOH
2154	Potassium hydrogen tartrate	$KHC_4H_4O_6$	868-14-4	188.177	wh cry			1.98	0.57^{20}	s acid, alk; i EtOH
2155	Potassium hydroxide	KOH	1310-58-3	56.105	wh rhomb cry; hyg	406	1327	2.044	121^{25}	s EtOH; s MeOH
2156	Potassium hypochlorite	KOCl	7778-66-7	90.550	exists only in aq soln					
2157	Potassium phosphinate	KH_2PO_2	7782-87-8	104.087	wh hyg cry	dec				vs H_2O; s EtOH

No.	Name	Formula	CAS Reg No.	Mol. weight	Physical form	mp/°C	bp/°C	Density g cm^{-3}	Solubility g/100 g H_2O	Qualitative solubility
2158	Potassium iodate	KIO_3	7758-05-6	214.001	wh monocl cry	560 dec		3.89	9.22^{25}	
2159	Potassium iodide	KI	7681-11-0	166.003	col cub cry	681	1323	3.12	148^{25}	sl EtOH
2160	Potassium iron(III) oxalate trihydrate	$K_3Fe(C_2O_4)_3 \cdot 3H_2O$		491.243	grn monocl cry	100	230 dec	2.133	4.7^{0}	i EtOH
2161	Potassium manganate	K_2MnO_4	10294-64-1	197.133	grn cry	190 dec				s H_2O; reac HCl
2162	Potassium metaarsenate	$KAsO_3$	19197-73-0	162.018	wh solid	660				
2163	Potassium metabisulfite	$K_2S_2O_5$	16731-55-8	222.324	wh powder	≈150 dec		2.3	49.5^{25}	reac acid; i EtOH
2164	Potassium metaborate	KBO_2	13709-94-9	81.908	wh hex cry	947		≈2.3		
2165	Potassium molybdate	K_2MoO_4	13446-49-6	238.14	wh hyg cry	919		2.3	183^{25}	i EtOH
2166	Potassium niobate	$KNbO_3$	12030-85-2	180.002	wh rhomb cry	≈1100		4.64		i H_2O
2167	Potassium nitrate	KNO_3	7757-79-1	101.103	col orth cry or powder	334	400 dec	2.105	38.3^{25}	i EtOH
2168	Potassium nitrite	KNO_2	7758-09-0	85.104	wh hyg cry	438	537 exp	1.915	312^{25}	sl EtOH
2169	Potassium oxalate	$K_2C_2O_4$	583-52-8	166.216	wh pwd					sl H_2O
2170	Potassium oxalate monohydrate	$K_2C_2O_4 \cdot H_2O$	6487-48-5	184.231	col cry	160 dec		2.13	36.4^{20}	
2171	Potassium oxide	K_2O	12136-45-7	94.196	gray cub cry	740		2.35		s H_2O, EtOH, eth
2172	Potassium perbromate	$KBrO_4$	22207-96-1	183.000	wh cry	275 dec			4.21^{25}	
2173	Potassium percarbonate monohydrate	$K_2C_2O_6 \cdot H_2O$	589-97-9	216.230	oran or blue pow				6.5^{20}	
2174	Potassium perchlorate	$KClO_4$	7778-74-7	138.549	col orth cry; hyg	525		2.52	2.08^{25}	
2175	Potassium periodate	KIO_4	7790-21-8	230.001	col tetr cry	582	exp	3.618	0.51^{25}	
2176	Potassium permanganate	$KMnO_4$	7722-64-7	158.034	purp orth cry	dec		2.7	7.60^{25}	reac EtOH
2177	Potassium peroxide	K_2O_2	17014-71-0	110.196	yel amorp solid	490				reac H_2O
2178	Potassium persulfate	$K_2S_2O_8$	7727-21-1	270.322	col cry	≈100 dec		2.48	4.7^{20}	
2179	Potassium phosphate	K_3PO_4	7778-53-2	212.266	wh orth cry; hyg	1340		2.564	106^{25}	i EtOH
2180	Potassium pyrophosphate	$K_4P_2O_7$	7320-34-5		wh cry	dec 1300				s H_2
2181	Potassium pyrophosphate trihydrate	$K_4P_2O_7 \cdot 3H_2O$	7790-67-2	384.383	col hyg cry	dec 300		2.33		vs H_2O; i EtOH
2182	Potassium pyrosulfate	$K_2S_2O_7$	7790-62-7	254.323	col needles	≈325		2.28		s H_2O
2183	Potassium selenate	K_2SeO_4	7790-59-2	221.16	wh powder			3.07	111[illegible]	
2184	Potassium selenide	K_2Se	1312-74-9	157.16	red cub cry; hyg	800		2.29		s H_2O
2185	Potassium selenite	K_2SeO_3	10431-47-7	205.16	wh hyg cry	875 dec			217^{25}	sl EtOH
2186	Potassium silver cyanide	$KAg(CN)_2$	506-61-6	199.000	wh cry					s H_2O
2187	Potassium sodium tartrate tetrahydrate	$KNaC_4H_4O_6 \cdot 4H_2O$	6381-59-5	282.220	wh cry	≈70 dec	anh at 130	1.79		vs H_2O; i EtOH
2188	Potassium stannate trihydrate	$K_2SnO_3 \cdot 3H_2O$	12142-33-5*	298.951	col cry			3.20		vs H_2O; i EtOH
2189	Potassium stearate	$KC_{18}H_{35}O_2$	593-29-3	322.568	wh pow					sl cold H_2O; s hot H_2O, EtOH
2190	Potassium sulfate	K_2SO_4	7778-80-5	174.260	wh orth cry	1069		2.66	12.0^{25}	i EtOH
2191	Potassium sulfide	K_2S	1312-73-8	110.262	red-yel cub cry; hyg	948		1.74		s H_2O, EtOH; i eth
2192	Potassium sulfide pentahydrate	$K_2S \cdot 5H_2O$	37248-34-3	200.338	col rhomb cry	60				vs H_2O, EtOH; i eth
2193	Potassium sulfite	K_2SO_3	10117-38-1	158.260	col hex cry				106^{25}	sl EtOH
2194	Potassium sulfite dihydrate	$K_2SO_3 \cdot 2H_2O$	7790-56-9	194.291	wh monocl cry	dec			107^{20}	sl EtOH; dec dil acid
2195	Potassium superoxide	KO_2	12030-88-5	71.097	yel tetr cry; hyg	380		2.16		reac H_2O
2196	Potassium tellurate(VI) trihydrate	$K_2TeO_4 \cdot 3H_2O$	15571-91-2*	323.84	wh cry powder					s H_2O
2197	Potassium tellurite	K_2TeO_3	7790-58-1	253.80	wh hyg cry	≈460 dec				vs H_2O
2198	Potassium tetraborate pentahydrate	$K_2B_4O_7 \cdot 5H_2O$	1332-77-0	323.513	wh cry powder				16.5^{30}	sl EtOH
2199	Potassium tetrachloroaurate dihydrate	$KAuCl_4 \cdot 2H_2O$	13682-61-6	413.908	yel monocl cry					s H_2O, EtOH, eth
2200	Potassium tetrachloroplatinate	K_2PtCl_4	10025-99-7	415.093	pink-red tetr cry	500 dec		3.38		s H_2O; i EtOH
2201	Potassium tetracyanocadmate	$K_2Cd(CN)_4$	14402-75-6	294.678	cub cry	≈450		1.85	25^{20}	sl EtOH
2202	Potassium tetracyanonickelate monohydrate	$K_2[Ni(CN)_4] \cdot H_2O$	14220-17-8*	258.975	red-oran cry	dec 100				
2203	Potassium tetracyanoplatinate(II) trihydrate	$K_2Pt(CN)_4 \cdot 3H_2O$	562-76-5*	431.397	col rhomb prisms					s H_2O
2204	Potassium tetracyanozincate	$K_2Zn(CN)_4$	14244-62-3	247.676	cry pow					vs H_2O
2205	Potassium tetraiodomercurate(II)	K_2HgI_4	7783-33-7	786.40	yel hyg cry			4.29		vs H_2O; s EtOH, eth, ace
2206	Potassium thiocyanate	$KSCN$	333-20-0	97.181	col tetr cry; hyg	173	500 dec	1.88	238^{25}	s EtOH
2207	Potassium thiosulfate	$K_2S_2O_3$	10294-66-3	190.325	col hyg cry				165^{25}	i EtOH
2208	Potassium titanate	K_2TiO_3	12030-97-6	174.062	wh orth cry	1515		3.1		reac H_2O
2209	Potassium triiodide monohydrate	$KI_3 \cdot H_2O$	7790-42-3	437.827	brn monocl cry; hyg	225 dec		3.5		s H_2O; reac EtOH, eth
2210	Potassium triiodozincate	$KZnI_3$	7790-43-4	485.221	hyg cry					vs H_2O
2211	Potassium thiocarbonate	K_2CS_3	26750-66-3	186.403	yel-red hyg cry					vs H_2O
2212	Potassium tungstate	K_2WO_4	7790-60-5	326.04	hyg cry	921		3.12		vs H_2O; i EtOH
2213	Potassium uranate	$K_2U_2O_7$	7790-63-8	666.251	oran cub cry			6.12		i H_2O; s acid

No.	Name	Formula	CAS Reg No.	Mol. weight	Physical form	mp/°C	bp/°C	Density g cm^{-3}	Solubility g/100 g H_2O	Qualitative solubility
2214	Potassium uranyl nitrate	$K(UO_2)(NO_3)_3$	18078-40-5	495.140	grn-yel cry pow					vs H_2O
2215	Potassium uranyl sulfate dihydrate	$K_2(UO_2)(SO_4)_2 \cdot 2H_2O$	27709-53-1	576.381	grn-yel cry pow	dec 120		3.36		vs H_2O
2216	Potassium zinc sulfate hexahydrate	$K_2Zn(SO_4)_2 \cdot 6H_2O$	13932-17-7	443.823	cry					s H_2O
2217	Potassium zirconium sulfate trihydrate	$K_4Zr(SO_4)_4 \cdot 3H_2O$	53608-79-0	685.914	wh cry pow					sl H_2O
2218	Praseodymium	Pr	7440-10-0	140.908	silv metal; hex	931	3520	6.77		
2219	Praseodymium boride	PrB_6	12008-27-4	205.774	blk cub cry	2610		4.84		
2220	Praseodymium nitride	PrN	25764-09-4	154.915	cub cry			7.46		
2221	Praseodymium silicide	$PrSi_2$	12066-83-0	197.079	tetr cry	1712		5.46		
2222	Praseodymium(II) iodide	PrI_2	65530-47-4	394.717	bronze solid	758				
2223	Praseodymium(III) bromate	$Pr(BrO_3)_3$	15162-93-3	524.615	grn cry					vs H_2O
2224	Praseodymium(III) bromide	$PrBr_3$	13536-53-3	380.620	grn hex cry; hyg	693		5.28		s H_2O
2225	Praseodymium(III) carbonate octahydrate	$Pr_2(CO_3)_3 \cdot 8H_2O$	14948-62-0	605.964	grn silky plates	dec 420 (anh)				i H_2O; s acid
2226	Praseodymium(III) chloride	$PrCl_3$	10361-79-2	247.267	grn hex needles; hyg	786		4.0	96.1^{25}	s EtOH
2227	Praseodymium(III) chloride heptahydrate	$PrCl_3 \cdot 7H_2O$	10025-90-8	373.374	grn cry	110 dec			96.1^{25}	s EtOH
2228	Praseodymium(III) fluoride	PrF_3	13709-46-1	197.903	grn hex cry	1399		6.3		
2229	Praseodymium(III) hydroxide	$Pr(OH)_3$	16469-16-2	191.930	grn solid	dec 220		3.7		i H_2O
2230	Praseodymium(III) iodide	PrI_3	13813-23-5	521.621	orth hyg cry	738		≈5.8		s H_2O
2231	Praseodymium(III) nitrate	$Pr(NO_3)_3$	10361-80-5	326.923	pale grn hyg cry				165^{25}	s EtOH
2232	Praseodymium(III) nitrate hexahydrate	$Pr(NO_3)_3 \cdot 6H_2O$	15878-77-0	435.014	grn needles				165^{25}	s EtOH, ace
2233	Praseodymium(III) oxide	Pr_2O_3	12036-32-7	329.813	wh hex cry	2183	3760	6.9		
2234	Praseodymium(III) perchlorate hexahydrate	$Pr(ClO_4)_3 \cdot 6H_2O$	13498-07-2*	547.351	hyg grn cry	dec 200				s H_2O, EtOH
2235	Praseodymium(III) sulfate octahydrate	$Pr_2(SO_4)_3 \cdot 8H_2O$	13510-41-3	714.125	grn monocl cry			2.83	17^{20}	s H_2O
2236	Praseodymium(III) sulfide	Pr_2S_3	12038-13-0	378.010	cub cry	1765		5.1		
2237	Praseodymium(III) telluride	Pr_2Te_3	12038-12-9	664.62	cub cry	1500		≈7.0		
2238	Praseodymium(IV) fluoride	PrF_4	15192-24-2	216.902	yel-wh solid	dec 90				
2239	Promethium	Pm	7440-12-2	145	silv metal; hex	1042	3000	7.26		
2240	Promethium(III) bromide	$PmBr_3$	14325-78-1	385	red cry	625				s H_2O
2241	Promethium(III) chloride	$PmCl_3$	13779-10-7	251	pale blue hyg cry	655				s H_2O
2242	Promethium(III) fluoride	PmF_3	13709-45-0	202	pink solid	1338				s H_2O
2243	Promethium(III) iodide	PmI_3	13818-73-0	526	red solid	695				
2244	Protactinium	Pa	7440-13-3	231.036	shiny metal; tetr or cub	1572		15.4		
2245	Protactinium(V) chloride	$PaCl_5$	13760-41-3	408.301	yel monocl cry	306		3.74		
2246	Radium	Ra	7440-14-4	226	wh metal; cub	696		5		
2247	Radium bromide	$RaBr_2$	10031-23-9	386	wh orth cry	728		5.79	70.6^{20}	s EtOH
2248	Radium carbonate	$RaCO_3$	7116-98-5	286	wh orth cry					i H_2O
2249	Radium chloride	$RaCl_2$	10025-66-8	297	wh orth cry	1000		4.9	24.5^{20}	s EtOH
2250	Radium fluoride	RaF_2	20610-49-5	264	wh cub cry			6.7		
2251	Radium nitrate	$Ra(NO_3)_2$	10213-12-4	350	cry				13.9	
2252	Radium sulfate	$RaSO_4$	7446-16-4	322	wh cry					i H_2O, acid
2253	Radon	Rn	10043-92-2	222	col gas	-71	-61.7	9.074 g/L		sl H_2O
2254	Rhenium	Re	7440-15-5	186.207	silv-gray metal	3185	5596	20.8		i HCl
2255	Perrhenic acid	$HReO_4$	13768-11-1	251.213	exists only in soln					vs H_2O, os
2256	Rhenium carbonyl	$Re_2(CO)_{10}$	14285-68-8	652.515	yel-wh cry	170 dec		2.87		s os
2257	Rhenium pentacarbonyl bromide	$Re(CO)_5Br$	14220-21-4	406.162	wh cry	90				
2258	Rhenium pentacarbonyl chloride	$Re(CO)_5Cl$	14099-01-5	361.711	wh cry		subl 140			
2259	Rhenium(III) bromide	$ReBr_3$	13569-49-8	425.919	red-brn monocl cry		500 subl	6.10		s ace, MeOH, EtOH
2260	Rhenium(III) chloride	$ReCl_3$	13569-63-6	292.566	red-blk hyg cry	500 dec		4.81		s H_2O
2261	Rhenium(III) iodide	ReI_3	15622-42-1	566.920	blk solid	dec				
2262	Rhenium(IV) chloride	$ReCl_4$	13569-71-6	328.019	purp-blk cry; hyg	300 dec		4.9		
2263	Rhenium(IV) fluoride	ReF_4	15192-42-4	262.201	blue tetr cry		>300 subl	7.49		
2264	Rhenium(IV) oxide	ReO_2	12036-09-8	218.206	gray orth cry	900 dec		11.4		
2265	Rhenium(IV) selenide	$ReSe_2$	12038-64-1	344.13	tricl cry					
2266	Rhenium(IV) silicide	$ReSi_2$	12038-66-3	242.378	refrac solid	2000				
2267	Rhenium(IV) sulfide	ReS_2	12038-63-0	250.337	tricl cry			7.6		
2268	Rhenium(IV) telluride	$ReTe_2$	12067-00-4	441.41	orth cry			8.50		
2269	Rhenium(V) bromide	$ReBr_5$	30937-53-2	585.727	brn solid	110 dec				
2270	Rhenium(V) chloride	$ReCl_5$	39368-69-9	363.472	brn-blk solid	220		4.9		reac H_2O
2271	Rhenium(V) fluoride	ReF_5	30937-52-1	281.199	yel-grn solid	48	221.3			
2272	Rhenium(V) oxide	Re_2O_5	12165-05-8	452.411	blue-blk tetr cry			≈7		

No.	Name	Formula	CAS Reg No.	Mol. weight	Physical form	mp/°C	bp/°C	Density g cm^{-3}	Solubility g/100 g H_2O	Qualitative solubility
2273	Rhenium(VI) chloride	$ReCl_6$	31234-26-1	398.925	red-grn solid	29				
2274	Rhenium(VI) dioxydifluoride	ReO_2F_2	81155-18-2	256.203	col cry	156				
2275	Rhenium(VI) fluoride	ReF_6	10049-17-9	300.197	yel liq or cub cry	18.5	33.8	4.06(cry)		s HNO_3
2276	Rhenium(VI) oxide	ReO_3	1314-28-9	234.205	red cub cry	400 dec		6.9		i H_2O, acid, alk
2277	Rhenium(VI) oxytetrachloride	$ReOCl_4$	13814-76-1	344.018	brn cry	29.3	223			reac H_2O
2278	Rhenium(VI) oxytetrafluoride	$ReOF_4$	17026-29-8	278.200	blue solid	108	171.7			
2279	Rhenium(VII) fluoride	ReF_7	17029-21-9	319.196	yel cub cry	48.3	73.7	4.32		
2280	Rhenium(VII) oxide	Re_2O_7	1314-68-7	484.410	yel hyg cry	327	360	6.10		s H_2O, EtOH, eth, diox, py
2281	Rhenium(VII) trioxychloride	ReO_3Cl	7791-09-5	269.658	col liq	4.5	128	3.87		reac H_2O
2282	Rhenium(VII) trioxyfluoride	ReO_3F	42246-24-2	253.203	yel solid	147	164			
2283	Rhenium(VII) dioxytrifluoride	ReO_2F_3	57246-89-6	275.201	yel solid	90	185.4			reac H_2O
2284	Rhenium(VII) oxypentafluoride	$ReOF_5$	23377-53-9	297.198	cream solid	43.8	73.0			
2285	Rhenium(VII) sulfide	Re_2S_7	12038-67-4	596.869	brn-blk tetr cry			4.87		i H_2O
2286	Rhodium	Rh	7440-16-6	102.906	silv-wh metal; cub	1964	3695	12.4		i acid, sl aqua regia
2287	Rhodium carbonyl	$Rh_6(CO)_{16}$	28407-51-4	1065.594	red-brn cry	220 dec				
2288	Rhodium carbonyl chloride	$[Rh(CO)_2Cl]_2$	14523-22-9	388.758	red-oran cry	124				s os
2289	Rhodium dodecacarbonyl	$Rh_4(CO)_{12}$	19584-30-6	747.743	red hyg cry	150 dec		2.52		reac H_2O
2290	Rhodium(III) bromide	$RhBr_3$	15608-29-4	342.618	dark brn plates	800 dec		5.56		s H_2O; i acid, os
2291	Rhodium(III) chloride	$RhCl_3$	10049-07-7	209.265	red monocl cry		717	5.38		i H_2O; s alk
2292	Rhodium(III) fluoride	RhF_3	60804-25-3	159.901	red hex cry			5.4		
2293	Rhodium(III) iodide	RhI_3	15492-38-3	483.619	blk monocl cry; hyg			6.4		
2294	Rhodium(III) nitrate	$Rh(NO_3)_3$	10139-58-9	288.921	hyg brn solid	600 dec				i H_2O
2295	Rhodium(III) nitrate dihydrate	$Rh(NO_3)_3 \cdot 2H_2O$	13465-43-5	324.951	blk solid	dec				i H_2O; s aqua regia
2296	Rhodium(III) oxide	Rh_2O_3	12036-35-0	253.809	gray hex cry	1100 dec		8.2		
2297	Rhodium(III) oxide pentahydrate	$Rh_2O_3 \cdot 5H_2O$	39373-27-8	309.010	yel pow	dec				sl H_2O; s acid
2298	Rhodium(III) sulfate	$Rh_2(SO_4)_3$	10489-46-0	493.999	red-yel solid	>500 dec				
2299	Rhodium(IV) oxide	RhO_2	12137-27-8	134.905	blk tetr cry			7.2		
2300	Rhodium(IV) oxide dihydrate	$RhO_2 \cdot 2H_2O$	[illegible]	170.9[illegible]	grn solid	dec		[illegible]		i H_2O, sol HCl, alk
2301	Rhodium(VI) fluoride	RhF_6	13693-07-7	216.896	blk cub cry	≈70		3.1		
2302	Rubidium	Rb	7440-17-7	85.468	soft silv metal; cub	39.30	688	1.53		reac H_2O
2303	Rubidium acetate	$RbC_2H_3O_2$	563-67-7	144.512	wh hyg cry	246				vs H_2O
2304	Rubidium aluminum sulfate	$RbAl(SO_4)_2$	13530-57-9	304.575	hex cry			≈3.1	1.60^{20}	i EtOH
2305	Rubidium aluminum sulfate dodecahydrate	$RbAl(SO_4)_2 \cdot 12H_2O$	7784-29-4	520.759	col cub cry	≈100 dec		≈1.9		s H_2O; i EtOH
2306	Rubidium azide	RbN_3	22756-36-1	127.488	tetr cry; exp	317		2.79	107^{16}	
2307	Rubidium bromate	$RbBrO_3$	13446-70-3	213.370	cub cry	430		3.68	2.95^{25}	
2308	Rubidium bromide	RbBr	7789-39-1	165.372	wh cub cry; hyg	692	1340	3.35	116^{25}	
2309	Rubidium carbonate	Rb_2CO_3	584-09-8	230.945	col monocl cry; hyg	837			223^{20}	
2310	Rubidium chlorate	$RbClO_3$	13446-71-4	168.919	col cry	324	dec 480	3.19	6.63^{25}	sl H_2O
2311	Rubidium chloride	RbCl	7791-11-9	120.921	wh cub cry; hyg	724	1390	2.76	93.9^{25}	sl EtOH
2312	Rubidium chromate	Rb_2CrO_4	13446-72-5	286.930	yel rhom cry			3.518	76.2^{25}	
2313	Rubidium dichromate	$Rb_2Cr_2O_7$	13446-73-6	386.924	red tricl or yel monocl cry			3.1		s H_2O
2314	Rubidium cyanide	RbCN	19073-56-4	111.486	wh cub cry			2.3		s H_2O; i EtOH, eth
2315	Rubidium fluoride	RbF	13446-74-7	104.466	wh cub cry; hyg	795	1410	3.2	300^{20}	i EtOH
2316	Rubidium fluoroborate	$RbBF_4$	18909-68-7	172.273	orth cry	612 dec		2.82		sl H_2O
2317	Rubidium formate	$RbCHO_2$	3495-35-0	130.486	wh hyg cry	dec				
2318	Rubidium hexafluorogermanate	Rb_2GeF_6	16962-48-4	357.57	wh cry	696				s H_2O
2319	Rubidium hydride	RbH	13446-75-8	86.476	wh cub cry; flam	≈170 dec		2.60		reac H_2O
2320	Rubidium hydrogen carbonate	$RbHCO_3$	19088-74-5	146.485	wh rhomb cry	175 dec			116^{20}	
2321	Rubidium hydrogen fluoride	$RbHF_2$	12280-64-7	124.473	tetr cry	188		3.3		
2322	Rubidium hydrogen sulfate	$RbHSO_4$	15587-72-1	182.539	col monocl cry	208		2.9		s H_2O
2323	Rubidium hydroxide	RbOH	1310-82-3	102.475	gray-wh orth cry; hyg	385		3.2	173^{30}	s EtOH
2324	Rubidium iodate	$RbIO_3$	13446-76-9	260.370	monocl or cub cry	dec		4.33	2.44^{25}	vs HCl
2325	Rubidium iodide	RbI	7790-29-6	212.372	wh cub cry	656	1300	3.55	165^{25}	s EtOH
2326	Rubidium molybdate	Rb_2MoO_4	13718-22-4	330.87	wh cry	958				s H_2O
2327	Rubidium nitrate	$RbNO_3$	13126-12-0	147.473	wh hex cry; hyg	310		3.11	65.0^{25}	vs H_2O
2328	Rubidium nitrite	$RbNO_2$	13825-25-7	131.474	wh cry	422				vs H_2O
2329	Rubidium oxide	Rb_2O	18088-11-4	186.935	yel-brn cub cry; hyg	400 dec		4.0		reac H_2O
2330	Rubidium perchlorate	$RbClO_4$	13510-42-4	184.919	wh hyg cry	550	dec >550	2.9	1.5^{25}	
2331	Rubidium permanganate	$RbMnO_4$	13465-49-1	204.404	dark purp cry	300 dec		3.24		sl H_2O
2332	Rubidium peroxide	Rb_2O_2	23611-30-5	202.935	wh orth cry	570		3.8		reac H_2O
2333	Rubidium selenide	Rb_2Se	31052-43-4	249.90	wh cub cry	733		3.22		reac H_2O

No.	Name	Formula	CAS Reg No.	Mol. weight	Physical form	mp/°C	bp/°C	Density g cm^{-3}	Solubility g/100 g H_2O	Qualitative solubility
2334	Rubidium sulfate	Rb_2SO_4	7488-54-2	266.999	wh orth cry	1066		3.61	50.8^{25}	
2335	Rubidium sulfide	Rb_2S	31083-74-6	203.001	wh cub cry	425		2.91		s H_2O
2336	Rubidium superoxide	RbO_2	12137-25-6	117.467	tetr cry	412		≈3.0		
2337	Ruthenium	Ru	7440-18-8	101.07	silv-wh metal; hex	2333	4150	12.1		i acid, aqua regia
2338	Ruthenium dodecacarbonyl	$Ru_3(CO)_{12}$	15243-33-1	639.33	oran cry	150 dec				
2339	Ruthenium pentacarbonyl	$Ru(CO)_5$	16406-48-7	241.12	col liq	-22	dec 50			i H_2O; s EtOH, bz, chl, hc
2340	Ruthenium nonacarbonyl	$Ru_2(CO)_9$	63128-11-0	454.23	stab below -40					s hex
2341	Ruthenium nitrosyl chloride monohydrate	$Ru(NO)Cl_3 \cdot H_2O$	18902-42-6	255.45	hyg red cry					
2342	Hexaammineruthenium(III) chloride	$Ru(NH_3)_6Cl_3$	14282-91-8	309.61	col monocl cry					s H_2O
2343	Ruthenium(III) bromide	$RuBr_3$	14014-88-1	340.78	brn hex cry	dec 500		5.3		i H_2O, acid, EtOH
2344	Ruthenium(III) chloride	$RuCl_3$	10049-08-8	207.43	blk-brn hex cry	≈500 dec		3.1		i H_2O; sl EtOH
2345	Ruthenium(III) fluoride	RuF_3	51621-05-7	158.07	brn rhomb cry	≈600 dec		5.36		i H_2O, dil acid
2346	Ruthenium(III) iodide	RuI_3	13896-65-6	481.78	blk hex cry	dec 300		6.0		sl H_2O
2347	Ruthenium(III) 2,4-pentanedioate	$Ru(CH_3COCHCOCH_3)_3$	14284-93-6	398.39	red-brn cry	230				
2348	Ruthenium(IV) fluoride	RuF_4	71500-16-8	177.06	yel-red cry					reac H_2O
2349	Ruthenium(IV) oxide	RuO_2	12036-10-1	133.07	gray-blk tetr cry	dec 1300		7.05		i H_2O, acid
2350	Ruthenium(V) fluoride	RuF_5	14521-18-7	196.06	grn monocl cry	86.5	227	3.90		
2351	Ruthenium(VI) fluoride	RuF_6	13693-08-8	215.06	dark brn orth cry	54	200 dec	3.54		reac H_2O
2352	Ruthenium(VIII) oxide	RuO_4	20427-56-9	165.07	yel monocl prisms	25.4	40	3.29	2.03^{20}	sl H_2O; vs ctc; reac EtOH
2353	Samarium	Sm	7440-19-9	150.36	silv metal; rhomb	1072	1794	7.52		
2354	Samarium boride	SmB_6	12008-30-9	215.23	refrac solid	2580		5.07		
2355	Samarium silicide	$SmSi_2$	12300-22-0	206.53	orth cry			5.14		
2356	Samarium(II) bromide	$SmBr_2$	50801-97-3	310.17	brn cry	669				reac H_2O
2357	Samarium(II) chloride	$SmCl_2$	13874-75-4	221.27	brn cry	855		3.69		reac H_2O
2358	Samarium(II) fluoride	SmF_2	15192-17-3	188.36	purp cry					reac H_2O
2359	Samarium(II) iodide	SmI_2	32248-43-4	404.17	grn cry	520				reac H_2O
2360	Samarium(III) acetate trihydrate	$Sm(C_2H_3O_2)_3 \cdot 3H_2O$	17829-86-6	381.54	hyg yel-wh solid			1.94		s H_2O
2361	Samarium(III) bromate nonahydrate	$Sm(BrO_3)_3 \cdot 9H_2O$	63427-22-5	696.20	pink hex cry	75 dec				vs H_2O; sl EtOH
2362	Samarium(III) bromide	$SmBr_3$	13759-87-0	390.07	yel cry	640				reac H_2O
2363	Samarium(III) carbonate	$Sm_2(CO_3)_3$	5895-47-6	480.75	wh-yel pow	dec >500				
2364	Samarium(III) chloride	$SmCl_3$	10361-82-7	256.72	yel cry	682		4.46	93.8^{25}	
2365	Samarium(III) chloride hexahydrate	$SmCl_3 \cdot 6H_2O$	13465-55-9	364.81	yel cry	dec		2.38	93.8^{25}	
2366	Samarium(III) fluoride	SmF_3	13765-24-7	207.36	wh cry	1306				reac H_2O
2367	Samarium(III) iodide	SmI_3	13813-25-7	531.07	oran cry	850				reac H_2O
2368	Samarium(III) nitrate	$Sm(NO_3)_3$	10361-83-8	336.38	yel-wh hyg solid				144^{25}	s EtOH
2369	Samarium(III) nitrate hexahydrate	$Sm(NO_3)_3 \cdot 6H_2O$	13759-83-6	444.47	pale yel cry	78				s H_2O, MeOH, ace
2370	Samarium(III) oxide	Sm_2O_3	12060-58-1	348.72	yel-wh cub cry	2269	3780	7.6		
2371	Samarium(III) sulfate octahydrate	$Sm_2(SO_4)_3 \cdot 8H_2O$	13465-58-2	733.03	yel cry			2.93	2.67^{20}	
2372	Samarium(III) sulfide	Sm_2S_3	12067-22-0	396.92	gray-brn cub cry	1720		5.87		
2373	Samarium(III) telluride	Sm_2Te_3	12040-00-5	683.52	orth cry			7.31		
2374	Scandium	Sc	7440-20-2	44.956	silv metal; hex	1541	2836	2.99		
2375	Scandium boride	ScB_2	12007-34-0	66.578	refrac solid	2250		3.17		
2376	Scandium bromide	$ScBr_3$	13465-59-3	284.668	wh hyg cry	969		9.33		s H_2O
2377	Scandium chloride	$ScCl_3$	10361-84-9	151.315	wh hyg cry	967		2.4		s H_2O; i EtOH
2378	Scandium fluoride	ScF_3	13709-47-2	101.951	wh powder	1552				sl H_2O
2379	Scandium hydroxide	$Sc(OH)_3$	17674-34-9	95.978	col amorp solid					i H_2O; s dil acid
2380	Scandium iodide	ScI_3	14474-33-0	425.669	hyg yel cry	953	subl			s H_2O, EtOH, CCD
2381	Scandium nitrate	$Sc(NO_3)_3$	13465-60-6	230.971	wh cry				169^{25}	s EtOH
2382	Scandium oxide	Sc_2O_3	12060-08-1	137.910	wh cub cry	2489		3.864		s conc acid
2383	Scandium sulfate pentahydrate	$Sc_2(SO_4)_3 \cdot 5H_2O$	15292-44-1	468.176	col cry	dec 110				vs H_2O
2384	Scandium sulfide	Sc_2S_3	12166-29-9	186.107	yel orth cry	1775		2.91		
2385	Scandium telluride	Sc_2Te_3	12166-44-8	472.71	blk hex cry			5.29		
2386	Selenium (gray)	Se	7782-49-2	78.96	gray metallic cry; hex	220.8	685	4.809		i H_2O, CS_2
2387	Selenium (α form)	Se	7782-49-2	78.96	red monocl cry	trans gray Se >120	685	4.39		i H_2O, EtOH; sl eth
2388	Selenium (vitreous)	Se	7782-49-2	78.96	blk amorp solid	trans gray Se 180	685	4.28		i H_2O; sl CS_2
2389	Selenic acid	H_2SeO_4	7783-08-6	144.97	wh hyg solid	58	260 dec	2.95		vs H_2O; reac EtOH
2390	Pentafluoroorthoselenic acid	$HOSeF_5$	38989-47-8	190.96	col solid	38	47			
2391	Selenous acid	H_2SeO_3	7783-00-8	128.97	wh hyg cry	70 dec		3.0		vs H_2O; s EtOH
2392	Selenium dioxide	SeO_2	7446-08-4	110.96	wh tetr needles or powder	340 tp	315 sp	3.95	264^{22}	s EtOH, MeOH; sl ace
2393	Selenium trioxide	SeO_3	13768-86-0	126.96	wh tetr cry; hyg	118	subl	3.44		s H_2O, os

No.	Name	Formula	CAS Reg No.	Mol. weight	Physical form	mp/°C	bp/°C	Density g cm^{-3}	Solubility g/100 g H_2O	Qualitative solubility
2394	Selenium bromide	Se_2Br_2	7789-52-8	317.73	red liq	5	225 dec	3.60		reac H_2O; s CS_2, chl
2395	Selenium chloride	Se_2Cl_2	10025-68-0	228.83	yel-brn oily liq	-85	127 dec	2.774		reac H_2O; s CS_2, bz, ctc, chl
2396	Selenium tetrabromide	$SeBr_4$	7789-65-3	398.58	oran-red cry	123				reac H_2O; s CS_2, chl
2397	Selenium tetrachloride	$SeCl_4$	10026-03-6	220.77	wh-yel cry	305 tp	191.4 sp	2.6		reac H_2O
2398	Selenium tetrafluoride	SeF_4	13465-66-2	154.95	col liq	-9.5	101.6	2.75		reac H_2O; vs EtOH, eth
2399	Selenium hexafluoride	SeF_6	7783-79-1	192.95	col gas	-34.6 tp	-46.6 sp	7.887 g/L		i H_2O
2400	Selenium chloride pentafluoride	SeF_5Cl	34979-62-9	209.41	col gas	-19	4.5			
2401	Selenium oxybromide	$SeOBr_2$	7789-51-7	254.77	red-yel solid	41.6	220 dec	3.38		reac H_2O; s CS_2, bz, ctc
2402	Selenium oxychloride	$SeOCl_2$	7791-23-3	165.87	col or yel liq	8.5	177	2.44		reac H_2O; s ctc, chl, bz, tol
2403	Selenium oxyfluoride	$SeOF_2$	7783-43-9	132.96	col liq	15	125	2.8		reac H_2O
2404	Selenium oxytetrafluoride	$SeOF_4$	53319-44-1	170.95	unstab col liq	12	65			
2405	Selenium dioxydifluoride	SeO_2F_2	14984-81-7	148.96	col gas	-99.5	-8.4	6.089 g/L		reac H_2O
2406	Selenium monosulfide	SeS	7446-34-6	111.03						
2407	Selenium disulfide	SeS_2	7488-56-4	143.09	red-yel cry	100				i H_2O; s acid
2408	Selenium sulfide (Se_2S_6)	Se_2S_6	75926-26-0	350.31	oran needles	121.5		2.44		s CS_2; sl bz
2409	Selenium sulfide (Se_4S_4)	Se_4S_4	75926-28-2	444.10	red cry	113 dec		3.29		s bz; sl CS_2
2410	Selenium sulfide (Se_6S_2)	Se_6S_2	75926-30-6	537.89	oran cry	121.5				s CS_2
2411	Silicon	Si	7440-21-3	28.086	gray cry or brn amorp solid	1414	3265	2.3296		i H_2O, acid; s alk
2412	Silane	SiH_4	7803-62-5	32.118	col gas; flam	-185	-111.9	1.313 g/L		reac H_2O; i EtOH, bz
2413	Disilane	Si_2H_6	1590-87-0	62.219	col gas; flam	-129.4	-14.8	2.543 g/L		reac H_2O, ctc, chl; s EtOH, bz
2414	Trisilane	Si_3H_8	7783-26-8	92.321	flam col liq	-117.4	52.9	0.739		reac H_2O
2415	Tetrasilane	Si_4H_{10}	7783-29-1	122.421	col liq; flam	-89.9	108.1	0.792		reac H_2O
2416	2-Silyltrisilane	Si_4H_{10}	13597-87-0	122.421	col liq	-99.4	101.7	0.792		reac H_2O
2417	Cyclopentasilane	Si_5H_{10}	289-22-5	150.507	col liq	-10.5	194.3	0.963		reac H_2O
2418	Pentasilane	Si_5H_{12}	14868-53-2	152.523	col liq	-72.8	153.2	0.827		reac H_2O
2419	2-Silyltetrasilane	Si_5H_{12}	14868-54-3	152.523	col liq	-109.9	146.2	0.820		reac H_2O
2420	2,2-Disilyltrisilane	Si_5H_{12}	15947-57-6	152.523	col liq	-57.8	134.3	0.815		reac H_2O
2421	Cyclohexasilane	Si_6H_{12}	291-59-8	180.608	col liq	16.5	226			reac H_2O
2422	Hexasilane	Si_6H_{14}	14693-61-9	182.624	col liq	-44.7	193.6	0.847		reac H_2O
2423	2-Silylpentasilane	Si_6H_{14}	14868-55-4	182.624	col liq	-78.4	185.2	0.840		
2424	3-Silylpentasilane	Si_6H_{14}	52988-75-7	182.624	col liq	-69	179.5	0.843		reac H_2O
2425	Heptasilane	Si_7H_{16}	14693-65-3	212.726	col liq	-30.1	226.8	0.859		reac H_2O
2426	Bromosilane	SiH_3Br	13465-73-1	111.014	col gas	-94	1.9	4.538 g/L		
2427	Dibromosilane	SiH_2Br_2	13768-94-0	189.910	liq	-70.1	66			
2428	Tribromosilane	$SiHBr_3$	7789-57-3	268.806	flam liq	-73	109	2.7		reac H_2O
2429	Tetrabromosilane	$SiBr_4$	7789-66-4	347.702	col fuming liq	5.39	154	2.8		reac H_2O
2430	Bromotrichlorosilane	$SiBrCl_3$	13465-74-2	214.349	col liq	-62	80.3	1.826		reac H_2O
2431	Dibromodichlorosilane	$SiBr_2Cl_2$	13465-75-3	258.800	col liq	-45.5	104	2.172		reac H_2O
2432	Tribromochlorosilane	$SiBr_3Cl$	13465-76-4	303.251	col liq	-20.8	127	2.497		reac H_2O
2433	Hexabromosilane	Si_2Br_6	13517-13-0	535.595	col cry	95	265			
2434	Octabromotrisilane	Si_3Br_8	54804-32-9	723.489	col liq	46				
2435	Chlorosilane	SiH_3Cl	13465-78-6	66.563	col gas	-118	-30.4	2.721 g/L		
2436	Dichlorosilane	SiH_2Cl_2	4109-96-0	101.008	col gas; flam	-122	8.3	4.129 g/L		reac H_2O
2437	Trichlorosilane	$SiHCl_3$	10025-78-2	135.453	fuming liq	-128.2	33	1.331		reac H_2O
2438	Tetrachlorosilane	$SiCl_4$	10026-04-7	169.898	col fuming liq	-68.74	57.65	1.5		reac H_2O
2439	Chlorotrifluorosilane	$SiClF_3$	14049-36-6	120.534	col gas	-138	-70.0	4.927 g/L		reac H_2O
2440	Dichlorodifluorosilane	$SiCl_2F_2$	18356-71-3	136.989	col gas	-44	-32	5.599 g/L		reac H_2O
2441	Trichlorofluorosilane	$SiCl_3F$	14965-52-7	153.443	col gas		12.25	6.272 g/L		
2442	Trichloroiodosilane	$SiCl_3I$	13465-85-5	261.349	col liq	-60	113.5			reac H_2O
2443	Hexachlorodisilane	Si_2Cl_6	13465-77-5	268.889	col liq	2.5	146			reac H_2O
2444	Octachlorotrisilane	Si_3Cl_8	13596-23-1	367.881	col liq	-67	216			
2445	Fluorosilane	SiH_3F	13537-33-2	50.108	col gas		-98.6	2.048 g/L		
2446	Difluorosilane	SiH_2F_2	13824-36-7	68.099	col gas	-122	-77.8	2.783 g/L		
2447	Trifluorosilane	$SiHF_3$	13465-71-9	86.089	col gas	-131	-95	3.519 g/L		
2448	Tetrafluorosilane	SiF_4	7783-61-1	104.080	col gas	-90.2	-86	4.254 g/L		reac H_2O
2449	Hexafluorodisilane	Si_2F_6	13830-68-7	170.161	col gas	-18.7 tp @ 780 mmHg	-19.1 sp			reac H_2O
2450	Octafluorotrisilane	Si_3F_8	14521-14-3	236.244	col liq	-1.2	42			
2451	Decafluorotetrasilane	Si_4F_{10}	14521-15-4	302.326	col cry	68	85.1			
2452	Iodosilane	SiH_3I	13598-42-0	158.014	col liq	-57	45.6			

No.	Name	Formula	CAS Reg No.	Mol. weight	Physical form	mp/°C	bp/°C	Density g cm^{-3}	Solubility g/100 g H_2O	Qualitative solubility
2453	Diiodosilane	SiH_2I_2	13760-02-6	283.911	col liq	-1	150			
2454	Triiodosilane	$SiHI_3$	13465-72-0	409.807	liq	8	220 dec			
2455	Tetraiodosilane	SiI_4	13465-84-4	535.704	wh powder	120.5	287.35	4.1		
2456	Hexaiododisilane	Si_2I_6	13510-43-5	817.598	pale yel cry	250				
2457	Disiloxane	$(SiH_3)_2O$	13597-73-4	78.218	gas	-144	-15.2	3.197 g/L		
2458	Hexachlorodisiloxane	$(SiCl_3)_2O$	14986-21-1	284.888	liq	-28	137			
2459	Methylsilane	SiH_3CH_3	992-94-9	46.145	col gas	-156.5	-57.5			
2460	Metasilicic acid	H_2SiO_3	7699-41-4	78.100	wh amorp powder					i H_2O; s HF
2461	Orthosilicic acid	H_4SiO_4	10193-36-9	96.116	exists only in soln					
2462	Fluorosilicic acid	H_2SiF_6	16961-83-4	144.092	stab only in aq soln					s H_2O
2463	Silicon carbide (hexagonal)	SiC	409-21-2	40.097	hard grn-black hex cry	2830		3.16		i H_2O, EtOH
2464	Silicon nitride (Si_3N_4)	Si_3N_4	12033-89-5	140.284	gray refrac solid; hex	1900		3.17		
2465	Silicon monoxide	SiO	10097-28-6	44.085	blk cub cry, stable >1200			2.18		
2466	Silicon dioxide (α-quartz)	SiO_2	14808-60-7	60.085	col hex cry	trans to beta quartz 573	2950	2.648		i H_2O, acid; s HF
2467	Silicon dioxide (β-quartz)	SiO_2	14808-60-7	60.085	col hex cry	trans to tridymite 867	2950	2.533^{600}		i H_2O, acid; s HF
2468	Silicon dioxide (tridymite)	SiO_2	15468-32-3	60.085	col hex cry	trans cristobalite 1470	2950	2.265		i H_2O, acid; s HF
2469	Silicon dioxide (cristobalite)	SiO_2	14464-46-1	60.085	col hex cry	1722	2950	2.334		i H_2O, acid; s HF
2470	Silicon dioxide (vitreous)	SiO_2	60676-86-0	60.085	col amorp solid	1713	2950	2.196		i H_2O, acid; s HF
2471	Silicon monosulfide	SiS	12504-41-5	60.151	yel-red hyg powder	1090	940	1.85		reac H_2O
2472	Silicon disulfide	SiS_2	13759-10-9	92.216	wh rhomb cry	1090	subl	2.04		reac H_2O, EtOH; i bz
2473	Silicon tetraacetate	$Si(C_2H_3O_2)_4$	562-90-3	264.262	wh hyg cry	110				reac H_2O; s ace, bz
2474	Silicon tetraboride	SiB_4	12007-81-7	71.330	gray refrac solid	1870 dec		2.4		
2475	Silicotungstic acid	$H_4SiO_4 \cdot (W_3O_9)_4$	12520-88-6	2878.17	hyg yel cry					vs H_2O, EtOH
2476	Silver	Ag	7440-22-4	107.868	silv metal; cub	961.78	2162	10.5		
2477	Silver azide	AgN_3	13863-88-2	149.888	orth cry; exp	exp ≈250		4.9	0.00081^{20}	
2478	Silver subfluoride	Ag_2F	1302-01-8	234.734	yel hex cry	100 dec		8.6		reac H_2O
2479	Silver(I) acetate	$AgC_2H_3O_2$	563-63-3	166.912	wh needles or powder	dec		3.26	1.04^{20}	
2480	Silver(I) acetylide	Ag_2C_2	7659-31-6	239.757	wh powder; exp					
2481	Silver(I) acetylide (AgC_2H)	AgC_2H	13092-75-6	132.897	wh powder; exp					
2482	Silver(I) arsenate	Ag_3AsO_4	13510-44-6	462.524	red cub cry	dec		6.657	0.00085	s NH_4OH
2483	Silver(I) benzoate	$Ag(C_6H_5CO_2)$	532-31-0	228.982	powder				30^{20}	
2484	Silver(I) bromate	$AgBrO_3$	7783-89-3	235.770	wh tetr cry	360 dec		5.21	0.193^{25}	
2485	Silver(I) bromide	AgBr	7785-23-1	187.772	yel cub cry	430	1502	6.47	0.000014^{25}	i H_2O, acid, EtOH
2486	Silver(I) carbonate	Ag_2CO_3	534-16-7	275.745	yel monocl cry	218		6.077	0.0036^{20}	s acid
2487	Silver(I) chlorate	$AgClO_3$	7783-92-8	191.319	wh tetr cry	230	270 dec	4.430	17.6^{25}	sl EtOH
2488	Silver(I) chloride	AgCl	7783-90-6	143.321	wh cub cry	455	1547	5.56	0.00019^{25}	
2489	Silver(I) chlorite	$AgClO_2$	7783-91-7	175.320	yel cry	105 exp			0.55^{25}	
2490	Silver(I) chromate	Ag_2CrO_4	7784-01-2	331.730	brn-red monocl cry			5.625	0.000014^{0}	
2491	Silver(I) citrate	$Ag_3C_6H_5O_7$	126-45-4	512.705	wh cry powder					i H_2O; s HNO_3
2492	Silver(I) cyanide	AgCN	506-64-9	133.886	wh-gray hex cry	320 dec		3.95	0.0000011	i EtOH, dil acid
2493	Silver(I) dichromate	$Ag_2Cr_2O_7$	7784-02-3	431.724	red cry			4.770		sl H_2O
2494	Silver(I) diethyldithiocarbamate	$Ag(C_2H_5)_2NCS_2$	1470-61-7	256.138	powder	173				s py
2495	Silver(I) fluoride	AgF	7775-41-9	126.866	yel-brn cub cry; hyg	435	1159	5.852	172^{20}	
2496	Silver(I) hexafluoroantimonate	$AgSbF_6$	26042-64-8	343.618	powder					
2497	Silver(I) hexafluoroarsenate	$AgAsF_6$	12005-82-2	296.780	powder					
2498	Silver(I) hexafluorophosphate	$AgPF_6$	26042-63-7	252.832	powder	102 dec				
2499	Silver(I) hydrogen fluoride	$AgHF_2$	12249-52-4	146.873	hyg cry	dec				
2500	Silver(I) iodate	$AgIO_3$	7783-97-3	282.770	wh orth cry	>200		5.53	0.053^{25}	
2501	Silver(I) iodide	AgI	7783-96-2	234.772	yel powder; hex	558	1506	5.68	0.000003	i acid
2502	Silver(I) lactate monohydrate	$AgC_3H_5O_3 \cdot H_2O$	128-00-7	214.954	gray cry powder					sl H_2O, EtOH
2503	Silver(I) metaphosphate	$AgPO_3$	13465-96-8	186.840	grn glass	490		6.37		i H_2O; s HNO_3, NH_4OH
2504	Silver(I) molybdate	Ag_2MoO_4	13765-74-7	375.67	yel cub cry	483		6.18		sl H_2O
2505	Silver(I) nitrate	$AgNO_3$	7761-88-8	169.873	col rhomb cry	210	440 dec	4.35	234^{25}	sl EtOH, ace
2506	Silver(I) nitrite	$AgNO_2$	7783-99-5	153.874	yel needles	140 dec		4.453	0.415^{25}	i EtOH; reac acid
2507	Silver(I) oxalate	$Ag_2C_2O_4$	533-51-7	303.755	wh cry powder	exp 140		5.03	0.0043^{20}	

No.	Name	Formula	CAS Reg No.	Mol. weight	Physical form	mp/°C	bp/°C	Density g cm⁻³	Solubility g/100 g H_2O	Qualitative solubility
2508	Silver(I) oxide	Ag_2O	20667-12-3	231.735	brn-blk cub cry	≈200 dec		7.2	0.0025	i EtOH; s acid, alk
2509	Silver(I) perchlorate	$AgClO_4$	7783-93-9	207.319	col cub cry; hyg	486 dec		2.806	558^{25}	s bz, py, os
2510	Silver(I) perchlorate monohydrate	$AgClO_4 \cdot H_2O$	14242-05-8	225.334	hyg wh cry	43 dec			558^{25}	
2511	Silver(I) permanganate	$AgMnO_4$	7783-98-4	226.804	viol monocl cry	dec		4.49	0.91^{18}	reac EtOH
2512	Silver(I) phosphate	Ag_3PO_4	7784-09-0	418.576	yel powder	849		6.37	0.0064	sl dil acid
2513	Silver(I) picrate monohydrate	$AgC_6H_2N_3O_7 \cdot H_2O$	146-84-9	353.979	yel cry					sl H_2O, EtOH; i chl, eth
2514	Silver(I) selenate	Ag_2SeO_4	7784-07-8	358.69	orth cry			5.72	0.118^{20}	
2515	Silver(I) selenide	Ag_2Se	1302-09-6	294.70	gray hex needles	880		8.216		i H_2O
2516	Silver(I) selenite	Ag_2SeO_3	7784-05-6	342.69	needles	530	>550 dec	5.930		sl H_2O; s acid
2517	Silver(I) sulfate	Ag_2SO_4	10294-26-5	311.799	col cry or powder	660		5.45	0.84^{25}	
2518	Silver(I) sulfide	Ag_2S	21548-73-2	247.801	gray-blk orth powder	825 (high press.)		7.23		i H_2O; s acid
2519	Silver(I) sulfite	Ag_2SO_3	13465-98-0	295.799	wh cry	100 dec			0.00046^{20}	s acid, NH_4OH
2520	Silver(I) telluride	Ag_2Te	12002-99-2	343.34	blk orth cry	955		8.4		
2521	Silver(I) tetraiodomercurate(II)	Ag_2HgI_4	7784-03-4	923.94	yel tetr cry	trans to red cub ≈40		6.1		i H_2O, dil acid
2522	Silver(I) thiocyanate	AgSCN	1701-93-5	165.951	wh powder	dec				i H_2O
2523	Silver(I) thiosulfate	$Ag_2S_2O_3$	23149-52-2	327.864	wh cry	dec				sl H_2O; s NH_4OH
2524	Silver(II) oxide	AgO	1301-96-8	123.867	gray powder; monocl or cub	>100 dec		7.5	0.0027^{25}	s alk; reac acid
2525	Silver(I) tungstate	Ag_2WO_4	13465-93-5	463.57	yel cry	620			0.015	s HNO_3, NH_4OH
2526	Silver(II) fluoride	AgF_2	7783-95-1	145.865	wh or gray hyg cry	690		4.58		reac H_2O
2527	Silver(II) oxide (Ag_2O_2)	Ag_2O_2	25455-73-6	247.735	gray-blk cub cry	>100		7.44		i H_2O; s acid, NH_4OH
2528	Sodium	Na	7440-23-5	22.990	soft silv met; cub	97.794	882.940	0.97		reac H_2O
2529	Sodium acetate	$NaC_2H_3O_2$	127-09-3	82.034	col cry	328.2		1.528	50.4^{25}	
2530	Sodium acetate trihydrate	$NaC_2H_3O_2 \cdot 3H_2O$	6131-90-4	136.079	col cry	58 dec		1.45	50.4^{25}	sl EtOH
2531	Sodium aluminate	$NaAlO_2$	1302-42-7	81.971	wh orth cry; hyg	1650		4.63		vs H_2O; i EtOH
2532	Sodium aluminum hydride	$NaAlH_4$	13770-96-2	54.004	wh hyg solid	174 dec		1.24		i eth; s thf
2533	Sodium aluminum sulfate dodecahydrate	$NaAl(SO_4)_2 \cdot 12H_2O$	10102-71-3	458.281	col cry	≈60		1.61	39.7[illegible]	i EtOH
2534	Sodium amide	$NaNH_2$	7782-92-5	39.013	wh-grn orth cry	210	500 dec	1.39		reac H_2O
2535	Sodium ammonium phosphate tetrahydrate	$NaNH_4HPO_4 \cdot 4H_2O$	13011-54-6	209.069	monocl cry	≈80 dec		1.54		s H_2O; i EtOH
2536	Sodium arsenate dodecahydrate	$Na_3AsO_4 \cdot 12H_2O$	7778-43-0	424.072	col monocl prism	86 dec				s H_2O; sl EtOH; i eth
2537	Sodium arsenite	$NaAsO_2$	7784-46-5	129.911	wh-gray hyg powder			1.87		vs H_2O; i EtOH
2538	Sodium azide	NaN_3	26628-22-8	65.010	col hex cry	300 dec		1.846	40.8^{20}	sl EtOH; i eth
2539	Sodium borohydride	$NaBH_4$	16940-66-2	37.833	wh cub cry; hyg	≈400 dec		1.07	55^{20}	reac EtOH
2540	Sodium bromate	$NaBrO_3$	7789-38-0	150.892	col cub cry	381		3.34	39.4^{25}	i EtOH
2541	Sodium bromide	NaBr	7647-15-6	102.894	wh cub cry	747	1390	3.200	94.6^{25}	s EtOH
2542	Sodium bromide dihydrate	$NaBr \cdot 2H_2O$	13466-08-5	138.925	wh cry	36 dec		2.18	94.6^{25}	sl EtOH
2543	Sodium carbonate	Na_2CO_3	497-19-8	105.989	wh hyg powder	856		2.54	30.7^{25}	i EtOH
2544	Sodium carbonate monohydrate	$Na_2CO_3 \cdot H_2O$	5968-11-6	124.005	col orth cry	100 dec		2.25	30.7^{25}	i EtOH
2545	Sodium carbonate decahydrate	$Na_2CO_3 \cdot 10H_2O$	6132-02-1	286.142	col cry	34 dec		1.46	30.7^{25}	i EtOH
2546	Sodium chlorate	$NaClO_3$	7775-09-9	106.441	col cub cry	248	dec 630	2.5	100^{25}	sl EtOH
2547	Sodium chloride	NaCl	7647-14-5	58.443	col cub cry	800.7	1465	2.17	36.0^{25}	sl EtOH
2548	Sodium chlorite	$NaClO_2$	7758-19-2	90.442	wh hyg cry	≈180 dec			64^{17}	
2549	Sodium chromate	Na_2CrO_4	7775-11-3	161.974	yel orth cry	794		2.72	87.6^{25}	sl EtOH
2550	Sodium chromate tetrahydrate	$Na_2CrO_4 \cdot 4H_2O$	10034-82-9	234.035	yel hyg cry	dec			87.6^{25}	sl EtOH
2551	Sodium citrate dihydrate	$Na_3C_6H_5O_7 \cdot 2H_2O$	6132-04-3	294.099	wh cry	150 dec				vs H_2O; i EtOH, eth
2552	Sodium citrate pentahydrate	$Na_3C_6H_5O_7 \cdot 5H_2O$	6858-44-2	348.145	hyg col cry	dec 150		1.86	92^{25}	vs H_2O; sl EtOH; i eth
2553	Sodium cyanate	NaCNO	917-61-3	65.007	col needles	550		1.89		s H_2O; sl EtOH; i eth
2554	Sodium cyanide	NaCN	143-33-9	49.008	wh cub cry; hyg	562		1.6	58.2^{20}	sl EtOH
2555	Sodium cyanoborohydride	$NaBH_3(CN)$	25895-60-7	62.843	wh hyg powder	240 dec		1.12		vs H_2O; s thf; sl EtOH; i bz, eth
2556	Sodium dichromate	$Na_2Cr_2O_7$	10588-01-9	261.968	red hyg cry	357	400 dec		187^{25}	
2557	Sodium dichromate dihydrate	$Na_2Cr_2O_7 \cdot 2H_2O$	7789-12-0	297.999	oran-red monocl cry	85 dec		2.35		vs H_2O; S HOAc
2558	Sodium dihydrogen phosphate	NaH_2PO_4	7558-80-7	119.977	col monocl cry	200 dec			94.9^{25}	
2559	Sodium dihydrogen phosphate monohydrate	$NaH_2PO_4 \cdot H_2O$	10049-21-5	137.993	wh hyg cry	100 dec			94.9^{25}	i EtOH
2560	Sodium dihydrogen phosphate dihydrate	$NaH_2PO_4 \cdot 2H_2O$	13472-35-0	156.008	col orth cry	60 dec		1.91	94.9^{25}	i EtOH
2561	Sodium dihydrogen hypophosphate hexahydrate	$Na_2H_2P_2O_6 \cdot 6H_2O$	7782-95-8	314.031	monocl plates	110 dec		1.849	2.0^{25}	i EtOH

No.	Name	Formula	CAS Reg No.	Mol. weight	Physical form	mp/°C	bp/°C	Density g cm⁻³	Solubility g/100 g H_2O	Qualitative solubility
2562	Sodium dihydrogen pyrophosphate	$Na_2H_2P_2O_7$	7758-16-9	221.939	wh powder	220 dec		≈1.9		s H_2O
2563	Sodium dithionate	$Na_2S_2O_4$	7775-14-6	174.108	gray-wh powder	52 dec			24.1^{20}	sl EtOH
2564	Sodium dithionate dihydrate	$Na_2S_2O_6 \cdot 2H_2O$	7631-94-9*	242.137	col orth cry	110 dec		2.19	15.1^{20}	i EtOH
2565	Sodium ethanolate	NaC_2H_5O	141-52-6	68.050	wh-yel hyg powder	260 dec				reac H_2O; s EtOH
2566	Sodium ferricyanide monohydrate	$Na_3Fe(CN)_6 \cdot H_2O$	14217-21-1*	298.933	red hyg cry					s H_2O; i EtOH
2567	Sodium ferrocyanide decahydrate	$Na_4Fe(CN)_6 \cdot 10H_2O$	13601-19-9	484.061	yel monocl cry	≈50 dec		1.46	20^{20}	i os
2568	Sodium fluoride	NaF	7681-49-4	41.988	col cub or tetr cry	996	1704	2.78	4.13^{25}	i EtOH
2569	Sodium fluorophosphate	Na_2PO_3F	10163-15-2	143.950	powder					
2570	Sodium formate	$NaCHO_2$	141-53-7	68.008	wh hyg cry	257.3	dec	1.92	94.9^{25}	sl EtOH
2571	Sodium germanate	Na_2GeO_3	12025-19-3	166.62	wh monocl hyg cry	1083		3.31		
2572	Sodium gold cyanide	$NaAu(CN)_2$	15280-09-8	271.991	wh-yel cry pow					s H_2O, NH_4OH
2573	Sodium gold thiosulfate dihydrate	$Na_3Au(S_2O_3)_2 \cdot 5H_2O$	10233-88-2	526.223	wh needles or prisms	dec 150		3.09		vs H_2O; i EtOH
2574	Sodium hexabromoplatinate(IV) hexahydrate	$Na_2PtBr_6 \cdot 6H_2O$	39277-13-9	828.579	cry					
2575	Sodium hexachloroiridate(IV) hexahydrate	$Na_2IrCl_6 \cdot 6H_2O$	19567-78-3	559.006	cry	600 dec				
2576	Sodium hexachloroplatinate(IV)	Na_2PtCl_6	16923-58-3	453.782	yel hyg cry				53^{16}	s EtOH
2577	Sodium hexachloroplatinate(IV) hexahydrate	$Na_2PtCl_6 \cdot 6H_2O$	16923-58-3	561.873	yel cry	110 dec		2.50	53^{16}	s EtOH; i eth
2578	Sodium hexafluoroaluminate	Na_3AlF_6	13775-53-6	209.941	col monocl cry; trans cub 560	1013		2.97		i H_2O
2579	Sodium hexafluoroantimonate	$NaSbF_6$	16925-25-0	258.740	wh cub cry			3.375	129^{20}	s EtOH, ace
2580	Sodium hexafluorophosphate monohydrate	$NaPF_6 \cdot H_2O$	20644-15-9	185.969	col orth cry			2.369	103^{0}	s EtOH, MeOH, ace
2581	Sodium hexafluorosilicate	Na_2SiF_6	16893-85-9	188.056	wh hex cry	847		2.7	0.67^{20}	i EtOH
2582	Sodium hexanitrocobaltate(III)	$Na_3Co(NO_2)_6$	14649-73-1	403.935	yel-brn cry powder					vs H_2O; sl EtOH
2583	Sodium hydride	NaH	7646-69-7	23.998	silv cub cry; flam	425 dec		1.39		reac H_2O, EtOH
2584	Sodium hydrogen arsenate	Na_2HAsO_4	7778-43-0	185.908	wh powder	≈195 dec			51^{20}	sl EtOH
2585	Sodium hydrogen arsenate heptahydrate	$Na_2HAsO_4 \cdot 7H_2O$	10048-95-0	312.014	wh monocl cry	≈50 dec		1.87	51^{20}	sl EtOH
2586	Sodium hydrogen carbonate	$NaHCO_3$	144-55-8	84.007	wh monocl cry	≈50 dec		2.20	10.3^{25}	i EtOH
2587	Sodium hydrogen fluoride	$NaHF_2$	1333-83-1	61.995	wh hex cry	>160 dec		2.08	3.25^{20}	
2588	Sodium hydrogen phosphate	Na_2HPO_4	7558-79-4	141.959	wh hyg powder			1.7	11.8^{25}	
2589	Sodium hydrogen phosphate heptahydrate	$Na_2HPO_4 \cdot 7H_2O$	7782-85-6	268.066	col cry			≈1.7	11.8^{25}	i EtOH
2590	Sodium hydrogen phosphate dodecahydrate	$Na_2HPO_4 \cdot 12H_2O$	10039-32-4	358.143	col cry	≈35 dec		≈1.5	11.8^{25}	i EtOH
2591	Sodium hydrogen sulfate	$NaHSO_4$	7681-38-1	120.061	wh hyg cry	≈315		2.43	28.5^{25}	
2592	Sodium hydrogen sulfate monohydrate	$NaHSO_4 \cdot H_2O$	10034-88-5	138.076	wh monocl cry			2.10	28.5^{25}	reac EtOH
2593	Sodium hydrogen sulfide	NaHS	16721-80-5	56.063	col rhomb cry	350		1.79		s H_2O, EtOh, eth
2594	Sodium hydrogen sulfide dihydrate	$NaHS \cdot 2H_2O$	16721-80-5	92.094	yel hyg needles	55 dec				vs H_2O, EtOH, eth
2595	Sodium hydrogen sulfite	$NaHSO_3$	7631-90-5	104.061	wh cry			1.48		s H_2O; sl EtOH
2596	Sodium hydroxide	NaOH	1310-73-2	39.997	wh orth cry; hyg	323	1388	2.13	100^{25}	s EtOH, MeOH
2597	Sodium hypochlorite	NaClO	7681-52-9	74.442	stab in aq soln	anh form exp			79.9^{25}	
2598	Sodium hypochlorite pentahydrate	$NaOCl \cdot 5H_2O$	10022-70-5	164.518	pale grn orth cry	18		1.6		s H_2O
2599	Sodium iodate	$NaIO_3$	7681-55-2	197.892	wh orth cry	422		4.28	9.47^{25}	i EtOH
2600	Sodium iodide	NaI	7681-82-5	149.894	wh cub cry; hyg	661	1304	3.67	184^{25}	s EtOH, ace
2601	Sodium iodide dihydrate	$NaI \cdot 2H_2O$	13517-06-1	185.925	hyg col monocl cry	69 dec		2.45	318^{0}	vs H_2O
2602	Sodium bismuthate	$NaBiO_3$	12232-99-4	279.968	yel-brn hyg cry					i cold H_2O, reac acid
2603	Sodium metabisulfite	$Na_2S_2O_5$	7681-57-4	190.107	wh cry				66.7^{25}	sl EtOH
2604	Sodium metaborate	$NaBO_2$	7775-19-1	65.800	wh hex cry	966	1434	2.46		s H_2O
2605	Sodium metasilicate	Na_2SiO_3	6834-92-0	122.064	wh amorp solid; hyg	1089		2.61		s cold H_2O; reac hot H_2O; i EtOH
2606	Sodium metasilicate pentahydrate	$Na_2SiO_3 \cdot 5H_2O$	13517-24-3	212.140	wh pow	72 dec				s H_2O
2607	Sodium molybdate	Na_2MoO_4	7631-95-0	205.92	col cub cry	687		≈3.5	65.0^{25}	
2608	Sodium molybdate dihydrate	$Na_2MoO_4 \cdot 2H_2O$	10102-40-6	241.95	cry powder	100 dec		≈3.5	65.0^{25}	
2609	Sodium molybdophosphate	$Na_3PO_4 \cdot 12MoO_3$	1313-30-0	1891.20	hyg solid			2.83		vs H_2O, EtOH
2610	Sodium niobate	$NaNbO_3$	12034-09-2	163.894	rhom cry	1422		4.55		i H_2O
2611	Sodium nitrate	$NaNO_3$	7631-99-4	84.995	col hex cry; hyg	306.5		2.261	91.2^{25}	sl EtOH, MeOH
2612	Sodium nitrite	$NaNO_2$	7632-00-0	68.996	wh orth cry; hyg	284	>320 dec	2.17	84.8^{25}	sl EtOH; reac acid
2613	Sodium nitroferricyanide dihydrate	$Na_2[Fe(CN)_5NO] \cdot 2H_2O$	13755-38-9	297.949	red cry			1.72	40^{16}	sl EtOH
2614	Sodium orthovanadate	Na_3VO_4	13721-39-6	183.909	col hex prisms	860				s H_2O; i EtOH
2615	Sodium oxalate	$Na_2C_2O_4$	62-76-0	133.999	wh powder	≈250 dec		2.34	3.61^{25}	i EtOH

No.	Name	Formula	CAS Reg No.	Mol. weight	Physical form	mp/°C	bp/°C	Density g cm^{-3}	Solubility g/100 g H_2O	Qualitative solubility
2616	Sodium oxide	Na_2O	1313-59-3	61.979	wh amorp powder	1134		2.27		reac H_2O
2617	Sodium perborate tetrahydrate	$NaBO_3 \cdot 4H_2O$	7632-04-4	153.861	wh cry	60 dec				reac H_2O
2618	Sodium perchlorate	$NaClO_4$	7601-89-0	122.441	wh orth cry; hyg	482 dec		2.52	205^{25}	
2619	Sodium perchlorate monohydrate	$NaClO_4 \cdot H_2O$	7791-07-3	140.456	wh hyg cry	≈130 dec		2.02	205^{25}	
2620	Sodium periodate	$NaIO_4$	7790-28-5	213.892	wh tetr cry	≈300 dec		3.86	14.4^{25}	s acid
2621	Sodium periodate trihydrate	$NaIO_4 \cdot 3H_2O$	13472-31-6	267.938	wh hex cry	175 dec		3.22	14.4^{25}	
2622	Sodium permanganate trihydrate	$NaMnO_4 \cdot 3H_2O$	10101-50-5*	195.972	red-blk hyg cry	170 dec		2.47	144^{20}	reac EtOH
2623	Sodium peroxide	Na_2O_2	1313-60-6	77.979	yel hyg powder	675		2.805		reac H_2O
2624	Sodium perrhenate	$NaReO_4$	13472-33-8	273.195	cry	300		5.39		
2625	Sodium persulfate	$Na_2S_2O_8$	7775-27-1	238.105	wh hyg cry					vs H_2O; reac EtOH
2626	Sodium phosphate	Na_3PO_4	7601-54-9	163.940	col cry	1583		2.54	14.5^{25}	s H_2O
2627	Sodium phosphate dodecahydrate	$Na_3PO_4 \cdot 12H_2O$	10101-89-0	380.124	col hex cry	≈75		1.62	14.4^{25}	i EtOH
2628	Sodium phosphate, chlorinated	$Na_3PO_4 \cdot NaOCl$	56802-99-4	238.383	wh cry				25^{25}	
2629	Sodium phosphide	Na_3P	12058-85-4	99.943	red solid	>650				reac H_2O
2630	Sodium phosphinate	NaH_2PO_2	7681-53-0	87.979	wh cry				100^{25}	
2631	Sodium phosphinate monohydrate	$NaH_2PO_2 \cdot H_2O$	10039-56-2	105.994	col hyg cry	310 dec			100^{25}	s EtOH
2632	Sodium phosphonate pentahydrate	$Na_2HPO_3 \cdot 5H_2O$	13517-23-2	216.036	wh hex plates	dec 200			429^{20}	vs H_2O; i EtOH
2633	Sodium pyrophosphate	$Na_4P_2O_7$	7722-88-5	265.902	col cry	988		2.53	7.09^{25}	
2634	Sodium selenate	Na_2SeO_4	13410-01-0	188.94	col orth cry				58.5^{25}	
2635	Sodium selenate decahydrate	$Na_2SeO_4 \cdot 10H_2O$	10102-23-5	369.09	wh cry			1.61	58.5^{25}	
2636	Sodium selenide	Na_2Se	1313-85-5	124.94	amorp solid	>875		2.62		reac H_2O
2637	Sodium selenite	Na_2SeO_3	10102-18-8	172.94	wh tetr cry				89.8^{25}	i EtOH
2638	Sodium selenite pentahydrate	$Na_2SeO_3 \cdot 5H_2O$	26970-82-1	184.054	wh tetr cry	dec				s H_2O; i EtOH
2639	Sodium stannate trihydrate	$Na_2SnO_3 \cdot 3H_2O$	12209-98-2	266.734	col hex cry	dec 140			61^{15}	vs H_2O; i EtOH, ace
2640	Sodium stearate	$NaC_{18}H_{35}O_2$	822-16-2	306.460	wh powder					sl H_2O, EtOH; vs hot H_2O
2641	Sodium succinate hexahydrate	$Na_2C_4H_4O_4 \cdot 6H_2O$	150-90-3	270.144	cry powder	120 dec			20	i EtOH
2642	Sodium sulfate	Na_2SO_4	7757-82-6	142.043	wh orth cry or powder	884		2.7	28.1^{25}	i EtOH
2643	Sodium sulfate heptahydrate	$Na_2SO_4 \cdot 7H_2O$	13472-39-1	268.152	wh cry	dec				vs H_2O
2644	Sodium sulfate decahydrate	$Na_2SO_4 \cdot 10H_2O$	7727-73-3	322.196	col monocl cry	32 dec		1.46	28.1^{25}	i EtOH
2645	Sodium sulfide	Na_2S	1313-82-2	78.045	wh cub cry; hyg	1172		1.856	20.6^{25}	sl EtOH; i eth
2646	Sodium sulfide pentahydrate	$Na_2S \cdot 5H_2O$	1313-83-3	168.121	col orth cry	120 dec		1.58	20.6^{25}	s EtOH; i eth
2647	Sodium sulfide nonahydrate	$Na_2S \cdot 9H_2O$	1313-84-4	240.183	wh-yel hyg cry	≈50 dec		1.43	20.6^{25}	sl EtOH; i eth
2648	Sodium sulfite	Na_2SO_3	7757-83-7	126.043	wh hex cry	911		2.63	30.7^{25}	i EtOH
2649	Sodium sulfite heptahydrate	$Na_2SO_3 \cdot 7H_2O$	10102-15-5	252.150	wh monocl cry; unstab			1.56	30.7^{25}	sl EtOH
2650	Sodium superoxide	NaO_2	12034-12-7	54.989	yel cub cry	552		2.2		reac H_2O
2651	Sodium tartrate dihydrate	$Na_2C_4H_4O_6 \cdot 2H_2O$	6106-24-7	230.082				1.545		s H_2O; i EtOH
2652	Sodium tellurate	Na_2TeO_4	10101-83-4	237.58	wh powder				0.8	
2653	Sodium tellurite	Na_2TeO_3	10102-20-2	221.58	wh rhomb prisms					sl H_2O
2654	Sodium tetraborate	$Na_2B_4O_7$	1330-43-4	201.220	col gl solid; hyg	743	1575	2.4	3.17^{25}	sl MeOH
2655	Sodium tetraborate tetrahydrate	$Na_2B_4O_7 \cdot 4H_2O$	12045-87-3	273.281	wh monocl cry			1.95	3.17^{25}	
2656	Sodium tetraborate pentahydrate	$Na_2B_4O_7 \cdot 5H_2O$	12045-88-4	291.296	hex cry	dec		1.88	3.17^{25}	
2657	Sodium tetraborate decahydrate	$Na_2B_4O_7 \cdot 10H_2O$	1303-96-4	381.373	wh monocl cry	75 dec		1.73	3.17^{25}	i EtOH
2658	Sodium tetrachloroaluminate	$NaAlCl_4$	7784-16-9	191.784	orth cry			2.01		s H_2O
2659	Sodium tetrachloroaurate(III) dihydrate	$NaAuCl_4 \cdot 2H_2O$	13874-02-7	397.800	oran-yel rhom cry	100 dec			150^{10}	s EtOH, eth
2660	Sodium tetrachloropalladate(II) trihydrate	$Na_2PdCl_4 \cdot 3H_2O$	13820-53-6	348.26	brn-red hyg cry					vs H_2O; s EtOH
2661	Sodium tetrachloroplatinate(II) tetrahydrate	$Na_2PtCl_4 \cdot 4H_2O$	10026-00-3	454.938	red prisms	100				s H_2O, EtOH
2662	Sodium tetrafluoroberyllate	Na_2BeF_4	13871-27-7	130.986	orth cry	575		2.47		sl H_2O
2663	Sodium tetrafluoroborate	$NaBF_4$	13755-29-8	109.795	wh orth prisms	384		2.47	108^{20}	sl EtOH
2664	Sodium thioantimonate nonahydrate	$Na_3SbS_4 \cdot 9H_2O$	10101-91-4	481.127	yel cry	dec 108		1.8	28^{20}	i EtOH
2665	Sodium thiocyanate	NaSCN	540-72-7	81.073	col hyg cry	287			151^{25}	
2666	Sodium thiophosphate dodecahydrate	$Na_3PO_3S \cdot 12H_2O$	10101-88-9	396.190	hex hyg leaflets	60				vs hot H_2O
2667	Sodium thiosulfate	$Na_2S_2O_3$	7772-98-7	158.108	col monocl cry	100 dec		1.69	76.4^{25}	i EtOH
2668	Sodium thiosulfate pentahydrate	$Na_2S_2O_3 \cdot 5H_2O$	10102-17-7	248.184	col cry	≈50 dec		1.69	76.4^{25}	i EtOH
2669	Sodium trimetaphosphate	$Na_3(PO_3)_3$	7785-84-4	305.885	wh cry			2.49	22	
2670	Sodium trimetaphosphate hexahydrate	$Na_3(PO_3)_3 \cdot 6H_2O$	7785-84-4	413.976	tricl-rhom hyg prisms	53		1.786	22	i EtOH
2671	Sodium tripolyphosphate	$Na_5P_3O_{10}$	7758-29-4	367.864	wh hyg powder	622			20^{25}	
2672	Sodium tungstate	Na_2WO_4	13472-45-2	293.82	wh rhom cry	695		4.18	74.2^{25}	
2673	Sodium tungstate dihydrate	$Na_2WO_4 \cdot 2H_2O$	10213-10-2	329.85	wh orth cry	100 dec		3.25	74.2^{25}	i EtOH
2674	Sodium uranate(VI) monohydrate	$Na_2U_2O_7 \cdot H_2O$	13721-34-1	652.049	yel powder					i H_2O; s acid

No.	Name	Formula	CAS Reg No.	Mol. weight	Physical form	mp/°C	bp/°C	Density g cm^{-3}	Solubility g/100 g H_2O	Qualitative solubility
2675	Sodium vanadate(V)	$NaVO_3$	13718-26-8	121.930	col monocl prisms	630			21^{25}	
2676	Sodium vanadate(V) tetrahydrate	$NaVO_3 \cdot 4H_2O$	13718-26-8	193.992	yel-wh cry powder				21^{25}	
2677	Strontium	Sr	7440-24-6	87.62	silv-wh metal; cub	777	1382	2.64		reac H_2O; s EtOH
2678	Strontium acetate	$Sr(C_2H_3O_2)_2$	543-94-2	205.71	col hyg cry	dec		2.1	40^{25}	vs H_2O
2679	Strontium arsenite tetrahydrate	$Sr(AsO_2)_2 \cdot 4H_2O$	10378-48-0	373.52	wh powder					sl H_2O, EtOH; sol dil acid
2680	Strontium bromate monohydrate	$Sr(BrO_3)_2 \cdot H_2O$	14519-18-7	361.44	yel hyg monocl cry	120 dec		3.773	39.0^{25}	
2681	Strontium bromide	$SrBr_2$	10476-81-0	247.43	wh tetr cry	657		4.216	107^{25}	
2682	Strontium bromide hexahydrate	$SrBr_2 \cdot 6H_2O$	7789-53-9	355.52	col hyg cry	88 dec			107^{25}	s EtOH; i eth
2683	Strontium carbide	SrC_2	12071-29-3	111.64	blk tetr cry	>1700		3.19		i H_2O
2684	Strontium carbonate	$SrCO_3$	1633-05-2	147.63	wh orth cry; hyg	1494		3.785	0.00034^{20}	s dil acid
2685	Strontium chlorate	$Sr(ClO_3)_2$	7791-10-8	254.52	col cry	120 dec		3.15	176^{25}	sl EtOH
2686	Strontium chloride	$SrCl_2$	10476-85-4	158.53	wh cub cry; hyg	874	1250	3.052	54.7^{25}	
2687	Strontium chloride hexahydrate	$SrCl_2 \cdot 6H_2O$	10025-70-4	266.62	col hyg cry	100 dec		1.96	54.7^{25}	s EtOH
2688	Strontium chromate	$SrCrO_4$	7789-06-2	203.61	yel monocl cry	dec		3.9	0.106^{20}	s dil acid
2689	Strontium cyanide dihydrate	$Sr(CN)_2 \cdot 4H_2O$	52870-08-3	211.72	wh hyg cry	dec				vs H_2O
2690	Strontium ferrocyanide pentadecahydrate	$SrFe(CN)_6 \cdot 15H_2O$	14654-44-5	569.80	yel monocl cry				50	
2691	Strontium fluoride	SrF_2	7783-48-4	125.62	wh cub cry or powder	1477	2460	4.24	0.021^{25}	s dil acid
2692	Strontium formate	$Sr(CHO_2)_2$	592-89-2	177.66	wh cry	71.9		2.693	9.1^{0}	
2693	Strontium formate dihydrate	$Sr(CHO_2)_2 \cdot 2H_2O$	6160-34-5	213.69	col rhom cry	100 dec		2.25	9.1^{37}	i EtOH, eth
2694	Strontium hexaboride	SrB_6	12046-54-7	152.49	blk cub cry	2235		3.39		i H_2O; s HNO_3
2695	Strontium hydride	SrH_2	13598-33-9	89.64	orth cry	1050		3.26		reac H_2O
2696	Strontium hydroxide	$Sr(OH)_2$	18480-07-4	121.64	col orth cry; hyg	535	710 dec	3.625	2.25^{25}	
2697	Strontium iodate	$Sr(IO_3)_2$	13470-01-4	437.43	tricl cry			5.045	0.165^{25}	
2698	Strontium iodide	SrI_2	10476-86-5	341.43	wh hyg cry	538	1773 dec	4.55	177^{25}	
2699	Strontium iodide hexahydrate	$SrI_2 \cdot 6H_2O$	73796-25-5	449.52	wh-yel hex cry; hyg	120 dec		4.4	177^{25}	s EtOH
2700	Strontium molybdate	$SrMoO_4$	13470-04-7	247.56	wh cry pow	1040		4.54		i H_2O
2701	Strontium niobate	$SrNb_2O_6$	12034-89-8	369.43	monocl cry	1225		5.11		i H_2O
2702	Strontium nitrate	$Sr(NO_3)_2$	10042-76-9	211.63	wh cub cry	570		2.99	80.2^{25}	sl EtOH, ace
2703	Strontium nitride	Sr_3N_2	12033-82-8	290.87	refrac solid	1200				reac H_2O; s HCl
2704	Strontium nitrite	$Sr(NO_2)_2$	13470-06-9	179.63	wh-yel hyg needles	240 dec		2.8	72.1^{30}	s H_2O
2705	Strontium orthosilicate	Sr_2SiO_4	13597-55-2	267.32	orth cry			4.5		
2706	Strontium oxalate monohydrate	$SrC_2O_4 \cdot H_2O$	814-95-9	193.65	cry pow	dec 150			0.00005^{20}	sl dil acid
2707	Strontium oxide	SrO	1314-11-0	103.62	col cub cry	2531		5.1		reac H_2O
2708	Strontium perchlorate	$Sr(ClO_4)_2$	13450-97-0	286.52	col hyg cry				306^{25}	s EtOH, MeOH
2709	Strontium permanganate trihydrate	$Sr(MnO_4)_2 \cdot 3H_2O$	14446-13-0	379.54	purp cub cry	175 dec		2.75	250^{18}	
2710	Strontium peroxide	SrO_2	1314-18-7	119.62	wh tetr cry; unstab	215 dec		4.78		reac H_2O
2711	Strontium phosphate	$Sr_3(PO_4)_2$	7446-28-8	452.80	wh powder				0.000011^{20}	s acid
2712	Strontium selenate	$SrSeO_4$	7446-21-1	230.58	orth cry			4.25	0.115^{20}	s hot HCl
2713	Strontium selenide	SrSe	1315-07-7	166.58	wh cub cry	1600		4.54		
2714	Strontium silicide	$SrSi_2$	12138-28-2	143.79	silv-gray cub cry	1100		3.35		
2715	Strontium sulfate	$SrSO_4$	7759-02-6	183.68	wh orth cry	1606		3.96	0.0135^{25}	i EtOH; sl acid
2716	Strontium sulfide	SrS	1314-96-1	119.69	gray cub cry	2226		3.70		sl H_2O; s acid
2717	Strontium sulfite	$SrSO_3$	13451-02-0	167.68	col cry	dec			0.0015^{25}	s H_2SO_4, HCl
2718	Strontium telluride	SrTe	12040-08-3	215.22	wh cub cry			4.83		
2719	Strontium thiosulfate pentahydrate	$SrS_2O_3 \cdot 5H_2O$	15123-90-7	289.82	monocl needles	100 dec		2.17	36.3^{25}	i EtOH
2720	Strontium titanate	$SrTiO_3$	12060-59-2	183.49	wh cub cry	2080		5.1		i H_2O
2721	Strontium tungstate	$SrWO_4$	13451-05-3	335.46	col tetr cry	dec		6.187	0.14^{15}	i EtOH
2722	Strontium zirconate	$SrZrO_3$	12036-39-4	226.84	col cry	2600				
2723	Sulfur (rhombic)	S	7704-34-9	32.065	yel orth cry	95.3 (trans to monocl)	444.61	2.07		i H_2O; sl EtOH, bz, eth; s CS_2
2724	Sulfur (monoclinic)	S	7704-34-9	32.065	yel monocl needles, stable 95.3-120	115.21	444.61	2.00		i H_2O; sl EtOH, bz, eth; s CS_2
2725	Sulfuric acid	H_2SO_4	7664-93-9	98.079	col oily liq	10.31	337	1.8302		vs H_2O
2726	Peroxysulfuric acid	H_2SO_5	7722-86-3	114.078	wh cry; unstab	45 dec				vs H_2O
2727	Nitrosylsulfuric acid	$HNOSO_4$	7782-78-7	127.077	prisms	73 dec				reac H_2O; s H_2SO_4
2728	Chlorosulfonic acid	$SO_2(OH)Cl$	7790-94-5	116.524	col-yel liq	-80	152	1.75		reac H_2O; s py
2729	Fluorosulfonic acid	$SO_2(OH)F$	7789-21-1	100.069	col liq	-89	163	1.726		reac H_2O
2730	Sulfurous acid	H_2SO_3	7782-99-2	82.079	exists only in aq soln					
2731	Sulfamic acid	H_2NSO_3H	5329-14-6	97.094	orth cry	≈205 dec		2.15	14.7^{0}	sl ace; i eth
2732	Sulfur dioxide	SO_2	7446-09-5	64.064	col gas	-75.5	-10.05	2.619 g/L		s H_2O, EtOh, eth, chl

No.	Name	Formula	CAS Reg No.	Mol. weight	Physical form	mp/°C	bp/°C	Density g cm⁻³	Solubility g/100 g H_2O	Qualitative solubility
2733	Sulfur trioxide	SO_3	7446-11-9	80.063	wh needles	62.2	subl			reac H_2O
2734	Sulfur trioxide (γ-form)	SO_3	7446-11-9	80.063	col solid or liq	16.8	44.5	1.90		reac H_2O
2735	Sulfur trioxide (β-form)	SO_3	7446-11-9	80.063	wh needles	30.5	44.5			reac H_2O
2736	Sulfur bromide ($SSBr_2$)	$SSBr_2$	13172-31-1	223.938	red oily liq	-46	>25 dec	2.63		reac H_2O
2737	Sulfur chloride ($SSCl_2$)	$SSCl_2$	10025-67-9	135.036	yel-red oily liq	-77	137	1.69		reac H_2O; s EtOH, bz, eth, ctc
2738	Sulfur fluoride (SSF_2)	SSF_2	16860-99-4	102.127	col gas	-164.6	-10.6	4.174 g/L		reac H_2O
2739	Sulfur fluoride (FSSF)	FSSF	13709-35-8	102.127	col gas	-133	15	4.174 g/L		reac H_2O
2740	Sulfur dichloride	SCl_2	10545-99-0	102.971	red visc liq	-122	59.6	1.62		reac H_2O
2741	Sulfur tetrafluoride	SF_4	7783-60-0	108.059	col gas	-125	-40.45	4.417 g/L		reac H_2O
2742	Sulfur hexafluoride	SF_6	2551-62-4	146.055	col gas	-49.596 tp	-63.8 sp	5.970 g/L		sl H_2O; s EtOH
2743	Sulfur bromide pentafluoride	SF_5Br	15607-89-3	206.961	col gas	-79	3.1	8.459 g/L		
2744	Sulfur chloride pentafluoride	SF_5Cl	13780-57-9	162.510	col gas	-64	-19.05	6.642 g/L		
2745	Sulfur decafluoride	S_2F_{10}	5714-22-7	254.114	liq	-52.7	30; dec 150	2.08		i H_2O
2746	Sulfuryl amide	$(NH_2)_2SO_2$	7803-58-9	96.109	orth plates	93	250 dec			vs H_2O; sl EtOH
2747	Sulfuryl chloride	SO_2Cl_2	7791-25-5	134.970	col liq	-51	69.4	1.680		reac H_2O; s bz, tol, eth
2748	Sulfuryl fluoride	SO_2F_2	2699-79-8	102.061	col gas	-135.8	-55.4	4.172 g/L		sl H_2O, EtOH, s tol, ctc
2749	Sulfuryl bromide fluoride	SO_2BrF	13536-61-3	162.966	col liq	-86	41			reac H_2O
2750	Sulfuryl chloride fluoride	SO_2ClF	13637-84-8	118.515	col gas	-124.7	7.1	1.62^{0}		reac H_2O
2751	Pyrosulfuryl chloride	$S_2O_5Cl_2$	7791-27-7	215.033	col fuming liq	-37	151	1.837		reac H_2O
2752	Thionyl bromide	$SOBr_2$	507-16-4	207.872	yel liq	-50	140			reac H_2O
2753	Thionyl chloride	$SOCl_2$	7719-09-7	118.970	yel fuming liq	-101	75.6	1.631		reac H_2O; s bz, ctc, chl
2754	Thionyl fluoride	SOF_2	7783-42-8	86.061	col gas	-129.5	-43.8	3.518 g/L		reac H_2O; s bz, eth
2755	Sulfur fluoride oxide (SOF_4)	SOF_4	13709-54-1	124.058	col gas	-99.6	-48.5	1.95^{-82}		reac H_2O
2756	Sulfur fluoride hypofluorite	F_5SOF	15179-32-5	162.054	col gas	-86	-35.1	6.624 g/L		
2757	Tetrasulfur tetranitride	S_4N_4	28950-34-7	184.287	yel-oran cry	178.2	subl			i H_2O; reac alk, acid
2758	Tantalum	Ta	7440-25-7	180.948	gray metal; cub	3017	5458	16.4		reac HF
2759	Tantalum aluminide	$TaAl_3$	12004-76-1	261.893	gray refrac powder	≈1400		7.02		i H_2O, acid, alk
2760	Tantalum boride (TaB)	TaB	12007-07-7	191.759	refrac orth cry	2040		14.2		
2761	Tantalum boride (TaB_2)	TaB_2	12007-35-1	202.570	blk hex cry	3100		11.2		i H_2O, acid, alk
2762	Tantalum carbide (TaC)	TaC	12070-06-3	192.959	gold-brown powder; cub	3880	4780	14.3		s HF-HNO_3 mixture
2763	Tantalum carbide (Ta_2C)	Ta_2C	12070-07-4	373.907	refrac hex cry	3327		15.1		
2764	Tantalum hydride	TaH	13981-95-8	181.956	gray metallic solid			15.1		i acid
2765	Tantalum nitride	TaN	12033-62-4	194.955	blk hex cry	3090		13.7		i H_2O; sl aqua regia; reac alk
2766	Tantalum silicide	$TaSi_2$	12039-79-1	237.119	gray powder	2200		9.14		
2767	Tantalum(III) bromide	$TaBr_3$	13842-73-4	420.660	gray-grn solid	dec 220				
2768	Tantalum(III) chloride	$TaCl_3$	13569-67-0	287.307	blk-grn solid	dec 440				s H_2O
2769	Tantalum(IV) bromide	$TaBr_4$	13842-76-7	500.564	dark blue solid	392		5.77		reac H_2O
2770	Tantalum(IV) chloride	$TaCl_4$	13569-72-7	322.760	dark grn solid	dec 300	subl	4.35		reac H_2O
2771	Tantalum(IV) iodide	TaI_4	14693-80-2	688.566	gray-blk solid	400 dec				reac H_2O
2772	Tantalum(IV) oxide	TaO_2	12036-14-5	212.947	tetr cry			10.0		
2773	Tantalum(IV) selenide	$TaSe_2$	12039-55-3	338.87	hex cry			6.7		
2774	Tantalum(IV) sulfide	TaS_2	12143-72-5	245.078	blk hex cry	>3000		6.86		i H_2O
2775	Tantalum(IV) telluride	$TaTe_2$	12067-66-2	436.15	monocl cry			9.4		
2776	Tantalum(V) bromide	$TaBr_5$	13451-11-1	580.468	yel cry powder	265.8	348.8	4.67		
2777	Tantalum(V) chloride	$TaCl_5$	7721-01-9	358.213	yel-wh monocl cry; hyg	216.6	239	3.68		reac H_2O; s EtOH
2778	Tantalum(V) fluoride	TaF_5	7783-71-3	275.940	wh monocl cry; hyg	96.9	229.5	4.74		s H_2O, eth; sl CS_2, ctc
2779	Tantalum(V) iodide	TaI_5	14693-81-3	815.470	blk hex cry; hyg	496	543	5.80		reac H_2O
2780	Tantalum(V) oxide	Ta_2O_5	1314-61-0	441.893	wh rhomb cry or powder	1875		8.24		i H_2O, EtOH, acid; s HF
2781	Technetium	Tc	7440-26-8	98	hex cry	2157	4265	11		
2782	Technetium(V) fluoride	TcF_5	31052-14-9	193	yel solid	50	dec			
2783	Technetium(VI) fluoride	TcF_6	13842-93-8	212	yel cub cry	37.4	55.3	3.0		
2784	Tellurium	Te	13494-80-9	127.60	gray-wh rhomb cry	449.51	988	6.232		i H_2O, bz, CS_2
2785	Telluric(VI) acid	H_6TeO_6	7803-68-1	229.64	wh monocl cry	136		3.07	50.1^{30}	
2786	Tellurous acid	H_2TeO_3	10049-23-7	177.61	wh cry	40 dec		3.0		sl H_2O; s dil acid, alk
2787	Tellurium dioxide	TeO_2	7446-07-3	159.60	wh orth cry	733	1245	5.9		i H_2O; s alk, acid
2788	Tellurium trioxide	TeO_3	13451-18-8	175.60	yel-oran cry	430		5.07		i H_2O

No.	Name	Formula	CAS Reg No.	Mol. weight	Physical form	mp/°C	bp/°C	Density g cm^{-3}	Solubility g/100 g H_2O	Qualitative solubility
2789	Tellurium dibromide	$TeBr_2$	7789-54-0	287.41	grn-brn hyg cry	210	339			reac H_2O; s eth; sl chl
2790	Tellurium dichloride	$TeCl_2$	10025-71-5	198.51	blk amorp solid; hyg	208	328	6.9		reac H_2O; i ctc
2791	Tellurium tetrabromide	$TeBr_4$	10031-27-3	447.22	yel-oran monocl cry	380	≈420 dec	4.3		reac H_2O; s eth
2792	Tellurium tetrachloride	$TeCl_4$	10026-07-0	269.41	wh monocl cry; hyg	224	387	3.0		reac H_2O; s EtOH, tol
2793	Tellurium tetrafluoride	TeF_4	15192-26-4	203.59	col cry	129	195 dec			reac H_2O
2794	Tellurium decafluoride	Te_2F_{10}	53214-07-6	445.18	col liq	-33.7	59			
2795	Tellurium tetraiodide	TeI_4	7790-48-9	635.22	blk orth cry	280		5.05		reac H_2O; sl ace
2796	Tellurium hexafluoride	TeF_6	7783-80-4	241.59	col gas	-37.6 tp	-38.9 sp	9.875 g/L		reac H_2O
2797	Terbium	Tb	7440-27-9	158.925	silv metal; hex	1359	3230	8.23		
2798	Terbium nitride	TbN	12033-64-6	172.932	cub cry			9.55		
2799	Terbium silicide	$TbSi_2$	12039-80-4	215.096	orth cry			6.66		
2800	Terbium(III) bromide	$TbBr_3$	14456-47-4	398.637	wh hex cry	830	1490			s H_2O
2801	Terbium(III) chloride	$TbCl_3$	10042-88-3	265.284	wh orth cry; hyg	582		4.35		s H_2O
2802	Terbium(III) chloride hexahydrate	$TbCl_3 \cdot 6H_2O$	13798-24-8	373.375	hyg cry			4.35		vs H_2O
2803	Terbium(III) iodide	TbI_3	13813-40-6	539.638	hex cry; hyg	955		≈5.2		s H_2O
2804	Terbium(III) nitrate	$Tb(NO_3)_3$	10043-27-3	344.940	pink hyg solid				157^{25}	s EtOH
2805	Terbium(III) nitrate hexahydrate	$Tb(NO_3)_3 \cdot 6H_2O$	13451-19-9	453.031	col needles	89				s H_2O, EtOH, ace
2806	Terbium(III) oxide	Tb_2O_3	12036-41-8	365.849	wh cub cry	2303		7.91		
2807	Terbium(III) sulfate octahydrate	$Tb_2(SO_4)_3 \cdot 8H_2O$	13842-67-6	750.161	wh cry	dec 360				sl H_2O
2808	Terbium(III) sulfide	Tb_2S_3	12138-11-3	414.046	cub cry			6.35		
2809	Terbium(III) fluoride	TbF_3	13708-63-9	215.920	wh solid	1175	2280			i H_2O
2810	Terbium(IV) fluoride	TbF_4	36781-15-4	234.919	wh monocl cry	dec 300				i H_2O
2811	Thallium	Tl	7440-28-0	204.383	soft blue-wh metal	304	1473	11.8		i H_2O; reac acid
2812	Thallium(I) acetate	$TlC_2H_3O_2$	563-68-8	263.427	hyg wh cry	131		3.68		s H_2O, EtOH
2813	Thallium(I) azide	TlN_3	13847-66-0	246.403	yel cry	334	exp			s H_2O
2814	Thallium(I) bromate	$TlBrO_3$	14550-84-6	332.285	col needles	120 dec			0.49^{30}	s EtOH
2815	Thallium(I) bromide	TlBr	7789-40-4	284.287	yel cub cry	460	819	7.5	0.059^{20}	
2816	Thallium(I) carbonate	Tl_2CO_3	6533-73-9	468.776	wh monocl cry	273		7.11	4.69^{20}	i EtOH
2817	Thallium(I) chlorate	$TlClO_3$	13453-30-0	287.834	col hex cry	dec 500		5.5	3.92^{20}	
2818	Thallium(I) chloride	TlCl	7791-12-0	239.836	wh cub cry	431	720	7.0	0.33^{20}	i EtOH
2819	Thallium(I) chromate	Tl_2CrO_4	13473-75-1	524.761	yel cry				0.003^{20}	sl acid, alk
2820	Thallium(I) cyanide	TlCN	13453-34-4	230.401	wh hex plates			6.523		s H_2O, acid, EtOH
2821	Thallium(I) ethanolate	TlC_2H_5O	20398-06-5	249.443	cloudy liq	-3	130 dec	3.49		reac H_2O
2822	Thallium(I) fluoride	TlF	7789-27-7	223.381	wh orth cry	326	826	8.36	245^{25}	
2823	Thallium(I) formate	$TlCHO_2$	992-98-3	249.401	hyg col needles	101		4.97		vs H_2O; s MeOH
2824	Thallium(I) hexafluorophosphate	$TlPF_6$	60969-19-9	349.347	wh cub cry			4.6		
2825	Thallium(I) hydroxide	TlOH	12026-06-1	221.390	yel needles	139 dec		7.44	34.3^{18}	
2826	Thallium(I) iodate	$TlIO_3$	14767-09-0	379.285	wh needles				0.058	sl HNO_3
2827	Thallium(I) iodide	TlI	7790-30-9	331.287	yel cry powder	441.7	824	7.1	0.0085^{20}	i EtOH
2828	Thallium(I) molybdate	Tl_2MoO_4	34128-09-1	568.71	yel-wh cub cry					i H_2O
2829	Thallium(I) nitrate	$TlNO_3$	10102-45-1	266.388	wh cry	206	450 dec	5.55	9.55^{20}	s H_2O; i EtOH
2830	Thallium(I) nitrite	$TlNO_2$	13826-63-6	250.389	yel cub cry	186		5.7	32.1^{25}	s H_2O
2831	Thallium(I) oxalate	$Tl_2C_2O_4$	30737-24-7	496.786	wh powder			6.31	1.83^{20}	
2832	Thallium(I) oxide	Tl_2O	1314-12-1	424.766	blk rhomb cry; hyg	579	≈1080	9.52		s H_2O, EtOH
2833	Thallium(I) perchlorate	$TlClO_4$	13453-40-2	303.834	col orth cry	501		4.89	19.7^{30}	
2834	Thallium(I) selenate	Tl_2SeO_4	7446-22-2	551.73	orth cry	>400		6.875	2.8^{20}	i EtOH, eth
2835	Thallium(I) selenide	Tl_2Se	15572-25-5	487.73	gray plates	340				i H_2O, acid
2836	Thallium(I) sulfate	Tl_2SO_4	7446-18-6	504.830	wh rhomb prisms	632		6.77	5.47^{25}	
2837	Thallium(I) sulfide	Tl_2S	1314-97-2	440.832	blue-blk cry	457	1367	8.39	0.02^{20}	sl alk; s acid
2838	Thallium(III) acetate	$Tl(C_2H_3O_2)_3$	2570-63-0	381.514	hyg wh platelets	182 dec				
2839	Thallium(III) bromide tetrahydrate	$TlBr_3 \cdot 4H_2O$	13701-90-1	516.157	yel orth cry			3.65		s H_2O, EtOH
2840	Thallium(III) chloride	$TlCl_3$	13453-32-2	310.742	monocl cry	155		4.7		vs H_2O, EtOH, eth
2841	Thallium(III) chloride tetrahydrate	$TlCl_3 \cdot 4H_2O$	13453-32-2*	382.804	orth cry			3.00		s H_2O
2842	Thallium(III) fluoride	TlF_3	7783-57-5	261.378	wh orth cry; hyg	550 dec		8.65		reac H_2O
2843	Thallium(III) nitrate	$Tl(NO_3)_3$	13746-98-0	390.398	col cry					reac H_2O
2844	Thallium(III) oxide	Tl_2O_3	1314-32-5	456.765	brn cub cry	834		10.2		i H_2O; reac acid
2845	Thallium(III) sulfate	$Tl_2(SO_4)_3$	16222-66-5	696.955	col leaflets					reac H_2O
2846	Thallium selenide	TlSe	12039-52-0	283.34	blk solid	330				i H_2O, acid
2847	Thorium	Th	7440-29-1	232.038	soft gray-wh metal; cub	1750	4788	11.7		s acid
2848	Thorium hydride	ThH_2	16689-88-6	234.054	tetr cry			9.5		
2849	Thorium boride	ThB_6	12229-63-9	296.904	refrac solid	2450		6.99		
2850	Thorium(IV) bromide	$ThBr_4$	13453-49-1	551.654	wh hyg cry	679			65^{20}	

No.	Name	Formula	CAS Reg No.	Mol. weight	Physical form	mp/°C	bp/°C	Density g cm^{-3}	Solubility g/100 g H_2O	Qualitative solubility
2851	Thorium carbide	ThC	12012-16-7	244.049	cub cry	2500		10.6		reac H_2O
2852	Thorium dicarbide	ThC_2	12071-31-7	256.059	yel monocl cry	≈2650		9.0		reac H_2O
2853	Thorium(IV) chloride	$ThCl_4$	10026-08-1	373.850	gray-wh tetr needles; hyg	770	921	4.59		s H_2O, EtOH
2854	Thorium(IV) fluoride	ThF_4	13709-59-6	308.032	wh monocl cry; hyg	1110	1680	6.1		
2855	Thorium(IV) iodide	ThI_4	7790-49-0	739.656	wh-yel monocl cry	566	837			
2856	Thorium(IV) nitrate	$Th(NO_3)_4$	13823-29-5	480.058	hyg wh plates	55 dec				vs H_2O, EtOH
2857	Thorium(IV) nitrate tetrahydrate	$Th(NO_3)_4 \cdot 4H_2O$	13470-07-0	552.119	wh hyg cry	500 dec			191^{20}	s EtOH
2858	Thorium nitride	ThN	12033-65-7	246.045	refrac cub cry	2820		11.6		reac H_2O
2859	Thorium(IV) oxide	ThO_2	1314-20-1	264.037	wh cub cry	3350	4400	10.0		i H_2O, alk; sl acid
2860	Thorium(IV) selenide	$ThSe_2$	60763-24-8	389.96	orth cry			8.5		
2861	Thorium orthosilicate	$ThSiO_4$	14553-44-7	324.122	brn tetr cry			6.7		
2862	Thorium silicide	$ThSi_2$	12067-54-8	288.209	tetr cry	1850		7.9		
2863	Thorium(IV) sulfate nonahydrate	$Th(SO_4)_2 \cdot 9H_2O$	10381-37-0	586.301	wh monocl cry	dec		2.8	1.2^{20}	
2864	Thorium(IV) sulfide	ThS_2	12138-07-7	296.168	dark brn cry	1905		7.30		i H_2O; s acid
2865	Thulium	Tm	7440-30-4	168.934	silv metal; hex	1545	1950	9.32		s dil acid
2866	Thulium(II) bromide	$TmBr_2$	64171-97-7	328.742	dark grn solid	619				
2867	Thulium(II) chloride	$TmCl_2$	22852-11-5	239.840	red or grn cry	718				reac H_2O
2868	Thulium(II) iodide	TmI_2	60864-26-8	422.743	blk hyg solid	756				reac H_2O
2869	Thulium(III) bromide	$TmBr_3$	14456-51-0	408.646	wh hyg cry	954				s H_2O
2870	Thulium(III) chloride	$TmCl_3$	13537-18-3	275.293	yel hyg cry	845				s H_2O
2871	Thulium(III) chloride heptahydrate	$TmCl_3 \cdot 7H_2O$	13778-39-7	401.400	hyg cry					s H_2O, EtOH
2872	Thulium(III) fluoride	TmF_3	13760-79-7	225.929	wh cry	1158				s H_2O
2873	Thulium(III) hydroxide	$Tm(OH)_3$	1311-33-7	219.956	wh or grn prec					i H_2O
2874	Thulium(III) iodide	TmI_3	13813-43-9	549.647	yel hyg cry	1021				
2875	Thulium(III) nitrate	$Tm(NO_3)_3$	14985-19-4	354.949	grn hyg solid				212^{25}	s EtOH
2876	Thulium(III) nitrate pentahydrate	$Tm(NO_3)_3 \cdot 5H_2O$	36548-87-5	445.025	grn hyg cry					s H_2O, EtOH, ace
2877	Thulium(III) oxalate hexahydrate	$Tm_2(C_2O_4)_3 \cdot 6H_2O$	26677-68-9	710.016	grn solid	dec 50				s alk oxalates
2878	Thulium(III) oxide	Tm_2O_3	12036-44-1	385.866	grn-wh cub cry	2341	3945	8.6		sl acid
2879	Tin (gray)	Sn	7440-31-5	118.710	cub cry	trans to wh Sn 13.2	2602	5.769		
2880	Tin (white)	Sn	7440-31-5	118.710	silv tetr cry	231.93	2602	7.287		
2881	Stannane	SnH_4	2406-52-2	122.742	unstab col gas	-146	-51.8	5.017 g/L		
2882	Methylstannane	SnH_3CH_3	1631-78-3	136.769	col gas		1.4	5.590 g/L		reac H_2O
2883	(Dimethylamino)trimethystannane	$Sn(CH_3)_3N(CH_3)_2$	993-50-0	207.890	liq	1	126	1.22		reac H_2O
2884	Tin monophosphide	SnP	25324-56-5	149.684	dull metallic solid	540				
2885	Tin triphosphide	Sn_4P_3	12286-33-8	567.761	wh cry	≈550		5.2		
2886	Tin(II) acetate	$Sn(C_2H_3O_2)_2$	638-39-1	236.799	wh orth cry	183	subl	2.31		s dil HCl
2887	Tin(II) bromide	$SnBr_2$	10031-24-0	278.518	yel powder	215	639	5.12	85^{0}	s EtOH, eth, ace
2888	Tin(II) chloride	$SnCl_2$	7772-99-8	189.616	wh orth cry	247.0	623	3.90	178^{10}	s EtOH, ace, eth; i xyl
2889	Tin(II) chloride dihydrate	$SnCl_2 \cdot 2H_2O$	10025-69-1	225.647	wh monocl cry	37 dec		2.71	178^{10}	s EtOH, NaOH; vs HCl
2890	Tin(II) fluoride	SnF_2	7783-47-3	156.707	wh monocl cry; hyg	215	850	4.57		s H_2O; i EtOH, eth, chl
2891	Tin(II) hexafluorozirconate	$SnZrF_6$	12419-43-1	323.924	cry			4.21		s H_2O
2892	Tin(II) hydroxide	$Sn(OH)_2$	12026-24-3	152.725	wh amorp solid					
2893	Tin(II) iodide	SnI_2	10294-70-9	372.519	red-oran powder	320	714	5.28	0.98^{20}	s bz, chl, CS_2
2894	Tin(II) oxalate	SnC_2O_4	814-94-8	206.729	wh powder	280 dec		3.56		i H_2O; s dil HCl
2895	Tin(II) oxide	SnO	21651-19-4	134.709	blue-blk tetr cry	1080 dec		6.45		i H_2O, EtOH; s acid
2896	Tin(II) pyrophosphate	$Sn_2P_2O_7$	15578-26-4	411.363	wh amorp powder	400 dec		4.009		i H_2O; s conc acid
2897	Tin(II) selenide	SnSe	1315-06-6	197.67	gray orth cry	861		6.18		i H_2O; s aqua regia
2898	Tin(II) sulfate	$SnSO_4$	7488-55-3	214.773	wh orth cry	378 dec		4.15	18.8^{19}	
2899	Tin(II) sulfide	SnS	1314-95-0	150.775	gray orth cry	881	1210	5.08		i H_2O; s conc acid
2900	Tin(II) tartrate	$SnC_4H_4O_6$	815-85-0	266.781	wh cry powder					s H_2O, dil HCl
2901	Tin(II) telluride	SnTe	12040-02-7	246.31	gray cub cry	806		6.5		
2902	Tin(IV) bromide	$SnBr_4$	7789-67-5	438.326	wh cry	29.1	205	3.34		vs H_2O; s EtOH
2903	Tin(IV) chloride	$SnCl_4$	7646-78-8	260.522	col fuming liq	-34.07	114.15	2.234		reac H_2O; s EtOH, ctc, bz, ace
2904	Tin(IV) chloride pentahydrate	$SnCl_4 \cdot 5H_2O$	10026-06-9	350.598	wh-yel cry	56 dec		2.04		vs H_2O; s EtOH
2905	Tin(IV) chromate	$Sn(CrO_4)_2$	38455-77-5	350.697	brn-yel cry powder	dec				s H_2O
2906	Tin(IV) fluoride	SnF_4	7783-62-2	194.704	wh tetr cry	442	705 subl	4.78		reac H_2O
2907	Tin(IV) iodide	SnI_4	7790-47-8	626.328	yel-brn cub cry	143	364.35	4.46		reac H_2O; s EtOH, bz, chl, eth
2908	Tin(IV) oxide	SnO_2	18282-10-5	150.709	gray tetr cry	1630		6.85		i H_2O, EtOH; s hot conc alk
2909	Tin(IV) selenide	$SnSe_2$	20770-09-6	276.63	red-brn cry	650		≈5.0		i H_2O; s alk, conc acid

No.	Name	Formula	CAS Reg No.	Mol. weight	Physical form	mp/°C	bp/°C	Density g cm⁻³	Solubility g/100 g H_2O	Qualitative solubility
2910	Tin(IV) selenite	$Sn(SeO_3)_2$	7446-25-5	372.63	cry powder					i H_2O; s hot HCl
2911	Tin(IV) sulfide	SnS_2	1315-01-1	182.840	gold-yel hex cry	600 dec		4.5		i H_2O; s alk, aqua regia
2912	Titanium	Ti	7440-32-6	47.867	gray metal; hex	1668	3287	4.506		
2913	Titanocene dichloride	$Ti(C_5H_5)_2Cl_2$	1271-19-8	248.959	red cry	289		1.60		sl H_2O, bz; s chl, EtOH, tol
2914	Titanium hydride	TiH_2	7704-98-5	49.883	gray-blk powder	≈450 dec		3.75		i H_2O
2915	Titanium boride	TiB_2	12045-63-5	69.489	gray refrac solid; hex	3225		4.38		
2916	Titanium carbide	TiC	12070-08-5	59.878	cub cry	3067		4.93		i H_2O; s HNO_3
2917	Titanium nitride	TiN	25583-20-4	61.874	yel-brn cub cry	2947		5.21		i H_2O; s aqua regia
2918	Titanium phosphide	TiP	12037-65-9	78.841	gray hex cry	1990		4.08		
2919	Titanium silicide	$TiSi_2$	12039-83-7	104.038	blk orth cry	1500		4.0		i H_2O, acid, alk; s HF
2920	Titanium(II) bromide	$TiBr_2$	13783-04-5	207.675	blk powder	dec 400		4.0		reac H_2O
2921	Titanium(II) chloride	$TiCl_2$	10049-06-6	118.773	blk hex cryc	1035	1500	3.13		reac H_2O; s EtOH; i chl, eth
2922	Titanium(II) iodide	TiI_2	13783-07-8	301.676	blk hex cry	dec 400		5.02		reac H_2O
2923	Titanium(II) oxide	TiO	12137-20-1	63.866	yel cub cry	1770	3227	4.95		
2924	Titanium(II) sulfide	TiS	12039-07-5	79.932	brn hex cry	1927		3.85		i H_2O; s conc acid
2925	Titanium(III) bromide	$TiBr_3$	13135-31-4	287.579	viol hex cry	dec 400				s H_2O
2926	Titanium(III) chloride	$TiCl_3$	7705-07-9	154.226	red-viol hex cry; hyg	425 dec	960	2.64		reac H_2O
2927	Titanium(III) fluoride	TiF_3	13470-08-1	104.862	viol hex cry	950 dec		2.98		i H_2O, dil acid, alk
2928	Titanium(III) iodide	TiI_3	13783-08-9	428.580	viol cry	dec 350				
2929	Titanium(III) oxide	Ti_2O_3	1344-54-3	143.732	blk hex cry	1842		4.486		s hot HF
2930	Titanium(III) sulfate	$Ti_2(SO_4)_3$	10343-61-0	383.922	grn cry					i H_2O, EtOH; s dil HCl
2931	Titanium(III) sulfide	Ti_2S_3	12039-16-6	191.929	blk hex cry			3.56		
2932	Titanium(III,IV) oxide	Ti_3O_5	12065-65-5	223.598	blk monocl cry	1777		4.24		
2933	Titanium(IV) bromide	$TiBr_4$	7789-68-6	367.483	yel-oran cub cry; hyg	38.3	233.5	3.37		reac H_2O
2934	Titanium(IV) chloride	$TiCl_4$	7550-45-0	189.679	col or yel liq	-24.12	136.45	1.73		reac H_2O; s EtOH
2935	Titanium(IV) fluoride	TiF_4	7783-63-3	123.861	wh hyg powder	377	subl 284	2.798		reac H_2O; s EtOH, py
2936	Titanium(IV) iodide	TiI_4	7720-83-4	555.485	red hyg powder	155	377	4.3		reac H_2O
2937	Titanium(IV) oxide (anatase)	TiO_2	1317-70-0	79.866	brn tetr cry	1560		3.9		
2938	Titanium(IV) oxide (brookite)	TiO_2	12188-41-9	79.866	wh orth cry			4.17		
2939	Titanium(IV) oxide (rutile)	TiO_2	1317-80-2	79.866	wh tetr cry	1843	≈3000	4.17		i H_2O, dil acid; s conc acid
2940	Titanium(IV) oxysulfate monohydrate	$TiOSO_4 \cdot H_2O$	13825-74-6*	177.944	col orth cry			2.71		reac H_2O
2941	Titanium(IV) sulfate	$Ti(SO_4)_2$	13693-11-3	239.992	wh-yel hyg cry	150 dec				s H_2O
2942	Titanium(IV) sulfide	TiS_2	12039-13-3	111.997	yel-brn hex cry; hyg			3.37		s H_2SO_4
2943	Tungsten	W	7440-33-7	183.84	gray-wh metal; cub	3422	5555	19.3		
2944	Tungstic acid	H_2WO_4	7783-03-1	249.85	yel amorp powder	100 dec		5.5		i H_2O, acid; s alk
2945	Tungsten boride (W_2B)	W_2B	12007-10-2	378.49	refrac blk powder	2670		16.0		i H_2O
2946	Tungsten boride (WB)	WB	12007-09-9	194.65	blk refrac powder	2665		15.2		i H_2O
2947	Tungsten boride (W_2B_5)	W_2B_5	12007-98-6	421.74	refrac solid	2370		11.0		i H_2O
2948	Tungsten carbide (W_2C)	W_2C	12070-13-2	379.69	refrac hex cry	≈2800		14.8		i H_2O
2949	Tungsten carbide (WC)	WC	12070-12-1	195.85	gray hex cry	2785		15.6		i H_2O; s HNO_3/HF
2950	Tungsten carbonyl	$W(CO)_6$	14040-11-0	351.90	wh cry	170 dec	subl	2.65		i H_2O; s os
2951	Tungsten nitride (WN_2)	WN_2	60922-26-1	211.85	hex cry	600 dec		7.7		
2952	Tungsten nitride (W_2N)	W_2N	12033-72-6	381.69	gray cub cry	dec		17.8		
2953	Tungsten silicide (WSi_2)	WSi_2	12039-88-2	240.01	blue-gray tetr cry	2160		9.3		i H_2O
2954	Tungsten silicide (W_5Si_3)	W_5Si_3	12039-95-1	1003.46	blue-gray refrac solid	2320		14.4		
2955	Tungsten(II) bromide	WBr_2	13470-10-5	343.65	yel powder	dec 400				i H_2O
2956	Tungsten(II) chloride	WCl_2	13470-12-7	254.75	gray solid	dec 500		5.44		sl H_2O
2957	Tungsten(II) iodide	WI_2	13470-17-2	437.65	oran-brn cry	dec 800		6.79		i H_2O
2958	Tungsten(III) bromide	WBr_3	15163-24-3	423.55	blk hex cry	dec 180				i H_2O
2959	Tungsten(III) chloride	WCl_3	20193-56-0	290.20	red solid	550 dec	subl			reac H_2O
2960	Tungsten(III) iodide	WI_3	15513-69-6	564.55	blk solid	dec r.t.				i H_2O; s ace; sl EtOH, chl
2961	Tungsten(IV) bromide	WBr_4	14055-81-3	503.46	blk orth cry		240 subl			reac H_2O
2962	Tungsten(IV) chloride	WCl_4	13470-13-8	325.65	blk hyg powder	450 dec		4.62		reac H_2O
2963	Tungsten(IV) fluoride	WF_4	13766-47-7	259.83	red-brn cry	dec 800				reac H_2O; s MeCN; i bz, tol, ctc

No.	Name	Formula	CAS Reg No.	Mol. weight	Physical form	mp/°C	bp/°C	Density g cm^{-3}	Solubility g/100 g H_2O	Qualitative solubility
2964	Tungsten(IV) iodide	WI_4	14055-84-6	691.46	blk cry	dec				reac H_2O; s EtOH; i eth chl
2965	Tungsten(IV) oxide	WO_2	12036-22-5	215.84	brn monocl cry	≈1500 dec	1730	10.8		i H_2O, os
2966	Tungsten(IV) selenide	WSe_2	12067-46-8	341.76	gray hex cry			9.2		
2967	Tungsten(IV) sulfide	WS_2	12138-09-9	247.97	gray hex cry	1250 dec		7.6		i H_2O, HCl, alk
2968	Tungsten(IV) telluride	WTe_2	12067-76-4	439.04	gray orth cry	1020		9.43		
2969	Tungsten(V) bromide	WBr_5	13470-11-6	583.36	brn-blk hyg solid	286	333			reac H_2O
2970	Tungsten(V) chloride	WCl_5	13470-14-9	361.11	blk-grn hyg cry	253	286	3.88		reac H_2O
2971	Tungsten(V) ethanolate	$W(C_2H_5O)_5$	62571-53-3	409.14	powder		105(0.05 mmHg)			s EtAc
2972	Tungsten(V) fluoride	WF_5	19357-83-6	278.83	yel solid	dec 20				reac H_2O
2973	Tungsten(V) oxytribromide	$WOBr_3$	20213-56-3	439.55	dark brn tetr cry			≈5.9		
2974	Tungsten(V) oxytrichloride	$WOCl_3$	14249-98-0	306.20	grn tetr cry			≈4.6		
2975	Tungsten(VI) bromide	WBr_6	13701-86-5	663.26	blue-blk cry	309		6.9		reac H_2O
2976	Tungsten(VI) chloride	WCl_6	13283-01-7	396.56	purp hex cry; hyg	282	337	3.52		reac H_2O; s EtOH, os
2977	Tungsten(VI) dioxydibromide	WO_2Br_2	13520-75-7	375.65	red cry		440 subl			
2978	Tungsten(VI) dioxydichloride	WO_2Cl_2	13520-76-8	286.75	yel orth cry	265		4.67		i H_2O
2979	Tungsten(VI) dioxydiiodide	WO_2I_2	14447-89-3	469.65	grn monocl cry	400 dec		6.39		
2980	Tungsten(VI) fluoride	WF_6	7783-82-6	297.83	yel liq or col gas	1.9	17.1	3.44		reac H_2O; vs ctc, cyhex
2981	Tungsten(VI) oxide	WO_3	1314-35-8	231.84	yel powder	1473	≈1700	7.2		i H_2O, os; sl acid; s alk
2982	Tungsten(VI) oxytetrabromide	$WOBr_4$	13520-77-9	519.46	red tetr cry	277	327	≈5.5		reac H_2O
2983	Tungsten(VI) oxytetrachloride	$WOCl_4$	13520-78-0	341.65	red hyg cry	210	230	11.92		reac H_2O; s bz, CS_2
2984	Tungsten(VI) oxytetrafluoride	WOF_4	13520-79-1	275.83	wh monocl cry	105	185.9	5.07		reac H_2O
2985	Tungsten(VI) sulfide	WS_3	12125-19-8	280.04	brn powder					sl H_2O; s alk
2986	Uranium	U	7440-61-1	238.029	silv-wh orth cry	1135	4131	19.1		
2987	Uranium boride (UB_2)	UB_2	12007-36-2	259.651	refrac solid	2430		12.7		
2988	Uranium boride (UB_4)	UB_4	12007-84-0	281.273	refrac solid	2530		9.32		i H_2O
2989	Uranium carbide (UC)	UC	12070-09-6	250.040	gray cub cry	2790				
2990	Uranium carbide (UC_2)	UC_2	12071-33-9	262.050	gray tetr cry	2350	4370	11.3		reac H_2O; sl EtOH
2991	Uranium carbide (U_2C_3)	U_2C_3	12076-62-9	512.090	gray cub cry	≈1700 dec		12.7		
2992	Uranium nitride (UN)	UN	25658-43-9	252.036	gray cub cry	2805		14.3		i H_2O
2993	Uranium nitride (U_2N_3)	U_2N_3	12033-83-9	518.078	cub cry	dec		11.3		
2994	Uranium(III) bromide	UBr_3	13470-19-4	477.741	red hyg cry	727				s H_2O
2995	Uranium(III) chloride	UCl_3	10025-93-1	344.388	grn hyg cry	837		5.51		vs H_2O; i bz, ctc
2996	Uranium(III) fluoride	UF_3	13775-06-9	295.024	blk hex cry	1495		8.9		i H_2O; s acid
2997	Uranium(III) hydride	UH_3	13598-56-6	241.053	gray-blk cub cry			11.1		
2998	Uranium(III) iodide	UI_3	13775-18-3	618.742	blk hyg cry	766				s H_2O
2999	Uranium(IV) bromide	UBr_4	13470-20-7	557.645	brn hyg cry	519				s H_2O, EtOH
3000	Uranium(IV) chloride	UCl_4	10026-10-5	379.841	grn octahed cry	590	791	4.72		reac H_2O; s EtOH
3001	Uranium(IV) fluoride	UF_4	10049-14-6	314.023	grn monocl cry	1036	1417	6.7	0.01^{25}	s conc acid, alk
3002	Uranium(IV) iodide	UI_4	13470-22-9	745.647	blk hyg cry	506				s H_2O, EtOH
3003	Uranium(IV) oxide	UO_2	1344-57-6	270.028	brn cub cry	2847		10.97		i H_2O, dil acid; s conc acid
3004	Uranium(IV,V) oxide	U_4O_9	12037-15-9	1096.111	cub cry			11.2		
3005	Uranium(V) bromide	UBr_5	13775-16-1	637.549	brn hyg cry					reac H_2O
3006	Uranium(V) chloride	UCl_5	13470-21-8	415.294	brn hyg cry	287				reac H_2O
3007	Uranium(V) fluoride	UF_5	13775-07-0	333.021	pale blue tetr cry; hyg	348		5.81		s H_2O
3008	Uranium(V,VI) oxide	U_3O_8	1344-59-8	842.082	grn-blk orth cry	1300 dec		8.38		
3009	Uranium(VI) chloride	UCl_6	13763-23-0	450.747	grn hex cry	177		3.6		
3010	Uranium(VI) fluoride	UF_6	7783-81-5	352.019	wh monocl solid	64.06 tp	56.5 sp	5.09		reac H_2O; s ctc, chl
3011	Uranium(VI) oxide	UO_3	1344-58-7	286.027	oran-yel cry			≈7.3		i H_2O; s acid
3012	Uranium(VI) oxide monohydrate	$UO_3 \cdot H_2O$	12326-21-5	304.043	yel orth cry	570 dec		7.05		
3013	Uranium peroxide dihydrate	$UO_4 \cdot 2H_2O$	19525-15-6	338.057	yel hyg cry	115 dec				i H_2O
3014	Uranyl acetate dihydrate	$UO_2(C_2H_3O_2)_2 \cdot 2H_2O$	6159-44-0	424.146	ye cry (HOAc)	80 dec		2.89		sl EtOH
3015	Uranyl chloride	UO_2Cl_2	7791-26-6	340.934	yel orth cry; hyg	577				vs H_2O; s EtOH, ace; i bz
3016	Uranyl fluoride	UO_2F_2	13536-84-0	308.025	yel hyg solid				64.4^{20}	i bz
3017	Uranyl hydrogen phosphate tetrahydrate	$UO_2HPO_4 \cdot 4H_2O$	18433-48-2	438.068	yel cry pow					i H_2O; s acid
3018	Uranyl nitrate	$UO_2(NO_3)_2$	10102-06-4	394.037	yel cry				127^{25}	s eth
3019	Uranyl nitrate hexahydrate	$UO_2(NO_3)_2 \cdot 6H_2O$	13520-83-7	502.129	yel orth cry; hyg	60	118 dec	2.81	127^{25}	s EtOH, eth
3020	Uranyl sulfate	UO_2SO_4	1314-64-3	366.090	yel cry					
3021	Uranyl sulfate trihydrate	$UO_2SO_4 \cdot 3H_2O$	20910-28-5	420.137	yel cry			3.28	152^{16}	sl EtOH
3022	Vanadium	V	7440-62-2	50.942	gray-wh metal; cub	1910	3407	6.0		i H_2O; s acid

No.	Name	Formula	CAS Reg No.	Mol. weight	Physical form	mp/°C	bp/°C	Density g cm⁻³	Solubility g/100 g H_2O	Qualitative solubility
3023	Vanadocene	$V(C_5H_5)_2$	1277-47-0	181.128	viol cry; hyg	167				s bz, thf
3024	Vanadocene dichloride	$V(C_5H_5)_2Cl_2$	12083-48-6	252.034	dark grn cry	205 dec				s H_2O, chl, EtOH
3025	Vanadium boride (VB)	VB	12045-27-1	61.753	refrac solid	2250				i H_2O
3026	Vanadium boride (VB_2)	VB_2	12007-37-3	72.564	refrac solid	2450				
3027	Vanadium carbide (VC)	VC	12070-10-9	62.953	refrac blk cry; cub	2810		5.77		i H_2O
3028	Vanadium carbide (V_2C)	V_2C	12012-17-8	113.894	hex cry	2167				
3029	Vanadium carbonyl	$V(CO)_6$	14024-00-1	219.002	blue-grn cry; flam	60 dec	subl			
3030	Vanadium nitride	VN	24646-85-3	64.949	blk powder; cub	2050		6.13		i H_2O; s aqua regia
3031	Vanadium silicide (VSi_2)	VSi_2	12039-87-1	107.113	metallic prisms			4.42		s HF
3032	Vanadium silicide (V_3Si)	V_3Si	12039-76-8	180.911	cub cry	1935		5.70		
3033	Vanadium(II) bromide	VBr_2	14890-41-6	210.750	oran-brn hex cry		800 subl	4.58		reac H_2O
3034	Vanadium(II) chloride	VCl_2	10580-52-6	121.848	grn hex plates	1350	910 subl	3.23		reac H_2O; s EtOH, eth
3035	Vanadium(II) fluoride	VF_2	13842-80-3	88.939	blue hyg cry	1490				reac H_2O
3036	Vanadium(II) iodide	VI_2	15513-84-5	304.751	red-viol hex cry		subl 800	5.44		reac H_2O
3037	Vanadium(II) oxide	VO	12035-98-2	66.941	gray-blk cry	1790		5.758		s acid
3038	Vanadium(II) sulfate heptahydrate	$VSO_4 \cdot 7H_2O$	36907-42-3	273.111	viol cry					
3039	Vanadium(III) bromide	VBr_3	13470-26-3	290.654	blk-grn hyg cry	dec 500	subl	4.00		reac H_2O
3040	Vanadium(III) chloride	VCl_3	7718-98-1	157.301	red-viol hex cry; hyg	500 dec		3.00		reac H_2O; s EtOH, eth
3041	Vanadium(III) fluoride	VF_3	10049-12-4	107.937	yel-grn hex cry	1395	subl	3.363		i H_2O, EtOH
3042	Vanadium(III) fluoride trihydrate	$VF_3 \cdot 3H_2O$	10049-12-4*	161.983	grn rhomb cry	≈100 dec				sl H_2O
3043	Vanadium(III) iodide	VI_3	15513-94-7	431.655	brn-blk rhomb cry; hyg	dec 300		5.21		reac H_2O
3044	Vanadium(III) oxide	V_2O_3	1314-34-7	149.881	blk powder	1957	≈3000	4.87		i H_2O
3045	Vanadium(III) 2,4-pentanedioate	$V(CH_3COCHCOCH_3)_3$	13476-99-8	348.266	brn cry	≈185	subl	≈1.0		s MeOH, ace, bz chl
3046	Vanadium(III) sulfate	$V_2(SO_4)_3$	13701-70-7	390.071	yel powder	≈400 dec				sl H_2O
3047	Vanadium(III) sulfide	V_2S_3	1315-03-3	198.078	grn-blk powder	dec		4.7		i H_2O; s hot HCl
3048	Vanadium(IV) bromide	VBr_4	13595-30-7	370.558	unstab purp cry	-23 dec				
3049	Vanadium(IV) chloride	VCl_4	7632-51-1	192.754	red-brn liq	-28	151	1.816		reac H_2O; s EtOH, eth
3050	Vanadium(IV) fluoride	VF_4	10049-16-8	126.936	grn hyg powder	325 dec	subl	3.15		vs H_2O
3051	Vanadium(IV) oxide	VO_2	12036-21-4	82.941	blue-blk powder	1967		4.339		i H_2O; s acid, alk
3052	Vanadium(V) fluoride	VF_5	7783-72-4	145.934	col liq	19.5	48.3	2.50		reac H_2O
3053	Vanadium(V) dioxide fluoride	VO_2F	14259-82-6	101.939	brn hyg cry	350 dec				reac H_2O
3054	Vanadium(V) dioxide chloride	VO_2Cl	13759-30-3	118.394	oran hyg cry	dec 180				s thf
3055	Vanadium(V) oxide	V_2O_5	1314-62-1	181.880	yel-brn orth cry	681	1750	3.35	0.07^{25}	s conc acid, alk; i EtOH
3056	Vanadium(V) sulfide	V_2S_5	12138-17-9	262.208	grn-blk pow	dec		3.0		i H_2O; s acid, alk
3057	Vanadyl bromide	VOBr	13520-88-2	146.845	viol cry	480 dec				
3058	Vanadyl chloride	VOCl	13520-87-1	102.394	brn orth cry		127	1.72		
3059	Vanadyl dibromide	$VOBr_2$	13520-89-3	226.749	yel-brn cry	180 dec				
3060	Vanadyl dichloride	$VOCl_2$	10213-09-9	137.847	grn hyg cry	380 dec		2.88		reac H_2O; s EtOH
3061	Vanadyl difluoride	VOF_2	13814-83-0	104.938	yel cry					
3062	Vanadyl selenite hydrate	$VOSeO_3 \cdot H_2O$	133578-89-9	211.92	grn tricl plates			3.506		
3063	Vanadyl sulfate dihydrate	$VOSO_4 \cdot 2H_2O$	27774-13-6	199.035	blue cry powder					s H_2O
3064	Vanadyl tribromide	$VOBr_3$	13520-90-6	306.653	deep red liq	-59	170			reac H_2O
3065	Vanadyl trichloride	$VOCl_3$	7727-18-6	173.300	fuming red-yel liq	-79	127	1.829		reac H_2O; s MeOH, eth, ace
3066	Vanadyl trifluoride	VOF_3	13709-31-4	123.936	yel hyg powder	300	480	2.459		reac H_2O
3067	Water	H_2O	7732-18-5	18.015	col liq	0.00	99.974	0.9970^{25}		vs EtOH, MeOH, ace
3068	Water-d_2	D_2O	7789-20-0	20.027	col liq	3.82	101.42	1.1044^{25}		
3069	Water-t_2	T_2O	14940-65-9	22.032	col liq	4.48	101.51	1.2138^{25}		
3070	Xenon	Xe	7440-63-3	131.293	col gas	-111.745 tp (81.6 kPa)	-108.09	5.366 g/L		sl H_2O
3071	Xenon trioxide	XeO_3	13776-58-4	179.291	col orth cry	exp ≈25		4.55		s H_2O
3072	Xenon tetroxide	XeO_4	12340-14-6	195.291	yel solid or col gas; exp	-35.9	≈0 dec			
3073	Xenon difluoride	XeF_2	13709-36-9	169.290	col tetr cry	129.03 tp	114.35 sp	4.32		sl H_2O
3074	Xenon tetrafluoride	XeF_4	13709-61-0	207.287	col monocl cry	117.10 tp	115.75 sp	4.04		reac H_2O
3075	Xenon hexafluoride	XeF_6	13693-09-9	245.283	col monocl cry	49.48	75.6	3.56		reac H_2O
3076	Xenon fluoride oxide	$XeOF_2$	13780-64-8	185.289	yel solid, stab <-25	exp ≈0				reac H_2O
3077	Xenon oxytetrafluoride	$XeOF_4$	13774-85-1	223.286	col liq	-46.2		3.17^{0}		reac H_2O
3078	Xenon dioxydifluoride	XeO_2F_2	13875-06-4	201.289	col orth cry	30.8 exp		4.10		
3079	Xenon difluoride trioxide	XeO_3F_2	15192-14-0	217.288	unstab at r.t.	-54.1	exp			
3080	Xenon pentafluoride hexafluoroarsenate	XeF_5AsF_6	20328-94-3	415.197	wh monocl cry	130.5		3.51		

No.	Name	Formula	CAS Reg No.	Mol. weight	Physical form	mp/°C	bp/°C	Density g cm^{-3}	Solubility g/100 g H_2O	Qualitative solubility
3081	Xenon pentafluoride hexafluororuthenate	XeF_5RuF_6	39796-98-0	441.35	grn orth cry	152		3.79		
3082	Xenon fluoride hexafluoroantimonate	XeF_3SbF_6	39797-63-2	424.039	yel-grn monocl cry	≈110		3.92		
3083	Xenon fluoride hexafluoroarsenate	$Xe_2F_3AsF_6$	50432-32-1	508.494	yel-grn monocl cry	99		3.62		reac H_2O
3084	Xenon fluoride hexafluororuthenate	$XeFRuF_6$	22527-13-5	365.35	yel-grn monocl cry	110		3.78		
3085	Xenon fluoride undecafluoroantimonate	$XeFSb_2F_{11}$	15364-10-0	602.794	yel monocl cry	63		3.69		
3086	Xenon trifluoride undecafluoroantimonate	$XeF_3Sb_2F_{11}$	35718-37-7	640.791	yel-grn tricl cry	82		3.98		
3087	Ytterbium	Yb	7440-64-4	173.04	silv metal; cub	824	1196	6.90		s dil acid
3088	Ytterbium silicide	$YbSi_2$	12039-89-3	229.21	hex cry			7.54		
3089	Ytterbium(II) bromide	$YbBr_2$	25502-05-0	332.85	yel cry	673				reac H_2O
3090	Ytterbium(II) chloride	$YbCl_2$	13874-77-6	243.95	grn cry	721		5.27		reac H_2O
3091	Ytterbium(II) fluoride	YbF_2	15192-18-4	211.04	gray solid	1407				i H_2O
3092	Ytterbium(II) iodide	YbI_2	19357-86-9	426.85	blk cry	772				reac H_2O
3093	Ytterbium(III) acetate tetrahydrate	$Yb(C_2H_3O_2)_3 \cdot 4H_2O$	15280-58-7	422.23	hyg col cry	dec 70		2.09		vs H_2O
3094	Ytterbium(III) bromide	$YbBr_3$	13759-89-2	412.75	col cry	956 dec				s H_2O
3095	Ytterbium(III) chloride	$YbCl_3$	10361-91-8	279.40	wh hyg powder	854				s H_2O
3096	Ytterbium(III) chloride hexahydrate	$YbCl_3 \cdot 6H_2O$	19423-87-1	387.49	grn hyg cry	150 dec		2.57		vs H_2O
3097	Ytterbium(III) fluoride	YbF_3	13760-80-0	230.04	wh cry	1157		8.2		i H_2O
3098	Ytterbium(III) iodide	YbI_3	13813-44-0	553.75	yel cry	dec 700				s H_2O
3099	Ytterbium(III) nitrate	$Yb(NO_3)_3$	13768-67-7	359.06	col hyg solid				239^{25}	s EtOH
3100	Ytterbium(III) oxide	Yb_2O_3	1314-37-0	394.08	col cub cry	2355	4070	9.2		s dil acid
3101	Ytterbium(III) sulfate octahydrate	$Yb_2(SO_4)_3 \cdot 8H_2O$	10034-98-7	778.39	col cry			3.3	38.4^{20}	
3102	Yttrium	Y	7440-65-5	88.906	silv metal; hex	1522	3345	4.47		reac H_2O; s dil acid
3103	Yttrium aluminum oxide	$Y_3Al_5O_{12}$	12005-21-9	593.619	grn cub cry			≈4.5		
3104	Yttrium antimonide	YSb	12186-97-9	210.666	cub cry	2310		5.97		
3105	Yttrium arsenide	YAs	12255-48-0	163.828	cub cry			5.59		
3106	Yttrium boride	YB_6	12008-32-1	153.772	refrac solid	2600		3.72		
3107	Yttrium bromide	YBr_3	13469-98-2	328.618	col hyg cry	904			83.3^{30}	
3108	Yttrium carbide	YC_2	12071-35-1	112.927	refrac solid	≈2400		4.13		
3109	Yttrium carbonate trihydrate	$Y_2(CO_3)_3 \cdot 3H_2O$	5970-44-5	411.885	red-brn powder					i H_2O; s dil acid
3110	Yttrium chloride	YCl_3	10361-92-9	195.265	wh monocl cry; hyg	721	1482	2.61	75.1^{20}	vs H_2O
3111	Yttrium chloride hexahydrate	$YCl_3 \cdot 6H_2O$	10025-94-2	303.356	hyg col cry	dec 100				vs H_2O; s EtOH
3112	Yttrium fluoride	YF_3	13709-49-4	145.901	wh hyg powder	1155		4.0		i H_2O
3113	Yttrium hydroxide	$Y(OH)_3$	16469-22-0	139.928	wh prec or pow	dec 190				i H_2O
3114	Yttrium iodide	YI_3	13470-38-7	469.619	hyg wh-yel cry	997				s H_2O, ace, EtOH
3115	Yttrium iron oxide	$Y_3Fe_5O_{12}$	12063-56-8	737.936	cub cry	1555				
3116	Yttrium nitrate	$Y(NO_3)_3$	10361-93-0	274.921	wh hyg solid				149^{25}	s EtOH
3117	Yttrium nitrate tetrahydrate	$Y(NO_3)_3 \cdot 4H_2O$	13773-69-8	346.982	red-wh prisms			2.68	149^{25}	
3118	Yttrium nitrate hexahydrate	$Y(NO_3)_3 \cdot 6H_2O$	13494-98-9	383.012	hyg cry				149^{25}	
3119	Yttrium oxide	Y_2O_3	1314-36-9	225.810	wh cry; cub	2439		5.03		s dil acid
3120	Yttrium phosphide	YP	12294-01-8	119.880	cub cry			≈4.4		
3121	Yttrium sulfate octahydrate	$Y_2(SO_4)_3 \cdot 8H_2O$	7446-33-5	610.122	red monocl cry			2.6	7.47^{16}	
3122	Yttrium sulfide	Y_2S_3	12039-19-9	274.007	yel cub cry	1925		3.87		
3123	Zinc	Zn	7440-66-6	65.409	blue-wh metal; hex	419.53	907	7.134		s acid, alk
3124	Zinc acetate dihydrate	$Zn(C_2H_3O_2)_2 \cdot 2H_2O$	5970-45-6	219.527	wh powder	237 dec		1.735	30.0^{20}	s EtOH
3125	Zinc ammonium sulfate	$Zn(NH_4)_2(SO_4)_2$	7783-24-6	293.611	wh cry				9.2^{20}	
3126	Zinc antimonide	ZnSb	12039-35-9	187.169	silv-wh orth cry	565		6.33		reac H_2O
3127	Zinc arsenate	$Zn_3(AsO_4)_2$	13464-44-3	474.065	wh powder				0.000078^{20}	s acid, alk
3128	Zinc arsenate octahydrate	$Zn_3(AsO_4)_2 \cdot 8H_2O$	13464-45-4	618.187	wh monocl cry			3.33	0.000078^{20}	s acid, alk
3129	Zinc arsenide	Zn_3As_2	12006-40-5	346.070	powder	1015		5.528		
3130	Zinc arsenite	$Zn(AsO_2)_2$	10326-24-6	279.250	col powder					i H_2O; s acid
3131	Zinc borate	$3ZnO \cdot 2B_2O_3$	27043-84-1	383.466	wh amorp powder			3.64		sl H_2O; s dil acid
3132	Zinc borate hemiheptahydrate	$2ZnO \cdot 3B_2O_3 \cdot 3.5H_2O$	12513-27-8	434.69	wh cry	980		4.22		i H_2O
3133	Zinc borate pentahydrate	$2ZnO \cdot 3B_2O_3 \cdot 5H_2O$	12536-65-1	461.753	wh powder			3.64	0.007^{25}	sl HCl
3134	Zinc bromate hexahydrate	$Zn(BrO_3)_2 \cdot 6H_2O$	13517-27-6	429.305	wh hyg solid	100		2.57		vs H_2O
3135	Zinc bromide	$ZnBr_2$	7699-45-8	225.217	wh hex cry; hyg	402	≈670	4.5	488^{25}	vs EtOH; s eth
3136	Zinc caprylate	$Zn(C_8H_{15}O_2)_2$	557-09-5	351.816	wh hyg cry	136				sl H_2O
3137	Zinc carbonate	$ZnCO_3$	3486-35-9	125.418	wh hex cry	140 dec		4.434	0.000091^{20}	s dil acid, alk
3138	Zinc carbonate hydroxide	$3Zn(OH)_2 \cdot 2ZnCO_3$	12070-69-8	549.107	wh powder					
3139	Zinc chlorate	$Zn(ClO_3)_2$	10361-95-2	232.311	yel hyg cry	60 dec		2.15	200^{20}	
3140	Zinc chloride	$ZnCl_2$	7646-85-7	136.315	wh hyg cry	290	732	2.907	408^{25}	vs H_2O; s EtOH, ace

No.	Name	Formula	CAS Reg No.	Mol. weight	Physical form	mp/°C	bp/°C	Density g cm^{-3}	Solubility g/100 g H_2O	Qualitative solubility
3141	Zinc chromate	$ZnCrO_4$	13530-65-9	181.403	yel prisms	316		3.40	3.08	s acid; i ace
3142	Zinc chromite	$ZnCr_2O_4$	12018-19-8	233.399	grn cub cry			5.29		
3143	Zinc citrate dihydrate	$Zn_3(C_6H_5O_7)_2 \cdot 2H_2O$	546-46-3	610.456	col powder					sl H_2O; s dil acid, alk
3144	Zinc cyanide	$Zn(CN)_2$	557-21-1	117.443	wh powder			1.852	0.00047^{20}	reac acid
3145	Zinc diethyl	$Zn(C_2H_5)_2$	557-20-0	123.531	col liq	-28	118	1.2065		reac H_2O; msc eth, peth, bz
3146	Zinc dithionate	ZnS_2O_4	7779-86-4	193.537	wh amorp solid	200 dec			40^{20}	
3147	Zinc fluoride	ZnF_2	7783-49-5	103.406	wh tetr needles; hyg	872	1500	4.9	1.55^{25}	sl H_2O
3148	Zinc fluoride tetrahydrate	$ZnF_2 \cdot 4H_2O$	13986-18-0	175.468	wh orth cry			2.30	1.55^{25}	
3149	Zinc fluoroborate hexahydrate	$Zn(BF_4)_2 \cdot 6H_2O$	27860-83-9	347.109	hex cry			2.12		vs H_2O; s EtOH
3150	Zinc formate dihydrate	$Zn(CHO_2)_2 \cdot 2H_2O$	5970-62-7	191.474	wh cry			2.207	5.2^{20}	i EtOH
3151	Zinc hexafluorosilicate hexahydrate	$ZnSiF_6 \cdot 6H_2O$	16871-71-9	315.576	wh cry					s H_2O
3152	Zinc hydroxide	$Zn(OH)_2$	20427-58-1	99.424	col orth cry	125 dec		3.05	0.000042^{20}	
3153	Zinc iodate	$Zn(IO_3)_2$	7790-37-6	415.214	wh cry powder				0.64^{25}	
3154	Zinc iodide	ZnI_2	10139-47-6	319.218	wh-yel hyg cry	450	625	4.74	438^{25}	vs H_2O; s EtOH, eth
3155	Zinc laurate	$Zn(C_{12}H_{23}O_2)_2$	2452-01-9	464.029	wh powder	128				sl H_2O
3156	Zinc molybdate	$ZnMoO_4$	13767-32-3	225.35	wh tetr cry	>700		4.3		i H_2O
3157	Zinc nitrate	$Zn(NO_3)_2$	7779-88-6	189.418	wh powder				120^{25}	
3158	Zinc nitrate hexahydrate	$Zn(NO_3)_2 \cdot 6H_2O$	10196-18-6	297.510	col orth cry	36 dec		2.067	120^{25}	vs EtOH
3159	Zinc nitride	Zn_3N_2	1313-49-1	224.240	blue-gray cub cry	700 dec		6.22		i H_2O
3160	Zinc nitrite	$Zn(NO_2)_2$	10102-02-0	157.420	hyg solid					reac H_2O
3161	Zinc oleate	$Zn(C_{18}H_{33}O_2)_2$	557-07-3	628.316	wh powder	70 dec				i H_2O; s EtOH, eth, bz
3162	Zinc oxalate	ZnC_2O_4	547-68-2	153.428	wh pwd				0.0026^{25}	
3163	Zinc oxalate dihydrate	$ZnC_2O_4 \cdot 2H_2O$	4255-07-6	189.458	wh powder	100 dec		2.56	0.0026^{25}	s dil acid
3164	Zinc tartrate dihydrate	$ZnC_4H_4O_6 \cdot 2H_2O$	22570-08-7	249.511	wh cry pow	dec 150			0.022^{20}	
3165	Zinc oxide	ZnO	1314-13-2	81.408	wh powder; hex	1974		5.6		i H_2O; s dil acid
3166	Zinc 2,4-pentanedioate	$Zn(CH_3COCHCOCH_3)_2$	14024-63-6	263.625	cry	137 dec				sl H_2O; s EtOH, DMSO
3167	Zinc pentanoate dihydrate	$Zn(C_5H_9O_2)_2 \cdot 2H_2O$	556-38-7	303.687	scales or powder					sl H_2O; reac acid; s EtOH
3168	Zinc perchlorate hexahydrate	$Zn(ClO_4)_2 \cdot 6H_2O$	10025-64-6	372.402	wh cub cry; hyg	106 dec		2.2	121.3^{25}	s EtOH
3169	Zinc permanganate hexahydrate	$Zn(MnO_4)_2 \cdot 6H_2O$	23414-72-4	411.372	blk orth cry; hyg			2.45		s H_2O; reac EtOH
3170	Zinc peroxide	ZnO_2	1314-22-3	97.408	yel-wh powder	>150 dec	212 exp	1.57		i H_2O; reac acid, EtOH, ace
3171	Zinc phosphate	$Zn_3(PO_4)_2$	7779-90-0	386.170	wh monocl cry	900		4.0		i H_2O
3172	Zinc phosphate tetrahydrate	$Zn_3(PO_4)_2 \cdot 4H_2O$	7543-51-3	458.231	col orth cry			3.04		i H_2O, EtOH; s dil acid, alk
3173	Zinc phosphide	Zn_3P_2	1314-84-7	258.175	gray tetr cry	1160		4.55		i H_2O, EtOH; reac acid; s bz
3174	Zinc pyrophosphate	$Zn_2P_2O_7$	7446-26-6	304.761	wh cry powder			3.75		i H_2O; s dil acid
3175	Zinc selenate pentahydrate	$ZnSeO_4 \cdot 5H_2O$	13597-54-1	298.44	tricl cry	50 dec		2.59	63.4^{25}	
3176	Zinc selenide	ZnSe	1315-09-9	144.37	yel-red cub cry	>1100	subl	5.65		i H_2O; s dil acid
3177	Zinc orthosilicate	Zn_2SiO_4	13597-65-4	222.902	wh hex cry	1509		4.1		i H_2O, dil acid
3178	Zinc selenite	$ZnSeO_3$	13597-46-1	192.37	wh powder	621				
3179	Zinc stearate	$Zn(C_{18}H_{35}O_2)_2$	557-05-1	632.348	wh powder	130		1.095		i H_2O, EtOH, eth; s bz
3180	Zinc sulfate	$ZnSO_4$	7733-02-0	161.472	col orth cry	680 dec		3.8	57.7^{25}	
3181	Zinc sulfate monohydrate	$ZnSO_4 \cdot H_2O$	7446-19-7	179.487	wh monocl cry	238 dec		3.20	57.7^{25}	i EtOH
3182	Zinc sulfate heptahydrate	$ZnSO_4 \cdot 7H_2O$	7446-20-0	287.578	col orth cry	100 dec		1.97	57.7^{25}	i EtOH
3183	Zinc sulfide (sphalerite)	ZnS	1314-98-3	97.474	gray-wh cub cry	trans wurtzite 1020		4.04		i H_2O, EtOH; s dil acid
3184	Zinc sulfide (wurtzite)	ZnS	1314-98-3	97.474	wh hex cry	1700	subl	4.09		i H_2O; s dil acid
3185	Zinc sulfite dihydrate	$ZnSO_3 \cdot 2H_2O$	7488-52-0	181.503	wh powder	200 dec			0.224^{25}	i EtOH
3186	Zinc telluride	ZnTe	1315-11-3	193.01	red cub cry	1239		5.9		i H_2O
3187	Zinc thiocyanate	$Zn(SCN)_2$	557-42-6	181.573	wh hyg cry					sl H_2O; s EtOH
3188	Zirconium	Zr	7440-67-7	91.224	gray-wh metal; hex	1854.7	4409	6.52		s hot conc acid
3189	Zirconocene dichloride	$Zr(C_5H_5)_2Cl_2$	1291-32-3	292.316	col cry	248	subl 150			
3190	Zirconium boride	ZrB_2	12045-64-6	112.846	gray refrac solid; hex	3050		6.17		
3191	Zirconium carbide	ZrC	12020-14-3	103.235	gray refrac solid; cub	3532		6.73		s HF
3192	Zirconium nitride	ZrN	25658-42-8	105.231	yel cub cry	2952		7.09		s conc HF; sl dil acid
3193	Zirconium phosphide	ZrP_2	12037-80-8	153.172	orth cry			≈5.1		
3194	Zirconium silicide	$ZrSi_2$	12039-90-6	147.395	gray powder	1620		4.88		i H_2O, aqua regia; s HF

No.	Name	Formula	CAS Reg No.	Mol. weight	Physical form	mp/°C	bp/°C	Density g cm⁻³	Solubility g/100 g H_2O	Qualitative solubility
3195	Zirconium(II) bromide	$ZrBr_2$	24621-17-8	251.032	blue-blk cry	dec 400				
3196	Zirconium(II) chloride	$ZrCl_2$	13762-26-0	162.130	blk cry	772		3.16		reac H_2O
3197	Zirconium(II) fluoride	ZrF_2	13842-94-9	129.221	blk cry	902				
3198	Zirconium(II) hydride	ZrH_2	7704-99-6	93.240	gray tetr cry	800 dec		5.6		i H_2O
3199	Zirconium(II) iodide	ZrI_2	15513-85-6	345.033	blk cry	827				
3200	Zirconium(III) bromide	$ZrBr_3$	24621-18-9	330.936	dark blue cry	dec 300				
3201	Zirconium(III) chloride	$ZrCl_3$	10241-03-9	197.583	dark blue cry	627		3.05		reac H_2O
3202	Zirconium(III) fluoride	ZrF_3	13814-22-7	148.219	blue-grn cry	927		4.26		i H_2O; s acid
3203	Zirconium(III) iodide	ZrI_3	13779-87-8	471.937	dark blue cry	727				
3204	Zirconium(IV) acetate hydroxide	$Zr(C_2H_3O_2)_2(OH)_2$	14311-93-4	243.327	wh amorp solid					s H_2O
3205	Zirconium(IV) ammonium carbonate dihydrate	$Zr(NH_4)_3OH(CO_3)_3 \cdot 2H_2O$	12616-24-9*	362.404	prisms; unstab					s H_2O
3206	Zirconium(IV) bromide	$ZrBr_4$	13777-25-8	410.840	wh cub cry	450 tp	360 sp	3.98		reac H_2O
3207	Zirconium(IV) chloride	$ZrCl_4$	10026-11-6	233.036	wh monocl cry; hyg	437 tp	331 sp	2.80		reac H_2O; s EtOH, eth
3208	Zirconium(IV) fluoride	ZrF_4	7783-64-4	167.218	wh monocl cry	910	912 sp	4.43	1.5^{25}	
3209	Zirconium(IV) hydroxide	$Zr(OH)_4$	14475-63-9	159.254	wh amorp powder	dec		3.25		i H_2O; s acid
3210	Zirconium(IV) iodide	ZrI_4	13986-26-0	598.842	yel-oran cub cry	500	431 sp	4.85		vs H_2O
3211	Zirconium(IV) nitrate pentahydrate	$Zr(NO_3)_4 \cdot 5H_2O$	13746-89-9	420.320	wh hyg cry	100 dec				vs H_2O; s EtOH
3212	Zirconium(IV) orthosilicate	$ZrSiO_4$	10101-52-7	183.308	wh tetr cry	1540 dec		4.6		i H_2O, acid
3213	Zirconium(IV) oxide	ZrO_2	1314-23-4	123.223	wh amorp powder	2710	4300	5.68		i H_2O; sl acid
3214	Zirconium(IV) pyrophosphate	ZrP_2O_7	13565-97-4	265.167	wh refrac solid	dec 1550				i H_2O, dil acid; s HF
3215	Zirconium(IV) sulfate	$Zr(SO_4)_2$	14644-61-2	283.349	wh hyg cry	410 dec		3.22		s H_2O; sl EtOH
3216	Zirconium(IV) sulfate tetrahydrate	$Zr(SO_4)_2 \cdot 4H_2O$	7446-31-3	355.411	wh tetr cry	100 dec		2.80		vs H_2O
3217	Zirconium(IV) sulfide	ZrS_2	12039-15-5	155.354	red-brn hex cry	1550		3.87		i H_2O
3218	Zirconium(IV) tungstate	$Zr(WO_4)_2$	16853-74-0	586.90	grn pow					
3219	Zirconyl chloride	$ZrOCl_2$	7699-43-6	178.129	wh solid	250 dec				s H_2O, EtOH
3220	Zirconyl chloride octahydrate	$ZrOCl_2 \cdot 8H_2O$	13520-92-8	322.252	tetr cry	400 dec		1.91		vs H_2O, EtOH

FORMULA INDEX OF INORGANIC COMPOUNDS

Formula	No.
Ac	1
$AcBr_3$	2
$AcCl_3$	3
AcF_3	4
AcI_3	5
Ac_2O_3	6
Ag	2476
$AgAsF_6$	2497
AgBr	2485
$AgBrO_3$	2484
AgCN	2492
AgC_2H	2481
$AgC_2H_3O_2$	2479
$AgC_3H_5O_3 \cdot H_2O$	2502
$Ag(C_2H_5)_2NCS_2$	2494
$AgC_6H_2N_3O_7 \cdot H_2O$	2513
$Ag(C_6H_5CO_2)$	2483
AgCl	2488
$AgClO_2$	2489
$AgClO_3$	2487
$AgClO_4$	2509
$AgClO_4 \cdot H_2O$	2510
AgF	2495
AgF_2	2526
$AgHF_2$	2499
AgI	2501
$AgIO_3$	2500
$AgMnO_4$	2511
$AgNO_2$	2506
$AgNO_3$	2505
AgN_3	2477
AgO	2524
$AgPF_6$	2498
$AgPO_3$	2503
AgSCN	2522
$AgSbF_6$	2496
Ag_2CO_3	2486
Ag_2C_2	2480
$Ag_2C_2O_4$	2507
Ag_2CrO_4	2490
$Ag_2Cr_2O_7$	2493
Ag_2F	2478
Ag_2HgI_4	2521
Ag_2MoO_4	2504
Ag_2O	2508
Ag_2O_2	2527
Ag_2S	2518
Ag_2SO_3	2519
Ag_2SO_4	2517
$Ag_2S_2O_3$	2523
Ag_2Se	2515
Ag_2SeO_3	2516
Ag_2SeO_4	2514
Ag_2Te	2520
Ag_2WO_4	2525
Ag_3AsO_4	2482
$Ag_3C_6H_5O_7$	2491
Ag_3PO_4	2512
Al	7
AlAs	13
AlB_2	26
$Al(BH_4)_3$	15
AlB_{12}	27
$AlBr_3$	17
$AlBr_3 \cdot 6H_2O$	18
$Al(BrO_3)_3 \cdot 9H_2O$	16
$Al(C_2H_5O)_3$	28
$Al(C_2H_3O_2)_3$	8
$Al(C_3H_5O_3)_3$	38
$Al(CH_3COCHCOCH_3)_3$	50
$AlC_{18}H_{35}O_2$	67
$Al(C_{15}H_{31}COO)_3$	49
$Al(C_{18}H_{33}O_2)_3$	43
$Al(C_{18}H_{35}O_2)_3$	65
$AlCl(C_2H_5)_2$	24
$AlCl(C_4H_9)_2$	25
$AlCl_2CH_3$	23
$AlCl_3$	21
$AlCl_3 \cdot 6H_2O$	22
$Al(ClO_3)_3 \cdot 9H_2O$	20
$Al(ClO_4)_3$	51
$Al(ClO_4)_3 \cdot 9H_2O$	52
AlF_3	29
$AlF_3 \cdot H_2O$	30
$AlF_3 \cdot 3H_2O$	31
AlH_3	33
$Al(H_2PO_2)_3$	57
AlI_3	36
$AlI_3 \cdot 6H_2O$	37
AlN	42
$AlNH_4(SO_4)_2$	10
$AlNH_4(SO_4)_2 \cdot 12H_2O$	11
$Al(NO_3)_3$	40
$Al(NO_3)_3 \cdot 9H_2O$	41
$Al(OH)(C_2H_3O_2)_2$	9
AlO(OH)	47
AlO(OH)	48
$Al(OH)_2(C_{18}H_{35}O_2)$	66
$Al(OH)_3$	34
AlP	58
$AlPO_4$	53
$AlPO_4 \cdot 2H_2O$	54
$Al(PO_3)_3$	56
$Al(SCN)_3$	72
AlSb	12
$Al_2(C_2O_4)_3 \cdot H_2O$	44
$Al_2(MoO_4)_3$	39
Al_2O_3	45
Al_2O_3	46
$2Al_2O_3 \cdot B_2O_3$	14
$Al_2O_3 \cdot 2SiO_2 \cdot 2H_2O$	64
$3Al_2O_3 \cdot 2SiO_2$	62
$Al_2(OH)_3PO_4$	55
$Al_2(OH)_5Cl \cdot 2H_2O$	35
Al_2S_3	70
$Al_2(SO_4)_3$	68
$Al_2(SO_4)_3 \cdot 18H_2O$	69
Al_2Se_3	59
Al_2SiO_5	60
Al_2SiO_5	61
Al_2SiO_5	63
$Al_2(SiF_6)_3 \cdot 9H_2O$	32
Al_2Te_3	71
Al_2TiO_5	73
Al_2Zr	74
Al_4C_3	19
Am	75
$AmBr_3$	77
$AmCl_3$	78
AmF_3	79
AmF_4	81
AmI_3	80
AmO_2	82
Am_2O_3	76
Ar	235
As	236
As	237
As	238
$AsBr_3$	247
$As(C_2H_5O)_3$	249
$AsCl_3$	248
$AsCl_5$	257
AsF_3	250
AsF_5	258
AsH_3	239
AsI_3	251
As_2H_4	240
As_2I_4	244
As_2O_3	252
As_2O_3	253
As_2O_5	259
As_2S_3	255
As_2S_5	261
As_2Se	245
As_2Se_3	254
As_2Se_5	260
As_2Te_3	256
As_4S_4	246
At	262
Au	1166
AuBr	1169
$AuBr_3$	1174
AuCN	1171
$Au(CN)_3 \cdot 3H_2O$	1176
AuCl	1170
$AuCl_3$	1175
AuF_3	1177
AuI	1172
AuI_3	1179
$Au(OH)_3$	1178
Au_2O_3	1180
Au_2S	1173
Au_2S_3	1183
Au_2Se_3	1182
$Au_2(SeO_4)_3$	1181
B	441
BAs	466
BBr_3	473
BCl_3	474
BF_3	475
$BF_3(C_2H_5)_2O$	476
BH_3CO	459
BI_3	477
BN	468
BP	470
B_2Br_4	456
B_2Cl_4	457
B_2F_4	458
B_2H_6	442
B_2O_3	469
B_2S_3	472
$B_3N_3H_6$	460
B_4C	467
B_4H_{10}	443
B_5H_9	444
B_5H_{11}	445
B_6H_{10}	446
B_6H_{12}	447
B_6Si	471
B_9H_{15}	448
$B_{10}H_{14}$	449
$B_{10}H_{16}$	450
$B_{12}H_{16}$	451
$B_{13}H_{19}$	452
$B_{14}H_{18}$	453
$B_{16}H_{20}$	454
$B_{18}H_{22}$	455
Ba	263
$BaAl_2O_4$	266
$BaAl_4$	267
$Ba(BO_2)_2 \cdot H_2O$	313
$Ba(BO_2)_2 \cdot 2H_2O$	314
BaB_6	297
$BaBiO_3$	269
$BaBr_2$	272
$BaBr_2 \cdot 2H_2O$	273
$Ba(BrO_3)_2$	270
$Ba(BrO_3)_2 \cdot H_2O$	271
$BaCO_3$	276
BaC_2	275
$Ba(CHO_2)_2$	296
$Ba(CN)_2$	289
BaC_2O_4	324
$BaC_2O_4 \cdot H_2O$	325
$Ba(CH_3COCHCOCH_3)_2 \cdot 8H_2O$	327
$BaC_4H_4O_6$	348
$Ba(C_2H_3O_2)_2$	264
$Ba(C_2H_3O_2)_2 \cdot H_2O$	265
$Ba(C_{18}H_{35}O_2)_2$	342
BaClF	281
$BaCl_2$	279
$BaCl_2 \cdot 2H_2O$	280
$Ba(ClO_3)_2$	277
$Ba(ClO_3)_2 \cdot H_2O$	278
$Ba(ClO_4)_2$	328
$Ba(ClO_4)_2 \cdot 3H_2O$	329
$BaCrO_4$	283
$BaCr_2O_7 \cdot 2H_2O$	290
$BaCuY_2O_5$	285

Formula	No.
BaF_2	295
$BaFe_{12}O_{19}$	293
$BaGeF_6$	298
$BaHPO_4$	301
BaH_2	300
$Ba(HS)_2$	302
$Ba(HS)_2 \cdot 4H_2O$	303
$Ba(H_2PO_2)_2 \cdot H_2O$	307
$BaHgI_4$	350
BaI_2	310
$BaI_2 \cdot 2H_2O$	311
$Ba(IO_3)_2$	308
$Ba(IO_3)_2 \cdot H_2O$	309
$BaK_2(CrO_4)_2$	333
$BaMnO_4$	312
$Ba(MnO_4)_2$	330
$BaMoO_4$	317
$Ba(NO_2)_2$	321
$Ba(NO_2)_2 \cdot H_2O$	322
$Ba(NO_3)_2$	319
$Ba(N_3)_2$	268
$Ba(NbO_3)_2$	318
BaO	326
$BaO \cdot ZrO_2 \cdot SiO_2$	364
BaO_2	331
$Ba(OH)_2$	304
$Ba(OH)_2 \cdot H_2O$	305
$Ba(OH)_2 \cdot 8H_2O$	306
$Ba(PO_3)_2$	315
$BaPbO_3$	332
$BaPt(CN)_4 \cdot 4H_2O$	349
BaS	346
$BaSO_3$	347
$BaSO_4$	345
$Ba(SCN)_2$	351
$Ba(SCN)_2 \cdot 2H_2O$	352
$Ba(SCN)_2 \cdot 3H_2O$	353
BaS_2O_3	354
$BaS_2O_3 \cdot H_2O$	355
$BaS_2O_6 \cdot 2H_2O$	292
$BaSe$	336
$BaSeO_3$	337
$BaSeO_4$	335
$BaSiF_6$	299
$BaSiO_3$	316
$BaSi_2$	338
$BaSi_2O_5$	291
$BaSnO_3$	340
$BaSnO_3 \cdot 3H_2O$	341
$BaSr(NbO_3)_4$	343
$BaTiO_3$	356
$BaTi(SiO_3)_3$	359
$BaTi_2O_5$	357
$BaTi_4O_9$	358
BaU_2O_7	361
$BaWO_4$	360
$BaZrO_3$	363
Ba_2CaWO_6	274
$Ba_2Cu_3YO_7$	286
$Ba_2Cu_4YO_8$	287
$Ba_2Fe(CN)_6 \cdot 6H_2O$	294
$Ba_2Na(NbO_3)_5$	339
$Ba_2P_2O_7$	334
Ba_2SrWO_6	344
$Ba_3(C_6H_5O_7)_2 \cdot H_2O$	284
$Ba_3(CrO_4)_2$	282
Ba_3N_2	320
$Ba_3(VO_4)_2$	323
$Ba_3Y_2WO_9$	362
$Ba_4Cu_7Y_2O_{15}$	288
Be	367
$BeAl_2O_4$	370
BeB_2	372
$Be(BH_4)_2$	376
BeB_6	373
$BeBr_2$	377
$BeCO_3 \cdot 4H_2O$	379
$Be(CHO_2)_2$	383
$BeC_2O_4 \cdot 3H_2O$	391
$Be(C_2H_3O_2)_2$	368
$Be(CH_3COCHCOCH_3)_2$	393
$BeCl_2$	381
$Be(ClO_4)_2 \cdot 4H_2O$	394
BeF_2	382
$BeHPO_4$	385
BeH_2	384
BeI_2	388
$Be(NO_3)_2 \cdot 3H_2O$	389
BeO	392
$Be(OH)_2$	386
$Be(OH)_2$	387
BeS	399
$BeSO_4$	396
$BeSO_4 \cdot 2H_2O$	397
$BeSO_4 \cdot 4H_2O$	398
$BeSeO_4 \cdot 4H_2O$	395
Be_2B	374
Be_2C	378
$Be_3Al_2(SiO_3)_6$	371
Be_3N_2	390
$Be_3(OH)_2(CO_3)_2$	380
Be_4B	375
$Be_4O(C_2H_3O_2)_6$	369
Bi	400
$BiAsO_4$	404
$BiBr_3$	431
$Bi(CH_3)_3$	436
$BiC_6H_5O_7$	407
$Bi(C_2H_3O_2)_3$	401
$Bi(CF_3COCHCOCF_3)_3$	411
$Bi(C_{18}H_{33}O_2)_3$	416
$BiCl_3$	432
BiF_3	433
BiF_5	434
BiH_3	408
BiK_4I_7	425
BiI_3	435
$Bi(NO_3)_3 \cdot 5H_2O$	414
$BiOBr$	420
$BiOC_2H_3O_2$	402
$BiOCl$	421
$BiOI$	422
$BiONO_3$	423
$Bi(OH)_3$	409
$BiPO_4$	424
$BiSb$	403
$BiVO_4$	439
$Bi_2(C_2O_4)_3$	417
Bi_2MoO_6	412
$Bi_2(MoO_4)_3$	413
$(BiO)_2CO_3$	405
Bi_2O_3	418
$Bi_2O_3 \cdot 2CrO_3$	406
$2Bi_2O_3 \cdot 3GeO_2$	410
$2Bi_2O_3 \cdot 3ZrO_2$	440
Bi_2O_4	419
Bi_2S_3	429
$Bi_2(SO_4)_3$	428
Bi_2Se_3	426
$Bi_2(SnO_3)_3 \cdot 5H_2O$	427
Bi_2Te_3	430
$Bi_2(WO_4)_3$	438
$Bi_4(TiO_4)_3$	437
$Bi_5O(OH)_9(NO_3)_4$	415
Bk	365
Bk	366
$BrCN$	665
$BrCl$	486
BrF	487
BrF_3	488
BrF_5	489
BrN_3	485
$BrOF_3$	490
BrO_2	481
BrO_2F	491
BrO_3F	492
Br_2	478
Br_2O	482
Br_2O_3	483
Br_2O_5	484
C	644
C	645
C	646
CO	650
$COBr_2$	660
$COCl_2$	661
COF_2	662
COS	656
$COSe$	657
CO_2	651
$CSSe$	658
$CSTe$	659
CS_2	653
CSe_2	655
C_2N_2	663
C_3O_2	652
C_3S_2	654
$(C_2H_5O)_2P(S)SNH_4$	105
C_{60}	647
$C_{60}F_{60}$	649
C_{70}	648
Ca	542
$Ca(AlH_4)_2$	637
$CaAl_2O_4$	546
$CaAsO_3$	549
$Ca(BO_2)_2$	597
$Ca(BO_2)_2 \cdot 7H_2O$	611
$CaB_4O_7 \cdot 6H_2O$	550
CaB_6	551
$CaBr_2$	554
$CaBr_2 \cdot 2H_2O$	555
$CaBr_2 \cdot 6H_2O$	556
$Ca(BrO_3)_2$	552
$Ca(BrO_3)_2 \cdot H_2O$	553
$CaCN_2$	572
$CaCO_3$	558
$CaCO_3$	559
$CaCO_3$	560
CaC_2	557
$Ca(CHO_2)_2$	581
$Ca(CN)_2$	573
CaC_2O_4	606
$CaC_2O_4 \cdot H_2O$	607
$CaC_4H_4O_4 \cdot 3H_2O$	629
$CaC_4H_4O_6 \cdot 4H_2O$	635
$Ca(C_2H_3O_2)_2$	543
$Ca(C_2H_3O_2)_2 \cdot H_2O$	544
$Ca(C_2H_3O_2)_2 \cdot 2H_2O$	545
$Ca(C_3H_5O_2)_2$	621
$Ca(CF_3COCHCOCF_3)_2$	583
$Ca(CH_3COCHCOCH_3)_2$	612
$Ca(C_8H_{15}O_2)_2$	576
$Ca(C_{16}H_{31}O_2)_2$	610
$Ca(C_{18}H_{33}O_2)_2$	605
$Ca(C_{18}H_{35}O_2)_2$	628
$CaCl_2$	563
$CaCl_2 \cdot H_2O$	564
$CaCl_2 \cdot 2H_2O$	565
$CaCl_2 \cdot 4H_2O$	566
$CaCl_2 \cdot 6H_2O$	567
$Ca(ClO_2)_2$	568
$Ca(ClO_3)_2$	561
$Ca(ClO_3)_2 \cdot 2H_2O$	562
$Ca(ClO_4)_2$	613
$Ca(ClO_4)_2 \cdot 4H_2O$	614
$CaCrO_4$	569
$CaCrO_4 \cdot 2H_2O$	570
$CaCr_2O_7 \cdot 3H_2O$	574
CaF_2	578
$CaHPO_3 \cdot H_2O$	619
$CaHPO_4$	586
$CaHPO_4 \cdot 2H_2O$	587
CaH_2	585
$Ca(HS)_2 \cdot 6H_2O$	589
$CaH_2(SO_3)_2$	588
$Ca(H_2PO_2)_2$	593
$Ca(H_2PO_4)_2 \cdot H_2O$	575
CaI_2	595
$CaI_2 \cdot 6H_2O$	596
$Ca(IO_3)_2$	594
$Ca(MnO_4)_2$	615
$CaMoO_4$	599
$Ca(NO_2)_2$	603
$Ca(NO_2)_2 \cdot H_2O$	604
$Ca(NO_3)_2$	600
$Ca(NO_3)_2 \cdot 4H_2O$	601
CaO	608
$2CaO \cdot 3B_2O_3 \cdot 5H_2O$	582
CaO_2	616
$Ca(OCl)_2$	592
$Ca(OH)_2$	590
$CaPO_3F \cdot 2H_2O$	580
CaS	633
$CaSO_3 \cdot 2H_2O$	634
$CaSO_4$	630
$CaSO_4 \cdot 0.5H_2O$	631

$CaSO_4 \cdot 2H_2O$ 632
$Ca(SCN)_2 \cdot 4H_2O$ 638
$CaS_2O_3 \cdot 6H_2O$ 639
$CaSe$ 624
$CaSeO_4 \cdot 2H_2O$ 623
$CaSi$ 625
$CaSiF_6 \cdot 2H_2O$ 584
$CaSiO_3$ 598
$CaSi_2$ 626
$CaSnO_3 \cdot 3H_2O$ 627
$CaTe$ 636
$CaTiO_3$ 640
$CaWO_4$ 641
$CaZrO_3$ 642
$Ca_2Fe(CN)_6 \cdot 12H_2O$ 577
$Ca_2P_2O_7$ 622
Ca_2PbO_4 620
$Ca_3Al_2O_6$ 547
$Ca_3(AsO_4)_2$ 548
$Ca_3(C_6H_5O_7)_2 \cdot 4H_2O$ 571
Ca_3N_2 602
Ca_3OSiO_4 609
Ca_3P_2 618
$Ca_3(PO_4)_2$ 617
$Ca_5(OH)(PO_4)_3$ 591
$Ca_5(PO_4)_3F$ 579
Cd 493
$Cd(BF_4)_2$ 539
$CdBr_2$ 500
$CdBr_2 \cdot 4H_2O$ 501
$CdCO_3$ 502
$Cd(CN)_2$ 508
CdC_2O_4 520
$CdC_2O_4 \cdot 3H_2O$ 521
$Cd(CH_3COCHCOCH_3)_2$ 523
$CdC_4H_4O_4$ 531
$Cd(C_2H_3O_2)_2$ 494
$Cd(C_2H_3O_2)_2 \cdot 2H_2O$ 495
$Cd(C_8H_{15}O_2)_2$ 510
$Cd(C_{18}H_{35}O_2)_2$ 530
$CdCl_2$ 504
$CdCl_2 \cdot H_2O$ 505
$CdCl_2 \cdot 2.5H_2O$ 506
$Cd(ClO_3)_2 \cdot 2H_2O$ 503
$Cd(ClO_4)_2 \cdot 6H_2O$ 524
$CdCrO_4$ 507
$CdCr_2O_7 \cdot H_2O$ 509
CdF_2 511
CdI_2 514
$Cd(IO_3)_2$ 513
$CdMoO_4$ 516
$Cd(NO_3)_2$ 518
$Cd(NO_3)_2 \cdot 4H_2O$ 519
$Cd(N_3)_2$ 498
CdO 522
$Cd(OH)_2$ 512
CdS 535
$CdSO_3$ 536
$CdSO_4$ 532
$CdSO_4 \cdot H_2O$ 533
$CdSO_4 \cdot 8H_2O$ 534
$CdSb$ 496
$CdSe$ 528
$CdSeO_3$ 529
$CdSeO_4 \cdot 2H_2O$ 527
$CdSiO_3$ 515
$CdTe$ 537
$CdTeO_3$ 538
$CdTiO_3$ 540
$CdWO_4$ 541
$Cd_2Nb_2O_7$ 517
Cd_3As_2 497
Cd_3P_2 526
$Cd_3(PO_4)_2$ 525
$Cd_5(BW_{12}O_{40}) \cdot 18H_2O$ 499
Ce 669
CeB_6 670
$CeBr_3$ 681
$CeBr_3 \cdot 7H_2O$ 682
CeC_2 671
$Ce(C_2H_3O_2)_3 \cdot 1.5H_2O$ 678
$Ce(C_2O_4)_3 \cdot 9H_2O$ 693
$Ce(CH_3COCHCOCH_3)_3 \cdot 3H_2O$ 695
$CeCl_3$ 685
$CeCl_3 \cdot 7H_2O$ 686
$Ce(ClO_4)_3 \cdot 6H_2O$ 696
CeF_3 687
CeF_4 704
CeH_2 675
CeH_3 688
CeI_2 676
CeI_3 690
$CeI_3 \cdot 9H_2O$ 691
CeN 673
$Ce(NO_3)_3 \cdot 6H_2O$ 692
CeO_2 706
$Ce(OH)_3$ 689
$Ce(OH)_4$ 705
CeS 677
$Ce(SO_4)_2 \cdot 4H_2O$ 707
$CeSi_2$ 674
Ce_2C_3 672
$Ce_2(CO_3)_3$ 683
$Ce_2(CO_3)_3 \cdot 5H_2O$ 684
Ce_2O_3 694
Ce_2S_3 700
$Ce_2(SO_4)_3$ 698
$Ce_2(SO_4)_3 \cdot 8H_2O$ 699
$Ce_2(SeO_4)_3$ 697
$Ce_2(WO_4)_3$ 701
Cf 643
$ClCN$ 666
ClF 753
ClF_3 754
ClF_5 755
$ClOClO_3$ 760
$ClOF_3$ 756
ClO_2 749
ClO_2F 757
ClO_2F_3 758
ClO_3F 759
Cl_2 744
Cl_2O 748
Cl_2O_3 750
Cl_2O_6 751
Cl_2O_7 752
Cm 1011
Co 826
$CoAl_2O_4$ 842
$CoAs$ 830
$CoAsS$ 829
$CoAs_2$ 831
$CoAs_3$ 832
CoB 833
$CoBr_2$ 845
$CoBr_2 \cdot 6H_2O$ 846
$Co(BrO_3)_2 \cdot 6H_2O$ 844
$CoCO_3$ 847
$2CoCO_3 \cdot 3Co(OH)_2 \cdot H_2O$ 848
$Co(CHO_2)_2 \cdot 2H_2O$ 862
$Co(CN)_2$ 856
$Co(CN)_2 \cdot 2H_2O$ 857
CoC_2O_4 878
$CoC_2O_4 \cdot 2H_2O$ 879
$Co(C_2H_3O_2)_2$ 840
$Co(C_2H_3O_2)_2 \cdot 4H_2O$ 841
$Co(C_2H_3O_2)_3$ 901
$Co(C_5H_5)_2$ 827
$Co(CH_3COCHCOCH_3)_2$ 881
$Co(CF_3COCHCOCF_3)_2$ 863
$Co(CH_3COCHCOCH_3)_3$ 910
$Co(C_{18}H_{33}O_2)_2$ 876
$Co(C_{18}H_{35}O_2)_2$ 890
$CoCl_2$ 850
$CoCl_2 \cdot 2H_2O$ 851
$CoCl_2 \cdot 6H_2O$ 852
$Co(ClO_3)_2 \cdot 6H_2O$ 849
$Co(ClO_4)_2$ 882
$Co(ClO_4)_2 \cdot 6H_2O$ 883
$CoCrO_4$ 853
$CoCr_2O_4$ 854
CoF_2 860
$CoF_2 \cdot 4H_2O$ 861
CoF_3 903
$CoFe_2O_4$ 858
CoI_2 868
$CoI_2 \cdot 2H_2O$ 869
$CoI_2 \cdot 6H_2O$ 870
$Co(IO_3)_2$ 867
$CoK_2(SO_4)_2 \cdot 6H_2O$ 885
$CoK_3(NO_2)_6 \cdot 1.5H_2O$ 911
$CoMoO_4$ 871
$CoMoO_4 \cdot H_2O$ 872
$CoNH_4PO_4$ 96
$CoNH_4PO_4 \cdot H_2O$ 97
$Co(NO_2)_2$ 875
$Co(NO_3)_2$ 873
$Co(NO_3)_2 \cdot 6H_2O$ 874
$Co(NO_3)_3$ 907
$Co(NH_3)_6Cl_3$ 905
CoO 880
$Co(OH)_2$ 865
$Co(OH)_2 \cdot H_2O$ 866
$Co(OH)_3$ 906
CoS 894
$CoSO_4$ 891
$CoSO_4 \cdot H_2O$ 892
$CoSO_4 \cdot 7H_2O$ 893
CoS_2 836
$Co(SCN)_2$ 896
$Co(SCN)_2 \cdot 3H_2O$ 897
$CoSb$ 828
$CoSe$ 887
$CoSeO_3 \cdot 2H_2O$ 888
$CoSeO_4 \cdot 5H_2O$ 886
$CoSiF_6 \cdot 6H_2O$ 864
$CoSi_2$ 839
$CoTe$ 895
$CoTiO_3$ 898
$CoWO_4$ 899
Co_2B 834
$Co_2(CO)_8$ 835
$Co_2F_6 \cdot 2H_2O$ 904
Co_2O_3 908
$Co_2O_3 \cdot H_2O$ 909
Co_2P 838
Co_2S_3 912
Co_2SiO_4 877
Co_2SnO_4 889
Co_2TiO_4 913
$Co_3(AsO_4)_2 \cdot 8H_2O$ 843
$Co_3(C_6H_5O_7)_2 \cdot 2H_2O$ 855
$Co_3[Fe(CN)_6]_2$ 859
Co_3O_4 900
$Co_3(PO_4)_2 \cdot 8H_2O$ 884
$Co_4(CO)_{12}$ 837
Cr 761
CrB 765
CrB_2 766
$CrBr_2$ 778
$CrBr_3$ 791
$CrBr_3(H_2O)_4 \cdot 2H_2O$ 792
$Cr(CHOO)_2 \cdot H_2O$ 782
$CrC_2O_4 \cdot H_2O$ 784
$Cr(C_2H_3O_2)_2(OH)$ 790
$Cr(C_2H_3O_2)_2 \cdot H_2O$ 777
$Cr(C_2H_3O_2)_3$ 787
$Cr(C_2H_3O_2)_3 \cdot H_2O$ 788
$Cr(C_2H_3O_2)_3 \cdot 6H_2O$ 789
$Cr(CO)_6$ 770
$Cr(CH_3COCHCOCH_3)_3$ 805
$CrCl_2$ 779
$[CrCl_2(H_2O)_4]Cl \cdot 2H_2O$ 795
$CrCl_3$ 794
$Cr(ClO_4)_3$ 806
$CrCl_4$ 816
CrF_2 781
CrF_3 796
$CrF_3 \cdot 3H_2O$ 797
CrF_4 817
CrF_5 819
CrF_6 821
$Cr(H_2O)_4Cl_2 \cdot 4H_2O$ 780
$Cr(H_2O)_6Br_3$ 793
$Cr(H_2O)_6F_3 \cdot 3H_2O$ 799
CrI_2 783
CrI_3 801
$CrK(SO_4)_2 \cdot 12H_2O$ 811
CrN 771
$Cr(NO_3)_3$ 802
$Cr(NO_3)_3 \cdot 9H_2O$ 803
$CrOF_4$ 823
$Cr(OH)SO_4$ 798
CrO_2 818

Formula	No.
CrO_2Cl_2	824
CrO_2F_2	825
CrO_3	822
$Cr(OH)_3 \cdot 3H_2O$	800
CrP	773
$CrPO_4$	807
$CrPO_4 \cdot 3.5H_2O$	808
$CrPO_4 \cdot 6H_2O$	809
$CrSO_4 \cdot 5H_2O$	785
$CrSb$	763
$CrSe$	774
$CrSi_2$	775
Cr_2As	764
Cr_2B	767
Cr_2N	772
Cr_2O_3	804
Cr_2O_5	820
Cr_2S_3	814
$Cr_2(SO_4)_3$	812
$Cr_2(SO_4)_3 \cdot 18H_2O$	813
Cr_2Te_3	815
Cr_3C_2	769
Cr_3O_4	786
Cr_3Si	776
Cr_5B_3	768
Cs	708
$CsAl(SO_4)_2 \cdot 12H_2O$	710
$CsBF_4$	722
$CsBO_2$	732
$CsBr$	714
$CsBrO_3$	713
$CsCHO_2$	723
$CsCN$	719
$Cs(C_2F_3O_2)$	743
$CsC_2H_3O_2$	709
$CsCl$	717
$CsClO_3$	716
$CsClO_4$	739
CsF	721
CsH	725
$CsHCO_3$	726
$CsHF_2$	727
$CsHSO_4$	728
CsI	731
$CsIBr_2$	720
$CsIO_3$	730
$CsIO_4$	740
$CsNH_2$	711
$CsNO_2$	735
$CsNO_3$	734
CsN_3	712
$CsOH$	729
CsO_2	737
Cs_2CO_3	715
Cs_2GeF_6	724
Cs_2MoO_4	733
Cs_2O	736
Cs_2O_3	738
Cs_2S	742
Cs_2SO_4	741
Cs_4CrO_4	718
Cu	914
$Cu(BF_4)_2$	1005
$Cu(BO_2)_2$	945
$CuBr$	922
$CuBr_2$	946
$CuCN$	924
$CuCO_3$	948
$CuCO_3 \cdot Cu(OH)_2$	949
CuC_2	941
$Cu(CHO_2)_2$	969
$Cu(CHO_2)_2 \cdot 4H_2O$	970
$CuC_2H_3O_2$	919
$Cu(C_2H_3O_2)_2$	937
$Cu(C_2H_3O_2)_2 \cdot CuO \cdot 6H_2O$	940
$Cu(C_2H_3O_2)_2 \cdot H_2O$	938
$Cu(C_2H_3O_2)_2 \cdot 3Cu(AsO_2)_2$	939
$Cu(CN)_2$	958
CuC_2O_4	982
$CuC_2O_4 \cdot 0.5H_2O$	983
$CuC_4H_4O_6 \cdot 3H_2O$	1002
$Cu(C_2H_5O)_2$	961
$Cu(C_4H_7O_2)_2 \cdot H_2O$	947
$Cu(CH_3COCHCOCH_3)_2$	985
$Cu(CF_3COCHCOCF_3)_2$	972
$Cu(CF_3COCHCOCH_3)_2$	1007
$Cu(C_2H_5CO_2CHCOCH_3)_2$	962
$CuC_{12}H_{22}O_{14}$	971
$Cu(C_8H_{15}O_2)_2$	963
$Cu(C_{10}H_{17}O_2)_2$	959
$CuC_{32}H_{16}N_8$	991
$Cu(C_{10}$[illegible]	981
$Cu(C_{18}H_{35}O_2)_2$	997
$CuCl$	923
$CuCl_2$	951
$CuCl_2 \cdot 2H_2O$	952
$CuCl_2 \cdot 3CuO \cdot 3.5H_2O$	986
$CuCl_2 \cdot 2NH_4Cl$	99
$CuCl_2 \cdot 2NH_4Cl \cdot 2H_2O$	100
$Cu(ClO_3)_2 \cdot 6H_2O$	950
$Cu(ClO_4)_2$	987
$Cu(ClO_4)_2 \cdot 6H_2O$	988
$CuCrO_4$	954
$CuCrO_4 \cdot 2Cu(OH)_2$	955
$CuCr_2O_4$	956
$CuCr_2O_7 \cdot 2H_2O$	960
CuF	925
CuF_2	967
$CuF_2 \cdot 2H_2O$	968
$CuFeS_2$	966
$CuFe_2O_4$	964
CuH	926
$CuHAsO_3$	943
CuI	927
$Cu(IO_3)_2$	975
$Cu(IO_3)_2 \cdot H_2O$	976
$CuMoO_4$	977
$Cu(NO_3)_2$	978
$Cu(NO_3)_2 \cdot 3H_2O$	979
$Cu(NO_3)_2 \cdot 6H_2O$	980
CuN_3	921
$Cu(N_3)_2$	944
CuO	984
$Cu(OH)_2$	974
CuP_2	917
CuS	1001
$CuSCN$	935
$CuSO_4$	998
$CuSO_4 \cdot 5H_2O$	999
$CuSe$	993
$CuSeO_3 \cdot 2H_2O$	994
$CuSeO_4 \cdot 5H_2O$	992
$CuSiF_6 \cdot 4H_2O$	973
$CuSiO_3 \cdot 2H_2O$	995
$CuSnO_3$	996
$CuTe$	1003
$CuTeO_3$	1004
$CuTiO_3$	1006
$Cu(VO_3)_2$	1010
$CuWO_4$	1008
$CuWO_4 \cdot 2H_2O$	1009
Cu_2C_2	920
$Cu_2C_6H_4O_7 \cdot 2.5H_2O$	957
$Cu_2Fe(CN)_6$	965
Cu_2HgI_4	928
Cu_2O	929
$Cu_2(OH)_3Cl$	953
Cu_2S	931
$Cu_2SO_3 \cdot 0.5H_2O$	932
$Cu_2SO_3 \cdot H_2O$	933
$Cu_2SO_3 \cdot CuSO_3 \cdot 2H_2O$	936
Cu_2Se	930
Cu_2Te	934
Cu_3As	915
$Cu_3(AsO_4)_2$	942
Cu_3N	916
$Cu_3(OH)_4SO_4$	1000
$Cu_3(PO_4)_2$	989
$Cu_3(PO_4)_2 \cdot 3H_2O$	990
Cu_5Si	918
DBr	1248
DCl	1251
DI	1255
DT	1246
D_2	1242
D_2O	3068
Dy	1012
DyB_4	1013
$DyBr_2$	1016
$DyBr_3$	1020
$Dy(C_2H_3O_2)_3 \cdot 4H_2O$	1019
$DyCl_2$	1017
$DyCl_3$	1022
$DyCl_3 \cdot 6H_2O$	1023
DyF_3	1024
DyH_3	1025
DyI_2	1018
DyI_3	1027
DyN	1014
$Dy(NO_3)_3 \cdot 5H_2O$	1028
$Dy(OH)_3$	1026
$DySi_2$	1015
$Dy_2(CO_3)_3 \cdot 4H_2O$	1021
Dy_2O_3	1029
Dy_2S_3	1031
$Dy_2(SO_4)_3 \cdot 8H_2O$	1030
Dy_2Te_3	1032
Er	1034
ErB_4	1035
$ErBr_3$	1037
$ErBr_3 \cdot 6H_2O$	1038
$Er(C_2H_3O_2)_3 \cdot 4H_2O$	1036
$ErCl_3$	1039
$ErCl_3 \cdot 6H_2O$	1040
ErF_3	1041
ErH_3	1042
ErI_3	1044
ErN	1046
$Er(NO_3)_3 \cdot 5H_2O$	1045
$Er(OH)_3$	1043
$ErSi_2$	1048
Er_2O_3	1047
Er_2S_3	1051
$Er_2(SO_4)_3$	1049
$Er_2(SO_4)_3 \cdot 8H_2O$	1050
Er_2Te_3	1052
Es	1033
Eu	1053
EuB_6	1054
$EuBr_2$	1057
$EuBr_3$	1065
$EuCl_2$	1058
$EuCl_3$	1066
$EuCl_3 \cdot 6H_2O$	1067
$Eu(ClO_4)_3 \cdot 6H_2O$	1073
EuF_2	1059
EuF_3	1068
EuI_2	1060
EuI_3	1069
EuN	1055
$Eu(NO_3)_3 \cdot 6H_2O$	1070
EuS	1063
$EuSO_4$	1062
$EuSe$	1061
$EuSi_2$	1056
$EuTe$	1064
$Eu_2(C_2O_4)_3$	1071
Eu_2O_3	1072
$Eu_2(SO_4)_3$	1074
$Eu_2(SO_4)_3 \cdot 8H_2O$	1075
FCN	667
FNO_3	1081
$FOClO_3$	1082
$FSSF$	2739
F_2	1077
F_2O	1078
F_2O_2	1079
F_2O_4	1080
F_5SOF	2756
Fe	1319
$Fe(AlO_2)_2$	1337
$FeAs$	1325
$FeAsO_4 \cdot 2H_2O$	1375
FeB	1326
$FeBr_2$	1340
$FeBr_2 \cdot 6H_2O$	1341
$FeBr_3$	1376
$FeCO_3$	1342
$FeC_2O_4 \cdot 2H_2O$	1356
$Fe(CHO_2)_3$	1385
$Fe(C_2H_3O_2)_2$	1335
$Fe(C_2H_3O_2)_2 \cdot 4H_2O$	1336
$FeC_4H_4O_6$	1367
$Fe(CO)_4H_2$	1321

Formula	No.	Formula	No.	Formula	No.	Formula	No.
$Fe(CO)_5$	1322	Fe_3O_4	1372	GeF_4	1160	H_2NSO_3H	2731
$FeC_6H_5O_7 \cdot 5H_2O$	1380	Fe_3P	1331	$GeHBr_3$	1147	H_2O	3067
$Fe(C_5H_5)_2$	1320	$Fe_3(PO_4)_2 \cdot 8H_2O$	1360	$GeHCl_3$	1148	H_2O_2	1256
$Fe(CH_3COCHCOCH_3)_2$	1358	$Fe_4[Fe(CN)_6]_3$	1382	GeH_2Br_2	1140	H_2PFO_3	2013
$Fe(CH_3COCHCOCH_3)_3$	1394	$Fe_4(P_2O_7)_3 \cdot 9H_2O$	1397	GeH_2Cl_2	1141	H_2PtCl_6	2063
$FeCl_2$	1343	Fm	1076	GeH_3Br	1137	$H_2PtCl_6 \cdot 6H_2O$	2083
$FeCl_2 \cdot 2H_2O$	1344	Fr	1083	GeH_3CH_3	1146	$H_2Pt(OH)_6$	2064
$FeCl_2 \cdot 4H_2O$	1345	Ga	1104	GeH_3Cl	1138	H_2S	1258
$Fe(ClO_4)_2$	1359	GaAs	1106	GeH_3F	1144	H_2SO_3	2730
$FeCl_3$	1377	$GaBr_3$	1114	GeH_3I	1145	H_2SO_4	2725
$FeCl_3 \cdot 6H_2O$	1378	$Ga(CH_3COCHCOCH_3)_3$	1124	GeH_4	1132	H_2SO_5	2726
$Fe(ClO_4)_3 \cdot 6H_2O$	1395	$GaCl_2$	1110	GeI_2	1153	H_2S_2	1259
$FeCr_2O_4$	1346	$GaCl_3$	1115	GeI_4	1161	H_2Se	1257
FeF_2	1347	$Ga(ClO_4)_3 \cdot 6H_2O$	1125	GeO	1154	H_2SeO_3	2391
$FeF_2 \cdot 4H_2O$	1348	GaF_3	1116	GeO_2	1163	H_2SeO_4	2389
FeF_3	1383	$GaF_3 \cdot 3H_2O$	1117	GeS	1156	H_2SiF_6	2462
$FeF_3 \cdot 3H_2O$	1384	GaH_3	1118	GeS_2	1165	H_2SiO_3	2460
$Fe(H_2PO_2)_3$	1398	GaI_3	1120	GeSe	1155	H_2Te	1260
FeI_2	1350	GaN	1107	$GeSe_2$	1164	H_2TeO_3	2786
$FeI_2 \cdot 4H_2O$	1351	$Ga(NO_3)_3$	1121	GeTe	1157	H_2WO_4	2944
$FeMoO_4$	1352	GaOOH	1123	Ge_2H_6	1133	H_3AsO_3	243
$Fe(NO_3)_2$	1353	$Ga(OH)_3$	1119	Ge_3H_8	1134	H_3AsO_4	241
$Fe(NO_3)_2 \cdot 6H_2O$	1354	GaP	1108	Ge_3N_4	1162	$H_3AsO_4 \cdot 0.5H_2O$	242
$Fe(NO_3)_3$	1389	GaS	1112	Ge_4H_{10}	1135	H_3BO_3	461
$Fe(NO_3)_3 \cdot 6H_2O$	1390	GaSb	1105	Ge_5H_{12}	1136	H_3PO_3	2006
$Fe(NO_3)_3 \cdot 9H_2O$	1391	GaSe	1111	$HAuBr_4 \cdot 5H_2O$	1167	H_3PO_4	2004
$Fe(NH_4)_3(C_6H_5O_7)_2$	1374	GaTe	1113	$HAuCl_4 \cdot 4H_2O$	1168	$H_3PW_{12}O_{40}$	2005
$FeNaP_2O_7$	1399	Ga_2O	1109	HBF_4	465	$H_3P(Mo_3O_{10})_4$	1790
FeO	1357	Ga_2O_3	1122	HBO_2	462	$H_4P_2O_6$	2009
$FeOH(C_2H_3O_2)_2$	1373	Ga_2S_3	1129	HBO_2	463	$H_4P_2O_7$	2010
FeO(OH)	1387	$Ga_2(SO_4)_3$	1127	HBO_2	464	H_4SiO_4	2461
$Fe(OH)_2$	1349	$Ga_2(SO_4)_3 \cdot 18H_2O$	1128	HBr	1247	$H_4SiO_4 \cdot (W_3O_9)_4$	2475
$Fe(OH)_3$	1386	Ga_2Se_3	1126	$HBrO_3$	479	H_6TeO_6	2785
FeP	1329	Ga_2Te_3	1130	HCN	1252	He	1208
$FePO_4 \cdot 2H_2O$	1396	Gd	1084	HCl	1249	Hf	1184
FeS	1365	GdB_6	1085	$HCl \cdot 2H_2O$	1250	HfB_2	1185
$FeSO_4$	1362	$GdBr_3$	1091	$HClO_3$	746	$HfBr_2$	1192
$FeSO_4 \cdot H_2O$	1363	$Gd(C_2H_3O_2)_2 \cdot 4H_2O$	1090	$HClO_4$	747	$HfBr_3$	1194
$FeSO_4 \cdot 7H_2O$	1364	$GdCl_3$	1092	HD	1244	$HfBr_4$	1197
FeS_2	1332	$GdCl_3 \cdot 6H_2O$	1093	HF	1253	HfC	1186
$Fe(SCN)_2 \cdot 3H_2O$	1369	GdF_3	1094	HI	1254	$Hf(C_5H_5)_2Cl_2$	1191
$Fe(SCN)_3$	1402	GdI_2	1088	HIO_3	1288	$HfCl_2$	1193
FeSe	1361	GdI_3	1095	$HIO_4 \cdot 2H_2O$	1289	$HfCl_3$	1195
FeSi	1333	GdN	1086	$HNOSO_4$	2727	$HfCl_4$	1198
$FeSi_2$	1334	$Gd(NO_3)_3 \cdot 5H_2O$	1096	HNO_2	1936	HfF_4	1199
$Fe(TaO_3)_2$	1366	$Gd(NO_3)_3 \cdot 6H_2O$	1097	HNO_3	1935	HfH_2	1187
FeTe	1368	GdSe	1089	HN_3	1235	HfI_3	1196
$FeTiO_3$	1370	$GdSi_2$	1087	HOBr	480	HfI_4	1200
$Fe(VO_3)_3$	1388	$Gd_2(C_2O_4)_3 \cdot 10H_2O$	1098	HOCl	745	HfN	1188
$FeWO_4$	1371	Gd_2O_3	1099	$HOSeF_5$	2390	$HfOCl_2 \cdot 8H_2O$	1202
Fe_2B	1327	Gd_2S_3	1102	HPF_6	2012	HfO_2	1201
$Fe_2(C_2O_4)_3$	1392	$Gd_2(SO_4)_3$	1100	HPH_2O_2	2007	HfP	1189
$Fe_2(CO)_9$	1323	$Gd_2(SO_4)_3 \cdot 8H_2O$	1101	HPO_2F_2	2011	HfS_2	1206
$Fe_2(CrO_4)_3$	1379	Gd_2Te_3	1103	HPO_3	2008	$Hf(SO_4)_2$	1205
$Fe_2(Cr_2O_7)_3$	1381	Ge	1131	$HReO_4$	2255	$HfSe_2$	1203
Fe_2O_3	1393	$GeBr_2$	1150	HT	1245	$HfSiO_4$	1204
Fe_2P	1330	$GeBr_4$	1158	H_2	1241	$HfSi_2$	1190
$Fe_2(SO_4)_3$	1400	$Ge(CH_3)_2Cl_2$	1143	H_2CrO_4	762	$HfTiO_4$	1207
$Fe_2(SO_4)_3 \cdot 9H_2O$	1401	$GeCl_2$	1151	$H_2MoO_4 \cdot H_2O$	1819	Hg	1735
Fe_2SiO_4	1355	$GeCl_3F$	1149	H_2NOH	1236	$HgBr_2$	1761
$Fe_3(AsO_4)_2$	1338	$GeCl_4$	1159	$H_2NOH \cdot HBr$	1237	$Hg(BrO_3)_2$	1760
$Fe_3(AsO_4)_2 \cdot 6H_2O$	1339	GeF_2	1152	$H_2NOH \cdot HCl$	1238	$Hg(CH_3)_2$	1736
Fe_3C	1328	GeF_2Cl_2	1142	$H_2NOH \cdot HClO_4$	1239	$Hg(CN)_2$	1765
$Fe_3(CO)_{12}$	1324	GeF_3Cl	1139	$2(H_2NOH) \cdot H_2SO_4$	1240	$Hg(CN)_2 \cdot HgO$	1779

$Hg(CNO)_2$ 1768
HgC_2O_4 1776
$Hg(C_2H_3O_2)_2$ 1757
$Hg(C_7H_5O_2)_2 \cdot H_2O$ 1759
$HgCl_2$ 1763
$Hg(ClO_3)_2$ 1762
$Hg(ClO_4)_2 \cdot 3H_2O$ 1780
$HgCrO_4$ 1764
$HgCr_2O_7$ 1766
HgF_2 1767
$HgHAsO_4$ 1769
HgI_2 1771
HgI_2 1772
$Hg(IO_3)_2$ 1770
$Hg(NH_2)Cl$ 1758
$Hg(NO_3)_2$ 1773
$Hg(NO_3)_2 \cdot H_2O$ 1774
$Hg(NO_3)_2 \cdot 2H_2O$ 1775
HgO 1777
HgS 1784
HgS 1785
$HgSO_4$ 1783
$Hg(SCN)_2$ 1787
$HgSe$ 1782
$HgTe$ 1786
$HgWO_4$ 1788
Hg_2Br_2 1739
$Hg_2(BrO_3)_2$ 1738
Hg_2CO_3 1740
$Hg_2C_2O_4$ 1750
$Hg_2(C_2H_3O_2)_2$ 1737
Hg_2Cl_2 1742
$Hg_2(ClO_3)_2$ 1741
$Hg_2(ClO_4)_2 \cdot 4H_2O$ 1752
Hg_2CrO_4 1743
Hg_2F_2 1744
Hg_2I_2 1746
$Hg_2(IO_3)_2$ 1745
$Hg_2(NO_2)_2$ 1749
$Hg_2(NO_3)_2$ 1747
$Hg_2(NO_3)_2 \cdot 2H_2O$ 1748
Hg_2O 1751
Hg_2S 1754
Hg_2SO_4 1753
$Hg_2(SCN)_2$ 1755
Hg_2WO_4 1756
$(Hg_3O_2)SO_4$ 1778
$Hg_3(PO_4)_2$ 1781
Ho 1209
$HoBr_3$ 1211
$Ho(C_2H_3O_2)_3$ 1210
$HoCl_3$ 1212
$HoCl_3 \cdot 6H_2O$ 1213
HoF_3 1214
HoI_3 1215
HoN 1217
$Ho(NO_3)_3 \cdot 5H_2O$ 1216
$HoSi_2$ 1220
$Ho_2(C_2O_4)_3 \cdot 10H_2O$ 1218
Ho_2O_3 1219
Ho_2S_3 1221
IBr 1294
ICN 668
ICl 1295
ICl_3 1296
IF 1297
IF_3 1298
IF_5 1299
IF_7 1300
IOF_3 1301
IOF_5 1302
IO_2F_3 1303
IO_3F 1304
I_2 1287
I_2O_4 1290
I_2O_5 1291
I_2O_6 1292
I_4O_9 1293
In 1261
$InAs$ 1263
$InBr$ 1266
$InBr_2$ 1269
$InBr_3$ 1272
$InCl$ 1267
$InCl_2$ 1270
$InCl_3$ 1273
$InCl_3 \cdot 4H_2O$ 1274
$In(ClO_4)_3 \cdot 8H_2O$ 1281
InF_3 1275
$InF_3 \cdot 3H_2O$ 1276
InI 1268
InI_3 1278
InN 1264
$In(NO_3)_3 \cdot 3H_2O$ 1279
$In(OH)_3$ 1277
InP 1265
$InPO_4$ 1282
InS 1271
$InSb$ 1262
In_2O_3 1280
In_2S_3 1285
$In_2(SO_4)_3$ 1284
In_2Se_3 1283
In_2Te_3 1286
Ir 1305
$IrBr_3$ 1307
$IrBr_3 \cdot 4H_2O$ 1308
$Ir(CH_3COCHCOCH_3)_3$ 1313
$IrCl_3$ 1309
$IrCl_4$ 1315
IrF_3 1310
IrF_6 1318
IrI_3 1311
IrO_2 1316
IrS_2 1317
Ir_2O_3 1312
Ir_2S_3 1314
$Ir_4(CO)_{12}$ 1306
K 2099
$KAg(CN)_2$ 2186
$KAl(SO_4)_2$ 2103
$KAl(SO_4)_2 \cdot 12H_2O$ 2104
$KAlSi_3O_8$ 2102
$KAsO_2$ 2107
$KAsO_3$ 2162
$KAu(CN)_2$ 2122
$KAuCl_4 \cdot 2H_2O$ 2199
KBF_4 2132
KBH_4 2109
KBO_2 2164
KBr 2111
$KBrO_3$ 2110
$KBrO_4$ 2172
$KCHO_2$ 2134
KCN 2121
$KCNO$ 2120
$KC_2H_3O_2$ 2100
$KC_{18}H_{35}O_2$ 2189
KCl 2115
$KClO_3$ 2114
$KClO_4$ 2174
$KCrO_3Cl$ 2116
KF 2130
$KF \cdot 2H_2O$ 2131
KH 2141
$KHCO_3$ 2143
$KHC_2O_4 \cdot 0.5H_2O$ 2146
$KHC_4H_4O_6$ 2154
KHF_2 2144
$KH(IO_3)_2$ 2145
KHS 2151
$KHS \cdot 0.5H_2O$ 2152
$KHSO_3$ 2153
$KHSO_4$ 2150
$KHSeO_3$ 2149
KH_2AsO_4 2124
KH_2PO_2 2157
KH_2PO_3 2126
KH_2PO_4 2125
KI 2159
KIO_3 2158
KIO_4 2175
$KI_3 \cdot H_2O$ 2209
$KMnO_4$ 2176
KNH_2 2105
KNO_2 2168
KNO_3 2167
KN_3 2108
$KNaC_4H_4O_6 \cdot 4H_2O$ 2187
$KNbO_3$ 2166
$KOCl$ 2156
KOH 2155
KO_2 2195
$KSCN$ 2206
$K(UO_2)(NO_3)_3$ 2214
$KZnI_3$ 2210
$K_2Al_2O_4 \cdot 3H_2O$ 2101
$K_2B_4O_7 \cdot 5H_2O$ 2198
K_2CO_3 2112
$K_2CO_3 \cdot 1.5H_2O$ 2113
K_2CS_3 2211
$K_2C_2O_4$ 2169
$K_2C_2O_4 \cdot H_2O$ 2170
$K_2C_2O_6 \cdot H_2O$ 2173
$K_2Cd(CN)_4$ 2201
$K_2Co(SeO_4)_2 \cdot 6H_2O$ 2119
K_2CrO_4 2117
$K_2Cr_2O_7$ 2123
K_2HAsO_4 2142
K_2HPO_3 2148
K_2HPO_4 2147
K_2HgI_4 2205
K_2MnF_6 2138
K_2MnO_4 2161
K_2MoO_4 2165
$K_2[Ni(CN)_4] \cdot H_2O$ 2202
K_2O 2171
K_2O_2 2177
K_2OsCl_6 2135
$K_2Pt(CN)_4 \cdot 3H_2O$ 2203
K_2PtCl_4 2200
K_2PtCl_6 2136
K_2S 2191
$K_2S \cdot 5H_2O$ 2192
K_2SO_3 2193
$K_2SO_3 \cdot 2H_2O$ 2194
K_2SO_4 2190
$K_2S_2O_3$ 2207
$K_2S_2O_5$ 2163
$K_2S_2O_6$ 2127
$K_2S_2O_7$ 2182
$K_2S_2O_8$ 2178
$K_2(SbC_4H_4O_6)_2 \cdot 3H_2O$ 213
K_2Se 2184
K_2SeO_3 2185
K_2SeO_4 2183
K_2SiF_6 2139
$K_2SnO_3 \cdot 3H_2O$ 2188
K_2TaF_7 2133
K_2TeO_3 2197
$K_2TeO_4 \cdot 3H_2O$ 2196
K_2TiO_3 2208
$K_2(UO_2)(SO_4)_2 \cdot 2H_2O$ 2215
$K_2U_2O_7$ 2213
K_2WO_4 2212
$K_2Zn(CN)_4$ 2204
$K_2Zn(SO_4)_2 \cdot 6H_2O$ 2216
K_2ZrF_6 2140
K_3AsO_4 2106
$K_3C_6H_5O_7 \cdot H_2O$ 2118
$K_3Co(CN)_6$ 2137
$K_3Cr(C_2O_4)_3 \cdot 3H_2O$ 810
$K_3Fe(CN)_6$ 2128
$K_3Fe(C_2O_4)_3 \cdot 3H_2O$ 2160
K_3PO_4 2179
$K_3Sb(C_2O_4)_3 \cdot 3H_2O$ 224
$K_4Fe(CN)_6 \cdot 3H_2O$ 2129
$K_4P_2O_7$ 2180
$K_4P_2O_7 \cdot 3H_2O$ 2181
$K_4Zr(SO_4)_4 \cdot 3H_2O$ 2217
Kr 1403
$KrFSb_2F_{11}$ 1405
KrF_2 1404
La 1406
$LaAlO_3$ 1407
LaB_6 1408
$LaBrO_3 \cdot 9H_2O$ 1409
$LaBr_3$ 1410
LaC_2 1411
$LaCl_3$ 1413
$LaCl_3 \cdot 7H_2O$ 1414
$La(ClO_4)_3 \cdot 6H_2O$ 1423
LaF_3 1415
LaH_3 1416
LaI_3 1419
$La(IO_3)_3$ 1418

LaN 1421
$La(NO_3)_3 \cdot 6H_2O$ 1420
$La(OH)_3$ 1417
LaS 1428
$LaSi_2$ 1424
$La_2(CO_3)_3 \cdot 8H_2O$ 1412
La_2O_3 1422
La_2S_3 1429
$La_2(SO_4)_3$ 1425
$La_2(SO_4)_3 \cdot 8H_2O$ 1426
$La_2(SO_4)_3 \cdot 9H_2O$ 1427
Li 1505
$LiAlH_4$ 1508
$LiAlSi_2O_6$ 1509
$LiAsF_6$ 1534
$LiBF_4$ 1574
$LiBH_4$ 1513
$LiBO_2$ 1547
$LiBO_2 \cdot 2H_2O$ 1548
$LiBr$ 1515
$LiBr \cdot H_2O$ 1516
$LiBrO_3$ 1514
$Li(CHO_2) \cdot H_2O$ 1532
$LiCN$ 1525
$LiC_2H_3O_2$ 1506
$LiC_2H_3O_2 \cdot 2H_2O$ 1507
$LiC_{18}H_{35}O_2$ 1566
$LiCl$ 1520
$LiCl \cdot H_2O$ 1521
$LiClO_3$ 1519
$LiClO_4$ 1561
$LiClO_4 \cdot 3H_2O$ 1562
$LiCoO_2$ 1524
LiD 1526
LiF 1531
$LiFeSi$ 1530
LiH 1538
LiH_2PO_4 1528
$LiHCO_3$ 1539
LiI 1544
$LiI \cdot 3H_2O$ 1545
$LiIO_3$ 1543
$LiN(C_3H_7)_2$ 1529
$LiNH_2$ 1510
$LiNO_2$ 1555
$LiNO_2 \cdot H_2O$ 1556
$LiNO_3$ 1553
LiN_3 1512
$LiNbO_3$ 1552
$LiOCl$ 1542
$LiOH$ 1540
$LiOH \cdot H_2O$ 1541
$LiPF_6$ 1535
$LiPO_3$ 1549
$LiSCN$ 1575
$LiSbF_6$ 1533
$LiSiO_4$ 1557
$LiTaO_3$ 1570
$LiVO_3$ 1578
$Li_2B_4O_7$ 1571
$Li_2B_4O_7 \cdot 5H_2O$ 1572
Li_2CO_3 1518
Li_2C_2 1517
$Li_2C_2O_4$ 1558
$Li_2CrO_4 \cdot 2H_2O$ 1522
$Li_2Cr_2O_7 \cdot 2H_2O$ 1527
Li_2MnO_3 1546
Li_2MoO_4 1551
Li_2O 1560
Li_2O_2 1563
$Li_2Pt(CN)_4 \cdot 5H_2O$ 1573
Li_2S 1569
Li_2SO_4 1567
$Li_2SO_4 \cdot H_2O$ 1568
$Li_2SeO_3 \cdot H_2O$ 1565
$Li_2SeO_4 \cdot H_2O$ 1564
Li_2SiF_6 1536
Li_2SiO_3 1550
Li_2SnF_6 1537
Li_2TiO_3 1576
Li_2WO_4 1577
Li_2ZrO_3 1579
Li_3AsO_4 1511
$Li_3C_6H_5O_7 \cdot 4H_2O$ 1523
Li_3N 1554
Li_3PO_4 1559
Lr 1430
Lu 1580
LuB_4 1581
$LuBr_3$ 1582
$LuCl_3$ 1583
$LuCl_3 \cdot 6H_2O$ 1584
$Lu(ClO_4)_3 \cdot 6H_2O$ 1591
LuF_3 1585
LuI_3 1586
LuN 1589
$Lu(NO_3)_3$ 1588
Lu_2O_3 1590
Lu_2S_3 1594
$Lu_2(SO_4)_3$ 1592
$Lu_2(SO_4)_3 \cdot 8H_2O$ 1593
Lu_2Te_3 1595
$Lu_3Fe_5O_{12}$ 1587
Md 1734
Mg 1596
$Mg(AlO_2)_2$ 1600
MgB_2 1606
$Mg(BO_2)_2 \cdot 8H_2O$ 1637
$Mg(BO_3)_2 \cdot 7H_2O$ 1651
MgB_6 1607
MgB_{12} 1608
$MgBr_2$ 1610
$MgBr_2 \cdot 6H_2O$ 1611
$Mg(BrO_3)_2 \cdot 6H_2O$ 1609
$MgCO_3$ 1612
$MgCO_3 \cdot 2H_2O$ 1613
$MgCO_3 \cdot 3H_2O$ 1614
$MgCO_3 \cdot 5H_2O$ 1615
$MgCO_3 \cdot Mg(OH)_2 \cdot 3H_2O$ 1618
$4MgCO_3 \cdot Mg(OH)_2 \cdot 4H_2O$ 1616
$4MgCO_3 \cdot Mg(OH)_2 \cdot 5H_2O$ 1617
$Mg(CHO_2)_2 \cdot 2H_2O$ 1627
MgC_2O_4 1648
$MgC_2O_4 \cdot 2H_2O$ 1649
$Mg(C_2H_3O_2)_2$ 1597
$Mg(C_2H_3O_2)_2 \cdot H_2O$ 1598
$Mg(C_2H_3O_2)_2 \cdot 4H_2O$ 1599
$MgCl_2$ 1620
$MgCl_2 \cdot 6H_2O$ 1621
$Mg(ClO_3)_2 \cdot 6H_2O$ 1619
$Mg(ClO_4)_2$ 1652
$Mg(ClO_4)_2 \cdot 6H_2O$ 1653
$MgCrO_4 \cdot 7H_2O$ 1622
$MgCr_2O_4$ 1623
MgF_2 1626
$MgHPO_4 \cdot 3H_2O$ 1631
MgH_2 1630
$Mg(H_2PO_4)_2 \cdot 2H_2O$ 1673
MgI_2 1634
$MgI_2 \cdot 6H_2O$ 1635
$MgI_2 \cdot 8H_2O$ 1636
$Mg(IO_3)_2 \cdot 4H_2O$ 1633
$Mg(MnO_4)_2 \cdot 6H_2O$ 1654
$MgMoO_4$ 1640
$MgNH_4PO_4 \cdot 6H_2O$ 1603
$Mg(NH_2)_2$ 1602
$Mg(NO_2)_2 \cdot 3H_2O$ 1645
$Mg(NO_3)_2$ 1641
$Mg(NO_3)_2 \cdot 2H_2O$ 1642
$Mg(NO_3)_2 \cdot 6H_2O$ 1643
MgO 1650
$MgO \cdot ZrO_2 \cdot SiO_2$ 1680
MgO_2 1655
$Mg(OH)_2$ 1632
MgS 1670
$MgSO_3 \cdot 3H_2O$ 1671
$MgSO_3 \cdot 6H_2O$ 1672
$MgSO_4$ 1667
$MgSO_4 \cdot H_2O$ 1668
$MgSO_4 \cdot 7H_2O$ 1669
$Mg(SCN)_2 \cdot 4H_2O$ 1674
$MgS_2O_3 \cdot 6H_2O$ 1675
$MgSe$ 1662
$MgSeO_3 \cdot 6H_2O$ 1663
$MgSeO_4 \cdot 6H_2O$ 1661
$MgSiF_6 \cdot 6H_2O$ 1664
$MgSiO_3$ 1638
$MgTiO_3$ 1639
$MgWO_4$ 1677
$MgZrO_3$ 1679
$Mg_2Al_3(AlSi_5O_{18})$ 1601
Mg_2Ge 1629
Mg_2GeO_4 1628
$Mg_2P_2O_7$ 1658
$Mg_2P_2O_7 \cdot 3H_2O$ 1659
Mg_2Si 1665
Mg_2SiO_4 1646
$Mg_2Si_3O_8$ 1676
Mg_2Sn 1666
Mg_2TiO_4 1647
$Mg_2V_2O_7$ 1678
Mg_3As_2 1605
$Mg_3(C_6H_5O_7)_2$ 1624
$Mg_3(C_6H_5O_7)_2 \cdot 14H_2O$ 1625
Mg_3N_2 1644
Mg_3P_2 1660
$Mg_3(PO_4)_2 \cdot 5H_2O$ 1656
$Mg_3(PO_4)_2 \cdot 8H_2O$ 1657
Mg_3Sb_2 1604
Mn 1681
MnB 1685
MnB_2 1686
$MnB_4O_7 \cdot 8H_2O$ 1724
$MnBr_2$ 1695
$MnBr_2 \cdot 4H_2O$ 1696
$MnCO_3$ 1697
$MnC_2O_4 \cdot 2H_2O$ 1712
$Mn(C_2H_3O_2)_2 \cdot 4H_2O$ 1694
$Mn(C_2H_3O_2)_3 \cdot 2H_2O$ 1728
$Mn(CO)_5Br$ 1690
$Mn(C_5H_5)_2$ 1682
$MnCl_2$ 1698
$MnCl_2 \cdot 4H_2O$ 1699
$Mn(ClO_4)_2 \cdot 6H_2O$ 1714
MnF_2 1701
MnF_3 1729
$Mn(H_2PO_2)_2 \cdot H_2O$ 1703
$Mn(H_2PO_4)_2 \cdot 2H_2O$ 1700
MnI_2 1704
$MnI_2 \cdot 4H_2O$ 1705
$MnMoO_4$ 1707
$Mn(NO_3)_2$ 1708
$Mn(NO_3)_2 \cdot 4H_2O$ 1709
$Mn(NO_3)_2 \cdot 6H_2O$ 1710
MnO 1713
MnO_2 1732
$MnO(OH)$ 1730
$Mn(OH)_2$ 1702
MnP 1691
MnS 1720
MnS 1721
MnS 1722
$MnSO_4$ 1717
$MnSO_4 \cdot H_2O$ 1718
$MnSO_4 \cdot 4H_2O$ 1719
$MnSb$ 1683
$MnSe$ 1716
$MnSiO_3$ 1706
$MnSi_2$ 1693
$MnTe$ 1723
$MnTiO_3$ 1725
$MnWO_4$ 1726
Mn_2B 1687
$Mn_2(CO)_{10}$ 1689
Mn_2O_3 1731
Mn_2O_7 1733
Mn_2P 1692
$Mn_2P_2O_7$ 1715
Mn_2Sb 1684
Mn_2SiO_4 1711
Mn_3C 1688
Mn_3O_4 1727
Mo 1789
$MoBr_2$ 1800
$MoBr_3$ 1803
$MoBr_4$ 1808
MoC 1793
$Mo(CO)_6$ 1795
$MoCl_2$ 1801
$MoCl_3$ 1804
$MoCl_4$ 1809
$MoCl_5$ 1816
MoF_3 1805

MoF_4 1810
MoF_5 1817
MoF_6 1822
MoI_2 1802
MoI_3 1806
MoI_4 1811
MoN 1796
$MoOCl_3$ 1818
$MoOCl_4$ 1825
$MoOF_4$ 1826
MoO_2 1812
MoO_2Cl_2 1820
MoO_2F_2 1821
MoO_3 1824
MoP 1798
$Mo(PO_3)_6$ 1823
MoS_2 1814
MoS_3 1827
$MoSe_2$ 1813
$MoSi_2$ 1799
$MoTe_2$ 1815
Mo_2B 1791
Mo_2B_5 1792
Mo_2C 1794
Mo_2N 1797
Mo_2O_3 1807
NBr_3 1943
$NClF_2$ 1947
NCl_3 1944
NF_3 1945
NHF_2 1950
NH_2COONH_4 90
NH_2Cl 1948
NH_2F 1949
NH_3 83
NH_4AlCl_4 188
NH_4BF_4 193
$NH_4B_5O_8 \cdot 4H_2O$ 164
NH_4Br 88
NH_4BrO_3 87
NH_4CHO_2 111
NH_4CN 101
$NH_4C_2H_3O_2$ 84
$NH_4C_3H_5O_3$ 149
$NH_4C_4H_5O_5$ 132
$NH_4C_4H_9CO_2$ 205
$NH_4C_6H_2N_3O_7$ 174
$NH_4C_7H_5O_2$ 86
$NH_4C_7H_5O_3$ 176
$NH_4C_8H_{15}O_2$ 89
$NH_4C_{15}H_{31}CO_2$ 163
$NH_4C_{18}H_{33}O_2$ 160
$NH_4C_{18}H_{35}O_2$ 179
$NH_4Ce(SO_4)_2 \cdot 4H_2O$ 680
NH_4Cl 93
NH_4ClO_3 92
NH_4ClO_4 167
$NH_4[Co(NH_3)_2(NO_2)_4]$ 902
$NH_4[Cr(NH_3)_2(SCN)_4] \cdot H_2O$ 194
$NH_4Cr(SO_4)_2 \cdot 12H_2O$ 95
NH_4F 109
$NH_4Fe(CrO_4)_2$ 146
$NH_4Fe(SO_4)_2 \cdot 12H_2O$ 148
NH_4HCO_3 129
$NH_4HC_2O_4 \cdot H_2O$ 133
$NH_4HC_4H_4O_6$ 140
NH_4HF_2 131
NH_4HS 138
NH_4HSO_3 139
NH_4HSO_4 137
NH_4HSeO_4 136
$NH_4H_2AsO_4$ 103
$NH_4H_2PO_2$ 142
$NH_4H_2PO_4$ 104
NH_4I 144
NH_4IO_3 143
$NH_4MgCl_3 \cdot 6H_2O$ 150
NH_4MnO_4 168
NH_4NH_2CSS 106
$NH_4NH_2SO_3$ 180
NH_4NO_2 158
NH_4NO_3 157
NH_4N_3 85
$NH_4NiCl_3 \cdot 6H_2O$ 1865
NH_4OH 141
NH_4PF_6 124
NH_4ReO_4 170
NH_4SCN 198
NH_4SO_3F 110
NH_4SbF_4 192
NH_4VO_3 153
NI_3 1946
NO 1938
$NOBr$ 1954
$NOCl$ 1955
NOF 1956
NOF_3 1957
NO_2 1939
NO_2Cl 1958
NO_2F 1959
NO_2NH_2 1934
N_2 1933
N_2F_2 1951
N_2F_2 1952
N_2F_4 1953
$(NH_2)_2SO_2$ 2746
N_2H_4 1222
$N_2H_4 \cdot CH_3COOH$ 1223
$N_2H_4 \cdot HBr$ 1226
$N_2H_4 \cdot HClO_4 \cdot 0.5H_2O$ 1232
$N_2H_4 \cdot HN_3$ 1224
$N_2H_4 \cdot HCl$ 1227
$N_2H_4 \cdot 2HNO_3$ 1231
$N_2H_4 \cdot 2HCl$ 1228
$N_2H_4 \cdot HI$ 1229
$N_2H_4 \cdot HNO_3$ 1230
$N_2H_4 \cdot H_2O$ 1225
$N_2H_4 \cdot H_2SO_4$ 1233
$(N_2H_4)_2 \cdot H_2SO_4$ 1234
$(NH_4)_2B_4O_7 \cdot 4H_2O$ 187
$(NH_4)_2CO_3$ 91
$(NH_4)_2C_2O_4$ 161
$(NH_4)_2C_2O_4 \cdot H_2O$ 162
$(NH_4)_2C_4H_4O_6$ 185
$(NH_4)_2Ce(NO_3)_6$ 702
$(NH_4)_2Ce(NO_3)_5 \cdot 4H_2O$ 679
$(NH_4)_2Co(SO_4)_2 \cdot 6H_2O$ 98
$(NH_4)_2CrO_4$ 94
$(NH_4)_2Cr_2O_7$ 102
$(NH_4)_2Fe(CN)_5NO$ 159
$(NH_4)_2Fe(SO_4)_2 \cdot 6H_2O$ 145
$(NH_4)_2GeF_6$ 123
$(NH_4)_2HAsO_4$ 128
$(NH_4)_2HC_6H_5O_7$ 130
$(NH_4)_2HPO_3 \cdot H_2O$ 135
$(NH_4)_2HPO_4$ 134
$(NH_4)_2HgCl_4 \cdot 2H_2O$ 151
$(NH_4)_2IrCl_6$ 116
$(NH_4)_2MoS_4$ 195
$(NH_4)_2Mo_2O_7$ 155
$(NH_4)_2OsBr_6$ 113
$(NH_4)_2OsCl_6$ 117
$(NH_4)_2PdCl_4$ 189
$(NH_4)_2PdCl_6$ 118
$(NH_4)_2PtBr_6$ 114
$(NH_4)_2PtCl_4$ 190
$(NH_4)_2PtCl_6$ 119
$(NH_4)_2RhCl_5 \cdot H_2O$ 165
$(NH_4)_2RuCl_6$ 120
$(NH_4)_2S$ 182
$(NH_4)_2S_x$ 175
$(NH_4)_2SO_3$ 183
$(NH_4)_2SO_3 \cdot H_2O$ 184
$(NH_4)_2SO_4$ 181
$(NH_4)_2S_2O_3$ 199
$(NH_4)_2S_2O_8$ 169
$(NH_4)_2SeO_3$ 178
$(NH_4)_2SeO_4$ 177
$(NH_4)_2SiF_6$ 125
$(NH_4)_2TaF_7$ 112
$(NH_4)_2TeO_4$ 186
$(NH_4)_2TiF$ 126
$(NH_4)_2TiO(C_2O_4)_2 \cdot H_2O$ 200
$(NH_4)_2U_2O_7$ 203
$(NH_4)_2WS_4$ 196
$(NH_4)_2ZnCl_4$ 191
$(NH_4)_2ZrF_6$ 127
N_2O 1937
N_2O_3 1940
N_2O_4 1941
N_2O_5 1942
N_3CN 664
$(NH_4)_3AlF_6$ 121
$(NH_4)_3Fe(CN)_6 \cdot 3H_2O$ 107
$(NH_4)_3Fe(C_2O_4)_3 \cdot 3H_2O$ 147
$(NH_4)_3GaF_6$ 122
$(NH_4)_3IrCl_6$ 115
$(NH_4)_3PO_4 \cdot 3H_2O$ 171
$(NH_4)_3PO_4 \cdot 12MoO_3$ 156
$(NH_4)_3PO_4 \cdot 12MoO_3 \cdot H_2O$ 172
$(NH_4)_3PO_4 \cdot 12WO_3 \cdot 2H_2O$ 173
$(NH_4)_3VS_4$ 197
$(NH_4)_3ZnCl_5$ 166
$(NH_4)_4Ce(SO_4)_4 \cdot 2H_2O$ 703
$(NH_4)_4Fe(CN)_6 \cdot 3H_2O$ 108
$(NH_4)_6Mo_7O_{24} \cdot 4H_2O$ 154
$(NH_4)_6W_7O_{24} \cdot 6H_2O$ 152
N_4Se_4 1960
$(NH_4)_{10}W_{12}O_{41}$ 201
$(NH_4)_{10}W_{12}O_{41} \cdot 5H_2O$ 202
Na 2528
$NaAlCl_4$ 2658
$NaAlH_4$ 2532
$NaAlO_2$ 2531
$NaAl(SO_4)_2 \cdot 12H_2O$ 2533
$NaAsO_2$ 2537
$NaAu(CN)_2$ 2572
$NaAuCl_4 \cdot 2H_2O$ 2659
$NaBF_4$ 2663
$NaBH_3(CN)$ 2555
$NaBH_4$ 2539
$NaBO_2$ 2604
$NaBO_3 \cdot 4H_2O$ 2617
$NaBiO_3$ 2602
$NaBr$ 2541
$NaBr \cdot 2H_2O$ 2542
$NaBrO_3$ 2540
$NaCHO_2$ 2570
$NaCN$ 2554
$NaCNO$ 2553
$NaC_2H_3O_2$ 2529
$NaC_2H_3O_2 \cdot 3H_2O$ 2530
NaC_2H_5O 2565
$NaC_{18}H_{35}O_2$ 2640
$NaCl$ 2547
$NaClO$ 2597
$NaClO_2$ 2548
$NaClO_3$ 2546
$NaClO_4$ 2618
$NaClO_4 \cdot H_2O$ 2619
NaF 2568
NaH 2583
$NaHCO_3$ 2586
$NaHF_2$ 2587
$NaHS$ 2593
$NaHS \cdot 2H_2O$ 2594
$NaHSO_3$ 2595
$NaHSO_4$ 2591
$NaHSO_4 \cdot H_2O$ 2592
NaH_2PO_2 2630
$NaH_2PO_2 \cdot H_2O$ 2631
NaH_2PO_4 2558
$NaH_2PO_4 \cdot H_2O$ 2559
$NaH_2PO_4 \cdot 2H_2O$ 2560
NaI 2600
$NaI \cdot 2H_2O$ 2601
$NaIO_3$ 2599
$NaIO_4$ 2620
$NaIO_4 \cdot 3H_2O$ 2621
$NaMnO_4 \cdot 3H_2O$ 2622
$NaNH_2$ 2534
$NaNH_4HPO_4 \cdot 4H_2O$ 2535
$NaNO_2$ 2612
$NaNO_3$ 2611
NaN_3 2538
$NaNbO_3$ 2610
$NaOCl \cdot 5H_2O$ 2598
$NaOH$ 2596
NaO_2 2650
$NaPF_6 \cdot H_2O$ 2580
$NaReO_4$ 2624
$NaSCN$ 2665
$NaSbF_6$ 2579

Formula	No.
$NaVO_3$	2675
$NaVO_3 \cdot 4H_2O$	2676
$Na_2B_4O_7$	2654
$Na_2B_4O_7 \cdot 4H_2O$	2655
$Na_2B_4O_7 \cdot 5H_2O$	2656
$Na_2B_4O_7 \cdot 10H_2O$	2657
Na_2BeF_4	2662
Na_2CO_3	2543
$Na_2CO_3 \cdot H_2O$	2544
$Na_2CO_3 \cdot 10H_2O$	2545
$Na_2C_2O_4$	2615
$Na_2C_4H_4O_4 \cdot 6H_2O$	2641
$Na_2C_4H_4O_6 \cdot 2H_2O$	2651
Na_2CrO_4	2549
$Na_2CrO_4 \cdot 4H_2O$	2550
$Na_2Cr_2O_7$	2556
$Na_2Cr_2O_7 \cdot 2H_2O$	2557
$Na_2[Fe(CN)_5NO] \cdot 2H_2O$	2613
Na_2GeO_3	2571
Na_2HAsO_4	2584
$Na_2HAsO_4 \cdot 7H_2O$	2585
$Na_2HPO_3 \cdot 5H_2O$	2632
Na_2HPO_4	2588
$Na_2HPO_4 \cdot 7H_2O$	2589
$Na_2HPO_4 \cdot 12H_2O$	2590
$Na_2H_2P_2O_6 \cdot 6H_2O$	2561
$Na_2H_2P_2O_7$	2562
$Na_2IrCl_6 \cdot 6H_2O$	2575
Na_2MoO_4	2607
$Na_2MoO_4 \cdot 2H_2O$	2608
Na_2O	2616
Na_2O_2	2623
Na_2PO_3F	2569
$Na_2PdCl_4 \cdot 3H_2O$	2660
$Na_2PtBr_6 \cdot 6H_2O$	2574
$Na_2PtCl_4 \cdot 4H_2O$	2661
Na_2PtCl_6	2576
$Na_2PtCl_6 \cdot 6H_2O$	2577
Na_2S	2645
$Na_2S \cdot 5H_2O$	2646
$Na_2S \cdot 9H_2O$	2647
Na_2SO_3	2648
$Na_2SO_3 \cdot 7H_2O$	2649
Na_2SO_4	2642
$Na_2SO_4 \cdot 7H_2O$	2643
$Na_2SO_4 \cdot 10H_2O$	2644
$Na_2S_2O_3$	2667
$Na_2S_2O_3 \cdot 5H_2O$	2668
$Na_2S_2O_4$	2563
$Na_2S_2O_5$	2603
$Na_2S_2O_6 \cdot 2H_2O$	2564
$Na_2S_2O_8$	2625
Na_2Se	2636
Na_2SeO_3	2637
$Na_2SeO_3 \cdot 5H_2O$	2638
Na_2SeO_4	2634
$Na_2SeO_4 \cdot 10H_2O$	2635
Na_2SiF_6	2581
Na_2SiO_3	2605
$Na_2SiO_3 \cdot 5H_2O$	2606
$Na_2SnO_3 \cdot 3H_2O$	2639
Na_2TeO_3	2653
Na_2TeO_4	2652
$Na_2U_2O_7 \cdot H_2O$	2674
Na_2WO_4	2672
$Na_2WO_4 \cdot 2H_2O$	2673
Na_3AlF_6	2578
$Na_3AsO_4 \cdot 12H_2O$	2536
$Na_3Au(S_2O_3)_2 \cdot 5H_2O$	2573
$Na_3C_6H_5O_7 \cdot 2H_2O$	2551
$Na_3C_6H_5O_7 \cdot 5H_2O$	2552
$Na_3Co(NO_2)_6$	2582
$Na_3Fe(CN)_6 \cdot H_2O$	2566
Na_3P	2629
$Na_3PO_3S \cdot 12H_2O$	2666
Na_3PO_4	2626
$Na_3PO_4 \cdot 12H_2O$	2627
$Na_3PO_4 \cdot 12MoO_3$	2609
$Na_3PO_4 \cdot NaOCl$	2628
$Na_3(PO_3)_3$	2669
$Na_3(PO_3)_3 \cdot 6H_2O$	2670
$Na_3SbS_4 \cdot 9H_2O$	2664
Na_3VO_4	2614
$Na_4Fe(CN)_6 \cdot 10H_2O$	2567
$Na_4P_2O_7$	2633
$Na_4Pb(S_2O_3)_3$	1484
$Na_5P_3O_{10}$	2671
Nb	1902
NbB	1904
NbB_2	1905
$NbBr_3$	1912
$NbBr_4$	1916
$NbBr_5$	1924
NbC	1906
$Nb(C_5H_5)_2Cl_2$	1903
$NbCl_3$	1913
$NbCl_4$	1917
$NbCl_5$	1925
NbF_3	1914
NbF_4	1918
NbF_5	1928
NbI_3	1915
NbI_4	1919
NbI_5	1929
NbN	1908
NbO	1911
$NbOBr_3$	1931
$NbOCl_3$	1932
NbO_2	1920
NbO_2F	1926
$Nb(OC_2H_5)_5$	1927
NbP	1909
NbS_2	1922
$NbSe_2$	1921
$NbSi_2$	1910
$NbTe_2$	1923
Nb_2C	1907
Nb_2O_5	1930
Nd	1828
NdB_6	1829
$NdBr_3$	1834
$Nd(BrO_3)_3 \cdot 9H_2O$	1833
$Nd(C_2H_3O_2)_2$	1831
$Nd(C_5H_5)_3$	1847
$NdCl_2$	1832
$NdCl_3$	1835
$NdCl_3 \cdot 6H_2O$	1836
NdF_3	1837
NdI_3	1839
NdN	1830
$Nd(NO_3)_3$	1840
$Nd(NO_3)_3 \cdot 6H_2O$	1841
$Nd(OH)_3$	1838
Nd_2O_3	1842
Nd_2S_3	1845
$Nd_2(SO_4)_3$	1843
$Nd_2(SO_4)_3 \cdot 8H_2O$	1844
Nd_2Te_3	1846
Ne	1848
Ni	1851
$NiAl$	1853
$NiAs$	1855
NiB	1856
$NiBr_2$	1869
$NiBr_2 \cdot 3H_2O$	1870
$NiCO_3$	1871
$Ni(CN)_2 \cdot 4H_2O$	1876
$NiC_2O_4 \cdot 2H_2O$	1887
$Ni(C_2H_3O_2)_2 \cdot 4H_2O$	1864
$Ni(CO)_4$	1859
$Ni(C_5H_5)_2$	1852
$NiCl_2$	1873
$NiCl_2 \cdot 6H_2O$	1874
$Ni(ClO_3)_2 \cdot 6H_2O$	1872
$Ni(ClO_4)_2 \cdot 6H_2O$	1889
$NiCrO_4$	1875
NiF_2	1877
$NiF_2 \cdot 4H_2O$	1878
NiI_2	1883
$NiI_2 \cdot 6H_2O$	1884
$Ni(IO_3)_2$	1881
$Ni(IO_3)_2 \cdot 4H_2O$	1882
$Ni(NH_4)_2(SO_4)_2$	1866
$Ni(NH_4)_2(SO_4)_2 \cdot 6H_2O$	1867
$Ni(NO_3)_2$	1885
$Ni(NO_3)_2 \cdot 6H_2O$	1886
NiO	1888
$Ni(OH)_2$	1879
$Ni(OH)_2 \cdot H_2O$	1880
NiS	1897
$NiSO_4$	1894
$NiSO_4 \cdot 6H_2O$	1895
$NiSO_4 \cdot 7H_2O$	1896
$Ni(SCN)_2$	1898
$NiSb$	1854
$NiSe$	1892
$NiSeO_4 \cdot 6H_2O$	1891
$NiSi_2$	1861
$NiSnO_3 \cdot 2H_2O$	1893
$NiTiO_3$	1899
Ni_2B	1857
Ni_2O_3	1901
Ni_2P	1860
Ni_2Si	1862
$Ni_3(AsO_4)_2 \cdot 8H_2O$	1868
Ni_3B	1858
$Ni_3(PO_4)_2 \cdot 8H_2O$	1890
Ni_3S_2	1863
Ni_3S_4	1900
No	1961
Np	1849
NpO_2	1850
O_2	1977
O_3	1978
Os	1962
$OsBr_3$	1968
$Os(CO)_5$	1965
$Os(C_5H_5)_2$	1963
$OsCl_2$	1967
$OsCl_3$	1969
$OsCl_4$	1970
OsF_4	1971
OsF_5	1973
OsF_6	1974
$OsOCl_4$	1975
OsO_2	1972
OsO_4	1976
$Os_2(CO)_9$	1966
$Os_3(CO)_{12}$	1964
P	1994
P	1995
P	1996
$PBrF_2$	2020
PBr_2F	2019
PBr_2F_3	2032
PBr_3	2018
PBr_4F	2031
PBr_5	2030
$PClF_2$	2023
$PClF_4$	2037
PCl_2F	2022
PCl_2F_3	2036
PCl_3	2021
PCl_3F_2	2035
PCl_4F	2034
PCl_5	2033
PF_3	2024
PF_5	2038
PH_3	1997
PH_4Cl	2002
PH_4I	2003
PI_3	2025
$POBrCl_2$	2046
$POBrClF$	2048
$POBrF_2$	2047
$POBr_2Cl$	2044
$POBr_2F$	2045
$POBr_3$	2043
$POClF_2$	2051
$POCl_2F$	2050
$POCl_3$	2049
POF_2H	2042
POF_3	2052
POI_3	2053
$PSBrF_2$	2056
$PSBr_2F$	2055
$PSBr_3$	2054
$PSClF_2$	2059
$PSCl_2F$	2058
$PSCl_3$	2057
PSF_3	2060
PSI_3	2061
P_2Cl_4	1999
P_2F_4	2000
P_2H_4	1998
P_2I_4	2001

Formula	No.
P_2O_3	2026
P_2O_5	2039
P_2S_3	2029
P_2S_5	2041
P_2Se_3	2028
P_2Se_5	2040
$(PNCl_2)_3$	2017
P_3N_5	2014
P_4O_6	2027
P_4S_3	2015
P_4S_7	2016
Pa	2244
$PaCl_5$	2245
Pb	1431
$Pb(AsO_2)_2$	1438
$Pb(BF_4)_2$	1456
$Pb(BO_2)_2 \cdot H_2O$	1440
$PbBr_2$	1442
$Pb(BrO_3)_2 \cdot H_2O$	1441
$PbBr_4$	1501
$PbCO_3$	1444
$Pb(CHO_2)_2$	1457
$Pb(CN)_2$	1453
PbC_2O_4	1469
$Pb(C_2H_3O_2)_2$	1433
$Pb(C_2H_3O_2)_2 \cdot 3H_2O$	1434
$Pb(C_2H_3O_2)_2 \cdot 2Pb(OH)_2$	1435
$Pb(C_3H_5O_3)_2$	1464
$Pb(C_4H_7O_2)_2$	1443
$Pb(C_2H_3O_2)_4$	1500
$Pb(CF_3COCHCOCF_3)_2$	1458
$Pb(CH_3COCHCOCH_3)_2$	1473
$Pb(C_7H_{15}CO_2)_2$	1454
$Pb(C_{18}H_{33}O_2)_2$	1468
$Pb(C_{18}H_{35}O_2)_2$	1485
PbClF	1448
$PbCl_2$	1447
$Pb(ClO_2)_2$	1449
$Pb(ClO_3)_2$	1446
$Pb(ClO_4)_2$	1474
$Pb(ClO_4)_2 \cdot 3H_2O$	1475
$PbCl_4$	1502
$PbCrO_4$	1450
$PbCrO_4 \cdot PbO$	1451
PbF_2	1455
PbF_4	1503
$PbHAsO_4$	1459
$PbHPO_4$	1460
PbH_4	1432
$Pb(H_2PO_2)_2$	1477
PbI_2	1463
$Pb(IO_3)_2$	1462
$PbMoO_4$	1465
$Pb(NO_3)_2$	1467
$Pb(N_3)_2$	1439
$Pb(NbO_3)_2$	1466
PbO	1470
PbO	1471
$3PbO \cdot H_2O$	1472
PbO_2	1504
$Pb(OH)_2$	1461
$Pb(OH)_2 \cdot 2PbCO_3$	1445
PbS	1487
$PbSO_3$	1488
$PbSO_4$	1486
$Pb(SCN)_2$	1491
PbS_2O_3	1492
PbSe	1482
$PbSeO_3$	1483
$PbSeO_4$	1481
$PbSiF_6 \cdot 2H_2O$	1480
$PbSiO_3$	1478
$Pb(TaO_3)_2$	1489
PbTe	1490
$PbTiO_3$	1493
$Pb(VO_3)_2$	1496
$PbWO_4$	1494
$PbWO_4$	1495
$PbZrO_3$	1497
Pb_2O_3	1498
Pb_2SiO_4	1479
$Pb_3(AsO_4)_2$	1437
$Pb_3(C_6H_5O_7)_2 \cdot 3H_2O$	1452
Pb_3O_4	1499
$Pb_3(PO_4)_2$	1476
$Pb_3(SbO_4)_2$	1436
Pd	1979
$PdBr_2$	1981
$Pd(CN)_2$	1984
$Pd(CH_3COCHCOCH_3)_2$	1989
$Pd(C_2H_3O_2)_2$	1980
$PdCl_2$	1982
$PdCl_2 \cdot 2H_2O$	1983
PdF_2	1985
PdI_2	1986
$Pd(NH_3)_2Cl_2$	1992
$Pd(NH_3)_2Cl_2$	1993
$Pd(NO_3)_2$	1987
PdO	1988
PdS	1991
$PdSO_4 \cdot 2H_2O$	1990
Pm	2239
$PmBr_3$	2240
$PmCl_3$	2241
PmF_3	2242
PmI_3	2243
Po	2096
$PoCl_4$	2097
PoO_2	2098
Pr	2218
PrB_6	2219
$PrBr_3$	2224
$Pr(BrO_3)_3$	2223
$PrCl_3$	2226
$PrCl_3 \cdot 7H_2O$	2227
$Pr(ClO_4)_3 \cdot 6H_2O$	2234
PrF_3	2228
PrF_4	2238
PrI_2	2222
PrI_3	2230
PrN	2220
$Pr(NO_3)_3$	2231
$Pr(NO_3)_3 \cdot 6H_2O$	2232
$Pr(OH)_3$	2229
$PrSi_2$	2221
$Pr_2(CO_3)_3 \cdot 8H_2O$	2225
Pr_2O_3	2233
Pr_2S_3	2236
$Pr_2(SO_4)_3 \cdot 8H_2O$	2235
Pr_2Te_3	2237
Pt	2062
$PtBr_2$	2065
$PtBr_3$	2071
$PtBr_4$	2073
$Pt(CN)_2$	2067
$PtCl_2$	2066
$PtCl_3$	2072
$PtCl_4$	2074
$PtCl_4 \cdot 5H_2O$	2075
PtF_4	2076
PtF_6	2080
PtI_2	2068
PtI_4	2077
$Pt(NH_3)_2Cl_2$	2081
$Pt(NH_3)_2Cl_2$	2082
PtO	2069
PtO_2	2078
PtS	2070
PtS_2	2079
PtSi	2084
Pu	2085
$PuBr_3$	2088
$PuCl_3$	2089
PuF_3	2090
PuF_4	2093
PuF_6	2095
PuI_3	2091
PuN	2086
PuO	2087
PuO_2	2094
Pu_2O_3	2092
Ra	2246
$RaBr_2$	2247
$RaCO_3$	2248
$RaCl_2$	2249
RaF_2	2250
$Ra(NO_3)_2$	2251
$RaSO_4$	2252
Rb	2302
$RbAl(SO_4)_2$	2304
$RbAl(SO_4)_2 \cdot 12H_2O$	2305
$RbBF_4$	2316
RbBr	2308
$RbBrO_3$	2307
$RbCHO_2$	2317
RbCN	2314
$RbC_2H_3O_2$	2303
RbCl	2311
$RbClO_3$	2310
$RbClO_4$	2330
RbF	2315
RbH	2319
$RbHCO_3$	2320
$RbHF_2$	2321
$RbHSO_4$	2322
RbI	2325
$RbIO_3$	2324
$RbMnO_4$	2331
$RbNO_2$	2328
$RbNO_3$	2327
RbN_3	2306
RbOH	2323
RbO_2	2336
Rb_2CO_3	2309
Rb_2CrO_4	2312
$Rb_2Cr_2O_7$	2313
Rb_2GeF_6	2318
Rb_2MoO_4	2326
Rb_2O	2329
Rb_2O_2	2332
Rb_2S	2335
Rb_2SO_4	2334
Rb_2Se	2333
Re	2254
$ReBr_3$	2259
$ReBr_5$	2269
$Re(CO)_5Br$	2257
$Re(CO)_5Cl$	2258
$ReCl_3$	2260
$ReCl_4$	2262
$ReCl_5$	2270
$ReCl_6$	2273
ReF_4	2263
ReF_5	2271
ReF_6	2275
ReF_7	2279
ReI_3	2261
$ReOCl_4$	2277
$ReOF_4$	2278
$ReOF_5$	2284
ReO_2	2264
ReO_2F_2	2274
ReO_2F_3	2283
ReO_3	2276
ReO_3Cl	2281
ReO_3F	2282
ReS_2	2267
$ReSe_2$	2265
$ReSi_2$	2266
$ReTe_2$	2268
$Re_2(CO)_{10}$	2256
Re_2O_5	2272
Re_2O_7	2280
Re_2S_7	2285
Rh	2286
$RhBr_3$	2290
$[Rh(CO)_2Cl]_2$	2288
$RhCl_3$	2291
RhF_3	2292
RhF_6	2301
RhI_3	2293
$Rh(NO_3)_3$	2294
$Rh(NO_3)_3 \cdot 2H_2O$	2295
RhO_2	2299
$RhO_2 \cdot 2H_2O$	2300
Rh_2O_3	2296
$Rh_2O_3 \cdot 5H_2O$	2297
$Rh_2(SO_4)_3$	2298
$Rh_4(CO)_{12}$	2289
$Rh_6(CO)_{16}$	2287
Rn	2253
Ru	2337
$RuBr_3$	2343
$Ru(CO)_5$	2339
$Ru(CH_3COCHCOCH_3)_3$	2347
$RuCl_3$	2344

Formula	No.
RuF_3	2345
RuF_4	2348
RuF_5	2350
RuF_6	2351
RuI_3	2346
$Ru(NO)Cl_3 \cdot H_2O$	2341
$Ru(NH_3)_6Cl_3$	2342
RuO_2	2349
RuO_4	2352
$Ru_2(CO)_9$	2340
$Ru_3(CO)_{12}$	2338
S	2723
S	2724
SCl_2	2740
SF_4	2741
SF_5Br	2743
SF_5Cl	2744
SF_6	2742
$SOBr_2$	2752
$SOCl_2$	2753
SOF_2	2754
SOF_4	2755
SO_2	2732
SO_2BrF	2749
SO_2Cl_2	2747
SO_2ClF	2750
SO_2F_2	2748
$SO_2(OH)Cl$	2728
$SO_2(OH)F$	2729
SO_3	2733
SO_3	2734
SO_3	2735
$SSBr_2$	2736
$SSCl_2$	2737
SSF_2	2738
S_2F_{10}	2745
$S_2O_5Cl_2$	2751
S_4N_4	2757
Sb	206
Sb	207
$SbAs$	212
$SbBr_3$	215
$Sb(CH_3)_3$	209
$Sb(CH_3)_5$	210
$Sb(C_2H_3O_2)_3$	214
$SbCl_2F_3$	232
$SbCl_3$	216
$SbCl_5$	230
SbF_3	217
SbF_5	231
SbH_3	208
$SbIS$	219
SbI_3	218
$SbOCl$	222
$SbPO_4$	223
$[Sb(CH_3)_2]_2$	211
Sb_2O_3	220
Sb_2O_3	221
Sb_2O_4	229
Sb_2O_5	233
Sb_2S_3	227
$Sb_2(SO_4)_3$	226
Sb_2S_5	234
Sb_2Se_3	225
Sb_2Te_3	228
Sc	2374
ScB_2	2375
$ScBr_3$	2376
$ScCl_3$	2377
ScF_3	2378
ScI_3	2380
$Sc(NO_3)_3$	2381
$Sc(OH)_3$	2379
Sc_2O_3	2382
Sc_2S_3	2384
$Sc_2(SO_4)_3 \cdot 5H_2O$	2383
Sc_2Te_3	2385
Se	2386
Se	2387
Se	2388
$SeBr_4$	2396
$SeCl_4$	2397
SeF_4	2398
SeF_5Cl	2400
SeF_6	2399
$SeOBr_2$	2401
$SeOCl_2$	2402
$SeOF_2$	2403
$SeOF_4$	2404
SeO_2	2392
SeO_2F_2	2405
SeO_3	2393
SeS	2406
SeS_2	2407
Se_2Br_2	2394
Se_2Cl_2	2395
Se_2S_6	2408
Se_4S_4	2409
Se_6S_2	2410
Si	2411
SiB_4	2474
$SiBrCl_3$	2430
$SiBr_2Cl_2$	2431
$SiBr_3Cl$	2432
$SiBr_4$	2429
SiC	2463
$Si(C_2H_3O_2)_4$	2473
$SiClF_3$	2439
$SiCl_2F_2$	2440
$SiCl_3F$	2441
$SiCl_3I$	2442
$SiCl_4$	2438
SiF_4	2448
$SiHBr_3$	2428
$SiHCl_3$	2437
$SiHF_3$	2447
$SiHI_3$	2454
SiH_2Br_2	2427
SiH_2Cl_2	2436
SiH_2F_2	2446
SiH_2I_2	2453
SiH_3Br	2426
SiH_3Cl	2435
SiH_3F	2445
SiH_3I	2452
SiH_3CH_3	2459
SiH_4	2412
SiI_4	2455
SiO	2465
SiO_2	2466
SiO_2	2467
SiO_2	2468
SiO_2	2469
SiO_2	2470
SiS	2471
SiS_2	2472
Si_2Br_6	2433
Si_2Cl_6	2443
$(SiCl_3)_2O$	2458
Si_2F_6	2449
Si_2H_6	2413
$(SiH_3)_2O$	2457
Si_2I_6	2456
Si_3Br_8	2434
Si_3Cl_8	2444
Si_3F_8	2450
Si_3H_8	2414
Si_3N_4	2464
Si_4F_{10}	2451
Si_4H_{10}	2415
Si_4H_{10}	2416
Si_5H_{10}	2417
Si_5H_{12}	2418
Si_5H_{12}	2419
Si_5H_{12}	2420
Si_6H_{12}	2421
Si_6H_{14}	2422
Si_6H_{14}	2423
Si_6H_{14}	2424
Si_7H_{16}	2425
Sm	2353
SmB_6	2354
$SmBr_2$	2356
$SmBr_3$	2362
$Sm(BrO_3)_3 \cdot 9H_2O$	2361
$Sm(C_2H_3O_2)_3 \cdot 3H_2O$	2360
$SmCl_2$	2357
$SmCl_3$	2364
$SmCl_3 \cdot 6H_2O$	2365
SmF_2	2358
SmF_3	2366
SmI_2	2359
SmI_3	2367
$Sm(NO_3)_3$	2368
$Sm(NO_3)_3 \cdot 6H_2O$	2369
$SmSi_2$	2355
$Sm_2(CO_3)_3$	2363
Sm_2O_3	2370
Sm_2S_3	2372
$Sm_2(SO_4)_3 \cdot 8H_2O$	2371
Sm_2Te_3	2373
Sn	2879
Sn	2880
$SnBr_2$	2887
$SnBr_4$	2902
SnC_2O_4	2894
$Sn(CH_3)_3N(CH_3)_2$	2883
$SnC_4H_4O_6$	2900
$Sn(C_2H_3O_2)_2$	2886
$SnCl_2$	2888
$SnCl_2 \cdot 2H_2O$	2889
$SnCl_4$	2903
$SnCl_4 \cdot 5H_2O$	2904
$Sn(CrO_4)_2$	2905
SnF_2	2890
SnF_4	2906
SnH_3CH_3	2882
SnH_4	2881
SnI_2	2893
SnI_4	2907
SnO	2895
SnO_2	2908
$Sn(OH)_2$	2892
SnP	2884
SnS	2899
$SnSO_4$	2898
SnS_2	2911
$SnSe$	2897
$SnSe_2$	2909
$Sn(SeO_3)_2$	2910
$SnTe$	2901
$SnZrF_6$	2891
$Sn_2P_2O_7$	2896
Sn_4P_3	2885
Sr	2677
$Sr(AsO_2)_2 \cdot 4H_2O$	2679
SrB_6	2694
$SrBr_2$	2681
$SrBr_2 \cdot 6H_2O$	2682
$Sr(BrO_3)_2 \cdot H_2O$	2680
$SrCO_3$	2684
SrC_2	2683
$Sr(CHO_2)_2$	2692
$Sr(CHO_2)_2 \cdot 2H_2O$	2693
$Sr(CN)_2 \cdot 4H_2O$	2689
$SrC_2O_4 \cdot H_2O$	2706
$Sr(C_2H_3O_2)_2$	2678
$SrCl_2$	2686
$SrCl_2 \cdot 6H_2O$	2687
$Sr(ClO_3)_2$	2685
$Sr(ClO_4)_2$	2708
$SrCrO_4$	2688
SrF_2	2691
$SrFe(CN)_6 \cdot 15H_2O$	2690
SrH_2	2695
SrI_2	2698
$SrI_2 \cdot 6H_2O$	2699
$Sr(IO_3)_2$	2697
$Sr(MnO_4)_2 \cdot 3H_2O$	2709
$SrMoO_4$	2700
$Sr(NO_2)_2$	2704
$Sr(NO_3)_2$	2702
$SrNb_2O_6$	2701
SrO	2707
SrO_2	2710
$Sr(OH)_2$	2696
SrS	2716
$SrSO_3$	2717
$SrSO_4$	2715
$SrS_2O_3 \cdot 5H_2O$	2719
$SrSe$	2713
$SrSeO_4$	2712
$SrSi_2$	2714
$SrTe$	2718
$SrTiO_3$	2720
$SrWO_4$	2721

$SrZrO_3$ 2722
Sr_2SiO_4 2705
Sr_3N_2 2703
$Sr_3(PO_4)_2$ 2711
T_2 1243
T_2O 3069
Ta 2758
$TaAl_3$ 2759
TaB 2760
TaB_2 2761
$TaBr_3$ 2767
$TaBr_4$ 2769
$TaBr_5$ 2776
TaC 2762
$TaCl_3$ 2768
$TaCl_4$ 2770
$TaCl_5$ 2777
TaF_5 2778
TaH 2764
TaI_4 2771
TaI_5 2779
TaN 2765
TaO_2 2772
TaS_2 2774
$TaSe_2$ 2773
$TaSi_2$ 2766
$TaTe_2$ 2775
Ta_2C 2763
Ta_2O_5 2780
Tb 2797
$TbBr_3$ 2800
$TbCl_3$ 2801
$TbCl_3 \cdot 6H_2O$ 2802
TbF_3 2809
TbF_4 2810
TbI_3 2803
TbN 2798
$Tb(NO_3)_3$ 2804
$Tb(NO_3)_3 \cdot 6H_2O$ 2805
$TbSi_2$ 2799
Tb_2O_3 2806
Tb_2S_3 2808
$Tb_2(SO_4)_3 \cdot 8H_2O$ 2807
Tc 2781
TcF_5 2782
TcF_6 2783
Te 2784
$TeBr_2$ 2789
$TeBr_4$ 2791
$TeCl_2$ 2790
$TeCl_4$ 2792
TeF_4 2793
TeF_6 2796
TeI_4 2795
TeO_2 2787
TeO_3 2788
Te_2F_{10} 2794
Th 2847
ThB_6 2849
$ThBr_4$ 2850
ThC 2851
ThC_2 2852
$ThCl_4$ 2853
ThF_4 2854
ThH_2 2848
ThI_4 2855
ThN 2858
$Th(NO_3)_4$ 2856
$Th(NO_3)_4 \cdot 4H_2O$ 2857
ThO_2 2859
ThS_2 2864
$Th(SO_4)_2 \cdot 9H_2O$ 2863
$ThSe_2$ 2860
$ThSiO_4$ 2861
$ThSi_2$ 2862
Ti 2912
TiI_3 2928
TiB_2 2915
$TiBr_2$ 2920
$TiBr_3$ 2925
$TiBr_4$ 2933
TiC 2916
$Ti(C_5H_5)_2Cl_2$ 2913
$TiCl_2$ 2921
$TiCl_3$ 2926
$TiCl_4$ 2934
TiF_3 2927
TiF_4 2935
TiH_2 2914
TiI_2 2922
TiI_4 2936
TiN 2917
TiO 2923
$TiOSO_4 \cdot H_2O$ 2940
TiO_2 2937
TiO_2 2938
TiO_2 2939
TiP 2918
TiS 2924
TiS_2 2942
$Ti(SO_4)_2$ 2941
$TiSi_2$ 2919
Ti_2O_3 2929
Ti_2S_3 2931
$Ti_2(SO_4)_3$ 2930
Ti_3O_5 2932
Tl 2811
$TlBr$ 2815
$TlBrO_3$ 2814
$TlBr_3 \cdot 4H_2O$ 2839
$TlCHO_2$ 2823
$TlCN$ 2820
$TlC_2H_3O_2$ 2812
TlC_2H_5O 2821
$Tl(C_2H_3O_2)_3$ 2838
$TlCl$ 2818
$TlClO_3$ 2817
$TlClO_4$ 2833
$TlCl_3$ 2840
$TlCl_3 \cdot 4H_2O$ 2841
TlF 2822
TlF_3 2842
TlI 2827
$TlIO_3$ 2826
$TlNO_2$ 2830
$TlNO_3$ 2829
TlN_3 2813
$Tl(NO_3)_3$ 2843
$TlOH$ 2825
$TlPF_6$ 2824
$TlSe$ 2846
Tl_2CO_3 2816
$Tl_2C_2O_4$ 2831
Tl_2CrO_4 2819
Tl_2MoO_4 2828
Tl_2O 2832
Tl_2O_3 2844
Tl_2S 2837
Tl_2SO_4 2836
$Tl_2(SO_4)_3$ 2845
Tl_2Se 2835
Tl_2SeO_4 2834
Tm 2865
$TmBr_2$ 2866
$TmBr_3$ 2869
$TmCl_2$ 2867
$TmCl_3$ 2870
$TmCl_3 \cdot 7H_2O$ 2871
TmF_3 2872
TmI_2 2868
TmI_3 2874
$Tm(NO_3)_3$ 2875
$Tm(NO_3)_3 \cdot 5H_2O$ 2876
$Tm(OH)_3$ 2873
$Tm_2(C_2O_4)_3 \cdot 6H_2O$ 2877
Tm_2O_3 2878
U 2986
UB_2 2987
UB_4 2988
UBr_3 2994
UBr_4 2999
UBr_5 3005
UC 2989
UC_2 2990
UCl_3 2995
UCl_4 3000
UCl_5 3006
UCl_6 3009
UF_3 2996
UF_4 3001
UF_5 3007
UF_6 3010
UH_3 2997
UI_3 2998
UI_4 3002
UN 2992
UO_2 3003
$UO_2(C_2H_3O_2)_2 \cdot 2H_2O$ 3014
UO_2Cl_2 3015
UO_2F_2 3016
$UO_2HPO_4 \cdot 4H_2O$ 3017
$UO_2(NO_3)_2$ 3018
$UO_2(NO_3)_2 \cdot 6H_2O$ 3019
$UO_2(NH_4)_3F_5$ 204
UO_2SO_4 3020
$UO_2SO_4 \cdot 3H_2O$ 3021
UO_3 3011
$UO_3 \cdot H_2O$ 3012
$UO_4 \cdot 2H_2O$ 3013
U_2C_3 2991
U_2N_3 2993
U_3O_8 3008
U_4O_9 3004
V 3022
VB 3025
VB_2 3026
VBr_2 3033
VBr_3 3039
VBr_4 3048
VC 3027
$V(CO)_6$ 3029
$V(C_5H_5)_2$ 3023
$V(C_5H_5)_2Cl_2$ 3024
$V(CH_3COCHCOCH_3)_3$ 3045
VCl_2 3034
VCl_3 3040
VCl_4 3049
VF_2 3035
VF_3 3041
$VF_3 \cdot 3H_2O$ 3042
VF_4 3050
VF_5 3052
VI_2 3036
VI_3 3043
VN 3030
VO 3037
$VOBr$ 3057
$VOBr_2$ 3059
$VOBr_3$ 3064
$VOCl$ 3058
$VOCl_2$ 3060
$VOCl_3$ 3065
VOF_2 3061
VOF_3 3066
$VOSO_4 \cdot 2H_2O$ 3063
$VOSeO_3 \cdot H_2O$ 3062
VO_2 3051
VO_2Cl 3054
VO_2F 3053
$VSO_4 \cdot 7H_2O$ 3038
VSi_2 3031
V_2C 3028
V_2O_3 3044
V_2O_5 3055
V_2S_3 3047
$V_2(SO_4)_3$ 3046
V_2S_5 3056
V_3Si 3032
W 2943
WB 2946
WBr_2 2955
WBr_3 2958
WBr_4 2961
WBr_5 2969
WBr_6 2975
WC 2949
$W(CO)_6$ 2950
$W(C_2H_5O)_5$ 2971
WCl_2 2956
WCl_3 2959
WCl_4 2962
WCl_5 2970
WCl_6 2976
WF_4 2963
WF_5 2972
WF_6 2980

WI_2 2957
WI_3 2960
WI_4 2964
WN_2 2951
$WOBr_3$ 2973
$WOBr_4$ 2982
$WOCl_3$ 2974
$WOCl_4$ 2983
WOF_4 2984
WO_2 2965
WO_2Br_2 2977
WO_2Cl_2 2978
WO_2I_2 2979
WO_3 2981
WS_2 2967
WS_3 2985
WSe_2 2966
WSi_2 2953
WTe_2 2968
W_2B 2945
W_2B_5 2947
W_2C 2948
W_2N 2952
W_5Si_3 2954
Xe 3070
$XeFRuF_6$ 3084
$XeFSb_2F_{11}$ 3085
XeF_2 3073
XeF_3SbF_6 3082
$XeF_3Sb_2F_{11}$ 3086
XeF_4 3074
XeF_5AsF_6 3080
XeF_5RuF_6 3081
XeF_6 3075
$XeOF_2$ 3076
$XeOF_4$ 3077
XeO_2F_2 3078
XeO_3 3071
XeO_3F_2 3079
XeO_4 3072
$Xe_2F_3AsF_6$ 3083
Y 3102
YAs 3105
YB_6 3106
YBr_3 3107
YC_2 3108
YCl_3 3110
$YCl_3 \cdot 6H_2O$ 3111
YF_3 3112
YI_3 3114
$Y(NO_3)_3$ 3116
$Y(NO_3)_3 \cdot 4H_2O$ 3117
$Y(NO_3)_3 \cdot 6H_2O$ 3118
$Y(OH)_3$ 3113
YP 3120
YSb 3104
$Y_2(CO_3)_3 \cdot 3H_2O$ 3109
Y_2O_3 3119
Y_2S_3 3122
$Y_2(SO_4)_3 \cdot 8H_2O$ 3121
$Y_3Al_5O_{12}$ 3103
$Y_3Fe_5O_{12}$ 3115
Yb 3087
$YbBr_2$ 3089
$YbBr_3$ 3094
$Yb(C_2H_3O_2)_3 \cdot 4H_2O$ 3093
$YbCl_2$ 3090
$YbCl_3$ 3095
$YbCl_3 \cdot 6H_2O$ 3096
YbF_2 3091
YbF_3 3097
YbI_2 3092
YbI_3 3098
$Yb(NO_3)_3$ 3099
$YbSi_2$ 3088
Yb_2O_3 3100
$Yb_2(SO_4)_3 \cdot 8H_2O$ 3101
Zn 3123
$Zn(AsO_2)_2$ 3130
$Zn(BF_4)_2 \cdot 6H_2O$ 3149
$ZnBr_2$ 3135
$Zn(BrO_3)_2 \cdot 6H_2O$ 3134
$ZnCO_3$ 3137
$Zn(CHO_2)_2 \cdot 2H_2O$ 3150
$Zn(CN)_2$ 3144
ZnC_2O_4 3162
$ZnC_2O_4 \cdot 2H_2O$ 3163
$ZnC_4H_4O_6 \cdot 2H_2O$ 3164
$Zn(C_2H_3O_2)_2 \cdot 2H_2O$ 3124
$Zn(C_2H_5)_2$ 3145
$Zn(C_8H_{15}O_2)_2$ 3136
$Zn(CH_3COCHCOCH_3)_2$ 3166
$Zn(C_5H_9O_2)_2 \cdot 2H_2O$ 3167
$Zn(C_{12}H_{23}O_2)_2$ 3155
$Zn(C_{18}H_{33}O_2)_2$ 3161
$Zn(C_{18}H_{35}O_2)_2$ 3179
$ZnCl_2$ 3140
$Zn(ClO_3)_2$ 3139
$Zn(ClO_4)_2 \cdot 6H_2O$ 3168
$ZnCrO_4$ 3141
$ZnCr_2O_4$ 3142
ZnF_2 3147
$ZnF_2 \cdot 4H_2O$ 3148
ZnI_2 3154
$Zn(IO_3)_2$ 3153
$Zn(MnO_4)_2 \cdot 6H_2O$ 3169
$ZnMoO_4$ 3156
$Zn(NH_4)_2(SO_4)_2$ 3125
$Zn(NO_2)_2$ 3160
$Zn(NO_3)_2$ 3157
$Zn(NO_3)_2 \cdot 6H_2O$ 3158
ZnO 3165
$3ZnO \cdot 2B_2O_3$ 3131
$2ZnO \cdot 3B_2O_3 \cdot 3.5H_2O$ 3132
$2ZnO \cdot 3B_2O_3 \cdot 5H_2O$ 3133
ZnO_2 3170
$Zn(OH)_2$ 3152
$3Zn(OH)_2 \cdot 2ZnCO_3$ 3138
ZnS 3183
ZnS 3184
$ZnSO_3 \cdot 2H_2O$ 3185
$ZnSO_4$ 3180
$ZnSO_4 \cdot H_2O$ 3181
$ZnSO_4 \cdot 7H_2O$ 3182
$Zn(SCN)_2$ 3187
ZnS_2O_4 3146
ZnSb 3126
ZnSe 3176
$ZnSeO_3$ 3178
$ZnSeO_4 \cdot 5H_2O$ 3175
$ZnSiF_6 \cdot 6H_2O$ 3151
ZnTe 3186
$Zn_2P_2O_7$ 3174
Zn_2SiO_4 3177
Zn_3As_2 3129
$Zn_3(AsO_4)_2$ 3127
$Zn_3(AsO_4)_2 \cdot 8H_2O$ 3128
$Zn_3(C_6H_5O_7)_2 \cdot 2H_2O$ 3143
Zn_3N_2 3159
Zn_3P_2 3173
$Zn_3(PO_4)_2$ 3171
$Zn_3(PO_4)_2 \cdot 4H_2O$ 3172
Zr 3188
ZrB_2 3190
$ZrBr_2$ 3195
$ZrBr_3$ 3200
$ZrBr_4$ 3206
ZrC 3191
$Zr(C_2H_3O_2)_2(OH)_2$ 3204
$Zr(C_5H_5)_2Cl_2$ 3189
$ZrCl_2$ 3196
$ZrCl_3$ 3201
$ZrCl_4$ 3207
ZrF_2 3197
ZrF_3 3202
ZrF_4 3208
ZrH_2 3198
ZrI_2 3199
ZrI_3 3203
ZrI_4 3210
ZrN 3192
$Zr(NH_4)_3OH(CO_3)_3 \cdot 2H_2O$ 3205
$Zr(NO_3)_4 \cdot 5H_2O$ 3211
$ZrOCl_2$ 3219
$ZrOCl_2 \cdot 8H_2O$ 3220
ZrO_2 3213
$Zr(OH)_4$ 3209
ZrP_2 3193
ZrP_2O_7 3214
ZrS_2 3217
$Zr(SO_4)_2$ 3215
$Zr(SO_4)_2 \cdot 4H_2O$ 3216
$ZrSiO_4$ 3212
$ZrSi_2$ 3194
$Zr(WO_4)_2$ 3218

PHYSICAL PROPERTIES OF THE RARE EARTH METALS

K.A. Gschneidner, Jr.

TABLE 1. Data for the Trivalent Ions of the Rare Earth Elements

Rare earth	Symbol	Atomic no.	Atomic wt.[a]	Electronic configuration for R^{3+}: No. 4f electrons	S	L	J	Spectroscopic ground state symbol
Scandium	Sc	21	44.955910	0	—	—	—	—
Yttrium	Y	39	88.90585	0	—	—	—	—
Lanthanum	La	57	138.9055	0	—	—	—	—
Cerium	Ce	58	140.115	1	1/2	3	5/2	$^2F_{5/2}$
Praseodymium	Pr	59	140.90765	2	1	5	4	$^3H_4$
Neodymium	Nd	60	144.24	3	3/2	6	9/2	$^4I_{9/2}$
Promethium	Pm	61	(145)	4	2	6	4	$^5I_4$
Samarium	Sm	62	150.36	5	5/2	5	5/2	$^6H_{5/2}$
Europium	Eu	63	151.965	6	3	3	0	$^7F_0$
Gadolinium	Gd	64	157.25	7	7/2	0	7/2	$^8S_{7/2}$
Terbium	Tb	65	158.92534	8	3	3	6	$^7F_6$
Dysprosium	Dy	66	162.50	9	5/2	5	15/2	$^6H_{15/2}$
Holmium	Ho	67	164.93032	10	2	6	8	$^5I_8$
Erbium	Er	68	167.26	11	3/2	6	15/2	$^4I_{15/2}$
Thulium	Tm	69	168.93421	12	1	5	6	$^3H_6$
Ytterbium	Yb	70	173.04	13	1/2	3	7/2	$^2F_{7/2}$
Lutetium	Lu	71	174.967	14	—	—	—	—

Note: For additional information, see Goldschmidt, Z.B., in *Handbook on the Physics and Chemistry of Rare Earths*, Vol. 1, Gschneidner, K.A., Jr. and Eyring, L., Eds., North-Holland Physics, Amsterdam, 1978; DeLaeter, J.R., and Heumann, K.G., *J. Phys. Chem. Ref. Data*, 20, 1313, 1991; *Pure Appl. Chem.*, 66, 2423, 1994.

[a] 1993 standard atomic weights.

TABLE 2. Crystallographic Data for the Rare Earth Metals at 24 °C (297 K) or Below

Rare earth metal	Crystal structure[a]	Lattice constants (Å): a_o	b_o	c_o	Metallic radius CN = 12 (Å)	Atomic volume (cm^3/mol)	Density (g/cm^3)
αSc	hcp	3.3088	—	5.2680	1.6406	15.039	2.989
αY	hcp	3.6482	—	5.7318	1.8012	19.893	4.469
αLa	dhcp	3.7740	—	12.171	1.8791	22.602	6.146
αCe[b]	fcc	4.85[b]	—	—	1.72[b]	17.2[b]	8.16[b]
βCe	dhcp	3.6810	—	11.857	1.8321	20.947	6.689
γCe[c]	fcc	5.1610	—	—	1.8247	20.696	6.770
αPr	dhcp	3.6721	—	11.8326	1.8279	20.803	6.773
αNd	dhcp	3.6582	—	11.7966	1.8214	20.583	7.008
αPm	dhcp	3.65	—	11.65	1.811	20.24	7.264
αSm	rhomb[d]	3.6290[d]	—	26.207	1.8041	20.000	7.520
Eu	bcc	4.5827	—	—	2.0418	28.979	5.244
αGd	hcp	3.6336	—	5.7810	1.8013	19.903	7.901
α′Tb[e]	ortho	3.605[e]	6.244[e]	5.706[e]	1.784[e]	19.34[e]	8.219[e]
αTb	hcp	3.6055	—	5.6966	1.7833	19.310	8.230
α′Dy[f]	ortho	3.595[f]	6.184[f]	5.678[f]	1.774[f]	19.00[f]	8.551[f]
αDy	hcp	3.5915	—	5.6501	1.7740	19.004	8.551
Ho	hcp	3.5778	—	5.6178	1.7661	18.752	8.795
Er	hcp	3.5592	—	5.5850	1.7566	18.449	9.066
Tm	hcp	3.5375	—	5.5540	1.7462	18.124	9.321
αYb[g]	hcp	3.8799[g]	—	6.3859[g]	1.9451[g]	25.067[g]	6.903[g]
βYb	fcc	5.4848	—	—	1.9392	24.841	6.966
Lu	hcp	3.5052	—	5.5494	1.7349	17.779	9.841

Note: For additional information, see Gschneidner, K.A., Jr. and Calderwood, F.W., in *Handbook on the Physics and Chemistry of Rare Earths*, Vol. 8, Gschneidner, K.A., Jr. and Eyring, L., Eds., North-Holland Physics, Amsterdam, 1986; Gschneidner, K.A., Jr., Pecharsky, V.K., Cho, Jaephil and Martin, S.W., *Scripta Mater.*, 34, 1717, 1996.

[a] hcp = hexagonal close-packed; $P6_3/mmc$, hP2, A3, Mg-type; dhcp = double-c hexagonal close-packed; $P6_3/mmc$, hP4, A3′, αLa-type; fcc = face-centered cubic; $Fm\bar{3}m$, cF4, A1, Cu-type; rhomb = rhombohedral; $R\bar{3}m$, hR3, αSm-type; bcc = body-centered cubic; $Im\bar{3}m$, cI2, A2, W-type; ortho = orthorhombic; Cmcm, oC4, α′ Dy-type.

[b] At 77 K (−196 °C).

[c] Equilibrium room temperature (standard state) phase.

[d] Rhombohedral is the primitive cell. Lattice parameters given are for the nonprimitive hexagonal cell.

[d] At 220 K (−53 °C).

[f] At 86 K (−187 °C).

[g] At 23 °C.

TABLE 3. Crystallographic Data for Rare Earth Metals at High Temperature

Rare earth metal	Structure	Lattice parameter (Å)	Temp. (°C)	Metallic radius CN = 8 (Å)	Metallic radius CN = 12 (Å)	Atomic volume (cm^3/mol)	Density (g/cm^3)
βSc	bcc	3.73 (est.)	1337	1.62	1.66	15.6	2.88
βY	bcc	4.10[a]	1478	1.78	1.83	20.8	4.28
βLa	fcc	5.303	325	—	1.875	22.45	6.187
γLa	bcc	4.26	887	1.84	1.90	23.3	5.97
δCe	bcc	4.12	757	1.78	1.84	21.1	6.65
βPr	bcc	4.13	821	1.79	1.84	21.2	6.64
βNd	bcc	4.13	883	1.79	1.84	21.2	6.80
βPm	bcc	4.10 (est.)	890	1.78	1.83	20.8	6.99
βSm	hcp	a = 3.6630 c = 5.8448	450[b]	—	1.8176	20.450	7.353
γSm	bcc	4.10 (est.)	922	1.77	1.82	20.8	7.25
βGd	bcc	4.06	1265	1.76	1.81	20.2	7.80
βTb	bcc	4.07[a]	1289	1.76	1.81	20.3	7.82
βDy	bcc	4.03[a]	1381	1.75	1.80	19.7	8.23
γYb	bcc	4.44	763[c]	1.92	1.98	26.4	6.57

Note: The rare earths Eu, Ho, Er, Tm, and Lu are monomorphic. For additional information, see Gschneidner, K.A., Jr. and Calderwood, F.W., in *Handbook on the Physics and Chemistry of Rare Earths*, Vol. 8, Gschneidner, K.A., Jr. and Eyring, L., Eds., North-Holland Physics, Amsterdam, 1986, 1.

[a] Determined by extrapolation to 0% solute of a vs. composition data for R-Mg alloys at 24 °C and corrected for thermal expansion to temperature given.

[b] The hcp phase was stabilized by impurities and the temperature of measurement was below the equilibrium transition temperature (see Table 4).

[c] The bcc phase was stabilized by impurities and the temperature of measurement was below the equilibrium transition temperature (see Table 4).

TABLE 4. High Temperature Transition Temperatures and Melting Point of Rare Earth Metals

Rare earth metal	Transition I (α – β)[a] Temp. (°C)	Transition I Phases	Transition II (β – γ)[a] Temp. (C°)	Transition II Phases	Melting point (C°)
Sc	1337	hcp ⇌ bcc	—	—	1541
Y	1478	hcp ⇌ bcc	—	—	1522
La[b]	310	dhcp ⇌ fcc	865	fcc ⇌ bcc	918
Ce[c,d]	139	dhcp ⇌ fcc (β - γ)	726	fcc ⇌ bcc (γ - δ)	798
Pr	795	dhcp ⇌ bcc	—	—	931
Nd	863	dhcp ⇌ bcc	—	—	1021
Pm	890	dhcp ⇌ bcc	—	—	1042
Sm[e]	734	rhom ⇌ hcp	922	hcp ⇌ bcc	1074
Eu	—	—	—	—	822
Gd	1235	hcp ⇌ bcc	—	—	1313
Tb	1289	hcp ⇌ bcc	—	—	1356
Dy	1381	hcp ⇌ bcc	—	—	1412
Ho	—	—	—	—	1474
Er	—	—	—	—	1529
Tm	—	—	—	—	1545
Yb	795	fcc ⇌ bcc (β - γ)	—	—	819
Lu	—	—	—	—	1663

Note: For additional information, see Gschneidner, K.A., Jr. and Calderwood, F.W., in *Handbook on the Physics and Chemistry of Rare Earths*, Vol. 8, Gschneidner, K.A., Jr. and Eyring, L., Eds., North-Holland Physics, Amsterdam, 1986; Gschneidner, K.A., Jr., Pecharsky, V.K., Cho, Jaephil and Martin, S.W., *Scripta Mater.*, 34, 1717, 1996.

[a] For all the transformations listed, unless otherwise noted.

[b] On cooling, fcc → dhcp (β → α), 260 °C.

[c] The β ⇌ γ equilibrium transition temperature is 10 ± 5 °C.

[d] On cooling, fcc → dhcp (γ → β), −16 °C.

[e] On cooling, hcp → rhomb (β → α), 727 °C.

TABLE 5. Low Temperature Transition Temperatures of the Rare Earth Metals

Rare earth metal	Cooling: Transformation	°C	K	Rare earth metal	Heating: Transformation	°C	K
Ce	γ → β[a]	−16	257	Ce	α → β	−148	125
	γ → α	−172	101		α → β + γ	−104	169
	β → α	−228	45		β → γ[a]	139	412
Tb	α → α′	−53	220	Yb	α → β	7	280
Dy	α → α′	−187	86				
Yb	β → α	−13	260				

Note: For additional information, see Beaudry, B.J. and Gschneidner, K.A., Jr., in *Handbook on the Physics and Chemistry of Rare Earths*, Vol. 1, Gschneidner, K.A., Jr. and Eyring, L., Eds., North-Holland Physics, Amsterdam, 1978, 173; Koskenmaki, D.C. and Gschneidner, K.A., Jr., 1978, in *Handbook on the Physics and Chemistry of Rare Earths*, Vol. 1, Gschneidner, K.A., Jr. and Eyring, L., Eds., North-Holland Physics, Amsterdam, 1978, 337; Gschneidner, K.A., Jr., Pecharsky, V.K., Cho, Jaephil and Martin, S.W., *Scripta Mater.*, 34, 1717, 1996.

[a] The β ⇌ γ equilibrium transition temperature is 10 ± 5 °C (283 ± 5K).

TABLE 6. Heat Capacity, Standard Entropy, Heats of Transformation, and Fusion of the Rare Earth Metals

Rare earth metal	Heat capacity at 298 K (J/mol K)	Standard entropy S°_{298} (J/mol K)	Heat of transformation (kJ/mol)				Heat of fusion (kJ/mol)
			trans. 1	ΔH_{tr}^{1}	trans. 2	ΔH_{tr}^{2}	
Sc	25.5	34.6	$\alpha \rightleftharpoons \beta$	4.00	—	—	14.1
Y	26.5	44.4	$\alpha \rightleftharpoons \beta$	4.99	—	—	11.4
La	27.1	56.9	$\alpha \rightleftharpoons \beta$	0.36	$\beta \rightleftharpoons \gamma$	3.12	6.20
Ce	26.9	72.0	$\beta \rightleftharpoons \gamma$	0.05	$\gamma \rightleftharpoons \delta$	2.99	5.46
Pr	27.2	73.2	$\alpha \rightleftharpoons \beta$	3.17	—	—	6.89
Nd	27.5	71.5	$\alpha \rightleftharpoons \beta$	3.03	—	—	7.14
Pm	27.3[a]	71.6[a]	$\alpha \rightleftharpoons \beta$	3.0[a]	—	—	7.7[a]
Sm	29.5	69.6	$\alpha \rightleftharpoons \beta$	0.2[a]	$\beta \rightleftharpoons \gamma$	3.11	8.62
Eu	27.7	77.8	—	—	—	—	9.21
Gd	37.0	68.1	$\alpha \rightleftharpoons \beta$	3.91	—	—	10.0
Tb	28.9	73.2	$\alpha \rightleftharpoons \beta$	5.02	—	—	10.79
Dy	27.7	75.6	$\alpha \rightleftharpoons \beta$	4.16	—	—	11.06
Ho	27.2	75.3	—	—	—	—	17.0[a]
Er	28.1	73.2	—	—	—	—	19.9
Tm	27.0	74.0	—	—	—	—	16.8
Yb	26.7	59.9	$\beta \rightleftharpoons \gamma$	1.75	—	—	7.66
Lu	26.9	51.0	—	—	—	—	22[a]

Note: For additional information, see Hultgren, R., Desai, P.D., Hawkins, D.T., Gleiser, M., Kelley, K.K., and Wagman, D.D., *Selected Values of the Thermodynamic Properties of the Elements,* ASM International, Metals Park, Ohio, 1973; Wagman, D.D., Evans, W.H., Parker, V.B., Schumm, R.H., Halow, I., Bailey, S.M., Churney, K.L., and Nuttall, R.L., *The NBS Tables of Chemical Thermodynamic Properties, J. Phys. Chem. Ref. Data,* Vol. 11, Suppl. 2, 1982; Amitin, E.B., Bessergenev, W.G., Kovalevskaya, Yu. A., and Paukov, I.E., *J. Chem. Thermodyn.*, 15, 181, 1983; Amitin, E.B., Bessergenev, W.G., Kovalevskaya, Yu. A., and Paukov, I.E., *J. Chem. Thermodyn.*, 15, 181, 1983.

[a] Estimated.

TABLE 7. Vapor Pressures, Boiling Points, and Heats of Sublimation of Rare Earth Metals

Rare earth metal	Temperature in °C[a] for a vapor pressure of				Boiling point[a] (°C)	Heat of sublimation at 25 °C (kJ/mol)
	10^{-8} atm (0.001 Pa)	10^{-6} atm (0.101 Pa)	10^{-4} atm (10.1Pa)	10^{-2} atm (1013 Pa)		
Sc	1036	1243	1533	1999	2836	377.8
Y	1222	1460	1812	2360	3345	424.7
La	1301	1566	1938	2506	3464	431.0
Ce	1290	1554	1926	2487	3443	422.6
Pr	1083	1333	1701	2305	3520	355.6
Nd	955	1175	1500	2029	3074	327.6
Pm	—	—	—	—	3000[b]	348[b]
Sm	508	642	835	1150	1794	206.7
Eu	399	515	685	964	1529	175.3
Gd	1167	1408	1760	2306	3273	397.5
Tb	1124	1354	1698	2237	3230	388.7
Dy	804	988	1252	1685	2567	290.4
Ho	845	1036	1313	1771	2700	300.8
Er	908	1113	1405	1896	2868	317.1
Tm	599	748	964	1300	1950	232.2
Yb	301	400	541	776	1196	152.1
Lu	1241	1483	1832	2387	3402	427.6

Note: For additional information, see Hultgren, R., Desai, P.D., Hawkins, D.T., Gleiser, M., Kelley, K.K., and Wagman, D.D., *Selected Values of the Thermodynamic Properties of the Elements,* ASM International, Metals Park, Ohio, 1973; Beaudry, B.J. and Gschneidner, K.A., Jr., in *Handbook on the Physics and Chemistry of Rare Earths,* Vol. 1, Gschneidner, K.A., Jr. and Eyring, L., Eds., North-Holland Physics, Amsterdam, 1978, 173.

[a] International Temperature Scale of 1990 (ITS-90) values.

[b] Estimated.

TABLE 8. Magnetic Properties of the Rare Earth Metals

Rare earth metal	$\chi_A \times 10^6$ at 298 K (emu/mol)	Effective magnetic moment: Paramagnetic at ~298 K		Effective magnetic moment: Ferromagnetic at ~0 K		Easy axis	Néel temp. T_N (K)		Curie temp. T_C (K)	θ_p (K)		
		Theory[a]	Obs.	Theory[b]	Obs.		Hex sites	Cubic sites		∥c	⊥c	Polycryst. or avg.
αSc	295.2	—	—	—	—	—	—	—	—	—	—	—
αY	187.7	—	—	—	—	—	—	—	—	—	—	—
αLa	95.9	—	—	—	—	—	—	—	—	—	—	—
βLa	105	—	—	—	—	—	—	—	—	—	—	—
γCe	2,270	2.54	2.52	2.14	—	—	—	14.4	—	—	—	−50
βCe	2,500	2.54	2.61	2.14	—	—	13.7	12.5	—	—	—	−41
αPr	5,530	3.58	3.56	3.20	2.7[c]	a	0.03	—	—	—	—	0
αNd	5,930	3.62	3.45	3.27	2.2[c]	b	19.9	7.5	—	0	5	3.3
αPm	—	2.68	—	2.40	—	—	—	—	—	—	—	—
αSm	1,278[d]	0.85	1.74	0.71	0.5[c]	a	109	14.0	—	—	—	—
Eu	30,900	7.94	8.48	7.0	5.9	<110>	—	90.4	—	—	—	100
αGd	185,000[e]	7.94	7.98	7.0	7.63	30° to c	—	—	293.4	317	317	317
αTb	170,000	9.72	9.77	—	—	—	230.0	—	—	195	239	224
α′Tb	—	—	—	9.0	9.34	b	—	—	219.5	—	—	—
αDy	98,000	10.64	10.83	—	—	—	180.2	—	—	121	169	153
α′Dy	—	—	—	10.0	10.33	a	—	—	90.5[g]	—	—	—
Ho	72,900	10.60	11.2	10.0	10.34	b	132	—	19.5	73.0	88.0	83.0
Er	48,000	9.58	9.9	9.0	9.1	30° to c	85	—	18.7	61.7	32.5	42.2
Tm	24,700	7.56	7.61	7.0	7.14	c	58	—	32.0	41.0	−17.0	2.3
βYb	67[d]	—	—	—	—	—	—	—	—	—	—	—
Lu	182.9	—	—	—	—	—	—	—	—	—	—	—

Note: For additional information, see McEwen, K.A., in *Handbook on the Physics and Chemistry of Rare Earths,* Vol. 1, Gschneidner, K.A., Jr. and Eyring, L., Eds., North-Holland Physics, Amsterdam, 1978, 411; Legvold, S., in *Ferromagnetic Materials,* Vol. 1, Wohlfarth, E.P., Ed., North-Holland Physics, Amsterdam, 1980, 183; Pecharsky, V.K., Gschneidner, K.A., Jr. and Fort, D., *Phys. Rev. B,* 47, 5063, 1993; Pecharsky, V.K., Gschneidner, K.A., Jr. and Fort, D., 1996, to be published; Steward, A.M. and Collocott, S.J., *J. Phys.: Condens. Matter,* 1, 677, 1988.

[a] $g[J(J+1)]^{1/2}$.
[b] gJ.
[c] At 38 T and 4.2 K.
[d] At 290 K.
[e] At 350 K.
[g] On cooling T_C = 89.6 K and on warming T_C = 91.5 K.

TABLE 9. Room Temperature Coefficient of Thermal Expansion, Thermal Conductivity, Electrical Resistance, and Hall Coefficient

Rare earth metal	Expansion ($\alpha_t \times 10^6$) (°C⁻¹)			Thermal conductivity (W/cm K)	Electrical resistance (μΩ·cm)			Hall coefficient ($R_i \times 10^{12}$) (V·cm/A·Oe)		
	α_a	α_c	α_{poly}		ρ_a	ρ_c	ρ_{poly}	R_a	R_c	R_{poly}
αSc	7.6	15.3	10.2	0.158	70.9	26.9	56.2[a]	—	—	−0.13
αY	6.0	19.7	10.6	0.172	72.5	35.5	59.6	−0.27	−1.6	—
aLa	4.5	27.2	12.1	0.134	—	—	61.5	—	—	−0.35
bCe	—	—	—	—	—	—	82.8	—	—	—
γCe	6.3	—	6.3	0.113	—	—	74.4	—	—	+1.81
αPr	4.5	11.2	6.7	0.125	—	—	70.0	—	—	+0.709
αNd	7.6	13.5	9.6	0.165	—	—	64.3	—	—	+0.971
αPm	9[b]	16[b]	11[b]	0.15[b]	—	—	75[b]	—	—	—
αSm	9.6	19.0	12.7	0.133	—	—	94.0	—	—	−0.21
Eu	35.0	—	35.0	0.139[b]	—	—	90.0	—	—	+24.4
αGd	9.1[c]	10.0[c]	9.4[c]	0.105	135.1	121.7	131.0	−10	−54	−4.48[d]
αTb	9.3	12.4	10.3	0.111	123.5	101.5	115.0	−1.0	−3.7	—
αDy	7.1	15.6	9.9	0.107	111.0	76.6	92.6	−0.3	−3.7	—
Ho	7.0	19.5	11.2	0.162	101.5	60.5	81.4	+0.2	−3.2	—
Er	7.9	20.9	12.2	0.145	94.5	60.3	86.0	+0.3	−3.6	—
Tm	8.8	22.2	13.3	0.169	88.0	47.2	67.6	—	—	−1.8
βYb	26.3	—	26.3	0.385	—	—	25.0	—	—	+3.77
Lu	4.8	20.0	9.9	0.164	76.6	34.7	58.2	+0.45	−2.6	−0.535

Note: For additional information, see Beaudry, B. J. and Gschneidner, K.A., Jr., in *Handbook on the Physics and Chemistry of Rare Earths,* Vol. 1, Gschneidner, K.A., Jr. and Eyring, L., Eds., North-Holland Physics, Amsterdam, 1978, 173; McEwen, K.A., in *Handbook on the Physics and Chemistry of Rare Earths,* Vol. 1, Gschneidner, K.A., Jr. and Eyring, L., Eds., North-Holland Physics, Amsterdam, 1978, 411.

[a] Calculated from single crystal values.
[b] Estimated.
[c] At 100 °C.
[d] At 77 °C.

TABLE 10. Electronic Specific Heat Constant (γ), Electron–Electron (Coulomb) Coupling Constant (μ*), Electron–Phonon Coupling Constant (λ), Debye Temperature at 0 K(θ_D), and Superconducting Transition Temperature

Rare earth metal	γ (mJ/mol·K²)	μ*	λ	θ_D (K) from Heat capacity	Elastic constants	Superconducting temperature (K)
αSc	10.334	0.16	0.30	345.3	—	0.050[a]
αY	7.878	0.15	0.30	244.4	258	1.3[b]
αLa	9.45	0.08	0.76	150	154	5.10
βLa	11.5	—	—	140	—	6.00
αCe	12.8	—	—	179	—	0.022[c]
αPr	20	—	1.07[d]	155[e]	153	—
αNd	f	—	0.86[d]	157[e]	163	—
αPm	—	—	—	159[e]	—	—
αSm	8.1 ± 1.5[g]	—	0.81[d]	162[e,f]	169	—
Eu	f	—	—	f	118	—
αGd	4.48	—	0.30	169	182	—
α′Tb	3.71	—	0.34[d]	169.6	177	—
α′Dy	4.9	—	0.32[d]	192	183	—
Ho	2.1	—	0.30[d]	175[e]	190	—
Er	8.7		0.33[d]	176.9	188	—
Tm	f	—	0.36[d]	179[e]	200	—
αYb	3.30	—	—	117.6	118	—
βYb	8.36	—	—	109	—	—
Lu	8.194	0.14	0.31	183.2	185	0.022[h]

Note: For additional information, see Sundström, L.J., in *Handbook on the Physics and Chemistry of Rare Earths,* Vol. 1, Gschneidner, K.A., Jr., and Eyring, L., Eds., North-Holland Physics, Amsterdam, 1978, 379; Scott, T., in *Handbook on the Physics and Chemistry of Rare Earths,* Vol. 1, Gschneidner, K.A., Jr. and Eyring, L., Eds., North-Holland Physics, Amsterdam, 1978, 591; Probst, C. and Wittig, J., in *Handbook on the Physics and Chemistry of Rare Earths,* Vol. 1, Gschneidner, K.A., Jr. and Eyring, L., Eds., North-Holland Physics, Amsterdam, 1978, 749; Tsang, T.-W.E., Gschneidner, K.A., Jr., Schmidt, F.A., and Thome, D.K., *Phys. Rev.,* B, 31, 235, 1985; Collocott, S.J., Hill, R.W. and Stewart, A.M., *J. Phys. F,* 18, L223, 1988; Hill, R.W. and Gschneidner, K.A., Jr., *J. Phys. F,* 18, 2545, 1988; Skriver, H.L. and Mertig, I., *Phys. Rev. B,* 41, 6553, 1990; Collocott, S.J. and Stewart, A.M., *J. Phys. Condens. Matter,* 4, 6743, 1992; Pecharsky, V.K., Gschneidner, K.A., Jr. and Fort, D., *Phys. Rev. B,* 47, 5063, 1993.

[a] At 18.6 GPa.
[b] At 11 GPa.
[c] At 2.2 GPa.
[d] Calculated value.
[e] Estimated.
[f] Heat capacity results have been reported, but the resultant γ and θ_D values are unreliable because of the presence of impurities and/or there was no reliable procedure or model to correct for the magnetic contribution to the heat capacity.
[g] Based on the values reported for the purer Sm sample (IV).
[h] At 4.5 GPa.

TABLE 11. Room Temperature Elastic Moduli and Mechanical Properties

Rare earth metal	Elastic moduli (GPa): Young's (elastic) modulus	Shear modulus	Bulk modulus	Poisson's ratio	Mechanical properties (MPa): Yield strength 0.2% offset	Ultimate tensile strength	Uniform elongation (%)	Reduction in area (%)	Recryst. temp. (°C)
Sc	74.4	29.1	56.6	0.279	173[a]	255[a]	5.0[a]	8.0[a]	550
Y	63.5	25.6	41.2	0.243	42	129	34.0	—	550
αLa	36.6	14.3	27.9	0.280	126[a]	130	7.9[a]	—	300
βCe	—	—	—	—	86	138	—	24.0	—
γCe	33.6	13.5	21.5	0.24	28	117	22.0	30.0	325
αPr	37.3	14.8	28.8	0.281	73	147	15.4	67.0	400
αNd	41.4	16.3	31.8	0.281	71	164	25.0	72.0	400
αPm	46[b]	18[b]	33[b]	0.28[b]	—	—	—	—	400[b]
αSm	49.7	19.5	37.8	0.274	68	156	17.0	29.5	440
Eu	18.2	7.9	8.3	0.152	—	—	—	—	300
αGd	54.8	21.8	37.9	0.259	15	118	37.0	56.0	500
αTb	55.7	22.1	38.7	0.261	—	—	—	—	500
αDy	61.4	24.7	40.5	0.247	43	139	30.0	30.0	550
Ho	64.8	26.3	40.2	0.231	—	—	—	—	520
Er	69.9	28.3	44.4	0.237	60	136	11.5	11.9	520
Tm	74.0	30.5	44.5	0.213	—	—	—	—	600
βYb	23.9	9.9	30.5	0.207	7	58	43.0	92.0	300
Lu	68.6	27.2	47.6	0.261	—	—	—	—	600

Note: For additional information, see Scott, T., in *Handbook on the Physics and Chemistry of Rare Earths,* Vol. 1, Gschneidner, K.A., Jr. and Eyring, L., Eds., North-Holland Physics, Amsterdam, 1978, 591.

[a] Value is questionable.
[b] Estimated.

TABLE 12. Liquid Metal Properties near the Melting Point

Rare earth metal	Density (g/cm³)	Surface tension (N/m)	Viscosity (centipoise)	Heat capacity (J/mol K)	Thermal conductivity (W/cm K)	Magnetic susceptibility χ × 10⁴ (emu/mol)	Electrical resistivity (μΩ·cm)	ΔV (l→s)[a] (%)	Spectral emittance at λ = 645 nm: ε (%)	Spectral emittance at λ = 645 nm: Temp. range (°C)
Sc	2.80	0.954	—	44.2[b]	—	—	—	—	—	—
Y	4.24	0.871	—	43.1	—	—	—	—	36.8	1522–1647
La	5.96	0.718	2.65	34.3	0.238	1.20	133	−0.6	25.4	920–1287
Ce	6.68	0.706	3.20	37.7	0.210	9.37	130	+1.1	32.2	877–1547
Pr	6.59	0.707	2.85	43.0	0.251	17.3	139	−0.02	28.4	931–1537
Nd	6.72	0.687	—	48.8	0.195	18.7	151	−0.9	39.4	1021–1567
Pm	6.9[b]	0.680[b]	—	50[b]	—	—	160[b]	—	—	—
Sm	7.16	0.431	—	50.2[b]	—	18.3	182	−3.6	43.7	1075
Eu	4.87	0.264	—	38.1	—	97	242	−4.8	—	—
Gd	7.4	0.664	—	37.2	0.149	67	195	−2.0	34.2	1313–1600
Tb	7.65	0.669	—	46.5	—	82	193	−3.1	—	—
Dy	8.2	0.648	—	49.9	0.187	95	210	−4.5	29.7	1412–1437
Ho	8.34	0.650	—	43.9	—	88	221	−7.4	—	—
Er	8.6	0.637	—	38.7	—	69	226	−9.0	37.2	1529–1587
Tm	9.0[b]	—	—	41.4	—	41	235[b]	−6.9	—	—
Yb	6.21	0.320	2.67	36.8	—	—	113	−5.1	—	—
Lu	9.3	0.940	—	47.9[b]	—	—	224	−3.6	—	—

Note: For additional information, see Van Zytveld, J., in *Handbook on the Physics and Chemistry of Rare Earths*, Vol. 12, Gschneidner, K.A., Jr. and Eyring, L., Eds., North-Holland Physics, Amsterdam, 1989, 357; Stretz, L.A. and Bautista, R.G., in *Temperature, Its Measurement and Control in Science and Industry*, Vol. 4, part I, H.H. Plumb, Ed., Instrument Society of America, Pittsburgh, 1972, 489; King, T.S., Baria, D.N., and Bautista, R.G., *Met. Trans. B*, 7, 411, 1976; Baria, D.N., King, T.S., and Bautista, R.G., *Met. Trans. B*, 7, 577, 1976.

[a] Volume change on freezing.

[b] Estimated.

TABLE 13. Ionization Potentials (Electronvolts)

Rare earth	I Neutral atom	II Singly ionized	III Doubly ionized	IV Triply ionized	V Quadruply ionized
Sc	6.56144	12.79967	24.75666	73.4894	91.65
Y	6.217	12.24	20.52	60.597	77.0
La	5.5770	11.060	19.1773	49.95	61.6
Ce	5.5387	10.85	20.198	36.758	65.55
Pr	5.464	10.55	21.624	38.98	57.53
Nd	5.5250	10.73	22.1	40.41	—
Pm	5.554	10.90	22.3	41.1	—
Sm	5.6437	11.07	23.4	41.4	—
Eu	5.6704	11.241	24.92	42.7	—
Gd	6.1500	12.09	20.63	44.0	—
Tb	5.8639	11.52	21.91	39.79	—
Dy	5.9389	11.67	22.8	41.47	—
Ho	6.0216	11.80	22.84	42.5	—
Er	6.1078	11.93	22.74	42.7	—
Tm	6.18431	12.05	23.68	42.7	—
Yb	6.25416	12.1761	25.05	43.56	—
Lu	5.42585	13.9	20.9594	45.25	66.8

Note: For references, see the table "Ionization Potentials of Atoms and Atomic Ions" in Section 10.

TABLE 14. Effective Ionic Radii (Å)[a]

Rare earth ion	R^{2+} CN = 6	R^{2+} CN = 8	R^{3+} CN = 6	R^{3+} CN = 8	R^{3+} CN = 12	R^{4+} CN = 6	R^{4+} CN = 8
Sc	—	—	0.745	0.87	1.116	—	—
Y	—	—	0.900	1.015	1.220	—	—
La	—	—	1.045	1.18	1.320	—	—
Ce	—	—	1.010	1.14	1.290	0.80	0.97
Pr	—	—	0.997	1.14	1.286	0.78	0.96
Nd	—	—	0.983	1.12	1.276	—	—
Pm	—	—	0.97	1.10	1.267	—	—
Sm	1.19	1.27	0.958	1.09	1.260	—	—
Eu	1.17	1.25	0.947	1.07	1.252	—	—
Gd	—	—	0.938	1.06	1.246	—	—
Tb	—	—	0.923	1.04	1.236	0.76	0.88
Dy	—	—	0.912	1.03	1.228	—	—
Ho	—	—	0.901	1.02	1.221	—	—
Er	—	—	0.890	1.00	1.214	—	—
Tm	—	—	0.880	0.99	1.207	—	—
Yb	1.00	1.07	0.868	0.98	1.199	—	—
Lu	—	—	0.861	0.97	1.194	—	—

Note: For additional information, see Shannon, R.D. and Prewitt, C.T., *Acta Cryst.*, 25, 925, 1969 and Shannon, R.D. and Prewitt, C.T., *Acta Cryst.*, 26, 1046, 1970.

[a] Radius of O^{2-} is 1.40 Å for a coordination number (CN) of 6.

MELTING, BOILING, TRIPLE, AND CRITICAL POINT TEMPERATURES OF THE ELEMENTS

This table summarizes the significant points on the phase diagrams for the elements for which data are available. Values are given for the solid–liquid–gas triple point t_{tp}, normal melting point t_m, normal boiling point t_b, and critical temperature t_c; all are on the ITS-90 scale. An "sp" notation indicates a sublimation point, where the vapor pressure of the solid phase reaches 101.325 kPa (1 atm). Transition temperatures between allotropic forms are included for several elements. The major data sources are listed below; values from Reference 1, which deals with reference points on the ITS-90 scale, were adopted when applicable.

References

1. Bedford, R. E., Bonnier, G., Maas, H., and Pavese, F., *Metrologia* 33, 133, 1996.
2. Dinsdale, A.T., SGTE Data for Pure Elements, *CALPHAD*, 15, 317–425, 1991.
3. Chase, M.W., Davies, C.A., Downey, J.R., Frurip, D.J., McDonald, R.A., and Syverud, A.N., *JANAF Thermochemical Tables, Third Edition, J. Phys. Chem. Ref. Data*, Vol. 14, Suppl. 1, 1985.
4. Gurvich, L.V., Veyts, I.V., and Alcock, C.B., *Thermodynamic Properties of Individual Substances, Fourth Edition,* Hemisphere Publishing Corp., New York, 1989.
5. Greenwood, N. N., and Earnshaw, A., *Chemistry of the Elements, Second Edition*, Butterworth-Heinemann, Oxford, 1997.

Element	t_{tp}/°C	t_m/°C	t_b/°C	t_c/°C
Actinium		1050	3198	
Aluminum		660.32	2519	6700
Americium		1176	2011	
Antimony		630.628	1587	
Argon	−189.36 (69 kPa)		−185.847	−122.28
Arsenic (gray)	817 (3.70 MPa)		616 sp	1400
Astatine		302		
Barium		727	1897	
Berkelium (β form)		986		
Beryllium		1287	2471	
Bismuth		271.406	1564	
Boron		2075	4000	
Bromine		−7.2	58.8	315
Cadmium		321.069	767	
Calcium		842	1484	
Californium		900		
Carbon (graphite)	4489 (10.3 MPa)		3825 sp	
Carbon (diamond)		4440 (12.4 GPa)		
Cerium		799	3443	
Cesium		28.5	671	1665
Chlorine		−101.5	−34.04	143.8
Chromium		1907	2671	
Cobalt		1495	2927	
Copper		1084.62	2562	
Curium		1345	~3100	
Dysprosium		1412	2567	
Einsteinium		860		
Erbium		1529	2868	
Europium		822	1529	
Fermium		1527		
Fluorine	−219.67		−188.12	−129.02
Francium		27		
Gadolinium		1313	3273	
Gallium	29.7666		2204	
Germanium		938.25	2833	
Gold		1064.18	2856	
Hafnium		2233	4603	
Helium			−268.93	−267.96
Holmium		1472	2700	
Hydrogen	−259.198 (7.2 kPa)	−259.1	−252.762	−240.18
Indium	156.5936	156.60	2072	
Iodine		113.7	184.4	546
Iridium		2446	4428	
Iron		1538	2861	
Krypton	−157.38 (73.2 kPa)		−153.34	−63.67
Lanthanum		920	3464	
Lawrencium		1627		
Lead		327.462	1749	

Element	t_{tp}/°C	t_m/°C	t_b/°C	t_c/°C
Lithium		180.50	1342	2950
Lutetium		1663	3402	
Magnesium		650	1090	
Manganese		1246	2061	
Mendelevium		827		
Mercury	−38.837	−38.8290	356.62	1491
Molybdenum		2623	4639	
Neodymium		1016	3074	
Neon	−248.609 (43 kPa)		−246.053	−228.7
Neptunium		644		
Nickel		1455	2913	
Niobium		2477	4744	
Nitrogen	−209.999	−210.0	−195.798	−146.94
Nobelium		827		
Osmium		3033	5012	
Oxygen		−218.79	−182.953	−118.56
Palladium		1554.8	2963	
Phosphorus (white)		44.15	280.5	721
Phosphorus (red)	590		431 sp	721
Phosphorus (black)		610		
Platinum		1768.2	3825	
Plutonium		640	3228	
Polonium		254	962	
Potassium		63.5	759	1950
Praseodymium		931	3520	
Promethium		1042	3000	
Protactinium		1572		
Radium		696		
Radon		−71	−61.7	104
Rhenium		3185	5596	
Rhodium		1964	3695	
Rubidium	39.26	39.30	688	1820
Ruthenium		2333	4150	
Samarium		1072	1794	
Scandium		1541	2836	
Selenium (vitreous)		180 (trans to gray)	685	1493
Selenium (gray)		220.8	685	1493
Silicon		1414	3265	
Silver		961.78	2162	
Sodium		97.794	882.940	2300
Strontium		777	1382	
Sulfur (rhombic)		95.3 (trans to monocl)	444.61	1041
Sulfur (monoclinic)		115.21	444.61	1041
Tantalum		3017	5458	
Technetium		2157	4265	
Tellurium		449.51	988	
Terbium		1359	3230	
Thallium		304	1473	
Thorium		1750	4788	
Thulium		1545	1950	
Tin (gray)		13.2 (trans to white)	2602	
Tin (white)		231.93	2602	
Titanium		1668	3287	
Tungsten		3422	5555	
Uranium		1135	4131	
Vanadium		1910	3407	
Xenon	−111.745 (81.6 kPa)		−108.09	16.58
Ytterbium		824	1196	
Yttrium		1522	3345	
Zinc		419.53	907	
Zirconium		1854.7	4409	

HEAT CAPACITY OF THE ELEMENTS AT 25 °C

This table gives the specific heat capacity (c_p) in J/g K and the molar heat capacity (C_p) in J/mol K at a temperature of 25 °C and a pressure of 100 kPa (1 bar or 0.987 standard atmospheres) for all the elements for which reliable data are available.

Name	c_p J/g K	C_p J/mol K
Actinium	0.120	27.2
Aluminum	0.897	24.20
Antimony	0.207	25.23
Argon	0.520	20.786
Arsenic	0.329	24.64
Barium	0.204	28.07
Beryllium	1.825	16.443
Bismuth	0.122	25.52
Boron	1.026	11.087
Bromine (Br_2)	0.474	75.69
Cadmium	0.232	26.020
Calcium	0.647	25.929
Carbon (graphite)	0.709	8.517
Cerium	0.192	26.94
Cesium	0.242	32.210
Chlorine (Cl_2)	0.479	33.949
Chromium	0.449	23.35
Cobalt	0.421	24.81
Copper	0.385	24.440
Dysprosium	0.173	28.16
Erbium	0.168	28.12
Europium	0.182	27.66
Fluorine (F_2)	0.824	31.304
Gadolinium	0.236	37.03
Gallium	0.373	26.03
Germanium	0.320	23.222
Gold	0.129	25.418
Hafnium	0.144	25.73
Helium	5.193	20.786
Holmium	0.165	27.15
Hydrogen (H_2)	14.304	28.836
Indium	0.233	26.74
Iodine (I_2)	0.214	54.43
Iridium	0.131	25.10
Iron	0.449	25.10
Krypton	0.248	20.786
Lanthanum	0.195	27.11
Lead	0.130	26.84
Lithium	3.582	24.860
Lutetium	0.154	26.86
Magnesium	1.023	24.869
Manganese	0.479	26.32
Mercury	0.140	27.983
Molybdenum	0.251	24.06
Neodymium	0.190	27.45
Neon	1.030	20.786
Nickel	0.444	26.07
Niobium	0.265	24.60
Nitrogen (N_2)	1.040	29.124
Osmium	0.130	24.7
Oxygen (O_2)	0.918	29.378
Palladium	0.246	25.98
Phosphorus (white)	0.769	23.824
Platinum	0.133	25.86
Potassium	0.757	29.600
Praseodymium	0.193	27.20
Radon	0.094	20.786
Rhenium	0.137	25.48
Rhodium	0.243	24.98
Rubidium	0.363	31.060
Ruthenium	0.238	24.06
Samarium	0.197	29.54
Scandium	0.568	25.52
Selenium	0.321	25.363
Silicon	0.712	19.99
Silver	0.235	25.350
Sodium	1.228	28.230
Strontium	0.306	26.79
Sulfur (rhombic)	0.708	22.70
Tantalum	0.140	25.36
Tellurium	0.202	25.73
Terbium	0.182	28.91
Thallium	0.129	26.32
Thorium	0.118	27.32
Thulium	0.160	27.03
Tin (white)	0.227	26.99
Titanium	0.523	25.060
Tungsten	0.132	24.27
Uranium	0.116	27.665
Vanadium	0.489	24.89
Xenon	0.158	20.786
Ytterbium	0.155	26.74
Yttrium	0.298	26.53
Zinc	0.388	25.390
Zirconium	0.278	25.36

VAPOR PRESSURE OF THE METALLIC ELEMENTS — EQUATIONS

C. B. Alcock

This table gives coefficients in an equation for the vapor pressure of 65 metallic elements in both the solid and liquid states. Vapor pressures in the range 10^{-10} to 10^2 Pa (10^{-15} to 10^{-3} atm) are covered. The equation is:

for p in atmospheres: $\log(p/\text{atm}) = A + BT^{-1} + C\log T + DT^{-3}$
for p in pascals: $\log(p/\text{Pa}) = 5.006 + A + BT^{-1} + C\log T + DT^{-3}$
for p in torr (mmHg): $\log(p/\text{torr}) = 2.881 + A + BT^{-1} + C\log T + DT^{-3}$

where T is the temperature in K.

This equation reproduces the observed vapor pressures to an accuracy of 5% or better. The metals are listed alphabetically by name, and the melting point is included.

The table following this one gives values of the vapor pressure at several temperatures in the 400 K to 2400 K range, as calculated from these equations.

Reprinted with permission of the publisher, Pergamon Press.

Reference

Alcock, C. B., Itkin, V. P., and Horrigan, M. K., *Canadian Metallurgical Quarterly*, 23, 309, 1984.

Element	Phase	*A*	*B*	*C*	*D*	Range/K	mp/K
Aluminum	Solid	9.459	-17342	-0.7927		298-mp	933
Aluminum	Liquid	5.911	-16211			mp-1800	
Americium	Solid	11.311	-15059	-1.3449		298-mp	1449
Barium	Solid	12.405	-9690	-2.2890		298-mp	1000
Barium	Liquid	4.007	-8163			mp-1200	
Beryllium	Solid	8.042	-17020	-0.4440		298-mp	1560
Beryllium	Liquid	5.786	-15731			mp-1800	
Cadmium	Solid	5.939	-5799			298-mp	594
Cadmium	Liquid	5.242	-5392			mp-650	
Calcium	Solid	10.127	-9517	-1.4030		298-mp	1115
Cerium	Solid	6.139	-21752			298-mp	1071
Cerium	Liquid	5.611	-21200			mp-2450	
Cesium	Solid	4.711	-3999			298-mp	302
Cesium	Liquid	4.165	-3830			mp-550	
Chromium	Solid	6.800	-20733	0.4391	-0.4094	298-2000	2180
Cobalt	Solid	10.976	-22576	-1.0280		298-mp	1768
Cobalt	Liquid	6.488	-20578			mp-2150	
Copper	Solid	9.123	-17748	-0.7317		298-mp	1358
Copper	Liquid	5.849	-16415			mp-1850	
Curium	Solid	8.369	-20364	-0.5770		298-mp	1618
Curium	Liquid	5.223	-18292			mp-2200	
Dysprosium	Solid	9.579	-15336	-1.1114		298-mp	1685
Erbium	Solid	9.916	-16642	-1.2154		298-mp	1802
Erbium	Liquid	4.668	-14380			mp-1900	
Europium	Solid	9.240	-9459	-1.1661		298-mp	1095
Gadolinium	Solid	8.344	-20861	-0.5775		298-mp	1586
Gadolinium	Liquid	5.557	-19389			mp-2250	
Gallium	Solid	6.657	-14208			298-mp	303
Gallium	Liquid	6.754	-13984	-0.3413		mp-1600	
Gold	Solid	9.152	-19343	-0.7479		298-mp	1337
Gold	Liquid	5.832	-18024			mp-2050	
Hafnium	Solid	9.445	-32482	-0.6735		298-mp	2506
Holmium	Solid	9.785	-15899	-1.1753		298-mp	1747
Indium	Solid	5.991	-12548			298-mp	430
Indium	Liquid	5.374	-12276			mp-1500	
Iridium	Solid	10.506	-35099	-0.7500		298-2500	2719
Iron	Solid	7.100	-21723	0.4536	-0.5846	298-mp	1811
Iron	Liquid	6.347	-19574			mp-2100	
Lanthanum	Solid	7.463	-22551	-0.3142		298-mp	1191
Lanthanum	Liquid	5.911	-21855			mp-2450	
Lead	Solid	5.643	-10143			298-mp	600
Lead	Liquid	4.911	-9701			mp-1200	
Lithium	Solid	5.667	-8310			298-mp	454
Lithium	Liquid	5.055	-8023			mp-1000	
Lutetium	Solid	8.793	-22423	-0.6200		298-mp	1936
Lutetium	Liquid	5.648	-20302			mp-2350	

Element	Phase	A	B	C	D	Range/K	mp/K
Magnesium	Solid	8.489	-7813	-0.8253		298-mp	923
Manganese	Solid	12.805	-15097	-1.7896		298-mp	1519
Mercury	Liquid	5.116	-3190			298-400	234
Molybdenum	Solid	11.529	-34626	-1.1331		298-2500	2895
Neodymium	Solid	8.996	-17264	-0.9519		298-mp	1294
Neodymium	Liquid	4.912	-15824			mp-2000	
Neptunium	Solid	19.643	-24886	-3.9991		298-mp	917
Neptunium	Liquid	10.076	-23378	-1.3250		mp-2500	
Nickel	Solid	10.557	-22606	-0.8717		298-mp	1728
Nickel	Liquid	6.666	-20765			mp-2150	
Niobium	Solid	8.882	-37818	-0.2575		298-2500	2750
Osmium	Solid	9.419	-41198	-0.3896		298-2500	3306
Palladium	Solid	9.502	-19813	-0.9258		298-mp	1828
Palladium	Liquid	5.426	-17899			mp 2100	
Platinum	Solid	4.882	-29387	1.1039	-0.4527	298-mp	2041
Platinum	Liquid	6.386	-26856			mp-2500	
Plutonium	Solid	26.160	-19162	-6.6675		298-600	913
Plutonium	Solid	18.858	-18460	1.1720		500-mp	
Plutonium	Liquid	3.666	-16658			mp-2450	
Potassium	Solid	4.961	-4646			298-mp	337
Potassium	Liquid	4.402	-4453			mp-600	
Praseodymium	Solid	8.859	-18720	-0.9512		298-mp	1204
Praseodymium	Liquid	4.772	-17315			mp-2200	
Protactinium	Solid	10.552	-34869	1.0075		298-mp	1845
Protactinium	Liquid	6.177	-32874			mp-2500	
Rhenium	Solid	11.543	-40726	-1.1629		298-2500	3459
Rhodium	Solid	10.168	-29010	-0.7068		298-mp	2236
Rhodium	Liquid	6.802	-26792			mp-2500	
Rubidium	Solid	4.857	-4215			298-mp	312
Rubidium	Liquid	4.312	-4040			mp-550	
Ruthenium	Solid	9.755	-34154	-0.4723		298-mp	2606
Samarium	Solid	9.988	-11034	-1.3287		298-mp	1347
Scandium	Solid	6.650	-19721	0.2885	-0.3663	298-mp	1814
Scandium	Liquid	5.795	-17681			mp-2000	
Silver	Solid	9.127	-14999	-0.7845		298-mp	1235
Silver	Liquid	5.752	-13827			mp-1600	
Sodium	Solid	5.298	-5603			298-mp	371
Sodium	Liquid	4.704	-5377			mp-700	
Strontium	Solid	9.226	-8572	-1.1926		298-mp	1050
Tantalum	Solid	16.807	-41346	-3.2152	0.7437	298-2500	3280
Terbium	Solid	9.510	-20457	-0.9247		298-mp	1629
Terbium	Liquid	5.411	-18639			mp-2200	
Thallium	Solid	5.971	-9447			298-mp	577
Thallium	Liquid	5.259	-9037			mp-1100	
Thorium	Solid	8.668	-31483	-0.5288		298-mp	2023
Thorium	Liquid	-18.453	-24569	6.6473		mp-2500	
Thulium	Solid	8.882	-12270	-0.9564		298-1400	1818
Tin	Solid	6.036	-15710			298-mp	505
Tin	Liquid	5.262	-15332			mp-1850	
Titanium	Solid	11.925	-24991	-1.3376		298-mp	1943
Titanium	Liquid	6.358	-22747			mp-2400	
Tungsten	Solid	2.945	-44094	1.3677		298-2350	3687
Tungsten	Solid	54.527	-57687	-12.2231		2200-2500	
Uranium	Solid	0.770	-27729	2.6982	-1.5471	298-mp	1408
Uranium	Liquid	20.735	-28776	-4.0962		mp-2500	
Vanadium	Solid	9.744	-27132	-0.5501		298-mp	2183
Vanadium	Liquid	6.929	-25011			mp-2500	
Ytterbium	Solid	9.111	-8111	-1.0849		298-900	1092
Yttrium	Solid	9.735	-22306	-0.8705		298-mp	1795
Yttrium	Liquid	5.795	-20341			mp-2300	
Zinc	Solid	6.102	-6776			298-mp	693
Zinc	Liquid	5.378	-6286			mp-750	
Zirconium	Solid	10.008	-31512	-0.7890		298-m.p	2127
Zirconium	Liquid	6.806	-30295			mp-2500	

VAPOR PRESSURE OF THE METALLIC ELEMENTS — DATA

The following values of the vapor pressure of metallic elements are calculated from the equations in the preceding table. All values are given in pascals. For conversion, note that 1 Pa = 7.50 μmHg = $9.87 \cdot 10^{-6}$ atm.

Metal	mp/K	Vapor Pressure in Pa										
		400 K	600 K	800 K	1000 K	1200 K	1400 K	1600 K	1800 K	2000 K	2200 K	2400 K
Aluminum	933			3.06×10^{-10}	5.08×10^{-6}	0.00256	0.218	6.10	81.4			
Americium	1449			3.88×10^{-7}	0.00167	0.423	21.35					
Barium	1000		7.97×10^{-6}	0.0450	7.11	162						
Beryllium	1560			3.04×10^{-10}	4.96×10^{-6}	0.00314	0.312	9.12	113			
Cadmium	594	0.000280	18.2									
Calcium	1115		2.36×10^{-5}	0.146	25.5							
Cerium	1071				2.47×10^{-11}	8.91×10^{-8}	2.97×10^{-5}	0.00233	0.0691	1.04	9.56	60.8
Cesium	302	0.394										
Chromium	2180				2.45×10^{-8}	7.59×10^{-5}	0.0239	1.80	52.1	774		
Cobalt	1768				2.09×10^{-10}	1.00×10^{-6}	0.000419	0.0379	1.15	16.0		
Copper	1358			6.60×10^{-11}	1.53×10^{-6}	0.00122	0.135	3.94	54.4			
Curium	1618				1.90×10^{-9}	4.24×10^{-6}	0.00103	0.0629	1.17	12.1	82.1	
Dysprosium	1685			1.54×10^{-8}	8.21×10^{-5}	0.0241	1.362	27.5				
Erbium	1802			3.90×10^{-10}	4.30×10^{-6}	0.00205	0.163	4.23	52.5			
Europium	1095		1.74×10^{-5}	0.109	19.4							
Gadolinium	1586				5.70×10^{-10}	1.54×10^{-6}	0.000429	0.0279	0.618	7.39	56.2	
Gallium	303			1.94×10^{-7}	0.000565	0.114	4.98	84.4				
Gold	1337				3.72×10^{-8}	5.44×10^{-5}	0.00920	0.374	6.68	67.0		
Hafnium	2506						1.35×10^{-11}	9.81×10^{-9}	1.63×10^{-6}	9.69×10^{-5}	0.00272	0.0437
Holmium	1747			3.20×10^{-9}	2.32×10^{-5}	0.00837	0.546	12.3				
Indium	430		8.31×10^{-11}	1.08×10^{-5}	0.0127	1.413	40.9					
Iridium	2719							1.48×10^{-9}	3.72×10^{-7}	3.06×10^{-5}	0.00112	0.0225
Iron	1811				5.54×10^{-9}	2.51×10^{-5}	0.0104	0.961	32.7	36.8		
Lanthanum	1191					5.09×10^{-8}	2.02×10^{-5}	0.00181	0.0596	0.976	9.61	64.7
Lead	601		5.54×10^{-7}	0.00618	1.64	68.1						
Lithium	454	7.90×10^{-11}	0.000489	1.08	109							
Lutetium	1936				3.28×10^{-11}	1.59×10^{-7}	6.79×10^{-5}	0.00628	0.211	3.18	26.7	
Magnesium	923	6.53×10^{-9}	0.0152	21.5								
Manganese	1519			5.55×10^{-7}	0.00221	0.524	24.9					
Mercury	234	140										
Molybdenum	2895							1.83×10^{-9}	4.07×10^{-7}	3.03×10^{-5}	0.00102	0.0189
Neodymium	1294			4.55×10^{-11}	7.62×10^{-7}	0.000483	0.0412	1.07	13.4	101		
Neptunium	917					3.31×10^{-9}	1.63×10^{-6}	0.000168	0.00604	0.105	1.06	7.28
Nickel	1728				2.19×10^{-10}	1.09×10^{-6}	0.000471	0.0438	1.37	19.5		
Niobium	2750							2.32×10^{-11}	9.54×10^{-9}	1.17×10^{-6}	5.98×10^{-5}	0.00158
Osmium	3306								1.85×10^{-10}	3.46×10^{-8}	2.49×10^{-6}	8.75×10^{-5}
Palladium	1828				8.27×10^{-9}	1.40×10^{-5}	0.00277	0.144	3.07	30.4		
Platinum	2041						2.34×10^{-8}	1.14×10^{-5}	0.00143	0.0689	0.153	1.59
Plutonium	913				1.03×10^{-8}	6.17×10^{-6}	0.000594	0.0182	0.262	2.20	12.6	53.8
Potassium	337	0.0188	96.9									
Praseodymium	1204				1.95×10^{-8}	2.16×10^{-5}	0.00257	0.0904	1.44	13.2	80.8	
Protactinium	1845							3.44×10^{-10}	8.06×10^{-8}	5.57×10^{-6}	0.000174	0.00306
Rhenium	3459								1.37×10^{-10}	2.22×10^{-8}	1.41×10^{-6}	4.45×10^{-5}
Rhodium	2236						1.69×10^{-8}	5.99×10^{-6}	0.000571	0.0217	0.422	4.41
Rubidium	312	0.165										
Ruthenium	2606							7.96×10^{-9}	1.77×10^{-6}	0.000133	0.00455	0.0858
Samarium	1347		8.17×10^{-8}	0.00221	0.942	51.0						
Scandium	1814				6.31×10^{-8}	0.000129	0.0300	1.80	43.6	91.3		
Silver	1235			1.27×10^{-7}	0.000603	0.165	7.61	131				
Sodium	371	0.000185	5.60									
Strontium	1050	4.99×10^{-11}	0.000429	1.134	121							
Tantalum	3280									3.36×10^{-10}	1.87×10^{-8}	5.21×10^{-7}
Terbium	1629				1.92×10^{-9}	4.18×10^{-6}	0.000988	0.0585	1.15	12.5	88.0	
Thallium	577		1.59×10^{-5}	0.0931	16.9							
Thorium	2023						3.33×10^{-11}	2.00×10^{-8}	2.89×10^{-6}	0.000154	0.00401	0.0610
Thulium	1818		6.03×10^{-10}	5.94×10^{-5}	0.0561	5.22	130					
Tin	505			1.26×10^{-9}	8.62×10^{-6}	0.00310	0.207	4.85	56.3			
Titanium	1943					9.69×10^{-9}	7.44×10^{-6}	0.00106	0.0493	0.978	10.6	76.9
Tungsten	3687									2.62×10^{-10}	3.01×10^{-8}	1.59×10^{-6}
Uranium	1408					9.47×10^{-10}	2.87×10^{-6}	4.27×10^{-6}	0.000263	0.00678	0.0933	0.803
Vanadium	2183					2.79×10^{-10}	4.35×10^{-7}	0.000107	0.00769	0.233	3.68	32.6
Ytterbium	1092	1.03×10^{-9}	0.00384	6.74								
Yttrium	1795				6.66×10^{-11}	2.96×10^{-7}	0.000117	0.0102	0.316	4.27	35.9	
Zinc	693	1.47×10^{-6}	0.653									
Zirconium	2127						1.05×10^{-10}	6.17×10^{-8}	8.68×10^{-6}	0.000450	0.0110	0.155

DENSITY OF MOLTEN ELEMENTS AND REPRESENTATIVE SALTS

This table lists the liquid density at the melting point, ρ_m, for elements that are solid at room temperature, as well as for some representative salts of these elements. Densities at higher temperatures (up to the t_{max} given in the last column) may be estimated from the equation

$$\rho(t) = \rho_m - k(t - t_m)$$

where t_m is the melting point and k is given in the fifth column of the table. If a value of t_{max} is not given, the equation should not be used to extrapolate more than about 20 °C beyond the melting point.

Data for the elements were selected from the primary literature; the assistance of Gernot Lang in compiling these data is gratefully acknowledged. The molten salt data were derived from Reference 1.

References

1. Janz, G. J., Thermodynamic and Transport Properties of Molten Salts: Correlation Equations for Critically Evaluated Density, Surface Tension, Electrical Conductance, and Viscosity Data, *J. Phys. Chem. Ref. Data*, 17, Suppl. 2, 1988.
2. Nasch, P. M., and Steinemann, S. G., *Phys. Chem. Liq.*, 29, 43, 1995.

Formula	Name	t_m/°C	ρ_m/g cm^{-3}	k/g cm^{-3} °C^{-1}	t_{max}
Ag	Silver	961.78	9.320	0.0009	1500
$AgBr$	Silver(I) bromide	430	5.577	0.001035	667
$AgCl$	Silver(I) chloride	455	4.83	0.00094	627
AgI	Silver(I) iodide	558	5.58	0.00101	802
$AgNO_3$	Silver(I) nitrate	210	3.970	0.001098	360
Ag_2SO_4	Silver(I) sulfate	660	4.84	0.001089	770
Al	Aluminum	660.32	2.375	0.000233	1340
$AlBr_3$	Aluminum bromide	97.5	2.647	0.002435	267
$AlCl_3$	Aluminum chloride	192.6	1.302	0.002711	296
AlI_3	Aluminum iodide	188.32	3.223	0.0025	240
As	Arsenic	817	5.22	0.000544	
Au	Gold	1064.18	17.31	0.001343	1200
B	Boron	2075	2.08		
Ba	Barium	727	3.338	0.000299	1550
$BaBr_2$	Barium bromide	857	3.991	0.000924	900
$BaCl_2$	Barium chloride	961	3.174	0.000681	1081
BaF_2	Barium fluoride	1368	4.14	0.000999	1727
BaI_2	Barium iodide	711	4.26	0.000977	975
Be	Beryllium	1287	1.690	0.00011	
$BeCl_2$	Beryllium chloride	415	1.54	0.0011	473
BeF_2	Beryllium fluoride	552	1.96	0.000015	850
Bi	Bismuth	271.406	10.05	0.00135	800
$BiBr_3$	Bismuth bromide	219	4.76	0.002637	927
$BiCl_3$	Bismuth chloride	234	3.916	0.0023	350
Ca	Calcium	842	1.378	0.000230	1484
$CaBr_2$	Calcium bromide	742	3.111	0.0005	791
$CaCl_2$	Calcium chloride	775	2.085	0.000422	950
CaF_2	Calcium fluoride	1418	2.52	0.000391	2027
CaI_2	Calcium iodide	783	3.443	0.000751	1028
Cd	Cadmium	321.069	7.996	0.001218	500
$CdBr_2$	Cadmium bromide	568	4.075	0.00108	720
$CdCl_2$	Cadmium chloride	568	3.392	0.00082	807
CdI_2	Cadmium iodide	388	4.396	0.001117	700
Ce	Cerium	799	6.55	0.000710	1460
$CeCl_3$	Cerium(III) chloride	807	3.25	0.00092	950
CeF_3	Cerium(III) fluoride	1430	4.659	0.000936	1927
Co	Cobalt	1495	7.75	0.00165	1580
Cr	Chromium	1907	6.3	0.0011	2100
Cs	Cesium	28.44	1.843	0.000556	510
$CsBr$	Cesium bromide	636	3.133	0.001223	860
$CsCl$	Cesium chloride	646	2.79	0.001065	906
CsF	Cesium fluoride	703	3.649	0.001282	912
CsI	Cesium iodide	632	3.197	0.001183	907
$CsNO_3$	Cesium nitrate	409	2.820	0.001166	491
Cs_2SO_4	Cesium sulfate	1005	3.1	0.00095	1530
Cu	Copper	1084.62	8.02	0.000609	1630
$CuCl$	Copper(I) chloride	423	3.692	0.00076	585

Formula	Name	t_m/°C	ρ_m/g cm^{-3}	k/g cm^{-3} °C^{-1}	t_{max}
Dy	Dysprosium	1411	8.37	0.00143	1540
$DyCl_3$	Dysprosium(III) chloride	718	3.62	0.00068	987
Er	Erbium	1529	8.86	0.00157	1700
Eu	Europium	822	5.13	0.0028	980
Fe	Iron	1538	6.98	0.000572	1680
$FeCl_2$	Iron(II) chloride	677	2.348	0.000555	877
Ga	Gallium	29.7666	6.08	0.00062	400
$GaBr_3$	Gallium(III) bromide	123	3.116	0.00246	135
$GaCl_3$	Gallium(III) chloride	77.9	2.053	0.002083	141
GaI_3	Gallium(III) iodide	212	3.630	0.002377	252
Gd	Gadolinium	1314	7.4		
$GdCl_3$	Gadolinium(III) chloride	602	3.56	0.000671	1007
GdI_3	Gadolinium(III) iodide	930	4.12	0.000908	1032
Ge	Germanium	938.25	5.60	0.00055	1600
Hf	Hafnium	2233	12		
$HgBr_2$	Mercury(II) bromide	241	5.126	0.003233	319
$HgCl_2$	Mercury(II) chloride	277	4.368	0.002862	304
HgI_2	Mercury(II) iodide	256	5.222	0.003235	354
Ho	Holmium	1472	8.34		
In	Indium	156.60	7.02	0.000836	500
$InBr_3$	Indium(III) bromide	420	3.121	0.0015	528
$InCl_3$	Indium(III) chloride	583	2.140	0.0021	666
InI_3	Indium(III) iodide	207	3.820	0.0015	360
Ir	Iridium	2446	19		
K	Potassium	63.38	0.828	0.000232	500
KBr	Potassium bromide	734	2.127	0.000825	930
KCl	Potassium chloride	771	1.527	0.000583	939
KF	Potassium fluoride	858	1.910	0.000651	1037
KI	Potassium iodide	681	2.448	0.000956	904
KNO_3	Potassium nitrate	334	1.865	0.000723	457
La	Lanthanum	920	5.94	0.00061	1600
$LaBr_3$	Lanthanum bromide	788	4.933	0.000096	912
$LaCl_3$	Lanthanum chloride	858	3.209	0.000777	973
LaF_3	Lanthanum fluoride	1493	4.589	0.000682	2177
LaI_3	Lanthanum iodide	778	4.29	0.001110	907
Li	Lithium	180.5	0.512	0.00052	285
LiBr	Lithium bromide	550	2.528	0.000652	739
LiCl	Lithium chloride	610	1.502	0.000432	781
LiF	Lithium fluoride	848.2	1.81	0.000490	1047
LiI	Lithium iodide	469	3.109	0.000917	667
$LiNO_3$	Lithium nitrate	253	1.781	0.000546	441
Li_2SO_4	Lithium sulfate	860	2.003	0.000407	1214
Lu	Lutetium	1663	9.3		
Mg	Magnesium	650	1.584	0.000234	900
$MgBr_2$	Magnesium bromide	711	2.62	0.000478	935
$MgCl_2$	Magnesium chloride	714	1.68	0.000271	826
MgI_2	Magnesium iodide	634	3.05	0.000651	888
Mn	Manganese	1246	5.95	0.00105	1590
$MnCl_2$	Manganese(II) chloride	650	2.353	0.000437	850
Mo	Molybdenum	2623	9.33		
Na	Sodium	97.794	0.927	0.00023	600
NaBr	Sodium bromide	747	2.342	0.000816	945
Na_2CO_3	Sodium carbonate	856	1.972	0.000448	1004
NaCl	Sodium chloride	800.7	1.556	0.000543	1027
NaF	Sodium fluoride	996	1.948	0.000636	1097
NaI	Sodium iodide	661	2.742	0.000949	912
$NaNO_3$	Sodium nitrate	306.5	1.90	0.000715	370
Na_2SO_4	Sodium sulfate	884	2.069	0.000483	1077
Nd	Neodymium	1016	6.89	0.00076	1350
Ni	Nickel	1455	7.81	0.000726	1700
$NiCl_2$	Nickel(II) chloride	1031	2.653	0.00066	1057
Os	Osmium	3033	20		
Pb	Lead	327.462	10.66	0.00122	700
$PbBr_2$	Lead(II) bromide	371	5.73	0.00165	600
$PbCl_2$	Lead(II) chloride	501	4.951	0.0015	710

Formula	Name	t_m/°C	ρ_m/g cm^{-3}	k/g cm^{-3} °C^{-1}	t_{max}
PbI_2	Lead(II) iodide	410	5.691	0.001594	697
Pd	Palladium	1554.8	10.38	0.001169	1700
Pr	Praseodymium	931	6.50	0.00093	1460
$PrCl_3$	Praseodymium chloride	786	3.23	0.00074	977
Pt	Platinum	1768.2	19.77	0.0024	2200
Pu	Plutonium	640	16.63	0.001419	950
Rb	Rubidium	39.31	1.46	0.000451	800
RbBr	Rubidium bromide	692	2.715	0.001072	907
Rb_2CO_3	Rubidium carbonate	837	2.84	0.000640	1007
RbCl	Rubidium chloride	724	2.248	0.000883	923
RbF	Rubidium fluoride	795	2.87	0.00102	1067
RbI	Rubidium iodide	656	2.904	0.001143	902
$RbNO_3$	Rubidium nitrate	310	2.519	0.001068	417
Rb_2SO_4	Rubidium sulfate	1066	2.56	0.000665	1545
Re	Rhenium	3185	18.9		
Rh	Rhodium	1964	10.7	0.000895	2200
Ru	Ruthenium	2334	10.65		
S	Sulfur	115.21	1.819	0.00080	160
Sb	Antimony	630.628	6.53	0.00067	745
$SbCl_3$	Antimony(III) chloride	73.4	2.681	0.002293	77
$SbCl_5$	Antimony(V) chloride	4	2.37	0.001869	77
SbI_3	Antimony(III) iodide	171	4.171	0.002483	322
Sc	Scandium	1541	2.80		
Se	Selenium	220.8	3.99		
Si	Silicon	1414	2.57	0.00036	1500
Sm	Samarium	1072	7.16		
Sn	Tin	231.93	6.99	0.000601	1200
$SnCl_2$	Tin(II) chloride	247	3.36	0.001253	480
$SnCl_4$	Tin(IV) chloride	−33	2.37	0.002687	138
Sr	Strontium	777	6.980		
$SrBr_2$	Strontium bromide	657	3.70	0.000745	1004
$SrCl_2$	Strontium chloride	874	2.727	0.000578	1037
SrF_2	Strontium fluoride	1477	3.470	0.000751	1927
SrI_2	Strontium iodide	538	4.085	0.000885	1026
Ta	Tantalum	3017	15		
$TaCl_5$	Tantalum(V) chloride	216.6	2.700	0.004316	457
Tb	Terbium	1359	7.65		
Te	Tellurium	449.51	5.70	0.00035	600
$ThCl_4$	Thorium chloride	770	3.363	0.0014	847
ThF_4	Thorium fluoride	1110	6.058	0.000759	1378
Ti	Titanium	1668	4.11		
$TiCl_4$	Titanium(IV) chloride	-25	1.807	0.001735	137
Tl	Thallium	304	11.22	0.00144	600
TlBr	Thallium(I) bromide	460	5.98	0.001755	647
TlCl	Thallium(I) chloride	431	5.628	0.0018	642
TlI	Thallium(I) iodide	441.8	6.15	0.001761	737
$TlNO_3$	Thallium(I) nitrate	206	4.91	0.001873	279
Tl_2SO_4	Thallium(I) sulfate	632	5.62	0.00130	927
Tm	Thulium	1545	8.56	0.00050	1675
U	Uranium	1135	17.3		
UCl_3	Uranium(III) chloride	837	4.84	0.007943	1057
UCl_4	Uranium(IV) chloride	590	3.572	0.001945	667
UF_4	Uranium(IV) fluoride	1036	6.485	0.000992	1341
V	Vanadium	1910	5.5		
W	Tungsten	3422	17.6		
Y	Yttrium	1526	4.24		
YCl_3	Yttrium chloride	721	2.510	0.0005	845
Yb	Ytterbium	824	6.21		
Zn	Zinc	419.53	6.57	0.0011	700
$ZnBr_2$	Zinc bromide	402	3.47	0.000959	602
$ZnCl_2$	Zinc chloride	290	2.54	0.00053	557
ZnI_2	Zinc iodide	450	3.878	0.00136	588
$ZnSO_4$	Zinc sulfate	680	3.14	0.00047	987
Zr	Zirconium	1854.7	5.8		
$ZrCl_4$	Zirconium chloride	437	1.643	0.007464	492

MAGNETIC SUSCEPTIBILITY OF THE ELEMENTS AND INORGANIC COMPOUNDS

When a material is placed in a magnetic field H, a magnetization (magnetic moment per unit volume) M is induced in the material which is related to H by $M = \kappa H$, where κ is called the volume susceptibility. Since H and M have the same dimensions, κ is dimensionless. A more useful parameter is the molar susceptibility χ_m, defined by

$$\chi_m = \kappa V_m = \kappa M / \rho$$

where V_m is the molar volume of the substance, M the molar mass, and ρ the mass density. When the cgs system is used, the customary units for χ_m are $cm^3\ mol^{-1}$; the corresponding SI units are $m^3\ mol^{-1}$.

Substances that have no unpaired electron orbital or spin angular momentum generally have negative values of χ_m and are called diamagnetic. Their molar susceptibility varies only slightly with temperature. Substances with unpaired electrons, which are termed paramagnetic, have positive χ_m and show a much stronger temperature dependence, varying roughly as $1/T$. The net susceptibility of a paramagnetic substance is the sum of the paramagnetic and diamagnetic contributions, but the former almost always dominates.

This table gives values of χ_m for the elements and selected inorganic compounds. All values refer to nominal room temperature (285 to 300 K) unless otherwise indicated. When the physical state (s = solid, l = liquid, g = gas, aq = aqueous solution) is not given, the most common crystalline form is understood. An entry of Ferro. indicates a ferromagnetic substance.

Substances are arranged in alphabetical order by the most common name, except that compounds such as hydrides, oxides, and acids are grouped with the parent element (the same ordering used in the table "Physical Constants of Inorganic Compounds").

In keeping with customary practice, the molar susceptibility is given here in units appropriate to the cgs system. These values should be multiplied by 4π to obtain values for use in SI equations (where the magnetic field strength H has units of A m^{-1}).

References

1. *Landolt-Börnstein, Numerical Data and Functional Relationships in Science and Technology, New Series,* II/16, *Diamagnetic Susceptibility,* Springer-Verlag, Heidelberg, 1986.
2. *Landolt-Börnstein, Numerical Data and Functional Relationships in Science and Technology, New Series,* III/19, Subvolumes a to i2, *Magnetic Properties of Metals,* Springer-Verlag, Heidelberg, 1986-1992.
3. *Landolt-Börnstein, Numerical Data and Functional Relationships in Science and Technology, New Series,* II/2, II/8, II/10, II/11,and II/12a, *Coordination and Organometallic Transition Metal Compounds,* Springer-Verlag, Heidelberg, 1966-1984.
4. *Tables de Constantes et Données Numérique, Volume 7, Relaxation Paramagnetique,* Masson, Paris, 1957.

Name	Formula	$\chi_m/10^{-6}$ cm³ mol⁻¹
Aluminum	Al	+16.5
Aluminum trifluoride	AlF_3	−13.9
Aluminum oxide	Al_2O_3	−37
Aluminum sulfate	$Al_2(SO_4)_3$	−93
Ammonia (g)	NH_3	−16.3
Ammonia (aq)	NH_3	−18.3
Ammonium acetate	$NH_4C_2H_3O_2$	−41.1
Ammonium bromide	NH_4Br	−47
Ammonium carbonate	$(NH_4)_2CO_3$	−42.5
Ammonium chlorate	NH_4ClO_3	−42.1
Ammonium chloride	NH_4Cl	−36.7
Ammonium fluoride	NH_4F	−23
Ammonium iodate	NH_4IO_3	−62.3
Ammonium iodide	NH_4I	−66
Ammonium nitrate	NH_4NO_3	−33
Ammonium sulfate	$(NH_4)_2SO_4$	−67
Ammonium thiocyanate	NH_4SCN	−48.1
Antimony	Sb	−99
Stibine (g)	SbH_3	−34.6
Antimony(III) bromide	$SbBr_3$	−111.4
Antimony(III) chloride	$SbCl_3$	−86.7
Antimony(III) fluoride	SbF_3	−46
Antimony(III) iodide	SbI_3	−147.2
Antimony(III) oxide	Sb_2O_3	−69.4
Antimony(III) sulfide	Sb_2S_3	−86
Antimony(V) chloride	$SbCl_5$	−120.5
Argon (g)	Ar	−19.32
Arsenic (gray)	As	−5.6
Arsenic (yellow)	As	−23.2
Arsine (g)	AsH_3	−35.2
Arsenic(III) bromide	$AsBr_3$	−106
Arsenic(III) chloride	$AsCl_3$	−72.5
Arsenic(III) iodide	AsI_3	−142.2
Arsenic(III) oxide	As_2O_3	−30.34
Arsenic(III) sulfide	As_2S_3	−70
Barium	Ba	+20.6
Barium bromide	$BaBr_2$	−92
Barium bromide dihydrate	$BaBr_2{\cdot}2H_2O$	−119.3
Barium carbonate	$BaCO_3$	−58.9
Barium chloride	$BaCl_2$	−72.6
Barium chloride dihydrate	$BaCl_2{\cdot}2H_2O$	−100
Barium fluoride	BaF_2	−51
Barium hydroxide	$Ba(OH)_2$	−53.2
Barium iodate	$Ba(IO_3)_2$	−122.5
Barium iodide	BaI_2	−124.4
Barium iodide dihydrate	$BaI_2{\cdot}2H_2O$	−163
Barium nitrate	$Ba(NO_3)_2$	−66.5
Barium oxide	BaO	−29.1
Barium peroxide	BaO_2	−40.6
Barium sulfate	$BaSO_4$	−65.8
Beryllium	Be	−9.0
Beryllium chloride	$BeCl_2$	−26.5
Beryllium hydroxide	$Be(OH)_2$	−23.1
Beryllium oxide	BeO	−11.9
Beryllium sulfate	$BeSO_4$	−37
Bismuth	Bi	−280.1
Bismuth tribromide	$BiBr_3$	−147
Bismuth trichloride	$BiCl_3$	−26.5
Bismuth fluoride	BiF_3	−61.2
Bismuth hydroxide	$Bi(OH)_3$	−65.8

Name	Formula	$\chi_m/10^{-6}$ cm³ mol⁻¹
Bismuth triiodide	BiI_3	–200.5
Bismuth nitrate pentahydrate	$Bi(NO_3)_3{\cdot}5H_2O$	–159
Bismuth oxide	Bi_2O_3	–83
Bismuth phosphate	$BiPO_4$	–77
Bismuth sulfate	$Bi_2(SO_4)_3$	–199
Bismuth sulfide	Bi_2S_3	–123
Boron	B	–6.7
Diborane (g)	B_2H_6	–21.0
Boric acid (orthoboric acid)	H_3BO_3	–34.1
Boron trichloride	BCl_3	–59.9
Boron oxide	B_2O_3	–38.7
Bromine (l)	Br_2	–56.4
Bromine (g)	Br_2	–73.5
Bromine trifluoride	BrF_3	–33.9
Bromine pentafluoride	BrF_5	–45.1
Cadmium	Cd	–19.7
Cadmium bromide	$CdBr_2$	–87.3
Cadmium bromide tetrahydrate	$CdBr_2{\cdot}4H_2O$	–131.5
Cadmium carbonate	$CdCO_3$	–46.7
Cadmium chloride	$CdCl_2$	–68.7
Cadmium chromate	$CdCrO_4$	–16.8
Cadmium cyanide	$Cd(CN)_2$	–54
Cadmium fluoride	CdF_2	–40.6
Cadmium hydroxide	$Cd(OH)_2$	–41
Cadmium iodate	$Cd(IO_3)_2$	–108.4
Cadmium iodide	CdI_2	–117.2
Cadmium nitrate	$Cd(NO_3)_2$	–55.1
Cadmium nitrate tetrahydrate	$Cd(NO_3)_2{\cdot}4H_2O$	–140
Cadmium oxide	CdO	–30
Cadmium sulfate	$CdSO_4$	–59.2
Cadmium sulfide	CdS	–50
Calcium	Ca	+40
Calcium bromide	$CaBr_2$	–73.8
Calcium carbonate	$CaCO_3$	–38.2
Calcium chloride	$CaCl_2$	–54.7
Calcium fluoride	CaF_2	–28
Calcium hydroxide	$Ca(OH)_2$	–22
Calcium iodate	$Ca(IO_3)_2$	–101.4
Calcium iodide	CaI_2	–109
Calcium oxide	CaO	–15.0
Calcium sulfate	$CaSO_4$	–49.7
Calcium sulfate dihydrate	$CaSO_4{\cdot}2H_2O$	–74
Carbon (diamond)	C	–5.9
Carbon (graphite)	C	–6.0
Carbon monoxide (g)	CO	–11.8
Carbon dioxide (g)	CO_2	–21.0
Cerium (β)	Ce	+2500
Cerium(II) sulfide	CeS	+2110
Cerium(III) chloride	$CeCl_3$	+2490
Cerium(III) fluoride	CeF_3	+2190
Cerium(III) sulfide	Ce_2S_3	+5080
Cerium(IV) oxide	CeO_2	+26
Cerium(IV) sulfate tetrahydrate	$Ce(SO_4)_2{\cdot}4H_2O$	–97
Cesium	Cs	+29
Cesium bromate	$CsBrO_3$	–75.1
Cesium bromide	CsBr	–67.2
Cesium carbonate	Cs_2CO_3	–103.6
Cesium chlorate	$CsClO_3$	–65
Cesium chloride	CsCl	–56.7
Cesium fluoride	CsF	–44.5
Cesium iodide	CsI	–82.6
Cesium superoxide	CsO_2	+1534
Cesium sulfate	Cs_2SO_4	–116
Chlorine (l)	Cl_2	–40.4
Chlorine trifluoride (g)	ClF_3	–26.5
Chromium	Cr	+167
Chromium(II) chloride	$CrCl_2$	+7230
Chromium(III) chloride	$CrCl_3$	+6350
Chromium(III) fluoride	CrF_3	+4370
Chromium(III) oxide	Cr_2O_3	+1960
Chromium(III) sulfate	$Cr_2(SO_4)_3$	+11800
Chromium(VI) oxide	CrO_3	+40
Cobalt	Co	Ferro.
Cobalt(II) bromide	$CoBr_2$	+13000
Cobalt(II) chloride	$CoCl_2$	+12660
Cobalt(II) chloride hexahydrate	$CoCl_2{\cdot}6H_2O$	+9710
Cobalt(II) cyanide	$Co(CN)_2$	+3825
Cobalt(II) fluoride	CoF_2	+9490
Cobalt(II) iodide	CoI_2	+10760
Cobalt(II) sulfate	$CoSO_4$	+10000
Cobalt(II) sulfide	CoS	+225
Cobalt(II,III) oxide	Co_3O_4	+7380
Cobalt(III) fluoride	CoF_3	+1900
Cobalt(III) oxide	Co_2O_3	+4560
Copper	Cu	–5.46
Copper(I) bromide	CuBr	49
Copper(I) chloride	CuCl	–40
Copper(I) cyanide	CuCN	–24
Copper(I) iodide	CuI	–63
Copper(I) oxide	Cu_2O	–20
Copper(II) bromide	$CuBr_2$	+685
Copper(II) chloride	$CuCl_2$	+1080
Copper(II) chloride dihydrate	$CuCl_2{\cdot}2H_2O$	+1420
Copper(II) fluoride	CuF_2	+1050
Copper(II) fluoride dihydrate	$CuF_2{\cdot}2H_2O$	+1600
Copper(II) hydroxide	$Cu(OH)_2$	+1170
Copper(II) nitrate trihydrate	$Cu(NO_3)_2{\cdot}3H_2O$	+1570
Copper(II) nitrate hexahydrate	$Cu(NO_3)_2{\cdot}6H_2O$	+1625
Copper(II) oxide	CuO	+238
Copper(II) sulfate	$CuSO_4$	+1330
Copper(II) sulfate pentahydrate	$CuSO_4{\cdot}5H_2O$	+1460
Copper(II) sulfide	CuS	–2.0
Dysprosium (α)	Dy	+98000
Dysprosium(III) oxide	Dy_2O_3	+89600
Dysprosium(III) sulfide	Dy_2S_3	+95200
Erbium	Er	+48000
Erbium oxide	Er_2O_3	+73920
Erbium sulfate octahydrate	$Er_2(SO_4)_3{\cdot}8H_2O$	+74600
Erbium sulfide	Er_2S_3	+77200
Europium	Eu	+30900
Europium(II) bromide	$EuBr_2$	+26800
Europium(II) chloride	$EuCl_2$	+26500
Europium(II) fluoride	EuF_2	+23750
Europium(II) iodide	EuI_2	+26000
Europium(II) sulfide	EuS	+23800
Europium(III) oxide	Eu_2O_3	+10100
Europium(III) sulfate	$Eu_2(SO_4)_3$	+10400
Fluorine	F_2	–9.63
Gadolinium (350 K)	Gd	+185000
Gadolinium(III) chloride	$GdCl_3$	+27930

Name	Formula	$\chi_m/10^{-6}$ cm³ mol⁻¹	Name	Formula	$\chi_m/10^{-6}$ cm³ mol⁻¹
Gadolinium(III) oxide	Gd_2O_3	+53200	Iron(II) chloride tetrahydrate	$FeCl_2{\cdot}4H_2O$	+12900
Gadolinium(III) sulfate octahydrate	$Gd_2(SO_4)_3{\cdot}8H_2O$	+53280	Iron(II) fluoride	FeF_2	+9500
Gadolinium(III) sulfide	Gd_2S_3	+55500	Iron(II) iodide	FeI_2	+13600
Gallium	Ga	−21.6	Iron(II) oxide	FeO	+7200
Gallium suboxide	Ga_2O	−34	Iron(II) sulfate	$FeSO_4$	+12400
Gallium(II) sulfide	GaS	−23	Iron(II) sulfate monohydrate	$FeSO_4{\cdot}H_2O$	+10500
Gallium(III) chloride	$GaCl_3$	−63	Iron(II) sulfate heptahydrate	$FeSO_4{\cdot}7H_2O$	+11200
Gallium(III) sulfide	Ga_2S_3	−80	Iron(II) sulfide	FeS	+1074
Germanium	Ge	−11.6	Iron(III) chloride	$FeCl_3$	+13450
Germane (g)	GeH_4	−29.7	Iron(III) chloride hexahydrate	$FeCl_3{\cdot}6H_2O$	+15250
Germanium(II) oxide	GeO	−28.8	Iron(III) fluoride	FeF_3	+13760
Germanium(II) sulfide	GeS	−40.9	Iron(III) fluoride trihydrate	$FeF_3{\cdot}3H_2O$	+7870
Germanium(IV) chloride	$GeCl_4$	−72	Iron(III) nitrate nonahydrate	$Fe(NO_3)_3{\cdot}9H_2O$	+15200
Germanium(IV) fluoride	GeF_4	−50	Krypton (g)	Kr	−29.0
Germanium(IV) iodide	GeI_4	−171	Lanthanum (α)	La	+95.9
Germanium(IV) oxide	GeO_2	−34.3	Lanthanum oxide	La_2O_3	−78
Germanium(IV) sulfide	GeS_2	−53.9	Lanthanum sulfate nonahydrate	$La_2(SO_4)_3{\cdot}9H_2O$	−262
Gold	Au	−28	Lanthanum sulfide	La_2S_3	−37
Gold(I) bromide	AuBr	−61	Lead	Pb	−23
Gold(I) chloride	AuCl	−67	Lead(II) acetate	$Pb(C_2H_3O_2)_2$	−89.1
Gold(I) iodide	AuI	−91	Lead(II) bromide	$PbBr_2$	−90.6
Gold(III) chloride	$AuCl_3$	−112	Lead(II) carbonate	$PbCO_3$	−61.2
Hafnium	Hf	+71	Lead(II) chloride	$PbCl_2$	−73.8
Hafnium oxide	HfO_2	−23	Lead(II) chromate	$PbCrO_4$	−18
Helium (g)	He	−2.02	Lead(II) fluoride	PbF_2	−58.1
Holmium	Ho	+72900	Lead(II) iodate	$Pb(IO_3)_2$	−131
Holmium oxide	Ho_2O_3	+88100	Lead(II) iodide	PbI_2	−126.5
Hydrazine (l)	N_2H_4	−201	Lead(II) nitrate	$Pb(NO_3)_2$	−74
Hydrogen (l, 20.3 K)	H_2	−5.44	Lead(II) oxide	PbO	−42
Hydrogen (g)	H_2	−3.99	Lead(II) phosphate	$Pb_3(PO_4)_2$	−182
Hydrogen chloride (l)	HCl	−22.6	Lead(II) sulfate	$PbSO_4$	−69.7
Hydrogen chloride (aq)	HCl	−22	Lead(II) sulfide	PbS	−83.6
Hydrogen fluoride (l)	HF	−8.6	Lithium	Li	+14.2
Hydrogen fluoride (aq)	HF	−9.3	Lithium bromide	LiBr	−34.3
Hydrogen iodide (s, 195 K)	HI	−47.3	Lithium carbonate	Li_2CO_3	−27
Hydrogen iodide (l, 233 K)	HI	−48.3	Lithium chloride	LiCl	−24.3
Hydrogen iodide (aq)	HI	−50.2	Lithium fluoride	LiF	−10.1
Hydrogen peroxide (l)	H_2O_2	−17.3	Lithium hydride	LiH	−4.6
Hydrogen sulfide (g)	H_2S	−25.5	Lithium hydroxide (aq)	LiOH	−12.3
Indium	In	−10.2	Lithium iodide	LiI	−50
Indium(I) chloride	InCl	−30	Lithium sulfate	Li_2SO_4	−41.6
Indium(II) chloride	$InCl_2$	−56	Lutetium	Lu	+182.9
Indium(II) sulfide	InS	−28	Magnesium	Mg	+13.1
Indium(III) bromide	$InBr_3$	−107	Magnesium bromide	$MgBr_2$	−72
Indium(III) chloride	$InCl_3$	−86	Magnesium carbonate	$MgCO_3$	−32.4
Indium(III) oxide	In_2O_3	−56	Magnesium chloride	$MgCl_2$	−47.4
Indium(III) sulfide	In_2S_3	−98	Magnesium fluoride	MgF_2	−22.7
Iodine	I_2	−90	Magnesium hydroxide	$Mg(OH)_2$	−22.1
Iodic acid	HIO_3	−48	Magnesium iodide	MgI_2	−111
Iodine pentoxide	I_2O_5	−79.4	Magnesium oxide	MgO	−10.2
Iodine chloride	ICl	−54.6	Magnesium sulfate	$MgSO_4$	−42
Iodine trichloride	ICl_3	−90.2	Magnesium sulfate monohydrate	$MgSO_4{\cdot}H_2O$	−61
Iodine pentafluoride	IF_5	−58.1	Magnesium sulfate heptahydrate	$MgSO_4{\cdot}7H_2O$	−135.7
Iridium	Ir	+25	Manganese	Mn	+511
Iridium(III) chloride	$IrCl_3$	−14.4	Manganese(II) bromide	$MnBr_2$	+13900
Iridium(IV) oxide	IrO_2	+224	Manganese(II) carbonate	$MnCO_3$	+11400
Iron	Fe	Ferro.	Manganese(II) chloride	$MnCl_2$	+14350
Iron(II) bromide	$FeBr_2$	+13600	Manganese(II) chloride tetrahydrate	$MnCl_2{\cdot}4H_2O$	+14600
Iron(II) carbonate	$FeCO_3$	+11300	Manganese(II) fluoride	MnF_2	+10700
Iron(II) chloride	$FeCl_2$	+14750	Manganese(II) hydroxide	$Mn(OH)_2$	+13500

Name	Formula	$\chi_m/10^{-6}$ cm³ mol⁻¹
Manganese(II) iodide	MnI_2	+14400
Manganese(II) oxide	MnO	+4850
Manganese(II) sulfate	$MnSO_4$	+13660
Manganese(II) sulfate monohydrate	$MnSO_4 \cdot H_2O$	+14200
Manganese(II) sulfate tetrahydrate	$MnSO_4 \cdot 4H_2O$	+14600
Manganese(II) sulfide (α form)	MnS	+5630
Manganese(II) sulfide (β form)	MnS	+3850
Manganese(II,III) oxide	Mn_3O_4	+12400
Manganese(III) fluoride	MnF_3	+10500
Manganese(III) oxide	Mn_2O_3	+14100
Manganese(IV) oxide	MnO_2	+2280
Mercury (s, 234 K)	Hg	−24.1
Mercury (l)	Hg	−33.5
Mercury(I) bromide	Hg_2Br_2	−105
Mercury(I) chloride	Hg_2Cl_2	−120
Mercury(I) fluoride	Hg_2F_2	−106
Mercury(I) iodide	Hg_2I_2	−166
Mercury(I) nitrate	$Hg_2(NO_3)_2$	−121
Mercury(I) oxide	Hg_2O	−76.3
Mercury(I) sulfate	Hg_2SO_4	−123
Mercury(II) bromide	$HgBr_2$	−94.2
Mercury(II) chloride	$HgCl_2$	−82
Mercury(II) cyanide	$Hg(CN)_2$	−67
Mercury(II) fluoride	HgF_2	−57.3
Mercury(II) iodide	HgI_2	−165
Mercury(II) nitrate	$Hg(NO_3)_2$	−74
Mercury(II) oxide	HgO	46
Mercury(II) sulfate	$HgSO_4$	−78.1
Mercury(II) sulfide	HgS	−55.4
Mercury(II) thiocyanate	$Hg(SCN)_2$	−96.5
Molybdenum	Mo	+72
Molybdenum(III) bromide	$MoBr_3$	+525
Molybdenum(III) chloride	$MoCl_3$	+43
Molybdenum(III) oxide	Mo_2O_3	−42.0
Molybdenum(IV) bromide	$MoBr_4$	+520
Molybdenum(IV) chloride	$MoCl_4$	+1750
Molybdenum(IV) oxide	MoO_2	+41
Molybdenum(V) chloride	$MoCl_5$	+990
Molybdenum(VI) fluoride	MoF_6	−26.0
Molybdenum(VI) oxide	MoO_3	+3
Neodymium (α)	Nd	+5930
Neodymium fluoride	NdF_3	+4980
Neodymium oxide	Nd_2O_3	+10200
Neodymium sulfate	$Nd_2(SO_4)_3$	+9990
Neodymium sulfide	Nd_2S_3	+5550
Neon (g)	Ne	−6.96
Neptunium	Np	+575
Nickel	Ni	Ferro.
Nickel(II) bromide	$NiBr_2$	+5600
Nickel(II) chloride	$NiCl_2$	+6145
Nickel(II) chloride hexahydrate	$NiCl_2 \cdot 6H_2O$	+4240
Nickel(II) fluoride	NiF_2	+2410
Nickel(II) hydroxide	$Ni(OH)_2$	+4500
Nickel(II) iodide	NiI_2	+3875
Nickel(II) nitrate hexahydrate	$Ni(NO_3)_2 \cdot 6H_2O$	+4300
Nickel(II) oxide	NiO	+660
Nickel(II) sulfate	$NiSO_4$	+4005
Nickel(II) sulfide	NiS	+190
Nickel(III) sulfide	Ni_3S_2	+1030
Niobium	Nb	+208
Niobium(V) oxide	Nb_2O_5	−10
Nitrogen (g)	N_2	−12.0
Nitric acid (l)	HNO_3	−19.9
Nitrous oxide (g)	N_2O	−18.9
Nitric oxide (s, 90 K)	NO	+19.8
Nitric oxide (l, 118 K)	NO	+114.2
Nitric oxide (g)	NO	+1461
Nitrogen dioxide (g, 408 K)	NO_2	+150
Nitrogen trioxide (g)	N_2O_3	−16
Nitrogen tetroxide (g)	N_2O_4	−23.0
Osmium	Os	+11
Oxygen (s, 54 K)	O_2	+10200
Oxygen (l, 90 K)	O_2	+7699
Oxygen (g)	O_2	+3415
Ozone (l)	O_3	+6.7
Palladium	Pd	+540
Palladium(II) chloride	$PdCl_2$	−38
Phosphorus (white)	P	−26.66
Phosphorus (red)	P	−20.77
Phosphine (g)	PH_3	−26.2
Phosphoric acid (aq)	H_3PO_4	−43.8
Phosphorous acid (aq)	H_3PO_3	−42.5
Phosphorus(III) chloride (l)	PCl_3	−63.4
Platinum	Pt	+193
Platinum(II) chloride	$PtCl_2$	−54
Platinum(III) chloride	$PtCl_3$	−66.7
Platinum(IV) chloride	$PtCl_4$	−93
Platinum(IV) fluoride	PtF_4	+445
Plutonium	Pu	+525
Plutonium(IV) fluoride	PuF_4	+1760
Plutonium(IV) oxide	PuO_2	+730
Plutonium(VI) fluoride	PuF_6	+173
Potassium	K	+20.8
Potassium bromate	$KBrO_3$	−52.6
Potassium bromide	KBr	−49.1
Potassium carbonate	K_2CO_3	−59
Potassium chlorate	$KClO_3$	−42.8
Potassium chloride	KCl	−38.8
Potassium chromate	K_2CrO_4	−3.9
Potassium cyanide	KCN	−37
Potassium ferricyanide	$K_3Fe(CN)_6$	+2290
Potassium ferrocyanide trihydrate	$K_4Fe(CN)_6 \cdot 3H_2O$	−172.3
Potassium fluoride	KF	−23.6
Potassium hydrogen sulfate	$KHSO_4$	−49.8
Potassium hydroxide (aq)	KOH	−22
Potassium iodate	KIO_3	−63.1
Potassium iodide	KI	−63.8
Potassium nitrate	KNO_3	−33.7
Potassium nitrite	KNO_2	−23.3
Potassium permanganate	$KMnO_4$	+20
Potassium sulfate	K_2SO_4	−67
Potassium sulfide	K_2S	−60
Potassium superoxide	KO_2	+3230
Potassium thiocyanate	$KSCN$	−48
Praseodymium (α)	Pr	+5530
Praseodymium chloride	$PrCl_3$	+44.5
Praseodymium oxide	Pr_2O_3	+8994
Praseodymium sulfide	Pr_2S_3	+10770
Protactinium	Pa	+277
Rhenium	Re	+67

Name	Formula	$\chi_m/10^{-6}$ cm^3 mol^{-1}
Rhenium(IV) oxide	ReO_2	+44
Rhenium(IV) sulfide	ReS_2	+38
Rhenium(V) chloride	$ReCl_5$	+1225
Rhenium(VI) oxide	ReO_3	+16
Rhenium(VII) oxide	Re_2O_7	−16
Rhodium	Rh	+102
Rhodium(III) chloride	$RhCl_3$	−7.5
Rhodium(III) oxide	Rh_2O_3	+104
Rubidium	Rb	+17
Rubidium bromide	RbBr	−56.4
Rubidium carbonate	Rb_2CO_3	−75.4
Rubidium chloride	RbCl	−46
Rubidium fluoride	RbF	−31.9
Rubidium iodide	RbI	−72.2
Rubidium nitrate	$RbNO_3$	−41
Rubidium sulfate	Rb_2SO_4	−88.4
Rubidium superoxide	RbO_2	+1527
Ruthenium	Ru	+39
Ruthenium(III) chloride	$RuCl_3$	+1998
Ruthenium(IV) oxide	RuO_2	+162
Samarium (α)	Sm	+1278
Samarium(II) bromide	$SmBr_2$	+5337
Samarium(III) bromide	$SmBr_3$	+972
Samarium(III) oxide	Sm_2O_3	+1988
Samarium(III) sulfate octahydrate	$Sm_2(SO_4)_3\cdot 8H_2O$	+1710
Samarium(III) sulfide	Sm_2S_3	+3300
Scandium (α)	Sc	+295.2
Selenium	Se	−25
Selenium dioxide	SeO_2	−27.2
Selenium bromide	Se_2Br_2	−113
Selenium chloride (l)	Se_2Cl_2	−94.8
Selenium hexafluoride (g)	SeF_6	−51
Silicon	Si	−3.12
Silane (g)	SiH_4	−20.4
Disilane (g)	Si_2H_6	−37.3
Tetramethylsilane (l)	$(CH_3)_4Si$	−74.80
Tetraethylsilane (l)	$(C_2H_5)_4Si$	−120.2
Tetrabromosilane (l)	$SiBr_4$	−126
Tetrachlorosilane (l)	$SiCl_4$	−87.5
Silicon carbide	SiC	−12.8
Silicon dioxide	SiO_2	−29.6
Silver	Ag	−19.5
Silver(I) bromide	AgBr	−61
Silver(I) carbonate	Ag_2CO_3	−80.90
Silver(I) chloride	AgCl	−49
Silver(I) chromate	Ag_2CrO_4	−40
Silver(I) cyanide	AgCN	−43.2
Silver(I) fluoride	AgF	−36.5
Silver(I) iodide	AgI	−80
Silver(I) nitrate	$AgNO_3$	−45.7
Silver(I) nitrite	$AgNO_2$	−42
Silver(I) oxide	Ag_2O	−134
Silver(I) phosphate	Ag_3PO_4	−120
Silver(I) sulfate	Ag_2SO_4	−92.90
Silver(I) thiocyanate	AgSCN	−61.8
Silver(II) oxide	AgO	−19.6
Sodium	Na	+16
Sodium acetate	$NaC_2H_3O_2$	−37.6
Sodium bromate	$NaBrO_3$	−44.2
Sodium bromide	NaBr	−41

Name	Formula	$\chi_m/10^{-6}$ cm^3 mol^{-1}
Sodium carbonate	Na_2CO_3	−41
Sodium chlorate	$NaClO_3$	−34.7
Sodium chloride	NaCl	−30.2
Sodium dichromate	$Na_2Cr_2O_7$	+55
Sodium fluoride	NaF	−15.6
Sodium hydrogen phosphate	Na_2HPO_4	−56.6
Sodium hydroxide (aq)	NaOH	−15.8
Sodium iodate	$NaIO_3$	−53
Sodium iodide	NaI	−57
Sodium nitrate	$NaNO_3$	−25.6
Sodium nitrite	$NaNO_2$	−14.5
Sodium oxide	Na_2O	−19.8
Sodium peroxide	Na_2O_2	−28.10
Sodium sulfate	Na_2SO_4	−52
Sodium sulfate decahydrate	$Na_2SO_4\cdot 10H_2O$	−184
Sodium sulfide	Na_2S	−39
Sodium tetraborate	$Na_2B_4O_7$	−85
Strontium	Sr	+92
Strontium bromide	$SrBr_2$	−86.6
Strontium bromide hexahydrate	$SrBr_2\cdot 6H_2O$	−160
Strontium carbonate	$SrCO_3$	−47
Strontium chlorate	$Sr(ClO_3)_2$	−73
Strontium chloride	$SrCl_2$	−61.5
Strontium chloride hexahydrate	$SrCl_2\cdot 6H_2O$	−145
Strontium chromate	$SrCrO_4$	−5.1
Strontium fluoride	SrF_2	−37.2
Strontium hydroxide	$Sr(OH)_2$	−40
Strontium iodate	$Sr(IO_3)_2$	−108
Strontium iodide	SrI_2	−112
Strontium nitrate	$Sr(NO_3)_2$	−57.2
Strontium oxide	SrO	−35
Strontium peroxide	SrO_2	−32.3
Strontium sulfate	$SrSO_4$	−57.9
Sulfur (rhombic)	S	−15.5
Sulfur (monoclinic)	S	−14.9
Sulfuric acid (l)	H_2SO_4	−39
Sulfur dioxide (g)	SO_2	−18.2
Sulfur trioxide (l)	SO_3	−28.54
Sulfur chloride (l)	$SSCl_2$	−62.2
Sulfur dichloride (l)	SCl_2	−49.4
Sulfur hexafluoride (g)	SF_6	−44
Thionyl chloride (l)	$SOCl_2$	−44.3
Tantalum	Ta	+154
Tantalum(V) chloride	$TaCl_5$	+140
Tantalum(V) oxide	Ta_2O_5	−32
Technetium	Tc	+115
Tellurium	Te	−38
Tellurium dibromide	$TeBr_2$	−106
Tellurium dichloride	$TeCl_2$	−94
Tellurium hexafluoride (g)	TeF_6	−66
Terbium (α)	Tb	+170000
Terbium oxide	Tb_2O_3	+78340
Thallium	Tl	−50
Thallium(I) bromate	$TlBrO_3$	−75.9
Thallium(I) bromide	TlBr	−63.9
Thallium(I) carbonate	Tl_2CO_3	−101.6
Thallium(I) chlorate	$TlClO_3$	−65.5
Thallium(I) chloride	TlCl	−57.8
Thallium(I) chromate	Tl_2CrO_4	−39.3
Thallium(I) cyanide	TlCN	−49

Name	Formula	$\chi_m/10^{-6}$ cm³ mol⁻¹
Thallium(I) fluoride	TlF	−44.4
Thallium(I) iodate	$TlIO_3$	−86.8
Thallium(I) iodide	TlI	−82.2
Thallium(I) nitrate	$TlNO_3$	−56.5
Thallium(I) nitrite	$TlNO_2$	−50.8
Thallium(I) sulfate	Tl_2SO_4	−112.6
Thallium(I) sulfide	Tl_2S	−88.8
Thorium	Th	+97
Thorium(IV) oxide	ThO_2	−16
Thulium	Tm	+24700
Thulium oxide	Tm_2O_3	+51444
Tin (gray)	Sn	−37.4
Tin(II) chloride	$SnCl_2$	−69
Tin(II) chloride dihydrate	$SnCl_2 \cdot 2H_2O$	−91.4
Tin(II) oxide	SnO	−19
Tin(IV) bromide	$SnBr_4$	−149
Tin(IV) chloride (l)	$SnCl_4$	−115
Tin(IV) oxide	SnO_2	−41
Titanium	Ti	+151
Titanium(II) bromide	$TiBr_2$	+720
Titanium(II) chloride	$TiCl_2$	+484
Titanium(II) iodide	TiI_2	+1790
Titanium(II) sulfide	TiS	+432
Titanium(III) bromide	$TiBr_3$	+660
Titanium(III) chloride	$TiCl_3$	+1110
Titanium(III) fluoride	TiF_3	+1300
Titanium(III) oxide	Ti_2O_3	+132
Titanium(IV) chloride	$TiCl_4$	−54
Titanium(IV) oxide	TiO_2	+5.9
Tungsten	W	+53
Tungsten carbide	WC	+10
Tungsten(II) chloride	WCl_2	−25
Tungsten(IV) oxide	WO_2	+57
Tungsten(IV) sulfide	WS_2	+5850
Tungsten(V) bromide	WBr_5	+270
Tungsten(V) chloride	WCl_5	+387
Tungsten(VI) chloride	WCl_6	−71
Tungsten(VI) fluoride (g)	WF_6	−53
Tungsten(VI) oxide	WO_3	−15.8
Uranium	U	+409
Uranium(III) bromide	UBr_3	+4740
Uranium(III) chloride	UCl_3	+3460
Uranium(III) hydride	UH_3	+6244
Uranium(III) iodide	UI_3	+4460
Uranium(IV) bromide	UBr_4	+3530
Uranium(IV) chloride	UCl_4	+3680
Uranium(IV) fluoride	UF_4	+3530
Uranium(IV) oxide	UO_2	+2360
Uranium(VI) fluoride	UF_6	+43
Uranium(VI) oxide	UO_3	+128
Vanadium	V	+285
Vanadium(II) bromide	VBr_2	+3230
Vanadium(II) chloride	VCl_2	+2410
Vanadium(III) bromide	VBr_3	+2910
Vanadium(III) chloride	VCl_3	+3030
Vanadium(III) fluoride	VF_3	+2757
Vanadium(III) oxide	V_2O_3	+1976
Vanadium(III) sulfide	V_2S_3	+1560
Vanadium(IV) chloride	VCl_4	+1215
Vanadium(IV) oxide	VO_2	+99
Vanadium(V) oxide	V_2O_5	+128
Water (s, 273 K)	H_2O	−12.63
Water (l, 293 K)	H_2O	−12.96
Water (l, 373 K)	H_2O	−13.09
Water (g, 373 K))	H_2O	−13.1
Xenon (g)	Xe	−45.5
Ytterbium (β)	Yb	+67
Yttrium (α)	Y	+187.7
Yttrium oxide	Y_2O_3	+44.4
Yttrium sulfide	Y_2S_3	+100
Zinc	Zn	−9.15
Zinc carbonate	$ZnCO_3$	−34
Zinc chloride	$ZnCl_2$	−55.33
Zinc cyanide	$Zn(CN)_2$	−46
Zinc fluoride	ZnF_2	−34.3
Zinc hydroxide	$Zn(OH)_2$	−67
Zinc iodide	ZnI_2	−108
Zinc oxide	ZnO	−27.2
Zinc phosphate	$Zn_3(PO_4)_2$	−141
Zinc sulfate	$ZnSO_4$	−47.8
Zinc sulfate monohydrate	$ZnSO_4 \cdot H_2O$	−63
Zinc sulfate heptahydrate	$ZnSO_4 \cdot 7H_2O$	−138
Zinc sulfide	ZnS	−25
Zirconium	Zr	+120
Zirconium carbide	ZrC	−26
Zirconium nitrate pentahydrate	$Zr(NO_3)_4 \cdot 5H_2O$	−77
Zirconium(IV) oxide	ZrO_2	−13.8

INDEX OF REFRACTION OF INORGANIC LIQUIDS

This table gives the index of refraction n of several inorganic substances in the liquid state at specified temperatures. The measurements refer to ambient atmospheric pressure except for substances whose normal boiling points are greater than the indicated temperature; in this case the pressure is the saturated vapor pressure of the substance. All values refer to a wavelength of 589 nm unless otherwise indicated. Entries are arranged in alphabetical order by chemical formula as normally written.

Data on the index of refraction at other temperatures and wavelengths may be found in Reference 1.

References

1. Wohlfarth, C., and Wohlfarth, B., *Landolt-Börnstein, Numerical Data and Functional Relationships in Science and Technology, New Series*, III/38A, Martienssen, W., Editor, Springer-Verlag, Heidelberg, 1996.
2. Francis, A.W., *J. Chem. Eng. Data*, 5, 534, 1960.

Formula	Name	t/°C	n
Ar	Argon	−188	1.2312
$AsCl_3$	Arsenic(III) chloride	16	1.604
BBr_3	Boron tribromide	16	1.312
BrF_3	Bromine trifluoride	25	1.4536
BrF_5	Bromine pentafluoride	25	1.3529
Br_2	Bromine	15	1.659
COS	Carbon oxysulfide	25	1.3506
CO_2	Carbon dioxide	24	1.6630
CS_2	Carbon disulfide	20	1.62774
C_3O_2	Carbon suboxide	0	1.453
Cl_2	Chlorine	20	1.3834
CrO_2Cl_2	Chromyl chloride	23	1.524
$Fe(CO)_5$	Iron pentacarbonyl	14	1.523
$GeBr_4$	Germanium(IV) bromide	26	1.6269
$GeCl_4$	Germanium(IV) chloride	25	1.4614
HBr	Hydrogen bromide	10	1.325
HCN	Hydrogen cyanide	20	1.26136
HCl	Hydrogen chloride	18	1.3287[a]
$HClO_4$	Perchloric acid	50	1.3819
HF	Hydrogen fluoride	25	1.1574
HI	Hydrogen iodide	16	1.466
HNO_3	Nitric acid	25	1.393
H_2	Hydrogen	−253	1.1096
H_2O	Water	20	1.33336
H_2O_2	Hydrogen peroxide	28	1.4061
H_2S	Hydrogen sulfide	−80	1.460
		20	1.3682
H_2SO_4	Sulfuric acid	20	1.4183
H_2S_2	Hydrogen disulfide	20	1.630
He	Helium	−269	1.02451[c]
Kr	Krypton	−157	1.3032[c]
NH_3	Ammonia	−77	1.3944[b]
		20	1.3327
NO	Nitric oxide	−90	1.330
N_2	Nitrogen	−196	1.19876[b]
N_2H_4	Hydrazine	22	1.470
N_2O	Nitrous oxide	25	1.238
O_2	Oxygen	−183	1.2243[c]
PBr_3	Phosphorus(III) bromide	25	1.687
PCl_3	Phosphorus(III) chloride	21	1.5122
PH_3	Phosphine	17	1.317
P_2O_3	Phosphorus(III) oxide	27	1.540
S	Sulfur	125	1.9170
SCl_2	Sulfur dichloride	14	1.557
SF_6	Sulfur hexafluoride	25	1.167
$SOCl_2$	Thionyl chloride	10	1.527
SO_2	Sulfur dioxide	25	1.3396
SO_2Cl_2	Sulfuryl chloride	12	1.444
SO_3	Sulfur trioxide	20	1.40965
$SSCl_2$	Sulfur chloride	20	1.671
$SbCl_5$	Antimony(V) chloride	22	1.5925
$SiBr_4$	Tetrabromosilane	31	1.5685
$SiCl_4$	Tetrachlorosilane	25	1.41156
$SnBr_4$	Tin(IV) bromide	31	1.6628
$SnCl_4$	Tin(IV) chloride	25	1.5086
$TiCl_4$	Titanium(IV) chloride	18	1.6076
Xe	Xenon	−112	1.3918[c]

[a] At 581 nm
[b] At 578 nm
[c] At 546 nm

PHYSICAL AND OPTICAL PROPERTIES OF MINERALS

The chemical formula, crystal system, density, hardness, and index of refraction of some common minerals are given in this table. Entries are arranged alphabetically by mineral name. The columns are:

- **Formula:** Chemical formula for a typical sample of the mineral. Composition often varies considerably with the origin of the sample.
- **Crystal system:** tricl = triclinic; monocl = monoclinic; orth = orthorhombic; tetr = tetragonal; hex = hexagonal; rhomb = rhombohedral; cub = cubic.
- **Density:** Typical density in g/cm³. Individual samples may vary by a few percent.
- **Hardness:** On the Mohs' scale (range of 1 to 10, with talc = 1 and diamond = 10).
- **Index of refraction:** Values are given for the three coordinate axes in the order of least, intermediate, and greatest index. For cubic crystals there is only a single value. See Reference 1 for details on the axis systems. Variations of several percent, depending on the origin and exact composition of the sample, are common.

References

1. Deer, W. A., Howie, R. A., and Zussman, J., *An Introduction to the Rock-Forming Minerals*, 2nd Edition, Longman Scientific & Technical, Harlow, Essex, 1992.
2. Carmichael, R. S., *Practical Handbook of Physical Properties of Rocks and Minerals*, CRC Press, Boca Raton, FL, 1989.
3. Donnay, J. D. H., and Ondik, H. M., *Crystal Data Determinative Tables, Third Edition, Volume 2, Inorganic Compounds*, Joint Committee on Powder Diffraction Standards, Swarthmore, PA, 1973.

Name	Formula	Crystal system	Density g/cm³	Hardness	Index of refraction		
					n_α	n_β	n_γ
Acanthite	Ag_2S	orth	7.2	2.3			
Actinolite	$Ca_2(Mg,Fe)_5Si_8O_{22}(OH,F)_2$	monocl	3.23	5.5	1.624	1.655	1.664
Aegirine	$NaFe(SiO_3)_2$	monocl	3.58	6	1.763	1.800	1.815
Akermanite	$Ca_2MgSi_2O_7$	tetr	2.94	5.5	1.632	1.640	
Alabandite	MnS	cub	4.0	3.8			
Albite	$NaAlSi_3O_8$	tricl	2.63	6.3	1.527	1.531	1.538
Allanite	$(Ca,Mn,Ce,La,Y,Th)_2(Fe,Ti)(Al,Fe)O{\cdot}OH(Si_2O_7)(SiO_4)$	monocl	3.8	5.8	1.75	1.78	1.80
Allemontite	$SbAs$	hex	6.0	3.5			
Almandine	$Fe_3Al_2Si_3O_{12}$	cub	4.32	6.8	1.830		
Altaite	$PbTe$	cub	8.16	3			
Aluminite	$Al_2(SO_4)(OH)_4{\cdot}7H_2O$	monocl	1.74	1.5	1.459	1.464	1.470
Alunite	$(K,Na)Al_3(SO_4)_2(OH)_6$	rhomb	2.8	3.8	1.572	1.592	
Alunogen	$Al_2(SO_4)_3{\cdot}18H_2O$	monocl	1.69	1.8	1.467	1.47	1.478
Amblygonite	$(Li,Na)Al(PO_4)(F,OH)$	tricl	3.1	5.8	1.591	1.604	1.613
Analcite	$NaAlSi_2O_6{\cdot}H_20$	cub	2.27	5.5	1.486		
Anatase	TiO_2	tetr	4.23	5.8	2.488	2.561	
Andalusite	Al_2OSiO_4	orth	3.15	7.5	1.635	1.639	1.644
Andesine	$NaAlSi_3O_8{\cdot}CaAl_2Si_2O_8$	tricl	2.67	6.3	1.550	1.553	1.557
Andorite	$PbAgSb_3S_6$	rhomb	5.35	3.3			
Andradite	$Ca_3(Fe,Ti)_2Si_3O_{12}$	cub	3.86	6.8	1.887		
Anglesite	$PbSO_4$	orth	6.29	2.8	1.877	1.883	1.894
Anhydrite	$CaSO_4$	orth	2.96	3.5	1.570	1.575	1.614
Ankerite	$Ca(Fe,Mg,Mn)(CO_3)_2$	rhomb	3.0	3.8	1.529	1.720	
Anorthite	$CaAl_2Si_2O_8$	tricl	2.76	6.3	1.577	1.585	1.590
Anorthoclase	$(Na,K)AlSi_3O_8$	tricl	2.58	6	1.523	1.528	1.529
Anthophyllite	$(Mg,Fe)_7Si_8O_{22}(OH,F)_2$	rhomb	3.21	5.8	1.645	1.658	1.668
Apatite	$Ca_5(PO_4)_3(OH,F,Cl)$	hex	3.2	5	1.645	1.648	
Apophyllite	$KFCa_4Si_8O_{20}{\cdot}8H_2O$	tetr	2.35	4.8	1.535	1.536	
Aragonite	$CaCO_3$	orth	2.83	3.5	1.531	1.680	1.686
Arcanite	K_2SO_4	orth	2.66		1.494	1.494	1.497
Argentite	Ag_2S	orth	7.2	2.3			
Arsenolite	As_2O_3	cub	3.86	1.5	1.755		
Arsenopyrite	$FeAsS$	monocl	6.1	5.8			
Atacamite	$Cu_2(OH)_3Cl$	rhomb	3.76	3.3	1.831	1.861	1.880
Augelite	$Al_2(PO_4)(OH)_3$	monocl	2.70	4.8	1.574	1.576	1.588
Augite	$(Ca,Mg,Fe,Ti,Al)_2(Si,Al)_2O_6$	monocl	3.38	6	1.703	1.707	1.738
Autunite	$Ca(UO_{22})(PO_4)_2{\cdot}10H20$	tetr	3.2	2.3	1.553	1.577	
Axinite	$(Ca,Mn,Fe)_3Al_2BO_3Si_4O_{12}(OH)$	tricl	3.31	6.8	1.684	1.691	1.694

Name	Formula	Crystal system	Density g/cm³	Hardness	Index of refraction n_α	n_β	n_γ
Azurite	$Cu_3(OH)_2(CO_3)_2$	monocl	3.77	3.8	1.730	1.758	1.838
Baddeleyite	ZrO_2	monocl	5.7	6.5	2.13	2.19	2.20
Barite	$BaSO_4$	orth	4.49	3.3	1.636	1.637	1.648
Benitoite	$BaTi(SiO_3)_3$	rhomb	3.65	6.3	1.757	1.804	
Bertrandite	$Be_4Si_2O_7(OH)_2$	rhomb	2.6	6	1.589	1.602	1.613
Beryl	$Be_3Al_2(SiO_3)_6$	hex	2.64	7.8	1.582	1.589	
Beryllonite	$NaBe(PO)_4$	monocl	2.81	5.8	1.552	1.558	1.561
Biotite	$K(Mg,Fe)_3AlSi_3O_{10}(OH,F)_2$	monocl	3.0	2.8	1.595	1.651	1.651
Bismuthinite	Bi_2S_3	orth	6.78	2			
Bixbyite	$(Mn,Fe)_2O_3$	cub	4.95	6.3			
Bloedite	$Na_2Mg(SO_4)_2{\cdot}4H_2O$	monocl	2.25	2.8	1.483	1.486	1.487
Boehmite	$AlO(OH)$	orth	3.44	3.8	1.64	1.65	1.66
Boracite	$Mg_3B_7O_{13}Cl$	rhomb	2.94	7.3	1.66	1.66	1.67
Borax	$Na_2B_4O_7{\cdot}10H_2O$	monocl	1.73	2.3	1.447	1.469	1.472
Bornite	Cu_5FeS_4	cub	5.07	3			
Boulangerite	$Pb_5Sb_4S_{11}$	monocl	6.1	2.8			
Bournonite	$PbCuSbS_3$	rhomb	5.83	2.8			
Braggite	PtS	tetr	10.2				
Braunite	$(Mn,Si)_2O_3$	tetr	4.78	6.3			
Bravoite	$(Ni,Fe)S_2$	cub	4.62	5.8			
Breithauptite	NiSb	hex	≈8.7	5.5			
Brochantite	$Cu_4(SO_4)(OH)_6$	monocl	3.79	3.8	1.728	1.771	1.800
Bromyrite	AgBr	cub	6.47	2.5	2.253		
Brookite	TiO_2	orth	4.23	5.8	2.583	2.584	2.700
Brucite	$Mg(OH)_2$	hex	2.37	2.5	1.575	1.59	
Bunsenite	NiO	cub	6.72	5.5			
Cacoxenite	$Fe_4(PO_4)_3(OH)_3{\cdot}12H_2O$	hex	2.3	3.5	1.580	1.646	
Calcite	$CaCO_3$	hex	2.71	3	1.486	1.658	
Caledonite	$Cu_2Pb_5(SO_4)_3(CO_3)(OH)_6$	rhomb	5.76	2.8	1.818	1.866	1.909
Calomel	Hg_2Cl_2	tetr	7.16	1.5	1.973	2.656	
Cancrinite	$(Na,Ca,K)_7[Al_6Si_6O_{24}](CO_3,SO_4,Cl,OH)_2{\cdot}H_2O$	hex	2.42	5.5	1.495	1.509	
Carnalite	$KMgCl_3{\cdot}6H_2O$	rhomb	1.60	2.5	1.466	1.475	1.494
Carnotite	$K_2(UO_2)_2(VO_4)_2{\cdot}3H_2O$	rhomb		1.5	1.75	1.92	1.95
Cassiterite	SnO_2	tetr	6.85	6.5	2.006	2.097	
Celestite	$SrSO_4$	orth	3.96	3.3	1.622	1.624	1.631
Celsian	$BaAl_2Si_2O_8$	monocl	3.25	6.3	1.583	1.588	1.594
Cerargyrite	AgCl	cub	5.56	2.5	2.071		
Cerussite	$PbCO_3$	orth	6.6	3.3	1.804	2.076	2.079
Cervantite	Sb_2O_4	orth	6.64	4.5			
Chabazite	$Ca[Al_2Si_4O_{12}]{\cdot}6H_2O$	trig	2.08	4.5	1.482		
Chalcanthite	$CuSO_4{\cdot}5H_2O$	tricl	2.29	2.5	1.514	1.537	1.543
Chalcocite	Cu_2S	orth	5.6	2.8			
Chalcopyrite	$CuFeS_2$	tetr	4.2	3.8			
Chiolite	$Na_5Al_3F_{14}$	tetr	3.00	3.8	1.342	1.349	
Chlorite	$(Mg,Al,Fe)_{12}(Si,Al)_8O_{20}(OH)_{16}$	monocl	3.0	2.5	1.61	1.62	1.62
Chloritoid	$FeAl_4O_2(SiO_4)_2(OH)_4$	monocl	3.66	6.5	1.717	1.721	1.726
Chondrodite	$Mg(OH,F)_2{\cdot}2Mg_2SiO_4$	monocl	3.21	6.5	1.604	1.615	1.634
Chromite	$FeCr_2O_4$	cub	5.0	5.5	2.16		
Chrysoberyl	$BeAl_2O_4$	orth	3.65	8.5	1.746	1.748	1.756
Chrysocolla	$CuSiO_3{\cdot}2H_2O$	rhomb	2.4	2	1.575	1.597	1.598
Cinnabar	HgS	hex	8.17	2.3	2.814	3.143	
Claudetite	As_2O_3	monocl	3.74	2.5	1.87	1.92	2.01
Clinohumite	$Mg(OH,F)_2{\cdot}4Mg_2SiO_4$	monocl	3.21	6	1.633	1.647	1.668
Clinozoisite	$Ca_2Al_3Si_3O_{12}(OH)$	monocl	3.30	6.5	1.693	1.700	1.712
Cobaltite	CoAsS	cub	≈6.1	5.5			
Colemanite	$Ca_2B_6O_{11}{\cdot}5H_2O$	monocl	2.42	4.5	1.586	1.592	1.614
Columbite	$(Fe,Mn)(Nb,Ta)_2O_6$	rhomb	5.20	6			
Connellite	$Cu_{19}(SO_4)Cl_4(OH)_{32}{\cdot}3H_2O$	hex	3.36	3	1.731	1.752	
Copiapite	$(Fe,Mg)Fe_4(SO_4)_6(OH)_2{\cdot}20H_2O$	tricl	2.13	2.8	1.52	1.54	1.59
Coquimbite	$Fe_2(SO_4)_3{\cdot}9H_2O$	hex	2.1	2.5	1.54	1.56	

Name	Formula	Crystal system	Density g/cm³	Hardness	Index of refraction n_α	n_β	n_γ
Cordierite	$Al_3(Mg,Fe)_2Si_5AlO_{18}$	rhomb	2.66	7	1.540	1.549	1.553
Corundum	Al_2O_3	hex	3.97	9	1.761	1.769	
Cotunnite	$PbCl_2$	orth	5.98	2.5	2.199	2.217	2.260
Covellite	CuS	hex	4.8	1.8			
Cristobalite	SiO_2	hex	2.33	6.5	1.484	1.487	
Crocoite	$PbCrO_4$	monocl	6.12	2.8	2.29	2.36	2.66
Cryolite	Na_3AlF_6	monocl	2.97	2.5	1.338	1.338	1.339
Cryolithionite	$Na_3Li_3Al_2F_{12}$	cub	2.77	2.8	1.340		
Cubanite	$CuFe_2S_3$	rhomb	4.11	3.5			
Cummingtonite	$(Mg,Fe)_7Si_8O_{22}(OH)_2$	monocl	3.4	5.5	1.650	1.660	1.676
Cuprite	Cu_2O	cub	6.0	3.8			
Danburite	$CaSi_2B_2O_8$	rhomb	3.0	7	1.63	1.63	1.63
Datolite	$CaBSiO_4(OH)$	monocl	2.98	5.3	1.624	1.652	1.668
Daubreelite	Cr_2FeS_4	cub	3.81				
Derbylite	$Fe_6Ti_6Sb_2O_{23}$	rhomb	4.53	5	2.45	2.45	2.51
Diamond	C	cub	3.51	10	2.418		
Diaspore	AlO(OH)	orth	3.4	6.8	1.694	1.715	1.741
Digenite	$Cu_{2-x}S$	cub	5.55	2.8			
Diopside	$CaMgSi_2O_6$	monocl	3.30	6	1.680	1.687	1.708
Dioptase	$CuSiO_2(OH)_2$	rhomb	3.5	5	1.65	1.70	
Dolomite	$CaMg(CO_3)_2$	rhomb	2.86	3.5	1.500	1.679	
Douglasite	$K_2FeCl_4{\cdot}2H_2O$	orth	2.16		1.488	1.500	
Dyscrasite	Ag_3Sb	rhomb	9.74	3.8			
Eddingtonite	$BaAl_2Si_3O_{10}{\cdot}4H_2O$	rhomb	2.8		1.541	1.553	1.557
Eglestonite	Hg_4OCl_2	cub	8.4	2.5	2.49		
Emplectite	$CuBiS_2$	rhomb	6.38	2			
Enargite	Cu_3AsS_4	rhomb	4.5	3			
Enstatite	$MgSiO_3$	monocl	3.19	5.5	1.656	1.662	1.669
Epidote	$Ca_2Al_2(Al,Fe)OH(SiO_4)_3$	monocl	3.44	6	1.733	1.755	1.765
Epsomite	$MgSO_4{\cdot}7H_2O$	orth	1.67	2.3	1.433	1.455	1.461
Erythrite	$(Co,Ni)_3(AsO_4)_2{\cdot}8H_2O$	monocl	3.06	2	1.626	1.661	1.699
Eucairite	CuAgSe	orth	7.7	2.5			
Euclasite	$BeAlSiO_4(OH)$	monocl	3.1	7.5	1.651	1.655	1.671
Eudialite	$(Na,Ca,Ce)_5(Fe,Mn)(Zr,Ti)(Si_3O_9)_2$ (OH,Cl)	hex	3.0	5.5	1.623	1.600	1.615
Eulytite	$Bi_4Si_3O_{12}$	cub	6.6	4.5	2.05		
Euxenite	$(Y,Ca,Ce,U,Th)(Nb,Ta,Ti)_2O_6$	rhomb	5.5	6	2.2		
Fayalite	Fe_2SiO_4	orth	4.30	6.5	1.827	1.869	1.879
Ferberite	$FeWO_4$	monocl	7.51	4.3			
Fergussonite	$(Y,Er,Ce,Fe)(Nb,Ta,Ti)O_4$	tetr	5.7	6	2.1		
Fluorite	CaF_2	cub	3.18	4	1.434		
Forsterite	Mg_2SiO_4	orth	3.21	7	1.635	1.651	1.670
Franklinite	$ZnFe_2O_4$	cub	5.21	6	2.36		
Gahnite	$ZnAl_2O_4$	cub	4.62	7.8	1.805		
Galaxite	$MnAl_2O_4$	cub	4.04	7.8	1.92		
Galena	PbS	cub	7.60	2.5	3.91		
Galenabismuthite	$PbBi_2S_4$	rhomb	7.04	3			
Ganomalite	$(Ca,Pb)_{10}(OH,Cl)_2(Si_2O_7)_3$	hex	5.6	3.5	1.910	1.945	
Gaylussite	$Na_2Ca(CO_3)_2{\cdot}5H_2O$	monocl	1.99	2.8	1.444	1.516	1.523
Gehlenite	$Ca_2Al_2SiO_7$	tetr	3.04	5.5	1.658	1.669	
Geikielite	$MgTiO_3$	hex	3.85	5.5	1.95	2.31	
Gibbsite	$Al(OH)_3$	monocl	2.42	3	1.57	1.57	1.59
Glauberite	$Na_2Ca(SO_4)_2$	monocl	2.80	2.8	1.515	1.535	1.536
Glauconite	$(K,Na,Ca)_{1.6}(Fe,Al,Mg)_{4.0}Si_{7.3}Al_{0.7}O_{20}(OH)_4$	monocl	2.7	2	1.60	1.63	1.63
Glaucophane	$Na_2Mg_3Al_2Si_8O_{22}(OH)_2$	monocl	3.19	6	1.634	1.645	1.648
Gmelinite	$(Ca,Na_2)[Al_2Si_4O_{12}]{\cdot}6H_2O$	hex	2.10	4.5	1.477	1.485	
Goethite	FeO(OH)	orth	4.3	5.3	2.268	2.401	2.457
Goslarite	$ZnSO_4{\cdot}7H_2O$	orth	1.97	2.3	1.457	1.480	1.484
Greenockite	CdS	hex	4.8	3.3	2.506	2.529	
Grossularite	$Ca_3Al_2Si_3O_{12}$	cub	3.59	6.8	1.734		
Gummite	$UO_3{\cdot}H_2O$	orth	7.05	3.8			

Name	Formula	Crystal system	Density g/cm³	Hardness	Index of refraction n_α	n_β	n_γ
Gypsum	$CaSO_4·2H_2O$	monocl	2.32	2	1.520	1.525	1.530
Halite	NaCl	cub	2.17	2	1.544		
Hambergite	$Be_2(OH)(BO_3)$	rhomb	2.36	7.5	1.56	1.59	1.63
Hanksite	$Na_{22}K(SO_4)_9(CO_3)_2Cl$	hex	2.56	3.3	1.461	1.481	
Harmotome	$Ba[Al_2Si_6O_{16}]·6H_2O$	monocl	2.44	4.5	1.506	1.507	1.511
Hausmannite	Mn_3O_4	tetr	4.84	5.5	2.15	2.46	
Haüyne	$(Na,Ca)_{4-8}Al_6Si_6O_{24}(SO_4,S)_{1-2}$	cub	2.47	5.8	1.502		
Hedenbergite	$CaFeSi_2O_6$	monocl	3.53	6	1.721	1.727	1.746
Helvite	$Mn_4Be_3Si_3O_{12}S$	cub	3.32	6	1.739		
Hematite	Fe_2O_3	hex	5.25	6	2.91	3.19	
Hemimorphite	$Zn_4Si_2O_7(OH)_2·H_2O$	rhomb	3.45	5	1.614	1.617	1.636
Hercynite	$Fe(AlO_2)_2$	cub	4.3	7.8	1.835		
Herderite	$CaBe(PO_4)(Fe,OH)$	monocl	2.98	5.3	1.592	1.612	1.621
Hessite	Ag_2Te	orth	8.4	2.5			
Heulandite	$(Ca,Na_2,K_2)[Al_2Si_7O_{18}]·6H_2O$	monocl	2.2	3.8	1.498	1.498	1.506
Hopeite	$Zn_3(PO_4)_2·4H_2O$	orth	3.0	3.2	1.58	1.59	1.59
Hornblende	$Ca_2(Mg,Fe)_4Al(Si_7AlO_{22})(OH)_2$	monocl	3.24	5.5	1.67	1.67	1.69
Huebnerite	$MnWO_4$	monocl	7.2	4.3	2.17	2.22	2.32
Humite	$Mg(OH,F)_2·3Mg_2SiO_4$	orth	3.3	6	1.625	1.636	1.657
Huntite	$Mg_3Ca(CO_3)_4$	trig	2.70				
Hydrogrossularite	$Ca_3Al_2Si_2O_8(SiO_4)_{1-m}(OH)_{4m}$	cub	3.4	6.8	1.70		
Hydromagnesite	$3MgCO_3·Mg(OH)_2·3H_2O$	monocl	2.24	3.5	1.523	1.527	1.545
Illite	$KAl_4[Si_7AlO_{20}](OH)_4$	monocl	2.8	1.5	1.56	1.59	1.59
Ilmenite	$FeTiO_3$	rhomb	4.72	5.5			
Iodyrite	AgI	hex	5.68	1.5	2.21	2.22	
Jacobsite	$MnFe_2O_4$	cub	4.87	7.8	2.3		
Jadeite	$NaAlSi_2O_6$	monocl	3.34	6	1.649	1.654	1.663
Jamesonite	$Pb_4FeSb_6S_{14}$	monocl	5.63	2.5			
Jarosite	$KFe_3(SO_4)_2(OH)_6$	rhomb	3.09	3	1.715	1.820	
Kainite	$KMg(SO_4)Cl·3H_2O$	monocl	2.15	2.8	1.494	1.505	1.516
Kaliophylite	$KAlSiO_4$	hex	2.61	6	1.532	1.537	
Kaolinite	$Al_4Si_4O_{10}(OH)_8$	tricl	2.65	2.3	1.549	1.564	1.565
Kernite	$Na_2B_4O_7·4H_2O$	monocl	1.95	2.5	1.454	1.472	1.488
Kieserite	$MgSO_4·H_2O$	monocl	2.57	3.5	1.520	1.533	1.584
Kyanite	Al_2OSiO_4	tricl	3.59	6.3	1.715	1.722	1.731
Lanarkite	$Pb_2(SO_4)O$	monocl	6.92	2.3	1.928	2.007	2.036
Lanthanite	$(La,Ce)_2(CO_3)_3·8H_2O$	rhomb	2.72	2.8	1.52	1.587	1.613
Laumontite	$Ca_4[Al_8Si_{16}O_{48}]·16H_2O$	monocl	2.3	3.3	1.508	1.517	1.519
Laurionite	Pb(OH)Cl	rhomb	6.24	3.3	2.08	2.12	2.16
Lawsonite	$CaAl_2(OH)_2Si_2O_7·H_2O$	rhomb	3.08	6	1.655	1.675	1.685
Lazulite	$(Mg,Fe)Al_2(PO_4)_2(OH)_2$	monocl	3.23	5.8	1.615	1.64	1.650
Lazurite	$Na_4SSi_3Al_3O_{12}$	cub	2.42	5.3	1.500		
Leadhillite	$Pb_4(SO_4)(CO_3)_2(OH)_2$	monocl	6.55	2.8	1.87	2.00	2.01
Lepidocrocite	FeO(OH)	orth	4.26	5	1.94	2.20	2.51
Lepidolite	$K_2(Li,Al)_{5-6}[Si_{6-7}Al_{2-1}O_{20}](OH,F)_4$	monocl	2.85	3.3	1.536	1.565	1.566
Leucite	$KAlSi_2O_6$	tetr	2.49	5.8	1.510		
Levyne	$(Ca,Na_2)Al_2Si_4O_{12}·6H_2O$	rhomb	2.10	4.5	1.496	1.501	
Litharge	PbO	tetr	9.35	2	2.535	2.665	
Loellingite	$FeAs_2$	rhomb	7.40	5.3			
Maghemite	Fe_2O_3	cub	4.88	7.8	2.63		
Magnesite	$MgCO_3$	hex	3.05	4	1.536	1.741	
Magnetite	Fe_3O_4	cub	5.17	6	2.42		
Malachite	$Cu_2(OH)_2(CO_3)$	monocl	4.05	3.8	1.655	1.875	1.909
Manganite	MnO(OH)	monocl	≈4.3	4	2.25	2.25	2.53
Manganosite	MnO	cub	5.37	5.5			
Marcasite	FeS_2	cub	5.02	6.3			
Marialite	$Na_4Al_3Si_9O_{24}Cl$	tetr	2.56	5.5	1.541	1.548	
Marshite	CuI	cub	5.67	2.5	2.346		
Mascagnite	$(NH_4)_2SO_4$	orth	1.77	2.3	1.520	1.523	1.533
Matlockite	PbClF	tetr	7.05	2.8	2.006	2.145	

Name	Formula	Crystal system	Density g/cm³	Hardness	Index of refraction n_α	n_β	n_γ
Meionite	$Ca_4Al_6Si_6O_{24}CO_3$	tetr	2.78	5.5	1.559	1.595	
Melanterite	$FeSO_4 \cdot 7H_2O$	monocl	1.89	2	1.47	1.48	1.49
Melilite	$(Ca,Na)_2(Mg,Fe,Al,Si)_3O_7$	tetr	3.00	5.5	1.639	1.645	
Mellite	$Al_2C_{12}O_{12} \cdot 18H_2O$	tetr	1.64	2.3	1.511	1.539	
Mendipite	$Pb_3O_2Cl_2$	rhomb	7.24	2.5	2.24	2.27	2.31
Mesolite	$Na_2Ca_2(Al_2Si_3O_{10})_3 \cdot 8H_2O$	orth	2.26	5	1.506		
Metacinnabar	HgS	cub	7.70	3			
Microcline	$KAlSi_3O_8$	monocl	2.56	6.3	1.522	1.526	1.530
Miersite	AgI	hex	5.68	2.5	2.20		
Millerite	NiS	hex	5.5	3.3			
Mimetite	$Pb_5(AsO_4,PO_4)_3Cl$	hex	7.24	3.8	2.128	2.147	
Minium	Pb_3O_4	tetr	8.9	2.5			
Mirabilite	$Na_2SO_4 \cdot 10H_2O$	monocl	1.46	1.8	1.394	1.396	1.398
Moissanite	SiC	hex	3.16	9.5	2.648	2.691	
Molybdenite	MoS_2	hex	5.06	1.3			
Monazite	$(Ce,La,Th)PO_4$	monocl	5.2	5	1.787	1.789	1.840
Monetite	$CaHPO_4$	tricl	2.92	3.5	1.587	1.61	1.640
Monticellite	$Ca(Mg,Fe)SiO_4$	orth	3.18	5.5	1.647	1.655	1.664
Montmorillonite	$(0.5Ca,Na)_{0.7}(Al,Mg,Fe)_4[(Si,Al)_8O_{20}](OH)_4 \cdot nH_2O$	monocl	2.5	1.5	1.55	1.57	1.57
Montroydite	HgO	orth	11.14	2.5	2.37	2.50	2.65
Mordenite	$(Na,K,Ca)[Al_2Si_{10}O_{24}] \cdot 7H_2O$	orth	2.13	3.5	1.478	1.480	1.482
Muscovite	$KAl_2Si_3AlO_{10}(OH,F)_2$	monocl	2.83	2.8	1.563	1.596	1.602
Nantokite	CuCl	cub	4.14	2.5	1.930		
Natrolite	$Na_2Al_2Si_3O_{10} \cdot 2H_2O$	orth	2.23	5	1.478	1.481	1.491
Nepheline	$Na_3KAl_4Si_4O_{16}$	hex	2.61	5.8	1.534	1.538	
Newberyite	$MgHPO_4 \cdot 3H_2O$	orth	2.13	3.3	1.514	1.517	1.533
Niccolite	NiAs	hex	7.77	5.3			
Norbergite	$Mg(OH,F)_2 \cdot Mg_2SiO_4$	orth	3.21	6.5	1.565	1.573	1.592
Nosean	$Na_8Al_6Si_6O_{24}SO_4$	cub	2.35	5.5	1.495		
Oldhamite	CaS	cub	2.59	4	2.137		
Oligoclase	$([NaSi]_{0.9\text{-}0.7}[CaAl]_{0.1\text{-}0.3})AlSi_2O_8$	tricl	2.64	6.3	1.539	1.543	1.547
Olivenite	$Cu_2(AsO_4)(OH)$	rhomb	4.2	3	1.77	1.80	1.85
Olivine	$(Mg,Fe)SiO_4$	rhomb	3.81	6.8	1.73	1.76	1.78
Opal	$SiO_2 \cdot nH_2O$	amorp	1.9	5	1.44		
Orpiment	As_2S_3	monocl	3.46	1.8	2.40	2.81	3.02
Orthoclase	$KAlSi_3O_8$	monocl	2.56	6	1.523	1.527	1.531
Orthopyroxene	$(Mg,Fe)SiO_3$	rhomb	3.6	5.5	1.709	1.712	1.723
Paragonite	$NaAl_2Si_3AlO_{10}(OH)_2$	monocl	2.85	2.5	1.572	1.602	1.605
Parisite	$(Ce,La,Na)FCO_3 \cdot CaCO_3$	hex	4.42	4.5	1.672	1.771	
Pectolite	$Ca_2NaH(SiO_3)_3$	tricl	2.88	4.8	1.603	1.610	1.639
Penfieldite	$Pb_4Cl_6(OH)_2$	hex	6.6		2.13	2.21	
Pentlandite	$(Fe,Ni)_9S_8$	cub	4.8	3.8			
Percylite	$PbCuCl_2(OH)_2$	cub		2.5	2.05		
Periclase	MgO	cub	3.6	5.5	1.735		
Perovskite	$CaTiO_3$	cub	3.98	5.5	2.34		
Petalite	$LiAlSi_4O_{10}$	monocl	2.42	6.5	1.506	1.511	1.519
Pharmacosiderite	$Fe_3(AsO_4)_2(OH)_3 \cdot 5H_2O$	cub	2.80	2.5	1.690		
Phenakite	Be_2SiO_4	rhomb	2.98	7.5	1.654	1.670	
Phillipsite	$K(Ca_{0.5},Na)_2[Al_3Si_5O_{16}] \cdot 6H_2O$	monocl	2.2	4.3	1.494	1.497	1.505
Phlogopite	$KMg_3AlSi_3O_{10}(OH,F)_2$	monocl	2.83	2.3	1.560	1.597	1.598
Phosgenite	$Pb_2(CO_3)Cl_2$	tetr	6.13	2.5	2.118	2.145	
Piemontite	$Ca_2(Mn,Fe,Al)_3O(Si_2O_7)(SiO_4)(OH)$	monocl	3.49	6	1.762	1.773	1.796
Pigeonite	$(Mg,Fe,Ca)(Mg,Fe)Si_2O_6$	monocl	3.38	6	1.702	1.703	1.728
Pollucite	$CsAlSi_2O_6$	tetr	2.9	6.5	1.517		
Polybasite	$(Ag,Cu)_{16}Sb_2S_{11}$	monocl	6.1	2.5			
Powellite	$Ca(Mo,W)O_4$	tetr	4.35	3.8	1.971	1.980	
Prehnite	$Ca_2Al_2Si_3O_{10}(OH)_2$	rhomb	2.93	6.3	1.622	1.628	1.648
Proustite	Ag_3AsS_3	rhomb	5.57	2.3	2.792	3.088	
Pseudobrookite	Fe_2TiO_5	rhomb	4.36	6	2.38	2.39	2.42
Psilomelane	$BaMn_9O_{16}(OH)_4$	rhomb	4.71	5.5			

Name	Formula	Crystal system	Density g/cm³	Hardness	Index of refraction n_α	n_β	n_γ
Pumpellyite	$Ca_2Al_2(Al,Fe,Mg)[Si_2(O,OH)_7] (SiO_4)(OH,O)_3$	monocl	3.21	5.5	1.688	1.695	1.705
Pyrargyrite	Ag_3SbS_3	rhomb	5.85	2.5	2.88	3.08	
Pyrite	FeS_2	cub	5.02	6.3			
Pyrochlore	$NaCaNb_2O_6F$	cub	5.3	5.3			
Pyrochroite	$Mn(OH)_2$	hex	3.26	2.5	1.68	1.72	
Pyrolusite	MnO_2	tetr	5.08	6.3			
Pyromorphite	$Pb_5(PO_4,AsO_4)_3Cl$	hex	7.04	3.8	2.048	2.058	
Pyrope	$Mg_3Al_2Si_3O_{12}$	cub	3.58	6.8	1.714		
Pyrophyllite	$Al_2Si_4O_{10}(OH)_2$	monocl	2.78	1.5	1.545	1.579	1.599
Pyrrhotite	Fe_7S_8	hex	4.62	4			
Quartz	SiO_2	hex	2.65	7	1.544	1.553	
Rammelsbergite	$NiAs_2$	orth	7.1	5.8			
Raspite	$PbWO_4$	monocl	8.46	2.8	1.27	1.27	1.30
Realgar	As_4S_4	monocl	3.5	1.8	2.538	2.684	2.704
Rhodochrosite	$MnCO_3$	hex	3.70	3.8	1.597	1.816	
Rhodonite	$(Mn,Fe,Ca)SiO_3$	orth	3.48	6	1.725	1.729	1.737
Riebeckite	$Na_2Fe_5(Si_8O_{22})(OH)_2$	monocl	3.3	5	1.675	1.683	1.694
Rutile	TiO_2	tetr	4.23	6.2	2.609	2.900	
Safflorite	$(Co,Fe)As_2$	rhomb	7.3	4.8			
Samarskite	$(Y,Er,Ce,U,Ca,Fe,Pb,Th) (Nb,Ta,Ti,Sn)_2O_6$	rhomb	5.69	5.5	2.200		
Sapphirine	$(Mg,Fe)_2Al_4O_6SiO_4$	monocl	3.49	7.5	1.709	1.712	1.715
Scapolite	$(Na,Ca)_4Al_3(Al,Si)_3Si_6O_{24} (Cl,F,OH,CO_3,SO_4)$	tetr	2.64	5.5	1.551	1.573	
Scheelite	$CaWO_4$	tetr	6.06	4.8	1.920	1.936	
Scolecite	$CaAl_2Si_3O_{10}\cdot 3H_2O$	monocl	2.27	5	1.510	1.518	1.519
Scorodite	$Fe(AsO_4)\cdot 2H_2O$	rhomb	3.28	3.8	1.784	1.795	1.814
Sellaite	MgF_2	tetr	3.15	5	1.378	1.390	
Senarmontite	Sb_2O_3	cub	5.58	2.3	2.087		
Serpentine	$Mg_3Si_2O_5(OH)_4$	monocl	2.55	3	1.55	1.56	1.56
Siderite	$FeCO_3$	hex	3.9	4.3	1.635	1.875	
Sillimanite	Al_2OSiO_4	rhomb	3.25	7	1.658	1.660	1.660
Skutterudite	$(Co,Ni)As_3$	cub	6.8	5.8			
Smithsonite	$ZnCO_3$	rhomb	4.4	4.3	1.621	1.848	
Sodalite	$Na_8Al_6Si_6O_{24}Cl_2$	cub	2.30	5.8	1.485		
Sperrylite	$PtAs_2$	cub	10.58	6.5			
Spessartite	$Mn_3Al_2Si_3O_{12}$	cub	4.19	6.8	1.800		
Sphalerite	ZnS	cub	4.0	3.8	2.369		
Sphene	$CaTiSiO_4(O,OH,F)$	monocl	3.50	5	1.90	1.95	2.03
Spinel	$MgAl_2O_4$	cub	3.55	7.8	1.719		
Spodumene	$LiAlSi_2O_6$	monocl	3.13	6.8	1.656	1.662	1.671
Stannite	Cu_2FeSn_4	tetr	4.4	4			
Staurolite	$(Fe,Mg,Zn)_2(Al,Fe,Ti)_9O_6 [(Si,Al)O_4]_4(O,OH)_2$	monocl	3.79	7.5	1.743	1.747	1.755
Stercorite	$Na(NH_4)H(PO_4)\cdot 4H_2O$	tricl	1.62	2	1.439	1.442	1.469
Stibiotantalite	$Sb(Ta,Nb)O_4$	rhomb	6.6	5.5	2.38	2.41	2.46
Stibnite	Sb_2S_3	orth	4.56	2			
Stilbite	$NaCa_2[Al_5Si_{13}O_{36}]\cdot 14H_2O$	monocl	2.2	3.8	1.492	1.499	1.503
Stilpnomelane	$(K,Na,Ca)_{0.6}(Fe,Mg)_6Si_8Al (O,OH)_{27}\cdot 2H_2O$	monocl	2.8	3.5	1.585	1.665	1.665
Stolzite	$PbWO_4$	tetr	8.2	2.8	2.19	2.27	
Strengite	$FePO_4\cdot 2H_2O$	orth	2.87	4	1.707	1.719	1.741
Strontianite	$SrCO_3$	orth	3.5	3.5	1.518	1.666	1.668
Struvite	$Mg(NH_4)(PO_4)\cdot 6H_2O$	rhomb	1.71	2	1.495	1.496	1.504
Sulfur	S	orth	2.07	2	1.958	2.038	2.245
Sylvanite	$(Ag,Au)Te_2$	monocl	8.16	1.8			
Sylvite	KCl	cub	1.99	2	1.490		
Talc	$Mg_3Si_4O_{10}(OH)_2$	monocl	2.71	1	1.545	1.592	1.595
Tantalite	$(Fe,Mn)(Ta,Nb)_2O_6$	rhomb	7.95	6.5	2.26	2.32	2.43
Tapiolite	$FeTa_2O_6$	tetr	7.9	6.3	2.27	2.42	
Tellurobismuthite	Bi_2Te_3	hex	7.74	1.8			
Terlinguaite	Hg_2OCl	monocl	8.73	2.5	2.35	2.64	2.66
Tetrahedrite	$(Cu,Fe)_{12}Sb_4S_{13}$	cub	4.9	3.8			
Thenardite	Na_2SO_4	orth	2.7	2.8	1.468	1.475	1.483

Name	Formula	Crystal system	Density g/cm³	Hardness	Index of refraction n_α	n_β	n_γ
Thermonatrite	$Na_2CO_3 \cdot H_2O$	orth	2.25	1.3	1.420	1.506	1.524
Thomsenolite	$NaCaAlF_6 \cdot H_2O$	monocl	2.98	2	1.407	1.414	1.415
Thorianite	ThO_2	cub	10.0	6.5	2.200		
Thorite	$ThSiO_4$	tetr	6.7	4.8	1.8		
Topaz	$Al_2SiO_4(OH,F)_2$	rhomb	3.53	8	1.618	1.620	1.627
Torbernite	$Cu(UO_2)_2(PO_4)_2 \cdot 8H_2O$	tetr	3.22	2.3	1.582	1.592	
Tourmaline	$Na(Mg,Fe,Mn,Li,Al)_3Al_6Si_6O_{18}(BO_3)_3$	rhomb	3.14	7	1.62	1.65	
Tremolite	$Ca_2Mg_5Si_8O_{22}(OH,F)_2$	monocl	3.0	5.5	1.599	1.612	1.622
Trevorite	$NiFe_2O_4$	cub	5.33	7.8	2.3		
Tridymite	SiO_2	hex	2.27	7	1.475	1.476	1.479
Triphyllite-Lithiophyllite	$Li(Fe,Mn)PO_4$	rhomb	3.46	4.5	1.68	1.68	1.69
Troegerite	$(UO_2)_3(AsO_4)_2 \cdot 12H_2O$	tetr		2.5	1.59	1.630	
Troilite	FeS	hex	4.7	4			
Trona	$Na_3H(CO_3)_2 \cdot 2H_2O$	monocl	2.14	2.8	1.412	1.492	1.540
Turquois	$Cu(Al,Fe)_6(PO_4)_4(OH)_8 \cdot 4H_2O$	tricl	2.9	5.3	1.70	1.73	1.75
Ullmannite	NiSbS	cub	6.65	5.3			
Uraninite	UO_2	cub	11.0	5.5			
Uvarovite	$Ca_3Cr_2Si_3O_{12}$	cub	3.83	6.8	1.865		
Valentinite	Sb_2O_3	orth	5.7	2.8	2.18	2.35	2.35
Vanadinite	$Pb_5(VO_4)_3Cl$	hex	6.8	2.9	2.350	2.416	
Variseite-Strengite	$(Al,Fe)(PO_4) \cdot 2H_2O$	rhomb	2.72	4	1.635	1.654	1.668
Vaterite	$CaCO_3$	hex	2.71		1.550	1.645	
Vermiculite	$(Mg,Ca)_{0.7}(Mg,Fe,Al)_6[(Al,Si)_8O_{20}](OH)_4 \cdot 8H_2O$	monocl	2.3	1.5	1.542	1.556	1.556
Vesuvianite	$Ca_{10}(Mg,Fe)_2Al_4(Si_2O_7)_2(SiO_4)_5(OH,F)_4$	tetr	3.33	6.5	1.72	1.73	
Villiaumite	NaF	cub	2.78	2.3	1.327		
Vivianite	$Fe_3(PO_4)_2 \cdot 8H_2O$	monocl	2.58	1.8	1.598	1.629	[illegible]
Wagnerite	$Mg_2(PO_4)F$	monocl	3.15	5.3	1.568	1.572	1.582
Wavellite	$Al_3(OH)_3(PO_4)_2 \cdot 5H_2O$	rhomb	2.36	3.6	1.527	1.535	1.553
Whewellite	$CaC_2O_4 \cdot H_2O$	cub	2.2	2.8	1.491	1.554	1.650
Willemite	Zn_2SiO_4	hex	4.1	5.5	1.691	1.719	
Witherite	$BaCO_3$	orth	4.29	3.5	1.529	1.676	1.677
Wolframite	$(Fe,Mn)WO_4$	monocl	7.3	4.3	2.26	2.32	2.42
Wollastonite	$CaSiO_3$	monocl	2.92	4.8	1.628	1.639	1.642
Wulfenite	$PbMoO_4$	tetr	6.7	2.9	2.283	2.403	
Wurtzite	ZnS	hex	4.09	3.8	2.356	2.378	
Xenotime	YPO_4	tetr	4.8	4.5	1.721	1.816	
Zeunerite	$Cu(UO_2)_2(AsO_4)_2 \cdot 10H_2O$	tetr			1.606		
Zincite	ZnO	hex	5.6	4	2.013	2.029	
Zircon	$ZrSiO_4$	tetr	4.6	7.5	1.94	1.99	
Zoisite	$Ca_2Al_3Si_3O_{12}(OH)$	rhomb	3.26	6	1.695	1.699	1.711

CRYSTALLOGRAPHIC DATA ON MINERALS

This table contains x-ray crystallographic data on about 400 common minerals, as well as selected crystalline elements. Entries are arranged alphabetically by mineral name. The columns are:

Name: Common name of the mineral.

Formula: Chemical formula for a typical sample of the mineral. Composition often varies considerably with the origin of the sample.

Crystal system: tricl = triclinic; monocl = monoclinic; orth = orthorhombic; tetr = tetragonal; hex = hexagonal; rhomb = rhombohedral; cubic = cubic.

Structure type: Prototype for the structural arrangement of the crystallographic cell.

Z: Number of formula units per the unit cell.

a, b, c: Lengths of the cell edges in Å (1Å = 10^{-8} cm).

α, β, γ: Angles between cell axes.

References

1. Robie, R.A., Bethke, P.M., and Beardsley, K.M., *U. S. Geological Survey Bulletin 1248*, U. S. Government Printing Office, Washington, D.C.
2. Donnay, J.D.H., and Ondik, H.M., *Crystal Data Determinative Tables, Third Edition, Volume 2, Inorganic Compounds*, Joint Committee on Powder Diffraction Standards, Swarthmore, PA, 1973.
3. Deer, W.A., Howie, R.A., and Zussman, J., *An Introduction to the Rock-Forming Minerals, 2nd Edition*, Longman Scientific & Technical, Harlow, Essex, 1992.

Name	Formula	Crystal system	Structure type	Z	a/Å	b/Å	c/Å	α	β	γ
Acanthite	Ag_2S	monocl		4	4.228	6.928	7.862		99.58°	
Acmite (Aegirine)	$NaFe(SiO_3)_2$	monocl	diopside	4	9.658	8.795	5.294		107.42°	
Akermanite	$Ca_2MgSi_2O_7$	tetr	melilite	2	7.8435		5.010			
Alabandite	MnS	cubic	rock salt	4	5.223					
Almandine (Almandite)	$Fe_3Al_2Si_3O_{12}$	cubic	garnet	8	11.526					
Altaite	PbTe	cubic	rock salt	4	6.4606					
Aluminum	Al	cubic	copper	4	4.049					
Alunite	$KAl_3(SO_4)_2(OH)_6$	rhomb		3	6.982		17.32			
Analcite	$NaAlSi_2O_6{\cdot}H_20$	cubic		16	13.733					
Anatase	TiO_2	tetr		4	3.785		9.514			
Andalusite	Al_2OSiO_4	orth		4	7.7959	7.8983	5.5583			
Andradite	$Ca_3Fe_2Si_3O_{12}$	cubic	garnet	8	12.048					
Anglesite	$PbSO_4$	orth	barite	4	8.480	5.398	6.958			
Anhydrite	$CaSO_4$	orth	anhydrite	4	6.991	6.996	6.238			
Annite	$KFe_3[AlSi_3O_{10}](OH)_2$	monocl	1M mica	2	10.29	9.33	5.39		105.1°	
Anorthite	$CaAl_2Si_2O_8$	tricl	primitive cell	8	8.177	12.877	14.169	93.17°	115.85°	91.22°
Anthophyllite	$Mg_7Si_8O_{22}(OH)_2$	orth		4	18.61	18.01	5.24			
Antimony	Sb	rhomb	arsenic	6	4.2996		11.2516			
Aragonite	$CaCO_3$	orth	aragonite	4	5.741	7.968	4.959			
Arcanite	K_2SO_4	orth	arcanite	4	5.772	10.072	7.483			
Argentite	Ag_2S	cubic		2	4.870					
Argentopyrite	$AgFe_2S_3$	orth		4	6.64	11.47	6.45			
Arsenic	As	rhomb	arsenic	6	3.760		10.555			
Arsenolite	As_2O_3	cubic	diamond	16	11.074					
Arsenopyrite	FeAsS	tricl		4	5.760	5.690	5.785	90.00°	112.23°	90.00°
Azurite	$Cu_3(OH)_2(CO_3)_2$	monocl		2	5.008	5.844	10.336		92.45°	
Baddeleyite	ZrO_2	monocl	baddeleyite	4	5.1454	5.2075	5.3107		99.23°	
Banalsite	$BaNa_2Al_4Si_4O_{16}$	orth		4	8.50	9.97	16.72			
Barite	$BaSO_4$	orth	barite	4	8.878	5.450	7.152			
Berlinite	$AlPO_4$	hex	α-quartz	3	4.942		10.97			
Beryl	$Be_3Al_2(SiO_3)_6$	hex	beryl	2	9.215		9.192			
Berzelianite	Cu_2Se	cubic		4	5.85					
Bismite	Bi_2O_3	monocl	pseudo-orth	4	7.48	8.14	5.83		112.9°	
Bismuth	Bi	rhomb	arsenic	6	4.5367		11.8383			
Bismuthinite	Bi_2S_3	orth	stibnite	4	11.150	11.300	3.981			
Bixbyite	Mn_2O_3	cubic	thallium trioxide	16	9.411					
Boehmite	AlO(OH)	orth	lepidocrocite	4	2.868	12.227	3.700			
Borax	$Na_2B_4O_7{\cdot}10H_2O$	monocl		4	11.858	10.674	12.197		106.68°	

Name	Formula	Crystal system	Structure type	Z	a/Å	b/Å	c/Å	α	β	γ
Bornite (metastable)	Cu_5FeS_4	cubic		8	10.94					
Breithauptite	NiSb	hex	niccolite	2	3.942		5.155			
Brochantite	$Cu_4SO_4(OH)_6$	monocl		4	13.066	9.85	6.022		103.27°	
Bromargyrite	AgBr	cubic	rock salt	4	5.7745					
Bromellite	BeO	hex	zincite	2	2.6979		4.3772			
Brookite	TiO_2	orth		8	5.456	9.182	5.143			
Brucite	$Mg(OH)_2$	hex	cadmium iodide	1	3.147		4.769			
Bunsenite	NiO	cubic	rock salt	4	4.177					
Bustamite	$CaMn(SiO_3)_2$	tricl		6	7.736	7.157	13.824	90.52°	94.58°	103.87°
Cadmium telluride	CdTe	cubic	sphalerite	4	6.4805					
Cadmoselite	CdSe	hex	zincite	2	4.2977		7.0021			
Calcite	$CaCO_3$	rhomb	calcite	6	4.9899		17.064			
Calomel	Hg_2Cl_2	tetr		4	4.478		10.910			
Carbonate-apatite	$Ca_{10}(PO_4)_6CO_3{\cdot}H_2O$	hex	apatite	1	9.436		6.883			
Cassiterite	SnO_2	tetr	rutile	2	4.738		3.188			
Cattierite	CoS_2	cubic	pyrite	4	5.5345					
Celestite	$SrSO_4$	orth	barite	4	8.359	5.352	6.866			
Celsian	$BaAl_2Si_2O_8$	monocl		8	8.627	13.045	14.408		115.20°	
Cerianite	CeO_2	cubic	fluorite	4	5.4110					
Cerussite	$PbCO_3$	orth	aragonite	4	6.152	8.436	5.195			
Cervantite	Sb_2O_4	orth		4	5.424	11.76	4.804			
Chalcanthite	$CuSO_4{\cdot}5H_2O$	tricl		2	6.1045	10.72	5.949	97.57°	107.28°	77.43°
Chalcocite	Cu_2S	orth		96	11.881	27.323	13.491			
Chalcopyrite	$CuFeS_2$	tetr		4	5.2988		10.434			
Chlorapatite	$Ca_5(PO_4)_3Cl$	hex	apatite	2	9.629		6.777			
Chlorargyrite	AgCl	cubic	rock salt	4	5.5491					
Chloritoid	$FeAl_4O_2(SiO_4)_2(OH)_4$	monocl		8	9.48	5.48	18.18		101.77°	
Chloromagnesite	$MgCl_2$	rhomb		3	3.632		17.795			
Chondrodite	$2Mg_2SiO_4{\cdot}MgF_2$	monocl		2	7.89	4.743	10.29		109.03°	
Chrysoberyl	$BeAl_2O_4$	orth	olivine	4	5.4756	9.4041	4.4267			
Cinnabar	HgS	hex	cinnabar	3	4.149		9.495			
Claudetite	As_2O_3	monocl		4	5.339	12.984	4.5405		94.27°	
Clausthalite	PbSe	cubic	rock salt	4	6.1255					
Clinoenstatite	$MgSiO_3$	monocl		8	9.620	8.825	5.188		108.33°	
Clinoferrosilite	$FeSiO_3$	monocl		8	9.7085	9.0872	5.2284		108.43°	
Clinohumite	$4Mg_2SiO_4{\cdot}MgF_2$	monocl		2	13.68	4.75	10.27		100.83°	
Clinozoisite	$Ca_2Al_3(SiO_4)_3OH$	monocl		2	8.887	5.581	10.14		115.93°	
Cobalt olivine	Co_2SiO_4	orth	olivine	4	4.782	10.301	6.003			
Cobalt oxide	CoO	cubic	rock salt	4	4.260					
Cobalt sulfide	CoS	cubic	sphalerite	4	5.339					
Cobalt titanate	$CoTiO_3$	rhomb	ilmenite	6	5.066		13.918			
Cobalticalcite	$CoCO_3$	rhomb	calcite	6	4.6581		14.958			
Cobaltite	CoAsS	cubic	NiSbS	4	5.60					
Coesite	SiO_2	monocl		16	7.152	12.379	7.152		120.00°	
Coffinite	$USiO_4$	tetr	zircon	4	6.995		6.263			
Colemanite	$Ca_2B_6O_{11}{\cdot}5H_2O$	monocl		4	8.743	11.264	6.102		110.12°	
Coloradoite	HgTe	cubic	sphalerite	4	6.4600					
Cooperite	PtS	tetr		2	3.4699		6.1098			
Copper	Cu	cubic	face-centered cubic	4	3.6150					
Corundum	Al_2O_3	rhomb	corundum	6	4.7591		12.9894			
Cotunnite	$PbCl_2$	orth		4	4.535	7.62	9.05			
Covellite	CuS	hex		6	3.792		16.34			
Cristobalite (α)	SiO_2	tetr		4	4.971		6.918			
Cristobalite (β)	SiO_2	cubic		8	7.1382					
Cryolite	Na_3AlF_6	monocl		2	5.40	5.60	7.776		90.18°	
Cubanite	$CuFe_2S_3$	orth		4	6.46	11.12	6.23			
Cummingtonite	$(Mg,Fe,Mn)_7(Si_4O_{11})_2(OH)_2$	monocl	tremolite	2	9.522	18.223	5.332		101.92°	
Cuprite	Cu_2O	cubic		2	4.2696					
Danburite	$CaB_2Si_2O_8$	orth		4	8.04	8.77	7.74			

Name	Formula	Crystal system	Structure type	Z	a/Å	b/Å	c/Å	α	β	γ
Datolite	$CaBSiO_4(OH)$	monocl		4	9.62	7.60	4.84		90.15°	
Daubreeite	$FeCr_2S_4$	cubic	spinel	8	9.966					
Diamond	C	cubic	diamond	8	3.5670					
Diaspore	$AlO(OH)$	orth		4	4.401	9.421	2.845			
Dickite	$Al_2Si_2O_5(OH)_4$	monocl		4	5.150	8.940	14.736		103.58°	
Digenite	$Cu_{1.79}S$	cubic	deformed fluorite	4	5.5695					
Diopside	$CaMg(SiO_3)_2$	monocl	diopside	4	9.743	8.923	5.251		105.93°	
Dioptase	$CuSiO_2(OH)_2$	rhomb	phenacite	18	14.61		7.80			
Dolerophanite	$Cu_2O(SO_4)$	monocl		4	8.334	6.312	7.628		108.4°	
Dolomite	$CaMg(CO_3)_2$	rhomb	calcite	3	4.8079		16.010			
Dravite	$NaMg_3Al_6B_3Si_6O_{27}(OH)_4$	rhomb	tourmaline	3	15.942		7.224			
Elbaite	$NaLiAl_{7.67}B_3Si_6O_{27}(OH)_4$	rhomb	tourmaline	3	15.842		7.009			
Enargite	Cu_3AsS_4	orth		2	6.426	7.422	6.144			
Enstatite	$MgSiO_3$	orth		16	8.829	18.22	5.192			
Epidote	$Ca_2Al_2(Al,Fe)OH(SiO_4)_3$	monocl		2	8.89	5.63	10.19		115.40°	
Epsomite	$MgSO_4{\cdot}7H_2O$	orth		4	11.86	11.99	6.858			
Eskolaite	Cr_2O_3	rhomb	corundum	6	4.9607		13.599			
Eucairite	AgCuSe	orth		10	4.105	20.35	6.31			
Euclase	$AlBeSiO_4(OH)$	monocl		4	4.763	14.29	4.618		100.25°	
Famatimite	Cu_3SbS_4	tetr		2	5.384		10.770			
Fayalite	Fe_2SiO_4	orth	olivine	4	4.817	10.477	6.105			
Fe-Cordierite	$Fe_2Al_3(AlSi_5O_{18})$	orth	cordierite	4	9.726	17.065	9.287			
Fe-Gehlenite	$Ca_2Fe_2SiO_7$	tetr	melilite	2	7.54		4.855			
Fe-Indialite	$Fe_2Al_3(AlSi_5O_{18})$	hex	beryl	2	9.860		9.285			
Fe-Leucite	$KFeSi_2O_6$	tetr		16	13.205		13.970			
Fe-Microcline	$KFeSi_3O_8$	tricl		4	8.68	13.10	7.340	90.75°	116.05°	86.23°
Fe-Sanidine	$KFeSi_3O_8$	monocl		4	8.689	13.12	7.319		116.10°	
Fe-Skutterudite	$FeAs_{2.95}$	cubic		8	8.1814					
Ferberite	$FeWO_4$	monocl	wolframite	2	4.732	5.708	4.965		90.00°	
Ferriannite	$KFe_3[FeSi_3O_{10}](OH)_2$	monocl		2	5.430	9.404	10.341		100.07°	
Ferroselite	$FeSe_2$	orth	marcasite	2	4.801	5.778	3.587			
Ferrotremolite	$Ca_2Fe_5[Si_8O_{22}](OH)_2$	monocl	tremolite	2	9.97	18.34	5.30		104.50°	
Fluor-edenite	$NaCa_2Mg_5[AlSi_7O_{22}]F_2$	monocl	tremolite	2	9.847	18.00	5.282		104.83°	
Fluor-humite	$3Mg_2SiO_4{\cdot}MgF_2$	orth		4	10.243	20.72	4.735			
Fluor-norbergite	$Mg_2SiO_4{\cdot}MgF_2$	orth		4	8.727	10.271	4.709			
Fluor-phlogopite	$KMg_3[AlSi_3O_{10}]F_2$	monocl	1M mica	2	5.299	9.188	10.135		99.92°	
Fluor-richterite	$Na_2CaMg_5[Si_8O_{22}]F_2$	monocl	tremolite	2	9.823	17.96	5.268		104.33°	
Fluor-tremolite	$Ca_2Mg_5[Si_8O_{22}]F_2$	monocl	tremolite	2	9.781	18.01	5.267		104.52°	
Fluorapatite	$Ca_5(PO_4)_3F$	hex	apatite	2	9.3684		6.8841			
Fluorite	CaF_2	cubic	fluorite	4	5.4638					
Forsterite	Mg_2SiO_4	orth	olivine	4	4.758	10.214	5.984			
Frohbergite	$FeTe_2$	orth	marcasite	2	5.265	6.265	3.869			
Gahnite	$ZnAl_2O_4$	cubic	spinel	8	8.0848					
Galaxite	$MnAl_2O_4$	cubic	spinel	8	8.258					
Galena	PbS	cubic	rock salt	4	5.9360					
Gallium oxide	Ga_2O_3	rhomb	corundum	6	4.9793		13.429			
Gehlenite	$Ca_2Al_2SiO_7$	tetr	melilite	2	7.690		5.0675			
Geikielite	$MgTiO_3$	rhomb	ilmenite	6	5.054		13.898			
Gerhardite	$Cu_2(NO_3)(OH)_3$	orth		4	6.075	13.812	5.592			
Gersdorfite	NiAsS	cubic		4	5.693					
Gibbsite	$Al(OH)_3$	monocl		8	9.719	5.0705	8.6412		94.57°	
Glauchroite	$CaMnSiO_4$	orth	olivine	4	4.944	11.19	6.529			
Glaucodot	(Co,Fe)AsS	orth		24	6.64	28.39	5.64			
Glaucophane I	$Na_2Mg_3Al_2[Si_8O_{22}](OH)_2$	monocl	tremolite	2	9.748	17.915	5.273		102.78°	
Glaucophane II	$Na_2Mg_3Al_2[Si_8O_{22}](OH)_2$	monocl	tremolite	2	9.663	17.696	5.277		103.67°	
Goethite	$FeO(OH)$	orth		4	4.596	9.957	3.021			
Gold	Au	cubic	face-centered cubic	4	4.0786					
Goldmanite	$Ca_3V_2Si_3O_{12}$	cubic	garnet	8	12.070					
Goslarite	$ZnSO_4{\cdot}7H_2O$	orth	epsomite	4	11.779	12.050	6.822			

Name	Formula	Crystal system	Structure type	Z	a/Å	b/Å	c/Å	α	β	γ
Graphite	C	hex	graphite	4	2.4612		6.7079			
Greenockite	CdS	hex	zincite	2	4.1354		6.7120			
Greigite	Fe_3S_4	cubic	spinel	8	9.876					
Grossularite	$Ca_3Al_2Si_3O_{12}$	cubic	garnet	8	11.851					
Grunerite	$Fe_7[Si_8O_{22}](OH)_2$	monocl	tremolite	2	9.572	18.44	5.342		101.77°	
Gudmundite	FeSbS	monocl		8	10.00	5.93	6.73		90.00°	
Gypsum	$CaSO_4{\cdot}2H_2O$	monocl		4	5.68	15.18	6.29		113.83°	
Hafnia	HfO_2	monocl	baddeleyite	4	5.1156	5.1722	5.2948		99.18°	
Halite	NaCl	cubic	rock salt	4	5.6402					
Hambergite	$Be_2(OH,F)BO_3$	orth		8	9.755	12.201	4.426			
Hardystonite	$Ca_2ZnSi_2O_7$	tetr	melilite	2	7.87		5.01			
Hauerite	MnS_2	cubic	pyrite	4	6.1014					
Hausmannite	Mn_3O_4	tetr		8	8.136		9.422			
Hawleyite	CdS	cubic	sphalerite	4	5.833					
Heazelwoodite	Ni_3S_2	rhomb		3	5.746		7.134			
Hedenbergite	$CaFe(SiO_3)_2$	monocl	diopside	4	9.854	9.024	5.263		104.23°	
Hematite	Fe_2O_3	rhomb	corundum	6	5.025		13.735			
Hemimorphite	$Zn_4(OH)_2Si_2O_7{\cdot}H_2O$	orth		2	8.370	10.719	5.120			
Hercynite	$Fe(AlO_2)_2$	cubic	spinel	8	8.150					
Herzenbergite	SnS	orth	germanium sulfide	4	4.328	11.190	3.978			
Hessite	Ag_2Te	monocl		4	8.13	4.48	8.09		111.9°	
Hexahydrite	$MgSO_4{\cdot}6H_2O$	monocl		8	10.110	7.212	24.41		98.30°	
High albite (Analbite)	$NaAlSi_3O_8$	tricl		4	8.160	12.870	7.106	93.54°	116.36°	90.19°
High argentite	Ag_2S	cubic		4	6.269					
High bornite	Cu_5FeS_4	cubic		1	5.50					
High carnegeite	$NaAlSiO_4$	cubic		4	7.325					
High chalcocite	Cu_2S	hex		2	3.961		6.722			
High clinoenstatite	$MgSiO_3$	tricl		8	10.000	8.934	5.170	88.27°	70.03°	91.01°
High digenite	Cu_2S	cubic		4	5.725					
High germania	GeO_2	hex	α–quartz	3	4.987		5.652			
High leucite	$KAlSi_2O_6$	cubic		16	13.43					
High naumanite	Ag_2Se	cubic		2	4.993					
High sanidine	$KAlSi_3O_8$	monocl		4	8.615	13.031	7.177		115.98°	
Huebnerite	$MnWO_4$	monocl	wolframite	2	4.834	5.758	4.999		91.18°	
Huntite	$Mg_3Ca(CO_3)_4$	rhomb	calcite	3	9.498		7.816			
Hydroxylapatite	$Ca_5(PO_4)_3OH$	hex	apatite	2	9.418		6.883			
Ice	H_2O	hex		4	4.5212		7.3666			
Ilmenite	$FeTiO_3$	rhomb	ilmenite	6	5.093		14.055			
Indialite (Cordierite)	$Mg_2Al_3(AlSi_5O_{18})$	hex	beryl	2	9.7698		9.3517			
Iodargyrite	AgI	hex	zincite	2	4.5955		7.5005			
Iron (α)	Fe	cubic	body-centered cubic	2	2.8664					
Jacobsite	$MnFe_2O_4$	cubic	spinel	8	8.499					
Jadeite	$NaAl(SiO_3)_2$	monocl	diopside	4	9.409	8.564	5.220		107.50°	
Jalpaite	$Ag_{1.55}Cu_{0.45}S$	tetr		16	8.673		11.756			
Johannsenite	$CaMn(SiO_3)_2$	monocl	diopside	4	9.83	9.04	5.27		105.00°	
Kaliophilite	$KAlSiO_4$	hex		54	26.930		8.522			
Kalsilite	$KAlSiO_4$	hex		2	5.1597		8.7032			
Kaolinite	$Al_2Si_2O_5(OH)_4$	tricl		2	5.155	8.959	7.407	91.68°	104.87°	89.93°
Karelianite	V_2O_3	rhomb	corundum	6	4.952		14.002			
Keatite	SiO_2	tetr		12	7.456		8.604			
Kernite	$Na_2B_4O_7{\cdot}4H_2O$	monocl		4	7.022	9.151	15.676		108.83°	
Kerschsteinite	$CaFeSiO_4$	orth	olivine	4	4.886	11.146	6.434			
Klockmannite	CuSe	hex	deformed covellite	78	14.206		17.25			
Knebelite	$MnFeSiO_4$	orth	olivine	4	4.854	10.602	6.162			
Kyanite	Al_2OSiO_4	tricl		4	7.123	7.848	5.564	89.92°	101.25°	105.97°
Larnite	Ca_2SiO_4	monocl		4	5.48	6.76	9.28		94.55°	
Laurite	RuS_2	cubic	pyrite	4	5.60					

Name	Formula	Crystal system	Structure type	Z	a/Å	b/Å	c/Å	α	β	γ
Lawrencite	$FeCl_2$	rhomb		3	3.593		17.58			
Lawsonite	$CaAl_2Si_2O_7(OH)_2 \cdot H_2O$	orth		4	8.787	5.836	13.123			
Lead	Pb	cubic	face-centered cubic	4	4.9505					
Leonhardtite	$MgSO_4 \cdot 4H_2O$	monocl		4	5.922	13.604	7.905		90.85°	
Lepidocrocite	FeO(OH)	orth		4	3.868	12.525	3.066			
Lepidolite	$K_2Al_3Li_2AlSi_7O_{20}(OH)_4$	monocl	2M2 mica	2	9.2	5.3	20.0		98.00°	
Leucite	$KAlSi_2O_6$	tetr		16	13.074		13.738			
Lime	CaO	cubic	rock salt	4	4.8108					
Lime olivine	Ca_2SiO_4	orth	olivine	4	5.091	11.371	6.782			
Linnaeite	Co_3S_4	cubic	spinel	8	9.401					
Litharge	PbO	tetr		2	3.9759		5.023			
Loellingite	$FeAs_2$	orth	marcasite	2	5.300	5.981	2.882			
Low albite	$NaAlSi_3O_8$	tricl		4	8.139	12.788	7.160	94.27°	116.57°	87.68°
Low bornite	Cu_5FeS_4	tetr		16	10.94		21.88			
Low cordierite	$Mg_2Al_3(AlSi_5O_{18})$	orth		4	9.721	17.062	9.339			
Low germania	GeO_2	tetr	rutile	2	4.3963		2.8626			
Low nepheline	$NaAlSiO_4$	hex		8	9.986		8.330			
Luzonite	Cu_3AsS_4	tetr		2	5.289		10.440			
Mackinawite	FeS	tetr		2	3.675		5.030			
Magnesioriebeckite	$Na_2Mg_3Fe_2[Si_8O_{22}](OH)_2$	monocl	tremolite	2	9.733	17.946	5.299		103.30°	
Magnesite	$MgCO_3$	rhomb	calcite	6	4.6330		15.016			
Magnetite	Fe_3O_4	cubic	spinel	8	8.3940					
Malachite	$Cu_2(OH)_2CO_3$	monocl		4	9.502	11.974	3.240		98.75°	
Maldonite	Au_2Bi	cubic		8	7.958					
Manganese sulfide (γ)	MnS	hex	zincite	2	3.976		6.432			
Manganese sulfide (β)	MnS	cubic	sphalerite	4	5.611					
Manganosite	MnO	cubic	rock salt	4	4.4448					
Marcasite	FeS_2	orth	marcasite	2	4.443	5.423	3.3876			
Margarite	$CaAl_2[AlSi_2O_{10}](OH)_2$	monocl	2M mica	4	5.13	8.92	19.50		95.00°	
Marialite	$Na_4Al_3Si_9O_{24}Cl$	tetr		2	12.064		7.514			
Marshite	CuI	cubic	sphalerite	4	6.0507					
Mascagnite	$(NH_4)_2SO_4$	orth	arcanite	4	7.782	5.993	10.636			
Massicot	PbO	orth		4	5.489	4.755	5.891			
Matlockite	PbClF	tetr		2	4.106		7.23			
Maucherite	$Ni_{11}As_8$	tetr		4	6.870		21.81			
Meionite	$Ca_4Al_6Si_6O_{24}CO_3$	tetr		2	12.174		7.652			
Melanophlogite	SiO_2	cubic	clathrate type	46	13.402					
Melanterite	$FeSO_4 \cdot 7H_2O$	monocl		4	14.072	6.503	11.041		105.57°	
Melonite	$NiTe_2$	hex	cadmium iodide	1	3.869		5.308			
Metacinnabar	HgS	cubic	sphalerite	4	5.8517					
Miargyrite	$AgSbS_2$	monocl		8	12.862	4.111	13.220		98.63°	
Microcline	$KAlSi_3O_8$	tricl		4	8.582	12.964	7.222	90.62°	115.92°	87.68°
Miersite	AgI	cubic	sphalerite	4	6.4963					
Millerite	NiS	rhomb		9	9.616		3.152			
Minium	Pb_3O_4	tetr		4	8.815		6.565			
Minnesotaite	$Fe_3Si_4O_{10}(OH)_2$	monocl		4	5.4	9.42	19.4		100.00°	
Mirabilite	$Na_2SO_4 \cdot 10H_2O$	monocl		4	11.51	10.38	12.83		107.75°	
Mn-Indialite	$Mn_2Al_3(AlSi_5O_{18})$	hex	beryl	2	9.925		9.297			
Molybdenite	MoS_2	hex	molybdenite	2	3.1604		12.295			
Molybdenum	Mo	cubic		2	3.1653					
Molybdite	MoO_3	orth		4	3.962	13.858	3.697			
Monteponite	CdO	cubic	rock salt	4	4.6953					
Monticellite	$CaMgSiO_4$	orth	olivine	4	4.827	11.084	6.376			
Montroydite	HgO	orth		4	6.608	5.518	3.519			
Mullite (2:1)	$2Al_2O_3 \cdot SiO_2$	orth		6	7.5788	7.6909	2.8883			
Mullite (3:2)	$3Al_2O_3 \cdot 2SiO_2$	orth		3	7.557	7.6876	2.8842			
Muscovite	$KAl_2AlSi_3O_{10}(OH)_2$	monocl	2M2 mica	4	5.203	8.995	20.030		94.47°	

Name	Formula	Crystal system	Structure type	Z	a/Å	b/Å	c/Å	α	β	γ
Nacrite	$Al_2Si_2O_5(OH)_4$	monocl		4	8.909	5.146	15.697		113.70°	
Nantokite	$CuCl$	cubic	sphalerite	4	5.416					
Natroalunite	$NaAl_3(SO_4)_2(OH)_6$	rhomb		3	6.974		16.69			
Natrolite	$Na_2Al_2Si_3O_{10}{\cdot}2H_2O$	orth		8	18.30	18.63	6.60			
Neighborite	$NaMgF_3$	orth	perovskite	4	5.363	7.676	5.503			
Ni-Skutterudite	$NiAs_{2.95}$	cubic		8	8.3300					
Niccolite	$NiAs$	hex	niccolite	2	3.618		5.034			
Nickel	Ni	cubic	face-centered cubic	4	3.5238					
Nickel carbonate	$NiCO_3$	rhomb	calcite	6	4.5975		14.723			
Nickel olivine	Ni_2SiO_4	orth	olivine	4	4.727	10.121	5.915			
Nickel selenide	$NiSe_2$	cubic	pyrite	4	5.9604					
Niter	KNO_3	orth	aragonite	4	6.431	9.164	5.414			
Norsethite	$BaMg(CO_3)_2$	rhomb	calcite	3	5.020		16.75			
Oldhamite	CaS	cubic	rock salt	4	5.689					
Orpiment	As_2S_3	monocl		4	11.49	9.59	4.25		90.45°	
Orthoclase	$KAlSi_3O_8$	monocl		4	8.562	12.996	7.193		116.02°	
Orthoferrosilite	$FeSiO_3$	orth	enstatite	16	9.080	18.431	5.238			
Otavite	$CdCO_3$	rhomb	calcite	6	4.9204		16.298			
Paracelsian	$BaAl_2Si_2O_8$	monocl		4	8.58	9.583	9.08		90.00°	
Paragonite	$NaAl_2AlSi_3O_{10}(OH)_2$	monocl	2M1 mica	4	5.13	8.89	19.32		95.17°	
Pararammelsbergite	$NiAs_2$	orth		8	5.75	5.82	11.428			
Paratellurite	TeO_2	tetr		4	4.810		7.613			
Parawollastonite	$CaSiO_3$	monocl		12	15.417	7.321	7.066		95.40°	
Pectolite	$Ca_2NaH(SiO_3)_3$	tricl		2	7.99	7.04	7.02	90.05°	95.27°	102.47°
Pentlandite	$Fe_{5.25}Ni_{3.75}S_8$	cubic		4	10.196					
Pentlandite	$Fe_{4.75}Ni_{5.25}S_8$	cubic		4	10.095					
Periclase	MgO	cubic	rock salt	4	4.2117					
Perovskite	$CaTiO_3$	orth	perovskite	4	5.3670	7.6438	5.4439			
Petalite	$LiAlSi_4O_{10}$	monocl		2	11.32	5.14	7.62		105.90°	
Petzite	Ag_3AuTe_2	cubic		8	10.38					
Phenacite	Be_2SiO_4	rhomb	phenacite	18	12.472		8.252			
Phlogopite	$KMg_3AlSi_3O_{10}(OH)_2$	monocl	1M mica	2	5.326	9.210	10.311		100.17°	
Picrochromite	$MgCr_2O_4$	cubic	spinel	8	8.333					
Piemontite	$Ca_2Al_{1.5}Mn_{1.5}(SiO_4)_3OH$	monocl		2	8.95	5.70	9.41		115.70°	
Platinum	Pt	cubic	face-centered cubic	4	3.9231					
Polymidite	Ni_3S_4	cubic	spinel	8	9.480					
Portlandite	$Ca(OH)_2$	hex	cadmium iodide	1	3.5933		4.9086			
Powellite	$CaMoO_4$	tetr	scheelite	4	5.226		11.43			
Protoenstatite	$MgSiO_3$	orth		8	9.25	8.74	5.32			
Proustite	Ag_3AsS_3	rhomb		6	10.816		8.6948			
Pseudowollastonite	$CaSiO_3$	tricl		24	6.90	11.78	19.65	90.00°	90.80°	90.00°
Pyrargyrite	Ag_3SbS_3	rhomb		6	11.052		8.7177			
Pyrite	FeS_2	cubic	pyrite	4	5.4175					
Pyrolusite	MnO_2	tetr	rutile	2	4.388		2.865			
Pyrope	$Mg_3Al_2Si_3O_{12}$	cubic	garnet	8	11.459					
Pyrophanite	$MnTiO_3$	rhomb	ilmenite	6	5.155		14.18			
Pyrophyllite	$Al_2Si_4O_{10}(OH)_2$	monocl	2M1 mica	4	5.14	8.90	18.55		99.92°	
Pyroxmangite	$MnFe(SiO_3)_2$	tricl		7	7.56	17.45	6.67	84.00°	94.30°	113.70°
Pyrrhotite	$Fe_{0.980}S$	hex	defect niccolite	2	3.446		5.848			
Pyrrhotite	$Fe_{0.885}S$	hex	defect niccolite	2	3.440		5.709			
Quartz (α)	SiO_2	hex		3	4.9136		5.4051			
Quartz (β)	SiO_2	hex		3	4.999		5.4592			
Rammelsbergite	$NiAs_2$	orth	marcasite	2	4.757	5.797	3.542			
Realgar	As_4S_4	monocl		16	9.29	13.53	6.57		106.55°	
Retgersite	$NiSO_4{\cdot}4H_2O$	tetr		4	6.782		18.28			
Rhodochrosite	$MnCO_3$	rhomb	calcite	6	4.7771		15.664			

Name	Formula	Crystal system	Structure type	Z	a/Å	b/Å	c/Å	α	β	γ
Rhodonite	$MnSiO_3$	tricl		10	7.682	11.818	6.707	92.36°	93.95°	105.66°
Riebeckite	$Na_2Fe_5FSi_8O_{22}(OH)_2$	monocl	tremolite	2	9.729	18.065	5.334		103.31°	
Rutile	TiO_2	tetr		2	4.5937		2.9618			
Safflorite	$Co_{0.5}Fe_{0.5}As_2$	orth	marcasite	2	5.231	5.953	2.962			
Sanmartinite	$ZnWO_4$	monocl	wolframite	2	4.691	5.720	4.925		89.36°	
Sapphirine	$Mg_2Al_4O_6SiO_4$	monocl		8	9.96	28.60	9.85		110.5°	
Scacchite	$MnCl_2$	rhomb		3	3.711		17.59			
Scheelite	$CaWO_4$	tetr	scheelite	4	5.242		11.372			
Schorl	$NaFe_3Al_6B_3Si_6O_{27}(OH)_4$	rhomb	tourmaline	3	16.032		7.149			
Selenium	Se	hex		3	4.3642		4.9588			
Selenolite	SeO_2	tetr		8	8.35		5.05			
Sellaite	MgF_2	tetr	rutile	2	4.621		3.050			
Senarmontite	Sb_2O_3	cubic	arsenic trioxide	16	11.152					
Shandite	$Ni_3Pb_2S_2$	rhomb		3	5.576		13.658			
Shortite	$Na_2Ca_2(CO_3)_3$	orth		2	4.961	11.03	7.12			
Siderite	$FeCO_3$	rhomb	calcite	6	4.6887		15.373			
Silicon	Si	cubic	diamond	8	5.4305					
Sillimanite	Al_2OSiO_4	orth		4	7.4843	7.6730	5.7711			
Silver	Ag	cubic	face-centered cubic	4	4.0862					
Silver telluride I	Ag_2Te	cubic		2	5.29					
Silver telluride II	Ag_2Te	cubic		4	6.585					
Smithsonite	$ZnCO_3$	rhomb	calcite	6	4.6528		15.025			
Soda niter	$NaNO_3$	rhomb	calcite	6	5.0696		16.829			
Sodium melilite	$NaCaAlSi_2O_7$	tetr	melilite	2	8.511		4.809			
Sperrylite	$PtAs_2$	cubic	pyrite	4	5.968					
Spessartite	$Mn_3Al_2Si_3O_{12}$	cubic	garnet	8	11.621					
Sphalerite	ZnS	cubic	sphalerite	4	5.4093					
Sphene	$CaTiSiO_5$	monocl		4	7.07	8.72	6.56		113.95°	
Spinel	$MgAl_2O_4$	cubic	spinel	8	8.080					
Spodumene	$LiAl(SiO_3)_2$	monocl	diopside	4	9.451	8.387	5.208		110.07°	
Spodumene (β)	$LiAl(SiO_3)_2$	tetr		4	7.5332		9.1540			
Staurolite	$Fe_2Al_9Si_4O_{22}(OH)_2$	monocl		2	7.90	16.65	5.63		90.00°	
Sternbergite	$AgFe_2S_3$	orth		8	11.60	12.675	6.63			
Stibnite	Sb_2S_3	orth	stibnite	4	11.229	11.310	3.8389			
Stilleite	ZnSe	cubic	sphalerite	4	5.6685					
Stishovite	SiO_2	tetr	rutile	2	4.1790		2.6649			
Stolzite	$PbWO_4$	tetr	scheelite	4	5.4616		12.046			
Stromeyerite	$Ag_{0.93}Cu_{1.07}S$	orth		4	4.066	6.628	7.972			
Strontianite	$SrCO_3$	orth	aragonite	4	6.029	8.414	5.107			
Sulfur (monoclinic)	S	monocl	S8 ring molecules	48	11.04	10.98	10.92		96.73°	
Sulfur (orthorhombic)	S	orth	S8 ring molecules	128	10.4646	12.8660	24.4860			
Sulfur (rhombohedral)	S	rhomb	S6 ring molecules	18	10.818		4.280			
Sylvite	KCl	cubic	rock salt	4	6.2931					
Syngenite	$K_2Ca(SO_4)_2{\cdot}H_2O$	monocl		2	9.775	7.156	6.251		104.00°	
Synthetic anorthite	$CaAl_2Si_2O_8$	hex		2	5.10		14.72			
Synthetic anorthite	$CaAl_2Si_2O_8$	orth		2	8.22	8.60	4.83			
Talc	$Mg_3Si_4O_{10}(OH)_2$	monocl	2M1 mica	4	5.287	9.158	18.95		99.50°	
Tantalum	Ta	cubic	tungsten	2	3.3058					
Teallite	$PbSnS_2$	orth	germanium sulfide	2	4.266	11.419	4.090			
Tellurite	TeO_2	orth	tellurite	8	5.607	12.034	5.463			
Tellurium	Te	hex	selenium	3	4.4570		5.9290			
Tellurobismuthite	Bi_2Te_3	rhomb		3	4.3835		30.487			
Tennantite	$Cu_{12}As_4S_{13}$	cubic	tetrahedrite	2	10.190					
Tenorite	CuO	monocl		4	4.684	3.425	5.129		99.47°	
Tephroite	Mn_2SiO_4	orth	olivine	4	4.871	10.636	6.232			

Name	Formula	Crystal system	Structure type	Z	a/Å	b/Å	c/Å	α	β	γ
Tetrahedrite	$Cu_{12}Sb_4S_{13}$	cubic	tetrahedrite	2	10.327					
Thenardite	Na_2SO_4	orth	thenardite	8	5.863	12.304	9.821			
Thorianite	ThO_2	cubic	fluorite	4	5.5952					
Thorite	$ThSiO_4$	tetr	zircon	4	7.143		6.327			
Tiemannite	HgSe	cubic	sphalerite	4	6.0853					
Tin	Sn	tetr		4	5.8315		3.1813			
Titanium	Ti	hex		2	2.953		4.729			
Titanium(III) oxide	Ti_2O_3	rhomb	corundum	6	5.149		13.642			
Topaz	$Al_2SiO_4(OH,F)_2$	orth		4	8.394	8.792	4.649			
Tremolite	$Ca_2Mg_5Si_8O_{22}(OH)_2$	monocl	tremolite	2	9.840	18.052	5.275		104.70°	
Trevorite	$NiFe_2O_4$	cubic	spinel	8	8.339					
Tridymite (β)	SiO_2	hex		4	5.0463		8.2563			
Trogtalite	$CoSe_2$	cubic	pyrite	4	5.8588					
Troilite	FeS	hex	niccolite	2	3.446		5.877			
Tschermakite	$CaAl_2SiO_6$	monocl	diopside	4	9.615	8.661	5.272		106.12°	
Tungsten	W	cubic		2	3.1653					
Tungstenite	WS_2	hex	molybdenite	2	3.154		12.362			
Turquois	$CuAl_6(PO_4)_4(OH)_8{\cdot}4H_2O$	tricl		1	7.424	7.629	9.910	68.61°	69.71°	65.08°
Umangite	Cu_3Se_2	tetr		2	6.402		4.276			
Uraninite	UO_2	cubic	fluorite	4	5.4682					
Ureyite	$NaCr(SiO_3)_2$	monocl	diopside	4	9.550	8.712	5.273		107.44°	
Uvarovite	$Ca_3Cr_2Si_3O_{12}$	cubic	garnet	8	11.999					
Uvite	$CaMg_4Al_5B_3Si_6O_{27}(OH)_4$	rhomb	tourmaline	3	15.86		7.19			
Vaesite	NiS_2	cubic	pyrite	4	5.6873					
Valentinite	Sb_2O_3	orth	antimony trioxide	4	4.914	12.468	5.421			
Vanthoffite	$MgSO_4{\cdot}3Na_2SO_4$	monocl		2	9.797	9.217	8.199		113.50°	
Vaterite	$CaCO_3$	hex		6	7.135		8.524			
Villiaumite	NaF	cubic	rock salt	4	4.6342					
Violarite	$FeNi_2S_4$	cubic	spinel	8	9.464					
Willemite	Zn_2SiO_4	rhomb	phenacite	18	13.94		9.309			
Witherite	$BaCO_3$	orth	aragonite	4	6.430	8.904	5.314			
Wolframite	$Fe_{0.5}Mn_{0.5}WO_4$	monocl	wolframite	2	4.782	5.731	4.982		90.57°	
Wollastonite	$CaSiO_3$	tricl		6	7.94	7.32	7.07	90.03°	95.37°	103.43°
Wulfenite	$PbMoO_4$	tetr	scheelite	4	5.435		12.110			
Wurtzite	ZnS	hex	zincite	2	3.8230		6.2565			
Wustite	$Fe_{0.953}O$	cubic	defect rock salt	4	4.3088					
Xenotime	YPO_4	tetr	zircon	4	6.885		5.982			
Zinc	Zn	hex	hexagonal close pack	2	2.665		4.947			
Zinc telluride	ZnTe	cubic	sphalerite	4	6.1020					
Zincite	ZnO	hex	zincite	2	3.2495		5.2069			
Zinkosite	$ZnSO_4$	orth	barite	4	8.588	6.740	4.770			
Zircon	$ZrSiO_4$	tetr	zircon	4	6.604		5.979			
Zoisite	$Ca_2Al_3(SiO_4)_3OH$	orth		4	16.15	5.581	10.06			

Section 5
Thermochemistry, Electrochemistry, and Kinetics

CODATA Key Values for Thermodynamics 5-1
Standard Thermodynamic Properties of Chemical Substances 5-4
Thermodynamic Properties as a Function of Temperature 5-43
Thermodynamic Properties of Aqueous Ions 5-66
Heat of Combustion 5-68
Energy Content of Fuels 5-69
Electrical Conductivity of Water 5-70
Electrical Conductivity of Aqueous Solutions 5-71
Standard KCl Solutions for Calibrating Conductivity Cells 5-72
Molar Conductivity of Aqueous HF, HCl, HBr, and HI 5-73
Equivalent Conductivity of Electrolytes in Aqueous Solution 5-74
Ionic Conductivity and Diffusion at Infinite Dilution 5-75
Activity Coefficients of Acids, Bases, and Salts 5-78
Mean Activity Coefficients of Electrolytes as a Function of Concentration 5-80
Enthalpy of Dilution of Acids 5-84
Enthalpy of Solution of Electrolytes 5-85
Enthalpy of Hydration of Gases 5-86
Chemical Kinetic Data for Stratospheric Modeling 5-90

CODATA KEY VALUES FOR THERMODYNAMICS

The Committee on Data for Science and Technology (CODATA) has conducted a project to establish internationally agreed values for the thermodynamic properties of key chemical substances. This table presents the final results of the project. Use of these recommended, internally consistent values is encouraged in the analysis of thermodynamic measurements, data reduction, and preparation of other thermodynamic tables.

The table includes the standard enthalpy of formation at 298.15 K, the entropy at 298.15 K, and the quantity $H°$ (298.15 K)–$H°$ (0). A value of 0 in the $\Delta_f H°$ column for an element indicates the reference state for that element. The standard state pressure is 100,000 Pa (1 bar). See the reference for information on the dependence of gas-phase entropy on the choice of standard state pressure.

Substances are listed in alphabetical order of their chemical formulas when written in the most common form.

The table is reprinted with permission of CODATA.

Reference

Cox, J. D., Wagman, D. D., and Medvedev, V. A., *CODATA Key Values for Thermodynamics,* Hemisphere Publishing Corp., New York, 1989.

Substance	State	$\Delta_f H°$ (298.15 K) kJ·mol^{-1}	$S°$ (298.15 K) J·K^{-1}·mol^{-1}	$H°$ (298.15 K)–$H°$ (0) kJ·mol^{-1}
Ag	cr	0	42.55 ± 0.20	5.745 ± 0.020
Ag	g	284.9 ± 0.8	172.997 ± 0.004	6.197 ± 0.001
Ag^{+}	aq	105.79 ± 0.08	73.45 ± 0.40	
AgCl	cr	−127.01 ± 0.05	96.25 ± 0.20	12.033 ± 0.020
Al	cr	0	28.30 ± 0.10	4.540 ± 0.020
Al	g	330.0 ± 4.0	164.554 ± 0.004	6.919 ± 0.001
Al^{+3}	aq	−538.4 ± 1.5	−325 ± 10	
AlF_3	cr	−1510.4 ± 1.3	66.5 ± 0.5	11.62 ± 0.04
Al_2O_3	cr, corundum	−1675.7 ± 1.3	50.92 ± 0.10	10.016 ± 0.020
Ar	g	0	154.846 ± 0.003	6.197 ± 0.001
B	cr, rhombic	0	5.90 ± 0.08	1.222 ± 0.008
B	g	565 ± 5	153.436 ± 0.015	6.316 ± 0.002
BF_3	g	−1136.0 ± 0.8	254.42 ± 0.20	11.650 ± 0.020
B_2O_3	cr	−1273.5 ± 1.4	53.97 ± 0.30	9.301 ± 0.040
Be	cr	0	9.50 ± 0.08	1.950 ± 0.020
Be	g	324 ± 5	136.275 ± 0.003	6.197 ± 0.001
BeO	cr	−609.4 ± 2.5	13.77 ± 0.04	2.837 ± 0.008
Br	g	111.87 ± 0.12	175.018 ± 0.004	6.197 ± 0.001
Br^{-}	aq	−121.41 ± 0.15	82.55 ± 0.20	
Br_2	l	0	152.21 ± 0.30	24.52 ± 0.01
Br_2	g	30.91 ± 0.11	245.468 ± 0.005	9.725 ± 0.001
C	cr, graphite	0	5.74 ± 0.10	1.050 ± 0.020
C	g	716.68 ± 0.45	158.100 ± 0.003	6.536 ± 0.001
CO	g	−110.53 ± 0.17	197.660 ± 0.004	8.671 ± 0.001
CO_2	g	−393.51 ± 0.13	213.785 ± 0.010	9.365 ± 0.003
CO_2	aq, undissoc.	−413.26 ± 0.20	119.36 ± 0.60	
CO_3^{-2}	aq	−675.23 ± 0.25	−50.0 ± 1.0	
Ca	cr	0	41.59 ± 0.40	5.736 ± 0.040
Ca	g	177.8 ± 0.8	154.887 ± 0.004	6.197 ± 0.001
Ca^{+2}	aq	−543.0 ± 1.0	−56.2 ± 1.0	
CaO	cr	−634.92 ± 0.90	38.1 ± 0.4	6.75 ± 0.06
Cd	cr	0	51.80 ± 0.15	6.247 ± 0.015
Cd	g	111.80 ± 0.20	167.749 ± 0.004	6.197 ± 0.001
Cd^{+2}	aq	−75.92 ± 0.60	−72.8 ± 1.5	
CdO	cr	−258.35 ± 0.40	54.8 ± 1.5	8.41 ± 0.08
$CdSO_4{\cdot}8/3H_2O$	cr	−1729.30 ± 0.80	229.65 ± 0.40	35.56 ± 0.04
Cl	g	121.301 ± 0.008	165.190 ± 0.004	6.272 ± 0.001
Cl^{-}	aq	−167.080 ± 0.10	56.60 ± 0.20	
ClO_4^{-}	aq	−128.10 ± 0.40	184.0 ± 1.5	
Cl_2	g	0	223.081 ± 0.010	9.181 ± 0.001
Cs	cr	0	85.23 ± 0.40	7.711 ± 0.020
Cs	g	76.5 ± 1.0	175.601 ± 0.003	6.197 ± 0.001
Cs^{+}	aq	−258.00 ± 0.50	132.1 ± 0.5	
Cu	cr	0	33.15 ± 0.08	5.004 ± 0.008

Substance	State	$\Delta_f H°$ (298.15 K) kJ·mol^{-1}	$S°$ (298.15 K) J·K^{-1}·mol^{-1}	$H°$ (298.15 K) − $H°$ (0) kJ·mol^{-1}
Cu	g	337.4 ± 1.2	166.398 ± 0.004	6.197 ± 0.001
Cu^{+2}	aq	64.9 ± 1.0	−98 ± 4	
$CuSO_4$	cr	−771.4 ± 1.2	109.2 ± 0.4	16.86 ± 0.08
F	g	79.38 ± 0.30	158.751 ± 0.004	6.518 ± 0.001
F^-	aq	−335.35 ± 0.65	−13.8 ± 0.8	
F_2	g	0	202.791 ± 0.005	8.825 ± 0.001
Ge	cr	0	31.09 ± 0.15	4.636 ± 0.020
Ge	g	372 ± 3	167.904 ± 0.005	7.398 ± 0.001
GeF_4	g	−1190.20 ± 0.50	301.9 ± 1.0	17.29 ± 0.10
GeO_2	cr, tetragonal	−580.0 ± 1.0	39.71 ± 0.15	7.230 ± 0.020
H	g	217.998 ± 0.006	114.717 ± 0.002	6.197 ± 0.001
H^+	aq	0	0	
HBr	g	−36.29 ± 0.16	198.700 ± 0.004	8.648 ± 0.001
HCO_3^-	aq	−689.93 ± 0.20	98.4 ± 0.5	
HCl	g	−92.31 ± 0.10	186.902 ± 0.005	8.640 ± 0.001
HF	g	−273.30 ± 0.70	173.779 ± 0.003	8.599 ± 0.001
HI	g	26.50 ± 0.10	206.590 ± 0.004	8.657 ± 0.001
HPO_4^{-2}	aq	−1299.0 ± 1.5	−33.5 ± 1.5	
HS^-	aq	−16.3 ± 1.5	67 ± 5	
HSO_4^-	aq	−886.9 ± 1.0	131.7 ± 3.0	
H_2	g	0	130.680 ± 0.003	8.468 ± 0.001
H_2O	l	−285.830 ± 0.040	69.95 ± 0.03	13.273 ± 0.020
H_2O	g	−241.826 ± 0.040	188.835 ± 0.010	9.905 ± 0.005
$H_2PO_4^-$	aq	−1302.6 ± 1.5	92.5 ± 1.5	
H_2S	g	−20.6 ± 0.5	205.81 ± 0.05	9.957 ± 0.010
H_2S	aq, undissoc.	−38.6 ± 1.5	126 ± 5	
H_3BO_3	cr	−1094.8 ± 0.8	89.95 ± 0.60	13.52 ± 0.04
H_3BO_3	aq, undissoc.	−1072.8 ± 0.8	162.4 ± 0.6	
He	g	0	126.153 ± 0.002	6.197 ± 0.001
Hg	l	0	75.90 ± 0.12	9.342 ± 0.008
Hg	g	61.38 ± 0.04	174.971 ± 0.005	6.197 ± 0.001
Hg^{+2}	aq	170.21 ± 0.20	−36.19 ± 0.80	
HgO	cr, red	−90.79 ± 0.12	70.25 ± 0.30	9.117 ± 0.025
Hg_2^{+2}	aq	166.87 ± 0.50	65.74 ± 0.80	
Hg_2Cl_2	cr	−265.37 ± 0.40	191.6 ± 0.8	23.35 ± 0.20
Hg_2SO_4	cr	−743.09 ± 0.40	200.70 ± 0.20	26.070 ± 0.030
I	g	106.76 ± 0.04	180.787 ± 0.004	6.197 ± 0.001
I^-	aq	−56.78 ± 0.05	106.45 ± 0.30	
I_2	cr	0	116.14 ± 0.30	13.196 ± 0.040
I_2	g	62.42 ± 0.08	260.687 ± 0.005	10.116 ± 0.001
K	cr	0	64.68 ± 0.20	7.088 ± 0.020
K	g	89.0 ± 0.8	160.341 ± 0.003	6.197 ± 0.001
K^+	aq	−252.14 ± 0.08	101.20 ± 0.20	
Kr	g	0	164.085 ± 0.003	6.197 ± 0.001
Li	cr	0	29.12 ± 0.20	4.632 ± 0.040
Li	g	159.3 ± 1.0	138.782 ± 0.010	6.197 ± 0.001
Li^+	aq	−278.47 ± 0.08	12.24 ± 0.15	
Mg	cr	0	32.67 ± 0.10	4.998 ± 0.030
Mg	g	147.1 ± 0.8	148.648 ± 0.003	6.197 ± 0.001
Mg^{+2}	aq	−467.0 ± 0.6	−137 ± 4	
MgF_2	cr	−1124.2 ± 1.2	57.2 ± 0.5	9.91 ± 0.06
MgO	cr	−601.60 ± 0.30	26.95 ± 0.15	5.160 ± 0.020
N	g	472.68 ± 0.40	153.301 ± 0.003	6.197 ± 0.001
NH_3	g	−45.94 ± 0.35	192.77 ± 0.05	10.043 ± 0.010
NH_4^+	aq	−133.26 ± 0.25	111.17 ± 0.40	
NO_3^-	aq	−206.85 ± 0.40	146.70 ± 0.40	
N_2	g	0	191.609 ± 0.004	8.670 ± 0.001
Na	cr	0	51.30 ± 0.20	6.460 ± 0.020
Na	g	107.5 ± 0.7	153.718 ± 0.003	6.197 ± 0.001
Na^+	aq	−240.34 ± 0.06	58.45 ± 0.15	

Substance	State	$\Delta_f H°$ (298.15 K) kJ·mol⁻¹	$S°$ (298.15 K) J·K⁻¹·mol⁻¹	$H°$ (298.15 K) − $H°$ (0) kJ·mol⁻¹
Ne	g	0	146.328 ± 0.003	6.197 ± 0.001
O	g	249.18 ± 0.10	161.059 ± 0.003	6.725 ± 0.001
OH^-	aq	−230.015 ± 0.040	−10.90 ± 0.20	
O_2	g	0	205.152 ± 0.005	8.680 ± 0.002
P	cr, white	0	41.09 ± 0.25	5.360 ± 0.015
P	g	316.5 ± 1.0	163.199 ± 0.003	6.197 ± 0.001
P_2	g	144.0 ± 2.0	218.123 ± 0.004	8.904 ± 0.001
P_4	g	58.9 ± 0.3	280.01 ± 0.50	14.10 ± 0.20
Pb	cr	0	64.80 ± 0.30	6.870 ± 0.030
Pb	g	195.2 ± 0.8	175.375 ± 0.005	6.197 ± 0.001
Pb^{+2}	aq	0.92 ± 0.25	18.5 ± 1.0	
$PbSO_4$	cr	−919.97 ± 0.40	148.50 ± 0.60	20.050 ± 0.040
Rb	cr	0	76.78 ± 0.30	7.489 ± 0.020
Rb	g	80.9 ± 0.8	170.094 ± 0.003	6.197 ± 0.001
Rb^+	aq	−251.12 ± 0.10	121.75 ± 0.25	
S	cr, rhombic	0	32.054 ± 0.050	4.412 ± 0.006
S	g	277.17 ± 0.15	167.829 ± 0.006	6.657 ± 0.001
SO_2	g	−296.81 ± 0.20	248.223 ± 0.050	10.549 ± 0.010
SO_4^{-2}	aq	−909.34 ± 0.40	18.50 ± 0.40	
S_2	g	128.60 ± 0.30	228.167 ± 0.010	9.132 ± 0.002
Si	cr	0	18.81 ± 0.08	3.217 ± 0.008
Si	g	450 ± 8	167.981 ± 0.004	7.550 ± 0.001
SiF_4	g	−1615.0 ± 0.8	282.76 ± 0.50	15.36 ± 0.05
SiO_2	cr, alpha quartz	−910.7 ± 1.0	41.46 ± 0.20	6.916 ± 0.020
Sn	cr, white	0	51.18 ± 0.08	6.323 ± 0.008
Sn	g	301.2 ± 1.5	168.492 ± 0.004	6.215 ± 0.001
Sn^{+2}	aq	−8.9 ± 1.0	−16.7 ± 4.0	
SnO	cr, tetragonal	−280.71 ± 0.20	57.17 ± 0.30	8.736 ± 0.020
SnO_2	cr, tetragonal	−577.63 ± 0.20	49.04 ± 0.10	8.384 ± 0.020
Th	cr	0	51.8 ± 0.5	6.35 ± 0.05
Th	g	602 ± 6	190.17 ± 0.05	6.197 ± 0.003
ThO_2	cr	−1226.4 ± 3.5	65.23 ± 0.20	10.560 ± 0.020
Ti	cr	0	30.72 ± 0.10	4.824 ± 0.015
Ti	g	473 ± 3	180.298 ± 0.010	7.539 ± 0.002
$TiCl_4$	g	−763.2 ± 3.0	353.2 ± 4.0	21.5 ± 0.5
TiO_2	cr, rutile	−944.0 ± 0.8	50.62 ± 0.30	8.68 ± 0.05
U	cr	0	50.20 ± 0.20	6.364 ± 0.020
U	g	533 ± 8	199.79 ± 0.10	6.499 ± 0.020
UO_2	cr	−1085.0 ± 1.0	77.03 ± 0.20	11.280 ± 0.020
UO_2^{+2}	aq	−1019.0 ± 1.5	−98.2 ± 3.0	
UO_3	cr, gamma	−1223.8 ± 1.2	96.11 ± 0.40	14.585 ± 0.050
U_3O_8	cr	−3574.8 ± 2.5	282.55 ± 0.50	42.74 ± 0.10
Xe	g	0	169.685 ± 0.003	6.197 ± 0.001
Zn	cr	0	41.63 ± 0.15	5.657 ± 0.020
Zn	g	130.40 ± 0.40	160.990 ± 0.004	6.197 ± 0.001
Zn^{+2}	aq	−153.39 ± 0.20	−109.8 ± 0.5	
ZnO	cr	−350.46 ± 0.27	43.65 ± 0.40	6.933 ± 0.040

STANDARD THERMODYNAMIC PROPERTIES OF CHEMICAL SUBSTANCES

This table gives the standard state chemical thermodynamic properties of about 2500 individual substances in the crystalline, liquid, and gaseous states. Substances are listed by molecular formula in a modified Hill order; all substances not containing carbon appear first, followed by those that contain carbon. The properties tabulated are:

$\Delta_f H°$	Standard molar enthalpy (heat) of formation at 298.15 K in kJ/mol
$\Delta_f G°$	Standard molar Gibbs energy of formation at 298.15 K in kJ/mol
$S°$	Standard molar entropy at 298.15 K in J/mol K
C_p	Molar heat capacity at constant pressure at 298.15 K in J/mol K

The standard state pressure is 100 kPa (1 bar). The standard states are defined for different phases by:

- The standard state of a pure gaseous substance is that of the substance as a (hypothetical) ideal gas at the standard state pressure.
- The standard state of a pure liquid substance is that of the liquid under the standard state pressure.
- The standard state of a pure crystalline substance is that of the crystalline substance under the standard state pressure.

An entry of 0.0 for $\Delta_f H°$ for an element indicates the reference state of that element. See References 1 and 2 for further information on reference states. A blank means no value is available.

The data are derived from the sources listed in the references, from other papers appearing in the *Journal of Physical and Chemical Reference Data*, and from the primary research literature. We are indebted to M. V. Korobov for providing data on fullerene compounds.

References

1. Cox, J. D., Wagman, D. D., and Medvedev, V. A., *CODATA Key Values for Thermodynamics*, Hemisphere Publishing Corp., New York, 1989.
2. Wagman, D. D., Evans, W. H., Parker, V. B., Schumm, R. H., Halow, I., Bailey, S. M., Churney, K. L., and Nuttall, R. L., *The NBS Tables of Chemical Thermodynamic Properties, J. Phys. Chem. Ref. Data*, Vol. 11, Suppl. 2, 1982.
3. Chase, M. W., Davies, C. A., Downey, J. R., Frurip, D. J., McDonald, R. A., and Syverud, A. N., *JANAF Thermochemical Tables, Third Edition, J. Phys. Chem. Ref. Data*, Vol. 14, Suppl. 1, 1985.
4. Chase, M. W., *NIST-JANAF Thermochemical Tables, Fourth Edition, J. Phys. Chem. Ref. Data*, Monograph 9, 1998.
5. Daubert, T. E., Danner, R. P., Sibul, H. M., and Stebbins, C. C., *Physical and Thermodynamic Properties of Pure Compounds: Data Compilation*, extant 1994 (core with 4 supplements), Taylor & Francis, Bristol, PA.
6. Pedley, J. B., Naylor, R. D., and Kirby, S. P., *Thermochemical Data of Organic Compounds, Second Edition*, Chapman & Hall, London, 1986.
7. Pedley, J. B., *Thermochemical Data and Structures of Organic Compounds*, Thermodynamic Research Center, Texas A & M University, College Station, TX, 1994.
8. Domalski, E. S., and Hearing, E. D., Heat Capacities and Entropies of Organic Compounds in the Condensed Phase, Volume III, *J. Phys. Chem. Ref. Data*, 25, 1–525, 1996.
9. Zabransky, M., Ruzicka , V., Majer, V., and Domalski, E. S., *Heat Capacity of Liquids, J. Phys. Chem. Ref. Data*, Monograph No. 6, 1996.
10. Gurvich, L. V., Veyts, I.V., and Alcock, C. B., *Thermodynamic Properties of Individual Substances, Fourth Edition, Vol. 1*, Hemisphere Publishing Corp., New York, 1989.
11. Gurvich, L. V., Veyts, I.V., and Alcock, C. B., *Thermodynamic Properties of Individual Substances, Fourth Edition, Vol. 3*, CRC Press, Boca Raton, FL, 1994.
12. *NIST Chemistry Webbook*, <webbook.nist.gov>

		Crystal				Liquid				Gas			
Molecular formula	**Name**	**$\Delta_f H°$ kJ/mol**	**$\Delta_f G°$ kJ/mol**	**$S°$ J/mol K**	**C_p J/mol K**	**$\Delta_f H°$ kJ/mol**	**$\Delta_f G°$ kJ/mol**	**$S°$ J/mol K**	**C_p J/mol K**	**$\Delta_f H°$ kJ/mol**	**$\Delta_f G°$ kJ/mol**	**$S°$ J/mol K**	**C_p J/mol K**
Substances not containing carbon:													
Ac	Actinium	0.0		56.5	27.2					406.0	366.0	188.1	20.8
Ag	Silver	0.0		42.6	25.4					284.9	246.0	173.0	20.8
AgBr	Silver(I) bromide	-100.4	-96.9	107.1	52.4								
$AgBrO_3$	Silver(I) bromate	-10.5	71.3	151.9									
AgCl	Silver(I) chloride	-127.0	-109.8	96.3	50.8								
$AgClO_3$	Silver(I) chlorate	-30.3	64.5	142.0									
$AgClO_4$	Silver(I) perchlorate	-31.1											
AgF	Silver(I) fluoride	-204.6											
AgF_2	Silver(II) fluoride	-360.0											
AgI	Silver(I) iodide	-61.8	-66.2	115.5	56.8								
$AgIO_3$	Silver(I) iodate	-171.1	-93.7	149.4	102.9								
$AgNO_3$	Silver(I) nitrate	-124.4	-33.4	140.9	93.1								
Ag_2	Disilver									410.0	358.8	257.1	37.0
Ag_2CrO_4	Silver(I) chromate	-731.7	-641.8	217.6	142.3								
Ag_2O	Silver(I) oxide	-31.1	-11.2	121.3	65.9								
Ag_2O_2	Silver(II) oxide	-24.3	27.6	117.0	88.0								
Ag_2O_3	Silver(III) oxide	33.9	121.4	100.0									
Ag_2O_4S	Silver(I) sulfate	-715.9	-618.4	200.4	131.4								
Ag_2S	Silver(I) sulfide (argentite)	-32.6	-40.7	144.0	76.5								
Al	Aluminum	0.0		28.3	24.2					330.0	289.4	164.6	21.4

		Crystal				Liquid				Gas			
Molecular formula	Name	$\Delta_f H°$ kJ/mol	$\Delta_f G°$ kJ/mol	$S°$ J/mol K	C_p J/mol K	$\Delta_f H°$ kJ/mol	$\Delta_f G°$ kJ/mol	$S°$ J/mol K	C_p J/mol K	$\Delta_f H°$ kJ/mol	$\Delta_f G°$ kJ/mol	$S°$ J/mol K	C_p J/mol K
AlB_3H_{12}	Aluminum borohydride					-16.3	145.0	289.1	194.6	13.0	147.0	379.2	
AlBr	Aluminum monobromide									-4.0	-42.0	239.5	35.6
$AlBr_3$	Aluminum bromide	-527.2		180.2	100.6					-425.1			
AlCl	Aluminum monochloride									-47.7	-74.1	228.1	35.0
$AlCl_2$	Aluminum dichloride									-331.0			
$AlCl_3$	Aluminum chloride	-704.2	-628.8	109.3	91.1					-583.2			
AlF	Aluminum monofluoride									-258.2	-283.7	215.0	31.9
AlF_3	Aluminum fluoride	-1510.4	-1431.1	66.5	75.1					-1204.6	-1188.2	277.1	62.6
AlF_4Na	Sodium tetrafluoroaluminate									-1869.0	-1827.5	345.7	105.9
AlH	Aluminum monohydride									259.2	231.2	187.9	29.4
AlH_3	Aluminum hydride	-46.0		30.0	40.2								
AlH_4K	Potassium aluminum hydride	-183.7											
AlH_4Li	Lithium aluminum hydride	-116.3	-44.7	78.7	83.2								
AlH_4Na	Sodium aluminum hydride	-115.5											
AlI	Aluminum monoiodide									65.5			36.0
AlI_3	Aluminum iodide	-302.9		195.9	98.7					-289.4		223.6	
AlN	Aluminum nitride	-318.0	-287.0	20.2	30.1								
AlO	Aluminum monoxide									91.2	65.3	218.4	30.9
AlO_4P	Aluminum phosphate	-1733.8	-1617.9	90.8	93.2								
AlP	Aluminum phosphide	-166.5											
AlS	Aluminum monosulfide									200.9	150.1	230.6	33.4
Al_2	Dialuminum									485.9	433.3	233.2	36.4
Al_2Br_6	Aluminum hexabromide									-970.7			
Al_2Cl_6	Aluminum hexachloride									-1290.8	-1220.4	490.0	
Al_2F_6	Aluminum hexafluoride									-2628.0			
Al_2I_6	Aluminum hexaiodide									-516.7			
Al_2O	Aluminum oxide (Al_2O)									-130.0	-159.0	259.4	45.7
Al_2O_3	Aluminum oxide (corundum)	-1675.7	-1582.3	50.9	79.0								
Al_2S_3	Aluminum sulfide	-724.0		116.9	105.1								
Am	Americium	0.0											
Ar	Argon									0.0		154.8	20.8
As	Arsenic (gray)	0.0		35.1	24.6					302.5	261.0	174.2	20.8
As	Arsenic (yellow)	14.6											
$AsBr_3$	Arsenic(III) bromide	-197.5								-130.0	-159.0	363.9	79.2
$AsCl_3$	Arsenic(III) chloride					-305.0	-259.4	216.3		-261.5	-248.9	327.2	75.7
AsF_3	Arsenic(III) fluoride					-821.3	-774.2	181.2	126.6	-785.8	-770.8	289.1	65.6
AsGa	Gallium arsenide	-71.0	-67.8	64.2	46.2								
AsH_3	Arsine									66.4	68.9	222.8	38.1
AsH_3O_4	Arsenic acid	-906.3											
AsI_3	Arsenic(III) iodide	-58.2	-59.4	213.1	105.8							388.3	80.6
AsIn	Indium arsenide	-58.6	-53.6	75.7	47.8								
AsO	Arsenic monoxide									70.0			
As_2	Diarsenic									222.2	171.9	239.4	35.0
As_2O_5	Arsenic(V) oxide	-924.9	-782.3	105.4	116.5								
As_2S_3	Arsenic(III) sulfide	-169.0	-168.6	163.6	116.3								
At	Astatine	0.0											
Au	Gold	0.0		47.4	25.4					366.1	326.3	180.5	20.8
AuBr	Gold(I) bromide	-14.0											
$AuBr_3$	Gold(III) bromide	-53.3											
AuCl	Gold(I) chloride	-34.7											
$AuCl_3$	Gold(III) chloride	-117.6											
AuF_3	Gold(III) fluoride	-363.6											
AuH	Gold hydride									295.0	265.7	211.2	29.2
AuI	Gold(I) iodide	0.0											
Au_2	Digold									515.1			36.9
B	Boron (β-rhombohedral)	0.0		5.9	11.1					565.0	521.0	153.4	20.8
BBr	Bromoborane(1)									238.1	195.4	225.0	32.9
BBr_3	Boron tribromide					-239.7	-238.5	229.7		-205.6	-232.5	324.2	67.8
BCl	Chloroborane(1)									149.5	120.9	213.2	31.7
BClO	Chloroxyborane									-314.0			
BCl_3	Boron trichloride					-427.2	-387.4	206.3	106.7	-403.8	-388.7	290.1	62.7
$BCsO_2$	Cesium metaborate	-972.0	-915.0	104.4	80.6								
BF	Fluoroborane(1)									-122.2	-149.8	200.5	29.6
BFO	Fluorooxyborane									-607.0			
BF_3	Boron trifluoride									-1136.0	-1119.4	254.4	
BF_3H_3N	Aminetrifluoroboron	-1353.9											

		Crystal				Liquid				Gas			
Molecular formula	Name	$\Delta_f H°$ kJ/mol	$\Delta_f G°$ kJ/mol	$S°$ J/mol K	C_p J/mol K	$\Delta_f H°$ kJ/mol	$\Delta_f G°$ kJ/mol	$S°$ J/mol K	C_p J/mol K	$\Delta_f H°$ kJ/mol	$\Delta_f G°$ kJ/mol	$S°$ J/mol K	C_p J/mol K
BF_3H_3P	Trihydro(phosphorus trifluoride)boron									-854.0			
BF_4Na	Sodium tetrafluoroborate	-1844.7	-1750.1	145.3	120.3								
BH	Borane(1)									442.7	412.7	171.8	29.2
BHO_2	Metaboric acid (β, monoclinic)	-794.3	-723.4	38.0						-561.9	-551.0	240.1	42.2
BH_3	Borane(3)									89.2	93.3	188.2	36.0
BH_3O_3	Boric acid	-1094.3	-968.9	90.0	86.1					-994.1			
BH_4K	Potassium borohydride	-227.4	-160.3	106.3	96.1								
BH_4Li	Lithium borohydride	-190.8	-125.0	75.9	82.6								
BH_4Na	Sodium borohydride	-188.6	-123.9	101.3	86.8								
BI_3	Boron triiodide									71.1	20.7	349.2	70.8
BKO_2	Potassium metaborate	-981.6	-923.4	80.0	66.7								
$BLiO_2$	Lithium metaborate	-1032.2	-976.1	51.5	59.8								
BN	Boron nitride	-254.4	-228.4	14.8	19.7					647.5	614.5	212.3	29.5
$BNaO_2$	Sodium metaborate	-977.0	-920.7	73.5	65.9								
BO	Boron monoxide									25.0	-4.0	203.5	29.2
BO_2	Boron dioxide									-300.4	-305.9	229.6	43.0
BO_2Rb	Rubidium metaborate	-971.0	-913.0	94.3	74.1								
BS	Boron monosulfide									342.0	288.8	216.2	30.0
B_2	Diboron									830.5	774.0	201.9	30.5
B_2Cl_4	Tetrachlorodiborane					-523.0	-464.8	262.3	137.7	-490.4	-460.6	357.4	95.4
B_2F_4	Tetrafluorodiborane									-1440.1	-1410.4	317.3	79.1
B_2H_6	Diborane									36.4	87.6	232.1	56.7
B_2O_2	Diboron dioxide									-454.8	-462.3	242.5	57.3
B_2O_3	Boron oxide	-1273.5	-1194.3	54.0	62.8					-843.8	-832.0	279.8	66.9
B_2S_3	Boron sulfide	-240.6		100.0	111.7					67.0			
$B_3H_6N_3$	Borazine					-541.0	-392.7	199.6					
B_4H_{10}	Tetraborane(10)									66.1	184.3	280.3	93.2
$B_4Na_2O_7$	Sodium tetraborate	-3291.1	-3096.0	189.5	186.8								
B_5H_9	Pentaborane(9)					42.7	171.8	184.2	151.1	73.2	173.6	280.6	99.6
B_5H_{11}	Pentaborane(11)					73.2				103.3	230.6	321.0	130.3
B_6H_{10}	Hexaborane(10)					56.3				94.6	211.3	296.8	125.7
B_9H_{15}	Nonaborane(15)									158.4	357.5	364.9	187.0
$B_{10}H_{14}$	Decaborane(14)									47.3	232.8	350.7	186.1
Ba	Barium	0.0		62.5	28.1					180.0	146.0	170.2	20.8
$BaBr_2$	Barium bromide	-757.3	-736.8	146.0									
$BaCl_2$	Barium chloride	-855.0	-806.7	123.7	75.1								
$BaCl_2H_4O_2$	Barium chloride dihydrate	-1456.9	-1293.2	203.0									
BaF_2	Barium fluoride	-1207.1	-1156.8	96.4	71.2								
BaH_2	Barium hydride	-177.0	-138.2	63.0	46.0								
BaH_2O_2	Barium hydroxide	-944.7											
BaI_2	Barium iodide	-602.1											
BaN_2O_4	Barium nitrite	-768.2											
BaN_2O_6	Barium nitrate	-988.0	-792.6	214.0	151.4								
BaO	Barium oxide	-548.0	-520.3	72.1	47.3					-112.0			
BaO_4S	Barium sulfate	-1473.2	-1362.2	132.2	101.8								
BaS	Barium sulfide	-460.0	-456.0	78.2	49.4								
Be	Beryllium	0.0		9.5	16.4					324.0	286.6	136.3	20.8
$BeBr_2$	Beryllium bromide	-353.5		108.0	69.4								
$BeCl_2$	Beryllium chloride	-490.4	-445.6	75.8	62.4								
BeF_2	Beryllium fluoride	-1026.8	-979.4	53.4	51.8								
BeH_2O_2	Beryllium hydroxide	-902.5	-815.0	45.5	62.1								
BeI_2	Beryllium iodide	-192.5		121.0	71.1								
BeO	Beryllium oxide	-609.4	-580.1	13.8	25.6								
BeO_4S	Beryllium sulfate	-1205.2	-1093.8	77.9	85.7								
BeS	Beryllium sulfide	-234.3		34.0	34.0								
Bi	Bismuth	0.0		56.7	25.5					207.1	168.2	187.0	20.8
BiClO	Bismuth oxychloride	-366.9	-322.1	120.5									
$BiCl_3$	Bismuth trichloride	-379.1	-315.0	177.0	105.0					-265.7	-256.0	358.9	79.7
BiH_3O_3	Bismuth hydroxide	-711.3											
BiI_3	Bismuth triiodide		-175.3										
Bi_2	Dibismuth									219.7			36.9
Bi_2O_3	Bismuth oxide	-573.9	-493.7	151.5	113.5								
$Bi_2O_{12}S_3$	Bismuth sulfate	-2544.3											
Bi_2S_3	Bismuth sulfide	-143.1	-140.6	200.4	122.2								
Bk	Berkelium	0.0											

		Crystal				Liquid				Gas			
Molecular formula	**Name**	**$\Delta_f H°$ kJ/mol**	**$\Delta_f G°$ kJ/mol**	**$S°$ J/mol K**	**C_p J/mol K**	**$\Delta_f H°$ kJ/mol**	**$\Delta_f G°$ kJ/mol**	**$S°$ J/mol K**	**C_p J/mol K**	**$\Delta_f H°$ kJ/mol**	**$\Delta_f G°$ kJ/mol**	**$S°$ J/mol K**	**C_p J/mol K**
Br	Bromine (atomic)									111.9	82.4	175.0	20.8
BrCl	Bromine chloride									14.6	-1.0	240.1	35.0
$BrCl_3Si$	Bromotrichlorosilane											350.1	90.9
BrCs	Cesium bromide	-405.8	-391.4	113.1	52.9								
BrCu	Copper(I) bromide	-104.6	-100.8	96.1	54.7								
BrF	Bromine fluoride									-93.8	-109.2	229.0	33.0
BrF_3	Bromine trifluoride					-300.8	-240.5	178.2	124.6	-255.6	-229.4	292.5	66.6
BrF_5	Bromine pentafluoride					-458.6	-351.8	225.1		-428.9	-350.6	320.2	99.6
BrGe	Germanium monobromide									235.6			37.1
$BrGeH_3$	Bromogermane											274.8	56.4
BrH	Hydrogen bromide									-36.3	-53.4	198.7	29.1
BrHSi	Bromosilylene									-464.4			
BrH_3Si	Bromosilane											262.4	52.8
BrH_4N	Ammonium bromide	-270.8	-175.2	113.0	96.0								
BrI	Iodine bromide									40.8	3.7	258.8	36.4
BrIn	Indium(I) bromide	-175.3	-169.0	113.0						-56.9	-94.3	259.5	36.7
BrK	Potassium bromide	-393.8	-380.7	95.9	52.3								
$BrKO_3$	Potassium bromate	-360.2	271.2	149.2	105.2								
$BrKO_4$	Potassium perbromate	-287.9	-174.4	170.1	120.2								
BrLi	Lithium bromide	-351.2	-342.0	74.3									
BrNO	Nitrosyl bromide									82.2	82.4	273.7	45.5
BrNa	Sodium bromide	-361.1	-349.0	86.8	51.4					-143.1	-177.1	241.2	36.3
$BrNaO_3$	Sodium bromate	-334.1	-242.6	128.9									
BrO	Bromine monoxide									125.8	109.6	233.0	34.2
BrO_2	Bromine dioxide									152.0	155.0	271.1	45.4
BrRb	Rubidium bromide	-394.6	-381.8	110.0	52.8								
BrSi	Bromosilyldyne									209.0			38.6
BrTl	Thallium(I) bromide	-173.2	-167.4	120.5						-37.7			
Br_2	Bromine					0.0		152.2	75.7	30.9	3.1	245.5	36.0
Br_2Ca	Calcium bromide	-682.8	-663.6	130.0									
Br_2Cd	Cadmium bromide	-316.2	-296.3	137.2	76.7								
Br_2Co	Cobalt(II) bromide	-220.9			79.5								
Br_2Cr	Chromium(II) bromide	-302.1											
Br_2Cu	Copper(II) bromide	-141.8											
Br_2Fe	Iron(II) bromide	-249.8	-238.1	140.6									
Br_2H_2Si	Dibromosilane											309.7	65.5
Br_2Hg	Mercury(II) bromide	-170.7	-153.1	172.0									
Br_2Hg_2	Mercury(I) bromide	-206.9	-181.1	218.0									
Br_2Mg	Magnesium bromide	-524.3	-503.8	117.2									
Br_2Mn	Manganese(II) bromide	-384.9											
Br_2Ni	Nickel(II) bromide	-212.1											
Br_2Pb	Lead(II) bromide	-278.7	-261.9	161.5	80.1								
Br_2Pt	Platinum(II) bromide	-82.0											
Br_2S_2	Sulfur bromide					-13.0							
Br_2Se	Selenium dibromide									-21.0			
Br_2Sn	Tin(II) bromide	-243.5											
Br_2Sr	Strontium bromide	-717.6	-697.1	135.1	75.3								
Br_2Ti	Titanium(II) bromide	-402.0											
Br_2Zn	Zinc bromide	-328.7	-312.1	138.5									
Br_3Ce	Cerium(III) bromide	-891.4											
Br_3ClSi	Tribromochlorosilane											377.1	95.3
Br_3Dy	Dysprosium(III) bromide	-836.2											
Br_3Fe	Iron(III) bromide	-268.2											
Br_3Ga	Gallium(III) bromide	-386.6	-359.8	180.0									
Br_3HSi	Tribromosilane					-355.6	-336.4	248.1		-317.6	-328.5	348.6	80.8
Br_3In	Indium(III) bromide	-428.9								-282.0			
Br_3OP	Phosphoric tribromide	-458.6										359.8	89.9
Br_3P	Phosphorus(III) bromide					-184.5	-175.7	240.2		-139.3	-162.8	348.1	76.0
Br_3Pt	Platinum(III) bromide	-120.9											
Br_3Re	Rhenium(III) bromide	-167.0											
Br_3Ru	Ruthenium(III) bromide	-138.0											
Br_3Sb	Antimony(III) bromide	-259.4	-239.3	207.1						-194.6	-223.9	372.9	80.2
Br_3Sc	Scandium bromide	-743.1											
Br_3Ti	Titanium(III) bromide	-548.5	-523.8	176.6	101.7								
Br_4Ge	Germanium(IV) bromide					-347.7	-331.4	280.7		-300.0	-318.0	396.2	101.8
Br_4Pa	Protactinium(IV) bromide	-824.0	-787.8	234.0									

		Crystal				Liquid				Gas			
Molecular formula	**Name**	**$\Delta_f H°$ kJ/mol**	**$\Delta_f G°$ kJ/mol**	**$S°$ J/mol K**	**C_p J/mol K**	**$\Delta_f H°$ kJ/mol**	**$\Delta_f G°$ kJ/mol**	**$S°$ J/mol K**	**C_p J/mol K**	**$\Delta_f H°$ kJ/mol**	**$\Delta_f G°$ kJ/mol**	**$S°$ J/mol K**	**C_p J/mol K**
Br_4Pt	Platinum(IV) bromide	-156.5											
Br_4Si	Tetrabromosilane					-457.3	-443.9	277.8		-415.5	-431.8	377.9	97.1
Br_4Sn	Tin(IV) bromide	-377.4	-350.2	264.4						-314.6	-331.4	411.9	103.4
Br_4Te	Tellurium tetrabromide	-190.4											
Br_4Ti	Titanium(IV) bromide	-616.7	-589.5	243.5	131.5					-549.4	-568.2	398.4	100.8
Br_4V	Vanadium(IV) bromide									-336.8			
Br_4Zr	Zirconium(IV) bromide	-760.7											
Br_5P	Phosphorus(V) bromide	-269.9											
Br_5Ta	Tantalum(V) bromide	-598.3											
Br_6W	Tungsten(VI) bromide	-348.5											
Ca	Calcium	0.0		41.6	25.9					177.8	144.0	154.9	20.8
$CaCl_2$	Calcium chloride	-795.4	-748.8	108.4	72.9								
CaF_2	Calcium fluoride	-1228.0	-1175.6	68.5	67.0								
CaH_2	Calcium hydride	-181.5	-142.5	41.4	41.0								
CaH_2O_2	Calcium hydroxide	-985.2	-897.5	83.4	87.5								
CaI_2	Calcium iodide	-533.5	-528.9	142.0									
CaN_2O_6	Calcium nitrate	-938.2	-742.8	193.2	149.4								
CaO	Calcium oxide	-634.9	-603.3	38.1	42.0								
CaO_4S	Calcium sulfate	-1434.5	-1322.0	106.5	99.7								
CaS	Calcium sulfide	-482.4	-477.4	56.5	47.4								
$Ca_3O_8P_2$	Calcium phosphate	-4120.8	-3884.7	236.0	227.8								
Cd	Cadmium	0.0		51.8	26.0					111.8		167.7	20.8
$CdCl_2$	Cadmium chloride	-391.5	-343.9	115.3	74.7								
CdF_2	Cadmium fluoride	-700.4	-647.7	77.4									
CdH_2O_2	Cadmium hydroxide	-560.7	-473.6	96.0									
CdI_2	Cadmium iodide	-203.3	-201.4	161.1	80.0								
CdO	Cadmium oxide	-258.4	-228.7	54.8	43.4								
CdO_4S	Cadmium sulfate	-933.3	-822.7	123.0	99.6								
CdS	Cadmium sulfide	-161.9	-156.5	64.9									
$CdTe$	Cadmium telluride	-92.5	-92.0	100.0									
Ce	Cerium (γ, fcc)	0.0		72.0	26.9					423.0	385.0	191.8	23.1
$CeCl_3$	Cerium(III) chloride	-1060.5	-984.8	151.0	87.4								
CeI_3	Cerium(III) iodide	-669.3											
CeO_2	Cerium(IV) oxide	-1088.7	-1024.6	62.3	61.6								
CeS	Cerium(II) sulfide	-459.4	-451.5	78.2	50.0								
Ce_2O_3	Cerium(III) oxide	-1796.2	-1706.2	150.6	114.6								
Cf	Californium	0.0											
Cl	Chlorine (atomic)									121.3	105.3	165.2	21.8
$ClCs$	Cesium chloride	-443.0	-414.5	101.2	52.5								
$ClCsO_4$	Cesium perchlorate	-443.1	-314.3	175.1	108.3								
$ClCu$	Copper(I) chloride	-137.2	-119.9	86.2	48.5								
ClF	Chlorine fluoride									-50.3	-51.8	217.9	32.1
$ClFO_3$	Perchloryl fluoride									-23.8	48.2	279.0	64.9
ClF_3	Chlorine trifluoride					-189.5				-163.2	-123.0	281.6	63.9
ClF_5S	Sulfur chloride pentafluoride					-1065.7							
$ClGe$	Germanium monochloride									155.2	124.2	247.0	36.9
$ClGeH_3$	Chlorogermane											263.7	54.7
ClH	Hydrogen chloride									-92.3	-95.3	186.9	29.1
$ClHO$	Hypochlorous acid									-78.7	-66.1	236.7	37.2
$ClHO_4$	Perchloric acid					-40.6							
ClH_3Si	Chlorosilane											250.7	51.0
ClH_4N	Ammonium chloride	-314.4	-202.9	94.6	84.1								
ClH_4NO_4	Ammonium perchlorate	-295.3	-88.8	186.2									
ClH_4P	Phosphonium chloride	-145.2											
ClI	Iodine chloride					-23.9	-13.6	135.1		17.8	-5.5	247.6	35.6
$ClIn$	Indium(I) chloride	-186.2								-75.0			
ClK	Potassium chloride	-436.5	-408.5	82.6	51.3					-214.6	-233.3	239.1	36.5
$ClKO_3$	Potassium chlorate	-397.7	-296.3	143.1	100.3								
$ClKO_4$	Potassium perchlorate	-432.8	-303.1	151.0	112.4								
$ClLi$	Lithium chloride	-408.6	-384.4	59.3	48.0								
$ClLiO_4$	Lithium perchlorate	-381.0											
$ClNO$	Nitrosyl chloride									51.7	66.1	261.7	44.7
$ClNO_2$	Nitryl chloride									12.6	54.4	272.2	53.2
$ClNa$	Sodium chloride	-411.2	-384.1	72.1	50.5								
$ClNaO_2$	Sodium chlorite	-307.0											
$ClNaO_3$	Sodium chlorate	-365.8	-262.3	123.4									

		Crystal				Liquid				Gas			
Molecular formula	Name	$\Delta_fH°$ kJ/mol	$\Delta_fG°$ kJ/mol	$S°$ J/mol K	C_p J/mol K	$\Delta_fH°$ kJ/mol	$\Delta_fG°$ kJ/mol	$S°$ J/mol K	C_p J/mol K	$\Delta_fH°$ kJ/mol	$\Delta_fG°$ kJ/mol	$S°$ J/mol K	C_p J/mol K
$ClNaO_4$	Sodium perchlorate	-383.3	-254.9	142.3									
ClO	Chlorine oxide									101.8	98.1	226.6	31.5
$ClOV$	Vanadyl chloride	-607.0	-556.0	75.0									
ClO_2	Chlorine dioxide									102.5	120.5	256.8	42.0
ClO_2	Chlorine superoxide (ClOO)									89.1	105.0	263.7	46.0
ClO_4Rb	Rubidium perchlorate	-437.2	-306.9	161.1									
$ClRb$	Rubidium chloride	-435.4	-407.8	95.9	52.4								
$ClSi$	Chlorosilylidyne									189.9			36.9
$ClTl$	Thallium(I) chloride	-204.1	-184.9	111.3	50.9					-67.8			
Cl_2	Chlorine									0.0		223.1	33.9
Cl_2Co	Cobalt(II) chloride	-312.5	-269.8	109.2	78.5								
Cl_2Cr	Chromium(II) chloride	-395.4	-356.0	115.3	71.2								
Cl_2CrO_2	Chromyl chloride					-579.5	-510.8	221.8		-538.1	-501.6	329.8	84.5
Cl_2Cu	Copper(II) chloride	-220.1	-175.7	108.1	71.9								
Cl_2Fe	Iron(II) chloride	-341.8	-302.3	118.0	76.7								
Cl_2H_2Si	Dichlorosilane											285.7	60.5
Cl_2Hg	Mercury(II) chloride	-224.3	-178.6	146.0									
Cl_2Hg_2	Mercury(I) chloride	-265.4	-210.7	191.6									
Cl_2Mg	Magnesium chloride	-641.3	-591.8	89.6	71.4								
Cl_2Mn	Manganese(II) chloride	-481.3	-440.5	118.2	72.9								
Cl_2Ni	Nickel(II) chloride	-305.3	-259.0	97.7	71.7								
Cl_2O	Chlorine monoxide									80.3	97.9	266.2	45.4
Cl_2OS	Thionyl chloride					-245.6			121.0	-212.5	-198.3	309.8	66.5
Cl_2O_2S	Sulfuryl chloride					-394.1			134.0	-364.0	-320.0	311.9	77.0
Cl_2O_2U	Uranyl chloride	-1243.9	-1146.4	150.5	107.9								
Cl_2Pb	Lead(II) chloride	-359.4	-314.1	136.0									
Cl_2Pt	Platinum(II) chloride	-123.4											
Cl_2S	Sulfur dichloride					-50.0							
Cl_2S_2	Sulfur chloride					[illegible]							
Cl_2Sn	Tin(II) chloride	-325.1											
Cl_2Sr	Strontium chloride	-828.9	-781.1	114.9	75.6								
Cl_2Ti	Titanium(II) chloride	-513.8	-464.4	87.4	69.8								
Cl_2Zn	Zinc chloride	-415.1	-369.4	111.5	71.3					-266.1			
Cl_2Zr	Zirconium(II) chloride	-502.0											
Cl_3Cr	Chromium(III) chloride	-556.5	-486.1	123.0	91.8								
Cl_3Dy	Dysprosium(III) chloride	-1000.0											
Cl_3Er	Erbium chloride	-998.7			100.0								
Cl_3Eu	Europium(III) chloride	-936.0											
Cl_3Fe	Iron(III) chloride	-399.5	-334.0	142.3	96.7								
Cl_3Ga	Gallium(III) chloride	-524.7	-454.8	142.0									
Cl_3Gd	Gadolinium(III) chloride	-1008.0			88.0								
Cl_3HSi	Trichlorosilane					-539.3	-482.5	227.6		-513.0	-482.0	313.9	75.8
Cl_3Ho	Holmium chloride	-1005.4			88.0								
Cl_3In	Indium(III) chloride	-537.2								-374.0			
Cl_3Ir	Iridium(III) chloride	-245.6											
Cl_3La	Lanthanum chloride	-1072.2			108.8								
Cl_3Lu	Lutetium chloride	-945.6								-649.0			
Cl_3N	Nitrogen trichloride					230.0							
Cl_3Nd	Neodymium chloride	-1041.0			113.0								
Cl_3OP	Phosphoric trichloride					-597.1	-520.8	222.5	138.8	-558.5	-512.9	325.5	84.9
Cl_3OV	Vanadyl trichloride					-734.7	-668.5	244.3		-695.6	-659.3	344.3	89.9
Cl_3Os	Osmium(III) chloride	-190.4											
Cl_3P	Phosphorus(III) chloride					-319.7	-272.3	217.1		-287.0	-267.8	311.8	71.8
Cl_3Pr	Praseodymium chloride	-1056.9			100.0								
Cl_3Pt	Platinum(III) chloride	-182.0											
Cl_3Re	Rhenium(III) chloride	-264.0	-188.0	123.8	92.4								
Cl_3Rh	Rhodium(III) chloride	-299.2											
Cl_3Ru	Ruthenium(III) chloride	-205.0											
Cl_3Sb	Antimony(III) chloride	-382.2	-323.7	184.1	107.9								
Cl_3Sc	Scandium chloride	-925.1											
Cl_3Sm	Samarium(III) chloride	-1025.9											
Cl_3Tb	Terbium chloride	-997.0											
Cl_3Ti	Titanium(III) chloride	-720.9	-653.5	139.7	97.2								
Cl_3Tl	Thallium(III) chloride	-315.1											
Cl_3Tm	Thulium chloride	-986.6											
Cl_3U	Uranium(III) chloride	-866.5	-799.1	159.0	102.5								

		Crystal				Liquid				Gas			
Molecular formula	**Name**	**$\Delta_f H°$ kJ/mol**	**$\Delta_f G°$ kJ/mol**	**$S°$ J/mol K**	**C_p J/mol K**	**$\Delta_f H°$ kJ/mol**	**$\Delta_f G°$ kJ/mol**	**$S°$ J/mol K**	**C_p J/mol K**	**$\Delta_f H°$ kJ/mol**	**$\Delta_f G°$ kJ/mol**	**$S°$ J/mol K**	**C_p J/mol K**
Cl_3V	Vanadium(III) chloride	-580.7	-511.2	131.0	93.2								
Cl_3Y	Yttrium chloride	-1000.0								-750.2			75.0
Cl_3Yb	Ytterbium(III) chloride	-959.8											
Cl_4Ge	Germanium(IV) chloride					-531.8	-462.7	245.6		-495.8	-457.3	347.7	96.1
Cl_4Hf	Hafnium(IV) chloride	-990.4	-901.3	190.8	120.5					-884.5			
Cl_4Pa	Protactinium(IV) chloride	-1043.0	-953.0	192.0									
Cl_4Pb	Lead(IV) chloride					-329.3							
Cl_4Pt	Platinum(IV) chloride	-231.8											
Cl_4Si	Tetrachlorosilane					-687.0	-619.8	239.7	145.3	-657.0	-617.0	330.7	90.3
Cl_4Sn	Tin(IV) chloride					-511.3	-440.1	258.6	165.3	-471.5	-432.2	365.8	98.3
Cl_4Te	Tellurium tetrachloride	-326.4			138.5								
Cl_4Th	Thorium(IV) chloride	-1186.2	-1094.1	190.4	120.3					-964.4	-932.0	390.7	107.5
Cl_4Ti	Titanium(IV) chloride					-804.2	-737.2	252.3	145.2	-763.2	-726.3	353.2	95.4
Cl_4U	Uranium(IV) chloride	-1019.2	-930.0	197.1	122.0					-809.6	-786.6	419.0	
Cl_4V	Vanadium(IV) chloride					-569.4	-503.7	255.0		-525.5	-492.0	362.4	96.2
Cl_4Zr	Zirconium(IV) chloride	-980.5	-889.9	181.6	119.8								
Cl_5Nb	Niobium(V) chloride	-797.5	-683.2	210.5	148.1					-703.7	-646.0	400.6	120.8
Cl_5P	Phosphorus(V) chloride	-443.5								-374.9	-305.0	364.6	112.8
Cl_5Pa	Protactinium(V) chloride	-1145.0	-1034.0	238.0									
Cl_5Ta	Tantalum(V) chloride	-859.0											
Cl_6U	Uranium(VI) chloride	-1092.0	-962.0	285.8	175.7					-1013.0	-928.0	431.0	
Cl_6W	Tungsten(VI) chloride	-602.5								-513.8			
Cm	Curium	0.0											
Co	Cobalt	0.0		30.0	24.8					424.7	380.3	179.5	23.0
CoF_2	Cobalt(II) fluoride	-692.0	-647.2	82.0	68.8								
CoH_2O_2	Cobalt(II) hydroxide	-539.7	-454.3	79.0									
CoI_2	Cobalt(II) iodide	-88.7											
CoN_2O_6	Cobalt(II) nitrate	-420.5											
CoO	Cobalt(II) oxide	-237.9	-214.2	53.0	55.2								
CoO_4S	Cobalt(II) sulfate	-888.3	-782.3	118.0									
CoS	Cobalt(II) sulfide	-82.8											
Co_2S_3	Cobalt(III) sulfide	-147.3											
Co_3O_4	Cobalt(II,III) oxide	-891.0	-774.0	102.5	123.4								
Cr	Chromium	0.0		23.8	23.4					396.6	351.8	174.5	20.8
CrF_2	Chromium(II) fluoride	-778.0											
CrF_3	Chromium(III) fluoride	-1159.0	-1088.0	93.9	78.7								
CrI_2	Chromium(II) iodide	-156.9											
CrI_3	Chromium(III) iodide	-205.0											
CrO_2	Chromium(IV) oxide	-598.0											
CrO_3	Chromium(VI) oxide									-292.9		266.2	56.0
CrO_4Pb	Lead(II) chromate	-930.9											
Cr_2FeO_4	Chromium iron oxide	-1444.7	-1343.8	146.0	133.6								
Cr_2O_3	Chromium(III) oxide	-1139.7	-1058.1	81.2	118.7								
Cr_3O_4	Chromium(II,III) oxide	-1531.0											
Cs	Cesium	0.0		85.2	32.2					76.5	49.6	175.6	20.8
CsF	Cesium fluoride	-553.5	-525.5	92.8	51.1								
CsF_2H	Cesium hydrogen fluoride	-923.8	-858.9	135.2	87.3								
CsH	Cesium hydride	-54.2											
CsHO	Cesium hydroxide	-416.2	-371.8	104.2	69.9					-256.0	-256.5	254.8	49.7
$CsHO_4S$	Cesium hydrogen sulfate	-1158.1											
CsH_2N	Cesium amide	-118.4											
CsI	Cesium iodide	-346.6	-340.6	123.1	52.8								
$CsNO_3$	Cesium nitrate	-506.0	-406.5	155.2									
CsO_2	Cesium superoxide	-286.2											
Cs_2O	Cesium oxide	-345.8	-308.1	146.9	76.0								
Cs_2O_3S	Cesium sulfite	-1134.7											
Cs_2O_4S	Cesium sulfate	-1443.0	-1323.6	211.9	134.9								
Cs_2S	Cesium sulfide	-359.8											
Cu	Copper	0.0		33.2	24.4					337.4	297.7	166.4	20.8
CuF_2	Copper(II) fluoride	-542.7											
CuH_2O_2	Copper(II) hydroxide	-449.8											
CuI	Copper(I) iodide	-67.8	-69.5	96.7	54.1								
CuN_2O_6	Copper(II) nitrate	-302.9											
CuO	Copper(II) oxide	-157.3	-129.7	42.6	42.3								
CuO_4S	Copper(II) sulfate	-771.4	-662.2	109.2									
CuO_4W	Copper(II) tungstate	-1105.0											

		Crystal				Liquid				Gas			
Molecular formula	**Name**	$\Delta_f H°$ kJ/mol	$\Delta_f G°$ kJ/mol	$S°$ J/mol K	C_p J/mol K	$\Delta_f H°$ kJ/mol	$\Delta_f G°$ kJ/mol	$S°$ J/mol K	C_p J/mol K	$\Delta_f H°$ kJ/mol	$\Delta_f G°$ kJ/mol	$S°$ J/mol K	C_p J/mol K
CuS	Copper(II) sulfide	-53.1	-53.6	66.5	47.8								
$CuSe$	Copper(II) selenide	-39.5											
Cu_2	Dicopper									484.2	431.9	241.6	36.6
Cu_2O	Copper(I) oxide	-168.6	-146.0	93.1	63.6								
Cu_2S	Copper(I) sulfide	-79.5	-86.2	120.9	76.3								
Dy	Dysprosium	0.0		75.6	27.7					290.4	254.4	196.6	20.8
DyI_3	Dysprosium(III) iodide	-620.5											
Dy_2O_3	Dysprosium(III) oxide	-1863.1	-1771.5	149.8	116.3								
Er	Erbium	0.0		73.2	28.1					317.1	280.7	195.6	20.8
ErF_3	Erbium fluoride	-1711.0											
Er_2O_3	Erbium oxide	-1897.9	-1808.7	155.6	108.5								
Es	Einsteinium	0.0											
Eu	Europium	0.0		77.8	27.7					175.3	142.2	188.8	20.8
Eu_2O_3	Europium(III) oxide	-1651.4	-1556.8	146.0	122.2								
Eu_3O_4	Europium(II,III) oxide	-2272.0	-2142.0	205.0									
F	Fluorine (atomic)									79.4	62.3	158.8	22.7
FGa	Gallium monofluoride									-251.9			33.3
FGe	Germanium monofluoride									-33.4			34.7
$FGeH_3$	Fluorogermane											252.8	51.6
FH	Hydrogen fluoride					-299.8				-273.3	-275.4	173.8	
FH_3Si	Fluorosilane											238.4	47.4
FH_4N	Ammonium fluoride	-464.0	-348.7	72.0	65.3								
FI	Iodine fluoride									-95.7	-118.5	236.2	33.4
FIn	Indium(I) fluoride									-203.4			
FK	Potassium fluoride	-567.3	-537.8	66.6	49.0								
FLi	Lithium fluoride	-616.0	-587.7	35.7	41.6								
FNO	Nitrosyl fluoride									-66.5	-51.0	248.1	41.3
FNO_2	Nitryl fluoride											260.4	49.8
FNS	Thionitrosyl fluoride (NSF)											259.8	44.1
FNa	Sodium fluoride	-576.6	-546.3	51.1	46.9								
FO	Fluorine oxide									109.0	105.3	216.4	32.0
FO_2	Fluorine superoxide (FOO)									25.4	39.4	259.5	44.5
FRb	Rubidium fluoride	-557.7											
FSi	Fluorosilylidyne									7.1	-24.3	225.8	32.6
FTl	Thallium(I) fluoride	-324.7								-182.4			
F_2	Fluorine									0.0		202.8	31.3
F_2Fe	Iron(II) fluoride	-711.3	-668.6	87.0	68.1								
F_2HK	Potassium hydrogen fluoride	-927.7	-859.7	104.3	76.9								
F_2HN	Difluoramine											252.8	43.4
F_2HNa	Sodium hydrogen fluoride	-920.3	-852.2	90.9	75.0								
F_2HRb	Rubidium hydrogen fluoride	-922.6	-855.6	120.1	79.4								
F_2Mg	Magnesium fluoride	-1124.2	-1071.1	57.2	61.6								
F_2N	Difluoroamidogen									43.1	57.8	249.9	41.0
F_2N_2	*cis*-Difluorodiazine									69.5			
F_2N_2	*trans*-Difluorodiazine									82.0			
F_2Ni	Nickel(II) fluoride	-651.4	-604.1	73.6	64.1								
F_2O	Fluorine monoxide									24.5	41.8	247.5	43.3
F_2OS	Thionyl fluoride											278.7	56.8
F_2O_2	Fluorine dioxide									19.2	58.2	277.2	62.1
F_2O_2S	Sulfuryl fluoride											284.0	66.0
F_2O_2U	Uranyl fluoride	-1653.5	-1557.4	135.6	103.2								
F_2Pb	Lead(II) fluoride	-664.0	-617.1	110.5									
F_2Si	Difluorosilylene									-619.0	-628.0	252.7	43.9
F_2Sr	Strontium fluoride	-1216.3	-1164.8	82.1	70.0								
F_2Zn	Zinc fluoride	-764.4	-713.3	73.7	65.7								
F_3Ga	Gallium(III) fluoride	-1163.0	-1085.3	84.0									
F_3Gd	Gadolinium(III) fluoride									-1297.0			
F_3HSi	Trifluorosilane											271.9	60.5
F_3Ho	Holmium fluoride	-1707.0											
F_3N	Nitrogen trifluoride									-132.1	-90.6	260.8	53.4
F_3Nd	Neodymium fluoride	-1657.0											
F_3OP	Phosphoric trifluoride									-1254.3	-1205.8	285.4	68.8
F_3P	Phosphorus(III) fluoride									-958.4	-936.9	273.1	58.7
F_3Sb	Antimony(III) fluoride	-915.5											
F_3Sc	Scandium fluoride	-1629.2	-1555.6	92.0						-1247.0	-1234.0	300.5	67.8
F_3Sm	Samarium(III) fluoride	-1778.0											

		Crystal				Liquid				Gas			
Molecular formula	**Name**	**$\Delta_f H°$ kJ/mol**	**$\Delta_f G°$ kJ/mol**	**$S°$ J/mol K**	**C_p J/mol K**	**$\Delta_f H°$ kJ/mol**	**$\Delta_f G°$ kJ/mol**	**$S°$ J/mol K**	**C_p J/mol K**	**$\Delta_f H°$ kJ/mol**	**$\Delta_f G°$ kJ/mol**	**$S°$ J/mol K**	**C_p J/mol K**
F_3Th	Thorium(III) fluoride									-1166.1	-1160.6	339.2	73.3
F_3U	Uranium(III) fluoride	-1502.1	-1433.4	123.4	95.1					-1058.5	-1051.9	331.9	74.3
F_3Y	Yttrium fluoride	-1718.8	-1644.7	100.0						-1288.7	-1277.8	311.8	70.3
F_4Ge	Germanium(IV) fluoride									-1190.2	-1150.0	301.9	
F_4Hf	Hafnium fluoride	-1930.5	-1830.4	113.0						-1669.8			
F_4N_2	Tetrafluorohydrazine									-8.4	79.9	301.2	79.2
F_4Pb	Lead(IV) fluoride	-941.8											
F_4S	Sulfur tetrafluoride									-763.2	-722.0	299.6	77.6
F_4Si	Tetrafluorosilane									-1615.0	-1572.8	282.8	73.6
F_4Th	Thorium(IV) fluoride	-2097.8	-2003.4	142.0	110.7					-1759.0	-1724.0	341.7	93.0
F_4U	Uranium(IV) fluoride	-1914.2	-1823.3	151.7	116.0					-1598.7	-1572.7	368.0	91.2
F_4V	Vanadium(IV) fluoride	-1403.3											
F_4Xe	Xenon tetrafluoride	-261.5											
F_4Zr	Zirconium(IV) fluoride	-1911.3	-1809.9	104.6	103.7								
F_5I	Iodine pentafluoride					-864.8				-822.5	-751.7	327.7	99.2
F_5Nb	Niobium(V) fluoride	-1813.8	-1699.0	160.2	134.7					-1739.7	-1673.6	321.9	97.1
F_5P	Phosphorus(V) fluoride									-1594.4	-1520.7	300.8	84.8
F_5Ta	Tantalum(V) fluoride	-1903.6											
F_5V	Vanadium(V) fluoride					-1480.3	-1373.1	175.7		-1433.9	-1369.8	320.9	98.6
$F_6H_8N_2Si$	Ammonium hexafluorosilicate	-2681.7	-2365.3	280.2	228.1								
F_6Ir	Iridium(VI) fluoride	-579.7	-461.6	247.7						-544.0	-460.0	357.8	121.1
F_6K_2Si	Potassium hexafluorosilicate	-2956.0	-2798.6	226.0									
F_6Mo	Molybdenum(VI) fluoride					-1585.5	-1473.0	259.7	169.8	-1557.7	-1472.2	350.5	120.6
F_6Na_2Si	Sodium hexafluorosilicate	-2909.6	-2754.2	207.1	187.1								
F_6Os	Osmium(VI) fluoride			246.0								358.1	120.8
F_6Pt	Platinum(VI) fluoride			235.6								348.3	122.8
F_6S	Sulfur hexafluoride									-1220.5	-1116.5	291.5	97.0
F_6Se	Selenium hexafluoride									-1117.0	-1017.0	313.9	110.5
F_6Si_2	Hexafluorodisilane	-2427.0	-2299.7	219.1	129.5					-2383.3	-2307.3	391.0	129.9
F_6Te	Tellurium hexafluoride									-1318.0			
F_6U	Uranium(VI) fluoride	-2197.0	-2068.5	227.6	166.8					-2147.4	-2063.7	377.9	129.6
F_6W	Tungsten(VI) fluoride					-1747.7	-1631.4	251.5		-1721.7	-1632.1	341.1	119.0
Fe	Iron	0.0		27.3	25.1					416.3	370.7	180.5	25.7
FeI_2	Iron(II) iodide	-113.0											
FeI_3	Iron(III) iodide									71.0			
$FeMoO_4$	Iron(II) molybdate	-1075.0	-975.0	129.3	118.5								
FeO	Iron(II) oxide	-272.0											
FeO_4S	Iron(II) sulfate	-928.4	-820.8	107.5	100.6								
FeO_4W	Iron(II) tungstate	-1155.0	-1054.0	131.8	114.6								
FeS	Iron(II) sulfide	-100.0	-100.4	60.3	50.5								
FeS_2	Iron disulfide	-178.2	-166.9	52.9	62.2								
Fe_2O_3	Iron(III) oxide	-824.2	-742.2	87.4	103.9								
Fe_2O_4Si	Iron(II) orthosilicate	-1479.9	-1379.0	145.2	132.9								
Fe_3O_4	Iron(II,III) oxide	-1118.4	-1015.4	146.4	143.4								
Fm	Fermium	0.0											
Fr	Francium	0.0		95.4									
Ga	Gallium	0.0	0.0	40.8	26.1	5.6				272.0	233.7	169.0	25.3
GaH_3O_3	Gallium(III) hydroxide	-964.4	-831.3	100.0									
GaI_3	Gallium(III) iodide	-238.9		205.0	100.0								
GaN	Gallium nitride	-110.5											
GaO	Gallium monoxide									279.5	253.5	231.1	32.1
GaP	Gallium phosphide	-88.0											
GaSb	Gallium antimonide	-41.8	-38.9	76.1	48.5								
Ga_2	Digallium									438.5			
Ga_2O	Gallium suboxide	-356.0											
Ga_2O_3	Gallium(III) oxide	-1089.1	-998.3	85.0	92.1								
Gd	Gadolinium	0.0		68.1	37.0					397.5	359.8	194.3	27.5
Gd_2O_3	Gadolinium(III) oxide	-1819.6			106.7								
Ge	Germanium	0.0		31.1	23.3					372.0	331.2	167.9	30.7
GeH_3I	Iodogermane											283.2	57.5
GeH_4	Germane									90.8	113.4	217.1	45.0
GeI_4	Germanium(IV) iodide	-141.8	-144.3	271.1						-56.9	-106.3	428.9	104.1
GeO	Germanium(II) oxide	-261.9	-237.2	50.0						-46.2	-73.2	224.3	30.9
GeO_2	Germanium(IV) oxide	-580.0	-521.4	39.7	52.1								
GeP	Germanium phosphide	-21.0	-17.0	63.0									
GeS	Germanium(II) sulfide	-69.0	-71.5	71.0						92.0	42.0	234.0	33.7

Molecular formula	Name	Crystal $\Delta_f H°$ kJ/mol	Crystal $\Delta_f G°$ kJ/mol	Crystal $S°$ J/mol K	Crystal C_p J/mol K	Liquid $\Delta_f H°$ kJ/mol	Liquid $\Delta_f G°$ kJ/mol	Liquid $S°$ J/mol K	Liquid C_p J/mol K	Gas $\Delta_f H°$ kJ/mol	Gas $\Delta_f G°$ kJ/mol	Gas $S°$ J/mol K	Gas C_p J/mol K
GeTe	Germanium(II) telluride	20.0											
Ge_2	Digermanium									473.1	416.3	252.8	35.6
Ge_2H_6	Digermane					137.3				162.3			
Ge_3H_8	Trigermane					193.7				226.8			
H	Hydrogen (atomic)									218.0	203.3	114.7	20.8
HI	Hydrogen iodide									26.5	1.7	206.6	29.2
HIO_3	Iodic acid	-230.1											
HK	Potassium hydride	-57.7											
HKO	Potassium hydroxide	-424.6	-379.4	81.2	68.9					-232.0	-229.7	238.3	49.2
HKO_4S	Potassium hydrogen sulfate	-1160.6	-1031.3	138.1									
HLi	Lithium hydride	-90.5	-68.3	20.0	27.9								
HLiO	Lithium hydroxide	-487.5	-441.5	42.8	49.6					-229.0	-234.2	214.4	46.0
HN	Imidogen									351.5	345.6	181.2	29.2
HNO_2	Nitrous acid									-79.5	-46.0	254.1	45.6
HNO_3	Nitric acid					-174.1	-80.7	155.6	109.9	-133.9	-73.5	266.9	54.1
HN_3	Hydrazoic acid					264.0	327.3	140.6		294.1	328.1	239.0	43.7
HNa	Sodium hydride	-56.3	-33.5	40.0	36.4								
HNaO	Sodium hydroxide	-425.8	-379.7	64.4	59.5					-191.0	-193.9	229.0	48.0
$HNaO_4S$	Sodium hydrogen sulfate	-1125.5	-992.8	113.0									
HNa_2O_4P	Sodium hydrogen phosphate	-1748.1	-1608.2	150.5	135.3								
HO	Hydroxyl									39.0	34.2	183.7	29.9
HORb	Rubidium hydroxide	-418.8	-373.9	94.0	69.0					-238.0	-239.1	248.5	49.5
HOTl	Thallium(I) hydroxide	-238.9	-195.8	88.0									
HO_2	Hydroperoxy									10.5	22.6	229.0	34.9
HO_3P	Metaphosphoric acid	-948.5											
HO_4RbS	Rubidium hydrogen sulfate	-1159.0											
HO_4Re	Perrhenic acid	-762.3	-656.4	158.2									
HRb	Rubidium hydride	-52.3											
HS	Mercapto									142.7	113.3	195.7	32.3
HSi	Silylidyne									361.0			
HTa_2	Tantalum hydride	-32.6	-69.0	79.1	90.8								
H_2	Hydrogen									0.0		130.7	28.8
H_2KN	Potassium amide	-128.9											
H_2KO_4P	Potassium dihydrogen phosphate	-1568.3	-1415.9	134.9	116.6								
H_2LiN	Lithium amide	-179.5											
H_2Mg	Magnesium hydride	-75.3	-35.9	31.1	35.4								
H_2MgO_2	Magnesium hydroxide	-924.5	-833.5	63.2	77.0								
H_2N	Amidogen									184.9	194.6	195.0	33.9
H_2NNa	Sodium amide	-123.8	-64.0	76.9	66.2								
H_2NRb	Rubidium amide	-113.0											
$H_2N_2O_2$	Nitramide	-89.5											
H_2NiO_2	Nickel(II) hydroxide	-529.7	-447.2	88.0									
H_2O	Water					-285.8	-237.1	70.0	75.3	-241.8	-228.6	188.8	33.6
H_2O_2	Hydrogen peroxide					-187.8	-120.4	109.6	89.1	-136.3	-105.6	232.7	43.1
H_2O_2Sn	Tin(II) hydroxide	-561.1	-491.6	155.0									
H_2O_2Sr	Strontium hydroxide	-959.0											
H_2O_2Zn	Zinc hydroxide	-641.9	-553.5	81.2									
H_2O_3Si	Metasilicic acid	-1188.7	-1092.4	134.0									
H_2O_4S	Sulfuric acid					-814.0	-690.0	156.9	138.9				
H_2O_4Se	Selenic acid	-530.1											
H_2S	Hydrogen sulfide									-20.6	-33.4	205.8	34.2
H_2S_2	Hydrogen disulfide					-18.1			84.1	15.5			51.5
H_2Se	Hydrogen selenide									29.7	15.9	219.0	34.7
H_2Sr	Strontium hydride	-180.3											
H_2Te	Hydrogen telluride									99.6			
H_2Th	Thorium hydride	-139.7	-100.0	50.7	36.7								
H_2Zr	Zirconium(II) hydride	-169.0	-128.8	35.0	31.0								
H_3ISi	Iodosilane											270.9	54.4
H_3N	Ammonia									-45.9	-16.4	192.8	35.1
H_3NO	Hydroxylamine	-114.2											
H_3O_2P	Phosphinic acid	-604.6				-595.4							
H_3O_3P	Phosphonic acid	-964.4											
H_3O_4P	Phosphoric acid	-1284.4	-1124.3	110.5	106.1	-1271.7	-1123.6	150.8	145.0				
H_3P	Phosphine									5.4	13.5	210.2	37.1
H_3Sb	Stibine									145.1	147.8	232.8	41.1
H_3U	Uranium(III) hydride	-127.2	-72.8	63.7	49.3								

		Crystal				Liquid				Gas			
Molecular formula	Name	$\Delta_fH°$ kJ/mol	$\Delta_fG°$ kJ/mol	$S°$ J/mol K	C_p J/mol K	$\Delta_fH°$ kJ/mol	$\Delta_fG°$ kJ/mol	$S°$ J/mol K	C_p J/mol K	$\Delta_fH°$ kJ/mol	$\Delta_fG°$ kJ/mol	$S°$ J/mol K	C_p J/mol K
H_4IN	Ammonium iodide	-201.4	-112.5	117.0									
H_4N_2	Hydrazine					50.6	149.3	121.2	98.9	95.4	159.4	238.5	48.4
$H_4N_2O_2$	Ammonium nitrite	-256.5											
$H_4N_2O_3$	Ammonium nitrate	-365.6	-183.9	151.1	139.3								
H_4N_4	Ammonium azide	115.5	274.2	112.5									
H_4O_4Si	Orthosilicic acid	-1481.1	-1332.9	192.0									
$H_4O_7P_2$	Diphosphoric acid	-2241.0				-2231.7							
H_4P_2	Diphosphine					-5.0				20.9			
H_4Si	Silane									34.3	56.9	204.6	42.8
H_4Sn	Stannane									162.8	188.3	227.7	49.0
H_5NO	Ammonium hydroxide					-361.2	-254.0	165.6	154.9				
H_5NO_3S	Ammonium hydrogen sulfite	-768.6											
H_5NO_4S	Ammonium hydrogen sulfate	-1027.0											
H_6Si_2	Disilane									80.3	127.3	272.7	80.8
$H_8N_2O_4S$	Ammonium sulfate	-1180.9	-901.7	220.1	187.5								
H_8Si_3	Trisilane					92.5				120.9			
$H_9N_2O_4P$	Ammonium hydrogen phosphate	-1566.9			188.0								
$H_{12}N_3O_4P$	Ammonium phosphate	-1671.9											
He	Helium									0.0		126.2	20.8
Hf	Hafnium	0.0		43.6	25.7					619.2	576.5	186.9	20.8
HfO_2	Hafnium oxide	-1144.7	-1088.2	59.3	60.3								
Hg	Mercury					0.0		75.9	28.0	61.4	31.8	175.0	20.8
HgI_2	Mercury(II) iodide	-105.4	-101.7	180.0									
HgO	Mercury(II) oxide	-90.8	-58.5	70.3	44.1								
HgO_4S	Mercury(II) sulfate	-707.5											
HgS	Mercury(II) sulfide (red)	-58.2	-50.6	82.4	48.4								
HgTe	Mercury(II) telluride	-42.0											
Hg_2	Dimercury									108.8	68.2	288.1	37.4
Hg_2I_2	Mercury(I) iodide	-121.3	-111.0	233.5									
Hg_2O_4S	Mercury(I) sulfate	-743.1	-625.8	200.7	132.0								
Ho	Holmium	0.0		75.3	27.2					300.8	264.8	195.6	20.8
Ho_2O_3	Holmium oxide	-1880.7	-1791.1	158.2	115.0								
I	Iodine (atomic)									106.8	70.2	180.8	20.8
IIn	Indium(I) iodide	-116.3	-120.5	130.0						7.5	-37.7	267.3	36.8
IK	Potassium iodide	-327.9	-324.9	106.3	52.9								
IKO_3	Potassium iodate	-501.4	-418.4	151.5	106.5								
IKO_4	Potassium periodate	-467.2	-361.4	175.7									
ILi	Lithium iodide	-270.4	-270.3	86.8	51.0								
INa	Sodium iodide	-287.8	-286.1	98.5	52.1								
$INaO_3$	Sodium iodate	-481.8			92.0								
$INaO_4$	Sodium periodate	-429.3	-323.0	163.0									
IO	Iodine monoxide									126.0	102.5	239.6	32.9
IRb	Rubidium iodide	-333.8	-328.9	118.4	53.2								
ITl	Thallium(I) iodide	-123.8	-125.4	127.6						7.1			
I_2	Iodine (rhombic)	0.0		116.1	54.4					62.4	19.3	260.7	36.9
I_2Mg	Magnesium iodide	-364.0	-358.2	129.7									
I_2Ni	Nickel(II) iodide	-78.2											
I_2Pb	Lead(II) iodide	-175.5	-173.6	174.9	77.4								
I_2Sn	Tin(II) iodide	-143.5											
I_2Sr	Strontium iodide	-558.1			81.6								
I_2Zn	Zinc iodide	-208.0	-209.0	161.1									
I_3In	Indium(III) iodide	-238.0								-120.5			
I_3La	Lanthanum iodide	-668.9											
I_3Lu	Lutetium iodide	-548.0											
I_3P	Phosphorus(III) iodide	-45.6										374.4	78.4
I_3Ru	Ruthenium(III) iodide	-65.7											
I_3Sb	Antimony(III) iodide	-100.4											
I_4Pt	Platinum(IV) iodide	-72.8											
I_4Si	Tetraiodosilane	-189.5											
I_4Sn	Tin(IV) iodide				84.9							446.1	105.4
I_4Ti	Titanium(IV) iodide	-375.7	-371.5	249.4	125.7					-277.8			
I_4V	Vanadium(IV) iodide									-122.6			
I_4Zr	Zirconium(IV) iodide	-481.6											
In	Indium	0.0		57.8	26.7					243.3	208.7	173.8	20.8
InO	Indium monoxide									387.0	364.4	236.5	32.6
InP	Indium phosphide	-88.7	-77.0	59.8	45.4								

Molecular formula	Name	Crystal $\Delta_fH°$ kJ/mol	Crystal $\Delta_fG°$ kJ/mol	Crystal $S°$ J/mol K	Crystal C_p J/mol K	Liquid $\Delta_fH°$ kJ/mol	Liquid $\Delta_fG°$ kJ/mol	Liquid $S°$ J/mol K	Liquid C_p J/mol K	Gas $\Delta_fH°$ kJ/mol	Gas $\Delta_fG°$ kJ/mol	Gas $S°$ J/mol K	Gas C_p J/mol K
InS	Indium(II) sulfide	-138.1	-131.8	67.0						238.0			
InSb	Indium antimonide	-30.5	-25.5	86.2	49.5					344.3			
In_2	Diindium									380.9			
In_2O_3	Indium(III) oxide	-925.8	-830.7	104.2	92.0								
In_2S_3	Indium(III) sulfide	-427.0	-412.5	163.6	118.0								
In_2Te_5	Indium(IV) telluride	-175.3											
Ir	Iridium	0.0		35.5	25.1					665.3	617.9	193.6	20.8
IrO_2	Iridium(IV) oxide	-274.1			57.3								
IrS_2	Iridium(IV) sulfide	-138.0											
Ir_2S_3	Iridium(III) sulfide	-234.0											
K	Potassium	0.0		64.7	29.6					89.0	60.5	160.3	20.8
$KMnO_4$	Potassium permanganate	-837.2	-737.6	171.7	117.6								
KNO_2	Potassium nitrite	-369.8	-306.6	152.1	107.4								
KNO_3	Potassium nitrate	-494.6	-394.9	133.1	96.4								
KNa	Potassium sodium					6.3							
KO_2	Potassium superoxide	-284.9	-239.4	116.7	77.5								
K_2	Dipotassium									123.7	87.5	249.7	37.9
K_2O	Potassium oxide	-361.5											
K_2O_2	Potassium peroxide	-494.1	-425.1	102.1									
K_2O_4S	Potassium sulfate	-1437.8	-1321.4	175.6	131.5								
K_2S	Potassium sulfide	-380.7	-364.0	105.0									
K_3O_4P	Potassium phosphate	-1950.2											
Kr	Krypton									0.0		164.1	20.8
La	Lanthanum	0.0		56.9	27.1					431.0	393.6	182.4	22.8
LaS	Lanthanum monosulfide	-456.0	-451.5	73.2	59.0								
La_2O_3	Lanthanum oxide	-1793.7	-1705.8	127.3	108.8								
Li	Lithium	0.0		29.1	24.8					159.3	126.6	138.8	20.8
$LiNO_2$	Lithium nitrite	-372.4	-302.0	96.0									
$LiNO_3$	Lithium nitrate	-483.1	-381.1	90.0									
Li_2	Dilithium									215.9	174.4	197.0	36.1
Li_2O	Lithium oxide	-597.9	-561.2	37.6	54.1								
Li_2O_2	Lithium peroxide	-634.3											
Li_2O_3Si	Lithium metasilicate	-1648.1	-1557.2	79.8	99.1								
Li_2O_4S	Lithium sulfate	-1436.5	-1321.7	115.1	117.6								
Li_2S	Lithium sulfide	-441.4											
Li_3O_4P	Lithium phosphate	-2095.8											
Lr	Lawrencium	0.0											
Lu	Lutetium	0.0		51.0	26.9					427.6	387.8	184.8	20.9
Lu_2O_3	Lutetium oxide	-1878.2	-1789.0	110.0	101.8								
Md	Mendelevium	0.0											
Mg	Magnesium	0.0		32.7	24.9					147.1	112.5	148.6	20.8
MgN_2O_6	Magnesium nitrate	-790.7	-589.4	164.0	141.9								
MgO	Magnesium oxide	-601.6	-569.3	27.0	37.2								
MgO_4S	Magnesium sulfate	-1284.9	-1170.6	91.6	96.5								
MgO_4Se	Magnesium selenate	-968.5											
MgS	Magnesium sulfide	-346.0	-341.8	50.3	45.6								
Mg_2	Dimagnesium									287.7			
Mg_2O_4Si	Magnesium orthosilicate	-2174.0	-2055.1	95.1	118.5								
Mn	Manganese	0.0		32.0	26.3					280.7	238.5	173.7	20.8
MnN_2O_6	Manganese(II) nitrate	-576.3											
$MnNaO_4$	Sodium permanganate	-1156.0											
MnO	Manganese(II) oxide	-385.2	-362.9	59.7	45.4								
MnO_2	Manganese(IV) oxide	-520.0	-465.1	53.1	54.1								
MnO_3Si	Manganese(II) metasilicate	-1320.9	-1240.5	89.1	86.4								
MnS	Manganese(II) sulfide (α form)	-214.2	-218.4	78.2	50.0								
MnSe	Manganese(II) selenide	-106.7	-111.7	90.8	51.0								
Mn_2O_3	Manganese(III) oxide	-959.0	-881.1	110.5	107.7								
Mn_2O_4Si	Manganese(II) orthosilicate	-1730.5	-1632.1	163.2	129.9								
Mn_3O_4	Manganese(II,III) oxide	-1387.8	-1283.2	155.6	139.7								
Mo	Molybdenum	0.0		28.7	24.1					658.1	612.5	182.0	20.8
$MoNa_2O_4$	Sodium molybdate	-1468.1	-1354.3	159.7	141.7								
MoO_2	Molybdenum(IV) oxide	-588.9	-533.0	46.3	56.0								
MoO_3	Molybdenum(VI) oxide	-745.1	-668.0	77.7	75.0								
MoO_4Pb	Lead(II) molybdate	-1051.9	-951.4	166.1	119.7								
MoS_2	Molybdenum(IV) sulfide	-235.1	-225.9	62.6	63.6								
Mo_3Si	Molybdenum silicide	-125.2	-125.7	106.3	93.1								

		Crystal				Liquid				Gas			
Molecular formula	**Name**	**$\Delta_f H°$ kJ/mol**	**$\Delta_f G°$ kJ/mol**	**$S°$ J/mol K**	**C_p J/mol K**	**$\Delta_f H°$ kJ/mol**	**$\Delta_f G°$ kJ/mol**	**$S°$ J/mol K**	**C_p J/mol K**	**$\Delta_f H°$ kJ/mol**	**$\Delta_f G°$ kJ/mol**	**$S°$ J/mol K**	**C_p J/mol K**
N	Nitrogen (atomic)									472.7	455.5	153.3	20.8
$NNaO_2$	Sodium nitrite	-358.7	-284.6	103.8									
$NNaO_3$	Sodium nitrate	-467.9	-367.0	116.5	92.9								
NO	Nitric oxide									91.3	87.6	210.8	29.9
NO_2	Nitrogen dioxide									33.2	51.3	240.1	37.2
NO_2Rb	Rubidium nitrite	-367.4	-306.2	172.0									
NO_3Rb	Rubidium nitrate	-495.1	-395.8	147.3	102.1								
NO_3Tl	Thallium(I) nitrate	-243.9	-152.4	160.7	99.5								
NP	Phosphorus nitride	-63.0								171.5	149.4	211.1	29.7
N_2	Nitrogen									0.0		191.6	29.1
N_2O	Nitrous oxide									81.6	103.7	220.0	38.6
N_2O_3	Nitrogen trioxide					50.3				86.6	142.4	314.7	72.7
N_2O_4	Nitrogen tetroxide					-19.5	97.5	209.2	142.7	11.1	99.8	304.4	79.2
N_2O_4Sr	Strontium nitrite	-762.3											
N_2O_5	Nitrogen pentoxide	-43.1	113.9	178.2	143.1					13.3	117.1	355.7	95.3
N_2O_6Pb	Lead(II) nitrate	-451.9											
N_2O_6Ra	Radium nitrate	-992.0	-796.1	222.0									
N_2O_6Sr	Strontium nitrate	-978.2	-780.0	194.6	149.9								
N_2O_6Zn	Zinc nitrate	-483.7											
N_3Na	Sodium azide	21.7	93.8	96.9	76.6								
N_4Si_3	Silicon nitride	-743.5	-642.6	101.3									
Na	Sodium	0.0		51.3	28.2					107.5	77.0	153.7	20.8
NaO_2	Sodium superoxide	-260.2	-218.4	115.9	72.1								
Na_2	Disodium									142.1	103.9	230.2	37.6
Na_2O	Sodium oxide	-414.2	-375.5	75.1	69.1								
Na_2O_2	Sodium peroxide	-510.9	-447.7	95.0	89.2								
Na_2O_3S	Sodium sulfite	-1100.8	-1012.5	145.9	120.3								
Na_2O_3Si	Sodium metasilicate	-1554.9	-1462.8	113.9									
Na_2O_4S	Sodium sulfate	-1387.1	-1270.2	149.6	128.2								
Na_2S	Sodium sulfide	-364.8	-349.8	83.7									
Nb	Niobium	0.0		36.4	24.6					725.9	681.1	186.3	30.2
NbO	Niobium(II) oxide	-405.8	-378.6	48.1	41.3								
NbO_2	Niobium(IV) oxide	-796.2	-740.5	54.5	57.5								
Nb_2O_5	Niobium(V) oxide	-1899.5	-1766.0	137.2	132.1								
Nd	Neodymium	0.0		71.5	27.5					327.6	292.4	189.4	22.1
Nd_2O_3	Neodymium oxide	-1807.9	-1720.8	158.6	111.3								
Ne	Neon									0.0		146.3	20.8
Ni	Nickel	0.0		29.9	26.1					429.7	384.5	182.2	23.4
NiO_4S	Nickel(II) sulfate	-872.9	-759.7	92.0	138.0								
NiS	Nickel(II) sulfide	-82.0	-79.5	53.0	47.1								
Ni_2O_3	Nickel(III) oxide	-489.5											
No	Nobelium	0.0											
O	Oxygen (atomic)									249.2	231.7	161.1	21.9
OP	Phosphorus monoxide									-28.5	-51.9	222.8	31.8
OPb	Lead(II) oxide (massicot)	-217.3	-187.9	68.7	45.8								
OPb	Lead(II) oxide (litharge)	-219.0	-188.9	66.5	45.8								
OPd	Palladium(II) oxide	-85.4			31.4					348.9	325.9	218.0	
ORa	Radium oxide	-523.0											
ORb_2	Rubidium oxide	-339.0											
ORh	Rhodium monoxide									385.0			
OS	Sulfur monoxide									6.3	-19.9	222.0	30.2
OSe	Selenium monoxide									53.4	26.8	234.0	31.3
OSi	Silicon monoxide									-99.6	-126.4	211.6	29.9
OSn	Tin(II) oxide	-280.7	-251.9	57.2	44.3					15.1	-8.4	232.1	31.6
OSr	Strontium oxide	-592.0	-561.9	54.4	45.0					1.5			
OTi	Titanium(II) oxide	-519.7	-495.0	50.0	40.0								
OTl_2	Thallium(I) oxide	-178.7	-147.3	126.0									
OU	Uranium(II) oxide									21.0			
OV	Vanadium(II) oxide	-431.8	-404.2	38.9	45.4								
OZn	Zinc oxide	-350.5	-320.5	43.7	40.3								
O_2	Oxygen									0.0		205.2	29.4
O_2P	Phosphorus dioxide									-279.9	-281.6	252.1	39.5
O_2Pb	Lead(IV) oxide	-277.4	-217.3	68.6	64.6								
O_2Rb	Rubidium superoxide	-278.7											
O_2Rb_2	Rubidium peroxide	-472.0											
O_2Ru	Ruthenium(IV) oxide	-305.0											

		Crystal				Liquid				Gas			
Molecular formula	Name	$\Delta_f H°$ kJ/mol	$\Delta_f G°$ kJ/mol	$S°$ J/mol K	C_p J/mol K	$\Delta_f H°$ kJ/mol	$\Delta_f G°$ kJ/mol	$S°$ J/mol K	C_p J/mol K	$\Delta_f H°$ kJ/mol	$\Delta_f G°$ kJ/mol	$S°$ J/mol K	C_p J/mol K
O_2S	Sulfur dioxide					-320.5				-296.8	-300.1	248.2	39.9
O_2Se	Selenium dioxide	-225.4											
O_2Si	Silicon dioxide (α-quartz)	-910.7	-856.3	41.5	44.4					-322.0			
O_2Sn	Tin(IV) oxide	-577.6	-515.8	49.0	52.6								
O_2Te	Tellurium dioxide	-322.6	-270.3	79.5									
O_2Th	Thorium(IV) oxide	-1226.4	-1169.2	65.2	61.8								
O_2Ti	Titanium(IV) oxide	-944.0	-888.8	50.6	55.0								
O_2U	Uranium(IV) oxide	-1085.0	-1031.8	77.0	63.6					-465.7	-471.5	274.6	51.4
O_2W	Tungsten(IV) oxide	-589.7	-533.9	50.5	56.1								
O_2Zr	Zirconium(IV) oxide	-1100.6	-1042.8	50.4	56.2								
O_3	Ozone									142.7	163.2	238.9	39.2
O_3PbS	Lead(II) sulfite	-669.9											
O_3PbSi	Lead(II) metasilicate	-1145.7	-1062.1	109.6	90.0								
O_3Pr_2	Praseodymium oxide	-1809.6			117.4								
O_3Rh_2	Rhodium(III) oxide	-343.0			103.8								
O_3S	Sulfur trioxide	-454.5	-374.2	70.7		-441.0	-373.8	113.8		-395.7	-371.1	256.8	50.7
O_3Sc_2	Scandium oxide	-1908.8	-1819.4	77.0	94.2								
O_3SiSr	Strontium metasilicate	-1633.9	-1549.7	96.7	88.5								
O_3Sm_2	Samarium(III) oxide	-1823.0	-1734.6	151.0	114.5								
O_3Tb_2	Terbium oxide	-1865.2			115.9								
O_3Ti_2	Titanium(III) oxide	-1520.9	-1434.2	78.8	97.4								
O_3Tm_2	Thulium oxide	-1888.7	-1794.5	139.7	116.7								
O_3U	Uranium(VI) oxide	-1223.8	-1145.7	96.1	81.7								
O_3V_2	Vanadium(III) oxide	-1218.8	-1139.3	98.3	103.2								
O_3W	Tungsten(VI) oxide	-842.9	-764.0	75.9	73.8								
O_3Y_2	Yttrium oxide	-1905.3	-1816.6	99.1	102.5								
O_3Yb_2	Ytterbium(III) oxide	-1814.6	-1726.7	133.1	115.4								
O_4Os	Osmium(VIII) oxide	-394.1	-304.9	143.9						-337.2	-292.8	293.8	74.1
O_4PbS	Lead(II) sulfate	-920.0	-813.0	148.5	103.2								
O_4PbSe	Lead(II) selenate	-609.2	-504.9	167.8									
O_4Pb_2Si	Lead(II) orthosilicate	-1363.1	-1252.6	186.6	137.2								
O_4Pb_3	Lead(II,II,IV) oxide	-718.4	-601.2	211.3	146.9								
O_4RaS	Radium sulfate	-1471.1	-1365.6	138.0									
O_4Rb_2S	Rubidium sulfate	-1435.6	-1316.9	197.4	134.1								
O_4Ru	Ruthenium(VIII) oxide	-239.3	-152.2	146.4									
O_4SSr	Strontium sulfate	-1453.1	-1340.9	117.0									
O_4STl_2	Thallium(I) sulfate	-931.8	-830.4	230.5									
O_4SZn	Zinc sulfate	-982.8	-871.5	110.5	99.2								
O_4SiSr_2	Strontium orthosilicate	-2304.5	-2191.1	153.1	134.3								
O_4SiZn_2	Zinc orthosilicate	-1636.7	-1523.2	131.4	123.3								
O_4SiZr	Zirconium(IV) orthosilicate	-2033.4	-1919.1	84.1	98.7								
O_4TiZr	Zirconium titanate	-2024.1	-1915.8	116.7	114.0								
O_5Sb_2	Antimony(V) oxide	-971.9	-829.2	125.1									
O_5Ta_2	Tantalum(V) oxide	-2046.0	-1911.2	143.1	135.1								
O_5Ti_3	Titanium(III,IV) oxide	-2459.4	-2317.4	129.3	154.8								
O_5V_2	Vanadium(V) oxide	-1550.6	-1419.5	131.0	127.7								
O_5V_3	Vanadium(III,IV) oxide	-1933.0	-1803.0	163.0									
O_7Re_2	Rhenium(VII) oxide	-1240.1	-1066.0	207.1	166.1					-1100.0	-994.0	452.0	
O_7U_3	Uranium(IV,VI) oxide	-3427.1	-3242.9	250.5	215.5								
O_8S_2Zr	Zirconium(IV) sulfate	-2217.1			172.0								
O_8U_3	Uranium(V,VI) oxide	-3574.8	-3369.5	282.6	238.4								
O_9U_4	Uranium(IV,V) oxide	-4510.4	-4275.1	334.1	293.3								
Os	Osmium	0.0		32.6	24.7					791.0	745.0	192.6	20.8
P	Phosphorus (white)	0.0		41.1	23.8					316.5	280.1	163.2	20.8
P	Phosphorus (red)	-17.6		22.8	21.2								
P	Phosphorus (black)	-39.3											
P_2	Diphosphorus									144.0	103.5	218.1	32.1
P_4	Tetraphosphorus									58.9	24.4	280.0	67.2
Pa	Protactinium	0.0		51.9						607.0	563.0	198.1	22.9
Pb	Lead	0.0		64.8	26.4					195.2	162.2	175.4	20.8
PbS	Lead(II) sulfide	-100.4	-98.7	91.2	49.5								
PbSe	Lead(II) selenide	-102.9	-101.7	102.5	50.2								
PbTe	Lead(II) telluride	-70.7	-69.5	110.0	50.5								
Pd	Palladium	0.0		37.6	26.0					378.2	339.7	167.1	20.8
PdS	Palladium(II) sulfide	-75.0	-67.0	46.0									
Pm	Promethium	0.0										187.1	24.3

		Crystal				Liquid				Gas			
Molecular formula	Name	$\Delta_f H°$ kJ/mol	$\Delta_f G°$ kJ/mol	$S°$ J/mol K	C_p J/mol K	$\Delta_f H°$ kJ/mol	$\Delta_f G°$ kJ/mol	$S°$ J/mol K	C_p J/mol K	$\Delta_f H°$ kJ/mol	$\Delta_f G°$ kJ/mol	$S°$ J/mol K	C_p J/mol K
Po	Polonium	0.0											
Pr	Praseodymium	0.0		73.2	27.2					355.6	320.9	189.8	21.4
Pt	Platinum	0.0		41.6	25.9					565.3	520.5	192.4	25.5
PtS	Platinum(II) sulfide	-81.6	-76.1	55.1	43.4								
PtS_2	Platinum(IV) sulfide	-108.8	-99.6	74.7	65.9								
Pu	Plutonium	0.0											
Ra	Radium	0.0		71.0						159.0	130.0	176.5	20.8
Rb	Rubidium	0.0		76.8	31.1					80.9	53.1	170.1	20.8
Re	Rhenium	0.0		36.9	25.5					769.9	724.6	188.9	20.8
Rh	Rhodium	0.0		31.5	25.0					556.9	510.8	185.8	21.0
Rn	Radon									0.0		176.2	20.8
Ru	Ruthenium	0.0		28.5	24.1					642.7	595.8	186.5	21.5
S	Sulfur (rhombic)	0.0		32.1	22.6					277.2	236.7	167.8	23.7
S	Sulfur (monoclinic)	0.3											
SSi	Silicon monosulfide									112.5	60.9	223.7	32.3
SSn	Tin(II) sulfide	-100.0	-98.3	77.0	49.3								
SSr	Strontium sulfide	-472.4	-467.8	68.2	48.7								
STl_2	Thallium(I) sulfide	-97.1	-93.7	151.0									
SZn	Zinc sulfide (wurtzite)	-192.6											
SZn	Zinc sulfide (sphalerite)	-206.0	-201.3	57.7	46.0								
S_2	Disulfur									128.6	79.7	228.2	32.5
Sb	Antimony	0.0		45.7	25.2					262.3	222.1	180.3	20.8
Sb_2	Diantimony									235.6	187.0	254.9	36.4
Sc	Scandium	0.0		34.6	25.5					377.8	336.0	174.8	22.1
Se	Selenium (gray)	0.0		42.4	25.4					227.1	187.0	176.7	20.8
Se	Selenium (α form)	6.7								227.1			
Se	Selenium (vitreous)	5.0								227.1			
SeSr	Strontium selenide	-385.8											
$SeTl_2$	Thallium(I) selenide	-59.0	-59.0	172.0									
SeZn	Zinc selenide	-163.0	-163.0	84.0									
Se_2	Diselenium									146.0	96.2	252.0	35.4
Si	Silicon	0.0		18.8	20.0					450.0	405.5	168.0	22.3
Si_2	Disilicon									594.0	536.0	229.9	34.4
Sm	Samarium	0.0		69.6	29.5					206.7	172.8	183.0	30.4
Sn	Tin (white)	0.0		51.2	27.0					301.2	266.2	168.5	21.3
Sn	Tin (gray)	-2.1	0.1	44.1	25.8								
Sr	Strontium	0.0		55.0	26.8					164.4	130.9	164.6	20.8
Ta	Tantalum	0.0		41.5	25.4					782.0	739.3	185.2	20.9
Tb	Terbium	0.0		73.2	28.9					388.7	349.7	203.6	24.6
Tc	Technetium	0.0								678.0		181.1	20.8
Te	Tellurium	0.0		49.7	25.7					196.7	157.1	182.7	20.8
Te_2	Ditellurium									168.2	118.0	268.1	36.7
Th	Thorium	0.0		51.8	27.3					602.0	560.7	190.2	20.8
Ti	Titanium	0.0		30.7	25.0					473.0	428.4	180.3	24.4
Tl	Thallium	0.0		64.2	26.3					182.2	147.4	181.0	20.8
Tm	Thulium	0.0		74.0	27.0					232.2	197.5	190.1	20.8
U	Uranium	0.0		50.2	27.7					533.0	488.4	199.8	23.7
V	Vanadium	0.0		28.9	24.9					514.2	754.4	182.3	26.0
W	Tungsten	0.0		32.6	24.3					849.4	807.1	174.0	21.3
Xe	Xenon									0.0		169.7	20.8
Y	Yttrium	0.0		44.4	26.5					421.3	381.1	179.5	25.9
Yb	Ytterbium	0.0		59.9	26.7					152.3	118.4	173.1	20.8
Zn	Zinc	0.0		41.6	25.4					130.4	94.8	161.0	20.8
Zr	Zirconium	0.0		39.0	25.4					608.8	566.5	181.4	26.7
Substances containing carbon:													
C	Carbon (graphite)	0.0		5.7	8.5					716.7	671.3	158.1	20.8
C	Carbon (diamond)	1.9	2.9	2.4	6.1								
CAgN	Silver(I) cyanide	146.0	156.9	107.2	66.7								
CAg_2O_3	Silver(I) carbonate	-505.8	-436.8	167.4	112.3								
$CBaO_3$	Barium carbonate	-1213.0	-1134.4	112.1	86.0								
$CBeO_3$	Beryllium carbonate	-1025.0		52.0	65.0								
$CBrClF_2$	Bromochlorodifluoromethane											318.5	74.6
$CBrCl_2F$	Bromodichlorofluoromethane											330.6	80.0
$CBrCl_3$	Bromotrichloromethane									-41.1			85.3
$CBrF_3$	Bromotrifluoromethane									-648.3			69.3

		Crystal				Liquid				Gas			
Molecular formula	**Name**	**$\Delta_f H°$ kJ/mol**	**$\Delta_f G°$ kJ/mol**	**$S°$ J/mol K**	**C_p J/mol K**	**$\Delta_f H°$ kJ/mol**	**$\Delta_f G°$ kJ/mol**	**$S°$ J/mol K**	**C_p J/mol K**	**$\Delta_f H°$ kJ/mol**	**$\Delta_f G°$ kJ/mol**	**$S°$ J/mol K**	**C_p J/mol K**
CBrN	Cyanogen bromide	140.5								186.2	165.3	248.3	46.9
$CBrN_3O_6$	Bromotrinitromethane					32.5				80.3			
CBr_2ClF	Dibromochlorofluoromethane											342.8	82.4
CBr_2Cl_2	Dibromodichloromethane											347.8	87.1
CBr_2F_2	Dibromodifluoromethane											325.3	77.0
CBr_2O	Carbonyl bromide					-127.2				-96.2	-110.9	309.1	61.8
CBr_3Cl	Tribromochloromethane											357.8	89.4
CBr_3F	Tribromofluoromethane											345.9	84.4
CBr_4	Tetrabromomethane	29.4	47.7	212.5	144.3					83.9	67.0	358.1	91.2
$CCaO_3$	Calcium carbonate (calcite)	-1207.6	-1129.1	91.7	83.5								
$CCaO_3$	Calcium carbonate (aragonite)	-1207.8	-1128.2	88.0	82.3								
$CCdO_3$	Cadmium carbonate	-750.6	-669.4	92.5									
CClFO	Carbonyl chloride fluoride											276.7	52.4
$CClF_3$	Chlorotrifluoromethane									-706.3			66.9
CClN	Cyanogen chloride					112.1				138.0	131.0	236.2	45.0
$CClN_3O_6$	Chlorotrinitromethane					-27.1				18.4			
CCl_2F_2	Dichlorodifluoromethane									-477.4	-439.4	300.8	72.3
CCl_2O	Carbonyl chloride									-219.1	-204.9	283.5	57.7
CCl_3	Trichloromethyl									59.0			
CCl_3F	Trichlorofluoromethane					-301.3	-236.8	225.4	121.6	-268.3			78.1
CCl_4	Tetrachloromethane					-128.2			130.7	-95.7			83.3
$CCoO_3$	Cobalt(II) carbonate	-713.0											
CCs_2O_3	Cesium carbonate	-1139.7	-1054.3	204.5	123.9								
CCuN	Copper(I) cyanide	96.2	111.3	84.5									
CFN	Cyanogen fluoride											224.7	41.8
CF_2O	Carbonyl fluoride									-639.8			46.8
CF_3	Trifluoromethyl									-477.0	-464.0	264.5	49.6
CF_3I	Trifluoroiodomethane									-587.8		307.4	70.0
CF_4	Tetrafluoromethane									-933.6		261.6	61.1
$CFeO_3$	Iron(II) carbonate	-740.6	-666.7	92.9	82.1								
CFe_3	Iron carbide	25.1	20.1	104.6	105.9								
CH	Methylidyne									595.8			
CHBrClF	Bromochlorofluoromethane											304.3	63.2
$CHBrCl_2$	Bromodichloromethane											316.4	67.4
$CHBrF_2$	Bromodifluoromethane									-424.9		295.1	58.7
$CHBr_2Cl$	Chlorodibromomethane											327.7	69.2
$CHBr_2F$	Dibromofluoromethane											316.8	65.1
$CHBr_3$	Tribromomethane					-22.3	-5.0	220.9	130.7	23.8	8.0	330.9	71.2
$CHClF_2$	Chlorodifluoromethane									-482.6		280.9	55.9
$CHCl_2F$	Dichlorofluoromethane											293.1	60.9
$CHCl_3$	Trichloromethane					-134.1	-73.7	201.7	114.2	-102.7	6.0	295.7	65.7
$CHCsO_3$	Cesium hydrogen carbonate	-966.1											
CHFO	Formyl fluoride											246.6	39.9
CHF_3	Trifluoromethane									695.4		259.7	51.0
CHI_3	Triiodomethane	-181.1								251.0		356.2	75.0
$CHKO_2$	Potassium formate	-679.7											
$CHKO_3$	Potassium hydrogen carbonate	-963.2	-863.5	115.5									
CHN	Hydrogen cyanide					108.9	125.0	112.8	70.6	135.1	124.7	201.8	35.9
CHNO	Isocyanic acid (HNCO)											238.0	44.9
CHNS	Isothiocyanic acid									127.6	113.0	247.8	46.9
CHN_3O_6	Trinitromethane					-32.8				-13.4		435.6	134.1
$CHNaO_2$	Sodium formate	-666.5	-599.9	103.8	82.7								
$CHNaO_3$	Sodium hydrogen carbonate	-950.8	-851.0	101.7	87.6								
CHO	Oxomethyl (HCO)									43.1	28.0	224.7	34.6
CH_2	Methylene									390.4	372.9	194.9	33.8
CH_2BrCl	Bromochloromethane											287.6	52.7
CH_2BrF	Bromofluoromethane											276.3	49.2
CH_2Br_2	Dibromomethane											293.2	54.7
CH_2ClF	Chlorofluoromethane											264.4	47.0
CH_2Cl_2	Dichloromethane					-124.2		177.8	101.2	-95.4		270.2	51.0
CH_2F_2	Difluoromethane									-452.3		246.7	42.9
CH_2I_2	Diiodomethane					68.5	90.4	174.1	134.0	119.5	95.8	309.7	57.7
CH_2N_2	Diazomethane											242.9	52.5
CH_2N_2	Cyanamide	58.8											
$CH_2N_2O_4$	Dinitromethane					-104.9				-61.5		358.1	86.4
CH_2O	Formaldehyde									-108.6	-102.5	218.8	35.4

		Crystal				Liquid				Gas			
Molecular formula	**Name**	$\Delta_fH°$ kJ/mol	$\Delta_fG°$ kJ/mol	$S°$ J/mol K	C_p J/mol K	$\Delta_fH°$ kJ/mol	$\Delta_fG°$ kJ/mol	$S°$ J/mol K	C_p J/mol K	$\Delta_fH°$ kJ/mol	$\Delta_fG°$ kJ/mol	$S°$ J/mol K	C_p J/mol K
$(CH_2O)_x$	Paraformaldehyde	-177.6											
CH_2O_2	Formic acid					-425.0	-361.4	129.0	99.0	-378.7			
CH_2S_3	Trithiocarbonic acid					24.0							
CH_3	Methyl									145.7	147.9	194.2	38.7
CH_3BO	Borane carbonyl									-111.2	-92.9	249.4	59.5
CH_3Br	Bromomethane					-59.8				-35.4	-26.3	246.4	42.4
CH_3Cl	Chloromethane									-81.9		234.6	40.8
CH_3Cl_3Si	Methyltrichlorosilane							262.8	163.1	-528.9		351.1	102.4
CH_3F	Fluoromethane											222.9	37.5
CH_3I	Iodomethane					-13.6		163.2	126.0	14.4		254.1	44.1
CH_3NO	Formamide					-254.0				-193.9			
CH_3NO_2	Nitromethane					-112.6	-14.4	171.8	106.6	-80.8		282.9	55.5
CH_3NO_2	Methyl nitrite									-66.1			
CH_3NO_3	Methyl nitrate					-156.3	-43.4	217.1	157.3	-122.0		305.8	76.6
CH_4	Methane									-74.6	-50.5	186.3	35.7
CH_4N_2	Ammonium cyanide	0.4			134.0								
CH_4N_2O	Urea	-333.1								-245.8			
CH_4N_2S	Thiourea	-89.1								22.9			
$CH_4N_4O_2$	Nitroguanidine	-92.4											
CH_4O	Methanol					-239.2	-166.6	126.8	81.1	-201.0	-162.3	239.9	44.1
CH_4S	Methanethiol					-46.7	-7.7	169.2	90.5	-22.9	-9.3	255.2	50.3
CH_5N	Methylamine					-47.3	35.7	150.2	102.1	-22.5	32.7	242.9	50.1
CH_5NO_3	Ammonium hydrogen carbonate	-849.4	-665.9	120.9									
CH_5N_3	Guanidine	-56.0											
CH_5N_3S	Hydrazinecarbothioamide	24.7											
$CH_5N_5O_2$	3-Amino-1-nitroguanidine	22.1											
CH_6ClN	Methylamine hydrochloride	-298.1											
CH_6N_2	Methylhydrazine					54.2	180.0	165.9	134.9	94.7	187.0	278.8	71.1
CH_6Si	Methylsilane											256.5	65.9
CHg_2O_3	Mercury(I) carbonate	-553.5	-468.1	180.0									
CIN	Cyanogen iodide	166.2	185.0	96.2						225.5	196.6	256.8	48.3
CI_4	Tetraiodomethane	-392.9								474.0		391.9	95.9
CKN	Potassium cyanide	-113.0	-101.9	128.5	66.3								
$CKNS$	Potassium thiocyanate	-200.2	-178.3	124.3	88.5								
CK_2O_3	Potassium carbonate	-1151.0	-1063.5	155.5	114.4								
CLi_2O_3	Lithium carbonate	-1215.9	-1132.1	90.4	99.1								
$CMgO_3$	Magnesium carbonate	-1095.8	-1012.1	65.7	75.5								
$CMnO_3$	Manganese(II) carbonate	-894.1	-816.7	85.8	81.5								
CN	Cyanide									437.6	407.5	202.6	29.2
$CNNa$	Sodium cyanide	-87.5	-76.4	115.6	70.4								
$CNNaO$	Sodium cyanate	-405.4	-358.1	96.7	86.6								
CN_4O_8	Tetranitromethane					38.4				82.4		503.7	176.1
CNa_2O_3	Sodium carbonate	-1130.7	-1044.4	135.0	112.3								
CO	Carbon monoxide									-110.5	-137.2	197.7	29.1
COS	Carbon oxysulfide									-142.0	-169.2	231.6	41.5
CO_2	Carbon dioxide									-393.5	-394.4	213.8	37.1
CO_3Pb	Lead(II) carbonate	-699.1	-625.5	131.0	87.4								
CO_3Rb_2	Rubidium carbonate	-1136.0	-1051.0	181.3	117.6								
CO_3Sr	Strontium carbonate	-1220.1	-1140.1	97.1	81.4								
CO_3Tl_2	Thallium(I) carbonate	-700.0	-614.6	155.2									
CO_3Zn	Zinc carbonate	-812.8	-731.5	82.4	79.7								
CS	Carbon monosulfide									280.3	228.8	210.6	29.8
CS_2	Carbon disulfide					89.0	64.6	151.3	76.4	116.7	67.1	237.8	45.4
CSe_2	Carbon diselenide					164.8							
CSi	Silicon carbide (cubic)	-65.3	-62.8	16.6	26.9								
CSi	Silicon carbide (hexagonal)	-62.8	-60.2	16.5	26.7								
C_2	Dicarbon									831.9	775.9	199.4	43.2
C_2BrF_5	Bromopentafluoroethane									-1064.4			
$C_2Br_2ClF_3$	1,2-Dibromo-1-chloro-1,2,2-trifluoroethane					-691.7				-656.6			
$C_2Br_2F_4$	1,2-Dibromotetrafluoroethane					-817.7				-789.1			
C_2Br_4	Tetrabromoethene											387.1	102.7
C_2Br_6	Hexabromoethane											441.9	139.3
C_2Ca	Calcium carbide	-59.8	-64.9	70.0	62.7								
C_2CaN_2	Calcium cyanide	-184.5											
C_2CaO_4	Calcium oxalate	-1360.6											

Molecular formula	Name	Crystal $\Delta_f H°$ kJ/mol	Crystal $\Delta_f G°$ kJ/mol	Crystal $S°$ J/mol K	Crystal C_p J/mol K	Liquid $\Delta_f H°$ kJ/mol	Liquid $\Delta_f G°$ kJ/mol	Liquid $S°$ J/mol K	Liquid C_p J/mol K	Gas $\Delta_f H°$ kJ/mol	Gas $\Delta_f G°$ kJ/mol	Gas $S°$ J/mol K	Gas C_p J/mol K
C_2ClF_3	Chlorotrifluoroethene					-522.7				-505.5	-523.8	322.1	83.9
C_2ClF_5	Chloropentafluoroethane									-1118.8			184.2
$C_2Cl_2F_4$	1,2-Dichloro-1,1,2,2-tetrafluoroethane					-960.2			111.7	-937.0			
$C_2Cl_2O_2$	Oxalyl chloride					-367.6				-335.8			
$C_2Cl_3F_3$	1,1,2-Trichloro-1,2,2-trifluoroethane					-745.0			170.1	-716.8			
C_2Cl_3N	Trichloroacetonitrile											336.6	96.1
C_2Cl_4	Tetrachloroethene					-50.6	3.0	266.9	143.4	-10.9			
$C_2Cl_4F_2$	1,1,1,2-Tetrachloro-2,2-difluoroethane									-489.9	-407.0	382.9	123.4
$C_2Cl_4F_2$	1,1,2,2-Tetrachloro-1,2-difluoroethane								173.6				
C_2Cl_4O	Trichloroacetyl chloride					-280.8				-239.8			
C_2Cl_6	Hexachloroethane	-202.8		237.3	198.2					-143.6			
C_2F_3N	Trifluoroacetonitrile									-497.9		298.1	77.9
C_2F_4	Tetrafluoroethene	-820.5								-658.9		300.1	80.5
C_2F_6	Hexafluoroethane									-1344.2		332.3	106.7
C_2HBr	Bromoacetylene											253.7	55.7
$C_2HBrClF_3$	1-Bromo-2-chloro-1,1,2-trifluoroethane					-675.3				-644.8			
$C_2HBrClF_3$	2-Bromo-2-chloro-1,1,1-trifluoroethane					-720.0				-690.4			
C_2HCl	Chloroacetylene											242.0	54.3
C_2HClF_2	1-Chloro-2,2-difluoroethene									-315.5	-289.1	303.0	72.1
C_2HCl_2F	1,1-Dichloro-2-fluoroethene											313.9	76.5
$C_2HCl_2F_3$	2,2-Dichloro-1,1,1-trifluoroethane											352.8	102.5
C_2HCl_3	Trichloroethene					-43.6		228.4	124.4	-9.0		324.8	80.3
C_2HCl_3O	Trichloroacetaldehyde					-234.5			151.0	-196.6			
C_2HCl_3O	Dichloroacetyl chloride					-280.4				-241.0			
$C_2HCl_3O_2$	Trichloroacetic acid	-503.3											
C_2HCl_5	Pentachloroethane					-187.6			173.8	-142.0			
C_2HF	Fluoroacetylene											231.7	52.4
C_2HF_3	Trifluoroethene									-490.5			
$C_2HF_3O_2$	Trifluoroacetic acid					-1069.9				-1031.4			
C_2HF_5	Pentafluoroethane									-1100.4			
C_2H_2	Acetylene									227.4	209.9	200.9	44.0
$C_2H_2BrF_3$	2-Bromo-1,1,1-trifluoroethane									-694.5			
$C_2H_2Br_2$	*cis*-1,2-Dibromoethene											311.3	68.8
$C_2H_2Br_2$	*trans*-1,2-Dibromoethene											313.5	70.3
$C_2H_2Br_2Cl_2$	1,2-Dibromo-1,2-dichloroethane									-36.9			
$C_2H_2Br_4$	1,1,2,2-Tetrabromoethane								165.7				
$C_2H_2ClF_3$	2-Chloro-1,1,1-trifluoroethane											326.5	89.1
$C_2H_2Cl_2$	1,1-Dichloroethene					-23.9	24.1	201.5	111.3	2.8	25.4	289.0	67.1
$C_2H_2Cl_2$	*cis*-1,2-Dichloroethene					-26.4		198.4	116.4	4.6		289.6	65.1
$C_2H_2Cl_2$	*trans*-1,2-Dichloroethene					-24.3	27.3	195.9	116.8	5.0	28.6	290.0	66.7
$C_2H_2Cl_2O$	Chloroacetyl chloride					-283.7				-244.8			
$C_2H_2Cl_2O_2$	Dichloroacetic acid					-496.3							
$C_2H_2Cl_3NO$	2,2,2-Trichloroacetamide	-358.0											
$C_2H_2Cl_4$	1,1,1,2-Tetrachloroethane											356.0	102.7
$C_2H_2Cl_4$	1,1,2,2-Tetrachloroethane					-195.0		246.9	162.3	-149.2		362.8	100.8
$C_2H_2F_2$	1,1-Difluoroethene									-335.0		266.2	60.1
$C_2H_2F_2$	*cis*-1,2-Difluoroethene											268.3	58.2
$C_2H_2F_3I$	1,1,1-Trifluoro-2-iodoethane									-644.5			
$C_2H_2I_2$	*cis*-1,2-Diiodoethene									-207.4			
C_2H_2O	Ketene					-67.9				-47.5	-48.3	247.6	51.8
$C_2H_2O_2$	Glyoxal									-212.0	-189.7	272.5	60.6
$C_2H_2O_4$	Oxalic acid	-829.9		109.8	91.0					-731.8	-662.7	320.6	86.2
$C_2H_2O_4Sr$	Strontium formate	-1393.3											
C_2H_2S	Thiirene									300.0	275.8	255.3	54.7
C_2H_3Br	Bromoethene									79.2	81.8	275.8	55.5
C_2H_3BrO	Acetyl bromide					-223.5				-190.4			
$C_2H_3BrO_2$	Bromoacetic acid									-383.5	-338.3	337.0	80.5
C_2H_3Cl	Chloroethene	-94.1			59.4	14.6				37.2	53.6	264.0	53.7
$C_2H_3ClF_2$	1-Chloro-1,1-difluoroethane											307.2	82.5
C_2H_3ClO	Acetyl chloride					-272.9	-208.0	200.8	117.0	-242.8	-205.8	295.1	67.8
$C_2H_3ClO_2$	Chloroacetic acid	-509.7								-427.6	-368.5	325.9	78.8

		Crystal				Liquid				Gas			
Molecular formula	**Name**	**$\Delta_f H°$ kJ/mol**	**$\Delta_f G°$ kJ/mol**	**$S°$ J/mol K**	**C_p J/mol K**	**$\Delta_f H°$ kJ/mol**	**$\Delta_f G°$ kJ/mol**	**$S°$ J/mol K**	**C_p J/mol K**	**$\Delta_f H°$ kJ/mol**	**$\Delta_f G°$ kJ/mol**	**$S°$ J/mol K**	**C_p J/mol K**
$C_2H_3Cl_2F$	1,1-Dichloro-1-fluoroethane											320.2	88.7
$C_2H_3Cl_3$	1,1,1-Trichloroethane					-177.4		227.4	144.3	-144.4		323.1	93.3
$C_2H_3Cl_3$	1,1,2-Trichloroethane					-190.8		232.6	150.9	-151.3		337.2	89.0
C_2H_3F	Fluoroethene									-138.8			
C_2H_3FO	Acetyl fluoride					-467.2				-442.1			
$C_2H_3F_3$	1,1,1-Trifluoroethane									-744.6		279.9	78.2
$C_2H_3F_3$	1,1,2-Trifluoroethane									-730.7			
$C_2H_3F_3O$	2,2,2-Trifluoroethanol					-932.4				-888.4			
C_2H_3I	Iodoethene											285.0	57.9
C_2H_3IO	Acetyl iodide					-163.5				-126.4			
$C_2H_3KO_2$	Potassium acetate	-723.0											
C_2H_3N	Acetonitrile					40.6	86.5	149.6	91.5	74.0	91.9	243.4	52.2
C_2H_3N	Isocyanomethane					130.8	159.5	159.0		163.5	165.7	246.9	52.9
C_2H_3NO	Methyl isocyanate					-92.0							
$C_2H_3NO_2$	Nitroethene									33.3		300.5	73.7
$C_2H_3NO_3$	Oxamic acid	-661.2								-552.3			
C_2H_3NS	Methyl isothiocyanate	79.4											
$C_2H_3NaO_2$	Sodium acetate	-708.8	-607.2	123.0	79.9								
C_2H_4	Ethylene									52.4	68.4	219.3	42.9
C_2H_4BrCl	1-Bromo-2-chloroethane								130.1				
$C_2H_4Br_2$	1,1-Dibromoethane					-66.2						327.7	80.8
$C_2H_4Br_2$	1,2-Dibromoethane					-79.2		223.3	136.0	-37.5			
C_2H_4ClF	1-Chloro-1-fluoroethane									-313.4			
$C_2H_4Cl_2$	1,1-Dichloroethane					-158.4	-73.8	211.8	126.3	-127.7	-70.8	305.1	76.2
$C_2H_4Cl_2$	1,2-Dichloroethane					-166.8			128.4	-126.4		308.4	78.7
$C_2H_4F_2$	1,1-Difluoroethane									-497.0		282.5	67.8
$C_2H_4I_2$	1,2-Diiodoethane	9.3								75.0			
$C_2H_4N_2O_2$	Oxamide	-504.4								-387.1			
$C_2H_4N_2O_2$	Ethanedial dioxime	-90.5											
$C_2H_4N_2O_4$	1,1-Dinitroethane					-148.2							
$C_2H_4N_2O_4$	1,2-Dinitroethane					-165.2							
$C_2H_4N_2S_2$	Ethanedithioamide	-20.8								83.0			
$C_2H_4N_4$	1*H*-1,2,4-Triazol-3-amine	76.8											
C_2H_4O	Acetaldehyde					-192.2	-127.6	160.2	89.0	-166.2	-133.0	263.8	55.3
C_2H_4O	Oxirane					-78.0	-11.8	153.9	88.0	-52.6	-13.0	242.5	47.9
C_2H_4OS	Thioacetic acid					-216.9				-175.1			
$C_2H_4O_2$	Acetic acid					-484.3	-389.9	159.8	123.3	-432.2	-374.2	283.5	63.4
$C_2H_4O_2$	Methyl formate					-386.1			119.1	-357.4		285.3	64.4
$C_2H_4O_3$	Peroxyacetic acid												82.4
$C_2H_4O_3$	Glycolic acid									-583.0	-504.9	318.6	87.1
C_2H_4S	Thiirane					51.6				82.0	96.8	255.2	53.3
C_2H_4Si	Ethynylsilane											269.4	72.6
C_2H_5Br	Bromoethane					-90.5	-25.8	198.7	100.8	-61.9	-23.9	286.7	64.5
C_2H_5Cl	Chloroethane					-136.8	-59.3	190.8	104.3	-112.1	-60.4	276.0	62.8
C_2H_5ClO	2-Chloroethanol					-295.4							
C_2H_5F	Fluoroethane											264.5	58.6
C_2H_5I	Iodoethane					-40.0	14.7	211.7	115.1	-8.1	19.2	306.0	66.9
C_2H_5N	Ethyleneimine					91.9				126.5			
C_2H_5NO	Acetamide	-317.0		115.0	91.3					-238.3			
C_2H_5NO	*N*-Methylformamide								123.8				
$C_2H_5NO_2$	Nitroethane					-143.9			134.4	-103.8		320.5	79.0
$C_2H_5NO_2$	Glycine	-528.5								-392.1			
$C_2H_5NO_3$	2-Nitroethanol					-350.7							
$C_2H_5NO_3$	Ethyl nitrate					-190.4				-154.1			
C_2H_5NS	Thioacetamide	-71.7								11.4			
C_2H_6	Ethane									-84.0	-32.0	229.2	52.5
C_2H_6Cd	Dimethyl cadmium					63.6	139.0	201.9	132.0	101.6	146.9	303.0	
C_2H_6Hg	Dimethyl mercury					59.8	140.3	209.0		94.4	146.1	306.0	83.3
$C_2H_6N_2O$	*N*-Methylurea	-332.8											
$C_2H_6N_4O_2$	1,2-Hydrazinedicarboxamide	-498.7											
$C_2H_6N_4O_2$	Oxalyl dihydrazide	-295.2											
C_2H_6O	Ethanol					-277.6	-174.8	160.7	112.3	-234.8	-167.9	281.6	65.6
C_2H_6O	Dimethyl ether					-203.3				-184.1	-112.6	266.4	64.4
C_2H_6OS	Dimethyl sulfoxide					-204.2	-99.9	188.3	153.0	-151.3			
$C_2H_6O_2$	Ethylene glycol					-460.0		163.2	148.6	-392.2		303.8	82.7
$C_2H_6O_2S$	Dimethyl sulfone	-450.1	-302.4	142.0						-373.1	-272.7	310.6	100.0

Molecular formula	Name	Crystal $\Delta_f H°$ kJ/mol	Crystal $\Delta_f G°$ kJ/mol	Crystal $S°$ J/mol K	Crystal C_p J/mol K	Liquid $\Delta_f H°$ kJ/mol	Liquid $\Delta_f G°$ kJ/mol	Liquid $S°$ J/mol K	Liquid C_p J/mol K	Gas $\Delta_f H°$ kJ/mol	Gas $\Delta_f G°$ kJ/mol	Gas $S°$ J/mol K	Gas C_p J/mol K
$C_2H_6O_3S$	Dimethyl sulfite					-523.6				-483.4			
$C_2H_6O_4S$	Dimethyl sulfate					-735.5				-687.0			
C_2H_6S	Ethanethiol					-73.6	-5.5	207.0	117.9	-46.1	-4.8	296.2	72.7
C_2H_6S	Dimethyl sulfide					-65.3		196.4	118.1	-37.4		286.0	74.1
$C_2H_6S_2$	1,2-Ethanedithiol					-54.3				-9.7			
$C_2H_6S_2$	Dimethyl disulfide					-62.6		235.4	146.1	-24.7			
C_2H_6Zn	Dimethyl zinc					23.4		201.6	129.2	53.0			
C_2H_7N	Ethylamine					-74.1			130.0	-47.5	36.3	283.8	71.5
C_2H_7N	Dimethylamine					-43.9	70.0	182.3	137.7	-18.8	68.5	273.1	70.7
C_2H_7NO	Ethanolamine								195.5				
C_2H_8ClN	Dimethylamine hydrochloride	-289.3											
$C_2H_8N_2$	1,2-Ethanediamine					-63.0			172.6	-18.0			
$C_2H_8N_2$	1,1-Dimethylhydrazine					48.9	206.4	198.0	164.1	84.1			
$C_2H_8N_2$	1,2-Dimethylhydrazine					52.7				92.2			
$C_2H_8N_2O_4$	Ammonium oxalate	-1123.0			226.0								
C_2HgO_4	Mercury(II) oxalate	-678.2											
C_2I_2	Diiodoacetylene											313.1	70.3
C_2I_4	Tetraiodoethene	305.0											
$C_2K_2O_4$	Potassium oxalate	-1346.0											
C_2MgO_4	Magnesium oxalate	-1269.0											
C_2N_2	Cyanogen					285.9				306.7		241.9	56.8
$C_2N_4O_6$	Trinitroacetonitrile					183.7							
$C_2Na_2O_4$	Sodium oxalate									1318.0			
C_2O_4Pb	Lead(II) oxalate	-851.4	-750.1	146.0	105.4								
C_3F_8	Perfluoropropane									-1783.2			
$C_3H_2N_2$	Malononitrile	186.4								265.5			
$C_3H_2O_2$	2-Propynoic acid					-193.2							
$C_3H_2O_3$	1,3-Dioxol-2-one					-459.9				-418.6			
$C_3H_3Cl_3$	1,2,3-Trichloropropene					-101.8							
$C_3H_3F_3$	3,3,3-Trifluoropropene									-614.2			
C_3H_3N	Acrylonitrile					147.1				180.6			
C_3H_3NO	Oxazole					-48.0				-15.5			
C_3H_3NO	Isoxazole					42.1				78.6			
C_3H_4	Allene									190.5			
C_3H_4	Propyne									184.9			
C_3H_4	Cyclopropene									277.1			
$C_3H_4Cl_2$	2,3-Dichloropropene					-73.3							
$C_3H_4Cl_4$	1,1,1,3-Tetrachloropropane					-208.7							
$C_3H_4Cl_4$	1,2,2,3-Tetrachloropropane					-251.8							
$C_3H_4F_4O$	2,2,3,3-Tetrafluoro-1-propanol					-1114.9				-1061.3			
$C_3H_4N_2$	1*H*-Pyrazole	105.4			81.0					179.4			
$C_3H_4N_2$	Imidazole	49.8								132.9			
C_3H_4O	Acrolein												71.3
$C_3H_4O_2$	1,2-Propanedione					-309.1				-271.0			
$C_3H_4O_2$	Acrylic acid					-383.8			145.7				
$C_3H_4O_2$	2-Oxetanone					-329.9		175.3	122.1	-282.9			
$C_3H_4O_3$	Ethylene carbonate					-682.8			133.9	-508.4			
C_3H_5Br	*cis*-1-Bromopropene					7.9				40.8			
C_3H_5Br	3-Bromopropene					12.2				45.2			
C_3H_5BrO	Bromoacetone									-181.0			
C_3H_5Cl	2-Chloropropene									-21.0			
C_3H_5Cl	3-Chloropropene								125.1				
C_3H_5ClO	Epichlorohydrin					-148.4			131.6	-107.8			
$C_3H_5ClO_2$	2-Chloropropanoic acid					-522.5				-475.8			
$C_3H_5ClO_2$	3-Chloropropanoic acid	-549.3											
$C_3H_5ClO_2$	Ethyl chloroformate					-505.3				-462.9			
$C_3H_5ClO_2$	Methyl chloroacetate					-487.0				-444.0			
$C_3H_5Cl_3$	1,2,3-Trichloropropane					-230.6			183.6	-182.9			
C_3H_5I	3-Iodopropene					53.7				91.5			
C_3H_5IO	Iodoacetone									-130.5			
$C_3H_5IO_2$	3-Iodopropanoic acid	-460.0											
C_3H_5N	Propanenitrile					15.5			119.3	51.7			
C_3H_5N	2-Propyn-1-amine					205.7							
C_3H_5N	Ethyl isocyanide					108.6				141.7			
C_3H_5NO	Acrylamide	-212.1			110.6	-224.0				-130.2			
$C_3H_5NO_3$	Nitroacetone					-278.6							

		Crystal				Liquid				Gas			
Molecular formula	**Name**	**$\Delta_f H°$ kJ/mol**	**$\Delta_f G°$ kJ/mol**	**$S°$ J/mol K**	**C_p J/mol K**	**$\Delta_f H°$ kJ/mol**	**$\Delta_f G°$ kJ/mol**	**$S°$ J/mol K**	**C_p J/mol K**	**$\Delta_f H°$ kJ/mol**	**$\Delta_f G°$ kJ/mol**	**$S°$ J/mol K**	**C_p J/mol K**
$C_3H_5NO_4$	Methyl nitroacetate					-464.0							
$C_3H_5N_3O_9$	Trinitroglycerol					-370.9				-279.1		545.9	234.2
C_3H_6	Propene					4.0				20.0			
C_3H_6	Cyclopropane					35.2				53.3	104.5	237.5	55.6
$C_3H_6Br_2$	1,2-Dibromopropane					-113.6				-71.6			
$C_3H_6Cl_2$	1,2-Dichloropropane, (±)					-198.8			149.1	-162.8			
$C_3H_6Cl_2$	1,3-Dichloropropane					-199.9				-159.2			
$C_3H_6Cl_2$	2,2-Dichloropropane					-205.8				-173.2			
$C_3H_6Cl_2O$	2,3-Dichloro-1-propanol					-381.5				-316.3			
$C_3H_6Cl_2O$	1,3-Dichloro-2-propanol					-385.3				-318.4			
$C_3H_6I_2$	1,2-Diiodopropane									35.6			
$C_3H_6I_2$	1,3-Diiodopropane					-9.0							
$C_3H_6N_2O_2$	Propanediamide	-546.1											
$C_3H_6N_2O_2$	*N*-(Aminocarbonyl)acetamide	-544.2								-441.2			
$C_3H_6N_2O_4$	1,1-Dinitropropane					-163.2				-100.7			
$C_3H_6N_2O_4$	1,3-Dinitropropane					-207.1							
$C_3H_6N_2O_4$	2,2-Dinitropropane					-181.2							
$C_3H_6N_6O_6$	Hexahydro-1,3,5-trinitro-1,3,5-triazine									192.0		482.4	230.2
C_3H_6O	Allyl alcohol					-171.8			138.9	-124.5			
C_3H_6O	Propanal					-215.6				-185.6		304.5	80.7
C_3H_6O	Acetone					-248.4		199.8	126.3	-217.1	-152.7	295.3	74.5
C_3H_6O	Methyloxirane					-123.0		196.5	120.4	-94.7		286.9	72.6
C_3H_6O	Oxetane					-110.8				-80.5			
$C_3H_6O_2$	Propanoic acid					-510.7		191.0	152.8	-455.7			
$C_3H_6O_2$	Ethyl formate								149.3				
$C_3H_6O_2$	Methyl acetate					-445.9			141.9	-413.3		324.4	86.0
$C_3H_6O_2$	1,3-Dioxolane					-333.5			118.0	-298.0			
$C_3H_6O_2S$	Thiolactic acid					-468.4							
$C_3H_6O_3$	1,3,5-Trioxane	-522.5		133.0	111.4					-465.9			
C_3H_6S	Thietane					24.7		184.9		60.6	107.1	285.0	68.3
C_3H_6S	Methylthiirane					11.3				45.8			
$C_3H_6S_2$	1,2-Dithiolane									0.0	47.7	313.5	86.5
$C_3H_6S_2$	1,3-Dithiolane									10.0	54.7	323.3	84.7
$C_3H_6S_3$	1,3,5-Trithiane									80.0	130.4	336.4	111.3
C_3H_7Br	1-Bromopropane					-121.9				-87.0			
C_3H_7Br	2-Bromopropane					-130.5				-99.4			
C_3H_7Cl	1-Chloropropane					-160.5				-131.9			
C_3H_7Cl	2-Chloropropane					-172.3				-144.9			
$C_3H_7ClO_2$	3-Chloro-1,2-propanediol					-525.3							
$C_3H_7ClO_2$	2-Chloro-1,3-propanediol					-517.5							
C_3H_7F	1-Fluoropropane									-285.9			
C_3H_7F	2-Fluoropropane									-293.5			
C_3H_7I	1-Iodopropane					-66.0				-30.0			
C_3H_7I	2-Iodopropane					-74.8				-40.3			
C_3H_7N	Allylamine					-10.0							
C_3H_7N	Cyclopropylamine					45.8		187.7	147.1	77.0			
C_3H_7NO	*N,N*-Dimethylformamide					-239.3			150.6	-192.4			
C_3H_7NO	Propanamide	-338.2								-259.0			
$C_3H_7NO_2$	1-Nitropropane					-167.2				-124.3		350.0	104.1
$C_3H_7NO_2$	2-Nitropropane					-180.3			170.3	-138.9			
$C_3H_7NO_2$	Ethyl carbamate	-517.1			156.4	-497.3				-446.3			
$C_3H_7NO_2$	*DL*-Alanine	-563.6											
$C_3H_7NO_2$	*D*-Alanine	-561.2											
$C_3H_7NO_2$	*L*-Alanine	-604.0								-465.9			
$C_3H_7NO_2$	β-Alanine	-558.0								-424.0			
$C_3H_7NO_2$	Sarcosine	-513.3								-367.3			
$C_3H_7NO_2S$	*L*-Cysteine	-534.1											
$C_3H_7NO_3$	Propyl nitrate					-214.5				-174.1		362.6	123.2
$C_3H_7NO_3$	Isopropyl nitrate					-229.7				-191.0			
$C_3H_7NO_3$	*DL*-Serine	-739.0											
$C_3H_7NO_3$	*L*-Serine	-732.7											
C_3H_8	Propane					-120.9				-103.8	-23.4	270.3	73.6
$C_3H_8N_2O$	*N*-Ethylurea	-357.8											
$C_3H_8N_2O$	*N,N*-Dimethylurea	-319.1											
$C_3H_8N_2O$	*N,N*-Dimethylurea	-312.1											

Molecular formula	Name	Crystal $\Delta_f H°$ kJ/mol	Crystal $\Delta_f G°$ kJ/mol	Crystal $S°$ J/mol K	Crystal C_p J/mol K	Liquid $\Delta_f H°$ kJ/mol	Liquid $\Delta_f G°$ kJ/mol	Liquid $S°$ J/mol K	Liquid C_p J/mol K	Gas $\Delta_f H°$ kJ/mol	Gas $\Delta_f G°$ kJ/mol	Gas $S°$ J/mol K	Gas C_p J/mol K
$C_3H_8N_2O_3$	Oxymethurea	-717.0											
C_3H_8O	1-Propanol					-302.6		193.6	143.9	-255.1		322.6	85.6
C_3H_8O	2-Propanol					-318.1		181.1	156.5	-272.6		309.2	89.3
C_3H_8O	Ethyl methyl ether									-216.4		309.2	93.3
$C_3H_8O_2$	1,2-Propylene glycol					-501.0			190.8	-429.8			
$C_3H_8O_2$	1,3-Propylene glycol					-480.8				-408.0			
$C_3H_8O_2$	Ethylene glycol monomethyl ether								171.1				
$C_3H_8O_2$	Dimethoxymethane					-377.8		244.0	162.0	-348.5			
$C_3H_8O_3$	Glycerol					-669.6		206.3	218.9	-577.9			
C_3H_8S	1-Propanethiol					-99.9		242.5	144.6	-67.8			
C_3H_8S	2-Propanethiol					-105.9		233.5	145.3	-76.2			
C_3H_8S	Ethyl methyl sulfide					-91.6		239.1	144.6	-59.6			
$C_3H_8S_2$	1,3-Propanedithiol					-79.4				-29.8			
C_3H_9Al	Trimethyl aluminum					-136.4	-9.9	209.4	155.6	-74.1			
C_3H_9B	Trimethylborane					-143.1	-32.1	238.9		-124.3	-35.9	314.7	88.5
$C_3H_9BO_3$	Trimethyl borate								189.9				
C_3H_9ClSi	Trimethylchlorosilane					-382.8	-246.4	278.2		-352.0	-243.5	309.1	
C_3H_9N	Propylamine					-101.5			164.1	-70.1	39.9	325.4	91.2
C_3H_9N	Isopropylamine					-112.3		218.3	163.8	-83.7	32.2	312.2	97.5
C_3H_9N	Trimethylamine					-45.7		208.5	137.9	-23.6		287.1	91.8
$C_3H_{10}ClN$	Propylamine hydrochloride	-354.7											
$C_3H_{10}ClN$	Trimethylamine hydrochloride	-282.9											
$C_3H_{10}N_2$	1,2-Propanediamine, (±)					-97.8				-53.6			
$C_3H_{10}Si$	Trimethylsilane											331.0	117.9
$C_3H_{12}BN$	Trimethylamine borane	-142.5	70.7	187.0									
$C_3H_{12}BN$	Aminetrimethylboron	-284.1	-79.3	218.0									
C_4Cl_6	Hexachloro-1,3-butadiene					-24.5							
C_4F_8	Perfluorocyclobutane									-1542.6			
C_4F_{10}	Perfluorobutane								127.2				
$C_4H_2N_2$	*trans*-2-Butenedinitrile	268.2								340.2			
$C_4H_2O_3$	Maleic anhydride	-469.8								-398.3			
$C_4H_2O_4$	2-Butynedioic acid	-577.3											
$C_4H_3NO_3$	2-Nitrofuran	-104.1								-28.8			
$C_4H_4BrNO_2$	*N*-Bromosuccinimide	-335.9											
$C_4H_4ClNO_2$	*N*-Chlorosuccinimide	-357.9											
$C_4H_4N_2$	Succinonitrile	139.7		191.6	145.6					209.7			
$C_4H_4N_2$	Pyrazine	139.8								196.1			
$C_4H_4N_2$	Pyrimidine					145.9				195.7			
$C_4H_4N_2$	Pyridazine					224.9				278.3			
$C_4H_4N_2O_2$	Uracil	-429.4			120.5					-302.9			
$C_4H_4N_2O_3$	Barbituric acid	-634.7											
C_4H_4O	Furan					-62.3		177.0	114.8	-34.8		267.2	65.4
$C_4H_4O_2$	Diketene					-233.1				-190.3			
$C_4H_4O_3$	Succinic anhydride	-608.6								-527.9			
$C_4H_4O_4$	Maleic acid	-789.4		160.8	137.0					-679.4			
$C_4H_4O_4$	Fumaric acid	-811.7		168.0	142.0					-675.8			
C_4H_4S	Thiophene					80.2		181.2	123.8	114.9	126.1	278.8	72.8
C_4H_5N	*trans*-2-Butenenitrile					95.1				134.3			
C_4H_5N	3-Butenenitrile					117.8				159.7			
C_4H_5N	2-Methylacrylonitrile								126.3				
C_4H_5N	Pyrrole					63.1		156.4	127.7	108.2			
C_4H_5N	Cyclopropanecarbonitrile					140.8				182.8			
$C_4H_5NO_2$	Succinimide	-459.0								-375.4			
C_4H_5NS	4-Methylthiazole					67.9				111.8			
$C_4H_5N_3O$	Cytosine	-221.3			132.6								
C_4H_6	1,2-Butadiene					138.6				162.3			
C_4H_6	1,3-Butadiene					88.5		199.0	123.6	110.0			
C_4H_6	1-Butyne					141.4				165.2			
C_4H_6	2-Butyne					119.1				145.7			
C_4H_6	Cyclobutene									156.7			
$C_4H_6N_2O_2$	2,5-Piperazinedione	-446.5											
C_4H_6O	Divinyl ether					-39.8				-13.6			
C_4H_6O	*trans*-2-Butenal					-138.7				-100.6			
$C_4H_6O_2$	*trans*-2-Butenoic acid												
$C_4H_6O_2$	Methacrylic acid								161.1				
$C_4H_6O_2$	Vinyl acetate					-349.2				-314.4			

		Crystal				Liquid				Gas			
Molecular formula	**Name**	**$\Delta_f H°$ kJ/mol**	**$\Delta_f G°$ kJ/mol**	**$S°$ J/mol K**	**C_p J/mol K**	**$\Delta_f H°$ kJ/mol**	**$\Delta_f G°$ kJ/mol**	**$S°$ J/mol K**	**C_p J/mol K**	**$\Delta_f H°$ kJ/mol**	**$\Delta_f G°$ kJ/mol**	**$S°$ J/mol K**	**C_p J/mol K**
$C_4H_6O_2$	Methyl acrylate					-362.2		239.5	158.8	-333.0			
$C_4H_6O_2$	γ-Butyrolactone					-420.9			141.4	-366.5			
$C_4H_6O_3$	Acetic anhydride					-624.4				-572.5			
$C_4H_6O_3$	Propylene carbonate					-613.2			218.6	-582.5			
$C_4H_6O_4$	Succinic acid	-940.5		167.3	153.1					-823.0			
$C_4H_6O_4$	Dimethyl oxalate	-756.3								-708.9			
C_4H_6S	2,3-Dihydrothiophene					52.9				90.7	133.5	303.5	79.8
C_4H_6S	2,5-Dihydrothiophene					47.0				86.9	131.6	297.1	83.3
C_4H_7ClO	2-Chloroethyl vinyl ether					-208.1				-170.1			
$C_4H_7ClO_2$	2-Chlorobutanoic acid					-575.5							
$C_4H_7ClO_2$	3-Chlorobutanoic acid					-556.3							
$C_4H_7ClO_2$	4-Chlorobutanoic acid					-566.3							
$C_4H_7ClO_2$	Propyl chlorocarbonate					-533.4				-492.7			
C_4H_7N	Butanenitrile					-5.8				33.6			
C_4H_7N	2-Methylpropanenitrile					-13.8				23.4			
C_4H_7NO	Acetone cyanohydrin					-120.9							
C_4H_7NO	2-Pyrrolidone					-286.2							
C_4H_7NO	2-Methyl-2-oxazoline					-169.5				-130.5			
$C_4H_7NO_4$	Iminodiacetic acid	-932.6											
$C_4H_7NO_4$	Ethyl nitroacetate					-487.1							
$C_4H_7NO_4$	*L*-Aspartic acid	-973.3											
$C_4H_7N_3O$	Creatinine	-238.5											
C_4H_8	1-Butene					-20.8		227.0	118.0	0.1			
C_4H_8	*cis*-2-Butene					-29.8		219.9	127.0	-7.1			
C_4H_8	*trans*-2-Butene					-33.3				-11.4			
C_4H_8	Isobutene					-37.5				-16.9			
C_4H_8	Cyclobutane					3.7				27.7			
C_4H_8	Methylcyclopropane					1.7							
$C_4H_8Br_2$	1,2-Dibromobutane					-142.1				-91.6			
$C_4H_8Br_2$	1,3-Dibromobutane					-148.0							
$C_4H_8Br_2$	1,4-Dibromobutane					-140.3				-87.8			
$C_4H_8Br_2$	2,3-Dibromobutane					-139.6				-102.0			
$C_4H_8Br_2$	1,2-Dibromo-2-methylpropane					-156.6				-113.3			
$C_4H_8Cl_2$	1,3-Dichlorobutane					-237.3				-195.0			
$C_4H_8Cl_2$	1,4-Dichlorobutane					-229.8				-183.4			
$C_4H_8Cl_2O$	Bis(2-chloroethyl) ether								220.9				
$C_4H_8I_2$	1,4-Diiodobutane					-30.0							
$C_4H_8N_2O_2$	Succinamide	-581.2											
$C_4H_8N_2O_2$	Dimethylglyoxime	-199.7											
$C_4H_8N_2O_3$	*L*-Asparagine	-789.4											
$C_4H_8N_2O_3$	*N*-Glycylglycine	-747.7											
$C_4H_8N_2O_4$	1,4-Dinitrobutane					-237.5							
$C_4H_8N_8O_8$	Cyclotetramethylenetetranitramine									187.9		568.8	275.5
C_4H_8O	Ethyl vinyl ether					-167.4				-140.8			
C_4H_8O	1,2-Epoxybutane					-168.9		230.9	147.0				
C_4H_8O	Butanal					-239.2		246.6	163.7	-204.8		343.7	103.4
C_4H_8O	Isobutanal					-247.3				-215.7			
C_4H_8O	2-Butanone					-273.3		239.1	158.7	-238.5		339.9	101.7
C_4H_8O	Tetrahydrofuran					-216.2		204.3	124.0	-184.1		302.4	76.3
C_4H_8OS	*S*-Ethyl thioacetate					-268.2				-228.1			
$C_4H_8O_2$	Butanoic acid					-533.8		222.2	178.6	-475.9			
$C_4H_8O_2$	2-Methylpropanoic acid								173.0				
$C_4H_8O_2$	Propyl formate					-500.3				-462.7			
$C_4H_8O_2$	Ethyl acetate					-479.3		257.7	170.7	-443.6			
$C_4H_8O_2$	Methyl propanoate								171.2				
$C_4H_8O_2$	1,3-Dioxane					-379.7			143.9	-340.6			
$C_4H_8O_2$	1,4-Dioxane					-353.9		270.2	152.1	-315.3			
$C_4H_8O_2$	2-Methyl-1,3-dioxolane					-386.9				-352.0			
$C_4H_8O_2S$	Sulfolane								180.0				
C_4H_8S	Tetrahydrothiophene					-72.9				-34.1	45.8	309.6	92.5
$C_4H_8S_2$	1,3-Dithiane									-10.0	72.4	333.5	110.4
$C_4H_8S_2$	1,4-Dithiane									0.0	84.5	326.2	109.7
C_4H_9Br	1-Bromobutane					-143.8				-107.1			
C_4H_9Br	2-Bromobutane, (±)					-154.9				-120.3			
C_4H_9Br	2-Bromo-2-methylpropane					-164.4				-132.4			
C_4H_9Cl	1-Chlorobutane					-188.1				-154.4			

Molecular formula	Name	Crystal $\Delta_f H°$ kJ/mol	Crystal $\Delta_f G°$ kJ/mol	Crystal $S°$ J/mol K	Crystal C_p J/mol K	Liquid $\Delta_f H°$ kJ/mol	Liquid $\Delta_f G°$ kJ/mol	Liquid $S°$ J/mol K	Liquid C_p J/mol K	Gas $\Delta_f H°$ kJ/mol	Gas $\Delta_f G°$ kJ/mol	Gas $S°$ J/mol K	Gas C_p J/mol K
C_4H_9Cl	2-Chlorobutane					-192.8				-161.1			
C_4H_9Cl	1-Chloro-2-methylpropane					-191.1				-159.3			
C_4H_9Cl	2-Chloro-2-methylpropane					-211.3				-182.2			
C_4H_9ClO	2-Chloroethyl ethyl ether					-335.6				-301.3			
C_4H_9I	1-Iodo-2-methylpropane								162.3				
C_4H_9I	2-Iodo-2-methylpropane					-107.5				-72.1			
C_4H_9N	Cyclobutanamine					5.6				41.2			
C_4H_9N	Pyrrolidine					-41.1		204.1	156.6	-3.6			
C_4H_9NO	Butanamide	-364.8								-282.0			
C_4H_9NO	*N*-Methylpropanamide								179.0				
C_4H_9NO	2-Methylpropanamide	-368.6								-282.6			
C_4H_9NO	*N,N*-Dimethylacetamide					-278.3			175.6	-228.0			
C_4H_9NO	Morpholine								164.8				
$C_4H_9NO_2$	1-Nitrobutane					-192.5				-143.9		369.9	115.1
$C_4H_9NO_2$	2-Nitroisobutane					-217.2				-177.1			
$C_4H_9NO_2$	Propyl carbamate	-552.6								-471.4			
$C_4H_9NO_2$	4-Aminobutanoic acid	-581.0								-411.0			
$C_4H_9NO_3$	3-Nitro-2-butanol					-390.0							
$C_4H_9NO_3$	2-Methyl-2-nitro-1-propanol	-410.1											
$C_4H_9NO_3$	*DL*-Threonine	-758.8											
$C_4H_9NO_3$	*L*-Threonine	-807.2											
$C_4H_9N_3O_2$	Creatine	-537.2											
C_4H_{10}	Butane					-147.3			140.9	-125.7			
C_4H_{10}	Isobutane					-154.2				-134.2			
$C_4H_{10}Hg$	Diethyl mercury					30.1			182.8	75.3			
$C_4H_{10}N_2$	Piperazine	-45.6											
$C_4H_{10}N_2O$	Trimethylurea	-330.5											
$C_4H_{10}N_2O_2$	*N*-Nitrodiethylamine					-106.2				-53.0			
$C_4H_{10}N_2O_4$	*L*-Asparagine, monohydrate	-1086.6											
$C_4H_{10}O$	1-Butanol					-327.3		225.8	177.2	-274.9			
$C_4H_{10}O$	2-Butanol					-342.6		214.9	196.9	-292.8		359.5	112.7
$C_4H_{10}O$	2-Methyl-1-propanol					-334.7		214.7	181.5	-283.8			
$C_4H_{10}O$	2-Methyl-2-propanol					-359.2		193.3	218.6	-312.5		326.7	113.6
$C_4H_{10}O$	Diethyl ether					-279.5		253.5	172.5	-252.1		342.7	119.5
$C_4H_{10}O$	Methyl propyl ether					-266.0		262.9	165.4	-238.1			
$C_4H_{10}O$	Isopropyl methyl ether					-278.8		253.8	161.9	-252.0			
$C_4H_{10}OS$	Diethyl sulfoxide					-268.0				-205.6			
$C_4H_{10}O_2$	1,2-Butanediol, (±)					-523.6							
$C_4H_{10}O_2$	1,3-Butanediol					-501.0				-433.2			
$C_4H_{10}O_2$	1,4-Butanediol					-505.3		223.4	200.1	-428.7			
$C_4H_{10}O_2$	2,3-Butanediol					-541.5			213.0	-482.3			
$C_4H_{10}O_2$	2-Methyl-1,2-propanediol					-539.7							
$C_4H_{10}O_2$	Ethylene glycol monoethyl ether								210.8				
$C_4H_{10}O_2$	Ethylene glycol dimethyl ether					-376.6			193.3				
$C_4H_{10}O_2$	Dimethylacetal					-420.6				-389.7			
$C_4H_{10}O_2$	*tert*-Butyl hydroperoxide					-293.6				-245.9			
$C_4H_{10}O_3$	Diethylene glycol					-628.5			244.8	-571.2			
$C_4H_{10}O_3S$	Diethyl sulfite					-600.7				-552.2			
$C_4H_{10}O_4S$	Diethyl sulfate					-813.2				-756.3			
$C_4H_{10}S$	1-Butanethiol					-124.7			171.2	-88.0			
$C_4H_{10}S$	2-Butanethiol					-131.0				-96.9			
$C_4H_{10}S$	2-Methyl-1-propanethiol					-132.0				-97.3			
$C_4H_{10}S$	2-Methyl-2-propanethiol					-140.5				-109.6			
$C_4H_{10}S$	Diethyl sulfide					-119.4		269.3	171.4	-83.5		368.1	117.0
$C_4H_{10}S$	Methyl propyl sulfide					-118.5		272.5	171.6	-82.2			
$C_4H_{10}S$	Isopropyl methyl sulfide					-124.7		263.1	172.4	-90.5			
$C_4H_{10}S_2$	1,4-Butanedithiol					-105.7				-50.6			
$C_4H_{10}S_2$	Diethyl disulfide					-120.1		269.3	171.4	-79.4			
$C_4H_{11}N$	Butylamine					-127.6			179.2	-91.9			
$C_4H_{11}N$	*sec*-Butylamine					-137.5				-104.6			
$C_4H_{11}N$	*tert*-Butylamine					-150.6			192.1	-121.0			
$C_4H_{11}N$	Isobutylamine					-132.6			183.2	-98.7			
$C_4H_{11}N$	Diethylamine					-103.7			169.2	-72.2			
$C_4H_{11}NO$	*N,N*-Dimethylethanolamine					-253.7				-203.6			
$C_4H_{11}NO_2$	Diethanolamine	-493.8			233.5					-397.1			
$C_4H_{11}NO_3$	Tris(hydroxymethyl)methylamine	-717.8											

		Crystal				Liquid				Gas			
Molecular formula	**Name**	**$\Delta_f H°$ kJ/mol**	**$\Delta_f G°$ kJ/mol**	**$S°$ J/mol K**	**C_p J/mol K**	**$\Delta_f H°$ kJ/mol**	**$\Delta_f G°$ kJ/mol**	**$S°$ J/mol K**	**C_p J/mol K**	**$\Delta_f H°$ kJ/mol**	**$\Delta_f G°$ kJ/mol**	**$S°$ J/mol K**	**C_p J/mol K**
$C_4H_{12}BrN$	Tetramethylammonium bromide	-251.0											
$C_4H_{12}ClN$	Diethylamine hydrochloride	-358.6											
$C_4H_{12}ClN$	Tetramethylammonium chloride	-276.4											
$C_4H_{12}IN$	Tetramethylammonium iodide	-203.9											
$C_4H_{12}N_2$	2-Methyl-1,2-propanediamine					-133.9				-90.3			
$C_4H_{12}Pb$	Tetramethyl lead					97.9				135.9			
$C_4H_{12}Si$	Tetramethylsilane					-264.0	-100.0	277.3	204.1	-239.1	-99.9	359.0	143.9
$C_4H_{12}Sn$	Tetramethylstannane					-52.3				-18.8			
$C_4H_{13}N_3$	Bis(2-aminoethyl)amine								254.0				
C_4N_2	2-Butynedinitrile					500.4				529.2			
C_4NiO_4	Nickel carbonyl					-633.0	-588.2	313.4	204.6	-602.9	-587.2	410.6	145.2
C_5FeO_5	Iron pentacarbonyl					-774.0	-705.3	338.1	240.6				
$C_5H_2F_6O_2$	Hexafluoroacetylacetone	-2286.7											
$C_5H_3NO_5$	5-Nitro-2-furancarboxylic acid	-516.8											
$C_5H_4N_4$	1*H*-Purine	169.4											
$C_5H_4N_4O$	Hypoxanthine	-110.8		145.6	134.5								
$C_5H_4N_4O_2$	Xanthine	-379.6		161.1	151.3								
$C_5H_4N_4O_3$	Uric acid	-618.8		173.2	166.1								
$C_5H_4O_2$	Furfural					-201.6			163.2	-151.0			
$C_5H_4O_3$	2-Furancarboxylic acid	-498.4								-390.0			
$C_5H_4O_3$	3-Methyl-2,5-furandione					-504.5				-447.2			
$C_5H_5F_3O_2$	1,1,1-Trifluoro-2,4-pentanedione					-1040.2				-993.3			
C_5H_5N	Pyridine					100.2			132.7	140.4			
C_5H_5NO	1*H*-Pyrrole-2-carboxaldehyde	-106.4											
$C_5H_5N_5$	Adenine	96.9			147.0					205.7			
$C_5H_5N_5O$	Guanine	-183.9											
C_5H_6	*cis*-3-Penten-1-yne					226.5							
C_5H_6	*trans*-3-Penten-1-yne					228.2							
C_5H_6	1,3-Cyclopentadiene					105.9				134.3			
$C_5H_6N_2O_2$	Thymine	-462.8			150.8					-328.7			
$C_5H_6O_2$	Furfuryl alcohol					-276.2			204.0	-211.8			
$C_5H_6O_4$	*trans*-1-Propene-1,2-dicarboxylic acid	-824.4											
C_5H_6S	2-Methylthiophene					44.6		218.5	149.8	83.5			
C_5H_6S	3-Methylthiophene					43.1				82.5			
C_5H_7N	*trans*-3-Pentenenitrile					80.9				125.7			
C_5H_7N	Cyclobutanecarbonitrile					103.0				147.4			
C_5H_7N	1-Methylpyrrole					62.4				103.1			
C_5H_7N	2-Methylpyrrole					23.3				74.0			
C_5H_7N	3-Methylpyrrole					20.5				70.2			
$C_5H_7NO_2$	Ethyl cyanoacetate								220.2				
C_5H_8	1,2-Pentadiene									140.7			
C_5H_8	*cis*-1,3-Pentadiene									81.4			
C_5H_8	*trans*-1,3-Pentadiene									76.1			
C_5H_8	1,4-Pentadiene									105.7			
C_5H_8	2,3-Pentadiene									133.1			
C_5H_8	3-Methyl-1,2-butadiene					101.2							
C_5H_8	2-Methyl-1,3-butadiene					48.2		229.3	152.6	75.5			
C_5H_8	Cyclopentene					4.3		201.2	122.4	34.0			
C_5H_8	Spiropentane					157.5		193.7	134.5	185.2			
C_5H_8	Methylenecyclobutane					93.8				121.6			
$C_5H_8N_4O_{12}$	Pentaerythritol tetranitrate	-538.6								-387.0		614.7	294.8
C_5H_8O	Cyclopentanone					-235.9				-192.1			
$C_5H_8O_2$	4-Pentenoic acid					-430.6							
$C_5H_8O_2$	Allyl acetate								184.1				
$C_5H_8O_2$	Ethyl acrylate					-370.6				-354.2			
$C_5H_8O_2$	Methyl *trans*-2-butenoate					-382.9				-341.9			
$C_5H_8O_2$	Methyl methacrylate								191.2				
$C_5H_8O_2$	2,4-Pentanedione					-423.8				-382.0			
$C_5H_8O_2$	Dihydro-4-methyl-2(3*H*)-furanone					-461.3				-406.5			
$C_5H_8O_2$	Tetrahydro-2*H*-pyran-2-one					-436.7				-379.6			
$C_5H_8O_3$	Methyl acetoacetate					-623.2							
$C_5H_8O_4$	Glutaric acid	-960.0											
$C_5H_9ClO_2$	Propyl chloroacetate					-515.5				-467.0			
C_5H_9N	Pentanenitrile					-33.1				10.5			
C_5H_9N	2,2-Dimethylpropanenitrile					-39.8		232.0	179.4	-2.3			

		Crystal				Liquid				Gas			
Molecular formula	**Name**	**$\Delta_f H°$ kJ/mol**	**$\Delta_f G°$ kJ/mol**	**$S°$ J/mol K**	**C_p J/mol K**	**$\Delta_f H°$ kJ/mol**	**$\Delta_f G°$ kJ/mol**	**$S°$ J/mol K**	**C_p J/mol K**	**$\Delta_f H°$ kJ/mol**	**$\Delta_f G°$ kJ/mol**	**$S°$ J/mol K**	**C_p J/mol K**
C_5H_9N	1,2,5,6-Tetrahydropyridine					33.5							
C_5H_9NO	2-Piperidinone	-306.6											
C_5H_9NO	*N*-Methyl-2-pyrrolidone					-262.2			307.8				
$C_5H_9NO_2$	*L*-Proline	-515.2								-366.2			
$C_5H_9NO_4$	*D*-Glutamic acid	-1005.3											
$C_5H_9NO_4$	*L*-Glutamic acid	-1009.7											
C_5H_{10}	1-Pentene					-46.9		262.6	154.0	-21.1			
C_5H_{10}	*cis*-2-Pentene					-53.7		258.6	151.7	-27.6			
C_5H_{10}	*trans*-2-Pentene					-58.2		256.5	157.0	-31.9			
C_5H_{10}	2-Methyl-1-butene					-61.1		254.0	157.2	-35.2			
C_5H_{10}	3-Methyl-1-butene					-51.5		253.3	156.1	-27.5			
C_5H_{10}	2-Methyl-2-butene					-68.6		251.0	152.8	-41.7			
C_5H_{10}	Cyclopentane					-105.1		204.5	128.8	-76.4			
C_5H_{10}	Methylcyclobutane					-44.5							
C_5H_{10}	Ethylcyclopropane					-24.8							
C_5H_{10}	1,1-Dimethylcyclopropane					-33.3				-8.2			
C_5H_{10}	*cis*-1,2-Dimethylcyclopropane					-26.3							
C_5H_{10}	*trans*-1,2-Dimethylcyclopropane					-30.7							
$C_5H_{10}Br_2$	2,3-Dibromo-2-methylbutane									-137.6			
$C_5H_{10}N_2O$	*N*-Nitrosopiperidine					-31.1				16.6			
$C_5H_{10}N_2O_2$	*N*-Nitropiperidine					-93.0				-44.5			
$C_5H_{10}N_2O_3$	*L*-Glutamine	-826.4											
$C_5H_{10}O$	Cyclopentanol					-300.1		204.1	182.5	-242.5		362.9	
$C_5H_{10}O$	Pentanal					-267.2				-228.4			
$C_5H_{10}O$	2-Pentanone					-297.3			184.1	-258.8			
$C_5H_{10}O$	3-Pentanone					-296.5		266.0	190.9	-257.9			
$C_5H_{10}O$	3-Methyl-2-butanone					-299.5		268.5	179.9	-262.6			
$C_5H_{10}O$	3,3-Dimethyloxetane					-182.2				-148.2			
$C_5H_{10}O$	Tetrahydropyran					[illegible]				-223.4			
$C_5H_{10}OS$	S-Propyl thioacetate					-294.5				-250.4			
$C_5H_{10}O_2$	Pentanoic acid					-559.4		259.8	210.3	-491.9			
$C_5H_{10}O_2$	2-Methylbutanoic acid					-554.5							
$C_5H_{10}O_2$	3-Methylbutanoic acid					-561.6				-510.0			
$C_5H_{10}O_2$	2,2-Dimethylpropanoic acid	-564.5								-491.3			
$C_5H_{10}O_2$	Butyl formate								200.2				
$C_5H_{10}O_2$	Propyl acetate								196.2				
$C_5H_{10}O_2$	Isopropyl acetate					-518.9			199.4	-481.6			
$C_5H_{10}O_2$	Ethyl propanoate					-502.7				-463.4			
$C_5H_{10}O_2$	Methyl butanoate								198.2				
$C_5H_{10}O_2$	(Ethoxymethyl)oxirane					-296.5							
$C_5H_{10}O_2$	4-Methyl-1,3-dioxane					-416.1				-376.9			
$C_5H_{10}O_2$	*cis*-1,2-Cyclopentanediol	-485.0											
$C_5H_{10}O_2$	*trans*-1,2-Cyclopentanediol	-490.1											
$C_5H_{10}O_2$	Tetrahydrofurfuryl alcohol					-435.7				-369.1			
$C_5H_{10}O_3$	Diethyl carbonate					-681.5				-637.9			
$C_5H_{10}O_3$	Ethylene glycol monomethyl ether acetate								310.0				
$C_5H_{10}O_3$	Ethyl lactate								254.0				
$C_5H_{10}O_4$	Glycerol 1-acetate, (*DL*)					-909.2							
$C_5H_{10}O_5$	*D*-Ribose	-1047.2											
$C_5H_{10}O_5$	*D*-Xylose	-1057.8											
$C_5H_{10}O_5$	α-*D*-Arabinopyranose	-1057.9											
$C_5H_{10}S$	Thiacyclohexane					-106.3		218.2	163.3	-63.5	53.1	323.0	109.7
$C_5H_{10}S$	Cyclopentanethiol					-89.5		256.9	165.2	-48.1			
$C_5H_{11}Br$	1-Bromopentane					-170.2				-128.9			
$C_5H_{11}Cl$	1-Chloropentane					-213.2				-174.9			
$C_5H_{11}Cl$	1-Chloro-3-methylbutane					-216.0				-179.7			
$C_5H_{11}Cl$	2-Chloro-2-methylbutane					-235.7				-202.2			
$C_5H_{11}Cl$	2-Chloro-3-methylbutane					-226.6				-185.1			
$C_5H_{11}N$	Cyclopentylamine					-95.1		241.0	181.2	-54.9			
$C_5H_{11}N$	Piperidine					-86.4		210.0	179.9	-47.1			
$C_5H_{11}NO$	Pentanamide	-379.5								-290.2			
$C_5H_{11}NO$	2,2-Dimethylpropanamide	-399.7								-313.1			
$C_5H_{11}NO_2$	1-Nitropentane					-215.4				-164.4		390.9	137.1
$C_5H_{11}NO_2$	*DL*-Valine	-628.9											
$C_5H_{11}NO_2$	*L*-Valine	-617.9								-455.1			

		Crystal				Liquid				Gas			
Molecular formula	**Name**	**$\Delta_f H°$ kJ/mol**	**$\Delta_f G°$ kJ/mol**	**$S°$ J/mol K**	**C_p J/mol K**	**$\Delta_f H°$ kJ/mol**	**$\Delta_f G°$ kJ/mol**	**$S°$ J/mol K**	**C_p J/mol K**	**$\Delta_f H°$ kJ/mol**	**$\Delta_f G°$ kJ/mol**	**$S°$ J/mol K**	**C_p J/mol K**
$C_5H_{11}NO_2$	5-Aminopentanoic acid	-604.1								-460.0			
$C_5H_{11}NO_2S$	*L*-Methionine	-577.5								-413.5			
$C_5H_{11}NO_4$	2-Ethyl-2-nitro-1,3-propanediol	-606.4											
C_5H_{12}	Pentane					-173.5			167.2	-146.9			
C_5H_{12}	Isopentane					-178.4		260.4	164.8	-153.6			
C_5H_{12}	Neopentane					-190.2				-168.0			
$C_5H_{12}N_2O$	Butylurea	-419.5											
$C_5H_{12}N_2O$	*tert*-Butylurea	-417.4											
$C_5H_{12}N_2O$	*N,N*-Diethylurea	-372.2											
$C_5H_{12}N_2O$	Tetramethylurea					-262.2							
$C_5H_{12}N_2S$	Tetramethylthiourea	-38.1								44.9			
$C_5H_{12}O$	1-Pentanol					-351.6			208.1	-294.6			
$C_5H_{12}O$	2-Pentanol					-365.2				-311.0			
$C_5H_{12}O$	3-Pentanol					-368.9			239.7	-314.9			
$C_5H_{12}O$	2-Methyl-1-butanol, (±)					-356.6				-301.4			
$C_5H_{12}O$	3-Methyl-1-butanol					-356.4				-300.7			
$C_5H_{12}O$	2-Methyl-2-butanol					-379.5			247.1	-329.3			
$C_5H_{12}O$	3-Methyl-2-butanol, (±)					-366.6				-313.5			
$C_5H_{12}O$	2,2-Dimethyl-1-propanol					-399.4							
$C_5H_{12}O$	Butyl methyl ether					-290.6		295.3	192.7	-258.1			
$C_5H_{12}O$	Methyl *tert*-butyl ether					-313.6		265.3	187.5	-283.7			
$C_5H_{12}O$	Ethyl propyl ether					-303.6		295.0	197.2	-272.0			
$C_5H_{12}O_2$	1,5-Pentanediol					-528.8				-450.8			
$C_5H_{12}O_2$	2,2-Dimethyl-1,3-propanediol	-551.2											
$C_5H_{12}O_2$	Diethoxymethane					-450.5				-414.7			
$C_5H_{12}O_2$	1,1-Dimethoxypropane					-443.6							
$C_5H_{12}O_2$	2,2-Dimethoxypropane					-459.4				-429.9			
$C_5H_{12}O_3$	Diethylene glycol monomethyl ether								271.1				
$C_5H_{12}O_3$	2-(Hydroxymethyl)-2-methyl-1,3-propanediol	-744.6											
$C_5H_{12}O_4$	Pentaerythritol	-920.6								-776.7			
$C_5H_{12}O_5$	Xylitol	-1118.5											
$C_5H_{12}S$	1-Pentanethiol					-151.3				-110.0			
$C_5H_{12}S$	2-Methyl-1-butanethiol, (+)					-154.4				-114.9			
$C_5H_{12}S$	3-Methyl-1-butanethiol					-154.4				-114.9			
$C_5H_{12}S$	2-Methyl-2-butanethiol					-162.8		290.1	198.1	-127.1			
$C_5H_{12}S$	3-Methyl-2-butanethiol					-158.8				-121.3			
$C_5H_{12}S$	2,2-Dimethyl-1-propanethiol					-165.4				-129.0			
$C_5H_{12}S$	Butyl methyl sulfide					-142.9		307.5	200.9	-102.4			
$C_5H_{12}S$	*tert*-Butyl methyl sulfide					-157.1		276.1	199.9	-121.3			
$C_5H_{12}S$	Ethyl propyl sulfide					-144.8		309.5	198.4	-104.8			
$C_5H_{12}S$	Ethyl isopropyl sulfide					-156.1				-118.3			
$C_5H_{13}N$	Pentylamine								218.0				
$C_5H_{14}N_2$	*N,N,N',N'*-Tetramethylmethanediamine					-51.1				-18.2			
C_6ClF_5	Chloropentafluorobenzene	-858.4								-809.3			
C_6Cl_6	Hexachlorobenzene	-127.6		260.2	201.2					-35.5			
C_6F_6	Hexafluorobenzene					-991.3		280.8	221.6	-955.4			
C_6F_{10}	Perfluorocyclohexene					-1963.5				-1932.7			
C_6F_{12}	Perfluorocyclohexane					-2406.3				-2370.4			
C_6HCl_5O	Pentachlorophenol	-292.5		253.2	202.0								
C_6HF_5	Pentafluorobenzene	-852.7				-841.8				-806.5			
C_6HF_5O	Pentafluorophenol	-1024.1				-1007.7							
$C_6H_2F_4$	1,2,4,5-Tetrafluorobenzene					-683.8							
$C_6H_3Cl_3$	1,2,3-Trichlorobenzene	-70.8								3.8			
$C_6H_3Cl_3$	1,2,4-Trichlorobenzene					-63.1				-8.1			
$C_6H_3Cl_3$	1,3,5-Trichlorobenzene	-78.4								-13.4			
$C_6H_3N_3O_6$	1,3,5-Trinitrobenzene	-37.0			214.6								
$C_6H_3N_3O_7$	2,4,6-Trinitrophenol	-217.9			239.7								
$C_6H_3N_3O_8$	2,4,6-Trinitro-1,3-benzenediol	-467.5											
$C_6H_4ClNO_2$	1-Chloro-4-nitrobenzene	-48.7			250.2								
$C_6H_4Cl_2$	*o*-Dichlorobenzene					-17.5			162.4	30.2			
$C_6H_4Cl_2$	*m*-Dichlorobenzene					-20.7				25.7			
$C_6H_4Cl_2$	*p*-Dichlorobenzene	-42.3		175.4	147.8					22.5			
$C_6H_4Cl_2O$	2,4-Dichlorophenol	-226.4								-156.3			
$C_6H_4F_2$	*o*-Difluorobenzene					-330.0		222.6	159.0	-293.8			
$C_6H_4F_2$	*m*-Difluorobenzene					-343.9		223.8	159.1	-309.2			

		Crystal				Liquid				Gas			
Molecular formula	**Name**	**$\Delta_f H°$ kJ/mol**	**$\Delta_f G°$ kJ/mol**	**$S°$ J/mol K**	**C_p J/mol K**	**$\Delta_f H°$ kJ/mol**	**$\Delta_f G°$ kJ/mol**	**$S°$ J/mol K**	**C_p J/mol K**	**$\Delta_f H°$ kJ/mol**	**$\Delta_f G°$ kJ/mol**	**$S°$ J/mol K**	**C_p J/mol K**
$C_6H_4F_2$	*p*-Difluorobenzene					-342.3			157.5	-306.7			
$C_6H_4N_2O_4$	1,2-Dinitrobenzene	-2.0			200.4								
$C_6H_4N_2O_4$	1,3-Dinitrobenzene	-27.0			197.5	-36.0							
$C_6H_4N_2O_4$	1,4-Dinitrobenzene	-38.0			200.0								
$C_6H_4N_2O_5$	2,4-Dinitrophenol	-232.7								-128.1			
$C_6H_4O_2$	*p*-Benzoquinone	-185.7			129.0					-122.9			
C_6H_5Br	Bromobenzene					60.9		219.2	154.3				
C_6H_5Cl	Chlorobenzene					11.1			150.1	52.0			
C_6H_5ClO	2-Chlorophenol								188.7				
C_6H_5ClO	3-Chlorophenol	-206.4				-189.3							
C_6H_5ClO	4-Chlorophenol	-197.7				-181.3							
$C_6H_5Cl_2N$	3,4-Dichloroaniline	-89.1											
C_6H_5F	Fluorobenzene					-150.6		205.9	146.4	-115.9			
C_6H_5I	Iodobenzene					117.2		205.4	158.7	164.9			
$C_6H_5NO_2$	Nitrobenzene					12.5			185.8	68.5		348.8	120.4
$C_6H_5NO_2$	3-Pyridinecarboxylic acid	-344.9								-221.5			
$C_6H_5NO_3$	2-Nitrophenol	-202.4											
$C_6H_5N_3$	1*H*-Benzotriazole	236.5								335.5			
$C_6H_5N_3O_4$	2,3-Dinitroaniline	-11.7											
$C_6H_5N_3O_4$	2,4-Dinitroaniline	-67.8											
$C_6H_5N_3O_4$	2,5-Dinitroaniline	-44.3											
$C_6H_5N_3O_4$	2,6-Dinitroaniline	-50.6											
$C_6H_5N_3O_4$	3,5-Dinitroaniline	-38.9											
C_6H_6	1,5-Hexadiyne					384.2							
C_6H_6	Benzene					49.1	124.5	173.4	136.0	82.9	129.7	269.2	82.4
C_6H_6ClN	2-Chloroaniline					-4.6							
C_6H_6ClN	3-Chloroaniline					-20.3			198.7				
C_6H_6ClN	4-Chloroaniline	-33.3			147.3								
$C_6H_6N_2O_2$	2-Nitroaniline	-26.1			166.0	0.4				63.8			
$C_6H_6N_2O_2$	3-Nitroaniline	-38.3			158.8	-14.4				58.4			
$C_6H_6N_2O_2$	4-Nitroaniline	-42.0			167.0	-20.7				58.8			
C_6H_6O	Phenol	-165.1		144.0	127.4					-96.4			
C_6H_6O	2-Vinylfuran					-10.3				27.8			
$C_6H_6O_2$	*p*-Hydroquinone	-364.5			136.0					-265.3			
$C_6H_6O_2$	Pyrocatechol	-354.1								-267.5			
$C_6H_6O_2$	Resorcinol	-368.0								-274.7			
$C_6H_6O_3$	1,2,3-Benzenetriol	-551.1								-434.2			
$C_6H_6O_3$	1,2,4-Benzenetriol	-563.8								-444.0			
$C_6H_6O_3$	1,3,5-Benzenetriol	-584.6								-452.9			
$C_6H_6O_3$	3,4-Dimethyl-2,5-furandione	-581.4											
$C_6H_6O_6$	*cis*-1-Propene-1,2,3-tricarboxylic acid	-1224.4											
$C_6H_6O_6$	*trans*-1-Propene-1,2,3-tricarboxylic acid	-1232.7											
C_6H_6S	Benzenethiol					63.7		222.8	173.2	111.3			
C_6H_7N	Aniline					31.6			191.9	87.5	-7.0	317.9	107.9
C_6H_7N	2-Methylpyridine					56.7			158.6	99.2			
C_6H_7N	3-Methylpyridine					61.9		216.3	158.7	106.4			
C_6H_7N	4-Methylpyridine					59.2		209.1	159.0	103.8			
C_6H_7N	1-Cyclopentenecarbonitrile					111.5				156.5			
$C_6H_8N_2$	Adiponitrile					85.1			128.7	149.5			
$C_6H_8N_2$	1,2-Benzenediamine	-0.3											
$C_6H_8N_2$	1,3-Benzenediamine	-7.8		154.5	159.6								
$C_6H_8N_2$	1,4-Benzenediamine	3.0											
$C_6H_8N_2$	Phenylhydrazine					141.0			217.0	202.9			
$C_6H_8N_2S$	Bis(2-cyanoethyl) sulfide					96.3							
$C_6H_8O_4$	Dimethyl maleate								263.2				
$C_6H_8O_6$	*L*-Ascorbic acid	-1164.6											
$C_6H_8O_7$	Citric acid	-1543.8											
$C_6H_9Cl_3O_2$	Butyl trichloroacetate					-545.8				-492.3			
$C_6H_9Cl_3O_2$	Isobutyl trichloroacetate					-553.4				-500.2			
C_6H_9N	Cyclopentanecarbonitrile					0.7				44.1			
C_6H_9N	2,4-Dimethylpyrrole	-422.3											
C_6H_9N	2,5-Dimethylpyrrole					-16.7				39.8			
$C_6H_9NO_3$	Triacetamide					-610.5				-550.1			
$C_6H_9NO_6$	Nitrilotriacetic acid	-1311.9											
$C_6H_9N_3O_2$	*L*-Histidine	-466.7											

		Crystal				Liquid				Gas			
Molecular formula	**Name**	**$\Delta_f H°$ kJ/mol**	**$\Delta_f G°$ kJ/mol**	**$S°$ J/mol K**	**C_p J/mol K**	**$\Delta_f H°$ kJ/mol**	**$\Delta_f G°$ kJ/mol**	**$S°$ J/mol K**	**C_p J/mol K**	**$\Delta_f H°$ kJ/mol**	**$\Delta_f G°$ kJ/mol**	**$S°$ J/mol K**	**C_p J/mol K**
C_6H_{10}	1,5-Hexadiene					54.1				84.2			
C_6H_{10}	3,3-Dimethyl-1-butyne					78.4							
C_6H_{10}	Cyclohexene					-38.5		214.6	148.3	-5.0			
C_6H_{10}	1-Methylcyclopentene					-36.4				-3.8			
C_6H_{10}	3-Methylcyclopentene					-23.7				7.4			
C_6H_{10}	4-Methylcyclopentene					-17.6				14.6			
$C_6H_{10}Cl_2O_2$	Butyl dichloroacetate					-550.1				-497.8			
$C_6H_{10}O$	Cyclohexanone					-271.2			182.2	-226.1			
$C_6H_{10}O$	2-Methylcyclopentanone					-265.2							
$C_6H_{10}O$	Mesityl oxide								212.5				
$C_6H_{10}O_2$	Ethyl *trans*-2-butenoate					-420.0				-375.6			
$C_6H_{10}O_2$	Methyl cyclobutanecarboxylate					-395.0				-350.2			
$C_6H_{10}O_3$	Ethyl acetoacetate								248.0				
$C_6H_{10}O_3$	Propanoic anhydride					-679.1				-626.5			
$C_6H_{10}O_4$	Adipic acid	-994.3											
$C_6H_{10}O_4$	Diethyl oxalate					-805.5				-742.0			
$C_6H_{10}O_4$	Ethylene glycol diacetate								310.0				
$C_6H_{11}Cl$	Chlorocyclohexane					-207.2				-163.7			
$C_6H_{11}ClO_2$	Ethyl 4-chlorobutanoate					-566.5				-513.8			
$C_6H_{11}ClO_2$	Propyl 3-chloropropanoate					-537.6				-485.7			
$C_6H_{11}ClO_2$	Butyl chloroacetate					-538.4				-487.4			
$C_6H_{11}NO$	Caprolactam	-329.4			156.8					-239.6			
$C_6H_{11}NO$	1-Methyl-2-piperidinone					-293.0							
C_6H_{12}	1-Hexene					-74.2		295.2	183.3	-43.5			
C_6H_{12}	*cis*-2-Hexene					-83.9				-52.3			
C_6H_{12}	*trans*-2-Hexene					-85.5				-53.9			
C_6H_{12}	*cis*-3-Hexene					-78.9				-47.6			
C_6H_{12}	*trans*-3-Hexene					-86.1				-54.4			
C_6H_{12}	2-Methyl-1-pentene					-90.0				-59.4			
C_6H_{12}	3-Methyl-1-pentene					-78.2				-49.5			
C_6H_{12}	4-Methyl-1-pentene					-80.0				-51.3			
C_6H_{12}	2-Methyl-2-pentene					-98.5				-66.9			
C_6H_{12}	3-Methyl-*cis*-2-pentene					-94.5				-62.3			
C_6H_{12}	3-Methyl-*trans*-2-pentene					-94.6				-63.1			
C_6H_{12}	4-Methyl-*cis*-2-pentene					-87.0				-57.5			
C_6H_{12}	4-Methyl-*trans*-2-pentene					-91.6				-61.5			
C_6H_{12}	2-Ethyl-1-butene					-87.1				-56.0			
C_6H_{12}	2,3-Dimethyl-1-butene					-93.2				-62.4			
C_6H_{12}	3,3-Dimethyl-1-butene					-87.5				-60.3			
C_6H_{12}	2,3-Dimethyl-2-butene					-101.4		270.2	174.7	-68.1			
C_6H_{12}	Cyclohexane					-156.4			154.9	-123.4			
C_6H_{12}	Methylcyclopentane					-137.9				-106.2			
C_6H_{12}	Ethylcyclobutane					-59.0				-27.5			
C_6H_{12}	1,1,2-Trimethylcyclopropane					-96.2							
$C_6H_{12}N_2O_4S_2$	*L*-Cystine	-1032.7											
$C_6H_{12}N_2S_4$	Thiram	40.2			301.7								
$C_6H_{12}O$	Butyl vinyl ether					-218.8			232.0	-182.6			
$C_6H_{12}O$	Hexanal							280.3	210.4				
$C_6H_{12}O$	2-Hexanone					-322.0			213.3	-278.9			
$C_6H_{12}O$	3-Hexanone					-320.2		305.3	216.9	-277.6			
$C_6H_{12}O$	4-Methyl-2-pentanone								213.3				
$C_6H_{12}O$	2-Methyl-3-pentanone					-325.9				-286.0			
$C_6H_{12}O$	3,3-Dimethyl-2-butanone					-328.6				-290.6			
$C_6H_{12}O$	Cyclohexanol					-348.2			208.2	-286.2			
$C_6H_{12}O$	*cis*-2-Methylcyclopentanol					-345.5							
$C_6H_{12}O_2$	Hexanoic acid					-583.8				-511.9			
$C_6H_{12}O_2$	Butyl acetate					-529.2			227.8	-485.3			
$C_6H_{12}O_2$	*tert*-Butyl acetate					-554.5			231.0	-516.5			
$C_6H_{12}O_2$	Isobutyl acetate								233.8				
$C_6H_{12}O_2$	Ethyl butanoate								228.0				
$C_6H_{12}O_2$	Methyl pentanoate					-514.2			229.3	-471.1			
$C_6H_{12}O_2$	Methyl 2,2-dimethylpropanoate					-530.0			257.9	-491.2			
$C_6H_{12}O_2$	Diacetone alcohol								221.3				
$C_6H_{12}O_3$	Ethylene glycol monoethyl ether acetate								376.0				
$C_6H_{12}O_3$	Paraldehyde					-673.1				-631.7			

		Crystal				Liquid				Gas			
Molecular formula	Name	$\Delta_f H°$ kJ/mol	$\Delta_f G°$ kJ/mol	$S°$ J/mol K	C_p J/mol K	$\Delta_f H°$ kJ/mol	$\Delta_f G°$ kJ/mol	$S°$ J/mol K	C_p J/mol K	$\Delta_f H°$ kJ/mol	$\Delta_f G°$ kJ/mol	$S°$ J/mol K	C_p J/mol K
$C_6H_{12}O_6$	β-*D*-Fructose	-1265.6											
$C_6H_{12}O_6$	*D*-Galactose	-1286.3											
$C_6H_{12}O_6$	α-*D*-Glucose	-1273.3											
$C_6H_{12}O_6$	*D*-Mannose	-1263.0											
$C_6H_{12}O_6$	*L*-Sorbose	-1271.5											
$C_6H_{12}S$	Thiepane									-65.8	79.4	363.5	131.3
$C_6H_{12}S$	Cyclohexanethiol					-140.7		255.6	192.6	-96.2			
$C_6H_{12}S$	Cyclopentyl methyl sulfide					-109.8				-64.7			
$C_6H_{13}Br$	1-Bromohexane					-194.2		453.0	204.0	-148.3			
$C_6H_{13}Cl$	2-Chlorohexane					-246.1				-204.3			
$C_6H_{13}N$	Cyclohexylamine					-147.6				-104.0			
$C_6H_{13}N$	2-Methylpiperidine, (±)					-124.9				-84.4			
$C_6H_{13}NO$	Hexanamide	-423.0								-324.2			
$C_6H_{13}NO$	*N*-Butylacetamide					-380.9				-305.9			
$C_6H_{13}NO_2$	*DL*-Leucine	-640.6											
$C_6H_{13}NO_2$	*D*-Leucine	-637.3											
$C_6H_{13}NO_2$	*L*-Leucine	-637.4			200.1					-486.8			
$C_6H_{13}NO_2$	*DL*-Isoleucine	-635.3											
$C_6H_{13}NO_2$	*L*-Isoleucine	-637.8											
$C_6H_{13}NO_2$	*L*-Norleucine	-639.1											
$C_6H_{13}NO_2$	6-Aminohexanoic acid	-637.3											
C_6H_{14}	Hexane					-198.7			195.6	-166.9			
C_6H_{14}	2-Methylpentane					-204.6		290.6	193.7	-174.6			
C_6H_{14}	3-Methylpentane					-202.4		292.5	190.7	-171.9			
C_6H_{14}	2,2-Dimethylbutane					-213.8		272.5	191.9	-185.9			
C_6H_{14}	2,3-Dimethylbutane					-207.4		287.8	189.7	-178.1			
$C_6H_{14}N_2$	Azopropane					11.5				51.3			
$C_6H_{14}N_2O_2$	*DL*-Lysine	-678.7											
$C_6H_{14}N_4O_2$	*D*-Arginine	-623.5		250.6	232.0								
$C_6H_{14}O$	1-Hexanol					-377.5		287.4	240.4	-315.9			
$C_6H_{14}O$	2-Hexanol					-392.0				-333.5			
$C_6H_{14}O$	3-Hexanol					-392.4			286.2				
$C_6H_{14}O$	2-Methyl-1-pentanol								248.0				
$C_6H_{14}O$	3-Methyl-2-pentanol								275.9				
$C_6H_{14}O$	4-Methyl-2-pentanol					-394.7			273.0				
$C_6H_{14}O$	2-Methyl-3-pentanol					-396.4							
$C_6H_{14}O$	3-Methyl-3-pentanol								293.4				
$C_6H_{14}O$	Dipropyl ether					-328.8		323.9	221.6	-293.0			
$C_6H_{14}O$	Diisopropyl ether					-351.5			216.8	-319.2			
$C_6H_{14}O$	Butyl ethyl ether								159.0				
$C_6H_{14}O$	*tert*-Butyl ethyl ether									-313.9			
$C_6H_{14}OS$	Dipropyl sulfoxide					-329.4				-254.9			
$C_6H_{14}O_2$	1,2-Hexanediol					-577.1				-490.1			
$C_6H_{14}O_2$	1,6-Hexanediol	-569.9				-548.6				-461.2			
$C_6H_{14}O_2$	2-Methyl-2,4-pentanediol								336.0				
$C_6H_{14}O_2$	Ethylene glycol monobutyl ether								281.0				
$C_6H_{14}O_2$	1,1-Diethoxyethane					-491.4				-453.5			
$C_6H_{14}O_2$	Ethylene glycol diethyl ether					-451.4			259.4	-408.1			
$C_6H_{14}O_3$	Diethylene glycol monoethyl ether								301.0				
$C_6H_{14}O_3$	Diethylene glycol dimethyl ether								274.1				
$C_6H_{14}O_3$	Trimethylolpropane	-750.9											
$C_6H_{14}O_4$	Triethylene glycol					-804.3				-725.0			
$C_6H_{14}O_4S$	Dipropyl sulfate					-859.0				-792.0			
$C_6H_{14}O_6$	Galactitol					-1317.0							
$C_6H_{14}O_6$	*D*-Mannitol					-1314.5							
$C_6H_{14}S$	1-Hexanethiol					-175.7				-129.9			
$C_6H_{14}S$	2-Methyl-2-pentanethiol					-188.3				-148.3			
$C_6H_{14}S$	2,3-Dimethyl-2-butanethiol					-187.1				-147.9			
$C_6H_{14}S$	Diisopropyl sulfide					-181.6		313.0	232.0	-142.0			
$C_6H_{14}S$	Butyl ethyl sulfide					-172.3				-127.8			
$C_6H_{14}S$	Methyl pentyl sulfide					-167.1				-121.8			
$C_6H_{14}S_2$	Dipropyl disulfide					-171.5				-118.3			
$C_6H_{15}B$	Triethylborane					-194.6	9.4	336.7	241.2	-157.7	16.1	437.8	
$C_6H_{15}N$	Dipropylamine					-156.1				-116.0			
$C_6H_{15}N$	Diisopropylamine					-178.5				-143.8			
$C_6H_{15}N$	Triethylamine					-127.7			219.9	-92.7			

Molecular formula	Name	Crystal $\Delta_f H°$ kJ/mol	Crystal $\Delta_f G°$ kJ/mol	Crystal $S°$ J/mol K	Crystal C_p J/mol K	Liquid $\Delta_f H°$ kJ/mol	Liquid $\Delta_f G°$ kJ/mol	Liquid $S°$ J/mol K	Liquid C_p J/mol K	Gas $\Delta_f H°$ kJ/mol	Gas $\Delta_f G°$ kJ/mol	Gas $S°$ J/mol K	Gas C_p J/mol K
$C_6H_{15}NO$	2-Diethylaminoethanol					-305.9							
$C_6H_{15}NO_3$	Triethanolamine	-664.2			389.0					-558.3			
$C_6H_{16}N_2$	1,6-Hexanediamine	-205.0											
$C_6H_{18}N_3OP$	Hexamethylphosphoric triamide								321.0				
$C_6H_{18}OSi_2$	Hexamethyldisiloxane					-815.0	-541.5	433.8	311.4	-777.7	-534.5	535.0	238.5
C_6MoO_6	Molybdenum hexacarbonyl	-982.8	-877.7	325.9	242.3					-912.1	-856.0	490.0	205.0
C_6N_4	Tetracyanoethene	623.8								705.0			
C_7F_8	Perfluorotoluene					-1311.1		355.5	262.3				
C_7F_{14}	Perfluoromethylcyclohexane					-2931.1			353.1	-2897.2			
C_7F_{16}	Perfluoroheptane					-3420.0		561.8	419.0	-3383.6			
$C_7H_3F_5$	2,3,4,5,6-Pentafluorotoluene					-883.8		306.4	225.8	-842.7			
$C_7H_4Cl_2O$	3-Chlorobenzoyl chloride					-189.7							
$C_7H_4N_2O_6$	3,5-Dinitrobenzoic acid	-409.8											
C_7H_5ClO	Benzoyl chloride					-158.0				-103.2			
$C_7H_5ClO_2$	2-Chlorobenzoic acid	-404.5								-325.0			
$C_7H_5ClO_2$	3-Chlorobenzoic acid	-424.3								-342.3			
$C_7H_5ClO_2$	4-Chlorobenzoic acid	-428.9			163.2					-341.0			
$C_7H_5F_3$	(Trifluoromethyl)benzene								188.4				
C_7H_5N	Benzonitrile					163.2		209.1	165.2	215.7			
C_7H_5NO	Benzoxazole	-24.2								44.8			
$C_7H_5NO_4$	2-Nitrobenzoic acid	-378.8											
$C_7H_5NO_4$	3-Nitrobenzoic acid	-394.7											
$C_7H_5NO_4$	4-Nitrobenzoic acid	-392.2											
$C_7H_5N_3O_6$	2,4,6-Trinitrotoluene	-63.2			243.3								
$C_7H_6N_2$	1*H*-Benzimidazole	79.5								181.7			
$C_7H_6N_2$	1*H*-Indazole	151.9								243.0			
$C_7H_6N_2O_4$	1-Methyl-2,4-dinitrobenzene	-66.4								33.2			
C_7H_6O	Benzaldehyde					-87.0		221.2	172.0	-36.7			
$C_7H_6O_2$	Benzoic acid	-385.2		167.6	146.8					-294.0			
$C_7H_6O_2$	Salicylaldehyde								222.0				
$C_7H_6O_2$	3-(2-Furanyl)-2-propenal	-182.0								-105.9			
$C_7H_6O_3$	2-Hydroxybenzoic acid	-589.9								-494.8			
C_7H_7Br	4-Bromotoluene					12.0							
C_7H_7Cl	2-Chlorotoluene								166.8				
C_7H_7Cl	(Chloromethyl)benzene					-32.5				18.9			
C_7H_7F	4-Fluorotoluene					-186.9			171.2	-147.4			
C_7H_7NO	Benzamide	-202.6								-100.9			
$C_7H_7NO_2$	Aniline-2-carboxylic acid	-401.1								-296.0			
$C_7H_7NO_2$	Aniline-3-carboxylic acid	-417.3								-283.6			
$C_7H_7NO_2$	Aniline-4-carboxylic acid	-410.0			177.8					-296.7			
$C_7H_7NO_2$	2-Nitrotoluene					-9.7							
$C_7H_7NO_2$	3-Nitrotoluene					-31.5							
$C_7H_7NO_2$	4-Nitrotoluene	-48.1			172.3					31.0			
$C_7H_7NO_2$	(Nitromethyl)benzene					-22.8				30.7			
$C_7H_7NO_2$	Salicylaldoxime	-183.7											
C_7H_8	Toluene					12.4			157.3	50.5			
$C_7H_8N_2O$	Phenylurea	-218.6											
C_7H_8O	*o*-Cresol	-204.6		165.4	154.6					-128.6			
C_7H_8O	*m*-Cresol					-194.0		212.6	224.9	-132.3			
C_7H_8O	*p*-Cresol	-199.3		167.3	150.2					-125.4			
C_7H_8O	Benzyl alcohol					-160.7		216.7	217.9	-100.4			
C_7H_8O	Anisole					-114.8				-67.9			
C_7H_9N	Benzylamine					34.2			207.2	94.4			
C_7H_9N	2-Methylaniline					-6.3				56.4	167.6	351.0	130.2
C_7H_9N	3-Methylaniline					-8.1				54.6	165.4	352.5	125.5
C_7H_9N	4-Methylaniline	-23.5								55.3	167.7	347.0	126.2
C_7H_9N	*N*-Methylaniline								207.1				
C_7H_9N	1-Cyclohexenecarbonitrile					48.1				101.6			
C_7H_9N	2,3-Dimethylpyridine					19.4		243.7	189.5	67.1			
C_7H_9N	2,4-Dimethylpyridine					16.1		248.5	184.8	63.6			
C_7H_9N	2,5-Dimethylpyridine					18.7		248.8	184.7	66.5			
C_7H_9N	2,6-Dimethylpyridine					12.7		244.2	185.2	58.1			
C_7H_9N	3,4-Dimethylpyridine					18.3		240.7	191.8	68.8			
C_7H_9N	3,5-Dimethylpyridine					22.5		241.7	184.5	72.0			
$C_7H_{10}O_2$	Ethyl 2-pentynoate					-301.8				-250.3			
$C_7H_{10}O_2$	Methyl 2-hexynoate					-242.7							

Molecular formula	Name	Crystal $\Delta_fH°$ kJ/mol	Crystal $\Delta_fG°$ kJ/mol	Crystal $S°$ J/mol K	Crystal C_p J/mol K	Liquid $\Delta_fH°$ kJ/mol	Liquid $\Delta_fG°$ kJ/mol	Liquid $S°$ J/mol K	Liquid C_p J/mol K	Gas $\Delta_fH°$ kJ/mol	Gas $\Delta_fG°$ kJ/mol	Gas $S°$ J/mol K	Gas C_p J/mol K
$C_7H_{11}Cl_3O_2$	Isopentyl trichloroacetate					-580.9				-523.1			
$C_7H_{11}N$	Cyclohexanecarbonitrile					-47.2				4.8			
C_7H_{12}	Bicyclo[2.2.1]heptane	-95.1			151.0					-54.8			
C_7H_{12}	1-Methylbicyclo(3,1,0)hexane					-33.2				1.7			
C_7H_{12}	Methylenecyclohexane					-61.3				-25.2			
C_7H_{12}	Vinylcyclopentane					-34.8							
C_7H_{12}	1-Ethylcyclopentene					-53.3				-19.8			
$C_7H_{12}O$	2-Methylenecyclohexanol					-277.6							
$C_7H_{12}O_2$	Butyl acrylate					-422.6			251.0	-375.3			
$C_7H_{12}O_4$	Diethyl malonate								285.0				
$C_7H_{13}ClO_2$	Butyl 2-chloropropanoate					-571.7				-517.3			
$C_7H_{13}ClO_2$	Isobutyl 2-chloropropanoate					-603.1				-549.6			
$C_7H_{13}ClO_2$	Butyl 3-chloropropanoate					-557.9				-502.3			
$C_7H_{13}ClO_2$	Isobutyl 3-chloropropanoate					-572.6				-517.3			
$C_7H_{13}ClO_2$	Propyl 2-chlorobutanoate					-630.7				-578.4			
$C_7H_{13}N$	Heptanenitrile					-82.8				-31.0			
C_7H_{14}	1-Heptene					-97.9		327.6	211.8	-62.3			
C_7H_{14}	*cis*-2-Heptene					-105.1							
C_7H_{14}	*trans*-2-Heptene					-109.5							
C_7H_{14}	*cis*-3-Heptene					-104.3							
C_7H_{14}	*trans*-3-Heptene					-109.3							
C_7H_{14}	5-Methyl-1-hexene					-100.0				-65.7			
C_7H_{14}	*cis*-3-Methyl-3-hexene					-115.9				-79.4			
C_7H_{14}	*trans*-3-Methyl-3-hexene					-112.7				-76.8			
C_7H_{14}	2,4-Dimethyl-1-pentene					-117.0				-83.8			
C_7H_{14}	4,4-Dimethyl-1-pentene					-110.6				-81.6			
C_7H_{14}	2,4-Dimethyl-2-pentene					-123.1				-88.7			
C_7H_{14}	*cis*-4,4-Dimethyl-2-pentene					-105.3				-72.6			
C_7H_{14}	*trans*-4,4-Dimethyl-2-pentene					-121.7				-88.8			
C_7H_{14}	2-Ethyl-3-methyl-1-butene					-114.1				-79.5			
C_7H_{14}	2,3,3-Trimethyl-1-butene					-117.7				-85.5			
C_7H_{14}	Cycloheptane					-156.6				-118.1			
C_7H_{14}	Methylcyclohexane					-190.1			184.8	-154.7			
C_7H_{14}	Ethylcyclopentane					-163.4		279.9		-126.9			
C_7H_{14}	1,1-Dimethylcyclopentane					-172.1				-138.2			
C_7H_{14}	*cis*-1,2-Dimethylcyclopentane					-165.3		269.2		-129.5			
C_7H_{14}	*trans*-1,2-Dimethylcyclopentane					-171.2				-136.6			
C_7H_{14}	*cis*-1,3-Dimethylcyclopentane					-170.1				-135.8			
C_7H_{14}	*trans*-1,3-Dimethylcyclopentane					-168.1				-133.6			
C_7H_{14}	1,1,2,2-Tetramethylcyclopropane					-119.8							
$C_7H_{14}Br_2$	1,2-Dibromoheptane					-212.3				-157.9			
$C_7H_{14}O$	1-Heptanal					-311.5		335.4	230.1	-263.8			
$C_7H_{14}O$	2-Heptanone								232.6				
$C_7H_{14}O$	3-Heptanone									-297.1			
$C_7H_{14}O$	4-Heptanone									-298.3			
$C_7H_{14}O$	2,2-Dimethyl-3-pentanone					-356.1				-313.6			
$C_7H_{14}O$	2,4-Dimethyl-3-pentanone					-352.9		318.0	233.7	-311.3			
$C_7H_{14}O$	*cis*-2-Methylcyclohexanol					-390.2				-327.0			
$C_7H_{14}O$	*trans*-2-Methylcyclohexanol, (±)					-415.7				-352.5			
$C_7H_{14}O$	*cis*-3-Methylcyclohexanol, (±)					-416.1				-350.9			
$C_7H_{14}O$	*trans*-3-Methylcyclohexanol, (±)					-394.4				-329.1			
$C_7H_{14}O$	*cis*-4-Methylcyclohexanol					-413.2				-347.5			
$C_7H_{14}O$	*trans*-4-Methylcyclohexanol					-433.3				-367.2			
$C_7H_{14}O_2$	Heptanoic acid					-610.2			265.4	-536.2			
$C_7H_{14}O_2$	Pentyl acetate								261.0				
$C_7H_{14}O_2$	Isopentyl acetate								248.5				
$C_7H_{14}O_2$	Ethyl pentanoate					-553.0				-505.9			
$C_7H_{14}O_2$	Ethyl 3-methylbutanoate					-571.0				-527.0			
$C_7H_{14}O_2$	Ethyl 2,2-dimethylpropanoate					-577.2				-536.0			
$C_7H_{14}O_2$	Methyl hexanoate					-540.2				-492.2			
$C_7H_{14}O_6$	α-Methylglucoside	-1233.3											
$C_7H_{15}Br$	1-Bromoheptane					-218.4				-167.8			
C_7H_{16}	Heptane					-224.2			224.7	-187.6			
C_7H_{16}	2-Methylhexane					-229.5		323.3	222.9	-194.5			
C_7H_{16}	3-Methylhexane					-226.4				-191.3			
C_7H_{16}	3-Ethylpentane					-224.9		314.5	219.6	-189.5			

		Crystal				Liquid				Gas			
Molecular formula	**Name**	**$\Delta_f H°$ kJ/mol**	**$\Delta_f G°$ kJ/mol**	**$S°$ J/mol K**	**C_p J/mol K**	**$\Delta_f H°$ kJ/mol**	**$\Delta_f G°$ kJ/mol**	**$S°$ J/mol K**	**C_p J/mol K**	**$\Delta_f H°$ kJ/mol**	**$\Delta_f G°$ kJ/mol**	**$S°$ J/mol K**	**C_p J/mol K**
C_7H_{16}	2,2-Dimethylpentane					-238.3		300.3	221.1	-205.7			
C_7H_{16}	2,3-Dimethylpentane					-233.1				-198.7			
C_7H_{16}	2,4-Dimethylpentane					-234.6		303.2	224.2	-201.6			
C_7H_{16}	3,3-Dimethylpentane					-234.2				-201.0			
C_7H_{16}	2,2,3-Trimethylbutane					-236.5		292.2	213.5	-204.4			
$C_7H_{16}O$	1-Heptanol					-403.3			272.1	-336.5			
$C_7H_{16}O$	*tert*-Butyl isopropyl ether					-392.8				-358.1			
$C_7H_{16}O_2$	1,7-Heptanediol					-574.2							
$C_7H_{16}O_2$	2,2-Diethoxypropane					-538.9				-506.9			
$C_7H_{16}S$	1-Heptanethiol					-200.5				-149.9			
$C_8H_4O_3$	Phthalic anhydride	-460.1		180.0	160.0					-371.4			
$C_8H_5NO_2$	1*H*-Indole-2,3-dione	-268.2											
$C_8H_6O_4$	Phthalic acid	-782.0		207.9	188.1								
$C_8H_6O_4$	Isophthalic acid	-803.0								-696.3			
$C_8H_6O_4$	Terephthalic acid	-816.1								-717.9			
C_8H_6S	Benzo[b]thiophene	100.6								166.3			
C_8H_7N	1*H*-Indole	86.6								156.5			
C_8H_8	Styrene					103.8			182.0	147.9			
C_8H_8O	Phenyl vinyl ether					-26.2				22.7			
C_8H_8O	Acetophenone					-142.5				-86.7			
$C_8H_8O_2$	*o*-Toluic acid	-416.5			174.9								
$C_8H_8O_2$	*m*-Toluic acid	-426.1			163.6								
$C_8H_8O_2$	*p*-Toluic acid	-429.2			169.0								
$C_8H_8O_2$	Methyl benzoate					-343.5			221.3	-287.9			
$C_8H_8O_3$	Methyl salicylate								249.0				
C_8H_9NO	Acetanilide	-209.4			179.3								
C_8H_{10}	1,7-Octadiyne					334.4							
C_8H_{10}	Ethylbenzene					-12.3			183.2	29.9			
C_8H_{10}	*o*-Xylene					-24.4			186.1	19.1			
C_8H_{10}	*m*-Xylene					-25.4			183.0	17.3			
C_8H_{10}	*p*-Xylene					-24.4			181.5	18.0			
$C_8H_{10}O$	2-Ethylphenol					-208.8				-145.2			
$C_8H_{10}O$	3-Ethylphenol					-214.3				-146.1			
$C_8H_{10}O$	4-Ethylphenol	-224.4			206.9					-144.1			
$C_8H_{10}O$	2,3-Xylenol	-241.1								-157.2			
$C_8H_{10}O$	2,4-Xylenol					-228.7				-163.8			
$C_8H_{10}O$	2,5-Xylenol	-246.6								-161.6			
$C_8H_{10}O$	2,6-Xylenol	-237.4								-162.1			
$C_8H_{10}O$	3,4-Xylenol	-242.3								-157.3			
$C_8H_{10}O$	3,5-Xylenol	-244.4								-162.4			
$C_8H_{10}O$	Benzeneethanol								252.6				
$C_8H_{10}O$	Ethoxybenzene					-152.6			228.5	-101.6			
$C_8H_{10}O_2$	1,2-Dimethoxybenzene					-290.3				-223.3			
$C_8H_{11}N$	*N*-Ethylaniline					8.2				56.3			
$C_8H_{11}N$	*N,N*-Dimethylaniline					46.0				100.5			
$C_8H_{11}N$	2,4-Dimethylaniline					-39.2							
$C_8H_{11}N$	2,5-Dimethylaniline					-38.9							
$C_8H_{11}N$	2,6-Dimethylaniline								238.9				
C_8H_{12}	1-Octen-3-yne					140.7							
C_8H_{12}	*cis*-1,2-Divinylcyclobutane					124.3				166.5			
C_8H_{12}	*trans*-1,2-Divinylcyclobutane					101.3				143.5			
$C_8H_{12}N_4$	2,2'-Azobis[isobutyronitrile]	246.0			237.6								
$C_8H_{12}O_2$	2,2,4,4-Tetramethyl-1,3-cyclobutanedione	-379.9								-307.6			
C_8H_{14}	Ethylidenecyclohexane					-103.5				-59.5			
C_8H_{14}	Allylcyclopentane					-64.5				-24.1			
$C_8H_{14}ClN_5$	Atrazine	-125.4											
$C_8H_{14}O_3$	Butanoic anhydride								283.7				
$C_8H_{15}ClO_2$	3-Methylbutyl 2-chloropropanoate					-627.3				-575.0			
$C_8H_{15}ClO_2$	3-Methylbutyl 3-chloropropanoate					-593.4				-539.4			
$C_8H_{15}N$	Octanenitrile					-107.3				-50.5			
C_8H_{16}	1-Octene					-124.5			241.0	-81.3			
C_8H_{16}	*cis*-2-Octene					-135.7			239.0				
C_8H_{16}	*trans*-2-Octene					-135.7			239.0				
C_8H_{16}	*cis*-2,2-Dimethyl-3-hexene					-126.4				-89.3			
C_8H_{16}	*trans*-2,2-Dimethyl-3-hexene					-144.9				-107.7			

Molecular formula	Name	Crystal $\Delta_f H°$ kJ/mol	Crystal $\Delta_f G°$ kJ/mol	Crystal $S°$ J/mol K	Crystal C_p J/mol K	Liquid $\Delta_f H°$ kJ/mol	Liquid $\Delta_f G°$ kJ/mol	Liquid $S°$ J/mol K	Liquid C_p J/mol K	Gas $\Delta_f H°$ kJ/mol	Gas $\Delta_f G°$ kJ/mol	Gas $S°$ J/mol K	Gas C_p J/mol K
C_8H_{16}	3-Ethyl-2-methyl-1-pentene					-137.9				-100.3			
C_8H_{16}	2,4,4-Trimethyl-1-pentene					-145.9				-110.5			
C_8H_{16}	2,4,4-Trimethyl-2-pentene					-142.4				-104.9			
C_8H_{16}	Cyclooctane					-167.7				-124.4			
C_8H_{16}	Ethylcyclohexane					-212.1		280.9	211.8	-171.5			
C_8H_{16}	1,1-Dimethylcyclohexane					-218.7		267.2	209.2	-180.9			
C_8H_{16}	*cis*-1,2-Dimethylcyclohexane					-211.8		274.1	210.2	-172.1			
C_8H_{16}	*trans*-1,2-Dimethylcyclohexane					-218.2		273.2	209.4	-179.9			
C_8H_{16}	*cis*-1,3-Dimethylcyclohexane					-222.9		272.6	209.4	-184.6			
C_8H_{16}	*trans*-1,3-Dimethylcyclohexane					-215.7		276.3	212.8	-176.5			
C_8H_{16}	*cis*-1,4-Dimethylcyclohexane					-215.6		271.1	212.1	-176.6			
C_8H_{16}	*trans*-1,4-Dimethylcyclohexane					-222.4		268.0	210.2	-184.5			
C_8H_{16}	Propylcyclopentane					-188.8		310.8	216.3	-147.7			
C_8H_{16}	1-Ethyl-1-methylcyclopentane					-193.8							
C_8H_{16}	*cis*-1-Ethyl-2-methylcyclopentane					-190.8							
C_8H_{16}	*trans*-1-Ethyl-2-methylcyclopentane					-195.1				-156.2			
C_8H_{16}	*cis*-1-Ethyl-3-methylcyclopentane					-194.4							
C_8H_{16}	*trans*-1-Ethyl-3-methylcyclopentane					-196.0							
$C_8H_{16}O$	Octanal									-291.9		365.4	
$C_8H_{16}O$	2-Ethylhexanal					-348.5				-299.6			
$C_8H_{16}O$	2-Octanone								273.3				
$C_8H_{16}O$	2,2,4-Trimethyl-3-pentanone					-381.6				-338.3			
$C_8H_{16}O_2$	Octanoic acid					-636.0			297.9	-554.3			
$C_8H_{16}O_2$	2-Ethylhexanoic acid					635.1				-559.5			
$C_8H_{16}O_2$	Hexyl acetate								282.8				
$C_8H_{16}O_2$	Isobutyl isobutanoate					-587.4				-542.9			
$C_8H_{16}O_2$	Propyl pentanoate					-583.0				-533.6			
$C_8H_{16}O_2$	Isopropyl pentanoate					-592.2				-544.9			
$C_8H_{16}O_2$	Methyl heptanoate					-567.1			285.1	-515.5			
$C_8H_{17}Br$	1-Bromooctane					-245.1				-189.3			
$C_8H_{17}Cl$	1-Chlorooctane					-291.3				-238.9			
$C_8H_{17}NO$	Octanamide	-473.2								-362.7			
C_8H_{18}	Octane					-250.1			254.6	-208.5			
C_8H_{18}	2-Methylheptane					-255.0		356.4	252.0	-215.3			
C_8H_{18}	3-Methylheptane, (*S*)					-252.3		362.6	250.2	-212.5			
C_8H_{18}	4-Methylheptane					-251.6			251.1	-211.9			
C_8H_{18}	3-Ethylhexane					-250.4				-210.7			
C_8H_{18}	2,2-Dimethylhexane					-261.9				-224.5			
C_8H_{18}	2,3-Dimethylhexane					-252.6				-213.8			
C_8H_{18}	2,4-Dimethylhexane					-257.0				-219.2			
C_8H_{18}	2,5-Dimethylhexane					-260.4			249.2	-222.5			
C_8H_{18}	3,3-Dimethylhexane					-257.5			246.6	-219.9			
C_8H_{18}	3,4-Dimethylhexane					-251.8				-212.8			
C_8H_{18}	3-Ethyl-2-methylpentane					-249.6				-211.0			
C_8H_{18}	3-Ethyl-3-methylpentane					-252.8				-214.8			
C_8H_{18}	2,2,3-Trimethylpentane					-256.9				-220.0			
C_8H_{18}	2,2,4-Trimethylpentane					-259.2			239.1	-224.0			
C_8H_{18}	2,3,3-Trimethylpentane					253.5			245.6	-216.3			
C_8H_{18}	2,3,4-Trimethylpentane					-255.0		329.3	247.3	-217.3			
C_8H_{18}	2,2,3,3-Tetramethylbutane	-269.0		273.7	239.2					-226.0			
$C_8H_{18}N_2$	Azobutane					-40.1				9.2			
$C_8H_{18}O$	1-Octanol					-426.5			305.2	-355.6			
$C_8H_{18}O$	2-Octanol								330.1				
$C_8H_{18}O$	2-Ethyl-1-hexanol					-432.8		347.0	317.5	-365.3			
$C_8H_{18}O$	Dibutyl ether					-377.9			278.2	-332.8			
$C_8H_{18}O$	Di-*sec*-butyl ether					-401.5				-360.6			
$C_8H_{18}O$	Di-*tert*-butyl ether					-399.6			276.1	-362.0			
$C_8H_{18}O$	*tert*-Butyl isobutyl ether					-409.1				-369.0			
$C_8H_{18}O_2$	1,8-Octanediol	-626.6											
$C_8H_{18}O_2$	2,5-Dimethyl-2,5-hexanediol	-681.7											
$C_8H_{18}O_3$	Diethylene glycol monobutyl ether								354.9				
$C_8H_{18}O_3$	Diethylene glycol diethyl ether								341.4				
$C_8H_{18}O_3S$	Dibutyl sulfite					-693.1				-625.3			
$C_8H_{18}O_5$	Tetraethylene glycol					-981.7			428.8	-883.0			
$C_8H_{18}S$	Dibutyl sulfide					-220.7		405.1	284.3	-167.7			
$C_8H_{18}S$	Di-*sec*-butyl sulfide					-220.7				-167.7			

		Crystal				Liquid				Gas			
Molecular formula	**Name**	**$\Delta_f H°$ kJ/mol**	**$\Delta_f G°$ kJ/mol**	**$S°$ J/mol K**	**C_p J/mol K**	**$\Delta_f H°$ kJ/mol**	**$\Delta_f G°$ kJ/mol**	**$S°$ J/mol K**	**C_p J/mol K**	**$\Delta_f H°$ kJ/mol**	**$\Delta_f G°$ kJ/mol**	**$S°$ J/mol K**	**C_p J/mol K**
$C_8H_{18}S$	Di-*tert*-butyl sulfide					-232.6				-188.8			
$C_8H_{18}S$	Diisobutyl sulfide					-229.2				-180.5			
$C_8H_{18}S_2$	Dibutyl disulfide					-222.9				-160.6			
$C_8H_{18}S_2$	Di-*tert*-butyl disulfide					-255.2				-201.0			
$C_8H_{19}N$	Dibutylamine					-206.0			292.9	-156.6			
$C_8H_{19}N$	Diisobutylamine					-218.5				-179.2			
$C_8H_{20}BrN$	Tetraethylammonium bromide	-342.7											
$C_8H_{20}O_4Si$	Ethyl silicate							533.1	364.4				
$C_8H_{20}Pb$	Tetraethyl lead					52.7		464.6	307.4	109.6			
$C_8H_{20}Si$	Tetraethylsilane								298.1				
$C_9H_6N_2O_2$	Toluene-2,4-diisocyanate								287.8				
C_9H_7N	Quinoline					141.2				200.5			
C_9H_7N	Isoquinoline					144.3		216.0	196.2	204.6			
C_9H_7NO	2-Quinolinol	-144.9								-25.5			
C_9H_7NO	8-Quinolinol	82.1											
C_9H_8	Indene					110.6		215.3	186.9	163.4			
$C_9H_8O_4$	2-(Acetyloxy)benzoic acid	-815.6											
C_9H_{10}	Cyclopropylbenzene					100.3				150.5			
C_9H_{10}	Indan					11.5		56.0	190.2	60.3			
$C_9H_{10}Cl_2N_2O$	Diuron	-329.0											
$C_9H_{10}N_2$	2,2'-Dipyrrolylmethane	126.2											
$C_9H_{10}O_2$	Ethyl benzoate								246.0				
$C_9H_{10}O_2$	Benzyl acetate								148.5				
$C_9H_{11}NO_2$	*L*-Phenylalanine	-466.9		213.6	203.0					-312.9			
$C_9H_{11}NO_3$	*L*-Tyrosine	-685.1		214.0	216.4								
C_9H_{12}	Propylbenzene					-38.3		287.8	214.7	7.9			
C_9H_{12}	Isopropylbenzene					-41.1			210.7	4.0			
C_9H_{12}	2-Ethyltoluene					-46.4				1.3			
C_9H_{12}	3-Ethyltoluene					-48.7				-1.8			
C_9H_{12}	4-Ethyltoluene					-49.8				-3.2			
C_9H_{12}	1,2,3-Trimethylbenzene					-58.5		267.9	216.4	-9.5			
C_9H_{12}	1,2,4-Trimethylbenzene					-61.8			215.0	-13.8			
C_9H_{12}	1,3,5-Trimethylbenzene					-63.4			209.3	-15.9			
$C_9H_{12}O$	2-Isopropylphenol					-233.7				-182.2			
$C_9H_{12}O$	3-Isopropylphenol					-252.5				-196.0			
$C_9H_{12}O$	4-Isopropylphenol	-270.0								-175.3			
$C_9H_{12}O_2$	Isopropylbenzene hydroperoxide					-148.3				-78.4			
$C_9H_{13}NO_2$	Ethyl 3,5-dimethylpyrrole-2-carboxylate	-474.5											
$C_9H_{13}NO_2$	Ethyl 2,4-dimethylpyrrole-3-carboxylate	-463.2											
$C_9H_{13}NO_2$	Ethyl 2,5-dimethylpyrrole-3-carboxylate	-478.7											
$C_9H_{13}NO_2$	Ethyl 4,5-dimethylpyrrole-3-carboxylate	-470.3											
$C_9H_{14}O$	Isophorone								253.5				
$C_9H_{14}O_6$	Triacetin					-1330.8		458.3	384.7	-1245.0			
$C_9H_{15}N$	3-Ethyl-2,4,5-trimethylpyrrole	-89.2											
C_9H_{16}	1-Nonyne					16.3				62.3			
$C_9H_{16}O_4$	Nonanedioic acid	-1054.3											
$C_9H_{17}NO$	2,2,6,6-Tetramethyl-4-piperidinone	-334.2								-273.4			
C_9H_{18}	Propylcyclohexane					-237.4		311.9	242.0	-192.3			
C_9H_{18}	1α,3α,5β-1,3,5-Trimethylcyclohexane									-212.1			
$C_9H_{18}O$	2-Nonanone					-397.2				-340.7			
$C_9H_{18}O$	5-Nonanone					-398.2		401.4	303.6	-344.9			
$C_9H_{18}O$	2,6-Dimethyl-4-heptanone					-408.5			297.3	-357.6			
$C_9H_{18}O_2$	Nonanoic acid					-659.7			362.4	-577.3			
$C_9H_{18}O_2$	Butyl pentanoate					-613.3				-560.2			
$C_9H_{18}O_2$	*sec*-Butyl pentanoate					-624.2				-573.2			
$C_9H_{18}O_2$	Isobutyl pentanoate					-620.0				-568.6			
$C_9H_{18}O_2$	Methyl octanoate					-590.3				-533.9			
$C_9H_{19}N$	*N*-Butylpiperidine					-171.8							
$C_9H_{19}N$	2,2,6,6-Tetramethylpiperidine					-206.9				-159.9			
C_9H_{20}	Nonane					-274.7			284.4	-228.2			
C_9H_{20}	2,2-Dimethylheptane					-288.1							
C_9H_{20}	2,2,3-Trimethylhexane					-282.7							

		Crystal				Liquid				Gas			
Molecular formula	**Name**	**$\Delta_f H°$ kJ/mol**	**$\Delta_f G°$ kJ/mol**	**$S°$ J/mol K**	**C_p J/mol K**	**$\Delta_f H°$ kJ/mol**	**$\Delta_f G°$ kJ/mol**	**$S°$ J/mol K**	**C_p J/mol K**	**$\Delta_f H°$ kJ/mol**	**$\Delta_f G°$ kJ/mol**	**$S°$ J/mol K**	**C_p J/mol K**
C_9H_{20}	2,2,4-Trimethylhexane					-282.8							
C_9H_{20}	2,2,5-Trimethylhexane					-293.3							
C_9H_{20}	2,3,3-Trimethylhexane					-281.1							
C_9H_{20}	2,3,5-Trimethylhexane					-284.0				-242.6			
C_9H_{20}	2,4,4-Trimethylhexane					-280.2							
C_9H_{20}	3,3,4-Trimethylhexane					-277.5							
C_9H_{20}	3,3-Diethylpentane					-275.4			278.2	-233.3			
C_9H_{20}	3-Ethyl-2,2-dimethylpentane					-272.7							
C_9H_{20}	3-Ethyl-2,4-dimethylpentane					-269.7							
C_9H_{20}	2,2,3,3-Tetramethylpentane					-278.3			271.5	-237.1			
C_9H_{20}	2,2,3,4-Tetramethylpentane					-277.7				-236.9			
C_9H_{20}	2,2,4,4-Tetramethylpentane					-280.0			266.3	-241.6			
C_9H_{20}	2,3,3,4-Tetramethylpentane					-277.9				-236.1			
$C_9H_{20}N_2O$	Tetraethylurea					-380.0				-316.4			
$C_9H_{20}O$	1-Nonanol					-453.4				-376.5			
$C_9H_{20}O_2$	1,9-Nonanediol	-657.6											
$C_9H_{21}N$	Tripropylamine					-207.1				-161.0			
$C_{10}H_6N_2$	2-Quinolinecarbonitrile	246.5											
$C_{10}H_6N_2$	3-Quinolinecarbonitrile	242.3											
$C_{10}H_6N_2O_4$	1,5-Dinitronaphthalene	29.8											
$C_{10}H_6N_2O_4$	1,8-Dinitronaphthalene	39.7											
$C_{10}H_7Cl$	1-Chloronaphthalene					54.6			212.6	119.8			
$C_{10}H_7Cl$	2-Chloronaphthalene	55.4								137.4			
$C_{10}H_7I$	1-Iodonaphthalene					161.5				233.8			
$C_{10}H_7I$	2-Iodonaphthalene	144.3								235.1			
$C_{10}H_7NO_2$	1-Nitronaphthalene	42.6								111.2			
$C_{10}H_8$	Naphthalene	78.5	201.6	167.4	165.7					150.6	224.1	333.1	131.9
$C_{10}H_8$	Azulene	212.3								289.1			
$C_{10}H_8O$	1-Naphthol	121.5			166.0					-30.4			149.4
$C_{10}H_8O$	2-Naphthol	-124.1		179.0	172.8					-29.9		366.6	147.8
$C_{10}H_9N$	1-Naphthylamine	67.8								132.8			
$C_{10}H_9N$	2-Naphthylamine	60.2								134.3			
$C_{10}H_{10}$	1,2-Dihydronaphthalene					71.6							
$C_{10}H_{10}$	1,4-Dihydronaphthalene					84.2							
$C_{10}H_{10}O$	1-Tetralone	-209.6											
$C_{10}H_{10}O_4$	Dimethyl phthalate								303.1				
$C_{10}H_{10}O_4$	Dimethyl isophthalate	-730.9											
$C_{10}H_{10}O_4$	Dimethyl terephthalate	-732.6			261.1								
$C_{10}H_{12}$	1,2,3,4-Tetrahydronaphthalene					-29.2			217.5	26.0			
$C_{10}H_{14}$	Butylbenzene					-63.2		321.2	243.4	-11.8			
$C_{10}H_{14}$	*sec*-Butylbenzene, (±)					-66.4				-18.4			
$C_{10}H_{14}$	*tert*-Butylbenzene					-71.9				-23.0			
$C_{10}H_{14}$	Isobutylbenzene					-69.8				-21.9			
$C_{10}H_{14}$	1-Isopropyl-2-methylbenzene					-73.3							
$C_{10}H_{14}$	1-Isopropyl-3-methylbenzene					-78.6							
$C_{10}H_{14}$	1-Isopropyl-4-methylbenzene					-78.0			236.4				
$C_{10}H_{14}$	*o*-Diethylbenzene					-68.5							
$C_{10}H_{14}$	*m*-Diethylbenzene					-73.5							
$C_{10}H_{14}$	*p*-Diethylbenzene					-72.8							
$C_{10}H_{14}$	3-Ethyl-1,2-dimethylbenzene					-80.5							
$C_{10}H_{14}$	4-Ethyl-1,2-dimethylbenzene					-86.0							
$C_{10}H_{14}$	2-Ethyl-1,3-dimethylbenzene					-80.1							
$C_{10}H_{14}$	2-Ethyl-1,4-dimethylbenzene					-84.8							
$C_{10}H_{14}$	1-Ethyl-2,4-dimethylbenzene					-84.1							
$C_{10}H_{14}$	1-Ethyl-3,5-dimethylbenzene					-87.8							
$C_{10}H_{14}$	1,2,4,5-Tetramethylbenzene	-119.9		245.6	215.1								
$C_{10}H_{14}O$	Thymol	-309.7								-218.5			
$C_{10}H_{16}$	Dipentene					-50.8			249.4	-2.6			
$C_{10}H_{16}$	*d*-Limonene					-54.5			249.0				
$C_{10}H_{16}$	α-Pinene					-16.4				28.3			
$C_{10}H_{16}$	β-Pinene					-7.7				38.7			
$C_{10}H_{16}$	α-Terpinene									-20.6			
$C_{10}H_{16}$	β-Myrcene					14.5							
$C_{10}H_{16}$	*cis, cis*-2,6-Dimethyl-2,4,6-octatriene					-24.0							
$C_{10}H_{16}N_2O_8$	Ethylenediaminetetraacetic acid	-1759.5											

		Crystal				Liquid				Gas			
Molecular formula	**Name**	**$\Delta_f H°$ kJ/mol**	**$\Delta_f G°$ kJ/mol**	**$S°$ J/mol K**	**C_p J/mol K**	**$\Delta_f H°$ kJ/mol**	**$\Delta_f G°$ kJ/mol**	**$S°$ J/mol K**	**C_p J/mol K**	**$\Delta_f H°$ kJ/mol**	**$\Delta_f G°$ kJ/mol**	**$S°$ J/mol K**	**C_p J/mol K**
$C_{10}H_{16}O$	Camphor, (±)	-319.4			271.2					-267.5			
$C_{10}H_{18}$	1,1'-Bicyclopentyl					-178.9							
$C_{10}H_{18}$	*cis*-Decahydronaphthalene					-219.4		265.0	232.0	-169.2			
$C_{10}H_{18}$	*trans*-Decahydronaphthalene					-230.6		264.9	228.5	-182.1			
$C_{10}H_{18}O_4$	Sebacic acid	-1082.6								-921.9			
$C_{10}H_{19}N$	Decanenitrile					-158.4				-91.5			
$C_{10}H_{20}$	1-Decene					-173.8		425.0	300.8	-123.3			
$C_{10}H_{20}$	*cis*-1,2-Di-*tert*-butylethene					-163.6							
$C_{10}H_{20}$	Butylcyclohexane					-263.1		345.0	271.0	-213.7			
$C_{10}H_{20}O_2$	Decanoic acid	-713.7				-684.3				-594.9			
$C_{10}H_{20}O_2$	Methyl nonanoate					-616.2				-554.2			
$C_{10}H_{21}NO_2$	1-Nitrodecane					-351.5							
$C_{10}H_{22}$	Decane					-300.9			314.4	-249.5			
$C_{10}H_{22}$	2-Methylnonane					-309.8		420.1	313.3	-260.2			
$C_{10}H_{22}$	5-Methylnonane					-307.9		423.8	314.4	-258.6			
$C_{10}H_{22}O$	1-Decanol					-478.1			370.6	-396.6			
$C_{10}H_{22}O$	Dipentyl ether								250.0				
$C_{10}H_{22}O$	Diisopentyl ether								379.0				
$C_{10}H_{22}O_2$	1,10-Decanediol	-678.9											
$C_{10}H_{22}O_2$	Ethylene glycol dibutyl ether								350.0				
$C_{10}H_{22}S$	1-Decanethiol	-309.9				-276.5		476.1	350.4	-211.5			
$C_{10}H_{22}S$	Dipentyl sulfide					-266.4				-204.9			
$C_{10}H_{22}S$	Diisopentyl sulfide					-281.8				-221.5			
$C_{10}H_{23}N$	Octyldimethylamine					-232.8							
$C_{11}H_8O_2$	1-Naphthalenecarboxylic acid	-333.5								-223.1			
$C_{11}H_8O_2$	2-Naphthalenecarboxylic acid	-346.1								-232.5			
$C_{11}H_{10}$	1-Methylnaphthalene					56.3		254.8	224.4				
$C_{11}H_{10}$	2-Methylnaphthalene	44.9		220.0	196.0					106.7			
$C_{11}H_{12}N_2O_2$	*L*-Tryptophan	-415.3		251.0	238.1								
$C_{11}H_{14}$	1,1-Dimethylindan					-53.6				-1.6			
$C_{11}H_{16}$	1-*tert*-Butyl-3-methylbenzene					-109.7							
$C_{11}H_{16}$	1-*tert*-Butyl-4-methylbenzene					-109.7				-57.0			
$C_{11}H_{16}$	Pentamethylbenzene	-144.6								-67.2			
$C_{11}H_{20}$	Spiro[5.5]undecane					-244.5				-188.3			
$C_{11}H_{22}$	1-Undecene								344.9				
$C_{11}H_{22}O_2$	Methyl decanoate					-640.5				-573.8			
$C_{11}H_{24}$	Undecane					-327.2			344.9	-270.8			
$C_{11}H_{24}O$	1-Undecanol					-504.8							
$C_{12}F_{27}N$	Tris(perfluorobutyl)amine								418.4				
$C_{12}H_8$	Acenaphthylene	186.7			166.4					259.7			
$C_{12}H_8N_2$	Phenazine	237.0								328.8			
$C_{12}H_8O$	Dibenzofuran	-5.3								83.4			
$C_{12}H_8S$	Dibenzothiophene	120.0								205.1			
$C_{12}H_8S_2$	Thianthrene	182.0								286.0			
$C_{12}H_9N$	Carbazole	101.7								200.7			
$C_{12}H_{10}$	Acenaphthene	70.3		188.9	190.4					156.0			
$C_{12}H_{10}$	Biphenyl	99.4		209.4	198.4					181.4			
$C_{12}H_{10}N_2O$	*trans*-Azoxybenzene	243.4								342.0			
$C_{12}H_{10}N_2O$	*N*-Nitrosodiphenylamine	227.2											
$C_{12}H_{10}O$	Diphenyl ether	-32.1		233.9	216.6	-14.9				52.0			
$C_{12}H_{10}O_2$	1-Naphthaleneacetic acid	-359.2											
$C_{12}H_{10}O_2$	2-Naphthaleneacetic acid	-371.9											
$C_{12}H_{11}N$	Diphenylamine	130.2								219.3			
$C_{12}H_{11}N$	2-Aminobiphenyl	93.8								184.4			
$C_{12}H_{11}N$	4-Aminobiphenyl	81.0											
$C_{12}H_{12}N_2$	*p*-Benzidine	70.7											
$C_{12}H_{14}O_4$	Diethyl phthalate					-776.6		425.1	366.1	-688.4			
$C_{12}H_{16}$	Cyclohexylbenzene					-76.6				-16.7			
$C_{12}H_{17}NO_4$	Diethyl 3,5-dimethylpyrrole-2,4-dicarboxylate	-916.7											
$C_{12}H_{18}$	3,9-Dodecadiyne					197.8							
$C_{12}H_{18}$	5,7-Dodecadiyne					181.5							
$C_{12}H_{18}$	1-*tert*-Butyl-3,5-dimethylbenzene					-146.5							
$C_{12}H_{18}$	Hexamethylbenzene	-162.4		306.3	245.6					-77.4			
$C_{12}H_{22}$	Cyclohexylcyclohexane					-273.7				-215.7			
$C_{12}H_{22}O_4$	Dodecanedioic acid	-1130.0								-976.9			

		Crystal				Liquid				Gas			
Molecular formula	**Name**	**$\Delta_f H°$ kJ/mol**	**$\Delta_f G°$ kJ/mol**	**$S°$ J/mol K**	**C_p J/mol K**	**$\Delta_f H°$ kJ/mol**	**$\Delta_f G°$ kJ/mol**	**$S°$ J/mol K**	**C_p J/mol K**	**$\Delta_f H°$ kJ/mol**	**$\Delta_f G°$ kJ/mol**	**$S°$ J/mol K**	**C_p J/mol K**
$C_{12}H_{22}O_{11}$	Sucrose	-2226.1											
$C_{12}H_{22}O_{11}$	β-*D*-Lactose	-2236.7											
$C_{12}H_{24}$	1-Dodecene					-226.2		484.8	360.7	-165.4			
$C_{12}H_{24}O_2$	Dodecanoic acid	-774.6			404.3	-737.9				-642.0			
$C_{12}H_{24}O_2$	Methyl undecanoate					-665.2				-593.8			
$C_{12}H_{24}O_{12}$	α-Lactose monohydrate	-2484.1											
$C_{12}H_{25}Br$	1-Bromododecane					-344.7				-269.9			
$C_{12}H_{25}Cl$	1-Chlorododecane					-392.3				-321.1			
$C_{12}H_{26}$	Dodecane					-350.9			375.8	-289.4			
$C_{12}H_{26}O$	1-Dodecanol					-528.5			438.1	-436.6			
$C_{12}H_{26}O_3$	Diethylene glycol dibutyl ether								452.0				
$C_{12}H_{27}N$	Tributylamine					-281.6							
$C_{12}H_{27}O_4P$	Tributyl phosphate								379.4				
$C_{13}H_8O_2$	Xanthone	-191.5											
$C_{13}H_9N$	Acridine	179.4								273.9			
$C_{13}H_9N$	Phenanthridine	141.9								240.5			
$C_{13}H_9N$	Benzo[f]quinoline	150.6								233.7			
$C_{13}H_{10}$	9*H*-Fluorene	89.9		207.3	203.1					175.0			173.1
$C_{13}H_{10}N_2$	9-Acridinamine	159.2											
$C_{13}H_{10}O$	Benzophenone	-34.5			224.8					54.9			
$C_{13}H_{11}N$	9-Methyl-9*H*-carbazole	105.5								201.0			
$C_{13}H_{12}$	Diphenylmethane	71.5		239.3		89.7				139.0			
$C_{13}H_{13}N$	*N*-Benzylaniline	101.4											
$C_{13}H_{14}N_2$	4,4′-Diaminodiphenylmethane				270.9								
$C_{13}H_{24}O_4$	Tridecanedioic acid	-1148.3											
$C_{13}H_{26}$	1-Tridecene								391.8				
$C_{13}H_{26}O_2$	Methyl dodecanoate					-693.0				-614.9			
$C_{13}H_{28}$	Tridecane								406.7				
[illegible]	[illegible]	[illegible]											
$C_{14}H_8O_2$	9,10-Anthracenedione	-188.5								-75.7			
$C_{14}H_8O_2$	9,10-Phenanthrenedione	-154.7								-46.6			
$C_{14}H_8O_4$	1,4-Dihydroxy-9,10-anthracenedione	-595.8								-471.7			
$C_{14}H_{10}$	Anthracene	129.2		207.5	210.5					230.9			
$C_{14}H_{10}$	Phenanthrene	116.2		215.1	220.6					207.5			
$C_{14}H_{10}$	Diphenylacetylene	312.4			225.9								
$C_{14}H_{10}O_2$	Benzil	-153.9								-55.5			
$C_{14}H_{10}O_4$	Benzoyl peroxide	-369.4								-281.7			
$C_{14}H_{12}$	*cis*-Stilbene					183.3				252.3			
$C_{14}H_{12}$	*trans*-Stilbene	136.9								236.1			
$C_{14}H_{14}$	1,1-Diphenylethane					48.7							
$C_{14}H_{14}$	1,2-Diphenylethane	51.5								142.9			
$C_{14}H_{22}$	1,3-Di-*tert*-butylbenzene					-188.8							
$C_{14}H_{22}$	1,4-Di-*tert*-butylbenzene	-212.0											
$C_{14}H_{23}N_3O_{10}$	Pentetic acid	-2225.2											
$C_{14}H_{27}N$	Tetradecanenitrile					-260.2				-174.9			
$C_{14}H_{28}O_2$	Tetradecanoic acid	-833.5			432.0	-788.8				-693.7			
$C_{14}H_{28}O_2$	Methyl tridecanoate					-717.9				-635.3			
$C_{14}H_{30}O$	1-Tetradecanol	-629.6			388.0	-580.6							
$C_{15}H_{16}O_2$	2,2-Bis(4-hydroxyphenyl)propane	-368.6											
$C_{15}H_{24}$	1,3-Di-*tert*-butyl-5-methylbenzene	-245.8											
$C_{15}H_{24}O$	2,6-Di-*tert*-butyl-4-methylphenol	-410.0								-296.9			
$C_{15}H_{30}$	Decylcyclopentane					-367.3							
$C_{15}H_{30}O_2$	Pentadecanoic acid	-861.7			443.3	-811.7				-699.0			
$C_{15}H_{30}O_2$	Methyl tetradecanoate					-743.9				-656.9			
$C_{15}H_{32}O$	1-Pentadecanol	-658.2											
$C_{16}H_{10}$	Fluoranthene	189.9		230.6	230.2					289.0			
$C_{16}H_{10}$	Pyrene	125.5		224.9	229.7					225.7			
$C_{16}H_{22}O_4$	Dibutyl phthalate					-842.6				-750.9			
$C_{16}H_{22}O_{11}$	α-*D*-Glucose pentaacetate	-2249.4											
$C_{16}H_{22}O_{11}$	β-*D*-Glucose pentaacetate	-2232.6											
$C_{16}H_{26}$	Decylbenzene					-218.3				-138.6			
$C_{16}H_{32}$	1-Hexadecene					-328.7		587.9	488.9	-248.4			
$C_{16}H_{32}O_2$	Hexadecanoic acid	-891.5		452.4	460.7	-838.1				-737.1			
$C_{16}H_{32}O_2$	Methyl pentadecanoate					-771.0				-680.0			
$C_{16}H_{33}Br$	1-Bromohexadecane					-444.5				-350.2			

		Crystal				Liquid				Gas			
Molecular formula	**Name**	**$\Delta_f H°$ kJ/mol**	**$\Delta_f G°$ kJ/mol**	**$S°$ J/mol K**	**C_p J/mol K**	**$\Delta_f H°$ kJ/mol**	**$\Delta_f G°$ kJ/mol**	**$S°$ J/mol K**	**C_p J/mol K**	**$\Delta_f H°$ kJ/mol**	**$\Delta_f G°$ kJ/mol**	**$S°$ J/mol K**	**C_p J/mol K**
$C_{16}H_{34}$	Hexadecane					-456.1			501.6	-374.8			
$C_{16}H_{34}O$	1-Hexadecanol	-686.5			422.0					-517.0			
$C_{16}H_{36}IN$	Tetrabutylammonium iodide	-498.6											
$C_{17}H_{34}O_2$	Heptadecanoic acid	-924.4			475.7	-865.6							
$C_{18}H_{12}$	Benz[a]anthracene	170.8								293.0			
$C_{18}H_{12}$	Chrysene	145.3								269.8			
$C_{18}H_{14}$	*o*-Terphenyl			298.8	274.8			337.1	369.1				
$C_{18}H_{14}$	*p*-Terphenyl	163.0		285.6	278.7					279.0			
$C_{18}H_{15}N$	Triphenylamine	234.7								326.8			
$C_{18}H_{15}O_4P$	Triphenyl phosphate			397.5	356.2								
$C_{18}H_{15}P$	Triphenylphosphine				312.5								
$C_{18}H_{30}$	1,3,5-Tri-*tert*-butylbenzene	-320.0											
$C_{18}H_{34}O_2$	Oleic acid								577.0				
$C_{18}H_{34}O_4$	Dibutyl sebacate								619.0				
$C_{18}H_{36}O_2$	Stearic acid	-947.7			501.5	-884.7				-781.2			
$C_{18}H_{37}Cl$	1-Chlorooctadecane					-544.1				-446.0			
$C_{18}H_{38}$	Octadecane	-567.4		480.2	485.6					-414.6			
$C_{18}H_{39}N$	Trihexylamine					-433.0							
$C_{19}H_{16}O$	Triphenylmethanol	-2.5											
$C_{19}H_{36}O_2$	Methyl oleate					-734.5				-649.9			
$C_{19}H_{36}O_2$	Methyl *trans*-9-octadecenoate					-737.0							
$C_{20}H_{12}$	Perylene	182.8		264.6	274.9								
$C_{20}H_{12}$	Benzo[a]pyrene												254.8
$C_{20}H_{14}O_4$	Diphenyl phthalate	-489.2											
$C_{20}H_{38}O_2$	Ethyl oleate					-775.8							
$C_{20}H_{38}O_2$	Ethyl *trans*-9-octadecenoate					-773.3							
$C_{20}H_{40}O_2$	Eicosanoic acid	-1011.9			545.1	-940.0				-812.4			
$C_{21}H_{21}O_4P$	Tri-*o*-cresyl phosphate			570.0	578.0								
$C_{22}H_{14}$	Dibenz[a,h]anthracene												283.9
$C_{22}H_{42}O_2$	*trans*-13-Docosenoic acid	-960.7											
$C_{22}H_{42}O_2$	Butyl oleate					-816.9							
$C_{22}H_{44}O_2$	Butyl stearate												
$C_{24}H_{38}O_4$	Bis(2-ethylhexyl) phthalate								704.7				
$C_{24}H_{51}N$	Trioctylamine					-585.0							
$C_{26}H_{18}$	9,10-Diphenylanthracene	308.7								465.6			
$C_{26}H_{54}$	5-Butyldocosane					-713.5				-587.6			
$C_{26}H_{54}$	11-Butyldocosane					-716.0				-593.4			
$C_{28}H_{18}$	9,9'-Bianthracene	326.2								454.3			
$C_{31}H_{64}$	11-Decylheneicosane					-848.0				-705.8			
$C_{32}H_{66}$	Dotriacontane	-968.3								-697.2			
C_{60}	Carbon (fullerene-C_{60})	2327.0	2302.0	426.0	520.0					2502.0	2442.0	544.0	512.0
C_{70}	Carbon (fullerene-C_{70})	2555.0	2537.0	464.0	650.0					2755.0	2692.0	614.0	585.0

THERMODYNAMIC PROPERTIES AS A FUNCTION OF TEMPERATURE

L. V. Gurvich, V. S. Iorish, V. S. Yungman, and O. V. Dorofeeva

The thermodynamic properties $C_p^\circ(T)$, $S^\circ(T)$, $H^\circ(T)$-$H^\circ(T_r)$, -$[G^\circ(T)$-$H^\circ(T_r)]/T$ and formation properties $\Delta_f H^\circ(T)$, $\Delta_f G^\circ(T)$, log $K_f^\circ(T)$ are tabulated as functions of temperature in the range 298.15 to 1500 K for 80 substances in the standard state. The reference temperature, T_r, is equal to 298.15 K. The standard state pressure is taken as 1 bar (100,000 Pa). The tables are presented in the JANAF Thermochemical Tables format (Reference 2). The numerical data are extracted from IVTANTHERMO databases except for C_2H_4O, C_3H_6O, C_6H_6, C_6H_6O, $C_{10}H_8$, and CH_5N, which are based upon TRC Tables. See the references for information on standard states and other details.

References

1. Gurvich, L. V., Veyts, I. V., and Alcock, C. B., Eds., *Thermodynamic Properties of Individual Substances, 4th ed.*, Hemisphere Publishing Corp., New York, 1989.
2. Chase, M. W., et al., *JANAF Thermochemical Tables, 3rd ed., J. Phys. Chem. Ref. Data*, 14, Suppl. 1, 1985.

Order of Listing of Tables

No.	Formula	Name	State
1	Ar	Argon	g
2	Br	Bromine	g
3	Br_2	Dibromine	g
4	BrH	Hydrogen bromide	g
5	C	Carbon (graphite)	cr
6	C	Carbon (diamond)	cr
7	C_2	Dicarbon	g
8	C_3	Tricarbon	g
9	CO	Carbon oxide	g
10	CO_2	Carbon dioxide	g
11	CH_4	Methane	g
12	C_2H_2	Acetylene	g
13	C_2H_4	Ethylene	g
14	C_2H_6	Ethane	g
15	C_3H_6	Cyclopropane	g
16	C_3H_8	Propane	g
17	C_6H_6	Benzene	l
18	C_6H_6	Benzene	g
19	$C_{10}H_8$	Naphthalene	cr, l
20	$C_{10}H_8$	Naphthalene	g
21	CH_2O	Formaldehyde	g
22	CH_4O	Methanol	g
23	C_2H_4O	Acetaldehyde	g
24	C_2H_6O	Ethanol	g
25	$C_2H_4O_2$	Acetic acid	g
26	C_3H_6O	Acetone	g
27	C_6H_6O	Phenol	g
28	CF_4	Carbon tetrafluoride	g
29	CHF_3	Trifluoromethane	g
30	$CClF_3$	Chlorotrifluoromethane	g
31	CCl_2F_2	Dichlorodifluoromethane	g
32	$CHClF_2$	Chlorodifluoromethane	g
33	CH_5N	Methylamine	g
34	Cl	Chlorine	g
35	Cl_2	Dichlorine	g
36	ClH	Hydrogen chloride	g
37	Cu	Copper	cr, l
38	Cu	Copper	g
39	CuO	Copper oxide	cr
40	Cu_2O	Dicopper oxide	cr
41	$CuCl_2$	Copper dichloride	cr, l
42	$CuCl_2$	Copper dichloride	g
43	F	Fluorine	g
44	F_2	Difluorine	g
45	FH	Hydrogen fluoride	g
46	Ge	Germanium	cr, l
47	Ge	Germanium	g
48	GeO_2	Germanium dioxide	cr, l
49	$GeCl_4$	Germanium tetrachloride	g
50	H	Hydrogen	g
51	H_2	Dihydrogen	g
52	HO	Hydroxyl	g
53	H_2O	Water	l
54	H_2O	Water	g
55	I	Iodine	g
56	I_2	Diiodine	cr, l
57	I_2	Diiodine	g
58	IH	Hydrogen iodide	g
59	K	Potassium	cr, l
60	K	Potassium	g
61	K_2O	Dipotassium oxide	cr, l
62	KOH	Potassium hydroxide	cr, l
63	KOH	Potassium hydroxide	g
64	KCl	Potassium chloride	cr, l
65	KCl	Potassium chloride	g
66	N_2	Dinitrogen	g
67	NO	Nitric oxide	g
68	NO_2	Nitrogen dioxide	g
69	NH_3	Ammonia	g
70	O	Oxygen	g
71	O_2	Dioxygen	g
72	S	Sulfur	cr, l
73	S	Sulfur	g
74	S_2	Disulfur	g
75	S_8	Octasulfur	g
76	SO_2	Sulfur dioxide	g
77	Si	Silicon	cr
78	Si	Silicon	g
79	SiO_2	Silicon dioxide	cr
80	$SiCl_4$	Silicon tetrachloride	g

T/K	J/K mol			kJ/mol			
	C_p°	S°	$-(G^\circ-H^\circ(T_r))/T$	$H^\circ-H^\circ(T_r)$	$\Delta_f H^\circ$	$\Delta_f G^\circ$	Log K_f
1. ARGON Ar (g)							
298.15	20.786	154.845	154.845	0.000	0.000	0.000	0.000
300	20.786	154.973	154.845	0.038	0.000	0.000	0.000
400	20.786	160.953	155.660	2.117	0.000	0.000	0.000
500	20.786	165.591	157.200	4.196	0.000	0.000	0.000
600	20.786	169.381	158.924	6.274	0.000	0.000	0.000
700	20.786	172.585	160.653	8.353	0.000	0.000	0.000
800	20.786	175.361	162.322	10.431	0.000	0.000	0.000
900	20.786	177.809	163.909	12.510	0.000	0.000	0.000
1000	20.786	179.999	165.410	14.589	0.000	0.000	0.000
1100	20.786	181.980	166.828	16.667	0.000	0.000	0.000
1200	20.786	183.789	168.167	18.746	0.000	0.000	0.000
1300	20.786	185.453	169.434	20.824	0.000	0.000	0.000
1400	20.786	186.993	170.634	22.903	0.000	0.000	0.000
1500	20.786	188.427	171.773	24.982	0.000	0.000	0.000
2. BROMINE Br (g)							
298.15	20.786	175.017	175.017	0.000	111.870	82.379	−14.432
300	20.786	175.146	175.018	0.038	111.838	82.196	−14.311
400	20.787	181.126	175.833	2.117	96.677	75.460	−9.854
500	20.798	185.765	177.373	4.196	96.910	70.129	−7.326
600	20.833	189.559	179.097	6.277	97.131	64.752	−5.637
700	20.908	192.776	180.827	8.364	97.348	59.338	−4.428
800	21.027	195.575	182.499	10.461	97.568	53.893	−3.519
900	21.184	198.061	184.093	12.571	97.796	48.420	−2.810
1000	21.365	200.302	185.604	14.698	98.036	42.921	−2.242
1100	21.559	202.347	187.034	16.844	98.291	37.397	−1.776
1200	21.752	204.231	188.390	19.010	98.560	31.850	−1.386
1300	21.937	205.980	189.676	21.195	98.844	26.279	−1.056
1400	22.107	207.612	190.900	23.397	99.141	20.686	−0.772
1500	22.258	209.142	192.065	25.615	99.449	15.072	−0.525
3. DIBROMINE Br_2 (g)							
298.15	36.057	245.467	245.467	0.000	30.910	3.105	−0.544
300	36.074	245.690	245.468	0.067	30.836	2.933	−0.511
332.25	36.340	249.387	245.671	1.235		pressure = 1 bar	
400	36.729	256.169	246.892	3.711	0.000	0.000	0.000
500	37.082	264.406	249.600	7.403	0.000	0.000	0.000
600	37.305	271.188	252.650	11.123	0.000	0.000	0.000
700	37.464	276.951	255.720	14.862	0.000	0.000	0.000
800	37.590	281.962	258.694	18.615	0.000	0.000	0.000
900	37.697	286.396	261.530	22.379	0.000	0.000	0.000
1000	37.793	290.373	264.219	26.154	0.000	0.000	0.000
1100	37.883	293.979	266.763	29.938	0.000	0.000	0.000
1200	37.970	297.279	269.170	33.730	0.000	0.000	0.000
1300	38.060	300.322	271.451	37.532	0.000	0.000	0.000
1400	38.158	303.146	273.615	41.343	0.000	0.000	0.000
1500	38.264	305.782	275.673	45.164	0.000	0.000	0.000
4. HYDROGEN BROMIDE HBr (g)							
298.15	29.141	198.697	198.697	0.000	−36.290	−53.360	9.348
300	29.141	198.878	198.698	0.054	−36.333	−53.466	9.309
400	29.220	207.269	199.842	2.971	−52.109	−55.940	7.305
500	29.454	213.811	202.005	5.903	−52.484	−56.854	5.939
600	29.872	219.216	204.436	8.868	−52.844	−57.694	5.023
700	30.431	223.861	206.886	11.882	−53.168	−58.476	4.363
800	31.063	227.965	209.269	14.957	−53.446	−59.214	3.866
900	31.709	231.661	211.555	18.095	−53.677	−59.921	3.478
1000	32.335	235.035	213.737	21.298	−53.864	−60.604	3.166
1100	32.919	238.145	215.816	24.561	−54.012	−61.271	2.909

	J/K·mol			kJ/mol			
T/K	C_p°	S°	$-(G^\circ-H^\circ(T_r))/T$	$H^\circ-H^\circ(T_r)$	$\Delta_f H^\circ$	$\Delta_f G^\circ$	Log K_f
1200	33.454	241.032	217.799	27.880	−54.129	−61.925	2.696
1300	33.938	243.729	219.691	31.250	−54.220	−62.571	2.514
1400	34.374	246.261	221.499	34.666	−54.291	−63.211	2.358
1500	34.766	248.646	223.230	38.123	−54.348	−63.846	2.223
5. CARBON (GRAPHITE) C (cr; graphite)							
298.15	8.536	5.740	5.740	0.000	0.000	0.000	0.000
300	8.610	5.793	5.740	0.016	0.000	0.000	0.000
400	11.974	8.757	6.122	1.054	0.000	0.000	0.000
500	14.537	11.715	6.946	2.385	0.000	0.000	0.000
600	16.607	14.555	7.979	3.945	0.000	0.000	0.000
700	18.306	17.247	9.113	5.694	0.000	0.000	0.000
800	19.699	19.785	10.290	7.596	0.000	0.000	0.000
900	20.832	22.173	11.479	9.625	0.000	0.000	0.000
1000	21.739	24.417	12.662	11.755	0.000	0.000	0.000
1100	22.452	26.524	13.827	13.966	0.000	0.000	0.000
1200	23.000	28.502	14.968	16.240	0.000	0.000	0.000
1300	23.409	30.360	16.082	18.562	0.000	0.000	0.000
1400	23.707	32.106	17.164	20.918	0.000	0.000	0.000
1500	23.919	33.749	18.216	23.300	0.000	0.000	0.000
6. CARBON (DIAMOND) C (cr; diamond)							
298.15	6.109	2.362	2.362	0.000	1.850	2.857	−0.501
300	6.201	2.400	2.362	0.011	1.846	2.863	−0.499
400	10.321	4.783	2.659	0.850	1.645	3.235	−0.422
500	13.404	7.431	3.347	2.042	1.507	3.649	−0.381
600	15.885	10.102	4.251	3.511	1.415	4.087	−0.356
700	17.930	12.709	5.274	5.205	1.361	4.537	−0.339
800	19.619	15.217	6.361	7.085	1.338	4.993	−0.326
900	21.006	17.611	7.479	9.118	1.343	5.450	−0.316
1000	22.129	19.884	8.607	11.277	1.372	5.905	−0.308
1100	23.020	22.037	9.731	13.536	1.420	6.356	−0.302
1200	23.709	24.071	10.842	15.874	1.484	6.802	−0.296
1300	24.222	25.990	11.934	18.272	1.561	7.242	−0.291
1400	24.585	27.799	13.003	20.714	1.646	7.675	−0.286
1500	24.824	29.504	14.047	23.185	1.735	8.103	−0.282
7. DICARBON C_2 (g)							
298.15	43.548	197.095	197.095	0.000	830.457	775.116	−135.795
300	43.575	197.365	197.096	0.081	830.506	774.772	−134.898
400	42.169	209.809	198.802	4.403	832.751	755.833	−98.700
500	39.529	218.924	201.959	8.483	834.170	736.423	−76.933
600	37.837	225.966	205.395	12.342	834.909	716.795	−62.402
700	36.984	231.726	208.758	16.078	835.148	697.085	−52.016
800	36.621	236.637	211.943	19.755	835.020	677.366	−44.227
900	36.524	240.943	214.931	23.411	834.618	657.681	−38.170
1000	36.569	244.793	217.728	27.065	834.012	638.052	−33.328
1100	36.696	248.284	220.349	30.728	833.252	618.492	−29.369
1200	36.874	251.484	222.812	34.406	832.383	599.006	−26.074
1300	37.089	254.444	225.133	38.104	831.437	579.596	−23.288
1400	37.329	257.201	227.326	41.824	830.445	560.261	−20.903
1500	37.589	259.785	229.405	45.570	829.427	540.997	−18.839
8. TRICARBON C_3 (g)							
298.15	42.202	237.611	237.611	0.000	839.958	774.249	−135.643
300	42.218	237.872	237.611	0.078	839.989	773.841	−134.736
400	43.383	250.164	239.280	4.354	841.149	751.592	−98.147
500	44.883	260.003	242.471	8.766	841.570	729.141	−76.172
600	46.406	268.322	246.104	13.331	841.453	706.659	−61.519
700	47.796	275.582	249.807	18.042	840.919	684.230	−51.057
800	48.997	282.045	253.440	22.884	840.053	661.901	−43.217

	J/K·mol			kJ/mol			
T/K	C_p°	S°	$-(G^\circ-H^\circ(T_r))/T$	$H^\circ-H^\circ(T_r)$	Δ_fH°	Δ_fG°	Log K_f
900	50.006	287.876	256.948	27.835	838.919	639.698	−37.127
1000	50.844	293.189	260.310	32.879	837.572	617.633	−32.261
1100	51.535	298.069	263.524	37.999	836.059	595.711	−28.288
1200	52.106	302.578	266.593	43.182	834.420	573.933	−24.982
1300	52.579	306.768	269.524	48.417	832.690	552.295	−22.191
1400	52.974	310.679	272.326	53.695	830.899	530.793	−19.804
1500	53.307	314.346	275.006	59.010	829.068	509.421	−17.739
9. CARBON OXIDE CO (g)							
298.15	29.141	197.658	197.658	0.000	−110.530	−137.168	24.031
300	29.142	197.838	197.659	0.054	−110.519	−137.333	23.912
400	29.340	206.243	198.803	2.976	−110.121	−146.341	19.110
500	29.792	212.834	200.973	5.930	−110.027	−155.412	16.236
600	30.440	218.321	203.419	8.941	−110.157	−164.480	14.319
700	31.170	223.067	205.895	12.021	−110.453	−173.513	12.948
800	31.898	227.277	208.309	15.175	−110.870	−182.494	11.915
900	32.573	231.074	210.631	18.399	−111.378	−191.417	11.109
1000	33.178	234.538	212.851	21.687	−111.952	−200.281	10.461
1100	33.709	237.726	214.969	25.032	−112.573	−209.084	9.928
1200	34.169	240.679	216.990	28.426	−113.228	−217.829	9.482
1300	34.568	243.430	218.920	31.864	−113.904	−226.518	9.101
1400	34.914	246.005	220.763	35.338	−114.594	−235.155	8.774
1500	35.213	248.424	222.527	38.845	−115.291	−243.742	8.488
10. CARBON DIOXIDE CO_2 (g)							
298.15	37.135	213.783	213.783	0.000	−393.510	−394.373	69.092
300	37.220	214.013	213.784	0.069	−393.511	−394.379	68.667
400	41.328	225.305	215.296	4.004	−393.586	−394.656	51.536
500	44.627	234.895	218.280	8.307	−393.672	−394.914	41.256
600	47.327	243.278	221.762	12.909	−393.791	−395.152	34.401
700	49.569	250.747	225.379	17.758	−393.946	−395.367	29.502
800	51.442	257.492	228.978	22.811	−394.133	−395.558	25.827
900	53.008	263.644	232.493	28.036	−394.343	−395.724	22.967
1000	54.320	269.299	235.895	33.404	−394.568	−395.865	20.678
1100	55.423	274.529	239.172	38.893	−394.801	−395.984	18.803
1200	56.354	279.393	242.324	44.483	−395.035	−396.081	17.241
1300	57.144	283.936	245.352	50.159	−395.265	−396.159	15.918
1400	57.818	288.196	248.261	55.908	−395.488	−396.219	14.783
1500	58.397	292.205	251.059	61.719	−395.702	−396.264	13.799
11. METHANE CH_4 (g)							
298.15	35.695	186.369	186.369	0.000	−74.600	−50.530	8.853
300	35.765	186.590	186.370	0.066	−74.656	−50.381	8.772
400	40.631	197.501	187.825	3.871	−77.703	−41.827	5.462
500	46.627	207.202	190.744	8.229	−80.520	−32.525	3.398
600	52.742	216.246	194.248	13.199	−82.969	−22.690	1.975
700	58.603	224.821	198.008	18.769	−85.023	−12.476	0.931
800	64.084	233.008	201.875	24.907	−86.693	−1.993	0.130
900	69.137	240.852	205.773	31.571	−88.006	8.677	−0.504
1000	73.746	248.379	209.660	38.719	−88.996	19.475	−1.017
1100	77.919	255.607	213.511	46.306	−89.698	30.358	−1.442
1200	81.682	262.551	217.310	54.289	−90.145	41.294	−1.797
1300	85.067	269.225	221.048	62.630	−90.367	52.258	−2.100
1400	88.112	275.643	224.720	71.291	−90.390	63.231	−2.359
1500	90.856	281.817	228.322	80.242	−90.237	74.200	−2.584
12. ACETYLENE C_2H_2 (g)							
298.15	44.036	200.927	200.927	0.000	227.400	209.879	−36.769
300	44.174	201.199	200.927	0.082	227.397	209.770	−36.524
400	50.388	214.814	202.741	4.829	227.161	203.928	−26.630
500	54.751	226.552	206.357	10.097	226.846	198.154	−20.701

	J/K·mol			kJ/mol			
T/K	C_p°	S°	$-(G^\circ-H^\circ(T_r))/T$	$H^\circ-H^\circ(T_r)$	Δ_fH°	Δ_fG°	Log K_f
600	58.121	236.842	210.598	15.747	226.445	192.452	−16.754
700	60.970	246.021	215.014	21.704	225.968	186.823	−13.941
800	63.511	254.331	219.418	27.931	225.436	181.267	−11.835
900	65.831	261.947	223.726	34.399	224.873	175.779	−10.202
1000	67.960	268.995	227.905	41.090	224.300	170.355	−8.898
1100	69.909	275.565	231.942	47.985	223.734	164.988	−7.835
1200	71.686	281.725	235.837	55.067	223.189	159.672	−6.950
1300	73.299	287.528	239.592	62.317	222.676	154.400	−6.204
1400	74.758	293.014	243.214	69.721	222.203	149.166	−5.565
1500	76.077	298.218	246.709	77.264	221.774	143.964	−5.013
13. ETHYLENE C_2H_4 (g)							
298.15	42.883	219.316	219.316	0.000	52.400	68.358	−11.976
300	43.059	219.582	219.317	0.079	52.341	68.457	−11.919
400	53.045	233.327	221.124	4.881	49.254	74.302	−9.703
500	62.479	246.198	224.864	10.667	46.533	80.887	−8.450
600	70.673	258.332	229.441	17.335	44.221	87.982	−7.659
700	77.733	269.770	234.393	24.764	42.278	95.434	−7.121
800	83.868	280.559	239.496	32.851	40.655	103.142	−6.734
900	89.234	290.754	244.630	41.512	39.310	111.036	−6.444
1000	93.939	300.405	249.730	50.675	38.205	119.067	−6.219
1100	98.061	309.556	254.756	60.280	37.310	127.198	−6.040
1200	101.670	318.247	259.688	70.271	36.596	135.402	−5.894
1300	104.829	326.512	264.513	80.599	36.041	143.660	−5.772
1400	107.594	334.384	269.225	91.223	35.623	151.955	−5.669
1500	110.018	341.892	273.821	102.107	35.327	160.275	−5.581
14. ETHANE C_2H_6 (g)							
298.15	52.487	229.161	229.161	0.000	−84.000	−32.015	5.609
300	52.711	229.487	229.162	0.097	−84.094	−31.692	5.518
400	65.459	246.378	231.379	5.999	−88.988	−13.473	1.759
500	77.941	262.344	235.989	13.177	−93.238	5.912	−0.618
600	89.188	277.568	241.660	21.545	−96.779	26.086	−2.271
700	99.136	292.080	247.835	30.972	−99.663	46.800	−3.492
800	107.936	305.904	254.236	41.334	−101.963	67.887	−4.433
900	115.709	319.075	260.715	52.525	−103.754	89.231	−5.179
1000	122.552	331.628	267.183	64.445	−105.105	110.750	−5.785
1100	128.553	343.597	273.590	77.007	−106.082	132.385	−6.286
1200	133.804	355.012	279.904	90.131	−106.741	154.096	−6.708
1300	138.391	365.908	286.103	103.746	−107.131	175.850	−7.066
1400	142.399	376.314	292.178	117.790	−107.292	197.625	−7.373
1500	145.905	386.260	298.121	132.209	−107.260	219.404	−7.640
15. CYCLOPROPANE C_3H_6 (g)							
298.15	55.571	237.488	237.488	0.000	53.300	104.514	−18.310
300	55.941	237.832	237.489	0.103	53.195	104.832	−18.253
400	76.052	256.695	239.924	6.708	47.967	122.857	−16.043
500	93.859	275.637	245.177	15.230	43.730	142.091	−14.844
600	108.542	294.092	251.801	25.374	40.405	162.089	−14.111
700	120.682	311.763	259.115	36.854	37.825	182.583	−13.624
800	130.910	328.564	266.755	49.447	35.854	203.404	−13.281
900	139.658	344.501	274.516	62.987	34.384	224.441	−13.026
1000	147.207	359.616	282.277	77.339	33.334	245.618	−12.830
1100	153.749	373.961	289.965	92.395	32.640	266.883	−12.673
1200	159.432	387.588	297.538	108.060	32.249	288.197	−12.545
1300	164.378	400.549	304.967	124.257	32.119	309.533	−12.437
1400	168.689	412.892	312.239	140.915	32.215	330.870	−12.345
1500	172.453	424.662	319.344	157.976	32.507	352.193	−12.264
16. PROPANE C_3H_8 (g)							
298.15	73.597	270.313	270.313	0.000	−103.847	−23.458	4.110

T/K	C_p°	S°	$-(G^\circ-H^\circ(T_r))/T$	$H^\circ-H^\circ(T_r)$	$\Delta_f H^\circ$	$\Delta_f G^\circ$	Log K_f
	J/K·mol			kJ/mol			
300	73.931	270.769	270.314	0.136	−103.972	−22.959	3.997
400	94.014	294.739	273.447	8.517	−110.33	15.029	−0.657
500	112.591	317.768	280.025	18.872	−115.658	34.507	−3.605
600	128.700	339.753	288.162	30.955	−119.973	64.961	−5.655
700	142.674	360.668	297.039	44.540	−123.384	96.065	−7.168
800	154.766	380.528	306.245	59.427	−126.016	127.603	−8.331
900	165.352	399.381	315.555	75.444	−127.982	159.430	−9.253
1000	174.598	417.293	324.841	92.452	−129.380	191.444	−10.000
1100	182.673	434.321	334.026	110.325	−130.296	223.574	−10.617
1200	189.745	450.526	343.064	128.954	−130.802	255.770	−11.133
1300	195.853	465.961	351.929	148.241	−130.961	287.993	−11.572
1400	201.209	480.675	360.604	168.100	−130.829	320.217	−11.947
1500	205.895	494.721	369.080	188.460	−130.445	352.422	−12.272
17. BENZENE C_6H_6 (l)							
298.15	135.950	173.450	173.450	0.000	49.080	124.521	−21.815
300	136.312	174.292	173.453	.252	49.077	124.989	−21.762
400	161.793	216.837	179.082	15.102	48.978	150.320	−19.630
500	207.599	257.048	190.639	33.204	50.330	175.559	−18.340
18. BENZENE C_6H_6 (g)							
298.15	82.430	269.190	269.190	0.000	82.880	129.750	−22.731
300	83.020	269.700	269.190	0.153	82.780	130.040	−22.641
400	113.510	297.840	272.823	10.007	77.780	146.570	−19.140
500	139.340	326.050	280.658	22.696	73.740	164.260	−17.160
600	160.090	353.360	290.517	37.706	70.490	182.680	−15.903
700	176.790	379.330	301.360	54.579	67.910	201.590	−15.042
800	190.460	403.860	312.658	72.962	65.910	220.820	−14.418
900	201.840	426.970	324.084	92.597	64.410	240.280	−13.945
1000	211.430	448.740	335.473	113.267	63.340	259.890	−13.575
1100	219.580	469.280	346.710	134.827	62.620	277.640	−13.184
1200	226.540	488.690	357.743	157.137	62.200	299.320	−13.029
1300	232.520	507.070	368.534	180.097	62.000	319.090	−12.821
1400	237.680	524.490	379.056	203.607	61.990	338.870	−12.643
1500	242.140	541.040	389.302	227.607	62.110	358.640	−12.489
19. NAPHTHALENE $C_{10}H_8$ (cr, l)							
298.15	165.720	167.390	167.390	0.000	78.530	201.585	−35.316
300	167.001	168.419	167.393	0.308	78.466	202.349	−35.232
353.43	208.722	198.948	169.833	10.290	96.099	224.543	−33.186
PHASE TRANSITION: $\Delta_{trs}H$ = 18.980 kJ/mol, $\Delta_{trs}S$ = 53.702 J/K·mol, cr–l							
353.43	217.200	252.650	169.833	29.270	96.099	224.543	−33.186
400	241.577	280.916	181.124	39.917	96.067	241.475	−31.533
470	276.409	322.712	199.114	58.091	97.012	266.859	−29.658
20. NAPHTHALENE $C_{10}H_8$ (g)							
298.15	131.920	333.150	333.150	0.000	150.580	224.100	−39.260
300	132.840	333.970	333.157	0.244	150.450	224.560	−39.098
400	180.070	378.800	338.950	15.940	144.190	250.270	−32.681
500	219.740	423.400	351.400	36.000	139.220	277.340	−28.973
600	251.530	466.380	367.007	59.624	135.350	305.330	−26.581
700	277.010	507.140	384.146	86.096	132.330	333.950	−24.919
800	297.730	545.520	401.935	114.868	130.050	362.920	−23.696
900	314.850	581.610	419.918	145.523	128.430	392.150	−22.759
1000	329.170	615.550	437.806	177.744	127.510	421.700	−22.027
1100	341.240	647.500	455.426	211.281	127.100	450.630	−21.398
1200	351.500	677.650	472.707	245.932	126.960	480.450	−20.913
1300	360.260	706.130	489.568	281.531	127.060	509.770	−20.482
1400	367.780	733.110	506.009	317.941	127.390	539.740	−20.137
1500	374.270	758.720	522.019	355.051	127.920	568.940	−19.812

	J/K·mol			kJ/mol			
T/K	C_p°	S°	$-(G^\circ-H^\circ(T_r))/T$	$H^\circ-H^\circ(T_r)$	$\Delta_f H^\circ$	$\Delta_f G^\circ$	Log K_f
21. FORMALDEHYDE H_2CO (g)							
298.15	35.387	218.760	218.760	0.000	−108.700	−102.667	17.987
300	35.443	218.979	218.761	0.066	−108.731	−102.630	17.869
400	39.240	229.665	220.192	3.789	−110.438	−100.340	13.103
500	43.736	238.900	223.028	7.936	−112.073	−97.623	10.198
600	48.181	247.270	226.381	12.534	−113.545	−94.592	8.235
700	52.280	255.011	229.924	17.560	−114.833	−91.328	6.815
800	55.941	262.236	233.517	22.975	−115.942	−87.893	5.739
900	59.156	269.014	237.088	28.734	−116.889	−84.328	4.894
1000	61.951	275.395	240.603	34.792	−117.696	−80.666	4.213
1100	64.368	281.416	244.042	41.111	−118.382	−76.929	3.653
1200	66.453	287.108	247.396	47.655	−118.966	−73.134	3.183
1300	68.251	292.500	250.660	54.392	−119.463	−69.294	2.784
1400	69.803	297.616	253.833	61.297	−119.887	−65.418	2.441
1500	71.146	302.479	256.915	68.346	−120.249	−61.514	2.142
22. METHANOL CH_3OH (g)							
298.15	44.101	239.865	239.865	0.000	−201.000	−162.298	28.434
300	44.219	240.139	239.866	0.082	−201.068	−162.057	28.216
400	51.713	253.845	241.685	4.864	−204.622	−148.509	19.393
500	59.800	266.257	245.374	10.442	−207.750	−134.109	14.010
600	67.294	277.835	249.830	16.803	−210.387	−119.125	10.371
700	73.958	288.719	254.616	23.873	−212.570	−103.737	7.741
800	79.838	298.987	259.526	31.569	−214.350	−88.063	5.750
900	85.025	308.696	264.455	39.817	−215.782	−72.188	4.190
1000	89.597	317.896	269.343	48.553	−216.916	−56.170	2.934
1100	93.624	326.629	274.158	57.718	−217.794	−40.050	1.902
1200	97.165	334.930	278.879	67.262	−218.457	−23.861	1.039
1300	100.277	342.833	283.497	77.137	−218.936	−7.624	0.306
1400	103.014	350.367	288.007	87.304	−219.261	8.644	−0.322
1500	105.422	357.558	292.405	97.729	−219.456	24.930	−0.868
23. ACETALDEHYDE C_2H_4O (g)							
298.15	55.318	263.840	263.840	0.000	−166.190	−133.010	23.302
300	55.510	264.180	263.837	0.103	−166.250	−132.800	23.122
400	66.282	281.620	266.147	6.189	−169.530	−121.130	15.818
500	76.675	297.540	270.850	13.345	−172.420	−108.700	11.356
600	85.942	312.360	276.550	21.486	−174.870	−95.720	8.334
700	94.035	326.230	282.667	30.494	−176.910	−82.350	6.145
800	101.070	339.260	288.938	40.258	−178.570	−68.730	4.487
900	107.190	351.520	295.189	50.698	−179.880	−54.920	3.187
1000	112.490	363.100	301.431	61.669	−180.850	−40.930	2.138
1100	117.080	374.040	307.537	73.153	−181.560	−27.010	1.283
1200	121.060	384.400	313.512	85.065	−182.070	−12.860	0.560
1300	124.500	394.230	319.350	97.344	−182.420	1.240	−0.050
1400	127.490	403.570	325.031	109.954	−182.640	15.470	−0.577
1500	130.090	412.460	330.571	122.834	−182.750	29.580	−1.030
24. ETHANOL C_2H_5OH (g)							
298.15	65.652	281.622	281.622	0.000	−234.800	−167.874	29.410
300	65.926	282.029	281.623	0.122	−234.897	−167.458	29.157
400	81.169	303.076	284.390	7.474	−239.826	−144.216	18.832
500	95.400	322.750	290.115	16.318	−243.940	−119.820	12.517
600	107.656	341.257	297.112	26.487	−247.260	−94.672	8.242
700	118.129	358.659	304.674	37.790	−249.895	−69.023	5.151
800	127.171	375.038	312.456	50.065	−251.951	−43.038	2.810
900	135.049	390.482	320.276	63.185	−253.515	−16.825	0.976
1000	141.934	405.075	328.033	77.042	−254.662	9.539	−0.498
1100	147.958	418.892	335.670	91.543	−255.454	36.000	−1.709
1200	153.232	431.997	343.156	106.609	−255.947	62.520	−2.721

T/K	J/K·mol C_p°	S°	$-(G^\circ-H^\circ(T_r))/T$	kJ/mol $H^\circ-H^\circ(T_r)$	$\Delta_f H^\circ$	$\Delta_f G^\circ$	Log K_f
1300	157.849	444.448	350.473	122.168	−256.184	89.070	−3.579
1400	161.896	456.298	357.612	138.160	−256.206	115.630	−4.314
1500	165.447	467.591	364.571	154.531	−256.044	142.185	−4.951
25. ACETIC ACID $C_2H_4O_2$ (g)							
298.15	63.438	283.470	283.470	0.000	−432.249	−374.254	65.567
300	63.739	283.863	283.471	0.118	−432.324	−373.893	65.100
400	79.665	304.404	286.164	7.296	−436.006	−353.840	46.206
500	93.926	323.751	291.765	15.993	−438.875	−332.950	34.783
600	106.181	341.988	298.631	26.014	−440.993	−311.554	27.123
700	116.627	359.162	306.064	37.169	−442.466	−289.856	21.629
800	125.501	375.331	313.722	49.287	−443.395	−267.985	17.497
900	132.989	390.558	321.422	62.223	−443.873	−246.026	14.279
1000	139.257	404.904	329.060	75.844	−443.982	−224.034	11.702
1100	144.462	418.429	336.576	90.039	−443.798	−202.046	9.594
1200	148.760	431.189	343.933	104.707	−443.385	−180.086	7.839
1300	152.302	443.240	351.113	119.765	−442.795	−158.167	6.355
1400	155.220	454.637	358.105	135.146	−442.071	−136.299	5.085
1500	157.631	465.432	364.903	150.793	−441.247	−114.486	3.987
26. ACETONE C_3H_6O (g)							
298.15	74.517	295.349	295.349	0.000	−217.150	−152.716	26.757
300	74.810	295.809	295.349	0.138	−217.233	−152.339	26.521
400	91.755	319.658	298.498	8.464	−222.212	−129.913	16.962
500	107.864	341.916	304.988	18.464	−226.522	−106.315	11.107
600	122.047	362.836	312.873	29.978	−230.120	−81.923	7.133
700	134.306	382.627	321.470	42.810	−233.049	−56.986	4.252
800	144.934	401.246	330.265	56.785	−235.350	−31.673	2.068
900	154.097	418.860	339.141	71.747	−237.149	−6.109	0.353
1000	162.046	435.513	347.950	87.563	−238.404	19.707	−1.030
1100	168.908	451.286	356.617	104.136	−239.283	45.396	−2.157
1200	174.891	466.265	365.155	121.332	−239.827	71.463	−3.110
1300	180.079	480.491	373.513	139.072	−240.120	97.362	−3.912
1400	184.556	493.963	381.596	157.314	−240.203	123.470	−4.607
1500	188.447	506.850	389.533	175.975	−240.120	149.369	−5.202
27. PHENOL C_6H_6O (g)							
298.15	103.220	314.810	314.810	0.000	−96.400	−32.630	5.720
300	103.860	315.450	314.810	0.192	−96.490	−32.230	5.610
400	135.790	349.820	319.278	12.217	−100.870	−10.180	1.330
500	161.910	383.040	328.736	27.152	−104.240	12.970	−1.360
600	182.480	414.450	340.430	44.412	−106.810	36.650	−3.190
700	198.840	443.860	353.134	63.508	−108.800	60.750	−4.530
800	212.140	471.310	366.211	84.079	−110.300	85.020	−5.550
900	223.190	496.950	379.327	105.861	−111.370	109.590	−6.360
1000	232.490	520.960	392.302	128.658	−111.990	134.280	−7.010
1100	240.410	543.500	405.033	152.314	−112.280	158.620	−7.530
1200	247.200	564.720	417.468	176.703	−112.390	183.350	−7.980
1300	253.060	584.740	429.568	201.723	−112.330	208.070	−8.360
1400	258.120	603.680	441.331	227.288	−112.120	233.050	−8.700
1500	262.520	621.650	452.767	253.325	−111.780	257.540	−8.970
28. CARBON TETRAFLUORIDE CF_4 (g)							
298.15	61.050	261.455	261.455	0.000	−933.200	−888.518	155.663
300	61.284	261.833	261.456	0.113	−933.219	−888.240	154.654
400	72.399	281.057	264.001	6.822	−933.986	−873.120	114.016
500	80.713	298.153	269.155	14.499	−934.372	−857.852	89.618
600	86.783	313.434	275.284	22.890	−934.490	−842.533	73.348
700	91.212	327.162	281.732	31.801	−934.431	−827.210	61.726
800	94.479	339.566	288.199	41.094	−934.261	−811.903	53.011
900	96.929	350.842	294.542	50.670	−934.024	−796.622	46.234

	J/K·mol			kJ/mol			
T/K	C_p°	S°	$-(G^\circ-H^\circ(T_r))/T$	$H^\circ-H^\circ(T_r)$	$\Delta_f H^\circ$	$\Delta_f G^\circ$	Log K_f
1000	98.798	361.156	300.695	60.460	−933.745	−781.369	40.814
1100	100.250	370.643	306.629	70.416	−933.442	−766.146	36.381
1200	101.396	379.417	312.334	80.500	−933.125	−750.952	32.688
1300	102.314	387.571	317.811	90.687	−932.800	−735.784	29.564
1400	103.059	395.181	323.069	100.957	−932.470	−720.641	26.887
1500	103.671	402.313	328.116	111.295	−932.137	−705.522	24.568
29. TRIFLUOROMETHANE CHF_3 (g)							
298.15	51.069	259.675	259.675	0.000	−696.700	−662.237	116.020
300	51.258	259.991	259.676	0.095	−696.735	−662.023	115.267
400	61.148	276.113	261.807	5.722	−698.427	−650.186	84.905
500	69.631	290.700	266.149	12.275	−699.715	−637.969	66.647
600	76.453	304.022	271.368	19.593	−700.634	−625.528	54.456
700	81.868	316.230	276.917	27.519	−701.253	−612.957	45.739
800	86.201	327.455	282.542	35.930	−701.636	−600.315	39.196
900	89.719	337.818	288.116	44.732	−701.832	−587.636	34.105
1000	92.617	347.426	293.572	53.854	−701.879	−574.944	30.032
1100	95.038	356.370	298.879	63.240	−701.805	−562.253	26.699
1200	97.084	364.730	304.022	72.849	−701.629	−549.574	23.922
1300	98.833	372.571	308.997	82.647	−701.368	−536.913	21.573
1400	100.344	379.952	313.804	92.607	−701.033	−524.274	19.561
1500	101.660	386.921	318.449	102.709	−700.635	−511.662	17.817
30. CHLOROTRIFLUOROMETHANE $CClF_3$ (g)							
298.15	66.886	285.419	285.419	0.000	−707.800	−667.238	116.896
300	67.111	285.834	285.421	0.124	−707.810	−666.986	116.131
400	77.528	306.646	288.187	7.383	−708.153	−653.316	85.313
500	85.013	324.797	293.734	15.532	−708.170	−639.599	66.818
600	90.329	340.794	300.271	24.314	−707.975	−625.901	54.489
700	94.132	355.020	307.096	33.547	−707.654	−612.246	45.686
800	96.899	367.780	313.897	43.106	−707.264	−598.642	39.087
900	98.951	379.317	320.536	52.903	−706.837	−585.090	33.957
1000	100.507	389.827	326.947	62.880	−706.396	−571.586	29.856
1100	101.708	399.465	333.108	72.993	−705.950	−558.126	26.503
1200	102.651	408.357	339.013	83.213	−705.505	−544.707	23.710
1300	103.404	416.604	344.668	93.517	−705.064	−531.326	21.349
1400	104.012	424.290	350.084	103.889	−704.628	−517.977	19.326
1500	104.512	431.484	355.273	114.316	−704.196	−504.660	17.574
31. DICHLORODIFLUOROMETHANE CCl_2F_2 (g)							
298.15	72.476	300.903	300.903	0.000	−486.000	−447.030	78.317
300	72.691	301.352	300.905	0.134	−486.002	−446.788	77.792
400	82.408	323.682	303.883	7.919	−485.945	−433.716	56.637
500	89.063	342.833	309.804	16.514	−485.618	−420.692	43.949
600	93.635	359.500	316.729	25.663	−485.136	−407.751	35.497
700	96.832	374.189	323.909	35.196	−484.576	−394.897	29.467
800	99.121	387.276	331.027	44.999	−483.984	−382.126	24.950
900	100.801	399.053	337.942	55.000	−483.388	−369.429	21.441
1000	102.062	409.742	344.596	65.146	−482.800	−356.799	18.637
1100	103.030	419.517	350.969	75.402	−482.226	−344.227	16.346
1200	103.786	428.515	357.061	85.745	−481.667	−331.706	14.439
1300	104.388	436.847	362.882	96.154	−481.121	−319.232	12.827
1400	104.874	444.602	368.445	106.618	−480.588	−306.799	11.447
1500	105.270	451.851	373.767	117.126	−480.065	−294.404	10.252
32. CHLORODIFLUOROMETHANE $CHClF_2$ (g)							
298.15	55.853	280.915	280.915	0.000	−475.000	−443.845	77.759
300	56.039	281.261	280.916	0.104	−475.028	−443.652	77.246
400	65.395	298.701	283.231	6.188	−476.390	−432.978	56.540
500	73.008	314.145	287.898	13.123	−477.398	−422.001	44.086
600	78.940	328.003	293.448	20.733	−478.103	−410.851	35.767

T/K	J/K·mol C_p°	S°	$-(G^\circ-H^\circ(T_r))/T$	kJ/mol $H^\circ-H^\circ(T_r)$	Δ_fH°	Δ_fG°	Log K_f
700	83.551	340.533	299.294	28.867	−478.574	−399.603	29.818
800	87.185	351.936	305.172	37.411	−478.870	−388.299	25.353
900	90.100	362.379	310.956	46.280	−479.031	−376.967	21.878
1000	92.475	371.999	316.586	55.413	−479.090	−365.622	19.098
1100	94.433	380.908	322.033	64.761	−479.068	−354.276	16.823
1200	96.066	389.196	327.289	74.289	−478.982	−342.935	14.927
1300	97.438	396.941	332.352	83.966	−478.843	−331.603	13.324
1400	98.601	404.206	337.228	93.769	−478.661	−320.283	11.950
1500	99.593	411.044	341.923	103.681	−478.443	−308.978	10.759
33. METHYLAMINE CH_5N (g)							
298.15	50.053	242.881	242.881	0.000	−22.529	32.734	−5.735
300	50.227	243.196	242.893	0.091	−22.614	33.077	−5.759
400	60.171	258.986	244.975	5.604	−26.846	52.294	−6.829
500	70.057	273.486	249.244	12.121	−30.431	72.510	−7.575
600	78.929	287.063	254.431	19.579	−33.364	93.382	−8.129
700	86.711	299.826	260.008	27.873	−35.712	114.702	−8.559
800	93.545	311.865	265.749	36.893	−37.548	136.316	−8.900
900	99.573	323.239	271.511	46.555	−38.949	158.138	−9.178
1000	104.886	334.006	277.220	56.786	−39.967	180.098	−9.407
1100	109.576	344.233	282.861	67.509	−40.681	201.822	−9.584
1200	113.708	353.944	288.374	78.685	−41.136	224.240	−9.761
1300	117.341	363.190	293.775	90.239	−41.376	246.364	−9.899
1400	120.542	372.012	299.061	102.131	−41.451	268.504	−10.018
1500	123.353	380.426	304.209	114.326	−41.381	290.639	−10.121
34. CHLORINE Cl (g)							
298.15	21.838	165.190	165.190	0.000	121.302	105.306	−18.449
300	21.852	165.325	165.190	0.040	121.311	105.207	−18.318
400	22.467	171.703	166.055	2.259	121.795	99.766	−13.028
500	22.744	176.752	167.708	4.522	122.272	94.203	−9.841
600	22.781	180.905	169.571	6.800	122.734	88.546	−7.709
700	22.692	184.411	171.448	9.074	123.172	82.813	−6.179
800	22.549	187.432	173.261	11.337	123.585	77.019	−5.029
900	22.389	190.079	174.986	13.584	123.971	71.175	−4.131
1000	22.233	192.430	176.615	15.815	124.334	65.289	−3.410
1100	22.089	194.542	178.150	18.031	124.675	59.368	−2.819
1200	21.959	196.458	179.597	20.233	124.996	53.416	−2.325
1300	21.843	198.211	180.963	22.423	125.299	47.439	−1.906
1400	21.742	199.826	182.253	24.602	125.587	41.439	−1.546
1500	21.652	201.323	183.475	26.772	125.861	35.418	−1.233
35. DICHLORINE Cl_2 (g)							
298.15	33.949	223.079	223.079	0.000	0.000	0.000	0.000
300	33.981	223.290	223.080	0.063	0.000	0.000	0.000
400	35.296	233.263	224.431	3.533	0.000	0.000	0.000
500	36.064	241.229	227.021	7.104	0.000	0.000	0.000
600	36.547	247.850	229.956	10.736	0.000	0.000	0.000
700	36.874	253.510	232.926	14.408	0.000	0.000	0.000
800	37.111	258.450	235.815	18.108	0.000	0.000	0.000
900	37.294	262.832	238.578	21.829	0.000	0.000	0.000
1000	37.442	266.769	241.203	25.566	0.000	0.000	0.000
1100	37.567	270.343	243.692	29.316	0.000	0.000	0.000
1200	37.678	273.617	246.052	33.079	0.000	0.000	0.000
1300	37.778	276.637	248.290	36.851	0.000	0.000	0.000
1400	37.872	279.440	250.416	40.634	0.000	0.000	0.000
1500	37.961	282.056	252.439	44.426	0.000	0.000	0.000
36. HYDROGEN CHLORIDE HCl (g)							
298.15	29.136	186.902	186.902	0.000	−92.310	−95.298	16.696
300	29.137	187.082	186.902	0.054	−92.314	−95.317	16.596

	J/K·mol			kJ/mol			
T/K	C_p°	S°	$-(G^\circ-H^\circ(T_r))/T$	$H^\circ-H^\circ(T_r)$	$\Delta_f H^\circ$	$\Delta_f G^\circ$	Log K_f
400	29.175	195.468	188.045	2.969	−92.587	−96.278	12.573
500	29.304	201.990	190.206	5.892	−92.911	−97.164	10.151
600	29.576	207.354	192.630	8.835	−93.249	−97.983	8.530
700	29.988	211.943	195.069	11.812	−93.577	−98.746	7.368
800	30.500	215.980	197.435	14.836	−93.879	−99.464	6.494
900	31.063	219.604	199.700	17.913	−94.149	−100.145	5.812
1000	31.639	222.907	201.858	21.049	−94.384	−100.798	5.265
1100	32.201	225.949	203.912	24.241	−94.587	−101.430	4.816
1200	32.734	228.774	205.867	27.488	−94.760	−102.044	4.442
1300	33.229	231.414	207.732	30.786	−94.908	−102.645	4.124
1400	33.684	233.893	209.513	34.132	−95.035	−103.235	3.852
1500	34.100	236.232	211.217	37.522	−95.146	−103.817	3.615
37. COPPER Cu (cr, l)							
298.15	24.440	33.150	33.150	0.000	0.000	0.000	0.000
300	24.460	33.301	33.150	0.045	0.000	0.000	0.000
400	25.339	40.467	34.122	2.538	0.000	0.000	0.000
500	25.966	46.192	35.982	5.105	0.000	0.000	0.000
600	26.479	50.973	38.093	7.728	0.000	0.000	0.000
700	26.953	55.090	40.234	10.399	0.000	0.000	0.000
800	27.448	58.721	42.322	13.119	0.000	0.000	0.000
900	28.014	61.986	44.328	15.891	0.000	0.000	0.000
1000	28.700	64.971	46.245	18.726	0.000	0.000	0.000
1100	29.553	67.745	48.075	21.637	0.000	0.000	0.000
1200	30.617	70.361	49.824	24.644	0.000	0.000	0.000
1300	31.940	72.862	51.501	27.769	0.000	0.000	0.000
1358	33.844	74.275	52.443	29.647	0.000	0.000	0.000
PHASE TRANSITION: $\Delta_{trs}H$ = 13.141 kJ/mol, $\Delta_{trs}S$ = 9.676 J/K·mol, cr–l							
1358	32.800	83.951	52.443	42.788	0.000	0.000	0.000
1400	32.800	84.950	53.403	44.166	0.000	0.000	0.000
1500	32.800	87.213	55.583	47.446	0.000	0.000	0.000
38. COPPER Cu (g)							
298.15	20.786	166.397	166.397	0.000	337.600	297.873	−52.185
300	20.786	166.525	166.397	0.038	337.594	297.626	−51.821
400	20.786	172.505	167.213	2.117	337.179	284.364	−37.134
500	20.786	177.143	168.752	4.196	336.691	271.215	−28.333
600	20.786	180.933	170.476	6.274	336.147	258.170	−22.475
700	20.786	184.137	172.205	8.353	335.554	245.221	−18.298
800	20.786	186.913	173.874	10.431	334.913	232.359	−15.171
900	20.786	189.361	175.461	12.510	334.219	219.581	−12.744
1000	20.786	191.551	176.963	14.589	333.463	206.883	−10.806
1100	20.788	193.532	178.380	16.667	332.631	194.265	−9.225
1200	20.793	195.341	179.719	18.746	331.703	181.726	−7.910
1300	20.803	197.006	180.986	20.826	330.657	169.270	−6.801
1400	20.823	198.548	182.186	22.907	316.342	157.305	−5.869
1500	20.856	199.986	183.325	24.991	315.146	145.987	−5.084
39. COPPER OXIDE CuO (cr)							
298.15	42.300	42.740	42.740	0.000	−162.000	−134.277	23.524
300	42.417	43.002	42.741	0.078	−161.994	−134.105	23.349
400	46.783	55.878	44.467	4.564	−161.487	−124.876	16.307
500	49.190	66.596	47.852	9.372	−160.775	−115.803	12.098
600	50.827	75.717	51.755	14.377	−159.973	−106.883	9.305
700	52.099	83.651	55.757	19.526	−159.124	−98.102	7.320
800	53.178	90.680	59.691	24.791	−158.247	−89.444	5.840
900	54.144	97.000	63.491	30.158	−157.356	−80.897	4.695
1000	55.040	102.751	67.134	35.617	−156.462	−72.450	3.784
1100	55.890	108.037	70.615	41.164	−155.582	−64.091	3.043
1200	56.709	112.936	73.941	46.794	−154.733	−55.812	2.429

	J/K·mol			kJ/mol			
T/K	C_p°	S°	$-(G^\circ-H^\circ(T_r))/T$	$H^\circ-H^\circ(T_r)$	$\Delta_f H^\circ$	$\Delta_f G^\circ$	Log K_f
1300	57.507	117.507	77.118	52.505	−153.940	−47.601	1.913
1400	58.288	121.797	80.158	58.295	−166.354	−39.043	1.457
1500	59.057	125.845	83.070	64.163	−165.589	−29.975	1.044
40. DICOPPER OXIDE Cu_2O (cr)							
298.15	62.600	92.550	92.550	0.000	−173.100	−150.344	26.339
300	62.721	92.938	92.551	0.116	−173.102	−150.203	26.152
400	67.587	111.712	95.078	6.654	−173.036	−142.572	18.618
500	70.784	127.155	99.995	13.580	−172.772	−134.984	14.101
600	73.323	140.291	105.643	20.789	−172.389	−127.460	11.096
700	75.552	151.764	111.429	28.235	−171.914	−120.009	8.955
800	77.616	161.989	117.121	35.894	−171.363	−112.631	7.354
900	79.584	171.245	122.629	43.755	−170.750	−105.325	6.113
1000	81.492	179.729	127.920	51.809	−170.097	−98.091	5.124
1100	83.360	187.584	132.992	60.052	−169.431	−90.922	4.317
1200	85.202	194.917	137.850	68.480	−168.791	−83.814	3.648
1300	87.026	201.808	142.507	77.092	−168.223	−76.756	3.084
1400	88.836	208.324	146.978	85.885	−194.030	−68.926	2.572
1500	90.636	214.515	151.276	94.858	−193.438	−60.010	2.090
41. COPPER DICHLORIDE $CuCl_2$ (cr, l)							
298.15	71.880	108.070	108.070	0.000	−218.000	−173.826	30.453
300	71.998	108.515	108.071	0.133	−217.975	−173.552	30.218
400	76.338	129.899	110.957	7.577	−216.494	−158.962	20.758
500	78.654	147.204	116.532	15.336	−214.873	−144.765	15.123
600	80.175	161.687	122.884	23.282	−213.182	−130.901	11.396
675	81.056	171.183	127.732	29.329	−211.185	−120.693	9.340
PHASE TRANSITION: $\Delta_{trs}H$ = 0.700 kJ/mol, $\Delta_{trs}S$ = 1.037 J/K·mol, crII–crI							
675	82.400	172.220	127.732	30.029	−211.185	−120.693	9.340
700	82.400	175.216	129.375	32.089	−210.719	−117.350	8.757
800	82.400	186.219	135.808	40.329	−208.898	−104.137	6.799
871	82.400	193.226	140.207	46.179	−192.649	−94.893	5.691
PHASE TRANSITION: $\Delta_{trs}H$ = 15.001 kJ/mol, $\Delta_{trs}S$ = 17.221 J/K·mol, crI–l							
871	100.000	210.447	140.207	61.180	−192.649	−94.893	5.691
900	100.000	213.723	142.523	64.080	−191.640	−91.655	5.319
1000	100.000	224.259	150.179	74.080	−188.212	−80.730	4.217
1100	100.000	233.790	157.353	84.080	−184.873	−70.144	3.331
1130.75	100.000	236.547	159.470	87.155	−183.867	−66.951	3.093
42. COPPER DICHLORIDE $CuCl_2$ (g)							
298.15	56.814	278.418	278.418	0.000	−43.268	−49.883	8.739
300	56.869	278.769	278.419	0.105	−43.271	−49.924	8.692
400	58.992	295.456	280.679	5.911	−43.428	−52.119	6.806
500	60.111	308.752	285.010	11.871	−43.606	−54.271	5.670
600	60.761	319.774	289.911	17.918	−43.814	−56.385	4.909
700	61.168	329.173	294.865	24.015	−44.060	−58.462	4.362
800	61.439	337.360	299.677	30.147	−44.349	−60.500	3.950
900	61.630	344.608	304.274	36.301	−44.688	−62.499	3.627
1000	61.776	351.109	308.638	42.471	−45.088	−64.457	3.367
1100	61.900	357.003	312.771	48.655	−45.566	−66.372	3.152
1200	62.022	362.394	316.685	54.851	−46.139	−68.239	2.970
1300	62.159	367.364	320.395	61.060	−46.829	−70.053	2.815
1400	62.325	371.976	323.916	67.284	−60.784	−71.404	2.664
1500	62.531	376.283	327.265	73.526	−61.613	−72.133	2.512
43. FLUORINE F (g)							
298.15	22.746	158.750	158.750	0.000	79.380	62.280	−10.911
300	22.742	158.891	158.750	0.042	79.393	62.173	−10.825
400	22.432	165.394	159.639	2.302	80.043	56.332	−7.356
500	22.100	170.363	161.307	4.528	80.587	50.340	−5.259

T/K	J/K·mol C_p°	S°	$-(G^\circ-H^\circ(T_r))/T$	kJ/mol $H^\circ-H^\circ(T_r)$	$\Delta_f H^\circ$	$\Delta_f G^\circ$	Log K_f
600	21.832	174.368	163.161	6.724	81.046	44.246	−3.852
700	21.629	177.717	165.008	8.897	81.442	38.081	−2.842
800	21.475	180.595	166.780	11.052	81.792	31.862	−2.080
900	21.357	183.117	168.458	13.193	82.106	25.601	−1.486
1000	21.266	185.362	170.039	15.324	82.391	19.308	−1.009
1100	21.194	187.386	171.525	17.447	82.654	12.986	−0.617
1200	21.137	189.227	172.925	19.563	82.897	6.642	−0.289
1300	21.091	190.917	174.245	21.675	83.123	0.278	−0.011
1400	21.054	192.479	175.492	23.782	83.335	−6.103	0.228
1500	21.022	193.930	176.673	25.886	83.533	−12.498	0.435
44. DIFLUORINE F_2 (g)							
298.15	31.304	202.790	202.790	0.000	0.000	0.000	0.000
300	31.337	202.984	202.790	0.058	0.000	0.000	0.000
400	32.995	212.233	204.040	3.277	0.000	0.000	0.000
500	34.258	219.739	206.453	6.643	0.000	0.000	0.000
600	35.171	226.070	209.208	10.117	0.000	0.000	0.000
700	35.839	231.545	212.017	13.669	0.000	0.000	0.000
800	36.343	236.365	214.765	17.279	0.000	0.000	0.000
900	36.740	240.669	217.409	20.934	0.000	0.000	0.000
1000	37.065	244.557	219.932	24.625	0.000	0.000	0.000
1100	37.342	248.103	222.334	28.346	0.000	0.000	0.000
1200	37.588	251.363	224.619	32.093	0.000	0.000	0.000
1300	37.811	254.381	226.794	35.863	0.000	0.000	0.000
1400	38.019	257.191	228.866	39.654	0.000	0.000	0.000
1500	38.214	259.820	230.843	43.466	0.000	0.000	0.000
45. HYDROGEN FLUORIDE HF (g)							
298.15	29.137	173.776	173.776	0.000	−273.300	−275.399	48.248
300	29.137	173.956	173.776	0.054	−273.302	−275.412	47.953
400	29.149	182.340	174.919	2.968	−273.450	−276.096	36.054
500	29.172	188.846	177.078	5.884	−273.679	−276.733	28.910
600	29.230	194.169	179.496	8.804	−273.961	−277.318	24.142
700	29.350	198.683	181.923	11.732	−274.277	−277.852	20.733
800	29.549	202.614	184.269	14.676	−274.614	−278.340	18.174
900	29.827	206.110	186.505	17.645	−274.961	−278.785	16.180
1000	30.169	209.270	188.626	20.644	−275.309	−279.191	14.583
1100	30.558	212.163	190.636	23.680	−275.652	−279.563	13.275
1200	30.974	214.840	192.543	26.756	−275.988	−279.904	12.184
1300	31.403	217.336	194.355	29.875	−276.315	−280.217	11.259
1400	31.831	219.679	196.081	33.037	−276.631	−280.505	10.466
1500	32.250	221.889	197.729	36.241	−276.937	−280.771	9.777
46. GERMANIUM Ge (cr, l)							
298.15	23.222	31.090	31.090	0.000	0.000	0.000	0.000
300	23.249	31.234	31.090	0.043	0.000	0.000	0.000
400	24.310	38.083	32.017	2.426	0.000	0.000	0.000
500	24.962	43.582	33.798	4.892	0.000	0.000	0.000
600	25.452	48.178	35.822	7.414	0.000	0.000	0.000
700	25.867	52.133	37.876	9.980	0.000	0.000	0.000
800	26.240	55.612	39.880	12.586	0.000	0.000	0.000
900	26.591	58.723	41.804	15.227	0.000	0.000	0.000
1000	26.926	61.542	43.639	17.903	0.000	0.000	0.000
1100	27.252	64.124	45.386	20.612	0.000	0.000	0.000
1200	27.571	66.509	47.048	23.353	0.000	0.000	0.000
1211.4	27.608	66.770	47.232	23.668	0.000	0.000	0.000
PHASE TRANSITION: $\Delta_{trs}H$ = 37.030 kJ/mol, $\Delta_{trs}S$ = 30.568 J/K·mol, cr–l							
1211.4	27.600	97.338	47.232	60.698	0.000	0.000	0.000
1300	27.600	99.286	50.714	63.143	0.000	0.000	0.000
1400	27.600	101.331	54.258	65.903	0.000	0.000	0.000
1500	27.600	103.236	57.460	68.663	0.000	0.000	0.000

T/K	J/K·mol C_p°	S°	$-(G^\circ-H^\circ(T_r))/T$	kJ/mol $H^\circ-H^\circ(T_r)$	$\Delta_f H^\circ$	$\Delta_f G^\circ$	Log K_f
47. GERMANIUM Ge (g)							
298.15	30.733	167.903	167.903	0.000	367.800	327.009	−57.290
300	30.757	168.094	167.904	0.057	367.814	326.756	−56.893
400	31.071	177.025	169.119	3.162	368.536	312.959	−40.868
500	30.360	183.893	171.415	6.239	369.147	298.991	−31.235
600	29.265	189.334	173.965	9.222	369.608	284.914	−24.804
700	28.102	193.758	176.487	12.090	369.910	270.773	−20.205
800	27.029	197.439	178.882	14.845	370.060	256.598	−16.754
900	26.108	200.567	181.122	17.501	370.073	242.414	−14.069
1000	25.349	203.277	183.205	20.072	369.969	228.234	−11.922
1100	24.741	205.664	185.141	22.575	369.763	214.069	−10.165
1200	24.264	207.795	186.941	25.025	369.471	199.928	−8.703
1300	23.898	209.722	188.621	27.432	332.088	188.521	−7.575
1400	23.624	211.483	190.192	29.807	331.704	177.492	−6.622
1500	23.426	213.105	191.666	32.159	331.296	166.491	−5.798
48. GERMANIUM DIOXIDE GeO_2 (cr, l)							
298.15	50.166	39.710	39.710	0.000	−580.200	−521.605	91.382
300	50.475	40.021	39.711	0.093	−580.204	−521.242	90.755
400	61.281	56.248	41.850	5.759	−579.893	−501.610	65.503
500	66.273	70.519	46.191	12.164	−579.013	−482.134	50.368
600	69.089	82.872	51.299	18.943	−577.915	−462.859	40.295
700	70.974	93.671	56.597	25.952	−576.729	−443.776	33.115
800	72.449	103.247	61.841	33.125	−575.498	−424.866	27.741
900	73.764	111.857	66.928	40.436	−574.235	−406.113	23.570
1000	75.049	119.696	71.819	47.877	−572.934	−387.502	20.241
1100	76.378	126.910	76.504	55.447	−571.582	−369.024	17.523
1200	77.796	133.616	80.987	63.155	−570.166	−350.671	15.264
1300	79.332	139.903	85.279	71.010	−605.685	−329.732	13.249
1308	79.460	140.390	85.615	71.646	−584.059	−328.034	13.100
PHASE TRANSITION: $\Delta_{trs}H$ = 21.500 kJ/mol, $\Delta_{trs}S$ = 16.437 J/K·mol, crII–crI							
1308	80.075	156.827	85.615	93.146	−584.059	−328.034	13.100
1388	81.297	161.617	89.858	99.601	−565.504	−312.415	11.757
PHASE TRANSITION: $\Delta_{trs}H$ = 17.200 kJ/mol, $\Delta_{trs}S$ = 12.392 J/K·mol, crI–l							
1388	78.500	174.009	89.858	116.801	−565.504	−312.415	11.757
1400	78.500	174.685	90.582	117.743	−565.328	−310.228	11.575
1500	78.500	180.100	96.372	125.593	−563.882	−292.057	10.170
49. GERMANIUM TETRACHLORIDE $GeCl_4$ (g)							
298.15	95.918	348.393	348.393	0.000	−500.000	−461.582	80.866
300	96.041	348.987	348.395	0.178	−499.991	−461.343	80.326
400	100.750	377.342	352.229	10.045	−499.447	−448.540	58.573
500	103.206	400.114	359.604	20.255	−498.845	−435.882	45.536
600	104.624	419.067	367.980	30.652	−498.234	−423.347	36.855
700	105.509	435.266	376.463	41.162	−497.634	−410.914	30.662
800	106.096	449.396	384.715	51.744	−497.057	−398.565	26.023
900	106.504	461.917	392.611	62.375	−496.509	−386.287	22.419
1000	106.799	473.155	400.113	73.041	−495.993	−374.068	19.539
1100	107.020	483.344	407.224	83.733	−495.512	−361.899	17.185
1200	107.189	492.664	413.961	94.444	−495.067	−349.772	15.225
1300	107.320	501.249	420.349	105.169	−531.677	−334.973	13.459
1400	107.425	509.206	426.416	115.907	−531.265	−319.857	11.934
1500	107.509	516.621	432.185	126.654	−530.861	−304.771	10.613
50. HYDROGEN H (g)							
298.15	20.786	114.716	114.716	0.000	217.998	203.276	−35.613
300	20.786	114.845	114.716	0.038	218.010	203.185	−35.377
400	20.786	120.824	115.532	2.117	218.635	198.149	−25.875
500	20.786	125.463	117.071	4.196	219.253	192.956	−20.158

T/K	J/K·mol $C_p°$	$S°$	$-(G°-H°(T_r))/T$	kJ/mol $H°-H°(T_r)$	$\Delta_f H°$	$\Delta_f G°$	Log K_f
600	20.786	129.252	118.795	6.274	219.867	187.639	−16.335
700	20.786	132.457	120.524	8.353	220.476	182.219	−13.597
800	20.786	135.232	122.193	10.431	221.079	176.712	−11.538
900	20.786	137.680	123.780	12.510	221.670	171.131	−9.932
1000	20.786	139.870	125.282	14.589	222.247	165.485	−8.644
1100	20.786	141.852	126.700	16.667	222.806	159.781	−7.587
1200	20.786	143.660	128.039	18.746	223.345	154.028	−6.705
1300	20.786	145.324	129.305	20.824	223.864	148.230	−5.956
1400	20.786	146.864	130.505	22.903	224.360	142.393	−5.313
1500	20.786	148.298	131.644	24.982	224.835	136.522	−4.754
51. DIHYDROGEN H_2 (g)							
298.15	28.836	130.680	130.680	0.000	0.000	0.000	0.000
300	28.849	130.858	130.680	0.053	0.000	0.000	0.000
400	29.181	139.217	131.818	2.960	0.000	0.000	0.000
500	29.260	145.738	133.974	5.882	0.000	0.000	0.000
600	29.327	151.078	136.393	8.811	0.000	0.000	0.000
700	29.440	155.607	138.822	11.749	0.000	0.000	0.000
800	29.623	159.549	141.172	14.702	0.000	0.000	0.000
900	29.880	163.052	143.412	17.676	0.000	0.000	0.000
1000	30.204	166.217	145.537	20.680	0.000	0.000	0.000
1100	30.580	169.113	147.550	23.719	0.000	0.000	0.000
1200	30.991	171.791	149.460	26.797	0.000	0.000	0.000
1300	31.422	174.288	151.275	29.918	0.000	0.000	0.000
1400	31.860	176.633	153.003	33.082	0.000	0.000	0.000
1500	32.296	178.846	154.653	36.290	0.000	0.000	0.000
52. HYDROXYL OH (g)							
298.15	29.886	183.737	183.737	0.000	39.349	34.631	−6.067
300	29.879	183.922	183.738	0.055	39.350	34.602	−6.025
400	29.604	192.476	184.906	3.028	39.384	33.012	−4.311
500	29.495	199.067	187.104	5.982	39.347	31.422	−3.283
600	29.513	204.445	189.560	8.931	39.252	29.845	−2.598
700	29.655	209.003	192.020	11.888	39.113	28.287	−2.111
800	29.914	212.979	194.396	14.866	38.945	26.752	−1.747
900	30.265	216.522	196.661	17.874	38.763	25.239	−1.465
1000	30.682	219.731	198.810	20.921	38.577	23.746	−1.240
1100	31.135	222.677	200.848	24.012	38.393	22.272	−1.058
1200	31.603	225.406	202.782	27.149	38.215	20.814	−0.906
1300	32.069	227.954	204.621	30.332	38.046	19.371	−0.778
1400	32.522	230.347	206.374	33.562	37.886	17.941	−0.669
1500	32.956	232.606	208.048	36.836	37.735	16.521	−0.575
53. WATER H_2O (l)							
298.15	75.300	69.950	69.950	0.000	−285.830	−237.141	41.546
300	75.281	70.416	69.951	0.139	−285.771	−236.839	41.237
373.21	76.079	86.896	71.715	5.666	−283.454	−225.160	31.513
54. WATER H_2O (g)							
298.15	33.598	188.832	188.832	0.000	−241.826	−228.582	40.046
300	33.606	189.040	188.833	0.062	−241.844	−228.500	39.785
400	34.283	198.791	190.158	3.453	−242.845	−223.900	29.238
500	35.259	206.542	192.685	6.929	−243.822	−219.050	22.884
600	36.371	213.067	195.552	10.509	−244.751	−214.008	18.631
700	37.557	218.762	198.469	14.205	−245.620	−208.814	15.582
800	38.800	223.858	201.329	18.023	−246.424	−203.501	13.287
900	40.084	228.501	204.094	21.966	−247.158	−198.091	11.497
1000	41.385	232.792	206.752	26.040	−247.820	−192.603	10.060
1100	42.675	236.797	209.303	30.243	−248.410	−187.052	8.882
1200	43.932	240.565	211.753	34.574	−248.933	−181.450	7.898
1300	45.138	244.129	214.108	39.028	−249.392	−175.807	7.064

	J/K·mol			kJ/mol			
T/K	C_p°	S°	$-(G^\circ-H^\circ(T_r))/T$	$H^\circ-H^\circ(T_r)$	$\Delta_f H^\circ$	$\Delta_f G^\circ$	Log K_f
1400	46.281	247.516	216.374	43.599	−249.792	−170.132	6.348
1500	47.356	250.746	218.559	48.282	−250.139	−164.429	5.726
55. IODINE I (g)							
298.15	20.786	180.787	180.787	0.000	106.760	70.172	−12.294
300	20.786	180.915	180.787	0.038	106.748	69.945	−12.178
400	20.786	186.895	181.602	2.117	97.974	58.060	−7.582
500	20.786	191.533	183.142	4.196	75.988	50.202	−5.244
600	20.786	195.323	184.866	6.274	76.190	45.025	−3.920
700	20.786	198.527	186.594	8.353	76.385	39.816	−2.971
800	20.787	201.303	188.263	10.432	76.574	34.579	−2.258
900	20.789	203.751	189.851	12.510	76.757	29.319	−1.702
1000	20.795	205.942	191.352	14.589	76.936	24.038	−1.256
1100	20.806	207.924	192.770	16.669	77.109	18.740	−0.890
1200	20.824	209.735	194.110	18.751	77.277	13.426	−0.584
1300	20.851	211.403	195.377	20.835	77.440	8.098	−0.325
1400	20.889	212.950	196.577	22.921	77.596	2.758	−0.103
1500	20.936	214.392	197.717	25.013	77.745	−2.592	0.090
56. DIIODINE I_2 (cr, l)							
298.15	54.440	116.139	116.139	0.000	0.000	0.000	0.000
300	54.518	116.476	116.140	0.101	0.000	0.000	0.000
386.75	61.531	131.039	117.884	5.088	0.000	0.000	0.000
PHASE TRANSITION: $\Delta_{trs}H$ = 15.665 kJ/mol, $\Delta_{trs}S$ = 40.504 J/K·mol, cr–l							
386.75	79.555	171.543	117.884	20.753	0.000	0.000	0.000
400	79.555	174.223	119.706	21.807	0.000	0.000	0.000
457.67	79.555	184.938	127.266	26.395	0.000	0.000	0.000
57. DIIODINE I_2 (g)							
298.15	36.887	260.685	260.685	0.000	62.420	19.324	−3.385
300	36.897	260.913	260.685	0.068	62.387	19.056	−3.318
400	37.256	271.584	262.138	3.778	44.391	5.447	−0.711
457.67	37.385	276.610	263.652	5.931		pressure = 1 bar	
500	37.464	279.921	264.891	7.515	0.000	0.000	0.000
600	37.613	286.765	267.983	11.269	0.000	0.000	0.000
700	37.735	292.573	271.092	15.037	0.000	0.000	0.000
800	37.847	297.619	274.099	18.816	0.000	0.000	0.000
900	37.956	302.083	276.965	22.606	0.000	0.000	0.000
1000	38.070	306.088	279.681	26.407	0.000	0.000	0.000
1100	38.196	309.722	282.249	30.220	0.000	0.000	0.000
1200	38.341	313.052	284.679	34.047	0.000	0.000	0.000
1300	38.514	316.127	286.981	37.890	0.000	0.000	0.000
1400	38.719	318.989	289.166	41.751	0.000	0.000	0.000
1500	38.959	321.668	291.245	45.635	0.000	0.000	0.000
58. HYDROGEN IODIDE HI (g)							
298.15	29.157	206.589	206.589	0.000	26.500	1.700	−0.298
300	29.158	206.769	206.589	0.054	26.477	1.546	−0.269
400	29.329	215.176	207.734	2.977	17.093	−6.289	0.821
500	29.738	221.760	209.904	5.928	−5.481	−9.946	1.039
600	30.351	227.233	212.348	8.931	−5.819	−10.806	0.941
700	31.070	231.965	214.820	12.002	−6.101	−11.614	0.867
800	31.807	236.162	217.230	15.145	−6.323	−12.386	0.809
900	32.511	239.950	219.548	18.362	−6.489	−13.133	0.762
1000	33.156	243.409	221.763	21.646	−6.608	−13.865	0.724
1100	33.735	246.597	223.878	24.991	−6.689	−14.586	0.693
1200	34.249	249.555	225.896	28.391	−6.741	−15.302	0.666
1300	34.703	252.314	227.823	31.839	−6.775	−16.014	0.643
1400	35.106	254.901	229.666	35.330	−6.797	−16.723	0.624
1500	35.463	257.336	231.430	38.858	−6.814	−17.432	0.607

	J/K·mol			kJ/mol			
T/K	C_p°	S°	$-(G^\circ-H^\circ(T_r))/T$	$H^\circ-H^\circ(T_r)$	$\Delta_f H^\circ$	$\Delta_f G^\circ$	Log K_f
59. POTASSIUM K (cr, l)							
298.15	29.600	64.680	64.680	0.000	0.000	0.000	0.000
300	29.671	64.863	64.681	0.055	0.000	0.000	0.000
336.86	32.130	68.422	64.896	1.188	0.000	0.000	0.000
PHASE TRANSITION: $\Delta_{trs}H$ = 2.321 kJ/mol, $\Delta_{trs}S$ = 6.891 J/K·mol, cr–l							
336.86	32.129	75.313	64.896	3.509	0.000	0.000	0.000
400	31.552	80.784	66.986	5.519	0.000	0.000	0.000
500	30.741	87.734	70.469	8.632	0.000	0.000	0.000
600	30.158	93.283	73.824	11.675	0.000	0.000	0.000
700	29.851	97.905	76.943	14.673	0.000	0.000	0.000
800	29.838	101.887	79.818	17.655	0.000	0.000	0.000
900	30.130	105.415	82.470	20.651	0.000	0.000	0.000
1000	30.730	108.618	84.927	23.691	0.000	0.000	0.000
1039.4	31.053	109.812	85.847	24.908	0.000	0.000	0.000
60. POTASSIUM K (g)							
298.15	20.786	160.340	160.340	0.000	89.000	60.479	−10.596
300	20.786	160.468	160.340	0.038	88.984	60.302	−10.499
400	20.786	166.448	161.155	2.117	85.598	51.332	−6.703
500	20.786	171.086	162.695	4.196	84.563	42.887	−4.480
600	20.786	174.876	164.419	6.274	83.599	34.643	−3.016
700	20.786	178.080	166.148	8.353	82.680	26.557	−1.982
800	20.786	180.856	167.817	10.431	81.776	18.601	−1.215
900	20.786	183.304	169.404	12.510	80.859	10.759	−0.624
1000	20.786	185.494	170.905	14.589	79.897	3.021	−0.158
1039.4	20.786	186.297	171.474	15.408		pressure = 1 bar	
1100	20.786	187.475	172.323	16.667	0.000	0.000	0.000
1200	20.786	189.284	173.662	18.746	0.000	0.000	0.000
1300	20.789	190.948	174.929	20.825	0.000	0.000	0.000
1400	20.793	192.489	176.129	22.904	0.000	0.000	0.000
1500	20.801	193.923	177.268	24.983	0.000	0.000	0.000
61. DIPOTASSIUM OXIDE K_2O (cr, l)							
298.15	72.000	96.000	96.000	0.000	−361.700	−321.171	56.267
300	72.130	96.446	96.001	0.133	−361.704	−320.920	55.876
400	79.154	118.158	98.914	7.698	−366.554	−306.416	40.013
500	86.178	136.575	104.647	15.964	−366.043	−291.423	30.444
590	92.500	151.348	110.662	24.005	−364.204	−278.079	24.619
PHASE TRANSITION: $\Delta_{trs}H$ = 0.700 kJ/mol, $\Delta_{trs}S$ = 1.186 J/K·mol, crIII–crII							
590	100.000	152.534	110.662	24.705	−364.204	−278.079	24.619
600	100.000	154.215	111.374	25.705	−363.968	−276.621	24.082
645	100.000	161.447	114.618	30.205	−358.901	−270.109	21.874
PHASE TRANSITION: $\Delta_{trs}H$ = 4.000 kJ/mol, $\Delta_{trs}S$ = 6.202 J/K·mol, crII–crI							
645	100.000	167.649	114.618	34.205	−358.901	−270.109	21.874
700	100.000	175.832	119.111	39.705	−357.592	−262.592	19.595
800	100.000	189.185	127.054	49.705	−355.224	−249.183	16.270
900	100.000	200.963	134.625	59.705	−352.919	−236.067	13.701
1000	100.000	211.499	141.794	69.705	−350.732	−223.202	11.659
1013	100.000	212.791	142.697	71.005	−323.459	−221.546	11.424
PHASE TRANSITION: $\Delta_{trs}H$ = 27.000 kJ/mol, $\Delta_{trs}S$ =26.654 J/K·mol, crI–l							
1013	100.000	239.444	142.697	98.005	−323.459	−221.546	11.424
1100	100.000	247.684	150.679	106.705	−479.439	−203.633	9.670
1200	100.000	256.385	159.131	116.705	−475.371	−178.740	7.780
1300	100.000	264.389	166.924	126.705	−471.321	−154.185	6.195
1400	100.000	271.800	174.154	136.705	−467.287	−129.941	4.848
1500	100.000	278.699	180.896	146.705	−463.268	−105.986	3.691

T/K	C_p°	S°	$-(G^\circ-H^\circ(T_r))/T$	$H^\circ-H^\circ(T_r)$	$\Delta_f H^\circ$	$\Delta_f G^\circ$	Log K_f
	J/K·mol			kJ/mol			
62. POTASSIUM HYDROXIDE KOH (cr, l)							
298.15	64.900	78.870	78.870	0.000	−424.580	−378.747	66.354
300	65.038	79.272	78.871	0.120	−424.569	−378.463	65.895
400	72.519	99.007	81.512	6.998	−426.094	−362.765	47.372
500	80.000	115.993	86.745	14.624	−424.572	−347.093	36.260
520	81.496	119.159	87.931	16.239	−417.725	−344.002	34.555
PHASE TRANSITION: $\Delta_{trs} H$ = 6.450 kJ/mol, $\Delta_{trs} S$ = 12.404 J/K·mol, crII–crI							
520	79.000	131.563	87.931	22.689	−417.725	−344.002	34.555
600	79.000	142.868	94.520	29.009	−416.274	−332.766	28.969
678	79.000	152.523	100.649	35.171	−405.464	−321.998	24.807
PHASE TRANSITION: $\Delta_{trs} H$ = 9.400 kJ/mol, $\Delta_{trs} S$ = 13.865 J/K·mol, crI–l							
678	83.000	166.388	100.649	44.571	−405.464	−321.998	24.807
700	83.000	169.038	102.757	46.397	−404.981	−319.297	23.826
800	83.000	180.121	111.750	54.697	−402.808	−307.206	20.058
900	83.000	189.897	119.901	62.997	−400.694	−295.383	17.143
1000	83.000	198.642	127.345	71.297	−398.668	−283.791	14.824
1100	83.000	206.553	134.192	79.597	−475.618	−267.780	12.716
1200	83.000	213.775	140.527	87.897	−472.711	−249.014	10.839
1300	83.000	220.418	146.421	96.197	−469.843	−230.490	9.261
1400	83.000	226.569	151.929	104.497	−467.011	−212.184	7.917
1500	83.000	232.296	157.098	112.797	−464.217	−194.080	6.758
63. POTASSIUM HYDROXIDE KOH (g)							
298.15	49.184	238.283	238.283	0.000	−227.989	−229.685	40.239
300	49.236	238.588	238.284	0.091	−228.007	−229.696	39.993
400	51.178	253.053	240.243	5.124	−231.377	−229.667	29.991
500	52.178	264.591	243.998	10.296	−232.309	−229.129	23.937
600	52.804	274.163	248.251	15.547	−233.145	−228.413	19.885
700	53.296	282.340	252.551	20.853	−233.934	−227.562	16.981
800	53.758	289.487	256.730	26.206	−234.708	−226.599	14.795
900	54.229	295.846	260.730	31.605	−235.495	−225.538	13.090
1000	54.713	301.585	264.533	37.052	−236.322	−224.388	11.721
1100	55.203	306.823	268.143	42.548	−316.077	−218.535	10.377
1200	55.686	311.647	271.570	48.092	−315.925	−209.674	9.127
1300	56.153	316.122	274.827	53.684	−315.764	−200.826	8.069
1400	56.598	320.300	277.927	59.322	−315.595	−191.991	7.163
1500	57.016	324.220	280.884	65.003	−315.420	−183.169	6.378
64. POTASSIUM CHLORIDE KCl (cr, l)							
298.15	51.300	82.570	82.570	0.000	−436.490	−408.568	71.579
300	51.333	82.887	82.571	0.095	−436.481	−408.395	71.107
400	52.977	97.886	84.605	5.312	−438.463	−398.651	52.058
500	54.448	109.867	88.498	10.685	−437.990	−388.749	40.612
600	55.885	119.921	92.919	16.201	−437.332	−378.960	32.991
700	57.425	128.649	97.413	21.865	−436.502	−369.295	27.557
800	59.205	136.430	101.812	27.694	−435.505	−359.760	23.490
900	61.361	143.523	106.058	33.719	−434.337	−350.360	20.334
1000	64.032	150.121	110.138	39.983	−432.981	−341.100	17.817
1044	65.405	152.908	111.882	42.830	−485.450	−336.720	16.847
PHASE TRANSITION: $\Delta_{trs} H$ = 26.320 kJ/mol, $\Delta_{trs} S$ = 25.210 J/K·mol, cr–l							
1044	72.000	178.118	111.882	69.150	−485.450	−336.720	16.847
1100	72.000	181.880	115.351	73.182	−483.633	−328.790	15.613
1200	72.000	188.145	121.160	80.382	−480.393	−314.856	13.705
1300	72.000	193.908	126.537	87.582	−477.158	−301.192	12.102
1400	72.000	199.244	131.542	94.782	−473.928	−287.778	10.737
1500	72.000	204.211	136.223	101.982	−470.704	−274.594	9.562

T/K	C_p° (J/K mol)	S° (J/K mol)	$-(G^\circ - H^\circ(T_r))/T$ (J/K mol)	$H^\circ - H^\circ(T_r)$ (kJ/mol)	$\Delta_f H^\circ$ (kJ/mol)	$\Delta_f G^\circ$ (kJ/mol)	Log K_f
65. POTASSIUM CHLORIDE KCl (g)							
298.15	36.505	239.091	239.091	0.000	−214.575	−233.320	40.876
300	36.518	239.317	239.092	0.068	−214.594	−233.436	40.644
400	37.066	249.904	240.532	3.749	−218.112	−239.107	31.224
500	37.384	258.212	243.267	7.473	−219.287	−244.219	25.513
600	37.597	265.048	246.344	11.222	−220.396	−249.100	21.686
700	37.769	270.857	249.441	14.991	−221.461	−253.799	18.938
800	37.907	275.910	252.441	18.775	−222.509	−258.347	16.868
900	38.041	280.382	255.302	22.572	−223.568	−262.764	15.250
1000	38.162	284.397	258.014	26.383	−224.667	−267.061	13.950
1100	38.279	288.039	260.581	30.205	−304.696	−266.627	12.661
1200	38.401	291.375	263.010	34.039	−304.821	−263.161	11.455
1300	38.518	294.454	265.312	37.885	−304.941	−259.684	10.434
1400	38.639	297.313	267.496	41.743	−305.053	−256.199	9.559
1500	38.761	299.983	269.574	45.613	−305.159	−252.706	8.800
66. DINITROGEN N_2 (g)							
298.15	29.124	191.608	191.608	0.000	0.000	0.000	0.000
300	29.125	191.788	191.608	0.054	0.000	0.000	0.000
400	29.249	200.180	192.752	2.971	0.000	0.000	0.000
500	29.580	206.738	194.916	5.911	0.000	0.000	0.000
600	30.109	212.175	197.352	8.894	0.000	0.000	0.000
700	30.754	216.864	199.812	11.936	0.000	0.000	0.000
800	31.433	221.015	202.208	15.046	0.000	0.000	0.000
900	32.090	224.756	204.509	18.222	0.000	0.000	0.000
1000	32.696	228.169	206.706	21.462	0.000	0.000	0.000
1100	33.241	231.311	208.802	24.759	0.000	0.000	0.000
1200	33.723	234.224	210.801	28.108	0.000	0.000	0.000
1300	34.147	236.941	212.708	31.502	0.000	0.000	0.000
1400	34.517	239.485	214.531	34.936	0.000	0.000	0.000
1500	34.842	241.878	216.275	38.404	0.000	0.000	0.000
67. NITRIC OXIDE NO (g)							
298.15	29.862	210.745	210.745	0.000	91.277	87.590	−15.345
300	29.858	210.930	210.746	0.055	91.278	87.567	−15.247
400	29.954	219.519	211.916	3.041	91.320	86.323	−11.272
500	30.493	226.255	214.133	6.061	91.340	85.071	−8.887
600	31.243	231.879	216.635	9.147	91.354	83.816	−7.297
700	32.031	236.754	219.168	12.310	91.369	82.558	−6.160
800	32.770	241.081	221.642	15.551	91.386	81.298	−5.308
900	33.425	244.979	224.022	18.862	91.405	80.036	−4.645
1000	33.990	248.531	226.298	22.233	91.426	78.772	−4.115
1100	34.473	251.794	228.469	25.657	91.445	77.505	−3.680
1200	34.883	254.811	230.540	29.125	91.464	76.237	−3.318
1300	35.234	257.618	232.516	32.632	91.481	74.967	−3.012
1400	35.533	260.240	234.404	36.170	91.495	73.697	−2.750
1500	35.792	262.700	236.209	39.737	91.506	72.425	−2.522
68. NITROGEN DIOXIDE NO_2 (g)							
298.15	37.178	240.166	240.166	0.000	34.193	52.316	−9.165
300	37.236	240.397	240.167	0.069	34.181	52.429	−9.129
400	40.513	251.554	241.666	3.955	33.637	58.600	−7.652
500	43.664	260.939	244.605	8.167	33.319	64.882	−6.778
600	46.383	269.147	248.026	12.673	33.174	71.211	−6.199
700	48.612	276.471	251.575	17.427	33.151	77.553	−5.787
800	50.405	283.083	255.107	22.381	33.213	83.893	−5.478
900	51.844	289.106	258.555	27.496	33.334	90.221	−5.236
1000	53.007	294.631	261.891	32.741	33.495	96.534	−5.042
1100	53.956	299.729	265.102	38.090	33.686	102.828	−4.883
1200	54.741	304.459	268.187	43.526	33.898	109.105	−4.749

T/K	C_p° (J/K mol)	S° (J/K mol)	$-(G^\circ-H^\circ(T_r))/T$ (J/K mol)	$H^\circ-H^\circ(T_r)$ (kJ/mol)	$\Delta_f H^\circ$ (kJ/mol)	$\Delta_f G^\circ$ (kJ/mol)	Log K_f
1300	55.399	308.867	271.148	49.034	34.124	115.363	−4.635
1400	55.960	312.994	273.992	54.603	34.360	121.603	−4.537
1500	56.446	316.871	276.722	60.224	34.604	127.827	−4.451
69. AMMONIA NH_3 (g)							
298.15	35.630	192.768	192.768	0.000	−45.940	−16.407	2.874
300	35.678	192.989	192.769	0.066	−45.981	−16.223	2.825
400	38.674	203.647	194.202	3.778	−48.087	−5.980	0.781
500	41.994	212.633	197.011	7.811	−49.908	4.764	−0.498
600	45.229	220.578	200.289	12.174	−51.430	15.846	−1.379
700	48.269	227.781	203.709	16.850	−52.682	27.161	−2.027
800	51.112	234.414	207.138	21.821	−53.695	38.639	−2.523
900	53.769	240.589	210.516	27.066	−54.499	50.231	−2.915
1000	56.244	246.384	213.816	32.569	−55.122	61.903	−3.233
1100	58.535	251.854	217.027	38.309	−55.589	73.629	−3.496
1200	60.644	257.039	220.147	44.270	−55.920	85.392	−3.717
1300	62.576	261.970	223.176	50.432	−56.136	97.177	−3.905
1400	64.339	266.673	226.117	56.779	−56.251	108.975	−4.066
1500	65.945	271.168	228.971	63.295	−56.282	120.779	−4.206
70. OXYGEN O (g)							
298.15	21.911	161.058	161.058	0.000	249.180	231.743	−40.600
300	21.901	161.194	161.059	0.041	249.193	231.635	−40.331
400	21.482	167.430	161.912	2.207	249.874	225.677	−29.470
500	21.257	172.197	163.511	4.343	250.481	219.556	−22.937
600	21.124	176.060	165.290	6.462	251.019	213.319	−18.571
700	21.040	179.310	167.067	8.570	251.500	206.997	−15.446
800	20.984	182.115	168.777	10.671	251.932	200.610	−13.098
900	20.944	184.584	170.399	12.767	252.325	194.171	−11.269
1000	20.915	186.789	171.930	14.860	252.686	187.689	−9.804
1100	20.893	188.782	173.372	16.950	253.022	181.173	−8.603
1200	20.877	190.599	174.733	19.039	253.335	174.628	−7.601
1300	20.864	192.270	176.019	21.126	253.630	168.057	−6.753
1400	20.853	193.815	177.236	23.212	253.908	161.463	−6.024
1500	20.845	195.254	178.389	25.296	254.171	154.851	−5.392
71. DIOXYGEN O_2 (g)							
298.15	29.378	205.148	205.148	0.000	0.000	0.000	0.000
300	29.387	205.330	205.148	0.054	0.000	0.000	0.000
400	30.109	213.873	206.308	3.026	0.000	0.000	0.000
500	31.094	220.695	208.525	6.085	0.000	0.000	0.000
600	32.095	226.454	211.045	9.245	0.000	0.000	0.000
700	32.987	231.470	213.612	12.500	0.000	0.000	0.000
800	33.741	235.925	216.128	15.838	0.000	0.000	0.000
900	34.365	239.937	218.554	19.244	0.000	0.000	0.000
1000	34.881	243.585	220.878	22.707	0.000	0.000	0.000
1100	35.314	246.930	223.096	26.217	0.000	0.000	0.000
1200	35.683	250.019	225.213	29.768	0.000	0.000	0.000
1300	36.006	252.888	227.233	33.352	0.000	0.000	0.000
1400	36.297	255.568	229.162	36.968	0.000	0.000	0.000
1500	36.567	258.081	231.007	40.611	0.000	0.000	0.000
72. SULFUR S (cr, l)							
298.15	22.690	32.070	32.070	0.000	0.000	0.000	0.000
300	22.737	32.210	32.070	0.042	0.000	0.000	0.000
368.3	24.237	37.030	32.554	1.649	0.000	0.000	0.000
PHASE TRANSITION: $\Delta_{trs}H$ = 0.401 kJ/mol, $\Delta_{trs}S$ = 1.089 J/K mol, crII–crI							
368.3	24.773	38.119	32.553	2.050	0.000	0.000	0.000
388.36	25.180	39.444	32.875	2.551	0.000	0.000	0.000
PHASE TRANSITION: $\Delta_{trs}H$ = 1.722 kJ/mol, $\Delta_{trs}S$ = 4.431 J/K mol, crI–l							

T/K	C_p°	S°	$-(G^\circ-H^\circ(T_r))/T$	$H^\circ-H^\circ(T_r)$	$\Delta_f H^\circ$	$\Delta_f G^\circ$	Log K_f
	J/K·mol			kJ/mol			
388.36	31.710	43.875	32.872	4.273	0.000	0.000	0.000
400	32.369	44.824	33.206	4.647	0.000	0.000	0.000
500	38.026	53.578	36.411	8.584	0.000	0.000	0.000
600	34.371	60.116	39.842	12.164	0.000	0.000	0.000
700	32.451	65.278	43.120	15.511	0.000	0.000	0.000
800	32.000	69.557	46.163	18.715	0.000	0.000	0.000
882.38	32.000	72.693	48.496	21.351	0.000	0.000	0.000
73. SULFUR S (g)							
298.15	23.673	167.828	167.828	0.000	277.180	236.704	−41.469
300	23.669	167.974	167.828	0.044	277.182	236.453	−41.170
400	23.233	174.730	168.752	2.391	274.924	222.962	−29.115
500	22.741	179.860	170.482	4.689	273.286	210.145	−21.953
600	22.338	183.969	172.398	6.942	271.958	197.646	−17.206
700	22.031	187.388	174.302	9.160	270.829	185.352	−13.831
800	21.800	190.314	176.125	11.351	269.816	173.210	11.309
900	21.624	192.871	177.847	13.522	215.723	162.258	−9.417
1000	21.489	195.142	179.465	15.677	216.018	156.301	−8.164
1100	21.386	197.185	180.985	17.821	216.284	150.317	−7.138
1200	21.307	199.043	182.413	19.955	216.525	144.309	−6.282
1300	21.249	200.746	183.759	22.083	216.743	138.282	−5.556
1400	21.209	202.319	185.029	24.206	216.940	132.239	−4.934
1500	21.186	203.781	186.231	26.325	217.119	126.182	−4.394
74. DISULFUR S_2 (g)							
298.15	32.505	228.165	228.165	0.000	128.600	79.696	−13.962
300	32.540	228.366	228.165	0.060	128.576	79.393	−13.823
400	34.108	237.956	229.462	3.398	122.703	63.380	−8.276
500	35.133	245.686	231.959	6.863	118.296	49.031	−5.122
600	35.815	252.156	234.800	10.413	114.685	35.530	−3.093
700	36.305	257.715	237.686	14.020	111.599	22.588	−1.685
800	36.697	262.589	240.501	17.671	108.841	10.060	−0.657
882.38	36.985	266.200	242.734	20.706	pressure = 1 bar		
900	37.045	266.932	243.201	21.358	0.000	0.000	0.000
1000	37.377	270.852	245.773	25.079	0.000	0.000	0.000
1100	37.704	274.430	248.218	28.833	0.000	0.000	0.000
1200	38.030	277.725	250.541	32.620	0.000	0.000	0.000
1300	38.353	280.781	252.751	36.439	0.000	0.000	0.000
1400	38.669	283.635	254.856	40.290	0.000	0.000	0.000
1500	38.976	286.314	256.865	44.173	0.000	0.000	0.000
75. OCTASULFUR S_8 (g)							
298.15	156.500	432.536	432.536	0.000	101.277	48.810	−8.551
300	156.768	433.505	432.539	0.290	101.231	48.484	−8.442
400	167.125	480.190	438.834	16.542	80.642	32.003	−4.179
500	173.181	518.176	451.022	33.577	66.185	21.409	−2.237
600	177.936	550.180	464.951	51.137	55.101	13.549	−1.180
700	182.441	577.948	479.152	69.157	46.349	7.343	−0.548
800	186.764	602.596	493.071	87.620	39.177	2.263	−0.148
900	190.595	624.821	506.495	106.494	−392.062	6.554	−0.380
1000	193.618	645.067	519.355	125.712	−387.728	50.614	−2.644
1100	195.684	663.625	531.639	145.185	−383.272	94.233	−4.475
1200	196.825	680.707	543.359	164.817	−378.786	137.444	−5.983
1300	197.195	696.480	554.539	184.524	−374.356	180.283	−7.244
1400	196.988	711.089	565.206	204.237	−370.048	222.785	−8.312
1500	196.396	724.662	575.389	223.909	−365.905	264.984	−9.227
76. SULFUR DIOXIDE SO_2 (g)							
298.15	39.842	248.219	248.219	0.000	−296.810	−300.090	52.574
300	39.909	248.466	248.220	0.074	−296.833	−300.110	52.253
400	43.427	260.435	249.828	4.243	−300.240	−300.935	39.298

	J/K·mol			kJ/mol			
T/K	C_p°	S°	$-(G^\circ-H^\circ(T_r))/T$	$H^\circ-H^\circ(T_r)$	$\Delta_f H^\circ$	$\Delta_f G^\circ$	Log K_f
500	46.490	270.465	252.978	8.744	−302.735	−300.831	31.427
600	48.938	279.167	256.634	13.520	−304.699	−300.258	26.139
700	50.829	286.859	260.413	18.513	−306.308	−299.386	22.340
800	52.282	293.746	264.157	23.671	−307.691	−298.302	19.477
900	53.407	299.971	267.796	28.958	−362.075	−295.987	17.178
1000	54.290	305.646	271.301	34.345	−362.012	−288.647	15.077
1100	54.993	310.855	274.664	39.810	−361.934	−281.314	13.358
1200	55.564	315.665	277.882	45.339	−361.849	−273.989	11.926
1300	56.033	320.131	280.963	50.920	−361.763	−266.671	10.715
1400	56.426	324.299	283.911	56.543	−361.680	−259.359	9.677
1500	56.759	328.203	286.735	62.203	−361.605	−252.053	8.777
77. SILICON Si (cr)							
298.15	19.789	18.810	18.810	0.000	0.000	0.000	0.000
300	19.855	18.933	18.810	0.037	0.000	0.000	0.000
400	22.301	25.023	19.624	2.160	0.000	0.000	0.000
500	23.610	30.152	21.231	4.461	0.000	0.000	0.000
600	24.472	34.537	23.092	6.867	0.000	0.000	0.000
700	25.124	38.361	25.006	9.348	0.000	0.000	0.000
800	25.662	41.752	26.891	11.888	0.000	0.000	0.000
900	26.135	44.802	28.715	14.478	0.000	0.000	0.000
1000	26.568	47.578	30.464	17.114	0.000	0.000	0.000
1100	26.974	50.130	32.138	19.791	0.000	0.000	0.000
1200	27.362	52.493	33.737	22.508	0.000	0.000	0.000
1300	27.737	54.698	35.265	25.263	0.000	0.000	0.000
1400	28.103	56.767	36.728	28.055	0.000	0.000	0.000
1500	28.462	58.719	38.130	30.883	0.000	0.000	0.000
78. SILICON Si (g)							
298.15	22.251	167.980	167.980	0.000	450.000	405.525	−71.045
300	22.234	168.117	167.980	0.041	450.004	405.249	−70.559
400	21.613	174.416	168.843	2.229	450.070	390.312	−50.969
500	21.316	179.204	170.456	4.374	449.913	375.388	−39.216
600	21.153	183.074	172.246	6.497	449.630	360.508	−31.385
700	21.057	186.327	174.032	8.607	449.259	345.682	−25.795
800	21.000	189.135	175.748	10.709	448.821	330.915	−21.606
900	20.971	191.606	177.375	12.808	448.329	316.205	−18.352
1000	20.968	193.815	178.911	14.904	447.791	301.553	−15.751
1100	20.989	195.815	180.358	17.002	447.211	286.957	−13.626
1200	21.033	197.643	181.723	19.103	446.595	272.416	−11.858
1300	21.099	199.329	183.014	21.209	445.946	257.927	−10.364
1400	21.183	200.895	184.236	23.323	445.268	243.489	−9.085
1500	21.282	202.360	185.396	25.446	444.563	229.101	−7.978
79. SILICON DIOXIDE SiO_2 (cr)							
298.15	44.602	41.460	41.460	0.000	−910.700	−856.288	150.016
300	44.712	41.736	41.461	0.083	−910.708	−855.951	149.032
400	53.477	55.744	43.311	4.973	−910.912	−837.651	109.385
500	60.533	68.505	47.094	10.705	−910.540	−819.369	85.598
600	64.452	79.919	51.633	16.971	−909.841	−801.197	69.749
700	68.234	90.114	56.414	23.590	−908.958	−783.157	58.439
800	76.224	99.674	61.226	30.758	−907.668	−765.265	49.966
848	82.967	104.298	63.533	34.569	−906.310	−756.747	46.613
PHASE TRANSITION: $\Delta_{trs}H$ = 0.411 kJ/mol, $\Delta_{trs}S$ = 0.484 J/K·mol, crII–crII′							
848	67.446	104.782	63.532	34.980	−906.310	−756.747	46.613
900	67.953	108.811	66.033	38.500	−905.922	−747.587	43.388
1000	68.941	116.021	70.676	45.345	−905.176	−730.034	38.133
1100	69.940	122.639	75.104	52.289	−904.420	−712.557	33.836
1200	70.947	128.768	79.323	59.333	−901.382	−695.148	30.259
PHASE TRANSITION: $\Delta_{trs}H$ = 2.261 kJ/mol, $\Delta_{trs}S$ = 1.883 J/K·mol, crII′–crI							

	J/K·mol			kJ/mol			
T/K	C_p°	S°	$-(G^\circ - H^\circ(T_r))/T$	$H^\circ - H^\circ(T_r)$	$\Delta_f H^\circ$	$\Delta_f G^\circ$	Log K_f
1200	71.199	130.651	79.323	61.594	−901.382	−695.148	30.259
1300	71.743	136.372	83.494	68.742	−900.574	−677.994	27.242
1400	72.249	141.707	87.463	75.941	−899.782	−660.903	24.658
1500	72.739	146.709	91.248	83.191	−899.004	−643.867	22.421
80. SILICON TETRACHLORIDE $SiCl_4$ (g)							
298.15	90.404	331.446	331.446	0.000	−662.200	−622.390	109.039
300	90.562	332.006	331.448	0.167	−662.195	−622.143	108.323
400	96.893	359.019	335.088	9.572	−661.853	−608.841	79.505
500	100.449	381.058	342.147	19.456	−661.413	−595.637	62.225
600	102.587	399.576	350.216	29.616	−660.924	−582.527	50.713
700	103.954	415.500	358.432	39.948	−660.417	−569.501	42.496
800	104.875	429.445	366.455	50.392	−659.912	−556.548	36.338
900	105.523	441.837	374.155	60.914	−659.422	−543.657	31.553
1000	105.995	452.981	381.490	71.491	−658.954	−530.819	27.727
1100	106.349	463.101	388.456	82.109	−658.515	−518.027	24.599
1200	106.620	472.366	395.068	92.758	−658.107	−505.274	21.994
1300	106.834	480.909	401.347	103.431	−657.735	−492.553	19.791
1400	107.003	488.833	407.316	114.123	−657.400	−479.860	17.904
1500	107.141	496.220	413.000	124.830	−657.104	−467.189	16.269

THERMODYNAMIC PROPERTIES OF AQUEOUS IONS

This table contains standard state thermodynamic properties of positive and negative ions in aqueous solution. It includes enthalpy and Gibbs energy of formation, entropy, and heat capacity, and thus serves as a companion to the preceding table, "Standard Thermodynamic Properties of Chemical Substances". The standard state is the hypothetical ideal solution with molality $m = 1$ mol/kg (mean ionic molality $m_{\pm}$ in the case of a species which is assumed to dissociate at infinite dilution). Further details on conventions may be found in Reference 1.

All values refer to standard conditions of 25 °C and 100 kPa pressure.

References

1. Wagman, D. D., Evans, W. H., Parker, V. B., Schumm, R. H., Halow, I., Bailey, S. M., Churney, K. L., and Nuttall, R. L., *The NBS Tables of Chemical Thermodynamic Properties, J. Phys. Chem. Ref. Data*, Vol. 11, Suppl. 2, 1982.
2. Zemaitis, J. F., Clark, D. M., Rafal, M., and Scrivner, N. C., *Handbook of Aqueous Electrolyte Thermodynamics*, American Institute of Chemical Engineers, New York, 1986.

Species	$\Delta_f H°$/ kJ mol⁻¹	$\Delta_f G°$/ kJ mol⁻¹	$S°$/ J mol⁻¹K⁻¹	C_p/ J mol⁻¹K⁻¹
Cations				
Ag^{+}	105.6	77.1	72.7	21.8
Al^{+3}	−531.0	−485.0	−321.7	
$AlOH^{+2}$		−694.1		
Ba^{+2}	−537.6	−560.8	9.6	
$BaOH^{+}$		−730.5		
Be^{+2}	−382.8	−379.7	−129.7	
Bi^{+3}		82.8		
$BiOH^{+2}$		−146.4		
Ca^{+2}	−542.8	−553.6	−53.1	
$CaOH^{+}$		−718.4		
Cd^{+2}	−75.9	−77.6	−73.2	
$CdOH^{+}$		−261.1		
Ce^{+3}	−696.2	−672.0	−205.0	
Ce^{+4}	−537.2	−503.8	−301.0	
Co^{+2}	−58.2	−54.4	−113.0	
Co^{+3}	92.0	134.0	−305.0	
Cr^{+2}	−143.5			
Cs^{+}	−258.3	−292.0	133.1	−10.5
Cu^{+}	71.7	50.0	40.6	
Cu^{+2}	64.8	65.5	−99.6	
Dy^{+3}	−699.0	−665.0	−231.0	21.0
Er^{+3}	−705.4	−669.1	−244.3	21.0
Eu^{+2}	−527.0	−540.2	−8.0	
Eu^{+3}	−605.0	−574.1	−222.0	8.0
Fe^{+2}	−89.1	−78.9	−137.7	
Fe^{+3}	−48.5	−4.7	−315.9	
$FeOH^{+}$	−324.7	−277.4	−29.0	
$FeOH^{+2}$	−290.8	−229.4	−142.0	
$Fe(OH)_2^{+}$		−438.0		
Ga^{+2}		−88.0		
Ga^{+3}	−211.7	−159.0	−331.0	
$GaOH^{+2}$		−380.3		
$Ga(OH)_2^{+}$		−597.4		
Gd^{+3}	−686.0	−661.0	−205.9	
H^{+}	0	0	0	0
Hg^{+2}	171.1	164.4	−32.2	
Hg_2^{+2}	172.4	153.5	84.5	
$HgOH^{+}$	−84.5	−52.3	71.0	
Ho^{+3}	−705.0	−673.7	−226.8	17.0
In^{+}		−12.1		
In^{+2}		−50.7		
In^{+3}	−105.0	−98.0	−151.0	
$InOH^{+2}$	−370.3	−313.0	−88.0	
$In(OH)_2^{+}$	−619.0	−525.0	25.0	
K^{+}	−252.4	−283.3	102.5	21.8
La^{+3}	−707.1	−683.7	−217.6	−13.0
Li^{+}	−278.5	−293.3	13.4	68.6
Lu^{+3}	−665.0	−628.0	−264.0	25.0
LuF^{+2}		−931.4		
Mg^{+2}	−466.9	−454.8	−138.1	
$MgOH^{+}$		−626.7		
Mn^{+2}	−220.8	−228.1	−73.6	50.0
$MnOH^{+}$	−450.6	−405.0	−17.0	
NH_4^{+}	−132.5	−79.3	113.4	79.9
$N_2H_5^{+}$	−7.5	82.5	151.0	70.3
Na^{+}	−240.1	−261.9	59.0	46.4
Nd^{+3}	−696.2	−671.6	−206.7	−21.0
Ni^{+2}	−54.0	−45.6	−128.9	
$NiOH^{+}$	−287.9	−227.6	−71.0	
PH_4^{+}		92.1		
Pa^{+4}	−619.0			
Pb^{+2}	−1.7	−24.4	10.5	
$PbOH^{+}$		−226.3		
Pd^{+2}	149.0	176.5	−184.0	
Po^{+2}		71.0		
Po^{+4}		293.0		
Pr^{+3}	−704.6	−679.1	−209.0	−29.0
Pt^{+2}		254.8		
Ra^{+2}	−527.6	−561.5	54.0	
Rb^{+}	−251.2	−284.0	121.5	
Re^{+}		−33.0		
Sc^{+3}	−614.2	−586.6	−255.0	
$ScOH^{+2}$	−861.5	−801.2	−134.0	
Sm^{+2}		−497.5		
Sm^{+3}	−691.6	−666.6	−211.7	−21.0
Sn^{+2}	−8.8	−27.2	−17.0	
$SnOH^{+}$	−286.2	−254.8	50.0	
Sr^{+2}	−545.8	−559.5	−32.6	
$SrOH^{+}$		−721.3		
Tb^{+3}	−682.8	−651.9	−226.0	17.0
$Te(OH)_3^{+}$	−608.4	−496.1	111.7	
Th^{+4}	−769.0	−705.1	−422.6	
$Th(OH)^{+3}$	−1030.1	−920.5	−343.0	
$Th(OH)_2^{+2}$	−1282.4	−1140.9	−218.0	
Tl^{+}	5.4	−32.4	125.5	

Species	$\Delta_f H°$/ kJ mol⁻¹	$\Delta_f G°$/ kJ mol⁻¹	$S°$/ J mol⁻¹K⁻¹	C_p/ J mol⁻¹K⁻¹
Tl^{+3}	196.6	214.6	−192.0	
$TlOH^{+2}$		−15.9		
$Tl(OH)_2^+$		−244.7		
Tm^{+3}	−697.9	−662.0	−243.0	25.0
U^{+3}	−489.1	−476.2	−188.0	
U^{+4}	−591.2	−531.9	−410.0	
Y^{+3}	−723.4	−693.8	−251.0	
$Y_2(OH)_2^{+4}$		−1780.3		
Yb^{+2}		−527.0		
Yb^{+3}	−674.5	−644.0	−238.0	25.0
$Y(OH)^{+2}$		−879.1		
Zn^{+2}	−153.9	−147.1	−112.1	46.0
$ZnOH^+$		−330.1		
Anions				
AlO_2^-	−930.9	−830.9	−36.8	
$Al(OH)_4^-$	−1502.5	−1305.3	102.9	
AsO_2^-	−429.0	−350.0	40.6	
AsO_4^{-3}	−888.1	−648.4	−162.8	
BF_4^-	−1574.9	−1486.9	180.0	
BH_4^-	48.2	114.4	110.5	
BO_2^-	−772.4	−678.9	−37.2	
$B_4O_7^{-2}$		−2604.8		
BeO_2^{-2}	−790.8	−640.1	−159.0	
Br^-	−121.6	−104.0	82.4	−141.8
BrO^-	−94.1	−33.4	42.0	
BrO_3^-	−67.1	18.6	161.7	
BrO_4^-	13.0	118.1	199.6	
$CHOO^-$	−425.6	−351.0	92.0	−87.9
CH_3COO^-	−486.0	−369.3	86.6	−6.3
$C_2O_4^{-2}$	−825.1	−673.9	45.6	
$C_2O_4H^-$	−818.4	−698.3	149.4	
Cl^-	−167.2	−131.2	56.5	−136.4
ClO^-	−107.1	−36.8	42.0	
ClO_2^-	−66.5	17.2	101.3	
ClO_3^-	−104.0	−8.0	162.3	
ClO_4^-	−129.3	−8.5	182.0	
CN^-	150.6	172.4	94.1	
CO_3^{-2}	−677.1	−527.8	−56.9	
CrO_4^{-2}	−881.2	−727.8	50.2	
$Cr_2O_7^{-2}$	−1490.3	−1301.1	261.9	
F^-	−332.6	−278.8	−13.8	−106.7
$Fe(CN)_6^{-3}$	561.9	729.4	270.3	
$Fe(CN)_6^{-4}$	455.6	695.1	95.0	
$HB_4O_7^-$		−2685.1		
HCO_3^-	−692.0	−586.8	91.2	

Species	$\Delta_f H°$/ kJ mol⁻¹	$\Delta_f G°$/ kJ mol⁻¹	$S°$/ J mol⁻¹K⁻¹	C_p/ J mol⁻¹K⁻¹
HF_2^-	−649.9	−578.1	92.5	
HPO_3F^-		−1198.2		
HPO_4^{-2}	−1292.1	−1089.2	−33.5	
$HP_2O_7^{-3}$	−2274.8	−1972.2	46.0	
HS^-	−17.6	12.1	62.8	
HSO_3^-	−626.2	−527.7	139.7	
HSO_4^-	−887.3	−755.9	131.8	−84.0
$HS_2O_4^-$		−614.5		
HSe^-	15.9	44.0	79.0	
$HSeO_3^-$	−514.6	−411.5	135.1	
$HSeO_4^-$	−581.6	−452.2	149.4	
$H_2AsO_3^-$	−714.8	−587.1	110.5	
$H_2AsO_4^-$	−909.6	−753.2	117.0	
$H_2PO_4^-$	−1296.3	−1130.2	90.4	
$H_2P_2O_7^{-2}$	−2278.6	−2010.2	163.0	
I^-	−55.2	−51.6	111.3	−142.3
IO^-	−107.5	−38.5	−5.4	
IO_3^-	−221.3	−128.0	118.4	
IO_4^-	−151.5	−58.5	222.0	
MnO_4^-	−541.4	−447.2	191.2	−82.0
MnO_4^{-2}	−653.0	−500.7	59.0	
MoO_4^{-2}	−997.9	−836.3	27.2	
NO_2^-	−104.6	−32.2	123.0	−97.5
NO_3^-	−207.4	−111.3	146.4	−86.6
N_3^-	275.1	348.2	107.9	
OCN^-	−146.0	−97.4	106.7	
OH^-	−230.0	−157.2	−10.8	−148.5
PO_4^{-3}	−1277.4	−1018.7	−220.5	
$P_2O_7^{-4}$	−2271.1	−1919.0	−117.0	
Re^-	46.0	10.1	230.0	
S^{-2}	33.1	85.8	−14.6	
SCN^-	76.4	92.7	144.3	−40.2
SO_3^{-2}	−635.5	−486.5	−29.0	
SO_4^{-2}	−909.3	−744.5	20.1	−293.0
S_2^{-2}	30.1	79.5	28.5	
$S_2O_3^{-2}$	−652.3	−522.5	67.0	
$S_2O_4^{-2}$	−753.5	−600.3	92.0	
$S_2O_8^{-2}$	−1344.7	−1114.9	244.3	
Se^{-2}		129.3		
SeO_3^{-2}	−509.2	−369.8	13.0	
SeO_4^{-2}	−599.1	−441.3	54.0	
VO_3^-	−888.3	−783.6	50.0	
VO_4^{-3}		−899.0		
WO_4^{-2}	−1075.7			

HEAT OF COMBUSTION

The heat of combustion of a substance at 25 °C can be calculated from the enthalpy of formation ($\Delta_f H°$) data in the table "Standard Thermodynamic Properties of Chemical Substances" in this section. We can write the general combustion reaction as

$$X + O_2 \rightarrow CO_2(g) + H_2O\ (l) + \text{other products}$$

For a compound containing only carbon, hydrogen, and oxygen, the reaction is simply

$$C_aH_bO_c + \left(a + \frac{1}{4}b - \frac{1}{2}c\right)O_2 \rightarrow a\ CO_2\ (g) + \frac{1}{2}b\ H_2O\ (l)$$

and the standard heat of combustion $\Delta_c H°$, which is defined as the negative of the enthalpy change for the reaction (i.e., the heat released in the combustion process), is given by

$$\Delta_c H° = -a\Delta_f H°(CO_2, g) - \frac{1}{2}b\Delta_f H°(H_2O, l) + \Delta_f H°(C_aH_bO_c)$$

$$= 393.51a + 142.915b + \Delta_f H°(C_aH_bO_c)$$

This equation applies if the reactants start in their standard states (25 °C and one atmosphere pressure) and the products return to the same conditions. The same equation applies to a compound containing another element if that element ends in its standard reference state (e.g., nitrogen, if the product is N_2); in general, however, the exact products containing the other elements must be known in order to calculate the heat of combustion.

The following table gives the standard heat of combustion calculated in this manner for a few representative substances.

Molecular formula	Name	$\Delta_c H°$/kJ mol^{-1}
Inorganic substances		
C	Carbon (graphite)	393.5
CO	Carbon monoxide (g)	283.0
H_2	Hydrogen (g)	285.8
H_3N	Ammonia (g)	382.8
H_4N_2	Hydrazine (g)	667.1
N_2O	Nitrous oxide (g)	82.1
Hydrocarbons		
CH_4	Methane (g)	890.8
C_2H_2	Acetylene (g)	1301.1
C_2H_4	Ethylene (g)	1411.2
C_2H_6	Ethane (g)	1560.7
C_3H_6	Propylene (g)	2058.0
C_3H_6	Cyclopropane (g)	2091.3
C_3H_8	Propane (g)	2219.2
C_4H_6	1,3-Butadiene (g)	2541.5
C_4H_{10}	Butane (g)	2877.6
C_5H_{12}	Pentane (l)	3509.0
C_6H_6	Benzene (l)	3267.6
C_6H_{12}	Cyclohexane (l)	3919.6
C_6H_{14}	Hexane (l)	4163.2
C_7H_8	Toluene (l)	3910.3
C_7H_{16}	Heptane (l)	4817.0
$C_{10}H_8$	Naphthalene (s)	5156.3
Alcohols and ethers		
CH_4O	Methanol (l)	726.1
C_2H_6O	Ethanol (l)	1366.8
C_2H_6O	Dimethyl ether (g)	1460.4
$C_2H_6O_2$	Ethylene glycol (l)	1189.2
C_3H_8O	1-Propanol (l)	2021.3
$C_3H_8O_3$	Glycerol (l)	1655.4
$C_4H_{10}O$	Diethyl ether (l)	2723.9
$C_5H_{12}O$	1-Pentanol (l)	3330.9
C_6H_6O	Phenol (s)	3053.5
Carbonyl compounds		
CH_2O	Formaldehyde (g)	570.7
C_2H_2O	Ketene (g)	1025.4
C_2H_4O	Acetaldehyde (l)	1166.9
C_3H_6O	Acetone (l)	1789.9
C_3H_6O	Propanal (l)	1822.7
C_4H_8O	2-Butanone (l)	2444.1
Acids and esters		
CH_2O_2	Formic acid (l)	254.6
$C_2H_4O_2$	Acetic acid (l)	874.2
$C_2H_4O_2$	Methyl formate (l)	972.6
$C_3H_6O_2$	Methyl acetate (l)	1592.2
$C_4H_8O_2$	Ethyl acetate (l)	2238.1
$C_6H_5NO_2$	Nicotinic acid (s)	2731.1
$C_7H_6O_2$	Benzoic acid (s)	3228.2
Nitrogen compounds		
CHN	Hydrogen cyanide (g)	671.5
CH_3NO_2	Nitromethane (l)	709.2
CH_4N_2O	Urea (s)	632.7
CH_5N	Methylamine (g)	1085.6
C_2H_3N	Acetonitrile (l)	1247.2
C_2H_5NO	Acetamide (s)	1184.6
C_3H_9N	Trimethylamine (g)	2443.1
C_5H_5N	Pyridine (l)	2782.3
C_6H_7N	Aniline (l)	3392.8

ENERGY CONTENT OF FUELS

Several fuels are compared in this table with respect to their energy content per unit mass and the amount of CO_2 released per unit of available energy. The energy content is taken to be the negative of the standard enthalpy of combustion (see the table "Heat of Combustion" in this section for more details). The energy is assumed to be released by combustion with oxygen at normal atmospheric pressure, with products of gaseous CO_2 and liquid H_2O at room temperature. This quantity is often called the "gross heat of combustion" to distinguish it from the "net heat of combustion," for which the water remains in the gas state. The latter quantity is typically 5% to 10% less than the values given here.

The energy content is given both in SI units of MJ/kg and conventional units of BTU/lb. Values for the fossil fuels and other materials are typical; individual samples show wide variations.

The last column gives the grams of carbon released as carbon dioxide per megajoule of energy. Examination of the table shows that the minimum CO_2 release occurs for fuels that have a high ratio of hydrogen to carbon. Furthermore, fuels containing oxygen have a lower energy content and higher CO_2 release than hydrocarbons with the same number of carbon atoms.

References

1. Domalski, E. S., Jobe, T. L., and Milne, T. A., *Thermodynamic Data for Biomass Conversion and Waste Incineration,* SERI/SP-271-2839, Solar Technical Information Program, U. S. Department of Energy, September 1986; see also NBSIR 78-1479, National Bureau of Standards, August 1978.
2. Green, D. W., and Ackers, D. E, *Perry's Chemical Engineers' Handbook, Eighth Edition,* McGraw-Hill, New York, 2007.
3. *Transportation Energy Data Book,* U. S. Department of Energy, May 2007, http://cta.ornl.gov/data/index.shtml.
4. *Chemical Composition of Natural Gas,* Union Gas Limited, http://www.uniongas.com/aboutus/aboutng/composition.asp.

Fuel	Energy content		g of C
	MJ/kg	10^3 BTU/lb	per MJ
Pure compounds			
Hydrogen	141.8	61.0	0.0
Methane	55.5	23.9	13.5
Ethane	51.9	22.3	15.4
Propane	50.3	21.7	16.2
Hexane	48.3	20.8	17.3
Heptane	48.1	20.7	17.5
Octane	47.9	20.6	17.6
Methanol	22.7	9.7	16.5
Ethanol	29.7	12.8	17.6
1-Propanol	33.6	14.5	17.8
1-Butanol	36.1	15.5	18.0
1-Octanol	40.7	17.5	18.1
Methyl *tert*-butyl ether	38.2	16.4	17.8

Fuel	Energy content		g of C
	MJ/kg	10^3 BTU/lb	per MJ
Fossil fuels			
Natural gas[a]	54.0	23.2	13.9
Gasoline	46.5	20.0	17.6
Kerosene	46.4	20.0	18.5
Fuel oil	40.9	17.6	21.3
Coal, high bituminous	36.3	15.6	23.5
Coal, low bituminous	28.9	12.4	26.3
Coal, anthracite	34.6	14.9	27.3
Other materials			
Wood, oak	18.9	8.1	25.3
Wood, locust	19.7	8.5	25.7
Wood, Ponderosa pine	20.0	8.6	24.6
Wood, redwood	20.7	8.9	24.4
Charcoal, wood	34.7	14.9	26.8
Newsprint	18.6	8.0	26.5
Cellulose	17.3	7.5	25.6
Grass (lawn clippings)	19.3	8.3	24.9

[a] Assumed to be 95% methane, 2.5% ethane, and 2.5% inert compounds; however, the actual composition varies widely (see Reference 4).

ELECTRICAL CONDUCTIVITY OF WATER

This table gives the electrical conductivity of highly purified water over a range of temperature and pressure. The first column of conductivity data refers to water at its own vapor pressure. Equations for calculating the conductivity at any temperature and pressure may be found in the reference.

Reference

Marshall, W. L., *J. Chem. Eng. Data* 32, 221, 1987.

Conductivity in μS/cm at the Indicated Pressure

t/°C	Sat. vapor	50 MPa	100 MPa	200 MPa	400 MPa	600 MPa
0	0.0115	0.0150	0.0189	0.0275	0.0458	0.0667
25	0.0550	0.0686	0.0836	0.117	0.194	0.291
100	0.765	0.942	1.13	1.53	2.45	3.51
200	2.99	4.08	5.22	7.65	13.1	19.5
300	2.41	4.87	7.80	14.1	28.9	46.5
400		1.17	4.91	14.3	39.2	71.3
600			0.134	4.65	33.8	85.7

ELECTRICAL CONDUCTIVITY OF AQUEOUS SOLUTIONS

The following table gives the electrical conductivity of aqueous solutions of some acids, bases, and salts as a function of concentration. All values refer to 20 °C. The conductivity κ (often called specific conductance in older literature) is the reciprocal of the resistivity. The molar conductivity Λ is related to this by $\Lambda = \kappa/c$, where c is the amount-of-substance concentration of the electrolyte. Thus if κ has units of millisiemens per centimeter (mS/cm), as in this table, and c is expressed in mol/L, then Λ has units of S cm² mol⁻¹. For these electrolytes the concentration c corresponding to the mass percent values given here can be found in the table "Concentrative Properties of Aqueous Solutions" in Section 8.

References

1. *CRC Handbook of Chemistry, and Physics, 70th Edition*, Weast, R. C., Ed., CRC Press, Boca Raton, FL, 1989, p. D-221.
2. Wolf, A. V., *Aqueous Solutions and Body Fluids*, Harper and Row, New York, 1966.

Electrical Conductivity κ in mS/cm for the Indicated Concentration in Mass Percent

Name	Formula	0.5%	1%	2%	5%	10%	15%	20%	25%	30%	40%	50%
Acetic acid	CH_3COOH	0.3	0.6	0.8	1.2	1.5	1.7	1.7	1.6	1.4	1.1	0.8
Ammonia	NH_3	0.5	0.7	1.0	1.1	1.0	0.7	0.5	0.4			
Ammonium chloride	NH_4Cl	10.5	20.4	40.3	95.3	180						
Ammonium sulfate	$(NH_4)_2SO_4$	7.4	14.2	25.7	57.4	105	147	185	215			
Barium chloride	$BaCl_2$	4.7	9.1	17.4	40.4	76.7	109.0	137.0				
Calcium chloride	$CaCl_2$	8.1	15.7	29.4	67.0	117	157	177	183	172	106	
Cesium chloride	$CsCl$	3.8	7.4	13.8	32.9	65.8	102	142				
Citric acid	$H_3C(OH)(COO)_3$	1.2	2.1	3.0	4.7	6.2	7.0	7.2	7.1			
Copper(II) sulfate	$CuSO_4$	2.9	5.4	9.3	19.0	32.2	42.3					
Formic acid	$HCOOH$	1.4	2.4	3.5	5.6	7.8	9.0	9.9	10.4	10.5	9.9	8.6
Hydrogen chloride	HCl	45.1	92.9	183								
Lithium chloride	$LiCl$	10.1	19.0	34.9	76.4	127	155	170	165	146		
Magnesium chloride	$MgCl_2$	8.6	16.6	31.2	66.9	108	129	134	122	98		
Magnesium sulfate	$MgSO_4$	4.1	7.6	13.3	27.4	42.7	54.2	51.1	44.1			
Manganese(II) sulfate	$MnSO_4$		6.2	10.6	21.6	34.5	43.7	47.6				
Nitric acid	HNO_3	28.4	56.1	108								
Oxalic acid	$H_2C_2O_4$	14.0	21.8	35.3	65.6							
Phosphoric acid	H_3PO_4	5.5	10.1	16.2	31.5	59.4	88.4	118	146	173	209	
Potassium bromide	KBr	5.2	10.2	19.5	47.7	95.6	144	194				
Potassium carbonate	K_2CO_3	7.0	13.6	25.4	58.0	109	152	188	223			
Potassium chloride	KCl	8.2	15.7	29.5	71.9	143	208					
Potassium dihydrogen phosphate	KH_2PO_4	3.0	5.9	11.0	25.0	44.6						
Potassium hydrogen carbonate	$KHCO_3$	4.6	8.9	17.0	38.8	72.4	101	128				
Potassium hydrogen phosphate	K_2HPO_4	5.2	9.9	18.3	40.3							
Potassium hydroxide	KOH	20.0	38.5	75.0	178							
Potassium iodide	KI	3.8	7.5	14.2	35.2	71.8	110	188	224			
Potassium nitrate	KNO_3	5.5	10.7	20.1	47.0	87.3	124	157	182			
Potassium permanganate	$KMnO_4$	3.5	6.9	13.0	30.5							
Potassium sulfate	K_2SO_4	5.8	11.2	21.0	48.0	88.6						
Silver(I) nitrate	$AgNO_3$	3.1	6.1	12.0	26.7	49.8	72.0	92.8	112	129	162	
Sodium acetate	$NaCH_3COO$	3.9	7.6	14.4	30.9	53.4	64.1	69.3	69.2	64.3		
Sodium bromide	$NaBr$	5.0	9.7	18.4	44.0	84.6	122	157	191	216		
Sodium carbonate	Na_2CO_3	7.0	13.1	23.3	47.0	74.4	88.6					
Sodium chloride	$NaCl$	8.2	16.0	30.2	70.1	126	171	204	222			
Sodium citrate	$Na_3C_6H_5O_7$		7.4	12.8	26.2	42.1	52.0	57.1	57.3	53.5		
Sodium dihydrogen phosphate	NaH_2PO_4	2.2	4.4	9.1	21.0	33.2	43.3	49.6	53.1	54.0	46.1	
Sodium hydrogen carbonate	$NaHCO_3$	4.2	8.2	15.0	31.4							
Sodium hydrogen phosphate	Na_2HPO_4	4.6	8.7	15.6	31.4							
Sodium hydroxide	$NaOH$	24.8	48.6	93.1	206							
Sodium nitrate	$NaNO_3$	5.4	10.6	20.4	46.2	82.6	111	134	152	165	178	
Sodium phosphate	Na_3PO_4	7.3	14.1	22.7	43.5							
Sodium sulfate	Na_2SO_4	5.9	11.2	19.8	42.7	71.3	91.1	109				
Sodium thiosulfate	$Na_2S_2O_3$	5.7	10.7	19.5	43.3	76.7	104	123	134	136	118	
Strontium chloride	$SrCl_2$	5.9	11.4	22.0	49.1	91.5	127	153	168	178		
Sulfuric acid	H_2SO_4	24.3	47.8	92	211							
Trichloroacetic acid	CCl_3COOH	10.3	19.6	37.2	84.7	148	193	221				
Zinc sulfate	$ZnSO_4$	2.8	5.4	10.0	20.5	33.7	43.3					

STANDARD KCl SOLUTIONS FOR CALIBRATING CONDUCTIVITY CELLS

This table presents recommended electrolytic conductivity (κ) values for aqueous potassium chloride solutions with molalities of 0.01 mol/kg, 0.1 mol/kg and 1.0 mol/kg at temperatures from 0 °C to 50 °C. The values, which are based on measurements at the National Institute of Standards and Technology, provide primary standards for the calibration of conductivity cells. The measurements at 0.01 and 0.1 molal are described in Reference 1, while those at 1.0 molal are in Reference 2. Temperatures are given on the ITS-90 scale. The uncertainty in the conductivity is about 0.03% for the 0.01 molal values and about 0.04% for the 0.1 and 1.0 molal values. The conductivity of water saturated with atmospheric CO_2 is given in the last column. These values were subtracted from the original measurements to give the values in the second, third, and fourth columns. All κ values are given in units of 10^{-4} S/m (numerically equal to μS/cm).

The assistance of Kenneth W. Pratt is appreciated.

References

1. Wu, Y. C., Koch, W. F., and Pratt, K. W., *J. Res. Natl. Inst. Stand. Technol.* 96, 191, 1991.
2. Wu, Y. C., Koch, W. F., Feng, D., Holland, L. A., Juhasz, E., Arvay, E., and Tomek, A., *J. Res. Natl. Inst. Stand. Technol.* 99, 241, 1994.
3. Pratt, K. W., Koch, W. F., Wu, Y. C., and Berezansky, P. A., *Pure Appl. Chem.* 73, 1783, 2001.

	10^4 κ/S m^{-1}			
t/°C	0.01 m KCl	0.1 m KCl	1.0 m KCl	H_2O (CO_2 sat.)
0	772.92	7 116.85	63 488	0.58
5	890.96	8 183.70	72 030	0.68
10	1 013.95	9 291.72	80 844	0.79
15	1 141.45	10 437.1	89 900	0.89
18	1 219.93	11 140.6	—	0.95
20	1 273.03	11 615.9	99 170	0.99
25	1 408.23	12 824.6	108 620	1.10
30	1 546.63	14 059.2	118 240	1.20
35	1 687.79	15 316.0	127 970	1.30
40	1 831.27	16 591.0	137 810	1.40
45	1 976.62	17 880.6	147 720	1.51
50	2 123.43	19 180.9	157 670	1.61

MOLAR CONDUCTIVITY OF AQUEOUS HF, HCl, HBr, AND HI

The molar conductivity Λ of an electrolyte solution is defined as the conductivity divided by amount-of-substance concentration. The customary unit is S cm² mol⁻¹ (i.e., Ω^{-1} cm² mol⁻¹). The first part of this table gives the molar conductivity of the hydrohalogen acids at 25 °C as a function of the concentration in mol/L. The second part gives the temperature dependence of Λ for HCl and HBr. More extensive tables and mathematical representations may be found in the reference.

Reference

Hamer, W. J., and DeWane, H. J., *Electrolytic Conductance and the Conductances of the Hydrohalogen Acids in Water*, Natl. Stand. Ref. Data Sys.- Natl. Bur. Standards (U.S.), No. 33, 1970.

c/mol L⁻¹	HF	HCl	HBr	HI
Inf. dil.	405.1	426.1	427.7	426.4
0.0001		424.5	425.9	424.6
0.0005		422.6	424.3	423.0
0.001		421.2	422.9	421.7
0.005	128.1	415.7	417.6	416.4
0.01	96.1	411.9	413.7	412.8
0.05	50.1	398.9	400.4	400.8
0.10	39.1	391.1	391.9	394.0
0.5	26.3	360.7	361.9	369.8
1.0	24.3	332.2	334.5	343.9
1.5		305.8	307.6	316.4
2.0		281.4	281.7	288.9
2.5		258.9	257.8	262.5
3.0		237.6	236.8	237.9
3.5		218.3	217.5	215.4
4.0		200.0	199.4	195.1
4.5		183.1	182.4	176.8
5.0		167.4	166.5	160.4
5.5		152.9	151.8	145.5
6.0		139.7	138.2	131.7
6.5		127.7	125.7	118.6
7.0		116.9	114.2	105.7
7.5		107.0	103.8	
8.0		98.2	94.4	
8.5		90.3	85.8	
9.0		83.1		
9.5		76.6		
10.0		70.7		

c/ mol L⁻¹	−20 °C	−10 °C	0 °C	10 °C	20 °C	30 °C	40 °C	50 °C
HCl								
0.5			228.7	283.0	336.4	386.8	436.9	482.4
1.0			211.7	261.6	312.2	359.0	402.9	445.3
1.5			196.2	241.5	287.5	331.1	371.6	410.8
2.0			182.0	222.7	262.9	303.3	342.4	378.2
2.5		131.7	168.5	205.1	239.8	277.0	315.2	347.6
3.0		120.8	154.6	188.5	219.3	253.3	289.3	319.0
3.5	85.5	111.3	139.6	172.2	201.6	232.9	263.9	292.1
4.0	79.3	102.7	129.2	158.1	185.6	214.2	242.2	268.2
4.5	73.7	94.9	119.5	145.4	170.6	196.6	222.5	246.7
5.0	68.5	87.8	110.3	133.5	156.6	180.2	204.1	226.5
5.5	63.6	81.1	101.7	122.5	143.6	165.0	187.1	207.7
6.0	58.9	74.9	93.7	112.3	131.5	151.0	171.3	190.3
6.5	54.4	69.1	86.2	103.0	120.4	138.2	156.9	174.3
7.0	50.2	63.7	79.3	94.4	110.2	126.4	143.3	159.7
7.5	46.3	58.6	73.0	86.5	100.9	115.7	131.6	146.2
8.0	42.7	54.0	67.1	79.4	92.4	106.1	120.6	134.0
8.5	39.4	49.8	61.7	72.9	84.7	97.3	110.7	123.0
9.0	36.4	45.9	56.8	67.1	77.8	89.4	101.7	112.9
9.5	33.6	42.3	52.3	61.8	71.5	82.3	93.6	103.9
10.0	31.2	39.1	48.2	57.0	65.8	75.9	86.3	95.7
10.5	28.9	36.1	44.5	52.7	60.7	70.1	79.6	88.4
11.0	26.8	33.4	41.1	48.8	56.1	64.9	73.6	81.7
11.5	24.9	31.0	38.0	45.3	51.9	60.1	68.0	75.6
12.0	23.1	28.7	35.3	42.0	48.0	55.6	62.8	70.0
12.5	21.4	26.7	32.7	39.0	44.4	51.4	57.9	64.8

c/ mol L⁻¹	−20 °C	−10 °C	0 °C	10 °C	20 °C	30 °C	40 °C	50 °C
HBr								
0.5			240.9	295.9	347.0	398.9	453.6	496.8
1.0			229.6	276.0	329.0	380.4	418.6	465.2
1.5			209.5	254.9	298.9	340.6	381.8	421.4
2.0		150.8	188.6	231.3	271.8	314.1	350.5	387.4
2.5		136.8	171.7	208.3	244.8	281.7	316.0	349.1
3.0		125.7	157.2	189.5	222.2	255.0	287.8	318.6
3.5		116.1	144.1	174.6	203.2	234.4	263.7	291.9
4.0	84.0	107.5	132.3	160.2	186.8	214.2	239.7	266.9
4.5	78.0	99.0	123.0	146.4	171.2	195.1	218.8	242.6
5.0	72.3	91.4	112.6	134.0	155.7	178.2	199.6	221.3
5.5	67.0	84.2	103.1	122.7	142.1	162.8	181.4	201.8
6.0	61.8	77.2	94.3	112.0	129.6	148.0	165.4	183.4
6.5	56.8	70.7	86.0	102.0	118.0	134.1	150.5	166.3
7.0	51.9	64.6	78.4	92.6	107.1	121.4	136.3	150.8

EQUIVALENT CONDUCTIVITY OF ELECTROLYTES IN AQUEOUS SOLUTION

Petr Vanýsek

This table gives the equivalent (molar) conductivity Λ at 25 °C for some common electrolytes in aqueous solution at concentrations up to 0.1 mol/L. The units of Λ are 10^{-4} m² S mol⁻¹.

For very dilute solutions, the equivalent conductivity for any electrolyte of concentration c can be approximately calculated using the Debye–Hückel–Onsager equation, which can be written for a symmetrical (equal charge on cation and anion) electrolyte as

$$\Lambda = \Lambda^\circ - (A + B\Lambda^\circ)c^{1/2}$$

For a solution at 25 °C and both cation and anion with charge |1|, the constants are A = 60.20 and B = 0.229. Λ° can be found from the next table, "Ionic Conductivity and Diffusion at Infinite Dilution." The equation is reliable for $c < 0.001$ mol/L; with higher concentration the error increases.

Compound	Infinite dilution	Concentration (mol/L)						
		0.0005	0.001	0.005	0.01	0.02	0.05	0.1
	Λ°	Λ (10^{-4} m² S mol⁻¹)						
$AgNO_3$	133.29	131.29	130.45	127.14	124.70	121.35	115.18	109.09
$1/2BaCl_2$	139.91	135.89	134.27	127.96	123.88	119.03	111.42	105.14
$1/2CaCl_2$	135.77	131.86	130.30	124.19	120.30	115.59	108.42	102.41
$1/2Ca(OH)_2$	258	—	—	233	226	214	—	—
$CuSO_4$	133.6	121.6	115.20	94.02	83.08	72.16	59.02	50.55
HCl	425.95	422.53	421.15	415.59	411.80	407.04	398.89	391.13
KBr	151.9	149.8	148.9	146.02	143.36	140.41	135.61	131.32
KCl	149.79	147.74	146.88	143.48	141.20	138.27	133.30	128.90
$KClO_4$	139.97	138.69	137.80	134.09	131.39	127.86	121.56	115.14
$1/3K_3Fe(CN)_6$	174.5	166.4	163.1	150.7	—	—	—	—
$1/4K_4Fe(CN)_6$	184	—	167.16	146.02	134.76	122.76	107.65	97.82
$KHCO_3$	117.94	116.04	115.28	112.18	110.03	107.17	—	—
KI	150.31	148.2	143.32	144.30	142.11	139.38	134.90	131.05
KIO_4	127.86	125.74	124.88	121.18	118.45	114.08	106.67	98.2
KNO_3	144.89	142.70	141.77	138.41	132.75	132.34	126.25	120.34
$KMnO_4$	134.8	132.7	131.9	—	126.5	—	—	113
KOH	271.5	—	234	230	228	—	219	213
$KReO_4$	128.20	126.03	125.12	121.31	118.49	114.49	106.40	97.40
$1/3LaCl_3$	145.9	139.6	137.0	127.5	121.8	115.3	106.2	99.1
$LiCl$	114.97	113.09	112.34	109.35	107.27	104.60	100.06	95.81
$LiClO_4$	105.93	104.13	103.39	100.52	98.56	96.13	92.15	88.52
$1/2MgCl_2$	129.34	125.55	124.15	118.25	114.49	109.99	103.03	97.05
NH_4Cl	149.6	147.5	146.7	143.9	141.21	138.25	133.22	128.69
$NaCl$	126.39	124.44	123.68	120.59	118.45	115.70	111.01	106.69
$NaClO_4$	117.42	115.58	114.82	111.70	109.54	106.91	102.35	98.38
NaI	126.88	125.30	124.19	121.19	119.18	116.64	112.73	108.73
$NaOOCCH_3$	91.0	89.2	88.5	85.68	83.72	81.20	76.88	72.76
$NaOH$	247.7	245.5	244.6	240.7	237.9	—	—	—
Na picrate	80.45	78.7	78.6	75.7	73.7	—	66.3	61.8
$1/2Na_2SO_4$	129.8	125.68	124.09	117.09	112.38	106.73	97.70	89.94
$1/2SrCl_2$	135.73	131.84	130.27	124.18	120.23	115.48	108.20	102.14
$ZnSO_4$	132.7	121.3	114.47	95.44	84.87	74.20	61.17	52.61

IONIC CONDUCTIVITY AND DIFFUSION AT INFINITE DILUTION

Petr Vanýsek

This table gives the molar (equivalent) conductivity λ for common ions at infinite dilution. All values refer to aqueous solutions at 25 °C. It also lists the diffusion coefficient D of the ion in dilute aqueous solution, which is related to λ through the equation

$$D = (RT/F^2)(\lambda/|z|)$$

where R is the molar gas constant, T the temperature, F the Faraday constant, and z the charge on the ion. The variation with temperature is fairly sharp; for typical ions, λ and D increase by 2 to 3% per degree as the temperature increases from 25 °C.

The diffusion coefficient for a salt, D_{salt}, may be calculated from the D_+ and D_- values of the constituent ions by the relation

$$D_{salt} = \frac{(z_+ + |z_-|)D_+D_-}{z_+D_+ + |z_-|D_-}$$

For solutions of simple, pure electrolytes (one positive and one negative ionic species), such as NaCl, equivalent ionic conductivity Λ°, which is the molar conductivity per unit concentration of charge, is defined as

$$\Lambda° = \Lambda_+ + \Lambda_-$$

where Λ_+ and Λ_- are equivalent ionic conductivities of the cation and anion. The more general formula is

$$\Lambda° = \nu_+\Lambda_+ + \nu_-\Lambda_-$$

where ν_+ and ν_- refer to the number of moles of cations and anions to which one mole of the electrolyte gives a rise in the solution.

References

1. Gray, D. E., Ed., *American Institute of Physics Handbook*, McGraw-Hill, New York, 1972, 2–226.
2. Robinson, R. A., and Stokes, R. H., *Electrolyte Solutions*, Butterworths, London, 1959.
3. Lobo, V. M. M., and Quaresma, J. L., *Handbook of Electrolyte Solutions*, Physical Science Data Series 41, Elsevier, Amsterdam, 1989.
4. Conway, B. E., *Electrochemical Data*, Elsevier, Amsterdam, 1952.
5. Milazzo, G., *Electrochemistry: Theoretical Principles and Practical Applications*, Elsevier, Amsterdam, 1963.

Ion	$\Lambda_\pm$ 10^{-4} m² S mol⁻¹	D 10^{-5} cm² s⁻¹
Inorganic Cations		
Ag^+	61.9	1.648
$1/3Al^{3+}$	61	0.541
$1/2Ba^{2+}$	63.6	0.847
$1/2Be^{2+}$	45	0.599
$1/2Ca^{2+}$	59.47	0.792
$1/2Cd^{2+}$	54	0.719
$1/3Ce^{3+}$	69.8	0.620
$1/2Co^{2+}$	55	0.732
$1/3[Co(NH_3)_6]^{3+}$	101.9	0.904
$1/3[Co(en)_3]^{3+}$	74.7	0.663
$1/6[Co_2(trien)_3]^{6+}$	69	0.306
$1/3Cr^{3+}$	67	0.595
Cs^+	77.2	2.056
$1/2Cu^{2+}$	53.6	0.714
D^+	249.9	6.655
$1/3Dy^{3+}$	65.6	0.582
$1/3Er^{3+}$	65.9	0.585
$1/3Eu^{3+}$	67.8	0.602
$1/2Fe^{2+}$	54	0.719
$1/3Fe^{3+}$	68	0.604
$1/3Gd^{3+}$	67.3	0.597
H^+	349.65	9.311
$1/2Hg^{2+}$	68.6	0.913
$1/2Hg^{2+}$	63.6	0.847
$1/3Ho^{3+}$	66.3	0.589
K^+	73.48	1.957
$1/3La^{3+}$	69.7	0.619
Li^+	38.66	1.029
$1/2Mg^{2+}$	53.0	0.706
$1/2Mn^{2+}$	53.5	0.712
NH_4^+	73.5	1.957
$N_2H_5^+$	59	1.571
Na^+	50.08	1.334
$1/3Nd^{3+}$	69.4	0.616
$1/2Ni^{2+}$	49.6	0.661
$1/4[Ni_2(trien)_3]^{4+}$	52	0.346
$1/2Pb^{2+}$	71	0.945
$1/3Pr^{3+}$	69.5	0.617
$1/2Ra^{2+}$	66.8	0.889
Rb^+	77.8	2.072
$1/3Sc^{3+}$	64.7	0.574
$1/3Sm^{3+}$	68.5	0.608
$1/2Sr^{2+}$	59.4	0.791
Tl^+	74.7	1.989
$1/3Tm^{3+}$	65.4	0.581
$1/2UO_2^{2+}$	32	0.426
$1/3Y^{3+}$	62	0.550
$1/3Yb^{3+}$	65.6	0.582
$1/2Zn^{2+}$	52.8	0.703
Inorganic Anions		
$Au(CN)_2^-$	50	1.331
$Au(CN)_4^-$	36	0.959
$B(C_6H_5)_4^-$	21	0.559
Br^-	78.1	2.080
Br_3^-	43	1.145
BrO_3^-	55.7	1.483
CN^-	78	2.077
CNO^-	64.6	1.720
$1/2CO_3^{2-}$	69.3	0.923
Cl^-	76.31	2.032
ClO_2^-	52	1.385
ClO_3^-	64.6	1.720
ClO_4^-	67.3	1.792
$1/3[Co(CN)_6]^{3-}$	98.9	0.878
$1/2CrO_4^{2-}$	85	1.132

Ion	$\Lambda_{\pm}$ 10^{-4} m² S mol⁻¹	D 10^{-5} cm² s⁻¹
F^-	55.4	1.475
$1/4[Fe(CN)_6]^{4-}$	110.4	0.735
$1/3[Fe(CN)_6]^{3-}$	100.9	0.896
$H_2AsO_4^-$	34	0.905
HCO_3^-	44.5	1.185
HF_2^-	75	1.997
$1/2HPO_4^{2-}$	57	0.759
$H_2PO_4^-$	36	0.959
$H_2PO_2^-$	46	1.225
HS^-	65	1.731
HSO_3^-	58	1.545
HSO_4^-	52	1.385
$H_2SbO_4^-$	31	0.825
I^-	76.8	2.045
IO_3^-	40.5	1.078
IO_4^-	54.5	1.451
MnO_4^-	61.3	1.632
$1/2MoO_4^{2-}$	74.5	1.984
$N(CN)_2^-$	54.5	1.451
NO_2^-	71.8	1.912
NO_3^-	71.42	1.902
$NH_2SO_3^-$	48.3	1.286
N_3^-	69	1.837
OCN^-	64.6	1.720
OD^-	119	3.169
OH^-	198	5.273
PF_6^-	56.9	1.515
$1/2PO_3F^{2-}$	63.3	0.843
$1/3PO_4^{3-}$	92.8	0.824
$1/4P_2O_7^{4-}$	96	0.639
$1/3P_3O_9^{3-}$	83.6	0.742
$1/5P_3O_{10}^{5-}$	109	0.581
ReO_4^-	54.9	1.462
SCN^-	66	1.758
$1/2SO_3^{2-}$	72	0.959
$1/2SO_4^{2-}$	80.0	1.065
$1/2S_2O_3^{2-}$	85.0	1.132
$1/2S_2O_4^{2-}$	66.5	0.885
$1/2S_2O_6^{2-}$	93	1.238
$1/2S_2O_8^{2-}$	86	1.145
$Sb(OH)_6^-$	31.9	0.849
$SeCN^-$	64.7	1.723
$1/2SeO_4^{2-}$	75.7	1.008
$1/2WO_4^{2-}$	69	0.919
Organic Cations		
Benzyltrimethylammonium⁺	34.6	0.921
Isobutylammonium⁺	38	1.012
Butyltrimethylammonium⁺	33.6	0.895
Decylpyridinium⁺	29.5	0.786
Decyltrimethylammonium⁺	24.4	0.650
Diethylammonium⁺	42.0	1.118
Dimethylammonium⁺	51.8	1.379
Dipropylammonium⁺	30.1	0.802
Dodecylammonium⁺	23.8	0.634
Dodecyltrimethylammonium⁺	22.6	0.602
Ethanolammonium⁺	42.2	1.124
Ethylammonium⁺	47.2	1.257
Ethyltrimethylammonium⁺	40.5	1.078
Hexadecyltrimethylammonium⁺	20.9	0.557
Hexyltrimethylammonium⁺	29.6	0.788
Histidyl⁺	23.0	0.612
Hydroxyethyltrimethylarsonium⁺	39.4	1.049
Methylammonium⁺	58.7	1.563
Octadecylpyridinium⁺	20	0.533
Octadecyltributylammonium⁺	16.6	0.442
Octadecyltriethylammonium⁺	17.9	0.477
Octadecyltrimethylammonium⁺	19.9	0.530
Octadecyltripropylammonium⁺	17.2	0.458
Octyltrimethylammonium⁺	26.5	0.706
Pentylammonium⁺	37	0.985
Piperidinium⁺	37.2	0.991
Propylammonium⁺	40.8	1.086
Pyrilammonium⁺	24.3	0.647
Tetrabutylammonium⁺	19.5	0.519
Tetradecyltrimethylammonium⁺	21.5	0.573
Tetraethylammonium⁺	32.6	0.868
Tetramethylammonium⁺	44.9	1.196
Tetraisopentylammonium⁺	17.9	0.477
Tetrapentylammmonium⁺	17.5	0.466
Tetrapropylammonium⁺	23.4	0.623
Triethylammonium⁺	34.3	0.913
Triethylsulfonium⁺	36.1	0.961
Trimethylammonium⁺	47.23	1.258
Trimethylhexylammonium⁺	34.6	0.921
Trimethylsulfonium⁺	51.4	1.369
Tripropylammonium⁺	26.1	0.695
Organic Anions		
Acetate⁻	40.9	1.089
p-Anisate⁻	29.0	0.772
1/2Azelate²⁻	40.6	0.541
Benzoate⁻	32.4	0.863
Bromoacetate⁻	39.2	1.044
Bromobenzoate⁻	30	0.799
Butyrate⁻	32.6	0.868
Chloroacetate⁻	39.8	1.060
m-Chlorobenzoate⁻	31	0.825
o-Chlorobenzoate⁻	30.2	0.804
1/3Citrate³⁻	70.2	0.623
Crotonate⁻	33.2	0.884
Cyanoacetate⁻	43.4	1.156
Cyclohexane carboxylate⁻	28.7	0.764
1/2 1,1-Cyclopropanedicarboxylate²⁻	53.4	0.711
Decylsulfate⁻	26	0.692
Dichloroacetate⁻	38.3	1.020
1/2Diethylbarbiturate²⁻	26.3	0.350
Dihydrogencitrate⁻	30	0.799
1/2Dimethylmalonate²⁻	49.4	0.658
3,5-Dinitrobenzoate⁻	28.3	0.754
Dodecylsulfate⁻	24	0.639
Ethylmalonate⁻	49.3	1.313
Ethylsulfate⁻	39.6	1.055
Fluoroacetate⁻	44.4	1.182
Fluorobenzoate⁻	33	0.879
Formate⁻	54.6	1.454
1/2Fumarate²⁻	61.8	0.823
1/2Glutarate²⁻	52.6	0.700
Hydrogenoxalate⁻	40.2	1.070
Isovalerate⁻	32.7	0.871
Iodoacetate⁻	40.6	1.081
Lactate⁻	38.8	1.033

Ion	$\Lambda_{\pm}$ 10^{-4} m^2 S mol^{-1}	D 10^{-5} cm^2 s^{-1}	Ion	$\Lambda_{\pm}$ 10^{-4} m^2 S mol^{-1}	D 10^{-5} cm^2 s^{-1}
1/2Malate^{2-}	58.8	0.783	Picrate^{-}	30.37	0.809
1/2Maleate^{2-}	61.9	0.824	Pivalate^{-}	31.9	0.849
1/2Malonate^{2-}	63.5	0.845	Propionate^{-}	35.8	0.953
Methylsulfate^{-}	48.8	1.299	Propylsulfate^{-}	37.1	0.988
Naphthylacetate^{-}	28.4	0.756	Salicylate^{-}	36	0.959
1/2Oxalate^{2-}	74.11	0.987	1/2Suberate^{2-}	36	0.479
Octylsulfate^{-}	29	0.772	1/2Succinate^{2-}	58.8	0.783
Phenylacetate^{-}	30.6	0.815	*p*-Sulfonate	29.3	0.780
1/2*o*-Phthalate^{2-}	52.3	0.696	1/2Tartarate^{2-}	59.6	0.794
1/2*m*-Phthalate^{2-}	54.7	0.728	Trichloroacetate^{-}	35	0.932

ACTIVITY COEFFICIENTS OF ACIDS, BASES, AND SALTS

Petr Vanýsek

This table gives mean activity coefficients at 25 °C for molalities in the range 0.1 to 1.0. See the following table for definitions, references, and data over a wider concentration range.

	0.1	**0.2**	**0.3**	**0.4**	**0.5**	**0.6**	**0.7**	**0.8**	**0.9**	**1.0**
$AgNO_3$	0.734	0.657	0.606	0.567	0.536	0.509	0.485	0.464	0.446	0.429
$AlCl_3$	0.337	0.305	0.302	0.313	0.331	0.356	0.388	0.429	0.479	0.539
$Al_2(SO_4)_3$	0.035	0.0225	0.0176	0.0153	0.0143	0.014	0.0142	0.0149	0.0159	0.0175
$BaCl_2$	0.500	0.444	0.419	0.405	0.397	0.391	0.391	0.391	0.392	0.395
$BeSO_4$	0.150	0.109	0.0885	0.0769	0.0692	0.0639	0.0600	0.0570	0.0546	0.0530
$CaCl_2$	0.518	0.472	0.455	0.448	0.448	0.453	0.460	0.470	0.484	0.500
$CdCl_2$	0.2280	0.1638	0.1329	0.1139	0.1006	0.0905	0.0827	0.0765	0.0713	0.0669
$Cd(NO_3)_2$	0.513	0.464	0.442	0.430	0.425	0.423	0.423	0.425	0.428	0.433
$CdSO_4$	0.150	0.103	0.0822	0.0699	0.0615	0.0553	0.0505	0.0468	0.0438	0.0415
$CoCl_2$	0.522	0.479	0.463	0.459	0.462	0.470	0.479	0.492	0.511	0.531
$CrCl_3$	0.331	0.298	0.294	0.300	0.314	0.335	0.362	0.397	0.436	0.481
$Cr(NO_3)_3$	0.319	0.285	0.279	0.281	0.291	0.304	0.322	0.344	0.371	0.401
$Cr_2(SO_4)_3$	0.0458	0.0300	0.0238	0.0207	0.0190	0.0182	0.0181	0.0185	0.0194	0.0208
CsBr	0.754	0.694	0.654	0.626	0.603	0.586	0.571	0.558	0.547	0.538
CsCl	0.756	0.694	0.656	0.628	0.606	0.589	0.575	0.563	0.553	0.544
CsI	0.754	0.692	0.651	0.621	0.599	0.581	0.567	0.554	0.543	0.533
$CsNO_3$	0.733	0.655	0.602	0.561	0.528	0.501	0.478	0.458	0.439	0.422
CsOH	0.795	0.761	0.744	0.739	0.739	0.742	0.748	0.754	0.762	0.771
CsOAc	0.799	0.771	0.761	0.759	0.762	0.768	0.776	0.783	0.792	0.802
Cs_2SO_4	0.456	0.382	0.338	0.311	0.291	0.274	0.262	0.251	0.242	0.235
$CuCl_2$	0.508	0.455	0.429	0.417	0.411	0.409	0.409	0.410	0.413	0.417
$Cu(NO_3)_2$	0.511	0.460	0.439	0.429	0.426	0.427	0.431	0.437	0.445	0.455
$CuSO_4$	0.150	0.104	0.0829	0.0704	0.0620	0.0559	0.0512	0.0475	0.0446	0.0423
$FeCl_2$	0.5185	0.473	0.454	0.448	0.450	0.454	0.463	0.473	0.488	0.506
HBr	0.805	0.782	0.777	0.781	0.789	0.801	0.815	0.832	0.850	0.871
HCl	0.796	0.767	0.756	0.755	0.757	0.763	0.772	0.783	0.795	0.809
$HClO_4$	0.803	0.778	0.768	0.766	0.769	0.776	0.785	0.795	0.808	0.823
HI	0.818	0.807	0.811	0.823	0.839	0.860	0.883	0.908	0.935	0.963
HNO_3	0.791	0.754	0.735	0.725	0.720	0.717	0.717	0.718	0.721	0.724
H_2SO_4	0.2655	0.2090	0.1826	—	0.1557	—	0.1417	—	—	0.1316
KBr	0.772	0.722	0.693	0.673	0.657	0.646	0.636	0.629	0.622	0.617
KCl	0.770	0.718	0.688	0.666	0.649	0.637	0.626	0.618	0.610	0.604
$KClO_3$	0.749	0.681	0.635	0.599	0.568	0.541	0.518	—	—	—
K_2CrO_4	0.456	0.382	0.340	0.313	0.292	0.276	0.263	0.253	0.243	0.235
KF	0.775	0.727	0.700	0.682	0.670	0.661	0.654	0.650	0.646	0.645
$K_3Fe(CN)_6$	0.268	0.212	0.184	0.167	0.155	0.146	0.140	0.135	0.131	0.128
$K_4Fe(CN)_6$	0.139	0.0993	0.0808	0.0693	0.0614	0.0556	0.0512	0.0479	0.0454	—
KH_2PO_4	0.731	0.653	0.602	0.561	0.529	0.501	0.477	0.456	0.438	0.421
KI	0.778	0.733	0.707	0.689	0.676	0.667	0.660	0.654	0.649	0.645
KNO_3	0.739	0.663	0.614	0.576	0.545	0.519	0.496	0.476	0.459	0.443
KOAc	0.796	0.766	0.754	0.750	0.751	0.754	0.759	0.766	0.774	0.783
KOH	0.798	0.760	0.742	0.734	0.732	0.733	0.736	0.742	0.749	0.756
KSCN	0.769	0.716	0.685	0.663	0.646	0.633	0.623	0.614	0.606	0.599
K_2SO_4	0.441	0.360	0.316	0.286	0.264	0.246	0.232	—	—	—
LiBr	0.796	0.766	0.756	0.752	0.753	0.758	0.767	0.777	0.789	0.803
LiCl	0.790	0.757	0.744	0.740	0.739	0.743	0.748	0.755	0.764	0.774
$LiClO_4$	0.812	0.794	0.792	0.798	0.808	0.820	0.834	0.852	0.869	0.887
LiI	0.815	0.802	0.804	0.813	0.824	0.838	0.852	0.870	0.888	0.910
$LiNO_3$	0.788	0.752	0.736	0.728	0.726	0.727	0.729	0.733	0.737	0.743
LiOH	0.760	0.702	0.665	0.638	0.617	0.599	0.585	0.573	0.563	0.554
LiOAc	0.784	0.742	0.721	0.709	0.700	0.691	0.689	0.688	0.688	0.689
Li_2SO_4	0.468	0.398	0.361	0.337	0.319	0.307	0.297	0.289	0.282	0.277
$MgCl_2$	0.529	0.489	0.477	0.475	0.481	0.491	0.506	0.522	0.544	0.570
$MgSO_4$	0.150	0.107	0.0874	0.0756	0.0675	0.0616	0.0571	0.0536	0.0508	0.0485

	0.1	0.2	0.3	0.4	0.5	0.6	0.7	0.8	0.9	1.0
$MnCl_2$	0.516	0.469	0.450	0.442	0.440	0.443	0.448	0.455	0.466	0.479
$MnSO_4$	0.150	0.105	0.0848	0.0725	0.0640	0.0578	0.0530	0.0493	0.0463	0.0439
NH_4Cl	0.770	0.718	0.687	0.665	0.649	0.636	0.625	0.617	0.609	0.603
NH_4NO_3	0.740	0.677	0.636	0.606	0.582	0.562	0.545	0.530	0.516	0.504
$(NH_4)_2SO_4$	0.439	0.356	0.311	0.280	0.257	0.240	0.226	0.214	0.205	0.196
NaBr	0.782	0.741	0.719	0.704	0.697	0.692	0.689	0.687	0.687	0.687
NaCl	0.778	0.735	0.710	0.693	0.681	0.673	0.667	0.662	0.659	0.657
$NaClO_3$	0.772	0.720	0.688	0.664	0.645	0.630	0.617	0.606	0.597	0.589
$NaClO_4$	0.775	0.729	0.701	0.683	0.668	0.656	0.648	0.641	0.635	0.629
Na_2CrO_4	0.464	0.394	0.353	0.327	0.307	0.292	0.280	0.269	0.261	0.253
NaF	0.765	0.710	0.676	0.651	0.632	0.616	0.603	0.592	0.582	0.573
NaH_2PO_4	0.744	0.675	0.629	0.593	0.563	0.539	0.517	0.499	0.483	0.468
NaI	0.787	0.751	0.735	0.727	0.723	0.723	0.724	0.727	0.731	0.736
$NaNO_3$	0.762	0.703	0.666	0.638	0.617	0.599	0.583	0.570	0.558	0.548
NaOAc	0.791	0.757	0.744	0.737	0.735	0.736	0.740	0.745	0.752	0.757
NaOH	0.766	0.727	0.708	0.697	0.690	0.685	0.681	0.679	0.678	0.678
NaSCN	0.787	0.750	—	0.720	0.715	0.712	0.710	0.710	0.711	0.712
Na_2SO_4	0.445	0.365	0.320	0.289	0.266	0.248	0.233	0.221	0.210	0.201
$NiCl_2$	0.522	0.479	0.463	0.460	0.464	0.471	0.482	0.496	0.515	0.563
$NiSO_4$	0.150	0.105	0.0841	0.0713	0.0627	0.0562	0.0515	0.0478	0.0448	0.0425
$Pb(NO_3)_2$	0.395	0.308	0.260	0.228	0.205	0.187	0.172	0.160	0.150	0.141
RbBr	0.763	0.706	0.673	0.650	0.632	0.617	0.605	0.595	0.586	0.578
RbCl	0.764	0.709	0.675	0.652	0.634	0.620	0.608	0.599	0.590	0.583
RbI	0.762	0.705	0.671	0.647	0.629	0.614	0.602	0.591	0.583	0.575
$RbNO_3$	0.734	0.658	0.606	0.565	0.534	0.508	0.485	0.465	0.446	0.430
RbOAc	0.796	0.767	0.756	0.753	0.755	0.759	0.766	0.773	0.782	0.792
Rb_2SO_4	0.451	0.374	0.331	0.301	0.279	0.263	0.249	0.238	0.228	0.219
$SrCl_2$	0.511	0.462	0.442	0.433	0.430	0.431	0.434	0.441	0.449	0.461
$TlClO_4$	0.730	0.652	0.599	0.559	0.527	—	—	—	—	—
$TlNO_3$	0.702	0.606	0.545	0.500	—	—	—	—	—	—
UO_2Cl_2	0.544	0.510	0.520	0.505	0.517	0.532	0.549	0.571	0.595	0.620
UO_2SO_4	0.150	0.102	0.0807	0.0689	0.0611	0.0566	0.0515	0.0483	0.0458	0.0439
$ZnCl_2$	0.515	0.462	0.432	0.411	0.394	0.380	0.369	0.357	0.348	0.339
$Zn(NO_3)_2$	0.531	0.489	0.474	0.469	0.473	0.480	0.489	0.501	0.518	0.535
$ZnSO_4$	0.150	0.140	0.0835	0.0714	0.0630	0.0569	0.0523	0.0487	0.0458	0.0435

MEAN ACTIVITY COEFFICIENTS OF ELECTROLYTES AS A FUNCTION OF CONCENTRATION

The mean activity coefficient γ of an electrolyte X_aY_b is defined as

$$\gamma = (\gamma_+^a \gamma_-^b)^{1/(a+b)}$$

where γ_+ and γ_- are activity coefficients of the individual ions (which cannot be directly measured). This table gives the mean activity coefficients of about 100 electrolytes in aqueous solution as a function of concentration, expressed in molality terms. All values refer to a temperature of 25 °C. Substances are arranged in alphabetical order by formula.

References

1. Hamer, W. J., and Wu, Y. C., *J. Phys. Chem. Ref. Data*, 1, 1047, 1972.
2. Staples, B. R., *J. Phys. Chem. Ref. Data*, 6, 385, 1977; 10, 767, 1981; 10, 779, 1981.
3. Goldberg, R. N. et al., *J. Phys. Chem. Ref. Data*, 7, 263, 1978; 8, 923, 1979; 8, 1005, 1979; 10, 1, 1981; 10, 671, 1981.

Mean Activity Coefficient at 25 °C

m/mol kg^{-1}	$AgNO_3$	$BaBr_2$	$BaCl_2$	BaI_2	$CaBr_2$	$CaCl_2$	CaI_2
0.001	0.964	0.881	0.887	0.890	0.890	0.888	0.890
0.002	0.950	0.850	0.849	0.853	0.853	0.851	0.853
0.005	0.924	0.785	0.782	0.792	0.791	0.787	0.791
0.010	0.896	0.727	0.721	0.737	0.735	0.727	0.736
0.020	0.859	0.661	0.653	0.678	0.674	0.664	0.677
0.050	0.794	0.573	0.559	0.600	0.594	0.577	0.600
0.100	0.732	0.517	0.492	0.551	0.540	0.517	0.552
0.200	0.656	0.463	0.436	0.520	0.502	0.469	0.524
0.500	0.536	0.435	0.391	0.536	0.500	0.444	0.554
1.000	0.430	0.470	0.393	0.664	0.604	0.495	0.729
2.000	0.316	0.654		1.242	1.125	0.784	
5.000	0.181				18.7	5.907	
10.000	0.108					43.1	
15.000	0.085						

m/mol kg^{-1}	$Cd(NO_2)_2$	$Cd(NO_3)_2$	$CoBr_2$	$CoCl_2$	CoI_2	$Co(NO_3)_2$	CsBr
0.001	0.881	0.888	0.890	0.889	0.887	0.888	0.965
0.002	0.837	0.851	0.854	0.852	0.849	0.850	0.951
0.005	0.759	0.787	0.794	0.789	0.783	0.786	0.925
0.010	0.681	0.728	0.740	0.732	0.724	0.728	0.898
0.020	0.589	0.664	0.681	0.670	0.661	0.663	0.864
0.050	0.451	0.576	0.605	0.586	0.582	0.576	0.806
0.100	0.344	0.515	0.556	0.528	0.540	0.516	0.752
0.200	0.247	0.465	0.523	0.483	0.527	0.469	0.691
0.500	0.148	0.428	0.538	0.465	0.596	0.446	0.605
1.000	0.098	0.437	0.685	0.532	0.845	0.492	0.540
2.000	0.069	0.517	1.421	0.864	2.287	0.722	0.485
5.000	0.054		13.9		55.3	3.338	0.454
10.000					196		

m/mol kg^{-1}	CsCl	CsF	CsI	$CsNO_3$	CsOH	Cs_2SO_4	$CuBr_2$
0.001	0.965	0.965	0.965	0.964	0.966	0.885	0.889
0.002	0.951	0.952	0.951	0.951	0.953	0.845	0.853
0.005	0.925	0.929	0.925	0.924	0.930	0.775	0.791
0.010	0.898	0.905	0.898	0.897	0.906	0.709	0.735
0.020	0.864	0.876	0.863	0.860	0.878	0.634	0.674
0.050	0.805	0.830	0.804	0.796	0.836	0.526	0.594
0.100	0.751	0.792	0.749	0.733	0.802	0.444	0.541
0.200	0.691	0.755	0.688	0.655	0.772	0.369	0.504
0.500	0.607	0.721	0.601	0.529	0.755	0.285	0.503
1.000	0.546	0.726	0.534	0.421	0.782	0.233	0.591
2.000	0.496	0.803	0.470				0.859
5.000	0.474						
10.000	0.508						

m/mol kg⁻¹	$CuCl_2$	$Cu(ClO_4)_2$	$Cu(NO_3)_2$	$FeCl_2$	HBr	HCl	$HClO_4$
0.001	0.887	0.890	0.888	0.888	0.966	0.965	0.966
0.002	0.849	0.854	0.851	0.850	0.953	0.952	0.953
0.005	0.783	0.795	0.787	0.785	0.930	0.929	0.929
0.010	0.722	0.741	0.729	0.725	0.907	0.905	0.906
0.020	0.654	0.685	0.664	0.659	0.879	0.876	0.878
0.050	0.561	0.613	0.577	0.570	0.837	0.832	0.836
0.100	0.495	0.572	0.516	0.509	0.806	0.797	0.803
0.200	0.441	0.553	0.466	0.462	0.783	0.768	0.776
0.500	0.401	0.617	0.431	0.443	0.790	0.759	0.769
1.000	0.405	0.892	0.456	0.500	0.872	0.811	0.826
2.000	0.453	2.445	0.615	0.782	1.167	1.009	1.055
5.000	0.601		2.083		3.800	2.380	3.100
10.000					33.4	10.4	30.8
15.000							323

m/mol kg⁻¹	HF	HI	HNO_3	H_2SO_4	KBr	KCNS	KCl
0.001	0.551	0.966	0.965	0.804	0.965	0.965	0.965
0.002	0.429	0.953	0.952	0.740	0.952	0.951	0.951
0.005	0.302	0.931	0.929	0.634	0.927	0.927	0.927
0.010	0.225	0.909	0.905	0.542	0.902	0.901	0.901
0.020	0.163	0.884	0.875	0.445	0.870	0.869	0.869
0.050	0.106	0.847	0.829	0.325	0.817	0.815	0.816
0.100	0.0766	0.823	0.792	0.251	0.771	0.768	0.768
0.200	0.0550	0.811	0.756	0.195	0.772	0.716	0.717
0.500	0.0352	0.845	0.725	0.146	0.658	0.647	0.649
1.000	0.0249	0.969	0.730	0.125	0.617	0.598	0.604
2.000	0.0175	1.363	0.788	0.119	0.593	0.556	0.573
5.000	0.0110	4.760	1.063	0.197	0.626	0.525	0.593
10.000	0.0085	49.100	1.644	0.527			
15.000	0.0077		2.212	1.077			
20.000	0.0075		2.607	1.701			

m/mol kg⁻¹	$KClO_3$	K_2CrO_4	KF	KH_2PO_4*	K_2HPO_4**	KI	KNO_3
0.001	0.965	0.886	0.965	0.964	0.886	0.965	0.964
0.002	0.951	0.847	0.952	0.950	0.847	0.952	0.950
0.005	0.926	0.779	0.927	0.924	0.779	0.927	0.924
0.010	0.899	0.715	0.902	0.896	0.715	0.902	0.896
0.020	0.865	0.643	0.870	0.859	0.643	0.871	0.860
0.050	0.805	0.539	0.818	0.793	0.538	0.820	0.797
0.100	0.749	0.460	0.773	0.730	0.457	0.776	0.735
0.200	0.681	0.385	0.726	0.652	0.379	0.731	0.662
0.500	0.569	0.296	0.670	0.529	0.283	0.676	0.546
1.000		0.239	0.645	0.422		0.646	0.444
2.000		0.199	0.658			0.638	0.332
5.000			0.871				
10.000			1.715				
15.000			3.120				

m/mol kg⁻¹	KOH	K_2SO_4	LiBr	LiCl	$LiClO_4$	LiI	$LiNO_3$
0.001	0.965	0.885	0.965	0.965	0.966	0.966	0.965
0.002	0.952	0.844	0.952	0.952	0.953	0.953	0.952
0.005	0.927	0.772	0.929	0.928	0.931	0.930	0.928
0.010	0.902	0.704	0.905	0.904	0.908	0.908	0.904
0.020	0.871	0.625	0.877	0.874	0.882	0.882	0.874
0.050	0.821	0.511	0.832	0.827	0.843	0.843	0.827
0.100	0.779	0.424	0.797	0.789	0.815	0.817	0.788
0.200	0.740	0.343	0.767	0.756	0.795	0.802	0.753
0.500	0.710	0.251	0.754	0.739	0.806	0.824	0.726
1.000	0.733		0.803	0.775	0.887	0.912	0.743
2.000	0.860		1.012	0.924	1.161	1.197	0.837
5.000	1.697		2.696	2.000			1.298

m/mol kg^{-1}	KOH	K_2SO_4	LiBr	LiCl	$LiClO_4$	LiI	$LiNO_3$
10.000	6.110		20.0	9.600			2.500
15.000	19.9		147	30.9			3.960
20.000	46.4		486				4.970

m/mol kg^{-1}	LiOH	Li_2SO_4	$MgBr_2$	$MgCl_2$	MgI_2	$MnBr_2$	$MnCl_2$
0.001	0.964	0.887	0.889	0.889	0.889	0.889	0.888
0.002	0.950	0.847	0.852	0.852	0.853	0.853	0.850
0.005	0.923	0.780	0.790	0.790	0.791	0.791	0.786
0.010	0.895	0.716	0.733	0.734	0.736	0.735	0.727
0.020	0.858	0.645	0.672	0.672	0.677	0.674	0.662
0.050	0.794	0.544	0.593	0.590	0.602	0.595	0.574
0.100	0.735	0.469	0.543	0.535	0.556	0.543	0.513
0.200	0.668	0.400	0.512	0.493	0.535	0.508	0.464
0.500	0.579	0.325	0.540	0.485	0.594	0.519	0.437
1.000	0.522	0.284	0.715	0.577	0.858	0.650	0.477
2.000	0.484	0.270	1.590	1.065	2.326	1.224	0.661
5.000	0.493		36.1	14.40	109.8	6.697	1.539

m/mol kg^{-1}	$Mn(ClO_4)_2$	NH_4Cl	NH_4ClO_4	$(NH_4)_2HPO_4$**	NH_4NO_3	NaBr	$NaBrO_3$
0.001	0.892	0.965	0.964	0.882	0.964	0.965	0.965
0.002	0.858	0.952	0.950	0.839	0.951	0.952	0.951
0.005	0.801	0.927	0.924	0.763	0.925	0.928	0.926
0.010	0.752	0.901	0.895	0.688	0.897	0.903	0.900
0.020	0.700	0.869	0.859	0.600	0.862	0.873	0.867
0.050	0.637	0.816	0.794	0.469	0.801	0.824	0.811
0.100	0.604	0.769	0.734	0.367	0.744	0.783	0.759
0.200	0.596	0.718	0.663	0.273	0.678	0.742	0.698
0.500	0.686	0.649	0.560	0.171	0.582	0.697	0.605
1.000	1.030	0.603	0.479	0.114	0.502	0.687	0.528
2.000	3.072	0.569	0.399	0.074	0.419	0.730	0.449
5.000		0.563			0.303	1.083	
10.000					0.220		
15.000					0.179		
20.000					0.154		

m/mol kg^{-1}	Na_2CO_3	NaCl	$NaClO_3$	$NaClO_4$	Na_2CrO_4	NaF	Na_2HPO_4*
0.001	0.887	0.965	0.965	0.965	0.887	0.965	0.887
0.002	0.847	0.952	0.952	0.952	0.849	0.951	0.848
0.005	0.780	0.928	0.927	0.928	0.783	0.926	0.780
0.010	0.716	0.903	0.902	0.903	0.722	0.901	0.717
0.020	0.644	0.872	0.870	0.872	0.653	0.868	0.644
0.050	0.541	0.822	0.818	0.821	0.554	0.813	0.539
0.100	0.462	0.779	0.771	0.777	0.479	0.764	0.456
0.200	0.385	0.734	0.719	0.729	0.406	0.710	0.373
0.500	0.292	0.681	0.646	0.668	0.318	0.633	0.266
1.000	0.229	0.657	0.590	0.630	0.261	0.573	0.191
2.000	0.182	0.668	0.537	0.608	0.231		0.133
5.000		0.874		0.648			

m/mol kg^{-1}	NaI	$NaNO_3$	NaOH	Na_2SO_3	Na_2SO_4	Na_2WO_4	$NiBr_2$
0.001	0.965	0.965	0.965	0.887	0.886	0.886	0.889
0.002	0.952	0.951	0.952	0.847	0.846	0.846	0.853
0.005	0.928	0.926	0.927	0.779	0.777	0.777	0.791
0.010	0.904	0.900	0.902	0.716	0.712	0.712	0.735
0.020	0.874	0.866	0.870	0.644	0.637	0.638	0.675
0.050	0.827	0.810	0.819	0.540	0.529	0.534	0.596
0.100	0.789	0.759	0.775	0.462	0.446	0.457	0.546
0.200	0.753	0.701	0.731	0.386	0.366	0.388	0.514
0.500	0.722	0.617	0.685	0.296	0.268	0.320	0.535
1.000	0.734	0.550	0.674	0.237	0.204	0.291	0.692
2.000	0.823	0.480	0.714	0.196	0.155	0.291	1.476

m/mol kg^{-1}	NaI	$NaNO_3$	NaOH	Na_2SO_3	Na_2SO_4	Na_2WO_4	$NiBr_2$
5.000	1.402	0.388	1.076				
10.000	4.011	0.329	3.258				
15.000			9.796				
20.000			19.410				

m/mol kg^{-1}	$NiCl_2$	$Ni(ClO_4)_2$	$Ni(NO_3)_2$	$Pb(ClO_4)_2$	$Pb(NO_3)_2$	RbBr	RbCl
0.001	0.889	0.891	0.889	0.889	0.882	0.965	0.965
0.002	0.852	0.855	0.851	0.851	0.840	0.951	0.951
0.005	0.789	0.797	0.787	0.787	0.764	0.926	0.926
0.010	0.732	0.745	0.730	0.729	0.690	0.900	0.900
0.020	0.669	0.690	0.666	0.666	0.604	0.866	0.867
0.050	0.584	0.621	0.581	0.580	0.476	0.811	0.811
0.100	0.527	0.582	0.524	0.522	0.379	0.760	0.761
0.200	0.482	0.567	0.481	0.476	0.291	0.705	0.707
0.500	0.465	0.639	0.467	0.458	0.195	0.630	0.633
1.000	0.538	0.946	0.528	0.516	0.136	0.578	0.583
2.000	0.915	2.812	0.797	0.799		0.535	0.546
5.000	4.785			4.043		0.514	0.544
10.000				33.8			

m/mol kg^{-1}	RbF	RbI	$RbNO_3$	Rb_2SO_4	$SrBr_2$	$SrCl_2$	SrI_2
0.001	0.965	0.965	0.964	0.886	0.889	0.888	0.890
0.002	0.952	0.951	0.950	0.845	0.852	0.850	0.854
0.005	0.927	0.926	0.924	0.776	0.790	0.785	0.793
0.010	0.902	0.900	0.896	0.710	0.734	0.725	0.740
0.020	0.871	0.866	0.859	0.635	0.673	0.659	0.681
0.050	0.821	0.810	0.795	0.526	0.591	0.569	0.606
0.100	0.780	0.759	0.733	0.448	0.536	0.506	0.557
0.200	0.739	0.703	0.657	0.365	0.492	0.455	0.526
0.500	0.701	0.627	0.536	0.274	0.476	0.421	0.542
1.000	0.697	0.574	0.430	0.217	0.545	0.451	0.686
2.000	0.724	0.532	0.320		0.921	0.650	
5.000		0.517					

m/mol kg^{-1}	UO_2Cl_2	$UO_2(NO_3)_2$	$ZnBr_2$	$ZnCl_2$	ZnI_2
0.001	0.888	0.888	0.890	0.887	0.893
0.002	0.851	0.849	0.854	0.847	0.859
0.005	0.787	0.784	0.794	0.781	0.804
0.010	0.729	0.726	0.741	0.719	0.757
0.020	0.666	0.663	0.683	0.652	0.708
0.050	0.583	0.583	0.606	0.561	0.644
0.100	0.529	0.535	0.553	0.499	0.601
0.200	0.493	0.509	0.515	0.447	0.574
0.500	0.501	0.532	0.516	0.384	0.635
1.000	0.601	0.673	0.558	0.330	0.836
2.000	0.948	1.223	0.578	0.283	1.062
5.000		3.020	0.788	0.342	1.546
10.000			2.317	0.876	4.698
15.000			5.381	1.914	
20.000			7.965	2.968	

* The anion is $H_2PO_4^-$.

** The anion is HPO_4^{-2}.

ENTHALPY OF DILUTION OF ACIDS

The quantity given in this table is $-\Delta_{dil}H$, the negative of the enthalpy (heat) of dilution to infinite dilution for aqueous solutions of several common acids; i.e., the negative of the enthalpy change when a solution of molality m at a temperature of 25 °C is diluted with an infinite amount of water. The tabulated numbers thus represent the heat produced (or, if the value is negative, the heat absorbed) when the acid is diluted. The initial molality m is given in the first column. The second column gives the dilution ratio, which is the number of moles of water that must be added to one mole of the acid to produce a solution of the molality in the first column.

Reference

Parker, V. B., *Thermal Properties of Aqueous Uni-Univalent Electrolytes*, Natl. Stand. Ref. Data Ser. - Natl. Bur. Stand. (U.S.) 2, U.S. Government Printing Office, 1965.

				$-\Delta_{dil}H$ in kJ/mol at 25 °C					
m	Dil. ratio	HF	HCl	$HClO_4$	HBr	HI	HNO_3	CH_2O_2	$C_2H_4O_2$
55.506	1.0		45.61		48.83		19.73	0.046	2.167
20	2.775	14.88	19.87	13.81	19.92	21.71	9.498	0.038	2.075
15	3.700	14.34	15.40	7.920	14.29	14.02	6.883	0.109	1.962
10	5.551	13.87	10.24	2.013	8.694	7.615	3.933	0.205	1.824
9	6.167	13.81	9.213	1.280	7.719	6.569	3.368	0.230	1.782
8	6.938	13.77	8.201	0.611	6.786	5.607	2.791	0.255	1.724
7	7.929	13.73	7.217	0.046	5.925	4.728	2.251	0.272	1.648
6	9.251	13.69	6.268	-0.351	5.004	3.975	1.749	0.280	1.540
5.5506	10	13.66	5.841	-0.490	4.590	3.577	1.540	0.285	1.477
5	11.10	13.62	5.318	-0.628	4.113	3.197	1.310	0.289	1.393
4.5	12.33	13.58	4.899	-0.732	3.711	2.828	1.109	0.289	1.310
4	13.88	13.53	4.402	-0.787	3.330	2.460	0.958	0.289	1.218
3.5	15.86	13.47	3.958	-0.820	2.966	2.105	0.791	0.289	1.121
3	18.50	13.45	3.506	-0.782	2.611	1.787	0.665	0.289	1.025
2.5	22.20	13.43	3.063	-0.724	2.301	1.527	0.582	0.285	0.912
2	27.75	13.40	2.623	-0.623	1.996	1.318	0.527	0.276	0.803
1.5	37.00	13.36	2.167	-0.431	1.665	1.125	0.506	0.259	0.678
1	55.51	13.30	1.695	-0.201	1.314	0.933	0.506	0.226	0.544
0.5551	100	13.22	1.234	0.050	0.983	0.736	0.502	0.184	0.423
0.5	111.0	13.20	1.172	0.075	0.941	0.711	0.498	0.176	0.406
0.2	277.5	13.09	0.761	0.247	0.649	0.536	0.439	0.146	0.331
0.1	555.1	12.80	0.556	0.272	0.498	0.439	0.372	0.134	0.289
0.0925	600	12.79	0.540	0.272	0.481	0.427	0.368	0.134	0.285
0.0793	700	12.70	0.502	0.272	0.452	0.402	0.351	0.134	0.285
0.0694	800	12.61	0.473	0.268	0.427	0.385	0.339	0.130	0.280
0.0617	900	12.50	0.448	0.264	0.406	0.368	0.326	0.126	0.276
0.05551	1000	12.42	0.427	0.259	0.385	0.351	0.318	0.121	0.272
0.05	1110	12.24	0.406	0.259	0.372	0.339	0.305	0.121	0.272
0.02775	2000	11.29	0.310	0.226	0.285	0.264	0.247	0.117	0.264
0.01850	3000	10.66	0.251	0.197	0.234	0.218	0.213	0.117	0.259
0.01388	4000	10.25	0.226	0.180	0.205	0.192	0.192	0.113	0.259
0.01110	5000	9.874	0.197	0.167	0.184	0.172	0.176	0.109	0.255
0.00555	10000	8.912	0.142	0.126	0.130	0.121	0.130	0.105	0.243
0.00278	20000	7.531	0.105	0.092	0.092	0.084	0.096	0.096	0.230
0.00111	50000	5.439	0.067	0.059	0.054	0.050	0.063	0.084	0.222
0.000555	100000	3.766	0.042	0.042	0.038	0.038	0.046	0.054	0.209
0.000111	500000	1.255	0.021	0.021	0.021	0.021	0.021	0.038	0.167
0	∞	0	0	0	0	0	0	0	0

ENTHALPY OF SOLUTION OF ELECTROLYTES

This table gives the molar enthalpy (heat) of solution at infinite dilution for some common uni-univalent electrolytes. This is the enthalpy change when 1 mol of solute in its standard state is dissolved in an infinite amount of water. Values are given in kilojoules per mole at 25 °C.

Reference

Parker, V. B., *Thermal Properties of Uni-Univalent Electrolytes*, Natl. Stand. Ref. Data Series — Natl. Bur. Stand.(U.S.), No.2, 1965.

Solute	State	$\Delta_{sol}H°$ kJ/mol	Solute	State	$\Delta_{sol}H°$ kJ/mol	Solute	State	$\Delta_{sol}H°$ kJ/mol
HF	g	–61.50	$LiBr \cdot 2H_2O$	c	–9.41	$KClO_3$	c	41.38
HCl	g	–74.84	$LiBrO_3$	c	1.42	$KClO_4$	c	51.04
$HClO_4$	l	–88.76	LiI	c	–63.30	KBr	c	19.87
$HClO_4 \cdot H_2O$	c	–32.95	$LiI \cdot H_2O$	c	–29.66	$KBrO_3$	c	41.13
HBr	g	–85.14	$LiI \cdot 2H_2O$	c	–14.77	KI	c	20.33
HI	g	–81.67	$LiI \cdot 3H_2O$	c	0.59	KIO_3	c	27.74
HIO_3	c	8.79	$LiNO_2$	c	–11.00	KNO_2	c	13.35
HNO_3	l	–33.28	$LiNO_2 \cdot H_2O$	c	7.03	KNO_3	c	34.89
HCOOH	l	–0.86	$LiNO_3$	c	–2.51	$KC_2H_3O_2$	c	–15.33
CH_3COOH	l	–1.51				KCN	c	11.72
			NaOH	c	–44.51	KCNO	c	20.25
NH_3	g	–30.50	$NaOH \cdot H_2O$	c	–21.41	KCNS	c	24.23
NH_4Cl	c	14.78	NaF	c	0.91	$KMnO_4$	c	43.56
NH_4ClO_4	c	33.47	NaCl	c	3.88			
NH_4Br	c	16.78	$NaClO_2$	c	0.33	RbOH	c	–62.34
NH_4I	c	13.72	$NaClO_2 \cdot 3H_2O$	c	28.58	$RbOH \cdot H_2O$	c	–17.99
NH_4IO_3	c	31.80	$NaClO_3$	c	21.72	$RbOH \cdot 2H_2O$	c	0.88
NH_4NO_2	c	19.25	$NaClO_4$	c	13.88	RbF	c	–26.11
NH_4NO_3	c	25.69	$NaClO_4 \cdot H_2O$	c	22.51	$RbF \cdot H_2O$	c	–0.42
$NH_4C_2H_3O_2$	c	–2.38	NaBr	c	–0.60	$RbF \cdot 1.5H_2O$	c	1.34
NH_4CN	c	17.57	$NaBr \cdot 2H_2O$	c	18.64	RbCl	c	17.28
NH_4CNS	c	22.59	$NaBrO_3$	c	26.90	$RbClO_3$	c	47.74
CH_3NH_3Cl	c	5.77	NaI	c	–7.53	$RbClO_4$	c	56.74
$(CH_3)_3NHCl$	c	1.46	$NaI \cdot 2H_2O$	c	16.13	RbBr	c	21.88
$N(CH_3)_4Cl$	c	4.08	$NaIO_3$	c	20.29	$RbBrO_3$	c	48.95
$N(CH_3)_4Br$	c	24.27	$NaNO_2$	c	13.89	RbI	c	25.10
$N(CH_3)_4I$	c	42.07	$NaNO_3$	c	20.50	$RbNO_3$	c	36.48
			$NaC_2H_3O_2$	c	–17.32			
$AgClO_4$	c	7.36	$NaC_2H_3O_2 \cdot 3H_2O$	c	19.66	CsOH	c	–71.55
$AgNO_2$	c	36.94	NaCN	c	1.21	$CsOH \cdot H_2O$	c	–20.50
$AgNO_3$	c	22.59	$NaCN \cdot 0.5H_2O$	c	3.31	CsF	c	–36.86
			$NaCN \cdot 2H_2O$	c	18.58	$CsF \cdot H_2O$	c	–10.46
LiOH	c	–23.56	NaCNO	c	19.20	$CsF \cdot 1.5H_2O$	c	–5.44
$LiOH \cdot H_2O$	c	–6.69	NaCNS	c	6.83	CsCl	c	17.78
LiF	c	4.73				$CsClO_4$	c	55.44
LiCl	c	–37.03	KOH	c	–57.61	CsBr	c	25.98
$LiCl \cdot H_2O$	c	–19.08	$KOH \cdot H_2O$	c	–14.64	$CsBrO_3$	c	50.46
$LiClO_4$	c	–26.55	$KOH \cdot 1.5H_2O$	c	–10.46	CsI	c	33.35
$LiClO_4 \cdot 3H_2O$	c	32.61	KF	c	–17.73	$CsNO_3$	c	40.00
LiBr	c	–48.83	$KF \cdot 2H_2O$	c	6.97			
$LiBr \cdot H_2O$	c	–23.26	KCl	c	17.22			

ENTHALPY OF HYDRATION OF GASES

The molar enthalpy of hydration $\Delta_{hyd}H^\infty$ is defined as the enthalpy change when one mole of an ideal gas is dissolved in an infinite amount of water. Another term for this quantity is enthalpy (heat) of solvation in water at infinite dilution. The enthalpy of hydration influences the distribution of a volatile compound between the aqueous solution phase and air and is thus important in fields such as environmental science, geochemistry, and chemical engineering. It is related to the enthalpy of solution of the liquid or solid phase, $\Delta_{sol}H^\infty$, by

$$\Delta_{hyd}H^\infty = \Delta_{sol}H^\infty - \Delta_{vap}H^0$$

where $\Delta_{vap}H^0$ is the molar enthalpy of vaporization. This table gives the molar enthalpy of hydration for a number of common substances.

References

1. Plyasunov, A. V., and Shock, E. L., *J. Chem. Eng. Data* [Hydrocarbons, alcohols, and ketones] 46, 1016, 2001.
2. Plyasunov, A. V., Plyasunova, N. V., and Shock, E. L., *J. Chem. Eng. Data* [Esters] 49, 1152, 2004.
3. Plyasunov, A. V., Plyasunova, N. V., and Shock, E. L., *J. Chem. Eng. Data* [Ethers, diethers, and polyethers] 51, 276, 2006.
4. Plyasunova, N. V., Plyasunov, A. V., and Shock, E. L., *J. Chem. Eng. Data* [Thiols, sulfides, and polysulfides] 50, 246, 2005.
5. Plyasunov, A. V., Plyasunova, N. V., and Shock, E. L., *J. Chem. Eng. Data* [Nitriles and dinitriles] 51, 1481, 2006.
6. Plyasunov, A. V., and Shock, E. L., *Geochim. Cosmochim. Acta* [Inorganics and halogen compounds; temperature dependence] 67, 4981, 2003.
7. Kühne, R., Ebert, R-U, and Schüürmann, G., *Environ. Sci. Technol.* [Compilation; temperature dependence] 39, 6705, 2005.
8. Mintz, C., Clark, M., Acree, W. E., and Abraham, M. H., *J. Chem. Inf. Model.* [Compilation and modelling] 47, 115, 2007.

Name	Mol. Form.	$\Delta_{hyd}H^\infty$/kJ mol^{-1} at 298.15 K
Acenaphthene	$C_{12}H_{10}$	−52.1
Acetic acid	$C_2H_4O_2$	−52.8
Acetone	C_3H_6O	−39.7
Acetonitrile	C_2H_3N	−34.9
Acetophenone	C_8H_8O	−53.3
Ammonia	H_3N	−35.4
Aniline	C_6H_7N	−56.5
Anisole	C_7H_8O	−41.4
Argon	Ar	−12.2
Benzaldehyde	C_7H_6O	−42.1
Benzene	C_6H_6	−28.1
Benzonitrile	C_7H_5N	−48.5
Benzyl alcohol	C_7H_8O	−66.9
Biphenyl	$C_{12}H_{10}$	−47.2
Bromobenzene	C_6H_5Br	−33.5
Bromodichloromethane	$CHBrCl_2$	−28.9
Bromoethane	C_2H_5Br	−29.5
Bromomethane	CH_3Br	−23.8
2-Bromo-2-methylpropane	C_4H_9Br	−25.4
1,3-Butadiene	C_4H_6	−31.4
Butane	C_4H_{10}	−24.8
1,4-Butanediamine	$C_4H_{12}N_2$	−91.6
1,4-Butanediol	$C_4H_{10}O_2$	−89.6
Butanenitrile	C_4H_7N	−42.1
1,2,3,4-Butanetetrol	$C_4H_{10}O_4$	−114
1-Butanethiol	$C_4H_{10}S$	−36.3
Butanoic acid	$C_4H_8O_2$	−59.5
1-Butanol	$C_4H_{10}O$	−61.9
2-Butanol	$C_4H_{10}O$	−62.7
2-Butanone	C_4H_8O	−41.9
1-Butene	C_4H_8	−24.1
2-Butoxyethanol	$C_6H_{14}O_2$	−73.6
Butyl acetate	$C_6H_{12}O_2$	−52.7
sec-Butyl acetate	$C_6H_{12}O_2$	−51.9
tert-Butyl acetate	$C_6H_{12}O_2$	−46.2
Butylamine	$C_4H_{11}N$	−59.2
sec-Butylamine	$C_4H_{11}N$	−57.1
tert-Butylamine	$C_4H_{11}N$	−59.0
Butylbenzene	$C_{10}H_{14}$	−38.5
Butyl butanoate	$C_8H_{16}O_2$	−63.5
Butyl ethyl ether	$C_6H_{14}O$	−48.4
tert-Butyl ethyl ether	$C_6H_{14}O$	−53.4
4-*tert*-Butylphenol	$C_{10}H_{14}O$	−63.8
Butyl propanoate	$C_7H_{14}O_2$	−57.8
1-Butyne	C_4H_6	−13.5
Carbon dioxide	CO_2	−17.9
Carbon monoxide	CO	−11.1
Chlorine	Cl_2	−23.4
Chlorine dioxide	ClO_2	−27.8
Chlorobenzene	C_6H_5Cl	−30.6
2-Chloro-1,1′-biphenyl	$C_{12}H_9Cl$	−42.8
1-Chlorobutane	C_4H_9Cl	−28.2
2-Chlorobutane	C_4H_9Cl	−34.6
Chlorodibromomethane	$CHBr_2Cl$	−33.3
Chlorodifluoromethane	$CHClF_2$	−22.8
Chloroethane	C_2H_5Cl	−22.0
Chlorofluoromethane	CH_2ClF	−21.7
1-Chlorohexane	$C_6H_{13}Cl$	−34.5
Chloromethane	CH_3Cl	−20.2
1-Chloropentane	$C_5H_{11}Cl$	−34.1
3-Chlorophenol	C_6H_5ClO	−50.3
1-Chloropropane	C_3H_7Cl	−27.0
2-Chloropyridine	C_5H_4ClN	−42.6
3-Chloropyridine	C_5H_4ClN	−46.2
2-Chlorotoluene	C_7H_7Cl	−38.3
3-Chlorotoluene	C_7H_7Cl	−37.0

Name	Mol. Form.	$\Delta_{hyd}H^{\circ}$/kJ mol^{-1} at 298.15 K
4-Chlorotoluene	C_7H_7Cl	−33.3
o-Cresol	C_7H_8O	−64.8
m-Cresol	C_7H_8O	−58.7
p-Cresol	C_7H_8O	−61.3
Cycloheptanol	$C_7H_{14}O$	−74.6
Cyclohexane	C_6H_{12}	−30.0
cis-1,2-Cyclohexanediol	$C_6H_{12}O_2$	−82.4
Cyclohexanol	$C_6H_{12}O$	−70.7
Cyclohexanone	$C_6H_{10}O$	−49.8
Cyclohexene	C_6H_{10}	−27.3
Cyclooctane	C_8H_{16}	−39.0
cis-Cyclooctene	C_8H_{14}	−45.5
Cyclopentane	C_5H_{10}	−30.3
Cyclopentanol	$C_5H_{10}O$	−58.5
Cyclopentanone	C_5H_8O	−44.3
Cyclopropane	C_3H_6	−15.4
Dibromomethane	CH_2Br_2	−33.0
Dibutylamine	$C_8H_{19}N$	−59.3
Dibutyl ether	$C_8H_{18}O$	−55.8
o-Dichlorobenzene	$C_6H_4Cl_2$	−37.3
m-Dichlorobenzene	$C_6H_4Cl_2$	−35.3
p-Dichlorobenzene	$C_6H_4Cl_2$	−28.4
2,3-Dichloro-1,1′-biphenyl	$C_{12}H_8Cl_2$	−45.6
2,4-Dichloro-1,1′-biphenyl	$C_{12}H_8Cl_2$	−43.0
2,4′-Dichloro-1,1′-biphenyl	$C_{12}H_8Cl_2$	−44.2
2,5-Dichlorobiphenyl	$C_{12}H_8Cl_2$	−45.6
Dichlorodifluoromethane	CCl_2F_2	−26.0
2,2-Dichloro-1,1-difluoro-1-methoxyethane	$C_3H_4Cl_2F_2O$	−30.4
1,1-Dichloroethane	$C_2H_4Cl_2$	−30.3
1,2-Dichloroethane	$C_2H_4Cl_2$	−27.9
1,1-Dichloroethene	$C_2H_2Cl_2$	−28.5
cis-1,2-Dichloroethene	$C_2H_2Cl_2$	−26.9
trans-1,2-Dichloroethene	$C_2H_2Cl_2$	−29.3
Dichloromethane	CH_2Cl_2	−30.3
1,2-Dichloropropane	$C_3H_6Cl_2$	−31.1
1,3-Dichloropropane	$C_3H_6Cl_2$	−29.7
1,2-Dichloro-1,1,2,2-tetrafluoroethane	$C_2Cl_2F_4$	−20.2
1,2-Diethoxyethane	$C_6H_{14}O_2$	−71.9
Diethylamine	$C_4H_{11}N$	−64.3
N,N-Diethylaniline	$C_{10}H_{15}N$	−45.7
p-Diethylbenzene	$C_{10}H_{14}$	−46.4
Diethylene glycol dimethyl ether	$C_6H_{14}O_3$	−96.2
Diethyl ether	$C_4H_{10}O$	−46.4
Diethyl sulfide	$C_4H_{10}S$	−40.2
1,1-Difluoroethane	$C_2H_4F_2$	−20.7
Difluoromethane	CH_2F_2	−17.2
Diiodomethane	CH_2I_2	−41.6
Diisopropyl ether	$C_6H_{14}O$	−51.7
1,2-Dimethoxyethane	$C_4H_{10}O_2$	−59.3
Dimethylamine	C_2H_7N	−53.1
2,4-Dimethylaniline	$C_8H_{11}N$	−58.7
2,5-Dimethylaniline	$C_8H_{11}N$	−61.5
2,6-Dimethylaniline	$C_8H_{11}N$	−60.5
N,N-Dimethylaniline	$C_8H_{11}N$	−49.6
2,3-Dimethylbutane	C_6H_{14}	−32.4
3,3-Dimethyl-2-butanone	$C_6H_{12}O$	−47.5
cis-1,2-Dimethylcyclohexane	C_8H_{16}	−38.3
trans-1,2-Dimethylcyclohexane	C_8H_{16}	−36.1
Dimethyl ether	C_2H_6O	−34.0
N,N-Dimethylformamide	C_3H_7NO	−62.9
2,4-Dimethyl-3-pentanone	$C_7H_{14}O$	−54.0
2,3-Dimethylpyridine	C_7H_9N	−57.7
2,4-Dimethylpyridine	C_7H_9N	−60.7
2,5-Dimethylpyridine	C_7H_9N	−54.9
2,6-Dimethylpyridine	C_7H_9N	−52.3
3,4-Dimethylpyridine	C_7H_9N	−50.5
3,5-Dimethylpyridine	C_7H_9N	−51.3
Dimethyl sulfide	C_2H_6S	−31.5
Dimethyl sulfoxide	C_2H_6OS	−71.9
2,5-Dimethyltetrahydrofuran	$C_6H_{12}O$	−56.3
1,4-Dioxane	$C_4H_8O_2$	−48.4
1,2-Dipropoxyethane	$C_8H_{18}O_2$	−76.8
Dipropylamine	$C_6H_{15}N$	−65.2
Dipropyl ether	$C_6H_{14}O$	−49.9
Dipropyl sulfide	$C_6H_{14}S$	−47.7
1-Dodecanol	$C_{12}H_{26}O$	−81.9
Ethane	C_2H_6	−17.9
1,2-Ethanediamine	$C_2H_8N_2$	−76.1
1,2-Ethanediol	$C_2H_6O_2$	−77.3
Ethanethiol	C_2H_6S	−28.9
Ethanol	C_2H_6O	−50.6
2-Ethoxyethanol	$C_4H_{10}O_2$	−66.4
1-Ethoxy-2-methoxyethane	$C_5H_{12}O_2$	−66.1
Ethyl acetate	$C_4H_8O_2$	−45.3
Ethylamine	C_2H_7N	−53.7
2-Ethylaniline	$C_8H_{11}N$	−59.7
4-Ethylaniline	$C_8H_{11}N$	−65.0
Ethylbenzene	C_8H_{10}	−39.4
Ethyl butanoate	$C_6H_{12}O_2$	−52.7
Ethylcyclohexane	C_8H_{16}	−36.8
Ethyl 2,2-dimethylpropanoate	$C_7H_{14}O_2$	−50.3
Ethylene	C_2H_4	−13.7
Ethyl formate	$C_3H_6O_2$	−38.1
Ethyl hexanoate	$C_8H_{16}O_2$	−60.2
Ethyl 3-methylbutanoate	$C_7H_{14}O_2$	−56.0
Ethyl 2-methylbutanoate	$C_7H_{14}O_2$	−55.4
Ethyl 2-methylpropanoate	$C_6H_{12}O_2$	−51.3
Ethyl pentanoate	$C_7H_{14}O_2$	−56.5
Ethyl propanoate	$C_5H_{10}O_2$	−49.5
2-Ethylpyridine	C_7H_9N	−55.7

Name	Mol. Form.	$\Delta_{hyd}H^\infty$/kJ mol^{-1} at 298.15 K
3-Ethylpyridine	C_7H_9N	−53.5
4-Ethylpyridine	C_7H_9N	−52.2
9*H*-Fluorene	$C_{13}H_{10}$	−42.7
Fluorobenzene	C_6H_5F	−29.3
Fluoromethane	CH_3F	−16.1
Glycerol	$C_3H_8O_3$	−103.5
Helium	He	−0.67
1,1,1,2,3,3,3-Heptafluoropropane	C_3HF_7	−24.8
Heptanal	$C_7H_{14}O$	−56.6
Heptane	C_7H_{16}	−34.0
1-Heptanol	$C_7H_{16}O$	−72.1
2-Heptanol	$C_7H_{16}O$	−72.6
4-Heptanol	$C_7H_{16}O$	−75.3
2-Heptanone	$C_7H_{14}O$	−54.9
4-Heptanone	$C_7H_{14}O$	−58.1
1,1,1,3,3,3-Hexafluoro-2-propanol	$C_3H_2F_6O$	−57.1
Hexanal	$C_6H_{12}O$	−55.2
Hexane	C_6H_{14}	−31.9
Hexanedinitrile	$C_6H_8N_2$	−66.6
1-Hexanol	$C_6H_{14}O$	−67.4
3-Hexanol	$C_6H_{14}O$	−69.6
2-Hexanone	$C_6H_{12}O$	−48.9
3-Hexanone	$C_6H_{12}O$	−46.0
1-Hexene	C_6H_{12}	−30.4
Hexyl acetate	$C_8H_{16}O_2$	−60.8
Hexylamine	$C_6H_{15}N$	−65.9
Hexylbenzene	$C_{12}H_{18}$	−52.7
Hydrogen	H_2	−0.402
Hydrogen selenide	H_2Se	−15.7
Hydrogen sulfide	H_2S	−18.0
3-Hydroxybenzaldehyde	$C_7H_6O_2$	−70.7
3-Hydroxybenzonitrile	C_7H_5NO	−70.7
4-Hydroxybenzonitrile	C_7H_5NO	−70.3
Iodoethane	C_2H_5I	−31.7
Iodomethane	CH_3I	−28.2
1-Iodopropane	C_3H_7I	−35.3
2-Iodopropane	C_3H_7I	−36.6
Isobutanal	C_4H_8O	−40.0
Isobutane	C_4H_{10}	−21.7
Isobutene	C_4H_8	−22.7
Isobutyl acetate	$C_6H_{12}O_2$	−51.8
Isobutyl formate	$C_5H_{10}O_2$	−43.0
Isobutyl isobutanoate	$C_8H_{16}O_2$	−55.3
Isobutyl propanoate	$C_7H_{14}O_2$	−54.7
Isoflurane	$C_3H_2ClF_5O$	−35.3
Isopentyl acetate	$C_7H_{14}O_2$	−53.8
Isopentyl formate	$C_6H_{12}O_2$	−47.7
Isophorone	$C_9H_{14}O$	−59.1
Isopropyl acetate	$C_5H_{10}O_2$	−46.8
Isopropylamine	C_3H_9N	−55.0
Isopropylbenzene	C_9H_{12}	−33.7

Name	Mol. Form.	$\Delta_{hyd}H^\infty$/kJ mol^{-1} at 298.15 K
Isopropyl formate	$C_4H_8O_2$	−43.0
1-Isopropyl-4-methylbenzene	$C_{10}H_{14}$	−34.6
Krypton	Kr	−15.6
Methane	CH_4	−12.0
Methanethiol	CH_4S	−24.4
Methanol	CH_4O	−52.0
2-Methoxyethanol	$C_3H_8O_2$	−60.4
2-Methoxy-2-methylbutane	$C_6H_{14}O$	−52.5
2-Methoxyphenol	$C_7H_8O_2$	−62.6
Methyl acetate	$C_3H_6O_2$	−40.1
Methylamine	CH_5N	−45.3
Methyl benzoate	$C_8H_8O_2$	−50.3
Methyl butanoate	$C_5H_{10}O_2$	−47.5
3-Methyl-1-butanol	$C_5H_{12}O$	−66.0
2-Methyl-2-butanol	$C_5H_{12}O$	−68.4
3-Methyl-2-butanone	$C_5H_{10}O$	−57.6
2-Methyl-2-butene	C_5H_{10}	−26.6
Methyl *tert*-butyl ether	$C_5H_{12}O$	−48.7
Methyl 2,2-dimethylpropanoate	$C_6H_{12}O_2$	−46.2
Methyl formate	$C_2H_4O_2$	−32.0
Methyl hexanoate	$C_7H_{14}O_2$	−54.7
Methyl isobutanoate	$C_5H_{10}O_2$	−46.0
4-Methylmorpholine	$C_5H_{11}NO$	−68.7
1-Methylnaphthalene	$C_{11}H_{10}$	−45.0
2-Methylnaphthalene	$C_{11}H_{10}$	−44.9
2-Methylpentane	C_6H_{14}	−30.5
3-Methylpentane	C_6H_{14}	−36.8
Methyl pentanoate	$C_6H_{12}O_2$	−50.4
4-Methyl-2-pentanol	$C_6H_{14}O$	−69.9
4-Methyl-2-pentanone	$C_6H_{12}O$	−44.6
1-Methylpiperidine	$C_6H_{13}N$	−65.8
2-Methylpropanenitrile	C_4H_7N	−40.0
Methyl propanoate	$C_4H_8O_2$	−44.5
2-Methyl-1-propanol	$C_4H_{10}O$	−60.2
2-Methyl-2-propanol	$C_4H_{10}O$	−62.9
Methyl propyl ether	$C_4H_{10}O$	−38.0
2-Methylpyridine	C_6H_7N	−50.3
3-Methylpyridine	C_6H_7N	−50.3
4-Methylpyridine	C_6H_7N	−51.8
N-Methylpyrrolidine	$C_5H_{11}N$	−63.4
2-Methyltetrahydrofuran	$C_5H_{10}O$	−51.4
Morpholine	C_4H_9NO	−69.5
Naphthalene	$C_{10}H_8$	−42.8
Neon	Ne	−3.90
Neopentane	C_5H_{12}	−23.4
Nitric oxide	NO	−11.9
Nitrobenzene	$C_6H_5NO_2$	−43.8
Nitroethane	$C_2H_5NO_2$	−32.5
Nitrogen	N_2	−1.04
Nitromethane	CH_3NO_2	−35.7
2-Nitrophenol	$C_6H_5NO_3$	−49.8

Name	Mol. Form.	$\Delta_{hyd}H^{\infty}$/kJ mol^{-1} at 298.15 K
3-Nitrophenol	$C_6H_5NO_3$	–67.7
4-Nitrophenol	$C_6H_5NO_3$	–68.6
1-Nitropropane	$C_3H_7NO_2$	–34.4
2-Nitropropane	$C_3H_7NO_2$	–34.1
2-Nitrotoluene	$C_7H_7NO_2$	–46.4
3-Nitrotoluene	$C_7H_7NO_2$	–38.5
Nitrous oxide	N_2O	–19.8
5-Nonanone	$C_9H_{18}O$	–62.8
2-Nonanone	$C_9H_{18}O$	–65.3
Octanal	$C_8H_{16}O$	–48.8
Octane	C_8H_{18}	–36.0
1-Octanol	$C_8H_{18}O$	–74.1
2-Octanone	$C_8H_{16}O$	–58.3
1-Octene	C_8H_{16}	39.2
Octylamine	$C_8H_{19}N$	–52.3
Oxygen	O_2	–1.20
Pentachlorobenzene	C_6HCl_5	–39.9
Pentafluoroethane	C_2HF_5	–21.5
2,2,3,3,3-Pentafluoro-1-propanol	$C_3H_3F_5O$	–51.9
Pentanal	$C_5H_{10}O$	–42.9
Pentane	C_5H_{12}	–28.3
1,5-Pentanediamine	$C_5H_{14}N_2$	–95.1
Pentanedinitrile	$C_5H_6N_2$	–63.5
Pentanenitrile	C_5H_9N	–45.6
1-Pentanol	$C_5H_{12}O$	–61.9
2-Pentanol	$C_5H_{12}O$	–63.3
3-Pentanol	$C_5H_{12}O$	–59.6
2-Pentanone	$C_5H_{10}O$	–45.3
3-Pentanone	$C_5H_{10}O$	–49.6
Pentyl acetate	$C_7H_{14}O_2$	–55.3
Pentylamine	$C_5H_{13}N$	–62.1
Pentylbenzene	$C_{11}H_{16}$	–49.5
Pentyl formate	$C_6H_{12}O_2$	–48.1
Perfluoropropene	C_3F_6	–17.4
Phenol	C_6H_6O	–57.7
1-Phenyl-1-propanone	$C_9H_{10}O$	–61.9
Piperidine	$C_5H_{11}N$	–65.4
Propanal	C_3H_6O	–39.4
Propanamide	C_3H_7NO	–73.4
Propane	C_3H_8	–20.4
1,3-Propanediamine	$C_3H_{10}N_2$	–85.6
1,3-Propanediol	$C_3H_8O_2$	–81.1
Propanenitrile	C_3H_5N	–39.5
1-Propanethiol	C_3H_8S	–30.2
Propanoic acid	$C_3H_6O_2$	–56.5
1-Propanol	C_3H_8O	–59.9
2-Propanol	C_3H_8O	–58.2
Propene	C_3H_6	–21.6
2-Propoxyethanol	$C_5H_{12}O_2$	–69.6
Propyl acetate	$C_5H_{10}O_2$	–48.7
Propylamine	C_3H_9N	–56.0
Propylbenzene	C_9H_{12}	–36.4
Propyl butanoate	$C_7H_{14}O_2$	–54.9
Propyl formate	$C_4H_8O_2$	–40.5
Propyl propanoate	$C_6H_{12}O_2$	–51.2
Propyne	C_3H_4	–15.6
Pyridine	C_5H_5N	–42.1
Quinoline	C_9H_7N	–58.2
Radon	Rn	–24.0
Styrene	C_8H_8	–28.4
Succinonitrile	$C_4H_4N_2$	–58.2
Sulfur hexafluoride	F_6S	–20.7
1,2,3,4-Tetrachlorobenzene	$C_6H_2Cl_4$	–35.0
1,1,2,2-Tetrachloroethane	$C_2H_2Cl_4$	–34.8
1,1,1,2-Tetrachloroethane	$C_2H_2Cl_4$	–36.2
Tetrachloroethene	C_2Cl_4	–41.5
Tetrachloromethane	CCl_4	–30.5
Tetraethylene glycol dimethyl ether	$C_{10}H_{22}O_5$	–126
1,1,1,2-Tetrafluoroethane	$C_2H_2F_4$	–22.2
Tetrafluoroethene	C_2F_4	–15.1
Tetrafluoromethane	CF_4	–13.5
2,2,3,3-Tetrafluoro-1-propanol	$C_3H_4F_4O$	–57.9
Tetrahydrofuran	C_4H_8O	–47.3
Tetrahydropyran	$C_5H_{10}O$	–48.9
Thiophene	C_4H_4S	–29.9
Toluene	C_7H_8	–32.4
Tribromomethane	$CHBr_3$	–35.8
1,2,3-Trichlorobenzene	$C_6H_3Cl_3$	–32.6
1,3,5-Trichlorobenzene	$C_6H_3Cl_3$	–34.2
1,1,1-Trichloroethane	$C_2H_3Cl_3$	–28.7
1,1,2-Trichloroethane	$C_2H_3Cl_3$	–32.5
Trichloroethene	C_2HCl_3	–32.2
Trichlorofluoromethane	CCl_3F	–19.8
Trichloromethane	$CHCl_3$	–33.5
1,1,2-Trichloro-1,2,2-trifluoroethane	$C_2Cl_3F_3$	–28.8
Triethylamine	$C_6H_{15}N$	–69.7
Triethylene glycol dimethyl ether	$C_8H_{18}O_4$	–102.4
2,2,2-Trifluoroethanol	$C_2H_3F_3O$	–50.2
Trifluoromethane	CHF_3	–22.6
1,1,1-Trifluoro-2-propanol	$C_3H_5F_3O$	–53.5
Trimethylamine	C_3H_9N	–52.7
1,2,3-Trimethylbenzene	C_9H_{12}	–37.4
1,2,4-Trimethylbenzene	C_9H_{12}	–36.6
1,3,5-Trimethylbenzene	C_9H_{12}	–39.1
2,2,4-Trimethylpentane	C_8H_{18}	–31.0
2,3,4-Trimethylpentane	C_8H_{18}	–38.5
Xenon	Xe	–19.4
o-Xylene	C_8H_{10}	–37.7
m-Xylene	C_8H_{10}	–38.6
p-Xylene	C_8H_{10}	–34.8

CHEMICAL KINETIC DATA FOR STRATOSPHERIC MODELING

The present compilation of kinetic data represents the 12th evaluation prepared by the NASA Panel for Data Evaluation. The Panel was established in 1977 by the NASA Upper Atmosphere Research Program Office for the purpose of providing a critical tabulation of the latest kinetic and photochemical data for use by modelers in computer simulations of stratospheric chemistry. The recommended rate data and cross sections are based on laboratory measurements. The major use of theoretical extrapolation of data is in connection with three-body reactions, in which the required pressure or temperature dependence is sometimes unavailable from laboratory measurements, and can be estimated by use of appropriate theoretical treatment. In the case of important rate constants for which no experimental data are available, the panel may provide estimates of rate constant parameters based on analogy to similar reactions for which data are available.

Rate constants are expressed in the form $k(T) = A \exp(-E/RT)$, where A is the pre-exponential factor, E the activation energy, R the gas constant, and T the absolute temperature. Uncertainties are expressed by the factor f, e.g., a value of 4.2×10^{-10} with $f = 2$ indicates that the true value is believed to lie between 2.1×10^{-10} and 8.4×10^{-10}. The value of f at other temperatures may be calculated from $f(298)$, given in the last column, by:

$$f(T) = f(298) \exp[(\Delta E/R)(1/T - 1/298)],$$

where $\Delta E/R$ is the uncertainty in E/R.

Table 1 covers rate constant data on second order reactions, grouped by class, while Table 2 covers association reactions. Relevant equilibrium constant data are given in Table 3. All concentrations are measured in molecules cm^{-3}. Notes on each reaction, as well as related photochemical data, may be found in the reference.

The assistance of Robert Hampson is gratefully acknowledged.

Reference

DeMore, W. B., Sander, S. P., Golden, D. M., Hampson, R. F., Kurylo, M. J., Howard, C. J., Ravishankara, A. R., Kolb, C. E., and Molina, M. J., *Chemical Kinetics and Photochemical Data for Use in Atmospheric Modeling. Evaluation Number 12*, Jet Propulsion Laboratory Publication 97-4, Pasadena CA, 1997.

The report is also available at the World Wide Web site < http://remus.jpl.nasa.gov/pub/jpl97>.

TABLE 1. Rate Constants for Second Order Reactions

Reaction	A cm^3 molecule^{-1} s^{-1}	E/R K	k (298 K) cm^3 molecule^{-1} s^{-1}	$f(298)$
O_x Reactions				
$O + O_3 \rightarrow O_2 + O_2$	8.0×10^{-12}	2060±250	8.0×10^{-15}	1.15
$O(^1D)$ Reactions				
$O(^1D) + O_2 \rightarrow O + O_2$	3.2×10^{-11}	−(70±100)	4.0×10^{-11}	1.2
$O(^1D) + O_3 \rightarrow O_2 + O_2$	1.2×10^{-10}	0±100	1.2×10^{-10}	1.3
$\rightarrow O_2 + O + O$	1.2×10^{-10}	0±100	1.2×10^{-10}	1.3
$O(^1D) + H_2 \rightarrow OH + H$	1.1×10^{-10}	0±100	1.1×10^{-10}	1.1
$O(^1D) + H_2O \rightarrow OH + OH$	2.2×10^{-10}	0±100	2.2×10^{-10}	1.2
$O(^1D) + N_2 \rightarrow O + N_2$	1.8×10^{-11}	−(110±100)	2.6×10^{-11}	1.2
$O(^1D) + N_2O \rightarrow N_2 + O_2$	4.9×10^{-11}	0±100	4.9×10^{-11}	1.3
$\rightarrow NO + NO$	6.7×10^{-11}	0±100	6.7×10^{-11}	1.3
$O(^1D) + NH_3 \rightarrow OH + NH_2$	2.5×10^{-10}	0±100	2.5×10^{-10}	1.3
$O(^1D) + CO_2 \rightarrow O + CO_2$	7.4×10^{-11}	−(120±100)	1.1×10^{-10}	1.2
$O(^1D) + CH_4 \rightarrow$ products	1.5×10^{-10}	0±100	1.5×10^{-10}	1.2
$O(^1D) + HCl \rightarrow$ products	1.5×10^{-10}	0±100	1.5×10^{-10}	1.2
$O(^1D) + HF \rightarrow OH + F$	1.4×10^{-10}	0±100	1.4×10^{-10}	2.0
$O(^1D) + HBr \rightarrow$ products	1.5×10^{-10}	0±100	1.5×10^{-10}	2.0
$O(^1D) + Cl_2 \rightarrow$ products	2.8×10^{-10}	0±100	2.8×10^{-10}	2.0
$O(^1D) + CCl_2O \rightarrow$ products	3.6×10^{-10}	0±100	3.6×10^{-10}	2.0
$O(^1D) + CClFO \rightarrow$ products	1.9×10^{-10}	0±100	1.9×10^{-10}	2.0
$O(^1D) + CF_2O \rightarrow$ products	7.4×10^{-11}	0±100	7.4×10^{-11}	2.0
$O(^1D) + CCl_4 \rightarrow$ products (CFC–10)	3.3×10^{-10}	0±100	3.3×10^{-10}	1.2
$O(^1D) + CH_3Br \rightarrow$ products	1.8×10^{-10}	0±100	1.8×10^{-10}	1.3
$O(^1D) + CH_2Br_2 \rightarrow$ products	2.7×10^{-10}	0±100	2.7×10^{-10}	1.3
$O(^1D) + CHBr_3 \rightarrow$ products	6.6×10^{-10}	0±100	6.6×10^{-10}	1.5
$O(^1D) + CH_3F \rightarrow$ products (HFC–41)	1.5×10^{-10}	0±100	1.5×10^{-10}	1.2
$O(^1D) + CH_2F_2 \rightarrow$ products (HFC–32)	5.1×10^{-11}	0±100	5.1×10^{-11}	1.3
$O(^1D) + CHF_3 \rightarrow$ products (HFC–23)	9.1×10^{-12}	0±100	9.1×10^{-12}	1.2
$O(^1D) + CHCl_2F \rightarrow$ products (HCFC–21)	1.9×10^{-10}	0±100	1.9×10^{-10}	1.3

Reaction	*A* cm³ molecule⁻¹ s⁻¹	*E/R* K	*k* (298 K) cm³ molecule⁻¹ s⁻¹	*f*(298)
$O(^1D) + CHClF_2 \rightarrow$ products (HCFC–22)	1.0×10^{-10}	0±100	1.0×10^{-10}	1.2
$O(^1D) + CCl_3F \rightarrow$ products (CFC–11)	2.3×10^{-10}	0±100	2.3×10^{-10}	1.2
$O(^1D) + CCl_2F_2 \rightarrow$ products (CFC–12)	1.4×10^{-10}	0±100	1.4×10^{-10}	1.3
$O(^1D) + CClF_3 \rightarrow$ products (CFC–13)	8.7×10^{-11}	0±100	8.7×10^{-11}	1.3
$O(^1D) + CClBrF_2 \rightarrow$ products (Halon–1211)	1.5×10^{-10}	0±100	1.5×10^{-10}	1.3
$O(^1D) + CBr_2F_2 \rightarrow$ products (Halon–1202)	2.2×10^{-10}	0±100	2.2×10^{-10}	1.3
$O(^1D) + CBrF_3 \rightarrow$ products (Halon–1301)	1.0×10^{-10}	0±100	1.0×10^{-10}	1.3
$O(^1D) + CF_4 \rightarrow CF_4 + O$ (CFC–14)	–	–	2.0×10^{-14}	1.5
$O(^1D) + CH_3CH_2F \rightarrow$ products (HFC–161)	2.6×10^{-10}	0±100	2.6×10^{-10}	1.3
$O(^1D) + CH_3CHF_2 \rightarrow$ products (HFC–152a)	2.0×10^{-10}	0±100	2.0×10^{-10}	1.3
$O(^1D) + CH_3CCl_2F \rightarrow$ products (HCFC–141b)	2.6×10^{-10}	0±100	2.6×10^{-10}	1.3
$O(^1D) + CH_3CClF_2 \rightarrow$ products (HCFC–142b)	2.2×10^{-10}	0±100	2.2×10^{-10}	1.3
$O(^1D) + CH_3CF_3 \rightarrow$ products (HFC–143a)	1.0×10^{-10}	0±100	1.0×10^{-10}	3.0
$O(^1D) + CH_2ClCClF_2 \rightarrow$ products (HCFC–132b)	1.6×10^{-10}	0±100	1.6×10^{-10}	2.0
$O(^1D) + CH_2ClCF_3 \rightarrow$ products (HCFC–133a)	1.2×10^{-10}	0±100	1.2×10^{-10}	1.3
$O(^1D) + CH_2FCF_3 \rightarrow$ products (HFC–134a)	4.9×10^{-11}	0±100	4.9×10^{-11}	1.3
$O(^1D) + CHCl_2CF_3 \rightarrow$ products (HCFC–123)	2.0×10^{-10}	0±100	2.0×10^{-10}	1.3
$O(^1D) + CHClFCF_3 \rightarrow$ products (HCFC–124)	8.6×10^{-11}	0±100	8.6×10^{-11}	1.3
$O(^1D) + CHF_2CF_3 \rightarrow$ products (HFC–125)	1.2×10^{-10}	0±100	1.2×10^{-10}	2.0
$O(^1D) + CCl_3CF_3 \rightarrow$ products (CFC–113a)	2×10^{-10}	0±100	2×10^{-10}	2.0
$O(^1D) + CCl_2FCClF_2 \rightarrow$ products (CFC–113)	2×10^{-10}	0±100	2×10^{-10}	2.0
$O(^1D) + CCl_2FCF_3 \rightarrow$ products (CFC 114a)	1×10^{-10}	0±100	1×10^{-10}	2.0
$O(^1D) + CClF_2CClF_2 \rightarrow$ products (CFC–114)	1.3×10^{-10}	0±100	1.3×10^{-10}	1.3
$O(^1D) + CClF_2CF_3 \rightarrow$ products (CFC–115)	5×10^{-11}	0±100	5×10^{-11}	1.3
$O(^1D) + CBrF_2CBrF_2 \rightarrow$ products (Halon–2402)	1.6×10^{-10}	0±100	1.6×10^{-10}	1.3
$O(^1D) + CF_3CF_3 \rightarrow O + CF_3CF_3$ (CFC–116)	–	–	1.5×10^{-13}	1.5
$O(^1D) + CHF_2CF_2CF_2CHF_2 \rightarrow$ products (HFC–338pcc)	1.8×10^{-11}	0±100	1.8×10^{-11}	1.5
$O(^1D) + c\text{–}C_4F_8 \rightarrow$ products	–	–	8×10^{-13}	1.3
$O(^1D) + CF_3CHFCHFCF_2CF_3 \rightarrow$ products (HFC–43–10mee)	2.1×10^{-10}	0±100	2.1×10^{-10}	4
$O(^1D) + C_5F_{12} \rightarrow$ products (CFC–41–12)	–	–	3.9×10^{-13}	2
$O(^1D) + C_6F_{14} \rightarrow$ products (CFC–51–14)	–	–	1×10^{-12}	2
$O(^1D) + 1,2\text{–}(CF_3)_2c\text{–}C_4F_6 \rightarrow$ products	–	–	2.8×10^{-13}	2
$O(^1D) + SF_6 \rightarrow$ products	–	–	1.8×10^{-14}	1.5
Singlet O_2 Reactions				
$O_2(^1\Delta) + O \rightarrow$ products	–	–	$<2\times10^{-16}$	–
$O_2(^1\Delta) + O_2 \rightarrow$ products	3.6×10^{-18}	220±100	1.7×10^{-18}	1.2
$O_2(^1\Delta) + O_3 \rightarrow O + 2O_2$	5.2×10^{-11}	2840±500	3.8×10^{-15}	1.2
$O_2(^1\Delta) + H_2O \rightarrow$ products	–	–	4.8×10^{-18}	1.5
$O_2(^1\Delta) + N \rightarrow NO + O$	–	–	$<9\times10^{-17}$	–
$O_2(^1\Delta) + N_2 \rightarrow$ products	–	–	$<10^{-20}$	–
$O_2(^1\Delta) + CO_2 \rightarrow$ products	–	–	$<2\times10^{-20}$	–
$O_2(^1\Sigma) + O \rightarrow$ products	–	–	8×10^{-14}	5.0
$O_2(^1\Sigma) + O_2 \rightarrow$ products	–	–	3.9×10^{-17}	1.5
$O_2(^1\Sigma) + O_3 \rightarrow$ products	2.2×10^{-11}	0±200	2.2×10^{-11}	1.2
$O_2(^1\Sigma) + H_2O \rightarrow$ products	–	–	5.4×10^{-12}	1.3
$O_2(^1\Sigma) + N \rightarrow$ products	–	–	$<10^{-13}$	–
$O_2(^1\Sigma) + N_2 \rightarrow$ products	2.1×10^{-15}	0±200	2.1×10^{-15}	1.2
$O_2(^1\Sigma) + CO_2 \rightarrow$ products	4.2×10^{-13}	0±200	4.2×10^{-13}	1.2
HO_x Reactions				
$O + OH \rightarrow O_2 + H$	2.2×10^{-11}	–(120±100)	3.3×10^{-11}	1.2
$O + HO_2 \rightarrow OH + O_2$	3.0×10^{-11}	–(200±100)	5.9×10^{-11}	1.2
$O + H_2O_2 \rightarrow OH + HO_2$	1.4×10^{-12}	2000±1000	1.7×10^{-15}	2.0
$H + O_3 \rightarrow OH + O_2$	1.4×10^{-10}	470±200	2.9×10^{-11}	1.25
$H + HO_2 \rightarrow$ products	8.1×10^{-11}	0±100	8.1×10^{-11}	1.3
$OH + O_3 \rightarrow HO_2 + O_2$	1.6×10^{-12}	940±300	6.8×10^{-14}	1.3
$OH + H_2 \rightarrow H_2O + H$	5.5×10^{-12}	2000±100	6.7×10^{-15}	1.1
$OH + HD \rightarrow$ products	5.0×10^{-12}	2130±200	4.0×10^{-15}	1.2
$OH + OH \rightarrow H_2O + O$	4.2×10^{-12}	240±240	1.9×10^{-12}	1.4

Reaction	A cm³ molecule⁻¹ s⁻¹	E/R K	k (298 K) cm³ molecule⁻¹ s⁻¹	f(298)
$OH + HO_2 \rightarrow H_2O + O_2$	4.8×10^{-11}	$-(250\pm200)$	1.1×10^{-10}	1.3
$OH + H_2O_2 \rightarrow H_2O + HO_2$	2.9×10^{-12}	160 ± 100	1.7×10^{-12}	1.2
$HO_2 + O_3 \rightarrow OH + 2O_2$	1.1×10^{-14}	$500\pm$	2.0×10^{-15}	1.3
$HO_2 + HO_2 \rightarrow H_2O_2 + O_2$	2.3×10^{-13}	$-(600\pm200)$	1.7×10^{-12}	1.3
$H_2O_2 + O_2$	$1.7\times10^{-33}[M]$	$-(1000\pm400)$	$4.9\times10^{-32}[M]$	1.3
NO_x Reactions				
$O + NO_2 \rightarrow NO + O_2$	6.5×10^{-12}	$-(120\pm120)$	9.7×10^{-12}	1.1
$O + NO_3 \rightarrow O_2 + NO_2$	1.0×10^{-11}	0 ± 150	1.0×10^{-11}	1.5
$O + N_2O_5 \rightarrow$ products			$<3.0\times10^{-16}$	
$O + HNO_3 \rightarrow OH + NO_3$			$<3.0\times10^{-17}$	
$O + HO_2NO_2 \rightarrow$ products	7.8×10^{-11}	3400 ± 750	8.6×10^{-16}	3.0
$H + NO_2 \rightarrow OH + NO$	4.0×10^{-10}	340 ± 300	1.3×10^{-10}	1.3
$OH + NO_3 \rightarrow$ products			2.2×10^{-11}	1.5
$OH + HONO \rightarrow H_2O + NO_2$	1.8×10^{-11}	$390\pm$	4.5×10^{-12}	1.5
$OH + HNO_3 \rightarrow H_2O + NO_3$	See reference	1.3		
$OH + HO_2NO_2 \rightarrow$ products	1.3×10^{-12}	$-(380\pm)$	4.6×10^{-12}	1.5
$OH + NH_3 \rightarrow H_2O + NH_2$	1.7×10^{-12}	710 ± 200	1.6×10^{-13}	1.2
$HO_2 + NO \rightarrow NO_2 + OH$	3.5×10^{-12}	$-(250\pm50)$	8.1×10^{-12}	1.15
$HO_2 + NO_2 \rightarrow HONO + O_2$	See reference			
$HO_2 + NO_3 \rightarrow$ products			3.5×10^{-12}	1.5
$HO_2 + NH_2 \rightarrow$ products			3.4×10^{-11}	2.0
$N + O_2 \rightarrow NO + O$	1.5×10^{-11}	3600 ± 400	8.5×10^{-17}	1.25
$N + O_3 \rightarrow NO + O_2$			$<2.0\times10^{-16}$	
$N + NO \rightarrow N_2 + O$	2.1×10^{-11}	$-(100\pm100)$	3.0×10^{-11}	1.3
$N + NO_2 \rightarrow N_2O + O$	5.8×10^{-12}	$-(220\pm100)$	1.2×10^{-11}	1.5
$NO + O_3 \rightarrow NO_2 + O_2$	2.0×10^{-12}	1400 ± 200	1.8×10^{-14}	1.1
$NO + NO_3 \rightarrow 2NO_2$	1.5×10^{-11}	$-(170\pm100)$	2.6×10^{-11}	1.3
$NO_2 + O_3 \rightarrow NO_3 + O_2$	1.2×10^{-13}	2450 ± 150	3.2×10^{-17}	1.15
$NO_2 + NO_3 \rightarrow NO + NO_2 + O_2$	See reference			
$NO_3 + NO_3 \rightarrow 2NO_2 + O_2$	8.5×10^{-13}	2450 ± 500	2.3×10^{-16}	1.5
$NH_2 + O_2 \rightarrow$ products			$<6.0\times10^{-21}$	
$NH_2 + O_3 \rightarrow$ products	4.3×10^{-12}	930 ± 500	1.9×10^{-13}	3.0
$NH_2 + NO \rightarrow$ products	4.0×10^{-12}	$-(450\pm150)$	1.8×10^{-11}	1.3
$NH_2 + NO_2 \rightarrow$ products	2.1×10^{-12}	$-(650\pm250)$	1.9×10^{-11}	3.0
$NH + NO \rightarrow$ products	4.9×10^{-11}	0 ± 300	4.9×10^{-11}	1.5
$NH + NO_2 \rightarrow$ products	3.5×10^{-13}	$-(1140\pm500)$	1.6×10^{-11}	2.0
$O_3 + HNO_2 \rightarrow O_2 + HNO_3$			$<5.0\times10^{-19}$	
$N_2O_5 + H_2O \rightarrow 2HNO_3$			$<2.0\times10^{-21}$	
$N_2(A,v) + O_2 \rightarrow$ products			2.5×10^{-12}, $v=0$	1.5
$N_2(A,v) + O_3 \rightarrow$ products			4.1×10^{-11}, $v=0$	2.0
Reactions of Organic Compounds				
$O + CH_3 \rightarrow$ products	1.1×10^{-10}	0 ± 250	1.1×10^{-10}	1.3
$O + HCN \rightarrow$ products	1.0×10^{-11}	4000 ± 1000	1.5×10^{-17}	10
$O + C_2H_2 \rightarrow$ products	3.0×10^{-11}	1600 ± 250	1.4×10^{-13}	1.3
$O + H_2CO \rightarrow$ products	3.4×10^{-11}	1600 ± 250	1.6×10^{-13}	1.25
$O + CH_3CHO \rightarrow CH_3CO + OH$	1.8×10^{-11}	1100 ± 200	4.5×10^{-13}	1.25
$O_3 + C_2H_2 \rightarrow$ products	1.0×10^{-14}	4100 ± 500	1.0×10^{-20}	3
$O_3 + C_2H_4 \rightarrow$ products	1.2×10^{-14}	2630 ± 100	1.7×10^{-18}	1.25
$O_3 + C_3H_6 \rightarrow$ products	6.5×10^{-15}	1900 ± 200	1.1×10^{-17}	1.2
$OH + CO \rightarrow$ products	1.5×10^{-13} x $(1+0.6P_{atm})$	0 ± 300	1.5×10^{-13} x $(1+0.6P_{atm})$	1.3
$OH + CH_4 \rightarrow CH_3 + H_2O$	2.45×10^{-12}	1775 ± 100	6.3×10^{-15}	1.1
$OH + {}^{13}CH_4 \rightarrow {}^{13}CH_3 + H_2O$	See reference			
$OH + CH_3D \rightarrow$ products	3.5×10^{-12}	1950 ± 200	5.0×10^{-15}	1.15
$OH + H_2CO \rightarrow H_2O + HCO$	1.0×10^{-11}	0 ± 200	1.0×10^{-11}	1.25
$OH + CH_3OH \rightarrow$ products	6.7×10^{-12}	600 ± 300	8.9×10^{-13}	1.2
$OH + CH_3OOH \rightarrow$ products	3.8×10^{-12}	$-(200\pm200)$	7.4×10^{-12}	1.5
$OH + HC(O)OH \rightarrow$ products	4.0×10^{-13}	0 ± 200	4.0×10^{-13}	1.3

Reaction	A cm^3 molecule^{-1} s^{-1}	E/R K	k (298 K) cm^3 molecule^{-1} s^{-1}	f(298)
$OH + HCN \rightarrow$ products	1.2×10^{-13}	400±150	3.1×10^{-14}	3
$OH + C_2H_6 \rightarrow H_2O + C_2H_5$	8.7×10^{-12}	1070±100	2.4×10^{-13}	1.1
$OH + C_3H_8 \rightarrow H_2O + C_3H_7$	1.0×10^{-11}	660±100	1.1×10^{-12}	1.2
$OH + CH_3CHO \rightarrow CH_3CO + H_2O$	5.6×10^{-12}	–(270±200)	1.4×10^{-11}	1.2
$OH + C_2H_5OH \rightarrow$ products	7.0×10^{-12}	235±100	3.2×10^{-12}	1.3
$OH + CH_3C(O)OH \rightarrow$ products	4.0×10^{-13}	–(200±400)	8.0×10^{-13}	1.3
$OH + CH_3C(O)CH_3 \rightarrow CH_3C(O)CH_2 + H_2O$	2.2×10^{-12}	685±100	2.2×10^{-13}	1.15
$OH + CH_3CN \rightarrow$ products	7.8×10^{-13}	1050±200	2.3×10^{-14}	1.5
$OH + CH_3ONO_2 \rightarrow$ products	5.0×10^{-13}	890±500	2.4×10^{-14}	3
$OH + CH_3C(O)O_2NO_2$ (PAN)$\rightarrow$ products			$<4\times10^{-14}$	
$OH + C_3H_7ONO_2 \rightarrow$ products	8.2×10^{-13}	450±300	1.8×10^{-13}	3
$HO_2 + CH_2O \rightarrow$ adduct	6.7×10^{-15}	–(600±600)	5.0×10^{-14}	5
$HO_2 + CH_3O_2 \rightarrow CH_3OOH + O_2$	3.8×10^{-13}	–(800±400)	5.6×10^{-12}	2
$HO_2 + C_2H_5O_2 \rightarrow C_2H_5OOH + O_2$	7.5×10^{-13}	–(700±250)	8.0×10^{-12}	1.5
$HO_2 + CH_3C(O)O_2 \rightarrow$ products	4.5×10^{-13}	–(1000±600)	1.3×10^{-11}	2
$NO_3 + CO \rightarrow$ products			$<4.0\times10^{-19}$	
$NO_3 + CH_2O \rightarrow$ products			5.8×10^{-16}	1.3
$NO_3 + CH_3CHO \rightarrow$ products	1.4×10^{-12}	1900±300	2.4×10^{-15}	1.3
$CH_3 + O_2 \rightarrow$ products			$<3.0\times10^{-16}$	
$CH_3 + O_3 \rightarrow$ products	5.4×10^{-12}	220±150	2.6×10^{-12}	2
$HCO + O_2 \rightarrow CO + HO_2$	3.5×10^{-12}	–(140±140)	5.5×10^{-12}	1.3
$CH_2OH + O_2 \rightarrow CH_2O + HO_2$	9.1×10^{-12}	0±200	9.1×10^{-12}	1.3
$CH_3O + O_2 \rightarrow CH_2O + HO_2$	3.9×10^{-14}	900±300	1.9×10^{-15}	1.5
$CH_3O + NO \rightarrow CH_2O + HNO$	See reference			
$CH_3O + NO_2 \rightarrow CH_2O + HONO$	1.1×10^{-11}	1200±600	2.0×10^{-13}	5
$CH_3O_2 + O_3 \rightarrow$ products			$<3.0\times10^{-17}$	
$CH_3O_2 + CH_3O_2 \rightarrow$ products	2.5×10^{-13}	–(190±190)	4.7×10^{-13}	1.5
$CH_3O_2 + NO \rightarrow CH_3O + NO_2$	3.0×10^{-12}	–(280±60)	7.7×10^{-12}	1.15
$CH_3O_2 + CH_3C(O)O_2 \rightarrow$ products	1.3×10^{-12}	–(640±200)	1.1×10^{-11}	1.5
$C_2H_5 + O_2 \rightarrow C_2H_4 + HO_2$			$<2.0\times10^{-14}$	
$C_2H_5O + O_2 \rightarrow CH_3CHO + HO_2$	6.3×10^{-14}	550±200	1.0×10^{-14}	1.5
$C_2H_5O_2 + C_2H_5O_2 \rightarrow$ products	6.8×10^{-14}	0±300	6.8×10^{-14}	2
$C_2H_5O_2 + NO \rightarrow$ products	2.6×10^{-12}	–(365±150)	8.7×10^{-12}	1.2
$CH_3C(O)O_2 + CH_3C(O)O_2 \rightarrow$ products	2.9×10^{-12}	–(500±150)	1.5×10^{-11}	1.5
$CH_3C(O)O_2 + NO \rightarrow$ products	5.3×10^{-12}	–(360±150)	1.8×10^{-11}	1.4
FO_x Reactions				
$O + FO \rightarrow F + O_2$	2.7×10^{-11}	0±250	2.7×10^{-11}	3.0
$O + FO_2 \rightarrow FO + O_2$	5.0×10^{-11}	0±250	5.0×10^{-11}	5.0
$OH + CH_3F \rightarrow CH_2F + H_2O$ (HFC–41)	3.0×10^{-12}	1500±300	2.0×10^{-14}	1.1
$OH + CH_2F_2 \rightarrow CHF_2 + H_2O$ (HFC–32)	1.9×10^{-12}	1550±200	1.0×10^{-14}	1.2
$OH + CHF_3 \rightarrow CF_3 + H_2O$ (HFC–23)	1.0×10^{-12}	2440±200	2.8×10^{-16}	1.3
$OH + CF_3OH \rightarrow CF_3O + H_2O$			$<2\times10^{-17}$	
$OH + CH_3CH_2F \rightarrow$ products (HFC–161)	7.0×10^{-12}	1100±300	1.7×10^{-13}	1.4
$OH + CH_3CHF_2 \rightarrow$ products (HFC–152a)	2.4×10^{-12}	1260±200	3.5×10^{-14}	1.2
$OH + CH_2FCH_2F \rightarrow CHFCH_2F$ (HFC–152) $+ H_2O$	1.7×10^{-11}	1500±500	1.1×10^{-13}	2.0
$OH + CH_3CF_3 \rightarrow CH_2CF_3 + H_2O$ (HFC–143a)	1.8×10^{-12}	2170±150	1.2×10^{-15}	1.1
$OH + CH_2FCHF_2 \rightarrow$ products (HFC–143)	4.0×10^{-12}	1650±300	1.6×10^{-14}	1.5
$OH + CH_2FCF_3 \rightarrow CHFCF_3 + H_2O$ (HFC–134a)	1.5×10^{-12}	1750±200	4.2×10^{-15}	1.1
$OH + CHF_2CHF_2 \rightarrow CF_2CHF_2$ (HFC–134) $+ H_2O$	1.6×10^{-12}	1680±300	5.7×10^{-15}	2.0
$OH + CHF_2CF_3 \rightarrow CF_2CF_3 + H_2O$ (HFC–125)	5.6×10^{-13}	1700±300	1.9×10^{-15}	1.3
$OH + CH_3OCHF_2 \rightarrow$ products (HFOC–152a)	6.0×10^{-12}	1530±150	3.5×10^{-14}	1.2
$OH + CF_3OCH_3 \rightarrow CF_3OCH_2 + H_2O$ (HFOC–143a)	1.5×10^{-12}	1450±150	1.2×10^{-14}	1.1
$OH + CF_2HOCF_2H \rightarrow CF_2OCF_2H$ (HFOC–134) $+ H_2O$	1.9×10^{-12}	2000±150	2.3×10^{-15}	1.2
$OH + CF_3OCHF_2 \rightarrow CF_3OCF_2 + H_2O$ (HFOC–125)	4.7×10^{-13}	2100±300	4.1×10^{-16}	1.2
$OH + CF_3CH_2CH_3 \rightarrow$ products (HFC–263fb)	–	–	4.2×10^{-14}	1.5
$OH + CH_2FCF_2CHF_2 \rightarrow$ products (HFC–245ca)	2.4×10^{-12}	1660±150	9.1×10^{-15}	1.3
$OH + CHF_2CHFCHF_2 \rightarrow$ products (HFC–245ea)	–	–	1.6×10^{-14}	2.0
$OH + CF_3CHFCH_2F \rightarrow$ products (HFC–245eb)	–	–	1.5×10^{-14}	2.0
$OH + CHF_2CH_2CF_3 \rightarrow$ products (HFC–245fa)	6.1×10^{-13}	1330±150	7.0×10^{-15}	1.2

Reaction	A cm³ molecule⁻¹ s⁻¹	E/R K	k (298 K) cm³ molecule⁻¹ s⁻¹	f(298)
$OH + CF_3CF_2CH_2F \rightarrow CF_3CF_2CHF$ (HFC–236cb) $+H_2O$	1.5×10^{-12}	1750±500	4.2×10^{-15}	2.0
$OH + CF_3CHFCHF_2 \rightarrow$ products (HFC–236ea)	1.1×10^{-12}	1590±150	5.3×10^{-15}	1.1
$OH + CF_3CH_2CF_3 \rightarrow CF_3CHCF_3$ (HFC–236fa) $+H_2O$	1.3×10^{-12}	2480±150	3.2×10^{-16}	1.1
$OH + CF_3CHFCF_3 \rightarrow CF_3CFCF_3+H_2O$ (HFC–227ea)	5.0×10^{-13}	1700±300	1.7×10^{-15}	1.1
$OH + CHF_2OCH_2CF_3 \rightarrow$ products (HFOC–245fa)	2.6×10^{-12}	1610±150	1.2×10^{-14}	2.0
$OH + CF_3CH_2CF_2CH_3 \rightarrow$ products (HFC–365mfc)	2.0×10^{-12}	1750±200	5.7×10^{-15}	1.3
$OH + CF_3CH_2CH_2CF_3 \rightarrow$ products (HFC–356mff)	3.0×10^{-12}	1800±300	7.1×10^{-15}	1.3
$OH + CF_3CF_2CH_2CH_2F \rightarrow$ products (HFC–356mcf)	1.7×10^{-12}	1110±200	4.2×10^{-14}	2.0
$OH + CHF_2CF_2CF_2CF_2H \rightarrow$ products (HFC–338pcc)	7.8×10^{-13}	1530±200	4.6×10^{-15}	1.5
$OH + CF_3CH_2CF_2CH_2CF_3 \rightarrow$ products (HFC–458mfcf)	1.2×10^{-12}	1830±200	2.6×10^{-15}	2.0
$OH + CF_3CHFCHFCF_2CF_3 \rightarrow$ products (HFC–43–10mee)	5.2×10^{-13}	1500±300	3.4×10^{-15}	1.3
$OH + CF_3CF_2CH_2CH_2CF_2CF_3 \rightarrow$ (HFC–55–10–mcff) products	–	–	8.3×10^{-15}	1.5
$F + O_3 \rightarrow FO + O_2$	2.2×10^{-11}	230±200	1.0×10^{-11}	1.5
$F + H_2 \rightarrow HF + H$	1.4×10^{-10}	500±200	2.6×10^{-11}	1.2
$F + H_2O \rightarrow HF + OH$	1.4×10^{-11}	0±200	1.4×10^{-11}	1.3
$F + HNO_3 \rightarrow HF + NO_3$	6.0×10^{-12}	–(400±200)	2.3×10^{-11}	1.3
$F + CH_4 \rightarrow HF + CH_3$	1.6×10^{-10}	260±200	6.7×10^{-11}	1.4
$FO + O_3 \rightarrow$ products			$<1\times10^{-14}$	
$FO + NO \rightarrow NO_2 + F$	8.2×10^{-12}	–(300±200)	2.2×10^{-11}	1.5
$FO + FO \rightarrow 2F + O_2$	1.0×10^{-11}	0±250	1.0×10^{-11}	1.5
$FO_2 + O_3 \rightarrow$ products			$<3.4\times10^{-16}$	
$FO_2 + NO \rightarrow FNO + O_2$	7.5×10^{-12}	690±400	7.5×10^{-13}	2.0
$FO_2 + NO_2 \rightarrow$ products	3.8×10^{-11}	2040±500	4.0×10^{-14}	2.0
$FO_2 + CO \rightarrow$ products			$<5.1\times10^{-16}$	
$FO_2 + CH_4 \rightarrow$ products			$<2\times10^{-16}$	
$CF_3O + O_2 \rightarrow FO_2 + CF_2O$	$<3\times10^{-11}$	5000	$<1.5\times10^{-18}$	
$CF_3O + O_3 \rightarrow CF_3O_2 + O_2$	2×10^{-12}	1400±600	1.8×10^{-14}	1.3
$CF_3O + H_2O \rightarrow OH + CF_3OH$	3×10^{-12}	>3600	$<2\times10^{-17}$	
$CF_3O + NO \rightarrow CF_2O + FNO$	3.7×10^{-11}	–(110±70)	5.4×10^{-11}	1.2
$CF_3O + NO_2 \rightarrow$ products	See reference			
$CF_3O + CO \rightarrow$ products			$<2\times10^{-15}$	
$CF_3O + CH_4 \rightarrow CH_3 + CF_3OH$	2.6×10^{-12}	1420±200	2.2×10^{-14}	1.1
$CF_3O + C_2H_6 \rightarrow C_2H_5 + CF_3OH$	4.9×10^{-12}	400±100	1.3×10^{-12}	1.2
$CF_3O_2 + O_3 \rightarrow CF_3O + 2O_2$			$<3\times10^{-15}$	
$CF_3O_2 + CO \rightarrow CF_3O + CO_2$			$<5\times10^{-16}$	
$CF_3O_2 + NO \rightarrow CF_3O + NO_2$	5.4×10^{-12}	–(320±150)	1.6×10^{-11}	1.1
ClO_x Reactions				
$O + ClO \rightarrow Cl + O_2$	3.0×10^{-11}	–(70±70)	3.8×10^{-11}	1.2
$O + OClO \rightarrow ClO + O_2$	2.4×10^{-12}	960±300	1.0×10^{-13}	2.0
$O + Cl_2O \rightarrow ClO + ClO$	2.7×10^{-11}	530±150	4.5×10^{-12}	1.3
$O + HCl \rightarrow OH + Cl$	1.0×10^{-11}	3300±350	1.5×10^{-16}	2.0
$O + HOCl \rightarrow OH + ClO$	1.7×10^{-13}	0±300	1.7×10^{-13}	3.0
$O + ClONO_2 \rightarrow$ products	2.9×10^{-12}	800±200	2.0×10^{-13}	1.5
$O_3 + OClO \rightarrow$ products	2.1×10^{-12}	4700±1000	3.0×10^{-19}	2.5
$O_3 + Cl_2O_2 \rightarrow$ products	–		$<1.0\times10^{-19}$	–
$OH + Cl_2 \rightarrow HOCl + Cl$	1.4×10^{-12}	900±400	6.7×10^{-14}	1.2
$OH + ClO \rightarrow$ products	1.1×10^{-11}	–(120±150)	1.7×10^{-11}	1.5
$OH + OClO \rightarrow HOCl + O_2$	4.5×10^{-13}	–(800±200)	6.8×10^{-12}	2.0
$OH + HCl \rightarrow H_2O + Cl$	2.6×10^{-12}	350±100	8.0×10^{-13}	1.2
$OH + HOCl \rightarrow H_2O + ClO$	3.0×10^{-12}	500±500	5.0×10^{-13}	3.0
$OH + ClNO_2 \rightarrow HOCl + NO_2$	2.4×10^{-12}	1250±300	3.6×10^{-14}	2.0
$OH + ClONO_2 \rightarrow$ products	1.2×10^{-12}	330±200	3.9×10^{-13}	1.5
$OH + CH_3Cl \rightarrow CH_2Cl + H_2O$	4.0×10^{-12}	1400±250	3.6×10^{-14}	1.2
$OH + CH_2Cl_2 \rightarrow CHCl_2 + H_2O$	3.8×10^{-12}	1050±150	1.1×10^{-13}	1.4
$OH + CHCl_3 \rightarrow CCl_3 + H_2O$	2.0×10^{-12}	900±150	1.0×10^{-13}	1.2
$OH + CCl_4 \rightarrow$ products	$\sim1.0\times10^{-12}$	>2300	$<5.0\times10^{-16}$	–
$OH + CFCl_3 \rightarrow$ products (CFC–11)	$\sim1.0\times10^{-12}$	>3700	$<5.0\times10^{-18}$	–
$OH + CF_2Cl_2 \rightarrow$ products (CFC–12)	$\sim1.0\times10^{-12}$	>3600	$<6.0\times10^{-18}$	–
$OH + CH_2ClF \rightarrow CHClF + H_2O$ (HCFC–31)	2.8×10^{-12}	1270±200	3.9×10^{-14}	1.2

Reaction	A cm³ molecule⁻¹ s⁻¹	E/R K	k (298 K) cm³ molecule⁻¹ s⁻¹	f(298)
$OH + CHFCl_2 \rightarrow CFCl_2 + H_2O$ (HCFC–21)	1.7×10^{-12}	1250±150	2.6×10^{-14}	1.2
$OH + CHF_2Cl \rightarrow CF_2Cl + H_2O$ (HCFC–22)	1.0×10^{-12}	1600±150	4.7×10^{-15}	1.1
$OH + CH_3OCl \rightarrow$ products	2.4×10^{-12}	360±200	7.2×10^{-13}	3.0
$OH + CH_3CCl_3 \rightarrow CH_2CCl_3 + H_2O$ (HCC–140)	1.8×10^{-12}	1550±150	1.0×10^{-14}	1.1
$OH + C_2HCl_3 \rightarrow$ products	4.9×10^{-13}	–(450±200)	2.2×10^{-12}	1.25
$OH + C_2Cl_4 \rightarrow$ products	9.4×10^{-12}	1200±200	1.7×10^{-13}	1.25
$OH + CCl_3CHO \rightarrow H_2O + CCl_3CO$	8.2×10^{-12}	600±300	1.1×10^{-12}	1.5
$OH + CH_3CFCl_2 \rightarrow CH_2CFCl_2 + H_2O$ (HCFC–141b)	1.7×10^{-12}	1700±150	5.7×10^{-15}	1.2
$OH + CH_3CF_2Cl \rightarrow CH_2CF_2Cl + H_2O$ (HCFC–142b)	1.3×10^{-12}	1800±150	3.1×10^{-15}	1.2
$OH + CH_2ClCF_2Cl \rightarrow CHClCF_2Cl$ (HCFC–132b) $+ H_2O$	3.6×10^{-12}	1600±400	1.7×10^{-14}	2.0
$OH + CHCl_2CF_2Cl \rightarrow CCl_2CF_2Cl$ (HCFC–122) $+ H_2O$	1.0×10^{-12}	900±150	4.9×10^{-14}	1.2
$OH + CHFClCFCl_2 \rightarrow CFClCFCl_2$ (HCFC–122a) $+ H_2O$	1.0×10^{-12}	1250±150	1.5×10^{-14}	1.1
$OH + CH_2ClCF_3 \rightarrow CHClCF_3 + H_2O$ (HCFC–133a)	5.2×10^{-13}	1100±300	1.3×10^{-14}	1.3
$OH + CHCl_2CF_3 \rightarrow CCl_2CF_3 + H_2O$ (HCFC–123)	7.0×10^{-13}	900±150	3.4×10^{-14}	1.2
$OH + CHFClCF_2Cl \rightarrow CFClCF_2Cl$ (HCFC–123a) $+ H_2O$	9.2×10^{-13}	1280±150	1.3×10^{-14}	1.2
$OH + CHFClCF_3 \rightarrow CFClCF_3 + H_2O$ (HCFC–124)	8.0×10^{-13}	1350±150	8.6×10^{-15}	1.2
$OH + CH_3CF_2CFCl_2 \rightarrow$ products (HCFC–243cc)	7.7×10^{-13}	1700±300	2.6×10^{-15}	2.0
$OH + CF_3CF_2CHCl_2 \rightarrow$ products (HCFC–225ca)	1.0×10^{-12}	1100±200	2.5×10^{-14}	1.3
$OH + CF_2ClCF_2CHFCl \rightarrow$ products (HCFC–225cb)	5.5×10^{-13}	1250±200	8.3×10^{-15}	1.3
$HO_2 + Cl \rightarrow HCl + O_2$	1.8×10^{-11}	–(170±200)	3.2×10^{-11}	1.5
$\rightarrow OH + ClO$	4.1×10^{-11}	450±200	9.1×10^{-12}	2.0
$HO_2 + ClO \rightarrow HOCl + O_2$	4.8×10^{-13}	–(700±)	5.0×10^{-12}	1.4
$H_2O + ClONO_2 \rightarrow$ products	–	–	$<2.0\times10^{-21}$	–
$NO + OClO \rightarrow NO_2 + ClO$	2.5×10^{-12}	600±300	3.4×10^{-13}	2.0
$NO + Cl_2O_2 \rightarrow$ products	–	–	$<2.0\times10^{-14}$	–
$NO_3 + HCl \rightarrow HNO_3 + Cl$	–	–	$<5.0\times10^{-17}$	–
$HO_2NO_2 + HCl \rightarrow$ products	–	–	$<1.0\times10^{-21}$	–
$Cl + O_3 \rightarrow ClO + O_2$	2.9×10^{-11}	260±100	1.2×10^{-11}	1.15
$Cl + H_2 \rightarrow HCl + H$	3.7×10^{-11}	2300±200	1.6×10^{-14}	1.25
$Cl + H_2O_2 \rightarrow HCl + HO_2$	1.1×10^{-11}	980±500	4.1×10^{-13}	1.5
$Cl + NO_3 \rightarrow ClO + NO_2$	2.4×10^{-11}	0±400	2.4×10^{-11}	1.5
$Cl + N_2O \rightarrow ClO + N_2$	See reference			
$Cl + HNO_3 \rightarrow$ products	–	–	$<2.0\times10^{-16}$	–
$Cl + CH_4 \rightarrow HCl + CH_3$	1.1×10^{-11}	1400±150	1.0×10^{-13}	1.1
$Cl + CH_3D \rightarrow$ products	–	–	7.4×10^{-14}	2.0
$Cl + H_2CO \rightarrow HCl + HCO$	8.1×10^{-11}	30±100	7.3×10^{-11}	1.15
$Cl + CH_3O_2 \rightarrow$ products	–	–	1.6×10^{-10}	1.5
$Cl + CH_3OH \rightarrow CH_2OH + HCl$	5.4×10^{-11}	0±250	5.4×10^{-11}	1.5
$Cl + C_2H_6 \rightarrow HCl + C_2H_5$	7.7×10^{-11}	90±90	5.7×10^{-11}	1.1
$Cl + C_2H_5O_2 \rightarrow ClO + C_2H_5O$	–	–	7.4×10^{-11}	2.0
$\rightarrow HCl + C_2H_4O_2$	–	–	7.7×10^{-11}	2.0
$Cl + CH_3CN \rightarrow$ products	1.6×10^{-11}	2140±300	1.2×10^{-14}	2.0
$Cl + CH_3CO_3NO_2 \rightarrow$ products	–	–	$<1\times10^{-14}$	
$Cl + C_3H_8 \rightarrow HCl + C_3H_7$	1.2×10^{-10}	–(40±250)	1.4×10^{-10}	1.3
$Cl + OClO \rightarrow ClO + ClO$	3.4×10^{-11}	–(160±200)	5.8×10^{-11}	1.25
$Cl + ClOO \rightarrow Cl_2 + O_2$	2.3×10^{-10}	0±250	2.3×10^{-10}	3.0
$\rightarrow ClO + ClO$	1.2×10^{-11}	0±250	1.2×10^{-11}	3.0
$Cl + Cl_2O \rightarrow Cl_2 + ClO$	6.2×10^{-11}	–(130±130)	9.6×10^{-11}	1.2
$Cl + Cl_2O_2 \rightarrow$ products	–	–	1.0×10^{-10}	2.0
$Cl + HOCl \rightarrow$ products	2.5×10^{-12}	130±250	1.6×10^{-12}	1.5
$Cl + ClNO \rightarrow NO + Cl_2$	5.8×10^{-11}	–(100±200)	8.1×10^{-11}	1.5
$Cl + ClONO_2 \rightarrow$ products	6.5×10^{-12}	–(135±50)	1.0×10^{-11}	1.2
$Cl + CH_3Cl \rightarrow CH_2Cl + HCl$	3.2×10^{-11}	1250±200	4.8×10^{-13}	1.2
$Cl + CH_2Cl_2 \rightarrow HCl + CHCl_2$	3.1×10^{-11}	1350±500	3.3×10^{-13}	1.5
$Cl + CHCl_3 \rightarrow HCl + CCl_3$	8.2×10^{-12}	1325±300	9.6×10^{-14}	1.3
$Cl + CH_3F \rightarrow HCl + CH_2F$ (HFC–41)	2.0×10^{-11}	1200±500	3.5×10^{-13}	1.3
$Cl + CH_2F_2 \rightarrow HCl + CHF_2$ (HFC–32)	1.2×10^{-11}	1630±500	5.0×10^{-14}	1.5
$Cl + CF_3H \rightarrow HCl + CF_3$ (HFC–23)	–	–	3.0×10^{-18}	5.0
$Cl + CH_2FCl \rightarrow HCl + CHFCl$ (HCFC–31)	1.2×10^{-11}	1390±500	1.1×10^{-13}	2.0
$Cl + CHFCl_2 \rightarrow HCl + CFCl_2$ (HCFC–21)	5.5×10^{-12}	1675±200	2.0×10^{-14}	1.3

Reaction	*A* cm^3 molecule^{-1} s^{-1}	*E*/*R* K	*k* (298 K) cm^3 molecule^{-1} s^{-1}	*f*(298)
$Cl + CHF_2Cl \rightarrow HCl + CF_2Cl$ (HCFC–22)	5.9×10^{-12}	2430±200	1.7×10^{-15}	1.3
$Cl + CH_3CCl_3 \rightarrow CH_2CCl_3 + HCl$	2.8×10^{-12}	1790±400	7.0×10^{-15}	2.0
$Cl + CH_3CH_2F \rightarrow HCl + CH_3CHF$ (HFC–161)	1.8×10^{-11}	290±500	6.8×10^{-12}	3.0
$\rightarrow HCl + CH_2CH_2F$	1.4×10^{-11}	880±500	7.3×10^{-13}	3.0
$Cl + CH_3CHF_2 \rightarrow HCl + CH_3CF_2$ (HFC–152a)	6.4×10^{-12}	950±500	2.6×10^{-13}	1.3
$\rightarrow HCl + CH_2CHF_2$	7.2×10^{-12}	2390±500	2.4×10^{-15}	3.0
$Cl + CH_2FCH_2F \rightarrow HCl + CHFCH_2F$ (HFC–152)	2.6×10^{-11}	1060±500	7.5×10^{-13}	3.0
$Cl + CH_3CFCl_2 \rightarrow HCl + CH_2CFCl_2$ (HCFC–141b)	1.8×10^{-12}	2000±300	2.2×10^{-15}	1.2
$Cl + CH_3CF_2Cl \rightarrow HCl + CH_2CF_2Cl$ (HCFC–142b)	1.4×10^{-12}	2420±500	4.2×10^{-16}	1.2
$Cl + CH_3CF_3 \rightarrow HCl + CH_2CF_3$ (HFC–143a)	1.2×10^{-11}	3880±500	2.6×10^{-17}	5.0
$Cl + CH_2FCHF_2 \rightarrow HCl + CH_2FCF_2$ (HFC–143)	5.5×10^{-12}	1610±500	2.5×10^{-14}	3.0
$\rightarrow HCl + CHFCHF_2$	7.7×10^{-12}	1720±500	2.4×10^{-14}	3.0
$Cl + CH_2ClCF_3 \rightarrow HCl + CHClCF_3$ (HCFC–133a)	1.8×10^{-12}	1710±500	5.9×10^{-15}	3.0
$Cl + CH_2FCF_3 \rightarrow HCl + CHFCF_3$ (HFC–134a)	–	–	1.5×10^{-15}	1.2
$Cl + CHF_2CHF_2 \rightarrow HCl + CF_2CHF_2$ (HCF–134)	7.5×10^{-12}	2430±500	2.2×10^{-15}	1.5
$Cl + CHCl_2CF_3 \rightarrow HCl + CCl_2CF_3$ (HCFC–123)	4.4×10^{-12}	1750±500	1.2×10^{-14}	1.3
$Cl + CHFClCF_3 \rightarrow HCl + CFClCF_3$ (HCFC–124)	1.1×10^{-12}	1800±500	2.7×10^{-15}	1.3
$Cl + CHF_2CF_3 \rightarrow HCl + CF_2CF_3$ (HFC–125)	–	–	2.4×10^{-16}	1.3
$ClO + O_3 \rightarrow ClOO + O_2$	–	–	$<1.4\times10^{-17}$	–
$\rightarrow OClO + O_2$	1.0×10^{-12}	>4000	$<1.0\times10^{-18}$	–
$ClO + H_2 \rightarrow$ products	$\sim1.0\times10^{-12}$	>4800	$<1.0\times10^{-19}$	–
$ClO + NO \rightarrow NO_2 + Cl$	6.4×10^{-12}	–(290±100)	1.7×10^{-11}	1.15
$ClO + NO_3 \rightarrow ClOO + NO_2$	4.7×10^{-13}	0±400	4.7×10^{-13}	1.5
$ClO + N_2O \rightarrow$ products	$\sim1.0\times10^{-12}$	>4300	$<6.0\times10^{-19}$	–
$ClO + CO \rightarrow$ products	$\sim1.0\times10^{-12}$	>3700	$<4.0\times10^{-18}$	–
$ClO + CH_4 \rightarrow$ products	$\sim1.0\times10^{-12}$	>3700	$<4.0\times10^{-18}$	–
$ClO + H_2CO \rightarrow$ products	$\sim1.0\times10^{-12}$	>2100	$<1.0\times10^{-15}$	–
$ClO + CH_3O_2 \rightarrow$ products	3.3×10^{-12}	115±115	2.2×10^{-12}	1.5
$ClO + ClO \rightarrow Cl_2 + O_2$	1.0×10^{-12}	1590±300	4.8×10^{-15}	1.5
$\rightarrow ClOO + Cl$	3.0×10^{-11}	2450±500	8.0×10^{-15}	1.5
$\rightarrow OClO + Cl$	3.5×10^{-13}	1370±300	3.5×10^{-15}	1.5
$HCl + ClONO_2 \rightarrow$ products	–	–	$<1.0\times10^{-20}$	–
$CH_2ClO + O_2 \rightarrow CHClO + HO_2$	–	–	6×10^{-14}	5
$CH_2ClO_2 + HO_2 \rightarrow CH_2ClO_2H + O_2$	3.3×10^{-13}	–(820±200)	5.2×10^{-12}	1.5
$CH_2ClO_2 + NO \rightarrow CH_2ClO + NO_2$	7×10^{-12}	–(300±200)	1.9×10^{-11}	1.5
$CCl_3O_2 + NO \rightarrow CCl_2O + NO_2 + Cl$	7.3×10^{-12}	–(270±200)	1.8×10^{-11}	1.3
$CCl_2FO_2 + NO \rightarrow CClFO + NO_2 + Cl$	4.5×10^{-12}	–(350±200)	1.5×10^{-11}	1.3
$CClF_2O_2 + NO \rightarrow CF_2O + NO_2 + Cl$	3.8×10^{-12}	–(400±200)	1.5×10^{-11}	1.2
BrO_x Reactions				
$O + BrO \rightarrow Br + O_2$	1.9×10^{-11}	–(230±150)	4.1×10^{-11}	1.5
$O + HBr \rightarrow OH + Br$	5.8×10^{-12}	1500±200	3.8×10^{-14}	1.3
$O + HOBr \rightarrow OH + BrO$	1.2×10^{-10}	430±300	2.8×10^{-11}	3.0
$OH + Br_2 \rightarrow HOBr + Br$	4.2×10^{-11}	0±600	4.2×10^{-11}	1.3
$OH + BrO \rightarrow$ products	–	–	7.5×10^{-11}	3.0
$OH + HBr \rightarrow H_2O + Br$	1.1×10^{-11}	0±250	1.1×10^{-11}	1.2
$OH + CH_3Br \rightarrow CH_2Br + H_2O$	4.0×10^{-12}	1470±150	2.9×10^{-14}	1.1
$OH + CH_2Br_2 \rightarrow CHBr_2 + H_2O$	2.4×10^{-12}	900±300	1.2×10^{-13}	1.1
$OH + CHBr_3 \rightarrow CBr_3 + H_2O$	1.6×10^{-12}	710±200	1.5×10^{-13}	2.0
$OH + CHF_2Br \rightarrow CF_2Br + H_2O$	1.1×10^{-12}	1400±200	1.0×10^{-14}	1.1
$OH + CH_2ClBr \rightarrow CHClBr + H_2O$	2.3×10^{-12}	930±150	1.0×10^{-13}	1.2
$OH + CF_2ClBr \rightarrow$ products	–	–	$<1.5\times10^{-16}$	–
$OH + CF_2Br_2 \rightarrow$ products	–	–	$<5.0\times10^{-16}$	–
$OH + CF_3Br \rightarrow$ products	–	–	$<1.2\times10^{-16}$	–
$OH + CH_2BrCF_3 \rightarrow CHBrCF_3 + H_2O$	1.4×10^{-12}	1340±200	1.6×10^{-14}	1.3
$OH + CHFBrCF_3 \rightarrow CFBrCF_3$	7.2×10^{-13}	1110±150	1.8×10^{-14}	1.5
$OH + CHClBrCF_3 \rightarrow CClBrCF_3 + H_2O$	1.3×10^{-12}	995±150	4.5×10^{-14}	1.5
$OH + CF_2BrCHFCl \rightarrow CF_2BrCFCl + H_2O$	9.3×10^{-13}	1250±150	1.4×10^{-14}	1.5
$OH + CF_2BrCF_2Br \rightarrow$ products	–	–	$<1.5\times10^{-16}$	–
$HO_2 + Br \rightarrow HBr + O_2$	1.5×10^{-11}	600±600	2.0×10^{-12}	2.0

Reaction	A cm^3 molecule^{-1} s^{-1}	E/R K	k (298 K) cm^3 molecule^{-1} s^{-1}	f(298)
HO_2 + BrO → products	3.4×10^{-12}	–(540±200)	2.1×10^{-11}	1.5
NO_3 + HBr → HNO_3 + Br	–	–	<1.0×10^{-16}	–
Cl + CH_2ClBr → HCl + CHClBr	4.3×10^{-11}	1370±500	4.3×10^{-13}	3.0
Cl + CH_3Br → HCl + CH_2Br	1.5×10^{-11}	1060±100	4.3×10^{-13}	1.2
Cl + CH_2Br_2 → HCl + $CHBr_2$	6.4×10^{-12}	810±100	4.2×10^{-13}	1.2
Br + O_3 → BrO + O_2	1.7×10^{-11}	800±200	1.2×10^{-12}	1.2
Br + H_2O_2 → HBr + HO_2	1.0×10^{-11}	>3000	<5.0×10^{-16}	–
Br + NO_3 → BrO + NO_2	–	–	1.6×10^{-11}	2.0
Br + H_2CO → HBr + HCO	1.7×10^{-11}	800±200	1.1×10^{-12}	1.3
Br + OClO → BrO + ClO	2.6×10^{-11}	1300±300	3.4×10^{-13}	2.0
Br + Cl_2O → BrCl + ClO	2.1×10^{-11}	470±150	4.3×10^{-12}	1.3
Br + Cl_2O_2 → products	–	–	3.0×10^{-12}	2.0
BrO + O_3 → products	~1.0×10^{-12}	>3200	<2.0×10^{-17}	–
BrO + NO → NO_2 + Br	8.8×10^{-12}	–(260±130)	2.1×10^{-11}	1.15
BrO + NO_3 → products	–	–	1.0×10^{-12}	3.0
BrO + ClO → Br + OClO	1.6×10^{-12}	–(430±200)	6.8×10^{-12}	1.25
→ Br + ClOO	2.9×10^{-12}	–(220±200)	6.1×10^{-12}	1.25
→ BrCl + O_2	5.8×10^{-13}	–(170±200)	1.0×10^{-12}	1.25
BrO + BrO → products	1.5×10^{-12}	–(230±150)	3.2×10^{-12}	1.15
CH_2BrO_2 + NO → CH_2O + NO_2 + Br	4×10^{-12}	–(300±200)	1.1×10^{-11}	1.5
IO_x Reactions				
O + I_2 → IO + I	1.4×10^{-10}	0±250	1.4×10^{-10}	1.4
O + IO → O_2 + I			1.2×10^{-10}	2.0
OH + I_2 → HOI + I			1.8×10^{-10}	2.0
OH + HI → H_2O + I			3.0×10^{-11}	2.0
OH + CH_3I → H_2O + CH_2I	3.1×10^{-12}	1120±500	7.2×10^{-14}	3.0
OH + CF_3I → HOI + CF_3			3.1×10^{-14}	5.0
HO_2 + I → HI + O_2	1.5×10^{-11}	1090±500	3.8×10^{-13}	2.0
HO_2 + IO → HOI + O_2			8.4×10^{-11}	1.5
NO_3 + HI → HNO_3 + I	See reference			
I + O_3 → IO + O_2	2.3×10^{-11}	870±200	1.2×10^{-12}	1.2
I + BrO → IO + Br	–	–	1.2×10^{-11}	2.0
IO + NO → I + NO_2	9.1×10^{-12}	–(240±150)	2.0×10^{-11}	1.2
IO + ClO → products	5.1×10^{-12}	–(280±200)	1.3×10^{-11}	2.0
IO + BrO → products	–	–	6.9×10^{-11}	1.5
IO + IO → products	1.5×10^{-11}	–(500±500)	8.0×10^{-11}	1.5
INO + INO → I_2 + 2NO	8.4×10^{-11}	2620±600	1.3×10^{-14}	2.5
INO_2 + INO_2 → I_2 + 2NO_2	2.9×10^{-11}	2600±1000	4.7×10^{-15}	3.0
SO_x Reactions				
O + SH → SO + H	–	–	1.6×10^{-10}	5.0
O + CS → CO + S	2.7×10^{-10}	760±250	2.1×10^{-11}	1.1
O + H_2S → OH + SH	9.2×10^{-12}	1800±550	2.2×10^{-14}	1.7
O + OCS → CO + SO	2.1×10^{-11}	2200±150	1.3×10^{-14}	1.2
O + CS_2 → CS + SO	3.2×10^{-11}	650±150	3.6×10^{-12}	1.2
O + CH_3SCH_3 → CH_3SO + CH_3	1.3×10^{-11}	–(410±100)	5.0×10^{-11}	1.1
O + CH_3SSCH_3 → CH_3SO + CH_3S	5.5×10^{-11}	–(250±100)	1.3×10^{-10}	1.3
O_3 + H_2S → products	–	–	<2.0×10^{-20}	–
O_3 + CH_3SCH_3 → products	–	–	<1.0×10^{-18}	–
O_3 + SO_2 → SO_3 + O_2	3.0×10^{-12}	>7000	<2.0×10^{-22}	–
OH + H_2S → SH + H_2O	6.0×10^{-12}	75±75	4.7×10^{-12}	1.2
OH + OCS → products	1.1×10^{-13}	1200±500	1.9×10^{-15}	2.0
OH + CS_2 → products	See reference	–	–	–
OH + CH_3SH → CH_3S + H_2O	9.9×10^{-12}	–(360±100)	3.3×10^{-11}	1.2
OH + CH_3SCH_3 → H_2O + CH_2SCH_3	1.2×10^{-11}	260±100	5.0×10^{-12}	1.15
OH + CH_3SSCH_3 → products	6.0×10^{-11}	–(400±200)	2.3×10^{-10}	1.2
OH + S → H + SO	–	–	6.6×10^{-11}	3.0
OH + SO → H + SO_2	–	–	8.6×10^{-11}	2.0
HO_2 + H_2S → products	–	–	<3.0×10^{-15}	–

Reaction	A cm³ molecule⁻¹ s⁻¹	E/R K	k (298 K) cm³ molecule⁻¹ s⁻¹	f(298)
$HO_2 + CH_3SH \rightarrow$ products	–	–	$<4.0\times10^{-15}$	–
$HO_2 + CH_3SCH_3 \rightarrow$ products	–	–	$<5.0\times10^{-15}$	–
$HO_2 + SO_2 \rightarrow$ products	–	–	$<1.0\times10^{-18}$	–
$NO_2 + SO_2 \rightarrow$ products	–	–	$<2.0\times10^{-26}$	–
$NO_3 + H_2S \rightarrow$ products	–	–	$<8.0\times10^{-16}$	–
$NO_3 + OCS \rightarrow$ products	–	–	$<1.0\times10^{-16}$	–
$NO_3 + CS_2 \rightarrow$ products	–	–	$<4.0\times10^{-16}$	–
$NO_3 + CH_3SH \rightarrow$ products	4.4×10^{-13}	–(210±210)	8.9×10^{-13}	1.25
$NO_3 + CH_3SCH_3 \rightarrow CH_3SCH_2 + HNO_3$	1.9×10^{-13}	–(500±200)	1.0×10^{-12}	1.2
$NO_3 + CH_3SSCH_3 \rightarrow$ products	1.3×10^{-12}	270±270	5.3×10^{-13}	1.4
$NO_3 + SO_2 \rightarrow$ products	–	–	$<7.0\times10^{-21}$	–
$N_2O_5 + CH_3SCH_3 \rightarrow$ products	–	–	$<1.0\times10^{-17}$	–
$CH_3O_2 + SO_2 \rightarrow$ products	–	–	$<5.0\times10^{-17}$	–
$F + CH_3SCH_3 \rightarrow$ products	–	–	2.4×10^{-10}	2.0
$Cl + H_2S \rightarrow HCl + SH$	3.7×10^{-11}	–(210±100)	7.4×10^{-11}	1.25
$Cl + OCS \rightarrow$ products	–	–	$<1.0\times10^{-16}$	–
$Cl + CS_2 \rightarrow$ products	–	–	$<4.0\times10^{-15}$	–
$Cl + CH_3SH \rightarrow CH_3S + HCl$	1.2×10^{-10}	–(150±50)	2.0×10^{-10}	1.25
$Cl + CH_3SCH_3 \rightarrow$ products	See reference	–	–	–
$ClO + OCS \rightarrow$ products	–	–	$<2.0\times10^{-16}$	–
$ClO + CH_3SCH_3 \rightarrow$ products	–	–	9.5×10^{-15}	2.0
$ClO + SO \rightarrow Cl + SO_2$	2.8×10^{-11}	0±50	2.8×10^{-11}	1.3
$ClO + SO_2 \rightarrow Cl + SO_3$	–	–	$<4.0\times10^{-18}$	–
$Br + H_2S \rightarrow HBr + SH$	1.4×10^{-11}	2750±300	1.4×10^{-15}	2.0
$Br + CH_3SH \rightarrow CH_3S + HBr$	9.2×10^{-12}	390±100	2.5×10^{-12}	2.0
$Br + CH_3SCH_3 \rightarrow$ products	See reference			
$BrO + CH_3SCH_3 \rightarrow$ products	1.5×10^{-14}	–(850±200)	2.6×10^{-13}	1.3
$BrO + SO \rightarrow Br + SO_2$			5.7×10^{-11}	1.4
$IO + CH_3SH \rightarrow$ products			6.6×10^{-16}	2.0
$IO + CH_3SCH_3 \rightarrow$ products			1.2×10^{-14}	1.5
$S + O_2 \rightarrow SO + O$	2.3×10^{-12}	0±200	2.3×10^{-12}	1.2
$S + O_3 \rightarrow SO + O_2$			1.2×10^{-11}	2.0
$SO + O_2 \rightarrow SO_2 + O$	2.6×10^{-13}	2400±500	8.4×10^{-17}	2.0
$SO + O_3 \rightarrow SO_2 + O_2$	3.6×10^{-12}	1100±200	9.0×10^{-14}	1.2
$SO + NO_2 \rightarrow SO_2 + NO$	1.4×10^{-11}	0±50	1.4×10^{-11}	1.2
$SO + OClO \rightarrow SO_2 + ClO$			1.9×10^{-12}	3.0
$SO_3 + H_2O \rightarrow$ products	See reference		–	–
$SO_3 + NO_2 \rightarrow$ products			1.0×10^{-19}	10.0
$SH + O_2 \rightarrow OH + SO$			$<4.0\times10^{-19}$	–
$SH + O_3 \rightarrow HSO + O_2$	9.0×10^{-12}	280±200	3.5×10^{-12}	1.3
$SH + H_2O_2 \rightarrow$ products			$<5.0\times10^{-15}$	–
$SH + NO_2 \rightarrow HSO + NO$	2.9×10^{-11}	–(240±50)	6.5×10^{-11}	1.2
$SH + Cl_2 \rightarrow ClSH + Cl$	1.7×10^{-11}	690±200	1.7×10^{-12}	2.0
$SH + BrCl \rightarrow$ products	2.3×10^{-11}	–(350±200)	7.4×10^{-11}	2.0
$SH + Br_2 \rightarrow BrSH + Br$	6.0×10^{-11}	–(160±160)	1.0×10^{-10}	2.0
$SH + F_2 \rightarrow FSH + F$	4.3×10^{-11}	1390±200	4.0×10^{-13}	2.0
$HSO + O_2 \rightarrow$ products			$<2.0\times10^{-17}$	–
$HSO + O_3 \rightarrow$ products			1.0×10^{-13}	1.3
$HSO + NO \rightarrow$ products			$<1.0\times10^{-15}$	–
$HSO + NO_2 \rightarrow HSO_2 + NO$			9.6×10^{-12}	2.0
$HSO_2 + O_2 \rightarrow HO_2 + SO_2$			3.0×10^{-13}	3.0
$HOSO_2 + O_2 \rightarrow HO_2 + SO_3$	1.3×10^{-12}	330±200	4.4×10^{-13}	1.2
$CS + O_2 \rightarrow OCS + O$			2.9×10^{-19}	2.0
$CS + O_3 \rightarrow OCS + O_2$			3.0×10^{-16}	3.0
$CS + NO_2 \rightarrow OCS + NO$			7.6×10^{-17}	3.0
$CH_3S + O_2 \rightarrow$ products			$<3.0\times10^{-18}$	–
$CH_3S + O_3 \rightarrow$ products	2.0×10^{-12}	–(290±100)	5.3×10^{-12}	1.15
$CH_3S + NO \rightarrow$ products			$<1.0\times10^{-13}$	–
$CH_3S + NO_2 \rightarrow CH_3SO + NO$	2.1×10^{-11}	–(320±100)	6.1×10^{-11}	1.15
$CH_2SH + O_2 \rightarrow$ products			6.5×10^{-12}	2.0

Reaction	*A* cm³ molecule⁻¹ s⁻¹	*E*/*R* K	*k* (298 K) cm³ molecule⁻¹ s⁻¹	*f*(298)
$CH_2SH + O_3 \rightarrow$ products			3.5×10^{-11}	2.0
$CH_2SH + NO \rightarrow$ products			1.9×10^{-11}	2.0
$CH_2SH + NO_2 \rightarrow$ products			5.2×10^{-11}	2.0
$CH_3SO + O_3 \rightarrow$ products			6.0×10^{-13}	1.5
$CH_3SO + NO_2 \rightarrow CH_3SO_2 + NO$			1.2×10^{-11}	1.4
$CH_3SOO + O_3 \rightarrow$ products			$<8.0\times10^{-13}$	–
$CH_3SOO + NO \rightarrow$ products	1.1×10^{-11}	0±100	1.1×10^{-11}	2.0
$CH_3SO_2 + NO_2 \rightarrow$ products	2.2×10^{-11}	0±100	2.2×10^{-11}	2.0
$CH_3SCH_2 + NO_3 \rightarrow$ products			3.0×10^{-10}	2.0
$CH_3SCH_2O_2 + NO \rightarrow CH_3SCH_2O + NO_2$			1.9×10^{-11}	2.0
$CH_3SS + O_3 \rightarrow$ products			4.6×10^{-13}	2.0
$CH_3SS + NO_2 \rightarrow$ products			1.8×10^{-11}	2.0
$CH_3SSO + NO_2 \rightarrow$ products			4.5×10^{-12}	2.0
Metal Reactions				
$Na + O_3 \rightarrow NaO + O_2$	1.0×10^{-9}	95±50	7.3×10^{-10}	1.2
$\rightarrow NaO_2 + O$	–	–	$<4.0\times10^{-11}$	–
$Na + N_2O \rightarrow NaO + N_2$	2.8×10^{-10}	1600±400	1.3×10^{-12}	1.2
$Na + Cl_2 \rightarrow NaCl + Cl$	7.3×10^{-10}	0±200	7.3×10^{-10}	1.3
$NaO + O \rightarrow Na + O_2$	3.7×10^{-10}	0±400	3.7×10^{-10}	3.0
$NaO + O_3 \rightarrow NaO_2 + O_2$	1.1×10^{-9}	570±300	1.6×10^{-10}	1.5
$\rightarrow Na + 2O_2$	6.0×10^{-11}	0±800	6.0×10^{-11}	3.0
$NaO + H_2 \rightarrow NaOH + H$	2.6×10^{-11}	0±600	2.6×10^{-11}	2.0
$NaO + H_2O \rightarrow NaOH + OH$	2.2×10^{-10}	0±400	2.2×10^{-10}	2.0
$NaO + NO \rightarrow Na + NO_2$	1.5×10^{-10}	0±400	1.5×10^{-10}	4.0
$NaO + HCl \rightarrow$ products	2.8×10^{-10}	0±400	2.8×10^{-10}	3.0
$NaO_2 + O \rightarrow NaO + O_2$	2.2×10^{-11}	0±600	2.2×10^{-11}	5.0
$NaO_2 + NO \rightarrow NaO + NO_2$	–	–	$<10^{-14}$	–
$NaO_2 + HCl \rightarrow$ products	2.3×10^{-10}	0±400	2.3×10^{-10}	3.0
$NaOH + HCl \rightarrow NaCl + H_2O$	2.8×10^{-10}	0±400	2.8×10^{-10}	3.0

TABLE 2. Rate Constants for Association Reactions

The values quoted are suitable for air as the third body, M. The integer in parentheses is the power of ten.

	Low pressure limit $k_0(T) = k_0(300)\ (T/300)^{-n}$ cm⁶ molecule⁻² s⁻¹		High pressure limit $k_\infty(T) = k_\infty(300)\ (T/300)^{-m}$ cm³ molecule⁻¹ s⁻¹	
Reaction	$k_0(300)$	*n*	$k_\infty(300)$	*m*
O_x Reactions				
$O + O_2 \rightarrow O_3$	(6.0±0.5) (–34)	2.3±0.5	—	—
$O(^1D)$ Reactions				
$O(^1D) + N_2 \rightarrow N_2O$	(3.5±3.0) (–37)	0.6	—	—
HO_x Reactions				
$H + O_2 \rightarrow HO_2$	(5.7±0.5) (–32)	1.6±0.5	(7.5±4.0) (–11)	0±1.0
$OH + OH \rightarrow H_2O_2$	(6.2±1.2) (–31)	1.0	(2.6±1.0) (–11)	0±0.5
NO_x Reactions				
$O + NO \rightarrow NO_2$	(9.0±2.0) (–32)	1.5±0.3	(3.0±1.0) (–11)	0±1.0
$O + NO_2 \rightarrow NO_3$	(9.0±1.0) (–32)	2.0±1.0	(2.2±0.3) (–11)	0±1.0
$OH + NO \rightarrow HONO$	(7.0±1.0) (–31)	2.6±0.3	(3.6±1.0) (–11)	0.1±0.5
$OH + NO_2 \rightarrow HNO_3$	(2.5±0.1) (–30)	4.4±0.3	(1.6±0.2) (–11)	1.7±0.2
$HO_2 + NO_2 \rightarrow HO_2NO_2$	(1.8±0.3) (–31)	3.2±0.4	(4.7±1.0) (–12)	1.4±1.4
$NO_2 + NO_3 \rightarrow N_2O_5$	(2.2±0.5) (–30)	3.9±1.0	(1.5±0.8) (–12)	0.7±0.4
$NO_3 \rightarrow NO + O_2$	See reference			
Hydrocarbon Reactions				
$CH_3 + O_2 \rightarrow CH_3O_2$	(4.5±1.5) (–31)	3.0±1.0	(1.8±0.2) (–12)	1.7±1.7
$C_2H_5 + O_2 \rightarrow C_2H_5O_2$	(1.5±1.0) (–28)	3.0±1.0	(8.0±1.0) (–12)	0±1.0

Reaction	Low pressure limit $k_0(T) = k_0(300)\ (T/300)^{-n}$ cm^6 molecule^{-2} s^{-1}		High pressure limit $k_\infty(T) = k_\infty(300)\ (T/300)^{-m}$ cm^3 molecule^{-1} s^{-1}	
	$k_0(300)$	n	$k_\infty(300)$	m
$OH + C_2H_2 \rightarrow HOCHCH$	(5.5±2.0) (−30)	0.0±0.2	(8.3±1.0) (−13)	−2
$OH + C_2H_4 \rightarrow HOCH_2CH_2$	(1.0±0.6) (−28)	0.8±2.0	(8.8±0.9) (−12)	0
$CH_3O + NO \rightarrow CH_3ONO$	(1.4±0.5) (−29)	3.8±1.0	(3.6±1.6) (−11)	0.6±1.0
$CH_3O + NO_2 \rightarrow CH_3ONO_2$	(1.1±0.4) (−28)	4.0±2.0	(1.6±0.5) (−11)	1.0±1.0
$C_2H_5O + NO \rightarrow C_2H_5ONO$	(2.8±1.0) (−27)	4.0±2.0	(5.0±1.0) (−11)	1.0±1.0
$C_2H_5O + NO_2 \rightarrow C_2H_5ONO_2$	(2.0±1.0) (−27)	4.0±2.0	(2.8±0.4) (−11)	1.0±1.0
$CH_3O_2 + NO_2 \rightarrow CH_3O_2NO_2$	(1.5±0.8) (−30)	4.0±2.0	(6.5±3.2) (−12)	2.0±2.0
$CH_3C(O)O_2 + NO_2 \rightarrow CH_3C(O)O_2NO_2$	(9.7±3.8) (−29)	5.6±2.8	(9.3±0.4)(−12)	1.5±0.3
FO_x Reactions				
$F + O_2 \rightarrow FO_2$	(4.4±0.4) (−33)	1.2±0.5	–	–
$F + NO \rightarrow FNO$	(1.8±0.3) (−31)	1.0±10	(2.8±1.4) (−10)	0.0±1.0
$F + NO_2 \rightarrow FNO_2$	(6.3±3.0) (−32)	2.0±2.0	(2.6±1.3) (−10)	0.0±1.0
$FO + NO_2 \rightarrow FONO_2$	(2.6±2.0) (−31)	1.3±1.3	(2.0±1.0) (−11)	1.5±1.5
$CF_3 + O_2 \rightarrow CF_3O_2$	(3.0±0.3) (−29)	4.0±2.0	(4.0±1.0) (−12)	1.0±1.0
$CF_3O + NO_2 \rightarrow CF_3ONO_2$	See reference			
$CF_3O_2 + NO_2CF_3O_2NO_2$	(2.2±0.5) (−29)	5.0±1.0	(6.0±1.0) (−12)	2.5±1.0
$CF_3O + CO \rightarrow CF_3OCO$	(2.5±0.2) (−31)	–	(6.8±0.4) (−14)	−1.2
$CF_3O \rightarrow CF_2O + F$	See reference			
ClO_x Reactions				
$Cl + O_2 \rightarrow ClOO$	(2.7±1.0) (−33)	1.5±0.5	–	–
$Cl + NO \rightarrow ClNO$	(9.0±2.0) (−32)	1.6±0.5	–	–
$Cl + NO_2$ $ClONO \rightarrow$	(1.3±0.2) (−30)	2.0±1.0	(1.0±0.5) (−10)	1.0±1.0
$ClNO_2$	(1.8±0.3) (−31)	2.0±1.0	(1.0±0.5) (−10)	1.0±1.0
$Cl + CO \rightarrow ClCO$	(1.3±0.5) (−33)	3.8±0.5	–	–
$Cl + C_2H_2 \rightarrow ClC_2H_2$	((5.9±1.0) (−30)	2.1±1.0	(2.1±0.4) (−10)	1.0±0.5
$Cl + C_2H_4 \rightarrow ClC_2H_4$	(1.6±1) (−29)	3.3±1.0	(3.1±2) (−10)	1.0±0.5
$Cl + C_2Cl_4 \rightarrow C_2Cl_5$	(1.4±0.6) (−28)	8.5±1.0	(4.0±1.0) (−11)	1.2±0.5
$ClO + NO_2 \rightarrow ClONO_2$	(1.8±0.3) (−31)	3.4±1.0	(1.5±0.7) (−11)	1.9±1.9
$OClO + NO_3 \rightarrow O_2ClONO_2$	See reference			
$ClO + ClO \rightarrow Cl_2O_2$	(2.2±0.4) (−32)	3.1±0.5	(3.5±2) (−12)	1.0±1.0
$ClO + OClO \rightarrow Cl_2O_3$	(6.2±1.0) (−32)	4.7±0.6	(2.4±1.2) (−11)	0±1.0
$OClO + O \rightarrow ClO_3$	(1.9±0.5) (−31)	1.1±1.0	(3.1±0.8) (−11)	0±1.0
$CH_2Cl + O_2 \rightarrow CH_2ClO_2$	(1.9±0.1) (−30)	3.2±0.2	(2.9±0.2) (−12)	1.2±0.6
$CHCl_2 + O_2 \rightarrow CHCl_2O_2$	(1.3±0.1) (−30)	4.0±0.2	(2.8±0.2) (−12)	1.4±0.6
$CCl_3 + O_2 \rightarrow CCl_3O_2$	(6.9±0.2) (−31)	6.4±0.3	(2.4±0.2) (−12)	2.1±0.6
$CFCl_2 + O_2 \rightarrow CFCl_2O_2$	(5.0±0.8) (−30)	4.0±2.0	(6.0±1.0) (−12)	1.0±1.0
$CF_2Cl + O_2 \rightarrow CF_2ClO_2$	(3.0±1.5) (−30)	4.0±2.0	(3±2) (−12)	1.0±1.0
$CCl_3O_2 + NO_2 \rightarrow CCl_3O_2NO_2$	(5.0±1.0) (−29)	5.0±1.0	(6.0±1.0) (−12)	2.5±1.0
$CFCl_2O_2 + NO_2 \rightarrow CFCl_2O_2NO_2$	(3.5±0.5) (−29)	5.0±1.0	(6.0±1.0) (−12)	2.5±1.0
$CF_2ClO_2 + NO_2 \rightarrow CF_2ClO_2NO_2$	(3.3±0.7) (−29)	6.7±1.3	(4.1±1.9) (−12)	2.8±0.7
BrO_x Reactions				
$Br + NO_2 \rightarrow BrNO_2$	(4.2±0.8) (−31)	2.4±0.5	(2.7±0.5) (−11)	0±1.0
$BrO + NO_2 \rightarrow BrONO_2$	(5.2±0.6) (−31)	3.2±0.8	(6.9±1.0) (−12)	2.9±1.0
IO_x Reactions				
$I + NO \rightarrow INO$	(1.8±0.5) (−32)	1.0±0.5	(1.7±1.0) (−11)	0±1.0
$I + NO_2 \rightarrow INO_2$	(3.0±1.5) (−31)	1.0±1.0	(6.6±5.0) (−11)	0±1.0
$IO + NO_2 \rightarrow IONO_2$	(5.9±2.0) (−31)	3.5±1.0	(9.0±1.0) (−12)	1.5±1.0
SO_x Reactions				
$HS + NO \rightarrow HSNO$	(2.4±0.4) (−31)	3.0±1.0	(2.7±0.5) (−11)	0
$CH_3S + NO \rightarrow CH_3SNO$	(3.2±0.4) (−29)	4.0±1.0	(3.9±0.6) (−11)	2.7±1.0
$O + SO_2 \rightarrow SO_3$	(1.3±)(−33)	−3.6±0.7		
$OH + SO_2 \rightarrow HOSO_2$	(3.0±1.0) (−31)	3.3±1.5	(1.5±0.5) (−12)	0
$CH_3SCH_2 + O_2 \rightarrow CH_3SCH_2O_2$	See reference			
$SO_3 + NH_3 \rightarrow H_3NSO_3$	(3.9±0.8) (−30)	3.0±3.0	(4.7±1.3) (−11)	0±1.0

Reaction	Low pressure limit $k_0(T) = k_0(300)\,(T/300)^{-n}$ cm^6 molecule^{-2} s^{-1}		High pressure limit $k_\infty(T) = k_\infty(300)\,(T/300)^{-m}$ cm^3 molecule^{-1} s^{-1}	
	$k_0(300)$	n	$k_\infty(300)$	m
Metal Reactions				
$Na + O_2 \rightarrow NaO_2$	(3.2±0.3) (−30)	1.4±0.3	(6.0±2.0) (−10)	0±1.0
$NaO + O_2 \rightarrow NaO_3$	(3.5±0.7) (−30)	2.0±2.0	(5.7±3.0) (−10)	0±1.0
$NaO + CO_2 \rightarrow NaCO_3$	(8.7±2.6) (−28)	2.0±2.0	(6.5±3.0) (−10)	0±1.0
$NaOH + CO_2 \rightarrow NaHCO_3$	(1.3±0.3) (−28)	2.0±2.0	(6.8±4.0) (−10)	0±1.0

TABLE 3. Equilibrium Constants

$K(T)$/cm^3 molecule^{-1} = $A \exp(B/T)$ [200 < T/K < 300]

Reaction	A/cm^3 molecule^{-1}	B/K	K (298 K)	f(298 K)
$HO_2 + NO_2 \rightarrow HO_2NO_2$	2.1×10^{-27}	10900±1000	1.6×10^{-11}	5
$NO + NO_2 \rightarrow N_2O_3$	3.3×10^{-27}	4667±100	2.1×10^{-20}	2
$NO_2 + NO_2 \rightarrow N_2O_4$	5.2×10^{-29}	6643±250	2.5×10^{-19}	2
$NO_2 + NO_3 \rightarrow N_2O_5$	2.7×10^{-27}	11000±500	2.9×10^{-11}	1.3
$CH_3O_2 + NO_2 \rightarrow CH_3O_2NO_2$	1.3×10^{-28}	11200±1000	2.7×10^{-12}	2
$CH_3C(O)O_2 + NO_2 \rightarrow CH_3C(O)O_2NO_2$	9.0×10^{-29}	14000±200	2.3×10^{-8}	2
$F + O_2 \rightarrow FOO$	3.2×10^{-25}	6100±1200	2.5×10^{-16}	1.0
$Cl + O_2 \rightarrow ClOO$	5.7×10^{-25}	2500±750	2.5×10^{-21}	2
$Cl + CO \rightarrow ClCO$	1.6×10^{-25}	4000±500	1.1×10^{-19}	5
$ClO + O_2 \rightarrow ClO{\cdot}O_2$	2.9×10^{-26}	<3700	$<7.2 \times 10^{-21}$	–
$ClO + ClO \rightarrow Cl_2O_2$	1.3×10^{-27}	8744±850	7.2×10^{-15}	1.5
$ClO + OClO \rightarrow Cl_2O_3$	1.1×10^{-24}	5455±300	9.8×10^{-17}	3
$OClO + NO_3 \rightarrow O_2ClONO_2$	1×10^{-28}	9300±1000	3.6×10^{-15}	5
$OH + CS_2 \rightarrow CS_2OH$	4.5×10^{-25}	5140±500	1.4×10^{-17}	1.4
$CH_3S + O_2 \rightarrow CH_3SO_2$	1.8×10^{-27}	5545±300	2.2×10^{-19}	1.4

Section 6
Fluid Properties

Thermophysical Properties of Water and Steam 6-1
Vapor Pressure and Other Saturation Properties of Water 6-5
Standard Density of Water 6-7
Fixed-Point Properties of H_2O and D_2O 6-9
Properties of Saturated Liquid D_2O 6-10
Properties of Ice and Supercooled Water 6-12
Vapor Pressure of Ice 6-13
Melting Point of Ice as a Function of Pressure 6-13
Permittivity (Dielectric Constant) of Water at Various Frequencies 6-14
Thermophysical Properties of Air 6-15
Thermophysical Properties of Fluids 6-21
Thermophysical Properties of Selected Fluids at Saturation 6-38
Virial Coefficients of Selected Gases 6-46
Van Der Waals Constants for Gases 6-55
Mean Free Path and Related Properties of Gases 6-56
Influence of Pressure on Freezing Points 6-57
Critical Constants of Organic Compounds 6-58
Critical Constants of Inorganic Compounds 6-82
Sublimation Pressure of Solids 6-85
Vapor Pressure 6-87
Vapor Pressure of Fluids at Temperatures Below 300 K 6-117
Vapor Pressure of Saturated Salt Solutions 6-125
IUPAC Recommended Data for Vapor Pressure Calibration 6-126
Enthalpy of Vaporization 6-127
Enthalpy of Fusion 6-145
Compressibility and Expansion Coefficients of Liquids 6-155
Temperature and Pressure Dependence of Liquid Density 6-157
Volumetric Properties of Aqueous Sodium Chloride Solutions 6-162
Properties of Cryogenic Fluids 6-163
Properties of Liquid Helium 6-164
Properties of Refrigerants 6-165
Properties of Gas Clathrate Hydrates 6-168
Ionic Liquids 6-173
Density and Specific Volume of Mercury 6-177
Thermal Properties of Mercury 6-178
Melting Curve of Mercury 6-179
Vapor Pressure of Mercury 6-180
Surface Tension of Common Liquids 6-181
Surface Tension of Aqueous Mixtures 6-185
Permittivity (Dielectric Constant) of Liquids 6-186
Permittivity (Dielectric Constant) of Gases 6-208
Azeotropic Data for Binary Mixtures 6-209
Viscosity of Gases 6-228
Viscosity of Liquids 6-229
Viscosity of Carbon Dioxide Along the Saturation Line 6-234
Viscosity and Density of Aqueous Hydroxide Solutions 6-235
Viscosity of Liquid Metals 6-236
Thermal Conductivity of Gases 6-238
Thermal Conductivity of Liquids 6-240
Diffusion in Gases 6-245
Diffusion of Gases in Water 6-247
Diffusion Coefficients in Liquids at Infinite Dilution 6-248

THERMOPHYSICAL PROPERTIES OF WATER AND STEAM

Eric W. Lemmon

These tables summarize the thermophysical properties of water and steam at equilibrium as accepted by the International Association for the Properties of Water and Steam (www.iapws.org) for general and scientific use. The thermodynamic properties are calculated from the equation of state of Wagner and Pruss (Ref. 6). The reference state for these tables is the liquid at the triple point, at which the internal energy and entropy are taken as zero. These tables refer to states at 1 bar (100 kPa) pressure with temperatures in °C in the first section; liquid and gaseous states at equilibrium as a function of temperature in the second section; and properties along isobars in the third section. The tabulated properties are pressure (P), density (ρ), enthalpy (H), entropy (S), isochoric heat capacity (C_v), isobaric heat capacity (C_p), speed of sound (u), viscosity (η), thermal conductivity (λ), and static dielectric constant (D). In the saturation tables, the first line of identical temperatures is for the liquid state and the second line is for the vapor state. A duplicate entry in the isobar section indicates a phase transition (liquid-vapor) at that temperature; property values are then given for both phases. These are identified by the high densities in the liquid and the low densities in the vapor. The temperature scale is ITS-90. Additional calculations at state points not listed below can be obtained by using the NIST Standard Reference Data program REFPROP (Ref. 5) or the water-specific program Steam (Ref. 2).

References

1. Fernandez, D.P., Goodwin, A.R.H., Lemmon, E.W., Levelt Sengers, J.M.H., and Williams, R.C., A Formulation for the Static Permittivity of Water and Steam at Temperatures from 238 K to 873 K at Pressures up to 1200 MPa, Including Derivatives and Debye-Hückel Coefficients, *J. Phys. Chem. Ref. Data*, 26, 1125, 1997.
2. Harvey, A.H., Peskin, A.P., and Klein, S.A., NIST Standard Reference Database 10: NIST/ASME Steam Properties, Version 2.22, National Institute of Standards and Technology, Standard Reference Data Program, Gaithersburg, Maryland, 2008 (www.nist.gov/srd/nist10.htm).
3. Huber, M.L., Perkins, R.A., Laesecke, A., Friend, D.G., Sengers, J.V., Assael, M.J., Metaxa, I.M., Vogel, E., Mares, R., and Miyagawa, K., New International Formulation for the Viscosity of Water, *J. Phys. Chem. Ref. Data*, 2009, in press.
4. Kestin, J., Sengers, J.V., Kamgar-Parsi, B., and Levelt Sengers, J.M.H., Thermophysical Properties of Fluid H_2O, *J. Phys. Chem. Ref. Data*, 13, 175, 1984.
5. Lemmon, E.W., Huber, M.L., and McLinden, M.O., NIST Standard Reference Database 23: Reference Fluid Thermodynamic and Transport Properties-REFPROP, Version 8.0, National Institute of Standards and Technology, Standard Reference Data Program, Gaithersburg, Maryland, 2007 (www.nist.gov/srd/nist23.htm).
6. Wagner, W. and Pruss, A., The IAPWS Formulation 1995 for the Thermodynamic Properties of Ordinary Water Substance for General and Scientific Use, *J. Phys. Chem. Ref. Data*, 31, 387, 2002.

T	P	ρ	H	S	C_v	C_p	u		η	λ
°C	MPa	kg m^{-3}	kJ kg^{-1}	kJ kg^{-1} K^{-1}	kJ kg^{-1} K^{-1}	kJ kg^{-1} K^{-1}	m s^{-1}	D	μPa s	mW m^{-1} K^{-1}
P = 0.1 MPa (1 bar)										
0.01	0.1	999.84	0.10186	0.000007	4.2170	4.2194	1402.4	87.899	1791.1	561.09
10	0.1	999.70	42.118	0.15108	4.1906	4.1952	1447.3	83.974	1305.9	580.05
20	0.1	998.21	84.006	0.29646	4.1567	4.1841	1482.3	80.223	1001.6	598.46
25	0.1	997.05	104.92	0.36720	4.1376	4.1813	1496.7	78.408	890.02	607.19
30	0.1	995.65	125.82	0.43673	4.1172	4.1798	1509.2	76.634	797.22	615.50
40	0.1	992.22	167.62	0.57237	4.0734	4.1794	1528.9	73.201	652.73	630.63
50	0.1	988.03	209.42	0.70377	4.0262	4.1813	1542.6	69.916	546.52	643.59
60	0.1	983.20	251.25	0.83125	3.9765	4.1850	1551.0	66.774	466.03	654.39
70	0.1	977.76	293.12	0.95509	3.9251	4.1901	1554.7	63.770	403.55	663.13
80	0.1	971.79	335.05	1.0755	3.8728	4.1968	1554.4	60.898	354.05	670.01
90	0.1	965.31	377.06	1.1928	3.8204	4.2052	1550.4	58.152	314.17	675.27
99.606	0.1	958.63	417.50	1.3028	3.7702	4.2152	1543.5	55.628	282.75	678.97
99.606	0.1	0.59034	2674.9	7.3588	1.5548	2.0784	471.99	1.0058	12.218	25.053
100	0.1	0.58967	2675.8	7.3610	1.5535	2.0766	472.28	1.0058	12.234	25.079
Saturation										
0.01	0.000612	999.79	0.000612	0.0	4.2174	4.2199	1402.3	87.895	1791.4	561.04
0.01	0.000612	0.0048546	2500.9	9.1555	1.4184	1.8844	409.00	1.00006	8.9458	17.071
10	0.0012282	999.65	42.021	0.15109	4.1910	4.1955	1447.1	83.971	1306.0	580.00
10	0.0012282	0.0094071	2519.2	8.8998	1.4269	1.8947	416.17	1.00012	9.2384	17.621
20	0.0023393	998.16	83.914	0.29648	4.1570	4.1844	1482.2	80.219	1001.6	598.42
20	0.0023393	0.017314	2537.4	8.6660	1.4359	1.9059	423.18	1.00021	9.5441	18.227
25	0.0031699	997.00	104.83	0.36722	4.1379	4.1816	1496.5	78.405	890.04	607.15
25	0.0031699	0.023075	2546.5	8.5566	1.4405	1.9118	426.63	1.00028	9.7009	18.550
30	0.0042470	995.61	125.73	0.43675	4.1175	4.1801	1509.0	76.630	797.22	615.46
30	0.0042470	0.030415	2555.5	8.4520	1.4452	1.9180	430.03	1.00036	9.8602	18.887
40	0.0073849	992.18	167.53	0.57240	4.0737	4.1796	1528.7	73.197	652.72	630.58

T °C	P MPa	ρ kg m^{-3}	H kJ kg^{-1}	S kJ kg^{-1} K^{-1}	C_v kJ kg^{-1} K^{-1}	C_p kJ kg^{-1} K^{-1}	u m s^{-1}	D	η μPa s	λ mW m^{-1} K^{-1}
40	0.0073849	0.051242	2573.5	8.2555	1.4552	1.9314	436.71	1.00059	10.185	19.599
50	0.012352	988.00	209.34	0.70381	4.0264	4.1815	1542.4	69.913	546.50	643.55
50	0.012352	0.083147	2591.3	8.0748	1.4663	1.9468	443.21	1.00094	10.516	20.365
60	0.019946	983.16	251.18	0.83129	3.9767	4.1851	1550.8	66.772	466.02	654.35
60	0.019946	0.13043	2608.8	7.9081	1.4789	1.9648	449.50	1.0014	10.854	21.187
70	0.031201	977.73	293.07	0.95513	3.9252	4.1902	1554.6	63.768	403.53	663.09
70	0.031201	0.19843	2626.1	7.7540	1.4937	1.9862	455.57	1.0021	11.195	22.068
80	0.047414	971.77	335.01	1.0756	3.8729	4.1969	1554.3	60.896	354.04	669.99
80	0.047414	0.29367	2643.0	7.6111	1.5111	2.0120	461.39	1.0030	11.539	23.011
90	0.070182	965.30	377.04	1.1929	3.8204	4.2053	1550.4	58.151	314.17	675.25
90	0.070182	0.42390	2659.5	7.4781	1.5316	2.0429	466.94	1.0043	11.885	24.019
100	0.10142	958.35	419.17	1.3072	3.7682	4.2157	1543.2	55.527	281.58	679.09
100	0.10142	0.59817	2675.6	7.3541	1.5558	2.0800	472.20	1.0059	12.232	25.096
100	0.10142	958.35	419.17	1.3072	3.7682	4.2157	1543.2	55.527	281.58	679.09
100	0.10142	0.59817	2675.6	7.3541	1.5558	2.0800	472.20	1.0059	12.232	25.096
120	0.19867	943.11	503.81	1.5279	3.6662	4.2435	1519.9	50.620	232.03	683.19
120	0.19867	1.1221	2705.9	7.1291	1.6177	2.1770	481.73	1.0105	12.927	27.467
140	0.36154	926.13	589.16	1.7392	3.5694	4.2826	1486.2	46.131	196.64	683.30
140	0.36154	1.9667	2733.4	6.9293	1.7002	2.3109	489.82	1.0177	13.618	30.140
160	0.61823	907.45	675.47	1.9426	3.4788	4.3354	1443.2	42.018	170.43	679.96
160	0.61823	3.2596	2757.4	6.7491	1.8044	2.4883	496.29	1.0282	14.304	33.131
180	1.0028	887.00	763.05	2.1392	3.3949	4.4050	1391.7	38.235	150.38	673.32
180	1.0028	5.1588	2777.2	6.5840	1.9279	2.7129	501.04	1.0431	14.985	36.449
200	1.5549	864.66	852.27	2.3305	3.3179	4.4958	1332.1	34.742	134.58	663.31
200	1.5549	7.8610	2792.0	6.4302	2.0666	2.9895	503.92	1.0636	15.666	40.113
220	2.3196	840.22	943.58	2.5177	3.2479	4.6146	1264.5	31.495	121.77	649.65
220	2.3196	11.615	2800.9	6.2840	2.2172	3.3289	504.77	1.0915	16.354	44.170
240	3.3469	813.37	1037.6	2.7020	3.1850	4.7719	1189.0	28.455	111.06	631.85
240	3.3469	16.749	2803.0	6.1423	2.3794	3.7537	503.32	1.1292	17.062	48.729
260	4.6923	783.63	1135.0	2.8849	3.1301	4.9856	1105.3	25.580	101.81	609.24
260	4.6923	23.712	2796.6	6.0016	2.5555	4.3075	499.21	1.1802	17.810	54.035
280	6.4166	750.28	1236.9	3.0685	3.0849	5.2889	1012.6	22.824	93.550	581.15
280	6.4166	33.165	2779.9	5.8579	2.7503	5.0731	491.93	1.2505	18.630	60.622
300	8.5879	712.14	1345.0	3.2552	3.0530	5.7504	909.40	20.135	85.855	547.43
300	8.5879	46.168	2749.6	5.7059	2.9708	6.2197	480.73	1.3504	19.580	69.667
320	11.284	667.09	1462.2	3.4494	3.0428	6.5373	793.16	17.440	78.310	509.21
320	11.284	64.638	2700.6	5.5372	3.2276	8.1589	464.43	1.5012	20.773	83.944
340	14.601	610.67	1594.5	3.6601	3.0781	8.2080	658.27	14.606	70.331	468.55
340	14.601	92.759	2621.8	5.3356	3.5430	12.236	440.72	1.7555	22.477	111.00
360	18.666	527.59	1761.7	3.9167	3.2972	15.004	479.74	11.225	60.306	425.77
360	18.666	143.90	2481.5	5.0536	4.0068	27.356	402.37	2.3096	25.638	181.77
373.95	22.064	322.00	2084.3	4.4070				5.3606	79.791	

T K	P MPa	ρ kg m^{-3}	H kJ kg^{-1}	S kJ kg^{-1} K^{-1}	C_v kJ kg^{-1} K^{-1}	C_p kJ kg^{-1} K^{-1}	u m s^{-1}	D	η μPa s	λ mW m^{-1} K^{-1}
P = 0.1 MPa (1 bar)										
273.16	0.1	999.84	0.10186	0.000007	4.2170	4.2194	1402.4	87.899	1791.1	561.09
280	0.1	999.91	28.894	0.10411	4.1998	4.2009	1434.3	85.192	1433.6	574.09
300	0.1	996.56	112.65	0.39306	4.1302	4.1806	1501.5	77.747	853.74	610.32
320	0.1	989.43	196.25	0.66281	4.0414	4.1805	1538.9	70.935	576.73	639.75
340	0.1	979.54	279.93	0.91646	3.9414	4.1883	1554.0	64.702	421.63	660.58
360	0.1	967.40	363.82	1.1562	3.8369	4.2023	1552.1	59.004	325.86	673.78
372.756	0.1	958.63	417.50	1.3028	3.7702	4.2152	1543.5	55.628	282.75	678.97
372.756	0.1	0.59034	2674.9	7.3588	1.5548	2.0784	471.99	1.0058	12.218	25.053
380	0.1	0.57824	2689.9	7.3986	1.5356	2.0507	477.08	1.0056	12.498	25.546

T K	*P* MPa	*ρ* kg m^{-3}	*H* kJ kg^{-1}	*S* kJ kg^{-1} K^{-1}	C_v kJ kg^{-1} K^{-1}	C_p kJ kg^{-1} K^{-1}	*u* m s^{-1}	*D*	*η* μPa s	*λ* mW m^{-1} K^{-1}
400	0.1	0.54761	2730.4	7.5025	1.5082	2.0078	490.31	1.0051	13.278	27.008
450	0.1	0.48458	2829.7	7.7365	1.4943	1.9752	520.60	1.0040	15.267	31.168
500	0.1	0.43514	2928.6	7.9447	1.5082	1.9813	548.31	1.0033	17.299	35.861
550	0.1	0.39507	3028.1	8.1344	1.5319	2.0010	574.19	1.0027	19.356	40.953
600	0.1	0.36185	3128.8	8.3096	1.5600	2.0268	598.61	1.0023	21.425	46.367
650	0.1	0.33384	3230.8	8.4730	1.5903	2.0557	621.79	1.0020	23.496	52.049
700	0.1	0.30988	3334.4	8.6264	1.6222	2.0867	643.92	1.0017	25.562	57.964
750	0.1	0.28915	3439.5	8.7715	1.6553	2.1191	665.11	1.0015	27.617	64.083
800	0.1	0.27102	3546.3	8.9093	1.6892	2.1525	685.47	1.0013	29.657	70.385
900	0.1	0.24085	3765.0	9.1668	1.7589	2.2216	724.03	1.0011	33.680	83.466
1000	0.1	0.21673	3990.7	9.4045	1.8297	2.2921	760.17	1.00088	37.615	97.085
1100	0.1	0.19701	4223.4	9.6263	1.9000	2.3621	794.33	1.00074	41.453	111.15
1200	0.1	0.18058	4463.0	9.8348	1.9682	2.4302	826.85	1.00063	45.192	125.58
***P* = 1 MPa (10 bar)**										
273.16	1.0	1000.3	1.0180	0.000066	4.2127	4.2150	1403.9	87.937	1789.1	561.59
280	1.0	1000.3	29.783	0.10407	4.1960	4.1973	1435.7	85.228	1432.5	574.54
300	1.0	996.96	113.48	0.39281	4.1272	4.1781	1503.0	77.781	853.66	610.73
320	1.0	989.82	197.03	0.66242	4.0390	4.1784	1540.5	70.967	576.89	640.16
340	1.0	979.93	280.67	0.91594	3.9395	4.1863	1555.7	64.734	421.86	661.02
360	1.0	967.81	364.52	1.1556	3.8353	4.2004	1553.9	59.035	326.10	674.24
380	1.0	953.74	448.73	1.3832	3.7315	4.2220	1538.4	53.827	262.82	681.49
400	1.0	937.87	533.47	1.6005	3.6315	4.2535	1511.3	49.065	218.82	684.10
450	1.0	890.39	749.20	2.1086	3.4076	4.3924	1400.6	38.814	153.23	674.64
453.028	1.0	887.13	762.52	2.1381	3.3954	4.4045	1392.0	38.258	150.49	673.37
453.028	1.0	5.1450	2777.1	6.5850	1.9271	2.7114	501.02	1.0430	14.981	36.428
500	1.0	4.5323	2891.2	6.8250	1.6699	2.2795	535.74	1.0344	17.054	38.799
550	1.0	4.0581	3001.8	7.0359	1.6159	2.1647	565.75	1.0282	19.215	42.798
600	1.0	3.6871	3109.0	7.2224	1.6098	2.1292	592.58	1.0236	21.349	47.636
650	1.0	3.3843	3215.2	7.3925	1.6227	2.1254	617.34	1.0202	23.462	53.002
700	1.0	3.1305	3321.7	7.5504	1.6447	2.1368	640.55	1.0174	25.555	58.735
750	1.0	2.9140	3429.0	7.6984	1.6715	2.1566	662.53	1.0153	27.628	64.745
800	1.0	2.7265	3537.5	7.8384	1.7014	2.1816	683.48	1.0135	29.680	70.983
900	1.0	2.4174	3758.5	8.0986	1.7663	2.2402	722.85	1.0108	33.716	84.000
1000	1.0	2.1723	3985.7	8.3380	1.8346	2.3048	759.50	1.0088	37.655	97.573
1100	1.0	1.9729	4219.5	8.5608	1.9034	2.3713	794.01	1.0074	41.494	111.57
1200	1.0	1.8074	4460.0	8.7699	1.9708	2.4371	826.77	1.0063	45.231	125.89
***P* = 10 MPa (100 bar)**										
273.16	10.0	1004.8	10.111	0.00049	4.1721	4.1726	1418.4	88.311	1770.0	566.56
280	10.0	1004.7	38.613	0.10355	4.1593	4.1622	1450.3	85.588	1422.2	579.06
300	10.0	1001.0	121.73	0.39029	4.0984	4.1536	1518.2	78.113	852.99	614.81
320	10.0	993.70	204.84	0.65846	4.0157	4.1580	1556.4	71.284	578.57	644.31
340	10.0	983.84	288.08	0.91079	3.9205	4.1672	1572.6	65.044	424.17	665.40
360	10.0	971.85	371.56	1.1493	3.8198	4.1810	1572.0	59.345	328.53	678.93
380	10.0	957.99	455.37	1.3759	3.7188	4.2013	1558.0	54.140	265.22	686.53
400	10.0	942.42	539.67	1.5921	3.6210	4.2302	1532.7	49.385	221.17	689.57
450	10.0	896.16	753.94	2.0967	3.4010	4.3553	1428.4	39.172	155.48	681.51
500	10.0	838.02	977.18	2.5669	3.2211	4.6022	1271.3	30.794	119.83	651.64
550	10.0	761.82	1218.8	3.0270	3.0865	5.1407	1054.6	23.531	96.080	592.51
584.147	10.0	688.42	1408.1	3.3606	3.0438	6.1237	847.33	18.660	81.718	526.84
584.147	10.0	55.463	2725.5	5.6160	3.1065	7.1408	472.51	1.4248	20.194	76.569
600	10.0	49.773	2820.0	5.7756	2.6239	5.1365	503.34	1.3649	21.017	71.115
650	10.0	40.479	3022.6	6.1009	2.1103	3.3968	562.10	1.2672	23.472	67.341
700	10.0	35.355	3177.4	6.3305	1.9338	2.8741	602.20	1.2145	25.773	69.297
750	10.0	31.810	3314.6	6.5200	1.8625	2.6452	634.58	1.1793	27.973	73.331

T K	P MPa	ρ kg m^{-3}	H kJ kg^{-1}	S kJ kg^{-1} K^{-1}	C_v kJ kg^{-1} K^{-1}	C_p kJ kg^{-1} K^{-1}	u m s^{-1}	D	η µPa s	λ mW m^{-1} K^{-1}
800	10.0	29.107	3443.7	6.6867	1.8367	2.5313	662.61	1.1536	30.101	78.476
900	10.0	25.123	3691.6	6.9787	1.8439	2.4458	710.98	1.1182	34.201	90.516
1000	10.0	22.241	3935.5	7.2357	1.8843	2.4397	753.03	1.0948	38.144	103.50
1100	10.0	20.017	4180.6	7.4693	1.9377	2.4661	791.02	1.0782	41.960	116.73
1200	10.0	18.230	4429.2	7.6855	1.9957	2.5070	826.16	1.0659	45.663	130.00
P = 100 MPa (1000 bar)										
273.16	100.0	1045.3	95.444	−0.0083717	3.8761	3.9053	1575.5	91.834	1660.1	612.20
280	100.0	1043.6	122.26	0.088571	3.8869	3.9328	1603.8	88.976	1367.4	621.51
300	100.0	1037.2	201.44	0.36171	3.8751	3.9798	1667.9	81.216	859.19	654.50
320	100.0	1028.9	281.30	0.61941	3.8289	4.0043	1707.7	74.222	599.18	684.86
340	100.0	1019.0	361.55	0.86265	3.7637	4.0194	1728.7	67.891	448.03	707.90
360	100.0	1007.8	442.05	1.0927	3.6883	4.0309	1735.4	62.150	352.49	724.05
380	100.0	995.37	522.79	1.3110	3.6089	4.0427	1730.9	56.937	288.37	734.91
400	100.0	981.82	603.78	1.5187	3.5293	4.0569	1717.3	52.201	243.33	741.80
450	100.0	943.51	807.84	1.9993	3.3430	4.1105	1652.8	42.149	175.71	745.40
500	100.0	899.21	1015.4	2.4366	3.1820	4.1968	1555.7	34.149	139.57	730.42
550	100.0	848.78	1228.2	2.8421	3.0452	4.3234	1435.5	27.667	117.36	696.87
600	100.0	791.49	1448.6	3.2256	2.9295	4.5019	1300.4	22.290	101.85	645.83
650	100.0	726.21	1679.5	3.5952	2.8329	4.7503	1158.6	17.717	89.647	580.63
700	100.0	651.77	1925.0	3.9589	2.7538	5.0832	1020.0	13.754	79.123	510.19
750	100.0	568.52	2188.5	4.3223	2.6866	5.4492	898.84	10.336	69.732	433.71
800	100.0	482.23	2466.5	4.6811	2.6169	5.6108	813.97	7.5622	61.842	351.53
900	100.0	343.61	3000.1	5.3104	2.4386	4.8879	765.30	4.2835	52.771	257.05
1000	100.0	265.45	3440.1	5.7749	2.2950	3.9788	792.50	2.9559	50.506	232.07
1100	100.0	220.62	3809.8	6.1276	2.2276	3.4715	832.67	2.3472	51.089	223.70
1200	100.0	191.53	4142.5	6.4172	2.2098	3.2098	872.28	2.0111	52.802	219.07

VAPOR PRESSURE AND OTHER SATURATION PROPERTIES OF WATER

Eric W. Lemmon

This table summarizes the vapor pressure, enthalpy (heat) of vaporization, and surface tension of water as accepted by the International Association for the Properties of Water and Steam (www.iapws.org) for general and scientific use. The vapor pressure and heat of vaporization are calculated from the equation of state of Wagner and Pruss (Ref. 1). The temperature scale is ITS-90. Additional calculations at state points not listed below can be obtained by using the NIST Standard Reference Data program REFPROP (www.nist.gov/srd/nist23.htm) or the water-specific program Steam (www.nist.gov/srd/nist10.htm).

References

1. Wagner, W. and Pruss, A., The IAPWS Formulation 1995 for the Thermodynamic Properties of Ordinary Water Substance for General and Scientific Use, *J. Phys. Chem. Ref. Data*, 31, 387, 2002.
2. International Association for the Properties of Water and Steam, Release on the surface tension of ordinary water substance, Physical Chemistry of Aqueous Systems: Proceedings of the 12th International Conference on the Properties of Water and Steam, Orlando, Florida, September 11–16, 1994, pp. A139–A142.

t °C	P kPa	$\Delta_{vap}H$ kJ kg^{-1}	Surf. Ten. mN m^{-1}
0.01	0.61165	2500.9	75.65
2	0.70599	2496.2	75.37
4	0.81355	2491.4	75.08
6	0.93536	2486.7	74.80
8	1.0730	2481.9	74.51
10	1.2282	2477.2	74.22
12	1.4028	2472.5	73.93
14	1.5990	2467.7	73.63
16	1.8188	2463.0	73.34
18	2.0647	2458.3	73.04
20	2.3393	2453.5	72.74
22	2.6453	2448.8	72.43
24	2.9858	2444.0	72.13
25	3.1699	2441.7	71.97
26	3.3639	2439.3	71.82
28	3.7831	2434.6	71.51
30	4.2470	2429.8	71.19
32	4.7596	2425.1	70.88
34	5.3251	2420.3	70.56
36	5.9479	2415.5	70.24
38	6.6328	2410.8	69.92
40	7.3849	2406.0	69.60
42	8.2096	2401.2	69.27
44	9.1124	2396.4	68.94
46	10.099	2391.6	68.61
48	11.177	2386.8	68.28
50	12.352	2381.9	67.94
52	13.631	2377.1	67.61
54	15.022	2372.3	67.27
56	16.533	2367.4	66.93
58	18.171	2362.5	66.58
60	19.946	2357.7	66.24
62	21.867	2352.8	65.89
64	23.943	2347.8	65.54
66	26.183	2342.9	65.19
68	28.599	2338.0	64.84
70	31.201	2333.0	64.48
72	34.000	2328.1	64.12
74	37.009	2323.1	63.76
76	40.239	2318.1	63.40
78	43.703	2313.0	63.04
80	47.414	2308.0	62.67
82	51.387	2302.9	62.31
84	55.635	2297.9	61.94
86	60.173	2292.8	61.56
88	65.017	2287.6	61.19
90	70.182	2282.5	60.82
92	75.684	2277.3	60.44
94	81.541	2272.1	60.06
96	87.771	2266.9	59.68
98	94.390	2261.7	59.30
100	101.42	2256.4	58.91
102	108.87	2251.1	58.53
104	116.78	2245.8	58.14
106	125.15	2240.4	57.75
108	134.01	2235.1	57.36
110	143.38	2229.6	56.96
112	153.28	2224.2	56.57
114	163.74	2218.7	56.17
116	174.77	2213.2	55.77
118	186.41	2207.7	55.37
120	198.67	2202.1	54.97
122	211.59	2196.5	54.56
124	225.18	2190.9	54.16
126	239.47	2185.2	53.75
128	254.50	2179.5	53.34
130	270.28	2173.7	52.93
132	286.85	2167.9	52.52
134	304.23	2162.1	52.11
136	322.45	2156.2	51.69
138	341.54	2150.3	51.27
140	361.54	2144.3	50.86
142	382.47	2138.3	50.44
144	404.37	2132.2	50.01
146	427.26	2126.1	49.59
148	451.18	2119.9	49.17
150	476.16	2113.7	48.74
152	502.25	2107.5	48.31
154	529.46	2101.2	47.89
156	557.84	2094.8	47.46
158	587.42	2088.4	47.02
160	618.23	2082.0	46.59

t °C	P kPa	$\Delta_{vap}H$ kJ kg^{-1}	Surf. Ten. mN m^{-1}
162	650.33	2075.5	46.16
164	683.73	2068.9	45.72
166	718.48	2062.3	45.28
168	754.62	2055.6	44.85
170	792.19	2048.8	44.41
172	831.22	2042.0	43.97
174	871.76	2035.1	43.52
176	913.84	2028.2	43.08
178	957.51	2021.2	42.64
180	1002.8	2014.2	42.19
182	1049.8	2007.0	41.74
184	1098.5	1999.8	41.30
186	1148.9	1992.6	40.85
188	1201.1	1985.3	40.40
190	1255.2	1977.9	39.95
192	1311.2	1970.4	39.49
194	1369.1	1962.8	39.04
196	1429.0	1955.2	38.59
198	1490.9	1947.5	38.13
200	1554.9	1939.7	37.67
202	1621.0	1931.9	37.22
204	1689.3	1923.9	36.76
206	1759.8	1915.9	36.30
208	1832.6	1907.8	35.84
210	1907.7	1899.6	35.38
212	1985.1	1891.4	34.92
214	2065.0	1883.0	34.46
216	2147.3	1874.6	33.99
218	2232.2	1866.0	33.53
220	2319.6	1857.4	33.07
222	2409.6	1848.6	32.60
224	2502.3	1839.8	32.14
226	2597.8	1830.9	31.67
228	2696.0	1821.8	31.20
230	2797.1	1812.7	30.74
232	2901.0	1803.5	30.27
234	3008.0	1794.1	29.80
236	3117.9	1784.7	29.33
238	3230.8	1775.1	28.86
240	3346.9	1765.4	28.39
242	3466.2	1755.6	27.92
244	3588.7	1745.7	27.45
246	3714.5	1735.6	26.98
248	3843.6	1725.5	26.51
250	3976.2	1715.2	26.04
252	4112.2	1704.7	25.57
254	4251.8	1694.2	25.10
256	4394.9	1683.5	24.63
258	4541.7	1672.6	24.16
260	4692.3	1661.6	23.69
262	4846.6	1650.5	23.22
264	5004.7	1639.2	22.75
266	5166.8	1627.8	22.28
268	5332.9	1616.2	21.81
270	5503.0	1604.4	21.34
272	5677.2	1592.5	20.87
274	5855.6	1580.4	20.40
276	6038.3	1568.1	19.93
278	6225.2	1555.6	19.46
280	6416.6	1543.0	18.99
282	6612.4	1530.1	18.53
284	6812.8	1517.1	18.06
286	7017.7	1503.8	17.59
288	7227.4	1490.4	17.13
290	7441.8	1476.7	16.66
292	7661.0	1462.7	16.20
294	7885.2	1448.6	15.74
296	8114.3	1434.2	15.28
298	8348.5	1419.5	14.82
300	8587.9	1404.6	14.36
302	8832.5	1389.4	13.90
304	9082.4	1374.0	13.45
306	9337.8	1358.2	12.99
308	9598.6	1342.1	12.54
310	9865.1	1325.7	12.09
312	10137.	1309.0	11.64
314	10415.	1291.9	11.19
316	10699.	1274.5	10.75
318	10989.	1256.6	10.30
320	11284.	1238.4	9.86
322	11586.	1219.7	9.43
324	11895.	1200.6	8.99
326	12209.	1180.9	8.56
328	12530.	1160.8	8.13
330	12858.	1140.2	7.70
332	13193.	1118.9	7.28
334	13534.	1097.1	6.86
336	13882.	1074.6	6.44
338	14238.	1051.3	6.03
340	14601.	1027.3	5.63
342	14971.	1002.5	5.22
344	15349.	976.7	4.83
346	15734.	949.9	4.43
348	16128.	922.0	4.05
350	16529.	892.7	3.67
352	16939.	862.1	3.29
354	17358.	829.8	2.93
356	17785.	795.5	2.57
358	18221.	759.0	2.22
360	18666.	719.8	1.88
362	19121.	677.3	1.55
364	19585.	630.5	1.23
366	20060.	578.2	0.93
368	20546.	517.8	0.65
370	21044.	443.8	0.39
372	21554.	340.3	0.16
373.95	22064.	0.0	0.0

STANDARD DENSITY OF WATER

This table gives the density ρ of water in the temperature range from 0 °C to 100 °C at a pressure of 101325 Pa (one standard atmosphere). Temperatures are given on the ITS-90 scale. From 0 °C to 40 °C the values are taken from the publication in Reference 1 and refer to standard mean ocean water (SMOW), free from dissolved salts and gases. SMOW is a standard water sample of high purity and known isotopic composition. Methods of correcting for different isotopic compositions are discussed in Ref. 2. The remaining values are calculated from the NIST REFPROP program, Ref. 3, which obtains thermodynamic properties from the equation of state of Wagner and Pruss given in Ref. 4.

References

1. Tanaka, M., Girard, G., Davis, R., Peuto, A., and Bignell, N., *Metrologia* 38, 301, 2001.
2. Marsh, K. N., Ed., *Recommended Reference Materials for the Realization of Physicochemical Properties,* Blackwell Scientific Publications, Oxford, 1987.
3. Lemmon, E.W., Huber, M.L., and McLinden, M.O., NIST Standard Reference Database 23: Reference Fluid Thermodynamic and Transport Properties-REFPROP, Version 8.0, National Institute of Standards and Technology, Standard Reference Data Program, Gaithersburg, Maryland, 2007 (www.nist.gov/srd/nist23.htm).
4. Wagner, W., and Pruss, A., *J. Phys. Chem. Ref. Data* 31, 387, 2002.

t/°C	ρ/g cm^{-3}	t/°C	ρ/g cm^{-3}	t/°C	ρ/g cm^{-3}	t/°C	ρ/g cm^{-3}	t/°C	ρ/g cm^{-3}
0.1	0.9998495	4.1	0.9999748	8.1	0.9998452	12.1	0.9994890	16.1	0.9989296
0.2	0.9998560	4.2	0.9999746	8.2	0.9998389	12.2	0.9994774	16.2	0.9989132
0.3	0.9998624	4.3	0.9999742	8.3	0.9998325	12.3	0.9994657	16.3	0.9988967
0.4	0.9998685	4.4	0.9999736	8.4	0.9998260	12.4	0.9994539	16.4	0.9988800
0.5	0.9998745	4.5	0.9999728	8.5	0.9998193	12.5	0.9994419	16.5	0.9988633
0.6	0.9998803	4.6	0.9999719	8.6	0.9998125	12.6	0.9994298	16.6	0.9988464
0.7	0.9998859	4.7	0.9999709	8.7	0.9998056	12.7	0.9994176	16.7	0.9988294
0.8	0.9998913	4.8	0.9999697	8.8	0.9997985	12.8	0.9994052	16.8	0.9988123
0.9	0.9998966	4.9	0.9999683	8.9	0.9997912	12.9	0.9993927	16.9	0.9987951
1.0	0.9999017	5.0	0.9999668	9.0	0.9997839	13.0	0.9993801	17.0	0.9987778
1.1	0.9999066	5.1	0.9999651	9.1	0.9997764	13.1	0.9993674	17.1	0.9987603
1.2	0.9999113	5.2	0.9999633	9.2	0.9997687	13.2	0.9993546	17.2	0.9987428
1.3	0.9999158	5.3	0.9999613	9.3	0.9997610	13.3	0.9993416	17.3	0.9987251
1.4	0.9999202	5.4	0.9999592	9.4	0.9997530	13.4	0.9993285	17.4	0.9987073
1.5	0.9999244	5.5	0.9999569	9.5	0.9997450	13.5	0.9993153	17.5	0.9986895
1.6	0.9999285	5.6	0.9999544	9.6	0.9997368	13.6	0.9993020	17.6	0.9986715
1.7	0.9999323	5.7	0.9999518	9.7	0.9997285	13.7	0.9992885	17.7	0.9986534
1.8	0.9999360	5.8	0.9999491	9.8	0.9997200	13.8	0.9992749	17.8	0.9986351
1.9	0.9999396	5.9	0.9999462	9.9	0.9997114	13.9	0.9992612	17.9	0.9986168
2.0	0.9999429	6.0	0.9999431	10.0	0.9997027	14.0	0.9992474	18.0	0.9985984
2.1	0.9999461	6.1	0.9999400	10.1	0.9996938	14.1	0.9992335	18.1	0.9985798
2.2	0.9999491	6.2	0.9999366	10.2	0.9996848	14.2	0.9992194	18.2	0.9985611
2.3	0.9999519	6.3	0.9999331	10.3	0.9996757	14.3	0.9992052	18.3	0.9985424
2.4	0.9999546	6.4	0.9999295	10.4	0.9996665	14.4	0.9991909	18.4	0.9985235
2.5	0.9999571	6.5	0.9999257	10.5	0.9996571	14.5	0.9991765	18.5	0.9985045
2.6	0.9999595	6.6	0.9999217	10.6	0.9996475	14.6	0.9991619	18.6	0.9984854
2.7	0.9999616	6.7	0.9999176	10.7	0.9996379	14.7	0.9991473	18.7	0.9984662
2.8	0.9999636	6.8	0.9999134	10.8	0.9996281	14.8	0.9991325	18.8	0.9984469
2.9	0.9999655	6.9	0.9999090	10.9	0.9996182	14.9	0.9991176	18.9	0.9984275
3.0	0.9999672	7.0	0.9999045	11.0	0.9996081	15.0	0.9991026	19.0	0.9984079
3.1	0.9999687	7.1	0.9998998	11.1	0.9995979	15.1	0.9990874	19.1	0.9983883
3.2	0.9999700	7.2	0.9998950	11.2	0.9995876	15.2	0.9990722	19.2	0.9983686
3.3	0.9999712	7.3	0.9998900	11.3	0.9995772	15.3	0.9990568	19.3	0.9983487
3.4	0.9999722	7.4	0.9998849	11.4	0.9995666	15.4	0.9990413	19.4	0.9983287
3.5	0.9999731	7.5	0.9998797	11.5	0.9995559	15.5	0.9990257	19.5	0.9983087
3.6	0.9999738	7.6	0.9998743	11.6	0.9995451	15.6	0.9990100	19.6	0.9982885
3.7	0.9999743	7.7	0.9998687	11.7	0.9995341	15.7	0.9989942	19.7	0.9982682
3.8	0.9999747	7.8	0.9998631	11.8	0.9995230	15.8	0.9989782	19.8	0.9982478
3.9	0.9999749	7.9	0.9998572	11.9	0.9995118	15.9	0.9989621	19.9	0.9982273
4.0	0.9999749	8.0	0.9998513	12.0	0.9995005	16.0	0.9989459	20.0	0.9982067

t/°C	ρ/g cm^{-3}	t/°C	ρ/g cm^{-3}	t/°C	ρ/g cm^{-3}	t/°C	ρ/g cm^{-3}	t/°C	ρ/g cm^{-3}
20.1	0.9981860	25.3	0.9969696	30.5	0.9954967	35.7	0.9937899	49.0	0.98848
20.2	0.9981652	25.4	0.9969436	30.6	0.9954660	35.8	0.9937549	50.0	0.98804
20.3	0.9981443	25.5	0.9969176	30.7	0.9954352	35.9	0.9937199	51.0	0.98758
20.4	0.9981233	25.6	0.9968914	30.8	0.9954044	36.0	0.9936847	52.0	0.98712
20.5	0.9981022	25.7	0.9968651	30.9	0.9953734	36.1	0.9936495	53.0	0.98665
20.6	0.9980810	25.8	0.9968387	31.0	0.9953424	36.2	0.9936142	54.0	0.98617
20.7	0.9980596	25.9	0.9968123	31.1	0.9953113	36.3	0.9935788	55.0	0.98569
20.8	0.9980382	26.0	0.9967857	31.2	0.9952801	36.4	0.9935434	56.0	0.98521
20.9	0.9980167	26.1	0.9967591	31.3	0.9952488	36.5	0.9935078	57.0	0.98471
21.0	0.9979950	26.2	0.9967324	31.4	0.9952175	36.6	0.9934722	58.0	0.98421
21.1	0.9979733	26.3	0.9967055	31.5	0.9951860	36.7	0.9934365	59.0	0.98371
21.2	0.9979514	26.4	0.9966786	31.6	0.9951545	36.8	0.9934007	60.0	0.98320
21.3	0.9979295	26.5	0.9966516	31.7	0.9951228	36.9	0.9933649	61.0	0.98268
21.4	0.9979074	26.6	0.9966245	31.8	0.9950911	37.0	0.9933290	62.0	0.98216
21.5	0.9978853	26.7	0.9965973	31.9	0.9950593	37.1	0.9932929	63.0	0.98163
21.6	0.9978630	26.8	0.9965700	32.0	0.9950275	37.2	0.9932569	64.0	0.98109
21.7	0.9978407	26.9	0.9965426	32.1	0.9949955	37.3	0.9932207	65.0	0.98055
21.8	0.9978182	27.0	0.9965151	32.2	0.9949635	37.4	0.9931844	66.0	0.98000
21.9	0.9977956	27.1	0.9964875	32.3	0.9949313	37.5	0.9931481	67.0	0.97945
22.0	0.9977730	27.2	0.9964599	32.4	0.9948991	37.6	0.9931117	68.0	0.97890
22.1	0.9977502	27.3	0.9964321	32.5	0.9948668	37.7	0.9930753	69.0	0.97833
22.2	0.9977273	27.4	0.9964043	32.6	0.9948344	37.8	0.9930387	70.0	0.97776
22.3	0.9977044	27.5	0.9963763	32.7	0.9948020	37.9	0.9930021	71.0	0.97719
22.4	0.9976813	27.6	0.9963483	32.8	0.9947694	38.0	0.9929654	72.0	0.97661
22.5	0.9976582	27.7	0.9963202	32.9	0.9947368	38.1	0.9929286	73.0	0.97603
22.6	0.9976349	27.8	0.9962920	33.0	0.9947041	38.2	0.9928917	74.0	0.97544
22.7	0.9976115	27.9	0.9962637	33.1	0.9946713	38.3	0.9928548	75.0	0.97484
22.8	0.9975881	28.0	0.9962353	33.2	0.9946384	38.4	0.9928178	76.0	0.97424
22.9	0.9975645	28.1	0.9962068	33.3	0.9946055	38.5	0.9927807	77.0	0.97364
23.0	0.9975408	28.2	0.9961783	33.4	0.9945724	38.6	0.9927435	78.0	0.97303
23.1	0.9975171	28.3	0.9961496	33.5	0.9945393	38.7	0.9927063	79.0	0.97241
23.2	0.9974932	28.4	0.9961208	33.6	0.9945061	38.8	0.9926689	80.0	0.97179
23.3	0.9974692	28.5	0.9960920	33.7	0.9944728	38.9	0.9926316	81.0	0.97116
23.4	0.9974452	28.6	0.9960631	33.8	0.9944394	39.0	0.9925941	82.0	0.97053
23.5	0.9974210	28.7	0.9960341	33.9	0.9944060	39.1	0.9925565	83.0	0.96990
23.6	0.9973968	28.8	0.9960050	34.0	0.9943724	39.2	0.9925189	84.0	0.96926
23.7	0.9973724	28.9	0.9959758	34.1	0.9943388	39.3	0.9924812	85.0	0.96861
23.8	0.9973480	29.0	0.9959465	34.2	0.9943051	39.4	0.9924434	86.0	0.96796
23.9	0.9973234	29.1	0.9959171	34.3	0.9942713	39.5	0.9924056	87.0	0.96731
24.0	0.9972988	29.2	0.9958876	34.4	0.9942375	39.6	0.9923677	88.0	0.96664
24.1	0.9972740	29.3	0.9958581	34.5	0.9942035	39.7	0.9923297	89.0	0.96598
24.2	0.9972492	29.4	0.9958285	34.6	0.9941695	39.8	0.9922916	90.0	0.96531
24.3	0.9972243	29.5	0.9957987	34.7	0.9941354	39.9	0.9922534	91.0	0.96463
24.4	0.9971992	29.6	0.9957689	34.8	0.9941012	40.0	0.9922152	92.0	0.96396
24.5	0.9971741	29.7	0.9957390	34.9	0.9940669	41.0	0.99183	93.0	0.96327
24.6	0.9971489	29.8	0.9957090	35.0	0.9940326	42.0	0.99144	94.0	0.96258
24.7	0.9971236	29.9	0.9956790	35.1	0.9939982	43.0	0.99104	95.0	0.96189
24.8	0.9970981	30.0	0.9956488	35.2	0.9939637	44.0	0.99063	96.0	0.96119
24.9	0.9970726	30.1	0.9956185	35.3	0.9939291	45.0	0.99021	97.0	0.96049
25.0	0.9970470	30.2	0.9955882	35.4	0.9938944	46.0	0.98979	98.0	0.95978
25.1	0.9970213	30.3	0.9955578	35.5	0.9938597	47.0	0.98936	99.0	0.95907
25.2	0.9969955	30.4	0.9955273	35.6	0.9938248	48.0	0.98893	99.974	0.95837

FIXED-POINT PROPERTIES OF H_2O AND D_2O

Allan H. Harvey

Temperatures are given on the ITS-90 scale.

References

1. International Association for the Properties of Water and Steam (IAPWS), *Guideline on the Use of Fundamental Physical Constants and Basic Constants of Water* (2001; 2008 update), available from http://www.iapws.org.
2. IAPWS, *Revised Release on the Pressure along the Melting and Sublimation Curves of Ordinary Water Substance* (2008), available from http://www.iapws.org.
3. IAPWS, *IAPWS Release on the Values of Temperature, Pressure, and Density of Ordinary and Heavy Water Substances at their Respective Critical Points* (1992), available from http://www.iapws.org.
4. Wagner, W., and Pruß, A., *J. Phys. Chem. Ref. Data* 31, 387, 2002.
5. Hill, P. G., MacMillan, R. D. C., and Lee, V., *J. Phys. Chem. Ref. Data* 11, 1, 1982.
6. Guildner, L. A., Johnson, D. P., and Jones, F. E., *J. Res. Nat. Bur. Stand.* 80A, 505, 1976.
7. Harvey, A. H., and Lemmon, E. W., *J. Phys. Chem. Ref. Data* 31, 173, 2002.
8. Harvey, A. H., Peskin, A. P., and Klein, S. A., NIST Standard Reference Database 10: *NIST/ASME Steam Properties*, Version 2.22, National Institute of Standards and Technology, Standard Reference Data Program, Gaithersburg, Maryland, 2008 (http://www.nist.gov/srd/nist10.htm).
9. Lemmon, E. W., Huber, M. L., and McLinden, M. O., NIST Standard Reference Database 23: *Reference Fluid Thermodynamic and Transport Properties-REFPROP*, Version 8.0, National Institute of Standards and Technology, Standard Reference Data Program, Gaithersburg, Maryland, 2007 (http://www.nist.gov/srd/nist23.htm).

	Unit	H_2O	D_2O
Molar mass	g mol^{-1}	18.015268	20.02751
Melting point (101.325 kPa)	°C	0.0025	3.81
Boiling point (101.325 kPa)	°C	99.974	101.40
Triple-point temperature	°C	0.01 (exact)	3.82
Triple-point pressure	Pa	611.657	661
Triple-point density (liq)	g cm^{-3}	0.99979	1.1055
Critical temperature	°C	373.946	370.697
Critical pressure	MPa	22.064	21.671
Critical density	g cm^{-3}	0.322	0.356
Maximum density (101.325 kPa)	g cm^{-3}	0.999975	1.1060
Temperature of maximum density	°C	3.98	11.2

PROPERTIES OF SATURATED LIQUID D_2O

Allan H. Harvey

Properties of saturated liquid heavy water, D_2O, are given in this table as a function of temperature from the melting point to the critical point. The vapor pressure was calculated from the formulation of Harvey and Lemmon (Ref. 2). The other properties were generated from the NIST REFPROP program (Ref. 1) and are consistent with formulations adopted for general and scientific use by The International Association for the Properties of Water and Steam (IAPWS). The background for the equation of state used for density and heat capacity is given by Hill et al. (Ref. 3), and the background for the transport property correlations is given by Matsunaga and Nagashima (Ref. 4). The unpublished surface tension correlation and the other IAPWS formulations may be found on the IAPWS Web site (http://www.iapws.org). The temperature scale is ITS-90. Additional calculations at state points not listed below can be obtained by using the REFPROP program (http://www.nist.gov/srd/nist23.htm). The properties are

P: vapor pressure
ρ: density
C_p: isobaric heat capacity
η: viscosity
λ: thermal conductivity
σ: surface tension

References

1. Lemmon, E.W., Huber, M.L., and McLinden, M.O., NIST Standard Reference Database 23: Reference Fluid Thermodynamic and Transport Properties-REFPROP, Version 8.0, National Institute of Standards and Technology, Standard Reference Data Program, Gaithersburg, Maryland, 2007 (www.nist.gov/srd/nist23.htm).
2. Harvey, A. H. and Lemmon, E. W., *J. Phys. Chem. Ref. Data* 31, 173, 2002.
3. Hill, P. G., MacMillan, R. D. C., and Lee, V., *J. Phys. Chem. Ref. Data* 11, 1, 1982.
4. Matsunaga, N. and Nagashima, A., *J. Phys. Chem. Ref. Data* 12, 933, 1983.

t/°C	P/kPa	ρ/kg m^{-3}	C_p/kJ kg^{-1} K^{-1}	η/mPa s	λ/W m^{-1} K^{-1}	σ/mN m^{-1}
3.82	0.661	1105.5	4.211	2.086	0.565	74.93
10	1.026	1106.0	4.231	1.679	0.575	74.06
20	1.999	1105.3	4.243	1.247	0.589	72.61
30	3.702	1103.2	4.240	0.972	0.600	71.09
40	6.550	1099.9	4.232	0.785	0.610	69.52
50	11.12	1095.6	4.220	0.651	0.618	67.89
60	18.20	1090.5	4.207	0.552	0.625	66.21
70	28.81	1084.6	4.194	0.476	0.629	64.47
80	44.24	1078.1	4.181	0.416	0.633	62.67
90	66.09	1071.0	4.170	0.368	0.635	60.82
100	96.30	1063.3	4.161	0.329	0.636	58.93
110	137.1	1055.1	4.155	0.296	0.636	56.98
120	191.3	1046.4	4.153	0.269	0.635	54.99
130	261.8	1037.2	4.155	0.247	0.632	52.95
140	352.0	1027.5	4.162	0.227	0.629	50.87
150	465.8	1017.2	4.174	0.210	0.625	48.75
160	607.3	1006.5	4.190	0.195	0.620	46.59
170	781.1	995.2	4.212	0.182	0.614	44.39
180	992.0	983.4	4.240	0.171	0.607	42.16
190	1246	971.1	4.273	0.161	0.600	39.90
200	1547	958.2	4.313	0.152	0.592	37.61
210	1903	944.7	4.360	0.144	0.583	35.29
220	2319	930.5	4.415	0.136	0.574	32.95
230	2802	915.7	4.479	0.130	0.563	30.59
240	3359	900.1	4.554	0.123	0.553	28.22
250	3998	883.7	4.643	0.118	0.541	25.84
260	4725	866.4	4.750	0.112	0.529	23.45
270	5550	848.0	4.880	0.107	0.516	21.07
280	6480	828.4	5.038	0.103	0.502	18.69
290	7525	807.5	5.237	0.098	0.488	16.33
300	8694	784.8	5.490	0.094	0.473	13.99
310	9998	760.1	5.823	0.089	0.458	11.68

t/°C	P/kPa	ρ/kg m^{-3}	C_p/kJ kg^{-1} K^{-1}	η/mPa s	λ/W m^{-1} K^{-1}	σ/mN m^{-1}
320	11449	732.7	6.281	0.085	0.442	9.428
330	13059	702.0	6.952	0.080	0.425	7.238
340	14845	666.5	8.041	0.075	0.408	5.141
350	16824	623.3	10.17	0.069	0.391	3.173
360	19024	565.1	16.41	0.062	0.382	1.405
370	21487	430.7	268.8	0.046	0.548	0.0467
370.697	21671	356.0				0

PROPERTIES OF ICE AND SUPERCOOLED WATER

Allan H. Harvey

The common form of ice at ambient pressure is hexagonal ice, designated as ice Ih (see phase diagram in Section 12). The data given here refer to that form, at standard atmospheric pressure (101.325 kPa). Data have been taken from the references indicated, which in most cases are formulations based on critical evaluation of available experimental data. Most properties are sensitive to the method of preparation of the sample, because air and other gases are sometimes occluded. For this reason, there is often disagreement among values in the literature. For all properties except the dielectric constant of ice, the cited reference contains information on the uncertainty of the property.

References

1. Wagner, W., and Pruß, A., *J. Phys. Chem. Ref. Data* 31, 387, 2002.
2. Feistel, R., and Wagner, W., *J. Phys. Chem. Ref. Data* 35, 1021, 2006.
3. International Association for the Properties of Water and Steam (IAPWS), *Revised Release on the Equation of State 2006 for H_2O Ice Ih* (2009), available from http://www.iapws.org.
4. Andersson, O., and Inaba, A., *Phys. Chem. Chem. Phys.* 7, 1441, 2005.
5. Slack, G.A., *Phys. Rev. B* 22, 3065, 1980.
6. Wörz, O., and Cole, R.H., *J. Chem. Phys.* 51, 1546, 1969.

Density of supercooled water (Ref. 1)

t/°C	ρ/g cm^{-3}
0	0.9998
−5	0.9993
−10	0.9981
−15	0.9963
−20	0.9936
−25	0.9896
−30	0.9838

Phase Transition Properties (Ref. 2)

$\Delta_{fus}H(0\ °C) = 333.4\ J\ g^{-1}$

$\Delta_{subl}H(0\ °C) = 2834\ J\ g^{-1}$

Thermophysical Properties of Ice I_h

ρ: mass density (Refs. 2, 3)

α_V: cubic expansion coefficient, $\alpha_V = -(1/V)(\partial V/\partial T)_p$ (Refs. 2, 3)

κ_s: isentropic compressibility, $\kappa_s = -(1/V)(\partial V/\partial p)_s$ (Refs. 2, 3)

c_p: specific heat capacity at constant pressure (Refs. 2, 3)

λ: thermal conductivity (Refs. 4, 5)

ε: static dielectric constant (relative permittivity) (Ref. 6)

t/°C	ρ/g cm^{-3}	$10^3\alpha_V$/K^{-1}	κ_s/GPa^{-1}	c_p/J g^{-1} K^{-1}	λ/W m^{-1} K^{-1}	ε
0	0.9167	0.160	0.114	2.10	2.16	91.2
−10	0.9182	0.155	0.113	2.02	2.26	95.1
−20	0.9196	0.150	0.111	1.95	2.38	99.4
−30	0.9209	0.144	0.110	1.88	2.50	104.1
−40	0.9222	0.138	0.108	1.80	2.63	109.2
−50	0.9235	0.132	0.107	1.73	2.77	115.0
−60	0.9247	0.126	0.106	1.66	2.93	121.4
−80	0.9269	0.111	0.103	1.52	3.27	136.6
−100	0.9288	0.095	0.101	1.38	3.69	
−120	0.9304	0.078	0.099	1.25	4.2	
−140	0.9317	0.060	0.097	1.11	4.9	
−160	0.9326	0.041	0.096	0.97	5.7	
−180	0.9332	0.025	0.095	0.82	7.0	
−200	0.9336	0.013	0.095	0.65	8.9	
−220	0.9337	0.0050	0.095	0.47	12.2	
−240	0.9338	0.0012	0.095	0.27	20	
−260	0.9338	0.00008	0.095	0.036		

VAPOR PRESSURE OF ICE

The values of the vapor (sublimation) pressure of ice Ih were calculated from the equation recommended by the International Association for the Properties of Water and Steam (IAPWS) in 2008. See Refs. 1 and 2 for details on the uncertainty in different temperature ranges. The first entry in the table is the triple point of water, whose pressure has an uncertainty of 0.010 Pa.

References

1. International Association for the Properties of Water and Steam (IAPWS), *Revised Release on the Pressure along the Melting and Sublimation Curves of Ordinary Water Substance* (2008), available from http://www.iapws.org.
2. Wagner, W., Feistel, R., and Riethmann, T., New Equations for the Melting Pressure and Sublimation Pressure of H_2O Ice Ih, *J. Phys. Chem. Ref. Data*, to be submitted.

t/°C	*p*/Pa	*t*/°C	*p*/Pa	*t*/°C	*p*/Pa	*t*/°C	*p*/Pa	*t*/°C	*p*/Pa
0.01	611.657	−10	259.87	−22	85.08	−50	3.938	−130	8.75×10^{-7}
0	611.15	−11	237.71	−24	69.89	−55	2.094	−140	3.62×10^{-8}
−1	562.66	−12	217.29	−26	57.24	−60	1.081	−150	9.00×10^{-10}
−2	517.70	−13	198.49	−28	46.72	−65	0.541	−160	1.18×10^{-11}
−3	476.04	−14	181.19	−30	38.01	−70	0.262	−170	6.80×10^{-14}
−4	437.45	−15	165.27	−32	30.81	−75	0.122	−180	1.32×10^{-16}
−5	401.74	−16	150.65	−34	24.89	−80	5.48×10^{-2}	−190	5.89×10^{-20}
−6	368.71	−17	137.22	−36	20.04	−90	9.68×10^{-3}	−200	3.31×10^{-24}
−7	338.17	−18	124.90	−38	16.07	−100	1.40×10^{-3}	−210	8.86×10^{-30}
−8	309.95	−19	113.60	−40	12.84	−110	1.61×10^{-4}	−220	2.06×10^{-37}
−9	283.91	−20	103.24	−45	7.203	−120	1.41×10^{-5}		

T/K	*p*/Pa	*T*/K	*p*/Pa	*T*/K	*p*/Pa	*T*/K	*p*/Pa	*T*/K	*p*/Pa
273.16	611.657	265	305.91	250	76.01	215	1.386	130	1.20×10^{-8}
273.15	611.15	264	280.18	248	62.39	210	0.702	120	2.48×10^{-10}
273	603.65	263	256.43	246	50.95	205	0.344	110	2.57×10^{-12}
272	555.70	262	234.54	244	41.51	200	0.163	100	1.09×10^{-14}
271	511.25	261	214.37	242	33.70	190	3.24×10^{-2}	90	1.39×10^{-17}
270	470.06	260	195.80	240	27.27	180	5.39×10^{-3}	80	3.50×10^{-21}
269	431.92	258	163.00	235	15.81	170	7.30×10^{-4}	70	8.58×10^{-26}
268	396.62	256	135.30	230	8.947	160	7.73×10^{-5}	60	6.51×10^{-32}
267	363.97	254	111.98	225	4.939	150	6.10×10^{-6}	50	1.93×10^{-40}
266	333.79	252	92.40	220	2.654	140	3.37×10^{-7}		

MELTING POINT OF ICE AS A FUNCTION OF PRESSURE

This table gives the melting temperature of ice at various pressures, calculated from the equation for the ice Ih – liquid water phase boundary recommended by the International Association for the Properties of Water and Steam (IAPWS) in 2008. See Refs. 1 and 2 for information on the solid/liquid transitions for high-pressure forms of ice. IAPWS gives the following locations for the triple points where equilibrium exists among two ice forms and liquid water:

ice I – ice III	208.566 MPa	−21.985 °C
ice III – ice V	350.1 MPa	−16.986 °C
ice V – ice VI	632.4 MPa	0.16 °C
ice VI – ice VII	2216 MPa	81.85 °C

References

1. International Association for the Properties of Water and Steam (IAPWS), *Revised Release on the Pressure along the Melting and Sublimation Curves of Ordinary Water Substance* (2008), available from http://www.iapws.org.
2. Wagner, W., Feistel, R., and Riethmann, T., New Equations for the Melting Pressure and Sublimation Pressure of H_2O Ice Ih, *J. Phys. Chem. Ref. Data*, to be submitted.

p/MPa	*t*/°C	*p*/MPa	*t*/°C	*p*/MPa	*t*/°C
0.000 612	0.01	20	-1.54	90	-7.91
0.1	0.0026	30	-2.36	100	-8.94
1	-0.064	40	-3.21	120	-11.09
2	-0.14	50	-4.09	140	-13.35
5	-0.37	60	-5.00	160	-15.73
10	-0.75	70	-5.94	180	-18.22
15	-1.14	80	-6.91	200	-20.83

PERMITTIVITY (DIELECTRIC CONSTANT) OF WATER AT VARIOUS FREQUENCIES

The permittivity of liquid water in the radiofrequency and microwave regions can be represented by the Debye equation (References 1 and 2):

$$\varepsilon' = \varepsilon_\infty + \frac{\varepsilon_s - \varepsilon_\infty}{1+\omega^2\tau^2}$$

$$\varepsilon'' = \frac{(\varepsilon_s - \varepsilon_\infty)\omega\tau}{1+\omega^2\tau^2}$$

where $\varepsilon = \varepsilon' + i\,\varepsilon''$ is the (complex) relative permittivity (i.e., the absolute permittivity divided by the permittivity of free space $\varepsilon_0 = 8.854{\cdot}10^{-12}$ F m^{-1}). Here ε_s is the static permittivity (see Reference 3 and the table "Thermophysical Properties of Water and Steam" in this Section); ε_∞ is a parameter describing the permittivity in the high frequency limit; τ is the relaxation time for molecular orientation; and $\omega = 2\pi f$ is the angular frequency. The values in this table have been calculated from parameters given in Reference 2:

	0 °C	25 °C	50 °C
ε_∞	5.7	5.2	4.0
τ/ps	17.67	8.27	4.75

Other useful quantities that can be calculated from the values in the table are the loss tangent:

$$\tan\delta = \varepsilon''/\varepsilon'$$

and the absorption coefficient α which describes the power attenuation per unit length ($P = P_0\, e^{-\alpha l}$):

$$\alpha = \frac{\pi f \varepsilon''}{c\sqrt{\varepsilon'}}$$

and c is the speed of light. The last equation is valid when $\varepsilon''/\varepsilon' << 1$.

References

1. Fernendez, D. P., Mulev, Y., Goodwin, A. R. H., and Levelt Sengers, J. M. H., *J. Phys. Chem. Ref. Data*, 24, 33, 1995.
2. Kaatze, U., *J. Chem. Eng. Data*, 34, 371, 1989.
3. Archer, D. G., and Wang, P., *J. Phys. Chem. Ref. Data*, 12, 817, 1983.

	0 °C		25 °C		50 °C	
Frequency	ε′	ε″	ε′	ε″	ε′	ε″
0	87.90	0.00	78.36	0.00	69.88	0.00
1 kHz	87.90	0.00	78.36	0.00	69.88	0.00
1 MHz	87.90	0.01	78.36	0.00	69.88	0.00
10 MHz	87.90	0.09	78.36	0.04	69.88	0.02
100 MHz	87.89	0.91	78.36	0.38	69.88	0.20
200 MHz	87.86	1.82	78.35	0.76	69.88	0.39
500 MHz	87.65	4.55	78.31	1.90	69.87	0.98
1 GHz	86.90	9.01	78.16	3.79	69.82	1.96
2 GHz	84.04	17.39	77.58	7.52	69.65	3.92
3 GHz	79.69	24.64	76.62	11.13	69.36	5.85
4 GHz	74.36	30.49	75.33	14.58	68.95	7.75
5 GHz	68.54	34.88	73.73	17.81	68.45	9.62
10 GHz	42.52	40.88	62.81	29.93	64.49	18.05
20 GHz	19.56	30.78	40.37	36.55	52.57	28.99
30 GHz	12.50	22.64	26.53	33.25	40.57	32.74
40 GHz	9.67	17.62	18.95	28.58	31.17	32.43
50 GHz	8.28	14.34	14.64	24.53	24.42	30.47

THERMOPHYSICAL PROPERTIES OF AIR

Eric W. Lemmon

These tables summarize the thermophysical properties of air in the liquid and gaseous states as calculated from the pseudo-pure fluid equation of state of Lemmon et al. (2000). The first table refers to liquid and gaseous air at equilibrium as a function of temperature. The tabulated properties are the bubble-point pressure (i.e., pressure at which boiling begins as the pressure of the liquid is lowered); the dew-point pressure (i.e., pressure at which condensation begins as the pressure of the gas is raised); density (ρ); enthalpy (H); entropy (S); isochoric heat capacity (C_v); isobaric heat capacity (C_p); speed of sound (u); viscosity (η); and thermal conductivity (λ). The first line of identical temperatures is the bubble-point (liquid) and the second line is the dew point (vapor). The normal boiling point of air, i.e., the temperature at which the bubble-point pressure reaches 1 standard atmosphere (1.01325 bar), is 78.90 K (−194.25 °C).

The second table gives the properties of air along various isobars. An entry with non-integer temperatures in the isobar section indicates a phase transition (liquid–vapor) at these temperatures; property values are then given for both phases. These are identified by the high densities in the liquid and the low densities in the vapor. Additional calculations at state points not listed below can be obtained by using the NIST program REFPROP (http://www.nist.gov/srd/nist23.htm).

References

1. Lemmon, E.W., Jacobsen, R.T, Penoncello, S.G., and Friend, D.G., Thermodynamic Properties of Air and Mixtures of Nitrogen, Argon, and Oxygen from 60 to 2000 K at Pressures to 2000 MPa, *J. Phys. Chem. Ref. Data*, 29, 331, 2000.
2. Lemmon, E.W. and Jacobsen, R.T, Viscosity and Thermal Conductivity Equations for Nitrogen, Oxygen, Argon, and Air, *Int. J. Thermophys.*, 25, 21, 2004.
3. Lemmon, E.W., Huber, M.L., McLinden, M.O., NIST Standard Reference Database 23: Reference Fluid Thermodynamic and Transport Properties-REFPROP, Version 8.0, National Institute of Standards and Technology, Standard Reference Data Program, Gaithersburg, Maryland, 2007 (www.nist.gov/srd/nist23.htm).

Thermophysical Properties of Air along the Boiling and Condensation Curves

T K	P MPa	ρ kg m^{-3}	H kJ kg^{-1}	S kJ kg^{-1} K^{-1}	C_v kJ kg^{-1} K^{-1}	C_p kJ kg^{-1} K^{-1}	u m s^{-1}	η μPa s	λ mW m^{-1} K^{-1}
59.75	0.005265	957.6	−36.66	−0.5306	1.174	1.901	1030.	376.6	171.4
59.75	0.002432	0.1421	185.5	3.340	0.7184	1.009	154.8	4.220	5.294
60	0.005546	956.5	−36.19	−0.5226	1.173	1.901	1028.	371.9	171.0
60	0.002584	0.1504	185.8	3.326	0.7186	1.009	155.1	4.238	5.320
62	0.008270	948.2	−32.38	−0.4603	1.157	1.901	1012.	336.9	167.8
62	0.004111	0.2318	187.7	3.225	0.7198	1.012	157.6	4.386	5.529
64	0.01200	939.9	−28.58	−0.3999	1.143	1.902	995.8	306.3	164.5
64	0.006325	0.3460	189.6	3.132	0.7212	1.015	160.0	4.532	5.739
66	0.01699	931.5	−24.77	−0.3414	1.129	1.903	979.1	279.4	161.3
66	0.009442	0.5018	191.5	3.047	0.7230	1.019	162.3	4.679	5.950
68	0.02352	923.0	−20.95	−0.2846	1.115	1.906	962.2	255.7	158.0
68	0.01371	0.7089	193.4	2.968	0.7252	1.024	164.5	4.825	6.162
70	0.03191	914.4	−17.13	−0.2293	1.102	1.908	945.1	234.8	154.7
70	0.01943	0.9785	195.2	2.896	0.7277	1.030	166.7	4.970	6.376
72	0.04250	905.7	−13.31	−0.1756	1.090	1.912	927.7	216.3	151.4
72	0.02692	1.322	197.0	2.828	0.7305	1.037	168.7	5.115	6.592
74	0.05566	897.0	−9.468	−0.1232	1.078	1.917	910.0	199.9	148.1
74	0.03655	1.753	198.7	2.766	0.7338	1.046	170.6	5.260	6.810
76	0.07179	888.1	−5.617	−0.07209	1.067	1.923	892.1	185.2	144.8
76	0.04870	2.285	200.4	2.708	0.7375	1.055	172.5	5.405	7.031
78	0.09129	879.1	−1.751	−0.02217	1.056	1.930	873.9	172.1	141.5
78	0.06381	2.933	202.0	2.653	0.7416	1.066	174.2	5.549	7.256
80	0.1146	870.0	2.132	0.02665	1.045	1.938	855.4	160.4	138.2
80	0.08232	3.711	203.6	2.602	0.7460	1.078	175.8	5.694	7.485
82	0.1422	860.7	6.036	0.07444	1.035	1.948	836.7	149.8	134.8
82	0.1047	4.635	205.1	2.554	0.7510	1.092	177.4	5.839	7.719
84	0.1745	851.3	9.962	0.1213	1.025	1.959	817.6	140.2	131.4
84	0.1315	5.724	206.5	2.509	0.7563	1.108	178.8	5.984	7.959
86	0.2121	841.7	13.91	0.1673	1.016	1.972	798.2	131.5	128.1
86	0.1631	6.993	207.8	2.466	0.7620	1.125	180.0	6.131	8.206
88	0.2553	832.0	17.90	0.2125	1.007	1.986	778.6	123.6	124.8
88	0.2002	8.464	209.1	2.425	0.7682	1.144	181.2	6.278	8.461

T K	P MPa	ρ kg m⁻³	H kJ kg⁻¹	S kJ kg⁻¹ K⁻¹	C_v kJ kg⁻¹ K⁻¹	C_p kJ kg⁻¹ K⁻¹	u m s⁻¹	η μPa s	λ mW m⁻¹ K⁻¹
90	0.3048	822.0	21.91	0.2569	0.9984	2.003	758.5	116.4	121.4
90	0.2432	10.16	210.3	2.386	0.7748	1.166	182.2	6.427	8.725
92	0.3609	811.8	25.97	0.3007	0.9902	2.022	738.2	109.7	118.0
92	0.2927	12.09	211.4	2.349	0.7817	1.190	183.1	6.578	9.001
94	0.4243	801.4	30.06	0.3439	0.9825	2.044	717.5	103.6	114.6
94	0.3493	14.29	212.3	2.313	0.7891	1.217	183.8	6.732	9.289
96	0.4954	790.7	34.21	0.3866	0.9752	2.069	696.5	97.88	111.2
96	0.4136	16.78	213.2	2.279	0.7969	1.248	184.5	6.889	9.593
98	0.5749	779.7	38.41	0.4288	0.9684	2.098	675.0	92.57	107.8
98	0.4861	19.60	214.0	2.246	0.8052	1.282	184.9	7.050	9.915
100	0.6631	768.4	42.66	0.4707	0.9619	2.131	653.3	87.61	104.4
100	0.5674	22.76	214.6	2.213	0.8138	1.320	185.3	7.215	10.26
102	0.7608	756.7	46.98	0.5122	0.9560	2.168	631.1	82.94	101.0
102	0.6582	26.32	215.1	2.182	0.8230	1.363	185.5	7.387	10.63
104	0.8684	744.6	51.38	0.5535	0.9505	2.212	608.5	78.53	97.62
104	0.7590	30.31	215.5	2.151	0.8326	1.413	185.6	7.566	11.02
106	0.9864	732.1	55.86	0.5947	0.9456	2.262	585.5	74.35	94.25
106	0.8706	34.78	215.7	2.120	0.8429	1.470	185.5	7.754	11.46
108	1.116	719.1	60.44	0.6358	0.9412	2.321	562.1	70.36	90.89
108	0.9934	39.79	215.8	2.089	0.8537	1.536	185.2	7.952	11.94
110	1.256	705.5	65.12	0.6769	0.9375	2.390	538.2	66.54	87.55
110	1.128	45.41	215.6	2.059	0.8653	1.614	184.9	8.163	12.47
112	1.409	691.2	69.93	0.7182	0.9345	2.472	513.9	62.87	84.24
112	1.276	51.73	215.2	2.028	0.8777	1.708	184.3	8.391	13.07
114	1.575	676.2	74.87	0.7598	0.9324	2.571	489.0	59.31	80.96
114	1.437	58.84	214.6	1.997	0.8912	1.821	183.6	8.637	13.76
116	1.755	660.3	79.98	0.8019	0.9312	2.693	463.5	55.85	77.72
116	1.612	66.88	213.8	1.965	0.9059	1.961	182.7	8.909	14.56
118	1.948	643.4	85.29	0.8447	0.9312	2.847	437.3	52.47	74.52
118	1.801	76.04	212.6	1.932	0.9220	2.139	181.7	9.210	15.50
120	2.156	625.1	90.83	0.8885	0.9327	3.048	410.2	49.13	71.36
120	2.007	86.55	211.0	1.898	0.9402	2.374	180.4	9.552	16.63
122	2.379	605.3	96.66	0.9338	0.9363	3.323	382.0	45.81	68.24
122	2.229	98.76	208.9	1.861	0.9608	2.694	179.1	9.946	18.04
124	2.617	583.3	102.9	0.9811	0.9427	3.723	352.3	42.46	65.17
124	2.468	113.2	206.3	1.821	0.9847	3.157	177.5	10.41	19.85
126	2.872	558.3	109.7	1.032	0.9537	4.367	320.4	39.01	62.18
126	2.727	130.6	202.9	1.777	1.013	3.882	175.8	10.98	22.29
128	3.143	528.3	117.3	1.088	0.9728	5.589	285.0	35.33	59.44
128	3.006	152.6	198.3	1.725	1.049	5.166	174.0	11.72	25.84
130	3.429	488.3	126.7	1.157	1.010	8.849	243.7	31.07	58.05
130	3.308	182.7	191.7	1.660	1.096	8.033	171.9	12.77	31.81
132	3.723	411.2	142.6	1.273	1.117	35.04	189.1	24.47	67.80
132	3.646	235.4	179.7	1.556	1.168	20.65	169.4	14.80	47.00
132.63	3.785	302.6	164.5	1.437				17.83	

Thermophysical Properties of Air along Various Isobars

T K	ρ kg m⁻³	H kJ kg⁻¹	S kJ kg⁻¹ K⁻¹	C_v kJ kg⁻¹ K⁻¹	C_p kJ kg⁻¹ K⁻¹	u m s⁻¹	η μPa s	λ mW m⁻¹ K⁻¹
P = 0.1 MPa (1 bar)								
60	956.7	−36.11	−0.5230	1.173	1.901	1029.	372.4	171.1
78.79	875.5	−0.2237	−0.002818	1.051	1.933	866.7	167.4	140.2
81.61	4.442	204.8	2.563	0.7500	1.089	177.1	5.811	7.673
100	3.557	224.3	2.779	0.7282	1.040	198.2	7.107	9.469

T K	ρ kg m^{-3}	H kJ kg^{-1}	S kJ kg^{-1} K^{-1}	C_v kJ kg^{-1} K^{-1}	C_p kJ kg^{-1} K^{-1}	u m s^{-1}	η μPa s	λ mW m^{-1} K^{-1}
120	2.938	244.9	2.966	0.7211	1.022	218.3	8.457	11.38
140	2.507	265.2	3.123	0.7184	1.014	236.4	9.750	13.24
160	2.188	285.5	3.258	0.7172	1.011	253.2	10.99	15.05
180	1.942	305.6	3.377	0.7166	1.008	268.8	12.18	16.80
200	1.746	325.8	3.483	0.7163	1.007	283.5	13.33	18.50
220	1.586	345.9	3.579	0.7163	1.006	297.4	14.44	20.16
240	1.453	366.0	3.667	0.7164	1.006	310.7	15.51	21.77
260	1.341	386.2	3.747	0.7168	1.006	323.4	16.55	23.35
280	1.245	406.3	3.822	0.7173	1.006	335.6	17.56	24.88
300	1.161	426.4	3.891	0.7181	1.007	347.4	18.54	26.38
320	1.089	446.5	3.956	0.7192	1.007	358.7	19.49	27.85
340	1.024	466.7	4.018	0.7206	1.009	369.6	20.41	29.29
360	0.9674	486.9	4.075	0.7223	1.010	380.3	21.32	30.71
380	0.9164	507.1	4.130	0.7243	1.012	390.5	22.20	32.09
400	0.8706	527.4	4.182	0.7266	1.014	400.5	23.06	33.45
500	0.6964	629.5	4.410	0.7426	1.030	446.4	27.09	39.94
600	0.5803	733.6	4.599	0.7641	1.051	487.1	30.77	46.01
700	0.4974	839.9	4.763	0.7879	1.075	523.9	34.18	51.76
800	0.4352	948.6	4.908	0.8117	1.099	557.8	37.37	57.25
900	0.3869	1060.	5.039	0.8340	1.121	589.6	40.39	62.54
1000	0.3482	1173.	5.158	0.8540	1.141	619.6	43.28	67.68

P = 0.5 MPa (5 bar)

T K	ρ kg m^{-3}	H kJ kg^{-1}	S kJ kg^{-1} K^{-1}	C_v kJ kg^{-1} K^{-1}	C_p kJ kg^{-1} K^{-1}	u m s^{-1}	η μPa s	λ mW m^{-1} K^{-1}
60	957.0	35.80	0.5248	1.173	1.900	1031.	374.6	171.4
80	870.9	2.387	0.02430	1.046	1.934	858.5	161.4	138.6
96.12	790.0	34.46	0.3892	0.9748	2.071	695.2	97.55	111.0
98.36	20.14	214.1	2.240	0.8067	1.288	185.0	7.079	9.974
100	19.65	216.2	2.261	0.7967	1.261	187.4	7.192	10.10
120	15.48	239.6	2.475	0.7461	1.115	212.4	8.542	11.81
140	12.94	261.4	2.643	0.7311	1.068	232.8	9.834	13.58
160	11.17	282.5	2.784	0.7245	1.045	250.9	11.07	15.32
180	9.842	303.3	2.906	0.7213	1.032	267.4	12.26	17.04
200	8.811	323.8	3.014	0.7195	1.025	282.6	13.41	18.71
220	7.981	344.3	3.112	0.7186	1.020	297.0	14.51	20.34
240	7.297	364.6	3.200	0.7182	1.017	310.6	15.58	21.94
260	6.724	385.0	3.282	0.7182	1.015	323.5	16.62	23.50
280	6.235	405.2	3.357	0.7185	1.014	335.9	17.62	25.02
300	5.813	425.5	3.427	0.7191	1.013	347.8	18.60	26.51
320	5.446	445.8	3.492	0.7200	1.013	359.3	19.54	27.97
340	5.123	466.0	3.553	0.7213	1.013	370.3	20.47	29.41
360	4.836	486.3	3.611	0.7229	1.014	381.0	21.37	30.81
380	4.580	506.6	3.666	0.7248	1.016	391.3	22.24	32.19
400	4.350	526.9	3.718	0.7271	1.018	401.3	23.10	33.55
500	3.477	629.3	3.947	0.7429	1.032	447.3	27.13	40.02
600	2.897	733.5	4.137	0.7643	1.053	488.0	30.80	46.07
700	2.483	840.0	4.301	0.7881	1.076	524.8	34.20	51.80
800	2.173	948.8	4.446	0.8119	1.100	558.7	37.40	57.29
900	1.932	1060.	4.577	0.8341	1.122	590.5	40.42	62.58
1000	1.739	1173.	4.696	0.8542	1.142	620.4	43.30	67.71

P = 1 MPa (10 bar)

T K	ρ kg m^{-3}	H kJ kg^{-1}	S kJ kg^{-1} K^{-1}	C_v kJ kg^{-1} K^{-1}	C_p kJ kg^{-1} K^{-1}	u m s^{-1}	η μPa s	λ mW m^{-1} K^{-1}
60	958.0	−35.42	−0.5271	1.174	1.898	1033.	377.2	171.7
80	872.2	2.720	0.02129	1.047	1.930	862.5	162.8	139.1
100	770.1	42.76	0.4673	0.9623	2.119	658.2	88.33	105.0
106.22	730.7	56.36	0.5992	0.9451	2.268	583.0	73.90	93.88
108.10	40.07	215.8	2.088	0.8543	1.540	185.2	7.963	11.96

T K	ρ kg m^{-3}	H kJ kg^{-1}	S kJ kg^{-1} K^{-1}	C_v kJ kg^{-1} K^{-1}	C_p kJ kg^{-1} K^{-1}	u m s^{-1}	η μPa s	λ mW m^{-1} K^{-1}
120	33.48	232.3	2.233	0.7844	1.285	204.0	8.718	12.59
140	27.02	256.3	2.419	0.7481	1.148	228.2	9.978	14.11
160	22.94	278.7	2.568	0.7341	1.093	248.1	11.20	15.74
180	20.03	300.2	2.695	0.7273	1.065	265.7	12.38	17.38
200	17.83	321.3	2.806	0.7236	1.048	281.7	13.51	19.00
220	16.09	342.2	2.906	0.7216	1.038	296.6	14.61	20.60
240	14.68	362.9	2.996	0.7204	1.031	310.6	15.67	22.17
260	13.50	383.5	3.078	0.7199	1.026	323.8	16.70	23.70
280	12.50	403.9	3.154	0.7199	1.023	336.4	17.70	25.21
300	11.64	424.4	3.224	0.7203	1.021	348.4	18.67	26.68
320	10.90	444.8	3.290	0.7211	1.020	360.0	19.62	28.13
340	10.25	465.2	3.352	0.7222	1.019	371.1	20.54	29.55
360	9.668	485.6	3.410	0.7237	1.019	381.9	21.43	30.95
380	9.153	506.0	3.465	0.7255	1.020	392.3	22.31	32.32
400	8.690	526.4	3.518	0.7277	1.022	402.3	23.16	33.67
500	6.943	629.1	3.747	0.7434	1.034	448.5	27.18	40.11
600	5.784	733.5	3.937	0.7646	1.054	489.2	30.84	46.15
700	4.957	840.0	4.101	0.7884	1.077	526.0	34.24	51.87
800	4.338	948.9	4.247	0.8121	1.100	559.9	37.43	57.35
900	3.857	1060.	4.378	0.8343	1.122	591.5	40.45	62.63
1000	3.472	1173.	4.497	0.8543	1.142	621.5	43.33	67.75
***P* = 2 MPa (20 bar)**								
60.11[a]	959.1	−34.44	−0.5282	1.175	1.895	1037.	380.5	172.2
80	874.6	3.390	0.01535	1.048	1.921	870.2	165.5	140.1
100	775.0	43.09	0.4576	0.9636	2.086	672.5	90.43	106.6
118.52	638.8	86.69	0.8559	0.9314	2.894	430.4	51.60	73.71
119.94	86.20	211.0	1.899	0.9396	2.365	180.5	9.540	16.59
120	86.03	211.2	1.900	0.9382	2.354	180.7	9.542	16.57
140	59.88	244.8	2.161	0.7883	1.387	218.3	10.44	15.68
160	48.61	270.5	2.333	0.7545	1.213	242.7	11.55	16.80
180	41.54	294.0	2.471	0.7396	1.139	262.8	12.67	18.21
200	36.52	316.3	2.589	0.7319	1.100	280.3	13.77	19.69
220	32.70	338.0	2.692	0.7275	1.076	296.2	14.84	21.19
240	29.66	359.4	2.785	0.7249	1.060	310.8	15.88	22.68
260	27.18	380.5	2.870	0.7235	1.050	324.6	16.89	24.16
280	25.11	401.4	2.947	0.7228	1.042	337.6	17.88	25.62
300	23.34	422.2	3.019	0.7227	1.037	349.9	18.84	27.06
320	21.82	442.9	3.086	0.7231	1.033	361.7	19.77	28.47
340	20.49	463.5	3.148	0.7240	1.031	373.0	20.68	29.87
360	19.32	484.1	3.207	0.7253	1.030	383.9	21.57	31.24
380	18.27	504.7	3.263	0.7269	1.029	394.4	22.44	32.59
400	17.34	525.3	3.315	0.7290	1.029	404.5	23.29	33.93
500	13.84	628.6	3.546	0.7442	1.039	450.8	27.28	40.31
600	11.52	733.4	3.737	0.7653	1.057	491.5	30.93	46.30
700	9.878	840.2	3.902	0.7889	1.079	528.3	34.32	52.00
800	8.646	949.3	4.047	0.8125	1.102	562.1	37.49	57.46
900	7.689	1061.	4.178	0.8346	1.123	593.7	40.50	62.73
1000	6.923	1174.	4.298	0.8546	1.143	623.6	43.38	67.84
***P* = 5 MPa (50 bar)**								
60.64[a]	961.4	−31.10	−0.5246	1.176	1.886	1048.	386.2	173.3
80	881.7	5.437	−0.001766	1.054	1.898	892.1	173.5	143.0
100	788.3	44.27	0.4311	0.9681	2.009	710.6	96.44	111.1
120	665.1	87.67	0.8256	0.9194	2.429	496.8	57.06	79.02
140	321.3	172.5	1.467	1.049	8.515	199.5	19.25	43.18

T K	ρ kg m^{-3}	H kJ kg^{-1}	S kJ kg^{-1} K^{-1}	C_v kJ kg^{-1} K^{-1}	C_p kJ kg^{-1} K^{-1}	u m s^{-1}	η μPa s	λ mW m^{-1} K^{-1}
160	151.4	240.7	1.930	0.8269	1.916	231.8	13.80	22.96
180	116.7	273.4	2.123	0.7785	1.453	258.2	14.14	22.02
200	97.93	300.6	2.267	0.7569	1.288	279.4	14.90	22.54
220	85.43	325.5	2.385	0.7451	1.205	297.7	15.77	23.49
240	76.25	349.0	2.488	0.7381	1.155	314.0	16.68	24.62
260	69.12	371.8	2.579	0.7338	1.123	328.8	17.60	25.83
280	63.36	394.0	2.661	0.7312	1.101	342.7	18.52	27.09
300	58.59	415.8	2.736	0.7297	1.085	355.6	19.42	28.39
320	54.55	437.4	2.806	0.7291	1.074	367.9	20.31	29.69
340	51.07	458.8	2.871	0.7292	1.065	379.6	21.18	30.98
360	48.05	480.1	2.932	0.7298	1.059	390.7	22.04	32.27
380	45.38	501.2	2.989	0.7310	1.055	401.4	22.88	33.55
400	43.01	522.3	3.043	0.7327	1.052	411.7	23.70	34.81
500	34.21	627.4	3.277	0.7466	1.052	458.3	27.61	40.97
600	28.48	733.2	3.470	0.7671	1.066	498.9	31.20	46.83
700	24.42	840.8	3.636	0.7903	1.085	535.4	34.55	52.43
800	21.39	950.4	3.782	0.8136	1.106	569.0	37.69	57.83
900	19.03	1062.	3.914	0.8356	1.127	600.3	40.68	63.04
1000	17.14	1176.	4.034	0.8554	1.146	629.9	43.54	68.12
P = 10 MPa (100 bar)								
61.52[a]	965.2	–25.55	–0.5187	1.177	1.871	1064.	395.1	175.0
80	892.4	8.950	–0.02831	1.063	1.868	925.2	186.8	147.5
100	806.9	46.78	0.3930	0.9767	1.924	763.5	105.8	117.8
120	706.1	86.65	0.7565	0.9192	2.094	591.1	67.00	89.01
140	573.7	132.1	1.106	0.8916	2.517	418.1	41.82	63.61
160	397.0	188.0	1.479	0.8787	2.882	297.6	24.99	43.38
180	273.3	238.0	1.774	0.8267	2.098	283.1	19.31	32.89
200	214.1	274.8	1.968	0.7908	1.646	296.3	18.16	29.63
220	179.9	305.4	2.114	0.7701	1.434	312.4	18.17	28.80
240	157.0	332.8	2.233	0.7574	1.317	328.0	18.60	28.89
260	140.2	358.3	2.336	0.7492	1.245	342.7	19.20	29.38
280	127.3	382.7	2.426	0.7439	1.196	356.5	19.90	30.15
300	116.9	406.3	2.507	0.7404	1.162	369.5	20.64	31.12
320	108.3	429.3	2.582	0.7383	1.138	381.8	21.40	32.15
340	101.0	451.8	2.650	0.7372	1.120	393.5	22.17	33.22
360	94.80	474.1	2.714	0.7369	1.106	404.6	22.95	34.33
380	89.36	496.1	2.773	0.7374	1.096	415.3	23.72	35.45
400	84.57	517.9	2.829	0.7384	1.088	425.6	24.49	36.58
500	67.06	625.7	3.070	0.7505	1.073	471.8	28.19	42.26
600	55.82	733.3	3.266	0.7699	1.080	511.9	31.67	47.84
700	47.90	842.0	3.433	0.7926	1.095	547.8	34.94	53.26
800	42.00	952.4	3.581	0.8155	1.113	580.8	38.04	58.52
900	37.42	1065.	3.713	0.8372	1.132	611.6	40.99	63.64
1000	33.75	1179.	3.833	0.8568	1.150	640.7	43.81	68.64
P = 20 MPa (200 bar)								
63.24[a]	972.3	–14.49	–0.5069	1.180	1.847	1094.	411.4	178.4
80	911.3	16.25	–0.07568	1.081	1.825	982.1	213.1	155.4
100	836.2	52.71	0.3311	0.9948	1.827	846.0	123.1	128.7
120	755.9	89.57	0.6670	0.9334	1.864	711.9	82.28	103.9
140	668.7	127.4	0.9589	0.8906	1.926	589.6	58.52	82.08
160	575.8	166.5	1.220	0.8606	1.977	491.8	43.00	65.18
180	484.9	206.0	1.452	0.8365	1.946	428.0	33.22	53.43
200	407.6	243.6	1.651	0.8143	1.808	397.2	27.80	46.02
220	348.5	278.1	1.815	0.7956	1.642	387.6	25.12	41.70

T K	ρ kg m^{-3}	H kJ kg^{-1}	S kJ kg^{-1} K^{-1}	C_v kJ kg^{-1} K^{-1}	C_p kJ kg^{-1} K^{-1}	u m s^{-1}	η μPa s	λ mW m^{-1} K^{-1}
240	304.6	309.5	1.952	0.7812	1.506	388.7	23.92	39.30
260	271.5	338.6	2.068	0.7705	1.405	394.7	23.48	37.98
280	245.7	365.9	2.169	0.7627	1.331	402.8	23.47	37.46
300	225.0	392.0	2.259	0.7572	1.277	412.0	23.70	37.57
320	208.0	417.1	2.340	0.7532	1.236	421.5	24.08	37.94
340	193.7	441.5	2.414	0.7506	1.204	431.1	24.56	38.47
360	181.5	465.3	2.482	0.7491	1.180	440.6	25.10	39.13
380	170.9	488.7	2.546	0.7485	1.161	450.0	25.68	39.88
400	161.7	511.7	2.605	0.7486	1.146	459.2	26.29	40.69
500	128.3	624.0	2.856	0.7575	1.108	501.8	29.49	45.27
600	107.0	734.4	3.057	0.7753	1.103	539.5	32.69	50.19
700	92.10	845.1	3.227	0.7969	1.112	573.7	35.78	55.18
800	80.99	957.0	3.377	0.8191	1.126	605.2	38.75	60.13
900	72.34	1070.	3.510	0.8402	1.142	634.8	41.61	65.02
1000	65.41	1185.	3.631	0.8594	1.157	662.8	44.36	69.84
P = 50 MPa (500 bar)								
68.21[a]	991.5	18.30	−0.4728	1.192	1.793	1173.	450.9	187.5
80	955.5	39.24	−0.1896	1.128	1.761	1112.	292.7	174.1
100	895.9	73.98	0.1982	1.043	1.715	1012.	171.6	152.8
120	837.8	107.9	0.5077	0.9811	1.680	920.9	118.9	133.2
140	781.1	141.2	0.7643	0.9342	1.649	839.4	90.60	115.9
160	726.2	173.9	0.9826	0.8983	1.619	770.2	72.86	101.0
180	673.9	205.9	1.171	0.8706	1.586	714.1	60.73	88.73
200	624.9	237.3	1.337	0.8488	1.549	670.6	52.14	79.05
220	579.8	267.9	1.482	0.8313	1.509	638.1	45.98	71.65
240	539.0	297.7	1.612	0.8171	1.467	614.7	41.60	66.11
260	502.4	326.6	1.728	0.8057	1.425	598.5	38.49	61.97
280	470.0	354.7	1.832	0.7964	1.385	588.0	36.32	58.99
300	441.2	382.0	1.926	0.7890	1.349	581.7	34.82	57.07
320	415.6	408.7	2.012	0.7832	1.316	578.4	33.83	55.77
340	392.9	434.7	2.091	0.7787	1.287	577.4	33.19	54.93
360	372.7	460.2	2.164	0.7754	1.262	578.1	32.82	54.41
380	354.6	485.2	2.232	0.7731	1.241	580.0	32.66	54.15
400	338.2	509.8	2.295	0.7717	1.223	582.9	32.64	54.07
500	276.1	629.0	2.561	0.7749	1.168	604.1	33.85	55.45
600	234.6	744.6	2.772	0.7890	1.148	629.5	36.00	58.34
700	204.6	859.2	2.948	0.8080	1.147	655.5	38.46	61.93
800	181.8	974.2	3.102	0.8285	1.154	681.0	41.00	65.86
900	163.8	1090.	3.238	0.8483	1.164	705.9	43.55	69.98
1000	149.2	1207.	3.362	0.8665	1.175	730.1	46.07	74.20

[a] Freezing point for the liquid state.

THERMOPHYSICAL PROPERTIES OF FLUIDS

Eric W. Lemmon

These tables give thermodynamic and transport properties of a variety of fluids, as generated from the equations of state presented in the references below. The properties tabulated are pressure (P), density (ρ), enthalpy (H), entropy (S), isochoric heat capacity (C_v), isobaric heat capacity (C_p), speed of sound (u), viscosity (η), thermal conductivity (λ), and static dielectric constant (D). All extensive properties are given on a mass basis. Not all properties are included for every substance. The references should be consulted for information on the uncertainties.

Values are given first along the saturation line. The first two points are the properties at the triple point. The final line gives the properties at the critical point. Two lines are given for each temperature (except at the critical point); the first line gives the values of the liquid phase (note the high density) and the second line gives the values of the vapor phase (at low densities). Following the saturation tables, values are given as a function of temperature for several isobars. A duplicate entry in the isobar section indicates a phase transition (liquid–vapor) at that temperature; property values are then given for both phases. The phase can be determined by noting the sharp decrease in density between two successive temperature entries; all lines above this point refer to the liquid phase, and all lines below refer to the gas phase. If there is no sharp discontinuity in density, all data in the table refer to the supercritical region (*i.e.*, the isobar is above the critical pressure). If the first temperature in the isobars is not an integer, this state point refers to the properties of the liquid at the melting line.

All temperatures are given on ITS-90, except those for oxygen (Ref. 16) and helium (Ref. 13), where the source equations of state still use the IPTS-68 temperature scale. Nitrogen, oxygen, and argon use a reference state based on zero enthalpy in the gas phase at 0 K; parahydrogen, helium, methane, and ethane use a reference state of zero enthalpy and entropy at the saturated liquid state at the normal boiling point; and propane and carbon dioxide use a reference state of 200 kJ kg^{-1} and 1 kJ kg^{-1} K^{-1}, respectively, at −40 °C. Additional calculations at state points not listed below and for fluids not contained here can be obtained at the NIST Chemistry WebBook website (webbook.nist.gov/chemistry/fluid/), or by using the NIST Standard Reference Data program REFPROP (Ref. 9).

References

1. Arp, V.D., McCarty, R.D., and Friend, D.G., Thermophysical Properties of Helium-4 from 0.8 to 1500 K with Pressures to 2000 MPa, NIST Technical Note 1334 (revised), 1998.
2. Buecker, D. and Wagner, W., A Reference Equation of State for the Thermodynamic Properties of Ethane for Temperatures from the Melting Line to 675 K and Pressures up to 900 MPa, *J. Phys. Chem. Ref. Data*, 35, 205, 2006.
3. Friend, D.G., Ely, J.F., and Ingham, H., Tables for the Thermophysical Properties of Methane, NIST Technical Note 1325, 1989.
4. Friend, D.G., Ingham, H., and Ely, J.F., Thermophysical Properties of Ethane, *J. Phys. Chem. Ref. Data*, 20, 275, 1991.
5. Hands, B.A. and Arp, V.D., A Correlation of Thermal Conductivity Data for Helium, *Cryogenics*, 21, 697, 1981.
6. Harvey, A.H. and Lemmon, E.W., Method for Estimating the Dielectric Constant of Natural Gas Mixtures, *Int. J. Thermophys.*, 26, 31, 2005.
7. Leachman, J.W., Jacobsen, R.T, Lemmon, E.W., and Penoncello, S.G., Fundamental Equations of State for Parahydrogen, Normal Hydrogen, and Orthohydrogen, submitted to *J. Phys. Chem. Ref. Data*, 2009.
8. Lemmon, E.W. and Jacobsen, R.T, Viscosity and Thermal Conductivity Equations for Nitrogen, Oxygen, Argon, and Air, *Int. J. Thermophys.*, 25, 21, 2004.
9. Lemmon, E.W., Huber, M.L., and McLinden, M.O., NIST Standard Reference Database 23: Reference Fluid Thermodynamic and Transport Properties-REFPROP, Version 8.0, National Institute of Standards and Technology, Standard Reference Data Program, Gaithersburg, Maryland, 2007 (www.nist.gov/srd/nist23.htm).
10. Lemmon, E.W., McLinden, M.O., and Friend, D.G., Thermophysical Properties of Fluid Systems, NIST Chemistry WebBook, NIST Standard Reference Database Number 69, Linstrom, P.J. and Mallard, W.G., Eds., National Institute of Standards and Technology, Gaithersburg, Maryland.
11. Lemmon, E.W., McLinden, M.O., and Wagner, W., Thermodynamic Properties of Propane. III. A Reference Equation of State for Temperatures from the Melting Line to 650 K and Pressures up to 1000 MPa, submitted to *J. Chem. Eng. Data*, 2009.
12. Marsh, K., Perkins, R., and Ramires, M.L.V., Measurement and Correlation of the Thermal Conductivity of Propane from 86 to 600 K at Pressures to 70 MPa, *J. Chem. Eng. Data*, 47, 932, 2002.
13. McCarty, R.D. and Arp, V.D., A New Wide Range Equation of State for Helium, *Adv. Cryo. Eng.*, 35, 1465, 1990.
14. McCarty, R.D. and Weber, L.A., Thermophysical Properties of Parahydrogen from the Freezing Liquid Line to 5000 R for Pressures to 10,000 psia, Natl. Bur. Stand., Tech. Note 617, 1972.
15. Quinones-Cisneros, S.E., Huber, M.L., and Deiters, U.K., Reference Correlation for the Viscosity of Methane, for submission to *J. Phys. Chem. Ref. Data*, 2009.
16. Schmidt, R. and Wagner, W., A New Form of the Equation of State for Pure Substances and its Application to Oxygen, *Fluid Phase Equilib.*, 19, 175, 1985.
17. Setzmann, U. and Wagner, W., A New Equation of State and Tables of Thermodynamic Properties for Methane Covering the Range from the Melting Line to 625 K at Pressures up to 1000 MPa, *J. Phys. Chem. Ref. Data*, 20, 1061, 1991.
18. Span, R. and Wagner, W., A New Equation of State for Carbon Dioxide Covering the Fluid Region from the Triple-Point Temperature to 1100 K at Pressures up to 800 MPa, *J. Phys. Chem. Ref. Data*, 25, 1509, 1996.
19. Span, R., Lemmon, E.W., Jacobsen, R.T, Wagner, W., and Yokozeki, A., A Reference Equation of State for the Thermodynamic Properties of Nitrogen for Temperatures from 63.151 to 1000 K and Pressures to 2200 MPa, *J. Phys. Chem. Ref. Data*, 29, 1361, 2000.
20. Tegeler, Ch., Span, R., and Wagner, W., A New Equation of State for Argon Covering the Fluid Region for Temperatures from the Melting Line to 700 K at Pressures up to 1000 MPa, *J. Phys. Chem. Ref. Data*, 28, 779, 1999.
21. Vogel, E., Kuechenmeister, C., Bich, E., and Laesecke, A., Reference Correlation of the Viscosity of Propane, *J. Phys. Chem. Ref. Data*, 27, 947, 1998.

Nitrogen (N_2)

T K	*P* MPa	*ρ* kg m⁻³	*H* kJ kg⁻¹	*S* kJ kg⁻¹ K⁻¹	C_v kJ kg⁻¹ K⁻¹	C_p kJ kg⁻¹ K⁻¹	*u* m s⁻¹	*D*	*η* μPa s	*λ* mW m⁻¹ K⁻¹
Saturation										
63.15	0.01252	867.2	−150.7	2.426	1.176	2.000	995.3	1.47003	311.6	173.2
63.15	0.01252	0.6743	64.78	5.838	0.7499	1.058	161.1	1.00032	4.376	5.621
70	0.03854	838.5	−137.0	2.632	1.130	2.014	925.7	1.45241	220.2	159.5
70	0.03854	1.896	71.10	5.605	0.7580	1.082	168.4	1.00089	4.883	6.355
80	0.1369	793.9	−116.6	2.903	1.069	2.056	824.4	1.42541	145.1	139.5
80	0.1369	6.089	79.10	5.349	0.7773	1.145	176.7	1.00286	5.652	7.506
90	0.3605	745.0	−95.52	3.147	1.020	2.141	719.0	1.39622	102.8	119.8
90	0.3605	15.08	84.97	5.153	0.8078	1.266	181.8	1.00710	6.482	8.868
100	0.7783	689.4	−73.21	3.376	0.9832	2.318	605.2	1.36351	75.76	100.1
100	0.7783	31.96	87.77	4.986	0.8548	1.503	183.3	1.01510	7.429	10.73
110	1.466	621.5	−48.49	3.601	0.9667	2.743	476.4	1.32430	55.99	80.44
110	1.466	62.58	85.84	4.823	0.9284	2.062	180.8	1.02974	8.626	13.83
120	2.511	523.4	−17.87	3.851	1.011	4.508	317.3	1.26895	38.43	61.01
120	2.511	125.1	74.17	4.618	1.099	4.631	172.6	1.06012	10.62	21.72
126.19	3.396	313.3	29.23	4.215				1.15559	18.30	
***P* = 0.1 MPa (1 bar)**										
63.17	0.1	867.3	−150.6	2.426	1.176	2.000	995.6	1.47007	311.6	173.3
77.24	0.1	806.6	−122.2	2.831	1.085	2.041	852.5	1.43304	161.4	145.0
77.24	0.1	4.556	77.07	5.412	0.7710	1.123	174.7	1.00214	5.435	7.174
80	0.1	4.379	80.15	5.451	0.7666	1.112	178.3	1.00206	5.623	7.443
100	0.1	3.437	101.9	5.694	0.7514	1.071	201.6	1.00162	6.958	9.381
120	0.1	2.840	123.2	5.888	0.7466	1.057	222.0	1.00133	8.244	11.27
140	0.1	2.424	144.2	6.050	0.7447	1.050	240.4	1.00114	9.480	13.11
160	0.1	2.116	165.2	6.190	0.7438	1.047	257.3	1.00099	10.67	14.89
180	0.1	1.878	186.1	6.313	0.7433	1.045	273.2	1.00088	11.81	16.61
200	0.1	1.688	207.0	6.423	0.7430	1.043	288.1	1.00079	12.91	18.28
220	0.1	1.534	227.9	6.523	0.7429	1.043	302.3	1.00072	13.97	19.90
240	0.1	1.405	248.7	6.613	0.7428	1.042	315.8	1.00066	15.00	21.48
260	0.1	1.297	269.5	6.697	0.7428	1.042	328.7	1.00061	15.99	23.01
280	0.1	1.204	290.4	6.774	0.7429	1.041	341.2	1.00057	16.96	24.51
300	0.1	1.123	311.2	6.846	0.7432	1.041	353.2	1.00053	17.89	25.97
320	0.1	1.053	332.0	6.913	0.7436	1.042	364.7	1.00049	18.80	27.39
340	0.1	0.9909	352.9	6.976	0.7441	1.042	375.9	1.00047	19.68	28.79
360	0.1	0.9357	373.7	7.036	0.7450	1.043	386.8	1.00044	20.55	30.15
400	0.1	0.8421	415.5	7.146	0.7475	1.045	407.5	1.00040	22.21	32.81
500	0.1	0.6736	520.5	7.380	0.7592	1.056	454.6	1.00032	26.06	39.04
600	0.1	0.5613	627.0	7.574	0.7781	1.075	496.3	1.00026	29.58	44.84
700	0.1	0.4811	735.6	7.741	0.8011	1.098	533.9	1.00023	32.83	50.31
800	0.1	0.4210	846.6	7.890	0.8254	1.122	568.4	1.00020	35.89	55.51
900	0.1	0.3742	960.0	8.023	0.8488	1.146	600.7	1.0018	38.78	60.52
1000	0.1	0.3368	1076.	8.145	0.8705	1.167	631.1	1.00016	41.54	65.36
***P* = 1 MPa (10 bar)**										
63.37	1.0	868.0	−149.5	2.427	1.177	1.995	998.9	1.47049	311.8	173.6
80	1.0	796.3	−116.0	2.896	1.071	2.044	832.3	1.42683	147.3	140.5
100	1.0	690.8	−73.18	3.373	0.9833	2.305	609.4	1.36432	76.26	100.6
103.75	1.0	665.8	−64.33	3.460	0.9739	2.431	559.2	1.34984	67.78	92.74
103.75	1.0	41.33	87.73	4.926	0.8786	1.652	182.8	1.01956	7.835	11.67
120	1.0	32.10	110.9	5.134	0.7980	1.297	208.9	1.01517	8.714	12.49
140	1.0	26.01	135.5	5.324	0.7701	1.177	232.8	1.01227	9.844	14.00
160	1.0	22.11	158.5	5.477	0.7588	1.127	252.9	1.01043	10.96	15.60
180	1.0	19.33	180.7	5.608	0.7531	1.101	270.6	1.00911	12.06	17.21
200	1.0	17.21	202.6	5.723	0.7500	1.085	286.9	1.00811	13.13	18.81
220	1.0	15.54	224.2	5.826	0.7480	1.075	301.9	1.00732	14.16	20.37
240	1.0	14.17	245.6	5.919	0.7468	1.067	316.1	1.00667	15.17	21.90
260	1.0	13.04	266.9	6.005	0.7460	1.062	329.5	1.00614	16.14	23.39

T K	P MPa	ρ kg m^{-3}	H kJ kg^{-1}	S kJ kg^{-1} K^{-1}	C_v kJ kg^{-1} K^{-1}	C_p kJ kg^{-1} K^{-1}	u m s^{-1}	D	η μPa s	λ mW m^{-1} K^{-1}
280	1.0	12.07	288.1	6.083	0.7456	1.059	342.4	1.00569	17.09	24.86
300	1.0	11.25	309.2	6.156	0.7454	1.056	354.6	1.00530	18.01	26.29
320	1.0	10.53	330.3	6.224	0.7455	1.054	366.4	1.00496	18.91	27.69
340	1.0	9.901	351.4	6.288	0.7458	1.053	377.8	1.00466	19.79	29.07
360	1.0	9.343	372.4	6.348	0.7464	1.052	388.8	1.00440	20.64	30.42
400	1.0	8.399	414.5	6.459	0.7486	1.052	409.7	1.00395	22.29	33.04
500	1.0	6.711	520.1	6.695	0.7600	1.061	457.0	1.00316	26.13	39.23
600	1.0	5.592	627.0	6.889	0.7786	1.078	498.7	1.00263	29.63	44.99
700	1.0	4.793	735.8	7.057	0.8015	1.100	536.2	1.00226	32.87	50.43
800	1.0	4.195	847.0	7.206	0.8257	1.124	570.7	1.00197	35.92	55.63
900	1.0	3.730	960.6	7.339	0.8492	1.147	602.9	1.00176	38.81	60.62
1000	1.0	3.358	1076.	7.461	0.8708	1.168	633.2	1.00158	41.57	65.45
P = 10 MPa (100 bar)										
65.32	10.0	875.0	−138.3	2.441	1.186	1.955	1031.	1.47447	314.0	177.1
80	10.0	818.4	−109.6	2.837	1.093	1.960	904.1	1.43984	169.0	150.4
100	10.0	733.6	−69.89	3.280	0.9997	2.022	734.2	1.38918	93.65	115.9
120	10.0	632.9	−27.82	3.663	0.9401	2.218	559.6	1.33060	58.57	84.11
140	10.0	499.8	20.66	4.036	0.9127	2.676	392.0	1.25566	36.11	57.60
160	10.0	344.2	76.44	4.408	0.8849	2.668	300.8	1.17162	22.40	40.20
180	10.0	248.7	122.6	4.681	0.8407	1.990	294.7	1.12203	18.50	32.06
200	10.0	199.4	158.4	4.870	0.8107	1.628	307.7	1.09701	17.70	29.38
220	10.0	169.4	188.9	5.015	0.7927	1.445	323.2	1.08198	17.74	28.72
240	10.0	148.8	216.7	5.136	0.7813	1.340	338.4	1.07172	18.13	28.87
260	10.0	133.4	242.8	5.240	0.7737	1.273	352.9	1.06415	18.67	29.39
280	10.0	121.4	267.7	5.333	0.7685	1.227	366.6	1.05827	19.29	30.21
300	10.0	111.7	291.9	5.417	0.7649	1.195	379.5	1.05351	19.96	31.14
320	10.0	103.6	315.6	5.493	0.7623	1.171	391.8	1.04958	20.66	32.13
340	10.0	96.80	338.8	5.563	0.7606	1.153	403.6	1.04625	21.37	33.16
360	10.0	90.89	361.7	5.629	0.7596	1.138	414.8	1.04338	22.08	34.21
400	10.0	81.19	406.8	5.748	0.7593	1.119	436.0	1.03869	23.51	36.35
500	10.0	64.50	517.5	5.995	0.7672	1.100	483.1	1.03065	26.99	41.73
600	10.0	53.74	627.6	6.195	0.7841	1.104	523.9	1.02549	30.28	47.00
700	10.0	46.15	738.7	6.367	0.8059	1.118	560.4	1.02187	33.40	52.10
800	10.0	40.49	851.4	6.517	0.8294	1.137	593.8	1.01917	36.35	57.04
900	10.0	36.08	966.1	6.652	0.8523	1.157	625.0	1.01708	39.17	61.85
1000	10.0	32.56	1083.	6.775	0.8735	1.176	654.4	1.01540	41.88	66.54

Oxygen (O_2)

T K	P MPa	ρ kg m^{-3}	H kJ kg^{-1}	S kJ kg^{-1} K^{-1}	C_v kJ kg^{-1} K^{-1}	C_p kJ kg^{-1} K^{-1}	u m s^{-1}	D	η μPa s	λ mW m^{-1} K^{-1}
Saturation										
54.36	0.0001463	1306.	−193.6	2.092	1.195	1.673	1123	1.56799	773.6	201.9
54.36	0.0001463	0.01036	49.11	6.557	0.6638	0.9260	140.3	1.00000	4.096	4.420
60	0.0007258	1282.	−184.2	2.257	1.089	1.673	1127.	1.55615	578.1	193.9
60	0.0007258	0.04659	54.19	6.230	0.6817	0.9475	147.0	1.00002	4.553	4.984
70	0.006262	1237.	−167.4	2.516	1.017	1.678	1066.	1.53399	371.8	179.7
70	0.006262	0.3457	63.09	5.809	0.7052	0.9780	158.1	1.00013	5.356	5.992
80	0.03012	1190.	−150.6	2.740	0.9697	1.682	987.4	1.51120	261.2	165.4
80	0.03012	1.468	71.69	5.519	0.6950	0.9743	168.4	1.00054	6.149	7.028
90	0.09935	1142.	−133.7	2.938	0.9296	1.699	905.9	1.48766	195.6	151.0
90	0.09935	4.387	79.55	5.308	0.6758	0.9705	177.3	1.00163	6.936	8.124
100	0.2540	1091.	−116.4	3.118	0.8949	1.738	822.2	1.46295	152.6	136.6
100	0.2540	10.42	86.16	5.144	0.6752	1.006	184.1	1.00387	7.728	9.336
110	0.5434	1035.	−98.64	3.286	0.8658	1.807	734.8	1.43650	121.5	121.9
110	0.5434	21.28	91.05	5.010	0.6988	1.101	188.1	1.00791	8.547	10.75
120	1.022	973.9	−79.90	3.444	0.8430	1.927	641.5	1.40746	97.43	107.2
120	1.022	39.31	93.75	4.892	0.7415	1.276	189.4	1.01465	9.427	12.51

T K	P MPa	ρ kg m^{-3}	H kJ kg^{-1}	S kJ kg^{-1} K^{-1}	C_v kJ kg^{-1} K^{-1}	C_p kJ kg^{-1} K^{-1}	u m s^{-1}	D	η μPa s	λ mW m^{-1} K^{-1}
130	1.749	902.5	−59.66	3.600	0.8293	2.153	539.5	1.37432	77.57	92.63
130	1.749	68.37	93.47	4.778	0.8002	1.600	187.8	1.02558	10.45	14.94
140	2.788	813.2	−36.70	3.761	0.8323	2.691	423.1	1.33363	60.22	78.22
140	2.788	116.8	88.47	4.655	0.8834	2.370	182.8	1.04395	11.82	18.98
150	4.219	675.5	−6.671	3.955	0.9057	5.464	273.8	1.27248	42.90	64.19
150	4.219	214.9	72.56	4.483	1.049	6.625	172.8	1.08192	14.72	29.67
154.58	5.043	436.1	32.42	4.201				1.17084	24.84	
P = 0.1 MPa (1 bar)										
54.37	0.1	1306.	−193.5	2.092	1.195	1.673	1124.	1.56802	773.8	201.9
60	0.1	1282.	−184.1	2.257	1.089	1.673	1128.	1.55621	578.5	194.0
80	0.1	1191.	−150.6	2.739	0.9699	1.681	987.7	1.51126	261.4	165.5
90.06	0.1	1142.	−133.6	2.939	0.9293	1.699	905.4	1.48751	195.3	151.0
90.06	0.1	4.413	79.60	5.307	0.6757	0.9705	177.4	1.00164	6.940	8.131
100	0.1	3.941	88.99	5.405	0.6527	0.9352	188.4	1.00146	7.712	9.085
120	0.1	3.252	107.6	5.575	0.6543	0.9280	207.3	1.00121	9.219	10.99
140	0.1	2.774	126.1	5.718	0.6530	0.9218	224.6	1.00103	10.67	12.86
160	0.1	2.420	144.5	5.841	0.6519	0.9180	240.5	1.00090	12.07	14.69
180	0.1	2.147	162.8	5.949	0.6513	0.9158	255.4	1.00080	13.41	16.48
200	0.1	1.930	181.1	6.045	0.6512	0.9146	269.3	1.00072	14.72	18.24
220	0.1	1.753	199.4	6.132	0.6516	0.9142	282.6	1.00065	15.98	19.95
240	0.1	1.606	217.7	6.212	0.6525	0.9146	295.2	1.00060	17.20	21.64
260	0.1	1.482	236.0	6.285	0.6539	0.9156	307.2	1.00055	18.38	23.28
280	0.1	1.376	254.3	6.353	0.6560	0.9174	318.7	1.00051	19.53	24.90
300	0.1	1.284	272.7	6.416	0.6587	0.9199	329.7	1.00048	20.65	26.49
320	0.1	1.203	291.1	6.476	0.6621	0.9231	340.3	1.00045	21.74	28.04
340	0.1	1.132	309.6	6.532	0.6661	0.9269	350.5	1.00042	22.80	29.58
360	0.1	1.069	328.2	6.585	0.6707	0.9313	360.4	1.00040	23.84	31.08
400	0.1	0.9622	365.7	6.684	0.6812	0.9416	379.0	1.00036	25.84	34.03
500	0.1	0.7696	461.3	6.897	0.7119	0.9722	421.3	1.00029	30.49	41.05
600	0.1	0.6413	560.1	7.077	0.7432	1.003	458.9	1.00024	34.73	47.66
700	0.1	0.5497	661.9	7.234	0.7710	1.031	493.3	1.00020	38.65	53.97
800	0.1	0.4809	766.2	7.373	0.7945	1.055	525.4	1.00018	42.33	60.02
900	0.1	0.4275	872.6	7.498	0.8140	1.074	555.6	1.00016	45.81	65.87
1000	0.1	0.3848	980.9	7.612	0.8301	1.090	584.3	1.00014	49.12	71.55
P = 1 MPa (10 bar)										
54.47	1.0	1307.	−192.8	2.093	1.190	1.669	1127.	1.56831	775.0	202.1
60	1.0	1283.	−183.6	2.254	1.089	1.671	1130.	1.55674	582.7	194.3
80	1.0	1192.	−150.1	2.736	0.9716	1.679	991.1	1.51204	263.6	166.0
100	1.0	1093.	−116.1	3.115	0.8964	1.731	826.8	1.46397	153.9	137.2
119.62	1.0	976.3	−80.64	3.438	0.8438	1.921	645.2	1.40862	98.25	107.8
119.62	1.0	38.46	93.70	4.896	0.7396	1.268	189.4	1.01433	9.392	12.43
120	1.0	38.25	94.18	4.900	0.7355	1.257	190.0	1.01425	9.421	12.45
140	1.0	30.39	116.8	5.075	0.6851	1.065	214.2	1.01131	10.89	13.90
160	1.0	25.65	137.5	5.213	0.6708	1.005	233.7	1.00954	12.29	15.54
180	1.0	22.33	157.2	5.329	0.6631	0.9743	250.8	1.00831	13.64	17.21
200	1.0	19.84	176.5	5.431	0.6590	0.9568	266.3	1.00738	14.94	18.88
220	1.0	17.88	195.5	5.521	0.6571	0.9462	280.6	1.00665	16.20	20.53
240	1.0	16.30	214.4	5.603	0.6566	0.9397	294.0	1.00606	17.41	22.15
260	1.0	14.98	233.1	5.678	0.6572	0.9359	306.6	1.00557	18.59	23.76
280	1.0	13.86	251.8	5.748	0.6586	0.9342	318.5	1.00515	19.73	25.34
300	1.0	12.91	270.5	5.812	0.6609	0.9340	329.9	1.00480	20.85	26.89
320	1.0	12.08	289.2	5.872	0.6640	0.9351	340.8	1.00449	21.93	28.43
340	1.0	11.35	307.9	5.929	0.6677	0.9373	351.2	1.00422	22.99	29.93
360	1.0	10.71	326.7	5.983	0.6720	0.9404	361.2	1.00398	24.02	31.42
400	1.0	9.623	364.5	6.082	0.6823	0.9487	380.1	1.00358	26.01	34.33
500	1.0	7.683	460.7	6.297	0.7126	0.9763	422.7	1.00286	30.63	41.29
600	1.0	6.398	559.8	6.478	0.7436	1.006	460.4	1.00238	34.85	47.86

T K	P MPa	ρ kg m^{-3}	H kJ kg^{-1}	S kJ kg^{-1} K^{-1}	C_v kJ kg^{-1} K^{-1}	C_p kJ kg^{-1} K^{-1}	u m s^{-1}	D	η μPa s	λ mW m^{-1} K^{-1}
700	1.0	5.483	661.8	6.635	0.7713	1.033	494.9	1.00204	38.77	54.14
800	1.0	4.798	766.2	6.774	0.7948	1.056	526.9	1.00179	42.43	60.17
900	1.0	4.265	872.8	6.900	0.8142	1.075	557.1	1.00159	45.90	66.00
1000	1.0	3.839	981.1	7.014	0.8303	1.091	585.8	1.00143	49.20	71.67
P = 10 MPa (100 bar)										
55.50	10.0	1312.	−185.4	2.103	1.149	1.640	1158.	1.57109	786.0	203.9
60	10.0	1294.	−178.0	2.231	1.092	1.653	1155.	1.56193	625.3	197.9
80	10.0	1207.	−144.8	2.708	0.9885	1.654	1023.	1.51940	285.2	171.5
100	10.0	1116.	−111.6	3.078	0.9136	1.672	877.1	1.47515	169.5	144.8
120	10.0	1015.	−77.51	3.389	0.8571	1.755	727.0	1.42679	112.4	118.1
140	10.0	892.5	−40.64	3.673	0.8191	1.966	564.5	1.36967	75.70	91.70
160	10.0	716.0	4.057	3.970	0.8137	2.659	379.7	1.29022	47.96	66.52
180	10.0	423.3	70.35	4.360	0.8161	3.312	250.9	1.16550	26.07	42.24
200	10.0	277.6	119.7	4.621	0.7506	1.894	259.0	1.10662	20.74	31.77
220	10.0	219.9	152.6	4.778	0.7152	1.462	278.2	1.08384	20.24	29.46
240	10.0	186.7	179.8	4.896	0.6973	1.280	295.8	1.07090	20.67	29.19
260	10.0	164.2	204.3	4.995	0.6878	1.183	311.6	1.06220	21.40	29.67
280	10.0	147.6	227.3	5.080	0.6830	1.124	326.1	1.05578	22.25	30.48
300	10.0	134.6	249.4	5.156	0.6809	1.086	339.4	1.05078	23.15	31.47
320	10.0	124.0	270.8	5.225	0.6809	1.060	351.7	1.04675	24.07	32.56
340	10.0	115.2	291.9	5.289	0.6823	1.042	363.3	1.04339	25.00	33.74
360	10.0	107.8	312.6	5.348	0.6848	1.030	374.2	1.04054	25.92	34.96
400	10.0	95.67	353.5	5.456	0.6923	1.017	394.5	1.03595	27.74	37.45
500	10.0	75.32	454.8	5.682	0.7187	1.015	438.7	1.02825	32.07	43.71
600	10.0	62.44	557.1	5.869	0.7477	1.031	476.8	1.02339	36.10	49.86
700	10.0	53.46	661.2	6.029	0.7743	1.050	511.2	1.02002	39.87	55.84
800	10.0	46.79	767.1	6.170	0.7970	1.069	543.1	1.01752	43.43	61.66
900	10.0	41.63	874.8	6.297	0.8159	1.085	572.9	1.01558	46.80	67.32
1000	10.0	37.51	984.0	6.412	0.8316	1.098	601.2	1.01404	50.03	72.85

Parahydrogen (H_2)

T K	P MPa	ρ kg m^{-3}	H kJ kg^{-1}	S kJ kg^{-1} K^{-1}	C_v kJ kg^{-1} K^{-1}	C_p kJ kg^{-1} K^{-1}	u m s^{-1}	D	η μPa s	λ mW m^{-1} K^{-1}
Saturation										
13.80	0.007041	76.98	−53.74	−3.084	5.131	6.924	1263.	1.25267	25.90	75.27
13.80	0.007041	0.1255	396.3	29.52	6.226	10.53	305.6	1.00038	0.6507	10.46
14	0.007884	76.82	−52.36	−2.986	5.158	6.981	1257.	1.25208	25.21	76.86
14	0.007884	0.1388	398.1	29.19	6.236	10.56	307.6	1.00042	0.6669	10.65
16	0.02155	75.13	−37.60	−2.013	5.334	7.659	1210.	1.24585	19.81	90.02
16	0.02155	0.3377	415.8	26.33	6.323	10.87	325.4	1.00102	0.8106	12.60
18	0.04815	73.25	−21.18	−1.068	5.472	8.513	1168.	1.23903	16.22	98.39
18	0.04815	0.6880	431.5	24.08	6.387	11.29	340.7	1.00207	0.9355	14.60
20	0.09341	71.14	−2.692	−.1281	5.637	9.569	1119.	1.23147	13.63	103.0
20	0.09341	1.244	444.5	22.23	6.450	11.92	353.5	1.00375	1.057	16.71
22	0.1635	68.74	18.32	0.8244	5.811	10.86	1059.	1.22301	11.65	104.9
22	0.1635	2.071	454.2	20.64	6.541	12.88	363.7	1.00625	1.182	18.98
24	0.2648	66.01	42.37	1.805	5.974	12.52	989.9	1.21345	10.06	104.7
24	0.2648	3.255	459.7	19.19	6.681	14.40	371.3	1.00984	1.315	21.54
26	0.4038	62.83	70.23	2.832	6.120	14.83	909.3	1.20242	8.725	102.6
26	0.4038	4.924	460.0	17.82	6.890	16.90	376.1	1.01491	1.463	24.53
28	0.5875	58.98	103.2	3.941	6.266	18.57	813.3	1.18922	7.533	98.79
28	0.5875	7.300	453.4	16.45	7.188	21.53	378.2	1.02216	1.634	28.26
30	0.8232	53.98	144.2	5.211	6.472	26.65	693.0	1.17223	6.390	92.78
30	0.8232	10.87	435.7	14.93	7.625	32.58	377.2	1.03313	1.865	33.47
32	1.120	45.90	204.1	6.945	7.010	68.19	522.6	1.14520	5.076	82.53
32	1.120	17.49	392.3	12.83	8.336	92.39	372.0	1.05368	2.337	43.17
32.94	1.286	31.32	295.6	9.625				1.09756	3.512	

T K	P MPa	ρ kg m⁻³	H kJ kg⁻¹	S kJ kg⁻¹ K⁻¹	C_v kJ kg⁻¹ K⁻¹	C_p kJ kg⁻¹ K⁻¹	u m s⁻¹	D	η μPa s	λ mW m⁻¹ K⁻¹
P = 0.1 MPa (1 bar)										
20	0.1	71.14	−2.629	−.1296	5.637	9.566	1119.	1.23150	13.63	103.1
20.23	0.1	70.88	−.4432	−.02096	5.657	9.702	1112.	1.23055	13.38	103.4
20.23	0.1	1.323	445.8	22.04	6.458	12.01	354.8	1.00399	1.071	16.95
40	0.1	0.6157	661.5	29.52	6.218	10.57	521.4	1.00186	2.039	31.85
60	0.1	0.4059	873.1	33.81	6.503	10.72	635.7	1.00122	2.865	45.53
80	0.1	0.3035	1096.	37.02	7.598	11.77	713.9	1.00091	3.566	61.36
100	0.1	0.2425	1348.	39.82	9.277	13.43	772.5	1.00073	4.190	80.43
120	0.1	0.2020	1633.	42.41	10.83	14.98	827.4	1.00061	4.763	101.4
140	0.1	0.1731	1943.	44.80	11.82	15.96	883.4	1.00052	5.301	119.7
160	0.1	0.1514	2267.	46.96	12.21	16.35	940.4	1.00046	5.814	134.8
180	0.1	0.1346	2594.	48.89	12.19	16.32	997.7	1.00041	6.306	146.5
200	0.1	0.1211	2918.	50.60	11.95	16.08	1054.	1.00037	6.780	155.6
220	0.1	0.1101	3237.	52.11	11.63	15.76	1110.	1.00033	7.239	163.4
240	0.1	0.1010	3549.	53.47	11.33	15.46	1163.	1.00030	7.684	170.6
260	0.1	0.09319	3856.	54.70	11.07	15.20	1214.	1.00028	8.118	177.7
280	0.1	0.08654	4157.	55.82	10.87	15.00	1263.	1.00026	8.540	184.9
300	0.1	0.08077	4456.	56.85	10.72	14.85	1310.	1.00024	8.953	192.3
320	0.1	0.07572	4752.	57.80	10.61	14.74	1355.	1.00023	9.357	200.0
340	0.1	0.07127	5045.	58.69	10.53	14.66	1398.	1.00022	9.752	208.0
360	0.1	0.06731	5338.	59.53	10.48	14.61	1439.	1.00020	10.14	216.4
400	0.1	0.06058	5921.	61.06	10.43	14.55	1518.	1.00018	10.89	233.1
500	0.1	0.04847	7374.	64.31	10.40	14.53	1698.	1.00015	12.67	275.0
600	0.1	0.04040	8828.	66.96	10.42	14.55	1859.	1.00012	14.34	317.6
700	0.1	0.03463	10290	69.20	10.48	14.60	2006.	1.00011	15.92	360.8
800	0.1	0.03030	11750	71.16	10.57	14.70	2142.	1.00009	17.43	404.7
900	0.1	0.02693	13230	72.90	10.71	14.83	2268.	1.00008	18.89	449.2
1000	0.1	0.02424	14720	74.47	10.87	14.99	2386.	1.00007	20.30	494.3
P = 1 MPa (10 bar)										
20	1.0	72.29	6.046	−.3233	5.634	9.187	1164.	1.23552	14.53	105.1
31.24	1.0	49.65	177.4	6.181	6.718	40.77	596.5	1.15769	5.627	87.27
31.24	1.0	14.31	414.3	13.76	8.017	52.90	374.7	1.04376	2.096	38.44
40	1.0	7.270	590.6	18.85	6.455	14.12	497.2	1.02207	2.228	37.28
60	1.0	4.228	838.4	23.90	6.588	11.70	634.2	1.01279	2.976	48.40
80	1.0	3.072	1075.	27.31	7.645	12.25	717.9	1.00929	3.648	63.68
100	1.0	2.429	1335.	30.19	9.307	13.72	778.4	1.00734	4.255	82.51
120	1.0	2.015	1624.	32.83	10.85	15.17	834.2	1.00609	4.817	103.1
140	1.0	1.723	1938.	35.24	11.84	16.09	890.7	1.00520	5.348	121.2
160	1.0	1.506	2264.	37.42	12.23	16.45	948.1	1.00455	5.854	136.1
180	1.0	1.338	2593.	39.36	12.20	16.40	1006.	1.00404	6.341	147.6
200	1.0	1.204	2918.	41.07	11.96	16.14	1062.	1.00364	6.812	156.7
220	1.0	1.095	3238.	42.60	11.64	15.81	1117.	1.00331	7.268	164.3
240	1.0	1.004	3551.	43.96	11.34	15.50	1171.	1.00303	7.711	171.5
260	1.0	0.9267	3858.	45.19	11.08	15.24	1222.	1.00280	8.143	178.6
280	1.0	0.8607	4161.	46.31	10.88	15.03	1271.	1.00260	8.563	185.7
300	1.0	0.8035	4460.	47.34	10.73	14.87	1317.	1.00243	8.975	193.0
320	1.0	0.7534	4756.	48.30	10.62	14.76	1362.	1.00228	9.377	200.7
340	1.0	0.7093	5050.	49.19	10.54	14.68	1405.	1.00215	9.771	208.7
360	1.0	0.6700	5343.	50.03	10.49	14.62	1446.	1.00203	10.16	217.0
400	1.0	0.6033	5927.	51.56	10.43	14.56	1525.	1.00183	10.91	233.7
500	1.0	0.4830	7381.	54.81	10.41	14.53	1704.	1.00146	12.68	275.5
600	1.0	0.4027	8835.	57.46	10.43	14.55	1865.	1.00122	14.35	318.0
700	1.0	0.3454	10290	59.71	10.48	14.61	2011.	1.00105	15.93	361.2
800	1.0	0.3023	11760	61.66	10.58	14.70	2147.	1.00092	17.44	405.0
900	1.0	0.2688	13230	63.40	10.71	14.83	2272.	1.00082	18.90	449.5
1000	1.0	0.2420	14720	64.97	10.87	15.00	2390.	1.00074	20.30	494.6

T K	P MPa	ρ kg m⁻³	H kJ kg⁻¹	S kJ kg⁻¹ K⁻¹	C_v kJ kg⁻¹ K⁻¹	C_p kJ kg⁻¹ K⁻¹	u m s⁻¹	D	η μPa s	λ mW m⁻¹ K⁻¹
P = 10 MPa (100 bar)										
20	10.0	79.91	98.92	−1.567	5.529	7.566	1467.	1.26244	23.45	120.3
40	10.0	63.19	310.5	5.532	6.665	13.45	1154.	1.20345	8.791	121.8
60	10.0	42.86	617.0	11.71	6.960	16.19	923.2	1.13508	5.594	100.6
80	10.0	29.87	933.6	16.27	7.952	15.47	884.9	1.09289	5.034	96.74
100	10.0	23.01	1244.	19.73	9.546	15.76	904.3	1.07104	5.169	106.3
120	10.0	18.87	1567.	22.67	11.05	16.56	942.4	1.05803	5.497	121.6
140	10.0	16.08	1905.	25.27	12.00	17.11	990.1	1.04932	5.888	136.6
160	10.0	14.06	2249.	27.57	12.37	17.22	1042.	1.04303	6.303	149.4
180	10.0	12.51	2592.	29.59	12.32	17.01	1096.	1.03824	6.727	159.4
200	10.0	11.28	2928.	31.36	12.07	16.63	1150.	1.03445	7.151	167.3
220	10.0	10.28	3257.	32.93	11.74	16.22	1203.	1.03138	7.571	173.9
240	10.0	9.453	3577.	34.32	11.43	15.84	1254.	1.02883	7.985	180.4
260	10.0	8.751	3891.	35.58	11.17	15.52	1303.	1.02667	8.393	186.8
280	10.0	8.149	4199	36.72	10.96	15.27	1350.	1.02483	8.795	193.3
300	10.0	7.626	4502.	37.77	10.80	15.08	1394.	1.02323	9.189	200.2
320	10.0	7.167	4802.	38.73	10.68	14.94	1437.	1.02183	9.577	207.4
340	10.0	6.762	5100.	39.64	10.60	14.84	1478.	1.02059	9.959	215.0
360	10.0	6.400	5396.	40.48	10.55	14.77	1518.	1.01948	10.33	223.0
400	10.0	5.784	5985.	42.03	10.49	14.68	1593.	1.01761	11.07	239.2
500	10.0	4.665	7447.	45.30	10.45	14.60	1764.	1.01420	12.81	280.0
600	10.0	3.911	8907.	47.96	10.46	14.59	1919.	1.01191	14.46	321.9
700	10.0	3.368	10370	50.21	10.51	14.63	2062.	1.01026	16.02	364.7
800	10.0	2.957	11830	52.17	10.60	14.72	2193.	1.00902	17.52	408.2
900	10.0	2.636	13310	53.91	10.73	14.84	2315.	1.00805	18.97	452.4
1000	10.0	2.378	14800	55.48	10.89	15.00	2430.	1.00727	20.37	497.3

Helium (He–4)

T K	P MPa	ρ kg m⁻³	H kJ kg⁻¹	S kJ kg⁻¹ K⁻¹	C_v kJ kg⁻¹ K⁻¹	C_p kJ kg⁻¹ K⁻¹	u m s⁻¹	D	η μPa s	λ mW m⁻¹ K⁻¹
Saturation										
2.177	0.004856	146.2	−7.501	−2.177	6.287	6.318	216.8	1.07842	3.596	13.52
2.177	0.004856	1.146	15.73	8.494	3.470	6.061	83.22	1.00045	0.5376	3.977
2.5	0.01000	145.0	−6.180	−1.620	2.521	2.700	216.0	1.07491	3.746	14.89
2.5	0.01000	2.116	16.97	7.641	3.486	6.258	87.62	1.00082	0.6382	4.765
3	0.02373	141.3	−4.869	−1.177	2.061	2.598	214.3	1.06975	3.694	16.55
3	0.02373	4.428	18.66	6.666	3.452	6.592	93.04	1.00173	0.7961	5.893
3.5	0.04663	136.1	−3.258	−.7324	2.332	3.414	202.4	1.06471	3.511	17.75
3.5	0.04663	8.022	19.95	5.898	3.378	7.122	96.94	1.00315	0.9640	7.060
4	0.08100	128.9	−1.164	−.2439	2.504	4.523	185.7	1.05941	3.278	18.47
4	0.08100	13.41	20.68	5.217	3.284	8.179	99.47	1.00529	1.151	8.356
4.5	0.1292	118.8	1.586	0.3106	2.601	6.776	164.0	1.05319	3.010	18.77
4.5	0.1292	21.76	20.52	4.519	3.179	10.96	100.8	1.00865	1.372	10.04
5	0.1945	100.8	5.854	1.081	2.709	20.24	133.0	1.04367	2.634	19.28
5	0.1945	38.21	18.18	3.547	3.053	29.09	101.6	1.01543	1.687	13.60
5.195	0.2275	69.64	11.87	2.183				1.02904	2.129	
P = 0.1 MPa (1 bar)										
20	0.1	2.408	108.5	13.94	3.121	5.250	264.3	1.00093	3.582	26.20
40	0.1	1.200	212.9	17.56	3.119	5.206	373.4	1.00047	5.542	40.44
60	0.1	0.8006	317.0	19.67	3.118	5.198	456.9	1.00031	7.116	52.55
80	0.1	0.6007	420.9	21.16	3.117	5.195	527.2	1.00023	8.503	63.52
100	0.1	0.4807	524.8	22.32	3.117	5.194	589.2	1.00019	9.778	73.71
120	0.1	0.4007	628.7	23.27	3.117	5.194	645.3	1.00016	10.79	83.33
140	0.1	0.3435	732.6	24.07	3.116	5.193	696.9	1.00013	11.94	92.50
160	0.1	0.3006	836.4	24.76	3.116	5.193	744.9	1.00012	13.05	101.3
180	0.1	0.2672	940.3	25.37	3.116	5.193	790.0	1.00010	14.11	109.8
200	0.1	0.2405	1044.	25.92	3.116	5.193	832.7	1.00009	15.14	118.0

T K	P MPa	ρ kg m^{-3}	H kJ kg^{-1}	S kJ kg^{-1} K^{-1}	C_v kJ kg^{-1} K^{-1}	C_p kJ kg^{-1} K^{-1}	u m s^{-1}	D	η μPa s	λ mW m^{-1} K^{-1}
220	0.1	0.2187	1148.	26.42	3.116	5.193	873.3	1.00008	16.14	126.0
240	0.1	0.2005	1252.	26.87	3.116	5.193	912.0	1.00008	17.12	133.7
260	0.1	0.1851	1356.	27.28	3.116	5.193	949.2	1.00007	18.08	141.3
280	0.1	0.1718	1460.	27.67	3.116	5.193	985.0	1.00007	19.01	148.7
300	0.1	0.1604	1563.	28.03	3.116	5.193	1020.	1.00006	19.93	156.0
320	0.1	0.1504	1667.	28.36	3.116	5.193	1053.	1.00006	20.83	163.1
340	0.1	0.1415	1771.	28.68	3.116	5.193	1085.	1.00005	21.72	170.1
360	0.1	0.1337	1875.	28.97	3.116	5.193	1117.	1.00005	22.59	177.0
400	0.1	0.1203	2083.	29.52	3.116	5.193	1177.	1.00005	24.29	190.4
500	0.1	0.09626	2602.	30.68	3.116	5.193	1316.	1.00004	28.36	222.3
600	0.1	0.08022	3121.	31.63	3.116	5.193	1442.	1.00003	32.22	252.4
700	0.1	0.06876	3641.	32.43	3.116	5.193	1557.	1.00003	35.89	281.1
800	0.1	0.06017	4160.	33.12	3.116	5.193	1664.	1.00002	39.43	308.5
900	0.1	0.05348	4679.	33.73	3.116	5.193	1765.	1.00002	42.85	335.0
1000	0.1	0.04814	5199.	34.28	3.116	5.193	1861.	1.00002	46.16	360.6
P = 1 MPa (10 bar)										
20	1.0	24.02	104.1	8.936	3.156	5.728	276.4	1.00938	3.933	29.04
40	1.0	11.72	213.2	12.73	3.143	5.317	384.9	1.00455	5.740	42.01
60	1.0	7.850	318.6	14.87	3.133	5.242	466.8	1.00305	7.269	53.72
80	1.0	5.912	423.1	16.37	3.128	5.217	535.8	1.00229	8.633	64.50
100	1.0	4.745	527.4	17.54	3.125	5.206	596.9	1.00184	9.892	74.60
120	1.0	3.963	631.4	18.48	3.123	5.200	652.2	1.00154	10.90	84.16
140	1.0	3.403	735.4	19.29	3.121	5.197	703.2	1.00132	12.04	93.29
160	1.0	2.982	839.3	19.98	3.120	5.195	750.7	1.00116	13.13	102.1
180	1.0	2.653	943.2	20.59	3.120	5.194	795.4	1.00103	14.19	110.5
200	1.0	2.390	1047.	21.14	3.119	5.194	837.7	1.00093	15.21	118.7
220	1.0	2.174	1151.	21.63	3.119	5.193	878.0	1.00084	16.20	126.7
240	1.0	1.994	1255.	22.09	3.118	5.193	916.5	1.00077	17.17	134.4
260	1.0	1.842	1359.	22.50	3.118	5.193	953.5	1.00071	18.12	142.0
280	1.0	1.711	1463.	22.89	3.118	5.192	989.1	1.00066	19.05	149.4
300	1.0	1.597	1566.	23.24	3.118	5.192	1023.	1.00062	19.96	156.6
320	1.0	1.498	1670.	23.58	3.117	5.192	1057.	1.00058	20.86	163.8
340	1.0	1.410	1774.	23.89	3.117	5.192	1089.	1.00055	21.74	170.7
360	1.0	1.332	1878.	24.19	3.117	5.192	1120.	1.00052	22.61	177.6
400	1.0	1.200	2086.	24.74	3.117	5.192	1180.	1.00047	24.32	191.0
500	1.0	0.9604	2605.	25.90	3.117	5.192	1319.	1.00037	28.38	222.9
600	1.0	0.8007	3124.	26.84	3.116	5.192	1444.	1.00031	32.23	253.0
700	1.0	0.6866	3643.	27.64	3.116	5.192	1559.	1.00027	35.91	281.6
800	1.0	0.6009	4163.	28.34	3.116	5.192	1666.	1.00023	39.44	309.1
900	1.0	0.5342	4682.	28.95	3.116	5.193	1767.	1.00021	42.86	335.5
1000	1.0	0.4809	5201.	29.50	3.116	5.193	1863.	1.00019	46.17	361.1
P = 10 MPa (100 bar)										
20	10.0	147.2	111.0	3.756	3.286	5.413	497.6	1.06056	6.928	58.72
40	10.0	91.33	225.2	7.703	3.305	5.721	512.2	1.03614	7.483	59.90
60	10.0	65.36	337.4	9.980	3.251	5.506	566.3	1.02560	8.638	68.12
80	10.0	51.09	446.0	11.54	3.215	5.376	620.9	1.01993	9.802	76.83
100	10.0	42.05	552.8	12.74	3.192	5.303	672.3	1.01637	10.93	85.42
120	10.0	35.77	658.4	13.70	3.176	5.261	720.4	1.01391	11.82	93.82
140	10.0	31.14	763.3	14.51	3.165	5.235	765.6	1.01210	12.88	102.0
160	10.0	27.58	867.8	15.21	3.157	5.219	808.5	1.01071	13.89	110.1
180	10.0	24.75	972.1	15.82	3.151	5.208	849.2	1.00961	14.87	118.0
200	10.0	22.46	1076.	16.37	3.146	5.201	888.2	1.00872	15.82	125.7
220	10.0	20.55	1180.	16.86	3.142	5.196	925.6	1.00798	16.75	133.3
240	10.0	18.94	1284.	17.31	3.139	5.193	961.6	1.00735	17.65	140.7
260	10.0	17.57	1388.	17.73	3.137	5.190	996.4	1.00682	18.54	148.0
280	10.0	16.38	1492.	18.11	3.135	5.188	1030.	1.00635	19.41	155.2
300	10.0	15.34	1595.	18.47	3.133	5.187	1063.	1.00595	20.26	162.3

T K	P MPa	ρ kg m⁻³	H kJ kg⁻¹	S kJ kg⁻¹ K⁻¹	C_v kJ kg⁻¹ K⁻¹	C_p kJ kg⁻¹ K⁻¹	u m s⁻¹	D	η µPa s	λ mW m⁻¹ K⁻¹
320	10.0	14.43	1699.	18.81	3.132	5.186	1094.	1.00560	21.14	169.2
340	10.0	13.62	1803.	19.12	3.130	5.186	1125.	1.00528	22.01	176.0
360	10.0	12.89	1907.	19.42	3.129	5.185	1155.	1.00500	22.86	182.8
400	10.0	11.65	2114.	19.96	3.127	5.185	1213.	1.00452	24.54	196.0
500	10.0	9.387	2632.	21.12	3.124	5.185	1347.	1.00364	28.56	227.6
600	10.0	7.860	3151.	22.07	3.122	5.186	1469.	1.00305	32.38	257.4
700	10.0	6.760	3670.	22.87	3.121	5.186	1582.	1.00262	36.04	285.9
800	10.0	5.930	4188.	23.56	3.120	5.187	1687.	1.00230	39.56	313.2
900	10.0	5.281	4707.	24.17	3.119	5.188	1786.	1.00205	42.96	339.6
1000	10.0	4.760	5226.	24.72	3.119	5.188	1880.	1.00185	46.26	365.1

Argon (Ar)

T K	P MPa	ρ kg m⁻³	H kJ kg⁻¹	S kJ kg⁻¹ K⁻¹	C_v kJ kg⁻¹ K⁻¹	C_p kJ kg⁻¹ K⁻¹	u m s⁻¹	D	η µPa s	λ mW m⁻¹ K⁻¹
Saturation										
83.81	0.06889	1417.	−121.4	1.329	0.5496	1.116	862.4	1.51232	290.2	133.6
83.81	0.06889	4.055	42.28	3.283	0.3247	0.5550	168.1	1.00126	6.856	5.359
90	0.1335	1379.	−114.5	1.409	0.5268	1.121	819.5	1.49650	240.0	124.5
90	0.1335	7.436	44.57	3.176	0.3309	0.5757	172.8	1.00231	7.413	5.835
100	0.3238	1314.	−103.1	1.528	0.4976	1.154	746.9	1.46993	181.3	110.2
100	0.3238	16.86	47.40	3.032	0.3445	0.6269	178.9	1.00525	8.349	6.689
110	0.6653	1243.	−91.13	1.639	0.4747	1.218	669.2	1.44136	140.4	96.41
110	0.6653	33.29	48.84	2.911	0.3633	0.7122	183.0	1.01039	9.366	7.732
120	1.213	1163.	−78.35	1.746	0.4576	1.332	584.2	1.40965	110.2	83.13
120	1.213	60.11	48.41	2.802	0.3893	0.8627	185.1	1.01884	10.54	9.154
130	2.025	1068.	−64.16	1.854	0.4492	1.564	487.9	1.37273	85.86	70.43
130	2.025	103.6	45.30	2.696	0.4275	1.172	184.8	1.03259	12.03	11.45
140	3.168	943.7	−47.16	1.971	0.4598	2.225	371.6	1.32519	63.62	58.06
140	3.168	178.9	37.47	2.576	0.4940	2.104	181.5	1.05677	14.32	16.39
150	4.735	680.4	−17.88	2.159	0.7060	23.58	174.7	1.22816	36.78	57.63
150	4.735	394.5	11.52	2.355	0.8218	35.47	157.0	1.12827	21.18	55.88
150.69	4.863	535.6	−4.332	2.248				1.17684	27.63	
***P* = 0.1 MPa (1 bar)**										
83.81	0.1	1417.	−121.4	1.329	0.5496	1.116	862.5	1.51233	290.2	133.6
87.18	0.1	1396.	−117.7	1.373	0.5366	1.117	839.2	1.50375	261.3	128.6
87.18	0.1	5.704	43.57	3.223	0.3279	0.5654	170.8	1.00178	7.157	5.614
100	0.1	4.915	50.69	3.299	0.3206	0.5470	184.2	1.00153	8.234	6.450
120	0.1	4.058	61.49	3.398	0.3161	0.5347	202.8	1.00126	9.878	7.735
140	0.1	3.461	72.12	3.480	0.3144	0.5293	219.6	1.00108	11.48	8.987
160	0.1	3.020	82.68	3.550	0.3135	0.5264	235.1	1.00094	13.03	10.21
180	0.1	2.680	93.19	3.612	0.3131	0.5247	249.6	1.00083	14.53	11.39
200	0.1	2.409	103.7	3.667	0.3128	0.5236	263.2	1.00075	16.00	12.54
220	0.1	2.189	114.1	3.717	0.3127	0.5229	276.2	1.00068	17.42	13.66
240	0.1	2.005	124.6	3.762	0.3125	0.5224	288.5	1.00062	18.80	14.74
260	0.1	1.850	135.0	3.804	0.3125	0.5220	300.3	1.00058	20.15	15.80
280	0.1	1.717	145.5	3.843	0.3124	0.5217	311.7	1.00053	21.46	16.83
300	0.1	1.603	155.9	3.879	0.3124	0.5215	322.7	1.00050	22.74	17.84
320	0.1	1.502	166.3	3.913	0.3124	0.5214	333.3	1.00047	23.99	18.82
340	0.1	1.414	176.8	3.944	0.3123	0.5212	343.5	1.00044	25.21	19.77
360	0.1	1.335	187.2	3.974	0.3123	0.5211	353.5	1.00042	26.40	20.71
400	0.1	1.201	208.0	4.029	0.3123	0.5209	372.7	1.00037	28.70	22.52
500	0.1	0.9608	260.1	4.145	0.3123	0.5207	416.6	1.00030	34.08	26.73
600	0.1	0.8006	312.2	4.240	0.3122	0.5206	456.4	1.00025	39.00	30.57
700	0.1	0.6862	364.2	4.320	0.3122	0.5205	493.0	1.00021	43.56	34.13
800	0.1	0.6004	416.3	4.390	0.3122	0.5205	527.0	1.00019	47.82	37.46
900	0.1	0.5337	468.3	4.451	0.3122	0.5204	558.9	1.00017	51.85	40.60
1000	0.1	0.4803	520.3	4.506	0.3122	0.5204	589.1	1.00015	55.69	43.58

T K	P MPa	ρ kg m^{-3}	H kJ kg^{-1}	S kJ kg^{-1} K^{-1}	C_v kJ kg^{-1} K^{-1}	C_p kJ kg^{-1} K^{-1}	u m s^{-1}	D	η μPa s	λ mW m^{-1} K^{-1}
P = 1 MPa (10 bar)										
84.04	1.0	1418.	−120.8	1.330	0.5498	1.113	865.3	1.51276	291.1	134.0
100	1.0	1316.	−102.8	1.525	0.4984	1.148	751.6	1.47104	183.1	110.9
116.60	1.0	1191.	−82.82	1.710	0.4627	1.285	614.1	1.42088	119.7	87.58
116.60	1.0	49.55	48.81	2.839	0.3795	0.8007	184.6	1.01550	10.12	8.608
120	1.0	47.20	51.45	2.861	0.3682	0.7559	189.3	1.01476	10.37	8.732
140	1.0	37.75	65.11	2.967	0.3375	0.6350	212.1	1.01179	11.87	9.704
160	1.0	31.93	77.31	3.048	0.3266	0.5907	230.5	1.00997	13.36	10.80
180	1.0	27.83	88.89	3.116	0.3214	0.5686	246.8	1.00868	14.83	11.90
200	1.0	24.74	100.1	3.176	0.3185	0.5556	261.6	1.00772	16.26	13.00
220	1.0	22.31	111.1	3.228	0.3168	0.5473	275.3	1.00696	17.65	14.07
240	1.0	20.33	122.0	3.275	0.3157	0.5417	288.3	1.00634	19.01	15.12
260	1.0	18.69	132.8	3.319	0.3149	0.5376	300.5	1.00583	20.34	16.15
280	1.0	17.30	143.5	3.358	0.3144	0.5346	312.2	1.00539	21.64	17.16
300	1.0	16.11	154.2	3.395	0.3140	0.5324	323.4	1.00502	22.90	18.14
320	1.0	15.08	164.8	3.429	0.3137	0.5306	334.2	1.00470	24.14	19.10
340	1.0	14.17	175.4	3.462	0.3135	0.5292	344.7	1.00441	25.35	20.04
360	1.0	13.37	186.0	3.492	0.3133	0.5280	354.7	1.00416	26.53	20.96
400	1.0	12.01	207.1	3.547	0.3131	0.5263	374.1	1.00374	28.82	22.75
500	1.0	9.593	259.6	3.664	0.3127	0.5239	418.3	1.00299	34.16	26.92
600	1.0	7.988	311.9	3.760	0.3126	0.5227	458.1	1.00249	39.06	30.73
700	1.0	6.846	364.1	3.840	0.3125	0.5220	494.6	1.00213	43.61	34.27
800	1.0	5.990	416.3	3.910	0.3124	0.5215	528.6	1.00186	47.87	37.58
900	1.0	5.325	468.4	3.971	0.3124	0.5212	560.5	1.00166	51.90	40.71
1000	1.0	4.793	520.6	4.026	0.3124	0.5210	590.7	1.00149	55.72	43.68
P = 10 MPa (100 bar)										
86.27	10.0	1428.	−114.2	1.333	0.5519	1.085	891.3	1.51685	300.0	137.1
100	10.0	1349.	−99.27	1.493	0.5089	1.093	805.9	1.48430	205.8	118.6
120	10.0	1222.	−76.82	1.698	0.4650	1.163	674.3	1.43296	130.0	93.55
140	10.0	1066.	−52.01	1.888	0.4374	1.349	527.4	1.37161	85.36	70.81
160	10.0	833.6	−20.15	2.100	0.4356	1.970	357.8	1.28395	50.84	49.64
180	10.0	491.9	26.15	2.373	0.4254	2.091	259.3	1.16159	27.88	30.35
200	10.0	337.7	57.55	2.539	0.3815	1.215	267.7	1.10910	23.43	23.57
220	10.0	270.9	78.52	2.639	0.3585	0.9257	283.7	1.08686	22.90	21.77
240	10.0	231.2	95.59	2.713	0.3457	0.7960	299.0	1.07382	23.22	21.31
260	10.0	204.0	110.7	2.774	0.3379	0.7240	313.1	1.06491	23.87	21.39
280	10.0	183.6	124.7	2.826	0.3327	0.6789	326.1	1.05829	24.68	21.73
300	10.0	167.6	138.0	2.872	0.3290	0.6481	338.4	1.05312	25.58	22.19
320	10.0	154.6	150.7	2.913	0.3264	0.6261	350.0	1.04892	26.53	22.78
340	10.0	143.7	163.1	2.950	0.3243	0.6095	361.0	1.04542	27.51	23.43
360	10.0	134.4	175.1	2.985	0.3228	0.5967	371.6	1.04245	28.49	24.11
400	10.0	119.4	198.6	3.046	0.3205	0.5784	391.5	1.03765	30.48	25.50
500	10.0	94.09	255.0	3.173	0.3174	0.5540	436.1	1.02957	35.35	29.04
600	10.0	78.03	309.8	3.272	0.3159	0.5423	475.7	1.02448	39.97	32.48
700	10.0	66.80	363.7	3.355	0.3151	0.5357	511.7	1.02093	44.34	35.76
800	10.0	58.47	417.0	3.427	0.3145	0.5317	545.1	1.01830	48.47	38.88
900	10.0	52.03	470.0	3.489	0.3142	0.5290	576.4	1.01628	52.40	41.86
1000	10.0	46.88	522.8	3.545	0.3139	0.5271	606.0	1.01466	56.16	44.71

Methane (CH_4)

T K	P MPa	ρ kg m^{-3}	H kJ kg^{-1}	S kJ kg^{-1} K^{-1}	C_v kJ kg^{-1} K^{-1}	C_p kJ kg^{-1} K^{-1}	u m s^{-1}	D	η μPa s	λ mW m^{-1} K^{-1}
Saturation										
90.69	0.01170	451.5	−71.82	−.7099	2.168	3.368	1539.	1.67721	193.4	211.2
90.69	0.01170	0.2507	472.4	5.291	1.574	2.110	249.1	1.00031	3.585	8.776
100	0.03438	438.9	−40.27	−.3793	2.114	3.408	1452.	1.65478	151.1	199.6
100	0.03438	0.6746	490.2	4.925	1.589	2.146	260.1	1.00082	3.914	9.910

T K	P MPa	ρ kg m^{-3}	H kJ kg^{-1}	S kJ kg^{-1} K^{-1}	C_v kJ kg^{-1} K^{-1}	C_p kJ kg^{-1} K^{-1}	u m s^{-1}	D	η μPa s	λ mW m^{-1} K^{-1}
110	0.08813	424.8	−5.813	−.05217	2.064	3.469	1355.	1.62995	121.0	186.1
110	0.08813	1.598	508.0	4.619	1.611	2.205	270.0	1.00196	4.273	11.22
120	0.1914	409.9	29.41	0.2521	2.020	3.549	1253.	1.60408	98.68	172.0
120	0.1914	3.262	524.0	4.374	1.639	2.293	277.8	1.00399	4.636	12.66
130	0.3673	394.0	65.63	0.5385	1.980	3.658	1148.	1.57683	81.29	157.7
130	0.3673	5.980	537.7	4.170	1.674	2.421	283.1	1.00733	5.008	14.28
140	0.6412	376.9	103.2	0.8116	1.945	3.813	1038.	1.54771	67.35	143.5
140	0.6412	10.15	548.3	3.991	1.717	2.611	285.9	1.01248	5.398	16.15
150	1.040	357.9	142.6	1.076	1.919	4.047	920.8	1.51599	55.99	129.2
150	1.040	16.33	555.2	3.827	1.773	2.908	286.0	1.02013	5.827	18.38
160	1.592	336.3	184.8	1.338	1.904	4.435	795.4	1.48045	46.55	115.0
160	1.592	25.38	557.1	3.664	1.847	3.419	283.0	1.03144	6.335	21.23
170	2.328	310.5	231.2	1.605	1.910	5.187	657.5	1.43870	38.45	100.5
170	2.328	38.97	551.5	3.490	1.956	4.459	276.7	1.04861	7.012	25.34
180	3.285	276.2	285.9	1.899	1.967	7.292	497.0	1.38453	30.92	85.50
180	3.285	61.38	532.8	3.271	2.140	7.574	266.0	1.07739	8.127	33.50
190	4.519	200.8	378.3	2.369	2.602	94.01	250.3	1.27031	20.34	94.07
190	4.519	125.2	459.0	2.794	2.855	140.8	238.5	1.16274	12.24	120.5
190.56	4.599	162.7	415.6	2.562				1.21521	15.91	
P = 0.1 MPa (1 bar)										
90.72	0.1	451.5	−71.60	−.7097	2.168	3.367	1539.	1.67726	193.6	211.2
100	0.1	438.9	−40.17	−.3798	2.114	3.408	1453.	1.65487	151.2	199.6
111.51	0.1	422.6	−.5573	−.004967	2.057	3.480	1340.	1.62612	117.2	184.0
111.51	0.1	1.795	510.6	4.579	1.615	2.216	271.3	1.00220	4.327	11.43
120	0.1	1.655	529.2	4.740	1.594	2.174	282.8	1.00202	4.668	12.35
140	0.1	1.403	572.1	5.071	1.574	2.129	307.6	1.00172	5.449	14.65
160	0.1	1.221	614.5	5.354	1.568	2.110	330.0	1.00149	6.213	16.98
180	0.1	1.081	656.6	5.602	1.568	2.104	350.7	1.00132	6.963	19.32
200	0.1	0.9709	698.7	5.824	1.574	2.106	370.0	1.00119	7.697	21.65
220	0.1	0.8812	740.9	6.025	1.587	2.115	388.0	1.00108	8.416	23.99
240	0.1	0.8069	783.4	6.209	1.607	2.133	404.9	1.00099	9.117	26.39
260	0.1	0.7442	826.3	6.381	1.634	2.159	420.8	1.00091	9.803	28.88
280	0.1	0.6906	869.8	6.542	1.670	2.194	435.7	1.00085	10.47	31.47
300	0.1	0.6443	914.1	6.695	1.713	2.236	449.7	1.00079	11.13	34.19
320	0.1	0.6038	959.3	6.841	1.763	2.285	463.1	1.00074	11.77	37.04
340	0.1	0.5681	1006.	6.981	1.819	2.340	475.7	1.00070	12.39	40.03
360	0.1	0.5364	1053.	7.117	1.880	2.401	487.8	1.00066	13.00	43.15
400	0.1	0.4826	1152.	7.376	2.013	2.534	510.6	1.00059	14.18	49.80
500	0.1	0.3859	1423.	7.981	2.381	2.901	561.9	1.00047	16.92	68.34
600	0.1	0.3215	1732.	8.543	2.754	3.273	608.0	1.00039	19.41	88.80
700	0.1	0.2756	2077.	9.074	3.110	3.629	650.8	1.00034	21.68	110.4
800	0.1	0.2411	2457.	9.581	3.441	3.959	690.9	1.00030	23.76	132.5
900	0.1	0.2143	2868.	10.06	3.744	4.263	729.0	1.00026	25.70	154.7
1000	0.1	0.1929	3308.	10.53	4.020	4.538	765.2	1.00024	27.49	176.7
P = 1 MPa (10 bar)										
90.95	1.0	451.8	−69.37	−.7070	2.169	3.363	1543.	1.67772	195.7	211.7
100	1.0	439.6	−38.76	−.3862	2.116	3.401	1460.	1.65604	153.4	200.5
120	1.0	410.8	30.49	0.2447	2.022	3.536	1262.	1.60560	99.86	173.0
140	1.0	377.5	103.5	0.8069	1.946	3.798	1044.	1.54879	67.81	144.0
149.14	1.0	359.6	139.2	1.054	1.920	4.023	931.2	1.51885	56.88	130.4
149.14	1.0	15.70	554.8	3.841	1.767	2.876	286.1	1.01935	5.788	18.17
160	1.0	13.97	584.2	4.031	1.682	2.588	305.2	1.01720	6.287	18.79
180	1.0	11.81	633.6	4.322	1.631	2.380	333.7	1.01453	7.089	20.65
200	1.0	10.33	680.2	4.568	1.613	2.289	357.8	1.01269	7.838	22.74
220	1.0	9.217	725.5	4.784	1.613	2.248	379.1	1.01133	8.558	24.90
240	1.0	8.344	770.3	4.978	1.626	2.234	398.2	1.01025	9.258	27.19
260	1.0	7.634	815.0	5.157	1.648	2.238	415.8	1.00938	9.939	29.60

T K	P MPa	ρ kg m^{-3}	H kJ kg^{-1}	S kJ kg^{-1} K^{-1}	C_v kJ kg^{-1} K^{-1}	C_p kJ kg^{-1} K^{-1}	u m s^{-1}	D	η μPa s	λ mW m^{-1} K^{-1}
280	1.0	7.043	859.9	5.324	1.681	2.258	432.0	1.00865	10.60	32.12
300	1.0	6.542	905.4	5.480	1.722	2.289	447.0	1.00803	11.25	34.79
320	1.0	6.110	951.5	5.629	1.770	2.330	461.1	1.00750	11.89	37.59
340	1.0	5.733	998.6	5.772	1.825	2.379	474.4	1.00704	12.51	40.54
360	1.0	5.402	1047.	5.910	1.885	2.434	487.0	1.00663	13.11	43.64
400	1.0	4.846	1147.	6.172	2.017	2.559	510.6	1.00595	14.28	50.23
500	1.0	3.860	1420.	6.781	2.383	2.915	563.0	1.00474	17.01	68.68
600	1.0	3.210	1730.	7.345	2.755	3.282	609.7	1.00394	19.48	89.08
700	1.0	2.750	2076.	7.878	3.110	3.635	652.8	1.00338	21.74	110.6
800	1.0	2.405	2456.	8.385	3.441	3.964	693.1	1.00296	23.82	132.7
900	1.0	2.137	2868.	8.870	3.744	4.266	731.2	1.00263	25.75	154.9
1000	1.0	1.924	3308.	9.334	4.020	4.541	767.5	1.00237	27.54	176.8
P = 10 MPa (100 bar)										
93.22	10.0	454.5	−47.07	−.6806	2.176	3.325	1582.	1.68223	212.8	216.9
100	10.0	446.0	−24.48	−.4466	2.139	3.344	1526.	1.66701	174.8	209.0
120	10.0	419.8	43.14	0.1696	2.047	3.424	1352.	1.62085	112.1	183.3
140	10.0	391.2	112.8	0.7062	1.969	3.552	1171.	1.57167	78.30	157.0
160	10.0	358.8	185.8	1.194	1.909	3.777	981.8	1.51718	56.70	131.5
180	10.0	319.6	265.3	1.661	1.873	4.226	780.8	1.45298	41.52	107.0
200	10.0	266.2	358.9	2.153	1.878	5.304	567.9	1.36870	29.55	83.93
220	10.0	187.6	482.7	2.742	1.903	6.713	404.4	1.25087	19.37	63.29
240	10.0	128.4	599.6	3.252	1.847	4.853	383.6	1.16707	14.89	49.74
260	10.0	101.0	683.8	3.589	1.803	3.730	402.8	1.12974	13.80	44.85
280	10.0	85.51	752.9	3.846	1.793	3.242	424.5	1.10902	13.61	43.81
300	10.0	75.18	815.1	4.060	1.807	3.002	444.5	1.09539	13.75	44.37
320	10.0	67.61	873.8	4.250	1.838	2.878	462.8	1.08549	14.04	45.79
340	10.0	61.75	930.6	4.422	1.880	2.817	479.5	1.07786	14.41	47.75
360	10.0	57.01	986.7	4.582	1.931	2.795	494.8	1.07173	14.82	50.09
400	10.0	49.74	1099.	4.877	2.051	2.819	522.6	1.06238	15.72	55.61
500	10.0	38.32	1391.	5.529	2.401	3.055	581.0	1.04783	18.06	72.56
600	10.0	31.47	1712.	6.113	2.766	3.370	630.6	1.03918	20.32	92.15
700	10.0	26.82	2066.	6.657	3.117	3.696	675.2	1.03334	22.44	113.2
800	10.0	23.41	2451.	7.171	3.445	4.008	716.3	1.02907	24.43	134.9
900	10.0	20.79	2866.	7.660	3.747	4.300	754.7	1.02581	26.28	156.8
1000	10.0	18.72	3310.	8.128	4.022	4.568	791.1	1.02323	28.02	178.5

Ethane (C_2H_6)

T K	P MPa	ρ kg m^{-3}	H kJ kg^{-1}	S kJ kg^{-1} K^{-1}	C_v kJ kg^{-1} K^{-1}	C_p kJ kg^{-1} K^{-1}	u m s^{-1}	D	η μPa s	λ mW m^{-1} K^{-1}
Saturation										
90.37	0.0000011	651.5	−219.2	−1.655	1.605	2.326	2009.	1.94483	1281.	255.6
90.37	0.0000011	0.0000457	375.6	4.927	0.8916	1.168	180.9	1.00000	3.043	2.908
100	0.0000111	640.9	−197.0	−1.422	1.541	2.283	1938.	1.92579	873.2	247.8
100	0.0000111	0.0004007	386.9	4.418	0.9107	1.187	189.9	1.00000	3.316	3.456
120	0.0003523	618.9	−151.5	−1.007	1.478	2.280	1794.	1.88657	485.8	229.9
120	0.0003523	0.01062	411.0	3.681	0.9528	1.230	206.9	1.00001	3.896	4.657
140	0.003814	596.6	−105.6	−.6532	1.450	2.311	1649.	1.84725	319.8	210.6
140	0.003814	0.09880	435.6	3.213	1.003	1.284	222.0	1.00011	4.496	5.966
160	0.02141	573.6	−58.95	−.3418	1.436	2.357	1501.	1.80737	231.2	190.8
160	0.02141	0.4890	460.3	2.904	1.048	1.338	235.1	1.00054	5.110	7.422
180	0.07864	549.5	−11.13	−.06082	1.434	2.421	1350.	1.76641	176.2	171.3
180	0.07864	1.625	484.2	2.691	1.098	1.409	245.5	1.00181	5.739	9.079
200	0.2172	524.0	38.30	0.1981	1.444	2.512	1196.	1.72364	138.3	152.6
200	0.2172	4.170	506.2	2.538	1.179	1.537	252.3	1.00465	6.391	11.01
220	0.4920	496.3	90.01	0.4419	1.468	2.645	1037.	1.67809	110.1	134.6
220	0.4920	9.017	525.5	2.421	1.280	1.720	254.6	1.01008	7.089	13.32
240	0.9668	465.3	145.0	0.6767	1.507	2.847	873.3	1.62822	87.80	117.5

T K	*P* MPa	*ρ* kg m⁻³	*H* kJ kg⁻¹	*S* kJ kg⁻¹ K⁻¹	C_v kJ kg⁻¹ K⁻¹	C_p kJ kg⁻¹ K⁻¹	*u* m s⁻¹	*D*	*η* μPa s	*λ* mW m⁻¹ K⁻¹
240	0.9668	17.43	540.9	2.326	1.388	1.976	252.1	1.01957	7.881	16.18
260	1.712	429.1	204.9	0.9095	1.566	3.195	700.5	1.57121	69.30	101.2
260	1.712	31.58	550.2	2.238	1.516	2.418	243.8	1.03567	8.876	20.02
280	2.807	382.7	273.1	1.152	1.654	3.987	512.4	1.50037	52.84	85.43
280	2.807	56.37	548.6	2.136	1.696	3.522	228.1	1.06440	10.37	26.33
300	4.357	303.5	364.4	1.451	1.912	10.02	274.9	1.38458	34.97	71.49
300	4.357	114.5	514.1	1.950	2.089	13.30	200.5	1.13418	14.02	47.46
305.32	4.872	206.2	439.0	1.690				1.25115	21.80	
***P* = 0.1 MPa (1 bar)**										
90.38	0.1	651.5	−219.0	−1.655	1.605	2.326	2009.	1.94485	1281.	255.6
100	0.1	641.0	−196.9	−1.422	1.541	2.283	1939.	1.92586	873.9	247.9
120	0.1	619.0	−151.4	−1.007	1.478	2.279	1795.	1.88665	486.1	230.0
140	0.1	596.6	−105.5	−.6535	1.450	2.311	1650.	1.84735	320.0	210.6
160	0.1	573.6	−58.85	−.3420	1.436	2.357	1502.	1.80747	231.3	190.9
180	0.1	549.5	−11.10	−.06091	1.434	2.421	1351.	1.76644	176.2	171.4
184.33	0.1	544.1	−.5954	−.003215	1.435	2.438	1317.	1.75734	166.9	167.2
184.33	0.1	2.030	489.1	2.654	1.113	1.432	247.3	1.00226	5.877	9.471
200	0.1	1.856	511.8	2.771	1.149	1.457	257.9	1.00207	6.368	10.72
220	0.1	1.676	541.3	2.912	1.200	1.499	270.3	1.00187	6.991	12.47
240	0.1	1.529	571.8	3.045	1.260	1.554	281.8	1.00170	7.608	14.39
260	0.1	1.407	603.5	3.172	1.328	1.617	292.5	1.00157	8.217	16.48
280	0.1	1.303	636.5	3.294	1.401	1.688	302.6	1.00145	8.817	18.76
300	0.1	1.214	671.1	3.413	1.479	1.764	312.2	1.00135	9.408	21.22
320	0.1	1.137	707.1	3.530	1.560	1.844	321.4	1.00127	9.989	23.85
340	0.1	1.069	744.8	3.644	1.644	1.927	330.3	1.00119	10.56	26.65
360	0.1	1.009	784.2	3.756	1.729	2.011	338.8	1.00112	11.12	29.61
400	0.1	0.9067	868.1	3.977	1.901	2.181	355.2	1.00101	12.21	35.96
500	0.1	0.7242	1107.	4.509	2.315	2.594	393.1	1.00081	14.78	53.77
600	0.1	0.6031	1386.	5.015	2.689	2.967	427.6	1.00067	17.13	73.34
700	0.1	0.5167	1699.	5.498	3.021	3.298	459.6	1.00058	19.32	93.86
800	0.1	0.4520	2044.	5.958	3.314	3.591	489.6	1.00050	21.36	114.8
900	0.1	0.4018	2416.	6.396	3.572	3.849	517.9	1.00045	23.28	135.9
1000	0.1	0.3616	2813.	6.814	3.799	4.076	544.8	1.00040	25.10	156.8
***P* = 1 MPa (10 bar)**										
90.53	1.0	651.7	−217.5	−1.653	1.605	2.324	2012.	1.94510	1283.	255.9
100	1.0	641.3	−195.7	−1.424	1.542	2.282	1942.	1.92644	880.2	248.3
120	1.0	619.4	−150.2	−1.010	1.480	2.278	1799.	1.88737	489.1	230.5
140	1.0	597.2	−104.4	−.6563	1.452	2.309	1654.	1.84822	322.0	211.2
160	1.0	574.3	−57.80	−.3452	1.437	2.354	1508.	1.80855	232.8	191.6
180	1.0	550.4	−10.13	−.06458	1.435	2.416	1358.	1.76780	177.5	172.1
200	1.0	524.9	39.03	0.1943	1.445	2.505	1204.	1.72518	139.3	153.3
220	1.0	497.1	90.36	0.4388	1.468	2.637	1044.	1.67944	110.7	135.2
240	1.0	465.4	145.0	0.6764	1.507	2.846	873.9	1.62834	87.85	117.6
241.10	1.0	463.5	148.1	0.6894	1.510	2.861	864.0	1.62531	86.70	116.6
241.10	1.0	18.04	541.6	2.321	1.395	1.994	251.8	1.02025	7.929	16.36
260	1.0	15.93	577.9	2.466	1.400	1.880	268.9	1.01786	8.485	17.92
280	1.0	14.32	615.3	2.605	1.449	1.869	283.8	1.01604	9.072	19.91
300	1.0	13.07	652.9	2.735	1.513	1.898	296.9	1.01463	9.652	22.18
320	1.0	12.05	691.3	2.859	1.585	1.948	308.7	1.01349	10.22	24.67
340	1.0	11.21	730.9	2.979	1.663	2.010	319.6	1.01254	10.79	27.36
360	1.0	10.49	771.8	3.095	1.745	2.079	329.9	1.01173	11.34	30.24
400	1.0	9.312	857.9	3.322	1.911	2.230	348.8	1.01041	12.42	36.46
500	1.0	7.323	1101.	3.862	2.321	2.620	390.2	1.00818	14.95	54.08
600	1.0	6.058	1381.	4.372	2.693	2.983	426.6	1.00677	17.29	73.56
700	1.0	5.173	1696.	4.857	3.023	3.309	459.7	1.00578	19.46	94.03
800	1.0	4.518	2042.	5.318	3.315	3.598	490.3	1.00505	21.48	115.0

T K	P MPa	ρ kg m⁻³	H kJ kg⁻¹	S kJ kg⁻¹ K⁻¹	C_v kJ kg⁻¹ K⁻¹	C_p kJ kg⁻¹ K⁻¹	u m s⁻¹	D	η μPa s	λ mW m⁻¹ K⁻¹
900	1.0	4.011	2415.	5.757	3.573	3.855	519.0	1.00448	23.39	136.0
1000	1.0	3.608	2812.	6.175	3.800	4.081	546.2	1.00403	25.20	156.9
P = 10 MPa (100 bar)										
91.96	10.0	653.3	−202.5	−1.640	1.605	2.310	2034.	1.94751	1310.	258.6
100	10.0	644.8	−184.1	−1.448	1.553	2.275	1976.	1.93215	947.3	252.5
120	10.0	623.6	−138.8	−1.035	1.490	2.266	1839.	1.89429	520.6	235.4
140	10.0	602.2	−93.25	−.6839	1.463	2.290	1701.	1.85661	342.0	216.9
160	10.0	580.4	−47.10	−.3758	1.449	2.326	1564.	1.81879	248.1	197.9
180	10.0	558.1	−.1206	−.09918	1.447	2.374	1425.	1.78049	190.5	179.3
200	10.0	534.8	47.98	0.1542	1.456	2.439	1287.	1.74125	151.3	161.4
220	10.0	510.2	97.60	0.3905	1.478	2.527	1147.	1.70050	122.4	144.4
240	10.0	483.8	149.3	0.6152	1.511	2.646	1008.	1.65744	99.96	128.5
260	10.0	454.6	203.7	0.8331	1.557	2.809	866.9	1.61088	81.74	113.6
280	10.0	421.4	262.1	1.049	1.616	3.044	724.4	1.55888	66.38	99.57
300	10.0	381.3	326.4	1.271	1.689	3.419	579.3	1.49786	52.87	86.39
320	10.0	328.8	400.9	1.511	1.783	4.109	434.8	1.42048	40.36	73.81
340	10.0	255.9	493.3	1.791	1.888	5.070	319.9	1.31781	28.77	62.21
360	10.0	187.0	591.3	2.071	1.941	4.452	286.8	1.22573	21.62	53.36
400	10.0	126.2	738.6	2.460	2.025	3.205	310.4	1.14850	17.94	48.70
500	10.0	80.57	1035.	3.123	2.369	2.933	377.8	1.09293	17.95	59.52
600	10.0	62.41	1338.	3.675	2.722	3.150	427.4	1.07144	19.50	76.88
700	10.0	51.79	1666.	4.180	3.043	3.415	467.8	1.05903	21.26	96.40
800	10.0	44.59	2021.	4.653	3.330	3.672	502.9	1.05068	23.01	116.8
900	10.0	39.30	2400.	5.100	3.584	3.909	534.3	1.04457	24.73	137.5
1000	10.0	35.20	2802.	5.523	3.809	4.122	563.2	1.03988	26.40	158.2

Propane (C_3H_8)

T K	P MPa	ρ kg m⁻³	H kJ kg⁻¹	S kJ kg⁻¹ K⁻¹	C_v kJ kg⁻¹ K⁻¹	C_p kJ kg⁻¹ K⁻¹	u m s⁻¹	D	η μPa s	λ mW m⁻¹ K⁻¹
Saturation										
85.53	0.17×10⁻⁹	733.1	−196.6	−1.396	1.355	1.916	2136.	2.08838	10780	207.9
85.53	0.17×10⁻⁹	0.11×10⁻⁷	366.3	5.186	0.6907	0.8792	143.3	1.00000	2.641	1.706
100	0.25×10⁻⁷	718.1	−168.8	−1.095	1.341	1.930	2038.	2.05908	3774.	203.2
100	0.25×10⁻⁷	0.0000013	379.4	4.387	0.7475	0.9361	153.7	1.00000	2.979	2.417
120	0.0000030	697.8	−129.9	−.7411	1.335	1.957	1901.	2.01930	1500.	194.4
120	0.0000030	0.0001310	398.9	3.666	0.8210	1.010	166.8	1.00000	3.467	3.500
140	0.0000790	677.6	−90.51	−.4372	1.335	1.988	1767.	1.98025	822.2	183.9
140	0.0000790	0.002993	419.7	3.207	0.8878	1.076	178.9	1.00000	3.970	4.699
160	0.0008502	657.3	−50.38	−.1693	1.342	2.025	1633.	1.94163	535.0	172.3
160	0.0008502	0.02821	441.8	2.907	0.9514	1.141	190.0	1.00003	4.483	6.013
180	0.005068	636.6	−9.433	0.07174	1.356	2.070	1499.	1.90310	382.2	160.2
180	0.005068	0.1499	464.9	2.707	1.017	1.209	200.1	1.00016	5.000	7.439
200	0.02019	615.4	32.53	0.2926	1.383	2.127	1366.	1.86434	288.0	147.8
200	0.02019	0.5417	488.6	2.573	1.088	1.287	208.7	1.00059	5.515	8.971
220	0.06057	593.4	75.80	0.4984	1.421	2.199	1233.	1.82499	224.2	135.7
220	0.06057	1.499	512.7	2.484	1.169	1.381	215.6	1.00164	6.026	10.61
240	0.1480	570.4	120.7	0.6932	1.469	2.289	1102.	1.78460	178.4	124.0
240	0.1480	3.438	536.6	2.426	1.259	1.494	220.1	1.00376	6.540	12.38
260	0.3107	545.8	167.7	0.8801	1.528	2.403	971.7	1.74256	143.9	112.9
260	0.3107	6.903	560.1	2.389	1.357	1.630	221.9	1.00755	7.073	14.32
280	0.5817	519.2	217.3	1.062	1.596	2.547	840.3	1.69800	117.1	102.5
280	0.5817	12.62	582.3	2.365	1.466	1.803	220.4	1.01383	7.655	16.54
300	0.9977	489.4	270.2	1.241	1.675	2.740	706.8	1.64961	95.28	92.86
300	0.9977	21.63	602.6	2.349	1.588	2.041	214.8	1.02381	8.340	19.24
320	1.599	454.9	327.3	1.421	1.764	3.028	569.1	1.59504	76.78	83.94
320	1.599	35.74	619.5	2.334	1.729	2.416	204.4	1.03957	9.230	22.78
340	2.431	411.8	390.9	1.608	1.872	3.585	422.5	1.52907	59.98	75.57

T K	P MPa	ρ kg m^{-3}	H kJ kg^{-1}	S kJ kg^{-1} K^{-1}	C_v kJ kg^{-1} K^{-1}	C_p kJ kg^{-1} K^{-1}	u m s^{-1}	D	η μPa s	λ mW m^{-1} K^{-1}
340	2.431	58.88	629.8	2.311	1.898	3.197	187.4	1.06579	10.57	28.20
360	3.555	345.6	468.2	1.820	2.049	5.984	251.0	1.43242	42.27	67.72
360	3.555	105.4	622.4	2.249	2.183	7.111	161.1	1.11995	13.38	41.36
369.89	4.251	220.5	555.2	2.052				1.26277	22.89	
P = 0.1 MPa (1 bar)										
85.53	0.1	733.1	−196.5	−1.396	1.355	1.916	2137.	2.08840	10790	207.9
100	0.1	718.2	−168.7	−1.096	1.341	1.930	2038.	2.05912	3778.	203.2
120	0.1	697.8	−129.8	−.7413	1.335	1.957	1902.	2.01936	1502.	194.5
140	0.1	677.6	−90.39	−.4374	1.335	1.988	1767.	1.98032	822.9	184.0
160	0.1	657.3	−50.27	−.1695	1.342	2.025	1633.	1.94172	535.5	172.4
180	0.1	636.7	−9.328	0.07149	1.357	2.070	1500.	1.90320	382.5	160.2
200	0.1	615.5	32.62	0.2924	1.383	2.127	1366.	1.86444	288.2	147.9
220	0.1	593.5	75.84	0.4983	1.421	2.198	1234.	1.82505	224.3	135.7
230.74	0.1	581.2	99.69	0.6042	1.446	2.245	1163.	1.80347	197.9	129.4
230.74	0.1	2.387	525.6	2.450	1.216	1.439	218.3	1.00261	6.301	11.54
240	0.1	2.284	539.0	2.507	1.249	1.467	222.9	1.00249	6.559	12.40
260	0.1	2.092	569.0	2.627	1.324	1.535	232.3	1.00228	7.110	14.32
280	0.1	1.932	600.5	2.744	1.405	1.610	241.1	1.00211	7.656	16.36
300	0.1	1.796	633.5	2.857	1.490	1.692	249.4	1.00196	8.196	18.51
320	0.1	1.679	668.2	2.969	1.578	1.777	257.3	1.00183	8.732	20.78
340	0.1	1.576	704.6	3.080	1.668	1.865	264.9	1.00172	9.262	23.16
360	0.1	1.486	742.8	3.189	1.758	1.954	272.3	1.00162	9.787	25.66
400	0.1	1.334	824.5	3.404	1.937	2.130	286.2	1.00145	10.82	30.99
500	0.1	1.064	1059.	3.925	2.356	2.547	318.3	1.00116	13.29	46.36
600	0.1	0.8852	1332.	4.422	2.722	2.912	347.4	1.00096	15.59	64.63
700	0.1	0.7581	1640.	4.896	3.038	3.228	374.2	1.00082	17.70	85.81
800	0.1	0.6631	1976.	5.345	3.313	3.502	399.3	1.00072	19.62	109.9
900	0.1	0.5892	2339.	5.772	3.553	3.743	422.8	1.00064	21.36	136.9
1000	0.1	0.5302	2724.	6.177	3.764	3.953	445.1	1.00058	22.93	166.8
P = 1 MPa (10 bar)										
85.62	1.0	733.3	−195.3	−1.396	1.356	1.916	2139.	2.08856	10840	208.1
100	1.0	718.5	−167.6	−1.097	1.342	1.930	2041.	2.05953	3816.	203.5
120	1.0	698.2	−128.8	−.7432	1.336	1.956	1905.	2.01987	1515.	194.8
140	1.0	678.1	−89.34	−.4394	1.336	1.987	1771.	1.98096	829.3	184.3
160	1.0	657.8	−49.24	−.1717	1.343	2.024	1638.	1.94249	539.4	172.8
180	1.0	637.3	−8.331	0.06918	1.358	2.068	1505.	1.90415	385.3	160.7
200	1.0	616.2	33.57	0.2899	1.384	2.124	1372.	1.86560	290.4	148.4
220	1.0	594.4	76.72	0.4955	1.422	2.194	1241.	1.82648	226.2	136.3
240	1.0	571.5	121.5	0.6901	1.470	2.284	1111.	1.78630	180.0	124.6
260	1.0	547.0	168.2	0.8771	1.528	2.395	980.1	1.74433	145.2	113.5
280	1.0	520.1	217.5	1.060	1.596	2.539	846.7	1.69945	117.9	102.9
300	1.0	489.5	270.2	1.241	1.675	2.740	706.9	1.64962	95.28	92.86
300.09	1.0	489.3	270.4	1.242	1.675	2.741	706.2	1.64937	95.19	92.82
300.09	1.0	21.68	602.7	2.349	1.588	2.042	214.8	1.02386	8.344	19.25
320	1.0	19.34	642.9	2.479	1.646	2.009	229.9	1.02125	8.877	21.49
340	1.0	17.61	683.2	2.601	1.717	2.032	242.4	1.01931	9.414	23.91
360	1.0	16.25	724.3	2.719	1.795	2.081	253.5	1.01779	9.948	26.46
400	1.0	14.17	810.1	2.944	1.958	2.211	272.6	1.01549	11.00	31.92
500	1.0	10.93	1050.	3.478	2.363	2.583	311.6	1.01191	13.48	47.56
600	1.0	8.964	1326.	3.980	2.725	2.933	344.1	1.00976	15.76	66.04
700	1.0	7.623	1635.	4.456	3.040	3.242	372.9	1.00830	17.86	87.38
800	1.0	6.641	1973.	4.907	3.315	3.512	399.1	1.00723	19.76	111.6
900	1.0	5.889	2336.	5.334	3.555	3.750	423.4	1.00641	21.49	138.7
1000	1.0	5.292	2722.	5.741	3.765	3.959	446.3	1.00576	23.04	168.7
P = 10 MPa (100 bar)										
86.45	10.0	735.2	−182.9	−1.394	1.362	1.914	2159.	2.09013	11310	209.8
100	10.0	721.5	−156.9	−1.115	1.350	1.927	2070.	2.06349	4212.	205.7

T K	P MPa	ρ kg m^{-3}	H kJ kg^{-1}	S kJ kg^{-1} K^{-1}	C_v kJ kg^{-1} K^{-1}	C_p kJ kg^{-1} K^{-1}	u m s^{-1}	D	η μPa s	λ mW m^{-1} K^{-1}
120	10.0	701.8	−118.1	−.7615	1.344	1.951	1939.	2.02490	1649.	197.6
140	10.0	682.2	−78.80	−.4586	1.345	1.979	1809.	1.98713	894.7	187.7
160	10.0	662.7	−38.90	−.1922	1.352	2.012	1682.	1.94997	579.3	176.7
180	10.0	643.1	1.726	0.04696	1.367	2.052	1556.	1.91318	413.7	165.2
200	10.0	623.2	43.23	0.2656	1.393	2.101	1432.	1.87653	312.8	153.5
220	10.0	602.8	85.84	0.4686	1.430	2.162	1310.	1.83978	245.3	142.0
240	10.0	581.9	129.8	0.6598	1.478	2.236	1192.	1.80267	197.1	130.9
260	10.0	560.1	175.4	0.8422	1.535	2.325	1077.	1.76484	161.3	120.5
280	10.0	537.1	222.9	1.018	1.600	2.429	963.4	1.72590	133.7	110.7
300	10.0	512.7	272.7	1.190	1.673	2.550	852.7	1.68529	111.7	101.8
320	10.0	486.1	325.0	1.359	1.751	2.692	744.0	1.64228	93.65	93.68
340	10.0	456.7	380.5	1.527	1.833	2.865	637.1	1.59580	78.30	86.37
360	10.0	423.2	439.9	1.697	1.921	3.085	532.1	1.54422	64.77	79.79
400	10.0	334.5	576.0	2.054	2.111	3.790	339.0	1.41464	41.09	68.25
500	10.0	143.3	947.1	2.887	2.438	3.221	275.9	1.16393	19.83	58.85
600	10.0	99.02	1263.	3.462	2.755	3.183	330.8	1.11103	19.67	73.95
700	10.0	78.82	1591.	3.967	3.058	3.386	372.8	1.08759	20.88	95.32
800	10.0	66.46	1940.	4.434	3.328	3.610	407.1	1.07350	22.27	120.1
900	10.0	57.89	2312.	4.871	3.565	3.821	436.7	1.06382	23.62	148.0
1000	10.0	51.49	2704.	5.284	3.774	4.013	463.1	1.05665	24.89	178.8

Carbon Dioxide (CO_2)

T K	P MPa	ρ kg m^{-3}	H kJ kg^{-1}	S kJ kg^{-1} K^{-1}	C_v kJ kg^{-1} K^{-1}	C_p kJ kg^{-1} K^{-1}	u m s^{-1}	D	η μPa s	λ mW m^{-1} K^{-1}
Saturation										
216.59	0.5180	1178.	80.04	0.5213	0.9747	1.953	975.8	1.75696	256.7	180.6
216.59	0.5180	13.76	430.4	2.139	0.6292	0.9087	222.8	1.00694	10.95	11.01
220	0.5991	1166.	86.73	0.5517	0.9698	1.962	951.2	1.74853	242.0	176.2
220	0.5991	15.82	431.6	2.119	0.6389	0.9303	223.1	1.00798	11.14	11.30
230	0.8929	1129.	106.6	0.6387	0.9567	1.997	879.1	1.72239	204.2	163.3
230	0.8929	23.27	434.6	2.065	0.6700	1.005	223.6	1.01178	11.69	12.22
240	1.282	1089.	126.8	0.7235	0.9454	2.051	806.4	1.69398	173.0	150.7
240	1.282	33.30	436.5	2.014	0.7053	1.103	223.0	1.01692	12.27	13.30
250	1.785	1046.	147.7	0.8068	0.9364	2.132	731.8	1.66298	146.7	138.5
250	1.785	46.64	437.0	1.964	0.7459	1.237	221.2	1.02382	12.90	14.61
260	2.419	998.9	169.4	0.8895	0.9323	2.255	652.6	1.62876	124.4	126.3
260	2.419	64.42	435.9	1.914	0.7943	1.429	218.2	1.03309	13.61	16.31
270	3.203	945.8	192.4	0.9732	0.9396	2.453	565.5	1.59026	105.0	114.3
270	3.203	88.37	432.6	1.863	0.8517	1.731	213.8	1.04574	14.47	18.69
280	4.161	883.6	217.3	1.060	0.9605	2.814	471.5	1.54547	87.73	102.0
280	4.161	121.7	425.9	1.805	0.9232	2.277	207.7	1.06363	15.60	22.47
290	5.318	804.7	245.6	1.154	0.9937	3.676	371.9	1.48961	71.41	89.55
290	5.318	172.0	413.8	1.734	1.026	3.614	199.4	1.09112	17.36	29.82
300	6.713	679.2	283.4	1.276	1.120	8.698	245.7	1.40346	53.11	80.59
300	6.713	268.6	387.1	1.622	1.248	11.92	185.3	1.14588	21.31	53.69
304.13	7.377	467.6	332.2	1.434				1.26600	33.04	
P = 0.1 MPa (1 bar)										
220	0.1	2.439	442.2	2.492	0.5791	0.7807	233.4	1.00122	11.06	10.90
240	0.1	2.228	458.0	2.561	0.5981	0.7962	243.2	1.00112	12.07	12.24
260	0.1	2.052	474.1	2.625	0.6184	0.8142	252.3	1.00103	13.06	13.68
280	0.1	1.902	490.6	2.686	0.6390	0.8333	261.0	1.00095	14.05	15.20
300	0.1	1.773	507.4	2.745	0.6593	0.8525	269.4	1.00089	15.02	16.79
320	0.1	1.661	524.7	2.800	0.6791	0.8715	277.4	1.00083	15.98	18.42
340	0.1	1.562	542.3	2.854	0.6982	0.8900	285.2	1.00078	16.93	20.09
360	0.1	1.474	560.3	2.905	0.7165	0.9079	292.8	1.00074	17.87	21.77
400	0.1	1.326	597.3	3.002	0.7510	0.9417	307.3	1.00066	19.70	25.14
500	0.1	1.059	695.2	3.221	0.8255	1.015	340.6	1.00053	24.02	33.49
600	0.1	0.8824	799.9	3.411	0.8867	1.076	370.8	1.00044	28.00	41.55

T K	P MPa	ρ kg m^{-3}	H kJ kg^{-1}	S kJ kg^{-1} K^{-1}	C_v kJ kg^{-1} K^{-1}	C_p kJ kg^{-1} K^{-1}	u m s^{-1}	D	η μPa s	λ mW m^{-1} K^{-1}
700	0.1	0.7562	910.2	3.581	0.9375	1.127	398.7	1.00038	31.68	49.30
800	0.1	0.6616	1025.	3.735	0.9800	1.169	424.7	1.00033	35.09	56.71
900	0.1	0.5880	1144.	3.874	1.015	1.205	449.2	1.00029	38.27	63.80
1000	0.1	0.5292	1266.	4.003	1.045	1.234	472.4	1.00027	41.26	70.57
P = 1 MPa (10 bar)										
216.70	1.0	1179.	80.37	0.5210	0.9751	1.950	977.8	1.75735	257.2	180.9
220	1.0	1167.	86.83	0.5506	0.9703	1.959	953.6	1.74911	242.8	176.5
233.03	1.0	1117.	112.7	0.6646	0.9530	2.011	857.2	1.71404	194.1	159.5
233.03	1.0	26.01	435.3	2.049	0.6803	1.032	223.5	1.01318	11.86	12.53
240	1.0	24.86	442.4	2.079	0.6733	0.9991	228.5	1.01259	12.20	12.94
260	1.0	22.21	461.7	2.157	0.6656	0.9450	241.2	1.01123	13.18	14.26
280	1.0	20.20	480.4	2.226	0.6709	0.9252	252.3	1.01019	14.15	15.71
300	1.0	18.58	498.8	2.289	0.6822	0.9209	262.4	1.00937	15.11	17.25
320	1.0	17.23	517.3	2.349	0.6960	0.9243	271.8	1.00868	16.07	18.84
340	1.0	16.09	535.8	2.405	0.7111	0.9320	280.6	1.00810	17.01	20.47
360	1.0	15.11	554.6	2.459	0.7266	0.9421	289.0	1.00760	17.94	22.12
400	1.0	13.48	592.7	2.559	0.7575	0.9655	304.7	1.00677	19.76	25.46
500	1.0	10.66	692.4	2.781	0.8282	1.027	339.8	1.00535	24.06	33.74
600	1.0	8.845	798.0	2.974	0.8881	1.083	370.9	1.00444	28.04	41.76
700	1.0	7.564	908.8	3.144	0.9384	1.132	399.2	1.00379	31.71	49.47
800	1.0	6.610	1024.	3.298	0.9805	1.173	425.5	1.00332	35.12	56.86
900	1.0	5.872	1143.	3.438	1.016	1.207	450.2	1.00295	38.30	63.93
1000	1.0	5.283	1265.	3.567	1.046	1.236	473.6	1.00265	41.28	70.69
P = 10 MPa (100 bar)										
218.60	10.0	1190.	86.78	0.5155	0.9877	1.902	1012.	1.76433	266.3	185.4
220	10.0	1186.	89.44	0.5277	0.9806	1.904	1003.	1.76113	260.2	183.6
240	10.0	1115.	127.9	0.6949	0.9536	1.949	870.9	1.71160	189.4	159.2
260	10.0	1035.	167.8	0.8545	0.9340	2.052	735.5	1.65372	139.5	136.0
280	10.0	938.2	210.8	1.014	0.9263	2.280	588.7	1.58337	102.2	113.1
300	10.0	801.6	261.8	1.189	0.9496	2.991	414.3	1.48612	71.03	88.91
320	10.0	448.3	362.9	1.514	1.058	7.617	219.1	1.25290	32.39	60.44
340	10.0	258.6	443.4	1.759	0.9025	2.402	238.1	1.13908	22.80	35.70
360	10.0	208.2	483.0	1.873	0.8526	1.707	257.8	1.11030	21.83	31.88
400	10.0	161.5	542.1	2.029	0.8287	1.333	286.9	1.08425	22.23	31.58
500	10.0	113.1	663.8	2.301	0.8552	1.162	337.4	1.05801	25.39	37.26
600	10.0	89.94	779.3	2.511	0.9017	1.156	375.4	1.04580	28.94	44.39
700	10.0	75.49	895.9	2.691	0.9465	1.178	407.4	1.03828	32.39	51.59
800	10.0	65.35	1015.	2.850	0.9859	1.205	435.8	1.03305	35.66	58.65
900	10.0	57.76	1137.	2.993	1.020	1.231	461.8	1.02916	38.75	65.48
1000	10.0	51.82	1261.	3.124	1.048	1.255	485.9	1.02613	41.66	72.05

THERMOPHYSICAL PROPERTIES OF SELECTED FLUIDS AT SATURATION

Eric W. Lemmon

These tables give thermodynamic and transport properties of a variety of fluids, as generated from the equations of state presented in the references below. The properties tabulated are pressure (P), density (ρ), enthalpy (H), entropy (S), isochoric heat capacity (C_v), isobaric heat capacity (C_p), speed of sound (u), viscosity (η), thermal conductivity (λ), and static dielectric constant (D). All extensive properties are given on a mass basis. Not all properties are included for every substance. The references should be consulted for information on the uncertainties. Values are given along the saturation line. The first two lines give the properties at the triple point. The final line gives the properties at the critical point. Two lines are given for each temperature (except at the critical point); the first line gives the values of the liquid phase (note the high density), and the second line gives the values of the vapor phase (at low densities). Ammonia uses a reference state based on zero energy and entropy at the triple point; benzene, carbon monoxide, ethylene, hydrogen sulfide, sulfur dioxide, and toluene use a reference state of zero enthalpy and entropy at the saturated liquid state at the normal boiling point; butane, isobutane, ethanol, propylene, R-134a, and sulfur hexafluoride use a reference state of 200 kJ kg^{-1} and 1 kJ kg^{-1} K^{-1} at –40 °C. Additional calculations at state points not listed below and for fluids not contained here can be obtained at the NIST Chemistry WebBook website (http://webbook.nist.gov/chemistry/fluid/), or by using the NIST Standard Reference Data program REFPROP (http://www.nist.gov/srd/nist23.htm).

References

1. Bücker, D., and Wagner, W., Reference Equations of State for the Thermodynamic Properties of Fluid Phase n-Butane and Isobutane, *J. Phys. Chem. Ref. Data* 35, 929, 2006.
2. Dillon, H. E., and Penoncello, S. G., A Fundamental Equation for Calculation of the Thermodynamic Properties of Ethanol, *Int. J. Thermophys.* 25, 321, 2004.
3. Fenghour, A., Wakeham, W. A., Vesovic, V., Watson, J. T. R., Millat, J., and Vogel, E., The Viscosity of Ammonia, *J. Phys. Chem. Ref. Data* 24, 1649, 1995.
4. Guder, C., and Wagner, W., A Reference Equation of State for the Thermodynamic Properties of Sulfur Hexafluoride (SF_6) for Temperatures from the Melting Line to 625 K and Pressures up to 150 MPa, *J. Phys. Chem. Ref. Data* 38, 33, 2009.
5. Harvey, A. H., and Lemmon, E. W., Method for Estimating the Dielectric Constant of Natural Gas Mixtures, *Int. J. Thermophys.* 26, 31, 2005.
6. Holland, P. M., Eaton, B. E., and Hanley, H. J. M., A Correlation of the Viscosity and Thermal Conductivity Data of Gaseous and Liquid Ethylene, *J. Phys. Chem. Ref. Data* 12, 917, 1983.
7. Huber, M. L., Lasescke, A., and Perkins, R. A., Model for the Viscosity and Thermal Conductivity of Refrigerants, Including a New Correlation for the Viscosity of R134a, *Ind. Eng. Chem. Res.* 42, 3163, 2003.
8. Kiselev, S. B., Ely, J. F., Abdulagatov, I. M., and Huber, M. L., Generalized SAFT-DFT/DMT Model for the Thermodynamic, Interfacial, and Transport Properties of Associating Fluids: Application for n-Alkanols, *Ind. Eng. Chem. Res.* 44, 6916, 2005.
9. Lemmon, E. W., and Span, R., Short Fundamental Equations of State for 20 Industrial Fluids, *J. Chem. Eng. Data* 51, 785, 2006.
10. Overhoff, U., Development of a New Equation of State for the Fluid Region of Propene for Temperatures from the Melting Line to 575 K with Pressures to 1000 MPa as Well as Software for the Computation of Thermodynamic Properties of Fluids, Ph.D. Dissertation, Ruhr University, Bochum, Germany, 2006.
11. Perkins, R. A, Ramires, M. L. V., Nieto de Castro, C. A., and Cusco, L., Measurement and Correlation of the Thermal Conductivity of Butane from 135 K to 600 K at Pressures to 70 MPa, *J. Chem. Eng. Data* 47, 1263, 2002.
12. Perkins, R. A., Laesecke, A., Howley, J., Ramires, M. L. V., Gurova, A. N., and Cusco, L., Experimental Thermal Conductivity Values for the IUPAC Round-Robin Sample of 1,1,1,2-Tetrafluoroethane (R134a), NISTIR, 2000.
13. Perkins, R. A., Measurement and Correlation of the Thermal Conductivity of Isobutane from 114 K to 600 K at Pressures to 70 MPa, *J. Chem. Eng. Data* 47, 1272, 2002.
14. Polt, A., Platzer, B., and Maurer, G., Parameter der thermischen Zustandsgleichung von Bender fuer 14 mehratomige reine Stoffe, *Chem. Tech.* (*Leipzig*) 44, 216, 1992.
15. Schmidt, K. A. G., Carroll, J. J., Quiñones-Cisneros, S. E., and Kvamme, B., Hydrogen Sulphide Viscosity Model, Proceedings of the 86th Annual GPA Convention, San Antonio, Texas, 2007.
16. Smukala, J., Span, R., and Wagner, W., New Equation of State for Ethylene Covering the Fluid Region for Temperatures from the Melting Line to 450 K at Pressures up to 300 MPa, *J. Phys. Chem. Ref. Data* 29, 1053, 2000.
17. Tillner-Roth, R. and Baehr, H. D., An International Standard Formulation for the Thermodynamic Properties of 1,1,1,2-Tetrafluoroethane (HFC-134a) for Temperatures from 170 K to 455 K and Pressures up to 70 MPa, *J. Phys. Chem. Ref. Data* 23, 657, 1994.
18. Tillner-Roth, R., Harms-Watzenberg, F., and Baehr, H. D., A New Fundamental Equation for Ammonia, *DKV-Tagungsbericht* 20, 167, 1993.
19. Tufeu, R., Ivanov, D. Y., Garrabos, Y., and Le Neindre, B., Thermal Conductivity of Ammonia in a Large Temperature and Pressure Range Including the Critical Region, *Ber. Bunsenges. Phys. Chem.* 88, 422, 1984.
20. Vogel, E., Kuechenmeister, C., and Bich, E., Viscosity Correlation for Isobutane over Wide Ranges of the Fluid Region, *Int. J. Thermophys.* 21, 343, 2000.
21. Vogel, E., Kuechenmeister, C., and Bich, E., Viscosity Correlation for n-Butane in the Fluid Region, *High Temp.-High Press.* 31, 173, 1999.

Ammonia (NH_3)

T K	P MPa	ρ kg m^{-3}	H kJ kg^{-1}	S kJ kg^{-1} K^{-1}	C_v kJ kg^{-1} K^{-1}	C_p kJ kg^{-1} K^{-1}	u m s^{-1}	η µPa s	λ mW m^{-1} K^{-1}
195.50	0.006091	732.9	0.008311	0.	2.934	4.202	2124.	559.6	819.0
195.50	0.006091	0.06409	1484.	7.593	1.557	2.063	354.1	6.840	19.64
200.00	0.008651	728.1	19.00	0.09601	2.926	4.227	2080.	507.3	803.1
200.00	0.008651	0.08908	1493.	7.465	1.565	2.075	357.9	6.952	19.68
220.00	0.03379	705.8	104.7	0.5042	2.889	4.342	1914.	346.7	733.2
220.00	0.03379	0.3188	1529.	6.978	1.620	2.160	373.4	7.485	20.13

T K	P MPa	ρ kg m⁻³	H kJ kg⁻¹	S kJ kg⁻¹ K⁻¹	C_v kJ kg⁻¹ K⁻¹	C_p kJ kg⁻¹ K⁻¹	u m s⁻¹	η μPa s	λ mW m⁻¹ K⁻¹
240.00	0.1022	681.8	192.7	0.8866	2.855	4.449	1767.	254.9	665.1
240.00	0.1022	0.8969	1562.	6.591	1.705	2.298	386.3	8.059	20.98
260.00	0.2553	656.2	282.8	1.246	2.821	4.548	1625.	197.3	600.1
260.00	0.2553	2.115	1590.	6.274	1.823	2.503	396.2	8.656	22.26
280.00	0.5509	629.1	375.0	1.586	2.790	4.656	1481.	158.1	538.5
280.00	0.5509	4.382	1612.	6.005	1.972	2.788	402.6	9.266	24.03
300.00	1.062	600.0	469.7	1.910	2.762	4.800	1333.	129.3	480.3
300.00	1.062	8.251	1628.	5.770	2.148	3.177	405.0	9.894	26.41
320.00	1.873	568.2	567.9	2.222	2.743	5.018	1178.	106.9	424.8
320.00	1.873	14.51	1634.	5.554	2.349	3.718	402.7	10.56	29.57
340.00	3.080	532.4	671.3	2.529	2.739	5.385	1012.	88.55	371.5
340.00	3.080	24.40	1629.	5.346	2.574	4.530	395.1	11.33	33.94
360.00	4.793	490.3	783.0	2.838	2.764	6.082	830.6	72.80	319.2
360.00	4.793	40.19	1608.	5.129	2.832	5.954	380.8	12.35	40.75
380.00	7.140	436.1	910.3	3.169	2.853	7.818	620.8	58.31	266.6
380.00	7.140	67.37	1558.	4.874	3.147	9.395	358.0	14.02	54.56
400.00	10.30	344.6	1085.	3.595	3.177	22.73	384.6	41.80	216.0
400.00	10.30	131.1	1432.	4.462	3.598	34.92	318.2	18.53	113.5
405.40	11.33	225.0	1262.	4.026				27.04	

Benzene (C_6H_6)

T K	P MPa	ρ kg m⁻³	H kJ kg⁻¹	S kJ kg⁻¹ K⁻¹	C_v kJ kg⁻¹ K⁻¹	C_p kJ kg⁻¹ K⁻¹	u m s⁻¹
278.70	0.004799	893.6	−131.6	−0.4170	1.226	1.703	1369.
278.70	0.004799	0.1623	315.2	1.186	0.8603	0.9684	182.1
280.00	0.005147	892.3	−129.4	−0.4091	1.219	1.699	1366.
280.00	0.005147	0.1733	316.4	1.183	0.8659	0.9741	182.5
300.00	0.01381	871.7	−95.46	−0.2922	1.191	1.706	1296.
300.00	0.01381	0.4357	336.2	1.147	0.9505	1.061	187.4
320.00	0.03203	850.3	−60.78	−0.1804	1.237	1.766	1204.
320.00	0.03203	0.9537	357.3	1.126	1.034	1.148	191.7
340.00	0.06613	828.3	−24.71	−0.07116	1.306	1.840	1108.
340.00	0.06613	1.872	379.5	1.118	1.116	1.236	195.3
360.00	0.1242	805.8	12.89	0.03606	1.377	1.915	1015.
360.00	0.1242	3.370	402.5	1.118	1.198	1.326	197.9
380.00	0.2161	782.7	51.99	0.1414	1.446	1.990	925.2
380.00	0.2161	5.662	426.3	1.127	1.280	1.421	199.5
400.00	0.3528	758.6	92.59	0.2451	1.509	2.064	839.4
400.00	0.3528	9.003	450.6	1.140	1.363	1.521	199.7
450.00	0.9724	692.4	201.1	0.4984	1.657	2.279	633.9
450.00	0.9724	24.23	511.9	1.189	1.573	1.816	193.2
500.00	2.165	609.2	322.2	0.7495	1.813	2.637	425.3
500.00	2.165	57.26	569.5	1.244	1.800	2.314	172.5
550.00	4.215	466.9	469.8	1.023	2.053	4.765	179.1
550.00	4.215	150.3	600.4	1.260	2.104	5.869	125.6
562.05	4.894	309.0	547.3	1.159			

Butane (C_4H_{10})

T K	P MPa	ρ kg m⁻³	H kJ kg⁻¹	S kJ kg⁻¹ K⁻¹	C_v kJ kg⁻¹ K⁻¹	C_p kJ kg⁻¹ K⁻¹	u m s⁻¹	D	η μPa s	λ mW m⁻¹ K⁻¹
134.90	0.6657·10⁻⁶	735.0	−89.82	−0.4651	1.441	1.973	1827.	2.03981	2304.	176.6
134.90	0.6657·10⁻⁶	0.3450E·10⁻⁴	406.1	3.211	0.9634	1.106	148.9	1.00000	3.321	4.855
150.00	0.8573·10⁻⁵	720.9	−59.93	−0.2550	1.442	1.986	1730.	2.01487	1370.	171.2
150.00	0.8573·10⁻⁵	0.3995E·10⁻³	423.2	2.966	1.015	1.158	156.5	1.00000	3.709	5.579
200.00	0.001939	674.0	41.02	0.3252	1.473	2.062	1438.	1.93301	496.5	149.4

T K	P MPa	ρ kg m^{-3}	H kJ kg^{-1}	S kJ kg^{-1} K^{-1}	C_v kJ kg^{-1} K^{-1}	C_p kJ kg^{-1} K^{-1}	u m s^{-1}	D	η μPa s	λ mW m^{-1} K^{-1}
200.00	0.001939	0.06792	484.8	2.544	1.174	1.319	178.9	1.00007	4.979	8.497
220.00	0.007805	654.8	82.75	0.5240	1.505	2.113	1326.	1.90007	373.4	139.9
220.00	0.007805	0.2496	511.3	2.472	1.244	1.392	186.4	1.00027	5.477	9.884
240.00	0.02409	635.1	125.6	0.7105	1.548	2.176	1216.	1.86666	290.9	130.4
240.00	0.02409	0.7120	538.6	2.431	1.322	1.476	192.9	1.00076	5.966	11.39
260.00	0.06098	614.6	170.0	0.8875	1.601	2.254	1106.	1.83250	232.3	121.2
260.00	0.06098	1.688	566.6	2.413	1.408	1.573	197.9	1.00180	6.450	13.03
280.00	0.1328	593.3	216.0	1.058	1.662	2.345	998.3	1.79722	189.0	112.3
280.00	0.1328	3.490	595.0	2.411	1.501	1.684	201.1	1.00373	6.935	14.82
300.00	0.2576	570.7	264.0	1.222	1.729	2.451	890.9	1.76038	155.6	103.9
300.00	0.2576	6.516	623.6	2.421	1.602	1.811	202.2	1.00697	7.436	16.78
320.00	0.4562	546.4	314.3	1.384	1.803	2.577	783.6	1.72136	129.1	96.10
320.00	0.4562	11.29	651.9	2.439	1.708	1.960	200.7	1.01209	7.978	19.00
340.00	0.7520	519.7	367.4	1.543	1.881	2.729	675.4	1.67929	107.2	88.88
340.00	0.7520	18.52	679.5	2.461	1.820	2.142	196.2	1.01990	8.601	21.58
360.00	1.170	489.6	423.8	1.702	1.965	2.926	564.3	1.63273	88.54	82.27
360.00	1.170	29.35	705.6	2.484	1.935	2.379	187.9	1.03167	9.369	24.72
380.00	1.740	454.2	484.3	1.862	2.059	3.221	447.1	1.57904	71.86	76.22
380.00	1.740	45.85	728.7	2.505	2.058	2.754	174.7	1.04981	10.41	28.81
400.00	2.495	408.5	551.2	2.029	2.173	3.838	318.4	1.51174	55.96	70.60
400.00	2.495	73.08	745.2	2.514	2.210	3.623	154.8	1.08025	12.03	35.03
420.00	3.490	327.8	635.2	2.227	2.357	8.852	165.6	1.39801	37.39	67.19
420.00	3.490	135.0	739.9	2.476	2.444	10.72	124.5	1.15196	15.96	53.10
425.13	3.796	228.0	693.9	2.363				1.26627	23.90	

Carbon monoxide (CO)

T K	P MPa	ρ kg m^{-3}	H kJ kg^{-1}	S kJ kg^{-1} K^{-1}	C_v kJ kg^{-1} K^{-1}	C_p kJ kg^{-1} K^{-1}	u m s^{-1}	η μPa s	λ mW m^{-1} K^{-1}
68.16	0.01554	849.5	−28.96	−0.3863	1.262	2.157	998.2	274.2	180.3
68.16	0.01554	0.7761	203.0	3.017	0.7529	1.063	167.2	4.637	6.687
70.00	0.02105	842.1	−24.99	−0.3290	1.243	2.150	980.5	252.2	175.5
70.00	0.02105	1.027	204.7	2.953	0.7553	1.069	169.2	4.777	6.884
80.00	0.08374	800.3	−3.524	−0.04334	1.154	2.142	883.4	178.0	151.4
80.00	0.08374	3.658	213.4	2.668	0.7747	1.120	178.4	5.608	7.982
90.00	0.2385	755.4	18.18	0.2099	1.088	2.188	783.6	139.5	129.2
90.00	0.2385	9.660	220.4	2.457	0.8073	1.216	184.7	6.618	9.129
100.00	0.5444	705.4	40.71	0.4428	1.037	2.306	678.7	114.0	108.3
100.00	0.5444	21.20	224.9	2.285	0.8544	1.389	187.5	7.924	10.37
110.00	1.067	647.4	64.89	0.6657	1.002	2.558	564.7	93.40	88.95
110.00	1.067	41.74	226.0	2.130	0.9190	1.725	186.7	9.722	11.83
120.00	1.877	574.6	92.30	0.8924	0.9951	3.202	433.1	73.95	71.30
120.00	1.877	78.77	221.2	1.967	1.016	2.601	182.0	12.51	14.10
130.00	3.065	456.2	129.3	1.169	1.110	8.070	254.0	51.35	53.11
130.00	3.065	164.8	201.1	1.721	1.235	9.853	171.9	18.94	21.85
132.86	3.494	303.9	164.7	1.429				32.12	

Ethanol (C_2H_6O)

T K	P MPa	ρ kg m^{-3}	H kJ kg^{-1}	S kJ kg^{-1} K^{-1}	C_v kJ kg^{-1} K^{-1}	C_p kJ kg^{-1} K^{-1}	u m s^{-1}	η μPa s	λ mW m^{-1} K^{-1}
250.00	$0.2701 \cdot 10^{-3}$	825.1	150.4	0.8103	1.664	2.032	1325.	3141.	178.1
250.00	$0.2701 \cdot 10^{-3}$	0.005988	1110.	4.647	1.278	1.459	226.9	7.272	14.94
260.00	$0.6105 \cdot 10^{-3}$	816.9	171.1	0.8917	1.764	2.124	1281.	2454.	175.0
260.00	$0.6105 \cdot 10^{-3}$	0.01302	1124.	4.558	1.306	1.487	231.0	7.586	15.46
280.00	0.002582	800.5	215.8	1.057	1.989	2.351	1203.	1564.	169.6
280.00	0.002582	0.05120	1154.	4.409	1.362	1.544	238.9	8.211	16.61

T K	P MPa	ρ kg m^{-3}	H kJ kg^{-1}	S kJ kg^{-1} K^{-1}	C_v kJ kg^{-1} K^{-1}	C_p kJ kg^{-1} K^{-1}	u m s^{-1}	η μPa s	λ mW m^{-1} K^{-1}
300.00	0.008841	783.8	265.3	1.228	2.213	2.597	1132.	1047.	164.7
300.00	0.008841	0.1641	1185.	4.293	1.420	1.605	246.2	8.829	17.90
320.00	0.02546	766.4	319.7	1.403	2.412	2.838	1065.	728.9	160.1
320.00	0.02546	0.4457	1216.	4.204	1.481	1.673	252.6	9.439	19.34
340.00	0.06354	747.7	378.7	1.582	2.582	3.064	997.9	524.9	155.7
340.00	0.06354	1.058	1247.	4.136	1.548	1.753	257.9	10.04	20.97
360.00	0.1409	727.5	442.2	1.763	2.719	3.274	930.2	388.5	151.3
360.00	0.1409	2.255	1277.	4.082	1.625	1.852	261.7	10.63	22.83
380.00	0.2833	705.3	509.8	1.945	2.827	3.471	860.3	293.8	146.9
380.00	0.2833	4.400	1305.	4.039	1.714	1.980	263.8	11.22	25.04
400.00	0.5245	680.6	581.2	2.127	2.910	3.665	787.2	225.9	142.5
400.00	0.5245	8.009	1331.	4.002	1.818	2.148	263.8	11.82	27.74
420.00	0.9063	653.0	656.6	2.310	2.971	3.868	709.9	175.7	138.1
420.00	0.9063	13.80	1353.	3.969	1.939	2.375	261.4	12.44	31.21
440.00	1.477	621.3	736.1	2.493	3.016	4.109	627.0	137.5	133.8
440.00	1.477	22.81	1371.	3.935	2.079	2.697	256.2	13.13	35.92
460.00	2.292	584.0	820.7	2.678	3.048	4.451	535.8	107.6	129.9
460.00	2.292	36.68	1382.	3.897	2.241	3.197	247.6	13.96	42.73
480.00	3.406	537.1	912.7	2.869	3.075	5.091	431.4	83.25	126.8
480.00	3.406	58.46	1383.	3.848	2.431	4.128	234.1	15.13	53.47
500.00	4.883	467.9	1020.	3.081	3.109	7.255	302.5	61.12	127.6
500.00	4.883	96.58	1365.	3.771	2.669	6.943	211.9	17.23	73.49
513.90	6.148	276.0	1192.	3.413				30.79	

Ethylene (C_2H_4)

T K	P MPa	ρ kg m^{-3}	H kJ kg^{-1}	S kJ kg^{-1} K^{-1}	C_v kJ kg^{-1} K^{-1}	C_p kJ kg^{-1} K^{-1}	u m s^{-1}	D	η μPa s	λ mW m^{-1} K^{-1}
103.99	$0.1220 \cdot 10^{-3}$	654.6	−158.1	−1.179	1.622	2.429	1767.	2.00849	685.7	270.6
103.99	$0.1220 \cdot 10^{-3}$	0.003958	409.4	4.279	0.8901	1.187	202.7	1.00000	0.7727	6.801
120.00	0.001368	634.2	−119.2	−0.8308	1.553	2.427	1660.	1.96600	427.5	248.8
120.00	0.001368	0.03852	428.3	3.731	0.8937	1.192	217.5	1.00004	3.339	6.780
140.00	0.01185	608.0	−70.81	−0.4581	1.465	2.408	1521.	1.91418	277.6	222.3
140.00	0.01185	0.2876	451.3	3.271	0.9064	1.212	233.9	1.00033	4.841	7.892
160.00	0.05623	580.9	−22.66	−0.1372	1.395	2.407	1377.	1.86255	200.1	197.5
160.00	0.05623	1.212	473.0	2.961	0.9336	1.260	247.4	1.00139	5.700	9.115
180.00	0.1818	552.2	25.87	0.1473	1.346	2.441	1227.	1.80986	153.0	174.6
180.00	0.1818	3.589	492.3	2.739	0.9783	1.347	257.0	1.00413	6.401	10.36
200.00	0.4555	521.2	75.68	0.4070	1.321	2.529	1070.	1.75467	120.5	153.7
200.00	0.4555	8.494	508.1	2.569	1.043	1.492	261.9	1.00979	7.133	11.86
220.00	0.9566	486.7	128.0	0.6516	1.320	2.700	903.5	1.69488	95.68	134.5
220.00	0.9566	17.45	519.1	2.429	1.132	1.738	261.5	1.02023	8.007	13.94
240.00	1.773	446.1	184.9	0.8911	1.344	3.043	724.4	1.62679	74.96	116.1
240.00	1.773	33.07	522.7	2.299	1.254	2.211	254.9	1.03865	9.162	17.14
260.00	3.003	393.5	250.4	1.141	1.407	3.946	524.1	1.54141	55.95	97.24
260.00	3.003	61.54	513.5	2.153	1.439	3.512	240.5	1.07303	10.96	23.22
265.00	3.390	376.5	269.4	1.210	1.437	4.473	467.2	1.51452	51.11	92.17
265.00	3.390	72.62	507.3	2.108	1.506	4.308	235.3	1.08667	11.63	26.05
270.00	3.812	356.4	290.3	1.284	1.484	5.409	405.0	1.48318	46.03	86.86
270.00	3.812	86.80	498.3	2.054	1.592	5.766	229.2	1.10433	12.51	30.45
275.00	4.275	330.8	314.6	1.368	1.566	7.588	334.7	1.44393	40.40	81.58
275.00	4.275	106.5	484.3	1.985	1.716	9.293	221.6	1.12922	13.76	39.17
280.00	4.784	290.7	347.8	1.481	1.778	19.56	246.7	1.38366	33.08	83.42
280.00	4.784	140.7	457.5	1.873	1.981	29.26	208.9	1.17365	16.18	71.72
282.35	5.042	214.2	399.4	1.661				1.27358	22.88	

Hydrogen sulfide (H_2S)

T K	*P* MPa	*ρ* kg m^{-3}	*H* kJ kg^{-1}	*S* kJ kg^{-1} K^{-1}	C_v kJ kg^{-1} K^{-1}	C_p kJ kg^{-1} K^{-1}	*u* m s^{-1}	*η* μPa s	*λ* mW m^{-1} K^{-1}
187.70	0.02326	992.3	–50.47	–0.2520	1.302	2.020	1438.	543.7	249.9
187.70	0.02326	0.5120	524.5	2.811	0.7437	0.9976	245.8	7.673	6.240
200.00	0.05034	971.5	–25.72	–0.1244	1.263	2.003	1373.	378.5	236.7
200.00	0.05034	1.046	535.5	2.682	0.7507	1.012	252.8	8.188	7.174
220.00	0.1437	936.5	14.28	0.06577	1.209	1.995	1268.	265.5	215.8
220.00	0.1437	2.758	552.2	2.511	0.7671	1.047	262.6	9.055	8.740
240.00	0.3377	899.8	54.38	0.2393	1.164	2.009	1162.	211.3	195.6
240.00	0.3377	6.093	566.9	2.375	0.7898	1.102	270.1	9.954	10.42
260.00	0.6875	860.6	95.09	0.4006	1.128	2.050	1054.	175.0	176.1
260.00	0.6875	11.87	579.1	2.262	0.8190	1.183	275.0	10.88	12.32
280.00	1.256	818.0	137.0	0.5533	1.099	2.127	942.1	146.2	157.6
280.00	1.256	21.16	588.0	2.164	0.8549	1.302	277.1	11.83	14.58
300.00	2.110	770.5	180.8	0.7008	1.078	2.262	825.0	121.9	139.7
300.00	2.110	35.48	592.7	2.074	0.8981	1.488	276.1	12.80	17.48
300.00	2.110	770.5	180.8	0.7008	1.078	2.262	825.0	121.9	139.7
300.00	2.110	35.48	592.7	2.074	0.8981	1.488	276.1	12.80	17.48
310.00	2.667	744.3	203.8	0.7738	1.070	2.365	763.8	110.9	131.0
310.00	2.667	45.26	592.9	2.029	0.9230	1.626	274.3	13.31	19.32
320.00	3.323	715.7	227.8	0.8472	1.064	2.509	700.2	100.6	122.5
320.00	3.323	57.40	591.4	1.983	0.9509	1.816	271.6	13.85	21.59
330.00	4.089	684.1	253.2	0.9218	1.062	2.719	633.5	90.67	114.0
330.00	4.089	72.67	587.6	1.935	0.9825	2.095	268.0	14.44	24.50
340.00	4.976	648.3	280.4	0.9989	1.064	3.055	562.6	81.07	105.6
340.00	4.976	92.34	580.8	1.883	1.020	2.548	263.4	15.13	28.43
350.00	5.997	605.8	310.4	1.081	1.074	3.678	485.6	71.43	97.33
350.00	5.997	118.9	569.3	1.821	1.065	3.409	257.7	16.04	34.18
360.00	7.171	551.1	345.6	1.175	1.099	5.271	398.9	61.01	89.85
360.00	7.171	158.3	549.8	1.742	1.126	5.653	250.8	17.54	43.94
370.00	8.529	457.9	397.2	1.309	1.176	18.59	292.8	46.39	88.51
370.00	8.529	238.3	506.6	1.604	1.225	23.66	242.8	21.88	72.03
373.10	9.000	347.3	450.5	1.449				31.95	

Isobutane (C_4H_{10})

T K	*P* MPa	*ρ* kg m^{-3}	*H* kJ kg^{-1}	*S* kJ kg^{-1} K^{-1}	C_v kJ kg^{-1} K^{-1}	C_p kJ kg^{-1} K^{-1}	*u* m s^{-1}	*D*	*η* μPa s	*λ* mW m^{-1} K^{-1}
113.73	$0.2289{\cdot}10^{-7}$	740.3	–112.4	–0.6798	1.174	1.689	2000.	2.10871	8767.	157.9
113.73	$0.2289{\cdot}10^{-7}$	$0.1407{\cdot}10^{-5}$	368.3	3.547	0.7366	0.8796	139.4	1.00000	2.848	2.272
150.00	$0.2388{\cdot}10^{-4}$	706.0	–49.02	–0.1971	1.251	1.805	1714.	2.03196	1748.	148.4
150.00	$0.2388{\cdot}10^{-4}$	0.001113	403.1	2.817	0.8944	1.037	157.8	1.00000	3.799	4.461
200.00	0.003814	657.7	45.22	0.3440	1.363	1.968	1389.	1.93554	547.1	129.1
200.00	0.003814	0.1338	459.6	2.416	1.095	1.241	179.4	1.00015	5.080	8.027
220.00	0.01402	637.7	85.31	0.5349	1.415	2.041	1269.	1.89795	395.7	120.6
220.00	0.01402	0.4500	484.5	2.349	1.180	1.330	186.5	1.00049	5.576	9.620
240.00	0.04022	616.9	127.0	0.7159	1.475	2.123	1151.	1.86024	298.3	112.1
240.00	0.04022	1.197	510.3	2.313	1.270	1.428	192.3	1.00130	6.063	11.30
260.00	0.09588	595.4	170.4	0.8893	1.541	2.215	1036.	1.82195	231.6	103.9
260.00	0.09588	2.685	536.7	2.298	1.366	1.538	196.3	1.00291	6.545	13.07
280.00	0.1988	572.6	215.8	1.057	1.613	2.320	922.8	1.78256	183.8	95.99
280.00	0.1988	5.316	563.5	2.299	1.469	1.664	198.3	1.00575	7.033	14.96
300.00	0.3700	548.3	263.5	1.220	1.690	2.442	810.3	1.74142	148.2	88.60
300.00	0.3700	9.610	590.4	2.310	1.578	1.810	197.7	1.01041	7.546	17.02
320.00	0.6333	521.8	313.9	1.381	1.773	2.589	697.4	1.69761	120.7	81.78
320.00	0.6333	16.27	616.7	2.328	1.694	1.985	194.2	1.01765	8.118	19.37
340.00	1.015	492.1	367.4	1.541	1.862	2.777	582.7	1.64968	98.39	75.56

T K	P MPa	ρ kg m^{-3}	H kJ kg^{-1}	S kJ kg^{-1} K^{-1}	C_v kJ kg^{-1} K^{-1}	C_p kJ kg^{-1} K^{-1}	u m s^{-1}	D	η µPa s	λ mW m^{-1} K^{-1}
340.00	1.015	26.36	642.0	2.349	1.814	2.209	186.9	1.02871	8.813	22.21
360.00	1.543	457.2	425.0	1.703	1.958	3.052	463.6	1.59499	79.37	69.94
360.00	1.543	41.79	664.7	2.368	1.939	2.548	175.0	1.04579	9.758	25.97
380.00	2.252	412.8	488.6	1.870	2.069	3.588	335.1	1.52752	61.79	64.93
380.00	2.252	66.87	681.9	2.379	2.101	3.298	156.7	1.07403	11.28	31.85
400.00	3.186	341.0	565.4	2.060	2.250	6.349	184.4	1.42315	42.50	61.90
400.00	3.186	118.4	683.3	2.355	2.354	7.555	128.9	1.13392	14.76	46.45
407.81	3.629	225.5	633.9	2.226				1.26683	24.42	

Propylene (C_3H_6)

T K	P MPa	ρ kg m^{-3}	H kJ kg^{-1}	S kJ kg^{-1} K^{-1}	C_v kJ kg^{-1} K^{-1}	C_p kJ kg^{-1} K^{-1}	u m s^{-1}	η µPa s	λ mW m^{-1} K^{-1}
87.95	$0.7421 \cdot 10^{-9}$	768.3	−200.6	−1.433	1.598	2.280	2130.		
87.95	$0.7421 \cdot 10^{-9}$	$0.4271E \cdot 10^{-7}$	386.1	5.238	0.7005	0.8981	149.3	2.589	3.100
100.00	$0.4108 \cdot 10^{-7}$	754.1	−173.8	−1.146	1.509	2.184	2027.	3817.	191.4
100.00	$0.4108 \cdot 10^{-7}$	$0.2079E \cdot 10^{-5}$	397.1	4.562	0.7310	0.9286	158.4	2.900	3.578
150.00	$0.4107 \cdot 10^{-3}$	697.6	−68.84	−0.2936	1.355	2.060	1669.	595.0	179.0
150.00	$0.4107 \cdot 10^{-3}$	0.01386	446.6	3.142	0.8573	1.056	190.9	4.250	5.894
200.00	0.02676	640.8	35.10	0.3038	1.345	2.117	1345.	265.1	160.5
200.00	0.02676	0.6864	501.0	2.633	1.007	1.217	215.6	5.647	8.829
220.00	0.07823	616.9	78.03	0.5079	1.366	2.174	1216.	208.4	151.9
220.00	0.07823	1.855	523.0	2.531	1.079	1.305	222.4	6.204	10.20
240.00	0.1872	591.8	122.3	0.6998	1.397	2.250	1087.	167.9	142.6
240.00	0.1872	4.178	544.6	2.459	1.159	1.413	226.7	6.773	11.74
260.00	0.3862	565.0	168.4	0.8829	1.439	2.352	957.0	137.5	132.8
260.00	0.3862	8.275	565.1	2.408	1.249	1.549	228.1	7.386	13.57
280.00	0.7128	535.8	216.9	1.060	1.489	2.488	825.1	113.8	122.6
280.00	0.7128	14.99	583.8	2.371	1.350	1.728	226.0	8.097	15.88
300.00	1.209	503.1	268.5	1.235	1.550	2.684	690.5	94.34	112.0
300.00	1.209	25.57	599.8	2.339	1.457	1.976	219.8	9.000	19.07
320.00	1.919	464.5	324.6	1.411	1.623	3.006	549.0	77.47	101.0
320.00	1.919	42.32	611.3	2.307	1.596	2.420	208.3	10.27	23.88
340.00	2.897	414.7	387.9	1.596	1.726	3.746	396.0	61.42	89.47
340.00	2.897	70.68	614.1	2.261	1.764	3.484	190.3	12.37	32.30
360.00	4.219	325.4	472.8	1.828	1.993	10.80	200.4	41.56	80.42
360.00	4.219	139.2	586.6	2.144	2.158	15.87	161.1	17.86	59.05
364.21	4.555	230.1	529.6	1.981				27.37	

R–134a (1,1,1,2–tetrafluoroethane, CF_3CH_2F)

T K	P MPa	ρ kg m^{-3}	H kJ kg^{-1}	S kJ kg^{-1} K^{-1}	C_v kJ kg^{-1} K^{-1}	C_p kJ kg^{-1} K^{-1}	u m s^{-1}	η µPa s	λ mW m^{-1} K^{-1}
169.85	$0.3896 \cdot 10^{-3}$	1591.	71.46	0.4126	0.7922	1.184	1120.	2154.	145.2
169.85	$0.3896 \cdot 10^{-3}$	0.02817	334.9	1.964	0.5030	0.5853	126.8	6.829	3.080
180.00	0.001128	1564.	83.48	0.4814	0.7912	1.187	1068.	1479.	139.1
180.00	0.001128	0.07701	340.9	1.911	0.5267	0.6097	130.1	7.232	3.893
200.00	0.006313	1510.	107.4	0.6073	0.8016	1.206	967.6	867.3	127.7
200.00	0.006313	0.3898	353.1	1.836	0.5732	0.6586	136.0	8.015	5.498
220.00	0.02443	1455.	131.8	0.7235	0.8193	1.233	869.9	582.2	117.2
220.00	0.02443	1.385	365.7	1.787	0.6204	0.7109	141.0	8.779	7.108
240.00	0.07248	1398.	156.8	0.8321	0.8403	1.267	775.0	420.2	107.3
240.00	0.07248	3.837	378.3	1.755	0.6700	0.7705	144.7	9.521	8.732
260.00	0.1768	1337.	182.6	0.9349	0.8631	1.308	682.1	316.6	97.92
260.00	0.1768	8.905	390.8	1.736	0.7234	0.8418	146.8	10.25	10.39
280.00	0.3727	1272.	209.3	1.033	0.8877	1.361	590.2	244.3	88.99
280.00	0.3727	18.23	402.5	1.724	0.7810	0.9296	146.6	10.98	12.12

T K	P MPa	ρ kg m^{-3}	H kJ kg^{-1}	S kJ kg^{-1} K^{-1}	C_v kJ kg^{-1} K^{-1}	C_p kJ kg^{-1} K^{-1}	u m s^{-1}	η μPa s	λ mW m^{-1} K^{-1}
300.00	0.7028	1200.	237.2	1.129	0.9144	1.432	497.9	190.5	80.34
300.00	0.7028	34.19	413.3	1.716	0.8426	1.044	143.9	11.77	14.01
320.00	1.217	1117.	266.8	1.223	0.9443	1.543	404.0	147.8	71.78
320.00	1.217	60.71	422.3	1.709	0.9093	1.211	137.9	12.73	16.30
340.00	1.972	1015.	298.9	1.318	0.9802	1.751	306.4	111.8	63.08
340.00	1.972	105.7	428.2	1.698	0.9852	1.524	127.6	14.16	19.71
360.00	3.040	870.1	336.1	1.421	1.039	2.437	196.0	78.15	54.06
360.00	3.040	193.6	427.1	1.674	1.085	2.606	111.2	17.14	27.37
374.21	4.059	511.9	389.6	1.562				34.69	

Sulfur hexafluoride (SF_6)

T K	P MPa	ρ kg m^{-3}	H kJ kg^{-1}	S kJ kg^{-1} K^{-1}	C_v kJ kg^{-1} K^{-1}	C_p kJ kg^{-1} K^{-1}	u m s^{-1}
223.56	0.2314	1845.	154.1	0.8175	0.5275	0.8371	552.3
223.56	0.2314	19.56	264.7	1.312	0.4833	0.5631	112.8
230.00	0.3013	1813.	159.5	0.8414	0.5407	0.8573	521.5
230.00	0.3013	25.14	267.4	1.310	0.4996	0.5851	112.8
240.00	0.4401	1761.	168.3	0.8784	0.5604	0.8902	474.7
240.00	0.4401	36.21	271.6	1.309	0.5249	0.6219	112.3
250.00	0.6220	1705.	177.4	0.9151	0.5796	0.9264	428.6
250.00	0.6220	50.87	275.6	1.308	0.5506	0.6636	111.0
260.00	0.8546	1646.	186.9	0.9517	0.5986	0.9681	382.8
260.00	0.8546	70.10	279.5	1.308	0.5762	0.7121	109.0
270.00	1.146	1580.	196.8	0.9884	0.6176	1.019	336.9
270.00	1.146	95.32	283.0	1.308	0.6026	0.7739	106.1
280.00	1.506	1507.	207.2	1.025	0.6372	1.087	290.2
280.00	1.506	128.7	286.1	1.307	0.6327	0.8623	102.1
290.00	1.942	1423.	218.3	1.063	0.6589	1.190	241.6
290.00	1.942	173.9	288.4	1.305	0.6650	0.9997	96.92
300.00	2.468	1319.	230.3	1.103	0.6863	1.380	189.8
300.00	2.468	238.4	289.6	1.300	0.7029	1.272	90.15
310.00	3.098	1175.	244.3	1.147	0.7308	1.961	131.5
310.00	3.098	344.1	288.2	1.289	0.7618	2.195	81.33
318.72	3.755	742.3	269.5	1.225			

Sulfur dioxide (SO_2)

T K	P MPa	ρ kg m^{-3}	H kJ kg^{-1}	S kJ kg^{-1} K^{-1}	C_v kJ kg^{-1} K^{-1}	C_p kJ kg^{-1} K^{-1}	u m s^{-1}
197.70	0.001660	1620.	−89.30	−0.3900	0.8776	1.376	1362.
197.70	0.001660	0.06483	355.0	1.857	0.4424	0.5746	182.2
200.00	0.002026	1615.	−86.14	−0.3741	0.8756	1.374	1350.
200.00	0.002026	0.07823	356.3	1.838	0.4443	0.5768	183.2
220.00	0.009334	1567.	−58.76	−0.2437	0.8578	1.364	1253.
220.00	0.009334	0.3289	367.1	1.692	0.4638	0.6006	191.1
240.00	0.03199	1519.	−31.52	−0.1252	0.8406	1.360	1160.
240.00	0.03199	1.041	377.7	1.580	0.4885	0.6330	198.1
260.00	0.08791	1469.	−4.277	−0.01632	0.8248	1.363	1070.
260.00	0.08791	2.677	387.6	1.491	0.5180	0.6750	203.9
280.00	0.2043	1417.	23.14	0.08495	0.8112	1.375	980.3
280.00	0.2043	5.897	396.6	1.419	0.5514	0.7274	208.4
300.00	0.4172	1362.	50.92	0.1802	0.7999	1.398	891.0
300.00	0.4172	11.57	404.6	1.359	0.5875	0.7915	211.3
320.00	0.7702	1304.	79.31	0.2710	0.7910	1.436	801.0
320.00	0.7702	20.83	411.1	1.308	0.6253	0.8712	212.4
340.00	1.313	1240.	108.6	0.3586	0.7849	1.495	709.5

T K	P MPa	ρ kg m⁻³	H kJ kg⁻¹	S kJ kg⁻¹ K⁻¹	C_v kJ kg⁻¹ K⁻¹	C_p kJ kg⁻¹ K⁻¹	u m s⁻¹
340.00	1.313	35.23	415.7	1.262	0.6648	0.9756	211.5
360.00	2.099	1168.	139.3	0.4445	0.7819	1.590	615.4
360.00	2.099	57.07	417.8	1.218	0.7065	1.126	208.2
380.00	3.190	1086.	172.2	0.5306	0.7832	1.759	516.7
380.00	3.190	90.35	416.1	1.173	0.7529	1.380	202.2
400.00	4.656	983.7	208.7	0.6205	0.7922	2.134	409.4
400.00	4.656	143.8	408.2	1.119	0.8091	1.942	192.9
420.00	6.590	832.0	254.3	0.7264	0.8262	3.867	279.4
420.00	6.590	248.6	385.7	1.039	0.8916	4.570	179.3
430.64	7.884	525.0	320.7	0.8778			

Toluene (CH_3–C_6H_5)

T K	P MPa	ρ kg m⁻³	H kJ kg⁻¹	S kJ kg⁻¹ K⁻¹	C_v kJ kg⁻¹ K⁻¹	C_p kJ kg⁻¹ K⁻¹	u m s⁻¹
178.00	$0.3939 \cdot 10^{-7}$	974.8	−344.9	−1.261	1.024	1.472	1888
178.00	$0.3939 \cdot 10^{-7}$	$0.2453 \cdot 10^{-5}$	147.1	1.503	0.5940	0.6843	136.0
180.00	$0.5534 \cdot 10^{-7}$	972.9	−342.0	−1.244	1.024	1.472	1877.
180.00	$0.5534 \cdot 10^{-7}$	$0.3407 \cdot 10^{-5}$	148.5	1.480	0.6004	0.6907	136.7
200.00	$0.1083 \cdot 10^{-5}$	953.5	−312.5	−1.089	1.039	1.479	1768.
200.00	$0.1083 \cdot 10^{-5}$	$0.6002 \cdot 10^{-4}$	163.0	1.288	0.6668	0.7570	143.1
220.00	$0.1148 \cdot 10^{-4}$	934.6	−282.7	−0.9471	1.068	1.504	1665.
220.00	$0.1148 \cdot 10^{-4}$	$0.5783 \cdot 10^{-3}$	178.8	1.151	0.7373	0.8275	149.3
240.00	$0.7754 \cdot 10^{-4}$	916.0	−252.2	−0.8147	1.108	1.542	1566.
240.00	$0.7754 \cdot 10^{-4}$	0.003581	196.1	1.053	0.8111	0.9014	155.1
260.00	$0.3731 \cdot 10^{-3}$	897.5	−220.9	−0.6895	1.156	1.590	1472.
260.00	$0.3731 \cdot 10^{-3}$	0.01591	214.8	0.9865	0.8876	0.9781	160.7
280.00	0.001383	879.0	−188.6	−0.5697	1.210	1.646	1381.
280.00	0.001383	0.05483	235.0	0.9433	0.9660	1.057	166.0
300.00	0.004177	860.4	−155.1	−0.4541	1.268	1.707	1294.
300.00	0.004177	0.1549	256.7	0.9184	1.046	1.138	171.0
320.00	0.01073	841.7	−120.3	−0.3418	1.329	1.773	1211.
320.00	0.01073	0.3745	279.7	0.9080	1.126	1.221	175.5
340.00	0.02417	822.6	−84.12	−0.2323	1.391	1.842	1130.
340.00	0.02417	0.7995	303.9	0.9090	1.208	1.305	179.4
360.00	0.04898	803.1	−46.54	−0.1250	1.455	1.915	1051.
360.00	0.04898	1.545	329.3	0.9190	1.289	1.391	182.7
380.00	0.09099	783.0	−7.483	−0.01956	1.518	1.989	974.1
380.00	0.09099	2.757	355.7	0.9361	1.370	1.479	185.1
400.00	0.1573	762.2	33.10	0.08428	1.581	2.066	898.1
400.00	0.1573	4.612	382.9	0.9587	1.450	1.570	186.4
450.00	0.4862	705.7	141.6	0.3385	1.737	2.272	709.6
450.00	0.4862	13.62	453.2	1.031	1.649	1.814	184.2
500.00	1.177	638.1	261.2	0.5882	1.887	2.529	516.5
500.00	1.177	33.64	523.7	1.113	1.846	2.132	170.9
550.00	2.428	544.8	395.8	0.8406	2.043	3.029	306.4
550.00	2.428	80.11	585.4	1.185	2.057	2.873	140.1
591.75	4.126	292.0	565.8	1.130			

VIRIAL COEFFICIENTS OF SELECTED GASES

Henry V. Kehiaian

This table gives second virial coefficients of about 110 inorganic and organic gases as a function of temperature. Selected data from the literature have been fitted by least squares to the equation

$$B/\text{cm}^3\text{mol}^{-1} = \sum_{i=1}^{n} a(i)[(T_0/T)-1]^{i-1}$$

where T_0 = 298.15 K. The table gives the coefficients $a(i)$ and values of B at fixed temperature increments, as calculated from this smoothing equation.

The equation may be used with the tabulated coefficients for interpolation within the indicated temperature range. It should not be used for extrapolation beyond this range.

Compounds are listed in the modified Hill order (see Introduction), with carbon-containing compounds following those compounds not containing carbon.

A useful compilation of virial coefficient data from the literature may be found in the reference.

Reference

J. H. Dymond and E. B. Smith, *The Virial Coefficients of Pure Gases and Mixtures, A Critical Compilation*, Oxford University Press, Oxford, 1980.

Compounds Not Containing Carbon

Mol. form.	Name	T/K	B/cm³ mol⁻¹
Ar	Argon	100	−184
		120	−131
		140	−98
	$a(1) = -16$	160	−76
	$a(2) = -60$	180	−60
	$a(3) = -9.7$	200	−48
	$a(4) = -1.5$	300	−16
		400	−1
		500	7
		600	12
		700	15
		800	18
		900	20
		1000	22
BF_3	Boron trifluoride	200	−338
		240	−202
		280	−129
	$a(1) = -106$	320	−85
	$a(2) = -330$	360	−56
	$a(3) = -251$	400	−37
	$a(4) = -80$	440	−23
ClH	Hydrogen chloride	190	−451
		230	−269
		270	−181
	$a(1) = -144$	310	−132
	$a(2) = -325$	350	−102
	$a(3) = -277$	390	−81
	$a(4) = -170$	430	−66
		470	−54
Cl_2	Chlorine	210	−508
		220	−483
		230	−457
	$a(1) = -303$	240	−432
	$a(2) = -555$	250	−407
	$a(3) = 9$	260	−383
	$a(4) = 329$	270	−360
	$a(5) = 68$	280	−339
		290	−318
		300	−299
		350	−221
		400	−166
		450	−126
		500	−97
		600	−59
		700	−36
		800	−22
		900	−12
F_2	Fluorine	80	−378
		110	−165
		140	−109
	$a(1) = 8.5$	170	−79
	$a(2) = -163.2$	200	−55
	$a(3) = 84.0$	230	−33
	$a(4) = -27.9$	260	−14
F_4Si	Silicon tetrafluoride	210	−268
		240	−213
		270	−170
	$a(1) = -138$	300	−136
	$a(2) = -312$	330	−108
		360	−84
		390	−64
		420	−47
		450	−32
F_5I	Iodine pentafluoride	320	−2540
		330	−2344
		340	−2172
	$a(1) = -3077$	350	−2021
	$a(2) = -8474$	360	−1890
	$a(3) = -9116$	370	−1775
		380	−1674
		390	−1587
		400	−1510
		410	−1443
F_5P	Phosphorus pentafluoride	320	−162
		340	−143
		360	−127
	$a(1) = -186$	380	−112
	$a(2) = -345$	400	−98
		420	−86
		440	−75
		460	−64
F_6Mo	Molybdenum hexafluoride	300	−896
		310	−810
		320	−737

Mol. form.	Name	T/K	B/cm^3 mol^{-1}
	$a(1) = -914$	330	−677
	$a(2) = -2922$	340	−627
	$a(3) = -4778$	350	−586
		360	−553
		370	−527
		380	−506
		390	−491
F_6S	Sulfur hexafluoride	200	−685
		250	−416
		300	−275
	$a(1) = -279$	350	−190
	$a(2) = -647$	400	−135
	$a(3) = -335$	450	−96
	$a(4) = -72$	500	−68
F_6U	Uranium hexafluoride	320	−1030
		340	−905
		360	−805
	$a(1) = -1204$	380	−724
	$a(2) = -2690$	400	−658
	$a(3) = -2144$	420	−604
		440	560
F_6W	Tungsten hexafluoride	320	−641
		340	−578
		360	−523
	$a(1) = -719$	380	−473
	$a(2) = -1143$	400	−428
		420	−387
		440	−350
		460	−317
H_2	Hydrogen	15	−230
		20	−151
		25	−108
	$a(1) = 15.4$	30	−82
	$a(2) = -9.0$	35	−64
	$a(3) = -0.21$	40	−52
		45	−42
		50	−35
		60	−24
		70	−16
		80	−11
		90	−7
		100	−3
		200	11
		300	15
		400	18
H_2O	Water	300	−1126
		320	−850
		340	−660
	$a(1) = -1158$	360	−526
	$a(2) = -5157$	380	−428
	$a(3) = -10301$	400	−356
	$a(4) = -10597$	420	−301
	$a(5) = -4415$	440	−258
		460	−224
		480	−197
		500	−175
		600	−104
		700	−67
		800	−44
		900	−30
		1000	−20
		1100	−14
		1200	−11
H_3N	Ammonia	290	−302
		300	−265
		310	−236
	$a(1) = -271$	320	−213
	$a(2) = -1022$	330	−194
	$a(3) = -2715$	340	−179
	$a(4) = -4189$	350	−166
		360	−154
		370	−144
		380	−135
		400	−118
		420	−101
H_3P	Phosphine	190	−457
		200	−404
		210	−364
	$a(1) = -146$	220	−332
	$a(2) = -733$	230	−305
	$a(3) = 1022$	240	−281
	$a(4) = -1220$	250	−258
		260	−235
		270	−213
		280	−190
		290	−166
He	Helium	2	−172
		6	−48
		10	−24
	$a(1) = 12.44$	14	−13
	$a(2) = -1.25$	18	−7
		22	−3
		26	−1
		30	1
		50	6
		70	8
		90	10
		110	10
		150	11
		250	12
		650	13
		700	13
Kr	Krypton	110	−363
		120	−307
		130	−263
	$a(1) = -51$	140	−229
	$a(2) = -118$	150	−201
	$a(3) = -29$	160	−178
	$a(4) = -5$	170	−159
		180	−143
		190	−129
		200	−117
		250	−75
		300	−51
		400	−23
		500	−8
		600	2
		700	8
NO	Nitric oxide	120	−232
		130	−176
		140	−138
	$a(1) = -12$	150	−113

Mol. form.	Name	T/K	B/cm³ mol⁻¹
	$a(2) = -119$	160	−96
	$a(3) = 89$	170	−83
	$a(4) = -73$	180	−73
		190	−65
		200	−58
		210	−52
		230	−42
		250	−32
		270	−24
N_2	Nitrogen	75	−274
		100	−161
		125	−104
	$a(1) = -4.3$	150	−71
	$a(2) = -55.7$	175	−49
	$a(3) = -11.8$	200	−34
		225	−24
		250	−15
		300	−4
		400	9
		500	16
		600	21
		700	24
N_2O	Nitrous oxide	240	−219
		260	−181
		280	−151
	$a(1) = -130$	300	−128
	$a(2) = -307$	320	−110
	$a(3) = -248$	340	−96
		360	−85
		380	−76
		400	−68
Ne	Neon	60	−25
		80	−13
		100	−6
	$a(1) = 10.8$	120	−1
	$a(2) = -7.5$	140	2
	$a(3) = -0.4$	160	4
		180	6
		200	7
		300	11
		400	13
		500	14
		600	15
O_2	Oxygen	90	−241
		110	−161
		130	−117
	$a(1) = -16$	150	−88
	$a(2) = -62$	170	−69
	$a(3) = -8$	190	−55
	$a(4) = -3$	210	−44
		230	−36
		250	−29
		270	−23
		290	−18
		310	−14
		330	−10
		350	−7
		400	−1
O_2S	Sulfur dioxide	290	−465
		320	−354
		350	−276

Mol. form.	Name	T/K	B/cm³ mol⁻¹
	$a(1) = -430$	380	−221
	$a(2) = -1193$	410	−181
	$a(3) = -1029$	440	−153
		470	−132
Xe	Xenon	160	−421
		170	−377
		180	−340
	$a(1) = -130$	190	−307
	$a(2) = -262$	200	−280
	$a(3) = -87$	210	−255
		220	−234
		230	−215
		240	−199
		250	−184
		300	−129
		350	−93
		400	−69
		500	−39
		600	−21
		650	−14

Compounds Containing Carbon

Mol. form.	Name	T/K	B/cm³ mol⁻¹
$CClF_3$	Chlorotrifluoromethane	240	−369
		290	−237
		340	−165
	$a(1) = -223$	390	−119
	$a(2) = -504$	440	−86
	$a(3) = -340$	490	−60
	$a(4) = -291$	540	−39
CCl_2F_2	Dichlorodifluoromethane	250	−769
		280	−570
		310	−441
	$a(1) = -486$	340	−353
	$a(2) = -1217$	370	−289
	$a(3) = -1188$	400	−241
	$a(4) = -698$	430	−204
		460	−174
CCl_3F	Trichlorofluoromethane	240	−1140
		280	−879
		320	−689
	$a(1) = -786$	360	−545
	$a(2) = -1428$	400	−431
	$a(3) = -142$	440	−340
		480	−265
CCl_4	Tetrachloromethane	320	−1345
		340	−1171
		360	−1040
	$a(1) = -1600$	380	−942
	$a(2) = -4059$	400	−868
	$a(3) = -4653$	420	−814
CF_4	Tetrafluoromethane	250	−137
		300	−87
		350	−55
	$a(1) = -88$	400	−32
	$a(2) = -238$	450	−16
	$a(3) = -70$	500	−4
		600	14
		700	25

Mol. form.	Name	T/K	B/cm³ mol⁻¹
		800	33
$CHClF_2$	Chlorodifluoromethane	300	−343
		325	−298
		350	−257
	a(1) = −347	375	−221
	a(2) = −575	400	−188
	a(3) = 187	425	−158
$CHCl_2F$	Dichlorofluoromethane	250	−728
		275	−634
		300	−557
	a(1) = −562	325	−491
	a(2) = −862	350	−434
		375	−385
		400	−343
		425	−305
		450	−271
$CHCl_3$	Trichloromethane	320	1001
		330	−926
		340	−858
	a(1) = −1193	350	−797
	a(2) = 2936	360	−740
	a(3) = −1751	370	−689
		380	−642
		390	−599
		400	−559
CHF_3	Trifluoromethane	200	−433
		220	−350
		240	−288
	a(1) = −177	260	−241
	a(2) = −399	280	−204
	a(3) = −250	300	−174
		320	−151
		340	−132
		360	−116
		380	−103
		400	−91
CH_2Cl_2	Dichloromethane	320	−706
		330	−634
		340	−574
	a(1) = −913	350	−524
	a(2) = −3371	360	−482
	a(3) = −5013	370	−447
		380	−420
		400	−380
		420	−357
CH_2F_2	Difluoromethane	280	−375
		290	−343
		300	−316
	a(1) = −321	310	−294
	a(2) = −754	320	−275
	a(3) = −1300	330	−260
		340	−248
		350	−238
CH_3Br	Bromomethane	280	−645
		290	−596
		300	−551
	a(1) = −559	310	−509
	a(2) = −1324	320	−469
		340	−396
		360	−332
		380	−274
CH_3Cl	Chloromethane	280	−466
		300	−402
		320	−348
	a(1) = −407	340	−304
	a(2) = −887	360	−266
	a(3) = −385	380	−234
		400	−206
		420	−182
		440	−161
		460	−142
		480	−126
		500	−112
		600	−58
CH_3F	Fluoromethane	280	−244
		300	−205
		320	−174
	a(1) = −209	340	−150
	a(2) = −525	360	−129
	a(3) = −365	380	−112
		400	−99
		420	−87
CH_3I	Iodomethane	310	−725
		320	−646
		330	−582
	a(1) = −844	340	−531
	a(2) = −3353	350	−492
	a(3) = −6590	360	−462
		370	−441
		380	−427
CH_4	Methane	110	−328
		120	−276
		130	−237
	a(1) = −43	140	−206
	a(2) = −114	150	−181
	a(3) = −19	160	−160
	a(4) = −7	170	−143
		180	−128
		190	−116
		200	−105
		250	−66
		300	−43
		350	−27
		400	−16
		500	0
		600	10
CH_4O	Methanol	320	−1431
		330	−1299
		340	−1174
	a(1) = −1752	350	−1056
	a(2) = −4694	360	−945
		370	−840
		380	−741
		390	−646
		400	−557
CH_5N	Methylamine	300	−451
		325	−367
		350	−304
	a(1) = −459	375	−257
	a(2) = −1191	400	−220
	a(3) = −995	425	−192
		450	−170

Mol. form.	Name	*T*/K	*B*/cm³ mol⁻¹
		500	−140
		550	−122
CO	Carbon monoxide	210	−36
		240	−24
		270	−15
	a(1) = −9	300	−8
	a(2) = −58	330	−3
	a(3) = −18	360	1
		420	7
		480	11
CO_2	Carbon dioxide	220	−244
		240	−204
		260	−172
	a(1) = −127	280	−146
	a(2) = −288	300	−126
	a(3) = −118	320	−108
		340	−94
		360	−81
		380	−71
		400	−62
		500	−30
		600	−13
		700	−1
		800	7
		900	12
		1000	16
		1100	19
CS_2	Carbon disulfide	280	−932
		310	−740
		340	−603
	a(1) = −807	370	−504
	a(2) = −1829	400	−431
	a(3) = −1371	430	−375
$C_2Cl_2F_4$	1,2-Dichloro-1,1,2,2-tetrafluoroethane	300	−801
		320	−695
		340	−608
	a(1) = −812	360	−536
	a(2) = −1773	380	−475
	a(3) = −963	400	−423
		420	−379
		440	−341
		460	−307
		480	−279
		500	−253
$C_2Cl_3F_3$	1,1,2-Trichloro-1,2,2-trifluoroethane	290	−1041
		310	−943
		330	−856
	a(1) = −999	350	−780
	a(2) = −1479	370	−712
		390	−651
		410	−596
		430	−546
		450	−500
C_2H_2	Ethyne	200	−573
		210	−500
		220	−440
	a(1) = −216	230	−390
	a(2) = −375	240	−349
	a(3) = −716	250	−315
		260	−287

Mol. form.	Name	*T*/K	*B*/cm³ mol⁻¹
		270	−263
C_2H_3N	Ethanenitrile	330	−3468
		340	−2971
		350	−2563
	a(1) = −5840	360	−2233
	a(2) = −29175	370	−1970
	a(3) = −47611	380	−1765
		390	−1610
		400	−1499
		410	−1425
C_2H_4	Ethene	240	−218
		270	−172
		300	−139
	a(1) = −140	330	−113
	a(2) = −296	360	−92
	a(3) = −101	390	−76
		420	−63
		450	−52
$C_2H_4Cl_2$	1,2-Dichloroethane	370	−812
		390	−716
		410	−635
	a(1) = −1362	430	−566
	a(2) = −3240	450	−508
	a(3) = −2100	470	−458
		490	−416
		510	−379
		530	−347
		550	−319
		570	−295
C_2H_4O	Ethanal	290	−1352
		320	−927
		350	−654
	a(1) = −1217	380	−482
	a(2) = −4647	410	−375
	a(3) = −5725	440	−314
		470	−283
$C_2H_4O_2$	Methyl methanoate	320	−821
		330	−744
		340	−677
	a(1) = −1035	350	−620
	a(2) = −3425	360	−571
	a(3) = −4203	370	−528
		380	−492
		390	−461
		400	−435
C_2H_5Cl	Chloroethane	320	−634
		360	−450
		400	−330
	a(1) = −777	440	−249
	a(2) = −2205	480	−195
	a(3) = −1764	520	−157
		560	−131
		600	−114
C_2H_6	Ethane	200	−409
		220	−337
		240	−284
	a(1) = −184	260	−242
	a(2) = −376	280	−209
	a(3) = −143	300	−181
	a(4) = −54	320	−159
		340	−140

Mol. form.	Name	T/K	B/cm³ mol⁻¹
		360	−123
		380	−109
		400	−96
		500	−52
		600	−24
C_2H_6O	Ethanol	320	−2710
		330	−2135
		340	−1676
	a(1) = −4475	350	−1317
	a(2) = −29719	360	−1043
	a(3) = −56716	370	−843
		380	−705
		390	−622
C_2H_6O	Dimethyl ether	275	−536
		280	−517
		285	−499
	a(1) = −455	290	−482
	a(2) = −965	295	−465
		300	−449
		305	−433
		310	−418
C_2H_7N	Dimethylamine	310	−606
		320	−563
		330	−523
	a(1) = −662	340	−487
	a(2) = −1504	350	−454
	a(3) = −667	360	−423
		370	−395
		380	−369
		390	−345
		400	−322
C_2H_7N	Ethylamine	300	−773
		310	−710
		320	−654
	a(1) = −785	330	−604
	a(2) = −2012	340	−558
	a(3) = −1397	350	−517
		360	−480
		370	−447
		380	−416
		390	−389
		400	−363
C_3H_6	Cyclopropane	300	−383
		310	−356
		320	−332
	a(1) = −388	330	−310
	a(2) = −861	340	−290
	a(3) = −538	350	−272
		360	−256
		370	−241
		380	−227
		390	−215
		400	−204
C_3H_6	Propene	280	−395
		300	−342
		320	−299
	a(1) = −347	340	−262
	a(2) = −727	360	−232
	a(3) = −325	380	−205
		400	−183
		420	−163

Mol. form.	Name	T/K	B/cm³ mol⁻¹
		440	−146
		460	−131
		480	−118
		500	−106
C_3H_6O	2-Propanone	300	−1996
		320	−1522
		340	−1198
	a(1) = −2051	360	−971
	a(2) = −8903	380	−806
	a(3) = −18056	400	−683
	a(4) = −16448	420	−586
		440	−506
		460	−437
		480	−375
C_3H_6O	Ethyl methanoate	330	−1003
		340	−916
		350	−839
	a(1) = −1371	360	−771
	a(2) = −4231	370	−712
	a(3) = −4312	380	−660
		390	−614
C_3H_6O	Methyl ethanoate	320	−1320
		330	−1186
		340	−1074
	a(1) = −1709	350	−980
	a(2) = −6348	360	−903
	a(3) = −9650	370	−840
		380	−789
		390	−749
C_3H_7Cl	1-Chloropropane	310	−1001
		340	−772
		370	−614
	a(1) = −1121	400	−501
	a(2) = −3271	430	−417
	a(3) = −3786	460	−352
	a(4) = −1974	490	−302
		520	−261
		550	−227
		580	−198
C_3H_8	Propane	240	−641
		260	−527
		280	−444
	a(1) = −386	300	−381
	a(2) = −844	320	−331
	a(3) = −720	340	−292
	a(4) = −574	360	−259
		380	−232
		400	−208
		440	−169
		480	−138
		520	−112
		560	−90
C_3H_8O	1-Propanol	380	−873
		385	−826
		390	−783
	a(1) = −2690	395	−744
	a(2) = −12040	400	−709
	a(3) = −16738	405	−679
		410	−651
		415	−627
		420	−606

Mol. form.	Name	T/K	B/cm³ mol⁻¹
C_3H_8O	2-Propanol	380	−821
		385	−766
		390	−717
	$a(1) = -3165$	395	−674
	$a(2) = -16092$	400	−636
	$a(3) = -24197$	405	−604
		410	−576
		415	−552
		420	−533
C_3H_9N	Trimethylamine	310	−675
		320	−628
		330	−585
	$a(1) = -737$	340	−547
	$a(2) = -1669$	350	−512
	$a(3) = -986$	360	−480
		370	−450
C_4H_8	1-Butene	300	−624
		320	−539
		340	−470
	$a(1) = -633$	360	−413
	$a(2) = -1442$	380	−366
	$a(3) = -932$	400	−327
		420	−294
C_4H_8O	2-Butanone	310	−2056
		320	−1878
		330	−1712
	$a(1) = -2282$	340	−1555
	$a(2) = -5907$	350	−1407
		360	−1267
		370	−1135
$C_4H_8O_2$	Propyl methanoate	330	−1496
		340	−1354
		350	−1231
	$a(1) = -2118$	360	−1126
	$a(2) = -7299$	370	−1035
	$a(3) = -8851$	380	−957
		390	−890
		400	−834
$C_4H_8O_2$	Ethyl ethanoate	330	−1543
		340	−1385
		350	−1254
	$a(1) = -2272$	360	−1144
	$a(2) = -8818$	370	−1055
	$a(3) = -13130$	380	−982
		390	−923
		400	−878
$C_4H_8O_2$	Methyl propanoate	330	−1588
		340	−1444
		350	−1319
	$a(1) = -2216$	360	−1211
	$a(2) = -7339$	370	−1117
	$a(3) = -8658$	380	−1037
		390	−968
		400	−908
C_4H_9Cl	1-Chlorobutane	330	−1224
		370	−898
		410	−691
	$a(1) = -1643$	450	−551
	$a(2) = -4897$	490	−449
	$a(3) = -6178$	530	−371
	$a(4) = -3718$	570	−309

Mol. form.	Name	T/K	B/cm³ mol⁻¹
C_4H_{10}	Butane	250	−1170
		280	−863
		310	−668
	$a(1) = -735$	340	−536
	$a(2) = -1835$	370	−442
	$a(3) = -1922$	400	−371
	$a(4) = -1330$	430	−315
		460	−270
		490	−232
		520	−199
		550	−171
C_4H_{10}	2-Methylpropane	270	−900
		300	−697
		330	−553
	$a(1) = -707$	360	−450
	$a(2) = -1719$	390	−374
	$a(3) = -1282$	420	−317
		450	−273
		480	−240
		510	−215
$C_4H_{10}O$	1-Butanol	350	−1693
		360	−1544
		370	−1402
	$a(1) = -2629$	380	−1268
	$a(2) = -6315$	390	−1141
		400	−1021
		420	−796
		440	−593
$C_4H_{10}O$	2-Methyl-1-propanol	390	−1076
		400	−979
		410	−887
	$a(1) = -2269$	420	−800
	$a(2) = -5065$	430	−716
		440	−636
$C_4H_{10}O$	2-Butanol	380	−1110
		390	−1005
		400	−906
	$a(1) = -2232$	410	−811
	$a(2) = -5209$	420	−721
$C_4H_{10}O$	2-Methyl-2-propanol	380	−924
		390	−827
		400	−736
	$a(1) = -1952$	410	−649
	$a(2) = -4775$	420	−567
$C_4H_{10}O$	Diethyl ether	280	−1550
		300	−1199
		320	−954
	$a(1) = -1226$	340	−776
	$a(2) = -4458$	360	−638
	$a(3) = -7746$	380	−525
	$a(4) = -10005$	400	−428
		420	−340
$C_4H_{11}N$	Diethylamine	320	−1228
		330	−1134
		340	−1056
	$a(1) = -1522$	350	−988
	$a(2) = -5204$	360	−926
	$a(3) = -15047$	370	−868
	$a(4) = -28835$	380	−812
		390	−755

Mol. form.	Name	T/K	B/cm³ mol⁻¹
		400	−697
C_5H_5N	Pyridine	350	−1257
		360	−1176
		370	−1099
	a(1) = −1765	380	−1026
	a(2) = −3431	390	−957
		400	−892
		420	−770
		440	−659
C_5H_{10}	Cyclopentane	300	−1049
		305	−1015
		310	−981
	a(1) = −1062	315	−949
	a(2) = −2116	320	−918
C_5H_{10}	1-Pentene	310	−966
		320	−898
		330	836
	a(1) = −1055	340	−780
	a(2) = −2377	350	−729
	a(3) = −1189	360	−681
		370	−638
		380	−598
		390	−561
		400	−527
		410	−495
$C_5H_{10}O$	2-Pentanone	330	−2850
		340	2420
		350	−2076
	a(1) = −4962	360	−1804
	a(2) = −26372	370	−1595
	a(3) = −46537	380	−1440
		390	−1332
C_5H_{12}	Pentane	300	−1234
		310	−1130
		320	−1038
	a(1) = −1254	330	−957
	a(2) = −3345	340	−884
	a(3) = −2726	350	−818
		400	−579
		450	−436
		500	−348
		550	−294
C_5H_{12}	2-Methylbutane	280	−1263
		290	−1166
		300	−1079
	a(1) = −1095	310	−1001
	a(2) = −2503	320	−931
	a(3) = −1534	330	−867
		340	−810
		350	−757
		400	−557
		450	−424
C_5H_{12}	2,2-Dimethylpropane	300	−916
		310	−843
		320	−780
	a(1) = −931	330	−724
	a(2) = −2387	340	−674
	a(3) = −2641	350	−629
	a(4) = −1810	360	−590
		370	−554
		380	−521

Mol. form.	Name	T/K	B/cm³ mol⁻¹
		390	−492
		400	−464
		450	−357
		500	−279
		550	−218
C_6H_6	Benzene	290	−1588
		300	−1454
		310	−1335
	a(1) = −1477	320	−1231
	a(2) = −3851	330	−1139
	a(3) = −3683	340	−1056
	a(4) = −1423	350	−983
		400	−712
		450	−542
		500	−429
		550	−349
		600	−291
C_6H_7N	2-Methylpyridine	360	−1656
		370	−1523
		380	−1404
	a(1) = −2940	390	−1297
	a(2) = −8813	400	−1202
	a(3) = −7809	410	−1117
		420	−1040
		430	−972
C_6H_7N	3-Methylpyridine	380	−1819
		390	−1612
		400	−1448
	a(1) = −6304	410	−1322
	a(2) = −30415	420	−1230
	a(3) = −44549	430	−1166
C_6H_7N	4-Methylpyridine	380	−1787
		390	−1578
		400	−1417
	a(1) = −6553	410	−1297
	a(2) = −32873	420	−1214
	a(3) = −49874	430	−1163
C_6H_{12}	Cyclohexane	300	−1698
		320	−1391
		340	−1170
	a(1) = −1733	360	−1007
	a(2) = −5618	380	−883
	a(3) = −9486	400	−786
	a(4) = −7936	420	−707
		440	−641
		460	−584
		480	−534
		500	−488
		520	−446
		540	−406
		560	−368
C_6H_{12}	Methylcyclopentane	305	−1447
		315	−1357
		325	−1272
	a(1) = −1512	335	−1192
	a(2) = −2910	345	−1117
C_6H_{14}	Hexane	300	−1920
		310	−1724
		320	−1561
	a(1) = −1961	330	−1424
	a(2) = −6691	340	−1309

Mol. form.	Name	T/K	B/cm³ mol⁻¹
	$a(3) = -13167$	350	−1209
	$a(4) = -15273$	360	−1123
		370	−1046
		380	−978
		390	−916
		400	−859
		410	−806
		430	−707
		450	−616
$C_6H_{15}N$	Triethylamine	330	−1562
		340	−1444
		350	−1340
	$a(1) = -2061$	360	−1249
	$a(2) = -5735$	370	−1169
	$a(3) = -5899$	380	−1099
		390	−1037
		400	−983
C_7H_8	Toluene	350	−1641
		360	−1511
		370	−1394
	$a(1) = -2620$	380	−1289
	$a(2) = -7548$	390	−1195
	$a(3) = -6349$	400	−1110
		410	−1034
		420	−965
		430	−903
C_7H_{14}	1-Heptene	340	−1781
		350	−1651
		360	−1532
	$a(1) = -2491$	370	−1424
	$a(2) = -6230$	380	−1324
	$a(3) = -3780$	390	−1233
		400	−1150
		410	−1073
C_7H_{16}	Heptane	300	−2782
		320	−2297
		340	−1928
	$a(1) = -2834$	360	−1641
	$a(2) = -8523$	380	−1415
	$a(3) = -10068$	400	−1233
	$a(4) = -5051$	420	−1085
		440	−963
		460	−862
		480	−775
		500	−702
		540	−583
		580	−490
		620	−416
		660	−355
		700	−304
C_8H_{10}	1,2-Dimethylbenzene	380	−2046
		390	−1848
		400	−1681
	$a(1) = -5632$	410	−1543
	$a(2) = -22873$	420	−1428
	$a(3) = -28900$	430	−1335
		440	−1261
C_8H_{10}	1,3-Dimethylbenzene	380	−2082
		390	−1865
		400	−1679
	$a(1) = -5808$	410	−1521
	$a(2) = -23244$	420	−1388
	$a(3) = -27607$	430	−1276
		440	−1184
C_8H_{10}	1,4-Dimethylbenzene	380	−2043
		390	−1851
		400	−1680
	$a(1) = -4921$	410	−1529
	$a(2) = -16843$	420	−1395
	$a(3) = -16159$	430	−1276
		440	−1171
C_8H_{16}	1-Octene	360	−2147
		370	−2000
		380	−1861
	$a(1) = -3273$	390	−1729
	$a(2) = -6557$	400	−1604
		410	−1485
C_8H_{18}	Octane	300	−4042
		350	−2511
		400	−1704
	$a(1) = -4123$	450	−1234
	$a(2) = -13120$	500	−936
	$a(3) = -16408$	550	−732
	$a(4) = -8580$	600	−583
		650	−468
		700	−375

VAN DER WAALS CONSTANTS FOR GASES

The van der Waals equation of state for a real gas is

$$(P + n^2a/V^2)(V - nb) = nRT$$

where P is the pressure, V the volume, T the temperature, n the amount of substance (in moles), and R the gas constant. The van der Waals constants a and b are characteristic of the substance and are independent of temperature. They are related to the critical temperature and pressure, T_c and P_c, by

$$a = 27R^2T_c^{\ 2}/64P_c \qquad b = RT_c/8P_c$$

This table gives values of a and b for some common gases. Most of the values have been calculated from the critical temperature and pressure values given in the table "Critical Constants" in this section. Van der Waals constants for other gases may easily be calculated from the data in that table.

To convert the van der Waals constants to SI units, note that 1 bar L²/mol² = 0.1 Pa m⁶/mol² and 1 L/mol = 0.001 m³/mol.

Reference

Reid, R. C, Prausnitz, J. M., and Poling, B. E., *The Properties of Gases and Liquids, Fourth Edition*, McGraw-Hill, New York, 1987.

Substance	*a* bar L²/mol²	*b* L/mol
Acetic acid	17.71	0.1065
Acetone	16.02	0.1124
Acetylene	4.516	0.0522
Ammonia	4.225	0.0371
Aniline	29.14	0.1486
Argon	1.355	0.0320
Benzene	18.82	0.1193
Bromine	9.75	0.0591
Butane	13.89	0.1164
1-Butanol	20.94	0.1326
2-Butanone	19.97	0.1326
Carbon dioxide	3.658	0.0429
Carbon disulfide	11.25	0.0726
Carbon monoxide	1.472	0.0395
Chlorine	6.343	0.0542
Chlorobenzene	25.80	0.1454
Chloroethane	11.66	0.0903
Chloromethane	7.566	0.0648
Cyclohexane	21.92	0.1411
Cyclopropane	8.34	0.0747
Decane	52.74	0.3043
1-Decanol	59.51	0.3086
Diethyl ether	17.46	0.1333
Dimethyl ether	8.690	0.0774
Dodecane	69.38	0.3758
1-Dodecanol	75.70	0.3750
Ethane	5.580	0.0651
Ethanol	12.56	0.0871
Ethylene	4.612	0.0582
Fluorine	1.171	0.0290
Furan	12.74	0.0926
Helium	0.0346	0.0238
Heptane	31.06	0.2049
1-Heptanol	38.17	0.2150
Hexane	24.84	0.1744
1-Hexanol	31.79	0.1856
Hydrazine	8.46	0.0462
Hydrogen	0.2452	0.0265
Hydrogen bromide	4.500	0.0442
Hydrogen chloride	3.700	0.0406
Hydrogen cyanide	11.29	0.0881
Hydrogen fluoride	9.565	0.0739
Hydrogen iodide	6.309	0.0530
Hydrogen sulfide	4.544	0.0434
Isobutane	13.32	0.1164
Krypton	5.193	0.0106
Methane	2.303	0.0431
Methanol	9.476	0.0659
Methylamine	7.106	0.0588
Neon	0.208	0.0167
Neopentane	17.17	0.1411
Nitric oxide	1.46	0.0289
Nitrogen	1.370	0.0387
Nitrogen dioxide	5.36	0.0443
Nitrogen trifluoride	3.58	0.0545
Nitrous oxide	3.852	0.0444
Octane	37.88	0.2374
1-Octanol	44.71	0.2442
Oxygen	1.382	0.0319
Ozone	3.570	0.0487
Pentane	19.09	0.1449
1-Pentanol	25.88	0.1568
Phenol	22.93	0.1177
Propane	9.39	0.0905
1-Propanol	16.26	0.1079
2-Propanol	15.82	0.1109
Propene	8.442	0.0824
Pyridine	19.77	0.1137
Pyrrole	18.82	0.1049
Silane	4.38	0.0579
Sulfur dioxide	6.865	0.0568
Sulfur hexafluoride	7.857	0.0879
Tetrachloromethane	20.01	0.1281
Tetrachlorosilane	20.96	0.1470
Tetrafluoroethylene	6.954	0.0809
Tetrafluoromethane	4.040	0.0633
Tetrafluorosilane	5.259	0.0724
Tetrahydrofuran	16.39	0.1082
Thiophene	17.21	0.1058
Toluene	24.86	0.1497
1,1,1-Trichloroethane	20.15	0.1317
Trichloromethane	15.34	0.1019
Trifluoromethane	5.378	0.0640
Trimethylamine	13.37	0.1101
Water	5.537	0.0305
Xenon	4.192	0.0516

MEAN FREE PATH AND RELATED PROPERTIES OF GASES

In the simplest version of the kinetic theory of gases, molecules are treated as hard spheres of diameter d which make binary collisions only. In this approximation the mean distance traveled by a molecule between successive collisions, the mean free path l, is related to the collision diameter by:

$$l = \frac{kT}{\pi\sqrt{2}Pd^2}$$

where P is the pressure, T the absolute temperature, and k the Boltzmann constant. At standard conditions (P = 100,000 Pa and T = 298.15 K) this relation becomes:

$$l = \frac{9.27 \cdot 10^{27}}{d^2}$$

where l and d are in meters.

Using the same model and the same standard pressure, the collision diameter can be calculated from the viscosity η by the kinetic theory relation:

$$\eta = \frac{2.67 \cdot 10^{-20}(MT)^{1/2}}{d^2}$$

where η is in units of µPa s and M is the molar mass in g/mol. Kinetic theory also gives a relation for the mean velocity $\bar{v}$ of molecules of mass m:

$$\bar{v} = \left(\frac{8kT}{\pi m}\right)^{1/2} = 145.5(T/M)^{1/2} \text{ m/s}$$

Finally, the mean time τ between collisions can be calculated from the relation $\tau\bar{v} = l$.

The table below gives values of l, $\bar{v}$, and τ for some common gases at 25 °C and atmospheric pressure, as well as the value of d, all calculated from measured gas viscosities (see References 2 and 3 and the table "Viscosity of Gases" in this section). It is seen from the above equations that the mean free path varies directly with T and inversely with P, while the mean velocity varies as the square root of T and, in this approximation, is independent of P.

A more accurate model, in which molecular interactions are described by a Lennard-Jones potential, gives mean free path values about 5% lower than this table (see Reference 4).

References

1. Reid, R. C., Prausnitz, J. M., and Poling, B. E., *The Properties of Gases and Liquids, Fourth Edition*, McGraw-Hill, New York, 1987.
2. Lide, D. R., and Kehiaian, H. V., *CRC Handbook of Thermophysical and Thermochemical Data*, CRC Press, Boca Raton, FL, 1994.
3. Vargaftik, N. B., *Tables of Thermophysical Properties of Liquids and Gases, Second Edition*, John Wiley, New York, 1975.
4. Kaye, G. W. C., and Laby, T. H., *Tables of Physical and Chemical Constants, 15th Edition*, Longman, London, 1986.

Gas	d/m	l/m	$\bar{v}$/m s^{-1}	τ/ps
Air	$3.66 \cdot 10^{-10}$	$6.91 \cdot 10^{-8}$	467	148
Ar	3.58	7.22	397	182
CO_2	4.53	4.51	379	119
H_2	2.71	12.6	1769	71
He	2.15	20.0	1256	159
Kr	4.08	5.58	274	203
N_2	3.70	6.76	475	142
NH_3	4.32	4.97	609	82
Ne	2.54	14.3	559	256
O_2	3.55	7.36	444	166
Xe	4.78	4.05	219	185

INFLUENCE OF PRESSURE ON FREEZING POINTS

This table illustrates the variation of the freezing point of representative types of liquids with pressure. Substances are listed in alphabetical order. Note that 1 MPa = 0.01 kbar = 9.87 atm.

References

1. Isaacs, N. S., *Liquid Phase High Pressure Chemistry*, John Wiley, New York, 1981.
2. Merrill, L., *J. Phys. Chem. Ref. Data*, 6, 1205, 1977; 11, 1005, 1982.

Substance	Molecular formula	Freezing point in °C at:		
		0.1 MPa	100 MPa	1000 MPa
Acetic acid	$C_2H_4O_2$	16.6	37	
Acetophenone	C_8H_8O	20.0	41.2	
Aniline	C_6H_7N	−6.0	13.5	140
Benzene	C_6H_6	5.5	33.4	
Benzonitrile	C_7H_5N	−12.8	7.6	
Benzyl alcohol	C_7H_8O	−15.2	0.2	
Bromobenzene	C_6H_5Br	−30.6	−12	108
Bromoethane	C_2H_5Br	−118.6	−108	
1-Bromonaphthalene	$C_{10}H_7Br$	−1.8	6.1	
1-Bromopropane	C_3H_7Br	−110	−98	
p-Bromotoluene	C_7H_7Br	28.0	56.7	
Butanoic acid	$C_4H_8O_2$	−5.7	13.8	
1-Butanol	$C_4H_{10}O$	−89.8	−77.2	
Carbon disulfide	CS_2	−111.5	−98	
Chlorobenzene	C_6H_5Cl	−45.2	−28	84
p-Chlorotoluene	C_7H_7Cl	6.9	33.1	
o-Cresol	C_7H_8O	29.8	47.7	
m-Cresol	C_7H_8O	11.8	25.6	
p-Cresol	C_7H_8O	35.8	56.2	
Cyclohexane	C_6H_{12}	6.6	32.5	
Cyclohexanol	$C_6H_{12}O$	25.5	62.3	
1,2-Dibromoethane	$C_2H_4Br_2$	9.9	34.0	
p-Dichlorobenzene	$C_6H_4Cl_2$	52.7	79.1	
Dichloromethane	CH_2Cl_2	−95.1	−83	
N,N-Dimethylaniline	$C_8H_{11}N$	2.5	26.3	
1,4-Dioxane	$C_4H_8O_2$	11	23	
Ethanol	C_2H_6O	−114.1	−108	
Formamide	CH_3NO	−15.5	10.8	
Formic acid	CH_2O_2	8.3	20.6	
Furan	C_4H_4O	−85.6	−73	
Hexamethyldisiloxane	$C_6H_{18}OSi_2$	−66	−37	
Menthol	$C_{10}H_{20}O$	42	60	
Methyl benzoate	$C_8H_8O_2$	−15	31.8	
2-Methyl-2-butanol	$C_5H_{12}O$	−8.8	13.4	
2-Methyl-2-propanol	$C_4H_{10}O$	25.4	58.1	
Naphthalene	$C_{10}H_8$	78.2	115.7	
Nitrobenzene	$C_6H_5NO_2$	5.7	13.5	
m-Nitrotoluene	$C_7H_7NO_2$	15.5	40.6	
Pentachloroethane	C_2HCl_5	−29.0	−6.3	
Potassium	K	63.7	78	170
Potassium chloride	ClK	771		945
Propanoic acid	$C_3H_6O_2$	−20.7	−1.2	
Silver chloride	AgCl	455		545
Sodium	Na	97.8	106	167
Sodium chloride	ClNa	800.7		997
Sodium fluoride	FNa	996		1115
Tetrachloromethane	CCl_4	−23.0	14.2	
Tribromomethane	$CHBr_3$	8.1	31.5	
Trichloromethane	$CHCl_3$	−63.6	−45.2	
Water	H_2O	0.0	−9.0	
o-Xylene	C_8H_{10}	−25.2	−3.5	
m-Xylene	C_8H_{10}	−47.8	−25.2	
p-Xylene	C_8H_{10}	13.2	46.0	

CRITICAL CONSTANTS OF ORGANIC COMPOUNDS

Chris D. Muzny, Vladimir Diky, Andrei Kazakov, Robert D. Chirico, and Michael Frenkel

The parameters of the liquid-gas critical point are important constants in determining the behavior of fluids. This table lists the critical temperature, pressure, and molar volume, as well as the normal boiling point, for over 850 organic substances. The properties and their units are:

- T_b: Normal boiling point in K at a pressure of 101.325 kPa (1 atmosphere); an "s" following the value indicates a sublimation point (temperature at which the solid is in equilibrium with the gas at a pressure of 101.325 kPa)
- T_c: Critical temperature in K
- P_c: Critical pressure in MPa
- V_c: Critical molar volume in $cm^3\ mol^{-1}$

The listed values are critically evaluated using NIST ThermoData Engine, TDE (Ref. 1), designed to implement the dynamic data evaluation concept (Refs. 2–5). This concept requires large electronic databases capable of storing essentially all relevant experimental data known to date with detailed descriptions of metadata and uncertainties. The combination of these electronic databases with expert-system software, designed to automatically generate recommended property values based on available experimental and predicted data, leads to the ability to produce critically evaluated data dynamically or "to order." The evaluated data have been generated only for compounds for which experimental data for critical properties are available. Group contribution methods such as Joback-Reed (Ref. 6), Constantinou-Gani (Ref. 7), Marrero-Pardillo (Ref. 8), and Wilson-Jasperson (Ref. 9) as well as quantitative structure-property relationship (QSPR) methods (Ref. 5) were used within the TDE environment to validate available experimental data. Each recommended value in the table is characterized with a combined expanded uncertainty (Ref. 10) (level of confidence, approximately 95%) listed in parentheses. The majority of the values of the normal boiling temperatures provided in the table are taken from the compilation presented in the 90th Edition of the CRC *Handbook of Chemistry and Physics*. Some additional values of the normal boiling temperature, along with the combined expanded uncertainties associated with them in parentheses, are generated with the use of TDE. Only references to original experimental data actually used by TDE to generate critically evaluated data are indicated for each compound. Compounds are listed alphabetically by name.

Name	Mol. Form.	T_b/K	T_c/K	P_c/MPa	V_c/cm³ mol⁻¹	Ref.
Acetaldehyde	C_2H_4O	293.3	462(8)	7.5(1)	154(5)	11–13
Acetic acid	$C_2H_4O_2$	391.1	593(2)	5.79(0.03)	171(2)	14–21
Acetic anhydride	$C_4H_6O_3$	412.7	606(1)	4.00(0.08)	294(12)	22
Acetone	C_3H_6O	329.20	508.1(0.2)	4.7(0.1)	221(20)	18, 23–31
Acetonitrile	C_2H_3N	354.80	545.47(0.07)	4.88(0.05)	173(59)	32–40
Acetophenone	C_8H_8O	475	709.5(0.7)	4.01(0.05)	373(40)	11, 41
Acetylene	C_2H_2	188.45 s	308.4(0.4)	6.24(0.04)	119(11)	42–48
Acrylonitrile	C_3H_3N	350.5	540(2)	4.6(0.1)	211(10)	49
Allene	C_3H_4	238.8	394(4)	6.5(0.7)	167(8)	50
Allyl alcohol	C_3H_6O	370.5	539.8(0.6)	5.76(0.04)	222(9)	51
Allylamine	C_3H_7N	326.5	540.0(0.7)	4.83(0.03)	217(11)	52
Allyl ethyl ether	$C_5H_{10}O$	340.8	518(10)		320(10)	21
2-Aminobiphenyl	$C_{12}H_{11}N$	572	838(2)	3.52(0.03)	548(99)	53
2-Aminoethanol	C_2H_7NO	443.4(0.2)	671(3)	8.0(0.5)	207(13)	41
2-(2-Aminoethoxy)ethanol	$C_4H_{11}NO_2$	494	721(4)	4.88(0.1)	333(19)	54
N-(2-Aminoethyl)ethanolamine	$C_4H_{12}N_2O$	512	739(2)	4.53(0.09)	340(17)	55
Amyl orthosilicate	$C_{20}H_{44}O_4Si$		714(14)			57
Aniline	C_6H_7N	457.32	704(7)	5.3(0.1)	291(3)	39, 40, 58–60
Anisole	C_7H_8O	426.9	646.1(0.2)	4.2(0.1)	355(12)	25, 39, 40, 49
Benzene	C_6H_6	353.24	562.0(0.1)	4.90(0.02)	257(11)	19, 27, 28, 34, 61–97
Benzeneacetic acid	$C_8H_8O_2$	538.7	766(8)	3.9(0.3)	372(16)	98
Benzenebutanoic acid	$C_{10}H_{12}O_2$	563	783(8)	3.2(0.2)	493(18)	98
Benzeneethanol	$C_8H_{10}O$	491.9	724(4)	4.0(0.2)	390(15)	99
Benzeneheptanoic acid	$C_{13}H_{18}O_2$	585(27)	798(8)	2.5(0.3)	662(21)	98
Benzenehexanoic acid	$C_{12}H_{16}O_2$	574(27)	794(8)	2.6(0.3)	611(20)	98
Benzenepentanoic acid	$C_{11}H_{14}O_2$	583(1)	790(8)	3.1(0.2)	526(19)	98
Benzenepropanoic acid	$C_9H_{10}O_2$	553.0	776(8)	3.5(0.2)	440(17)	98
Benzo[b]thiophene	C_8H_6S	494	764(2)	4.68(0.04)	359(59)	100
Benzonitrile	C_7H_5N	464.3	691(9)	4.22(0.04)	348(7)	39, 40, 101, 102
Benzophenone	$C_{13}H_{10}O$	578.6	830(2)	3.0(0.1)	568(44)	101
Benzyl alcohol	C_7H_8O	478.46	715(3)	4.3(0.2)	333(13)	103
[1,1'-Bicyclohexyl]-2-one	$C_{12}H_{20}O$	537	787(70)		584(20)	104
1,1'-Bicyclopentyl	$C_{10}H_{18}$	463.61(0.03)	690(2)	3.27(0.03)	497(41)	105

Name	Mol. Form.	T_b/K	T_c/K	P_c/MPa	V_c/cm³ mol⁻¹	Ref.
Biphenyl	$C_{12}H_{10}$	529.3	773(5)	3.43(0.06)	481(69)	39, 106–109
Bis(2-aminoethyl)amine	$C_4H_{13}N_3$	480	710(2)	4.43(0.07)	350(21)	55
1,1-Bis(difluoromethoxy)-1,2,2,2-tetrafluoroethane	$C_4H_2F_8O_2$	319.78(0.08)	450(1)	2.40(0.08)	410(4)	110
Bis(difluoromethyl) ether	$C_2H_2F_4O$	275	420.2(0.1)	4.16(0.06)	223(13)	111
Bis(2-ethylhexyl) phthalate	$C_{24}H_{38}O_4$	657	835(9)	1.1(0.2)	1495(27)	112
Bis(2-hydroxyethyl)methylamine	$C_5H_{13}NO_2$	520	742(4)	4.2(0.4)	404(17)	54
Bis(2,2,2-trifluoroethyl) ether	$C_4H_4F_6O$	336.91	476.31(0.09)	2.78(0.05)	365(5)	113
Bis(trimethylsilyl)methane	$C_7H_{20}Si_2$	406	573.9(0.3)			114
Bromochlorodifluoromethane	$CBrClF_2$	269.5	428(12)	4.31(0.05)	229(16)	115
Bromodifluoromethane	$CHBrF_2$	258.6	412.0(0.3)	5.2(0.1)	173(18)	116
Bromoethane	C_2H_5Br	311.7	503.9(0.4)	6.2(0.1)	214(10)	117, 118
1-Bromo-2-fluorobenzene	C_6H_4BrF	427	669.6(0.6)	4.3(0.6)	342(18)	119
1-Bromo-3-fluorobenzene	C_6H_4BrF	423	652.0(0.4)	4.2(0.6)	337(18)	119
1-Bromo-4-fluorobenzene	C_6H_4BrF	424.7	654.8(0.4)	4.2(0.2)	338(18)	119
1-Bromopropane	C_3H_7Br	344.3	536.9(0.1)	4.33(0.06)	271(6)	120
Bromotrifluoromethane	$CBrF_3$	215.4	340.06(0.05)	3.96(0.01)	199(6)	121
1-Bromo-2-(trifluoromethyl)benzene	$C_7H_4BrF_3$	440.7	656.5(0.4)	3.3(0.8)	415(24)	119
1-Bromo-3-(trifluoromethyl)benzene	$C_7H_4BrF_3$	424.7	627.1(0.4)	3.2(0.7)	413(24)	119
1-Bromo-4-(trifluoromethyl)benzene	$C_7H_4BrF_3$	433	629.8(0.4)	3.2(0.8)	413(24)	119
1,3-Butadiene	C_4H_6	268.74	425(1)	4.35(0.07)	221(23)	122, 123
Butanal	C_4H_8O	348.0	537(2)		258(9)	104, 124
Butane	C_4H_{10}	272.7	425.2(0.1)	3.79(0.01)	257(4)	29, 125–139
1,4-Butanediamine	$C_4H_{12}N_2$	431.7	651(7)	4.5(0.5)	317(14)	140
1,2-Butanediol	$C_4H_{10}O_2$	463.7	680(2)	5.4(0.1)	298(12)	141
1,3-Butanediol	$C_4H_{10}O_2$	480.7	679(17)	4.7(0.1)	302(67)	54, 141
1,4-Butanediol	$C_4H_{10}O_2$	508	724(4)	5.5(0.2)	307(14)	49, 55
Butanenitrile	C_4H_7N	390.8	585.40(0.07)	3.82(0.05)	265(6)	35, 39, 40, 142
1-Butanethiol	$C_4H_{10}S$	371.7	570.1(0.6)	4.01(0.02)	324(12)	143, 144
Butanoic acid	$C_4H_8O_2$	436.90	623(6)	4.0(0.3)	292(10)	145–147
1-Butanol	$C_4H_{10}O$	390.88	563.0(0.4)	4.43(0.07)	280(14)	34, 80, 148–155
2-Butanol	$C_4H_{10}O$	372.66	535(4)	4.2(0.1)	269(4)	148, 149, 153, 156
2-Butanone	C_4H_8O	352.74	537(1)	4.18(0.02)	274(30)	25, 30, 33, 34, 38, 157
1-Butene	C_4H_8	266.89	419.3(0.1)	4.00(0.05)	236(14)	123, 158–161
cis-2-Butene	C_4H_8	276.86	435.7(0.2)	4.23(0.02)	235(4)	86, 123, 158
trans-2-Butene	C_4H_8	274.03	428.6(0.1)	4.03(0.02)	238(4)	86, 123, 158
2-Butoxyethanol	$C_6H_{14}O_2$	441.6	633.9(1)	3.3(0.1)	424(15)	11, 41
1-*tert*-Butoxy-2-ethoxyethane	$C_8H_{18}O_2$	421.2	585(3)	2.5(0.4)	546(14)	162
2-Butoxyethyl acetate	$C_8H_{16}O_3$	465	640(2)	2.7(0.2)	551(21)	144, 163
1-*tert*-Butoxy-2-methoxyethane	$C_7H_{16}O_2$	404(15)	574(1)	2.8(0.7)	480(13)	162
1-Butoxy-2-propanol	$C_7H_{16}O_2$	444.7	625(1)	2.7(0.1)	479(20)	32
Butyl acetate	$C_6H_{12}O_2$	399.3	578(10)	3.16(0.06)	403(6)	162, 164–167
sec-Butyl acetate	$C_6H_{12}O_2$	385	571.1(0.5)	3.01(0.1)	398(14)	164, 165
tert-Butyl acetate	$C_6H_{12}O_2$	368.3	541(4)	3.0(0.1)	399(7)	54
Butyl acrylate	$C_7H_{12}O_2$	418	597.4(0.6)	2.76(0.03)	445(7)	51
Butylamine	$C_4H_{11}N$	350.15	531.9(0.2)	4.20(0.04)	291(13)	168
sec-Butylamine	$C_4H_{11}N$	335.88	514.3(0.2)	4.0(0.2)	284(11)	168
tert-Butylamine	$C_4H_{11}N$	317.19	483.7(0.6)	3.85(0.06)	293(23)	75
Butylbenzene	$C_{10}H_{14}$	456.46	660.5(0.1)	2.89(0.03)	498(18)	79, 86, 88
sec-Butylbenzene	$C_{10}H_{14}$	446.5	652(1)	2.94(0.03)	488(39)	58
tert-Butylbenzene	$C_{10}H_{14}$	442.3	648(1)	3.00(0.03)	474(30)	58, 64
Butyl benzoate	$C_{11}H_{14}O_2$	522(2)	725(14)	2.4(0.3)	594(10)	169
Butyl butanoate	$C_8H_{16}O_2$	439	612(3)	2.4(0.2)	550(9)	162
Butylcyclohexane	$C_{10}H_{20}$	453	653.1(0.4)	2.57(0.07)	547(14)	170, 171
tert-Butylcyclohexane	$C_{10}H_{20}$	444.7	652.0(0.4)	2.82(0.09)	537(15)	170
tert-Butyl ethyl ether	$C_6H_{14}O$	345.8	509(2)	3.0(0.2)	394(4)	172
Butyl methyl ether	$C_5H_{12}O$	343.31	512.7(0.1)	3.37(0.02)	340(2)	25, 173, 174
Butyl propanoate	$C_7H_{14}O_2$	420.0	594(1)	2.8(0.2)	464(10)	162
Butyl vinyl ether	$C_6H_{12}O$	367	540(1)	3.12(0.05)	379(10)	175
γ-Butyrolactone	$C_4H_6O_2$	477	731(1)	5(1)	246(18)	49
Chlorobenzene	C_6H_5Cl	404.87	632.4(0.1)	4.5(0.1)		19, 33, 34

Name	Mol. Form.	T_b/K	T_c/K	P_c/MPa	V_c/cm³ mol⁻¹	Ref.
1-Chlorobutane	C_4H_9Cl	351.6	539.2(0.6)	4.1(0.2)	303(11)	119
2-Chlorobutane	C_4H_9Cl	341.4	518.6(0.6)	3.4(0.2)	307(14)	119
1-Chloro-2,4-difluorobenzene	$C_6H_3ClF_2$	400	609.6(0.4)	4.0(0.7)	333(16)	119
1-Chloro-2,5-difluorobenzene	$C_6H_3ClF_2$	401	612.5(0.4)	4.0(0.7)	333(16)	119
1-Chloro-3,4-difluorobenzene	$C_6H_3ClF_2$	400	609.2(0.4)	4.0(0.6)	333(16)	119
1-Chloro-3,5-difluorobenzene	$C_6H_3ClF_2$	391.7	592.0(0.4)	3.9(0.7)	327(16)	119
1-Chloro-1,1-difluoroethane	$C_2H_3ClF_2$	264.1	410.31(0.05)	4.06(0.03)	230(6)	113, 176–178
1-Chloro-2,2-difluoroethene	C_2HClF_2	254.7	400.5(0.7)	4.54(0.07)	197(6)	179
Chlorodifluoromethane	$CHClF_2$	232.5	369.30(0.05)	4.98(0.01)	165(2)	180–191
2-Chloro-2-(difluoromethoxy)-1,1,1-trifluoroethane	$C_3H_2ClF_5O$	322.4(0.1)	467.8(0.6)	3.05(0.03)	316(24)	192
Chloroethane	C_2H_5Cl	285.5	460.3(0.4)	5.24(0.04)	198(11)	193
Chloroethene	C_2H_3Cl	259.4	425(5)	5.60(0.03)	171(9)	194
1-Chloro-2-fluorobenzene	C_6H_4ClF	410.8	633.8(0.4)	4.3(0.6)	319(21)	119
1-Chloro-3-fluorobenzene	C_6H_4ClF	400.8	615.9(0.4)	4.2(0.6)	324(21)	119
1-Chloro-4-fluorobenzene	C_6H_4ClF	403	620.1(0.4)	4.2(0.4)	322(18)	119
1-Chloroheptane	$C_7H_{15}Cl$	433.6	614(8)	3.1(0.6)	492(14)	119
1-Chlorohexane	$C_6H_{13}Cl$	408.3	599(3)	3.3(0.3)	422(12)	119
Chloromethane	CH_3Cl	249.06	416.24(0.04)	6.72(0.03)	136(2)	195, 196
2-Chloro-2-methylbutane	$C_5H_{11}Cl$	358.8	509.1(0.6)	3.2(0.5)	397(15)	119
3-Chloro-3-methylpentane	$C_6H_{13}Cl$	389	528(3)		414(14)	119
2-Chloro-2-methylpropane	C_4H_9Cl	324.1	497.8(0.1)	3.7(0.4)	308(13)	34
1-Chlorooctane	$C_8H_{17}Cl$	456.7	643(2)	2.5(0.4)	543(13)	119
Chloropentafluoroacetone	C_3ClF_5O	281	410.6(0.1)	2.89(0.01)	277(21)	197
Chloropentafluorobenzene	C_6ClF_5	391.11	570(1)	3.2(0.2)	367(25)	198
Chloropentafluoroethane	C_2ClF_5	234.1	353.0(0.2)	3.141(0.01)	255(4)	199, 200
1-Chloropentane	$C_5H_{11}Cl$	381.6	571.2(0.4)	3.3(0.2)	361(12)	119
1-Chloropropane	C_3H_7Cl	319.7	503.3(0.4)	4.56(0.04)	268(24)	118, 193, 201, 202
2-Chloropropane	C_3H_7Cl	308.9	482.4(0.4)	4.25(0.04)	245(16)	201, 202
1-Chloro-1,2,2,2-tetrafluoroethane	C_2HClF_4	261	395.43(0.06)	3.62(0.01)	244(4)	203, 204
4-Chlorotoluene	C_7H_7Cl	435.0(0.1)	615.9(0.5)	2.33(0.09)	377(16)	88
2-Chloro-1,1,1-trifluoroethane	$C_2H_2ClF_3$	279.3	425.0(0.2)	4.02(0.02)	232(6)	205
Chlorotrifluoroethene	C_2ClF_3	245.4	380.1(0.1)	3.95(0.03)	214(12)	206, 207
2-Chloro-1,1,2-trifluoroethyl difluoromethyl ether	$C_3H_2ClF_5O$	329.92(0.07)	475.0(0.6)	2.98(0.03)	343(25)	192
Chlorotrifluoromethane	$CClF_3$	191.8	301.9(0.2)	3.89(0.01)	180.3(1)	191, 208–216
m-Cresol	C_7H_8O	475.42	705.8(0.4)	4.4(0.2)	337(12)	39, 40, 217, 218
o-Cresol	C_7H_8O	464.19	697.6(0.2)	4.2(0.2)	336(12)	217, 218
p-Cresol	C_7H_8O	475.13	704.6(0.3)	4.1(0.2)	349(13)	217, 218
Cyanogen	C_2N_2	252.1	397(3)	6.2(0.4)	149(8)	219
Cycloheptane	C_7H_{14}	391.6	604.2(0.1)	3.85(0.04)	361(12)	34, 220, 221
Cyclohexane	C_6H_{12}	353.88	553.4(0.3)	4.07(0.01)	307(12)	19, 34, 78, 82, 84, 88, 90, 144, 163, 170, 171, 220, 222–231
Cyclohexanol	$C_6H_{12}O$	433.99	647.1(0.3)	4.3(0.1)	334(33)	49, 232, 233
Cyclohexanone	$C_6H_{10}O$	428.58	665(1)	4.61(0.09)	354(12)	162, 166
Cyclohexene	C_6H_{10}	356.13	560.45(0.05)	4.43(0.08)	290(20)	86, 201, 202
Cyclohexylamine	$C_6H_{13}N$	407	626.8(0.9)	3.9(0.6)	349(15)	170
Cyclooctane	C_8H_{16}	422	647.2(0.4)	3.55(0.06)	417(28)	34, 170, 220, 221
Cyclopentane	C_5H_{10}	322.5	511.7(0.2)	4.51(0.07)	264(10)	34, 78, 90, 234–236
Cyclopentanol	$C_5H_{10}O$	413.57	619(1)	4.9(0.1)	288(19)	233
Cyclopentanone	C_5H_8O	403.72	624(2)	4.59(0.05)	276(17)	233
Cyclopentene	C_5H_8	317.4	506.1(0.2)	4.78(0.05)	252(16)	90, 144, 163, 201, 202
Cyclopropane	C_3H_6	240.34	398.2(0.4)	5.58(0.02)	164(9)	237, 238
Decafluorobiphenyl	$C_{12}F_{10}$	477(6)	640(4)	2.3(0.3)	641(40)	239
cis-Decahydronaphthalene	$C_{10}H_{18}$	469.0	702(1)	3.2(0.3)	492(19)	241
trans-Decahydronaphthalene	$C_{10}H_{18}$	460.5	687(1)	3.1(0.1)	499(19)	241
Decamethylcyclopentasiloxane	$C_{10}H_{30}O_5Si_5$	489(2)	617.4(0.3)	1.04(0.02)	1201(8)	34, 114
Decanal	$C_{10}H_{20}O$	481.7	674(1)	2.6(0.3)	601(14)	163, 242
Decane	$C_{10}H_{22}$	447.30	618.1(0.9)	2.10(0.03)	621(35)	16, 33, 65, 78, 86, 131, 243–252
1,10-Decanediamine	$C_{10}H_{24}N_2$	535(12)	736(8)	2.4(0.3)	654(28)	140

Name	Mol. Form.	T_b/K	T_c/K	P_c/MPa	V_c/cm³ mol⁻¹	Ref.
Decanedioic acid	$C_{10}H_{18}O_4$	627.0	845(13)	2.5(0.1)	724(21)	253
Decanoic acid	$C_{10}H_{20}O_2$	541.9	724(5)		638(24)	145, 146
1-Decanol	$C_{10}H_{22}O$	504.3	690(10)	2.3(0.1)	624(87)	149, 152, 254, 255
2-Decanol	$C_{10}H_{22}O$	484	668.5(0.3)	2.3(0.5)	646(13)	255
3-Decanol	$C_{10}H_{22}O$	486	666.1(0.3)	2.3(0.3)	643(13)	255
4-Decanol	$C_{10}H_{22}O$	483.7	663.7(0.3)	2.3(0.1)	643(13)	255
5-Decanol	$C_{10}H_{22}O$	474	663.2(0.4)	2.3(0.4)	646(13)	255
2-Decanone	$C_{10}H_{20}O$	483	671.8(0.5)	2.2(0.3)	625(25)	256
3-Decanone	$C_{10}H_{20}O$	476	668(1)	2.2(0.2)	628(15)	256
4-Decanone	$C_{10}H_{20}O$	479.7	662.9(0.5)	2.2(0.2)	636(18)	256
5-Decanone	$C_{10}H_{20}O$	477	661.0(0.4)	2.2(0.2)	628(25)	256
1-Decene	$C_{10}H_{20}$	443.7	616.0(0.3)	2.16(0.05)	594(3)	257
Decylbenzene	$C_{16}H_{26}$	566	752(8)	1.72(0.1)	879(29)	258
Dibenzofuran	$C_{12}H_8O$	560	824(2)	3.37(0.03)	494(32)	259
Dibenzothiophene	$C_{12}H_8S$	605.7	897(2)	3.9(0.2)		260
1,2-Dibromo-1-chloro-1,2,2-trifluoroethane	$C_2Br_2ClF_3$	366	560.6(0.2)	3.61(0.02)	368(4)	261
1,4-Dibromooctafluorobutane	$C_4Br_2F_8$	370	532(2)	2.4(0.3)	452(29)	198
Dibutylamine	$C_8H_{19}N$	432.8	607.5(0.2)	3.11(0.03)	532(21)	168
1,4-Di-*tert*-butylbenzene	$C_{14}H_{22}$	510.45(0.02)	708(2)	2.23(0.01)	732(70)	232
Dibutyl ether	$C_8H_{18}O$	413.43	584.1(0.2)	2.4(0.2)	521(12)	262
Dibutyl phthalate	$C_{16}H_{22}O_4$	613	797(9)	1.6(0.3)	954(18)	112
m-Dichlorobenzene	$C_6H_4Cl_2$	446	685.7(0.4)	4.2(0.2)	366(22)	119
1,4-Dichlorobenzene	$C_6H_4Cl_2$	447.04(0.05)	669(5)	3.54(0.07)	364(22)	263
Dichlorodiethylsilane	$C_4H_{10}Cl_2Si$	403.6(0.4)	595.7(0.6)	3.06(0.03)	455(4)	264
Dichlorodifluoromethane	CCl_2F_2	243.4	384.9(0.2)	4.12(0.01)	218(36)	191, 216, 265
Dichlorodimethylsilane	$C_2H_6Cl_2Si$	343.5	520.3(0.6)	3.49(0.03)	350(5)	266
1,1-Dichloroethane	$C_2H_4Cl_2$	330.5	523.4(0.1)	5.1(0.5)	248(12)	267
1,2-Dichloroethane	$C_2H_4Cl_2$	356.7	561.5(0.4)	5.4(0.1)	225(8)	33, 34, 268, 269
cis-1,2-Dichloroethene	$C_2H_2Cl_2$	333.3	535.8(0.4)	5.4(0.3)	220(15)	34
trans-1,2-Dichloroethene	$C_2H_2Cl_2$	321.9	515.5(0.2)	5.3(0.2)	216(14)	33, 34
1,1-Dichloro-1-fluoroethane	$C_2H_3Cl_2F$	305.2	477.3(0.1)	4.20(0.02)	253.7(0.6)	178, 270
Dichlorofluoromethane	$CHCl_2F$	282.1	451.6(0.4)	5.20(0.01)	196(1)	271
1,2-Dichlorohexafluoropropane	$C_3Cl_2F_6$	306.9(0.2)	451.8(0.1)	2.63(0.07)	365(48)	272
Dichloromethane	CH_2Cl_2	313	508.0(0.2)	6.35(0.05)	177(13)	273
1,2-Dichloropropane	$C_3H_6Cl_2$	369.6	578(2)	4.63(0.06)	292(6)	119, 232
1,3-Dichloropropane	$C_3H_6Cl_2$	394.1	615(3)	4.7(0.5)	299(16)	119
1,1-Dichloro-1,2,2,2-tetrafluoroethane	$C_2Cl_2F_4$	276.6	418.6(0.8)	3.31(0.03)	294(8)	179
1,2-Dichloro-1,1,2,2-tetrafluoroethane	$C_2Cl_2F_4$	276.7	418.74(0.06)	3.25(0.02)	295(3)	191, 274
1,2-Dichloro-1,1,2-trifluoroethane	$C_2HCl_2F_3$	302.7	461.6(0.1)	3.77(0.08)	283.2(0.5)	178
2,2-Dichloro-1,1,1-trifluoroethane	$C_2HCl_2F_3$	300.97	456.8(0.2)	3.67(0.01)	278(2)	182, 275–280
Didecyl phthalate	$C_{28}H_{46}O_4$	736(4)	870(10)	0.94(0.05)	1807(27)	112
1,1-Diethoxyethane	$C_6H_{14}O_2$	375.40	539.7(0.4)	3.22(0.08)	426(12)	166
1,2-Diethoxyethane	$C_6H_{14}O_2$	394.4	542(3)	2.14(0.02)	432(11)	162
Diethoxymethane	$C_5H_{12}O_2$	361	532(1)	3.4(0.5)	370(10)	162
Diethylamine	$C_4H_{11}N$	328.7	499.5(0.4)	3.75(0.02)	304(32)	38, 117, 133, 281, 282
p-Diethylbenzene	$C_{10}H_{14}$	456.9	657.90(0.05)	2.80(0.08)	494(12)	79, 86
Diethylene glycol	$C_4H_{10}O_3$	519.0	753(4)	4.8(0.2)	325(19)	283
Diethylene glycol diethyl ether	$C_8H_{18}O_3$	461	612(10)	2.4(0.7)	587(18)	162
Diethylene glycol dimethyl ether	$C_6H_{14}O_3$	435	617(4)	3.0(0.6)	450(16)	162
Diethylene glycol monobutyl ether	$C_8H_{18}O_3$	504	692(3)	2.8(0.6)	546(26)	41
Diethylene glycol monobutyl ether acetate	$C_{10}H_{20}O_4$	518	694(2)	2.15(0.05)	627(20)	55
Diethylene glycol monoethyl ether	$C_6H_{14}O_3$	469	670(4)	3.2(0.1)	427(23)	49
Diethylene glycol monoethyl ether acetate	$C_8H_{16}O_4$	491.7	670(12)	2.50(0.06)	524(18)	49, 55
Diethylene glycol monomethyl ether	$C_5H_{12}O_3$	466	672(2)	3.7(0.2)	378(25)	49
Diethylene glycol monopropyl ether	$C_7H_{16}O_3$	486	680(2)	3.05(0.07)	495(18)	41, 104
Diethyl ether	$C_4H_{10}O$	307.7	466.8(0.3)	3.64(0.01)	280(5)	19, 77, 95, 155, 196, 284–301
Diethyl oxalate	$C_6H_{10}O_4$	458.9	618(2)	2.14(0.02)	464(37)	101
Diethyl phthalate	$C_{12}H_{14}O_4$	568	776(9)	2.2(0.2)	687(15)	112
Diethyl succinate	$C_8H_{14}O_4$	490.9	663(30)	2.26(0.02)	567(44)	101, 302
Diethyl sulfide	$C_4H_{10}S$	365.3	557.5(1)	4.0(0.1)	322(8)	32, 303, 304

Name	Mol. Form.	T_b/K	T_c/K	P_c/MPa	V_c/cm³ mol⁻¹	Ref.
m-Difluorobenzene	$C_6H_4F_2$	355.8	548.4(0.4)	4.20(0.07)	289(21)	119
o-Difluorobenzene	$C_6H_4F_2$	367	566.0(0.4)	4.28(0.07)	290(21)	119
p-Difluorobenzene	$C_6H_4F_2$	362	556.9(0.4)	4.28(0.07)	297(22)	119
1,1-Difluoroethane	$C_2H_4F_2$	249.10	386.4(0.1)	4.52(0.01)	178(2)	178, 180, 185, 305–307
1,1-Difluoroethene	$C_2H_2F_2$	187.5	302.9(0.6)	4.48(0.05)	155(4)	179, 215
2,2-Difluoroethylbis(trifluoromethyl)amine	$C_4H_3F_8N$	324.50	460.20(0.09)	2.64(0.01)	375(1)	308
Difluoromethane	CH_2F_2	221.6	351.28(0.03)	5.79(0.01)	121(4)	306, 309–316
3-Difluoromethoxy-1,1,1,2,2-pentafluoropropane	$C_4H_3F_7O$	319.09	455.1(0.1)	2.77(0.02)	363(1)	113
2-(Difluoromethoxy)-1,1,1-trifluoroethane	$C_3H_3F_5O$	302.15	444.9(0.3)	3.43(0.01)	291(19)	113
2,4-Difluorotoluene	$C_7H_6F_2$	390	581.4(0.4)	3.7(0.4)	340(21)	119
2,5-Difluorotoluene	$C_7H_6F_2$	391	587.8(0.4)	3.8(0.5)	341(21)	119
2,6-Difluorotoluene	$C_7H_6F_2$	385	581.8(0.4)	3.7(0.4)	341(21)	119
3,4-Difluorotoluene	$C_7H_6F_2$	385	598.5(0.5)	3.8(0.6)	342(22)	119
Diheptyl phthalate	$C_{22}H_{34}O_4$	633	830(9)	1.24(0.08)	1153(22)	112
Dihexyl phthalate	$C_{20}H_{30}O_4$	652(5)	817(9)	1.3(0.1)	1061(22)	112
3,4-Dihydro-2*H*-pyran	C_5H_8O	359	561(2)	4.63(0.08)	268(34)	75
Diisobutylamine	$C_8H_{19}N$	412.8	584.4(0.2)	3.20(0.06)	518(22)	168
Diisopropylamine	$C_6H_{15}N$	357.1	523.1(0.2)	3.02(0.02)	407(18)	168
1,4-Diisopropylbenzene	$C_{12}H_{18}$	483.5	675(1)	2.30(0.04)	610(65)	317
Diisopropyl ether	$C_6H_{14}O$	341.6	500.2(0.7)	2.85(0.04)	386(5)	25, 290, 318, 319
1,2-Dimethoxyethane	$C_4H_{10}O_2$	357.7	539(4)	3.91(0.05)	305(9)	162, 166, 175, 290
Dimethoxymethane	$C_3H_8O_2$	315	488(11)	4.0(0.2)	259(10)	75, 162, 320
1,2-Dimethoxypropane	$C_5H_{12}O_2$	369	543(1)	3.4(0.6)	356(12)	162
2,2-Dimethoxypropane	$C_5H_{12}O_2$	356	510(3)		360(13)	162
Dimethyl adipate	$C_8H_{14}O_4$	504(3)	692(14)	2.5(0.5)	561(13)	321
Dimethylamine	C_2H_7N	280.03	437.5(0.4)	5.34(0.05)	188(13)	193, 322, 323
N,*N*-Dimethylaniline	$C_8H_{11}N$	467.30	687.7(0.6)	3.63(0.09)	407(15)	39, 40
2,2-Dimethylbutane	C_6H_{14}	322.88	489.1(0.5)	3.10(0.01)	364(1)	86, 324–328
2,3-Dimethylbutane	C_6H_{14}	331.08	500.2(0.3)	3.13(0.01)	358(1)	19, 86, 252, 324–330
3,3-Dimethyl-2-butanone	$C_6H_{12}O$	379.3	570.9(0.3)	3.67(0.03)	383(6)	164, 165
2,3-Dimethyl-1-butene	C_6H_{12}	328.8	497.7(0.9)	3.31(0.01)	346(9)	170
3,3-Dimethyl-1-butene	C_6H_{12}	314.4	477.4(0.9)	3.18(0.02)	348(10)	170
2,3-Dimethyl-2-butene	C_6H_{12}	346.5	521.0(0.9)	3.4(0.1)	344(7)	170
Dimethyl carbonate	$C_3H_6O_3$	363.7	557(1)	4.8(0.2)		331, 332
cis-1,3-Dimethylcyclohexane	C_8H_{16}	393.3	587.7(0.5)	2.88(0.01)	429(10)	64
cis-1,4-Dimethylcyclohexane	C_8H_{16}	397.43(0.02)	603.2(0.3)	3.44(0.02)	434(7)	164, 165
trans-1,4-Dimethylcyclohexane	C_8H_{16}	392.6	588(2)	3.04(0.01)	439(18)	74
Dimethyl disulfide	$C_2H_6S_2$	382.89	608(4)	5.1(0.1)	266(8)	54
Dimethyl ether	C_2H_6O	248.4	400.1(0.8)	5.31(0.03)	171(3)	125, 174, 186, 333–343
N,*N*-Dimethylformamide	C_3H_7NO	426	649.6(0.8)	4.4(0.1)	262(9)	11, 344
Dimethyl glutarate	$C_7H_{12}O_4$	487(3)	682(14)	2.8(0.4)	488(14)	321
2,2-Dimethylheptane	C_9H_{20}	405.9	576.7(0.5)	2.35(0.07)	546(12)	345
2,2-Dimethylhexane	C_8H_{18}	380.01	549.9(0.4)	2.53(0.03)	481(10)	346
2,3-Dimethylhexane	C_8H_{18}	388.77	563.5(0.4)	2.63(0.02)	466(16)	346
2,4-Dimethylhexane	C_8H_{18}	382.7	553(3)	2.55(0.02)	480(39)	330, 346
2,5-Dimethylhexane	C_8H_{18}	382.27	550.0(0.3)	2.49(0.02)	485(20)	19, 346
3,3-Dimethylhexane	C_8H_{18}	385.12	562.0(0.4)	2.65(0.02)	450(17)	346
3,4-Dimethylhexane	C_8H_{18}	390.88	568.8(0.4)	2.69(0.02)	467(21)	346
Dimethyl malonate	$C_5H_8O_4$	454.6	647(1)	3.5(0.1)	368(13)	317, 321
2,7-Dimethylnaphthalene	$C_{12}H_{12}$	538	775(2)	3.02(0.03)	515(45)	347
Dimethyl octanedioate	$C_{10}H_{18}O_4$	533(4)	723(14)	2.3(0.3)	672(14)	321
Dimethyl oxalate	$C_4H_6O_4$	436.6(0.6)	632(14)	4.0(0.3)	315(12)	321, 348
2,2-Dimethyloxirane	C_4H_8O	325	500(4)	4.4(0.1)	327(31)	54
2,2-Dimethylpentane	C_7H_{16}	352.4	520.6(0.6)	2.77(0.04)	409(12)	346, 349
2,3-Dimethylpentane	C_7H_{16}	362.93	537.5(0.6)	2.92(0.08)	394(23)	346, 349, 350
2,4-Dimethylpentane	C_7H_{16}	353.64	520.0(0.7)	2.74(0.06)	415(19)	346, 349
3,3-Dimethylpentane	C_7H_{16}	359.21	536.4(0.4)	2.94(0.02)	413(24)	346
2,3-Dimethyl-1-pentene	C_7H_{14}	357.5	534(4)	2.9(0.8)	400(11)	170
4,4-Dimethyl-1-pentene	C_7H_{14}	345.7	516(4)	2.91(0.01)	406(11)	170
Dimethyl phthalate	$C_{10}H_{10}O_4$	556.9	772(9)	2.76(0.08)	557(17)	112
Dimethyl pimelate	$C_9H_{16}O_4$	518(2)	711(14)	2.4(0.1)	608(14)	321

Name	Mol. Form.	T_b/K	T_c/K	P_c/MPa	V_c/cm³ mol⁻¹	Ref.
2,3-Dimethylpyridine	C_7H_9N	434.27	655.5(0.3)	4.03(0.01)	337(28)	86, 351
2,4-Dimethylpyridine	C_7H_9N	431.53	647.1(0.9)	3.83(0.02)	363(34)	90, 352
2,5-Dimethylpyridine	C_7H_9N	430.13	644.2(0.3)	4.11(0.04)	371(45)	86
2,6-Dimethylpyridine	C_7H_9N	417.16	623.8(0.2)	3.80(0.04)	353(43)	90
3,4-Dimethylpyridine	C_7H_9N	452.25	683.8(0.4)	4.06(0.01)	353(13)	86, 351, 352
3,5-Dimethylpyridine	C_7H_9N	444.99	667.3(0.3)	3.84(0.02)	369(12)	86, 351
2,6-Dimethylquinoline	$C_{11}H_{11}N$	541.22(0.02)	786(2)	3.27(0.02)	505(24)	529
Dimethyl sebacate	$C_{12}H_{22}O_4$	562(3)	742(14)	2.1(0.2)	695(14)	321
Dimethyl succinate	$C_6H_{10}O_4$	469.2(0.7)	662(14)	3.5(0.2)	426(16)	321
Dimethyl sulfide	C_2H_6S	310.48	503.0(0.3)	5.40(0.06)	201(9)	52, 117, 304
Dimethyl sulfoxide	C_2H_6OS	465.1(0.7)	707(1)	4.6(0.7)	228(7)	23
1,3-Dimethyl-1,1,3,3-tetraphenyldisiloxane	$C_{26}H_{26}OSi_2$	701(2)	893(9)	1.38(0.1)	1300(89)	353
1,3-Dimethyltricyclo[3.3.1.1^{3,7}]decane	$C_{12}H_{20}$	476.53	708(2)	2.86(0.01)	595(115)	141
Dinonyl phthalate	$C_{26}H_{42}O_4$	686	858(9)	1.0(0.3)	1652(26)	112
Dioctyl phthalate	$C_{24}H_{38}O_4$	688(4)	840(9)	1.1(0.1)	1510(22)	112
1,4-Dioxane	$C_4H_8O_2$	374.7	587.3(0.1)	5.2(0.2)	251(5)	34, 290, 354
Dipentyl phthalate	$C_{18}H_{26}O_4$	614(40)	811(9)		957(21)	112
Diphenyl ether	$C_{12}H_{10}O$	531.2	766.9(0.8)	3.10(0.04)	526(23)	25, 355
Diphenylmethane	$C_{13}H_{12}$	538.2	776(9)	3.02(0.07)		39, 302, 356, 357
1,3-Diphenyltetramethyldisiloxane	$C_{16}H_{22}OSi_2$		750(8)			353
Dipropylamine	$C_6H_{15}N$	382.5	555.8(0.1)	3.6(0.1)	414(15)	168
Dipropylene glycol	$C_6H_{14}O_3$	503.7	705(4)	3.4(0.1)	444(18)	54
Dipropyl ether	$C_6H_{14}O$	363.23	531(2)	2.92(0.05)	402(8)	25, 52
Dipropyl phthalate	$C_{14}H_{18}O_4$	577.7	784(9)	1.9(0.1)	816(20)	112
Diundecyl phthalate	$C_{30}H_{50}O_4$	711(25)	886(10)	0.89(0.1)	1590(25)	112
Docosane	$C_{22}H_{46}$	641.8	786(6)	1.0(0.1)	1434(50)	62, 358, 359
Docosanoic acid	$C_{22}H_{44}O_2$	693(3)	837(8)	1.11(0.08)	1485(29)	360
1-Docosanol	$C_{22}H_{46}O$	680(12)	827(8)		1243(20)	361
1,2,2,3,3,4,4,5,5,6,6,7-Dodecafluoro-1-heptanol	$C_7H_4F_{12}O$	444.7(0.7)	589(5)	2.0(0.2)	620(34)	362
Dodecane	$C_{12}H_{26}$	489.47	658.8(0.9)	1.80(0.09)	747(18)	34, 86, 118, 243, 244, 363, 364
1,12-Dodecanediamine	$C_{12}H_{28}N_2$	572(13)	767(8)	2.0(0.3)	765(37)	140
Dodecanedioic acid	$C_{12}H_{22}O_4$	621(10)	859(13)	2.1(0.2)	730(19)	253
1-Dodecanethiol	$C_{12}H_{26}S$	549.9	734(4)	1.81(0.1)	726(25)	99
Dodecanoic acid	$C_{12}H_{24}O_2$	572(1)	743(7)	1.9(0.2)	787(19)	360
1-Dodecanol	$C_{12}H_{26}O$	533	719.4(0.6)	2.02(0.05)	805(15)	149
2-Dodecanone	$C_{12}H_{24}O$	519.7	702(4)	1.9(0.4)	742(20)	256
3-Dodecanone	$C_{12}H_{24}O$	523(3)	701(2)	1.9(0.2)	680(19)	256
4-Dodecanone	$C_{12}H_{24}O$	528(5)	697(2)	1.9(0.4)	672(19)	256
5-Dodecanone	$C_{12}H_{24}O$	521(4)	695(5)	1.9(0.4)	678(19)	256
6-Dodecanone	$C_{12}H_{24}O$	522(4)	694(2)	1.9(0.3)	677(19)	256
1-Dodecene	$C_{12}H_{24}$	487.0	657.6(0.6)	1.88(0.01)	710(15)	257
Eicosane	$C_{20}H_{42}$	616	768(6)	1.08(0.05)	1325(48)	358, 359
Eicosanoic acid	$C_{20}H_{40}O_2$	601	820(8)	1.2(0.1)	1346(27)	360
1-Eicosanol	$C_{20}H_{42}O$	629	808(8)	1.1(0.2)	1130(18)	361
1-Eicosene	$C_{20}H_{40}$	620(15)	772(15)	1.1(0.3)	1213(36)	365
Ethane	C_2H_6	184.6	305.36(0.04)	4.88(0.01)	146(3)	44, 48, 92, 131, 132, 134, 214, 215, 229, 246, 366–393
1,2-Ethanediamine	$C_2H_8N_2$	390	613.1(0.3)	6.71(0.04)	204(13)	49
1,2-Ethanediol	$C_2H_6O_2$	470.5	719(5)	8.1(0.4)	180(11)	11, 32, 41, 394
1,1-Ethanediol, diacetate	$C_6H_{10}O_4$	442	618(4)	2.9(0.1)	457(13)	54
Ethanethiol	C_2H_6S	308.2	498.7(0.3)	5.53(0.08)	208(10)	304
Ethanol	C_2H_6O	351.44	515(1)	6.25(0.04)	169(4)	19, 25, 34, 80, 85, 128, 148–150, 152, 222, 395–409
Ethoxybenzene	$C_8H_{10}O$	442.96	647(2)	3.45(0.05)	407(16)	39, 40
2-Ethoxyethyl acetate	$C_6H_{12}O_3$	429.6	609(2)	3.07(0.03)	443(38)	141, 144, 163
2-Ethoxy-2-methylbutane	$C_7H_{16}O$	374.6	546(2)	2.83(0.09)	448(29)	410
1-Ethoxy-1,1,2,2,3,3,4,4,4-nonafluorobutane	$C_6H_5F_9O$	350.04	482.0(0.1)	1.98(0.01)	518(2)	113
Ethyl acetate	$C_4H_8O_2$	350.26	523.27(0.07)	3.88(0.02)	288(19)	19, 23, 400, 411–413
Ethylamine	C_2H_7N	289.7	456.5(0.9)	5.6(0.1)	183(22)	193, 414

Name	Mol. Form.	T_b/K	T_c/K	P_c/MPa	V_c/cm³ mol⁻¹	Ref.
Ethylbenzene	C_8H_{10}	409.34	617.1(0.1)	3.61(0.01)	365(61)	68, 78, 79, 86, 88, 172, 345
Ethyl benzoate	$C_9H_{10}O_2$	485.63(0.03)	700(14)	3.01(0.05)	470(12)	169
Ethyl butanoate	$C_6H_{12}O_2$	394.5	566.1(0.1)	3.2(0.3)	421(30)	415, 416
Ethyl *trans*-2-butenoate	$C_6H_{10}O_2$	411	599(10)		382(7)	21
Ethylcyclohexane	C_8H_{16}	405.1	606.9(0.4)	3.27(0.04)	431(12)	170, 171
Ethylcyclopentane	C_7H_{14}	376.7	569.48(0.05)	3.40(0.07)	377(3)	236
Ethyl 2,2-dimethylpropanoate	$C_7H_{14}O_2$	391.6	566(2)	2.88(0.02)	461(8)	417
Ethylene	C_2H_4	169.38	282.35(0.03)	5.06(0.01)	130.9(0.2)	61, 246, 371, 418–436
Ethyl 3-ethoxypropanoate	$C_7H_{14}O_3$	439	621(3)	2.7(0.1)	478(19)	11, 41
Ethyl formate	$C_3H_6O_2$	327.6	508.5(0.5)	4.78(0.03)		413, 415
Ethyl heptanoate	$C_9H_{18}O_2$	460	634(1)	2.3(0.2)	587(8)	162
3-Ethylhexane	C_8H_{18}	391.8	565.5(0.4)	2.61(0.02)	450(15)	346
Ethyl hexanoate	$C_8H_{16}O_2$	440	615(1)	2.6(0.2)	528(8)	162
2-Ethylhexanoic acid	$C_8H_{16}O_2$	501	674(1)	2.75(0.06)	543(8)	144, 163
2-Ethyl-1-hexanol	$C_8H_{18}O$	457.8	640.2(0.3)	3.0(0.2)	508(29)	437
2-Ethylhexyl acetate	$C_{10}H_{20}O_2$	472	642(2)	2.02(0.01)	644(10)	410
Ethyl 3-methylbutanoate	$C_7H_{14}O_2$	408.2	584(6)		463(9)	147, 162
Ethyl methyl ether	C_3H_8O	280.6	437.8(0.2)	4.39(0.06)	219(5)	174, 304, 438
3-Ethyl-2-methylpentane	C_8H_{18}	388.81	567.1(0.4)	2.70(0.02)	442(22)	346
3-Ethyl-3-methylpentane	C_8H_{18}	391.42	576.5(0.4)	2.77(0.01)	463(13)	346
Ethyl 2-methylpropanoate	$C_6H_{12}O_2$	383.3	554(4)	3.1(0.3)	421(76)	415
Ethyl methyl sulfide	C_3H_8S	339.9	533(10)	4.62(0.03)	260(6)	303
Ethyl nonanoate	$C_{11}H_{22}O_2$	500.2	664(1)	2.0(0.4)	715(8)	162
Ethyl octanoate	$C_{10}H_{20}O_2$	481.7	652(12)		657(8)	147, 162
3-Ethylpentane	C_7H_{16}	366.7	540.7(0.4)	2.90(0.03)	412(14)	346, 349
Ethyl pentanoate	$C_7H_{14}O_2$	419.3	593(1)	2.8(0.4)	466(10)	162
2-Ethylphenol	$C_8H_{10}O$	477.7	703(1)	3.7(0.3)	388(15)	218
3-Ethylphenol	$C_8H_{10}O$	491.6	716(1)	3.8(0.3)	393(15)	218
4-Ethylphenol	$C_8H_{10}O$	491.1	716(1)	3.1(0.6)	395(15)	218
Ethyl propanoate	$C_5H_{10}O_2$	372.3	547(1)	3.37(0.05)	343(23)	19, 64, 162, 413, 415
Ethyl propyl ether	$C_5H_{12}O$	336.36	500.2(0.4)	3.37(0.01)	343(44)	25, 304
S-Ethyl thioacetate	C_4H_8OS	389.6	590.5(0.2)	4.1(0.1)	320(10)	49
4-Ethyltoluene	C_9H_{12}	435	640.2(0.5)	3.23(0.04)	446(12)	439, 440
Ethyl vinyl ether	C_4H_8O	308.7	475(2)	4.06(0.04)	262(8)	290
Fluorobenzene	C_6H_5F	357.88	560.10(0.07)	4.55(0.01)	272(9)	86, 441
Fluoroethane	C_2H_5F	235.5	375.2(0.2)	5.02(0.01)	164(3)	206, 442, 443
Fluoromethane	CH_3F	194.8	317.42(0.01)	5.88(0.01)	112.4(0.1)	444, 445
2-Fluorotoluene	C_7H_7F	388	591.2(0.4)	3.9(0.3)	323(47)	119
3-Fluorotoluene	C_7H_7F	388	591.8(0.4)	3.9(0.3)	332(17)	119
4-Fluorotoluene	C_7H_7F	389.8	592.1(0.8)	3.85(0.01)	332(17)	119
Formic acid	CH_2O_2	374	588(10)		115.9(0.1)	146
Furan	C_4H_4O	304.7	490.2(0.2)	5.43(0.08)	218(3)	241, 290
Glycerol	$C_3H_8O_3$	563	850(9)	7.6(0.8)	251(15)	394
Heneicosane	$C_{21}H_{44}$	629.7	778(8)	1.0(0.1)	1366(48)	359
Heptadecane	$C_{17}H_{36}$	575.2	736(1)	1.33(0.07)	1081(28)	359
Heptadecanoic acid	$C_{17}H_{34}O_2$	635(4)	792(8)	1.4(0.1)	1130(24)	360
1-Heptadecanol	$C_{17}H_{36}O$	597	780(8)	1.4(0.1)	1097(18)	361
1-Heptadecene	$C_{17}H_{34}$	574(1)	734(7)	1.34(0.08)	1053(35)	365
2,2,3,3,5,5,6-Heptafluoro-1,4-dioxane	$C_4HF_7O_2$	312.5(0.1)	453(1)	2.86(0.06)	359(4)	110
1,1,1,2,2,3,3-Heptafluoropentan-4-one	$C_5H_3F_7O$	337.40	476.55(0.08)	2.57(0.01)	394.2(0.7)	308
1,1,1,2,3,3,3-Heptafluoropropane	C_3HF_7	257.65	375.0(0.1)	2.93(0.01)	299(7)	446–449
1,1,1,2,4,4,4-Heptafluoro-2-trifluoromethoxybutane	$C_5H_2F_{10}O$	322.73(0.09)	447(1)	2.15(0.06)	465(5)	110
1,1,1,2,2,3,3-Heptafluoro-3-(trifluoromethoxy) propane	$C_4F_{10}O$	280.0(0.4)	391.7(0.7)	1.89(0.05)	431(31)	450
2,2,4,4,6,8,8-Heptamethylnonane	$C_{16}H_{34}$	519.5	692(4)	1.53(0.01)	957(27)	451
1,1,1,3,5,5,5-Heptamethyltrisiloxane	$C_7H_{22}O_2Si_3$	415.8(0.2)	553.4(0.6)	1.48(0.02)	828(3)	452
Heptanal	$C_7H_{14}O$	426.0	616.8(0.4)	3.2(0.2)	434(7)	242
2-Heptanamine	$C_7H_{17}N$	415	598(2)	2.9(0.3)	455(18)	170

Name	Mol. Form.	T_b/K	T_c/K	P_c/MPa	V_c/cm³ mol⁻¹	Ref.
Heptane	C_7H_{16}	371.6	540.1(0.2)	2.74(0.01)	428(15)	34, 36, 62, 75, 78, 86, 131, 133–135, 157, 243, 244, 254, 286, 324, 346, 359, 394, 432, 453– 463
Heptanedioic acid	$C_7H_{12}O_4$	615.1	842(13)	3.3(0.2)	463(15)	253
Heptanoic acid	$C_7H_{14}O_2$	495.4	678(2)	3.0(0.3)	476(5)	145, 146, 464
1-Heptanol	$C_7H_{16}O$	449.60	632.4(0.6)	3.1(0.2)	430(9)	149, 254, 465
2-Heptanol	$C_7H_{16}O$	432	608.4(0.6)	3.0(0.1)	442(2)	149, 465
3-Heptanol	$C_7H_{16}O$	430(1)	605.4(0.3)	3.1(0.4)	451(3)	465
4-Heptanol	$C_7H_{16}O$	429	602.6(0.3)	3.1(0.6)	455(4)	465
2-Heptanone	$C_7H_{14}O$	424.20	611.4(0.2)	2.98(0.04)	436(4)	25, 172, 466
3-Heptanone	$C_7H_{14}O$	420	606.6(0.2)	3.0(0.1)	433(5)	466
4-Heptanone	$C_7H_{14}O$	417	602.0(0.2)	3.0(0.3)	434(5)	466
1-Heptene	C_7H_{14}	366.79	537.3(0.3)	2.85(0.02)	409(2)	51, 52, 86, 124, 257, 467
cis-2-Heptene	C_7H_{14}	371.6	548.5(0.6)	3.0(0.3)	410(7)	170
trans-2-Heptene	C_7H_{14}	371	542.8(0.4)	3.0(0.2)	410(7)	170
trans-3-Heptene	C_7H_{11}	368.9	538.6(0.7)	3.0(0.2)	411(7)	170
Heptylbenzene	$C_{13}H_{20}$	513	708(7)	2.1(0.2)	680(16)	258
Heptyl orthosilicate	$C_{28}H_{60}O_4Si$		778(16)			57
Hexacosane	$C_{26}H_{54}$	688(11)	816(8)	0.8(0.2)	1740(59)	358
Hexadecane	$C_{16}H_{34}$	560.01	722.2(0.8)	1.4(0.2)	1009(53)	34, 244, 245
Hexadecanoic acid	$C_{16}H_{32}O_2$	624.7	785(8)	1.5(0.2)	1059(23)	360
1-Hexadecanol	$C_{16}H_{34}O$	585	770(8)	1.47(0.1)	1019(17)	361
1-Hexadecene	$C_{16}H_{32}$	557.9(0.7)	718(7)	1.4(0.2)	986(33)	365
Hexaethyldisiloxane	$C_{12}H_{30}OSi_2$	525(6)	692.9(0.1)		955(2)	34
Hexafluoroacetylacetone	$C_5H_2F_6O_2$	327.30	485.1(0.5)	2.9(0.2)	313(49)	468
Hexafluorobenzene	C_6F_6	353.41	516.4(0.5)	3.28(0.01)	337(4)	74, 198, 286, 469–473
2,2,4,4,5,5-Hexafluoro-1,3-dioxolane	$C_3F_6O_2$	251.0(0.2)	368.1(0.7)	2.72(0.04)	293(30)	116
Hexafluoroethane	C_2F_6	195.1	292.9(0.2)	3.03(0.01)	223(3)	172, 474–477
1,1,1,3,3,3-Hexafluoro-2-methoxy-2-(trifluoromethyl)propane	$C_5H_3F_9O$	327(1)	463(1)	2.37(0.07)	448(5)	110
1,1,1,2,3,3-Hexafluoro-3-(2,2,3,3,3-pentafluoropropoxy)propane	$C_6H_3F_{11}O$	360.64	486.48(0.07)	1.95(0.01)	529(2)	113
1,1,1,2,3,3-Hexafluoropropane	$C_3H_2F_6$	277.65	412.40(0.06)	3.42(0.01)	270(5)	203, 447, 448, 478, 479
1,1,1,3,3,3-Hexafluoropropane	$C_3H_2F_6$	272.2	398.07(0.06)	3.18(0.01)	262(18)	203
1,1,1,2,3,3-Hexafluoro-3-(2,2,3,3-tetrafluoropropoxy)propane	$C_6H_4F_{10}O$	379.07	516.2(0.3)	2.2(0.2)	543(34)	113
1,1,1,2,3,3-Hexafluoro-3-(2,2,2-trifluoroethoxy) propane	$C_5H_3F_9O$	345.87	475.74(0.09)	2.23(0.02)	455(2)	113
Hexamethylbenzene	$C_{12}H_{18}$	536.6	758(2)	2.6(0.4)	581(15)	25
1,1,1,5,5,5-Hexamethyl-3,3-bis[(trimethylsilyl) oxy]trisiloxane	$C_{12}H_{36}O_4Si_5$	494.6(0.2)	622.6(0.2)	1.03(0.02)	1323(90)	34
Hexamethyldisiloxane	$C_6H_{18}OSi_2$	373.64(0.02)	518.7(0.6)	1.95(0.02)	629(15)	480
2,6,10,15,19,23-Hexamethyltetracosane	$C_{30}H_{62}$	694(3)	796(2)	0.60(0.04)	2060(70)	481
Hexanal	$C_6H_{12}O$	404	592(3)	3.4(0.2)	378(7)	163, 242
Hexane	C_6H_{14}	341.88	507.5(0.1)	3.03(0.01)	366.0(0.8)	19, 29, 33, 34, 38, 52, 77, 78, 84, 86, 96, 106, 124, 131, 133, 134, 243, 244, 254, 281, 286, 318, 324–328, 401, 456, 458, 470, 482–488
1,6-Hexanediamine	$C_6H_{16}N_2$	477.7	685(7)	3.6(0.5)	446(17)	140
1,6-Hexanedioic acid	$C_6H_{10}O_4$	610.5	841(13)	3.8(0.3)	449(17)	253
1,6-Hexanediol	$C_6H_{14}O_2$	481	741(10)	4.1(0.1)	404(13)	489
Hexanenitrile	$C_6H_{11}N$	436.80	633.8(0.2)	2.99(0.06)	378(8)	35
Hexanoic acid	$C_6H_{12}O_2$	478.4	661(7)		413(15)	145, 146, 360, 464
1-Hexanol	$C_6H_{14}O$	430.8	611.0(0.4)	3.40(0.09)	381(30)	34, 149, 151, 152, 254, 255, 456
2-Hexanol	$C_6H_{14}O$	413	585.9(0.5)	3.3(0.3)	406(8)	149, 255, 437
3-Hexanol	$C_6H_{14}O$	408	582.4(0.4)	3.3(0.1)	378(14)	64, 254, 255
2-Hexanone	$C_6H_{12}O$	400.8	586.7(0.5)	3.31(0.04)	377(4)	25, 172, 466
3-Hexanone	$C_6H_{12}O$	396.7	583.1(0.5)	3.32(0.01)	378(4)	25, 466
Hexatriacontane	$C_{36}H_{74}$	777(7)	872(9)	0.47(0.07)	2711(2)	358

Name	Mol. Form.	T_b/K	T_c/K	P_c/MPa	V_c/cm³ mol⁻¹	Ref.
1-Hexene	C_6H_{12}	336.63	504.1(0.9)	3.20(0.03)	381(9)	86, 201, 202, 252, 257, 490
cis-2-Hexene	C_6H_{12}	342.0	513.4(0.9)	3.34(0.06)	347(7)	170
trans-2-Hexene	C_6H_{12}	341.1	509.0(0.7)	3.16(0.01)	353(7)	170
cis-3-Hexene	C_6H_{12}	339.6	510(1)	3.29(0.01)	351(7)	170
trans-3-Hexene	C_6H_{12}	340.3	507(2)	3.18(0.01)	352(7)	170
5-Hexen-2-one	$C_6H_{10}O$	402.27(0.02)	593.5(0.6)	3.51(0.04)	359(10)	51
Hexyl acetate	$C_8H_{16}O_2$	444.7	618(1)	2.5(0.1)	526(8)	162
Hexylamine	$C_6H_{15}N$	406.0	592.3(0.7)	3.4(0.3)	402(15)	170
Hexylbenzene	$C_{12}H_{18}$	499.3	695(7)	2.4(0.2)	620(14)	258
Hexyl benzoate	$C_{13}H_{18}O_2$	550(26)	748(14)	2.0(0.3)	658(14)	169
Indan	C_9H_{10}	451.12	684.8(0.4)	3.95(0.03)	385(22)	25
Isobutanal	C_4H_8O	337.7	543.6(0.6)	5.12(0.08)	283(10)	164, 165
Isobutane	C_4H_{10}	261.42	407.84(0.07)	3.64(0.02)	256(7)	125, 132, 491, 492
Isobutene	C_4H_8	266.3	418.0(0.3)	4.00(0.04)	240(3)	123, 158, 493, 494
Isobutyl acetate	$C_6H_{12}O_2$	389.7	562(2)	2.97(0.06)	369(37)	166, 201, 202, 415
Isobutylbenzene	$C_{10}H_{14}$	445.94	650(3)	3.0(0.2)	493(15)	96
Isobutyl butanoate	$C_8H_{16}O_2$	430.1	611(6)	2.5(0.3)	524(9)	147
Isobutylcyclohexane	$C_{10}H_{20}$	444.5	642.1(0.6)	2.61(0.07)	550(14)	170
Isobutyl formate	$C_5H_{10}O_2$	371.4	551(4)	3.9(0.4)	359(25)	415
Isobutyl isobutanoate	$C_8H_{16}O_2$	421.8	602(6)		530(9)	147
Isobutyl 3-methylbutanoate	$C_9H_{18}O_2$	441.7	621(6)		581(9)	147
Isobutyl propanoate	$C_7H_{14}O_2$	410	586(8)		462(9)	162, 167
Isopentane	C_5H_{12}	301.03	460.37(0.09)	3.35(0.06)	313(17)	19, 86, 96, 127, 495
Isopentyl acetate	$C_7H_{14}O_2$	415.7	586.1(0.4)	2.76(0.07)	464(9)	166
Isopentyl butanoate	$C_9H_{18}O_2$	452	619(6)		595(10)	147
Isopentyl nitrite	$C_5H_{11}NO_2$	372(2)	626(16)	5.07(0.04)	386.2(0.1)	52
Isopentyl propanoate	$C_8H_{16}O_2$	433.4	611(6)		523(9)	147
Isopropyl acetate	$C_5H_{10}O_2$	361.8	531.1(0.6)	3.31(0.04)	343(4)	64, 166, 172, 411, 490
Isopropylamine	C_3H_9N	304.91	472.2(0.9)	4.55(0.07)	231(5)	75, 170
Isopropylbenzene	C_9H_{12}	425.56	631(1)	3.2(0.1)	423(5)	79, 90, 96, 172
Isopropylcyclohexane	C_9H_{18}	428.0	632.2(0.4)	3.1(0.2)	484(14)	170
Isopropyl formate	$C_4H_8O_2$	341.4	534.6(0.5)	3.95(0.03)	294(11)	164, 165
1-Isopropyl-4-methylbenzene	$C_{10}H_{14}$	450.3	654(8)	2.8(0.1)	495(16)	96, 147
(1*S*,2*R*,5*S*)-2-Isopropyl-5-methylcyclohexanol	$C_{10}H_{20}O$	490(2)	694(5)	2.7(0.6)	539(14)	302
Isopropyl methyl ether	$C_4H_{10}O$	303.92	464.4(0.2)	3.76(0.01)	287(11)	25
Isoquinoline	C_9H_7N	516.37	803(8)	5.07(0.03)	380(17)	218
d-Limonene	$C_{10}H_{16}$	451	653(2)	2.81(0.02)	498(11)	496
Mesityl oxide	$C_6H_{10}O$	403	605(2)	3.85(0.02)	353(27)	497
Methane	CH_4	111.67	190.56(0.02)	4.60(0.01)	99(3)	132, 134, 498–508
Methane-d_4	CD_4		189.2(0.6)		98(3)	507
Methanethiol	CH_4S	279.1	469.9(0.3)	7.24(0.09)	148(4)	304
Methanol	CH_4O	337.8	512.7(0.6)	8.01(0.03)	117(4)	19, 38, 67, 80, 85, 87, 131, 148, 152, 155, 196, 405, 458, 482, 509–516
1-Methoxy-2,4-dimethylbenzene	$C_9H_{12}O$	465	682(4)	3.2(0.7)	451(18)	162
2-Methoxy-1,4-dimethylbenzene	$C_9H_{12}O$	467	677(1)	3.2(0.7)	451(15)	162
2-Methoxyethanol	$C_3H_8O_2$	397.3	598(1)	5.28(0.08)	263(6)	49
2-Methoxyethyl acetate	$C_5H_{10}O_3$	416	603(3)	3.6(0.5)	368(15)	162
4-Methoxy-1,1,1,2,2,3,3-heptafluorobutane	$C_5H_5F_7O$	344.13	481.5(0.2)	2.38(0.01)	431(2)	113
1-Methoxy-1,1,2,2,3,3-hexafluoropropane	$C_4H_4F_6O$	341.02	487.0(0.3)	2.9(0.1)	370(25)	113
2-Methoxy-2-methylbutane	$C_6H_{14}O$	359.3	536(2)	3.23(0.09)	372(19)	172, 410, 517
5-Methoxy-1,1,2,2,3,3,4,4-octafluoropentane	$C_6H_6F_8O$	395.83	546.1(0.3)	2.40(0.07)	493(30)	113
1-Methoxy-2-propanol	$C_4H_{10}O_2$	392	579.8(0.3)	4.11(0.04)	304(12)	49
2-Methoxypropene	C_4H_8O	308.9(0.3)	478.5(0.6)	4.2(0.3)	257(13)	518
Methyl acetate	$C_3H_6O_2$	330.02	506.7(0.4)	4.73(0.07)	227(22)	19, 411–413
Methylamine	CH_5N	266.83	430.6(0.6)	7.61(0.09)	139(1)	130, 193, 322, 323
N-Methylaniline	C_7H_9N	469.4	702(5)	5.2(0.6)	347(16)	117
2-Methylaniline	C_7H_9N	473.5	710(1)	3.6(0.1)	377(52)	519
3-Methylaniline	C_7H_9N	476.5	709(10)	4.6(0.5)	346(17)	106
4-Methylaniline	C_7H_9N	473.6	667(10)	3.3(0.7)	334(16)	106
2-Methylanisole	$C_8H_{10}O$	444	662(1)	3.6(0.3)	397(18)	162

Name	Mol. Form.	T_b/K	T_c/K	P_c/MPa	V_c/cm³ mol⁻¹	Ref.
3-Methylanisole	$C_8H_{10}O$	448.7	665(1)	3.6(0.3)	449(22)	162
4-Methylanisole	$C_8H_{10}O$	448.7	667(1)	3.6(0.4)	396(15)	162
α-Methylbenzenemethanol	$C_8H_{10}O$	478	699(5)	3.8(0.7)	399(16)	32
Methyl benzoate	$C_8H_8O_2$	472	702(1)	3.8(0.1)	408(11)	517
2-Methylbutanal	$C_5H_{10}O$	363(2)	531.6(0.1)	4.04(0.03)	318(10)	124
Methyl butanoate	$C_5H_{10}O_2$	376.0	554.4(0.1)	3.49(0.08)	341(20)	19, 413, 416
3-Methylbutanoic acid	$C_5H_{10}O_2$	449.7	629(1)	3.4(0.2)	355(10)	146
2-Methyl-1-butanol	$C_5H_{12}O$	400.7	575.4(0.5)	3.9(0.1)	342(8)	254
2-Methyl-2-butanol	$C_5H_{12}O$	375.6	544(1)	3.71(0.05)	326(9)	147, 254
3-Methyl-1-butanol	$C_5H_{12}O$	404.3	579(2)	3.9(0.3)	335(7)	21, 91, 222, 254, 511, 520
3-Methyl-2-butanol	$C_5H_{12}O$	386.1	556.1(0.5)	3.9(0.4)	336(9)	254
3-Methyl-2-butanone	$C_5H_{10}O$	367.48	553.1(0.3)	3.83(0.1)	321(33)	157, 166
2-Methyl-2-butene	C_5H_{10}	311.71	470(1)	3.4(0.1)	299(7)	521
3-Methyl-1-butene	C_5H_{10}	293.3	452.7(0.5)	3.51(0.04)	305(8)	490
Methyl *tert*-butyl ether	$C_5H_{12}O$	328.2	497.0(0.6)	3.41(0.05)	335(10)	25, 440
Methylcyclohexane	C_7H_{14}	374.08	572.3(0.2)	3.48(0.09)	368(3)	34, 74, 78, 86, 88, 171, 235, 236
Methylcyclopentane	C_6H_{12}	345.0	532.78(0.05)	3.79(0.05)	322(2)	78, 235, 236
2-Methylcyclopentanone	$C_6H_{10}O$	412.7	631(2)	4.0(0.6)	328(17)	162
2-Methyl-*N,N*-dimethylaniline	$C_9H_{13}N$	467.3	668.0(0.7)	3.12(0.08)	466(15)	39, 40
Methyl dodecanoate	$C_{13}H_{26}O_2$	540	712(5)	1.4(0.4)	842(9)	218
1,1'-Methylenebis[(1-methylethyl)benzene]	$C_{19}H_{24}$	592(36)	795(8)	1.6(0.1)	871(30)	394
Methyl formate	$C_2H_4O_2$	304.9	487.16(0.1)	6.01(0.01)	172(6)	19, 412, 413
2-Methylfuran	C_5H_6O	337.9	528(3)	4.77(0.08)	252(3)	290
2-Methylheptane	C_8H_{18}	390.81	559.6(0.1)	2.50(0.02)	487(12)	86, 346, 522
3-Methylheptane	C_8H_{18}	392.1	563.7(0.4)	2.54(0.02)	463(12)	346
4-Methylheptane	C_8H_{18}	390.87	561.7(0.4)	2.54(0.02)	480(14)	346
Methyl heptanoate	$C_8H_{16}O_2$	447	628(2)	2.6(0.4)	521(8)	162
4-Methyl-3-heptanol	$C_8H_{18}O$	443	623.5(0.7)	2.8(0.4)	505(13)	437
5-Methyl-3-heptanol	$C_8H_{18}O$	445	621.2(0.3)	2.8(0.3)	493(13)	437
2-Methyl-3-heptanone	$C_8H_{16}O$	431	615(1)	2.7(0.3)	487(11)	162
5-Methyl-3-heptanone	$C_8H_{16}O$	434	619(4)	2.7(0.7)	484(11)	162
2-Methyl-1-heptene	C_8H_{16}	392.5	567.5(0.9)	2.6(0.2)	466(10)	170
2-Methyl-2-heptene	C_8H_{16}	395.8	569(1)	2.6(0.4)	465(11)	170
2-Methylhexane	C_7H_{16}	363.19	530.4(0.1)	2.73(0.03)	420(15)	86, 346, 349, 522
3-Methylhexane	C_7H_{16}	365.0(0.1)	535.4(0.5)	2.82(0.06)	405(19)	346, 349
2-Methyl-3-hexanone	$C_7H_{14}O$	408	593(1)	2.9(0.4)	428(13)	162
5-Methyl-2-hexanone	$C_7H_{14}O$	417	604(1)	2.9(0.3)	434(13)	162
2-Methyl-1-hexene	C_7H_{14}	365	542(1)	2.9(0.3)	407(8)	170
5-Methyl-1-hexene	C_7H_{14}	358.5	528.7(0.4)	2.9(0.2)	410(9)	170
N-Methylhexylamine	$C_7H_{17}N$	418(8)	592(1)	2.8(0.8)	458(20)	170
Methyl isobutanoate	$C_5H_{10}O_2$	365.7	540.7(0.5)	3.43(0.01)	341(42)	19, 413
Methyl methacrylate	$C_5H_8O_2$	373.76(0.07)	540.3(0.6)	2.97(0.06)	320(6)	51
1-Methylnaphthalene	$C_{11}H_{10}$	517.9	771(5)	3.56(0.07)	479(22)	86, 218, 523
2-Methylnaphthalene	$C_{11}H_{10}$	514.3	761(3)	3.37(0.06)	464(20)	218
2-Methyloctane	C_9H_{20}	416.4	582.8(0.2)	2.30(0.02)	547(17)	345, 522
Methyloxirane	C_3H_6O	308	488.11(0.08)	5.44(0.02)		290, 524
Methyl pentafluoroethyl ether	$C_3H_3F_5O$	278.74	406.81(0.05)	2.89(0.01)	301(5)	176, 448, 525
2-Methylpentane	C_6H_{14}	333.41	497.9(0.2)	3.03(0.01)	371(2)	86, 127, 252, 325–328, 487, 522
3-Methylpentane	C_6H_{14}	336.42	504.6(0.2)	3.12(0.01)	368.7(0.3)	252, 324–328, 526
Methyl pentanoate	$C_6H_{12}O_2$	400.6	588.9(0.3)	3.20(0.05)	398(6)	164, 165
2-Methyl-1-pentanol	$C_6H_{14}O$	422	604.4(0.5)	3.4(0.2)	410(8)	254
2-Methyl-2-pentanol	$C_6H_{14}O$	394.3	559.5(0.7)	3.6(0.4)	410(11)	437
2-Methyl-3-pentanol	$C_6H_{14}O$	399.7	576(1)	3.5(0.1)	380(9)	254
3-Methyl-3-pentanol	$C_6H_{14}O$	395.6	575.6(0.6)	3.5(0.2)	376(10)	254
4-Methyl-1-pentanol	$C_6H_{14}O$	425.1	603.5(0.7)	3.4(0.4)	406(7)	437
4-Methyl-2-pentanol	$C_6H_{14}O$	404.8	574.4(0.5)		389(9)	437
4-Methyl-2-pentanone	$C_6H_{12}O$	389.7	575.4(1)	3.4(0.1)	378(12)	166, 527
2-Methyl-2-pentene	C_6H_{12}	340.5	509.3(0.5)	3.26(0.01)	348(7)	170
4-Methyl-1-pentene	C_6H_{12}	327.1	493.1(0.5)	3.18(0.07)		170, 497

Name	Mol. Form.	T_b/K	T_c/K	P_c/MPa	V_c/cm³ mol⁻¹	Ref.
4-Methyl-*cis*-2-pentene	C_6H_{12}	329.5	496.3(0.7)	3.24(0.01)	350(7)	170
Methyl pentyl ether	$C_6H_{14}O$	372	546.5(0.2)	3.04(0.1)	395(14)	173, 174
2-Methyl-1,3-propanediol	$C_4H_{10}O_2$	484.8	708(2)	5.4(0.4)	300(12)	55
Methyl propanoate	$C_4H_8O_2$	353.0	530.57(0.1)	4.0(0.2)		19, 412, 413, 415
2-Methylpropanoic acid	$C_4H_8O_2$	427.60	605(2)	3.7(0.3)	296(10)	146
2-Methyl-1-propanol	$C_4H_{10}O$	381.04	548(2)	4.30(0.04)	274(17)	25, 91, 148, 153, 155
2-Methyl-2-propanol	$C_4H_{10}O$	355.6	506.2(0.1)	3.98(0.07)	283(4)	153
Methyl propyl ether	$C_4H_{10}O$	312.3	476.2(0.2)	3.80(0.01)	281(7)	25
2-Methylpyridine	C_6H_7N	402.53	622(1)	4.62(0.04)		75, 528
3-Methylpyridine	C_6H_7N	417.29	644.8(0.6)	4.63(0.03)		90, 528
4-Methylpyridine	C_6H_7N	418.51	645.8(0.5)	4.68(0.04)		75, 90
N-Methyl-2-pyrrolidinone	C_5H_9NO	475	721.7(0.4)	4.5(0.4)	330(12)	144, 242
2-Methylquinoline	$C_{10}H_9N$	519.7	778(2)	3.91(0.02)	447(49)	530
8-Methylquinoline	$C_{10}H_9N$	520.7	787(2)	4.22(0.02)		530
Methyl salicylate	$C_8H_8O_3$	496.1	709(30)	4.4(0.7)	436(17)	302
2-Methyltetrahydrofuran	$C_5H_{10}O$	351	537(2)	3.74(0.06)	292(4)	290
(Methylthio)benzene	C_7H_8S	467.5	706(4)	4.1(0.1)	374(13)	99
Methyl trifluoromethyl ether	$C_2H_3F_3O$	249.15	377.92(0.06)	3.64(0.03)	219(2)	447, 525
Methyltris(trimethylsiloxy)silane	$C_{10}H_{30}O_3Si_4$	464.3(0.2)	597.4(0.2)	1.23(0.02)	1089(74)	34
4-Morpholinecarboxaldehyde	$C_5H_9NO_2$	512	779(4)	5.0(0.4)	326(14)	54
Naphthalene	$C_{10}H_8$	491.1	748.3(0.4)	4.06(0.04)	408(21)	79, 82, 86, 241, 251, 355, 451, 531
Neopentane	C_5H_{12}	282.63	433.71(0.01)	3.20(0.01)	311.6(0.7)	532
Nitromethane	CH_3NO_2	374.34	588(3)	6.0(0.2)	175(2)	533, 534
Nonadecane	$C_{19}H_{40}$	603.1	756(5)	1.16(0.07)	1216(43)	358, 359
1-Nonadecene	$C_{19}H_{38}$	604(17)	755(8)	1.2(0.2)	1196(36)	365
1,1,1,2,2,3,3,4,4-Nonafluorohexan-5-one	$C_6H_3F_9O$	360.47	498.97(0.08)	2.20(0.02)	504(2)	308
Nonanal	$C_9H_{18}O$	464	658(2)	2.7(0.1)	546(10)	242
Nonane	C_9H_{20}	423.97	594.2(0.5)	2.29(0.05)	547(23)	34, 36, 78, 86, 131, 243–245, 247, 249, 251, 454, 456, 535, 536
1,9-Nonanediamine	$C_9H_{22}N_2$	531.8	726(7)	2.6(0.3)	600(23)	140
Nonanedioic acid	$C_9H_{16}O_4$	630.2	844(13)	2.7(0.2)	586(17)	253
Nonanoic acid	$C_9H_{18}O_2$	527.7	712(3)		592(16)	146
1-Nonanol	$C_9H_{20}O$	486.52	670.6(0.5)	2.54(0.07)	555(75)	149, 152, 254, 255
2-Nonanol	$C_9H_{20}O$	466.7	649(1)	2.53(0.1)	575(11)	149, 255
3-Nonanol	$C_9H_{20}O$	468	648.0(0.3)	2.5(0.3)	577(12)	255
4-Nonanol	$C_9H_{20}O$	465.7	645.1(0.3)	2.5(0.3)	577(12)	255
2-Nonanone	$C_9H_{18}O$	468.5	652.1(0.7)	2.5(0.1)	560(8)	49, 466
3-Nonanone	$C_9H_{18}O$	463	648(4)	2.4(0.6)	560(7)	466
4-Nonanone	$C_9H_{18}O$	460.7	643.7(0.3)	2.4(0.3)	560(7)	466
5-Nonanone	$C_9H_{18}O$	461.60	641.4(0.3)	2.35(0.02)	560(7)	466
1-Nonene	C_9H_{18}	420.1	594(1)	2.38(0.01)	529(2)	257
Octacosane	$C_{28}H_{58}$	707(3)	824(8)	0.8(0.1)	1916(65)	358
Octadecane	$C_{18}H_{38}$	589.5	748(1)	1.3(0.1)	1167(41)	244
1-Octadecanol	$C_{18}H_{38}O$	608	790(8)	1.28(0.1)	1157(18)	361
1-Octadecene	$C_{18}H_{36}$	588.7(1)	748(8)	1.3(0.1)	1119(34)	365
1,1,1,2,2,3,3,4-Octafluorobutane	$C_4H_2F_8$	300.62(0.02)	432.0(0.1)	2.80(0.02)	360(22)	537
1,2,2,3,3,4,4,5-Octafluoro-1-pentanol	$C_5H_4F_8O$	413.1(0.3)	571(1)	2.9(0.1)	440(24)	362
Octafluorotetrahydrofuran	C_4F_8O	272.3(0.5)	399.6(0.7)	2.68(0.09)	350(30)	450
Octamethylcyclotetrasiloxane	$C_8H_{24}O_4Si_4$	448.5(0.2)	585.8(0.9)	1.33(0.01)	1006(67)	114, 538
Octamethyltrisiloxane	$C_8H_{24}O_2Si_3$	425.7(0.1)	564.1(0.2)	1.42(0.01)	882(16)	539
Octanal	$C_8H_{16}O$	444	639.3(0.3)	3.0(0.3)	489(7)	163, 242
Octane	C_8H_{18}	398.82	568.7(0.1)	2.48(0.01)	490(22)	11, 19, 33, 34, 36, 78, 83, 86, 91, 96, 131, 133, 134, 243–245, 247, 249, 318, 319, 324, 346, 465, 469, 535, 540, 541
1,8-Octanediamine	$C_8H_{20}N_2$	498.7	712(7)	2.8(0.3)	547(20)	140
Octanedioic acid	$C_8H_{14}O_4$	618.6	843(13)	3.0(0.2)	520(16)	253
Octanenitrile	$C_8H_{15}N$	478.40	674.4(0.4)	2.85(0.03)	494(10)	35
Octanoic acid	$C_8H_{16}O_2$	512	694(1)	2.9(0.3)	522(19)	145, 146, 360, 464

Name	Mol. Form.	T_b/K	T_c/K	P_c/MPa	V_c/cm³ mol⁻¹	Ref.
1-Octanol	$C_8H_{18}O$	468.31	651(2)	2.80(0.07)	490(47)	25, 149, 152, 254, 255
2-Octanol	$C_8H_{18}O$	452.5	629.5(0.9)	2.75(0.04)	519(10)	149, 254, 255
3-Octanol	$C_8H_{18}O$	444	628.4(0.3)	2.8(0.4)	515(10)	255
4-Octanol	$C_8H_{18}O$	449.5	625.1(0.3)	2.8(0.3)	516(10)	255
2-Octanone	$C_8H_{16}O$	445.7	632.7(0.2)	2.7(0.5)	497(6)	466
3-Octanone	$C_8H_{16}O$	440.7	627.7(0.2)	2.7(0.3)	497(6)	466
4-Octanone	$C_8H_{16}O$	436	623.8(0.2)	2.7(0.3)	497(6)	466
1-Octene	C_8H_{16}	394.44	566.58(0.05)	2.68(0.02)	464(2)	86, 257
trans-2-Octene	C_8H_{16}	398	569.8(0.4)	2.58(0.09)	471(9)	170
trans-4-Octene	C_8H_{16}	395.5	566(1)	2.55(0.06)	472(9)	170
Octylamine	$C_8H_{19}N$	452.8	641(1)	2.82(0.03)	494(41)	542
Octylbenzene	$C_{14}H_{22}$	537	725(7)	2.0(0.2)	746(17)	258
Octyl orthosilicate	$C_{32}H_{68}O_4Si$		812(16)			57
Oxazole	C_3H_3NO	342.7	551(4)	6.8(0.2)	185(23)	54
Oxirane	C_2H_4O	283.8	469(1)	7.2(0.2)	138(4)	543, 544
Paraldehyde	$C_6H_{12}O_3$	397.5	563(10)		410(15)	12
Pentacene	$C_{22}H_{14}$		1115(17)		806(23)	545
1*H*-Pentadecafluoroheptane	C_7HF_{15}	369.2	495.8(0.7)	1.7(0.5)	644(38)	546
Pentadecane	$C_{15}H_{32}$	543.8	707(2)	1.54(0.09)	938(36)	243, 244, 245, 247, 249, 363, 547
Pentadecanoic acid	$C_{15}H_{30}O_2$	612(4)	777(8)	1.6(0.2)	1002(22)	360
1-Pentadecanol	$C_{15}H_{32}O$	573	757(8)	1.6(0.2)	961(16)	361
1-Pentadecene	$C_{15}H_{30}$	541.55(0.07)	705(7)	1.56(0.05)	933(30)	365
Pentafluorobenzene	C_6HF_5	358.89	530.93(0.05)	3.53(0.01)	322(22)	472, 548
3,3,4,4,4-Pentafluoro-2-butanone	$C_4H_3F_5O$	314.36(0.04)	453(1)	2.90(0.06)	333(4)	110
Pentafluoroethane	C_2HF_5	224.65	339.2(0.2)	3.63(0.01)	210(3)	180, 309, 312, 315, 447, 448, 549–554
1,1,1,2,2-Pentafluoropentan-3-one	$C_5H_5F_5O$	335.24	475.5(0.1)	2.64(0.01)	356(1)	308
1,1,1,2,2-Pentafluoropropane	$C_3H_3F_5$	255.8	380.1(0.4)	3.14(0.02)	273(3)	555
1,1,1,3,3-Pentafluoropropane	$C_3H_3F_5$	288.5	427.20(0.07)	3.66(0.02)	262(14)	203
1,1,2,2,3-Pentafluoropropane	$C_3H_3F_5$	298.2	447.57(0.06)	3.96(0.02)	258(13)	203
1,1,1,2,2-Pentafluoro-3-(1,1,2,2-tetrafluoroethoxy) propane	$C_5H_3F_9O$	343.4	473.0(0.1)	2.24(0.01)	457(2)	113
Pentafluoro(trifluoromethoxy)ethane	C_3F_8O	249.5(0.3)	356.8(0.1)	2.4(0.5)	319(5)	442
Pentafluoro(trifluoromethyl)sulfur	CF_8S	252.4(0.2)	381.2(0.1)	3.4(0.1)	284(4)	442
Pentanal	$C_5H_{10}O$	376	567(3)	3.1(0.3)	313(11)	104, 124
Pentane	C_5H_{12}	309.21	469.7(0.1)	3.37(0.01)	310(1)	19, 34, 36, 52, 64, 77, 84–86, 127, 131, 133, 134, 201, 202, 243, 244, 254, 286, 324, 359, 394, 456, 490, 556–562
Pentanedioic acid	$C_5H_8O_4$	576.0	840(13)	4.3(0.5)	343(13)	253
Pentanenitrile	C_5H_9N	414.5	610.3(0.2)	3.58(0.05)	320(8)	35
Pentanoic acid	$C_5H_{10}O_2$	459.3	639(2)	3.6(0.1)	347(15)	11, 41, 145, 146, 464
1-Pentanol	$C_5H_{12}O$	411.13	587.9(0.4)	3.9(0.3)	331(9)	25, 34, 149, 329, 456, 465, 563
2-Pentanol	$C_5H_{12}O$	392.5	560.4(0.2)	4.2(0.4)	340(3)	149, 254, 465
3-Pentanol	$C_5H_{12}O$	389.40	559.6(0.3)	4.9(0.9)	325(2)	465
2-Pentanone	$C_5H_{10}O$	375.41	561.0(0.2)	3.70(0.06)	324(4)	25, 172
3-Pentanone	$C_5H_{10}O$	374.9	561.4(0.2)	3.73(0.07)	319(11)	25
1-Pentene	C_5H_{10}	303.11	464.74(0.05)	3.55(0.02)	301.0(0.1)	86, 470, 559
cis-2-Pentene	C_5H_{10}	310.08	474.9(0.4)	3.69(0.02)	301(7)	564
Pentyl acetate	$C_7H_{14}O_2$	422.4	600(2)	2.79(0.03)		141, 162, 166
Pentylbenzene	$C_{11}H_{16}$	478.6	675(7)	2.6(0.3)	559(12)	258
Pentyl benzoate	$C_{12}H_{16}O_2$	533(3)	736(14)	2.2(0.2)	661(14)	169
Pentyl formate	$C_6H_{12}O_2$	403.6	576(4)	3.5(0.8)	453(35)	415
Perfluoroacetone	C_3F_6O	245.8	357.2(0.1)	2.85(0.01)	329(7)	197, 470
Perfluorobutane	C_4F_{10}	271.3	386.3(0.2)	2.33(0.02)	380(18)	565, 566
Perfluorocyclobutane	C_4F_8	267.3	388.4(0.1)	2.78(0.01)	316(8)	265, 470, 567, 568
Perfluorocyclohexane	C_6F_{12}	325.95 s	457.1(0.5)	2.24(0.03)	424(32)	569
Perfluorocyclohexene	C_6F_{10}	325.2	461.7(0.7)	2.6(0.1)	434(28)	546
Perfluorodecane	$C_{10}F_{22}$	417.4	542.4(0.4)	1.45(0.03)	892(7)	240, 570

Name	Mol. Form.	T_b/K	T_c/K	P_c/MPa	V_c/cm³ mol⁻¹	Ref.
Perfluorodimethoxymethane	$C_3F_8O_2$	263	372.4(0.2)	2.34(0.01)	370(30)	116, 571
Perfluoro-2,3-dimethylbutane	C_6F_{14}	333.0	463.0(0.1)	1.95(0.03)	523(18)	240, 572
Perfluoroethyl ethyl ether	$C_4H_5F_5O$	301.26	431.23(0.08)	2.53(0.01)	366(3)	176
Perfluoroethyl 2,2,2-trifluoroethyl ether	$C_4H_2F_8O$	301.04	421.68(0.08)	2.33(0.01)	409(3)	176
Perfluoroheptane	C_7F_{16}	355.7	477(3)	1.63(0.01)	603(14)	232, 566, 570, 573–576
Perfluoro-1-heptene	C_7F_{14}	354.2	478.2(0.7)	1.7(0.3)	555(34)	546
Perfluorohexane	C_6F_{14}	330.3	451(3)	1.88(0.02)	552(77)	198, 470, 546, 570, 572, 573, 577
1*H*-Perfluorohexane	C_6HF_{13}	345(4)	471.8(0.7)	2.0(0.3)	504(32)	546
Perfluoro-1-hexene	C_6F_{12}	330.2	454.3(0.7)		462(31)	546
Perfluoroisobutane	C_4F_{10}	273	395.4(0.7)		396(22)	578
Perfluoroisopropyl methyl ether	$C_4H_3F_7O$	302.49	433.30(0.08)	2.55(0.01)	369(3)	176
Perfluoromethylcyclohexane	C_7F_{14}	349.5	486.5(1)	2.02(0.01)	561(4)	325, 326, 327, 566, 569
Perfluoromethylcyclopentane	C_6F_{12}	321.58(0.01)	451.43(0.05)	2.17(0.01)	419(37)	579
Perfluoro-2-methylpentane	C_6F_{14}	330.8	454.6(0.2)	1.87(0.02)	585(10)	572, 580
Perfluoro-3-methylpentane	C_6F_{14}	331.6	450(1)	1.69(0.01)	511(36)	572
Perfluoronaphthalene	$C_{10}F_8$	482	673(1)	2.9(0.6)	464(29)	546
Perfluorononane	C_9F_{20}	390.76	524.0(0.1)	1.56(0.04)	846(48)	570
Perfluorooctane	C_8F_{18}	379.1	502.3(0.1)	1.66(0.02)	738(17)	34, 240, 570, 573
Perfluorooxetane	C_3F_6O	244.8	361.8(0.5)	3.10(0.02)	274(24)	116, 571
1*H*-Perfluoropentane	C_5HF_{11}	319(2)	443.9(0.7)	2.2(0.3)	413(28)	546
Perfluoropentane	C_5F_{12}	302.4	421.8(0.1)	2.04(0.02)	463(7)	570, 573
Perfluoropropane	C_3F_8	236.6	345.03(0.08)	2.67(0.01)	301(12)	187, 305, 470, 581
Perfluoropropyl methyl ether	$C_4H_3F_7O$	307.38	437.7(0.1)	2.48(0.01)	382(3)	176
Perfluorotoluene	C_7F_8	376.70	534.4(0.2)	2.70(0.02)	423(26)	25
Perfluorotributylamine	$C_{12}F_{27}N$	451(2)	566(4)	1.24(0.09)	1196(66)	582
Perfluorovaleric acid	$C_5HF_9O_2$	415.3(0.3)	545.5(0.2)	2.10(0.05)	496(29)	583
Phenol	C_6H_6O	455.02	694.3(0.1)	5.5(0.2)	283(10)	218, 584
Phenyl acetate	$C_8H_8O_2$	469	686(2)	3.60(0.06)	407(12)	55
4-Phenyl-1-butanol	$C_{10}H_{14}O$	537(3)	746(14)	3.1(0.2)	493(19)	584
Phenyl isocyanate	C_7H_5NO	436	657(10)	3.6(0.1)	342(15)	54
3-Phenyl-1-propanol	$C_9H_{12}O$	515(4)	732(14)	3.4(0.5)	450(14)	584
(1*S*)-(-)-α-Pinene	$C_{10}H_{16}$	429.1(0.4)	644(2)	3.4(0.2)	472(10)	496
Piperazine	$C_4H_{10}N_2$	421.8	660(5)	5.4(0.4)	283(16)	331, 489
Piperidine	$C_5H_{11}N$	379.37	594.14(0.05)	4.7(0.1)	294(10)	39, 218
Propanal	C_3H_6O	321	503.7(0.8)	5.04(0.03)	218(9)	104, 124, 163, 242
Propane	C_3H_8	231.1	369.9(0.1)	4.25(0.01)	199(6)	29, 125, 127, 131–134, 252, 322, 323, 336, 371, 387, 470, 486, 509, 540, 585–596
1,3-Propanediamine	$C_3H_{10}N_2$	413.2	632(7)	5.7(0.6)	257(13)	140
1,2-Propanediol	$C_3H_8O_2$	460.8	676(1)	5.9(0.2)	237(10)	32
1,3-Propanediol	$C_3H_8O_2$	487.6	718(2)	6.7(0.2)	255(12)	55
Propanenitrile	C_3H_5N	370.29	561.3(0.2)	4.26(0.07)	211(7)	35
1-Propanethiol	C_3H_8S	341.0	536.6(0.6)	4.7(0.1)	286(11)	143, 144
Propanoic acid	$C_3H_6O_2$	414.30	603(3)	4.5(0.7)	232(12)	39, 145, 146, 493, 597, 598
1-Propanol	C_3H_8O	370.4	536.8(0.2)	5.1(0.1)	220(22)	19, 34, 80, 148–150, 152, 153, 599, 600
2-Propanol	C_3H_8O	355.5	508.3(0.2)	4.7(0.1)	226(1)	25, 91, 148, 149, 153, 324, 440, 520, 601, 602
Propene	C_3H_6	225.46	364.9(0.5)	4.59(0.02)	184(11)	67, 123, 172, 181, 427, 603–610
2-Propoxyethanol	$C_5H_{12}O_2$	423.0	614.7(0.7)	3.65(0.09)	364(13)	11, 41
1-Propoxy-2-propanol	$C_6H_{14}O_2$	423	605(1)	3.1(0.1)	417(16)	32
Propyl acetate	$C_5H_{10}O_2$	374.69	549.69(0.08)	3.37(0.07)	346(24)	19, 411, 413, 415
Propylamine	C_3H_9N	320.37	499.2(0.4)	4.77(0.06)	230(12)	170, 193
Propylbenzene	C_9H_{12}	432.39	638.3(0.1)	3.20(0.02)	441(6)	79, 86, 88
Propyl benzoate	$C_{10}H_{12}O_2$	504(2)	710(14)	2.6(0.3)	530(10)	169
Propyl butanoate	$C_7H_{14}O_2$	416.2	593(1)	2.72(0.06)	463(10)	162, 166
Propylcyclohexane	C_9H_{18}	429	630.8(0.9)	2.87(0.04)	489(13)	170, 171
Propylene carbonate	$C_4H_6O_3$	515	763(2)	4.1(0.2)	256.5(0.5)	55
1,2-Propylene glycol 1-*tert*-butyl ether	$C_7H_{16}O_2$	425.2	601(4)	2.7(0.1)	468(15)	99

Name	Mol. Form.	T_b/K	T_c/K	P_c/MPa	V_c/cm³ mol⁻¹	Ref.
1,2-Propylene glycol monomethyl ether acetate	$C_6H_{12}O_3$	420	598(1)	3.1(0.2)	432(16)	41, 144
Propyl formate	$C_4H_8O_2$	354.1	538.1(0.1)	4.07(0.02)	281(31)	19, 412, 413, 415
Propyl isobutanoate	$C_7H_{14}O_2$	408.6	582(11)		463(9)	162, 167
Propyl 3-methylbutanoate	$C_8H_{16}O_2$	429.1	609(6)		523(9)	147
Propyl propanoate	$C_6H_{12}O_2$	395.7	569(3)	3.1(0.1)	403(6)	162, 166
Propyne	C_3H_4	250.0	402(2)	5.63(0.06)	160(10)	605, 611
Pyrazine	$C_4H_4N_2$	388	627(1)	6.49(0.03)	225(16)	612
Pyridine	C_5H_5N	388.38	619(2)	5.63(0.07)	248(12)	16, 17, 60, 90, 246, 290, 520, 613
Pyrrole	C_4H_5N	402.94	639.7(0.2)	8.0(0.2)	222(15)	241
Pyrrolidine	C_4H_9N	359.71	568.6(0.2)	5.69(0.08)	259(3)	241, 290
Quinoline	C_9H_7N	510.31	782(3)	4.75(0.1)	382(17)	218
Resorcinol	$C_6H_6O_2$	549.7	836(10)	6.3(0.3)	292(10)	489
Stearic acid	$C_{18}H_{36}O_2$	623	803(8)	1.3(0.2)	1251(27)	360
Styrene	C_8H_8	418	635(2)	3.9(0.2)	357(15)	481
Succinic acid	$C_4H_6O_4$	508.2	851(20)		308(21)	253
m-Terphenyl	$C_{10}H_{11}$	636	883(7)	2.2(0.2)	747(37)	108, 358
o-Terphenyl	$C_{18}H_{14}$	605	857(6)	2.9(0.1)	737(37)	108
p-Terphenyl	$C_{18}H_{14}$	649	913(22)	2.5(0.5)	713(37)	108, 614
Tetrabutyl silicate	$C_{16}H_{36}O_4Si$		682(14)			57
Tetrachloromethane	CCl_4	350.0	556.5(0.3)	4.57(0.07)	276(9)	19, 27, 469, 615–619
Tetracosane	$C_{24}H_{50}$	664.5	800(5)	0.9(0.1)	1585(55)	358, 359
Tetradecamethylcycloheptasiloxane	$C_{14}H_{42}O_7Si_7$	548.4(0.2)	683.2(0.2)	0.99(0.02)	1634(110)	34
Tetradecane	$C_{14}H_{30}$	526.73	693(1)	1.56(0.08)	870(49)	34, 244
Tetradecanedioic acid	$C_{14}H_{26}O_4$	639(10)		1.9(0.2)		253
Tetradecanoic acid	$C_{14}H_{28}O_2$	599(1)	763(8)	1.6(0.2)	921(20)	360
1-Tetradecanol	$C_{14}H_{30}O$	560	743(7)	1.70(0.04)	887(15)	361
2-Tetradecanone	$C_{14}H_{28}O$	562(6)	728(9)		896(26)	256
3-Tetradecanone	$C_{14}H_{28}O$	552(7)	727(6)		896(26)	256
4-Tetradecanone	$C_{14}H_{28}O$	552(7)	725(6)		900(27)	256
7-Tetradecanone	$C_{14}H_{28}O$	552(8)	723(8)		904(27)	256
1-Tetradecene	$C_{14}H_{28}$	524.3(0.1)	691(7)	1.58(0.07)	851(24)	365
Tetradecyl orthosilicate	$C_{40}H_{84}O_4Si$		849(16)			57
Tetraethoxysilane	$C_8H_{20}O_4Si$	441(1)	587(12)	2.0(0.4)	701.3(0.2)	57
Tetraethylene glycol	$C_8H_{18}O_5$	601	800(30)	2.8(0.7)	608(31)	283
Tetraethylsilane	$C_8H_{20}Si$	427.9	606(2)	2.297(0.01)	596.4(0.2)	620
1,2,3,4-Tetrafluorobenzene	$C_6H_2F_4$	367.5	550.8(0.2)	3.791(0.01)	312(22)	25
1,2,3,5-Tetrafluorobenzene	$C_6H_2F_4$	357.6	535.2(0.2)	3.75(0.01)	311(22)	25
1,2,4,5-Tetrafluorobenzene	$C_6H_2F_4$	363.4	543.3(0.2)	3.80(0.01)	309(22)	25
1,1,2,2-Tetrafluoro-2-(2,2-difluoromethoxy) ethane	$C_4H_4F_6O$	352.13	501.08(0.08)	3.09(0.02)	356(1)	113
1,1,1,2-Tetrafluoroethane	$C_2H_2F_4$	246.6	374.2(0.2)	4.06(0.01)	200(2)	113, 182, 183, 188, 278–280, 305, 313, 315, 447, 448, 478, 621–631
1,1,2,2-Tetrafluoroethane	$C_2H_2F_4$	253.3	391.75(0.08)	4.61(0.01)	192(1)	178, 632
Tetrafluoroethene	C_2F_4	197.3	307(1)	3.94(0.05)	183(3)	633, 634
1,2,2,2-Tetrafluoroethyl difluoromethyl ether	$C_3H_2F_6O$	296.50	428.95(0.08)	3.05(0.01)	315(2)	176
1,1,2,2-Tetrafluoroethyl 1,1,1-trifluoroethyl ether	$C_4H_3F_7O$	329.37	463.89(0.07)	2.71(0.01)	373(1)	113
Tetrafluoromethane	CF_4	145.2	227.54(0.03)	3.73(0.03)	140(1)	506, 635
1,1,2,2-Tetrafluoro-3-methoxypropane	$C_4H_6F_4O$	347.4(0.1)	505.4(0.1)	3.28(0.01)	331(1)	113
1,2,2,3-Tetrafluoro-1-propanol	$C_3H_4F_4O$	386.4(0.4)	554(2)	3.3(0.2)	280(15)	362
1,1,2,2-Tetrafluoro-3-(1,1,2,2-tetrafluoroethoxy) propane	$C_5H_4F_8O$	366.32	510.07(0.08)	2.58(0.01)	440(2)	113
1,1,1,2-Tetrafluoro-2-(trifluoromethoxy)ethane	C_3HF_7O	264(2)	377.26(0.06)	2.62(0.01)	321(2)	525
3,4,4,4-Tetrafluoro-3-(trifluoromethyl)-2-butanone	$C_5H_3F_7O$	328.76(0.05)	468(1)	2.50(0.06)	409(5)	110
4,4,5,5-Tetrafluoro-2-(trifluoromethyl)-1,3-dioxolane	$C_4HF_7O_2$	304.6(0.1)	435(1)	2.62(0.07)	376(4)	110
Tetrahexoxysilane	$C_{24}H_{52}O_4Si$		757(16)			57
Tetrahydrofuran	C_4H_8O	338	540(1)	5.29(0.06)	223(2)	23, 241, 290
1,2,3,4-Tetrahydronaphthalene	$C_{10}H_{12}$	480.8	720(1)	3.6(0.1)	431(40)	144, 242, 636
Tetrahydropyran	$C_5H_{10}O$	361	572.0(0.3)	4.8(0.2)	278(18)	75

Name	Mol. Form.	T_b/K	T_c/K	P_c/MPa	V_c/cm³ mol⁻¹	Ref.
Tetrahydrothiophene	C_4H_8S	394.3	632.0(0.2)	5.4(0.6)	276(19)	143, 241
1,2,4,5-Tetraisopropylbenzene	$C_{18}H_{30}$	532	703(1)	1.65(0.02)	983(83)	317
Tetramethoxysilane	$C_4H_{12}O_4Si$	393.3(0.7)	558(12)	2.8(0.6)	464(22)	57
1,2,4,5-Tetramethylbenzene	$C_{10}H_{14}$	470.0	676(2)	2.9(0.3)	489(11)	39
2,2,3,3-Tetramethylhexane	$C_{10}H_{22}$	433.5	623.0(0.5)	2.51(0.06)	574(14)	536
2,2,5,5-Tetramethylhexane	$C_{10}H_{22}$	410.6	581.4(0.5)	2.19(0.01)	600(14)	536
2,2,3,3-Tetramethylpentane	C_9H_{20}	413.4	607.5(0.5)	2.74(0.03)	514(15)	536
2,2,3,4-Tetramethylpentane	C_9H_{20}	406.2	592.6(0.5)	2.60(0.03)	517(17)	536
2,2,4,4-Tetramethylpentane	C_9H_{20}	395.44	574.6(0.5)	2.49(0.01)	532(16)	536
2,3,3,4-Tetramethylpentane	C_9H_{20}	414.7	607.5(0.5)	2.72(0.04)	517(17)	536
Tetramethylsilane	$C_4H_{12}Si$	299.8	449(2)	2.82(0.01)	362(7)	637–639
Tetramethylstannane	$C_4H_{12}Sn$	351	521.77(0.05)	2.98(0.01)		640, 641
Tetranonoxysilane	$C_{36}H_{76}O_4Si$		830(16)			57
Tetrapropyl silicate	$C_{12}H_{28}O_4Si$		649(12)			57
Thiacyclohexane	$C_5H_{10}S$	414.84(0.04)	684(44)	6.50(0.07)	284(11)	104
Thiobis(trifluoromethane)	C_2F_6S	251.3(0.3)	376.8(0.1)	3.2(0.5)	216(19)	442
Thiophene	C_4H_4S	357.2	579.4(0.2)	5.7(0.2)	230(3)	241, 290
Thymol	$C_{10}H_{14}O$	505.7	698(10)		528(21)	302
Toluene	C_7H_8	383.78	591.9(0.2)	4.13(0.02)	314(7)	34, 63, 68, 74, 78, 79, 84, 86, 88, 90, 96, 163, 172, 258, 481, 636, 642–646
Triacontane	$C_{30}H_{62}$	725.1	843(8)	0.6(0.1)	2055(69)	358
Tribromomethane	$CHBr_3$	422.4(0.2)	682(1)	5.8(0.2)	261(12)	647
Trichloroacetyl chloride	C_2Cl_4O	391.36(0.02)	604(2)	4.21(0.03)	331(54)	410
Trichloroethylsilane	$C_2H_5Cl_3Si$	371.8(0.4)	559.9(0.6)	3.34(0.04)	403(5)	266
Trichlorofluoromethane	CCl_3F	296.9	471.1(0.2)	4.40(0.03)	248.0(0.9)	33, 34, 191, 271, 571
Trichloromethane	$CHCl_3$	334.32	536.0(0.4)	5.5(0.2)	237(7)	18, 27, 28, 117
Trichloromethylsilane	CH_3Cl_3Si	338.8	517.7(0.3)	3.52(0.03)	329(9)	264, 648
1,3,5-Trichloro-2,4,6-trifluorobenzene	$C_6Cl_3F_3$	471.6	684.7(0.4)	3.3(0.1)	443(27)	25
1,1,2-Trichloro-1,2,2-trifluoroethane	$C_2Cl_3F_3$	320.9	487.4(0.2)	3.40(0.02)	325(1)	191, 265, 271, 649
Tricosane	$C_{23}H_{48}$	653	790(8)	0.9(0.1)	1527(53)	359
Tridecane	$C_{13}H_{28}$	508.62	676(1)	1.68(0.04)	824(30)	34, 244, 249
1-Tridecanol	$C_{13}H_{28}O$	547	732(7)	1.8(0.2)	828(14)	361
2-Tridecanone	$C_{13}H_{26}O$	536	717(6)	1.8(0.2)	820(24)	256
3-Tridecanone	$C_{13}H_{26}O$	539(7)	716(5)	1.7(0.5)	823(24)	256
4-Tridecanone	$C_{13}H_{26}O$	539(7)	712(6)	1.7(0.5)	823(24)	256
5-Tridecanone	$C_{13}H_{26}O$	539(8)	710(8)	1.7(0.5)	826(17)	256
6-Tridecanone	$C_{13}H_{26}O$	539(7)	709(5)	1.7(0.5)	826(24)	256
7-Tridecanone	$C_{13}H_{26}O$	534	708(5)	1.7(0.5)	830(24)	256
1-Tridecene	$C_{13}H_{26}$	505.9(0.1)	673(7)	1.74(0.05)	770(17)	365
Tridecylbenzene	$C_{19}H_{32}$	619	790(8)	1.5(0.1)	1079(43)	258
Triethylamine	$C_6H_{15}N$	362	535.6(0.3)	3.1(0.3)	392(25)	117, 286
1,3,5-Triethylbenzene	$C_{12}H_{18}$	488.92(0.02)	679(2)	2.32(0.01)	624(60)	332
Triethylene glycol	$C_6H_{14}O_4$	558	775(30)	3.3(0.2)	454(25)	283
Trifluoroacetonitrile	C_2F_3N	204.4	311.1(0.4)	3.61(0.04)	202(4)	470
1,2,3-Trifluorobenzene	$C_6H_3F_3$	368	560.3(0.4)	4.1(0.4)	296(20)	119
1,2,4-Trifluorobenzene	$C_6H_3F_3$	363	551.1(0.4)	4.1(0.6)	297(20)	119
1,3,5-Trifluorobenzene	$C_6H_3F_3$	348.7	530.9(0.4)	3.8(0.2)	300(20)	119
1,1,1-Trifluoroethane	$C_2H_3F_3$	225.90	345.89(0.07)	3.77(0.01)	195(15)	179, 203, 309, 478, 479, 549, 624, 650–653
2,2,2-Trifluoroethanol	$C_2H_3F_3O$	346.97(0.06)	498.57(0.05)	4.81(0.01)	211(12)	654, 655
2,2,2-Trifluoroethyl methyl ether	$C_3H_5F_3O$	304.77	448.98(0.08)	3.51(0.06)	277(3)	176
Trifluoroiodomethane	CF_3I	250.7	396.44(0.06)	3.95(0.01)	231(3)	656–659
Trifluoromethane	CHF_3	191.1	299.00(0.02)	4.82(0.01)	133(1)	310, 606, 660–664
Trifluoromethyl difluoromethyl ether	C_2HF_5O	235	354.49(0.06)	3.36(0.02)	226(20)	203
Trifluoromethyl 1,1,2,2-tetrafluoroethyl ether	C_3HF_7O	270	387.8(0.5)	2.65(0.01)	341(22)	116, 571
3,3,3-Trifluoropropene	$C_3H_3F_3$	256	378.6(0.5)	3.61(0.08)	229(14)	440
Trimethylamine	C_3H_9N	276.02	433.0(0.6)	4.08(0.04)	254(6)	322, 323, 665
1,2,3-Trimethylbenzene	C_9H_{12}	449.27	664.4(0.1)	3.45(0.03)	423(11)	79
1,2,4-Trimethylbenzene	C_9H_{12}	442.53	649.1(0.1)	3.3(0.1)	436(12)	79, 86, 96, 251
1,3,5-Trimethylbenzene	C_9H_{12}	437.89	637.31(0.1)	3.13(0.05)	435(12)	79

Name	Mol. Form.	T_b/K	T_c/K	P_c/MPa	V_c/cm^3 mol^{-1}	Ref.
3,7,7-Trimethyl-bicyclo[4.1.0]hept-3-ene	$C_{10}H_{16}$	445(2)	658(2)	2.9(0.5)	487(10)	496
2,2,3-Trimethylbutane	C_7H_{16}	354.01	531.3(0.5)	2.96(0.03)	401(13)	346, 349
Trimethylchlorosilane	C_3H_9ClSi	333	497.7(0.6)	3.20(0.03)	366(6)	266
1α,3α,5β-1,3,5-Trimethylcyclohexane	C_9H_{18}	413.7	602(2)	2.6(0.3)	494(14)	74
3,3,5-Trimethylheptane	$C_{10}H_{22}$	428.9	609.5(0.5)	2.32(0.05)	583(18)	536
2,2,5-Trimethylhexane	C_9H_{20}	397.24	570(2)	2.46(0.03)	547(18)	350
2,2,3-Trimethylpentane	C_8H_{18}	383	563.5(0.4)	2.73(0.02)	442(16)	346
2,2,4-Trimethylpentane	C_8H_{18}	372.37	543.9(0.4)	2.57(0.02)	475(20)	33, 34, 86, 159, 346, 666, 667
2,3,3-Trimethylpentane	C_8H_{18}	388.0	573.5(0.4)	2.82(0.03)	454(14)	346
2,3,4-Trimethylpentane	C_8H_{18}	386.7	566.4(0.4)	2.72(0.02)	462(12)	346
cis-Tri(methylphenyl)trisiloxane	$C_{21}H_{24}O_3Si_3$		824(8)			353
trans-2,4,6-Trimethyl-2,4,6-triphenylcyclotrisiloxane	$C_{21}H_{24}O_3Si_3$		839(8)			353
Undecafluorocyclohexane	C_6HF_{11}	335.2	477.7(0.7)			546
Undecane	$C_{11}H_{24}$	469.1	638.8(0.2)	2.01(0.03)	683(20)	34, 86, 131, 243, 249, 363, 535
Undecanoic acid	$C_{11}H_{22}O_2$	553	728(7)	2.1(0.2)	741(20)	360
1-Undecanol	$C_{11}H_{24}O$	518	703.0(0.6)	2.15(0.07)	707(12)	149
2-Undecanone	$C_{11}H_{22}O$	504.7	688(2)	2.08(0.01)	692(20)	256
3-Undecanone	$C_{11}H_{22}O$	500	685(2)	2.0(0.4)	692(20)	256
4-Undecanone	$C_{11}H_{22}O$	501(3)	681(2)	2.0(0.2)	692(20)	256
5-Undecanone	$C_{11}H_{22}O$	500	679(2)	2.0(0.2)	692(20)	256
6-Undecanone	$C_{11}H_{22}O$	501	678(2)	2.02(0.01)	692(20)	256
Undecylbenzene	$C_{17}H_{28}$	589	763(8)	1.6(0.1)	946(35)	258
Vinyl acetate	$C_4H_6O_2$	346.0	519.2(0.2)	4.17(0.03)	269(7)	156, 440
m-Xylene	C_8H_{10}	412.27	616.9(0.3)	3.54(0.01)	377(7)	34, 68, 79, 88, 90, 106, 668
o-Xylene	C_8H_{10}	417.7	630.26(0.1)	3.74(0.01)	372(40)	34, 68, 78, 79, 88, 669
p-Xylene	C_8H_{10}	411.52	616.17(0.09)	3.55(0.02)	372(35)	34, 68, 74, 79, 88, 90, 669
2,3-Xylenol	$C_8H_{10}O$	490.1	723(1)	4.1(0.3)	397(15)	218
2,4-Xylenol	$C_8H_{10}O$	484.13	708(1)	3.5(0.3)	389(15)	218
2,5-Xylenol	$C_8H_{10}O$	484.3	707(1)	3.9(0.1)	397(15)	25
2,6-Xylenol	$C_8H_{10}O$	474.22	701(1)	3.8(0.1)	396(15)	218
3,4-Xylenol	$C_8H_{10}O$	500	730(1)	4.9(0.5)	388(15)	218
3,5-Xylenol	$C_8H_{10}O$	494.89	716(1)	3.8(0.2)	396(15)	218

References

1. Frenkel, M., Chirico, R. D., Diky, V. V., Kazakov, A., and Muzny, C. D., ThermoData Engine. NIST Standard Reference Database 103b, Version 4.0 (Pure Compounds, Binary Mixtures, and Chemical Reactions, TDE-SOURCE Version 4.3), National Institute of Standards and Technology, Gaithersburg, MD – Boulder, CO, 2009, http://www.nist.gov/srd/nist103b.htm
2. Frenkel, M., Chirico, R. D., Diky, V., Yan, X., Dong, Q., and Muzny, C., *J. Chem. Inf. Model.* 45, 816, 2005.
3. Diky, V., Muzny, C. D., Lemmon, E. W., Chirico, R. D., and Frenkel, M., *J. Chem. Inf. Model.* 47, 1713, 2007.
4. Diky, V., Chirico, R. D., Kazakov, A. F., Muzny, C., and Frenkel, M., *J. Chem. Inf. Model.* 49, 503, 2009.
5. Diky, V., Chirico, R. D., Kazakov, A. F., Muzny, C., and Frenkel, M., *J. Chem. Inf. Model.* 49, 2883, 2009.
6. Joback, K. G., and Reid, R. C., *Chem. Eng. Commun.* 57, 233, 1987.
7. Constantinou, L., and Gani, R., *AIChE J.* 40, 1697, 1994.
8. Marrero-Morejon, J., and Pardillo-Fontdevila, E., *AIChE J.* 45, 615, 1999.
9. Wilson, G. M., and Jasperson, L. V., AIChE Meeting, New Orleans, LA, 1996.
10. Chirico, R. D., Frenkel, M., Diky, V. V., Marsh, K. N., and Wilhoit, R. C., *J. Chem. Eng. Data* 48, 1344, 2003.
11. Teja, A. S., and Anselme, M. J., *AIChE Symp. Ser.* 86 (279), 115, 1990.
12. Hollmann, R., *Z. Phys. Chem., Stoechiom. Verwandtschaftsl.* 43, 129, 1903.
13. Van der Waals, J. D., *Continuity of Gas and Liquid Data*, 1st edition, Leipzig, p. 168, 1881.
14. Vandana, V., and Teja, A. S., *Fluid Phase Equilib.* 103, 113, 1995.
15. Ambrose, D., Ellender, J. H., Sprake, C. H. S., and Townsend, R., *J. Chem. Thermodyn.* 9, 735, 1977.
16. Kreglewski, A., *Rocz. Chem.* 31, 1001, 1957.
17. Swietoslawski, W., and Kreglewski, A., *Bull. Acad. Pol. Sci., Cl. 3* 2, 77, 1954.
18. Swietoslawski, W., and Kreglewski, A., *Bull. Acad. Pol. Sci., Cl. 3* 2, 187, 1954.
19. Young, S., *Sci. Proc. R. Dublin Soc.* 12, 374, 1910.
20. Young, S., *J. Chem. Soc.* 59, 903, 1891.
21. Pawlewski, B., *Ber. Dtsch. Chem. Ges.* 16, 2633, 1883.
22. Ambrose, D., and Ghiassee, N. B., *J. Chem. Thermodyn.* 19, 911, 1987.
23. Sassa, Y., Konishi, R., and Katayama, T., *J. Chem. Eng. Data* 19, 44, 1974.
24. Ambrose, D., Sprake, C. H. S., and Townsend, R., *J. Chem. Thermodyn.* 6, 693, 1974.
25. Ambrose, D., Broderick, B. E., and Townsend, R., *J. Appl. Chem. Biotechnol.* 24, 359, 1974.
26. Campbell, A. N., and Musbally, G. M., *Can. J. Chem.* 48, 3173, 1970.
27. Campbell, A. N., and Chatterjee, R. M., *Can. J. Chem.* 47, 3893, 1969.
28. Campbell, A. N., and Chatterjee, R. M., *Can. J. Chem.* 46, 575, 1968.
29. Kay, W. B., *J. Phys. Chem.* 68, 827, 1964.
30. Rosenbaum, M., M.S. Thesis, Univ. Texas, Austin, TX, 1951.
31. Kuenen, J. P., and Robson, W. G., *Philos. Mag.* 3, 622, 1902.
32. VonNiederhausern, D. M., Wilson, L. C., Giles, N. F., and Wilson, G. M., *J. Chem. Eng. Data* 45, 154, 2000.
33. Christou, G., Young, C. L., and Svejda, P., *Ber. Bunsen-Ges. Phys. Chem.* 95, 510, 1991.

34. Christou, G., Ph. D. Dissertation, Univ. Melbourne, 1988.
35. Castillo-Lopez, N., and Trejo Rodriguez, A., *J. Chem. Thermodyn.* 19, 671, 1987.
36. Trejo Rodriguez, A., and McLure, I. A., *Fluid Phase Equilib.* 12, 297, 1983.
37. Trejo Rodriguez, A., and McLure, I. A., *J. Chem. Thermodyn.* 11, 1113, 1979.
38. Khera, R., Ph.D. Thesis, Ohio State Univ., Columbus, OH, 1968.
39. Guye, P. A., and Mallet, E., *Arch. Sci. Phys. Nat.* 13, 274, 1902.
40. Guye, P. A., and Mallet, E., *C. R. Hebd. Seances Acad. Sci.* 133, 168, 1902.
41. Teja, A. S., and Rosenthal, D. J., *Experimental Results for Phase Equilibria and Pure Component Properties*, DIPPR DATA Series No. 1, p. 96, 1991.
42. Goloborod'ko, N. P., and Khodeeva, S. M., *Russ. J. Phys. Chem.* (*Engl. Transl.*) 46, 235, 1972.
43. Mislavskaya, V. S., and Khodeeva, S. M., *Zh. Fiz. Khim.* 43, 2367, 1969.
44. Khodeeva, S. M., *Russ. J. Phys. Chem.* (*Engl. Transl.*) 40, 1061, 1966.
45. Ambrose, D., and Townsend, R., *Trans. Faraday Soc.* 60, 1025, 1964.
46. Ambrose, D., *Trans. Faraday Soc.* 52, 772, 1956.
47. McIntosh, D., *J. Phys. Chem.* 11, 306, 1907.
48. Kuenen, J. P., *Philos. Mag.* 44, 174, 1897.
49. Wilson, L. C., Wilson, H. L., Wilding, W. V., and Wilson, G. M., *J. Chem. Eng. Data* 41, 1252, 1996.
50. Lespieau, R., and Chavanne, G., *C. R. Hebd. Seances Acad. Sci.* 140, 1035, 1905.
51. Wang, X., Jia, Q., Gao, J., Xia, S., and Ma, P. S., *J. Chem. Ind. Eng.* (*China*) 56, 1385, 2005.
52. Liang, Y.-H., Ma, P. S., and Zhang, H., *J. Chem. Ind. Eng.* (*China*) 51, 243, 2000.
53. Steele, W. V., Chirico, R. D., Knipmeyer, S. E., and Nguyen, A., *J. Chem. Thermodyn.* 23, 957, 1991.
54. VonNiederhausern, D. M., Wilson, G. M., and Giles, N. F., *J. Chem. Eng. Data* 51, 1990, 2006.
55. Wilson, G. M., VonNiederhausern, D. M., and Giles, N. F., *J. Chem. Eng. Data* 47, 761, 2002.
56. Chashkin, Yu. R., Gorbunova, V. G., and Voronel, A. V., *Zh. Exp. Teor. Fiz.* 49, 432, 1965.
57. Nikitin, E. D., and Popov, A. P., *J. Chem. Eng. Data* 53, 1371, 2008.
58. Steele, W. V., Chirico, R. D., Knipmeyer, S. E., and Nguyen, A., J. Chem. Eng. Data 47, 648 - 666, 2002.
59. Lagutkin, O. D., and Kuropatkin, E. I., *Zh. Fiz. Khim.* 55, 1329, 1981.
60. Livingston, J., Morgan, R., and Higgins, E., *Z. Phys. Chem., Stoechiom. Verwandtschaftsl.* 64, 170, 1908.
61. Liu, T., Fu, J., Wang, K., Gao, Y., and Yuan, W., *J. Chem. Eng. Data* 46, 809, 2001.
62. Nikitin, E. D., Pavlov, P. A., and Skutin, M., *Fluid Phase Equilib.* 161, 119, 1999.
63. Chirico, R. D., and Steele, W. V., *Ind. Eng. Chem. Res.* 33, 157, 1994.
64. Zhang, J., Zhao, X., and Ma, P., *Huagong Xuebao* 43, 105, 1992.
65. Knipmeyer, S. E., Archer, D. G., Chirico, R. D., Gammon, B. E., Hossenlopp, I. A., Nguyen, A., Smith, N. K., Steele, W. V., and Strube, M. M., *Fluid Phase Equilib.* 52, 185, 1989.
66. Goodwin, R. D., *J. Phys. Chem. Ref. Data* 17, 1541, 1988.
67. Brunner, E., *J. Chem. Thermodyn.* 20, 1397, 1988.
68. Ambrose, D., *J. Chem. Thermodyn.* 19, 1007, 1987.
69. Kay, W. B., and Kreglewski, A., *Fluid Phase Equilib.* 11, 251, 1983.
70. Hales, J. L., and Gundry, H. A., *J. Phys. E* 16, 91, 1983.
71. Hugill, J. A., and McGlashan, M. L., *J. Chem. Thermodyn.* 13, 429, 1981.
72. Ewing, M. B., McGlashan, M. L., and Tzias, P., *J. Chem. Thermodyn.* 13, 527, 1981.
73. Akhundov, T. S., and Abdullaev, F. G., *Izv. Vyssh. Uchebn. Zaved., Neft Gaz* 20, 73, 1977.
74. Powell, R. J., Swinton, F. L., and Young, C. L., *J. Chem. Thermodyn.* 2, 105, 1970.
75. Kobe, K. A., and Mathews, J. F., *J. Chem. Eng. Data* 15, 182, 1970.
76. Artyukhovskaya, L. M., Shimanskaya, E. T., and Shimanskii, Yu. I., *Ukr. Fiz. Zh.* (*Ukr. Ed.*) 15, 1974, 1970.
77. Skripov, V. P., and Sinitsyn, E. N., *Zh. Fiz. Khim.* 42, 309, 1968.
78. Kay, W. B., and Hissong, D. W., *Proc. - Am. Pet. Inst., Div. Refin.* 47, 653, 1967.
79. Ambrose, D., Broderick, B. E., and Townsend, R., *J. Chem. Soc. A* 633, 1967.
80. Skaates, J. M., and Kay, W. B., *Chem. Eng. Sci.* 19, 431, 1964.
81. Makhan'ko, I. G., and Nozdrev, V. F., *Akust. Zh.* 10, 249, 1964.
82. Cheng, D. C. H., *Chem. Eng. Sci.* 18, 715, 1963.
83. Connolly, J. F., and Kandalic, G. A., *J. Chem. Eng. Data* 7, 137, 1962.
84. Partington, E. J., Rowlinson, J. S., and Weston, J. F., *Trans. Faraday Soc.* 56, 479, 1960.
85. McCracken, P. G., Storvick, T. S., and Smith, J. M., *J. Chem. Eng. Data* 5, 130, 1960.
86. Ambrose, D., Cox, J. D., and Townsend, R., *Trans. Faraday Soc.* 56, 1452, 1960.
87. Krichevskii, I. R., Khazanova, N. E., and Linshits, L. R., *Tr. GIAP* No. 9, 40, 1959.
88. Simon, M., *Bull. Soc. Chim. Belg.* 66, 375, 1957.
89. Krichevskii, I. R., Khazanova, N. E., and Linshits, L. R., *Zh. Fiz. Khim.* 31, 2711, 1957.
90. Ambrose, D., and Grant, D. G., *Trans. Faraday Soc.* 53, 771, 1957.
91. Kreglewski, A., *Rocz. Chem.* 29, 754, 1955.
92. Kay, W. B., and Nevens, T. D., *Chem. Eng. Prog., Symp. Ser.* 48, 108, 1952.
93. Bender, P., Furukawa, G. T., and Hyndman, J. R., *Ind. Eng. Chem.* 44, 387, 1952.
94. Gornowski, E. J., Amick, E. H., and Hixson, A. N., *Ind. Eng. Chem.* 39, 1348, 1947.
95. Schamhardt, H. O., Thesis, Amsterdam, The Netherlands, 1908.
96. Altschul, M., *Z. Phys. Chem., Stoechiom. Verwandtschaftsl.* 11, 577, 1893.
97. Young, S., *J. Chem. Soc., Trans.* 55, 486, 1889.
98. Nikitin, E. D., Popov, A. P., and Yatluk, Y. G., *J. Chem. Eng. Data* 51, 1335, 2006.
99. VonNiederhausern, D. M., Wilson, G. M., and Giles, N. F., *J. Chem. Eng. Data* 51, 1982, 2006.
100. Chirico, R. D., Knipmeyer, S. E., Nguyen, A., and Steele, W. V., *J. Chem. Thermodyn.* 23, 759, 1991.
101. Steele, W. V., Chirico, R. D., Hossenlopp, I. A., Knipmeyer, S. E., Nguyen, A., and Smith, N. K., *Experimental Results for DIPPR 1990-91 Projects on Phase Equilibria and Pure Component Properties*, DIPPR Data Ser. No. 2, p. 188, 1994.
102. Guseinov, S. O., Naziev, Y. M., Farzaliev, B. I., and Movsunov, T. G., *Izv. Vyssh. Uchebn. Zaved., Neft Gaz* 21, 48, 1978.
103. Ambrose, D., and Ghiassee, N. B., *J. Chem. Thermodyn.* 22, 307, 1990.
104. Anselme, M. J., and Teja, A. S., *AIChE Symp. Ser.* 86 (279), 128, 1990.
105. Chirico, R. D., and Steele, W. V., *J. Chem. Thermodyn.* 36, 633, 2004.
106. Glaser, F., and Ruland, H., *Chem.-Ing.-Tech.* 29, 772, 1957.
107. Chirico, R. D., Knipmeyer, S. E., Nguyen, A., and Steele, W. V., *J. Chem. Thermodyn.* 21, 1307, 1989.
108. Reiter, R. W., *NASA Document* N63, 1963.
109. Ellard, J. A., and Yanko, W. H., *U. S. A. E. C. Rep.* IDO-11008, 1963.
110. Sako, T., Yasumoto, M., Nakazawa, N., and Kamizawa, C., *J. Chem. Eng. Data* 46, 1078, 2001.
111. Defibaugh, D. R., Gillis, K. A., Moldover, M. R., Morrison, G., and Schmidt, J. W., *Fluid Phase Equilib.* 81, 285, 1992.
112. Nikitin, E. D., Popov, A. P., and Yatluk, Y. G., *J. Chem. Eng. Data* 51, 1326, 2006.
113. Yasumoto, M., Yamada, Y., Murata, J., Urata, S., and Otake, K., *J. Chem. Eng. Data* 48, 1368, 2003.
114. McLure, I. A., and Neville, J. F., *J. Chem. Thermodyn.* 14, 385, 1982.
115. Badylkes, S., *Kholod. Tekh.* 43, 18, 1966.
116. Salvi-Narkhede, M., Wang, B. -H., Adcock, J. I., and Van Hook, W. A., *J. Chem. Thermodyn.* 24, 1065, 1992.
117. Herz, W., and Neukirch, E., *Z. Phys. Chem., Stoechiom. Verwandtschaftsl.* 104, 433, 1923.
118. Adamenko, I. I., and Chernyavskaya, I. A., *Ukr. Fiz. Zh.* 11, 336, 1966.
119. Morton, D. W., Lui, M. P. W., Tran, C. A., and Young, C. L., *J. Chem. Eng. Data* 45, 437, 2000.
120. Li, Y., Ma, P., and Ruan, Y., *Shiyou Huagong* 22, 322, 1993.
121. Higashi, Y., Uematsu, M., and Watanabe, K., *Bull. JSME* 28, 2660, 1985.

122. Scott, R. B., Meyers, C. H., Rands, R. D., Brickwedde, F. G., and Bekkedahl, N., *J. Res. Natl. Bur. Stand.* (*U. S.*) 35, 39, 1945.
123. Cragoe, C. S., *Natl. Bur. Stand.* (*U. S.*) LC-736, 1943.
124. Ma, P. S., Gao, J., and Xia, S., *Chin. J. Chem. Eng.* 10, 473, 2002.
125. Yasumoto, M., Uchida, Y., Ochi, K., Furuya, T., and Otake, K., *J. Chem. Eng. Data* 50, 596, 2005.
126. Warowny, W., *J. Chem. Eng. Data* 41, 689, 1996.
127. Holcomb, C. D., Magee, J. W., and Haynes, W. M., *Research Report RR-147*, Gas Processors Association Project 916, Tulsa, OK, 1995.
128. Deak, A., Victorov, A. I., and De Loos, T. W., *Fluid Phase Equilib.* 107, 277, 1995.
129. Vasserman, A. A., Khasilev, I. P., and Cymarnyi, V. A., *Deposited Doc. VNIIKI*, Doc. No. 604-kk, 1989.
130. Li, L., and Kiran, E., *J. Chem. Eng. Data* 33, 342, 1988.
131. Brunner, E., *J. Chem. Thermodyn.* 20, 273, 1988.
132. Younglove, B. A., and Ely, J. F., *J. Phys. Chem. Ref. Data* 16, 577, 1987.
133. Kreglewski, A., and Kay, W. B., *J. Phys. Chem.* 73, 3359, 1969.
134. Golubev, I. F., and Agaev, N. A., *Dokl. Akad. Nauk SSSR* 151(4), 875, 1963.
135. Kay, W. B., *Ind. Eng. Chem.* 33, 590, 1941.
136. Sage, B. H., Hicks, B. L., and Lacey, W. N., *Ind. Eng. Chem.* 32, 1085, 1940.
137. Kay, W. B., *Ind. Eng. Chem.* 32, 353, 1940.
138. Beattie, J. A., Simard, G. L., and Su, G.-J., *J. Am. Chem. Soc.* 61, 24, 1939.
139. Kuenen, J. P., *Commun. Kamerlingh Onnes Lab., Univ. Leiden*, No. 125, 1, 1911.
140. Nikitin, E. D., Popov, A. P., and Yatluk, Y. G., *J. Chem. Eng. Data* 51, 609, 2006.
141. Steele, W. V., Chirico, R. D., Knipmeyer, S. E., and Nguyen, A., *J. Chem. Eng. Data* 41, 1255, 1996.
142. Eliosa, G., Murrieta-Guevara, F., Reza, J., and Trejo Rodriguez, A., *Fluid Phase Equilib.* 61, 99, 1990.
143. Tsonopoulos, C., and Ambrose, D., *J. Chem. Eng. Data* 46, 480, 2001.
144. Teja, A. S., and Anselme, M. J., *AIChE Symp. Ser.* 86 (279), 122, 1990.
145. Gude, M. T., Mendez-Santiago, J., and Teja, A. S., *J. Chem. Eng. Data* 42, 278, 1997.
146. Ambrose, D., and Ghiassee, N. B., *J. Chem. Thermodyn.* 19, 505, 1987.
147. Brown, J. C., *J. Chem. Soc., Trans.* 89, 311, 1906.
148. Lydersen, A. L., and Tsochev, V., *Chem. Eng. Technol.* 13, 125, 1990.
149. Rosenthal, D. J., and Teja, A. S., *Ind. Eng. Chem. Res.* 28, 1693, 1989.
150. Christou, G., and Young, C. L., *Int. DATA Ser.*, Sel. Data Mixtures, Ser. A, 14(4), 245, 1986.
151. Naumova, A. A., Tyvina, T. N., and Fokina, V. V., *Zh. Prikl. Khim.* 1980, 1667, 1980.
152. Efremov, Yu. V., *Zh. Fiz. Khim.* 40, 1240, 1966.
153. Ambrose, D., and Townsend, R., *J. Chem. Soc.* 54, 3614, 1963.
154. Singh, R., and Shemilt, L. W., *J. Chem. Phys.* 23, 1370, 1955.
155. Kay, W. B., and Donham, W. E., *Chem. Eng. Sci.* 4, 1, 1955.
156. Stevens, R. M. M., Van Roermund, J. C., Jager, M. D., De Loos, T. W., and De Swaan Arons, J., *Fluid Phase Equilib.*, 138, 159, 1997.
157. Kobe, K. A., Crawford, H. R., and Stephenson, R. W., *Ind. Eng. Chem.* 47, 1767, 1955.
158. Ihmels, E. C., Fischer, K., and Gmehling, J., *Fluid Phase Equilib.* 228-229, 155, 2005.
159. Li, J., Qin, Z., Wang, G., Dong, M., and Wang, J., *J. Chem. Eng. Data* 52, 1736, 2007.
160. Beattie, J. A., and Marple, S., *J. Am. Chem. Soc.* 72, 1449, 1950.
161. Olds, R. H., Sage, B. H., and Lacey, W. N., *Ind. Eng. Chem.* 38, 301, 1946.
162. Morton, D. W., Lui, M., and Young, C. L., *J. Chem. Thermodyn.* 31, 675, 1999.
163. Teja, A. S., and Rosenthal, D. J., *AIChE Symp. Ser.* 86 (279), 133, 1990.
164. Ma, F., Wang, J., and Ruan, Y., *J. Chem. Eng. Chin. Univ.* 9, 62, 1995.
165. Ma, P., and Ruan, Y., *Gaoxiao Huaxue Gongcheng Xuebao* 9, 62, 1995.
166. Quadri, S. K., and Kudchadker, A. P., *J. Chem. Thermodyn.* 23, 129, 1991.
167. Pawlewski, B., *Ber. Dtsch. Chem. Ges.* 15, 2460, 1882.
168. Toczylkin. L. S., and Young, C. L., *J. Chem. Thermodyn.* 12, 365, 1980.
169. Nikitin, E. D., and Popov, A. P., *J. Chem. Eng. Data* 52, 1336, 2007.
170. Morton, D. W., Lui, M. P. W., Tran, C. A., and Young, C. L., *J. Chem. Eng. Data* 49, 283, 2004.
171. Nikitin, E. D., Popov, A. P., and Bogatishcheva, N. S., *J. Chem. Eng. Data* 48, 1137, 2003.
172. Wilson, L. C., Wilding, W. V., Wilson, H. L., and Wilson, G. M., *J. Chem. Eng. Data* 40, 765, 1995.
173. Zawisza, A. C., and Glowka, S., *Bull. Acad. Pol. Sci., Ser. Sci. Chim.* 19, 191, 1971.
174. Osipiuk, B., and Stryjek, R., *Bull. Acad. Pol. Sci., Ser. Sci. Chim.* 18, 289, 1970.
175. Steele, W. V., Chirico, R. D., Knipmeyer, S. E., Nguyen, A., and Smith, N. K., *J. Chem. Eng. Data* 41, 1285, 1996.
176. Sako, T., Sato, M., Nakazawa, N., Oowa, M., Yasumoto, M., Ito, H., and Yamashita, S., *J. Chem. Eng. Data* 41, 802, 1996.
177. Tanikawa, S., Tatoh, J., Maezawa, Y., Sato, H., and Watanabe, K., *J. Chem. Eng. Data* 37, 74, 1992.
178. Chae, H. B., Schmidt, J. W., and Moldover, M. R., *J. Phys. Chem.* 94, 8840, 1990.
179. Mears, W. H., Stahl, R. F., Orfeo, S. R., Shair, R. C., Kells, L. F., Thompson, W., and McCann, H., *Ind. Eng. Chem.* 47, 1449, 1955.
180. Yata, J., Hori, M., Kawakatsu, H., and Minamiyama, T., *Int. J. Thermophys.* 17, 65, 1996.
181. Zhao, X., and Ma, P. S., *Chin. J. Chem. Eng.* 3, 233, 1995.
182. Nishiumi, H., Kohmatsu, S., Yokoyama, T., and Konda, A., *Fluid Phase Equilib.* 104, 131, 1995.
183. Economou, I. G., Peters, C. J., Florusse, L. J., and De Swaan Arons, J., *Fluid Phase Equilib.* 111, 239, 1995.
184. Nishiumi, H., Komatsu, M., Yokoyama, T., and Kohmatsu, S., *Fluid Phase Equilib.* 83, 109, 1993.
185. Wang, J., Liu, Z. G., Tan, L. C., and Yin, J. M., *Fluid Phase Equilib.* 80, 203, 1992.
186. Noles, J. R., and Zollweg, J. A., *J. Chem. Eng. Data* 37, 306, 1992.
187. Leu, A. D., and Robinson, D. B., *J. Chem. Eng. Data* 37, 7, 1992.
188. Goodwin, A. R. H., Defibaugh, D. R., and Weber, L. A., *Int. J. Thermophys.* 13, 837, 1992.
189. Zhimai, H., and Jianfen, H., *Gongcheng Rewuli Xuebao* 10, 233, 1989.
190. He, Z., Zhang, Y., and Hong, J., *The Second Asian Thermophysical Properties Conference*, Hunan University of Science and Technology, Guangzhou, China, p. 519, 1989.
191. *Chemicals and Plastics Physical Properties*, Union Carbide Corp. (Product Bulletin), 1968.
192. Ambrose, D., and Ghiassee, N. B., *J. Chem. Thermodyn.* 20, 765, 1988.
193. Berthoud, A., *J. Chim. Phys. Phys.-Chim. Biol.* 15, 3, 1917.
194. Cullick, A. S., and Ely, J. F., *J. Chem. Eng. Data* 27, 276, 1982.
195. Mansoorian, H., Hall, K. R., Holste, J. C., and Eubank, P. T., *J. Chem. Thermodyn.* 13, 1001, 1981.
196. Centnerszwer, M., *Z. Phys. Chem., Stoechiom. Verwandtschaftsl.* 49, 199, 1904.
197. Murphy, K. P., *J. Chem. Eng. Data* 9, 259, 1964.
198. Skripov, V. P., and Muratov, G. N., *Russ. J. Phys. Chem.* (*Engl. Transl.*) 51, 806, 1977.
199. Yada, N., Uematsu, M., and Watanabe, K., *Nippon Kikai Gakkai Ronbunshu, B-hen* 55, 2426, 1989.
200. Mears, W. H., Rosenthal. E., and Sinka, J. V., *J. Chem. Eng. Data* 11, 338, 1966.
201. Ma, P., Fang, Z., Zhang, J., and Ruan, Y., *J. Chem. Eng. Chin. Univ.* 6, 112, 1992.
202. Ma, P., Fang, Z., Zhang, J., and Ruan, Y., *Gaoxiao Huaxue Gongcheng Xuebao* 6, 112, 1992.
203. Schmidt, J. W., Carrillo-Nava, E., and Moldover, M. R., *Fluid Phase Equilib.* 122, 187, 1996.
204. Fukushima, M., and Watanabe, N., *Nippon Reito Kyokai Ronbunshu* 10, 75, 1993.
205. Liu, Z., Liang, D., He, M., Ju, B., and Yin, J., *Gongchengrewuli Xuebao* 18, 261, 1997.
206. Booth, H. S., and Swinehart, C. F., *J. Am. Chem. Soc.* 57, 1337, 1935.
207. Oliver, G. D., Grisard, J. W., and Cunningham, C. W., *J. Am. Chem. Soc.* 73, 5719, 1951.
208. Weber, L. A., *J. Chem. Eng. Data* 34, 171, 1989.
209. Shavandrin, A. M., and Li, S. A., *Inzh.-Fiz. Zh.* 37, 830, 1979.

210. Vitkalov, V. S., Kolpakov, Y. D., and Skripov, V. P., *Zh. Fiz. Khim.* 50, 2336, 1976.
211. Oguchi, K., Tanishita, I., Watanabe, K., Yamaguchi, T., and Sasayama, A., *Bull. JSME* 18, 1456, 1975.
212. Muratov, G. N., and Skripov, V. P., *Zh. Fiz. Khim.* 49, 2148, 1975.
213. Levelt Sengers, J. M. H., Straub, J., and Vincentini-Missoni, M., *J. Chem. Phys.* 54, 5034, 1971.
214. Tsiklis, D. S., and Prokhorov, V. M., *Dokl. Akad. Nauk SSSR* 174, 470, 1967.
215. Tsiklis, D. S., and Prokhorov, V. M., *Zh. Fiz. Khim.* 41, 2195, 1967.
216. Michels, A., Wassenaar, T., Wolkers, G. J., Prins, Chr., and van de Klundert, L., *J. Chem. Eng. Data* 11, 449, 1966.
217. Delaunois, C., *Ann. Mines Belg.* No. 1, 9, 1968.
218. Ambrose, D., *Trans. Faraday Soc.* 59, 1988, 1963.
219. Dewar, J., *Philos. Mag.* 18, 210, 1884.
220. Young, C. L., *Aust. J. Chem.* 25, 1625, 1972.
221. Hicks, C. P., and Young, C. L., *Trans. Faraday Soc.* 67, 1605, 1971.
222. Fischer, R., and Reichel, T., *Mikrochem. Ver. Mikrochim. Acta* 31, 102, 1943.
223. Zhang, R., Qin, Z., Wang, G., Dong, M., Hou, X., and Wang, J., *J. Chem. Eng. Data* 50, 1414, 2005.
224. Hugill, J. A., and McGlashan, M. L., *J. Chem. Thermodyn.* 10, 95, 1978.
225. Naziev, Y. M., Abasov, A. A., Nurberdiev, A. A., and Shakhverdiev, A. N., *Zh. Fiz. Khim.* 68, 434, 1974.
226. Krichevskii, I. R., and Sorina, G. A., *Zh. Fiz. Khim.* 34, 1420, 1960.
227. Richardson, M. J., and Rowlinson, J. S., *Trans. Faraday Soc.* 53, 1586, 1959.
228. Reamer, H. H., Sage, B. H., and Lacey, W. N., *Chem. Eng. Data Ser.* 3, 240, 1958.
229. Kay, W. B., and Albert, R. E., *Ind. Eng. Chem.* 48, 422, 1956.
230. Rotinyantz, L., and Nagornov, N. N., *Z. Phys. Chem., Abt. A* 169, 20, 1934.
231. Young, S., and Fortey, E. C., *J. Chem. Soc., Trans.* 75, 873, 1899.
232. Steele, W. V., Chirico, R. D., Knipmeyer, S. E., and Nguyen, A., *J. Chem. Eng. Data* 42, 1021, 1997.
233. Ambrose, D., and Ghiassee, N. B., *J. Chem. Thermodyn.* 19, 903, 1987.
234. Alekhin, O. D., Krupskii, N. P., and Minchenko, Y. B., *Ukr. Fiz. Zh. (Ukr. Ed.)* 15, 509, 1970.
235. Kudchadker, A. P., Alani, G. H., and Zwolinski, B. J., *Chem. Rev.* 68, 659, 1968.
236. Kay, W. B., *J. Am. Chem. Soc.* 69, 1273, 1947.
237. Lin, D. C. K., Silberberg, I. H., and McKetta, J. J., *J. Chem. Eng. Data* 15, 483, 1970.
238. Booth, H. S., and Morris, W. C., *J. Phys. Chem.* 62, 875, 1958.
239. Grzyll, L. R., Ramos, C., and Back, D. D., *J. Chem. Eng. Data* 41, 446, 1996.
240. Ermakov, G. V., and Skripov, V. P., *Zh. Fiz. Khim.* 43, 1308, 1969.
241. Cheng, D. C. H., McCoubrey, J. C., and Phillips, D. G., *Trans. Faraday Soc.* 58, 224, 1962.
242. Gude, M. T., and Teja, A. S., *Experimental Results for DIPPR 1990-91 Projects on Phase Equilibria and Pure Component Properties*, DIPPR Data Series No. 2, p. 174, 1994.
243. Anselme, M. J., Gude, M., and Teja, A. S., *Fluid Phase Equilib.* 57, 317, 1990.
244. Rosenthal, D. J., and Teja, A. S., *AIChE J.* 35, 1829, 1989.
245. Smith, R. L., Teja, A. S., and Kay, W. B., *AIChE J.* 33, 232, 1987.
246. Brunner, E., *J. Chem. Thermodyn.* 19, 823, 1987.
247. Smith, R. L., Anselme, M., and Teja, A. S., *Proc. World Congress III Chem. Eng.*, Tokyo, Vol. II, p. 135, 1986.
248. Gehrig, M., and Lentz, H., *J. Chem. Thermodyn.* 15, 1159, 1983.
249. Mogollon, E., Kay, W. B., and Teja, A. S., *Ind. Eng. Chem. Fundam.* 21, 173, 1982.
250. Cholpan, P. F., Sperkach, V. S., and Garkusha, L. N., *Fiz. Zhidk. Sostoyaniya* 9, 79, 1981.
251. Kay, W. B., and Pak, S. C., *J. Chem. Thermodyn.* 12, 673, 1980.
252. Chun, S. W., Ph.D. Thesis, Ohio State Univ., Columbus, OH, 1964.
253. Nikitin, E. D., Popov, A. P., Bogatishcheva, N. S., and Yatluk, Y. G., *J. Chem. Eng. Data* 49, 1515, 2004.
254. Quadri, S. K., Khilar, K. C., Kudchadker, A. P., and Patni, M. J., *J. Chem. Thermodyn.* 23, 67, 1991.
255. Anselme, M. J., and Teja, A. S., *Fluid Phase Equilib.* 40, 127, 1988.
256. Pulliam, M. K., Gude, M. T., and Teja, A. S., *J. Chem. Eng. Data* 40, 455, 1995.
257. Gude, M. T., Rosenthal, D. J., and Teja, A. S., *Fluid Phase Equilib.* 70, 55, 1991.
258. Nikitin, E. D., Popov, A. P., Bogatishcheva, N. S., and Yatluk, Y. G., *J. Chem. Eng. Data* 47, 1012, 2002.
259. Chirico, R. D., Gammon, B. E., Knipmeyer, S. E., Nguyen, A., Strube, M. M., Tsonopoulos, C., and Steele, W. V., *J. Chem. Thermodyn.* 22, 1075, 1990.
260. Chirico, R. D., Knipmeyer, S. E., Nguyen, A., and Steele, W. V., *J. Chem. Thermodyn.* 23, 431, 1991.
261. Nisel'son, L. A., Tret'yakova, K. V., Yatko, M. E., Tsirut, E. K., and Antonova, N. P., *Thermophysical Properties of Matter and Substances*, Vol. 4, Rabinovich, V. A., Ed., Amerind Pub., New Delhi, p. 132, 1975.
262. Toczylkin. L. S., and Young, C. L., *J. Chem. Thermodyn.* 12, 355, 1980.
263. Golik, A. Z., and Ravikovich, S. D., *Zh. Fiz. Khim.* 23, 86, 1949.
264. Stepanov, N. G., *Russ. J. Phys. Chem. (Engl. Transl.)* 46, 464, 1972.
265. Krauss, R., and Stephan, K., *J. Phys. Chem. Ref. Data* 18, 43, 1989.
266. Stepanov, N. G., and Nozdrev, V. F., *Russ. J. Phys. Chem. (Engl. Transl.)* 42, 1300, 1968.
267. Garcia-Sanchez, F., and Trejo Rodriguez, A., *J. Chem. Thermodyn.* 19, 359, 1987.
268. Garcia-Sanchez, F., and Trejo Rodriguez, A., *J. Chem. Thermodyn.* 17, 981, 1985.
269. Hojendahl, K., *Mat.-Fys. Medd. - K. Dan. Vidensk. Selsk.* 24, 1, 1946.
270. Duarte-Garza, H. A., Hwang, C.-A., Kellerman, S. A., Miller, R. C., Hall, K. R., Holste, J. C., Marsh, K. N., and Gammon, B. E., *J. Chem. Eng. Data* 42, 497, 1997.
271. Benning, A. F., and McHarness, R. C., *Ind. Eng. Chem.* 32, 814, 1940.
272. Gorchakovskii, V. K., Zadov, V. E., and Podvezennyi, V. N., *Inzh.-Fiz. Zh.* 59, 122, 1990.
273. Garcia-Sanchez, F., Romero-Martinez, A., and Trejo Rodriguez, A., *J. Chem. Thermodyn.* 21, 823, 1989.
274. Higashi, Y., Uematsu, M., and Watanabe, K., *Bull. JSME* 28, 2968, 1985.
275. Piao, C. C., Sato, H., and Watanabe, K., *J. Chem. Eng. Data* 36, 398, 1991.
276. Weber, L. A., and Levelt Sengers, J. M. H., *Fluid Phase Equilib.* 55, 241, 1990.
277. Tanikawa, S., Kabata, Y., Sato, H., and Watanabe, K., *J. Chem. Eng. Data* 35, 381, 1990.
278. Fukushima, M., Watanabe, N., and Kamimura, T., *Nippon Reito Kyokai Ronbunshu* 7, 243, 1990.
279. Fukushima, M., Watanabe, N., and Kamimura, T., *Nippon Reito Kyokai Ronbunshu* 7, 189, 1990.
280. Yamashita, T., Kubota, H., Tanaka, Y., Makita, T., and Kashiwagi, H., *Proc.10th Symp. Thermophys. Prop.*, Japan, pp. 75-78, 1989.
281. Mandlekar, A. V., Kay, W. B., Smith, R. L., and Teja, A. S., *Fluid Phase Equilib.* 23, 79, 1985.
282. Herz, W., *Z. Anorg. Allg. Chem.* 149, 230, 1925.
283. Nikitin, E. D., Pavlov, P. A., and Popov, A. P., *J. Chem. Thermodyn.* 27, 43, 1995.
284. Ratzsh, M. T., *Z. Phys. Chem. Leipzig* 243, 212, 1970.
285. Schmidt, G. C., *Justus Liebigs Ann. Chem.* 266, 266, 1891.
286. Young, C. L., *Int. DATA Ser., Sel. Data Mixtures, Ser. A*, No. 1, 66, 1975.
287. Ambrose, D., Sprake, C. H. S., and Townsend, R., *J. Chem. Thermodyn.* 4, 247, 1972.
288. Zawisza, A. C., *Bull. Acad. Pol. Sci., Ser. Sci. Chim.* 15, 291, 1967.
289. Stryjek, R., and Kreglewski, A., *Bull. Acad. Pol. Sci., Ser. Sci. Chim.* 13, 201, 1965.
290. Kobe, K. A., Ravicz, A. E., and Vohra, S. P., *J. Chem. Eng. Data* 1, 50, 1956.
291. Schroeer, E., *Z. Phys. Chem., Abt. A* 140, 379, 1929.
292. Schroeer, E., *Z. Phys. Chem., Abt. A* 140, 241, 1929.
293. Wilip, J., *Eesti Vabariigi Tartu Ulik. Toim. A* 6 (2), 1924.
294. Audant, *C. R. Hebd. Seances Acad. Sci.* 170, 1573, 1920.
295. Prins, A., and Scheffer, F. E. C., *J. Phys. Chem.* 84, 827, 1913.

296. Travers, M. W., and Usher, F. L., *Z. Phys. Chem., Stoechiom. Verwandtschaftsl.* 57, 365, 1906.
297. Centerszwer, M., and Pakalneet, A., *Z. Phys. Chem., Stoechiom. Verwandtschaftsl.* 55, 303, 1906.
298. Smits, A., *Z. Phys. Chem., Stoechiom. Verwandtschaftsl.* 52, 587, 1905.
299. Galitzine, B., and Wilip, J., *Bull. Acad. Pet.* 11, No. 3, 117, 1901.
300. De Vries, E. C., *Arch. Neerl. Sci. Exactes Nat.* 28, 215, 1895.
301. Ramsay, W., and Young, S., *Philos. Trans. R. Soc. London, A* 178, 57, 1887.
302. Radice, G., Ph. D. Thesis, Univ. of Geneve, 1899.
303. Vespigniani, G. R., *Gazz. Chim. Ital.* 33, 73, 1903.
304. Berthoud, A., and Brum, R., *J. Chim. Phys. Phys.-Chim. Biol.* 21, 143, 1924.
305. Grebenkov, A. J., Zhelezny, V. P., Klepatsky, P. M., Beljajeva, O. V., Chernjak, Y. A., Kotelevsky, Y. G., and Timofejev, B. D., *Int. J. Thermophys.* 17, 535, 1996.
306. Holcomb, C. D., Niesen, V. G., Van Poolen, L. J., and Outcalt, S. L., *Fluid Phase Equilib.* 91, 145, 1993.
307. Higashi, Y., Ashizawa, M., Kabata, Y., Majima, T., Uematsu, M., and Watanabe, K., *JSME Int. J.* 30, 1106, 1987.
308. Otake, K., Yasumoto, M., Yamada, Y., Murata, J., and Urata, S., *J. Chem. Eng. Data* 48, 1380, 2003.
309. Pitschmann, M., and Straub, J., *Int. J. Thermophys.* 23, 877, 2002.
310. Diefenbacher, A., and Tuerk, M., *J. Chem. Thermodyn.* 31, 905, 1999.
311. Shi, L., Zhu, M., Han, L., Duan, Y., Sun, L;, and Fu, Y.-D., *Science in China, Ser. E* 41, 435, 1998.
312. Kuwabara, S., Aoyama, H., Sato, H., and Watanabe, K., *J. Chem. Eng. Data* 40, 112, 1995.
313. Higashi, Y., *Int. J. Thermophys.* 16, 1175, 1995.
314. Fu, Y. D., Han, L. -Z., and Zhu, M. -S., *Fluid Phase Equilib.* 111, 273, 1995.
315. Higashi, Y., *Int. J. Refrig.* 17, 524, 1994.
316. Malbrunot, P. F., Meunier, P. A., Scatena, G. M., Mears, W. H., Murphy, K. P., and Sinka, J. V., *J. Chem. Eng. Data* 13, 16, 1968.
317. Steele, W. V., Chirico, R. D., Cowell, A. B., Knipmeyer, S. E., and Nguyen, A., *J. Chem. Eng. Data* 47, 725, 2002.
318. Young, C. L., *Int. DATA Ser., Sel. Data Mixtures, Ser. A*, No. 1, 159, 1975.
319. Durig, J. R., and Li, Y. S., *J. Chem. Phys.* 63, 4110, 1975.
320. Bourgou, A., *Bull. Soc. Chim. Belg.* 33, 101, 1924.
321. Verevkin, S. P., Kozlova, S. A., Emel'yanenko, V. N., Nikitin, E. D., Popov, A. P., and Krasnykh, E. L., *J. Chem. Eng. Data* 51, 1896, 2006.
322. Kay, W. B., and Young, C. L., *Int. DATA Ser., Sel. Data Mixtures, Ser. A*, No. 2, 154, 1974.
323. Weaver, D. L., M.S. Thesis, Ohio State Univ., Columbus, OH, 1973.
324. Young, C. L., *Int. DATA Ser., Sel. Data Mixtures, Ser. A*, No. 1, 47, 1974.
325. Genco, J. M., Teja, A. S., and Kay, W. B., *J. Chem. Eng. Data* 25, 350, 1980.
326. Kay, W. B., and Young, C. L., *Int. DATA Ser., Sel. Data Mixtures, Ser. A*, No. 1, 52, 1975.
327. Genco, J. M., Ph.D. Thesis, Ohio State Univ., Columbus, OH, 1965.
328. Kay, W. B., *J. Am. Chem. Soc.* 68, 1336, 1946.
329. Quadri, S. K., and Kudchadker, A. P., *J. Chem. Thermodyn.* 24, 473, 1992.
330. Young, S., and Fortey, E. C., *J. Chem. Soc.* 35, 1126, 1879.
331. Steele, W. V., Chirico, R. D., Knipmeyer, S. E., Nguyen, A., and Smith, N. K., *J. Chem. Eng. Data* 42, 1037, 1997.
332. Steele, W. V., Chirico, R. D., Knipmeyer, S. E., and Nguyen, A., *J. Chem. Eng. Data* 42, 1008, 1997.
333. Ihmels, E. C. C., and Lemmon, E. W., *Fluid Phase Equilib.* 260, 36, 2007.
334. Wu, J., Liu, Z., Wang, B., and Pan, J., *J. Chem. Eng. Data* 49, 704, 2004.
335. Noles, J. R., and Zollweg, J. A., *Fluid Phase Equilib.* 66, 275, 1991.
336. Glowka, S., *Bull. Acad. Pol. Sci., Ser. Sci. Chim.* 20, 163, 1972.
337. Zawisza, A. C., and Glowka, S., *Bull. Acad. Pol. Sci., Ser. Sci. Chim.* 18, 549, 1970.
338. Edwards, J., and Maass, O., *Can. J. Res., Sect. A* 12, 357, 1935.
339. Tapp, J. S., Steacie, E. W. R., and Maass, O., *Can. J. Res.* 9, 217, 1933.
340. Cardoso, E., and Coppola, A. A., *J. Chim. Phys. Phys.-Chim. Biol.* 20, 337, 1923.
341. Cardoso, E., and Bruno, A., *J. Chim. Phys. Phys.-Chim. Biol.* 20, 347, 1923.
342. Briner, E., and Cardoso, E., *J. Chim. Phys. Phys.-Chim. Biol.* 6, 641, 1908.
343. Briner, E., and Cardoso, E., *C. R. Hebd. Seances Acad. Sci.* 144, 911, 1907.
344. Bogoslovskii, V. E., Mikhalyuk, G. I., and Shamolin, A. I., *Zh. Prikl. Khim. (Leningrad)* 45, 1154, 1972.
345. Kay, W. B., and Hissong, D. W., *Proc. - Am. Pet. Inst., Div. Refin.* 49, 13, 1969.
346. McMicking, J. H., and Kay, W. B., *Proc., Am. Pet. Inst., Sect. 3* 45, 75, 1965.
347. Chirico, R. D., Knipmeyer, S. E., Nguyen, A., and Steele, W. V., *J. Chem. Thermodyn.* 25, 1461, 1993.
348. Stern, S. A., and Kay, W. B., *J. Phys. Chem.* 61, 374, 1957.
349. Edgar, G., and Calingaert, G., *J. Am. Chem. Soc.* 51, 1540, 1929.
350. Francis, A. W., *Ind. Eng. Chem.* 49, 1779, 1957.
351. Cox, J. D., *Trans. Faraday Soc.* 56, 959, 1960.
352. Steele, W. V., Chirico, R. D., Nguyen, A., and Knipmeyer, S. E., *J. Chem. Thermodyn.* 27, 311, 1995.
353. Nikitin, E. D., Pavlov, P. A., and Popov, A. P., *J. Chem. Thermodyn.* 26, 1047, 1994.
354. Cristou, G., Young, C. L., and Svejda, P., *Fluid Phase Equilib.* 67, 45, 1991.
355. Zhuravlev, D. I., *Zh. Fiz. Khim.* 9, 875, 1937.
356. Chirico, R. D., and Steele, W. V., *J. Chem. Eng. Data* 50, 1052, 2005.
357. Smith, R. L., Ph. D. Dissertation, Georgia Institute of Technology 1985.
358. Nikitin, E. D., Pavlov, P. A., and Popov, A. P., *Fluid Phase Equilib.* 141, 155, 1997.
359. Nikitin, E. D., Pavlov, P. A., and Bessonova, N. V., *J. Chem. Thermodyn.* 26, 177, 1994.
360. Nikitin, E. D., Pavlov, P. A., and Popov, A. P., *Fluid Phase Equilib.* 189, 151, 2001.
361. Nikitin, E. D., Pavlov, P. A., and Popov, A. P., *Fluid Phase Equilib.* 149, 223, 1998.
362. Sinicyn, E. N., Mikhalevich, L. A., and Yankovskaya, O. P., *Deposited Doc. VINITI*, Doc. No. 2510-V90, 1990.
363. Teja, A. S., Gude, M., and Rosenthal, D. J., *Fluid Phase Equilib.* 52, 193, 1989.
364. Beale, E. S., and Docksey, P., *J. Inst. Pet.* 21, 860, 1935.
365. Nikitin, E. D., and Popov, A. P., *Fluid Phase Equilib.* 166, 237, 1999.
366. Horstman, S., Fischer, K., Gmehling, J., and Kolar, P., *J. Chem. Thermodyn.* 32, 451, 2000.
367. Colgate, S. O., Sivaraman, A., and Dejsupa, C., *Fluid Phase Equilib.* 76, 175, 1992.
368. Friend, D. G., Ingham, H., and Ely, J. F., *J. Phys. Chem. Ref. Data* 20, 275, 1991.
369. Jangkamolkulchai, A., and Luks, K. D., *J. Chem. Eng. Data* 34, 92, 1989.
370. Calado, J. C. G., Chang, E., Clancy, P., and Streett, W. B., *J. Phys. Chem.* 91, 3914, 1987.
371. Brunner, E., *J. Chem. Thermodyn.* 17, 871, 1985.
372. Morrison, G., and Kincaid, J. M., *AIChE J.* 30, 257, 1984.
373. Sychev, V. V., Vasserman, A. A., Kozlov, A. D., Zagoruchenko, V. A., Spiridonov, G. A., and Tsymarny, V. A., *Thermodynamic Properties of Ethane*, Standards Publishing House, Moscow, 1982.
374. Morrison, G., *J. Phys. Chem.* 85, 759, 1981.
375. Bulavin, L. A., and Shimanskii, Yu. I., *Zh. Eksp. Teor. Fiz.* 29, 482, 1979.
376. Strumpf, H. J., Collings, A. F., and Pings, C. J., *J. Chem. Phys.* 60, 3109, 1974.
377. Burton, M., and Balzarini, D., *Can. J. Phys.* 52, 2011, 1974.
378. Douslin, D. R., and Harrison, R. H., *J. Chem. Thermodyn.* 5, 491, 1973.
379. Berestov, A. T., Giterman, M. S., and Shmakov, N. G., *Sov. Phys. - JETP (Engl. Transl.)* 37, 1128, 1973.
380. Efremova, G. D., and Shvarts, A. V., *Russ. J. Phys. Chem. (Engl. Transl.)* 46, 237, 1972.

381. Miniovich, V. M., and Sorina, G. A., *Russ. J. Phys. Chem. (Engl. Transl.)* 45, 306, 1971.
382. Khazanova, N. E., and Sominskaya, E. E., *Russ. J. Phys. Chem. (Engl. Transl.)* 45, 88, 1971.
383. Bulavin, L. A., Ostanevich, Yu. M., Simkina, A. P., and Stelkov, A. V., *Ukr. Fiz. Zh. (Ukr. Ed.)* 16, 90, 1971.
384. Chashkin, Yu. R., Smirnov, V. A., and Voronel, A. V., *Teplofiz. Svoistva Veshchestv Mater.* 2, 139, 1970.
385. Sliwinski, P., *Z. Phys. Chem. (Munich)* 63, 263, 1969.
386. Khazanova, N. E., Lesnevskaya, L. S., and Zakharova, A. V., *Khim. Prom-st. (Moscow)* 42, 364, 1966.
387. Matschke, D. E., and Thodos, G., *J. Chem. Eng. Data* 7, 232, 1962.
388. Schmidt, E. , and Thomas, W., *Forsch. Geb. Ingenieurw.* 20B, 161, 1954.
389. Kay, W. B., and Brice, D. B., *Ind. Eng. Chem.* 45, 615, 1953.
390. Murray, F. E., and Mason, S. G., *Can. J. Chem.* 30, 550, 1952.
391. Mason, S. G., Naldrett, S. N., and Maass, O., *Can. J. Res., Sect. B* 18, 103, 1940.
392. Beattie, J. A., Su, G.-J., and Simard, G. L., *J. Am. Chem. Soc.* 61, 924, 1939.
393. Price, T. W., *J. Chem. Soc.* 107, 188, 1915.
394. Nikitin, E. D., Pavlov, P. A., and Skripov, P. V., *J. Chem. Thermodyn.* 25, 869, 1993.
395. Marshall, W. L., and Jones, E. V., *J. Inorg. Nucl. Chem.* 36, 2319, 1974.
396. Sajotschewsky, W., *Beibl. Ann. Phys.* 3, 741, 1879.
397. Mocharnyuk, R. F., *Zh. Obshch. Khim.* 30, 1098, 1960.
398. Golik, A. Z., Ravikovich, S. D., and Orishchenko, A. V., *Ukr. Khim. Zh. (Russ. Ed.)* 21, 167, 1955.
399. Polikhronidi, N. G., Abdulagatov, I. M., Stepanov, G. V., and Batyrova, R. G., *J. Supercrit. Fluids* 43, 1, 2007.
400. Hu, T., Qin, Z., Wang, G., Hou, X., and Wang, J., *J. Chem. Eng. Data* 49, 1809, 2004.
401. Sauermann, P., Holzapfel, K., Oprzynski, J., Kohler, F., Poot, W., and De Loos, T. W., *Fluid Phase Equilib.* 112, 249, 1995.
402. Mousa, A. H. N., *J. Chem. Eng. Jpn.* 20, 635, 1987.
403. Wilson, K. S., Lindley, D. D., Kay, W. B., and Hershey, H. C., *J. Chem. Eng. Data* 29, 243, 1984.
404. Hentze, G., *Thermochim. Acta* 20, 27, 1977.
405. Nozdrev, V. F., *Akust. Zh.* 2, 209, 1956.
406. Griswold, J., Haney, J. D., and Klein, V. A., *Ind. Eng. Chem.* 35, 701, 1943.
407. Battelli, A., *Mem. Torino, Ser. 2* 44, 57, 1893.
408. Ramsay, W., and Young, S., *Philos. Trans. R. Soc. London* 177, 123, 1886.
409. Strauss, O, *Beibl. Ann. Phys.* 6, 282, 1882.
410. Steele, W. V., Chirico, R. D., Knipmeyer, S. E., and Nguyen, A., *Experimental Results for DIPPR 1990-91 Projects on Phase Equilibria and Pure Component Properties*, DIPPR Data Series No. 2, 154, 1994.
411. Ambrose, D., Ellender, J. H., Gundry, H. A., Lee, D. A., and Townsend, R., *J. Chem. Thermodyn.* 13, 795, 1981.
412. Lambert, J. D., Clarke, J. S., Duke, J. F., Hicks, C. L., Lawrence, S. D., Morris, D. M., and Shone, M. G. T., *Proc. R. Soc. London, A* 249, 414, 1959.
413. Young, S., and Thomas, G. L., *J. Chem. Soc.* 63, 1191, 1893.
414. Pohland, E., and Mehl, W., *Z. Phys. Chem., Abt. A* 164, 48, 1933.
415. Nadezhdin, A., *Rep. Phys.* 23, 708, 1887.
416. Guseinov, K. D., and Zhabbarov, O., *Izv. Vyssh. Uchebn. Zaved. Neft Gaz* 2, 76, 1975.
417. Steele, W. V., Chirico, R. D., Cowell, A. B., Knipmeyer, S. E., and Nguyen, A., *J. Chem. Eng. Data* 47, 700, 2002.
418. Nowak, P., Kleinrahm, R., and Wagner, W., *J. Chem. Thermodyn.* 28, 1441, 1996.
419. Hasch, B. M., and McHugh, M. A., *Fluid Phase Equilib.* 64, 251, 1991.
420. Jahangiri, M., Jacobsen, R. T., Stewart, R. B., and McCarty, R. D., *J. Phys. Chem. Ref. Data* 15, 593, 1986.
421. Younglove, B. A., *J. Phys. Chem. Ref. Data*, Vol. 11, Suppl. No. 1, Am. Chem. Soc., Washington, DC, 1982.
422. McCarty, R. D., and Jacobsen, R. T., *NBS Tech. Note (U. S.)* 1045, 1981.
423. Thomas, W., and Zander, M., *Int. J. Thermophys* 1, 383, 1980.
424. Hastings, J. R., Levelt Sengers, J. M. H., and Balfour, F. W., *J. Chem. Thermodyn.* 12, 1009, 1980.
425. Hastings, J. R., and Levelt Sengers, J. M. H., *Proc. 7th Symp. Thermophys. Prop.*, Cezairliyan, A., Ed., ASME, New York, p. 794, 1977.
426. Douslin, D. R., and Harrison, R. H., *J. Chem. Thermodyn.* 8, 301, 1976.
427. Bender, E., *Cryogenics* 15, 667, 1975.
428. Moldover, M. R., *J. Chem. Phys.* 61, 1766, 1974.
429. Angus, S., Armstrong, B., and de Reuck, K. M., *International Thermodynamic Tables of the Fluid State - 2 Ethylene*, Butterworths, London, 1974.
430. Zernov, V. S., Kogan, V. B., and Lyubetskii, S. G., *Zh. Prikl. Khim. (Leningrad)* 44, 1819, 1971.
431. Shim, J., and Kohn, J. P., *J. Chem. Eng. Data* 9, 1, 1964.
432. Kay, W. B., *Ind. Eng. Chem.* 40, 1459, 1948.
433. Diepen, G. A. M., and Scheffer, F. E. C., *J. Am. Chem. Soc.* 70, 4081, 1948.
434. Naldrett, S. N., and Maass, O., *Can. J. Res., Sect. B* 18, 118, 1940.
435. Dacey, J. R., McIntosh, R. L., and Maass, O., *Can. J. Res., Sect. B* 17, 206, 1939.
436. Maass, O., and Geddes, A. L., *Philos. Trans. R. Soc. London, A* 236, 303, 1937.
437. Lawrenson, I. J., and Lee, D. A., *J. Chem. Thermodyn.* 10, 1111, 1978.
438. Zawisza, A. C., and Glowka, S., *Bull. Acad. Pol. Sci., Ser. Sci. Chim.* 18, 555, 1970.
439. Lyons, R. L., M. S. Thesis, Pennsylvania State Univ., University Park, PA, 1985.
440. Daubert, T. E., Jalowka, J. W., and Goren, V., *AIChE Symp. Ser.* 83 (256), 128, 1987.
441. Douslin, D. R., Moore, R. T., Dawson, J. P., and Waddington, G., *J. Am. Chem. Soc.* 80, 2031, 1958.
442. Beyerlein, A. L., DesMarteau, D. D., Kul, I., and Zhao, G., *Fluid Phase Equilib.* 150, 287, 1998.
443. Parthasarathy, S., *Proc. - Indian Acad. Sci., Sect. A* 2, 497, 1935.
444. Bominaar, S. A. R. C., Trappeniers, N. J., and Biswas, S. N., *J. Phys. Chem.* 94, 1097, 1990.
445. Bominaar, S. A. R. C., Biswas, S. N., Trappeniers, N. J., and Ten Seldam, C. A., *J. Chem. Thermodyn.* 19, 959, 1987.
446. Froba, A. P., Botero, C., and Leipertz, A., *Int. J. Thermophys.* 27, 1609, 2006.
447. Uchida, Y., Yasumoto, M., Yamada, Y., Ochi, K., Furuya, T., and Otake, K., *J. Chem. Eng. Data* 49, 1615, 2004.
448. Otake, K., Uchida, Y., Yasumoto, M., Yamada, Y., Furuya, T., and Ochi, K., *J. Chem. Eng. Data* 49, 1643, 2004.
449. Hu, P. , and Chen, Z. S., *Fluid Phase Equilib.* 221, 7, 2004.
450. Salvi-Narkhede, M., Adcock, J. L., Gakh, A., and Van Hook, W. A., *J. Chem. Thermodyn.* 25, 643, 1993.
451. Ambrose, D., and Ghiassee, N. B., *J. Chem. Thermodyn.* 20, 1231, 1988.
452. Myers, J. E., Hershey, H. C., and Kay, W. B., *J. Chem. Thermodyn.* 11, 1019, 1979.
453. Golik, A. Z., and Adamenko, I. I., *Ukr. Fiz. Zh. (Ukr. Ed.)* 10, 443, 1965.
454. Golik, A. Z., and Ivanova, I. I., *Zh. Fiz. Khim.* 36, 1768, 1962.
455. Kay, W. B., *Ind. Eng. Chem.* 30, 459, 1938.
456. Christou, G., Sadus, R. J., and Young, C. L., *Fluid Phase Equilib.* 67, 259, 1991.
457. Kurumov, D. S., Grigor'ev, B. A., and Vasil'ev, Yu. L., *Teplofiz. Svoistva Veshchestv Mater.* No. 27, 101, 1989.
458. De Loos, T. W., Poot, W., and De Swaan Arons, J., *Fluid Phase Equilib.* 42, 209, 1988.
459. Artyukhovskaya, L. M., Shimanskaya, E. T., and Shimanskii, Yu. I., *Opt. Spektrosk.* 37, 935, 1974.
460. Artyukhovskaya, L. M., Shimanskaya, E. T., and Shimanskii, Yu. I., *Sov. Phys. - JETP (Engl. Transl.)* 37, 848, 1973.
461. Artyukhovskaya, L. M., Shimanskaya, E. T., and Shimanskii, Yu. I., *Zh. Eksp. Teor. Fiz.* 63, 2159, 1972.
462. Smith, L. B., Beattie, J. A., and Kay, W. C., *J. Am. Chem. Soc.* 59, 1587, 1937.
463. Beattie, J. A., and Kay, W. C., *J. Am. Chem. Soc.* 59, 1586, 1937.
464. Rosenthal, D. J., Gude, M. T., Teja, A. S., and Mendez-Santiago, J., *Fluid Phase Equilib.* 135, 89, 1997.

465. Smith, R. L., Anselme, M. J., and Teja, A. S., *Fluid Phase Equilib.* 31, 161, 1986.
466. Pulliam, M. K., Gude, M. T., and Teja, A. S., *Experimental Results for DIPPR 1990-91 Projects on Phase Equilibria and Pure Component Properties*, DIPPR Data Ser. No. 2, p. 184, 1994.
467. Naziev, Y. M., and Abasov, A. A., *Izv. Vyssh. Uchebn. Zaved., Neft Gaz* 12, 81, 1969.
468. Mousa, A. H. N., *J. Chem. Eng. Data* 26, 248, 1981.
469. Hicks, C. P., and Young, C. L., *Chem. Rev.* 75, 119, 1975.
470. Mousa, A. H. N., Kay, W. B., and Kreglewski, A., *J. Chem. Thermodyn.* 4, 301, 1972.
471. Douslin, D. R., Harrison, R. H., and Moore, R. T., *J. Chem. Thermodyn.* 1, 305, 1969.
472. Evans, F. D., and Tiley, P. F., *J. Chem. Soc. B* 134, 1966.
473. Counsell, J. F., Green, J. H. S., Hales, J. L., and Martin, J. F., *Trans. Faraday Soc.* 61, 212, 1965.
474. Saikawa, K., Kijima, J., Uematsu, M., and Watanabe, K., *J. Chem. Eng. Data* 24, 165, 1979.
475. Kijima, J., Saikawa, K., Watanabe, K., Oguchi, K., and Tanishita, I., *Proc. 7th Symp. Thermophys. Prop.*, Cezairliyan, A., Ed., ASME, New York, p. 480, 1977.
476. Kim, K. Y., Ph.D. Dissertation, Univ. Michigan, Ann Arbor, MI, 1974.
477. Swarts, F., *Bull. Soc. Chim. Belg.* 42, 114, 1933.
478. Aoyama, H., Kishizawa, G., Sato, H., and Watanabe, K., *J. Chem. Eng. Data* 41, 1046, 1996.
479. Aoyama, H., Sato, H., and Watanabe, K., *Sixteenth Japan Symposium on Thermophysical Properties*, Hiroshima, p. 173, 1995.
480. McLure, I. A., and Dickinson, E., *J. Chem. Thermodyn.* 8, 93, 1976.
481. VonNiederhausern, D. M., Wilson, G. M., and Giles, N. F., *J. Chem. Eng. Data* 45, 157, 2000.
482. Liu, J., Qin, Z., Wang, G., Hou, X., and Wang, J., *J. Chem. Eng. Data* 48, 1610, 2003.
483. Gude, M. T., and Teja, A. S., *Fluid Phase Equilib.* 83, 139, 1993.
484. Grigor'ev, B. A., Rastorguev, Yu. L., Gerasimov, A. A., Kurumov, D. S., and Plotnikov, S. A., *Int. J. Thermophys.* 9, 439, 1988.
485. Zawisza, A., *J. Chem. Thermodyn.* 17, 941, 1985.
486. Mousa, A. H. N., *J. Chem. Thermodyn.* 9, 1063, 1977.
487. Young, C. L., *Int. DATA Ser., Sel. Data Mixtures, Ser. A*, No. 1, 157, 1975.
488. Nichols, W. B., Reamer, H. H., and Sage, B. H., *AIChE J.* 3, 262, 1957.
489. VonNiederhausern, D. M., Wilson, G. M., and Giles, N. F., *J. Chem. Eng. Data* 51, 1986, 2006.
490. Ma, P., Ma, Y., and Zhang, J., *Gaoxiao Huaxue Gongcheng Xuebao* 5, 175, 1991.
491. Masui, G., Honda, Y., and Uematsu, M., *J. Chem. Thermodyn.* 38, 1711, 2006.
492. Goodwin, R. D., and Haynes, W. M., *NBS Tech. Note (U. S.)* No. 1051, 1982.
493. Pryanikova, R. O., Plenkina, R. M., Kuzyakina, N. V., and Markina, I. A., *Khim. Prom-st (Moscow)*, 13. 1987.
494. Beattie, J. A., Ingersoll, H. G., and Stockmayer, W. H., *J. Am. Chem. Soc.* 64, 546, 1942.
495. Vohra, S. P., and Kobe, K. A., *J. Chem. Eng. Data* 4, 329, 1959.
496. Smith, R. L., Negishi, E., Arai, K., and Saito, S., *J. Chem. Eng. Jpn.* 23, 99, 1990.
497. Steele, W. V., Chirico, R. D., Cowell, A. B., Knipmeyer, S. E., and Nguyen, A., *J. Chem. Eng. Data* 42, 1053, 1997.
498. Setzmann, U., and Wagner, W., *J. Phys. Chem. Ref. Data* 20, 1061, 1991.
499. Friend, D. G., Ely, J. F., and Ingham, H., *J. Phys. Chem. Ref. Data* 18, 583, 1989.
500. Kleinrahm, R., and Wagner, W., *J. Chem. Thermodyn.* 18, 739, 1986.
501. Calado, J. C. G., Dieters, U., and Strett, W. B., *J. Chem. Soc., Faraday Trans. 1* 77, 2503, 1981.
502. Angus, S., Armstrong, B., and de Reuck, K. M., *International Thermodynamic Tables of the Fluid State - 5 Methane*, Pergamon, Oxford, 1978.
503. Goodwin, R. D., *NBS Tech. Note (U. S.)* No. 653, 1974.
504. Gielen, H., Jansoone, F., and Verbeke, O. B., *J. Chem. Phys.* 59, 5763, 1973.
505. Jansoone, V., Gielen, H., De Boelpaep, J., and Verbeke, O. B., *Physica (Amsterdam)* 46, 213, 1970.
506. Terry, M. J., Lynch, J. T., Bunclark, M., Mansell, K. R., and Staveley, L. A. K., *J. Chem. Thermodyn.* 1, 413, 1969.
507. Grigor, A. F., and Steele, W. A., *J. Chem. Phys.* 48, 1032, 1968.
508. Keyes, F. G., Taylor, R. S., and Smith, L. B., *J. Math. Phys. (Cambridge, Mass.)* 1, 211, 1922.
509. Kuenen, J. P., *Philos. Mag.* 6, 637, 1903.
510. Crismer, L., *Bull. Soc. Chim. Belg.* 18, 18, 1904.
511. Schmidt, G. C., *Z. Phys. Chem., Stoechiom. Verwandtschaftsl.* 8, 628, 1891.
512. Polikhronidi, N. G., Radzhabova, L. M., Rasulov, A. R., and Stepanov, G. V., *High Temp. (Engl. Transl.)* 44, 512, 2006.
513. Francesconi, Artur Zaghini, Lentz, H., and Franck, E. U., *J. Phys. Chem.* 85, 3303, 1981.
514. Swami, D. R., Kumarkrishna Rao, V. N., and Narasinga Rao, N., *Trans., Indian Inst. Chem. Eng.* 9, 32, 1956.
515. Salzwedel, E., *Ann. Phys. (Leipzig)* 5, 853, 1930.
516. Ramsay, W., and Young, S., *Philos. Trans. R. Soc. London, A* 178, 313, 1887.
517. Steele, W. V., Chirico, R. D., Cowell, A. B., Knipmeyer, S. E., and Nguyen, A., *J. Chem. Eng. Data* 47, 667, 2002.
518. Gurarii, L. L., Kuleshov, G. G., Baglai, A. K., and Petrashkevich, R. I., *Khim.-Farm. Zh.* 21, 247, 1987.
519. Guseinov, S. O., Farzaliev, B. I., and Naziev, Y. M., *Izv. Vyssh. Uchebn. Zaved., Neft Gaz* 22, 52, 1979.
520. Kreglewski, A., *Bull. Acad. Pol. Sci., Cl. 3* 2, 191, 1954.
521. Kiyama, R., Suzuki, K., and Ikegami, T., *Rev. Phys. Chem. Jpn.* 21, 50, 1951.
522. Abara, J. A., Jennings, D. W., Kay, W. B., and Teja, A. S., *J. Chem. Eng. Data* 33, 242, 1988.
523. Wilson, G. M., Johnston, R. H., Hwang, S.-C., and Tsonopoulos, C., *Ind. Eng. Chem. Process Des. Dev.* 20, 94, 1981.
524. Rutenberg, [illegible], and Shakhova, S. F., *Russ. J. Phys. Chem. (Engl. Transl.)* 47, 124, 1973.
525. Yasumoto, M., Uchida, Y., Ochi, K., Furuya, T., Shono, A., and Otake, K., *J. Chem. Eng. Data* 52, 1726, 2007.
526. Day, H. O., and Felsing, W. A., *J. Am. Chem. Soc.* 74, 1951, 1952.
527. Ambrose, D., and Ghiassee, N. B., *J. Chem. Thermodyn.* 20, 767, 1988.
528. Chirico, R. D., Knipmeyer, S. E., Nguyen, A., and Steele, W. V., *J. Chem. Thermodyn.* 31, 339, 1999.
529. Chirico, R. D., Johnson, R. D. I., and Steele, W. V., *J. Chem. Thermodyn.* 39, 698, 2007.
530. Chirico, R. D., and Steele, W. V., *J. Chem. Eng. Data* 50, 697, 2005.
531. Schroeer, E., *Z. Phys. Chem., Abt. B* 49, 271, 1941.
532. Dawson, P. P., Silberberg, I. H., and McKetta, J. J., *J. Chem. Eng. Data* 18, 7, 1973.
533. Ambrose, D., Counsell, J. F., and Hicks, C. P., *J. Chem. Thermodyn.* 10, 771, 1978.
534. Griffin, D. N., *J. Am. Chem. Soc.* 71, 1423, 1949.
535. Matzik, I., and Schneider, G. M., *Ber. Bunsen-Ges. Phys. Chem.* 89, 551, 1985.
536. Ambrose, D., and Townsend, R., *Trans. Faraday Soc.* 64, 2622, 1968.
537. Defibaugh, D. R., Carrillo-Nava, E., Hurly, J. J., Moldover, M. R., Schmidt, J. W., and Weber, L. A., *J. Chem. Eng. Data* 42, 488, 1997.
538. Young, C. L., *J. Chem. Thermodyn.* 4, 65, 1972.
539. Lindley, D. D., and Hershey, H. C., *Fluid Phase Equilib.* 55, 109, 1990.
540. Kreglewski, A., *Bull. Acad. Pol. Sci., Cl. 3* 5, 323, 1957.
541. Young, S., *J. Chem. Soc.* 77, 1145, 1900.
542. Steele, W. V., Chirico, R. D., Knipmeyer, S. E., Nguyen, A., Smith, N. K., and Tasker, I. R., *J. Chem. Eng. Data* 41, 1269, 1996.
543. Post, R. G., *Unpublished Rep., Chem. Eng. No. 362,* Univ. Texas, Austin, TX, 1950.
544. Hess, L. G., and Tilton, V. V., *Ind. Eng. Chem.* 42, 1251, 1950.
545. Thodos, G., *AIChE J.* 3, 428, 1957.
546. Cheng, D. C. H., and McCoubrey, J. C., *J. Chem. Soc.* 4993, 1963.
547. Teja, A. S., and Smith, R. L., *AIChE J.* 33, 1560, 1987.
548. Ambrose, D., and Sprake, C. H. S., *J. Chem. Soc. A* 1263, 1971.
549. Yata, J., Hori, M., Kohno, K., and Minamiyama, T., *High Temp. - High Press.* 29, 19, 1997.

550. Duarte-Garza, H. A., Stouffer, C. E., Hall, K. R., Hall, K. R., Holste, J. C., Marsh, K. N., and Gammon, B. E., *J. Chem. Eng. Data* 42, 745, 1997.
551. Ye, F., Sato, H., and Watanabe, K., *J. Chem. Eng. Data* 40, 148, 1995.
552. Tsvetkov, O. B., Kletskii, A. V., Laptev, Yu. A., Asambaev, A. J., and Zausaev, I. A., *Int. J. Thermophys.* 16, 1185, 1995.
553. Sagawa, T., Sato, H., and Watanabe, K., *High Temp. - High Press.* 26, 193, 1994.
554. Wilson, L. C., Wilding, W. V., Wilson, G. M., Rowley, R. L., Felix, V. M., and Chisolm-Carter, T., *Fluid Phase Equilib.* 80, 167, 1992.
555. Shank, R. L., *J. Chem. Eng. Data* 12, 474, 1967.
556. Gude, M. T., and Teja, A. S., *AIChE Symp. Ser.* 90 (298), 14, 1994.
557. Grigor'ev, B. A., Rastorguev, Yu. L., Kurumov, D. S., Gerasimov, A. A., Kharin, V. E., and Plotnikov, S. A., *Int. J. Thermophys.* 11, 487, 1990.
558. Kratzke, H., *AIChE J.* 31, 693, 1985.
559. Wolfe, D., Kay, W. B., and Teja, A. S., *J. Chem. Eng. Data* 28, 319, 1983.
560. Artyukhovskaya, L. M., Shimanskaya, E. T., and Shimanskii, Yu. I., *Sov. Phys. - JETP* (*Engl. Transl.*) 59, 375, 1970.
561. Beattie, J. A., Levine, S. W., and Douslin, D. R., *J. Am. Chem. Soc.* 73, 4431, 1951.
562. Sage, B. H., and Lacey, W. N., *Ind. Eng. Chem.* 32, 992, 1940.
563. Quadri, S. K., and Kudchadker, A. P., *AIChE Symp. Ser.* 298, 1, 1994.
564. Lenoir, J. M., Rebert, C. J., and Hipkin, H. G., *J. Chem. Eng. Data* 16, 401, 1971.
565. Brown, J. A., and Mears, W. H., *J. Phys. Chem.* 62, 960, 1958.
566. Fowler, R. D., Hamilton, J. M., Kasper, J. S., Weber, C. E., Burford, W. B., and Anderson, H. C., *Ind. Eng. Chem.* 39, 375, 1947.
567. Martin, J. J., *J. Chem. Eng. Data* 7, 68, 1962.
568. Douslin, D. R., Moore, R. T., and Waddington, G., *J. Phys. Chem.* 63, 1959, 1959.
569. Rowlinson, J. S., and Thacker, R., *Trans. Faraday Soc.* 53, 1, 1957.
570. Ermakov, G. V., and Skripov, V. P., *Russ. J. Phys. Chem.* (*Engl. Transl.*) 41, 39, 1967.
571. Wang, B. -H., Adcock, J. L., Mathur, S. B., and Van Hook, W. A., *J. Chem. Thermodyn.* 23, 699, 1991.
572. Taylor, Z. L., and Reed, T. M., *AIChE J.* 16, 738, 1970.
573. Vandana, V., Rosenthal, D. J., and Teja, A. S., *Fluid Phase Equilib.* 99, 209, 1994.
574. Milton, H. T., and Oliver, G. D., *J. Am. Chem. Soc.* 74, 3951, 1952.
575. Oliver, G. D., and Grisard, J. W., *J. Am. Chem. Soc.* 73, 1688, 1951.
576. Oliver, G. D., Blumkin, S., and Cunningham, C. W., *J. Am. Chem. Soc.* 73, 5722, 1951.
577. Mousa, A. H. N., *J. Chem. Eng. Data* 23, 133, 1978.
578. McLure, I. A., Trejo Rodriguez, A., and Soares, V. A. M., *J. Chem. Thermodyn.* 14, 402, 1982.
579. Ewing, M. B., and Sanchez Ochoa, J. C., *J. Chem. Thermodyn.* 30, 189, 1998.
580. Ernst, G., Gurtner, J., and Wirbser, H., *J. Chem. Thermodyn.* 29, 1125, 1997.
581. Brown, J. A., *J. Chem. Eng. Data* 8, 106, 1963.
582. Young, C. L., *Int. DATA Ser., Sel. Data Mixtures, Ser. A*, No. 4, 291, 1985.
583. Sinitsyn, E. N., Mikhalevich, L. A., Biryukova, L. V., Danilov, N. N., Muratov, G. N., and Fedorov, A. P., *Deposited Doc. VINITI*, Doc. No. 1516-80, 1980.
584. Nikitin, E. D., Popov, A. P., and Yatluk, Y. G., *J. Chem. Eng. Data* 52, 315, 2007.
585. Honda, Y., Sato, T,, and Uematsu, M., *J. Chem. Thermodyn.* 40, 208, 2008.
586. Jou, F.-Y., Carroll, J. J., and Mather, A. E., *Fluid Phase Equilib.* 109, 235, 1995.
587. Sychev, V. V., Vasserman, A. A., Kozlov, A. D., and Tsymarny, V. A., *Thermodynamic Properties of Propane*, Standards Publishing House, Moscow, 1989.
588. Goodwin, R. D., and Haynes, W. M., *NBS Monogr.* (*U. S.*) No. 170, 1982.
589. Barber, J. R., Kay, W. B., and Teja, A. S., *AIChE J.* 28, 134, 1982.
590. Yesavage, V. F., Katz, D. L., and Powers, J. E., *J. Chem. Eng. Data* 14, 197, 1969.
591. Clegg, H. P., and Rowlinson, J. S., *Trans. Faraday Soc.* 51, 1333, 1955.
592. Kay, W. B., and Rambosek, G. M., *Ind. Eng. Chem.* 45, 221, 1953.
593. Meyers, C. H., *J. Res. Natl. Bur. Stand.* (*U. S.*) 29, 157, 1942.
594. Meyer, R. E., Ph.D. Thesis, Penn. State Univ., University Park, PA, 1941.
595. Deschner, W. W., and Brown, G. G., *Ind. Eng. Chem.* 32, 836, 1940.
596. Beattie, J. A., Poffenberger, N., and Hadlock, C., *J. Chem. Phys.* 3, 96, 1935.
597. Efremova, G. D., and Sokolova, E. S., *Russ. J. Phys. Chem.* (*Engl. Transl.*) 46, 1084, 1972.
598. Anonymous, B., *International Critical Tables of Numerical Data, Phys., Chem. Technol.*, Vol. III, Washburn, E. W., Ed., McGraw-Hill, New York, 1928.
599. Kuenen, J. P., and Robson, W. G., *Philos. Mag.* 4, 116, 1902.
600. Ramsay, W., and Young, S., *Philos. Trans. R. Soc. London, A* 180, 137, 1889.
601. Oh, B. C., Lee, S., Seo, J., and Kim, H., *J. Chem. Eng. Data* 49, 221, 2004.
602. Ambrose, D., Counsell, J. F., Lawrenson, I. J., and Lewis, G. B., *J. Chem. Thermodyn.* 10, 1033, 1978.
603. Seibert, F. M., and Burrell, G. A., *J. Am. Chem. Soc.* 37, 2683, 1915.
604. Lu, H., Newitt, D. M., and Ruhemann, M., *Proc. R. Soc. London, A* 178, 506, 1941.
605. Maass, O., and Wright, C. H., *J. Am. Chem. Soc.* 43, 1098, 1921.
606. Ohgaki, K., Umezono, S., and Katayama, T., *J. Supercrit. Fluids* 3, 78, 1990.
607. Marchman, H., Prengle, H. W., and Motard, R. L., *Ind. Eng. Chem.* 41, 2658, 1949.
608. Farrington, P. S., and Sage, B. H., *Ind. Eng. Chem.* 41, 1734, 1949.
609. Vaughan, W. E., and Graves, N. R., *Ind. Eng. Chem.* 32, 1252, 1940.
610. Winkler, C. A., and Maass, O., *Can. J. Res.* 9, 613, 1933.
611. Vohra, S. P., Kang, T.-L., Kobe, K. A., and McKetta, J. J., *J. Chem. Eng. Data* 7, 150, 1962.
612. Steele, W. V., Chirico, R. D., Knipmeyer, S. E., and Nguyen, A., *J. Chem. Eng. Data* 47, 689, 2002.
613. Chirico, R. D., Steele, W. V., Nguyen, A., Klots, T. D., and Knipmeyer, S. E., *J. Chem. Thermodyn.* 28, 797, 1996.
614. Mandel, H., and Ewbank, N., *Atomics International* NAA-S-R-5129, 1960.
615. Gallant, R. W., *Hydrocarbon Process.* 45, 161, 1966.
616. Altunin, V. V., Geller, V. Z., Kremenvskaya, E. A., Perel'shtein, I. I., and Petrov, E. K., *Thermophysical Properties of Freons, Methane Ser.*, Part 2, Vol. 9, NSRDS-USSR, Selover, T. B., Ed., Hemisphere, New York, 1987.
617. Toczylkin. L. S., and Young, C. L., *Aust. J. Chem.* 30, 1591, 1977.
618. Kordes, E., *Z. Elektrochem.* 58, 76, 1954.
619. Lewis, D. T., *J. Appl. Chem.* 3, 154, 1953.
620. Steele, W. V., Chirico, R. D., Nguyen, A., Hossenlopp, I. A., and Smith, N. K., *DIPPR Data Ser.* 1, 101, 1991.
621. Poot, W., and De Loos, T. W., *Fluid Phase Equilib.* 222-223, 255, 2004.
622. Poot, W., and De Loos, T. W., *Fluid Phase Equilib.* 210, 69, 2003.
623. Yata, J., Hori, M., Niki, M., Isono, Y., and Yanagitani, Y., *Fluid Phase Equilib.* 174, 221, 2000.
624. Fujiwara, K., Nakamura, S., and Noguchi, M., *J. Chem. Eng. Data* 43, 55, 1998.
625. Morrison, G., and Ward, D., *Fluid Phase Equilib.* 62, 65, 1991.
626. Piao, C. C., Sato, H., and Watanabe, K., *ASHRAE Trans.* 96, 132, 1990.
627. Piao, C. C., Sato, H., and Watanabe, K., *ASHRAE Trans.* 41, 132, 1989.
628. Kubota, H., Yamashita, T., Tanaka, Y., and Makita, T., *Int. J. Thermophys.* 10, 629, 1989.
629. Kabata, Y., Tanikawa, S., Uematsu, M., and Watanabe, K., *Int. J. Thermophys.* 10, 605, 1989.
630. Basu, R. S., and Wilson, D. P., *Int. J. Thermophys.* 10, 591, 1989.
631. Wilson, D. P., and Basu, R. S., *ASHRAE Trans.* 94, 2095, 1988.
632. Tatoh, J., Kuwabara, S., Sato, H., and Watanabe, K., *J. Chem. Eng. Data* 38, 116, 1993.
633. Lebedeva, E. S., and Khodeeva, S. M., *Zh. Fiz. Khim.* 41, 2081, 1967.
634. Renfrew, M. M., and Lewis, E. E., *Ind. Eng. Chem.* 38, 870, 1946.

635. Chari, N. C., Ph.D. Dissertation, Univ. Michigan, Ann Arbor, MI, 1960.
636. Steele, W. V., Chirico, R. D., Knipmeyer, S. E., and Smith, N. K., *Report*, NIPPR-360, NTIS Order No. DE89000709, Dec. 1988.
637. McGlashan, M. L., and McKinnon, I. R., *J. Chem. Thermodyn.* 9, 1205, 1977.
638. Cipollint, N. E., and Allen, A. O., *J. Chem. Phys.* 67, 131, 1977.
639. Hicks, C. P., and Young, C. L., *J. Chem. Soc., Faraday Trans. 1* 72, 122, 1976.
640. Hugill, J. A., and McGlashan, M. L., *J. Chem. Thermodyn.* 10, 85, 1978.
641. Bendtsen, J., *J. Raman Spectrosc.* 6, 306, 1977.
642. Abdulagatov, I. M., Polikhronidi, N. G., Bruno, T. J., Batyrova, R. G., and Stepanov, G. V., *Fluid Phase Equilib.* 263, 71, 2008.
643. Polikhronidi, N. G., Abdulagatov, I. M., Magee, J. W., and Batyrova, R. G., *J. Chem. Eng. Data* 46, 1064, 2001.
644. Goodwin, R. D., *J. Phys. Chem. Ref. Data* 18, 1565, 1989.
645. Akhundov, T. S., and Abdullaev, F. G., *Izv. Vyssh. Uchebn. Zaved., Neft Gaz* 12, 44, 1969.
646. Krase, N. W., and Goodman, J. B., *Ind. Eng. Chem.* 22, 13, 1930.
647. Buchowski, H., Janaszewski, B., and Toporek, J., *Bull. Acad. Pol. Sci., Ser. Sci. Chim.* 14, 403, 1966.
648. Sokolova, T. D., Prokof'eva, N. K., and Nisel'son, L. A., *Russ. J. Phys. Chem.* (*Engl. Transl.*) 47, 154, 1973.
649. Mastroianni, M. J., Stahl, R. F., and Sheldon, P. N., *J. Chem. Eng. Data* 23, 113, 1978.
650. Weber, L. A., and Defibaugh, D. R., *J. Chem. Eng. Data* 41, 1477, 1996.
651. Higashi, Y., and Ikeda, T., *Fluid Phase Equilib.* 125, 139, 1996.
652. Wang, H., Ma, Y., Lu, C., and Tian, Y., *J. Eng. Thermophys.* 14, 122, 1993.
653. Fukushima, M., *Nippon Reito Kyokai Ronbunshu* 10, 87, 1993.
654. Bier, K., Turk, M., and Zhai, J., *Vapour Pressure of Trifluoroethanol*, Insitut für Technische Thermodynamik und Kaltetechnik, Universität Karlsruhe (TH), D 7500 Karlsruhe l, FRG, 1991.
655. Bier, K., Tuerk, M., and Zhai, J., *Proc. Int. Inst. Ref., Comm. B1 Meet.*, Herzlia, Israel, pp. 129-139, 1990.
656. Zhang, C., Duan, Y. -Y., Shi, L., Zhu, M. -S., and Han, L. -Z., *J. Tsinghua Univ.* (*Sci. & Technol.*) 40, 77, 2000.
657. Duan, Y. -Y., Shi, L., Zhu, M. -S., and Han, L. -Z., *J. Tsinghua Univ.* (*Sci. & Technol.*) 40, 60, 2000.
658. Duan, Y. -Y., Shi, L., Sun, L. -Q., Zhu, M. -S., and Han, L. -Z., *Int. J. Thermophys.* 21, 393, 2000.
659. Duan, Y. -Y., Shi, L., Zhu, M. -S., and Han, L. -Z., *J. Chem. Eng. Data* 44, 501, 1999.
660. Khodeeva, S. M., and Gubochkin, I. V., *Russ. J. Phys. Chem.* (*Engl. Transl.*) 51, 998, 1977.
661. Diefenbacher, A., Crone, M., and Turk, M., *J. Chem. Thermodyn.* 30, 481, 1998.
662. Hori, K., Okazaki, S., Uematsu, M., and Watanabe, K., *Proc. 8th Symp. Thermophys. Prop.*, Vol. II, Sengers, J. V., Ed., ASME, New York, pp. 370-376, 1982.
663. Wagner, W., *Kaeltetech.-Klim.* 20, 238, 1968.
664. Hou, Y.-C., and Martin, J. J., *AIChE J.* 5, 125, 1959.
665. Day, H. O., and Felsing, W. A., *J. Am. Chem. Soc.* 72, 1698, 1950.
666. Kay, W. B., and Warzel, F. M., *Ind. Eng. Chem.* 43, 1150, 1951.
667. Beattie, J. A., and Edwards, D. G., *J. Am. Chem. Soc.* 70, 3382, 1948.
668. Akhundov, T. S., and Asadullaeva, N. N., *Izv. Vyssh. Uchebn. Zaved., Neft Gaz* 11, 83, 1968.
669. Akhundov, T. S., and Imanov, Sh. Yu., *Teplofiz. Svoistva Zhidk.* 48, 1970.

CRITICAL CONSTANTS OF INORGANIC COMPOUNDS

The parameters of the liquid-gas critical point are important constants in determining the behavior of fluids. This table lists the critical temperature, pressure, and molar volume, as well as the normal boiling point, for over 140 inorganic substances. The properties and their units are:

- T_b: Normal boiling point in K at a pressure of 101.325 kPa (1 atmosphere); an "s" following the value indicates a sublimation point (temperature at which the solid is in equilibrium with the gas at a pressure of 101.325 kPa)
- T_c: Critical temperature in K
- P_c: Critical pressure in MPa
- V_c: Critical molar volume in cm^3 mol^{-1}

The number of digits given for T_b, T_c, and P_c indicates the estimated accuracy of these quantities; however, values of T_c greater than 750 K may be in error by 10 K or more. Although most V_c values are given to three figures, they cannot be assumed accurate to better than a few percent. All values are experimentally determined except for a few values, indicated by an asterisk*, that are based on extrapolations. Methods of measurement are described and critiqued in Reference 1. Compounds are listed alphabetically by name.

References

1. Ambrose, D., and Young, C. L., *J. Chem. Eng. Data* 40, 345, 1995.
2. Morel, V., Bultel, A., and Chéron, B. G., *Int. J. Thermophys.* 30, 1853, 2009.
3. Ambrose, D., "Vapor-Liquid Constants of Fluids," in *Handbook of the Thermodynamics of Organic Compounds*, Stevenson, R. M., and Malanowski, S., Eds., Elsevier, New York, 1987.
4. Sato, M., Masui, G., and Uematsu, M., *J. Chem. Thermodyn.* 37, 931, 2005.
5. Velasco, S., Roman, F. L., White, J. A., and Mulero, A., *Fluid Phase Equilib.* 244, 11, 2006.
6. Nowak, P., Tielkes, T., Kleinrahm, R., and Wagner, W., *J. Chem. Thermodyn.* 29, 885, 1997.
7. Lemmon, E. W., and Span, R., *J. Chem. Eng. Data* 51, 785, 2006.
8. Goodwin, R. D., *J. Phys. Chem. Ref. Data* 14, 849, 1985.
9. Vargaftik, N. B., *Int. J. Thermophys.* 11, 467, 1990.
10. Nikitin, E. D., Pavlov, P. A., Popov, A. P., and Nikitina, H. E., *J. Chem. Thermodyn.* 27, 945, 1995.
11. Dillon, I. G., Nelson, P. A., and Swanson, B. S., *J. Chem. Phys.* 44, 4229, 1966.
12. Huber, M. L., Laesecke, A., and Friend, D. G., *The Vapor Pressure of Mercury*, NISTIR 6643, National Institute of Standards and Technology, Boulder, CO, March 2006; *Ind. Eng. Chem. Res.* 45, 7351, 2006.
13. Rau, H., Kutty, T. R. N., and Guedes de Carvalho, J. R. F., *J. Chem. Thermodyn.* 5, 291, 1973.
14. Funke, M., Kleinrahm, R., and Wagner, W., *J. Chem. Thermodyn.* 34, 717, 2002.
15. Sifner, O., and Klomfar, J., *J. Phys. Chem. Ref. Data* 23, 63, 1994.

Name	Formula	T_b/K	T_c/K	P_c/MPa	V_c/cm^3 mol^{-1}	Ref.
Aluminum	Al	2792	6700*			2
Aluminum bromide	$AlBr_3$	528	763	2.89	310	3
Aluminum chloride	$AlCl_3$	453 s	620	2.63	257	3
Aluminum iodide	AlI_3	655	983		408	3
Ammonia	NH_3	239.82	405.56	11.357	69.8	3,4
Ammonium chloride	NH_4Cl	611 s	1155	163.5		3
Antimony(III) bromide	$SbBr_3$	561	904		300	3
Antimony(III) chloride	$SbCl_3$	493.5	794		272	3
Antimony(III) iodide	SbI_3	673	1102			3
Argon	Ar	87.30	150.87	4.898	75	3
Arsenic	As	889 s	1673		35	3
Arsenic(III) chloride	$AsCl_3$	403	654		252	3
Arsine	AsH_3	210.7	373.1			3
Beryllium	Be	2744	5205*			5
Bismuth	Bi	1837	4620*			5
Bismuth tribromide	$BiBr_3$	735	1220		301	3
Bismuth trichloride	$BiCl_3$	714	1179	12.0	261	3
Boron tribromide	BBr_3	364.4	581		272	3
Boron trichloride	BCl_3	285.80	455	3.87	239	3
Boron trifluoride	BF_3	173.3	260.8	4.98	115	3
Boron triiodide	BI_3	482.7	773		356	3
Bromine	Br_2	332.0	588	10.34	127	3
Carbon dioxide	CO_2	194.6 s	304.13	7.375	94	6
Carbon disulfide	CS_2	319	552	7.90	173	3
Carbon monoxide	CO	81.7	132.86	3.494	93	3,7,8
Carbon oxysulfide	COS	223	375	5.88	137	3,7
Cesium	Cs	944	1938	9.4	341	9
Chlorine	Cl_2	239.11	416.9	7.991	123	3
Chlorine pentafluoride	ClF_5	260.1	416	5.27	233	3

Name	Formula	T_b/K	T_c/K	P_c/MPa	V_c/cm³ mol⁻¹	Ref.
Chlorotrifluorosilane	$SiClF_3$	203.2	307.7	3.46		3
Diborane	B_2H_6	180.8	289.8	4.05		3
Dichlorodifluorosilane	$SiCl_2F_2$	241	369.0	3.5		3
Difluoramine	NHF_2	250	403			3
cis-Difluorodiazine	N_2F_2	167.40	272	7.09		3
trans-Difluorodiazine	N_2F_2	161.70	260	5.57		3
Fluorine	F_2	85.03	144.13	5.172	66	3
Fluorine monoxide	F_2O	128.8	215			3
Gallium(III) bromide	$GaBr_3$	552	806.7		303	3
Gallium(III) chloride	$GaCl_3$	474	694		263	3
Gallium(III) iodide	GaI_3	613	951		395	3
Germane	GeH_4	185.1	312.2	4.95	147	3
Germanium	Ge	3106	9802*			5
Germanium(IV) bromide	$GeBr_4$	459.50	718		392	3
Germanium(IV) chloride	$GeCl_4$	359.70	553.2	3.861	330	3
Germanium(IV) iodide	GeI_4	621	973		500	3
Hafnium(IV) bromide	$HfBr_4$	596 s	746		415	3
Hafnium(IV) chloride	$HfCl_4$	590 s	725.7	5.42	314	3
Hafnium(IV) iodide	HfI_4	667 s	916		528	3
Helium	He	4.22	5.19	0.227	57	3
Hydrazine	N_2H_4	386.70	653	14.7		3
Hydrogen	H_2	20.28	32.97	1.293	65	3
Hydrogen bromide	HBr	206.77	363.2	8.55		3
Hydrogen chloride	HCl	188	324.7	8.31	81	3
Hydrogen fluoride	HF	293	461	6.48	69	3
Hydrogen iodide	HI	237.60	424.0	8.31		3
Hydrogen peroxide	H_2O_2	423.4	728*	22*		10
Hydrogen selenide	H_2Se	231.90	411	8.92		3
Hydrogen sulfide	H_2S	213.60	373.1	9.00	99	3,7
Iodine	I_2	457.6	819		155	3
Iodine bromide	IBr	389	719		139	3
Iron	Fe	3134	9340*			5
Krypton	Kr	119.93	209.48	5.525	91	3,7
Lithium	Li	1615	3223*	67*	66*	11
Manganese	Mn	2334	4325*			5
Mercury	Hg	629.77	1764	167	43	3,12
Mercury(II) bromide	$HgBr_2$	591	1012			3
Mercury(II) chloride	$HgCl_2$	577	973		174	3
Mercury(II) iodide	HgI_2	624	1072			3
Molybdenum(V) chloride	$MoCl_5$	541	850		369	3
Molybdenum(VI) fluoride	MoF_6	307.2	473	4.75	226	3
Neon	Ne	27.10	44.4	2.76	42	3
Niobium(V) chloride	$NbCl_5$	520.6	803.5	4.88	397	3
Niobium(V) fluoride	NbF_5	507	737	6.28	155	3
Nitric oxide	NO	121.41	180	6.48	58	3
Nitrogen	N_2	77.35	126.21	3.39	90	3
Nitrogen chloride difluoride	$NClF_2$	206	337.5	5.15		3
Nitrogen tetroxide	N_2O_4	294.30	431	10.1	167	3
Nitrogen trifluoride	NF_3	144.40	234.0	4.46	126	3
Nitrosyl chloride	NOCl	267.7	440			3
Nitrous oxide	N_2O	184.67	309.52	7.245	97	3,7
Nitryl fluoride	NO_2F	200.8	349.5			3
Osmium(VIII) oxide	OsO_4	404.4	678			3
Oxygen	O_2	90.20	154.59	5.043	73	3
Ozone	O_3	161.80	261.1	5.57	89	3
Perchloryl fluoride	ClO_3F	226.40	368.4	5.37	161	3
Phosphine	PH_3	185.40	324.5	6.54		3
Phosphonium chloride	PH_4Cl	246 s	322.3	7.37		3
Phosphorothioc chloride difluoride	$PSClF_2$	279.5	439.2	4.14		3
Phosphorothioc trifluoride	PSF_3	220.90	346.0	3.82		3

Name	Formula	T_b/K	T_c/K	P_c/MPa	V_c/cm³ mol⁻¹	Ref.
Phosphorus	P	553.7	994			3
Phosphorus(III) bromide	PBr_3	446.4	711		300	3
Phosphorus(III) chloride	PCl_3	349.3	563		264	3
Phosphorus(V) chloride	PCl_5	433 s	646			3
Phosphorus(III) chloride difluoride	$PClF_2$	225.9	362.4	4.52		3
Phosphorus(III) dichloride fluoride	PCl_2F	287.00	463.0	4.96		3
Phosphorus(III) fluoride	PF_3	171.4	271.2	4.33		3
Potassium	K	1032	2223*	16*	209*	11
Radon	Rn	211.5	377	6.28		3
Rhenium(VII) oxide	Re_2O_7	633	942		334	3
Rhenium(VI) oxytetrachloride	$ReOCl_4$	496	781		362	3
Rubidium	Rb	961	2093*	16*	247*	11
Selenium	Se	958	1766	27.2		3
Selenium hexafluoride	SeF_6	226.55 s	345.5			3
Selenium oxychloride	$SeOCl_2$	450	730	7.09	235	3
Silver	Ag	2435	6410*			5
Sodium	Na	1156	2573*	35*	116*	11
Sulfur	S	717.76	1314	20.7	57.0	3,13
Sulfur chloride pentafluoride	SF_5Cl	254.10	390.9			3
Sulfur dioxide	SO_2	263.10	430.64	7.884	122	3,7
Sulfur hexafluoride	SF_6	209.35 s	318.723	3.77	197	3,14
Sulfur tetrafluoride	SF_4	232.70	364			3
Sulfur trioxide	SO_3	317.7	491.0	8.2	127	3
Tantalum(V) bromide	$TaBr_5$	622	974		461	3
Tantalum(V) chloride	$TaCl_5$	512.50	767		402	3
Tellurium	Te	1261	2329*			5
Tellurium hexafluoride	TeF_6	234.25 s	356			3
Tellurium tetrachloride	$TeCl_4$	660	1002	8.56	310	3
Tetrabromosilane	$SiBr_4$	427	663		382	3
Tetrachlorosilane	$SiCl_4$	330.80	508.1	3.593	326	3
Tetrafluorohydrazine	N_2F_4	199	309	3.75		3
Tetrafluorosilane	SiF_4	187	259.0	3.72		3
Tetraiodosilane	SiI_4	560.50	944		558	3
Tin(IV) bromide	$SnBr_4$	478	744		417	3
Tin(IV) chloride	$SnCl_4$	387.30	591.9	3.75	351	3
Tin(IV) iodide	SnI_4	637.50	968		531	3
Titanium(IV) bromide	$TiBr_4$	506.7	795.7		391	3
Titanium(IV) chloride	$TiCl_4$	409.60	638	4.66	339	3
Titanium(IV) iodide	TiI_4	650	1040		505	3
Tribromosilane	$SiHBr_3$	382	610.0		305	3
Trichlorofluorosilane	$SiCl_3F$	285.40	438.6	3.58		3
Trichlorosilane	$SiHCl_3$	306	479		268	3
Trifluoramine oxide	NOF_3	185.7	303	6.43	147	3
Tungsten(VI) chloride	WCl_6	610	923		422	3
Tungsten(VI) fluoride	WF_6	290.3	444	4.34	233	3
Tungsten(VI) oxytetrachloride	$WOCl_4$	503	782		338	3
Uranium(VI) fluoride	UF_6	329.65 s	505.8	4.66	250	3
Vanadyl chloride	$VOCl$	400	636		171	3
Water	H_2O	373.12	647.10	22.06	56	3
Xenon	Xe	165.03	289.733	5.842	118	7,15
Xenon difluoride	XeF_2	387.50 s	631	9.32	148	3
Xenon tetrafluoride	XeF_4	388.90 s	612	7.04	188	3
Zirconium(IV) bromide	$ZrBr_4$	633 s	805		424	3
Zirconium(IV) chloride	$ZrCl_4$	604 s	778	5.77	319	3
Zirconium(IV) iodide	ZrI_4	704 s	960		530	3

SUBLIMATION PRESSURE OF SOLIDS

This table gives the sublimation (vapor) pressure of some representative solids as a function of temperature. Entries include simple inorganic and organic substances in their solid phase below room temperature, as well as polycyclic organic compounds which show measurable sublimation pressure only at elevated temperatures. Substances are listed by molecular formula in the Hill order. Values marked by * represent the solid–liquid–gas triple point. Note that some pressure values are in pascals (Pa) and others are in kilopascals (kPa). For conversion, 1 kPa = 7.506 mmHg = 0.0098692 atm.

References

1. Lide, D. R. and Kehiaian, H. V., *CRC Handbook of Thermophysical and Thermochemical Data*, CRC Press, Boca Raton, FL, 1994.
2. *TRC Thermodynamic Tables*, Thermodynamic Research Center, Texas A&M University, College Station, TX.
3. Oja, V. and Suuberg, E. M., *J. Chem. Eng. Data*, 43, 486, 1998.

Ar	*T*/K	55	60	65	70	75	80	83.81*	
Argon	*p*/kPa	0.2	0.8	2.8	7.7	18.7	40.7	68.8*	
BrH	*T*/K	135	140	150	160	170	180	185.1*	
Hydrogen bromide	*p*/kPa	0.1	0.3	1.1	3.3	8.7	20.1	27.4*	
Br_2	*T*/K	170	180	190	200	210	220	230	240*
Bromine	*p*/Pa	0.069	0.416	2.04	8.45	30.3	96.0	273	710*
ClH	*T*/K	120	130	140	150	155	159.0*		
Hydrogen chloride	*p*/kPa	0.1	0.5	1.9	5.8	9.5	13.5*		
Cl_2	*T*/K	120	130	140	150	160	170*		
Chlorine	*p*/Pa	0.144	1.52	11.2	63.1	283	1054*		
F_4Si	*T*/K	130	140	150	160	170	175	180	186.3*
Tetrafluorosilane	*p*/kPa	0.2	0.9	3.9	14.0	43.8	74.2	122.4	220.8*
F_6S	*T*/K	150	165	180	190	200	210	220	223.1*
Sulfur hexafluoride	*p*/kPa	0.4	2.6	11.3	25.9	54.5	106.1	195.1	232.7*
HI	*T*/K	160	170	180	190	200	210	220	222.4*
Hydrogen iodide	*p*/kPa	0.2	0.8	2.2	5.3	11.7	23.6	44.1	49.3*
H_2O	*T*/K	190	210	225	240	250	260	270	273.16*
Water	*p*/Pa	0.032	0.702	4.942	27.28	76.04	195.8	470.1	611.66*
H_2S	*T*/K	140	150	160	165	170	175	180	187.6*
Hydrogen sulfide	*p*/kPa	0.2	0.6	1.9	3.2	5.2	8.3	12.7	22.7*
H_3N	*T*/K	160	170	180	190	195	195.4*		
Ammonia	*p*/kPa	0.1	0.4	1.2	3.5	5.8	6.12*		
I_2	*T*/K	240	250	260	270	280	290	300	310*
Iodine	*p*/Pa	0.081	0.297	0.971	2.89	7.92	20.1	47.9	107*
Kr	*T*/K	80	90	95	100	105	110	115.8*	
Krypton	*p*/kPa	0.4	2.7	6.0	12.1	22.8	40.4	73.1*	
NO	*T*/K	85	90	95	100	105	109.5*		
Nitric oxide	*p*/kPa	0.1	0.4	1.3	3.8	10.0	21.9*		
Xe	*T*/K	110	120	130	140	150	155	160	161.4*
Xenon	*p*/kPa	0.3	1.5	4.9	14.0	34.2	51.1	74.2	81.7*
CHN	*T*/K	200	210	220	230	240	250	255	259.83*
Hydrogen cyanide	*p*/kPa	0.2	0.4	1.0	2.2	4.8	9.7	13.6	18.62*
CH_4	*T*/K	65	70	75	80	85	90.69*		
Methane	*p*/kPa	0.1	0.3	0.8	2.1	4.9	11.70*		
CO	*T*/K	50	55	60	65	68.13*			
Carbon monoxide	*p*/kPa	0.1	0.6	2.6	8.2	15.4*			
CO_2	*T*/K	130	140	155	170	185	194.7	205	216.58*
Carbon dioxide	*p*/kPa	0.032	0.187	1.674	9.987	44.02	101.3	227.1	518.0*
C_2Cl_6	*T*/K	275	300	325	350	375	400	425	459.9*
Hexachloroethane	*p*/Pa	0.004	0.056	0.383	1.62	5.30	14.8	36.4	107.4*
C_2H_2	*T*/K	130	140	150	160	170	180	190	192.4*
Acetylene	*p*/kPa	0.2	0.7	2.6	7.8	20.6	49.0	106.3	126.0*
$C_2H_4O_2$	*T*/K	250	260	270	280	289.7*			
Acetic acid	*p*/kPa	0.092	0.199	0.406	0.79	1.29*			
C_5H_{12}	*T*/K	200	210	220	230	240	250	255	256.58*
Neopentane	*p*/kPa	0.7	1.6	3.6	7.3	13.9	24.8	32.4	35.8*

Compound									
$C_6H_6Cl_6$	T/K	300	320	330	340	350	360	370	380
1,2,3,4,5,6–Hexa–chlorocyclohexane (Lindane)	p/Pa	0.01	0.13	0.39	1.04	2.66	6.42	14.8	32.7
$C_6H_6O_2$	T/K	330	340	350	360	370	380		
Resorcinol	p/Pa	1.03	2.78	7.09	17.2	39.6	87.6		
$C_6H_6O_2$	T/K	350	360	370	380	390	400		
p–Hydroquinone	p/Pa	1.20	3.18	7.96	19.0	43.4	95.1		
$C_{10}H_8$	T/K	250	270	280	290	300	310	330	353.43*
Naphthalene	p/Pa	0.036	0.514	1.662	4.918	13.43	34.15	182.9	999.6*
$C_{12}H_8N_2$	T/K	290	300	310	320				
Phenazine	p/Pa	0.0013	0.0046	0.0150	0.0448				
$C_{12}H_8O$	T/K	300	310	320	330	340	350		
Dibenzofuran	p/Pa	0.408	1.21	3.35	8.71	21.4	50.0		
$C_{12}H_9N$	T/K	350	355	360					
Carbazole	p/Pa	0.086	0.140	0.245					
$C_{13}H_7NO_2$	T/K	330	340	350	360	370	380		
Benz[g]isoquinoline–5,10–dione	p/Pa	0.006	0.018	0.053	0.148	0.394	0.994		
$C_{13}H_8O$	T/K	330	340	350					
1H–Phenalen–1–one	p/Pa	0.040	0.113	0.302					
$C_{13}H_8O_2$	T/K	400	410	420	430				
3–Hydroxy–1H–phenalen–1–one	p/Pa	0.006	0.018	0.053	0.144				
$C_{13}H_9N$	T/K	290	300	310	320				
Acridine	p/Pa	0.0024	0.0085	0.0278	0.0845				
$C_{13}H_9N$	T/K	310	320	330	340				
Phenanthridine	p/Pa	0.020	0.066	0.206	0.603				
$C_{14}H_{10}$	T/K	320	330	340	350	360	370	380	390
Anthracene	p/Pa	0.014	0.043	0.125	0.342	1.01	2.38	5.35	11.5
$C_{14}H_{10}$	T/K	300	310	320	330	340	350	360	
Phenanthrene	p/Pa	0.025	0.085	0.270	0.796	2.02	4.89	11.2	
$C_{16}H_{10}$	T/K	320	330	340	350	360	370	380	390
Pyrene	p/Pa	0.008	0.024	0.073	0.208	0.556	1.32	2.86	6.30
$C_{16}H_{10}O$	T/K	360	370	380	390	400			
1–Pyrenol	p/Pa	0.005	0.016	0.047	0.135	0.364			
$C_{16}H_{12}S$	T/K	330	340	350	360	370	380	390	
Benzo[b]naphtho–(2,1–d)thiophene	p/Pa	0.001	0.004	0.012	0.036	0.098	0.255	0.631	
$C_{17}H_{12}$	T/K	340	350	360	370	380	390	400	
11*H*–Benzo[b]fluorene	p/Pa	0.003	0.009	0.029	0.085	0.235	0.619	1.55	
$C_{18}H_{10}O_4$	T/K	420	430	440	450				
6,11–Dihydroxy–5,12–naphthacenedione	p/Pa	0.008	0.022	0.055	0.131				
$C_{18}H_{12}$	T/K	390	400	410	420				
Chrysene	p/Pa	0.087	0.221	0.539	1.26				
$C_{18}H_{12}$	T/K	390	400	410	420	430	440	450	460
Naphthacene	p/Pa	0.005	0.014	0.035	0.084	0.194	0.432	0.928	1.929
$C_{20}H_{12}$	T/K	390	400	410	420	430			
Perylene	p/Pa	0.006	0.015	0.040	0.102	0.246			
$C_{22}H_{14}$	T/K	450	460	470	480	490			
Pentacene	p/Pa	0.002	0.006	0.013	0.031	0.069			
$C_{24}H_{12}$	T/K	430	440	450	460	470	480	490	500
Coronene	p/Pa	0.004	0.010	0.021	0.046	0.097	0.197	0.389	0.747

VAPOR PRESSURE

This table gives vapor pressure data for about 1800 inorganic and organic substances. In order to accommodate elements and compounds ranging from refractory to highly volatile in a single table, the temperature at which the vapor pressure reaches specified pressure values is listed. The pressure values run in decade steps from 1 Pa (about 7.5 μm Hg) to 100 kPa (about 750 mm Hg). All temperatures are given in °C.

The data used in preparing the table came from a large number of sources; the main references used for each substance are indicated in the last column. Since the data were refit in most cases, values appearing in this table may not be identical with values in the source cited. The temperature entry in the 100 kPa column is close to, but not identical with, the normal boiling point (which is defined as the temperature at which the vapor pressure reaches 101.325 kPa). Although some temperatures are quoted to 0.1 °C, uncertainties of several degrees should generally be assumed. Values followed by an "e" were obtained by extrapolating (usually with an Antoine equation) beyond the region for which experimental measurements were available and are thus subject to even greater uncertainty.

Compounds are listed by molecular formula following the Hill convention. Substances not containing carbon are listed first, followed by those that contain carbon. To locate an organic compound by name or CAS Registry Number when the molecular formula is not known, use the table "Physical Constants of Organic Compounds" in Section 3 and its indexes to determine the molecular formula. The indexes to "Physical Constants of Inorganic Compounds" in Section 4 can be used in a similar way.

More extensive and detailed vapor pressure data on selected important substances appear in other tables in this section of the *Handbook*. These substances are flagged by a symbol following the name as follows:

- * See "Vapor Pressure of Fluids at Temperatures below 300 K"
- ** See "IUPAC Recommended Data for Vapor Pressure Calibration"
- *** See "Vapor Pressure of Ice" and "Vapor Pressure and other Saturation Properties of Water"

The following notations appear after individual temperature entries:

- s — Indicates the substance is a solid at this temperature.
- e — Indicates an extrapolation beyond the region where experimental measurements exist.
- i — Indicates the value was calculated from ideal gas thermodynamic functions, such as those in the *JANAF Thermochemical Tables* (see Reference 8).

References

1. Lide, D.R., and Kehiaian, H.V., CRC Handbook of *Thermophysical and Thermochemical Data*, CRC Press, Boca Raton, FL, 1994.
2. Stull, D., in *American Institute of Physics Handbook, Third Edition*, Gray, D.E., Ed., McGraw Hill, New York, 1972.
3. Hultgren, R., Desai, P.D., Hawkins, D.T., Gleiser, M., Kelley, K.K., and Wagman, D.D., *Selected Values of Thermodynamic Properties of the Elements*, American Society for Metals, Metals Park, OH, 1973.
4. Stull, D., *Ind. Eng. Chem.*, 39, 517, 1947.
5. *TRCVP, Vapor Pressure Database, Version 2.2P*, Thermodynamic Research Center, Texas A&M University, College Station, TX.
6. *TRC Thermodynamic Tables*, Thermodynamic Research Center, Texas A&M University, College Station, TX.
7. Ohe, S., *Computer Aided Data Book of Vapor Pressure*, Data Book Publishing Co., Tokyo, 1976.
8. Chase, M.W., Davies, C.A., Downey, J.R., Frurip, D.J., McDonald, R.A., and Syverud, A.N., *JANAF Thermochemical Tables, Third Edition, J. Phys. Chem. Ref. Data*, Vol. 14, Suppl. 1, 1985.
9. Barin, I., *Thermochemical Data of Pure Substances*, VCH Publishers, New York, 1993.
10. Jacobsen, R.T., et al, *International Thermodynamic Tables of the Fluid State, No. 10. Ethylene*, Blackwell Scientific Publications, Oxford, 1988.
11. Wakeham, W.A., *International Thermodynamic Tables of the Fluid State, No. 12. Methanol*, Blackwell Scientific Publications, Oxford, 1993.
12. Janz, G.J., *Molten Salts Handbook*, Academic Press, New York, 1967.
13. Ohse, R.W. *Handbook of Thermodynamic and Transport Properties of Alkali Metals*, Blackwell Scientific Publications, Oxford, 1994.
14. Gschneidner, K.A., in *CRC Handbook of Chemistry and Physics, 77th Edition*, p. 4–112, CRC Press, Boca Raton, FL, 1996.
15. Leider, H.R., Krikorian, O.H., and Young, D.A., *Carbon*, 11, 555, 1973.
16. Ruzicka, K., and Majer, V., *J. Phys. Chem. Ref. Data*, 23, 1, 1994.
17. Tillner–Roth, R., and Baehr, H.D., *J. Phys. Chem. Ref. Data*, 23, 657, 1994.
18. Younglove, B.A., and McLinden, M.O., *J. Phys. Chem. Ref. Data*, 23, 731, 1994.
19. Outcalt, S.L., and McLinden, M.O., *J. Phys. Chem. Ref. Data*, 25, 605, 1996.
20. Weber, L.A., and Defibaugh, D.R., *J. Chem. Eng. Data*, 41, 382, 1996.
21. Rodrigues, M.F., and Bernardo-Gil, M.G., *J. Chem. Eng. Data*, 41, 581, 1996.
22. Piacente, V., Gigli, G., Scardala, P., and Giustini, A., *J. Phys. Chem.*, 100, 9815, 1996.
23. Barton, J.L., and Bloom, H., *J. Phys. Chem.*, 60, 1413, 1956.
24. Sense, K.A., Alexander, C.A., Bowman, R.E., and Filbert, R.B., *J. Phys. Chem.*, 61, 337, 1957.
25. Ewing, C.T., and Stern, K.H., *J. Phys. Chem.* 78, 1998, 1974.
26. Cady, G.H., and Hargreaves, G.B., *J. Chem. Soc.*, 1563, 1961; 1568, 1961.
27. Skudlarski, K., Dudek, J., and Kapala, J., *J. Chem. Thermodynamics*, 19, 857, 1987.
28. Wagner, W., and de Reuck, K.M., *International Thermodynamic Tables of the Fluid State, No. 9. Oxygen*, Blackwell Scientific Publications, Oxford, 1987.
29. Marsh, K.N., Ed., *Recommended Reference Materials for the Realization of Physicochemical Properties*, Blackwell Scientific Publications, Oxford, 1987.
30. Alcock, C.B., Itkin, V.P., and Horrigan, M.K., *Canadian Metallurgical Quarterly*, 23, 309, 1984.
31. Stewart, R.B., and Jacobsen, R.T., *J. Phys. Chem. Ref. Data*, 18, 639, 1989.
32. Sifner, O., and Klomfar, J., *J. Phys. Chem. Ref. Data*, 23, 63, 1994.
33. Bah, A., and Dupont–Pavlovsky, N., *J. Chem. Eng. Data*, 40, 869, 1995.
34. Behrens, R.G., and Rosenblatt, G., *J. Chem. Thermodynamics*, 4, 175, 1972.
35. Behrens, R.G., and Rosenblatt, G., *J. Chem. Thermodynamics*, 5, 173, 1973.
36. Haar, L., Gallagher, J.S., and Kell, G.S., *NBS/NRC Steam Tables*, Hemisphere Publishing Corp., New York, 1984.
37. Wagner, W., Saul, A., and Pruss, A., *J. Phys. Chem. Ref. Data*, 23, 515. 1994.
38. Behrens, R.G., Lemons, R.S., and Rosenblatt, G., *J. Chem. Thermodynamics*, 6, 457, 1974.
39. Boublik, T., Fried, V., and Hala, E., *The Vapor Pressure of Pure Substances, Second Edition*, Elsevier, Amsterdam, 1984.
40. Goodwin, R.D., *J. Phys. Chem. Ref. Data*, 14, 849, 1985.
41. Younglove, B.A., and Ely, J.F., *J. Phys. Chem. Ref. Data*, 16, 577, 1987.

Mol. form.	Name	Temperature in °C for the indicated pressure						Ref.
		1 Pa	10 Pa	100 Pa	1 kPa	10 kPa	100 kPa	
Substances not containing carbon:								
Ag	Silver	1010	1140	1302	1509	1782	2160	2
AgBr	Silver(I) bromide	569 i	656 i	765 i	905 i	1093 i	1359 i	9
AgCl	Silver(I) chloride	670	769	873	1052	1264	1561	4
AgI	Silver(I) iodide	594	686	803	959	1177	1503	4
Al	Aluminum	1209	1359	1544	1781	2091	2517	2
AlB_3H_{12}	Aluminum borohydride				–46.8	–9.4	45.5	4
$AlCl_3$	Aluminum trichloride	58.4 s	76.5 s	97.1 s	120.7 s	148.2 s	180.5 s	4
AlF_3	Aluminum trifluoride	744 s	819 s	906 s	1008 s	1130 s	1276 s	8
AlI_3	Aluminum triiodide				218	285	385	4
Al_2O_3	Aluminum oxide			2122	2351	2629	2975	4
Ar	Argon*		–226.4 s	–220.3 s	–212.4 s	–201.7 s	–186.0	1,5,31
As	Arsenic	280 s	323 s	373 s	433 s	508 s	601 s	3
$AsCl_3$	Arsenic(III) chloride			–8 e	21.3	63.1	129.4	1
AsF_3	Arsenic(III) fluoride					8.1	56.0	4
AsI_3	Arsenic(III) iodide				187	261	367 e	7
As_2O_3	Arsenic(III) oxide (arsenolite)	133.7 s	163.0 s	196.8 s	236.2 s	283.0		34
At	Astatine	88 s	119 s	156 s	202 s	258 s	334	2
Au	Gold	1373	1541	1748	2008	2347	2805	2
B	Boron	2075	2289	2549	2868	3272	3799	2
BBr_3	Boron tribromide			–45 e	–15 e	27.5	90.4	1
BCl_3	Boron trichloride*			–94.0	–70.5	–37.4	12.3	4
BF_3	Boron trifluoride*	–173.9 s	–166.0 s	–156.0 s	–143.0 s	–125.9	–101.1	4
B_2F_4	Tetrafluorodiborane						–34	1
B_2H_6	Diborane			–162 e	–147.0	–125.8	–92.6	1
B_5H_9	Pentaborane(9)				–34.8	3.8	57.6	4
Ba	Barium	638 s	765	912	1115	1413	1897	9
Be	Beryllium	1189 s	1335	1518	1750	2054	2469	2
$BeBr_2$	Beryllium bromide	203 s	240 s	283 s	335 s	397 s	473 s	4
$BeCl_2$	Beryllium chloride	196 s	237 s	284 s	339 s	402 s	487	4
BeF_2	Beryllium fluoride		686 e	767 e	869	999	1172 e	7
BeI_2	Beryllium iodide	188 s	229 s	276 s	333 s	402 s	487	4
Bi	Bismuth	668	768	892	1052	1265	1562	2
$BiBr_3$	Bismuth tribromide			217 s	273 i	348 i	455 i	4,9
$BiCl_3$	Bismuth trichloride				248.9	328.6	438.7	1,4
BrCs	Cesium bromide	531 s	601 s	701 i	834 i	1019 i	1293 e	9
BrH	Hydrogen bromide*		–153.3 s	–140.4 s	–123.8 s	–101.5 s	–67.0	5
BrH_3Si	Bromosilane				–81.0	–47.3	2.2	4
BrH_4N	Ammonium bromide	121 s	154 s	195 s	246 s	310.4 s	395.1 s	5
BrK	Potassium bromide	597 s	674 s	773				25
BrLi	Lithium bromide		630	733	868	1049	1308	4
BrNa	Sodium bromide			791	931	1120	1389	4
BrRb	Rubidium bromide			766	903	1087	1350	4
BrTl	Thallium(I) bromide				509	635	817	4
Br_2	Bromine*	–87.7 s	–71.8 s	–52.7 s	–29.3 s	2.5	58.4	1
Br_2Cd	Cadmium bromide	373 s	435 s	509 s				27
Br_2Hg	Mercury(II) bromide	71 s	98 s	132 s	174 s	227 s	318	4
Br_2OS	Thionyl bromide	–49 e	–29 e	–5 e	27.8	72.9	139.6	5
Br_2Pb	Lead(II) bromide	374	431	502	597	726	914	4
Br_2S_2	Sulfur bromide	–7 e	15 e	42 e	78.4	128.1	200.9	5
Br_3In	Indium(III) bromide			304.6 s	328.7 s	364.8 s		1
Br_3OP	Phosphorus(V) oxybromide				64 e	115.5	191.4	5
Br_3P	Phosphorus(III) bromide		–23 e	5 e	42.3	94.6	172.6	5
Br_3Sb	Antimony(III) bromide				136.5	196.9	286.5	1
Br_4Ge	Germanium(IV) bromide				51	105	188	4
Br_4Sn	Tin(IV) bromide				67	122	204	4
Br_4Zr	Zirconium(IV) bromide	136 s	167 s	203 s	245 s	295 s	356 s	4
Br_5P	Phosphorus(V) bromide		–19 s	4 s	31 s	65.5 s	110.1	5
Ca	Calcium	591 s	683 s	798 s	954	1170	1482	2
Cd	Cadmium	257 s	310 s	381	472	594	767	2

Mol. form.	Name	Temperature in °C for the indicated pressure 1 Pa	10 Pa	100 Pa	1 kPa	10 kPa	100 kPa	Ref.
$CdCl_2$	Cadmium chloride	412 s	471 s	541 s	634	768	959	23, 27
CdF_2	Cadmium fluoride				1257	1461	1742	4
CdI_2	Cadmium iodide	296 s	344 s	406	498	622	795	4,27
CdO	Cadmium oxide	770 s	866 s	983 s	1128 s	1314 s	1558 s	4
Ce	Cerium	1719	1921	2169	2481	2886	3432	14
ClCs	Cesium chloride			730	864	1043	1297	4
ClCu	Copper(I) chloride		459	543	675	914	1477	4
ClF	Chlorine fluoride*				−144.4	−122.6	−90.2	5
ClF_2P	Phosphorus(III) chloride difluoride				−119.5	−91.1	−47.6	5
ClF_3	Chlorine trifluoride				−63.7	−33.0	11.4	5
ClF_5	Chlorine pentafluoride				−88 e	−59	−14	7
ClH	Hydrogen chloride*				−138.2 s	−118.0	−85.2	1,5
$ClHO_3S$	Chlorosulfonic acid	−40 e	−20 e	5 e	38.7	85.0	153.6	5
ClH_4N	Ammonium chloride	91 s	121 s	159 s	204.7 s	263.1 s	339.5 s	5
ClK	Potassium chloride	625 s	704 s	804	915	1137	1411	23,25
ClLi	Lithium chloride		649 i	761 i	905 i	1101 i	1381 i	8
ClNO	Nitrosyl chloride		−116 s	−100 s	−78.7 s	−50.2	−5.7	5
$ClNO_2$	Nitryl chloride	−121 e	−113 e	−102 e	−86.1	−60.9	−15.7	5
ClNa	Sodium chloride	653 s	733 s	835	987	1182	1461	23,25
ClO_2	Chlorine dioxide*					−34.3	10.5	5
ClRb	Rubidium chloride			777	916	1105	1379	4
ClTl	Thallium(I) chloride				504	626	806	4
Cl_2	Chlorine*	−145 s	−133.7 s	−120.2 s	−103.6 s	−76.1	−34.2	1
Cl_2Co	Cobalt(II) chloride					818	1048	4
Cl_2FP	Phosphorus(III) dichloride fluoride				−71.1	−37.4	13.5	5
Cl_2F_3P	Phosphorus(V) dichloride trifluoride		−120 e	−101 e	−77.1	−44.3	3 e	7
Cl_2Fe	Iron(II) chloride				685	821	1025	4
Cl_2Hg	Mercury(II) chloride	64.4 s	94.7 s	130.8 s	174.5 s	228.5 s	304.0	4
Cl_2Mg	Magnesium chloride			762	908	1111	1414	4
Cl_2Mn	Manganese(II) chloride				760	933	1189	4
Cl_2Ni	Nickel(II) chloride	534 s	592 s	662 s	747 s	852 s	985 s	4
Cl_2OS	Thionyl chloride	−99 e	−81 e	−58 e	−27.1	14.6	75.2	5
Cl_2O_2S	Sulfuryl chloride				−27 e	11.8	69.0	5
Cl_2Pb	Lead(II) chloride			541 e	637	765	949	23
Cl_2S	Sulfur dichloride	−76 e	−61 e	−41 e	−16.7	15.3	58.7	5
Cl_2S_2	Sulfur chloride	−55 e	−36 e	−12 e	21.0	67.2	137.1	5
Cl_2Sn	Tin(II) chloride		253	308	381	479	622	4
Cl_2Zn	Zinc chloride	305 i	356 i	419 i	497 i	596 i	726 i	4,9,12
Cl_3Fe	Iron(III) chloride	118 s	153 s	190 s	229 s	268 s	319	4
Cl_3HSi	Trichlorosilane			−81 e	−56 e	−21 e	31.6	7
Cl_3N	Nitrogen trichloride				−25 e	13.2	70.6	5
Cl_3OP	Phosphorus(V) oxychloride					39.9	105.0	5
Cl_3P	Phosphorus(III) chloride	−93 e	−77 e	−55 e	−26.0	14.5	75.7	5
Cl_4Po	Polonium(IV) chloride					300.6	389.4	5
Cl_4Se	Selenium tetrachloride	23 s	45 s	71 s	102 s	141.4 s	191.1 s	5
Cl_4Si	Tetrachlorosilane*				−39 e	0 e	57.3	1
Cl_4Te	Tellurium tetrachloride				237 e	299.4	387.8	5
Cl_4Zr	Zirconium(IV) chloride	117 s	146 s	181 s	222 s	272 s	336 s	9
Cl_5P	Phosphorus(V) chloride	−2 s	19 s	44 s	74 s	111.4 s	158.9 s	5
Co	Cobalt	1517	1687	1892	2150	2482	2925	2
Cr	Chromium	1383 s	1534 s	1718 s	1950	2257	2669	2
Cs	Cesium	144.5	195.6	260.9	350.0	477.1	667.0	13,30
CsF	Cesium fluoride				825	999	1249	4
CsI	Cesium iodide	523 s	595 s	692	854	1029	1278	4,25
Cu	Copper	1236	1388	1577	1816	2131	2563	2
CuI	Copper(I) iodide				636	864	1331	4
Dy	Dysprosium	1105 s	1250 s	1431 i	1681 i	2031 i	2558 i	3
Er	Erbium	1231 s	1390 s	1612 i	1890 i	2279 i	2859 i	3

Mol. form.	Name	Temperature in °C for the indicated pressure 1 Pa	10 Pa	100 Pa	1 kPa	10 kPa	100 kPa	Ref.
Eu	Europium	590 s	684 s	799 s	961	1179	1523	14
FH	Hydrogen fluoride*				−71.1	−33.7	19.2	1,5
FHO_3S	Fluorosulfonic acid	−14 e	4 e	28 e	59.1	101.3	162.2	5
FK	Potassium fluoride			869	1017	1216	1499	4
FLi	Lithium fluoride	801 s	896	1024	1188	1395	1672	4,12,25
FNO	Nitrosyl fluoride			−131 e	−116.1	−94.3	−60.1	5
FNO_2	Nitryl fluoride		−156 e	−144 e	−128.1	−106.0	−72.6	5
FNO_3	Fluorine nitrate	−160 e	−149 e	−135 e	−115.1	−87.4	−45.0	5
FNa	Sodium fluoride		920 s	1058	1218	1426	1702	4,12,24
FRb	Rubidium fluoride			910	1001	1145	1409	4,12
F_2	Fluorine*	−235 s	−229.5 s	−222.9 s	−214.8	−204.3	−188.3	1,5
F_2O	Fluorine monoxide*	−211.7	−204.7	−195.9	−184.2	−168.2	−144.9	5
F_2OS	Thionyl fluoride			−124 e	−106.5	−81.5	−44.1	5
F_2O_2Re	Rhenium(VI) dioxydifluoride				89.2	131.9	185 e	26
F_2Pb	Lead(II) fluoride				865	1054	1292	4
F_2Xe	Xenon difluoride			2.9 s	31.8 s	67.9 s	114 s	1,5
F_2Zn	Zinc fluoride	731 s	813 s	911 i	1048 i	1237 i	1503 i	9
F_3N	Nitrogen trifluoride*	−201 e	−194 e	−185 e	−172.8	−155.5	−129.2	5
F_3OP	Phosphorus(V) oxyfluoride	−124 s	−113 s	−100 s	−83.7 s	−64.1 s	−39.7 s	5
F_3P	Phosphorus(III) fluoride*				−152 e	−132.6	−101.4	5
F_4MoO	Molybdenum(VI) oxytetrafluoride	−21 s	3 s	33 s	69.3 s	117.3	184.1	26
F_4ORe	Rhenium(VI) oxytetrafluoride	5 s	26 s	50.7 s	80.1 s	117.1	171.2	26
F_4OW	Tungsten(VI) oxytetrafluoride	2 s	25 s	52.1 s	84.3 s	126.7	185.4	26
F_4S	Sulfur tetrafluoride				−110.0	−82.1	−40.3	5
F_4Se	Selenium tetrafluoride				13.6	51.6	104.7	5
F_4Si	Tetrafluorosilane*	−166 s	−157 s	−145.6 s	−132.3 s	−115.7 s	−94.9 s	4,7
F_5Mo	Molybdenum(V) fluoride				86.6	140.3	213 e	26
F_5Nb	Niobium(V) fluoride				80	140	224	4
F_5ORe	Rhenium(VII) oxypentafluoride	−103 s	−84 s	−59 s	−28 s	13.7 s	72.8	26
F_5Os	Osmium(V) fluoride			74.1	113.2	162.3	226 e	26
F_5P	Phosphorus(V) fluoride	−157 s	−148 s	−137 s	−124.5 s	−108.6 s	−84.8	5
F_5Re	Rhenium(V) fluoride			58.8	99.5	152 e	221 e	26
F_5Ta	Tantalum(V) fluoride					119	229	4
F_6Ir	Iridium(VI) fluoride	−88 s	−71 s	−51 s	−27 s	3.8 s	53.1	26
F_6Mo	Molybdenum(VI) fluoride	−98 s	−82 s	−64 s	−41.2 s	−13.4 s	33.5	26
F_6Os	Osmium(VI) fluoride	−89 s	−73 s	−54 s	−30.6 s	−1.7 s	47.4	26
F_6Re	Rhenium(VI) fluoride	−97 s	−82 s	−63 s	−40.2 s	−11.9 s	33.4	26
F_6S	Sulfur hexafluoride*	−158 s	−147 s	−133.6 s	−116.6 s	−94.4 s	−64.1 s	5
F_6Se	Selenium hexafluoride	−143 s	−132 s	−118 s	−100.7 s	−77.8 s	−46.5 s	5
F_6Te	Tellurium hexafluoride	−142 s	−130 s	−115 s	−96 s	−71.8 s	−39.1 s	5
F_6W	Tungsten(VI) fluoride	−107 s	−92 s	−74 s	−52.1 s	−24.8 s	16.9	26
$F_{10}S_2$	Sulfur decafluoride					−22.0	28.5	5
Fe	Iron	1455 s	1617	1818	2073	2406	2859	2
Fr	Francium	131 e	181 e	246 e	335 e	465 e	673 e	2
Ga	Gallium	1037	1175	1347	1565	1852	2245	2
Gd	Gadolinium	1563 i	1755 i	1994 i	2300 i	2703 i	3262 i	3
Ge	Germanium	1371	1541	1750	2014	2360	2831	2
HI	Hydrogen iodide*	−146 s	−135.2 s	−120.8 s	−101.9 s	−75.9 s	−35.9	5
HKO	Potassium hydroxide	520 e	601 e	704	842	1035	1325	4
HNO_3	Nitric acid			−37 e	−9 e	28.4	82.2	5
HN_3	Hydrazoic acid			−79 e	−54 e	−18.0	35.7	5
HNaO	Sodium hydroxide	513	605	722	874	1080	1377	4
H_2	Hydrogen*					−258.6	−252.8	1
H_2I_2Si	Diiodosilane				11.8	70.5	149.4	4
H_2O	Water***	−60.7 s	−42.2 s	−20.3 s	7.0	45.8	99.6	36,37
H_2O_2	Hydrogen peroxide			13 e	45 e	89.0	149.8	5
H_2O_4S	Sulfuric acid	72	103	140	187	248	330	4
H_2S	Hydrogen sulfide*		−149 s	−136 s	−118.9 s	−95.9 s	−60.5	1,5
H_2S_2	Hydrogen disulfide				−27 e	12.2	70.7	5

Mol. form.	Name	Temperature in °C for the indicated pressure 1 Pa	10 Pa	100 Pa	1 kPa	10 kPa	100 kPa	Ref.
H_2Se	Hydrogen selenide	−145 s	−134 s	−120 s	−102.8 s	−78.9 s	−41.5	5
H_2Te	Hydrogen telluride					−46.6	−2.3	5
H_3ISi	Iodosilane				−47.7	−10.1	45.2	4
H_3N	Ammonia*	−139 s	−127 s	−112 s	−94.5 s	−71.3	−33.6	1,5,6
H_3NO	Hydroxylamine				43.7	73.3	109.8	4
H_3P	Phosphine*	−182 s	−173 s	−161 s	−145 s	−122.7	−88.0	5
H_4IN	Ammonium iodide	125 s	159 s	201 s	253 s	318.4 s	405.2 s	5
H_4N_2	Hydrazine				14.7	55.6	113 e	5
H_4Si	Silane*			−181	−165.4	−143.7	−111.8	4
He	Helium*					−270.6	−268.9	2
Hf	Hafnium	2416	2681	3004	3406	3921	4603	9
Hg	Mercury**	42.0	76.6	120.0	175.6	250.3	355.9	29,30
HgI_2	Mercury(II) iodide	85.1 s	115.6 s	152.4 s	197.8 s	255.1 s	353.6	4
Ho	Holmium	1159 s	1311 s	1502 i	1767 i	2137 i	2691 i	3
IK	Potassium iodide			731	866	1052	1322	4
ILi	Lithium iodide	545	619	710	824	972	1170	4
INa	Sodium iodide			753	883	1058	1301	4
IRb	Rubidium iodide			733	866	1045	1302	4
ITl	Thallium(I) iodide				520	644	821	4
I_2	Iodine (rhombic)	−12.8 s	9.3 s	35.9 s	68.7 s	108 s	184.0	1,2
I_2Pb	Lead(II) iodide			470	558	682	869	4
I_2Zn	Zinc iodide	301 s	351 s	409 s	488 i	598 i	750 i	9
I_3Sb	Antimony(III) iodide				214.9	292.0	401.2	4
I_4Sn	Tin(IV) iodide				167.1	242.7	347.7	4
I_4Zr	Zirconium(IV) iodide	187 s	220 s	259 s	305 s	361 s	430 s	4
In	Indium	923	1052	1212	1417	1689	2067	2
Ir	Iridium	2440 s	2684	2979	3341	3796	4386	2
K	Potassium	200.2	256.5	328	424	559	756.2	13,30
Kr	Krypton*	−214.0 s	−208.0 s	−199.4 s	−188.9 s	−174.6 s	−153.6	5
La	Lanthanum	1732 i	1935 i	2185 i	2499 i	2905 i	3453 i	3
Li	Lithium	524.3	612.3	722.1	871.2	1064.3	1337.1	13,30
Lu	Lutetium	1633 s	1829.8	2072.8	2380 i	2799 i	3390 i	3
Mg	Magnesium	428 s	500 s	588 s	698	859	1088	2
Mn	Manganese	955 s	1074 s	1220 s	1418	1682	2060	2
Mo	Molybdenum	2469 s	2721	3039	3434	3939	4606	2
MoO_3	Molybdenum(VI) oxide				801	935	1151	4
NO	Nitric oxide*	−201 s	−195 s	−188 s	−179.3 s	−168.1 s	−151.9	5
N_2	Nitrogen*	−236 s	−232 s	−226.8 s	−220.2 s	−211.1 s	−195.9	1,5
N_2O	Nitrous oxide*	−167 s	−157 s	−145.4 s	−131.1 s	−112.9 s	−88.7	5
N_2O_4	Nitrogen tetroxide	−92 s	−78 s	−61 s	−41.1 s	−16.6 s	28.7	5
N_2O_5	Nitrogen pentoxide	−71 s	−56 s	−40 s	−19.9 s	3.9 s	33.2	5
Na	Sodium	280.6	344.2	424.3	529	673	880.2	13,30
Nb	Niobium	2669	2934	3251	3637	4120	4740	2
Nd	Neodymium	1322.3	1501.2	1725.3	2023 i	2442 i	3063 i	3
Ne	Neon*	−261 s	−260 s	−258 s	−255 s	−252 s	−246.1	2
Ni	Nickel	1510	1677	1881	2137	2468	2911	2
OPb	Lead(II) oxide	724	816	928	1065	1241	1471	4
OSr	Strontium oxide	1789 s	1903 s	2047 s	2235 s	2488 s		4
O_2	Oxygen*				−211.9	−200.5	−183.1	1,28
O_2S	Sulfur dioxide*			−98 s	−80 s	−52.2	−10.3	1,5
O_2Se	Selenium dioxide	124.5 s	153.9 s	188 s	228 s	275 s	315 s	38
O_2Si	Silicon dioxide	1966 i	2149 i	2368 i				8
O_3	Ozone*	−189 e	−182 e	−172 e	−158 e	−139.7	−111.5	5
O_3P_2	Phosphorus(III) oxide				47.3	100.3	172.8	4
O_3S	Sulfur trioxide				−20 s	6.6 s	44.5	5
O_3Sb_2	Antimony(III) oxide (valentinite)	426.1 s	478 s	539 s	610 s	907	1420	4,35
O_5P_2	Phosphorus(V) oxide	285 s	328 s	377.5 s	434.4 s	500.5 s	591	4
O_7Re_2	Rhenium(VII) oxide	147 s	176 s	208 s	244 s	284 s	362	4
Os	Osmium	2887 s	3150	3478	3875	4365	4983	2

Mol. form.	Name	Temperature in °C for the indicated pressure						Ref.
		1 Pa	10 Pa	100 Pa	1 kPa	10 kPa	100 kPa	
P	Phosphorus (white)	6 s	34 s	69	115	180	276	3,9
P	Phosphorus (red)	182 s	216 s	256 s	303 s	362 s	431 s	2,3
Pb	Lead	705	815	956	1139	1387	1754	2
PbS	Lead(II) sulfide	656 s	741 s	838 s	953 s	1088 s	1280	4
Pd	Palladium	1448 s	1624	1844	2122	2480	2961	2
Po	Polonium				573 e	730.2	963.3	5
Pr	Praseodymium	1497.7	1699.4	1954 i	2298 i	2781 i	3506 i	3
Pt	Platinum	2057	2277 e	2542	2870	3283	3821	2
Pu	Plutonium	1483	1680	1925	2238	2653	3226	2
Ra	Radium	546 s	633 s	764	936	1173	1526	2
Rb	Rubidium	160.4	212.5	278.9	368	496.1	685.3	13,30
Re	Rhenium	3030 s	3341	3736	4227	4854	5681	2
Rh	Rhodium	2015	2223	2476	2790	3132	3724	2
Rn	Radon*	−163 s	−152 s	−139 s	−121.4 s	−97.6 s	−62.3	5
Ru	Ruthenium	2315 s	2538	2814	3151	3572	4115	2
S	Sulfur	102 s	135	176	235	318	444	3
Sb	Antimony	534 s	603 s	738	946	1218	1585	2,3
Sc	Scandium	1372 s	1531 s	1733 i	1993 i	2340 i	2828 i	3
Se	Selenium	227	279	344	431	540	685	3
Si	Silicon	1635	1829	2066	2363	2748	3264	2
Sm	Samarium	728 s	833 s	967 s	1148 i	1402 i	1788 i	3
Sn	Tin	1224	1384	1582	1834	2165	2620	2
Sr	Strontium	523 s	609 s	717 s	866	1072	1373	2
Ta	Tantalum	3024	3324	3684	4122	4666	5361	2
Tb	Terbium	1516.1	1706.1	1928 i	2232 i	2640 i	3218 i	3
Tc	Technetium	2454 e	2725 e	3051 e	3453 e	3961 e	4621 e	2
Te	Tellurium			502 e	615 e	768.8	992.4	5
Th	Thorium	2360	2634	2975	3410	3986	4782	2
Ti	Titanium	1709	1898	2130 e	2419	2791	3285	2
Tl	Thallium	609	704	824	979	1188	1485	2
Tm	Thulium	844 s	962 s	1108 s	1297 s	1548 i	1944 i	3
U	Uranium	2052	2291	2586	2961	3454	4129	2
V	Vanadium	1828 s	2016	2250	2541	2914	3406	2
W	Tungsten	3204 s	3500	3864	4306	4854	5550	2
Xe	Xenon*	−190 s	−181 s	−170 s	−155.8 s	−136.6 s	−108.4	5,32
Y	Yttrium	1610.1	1802.3	2047 i	2354 i	2763 i	3334 i	3
Yb	Ytterbium	463 s	540 s	637 s	774 s	993 i	1192 i	3
Zn	Zinc	337 s	397 s	477	579	717	912 e	2
Zr	Zirconium	2366	2618	2924	3302	3780	4405	2
Substances containing carbon:								
C	Carbon (graphite)		2566 s	2775 s	3016 s	3299 s	3635 s	15
$CBrClF_2$	Bromochloro-difluoromethane	−136 e	−123 e	−106 e	−83.4	−51.8	−4.3	1
$CBrCl_3$	Bromotrichloromethane				−6 e	38.9	104.4	5
$CBrF_3$	Bromotrifluoromethane*	−168 e	−156 e	−142 e	−122.8	−96.6	−58.1	5
CBrN	Cyanogen bromide				−13 s	17.7 s	61.0	1
CBr_2F_2	Dibromodifluoromethane		−110 e	−91 e	−66 e	−30 e	22.5	1
CBr_4	Tetrabromomethane			25.6 s	65.8 s	111.6	188.9	5
$CClF_3$	Chlorotrifluoromethane	−176 e	−167 e	−155 e	−139 e	−116 e	−81.7	5
CClN	Cyanogen chloride		−94.6 s	−78.1 s	−57 s	−29 s	13.0	5
CCl_2F_2	Dichlorodifluoromethane*	−150 e	−138 e	−122 e	−101.8	−73.1	−30.0	5
CCl_2O	Carbonyl chloride	−127 e	−113 e	−96 e	−73 e	−40.6	7.2	5
CCl_3F	Trichlorofluoromethane*		−107 e	−89 e	−63 e	−28.5	23.3	1,5
CCl_3NO_2	Trichloronitromethane		−59 e	−30 e	4.4	47.8	112.0	5
CCl_4	Tetrachloromethane*	−79.4 s	−70.8 s	−53.5 s	−24.4 s	15.8	76.2	1,5
CFN	Cyanogen fluoride		−135 s	−121.2 s	−104.1 s	−82.8 s	−46.2	1,5
CF_4	Tetrafluoromethane*	−199.9 s	−193 s	−183.9 s	−171.6	−153.9	−128.3	1,5
$CHBrF_2$	Bromodifluoromethane		−128 s	−111.4 s	−89.7 s	−59.7 s	−16 s	5
$CHBr_3$	Tribromomethane				30.5	78.3	148.8	1
$CHClF_2$	Chlorodifluoromethane*	−152 e	−141 e	−126 e	−107.1	−80.5	−41.1	5
$CHCl_2F$	Dichlorofluoromethane	−76 e	−70 e	−61 e	−49 e	−28.7	8.6	1

Mol. form.	Name	Temperature in °C for the indicated pressure						Ref.
		1 Pa	10 Pa	100 Pa	1 kPa	10 kPa	100 kPa	
$CHCl_3$	Trichloromethane*			−61 e	−34 e	4.3	60.8	1
CHF_3	Trifluoromethane*			−152 e	−136 e	−114.4	−82.3	1
CHI_3	Triiodomethane	51.1 s	82.7 s	121 e			218.0	5
CHN	Hydrogen cyanide*			−77 s	−52.6 s	−22.7 s	25.4	1,5
CHNO	Cyanic acid			−81.1	−56.8	−23.9	23 e	5
CH_2BrCl	Bromochloromethane	−83 e	−69 e	−50 e	−25 e	11.4	67.7	1
CH_2Br_2	Dibromomethane			−37 e	−7 e	35.2	96.5	5
CH_2ClF	Chlorofluoromethane		−124 e	−108 e	−86.2	−55.7	−9.4	5
CH_2Cl_2	Dichloromethane*		−92 e	−73 e	−48 e	−12.5	39.3	1
CH_2F_2	Difluoromethane*	−156.7	−145.8	−131.9	−113.6	−88.6	−51.9	1
CH_2I_2	Diiodomethane			17 e	55 e	106.1	181.6	5
CH_2O	Formaldehyde*				−91 e	−61.7	−19.3	1
CH_2O_2	Formic acid	−56 s	−40.4 s	−22.3 s	−0.8 s	37.0	100.2	1,5
CH_3AsF_2	Methyldifluoroarsine				−15 e	22.1	76.1	5
CH_3BO	Borane carbonyl				−124	−99	−64	4
CH_3Br	Bromomethane				77 e	−44.3	3.3	1
CH_3Cl	Chloromethane*	−140.2 s	−128.6 s	−114.7 s	−96 e	−67.1	−24.4	1,33
CH_3Cl_3Si	Methyltrichlorosilane		−83 e	−61 e	−33 e	7 e	65.7	1
CH_3F	Fluoromethane*				−130 e	−111 e	−78.6	1
CH_3I	Iodomethane				−49 e	−12.4	42.1	1
CH_3NO	Formamide		22 e	53 e	93 e	145.0	218 e	5
CH_3NO_2	Nitromethane				−2 e	40 e	100.8	1
CH_3NO_3	Methyl nitrate		−75 e	−55 e	−27 e	9.8	63 e	5
CH_4	Methane*	−220 s	−214.2 s	−206.8 s	−197 s	−183.6 s	−161.7	5,41
CH_4Cl_2Si	Dichloromethylsilane			−77 e	−51 e	−14 e	40.5	1
CH_4O	Methanol*	−87 e	−69 e	−47.5	−20.4	15.2	64.2	11
CH_4S	Methanethiol		−115 e	−97 e	−74 e	−41.7	5.7	1
CH_5ClSi	Chloromethylsilane	−129 e	−115 e	−97.9	−74.4	−41.5	8.3	5
CH_5N	Methylamine				−76.7	−48.1	−6.6	1
CH_6N_2	Methylhydrazine			−31 e	−4.7	32.9	91 e	1
CH_6OSi	Methyl silyl ether				−90.2	−61.8	−18 e	1
CH_6Si	Methylsilane			−144 e	−124.6	−97.5	−57.5	5
CIN	Cyanogen iodide						153.8	5
CNNa	Sodium cyanide		672 e	798	961	1182	1497	4
CN_4O_8	Tetranitromethane				18.0	61.8	124 e	5
CO	Carbon monoxide*			−223 s	−216.5 s	−207.2 s	−191.7	40
COS	Carbon oxysulfide*			−136 e	−117 e	−90.0	−50.4	1
COSe	Carbon oxyselenide			−120	−98	−67	−22	4
CO_2	Carbon dioxide*	−159.1 s	−148.9 s	−136.7 s	−121.6 s	−103.1 s	−78.6 s	5
CS_2	Carbon disulfide		−96 e	−76 e	−49 e	−10.9	45.9	1
CSe_2	Carbon diselenide			−24 e	9.4	56.2	127 e	1
$C_2Br_2ClF_3$	1,2-Dibromo-1-chloro-1,2,2-trifluoroethane						92.3	5
$C_2Br_2F_4$	1,2-Dibromotetrafluoroethane		−97 e	−75 e	−46 e	−7.2	47.1	5
C_2Br_4	Tetrabromoethylene		−54.5 s	−31.7 s	−3.5 s	32.2 s	226.0	5
C_2ClF_3	Chlorotrifluoroethylene	−146 e	−134 e	−119 e	−99 e	−71 e	−28.4	1
C_2ClF_5	Chloropentafluoroethane					−80.3	−39.4	1
$C_2Cl_2F_4$	1,1-Dichlorotetrafluoroethane					−45.4	2.7	5
$C_2Cl_2F_4$	1,2-Dichlorotetrafluoroethane				−76.8	−44.9	3.2	5
$C_2Cl_3F_3$	1,1,1-Trichlorotrifluoroethane						45.6	1,5
$C_2Cl_3F_3$	1,1,2-Trichlorotrifluoroethane					−8.2	47.3	1,5
C_2Cl_3N	Trichloroacetonitrile				−16 e	25.3	85.1	1
C_2Cl_4	Tetrachloroethylene			−22 e	10 e	54.4	120.7	1
$C_2Cl_4F_2$	1,1,1,2-Tetrachloro-2,2-difluoroethane				−7 e	31.0	91.1	5
$C_2Cl_4F_2$	1,1,2,2-Tetrachloro-1,2-difluoroethane					32.3	92.5	1
C_2Cl_4O	Trichloroacetyl chloride			−25 e	7 e	51.7	117.8	1,5
C_2Cl_6	Hexachloroethane	−7.6 s	9.9 s	33.6 s	67.7 s	116.9 s	184.2 s	5
C_2F_3N	Trifluoroacetonitrile				−126.1	−102.5	−67.8	1
C_2F_4	Tetrafluoroethylene				−132.3	−109.7	−75.8	1
$C_2F_4N_2O_4$	1,1,2,2-Tetrafluoro-1,2-dinitroethane				−30 e	6.4	59.5	5

Mol. form.	Name	Temperature in °C for the indicated pressure						Ref.
		1 Pa	10 Pa	100 Pa	1 kPa	10 kPa	100 kPa	
C_2F_6	Hexafluoroethane**			–155.2 s	–137.5 s	–113.4 s	–78.4 s	1,5
$C_2HBrClF_3$	2-Bromo-2-chloro-1,1,1-trifluoroethane				–41.4	–4.8	49.8	1
C_2HBr_3O	Tribromoacetaldehyde			15.0	52.7	103.0	173.5	5
C_2HClF_4	1-Chloro-1,1,2,2-tetrafluoroethane			–110 e	–87.6	–57.0	–12.1	5
$C_2HCl_2F_3$	2,2-Dichloro-1,1,1-trifluoroethane		–101.0	–82.2	–57.4	–23.3	26.7	18
C_2HCl_3	Trichloroethylene	–74 e	–59 e	–39 e	–12 e	26.7	86.8	1
C_2HCl_3O	Trichloroacetaldehyde			–41.6	–9.8	33.8	97.4	5
$C_2HCl_3O_2$	Trichloroacetic acid				83.8	130.0	197.2	1,5
C_2HCl_5	Pentachloroethane		–23 e	3 e	37.4	86.0	159.4	1
$C_2HF_3O_2$	Trifluoroacetic acid					16.8	71.4	1,5
C_2HF_5O	Trifluoromethyl difluoromethyl ether	–147 e	–136 e	–121 e	–102 e	–75.0	–35.4	20
C_2H_2	Acetylene*			–146.6 s	–130.7 s	–110.6 s	–84.8 s	5
$C_2H_2Br_2$	*cis*-1,2-Dibromoethylene		–45 e	–21 e	10 e	52.2	114.8	1
$C_2H_2Br_2$	*trans*-1,2–Dibromoethylene				–4 e	42.2	107.4	5
$C_2H_2Br_2Cl_2$	1,2-Dibromo-1,1-dichloroethane					103.6	177.8	5
$C_2H_2Br_2Cl_2$	1,2-Dibromo-1,2-dichloroethane		–11 e	22 e	64.1	119 e	193 e	5
$C_2H_2Br_4$	1,1,2,2-Tetrabromoethane	14 e	38 e	69 e	109 e	163.7	242.9	5
$C_2H_2Cl_2$	1,1-Dichloroethylene	–116 e	–101 e	–82 e	–57 e	–21.4	31.2	1
$C_2H_2Cl_2$	*cis*-1,2-Dichloroethylene			–62 e	–34 e	3.8	60.3	1
$C_2H_2Cl_2$	*trans*-1,2-Dichloroethylene				–44 e	–7.5	47.3	1
$C_2H_2Cl_2F_2$	1,2-Dichloro-1,1-difluoroethane	–101 e	–87 e	–68 e	–42.2	–6.8	46.3	5
$C_2H_2Cl_2O$	Chloroacetyl chloride			–23.7	5.6	46.1	105.6	5
$C_2H_2Cl_4$	1,1,1,2-Tetrachloroethane	–58 e	–40 e	–15 e	17 e	62.2	129.7	1
$C_2H_2Cl_4$	1,1,2,2-Tetrachloroethane		–22 e	1 e	32.4	76.9	144.7	1
$C_2H_2F_4$	1,1,1,2-Tetrafluoroethane				–94.3	–66.8	–26.4	17
$C_2H_2F_4$	1,1,2,2-Tetrafluoroethane				–96.0	–66.9	–23.3	5
C_2H_2O	Ketene		–151 e	–135 e	–115 e	–88.2	–50.0	1
C_2H_3Br	Bromoethylene	–124 e	–110 e	–92 e	–68 e	–34.5	15.4	5
C_2H_3BrO	Acetyl bromide	–78 e	–65 e	–49 e	–25 e	13.9	84 e	5
$C_2H_3Br_3$	1,1,2-Tribromoethane	–18 e	4 e	32 e	68 e	117.1	188.4	5
C_2H_3Cl	Chloroethylene	–139 e	–127 e	–110 e	–89 e	–59.0	–14.1	1
$C_2H_3ClF_2$	1-Chloro-1,1-difluoroethane		–123 e	–107 e	–85.3	–55.4	–10.5	5
C_2H_3ClO	Acetyl chloride	–100 e	–85 e	–66 e	–40 e	–3.6	50.4	1
$C_2H_3ClO_2$	Chloroacetic acid				78.4	123.9	188.9	1
$C_2H_3Cl_2F$	1,1-Dichloro-1-fluoroethane		–101 e	–83 e	–57.9	–22.7	31.4	5
$C_2H_3Cl_2F$	1,2-Dichloro-1-fluoroethane			–50 e	–23.8	14.1	73.4	5
$C_2H_3Cl_3$	1,1,1-Trichloroethane				-25.3	14.2	73.7	5
$C_2H_3Cl_3$	1,1,2-Trichloroethane			–23 e	7 e	49.9	113.4	1
C_2H_3F	Fluoroethylene			–153.3	–135.2	–109.9	–72.2	5
C_2H_3FO	Acetyl fluoride					–64.1	17.0	5
$C_2H_3F_3$	1,1,1-Trifluoroethane				–113 e	–86.6	–47.8	1
$C_2H_3F_3O$	2,2,2-Trifluoroethanol			–33 e	–8 e	26.0	74 e	5
C_2H_3I	Iodoethylene				–41 e	–3 e	55.6	5
C_2H_3IO	Acetyl iodide				–0.6	47 e	107.0	5
C_2H_3N	Acetonitrile				–20 e	21.4	81.2	1
C_2H_3NO	Methylisocyanate				–43.5	–10.2	38.8	1
C_2H_3NS	Methyl thiocyanate			–18.4	16.2	63.5	132.5	5
C_2H_4	Ethylene*				–155.6	–135.1	–104.0	1,10
C_2H_4BrCl	1-Bromo-2-chloroethane				–0.4	41.7	105.7	6
$C_2H_4Br_2$	1,1-Dibromoethane		–49 e	–26 e	5 e	46.4	107.6	5
$C_2H_4Br_2$	1,2-Dibromoethane				18 e	62.2	130.9	1
C_2H_4ClF	1-Chloro-1-fluoroethane				–69.9	–36.1	15.8	5
$C_2H_4Cl_2$	1,1-Dichloroethane		–84 e	–64 e	–36.7	1.0	56.9	1
$C_2H_4Cl_2$	1,2-Dichloroethane				–16.4	23.7	83.1	1
$C_2H_4F_2$	1,1-Difluoroethane			–115.2	–94.6	–66.1	–24.3	19
$C_2H_4N_2O_6$	Ethylene glycol dinitrate	4 e	25.6	51.0	81 e	117 e	162 e	5
C_2H_4O	Acetaldehyde		–105 e	–87 e	–62.8	–29.4	20.0	5
C_2H_4O	Ethylene oxide		–111 e	–93 e	–70 e	–37.0	10.2	1
$C_2H_4O_2$	Acetic acid	–42.8 s	–26.7 s	–8 s	14.2 s	55.9	117.5	1,5
$C_2H_4O_2$	Methyl formate		–95 e	–76 e	–51.8	–18.1	31.4	5

Mol. form.	Name	Temperature in °C for the indicated pressure 1 Pa	10 Pa	100 Pa	1 kPa	10 kPa	100 kPa	Ref.
$C_2H_4O_3$	Peroxyacetic acid				14.4	55.3	109.7	5
$C_2H_4O_3$	Glycolic acid						99.9	5
$C_2H_5AsF_2$	Ethyldifluoroarsine			−36 e	−6.0	35.0	93.1	5
C_2H_5Br	Bromoethane	−111 e	−96 e	−77 e	−51.3	−15.5	38.0	5
C_2H_5Cl	Chloroethane	−126 e	−112 e	−94 e	−70 e	−37.0	12.0	1
C_2H_5ClO	2-Chloroethanol	−61 e	−39 e	−12 e	23 e	67.1	127.3	5
C_2H_5ClO	Chloromethyl methyl ether	−96 e	−80 e	−59 e	−32 e	6 e	61 e	5
$C_2H_5Cl_3OSi$	Trichloroethoxysilane	−78 e	−60 e	−36.0	−4.6	38.7	102.0	5
$C_2H_5Cl_3Si$	Trichloroethylsilane	−79 e	−61 e	−38 e	−8 e	34.9	98.7	5
C_2H_5F	Fluoroethane		−142 e	−127 e	−106.3	−78.7	−37.9	1
C_2H_5FO	2-Fluoroethanol			−22 e	8.3	47.5	99 e	5
C_2H_5I	Iodoethane	−94 e	−78 e	−56 e	−27.9	11.9	71.9	5
C_2H_5N	Ethyleneimine		−74 e	−55 e	−30 e	4.1	55 e	5
C_2H_5NO	Acetamide	16.7 s	39.1 s	65.2 s	102.8	150.8	218.2	5
C_2H_5NO	*N*-Methylformamide		13 e	41 e	78 e	127.9	199.1	1
$C_2H_5NO_2$	Nitroethane	−61 e	−44 e	−21 e	8.3	50.1	113.5	5
$C_2H_5NO_3$	Ethyl nitrate	−81 e	−63 e	−41 e	−12 e	28.2	87 e	1
C_2H_6	Ethane*	−183.3 s	−173.2	−161.3	−145.3	−122.8	−88.8	41
$C_2H_6Cl_2Si$	Dichlorodimethylsilane					11.1	70.1	5
C_2H_6Hg	Dimethyl mercury				−13.5	29.0	92.1	5
$C_2H_6N_2O$	*N*-Nitrosodimethylamine				30.7	80.5	149.8	5
C_2H_6O	Ethanol	−73 e	−56 e	−34 e	−7 e	29.2	78.0	1,5
C_2H_6O	Dimethyl ether*		−135 e	−118 e	−96.8	−67.6	−25.1	1,5
C_2H_6OS	Dimethyl sulfoxide			27.4	65.0	115.9	188.6	1
$C_2H_6O_2$	Ethylene glycol	2 e	24 e	51.1	86.1	132.5	196.9	1
$C_2H_6O_2$	Ethyl hydroperoxide	−70 e	−49 e	−25 e	6.8	47.0	101 e	5
$C_2H_6O_2S$	Dimethyl sulfone				109 e	166.8	248.9	5
C_2H_6S	Ethanethiol	−112 e	−97 e	−78 e	−53 e	−18 e	34.7	1
C_2H_6S	Dimethyl sulfide		−96 e	−77 e	−51.2	−16.0	37.0	1,5
$C_2H_6S_2$	Dimethyl disulfide	−71 e	−53 e	−29 e	1.7	45.0	109.3	5
$C_2H_7BO_2$	Dimethoxyborane	−116 e	−101.9	−83.5	−59.2	−25.4	25 e	5
C_2H_7N	Ethylamine			−71 e	−53 e	−27 e	16.4	1
C_2H_7N	Dimethylamine			−88 e	−66.9	−37.2	6.6	1
C_2H_7NO	Ethanolamine		11 e	35 e	66.2	109.0	170.6	1
$C_2H_8N_2$	1,2-Ethanediamine				17.0	57.5	116.6	1,5
$C_2H_8N_2$	1,1-Dimethylhydrazine			−52 e	−25.6	10.5	63 e	5
$C_2H_8N_2$	1,2-Dimethylhydrazine		−49 e	−33 e	−9 e	26.4	88 e	1
C_2N_2	Cyanogen	−127 s	−114.1 s	−98.5 s	−79.2 s	−54.9 s	−21.4	5
C_3ClF_5O	Chloropentafluoroacetone	−122 e	−109 e	−93 e	−71 e	−39.4	7.4	5
C_3Cl_6	Hexachloropropene	−12 e	11 e	40 e	79 e	132.8	213.6	5
C_3F_6	Perfluoropropene	−150 e	−138 e	−122 e	−101 e	−72 e	−30.6	5
C_3F_6O	Perfluoroacetone			−113 e	−94 e	−67.8	−27.6	5
C_3F_8	Perfluoropropane		−139 e	−124 e	−105 e	−77.5	−37.0	1
C_3HN	Cyanoacetylene			−58.7 s	−35.6 s	−7 s	42.0	5
$C_3H_2F_6O$	1,1,1,3,3,3-Hexafluoro-2-propanol					12.7	57.1	5
$C_3H_3F_5$	1,1,1,2,2–Pentafluoropropane					−60 e	−17.9	5
C_3H_3N	2-Propenenitrile		−72 e	−50 e	−22 e	17.7	77.0	1
C_3H_3NS	Thiazole					54.4	117.8	5
C_3H_4	Allene*		−129 e	−118 e	−101.4	−76.7	−34.7	5
C_3H_4	Propyne				−94 e	−65.3	−23.2	1
$C_3H_4ClF_3$	3-Chloro-1,1,1-trifluoropropane	−102 e	−87 e	−68 e	−43 e	−8 e	45.3	5
$C_3H_4Cl_2O$	1,1-Dichloroacetone				1 e	47.8	118.0	5
$C_3H_4Cl_2O_2$	Methyl dichloroacetate	−44 e	−25 e	0 e	33 e	77.7	142.3	5
$C_3H_4Cl_4$	1,1,1,2-Tetrachloropropane	−48 e	−28 e	−2 e	32 e	79.1	149.5	5
$C_3H_4F_4O$	2,2,3,3-Tetrafluoro-1-propanol			−10 e	17 e	53.9	107.2	5
C_3H_4O	Acrolein		−87 e	−67 e	−40 e	−3.0	52.8	1
$C_3H_4O_2$	Propenoic acid				35 e	78.0	140.7	1
$C_3H_4O_2$	Vinyl formate			−58 e	−34 e	−1.6	46.2	1
$C_3H_4O_2$	2-Oxetanone		−21 e	8 e	45.5	93.8	159.3	5
$C_3H_4O_3$	Ethylene carbonate	12.7 s	37 e				247	5

Mol. form.	Name	Temperature in °C for the indicated pressure						Ref.
		1 Pa	10 Pa	100 Pa	1 kPa	10 kPa	100 kPa	
C_3H_5Br	*cis*-1-Bromopropene	−100 e	−84 e	−64 e	−37 e	1.0	57.4	5
C_3H_5Br	2-Bromopropene	−112 e	−95 e	−75 e	−47 e	−9 e	48.0	5
C_3H_5Br	3-Bromopropene	−98 e	−80 e	−58 e	−28 e	12 e	69.6	5
C_3H_5Cl	*cis*-1-Chloropropene	−114 e	−100 e	−81 e	−55 e	−20.1	32.4	5
C_3H_5Cl	*trans*-1-Chloropropene		−97 e	−77 e	−52 e	−16.2	37.0	5
C_3H_5Cl	2-Chloropropene	−120 e	−106 e	−87 e	−63 e	−28.7	22.3	5
C_3H_5Cl	3-Chloropropene	−107 e	−92 e	−72.4	−46.3	−9.8	44.6	5
C_3H_5ClO	Epichlorohydrin			−21 e	11 e	53.8	115.5	5
$C_3H_5ClO_2$	Methyl chloroacetate		−28 e	−5 e	25 e	66.9	129.1	5
$C_3H_5Cl_3$	1,1,3-Trichloropropane	−51 e	−31 e	−5 e	28 e	75.3	145.1	5
$C_3H_5Cl_3$	1,2,3-Trichloropropane			2 e	37 e	84.9	156.3	5
$C_3H_5Cl_3Si$	Trichloro-2-propenylsilane					53.0	116.5	5
C_3H_5I	3-Iodopropene	−80 e	−62 e	−39 e	−8 e	36 e	101.5	5
C_3H_5N	Propanenitrile	−69.4	−55.3	−36.0	−7.9	35.2	97.4	1,5
C_3H_5NO	Acrylamide			109.6	161 e			5
C_3H_5NO	3-Hydroxypropanenitrile	−11 e	18 e	53 e	96.1	150.3	220.8	5
C_3H_5NS	Ethyl thiocyanate	−39 e	−20 e	4 e	35 e	79.1	143.4	5
C_3H_5NS	Ethyl isothiocyanate				17.4	66 e	136 e	5
$C_3H_5N_3O_9$	Trinitroglycerol	48.6	75.7	118 e	191 e	353 e	1007 e	5
C_3H_6	Propene*	−160.6	−149.0	−134.3	−114.9	−88.2	−47.9	1,5
C_3H_6	Cyclopropane			−124 e	−104 e	−75.7	−33.1	1
C_3H_6BrCl	1-Bromo-3-chloropropane	−51 e	−31 e	−6 e	28 e	74.1	142.9	5
$C_3H_6Br_2$	1,2-Dibromopropane	−46 e	−26 e	−2 e	31 e	75.3	139.5	5
$C_3H_6Br_2$	1,3-Dibromopropane	−30 e	−9 e	17 e	52 e	98.7	166.8	5
$C_3H_6Cl_2$	1,1-Dichloropropane				−14 e	27.0	87.7	5
$C_3H_6Cl_2$	1,2-Dichloropropane	−78 e	−61 e	−38.1	−8.1	33.7	95.9	5
$C_3H_6Cl_2$	1,3-Dichloropropane	−65 e	−46 e	−22 e	10 e	54.0	119.9	5
$C_3H_6Cl_2$	2,2-Dichloropropane				−28 e	10.8	68.9	5
$C_3H_6Cl_2O$	1,3-Dichloro-2-propanol			21.8	59.0	107.6	173.9	5
$C_3H_6N_2O_4$	1,1-Dinitropropane	−9 e	12 e	39 e	73.2	120 e	187 e	5
C_3H_6O	Allyl alcohol	−63 e	−48 e	−21.9	6.8	44.5	96.2	5
C_3H_6O	Methyl vinyl ether			−114 e	−89 e	−52.7	4.6	1
C_3H_6O	Propanal			−69 e	−42 e	−6 e	47.7	1
C_3H_6O	Acetone	−95	−81.8	−62.8	−35.6	1.3	55.7	1,5
C_3H_6O	Methyloxirane	−109 e	−95 e	−76 e	−51.5	−17.2	33.9	5
$C_3H_6O_2$	Propanoic acid			0 e	35.1	79.9	140.8	1,5
$C_3H_6O_2$	Ethyl formate		−80 e	−61 e	−35 e	1 e	54.0	1
$C_3H_6O_2$	Methyl acetate	−95 e	−79 e	−59 e	−33 e	3.3	56.6	1
$C_3H_6O_2$	1,3-Dioxolane		−72 e	−50 e	−22 e	17.0	75.3	1
$C_3H_6O_3$	1,3,5-Trioxane					53 e	113.7	1
C_3H_6S	Thietane		−62 e	−40 e	−9 e	32.5	94.5	5
C_3H_7Br	1-Bromopropane	−95 e	−78 e	−57 e	−28 e	11.6	70.6	1
C_3H_7Br	2-Bromopropane		−84 e	−65 e	−39.6	−1.7	59.1	1,5
C_3H_7Cl	1-Chloropropane	−106 e	−90 e	−71 e	−44.5	−8.1	46.2	1
C_3H_7Cl	2-Chloropropane		−91 e	−74 e	−51.1	−17.8	35.4	1,5
C_3H_7ClO	2-Chloro-1-propanol				23 e	63.8	125.7	5
C_3H_7F	1-Fluoropropane	−133 e	−120 e	−103 e	−80.7	−49.4	−2.8	5
C_3H_7I	1-Iodopropane	−78 e	−60 e	−37 e	−6 e	36.9	102.0	5
C_3H_7I	2-Iodopropane	−89 e	−71 e	−47 e	−16.3	26.5	89.2	5
C_3H_7N	Allylamine		−88 e	−65 e	−37 e	0.4	52 e	5
C_3H_7NO	*N*,*N*-Dimethylformamide	−39 e	−20 e	5 e	38.0	83.9	152.6	1
C_3H_7NO	*N*-Methylacetamide	−13.3 s	13 s	43 e	83.8	136.1	206.3	5
$C_3H_7NO_2$	1-Nitropropane	−56 e	−37 e	−13 e	20 e	64.8	130.8	1
$C_3H_7NO_2$	2-Nitropropane		−48 e	−22 e	10.7	55.6	119.8	1
$C_3H_7NO_3$	Propyl nitrate			−23.9	6.1	48.1	111 e	5
C_3H_8	Propane*	−156.9	−145.6	−130.9	−111.4	−83.8	−42.3	1,41
C_3H_8O	1-Propanol	−54 e	−38 e	−16 e	10 e	47 e	96.9	1,5
C_3H_8O	2-Propanol	−65 e	−49 e	−28 e	−1.3	33.6	82.0	1,5
C_3H_8O	Ethyl methyl ether	−98 e	−89 e	−77 e	−60 e	−34.8	7.0	5
$C_3H_8O_2$	1,2-Propylene glycol	−11 e	13 e	42 e	78 e	125.0	187.2	5

Mol. form.	Name	Temperature in °C for the indicated pressure						Ref.
		1 Pa	10 Pa	100 Pa	1 kPa	10 kPa	100 kPa	
$C_3H_8O_2$	1,3-Propylene glycol	4 e	30 e	62 e	101 e	149.9	214.0	5
$C_3H_8O_2$	Ethylene glycol monomethyl ether	−57 e	−37 e	−12 e	21 e	63.8	124.3	1
$C_3H_8O_2$	Dimethoxymethane	−93 c	−81 e	−64 e	−42 e	−9.3	41.7	5
$C_3H_8O_3$	Glycerol	96 e	113 e	136 e	168 e	213.4	287 e	1
C_3H_8S	1-Propanethiol	−94 e	−78 e	−57 e	−29.1	9.6	67.4	1,5
C_3H_8S	2-Propanethiol	−102 e	−87 e	−67 e	−41 e	−3 e	52.2	1
C_3H_8S	Ethyl methyl sulfide	−94 e	−78 e	−57 e	−29.7	8.8	66.3	1
$C_3H_8S_2$	1,3–Propanedithiol	−53 e	−28 e	3 e	43 e	97 e	172.4	5
C_3H_9As	Trimethylarsine			−74 e	−45 e	−5.4	52.0	5
$C_3H_9BO_3$	Trimethyl borate				−14 e	15.6	67.9	5
C_3H_9BS	Methyl dimethylthioborane			−62 e	−30.4	11.4	70.7	5
C_3H_9ClSi	Trimethylchlorosilane				−37.8	0.4	57.3	5
C_3H_9N	Propylamine		−81 e	−63 e	−38.3	−4.1	46.9	1,5
C_3H_9N	Isopropylamine		−91 e	−74 e	−50.4	−17.6	31.5	1,5
C_3H_9N	Trimethylamine		−114 e	−97 e	−75.0	−43.8	2.6	1,5
C_3H_9NO	1-Amino-2-propanol			18 e	53.2	98.2	157.9	5
$C_3H_9O_4P$	Trimethyl phosphate	−31 e	−7 e	23.6	62.8	116.0	192.0	5
C_3H_9P	Trimethylphosphine			−81 e	−53 e	−15.0	37.1	5
C_3H_9Sb	Trimethylstibine			−56 e	−23.8	19 e	80 e	5
$C_3H_{10}N_2$	1,2-Propanediamine		−35.4	−12.0	18.8	61 e	119 e	5
C_3N_2O	Carbonyl dicyanide				−21.7	15.3	65.2	5
C_4Cl_6	Hexachloro-1,3-butadiene	−1 e	22 e	50 e	86.7	137.0	209.7	5
$C_4F_6O_3$	Trifluoroacetic acid anhydride			−63 e	−39 e	−7.1	38.8	5
C_4F_8	Perfluorocyclobutane						−6.2	1
C_4F_{10}	Perfluorobutane		−122 e	−105 e	−82 e	−49.8	−2.5	1,5
$C_4H_2Cl_2O_2$	*trans*-2-Butenedioyl dichloride			8.0	45.6	94.3	159.8	5
$C_4H_2Cl_2S$	2,5-Dichlorothiophene			−20 e	22 e	81.4	171 e	5
$C_4H_2O_3$	Maleic anhydride				73.7	127.9	201.7	5
C_4H_3ClS	2-Chlorothiophene		−62 e	−35 e	2 e	51.8	123 e	5
C_4H_3IS	2-Iodothiophene			−25 e	23 e	94.9	181.0	5
C_4H_4	1-Buten-3-yne			−96.1	−73.4	−41.8	4.9	5
$C_4H_4N_2$	Succinonitrile	24.8 s					266.0	5
C_4H_4O	Furan			−78 e	−54 e	−20 e	31.0	1
$C_4H_4O_2$	Diketene				19.3	63.3	126 e	5
$C_4H_4O_3$	Succinic anhydride				121 e	180.8	260.8	5
$C_4H_4O_4$	Fumaric acid	123.9 s	150 s	180 s				5
C_4H_4S	Thiophene				−17 e	23.7	83.7	5
C_4H_5Cl	2-Chloro-1,3-butadiene	−113 e	−95 e	−71 e	−41 e	0.3	59.0	5
C_4H_5ClO	2-Methyl-2-propenoyl chloride		−57 e	−35 e	−5 e	36.4	98.2	5
$C_4H_5Cl_3O_2$	Ethyl trichloroacetate			15.3	51.9	100.1	166.6	5
C_4H_5N	3-Butenenitrile	−67 e	−48 e	−23.1	9.3	53.7	118.4	5
C_4H_5N	Methylacrylonitrile				−12 e	29.0	89.8	5
C_4H_5N	Pyrrole			−8 e	24 e	66.7	129.4	1
$C_4H_5NO_2$	Methyl cyanoacetate	−3 e	19 e	48 e	84 e	134.0	204.6	5
C_4H_5NS	Allyl isothiocyanate	−45 e	−27 e	−3 e	32.1	89 e	198 e	5
C_4H_5NS	4-Methylthiazole						67.0	5
C_4H_6	1,2-Butadiene	−132 e	−117 e	−98 e	−72.8	−38.9	10.5	5
C_4H_6	1,3-Butadiene*			−106 e	−83 e	−51.9	−4.7	1
C_4H_6	1-Butyne	−125 e	−111 e	−94 e	−71.2	−39.4	7.8	1
C_4H_6	2-Butyne		−89.2 s	−73.8 s	−53.5 s	−23.9	26.6	5
$C_4H_6Cl_2O_2$	Ethyl dichloroacetate			2.6	40.1	89.1	156.3	5
C_4H_6O	Divinyl ether		−99 e	−80 e	−56 e	−22.1	28.0	5
C_4H_6O	*trans*-2-Butenal	−74 e	−56 e	−33 e	−3 e	39.7	102.4	5
C_4H_6O	3-Buten-2-one					21 e	81.0	5
C_4H_6O	Cyclobutanone			−34 e	−4 e	37.1	97 e	5
$C_4H_6O_2$	*cis*-Crotonic acid			30 e	63 e	106.7	168.9	5
$C_4H_6O_2$	*trans*-Crotonic acid				74 e	120.8	184.9	5
$C_4H_6O_2$	3-Butenoic acid	−19 e	2 e	27 e	61 e	105.6	168.6	5
$C_4H_6O_2$	Methacrylic acid			22 e	56 e	99.9	161.5	5
$C_4H_6O_2$	Vinyl acetate	−88 e	−71 e	−50 e	−22 e	16.2	72.2	1

Mol. form.	Name	Temperature in °C for the indicated pressure						Ref.
		1 Pa	10 Pa	100 Pa	1 kPa	10 kPa	100 kPa	
$C_4H_6O_2$	Methyl acrylate		–71 e	–48 e	–18 e	22 e	79.9	5
$C_4H_6O_2$	2,3-Butanedione					30.7	84.8	5
$C_4H_6O_2$	gamma-Butyrolactone		–17 e	24 e	72 e	130.2	203 e	5
$C_4H_6O_3$	Acetic anhydride	–44 e	–25 e	–1 e	31 e	75.1	139.7	1
$C_4H_6O_3$	Propylene carbonate	–40 e	–5 e	43 e	112 e	220 e	410 e	5
$C_4H_6O_4$	Dimethyl oxalate				50.5	98.1	163.0	5
C_4H_7Br	*trans*-1-Bromo-1-butene	–87 e	–68 e	–43.3	–11.4	31.9	94.4	5
C_4H_7Br	2-Bromo-1-butene	–87 e	–70 e	–48 e	–20 e	20.7	80.6	5
C_4H_7Br	*cis*-2-Bromo-2-butene	–90 e	–72 e	–49.0	–18.5	23.5	85.2	5
C_4H_7Br	*trans*-2-Bromo-2-butene	–86 e	–67 e	–43.4	–12.0	31.0	93.5	5
$C_4H_7Br_3$	1,2,3-Tribromobutane	0 e	23 e	53 e	91 e	143.7	219.5	5
$C_4H_7Br_3$	1,2,4-Tribromobutane	–3 e	20 e	49 e	87 e	139.4	214.5	5
C_4H_7Cl	3-Chloro-1-butene			–64 e	–36 e	4 e	63.6	5
C_4H_7Cl	*cis*-2-Chloro-2-butene	–100 e	–83 e	–62 e	–34 e	6 e	66.4	5
C_4H_7Cl	*trans*-2-Chloro-2-butene	–102 e	–86 e	–65 e	–37 e	3 e	62.2	5
C_4H_7Cl	3-Chloro-2-methylpropene		–75 e	–54 e	–25 e	13.8	71.5	5
$C_4H_7ClO_2$	Ethyl chloroacetate			–2.6	32.6	79.1	143.8	5
C_4H_7N	Butanenitrile	–67 e	–48 e	–24 e	8 e	52.3	117.2	1
C_4H_8	1-Butene	–139.0	–125.2	–107.8	–85.3	–53.7	–6.6	1,5
C_4H_8	*cis*-2-Butene	–131.2	–117.4	–99.8	–76.7	–44.8	3.4	1,5
C_4H_8	*trans*-2-Butene			–102 e	–80 e	–47.6	0.6	1
C_4H_8	Isobutene	–139.1	–125.5	–108.2	–85.5	–54.5	–7.3	1,5
C_4H_8	Cyclobutane				–71.8	–38.1	12.1	5
C_4H_8	Methylcyclopropane	–130 e	–116 e	–99.3	–76.3	–44.2	4.2	5
$C_4H_8Br_2$	1,2-Dibromobutane	–54 e	–30 e	0.4	39.6	92.1	166.1	5
$C_4H_8Br_2$	1,4-Dibromobutane	–13 e	9 e	37 e	74 e	124.0	196.5	5
$C_4H_8Cl_2$	1,1-Dichlorobutane			–25 e	6 e	49.3	113.4	5
$C_4H_8Cl_2$	1,2-Dichlorobutane			–28.4	5.8	53.1	123.1	5
$C_4H_8Cl_2$	1,4-Dichlorobutane		–26 e	0 e	35 e	82.4	153.4	5
$C_4H_8Cl_2$	2,2-Dichlorobutane		–58 e	–35 e	–5 e	37.8	102.1	5
$C_4H_8Cl_2O$	Bis(2-chloroethyl) ether	–32 e	–9 e	19.8	56.9	106.9	177.9	5
C_4H_8O	Ethyl vinyl ether		–102 e	–81 e	–53.1	–16.5	34.7	5
C_4H_8O	1,2-Epoxybutane	–135 e	–114 e	–87 e	–53 e	–5.5	62.1	5
C_4H_8O	Butanal	–88 e	–72 e	–50 e	–22 e	16.6	74.5	1,5
C_4H_8O	Isobutanal			–56 e	–29 e	8 e	63.8	1
C_4H_8O	2-Butanone	–85 e	–68 e	–46 e	–18.1	21.2	79.2	1
C_4H_8O	Tetrahydrofuran	–94 e	–78 e	–57.3	–29.8	9 e	65.6	1
$C_4H_8O_2$	Butanoic acid			12.9	52.2	101.4	163.3	1,5
$C_4H_8O_2$	2-Methylpropanoic acid	–30.1	–8.2	18.1	50.5	92.9	154.0	5
$C_4H_8O_2$	Propyl formate	–78 e	–62 e	–42 e	–15.1	23.0	80.4	1,5
$C_4H_8O_2$	Isopropyl formate	–80 e	–65 e	–47 e	–22.2	13.2	67.7	5
$C_4H_8O_2$	Ethyl acetate	–83 e	–66 e	–45 e	–18 e	20.4	76.8	1
$C_4H_8O_2$	Methyl propanoate	–80 e	–64 e	–43 e	–15.8	22.2	79.0	1
$C_4H_8O_2$	*cis*-2-Butene-1,4-diol	17 e	44 e	77 e	117.4	168.5	234.9	5
$C_4H_8O_2$	1,3-Dioxane			–37 e	–3 e	43.4	106.0	5
$C_4H_8O_2$	1,4-Dioxane					39.6	101.0	1
$C_4H_8O_2S$	Sulfolane		49 e	87 e	135 e	198.0	283.5	5
C_4H_8S	Tetrahydrothiophene	–66 e	–47 e	–23 e	9.4	54.1	120.5	1
C_4H_9Br	1-Bromobutane	–68.4	–53.9	–34.1	–5.4	37.6	101.1	1,5
C_4H_9Br	2-Bromobutane	–86 e	–68 e	–46 e	–16 e	26.6	90.7	5
C_4H_9Br	1-Bromo-2-methylpropane	–85 e	–68 e	–46 e	–16 e	26.8	91.1	5
C_4H_9Br	2-Bromo-2-methylpropane					11.7	72.4	1,5
C_4H_9Cl	1-Chlorobutane	–87 e	–71 e	–49 e	–21 e	18.4	78.1	1
C_4H_9Cl	2-Chlorobutane	–96 e	–80 e	–59 e	–31.0	8.5	67.9	1
C_4H_9Cl	1-Chloro-2-methylpropane	–94 e	–78 e	–56.6	–28.7	10.2	68.5	5
C_4H_9Cl	2-Chloro-2-methylpropane					–4.2	50.3	5
$C_4H_9Cl_3Si$	Butyltrichlorosilane					77.2	148.4	5
C_4H_9F	1-Fluorobutane	–114 e	–99 e	–80 e	–55 e	–20.0	32.1	5
C_4H_9F	2-Fluorobutane	–117 e	–103 e	–85 e	–60.7	–26.7	24.7	5
C_4H_9I	1-Iodobutane	–62 e	–43 e	–19 e	14 e	60.5	130.0	5

Mol. form.	Name	Temperature in °C for the indicated pressure						Ref.
		1 Pa	10 Pa	100 Pa	1 kPa	10 kPa	100 kPa	
C_4H_9I	2-Iodobutane	−70 e	−51 e	−27 e	5 e	50 e	119.5	5
C_4H_9I	1-Iodo-2-methylpropane		−47 e	−21.4	12.0	56.8	120.0	5
C_4H_9I	2-Iodo-2-methylpropane	−75.1 s	−58.8 s	−39.5 s	-5.2	41 e	100.0	5
C_4H_9N	Pyrrolidine		−59 e	−38 e	−10 e	28.5	86.2	1
C_4H_9NO	*N*-Methylpropanamide				81.1	105 e		5
C_4H_9NO	*N,N*-Dimethylacetamide	−8 e	8 e	28.0	56.4	98.2	165.7	1
C_4H_9NO	2-Butanone oxime		−18 e	7 e	38.9	81.9	142.9	5
C_4H_9NO	Morpholine				21 e	64.5	128.5	1
$C_4H_9NO_3$	Isobutyl nitrate			−18 e	15.1	59.2	123.0	5
C_4H_{10}	Butane*	−134.3	−121.0	−103.9	−81.1	−49.1	−0.8	1,41
C_4H_{10}	Isobutane*		−129.0	−113.0	−90.9	−59.4	−12.0	1,41
$C_4H_{10}O$	1-Butanol	−37 e	−20 e	0 e	28 e	64 e	117.4	1
$C_4H_{10}O$	2-Butanol	−50 e	−34 e	−14 e	12.6	48.2	99.2	1,5
$C_4H_{10}O$	2-Methyl-1-propanol	−39 e	−24 e	−5 e	20.9	56.0	107.6	1,5
$C_4H_{10}O$	2-Methyl-2-propanol					34.4	82.1	1,5
$C_4H_{10}O$	Diethyl ether	−111 e	−96 e	−77 e	−52.6	−17.8	34.1	1
$C_4H_{10}O$	Methyl propyl ether				−40 e	−11.3	38.7	5
$C_4H_{10}O$	Isopropyl methyl ether				−56 e	−21.2	30.4	5
$C_4H_{10}O_2$	1,3-Butanediol	−4 e	23 e	55 e	94 e	142.9	206.1	5
$C_4H_{10}O_2$	1,4-Butanediol		45 e	77 e	116 e	164.7	227.6	5
$C_4H_{10}O_2$	2,3-Butanediol		15 e	43 e	77 e	121.2	180.3	5
$C_4H_{10}O_2$	Ethylene glycol monoethyl ether	−49 e	−29 e	−3 e	30 e	73.6	135.3	1
$C_4H_{10}O_2$	Ethylene glycol dimethyl ether			−44 e	−15 e	25.2	85.2	1
$C_4H_{10}O_2$	Dimethylacetal	−89 e	−74 e	−55 e	−29 e	7.7	64.1	5
$C_4H_{10}O_2$	Diethylperoxide				−39 e	3.6	65.0	5
$C_4H_{10}O_2S$	Bis(2-hydroxyethyl) sulfide			31 e	114.2		282.0	5
$C_4H_{10}O_3$	Diethylene glycol	35 e	58 e	86 e	123 e	173.6	245.2	1
$C_4H_{10}O_4S$	Diethyl sulfate		3 e	36 e	79 e	134 e	208.3	5
$C_4H_{10}S$	1-Butanethiol	−77 e	−59 e	−37 e	−6 e	35.4	98.0	5
$C_4H_{10}S$	2-Butanethiol	−86 e	−69 e	−47 e	−17 e	23.4	84.5	5
$C_4H_{10}S$	2-Methyl-1-propanethiol		−66 e	−44 e	−15 e	26.5	88.1	5
$C_4H_{10}S$	2-Methyl-2-propanethiol					5.8	63.8	5
$C_4H_{10}S$	Diethyl sulfide	−80 e	−62 e	−40 e	−10.8	30.3	91.7	1
$C_4H_{10}S$	Methyl propyl sulfide	−78 e	−61 e	−38 e	−8 e	33.1	95.1	5
$C_4H_{10}S$	Isopropyl methyl sulfide	−85 e	−68 e	−46 e	−17 e	23.4	84.3	5
$C_4H_{10}S_2$	1,4-Butanedithiol	−17 e	5 e	32 e	69.1	119.9	195.1	5
$C_4H_{10}S_2$	Diethyl disulfide	−46 e	−26 e	0 e	35 e	82.4	153.5	5
$C_4H_{11}N$	Butylamine			−46 e	−18.1	20.0	75.9	5
$C_4H_{11}N$	*sec*-Butylamine			−55 e	−29.1	7.5	62.3	5
$C_4H_{11}N$	*tert*-Butylamine			−67 e	−42.4	−8.1	43.7	5
$C_4H_{11}N$	Isobutylamine	−85 e	−70 e	−50 e	−24.5	12.0	67.3	5
$C_4H_{11}N$	Diethylamine			−46 e	−26 e	5 e	55.2	1
$C_4H_{11}NO$	*N,N*-Dimethylethanolamine	−52 e	−31 e	−6 e	27 e	70.9	133 e	5
$C_4H_{11}NO_2$	Diethanolamine	53 e	77 e	107 e	146 e	197.3	268 e	5
$C_4H_{12}BN$	(Dimethylamino)dimethyl-borane		−81 e	−60.1	−31.9	7.0	64.2	5
$C_4H_{12}Cl_2OSi_2$	1,3-Dichloro-1,1,3,3-tetramethyldisiloxane		−33 e	−9 e	23.8	69.1	136.5	5
$C_4H_{12}O_4Si$	Tetramethyl silicate				14.4	59.3	119.7	5
$C_4H_{12}Si$	Tetramethylsilane			−83 e	−59 e	−25 e	26.7	5
$C_4H_{12}Sn$	Tetramethylstannane			−55.0	−25.6	16.6	77.7	5
$C_4H_{13}N_3$	Diethylenetriamine	−10 e	13 e	43 e	80 e	129.6	198 e	5
C_4NiO_4	Nickel carbonyl					−12	42	4
C_5F_{12}	Perfluoropentane				−54.7	−20.9	28.6	5
C_5FeO_5	Iron pentacarbonyl				0	44	105	4
C_5H_4ClN	2-Chloropyridine			7.4	45.8	97.3	169.9	5
$C_5H_4O_2$	Furfural	−26 e	−8 e	16 e	47 e	92.4	161.4	1
C_5H_5N	Pyridine			−23 e	8 e	51.0	114.9	1
C_5H_6	1,3-Cyclopentadiene			−77 e	−51 e	−14 e	39.8	5
$C_5H_6N_2$	Pentanedinitrile	24.1	52 e	85 e	126 e	178 e	245 e	5
C_5H_6O	2-Methylfuran			−66 e	−35 e	6 e	64.5	1

Mol. form.	Name	Temperature in °C for the indicated pressure						Ref.
		1 Pa	10 Pa	100 Pa	1 kPa	10 kPa	100 kPa	
$C_5H_6O_2$	Furfuryl alcohol	−30 e	−5 e	25 e	62.6	109.3	169.7	5
C_5H_6S	2-Methylthiophene		−58 e	−32 e	2 e	47.9	112.2	1
C_5H_6S	3-Methylthiophene		−53 e	−28 e	6 e	50.6	115.1	1
C_5H_7N	1-Methylpyrrole				8 e	49.9	112.3	5
$C_5H_7NO_2$	Ethyl cyanoacetate	16 e	39 e	67.0	102.1	146.7	205.6	5
C_5H_8	1,2-Pentadiene	−109 e	−93 e	−73 e	−46.1	−9.7	44.5	5
C_5H_8	*cis*-1,3-Pentadiene	−109 e	−93 e	−73 e	−47.0	−10.5	43.7	1,5
C_5H_8	*trans*-1,3-Pentadiene			−75 e	−49.0	−13 e	42 e	1
C_5H_8	1,4-Pentadiene	−120 e	−105 e	−86 e	−60.9	−26.2	25.6	5
C_5H_8	2,3-Pentadiene	−106 e	−90 e	−70 e	−42.9	−6.3	47.9	5
C_5H_8	3-Methyl-1,2-butadiene	−111 e	−95 e	−75 e	−49.2	−13.1	40.4	5
C_5H_8	2-Methyl-1,3-butadiene	−115 e	−100 e	−81 e	−55.4	−19.7	33.7	1,5
C_5H_8	1-Pentyne			−75 e	−49.1	−13.5	39.9	5
C_5H_8	2-Pentyne	−100 e	−85 e	−65 e	−37.9	−0.5	55.7	5
C_5H_8	3-Methyl-1-butyne			−82 e	−57.5	−23.1	28.6	5
C_5H_8	Cyclopentene	−109 e	−94 e	−74 e	−48 e	−11.1	43.8	5
C_5H_8	Spiropentane	−110 e	−95 e	−76 e	−51 e	−15 e	38.6	5
C_5H_8O	3-Methyl-3-buten-2-one			−35 e	−5 e	36.0	97.3	5
C_5H_8O	Cyclopropyl methyl ketone		−57 e	−31 e	3 e	49 e	112 e	5
C_5H_8O	Cyclopentanone		−39 e	−14 e	19 e	64 e	130.3	1
C_5H_8O	3,4-Dihydro-2H-pyran				−22 e	22.0	84.9	5
$C_5H_8O_2$	4-Pentenoic acid	0 e	19 e	44 e	77 e	122.0	187.5	5
$C_5H_8O_2$	Vinyl propanoate					31.2	94 e	5
$C_5H_8O_2$	Ethyl acrylate		−55 e	−32.7	−2.8	38.5	99.2	5
$C_5H_8O_2$	Methyl methacrylate			−31 e	−1 e	39.7	100.0	1
$C_5H_8O_2$	2,4-Pentanedione			−5 e	24.7	67.8	137.4	1
$C_5H_8O_2$	Tetrahydro-2*H*-pyran-2-one		5 e	35.1	74.4	128.3	207.0	5
$C_5H_8O_3$	Methyl acetoacetate				50.1	101.1	171.3	5
$C_5H_8O_4$	Glutaric acid		121 e	153.2	191.9	240.3	302.5	5
$C_5H_8O_4$	Dimethyl malonate	−22 e	1 e	30.0	66.7	114.7	180.2	5
$C_5H_9ClO_2$	Ethyl 2-chloropropanoate			1.4	36.4	82.5	146.0	5
$C_5H_9ClO_2$	Isopropyl chloroacetate			−2 e	35.0	83.3	148.1	5
C_5H_9N	Pentanenitrile	−54 e	−34 e	−8 e	26 e	72.2	140.9	1
C_5H_9N	2,2-Dimethylpropanenitrile					41.1	104.8	5
C_5H_9NO	*N*-Methyl-2-pyrrolidone	1 e	24 e	53.1	92.3	147.2	229 e	5
C_5H_{10}	1-Pentene	−118.9	−103.4	−84.0	−58.8	−23.3	29.6	1,5
C_5H_{10}	*cis*-2-Pentene	−113.8	−98.1	−78.4	−52.7	−16.8	36.6	1,5
C_5H_{10}	*trans*-2-Pentene	−114.5	−98.9	−79.1	−53.3	−17.5	36.0	1,5
C_5H_{10}	2-Methyl-1-butene	−117.7	−102.2	−82.7	−57.2	−21.9	30.8	1,5
C_5H_{10}	3-Methyl-1-butene	−125.0	−110.1	−91.2	−66.7	−32.1	19.7	1,5
C_5H_{10}	2-Methyl-2-butene	−113.4	−97.6	−77.7	−51.6	−15.8	38.2	1,5
C_5H_{10}	Cyclopentane			−77.0	−45.4	−7.1	48.8	5
C_5H_{10}	Ethylcyclopropane	−118 e	−102 e	−83 e	−57 e	−20 e	35.5	5
C_5H_{10}	*cis*-1,2-Dimethylcyclo-propane	−118 e	−103 e	−83 e	−57 e	−20 e	36.6	5
C_5H_{10}	*trans*-1,2-Dimethylcyclo-propane	−122 e	−108 e	−89 e	−63 e	−27 e	27.8	5
$C_5H_{10}Br_2$	1,5-Dibromopentane	1 e	25 e	54 e	93 e	145.6	221.8	5
$C_5H_{10}Cl_2$	1,2-Dichloropentane				30 e	77.4	147.8	5
$C_5H_{10}Cl_2$	1,5-Dichloropentane	−31 e	−10 e	17 e	54 e	104.1	178.9	5
$C_5H_{10}N_2$	3-(Dimethylamino)-propanenitrile				51.1	101.8	171.4	5
$C_5H_{10}O$	Cyclopentanol		−13 e	11.5	42.2	82.5	140.0	5
$C_5H_{10}O$	Allyl ethyl ether			−56 e	−28.7	9.8	67.2	5
$C_5H_{10}O$	Pentanal	−71 e	−53 e	−31 e	−1 e	40.8	102.6	5
$C_5H_{10}O$	2-Pentanone				−1 e	40.3	101.9	1,5
$C_5H_{10}O$	3-Pentanone			−31 e	−1 e	40 e	101.6	1
$C_5H_{10}O$	3-Methyl-2-butanone	−69 e	−54 e	−34 e	−6.9	32.2	94.0	1,5
$C_5H_{10}O$	Tetrahydropyran				−15 e	26.0	88 e	5
$C_5H_{10}O$	2-Methyltetrahydrofuran				−20 e	19.7	79.8	5
$C_5H_{10}O_2$	Pentanoic acid	−7.4	15.3	42.7	76.3	122.1	185.7	5
$C_5H_{10}O_2$	2-Methylbutanoic acid	−10 e	10 e	36 e	69 e	112.8	175.2	5
$C_5H_{10}O_2$	3-Methylbutanoic acid	−15.8	4 e	30.0	64.7	110.6	176.1	5

Mol. form.	Name	1 Pa	10 Pa	100 Pa	1 kPa	10 kPa	100 kPa	Ref.
		Temperature in °C for the indicated pressure						
$C_5H_{10}O_2$	Butyl formate			–29 e	2 e	44.4	105.7	5
$C_5H_{10}O_2$	Isobutyl formate	–69 e	–53 e	–31 e	–3 e	37.4	97.6	5
$C_5H_{10}O_2$	Propyl acetate	–69 e	–51 e	–29 e	0 e	40.9	101.2	1
$C_5H_{10}O_2$	Isopropyl acetate		–61 e	–40 e	–11 e	29.8	88.2	5
$C_5H_{10}O_2$	Ethyl propanoate	–69 e	–52 e	–30 e	–1 e	38.9	98.7	1
$C_5H_{10}O_2$	Methyl butanoate	–68 e	–50 e	–28 e	0.9	41.7	102.3	5
$C_5H_{10}O_2$	Methyl isobutanoate	–83 e	–65 e	–41 e	–11 e	31 e	92.1	5
$C_5H_{10}O_2$	Tetrahydrofurfuryl alcohol	–40 e	–16 e	15 e	55 e	106 e	176.8	5
$C_5H_{10}O_3$	Diethyl carbonate		–42 e	–17 e	17 e	61.6	125.9	5
$C_5H_{10}O_3$	Ethylene glycol monomethyl ether acetate	–47 e	–26 e	0 e	34 e	79.4	144.1	5
$C_5H_{10}S$	Thiacyclohexane				24 e	71.1	141.2	5
$C_5H_{10}S$	Cyclopentanethiol				18 e	64 e	131.7	5
$C_5H_{11}Br$	1-Bromopentane	–60 e	–41 e	–16 e	16 e	61.5	129.1	5
$C_5H_{11}Br$	2-Bromopentane	–69 e	–51 e	–27 e	5 e	49.7	116.9	5
$C_5H_{11}Br$	3-Bromopentane	–68 e	–50 e	–26 e	6 e	50.8	118.1	5
$C_5H_{11}Br$	1-Bromo-3-methylbutane	–67 e	–49 e	25 e	8 e	52.4	119.9	5
$C_5H_{11}Cl$	1-Chloropentane	–73 e	–55 e	–32 e	–1 e	42.5	107.9	5
$C_5H_{11}Cl$	2-Chloropentane	–80 e	–62 e	–39 e	–9 e	33.2	96.1	5
$C_5H_{11}Cl$	3-Chloropentane	–77 e	–60 e	–37 e	–7 e	34.9	97.3	5
$C_5H_{11}Cl$	2-Chloro-2-methylbutane			–52 e	–21 e	21.8	85.2	5
$C_5H_{11}Cl$	1-Chloro-2,2-dimethyl-propane				–17 e	23.5	83.9	5
$C_5H_{11}F$	1-Fluoropentane	–97 e	–80 e	–60 e	–32 e	5.7	62.4	5
$C_5H_{11}I$	1-Iodopentane	–47 e	–27 e	–1 e	34 e	83.0	156.5	5
$C_5H_{11}I$	1-Iodo-3-methylbutane		–34 e	–6.6	28.8	77.3	147.8	5
$C_5H_{11}N$	Cyclopentylamine	–66 e	–48 e	–26 e	4 e	45.8	108 e	5
$C_5H_{11}N$	Piperidine				2 e	43.3	105.8	5
$C_5H_{11}N$	*N*-Methylpyrrolidine				–23 e	18.5	78 e	5
$C_5H_{11}NO_3$	3-Methylbutyl nitrate		–26 e	1.0	35.5	81.7	147.0	5
C_5H_{12}	Pentane**	–115.5	–99.8	–80.0	–54.0	–18.1	35.7	16
C_5H_{12}	Isopentane	–119 e	–105 e	–86 e	–61 e	–26 e	27.5	1
C_5H_{12}	Neopentane*		–107.5 s	–90.8 s	–68.8 s	–38.5 s	9.2	1,5
$C_5H_{12}N_2O$	Tetramethylurea			20.7	58.0	106.7	179.5	5
$C_5H_{12}O$	1-Pentanol	–27 e	–10 e	12 e	41 e	79.8	137.4	5
$C_5H_{12}O$	2-Pentanol	–35 e	–19 e	1 e	28.0	64.9	118.7	1
$C_5H_{12}O$	3-Pentanol	–41 e	–25 e	–4 e	24 e	61.1	114.9	5
$C_5H_{12}O$	2-Methyl-1-butanol	–27 e	–11 e	9 e	36.2	73.4	128.3	1
$C_5H_{12}O$	3-Methyl-1-butanol	–22 e	–7 e	13 e	39.1	75.7	130.1	5
$C_5H_{12}O$	2-Methyl-2-butanol			–5 e	17.7	50.6	101.7	1,5
$C_5H_{12}O$	3-Methyl-2-butanol			–3 e	22.7	58.2	111.1	5
$C_5H_{12}O$	2,2-Dimethyl-1-propanol					59.2	112.7	5
$C_5H_{12}O$	Butyl methyl ether			–54 e	–27 e	12 e	69.8	1
$C_5H_{12}O$	Methyl *tert*-butyl ether			–66 e	–39 e	–2 e	54.8	1
$C_5H_{12}O$	Ethyl propyl ether	–92 e	–77 e	–57 e	–30.5	6.7	63.4	1,5
$C_5H_{12}O_2$	1,5-Pentanediol	25 e	52 e	85 e	125 e	175.1	238.9	5
$C_5H_{12}O_2$	Ethylene glycol monopropyl ether				40 e	85.6	149.3	5
$C_5H_{12}O_2$	Diethoxymethane		–65 e	–43 e	–14 e	27.3	87.7	5
$C_5H_{12}O_3$	Diethylene glycol monomethyl ether		12 e	40 e	76 e	124.2	193.7	1
$C_5H_{12}S$	1-Pentanethiol	–60 e	–41 e	–17 e	15 e	60 e	126.2	1
$C_5H_{12}S$	2-Pentanethiol	–70 e	–52 e	–28 e	3 e	46.6	111.9	5
$C_5H_{12}S$	3-Pentanethiol	–70 e	–51 e	–28 e	4 e	47.7	113.4	5
$C_5H_{12}S$	2-Methyl-1-butanethiol				8.0	52.3	118.5	5
$C_5H_{12}S$	3-Methyl-1-butanethiol				7.8	51.9	117.9	5
$C_5H_{12}S$	2-Methyl-2-butanethiol				–8.0	34.6	98.7	5
$C_5H_{12}S$	Butyl methyl sulfide		–43 e	–19 e	13 e	57 e	123.0	1
$C_5H_{12}S$	*tert*-Butyl methyl sulfide				–7.8	34.7	98.4	5
$C_5H_{12}S$	Ethyl propyl sulfide	–64 e	–46 e	–23 e	9 e	52.7	118.0	5
$C_5H_{12}S$	Ethyl isopropyl sulfide	–72 e	–54 e	–31 e	0 e	42.7	106.9	5
$C_5H_{13}N$	Pentylamine		–52 e	–29 e	1 e	42.8	104.0	5
C_6BrF_5	Bromopentafluorobenzene			–10 e	23 e	68 e	136.0	5
C_6ClF_5	Chloropentafluorobenzene		–44 e	–21 e	11 e	53.8	117.6	1

Mol. form.	Name	Temperature in °C for the indicated pressure						Ref.
		1 Pa	10 Pa	100 Pa	1 kPa	10 kPa	100 kPa	
$C_6Cl_3F_3$	1,3,5-Trichloro-2,4,6-trifluorobenzene	−19 e	4 e	32 e	70 e	121.7	197.9	1
C_6F_6	Hexafluorobenzene		−56.9 s	−36 s	−11.5 s	22.6	79.9	1,5
C_6F_{12}	Perfluorocyclohexane				−46.2 s	−7.6 s	48.9 s	5
C_6F_{14}	Perfluorohexane		−75 e	−57 e	−32 e	2.8	56.8	5
C_6F_{14}	Perfluoro-2-methylpentane				−33 e	2.9	57.1	5
C_6F_{14}	Perfluoro-3-methylpentane	−95 e	−80 e	−60 e	−34 e	2.8	57.9	5
C_6F_{14}	Perfluoro-2,3-dimethylbutane					4.3	59.3	5
C_6HF_5	Pentafluorobenzene			−41 e	−13 e	27 e	85.3	5
C_6HF_5O	Pentafluorophenol				39 e	82 e	145.2	5
$C_6H_2F_4$	1,2,3,4-Tetrafluorobenzene			−36 e	−7 e	33.8	94.0	1
$C_6H_2F_4$	1,2,3,5-Tetrafluorobenzene			−43 e	−14 e	25.5	84.1	1
$C_6H_2F_4$	1,2,4,5-Tetrafluorobenzene					30.7	89.9	1
$C_6H_3Cl_3O$	2,4,6-Trichlorophenol			71.8	114.0	169.5	245.7	5
$C_6H_3F_3$	1,3,5-Trifluorobenzene					18.2	75.0	5
$C_6H_4Br_2$	*m*-Dibromobenzene	−7 e	16 e	44 e	83 e	137.0	218.2	5
$C_6H_4ClNO_2$	1-Chloro-4-nitrobenzene	15.4 s	35.8 s		97 e	156.0	238 e	5
$C_6H_4Cl_2$	*o*-Dichlorobenzene		−13 e	16.3	53.9	104.6	180.0	1,5
$C_6H_4Cl_2$	*m*-Dichlorobenzene		−22 e	8.0	46.7	97.8	172.5	1,5
$C_6H_4Cl_2$	*p*-Dichlorobenzene	−45.5 s	−21.8 s	8 s	46.7 s	99.0	173.6	1,5
$C_6H_4O_2$	*p*-Benzoquinone	−4.1 s	17.8 s	43.5 s	74.3 s	111.6 s		5
$C_6H_5AsCl_2$	Dichlorophenylarsine	6.9	35.2	70 e	113 e	170 e	245 e	5
C_6H_5Br	Bromobenzene		−25 e	1 e	34.9	83.1	155.4	1
C_6H_5Cl	Chlorobenzene		−43 e	−17 e	16.8	62.9	131.3	1,5
C_6H_5ClO	*o*-Chlorophenol				45.8	97.9	173.9	5
C_6H_5ClO	*m*-Chlorophenol			39.7	80.2	135.1	213.4	5
C_6H_5ClO	*p*-Chlorophenol			45.0	86.5	142.0	219.9	5
$C_6H_5Cl_3Si$	Trichlorophenylsilane			33 e	70.2	122.6	201 e	5
C_6H_5F	Fluorobenzene				−16.9	24.2	84.4	1
C_6H_5I	Iodobenzene	−30 e	−7 e	20.9	58.5	110.6	187.8	1
$C_6H_5NO_2$	Nitrobenzene		10 e	40 e	78 e	132 e	210.3	1
$C_6H_5NO_3$	*p*-Nitrophenol	72.6 s	97.4 s					5
C_6H_6	1,5-Hexadien-3-yne	−82 e	−66 e	−44.3	−16.0	23.7	83.6	5
C_6H_6	Benzene**			−40 s	−15.1 s	20.0	79.7	1,5
C_6H_6ClN	*o*-Chloroaniline		10 e	39.0	75.2	131.4	208.3	5
C_6H_6ClN	*m*-Chloroaniline	−5 e	19.7	49.4	94.2	162 e	1069 e	5
$C_6H_6N_2O_2$	*p*-Nitroaniline	87.8 s			192.0	252.6	331.2	5
C_6H_6O	Phenol	−9.7 s	9.6 s	34.1 s	68.9	113.7	181.4	1,5
$C_6H_6O_3$	1,2,3-Benzenetriol				162.0	222.8	308.3	5
C_6H_6S	Benzenethiol		−15 e	12 e	47 e	96.0	168.6	5
C_6H_7N	Aniline		−2.5	26.7	63.5	112.5	183.5	1,5
C_6H_7N	2-Methylpyridine	−56.5	−37.8	−13.9	18.3	62.9	129.0	1,5
C_6H_7N	3-Methylpyridine			−5 e	28.8	75.2	143.7	1
C_6H_7N	4-Methylpyridine	−58.2 s	−43.1 s	−3.9 s	29.6	76.1	144.9	1,5
C_6H_8	*cis*-1,3,5-Hexatriene					21 e	78 e	5
C_6H_8	1,3-Cyclohexadiene	−88 e	−71 e	−50 e	−21 e	19 e	79.9	5
C_6H_8	1,4-Cyclohexadiene				−15 e	27.3	85.0	5
$C_6H_8N_2$	Adiponitrile	30 e	61 e	100 e	148.6	211.8	297 e	5
$C_6H_8N_2$	*m*-Phenylenediamine			94.5	140.2	200.8	285.0	5
$C_6H_8N_2$	Phenylhydrazine		38 e	69 e	109 e	163.9	242.5	5
$C_6H_8O_4$	Dimethyl maleate		5 e	36 e	76 e	127.3	197 e	5
C_6H_8S	2,5-Dimethylthiophene		−43 e	−16 e	20 e	67.5	134.8	5
C_6H_{10}	*trans*-1,3-Hexadiene	−86 e	−70 e	−51 e	−24 e	14 e	72 e	5
C_6H_{10}	*trans*-1,4-Hexadiene	−98 e	−81 e	−60 e	−33 e	7 e	65 e	5
C_6H_{10}	1,5-Hexadiene	−99 e	−84 e	−64 e	−37 e	0.9	59.2	5
C_6H_{10}	*cis,cis*-2,4-Hexadiene					18 e	79.6	5
C_6H_{10}	*trans,cis*-2,4-Hexadiene	−89 e	−73 e	−52 e	−23 e	18 e	79.6	5
C_6H_{10}	*trans,trans*-2,4-Hexadiene				−23 e	18 e	79.6	5
C_6H_{10}	*trans*-2-Methyl-1,3-pentadiene	−92 e	−75 e	−54 e	−26 e	14 e	75.6	5
C_6H_{10}	2,3-Dimethyl-1,3-butadiene			−59 e	−30 e	9.7	68.1	5
C_6H_{10}	1-Hexyne	−91 e	−75 e	−54 e	−26 e	12.8	71.0	5

Mol. form.	Name	Temperature in °C for the indicated pressure						Ref.
		1 Pa	10 Pa	100 Pa	1 kPa	10 kPa	100 kPa	
C_6H_{10}	2-Hexyne	−84 e	−67 e	−46 e	−17 e	23.6	84.1	5
C_6H_{10}	3-Hexyne	−86 e	−69 e	−48 e	−19.1	21.0	81.0	1,5
C_6H_{10}	4-Methyl-1-pentyne	−97 e	−81 e	−61 e	−34 e	4.1	60.7	5
C_6H_{10}	4-Methyl-2-pentyne	−91 e	−74 e	−54 e	−26 e	13.8	72.7	5
C_6H_{10}	Cyclohexene	−87 e	−70 e	−49 e	−19 e	21 e	82.6	1
$C_6H_{10}Cl_2$	1,1-Dichlorocyclohexane	−39 e	−19 e	8 e	43 e	93.5	170.5	5
$C_6H_{10}Cl_2$	*cis*-1,2-Dichlorocyclohexane			27 e	69 e	125.7	206.2	5
$C_6H_{10}O$	4-Methyl-4-penten-2-one	−59 e	−41 e	−17 e	14 e	57.0	121.0	5
$C_6H_{10}O$	Cyclohexanone		−25 e	1 e	36 e	84 e	155.2	1
$C_6H_{10}O$	Mesityl oxide	−56 e	−37 e	−13 e	19 e	63.5	129.3	5
$C_6H_{10}O_2$	Vinyl butanoate					53 e	114.5	5
$C_6H_{10}O_2$	Ethyl methacrylate				8 e	53.2	116.8	5
$C_6H_{10}O_2$	Allyl glycidyl ether				40.1	85.7	152.8	5
$C_6H_{10}O_3$	Ethyl acetoacetate	−25 e	−3 e	25.7	62.3	111.3	180.2	5
$C_6H_{10}O_3$	Propanoic anhydride	−32 e	−15 e	6 e	36 e	77.6	142.9	5
$C_6H_{10}O_4$	Diethyl oxalate	−5 e	18 e	44.9	79.4	124.3	185.2	5
$C_6H_{10}O_4$	Dimethyl succinate			30 e	70.4	123.3	195.4	5
$C_6H_{10}O_4$	Ethylene glycol diacetate	−17 e	6 e	35.0	71.9	121.1	190.0	5
$C_6H_{10}S$	Diallylsulfide	−58 e	−38 e	−12.4	21.7	68.8	138.1	5
$C_6H_{11}Cl$	Chlorocyclohexane		−35 e	−9 e	25 e	71.6	142.1	5
$C_6H_{11}N$	Hexanenitrile	−40 e	−19 e	8 e	43 e	91.5	163.2	1,5
$C_6H_{11}N$	4-Methylpentanenitrile		−50 e	−20 e	20 e	75.2	155.2	5
$C_6H_{11}NO$	Caprolactam	36.8 s	58.9 s	86.6 s			270	5
C_6H_{12}	1-Hexene	−99.8	−82.8	−61.4	−33.7	5.2	63.1	1,5
C_6H_{12}	*cis*-2-Hexene	−97 e	−80 e	−58 e	−30 e	9.9	68.5	5
C_6H_{12}	*trans*-2-Hexene	91 e	−78 e	57 e	−30 e	9.3	67.5	5
C_6H_{12}	*cis*-3-Hexene	−96 e	−79 e	−59 e	−30.8	7.9	66.0	5
C_6H_{12}	*trans*-3-Hexene	−95 e	−79 e	−58 e	−30.0	8.8	66.7	5
C_6H_{12}	2-Methyl-1-pentene	−98 e	−82 e	−62 e	−34.2	4.1	61.7	5
C_6H_{12}	3-Methyl-1-pentene	−104 e	−88 e	−68 e	−41.5	−3.6	53.8	5
C_6H_{12}	4-Methyl-1-pentene	−105 e	−89 e	−69 e	−41.6	−3.6	53.5	5
C_6H_{12}	2-Methyl-2-pentene	−95 e	−78 e	−58 e	−30 e	9.0	66.9	5
C_6H_{12}	3-Methyl-*cis*-2-pentene	−95 e	−79 e	−58 e	−30 e	8.9	67.3	5
C_6H_{12}	3-Methyl-*trans*-2-pentene	−93 e	−77 e	−55 e	−27.4	11.7	70.0	5
C_6H_{12}	4-Methyl-*cis*-2-pentene	−102 e	−86 e	−66 e	−38.7	−0.9	56.0	5
C_6H_{12}	4-Methyl-*trans*-2-pentene	−100 e	−84 e	−64 e	−36.8	1.2	58.2	5
C_6H_{12}	2-Ethyl-1-butene	−98 e	−81 e	−60 e	−32 e	6.6	64.3	5
C_6H_{12}	2,3-Dimethyl-1-butene	−103 e	−87 e	−67 e	−39.9	−1.9	55.2	5
C_6H_{12}	3,3-Dimethyl-1-butene	−110 e	−95 e	−76 e	−50.8	−14.5	40.8	5
C_6H_{12}	2,3-Dimethyl-2-butene		−75 e	−54 e	−25 e	14 e	72.9	1
C_6H_{12}	Cyclohexane	−85.6 s	−68.9 s	−47.6 s	−19.8 s	19.3	80.4	1,5
C_6H_{12}	Methylcyclopentane	−97 e	−80 e	−58 e	−28.8	11.6	71.4	1,5
C_6H_{12}	Ethylcyclobutane	−99 e	−82 e	−61 e	−32 e	9 e	70.2	5
C_6H_{12}	Isopropylcyclopropane	−104 e	−88 e	−68 e	−40 e	−1 e	57.9	5
C_6H_{12}	1-Ethyl-1-methylcyclopropane	−105 e	−89 e	−69 e	−41 e	−3 e	56.3	5
C_6H_{12}	1,1,2-Trimethylcyclopropane	−109 e	−94 e	−73 e	−46 e	−7 e	52.0	5
$C_6H_{12}Cl_2$	1,2-Dichlorohexane				49 e	98.1	171.7	5
$C_6H_{12}Cl_2O$	2,2′-Dichlorodiisopropyl ether		−1 e	27.3	63.4	112.3	182.1	5
$C_6H_{12}O$	Butyl vinyl ether	−87 e	−67 e	−42 e	−9.3	33.6	93.2	5
$C_6H_{12}O$	Isobutyl vinyl ether	−87 e	−68 e	−44 e	−13 e	26.5	80.7	5
$C_6H_{12}O$	Hexanal	−56 e	−37 e	−13 e	19 e	62.6	127.8	5
$C_6H_{12}O$	2-Hexanone	−43 e	−21 e	4.2	34.5	61.9	127.2	1,5
$C_6H_{12}O$	3-Hexanone		−40 e	−16 e	15 e	58.5	123.1	1
$C_6H_{12}O$	3-Methyl-2-pentanone				8.5	52.7	117.0	5
$C_6H_{12}O$	4-Methyl-2-pentanone	−61 e	−43 e	−21 e	9 e	51.5	116.1	5
$C_6H_{12}O$	2-Methyl-3-pentanone					50.2	113.0	5
$C_6H_{12}O$	3,3-Dimethyl-2-butanone			−30 e	0 e	42.5	105.7	1
$C_6H_{12}O$	Cyclohexanol			34 e	61 e	99.2	160.7	1
$C_6H_{12}O_2$	Hexanoic acid		33 e	59 e	93 e	139.3	204.5	1
$C_6H_{12}O_2$	4-Methylpentanoic acid	36 e	49 e	67.1	92.9	133.6	206.8	5

Mol. form.	Name	Temperature in °C for the indicated pressure						Ref.
		1 Pa	10 Pa	100 Pa	1 kPa	10 kPa	100 kPa	
$C_6H_{12}O_2$	Diethylacetic acid	–9 e	16 e	46 e	83 e	130.7	192.5	5
$C_6H_{12}O_2$	Isopentyl formate	–60 e	–41 e	–17 e	15 e	59.1	124 e	5
$C_6H_{12}O_2$	Butyl acetate	–63 e	–43 e	–19 e	14 e	61.0	125.6	1,5
$C_6H_{12}O_2$	Isobutyl acetate	–63 e	–45 e	–21 e	10 e	53.4	116 e	5
$C_6H_{12}O_2$	Propyl propanoate	–62 e	–42 e	–18 e	14 e	58.3	122.0	5
$C_6H_{12}O_2$	Ethyl butanoate	–49 e	–34 e	–14 e	14.3	55.2	121.1	5
$C_6H_{12}O_2$	Ethyl 2-methylpropanoate	–65 e	–47 e	–24.6	5.4	47.3	109.8	5
$C_6H_{12}O_2$	Methyl pentanoate				19.2	63.7	127.4	5
$C_6H_{12}O_2$	Methyl isopentanoate					53.3	116.3	5
$C_6H_{12}O_2$	Diacetone alcohol	–41 e	–17 e	13 e	50.1	98.5	164 e	5
$C_6H_{12}O_3$	Ethylene glycol monoethyl ether acetate	–25 e	–8 e	14 e	44.6	88.0	155.6	5
$C_6H_{12}O_3$	Paraldehyde				17 e	62.2	124 e	5
$C_6H_{12}S$	Cyclohexanethiol					84.8	158.3	5
$C_6H_{12}S$	*cis*-Tetrahydro-2,5-dimethylthiophene	–53 e	–34 e	–8 e	25 e	72.0	142.1	5
$C_6H_{12}S$	Tetrahydro-3-methyl-2H-thiopyran	–48 e	–27 e	0 e	35 e	84.1	157.5	5
$C_6H_{13}Br$	1-Bromohexane	–45 e	–25 e	2 e	36 e	83.7	154.8	5
$C_6H_{13}Cl$	1-Chlorohexane	–55 e	–36 e	–11 e	21 e	66.7	134.6	5
$C_6H_{13}F$	1-Fluorohexane	–80 e	–62 e	–40 e	–11 e	30.4	91.1	5
$C_6H_{13}I$	1-Iodohexane	–33 e	–11 e	16 e	53 e	104.0	180.8	5
$C_6H_{13}N$	Cyclohexylamine			–9 e	22 e	66.6	133.5	1
C_6H_{14}	Hexane	–96.4 s	–79.2	–57.6	–29.3	9.8	68.3	16
C_6H_{14}	2-Methylpentane	–100 e	–84 e	–64 e	–36 e	2 e	59.9	1
C_6H_{14}	3-Methylpentane	–99 e	–83 e	–62 e	–34.3	4.6	62.9	1
C_6H_{14}	2,2-Dimethylbutane		–90 e	–71.5	–45.5	–7.7	49.4	1
C_6H_{14}	2,3-Dimethylbutane	–103 e	–87 e	–66 e	–39.0	–0.4	57.6	1
$C_6H_{14}O$	1-Hexanol		5 e	28 e	56.8	97.3	157.1	1
$C_6H_{14}O$	2-Hexanol	–28 e	–10 e	12 e	41.4	81.5	139.6	1
$C_6H_{14}O$	3-Hexanol	–43 e	–23 e	1 e	33 e	75.4	135.1	1
$C_6H_{14}O$	2-Methyl-1-pentanol			14 e	45.9	88.3	147.6	5
$C_6H_{14}O$	4-Methyl-1-pentanol			24 e	53 e	92.4	151.4	5
$C_6H_{14}O$	2-Methyl-2-pentanol	–29 e	–15 e	3 e	27.1	63.0	120.9	5
$C_6H_{14}O$	3-Methyl-2-pentanol				36.5	76.1	133.8	5
$C_6H_{14}O$	4-Methyl-2-pentanol	–43 e	–24 e	0 e	30 e	71.9	131.3	5
$C_6H_{14}O$	2-Methyl-3-pentanol				29.8	68.8	126.0	5
$C_6H_{14}O$	3-Methyl-3-pentanol		–23 e	–4 e	22.9	61.1	121.1	5
$C_6H_{14}O$	2-Ethyl-1-butanol		–5 e	17 e	46 e	85.7	146.1	5
$C_6H_{14}O$	3,3-Dimethyl-1-butanol	–37 e	–16 e	9 e	42 e	84.3	142.5	5
$C_6H_{14}O$	2,3-Dimethyl-2-butanol			–5 e	23 e	61.3	118.2	5
$C_6H_{14}O$	Dipropyl ether	–80 e	–63 e	–41 e	–12 e	28.8	89.7	1
$C_6H_{14}O$	Diisopropyl ether		–76 e	–55 e	–28 e	11 e	68.1	1
$C_6H_{14}O$	Butyl ethyl ether	–78 e	–61 e	–39 e	–10 e	31.0	91.9	1
$C_6H_{14}O$	*tert*-Butyl ethyl ether	–90 e	–74 e	–53 e	–24.6	14.4	72.6	5
$C_6H_{14}O_2$	2-Methyl-2,4-pentanediol	–8 e	17 e	48 e	86 e	134.4	197.5	5
$C_6H_{14}O_2$	Ethylene glycol monobutyl ether	–31 e	–8 e	20 e	55 e	103.2	170.2	5
$C_6H_{14}O_2$	1,1-Diethoxyethane	–68 e	–49 e	–26 e	3.7	44.2	101.9	5
$C_6H_{14}O_2$	Ethylene glycol diethyl ether		–59 e	–35.3	–2.8	44.4	118.8	5
$C_6H_{14}O_3$	1,2,6-Hexanetriol	92 e	114.8	146.0	191 e			5
$C_6H_{14}O_3$	Dipropylene glycol				110 e	162.6	231.4	5
$C_6H_{14}O_3$	Diethylene glycol monoethyl ether			40 e	80.3	132.4	201.4	5
$C_6H_{14}O_3$	Diethylene glycol dimethyl ether	–42 e	–20 e	8.3	44.3	92.3	159.4	5
$C_6H_{14}O_3$	Trimethylolpropane	73 e	98 e	128 e	167.8	220.5	295 e	5
$C_6H_{14}O_4$	Triethylene glycol	44 e	74 e	109.0	152.6	207.2	277.9	5
$C_6H_{14}S$	1-Hexanethiol	–45 e	–25 e	1 e	35 e	81.7	152.2	5
$C_6H_{14}S$	2-Hexanethiol	–50 e	–32 e	–8 e	25 e	69.9	138.4	5
$C_6H_{14}S$	Dipropyl sulfide	–50 e	–30 e	–6 e	28 e	73.6	142.4	5
$C_6H_{14}S$	Diisopropyl sulfide	–65 e	–47 e	–23 e	9 e	53.1	119.6	5
$C_6H_{14}S$	Isopropyl propyl sulfide				18.5	63.8	131.6	5
$C_6H_{14}S$	Butyl ethyl sulfide	–49 e	–30 e	–5 e	29 e	74.8	143.8	5
$C_6H_{15}N$	Hexylamine			–10 e	22 e	66.0	130.6	5
$C_6H_{15}N$	Butylethylamine				6.1	47.7	107.0	5

Mol. form.	Name	Temperature in °C for the indicated pressure						Ref.
		1 Pa	10 Pa	100 Pa	1 kPa	10 kPa	100 kPa	
$C_6H_{15}N$	Dipropylamine		−48 e	−25 e	6 e	47.5	108.8	5
$C_6H_{15}N$	Diisopropylamine			−47 e	−17.5	23.5	84.0	5
$C_6H_{15}N$	Triethylamine	−58 e	−45 e	−29 e	−5 e	29.9	88.5	1
$C_6H_{15}NO$	2-Diethylaminoethanol					97 e	160.6	5
$C_6H_{15}NO_3$	Triethanolamine	75 e	108 e	148 e	196 e	256.7	334 e	5
$C_6H_{15}O_4P$	Triethyl phosphate			34	76	132	211	4
$C_6H_{16}N_2$	Hexamethylenediamine				76.0	128.2	199.0	5
$C_6H_{16}O_2Si$	Diethoxydimethylsilane	−62 e	−44 e	−21.2	9.1	51.0	113.0	5
$C_6H_{18}Cl_2O_2Si_3$	1,5-Dichloro-1,1,3,3,5,5-hexamethyltrisiloxane	−29 e	−7 e	22.2	59.7	110.5	183.4	5
$C_6H_{18}OSi_2$	Hexamethyldisiloxane		−56 e	−34 e	−5 e	37.1	100.1	5
C_6MoO_6	Molybdenum hexacarbonyl		17.4 s	42.8 s	73.1 s	109.9 s	155.4 s	5
C_7F_{14}	Perfluoromethylcyclohexane				−21 e	18 e	75.9	1
C_7F_{16}	Perfluoroheptane		−62 e	−41 e	−14 e	24.7	82.1	1
C_7HF_{15}	1H-Pentadecafluoroheptane				−7 e	35.9	96.0	5
$C_7H_3ClF_3NO_2$	1-Chloro-2-nitro-4-trifluoromethyl)benzene	3 e	26 e	55 e	92.8	145.2	222.0	5
$C_7H_3F_5$	2,3,4,5,6-Pentafluorotoluene			−20 e	11 e	53.6	117.0	5
$C_7H_4ClF_3$	1-Chloro-2-(trifluoromethyl) benzene			1 e	34.5	81.8	151.8	5
$C_7H_4ClF_3$	1-Chloro-3-(trifluoromethyl) benzene	−53 e	−34 e	−9 e	24.2	69.8	137.2	5
$C_7H_4ClF_3$	1-Chloro-4-(trifluoromethyl) benzene			−9 e	24.2	70.4	138.1	5
$C_7H_4Cl_2O$	*o*-Chlorobenzoyl chloride				93 e	149 e	237.0	5
$C_7H_4Cl_2O$	*m*-Chlorobenzoyl chloride				87.8	147 e	225.0	5
$C_7H_4F_3NO_2$	1-Nitro-3-(trifluoromethyl) benzene		11 e	39 e	76.2	127.3	202.2	5
$C_7H_4F_4$	1-Fluoro-4-(trifluoromethyl) benzene			−38 e	−6 e	38.6	102.3	5
C_7H_5BrO	Benzoyl bromide	−15 e	11 e	42.6	83.9	139.5	218.0	5
C_7H_5ClO	Benzoyl chloride			27.5	67.0	120.4	196.7	5
$C_7H_5Cl_3$	(Trichloromethyl)benzene		9 e	40.6	81.5	136.2	213.0	5
$C_7H_5F_3$	(Trifluoromethyl)benzene				−3 e	39 e	101.6	5
C_7H_5N	Benzonitrile		−6 e	23.9	63.1	115.7	190.0	5
C_7H_5NS	Phenyl isothiocyanate				79.4	105 e	117 e	5
$C_7H_6Cl_2$	2,4-Dichlorotoluene		6 e	33 e	68.3	119.5	199.1	5
$C_7H_6Cl_2$	3,4-Dichlorotoluene	−13 e	9 e	38 e	76 e	129.3	208.4	5
$C_7H_6Cl_2$	(Dichloromethyl)benzene			31	72	130	213	4
C_7H_6O	Benzaldehyde		−9 e	19 e	54.6	104.6	178.3	1
$C_7H_6O_2$	Salicylaldehyde		−1 e	29 e	68 e	120.7	196.2	5
C_7H_7Br	*o*-Bromotoluene		−10 e	17 e	54 e	104.8	181.1	5
C_7H_7Br	*m*-Bromotoluene	−34 e	−11 e	19.4	58.1	109.9	183.1	5
C_7H_7Br	*p*-Bromotoluene				57 e	107.8	183.8	5
C_7H_7Br	(Bromomethyl)benzene			25.4	66.8	121.7	198.3	5
C_7H_7Cl	*o*-Chlorotoluene		−24 e	3 e	38 e	86.3	158.7	1,5
C_7H_7Cl	*m*-Chlorotoluene	−41 e	−21 e	6 e	41 e	89 e	161.8	5
C_7H_7Cl	*p*-Chlorotoluene				40 e	88.9	161.5	1,5
C_7H_7Cl	(Chloromethyl)benzene	−34 e	−11 e	17.7	55.4	106.3	178.9	5
C_7H_7ClO	1-Chloro-2-methoxy-benzene	−22 e	2 e	33 e	72 e	125.2	201 e	5
C_7H_7F	*o*-Fluorotoluene		−50 e	−26 e	5 e	49.0	113.9	5
C_7H_7F	*m*-Fluorotoluene	−67 e	−48 e	−25 e	7 e	51.0	116.1	5
C_7H_7F	*p*-Fluorotoluene		−48 e	−24 e	7 e	51 e	116.2	5
$C_7H_7NO_2$	*o*-Nitrotoluene	23 e	40 e	62 e	94 e	141.9	221.9	5
$C_7H_7NO_2$	*m*-Nitrotoluene			45 e	89.7	148.7	231.3	5
$C_7H_7NO_3$	2-Nitroanisole	15 e	45 e	82 e	129 e	189.4	271.8	5
C_7H_8	Toluene	−78.1	−57.1	−31.3	1.5	45.2	110.1	5
C_7H_8	Bicyclo[2.2.1]hepta-2,5-diene				−15 e	27.4	91 e	5
$C_7H_8Cl_2Si$	Dichloromethylphenylsilane			32.4	71.8	126.0	205.0	5
C_7H_8O	*o*-Cresol	−6.4 s	12.8 s	40.2	72.3	120.3	190.5	1,5
C_7H_8O	*m*-Cresol	20.8	33.6	52.4	82.6	130.6	201.8	1,5
C_7H_8O	*p*-Cresol	−0.2 s	20.7 s	52.7	83.1	130.7	201.5	1,5
C_7H_8O	Benzyl alcohol	8 e	28 e	54 e	88 e	134.7	204.9	1
C_7H_8O	Anisole		−21 e	4 e	38 e	84 e	153.2	1,5
C_7H_8S	3-Methylbenzenethiol		0 e	29 e	66 e	117.9	194.6	5
C_7H_9N	Benzylamine			25.6	62.6	112.7	183.9	5

Mol. form.	Name	Temperature in °C for the indicated pressure						Ref.
		1 Pa	10 Pa	100 Pa	1 kPa	10 kPa	100 kPa	
C_7H_9N	*o*-Methylaniline	1.0	18.8	42.6	76.1	125.6	199.9	1,5
C_7H_9N	*m*-Methylaniline	3.8	22.0	46.2	80.1	128.8	202.9	1,5
C_7H_9N	*p*-Methylaniline				77.1	126.2	199.9	5
C_7H_9N	*N*-Methylaniline	−16 e	6 e	34 e	70.3	121.1	195.8	1
C_7H_9N	2-Ethylpyridine	−46 e	−26 e	−1 e	33 e	79.3	149.0	5
C_7H_9N	3-Ethylpyridine	−38 e	−17 e	9 e	44 e	92.7	166.5	5
C_7H_9N	4-Ethylpyridine	−35 e	−15 e	11 e	46 e	94.4	168.6	5
C_7H_9N	2,3-Dimethylpyridine				42 e	89.9	160.6	5
C_7H_9N	2,4-Dimethylpyridine		−25 e	3.7	40.0	87.5	157.9	1,5
C_7H_9N	2,5-Dimethylpyridine			4 e	39 e	86.2	156.6	1
C_7H_9N	2,6-Dimethylpyridine			−3 e	29.9	75.8	143.6	1
C_7H_9N	3,4-Dimethylpyridine		−9 e	19 e	55 e	104.8	178.6	5
C_7H_9N	3,5-Dimethylpyridine			11 e	48 e	98 e	171.5	1
$C_7H_{10}N_2$	Toluene-2,4-diamine			100.4	145.3	202.9	279.5	5
C_7H_{12}	1-Heptyne	−75 e	−57 e	−35 e	−5 e	37.1	99.5	5
C_7H_{12}	2-Heptyne		−51 e	−27 e	4 e	46.9	111.5	5
C_7H_{12}	3-Heptyne	−71 e	−53 e	−31 e	0 e	42.7	106.4	5
C_7H_{12}	5-Methyl-1-hexyne	−80 e	−62 e	−40 e	−11 e	30.1	91.4	5
C_7H_{12}	5-Methyl-2-hexyne	−75 e	−57 e	−34 e	−4 e	38.6	102.0	5
C_7H_{12}	2-Methyl-3-hexyne	−78 e	−61 e	−39 e	−9 e	32.6	94.8	5
C_7H_{12}	4,4-Dimethyl-1-pentyne		−73 e	−52 e	−24 e	15.9	75.6	5
C_7H_{12}	4,4-Dimethyl-2-pentyne		−70 e	−48 e	−19 e	21.4	82.6	5
C_7H_{12}	Bicyclo[4.1.0]heptane					49.9	116.3	5
C_7H_{12}	Cycloheptene			−30.0	3.4	47.5	108 e	5
C_7H_{12}	1-Methylbicyclo(3,1,0)hexane					29.8	92.6	5
C_7H_{12}	Methylenecyclohexane	−76 e	−58 e	−35 e	−5 e	38 e	103.0	5
C_7H_{12}	1-Methylcyclohexene	−72 e	−53 e	−30 e	1 e	45 e	109.8	5
C_7H_{12}	4-Methylcyclohexene	−76 e	−59 e	−36 e	−5 e	37.9	102.3	5
C_7H_{12}	1-Ethylcyclopentene	−75 e	−57 e	−34 e	−3 e	40.7	105.8	5
C_7H_{12}	1,2-Dimethylcyclopentene	−75 e	−57 e	−34 e	−3 e	40.2	105.3	5
C_7H_{12}	1,5-Dimethylcyclopentene	−77 e	−59 e	−36 e	−5.5	37.3	101.5	5
$C_7H_{12}O$	Cycloheptanone			18 e	53.7	104.0	178.7	5
$C_7H_{12}O_2$	Butyl acrylate	−52 e	−31 e	−4.5	30.4	78.0	146.9	5
$C_7H_{12}O_2$	Propyl methacrylate				26 e	73.8	139.7	5
$C_7H_{12}O_3$	Ethyl levulinate		17 e	45.3	82.6	133.2	205.7	5
$C_7H_{12}O_4$	Diethyl malonate	−23 e	4 e	36.0	76.4	128.5	198.3	5
$C_7H_{12}O_4$	Dimethyl glutarate	−11 e	15 e	47 e	87.7	139.8	209.5	5
$C_7H_{13}ClO$	Heptanoyl chloride	−17 e	4 e	29.4	59.7	96.9	144.0	5
C_7H_{14}	1-Heptene	−82.1	−63.8	−40.6	−10.7	31.1	93.2	1,5
C_7H_{14}	*cis*-2-Heptene	−79 e	−61 e	−38 e	−8 e	34.3	98.0	5
C_7H_{14}	*trans*-2-Heptene	−79 e	−61 e	−39 e	−8 e	34.0	97.5	5
C_7H_{14}	*cis*-3-Heptene	−80 e	−62 e	−40 e	−10 e	32.3	95.3	5
C_7H_{14}	*trans*-3-Heptene	−80 e	−62 e	−40 e	−10 e	32.2	95.2	5
C_7H_{14}	2-Methyl-1-hexene	−81 e	−64 e	−42 e	−12 e	29.3	91.6	5
C_7H_{14}	4-Methyl-1-hexene	−84 e	-67 e	−45 e	−16 e	25.3	86.3	5
C_7H_{14}	2-Methyl-2-hexene	−80 e	−63 e	−40 e	−10 e	32.0	95.0	5
C_7H_{14}	*cis*-3-Methyl-2-hexene	−79 e	−62 e	−39 e	−9 e	33.4	96.8	5
C_7H_{14}	*trans*-4-Methyl-2-hexene	−83 e	−66 e	−44 e	−15 e	25.9	87.1	5
C_7H_{14}	*trans*-5-Methyl-2-hexene	−83 e	−66 e	−44 e	−15 e	26.3	87.7	5
C_7H_{14}	*trans*-2-Methyl-3-hexene	−84 e	−67 e	−45 e	−16 e	24.6	85.5	5
C_7H_{14}	3-Ethyl-1-pentene	−85 e	−68 e	−46 e	−17 e	23.2	83.7	5
C_7H_{14}	2,3-Dimethyl-1-pentene	−85 e	−68 e	−46 e	−17 e	23.4	83.8	5
C_7H_{14}	2,4-Dimethyl-1-pentene	−88 e	−71 e	−50 e	−21 e	20.0	81.2	5
C_7H_{14}	3,3-Dimethyl-1-pentene	−87 e	−71 e	−50 e	−21 e	18.1	77.1	5
C_7H_{14}	4,4-Dimethyl-1-pentene	−94 e	−78 e	−57 e	−28 e	11.5	72.1	5
C_7H_{14}	2,3-Dimethyl-2-pentene	−79 e	−62 e	−39 e	−9 e	33.5	96.9	5
C_7H_{14}	2,4-Dimethyl-2-pentene	−84 e	−68 e	−46 e	−18 e	22.6	82.9	5
C_7H_{14}	*cis*-3,4-Dimethyl-2-pentene	−83 e	−65 e	−43 e	−14 e	27.2	88.8	5
C_7H_{14}	*trans*-3,4-Dimethyl-2-pentene	−82 e	−64 e	−42 e	−13 e	29.0	91.1	5
C_7H_{14}	*cis*-4,4-Dimethyl-2-pentene	−90 e	−73 e	−51 e	−22 e	18.6	80.0	5

Mol. form.	Name	Temperature in °C for the indicated pressure 1 Pa	10 Pa	100 Pa	1 kPa	10 kPa	100 kPa	Ref.
C_7H_{14}	*trans*-4,4-Dimethyl-2-pentene	−90 e	−73 e	−52 e	−23 e	16.6	76.3	5
C_7H_{14}	2,3,3-Trimethyl-1-butene	−91 e	−75 e	−53 e	−24.2	16.3	77.5	5
C_7H_{14}	Cycloheptane				6 e	51.1	118.4	1
C_7H_{14}	Methylcyclohexane	−79 e	−62 e	−39 e	−7.9	35.5	100.5	1
C_7H_{14}	Ethylcyclopentane	−76 e	−59 e	−35 e	−5 e	38.4	103.0	5
C_7H_{14}	1,1-Dimethylcyclopentane		−69 e	−47 e	−17 e	24.8	87.4	5
C_7H_{14}	*cis*-1,2-Dimethylcyclopentane			−38 e	−8 e	34.9	99.0	5
C_7H_{14}	*trans*-1,2-Dimethylcyclopentane	−83 e	−66 e	−43 e	−13 e	28.4	91.4	5
C_7H_{14}	*cis*-1,3-Dimethylcyclopentane	−84 e	−66 e	−44 e	−14 e	28.2	91.1	5
C_7H_{14}	*trans*-1,3-Dimethylcyclopentane	−84 e	−67 e	−44 e	−14 e	27.4	90.3	5
$C_7H_{14}O$	1-Heptanal	−41 e	−21 e	4 e	37 e	83.7	152.3	5
$C_7H_{14}O$	2-Heptanone		−22 e	3 e	36 e	82.2	150.6	1
$C_7H_{14}O$	3-Heptanone		−28 e	0 e	36 e	83.2	147.0	5
$C_7H_{14}O$	4-Heptanone	−27 e	−6 e	18.8	50.2	90.3	143.4	5
$C_7H_{14}O$	5-Methyl-2-hexanone		−27 e	−2 e	31.0	76.6	144.4	5
$C_7H_{14}O$	2,4-Dimethyl-3-pentanone	−61 e	−42 e	−18 e	14 e	58.5	124.8	1
$C_7H_{14}O_2$	Heptanoic acid	24 e	46 e	72 e	107 e	154.6	222.6	5
$C_7H_{14}O_2$	Pentyl acetate	−58 e	−39 e	−14 e	20 e	70.1	149 e	5
$C_7H_{14}O_2$	Isopentyl acetate	−51 e	−30 e	−4 e	30.3	76.2	141.4	5
$C_7H_{14}O_2$	Isobutyl propanoate	−35 e	−19 e	2 e	31 e	72.0	136.1	5
$C_7H_{14}O_2$	Propyl butanoate	−35 e	−19 e	3 e	32.0	74.9	142.8	5
$C_7H_{14}O_2$	Propyl isobutanoate		−28 e	−5.7	24.5	67.5	133.3	5
$C_7H_{14}O_2$	Isopropyl isobutanoate		−44 e	−19.7	12.2	56.0	120.1	5
$C_7H_{14}O_2$	Ethyl 3-methylbutanoate	−57 e	−36 e	−10 e	23.9	69.5	134.4	5
$C_7H_{14}O_2$	Methyl hexanoate	−47 e	−26 e	2 e	36.6	83.3	149 e	5
$C_7H_{14}O_2$	4-Methoxy-4-methyl-2-pentanone				43 e	89.8	160 e	5
$C_7H_{15}Br$	1-Bromoheptane	−30 e	−9 e	18 e	54 e	104.4	178.4	5
$C_7H_{15}Cl$	1-Chloroheptane	−39 e	−19 e	7 e	41 e	88.6	159.9	5
$C_7H_{15}F$	1-Fluoroheptane	−64 e	−45 e	−22 e	10 e	53.3	117.4	5
$C_7H_{15}I$	1-Iodoheptane	−19 e	3 e	32 e	71 e	123.8	203.4	5
C_7H_{16}	Heptane	−78.6	−60.2	−37.0	−6.6	35.4	98.0	16
C_7H_{16}	2-Methylhexane	−82 e	−65 e	−43 e	−13 e	27.8	89.7	1
C_7H_{16}	3-Methylhexane	−81 e	−64 e	−42 e	−12 e	29.2	91.5	1
C_7H_{16}	3-Ethylpentane	−81 e	−63 e	−41 e	−11 e	30.5	93.1	1
C_7H_{16}	2,2-Dimethylpentane	−90 e	−73 e	−52 e	−22.9	17.6	78.8	1
C_7H_{16}	2,3-Dimethylpentane	−87 e	−68.4	−45.3	−14.9	26.8	89.3	5
C_7H_{16}	2,4-Dimethylpentane	−89 e	−72 e	−50 e	−21.3	19.2	80.1	1
C_7H_{16}	3,3-Dimethylpentane	−88 e	−71 e	−49 e	−18.8	22.9	85.6	1
C_7H_{16}	2,2,3-Trimethylbutane				−23.2	18.1	80.4	5
$C_7H_{16}O$	1-Heptanol		17 e	40 e	70.1	112.5	176 e	1
$C_7H_{16}O$	2-Heptanol	−9 e	7 e	27 e	55.0	95.2	158.7	5
$C_7H_{16}O$	3-Heptanol	−8 e	7 e	27 e	54.5	93.9	156.3	5
$C_7H_{16}O$	4-Heptanol	−16 e	1 e	22 e	51 e	91.9	154.6	5
$C_7H_{16}O$	2,2-Dimethyl-3-pentanol			9 e	35 e	73.1	135.5	5
$C_7H_{16}S$	1-Heptanethiol	−30 e	−9 e	18 e	53 e	102.7	176.4	5
$C_7H_{17}N$	Heptylamine			5 e	39 e	86.7	156.4	5
$C_7H_{18}N_2$	*N,N*-Diethyl-1,3-propanediamine				50.1	99.9	167.7	5
C_8F_{18}	Perfluorooctane				5 e	45.0	105.6	5
$C_8H_4O_3$	Phthalic anhydride	48.2 s	72.4 s			192.7	284.2	5
C_8H_6O	Benzofuran		−16 e	12 e	47.9	97.7	170.7	5
C_8H_7Cl	*o*-Chlorostyrene	−33 e	−10 e	20 e	58 e	110.8	188 e	5
C_8H_7N	2-Methylbenzonitrile		1 e	32.1	72.2	126.6	204.7	5
C_8H_7N	4-Methylbenzonitrile			40.1	78.7	134.3	221.3	5
C_8H_7N	Benzeneacetonitrile	−3 e	23 e	55.3	97.4	153.7	233.1	5
C_8H_7N	Indole	20.6 s	44.5 s				254.0	5
$C_8H_7NO_4$	Methyl 2-nitrobenzoate	17 e	49 e	89 e	140 e	208 e	302 e	5
C_8H_8	Styrene		−31 e	−5 e	28.6	75.4	144.7	1
C_8H_8	1,3,5,7-Cyclooctatetraene				24.3	71.0	140.1	5
C_8H_8O	Acetophenone			36 e	73 e	125.3	201.5	5
$C_8H_8O_2$	Phenyl acetate		3 e	33.1	72.2	123.9	195.5	5

Mol. form.	Name	Temperature in °C for the indicated pressure						Ref.
		1 Pa	10 Pa	100 Pa	1 kPa	10 kPa	100 kPa	
$C_8H_8O_2$	Methyl benzoate		−1 e	29 e	68 e	121.2	198.9	5
$C_8H_8O_2$	4-Methoxybenzaldehyde	9 e	35 e	68.1	110.8	167.9	248.5	5
$C_8H_8O_3$	Methyl salicylate	−1 e	22 e	51 e	88.8	141.8	219.9	5
C_8H_9Cl	1-Chloro-2-ethylbenzene	−30 e	−9 e	18 e	54 e	103.7	177.9	5
C_8H_9Cl	1-Chloro-4-ethylbenzene	−27 e	−6 e	22 e	58 e	108.7	183.9	5
$C_8H_9NO_2$	1-Ethyl-4-nitrobenzene	10 e	36 e	69 e	111.6	168 e	245 e	5
C_8H_{10}	Ethylbenzene	−56.2	−36.8	−12.0	21.1	67.1	135.7	1
C_8H_{10}	*o*-Xylene			−7 e	27 e	74.2	143.9	1
C_8H_{10}	*m*-Xylene		−35 e	−10 e	23.4	69.8	138.7	1
C_8H_{10}	*p*-Xylene				22.4	68.9	137.9	1
$C_8H_{10}O$	*o*-Ethylphenol		16.9	44.5	81.1	130.9	204.0	5
$C_8H_{10}O$	*m*-Ethylphenol	5.6	29.2	57.5	91.9	144.8	217.9	5
$C_8H_{10}O$	*p*-Ethylphenol			60 e	95.5	144.6	217.5	5
$C_8H_{10}O$	2,3-Xylenol	14.3 s	34.3 s	57.2 s	91.4	141.7	216.4	1,5
$C_8H_{10}O$	2,4-Xylenol			50.2	85.5	137.2	210.5	1,5
$C_8H_{10}O$	2,5-Xylenol	13.4 s	33.2 s	55.9 s	87.4	137.0	210.6	5
$C_8H_{10}O$	2,6-Xylenol	−3.1 s	16.7 s	39.6 s	75.3	125.9	200.6	1,5
$C_8H_{10}O$	3,4-Xylenol	19.7 s	40.2 s	63.7 s	102.1	152.3	226.4	1,5
$C_8H_{10}O$	3,5-Xylenol	16.5 s	37.2 s	61.1 s	98.0	147.9	221.3	1,5
$C_8H_{10}O$	Benzeneethanol	2 e	25 e	54 e	92 e	143.6	217.7	5
$C_8H_{10}O$	Phenetole		−9 e	17 e	51 e	99 e	169.3	5
$C_8H_{10}O_2$	2-Phenoxyethanol	21 e	46 e	75.9	115.4	168.7	244.8	5
$C_8H_{10}O_2$	1,3-Dimethoxybenzene	18 e	34 e	56 e	86.7	135.5	223 e	5
$C_8H_{11}N$	*p*-Ethylaniline	−2 e	21 e	49 e	87 e	139.4	216.7	5
$C_8H_{11}N$	*N*-Ethylaniline	−15 e	8 e	38 e	76.4	128.8	204.2	5
$C_8H_{11}N$	*N,N*-Dimethylaniline			28 e	66 e	118.1	193.6	1
$C_8H_{11}N$	2,4-Xylidine	−2 e	21 e	51 e	88 e	139.1	210.9	5
$C_8H_{11}N$	2,6-Xylidine			37 e	80 e	137.7	217.7	5
$C_8H_{11}N$	5-Ethyl-2-picoline	−33 e	−9.3	20 e			178.0	5
$C_8H_{11}NO$	*o*-Phenetidine	0 e	27 e	60 e	102.2	156.0	228.1	5
C_8H_{12}	1,5-Cyclooctadiene		−37 e	−8 e	30 e	80.2	150 e	5
C_8H_{12}	4-Vinylcyclohexene	−62 e	−43 e	−19 e	14.1	59.9	129 e	5
$C_8H_{12}O_4$	Diethyl maleate	−6 e	20 e	52.2	93.5	148.4	224.8	5
C_8H_{14}	2,5-Dimethyl-1,5-hexadiene	−38 e	−26 e	−10 e	14 e	50.8	115.1	5
C_8H_{14}	1-Octyne	−59 e	−40 e	−16 e	16 e	60.3	125.8	1
C_8H_{14}	2-Octyne	−52 e	−33 e	−8 e	25 e	70.6	137.8	1
C_8H_{14}	3-Octyne	−55 e	−35 e	−11 e	22 e	66.8	132.8	1
C_8H_{14}	4-Octyne	−56 e	−36 e	−12 e	21 e	65.6	131.4	1
C_8H_{14}	1-Ethylcyclohexene	−55 e	−35 e	−11 e	22 e	68 e	136.5	5
$C_8H_{14}O_2$	Cyclohexyl acetate					103.1	172.9	5
$C_8H_{14}O_2$	Butyl methacrylate				47 e	93.3	159.0	5
$C_8H_{14}O_3$	Butanoic anhydride	−28 e	−2 e	30 e	71 e	123.8	196.5	5
$C_8H_{14}O_4$	Ethyl succinate	−6 e	20 e	51.0	91.1	143.7	216.1	5
$C_8H_{14}O_4$	Dipropyl oxalate	−4 e	20 e	49.9	88.6	140.4	213.0	5
$C_8H_{14}O_4$	Dimethyl adipate		28 e	61 e	103 e	156.1	227.3	5
$C_8H_{15}Br$	(2-Bromoethyl)cyclohexane	−14 e	8 e	36.9	75.3	129.7	212.5	5
$C_8H_{15}ClO$	Octanoyl chloride	1 e	22 e	46 e	74.7	109 e	150 e	5
$C_8H_{15}N$	Octanenitrile	−15 e	8 e	37 e	75 e	127.7	204.4	5
C_8H_{16}	1-Octene	−65.7	−46.1	−21.4	10.5	54.9	120.9	1,5
C_8H_{16}	*cis*-2-Octene	−59 e	−41 e	−17 e	15 e	59 e	125.2	5
C_8H_{16}	*trans*-2-Octene	−59 e	−41 e	−17 e	14 e	59 e	124.5	5
C_8H_{16}	*cis*-3-Octene	−65 e	−46 e	−22 e	10 e	55.1	122.4	5
C_8H_{16}	*trans*-3-Octene	−61 e	−43 e	−19 e	13 e	57 e	122.8	5
C_8H_{16}	*cis*-4-Octene	−63 e	−44 e	−20 e	11 e	56 e	122.1	5
C_8H_{16}	*trans*-4-Octene	−65 e	−46 e	−22 e	10 e	54.6	121.8	5
C_8H_{16}	2-Methyl-1-heptene	−66 e	−48 e	−24 e	8 e	52.3	118.7	5
C_8H_{16}	2,2-Dimethyl-*cis*-3-hexene	−74 e	−56 e	−33 e	−3 e	40.1	105.0	5
C_8H_{16}	2,3-Dimethyl-2-hexene	−65 e	−47 e	−23 e	10 e	54.3	121.3	5
C_8H_{16}	2,3,3-Trimethyl-1-pentene		−53 e	−30 e	1 e	43.8	107.9	5
C_8H_{16}	2,4,4-Trimethyl-1-pentene	−79 e	−61 e	−38 e	−7 e	36.2	101.0	5

Mol. form.	Name	Temperature in °C for the indicated pressure						Ref.
		1 Pa	10 Pa	100 Pa	1 kPa	10 kPa	100 kPa	
C_8H_{16}	2,3,4-Trimethyl-2-pentene	−68 e	−49 e	−26 e	6 e	50.0	115.8	5
C_8H_{16}	2,4,4-Trimethyl-2-pentene	−73 e	−56 e	−33 e	−2 e	40.4	104.5	5
C_8H_{16}	Cyclooctane				30 e	78 e	150.7	1
C_8H_{16}	Ethylcyclohexane	−61 e	−42 e	−17 e	15.8	61.9	131.3	5
C_8H_{16}	1,1-Dimethylcyclohexane			−27 e	5 e	50.6	119.1	5
C_8H_{16}	*cis*-1,2-Dimethylcyclohexane		−44 e	−20 e	14 e	59.7	129.2	5
C_8H_{16}	*trans*-1,2-Dimethylcyclohexane	−68 e	−49 e	−25 e	8 e	53.9	122.9	5
C_8H_{16}	*cis*-1,3-Dimethylcyclohexane	−68 e	−48 e	−23 e	10 e	55.6	123.1	5
C_8H_{16}	*trans*-1,3-Dimethylcyclohexane	−62 e	−45 e	−23 e	8 e	51.5	120.9	5
C_8H_{16}	*cis*-1,4-Dimethylcyclohexane	−66 e	−47 e	−23 e	10 e	55.3	123.8	5
C_8H_{16}	*trans*-1,4-Dimethylcyclohexane			−27 e	5 e	50.6	118.9	5
C_8H_{16}	Propylcyclopentane	−60 e	−41 e	−16 e	16.5	62.1	130.5	5
C_8H_{16}	Isopropylcyclopentane	−65 e	−46 e	−21 e	12 e	57.3	125.9	5
C_8H_{16}	1-Ethyl-1-methylcyclopentane	−67 e	−49 e	−24 e	8 e	53.2	121.0	5
C_8H_{16}	*cis*-1-Ethyl-2-methylcyclopentane	−63 e	−44 e	−19 e	13.3	59.1	127.6	5
C_8H_{16}	1,1,2-Trimethylcyclopentane				2 e	46.2	113.2	5
C_8H_{16}	1,1,3-Trimethylcyclopentane	−77 e	−59 e	−36 e	−5 e	38.7	104.4	5
C_8H_{16}	1′,2′,4a-1,2,4-Trimethylcyclo-pentane	−70 e	−52 e	−28 e	4 e	48.9	116.2	5
C_8H_{16}	1′,2a,4′-1,2,4-Trimethylcyclo-pentane	−74 e	−56 e	−33 e	−1 e	42.8	108.8	5
$C_8H_{16}O$	1-Propylcyclopentanol	9 e	24 e	43 e	69.0	108.4	173.5	5
$C_8H_{16}O$	Octanal			6 e	45.7	97.8	170.2	5
$C_8H_{16}O$	2-Octanone		−3 e	23 e	57 e	103.8	172.1	5
$C_8H_{16}O$	3-Octanone			8 e	47.7	97 e	161 e	5
$C_8H_{16}O$	2,2,4-Trimethyl-3-pentanone			11.3	42.1	81.7	134.6	5
$C_8H_{16}O_2$	Octanoic acid	37 e	58 e	85 e	120 e	165.5	238.4	1,5
$C_8H_{16}O_2$	2-Ethylhexanoic acid				108 e	159.6	226.6	5
$C_8H_{16}O_2$	Hexyl acetate	−37 e	−13 e	16 e	52.8	100.4	164 e	5
$C_8H_{16}O_2$	Isopentyl propanoate			3.1	40.7	90.6	159.8	5
$C_8H_{16}O_2$	Isobutyl isobutanoate	−47 e	−26 e	0.4	34.8	81.1	147.0	5
$C_8H_{16}O_2$	Propyl 3-methylbutanoate			1.8	38.9	87.9	155.6	5
$C_8H_{16}O_2$	Ethyl hexanoate	−31 e	−9 e	18.7	53.9	100.7	166.2	5
$C_8H_{16}O_2$	Methyl heptanoate	−30 e	−9 e	19 e	54.2	102.4	172 e	5
$C_8H_{16}O_4$	Diethylene glycol monoethyl ether acetate	−16 e	10.6	43.9	86.2	141.3	216.6	5
$C_8H_{17}Br$	1-Bromooctane	−17 e	6 e	34 e	72 e	123.8	200.3	5
$C_8H_{17}Cl$	1-Chlorooctane	−25 e	−4 e	23 e	59 e	108.8	182.9	5
$C_8H_{17}Cl$	3-(Chloromethyl)heptane					100.3	172.4	5
$C_8H_{17}F$	1-Fluorooctane				29 e	74.6	141.8	5
$C_8H_{17}I$	1-Iodooctane	−6 e	18 e	48 e	87 e	142.5	224.5	5
C_8H_{18}	Octane		−42.6	−17.9	14.4	58.9	125.3	16
C_8H_{18}	2-Methylheptane	−69 e	−49.1	−24.5	7.6	51.6	117.2	1,5
C_8H_{18}	3-Methylheptane	−67 e	−48.1	−23.6	8.5	52.7	118.5	1,5
C_8H_{18}	4-Methylheptane	−65 e	−47 e	−24 e	7.8	51.6	117.2	5
C_8H_{18}	3-Ethylhexane				8 e	52.1	118.1	5
C_8H_{18}	2,2-Dimethylhexane	−73 e	−55 e	−32 e	−1.5	41.6	106.4	5
C_8H_{18}	2,3-Dimethylhexane				5 e	49.2	115.1	5
C_8H_{18}	2,4-Dimethylhexane				0.6	43.9	109.0	5
C_8H_{18}	2,5-Dimethylhexane	−71 e	−53 e	−30 e	0.7	43.8	108.6	5
C_8H_{18}	3,3-Dimethylhexane	−72 e	−54 e	−30 e	1.4	45.4	111.5	5
C_8H_{18}	3,4-Dimethylhexane				7 e	50.9	117.3	5
C_8H_{18}	3-Ethyl-2-methylpentane	−69 e	−50 e	−27 e	5 e	48.9	115.2	5
C_8H_{18}	3-Ethyl-3-methylpentane	−70 e	−51 e	−27 e	5 e	50.2	117.8	5
C_8H_{18}	2,2,3-Trimethylpentane	−74 e	−56 e	−32 e	−0.8	43.1	109.4	5
C_8H_{18}	2,2,4-Trimethylpentane	−81.9	−63.4	−39.8	−8.9	34.0	98.8	5
C_8H_{18}	2,3,3-Trimethylpentane	−72 e	−54 e	−30 e	2.1	46.9	114.3	5
C_8H_{18}	2,3,4-Trimethylpentane	−74 e	−54.5	−30.0	2.2	46.7	113.1	1,5
C_8H_{18}	2,2,3,3-Tetramethylbutane	−62.5 s	−44 s	−20.9 s	8.9 s	48.8 s	105.8	5
$C_8H_{18}O$	1-Octanol	12 e	30 e	53 e	84 e	128.2	194.8	1,39
$C_8H_{18}O$	2-Octanol			40 e	69.9	112.5	179.4	1,39
$C_8H_{18}O$	3-Octanol	12 e	24 e	40 e	64 e	102.8	174.1	1

Mol. form.	Name	Temperature in °C for the indicated pressure 1 Pa	10 Pa	100 Pa	1 kPa	10 kPa	100 kPa	Ref.
$C_8H_{18}O$	4-Octanol			40 e	66.9	107.3	176.0	1,39
$C_8H_{18}O$	4-Methyl-3-heptanol	−52 e	−28 e	1 e	39 e	87.6	155.0	5
$C_8H_{18}O$	5-Methyl-3-heptanol	−35 e	−16 e	8 e	40 e	84.8	153.0	5
$C_8H_{18}O$	4-Methyl-4-heptanol	−17 e	1 e	24 e	55 e	97.2	160.7	5
$C_8H_{18}O$	2-Ethyl-1-hexanol			45 e	75 e	118.3	184.2	1
$C_8H_{18}O$	2-Ethyl-2-hexanol	−13 e	4 e	26 e	55 e	96.3	160.3	5
$C_8H_{18}O$	2,4,4-Trimethyl-2-pentanol		−7 e	13 e	40 e	79.8	146.1	5
$C_8H_{18}O$	2,2,4-Trimethyl-3-pentanol	−2 e	9 e	24 e	47 e	82.6	150.4	5
$C_8H_{18}O$	Dibutyl ether	−55 e	−35 e	−8 e	26 e	73.0	141.2	5
$C_8H_{18}O$	Di-*sec*-butyl ether			−19 e	12.1	55.4	120.6	5
$C_8H_{18}O$	Di-*tert*-butyl ether			−33 e	−2 e	41.7	106.8	1
$C_8H_{18}O_2$	Ethylene glycol monohexyl ether	−13 e	14 e	46 e	86 e	137.7	206.9	5
$C_8H_{18}O_2$	1,2-Dipropoxyethane			−44.2	−2.0	63.6	179.2	5
$C_8H_{18}O_2$	Di-*tert*-butyl peroxide			−26 e	4.3	46.6	110.5	5
$C_8H_{18}O_3$	Diethylene glycol monobutyl ether	14 e	37 e	66.8	104.9	153 e	230.4	5
$C_8H_{18}O_3$	Diethylene glycol diethyl ether	−32 e	−7 e	25 e	64.9	117.1	189 e	5
$C_8H_{18}O_5$	Tetraethylene glycol	89 e	117 e	151.1	192.2	242.9	307.3	5
$C_8H_{18}S$	1-Octanethiol	−15 e	6 e	34 e	71 e	122.1	198.5	5
$C_8H_{18}S$	Dibutyl sulfide	−22 e	0 e	27 e	63 e	113.5	188.4	5
$C_8H_{19}N$	Dibutylamine	−37 e	−16 e	10 e	44 e	90.8	159.1	5
$C_8H_{19}N$	Diisobutylamine	−57 e	−36 e	−9.0	25.5	72.2	139.0	5
$C_8H_{20}O_4Si$	Ethyl silicate	−77 e	−52 e	−21 e	21.6	80.5	164.1	5
$C_8H_{20}Si$	Tetraethylsilane			−6.5	30.5	80.6	152.6	5
C_9F_{20}	Perfluorononane					40 e	114.7	5
$C_9H_6N_2O_2$	Toluene-2,4-diisocyanate		39 e	72 e	113.9	169.7	247 e	5
C_9H_7N	Quinoline	−1.3	23.7	55.4	96.8	153.4	236.5	1,5
C_9H_7N	Isoquinoline		30.2	60.7	101.3	157.9	242.7	1,5
C_9H_8	Indene			12 e	53.0	106.8	181.0	5
C_9H_{10}	*cis*-1-Propenylbenzene	−38 e	−15.4	13.3	51.4	103.7	178.4	5
C_9H_{10}	*trans*-1-Propenylbenzene		−16 e	13.3	51.6	103.7	178.4	5
C_9H_{10}	Isopropenylbenzene			3.2	41.5	92.8	164.9	5
C_9H_{10}	Indan	−33 e	−12 e	16 e	52 e	102.3	177.5	1
$C_9H_{10}O$	2,4-Dimethylbenzaldehyde	−3 e	23 e	54 e	93.2	144.6	214.5	5
$C_9H_{10}O_2$	Ethyl benzoate	−18 e	8 e	39 e	80.1	135.1	212.8	5
$C_9H_{10}O_2$	Benzyl acetate	−11 e	15 e	46.6	86.9	139.5	211 e	5
$C_9H_{11}Br$	1-Bromo-4-isopropylbenzene	−8 e	15 e	45 e	84 e	138.1	218.5	5
$C_9H_{11}Cl$	1-Chloro-2-isopropylbenzene	−23 e	−1 e	27 e	64 e	114.6	190.5	5
$C_9H_{11}Cl$	1-Chloro-4-isopropylbenzene		3 e	31 e	69 e	120.5	197.8	5
C_9H_{12}	Propylbenzene	−43 e	−23 e	4 e	38 e	86.7	158.8	1
C_9H_{12}	Isopropylbenzene	−46 e	−26 e	−1 e	33 e	80.9	152.0	1
C_9H_{12}	*o*-Ethyltoluene	−40 e	−19 e	8 e	43 e	92.1	164.7	5
C_9H_{12}	*m*-Ethyltoluene	−42 e	−21 e	5 e	40.4	88.9	160.8	5
C_9H_{12}	*p*-Ethyltoluene	−41 e	−21 e	6 e	41 e	89.2	161.5	5
C_9H_{12}	1,2,3-Trimethylbenzene		−12 e	15 e	52 e	101.5	175.6	1
C_9H_{12}	1,2,4-Trimethylbenzene	−37 e	−16 e	11 e	47 e	95.9	168.9	1
C_9H_{12}	1,3,5-Trimethylbenzene	−39 e	−18 e	9 e	43.7	92.4	164.3	1
$C_9H_{12}O$	Benzyl ethyl ether		−10 e	20.4	59.3	111.3	184.5	5
$C_9H_{12}O$	Phenyl propyl ether		−10 e	21 e	61 e	113.9	189.3	5
$C_9H_{12}O$	Phenyl isopropyl ether	−20 e	−1 e	23 e	56 e	103.7	176.9	5
$C_9H_{13}N$	2,4,6-Trimethylaniline	12 e	36 e	66 e	104.1	154.9	226 e	5
$C_9H_{13}N$	*N,N*-Dimethyl-*o*-toluidine	−25 e	−3 e	24.4	60.6	110.7	184.5	5
$C_9H_{13}N$	Amphetamine			33 e	70.1	118 e	202.0	5
$C_9H_{14}O$	Isophorone		1 e	33.1	75.1	132.4	215.1	5
$C_9H_{14}O_6$	Triacetin	37.6	62 e	90 e	124 e	165 e	214 e	5
$C_9H_{16}O_4$	Diethyl glutarate	−1 e	26 e	60.2	103.3	159.6	236.5	5
$C_9H_{17}N$	Nonanenitrile	−3 e	21 e	50.9	90.7	145.4	225.1	5
C_9H_{18}	1-Nonene	−50.1	−29.4	−3.3	30.4	77.1	146.4	1,5
C_9H_{18}	2-Methyl-1-octene	−53 e	−34 e	−9 e	25 e	72 e	144.1	5
C_9H_{18}	Butylcyclopentane	−45 e	−24 e	1 e	36 e	84 e	156.1	5
C_9H_{18}	Propylcyclohexane	−46 e	−26 e	0 e	35.1	83.6	156.2	5

Mol. form.	Name	Temperature in °C for the indicated pressure 1 Pa	10 Pa	100 Pa	1 kPa	10 kPa	100 kPa	Ref.
C_9H_{18}	Isopropylcyclohexane	−48 e	−28 e	−2 e	33 e	81.3	154.0	5
C_9H_{18}	*trans*-1-Ethyl-4-methylcyclo-hexane	−53 e	−33 e	−8 e	25 e	71.8	141.5	5
C_9H_{18}	1,1,2-Trimethylcyclohexane			−12 e	23 e	71.5	145.5	5
C_9H_{18}	1,1,3-Trimethylcyclohexane	−60 e	−41 e	−16 e	18 e	65.2	136.1	5
C_9H_{18}	1′,2a,4a-1,2,4-Trimethylcyclo-hexane	−71 e	−50 e	−22 e	15 e	65.7	140.7	5
C_9H_{18}	1′,3′,5′-1,3,5-Trimethylcyclo-hexane	−72 e	−50 e	−22 e	14 e	65.1	140.0	5
C_9H_{18}	Isobutylcyclopentane	−105 e	−88 e	−64 e	−28 e	31 e	147.0	5
C_9H_{18}	*cis*-1-Methyl-2-propylcyclo-pentane	−52 e	−33 e	−7 e	28 e	77 e	152.0	5
C_9H_{18}	*trans*-1-Methyl-2-propylcyclo-pentane	−56 e	−36 e	−11 e	23 e	72 e	145.8	5
C_9H_{18}	1,1,3,3-Tetramethylcyclo-pentane	−72 e	−54 e	−30 e	2 e	47 e	117.4	5
$C_9H_{18}O$	Nonanal		−3 e	27.4	65.5	115.6	184.6	5
$C_9H_{18}O$	2-Nonanone		8 e	35 e	71 e	121.0	194.0	5
$C_9H_{18}O$	5-Nonanone			−1 e	39.1	94 e	188 e	5
$C_9H_{18}O$	2,6–Dimethyl-4-heptanone	−32 e	−12 e	14 e	48 e	96.2	167.7	5
$C_9H_{18}O_2$	Nonanoic acid	48 e	69 e	97 e	133 e	182.7	255.1	5
$C_9H_{18}O_2$	Heptyl acetate	−16 e	6 e	31 e	70 e	119.9	191.9	5
$C_9H_{18}O_2$	Isopentyl butanoate				55 e	105.6	178.4	5
$C_9H_{18}O_2$	Isobutyl 3-methylbutanoate			11.3	48.3	97.9	168.3	5
$C_9H_{18}O_2$	Propyl hexanoate	−26 e	−2 e	28 e	65.1	113.4	178 e	5
$C_9H_{18}O_2$	Methyl octanoate	26 e	−9 e	13 e	40 e	76 e	127.9	5
$C_9H_{19}Cl$	1-Chlorononane	−11 e	11 e	39 e	76 e	127.8	204.7	5
C_9H_{20}	Nonane	−46.8	−26.0	0.0	34.0	80.8	150.3	16
C_9H_{20}	2-Methyloctane	−49 e	−30 e	−5 e	28 e	73.9	142.8	5
C_9H_{20}	3-Methyloctane	−49 e	−29 e	−5 e	29 e	74.7	143.7	5
C_9H_{20}	4-Methyloctane	−50 e	−30 e	−6 e	27 e	73.2	141.9	5
C_9H_{20}	2,2-Dimethylheptane	−58 e	−39 e	−15 e	18 e	63.6	132.3	5
C_9H_{20}	2,3-Dimethylheptane	−53 e	−33 e	−9 e	25 e	70.8	140.0	5
C_9H_{20}	2,6-Dimethylheptane	−55 e	−36 e	−12 e	21 e	66.4	134.7	5
C_9H_{20}	3-Ethyl-4-methylhexane			−9 e	24 e	70.6	139.9	5
C_9H_{20}	2,2,4-Trimethylhexane	−66.1	−46.4	−21.3	11.8	57.7	126.0	5
C_9H_{20}	2,2,5-Trimethylhexane	−65.1	−45.8	−21.2	11.2	56.2	123.7	1,5
C_9H_{20}	2,3,3-Trimethylhexane	−58 e	−38 e	−13 e	20 e	66.7	137.2	5
C_9H_{20}	2,3,5-Trimethylhexane	−60 e	−41 e	−16 e	17 e	62.3	130.9	5
C_9H_{20}	2,4,4-Trimethylhexane	−62 e	−43 e	−18 e	15 e	61.0	130.2	5
C_9H_{20}	3,3,4-Trimethylhexane	−53 e	−33 e	−7 e	28 e	76.3	148.9	5
C_9H_{20}	3,3-Diethylpentane			−9 e	26 e	73.7	145.7	1
C_9H_{20}	3-Ethyl-2,4-dimethylpentane	−58 e	−38 e	−13 e	20 e	66.7	136.2	5
C_9H_{20}	2,2,3,3-Tetramethylpentane				21 e	68.5	139.8	1
C_9H_{20}	2,2,3,4-Tetramethylpentane	−61 e	−42 e	−17 e	16 e	62.5	132.6	1
C_9H_{20}	2,2,4,4-Tetramethylpentane		−49 e	−25 e	8 e	53.2	121.8	1
C_9H_{20}	2,3,3,4-Tetramethylpentane	−57 e	−37 e	−12 e	22 e	69.7	141.1	1
$C_9H_{20}O$	1-Nonanol		40 e	64 e	96.9	141.0	213.0	5,39
$C_9H_{20}O$	3-Nonanol		24 e	47 e	78 e	123.0	194.2	5
$C_9H_{20}O$	4-Nonanol			45 e	76.4	121.3	192.0	5
$C_9H_{20}O$	5-Nonanol	13 e	31 e	54 e	84.5	128.1	194.7	5
$C_9H_{20}O$	2,2,4,4-Tetramethyl-3-pentanol				58	100	167	5
$C_9H_{20}S$	1-Nonanethiol	−2 e	21 e	49 e	87 e	140.4	219.2	5
$C_9H_{21}BO_3$	Triisopropyl borate					73.1	139.0	5
$C_9H_{21}N$	Nonylamine		9 e	37 e	75 e	126.2	202.1	5
$C_9H_{21}N$	Tripropylamine	−39 e	−18 e	8 e	42 e	88.2	156.0	5
$C_{10}F_8$	Perfluoronaphthalene	5.2 s	25.1 s	48.1 s				5
$C_{10}F_{22}$	Perfluorodecane					52 e	132.9	5
$C_{10}H_7Br$	1-Bromonaphthalene	17 e	45 e	80.3	126.7	189.8	280.5	5
$C_{10}H_7Cl$	1-Chloronaphthalene	14 e	39 e	70.5	112.8	171.6	258.6	5
$C_{10}H_8$	Naphthalene**	3.2 s	24.1 s	49.3 s	80.7	135.6	217.5	1,5
$C_{10}H_8$	Azulene	24.1 s	46 s	71.5 s	103.3	162.6	244.0	5
$C_{10}H_8O$	1-Naphthol				137.2	196.7	281.8	5
$C_{10}H_8O$	2-Naphthol				140.7	200.5	286.8	5
$C_{10}H_9N$	1-Naphthalenamine		62 e	99.0	146.9	210.7	300.1	5
$C_{10}H_9N$	2-Naphthalenamine	36.3 s	65.9 s	103 s	150.9	215.1	305.5	5

Mol. form.	Name	Temperature in °C for the indicated pressure						Ref.
		1 Pa	10 Pa	100 Pa	1 kPa	10 kPa	100 kPa	
$C_{10}H_9N$	2-Methylquinoline	5.3	31.9	63.8	102.9	165.8	247.2	5
$C_{10}H_9N$	4-Methylquinoline	29 e	54 e	85 e	127 e	183.0	265.1	5
$C_{10}H_9N$	6-Methylquinoline	27 e	51 e	81 e	122 e	179.2	264.5	5
$C_{10}H_9N$	8-Methylquinoline	15 e	40 e	70 e	111 e	166.1	247.3	5
$C_{10}H_{10}$	*m*-Divinylbenzene	−29 e	−4 e	27.1	67.6	122.1	199 e	5
$C_{10}H_{10}O_4$	Dimethyl phthalate	27 e	56 e	92.7	137.8	195.8	272.7	5
$C_{10}H_{10}O_4$	Dimethyl isophthalate			85 e	129.5	189.2	273 e	5
$C_{10}H_{10}O_4$	Dimethyl terephthalate	56.6 s	79.4 s	106.1 s	137.9 s	197.9	282 e	5
$C_{10}H_{12}$	1,2,3,4-Tetrahydronaphthalene	−21 e	3 e	33.2	74.1	127.4	207.8	5
$C_{10}H_{12}$	2-Ethylstyrene	−31 e	−8 e	21 e	60 e	111.7	187 e	5
$C_{10}H_{12}$	3-Ethylstyrene	−28 e	−5.3	24.1	62.6	116 e	193 e	5
$C_{10}H_{12}$	4-Ethylstyrene	−31 e	−8.2	21.3	60.5	115 e	196 e	5
$C_{10}H_{12}O$	Estragole			48.5	88.0	140.7	214.6	5
$C_{10}H_{12}O$	4-Isopropylbenzaldehyde			54.1	96.0	152.2	231.5	5
$C_{10}H_{12}O_2$	4-Allyl-2-methoxyphenol	9 e	37 e	72 e	115.9	173.8	252.9	5
$C_{10}H_{12}O_2$	2-Phenylethyl acetate	−4 e	22 e	54 e	96 e	152.3	232.0	5
$C_{10}H_{12}O_2$	Propyl benzoate	−8 e	18 e	50.2	92.3	149.2	230.5	5
$C_{10}H_{12}O_2$	Ethyl phenylacetate	−9 e	19 e	52 e	95 e	150.2	225 e	5
$C_{10}H_{12}O_2$	Isoeugenol				125 e	185.3	267.1	5
$C_{10}H_{14}$	Butylbenzene	−28 e	−7 e	21 e	56.9	107.6	182.8	1,5
$C_{10}H_{14}$	*sec*-Butylbenzene	−35 e	−14 e	13 e	48 e	98.3	172.8	5
$C_{10}H_{14}$	*tert*-Butylbenzene	−37 e	−16 e	10 e	46 e	94.9	168.6	5
$C_{10}H_{14}$	Isobutylbenzene	−36 e	−15 e	12 e	47.9	97.8	172.3	5
$C_{10}H_{14}$	*o*-Cymene	−39 e	−16 e	13 e	51 e	103.1	177.8	5
$C_{10}H_{14}$	*m*-Cymene	−34 e	−13 e	14 e	50 e	99.9	174.6	5
$C_{10}H_{14}$	*p*-Cymene	−33 e	−12 e	16 e	52 e	102.2	176.6	5
$C_{10}H_{14}$	*o*-Diethylbenzene	−28 e	−6 e	21 e	58 e	107.9	182.9	5
$C_{10}H_{14}$	*m*-Diethylbenzene	−28 e	−7 e	20 e	56 e	106.2	180.6	5
$C_{10}H_{14}$	*p*-Diethylbenzene	−28 e	−6 e	21 e	57 e	108.1	183.3	5
$C_{10}H_{14}$	3-Ethyl-1,2-dimethylbenzene	−22 e	0 e	28 e	66 e	117.2	193.4	5
$C_{10}H_{14}$	4-Ethyl-1,2-dimethylbenzene	−24 e	−2 e	26 e	63 e	113.6	189.2	5
$C_{10}H_{14}$	2-Ethyl-1,3-dimethylbenzene	−24 e	−2 e	26 e	63 e	113.7	189.5	5
$C_{10}H_{14}$	2-Ethyl-1,4-dimethylbenzene	−27 e	−5 e	23 e	60 e	110.6	186.4	5
$C_{10}H_{14}$	1-Ethyl-2,4-dimethylbenzene	−25 e	−4 e	24 e	61 e	112.2	187.9	5
$C_{10}H_{14}$	1-Ethyl-3,5-dimethylbenzene	−28 e	−6 e	21 e	58 e	108.3	183.2	5
$C_{10}H_{14}$	1-Methyl-2-propylbenzene	−27 e	−6 e	22 e	58.2	108.9	184.3	5
$C_{10}H_{14}$	1-Methyl-3-propylbenzene	−29 e	−8 e	20 e	56.1	106.5	181.3	5
$C_{10}H_{14}$	1-Methyl-4-propylbenzene	−29 e	−7 e	20 e	56.6	107.4	182.8	5
$C_{10}H_{14}$	1,2,3,4-Tetramethylbenzene	−16 e	7 e	36 e	74 e	126.6	204.5	5
$C_{10}H_{14}$	1,2,3,5-Tetramethylbenzene	−19 e	3 e	32 e	69 e	120.9	197.5	5
$C_{10}H_{14}$	1,2,4,5-Tetramethylbenzene					119.9	196.3	5
$C_{10}H_{14}O$	2-Butylphenol	7 e	31 e	61 e	101 e	155.2	234.4	5
$C_{10}H_{14}O$	Butyl phenyl ether	−16 e	8 e	38 e	77 e	131.3	209.7	5
$C_{10}H_{14}O$	Thymol	18.9 s	37.9 s	59.5	101.2	155.0	230.4	5
$C_{10}H_{15}N$	2-Methyl-5-isopropylaniline	19 e	43 e	72 e	107.4	150 e	204 e	5
$C_{10}H_{15}N$	*N*-Butylaniline	11 e	35 e	66 e	106 e	160.9	241.0	5
$C_{10}H_{15}N$	*N,N*-Diethylaniline	−11 e	14 e	44.3	84.2	138.4	216.3	5
$C_{10}H_{16}$	Dipentene	−42 e	−19 e	10.6	48.7	100.2	173.9	5
$C_{10}H_{16}$	*d*-Limonene	−45 e	−21 e	9.1	48.0	100.4	174.5	5
$C_{10}H_{16}$	*l*-Limonene	−33 e	−12 e	16 e	52.0	102.3	177.0	21
$C_{10}H_{16}$	β-Myrcene			9.4	47.3	98.3	171.0	5
$C_{10}H_{16}$	α-Pinene	−48 e	−27 e	−1 e	33.6	82.2	155.1	21
$C_{10}H_{16}$	β-Pinene	−43 e	−22 e	5.0	40.6	90.5	165.5	21
$C_{10}H_{16}$	Camphene					90.7	160.1	4
$C_{10}H_{16}$	Terpinolene			26.5	64.9	115.4	184.6	5
$C_{10}H_{16}$	β-Phellandrene			16 e	53.2	104 e	171.0	5
$C_{10}H_{16}O$	(+)-Camphor	−15.8 s	10 s	41.5 s	80.8 s	131.4 s	207.6	5
$C_{10}H_{16}O$	Pulegone	37 e	49.1	66.4	92.2	135.1	220.2	5
$C_{10}H_{18}$	1-Decyne	−34 e	−13 e	14 e	51 e	100.3	173.5	5
$C_{10}H_{18}$	*cis*-Decahydronaphthalene	−26 e	−4 e	24 e	62.4	115.5	195.3	1

Mol. form.	Name	Temperature in °C for the indicated pressure 1 Pa	10 Pa	100 Pa	1 kPa	10 kPa	100 kPa	Ref.
$C_{10}H_{18}$	*trans*-Decahydronaphthalene		–10 e	18 e	55.3	107.9	186.8	1
$C_{10}H_{18}O$	α-Terpineol			48	89	142	217	4
$C_{10}H_{18}O$	Eucalyptol			10.6	48.5	100.3	175.4	5
$C_{10}H_{18}O$	*trans*-Geraniol	4 e	31 e	63.2	104.3	157.7	229.6	5
$C_{10}H_{18}O_4$	Sebacic acid	125.9 s						5
$C_{10}H_{18}O_4$	Dipropyl succinate	11 e	38 e	72.1	115.4	172.3	250.4	5
$C_{10}H_{18}O_4$	Diethyl adipate	4 e	35 e	72 e	116.6	171.2	239.5	5
$C_{10}H_{19}N$	Decanenitrile	13 e	36 e	66 e	105.8	160.6	241.6	5
$C_{10}H_{20}$	1-Decene	–35.5	–13.7	13.7	49.0	97.9	170.1	1,5
$C_{10}H_{20}$	Cyclodecane			29 e	68 e	121.3	201.8	1
$C_{10}H_{20}$	Butylcyclohexane	–31 e	–9 e	18 e	54 e	104.7	180.4	5
$C_{10}H_{20}$	Isobutylcyclohexane	–37 e	–16 e	10 e	46 e	95.9	170.8	5
$C_{10}H_{20}$	*tert*-Butylcyclohexane	–39 e	–18 e	9 e	45 e	95.3	171.1	5
$C_{10}H_{20}O$	Decanal		16 e	47.2	86.3	137.7	208.0	5
$C_{10}H_{20}O_2$	Decanoic acid	58 e	80 e	108 e	145 e	195.2	269.5	5
$C_{10}H_{20}O_2$	Octyl acetate	–26 e	–3 e	27 e	66.3	120.0	198.2	5
$C_{10}H_{20}O_2$	2-Ethylhexyl acetate	–11 e	5 e	26 e	57.6	107.1	197.2	5
$C_{10}H_{20}O_2$	Isopentyl isopentanoate			22 e	62.8	116.9	193.6	5
$C_{10}H_{20}O_2$	Ethyl octanoate	–17 e	9 e	41 e	81.4	133.2	203 e	5
$C_{10}H_{20}O_4$	Diethylene glycol monobutyl ether acetate	6 e	34 e	69 e	112.6	169.2	245.4	5
$C_{10}H_{21}Br$	1-Bromodecane	9 e	33 e	63 e	104 e	159.2	240.0	5
$C_{10}H_{21}Cl$	1-Chlorodecane	2 e	25 e	54 e	92 e	145.7	225.3	5
$C_{10}H_{21}F$	1-Fluorodecane	–22 e	0 e	27 e	64 e	113.3	185.7	5
$C_{10}H_{22}$	Decane		–10.6	16.7	52.3	101.1	173.7	16
$C_{10}H_{22}$	2-Methylnonane	–34 e	–14 e	12 e	47 e	94.8	166.5	5
$C_{10}H_{22}$	3-Methylnonane	–34 e	–14 e	12 e	47 e	95.1	167.3	5
$C_{10}H_{22}$	4-Methylnonane	–36 e	–16 e	10 e	45 e	93.1	165.2	5
$C_{10}H_{22}$	5-Methylnonane	–36 e	–16 e	10 e	45 e	92.6	164.6	5
$C_{10}H_{22}$	2,4-Dimethyloctane				38 e	84.9	155.4	5
$C_{10}H_{22}$	2,7-Dimethyloctane	–39 e	–19 e	7 e	41 e	88.4	159.4	5
$C_{10}H_{22}$	2,2,6-Trimethylheptane	–46 e	–27 e	–2 e	32 e	78.5	148.4	5
$C_{10}H_{22}$	3,3,5-Trimethylheptane			0 e	35 e	82.7	155.2	5
$C_{10}H_{22}$	2,2,3,3-Tetramethylhexane	–46 e	–25 e	1 e	36 e	85.6	159.8	5
$C_{10}H_{22}$	2,2,5,5-Tetramethylhexane			–10 e	22 e	68.3	137.0	5
$C_{10}H_{22}$	2,4-Dimethyl-3-isopropylpentane	–46 e	–26 e	0 e	35 e	83.2	156.5	5
$C_{10}H_{22}$	2,2,3,3,4-Pentamethylpentane		–24 e	3 e	39 e	89.1	165.5	5
$C_{10}H_{22}$	2,2,3,4,4-Pentamethylpentane		–29 e	–3 e	33 e	82.8	158.7	5
$C_{10}H_{22}O$	1-Decanol	30 e	50 e	75 e	109 e	157.3	230.6	1,39
$C_{10}H_{22}O$	4-Decanol	18 e	37 e	61 e	93 e	139 e	210 e	5
$C_{10}H_{22}O$	Dipentyl ether	–31 e	–8 e	22 e	60 e	111.6	186.2	5
$C_{10}H_{22}O$	Diisopentyl ether			14.0	51.5	101.8	172.8	5
$C_{10}H_{22}O_2$	Ethylene glycol dibutyl ether	0 e	20 e	44 e	78.4	127.1	202.9	5
$C_{10}H_{22}O_5$	Tetraethylene glycol dimethyl ether				138 e	200.9	275.3	5
$C_{10}H_{22}S$	1-Decanethiol	11 e	34 e	64 e	103 e	157.5	238.6	5
$C_{10}H_{22}S$	Diisopentylsulfide			7 e	82 e	118 e	139 e	5
$C_{10}H_{23}N$	Dipentylamine				77 e	127.7	202.0	5
$C_{10}H_{30}O_3Si_4$	Decamethyltetrasiloxane	–31 e	–6 e	26 e	66.8	118.8	193.9	5
$C_{10}H_{30}O_5Si_5$	Decamethylcyclopentasiloxane	–2 e	19 e	46 e	82 e	132.9	210.4	5
$C_{11}H_8O_2$	1-Naphthalenecarboxylic acid				191.9	239.3	299.6	5
$C_{11}H_8O_2$	2-Naphthalenecarboxylic acid				197.9	246.0	308.1	5
$C_{11}H_{10}$	1-Methylnaphthalene	5 e	29 e	60 e	102 e	159.1	244.1	1
$C_{11}H_{10}$	2-Methylnaphthalene			57 e	99 e	156.0	240.5	1
$C_{11}H_{12}O_2$	Ethyl *trans*-cinnamate			79	125	187	271	4
$C_{11}H_{12}O_3$	Myristicin	23 e	53 e	88.9	135.2	196.0	279.4	5
$C_{11}H_{14}$	4-Isopropylstyrene	–25 e	–1 e	30.2	70.3	124.5	202.1	5
$C_{11}H_{14}$	1,2,3,4-Tetrahydro-5-methylnaphthalene	9 e	31 e	60 e	99 e	153.1	233.8	5
$C_{11}H_{14}$	1,2,3,4-Tetrahydro-6-methylnaphthalene	17 e	36 e	62 e	97 e	147.8	228.5	5
$C_{11}H_{14}O_2$	Butyl benzoate	6 e	34 e	67.9	110.3	165 e	237 e	5
$C_{11}H_{16}$	Pentylbenzene	–14 e	8 e	37 e	74 e	126.7	204.9	5

Mol. form.	Name	1 Pa	10 Pa	100 Pa	1 kPa	10 kPa	100 kPa	Ref.
		Temperature in °C for the indicated pressure						
$C_{11}H_{16}$	*p-tert*-Butyltoluene	−24 e	−2 e	27 e	64.1	115.5	190.8	5
$C_{11}H_{16}$	1,3-Diethyl-5-methylbenzene	−26 e	−1 e	29.5	69.5	123.5	200.2	5
$C_{11}H_{16}$	2-Ethyl-1,3,5-trimethylbenzene		6 e	36 e	75.7	129.6	207.6	5
$C_{11}H_{16}$	1-Ethyl-2,4,5-trimethylbenzene	−13 e	11 e	40 e	79.4	132.1	207.7	5
$C_{11}H_{20}$	1-Undecyne	−22 e	0 e	29 e	67 e	118.5	194.5	5
$C_{11}H_{20}$	2-Undecyne	−17 e	6 e	35 e	74 e	127.4	205.4	5
$C_{11}H_{20}O_2$	10-Undecenoic acid	35 e	67 e	105 e	150.0	205.4	274.5	5
$C_{11}H_{20}O_4$	Ethyl diethylmalonate			74 e	105 e	149.4	219 e	5
$C_{11}H_{21}N$	Undecanenitrile			78.6	120.3	177.3	259.9	5
$C_{11}H_{22}$	1-Undecene	−21.6	1.2	29.7	66.4	117.1	192.2	5
$C_{11}H_{22}$	*cis*-2-Undecene	−14 e	7 e	34 e	70.2	120.6	196 e	5
$C_{11}H_{22}$	*trans*-2-Undecene	−14 e	7 e	33 e	69.3	119.6	195 e	5
$C_{11}H_{22}$	*cis*-4-Undecene	−19 e	3 e	30 e	66.6	117.1	192 e	5
$C_{11}H_{22}$	*trans*-4-Undecene	−17 e	4 e	31 e	67.1	117.4	193 e	5
$C_{11}H_{22}$	*cis*-5-Undecene	−19 e	2 e	30 e	66.2	116.7	191 e	5
$C_{11}H_{22}$	*trans*-5-Undecene	−18 e	3 e	31 e	67.0	117.4	192 e	5
$C_{11}H_{22}$	Pentylcyclohexane	−17 e	6 e	34 e	72 e	124.2	202.7	5
$C_{11}H_{22}$	Hexylcyclopentane	−15 e	7 e	36 e	73 e	125.0	202.5	5
$C_{11}H_{22}O$	2-Undecanone	17 e	37 e	64.3	103.0	153.6	232.6	1,5
$C_{11}H_{22}O$	6-Undecanone		28 e	57 e	95 e	148.4	226.9	1
$C_{11}H_{22}O_2$	Undecanoic acid	68 e	90 e	118 e	156 e	207.2	283.6	5
$C_{11}H_{22}O_2$	Heptyl butanoate	2 e	29 e	62 e	102.6	155.1	224.7	5
$C_{11}H_{22}O_2$	Propyl octanoate	−2 e	23 e	55 e	94.0	145.2	215 e	5
$C_{11}H_{22}O_2$	Methyl decanoate	10 e	33 e	62 e	100.9	154.0	232 e	5
$C_{11}H_{24}$	Undecane	−18.4	4.3	32.6	69.5	120.2	195.4	16
$C_{11}H_{24}$	2-Methyldecane	−20 e	1 e	28 e	64 e	114.0	188.7	5
$C_{11}H_{24}$	3-Methyldecane	−35 e	−10 e	22 e	61.9	115.6	190.4	5
$C_{11}H_{24}$	4-Methyldecane	−38 e	−12 e	20 e	60.8	113.9	186.4	5
$C_{11}H_{24}$	2,4,7-Trimethyloctane				43 e	94 e	170.4	5
$C_{11}H_{24}O$	1-Undecanol	52.2	80.0	82 e	118 e	167.6	244.1	5
$C_{11}H_{24}S$	1-Undecanethiol	23 e	47 e	77 e	118 e	173.6	256.8	5
$C_{12}F_{27}N$	Trinonafluorobutylamine		3 e	29.0	63.3	109.9	176.8	5
$C_{12}H_8$	Acenaphthylene	24 s	49.8 s	80.6 s				5
$C_{12}H_9N$	Carbazole					254.7	354.0	5
$C_{12}H_{10}$	Acenaphthene				126.2	187 e	276 e	1
$C_{12}H_{10}$	Biphenyl			69.0	111.1	169.5	254.7	1
$C_{12}H_{10}N_2$	Azobenzene			98.1	144.8	206.7	292.7	4
$C_{12}H_{10}O$	Diphenyl ether		44 e	75 e	116 e	173 e	257.4	5
$C_{12}H_{10}O$	1-Acetonaphthone	37 e	69 e	107.0	154.6	215.2	294.9	5
$C_{12}H_{10}O$	2-Acetonaphthone	48.3 s		118.7	163.0	221.1	300.3	5
$C_{12}H_{10}S$	Diphenyl sulfide	20 e	51 e	88.7	137.5	202.2	291.8	5
$C_{12}H_{11}N$	Diphenylamine	48 s		102.8	150.5	213.7	301.4	5
$C_{12}H_{12}$	1-Ethylnaphthalene	16 e	41 e	72 e	114 e	171.8	257.7	5
$C_{12}H_{12}$	2-Ethylnaphthalene	14 e	39 e	71 e	113 e	171.2	257.3	5
$C_{12}H_{12}$	1,2-Dimethylnaphthalene	26 e	51 e	82 e	123 e	180.5	265.7	5
$C_{12}H_{12}$	2,7-Dimethylnaphthalene	31.5 s	53.1 s	78.8 s	115.9	175 e	260 e	5
$C_{12}H_{14}O_4$	Diethyl phthalate	12 e	51 e	96 e	150.5	215.9	296.2	5
$C_{12}H_{16}$	*p*-Isopropenylisopropylbenzene	−11 e	15 e	46 e	87 e	142.4	221 e	5
$C_{12}H_{16}$	Cyclohexylbenzene		28 e	58 e	98 e	154.7	239.5	5
$C_{12}H_{16}O_2$	3-Methylbutyl benzoate			66 e	115.0	177.7	261.4	5
$C_{12}H_{18}$	Hexylbenzene	−2 e	22 e	51 e	90 e	144.5	225.5	5
$C_{12}H_{18}$	1,2-Diisopropylbenzene	−14 e	9 e	37 e	74 e	125.9	203.2	5
$C_{12}H_{18}$	1,3-Diisopropylbenzene	−14 e	8 e	36 e	74 e	125.5	202.6	5
$C_{12}H_{18}$	1,4-Diisopropylbenzene	−6 e	18 e	49 e	90 e	148.8	238 e	5
$C_{12}H_{18}$	Hexamethylbenzene	46.3 s	72.5 s	81.7 s	121.8 s	178.3	263.7	5
$C_{12}H_{18}$	1,5,9-Cyclododecatriene	−14 e	11 e	44 e	87 e	145.0	229.8	5
$C_{12}H_{20}O_2$	Geranyl acetate			67.7	110.8	166.9	242.9	5
$C_{12}H_{20}O_4$	Dibutyl maleate	12.3	50.4	94.0	144.2	203 e	272 e	5
$C_{12}H_{22}$	1-Dodecyne	−11 e	13 e	43 e	82 e	135.8	214.4	5
$C_{12}H_{22}$	Cyclohexylcyclohexane		20 e	53.1	96.0	154.1	237.2	5

Mol. form.	Name	Temperature in °C for the indicated pressure						Ref.
		1 Pa	10 Pa	100 Pa	1 kPa	10 kPa	100 kPa	
$C_{12}H_{22}O_2$	Methyl 10-undecenoate	10 e	38 e	73 e	116 e	172.2	247.1	5
$C_{12}H_{22}O_4$	Dimethyl sebacate		53 e	97	150	214	293	4
$C_{12}H_{23}N$	Dodecanenitrile	36 e	60 e	92 e	133 e	190.5	275.5	5
$C_{12}H_{24}$	1-Dodecene	−8.3	15.2	44.8	82.9	135.4	212.8	5
$C_{12}H_{24}$	Hexylcyclohexane	−3 e	20 e	50 e	89 e	143.1	224.2	5
$C_{12}H_{24}$	Heptylcyclopentane	−1 e	22 e	51 e	90 e	143.5	223.5	5
$C_{12}H_{24}O$	Dodecanal			70 e	116.2	175.9	256.6	5
$C_{12}H_{24}O_2$	Dodecanoic acid	78 e	100 e	128 e	166 e	219.1	298.1	5
$C_{12}H_{24}O_2$	Decyl acetate	12 e	40 e	74 e	115.1	168.1	238 e	5
$C_{12}H_{24}O_2$	Ethyl decanoate	8 e	35 e	69 e	111.8	166.1	238 e	5
$C_{12}H_{25}Br$	1-Bromododecane	31 e	57 e	90 e	132 e	190.8	275.3	5
$C_{12}H_{25}Cl$	1-Chlorododecane	27 e	51 e	81 e	122 e	178.7	262.6	5
$C_{12}H_{26}$	Dodecane	−5.4	18.2	47.6	85.8	138.2	215.8	16
$C_{12}H_{26}O$	1-Dodecanol				133 e	185.0	264.1	1
$C_{12}H_{26}O_3$	Diethylene glycol dibutyl ether	5 e	34.4	70.2	115.3	174.1	253.8	5
$C_{12}H_{27}N$	Tributylamine	−26 e	1 e	35 e	77.7	134.5	213.4	5
$C_{12}H_{27}N$	Triisobutylamine		1 e	28.9	64.9	112.5	178.5	5
$C_{12}H_{27}O_4P$	Tributyl phosphate					205 e	288.3	5
$C_{12}H_{36}O_6Si_6$	Dodecamethylcyclohexasiloxane	18 e	41 e	69 e	108 e	162.2	244.7	5
$C_{13}H_9N$	Acridine			124.4	176.2	246.0	345.4	5
$C_{13}H_9N$	Phenanthridine	79 s						5
$C_{13}H_{10}$	Fluorene	48.4 s			137.4	205.4	295 e	5
$C_{13}H_{10}O_2$	Phenyl benzoate			102.3	151.4	217.9	313.3	5
$C_{13}H_{10}O_3$	Phenyl salicylate				166.0	224.8	312.4	5
$C_{13}H_{12}$	Diphenylmethane		45 e	77 e	119.3	177.7	263.6	1,5
$C_{13}H_{13}N$	Methyldiphenylamine	35 e	63 e	98.4	143.1	201.6	281.6	5
$C_{13}H_{14}$	1-Isopropylnaphthalene	27 e	51 e	82 e	123.2	180.8	267.3	5
$C_{13}H_{20}$	Heptylbenzene	12 e	36 e	66 e	107 e	162.7	246.2	5
$C_{13}H_{24}O_2$	Ethyl 10-undecenoate	32 e	55 e	86 e	125.2	179.5	258.4	5
$C_{13}H_{26}$	1-Tridecene	4.1	28.5	59.0	98.3	152.5	232.3	5
$C_{13}H_{26}$	Heptylcyclohexane	11 e	34 e	65 e	105 e	160.9	244.3	5
$C_{13}H_{26}$	Octylcyclopentane	13 e	36 e	66 e	106 e	160.9	243.1	5
$C_{13}H_{26}O_2$	Tridecanoic acid	87 e	109 e	138 e	176 e	230.3	311.5	5
$C_{13}H_{26}O_2$	Methyl dodecanoate	38 e	61 e	90 e	130 e	184.9	269 e	5
$C_{13}H_{28}$	Tridecane	7.2	31.5	61.8	101.1	155.1	234.9	16
$C_{13}H_{28}O$	1-Tridecanol	71.6	101.0	103 e	140 e	192.3	273.1	5
$C_{14}H_{10}$	Anthracene	89.2 s	125.9 s	151.5 s	165 s	238.8	340.2	1,5
$C_{14}H_{10}$	Phenanthrene	53 s	83 s	120.8	170.4	238.4	337.7	5
$C_{14}H_{10}O_2$	Benzil			123	175	246	346	4
$C_{14}H_{12}$	*cis*-Stilbene	26 e	54 e	88 e	130.4	183 e	253 e	5
$C_{14}H_{12}$	*trans*-Stilbene				155.6	218.1	305.8	5
$C_{14}H_{12}O_2$	Benzoin				181	248	342	4
$C_{14}H_{14}$	1,1-Diphenylethane	19 e	47 e	82.0	125.3	181 e	254 e	5
$C_{14}H_{15}N$	Dibenzylamine	48 e	77 e	113.1	158.9	218.5	299.4	5
$C_{14}H_{16}$	1-Butylnaphthalene	67 e	82 e	103 e	135 e	186.7	288.6	5
$C_{14}H_{16}$	2-Butylnaphthalene	44 e	67 e	98 e	139 e	197.5	287.4	5
$C_{14}H_{22}$	Octylbenzene	20.1	46.2	79.1	121.9	178.1	263.8	5
$C_{14}H_{26}O_4$	Diethyl sebacate		83 e	120	166	225	305	4
$C_{14}H_{27}N$	Tetradecanenitrile	52 e	79 e	114.0	159.0	219.7	306.3	5
$C_{14}H_{28}$	1-Tetradecene	16.1	41.3	72.7	113.2	168.7	250.6	5
$C_{14}H_{28}$	Octylcyclohexane	16.9	44.3	77.8	120.0	177.6	263.2	5
$C_{14}H_{28}$	Nonylcyclopentane	25 e	49 e	80 e	120 e	177.2	261.5	5
$C_{14}H_{28}O_2$	Tetradecanoic acid	96 e	118 e	147 e	186 e	241.3	325.6	5
$C_{14}H_{30}$	Tetradecane	19.1	44.1	75.3	115.7	171.1	253.0	16
$C_{14}H_{30}O$	1-Tetradecanol	80.0	110.5	149.6	152 e	205.3	286.7	5
$C_{14}H_{31}N$	Tetradecylamine			104 e	147 e	206.1	290.9	5
$C_{14}H_{42}O_5Si_6$	Tetradecamethylhexasiloxane	6 e	36 e	72 e	117 e	176.0	259.1	5
$C_{15}H_{18}$	1-Pentylnaphthalene	34 e	62 e	96 e	141.3	202.2	289 e	5
$C_{15}H_{24}$	Nonylbenzene	33.0	58.9	92.0	135.4	193.7	281.4	5
$C_{15}H_{30}$	Nonylcyclohexane	35 e	60 e	92 e	134 e	193.4	280.9	5

Mol. form.	Name	Temperature in °C for the indicated pressure						Ref.
		1 Pa	10 Pa	100 Pa	1 kPa	10 kPa	100 kPa	
$C_{15}H_{30}$	Decylcyclopentane	37 e	61 e	93 e	134 e	192.5	278.8	5
$C_{15}H_{30}O_2$	Methyl tetradecanoate		75 e	110	155	214	295	4
$C_{15}H_{32}$	Pentadecane	30.5	56.1	88.1	129.6	186.3	270.1	16
$C_{16}H_{22}O_4$	Dibutyl phthalate		104.0	142.7	191.5	254.5	339.4	4
$C_{16}H_{32}$	1-Hexadecene	38.4	65.0	98.1	140.5	198.8	284.3	5
$C_{16}H_{32}O_2$	Hexadecanoic acid		136 e	165 e	205 e	261.9	350.2	5
$C_{16}H_{34}$	Hexadecane	41.1	67.4	100.3	142.7	200.7	286.3	16
$C_{16}H_{34}O$	1-Hexadecanol	99.5	130.6	171.9	175 e	229.0	311.7	5
$C_{16}H_{35}N$	Hexadecylamine	63 e	91 e	126 e	171 e	232.6	320.5	5
$C_{17}H_{10}O$	Benzanthrone		184 e	229.3	290.3	377.2	511 e	5
$C_{17}H_{34}O_2$	Methyl hexadecanoate	65 e	93	129	177			4
$C_{17}H_{36}$	Heptadecane	51.5	78.5	112.0	155.3	214.5	302 e	16
$C_{17}H_{36}O$	1-Heptadecanol	94 e	117 e	146 e	185 e	240.1	323.3	5
$C_{18}H_{14}$	*o*-Terphenyl	66 e	94 e	129 e	176 e	241.3	336.3	5
$C_{18}H_{14}$	*m*-Terphenyl	87 e	118 e	156 e	206.6	275.3	374.6	5
$C_{18}H_{14}$	*p*-Terphenyl	127.1 s	154.7 s		217.2	284.0	383.0	5
$C_{18}H_{30}$	Hexaethylbenzene				144.1	206.8	297.5	5
$C_{18}H_{34}O_2$	Oleic acid	94 e	126 e	165.5	214.5	277.0	359.7	5
$C_{18}H_{34}O_2$	Elaidic acid		124 e	166	216	280	361	4
$C_{18}H_{36}O$	Stearaldehyde			142 e	186 e	246.9	336.7	5
$C_{18}H_{36}O_2$	Stearic acid		153 e	183 e	223 e	281.6	374.5	5
$C_{18}H_{38}$	Octadecane	61.5	89.0	123.1	167.3	227.6	316 e	16
$C_{18}H_{38}O$	1-Octadecanol	106 e	130 e	160 e	200.5	257.3	343.0	5
$C_{19}H_{16}$	Triphenylmethane	81 s		112 e	175 e	254.6	360.0	5
$C_{19}H_{36}O_2$	Methyl oleate	85 e	114 e	149.7	195.6	256 e	340 e	5
$C_{19}H_{40}$	Nonadecane	71.1	99.1	133.8	178.8	240.1	330 e	16
$C_{20}H_{42}$	Eicosane	80.4	108.9	144.2	189.8	252.1	344 e	16
$C_{20}H_{42}O$	1-Eicosanol	119 e	143 e	173 e	213 e	270.0	355.1	5
$C_{20}H_{60}O_8Si_9$	Eicosamethylnonasiloxane			141 e	183.1	236.7	307.1	5
$C_{21}H_{21}O_4P$	Tri-*o*-cresyl phosphate	119.0	156.1	201.0	256.3	326.3	418 e	5
$C_{21}H_{21}O_4P$	Tri-*m*-cresyl phosphate	147.8	177.3	211.4	251.3	298 e	355 e	5
$C_{21}H_{21}O_4P$	Tri-*p*-cresyl phosphate	140.6	174 e	214 e	262 e	320 e	392 e	5
$C_{21}H_{44}$	Heneicosane	82.3	113.5	152.2	201.6	263.8	355.9	5
$C_{22}H_{42}O_2$	Brassidic acid	134 e	166 e	203.6	249.8	307.6	382.0	5
$C_{22}H_{42}O_2$	Erucic acid	126 e	160 e	199.4	247.4	306.5	381.1	5
$C_{22}H_{42}O_2$	Butyl oleate	95.5	124.2	158 e	198 e	245 e	304 e	5
$C_{22}H_{44}O_2$	Behenic acid	145.4	176.5	213.7	259.3	316.2	390 e	5
$C_{22}H_{44}O_2$	Butyl stearate	99.6	128 e	162 e	201 e	249 e	307 e	5
$C_{22}H_{46}$	Docosane	83.5	115.0	154.0	203.6	274.8	368.0	5
$C_{23}H_{48}$	Tricosane	102.9	135.1	174.8	221 e	285.3	379.5	5
$C_{24}H_{38}O_4$	Dioctyl phthalate	130 e	163.7	203.8	252 e	311 e	385 e	5
$C_{24}H_{38}O_4$	Bis(2-ethylhexyl) phthalate	122.0	153.2	189.2	231.3	281.1	341.1	5
$C_{24}H_{50}$	Tetracosane	115.0	148.1	188.5	239.1	295.4	390.6	5
$C_{25}H_{52}$	Pentacosane	119.7	152.7	193.2	244.4	305.0	401.1	5
$C_{26}H_{54}$	Hexacosane	125.1	158.8	200.1	252.1	314.3	411.3	5
$C_{27}H_{56}$	Heptacosane	136.7	168.8	206.5	255.8	323.3	421.2	5
$C_{28}H_{58}$	Octacosane	136.5	169.8	210.9	263.1	332.0	430.6	5
$C_{29}H_{60}$	Nonacosane	148.2	182.8	221.2	271.5	340.2	439.7	5
$C_{30}H_{62}$	Squalane	66 e	84 e	105.8	131.9	163.7	203.2	5
C_{70}	Carbon (fullerene-C_{70})	598 s	662 s					22

VAPOR PRESSURE OF FLUIDS AT TEMPERATURES BELOW 300 K

This table gives vapor pressures of 67 important fluids in the temperature range 2 to 300 K. Helium (^4He), hydrogen (H_2), and neon (Ne) are covered on this page. The remaining fluids are listed on subsequent pages by molecular formula in the Hill order (see Introduction). The data have been taken from evaluated sources; references are listed at the end of the table.

Pressures are given in kilopascals (kPa). Note that:

1 kPa = 7.50062 Torr
100 kPa = 1 bar
101.325 kPa = 1 atm

"s" following an entry indicates that the compound is solid at that temperature.

Helium		Hydrogen		Neon	
T/K	*P*/kPa	*T*/K	*P*/kPa	*T*/K	*P*/kPa
2.2	5.3	14.0	7.90	25.0	51.3
2.3	6.7	14.5	10.38	26.0	71.8
2.4	8.3	15.0	13.43	27.0	98.5
2.5	10.2	15.5	17.12	28.0	132.1
2.6	12.4	16.0	21.53	29.0	173.5
2.7	14.8	16.5	26.71	30.0	223.8
2.8	17.5	17.0	32.84	31.0	284.0
2.9	20.6	17.5	39.92	32.0	355.2
3.0	24.0	18.0	48.08	33.0	438.6
3.1	27.8	18.5	57.39	34.0	535.2
3.2	32.0	19.0	67.96	35.0	646.2
3.3	36.5	19.5	79.89	36.0	772.8
3.4	41.5	20.0	93.26	37.0	916.4
3.5	47.0	20.5	108.2	38.0	1078
3.6	52.9	21.0	124.7	39.0	1260
3.7	59.3	21.5	143.1	40.0	1462
3.8	66.1	22.0	163.2	41.0	1688
3.9	73.5	22.5	185.3	42.0	1939
4.0	81.5	23.0	209.4	43.0	2216
4.1	90.0	23.5	235.7	44.0	2522

Helium		Hydrogen		Neon	
T/K	*P*/kPa	*T*/K	*P*/kPa	*T*/K	*P*/kPa
4.2	99.0	24.0	264.2		
4.3	108.7	24.5	295.1		
4.4	119.0	25.0	328.5		
4.5	129.9	25.5	364.3		
4.6	141.6	26.0	402.9		
4.7	153.9	26.5	444.3		
4.8	167.0	27.0	488.5		
4.9	180.8	27.5	535.7		
5.0	195.4	28.0	586.1		
5.1	210.9	28.5	639.7		
		29.0	696.7		
		29.5	757.3		
		30.0	821.4		
		30.5	889.5		
		31.0	961.5		
		31.5	1038.0		
		32.0	1119.0		
		32.5	1204.0		
Ref.	17,18		1		13

T/K	Ar Argon	BCl_3 Boron trichloride	BF_3 Boron trifluoride	BrH Hydrogen bromide	Br_2 Bromine	ClF Chlorine fluoride	ClH Hydrogen chloride
50	0.1 s						
55	0.2 s						
60	0.8 s						
65	2.8 s						
70	7.7 s						
75	18.7 s						
80	40.7 s						
85	79.0						
90	134						
95	213						
100	324						
105	473						
110	666						
115	910					0.1	
120	1214					0.3	0.1 s
125	1584					0.6	0.3 s
130	2027					1.2	0.5 s
135	2553			0.1 s		2.1	1.0 s
140	3170			0.3 s		3.6	1.9 s
145	3892		7.7	0.6 s		6.0	3.4 s
150	4736		13.4	1.1 s		9.5	5.8 s
155			22.3	1.9 s		14.6	9.5 s
160			35.2	3.3 s		21.8	14.7
165			53.7	5.4 s		31.7	22.0
170			79.1	8.7 s		44.8	31.9
175			113	13.4 s		62.0	45.1

T/K	Ar Argon	BCl_3 Boron trichloride	BF_3 Boron trifluoride	BrH Hydrogen bromide	Br_2 Bromine	ClF Chlorine fluoride	ClH Hydrogen chloride
180		0.1	157	20.1 s		84.2	62.5
185		0.2	214	29.5 s		112	84.7
190		0.3	285	37.9		147	113
195		0.5	372	51.8		190	148
200		0.8	479	69.5		242	190
205		1.2	608	91.8		304	242
210		1.8	762	119		378	304
215		2.6	944	153		464	377
220		3.8	1160	194	0.1 s	564	463
225		5.2	1413	242	0.2 s	680	563
230		7.2	1709	299	0.3 s	812	678
235		9.7	2056	366	0.4 s	961	811
240		12.9	2460	443	0.7 s	1130	961
245		17.0	2913	532	1.1 s	1319	1132
250		22.0	3481	633	1.7 s	1529	1325
255		28.1	4123	748	2.6 s	1762	1542
260		35.6	4874	878	3.8 s	2019	1784
265		44.5		1023	5.5 s	2301	2054
270		55.1		1185	7.3	2608	2354
275		67.6		1364	9.5	2941	2686
280		82.2		1562	12.3	3303	3053
285		99.1		1780	15.6	3693	3457
290		119		2018	19.7	4111	3901
295		141		2278	24.6	4560	4388
300		166		2561	30.5	5039	4921
Ref.	8,15	12	12	12	12	12	12

T/K	ClO_2 Chlorine dioxide	Cl_2 Chlorine	Cl_4Si Silicon tetrachloride	FH Hydrogen fluoride	F_2 Fluorine	F_2O Difluorine oxide	F_3N Nitrogen trifluoride
50							
55					0.4		
60					1.5		
65					4.8		
70					12.3		
75					27.6	0.1	
80					55.3	0.2	
85					101	0.5	0.1
90					172	1.2	0.2
95					276	2.6	0.4
100					420	5.3	0.9
105					615	10.1	2.0
110					870	18.0	4.0
115					1196	30.5	7.3
120					1605	49.3	12.8
125					2108	76.7	21.1
130					2721	115	33.5
135					3458	168	51.1
140					4339	237	75.4
145						328	108
150						444	150
155						588	205
160						766	273
165						981	357
170						1238	459
175		1.8				1541	581
180		2.8				1895	726
185		4.2				2303	896
190		6.1		0.3		2771	1092
195	0.1	8.7		0.5		3302	1319
200	0.3	12.3		0.8		3899	1578
205	0.5	16.9		1.2		4567	1871
210	0.9	22.9	0.1	1.7		5308	2203

T/K	ClO_2 Chlorine dioxide	Cl_2 Chlorine	Cl_4Si Silicon tetrachloride	FH Hydrogen fluoride	F_2 Fluorine	F_2O Difluorine oxide	F_3N Nitrogen trifluoride
215	1.4	30.5	0.2	2.3			2577
220	2.3	40.1	0.3	3.2			2995
225	3.5	51.9	0.5	4.4			3464
230	5.3	66.4	0.7	5.9			3991
235	7.6	84.0	1.0	7.9			
240	10.8	105	1.5	10.3			
245	14.9	130	2.0	13.4			
250	20.1	160	2.8	17.2			
255	26.6	194	3.8	21.8			
260	34.6	234	5.0	27.4			
265	44.4	280	6.6	34.2			
270	56.1	332	8.6	42.2			
275	69.9	392	11.1	51.8			
280	86.2	459	14.2	63.1			
285	105	535	17.9	76.3			
290	127	619	22.3	91.7			
295	151	714	27.7	110			
300	179	818	34.0	130			
Ref.	12	5	12	12	12	12	1

T/K	F_3P Phosphorus trifluoride	F_4Si Silicon tetrafluoride	F_6S Sulfur hexafluoride	HI Hydrogen iodide	H_2S Hydrogen sulfide	H_3N Ammonia	H_3P Phosphine
105	0.1						
110	0.2						0.1
115	0.5						0.2
120	1.0						0.4
125	1.9	0.1 s					0.7
130	3.5	0.2 s					1.3
135	5.9	0.4 s			0.1 s		2.3
140	9.5	0.9 s	0.1 s		0.2 s		3.9
145	14.9	1.9 s	0.2 s		0.3 s		6.2
150	22.5	3.8 s	0.4 s		0.6 s		9.6
155	33.1	7.5 s	0.8 s	0.1 s	1.1 s		14.5
160	47.3	14.0 s	1.5 s	0.2 s	1.9 s	0.1 s	21.1
165	66.0	25.2 s	2.6 s	0.4 s	3.2 s	0.2 s	30.0
170	90.1	43.8 s	4.4 s	0.8 s	5.2 s	0.3 s	41.6
175	121	74.2 s	7.1 s	1.3 s	8.3 s	0.6 s	56.6
180	159	122 s	11.3 s	2.2 s	12.7 s	1.2 s	75.6
185	206	197 s	17.3 s	3.4 s	18.9 s	2.1 s	99.2
190	262	280	25.9 s	5.3 s	26.6	3.5 s	128
195	330	376	38.0 s	8.0 s	36.7	5.8 s	163
200	410	488	54.4 s	11.7 s	49.8	8.7	205
205	503	618	76.6 s	16.8 s	66.4	12.6	254
210	611	766	106 s	23.6 s	87.1	17.9	312
215	736	932	145 s	32.5 s	113	24.9	379
220	877	1117	195 s	44.0 s	144	34.1	456
225	1037	1324	249	56.2	182	45.9	544
230	1217	1555	305	71.4	227	60.8	644
235	1418	1816	371	89.7	281	79.6	756
240	1640	2111	448	112	344	103	881
245	1885	2449	536	137	416	131	1019
250	2154	2841	636	168	500	165	1172
255	2448	3301	750	203	597	207	1341
260	2767		878	244	706	256	1525
265	3112		1021	290	830	313	1725
270			1181	343	969	381	1942
275			1358	404	1124	460	2176
280			1554	472	1297	552	2428
285			1768	548	1488	655	2699
290			2003	633	1698	774	2987
295			2258	727	1929	909	3295
300			2534	831	2181	1062	3621
Ref.	12	12	12,15	12	12,15	11	12

T/K	H_4Si Silane	Kr Krypton	NO Nitric oxide	N_2 Nitrogen	N_2O Nitrous oxide	O_2 Oxygen	O_2S Sulfur dioxide
50				0.4 s			
55				1.8 s		0.2	
60				6.3 s		0.7	
65				17.4		2.3	
70				38.6		6.3	
75		0.1 s		76.1		14.5	
80		0.4 s		137		30.1	
85		1.1 s	0.1 s	229		56.8	
90		2.7 s	0.4 s	361		99.3	
95	0.1	6.0 s	1.3 s	541		163	
100	0.2	12.1 s	3.8 s	779		254	
105	0.4	22.8 s	10.0 s	1084		379	
110	1.0	40.4 s	23.5	1467		543	
115	1.9	68.0 s	46.8	1939	0.1	756	
120	3.5	103	86.5	2513	0.1	1022	
125	6.1	150	151	3209	0.3	1351	
130	10.0	211	248		0.7	1749	
135	15.8	290	391		1.3	2225	
140	24.1	388	592		2.5	2788	
145	35.3	509	867		4.3	3448	
150	50.3	655	1231		7.1	4219	
155	69.8	830	1703		11.4		
160	94.6	1037	2302		17.6		
165	126	1278	3050		26.4		
170	164	1557	3971		38.5		0.1
175	210	1877	5089		54.7		0.2
180	265	2241	6433		75.9		0.3
185	331	2655			103		0.5
190	408	3120			138		0.8
195	498	3641			181		1.3
200	602	4223			234		2.0
205	722	4870			298		3.0
210	859				374		4.4
215	1017				465		6.3
220	1196				571		9.0
225	1398				694		12.6
230	1628				835		17.3
235	1888				996		23.3
240	2180				1179		31.1
245	2509				1385		40.9
250	2880				1615		53.2
255	3296				1870		68.3
260	3763				2152		86.7
265	4288				2462		109
270					2802		136
275					3172		168
280					3573		205
285					4006		249
290					4473		300
295					4973		359
300					5508		426
Ref.	12	13, 15	12, 15	1	12	3	12

T/K	O_3 Ozone	Rn Radon	Xe Xenon	$CBrF_3$ Bromotri-fluoromethane	$CClF_3$ Chlorotri-fluoromethane	CCl_2F_2 Dichlorodi-fluoromethane	CCl_3F Trichloro-fluoromethane
100	0.1		0.1 s				
105	0.2		0.1 s				
110	0.4		0.3 s				
115	1.0		0.7 s		0.1		
120	2.0		1.5 s		0.2		
125	3.8		2.7 s		0.3		
130	6.8	0.1	4.9 s		0.6		

T/K	O_3 Ozone	Rn Radon	Xe Xenon	$CBrF_3$ Bromotri-fluoromethane	$CClF_3$ Chlorotri-fluoromethane	CCl_2F_2 Dichlorodi-fluoromethane	CCl_3F Trichloro-fluoromethane
135	11.5	0.3	8.5 s	0.1	1.1		
140	18.7	0.5	14.0 s	0.3	2.0		
145	29.1	0.9	22.2 s	0.5	3.3		
150	43.7	1.5	34.2 s	0.9	5.3		
155	63.6	2.4	51.1 s	1.5	8.3	0.1	
160	89.9	3.8	74.2 s	2.5	12.6	0.3	
165	124	5.8	101	3.9	18.6	0.5	
170	168	8.6	134	5.9	26.8	0.8	
175	222	12.5	173	8.8	37.6	1.3	
180	289	17.7	222	12.8	51.7	2.1	
185	367	24.5	280	18.1	69.7	3.2	
190	468	33.2	348	25.1	92.3	4.8	0.2
195	584	44.4	428	34.1	120	6.9	0.3
200	721	58.2	521	45.6	155	9.9	0.4
205	881	75.3	628	60.0	196	13.7	0.6
210	1068	96	750	77.8	246	18.8	1.0
215	1285	121	889	99.5	304	25.2	1.4
220	1536	151	1045	126	372	33.3	2.0
225	1824	185	1220	157	451	43.3	2.9
230	2155		1416	194	542	55.5	4.1
235	2534		1633	237	646	70.4	5.6
240	2968		1872	287	763	88.1	7.6
245	3464		2136	344	896	109	10.1
250	4031		2425	410	1044	134	13.3
255	4678		2742	485	1210	163	17.2
260	5417		3087	570	1394	196	22.1
265			3462	665	1598	234	28.0
270			3869	771	1823	278	35.1
275			4310	889	2071	327	43.7
280			4786	1021	2343	383	53.8
285			5299	1166	2641	445	65.7
290				1325	2968	515	79.6
295				1501	3325	593	95.6
300				1692	3716	679	114.1
Ref.	12	15	12,13	12	12	12	12

T/K	CCl_4 Tetrachloro-methane	CF_4 Tetrafluoro-methane	CO Carbon monoxide	COS Carbon oxysulfide	CO_2 Carbon dioxide	$CHClF_2$ Chlorodifluoro-methane	$CHCl_3$ Trichloro-methane
50			0.1 s				
55			0.6 s				
60			2.6 s				
65			8.2 s				
70			21.0				
75			44.4				
80			83.7				
85			147				
90		0.1	239				
95		0.3	371				
100		0.8	545				
105		1.7	771				
110		3.4	1067				
115		6.5	1428				
120		11.5	1877				
125		19.3	2400				
130		30.8	3064				
135		47.4			0.1 s		
140		70.2		0.1	0.2 s		
145		101		0.2	0.4 s		
150		141		0.4	0.8 s	0.1	
155		191		0.8	1.7 s	0.3	

T/K	CCl_4 Tetrachloro-methane	CF_4 Tetrafluoro-methane	CO Carbon monoxide	COS Carbon oxysulfide	CO_2 Carbon dioxide	$CHClF_2$ Chlorodifluoro-methane	$CHCl_3$ Trichloro-methane
160		254		1.3	3.1 s	0.5	
165		332		2.2	5.7 s	0.8	
170		425		3.4	9.9 s	1.4	
175		537		5.2	16.8 s	2.3	
180		669		7.8	27.6 s	3.6	
185		824		11.3	44.0 s	5.5	
190		1005		15.9	68.4 s	8.1	
195		1216		22.1	104 s	11.8	
200		1460		30.0	155 s	16.7	
205		1743		40.1	227 s	23.1	
210		2073		52.7	327 s	31.5	
215		2457		68.2	465 s	42.1	0.1
220		2907		87.2	599	55.3	0.2
225		3438		110	734	71.7	0.3
230				137	893	91.6	0.4
235				169	1075	116	0.7
240				207	1283	144	1.0
245				250	1519	178	1.4
250				301	1785	218	2.0
255	1.5			358	2085	264	2.7
260	2.1			423	2419	317	3.7
265	2.8			497	2790	377	5.0
270	3.7			580	3203	446	6.6
275	4.9			673	3658	525	8.7
280	6.4			777	4161	613	11.3
285	8.2			892	4714	711	14.4
290	10.5			1019	5318	821	18.3
295	13.2			1159	5984	944	22.9
300	16.5			1313	6713	1080	28.5
Ref.	12	12	9	12	6, 19	12	12

T/K	CHF_3 Trifluoro-methane	CHN Hydrogen cyanide	CH_2Cl_2 Dichloro-methane	CH_2F_2 Difluoro-methane	CH_2O Formaldehyde	CH_3Cl Chloromethane	CH_3F Fluoromethane
120	0.1						
125	0.2						
130	0.4						
135	0.7						0.6
140	1.4			0.1			1.2
145	2.5			0.2			2.1
150	4.3			0.3			3.6
155	7.1			0.6			5.9
160	11.1			1.0			9.3
165	17.0			1.7			14.1
170	25.3			2.8			20.9
175	36.5			4.4			29.9
180	51.4			6.8			42.0
185	70.9			10.2	1.3	2.1	57.6
190	95.8			14.8	2.0	3.1	77.4
195	127			21.2	3.0	4.6	102
200	166	0.1 s	0.1	29.5	4.4	6.7	133
205	214	0.2 s	0.2	40.5	6.4	9.5	171
210	271	0.4 s	0.3	54.5	9.1	13.1	216
215	340	0.6 s	0.4	72.1	12.7	17.9	270
220	421	1 s	0.6	94.1	17.4	24.0	333
225	516	1.5 s	0.9	121	23.4	31.8	408
230	626	2.2 s	1.4	154	31.0	41.4	495
235	754	3.3 s	2.0	193	40.6	53.3	595
240	900	4.7 s	2.8	240	52.5	67.7	711
245	1067	6.8 s	3.8	295	67.0	85.1	843
250	1257	9.7 s	5.3	360	84.6	106	993
255	1472	13.6 s	7.1	434	106	131	1163
260	1713	18.8	9.5	521	131	159	1355

T/K	CHF_3 Trifluoro-methane	CHN Hydrogen cyanide	CH_2Cl_2 Dichloro-methane	CH_2F_2 Difluoro-methane	CH_2O Formaldehyde	CH_3Cl Chloromethane	CH_3F Fluoromethane
265	1984	24.1	12.4	620	161	193	1571
270	2287	30.5	16.1	732	196	232	1813
275	2624	38.3	20.7	860	236	277	2084
280	3000	47.7	26.3	1004	283	327	2387
285	3418	58.8	33.0	1165	337	385	2724
290	3881	72.1	41.1	1346	399	450	3099
295	4393	87.6	50.8	1547	470	524	3516
300		105.9	62.1	1770	549	606	3978
Ref.	12	12,16	12	12	12	12	12

T/K	CH_4 Methane	CH_4O Methanol	C_2H_2 Acetylene	C_2H_4 Ethylene	C_2H_6 Ethane	C_2H_6O Dimethyl ether	C_3H_4 Propadiene
65	0.1						
70	0.3						
75	0.8						
80	2.1						
85	4.9						
90	10.6						
95	20.0						
100	34.5						
105	57.0						
110	88.4			0.3			
115	133			0.8	0.1		
120	192			1.4	0.4		
125	269			2.7	0.7		
130	368		0.1 s	4.5	1.3		
135	491		0.3 s	7.7	2.2		
140	642		0.7 s	11.9	3.8		
145	824		1.3 s	18.3	6.0		
150	1041		2.6 s	27.5	9.7		0.1
155	1297		4.6 s	39.9	15.0	0.1	0.2
160	1594		7.8 s	56.4	21.5	0.2	0.3
165	1937		12.8 s	77.9	31.0	0.3	0.6
170	2331		20.6 s	105	42.9	0.5	1.0
175	2779		32.2 s	140	59.0	0.9	1.7
180	3288		49.0 s	182	78.7	1.4	2.7
185	3865		72.9 s	234	104	2.1	4.1
190	4520		106 s	296	135	3.2	6.1
195			146	369	172	4.7	8.9
200			190	456	217	6.8	12.5
205			244	557	271	9.6	17.4
210			309	673	334	13.3	23.7
215			385	806	407	18.1	31.6
220			475	958	492	24.3	41.4
225			579	1128	590	32.1	53.5
230		0.1	699	1321	700	41.9	68.2
235		0.2	837	1535	826	53.9	85.8
240		0.4	993	1774	967	68.6	107
245		0.5	1170	2039	1125	86.3	131
250		0.8	1370	2331	1301	108	160
255		1.2	1593	2652	1496	133	193
260		1.7	1843	3005	1712	162	230
265		2.4	2121	3391	1949	197	273
270		3.3	2429	3813	2210	237	322
275		4.5	2771	4275	2495	283	376
280		6.2	3150		2806	335	438
285		8.3	3567		3146	395	506
290		11	4028		3515	463	582
295		14.4	4535		3917	538	666
300		18.7	5093		4355	623	759
Ref.	2,16	12	12,16	4	2	12	12

T/K	C_3H_6 Propylene	C_3H_8 Propane	C_4H_6 Buta-1,3-diene	C_4H_{10} Butane	C_4H_{10} Isobutane	C_5H_{12} Pentane	C_5H_{12} Neopentane
140	0.1						
145	0.2						
150	0.4						
155	0.7						
160	1.2	0.8			0.1		
165	2.0	1.4			0.1		
170	3.1	2.2	0.1	0.1	0.3		
175	4.7	3.3	0.2	0.2	0.4		
180	7.0	5.0	0.4	0.3	0.7		
185	10.1	7.3	0.6	0.5	1.1		0.1 s
190	14.2	10.5	1.0	0.8	1.7		0.2 s
195	19.7	15.0	1.5	1.3	2.5		0.4 s
200	26.9	20.1	2.3	1.9	3.7		0.7 s
205	35.9	27.0	3.4	2.8	5.3		1.1 s
210	47.3	36.0	4.8	4.0	7.4		1.6 s
215	61.3	47.0	6.7	5.7	10.2		2.4 s
220	78.5	60.0	9.2	7.8	13.8	1.0	3.6 s
225	99.2	77.0	12.5	10.6	18.3	1.5	5.2 s
230	124	97.0	16.7	14.1	24.0	2.1	7.3 s
235	153	120	21.9	18.5	31.1	3.0	10.2 s
240	188	148	28.4	24.1	39.8	4.2	13.9 s
245	228	180	36.3	30.9	50.3	5.7	18.7 s
250	274	218	46.0	39.1	62.9	7.6	24.8 s
255	327	261	57.6	49.1	77.8	10.0	32.4 s
260	387	311	71.3	61.0	95.4	13.0	41.6
265	456	367	87.6	75.0	116	16.6	51.4
270	533	431	107	91.5	140	21.1	63.0
275	619	502	129	111	167	26.6	76.6
280	715	582	154	133	198	33.1	92.3
285	822	671	184	159	234	40.8	111
290	940	769	217	188	274	50.0	131
295	1069	878	255	221	319	60.7	155
300	1212	998	297	258	370	73.2	182
Ref.	7	2	12	2	2	14	12,16

References

1. B. A. Younglove, Thermophysical properties of fluids. I. Ethylene, parahydrogen, nitrogen trifluoride, and oxygen, *J. Phys. Chem. Ref. Data*, 11, Supp. 1, 1982.
2. B. A. Younglove and J. F. Ely, Thermophysical properties of fluids. II. Methane, ethane, propane, isobutane, and normal butane, *J. Phys. Chem. Ref. Data*, 16, 577, 1987.
3. W. Wagner, et al., *International Tables for the Fluid State: Oxygen*, Blackwell Scientific Publications, Oxford, 1987.
4. R. T. Jacobsen, et al., *International Tables for the Fluid State: Ethylene*, Blackwell Scientific Publications, Oxford, 1988.
5. S. Angus, et al., *International Tables for the Fluid State: Chlorine*, Pergamon Press, Oxford, 1985.
6. S. Angus, et al., *International Tables for the Fluid State: Carbon Dioxide*, Pergamon Press, Oxford, 1976.
7. S. Angus, et al., *International Tables for the Fluid State: Propylene*, Pergamon Press, Oxford, 1980.
8. R. B. Stewart and R. T. Jacobsen, Thermophysical properties of argon, *J. Phys. Chem. Ref. Data*, 18, 639, 1989.
9. R. D. Goodwin, Carbon monoxide thermophysical properties, *J. Phys. Chem. Ref. Data*, 14, 849, 1985.
10. R. D. Goodwin, Methanol thermophysical properties, *J. Phys. Chem. Ref. Data*, 16, 799, 1987.
11. L. Haar, Thermodynamic properties of ammonia, *J. Phys. Chem. Ref. Data*, 7, 635, 1978.
12. DIPPR Data Compilation of Pure Compound Properties, Design Institute for Physical Properties Data, American Institute of Chemical Engineers, 1987.
13. V. A. Rabinovich, et al., *Thermophysical Properties of Neon, Argon, Krypton, and Xenon*, Hemisphere Publishing Corp., New York, 1987.
14. K. N. Marsh, *Recommended Reference Methods for the Realization of Physicochemical Properties*, Blackwell Scientific Publications, Oxford, 1987.
15. TRC Thermodynamic Tables: Non-Hydrocarbons, Thermodynamic Research Center, Texas A & M University, College Station, Texas, 1985.
16. R. M. Stevenson and S. Malanowski, *Handbook of the Thermodynamics of Organic Compounds*, Elsevier, New York, 1987.
17. S. Angus and K. M. de Reuck, *International Tables of the Fluid State: Helium-4*, Pergamon Press, Oxford, 1977.
18. R. D. McCarty, *J. Phys. Chem. Ref. Data*, 2, 923, 1973.
19. R. Span and W. Wagner, *J. Phys. Chem. Ref. Data*, 25, 1509, 1996.

VAPOR PRESSURE OF SATURATED SALT SOLUTIONS

This table gives the vapor pressure of water above saturated solutions of some common salts at ambient temperatures. Data on pure water are given on the last line for comparison.

The references provide additional information on water activity, osmotic coefficient, and enthalpy of vaporization.

References

1. Apelblat, A., *J. Chem. Thermodynamics*, 24, 619, 1992.
2. Apelblat, A., *J. Chem. Thermodynamics*, 25, 63, 1993.
3. Apelblat, A., *J. Chem. Thermodynamics*, 25, 1513, 1993.
4. Apelblat, A. and Korin, E., *J. Chem. Thermodynamics*, 30, 59, 1998.

Vapor Pressure in kPa

Salt	10 °C	15 °C	20 °C	25 °C	30 °C	35 °C	40 °C	Ref.
$BaCl_2$	0.971	1.443	2.073	2.887	3.903	5.133	6.576	1
$Ca(NO_3)_2$	0.701	1.015	1.381	1.772	2.154	2.487		1
$CuSO_4$	1.113	1.574	2.189	2.996	4.037	5.363		3
$FeSO_4$	0.978	1.516	2.208	3.035	3.950	4.884		3
KBr	0.953	1.338	1.853	2.533	3.419	4.563		3
KIO_3	1.100	1.564	2.177	2.970	3.979	5.236	6.778	4
K_2CO_3	0.541	0.802	1.134	1.536	1.997	2.499	3.016	1
LiCl	0.128	0.193	0.279	0.384				2
$Mg(NO_3)_2$	0.726	0.999	1.339	1.749	2.231	2.782	3.397	1
$MnCl_2$	0.697	1.064	1.515	2.020	2.535	3.002		3
NH_4Cl	0.971	1.328	1.836	2.481				2
NH_4NO_3	0.853	1.152	1.524	1.972				2
$(NH_4)_2SO_4$	0.901	1.319	1.871	2.573	3.439	4.474		3
NaBr	0.722	1.004	1.376	1.858	2.475	3.255	4.229	4
NaCl	0.921	1.285	1.768	2.401	3.218	4.262	5.581	4
$NaNO_2$	0.703	0.994	1.381	1.888	2.540	3.368	4.403	4
$NaNO_3$	0.884	1.244	1.719	2.335	3.121	4.109	5.333	4
RbCl	0.862	1.215	1.684	2.298	3.088	4.089	5.343	4
$ZnSO_4$	0.945	1.401	1.986	2.698	3.523	4.431	5.382	1
Water	1.228	1.706	2.339	3.169	4.246	5.627	7.381	

IUPAC RECOMMENDED DATA FOR VAPOR PRESSURE CALIBRATION

These precise vapor pressure values are recommended as secondary standards. Values are given in kPa (1 kPa = 0.0098692 atm = 7.5006 Torr). Reprinted by permission of IUPAC.

References

1. Marsh, K.N., Ed., *Recommended Reference Materials for the Realization of Physicochemical Properties*, Blackwell Scientific Publications, Oxford, 1987.
2. Ruzicka, K., Fulem, M., and Ruzicka, V., "Recommended Vapor Pressure of Solid Naphthalene," *J. Chem. Eng. Data*, 50, 1956–1970, 2005.

T/K	CO_2(s)	H_2O(s)	$C_{10}H_8$(s)	*n*–C_5H_{12}	C_6H_6	C_6F_6	H_2O	Hg
180	27.62		$3.04 \cdot 10^{-11}$					
190	68.44		$4.33 \cdot 10^{-10}$					
200	155.11	0.0002	$4.69 \cdot 10^{-9}$					
210	327.17	0.0007	$4.02 \cdot 10^{-8}$					
220		0.0026	$2.82 \cdot 10^{-7}$					
230		0.0089	$1.66 \cdot 10^{-6}$					
240		0.0273	$8.37 \cdot 10^{-6}$					
250		0.0760	$3.69 \cdot 10^{-5}$	7.60				
260		0.1958	$1.446 \cdot 10^{-4}$	12.98				
270		0.4701	$5.09 \cdot 10^{-4}$	21.15			0.485	
280			0.00163	33.11	5.148	4.322	0.991	
290			0.004798	50.01	8.606	7.463	1.919	
300			0.01308	73.17	13.816	12.328	3.535	
310			0.03328	104.07	21.389	19.576	6.228	
320			0.07956	144.3	32.054	30.009	10.540	
330			0.1797	195.7	46.656	44.578	17.202	
340			0.3854	260.1	66.152	64.380	27.167	
350			0.7884	339.4	91.609	90.664	41.647	
360				435.9	124.192	124.816	62.139	
370				551.5	165.2	168.4	90.453	
380				688.8	215.9	223.0	128.74	
390				850.2	277.7	290.4	179.48	
400				1038	353.2	372.6	245.54	0.138
410				1256	441.0	471.5	330.15	0.215
420				1507	545.5	589.3	436.89	0.329
430				1793	667.6	728.3	569.73	0.493
440				2120	808.8	890.9	732.99	0.724
450				2490	971.1	1080	931.36	1.045
460				2910	1156	1297	1169.9	1.485
470					1366	1547	1453.9	2.078
480					1602	1833	1789.0	2.866
490					1868	2159	2181.4	3.899
500					2164	2530	2637.3	5.239
510					2494	2954	3163.3	6.955
520					2861		3766.4	9.131
530					3267		4453.9	11.861
540					3717		5233.5	15.256
550					4216		6113.4	19.438
560					4770		7102.0	25.547
570							8208.6	30.74
580							9443.0	38.19
590							10816	47.09
600							12339	57.64

ENTHALPY OF VAPORIZATION

The molar enthalpy (heat) of vaporization $\Delta_{vap}H$, which is defined as the enthalpy change in the conversion of one mole of liquid to gas at constant temperature, is tabulated here for about 950 inorganic and organic compounds. Values are given, when available, both at the normal boiling point t_b, referred to a pressure of 101.325 kPa (760 mmHg), and at 25 °C.

The values in this table were measured either by calorimetric techniques or by application of the Claperyon equation to the variation of vapor pressure with temperature. See Reference 1 for a discussion of the accuracy of different experimental techniques and methods of estimating enthalpy of vaporization at other temperatures. Several of the references present empirical techniques for correlating enthalpy of vaporization with molecular structure.

Compounds are listed by systematic name, with compounds not containing carbon preceding those that do contain carbon. To locate a compound by molecular formula or CAS Registry Number, use the indexes to the table "Physical Constants of Organic Compounds" in Section 3, which point to the entry in that table from which the name can be determined.

References

1. Majer, V., and Svoboda, V., *Enthalpies of Vaporization of Organic Compounds*, Blackwell Scientific Publications, Oxford, 1985.
2. Chase, M. W., Davies, C. A., Downey, J. R., Frurip, D. J., McDonald, R. A., and Syverud, A. N., *JANAF Thermochemical Tables, Third Edition*, *J. Phys. Chem. Ref. Data*, 14, Suppl. 1, 1985.
3. *Landolt-Börnstein, Numerical Data and Functional Relationships in Science and Technology, Sixth Edition*, II/4, *Caloric Quantities of State*, Springer-Verlag, Heidelberg, 1961.
4. Daubert, T. E., Danner, R. P., Sibul, H. M., and Stebbins, C. C., *Physical and Thermodynamic Properties of Pure Compounds: Data Compilation*, extant 1994 (core with 4 supplements), Taylor & Francis, Bristol, PA.
5. Ruzicka, K., and Majer, V., Simultaneous Treatment of Vapor Pressures and Related Thermal Data Between the Triple and Normal Boiling Temperatures for *n*-Alkanes C5–C20, *J. Phys. Chem. Ref. Data*, 23, 1, 1994.
6. Verevkin, S. P., Thermochemistry of Amines: Experimental Standard Molar Enthalpies of Formation of Some Aliphatic and Aromatic Amines, *J. Chem. Thermodynamics*, 29, 891, 1997.
7. Cady, G. H., and Hargreaves, G. B., The Vapor Pressure of Some Heavy Transition Metal Hexafluorides, *J. Chem. Soc.*, 1563, 1578, 1961.
8. Steele, W. V., Chirico, R. D., Knipmeyer, S. E., and Nguyen, A., *J. Chem. Eng. Data* 41, 1255, 1996.
9. Nichols, G., et al., *J. Chem. Eng. Data* 51, 475, 2006.
10. Dias, A. M. A., et al., *J. Chem. Eng. Data* 50, 1328, 2005.
11. Umnahanant, P., et al., *J. Chem. Eng. Data* 51, 2246, 2006.
12. Raganov, G. N., Pisarev, P. N., and Emel'yanenko, V. N., *J. Chem. Eng. Data* 50, 1114, 2005.
13. Verevkin, S. P., et al., *J. Chem. Eng. Data* 51, 1896, 2006.
14. Verevkin, S. P., *J. Chem. Thermodynamics* 38, 1111, 2006.

Name	Mol. Form.	t_b °C	$\Delta_{vap}H(t_b)$ kJ/mol	$\Delta_{vap}H$(25 °C) kJ/mol
Compounds not containing carbon				
Aluminum	Al	2519	294	
Aluminum borohydride	AlB_3H_{12}	44.5	30	
Aluminum bromide	$AlBr_3$	255	23.5	
Aluminum iodide	AlI_3	382	32.2	
Ammonia	H_3N	–33.33	23.33	19.86
Antimony(III) bromide	Br_3Sb	288	59	
Antimony(III) chloride	Cl_3Sb	220.3	45.19	
Antimony(III) iodide	I_3Sb	400	68.6	
Argon	Ar	–185.85	6.43	
Arsenic(III) bromide	$AsBr_3$	221	41.8	
Arsenic(III) chloride	$AsCl_3$	130	35.01	
Arsenic(III) fluoride	AsF_3	57.13	29.7	
Arsenic(V) fluoride	AsF_5	–52.8	20.8	
Arsenic(III) iodide	AsI_3	424	59.3	
Arsine	AsH_3	–62.5	16.69	
Barium	Ba	1897	140	
Beryllium chloride	$BeCl_2$	482	105	
Beryllium iodide	BeI_2	590	70.5	
Bismuth	Bi	1564	151	
Bismuth tribromide	$BiBr_3$	462	75.4	
Bismuth trichloride	$BiCl_3$	441	72.61	
Boron	B	4000	480	
Boron tribromide	BBr_3	91.3	30.5	
Boron trichloride	BCl_3	12.5	23.77	23.1
Boron trifluoride	BF_3	–99.9	19.33	
Boron triiodide	BI_3	209.5	40.5	

Name	Mol. Form.	t_b °C	$\Delta_{vap}H(t_b)$ kJ/mol	$\Delta_{vap}H(25\ °C)$ kJ/mol
Bromine	Br_2	58.8	29.96	30.91
Bromine fluoride	BrF	20	25.1	
Bromine pentafluoride	BrF_5	41.3	30.6	
Bromine trifluoride	BrF_3	125.8	47.57	
Bromosilane	BrH_3Si	1.9	24.4	
Cadmium	Cd	767	99.87	
Cadmium bromide	Br_2Cd	863	115	
Cadmium chloride	$CdCl_2$	964	124.3	
Cadmium fluoride	CdF_2	1750	214	
Cadmium iodide	CdI_2	744	115	
Chlorine	Cl_2	−34.04	20.41	17.65
Chlorine dioxide	ClO_2	11	30	
Chlorine fluoride	ClF	−101.1	24	
Chlorine monoxide	Cl_2O	2.2	25.9	
Chlorine trifluoride	ClF_3	11.75	27.53	
Chlorosilane	ClH_3Si	−30.4	21	
Chlorotrifluorosilane	ClF_3Si	−70.0	18.7	
Chromium(II) chloride	Cl_2Cr	1120	197	
Chromium(VI) dichloride dioxide	Cl_2CrO_2	117	35.1	
Diborane	B_2H_6	−92.49	14.28	
Dibromosilane	Br_2H_2Si	66	31	
Dichlorodifluorosilane	Cl_2F_2Si	−32	21.2	
Dichlorosilane	Cl_2H_2Si	8.3	25	24.2
Difluorine dioxide	F_2O_2	−57	19.1	
Difluorosilane	F_2H_2Si	−77.8	16.3	
Digermane	Ge_2H_6	29	25.1	
Diphosphine	H_4P_2	63.5	28.8	
Disilane	H_6Si_2	−14.8	21.2	
Fluorine	F_2	−188.12	6.62	
Fluorine monoxide	F_2O	−144.3	11.09	
Fluorosilane	FH_3Si	−98.6	18.8	
Gallium	Ga	2204	254	
Gallium(III) bromide	Br_3Ga	279	38.9	
Gallium(III) chloride	Cl_3Ga	201	23.9	
Gallium(III) iodide	GaI_3	340	56.5	
Germane	GeH_4	−88.1	14.06	
Germanium	Ge	2833	334	
Germanium(IV) bromide	Br_4Ge	186.35	41.4	
Germanium(IV) chloride	Cl_4Ge	86.55	27.9	
Gold	Au	2856	324	
Helium	He	−268.93	0.08	
Hydrazine	H_4N_2	113.55	41.8	44.7
Hydrazoic acid	HN_3	35.7	30.5	
Hydrogen	H_2	−252.76	0.90	
Hydrogen bromide	BrH	−66.38		12.69
Hydrogen chloride	ClH	−85	16.15	9.08
Hydrogen disulfide	H_2S_2	70.7		33.78
Hydrogen iodide	HI	−35.55	19.76	17.36
Hydrogen peroxide	H_2O_2	150.2		51.6
Hydrogen selenide	H_2Se	−41.25	19.7	
Hydrogen sulfide	H_2S	−59.55	18.67	14.08
Hydrogen telluride	H_2Te	−2	19.2	
Indium(I) bromide	$BrIn$	656	92	
Indium(I) iodide	IIn	712	90.8	
Iodine	I_2	184.4	41.57	

Name	Mol. Form.	t_b °C	$\Delta_{vap}H(t_b)$ kJ/mol	$\Delta_{vap}H(25\,°C)$ kJ/mol
Iodine pentafluoride	F_5I	100.5	41.3	
Iridium(VI) fluoride	F_6Ir	53.6	30.9	
Krypton	Kr	–153.34	9.08	
Lead	Pb	1749	179.5	
Lead(II) bromide	Br_2Pb	892	133	
Lead(II) chloride	Cl_2Pb	951	127	
Lead(II) fluoride	F_2Pb	1293	160.4	
Lead(II) iodide	I_2Pb	872	104	
Lithium fluoride	FLi	1673	147	
Lithium hydroxide	HLiO	1626	188	
Mercury	Hg	356.62	59.11	
Mercury(II) bromide	Br_2Hg	318	58.89	
Mercury(II) chloride	Cl_2Hg	304	58.9	
Mercury(II) iodide	HgI_2	351	59.2	
Molybdenum(V) chloride	Cl_5Mo	268	62.8	
Molybdenum(V) fluoride	F_5Mo	213.6	51.8	
Molybdenum(VI) fluoride	F_6Mo	34.0	29.0	
Molybdenum(VI) oxide	MoO_3	1155	138	
Molybdenum(VI) oxytetrafluoride	F_4MoO	186.0	50.6	
Neon	Ne	–246.05	1.71	
Niobium(V) chloride	Cl_5Nb	247.4	52.7	
Niobium(V) fluoride	F_5Nb	234	52.3	
Nitric acid	HNO_3	83		39.1
Nitric oxide	NO	–151.74	13.83	
Nitrogen	N_2	–195.79	5.57	
Nitrogen tetroxide	N_2O_4	21.15	38.12	
Nitrogen trifluoride	F_3N	–128.75	11.56	
Nitrosyl chloride	ClNO	–5.5	25.78	
Nitrosyl fluoride	FNO	–59.9	19.28	
Nitrous oxide	N_2O	–88.48	16.53	
Nitryl chloride	$ClNO_2$	–15	25.7	
Nitryl fluoride	FNO_2	–72.4	18.05	
Osmium(V) fluoride	F_5Os	233	65.6	
Osmium(VI) fluoride	F_6Os	47.5	28.1	
Oxygen	O_2	–182.95	6.82	
Pentaborane(11)	B_5H_{11}	65	31.8	
Perchloryl fluoride	$ClFO_3$	–46.75	19.33	
Phosphine	H_3P	–87.75	14.6	
Phosphorothioc trifluoride	F_3PS	–52.25	19.6	
Phosphorus	P	280.5	12.4	14.2
Phosphorus(III) bromide	Br_3P	173.2	38.8	
Phosphorus(III) chloride	Cl_3P	76	30.5	32.1
Phosphorus(III) chloride difluoride	ClF_2P	–47.3	17.6	
Phosphorus(III) dichloride fluoride	Cl_2FP	13.85	24.9	
Phosphorus(III) fluoride	F_3P	–101.8	16.5	
Phosphorus(V) fluoride	F_5P	–84.6	17.2	
Phosphorus(III) iodide	I_3P	227	43.9	
Phosphoryl bromide	Br_3OP	191.7	38	
Phosphoryl chloride	Cl_3OP	105.5	34.35	38.6
Rhenium(VII) dioxytrifluoride	F_3O_2Re	185.4	65.7	
Rhenium(V) fluoride	F_5Re	221.3	58.1	
Rhenium(VI) fluoride	F_6Re	33.8	28.7	
Rhenium(VI) oxytetrafluoride	F_4ORe	171.7	61.0	
Selenium	Se	685	95.48	
Selenium tetrafluoride	F_4Se	101.6	47.2	

Name	Mol. Form.	t_b °C	$\Delta_{vap}H(t_b)$ kJ/mol	$\Delta_{vap}H$(25 °C) kJ/mol
Silane	H_4Si	−111.9	12.1	
Silver(I) bromide	AgBr	1502	198	
Silver(I) chloride	AgCl	1547	199	
Silver(I) iodide	AgI	1506	143.9	
Sodium hydroxide	HNaO	1388	175	
Stannane	H_4Sn	−51.8	19.05	
Stibine	H_3Sb	−17	21.3	
Sulfur	S	444.61	45	
Sulfur dioxide	O_2S	−10.05	24.94	22.92
Sulfur hexafluoride	F_6S			8.99
Sulfur tetrafluoride	F_4S	−40.45	26.44	
Sulfur trioxide	O_3S	44.5	40.69	43.14
Sulfuryl chloride	Cl_2O_2S	69.4	31.4	30.1
Tantalum(V) bromide	Br_5Ta	348.8	62.3	
Tantalum(V) chloride	Cl_5Ta	239	54.8	
Tantalum(V) fluoride	F_5Ta	229.5	56.9	
Tellurium	Te	988	114.1	
Tellurium tetrachloride	Cl_4Te	387	77	
Tetraborane(10)	B_4H_{10}	18	27.1	
Tetrabromosilane	Br_4Si	154	37.9	
Tetrachlorosilane	Cl_4Si	57.65	28.7	29.7
Tetrafluorodiborane	B_2F_4	−34.0	28	
Tetrafluorohydrazine	F_4N_2	−74	13.27	
Tetraiodosilane	I_4Si	287.35	50.2	
Thallium(I) bromide	BrTl	819	99.56	
Thallium(I) chloride	ClTl	720	102.2	
Thallium(I) iodide	ITl	824	104.7	
Thallium(I) sulfide	STl_2	1367	154	
Thionitrosyl fluoride (NSF)	FNS	4.8	22.2	
Thionyl chloride	Cl_2OS	75.6	31.7	31
Thionyl fluoride	F_2OS	−43.8	21.8	
Thorium(IV) chloride	Cl_4Th	921	146.4	
Thorium(IV) fluoride	F_4Th	1680	258	
Tin(II) bromide	Br_2Sn	639	102	
Tin(IV) bromide	Br_4Sn	205	43.5	
Tin(II) chloride	Cl_2Sn	623	86.8	
Tin(IV) chloride	Cl_4Sn	114.15	34.9	
Tin(II) iodide	I_2Sn	714	105	
Tin(IV) iodide	I_4Sn	364.35	56.9	
Titanium(IV) bromide	Br_4Ti	233.5	44.37	
Titanium(II) chloride	Cl_2Ti	1500	232	
Titanium(III) chloride	Cl_3Ti	960	124	
Titanium(IV) chloride	Cl_4Ti	136.45	36.2	
Titanium(IV) iodide	I_4Ti	377	58.4	
Tribromosilane	Br_3HSi	109	34.8	
Trichlorosilane	Cl_3HSi	33		25.7
Trifluorosilane	F_3HSi	−95	16.2	
Trigermane	Ge_3H_8	110.5	32.2	
Trisilane	H_8Si_3	52.9	28.5	
Tungsten(VI) chloride	Cl_6W	337	52.7	
Tungsten(VI) fluoride	F_6W	17.1	26.5	
Tungsten(VI) oxytetrachloride	Cl_4OW	230	67.8	
Tungsten(VI) oxytetrafluoride	F_4OW	185.9	59.5	
Vanadium(IV) chloride	Cl_4V	151	41.4	42.5
Vanadium(V) fluoride	F_5V	48.3	44.52	

Name	Mol. Form.	t_b °C	$\Delta_{vap}H(t_b)$ kJ/mol	$\Delta_{vap}H(25\ °C)$ kJ/mol
Vanadyl trichloride	Cl_3OV	127	36.78	
Water	H_2O	99.97	40.65	43.98
Xenon	Xe	–108.09	12.57	
Zinc bromide	Br_2Zn	670	118	
Zinc chloride	Cl_2Zn	732	126	
Zinc fluoride	F_2Zn	1500	190.1	
Compounds containing carbon				
Acetaldehyde	C_2H_4O	20.1	25.76	25.47
Acetic acid	$C_2H_4O_2$	117.9	23.70	23.36
Acetic anhydride	$C_4H_6O_3$	139.5	38.2	
Acetone	C_3H_6O	56.05	29.10	30.99
Acetonitrile	C_2H_3N	81.65	29.75	32.94
Acetophenone	C_8H_8O	202	43.98	55.40
Acrolein	C_3H_4O	52.6	28.3	
Acrylonitrile	C_3H_3N	77.3	32.6	
Allyl acetate	$C_5H_8O_2$	103.5	36.3	
Allyl alcohol	C_3H_6O	97.4	40.0	
2-Amino-2-methyl-1-propanol	$C_4H_{11}NO$	165.5	50.6	
Aniline	C_6H_7N	184.17	42.44	55.83
Anisole	C_7H_8O	153.7	38.97	46.90
Azobutane	$C_8H_{18}N_2$			49.31
Azopropane	$C_6H_{14}N_2$	114		39.88
Benzaldehyde	C_7H_6O	178.8	42.5	
Benzene	C_6H_6	80.09	30.72	33.83
Benzenethiol	C_6H_6S	169.1	39.93	47.56
Benzonitrile	C_7H_5N	191.1	45.9	
Benzyl acetate	$C_9H_{10}O_2$	213	49.4	
Benzyl alcohol	C_7H_8O	205.31	50.48	
Benzylamine	C_7H_9N	185		60.16
N-Benzylaniline	$C_{13}H_{13}N$	306.5		79.6
Benzyl benzoate	$C_{14}H_{12}O_2$	323.5	53.6	
Bis(2-chloroethyl) ether	$C_4H_8Cl_2O$	178.5	45.2	
Bis(ethoxymethyl) ether	$C_6H_{14}O_3$	140.6	36.17	44.69
Bromobenzene	C_6H_5Br	156.06		44.54
1-Bromobutane	C_4H_9Br	101.6	32.51	36.64
2-Bromobutane	C_4H_9Br	91.3	30.77	34.41
Bromochloromethane	CH_2BrCl	68.0	30.0	
2-Bromo-2-chloro-1,1,1-trifluoroethane	$C_2HBrClF_3$	50.2	28.08	29.61
Bromoethane	C_2H_5Br	38.5	27.04	28.03
Bromoethene	C_2H_3Br	15.8	23.4	
1-Bromoheptane	$C_7H_{15}Br$	178.9		50.60
1-Bromohexane	$C_6H_{13}Br$	155.3		45.89
Bromomethane	CH_3Br	3.5	23.91	22.81
1-Bromo-2-methylpropane	C_4H_9Br	91.1	31.33	34.82
2-Bromo-2-methylpropane	C_4H_9Br	73.3	29.23	31.81
1-Bromonaphthalene	$C_{10}H_7Br$	281	39.3	
1-Bromooctane	$C_8H_{17}Br$	200.8		55.77
1-Bromopentane	$C_5H_{11}Br$	129.8	35.01	41.28
1-Bromopropane	C_3H_7Br	71.1	29.84	32.01
2-Bromopropane	C_3H_7Br	59.5	28.33	30.17
3-Bromopropene	C_3H_5Br	70.1	30.24	32.73
1,2-Butadiene	C_4H_6	10.9	24.02	23.21
1,3-Butadiene	C_4H_6	–4.41	22.47	20.86
Butanal	C_4H_8O	74.8	31.5	

Name	Mol. Form.	t_b °C	$\Delta_{vap}H(t_b)$ kJ/mol	$\Delta_{vap}H$(25 °C) kJ/mol
Butane	C_4H_{10}	–0.5	22.44	21.02
1,2-Butanediol	$C_4H_{10}O_2$	190.5	52.84	71.55
1,3-Butanediol	$C_4H_{10}O_2$	207.5	54.31	74.46
1,4-Butanediol	$C_4H_{10}O_2$	235		77.1
1,4-Butanedithiol	$C_4H_{10}S_2$	195.5		55.10
Butanenitrile	C_4H_7N	117.6	33.68	39.33
1-Butanethiol	$C_4H_{10}S$	98.5	32.23	36.63
2-Butanethiol	$C_4H_{10}S$	85.0	30.59	33.99
Butanoic acid	$C_4H_8O_2$	163.75		40.45
Butanoic anhydride	$C_8H_{14}O_3$	200	50.0	
1-Butanol	$C_4H_{10}O$	117.73	43.29	52.35
2-Butanol	$C_4H_{10}O$	99.51	40.75	49.72
2-Butanone	C_4H_8O	79.59	31.30	34.79
1-Butene	C_4H_8	–6.26	22.07	20.22
cis-2-Butene	C_4H_8	3.71	23.34	22.16
trans-2-Butene	C_4H_8	0.88	22.72	21.40
2-Butoxyethanol	$C_6H_{14}O_2$	168.4		56.59
Butyl acetate	$C_6H_{12}O_2$	126.1	36.28	43.86
tert-Butyl acetate	$C_6H_{12}O_2$	95.1	33.07	38.03
Butylamine	$C_4H_{11}N$	77.00	31.81	35.72
sec-Butylamine	$C_4H_{11}N$	62.73	29.92	32.85
tert-Butylamine	$C_4H_{11}N$	44.04	28.27	29.64
Butylbenzene	$C_{10}H_{14}$	183.31	38.87	50.8
sec-Butylbenzene	$C_{10}H_{14}$	173.3		48.1
tert-Butylbenzene	$C_{10}H_{14}$	169.1		47.6
Butylcyclohexane	$C_{10}H_{20}$	180		49.36
Butylcyclopentane	C_9H_{18}	156.6	36.16	45.89
Butylethylamine	$C_6H_{15}N$	107.5	33.97	40.15
Butyl ethyl ether	$C_6H_{14}O$	92.3	31.63	36.32
Butyl ethyl sulfide	$C_6H_{14}S$	144.3	37.01	44.51
Butyl formate	$C_5H_{10}O_2$	106.1	36.58	41.11
tert-Butyl isobutyl ether	$C_8H_{18}O$	112.0	33.11	40.5
Butyl methyl ether	$C_5H_{12}O$	70.16	29.55	32.37
sec-Butyl methyl ether	$C_5H_{12}O$	59.1	28.09	30.23
Butyl methyl sulfide	$C_5H_{12}S$	123.4	34.47	40.46
tert-Butyl methyl sulfide	$C_5H_{12}S$	98.9	31.47	35.84
Butyl propyl ether	$C_7H_{16}O$	118.1	33.72	40.22
Butyl vinyl ether	$C_6H_{12}O$	94	31.58	36.17
1-Butyne	C_4H_6	8.08	24.52	23.35
2-Butyne-1,4-diol	$C_4H_6O_2$	238		81.5
γ-Butyrolactone	$C_4H_6O_2$	204	52.2	
Camphor, (+)	$C_{10}H_{16}O$	207.4	59.5	
Carbon disulfide	CS_2	46	26.74	27.51
Carbon monoxide	CO	–191.5	6.04	
2-Chloroaniline	C_6H_6ClN	208.8	44.4	
Chlorobenzene	C_6H_5Cl	131.72	35.19	40.97
1-Chlorobutane	C_4H_9Cl	78.4	30.39	33.51
2-Chlorobutane	C_4H_9Cl	68.2	29.17	31.53
Chlorodifluoromethane	$CHClF_2$	–40.7	20.2	
Chloroethane	C_2H_5Cl	12.3	24.65	
2-Chloroethanol	C_2H_5ClO	128.6	41.4	
Chloroethene	C_2H_3Cl	–13.8	20.8	
1-Chloroheptane	$C_7H_{15}Cl$	160.4		47.66
1-Chlorohexane	$C_6H_{13}Cl$	135.1	35.67	42.83
Chloromethane	CH_3Cl	–24.09	21.40	18.92

Name	Mol. Form.	t_b °C	$\Delta_{vap}H(t_b)$ kJ/mol	$\Delta_{vap}H$(25 °C) kJ/mol
1-Chloro-3-methylbutane	$C_5H_{11}Cl$	98.9	32.02	36.24
1-Chloro-2-methylpropane	C_4H_9Cl	68.5	29.22	31.67
2-Chloro-2-methylpropane	C_4H_9Cl	50.9	27.55	28.98
1-Chloronaphthalene	$C_{10}H_7Cl$	259	52.1	
1-Chlorooctane	$C_8H_{17}Cl$	183.5		52.42
Chloropentafluorobenzene	C_6ClF_5	117.96	34.76	41.07
Chloropentafluoroethane	C_2ClF_5	−39.1	19.41	
1-Chloropentane	$C_5H_{11}Cl$	108.4	33.15	38.24
2-Chloropentane	$C_5H_{11}Cl$	97.0	31.79	36.03
1-Chloropropane	C_3H_7Cl	46.5	27.18	28.35
2-Chloropropane	C_3H_7Cl	35.7	26.30	26.90
3-Chloropropene	C_3H_5Cl	45.1	29.0	
2-Chlorotoluene	C_7H_7Cl	159.0	37.5	
4-Chlorotoluene	C_7H_7Cl	162.4	38.7	
Chlorotrifluoromethane	$CClF_3$	−81.37	15.8	
Cholesterol	$C_{27}H_{46}O$	360 dec		148.0
o-Cresol	C_7H_8O	191.04	45.19	
m-Cresol	C_7H_8O	202.27	47.40	61.71
p-Cresol	C_7H_8O	201.98	47.45	
Cyanogen	C_2N_2	−21.1	23.33	19.75
Cyclobutane	C_4H_8	12.6	24.19	23.51
Cyclobutanecarbonitrile	C_5H_7N	149.6	36.88	44.34
Cyclohexane	C_6H_{12}	80.73	29.97	33.01
Cyclohexanecarbonitrile	$C_7H_{11}N$	184		51.92
Cyclohexanethiol	$C_6H_{12}S$	158.8	37.06	44.57
Cyclohexanol	$C_6H_{12}O$	160.84		62.01
Cyclohexanone	$C_6H_{10}O$	155.43		45.06
Cyclohexene	C_6H_{10}	82.98	30.46	33.47
1-Cyclohexenecarbonitrile	C_7H_9N			53.55
Cyclohexylamine	$C_6H_{13}N$	134	36.14	43.67
Cyclohexylbenzene	$C_{12}H_{16}$	240.1		60.8
Cyclohexylcyclohexane	$C_{12}H_{22}$	238		57.98
Cyclopentane	C_5H_{10}	49.3	27.30	28.52
Cyclopentanecarbonitrile	C_6H_9N	170		43.43
Cyclopentanethiol	$C_5H_{10}S$	132.1	35.32	41.42
Cyclopentanol	$C_5H_{10}O$	140.42		57.05
Cyclopentanone	C_5H_8O	130.57	36.35	42.72
1-Cyclopentenecarbonitrile	C_6H_7N			44.98
Cyclopropane	C_3H_6	−32.81	20.05	16.93
Cyclopropanecarbonitrile	C_4H_5N	135.1	35.55	41.94
Cyclopropylbenzene	C_9H_{10}	173.6		50.22
Cyclopropyl methyl ketone	C_5H_8O	111.3	34.07	39.41
cis-Decahydronaphthalene	$C_{10}H_{18}$	195.8	41.0	
trans-Decahydronaphthalene	$C_{10}H_{18}$	187.3	40.2	
Decane	$C_{10}H_{22}$	174.15	39.58	51.42
1,10-Decanediol	$C_{10}H_{22}O_2$			120.0
Decanenitrile	$C_{10}H_{19}N$	243		66.84
1-Decanethiol	$C_{10}H_{22}S$	240.6		65.48
1-Decanol	$C_{10}H_{22}O$	231.1		81.50
1-Decene	$C_{10}H_{20}$	170.5		50.43
Decylbenzene	$C_{16}H_{26}$	293		78.2
1,4-Dibromobutane	$C_4H_8Br_2$	197		53.09
1,2-Dibromo-1-chloro-1,2,2-trifluoroethane	$C_2Br_2ClF_3$	93	31.17	35.04
1,2-Dibromoethane	$C_2H_4Br_2$	131.6	34.77	41.73
Dibromomethane	CH_2Br_2	97	32.92	36.97

Name	Mol. Form.	t_b °C	$\Delta_{vap}H(t_b)$ kJ/mol	$\Delta_{vap}H(25\,°C)$ kJ/mol
1,2-Dibromopropane	$C_3H_6Br_2$	141.9	35.61	41.67
1,3-Dibromopropane	$C_3H_6Br_2$	167.3		47.45
1,2-Dibromotetrafluoroethane	$C_2Br_2F_4$	47.35	27.03	28.39
Dibutylamine	$C_8H_{19}N$	159.6	38.44	49.45
Dibutyl ether	$C_8H_{18}O$	140.28	36.49	44.97
Di-*sec*-butyl ether	$C_8H_{18}O$	121.1	34.06	40.84
Di-*tert*-butyl ether	$C_8H_{18}O$	107.23	32.15	37.61
Dibutyl phthalate	$C_{16}H_{22}O_4$	340	79.2	
Dibutyl sulfide	$C_8H_{18}S$	185		52.96
Di-*tert*-butyl sulfide	$C_8H_{18}S$	149.1	33.26	43.76
o-Dichlorobenzene	$C_6H_4Cl_2$	180	39.66	50.21
m-Dichlorobenzene	$C_6H_4Cl_2$	173	38.62	48.58
p-Dichlorobenzene	$C_6H_4Cl_2$	174	38.79	49.0
1,2-Dichlorobutane	$C_4H_8Cl_2$	124.1	33.90	39.58
1,4-Dichlorobutane	$C_4H_8Cl_2$	161		46.36
Dichlorodifluoromethane	CCl_2F_2	−29.8	20.1	
1,1-Dichloroethane	$C_2H_4Cl_2$	57.3	28.85	30.62
1,2-Dichloroethane	$C_2H_4Cl_2$	83.5	31.98	35.16
1,1-Dichloroethene	$C_2H_2Cl_2$	31.6	26.14	26.48
cis-1,2-Dichloroethene	$C_2H_2Cl_2$	60.1	30.2	
trans-1,2-Dichloroethene	$C_2H_2Cl_2$	48.7	28.9	
1,1-Dichloro-1-fluoroethane	$C_2H_3Cl_2F$	32.0	26.06	26.48
Dichlorofluoromethane	$CHCl_2F$	8.9	25.2	
1,2-Dichloro-1,1,2,3,3,3-hexafluoropropane	$C_3Cl_2F_6$	34.1	26.28	26.93
1,2-Dichlorohexane	$C_6H_{12}Cl_2$	173		48.16
Dichloromethane	CH_2Cl_2	40	28.06	28.82
1,2-Dichloropentane	$C_5H_{10}Cl_2$	148.3	36.45	43.89
1,5-Dichloropentane	$C_5H_{10}Cl_2$	179		50.71
1,3-Dichloropropane	$C_3H_6Cl_2$	120.9	35.18	40.75
1,2-Dichloro-1,1,2,2-tetrafluoroethane	$C_2Cl_2F_4$	3.5	23.3	
Dicyclopropyl ketone	$C_7H_{10}O$	161		53.70
Diethanolamine	$C_4H_{11}NO_2$	268.8	65.2	
1,1-Diethoxyethane	$C_6H_{14}O_2$	102.25	36.28	43.20
1,2-Diethoxyethane	$C_6H_{14}O_2$	121.2	36.28	43.20
Diethoxymethane	$C_5H_{12}O_2$	88	31.33	35.65
Diethylamine	$C_4H_{11}N$	55.5	29.06	31.31
Diethyl carbonate	$C_5H_{10}O_3$	126		43.60
Diethyl disulfide	$C_4H_{10}S_2$	154.0	37.58	45.18
Diethylene glycol	$C_4H_{10}O_3$	245.8	52.3	
Diethylene glycol diethyl ether	$C_8H_{18}O_3$	188		58.40
Diethylene glycol dimethyl ether	$C_6H_{14}O_3$	162	36.17	44.69
Diethylene glycol monoethyl ether	$C_6H_{14}O_3$	196	47.5	
Diethylene glycol monomethyl ether	$C_5H_{12}O_3$	193	46.6	
Diethyl ether	$C_4H_{10}O$	34.5	26.52	27.10
Diethyl malonate	$C_7H_{12}O_4$	200	54.8	
Diethyl oxalate	$C_6H_{10}O_4$	185.7	42.0	
3,3-Diethylpentane	C_9H_{20}	146.3	34.61	42.0
Diethyl sulfide	$C_4H_{10}S$	92.1	31.77	35.80
o-Difluorobenzene	$C_6H_4F_2$	94	32.21	36.18
m-Difluorobenzene	$C_6H_4F_2$	82.6	31.10	34.59
p-Difluorobenzene	$C_6H_4F_2$	89	31.77	35.54
1,1-Difluoroethane	$C_2H_4F_2$	−24.05	21.56	19.08
2,3-Dihydrothiophene	C_4H_6S	112.1	33.24	37.74
2,5-Dihydrothiophene	C_4H_6S	122.4	34.83	39.95
Diiodomethane	CH_2I_2	182	42.5	

Name	Mol. Form.	t_b °C	$\Delta_{vap}H(t_b)$ kJ/mol	$\Delta_{vap}H(25\ °C)$ kJ/mol
Diisobutyl sulfide	$C_8H_{18}S$	171		48.71
Diisopentyl ether	$C_{10}H_{22}O$	172.5	35.1	
Diisopropylamine	$C_6H_{15}N$	83.9	30.40	34.61
Diisopropyl ether	$C_6H_{14}O$	68.4	29.10	32.12
Diisopropyl sulfide	$C_6H_{14}S$	120.0	33.80	39.60
Diketene	$C_4H_4O_2$	126.1	36.80	42.89
1,2-Dimethoxyethane	$C_4H_{10}O_2$	84.5	32.42	36.39
N,N-Dimethylacetamide	C_4H_9NO	165		50.24
Dimethylamine	C_2H_7N	6.88	26.40	25.05
2,4-Dimethylaniline	$C_8H_{11}N$	214		61.3
2,5-Dimethylaniline	$C_8H_{11}N$	214		61.7
N,N-Dimethylaniline	$C_8H_{11}N$	194.15		52.83
2,2-Dimethylbutane	C_6H_{14}	49.73	26.31	27.68
2,3-Dimethylbutane	C_6H_{14}	57.93	27.38	29.12
2,3-Dimethyl-2-butanethiol	$C_6H_{14}S$	126.1		39.3
3,3-Dimethyl-2-butanone	$C_6H_{12}O$	106.1	33.39	37.91
2,3-Dimethyl-1-butene	C_6H_{12}	55.6		29.18
3,3-Dimethyl-1-butene	C_6H_{12}	41.2		26.61
2,3-Dimethyl-2-butene	C_6H_{12}	73.3	29.64	32.51
1,1-Dimethylcyclohexane	C_8H_{16}	119.6	32.51	37.92
cis-1,2-Dimethylcyclohexane	C_8H_{16}	129.8	33.47	39.70
trans-1,2-Dimethylcyclohexane	C_8H_{16}	123.5	32.96	38.36
cis-1,3-Dimethylcyclohexane	C_8H_{16}	120.1	32.91	38.26
trans-1,3-Dimethylcyclohexane	C_8H_{16}	124.5	33.39	39.16
cis-1,4-Dimethylcyclohexane	C_8H_{16}	124.4	33.28	39.02
trans-1,4-Dimethylcyclohexane	C_8H_{16}	119.4	32.56	37.90
cis-1,3-Dimethylcyclopentane	C_7H_{14}	90.8	30.40	34.20
Dimethyl decanedioate	$C_{12}H_{22}O_4$			86.4
Dimethyl disulfide	$C_2H_6S_2$	109.74	33.78	37.86
Dimethyl ether	C_2H_6O	–24.8	21.51	18.51
N,N-Dimethylformamide	C_3H_7NO	153		46.89
Dimethyl glutarate	$C_7H_{12}O_4$	214		65.7
Dimethyl heptanedioate	$C_9H_{16}O_4$			73.5
2,6-Dimethyl-4-heptanol	$C_9H_{20}O$	174.5		65.17
2,6-Dimethyl-4-heptanone	$C_9H_{18}O$	169.4		50.92
2,2-Dimethylhexane	C_8H_{18}	106.86	32.07	37.28
2,3-Dimethylhexane	C_8H_{18}	115.62	33.17	38.78
2,4-Dimethylhexane	C_8H_{18}	109.5	32.51	37.76
2,5-Dimethylhexane	C_8H_{18}	109.12	32.54	37.85
3,3-Dimethylhexane	C_8H_{18}	111.97	32.31	37.53
3,4-Dimethylhexane	C_8H_{18}	117.73	33.24	38.97
Dimethyl 1,6-hexanedioate	$C_8H_{14}O_4$			69.0
2,5-Dimethyl-2,5-hexanediol	$C_8H_{18}O_2$	214		85.2
cis-2,2-Dimethyl-3-hexene	C_8H_{16}	105.5		36.86
trans-2,2-Dimethyl-3-hexene	C_8H_{16}	100.8		37.03
2,5-Dimethyl-3-hexyne-2,5-diol	$C_8H_{14}O_2$	205		82.8
1,1-Dimethylhydrazine	$C_2H_8N_2$	63.9	32.55	35.0
Dimethyl malonate	$C_5H_8O_4$	181.4		57.5
Dimethyl nonanedioate	$C_{11}H_{20}O_4$			82.3
2,4-Dimethyloctane	$C_{10}H_{22}$	156	36.47	47.13
Dimethyl octanedioate	$C_{10}H_{18}O_4$	268		78.1
Dimethyl oxalate	$C_4H_6O_4$	163.5		54.7
3,3-Dimethyloxetane	$C_5H_{10}O$	80.6	30.85	33.94
2,2-Dimethylpentane	C_7H_{16}	79.2	29.23	32.42
2,3-Dimethylpentane	C_7H_{16}	89.78	30.46	34.26

Name	Mol. Form.	t_b °C	$\Delta_{vap}H(t_b)$ kJ/mol	$\Delta_{vap}H(25\ °C)$ kJ/mol
2,4-Dimethylpentane	C_7H_{16}	80.49	29.55	32.88
3,3-Dimethylpentane	C_7H_{16}	86.06	29.62	33.03
2,2-Dimethyl-3-pentanone	$C_7H_{14}O$	125.6	36.09	42.34
2,4-Dimethyl-3-pentanone	$C_7H_{14}O$	125.4	34.64	41.51
2,4-Dimethyl-1-pentene	C_7H_{14}	81.6		33.03
4,4-Dimethyl-1-pentene	C_7H_{14}	72.5		31.13
2,4-Dimethyl-2-pentene	C_7H_{14}	83.4		34.19
cis-4,4-Dimethyl-2-pentene	C_7H_{14}	80.4		32.56
trans-4,4-Dimethyl-2-pentene	C_7H_{14}	76.7		32.81
2,2-Dimethylpropanenitrile	C_5H_9N	106.1	32.40	37.35
2,2-Dimethyl-1-propanethiol	$C_5H_{12}S$	103.7		36.4
2,3-Dimethylpyridine	C_7H_9N	161.12	39.08	47.82
2,4-Dimethylpyridine	C_7H_9N	158.38	38.53	47.49
2,5-Dimethylpyridine	C_7H_9N	156.98	38.68	47.04
2,6-Dimethylpyridine	C_7H_9N	144.01	37.46	45.34
3,4-Dimethylpyridine	C_7H_9N	179.10	39.99	50.50
3,5-Dimethylpyridine	C_7H_9N	171.84	39.46	49.33
Dimethyl succinate	$C_6H_{10}O_4$	196.4		61.0
Dimethyl sulfide	C_2H_6S	37.33	27.0	27.65
Dimethyl sulfoxide	C_2H_6OS	189	43.1	
1,3-Dioxane	$C_4H_8O_2$	106.1	34.37	39.09
1,4-Dioxane	$C_4H_8O_2$	101.5	34.16	38.60
Diphenyl ether	$C_{12}H_{10}O$	258.0	48.2	
1,2-Dipropoxyethane	$C_8H_{18}O_2$	163.2		50.62
Dipropylamine	$C_6H_{15}N$	109.3	33.47	40.04
Dipropyl ether	$C_6H_{14}O$	90.08	31.31	35.69
Dipropyl sulfide	$C_6H_{14}S$	142.9	36.60	44.21
1-Docosanol	$C_{22}H_{46}O$			135.9
Dodecane	$C_{12}H_{26}$	216.32	44.09	61.52
1,12-Dodecanediol	$C_{12}H_{26}O_2$			135
Dodecanenitrile	$C_{12}H_{23}N$	277		76.12
1-Dodecanol	$C_{12}H_{26}O$	260		90.8
1-Dodecene	$C_{12}H_{24}$	213.8		60.78
Dodecylbenzene	$C_{18}H_{30}$	328		86.6
Eicosane	$C_{20}H_{42}$	343	58.49	101.81
1-Eicosanol	$C_{20}H_{42}O$	356		125.9
1,2-Epoxybutane	C_4H_8O	63.4	30.3	
Ethane	C_2H_6	–88.6	14.69	5.16
1,2-Ethanediamine	$C_2H_8N_2$	117	37.98	44.98
1,2-Ethanediol	$C_2H_6O_2$	197.3	50.5	63.9
1,2-Ethanediol, diacetate	$C_6H_{10}O_4$	190		61.44
1,2-Ethanedithiol	$C_2H_6S_2$	146.1	37.93	44.68
Ethanethiol	C_2H_6S	35.0	26.79	27.30
Ethanol	C_2H_6O	78.29	38.56	42.32
Ethanolamine	C_2H_7NO	171	49.83	
Ethoxybenzene	$C_8H_{10}O$	169.81		51.04
2-Ethoxyethanol	$C_4H_{10}O_2$	135	39.22	48.21
2-Ethoxyethyl acetate	$C_6H_{12}O_3$	156.4	40.76	52.61
1-Ethoxy-2-methoxyethane	$C_5H_{12}O_2$	103.5	34.33	39.83
N-Ethylacetamide	C_4H_9NO	205		64.89
Ethyl acetate	$C_4H_8O_2$	77.11	31.94	35.60
Ethyl acrylate	$C_5H_8O_2$	99.4	34.7	
N-Ethylaniline	$C_8H_{11}N$	203.0		58.3
Ethylbenzene	C_8H_{10}	136.16	35.57	42.24
Ethyl butanoate	$C_6H_{12}O_2$	121.3	35.47	42.68

Name	Mol. Form.	t_b °C	$\Delta_{vap}H(t_b)$ kJ/mol	$\Delta_{vap}H(25\,°C)$ kJ/mol
2-Ethyl-1-butanol	$C_6H_{14}O$	147	43.2	
2-Ethyl-1-butene	C_6H_{12}	64.7		31.13
Ethyl chloroacetate	$C_4H_7ClO_2$	144.3	40.43	49.47
Ethylcyclobutane	C_6H_{12}	70.8	28.67	31.24
Ethylcyclohexane	C_8H_{16}	131.9	34.04	40.56
Ethylcyclopentane	C_7H_{14}	103.5	31.96	36.40
Ethyl dichloroacetate	$C_4H_6Cl_2O_2$	155		50.60
Ethyl 2,2-dimethylpropanoate	$C_7H_{14}O_2$	118.4	34.51	41.25
Ethylene	C_2H_4	−103.77	13.53	
N-Ethylformamide	C_3H_7NO	198		58.44
Ethyl formate	$C_3H_6O_2$	54.4	29.91	31.96
3-Ethylhexane	C_8H_{18}	118.6	33.59	39.64
Ethyl hexanoate	$C_8H_{16}O_2$	167		51.72
2-Ethylhexanoic acid	$C_8H_{16}O_2$	228		75.60
2-Ethyl-1-hexanol	$C_8H_{18}O$	184.6	54.2	68.51
2-Ethylhexyl acetate	$C_{10}H_{20}O_2$	199	43.5	
2-Ethylhexylamine	$C_8H_{19}N$	169.2	40.0	
Ethylisopropylamine	$C_5H_{13}N$	69.6	29.94	33.13
Ethyl isopropyl ether	$C_5H_{12}O$	54.1	28.21	30.08
Ethyl isopropyl sulfide	$C_5H_{12}S$	107.5	32.74	37.78
Ethyl 3-methylbutanoate	$C_7H_{14}O_2$	135.0	37.0	
2-Ethyl-3-methyl-1-butene	C_7H_{14}	89		34.35
1-Ethyl-1-methylcyclopentane	C_8H_{16}	121.6	33.20	38.85
3-Ethyl-2-methylpentane	C_8H_{18}	115.66	32.93	38.52
3-Ethyl-3-methylpentane	C_8H_{18}	118.27	32.78	37.99
3-Ethyl-2-methyl-1-pentene	C_8H_{16}	109.5		37.27
Ethyl 2-methylpropanoate	$C_6H_{12}O_2$	110.1	33.67	39.83
Ethyl methyl sulfide	C_3H_8S	66.7	29.53	31.85
3-Ethylpentane	C_7H_{16}	93.5	31.12	35.22
Ethyl pentanoate	$C_7H_{14}O_2$	146.1	36.96	47.01
3-Ethyl-3-pentanol	$C_7H_{16}O$	142		57.34
Ethyl pentyl ether	$C_7H_{16}O$	117.6	34.41	41.01
Ethyl propanoate	$C_5H_{10}O_2$	99.1	33.88	39.21
Ethyl propyl ether	$C_5H_{12}O$	63.21	28.94	31.43
Ethyl propyl sulfide	$C_5H_{12}S$	118.6	34.24	39.97
Ethyl trichloroacetate	$C_4H_5Cl_3O_2$	167.5		50.97
Ethyl vinyl ether	C_4H_8O	35.5	26.2	
Fluorobenzene	C_6H_5F	84.73	31.19	34.58
1-Fluorooctane	$C_8H_{17}F$	142.3	40.43	49.65
2-Fluorotoluene	C_7H_7F	115	35.4	
4-Fluorotoluene	C_7H_7F	116.6	34.08	39.42
Formamide	CH_3NO	220		60.15
Formic acid	CH_2O_2	101	22.69	20.10
Furan	C_4H_4O	31.5	27.10	27.45
Furfural	$C_5H_4O_2$	161.7	43.2	
Furfuryl alcohol	$C_5H_6O_2$	171	53.6	
Glycerol	$C_3H_8O_3$	290	61.0	
Glycerol triacetate	$C_9H_{14}O_6$	259		85.74
Heptadecane	$C_{17}H_{36}$	302.0	53.58	86.47
1-Heptadecanol	$C_{17}H_{36}O$	324		112.5
6-Heptadecanol	$C_{17}H_{36}O$			108.6
7-Heptadecanol	$C_{17}H_{36}O$			108.2
9-Heptadecanol	$C_{17}H_{36}O$			108.5
Heptane	C_7H_{16}	98.4	31.77	36.57
1,7-Heptanediol	$C_7H_{16}O_2$	262		97.9

Name	Mol. Form.	t_b °C	$\Delta_{vap}H(t_b)$ kJ/mol	$\Delta_{vap}H$(25 °C) kJ/mol
1-Heptanol	$C_7H_{16}O$	176.45		66.81
3-Heptanol	$C_7H_{16}O$	157	42.5	
2-Heptanone	$C_7H_{14}O$	151.05		47.24
1-Heptene	C_7H_{14}	93.64		35.49
cis-2-Heptene	C_7H_{14}	98.4		36.26
trans-2-Heptene	C_7H_{14}	98		36.27
cis-3-Heptene	C_7H_{14}	95.8		35.81
trans-3-Heptene	C_7H_{14}	95.7		35.84
Heptylamine	$C_7H_{17}N$	156		49.96
Heptylbenzene	$C_{13}H_{20}$	240		64.2
1-Hexacosanol	$C_{26}H_{54}O$			153.7
Hexadecane	$C_{16}H_{34}$	286.86	51.84	81.35
1,16-Hexadecanediol	$C_{16}H_{34}O_2$			163
1-Hexadecanol	$C_{16}H_{34}O$	312		107.7
1-Hexadecene	$C_{16}H_{32}$	284.9		80.25
Hexadecylbenzene	$C_{22}H_{38}$	385		104.8
Hexafluoroacetylacetone	$C_5H_2F_6O_2$	54.15	27.05	30.58
Hexafluorobenzene	C_6F_6	80.32	31.66	35.71
Hexafluoroethane	C_2F_6	–78.1	16.15	
Hexane	C_6H_{14}	68.73	28.85	31.56
1,6-Hexanediol	$C_6H_{14}O_2$	208		90.2
Hexanenitrile	$C_6H_{11}N$	163.65		47.91
1-Hexanol	$C_6H_{14}O$	157.6	44.50	61.61
2-Hexanol	$C_6H_{14}O$	140	41.01	58.46
2-Hexanone	$C_6H_{12}O$	127.6	36.35	43.14
3-Hexanone	$C_6H_{12}O$	123.5	35.36	42.47
1-Hexene	C_6H_{12}	63.48		30.61
cis-2-Hexene	C_6H_{12}	68.8		32.19
trans-2-Hexene	C_6H_{12}	67.9		31.60
cis-3-Hexene	C_6H_{12}	66.4		31.23
trans-3-Hexene	C_6H_{12}	67.1		31.55
Hexylamine	$C_6H_{15}N$	132.8	36.54	45.10
Hexylbenzene	$C_{12}H_{18}$	226.1		60.4
Hexyl methyl ether	$C_7H_{16}O$	126.1	34.93	42.07
Indan	C_9H_{10}	177.97	39.63	48.79
Iodobenzene	C_6H_5I	188.4	39.5	
1-Iodobutane	C_4H_9I	130.5	34.66	40.63
2-Iodobutane	C_4H_9I	120.1	33.27	38.46
Iodoethane	C_2H_5I	72.3	29.44	31.93
1-Iodohexane	$C_6H_{13}I$	181.3		49.75
Iodomethane	CH_3I	42.43	27.34	27.97
1-Iodo-2-methylpropane	C_4H_9I	121.1	33.54	38.83
2-Iodo-2-methylpropane	C_4H_9I	100.1	31.43	35.41
1-Iodopentane	$C_5H_{11}I$	157.0		45.27
1-Iodopropane	C_3H_7I	102.5	32.08	36.25
2-Iodopropane	C_3H_7I	89.5	30.68	34.06
Isobutane	C_4H_{10}	–11.73	21.30	19.23
Isobutyl acetate	$C_6H_{12}O_2$	116.5	35.9	
Isobutylamine	$C_4H_{11}N$	67.75	30.61	33.85
Isobutylbenzene	$C_{10}H_{14}$	172.79		48.0
Isobutyl formate	$C_5H_{10}O_2$	98.2	33.6	
Isobutyl isobutanoate	$C_8H_{16}O_2$	148.6	38.2	
Isobutyl methyl ether	$C_5H_{12}O$	58.6	28.02	30.13
Isopentane	C_5H_{12}	27.88	24.69	24.85
Isopentyl acetate	$C_7H_{14}O_2$	142.5	37.5	

Name	Mol. Form.	t_b °C	$\Delta_{vap}H(t_b)$ kJ/mol	$\Delta_{vap}H$(25 °C) kJ/mol
Isopentyl isopentanoate	$C_{10}H_{20}O_2$	190.4	45.9	
Isopropyl acetate	$C_5H_{10}O_2$	88.7	32.93	37.20
Isopropylamine	C_3H_9N	31.76	27.83	28.36
Isopropylbenzene	C_9H_{12}	152.41		45.13
Isopropylcyclohexane	C_9H_{18}	154.8		44.02
Isopropylcyclopentane	C_8H_{16}	126.5	33.56	39.44
Isopropylmethylamine	$C_4H_{11}N$	50.4	28.71	30.69
1-Isopropyl-4-methylbenzene	$C_{10}H_{14}$	177.1	38.2	
Isopropyl methyl ether	$C_4H_{10}O$	30.77	26.05	26.41
Isopropyl methyl sulfide	$C_4H_{10}S$	84.8	30.71	34.15
Isopropylpropylamine	$C_6H_{15}N$	96.9	32.14	37.23
Isopropyl propyl sulfide	$C_6H_{14}S$	132.1	35.11	41.78
Isoquinoline	C_9H_7N	243.22	49.0	60.26
Mesityl oxide	$C_6H_{10}O$	130	36.1	
Methane	CH_4	–161.48	8.19	
Methanol	CH_4O	64.6	35.21	37.43
2-Methoxyethanol	$C_3H_8O_2$	124.1	37.54	45.17
2-Methoxyethyl acetate	$C_5H_{10}O_3$	143	43.9	
Methyl acetate	$C_3H_6O_2$	56.87	30.32	32.29
Methyl acrylate	$C_4H_6O_2$	80.7	33.1	
2-Methylacrylonitrile	C_4H_5N	90.3	31.8	
Methylamine	CH_5N	–6.32	25.60	23.37
2-Methylaniline	C_7H_9N	200.3	44.6	
3-Methylaniline	C_7H_9N	203.3	44.9	
4-Methylaniline	C_7H_9N	200.4	44.3	
Methyl benzoate	$C_8H_8O_2$	199		55.57
1-Methylbicyclo[3,1,0]hexane	C_7H_{12}	93.1	31.07	34.77
3-Methylbutanenitrile	C_5H_9N	127.5	35.10	41.64
2-Methyl-1-butanethiol	$C_5H_{12}S$	119.1	33.79	39.45
2-Methyl-2-butanethiol	$C_5H_{12}S$	99.1	31.37	35.67
3-Methyl-2-butanethiol	$C_5H_{12}S$	109.8		37.5
Methyl butanoate	$C_5H_{10}O_2$	102.8	33.79	39.28
2-Methylbutanoic acid	$C_5H_{10}O_2$	177		46.91
2-Methyl-1-butanol	$C_5H_{12}O$	127.5		55.16
3-Methyl-1-butanol	$C_5H_{12}O$	131.1	44.07	55.61
2-Methyl-2-butanol	$C_5H_{12}O$	102.4	39.04	50.10
3-Methyl-2-butanol	$C_5H_{12}O$	112.9		53.0
3-Methyl-2-butanone	$C_5H_{10}O$	94.33	32.35	36.78
2-Methyl-1-butene	C_5H_{10}	31.2	25.50	25.92
3-Methyl-1-butene	C_5H_{10}	20.1		23.77
2-Methyl-2-butene	C_5H_{10}	38.56	26.31	27.06
(1-Methylbutyl)benzene	$C_{11}H_{16}$	199		53.0
Methyl *tert*-butyl ether	$C_5H_{12}O$	55.0	27.94	29.82
Methyl chloroacetate	$C_3H_5ClO_2$	129.5	39.23	46.73
Methyl cyanoacetate	$C_4H_5NO_2$	200.5	48.2	
Methyl cyclobutanecarboxylate	$C_6H_{10}O_2$	135.5	37.13	44.72
Methylcyclohexane	C_7H_{14}	100.93	31.27	35.36
1-Methylcyclohexanol	$C_7H_{14}O$	155	79.0	
cis-2-Methylcyclohexanol	$C_7H_{14}O$	165	48.5	
trans-2-Methylcyclohexanol	$C_7H_{14}O$	167.5	53.0	
Methylcyclopentane	C_6H_{12}	71.8	29.08	31.64
Methyl cyclopropanecarboxylate	$C_5H_8O_2$	114.9	35.25	41.27
2-Methyldecane	$C_{11}H_{24}$	189.3	40.25	54.28
4-Methyldecane	$C_{11}H_{24}$	187	40.70	53.76
Methyl dichloroacetate	$C_3H_4Cl_2O_2$	142.9	39.28	47.72

Name	Mol. Form.	t_b °C	$\Delta_{vap}H(t_b)$ kJ/mol	$\Delta_{vap}H(25\ °C)$ kJ/mol
Methyl 2,2-dimethylpropanoate	$C_6H_{12}O_2$	101.1	33.42	38.76
Methyl dodecanoate	$C_{13}H_{26}O_2$	267		77.17
N-Methylformamide	C_2H_5NO	199.51		56.19
Methyl formate	$C_2H_4O_2$	31.7	27.92	28.35
2-Methylheptane	C_8H_{18}	117.66	33.26	39.67
3-Methylheptane	C_8H_{18}	118.9	33.66	39.83
4-Methylheptane	C_8H_{18}	117.72	33.35	39.69
Methyl heptanoate	$C_8H_{16}O_2$	174		51.62
2-Methyl-2-heptanol	$C_8H_{18}O$	156		62.87
2-Methylhexane	C_7H_{16}	90.04	30.62	34.87
3-Methylhexane	C_7H_{16}	92	30.9	
Methyl hexanoate	$C_7H_{14}O_2$	149.5	38.55	48.04
2-Methyl-2-hexanol	$C_7H_{16}O$	143		58.57
5-Methyl-3-hexanol	$C_7H_{16}O$	147		59.82
cis-3-Methyl-3-hexene	C_7H_{14}	95.4		36.31
trans-3-Methyl-3-hexene	C_7H_{14}	93.5		35.70
Methylhydrazine	CH_6N_2	87.5	36.12	40.37
Methyl isobutanoate	$C_5H_{10}O_2$	92.5	32.61	37.32
Methyl methacrylate	$C_5H_8O_2$	100.5	36.0	
1-Methylnaphthalene	$C_{11}H_{10}$	244.7	45.5	
2-Methylnonane	$C_{10}H_{22}$	167.1	38.23	49.63
3-Methylnonane	$C_{10}H_{22}$	167.9	38.26	49.71
5-Methylnonane	$C_{10}H_{22}$	165.1	38.14	49.34
Methyl octanoate	$C_9H_{18}O_2$	192.9		56.41
Methyloxirane	C_3H_6O	35	27.35	27.89
2-Methylpentane	C_6H_{14}	60.26	27.79	29.89
3-Methylpentane	C_6H_{14}	63.27	28.06	30.28
2-Methyl-2,4-pentanediol	$C_6H_{14}O_2$	197.1	57.3	
Methyl pentanoate	$C_6H_{12}O_2$	127.4	35.36	43.10
2-Methyl-1-pentanol	$C_6H_{14}O$	149	50.2	
4-Methyl-1-pentanol	$C_6H_{14}O$	151.9	44.46	60.47
2-Methyl-2-pentanol	$C_6H_{14}O$	121.1	39.59	54.77
4-Methyl-2-pentanol	$C_6H_{14}O$	131.6	44.2	
3-Methyl-2-pentanone	$C_6H_{12}O$	117.5	34.16	40.53
4-Methyl-2-pentanone	$C_6H_{12}O$	116.5	34.49	40.61
2-Methyl-3-pentanone	$C_6H_{12}O$	113.5	33.84	39.79
2-Methyl-1-pentene	C_6H_{12}	62.1		30.48
3-Methyl-1-pentene	C_6H_{12}	54.2		28.62
4-Methyl-1-pentene	C_6H_{12}	53.9		28.71
2-Methyl-2-pentene	C_6H_{12}	67.3		31.60
3-Methyl-*cis*-2-pentene	C_6H_{12}	67.7		32.09
3-Methyl-*trans*-2-pentene	C_6H_{12}	70.4		31.35
4-Methyl-*cis*-2-pentene	C_6H_{12}	56.3		29.48
4-Methyl-*trans*-2-pentene	C_6H_{12}	58.6		29.97
Methyl pentyl ether	$C_6H_{14}O$	99	32.02	36.85
Methyl pentyl sulfide	$C_6H_{14}S$	145.1	37.41	45.24
2-Methylpropanenitrile	C_4H_7N	103.9	32.39	37.13
2-Methyl-1-propanethiol	$C_4H_{10}S$	88.5	31.01	34.63
2-Methyl-2-propanethiol	$C_4H_{10}S$	64.2	28.45	30.78
Methyl propanoate	$C_4H_8O_2$	79.8	32.24	35.85
2-Methylpropanoic acid	$C_4H_8O_2$	154.45		35.30
2-Methyl-1-propanol	$C_4H_{10}O$	107.89	41.82	50.82
2-Methyl-2-propanol	$C_4H_{10}O$	82.4	39.07	46.69
Methyl propyl ether	$C_4H_{10}O$	39.1	26.75	27.60
Methyl propyl sulfide	$C_4H_{10}S$	95.6	32.08	36.24

Name	Mol. Form.	t_b °C	$\Delta_{vap}H(t_b)$ kJ/mol	$\Delta_{vap}H(25\ °C)$ kJ/mol
2-Methylpyridine	C_6H_7N	129.38	36.17	42.48
3-Methylpyridine	C_6H_7N	144.14	37.35	44.44
4-Methylpyridine	C_6H_7N	145.36	37.51	44.56
2-Methylquinoline	$C_{10}H_9N$	246.5		66.1
4-Methylquinoline	$C_{10}H_9N$	262		67.6
6-Methylquinoline	$C_{10}H_9N$	258.6		67.7
8-Methylquinoline	$C_{10}H_9N$	247.5		65.7
Methyl salicylate	$C_8H_8O_3$	222.9	46.7	
4-Methylthiazole	C_4H_5NS	133.3	37.58	43.85
2-Methylthiophene	C_5H_6S	112.6	33.90	38.87
3-Methylthiophene	C_5H_6S	115.5	34.24	39.43
Methyl trichloroacetate	$C_3H_3Cl_3O_2$	153.8		48.33
Molybdenum carbonyl	C_6MoO_6	701	72.51	
Morpholine	C_4H_9NO	128	37.1	
Naphthalene	$C_{10}H_8$	217.9	43.2	
Neopentane	C_5H_{12}	9.48	22.74	21.84
Nitrobenzene	$C_6H_5NO_2$	210.8		55.01
Nitroethane	$C_2H_5NO_2$	114.0	38.0	
Nitromethane	CH_3NO_2	101.19	33.99	38.27
1-Nitropropane	$C_3H_7NO_2$	131.1	38.5	
2-Nitropropane	$C_3H_7NO_2$	120.2	36.8	
Nonadecane	$C_{19}H_{40}$	329.9	56.93	96.4
Nonane	C_9H_{20}	150.82	37.18	46.55
1,9-Nonanediol	$C_9H_{20}O_2$			112.5
1-Nonanol	$C_9H_{20}O$	213.37		76.86
2-Nonanone	$C_9H_{18}O$	195.3		56.44
5-Nonanone	$C_9H_{18}O$	188.45		53.30
Nonylbenzene	$C_{15}H_{24}$	280.5		74.1
Octadecane	$C_{18}H_{38}$	316.3	55.23	91.44
1-Octadecanol	$C_{18}H_{38}O$	335		116.8
cis-9-Octadecenoic acid	$C_{18}H_{34}O_2$	360	67.4	
Octane	C_8H_{18}	125.67	34.41	41.49
1,8-Octanediol	$C_8H_{18}O_2$			104.9
Octanenitrile	$C_8H_{15}N$	205.25		56.80
Octanoic acid	$C_8H_{16}O_2$	239	58.5	
1-Octanol	$C_8H_{18}O$	195.16		70.98
2-Octanol	$C_8H_{18}O$	179.3	44.4	
1-Octene	C_8H_{16}	121.29	34.07	40.34
Octylbenzene	$C_{14}H_{22}$	264		69.1
1-Octyne	C_8H_{14}	126.3	35.83	42.30
2-Octyne	C_8H_{14}	137.6	37.26	44.49
3-Octyne	C_8H_{14}	133.1	36.94	43.92
4-Octyne	C_8H_{14}	131.6	36.0	42.73
Oxetane	C_3H_6O	47.6	28.67	29.85
2-Oxetanone	$C_3H_4O_2$	162		47.03
Oxirane	C_2H_4O	10.6	25.54	24.75
Pentachloroethane	C_2HCl_5	162.0	36.9	
Pentadecane	$C_{15}H_{32}$	270.6	50.08	76.77
1,15-Pentadecanediol	$C_{15}H_{32}O_2$			139
1-Pentadecanol	$C_{15}H_{32}O$	300		103.5
Pentadecylbenzene	$C_{21}H_{36}$	373		100.3
Pentafluorobenzene	C_6HF_5	85.74	32.15	36.27
2,3,4,5,6-Pentafluorotoluene	$C_7H_3F_5$	117.5	34.75	41.12
2,2,4,6,6-Pentamethylheptane	$C_{12}H_{26}$	177.8		48.97
Pentane	C_5H_{12}	36.06	25.79	26.43

Name	Mol. Form.	t_b °C	$\Delta_{vap}H(t_b)$ kJ/mol	$\Delta_{vap}H$(25 °C) kJ/mol
1,5-Pentanediol	$C_5H_{12}O_2$	239	60.7	83.0
2,4-Pentanedione	$C_5H_8O_2$	138	34.30	41.77
Pentanenitrile	C_5H_9N	141.3	36.09	43.60
1-Pentanethiol	$C_5H_{12}S$	126.6	34.88	41.24
Pentanoic acid	$C_5H_{10}O_2$	186.1	44.1	
1-Pentanol	$C_5H_{12}O$	137.98	44.36	57.02
2-Pentanol	$C_5H_{12}O$	119.3	41.40	54.21
3-Pentanol	$C_5H_{12}O$	116.25		54.0
2-Pentanone	$C_5H_{10}O$	102.26	33.44	38.40
3-Pentanone	$C_5H_{10}O$	101.7	33.45	38.52
1-Pentene	C_5H_{10}	29.96	25.20	25.47
cis-2-Pentene	C_5H_{10}	36.93		26.86
trans-2-Pentene	C_5H_{10}	36.34		26.76
trans-3-Pentenenitrile	C_5H_7N	144	37.09	44.77
Pentyl acetate	$C_7H_{14}O_2$	149.2	38.42	48.56
Pentylamine	$C_5H_{13}N$	104.3	34.01	40.08
Pentylbenzene	$C_{11}H_{16}$	205.4		55.1
Pentylcyclohexane	$C_{11}H_{22}$	203.7		53.88
Perfluorobutane	C_4F_{10}	-1.9	22.9	
Perfluorocyclobutane	C_4F_8	-5.91	23.2	
Perfluorodecalin	$C_{10}F_{18}$	142.02		41.54
Perfluorohexane	C_6F_{14}	57.14		32.47
Perfluorononane	C_9F_{20}	117.61		45.27
Perfluorooctane	C_8F_{18}	105.9	33.38	41.13
Perfluorotoluene	C_7F_8	103.55		40.52
Phenanthrene	$C_{14}H_{10}$	340		75.50
Phenol	C_6H_6O	181.87	45.69	57.82
Piperidine	$C_5H_{11}N$	106.22		39.29
Propanal	C_3H_6O	48	28.31	29.62
Propane	C_3H_8	-42.1	19.04	14.79
1,3-Propanediamine	$C_3H_{10}N_2$	139.8	40.85	50.16
1,2-Propanediol	$C_3H_8O_2$	187.6	52.4	
1,3-Propanediol	$C_3H_8O_2$	214.4	57.9	69.8
1,3-Propanedithiol	$C_3H_8S_2$	172.9		49.66
Propanenitrile	C_3H_5N	97.14	31.81	36.03
1-Propanethiol	C_3H_8S	67.8	29.54	31.89
2-Propanethiol	C_3H_8S	52.6	27.91	29.45
Propanoic acid	$C_3H_6O_2$	141.15		32.14
Propanoic anhydride	$C_6H_{10}O_3$	170	41.7	
1-Propanol	C_3H_8O	97.2	41.44	47.45
2-Propanol	C_3H_8O	82.3	39.85	45.39
Propene	C_3H_6	-47.69	18.42	14.24
2-Propoxyethanol	$C_5H_{12}O_2$	149.8	41.40	52.12
Propyl acetate	$C_5H_{10}O_2$	101.3	33.92	39.72
Propylamine	C_3H_9N	47.22	29.55	31.27
Propylbenzene	C_9H_{12}	159.24		46.22
Propylcyclohexane	C_9H_{18}	156		45.08
Propylcyclopentane	C_8H_{16}	131	34.70	41.08
Propyl formate	$C_4H_8O_2$	80.9	33.61	37.53
Propyl propanoate	$C_6H_{12}O_2$	122.5	35.54	43.45
Pyridazine	$C_4H_4N_2$	208		53.47
Pyridine	C_5H_5N	115.23	35.09	40.21
Pyrimidine	$C_4H_4N_2$	123.8	43.09	49.79
Pyrrole	C_4H_5N	129.79	38.75	45.09
Pyrrolidine	C_4H_9N	86.56	33.01	37.52

Name	Mol. Form.	t_b °C	$\Delta_{vap}H(t_b)$ kJ/mol	$\Delta_{vap}H(25\,°C)$ kJ/mol
Quinoline	C_9H_7N	237.16	49.7	59.30
Salicylaldehyde	$C_7H_6O_2$	197	38.2	
Spiro[2.2]pentane	C_5H_8	39	26.76	27.49
Styrene	C_8H_8	145	38.7	
Succinonitrile	$C_4H_4N_2$	266	48.5	
1,1,2,2-Tetrabromoethane	$C_2H_2Br_4$	243.5	48.7	
1,1,2,2-Tetrachloroethane	$C_2H_2Cl_4$	145.2	37.64	45.71
Tetrachloroethene	C_2Cl_4	121.3	34.68	39.68
Tetrachloromethane	CCl_4	76.8	29.82	32.43
Tetradecane	$C_{14}H_{30}$	253.58	48.16	71.73
1,14-Tetradecanediol	$C_{14}H_{30}O_2$			149.7
Tetradecanenitrile	$C_{14}H_{27}N$			85.29
1-Tetradecanol	$C_{14}H_{30}O$	287		98.9
Tetradecylbenzene	$C_{20}H_{34}$	359		95.8
Tetrahydrofuran	C_4H_8O	65	29.81	31.99
Tetrahydrofurfuryl alcohol	$C_5H_{10}O_2$	178	45.2	
1,2,3,4-Tetrahydronaphthalene	$C_{10}H_{12}$	207.6	43.9	
Tetrahydropyran	$C_5H_{10}O$	88	31.17	34.58
Tetrahydrothiophene	C_4H_8S	121.1	34.66	39.43
2,2,3,3-Tetramethylbutane	C_8H_{18}	106.45		42.90
2,2,4,4-Tetramethylpentane	C_9H_{20}	122.29	32.51	38.49
Tetranitromethane	CN_4O_8	126.1	40.74	49.93
Thiacyclohexane	$C_5H_{10}S$	141.8	35.96	42.58
Thietane	C_3H_6S	95.0	32.32	35.97
Thiophene	C_4H_4S	84.0	31.48	34.70
Toluene	C_7H_8	110.63	33.18	38.01
Triacetamide	$C_6H_9NO_3$			60.41
Tribromomethane	$CHBr_3$	149.1	39.66	46.05
Tributylamine	$C_{12}H_{27}N$	216.5	46.9	
Tributyl borate	$C_{12}H_{27}BO_3$	234	56.1	
1,1,1-Trichloroethane	$C_2H_3Cl_3$	74.09	29.86	32.50
1,1,2-Trichloroethane	$C_2H_3Cl_3$	113.8	34.82	40.24
Trichloroethene	C_2HCl_3	87.21	31.40	34.54
Trichlorofluoromethane	CCl_3F	23.7	25.1	
Trichloromethane	$CHCl_3$	61.17	29.24	31.28
1,2,3-Trichloropropane	$C_3H_5Cl_3$	157	37.1	
1,1,1-Trichloro-2,2,2-trifluoroethane	$C_2Cl_3F_3$	45.5	26.85	28.08
1,1,2-Trichloro-1,2,2-trifluoroethane	$C_2Cl_3F_3$	47.7	27.04	28.40
Tridecane	$C_{13}H_{28}$	235.47	46.20	66.68
1,13-Tridecanediol	$C_{13}H_{28}O_2$			133
1-Tridecanol	$C_{13}H_{28}O$	274		94.7
Tridecylbenzene	$C_{19}H_{32}$	346		91.8
Triethylamine	$C_6H_{15}N$	89	31.01	34.84
Triethylene glycol	$C_6H_{14}O_4$	285	71.4	
Trifluoroacetic acid	$C_2HF_3O_2$	73	33.3	
1,1,1-Trifluoroethane	$C_2H_3F_3$	-47.25	18.99	
(Trifluoromethyl)benzene	$C_7H_5F_3$	102.1	32.63	37.60
Trimethylamine	C_3H_9N	2.87	22.94	21.66
1,2,3-Trimethylbenzene	C_9H_{12}	176.12		49.05
1,2,4-Trimethylbenzene	C_9H_{12}	169.38		47.93
1,3,5-Trimethylbenzene	C_9H_{12}	164.74		47.50
2,2,3-Trimethylbutane	C_7H_{16}	80.86	28.90	32.05
2,3,3-Trimethyl-1-butene	C_7H_{14}	77.9		32.09
2,2,5-Trimethylhexane	C_9H_{20}	124.09	33.65	40.16
2,3,5-Trimethylhexane	C_9H_{20}	131.4	34.43	41.41

Name	Mol. Form.	t_b °C	$\Delta_{vap}H(t_b)$ kJ/mol	$\Delta_{vap}H(25\ °C)$ kJ/mol
3,5,5-Trimethyl-1-hexanol	$C_9H_{20}O$	194		67.86
2,4,7-Trimethyloctane	$C_{11}H_{24}$	168.1	38.22	49.91
2,2,3-Trimethylpentane	C_8H_{18}	110	31.94	36.91
2,2,4-Trimethylpentane	C_8H_{18}	99.22	30.79	35.14
2,3,3-Trimethylpentane	C_8H_{18}	114.8	32.12	37.27
2,3,4-Trimethylpentane	C_8H_{18}	113.5	32.36	37.75
2,2,4-Trimethyl-3-pentanone	$C_8H_{16}O$	135.1	35.64	43.30
2,4,4-Trimethyl-1-pentene	C_8H_{16}	101.4		35.59
2,4,4-Trimethyl-2-pentene	C_8H_{16}	104.9		37.23
2,3,6-Trimethylpyridine	$C_8H_{11}N$	171.6	39.95	50.61
2,4,6-Trimethylpyridine	$C_8H_{11}N$	170.6	39.87	50.33
Tris(perfluorobutyl)amine	$C_{12}F_{27}N$	178	46.4	
Undecane	$C_{11}H_{24}$	195.9	41.91	56.58
1,11-Undecanediol	$C_{11}H_{24}O_2$			132
Undecanenitrile	$C_{11}H_{21}N$	253		71.14
1-Undecanol	$C_{11}H_{24}O$	245		85.8
Undecylbenzene	$C_{17}H_{28}$	316		82.4
Vinyl acetate	$C_4H_6O_2$	72.8	34.6	
o-Xylene	C_8H_{10}	144.5	36.24	43.43
m-Xylene	C_8H_{10}	139.07	35.66	42.65
p-Xylene	C_8H_{10}	138.23	35.67	42.40
2,4-Xylenol	$C_8H_{10}O$	210.98		64.96
2,5-Xylenol	$C_8H_{10}O$	211.1	46.9	
2,6-Xylenol	$C_8H_{10}O$	201.07		75.31
3,4-Xylenol	$C_8H_{10}O$	227		85.03
3,5-Xylenol	$C_8H_{10}O$	221.74		82.01

ENTHALPY OF FUSION

This table lists the molar enthalpy (heat) of fusion, $\Delta_{fus}H$, of over 1100 inorganic and organic compounds. All values refer to the enthalpy change at equilibrium between the liquid phase and the most stable solid phase at the phase transition temperature. Most values of $\Delta_{fus}H$ are given at the normal melting point t_m. However, a "t" following the entry in the melting point column indicates a triple–point temperature, where the solid, liquid, and gas phases are in equilibrium. Temperatures are given on the ITS–90 scale.

A * following an entry indicates that the value includes the enthalpy of transition between crystalline phases whose transformation occurs within 1 °C of the melting point.

Substances are listed by name, either an IUPAC systematic name or, in the case of drugs and other complex compounds, a common synonym. Inorganic compounds, including metal salts of organic acids, are listed first, followed by organic compounds. The molecular formula in the Hill convention is included.

References

1. Chase, M. W., Davies, C. A., Downey, J. R., Frurip, D. J., McDonald, R. A., and Syverud, A. N., *JANAF Thermochemical Tables, Third Edition, J. Phys. Chem. Ref. Data,* Vol. 14, Suppl. 1, 1985.
2. Chase, M. W., *NIST–JANAF Thermochemical Tables, Fourth Edition, J. Phys. Chem. Ref. Data,* Monograph No. 9, 1998.
3. Gurvich, L. V., Veyts, I. V., and Alcock, C. B., *Thermodynamic Properties of Individual Substances, Fourth Edition;* Vol. 2, Hemisphere Publishing Corp., New York, 1991; Vol. 3, CRC Press, Boca Raton, FL, 1994.
4. Dinsdale, A. T., "SGTE Data for Pure Elements", *CALPHAD,* 15, 317–425, 1991.
5. *Landolt–Börnstein, Numerical Data and Functional Relationships in Science and Technology, New Series,* IV/8A, "Enthalpies of Fusion and Transition of Organic Compounds," Springer–Verlag, Heidelberg, 1995.
6. *Landolt–Börnstein, Numerical Data and Functional Relationships in Science and Technology, New Series,* IV/19A, "Thermodynamic Properties of Inorganic Materials compiled by SGTE," Springer–Verlag, Heidelberg; Part 1, 1999; Part 2; 1999; Part 3, 2000; Part 4, 2001.
7. Janz, G. J., et al., *Physical Properties Data Compilations Relevant to Energy Storage. II. Molten Salts,* Nat. Stand. Ref. Data Sys.– Nat. Bur. Standards (U.S.), No. 61, Part 2, 1979.
8. Dirand, M., Bouroukba, M., Chevallier, V., Petitjean, D., Behar, E., and Ruffier–Meray, V., "Normal Alkanes, Multialkane Synthetic Model Mixtures, and Real Petroleum Waxes: Crystallographic Structures, Thermodynamic Properties, and Crystallization," *J. Chem. Eng. Data,* 47, 115–143, 2002.
9. Linstrom, P. J., and Mallard, W. G., Editors, *NIST Chemistry WebBook,* NIST Standard Reference Database No. 69, June 2005, National Institute of Standards and Technology, Gaithersburg, MD, <http://webbook.nist.gov>.
10. Thermodynamic Research Center, National Institute of Standards and Technology, *TRC Thermodynamic Tables,* <http://trc.nist.gov>.
11. Sangster, J., "Phase Diagrams and Thermodynamic Properties of Binary Systems of Drugs," *J. Phys. Chem. Ref. Data* 28, 889, 1999.

Name	Molecular formula	t_m/°C	$\Delta_{fus}H$/ kJ mol^{-1}
Inorganic compounds (including salts of organic acids)			
Actinium	Ac	1050	12.0
Aluminum	Al	660.32	10.71
Aluminum bromide	$AlBr_3$	97.5	11.25
Aluminum chloride	$AlCl_3$	192.6	35.35
Aluminum fluoride	AlF_3	2250 t	98
Aluminum iodide	AlI_3	188.28	15.90
Aluminum oxide (α)	Al_2O_3	2054	111.1
Aluminum sulfide	Al_2S_3	1100	66
Americium	Am	1176	14.39
Ammonia	H_3N	−77.73	5.66
Ammonium chloride	ClH_4N	520.1	10.6
Ammonium fluoride	FH_4N	238	12.6
Ammonium iodide	H_4IN	551	21
Ammonium nitrate	$H_4N_2O_3$	169.7	5.86
Antimony (gray)	Sb	630.628	19.79
Antimony(III) bromide	Br_3Sb	97	14.6
Antimony(III) chloride	Cl_3Sb	73.4	12.97
Antimony(III) fluoride	F_3Sb	287	22.8
Antimony(III) iodide	I_3Sb	171	22.8
Antimony(III) oxide (valentinite)	O_3Sb_2	655	54
Antimony(III) sulfide	S_3Sb_2	550	47.9
Argon	Ar	−189.36	1.18
Arsenic (gray)	As	817	24.44
Arsenic(III) bromide	$AsBr_3$	31.1	11.7
Arsenic(III) chloride	$AsCl_3$	−16	10.1
Arsenic(III) fluoride	AsF_3	−5.9	10.4
Arsenic(III) iodide	AsI_3	141	21.8
Arsenic(III) oxide (claudetite)	As_2O_3	314	18
Arsenic(V) oxide	As_2O_5	730	60
Arsenic(III) selenide	As_2Se_3	377	40.8
Arsenic(III) sulfide	As_2S_3	312	28.7
Arsenic sulfide	As_4S_4	307	25.4
Arsenic(III) telluride	As_2Te_3	375	46.0
Barium	Ba	727	7.12
Barium bromide	$BaBr_2$	857	32.2
Barium carbonate	$CBaO_3$	1555 (high pres.)	40
Barium chloride	$BaCl_2$	961	15.85
Barium fluoride	BaF_2	1368	23.36
Barium hydride	BaH_2	1200	25
Barium hydroxide	BaH_2O_2	408	16
Barium iodide	BaI_2	711	26.5
Barium oxide	BaO	1973	46
Barium sulfate	BaO_4S	1580	40
Barium sulfide	BaS	2227	63
Beryllium	Be	1287	7.895
Beryllium bromide	$BeBr_2$	508	18
Beryllium carbide	CBe_2	2127	75.3
Beryllium chloride	$BeCl_2$	415	8.66
Beryllium fluoride	BeF_2	552	4.77
Beryllium iodide	BeI_2	480	20.92
Beryllium nitride	Be_3N_2	2200	111
Beryllium oxide	BeO	2578	86
Beryllium sulfate	BeO_4S	1127	6
Bismuth	Bi	271.406	11.106
Bismuth oxide	Bi_2O_3	825	14.7

Name	Molecular formula	t_m/°C	$\Delta_{fus}H$/ kJ mol⁻¹
Bismuth sulfide	Bi_2S_3	777	78.2
Bismuth tribromide	$BiBr_3$	219	21.7
Bismuth trichloride	$BiCl_3$	234	23.6
Bismuth trifluoride	BiF_3	649	21.6
Bismuth triiodide	BiI_3	408.6	39.1
Boric acid	BH_3O_3	170.9	22.3
Boron	B	2075	50.2
Boron nitride	BN	2967	81
Boron oxide	B_2O_3	450	24.56
Boron sulfide	B_2S_3	563	48.12
Boron trichloride	BCl_3	−107.3	2.10
Boron trifluoride	BF_3	−126.8	4.20
Bromine	Br_2	−7.2	10.57
Bromine pentafluoride	BrF_5	−60.5	5.67
Cadmium	Cd	321.069	6.21
Cadmium bromide	Br_2Cd	568	33.35
Cadmium chloride	$CdCl_2$	568	48.58
Cadmium fluoride	CdF_2	1075	22.6
Cadmium iodide	CdI_2	388	15.3
Cadmium nitrate	CdN_2O_6	360	18.3
Calcium	Ca	842	8.54
Calcium bromide	Br_2Ca	742	29.1
Calcium carbonate (calcite)	$CCaO_3$	800	36
Calcium chloride	$CaCl_2$	775	28.05
Calcium fluoride	CaF_2	1418	30
Calcium hydride	CaH_2	1000	6.7
Calcium iodide	CaI_2	783	41.8
Calcium nitrate	CaN_2O_6	561	23.4
Calcium oxide	CaO	2613	80
Calcium sulfate	CaO_4S	1460	28
Calcium sulfide	CaS	2524	70
Carbon (graphite)	C	4489	117.4
Cerium	Ce	799	5.460
Cerium(III) bromide	Br_3Ce	732	51.9
Cerium(III) chloride	$CeCl_3$	807	53.1
Cerium(III) fluoride	CeF_3	1430	55.6
Cerium(III) iodide	CeI_3	760	51.0
Cerium(III) oxide	Ce_2O_3	2250	120
Cerium(IV) oxide	CeO_2	2480	80
Cesium	Cs	28.5	2.09
Cesium carbonate	CCs_2O_3	793	31
Cesium chloride	ClCs	646	20.4
Cesium chromate	$CrCs_2O_4$	963	35.3
Cesium fluoride	CsF	703	21.7
Cesium hydride	CsH	528	15
Cesium hydroxide	CsHO	342.3	7.78
Cesium iodide	CsI	632	25.7
Cesium metaborate	$BCsO_2$	732	27
Cesium molybdate	Cs_2MoO_4	956.3	31.8
Cesium nitrate	$CsNO_3$	409	13.8
Cesium nitrite	$CsNO_2$	406	10.9
Cesium oxide	Cs_2O	495	20
Cesium peroxide	Cs_2O_2	594	22
Cesium sulfate	Cs_2O_4S	1005	35.7
Chlorine	Cl_2	−101.5	6.40
Chromium	Cr	1907	21.00
Chromium(II) bromide	Br_2Cr	842	45
Chromium(III) bromide	Br_3Cr	812	60
Chromium(II) chloride	Cl_2Cr	824	45.0
Chromium(III) chloride	Cl_3Cr	827	60
Chromium(II) fluoride	CrF_2	894	34
Chromium(III) fluoride	CrF_3	1425	66
Chromium(II) iodide	CrI_2	867	46
Chromium(III) iodide	CrI_3	857	61
Chromium(III) oxide	Cr_2O_3	2432	125
Chromium(VI) oxide	CrO_3	197	14.2
Chromium(II) sulfide	CrS	1567	25.5
Cobalt	Co	1495	16.20
Cobalt(II) bromide	Br_2Co	678	43
Cobalt(II) chloride	Cl_2Co	737	46.0
Cobalt(II) fluoride	CoF_2	1127	58.1
Cobalt(II) iodide	CoI_2	520	35
Cobalt(II) selenite	CoO_3Se	659	16.3
Cobalt(II) sulfide	CoS	1117	30
Copper	Cu	1084.62	13.26
Copper(I) bromide	BrCu	483	5.1
Copper(I) chloride	ClCu	423	7.08
Copper(II) chloride	Cl_2Cu	598	15.0
Copper(II) fluoride	CuF_2	836	55
Copper(I) iodide	CuI	591	7.93
Copper(I) oxide	Cu_2O	1244	65.6
Copper(II) oxide	CuO	1227	49
Copper(I) sulfide	Cu_2S	1129	9.62
Curium	Cm	1345	14.64
Decaborane(14)	$B_{10}H_{14}$	98.78	21.97
Dysprosium	Dy	1412	11.35
Dysprosium(III) fluoride	DyF_3	1157	58.6
Dysprosium(III) oxide	Dy_2O_3	2408	120
Einsteinium	Es	860	9.41
Erbium	Er	1529	19.90
Erbium chloride	Cl_3Er	776	32.6
Erbium fluoride	ErF_3	1146	28.2
Erbium oxide	Er_2O_3	2418	130
Europium	Eu	822	9.21
Europium(II) bromide	Br_2Eu	683	25.1
Europium(III) chloride	Cl_3Eu	623	33.1
Europium(III) fluoride	EuF_3	647	6.40
Europium (II) oxide	EuO	1967	40
Europium(III) oxide	Eu_2O_3	2350	117
Fluorine	F_2	−219.67	0.51
Gadolinium	Gd	1313	9.67
Gadolinium(III) bromide	Br_3Gd	785	38.1
Gadolinium(III) chloride	Cl_3Gd	602	40.6
Gadolinium(III) fluoride	F_3Gd	1232	52.4
Gadolinium(III) iodide	GdI_3	930	54.0
Gadolinium(III) oxide	Gd_2O_3	2425	60
Gallium	Ga	29.7666	5.585
Gallium antimonide	GaSb	712	25.1
Gallium arsenide	AsGa	1238	87.64
Gallium(III) bromide	Br_3Ga	123	11.7
Gallium(III) chloride	Cl_3Ga	77.9	11.51
Gallium(III) iodide	GaI_3	212	12.9
Gallium(III) oxide	Ga_2O_3	1807	100
Germanium	Ge	938.25	36.94
Germanium(IV) bromide	Br_4Ge	26.1	12
Germanium(II) iodide	GeI_2	428	33.3
Germanium(IV) iodide	GeI_4	146	19.1
Germanium(IV) oxide	GeO_2	1116	12.6
Germanium(II) selenide	GeSe	675	24.7
Germanium(II) sulfide	GeS	658	21.3

Name	Molecular formula	t_m/°C	$\Delta_{fus}H$/ kJ mol^{-1}
Germanium(IV) sulfide	GeS_2	840	16.3
Germanium(II) telluride	GeTe	724	47.3
Gold	Au	1064.18	12.55
Hafnium	Hf	2233	27.20
Hafnium nitride	HfN	3310	62.8
Hafnium(IV) oxide	HfO_2	2800	96
Holmium	Ho	1472	11.76
Holmium bromide	Br_3Ho	919	50.1
Holmium chloride	Cl_3Ho	720	30.5
Holmium fluoride	F_3Ho	1143	56.3
Holmium oxide	Ho_2O_3	2415	130
Hydrazine	H_4N_2	1.54	12.66
Hydrogen	H_2	−259.198 t	0.12
Hydrogen bromide	BrH	−86.80	2.41
Hydrogen chloride	ClH	−114.17	2.00
Hydrogen fluoride	FH	−83.36	4.58
Hydrogen iodide	HI	−50.76	2.87
Hydrogen peroxide	H_2O_2	−0.43	12.50
Hydrogen sulfide	H_2S	−85.5	2.38
Indium	In	156.60	3.291
Indium antimonide	InSb	524	47.7
Indium arsenide	AsIn	942	77.0
Indium(I) bromide	BrIn	285	24.3
Indium(III) bromide	Br_3In	420	26
Indium(I) chloride	ClIn	225	9.20
Indium(III) chloride	Cl_3In	583	27
Indium(III) fluoride	F_3In	1172	64
Indium(I) iodide	IIn	364.4	17.26
Indium(II) iodide	I_2In	155	1.29
Indium(III) iodide	I_3In	207	18.48
Indium(III) oxide	In_2O_3	1912	105
Indium(II) sulfide	InS	692	36.0
Iodine	I_2	113.7	15.52
Iodine chloride	ClI	27.38	11.6
Iridium	Ir	2446	41.12
Iridium(VI) fluoride	F_6Ir	44	8.40
Iron	Fe	1538	13.81
Iron boride (FeB)	BFe	1658	62.66
Iron(II) bromide	Br_2Fe	691	43.0
Iron(II) chloride	Cl_2Fe	677	42.83
Iron(III) chloride	Cl_3Fe	307.6	40
Iron(II) fluoride	F_2Fe	1100	50
Iron(III) fluoride	F_3Fe	367	0.58
Iron(II) iodide	FeI_2	594	39
Iron(II) oxide	FeO	1377	24.1
Iron(II,III) oxide	Fe_3O_4	1597	138
Iron(III) oxide	Fe_2O_3	1539	87
Iron sodium oxide	$FeNaO_2$	1347	49.4
Iron(II) sulfide	FeS	1188	31.5
Krypton	Kr	−157.38	1.64
Lanthanum	La	920	6.20
Lanthanum bromide	Br_3La	788	54.0
Lanthanum chloride	Cl_3La	858	54.4
Lanthanum fluoride	F_3La	1493	50.2
Lanthanum iodide	I_3La	778	56.1
Lead	Pb	327.462	4.774
Lead(II) bromide	Br_2Pb	371	16.44
Lead(II) chloride	Cl_2Pb	501	21.88
Lead(II) fluoride	F_2Pb	830	14.7
Lead(II) iodide	I_2Pb	410	23.4
Lead(II) oxide (massicot)	OPb	887	25.6
Lead(II) sulfate	O_4PbS	1087	40.2
Lead(II) sulfide	PbS	1113	49.4
Lithium	Li	180.50	3.00
Lithium aluminate	$AlLiO_2$	1610	87.9
Lithium bromide	BrLi	550	17.66
Lithium carbonate	CLi_2O_3	732	44.8
Lithium chloride	ClLi	610	19.8
Lithium chromate	$CrLi_2O_4$	482	30.5
Lithium fluoride	FLi	848.2	27.09
Lithium hexafluoroaluminate	AlF_6Li_3	785	86.19
Lithium hydride	HLi	692	21.8
Lithium hydride-*d*	DLi	694	22
Lithium hydroxide	HLiO	473	20.9
Lithium iodide	ILi	469	14.6
Lithium metasilicate	Li_2O_3Si	1201	28
Lithium nitrate	$LiNO_3$	253	26.7
Lithium nitrite	$LiNO_2$	222	9.2
Lithium oxide	Li_2O	1438	35.6
Lithium perchlorate	$ClLiO_4$	236	29.3
Lithium sulfate	Li_2O_4S	860	9.00
Lutetium	Lu	1663	18.65
Lutetium oxide	Lu_2O_3	2490	133
Magnesium	Mg	650	8.48
Magnesium bromide	Br_2Mg	711	39.3
Magnesium carbonate	$CMgO_3$	990	59
Magnesium chloride	Cl_2Mg	714	43.1
Magnesium fluoride	F_2Mg	1263	58.7
Magnesium hydride	H_2Mg	327	14
Magnesium iodide	I_2Mg	634	26
Magnesium orthosilicate	Mg_2O_4Si	1897	71
Magnesium oxide	MgO	2825	77
Magnesium phosphate	$Mg_3O_8P_2$	1348	121
Magnesium sulfate	MgO_4S	1137	14.6
Magnesium sulfide	MgS	2226	63
Magnesium tetraboride	B_4Mg	727	0.0
Manganese	Mn	1246	12.91
Manganese(II) bromide	Br_2Mn	698	33.5
Manganese(II) chloride	Cl_2Mn	650	30.7
Manganese(II) fluoride	F_2Mn	900	30
Manganese(II) iodide	I_2Mn	638	41.8
Manganese(II) oxide	MnO	1842	43.9
Manganese(II) sulfide (α form)	MnS	1530	26.1
Mercury	Hg	−38.829	2.295
Mercury(II) bromide	Br_2Hg	241	17.9
Mercury(II) chloride	Cl_2Hg	277	19.41
Mercury(II) fluoride	F_2Hg	645	23.0
Mercury(I) iodide	Hg_2I_2	290	31.4
Mercury(II) iodide (yellow)	HgI_2	256	15.6
Mercury(II) sulfide (black)	HgS	820	40
Metaboric acid (γ form)	BHO_2	236	14.3
Molybdenum	Mo	2623	37.48
Molybdenum boride (Mo_2B_5)	B_5Mo_2	2210	226
Molybdenum(IV) chloride	Cl_4Mo	317	16.7
Molybdenum(V) chloride	Cl_5Mo	194	19
Molybdenum(VI) dioxydichloride	Cl_2MoO_2	176	17.0
Molybdenum(V) fluoride	F_5Mo	45.67	6.1
Molybdenum(VI) fluoride	F_6Mo	17.5	4.33
Molybdenum monoboride	BMo	2600	55.23
Molybdenum(VI) oxide	MoO_3	802	48.7

Name	Molecular formula	t_m/°C	$\Delta_{fus}H$/ kJ mol⁻¹
Molybdenum(VI) oxytetrachloride	Cl_4MoO	105	14.3
Molybdenum(VI) oxytetrafluoride	F_4MoO	97.2	4
Molybdenum(V) oxytrichloride	Cl_3MoO	310	22
Molybdenum(III) sulfide	Mo_2S_3	1807	0.13
Neodymium	Nd	1016	7.14
Neodymium(III) bromide	Br_3Nd	682	45.3
Neodymium(III) chloride	Cl_3Nd	759	48.5
Neodymium(III) fluoride	F_3Nd	1377	54.8
Neodymium(III) iodide	I_3Nd	787	41.5
Neon	Ne	−248.609	0.328
Neptunium	Np	644	3.20
Nickel	Ni	1455	17.48
Nickel boride (Ni_2B)	BNi_2	1125	42.15
Nickel boride (Ni_3B)	BNi_3	1166	72.28
Nickel(II) bromide	Br_2Ni	963	56
Nickel(II) chloride	Cl_2Ni	1031	77.9
Nickel(II) fluoride	F_2Ni	1380	69
Nickel(II) iodide	I_2Ni	800	48
Nickel(II) oxide	NiO	1957	50.7
Nickel(II) sulfide	NiS	976	30.1
Nickel disulfide	NiS_2	1007	65.7
Nickel subsulfide	Ni_3S_2	789	19.7
Niobium	Nb	2477	30
Niobium(V) bromide	Br_5Nb	254	24.0
Niobium(V) chloride	Cl_5Nb	205.8	33.9
Niobium(V) fluoride	F_5Nb	80	12.2
Niobium(V) iodide	I_5Nb	327	37.7
Niobium nitride	NNb	2050	46.0
Niobium(II) oxide	NbO	1937	85.4
Niobium(IV) oxide	NbO_2	1901	92
Niobium(V) oxide	Nb_2O_5	1512	104.3
Nitric acid	HNO_3	−41.6	10.5
Nitric oxide	NO	−163.6	2.30
Nitrogen	N_2	−210.0	0.71
Nitrogen tetroxide	N_2O_4	−9.3	14.65
Nitrous oxide	N_2O	−90.8	6.54
Osmium	Os	3033	57.85
Osmium(VIII) oxide	O_4Os	40.6	14.3
Oxygen	O_2	−218.79	0.44
Palladium	Pd	1554.8	16.74
Palladium(II) chloride	Cl_2Pd	679	18.41
Phosphinic acid	H_3O_2P	26.5	9.7
Phosphonic acid	H_3O_3P	74.4	12.8
Phosphoric acid	H_3O_4P	42.4	13.4
Phosphorus (white)	P	44.15	0.659
Phosphorus (red)	P	579.2	18.54
Phosphorus(III) chloride	Cl_3P	−93	7.10
Phosphorus heptasulfide	P_4S_7	308	36.6
Phosphorus(V) oxide	O_5P_2	562	27.2
Phosphorus sesquisulfide	P_4S_3	173	20.1
Phosphoryl chloride	Cl_3OP	1.18	13.1
Platinum	Pt	1768.2	22.175
Plutonium	Pu	640	2.824
Plutonium(III) bromide	Br_3Pu	681	58.6
Plutonium(III) chloride	Cl_3Pu	760	63.6
Plutonium(III) fluoride	F_3Pu	1396	59.8
Plutonium(IV) fluoride	F_4Pu	1037	42.7
Plutonium(VI) fluoride	F_6Pu	51.6	18.6
Plutonium(III) iodide	I_3Pu	777	50.2
Plutonium(III) oxide	O_3Pu_2	2085	113
Plutonium(IV) oxide	O_2Pu	2390	67
Polonium	Po	254	10.0
Potassium	K	63.5	2.335
Potassium aluminate	$AlKO_2$	1713	82
Potassium bromide	BrK	734	25.52
Potassium carbonate	CK_2O_3	899	27.6
Potassium chloride	ClK	771	26.28
Potassium chromate	CrK_2O_4	974	33.0
Potassium cyanide	CKN	622	14.6
Potassium fluoride	FK	858	27.2
Potassium fluoroborate	BF_4K	570	17.66
Potassium hydride	HK	619	21
Potassium hydrogen fluoride	F_2HK	238.8	6.62
Potassium hydroxide	HKO	406	7.90
Potassium iodide	IK	681	24.0
Potassium metaborate	BKO_2	947	31.38
Potassium nitrate	KNO_3	334	9.6
Potassium nitrite	KNO_2	438	16.7
Potassium oxide	K_2O	740	27
Potassium peroxide	K_2O_2	545	20.5
Potassium sulfate	K_2O_4S	1069	36.6
Potassium sulfide	K_2S	948	16.15
Potassium superoxide	KO_2	535	20.6
Praseodymium	Pr	931	6.89
Praseodymium(III) bromide	Br_3Pr	693	47.3
Praseodymium(III) chloride	Cl_3Pr	786	50.6
Praseodymium(III) fluoride	F_3Pr	1399	57.3
Praseodymium(III) iodide	I_3Pr	738	53.1
Protactinium	Pa	1572	12.34
Radium	Ra	696	7.7
Rhenium	Re	3185	34.08
Rhenium(VII) oxide	O_7Re_2	327	65.7
Rhodium	Rh	1964	26.59
Rubidium	Rb	39.30	2.19
Rubidium bromide	BrRb	692	23.3
Rubidium carbonate	CO_3Rb_2	873	30
Rubidium chloride	ClRb	724	24.4
Rubidium fluoride	FRb	795	25.8
Rubidium hydride	HRb	585	22
Rubidium hydroxide	HORb	385	8.0
Rubidium iodide	IRb	656	22.1
Rubidium metaborate	BO_2Rb	860	31
Rubidium nitrate	NO_3Rb	310	4.6
Rubidium nitrite	NO_2Rb	422	12.1
Rubidium oxide	ORb_2	505	20
Rubidium peroxide	O_2Rb_2	570	21
Rubidium sulfate	O_4Rb_2S	1066	37.3
Rubidium superoxide	O_2Rb	540	21
Ruthenium	Ru	2333	38.59
Ruthenium(V) fluoride	F_5Ru	101	74.5
Samarium	Sm	1072	8.62
Samarium(III) oxide	O_3Sm_2	2335	119
Scandium	Sc	1541	14.10
Scandium chloride	Cl_3Sc	967	67.4
Scandium fluoride	F_3Sc	1552	62.6
Scandium oxide	O_3Sc_2	2489	127
Selenium (gray)	Se	220.8	6.69
Selenium dioxide	O_2Se	360	17.6

Name	Molecular formula	t_m/°C	$\Delta_{fus}H$/ kJ mol^{-1}
Silicon	Si	1414	50.21
Silicon dioxide (cristobalite)	O_2Si	1722	9.6
Silicon monosulfide	SSi	1090	31
Silver	Ag	961.78	11.30
Silver(I) bromide	$AgBr$	430	9.163
Silver(I) chloride	$AgCl$	455	13.054
Silver(I) iodide	AgI	558	9.414
Silver(I) nitrate	$AgNO_3$	210	11.72
Silver(I) oxide	Ag_2O	827	15
Silver(I) sulfate	Ag_2O_4S	660	17.99
Silver(I) sulfide	Ag_2S	836	7.9
Sodium	Na	97.794	2.60
Sodium bromate	$BrNaO_3$	381	28.11
Sodium bromide	$BrNa$	747	26.23
Sodium carbonate	CNa_2O_3	856	29.7
Sodium chlorate	$ClNaO_3$	248	22.6
Sodium chloride	$ClNa$	800.7	28.16
Sodium chromate	$CrNa_2O_4$	794	24.7
Sodium cyanide	$CNNa$	562	8.79
Sodium fluoride	FNa	996	33.35
Sodium formate	$CHNaO_2$	257.3	17.7
Sodium hexafluoroaluminate	AlF_6Na_3	1013	114.4
Sodium hexafluorosilicate	F_6Na_2Si	847	99.6
Sodium hydride	HNa	638	26
Sodium hydroxide	$HNaO$	323	6.60
Sodium iodate	$INaO_3$	422	35.1
Sodium iodide	INa	661	23.7
Sodium metaborate	$BNaO_2$	966	36.2
Sodium metasilicate	Na_2O_3Si	1089	51.8
Sodium nitrate	$NNaO_3$	306.5	15.5
Sodium nitrite	$NNaO_2$	284	14.9
Sodium oxide	Na_2O	1134	47.7
Sodium peroxide	Na_2O_2	675	24.5
Sodium sulfate	Na_2O_4S	884	23.85
Sodium sulfide	Na_2S	1172	19
Sodium sulfite	Na_2O_3S	911	25.9
Strontium	Sr	777	7.43
Strontium bromide	Br_2Sr	657	10.5
Strontium carbonate	CO_3Sr	1494	40
Strontium chloride	Cl_2Sr	874	16.22
Strontium fluoride	F_2Sr	1477	29.7
Strontium hydride	H_2Sr	1050	23
Strontium hydroxide	H_2O_2Sr	535	23
Strontium iodide	I_2Sr	538	19.7
Strontium nitrate	N_2O_6Sr	570	44.6
Strontium oxide	OSr	2531	81
Strontium sulfate	O_4SSr	1606	36
Strontium sulfide	SSr	2226	63
Sulfur (monoclinic)	S	115.21	1.721
Sulfur hexafluoride	F_6S	−49.596	5.02
Sulfuric acid	H_2O_4S	10.31	10.71
Sulfur trioxide (γ-form)	O_3S	16.8	8.60
Tantalum	Ta	3017	36.57
Tantalum boride (TaB_2)	B_2Ta	3100	83.68
Tantalum(V) bromide	Br_5Ta	240	37.7
Tantalum(V) chloride	Cl_5Ta	216.6	35.1
Tantalum(V) fluoride	F_5Ta	96.9	12
Tantalum(V) iodide	I_5Ta	496	7.74
Tantalum nitride (TaN)	NTa	3090	6.7
Tantalum nitride (Ta_2N)	NTa_2	2727	92.0
Tantalum(V) oxide	O_5Ta_2	1875	120
Technetium	Tc	2157	33.29
Tellurium	Te	449.51	17.38
Tellurium dioxide	O_2Te	733	28.9
Tellurium tetrabromide	Br_4Te	380	24.7
Tellurium tetrachloride	Cl_4Te	224	18.9
Terbium	Tb	1359	10.15
Terbium(III) bromide	Br_3Tb	830	31.5
Terbium(III) chloride	Cl_3Tb	582	19.5
Tetrachlorosilane	Cl_4Si	−68.74	7.60
Tetraiodosilane	I_4Si	120.5	19.7
Thallium	Tl	304	4.142
Thallium(I) bromide	$BrTl$	460	16.4
Thallium(I) carbonate	CO_3Tl_2	273	18
Thallium(I) chloride	$ClTl$	431	15.56
Thallium(I) fluoride	FTl	326	13.87
Thallium(I) formate	CHO_2Tl	101	10.9
Thallium(I) iodide	ITl	441.7	14.7
Thallium(I) nitrate	NO_3Tl	206	9.6
Thallium(I) oxide	OTl_2	579	30.3
Thallium(III) oxide	O_3Tl_2	834	53
Thallium(I) sulfate	O_4STl_2	632	23.8
Thallium(I) sulfide	STl_2	457	23.0
Thorium	Th	1750	13.81
Thorium(IV) bromide	Br_4Th	679	54.4
Thorium(IV) chloride	Cl_4Th	770	43.9
Thorium(IV) fluoride	F_4Th	1110	41.8
Thorium(IV) iodide	I_4Th	566	48.1
Thorium(IV) oxide	O_2Th	3350	90
Thulium	Tm	1545	16.84
Thulium(III) chloride	Cl_3Tm	845	34.9
Thulium(III) fluoride	F_3Tm	1158	28.9
Tin (white)	Sn	231.93	7.15
Tin(II) bromide	Br_2Sn	232	18.0
Tin(IV) bromide	Br_4Sn	29.1	12.2
Tin(II) chloride	Cl_2Sn	247.0	14.52
Tin(IV) chloride	Cl_4Sn	−34.07	9.20
Tin(II) fluoride	F_2Sn	215	10.5
Tin(IV) fluoride	F_4Sn	442	27.6
Tin(II) iodide	I_2Sn	320	18.0
Tin(IV) iodide	I_4Sn	402	0.16
Tin(II) oxide	OSn	977	27.7
Tin(IV) oxide	O_2Sn	1630	23.4
Tin(II) sulfide	SSn	881	31.6
Tin(II) telluride	$SnTe$	806	45.2
Titanium	Ti	1668	14.15
Titanium boride	B_2Ti	2920	100.4
Titanium(IV) bromide	Br_4Ti	38.3	12.9
Titanium(II) chloride	Cl_2Ti	1035	34.3
Titanium(IV) chloride	Cl_4Ti	−24.12	9.97
Titanium(IV) fluoride	F_4Ti	377	41
Titanium(IV) iodide	I_4Ti	155	19.8
Titanium nitride	NTi	2947	66.9
Titanium(III) oxide	O_3Ti_2	1842	104.6
Titanium(IV) oxide (rutile)	O_2Ti	1912	68
Titanium(II) sulfide	STi	1927	32
Tungsten	W	3422	52.31
Tungsten boride (WB)	BW	2800	80
Tungsten boride (W_2B)	BW_2	2740	117
Tungsten boride (W_2B_5)	B_5W_2	2370	240

Name	Molecular formula	t_m/°C	$\Delta_{fus}H$/ kJ mol^{-1}
Tungsten(V) bromide	Br_5W	286	17.2
Tungsten(V) chloride	Cl_5W	253	20.6
Tungsten(VI) chloride	Cl_6W	282	6.69
Tungsten(VI) fluoride	F_6W	1.9	4.10
Tungsten(VI) oxide	O_3W	1473	73
Tungsten(VI) oxytetrachloride	Cl_4OW	210	18.8
Tungsten(VI) oxytetrafluoride	F_4OW	105	6
Uranium	U	1135	9.14
Uranium(III) bromide	Br_3U	727	43.9
Uranium(IV) bromide	Br_4U	519	55.2
Uranium(IV) chloride	Cl_4U	590	44.8
Uranium(III) fluoride	F_3U	1495	36.8
Uranium(IV) fluoride	F_4U	1036	47
Uranium(V) fluoride	F_5U	348	35
Uranium(VI) fluoride	F_6U	64.06	19.2
Uranium(IV) iodide	I_4U	506	42.1
Uranium(IV) oxide	O_2U	2847	74.2
Uranyl chloride	Cl_2O_2U	577	44.06
Vanadium	V	1910	21.5
Vanadium(II) chloride	Cl_2V	1350	35.0
Vanadium(IV) chloride	Cl_4V	−28	2.30
Vanadium(II) fluoride	F_2V	1490	44
Vanadium(III) fluoride	F_3V	1395	57
Vanadium(V) fluoride	F_5V	19.5	49.96
Vanadium(II) oxide	OV	1790	50
Vanadium(III) oxide	O_3V_2	1957	140
Vanadium(IV) oxide	O_2V	1545	56.0
Vanadium(V) oxide	O_5V_2	681	64
Water	H_2O	0.00	6.01
Xenon	Xe	−111.745 t	2.27
Xenon difluoride	F_2Xe	129.03	16.8
Xenon tetrafluoride	F_4Xe	117.1	16.3
Xenon hexafluoride	F_6Xe	49.48	5.74
Ytterbium	Yb	824	7.66
Ytterbium(III) chloride	Cl_3Yb	854	35.4
Yttrium	Y	1522	11.39
Yttrium chloride	Cl_3Y	721	31.5
Yttrium fluoride	F_3Y	1155	27.9
Yttrium oxide	O_3Y_2	2439	81
Zinc	Zn	419.53	7.068
Zinc bromide	Br_2Zn	402	15.7
Zinc chloride	Cl_2Zn	325	10.30
Zinc fluoride	F_2Zn	872	40
Zinc iodide	I_2Zn	450	17
Zinc oxide	OZn	1974	70
Zinc phosphide (ZnP_2)	P_2Zn	980	92.9
Zinc selenite	O_3SeZn	621	46.4
Zinc sulfide (wurtzite)	SZn	1827	30
Zinc telluride	TeZn	1295	63
Zirconium	Zr	1854.7	21.00
Zirconium boride	B_2Zr	3050	104.6
Zirconium(II) bromide	Br_2Zr	827	28
Zirconium(III) bromide	Br_3Zr	727	33
Zirconium(IV) bromide	Br_4Zr	450	
Zirconium(II) chloride	Cl_2Zr	722	27.0
Zirconium(III) chloride	Cl_3Zr	627	30
Zirconium(IV) chloride	Cl_4Zr	437	29
Zirconium(II) fluoride	F_2Zr	902	37.7
Zirconium(III) fluoride	F_3Zr	927	50
Zirconium(IV) fluoride	F_4Zr	910	61

Name	Molecular formula	t_m/°C	$\Delta_{fus}H$/ kJ mol^{-1}
Zirconium(II) iodide	I_2Zr	827	28
Zirconium(III) iodide	I_3Zr	727	33
Zirconium(IV) iodide	I_4Zr	500	32
Zirconium nitride	NZr	2952	67.4
Zirconium(IV) oxide	O_2Zr	2710	90
Zirconium(IV) sulfide	S_2Zr	1550	45
Organic compounds			
Acenaphthene	$C_{12}H_{10}$	93.4	21.49
Acenaphthylene	$C_{12}H_8$	91.8	6.9
Acetaldehyde	C_2H_4O	−123.37	2.31
Acetamide	C_2H_5NO	80.16	15.59
Acetaminophen	$C_8H_9NO_2$	169.3	30.5
Acetanilide	C_8H_9NO	114.3	21.3
Acetic acid	$C_2H_4O_2$	16.64	11.73
Acetic anhydride	$C_4H_6O_3$	−74.1	10.5
Acetone	C_3H_6O	−94.7	5.77
Acetonitrile	C_2H_3N	−43.82	8.16
Acrylic acid	$C_3H_4O_2$	12.5	9.51
Acrylonitrile	C_3H_3N	−83.48	6.23
Allene	C_3H_4	−136.6	4.40
Allobarbital	$C_{10}H_{12}N_2O_3$	172	32.3
2-Aminobenzoic acid	$C_7H_7NO_2$	146	20.5
4-Aminobenzoic acid	$C_7H_7NO_2$	188.2	22.5
3-Amino-1-propanol	C_3H_9NO	12.4	19.7
Aminopyrine	$C_{13}H_{17}N_3O$	107.5	27.6
Ampyrone	$C_{11}H_{13}N_3O$	109	24.9
Aniline	C_6H_7N	−6.02	10.54
Anisole	C_7H_8O	−37.13	12.9
Anthracene	$C_{14}H_{10}$	215.76	29.4
Antipyrine	$C_{11}H_{12}N_2O$	112	27.3
trans-Azobenzene	$C_{12}H_{10}N_2$	67.88	22.52
trans-Azoxybenzene	$C_{12}H_{10}N_2O$	34.6	17.9
Barbital	$C_8H_{12}N_2O_3$	190	24.7
Benzaldehyde	C_7H_6O	−57.1	9.32
Benzamide	C_7H_7NO	127.3	19.5
Benz[a]anthracene	$C_{18}H_{12}$	160.5	21.4
Benzene	C_6H_6	5.49	9.87
Benzeneacetic acid	$C_8H_8O_2$	76.5	16.3
1,2-Benzenediamine	$C_6H_8N_2$	102.1	23.1
1,3-Benzenediamine	$C_6H_8N_2$	66.0	15.57
1,4-Benzenediamine	$C_6H_8N_2$	141.1	23.8
Benzenethiol	C_6H_6S	−14.93	11.48
p-Benzidine	$C_{12}H_{12}N_2$	127	19.1
Benzil	$C_{14}H_{10}O_2$	94.87	23.5
Benzocaine	$C_9H_{11}NO_2$	89.7	22.3
Benzoic acid	$C_7H_6O_2$	122.35	18.02
Benzonitrile	C_7H_5N	−13.99	9.1
Benzo[c]phenanthrene	$C_{18}H_{12}$	68	16.3
Benzophenone	$C_{13}H_{10}O$	47.9	18.19
Benzo[a]pyrene	$C_{20}H_{12}$	181.1	17.3
Benzo[e]pyrene	$C_{20}H_{12}$	181.4	16.6
p-Benzoquinone	$C_6H_4O_2$	115	18.5
Benzoyl chloride	C_7H_5ClO	−0.4	19.2
Benzyl alcohol	C_7H_8O	−15.4	8.97
2,2′-Binaphthalene	$C_{20}H_{14}$	187.9	38.9
Biphenyl	$C_{12}H_{10}$	68.93	18.57
Bromobenzene	C_6H_5Br	−30.72	10.70
1-Bromobutane	C_4H_9Br	−112.6	9.23
2-Bromobutane	C_4H_9Br	−112.65	6.89

Name	Molecular formula	t_m/°C	$\Delta_{fus}H$/ kJ mol^{-1}
Bromoethane	C_2H_5Br	−118.6	7.47
Bromoethene	C_2H_3Br	−139.54	5.12
1-Bromoheptane	$C_7H_{15}Br$	−56.1	21.8
1-Bromohexane	$C_6H_{13}Br$	−83.7	18.1
Bromomethane	CH_3Br	−93.68	5.98
1-Bromonaphthalene	$C_{10}H_7Br$	6.1	15.2
2-Bromonaphthalene	$C_{10}H_7Br$	55.9	14.4
1-Bromooctane	$C_8H_{17}Br$	−55.0	24.7
1-Bromopentane	$C_5H_{11}Br$	−88.0	14.37
1-Bromopropane	C_3H_7Br	−110.3	6.44
2-Bromopropane	C_3H_7Br	−89.0	6.53
Bromotrichloromethane	$CBrCl_3$	−5.65	2.53
1,2-Butadiene	C_4H_6	−136.2	6.96
1,3-Butadiene	C_4H_6	−108.91	7.98
Butanal	C_4H_8O	−96.86	10.77
Butane	C_4H_{10}	−138.3	4.66
1,4-Butanediol	$C_4H_{10}O_2$	20.4	18.70
1-Butanethiol	$C_4H_{10}S$	−115.7	10.46
Butanoic acid	$C_4H_8O_2$	−5.1	11.59
1-Butanol	$C_4H_{10}O$	−88.6	9.37
2-Butanol	$C_4H_{10}O$	−88.5	5.97
2-Butanone	C_4H_8O	−86.64	8.39
1-Butene	C_4H_8	−185.34	3.96
cis-2-Butene	C_4H_8	−138.88	7.31
trans-2-Butene	C_4H_8	−105.52	9.76
cis-2-Butenoic acid	$C_4H_6O_2$	15	12.6
trans-2-Butenoic acid	$C_4H_6O_2$	71.5	13.0
tert-Butylamine	$C_4H_{11}N$	−66.94	0.882
Butylbenzene	$C_{10}H_{14}$	−87.85	11.22
Butylcyclohexane	$C_{10}H_{20}$	−74.73	14.16
Butyl methyl ether	$C_5H_{12}O$	−115.7	10.85
1-Butyne	C_4H_6	−125.7	6.03
2-Butyne	C_4H_6	−32.2	9.23
γ-Butyrolactone	$C_4H_6O_2$	−43.61	9.57
Caffeine	$C_8H_{10}N_4O_2$	236.3	22.0
Carbazole	$C_{12}H_9N$	246.3	24.1
Carbon dioxide	CO_2	−56.558	9.02
Carbon diselenide	CSe_2	−43.7	6.36
Carbon disulfide	CS_2	−112.1	4.39
Carbon monoxide	CO	−205.02	0.833
Carbon oxysulfide	COS	−138.8	4.73
Carbonyl chloride	CCl_2O	−127.78	5.74
Chloroacetic acid	$C_2H_3ClO_2$	63	12.28
2-Chloroaniline	C_6H_6ClN	−1.9	11.9
3-Chloroaniline	C_6H_6ClN	−10.28	10.15
4-Chloroaniline	C_6H_6ClN	70.5	20.0
Chlorobenzene	C_6H_5Cl	−45.31	9.6
2-Chlorobenzoic acid	$C_7H_5ClO_2$	140.2	25.6
Chlorocyclohexane	$C_6H_{11}Cl$	−43.81	2.043
Chlorodifluoromethane	$CHClF_2$	−157.42	4.12
Chloroethane	C_2H_5Cl	−138.4	4.45
Chloroethene	C_2H_3Cl	−153.84	4.92
Chloromethane	CH_3Cl	−97.7	6.43
2-Chloro-2-methylpropane	C_4H_9Cl	−25.60	2.07
1-Chloronaphthalene	$C_{10}H_7Cl$	−2.5	12.9
2-Chloronaphthalene	$C_{10}H_7Cl$	58.0	14.0
1-Chloro-2-nitrobenzene	$C_6H_4ClNO_2$	32.1	17.9
1-Chloro-3-nitrobenzene	$C_6H_4ClNO_2$	44.4	19.4
1-Chloro-4-nitrobenzene	$C_6H_4ClNO_2$	82	14.1
Chloropentafluoroethane	C_2ClF_5	−99.4	1.86

Name	Molecular formula	t_m/°C	$\Delta_{fus}H$/ kJ mol^{-1}
2-Chlorophenol	C_6H_5ClO	9.4	13.0
3-Chlorophenol	C_6H_5ClO	32.6	14.9
4-Chlorophenol	C_6H_5ClO	42.8	14.1
1-Chloropropane	C_3H_7Cl	−122.9	5.54
2-Chloropropane	C_3H_7Cl	−117.18	7.39
2-Chlorotoluene	C_7H_7Cl	−35.8	9.6
Chlorotrifluoroethene	C_2ClF_3	−158.2	5.55
Chrysene	$C_{18}H_{12}$	255.5	26.2
Coronene	$C_{24}H_{12}$	437.4	19.2
o-Cresol	C_7H_8O	31.03	15.82
m-Cresol	C_7H_8O	12.24	10.71
p-Cresol	C_7H_8O	34.77	12.71
Cyanamide	CH_2N_2	45.56	7.27
Cyanogen	C_2N_2	−27.83	8.11
Cyclobutane	C_4H_8	−90.7	1.09
Cycloheptane	C_7H_{14}	−8.46	1.88
Cycloheptanol	$C_7H_{14}O$	7.2	1.60
Cyclohexane	C_6H_{12}	6.59	2.68
Cyclohexanol	$C_6H_{12}O$	25.93	1.78
Cyclohexanone	$C_6H_{10}O$	−27.9	1.328
Cyclohexene	C_6H_{10}	−103.5	3.29
Cyclohexylamine	$C_6H_{13}N$	−17.8	17.5
Cyclohexylbenzene	$C_{12}H_{16}$	7.07	15.6
Cyclooctane	C_8H_{16}	14.59	2.41
Cyclopentane	C_5H_{10}	−93.4	0.61
Cyclopentanol	$C_5H_{10}O$	−17.5	1.535
Cyclopentene	C_5H_8	−135.0	3.36
Cyclopentylamine	$C_5H_{11}N$	−82.7	8.31
Cyclopropane	C_3H_6	−127.58	5.44
Cyclopropylamine	C_3H_7N	−35.39	13.18
cis-Decahydronaphthalene	$C_{10}H_{18}$	−42.9	9.49
trans-Decahydronaphthalene	$C_{10}H_{18}$	−30.4	14.41
Decanal	$C_{10}H_{20}O$	−4.0	34.5
Decane	$C_{10}H_{22}$	−29.6	28.72
Decanoic acid	$C_{10}H_{20}O_2$	31.4	27.8
1-Decanol	$C_{10}H_{22}O$	6.9	43
1-Decene	$C_{10}H_{20}$	−66.3	13.81
1,2-Dibromoethane	$C_2H_4Br_2$	9.84	10.89
1,2-Dibromopropane	$C_3H_6Br_2$	−55.49	8.94
1,3-Dibromopropane	$C_3H_6Br_2$	−34.5	14.6
1,2-Dibromotetrafluoroethane	$C_2Br_2F_4$	−110.32	7.04
o-Dichlorobenzene	$C_6H_4Cl_2$	−17.0	12.4
m-Dichlorobenzene	$C_6H_4Cl_2$	−24.8	12.6
p-Dichlorobenzene	$C_6H_4Cl_2$	53.09	18.19
1,1-Dichloroethane	$C_2H_4Cl_2$	−96.9	7.87
1,2-Dichloroethane	$C_2H_4Cl_2$	−35.7	8.84
1,1-Dichloroethene	$C_2H_2Cl_2$	−122.56	6.51
cis-1,2-Dichloroethene	$C_2H_2Cl_2$	−80.0	7.2
Dichloromethane	CH_2Cl_2	−97.2	4.60
1,2-Dichloropropane	$C_3H_6Cl_2$	−100.53	6.40
2,2-Dichloropropane	$C_3H_6Cl_2$	−33.9	2.30
1,2-Dichloro-1,1,2,2-tetrafluoroethane	$C_2Cl_2F_4$	−92.53	1.51
Diethyl ether	$C_4H_{10}O$	−116.2	7.19
3,3-Diethylpentane	C_9H_{20}	−33.1	10.09
Diethyl sulfide	$C_4H_{10}S$	−103.91	10.90
o-Difluorobenzene	$C_6H_4F_2$	−47.1	11.05
m-Difluorobenzene	$C_6H_4F_2$	−69.12	8.58
Diisopropyl ether	$C_6H_{14}O$	−85.4	12.04
1,2-Dimethoxyethane	$C_4H_{10}O_2$	−69.20	12.6
Dimethoxymethane	$C_3H_8O_2$	−105.1	8.33

Name	Molecular formula	t_m/°C	$\Delta_{fus}H$/ kJ mol^{-1}
Dimethylamine	C_2H_7N	−92.18	5.94
2,2-Dimethylbutane	C_6H_{14}	−98.8	0.58
2,3-Dimethylbutane	C_6H_{14}	−128.10	0.79
2,3-Dimethyl-2-butene	C_6H_{12}	−74.19	6.45
1,1-Dimethylcyclohexane	C_8H_{16}	−33.3	2.07
cis-1,2-Dimethylcyclohexane	C_8H_{16}	−49.8	1.64
trans-1,2-Dimethylcyclohexane	C_8H_{16}	−88.15	10.49
cis-1,3-Dimethylcyclohexane	C_8H_{16}	−75.53	10.82
trans-1,3-Dimethylcyclohexane	C_8H_{16}	−90.07	9.87
cis-1,4-Dimethylcyclohexane	C_8H_{16}	−87.39	9.31
trans-1,4-Dimethylcyclohexane	C_8H_{16}	−36.93	12.33
Dimethyl disulfide	$C_2H_6S_2$	−84.67	9.19
Dimethyl ether	C_2H_6O	−141.5	4.94
N,N-Dimethylformamide	C_3H_7NO	−60.48	7.90
1,1-Dimethylhydrazine	$C_2H_8N_2$	−57.20	10.07
1,2-Dimethylhydrazine	$C_2H_8N_2$	−8.9	13.64
Dimethyl oxalate	$C_4H_6O_4$	54.8	21.1
2,2-Dimethylpentane	C_7H_{16}	−123.7	5.82
2,4-Dimethylpentane	C_7H_{16}	−119.2	6.85
3,3-Dimethylpentane	C_7H_{16}	−134.4	6.85
Dimethyl sulfide	C_2H_6S	−98.24	7.99
Dimethyl sulfone	$C_2H_6O_2S$	108.9	18.30
Dimethyl sulfoxide	C_2H_6OS	17.89	14.37
N,N-Dimethylurea	$C_3H_8N_2O$	182.1	23.0
N,N'-Dimethylurea	$C_3H_8N_2O$	106.6	13.0
Dimethyl zinc	C_2H_6Zn	−43.0	6.83
1,4-Dioxane	$C_4H_8O_2$	11.85	12.84
1,3-Dioxolane	$C_3H_6O_2$	−97.22	6.57
Diphenylamine	$C_{12}H_{11}N$	53.2	18.5
Diphenyl ether	$C_{12}H_{10}O$	26.864	17.22
Diphenylmethane	$C_{13}H_{12}$	25.4	18.6
Dipropyl ether	$C_6H_{14}O$	−114.8	10.8
Divinyl ether	C_4H_6O	−100.6	7.9
Docosane	$C_{22}H_{46}$	43.6	48.8
Dodecane	$C_{12}H_{26}$	−9.57	36.8
Dodecanoic acid	$C_{12}H_{24}O_2$	43.8	36.3
1-Dodecanol	$C_{12}H_{26}O$	23.9	40.2
1-Dodecene	$C_{12}H_{24}$	−35.2	19.9
Dotriacontane	$C_{32}H_{66}$	69.4	75.8
Eicosane	$C_{20}H_{42}$	36.6	69.9
1-Eicosanol	$C_{20}H_{42}O$	65.4	42
Estradiol benzoate	$C_{25}H_{28}O_3$	193	41.8
Ethane	C_2H_6	−182.79	2.72*
1,2-Ethanediamine	$C_2H_8N_2$	11.14	22.58
1,2-Ethanediol	$C_2H_6O_2$	−12.69	9.96
Ethanethiol	C_2H_6S	−147.88	4.98
Ethanol	C_2H_6O	−114.14	4.931
Ethinylestradiol	$C_{20}H_{24}O_2$	183.5	27.9
Ethyl acetate	$C_4H_8O_2$	−83.8	10.48
Ethylbenzene	C_8H_{10}	−94.96	9.18
Ethylcyclohexane	C_8H_{16}	−111.3	8.33
Ethylene	C_2H_4	−169.15	3.35
Ethyl methyl sulfide	C_3H_8S	−105.93	9.76
3-Ethylpentane	C_7H_{16}	−118.55	9.55
2-Ethyltoluene	C_9H_{12}	−79.83	9.96
3-Ethyltoluene	C_9H_{12}	−95.6	7.6
4-Ethyltoluene	C_9H_{12}	−62.35	12.7
Fluoranthene	$C_{16}H_{10}$	110.19	18.69
9*H*-Fluorene	$C_{13}H_{10}$	114.77	19.58
Fluorobenzene	C_6H_5F	−42.18	11.31
Formamide	CH_3NO	2.49	8.44
Formic acid	CH_2O_2	8.3	12.68
Furan	C_4H_4O	−85.61	3.80
Furfural	$C_5H_4O_2$	−38.1	14.37
Furfuryl alcohol	$C_5H_6O_2$	−14.6	13.13
Glycerol	$C_3H_8O_3$	18.1	18.3
Heneicosane	$C_{21}H_{44}$	40.01	45.21
Heptacosane	$C_{27}H_{56}$	59.23	61.9
Heptadecane	$C_{17}H_{36}$	22.0	40.16
Heptanal	$C_7H_{14}O$	−43.4	23.2
Heptane	C_7H_{16}	−90.55	14.03
Heptanoic acid	$C_7H_{14}O_2$	−7.17	15.13
1-Heptanol	$C_7H_{16}O$	−33.2	18.17
1-Heptene	C_7H_{14}	−118.9	12.41
Hexachlorobenzene	C_6Cl_6	228.83	25.2
Hexachloroethane	C_2Cl_6	186.8t	9.75
Hexacontane	$C_{60}H_{122}$	99.3	193.2
Hexacosane	$C_{26}H_{54}$	56.1	60.0
Hexadecane	$C_{16}H_{34}$	18.12	53.36
Hexadecanoic acid	$C_{16}H_{32}O_2$	62.5	53.7
1-Hexadecanol	$C_{16}H_{34}O$	49.2	33.6
Hexafluorobenzene	C_6F_6	5.03	11.59
Hexafluoroethane	C_2F_6	−100.05	2.69
Hexamethylbenzene	$C_{12}H_{18}$	165.5	20.6
Hexanal	$C_6H_{12}O$	−56	13.3
Hexane	C_6H_{14}	−95.35	13.08
1,6-Hexanedioic acid	$C_6H_{10}O_4$	152.5	36.3
1,6-Hexanediol	$C_6H_{14}O_2$	41.5	22.2
1-Hexanol	$C_6H_{14}O$	−47.4	15.38
2-Hexanone	$C_6H_{12}O$	−55.5	14.9
3-Hexanone	$C_6H_{12}O$	−55.4	13.49
Hexatetracontane	$C_{46}H_{94}$	87.6	176.0
Hexatriacontane	$C_{36}H_{74}$	75.8	87.7
1-Hexene	C_6H_{12}	−139.76	9.35
cis-2-Hexene	C_6H_{12}	−141.11	8.88
Hydrogen cyanide	CHN	−13.29	8.41
p-Hydroquinone	$C_6H_6O_2$	172.4	26.8
2-Hydroxybenzoic acid	$C_7H_6O_3$	159.0	14.2
Imidazole	$C_3H_4N_2$	89.5	12.82
Indan	C_9H_{10}	−51.38	8.60
Indene	C_9H_8	−1.5	10.20
Indomethacin	$C_{19}H_{16}ClNO_4$	160	36.9
Iodobenzene	C_6H_5I	−31.3	9.75
Isobutane	C_4H_{10}	−159.4	4.54
Isobutene	C_4H_8	−140.7	5.92
Isopentane	C_5H_{12}	−159.77	5.15
Isopropylamine	C_3H_9N	−95.13	7.33
Isopropylbenzene	C_9H_{12}	−96.02	7.33
1-Isopropyl-4-methylbenzene	$C_{10}H_{14}$	−67.94	9.66
Isoquinoline	C_9H_7N	26.47	13.54
Khellin	$C_{14}H_{12}O_5$	154	32.3
Maleic anhydride	$C_4H_2O_3$	52.56	13.60
Methane	CH_4	−182.47	0.94
Methanethiol	CH_4S	−123	5.91
Methanol	CH_4O	−97.53	3.215
Methyl acetate	$C_3H_6O_2$	−98.25	7.49
Methylamine	CH_5N	−93.5	6.13
2-Methylaniline	C_7H_9N	−14.41	11.66
3-Methylaniline	C_7H_9N	−31.3	7.9
4-Methylaniline	C_7H_9N	43.6	18.9

Name	Molecular formula	t_m/°C	$\Delta_{fus}H$/ kJ mol^{-1}
Methyl benzoate	$C_8H_8O_2$	−12.4	9.74
2-Methyl-1,3-butadiene	C_5H_8	−145.9	4.93
2-Methyl-2-butanol	$C_5H_{12}O$	−9.1	4.46
3-Methyl-2-butanone	$C_5H_{10}O$	−93.1	9.34
2-Methyl-1-butene	C_5H_{10}	−137.53	7.91
3-Methyl-1-butene	C_5H_{10}	−168.43	5.36
2-Methyl-2-butene	C_5H_{10}	−133.72	7.60
Methyl *tert*-butyl ether	$C_5H_{12}O$	−108.6	7.60
Methylcyclohexane	C_7H_{14}	−126.6	6.75
Methylcyclopentane	C_6H_{12}	−142.42	6.93
Methylcyclopropane	C_4H_8	−177.6	2.8
2-Methylfuran	C_5H_6O	−91.3	8.55
2-Methylheptane	C_8H_{18}	−109.02	11.92
3-Methylheptane	C_8H_{18}	−120.48	11.69
4-Methylheptane	C_8H_{18}	−121.0	10.8
2-Methylhexane	C_7H_{16}	−118.2	9.19
Methylhydrazine	CH_6N_2	−52.36	10.42
Methyl methacrylate	$C_5H_8O_2$	−47.55	14.4
1-Methylnaphthalene	$C_{11}H_{10}$	−30.43	6.95
2-Methylnaphthalene	$C_{11}H_{10}$	34.6	12.13
Methyl nitrate	CH_3NO_3	−83.0	8.24
Methyloxirane	C_3H_6O	−111.9	6.53
2-Methylpentane	C_6H_{14}	−153.6	6.27
3-Methylpentane	C_6H_{14}	−162.90	5.30
2-Methyl-1-propanol	$C_4H_{10}O$	−101.9	6.32
2-Methyl-2-propanol	$C_4H_{10}O$	25.69	6.70
2-Methylpyridine	C_6H_7N	−66.68	9.72
3-Methylpyridine	C_6H_7N	−18.14	14.18
4-Methylpyridine	C_6H_7N	3.67	12.58
N-Methylurea	$C_2H_6N_2O$	104.9	14.0
Morpholine	C_4H_9NO	−4.8	14.5
Naphthalene	$C_{10}H_8$	80.26	19.01
1-Naphthol	$C_{10}H_8O$	95.0	23.1
2-Naphthol	$C_{10}H_8O$	121.5	18.1
Neopentane	C_5H_{12}	−16.4	3.10
Niacinamide	$C_6H_6N_2O$	130	23.2
2-Nitroaniline	$C_6H_6N_2O_2$	71.0	16.1
3-Nitroaniline	$C_6H_6N_2O_2$	113.4	23.6
4-Nitroaniline	$C_6H_6N_2O_2$	147.5	21.2
Nitrobenzene	$C_6H_5NO_2$	5.7	12.12
Nitroethane	$C_2H_5NO_2$	−89.5	9.85
Nitromethane	CH_3NO_2	−28.38	9.70
2-Nitrophenol	$C_6H_5NO_3$	44.8	17.7
3-Nitrophenol	$C_6H_5NO_3$	96.8	20.6
4-Nitrophenol	$C_6H_5NO_3$	113.6	18.8
Nitrosobenzene	C_6H_5NO	67	31.0
4-Nitrotoluene	$C_7H_7NO_2$	51.63	16.81
Nonacosane	$C_{29}H_{60}$	63.7	66.9
Nonadecane	$C_{19}H_{40}$	32.0	45.8
Nonanal	$C_9H_{18}O$	−19.3	30.5
Nonane	C_9H_{20}	−53.46	15.47
Nonanoic acid	$C_9H_{18}O_2$	12.4	19.82
5-Nonanone	$C_9H_{18}O$	−3.8	24.93
Octacosane	$C_{28}H_{58}$	61.1	65.1
Octadecane	$C_{18}H_{38}$	28.2	61.7
1-Octadecanol	$C_{18}H_{38}O$	57.9	45
Octane	C_8H_{18}	−56.82	20.73
Octanoic acid	$C_8H_{16}O_2$	16.5	21.35
1-Octanol	$C_8H_{18}O$	−14.8	23.7
Octatriacontane	$C_{38}H_{78}$	78.6	133.2
1-Octene	C_8H_{16}	−101.7	15.31
2-Oxepanone	$C_6H_{10}O_2$	−1.0	13.83
Oxetane	C_3H_6O	−97	6.5
Oxirane	C_2H_4O	−112.5	5.17
4-Oxopentanoic acid	$C_5H_8O_3$	33	9.22
Paraldehyde	$C_6H_{12}O_3$	12.6	13.5
Pentachloroethane	C_2HCl_5	−28.78	11.3
Pentacontane	$C_{50}H_{102}$	92.1	162.4
Pentacosane	$C_{25}H_{52}$	53.93	56.9
Pentadecane	$C_{15}H_{32}$	9.95	34.6
cis-1,3-Pentadiene	C_5H_8	−140.8	5.64
trans-1,3-Pentadiene	C_5H_8	−87.4	7.14
1,4-Pentadiene	C_5H_8	−148.2	6.12
Pentaerythritol	$C_5H_{12}O_4$	258	4.8
Pentafluorobenzene	C_6HF_5	−47.4	10.87
Pentafluorophenol	C_6HF_5O	37.5	16.41
2,3,4,5,6-Pentafluorotoluene	$C_7H_3F_5$	−29.78	13.1
Pentane	C_5H_{12}	−129.67	8.40
Pentanedioic acid	$C_5H_8O_4$	97.8	20.3
Pentanenitrile	C_5H_9N	−96.2	9
1-Pentanethiol	$C_5H_{12}S$	−75.65	17.53
Pentanoic acid	$C_5H_{10}O_2$	−33.6	14.16
1-Pentanol	$C_5H_{12}O$	−77.6	10.50
2-Pentanone	$C_5H_{10}O$	−76.8	10.63
3-Pentanone	$C_5H_{10}O$	−39	11.59
Pentatriacontane	$C_{35}H_{72}$	74.6	86.3
1-Pentene	C_5H_{10}	−165.12	5.94
cis-2-Pentene	C_5H_{10}	−151.36	7.11
trans-2-Pentene	C_5H_{10}	−140.21	8.35
Perfluoroacetone	C_3F_6O	−125.45	8.38
Perfluorobutane	C_4F_{10}	−129.1	7.66
Perfluorocyclobutane	C_4F_8	−40.19	2.77
Perfluoroheptane	C_7F_{16}	−51.2	6.95
Perfluorohexane	C_6F_{14}	−88.2	6.84
Perfluoropropane	C_3F_8	−147.70	0.477
Perfluorotoluene	C_7F_8	−65.49	11.54
Perylene	$C_{20}H_{12}$	277.76	31.9
Phenacetin	$C_{10}H_{13}NO_2$	134	33.0
Phenanthrene	$C_{14}H_{10}$	99.24	16.46
Phenobarbital	$C_{12}H_{12}N_2O_3$	174.0	27.8
Phenol	C_6H_6O	40.89	11.51
α-Phenylbenzeneacetic acid	$C_{14}H_{12}O_2$	147.29	31.3
Phenylbutazone	$C_{19}H_{20}N_2O_2$	105	27.7
Phenylhydrazine	$C_6H_8N_2$	20.6	14.05
Piperidine	$C_5H_{11}N$	−11.02	14.85
Potassium acetate	$C_2H_3KO_2$	309	7.65
Propane	C_3H_8	−187.63	3.50
1,3-Propanediol	$C_3H_8O_2$	−27.7	7.1
Propanenitrile	C_3H_5N	−92.78	5.03
1-Propanethiol	C_3H_8S	−113.13	5.48
2-Propanethiol	C_3H_8S	−130.5	5.74
Propanoic acid	$C_3H_6O_2$	−20.5	10.66
1-Propanol	C_3H_8O	−124.39	5.37
2-Propanol	C_3H_8O	−87.9	5.41
Propene	C_3H_6	−185.24	3.003
Propylamine	C_3H_9N	−84.75	10.97
Propylbenzene	C_9H_{12}	−99.6	9.27
Propylcyclohexane	C_9H_{18}	−94.9	10.37
Pyrazine	$C_4H_4N_2$	51.0	12.9
1*H*-Pyrazole	$C_3H_4N_2$	70.7	14.0

Name	Molecular formula	t_m/°C	$\Delta_{fus}H$/ kJ mol⁻¹
Pyrene	$C_{16}H_{10}$	150.62	17.36
Pyridine	C_5H_5N	−41.70	8.28
Pyrocatechol	$C_6H_6O_2$	104.6	22.8
Pyrrole	C_4H_5N	−23.39	7.91
Pyrrolidine	C_4H_9N	−57.79	8.58
Quinoline	C_9H_7N	−14.78	10.66
Resorcinol	$C_6H_6O_2$	109.4	20.4
Sebacic acid	$C_{10}H_{18}O_4$	130.9	40.8
Sodium acetate	$C_2H_3NaO_2$	328.2	17.9
Sodium hydrogen carbonate	$CHNaO_3$	527	25
Spiro[2.2]pentane	C_5H_8	−107.0	6.43
Stearic acid	$C_{18}H_{36}O_2$	69.3	61.2
trans-Stilbene	$C_{14}H_{12}$	124.2	27.7
Styrene	C_8H_8	−30.65	10.9
Succinic acid	$C_4H_6O_4$	187.9	32.4
Succinic anhydride	$C_4H_4O_3$	119	20.4
Succinonitrile	$C_4H_4N_2$	58.06	3.70
Sulfacetamide	$C_8H_{10}N_2O_3S$	183	22.4
Sulfadiazine	$C_{10}H_{10}N_4O_2S$	258	42.6
Sulfamerazine	$C_{11}H_{12}N_4O_2S$	236	38.7
Sulfamethoxazole	$C_{10}H_{11}N_3O_3S$	170	32.2
Sulfamethoxypyridazine	$C_{11}H_{12}N_4O_3S$	182.5	31.3
Sulfapyridine	$C_{11}H_{11}N_3O_2S$	192	34.4
Sulfathiazole	$C_9H_9N_3O_2S_2$	202	26.4
Sulfisoxazole	$C_{11}H_{13}N_3O_3S$	196	30.2
o-Terphenyl	$C_{18}H_{14}$	56.20	17.19
p-Terphenyl	$C_{18}H_{14}$	213.9	35.3
Tetrabromomethane	CBr_4	92.3	3.76
1,1,2,2-Tetrachloro-1,2-difluoroethane	$C_2Cl_4F_2$	24.8	3.67
1,1,2,2-Tetrachloroethane	$C_2H_2Cl_4$	−42.4	9.17
Tetrachloroethene	C_2Cl_4	−22.3	10.88
Tetrachloromethane	CCl_4	−22.62	2.56
Tetracontane	$C_{40}H_{82}$	81.5	135.5
Tetracosane	$C_{24}H_{50}$	50.4	54.4
Tetradecane	$C_{14}H_{30}$	5.82	45.07
Tetradecanoic acid	$C_{14}H_{28}O_2$	54.2	45.1
1-Tetradecanol	$C_{14}H_{30}O$	38.2	25.1*
1,2,3,5-Tetrafluorobenzene	$C_6H_2F_4$	−46.25	6.36
1,2,4,5-Tetrafluorobenzene	$C_6H_2F_4$	3.88	15.05
Tetrafluoroethene	C_2F_4	−131.15	7.72
Tetrafluoromethane	CF_4	−183.60	0.704
Tetrahydrofuran	C_4H_8O	−108.44	8.54
Tetrahydropyran	$C_5H_{10}O$	−49.1	1.8
Tetrahydrothiophene	C_4H_8S	−96.2	7.35
1,2,4,5-Tetramethylbenzene	$C_{10}H_{14}$	79.3	21
Tetramethyl lead	$C_4H_{12}Pb$	−30.2	10.80
2,2,3,3-Tetramethylpentane	C_9H_{20}	−9.75	2.33
2,2,4,4-Tetramethylpentane	C_9H_{20}	−66.54	9.74
Tetramethylsilane	$C_4H_{12}Si$	−99.06	6.87
Tetramethylstannane	$C_4H_{12}Sn$	−55.1	9.30
Tetratetracontane	$C_{44}H_{90}$	85.6	149.6
Tetratriacontane	$C_{34}H_{70}$	72.5	79.4
1*H*-Tetrazole	CH_2N_4	157.3	18.2
Thiazole	C_3H_3NS	−33.62	9.57
Thietane	C_3H_6S	−73.24	8.25
Thiophene	C_4H_4S	−38.21	5.07
Thiourea	CH_4N_2S	178	14.0
Thymol	$C_{10}H_{14}O$	49.5	21.3
Toluene	C_7H_8	−94.95	6.64
o-Toluic acid	$C_8H_8O_2$	103.5	19.5
m-Toluic acid	$C_8H_8O_2$	109.9	15.7
p-Toluic acid	$C_8H_8O_2$	179.6	22.7
Triacontane	$C_{30}H_{62}$	65.1	68.3
1,3,5-Triazine	$C_3H_3N_3$	80.3	14.56
Tribromomethane	$CHBr_3$	8.69	11.05
Trichloroacetic acid	$C_2HCl_3O_2$	59.2	5.90
1,2,3-Trichlorobenzene	$C_6H_3Cl_3$	51.3	17.9
1,2,4-Trichlorobenzene	$C_6H_3Cl_3$	16.92	16.4
1,3,5-Trichlorobenzene	$C_6H_3Cl_3$	62.8	18.1
1,1,1-Trichloroethane	$C_2H_3Cl_3$	−30.01	2.35
1,1,2-Trichloroethane	$C_2H_3Cl_3$	−36.3	11.46
Trichloroethene	C_2HCl_3	−84.7	8.45
Trichlorofluoromethane	CCl_3F	−110.44	6.89
Trichloromethane	$CHCl_3$	−63.41	9.5
1,1,2-Trichloro-1,2,2-trifluoroethane	$C_2Cl_3F_3$	−36.22	2.47
Tricosane	$C_{23}H_{48}$	47.76	50.86
Tridecane	$C_{13}H_{28}$	−5.4	28.50
1-Tridecanol	$C_{13}H_{28}O$	31.7	41.4
1,1,1-Trifluoroethane	$C_2H_3F_3$	−111.3	6.19
Trifluoromethane	CHF_3	−155.2	4.06
Triiodomethane	CHI_3	121.2	16.44
Trimethoprim	$C_{14}H_{18}N_4O_3$	199	49.4
Trimethylamine	C_3H_9N	−117.1	7
1,2,3-Trimethylbenzene	C_9H_{12}	−25.4	8.18
1,2,4-Trimethylbenzene	C_9H_{12}	−43.77	13.19
1,3,5-Trimethylbenzene	C_9H_{12}	−44.72	9.51
2,2,3-Trimethylbutane	C_7H_{16}	−24.6	2.26
2,2,4-Trimethylpentane	C_8H_{18}	−107.3	9.20
1,3,5-Trinitrobenzene	$C_6H_3N_3O_6$	122.9	15.4
Trinitroglycerol	$C_3H_5N_3O_9$	13.5	21.87
2,4,6-Trinitrotoluene	$C_7H_5N_3O_6$	80.5	22.9
1,3,5-Trioxane	$C_3H_6O_3$	60.29	15.11
Triphenylamine	$C_{18}H_{15}N$	126.5	24.9
Triphenylene	$C_{18}H_{12}$	197.8	24.74
Tritriacontane	$C_{33}H_{68}$	71.2	79.5
Undecane	$C_{11}H_{24}$	−25.5	22.2
Urea	CH_4N_2O	133.3	13.9
o-Xylene	C_8H_{10}	−25.2	13.6
m-Xylene	C_8H_{10}	−47.8	11.6
p-Xylene	C_8H_{10}	13.25	17.12
2,3-Xylenol	$C_8H_{10}O$	72.5	21.0
2,5-Xylenol	$C_8H_{10}O$	74.8	23.4
2,6-Xylenol	$C_8H_{10}O$	45.8	18.9
3,4-Xylenol	$C_8H_{10}O$	65.1	18.1
3,5-Xylenol	$C_8H_{10}O$	63.4	17.4

COMPRESSIBILITY AND EXPANSION COEFFICIENTS OF LIQUIDS

This table gives data on the variation of the density of some common liquids with pressure and temperature. The pressure dependence is described to first order by the isothermal compressibility coefficient κ defined as

$$\kappa_T = -(1/V)\,(\partial V/\partial P)_T$$

where V is the volume, and the temperature dependence by the cubic expansion coefficient α,

$$\alpha_V = (1/V)\,(\partial V/\partial T)_P$$

Substances are listed by molecular formula in the Hill order. More precise data on the variation of density with temperature over a wide temperature range can be found in Reference 1.

References

1. Lide, D. R., and Kehiaian, H. V., *CRC Handbook of Thermophysical and Thermochemical Data*, CRC Press, Boca Raton, FL, 1994.
2. Le Neindre, B., *Effets des Hautes et Très Hautes Pressions*, in *Techniques de l'Ingénieur*, Paris, 1991.
3. *Landolt-Börnstein, Numerical Data and Functional Relationships in Science and Technology, New Series*, IV/4, *High-Pressure Properties of Matter*, Springer-Verlag, Heidelberg, 1980.
4. Riddick, J.A., Bunger, W.B., and Sakano, T.K., *Organic Solvents, Fourth Edition*, John Wiley & Sons, New York, 1986.
5. Isaacs, N. S., *Liquid Phase High Pressure Chemistry*, John Wiley, New York, 1981.

Molecular formula	Name	Isothermal compressibility		Cubic expansion coefficient	
		t/°C	$\kappa_T \times 10^4$/MPa^{-1}	t/°C	$\alpha_V \times 10^3$/°C^{-1}
Cl_3P	Phosphorus trichloride	20	9.45	20	1.9
H_2O	Water	20	4.591	20	0.206
		25	4.524	25	0.256
		30	4.475	30	0.302
Hg	Mercury	20	0.401	20	0.1811
CCl_4	Tetrachloromethane	20	10.50	20	1.14
		40	12.20	40	1.21
		70	15.6	70	1.33
$CHBr_3$	Tribromomethane	50	8.76	25	0.91
$CHCl_3$	Trichloromethane	20	9.96	20	1.21
		50	12.9	50	1.33
CH_2Br_2	Dibromomethane	27	6.85		
CH_2Cl_2	Dichloromethane	25	10.3	25	1.39
CH_3I	Iodomethane	27	10.3	25	1.26
CH_4O	Methanol	20	12.14	20	1.49
		40	13.83	40	1.59
CS_2	Carbon disulfide	20	9.38	20	1.12
		40	10.6	35	1.16
C_2Cl_4	Tetrachloroethylene	25	7.56	25	1.02
C_2HCl_3	Trichloroethylene	25	8.57	25	1.17
$C_2H_2Cl_2$	*trans*-1,2-Dichloroethylene	25	11.2	25	1.36
$C_2H_4Cl_2$	1,1-Dichloroethane	20	7.97	25	0.93
$C_2H_4Cl_2$	1,2-Dichloroethane	30	8.46	20	1.14
$C_2H_4O_2$	Acetic acid	20	9.08	20	1.08
		80	13.7	80	1.38
C_2H_5Br	Bromoethane	20	11.53	20	1.31
C_2H_5I	Iodoethane	20	9.82	25	1.17
C_2H_6O	Ethanol	20	11.19	20	1.40
		70	15.93	70	1.67
$C_2H_6O_2$	Ethylene glycol	20	3.64	20	0.626
C_3H_6O	Acetone	20	12.62	20	1.46
		40	15.6	40	1.57
C_3H_7Br	1-Bromopropane	0	10.22	25	1.2
C_3H_7Cl	1-Chloropropane	0	12.09	20	1.4
C_3H_7I	1-Iodopropane	0	10.22	25	1.09
C_3H_8O	1-Propanol	0	8.43	0	1.22
C_3H_8O	2-Propanol	40	13.32	40	1.55
$C_3H_8O_2$	1,2-Propanediol	0	4.45	20	0.695
$C_3H_8O_2$	1,3-Propanediol	0	4.09	20	0.61
$C_3H_8O_3$	Glycerol	0	2.54	20	0.520

Molecular formula	Name	Isothermal compressibility t/°C	$\kappa_T \times 10^4$/MPa^{-1}	Cubic expansion coefficient t/°C	$\alpha_V \times 10^3$/°C^{-1}
$C_4H_8O_2$	Ethyl acetate	20	11.32	20	1.35
		60	16.2	60	1.54
C_4H_9Br	1-Bromobutane	25	10.26	20	1.13
C_4H_9I	1-Iodobutane	0	7.73	25	1.02
$C_4H_{10}O$	1-Butanol	0	8.10	0	1.12
$C_4H_{10}O$	Diethyl ether	20	18.65	20	1.65
		30	20.85	30	1.72
$C_4H_{10}O_3$	Diethylene glycol	0	3.34	20	0.635
C_5H_{10}	Cyclopentane	20	13.31	20	1.35
$C_5H_{11}Br$	1-Bromopentane	0	8.42	25	1.04
$C_5H_{11}I$	1-Iodopentane	0	7.56		
C_5H_{12}	Pentane	25	21.80	25	1.64
$C_5H_{12}O$	1-Pentanol	0	7.71	0	1.02
C_6H_5Br	Bromobenzene	20	6.46	20	0.86
C_6H_5Cl	Chlorobenzene	20	7.45	20	0.94
$C_6H_5NO_2$	Nitrobenzene	20	4.93	25	0.833
C_6H_6	Benzene	25	9.66	25	1.14
		45	11.28	45	1.21
C_6H_6O	Phenol	60	6.05	60	0.82
C_6H_7N	Aniline	20	4.53	20	0.81
		80	6.32	80	0.91
C_6H_{12}	Cyclohexane	20	11.30	20	1.15
		60	15.2	60	1.29
C_6H_{14}	Hexane	25	16.69	25	1.41
		45	20.27	45	1.52
C_6H_{14}	2-Methylpentane	0	13.97	25	1.43
C_6H_{14}	3-Methylpentane	0	14.57	25	1.40
C_6H_{14}	2,3-Dimethylbutane	20	17.97	25	1.39
$C_6H_{14}O$	1-Hexanol	25	8.24	25	1.03
$C_6H_{15}NO_3$	Triethanolamine	0	3.61	55	0.53
C_7H_8	Toluene	20	8.96	20	1.05
		50	11.0	50	1.13
C_7H_8O	Anisole	20	6.60	20	0.951
C_7H_{14}	Cycloheptane	20	9.22		
C_7H_{16}	Heptane	25	14.38	25	1.26
C_8H_{10}	*o*-Xylene	25	8.10	25	0.96
C_8H_{10}	*m*-Xylene	20	8.46	20	0.99
C_8H_{10}	*p*-Xylene	25	8.59	25	1.00
C_8H_{16}	Cyclooctane	20	8.03		
C_8H_{18}	Octane	25	12.82	25	1.16
		45	15.06	45	1.23
$C_8H_{18}O$	1-Octanol	25	7.64	25	0.827
C_9H_{12}	Mesitylene	25	8.14	25	0.94
$C_9H_{14}O_6$	Triacetin	0	4.49	25	0.94
C_9H_{20}	Nonane	25	11.75	25	1.08
$C_{10}H_{22}$	Decane	25	10.94	25	1.02
$C_{11}H_{24}$	Undecane	25	10.31	25	0.97
$C_{12}H_{26}$	Dodecane	25	9.88	25	0.93
$C_{13}H_{28}$	Tridecane	25	9.48	25	0.90
$C_{14}H_{30}$	Tetradecane	25	9.10	25	0.87
$C_{15}H_{32}$	Pentadecane	25	8.82		
$C_{16}H_{22}O_4$	Butyl phthalate	0	5.0	25	0.86
$C_{16}H_{34}$	Hexadecane	25	8.57		
		45	9.78		
$C_{19}H_{36}O_2$	Methyl oleate	0	6.18	60	0.85

TEMPERATURE AND PRESSURE DEPENDENCE OF LIQUID DENSITY

Ivan Cibulka

This table records parameters of the Tait equation (Refs. 1,2) that gives the ratio between the density at pressure P, $\rho(T,P)$, relative to the density at a reference pressure, $\rho(T,P_{ref})$, at the same temperature T:

$$\frac{\rho(T,P)}{\rho(T,P_{\mathrm{ref}})}=\frac{1}{1-C(T)\ln\left\{\dfrac{B(T)+P}{B(T)+P_{\mathrm{ref}}}\right\}},$$

$$C(T)=a_1+b_1(T/\mathrm{K})+c_1(T/\mathrm{K})^2, \tag{1}$$

$$B(T)/\mathrm{MPa}=a_2+a_3(T/\mathrm{K})+a_4(T/\mathrm{K})^2+a_5(T/\mathrm{K})^3+a_6(T/\mathrm{K})^4$$

Parameters a_i (i = 1, ..., 6) and b_1, c_1 were adjusted to selected experimental data using the weighted least-squares method. Parameters b_1 and c_1 are zero for most substances, and therefore their values are given in footnotes for those few substances where the statistical significance was justified. The reference pressure is P_{ref} = 0.101325 MPa at temperatures either at or below the normal boiling point temperature (T_{nbp}) and $P_{ref} = P_{sat}(T)$ (saturated vapor pressure) at temperatures $T > T_{nbp}$. Ranges of validity of the equation (T_{min}, T_{max}, P_{max}) are derived from ranges of experimental data; the minimal pressure of validity is taken as P_{ref}, *i.e.*, interpolation between P_{ref} and lowest experimental pressure is allowed. The upper limit of application is the freezing line (if not limited by the ranges of validity). To avoid any large-scale extrapolation, the validity ranges are rectangular areas ($T_{max} - T_{min}$)P_{max}. If in particular temperature interval(s) the maximum experimental pressure exceeded the given value of P_{max}, then the maximum pressure given in the table is denoted by (r) which means that the validity range given in the table is a rectangular subset of the non-rectangular experimental T, P range. In a few cases P_{max} is given as a ratio where the first value corresponds to T_{min} and the second one to T_{max}, *i.e.*, the validity range has approximately a trapezoidal shape. Values of parameters were taken from the papers (Refs. 3–10) where detailed information on the fits, experimental data, and application ranges is available. The numerical values of the parameters are different from those reported in papers (Refs. 3–10) since the forms of polynomials $C(T)$ and $B(T)$ differ. Besides, the parameters recorded in the table below must not necessarily correspond to those in Refs. 3–10 since some fits were updated using newly published experimental data.

In the second line for each substance, the parameters of the smoothing function for the density at reference pressure ρ (T,P_{ref}) are given. The functions are either the polynomial expansion,

$$\rho(T,P_{\mathrm{ref}})/(\mathrm{kg\,m^{-3}})=\sum_{i=1}^{N_p}a_i(T/\mathrm{K})^{(i-1)} \tag{2}$$

or the expansion,

$$\rho(T,P_{\mathrm{ref}})=\rho_c\left[1+\sum_{i=1}^{N_p}a_i(1-T_r)^{(i/3)}\right],\quad T_r=T/T_c \tag{3}$$

where N_p is the number of adjustable parameters. Values of the critical density ρ_c and the critical temperature T_c used for the fits using Eq. 3 are also recorded in the table. Parameters were mostly taken from Refs. 3–10, those for 1-alkanols C_1 to C_{10} and n-alkanes C_5 to C_{16} are from Ref. 11. Data used for the fits were predominantly recommended values published in the TRC Thermodynamic Tables (Refs. 12,13), sometimes combined with the original experimental data or, in a few cases, the original experimental data were correlated.

RMSD is a relative root-mean-square deviation (in per cent) between experimental values of density and those calculated from the particular function (Tait Eq. 1 or Eqs. 2, 3);

$$RMSD/\%=100\left\{\frac{1}{N}\sum_{i=1}^{N}\left(\frac{\rho_{\mathrm{exp}}-\rho_{\mathrm{calc}}}{\rho_{\mathrm{exp}}}\right)^2\right\}^{1/2}$$

where N is the number of experimental values included in the fit.

If the maximum temperature T_{max} of validity of the Tait equation is greater than the normal boiling point temperature T_{nbp}, then parameters of the Wagner equation in the form of either

$$P_{\mathrm{sat}}(T)=P_c\exp\left[\frac{a_1(1-T_r)+a_2(1-T_r)^{1.5}+a_3(1-T_r)^{2.5}+a_4(1-T_r)^5}{T_r}\right] \tag{4}$$

or

$$P_{\mathrm{sat}}(T)=P_c\exp\left[\frac{a_1(1-T_r)+a_2(1-T_r)^{1.5}+a_3(1-T_r)^3+a_4(1-T_r)^6}{T_r}\right] \tag{5}$$

where $T_r = T/T_c$ are recorded in the third line for each substance. Values of the critical pressure P_c and critical temperature T_c used in Eqs. 4 and 5 are also recorded in the table (values of the critical temperature may differ a little from those recorded for the function, Eq. 3). Parameters of Eqs. 4 and 5 were taken mostly from the papers by McGarry (Ref. 14) and Ambrose and Walton (Ref. 15); in a few cases, the fits were performed using original experimental data or in combination with the recommended values from the TRC Thermodynamic Tables (Refs. 12,13).

The two right-hand-most columns gives values of the isothermal compressibility coefficient, $\kappa_T = -(1/V)(\partial V/\partial P)_T = (1/\rho)(\partial\rho/\partial P)_T$, and the isobaric cubic expansion coefficient, $\alpha_p = (1/V)(\partial V/\partial T)_p = -(1/\rho)(\partial\rho/\partial T)_p$, calculated for T = 298.15 K and P = 0.101325 MPa from the Tait Eq. 1 and from the ρ (T, P_{ref}) equation, respectively. In a very few cases when the lower temperature limit of the Tait equation T_{min} is greater than 298.15 K, the extrapolated values of isothermal compressibility are given.

References

1. Tait, P. G., in *Physics and Chemistry of the Voyage of H.M.S. Challenger*, Vol. II, Part IV, Thomson, C. W., and Murray, J., Eds., H.M.S.O., London, 1889.
2. Tamman, G., *Z. Phys. Chem.* 17, 620, 1895.

3. Cibulka, I., and Ziková, M., *J. Chem. Eng. Data* 39, 876, 1994.
4. Cibulka, I., and Hnědkovský, L., *J. Chem. Eng. Data* 41, 657, 1996.
5. Cibulka, I., Hnědkovský, L., and Takagi, T., *J. Chem. Eng. Data* 42, 2, 1997.
6. Cibulka, I., Hnědkovský, L., and Takagi, T., *J. Chem. Eng. Data* 42, 415, 1997.
7. Cibulka, I., and Takagi, T., *J. Chem. Eng. Data* 44, 411, 1999.
8. Cibulka, I., and Takagi, T., *J. Chem. Eng. Data* 44, 1105, 1999.
9. Cibulka, I., Takagi, T., and Růžička, K., *J. Chem. Eng. Data* 46, 2, 2001.
10. Cibulka, I., and Takagi, T., *J. Chem. Eng. Data* 47, 1037, 2002.
11. Cibulka, I., *Fluid Phase Equilib.* 89, 1, 1993.
12. TRC Thermodynamic Tables, Hydrocarbons, Thermodynamics Research Center (TRC), NIST, Thermophysical Properties Division, Boulder, Colorado.
13. TRC Thermodynamic Tables, Non-Hydrocarbons, Thermodynamics Research Center (TRC), NIST, Thermophysical Properties Division, Boulder, Colorado.
14. McGarry, J., *Ind. Eng. Chem., Process Des. Develop.* 22, 313, 1983.
15. Ambrose, D., and Walton, J., *Pure Appl. Chem.* 61, 1395, 1989.

Eq.	a_1	a_2	a_3	a_4	a_5	a_6	T_{min}/T_{max} K	P_{max} MPa	T_c K	P_c MPa	ρ_c kg m^{-3}	RMSD %	T_{nbp} K	κ_T GPa^{-1}	α_P kK^{-1}
CCl_4 Tetrachloromethane (Ref. 9)															
1	$9.33340\cdot10^{-2}$	$1.11363\cdot10^{3}$	−8.68453	$2.80698\cdot10^{-2}$	$-4.22880\cdot10^{-5}$	$2.37923\cdot10^{-8}$	273/413	51/388				0.040	349.9	1.074	1.209
3	1.58994	2.51946	−5.82313	6.96793	−2.51359		253/554		556.40		557.33	0.042			
5	−7.07139	1.71497	−2.89930	−2.49466			250/556		556.40	4.551					
$CHBr_3$ Tribromomethane (Ref. 9)															
1	$1.03492\cdot10^{-1}$	$2.64208\cdot10^{2}$	$-4.57399\cdot10^{-1}$				323/368	150/343				0.058	422.3	0.809	0.907
2	$3.55953\cdot10^{3}$	−1.96212	$-1.08712\cdot10^{-3}$				283/403					0.001			
$CHCl_3$ Trichloromethane (Ref. 9)															
1	$9.57210\cdot10^{-2}$	$4.79593\cdot10^{2}$	−1.84011	$1.81340\cdot10^{-3}$			273/348	100(r)				0.031	334.4	1.037	1.274
3	3.56339	−3.86051	3.35636				213/333		536.40		499.49	0.043			
5	−6.95546	1.16625	−2.13970	−3.44421			215/536		536.40	5.366					
CH_2Cl_2 Dichloromethane (Ref. 9)															
1	$9.76370\cdot10^{-2}$	$5.24365\cdot10^{2}$	−2.06633	$2.09494\cdot10^{-3}$			293/423	100(r)				0.091	313.4	1.032	1.428
3	3.00368	−2.19763	2.34269				178/383		510.00		440.07	0.014			
5	−7.35739	2.17546	−4.07038	3.50701			233/510		510.00	6.300					
CH_3I Iodomethane (Ref. 9)															
1	$9.54770\cdot10^{-2}$	$5.36810\cdot10^{2}$	−2.25115	$2.53188\cdot10^{-3}$			253/313	160				0.038	315.6	1.052	1.255
2	$3.48981\cdot10^{3}$	−7.47709	$1.83592\cdot10^{-2}$	$-2.36742\cdot10^{-5}$			213/313					0.011			
CH_4O Methanol (Ref. 3)															
1[a]	$1.15068\cdot10^{-1}$	$6.49718\cdot10^{2}$	−4.34583	$1.30722\cdot10^{-2}$	$-2.00292\cdot10^{-5}$	$1.20566\cdot10^{-8}$	183/483	104(r)				0.059	337.7	1.231	1.201
3	2.62781	−4.04742	$1.58343\cdot10$	$-2.25066\cdot10$	$1.09160\cdot10$	$3.04774\cdot10^{-1}$	175/509		512.60		272.00	0.180			
4	−8.63571	1.17982	−2.47900	−1.02400			175/513		512.64	8.092					
C_2Cl_4 Tetrachloroethylene (Ref. 9)															
1	$1.01727\cdot10^{-1}$	$1.36610\cdot10^{2}$					298/298	101				0.007	394.5	0.744	1.012
3	1.10232	1.45142					298/313		620.20		571.84	0.019			
C_2HCl_3 Trichloroethylene (Ref. 9)															
1	$1.00987\cdot10^{-1}$	$1.16005\cdot10^{2}$					298/298	101				0.010	360.4	0.870	1.142
3	1.30631	1.33512					291/315		571.00		513.24	0.008			
$C_2H_4Cl_2$ 1,1-Dichloroethane (Ref. 9)															
1	$9.79930\cdot10^{-2}$	$4.40906\cdot10^{2}$	−1.67041	$1.60492\cdot10^{-3}$			298/398	101(r)				0.028	330.4	1.144	1.320
3	2.81773	−2.41755	2.40906				263/398		523.00		419.32	0.084			
4	−9.51254	7.49332	−7.87028	$6.32540\cdot10^{-1}$			320/450		523.00	5.070					
$C_2H_4Cl_2$ 1,2-Dichloroethane (Ref. 9)															
1	$9.60440\cdot10^{-2}$	$5.39476\cdot10^{2}$	−1.94358	$1.79758\cdot10^{-3}$			278/398	101(r)				0.027	356.7	0.801	1.158
3	2.77660	−2.33774	2.31530				263/398		561.60		439.82	0.063			
5	−7.36864	1.76727	−3.34295	−1.43530			260/566		566.00	5.362					
$C_2H_4O_2$ Acetic acid (Ref. 5)															
1	$9.44550\cdot10^{-2}$	$7.22131\cdot10^{2}$	−3.33202	$4.23588\cdot10^{-3}$			293/328	25/253				0.040	391.1	0.897	1.085
3	4.74861	$-1.76644\cdot10$	$4.44807\cdot10$	$-4.82535\cdot10$	$1.98308\cdot10$		293/493		592.71		351.19	0.003			
C_2H_5Br Bromoethane (Ref. 9)															
1	$9.43490\cdot10^{-2}$	$7.70215\cdot10^{2}$	−4.06470	$5.75736\cdot10^{-3}$			253/313	157(r)				0.055	311.6	1.344	1.475
3	1.50516	1.35063					253/313		503.90		506.82	0.058			
5	−9.14807	5.49831	−6.68657	6.27287			301/504		503.80	6.232					
C_2H_6O Ethanol (Ref. 3)															
1[b]	$1.09012\cdot10^{-1}$	$5.22523\cdot10^{2}$	−2.99367	$8.25136\cdot10^{-3}$	$-1.31014\cdot10^{-5}$	$8.74008\cdot10^{-9}$	193/480	200(r)				0.055	351.4	1.117	1.097
3[c]	$-9.92640\cdot10^{-1}$	$3.80287\cdot10$	$-1.81117\cdot10^{2}$	$4.45905\cdot10^{2}$	$-5.88518\cdot10^{2}$	$3.93920\cdot10^{2}$	159/508		513.88		276.00	0.160			
4	−8.68587	1.17831	−4.87620	1.58800			159/514		513.92	6.132					
$C_2H_6O_2$ Ethylene glycol (Ref. 6)															
1	$9.50140\cdot10^{-2}$	$6.74427\cdot10^{2}$	−1.77594	$1.27583\cdot10^{-3}$			298/378	100(r)				0.031	470.5	0.368	0.639
3	1.77482	1.11208					298/378		790.00		333.70	0.022			
C_3H_6O Acetone (Ref. 5)															
1	$9.92390\cdot10^{-2}$	$6.22264\cdot10^{2}$	−2.94315	$3.73329\cdot10^{-3}$			278/323	392				0.032	329.2	1.293	1.457
3	1.83805	$5.82830\cdot10^{-1}$	$-1.00291\cdot10^{-1}$	$5.70410\cdot10^{-1}$			179/506		508.10		277.90	0.025			

Eq.	a_1	a_2	a_3	a_4	a_5	a_6	T_{min}/T_{max} K	P_{max} MPa	T_c K	P_c MPa	ρ_c kg m^{-3}	RMSD %	T_{nbp} K	κ_T GPa^{-1}	α_P kK^{-1}
C_3H_7Br 1-Bromopropane (Ref. 9)															
1	$8.88710 \cdot 10^{-2}$	$4.24716 \cdot 10^{2}$	−1.81352	$2.55049 \cdot 10^{-3}$	$-1.23227 \cdot 10^{-6}$		280/500	100				0.094	344.0	1.137	1.251
3	1.65835	1.17536					280/340		536.00		455.53	0.009			
4	−9.00919	7.16793	7.05282	1.00727			300/536		536.00	4.750					
C_3H_7Cl 1-Chloropropane (Ref. 9)															
1	$1.07245 \cdot 10^{-1}$	$2.78606 \cdot 10^{2}$	$-5.17553 \cdot 10^{-2}$	$-3.83546 \cdot 10^{-3}$	$5.85824 \cdot 10^{-6}$		273/368	98(r)				0.078	319.6	1.382	1.462
3	4.15272	−5.73954	4.74431				253/343		503.00		309.22	0.015			
5	−7.55764	2.60153	−5.06041	3.31163			248/503		503.00	4.580					
C_3H_7I 1-Iodopropane (Ref. 9)															
1	$1.01044 \cdot 10^{-1}$	$3.01114 \cdot 10^{2}$	$-6.61560 \cdot 10^{-1}$	$8.74342 \cdot 10^{-5}$			273/368	1177				0.070	375.7	0.904	1.072
2	$2.22323 \cdot 10^{3}$	−1.38867	$-7.96985 \cdot 10^{-4}$				288/361					0.196			
C_3H_8O 1-Propanol (Ref. 3)															
1[d]	$9.60290 \cdot 10^{-2}$	$3.10861 \cdot 10^{2}$	$-8.35531 \cdot 10^{-1}$	$4.27987 \cdot 10^{-5}$	$7.78234 \cdot 10^{-7}$		170/524	49(r)				0.109	370.3	1.007	0.999
3	$9.40534 \cdot 10^{-1}$	$1.29442 \cdot 10$	$-5.39519 \cdot 10$	$1.13639 \cdot 10^{2}$	$-1.13866 \cdot 10^{2}$	$4.33832 \cdot 10$	153/530		536.74		274.00	0.160			
4	−8.53706	1.96214	−7.69180	2.94500			147/537		536.78	5.168					
C_3H_8O 2-Propanol (Ref. 6)															
1	$8.90020 \cdot 10^{-2}$	$1.87411 \cdot 10^{2}$	$-2.36391 \cdot 10^{-1}$	$-4.24787 \cdot 10^{-4}$			273/400	50(r)				0.054	355.4	1.123	1.099
3	$6.24542 \cdot 10^{-1}$	6.18938	−6.79039	2.49009			243/430		508.30		273.16	0.093			
5	−8.16927	$-9.43213 \cdot 10^{-2}$	−8.10040	7.85000			250/508		508.30	4.742					
$C_3H_8O_2$ 1,2-Propanediol (Ref. 6)															
1	$8.38940 \cdot 10^{-2}$	$9.09931 \cdot 10^{2}$	−4.03075	$5.24295 \cdot 10^{-3}$			273/368	200				0.019	460.2	0.481	0.714
2	$1.18071 \cdot 10^{3}$	$-2.50645 \cdot 10^{-1}$	$-0.05917 \cdot 10^{-4}$				283/363					0.017			
$C_3H_8O_2$ 1,3-Propanediol (Ref. 6)															
1	$9.10310 \cdot 10^{-2}$	$5.95160 \cdot 10^{2}$	−1.64714	$1.36994 \cdot 10^{-3}$			273/368	200				0.067	487.6	0.403	0.593
3	3.64841	−2.06696	1.22738				283/363		685.70		318.39	0.002			
$C_3H_8O_3$ Glycerol (Ref. 6)															
1	$1.14255 \cdot 10^{-1}$	$9.25959 \cdot 10^{2}$	−1.58140	$4.48909 \cdot 10^{-4}$			223/368	686				0.061	563.1	0.231	0.486
3	6.94960	−9.25236	5.43535				214/364		800.00		351.51	0.069			
$C_4H_8O_2$ Ethyl acetate (Ref. 5)															
1	$8.94090 \cdot 10^{-2}$	$1.05394 \cdot 10^{3}$	−7.58700	$1.98689 \cdot 10^{-2}$	$-1.82587 \cdot 10^{-5}$		253/343	49(r)				0.106	350.3	1.204	1.425
3	1.69043	2.64150	−4.70476	3.52368			253/473		523.20		308.06	0.015			
C_4H_9Br 1-Bromobutane (Ref. 9)															
1	$9.76600 \cdot 10^{-2}$	$4.97301 \cdot 10^{2}$	−2.02541	$2.24791 \cdot 10^{-3}$			273/368	1177				0.074	374.8	1.046	1.119
3	1.59525	1.23746					293/323		572.00		421.60	0.003			
$C_4H_{10}O$ 1-Butanol (Ref. 3)															
1[e]	$1.01542 \cdot 10^{-1}$	$5.01888 \cdot 10^{2}$	−2.41576	$5.35877 \cdot 10^{-3}$	$-7.31982 \cdot 10^{-6}$	$4.60628 \cdot 10^{-9}$	195/524	49(r)				0.053	390.9	0.914	0.949
3	−3.57379	$5.02450 \cdot 10$	$-1.75934 \cdot 10^{2}$	$3.08588 \cdot 10^{2}$	$-2.65628 \cdot 10^{2}$	$8.94997 \cdot 10$	186/559		563.01		271.00	0.352			
4	−8.40615	2.23010	−8.24860	$-7.11000 \cdot 10^{-1}$			185/563		563.05	4.424					
$C_4H_{10}O$ Diethyl ether (Ref. 5)															
1	$9.70250 \cdot 10^{-2}$	$3.95413 \cdot 10^{2}$	−1.68339	$1.82398 \cdot 10^{-3}$			293/353	981				0.239	307.6	1.740	1.657
3	3.31808	−8.55179	$2.09100 \cdot 10$	$-2.10672 \cdot 10$	8.26525		140/430		466.74		264.72	0.027			
5	−7.29916	1.24828	−2.91931	−3.36740			250/467		466.74	3.646					
C_5H_{10} Cyclopentane (Ref. 8)															
1	$8.85580 \cdot 10^{-2}$	$8.94968 \cdot 10^{2}$	−7.76576	$3.01636 \cdot 10^{-2}$	$-5.82645 \cdot 10^{-5}$	$4.43996 \cdot 10^{-8}$	193/353	48/196				0.050	322.4	1.308	1.325
3	1.68431	$6.19856 \cdot 10^{-1}$	$4.90670 \cdot 10^{-2}$	$2.19121 \cdot 10^{-1}$			179/508		511.70		275.04	0.031			
5	−6.51809	$3.84422 \cdot 10^{-1}$	−1.11706	−4.50275			289/512		511.60	4.509					
$C_5H_{11}Br$ 1-Bromopentane (Ref. 9)															
1	$8.98660 \cdot 10^{-2}$	$5.28042 \cdot 10^{2}$	−2.41620	$3.92466 \cdot 10^{-3}$	$-2.31231 \cdot 10^{-6}$		283/523	98				0.061	402.7	0.943	1.076
2	$1.60209 \cdot 10^{3}$	−1.30486					283/363					0.006			
4	−8.59341	3.46288	−3.06661	−4.07997			294/606		605.90	3.900					
C_5H_{12} Pentane (Ref. 4)															
1[f]	$1.00620 \cdot 10^{-1}$	$6.52014 \cdot 10^{2}$	−4.94167	$1.54854 \cdot 10^{-2}$	$-2.36811 \cdot 10^{-5}$	$1.44530 \cdot 10^{-8}$	173/373	87/285				0.119	309.2	2.133	1.610
3	1.17756	3.89157	−5.50896	3.29181			143/444		469.80		232.00	0.154			
4	−7.30698	1.75845	−2.16290	−2.91300			143/470		469.80	3.375					
$C_5H_{12}O$ 1-Pentanol (Ref. 3)															
1	$9.52000 \cdot 10^{-2}$	$3.00019 \cdot 10^{2}$	$-5.62481 \cdot 10^{-1}$	$-5.98193 \cdot 10^{-4}$	$1.15744 \cdot 10^{-6}$		233/550	59(r)				0.149	411.2	0.866	0.905
3	−2.61129	$4.23375 \cdot 10$	$-1.49640 \cdot 10^{2}$	$2.62210 \cdot 10^{2}$	$-2.23090 \cdot 10^{2}$	$7.37879 \cdot 10$	195/585		588.11		270.00	0.353			
4	−8.98005	3.91624	−9.90810	−2.19100			196/588		588.15	3.909					
C_6H_5Br Bromobenzene (Ref. 9)															
1	$9.73880 \cdot 10^{-2}$	$4.87590 \cdot 10^{2}$	−1.49313	$1.16283 \cdot 10^{-3}$			278/358	100(r)				0.043	429.2	0.668	0.901
3	2.84948	−2.30055	2.31394				298/358		670.00		484.60	0.002			
C_6H_5Cl Chlorobenzene (Ref. 9)															
1[g]	$1.06896 \cdot 10^{-1}$	$5.95178 \cdot 10^{2}$	−2.52143	$4.02563 \cdot 10^{-3}$	$-3.10524 \cdot 10^{-6}$	$1.06813 \cdot 10^{-9}$	278/583	50(r)				0.057	404.9	0.752	0.969
3	$1.17137 \cdot 10^{-1}$	$1.23712 \cdot 10$	$-2.84464 \cdot 10$	$2.94048 \cdot 10$	$-1.07075 \cdot 10$		253/630		632.40		365.45	0.008			
5	−7.58700	2.26551	−4.09118	$1.70377 \cdot 10^{-1}$			335/632		632.40	4.519					

Eq.	a_1	a_2	a_3	a_4	a_5	a_6	T_{min}/T_{max} K	P_{max} MPa	T_c K	P_c MPa	ρ_c kg m^{-3}	RMSD %	T_{nbp} K	κ_T GPa^{-1}	α_P kK^{-1}
$C_6H_5NO_2$ Nitrobenzene (Ref. 10)															
1	$9.31940\cdot10^{-2}$	$6.10698\cdot10^{2}$	−1.89353	$1.56305\cdot10^{-3}$			293/358	100				0.005	484.0	0.503	0.826
3	4.47929	−5.72846	4.39478				273/373		718.00		362.09	0.027			
C_6H_6 Benzene (Ref. 7)															
1	$9.36560\cdot10^{-2}$	$3.28066\cdot10^{2}$	$2.79270\cdot10^{-2}$	$-6.54288\cdot10^{-3}$	$1.66249\cdot10^{-5}$	$-1.24527\cdot10^{-8}$	283/499	58(r)				0.079	353.2	0.965	1.222
3	1.81679	$-1.44828\cdot10^{-1}$	3.65168	−5.85231	3.50125		279/561		562.16		301.60	0.024			
4	−7.01433	1.55256	−1.84790	−3.71300			288/562		562.16	4.898					
C_6H_7N Aniline (Ref. 10)															
1	$9.43630\cdot10^{-2}$	$7.27479\cdot10^{2}$	−2.43017	$2.23750\cdot10^{-3}$			298/358	100(r)				0.043	457.3	0.467	0.830
3	2.71043	−3.14236	1.48381·10	−3.89610·10	4.62777·10	−1.91216·10	263/699		699.00		332.60	0.093			
C_6H_{12} Cyclohexane (Ref. 8)															
1	$8.51590\cdot10^{-2}$	$6.06614\cdot10^{2}$	−3.64438	$9.15077\cdot10^{-3}$	$-1.15323\cdot10^{-5}$	$5.95736\cdot10^{-9}$	287/523	81(r)				0.087	353.9	1.135	1.207
3	1.62642	$9.95043\cdot10^{-1}$	$-2.55766\cdot10^{-1}$	$-8.79980\cdot10^{-2}$	$4.74534\cdot10^{-1}$		273/553		553.50		273.25	0.038			
5	−6.96009	1.31328	−2.75683	−2.45491			293/554		553.64	4.075					
C_6H_{14} Hexane (Ref. 4)															
1[h]	$1.05863\cdot10^{-1}$	$4.87649\cdot10^{2}$	−2.71631	$5.75951\cdot10^{-3}$	$-5.74075\cdot10^{-6}$	$2.33450\cdot10^{-9}$	223/498	200(r)				0.112	341.9	1.645	1.387
3	1.59756	1.84266	−1.72631	$4.94308\cdot10^{-1}$	$6.46314\cdot10^{-1}$		183/507		507.90		234.00	0.124			
4	−7.53998	1.83759	−2.54380	−3.16300			178/508		507.90	3.035					
C_6H_{14} 2-Methylpentane (Ref. 8)															
1	$8.95370\cdot10^{-2}$	$3.60920\cdot10^{2}$	−1.70650	$2.61411\cdot10^{-3}$	$-1.30788\cdot10^{-6}$		273/473	32(r)				0.272	333.4	1.793	1.430
3	5.72227	−2.82931·10	7.80587·10	−9.08956·10	3.90010·10		273/473		497.50		234.82	0.090			
5	−7.28750	1.29015	−2.97853	−2.17234			240/498		498.10	3.033					
C_6H_{14} 3-Methylpentane (Ref. 8)															
1	$8.79150\cdot10^{-2}$	$2.68648\cdot10^{2}$	−1.00644	$9.29350\cdot10^{-4}$			293/473	32(r)				0.086	336.4	1.714	1.350
3	4.03149	−1.18248·10	2.42559·10	−1.83295·10	4.33405		293/473		504.50		234.82	0.143			
5	−7.27084	1.26113	−2.81741	−2.17642			235/504		504.40	3.122					
C_6H_{14} 2,3-Dimethylbutane (Ref. 8)															
1[i]	$6.09940\cdot10^{-2}$	$2.88312\cdot10^{2}$	−1.38880	$2.23422\cdot10^{-3}$	$-1.24468\cdot10^{-6}$		208/473	32(r)				0.072	331.2	1.831	1.364
3	4.23996	−1.47401·10	3.56742·10	−3.61641·10	1.37668·10		208/463		499.98		240.72	0.078			
5	−7.27870	1.56349	−3.05387	−1.57752			235/500		500.30	3.146					
$C_6H_{14}O$ 1-Hexanol (Ref. 3)															
1	$9.45430\cdot10^{-2}$	$3.56654\cdot10^{2}$	$-9.63938\cdot10^{-1}$	$3.19317\cdot10^{-4}$	$7.27932\cdot10^{-7}$	$-3.68450\cdot10^{-10}$	298/503	400				0.045	430.5	0.828	0.878
3	$-1.55309\cdot10^{-1}$	1.77262·10	−5.76007·10	9.99026·10	−8.65883·10	2.96281·10	223/607		610.70		268.00	0.191			
4	−9.49034	5.13288	−1.05817·10	−5.15400			226/611		610.70	3.470					
C_7H_8 Toluene (Ref. 7)															
1[j]	$1.08655\cdot10^{-1}$	$6.58594\cdot10^{2}$	−3.36063	$6.79417\cdot10^{-3}$	$-6.73894\cdot10^{-6}$	$2.80015\cdot10^{-9}$	179/583	50(r)				0.052	383.8	0.900	1.080
3	2.33057	−3.10784	1.04839·10	−1.28168·10	6.29458	$-2.64550\cdot10^{-1}$	178/588		591.79		291.59	0.041			
5	−7.28607	1.38091	−2.83433	−2.79168			309/592		591.72	4.106					
C_7H_8O Anisole (Ref. 5)															
1	$9.57370\cdot10^{-2}$	$3.76997\cdot10^{2}$	$-7.76199\cdot10^{-1}$				298/353	196				0.032	426.9	0.657	0.949
3	2.32023	−1.28589	1.68888				273/353		645.60		335.84	0.031			
C_7H_{14} Cycloheptane (Ref. 8)															
1	$9.09350\cdot10^{-2}$	$3.25145\cdot10^{2}$	$-8.81409\cdot10^{-1}$	$4.29576\cdot10^{-4}$			294/393	40(r)				0.041	391.6	0.904	1.062
2	$1.06207\cdot10^{3}$	$-8.56800\cdot10^{-1}$					298/353					0.003			
C_7H_{16} Heptane (Ref. 4)															
1[k]	$8.14360\cdot10^{-2}$	$5.31891\cdot10^{2}$	−3.38490	$9.42976\cdot10^{-3}$	$-1.36183\cdot10^{-5}$	$8.10257\cdot10^{-9}$	198/511	150(r)				0.153	371.6	1.429	1.251
3	1.33159	3.30092	−4.50961	2.76549			183/538		540.11		236.00	0.377			
4	−7.77404	1.85614	−2.82980	−3.50700			183/540		540.15	2.735					
C_8H_{10} o-Xylene (Ref. 7)															
1	$8.02720\cdot10^{-2}$	$4.52200\cdot10^{2}$	−1.81150	$2.45174\cdot10^{-3}$	$-1.15349\cdot10^{-6}$		257/598	51(r)				0.082	417.6	0.806	0.944
3	1.93903	$6.93214\cdot10^{-1}$	$-5.52518\cdot10^{-1}$	$7.42481\cdot10^{-1}$			248/628		630.30		287.72	0.006			
5	−7.53357	1.40968	−3.10985	−2.85992			337/630		630.25	3.733					
C_8H_{10} m-Xylene (Ref. 7)															
1	$8.13550\cdot10^{-2}$	$5.95204\cdot10^{2}$	−3.17132	$6.99286\cdot10^{-3}$	$-7.69418\cdot10^{-6}$	$3.47267\cdot10^{-9}$	230/598	20(r)				0.054	412.3	0.857	0.987
3	1.96653	$7.70424\cdot10^{-1}$	$-8.14622\cdot10^{-1}$	$9.44937\cdot10^{-1}$			225/613		617.05		282.36	0.007			
5	−7.59222	1.39441	−3.22746	−2.40376			332/617		616.97	3.537					
C_8H_{10} p-Xylene (Ref. 7)															
1	$8.45420\cdot10^{-2}$	$4.38870\cdot10^{2}$	−1.77395	$2.41453\cdot10^{-3}$	$-1.13719\cdot10^{-6}$		288/598	51(r)				0.076	411.5	0.894	1.003
3	2.01680	$6.60504\cdot10^{-1}$	$-7.70676\cdot10^{-1}$	$9.97453\cdot10^{-1}$			286/603		616.20		280.13	0.005			
5	−7.63495	1.50724	−3.19678	−2.78710			331/616		616.15	3.513					
C_8H_{16} Cyclooctane (Ref. 8)															
1	$8.66040\cdot10^{-2}$	$3.85135\cdot10^{2}$	−1.19668	$9.00890\cdot10^{-4}$			314/394	40				0.026	422.2	0.798	0.960
3	1.25843	1.53566					293/394		647.20		273.70	0.040			

Eq.	a_1	a_2	a_3	a_4	a_5	a_6	T_{min}/T_{max} K	P_{max} MPa	T_c K	P_c MPa	ρ_c kg m^{-3}	RMSD %	T_{nbp} K	κ_T GPa^{-1}	α_P kK^{-1}
C_8H_{18} Octane (Ref. 4)															
1[l]	$1.09943\cdot10^{-1}$	$1.17639\cdot10^{3}$	$-1.02441\cdot10$	$3.73444\cdot10^{-2}$	$-6.42362\cdot10^{-5}$	$4.23381\cdot10^{-8}$	248/393	108(r)				0.069	398.8	1.255	1.158
3	1.96977	−1.10062	6.36417	−8.69348	4.42005		216/563		568.91		237.00	0.221			
4	−8.04937	2.03865	−3.31200	−3.64800			216/569		568.95	2.490					
$C_8H_{18}O$ 1-Octanol (Ref. 3)															
1	$9.39730\cdot10^{-2}$	$5.22871\cdot10^{2}$	−1.99322	$2.56178\cdot10^{-3}$	$-1.13255\cdot10^{-6}$		283/623	79(r)				0.177	468.3	0.743	0.844
3	−3.53373	$5.48581\cdot10$	$-1.99653\cdot10^{2}$	$3.45797\cdot10^{2}$	$-2.85030\cdot10^{2}$	$9.05445\cdot10$	258/643		652.50		266.00	0.262			
4	$-1.00144\cdot10$	5.90629	$-1.04026\cdot10$	−9.04800			258/653		652.50	2.860					
C_9H_{12} Mesitylene (Ref. 7)															
1	$8.84360\cdot10^{-2}$	$4.00368\cdot10^{2}$	−1.18616	$7.38886\cdot10^{-4}$			238/362	200				0.056	437.9	0.786	0.944
3	$1.23001\cdot10$	$-3.23171\cdot10$	$3.37700\cdot10$	$-1.06994\cdot10$			238/353		637.25		277.59	0.012			
C_9H_{20} Nonane (Ref. 4)															
1[m]	$1.04176\cdot10^{-1}$	$6.60874\cdot10^{2}$	−3.94793	$9.86964\cdot10^{-3}$	$-1.23014\cdot10^{-5}$	$6.27200\cdot10^{-9}$	248/511	65(r)				0.064	424.0	1.125	1.088
3	1.92778	$9.30219\cdot10^{-1}$	−1.33413	1.39282			223/511		594.90		238.00	0.078			
4	−8.32886	2.25707	−3.82570	−3.73200			220/595		594.90	2.290					
$C_{10}H_{22}$ Decane (Ref. 4)															
1[n]	$7.76110\cdot10^{-2}$	$5.12895\cdot10^{2}$	−2.98749	$7.58178\cdot10^{-3}$	$-9.71758\cdot10^{-6}$	$5.01670\cdot10^{-9}$	248/503	200(r)				0.079	447.3	1.096	1.042
3	$3.29139\cdot10^{-1}$	7.36434	−9.98510	5.28361			243/511		617.61		239.00	0.108			
4	−8.60643	2.44659	−4.29250	−3.90800			244/618		617.65	2.105					
$C_{11}H_{24}$ Undecane (Ref. 4)															
1[o]	$9.36240\cdot10^{-2}$	$9.78153\cdot10^{3}$	$-1.13416\cdot10^{2}$	$4.97036\cdot10^{-1}$	$-9.65143\cdot10^{-4}$	$6.97765\cdot10^{-7}$	258/423	50/500				0.044	469.1	1.078	0.998
3	−2.75532·10 1	9.28503	$-1.19923\cdot10$	5.96957			258/473		638.81		240.00	0.069			
$C_{12}H_{26}$ Dodecane (Ref. 4)															
1	9.0545010^{-2}	$9.00366\cdot10^{3}$	$-1.03134\cdot10^{2}$	$4.49084\cdot10^{-1}$	$-8.70464\cdot10^{-4}$	$6.31007\cdot10^{-7}$	268/393	20/442				0.083	489.5	0.998	0.977
3	$-3.04946\cdot10^{-2}$	8.32535	$-1.08268\cdot10$	5.55178			263/483		658.60		240.00	0.055			
$C_{13}H_{28}$ Tridecane (Ref. 4)															
1	$8.79880\cdot10^{-2}$	$3.53129\cdot10^{2}$	−1.17372	$1.00334\cdot10^{-3}$			303/473	500				0.080	508.6	0.951	0.947
3	2.52779	−1.64864	2.10099				265/473		676.00		240.00	0.139			
$C_{14}H_{30}$ Tetradecane (Ref. 4)															
1	$9.01310\cdot10^{-2}$	$2.44902\cdot10^{2}$	$-4.92510\cdot10^{-1}$				293/358	50/367				0.040	526.7	0.918	0.924
3	3.50992	−4.09768	3.61064				279/372		693.00		241.00	0.050			
C\` $_{15}H_{32}$ Pentadecane (Ref. 4)															
1	$8.85030\cdot10^{-2}$	$4.04942\cdot10^{2}$	−1.37693	$1.25370\cdot10^{-3}$			311/408	69/655				0.048	543.8	0.835	0.882
3	1.08801	1.82524					288/408		708.00		241.00	0.038			
$C_{16}H_{34}$ Hexadecane (Ref. 4)															
1	$9.05120\cdot10^{-2}$	$-6.87292\cdot10^{2}$	8.07128	$-2.59370\cdot10^{-2}$	$2.60661\cdot10^{-5}$		293/393	10/451				0.047	560.0	0.866	0.889
3	2.61306	−1.99964	2.40011				293/490		722.00		241.00	0.035			

[a] $b_1 = -5.322\cdot10^{-5}$
[b] $b_1 = -4.588\cdot10^{-5}$
[c] Additional term $a_7(1 - T_r)^{(7/3)}$ is included in Eq. 3 with $a_7 = -1.04344\ 10^2$
[d] $b_1 = -3.058\cdot10^{-5}$
[e] $b_1 = -3.268\cdot10^{-5}$
[f] $b_1 = -3.788\cdot10^{-5}$
[g] $b_1 = -3.690\cdot10^{-5}$
[h] $b_1 = -4.522\cdot10^{-5}$
[i] $b_1 = 4.085\cdot10^{-5}$
[j] $b_1 = -5.004\cdot10^{-5}$
[k] $b_1 = 7.924\cdot10^{-5}$, $c_1 = -1.513\cdot10^{-7}$
[l] $b_1 = -5.741\cdot10^{-5}$
[m] $b_1 = -2.960\cdot10^{-5}$
[n] $b_1 = 2.778\cdot10^{-5}$
[o] $b_1 = -1.105\cdot10^{-5}$

VOLUMETRIC PROPERTIES OF AQUEOUS SODIUM CHLORIDE SOLUTIONS

This table gives the following properties of aqueous solutions of NaCl as a function of temperature and concentration:

Specific volume v (reciprocal of density) in cm^3/g
Isothermal compressibility $\kappa_T = -(1/v)(\partial v/\partial P)_T$ in GPa^{-1}
Cubic expansion coefficient $\alpha_v = (1/v)(\partial v/\partial T)_P$ in kK^{-1}

All data refer to a pressure of 100 kPa (1 bar). The reference gives properties over a wider range of temperature and pressure.

Reference

Rogers, P. S. Z., and Pitzer, K. S., *J. Phys. Chem. Ref. Data*, 11, 15, 1982.

	Molality in mol/kg								
t/°C	0.100	0.250	0.500	0.750	1.000	2.000	3.000	4.000	5.000
Specific volume v in cm^3/g									
0	0.995732	0.989259	0.978889	0.968991	0.959525	0.925426	0.896292	0.870996	0.848646
10	0.995998	0.989781	0.979804	0.970256	0.961101	0.927905	0.899262	0.874201	0.851958
20	0.997620	0.991564	0.981833	0.972505	0.963544	0.930909	0.902565	0.877643	0.855469
25	0.998834	0.992832	0.983185	0.973932	0.965038	0.932590	0.904339	0.879457	0.857301
30	1.000279	0.994319	0.984735	0.975539	0.966694	0.934382	0.906194	0.881334	0.859185
40	1.003796	0.997883	0.988374	0.979243	0.970455	0.938287	0.910145	0.885276	0.863108
50	1.008064	1.002161	0.992668	0.983551	0.974772	0.942603	0.914411	0.889473	0.867241
60	1.0130	1.0071	0.9976	0.9885	0.9797	0.9474	0.9191	0.8940	0.8716
70	1.0186	1.0127	1.0031	0.9939	0.9851	0.9526	0.9240	0.8987	0.8762
80	1.0249	1.0188	1.0092	0.9999	0.9909	0.9581	0.9293	0.9037	0.8809
90	1.0317	1.0256	1.0157	1.0063	0.9972	0.9640	0.9348	0.9089	0.8858
100	1.0391	1.0329	1.0228	1.0133	1.0040	0.9703	0.9406	0.9144	0.8910
Isothermal Compressibility κ_T in GPa^{-1}									
0	0.503	0.492	0.475	0.459	0.443	0.389	0.346	0.315	0.294
10	0.472	0.463	0.449	0.436	0.423	0.377	0.341	0.313	0.294
20	0.453	0.446	0.433	0.422	0.411	0.371	0.338	0.313	0.294
25	0.447	0.440	0.428	0.417	0.407	0.369	0.337	0.313	0.294
30	0.443	0.436	0.425	0.414	0.404	0.367	0.337	0.313	0.294
40	0.438	0.432	0.421	0.411	0.401	0.367	0.338	0.315	0.296
50	0.438	0.431	0.421	0.411	0.402	0.369	0.340	0.317	0.299
60	0.44	0.44	0.43	0.42	0.41	0.38	0.35	0.32	0.30
70	0.45	0.44	0.43	0.42	0.42	0.38	0.36	0.33	0.31
80	0.46	0.45	0.44	0.43	0.43	0.39	0.37	0.34	0.32
90	0.47	0.47	0.46	0.45	0.44	0.41	0.38	0.35	0.33
100	0.49	0.48	0.47	0.46	0.45	0.42	0.39	0.37	0.34
Cubic expansion coefficient α_v in kK^{-1}									
0	−0.058	−0.026	0.024	0.069	0.110	0.237	0.313	0.355	
10	0.102	0.123	0.156	0.186	0.213	0.297	0.349	0.380	
20	0.218	0.232	0.254	0.274	0.292	0.349	0.384	0.406	
25	0.267	0.278	0.296	0.312	0.327	0.373	0.401	0.420	
30	0.311	0.320	0.334	0.347	0.359	0.395	0.418	0.433	
40	0.389	0.394	0.402	0.410	0.417	0.438	0.451	0.460	
50	0.458	0.460	0.464	0.467	0.470	0.479	0.484	0.486	
60	0.52	0.52	0.52	0.52	0.52	0.52	0.52	0.52	
70	0.58	0.58	0.58	0.57	0.57	0.56	0.55	0.54	
80	0.64	0.63	0.63	0.62	0.61	0.60	0.58	0.56	
90	0.69	0.68	0.67	0.67	0.66	0.63	0.61	0.59	
100	0.74	0.73	0.72	0.71	0.70	0.66	0.64	0.61	

PROPERTIES OF CRYOGENIC FLUIDS

This table gives physical and thermodynamic properties of eight cryogenic fluids. The properties are:

M	Molar mass in grams per mole
T_t	Triple point temperature in kelvins
P_t	Triple point pressure in kilopascals
ρ_t (l)	Liquid density at the triple point in grams per milliliter
$\Delta_{fus}H$ @ T_t	Enthalpy of fusion at the triple point in joules per gram
T_b	Normal boiling point in kelvins at a pressure of 101325 pascals (760 mmHg)
$\Delta_{vap}H$ @ T_b	Enthalpy of vaporization at the normal boiling point in joules per gram
ρ (l) @ T_b	Liquid density at the normal boiling point in grams per milliliter
ρ (g) @ T_b	Vapor density at the normal boiling point in grams per liter
C_p (l) @ T_b	Liquid heat capacity at constant pressure at the normal boiling point in joules per gram kelvin
C_p (g) @ T_b	Vapor heat capacity at constant pressure at the normal boiling point in joules per gram kelvin
T_c	Critical temperature in kelvins
P_c	Critical pressure in megapascals
ρ_c	Critical density in grams per milliliter

In the case of air, the value given for the triple point temperature is the incipient solidification temperature, and the normal boiling point value is the incipient boiling (bubble) point. See Reference 3 for more details.

References

1. Younglove, B. A., *J. Phys. Chem. Ref. Data*, 11, Suppl. 1, 1982.
2. Daubert, T. E., Danner, R. P., Sibul, H. M., and Stebbins, C. C., *Physical and Thermodynamic Properties of Pure Compounds: Data Compilation*, extant 1994 (core with 4 supplements), Taylor & Francis, Bristol, PA (also available as database).
3. Sytchev, V. V., et al., *Thermodynamic Properties of Air*, Hemisphere Publishing, New York, 1987.
4. Jacobsen, R. T., Stewart, R. B., and Jahangiri, M., *J. Phys. Chem. Ref. Data*, 15, 735, 1986. [Nitrogen]
5. Stewart, R. B., Jacobsen, R. T., and Wagner, W., *J. Phys. Chem. Ref. Data*, 20, 917, 1991. [Oxygen]
6. McCarty, R. D., *J. Phys. Chem. Ref. Data*, 2, 923, 1973. [Helium] Also, Donnelly, R. J., private communication.
7. Stewart, R. B. and Jacobsen, R. T., *J. Phys. Chem. Ref. Data*, 18, 639, 1989. [Argon]
8. Setzmann, U. and Wagner, W., *J. Phys. Chem. Ref. Data*, 20, 1061, 1991. [Methane]
9. Vargaftik, N. B., *Thermophysical Properties of Liquids and Gases*, 2nd ed., John Wiley, New York, 1975.

Property	Units	Air	N_2	O_2	H_2	He	Ne	Ar	Kr	Xe	CH_4
M	g/mol	28.96	28.014	31.999	2.0159	4.0026	20.180	39.948	83.800	131.290	16.043
T_t	K	59.75	63.15	54.3584	13.8		24.5561	83.8058	115.8	161.4	90.694
P_t	kPa		12.463	0.14633	7.042		50	68.95	72.92	81.59	11.696
ρ_t (l)	g/mL	0.959	0.870	1.306	0.0770		1.251	1.417	2.449	2.978	0.4515
$\Delta_{fus}H$ @ T_t	J/g		25.3	13.7	59.5		16.8	28.0	16.3	13.8	58.41
T_b	K	78.67	77.35	90.188	20.28	4.2221	27.07	87.293	119.92	165.10	111.668
$\Delta_{vap}H$ @ T_b	J/g	198.7	198.8	213.1	445	20.7	84.8	161.0	108.4	96.1	510.83
ρ (l) @ T_b	g/mL	0.8754	0.807	1.141	0.0708	0.124901	1.204	1.396	2.418	2.953	0.4224
ρ (g) @ T_b	g/L	3.199	4.622	4.467	1.3390	16.89	9.51	5.79	8.94		1.816
C_p (l) @ T_b	J/g K	1.865	2.042	1.699	9.668	4.545	1.877	1.078	0.533	0.340	3.481
C_p (g) @ T_b	J/g K		1.341	0.980	12.24	9.78		0.570	0.248	0.158	2.218
T_c	K	132.5	126.20	154.581	32.98	5.1953	44.40	150.663	209.40	289.73	190.56
P_c	MPa	3.766	3.390	5.043	1.293	0.227460	2.760	4.860	5.500	5.840	4.592
ρ_c	g/mL	0.316	0.313	0.436	0.031	0.06964	0.484	0.531	0.919	1.110	0.1627

PROPERTIES OF LIQUID HELIUM

The following data were obtained by a critical evaluation of all existing experimental measurements on liquid helium, using a fitting procedure described in the reference. All values refer to liquid helium at saturated vapor pressure; temperatures are on the ITS-90 scale. Several properties show a singularity at the lambda point (2.1768 K).

p: vapor pressure
ρ: density
C_s: molar heat capacity
$\Delta_{vap}H$: molar enthalpy of vaporization
ε: relative permittivity (dielectric constant)
σ: surface tension
α_V: cubic expansion coefficient
η: viscosity
λ: thermal conductivity

Reference

Donnelly, R. J., and Barenghi, C. F., *J. Phys. Chem. Ref. Data*, 27, 1217, 1998.

T/K	p/kPa	ρ/g cm^{-3}	C_s/J mol^{-1} K^{-1}	$\Delta_{vap}H$/J mol^{-1}	ε	σ/mN m^{-1}	$10^3\alpha_V$/K^{-1}	η/μPa s	λ/W cm^{-1} K^{-1}
0.0		0.1451397	0	59.83	1.057255		0.000		
0.5		0.1451377	0.010	70.24	1.057254	0.3530	0.107		
1.0	0.01558	0.1451183	0.415	80.33	1.057246	0.3471	0.309	3.873	
1.5	0.4715	0.1451646	4.468	89.35	1.057265	0.3322	−2.36	1.346	
2.0	3.130	0.1456217	21.28	93.07	1.057449	0.3021	−12.2	1.468	
2.5	10.23	0.1448402	9.083	92.50	1.057135	0.2623	39.4	3.259	0.1497
3.0	24.05	0.1412269	9.944	94.11	1.055683	0.2161	61.5	3.517	0.1717
3.5	47.05	0.1360736	12.37	92.84	1.053615	0.1626	88.7	3.509	0.1868
4.0	81.62	0.1289745	15.96	87.00	1.050770	0.1095	129	3.319	0.1965
4.5	130.3	0.1188552	21.8	75.86	1.046725	0.0609	211		
5.0	196.0		44.7	47.67		0.0157			

PROPERTIES OF REFRIGERANTS

This table gives physical properties of compounds that have been used as working fluids in traditional refrigeration systems or are under consideration as replacements in newer systems. Some are also used as solvents and blowing agents. Many of the compounds listed are believed to be less harmful to the environment than the traditional halocarbon refrigerants.

Compounds are listed by their ASHRAE standard refrigerant designations (Reference 1), which appear in the first column. These codes are often prefixed by symbols such as CFC- (for chlorofluorocarbon), HCFC- (for hydrochlorofluorocarbon), or simply R- (for refrigerant). The "R" number assigned to refrigerants is specified by ANSI/ASHRAE Standard 34. This system is most useful for the hydrocarbons and halocarbons with one to three carbons; for such molecules the chemical composition can be determined from the number and vice versa. The first digit on the far right is the number of fluorine atoms in the compound. The second digit from the right is one more than the number of hydrogen atoms. The third digit from the right is one less than the number of carbon atoms; for single-carbon compounds, this digit is omitted. The fourth digit from the right is equal to the number of unsaturated carbon–carbon bonds; for saturated compounds, this digit is omitted. The number of bromine and iodine atoms is indicated, if needed, by appending "Bn" or "In" to the digits specified by the above rules, where "n" is the number of bromine or iodine atoms. All atoms not specified by the above are assumed to be chlorine. Appended lowercase letter(s) designate different isomers. Additional rules are used to specify cyclic compounds, ethers, inorganic fluids (R700- and R7000-series), miscellaneous organic compounds (R600-series), and blends (R400- and R500-series).

The properties tabulated are

- t_m normal melting point in °C
- t_b normal boiling point in °C (at 101.325 kPa or 760 mmHg)
- t_c critical temperature in °C
- TLV Threshold Limit Value, which is the maximum safe concentration in air in the workplace, expressed as the time-weighted average (TWA) in parts per million by volume, over an 8-hr workday and 40-hr workweek.

Many of the critical temperatures and normal boiling points have been calculated from the equations of state described in References 9 and 10. These values differ slightly in some cases (generally no more than about 0.1 °C) from values elsewhere in this book, but the differences are probably within the experimental uncertainty. Further references and additional data on the critical properties may be found in the table "Critical Constants" in this Section.

REFERENCES

1. ASHRAE (2007). ANSI/ASHRAE Standard 34-2007 Designation and Safety Classification of Refrigerants, American Society of Heating, Refrigerating and Air-Conditioning Engineers, Atlanta, GA.
2. *ASHRAE Fundamentals Handbook 2001*, Chapter 19, Refrigerants, American Society of Heating, Refrigerating, and Air-Conditioning Engineers, Atlanta, GA, 2001.
3. Platzer, B., Polt, A., and Mauer, G., *Thermophysical Properties of Refrigerants*, Springer, Berlin, 1990.
4. Sako, T., Sato, M., Nakazawa, N., Oowa, M., Yasumoto, M., Ito, H., and Yamashita, S., *J. Chem. Eng. Data* 41, 802, 1996.
5. Schmidt, J. W., Carrillo-Nava, E., and Moldover, M. R., *Fluid Phase Equilib.* 122, 187, 1996.
6. Salvi-Narkhede, M., Wang, B-H., Adcock, J. L., and Van Hook, W. A., *J. Chem. Thermodyn.* 24, 1065, 1992.
7. Fialho, P. S., and Nieto de Castro, C. A., *Int. J. Thermophys.* 21, 385, 2000.
8. Daubert, T. E., Danner, R. P., Sibul, H. M., and Stebbins, C. C., *Physical and Thermodynamic Properties of Pure Compounds: Data Compilation*, extant 2002 (core with supplements), Taylor & Francis, Bristol, PA.
9. McLinden, M.O., Lemmon, E.W., and Huber, M.L., The REFPROP Database for the Thermophysical Properties of Refrigerants, 21st International Congress of Refrigeration, Washington D.C., International Institute of Refrigeration, Paper ICR0443, 2003.
10. Lemmon, E.W., Huber, M.L., and McLinden, M.O., NIST Standard Reference Database 23: Reference Fluid Thermodynamic and Transport Properties - REFPROP, Version 8.0, National Institute of Standards and Technology, Standard Reference Data Program, Gaithersburg, MD, 2007 (www.nist.gov/srd/nist23.htm).

Code	Name	Molecular Formula	CAS Reg. No.	t_m/°C	t_b/°C	t_c/°C	TLV/ ppm
10	Tetrachloromethane	CCl_4	56-23-5	−22.62	76.8	283.4	5
11	Trichlorofluoromethane	CCl_3F	75-69-4	−110.47	23.71	197.96	1000
12	Dichlorodifluoromethane	CCl_2F_2	75-71-8	−157.05	−29.75	111.97	1000
12B1	Bromochlorodifluoromethane	$CBrClF_2$	353-59-3	−159.5	−3.7	153.73	
12B2	Dibromodifluoromethane	CBr_2F_2	75-61-6	−110.1	22.76	198.1	100
13	Chlorotrifluoromethane	$CClF_3$	75-72-9	−181.15	−81.48	28.85	
13B1	Bromotrifluoromethane	$CBrF_3$	75-63-8	−172	−57.8	67.0	1000
14	Tetrafluoromethane	CF_4	75-73-0	−183.61	−128.05	−45.64	
20	Trichloromethane	$CHCl_3$	67-66-3	−63.41	61.17	263.2	10
21	Dichlorofluoromethane	$CHCl_2F$	75-43-4	−130.35	8.86	178.33	10
22	Chlorodifluoromethane	$CHClF_2$	75-45-6	−157.42	−40.81	96.15	1000
22B1	Bromodifluoromethane	$CHBrF_2$	1511-62-2	−145	−14.6	138.83	
23	Trifluoromethane	CHF_3	75-46-7	−155.13	−82.02	26.14	
30	Dichloromethane	CH_2Cl_2	75-09-2	−97.2	40	237	50

Code	Name	Molecular Formula	CAS Reg. No.	t_m/°C	t_b/°C	t_c/°C	TLV/ ppm
31	Chlorofluoromethane	CH_2ClF	593-70-4	–135.1	–9.1	154	
32	Difluoromethane	CH_2F_2	75-10-5	–136.81	–51.65	78.11	
40	Chloromethane	CH_3Cl	74-87-3	–97.7	–24.09	143.10	50
41	Fluoromethane	CH_3F	593-53-3	–143.33	–78.31	44.13	
50	Methane	CH_4	74-82-8	–182.43	–161.48	–82.59	1000
110	Hexachloroethane	C_2Cl_6	67-72-1	186.8	184.7 sp	422	1
111	Pentachlorofluoroethane	C_2Cl_5F	354-56-3	101.3	138		
112	1,1,2,2-Tetrachloro-1,2-difluoroethane	$C_2Cl_4F_2$	76-12-0	24.8	92.8	278	50
112a	1,1,1,2-Tetrachloro-2,2-difluoroethane	$C_2Cl_4F_2$	76-11-9	41.0	92.8		100
113	1,1,2-Trichloro-1,2,2-trifluoroethane	$C_2Cl_3F_3$	76-13-1	–36.22	47.59	214.06	1000
113a	1,1,1-Trichloro-2,2,2-trifluoroethane	$C_2Cl_3F_3$	354-58-5	14.37	45.5	209.7	
114	1,2-Dichloro-1,1,2,2-tetrafluoroethane	$C_2Cl_2F_4$	76-14-2	–92.52	3.59	145.68	1000
114a	1,1-Dichloro-1,2,2,2-tetrafluoroethane	$C_2Cl_2F_4$	374-07-2	–56.6	3.4	145.4	
114B2	1,2-Dibromotetrafluoroethane	$C_2Br_2F_4$	124-73-2	–110.32	47.35	214.6	
115	Chloropentafluoroethane	C_2ClF_5	76-15-3	–99.4	–39.22	79.95	1000
116	Hexafluoroethane	C_2F_6	76-16-4	–100.05	–78.09	19.88	
120	Pentachloroethane	C_2HCl_5	76-01-7	–28.78	162.0		
121	1,1,2,2-Tetrachloro-1-fluoroethane	C_2HCl_4F	354-14-3	–82.6	116.7		
121a	1,1,1,2-Tetrachloro-2-fluoroethane	C_2HCl_4F	354-11-0	–95.3	117.1		
122	1,2,2-Trichloro-1,1-difluoroethane	$C_2HCl_3F_2$	354-21-2	–140	71.9		
122a	1,2,2-Trichloro-1,2-difluoroethane	$C_2HCl_3F_2$	354-15-4	–174	72.5		
122b	1,1,1-Trichloro-2,2-difluoroethane	$C_2HCl_3F_2$	354-12-1		73		
123	2,2-Dichloro-1,1,1-trifluoroethane	$C_2HCl_2F_3$	306-83-2	–107.15	27.82	183.68	
123a	1,2-Dichloro-1,1,2-trifluoroethane	$C_2HCl_2F_3$	354-23-4	–78	29.5	188.4	
124	1-Chloro-1,2,2,2-tetrafluoroethane	C_2HClF_4	2837-89-0	–199.15	–11.96	122.28	
124a	1-Chloro-1,1,2,2-tetrafluoroethane	C_2HClF_4	354-25-6	–117	–11.7	126.7	
125	Pentafluoroethane	C_2HF_5	354-33-6	–100.63	–48.09	66.02	
E125	Trifluoromethyl difluoromethyl ether	C_2HF_5O	3822-68-2	–157	–38	80.8	
130	1,1,2,2-Tetrachloroethane	$C_2H_2Cl_4$	79-34-5	–42.4	145.2	388.00	1
131	1,1,2-Trichloro-2-fluoroethane	$C_2H_2Cl_3F$	359-28-4		102.4		
132	1,2-Dichloro-1,2-difluoroethane	$C_2H_2Cl_2F_2$	431-06-1	–101.2	59.6		
132b	1,2-Dichloro-1,1-difluoroethane	$C_2H_2Cl_2F_2$	1649-08-7	–101.2	46.2		
133	1-Chloro-1,2,2-trifluoroethane	$C_2H_2ClF_3$	431-07-2		17.3		
133a	2-Chloro-1,1,1-trifluoroethane	$C_2H_2ClF_3$	75-88-7	–105.5	6.1	151.86	
133b	1-Chloro-1,1,2-trifluoroethane	$C_2H_2ClF_3$	421-04-5		12		
134	1,1,2,2-Tetrafluoroethane	$C_2H_2F_4$	359-35-3	–89	–19.9	118.59	
134a	1,1,1,2-Tetrafluoroethane	$C_2H_2F_4$	811-97-2	–103.3	–26.07	101.06	
E134	Bis(difluoromethyl) ether	$C_2H_2F_4O$	1691-17-4		2	147.10	
140	1,1,2-Trichloroethane	$C_2H_3Cl_3$	79-00-5	–36.3	113.8	329	10
140a	1,1,1-Trichloroethane	$C_2H_3Cl_3$	71-55-6	–30.01	74.09	272	350
141	1,2-Dichloro-1-fluoroethane	$C_2H_3Cl_2F$	430-57-9	–60	73.8		
141b	1,1-Dichloro-1-fluoroethane	$C_2H_3Cl_2F$	1717-00-6	–103.47	32.05	204.35	
142	1-Chloro-2,2-difluoroethane	$C_2H_3ClF_2$	338-65-8		35.1		
142b	1-Chloro-1,1-difluoroethane	$C_2H_3ClF_2$	75-68-3	–130.43	–9.12	137.11	
143	1,1,2-Trifluoroethane	$C_2H_3F_3$	430-66-0	–84	3.7	156.6	
143a	1,1,1-Trifluoroethane	$C_2H_3F_3$	420-46-2	–111.81	–47.24	72.71	
143m	Methyl trifluoromethyl ether	$C_2H_3F_3O$	421-14-7	–149	–23.66	104.77	
E143a	2,2,2-Trifluoroethyl methyl ether	$C_3H_5F_3O$	460-43-5		31.62	175.83	
150	1,2-Dichloroethane	$C_2H_4Cl_2$	107-06-2	–35.7	83.5	288	10
150a	1,1-Dichloroethane	$C_2H_4Cl_2$	75-34-3	–96.9	57.3	250	100
151	1-Chloro-2-fluoroethane	C_2H_4ClF	762-50-5		52.8		

Code	Name	Molecular Formula	CAS Reg. No.	t_m/°C	t_b/°C	t_c/°C	TLV/ ppm
151a	1-Chloro-1-fluoroethane	C_2H_4ClF	1615-75-4		16.2		
152	1,2-Difluoroethane	$C_2H_4F_2$	624-72-6		26		
152a	1,1 Difluoroethane	$C_2H_4F_2$	75-37-6	−118.59	−24.02	113.26	
160	Chloroethane	C_2H_5Cl	75-00-3	−138.4	12.3	187.2	100
161	Fluoroethane	C_2H_5F	353-36-6	−143.2	−37.7	102.16	
170	Ethane	C_2H_6	74-84-0	−182.77	−88.58	32.17	1000
E170	Dimethyl ether	C_2H_6O	115-10-6	−141.50	−24.81	127.15	
216ca	1,3-Dichloro-1,1,2,2,3,3-hexafluoropropane	$C_3Cl_2F_6$	662-01-1	−125.4	35.7	180	
218	Perfluoropropane	C_3F_8	76-19-7	−147.70	−36.79	71.87	
227ca2	Trifluoromethyl 1,1,2,2-tetrafluoroethyl ether	C_3HF_7O	2356-61-8	−141	−3	114.63	
227ea	1,1,1,2,3,3,3-Heptafluoropropane	C_3HF_7	431-89-0	−126.80	−16.34	101.75	
227me	Trifluoromethyl 1,2,2,2-tetrafluoroethyl ether	C_3HF_7O	2356-62-9		−9.6		
236ea	1,1,1,2,3,3-Hexafluoropropane	$C_3H_2F_6$	431-63-0		6.20	139.29	
236fa	1,1,1,3,3,3-Hexafluoropropane	$C_3H_2F_6$	690-39-1	−93.63	−1.44	124.92	
236me	1,2,2,2-Tetrafluoroethyl difluoromethyl ether	$C_3H_2F_6O$	57041-67-5		23.35	155.80	
245ca	1,1,2,2,3-Pentafluoropropane	$C_3H_3F_5$	679-86-7		25.13	174.42	
245cb	1,1,1,2,2-Pentafluoropropane	$C_3H_3F_5$	1814-88-6		−17.4	106.96	
245fa	1,1,1,3,3-Pentafluoropropane	$C_3H_3F_5$	460-73-1	−102.10	15.14	154.01	
245mc	Methyl pentafluoroethyl ether	$C_3H_3F_5O$	22410-44-2		5.59	133.65	
245mf	Difluoromethyl 2,2,2-trifluoroethyl ether	$C_3H_3F_5O$	1885-48-9		29.24	170.84	
245qc	Difluoromethyl 1,1,2-trifluoroethyl ether	$C_3H_3F_5O$	69948-24-9		43.1		
254pc	Methyl 1,1,2,2-tetrafluoroethyl ether	$C_3H_4F_4O$	425-88-7	−107	37.1		
290	Propane	C_3H_8	74-98-6	−187.62	−42.11	96.74	1000
C316	1,2-Dichloro-1,2,3,3,4,4-hexafluorocyclobutane	$C_4Cl_2F_6$	356-18-3	−24.2	59.5	224	
C317	1-Chloro-1,2,2,3,3,4,4-heptafluorocyclobutane	C_4ClF_7	377-41-3	−39.1	25		
C318	Perfluorocyclobutane	C_4F_8	115-25-3	−40.19	−5.9	115.31	
347mcc	Perfluoropropyl methyl ether	$C_4H_3F_7O$	375-03-1		34.23	164.55	
347mmy	Perfluoroisopropyl methyl ether	$C_4H_3F_7O$	22052-84-2		29.34	160.15	
600	Butane	C_4H_{10}	106-97-8	−138.24	−0.49	151.98	1000
600a	Isobutane	C_4H_{10}	75-28-5	−159.38	−11.75	134.66	1000
610	Diethyl ether	$C_4H_{10}O$	60-29-7	−116.2	34.5	193.5	400
611	Methyl formate	$C_2H_4O_2$	107-31-3	−99	31.7	214.0	100
717	Ammonia	H_3N	7664-41-7	−77.65	−33.33	132.25	25
744	Carbon dioxide	CO_2	124-38-9	−56.56	−78.46 sp	30.98	5000
764	Sulfur dioxide	O_2S	7446-09-5	−75.45	−10.02	157.49	2
1112a	1,1-Dichloro-2,2-difluoroethene	$C_2Cl_2F_2$	79-35-6	−116	19		
1113	Chlorotrifluoroethene	C_2ClF_3	79-38-9	−158.2	−27.8	106	
1114	Tetrafluoroethene	C_2F_4	116-14-3	−131.15	−75.9	33.3	2
1120	Trichloroethene	C_2HCl_3	79-01-6	−84.7	87.21	271.0	10
1130	*trans*-1,2-Dichloroethene	$C_2H_2Cl_2$	156-60-5	−49.8	48.7	243.3	200
1132a	1,1-Difluoroethene	$C_2H_2F_2$	75-38-7	−144	−85.7	29.7	500
1140	Chloroethene	C_2H_3Cl	75-01-4	−153.84	−13.8	159	1
1141	Fluoroethene	C_2H_3F	75-02-5	−160.5	−72	54.7	1
1150	Ethylene	C_2H_4	74-85-1	−169.15	−103.77	9.20	200
1234yf	2,3,3,3-Tetrafluoroprop-l-ene	$C_3H_2F_4$	754-12-1		−29.5	94.7	
1270	Propene	C_3H_6	115-07-1	−185.19	−47.62	91.06	500

PROPERTIES OF GAS CLATHRATE HYDRATES

Carolyn A. Koh and E. Dendy Sloan

Gas clathrate hydrates (also known as gas hydrates) are crystalline inclusion compounds composed of hydrogen-bonded water cavities (host) which encage small gas (guest) molecules. Generally, a maximum of one guest molecule occupies each water cavity. Typical guest molecules that form gas hydrates are methane, ethane, carbon dioxide, and propane (see gas hydrate phase equilibria data in Table II). The structural and physical properties of gas hydrates are given in Tables Ia and Ib. Data have been taken from the references indicated.

Table Ia. Gas Hydrate Structural Properties (Ref. 1)

Structure	**sI**		**sII**		**sH**		
Crystal system	Cubic		Cubic		Hexagonal		
Space group	Pm3n (No. 223)[b]		Fd3m (No. 227)[b]		P6/mmm (No. 191)[b]		
Lattice description	Primitive		Face centered		Hexagonal		
Lattice parameters[a]	a = 12 Å $\alpha = \beta = \gamma = 90°$		a = 17.3 Å $\alpha = \beta = \gamma = 90°$		a = 12.2 Å, c = 10.1 Å $\alpha = \beta = 90°$, $\gamma = 120°$		
Ideal unit cell formula	$6(5^{12}6^2)\cdot 2(5^{12})\cdot 46H_2O$		$8(5^{12}6^4)\cdot 16(5^{12})\cdot 136H_2O$		$1(5^{12}6^8)\cdot 3(5^{12})\cdot 2(4^35^66^3)\cdot 34H_2O$		
Cavity	Small	Large	Small	Large	Small	Medium	Large
Description	5^{12}	$5^{12}6^2$	5^{12}	$5^{12}6^4$	5^{12}	$4^35^66^3$	$5^{12}6^8$
Number of cavities/unit cell	2	6	16	8	3	2	1
Average cavity radius[c] **(Å)**	3.95	4.33	3.91	4.73	3.94[d]	4.04[d]	5.79[d]
H_2O molecules/cavity[e]	20	24	20	28	20	20	36

[a] Lattice parameters are a function of temperature, pressure, and guest composition. Typical average values given.
[b] Space group reference numbers from the International Tables of Crystallography.
[c] The average cavity radius will vary with temperature, pressure, and guest composition.
[d] From the atomic coordinates measured using single crystal x-ray diffraction on 2,2-dimethylpentane·5(Xe,H_2S)·34H_2O at 173 K (Ref. 2). The Rietveld refinement package, GSAS was used to determine the atomic distances for each cage oxygen to the cage center.
[e] Number of oxygen atoms at the periphery of each cavity.

Table Ib. Physical Properties of sI, sII Hydrates Compared to Ice, Ih (Ref. 1,3,4,5)

Property	**Ice**	**sI**	**sII**
Dielectric constant at 273 K	94	~58	~58
H_2O reorientation time at 273 K (μs)	21	~10	~10
H_2O diffusion jump time (μs)	2.7	>200	>200
Isothermal Young's modulus at 268 K (10^9 Pa)	9.5	8.4[est]	8.2[est]
Poisson's ratio	0.3301[f]	0.31403[f]	0.31119[f]
Bulk modulus (GPa)	9.097[f]	8.762[f]	8.482[f]
Shear modulus (GPa)	3.488[f]	3.574[f]	3.6663[f]
Compressional velocity, V_p (m/s)	3870.1[f]	3778[f]	3821.8[f]
Shear velocity, V_s (m/s)	1949[f]	1963.6	2001.14[g]
Linear thermal expansion at 200 K (K^{-1})	56×10^{-6}	77×10^{-6}	52×10^{-6}
Thermal conductivity (W $m^{-1}K^{-1}$) at 263 K	2.18±0.01[h]	0.51±0.01[h]	0.50±0.01[h]
Adiabatic bulk compression at 273 K (GPa)	12	14[est]	14[est]
Heat capacity (J $kg^{-1}K^{-1}$)	1700±200[h]	2080	2130±40[h]
Refractive index (632.8 nm, −3 °C)	1.3082 (Ref. 9)	1.346 (Ref. 9)	1.350 (Ref. 9)
Density (g/cm^3)	0.91[j]	0.94	1.291[k]

[f] At 253–268 K, 22.4–32.8 MPa (ice, Ih), 258–288 K, 27.1–62.1 MPa (CH_4, sI), 258–288 K, 30.5–91.6 MPa (CH_4–C_2H_6, sII), Ref. 6.
[g] At 258–288 K, 26.6–62.1 MPa, Ref. 7.
[h] At 248–268 K (ice, Ih), 253–288 K (CH_4, sI), 248–265.5 K (THF, sII), Ref. 8.
[j] Fractional occupancy (calculated from a theoretical model) in small (S) and large (L) cavities: sI = CH_4: 0.87 (S) and CH_4: 0.973 (L); sII = CH_4: 0.672 (S), 0.057 (L); C_2H_6: 0.096 (L) only; C_3H_8: 0.84 (L) only.
[k] Calculated for 2,2-dimethylpentane·5(Xe,H_2S)·34H_2O, Ref. 2; est = estimated.

References for Table I

1. Sloan, E.D. and Koh, C.A., *Clathrate Hydrates of Natural Gases*, 3rd Edition, CRC Press, 2008.
2. Udachin, K.A., Ratcliffe, C.I., Enright, G.D., and Ripmeester, J.A., *Supramol. Chem.*, 8, 173, 1997.
3. Davidson, D.W., *Natural Gas Hydrates* (Cox, J.L., Ed.) Butterworths, Boston, 1, 1983.
4. Davidson, D.W., Handa, Y.P., and Ripmeester, J.A., *J. Phys. Chem.*, 90, 6549, 1986.
5. Ripmeester, J.A., Ratcliffe, C.I., Klug, D.D., and Tse, J.S., in *Proc. First International Conference on Natural Gas Hydrates*, (Sloan, E.D., Happel, J., and Hnatow, M.A., eds.) *Annals of the New York Academy of Sciences*, 715, 161, 1994.
6. Helgerud, M.B., Circone, S., Stern, L., Kirby, S., and Lorenson, T.D., in *Proc. Fourth International Conference on Gas Hydrates*, Yokohama May 19–23, 2002, 716, 2002.
7. Helgerud, M.B., Waite, W.F., Kirby, S.H., and Nur, A., *Can. J. Phys.*, 81, 47, 2003.
8. Waite, W.F., Gilbert, L.Y., Winters, W.J., and Mason, D.H., in *Proc. Fifth International Conference on Gas Hydrates*, Trondheim, Norway, June 13–16, Paper 5042, 2005.
9. Bylov, M. and Rasmussen, P., *Chem. Eng. Sci.*, 52, 3295, 1997.

Table II: Phase Equilibria Data of Gas Clathrate Hydrates

This table gives measured phase equilibria data of sI and sII gas clathrate hydrates (see Table I for gas hydrate structure and physical property data). The temperature and pressure conditions at which gas hydrates are stable are listed here for typical guest molecules (Tables IIa–d). For example, data for methane hydrate show that at 277.1 K methane hydrate will dissociate at pressures below 3.81 MPa.

Table IIa. Methane Hydrate (Ref. 1)

I–H–V

T (K)	*P* (MPa)	*T* (K)	*P* (MPa)	*T* (K)	*P* (MPa)	*T* (K)	*P* (MPa)	
262.4	1.79	266.5	2.08	268.6	2.22	270.9	2.39	
264.2	1.90							

L_W–H–V

T (K)	*P* (MPa)	*T* (K)	*P* (MPa)	*T* (K)	*P* (MPa)	*T* (K)	*P* (MPa)	
273.7	2.77	275.9	3.43	280.4	5.35	282.6	6.77	
274.3	2.90	277.1	3.81	280.9	5.71	284.3	8.12	
275.4	3.24	279.3	4.77	281.5	6.06	285.9	9.78	
275.9	3.42							Ref. 2

L_W-H-V

T (K)	*P* (MPa)	*T* (K)	*P* (MPa)	*T* (K)	*P* (MPa)	*T* (K)	*P* (MPa)	
295.7	33.99	295.9	35.30	301.0	64.81	302.0	77.50	Ref. 3

L_W–H–V

T (K)	*P* (MPa)	*T* (K)	*P* (MPa)	*T* (K)	*P* (MPa)	*T* (K)	*P* (MPa)	
285.7	9.62	285.7	9.62	295.9	34.75	300.9	62.40	
286.3	10.31	289.0	13.96	298.7	48.68	301.6	68.09	
286.1	10.10	292.1	21.13					Ref. 4

L_W-H-V

T (K)	*P* (MPa)	*T* (K)	*P* (MPa)	*T* (K)	*P* (MPa)	*T* (K)	*P* (MPa)	
275.4	2.87	277.2	3.90	279.2	4.90	281.2	6.10	
276.2	3.37	278.2	4.50					Ref. 5

I–H–V

T (K)	*P* (MPa)	*T* (K)	*P* (MPa)	*T* (K)	*P* (MPa)	*T* (K)	*P* (MPa)	
190.2	0.08251	208.2	0.222	243.2	0.9550	262.4	1.798	
198.2	0.1314	218.2	0.3571					Ref. 6

Table IIb. Ethane Hydrate (Ref. 1)

T (K)	P (kPa)	Phases	T (K)	P (kPa)	Phases	
260.8	294	I–H–V	285.8	2537	L_w–H–V	
260.9	290	I–H–V	287.0	3054	L_W–H–V	
269.3	441	I–H–V	287.7	4909	L_W–H–L_E	
273.4	545	L_W–H–V	287.8	3413	L_W–H–L_E	
275.4	669	L_W–H–V	287.8	4289	L_W–H–L_E	
277.6	876	L_W–H–V	288.1	3716	L_W–H–L_E	
279.1	1048	L_W–H–V	288.1	6840	L_W–H–L_E	
219.7	1131	L_W–H–V	288.2	4944	L_W–H–L_E	
281.1	1317	L_W–H–V	288.2	5082	L_W–H–L_E	
282.8	1641	L_W–H–V	288.3	4358	L_W–H–L_E	
284.4	2137	L_W–H–V	288.4	6840	L_W–H–L_E	Ref. 7
284.6	2055	L_W–H–V				

I–H–V

T (K)	P (kPa)	T (K)	P (kPa)	T (K)	P (kPa)	T (K)	P (kPa)
263.6	313	266.5	357	269.3	405	272.0	457

L_W–H–V

T (K)	P (kPa)	T (K)	P (kPa)	T (K)	P (kPa)	T (K)	P (kPa)	
273.7	510	278.7	931	280.4	1165	283.2	1689	
273.7	503	278.7	931	280.9	1255	284.3	1986	
274.8	579	279.3	1007	281.5	1345	285.4	2303	
275.9	662	279.8	1083	282.1	1448	285.4	2310	
277.6	814	280.4	1165	282.6	1558	286.5	2730	Ref. 2

L_W–H–V

T (K)	P (kPa)	T (K)	P (kPa)	T (K)	P (kPa)	T (K)	P (kPa)	
277.5	780	279.9	1040	283.3	1660	286.5	2620	
278.1	840	281.5	1380	284.5	2100			Ref. 8

Table IIc. Propane Hydrate (Ref. 1)

I–H–V

T (K)	P (kPa)	T (K)	P (kPa)	T (K)	P (kPa)	T (K)	P (kPa)
261.2	100	267.4	132	269.8	149	272.9	172
264.2	115	267.6	135	272.2	167		

L_W–H–V

T (K)	P (kPa)	T (K)	P (kPa)	T (K)	P (kPa)	T (K)	P (kPa)	
273.7	183	274.8	232	275.9	301	277.1	386	
273.7	183	275.4	270					Ref. 2

I–H–V

T (K)	P (kPa)	T (K)	P (kPa)	T (K)	P (kPa)	T (K)	P (kPa)	
247.9	48.2	251.6	58.3	258.2	81.1	260.9	94.5	
251.4	58.3	255.4	69.6	260.8	90.5	262.1	99.4	Ref. 9

Feed composition: x_{H_2O} = 0.9503, $x_{C_3H_8}$ = 0.0407
Q_2 at T = 278.62, P = 0.6 MPa

L_W–H–V		L_W–H–L $x_{C_3H_8}$		
T (K)	P (MPa)	T (K)	P (MPa)	
276.77	0.368	278.71	0.643	
277.01	0.377	278.75	0.893	
277.22	0.405	278.75	1.393	
277.36	0.425	278.75	1.891	
277.44	0.433	278.78	1.893	
277.87	0.473	278.80	2.391	
278.01	0.527	278.80	2.891	
278.22	0.483	278.79	2.893	
278.55	0.547	278.75	3.891	
		278.77	3.391	
		278.81	4.391	
		278.79	5.892	
		278.86	6.392	
		278.88	6.892	
		278.80	8.393	
		278.84	8.893	
		278.89	9.893	Ref. 10

Table IId. Carbon Dioxide Hydrate (Ref. 1)

L_W–H–V					
T (K)	P (MPa)	T (K)	P (MPa)	T (K)	P (MPa)
279.6	2.74	282.1	4.01	282.8	4.36

L_W–H–L_{CO_2}								
T (K)	P (MPa)	T (K)	P (MPa)	T (K)	P (MPa)	T (K)	P (MPa)	
282.9	5.03	283.1	6.47	283.6	11.98	283.9	14.36	
282.9	5.62	283.2	9.01					Ref. 11

Overall feed composition:
x_{H_2O} = 0.8668, x_{CO_2} = 0.1332
Q_2 at 283.27 K and 4.48 MPa

L_W–H–V		L_W–H–L_{CO_2}		
T (K)	P (MPa)	T (K)	P (MPa)	
276.52	1.82	283.33	5.97	
277.85	1.95	283.36	7.35	
278.52	2.21			
279.49	2.62			
280.44	2.88			
281.49	3.35			
281.97	3.68			
282.00	3.69			
282.45	3.85			
282.50	4.01			Ref. 12

References for Table II

1. Sloan, E.D. and Koh, C.A., *Clathrate Hydrates of Natural Gases*, 3rd Edition, CRC Press, 2008.
2. Deaton, W.M. and Frost, E.M., Jr., *Gas Hydrates and Their Relation to the Operation of Natural-Gas Pipe Lines*, U.S. Bureau of Mines Monograph 8, p. 101, 1946.
3. Kobayashi, R. and Katz, D.L., *Trans AIME*, **186**, 66, 1949.
4. McLeod, H.O. and Campbell, J.M., *J. Petl Tech.*, **222**, 590, 1961.
5. Thakore, J.L. and Holder, G.D., *Ind. Eng Chem. Res.*, **26**, 462, 1987.
6. Makogon, T.Y. and Sloan, E.D., *J. Chem. Eng. Data*, **39**, 351, 1994.
7. Roberts, O.L., Brownscombe, E.R., and Howe, L.S., *Oil Gas J.*, **39**, 37, 1940.
8. Holder, G.D. and Grigoriou, G.C., *J. Chem. Thermodyn.*, **12**, 1093, 1980.
9. Holder, G.D. and Godbole, S.P., *AIChE J.*, **28**, 930, 1982.
10. Mooijer-van den Heuvel, M.M., Peters, C.J., and de Swaan Arons, J., *Fluid Phase Equilib.*, **193**, 245, 2002.
11. Ng, H.-J. and Robinson, D.B., *Fluid Phase Equilib.*, **21**, 145, 1985.
12. Mooijer-van den Heuvel, M.M., Witteman, R., and Peters, C.J., *Fluid Phase Equilib.*, **182**, 97, 2001.

IONIC LIQUIDS

Ionic liquids are a class of organic salts with relatively low melting points. The term usually implies a melting point of 100 °C or lower, and many are liquid at room temperature. They offer several advantages as solvents, such as very low vapor pressure, good thermal stability, and nonflammable behavior. For these reasons they are attractive as constituents of environmentally friendly chemical processes.

This table lists some of the ionic liquids that have been studied. The following properties are given:

Mol. Form. — molecular formula in the Hill convention
CASRN — Chemical Abstracts Service Registry Number
Mol. Wt. — molecular weight (relative molar mass)
t_m — normal melting point in °C; the notation "gl" indicates a glass–liquid transition, rather than a crystal–liquid transition
ρ — density in g/cm³. The superscript indicates the temperature in °C; if there is no superscript, room temperature can be assumed.
η — viscosity in mPa s. The superscript indicates the temperature in °C; if there is no superscript, room temperature can be assumed.

The phase behavior of ionic liquids can be complicated. Some are crystalline at low temperatures and show a sharp transition from crystal to liquid state (a true melting point) as the temperature is raised, but others exist as a glass at low temperatures and convert to a liquid at the glass–liquid transition temperature, denoted by a small change in heat capacity. Still others are glasses at very low temperatures, transform to crystals as the temperature is raised, and finally become liquid at a still higher temperature. See Reference 3 for a discussion of the types of phase behavior.

References

1. *Ionic Liquids Database-(ILThermo),* NIST Standard Reference Database #147, 2006, National Institute of Standards and Technology, Gaithersburg, MD, <ilthermo.boulder.nist.gov>.
2. *Ionic Liquids — New Materials for New Applications,* 2006, < ildb.merck.de/ionicliquids/en/startpage.htm >.
3. Crosthwaite, J. M., Muldoon, M. J., Dixon, J. K., Anderson, J. L., and Brennecke, J. F., *J. Chem. Thermodynamics* 37, 559, 2005.
4. Sun, J., Forsyth, M., and MacFarlane, D. R., *J. Phys. Chem.* B 102, 8858, 1998.
5. Fredlake, C. P., Crosthwaite, J. M., Hert, D. G., Aki, S. N. V. K., and Brennecke, J. F., *J. Chem. Eng. Data* 49, 954, 2004.

Name	Mol. Form.	CASRN	Mol. Wt	t_m/°C	ρ/g cm⁻³	η/mPa s
1-Benzyl-3-methylimidazolium tetrafluoroborate	$C_{11}H_{13}BF_4N_2$	500996-04-3	260.039	63		
1-Butyl-4-(dimethylamino)pyridinium bromide	$C_{11}H_{19}BrN_2$		259.186	222		
1-Butyl-2,3-dimethylimidazolium chloride	$C_9H_{17}ClN_2$	98892-75-2	188.697	99		
1-Butyl-2,3-dimethylimidazolium hexafluorophosphate	$C_9H_{17}F_6N_2P$	227617-70-1	298.208	–58 gl	1.2416^{23}	
1-Butyl-2,3-dimethylimidazolium iodide	$C_9H_{17}IN_2$	108203-70-9	280.148	97		
1-Butyl-2,3-dimethylimidazolium octylsulfate	$C_{17}H_{34}N_2O_4S$		362.528	90		
1-Butyl-2,3-dimethylimidazolium tetrafluoroborate	$C_9H_{17}N_2BF_4$	402846-78-0	240.049	37	1.0762^{40}	
1-Butyl-2,3-dimethylimidazolium trifluoromethanesulfonate	$C_{10}H_{18}F_3N_2O_3S$		303.321	41		
1-Butyl-3,5-dimethylpyridinium bromide	$C_{11}H_{18}BrN$		244.172	95		
1-Butyl-3,4-dimethylpyridinium chloride	$C_{11}H_{18}ClN$		199.721	72		
1-Butyl-3,5-dimethylpyridinium chloride	$C_{11}H_{18}ClN$		199.721	100		
1-Butyl-3-methylimidazolium acetate	$C_{10}H_{18}N_2O_2$		198.262			440^{25}
1-Butyl-3-methylimidazolium bis(trifluoromethylsulfonyl)imide	$C_{10}H_{15}F_6N_3O_4S_2$	174899-83-3	419.364	–2	1.4370^{25}	70^{25}
1-Butyl-3-methylimidazolium bromide	$C_8H_{15}BrN_2$	85100-77-2	219.122	78		
1-Butyl-3-methylimidazolium chloride	$C_8H_{15}ClN_2$	79917-90-1	174.671	67	1.08^{25}	
1-Butyl-3-methylimidazolium dicyanamide	$C_{10}H_{15}N_5$	448245-52-1	205.260	–6	1.0580^{24}	
1-Butyl-3-methylimidazolium (diethylene glycol monomethyl ether)sulfate	$C_{13}H_{27}N_2O_6S$		339.427	–62 gl		1033^{25}
1-Butyl-3-methylimidazolium hexafluorophosphate	$C_8H_{15}F_6N_2P$	174501-64-5	284.182	11	1.367^{20}	382^{20}
1-Butyl-3-methylimidazolium iodide	$C_8H_{15}IN_2$	65039-05-6	266.122	–72	1.44^{25}	1100^{25}
1-Butyl-3-methylimidazolium methide	$C_{12}H_{15}F_9N_2O_6S_2$	731774-32-6	550.437		1.57^{25}	
1-Butyl-3-methylimidazolium methylsulfate	$C_9H_{18}N_2O_4S$	401788-98-5	250.315	13	1.212^{25}	
1-Butyl-3-methylimidazolium nitrate	$C_8H_{15}N_3O_3$	179075-88-8	201.223		1.15^{40}	67^{35}
1-Butyl-3-methylimidazolium octylsulfate	$C_{16}H_{32}N_2O_4S$	445473-58-5	348.501	31	1.07	
1-Butyl-3-methylimidazolium tetrafluoroborate	$C_8H_{15}BF_4N_2$	174501-65-6	226.023	–85 gl	1.2048^{22}	120^{25}
1-Butyl-3-methylimidazolium tosylate	$C_{15}H_{22}N_2O_3S$	410522-18-8	310.412	72		
1-Butyl-3-methylimidazolium trifluoroacetate	$C_{10}H_{15}F_3N_2O_2$	174899-94-6	252.233			70^{25}
1-Butyl-3-methylimidazolium trifluoromethanesulfonate	$C_9H_{15}F_3N_2O_3S$	174899-66-2	288.286	13	1.3013^{23}	
1-Butyl-3-methylpyridinium bis(trifluoromethylsulfonyl)imide	$C_{12}H_{16}F_6N_2O_4S_2$		430.386	–84 gl		63^{25}
1-Butyl-3-methylpyridinium bromide	$C_{10}H_{16}BrN$		230.145	–36 gl		
1-Butyl-3-methylpyridinium chloride	$C_{10}H_{16}ClN$	125652-55-3	185.694	117		
1-Butyl-3-methylpyridinium hexafluorophosphate	$C_{10}H_{16}F_6NP$		295.205	46		

Name	Mol. Form.	CASRN	Mol. Wt	t_m/°C	ρ/g cm^{-3}	η/mPa s
1-Butyl-4-methylpyridinium hexafluorophosphate	$C_{10}H_{16}F_6NP$	401788-99-6	295.205	44		
1-Butyl-3-methylpyridinium methylsulfate	$C_{11}H_{19}NO_4S$		261.339	<−50	1.19	
1-Butyl-4-methylpyridinium tetrafluorborate	$C_{10}H_{16}BF_4N$	343952-33-0	237.046		1.1842^{25}	
1-Butyl-3-methylpyridinium tetrafluoroborate	$C_{10}H_{16}BF_4N$		237.046	−76 gl		177^{25}
1-Butyl-1-methylpyrrolidinium bis(trifluoromethylsulfonyl)imide	$C_{11}H_{20}F_6N_2O_4S_2$	223437-11-4	422.408		1.394^{25}	
1-Butyl-1-methylpyrrolidinium dicyanamide	$C_{11}H_{20}N_4$	370865-80-8	208.304	<−50	1.02	
1-Butyl-1-methylpyrrolidinium hexafluorophosphate	$C_9H_{20}F_6NP$	330671-29-9	287.226	85		
1-Butyl-1-methylpyrrolidinium methylsulfate	$C_{10}H_{23}NO_4S$		253.360	10	1.17	
1-Butyl-1-methylpyrrolidinium trifluoroacetate	$C_{11}H_{20}F_3NO_2$		255.278	31		
1-Butyl-1-methylpyrrolidinium trifluoromethanesulfonate	$C_{10}H_{20}F_3NO_3S$	367522-96-1	291.331	3	1.25	
1-Butyl-1-methylpyrrolidinium tris(pentafluoroethyl)trifluorophosphate	$C_{15}H_{20}F_{18}NP$	851856-47-8	587.272	4	1.59	
1-Butylnicotinic acid	$C_{16}H_{22}F_6N_2O_6S_2$		516.475	15		531^{25}
1-Butylpyridinium bromide	$C_9H_{14}BrN$	874-80-6	216.118	105		
1-Butylpyridinium chloride	$C_9H_{14}ClN$	1124-64-7	171.667	127		
1-Butylpyridinium hexafluorophosphate	$C_9H_{14}F_6NP$	186088-50-6	281.178	75		
1-Butylpyridinium methylsulfate	$C_{10}H_{17}NO_4S$		247.312	<−50	1.22	
1-Butylpyridinium tetrafluoroborate	$C_9H_{14}BF_4N$	203389-28-0	223.019		1.214^{25}	
1-Butylpyridinium trifluoromethanesulfonate	$C_{10}H_{14}F_3NO_3S$	390423-43-5	285.283	35		
Cocosalky pentaethoxi methylammonium methylsulfate	$C_{23}H_{51}NO_9S$		517.718			2800^{25}
1-Decyl-3-methylimidazolium bromide	$C_{14}H_{27}BrN_2$	188589-32-4	303.281	16	112^{20}	
1-Decyl-3-methylimidazolium chloride	$C_{14}H_{27}ClN_2$	171058-18-7	258.830	38		
1,1-Dibutylpyrrolidinium bis(trifluoromethylsulfonyl)imide	$C_{14}H_{26}F_6N_2O_4S_2$		464.487	41		
N,*N*-Diethyl-*N*-(1-methylethyl)-2-propanaminium bis(trifluoromethylsulfonyl)imide	$C_{12}H_{24}F_6N_2O_4S_2$	210230-41-4	438.450	148		
1,3-Dimethylimidazolium bis(trifluoromethylsulfonyl)imide	$C_7H_9F_6N_3O_4S_2$	174899-81-1	377.284		1.570^{25}	
1,3-Dimethylimidazolium chloride	$C_5H_9ClN_2$	79917-88-7	132.591	126		
1,3-Dimethylimidazolium dimethylphosphate	$C_7H_{15}N_2O_4P$	654058-04-5	222.178		1.253^{30}	
1,3-Dimethylimidazolium methoxyethylsulfate	$C_8H_{16}N_2O_5S$	790663-78-4	252.288		1.314^{25}	
1,3-Dimethylimidazolium methylsulfate	$C_6H_{12}N_2O_4S$	97345-90-9	208.235			
1,3-Dimethylimidazolium trifluoromethylsulfonate	$C_6H_9F_3N_2O_3S$	121091-30-3	246.206	43		
1,2-Dimethyl-3-propylimidazolium bis(trifluoromethylsulfonyl)imide	$C_{10}H_{15}F_6N_3O_4S_2$	169051-76-7	419.364	11	1.457^{22}	90.0^{25}
1,1-Dimethylpyrrolidinium tris(pentafluoroethyl)trifluorophosphate	$C_{12}H_{14}F_{18}NP$		545.191	107	1.81	
1-Dodecyl-3-methylimidazolium chloride	$C_{16}H_{31}ClN_2$	114569-84-5	286.883	97		
1-Dodecyl-3-methylimidazolium hexafluorophosphate	$C_{16}H_{31}F_6N_2P$	219947-93-0	396.394			
1-Dodecyl-3-methylimidazolium tetrafluoroborate	$C_{16}H_{31}BF_4N_2$	244193-59-7	338.235			
N-Ethyl-*N*,*N*-bis(1-methylethyl)-1-heptanaminium bis(trifluoromethylsulfonyl)imide	$C_{17}H_{34}F_6N_2O_4S_2$	210230-53-8	508.583	−82 gl	1.27^{20}	362^{25}
1-Ethyl-2,3-dimethylimidazolium bis(trifluoromethylsulfonyl)imide	$C_9H_{13}F_6N_3O_4S_2$	174899-90-2	405.337	25	1.4913^{23}	
1-Ethyl-2,3-dimethylimidazolium bromide	$C_7H_{13}BrN_2$	98892-76-3	205.095	138		
1-Ethyl-2,3-dimethylimidazolium methylsulfate	$C_8H_{16}N_2O_4S$		236.289	46		
1-Ethyl-2,3-dimethylimidazolium tetrafluoroborate	$C_7H_{13}BF_4N_2$	307492-75-7	211.996	94		
1-Ethyl-2,3-dimethylimidazolium trifluoromethanesulfonate	$C_8H_{13}F_3N_2O_3S$	174899-72-0	274.260	110		
1-Ethyl-3-methylimidazolium bis(pentafluroethyl)phosphinate	$C_{10}H_{11}F_{10}N_2O_2P$		412.164	20	1.53	
1-Ethyl-3-methylimidazolium bis(trifluoromethylsulfonyl)imide	$C_8H_{11}F_6N_3O_4S_2$	174899-82-2	391.311	−17	1.5213^{23}	32^{25}
1-Ethyl-3-methylimidazolium bromide	$C_6H_{11}BrN_2$	65039-08-9	191.068	77		
1-Ethyl-3-methylimidazolium chloride	$C_6H_{11}ClN_2$	65039-09-0	146.617	85		
1-Ethyl-3-methylimidazolium ethylsulfate	$C_8H_{16}N_2O_4S$	342573-75-5	236.289		1.2388^{25}	100^{25}
1-Ethyl-3-methylimidazolium hexafluorophosphate	$C_6H_{11}F_6N_2P$	155371-19-0	256.128	59.6		
1-Ethyl-3-methylimidazolium methanesulfonate	$C_7H_{14}N_2O_3S$	145022-45-3	206.262		1.2437^{25}	
1-Ethyl-3-methylimidazolium octyl sulfate	$C_{14}H_{28}N_2O_4S$	790663-79-5	320.448			
1-Ethyl-3-methylimidazolium tetrafluoroborate	$C_6H_{11}BF_4N_2$	143314-16-3	197.969	14	1.2526^{60}	36.1^{25}
1-Ethyl-3-methylimidazolium thiocyanate	$C_7H_{11}N_3S$	331717-63-6	169.247	<−50	1.11	
1-Ethyl-3-methylimidazolium tosylate	$C_{13}H_{18}N_2O_3S$	328090-25-1	282.358	56		
1-Ethyl-3-methylimidazolium trifluoromethanesulfonate	$C_7H_{11}F_3N_2O_3S$	145022-44-2	260.233		1.385^{25}	
N-Ethyl-*N*-methyl-*N*-(1-methylethyl)-2-propanaminium bis(trifluoromethylsulfonyl)imide	$C_{11}H_{22}F_6N_2O_4S_2$	210230-42-5	424.424	140		

Name	Mol. Form.	CASRN	Mol. Wt	t_m/°C	ρ/g cm^{-3}	η/mPa s
1-Ethyl-3-methylpyridinium ethylsulfate	$C_{10}H_{17}NO_4S$		247.312	−71 gl		150^{25}
1-Ethyl-1-methylpyrrolidinium methylsulfate	$C_8H_{19}NO_4S$		225.307	23	1.23	
1-Ethyl-1-methylpyrrolidinium tetrafluoroborate	$C_7H_{16}BF_4N$	15302-90-6	201.014	69		
1-Ethylnicotinic acid ethyl ester ethylsulfate	$C_{12}H_{19}NO_6S$		305.347	−44 gl		3200^{25}
1-Ethylpyridinium bromide	$C_7H_{10}BrN$	1906-79-2	188.065	122		
1-Ethylpyridinium chloride	$C_7H_{10}ClN$	2294-38-4	143.614	119		
1-Ethylpyridinium ethylsulfate	$C_9H_{15}NO_4S$		233.285			137^{25}
1-Ethylpyridinium tetrafluoroborate	$C_7H_{10}BF_4N$	350-48-1	194.966		1.302^{20}	
1-Ethylpyridinium trifluoroacetate	$C_9H_{10}F_3NO_2$	474461-33-1	221.176		1.273^{20}	
1-Ethylpyridinium bis(trifluoromethylsulfonyl)imide	$C_9H_{10}F_6N_2O_4S_2$	712354-97-7	388.306		1.536^{25}	
1-Heptyl-3-methylimidazolium hexafluorophosphate	$C_{11}H_{21}F_6N_2P$	357915-04-9	326.262	−84 gl	1.274^{21}	
1-Hexadecyl-2,3-dimethylimidazolium chloride	$C_{21}H_{41}ClN_2$		357.017	210		
1-Hexadecyl-3-methylimidazolium chloride	$C_{20}H_{39}ClN_2$	61546-01-8	342.990	222		
1-Hexyl-4-(dimethylamino)pyridinium bis(trifluoromethylsulfonyl)imide	$C_{15}H_{23}F_6N_3O_4S_2$		487.482	−69		111^{25}
1-Hexyl-4-(dimethylamino)pyridinium bromide	$C_{13}H_{23}BrN_2$		287.239	196		
1-Hexyl-2,3-dimethylimidazolium bis(trifluoromethylsulfonyl)imide	$C_{13}H_{21}F_6N_3O_4S_2$	384347-22-2	461.444	−5	1.361^{25}	131^{25}
1-Hexyl-2,3-dimethylimidazolium chloride	$C_{11}H_{21}ClN_2$	455270-59-4	216.751	46		
1-Hexyl-2,3-dimethylimidazolium tetrafluoroborate	$C_{11}H_{21}BF_4N_2$	384347-21-1	268.103	14	1.15	
1-Hexyl-3,5-dimethylpyridinium bis(trifluoromethylsulfonyl)imide	$C_{15}H_{22}F_6N_2O_4S_2$		472.467	10		104^{25}
1-Hexyl-2-ethyl-3,5-dimethylpyridinium bis(trifluoromethylsulfonyl)-imide	$C_{17}H_{26}F_6N_2O_4S_2$		500.519	−66 gl		245^{25}
1-Hexyl-3-methyl-4-(dimethylamino)pyridinium bis(trifluoromethyl-sulfonyl)imide	$C_{16}H_{25}F_6N_3O_4S_2$		501.508	−2		112^{25}
1-Hexyl-3-methyl-4-(dimethylamino)pyridinium bromide	$C_{14}H_{25}BrN_2$		301.266	119		
1-Hexyl-3-methylimidazolium bis(trifluoromethylsulfonyl)imide	$C_{12}H_{19}F_6N_3O_4S_2$	382150-50-7	447.417	−7	1.3708^{25}	68^{25}
1-Hexyl-3-methylimidazolium bromide	$C_{10}H_{19}BrN_2$		247.175	−49 gl		
1-Hexyl-3-methylimidazolium chloride	$C_{10}H_{19}ClN_2$	171058-17-6	202.724	−75 gl	1.0400^{25}	3400000^{40}
1-Hexyl-3-methylimidazolium hexafluorophosphate	$C_{10}H_{19}F_6N_2P$	304680-35-1	312.235	−79 gl	1.294^{25}	585^{25}
1-Hexyl-3-methylimidazolium tetrafluoroborate	$C_{10}H_{19}BF_4N_2$	244193-50-8	254.076	−79 gl	1.136^{37}	
1-Hexyl-3-methylimidazolium trifluoromethanesulfonate	$C_{11}H_{19}F_3N_2O_3S$	460345-16-8	316.340	28	1.24	
1-Hexyl-4-(4-methylpiperidino)pyridinium bis(trifluoromethylsulfonyl)-imide	$C_{19}H_{29}F_6N_3O_4S_2$		541.571	37		
1-Hexyl-4-(4-methylpiperidino)pyridinium bromide	$C_{17}H_{29}BrN_2$		341.329	33 gl		
1-Hexyl-3-methylpyridinium bis(trifluoromethylsulfonyl)imide	$C_{14}H_{20}F_6N_2O_4S_2$		458.440	−82 gl		85^{25}
1-Hexyl-3-methylpyridinium bromide	$C_{12}H_{20}BrN$		258.198	−37 gl		
1-Hexyl-2-propyl-3,5-diethylpyridinium bis(trifluoromethylsulfonyl)-imide	$C_{20}H_{32}F_6N_2O_4S_2$		542.599	−67 gl		206^{25}
1-Hexylpyridinium bis(trifluoromethylsulfonyl)imide	$C_{13}H_{18}F_6N_2O_4S_2$		444.413	0		80^{25}
1-Hexylpyridinium bromide	$C_{11}H_{18}BrN$	74440-81-6	244.172	46		
1-Hexylpyridinium hexafluorophosphate	$C_{11}H_{18}F_6NP$		309.232	48		
1-Hexylpyridinium trifluoromethanesulfonate	$C_{12}H_{18}F_3NO_3S$		313.336	62		
1-Methylimidazolium tetrafluoroborate	$C_4H_7BF_4N_2$	151200-14-5	169.917	63		
1-Methylimidazolium tosylate	$C_{11}H_{14}N_2O_3S$	63458-90-2	254.305	89		
1-Methyl-3-octadecylimidazolium bis(trifluoromethylsulfonyl)imide	$C_{24}H_{43}F_6N_3O_4S_2$	404001-51-0	615.736	52		
1-Methyl-3-octadecylimidazolium tris(pentafluoroethyl)trifluoro-phosphate	$C_{28}H_{43}F_{18}N_2P$		780.599	43		
1-Methyl-3-octylimidazolium bromide	$C_{12}H_{23}BrN_2$	61545-99-1	275.228			
1-Methyl-3-octylimidazolium chloride	$C_{12}H_{23}ClN_2$	64697-40-1	230.777	−87 gl	1.0088^{25}	4100000^{40}
1-Methyl-3-octylimidazolium 2-(2-methoxyethoxy)ethyl sulfate	$C_{17}H_{34}N_2O_6S$	595565-55-2	394.526			
1-Methyl-3-octylimidazolium nitrate	$C_{12}H_{23}N_3O_3$	203389-27-9	257.329			1240^{20}
1-Methyl-3-octylimidazolium octylsulfate	$C_{20}H_{40}N_2O_4S$		404.608	76		
1-Methyl-3-octylimidazolium tetrafluoroborate	$C_{12}H_{23}BF_4N_2$	244193-52-0	282.129	−81 gl	1.104^{25}	341^{25}
1-Methyl-3-octylimidazolium trifluoromethanesulfonate	$C_{13}H_{23}F_3N_2O_3S$	403842-84-2	344.393	14	1.19	
1-Methyl-3-octylpyridinium bis(trifluoromethylsulfonyl)imide	$C_{16}H_{24}F_6N_2O_4S_2$		486.493	−80		112^{25}
1-Methyl-3-octylpyrrolidinium bis(trifluoromethylsulfonyl)imide	$C_{15}H_{28}F_6N_2O_4S_2$		478.514	−12	1.29	
1-Methyl-3-pentylimidazolium hexafluorophosphate	$C_9H_{17}F_6N_2P$	280779-52-4	298.208	−80 gl	1.333^{21}	
1-Methyl-3-propylimidazolium chloride	$C_7H_{13}ClN_2$	79917-89-8	160.644	62		

Name	Mol. Form.	CASRN	Mol. Wt	t_m/°C	ρ/g cm^{-3}	η/mPa s
1-Methyl-1-propylpyrrolidinium bis(trifluoromethylsulfonyl)imide	$C_{10}H_{18}F_6N_2O_4S_2$	223437-05-6	408.381			
3-Methyl-1-tetradecylimidazolium chloride	$C_{18}H_{35}ClN_2$	171058-21-2	314.937	195		
1-Methyl-3-tetradecylimidazolium tetrafluoroborate	$C_{18}H_{35}BF_4N_2$	244193-61-1	366.289	116		
Methyltrioctylammonium trifluoroacetate	$C_{27}H_{54}F_3NO_2$	121107-16-2	481.719	<−50	0.97	
Methyltrioctylammonium trifluoromethanesulfonate	$C_{26}H_{54}F_3NO_3S$	121107-18-4	517.772	56		
1-Nonyl-3-methylimidazolium hexafluorophosphate	$C_{13}H_{25}F_6N_2P$	343952-29-4	354.315	14	1.20^{21}	
1-Octyl-3-methylimidazolium bis(trifluoromethylsulfonyl)imide	$C_{14}H_{23}F_6N_3O_4S_2$	178631-04-4	475.471	−84 gl	1.325^{25}	
1-Octyl-3-methylimidazolium hexafluorophosphate	$C_{12}H_{23}F_6N_2P$	304680-36-2	340.288	−82 gl	1.237^{25}	734^{25}
1-Octylpyridinium chloride	$C_{13}H_{22}ClN$	4086-73-1	227.774	46		
1-Pentyl-3-methylimidazolium bis(trifluoromethylsulfonyl)imide	$C_{10}H_{15}F_6N_3O_4S_2$	280779-53-5	419.364			
1-(3,4,5,6-Perfluorohexyl)-3-methylimidazolium bis(trifluoromethyl-sulfonyl)imide	$C_{12}H_{10}F_{15}N_3O<s$		609.331	−56		
1-Propyl-3-methylimidazolium bromide	$C_7H_{13}BrN_2$		205.095	37		
1-Propyl-3-methylimidazolium tetrafluoroborate	$C_7H_{13}BF_4N_2$		211.996	−17	1.240^{25}	103^{25}
Pyridinium ethoxyethylsulfate	$C_9H_{15}NO_5S$	630393-27-0	249.284		1.281^{25}	
Tetrabutylammonium bis(trifluoromethylsulfonyl)imide	$C_{18}H_{35}F_6N_2O_4S_2$	210230-40-3	522.610	92		
Tetrabutylammonium docusate	$C_{36}H_{73}NO_7S$		664.033	−62 gl		12000^{25}
Tetrabutylammonium tris(pentafluoroethyl)trifluorophosphate	$C_{22}H_{36}F_{18}NP$		687.473	58		
Tetrabutylammonium tris(trifluoromethylsulfonyl)methide	$C_{20}H_{36}F_9NO_6S_3$	196958-57-3	653.684	96		
Tetramethylammonium tris(pentafluoroethyl)trifluorophosphate	$C_{14}H_{20}F_{18}NP$	394692-80-9	575.261	97		
N,N,N-Tributyl-1-heptanaminium bis(trifluoromethylsulfonyl)imide	$C_{21}H_{42}F_6N_2O_4S_2$	210230-50-5	564.689	−67 gl	1.17^{20}	606^{25}
N,N,N-Tributyl-1-heptanaminium trifluoromethanesulfonate	$C_{20}H_{42}F_3NO_3S$	210230-54-9	433.612	−55 gl		
N,N,N-Tributyl-1-hexanaminium bis(trifluoromethylsulfonyl)imide	$C_{20}H_{40}F_6N_2O_4S_2$	210230-49-2	550.663	26	1.15^{20}	595^{25}
Tributylmethylammonium bis(trifluoromethylsulfonyl)imide	$C_{15}H_{30}F_6N_2O_4S_2$	405514-94-5	480.530		1.266^{24}	
N,N,N-Tributyl-1-octanaminium bis(trifluoromethylsulfonyl)imide	$C_{22}H_{44}F_6N_2O_4S_2$	210230-51-6	578.715	−63 gl	1.12^{20}	574^{25}
N,N,N-Tributyl-1-octanaminium trifluoromethanesulfonate	$C_{21}H_{44}F_3NO_3S$	210230-58-3	447.639	−57 gl	1.02^{20}	2000^{25}
N,N,N-Triethylethanaminium bis(trifluoromethylsulfonyl)imide	$C_{10}H_{20}F_6N_2O_4S_2$	161401-26-9	410.397			
N,N,N-Triethyl-1-heptanaminium bis(trifluoromethylsulfonyl)imide	$C_{15}H_{30}F_6N_2O_4S_2$	210230-47-0	480.530	−79 gl	1.26^{20}	76^{25}
N,N,N-Triethyl-1-hexanaminium bis(trifluoromethylsulfonyl)imide	$C_{14}H_{28}F_6N_2O_4S_2$	210230-46-9	466.503	20	1.27^{20}	167^{25}
N,N,N-Triethyl-1-octanaminium bis(trifluoromethylsulfonyl)imide	$C_{16}H_{32}F_6N_2O_4S_2$	210230-48-1	494.556	−74 gl	1.25^{20}	202^{25}
Trihexyl(tetradecyl)phosphonium acetate	$C_{34}H_{71}O_2P$	460092-04-0	542.901		0.890^{25}	
Trihexyl(tetradecyl)phosphonium bis(trifluoromethylsulfonyl)imide	$C_{34}H_{68}F_6NO_4PS_2$	460092-03-9	764.002		1.067^{25}	
Trihexyl(tetradecyl)phosphonium chloride	$C_{32}H_{68}ClP$	258864-54-9	519.309		0.89^{25}	
Trihexyltetradecylphosphonium hexafluorophosphate	$C_{32}H_{68}F_6P_2$	374683-44-0	628.820	39		
Trihexyltetradecylphosphonium tetrafluoroborate	$C_{32}H_{68}BF_4P$	374683-55-3	570.661	25	0.94	
Trihexyl(tetradecyl)phosphonium tris(pentafluoroethyl)trifluoro-phosphate	$C_{38}H_{68}F_{18}P_2$		928.866	<−50	1.18	
Triisobutylmethylphosphonium *p*-toluenesulfonate	$C_{20}H_{37}O_3PS$	344774-05-6	388.545		1.069^{25}	
N,N,N-Trimethyl-1-heptanaminium bis(trifluoromethylsulfonyl)imide	$C_{12}H_{24}F_6N_2O_4S_2$	210230-44-7	438.450	−73 gl	1.28^{20}	153^{25}
N,N,N-Trimethyl-1-hexanaminium bis(trifluoromethylsulfonyl)imide	$C_{11}H_{22}F_6N_2O_4S_2$	210230-43-6	424.424	−74 gl	1.33^{20}	153^{25}
N,N,N-Trimethylmethanaminium bis(trifluoromethylsulfonyl)imide	$C_6H_{12}F_6N_2O_4S_2$	161401-25-8	354.290	133		
N,N,N-Trimethyl-1-octanaminium bis(trifluoromethylsulfonyl)imide	$C_{13}H_{26}F_6N_2O_4S_2$	210230-45-8	452.476	−73 gl	1.27^{20}	181^{25}
N,N,N-Tripropyl-1-propanaminium bis(trifluoromethylsulfonyl)imide	$C_{14}H_{28}F_6N_2O_4S_2$	210230-39-0	466.503	105		

DENSITY AND SPECIFIC VOLUME OF MERCURY

The data in this table have been adjusted to the ITS-90 temperature scale. The uncertainty in density values is 0.0003 g/mL between −20 and −10 °C; 0.0001 or less between −10 and 200 °C; and 0.0002 between 200 and 300 °C.

Reference

Ambrose, D., *Metrologia*, 27, 245, 1990.

t/°C	ρ/(g/mL)	*v*/(mL/kg)
−20	13.64461	73.2890
−19	13.64212	73.3024
−18	13.63964	73.3157
−17	13.63716	73.3291
−16	13.63468	73.3424
−15	13.63220	73.3558
−14	13.62972	73.3691
−13	13.62724	73.3824
−12	13.62476	73.3958
−11	13.62228	73.4091
−10	13.61981	73.4225
−9	13.61733	73.4358
−8	13.61485	73.4492
−7	13.61238	73.4625
−6	13.60991	73.4759
−5	13.60743	73.4892
−4	13.60496	73.5026
−3	13.60249	73.5160
−2	13.60002	73.5293
−1	13.59755	73.5427
0	13.59508	73.5560
1	13.59261	73.5694
2	13.59014	73.5827
3	13.58768	73.5961
4	13.58521	73.6095
5	13.58275	73.6228
6	13.58028	73.6362
7	13.57782	73.6495
8	13.57535	73.6629
9	13.57289	73.6763
10	13.57043	73.6896
11	13.56797	73.7030
12	13.56551	73.7164
13	13.56305	73.7297
14	13.56059	73.7431
15	13.55813	73.7565
16	13.55567	73.7698
17	13.55322	73.7832
18	13.55076	73.7966
19	13.54831	73.8100
20	13.54585	73.8233
21	13.54340	73.8367
22	13.54094	73.8501
23	13.53849	73.8635
24	13.53604	73.8769
25	13.53359	73.8902
26	13.53114	73.9036
27	13.52869	73.9170
28	13.52624	73.9304
29	13.52379	73.9438
30	13.52134	73.9572
31	13.51889	73.9705
32	13.51645	73.9839
33	13.51400	73.9973
34	13.51156	74.0107
35	13.50911	74.0241
36	13.50667	74.0375
37	13.50422	74.0509
38	13.50178	74.0643
39	13.49934	74.0777
40	13.49690	74.0911
41	13.49446	74.1045
42	13.49202	74.1179
43	13.48958	74.1313
44	13.48714	74.1447
45	13.48470	74.1581
46	13.48226	74.1715
47	13.47982	74.1850
48	13.47739	74.1984
49	13.47495	74.2118
50	13.47251	74.2252
51	13.47008	74.2386
52	13.46765	74.2520
53	13.46521	74.2655
54	13.46278	74.2789
55	13.46035	74.2923
56	13.45791	74.3057
57	13.45548	74.3192
58	13.45305	74.3326
59	13.45062	74.3460
60	13.44819	74.3594
61	13.44576	74.3729
62	13.44333	74.3863
63	13.44090	74.3998
64	13.43848	74.4132
65	13.43605	74.4266
66	13.43362	74.4401
67	13.43120	74.4535
68	13.42877	74.4670
69	13.42635	74.4804
70	13.42392	74.4939
71	13.42150	74.5073
72	13.41908	74.5208
73	13.41665	74.5342
74	13.41423	74.5477
75	13.41181	74.5612
76	13.40939	74.5746
77	13.40697	74.5881
78	13.40455	74.6016
79	13.40213	74.6150
80	13.39971	74.6285
81	13.39729	74.6420
82	13.39487	74.6554
83	13.39245	74.6689
84	13.39003	74.6824
85	13.38762	74.6959
86	13.38520	74.7094
87	13.38278	74.7229
88	13.38037	74.7364
89	13.37795	74.7498
90	13.37554	74.7633
91	13.37313	74.7768
92	13.37071	74.7903
93	13.36830	74.8038
94	13.36589	74.8173
95	13.36347	74.8308
96	13.36106	74.8443
97	13.35865	74.8579
98	13.35624	74.8714
99	13.35383	74.8849
100	13.35142	74.8984
110	13.3273	75.0337
120	13.3033	75.1693
130	13.2793	75.3052
140	13.2553	75.4413
150	13.2314	75.5778
160	13.2075	75.7147
170	13.1836	75.8519
180	13.1597	75.9895
190	13.1359	76.1274
200	13.1120	76.2659
210	13.0882	76.4047
220	13.0644	76.5440
230	13.0406	76.6838
240	13.0167	76.8241
250	12.9929	76.9650
260	12.9691	77.1064
270	12.9453	77.2484
280	12.9214	77.3909
290	12.8975	77.5341
300	12.8736	77.6779

THERMAL PROPERTIES OF MERCURY

Lev R. Fokin

The first of these tables gives the molar heat capacity at constant pressure of liquid and gaseous mercury as a function of temperature. To convert to specific heat in units of J/g K, divide these values by 200.59, the atomic weight of mercury.

Reference

Douglas, T. B., Ball, A. T., and Ginnings, D. C., *J. Res. Natl. Bur. Stands.*, 46, 334, 1951.

	C_p/(J/mol K)			C_p/(J/mol K)			C_p/(J/mol K)	
t/°C	Liquid	Gas	t/°C	Liquid	Gas	t/°C	Liquid	Gas
−38.84	28.2746	20.786	140	27.3675	20.786	340	27.1500	20.836
−20	28.1466	20.786	160	27.3090	20.786	356.73	27.1677	20.849
0	28.0190	20.786	180	27.2588	20.790	360	27.1709	20.853
20	27.9002	20.786	200	27.2169	20.790	380	27.1981	20.870
25	27.8717	20.786	220	27.1834	20.794	400	27.2324	20.891
40	27.7897	20.786	240	27.1583	20.794	420	27.2738	20.916
60	27.6880	20.786	260	27.1412	20.799	440	27.3207	20.941
80	27.5952	20.786	280	27.1320	20.807	460	27.3742	20.974
100	27.5106	20.786	300	27.1303	20.815	480	27.4332	21.008
120	27.4349	20.786	320	27.1366	20.824	500	27.4985	21.046

The second table gives the molar heat capacity of solid mercury in its rhombohedral (α-mercury) form.

References

1. Busey and Giaque, *J. Am. Chem. Soc.*, 75, 806, 1953.
2. Amitin, Lebedeva, and Paukov, *Rus. J. Phys. Chem.*, 2666, 1979.

t/°C	C_p/(J/mol K)	t/°C	C_p/(J/mol K)	t/°C	C_p/(J/mol K)	t/°C	C_p/(J/mol K)
−268.99	0.99*	−248.15	12.74	−193.15	23.16	−93.15	26.69
−268.99	0.97**	−243.15	14.78	−183.15	23.76	−73.15	27.28
−268.15	1.6	−233.15	17.90	−173.15	24.24	−53.15	27.96
−263.15	4.6	−223.15	19.94	−153.15	25.00	−38.87	28.5
−258.15	7.6	−213.15	21.40	−133.15	25.61		
−253.15	10.33	−203.15	22.42	−113.15	26.15		

* Superconducting state
**Normal state

The final table gives the cubic thermal expansion coefficient α, the isothermal compressibility coefficient κ_T, and the speed of sound u for liquid mercury as a function of temperature. These properties are defined as follows:

$$\alpha = \frac{1}{v}\left(\frac{\partial v}{\partial T}\right)_p \qquad \kappa_T = -\frac{1}{v}\left(\frac{\partial v}{\partial P}\right)_T \qquad u^2 = \left(\frac{\partial P}{\partial \rho}\right)_s \qquad \rho = v^{-1}$$

where v is the specific volume (given in the table on the preceding page).

Reference

Vukalovich, M. P., et al., *Thermophysical Properties of Mercury*, Moscow Standard Press, 1971.

		$\kappa_T \times 10^6$/bar^{-1}					$\kappa_T \times 10^6$/bar^{-1}		
t/°C	$\alpha \times 10^4$/K^{-1}	At 1 bar	At 1000 bar	u/m s^{-1}	t/°C	$\alpha \times 10^4$/K^{-1}	At 1 bar	At 1000 bar	u/m s^{-1}
−20	1.818	3.83		1470	120	1.8058	4.513	4.33	1404.7
0	1.8144	3.918	3.78	1460.8	140	1.8074	4.622		1395.4
20	1.8110	4.013	3.87	1451.4	160	1.8100	4.731	4.53	1386.1
40	1.8083	4.109	3.96	1442.0	180	1.8136	4.844		1376.7
60	1.8064	4.207		1432.7	200	1.818	4.96		1367
80	1.8053	4.308	4.14	1423.4	250	1.834	5.26		1344
100	1.8051	4.410		1414.1	300	1.856	5.59		1321

MELTING CURVE OF MERCURY

The solid–liquid phase boundary of mercury provides a convenient means of calibrating pressures up to 1 GPa in the neighborhood of room temperature. The best representation of this curve is given by:

$$p/\text{GPa} = 1.932845 \cdot 10^{-2}d + 1.8333 \cdot 10^{-6}d^2 + 5.9791 \cdot 10^{-8}d^3$$

where $d = t/°C - 38.8344$. Temperature is on the ITS-90 scale, and the relation is valid for pressures up to about 1.2 GPa. The following table is calculated from this equation.

p vs. t		t vs. p	
t/°C	p/GPa	p/GPa	t/°C
−38.83	0.000	0.010	−38.32
−35.00	0.074	0.020	−37.80
−30.00	0.171	0.050	−36.25
−25.00	0.268	0.100	−33.66
−20.00	0.365	0.200	−28.50
−15.00	0.463	0.300	−23.35
−10.00	0.560	0.400	−18.21
−5.00	0.658	0.500	−13.08
0.00	0.757	0.600	−7.97
5.00	0.856	0.700	−2.88
10.00	0.955	0.800	2.18
15.00	1.055	0.900	7.23
20.00	1.156	1.000	12.24
25.00	1.257	1.100	17.23
30.00	1.359	1.200	22.19

Reference

Molinar, G. F., Bean, V., Houck, J., and Welch, B., *Metrologia* 16, 21, 1980; 28, 353, 1991.

VAPOR PRESSURE OF MERCURY

The following table gives the vapor pressure of mercury in kilopascals (100 kPa = 1 bar) from the triple point (234.3156 K) to the critical point (1764 K). The data are generated from the formulation of Huber, Laesecke, and Friend in Reference 1, which is based on a critical evaluation of all the published data on mercury vapor pressure and related thermodynamic properties. The estimated uncertainty in the vapor pressure is:

−38 to −10 °C	3%
0 to 130 °C	1%
140 to 350 °C	0.15%
360 to 620 °C	0.5%
630 to 1491 °C	5%

Most of the entries in this table carry one significant figure beyond the estimated accuracy.

Note that the table refers to mercury vapor in equilibrium with liquid mercury, in the absence of air or other gases.

References

1. Huber, M. L., Laesecke, A., and Friend, D. G., *The Vapor Pressure of Mercury*, NISTIR 6643, National Institute of Standards and Technology, Boulder, CO, March 2006.
2. Huber, M. L., Laesecke, A., and Friend, D. G., *Ind. Eng. Chem. Res.* 45, 7351, 2006.
3. Vargaftik, N. B., Vinogradov, Y. K., and Yargin, V. S., *Handbook of Physical Properties of Liquids and Gases, Third Edition*, Begell House, New York, 1996.

t/°C	*p*/kPa	*t*/°C	*p*/kPa	*t*/°C	*p*/kPa	*t*/°C	*p*/kPa
−38.83	2.985×10^{-7}	38	0.0007350	300	32.965	720	6254
−30	9.451×10^{-7}	39	0.0007929	310	40.856	730	6718
−20	3.160×10^{-6}	40	0.0008551	320	50.260	740	7205
−10	9.625×10^{-6}	41	0.0009216	330	61.396	750	7718
0	2.699×10^{-5}	42	0.0009928	340	74.498	760	8258
1	2.979×10^{-5}	43	0.001069	350	89.823	770	8824
2	3.287×10^{-5}	44	0.001151	360	107.65	780	9417
3	3.623×10^{-5}	45	0.001238	370	128.26	790	10040
4	3.991×10^{-5}	46	0.001331	380	151.99	800	10690
5	4.393×10^{-5}	47	0.001430	390	179.17	810	11370
6	4.833×10^{-5}	48	0.001537	400	210.15	820	12080
7	5.312×10^{-5}	49	0.001650	410	245.32	830	12820
8	5.836×10^{-5}	50	0.001771	420	285.07	840	13600
9	6.406×10^{-5}	55	0.002506	430	329.82	850	14410
10	7.028×10^{-5}	60	0.003508	440	380.00	860	15250
11	7.705×10^{-5}	65	0.004862	450	436.07	870	16120
12	8.441×10^{-5}	70	0.006673	460	498.51	880	17030
13	9.242×10^{-5}	75	0.009075	470	567.81	890	17980
14	0.0001011	80	0.01223	480	644.46	900	18960
15	0.0001106	85	0.01635	490	729.01	910	19980
16	0.0001208	90	0.02167	500	821.99	920	21040
17	0.0001320	95	0.02850	510	923.96	930	22140
18	0.0001440	100	0.03721	520	1035.5	940	23270
19	0.0001571	110	0.06209	530	1157.2	950	24450
20	0.0001713	120	0.1009	540	1289.6	960	25670
21	0.0001866	130	0.1599	550	1433.3	970	26930
22	0.0002032	140	0.2478	560	1589.1	980	28230
23	0.0002211	150	0.3759	570	1757.4	990	29580
24	0.0002404	160	0.5592	580	1939	1000	30970
25	0.0002613	170	0.8168	590	2135	1050	38600
26	0.0002839	180	1.1728	600	2345	1100	47450
27	0.0003082	190	1.6573	610	2570	1150	57590
28	0.0003344	200	2.3071	620	2811	1200	69100
29	0.0003627	210	3.1670	630	3069	1250	82100
30	0.0003931	220	4.2906	640	3344	1300	96600
31	0.0004259	230	5.7414	650	3637	1350	112700
32	0.0004611	240	7.5939	660	3949	1400	130000
33	0.0004990	250	9.9347	670	4281	1450	150000
34	0.0005398	260	12.863	680	4632	1491	167000
35	0.0005835	270	16.494	690	5005		
36	0.0006305	280	20.955	700	5399		
37	0.0006809	290	26.392	710	5815		

SURFACE TENSION OF COMMON LIQUIDS

The surface tension γ of about 200 liquids is tabulated here as a function of temperature. Values of γ are given in units of millinewtons per meter (mN/m), which is equivalent to dyn/cm in c.g.s. units. The values refer to a nominal pressure of one atmosphere (about 100 kPa) except in cases where the indicated temperature is above the normal boiling point of the substance; in those cases, the applicable pressure is the saturation vapor pressure at the temperature in question.

The uncertainty of the values is 0.1 to 0.2 mN/m or less in most cases. Values at temperatures between the points tabulated can be obtained by linear interpolation to a good approximation.

Substances are listed by molecular formula in the modified Hill order, with substances not containing carbon appearing before those that do contain carbon. A more extensive compilation of surface tension may be found in Reference 1.

References

1. Jasper, J. J., *J. Phys. Chem. Ref. Data*, 1, 841, 1972.
2. Kahl, H., Wadewitz, T., and Winkelmann, J., *J. Chem. Eng. Data*, 48, 580, 2003.

		γ in mN/m				
Mol. formula	**Name**	**10 °C**	**25 °C**	**50 °C**	**75 °C**	**100 °C**
Br_2	Bromine	43.68	40.95	36.40		
Cl_2O_2S	Sulfuryl chloride		28.78			
Cl_3OP	Phosphoryl chloride		32.03	28.85	25.66	
Cl_3P	Phosphorus trichloride		27.98	24.81		
Cl_4Si	Silicon tetrachloride	19.78	18.29	15.80		
H_2O	Water	74.23	71.99	67.94	63.57	58.91
H_4N_2	Hydrazine		66.39			
Hg	Mercury	488.55	485.48	480.36	475.23	470.11
CCl_4	Tetrachloromethane		26.43	23.37	20.31	17.25
CS_2	Carbon disulfide	33.81	31.58	27.87		
$CHBr_3$	Tribromomethane		44.87	41.60	38.33	
$CHCl_3$	Trichloromethane		26.67	23.44	20.20	
CH_2Br_2	Dibromomethane		39.05	35.33	31.61	
CH_2Cl_2	Dichloromethane		27.20			
CH_2O_2	Formic acid		37.13	34.38	31.64	
CH_3I	Iodomethane	32.19	30.34			
CH_3NO	Formamide		57.03	54.92	52.82	50.71
CH_3NO_2	Nitromethane	39.04	36.53	32.33		
CH_4O	Methanol	23.23	22.07	20.14		
CH_5N	Methylamine		19.15			
C_2HCl_5	Pentachloroethane		34.15	31.20	28.26	
$C_2HF_3O_2$	Trifluoroacetic acid		13.53	11.42		
$C_2H_2Cl_4$	1,1,2,2-Tetrachloroethane		35.58	32.41	29.24	26.07
$C_2H_3Cl_3$	1,1,1-Trichloroethane		25.18	22.07		
$C_2H_3Cl_3$	1,1,2-Trichloroethane		34.02	30.65	27.27	23.89
C_2H_3N	Acetonitrile		28.66	25.51		
$C_2H_4Br_2$	1,2-Dibromoethane		39.55	36.25	32.95	
$C_2H_4Cl_2$	1,1-Dichloroethane		24.07			
$C_2H_4Cl_2$	1,2-Dichloroethane		31.86	28.29	24.72	
C_2H_4O	Acetaldehyde	22.54	20.50	17.10		
$C_2H_4O_2$	Acetic acid		27.10	24.61	22.13	
$C_2H_4O_2$	Methyl formate	26.72	24.36	20.43	16.50	12.57
C_2H_5Br	Bromoethane	25.36	23.62			
C_2H_5I	Iodoethane	30.38	28.46	25.24		
$C_2H_5NO_2$	Nitroethane	34.02	32.13	29.00		
C_2H_6O	Ethanol	23.22	21.97	19.89		
C_2H_6OS	Dimethyl sulfoxide		42.92	40.06		
$C_2H_6O_2$	Ethylene glycol		47.99	45.76	43.54	41.31
C_2H_6S	Dimethyl sulfide	25.27	24.06			
C_2H_6S	Ethanethiol		23.08			
$C_2H_6S_2$	Dimethyl disulfide		33.39	30.04		
C_2H_7N	Dimethylamine		26.34			
C_2H_7N	Ethylamine		19.20			
C_2H_7NO	Ethanolamine		48.32	45.53	42.73	
C_3H_5Br	3-Bromopropene		26.31	23.17		

Mol. formula	Name	γ in mN/m				
		10 °C	25 °C	50 °C	75 °C	100 °C
C_3H_5Cl	3-Chloropropene		23.14			
C_3H_5ClO	Epichlorohydrin	38.40	36.36	32.96	29.56	26.16
C_3H_5N	Propanenitrile		26.75	23.87		
$C_3H_6Cl_2$	1,2-Dichloropropane		28.32	25.22	22.12	
C_3H_6O	Acetone	24.57	22.72	19.65		
C_3H_6O	Allyl alcohol	26.63	25.28	23.02	20.77	
$C_3H_6O_2$	Ethyl formate	25.16	23.18			
$C_3H_6O_2$	Methyl acetate	26.66	24.73	21.51		
$C_3H_6O_2$	Propanoic acid		26.20	23.72	21.23	
C_3H_7Br	1-Bromopropane	27.08	25.26	22.21		
C_3H_7Br	2-Bromopropane	25.03	23.25	20.30		
C_3H_7Cl	1-Chloropropane	23.16	21.30			
C_3H_7Cl	2-Chloropropane	20.49	19.16			
C_3H_7NO	*N,N*-Dimethylformamide	37.56	35.74	32.70	29.66	26.62
$C_3H_7NO_2$	2-Nitropropane	31.02	29.29	26.39		
C_3H_8O	1-Propanol	24.48	23.32	21.38	19.43	
C_3H_8O	2-Propanol	22.11	20.93	18.96	16.98	
$C_3H_8O_2$	2-Methoxyethanol	32.32	30.84	28.38	25.92	23.46
C_3H_8S	1-Propanethiol		24.20	21.02		
C_3H_8S	2-Propanethiol		21.33	18.39		
C_3H_9N	Propylamine		21.75			
C_3H_9N	Trimethylamine		13.41			
$C_4H_4N_2$	Pyridazine	49.51	47.96	45.37	42.78	40.19
$C_4H_4N_2$	Pyrimidine		30.33	27.80	25.28	22.75
C_4H_4S	Thiophene		30.68	27.36		
C_4H_5N	Pyrrole	38.71	37.06	34.31		
$C_4H_6O_3$	Acetic anhydride	34.08	31.93	28.34	24.75	21.16
C_4H_7N	Butanenitrile		26.92	24.33	21.73	
C_4H_8O	2-Butanone		23.97	21.16		
$C_4H_8O_2$	1,4-Dioxane		32.75	29.28	25.80	22.32
$C_4H_8O_2$	Ethyl acetate	25.13	23.39	20.49	17.58	14.68
$C_4H_8O_2$	Methyl propanoate	26.32	24.44	21.29		
$C_4H_8O_2$	Butanoic acid		26.05	23.75	21.45	
C_4H_9Br	1-Bromobutane	27.58	25.90	23.08	20.27	17.45
C_4H_9Cl	1-Chlorobutane	24.85	23.18	20.39		
C_4H_9I	1-Iodobutane	29.79	28.24	25.67	23.09	20.51
C_4H_9N	Pyrrolidine	30.58	29.23	26.98		
$C_4H_{10}O$	1-Butanol	26.28	24.93	22.69	20.44	18.20
$C_4H_{10}O$	2-Butanol	23.74	22.54	20.56	18.57	16.58
$C_4H_{10}O$	2-Methyl-2-propanol		19.96	17.71		
$C_4H_{10}O$	Diethyl ether		16.65			
$C_4H_{10}O_2$	2-Ethoxyethanol		28.35	26.11	23.86	21.62
$C_4H_{10}O_3$	Diethylene glycol		44.77	42.57	40.37	38.17
$C_4H_{10}S$	Diethyl sulfide	26.22	24.57	21.80		
$C_4H_{11}N$	Butylamine		23.44	20.63		
$C_4H_{11}N$	Isobutylamine		21.75	19.02		
$C_4H_{11}N$	*tert*-Butylamine		16.87			
$C_4H_{11}N$	Diethylamine		19.85			
$C_5H_4O_2$	Furfural	45.08	43.09	39.78	36.46	33.14
C_5H_5N	Pyridine		36.56	33.29	30.03	
C_5H_8	Cyclopentene	24.45	22.20			
C_5H_8O	Cyclopentanone	34.45	32.80	30.05	27.30	24.55
C_5H_9NO	*N*-Methyl-2-pyrrolidinone	41.94	40.21	37.33	34.45	31.57
C_5H_{10}	1-Pentene	17.10	15.45			
C_5H_{10}	2-Methyl-2-butene	18.61	17.15			
C_5H_{10}	Cyclopentane	24.07	21.88	18.22		
$C_5H_{10}O$	2-Pentanone		23.25	21.62		
$C_5H_{10}O$	3-Pentanone		24.74	22.13		
$C_5H_{10}O$	Pentanal	26.95	25.44	22.91		
$C_5H_{10}O_2$	Butyl formate	26.05	24.52	21.95	19.39	16.82

Mol. formula	Name	γ in mN/m 10 °C	25 °C	50 °C	75 °C	100 °C
$C_5H_{10}O_2$	Propyl acetate	25.48	23.80	21.00	18.20	15.40
$C_5H_{10}O_2$	Isopropyl acetate	23.37	21.76	19.08	16.40	
$C_5H_{10}O_2$	Ethyl propanoate	25.55	23.80	20.88	17.96	
$C_5H_{10}O_2$	Methyl butanoate	26.34	24.62	21.76	18.89	16.03
$C_5H_{11}Cl$	1-Chloropentane	26.01	24.40	21.71	19.02	16.33
$C_5H_{11}N$	Piperidine	30.64	28.91	26.03	23.14	20.26
C_5H_{12}	Pentane	17.15	15.49			
$C_5H_{12}O$	1-Pentanol	26.67	25.36	23.17	20.99	18.80
$C_5H_{12}O$	2-Pentanol	24.96	23.45	20.94	18.43	15.92
$C_5H_{12}O$	3-Methyl-1-butanol	24.94	23.71	21.66	19.61	17.56
$C_5H_{13}N$	Pentylamine		24.69	22.14	19.58	
$C_6H_4Cl_2$	*m*-Dichlorobenzene	37.15	35.43	32.57	29.70	26.83
C_6H_5Br	Bromobenzene	36.98	35.24	32.34	29.44	26.54
C_6H_5Cl	Chlorobenzene	34.78	32.99	30.02	27.04	24.06
C_6H_5ClO	*o*-Chlorophenol		39.70	36.89	34.09	31.28
C_6H_5ClO	*m*-Chlorophenol		41.18	38.66	36.13	33.61
C_6H_5F	Fluorobenzene	28.47	26.66	23.65	20.64	
C_6H_5I	Iodobenzene	40.40	38.71	35.91	33.10	30.29
$C_6H_5NO_2$	Nitrobenzene			40.56	37.66	34.77
C_6H_6	Benzene		28.22	25.00	21.77	
C_6H_6O	Phenol			38.20	35.53	32.86
C_6H_7N	Aniline		42.12	39.41	36.69	
C_6H_7N	2-Methylpyridine		33.00	29.90	26.79	
$C_6H_8N_2$	Adiponitrile		45.45	43.02	40.58	
C_6H_{10}	Cyclohexene	28.01	26.17	23.12		
$C_6H_{10}O$	Cyclohexanone	36.43	34.57	31.46	28.36	25.25
$C_6H_{11}N$	Hexanenitrile		27.07	25.11	22.81	
C_6H_{12}	Cyclohexane	25.91	24.16	21.26	15.44	
C_6H_{12}	Methylcyclopentane	23.47	21.72	18.82		
C_6H_{12}	1-Hexene	19.44	17.90	15.33		
$C_6H_{12}O$	Cyclohexanol		32.92	30.50	28.09	25.67
$C_6H_{12}O$	2-Hexanone		25.45	22.72		
$C_6H_{12}O_2$	Butyl acetate	26.48	24.88	22.21	19.54	16.87
$C_6H_{12}O_2$	Isobutyl acetate	24.58	23.06	20.53	17.99	15.46
$C_6H_{12}O_2$	Ethyl butanoate	25.51	23.94	21.33	18.71	16.10
$C_6H_{12}O_3$	Paraldehyde	27.22	25.63	22.97	20.32	17.66
$C_6H_{13}Cl$	1-Chlorohexane	27.28	25.73	23.13	20.54	17.94
$C_6H_{13}N$	Cyclohexylamine		31.22	28.25	25.28	
C_6H_{14}	Hexane	19.42	17.89	15.33		
C_6H_{14}	2-Methylpentane	18.37	16.88	14.39		
C_6H_{14}	3-Methylpentane	19.20	17.61	14.96		
$C_6H_{14}O$	Diisopropyl ether		17.27	14.65		
$C_6H_{14}O$	1-Hexanol		25.81	23.81	21.80	19.80
$C_6H_{14}O_2$	1,1-Diethoxyethane		20.89	18.31	15.74	
$C_6H_{14}O_2$	2-Butoxyethanol	27.36	26.14	24.10	22.06	20.02
$C_6H_{15}N$	Triethylamine		20.22	17.74		
$C_6H_{15}N$	Dipropylamine		22.31	19.75	17.20	
$C_6H_{15}N$	Diisopropylamine		19.14	16.45		
C_7H_5N	Benzonitrile		38.79	35.90	33.00	
C_7H_6O	Benzaldehyde	39.63	38.00	35.27	32.55	29.82
C_7H_8	Toluene	29.46	27.73	24.85	21.98	19.10
C_7H_8O	*o*-Cresol		36.90	34.38	31.85	29.32
C_7H_8O	*m*-Cresol		35.69	33.38	31.07	28.76
C_7H_8O	Benzyl alcohol				27.89	24.44
C_7H_8O	Anisole		35.10	32.09	29.08	
C_7H_9N	*N*-Methylaniline		36.90	34.47	32.05	
C_7H_9N	2,3-Dimethylpyridine		32.71	30.04	27.36	
C_7H_9N	Benzylamine		39.30	36.27	33.23	
C_7H_{14}	Methylcyclohexane	24.98	23.29	20.46		
C_7H_{14}	1-Heptene	21.29	19.80	17.33	14.85	

Mol. formula	Name	γ in mN/m 10 °C	25 °C	50 °C	75 °C	100 °C
$C_7H_{14}O$	2-Heptanone		26.12	23.48		
$C_7H_{14}O_2$	Pentyl acetate	26.67	25.17	22.69	20.20	17.72
$C_7H_{14}O_2$	Heptanoic acid		27.76	25.64		
C_7H_{16}	Heptane	21.14	19.66	17.19	14.73	
C_7H_{16}	3-Methylhexane	20.76	19.31	16.88	14.46	
C_8H_8O	Acetophenone		39.04	36.15	33.27	
$C_8H_8O_2$	Methyl benzoate		37.17	34.25	31.32	
$C_8H_8O_3$	Methyl salicylate	40.98	39.22	36.28	33.35	30.41
C_8H_{10}	Ethylbenzene	30.39	28.75	26.01	23.28	20.54
C_8H_{10}	*o*-Xylene	31.41	29.76	27.01	24.25	21.50
C_8H_{10}	*m*-Xylene	30.13	28.47	25.71	22.95	20.19
C_8H_{10}	*p*-Xylene		28.01	25.32	22.64	19.95
$C_8H_{10}O$	Phenetole		32.41	29.65	26.89	
$C_8H_{11}N$	*N,N*-Dimethylaniline		35.52	32.90	30.27	
$C_8H_{11}N$	*N*-Ethylaniline		36.33	33.65	30.98	
C_8H_{16}	Ethylcyclohexane	26.73	25.15	22.51		
C_8H_{18}	Octane	22.57	21.14	18.77	16.39	14.01
C_8H_{18}	2,5-Dimethylhexane	20.77	19.40	17.12	14.84	12.56
$C_8H_{18}O$	1-Octanol	28.30	27.10	25.12		
$C_8H_{19}N$	Dibutylamine		24.12	21.74	19.36	
$C_8H_{19}N$	Diisobutylamine		21.72	19.44	17.16	
C_9H_7N	Quinoline	44.19	42.59	39.94	37.28	34.62
C_9H_{12}	Cumene	29.27	27.69	25.05	22.42	19.78
C_9H_{12}	1,2,4-Trimethylbenzene	30.74	29.20	26.64	24.07	21.51
C_9H_{12}	Mesitylene	28.89	27.55	25.31	23.07	20.82
$C_9H_{18}O$	5-Nonanone		26.28	23.85		
C_9H_{20}	Nonane	23.79	22.38	20.05	17.71	15.37
$C_9H_{20}O$	1-Nonanol	29.03	27.89	26.00	24.10	22.20
$C_{10}H_{12}$	1,2,3,4-Tetrahydronaphthalene		33.17	30.78	28.40	
$C_{10}H_{22}$	Decane	24.75	23.37	21.07	18.77	16.47
$C_{10}H_{22}O$	1-Decanol	29.61	28.51	26.68	24.85	23.02
$C_{11}H_{24}$	Undecane	25.56	24.21	21.96	19.70	17.45
$C_{12}H_{10}O$	Diphenyl ether		26.75	24.80		
$C_{12}H_{27}N$	Tributylamine		24.39	22.32	20.24	
$C_{13}H_{28}$	Tridecane	26.86	25.55	23.37	21.19	19.01
$C_{14}H_{12}O_2$	Benzyl benzoate	44.47	42.82	40.06	37.31	34.55
$C_{14}H_{30}$	Tetradecane	27.43	26.13	23.96	21.78	19.61
$C_{16}H_{34}$	Hexadecane		27.05	24.91	22.78	20.64
$C_{18}H_{38}$	Octadecane		27.87	25.77	23.66	21.55

SURFACE TENSION OF AQUEOUS MIXTURES

The composition dependence of the surface tension of binary mixtures of several compounds with water is given in this table. The data are tabulated as a function of the mass percent of the non-aqueous component. Data for methanol, ethanol, 1-propanol, and 2-propanol are taken from Reference 1, which also gives values at other temperatures.

References

1. Vazquez, G., Alvarez, E., and Navaza, J. M., *J. Chem. Eng. Data*, 40, 611, 1995.
2. *Landolt-Börnstein, Numerical Data and Functional Relationships in Science and Technology, New Series*, IV/16, *Surface Tension*, Springer-Verlag, Heidelberg, 1997.

Surface Tension in mN/m for the Specified Mass %

Compound	t/°C	0%	10%	20%	30%	40%	50%	60%	70%	80%	90%	100%
Acetic acid	30	71.2	51.4	43.3	41.2	38.2	37.4	36.1	33.5	31.5	30.2	26.3
Acetone	25	72.0	44.9	40.5	36.7	33.0	30.1	29.4	29.4	27.6	24.5	23.1
Acetonitrile	20	72.8	48.5	40.2	34.1	31.6	30.6	30.0	29.6	29.1	28.7	28.4
1,2-Butanediol	25	72.0	66.1	60.4	55.1	50.1	45.6	43.3	41.9	40.8	39.2	35.8
1,3-Butanediol	30	71.2	58.1	51.6	48.7	45.8	43.9	42.4	41.2	40.0	39.0	37.0
1,4-Butanediol	30	71.2	61.2	56.9	54.2	52.0	50.7	49.5	47.9	46.6	45.2	13.8
Butanoic acid	30	71.2	42.4	37.5	35.5	34.8	32.2	30.8	29.2	27.4	26.3	25.5
2-Butanone	20	72.8	41.6	32.2			25.2					24.6
γ-Butyrolactone	30	71.2	64	58	53	50	48	46	45	44	42.8	42.7
Chloroacetic acid	25	72.0	59.8	53.6	51.3	49.7	48.3	47.5	46.1			
Diethanolamine	25	72.0	66.8	63.2	60.7	58.8	57.2	55.7	54.3	52.7	50.6	47.2
N,N-Dimethylacetamide	25	72.0	72.0	72.0	72.4	73.5	74.9	75.4	73.0	65.7	54.7	36.4
N,N-Dimethylformamide	25	72.0	65.4	59.2	53.8	49.6	47.3	46.9	44.9	42.3	38.4	35.2
1,4-Dioxane	25	72.0					41.2	39.6	37.9	36.2	34.5	33.7
Ethanol	25	72.01	47.53	37.97	32.98	30.16	27.96	26.23	25.01	23.82	22.72	21.82
Ethylene glycol	20	72.8	68.5	61.9	61.9		57.0					18.2
Formic acid	20	72.8	66	60	55.7	52.2	50.3	48.8	47.1	44.7	40.9	38.0
Glycerol	25	72.0	70.5	69.5	68.5	67.9	67.4	66.9	66.5	65.7	64.5	62.5
Methanol	25	72.01	56.18	47.21	41.09	36.51	32.86	29.83	27.48	25.54	23.93	22.51
Morpholine	20	72.8	65.1	60.7	58.9	56.7	53.0	49.6	47.0	43.7	41.8	38.7
Nitric acid	20	72.8	71.9	70.7	68.9	66.6	63.8	60.6	56.8	52.6	47.9	42.6
Propanoic acid	30	71.2	46.6	42.2	37.7	35.6	33.1	31.7	30.2	28.2	27.4	25.8
1-Propanol	25	72.01	34.32	27.84	25.98	25.26	24.80	24.49	24.08	23.86	23.59	23.28
2-Propanol	25	72.01	40.42	30.57	26.82	25.27	24.26	23.51	22.68	22.14	21.69	21.22
1,2-Propylene glycol	30	71.2	60.5	54.9	50.7	47.2	44.5	41.5	38.6	37.6	36.3	35.5
1,3-Propylene glycol	30	71.2	62.6	58.8	55.7	53.8	52.8	51.7	50.8	49.6	48.2	47.0
Pyridine	25	72.0	52.8	51.2	48.0	46.8	46.6	45.8	45.0	43.6	40.9	37.0
Sulfolane	20	72.8					62.5	61.6	59.6	57.1	54.9	50.9
Sulfuric acid	50	67.9	73.5	75.1	73.6	71.2	68.0	64.1	60.0	56.4	53.6	51.7
Trichloroacetaldehyde	25	72.0	56.7	51.0	46.7	44.1	43.0	42.5	41.5	38.9	34.7	29.4
Trichloroacetic acid	25	72.0	55.8	46.5	42.8	41.6	40.6	39.4	38.3	37.4	36.5	

PERMITTIVITY (DIELECTRIC CONSTANT) OF LIQUIDS

Christian Wohlfarth

The permittivity of a substance (often called the dielectric constant) is the ratio of the electric displacement D to the electric field strength E when an external field is applied to the substance. The quantity tabulated here is the relative permittivity, which is the ratio of the actual permittivity to the permittivity of a vacuum; it is a dimensionless number.

The table gives the static relative permittivity ε_r, i.e., the relative permittivity measured in static fields or at low frequencies where no relaxation effects occur. The fourth column of the table lists the value of ε_r at the temperature specified in the third column, usually 293.15 or 298.15 K. Otherwise, the temperature closest to 293.15 K was chosen, or (as it is the case for many of the substances included here) ε_r is given at the only temperature for which data are available.

The static permittivity refers to nominal atmospheric pressure as long as the corresponding temperature is below the normal boiling point. Otherwise, at temperatures above the normal boiling point, the pressure is understood to be the saturated vapor pressure of the substance considered.

For substances where information on the temperature dependence of the permittivity is available, the table gives the coefficients of a simple polynomial fitting of permittivity to temperature with an equation of the form

$$\varepsilon_r(T) = a + bT + cT^2 + dT^3$$

where T is the absolute temperature in K. Since the parameter d was used in only a few cases where the quadratic fit was not satisfactory, only a, b, and c are listed as columns in the table, while the d values are given at the end of this introduction. For all other substances, d = 0. The temperature range of the fit is given in the last column. The coefficients of the fitting equation can be used to calculate dielectric constants within the fitted temperature range but should not be used for extrapolation outside this range. The user who needs dielectric constant data with more accuracy than can be provided by this equation is referred to Reference 1, which gives the original data together with their literature source.

Substances are listed by molecular formula in modified Hill order, with substances not containing carbon preceding those that do contain carbon.

* Indicates that the isomer was not specified in the original reference.

** Indicates a compound for which the cubic term is needed:

Ethanol	d = −0.15512E-05
N-Methylacetamide	d = −0.12998E-04
1,2-Propylene glycol	d = −0.32544E-05
1-Butanol	d = −0.48841E-06
2-Butanol	d = −0.89512E-06
2-Methyl-1-propanol	d = −0.45229E-06
2-Methyl-2-propanol	d = −0.25968E-05
N-Butylacetamide	d = −0.48716E-05

References

1. Wohlfarth, Ch., Static Dielectric Constants of Pure Liquids and Binary Liquid Mixtures, *Landolt-Börnstein, Numerical Data and Functional Relationships in Science and Technology, New Series*, Editor in Chief, O. Madelung, Group IV, Macroscopic and Technical Properties of Matter, Volume 6, Springer-Verlag, Berlin, 1991.
2. Marsh, K. N., Ed., *Recommended Reference Materials for the Realization of Physicochemical Properties*, Blackwell Scientific Publications, Oxford, 1987.

Mol. form.	Name	T/K	ε_r	a	b	c	Range/K
$AlBr_3$	Aluminum tribromide	373.2	3.38				
Ar	Argon	140.00	1.3247	0.12408E+01	0.68755E-02	−0.45344E-04	87-149
AsH_3	Arsine	200.9	2.40	0.37674E+01	−0.97454E-02	0.14537E-04	157-201
BBr_3	Boron tribromide	273.2	2.58				
B_2H_6	Diborane	180.66	1.8725	0.23848E+01	−0.29501E−02	0.64189E-06	108-181
B_5H_9	Pentaborane(9)	298.2	21.1	0.40952E+03	−0.24414E+01	0.38225E-02	226-298
BrF_3	Bromine trifluoride	298.2	106.8				
BrF_5	Bromine pentafluoride	297.7	7.91	0.11428E+02	−0.11822E-01		262-298
BrH	Hydrogen bromide	186.8	8.23				
BrNO	Nitrosyl bromide	288.4	13.4				
Br_2	Bromine	297.9	3.1484	0.32701E+01	−0.12535E-03		273-327
Br_2OS	Thionyl bromide	293.2	9.06				
Br_3OV	Vanadyl tribromide	298.2	3.6	0.61112E+01	−0.84211E-02		203-298
Br_4Ge	Germanium(IV) bromide	299.9	2.955	0.34450E+01	−0.16083E-02		300-316
Br_4Sn	Tin(IV) bromide	303.45	3.169	0.50001E+01	−0.60383E-02		304-316
$ClFO_3$	Perchloryl fluoride	150.2	2.194	0.23808E+01	−0.38629E-03	−0.57143E-05	125-150
ClF_3	Chlorine trifluoride	293.2	4.394	0.96716E+01	−0.18000E-01		273-313
ClF_5	Chlorine pentafluoride	193.2	4.28	0.78192E+01	−0.20860E-01	0.13132E-04	193-256
ClH	Hydrogen chloride	158.9	14.3	0.47316E+02	−0.28455E+00	0.48650E-03	159-258
ClNO	Nitrosyl chloride	285.2	18.2				
Cl_2	Chlorine	208.0	2.147	0.29440E+01	−0.44649E-02	0.30388E-05	208-240
Cl_2F_3P	Phosphorus(V) dichloride trifluoride	228.63	2.8129	0.46501E+01	−0.80358E-02		172-229
Cl_2OS	Thionyl chloride	298.2	8.675				
Cl_2OSe	Selenium oxychloride	293.2	46.2				
Cl_2O_2S	Sulfuryl chloride	293.2	9.1				

Mol. form.	Name	*T*/K	ε_r	*a*	*b*	*c*	Range/K
Cl_2S	Sulfur dichloride	298.2	2.915				
Cl_2S_2	Sulfur chloride	288.2	4.79				
Cl_3F_2P	Phosphorus(V) trichloride difluoride	268.0	2.3752	0.28905E+01	−0.19228E-02		215-268
Cl_3OP	Phosphorus(V) oxychloride	293.2	14.1				
Cl_3OV	Vanadyl trichloride	298.2	3.4				
Cl_3P	Phosphorus(III) chloride	290.2	3.498	0.59098E+01	−0.83322E-02		290-333
Cl_3PS	Phosphorus(V) sulfide trichloride	298.2	4.94				
Cl_4FP	Phosphorus(V) tetrachloride fluoride	272.64	2.6499	0.33503E+01	−0.29651E-02		244-273
Cl_4Ge	Germanium(IV) chloride	273.2	2.463	−0.55078E+01	0.64881E-01	−0.13091E-03	246-273
Cl_4Pb	Lead(IV) chloride	293.2	2.78				
Cl_4Si	Tetrachlorosilane	273.2	2.248	0.58041E+01	−0.27129E-01	0.51678E-04	207-273
Cl_4Sn	Tin(IV) chloride	273.2	3.014	0.43951E+01	−0.48805E-02		234-273
Cl_4Ti	Titanium(IV) chloride	257.4	2.843	0.33668E+01	−0.19675E-02		237-257
Cl_4V	Vanadium(IV) chloride	298.2	3.05				
Cl_5P	Phosphorus(V) chloride	433.2	2.85				
Cl_5Sb	Antimony(V) chloride	293.0	3.222	0.45413E+01	−0.45078E-02		276-320
FH	Hydrogen fluoride	273.2	83.6	0.50352E+03	−0.19297E+01	0.14372E-02	200-273
F_2	Fluorine	53.48	1.4913	0.14144E+01	0.26387E-02	−0.28356E-04	54-144
F_5I	Iodine pentafluoride	293.2	37.13	0.95184E+02	−0.19800E+00		273-313
F_6S	Sulfur hexafluoride	223.2	1.81				
F_6Xe	Xenon hexafluoride	328.2	4.10				
F_7I	Iodine heptafluoride	298.2	1.75				
$F_{10}S_2$	Sulfur decafluoride	293.2	2.0202				
HI	Hydrogen iodide	220.2	3.87	0.51557E+03	−0.44552E+01	0.96795E-02	220-236
H_2	Hydrogen	13.52	1.2792	0.13327E+01	−0.51946E-02		14-19
H_2O	Water	293.2	80.100	0.24921E+03	−0.79069E+00	0.72997E-03	273-372
H_2O_2	Hydrogen peroxide	290.2	74.6	0.48511E+03	−0.23145E+01	0.31020E-02	233-303
H_2S	Hydrogen sulfide	283.2	5.93	0.14736E+02	−0.33675E-01	0.96740E-05	212-363
H_3N	Ammonia	293.2	16.61	0.66756E+02	−0.24696E+00	0.25913E-03	238-323
H_4N_2	Hydrazine	298.2	51.7	0.22061E+03	−0.89633E+00	0.11066E-02	278-323
He	Helium	2.055	1.0555	0.10640E+01	−0.35584E-02		2-4
I_2	Iodine	391.25	11.08	0.64730E+02	−0.29266E+00	0.39759E-03	391-441
Kr	Krypton	119.80	1.664				
Mn_2O_7	Manganese(VII) oxide	293.2	3.28	0.37655E+01	−0.16463E-02		283-312
NO	Nitric oxide	1.997					
N_2	Nitrogen	63.15	1.4680	0.12550E+01	0.67949E-02	−0.56704E-04	63-126
N_2O_3	Nitrogen trioxide	203.2	31.13	0.92287E+02	−0.43306E+00	0.65000E-03	203-243
N_2O_4	Nitrogen tetroxide	293.2	2.44	0.28212E+01	−0.13000E-02		253-293
Ne	Neon	26.11	1.1907	0.12667E+01	−0.29064E-02		26-29
O_2	Oxygen	54.478	1.5684	0.15434E+01	0.14615E-02	−0.21964E-04	55-154
O_2S	Sulfur dioxide	298.2	16.3	0.52045E+02	−0.16125E+00	0.11042E-03	213-449
O_3	Ozone	90.2	4.75	0.86344E+01	−0.54807E-01	0.12596E-03	90-185
O_3S	Sulfur trioxide	291.2	3.11				
P	Phosphorus	307.2	4.096	0.79018E+00	0.23911E-01	−0.42826E-04	307-358
S	Sulfur	407.2	3.4991	0.51651E+01	−0.77381E-02	0.89120E-05	407-479
Se	Selenium	510.65	5.44	0.67569E+01	−0.25829E-02		511-575
Xe	Xenon	161.35	1.880				
$CBrClF_2$	Bromochlorodifluoromethane	123.2	3.920	0.52442E+01	−0.11000E-01		123-223
$CBrCl_3$	Bromotrichloromethane	293.2	2.405	0.29249E+01	−0.17650E-02		273-333
$CBrF_3$	Bromotrifluoromethane	123.2	3.730	0.54154E+01	−0.13680E-01		123-173
CBr_2Cl_2	Dibromodichloromethane	298.2	2.542	0.32330E+01	−0.23162E-02		298-333
CBr_2F_2	Dibromodifluoromethane	273.2	2.939	0.67296E+01	−0.22133E-01	0.30213E-04	139-273
CBr_3Cl	Tribromochloromethane	333.2	2.601				
CBr_3F	Tribromofluoromethane	293.2	3.00	0.53203E+01	−0.11061E-01	0.10688E-04	206-323
CBr_3NO_2	Tribromonitromethane	298.2	9.034	0.16079E+02	−0.23630E-01		298-328
$CClF_3$	Chlorotrifluoromethane	123.2	3.010	0.43677E+01	−0.11020E-01		123-173
CCl_2F_2	Dichlorodifluoromethane	123.2	3.500	0.46984E+01	−0.97600E-02		123-223
CCl_2O	Carbonyl chloride	295.2	4.30				
CCl_3D	Trichloromethane-*d*	298.2	4.67				
CCl_3F	Trichlorofluoromethane	293.2	3.00	0.53203E+01	−0.11061E-01	0.10688E-04	206-323
CCl_3NO_2	Trichloronitromethane	293.2	7.319	0.14403E+02	−0.24178E-01		276-333
CCl_4	Tetrachloromethane	293.2	2.2379	0.28280E+01	−0.20339E-02	0.71795E-07	283-333
CF_4	Tetrafluoromethane	126.3	1.685	0.20350E+01	−0.27616E-02		126-142

Mol. form.	Name	*T*/K	ε_r	*a*	*b*	*c*	Range/K
$CHBr_3$	Tribromomethane	283.2	4.404	0.71707E+01	−0.98000E-02		283-343
$CHCl_3$	Trichloromethane	293.2	4.8069	0.15115E+02	−0.51830E-01	0.56803E-04	218-323
CHF_3	Trifluoromethane	294.0	5.2	0.11442E+03	−0.75600E+00	0.13562E-02	130-263
CHN	Hydrogen cyanide	293.2	114.9	0.37331E+04	−0.23180E+02	0.36963E-01	258-299
CH_2Br_2	Dibromomethane	283.2	7.77	0.18060E+02	−0.36333E-01		283-313
CH_2Cl_2	Dichloromethane	298.0	8.93	0.40452E+02	−0.17748E+00	0.23942E-03	184-306
CH_2F_2	Difluoromethane	152.2	53.74	0.19428E+03	−0.12939E+01	0.24280E-02	152-224
CH_2I_2	Diiodomethane	298.2	5.32				
CH_2O_2	Formic acid	298.2	51.1	0.14040E+03	−0.24673E+00	−0.17151E-03	287-358
CH_3Br	Bromomethane	275.7	9.71	0.40580E+02	−0.18418E+00	0.26219E-03	195-276
CH_3Cl	Chloromethane	295.2	10.0	0.42775E+02	−0.16175E+00	0.17108E-03	190-392
CH_3ClO_2S	Methanesulfonyl chloride	293.2	34.0	0.10384E+03	−0.33838E+00	0.34156E-03	293-373
CH_3DO	Methan-d_1-ol	297.5	31.68	0.20839E+03	−0.10318E+01	0.14740E-02	176-298
CH_3F	Fluoromethane	131.0	51.0	0.11338E+03	−0.63979E+00	0.96983E-03	150-299
CH_3I	Iodomethane	293.2	6.97	0.24264E+02	−0.93914E-01	0.11926E-03	223-303
CH_3NO	Formamide	293.2	111.0	0.26076E+03	−0.61145E+00	0.34296E-03	278-333
CH_3NO_2	Nitromethane	293.2	37.27	0.11227E+03	−0.35591E+00	0.34206E-03	288-343
CH_3NO_2	Methyl nitrite	200.0	20.77	0.11071E+03	−0.73428E+00	0.14054E-02	110-260
CH_3NO_3	Methyl nitrate	293.2	23.9				
CH_4	Methane	91.0	1.6761	0.15996E+01	0.27434E-02	−0.22086E-04	91-184
CH_4O	Methanol	293.2	33.0	0.19341E+03	−0.92211E+00	0.12839E-02	177-293
CH_5N	Methylamine	215.2	16.7	0.34398E+02	−0.73630E-01	−0.41279E-04	198-258
CN_4O_8	Tetranitromethane	293.2	2.317				
COS	Carbon oxysulfide	185.0	4.47	0.84702E+01	−0.21488E-01		143-185
COSe	Carbon oxyselenide	283.2	3.47	0.48740E+01	−0.49425E-02		219-283
CO_2	Carbon dioxide	295.0	1.4492	0.79062E+00	0.10639E-01	−0.28510E-04	220-300
CS_2	Carbon disulfide	293.2	2.6320	0.45024E+01	−0.12054E-01	0.19147E-04	154-319
$C_2Br_2F_4$	1,2-Dibromotetrafluoroethane	298.2	2.34				
$C_2Cl_2F_4$	1,2-Dichlorotetrafluoroethane	273.2	2.4842	0.36663E+01	−0.42271E-02	−0.36255E-06	193-273
$C_2Cl_2O_2$	Oxalyl chloride	294.35	3.470				
C_2Cl_3N	Trichloroacetonitrile	292.2	7.85				
C_2Cl_4	Tetrachloroethylene	303.2	2.268				
$C_2Cl_4F_2$	1,1,2,2-Tetrachloro-1,2-difluoroethane	308.2	2.52				
C_2HBr_3O	Tribromoacetaldehyde	293.2	7.6				
C_2HCl_3	Trichloroethylene	301.5	3.390	0.58319E+01	−0.80828E-02		302-338
$C_2HCl_3F_2$	1,2,2-Trichloro-1,1-difluoroethane	303.2	4.01	0.75423E+01	−0.11667E-01		303-333
C_2HCl_3O	Trichloroacetaldehyde	298.2	6.8				
$C_2HCl_3O_2$	Trichloroacetic acid	333.2	4.34	0.13412E+01	0.90000E-02	−0.24130E-14	333-393
C_2HCl_5	Pentachloroethane	298.2	3.716	0.65972E+01	−0.96800E-02		298-338
$C_2HF_3O_2$	Trifluoroacetic acid	293.2	8.42	0.21652E+02	−0.68146E-01	0.78571E-04	263-323
C_2H_2	Acetylene	195.0	2.4841				
$C_2H_2Br_2$	*cis*-1,2-Dibromoethylene	298.2	7.08				
$C_2H_2Br_2$	*trans*-1,2-Dibromoethylene	298.2	2.88				
$C_2H_2Br_4$	1,1,2,2-Tetrabromoethane	303.2	6.72	0.16246E+02	−0.31500E-01		303-333
$C_2H_2Cl_2$	1,1-Dichloroethylene	293.2	4.60				
$C_2H_2Cl_2$	*cis*-1,2-Dichloroethylene	298.2	9.20				
$C_2H_2Cl_2$	*trans*-1,2-Dichloroethylene	293.2	2.14				
$C_2H_2Cl_2O_2$	Dichloroacetic acid	293.2	8.33	0.11014E+02	−0.10859E-01	0.49242E-05	284-363
$C_2H_2Cl_4$	1,1,1,2-Tetrachloroethane	207.2	9.22	0.19606E+02	−0.49847E-01		207-233
$C_2H_2Cl_4$	1,1,2,2-Tetrachloroethane	293.2	8.50				
$C_2H_2I_2$	*cis*-1,2-Diiodoethylene	345.65	4.46				
C_2H_3ClO	Acetyl chloride	295.2	15.8				
$C_2H_3ClO_2$	Chloroacetic acid	338.2	12.35	0.17310E+02	−0.14674E-01		338-393
$C_2H_3Cl_2NO_2$	1,1-Dichloro-1-nitroethane	303.2	16.3	0.37576E+02	−0.70400E-01		303-333
$C_2H_3Cl_3$	1,1,1-Trichloroethane	293.2	7.243	0.27705E+02	−0.10621E+00	0.12424E-03	258-318
$C_2H_3Cl_3$	1,1,2-Trichloroethane	298.2	7.1937	0.17147E+02	−0.33371E-01		288-318
$C_2H_3F_3O$	2,2,2-Trifluoroethanol	293.2	27.68	0.90593E+02	−0.21421E+00		293-318
C_2H_3N	Acetonitrile	293.2	36.64	0.29724E+03	−0.15508E+01	0.22591E-02	288-333
C_2H_3NO	Methyl isocyanate	288.7	21.75				
C_2H_4	Ethylene	270.0	1.4833	0.13546E+01	0.62614E-02	−0.21374E-04	200-270
C_2H_4BrCl	1-Bromo-2-chloroethane	283.2	7.41	0.19493E+02	−0.59054E-01	0.58036E-04	263-363
$C_2H_4Br_2$	1,2-Dibromoethane	293.2	4.9612	0.67142E+01	−0.59800E-02		293-313
$C_2H_4Cl_2$	1,1-Dichloroethane	298.2	10.10	0.24429E+02	−0.48000E-01		288-318

Mol. form.	Name	*T*/K	ε_r	*a*	*b*	*c*	Range/K
$C_2H_4Cl_2$	1,2-Dichloroethane	293.2	10.42	0.24404E+02	−0.47892E-01		293-343
$C_2H_4Cl_2O$	Bis(chloromethyl) ether	293.2	3.51				
$C_2H_4N_2O_6$	Ethylene glycol dinitrate	293.2	28.26				
C_2H_4O	Acetaldehyde	291.2	21.0				
C_2H_4O	Ethylene oxide	293.2	12.42	0.52661E+02	−0.21337E+00	0.25947E-03	293-243
C_2H_4OS	Thioacetic acid	298.2	14.30				
$C_2H_4O_2$	Acetic acid	293.2	6.20	−0.15731E+02	0.12662E+00	−0.17738E-03	293-363
$C_2H_4O_2$	Methyl formate	288.2	9.20	0.19699E+02	−0.36429E-01		288-302
$C_2H_4O_3S$	Ethylene glycol sulfite	298.2	39.6	0.85483E+02	−0.15400E+00		298-328
C_2H_5Br	Bromoethane	298.2	9.01	0.28473E+02	−0.85495E-01	0.67971E-04	243-308
C_2H_5Cl	Chloroethane	293.2	9.45	0.60693E+02	−0.31290E+00	0.47154E-03	237-293
C_2H_5ClO	2-Chloroethanol	293.2	25.80	0.11155E+03	−0.30149E+00		140-175
C_2H_5I	Iodoethane	293.2	7.82	0.25598E+02	−0.94367E-01	0.11424E-03	183-343
C_2H_5N	Ethyleneimine	298.2	18.3	0.61405E+02	−0.14474E+00		273-298
C_2H_5NO	Acetamide	363.7	67.6	−0.20055E+03	0.15515E+01	−0.22392E-02	364-448
C_2H_5NO	*N*-Methylformamide	293.2	189.0	0.10383E+04	−0.43165E+01	0.48398E-02	276-353
C_2H_5NO	Acetaldoxime	298.2	4.70				
$C_2H_5NO_2$	Nitroethane	288.2	29.11	0.57406E+02	−0.97657E-01		276-333
$C_2H_5NO_2$	Methyl carbamate	328.2	18.48	0.36773E+02	−0.55700E-01		328-368
$C_2H_5NO_3$	Ethyl nitrate	293.2	19.7				
C_2H_6	Ethane	95.0	1.9356	0.20185E+01	−0.51493E-03	−0.48148E-05	95-295
C_2H_6O	Ethanol**	293.2	25.3	0.15145E+03	0.87020E+00	0.19570E-02	163-523
C_2H_6O	Dimethyl ether	258.0	6.18	0.22389E+02	−0.86524E-01	0.91291E-04	155-258
C_2H_6OS	Dimethyl sulfoxide	293.2	47.24	0.38478E+02	0.16939E+00	−0.47423E-03	288-343
$C_2H_6O_2$	Ethylene glycol	293.2	41.4	0.14355E+03	−0.48573E+00	0.46703E-03	293-423
$C_2H_6O_2S$	Dimethyl sulfone	383.2	47.39	0.10830E+03	−0.15900E+00		383-398
$C_2H_6O_4S$	Dimethyl sulfate	298.2	55.0				
C_2H_6S	Ethanethiol	298.2	6.667				
C_2H_6S	Dimethyl sulfide	294.2	6.70				
$C_2H_6S_2$	1,2-Ethanedithiol	293.2	7.26	0.11228E+02	−0.13500E-01		293-333
$C_2H_6S_2$	Dimethyl disulfide	298.2	9.6	0.19109E+02	−0.32000E-01		298-323
C_2H_7N	Ethylamine	273.2	8.7	0.30163E+02	−0.79000E-01		233-273
C_2H_7NO	Ethanolamine	293.2	31.94	0.14890E+03	−0.62491E+00	0.77143E-03	253-293
$C_2H_8N_2$	1,2-Ethanediamine	293.2	13.82	0.48922E+02	−0.17021E+00	0.17262E-03	273-333
C_3Cl_6O	Hexachloroacetone	291.9	3.925	0.76423E+01	−0.15838E-01	0.10618E-04	269-303
C_3F_6O	Perfluoroacetone	202.2	2.104	0.34809E+01	−0.92883E-02	0.12282E-04	151-238
C_3HN	Cyanoacetylene	291.9	72.3	0.91803E+03	−0.49149E+01	0.69104E-02	281-314
$C_3H_2F_6O$	1,1,1,3,3,3-Hexafluoro-2-propanol	293.2	16.70				
$C_3H_3ClO_3$	4-Chloro-1,3-dioxolan-2-one	313.2	62.0				
C_3H_3N	Acrylonitrile	293.2	33.0	0.11109E+03	−0.36806E+00	0.34879E-03	233-413
$C_3H_3NO_2$	Cyanoacetic acid	277.2	33.4				
C_3H_4	Allene	269.0	2.025	0.26049E+01	−0.44147E-03	−0.63420E-05	156-269
C_3H_4	Propyne	246.0	3.218	0.60871E+01	−0.11730E-01		185-246
$C_3H_4ClF_3$	3-Chloro-1,1,1-trifluoropropane	295.2	7.32	0.22361E+02	−0.68840E-01	0.60594E-04	275-313
C_3H_4ClNO	2-Chloroethyl isocyanate	288.2	29.1	0.64311E+02	−0.12217E+00		288-403
$C_3H_4Cl_2O$	1,1-Dichloroacetone	293.2	14.6				
$C_3H_4F_4O$	2,2,3,3-Tetrafluoro-1-propanol	298.2	21.03				
C_3H_4O	Propargyl alcohol	293.2	20.8	0.99895E+02	−0.38911E+00	0.40776E-03	213-293
$C_3H_4O_3$	Ethylene carbonate	313.2	89.78	0.20746E+03	−0.37610E+00		313-343
C_3H_5Br	3-Bromopropene	293.2	7.0				
$C_3H_5BrO_2$	2-Bromopropanoic acid	294.2	11.0				
$C_3H_5Br_3$	1,2,3-Tribromopropane	303.2	6.00	0.11024E+02	−0.16596E-01		303-358
C_3H_5Cl	2-Chloropropene	299.25	8.92				
C_3H_5Cl	3-Chloropropene	293.2	8.2				
$C_3H_5ClN_2O_6$	3-Chloro-1,2-propanediol dinitrate	293.2	17.50				
C_3H_5ClO	Epichlorohydrin	293.2	22.6				
$C_3H_5ClO_2$	Ethyl chloroformate	308.7	9.736	0.15356E+02	−0.18250E-01		309-349
$C_3H_5ClO_2$	Methyl chloroacetate	293.2	12.0				
$C_3H_5Cl_3$	1,2,3-Trichloropropane	293.2	7.5				
C_3H_5I	3-Iodopropene	292.2	6.1				
C_3H_5N	Propanenitrile	293.2	29.7	0.82222E+02	−0.22937E+00	0.17424E-03	213-473
C_3H_5NO	Ethyl isocyanate	293.2	19.7				
C_3H_5NS	Ethyl isothiocyanate	293.2	19.6				

Mol. form.	Name	T/K	ε_r	*a*	*b*	*c*	Range/K
$C_3H_5N_3O_9$	Trinitroglycerol	293.2	19.25				
C_3H_6	Propene	220.0	2.1365	0.29623E+01	−0.37564E-02		220-250
$C_3H_6Br_2$	1,2-Dibromopropane	283.2	4.60	0.54973E+01	−0.31695E-02		283-333
$C_3H_6Br_2$	1,3-Dibromopropane	293.2	9.482	0.29193E+02	−0.94450E-01	0.92800E-04	293-368
$C_3H_6ClNO_2$	2-Chloro-2-nitropropane	250.4	31.90				
$C_3H_6Cl_2$	1,2-Dichloropropane	293.2	8.37	0.18915E+02	−0.35907E-01		281-323
$C_3H_6Cl_2$	1,3-Dichloropropane	303.2	10.27	0.21609E+02	−0.37333E-01		303-333
$C_3H_6Cl_2$	2,2-Dichloropropane	293.2	11.37	0.32421E+02	−0.72188E-01		245-293
$C_3H_6N_2O_4$	2,2-Dinitropropane	325.1	42.4				
C_3H_6O	Allyl alcohol	293.2	19.7	0.62714E+02	−0.14771E+00	0.37879E-05	213-303
C_3H_6O	Propanal	290.2	18.5				
C_3H_6O	Acetone	293.2	21.01	0.88157E+02	−0.34300E+00	0.38925E-03	273-323
$C_3H_6O_2$	Propanoic acid	298.2	3.44	0.18793E+01	0.46841E-02	0.19983E-05	289-408
$C_3H_6O_2$	Ethyl formate	288.2	8.57	0.15884E+02	−0.25333E-01		288-318
$C_3H_6O_2$	Methyl acetate	288.2	7.07	0.13190E+02	−0.21226E-01		276-318
$C_3H_6O_3$	3-Hydroxypropanoic acid	296.2	30.0				
$C_3H_6O_3$	Dimethyl carbonate	298.2	3.087				
$C_3H_6O_3$	1,3,5-Trioxane	338.2	15.55				
C_3H_7Br	1-Bromopropane	293.2	8.09	0.17769E+02	−0.32599E-01		274-328
C_3H_7Br	2-Bromopropane	293.2	9.46	0.26195E+02	−0.72995E-01	0.55454E-04	186-328
C_3H_7Cl	1-Chloropropane	293.2	8.588	0.21214E+02	−0.43130E-01		273-313
C_3H_7ClO	3-Chloro-1-propanol	215.2	36.0	0.12436E+03	−0.60841E+00	0.92060E-03	145-215
C_3H_7ClO	1-Chloro-2-propanol	153.2	59.0	−0.19169E+02	0.13605E+01	−0.55567E-02	153-177
$C_3H_7ClO_2$	3-Chloro-1,2-propanediol	293.2	31.0				
C_3H_7I	1-Iodopropane	293.2	7.07	0.13744E+02	−0.22745E-01		293-323
C_3H_7I	2-Iodopropane	298.2	8.19				
C_3H_7NO	*N*-Ethylformamide	298.2	102.7	0.64764E+03	−0.28499E+01	0.34286E-02	298-338
C_3H_7NO	*N,N*-Dimethylformamide	293.2	38.25	0.15364E+03	−0.60367E+00	0.71505E-03	213-353
C_3H_7NO	*N*-Methylacetamide**	303.2	179.0	0.15975E+04	−0.90451E+01	0.18345E-01	303-473
$C_3H_7NO_2$	1-Nitropropane	288.2	24.70	0.94999E+02	−0.38358E+00	0.48480E-03	276-333
$C_3H_7NO_2$	2-Nitropropane	288.2	26.74	0.60138E+02	−0.11566E+00		276-303
$C_3H_7NO_2$	Propyl nitrite	250.0	12.35	0.70552E+02	−0.40362E+00	0.66687E-03	110-310
$C_3H_7NO_2$	Isopropyl nitrite	260.0	13.92	0.74578E+02	−0.38283E+00	0.57071E-03	150-300
$C_3H_7NO_2$	Ethyl carbamate	328.2	14.14	0.32431E+02	−0.65097E-01	0.28571E-04	328-368
C_3H_8	Propane	293.19	1.6678	0.22883E+01	−0.23276E-02	0.84710E-06	90-300
C_3H_8O	1-Propanol	293.2	20.8	0.98045E+02	−0.36860E+00	0.36422E-03	193-493
C_3H_8O	2-Propanol	293.2	20.18	0.10416E+03	−0.41011E+00	0.42049E-03	193-493
$C_3H_8O_2$	1,2-Propylene glycol**	303.2	27.5	0.24546E+03	−0.15738E+01	0.38068E-02	193-403
$C_3H_8O_2$	1,3-Propylene glycol	293.2	35.1	0.11365E+03	−0.36680E+00	0.33766E-03	288-328
$C_3H_8O_2$	Ethylene glycol monomethyl ether	298.2	17.2	0.11803E+03	−0.58000E+00	0.81001E-03	254-318
$C_3H_8O_2$	Dimethoxymethane	293.2	2.644	0.25877E+01	−0.93019E-03	0.38472E-05	171-293
$C_3H_8O_3$	Glycerol	293.2	46.53	0.77503E+02	−0.37984E-01	−0.23107E-03	288-343
C_3H_8S	1-Propanethiol	288.2	5.937	0.11602E+02	−0.19580E-01		273-318
C_3H_8S	2-Propanethiol	298.2	5.952				
$C_3H_8S_2$	1,2-Propanedithiol	293.2	7.24	0.14667E+02	−0.32660E-01	0.25000E-04	293-333
$C_3H_8S_2$	1,3-Propanedithiol	303.2	8.11	0.66607E+01	0.31310E-01	−0.87500E-04	303-343
$C_3H_9BO_3$	Trimethyl borate	293.2	2.2762				
C_3H_9ClSi	Trimethylchlorosilane	273.2	10.21	−0.19492E+02	0.29806E+00	−0.69284E-03	223-273
C_3H_9N	Propylamine	296.2	5.08	0.17719E+02	−0.59022E-01	0.54780E-04	204-296
C_3H_9N	Isopropylamine	293.2	5.6268	0.40429E+02	−0.21441E+00	0.32634E-03	213-298
C_3H_9N	Trimethylamine	298.2	2.440	0.39745E+01	−0.51331E-02		273-298
$C_3H_9O_4P$	Trimethyl phosphate	293.2	20.6				
C_4Cl_6	Hexachloro-1,3-butadiene	293.2	2.55				
$C_4Cl_6O_3$	Trichloroacetic anhydride	298.2	5.0				
$C_4F_6O_3$	Trifluoroacetic acid anhydride	298.2	2.7				
$C_4H_2Cl_4O_3$	Dichloroacetic anhydride	298.2	15.8				
$C_4H_2O_3$	Maleic anhydride	326.2	52.75				
$C_4H_3F_7O$	2,2,3,3,4,4,4-Heptafluoro-1-butanol	298.2	14.4				
$C_4H_4N_2$	Succinonitrile	298.2	62.6	0.17724E+03	−0.54654E+00	0.54046E-03	236-351
$C_4H_4N_2$	Pyrazine	323.2	2.80				
C_4H_4O	Furan	277.1	2.88	0.13636E+01	0.12864E-01	−0.22701E-04	188-277
C_4H_4S	Thiophene	293.2	2.739	0.32941E+01	−0.19019E-02		253-293
C_4H_5Cl	2-Chloro-1,3-butadiene	293.2	4.914				

Mol. form.	Name	T/K	ε_r	a	b	c	Range/K
$C_4H_5Cl_3O_2$	Ethyl trichloroacetate	293.2	8.428				
C_4H_5N	Pyrrole	293.0	8.00	0.12672E+02	−0.14075E-01	−0.62671E-05	293-357
C_4H_5NO	Allyl isocynate	288.2	15.15	0.34299E+02	−0.66444E-01		288-333
C_4H_6	1,3-Butadiene	265.0	2.050	0.27674E+01	−0.26738E-02		185-265
C_4H_6O	Divinyl ether	288.2	3.94				
C_4H_6O	Ethoxyacetylene	298.2	8.05				
C_4H_6O	Cyclobutanone	298.2	14.27	0.43974E+02	−0.15712E+00	0.19264E-03	220-317
$C_4H_6O_2$	Methyl acrylate	303.2	7.03	0.11968E+02	−0.16500E-01		303-333
$C_4H_6O_2$	2,3-Butanedione	298.2	4.04	0.46907E+01	−0.22302E-02		278-348
$C_4H_6O_2$	γ-Butyrolactone	293.2	39.0				
$C_4H_6O_3$	Acetic anhydride	293.2	22.45				
$C_4H_6O_3$	Propylene carbonate	293.0	66.14	0.15940E+03	−0.39530E+00	0.26284E-03	273-333
C_4H_7Br	*cis*-2-Bromo-2-butene	293.2	5.38				
C_4H_7Br	*trans*-2-Bromo-2-butene	293.2	6.76				
$C_4H_7BrO_2$	2-Bromobutanoic acid	293.2	7.2				
$C_4H_7BrO_2$	Ethyl bromoacetate	303.2	9.75	0.15627E+02	−0.19600E-01		303-333
$C_4H_7BrO_2$	Methyl 3-bromopropanoate	303.2	5.81	0.36001E+01	0.72500E-02		303-343
$C_4H_7ClO_2$	Propyl chlorocarbonate	293.2	11.2				
$C_4H_7ClO_2$	Methyl 2-chloropropanoate	303.2	11.45	0.22449E+02	−0.36250E-01		303-343
C_4H_7N	Butanenitrile	293.2	24.83	0.53884E+02	−0.99257E-01		293-333
C_4H_7N	2-Methylpropanenitrile	293.2	24.42	0.52554E+02	−0.96000E-01		293-313
C_4H_7NO	2-Pyrrolidone	298.2	28.18	0.11054E+03	−0.47945E+00	0.68182E-03	298-338
C_4H_8	1-Butene	220.0	2.2195	0.29354E+01	−0.32580E-02		220-250
C_4H_8	*cis*-2-Butene	296.0	1.960	0.28802E+01	−0.31064E-02		197-296
C_4H_8	Isobutene	288.7	2.1225	0.33701E+01	−0.43295E-02		220-289
$C_4H_8Br_2$	1,2-Dibromobutane	293.2	4.74	0.11199E+03	−0.63334E+00	0.91250E-03	293-333
$C_4H_8Br_2$	1,3-Dibromobutane	293.2	9.14	0.34031E+02	−0.13254E+00	0.16250E-03	293-333
$C_4H_8Br_2$	1,4-Dibromobutane	303.2	8.68	0.20944E+02	−0.55620E-01	0.50000E-04	303-333
$C_4H_8Br_2$	2,3-Dibromobutane	298.2	6.245	0.23849E+02	−0.96300E-01	0.12500E-03	293-333
$C_4H_8Br_2$	1,2-Dibromo-2-methylpropane	293.2	4.1				
$C_4H_8Cl_2$	1,2-Dichlorobutane	293.2	7.74	0.31925E+02	−0.13232E+00	0.17007E-03	293-356
$C_4H_8Cl_2$	1,4-Dichlorobutane	308.2	9.30	0.59766E+01	0.49300E-01	−0.12500E-03	308-338
$C_4H_8Cl_2$	1,2-Dichloro-2-methylpropane	296.0	7.15	0.39429E+02	−0.20028E+00	0.30917E-03	165-296
$C_4H_8Cl_2O$	Bis(2-chloroethyl) ether	293.2	21.20				
C_4H_8O	Butanal	298.2	13.45				
C_4H_8O	2-Butanone	293.2	18.56	0.15457E+02	0.90152E-01	−0.27100E-03	293-333
C_4H_8O	Tetrahydrofuran	295.2	7.52	0.30739E+02	−0.12946E+00	0.17195E-03	224-295
$C_4H_8O_2$	Butanoic acid	287.2	2.98	0.15010E+01	0.50046E-02		287-403
$C_4H_8O_2$	2-Methylpropanoic acid	293.2	2.58				
$C_4H_8O_2$	Propyl formate	303.2	6.92				
$C_4H_8O_2$	Ethyl acetate	293.2	6.0814	0.15646E+02	−0.44066E-01	0.39137E-04	293-433
$C_4H_8O_2$	Methyl propanoate	293.2	6.200	0.12798E+02	−0.22540E-01		293-333
$C_4H_8O_2$	1,4-Dioxane	293.2	2.2189	0.27299E+01	−0.17440E-02		293-313
$C_4H_8O_3$	2-Hydroxybutanoic acid	296.2	37.7				
$C_4H_8O_3$	3-Hydroxybutanoic acid	296.2	31.5				
$C_4H_8O_3$	Ethyl methyl carbonate	293.2	2.985				
$C_4H_8O_3$	Ethylene glycol monoacetate	303.2	12.95				
C_4H_9Br	1-Bromobutane	283.2	7.315	0.22542E+02	−0.79306E-01	0.89867E-04	183-363
C_4H_9Br	2-Bromobutane	298.2	8.64	0.18461E+02	−0.32933E-01		274-328
C_4H_9Br	1-Bromo-2-methylpropane	273.2	7.70	0.37558E+02	−0.20571E+00	0.35496E-03	112-273
C_4H_9Br	2-Bromo-2-methylpropane	293.0	10.98	0.35085E+02	−0.14075E+00	0.19960E-03	258-293
C_4H_9Cl	1-Chlorobutane	293.2	7.276	0.13565E+02	−0.10161E-01	−0.38750E-04	273-323
C_4H_9Cl	2-Chlorobutane	293.2	8.564	0.30376E+02	−0.11377E+00	0.13429E-03	273-323
C_4H_9Cl	1-Chloro-2-methylpropane	293.2	7.027	0.14945E+02	−0.33747E-01	0.23036E-04	273-323
C_4H_9Cl	2-Chloro-2-methylpropane	293.2	9.663	0.35077E+02	−0.12867E+00	0.14304E-03	273-323
C_4H_9I	1-Iodobutane	293.2	6.27	0.16493E+02	−0.50262E-01	0.52485E-04	293-323
C_4H_9I	2-Iodobutane	293.2	7.873	0.10883E+02	−0.14680E-02	−0.30000E-04	293-323
C_4H_9I	2-Iodo-2-methylpropane	283.2	6.65	0.76780E+01	0.69900E-02	−0.37500E-04	283-323
C_4H_9N	Pyrrolidine	293.0	8.30	0.38191E+02	−0.15462E+00	0.17941E-03	274-333
C_4H_9NO	*N*-Methylpropanamide	293.2	170.0				
C_4H_9NO	*N*-Ethylacetamide	293.2	135.0	0.74494E+03	−0.31400E+01	0.36131E-02	213-353
C_4H_9NO	*N*,*N*-Dimethylacetamide	294.2	38.85	0.15420E+03	−0.57506E+00	0.61911E-03	294-433
C_4H_9NO	2-Butanone oxime	293.2	3.4				

Mol. form.	Name	T/K	ε_r	a	b	c	Range/K
C_4H_9NO	Morpholine	298.2	7.42				
$C_4H_9NO_2$	*tert*-Butyl nitrite	298.2	11.47				
$C_4H_9NO_2$	Propyl carbamate	338.2	12.06	0.24356E+02	−0.36400E-01		338-378
$C_4H_9NO_2$	Ethyl-*N*-methyl carbamate	298.2	21.10	0.11477E+03	−0.47568E+00	0.54127E-03	298-373
$C_4H_9NO_2$	*N*-Acetylethanolamine	298.2	96.6	0.37016E+03	−0.13113E+01	0.13214E-02	298-348
$C_4H_9NO_3$	Butyl nitrate	293.2	13.10				
C_4H_{10}	Butane	295.0	1.7697	0.22379E+01	−0.13884E-02	−0.66711E-06	135-303
C_4H_{10}	Isobutane	295.0	1.7518	0.23295E+01	−0.19953E-02	0.14197E-06	115-303
$C_4H_{10}O$	1-Butanol**	293.2	17.84	0.10578E+03	−0.50587E+00	0.84733E-03	193-553
$C_4H_{10}O$	2-Butanol**	293.2	17.26	0.13850E+03	−0.75146E+00	0.14086E-02	172-533
$C_4H_{10}O$	2-Methyl-1-propanol**	293.2	17.93	0.10762E+03	−0.51398E+00	0.83702E-03	173-533
$C_4H_{10}O$	2-Methyl-2-propanol**	298.2	12.47	0.22541E+03	−0.14990E+01	0.34050E-02	298-503
$C_4H_{10}O$	Diethyl ether	293.2	4.2666	0.79725E+01	−0.12519E-01		283-301
$C_4H_{10}O_2$	1,2-Butanediol	298.2	22.4	0.63702E+02	−0.13807E+00		278-323
$C_4H_{10}O_2$	1,3-Butanediol	298.2	28.8	0.72883E+02	−0.14770E+00		278-323
$C_4H_{10}O_2$	1,4-Butanediol	298.2	31.9	0.13079E+03	−0.46985E+00	0.46320E-03	288-328
$C_4H_{10}O_2$	Ethylene glycol monoethyl ether	298.2	13.38				
$C_4H_{10}O_2$	Ethylene glycol dimethyl ether	296.7	7.30	0.48832E+02	−0.24218E+00	0.34413E-03	256-318
$C_4H_{10}O_2S$	Bis(2-hydroxyethyl) sulfide	293.2	28.61	0.13128E+03	−0.52719E+00	0.60465E-03	253-333
$C_4H_{10}O_3$	Diethylene glycol	293.2	31.82	0.13973E+03	−0.54725E+00	0.61149E-03	288-343
$C_4H_{10}O_3S$	Diethyl sulfite	293.2	15.6				
$C_4H_{10}O_4$	1,2,3,4-Butanetetrol	393.2	28.2				
$C_4H_{10}O_4S$	Diethyl sulfate	293.2	29.2				
$C_4H_{10}S$	1-Butanethiol	288.2	5.204	0.11201E+02	−0.20767E-01		273-318
$C_4H_{10}S$	2-Butanethiol	288.2	5.645	0.10866E+02	−0.17993E-01		273-318
$C_4H_{10}S$	2-Methyl-1-propanethiol	298.2	4.961				
$C_4H_{10}S$	2-Methyl-2-propanethiol	293.2	5.475	0.10597E+02	−0.17500E-01		283-313
$C_4H_{10}S$	Diethyl sulfide	298.2	5.723				
$C_4H_{11}N$	Butylamine	293.2	4.71	0.13322E+02	−0.44176E-01	0.50250E-04	223-333
$C_4H_{11}N$	Diethylamine	293.2	3.680	0.26462E+02	−0.13750E+00	0.20373E-03	243-323
$C_4H_{11}NO_2$	Diethanolamine	293.2	25.75	0.73435E+02	−0.21377E+00	0.17500E-03	273-323
$C_4H_{12}O_2Si$	Dimethoxydimethylsilane	298.2	3.663				
$C_4H_{12}O_3Si$	Trimethoxymethylsilane	298.2	4.9				
$C_4H_{12}O_4Si$	Tetramethyl silicate	293.2	6.0				
$C_4H_{12}Si$	Diethylsilane	293.2	2.544				
$C_4H_{12}Si$	Tetramethylsilane	293.2	1.921				
$C_4H_{13}N_3$	Diethylenetriamine	293.2	12.62	0.57840E+02	−0.23873E+00	0.28841E-03	213-333
C_5FeO_5	Iron pentacarbonyl	293.2	2.602				
C_5H_4BrN	2-Bromopyridine	298.2	23.18	0.73391E+02	−0.23678E+00	0.22930E-03	298-398
C_5H_4ClN	2-Chloropyridine	298.2	27.32	0.98702E+02	−0.34237E+00	0.34502E-03	298-398
$C_5H_4F_8O$	2,2,3,3,4,4,5,5-Octafluoro-1-pentanol	298.2	15.30				
$C_5H_4O_2$	Furfural	293.2	42.1				
C_5H_5N	Pyridine	293.2	13.260	0.43991E+02	−0.15150E+00	0.15925E-03	293-323
C_5H_5NO	Pyridine-1-oxide	343.0	35.94	0.20878E+02	0.16450E+00	−0.35269E-03	343-398
C_5H_6O	2-Methylfuran	293.2	2.76				
$C_5H_6O_2$	Furfuryl alcohol	298.2	16.85				
$C_5H_7Cl_3O_2$	Propyl trichloroacetate	298.2	8.32				
$C_5H_7NO_2$	Ethyl cyanoacetate	263.2	31.62				
C_5H_8	1,3-Pentadiene*	298.2	2.319				
C_5H_8	1,4-Pentadiene	294.0	2.054	0.29994E+01	−0.34578E-02	0.85300E-06	178-294
C_5H_8	2-Methyl-1,3-butadiene	293.2	2.098	0.28170E+01	−0.23147E-02	−0.43975E-06	198-293
C_5H_8	Cyclopentene	295.0	2.083	0.28177E+01	−0.27597E-02	0.89346E-06	171-319
C_5H_8O	Cyclopentanone	298.2	13.58	0.24083E+02	−0.30286E-01	−0.16802E-04	219-298
$C_5H_8O_2$	Ethyl acrylate	303.2	6.05	0.47827E+02	−0.24394E+00	0.35000E-03	303-343
$C_5H_8O_2$	Methyl *trans*-2-butenoate	293.2	6.6645				
$C_5H_8O_2$	Methyl methacrylate	303.2	6.32	0.32098E+02	−0.14568E+00	0.20000E-03	303-343
$C_5H_8O_2$	2,4-Pentanedione	303.2	26.524				
$C_5H_8O_4$	Dimethyl malonate	293.2	9.82	0.26470E+02	−0.76656E-01	0.67888E-04	293-433
$C_5H_9BrO_2$	Ethyl 2-bromopropanoate	293.2	9.4				
$C_5H_9ClO_2$	Isobutyl chlorocarbonate	293.2	9.1				
$C_5H_9ClO_2$	Ethyl 2-chloropropanoate	303.2	11.95	0.25965E+02	−0.46250E-01		303-343
$C_5H_9ClO_2$	Ethyl 3-chloropropanoate	303.2	10.19	0.21951E+02	−0.38750E-01		303-343
$C_5H_9ClO_2$	Methyl 4-chlorobutanoate	303.2	9.51	0.17127E+02	−0.25000E-01		303-343

Mol. form.	Name	*T*/K	ε_r	*a*	*b*	*c*	Range/K
C_5H_9N	Pentanenitrile	293.2	20.04	0.55793E+02	−0.15750E+00	0.12432E-03	183-333
C_5H_9N	2,2-Dimethylpropanenitrile	293.2	21.1	0.58418E+02	−0.16884E+00	0.14131E-03	293-453
C_5H_9NO	Isobutyl isocyanate	293.2	11.638	0.38026E+02	−0.12714E+00	0.12679E-03	293-353
C_5H_9NO	N-Methyl-2-pyrrolidone	293.2	32.55				
C_5H_{10}	1-Pentene	293.2	2.011	−0.11438E+01	0.25420E-01	−0.50000E-04	273-293
C_5H_{10}	2-Methyl-1-butene	293.2	2.180				
C_5H_{10}	2-Methyl-2-butene	296.0	1.979	0.26064E+01	−0.19578E-02	−0.53908E-06	225-296
C_5H_{10}	Cyclopentane	293.2	1.9687	0.24287E+01	−0.15304E-02	−0.13095E-06	278-313
C_5H_{10}	Ethylcyclopropane	293.2	1.933				
$C_5H_{10}Br_2$	1,2-Dibromopentane	298.2	4.39				
$C_5H_{10}Br_2$	1,4-Dibromopentane	293.2	9.05	0.26443E+02	−0.88640E-01	0.10000E-03	293-333
$C_5H_{10}Br_2$	1,5-Dibromopentane	303.2	9.14	0.38192E+02	−0.15648E+00	0.20000E-03	303-333
$C_5H_{10}Cl_2$	1,2-Dichloropentane	293.2	6.89	0.19016E+02	−0.57954E-01	0.56801E-04	293-356
$C_5H_{10}Cl_2$	1,5-Dichloropentane	298.2	9.92				
$C_5H_{10}O$	Cyclopentanol	288.2	18.5	0.10565E+03	−0.44244E+00	0.48657E-03	258-323
$C_5H_{10}O$	Pentanal	293.2	10.00				
$C_5H_{10}O$	2,2-Dimethylpropanal	293.2	9.051	0.18645E+02	−0.32395E-01	−0.16157E-05	280-333
$C_5H_{10}O$	2-Pentanone	293.2	15.45	0.40893E+02	−0.10423E+00	0.60557E-04	204-353
$C_5H_{10}O$	3-Pentanone	293.2	17.00	0.12690E+02	0.95177E-01	−0.27321E-03	233-353
$C_5H_{10}O$	3-Methyl-2-butanone	293.2	10.37	0.30695E+02	−0.10962E+00	0.13810E-03	293-328
$C_5H_{10}O$	Tetrahydropyran	293.2	5.66	0.19793E+02	−0.76071E-01	0.94852E-04	234-333
$C_5H_{10}O$	2-Methyltetrahydrofuran	298.2	6.97				
$C_5H_{10}O_2$	Pentanoic acid	294.4	2.661	0.33491E+01	−0.75156E-02	0.17820E-04	250-344
$C_5H_{10}O_2$	Butyl formate	303.2	6.10	0.21532E+02	−0.84106E-01	0.10952E-03	288-323
$C_5H_{10}O_2$	Isobutyl formate	293.2	6.41				
$C_5H_{10}O_2$	Propyl acetate	293.2	5.62	0.17677E+02	−0.61404E-01	0.69196E-04	253-353
$C_5H_{10}O_2$	Ethyl propanoate	293.2	5.76				
$C_5H_{10}O_2$	Methyl butanoate	301.2	5.48	0.38604E+02	−0.19171E+00	0.27128E-03	301-343
$C_5H_{10}O_2$	Tetrahydrofurfuryl alcohol	303.2	13.48				
$C_5H_{10}O_2S$	3-Methyl sulfolane	298.2	29.4	0.53158E+02	−0.93730E-01	0.47275E-04	298-398
$C_5H_{10}O_3$	Diethyl carbonate	297.2	2.820				
$C_5H_{10}O_3$	Ethyl lactate	303.2	15.4	0.31225E+02	−0.43531E-01	−0.28571E-04	273-373
$C_5H_{10}O_4$	1,2,3-Propanetriol-1-acetate	242.2	38.57	0.10653E+03	−0.26439E+00	−0.62371E-04	215-242
$C_5H_{11}Br$	2-Bromo-2-methylbutane	298.2	9.21				
$C_5H_{11}Br$	1-Bromopentane	299.2	6.31	0.20954E+02	−0.78743E-01	0.98908E-04	183-328
$C_5H_{11}Br$	3-Bromopentane	298.2	8.37				
$C_5H_{11}Br$	1-Bromo-3-methylbutane	291.5	6.33	0.27743E+02	−0.13927E+00	0.22627E-03	123-292
$C_5H_{11}Cl$	1-Chloropentane	293.2	6.654	0.18626E+02	−0.54719E-01	0.47143E-04	273-323
$C_5H_{11}Cl$	1-Chloro-3-methylbutane	292.0	6.10	0.22228E+02	−0.93189E-01	0.12991E-03	171-297
$C_5H_{11}Cl$	2-Chloro-2-methylbutane	222.75	12.31	0.55104E+02	−0.29866E+00	0.47840E-03	201-223
$C_5H_{11}F$	1-Fluoropentane	293.2	3.931				
$C_5H_{11}I$	1-Iodopentane	293.2	5.78	0.15753E+02	−0.50543E-01	0.56401E-04	293-323
$C_5H_{11}I$	3-Iodopentane	293.2	7.432				
$C_5H_{11}I$	1-Iodo-3-methylbutane	292.2	5.6				
$C_5H_{11}I$	2-Iodo-2-methylbutane	293.2	8.192				
$C_5H_{11}N$	Piperidine	293.0	4.33	0.82317E+01	−0.11229E-01	−0.71429E-05	293-333
$C_5H_{11}N$	*N*-Methylpyrrolidine	298.2	32.2				
$C_5H_{11}NO$	2,2-Dimethylpropanamide	298.2	20.13	0.10400E+03	−0.46017E+00	0.60000E-03	298-328
$C_5H_{11}NO$	*N,N*-Diethylformamide	293.2	29.6				
$C_5H_{11}NO$	2-Pentanone oxime	293.2	3.3				
$C_5H_{11}NO_2$	Pentyl nitrite	298.2	7.21				
C_5H_{12}	Pentane	293.2	1.8371				
C_5H_{12}	Isopentane	293.2	1.845	0.22384E+01	−0.12985E-02	−0.16182E-06	143-293
C_5H_{12}	Neopentane	296.0	1.769	0.10949E+02	−0.63057E-01	0.10835E-03	251-296
$C_5H_{12}N_2O$	Tetramethylurea	293.2	23.10				
$C_5H_{12}O$	1-Pentanol	298.2	15.13	0.73397E+02	−0.28165E+00	0.28427E-03	213-513
$C_5H_{12}O$	2-Pentanol	298.2	13.71	0.16437E+03	−0.86506E+00	0.11955E-02	273-323
$C_5H_{12}O$	3-Pentanol	298.2	13.35	0.12838E+03	−0.60980E+00	0.75000E-03	288-318
$C_5H_{12}O$	2-Methyl-1-butanol	298.2	15.63	0.14020E+02	0.13948E+00	−0.45000E-03	288-318
$C_5H_{12}O$	3-Methyl-1-butanol	293.2	15.63	0.79733E+02	−0.31272E+00	0.32014E-03	173-513
$C_5H_{12}O$	2-Methyl-2-butanol	298.2	5.78	0.11662E+03	−0.69756E+00	0.10920E-02	268-318
$C_5H_{12}O$	3-Methyl-2-butanol	298.2	12.1				
$C_5H_{12}O$	2,2-Dimethyl-1-propanol	333.2	8.35	0.92350E+02	−0.41870E+00	0.50000E-03	333-373

Mol. form.	Name	T/K	ε_r	a	b	c	Range/K
$C_5H_{12}O_2$	1,2-Pentanediol	296.8	17.31	0.18436E+03	−0.10682E+01	0.17037E-02	197-297
$C_5H_{12}O_2$	1,4-Pentanediol	295.7	26.74	0.13568E+03	−0.59198E+00	0.75398E-03	193-318
$C_5H_{12}O_2$	1,5-Pentanediol	293.2	26.2	0.11858E+03	−0.45920E+00	0.49341E-03	243-343
$C_5H_{12}O_2$	2,3-Pentanediol	296.9	17.37	0.95876E+02	−0.46463E+00	0.67434E-03	238-297
$C_5H_{12}O_2$	2,4-Pentanediol	294.2	24.69	0.11914E+03	−0.52569E+00	0.69607E-03	224-294
$C_5H_{12}O_2$	Diethoxymethane	293.2	2.527	0.25294E+01	0.73988E-04	−0.28331E-06	227-293
$C_5H_{12}O_4$	Tetramethoxymethane	293.2	2.40				
$C_5H_{12}O_5$	Xylitol	293.2	40.0				
$C_5H_{12}S$	1-Pentanethiol	293.2	4.847	0.71131E+01	−0.30228E-02	−0.16414E-04	273-333
$C_5H_{12}S$	2-Methyl-2-butanethiol	293.2	5.087	0.15116E+02	−0.50700E-01	0.56250E-04	273-333
$C_5H_{12}S_4$	Tetrakis(methylthio)methane	343.2	2.818				
$C_5H_{13}N$	Pentylamine	293.2	4.27	0.11274E+02	−0.34965E-01	0.37706E-04	223-353
$C_5H_{13}N_3$	1,1,3,3-Tetramethylguanidine	298.2	11.5				
$C_5H_{14}OSi$	Ethoxytrimethylsilane	298.2	3.013				
C_6F_6	Hexafluorobenzene	298.2	2.029	0.24041E+01	−0.83086E-03	−0.14286E-05	298-338
C_6F_{14}	Perfluorohexane	298.2	1.76				
$C_6H_3N_3O_7$	2,4,6-Trinitrophenol	294.2	4.0				
C_6H_4BrF	1-Bromo-2-fluorobenzene	298.2	4.72				
C_6H_4BrF	1-Bromo-3-fluorobenzene	298.2	4.85				
C_6H_4BrF	1-Bromo-4-fluorobenzene	298.2	2.60				
$C_6H_4BrNO_2$	1-Bromo-3-nitrobenzene	328.2	20.2	0.81413E+02	−0.27645E+00	0.27367E-03	328-413
$C_6H_4Br_2$	*o*-Dibromobenzene	293.2	7.86	−0.81849E-02	0.62671E-01	−0.12222E-03	293-353
$C_6H_4Br_2$	*m*-Dibromobenzene	293.2	4.81	0.93214E+01	−0.20273E-01	0.16667E-04	293-353
$C_6H_4Br_2$	p-Dibromobenzene	368.2	2.57				
C_6H_4ClF	1-Chloro-2-fluorobenzene	298.2	6.10				
C_6H_4ClF	1-Chloro-3-fluorobenzene	298.2	4.96				
C_6H_4ClF	1-Chloro-4-fluorobenzene	298.2	3.34				
$C_6H_4ClNO_2$	1-Chloro-2-nitrobenzene	323.2	37.7	0.16800E+03	−0.59708E+00	0.59957E-03	323-436
$C_6H_4ClNO_2$	1-Chloro-3-nitrobenzene	323.2	20.9	0.77193E+02	−0.25118E+00	0.23798E-03	323-433
$C_6H_4ClNO_2$	1-Chloro-4-nitrobenzene	393.2	8.09				
$C_6H_4Cl_2$	*o*-Dichlorobenzene	293.2	10.12	0.13629E+02	0.10622E-02	−0.44444E-04	293-353
$C_6H_4Cl_2$	*m*-Dichlorobenzene	293.2	5.02	0.77565E+01	−0.93333E-02	−0.26880E-14	293-353
$C_6H_4Cl_2$	*p*-Dichlorobenzene	328.2	2.3943	0.26999E+01	−0.35325E-03	−0.17619E-05	328-363
C_6H_4FI	1-Fluoro-2-iodobenzene	298.2	8.22				
C_6H_4FI	1-Fluoro-4-iodobenzene	298.2	3.12				
$C_6H_4F_2$	*o*-Difluorobenzene	301.2	13.38	0.59107E+02	−0.23611E+00	0.27987E-03	273-323
$C_6H_4F_2$	*m*-Difluorobenzene	301.2	5.01	0.14448E+02	−0.46982E-01	0.51948E-04	273-323
$C_6H_4I_2$	*o*-Diiodobenzene	323.2	5.41	0.31150E+02	−0.14428E+00	0.20000E-03	323-353
$C_6H_4I_2$	*m*-Diiodobenzene	323.2	4.11				
$C_6H_4I_2$	*p*-Diiodobenzene	393.2	2.88				
$C_6H_4N_2$	2-Pyridinecarbonitrile	303.2	93.77	0.45596E+03	−0.17746E+01	0.19105E-02	303-398
$C_6H_4N_2$	3-Pyridinecarbonitrile	323.2	20.54	0.60484E+02	−0.17280E+00	0.15218E-03	323-398
$C_6H_4N_2$	4-Pyridinecarbonitrile	353.2	5.23	0.12533E+02	−0.30115E-01	0.26674E-04	353-398
$C_6H_4N_2O_4$	1,3-Dinitrobenzene	365.2	22.9	0.10406E+03	−0.34133E+00	0.32609E-03	365-413
C_6H_5Br	Bromobenzene	293.2	5.45	0.94100E+01	−0.12537E-01	−0.31127E-05	234-333
C_6H_5Cl	Chlorobenzene	293.2	5.6895	0.19471E+02	−0.70786E-01	0.82466E-04	293-430
C_6H_5ClO	*o*-Chlorophenol	296.2	7.40	0.29755E+02	−0.11256E+00	0.12390E-03	296-448
C_6H_5ClO	*m*-Chlorophenol	293.2	6.255				
C_6H_5ClO	*p*-Chlorophenol	314.2	11.18	0.31997E+02	−0.94241E-01	0.88392E-04	314-453
$C_6H_5ClO_2S$	Benzenesulfonyl chloride	323.2	28.90	0.83886E+02	−0.23405E+00	0.19713E-03	323-473
C_6H_5ClS	4-Chlorobenzenethiol	338.2	3.59				
C_6H_5F	Fluorobenzene	293.2	5.465				
C_6H_5I	Iodobenzene	293.2	4.59	0.89442E+01	−0.20008E-01	0.17641E-04	243-323
C_6H_5NOS	*N*-Sulfinylaniline	298.2	6.97				
$C_6H_5NO_2$	Nitrobenzene	293.0	35.6	0.11212E+03	−0.35211E+00	0.31128E-03	279-533
$C_6H_5NO_3$	*o*-Nitrophenol	323.2	16.50	0.33827E+02	−0.62123E-01	0.26774E-04	323-453
$C_6H_5NO_3$	*m*-Nitrophenol	373.2	35.45	0.18967E+03	−0.66144E+00	0.66532E-03	373-458
$C_6H_5NO_3$	*p*-Nitrophenol	393.2	42.20	0.22901E+03	−0.74264E+00	0.68006E-03	393-463
C_6H_6	Benzene	293.2	2.2825	0.26706E+01	−0.91648E-03	−0.14257E-05	293-513
C_6H_6BrN	*m*-Bromoaniline	293.2	13.0				
C_6H_6ClN	*o*-Chloroaniline	293.2	13.40				
C_6H_6ClN	*m*-Chloroaniline	293.2	13.3				
$C_6H_6N_2O_2$	*o*-Nitroaniline	353.0	47.3	0.18900E+03	−0.56977E+00	0.47484E-03	353-468

Mol. form.	Name	T/K	ε_r	a	b	c	Range/K
$C_6H_6N_2O_2$	*m*-Nitroaniline	398.0	35.6	0.20352E+03	−0.66582E+00	0.61310E-03	398-468
$C_6H_6N_2O_2$	*p*-Nitroaniline	428.0	78.5	0.48673E+03	−0.15040E+01	0.12857E-02	428-468
C_6H_6O	Phenol	303.2	12.40	0.63391E+02	−0.24988E+00	0.26930E-03	303-433
$C_6H_6O_2$	Pyrocatechol	388.2	17.57	0.74930E+02	−0.22142E+00	0.18919E-03	388-463
$C_6H_6O_2$	Resorcinol	393.2	13.55	0.30252E+02	−0.56443E-01	0.35578E-04	393-463
C_6H_6S	Benzenethiol	303.2	4.26	0.57155E+01	−0.70336E-02	0.73617E-05	303-358
C_6H_7N	Aniline	293.2	7.06	0.89534E+01	0.38990E-02	−0.36310E-04	293-413
C_6H_7N	2-Methylpyridine	293.2	10.18	0.34560E+02	−0.11980E+00	0.12500E-03	293-333
C_6H_7N	3-Methylpyridine	303.0	11.10	0.19643E+03	−0.11167E+01	0.16667E-02	303-333
C_6H_7N	4-Methylpyridine	293.0	12.2	0.33765E+02	−0.10113E+00	0.93860E-04	274-333
C_6H_7NO	2-Methylpyridine-1-oxide	323.2	36.4	0.11705E+03	−0.35301E+00	0.32000E-03	323-398
C_6H_7NO	3-Methylpyridine-1-oxide	318.2	28.26	0.59851E+02	−0.12682E+00	0.86622E-04	318-398
C_6H_8	1,3-Cyclohexadiene	184.2	2.68				
C_6H_8	1,4-Cyclohexadiene	296.0	2.211	0.27459E+01	−0.16975E-02	−0.36461E-06	232-356
$C_6H_8N_2$	Phenylhydrazine	293.2	7.15				
$C_6H_8N_2$	2,5-Dimethylpyrazine	293.2	2.436				
$C_6H_8N_2$	2,6-Dimethylpyrazine	308.2	2.653				
$C_6H_8O_2$	1,4-Cyclohexanedione	351.2	4.40				
$C_6H_9Cl_3O_2$	Butyl trichloroacetate	293.2	7.480				
$C_6H_9Cl_3O_2$	Isobutyl trichloroacetate	293.2	7.667				
C_6H_9N	Cyclopentanecarbonitrile	293.2	22.68	0.69830E+02	−0.25303E+00	0.31491E-03	201-293
C_6H_{10}	1,5-Hexadiene	291.0	2.125	0.30014E+01	−0.28668E-02	−0.31026E-06	151-294
C_6H_{10}	*cis,cis*-2,4-Hexadiene	297.0	2.163	0.27284E+01	−0.17178E-02	−0.62926E-06	234-351
C_6H_{10}	*trans,trans*-2,4-Hexadiene	297.0	2.123	0.26774E+01	−0.16977E-02	−0.55637E-06	232-353
C_6H_{10}	2-Methyl-1,3-pentadiene*	298.2	2.422				
C_6H_{10}	3-Methyl-1,3-pentadiene	298.2	2.426				
C_6H_{10}	4-Methyl-1,3-pentadiene	293.2	2.599	0.51328E+01	−0.12774E-01	0.14215E-04	198-323
C_6H_{10}	2,3-Dimethyl-1,3-butadiene	293.2	2.102	0.26258E+01	−0.17990E-02	0.12035E-06	223-323
C_6H_{10}	1-Hexyne	296.0	2.621	0.58591E+01	−0.17099E-01	0.20856E-04	184-296
C_6H_{10}	Cyclohexene	293.2	2.2176	0.30598E+01	−0.39841E-02	0.37554E-05	141-313
$C_6H_{10}O$	Butoxyacetylene	298.2	6.62				
$C_6H_{10}O$	Cyclohexanone	293.0	16.1	0.41577E+02	−0.11463E+00	0.92454E-04	253-423
$C_6H_{10}O$	Mesityl oxide	273.2	15.6				
$C_6H_{10}O_2$	Ethyl 2-butenoate	293.2	5.4				
$C_6H_{10}O_2$	Ethyl methacrylate	303.2	5.68	0.40962E+02	−0.20520E+00	0.29286E-03	303-343
$C_6H_{10}O_3$	Ethyl acetoacetate	293.2	14.0				
$C_6H_{10}O_3$	Propanoic anhydride	293.2	18.30				
$C_6H_{10}O_4$	Monomethyl glutarate	293.2	8.37	0.16779E+02	−0.39839E-01	0.38095E-04	293-363
$C_6H_{10}O_4$	Diethyl oxalate	293.2	8.266	0.21938E+02	−0.66226E-01	0.66800E-04	293-368
$C_6H_{10}O_4$	Dimethyl succinate	293.2	7.19	0.13551E+02	−0.23109E-01	0.55440E-05	293-433
$C_6H_{10}O_4$	Ethylene glycol diacetate	290.2	7.7	0.25093E+02	−0.95171E-01	0.12224E-03	223-290
$C_6H_{11}Br$	Bromocyclohexane	303.2	8.0026				
$C_6H_{11}BrO_2$	Ethyl 2-bromobutanoate	303.2	8.57	0.49005E+02	−0.23193E+00	0.32500E-03	303-333
$C_6H_{11}BrO_2$	Ethyl 2-bromo-2-methylpropanoate	303.2	8.55	0.77044E+02	−0.40784E+00	0.60000E-03	303-333
$C_6H_{11}Cl$	Chlorocyclohexane	303.2	7.9505				
$C_6H_{11}N$	Hexanenitrile	298.2	17.26				
$C_6H_{11}N$	4-Methylpentanenitrile	295.2	17.5				
$C_6H_{11}NO$	Cyclohexanone oxime	362.2	3.04				
C_6H_{12}	1-Hexene	294.0	2.077	0.31476E+01	−0.50003E-02	0.46673E-05	149-294
C_6H_{12}	*trans*-2-Hexene	295.0	1.978	0.24338E+01	−0.11323E-02	−0.13720E-05	157-295
C_6H_{12}	*cis*-3-Hexene	296.0	2.069	0.30691E+01	−0.45458E-02	0.39898E-05	155-296
C_6H_{12}	*trans*-3-Hexene	293.2	1.954				
C_6H_{12}	Cyclohexane	293.2	2.0243	0.24293E+01	−0.12095E-02	−0.58741E-06	283-333
C_6H_{12}	Methylcyclopentane	293.2	1.9853	0.21587E+01	−0.22450E-03	−0.12500E-05	293-323
C_6H_{12}	Ethylcyclobutane	293.2	1.965				
$C_6H_{12}Br_2$	1,6-Dibromohexane	298.2	8.52	−0.55185E+01	0.11746E+00	−0.23658E-03	274-328
$C_6H_{12}Br_2$	3,4-Dibromohexane	298.2	6.732				
$C_6H_{12}Cl_2$	1,6-Dichlorohexane	308.2	8.60	0.11277E+02	0.67200E-02	−0.50000E-04	308-338
$C_6H_{12}O$	1-Methylcyclopentanol	310.1	7.11	0.75444E+02	−0.36617E+00	0.47021E-03	310-333
$C_6H_{12}O$	Isobutyl vinyl ether	293.2	3.34	0.48060E+01	−0.50000E-02	−0.41495E-14	293-323
$C_6H_{12}O$	2-Hexanone	293.2	14.56	0.70378E+02	−0.29385E+00	0.35289E-03	243-293
$C_6H_{12}O$	4-Methyl-2-pentanone	293.2	13.11	0.36341E+02	−0.97119E-01	0.61896E-04	204-373
$C_6H_{12}O$	3,3-Dimethyl-2-butanone	293.2	12.73	0.66857E+02	−0.28552E+00	0.34422E-03	243-293

Mol. form.	Name	T/K	ε_r	*a*	*b*	*c*	Range/K
$C_6H_{12}O$	Cyclohexanol	293.2	16.40	0.10173E+03	−0.43072E+00	0.47926E-03	293-423
$C_6H_{12}O_2$	Hexanoic acid	298.2	2.600	0.21730E+01	0.14840E-02	−0.16526E-06	298-433
$C_6H_{12}O_2$	2-Ethylbutanoic acid	296.2	2.72				
$C_6H_{12}O_2$	*tert*-Butylacetic acid	296.2	2.85				
$C_6H_{12}O_2$	Pentyl formate	292.2	5.7				
$C_6H_{12}O_2$	Isopentyl formate	288.2	5.44	0.29257E+02	−0.14028E+00	0.20000E-03	288-323
$C_6H_{12}O_2$	Butyl acetate	293.2	5.07	0.13825E+02	−0.43994E-01	0.48214E-04	253-353
$C_6H_{12}O_2$	*sec*-Butyl acetate	293.2	5.135	0.12427E+02	−0.32035E-01	0.24286E-04	273-323
$C_6H_{12}O_2$	*tert*-Butyl acetate	293.2	5.672	0.55435E+02	−0.30494E+00	0.46107E-03	273-323
$C_6H_{12}O_2$	Isobutyl acetate	293.2	5.068	0.14323E+02	−0.46048E-01	0.49286E-04	273-323
$C_6H_{12}O_2$	Propyl propanoate	293.2	5.249				
$C_6H_{12}O_2$	Ethyl butanoate	301.2	5.18	0.48698E+02	−0.25660E+00	0.37237E-03	301-343
$C_6H_{12}O_2$	Methyl pentanoate	293.2	4.992				
$C_6H_{12}O_2$	Diacetone alcohol	298.2	18.2				
$C_6H_{12}O_3$	Ethylene glycol monoethyl ether acetate	303.2	7.567	0.23290E+02	−0.71566E-01	0.65000E-04	303-323
$C_6H_{12}S$	Cyclohexanethiol	298.2	5.420				
$C_6H_{13}Br$	1-Bromohexane	298.2	5.82	0.15233E+02	−0.44385E-01	0.43039E-04	274-328
$C_6H_{13}Cl$	1-Chlorohexane	293.2	6.104	0.15994E+02	−0.43647E-01	0.33393E-04	273-323
$C_6H_{13}ClO$	6-Chloro-1-hexanol	242.2	21.6	−0.73364E+01	0.46377E+00	−0.14202E-02	195-242
$C_6H_{13}I$	1-Iodohexane	293.3	5.35	0.16685E+02	−0.61309E-01	0.77262E-04	293-323
$C_6H_{13}N$	Cyclohexylamine	293.2	4.547				
$C_6H_{13}NO$	*N*-Propylpropanamide	298.2	118.1	0.58846E+03	−0.22012E+01	0.20870E-02	298-328
$C_6H_{13}NO$	*N*-Butylacetamide**	293.2	104.0	0.70739E+03	−0.37369E+01	0.71585E-02	253-493
$C_6H_{13}NO$	*N,N*-Diethylacetamide	293.2	32.1				
C_6H_{14}	Hexane	293.2	1.8865	0.19768E+01	0.70933E-03	−0.34470E-05	293-473
C_6H_{14}	2-Methylpentane	293.2	1.886	0.20745E+01	0.50871E-03	−0.39286E-05	273-323
C_6H_{14}	3-Methylpentane	293.2	1.886	0.24739E+01	−0.23190E-02	0.10714E-05	273-323
C_6H_{14}	2,2-Dimethylbutane	293.2	1.869	0.22740E+01	−0.96229E-03	−0.14286E-05	273-313
C_6H_{14}	2,3-Dimethylbutane	293.2	1.889	0.24305E+01	−0.20081E-02	0.53571E-06	273-323
$C_6H_{14}O$	1-Hexanol	293.2	13.03	0.62744E+02	−0.24214E+00	0.24704E-03	233-513
$C_6H_{14}O$	2-Hexanol	298.2	11.06				
$C_6H_{14}O$	3-Hexanol	298.2	9.66				
$C_6H_{14}O$	3-Methyl-1-pentanol	298.2	15.2				
$C_6H_{14}O$	3-Methyl-3-pentanol	293.2	4.322				
$C_6H_{14}O$	2-Ethyl-1-butanol	362.2	6.19				
$C_6H_{14}O$	2,2-Dimethyl-1-butanol	293.2	10.5	0.14054E+03	−0.72925E+00	0.97821E-03	243-393
$C_6H_{14}O$	Dipropyl ether	297.0	3.38	0.14600E+02	−0.72670E-01	0.11742E-03	161-297
$C_6H_{14}O$	Diisopropyl ether	303.2	3.805				
$C_6H_{14}OS$	Dipropyl sulfoxide	303.2	30.37	0.84868E+02	−0.23486E+00	0.18198E-03	303-373
$C_6H_{14}O_2$	2-Methyl-2,4-pentanediol	293.2	25.86	0.14531E+03	−0.65285E+00	0.83503E-03	203-333
$C_6H_{14}O_2$	Ethylene glycol diethyl ether	293.2	3.90	0.99099E+01	−0.33403E-01	0.44048E-04	223-303
$C_6H_{14}O_2S$	Dipropyl sulfone	303.2	32.62	0.70195E+02	−0.15008E+00	0.86506E-04	303-398
$C_6H_{14}O_3$	1,2,6-Hexanetriol	285.3	31.5	0.26127E+03	−0.14552E+01	0.22765E-02	261-285
$C_6H_{14}O_3$	Diethylene glycol dimethyl ether	298.2	7.23	0.28291E+02	−0.11236E+00	0.14000E-03	298-333
$C_6H_{14}O_4$	Triethylene glycol	293.2	23.69	0.91845E+02	−0.33827E+00	0.36062E-03	253-333
$C_6H_{14}O_6$	*D*-Glucitol	353.2	35.5				
$C_6H_{14}O_6$	*D*-Mannitol	443.2	24.6				
$C_6H_{14}S$	1-Hexanethiol	293.2	4.436	0.11774E+02	−0.37298E-01	0.41875E-04	273-333
$C_6H_{15}B$	Triethylborane	293.2	1.974				
$C_6H_{15}N$	Hexylamine	293.2	4.08	0.80244E+01	−0.16627E-01	0.10874E-04	253-373
$C_6H_{15}N$	Dipropylamine	293.2	2.923	0.11376E+02	−0.49796E-01	0.71792E-04	243-323
$C_6H_{15}N$	Triethylamine	293.2	2.418	0.29205E+01	−0.14007E-02	−0.13469E-05	233-323
$C_6H_{15}OP$	Triethylphosphine oxide	323.2	35.5				
$C_6H_{15}O_4P$	Triethyl phosphate	298.2	13.20	0.61230E+02	−0.26047E+00	0.33333E-03	298-333
$C_6H_{15}PS$	Triethylphosphine sulfide	371.2	39.0				
$C_6H_{16}O_2Si$	Diethoxydimethylsilane	298.2	3.216				
$C_6H_{16}Si$	Triethylsilane	293.2	2.323				
$C_6H_{18}N_3OP$	Hexamethylphosphoric triamide	293.2	31.3	0.95666E+02	−0.29769E+00	0.26407E-03	283-363
$C_6H_{18}N_4$	*N,N′*-Bis(2-aminoethyl)-1,2-ethanediamine	293.2	10.76	0.50699E+02	−0.21730E+00	0.27582E-03	213-333
$C_6H_{18}OSi_2$	Hexamethyldisiloxane	293.2	2.179	0.34537E+01	−0.61530E-02	0.61544E-05	213-313
$C_6H_{18}O_3Si_3$	Hexamethylcyclotrisiloxane	343.2	2.139				

Mol. form.	Name	T/K	ε_r	*a*	*b*	*c*	Range/K
$C_6H_{19}NSi_2$	Hexamethyldisilazane	294.2	2.273	0.23358E+01	0.16127E-02	−0.62078E-05	294-333
C_7F_{14}	Perfluoromethylcyclohexane	298.2	1.82				
C_7F_{16}	Perfluoroheptane	289.2	1.847				
$C_7H_3Cl_5$	2,3,4,5,6-Pentachlorotoluene	293.2	4.8				
C_7H_4ClNO	4-Chlorophenyl isocyanate	288.2	3.177	0.40896E+01	−0.31667E-02		288-348
C_7H_5BrO	Benzoyl bromide	293.2	21.33	0.84231E+02	−0.31089E+00	0.32857E-03	283-313
C_7H_5ClO	Benzoyl chloride	293.2	23.0				
C_7H_5FO	Benzoyl fluoride	293.2	22.7				
$C_7H_5F_3$	(Trifluoromethyl)benzene	298.2	9.22				
C_7H_5N	Benzonitrile	293.2	25.9	0.57605E+02	−0.13354E+00	0.87767E-04	273-453
C_7H_5NO	Phenyl isocyanate	293.2	8.940	0.17541E+02	−0.29790E-01	0.15476E-05	293-353
$C_7H_6ClNO_2$	4-Chloro-3-nitrotoluene	301.2	28.07				
$C_7H_6Cl_2$	2,4-Dichlorotoluene	301.2	5.68				
$C_7H_6Cl_2$	2,6-Dichlorotoluene	301.2	3.36				
$C_7H_6Cl_2$	3,4-Dichlorotoluene	301.2	9.39				
$C_7H_6Cl_2$	(Dichloromethyl)benzene	293.2	6.9				
C_7H_6O	Benzaldehyde	293.2	17.85	0.35046E+02	−0.61271E-01	0.16222E-04	301-346
$C_7H_6O_2$	Salicylaldehyde	293.2	18.35	0.51315E+02	−0.15379E+00	0.14111E-03	289-453
C_7H_7Br	*o*-Bromotoluene	293.2	4.641	0.10229E+02	−0.25050E-01	0.20357E-04	273-323
C_7H_7Br	*m*-Bromotoluene	293.2	5.566	0.11522E+02	−0.24946E-01	0.15714E-04	273-323
C_7H_7Br	*p*-Bromotoluene	293.2	5.503	0.10014E+02	−0.13918E-01	−0.50000E-05	273-293
C_7H_7Br	(Bromomethyl)benzene	293.2	6.658	0.18482E+02	−0.57207E-01	0.57321E-04	273-323
C_7H_7BrO	*o*-Bromoanisole	303.2	8.96	0.12023E+02	−0.59116E-02	−0.13787E-04	303-358
C_7H_7BrO	*p*-Bromoanisole	303.2	7.40	0.74367E+01	0.12648E-01	−0.42128E-04	303-358
C_7H_7Cl	*o*-Chlorotoluene	293.2	4.721	0.11507E+02	−0.31148E-01	0.27143E-04	273-323
C_7H_7Cl	*m*-Chlorotoluene	293.2	5.763	0.13921E+02	−0.37186E-01	0.31786E-04	273-323
C_7H_7Cl	*p*-Chlorotoluene	293.2	6.25	0.20265E+01	0.40060E-01	−0.87500E-04	293-333
C_7H_7Cl	(Chloromethyl)benzene	293.2	6.854	0.17108E+02	−0.45285E-01	0.35000E-04	273-323
C_7H_7ClO	*p*-Chloroanisole	293.2	7.84	0.64019E+01	0.30560E-01	−0.87500E-04	293-333
$C_7H_7ClO_2S$	*p*-Toluenesulfonyl chloride	343.2	22.6				
$C_7H_7ClO_3S$	4-Methoxybenzenesulfonyl chloride	314.2	27.2				
C_7H_7F	*o*-Fluorotoluene	298.2	4.23				
C_7H_7F	*m*-Fluorotoluene	298.2	5.41				
C_7H_7F	*p*-Fluorotoluene	298.2	5.88				
C_7H_7I	*p*-Iodotoluene	308.2	4.4				
C_7H_7N	2-Vinylpyridine	293.2	9.126				
C_7H_7N	4-Vinylpyridine	293.2	10.50				
$C_7H_7NO_2$	Benzyl nitrite	298.2	7.78				
$C_7H_7NO_2$	*o*-Nitrotoluene	293.0	26.26	0.10420E+03	−0.41726E+00	0.51607E-03	273-323
$C_7H_7NO_2$	*m*-Nitrotoluene	303.2	24.95	0.62492E+02	−0.16235E+00	0.12844E-03	303-403
$C_7H_7NO_2$	*p*-Nitrotoluene	331.2	22.2				
$C_7H_7NO_2S$	4-Nitrothioanisole	346.0	21.7				
$C_7H_7NO_3$	2-Nitroanisole	293.2	45.75	0.16684E+03	−0.58196E+00	0.57382E-03	293-423
$C_7H_7NO_3$	3-Nitroanisole	318.2	25.7	0.65402E+02	−0.16460E+00	0.12560E-03	318-443
$C_7H_7NO_3$	4-Nitroanisole	338.2	26.95	0.59811E+02	−0.10955E+00	0.36042E-04	338-443
C_7H_8	Toluene	296.35	2.379	0.32584E+01	−0.34410E-02	0.15937E-05	207-316
C_7H_8O	*o*-Cresol	298.2	6.76	0.21633E+02	−0.71069E-01	0.70590E-04	298-453
C_7H_8O	*m*-Cresol	298.2	12.44	0.81716E+02	−0.35039E+00	0.39878E-03	274-463
C_7H_8O	*p*-Cresol	298.2	13.05	0.70253E+02	−0.28870E+00	0.31979E-03	298-453
C_7H_8O	Benzyl alcohol	303.2	11.916	0.13661E+03	−0.72127E+00	0.10225E-02	303-333
C_7H_8O	Anisole	294.2	4.30	0.10887E+02	−0.32372E-01	0.33629E-04	294-413
$C_7H_8O_2$	2-Methoxyphenol	298.2	11.95	0.31751E+02	−0.88173E-01	0.72953E-04	291-448
$C_7H_8O_2$	3-Methoxyphenol	298.2	11.59	0.37279E+02	−0.12113E+00	0.11698E-03	298-433
$C_7H_8O_2$	4-Methoxyphenol	333.7	11.05	0.39483E+02	−0.12142E+00	0.10841E-03	334-453
$C_7H_8O_2S$	Ethyl thiophene-2-carboxylate	293.2	6.18				
$C_7H_8O_2S$	Methyl phenyl sulfone	373.2	37.9				
C_7H_8S	Benzenemethanethiol	298.2	4.705	0.16628E+02	−0.68276E-01	0.94636E-04	298-358
C_7H_8S	4-Methylbenzenethiol	323.2	4.74	0.87052E+01	−0.15347E-01	0.95238E-05	323-358
C_7H_8S	(Methylthio)benzene	303.2	4.88	0.21841E+02	−0.97630E-01	0.13750E-03	303-343
C_7H_9N	Benzylamine	293.2	5.18				
C_7H_9N	*o*-Methylaniline	298.2	6.138	0.10988E+02	−0.18976E-01	0.91958E-05	298-398
C_7H_9N	*m*-Methylaniline	298.2	5.816	0.13477E+02	−0.35551E-01	0.33135E-04	298-398
C_7H_9N	*p*-Methylaniline	333.2	5.058	0.78897E+01	−0.10196E-01	0.51190E-05	333-403

Mol. form.	Name	T/K	ε_r	a	b	c	Range/K
C_7H_9N	*N*-Methylaniline	293.2	5.96				
C_7H_9N	2-Ethylpyridine	293.2	8.33	0.36397E+02	−0.15070E+00	0.18750E-03	293-333
C_7H_9N	4-Ethylpyridine	293.2	10.98	−0.73831E+01	0.14326E+00	−0.27500E-03	293-333
C_7H_9N	2,4-Dimethylpyridine	293.2	9.60	0.25895E+02	−0.73900E-01	0.62500E-04	293-333
C_7H_9N	2,6-Dimethylpyridine	293.2	7.33	0.17714E+02	−0.39080E-01	0.12500E-04	293-333
C_7H_9NO	2,6-Dimethylpyridine-1-oxide	298.2	46.11	0.22765E+03	−0.90760E+00	0.10011E-02	298-398
C_7H_9NO	*o*-Methoxyaniline	303.2	5.230	0.79911E+01	−0.92183E-02	0.37879E-06	303-393
C_7H_9NO	*m*-Methoxyaniline	298.2	8.76	0.28179E+02	−0.97840E-01	0.11027E-03	289-393
C_7H_9NO	*p*-Methoxyaniline	333.2	7.85	0.30149E+02	−0.10523E+00	0.11467E-03	333-453
$C_7H_{10}N_2$	1-Methyl-1-phenylhydrazine	292.2	7.3				
$C_7H_{11}Cl_3O_2$	Isopentyl trichloroacetate	293.2	7.287				
C_7H_{12}	1,6-Heptadiene	293.0	2.161	0.30815E+01	−0.36095E-02	0.16354E-05	184-293
C_7H_{12}	Cycloheptene	295.0	2.265	0.32309E+01	−0.42373E-02	0.32572E-05	227-363
$C_7H_{12}O$	Cycloheptanone	298.2	13.16	0.17511E+03	−0.11221E+01	0.19417E-02	258-298
$C_7H_{12}O$	2-Methylcyclohexanone	293.2	14.0				
$C_7H_{12}O$	3-Methylcyclohexanone	293.2	12.4				
$C_7H_{12}O$	4-Methylcyclohexanone	293.2	12.35				
$C_7H_{12}O_2$	Cyclohexanecarboxylic acid	304.2	2.67				
$C_7H_{12}O_2$	Cyclohexyl formate	293.2	6.47				
$C_7H_{12}O_2$	Butyl acrylate	301.2	5.25	0.38296E+02	−0.19109E+00	0.27006E-03	301-343
$C_7H_{12}O_4$	Monomethyl adipate	293.2	6.69	0.11962E+02	−0.23973E-01	0.20608E-04	293-433
$C_7H_{12}O_4$	Diethyl malonate	304.2	7.550	0.14809E+02	−0.31207E-01	0.24066E-04	304-393
$C_7H_{12}O_4$	Dimethyl glutarate	293.2	7.87	0.20697E+02	−0.57794E-01	0.48405E-04	293-433
$C_7H_{12}O_5$	1,2,3-Propanetriol-1,3-diacetate	288.2	9.80	0.28321E+02	−0.89073E-01	0.86891E-04	258-374
C_7H_{14}	1-Heptene	293.2	2.092	0.21755E+01	0.13896E-02	−0.57049E-05	273-323
C_7H_{14}	2-Methyl-2-hexene	293.2	2.962				
C_7H_{14}	3-Ethyl-2-pentene	293.2	2.051				
C_7H_{14}	Cycloheptane	293.2	2.0784	0.25136E+01	−0.15089E-02	0.84915E-07	278-333
C_7H_{14}	Methylcyclohexane	293.2	2.024				
$C_7H_{14}Br_2$	1,2-Dibromoheptane	298.2	3.77				
$C_7H_{14}Br_2$	2,3-Dibromoheptane	298.2	5.08				
$C_7H_{14}Br_2$	3,4-Dibromoheptane	298.2	4.70				
$C_7H_{14}Cl_2$	1,7-Dichloroheptane	298.2	8.34				
$C_7H_{14}O$	1-Heptanal	295.2	9.07				
$C_7H_{14}O$	2-Heptanone	293.2	11.95	0.38348E+02	−0.12531E+00	0.12005E-03	253-413
$C_7H_{14}O$	3-Heptanone	293.2	12.7				
$C_7H_{14}O$	4-Heptanone	293.2	12.60	0.41520E+02	−0.13839E+00	0.13497E-03	253-393
$C_7H_{14}O$	5-Methyl-2-hexanone	293.2	13.53	0.52353E+02	−0.17695E+00	0.15195E-03	293-333
$C_7H_{14}O$	Cyclohexanemethanol	333.2	9.70	0.10164E+03	−0.45839E+00	0.54762E-03	333-368
$C_7H_{14}O$	2-Methylcyclohexanol*	293.2	9.375	0.17315E+03	−0.98794E+00	0.14634E-02	273-323
$C_7H_{14}O$	3-Methylcyclohexanol*	293.2	13.79	0.65896E+02	−0.21954E+00	0.14107E-03	273-323
$C_7H_{14}O$	4-Methylcyclohexanol*	293.2	13.45	0.65021E+02	−0.22896E+00	0.17946E-03	273-323
$C_7H_{14}O_2$	Heptanoic acid	288.2	3.04	0.36423E+01	−0.31996E-02	0.39362E-05	288-423
$C_7H_{14}O_2$	Pentyl acetate	293.2	4.79	0.12091E+02	−0.36536E-01	0.39732E-04	253-353
$C_7H_{14}O_2$	Isopentyl acetate	293.2	4.72				
$C_7H_{14}O_2$	Butyl propanoate	293.2	4.838				
$C_7H_{14}O_2$	Propyl butanoate	293.2	4.3				
$C_7H_{14}O_2$	Ethyl pentanoate	291.2	4.71				
$C_7H_{14}O_2$	Ethyl 3-methylbutanoate	293.2	4.71				
$C_7H_{14}O_2$	Methyl hexanoate	293.2	4.615				
$C_7H_{15}Br$	1-Bromoheptane	303.2	5.255	0.15289E+02	−0.50621E-01	0.57753E-04	203-343
$C_7H_{15}Br$	2-Bromoheptane	295.2	6.46				
$C_7H_{15}Br$	4-Bromoheptane	295.2	6.81				
$C_7H_{15}Cl$	1-Chloroheptane	293.2	5.521	0.14279E+02	−0.39431E-01	0.32321E-04	273-323
$C_7H_{15}Cl$	2-Chloroheptane	295.2	6.52				
$C_7H_{15}Cl$	3-Chloroheptane	295.2	6.70				
$C_7H_{15}Cl$	4-Chloroheptane	295.2	6.54				
$C_7H_{15}I$	1-Iodoheptane	298.2	4.92	0.11856E+02	−0.33493E-01	0.34368E-04	294-323
$C_7H_{15}I$	3-Iodoheptane	295.2	6.39				
C_7H_{16}	Heptane	293.2	1.9209	0.24740E+01	−0.22577E-02	0.12428E-05	273-373
C_7H_{16}	2-Methylhexane	293.2	1.9221	0.24759E+01	−0.22535E-02	0.12500E-05	293-323
C_7H_{16}	3-Methylhexane	293.2	1.920	0.27089E+01	−0.37908E-02	0.37500E-05	273-323
C_7H_{16}	3-Ethylpentane	293.2	1.942	0.23771E+01	−0.15140E-02	0.10093E-06	163-363

Mol. form.	Name	T/K	ε_r	a	b	c	Range/K
C_7H_{16}	2,2-Dimethylpentane	293.2	1.915	0.23414E+01	−0.14362E-02	−0.51322E-07	153-353
C_7H_{16}	2,3-Dimethylpentane	293.2	1.929	0.25637E+01	−0.26328E-02	0.16071E-05	273-323
C_7H_{16}	2,4-Dimethylpentane	293.2	1.902	0.23979E+01	−0.17436E-02	0.17857E-06	273-323
C_7H_{16}	3,3-Dimethylpentane	291.3	1.9419	0.24007E+01	−0.16802E-02	0.36069E-06	291-322
C_7H_{16}	2,2,3-Trimethylbutane	293.2	1.930				
$C_7H_{16}O$	1-Heptanol	293.2	11.75	0.60662E+02	−0.24049E+00	0.25155E-03	239-513
$C_7H_{16}O$	2-Heptanol	293.7	9.72	0.10050E+03	−0.49793E+00	0.64504E-03	207-365
$C_7H_{16}O$	3-Heptanol	296.1	7.07	0.19586E+03	−0.11465E+01	0.17175E-02	248-349
$C_7H_{16}O$	4-Heptanol	296.2	6.18	0.28995E+03	−0.18499E+01	0.30109E-02	270-301
$C_7H_{16}O$	2-Methyl-2-hexanol	297.0	3.257				
$C_7H_{16}O$	3-Methyl-2-hexanol	297.2	4.990	0.59724E+02	−0.32417E+00	0.47058E-03	244-372
$C_7H_{16}O$	3-Methyl-3-hexanol	298.2	3.248				
$C_7H_{16}O$	3-Ethyl-3-pentanol	293.2	3.158				
$C_7H_{16}O$	2,2-Dimethyl-1-pentanol	293.2	6.020	0.37318E+02	−0.17095E+00	0.22022E-03	283-393
$C_7H_{16}O$	Ethyl pentyl ether	296.2	3.6				
$C_7H_{16}O$	Ethyl isopentyl ether	293.2	3.955	0.66541E+01	−0.55450E-02	−0.12500E-04	293-323
$C_7H_{16}O_3$	Triethoxymethane	293.2	4.779				
$C_7H_{16}S$	1-Heptanethiol	293.2	4.194	0.71333E+01	−0.97320E-02	−0.12500E-05	273-333
$C_7H_{17}N$	Heptylamine	293.2	3.81	0.87794E+01	−0.24363E-01	0.25325E-04	253-373
$C_7H_{18}O_3Si$	Triethoxymethylsilane	298.2	3.845				
$C_8H_4F_6$	1,3-Bis(trifluoromethyl)benzene	303.2	5.98				
C_8H_6	Phenylacetylene	298.2	2.98				
$C_8H_6Cl_2$	2,5-Dichlorostyrene	298.2	2.58				
$C_8H_6Cl_4$	1,2,3,4-Tetrachloro-5,6-dimethylbenzene	293.2	8.0				
$C_8H_6Cl_4$	1,2,3,5-Tetrachloro-4,6-dimethylbenzene	293.2	5.4				
C_8H_6O	Phenoxyacetylene	298.2	4.76				
C_8H_7N	Benzeneacetonitrile	299.2	17.87	0.82175E+02	−0.37416E+00	0.53220E-03	299-343
$C_8H_7NO_2$	4-Methoxyphenyl isocyanate	333.2	10.26	0.20780E+02	−0.31571E-01		333-403
$C_8H_7NO_4$	Methyl 2-nitrobenzoate	300.1	27.76				
C_8H_8	Styrene	293.2	2.4737	0.44473E+01	−0.11422E-01	0.16000E-04	293-313
C_8H_8O	Acetophenone	298.2	17.44	0.26099E+02	0.64048E-02	−0.11905E-03	298-333
$C_8H_8O_2$	Benzeneacetic acid	353.2	3.47	0.24104E+01	0.30000E-02		353-393
$C_8H_8O_2$	Benzyl formate	303.2	6.34	0.26162E+02	−0.11026E+00	0.14787E-03	303-358
$C_8H_8O_2$	Phenyl acetate	298.2	5.403	0.11327E+02	−0.26707E-01	0.22938E-04	298-404
$C_8H_8O_2$	Methyl benzoate	302.7	6.642	0.17486E+02	−0.51027E-01	0.50222E-04	303-393
$C_8H_8O_2$	(Hydroxyacetyl)benzene	298.2	21.33	0.42286E+02	−0.69215E-01	−0.35714E-05	298-368
$C_8H_8O_2$	4-Methoxybenzaldehyde	303.2	22.0				
$C_8H_8O_3$	Methyl salicylate	314.4	8.80	0.20501E+02	−0.39045E-01	0.68298E-05	223-398
C_8H_9Br	1-Bromo-2-ethylbenzene	298.2	5.55				
C_8H_9Br	1-Bromo-3-ethylbenzene	298.2	5.56				
C_8H_9Br	1-Bromo-4-ethylbenzene	298.2	5.42				
C_8H_9BrO	1-Bromo-2-ethoxybenzene	313.2	7.04	0.23146E+02	−0.75753E-01	0.77778E-04	313-358
C_8H_9Cl	1-Chloro-2-ethylbenzene	298.2	4.36				
C_8H_9Cl	1-Chloro-3-ethylbenzene	298.2	5.18				
C_8H_9Cl	1-Chloro-4-ethylbenzene	298.2	5.16				
$C_8H_9NO_2$	1-Ethyl-2-nitrobenzene	273.4	21.9				
$C_8H_9NO_2$	Methyl 2-aminobenzoate	298.2	21.9				
$C_8H_9NO_2$	Ethyl 4-pyridinecarboxylate	293.2	8.95				
C_8H_{10}	Ethylbenzene	293.2	2.4463	0.35969E+01	−0.53169E-02	0.47500E-05	293-323
C_8H_{10}	*o*-Xylene	293.2	2.562	0.36163E+01	−0.40177E-02	0.14286E-05	273-323
C_8H_{10}	*m*-Xylene	293.2	2.359	0.28421E+01	−0.10191E-02	−0.21429E-05	273-323
C_8H_{10}	*p*-Xylene	293.2	2.2735	0.23140E+01	0.97221E-03	−0.37500E-05	293-363
$C_8H_{10}O$	2,3-Xylenol	343.2	4.81	0.14399E+02	−0.41438E-01	0.39244E-04	343-433
$C_8H_{10}O$	2,4-Xylenol	303.2	5.060	0.22125E+02	−0.85543E-01	0.96548E-04	303-363
$C_8H_{10}O$	2,5-Xylenol	338.2	5.36	0.18049E+02	−0.54991E-01	0.51656E-04	338-455
$C_8H_{10}O$	2,6-Xylenol	313.2	4.90	0.12284E+02	−0.32996E-01	0.29867E-04	313-453
$C_8H_{10}O$	3,4-Xylenol	333.2	9.02	0.54423E+02	−0.21153E+00	0.22508E-03	333-453
$C_8H_{10}O$	3,5-Xylenol	323.2	9.06	0.54251E+02	−0.21647E+00	0.23542E-03	323-453
$C_8H_{10}O$	Benzeneethanol	293.2	12.31	0.12170E+03	−0.63124E+00	0.87776E-03	278-333
$C_8H_{10}O$	1-Phenylethanol	293.2	8.77	0.32971E+02	−0.12042E+00	0.12809E-03	293-423
$C_8H_{10}O$	Phenetole	293.2	4.216	−0.15043E+02	0.13752E+00	−0.24500E-03	293-313

Mol. form.	Name	T/K	ε_r	a	b	c	Range/K
$C_8H_{10}O$	2-Methylanisole	293.2	3.502	0.50825E+01	−0.62297E-02	0.28571E-05	293-333
$C_8H_{10}O$	3-Methylanisole	293.2	3.967	0.12830E+02	−0.49701E-01	0.66429E-04	293-333
$C_8H_{10}O$	4-Methylanisole	293.2	3.914	0.86608E+01	−0.23510E-01	0.25000E-04	293-333
$C_8H_{10}O_2$	1,2-Dimethoxybenzene	293.2	4.45	0.74604E+01	−0.13445E-01	0.10737E-04	293-443
$C_8H_{10}O_2$	1,3-Dimethoxybenzene	298.2	5.363	0.11911E+02	−0.30804E-01	0.29643E-04	298-358
$C_8H_{10}O_2$	1,4-Dimethoxybenzene	333.7	5.60	0.11289E+02	−0.20765E-01	0.11987E-04	334-463
$C_8H_{10}O_2S$	Ethyl phenyl sulfone	348.2	39.0				
$C_8H_{10}S$	(Ethylthio)benzene	298.2	4.95				
$C_8H_{11}N$	*p*-Ethylaniline	298.2	4.84				
$C_8H_{11}N$	*N*-Ethylaniline	293.2	5.87				
$C_8H_{11}N$	*N,N*-Dimethylaniline	298.2	4.90	0.84052E+01	−0.13549E-01	0.62835E-05	289-453
$C_8H_{11}N$	2,4,6-Trimethylpyridine	298.2	7.807	0.20990E+02	−0.57419E-01	0.44286E-04	298-358
$C_8H_{11}NO$	4-Ethoxyaniline	298.2	7.43				
$C_8H_{12}N_2O_2$	Hexamethylene diisocyanate	288.2	14.41	0.26715E+02	−0.42696E-01		288-403
$C_8H_{12}O_4$	Diethyl maleate	298.2	7.560	0.13953E+02	−0.21969E-01	0.17817E-05	298-343
$C_8H_{12}O_4$	Diethyl fumarate	296.2	6.56				
C_8H_{14}	1,7-Octadiene	293.0	2.186	0.28376E+01	−0.17442E-02	−0.16141E-05	214-293
C_8H_{14}	*cis*-Cyclooctene	296.0	2.306	0.31115E+01	−0.32058E-02	0.16713E-05	269-406
C_8H_{14}	1,2-Dimethylcyclohexene	296.0	2.144	0.26443E+01	−0.17973E-02	0.35815E-06	211-374
C_8H_{14}	1,3-Dimethylcyclohexene	296.0	2.182	0.29951E+01	−0.34615E-02	0.24026E-05	213-373
$C_8H_{14}O_2$	Methyl cyclohexanecarboxylate	293.2	4.87				
$C_8H_{14}O_2$	Cyclohexyl acetate	293.2	5.08				
$C_8H_{14}O_3$	Butanoic anhydride	293.2	12.8				
$C_8H_{14}O_3$	2-Methylpropanoic anhydride	292.2	13.6				
$C_8H_{14}O_4$	Diisopropyl oxalate	293.2	6.403	0.10709E+02	−0.16328E-01	0.56000E-05	293-368
$C_8H_{14}O_4$	Diethyl succinate	293.2	6.098	0.80213E+01	0.11810E-02	−0.26400E-04	293-343
$C_8H_{14}O_4$	Dimethyl adipate	293.2	6.84	0.11739E+02	−0.17281E-01	0.11447E-05	293-433
$C_8H_{15}N$	Octanenitrile	293.2	13.90				
C_8H_{16}	1-Octene	293.2	2.113	0.24348E+01	0.34200E-03	−0.50000E-05	273-323
C_8H_{16}	*cis*-3-Octene	298.2	2.062				
C_8H_{16}	*trans*-3-Octene	298.2	2.002				
C_8H_{16}	*cis*-4-Octene	298.2	2.053				
C_8H_{16}	*trans*-4-Octene	298.2	2.004				
C_8H_{16}	3-Methyl-2-heptene*	293.2	2.436				
C_8H_{16}	2,5-Dimethyl-2-hexene	293.2	2.431				
C_8H_{16}	2,4,4-Trimethyl-1-pentene	298.2	2.0908				
C_8H_{16}	Cyclooctane	295.0	2.116	0.25036E+01	−0.12460E-02	−0.23175E-06	295-411
$C_8H_{16}Br_2$	1,8-Dibromooctane	298.2	7.43	0.94117E+00	0.61520E-01	−0.13333E-03	298-328
$C_8H_{16}Cl_2$	1,8-Dichlorooctane	298.2	7.64				
$C_8H_{16}O$	2-Octanone	293.2	9.51	−0.16219E+02	0.18799E+00	−0.34156E-03	293-333
$C_8H_{16}O$	3-Octanone	303.2	10.50				
$C_8H_{16}O_2$	Octanoic acid	288.2	2.85	0.29391E+01	−0.38721E-03		288-423
$C_8H_{16}O_2$	2-Ethylhexanoic acid	296.2	2.64				
$C_8H_{16}O_2$	Hexyl acetate	293.2	4.42				
$C_8H_{16}O_2$	Pentyl propanoate	293.2	4.552				
$C_8H_{16}O_2$	Isopentyl propanoate	273.2	5.21	0.17665E+02	−0.71718E-01	0.95635E-04	273-373
$C_8H_{16}O_2$	Butyl butanoate	298.2	4.39	0.79684E+01	−0.12000E-01	0.15266E-13	298-318
$C_8H_{16}O_2$	Propyl pentanoate	292.2	4.0				
$C_8H_{16}O_2$	Ethyl hexanoate	293.2	4.45	0.11007E+02	−0.32800E-01	0.35714E-04	253-353
$C_8H_{16}O_2$	Methyl heptanoate	293.2	4.355				
$C_8H_{16}O_3$	Isopentyl lactate	273.2	11.2	0.48649E+02	−0.21253E+00	0.27619E-03	273-373
$C_8H_{17}Br$	1-Bromooctane	293.2	5.0957	0.12404E+02	−0.35050E-01	0.34542E-04	283-353
$C_8H_{17}Br$	2-Bromooctane	293.2	5.44				
$C_8H_{17}Cl$	1-Chlorooctane	298.2	5.05	0.11346E+02	−0.25120E-01	0.13450E-04	274-328
$C_8H_{17}Cl$	2-Chlorooctane	293.2	5.42				
$C_8H_{17}F$	1-Fluorooctane	293.2	3.89				
$C_8H_{17}I$	1-Iodooctane	293.2	4.67	0.12452E+02	−0.41229E-01	0.50108E-04	233-313
$C_8H_{17}NO_2$	1-Nitrooctane	293.2	11.46				
C_8H_{18}	Octane	293.2	1.948	0.22590E+01	−0.84212E-03	−0.75758E-06	233-393
C_8H_{18}	2-Methylheptane	293.2	1.9519				
C_8H_{18}	3-Ethylhexane	293.2	1.9617				
C_8H_{18}	2,2-Dimethylhexane	293.2	1.9498				
C_8H_{18}	2,5-Dimethylhexane	293.95	1.9619	0.25821E+01	−0.26804E-02	0.19404E-05	294-324

Mol. form.	Name	*T*/K	ε_r	*a*	*b*	*c*	Range/K
C_8H_{18}	3,3-Dimethylhexane	293.2	1.9645				
C_8H_{18}	3,4-Dimethylhexane	292.1	1.9814	0.26849E+01	–0.33712E-02	0.32949E-05	292-324
C_8H_{18}	3-Ethyl-3-methylpentane	291.49	1.9869	0.25983E+01	–0.28027E-02	0.24195E-05	292-324
C_8H_{18}	2,2,3-Trimethylpentane	293.2	1.960				
C_8H_{18}	2,2,4-Trimethylpentane	293.2	1.943	0.23677E+01	–0.14768E-02	0.94261E-07	173-373
C_8H_{18}	2,3,3-Trimethylpentane	293.2	1.9780				
C_8H_{18}	2,3,4-Trimethylpentane	293.2	1.9738				
$C_8H_{18}O$	1-Octanol	293.2	10.30	0.51647E+02	–0.20371E+00	0.21320E-03	258-513
$C_8H_{18}O$	2-Octanol	293.2	8.13	0.63760E+02	–0.27643E+00	0.31075E-03	213-513
$C_8H_{18}O$	3-Octanol	293.2	5.55	0.12505E+03	–0.70646E+00	0.10245E-02	223-383
$C_8H_{18}O$	4-Octanol	293.2	4.48	0.51049E+02	–0.26664E+00	0.37280E-03	243-403
$C_8H_{18}O$	2-Methyl-1-heptanol	293.1	5.16	0.61698E+02	–0.33647E+00	0.49066E-03	236-328
$C_8H_{18}O$	3-Methyl-1-heptanol	290.3	2.884	0.84687E+01	–0.33712E-01	0.49793E-04	241-316
$C_8H_{18}O$	4-Methyl-1-heptanol	290.6	4.63	0.48612E+02	–0.26773E+00	0.39972E-03	237-332
$C_8H_{18}O$	5-Methyl-1-heptanol	290.4	7.68	0.54581E+02	–0.24772E+00	0.29734E-03	235-328
$C_8H_{18}O$	6-Methyl-1-heptanol	290.3	10.54	0.57997E+02	–0.23517E+00	0.24663E-03	265-328
$C_8H_{18}O$	2-Methyl-2-heptanol	292.2	3.43				
$C_8H_{18}O$	3-Methyl-2-heptanol	289.6	7.47	0.39178E+02	–0.17976E+00	0.24218E-03	229-329
$C_8H_{18}O$	4-Methyl-2-heptanol	290.0	3.59	0.39715E+02	–0.23115E+00	0.36771E-03	240-333
$C_8H_{18}O$	5-Methyl-2-heptanol	278.5	7.5	0.68568E+02	–0.40706E+00	0.67433E-03	230-279
$C_8H_{18}O$	6-Methyl-2-heptanol	290.1	6.41	0.77520E+02	–0.41724E+00	0.59448E-03	239-329
$C_8H_{18}O$	2-Methyl-3-heptanol	293.2	3.260	–0.59739E+01	0.56700E-01	–0.83125E-04	343-403
$C_8H_{18}O$	3-Methyl-3-heptanol	293.2	3.013	–0.38440E+01	0.42327E-01	–0.61250E-04	343-403
$C_8H_{18}O$	4-Methyl-3-heptanol	293.2	3.312	–0.48003E+01	0.50740E-01	–0.75000E-04	343-403
$C_8H_{18}O$	5-Methyl-3-heptanol	293.2	3.832	0.61967E+01	–0.63750E-02		343-383
$C_8H_{18}O$	6-Methyl-3-heptanol	293.2	4.992	0.23037E+02	–0.98029E-01	0.12479E-03	283-383
$C_8H_{18}O$	2-Methyl-4-heptanol	296.3	3.338	0.42102E+00	0.10427E-01	–0.20438E-05	230-333
$C_8H_{18}O$	3-Methyl-4-heptanol	290.0	7.46	0.33354E+02	–0.14077E+00	0.17750E-03	230-330
$C_8H_{18}O$	4-Methyl-4-heptanol	296.2	2.902				
$C_8H_{18}O$	2-Ethyl-1-hexanol	298.2	7.58	0.86074E+02	–0.42636E+00	0.55078E-03	208-318
$C_8H_{18}O$	2,2-Dimethyl-1-hexanol	293.2	4.50	0.91244E+01	–0.21785E-01	0.21018E-04	283-393
$C_8H_{18}O$	Dibutyl ether	293.2	3.0830	0.65383E+01	–0.16172E-01	0.14969E-04	293-314
$C_8H_{18}OS$	Dibutyl sulfoxide	313.2	24.73	0.67156E+02	–0.16448E+00	0.92275E-04	313-393
$C_8H_{18}O_2$	2-Ethyl-1,3-hexanediol	293.2	18.73	0.57919E+02	–0.17128E+00	0.12949E-03	233-333
$C_8H_{18}O_2S$	Dibutyl sulfone	323.2	25.72	0.66248E+02	–0.16417E+00	0.12001E-03	323-398
$C_8H_{18}O_4$	Triethylene glycol dimethyl ether	298.2	7.62				
$C_8H_{18}O_5$	Tetraethylene glycol	293.2	20.44	0.83547E+02	–0.31691E+00	0.34689E-03	253-333
$C_8H_{18}S$	1-Octanethiol	293.2	3.949	0.63667E+01	–0.87920E-02	0.18750E-05	273-333
$C_8H_{18}S$	Dibutyl sulfide	298.2	4.29				
$C_8H_{19}N$	Octylamine	293.2	3.58	0.77931E+01	–0.20015E-01	0.19347E-04	273-373
$C_8H_{19}N$	Dibutylamine	293.2	2.765	0.52504E+01	–0.10538E-01	0.71485E-05	243-323
$C_8H_{20}O_4Si$	Ethyl silicate	293.2	2.50				
$C_8H_{20}Si$	Tetraethylsilane	293.2	2.090				
$C_8H_{20}Sn$	Tetraethylstannane	293.2	2.241				
$C_8H_{23}N_5$	Tetraethylenepentamine	293.2	9.40	0.40553E+02	–0.16681E+00	0.20659E-03	213-333
$C_8H_{24}O_4Si_4$	Octamethylcyclotetrasiloxane	296.2	2.390	0.36286E+01	–0.56885E-02	0.50874E-05	296-333
$C_9H_6N_2O_2$	Toluene-2,4-diisocyanate	293.2	8.433	0.22174E+02	–0.66982E-01	0.68571E-04	293-353
$C_9H_6O_2$	2*H*-1-Benzopyran-2-one	343.2	34.04	0.11311E+03	–0.33804E+00	0.31324E-03	343-423
C_9H_7N	Quinoline	293.2	9.16	0.33432E+02	–0.13497E+00	0.17788E-03	258-323
C_9H_7N	Isoquinoline	298.2	11.0	0.14412E+03	–0.79935E+00	0.11839E-02	298-323
C_9H_8O	Cinnamaldehyde	305.8	17.72	0.41837E+02	–0.11060E+00	0.10401E-03	306-354
$C_9H_8O_4$	2-(Acetyloxy)benzoic acid	333.2	6.55	0.69994E+01	–0.14553E-02		333-416
C_9H_{10}	1-Propenylbenzene	293.2	2.73				
C_9H_{10}	Allylbenzene	293.2	2.63				
C_9H_{10}	Isopropenylbenzene	293.2	2.28				
$C_9H_{10}OS$	4-Acetylthioanisole	355.2	11.34				
$C_9H_{10}O_2$	Ethyl benzoate	293.2	6.20	0.18216E+02	–0.62361E-01	0.72884E-04	288-343
$C_9H_{10}O_2$	Methyl 4-methylbenzoate	306.2	4.3				
$C_9H_{10}O_2$	Benzyl acetate	303.2	5.34	0.11727E+02	–0.30869E-01	0.32340E-04	303-358
$C_9H_{10}O_2$	Phenyl propanoate	293.2	4.77				
$C_9H_{10}O_2$	4-Acetylanisole	313.2	17.3				
$C_9H_{10}O_3$	Ethyl salicylate	308.2	8.48	0.18910E+02	–0.35623E-01	0.46529E-05	225-321
$C_9H_{10}O_3$	Methyl 2-methoxybenzoate	294.2	7.7				

Mol. form.	Name	T/K	ε_r	a	b	c	Range/K
$C_9H_{11}Br$	(3-Bromopropyl)benzene	302.2	5.41	0.11360E+02	−0.27471E-01	0.25775E-04	302-358
$C_9H_{11}NO$	*N*-Ethylbenzamide	352.7	42.6	−0.20109E+03	0.17866E+01	−0.31065E-02	353-389
$C_9H_{11}NO$	*N,N*-Dimethylbenzamide	318.2	20.77	0.76725E+02	−0.26908E+00	0.29409E-03	318-443
$C_9H_{11}NO_2$	Ethyl 2-aminobenzoate	298.2	4.14				
C_9H_{12}	Propylbenzene	293.2	2.370	0.26933E+01	0.21679E-03	−0.44643E-05	273-323
C_9H_{12}	Isopropylbenzene	293.2	2.381	0.31149E+01	−0.30801E-02	0.19643E-05	273-323
C_9H_{12}	*o*-Ethyltoluene	293.2	2.595				
C_9H_{12}	*m*-Ethyltoluene	293.2	2.365				
C_9H_{12}	*p*-Ethyltoluene	293.2	2.265				
C_9H_{12}	1,2,3-Trimethylbenzene	293.2	2.656	0.76006E+01	−0.29118E-01	0.41786E-04	273-323
C_9H_{12}	1,2,4-Trimethylbenzene	293.2	2.377	0.31517E+01	−0.30634E-02	0.14286E-05	273-323
C_9H_{12}	1,3,5-Trimethylbenzene	293.2	2.279	0.38998E+01	−0.88072E-02	0.11149E-04	288-358
$C_9H_{12}O$	Benzenepropanol	293.2	11.97	0.94482E+02	−0.45540E+00	0.59307E-03	213-303
$C_9H_{12}O$	α-Ethylbenzenemethanol	293.2	6.68	0.44520E+02	−0.21505E+00	0.29443E-03	233-373
$C_9H_{12}O$	α,α-Dimethylbenzenemethanol	303.2	5.61	0.57072E+01	0.86568E-02	−0.29580E-04	303-373
$C_9H_{12}O$	1-Phenyl-2-propanol	293.2	9.35	0.10762E+03	−0.56026E+00	0.76915E-03	233-373
$C_9H_{12}O$	Benzyl ethyl ether	298.2	3.90				
$C_9H_{12}O$	2,6-Dimethylanisole	293.2	3.780	0.76700E+01	−0.18298E-01	0.17143E-04	293-333
$C_9H_{12}O$	3,5-Dimethylanisole	293.2	3.711	0.54981E+01	−0.56651E-02	−0.14286E-05	293-333
$C_9H_{12}O_2S$	Butyl thiophene-2-carboxylate	293.2	6.40				
$C_9H_{12}S$	Benzenepropanethiol	303.2	4.36	0.82411E+01	−0.15034E-01	0.73617E-05	303-358
$C_9H_{13}N$	Benzylethylamine	293.2	4.3				
$C_9H_{13}N$	*N*-Propylaniline	293.2	5.48				
$C_9H_{13}N$	2-Methyl-*N,N*-dimethylaniline	293.2	3.4				
$C_9H_{13}N$	4-Methyl-*N,N*-dimethylaniline	293.2	3.9				
$C_9H_{14}OSi$	Trimethylphenoxysilane	298.2	3.3953				
$C_9H_{14}O_6$	Triacetin	293.6	7.11	0.17819E+02	−0.53656E-01	0.57759E-04	219-304
$C_9H_{14}Si$	Trimethylphenylsilane	298.2	2.3533	0.21463E+01	0.32711E-02	−0.86264E-05	288-323
$C_9H_{16}O_2$	2-Nonenoic acid	296.2	2.5				
$C_9H_{16}O_2$	Cyclohexyl propanoate	293.2	4.82				
$C_9H_{16}O_2$	Ethyl cyclohexanecarboxylate	293.2	4.64				
$C_9H_{16}O_4$	Diethyl glutarate	303.2	6.659				
$C_9H_{17}N$	Nonanenitrile	293.2	12.08				
C_9H_{18}	1-Nonene	293.2	2.180	0.22710E+01	0.15797E-02	−0.64286E-05	273-323
$C_9H_{18}Br_2$	1,9-Dibromononane	293.2	7.153	0.18931E+02	−0.57764E-01	0.60000E-04	293-343
$C_9H_{18}O$	2-Nonanone	295.2	9.14				
$C_9H_{18}O$	5-Nonanone	293.2	10.6				
$C_9H_{18}O$	Di-*tert*-butyl ketone	287.65	10.0				
$C_9H_{18}O$	2,6-Dimethyl-4-heptanone	293.2	9.91	0.33178E+02	−0.11290E+00	0.11454E-03	273-393
$C_9H_{18}O_2$	Nonanoic acid	294.9	2.475	0.25039E+01	0.67274E-03	−0.24180E-05	295-365
$C_9H_{18}O_2$	2-Methyloctanoic acid	293.2	2.39				
$C_9H_{18}O_2$	2-Ethylheptanoic acid	293.2	1.98				
$C_9H_{18}O_2$	Heptyl acetate	293.2	4.2				
$C_9H_{18}O_2$	Pentyl butanoate	301.2	4.08	0.59029E+01	−0.49905E-02	−0.34292E-05	301-343
$C_9H_{18}O_2$	Isopentyl butanoate	293.2	4.0				
$C_9H_{18}O_2$	Isobutyl pentanoate	292.2	3.8				
$C_9H_{18}O_2$	Methyl octanoate	293.2	4.101				
$C_9H_{19}Br$	1-Bromononane	298.2	4.74	0.79870E+01	−0.10488E-01	−0.13450E-05	274-328
$C_9H_{19}Cl$	1-Chlorononane	293.2	4.803	0.95528E+01	−0.16200E-01	−0.16365E-13	293-323
$C_9H_{19}NO$	*N,N*-Dibutylformamide	293.2	18.4				
C_9H_{20}	Nonane	293.2	1.9722	0.23894E+01	−0.14830E-02	0.14881E-06	253-393
C_9H_{20}	2-Methyloctane	293.2	1.967				
C_9H_{20}	4-Methyloctane	293.2	1.967				
C_9H_{20}	2,4-Dimethylheptane	293.2	1.89				
C_9H_{20}	2,5-Dimethylheptane	293.2	1.89				
C_9H_{20}	2,6-Dimethylheptane	293.2	1.987				
$C_9H_{20}N_2O$	Tetraethylurea	296.8	14.29	0.52820E+02	−0.18790E+00	0.19580E-03	205-411
$C_9H_{20}O$	1-Nonanol	293.2	8.83	0.97467E+02	−0.51103E+00	0.71429E-03	288-343
$C_9H_{20}O$	2-Nonanol	298.2	6.66	0.10136E+03	−0.55612E+00	0.80000E-03	288-308
$C_9H_{20}O$	3-Nonanol	298.2	4.49	0.55214E+02	−0.31920E+00	0.50000E-03	288-308
$C_9H_{20}O$	4-Nonanol	298.2	3.69	0.27954E+01	0.30000E-02	−0.52375E-13	288-308
$C_9H_{20}O$	5-Nonanol	298.2	3.54	−0.25463E+01	0.35320E-01	−0.50000E-04	288-308
$C_9H_{21}B$	Tripropylborane	293.2	2.026				

Mol. form.	Name	T/K	ε_r	a	b	c	Range/K
$C_9H_{21}N$	Nonylamine	293.2	3.42	0.53575E+01	−0.71982E-02	0.19481E-05	293-373
$C_9H_{21}N$	Tripropylamine	293.2	2.380	0.33380E+01	−0.86332E-02	0.18322E-04	243-293
$C_9H_{21}O_4P$	Tripropyl phosphate	293.2	10.93	0.33166E+02	−0.10514E+00	0.10000E-03	293-373
$C_{10}H_7Br$	1-Bromonaphthalene	298.2	4.768	0.10561E+02	−0.27671E-01	0.27655E-04	293-323
$C_{10}H_7Cl$	1-Chloronaphthalene	298.2	5.04	0.84861E+01	−0.12357E-01	0.26899E-05	274-328
$C_{10}H_7NO_2$	1-Nitronaphthalene	333.2	19.68	0.36267E+02	−0.41283E-01	−0.25595E-04	333-403
$C_{10}H_8$	Naphthalene	363.2	2.54				
$C_{10}H_8O$	1-Naphthol	373.0	5.03	0.16489E+02	−0.46700E-01	0.42857E-04	373-453
$C_{10}H_8O$	2-Naphthol	413.0	4.95	0.92865E+01	−0.10500E-01	0.42501E-15	413-453
$C_{10}H_9N$	1-Naphthylamine	333.2	5.20	0.10577E+02	−0.22114E-01	0.17857E-04	333-453
$C_{10}H_9N$	2-Naphthylamine	393.0	5.26	0.19722E+02	−0.60679E-01	0.60714E-04	393-473
$C_{10}H_9N$	2-Methylquinoline	293.2	7.24	0.11688E+02	−0.78400E-02	−0.25000E-04	293-333
$C_{10}H_9N$	4-Methylquinoline	293.2	9.31	0.17788E+02	−0.32580E-01	0.12500E-04	293-333
$C_{10}H_9N$	6-Methylquinoline	293.2	8.48	0.21696E+02	−0.63400E-01	0.62500E-04	293-333
$C_{10}H_9N$	8-Methylquinoline	293.2	6.58	0.19356E+02	−0.61900E-01	0.62500E-04	293-333
$C_{10}H_{10}O_4$	Methyl 2-(acetyloxy)benzoate	328.9	5.31	0.19579E+02	−0.69970E-01	0.80889E-04	329-371
$C_{10}H_{10}O_4$	Dimethyl phthalate	293.2	8.66				
$C_{10}H_{12}$	1,2,3,4-Tetrahydronaphthalene	298.2	2.771	0.29172E+01	0.12832E-02	−0.59453E-05	298-343
$C_{10}H_{12}$	4-Ethylstyrene	298.2	3.350				
$C_{10}H_{12}$	Dicyclopentadiene	313.2	2.43	0.30564E+01	−0.20000E-02	0.82443E-15	313-373
$C_{10}H_{12}O$	Tetrahydro-2-naphthol*	293.2	11.70	0.98978E+02	−0.48267E+00	0.63008E-03	293-363
$C_{10}H_{12}O$	4-Isopropylbenzaldehyde	288.2	10.68				
$C_{10}H_{12}O_2$	4-Allyl-2-methoxyphenol	293.2	9.55	0.52377E+02	−0.24380E+00	0.33333E-03	273-323
$C_{10}H_{12}O_2$	2-Phenylethyl acetate	297.2	4.93				
$C_{10}H_{12}O_2$	Benzyl propanoate	303.0	5.11	0.42301E+01	0.13962E-01	−0.36426E-04	303-358
$C_{10}H_{12}O_2$	Phenyl butanoate	293.2	4.48				
$C_{10}H_{12}O_2$	Propyl benzoate	303.2	5.78	0.10927E+02	−0.20535E-01	0.11745E-04	303-358
$C_{10}H_{12}O_2$	Ethyl phenylacetate	293.2	5.320				
$C_{10}H_{14}$	Butylbenzene	293.2	2.359				
$C_{10}H_{14}$	*sec*-Butylbenzene	293.2	2.357	0.28348E+01	−0.68586E-03	−0.32143E-05	273-323
$C_{10}H_{14}$	*tert*-Butylbenzene	293.2	2.359	0.27924E+01	−0.38350E-03	−0.37500E-05	273-323
$C_{10}H_{14}$	Isobutylbenzene	293.2	2.318	0.28055E+01	−0.92614E-03	−0.25000E-05	273-323
$C_{10}H_{14}$	1-Isopropyl-4-methylbenzene	298.2	2.2322	0.25266E+01	−0.25121E-03	−0.24867E-05	277-333
$C_{10}H_{14}$	*o*-Diethylbenzene	293.2	2.594				
$C_{10}H_{14}$	*m*-Diethylbenzene	293.2	2.369				
$C_{10}H_{14}$	*p*-Diethylbenzene	293.2	2.259				
$C_{10}H_{14}$	1-Ethyl-3,5-dimethylbenzene	293.2	2.275				
$C_{10}H_{14}$	1,2,3,4-Tetramethylbenzene	296.0	2.538	0.33822E+01	−0.33630E-02	0.17475E-05	273-412
$C_{10}H_{14}$	1,2,4,5-Tetramethylbenzene	356.0	2.223	0.26834E+01	−0.10327E-02	−0.73533E-06	356-430
$C_{10}H_{14}N_2$	*L*-Nicotine	293.2	8.937	0.21347E+02	−0.57177E-01	0.50655E-04	293-363
$C_{10}H_{14}O$	1-Phenyl-2-methyl-2-propanol	298.2	5.71	0.21922E+02	−0.84231E-01	0.99475E-04	298-423
$C_{10}H_{14}O$	Butyl phenyl ether	293.2	3.734				
$C_{10}H_{14}O$	Thymol	333.2	4.259				
$C_{10}H_{15}N$	*N,N*-Diethylaniline	303.2	5.15	0.50773E+01	0.15399E-01	−0.50000E-04	303-328
$C_{10}H_{16}$	γ-Terpinene	298.2	2.2738				
$C_{10}H_{16}$	*d*-Limonene	298.2	2.3746				
$C_{10}H_{16}$	*l*-Limonene	298.2	2.3738				
$C_{10}H_{16}$	Terpinolene	298.2	2.2918				
$C_{10}H_{16}$	α-Pinene	298.2	2.1787				
$C_{10}H_{16}$	β-Pinene	298.2	2.4970				
$C_{10}H_{16}$	α-Terpinene	298.2	2.4526				
$C_{10}H_{16}$	β-Myrcene	298.2	2.3				
$C_{10}H_{16}O$	Carvenone	293.2	18.8				
$C_{10}H_{16}O$	*d*-Fenchone	294.2	12.8				
$C_{10}H_{17}Cl$	2-Chlorobornane	368.2	5.21				
$C_{10}H_{18}$	Pinane	298.2	2.1456				
$C_{10}H_{18}$	*cis*-Decahydronaphthalene	293.2	2.219	0.25410E+01	−0.11420E-02	0.15092E-06	293-373
$C_{10}H_{18}$	*trans*-Decahydronaphthalene	293.2	2.184	0.26615E+01	−0.21241E-02	0.16864E-05	293-373
$C_{10}H_{18}O$	Eucalyptol	298.2	4.57				
$C_{10}H_{18}O_2$	Cyclohexyl butanoate	293.2	4.58				
$C_{10}H_{18}O_4$	Diethyl adipate	293.2	6.109	0.14824E+02	−0.40749E-01	0.37600E-04	293-343
$C_{10}H_{20}$	1-Decene	293.2	2.136	0.19091E+01	0.33442E-02	−0.87500E-05	273-323
$C_{10}H_{20}$	*cis*-5-Decene	298.2	2.071				

Mol. form.	Name	T/K	ε_r	a	b	c	Range/K
$C_{10}H_{20}$	*trans*-5-Decene	298.2	2.030				
$C_{10}H_{20}$	5-Methyl-4-nonene	293.2	2.175				
$C_{10}H_{20}$	2,4,6-Trimethyl-3-heptene	293.2	2.293				
$C_{10}H_{20}Br_2$	1,10-Dibromodecane	303.2	6.56	0.17350E+02	−0.50328E-01	0.48633E-04	303-368
$C_{10}H_{20}Cl_2$	1,10-Dichlorodecane	308.2	6.68	−0.57423E+01	0.94220E-01	−0.17500E-03	308-338
$C_{10}H_{20}O$	2-Decanone	287.2	8.3				
$C_{10}H_{20}O$	Menthol	309.3	3.90	0.68202E+01	−0.15894E-01	0.20837E-04	309-358
$C_{10}H_{20}O_2$	2,2-Dimethyloctanoic acid	296.2	2.8				
$C_{10}H_{20}O_2$	Octyl acetate	288.2	4.18	−0.34691E+01	0.58106E-01	−0.10952E-03	288-323
$C_{10}H_{20}O_2$	2-Methylheptyl acetate	288.2	4.27	0.23285E+02	−0.11538E+00	0.17143E-03	288-323
$C_{10}H_{20}O_2$	Pentyl pentanoate	305.6	4.076	0.77641E+01	−0.14335E-01	0.73740E-05	306-393
$C_{10}H_{20}O_2$	Isopentyl pentanoate	292.2	3.6				
$C_{10}H_{20}O_2$	Isopentyl isopentanoate	288.2	4.39	0.14698E+02	−0.57726E-01	0.76190E-04	288-323
$C_{10}H_{20}O_2$	Methyl nonanoate	293.2	3.943				
$C_{10}H_{21}Br$	1-Bromodecane	298.2	4.44	0.11202E+02	−0.33491E-01	0.36314E-04	274-328
$C_{10}H_{21}Cl$	1-Chlorodecane	293.2	4.581	0.68741E+01	−0.12210E-02	−0.22500E-04	293-323
$C_{10}H_{21}NO$	*N,N*-Dibutylacetamide	293.2	19.1				
$C_{10}H_{22}$	Decane	293.2	1.9853	0.24054E+01	−0.15445E-02	0.44643E-06	253-393
$C_{10}H_{22}$	2,7-Dimethyloctane	293.2	1.98				
$C_{10}H_{22}$	4-Propylheptane	293.2	1.9955				
$C_{10}H_{22}O$	1-Decanol	293.2	7.93	0.47195E+02	−0.20740E+00	0.24942E-03	293-343
$C_{10}H_{22}O$	2-Decanol	298.2	5.82	0.13621E+03	−0.81000E+00	0.12500E-02	288-308
$C_{10}H_{22}O$	3-Decanol	298.2	4.05	0.52090E+02	−0.31020E+00	0.50000E-03	288-308
$C_{10}H_{22}O$	4-Decanol	298.2	3.42	−0.11260E+02	0.93960E-01	−0.15000E-03	288-308
$C_{10}H_{22}O$	5-Decanol	298.2	3.24	−0.25832E+01	0.31456E-01	−0.40000E-04	288-308
$C_{10}H_{22}O$	2,2-Dimethyl-1-octanol	293.2	7.86	0.69536E+02	−0.34596E+00	0.46250E-03	293-333
$C_{10}H_{22}O$	Dipentyl ether	298.2	2.798				
$C_{10}H_{22}O$	Diisopentyl ether	293.2	2.817	0.44690E+01	−0.63710E-02	0.25000E-05	293-323
$C_{10}H_{22}OS$	Dipentyl sulfoxide	348.2	18.8				
$C_{10}H_{22}O_5$	Tetraethylene glycol dimethyl ether	298.2	7.68				
$C_{10}H_{22}S$	Dipentyl sulfide	298.2	3.826				
$C_{10}H_{23}N$	Decylamine	293.2	3.31	0.61497E+01	−0.12801E-01	0.10606E-04	293-373
$C_{10}H_{30}O_3Si_4$	Decamethyltetrasiloxane	293.2	2.370				
$C_{10}H_{30}O_5Si_5$	Decamethylcyclopentasiloxane	293.2	2.50				
$C_{11}H_{10}$	1-Methylnaphthalene	293.2	2.915	0.45126E+01	−0.76480E-02	0.75000E-05	293-333
$C_{11}H_{10}$	2-Methylnaphthalene	313.2	2.747				
$C_{11}H_{10}O$	1-Methoxynaphthalene	293.2	4.020	0.71885E+01	−0.14838E-01	0.13750E-04	293-333
$C_{11}H_{10}O$	2-Methoxynaphthalene	353.2	3.563	0.56702E+01	−0.69754E-02	0.28571E-05	353-373
$C_{11}H_{12}O_2$	Ethyl *trans*-cinnamate	293.2	5.63				
$C_{11}H_{12}O_3$	Ethyl benzoylacetate	303.2	13.50	0.93644E+01	0.74280E-01	−0.20000E-03	303-323
$C_{11}H_{14}O_2$	Benzyl butanoate	301.2	4.55				
$C_{11}H_{14}O_2$	Phenyl pentanoate	293.2	4.30				
$C_{11}H_{14}O_2$	Butyl benzoate	303.2	5.52	0.77854E+01	−0.34972E-02	−0.13149E-04	303-358
$C_{11}H_{14}O_2$	Isobutyl benzoate	291.2	5.39				
$C_{11}H_{16}$	1,3-Diethyl-5-methylbenzene	293.2	2.264				
$C_{11}H_{16}$	Pentamethylbenzene	334.0	2.358	0.30196E+01	−0.22619E-02	0.83831E-06	334-413
$C_{11}H_{22}$	1-Undecene	293.2	2.137	0.22132E+01	0.13121E-02	−0.53571E-05	273-323
$C_{11}H_{22}O$	2-Undecanone	285.3	8.3				
$C_{11}H_{22}O_2$	Nonyl acetate	293.2	3.87				
$C_{11}H_{22}O_2$	Pentyl hexanoate	288.2	4.22	0.83503E+01	−0.18449E-01	0.14286E-04	288-323
$C_{11}H_{23}Br$	1-Bromoundecane	272.6	4.61				
$C_{11}H_{24}$	Undecane	293.2	1.9972	0.23637E+01	−0.12500E-02	−0.85869E-16	283-363
$C_{11}H_{24}O$	1-Undecanol	313.2	5.98				
$C_{11}H_{25}N$	Undecylamine	293.2	3.25	0.54945E+01	−0.96161E-02	0.66017E-05	293-373
$C_{12}F_{27}N$	Tris(perfluorobutyl)amine	293.2	2.15				
$C_{12}H_8O$	Dibenzofuran	373.2	3.00				
$C_{12}H_{10}$	Biphenyl	348.2	2.53	0.26869E+01	0.63072E-03	−0.30995E-05	348-428
$C_{12}H_{10}N_2O$	*trans*-Azoxybenzene	311.2	5.2				
$C_{12}H_{10}O$	Diphenyl ether	283.2	3.726				
$C_{12}H_{10}O$	2-Acetonaphthone	333.2	13.03	0.14538E+03	−0.73040E+00	0.10000E-02	333-363
$C_{12}H_{10}OS$	Diphenyl sulfoxide	344.7	16.6				
$C_{12}H_{10}O_2S$	Diphenyl sulfone	406.2	21.1				
$C_{12}H_{10}S$	Diphenyl sulfide	298.2	5.43				

Mol. form.	Name	*T*/K	ε_r	*a*	*b*	*c*	Range/K
$C_{12}H_{11}N$	Diphenylamine	323.2	3.73				
$C_{12}H_{11}NO$	*N*-1-Naphthylenylacetamide	433.2	24.3	0.84739E+02	−0.12391E+00	−0.35714E-04	433-533
$C_{12}H_{12}$	1,6-Dimethylnaphthalene	293.2	2.7250				
$C_{12}H_{12}O$	1-Ethoxynaphthalene	292.2	3.3				
$C_{12}H_{14}O_2$	Propyl cinnamate	293.2	5.45				
$C_{12}H_{14}O_4$	Diethyl phthalate	293.2	7.86				
$C_{12}H_{16}O$	2-Cyclohexylphenol	328.2	3.97				
$C_{12}H_{16}O$	4-Cyclohexylphenol	404.2	4.42				
$C_{12}H_{16}O_2$	Pentyl benzoate	293.2	5.07				
$C_{12}H_{16}O_3$	Pentyl salicylate	301.2	6.25				
$C_{12}H_{16}O_3$	Isopentyl salicylate	293.12	7.26	0.13129E+02	−0.19190E-01	−0.36060E-05	225-397
$C_{12}H_{17}NO$	*N*-Butyl-*N*-phenylacetamide	298.2	11.66				
$C_{12}H_{18}$	Hexylbenzene	293.2	2.3				
$C_{12}H_{18}$	1,3,5-Triethylbenzene	293.2	2.256				
$C_{12}H_{18}$	Hexamethylbenzene	449.0	2.172	0.35710E+01	−0.46912E-02	0.35088E-05	449-489
$C_{12}H_{20}O_2$	*l*-Bornyl acetate	303.2	4.46	0.60791E+01	0.98200E-02	−0.50000E-04	303-323
$C_{12}H_{22}O$	Dicyclohexyl ether	293.2	3.45	0.95324E+01	−0.31740E-01	0.37500E-04	293-333
$C_{12}H_{22}O$	Cyclododecanone	303.2	11.4	0.39327E+02	−0.13248E+00	0.13298E-03	303-423
$C_{12}H_{22}O_6$	Dibutyl tartrate	314.2	9.4				
$C_{12}H_{24}$	1-Dodecene	293.2	2.152	0.22581E+01	0.11106E-02	−0.50000E-05	273-323
$C_{12}H_{24}O_2$	Decyl acetate	293.2	3.75				
$C_{12}H_{24}O_2$	Ethyl decanoate	293.2	3.75	0.70969E+01	−0.15080E-01	0.12500E-04	293-353
$C_{12}H_{24}O_2$	Methyl undecanoate	293.2	3.671				
$C_{12}H_{25}Br$	1-Bromododecane	298.2	4.07	0.86103E+01	−0.20891E-01	0.18994E-04	274-328
$C_{12}H_{25}Cl$	1-Chlorododecane	298.2	4.17	0.10002E+02	−0.27798E-01	0.27559E-04	274-328
$C_{12}H_{25}I$	1-Iodododecane	298.2	3.91	0.34641E+01	0.97404E-02	−0.27602E-04	293-323
$C_{12}H_{26}$	Dodecane	293.2	2.0120	0.23697E+01	−0.12200E-02	−0.36375E-16	283-363
$C_{12}H_{26}O$	1-Dodecanol	303.2	5.82	0.18518E+02	−0.44859E-01	0.99900E-05	303-358
$C_{12}H_{26}O$	2-Butyl-1-octanol	363.2	3.28				
$C_{12}H_{27}BO_3$	Tributyl borate	293.2	2.23				
$C_{12}H_{27}N$	Dodecylamine	303.2	3.07	0.27999E+01	0.44810E-02	−0.11905E-04	303-373
$C_{12}H_{27}N$	Tributylamine	293.2	2.340	0.19846E+01	0.28108E-02	−0.54545E-05	233-293
$C_{12}H_{27}O_4P$	Tributyl phosphate	293.2	8.34	0.26304E+02	−0.88480E-01	0.92857E-04	293-373
$C_{12}H_{28}O_4Si$	Tetrapropoxysilane	298.2	3.21				
$C_{12}H_{28}Sn$	Tetrapropylstannane	293.2	2.267				
$C_{12}H_{30}OSi_2$	Hexaethyldisiloxane	298.2	2.259	0.36559E+01	−0.72406E-02	0.85714E-05	298-333
$C_{13}H_{10}O$	Benzophenone	300.2	12.62	0.34130E+02	−0.10249E+00	0.10268E-03	300-420
$C_{13}H_{10}O_3$	Phenyl salicylate	290.2	6.92	0.26545E+02	−0.11180E+00	0.15220E-03	290-358
$C_{13}H_{12}$	Diphenylmethane	303.2	2.540	0.30638E+01	−0.17286E-02		303-333
$C_{13}H_{12}O$	Benzyl phenyl ether	313.2	3.748				
$C_{13}H_{18}O_2$	Hexyl benzoate	293.2	4.80				
$C_{13}H_{20}$	Heptylbenzene	293.2	2.26				
$C_{13}H_{20}O$	α-Ionone*	292.4	10.78				
$C_{13}H_{20}O$	β-Ionone*	297.65	11.66				
$C_{13}H_{24}O_4$	Diethyl nonanedioate	303.2	5.133				
$C_{13}H_{26}$	1-Tridecene	293.2	2.139	0.14154E+01	0.66514E-02	−0.14286E-04	273-323
$C_{13}H_{26}O$	7-Tridecanone	303.2	7.6				
$C_{13}H_{26}O_2$	Ethyl undecanoate	293.2	3.55				
$C_{13}H_{26}O_2$	Methyl dodecanoate	293.2	3.539				
$C_{13}H_{27}Br$	1-Bromotridecane	281.15	4.19				
$C_{13}H_{28}$	Tridecane	293.2	2.0213	0.23731E+01	−0.12000E-02	−0.21841E-15	283-363
$C_{13}H_{28}$	5-Butylnonane	293.2	2.0319				
$C_{13}H_{28}O$	1-Tridecanol	333.2	4.02				
$C_{14}H_{10}$	Anthracene	502.0	2.649	0.20571E+02	−0.69169E-01	0.66667E-04	502-516
$C_{14}H_{10}$	Phenanthrene	383.2	2.72				
$C_{14}H_{10}O_2$	Benzil	368.2	13.04	−0.23599E+02	0.22715E+00	−0.34667E-03	368-393
$C_{14}H_{12}O_2$	Benzyl benzoate	303.2	5.26	0.76856E+01	−0.80000E-02	−0.80361E-15	303-358
$C_{14}H_{12}O_3$	Benzyl salicylate	301.2	4.12				
$C_{14}H_{14}$	1,2-Diphenylethane	331.2	2.47	0.31178E+01	−0.21572E-02	0.59800E-06	331-451
$C_{14}H_{14}O$	Dibenzyl ether	293.2	3.821	0.80154E+01	−0.20536E-01	0.21250E-04	293-333
$C_{14}H_{15}N$	Dibenzylamine	293.2	3.446				
$C_{14}H_{16}O_2Si$	Dimethyldiphenoxysilane	298.2	3.500	0.51669E+01	−0.77001E-02	0.70156E-05	283-353
$C_{14}H_{18}O_2$	Pentyl cinnamate	293.2	4.89				

Mol. form.	Name	*T*/K	ε_r	*a*	*b*	*c*	Range/K
$C_{14}H_{22}$	Octylbenzene	293.2	2.26				
$C_{14}H_{26}O_4$	Diisobutyl adipate	293.2	5.19				
$C_{14}H_{26}O_4$	Diethyl sebacate	303.2	4.995	0.39143E+02	−0.20965E+00	0.32000E-03	303-313
$C_{14}H_{28}O_2$	Dodecyl acetate	293.2	3.6				
$C_{14}H_{28}O_2$	Ethyl laurate	273.2	3.94				
$C_{14}H_{28}O_2$	Methyl tridecanoate	293.2	3.442				
$C_{14}H_{29}Br$	1-Bromotetradecane	293.2	3.84	0.10058E+02	−0.33905E-01	0.43528E-04	274-328
$C_{14}H_{30}$	Tetradecane	293.2	2.0343	0.23832E+01	−0.11900E-02	−0.51229E-16	283-363
$C_{14}H_{30}O$	1-Tetradecanol	318.2	4.42	0.12272E+02	−0.24667E-01	−0.13168E-13	318-358
$C_{14}H_{31}N$	Tetradecylamine	312.55	2.90				
$C_{15}H_{12}O_4$	Phenyl 2-(acetyloxy)benzoate	384.2	4.33				
$C_{15}H_{26}O_6$	Tributyrin	282.8	5.72	0.13152E+02	−0.36684E-01	0.36795E-04	199-283
$C_{15}H_{30}O_2$	Methyl tetradecanoate	293.2	3.352				
$C_{15}H_{31}Br$	1-Bromopentadecane	293.35	3.88				
$C_{15}H_{32}$	Pentadecane	293.2	2.0391	0.23792E+01	−0.11600E-02	−0.71069E-16	283-363
$C_{15}H_{32}O$	1-Pentadecanol	333.2	3.70				
$C_{15}H_{33}N$	Pentadecylamine	313.25	2.85				
$C_{15}H_{33}N$	Triisopentylamine	294.2	2.29				
$C_{16}H_{22}O_4$	Dibutyl phthalate	293.2	6.58	0.12444E+02	−0.20000E-01		293-333
$C_{16}H_{32}O_2$	Hexadecanoic acid	338.2	2.417				
$C_{16}H_{32}O_2$	Ethyl myristate	293.2	3.50	0.52642E+01	−0.60000E-02	−0.47358E-15	293-353
$C_{16}H_{32}O_2$	Methyl pentadecanoate	293.2	3.296				
$C_{16}H_{33}Br$	1-Bromohexadecane	298.2	3.68	0.58668E+01	−0.73333E-02	−0.52666E-14	298-328
$C_{16}H_{33}I$	1-Iodohexadecane	293.2	3.57	0.79531E+01	−0.22859E-01	0.26955E-04	293-323
$C_{16}H_{34}$	Hexadecane	293.2	2.0460	0.23861E+01	−0.11600E-02	0.25555E-15	293-363
$C_{16}H_{34}O$	1-Hexadecanol	333.2	3.69	0.85935E+01	−0.14714E-01	−0.45533E-13	333-363
$C_{16}H_{35}N$	Hexadecylamine	328.35	2.71				
$C_{16}H_{36}Sn$	Tetrabutylstannane	293.2	9.74	0.56115E+02	−0.24812E+00	0.30682E-03	293-313
$C_{17}H_{12}O_3$	2-Naphthyl salicylate	293.0	6.30	0.11229E+02	−0.18857E-01	0.70332E-05	293-353
$C_{17}H_{34}O$	9-Heptadecanone	328.2	5.43	0.44176E+02	−0.21183E+00	0.28571E-03	328-363
$C_{17}H_{34}O_2$	Methyl palmitate	313.2	3.124				
$C_{17}H_{36}$	Heptadecane	293.2	2.0578	0.23627E+01	−0.10400E-02	−0.10397E-12	293-308
$C_{17}H_{36}O$	1-Heptadecanol	333.2	3.41				
$C_{18}H_{26}O_4$	Dipentyl phthalate	293.2	6.00				
$C_{18}H_{28}O_2$	Phenyl laurate	293.2	3.28				
$C_{18}H_{30}O_2$	Linolenic acid	293.2	2.825	0.33867E+01	−0.19181E-02		274-368
$C_{18}H_{30}O_4$	Dicyclohexyl adipate	308.2	4.84				
$C_{18}H_{32}O_2$	Linoleic acid	293.2	2.754	0.32073E+01	−0.15477E-02		275-368
$C_{18}H_{34}O_2$	Oleic acid	293.2	2.336	0.25385E+01	−0.69448E-03		275-368
$C_{18}H_{34}O_4$	Dibutyl sebacate	293.2	4.54				
$C_{18}H_{36}O_2$	Stearic acid	293.2	2.314	0.27159E+01	−0.13300E-02		293-373
$C_{18}H_{36}O_2$	Hexadecyl acetate	308.2	3.19	0.47310E+01	−0.50000E-02	0.41338E-14	308-348
$C_{18}H_{36}O_2$	Ethyl palmitate	303.2	3.07	0.57938E+01	−0.12294E-01	0.10919E-04	303-455
$C_{18}H_{36}O_2$	Methyl heptadecanoate	313.2	3.07				
$C_{18}H_{37}Br$	1-Bromooctadecane	303.35	3.53	0.46790E+01	−0.30355E-02	−0.24798E-05	303-332
$C_{18}H_{38}O$	1-Octadecanol	333.2	3.38	0.73784E+01	−0.12000E-01	−0.22871E-13	333-363
$C_{18}H_{39}BO_3$	Trihexyl borate	293.2	2.22				
$C_{18}H_{39}N$	Octadecylamine	326.35	2.67				
$C_{19}H_{16}$	Triphenylmethane	367.2	2.46	0.40201E+01	−0.66507E-02	0.65329E-05	367-448
$C_{19}H_{18}O_3Si$	Methyltriphenoxysilane	298.2	3.628				
$C_{19}H_{32}O_2$	Methyl linolenate	293.2	3.355				
$C_{19}H_{34}O_2$	Methyl linoleate	293.2	3.466				
$C_{19}H_{36}O_2$	Methyl oleate	293.2	3.211				
$C_{19}H_{38}O$	10-Nonadecanone	353.2	5.37				
$C_{19}H_{38}O_2$	Methyl stearate	313.2	3.021				
$C_{19}H_{40}$	Nonadecane	293.2	2.0706				
$C_{20}H_{30}O_4$	Dihexyl phthalate	293.2	5.62				
$C_{20}H_{38}O_2$	Ethyl oleate	301.2	3.17	0.57033E+01	−0.11223E-01	0.93447E-05	301-423
$C_{20}H_{40}O_2$	Octadecyl acetate	308.2	3.07	0.44569E+01	−0.45000E-02	0.33923E-14	308-348
$C_{20}H_{40}O_2$	Ethyl stearate	313.2	2.958	0.70930E+01	−0.19081E-01	0.19555E-04	331-440
$C_{20}H_{40}O_2$	Methyl nonadecanoate	313.2	2.982				
$C_{20}H_{42}O$	1-Eicosanol	338.2	3.13	0.21700E+01	0.12497E-01	−0.28571E-04	338-363
$C_{20}H_{42}O$	Didecyl ether	293.2	2.644	0.41465E+01	−0.62240E-02	0.37500E-05	293-333

Mol. form.	Name	*T*/K	ε_r	*a*	*b*	*c*	Range/K
$C_{20}H_{60}O_8Si_9$	Eicosamethylnonasiloxane	293.2	2.645	0.57840E+01	−0.16568E-01	0.20000E-04	293-323
$C_{21}H_{21}O_4P$	Tricresyl phosphate*	298.2	6.7				
$C_{21}H_{38}O_6$	1,2,3-Propanetriyl hexanoate	293.2	4.476				
$C_{22}H_{42}O_2$	Butyl oleate	298.2	4.00				
$C_{22}H_{44}O_2$	Butyl stearate	298.2	3.120	0.73894E+02	−0.46261E+00	0.75500E-03	298-343
$C_{22}H_{46}$	Docosane	293.2	2.0840				
$C_{22}H_{46}O$	1-Docosanol	348.2	2.94	0.82062E+01	−0.25069E-01	0.28571E-04	348-373
$C_{24}H_{20}O_4Si$	Tetraphenoxysilane	333.2	3.4915				
$C_{24}H_{38}O_4$	Dioctyl phthalate	293.2	5.22				
$C_{26}H_{50}O_4$	Dioctyl sebacate	299.2	4.01				
$C_{27}H_{50}O_6$	1,2,3-Propanetriyl octanoate	293.2	3.931				
$C_{30}H_{58}O_4$	Ethylene glycol ditetradecanoate	343.2	2.98				
$C_{30}H_{62}$	Triacontane	373.2	1.9112				
$C_{30}H_{62}$	2,6,10,15,19,23-Hexamethyltetracosane	373.2	1.9106				
$C_{34}H_{66}O_4$	Ethylene glycol dipalmitate	348.2	2.89				
$C_{34}H_{68}O_2$	Hexadecyl stearate	333.2	2.61				
$C_{38}H_{74}O_4$	Ethylene glycol distearate	353.2	2.79				
$C_{39}H_{74}O_6$	Glycerol trilaurate	313.2	3.287				
$C_{51}H_{98}O_6$	Glycerol tripalmitate	328.2	2.901	−0.29131E+01	0.32206E-01	−0.44154E-04	328-393
$C_{57}H_{104}O_6$	Glycerol trioleate	293.2	3.109				
$C_{57}H_{104}O_6$	Glycerol trielaidate	313.2	2.980				
$C_{57}H_{110}O_6$	Glycerol tristearate	353.2	2.740				

* Isomer was not specified in the original reference.
** Cubic term is needed; see Introduction.

PERMITTIVITY (DIELECTRIC CONSTANT) OF GASES

This table gives the relative permittivity ε (often called the dielectric constant) of some common gases at a temperature of 20 °C and pressure of one atmosphere (101.325 kPa). Values of the permanent dipole moment μ in Debye Units (1 D = 3.33564 × 10^{-30} C m) are also included.

The density dependence of the permittivity is given by the equation

$$\frac{\varepsilon-1}{\varepsilon-2}=\rho_{\mathrm{m}}\left(\frac{4\pi N\alpha}{3}+\frac{4\pi N\mu^2}{9kT}\right)$$

where ρ_m is the molar density, N is Avogadro's number, k is the Boltzmann constant, T is the temperature, and α is the molecular polarizability. Therefore, in regions where the gas can be considered ideal, ε – 1 is approximately proportional to the pressure at constant temperature. For nonpolar gases (μ = 0), ε –1 is inversely proportional to temperature at constant pressure.

The number of significant figures indicates the accuracy of the values given. The values of ε for air, Ar, H_2, He, N_2, O_2, and CO_2 are recommended as reference values; these are accurate to 1 ppm or better.

The second part of the table gives the permittivity of water vapor in equilibrium with liquid water as a function of temperature (derived from Reference 4).

References

1. A. A. Maryott and F. Buckley, *Table of Dielectric Constants and Electric Dipole Moments of Substances in the Gaseous State*, National Bureau of Standards Circular 537, 1953.
2. B. A. Younglove, *J. Phys. Chem. Ref. Data*, 11, Suppl. 1, 1982; 16, 577, 1987 (for data on N_2, H_2, O_2, and hydrocarbons over a range of pressure and temperature).
3. *Landolt-Börnstein, Numerical Data and Functional Relationships in Science and Technology*, New Series, Group IV, Vol. 4, Springer-Verlag, Heidelberg, 1980 (for data at high pressures).
4. G. Birnbaum and S. K. Chatterjee, *J. Appl. Phys.*, 23, 220, 1952 (for data on water vapor).

Mol. form.	Name	ε	μ/D
Compounds not containing carbon			
	Air (dry, CO_2 free)	1.0005364	
Ar	Argon	1.0005172	0
BF_3	Boron trifluoride	1.0011	0
BrH	Hydrogen bromide	1.00279	0.827
ClH	Hydrogen chloride	1.00390	1.109
F_3N	Nitrogen trifluoride	1.0013	0.235
F_6S	Sulfur hexafluoride	1.00200	0
HI	Hydrogen iodide	1.00214	0.448
H_2	Hydrogen	1.0002538	0
H_2S	Hydrogen sulfide	1.00344	0.97
H_3N	Ammonia	1.00622	1.471
He	Helium	1.0000650	0
Kr	Krypton	1.00078	0
NO	Nitric oxide	1.00060	0.159
N_2	Nitrogen	1.0005480	0
N_2O	Nitrous oxide	1.00104	0.161
Ne	Neon	1.00013	0
O_2	Oxygen	1.0004947	0
O_2S	Sulfur dioxide	1.00825	1.633
O_3	Ozone	1.0017	0.534
Xe	Xenon	1.00126	0

Mol. form.	Name	ε	μ/D
Compounds containing carbon			
CF_4	Tetrafluoromethane	1.00121	0
CO	Carbon monoxide	1.00065	0.110
CO_2	Carbon dioxide	1.000922	0
CH_3Br	Bromomethane	1.01028	1.822
CH_3Cl	Chloromethane	1.01080	1.892
CH_3F	Fluoromethane	1.00973	1.858
CH_3I	Iodomethane	1.00914	1.62
CH_4	Methane	1.00081	0
C_2H_2	Acetylene	1.00124	0
C_2H_3Cl	Chloroethylene	1.0075	1.45
C_2H_4	Ethylene	1.00134	0
C_2H_5Cl	Chloroethane	1.01325	2.05
C_2H_6	Ethane	1.00140	0
C_2H_6O	Dimethyl ether	1.0062	1.30
C_3H_6	Propene	1.00228	0.366
C_3H_6	Cyclopropane	1.00178	0
C_3H_8	Propane	1.00200	0.084
C_4H_{10}	Butane	1.00258	0
C_4H_{10}	Isobutane	1.00260	0.132

Permittivity of Saturated Water Vapor

t/°C	ε	*t*/°C	ε
0	1.00006	60	1.00143
10	1.00012	70	1.00211
20	1.00021	80	1.00304
30	1.00036	90	1.00428
40	1.00059	100	1.00589
50	1.00094		

AZEOTROPIC DATA FOR BINARY MIXTURES

J. Gmehling, J. Menke, J. Krafczyk, K. Fischer, J.-C. Fontaine, and H. V. Kehiaian

Binary homogeneous (single-phase) liquid mixtures having an extremum (maximum or minimum) vapor pressure P at constant temperature T, as a function of composition, are called azeotropic mixtures, or simply azeotropes. The composition is usually expressed as mole fractions, where x_1 for component 1 in the liquid phase and y_1 for component 1 in the vapor phase are identical. Mixtures that do not show a maximum or minimum are called zeotropic. A maximum (minimum) of the $P(x_1)$ or $P(y_1)$ curves corresponds to a minimum (maximum) of the boiling temperature T at constant P, plotted as a function of x_1 or y_1 [see $T(x_1)$ and $T(y_1)$ curves, Types I and III, in Fig. 1]. Azeotropes in which the pressure is a maximum (temperature is a minimum) are often called positive azeotropes, while pressure-minimum (temperature-maximum) azeotropes are called negative azeotropes. The coordinates of an azeotropic point are the azeotropic temperature T_{Az}, pressure P_{Az}, and the vapor-phase composition $y_{1,Az}$, which is the same as the liquid-phase composition $x_{1,Az}$.

In the two-phase liquid-liquid region of partially miscible (heterogeneous) mixtures, the vapor pressure at constant T (or the boiling temperature at constant P) is independent of the global composition x_1 of the two coexisting liquid phases between the equilibrium compositions $x_1'$and x_1'' ($x_1' < x_1''$).

The constant vapor pressure (boiling temperature) above the two-phase region of certain partially miscible mixtures is usually larger (smaller) than the vapor pressure (boiling temperature) at any other liquid-phase composition in the homogeneous region. In this case, the vapor-phase composition is inside the miscibility gap. Mixtures of this type are called heteroazeotropic mixtures, or simply heteroazeotropes. (Fig. 1, Type II), as opposed to the other types of azeotropes, called homoazeotropes.

Only in a few cases partially miscible mixtures present a positive or negative azeotropic point in the single-phase region, outside the miscibility gap, similar to the azeotropic points of homogeneous mixtures (Fig. 1, Types IV and VI).

A few binary mixtures, for example the system perfluorobenzene + benzene, may present two azeotropic points at constant temperature (pressure), a positive and a negative one. They are called double azeotropic mixtures, or simply double azeotropes. (Fig. 1, Type V).

The knowledge of the occurrence of azeotropic points in binary and higher systems is of special importance for the design of distillation processes. The number of theoretical stages of a distillation column required for the separation depends on the separation factor α_{12}, i.e., the ratio of the K_i-factors ($K_i = y_i/x_i$) of the components i (i = 1, 2). The required separation factor can be calculated with the following simplified relation (Reference 1):

$$\alpha_{12} = K_1/K_2 = (y_1/x_1)/\,(y_2/x_2) = (\gamma_1 P_1^s)/\,(\gamma_2 P_2^s) \qquad (1)$$

where γ_i is the activity coefficient of component i in the liquid phase and P_i^s is the vapor pressure of the pure component i.

In distillation processes, only the difference between the separation factor and unity (α_{12} – 1) can be exploited for the separation. If the separation factor is close to unity, a large number of theoretical stages is required for the separation. If the binary system to be separated shows an azeotropic point (α_{12} = 1), the separation is impossible by ordinary distillation, even with an infinitely large number of stages.

Following eq. (1) azeotropic behavior will always occur in homogeneous binary systems when the vapor pressure ratio P_1^s/P_2^s is equal to the ratio of the activity coefficients γ_2/γ_1.

Various thermodynamic methods based on g^E-models (Wilson, NRTL, UNIQUAC) or group contribution methods (UNIFAC, modified UNIFAC, ASOG, PSRK) can be used for either calculating or predicting the required activity coefficients for the components under given conditions of temperature and composition (Reference 2).

Because of the importance of azeotropic data for the design of distillation processes, compilations have been available in book form for quite some time (References 3-7). The most recent printed data collection was published in 1994 (Reference 8). A revised and extended version appeared in 2004 (Reference 9).

A collection of approximately 47,400 zeotropic and azeotropic data sets, compiled from 6600 references, are stored in a comprehensive computerized data bank (Reference 10). The references from the above-mentioned compilations and from the vapor-liquid equilibrium part of the Dortmund Data Bank (Reference 11) were supplemented by references found from CAS online searches, private communications, data from industry, etc.. Over 24,000 zeotropic data and over 20,000 azeotropic data are available for binary systems. Nearly 90% of the binary azeotropic data show a pressure maximum. In most cases (ca. 90%) these are homogeneous azeotropes, and in approximately 7–8% of the cases heterogeneous azeotropes are reported. Less than 10% of the data stored show a pressure minimum. Approximately 21,000 of the data sets stored were published after 1970.

The table below provides information about azeotropes for 808 selected binary systems. Compounds are listed in the modified Hill order, with carbon-containing compounds following those compounds not containing carbon. In columns 1 and 2 are the molecular formulas of components 1 and 2 written in the Hill convention. In column 3 the names of the components are given, either a systematic IUPAC name or a name in ubiquitous use. Columns 4, 5, and 6 contain the azeotropic coordinates of the mixtures: temperature T_{Az}, pressure P_{Az}, and vapor-phase composition $y_{1,Az}$. The explanation of the type of azeotrope (column 7) is given by the following codes:

O: homogeneous azeotrope in a completely miscible system
L: homogeneous azeotrope in a partially miscible system
E: heterogeneous azeotrope
X: pressure maximum
N: pressure minimum
D: double azeotrope
C: system contains a supercritical compound

References

1. Gmehling, J. and Brehm, A., *Grundoperationen*, Thieme-Verlag, Stuttgart, 1996.
2. Gmehling, J. and Kolbe, B., *Thermodynamik*, VCH-Verlag, Weinheim, 1992.
3. Lecat, M., *Doctoral Dissertation*, 1908.

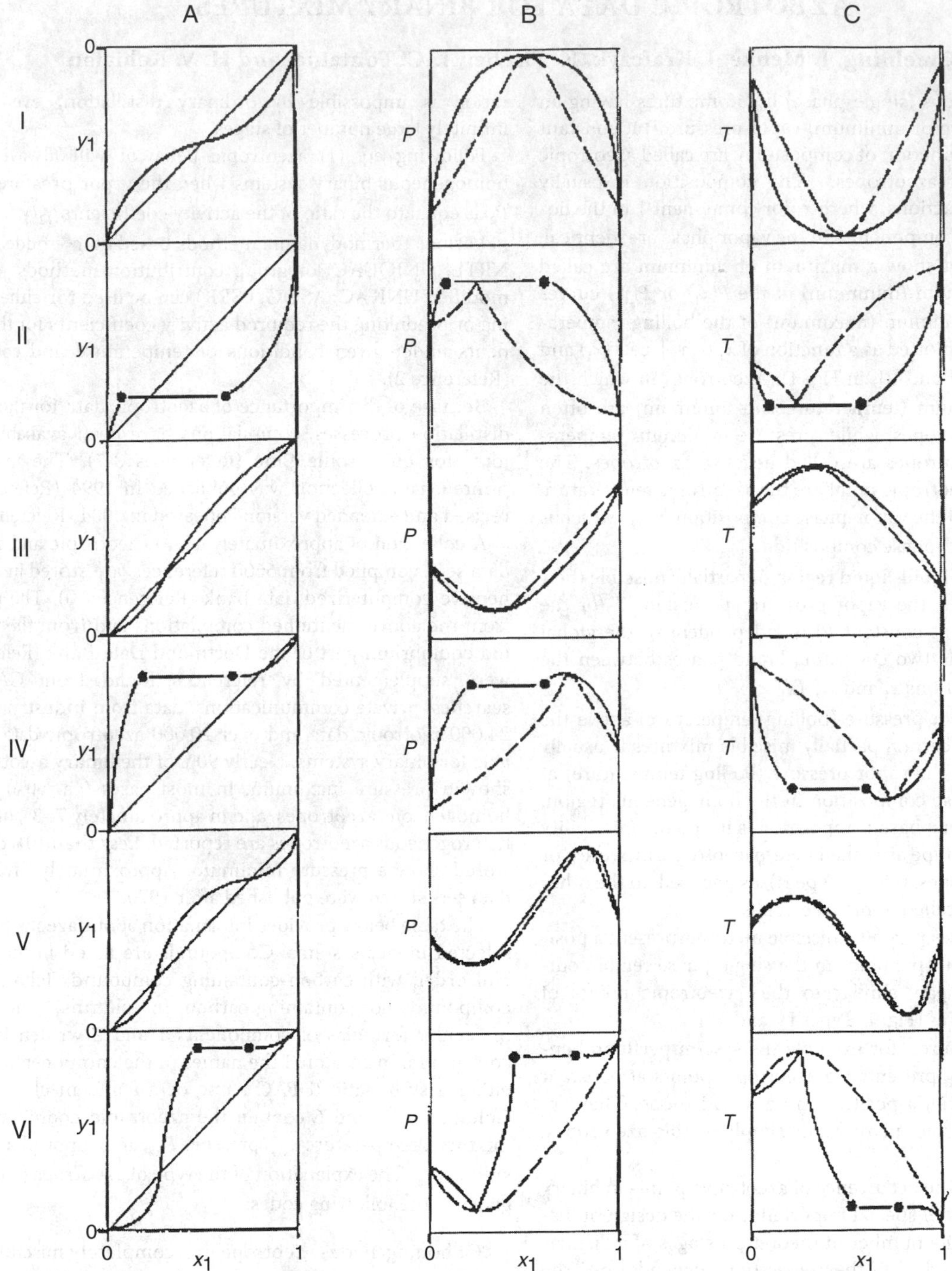

Figure 1 Different types of binary azeotropic systems: I - homogeneous pressure-maximum azeotrope in a completely miscible system (OX); II - heterogeneous pressure-maximum azeotrope (EX); III - homogeneous pressure-minimum azeotrope in a completely miscible system (ON); IV - homogeneous pressure-maximum azeotrope in a partially miscible system (LX); V - D: double azeotrope (OND, OXD); VI - homogeneous pressure-minimum azeotrope in a partially miscible system (LN). A - $y_1(x_1)$; B - $P(x_1)$ and $P(y_1)$; C - $T(x_1)$ and $T(y_1)$. Continuous line - (x_1); Dashed line - (y_1).

4. Lecat, M., *L'Azeotropisme*, Monograph, L'Auteur, Brussel, 1918.
5. Lecat, M., *Tables Azeotropiques*, Monograph, Lamertin, Brussel 1949.
6. Ogorodnikov, S. K., Lesteva, T. M., and Kogan V. B., *Azeotropic Mixtures*, Khimia, Leningrad, 1971.
7. Horsley, L. H., *Azeotropic Data III*, American Chemical Society, Washington, 1973.
8. Gmehling, J., Menke, J., Krafczyk, J., and Fischer, K., *Azeotropic Data*, 2 Volumes, VCH Verlag, Weinheim, 1994.
9. Gmehling, J., Menke, J., Krafczyk, J., and Fischer, K., *Azeotropic Data*, 2nd Ed., 3 Volumes, VCH Verlag, Weinheim, 2004.
10. Gmehling, J., Menke, J., Krafczyk, J., and Fischer, K., *A Data Bank for Azeotropic Data, Status and Applications, Fluid Phase Equilib.* 103, 51, 1995.
11. Dortmund Data Bank, www.ddbst.de

Molecular formula						
Comp. 1	**Comp. 2**	**Name**	T_{Az}/K	$y_{1,Az}$	P_{Az}/kPa	**Type**
$AlCl_3$		**Aluminum chloride**				
	Cl_3OP	Phosphoryl trichloride	660.15	0.5150	101.33	ONC
ClH		**Hydrogen chloride**				
	H_2O	Water	389.34	0.1083	133.32	ONC
Cl_2OS		**Thionyl chloride**				
	Cl_3P	Phosphorus(III) chloride	345.85	0.4200	101.33	OX
Cl_2O_2S		**Sulfuryl chloride**				
	Cl_3P	Phosphorus(III) chloride	364.15	0.5000	101.33	ON
Cl_3OP		**Phosphoryl trichloride**				
	Cl_5Nb	Niobium(V) chloride	536.15	0.4020	101.33	ON
	Cl_5Ta	Tantalum(V) chloride	558.85	0.4650	101.33	ON
Cl_4Ge		**Germanium(IV) chloride**				
	$C_2H_4Cl_2$	1,2-Dichloroethane	350.75	0.4630	101.33	OX
Cl_4Si		**Tetrachlorosilane**				
	C_2H_3N	Acetonitrile	321.05	0.6900	101.33	EX
Cl_5Mo		**Molybdenum(V) chloride**				
	Cl_6W	Tungsten(VI) chloride	274.70	0.9750	101.33	OX
FH		**Hydrogen fluoride**				
	H_2O	Water	382.15	0.3508	101.33	ON
	CCl_3F	Trichlorofluoromethane	283.15	0.7840	129.45	EX
HNO_3		**Nitric acid**				
	H_2O	Water	393.20	0.3820	101.33	ON
H_2O		**Water**				
	$CHCl_3$	Trichloromethane	329.27	0.1603	101.33	EX
	CH_2O	Formaldehyde	366.76	0.9300	53.33	OX
	CH_2O_2	Formic acid	380.35	0.4272	101.33	ON
	CH_3NO_2	Nitromethane	356.90	0.5160	101.33	EX
	C_2HCl_3	Trichloroethene	346.55	0.3560	101.33	EX
	C_2H_3N	Acetonitrile	349.95	0.3100	101.33	OX
	$C_2H_4Cl_2$	1,2-Dichloroethane	345.43	0.3570	101.33	EX
	C_2H_6O	Ethanol	351.25	0.1030	101.33	OX
	$C_2H_8N_2$	1,2-Ethanediamine	391.85	0.4450	101.33	ON
	C_3H_3N	Acrylonitrile	344.05	0.2850	101.33	EX
	C_3H_4O	Acrolein	325.45	0.0730	101.33	LX
	C_3H_6O	Propanal	320.65	0.0600	101.33	LX
	C_3H_6O	Allyl alcohol	361.15	0.5562	101.33	OX
	$C_3H_6O_2$	Methyl acetate	330.05	0.1060	103.62	LX
	$C_3H_6O_2$	1,3-Dioxolane	344.95	0.2520	101.30	OX
	$C_3H_6O_2$	Ethyl formate	325.75	0.0700	101.33	EX
	$C_3H_6O_2$	Propanoic acid	373.05	0.9500	101.33	OX
	C_3H_7Br	1-Bromopropane	336.35	0.2210	101.33	EX
	C_3H_8O	1-Propanol	360.80	0.5680	101.33	OX
	C_3H_8O	2-Propanol	353.70	0.3260	101.33	OX
	$C_3H_8O_2$	2-Methoxyethanol	372.65	0.9441	99.99	OX
	$C_3H_8O_2$	Dimethoxymethane	315.05	0.0269	101.38	LX
	C_4H_5N	*cis*-2-Butenenitrile	358.45	0.3832	101.33	EX
	C_4H_5N	*trans*-2-Butenenitrile	363.05	0.6843	101.33	EX
	C_4H_5N	Pyrrole	348.15	0.7514	50.13	EX
	$C_4H_6O_2$	Methacrylic acid	372.25	0.9464	98.93	OX
	C_4H_8O	2-Butanone	346.54	0.3480	101.33	LX
	C_4H_8O	Tetrahydrofuran	336.67	0.1828	101.33	OX
	C_4H_8O	Isobutanal	332.80	0.1698	100.99	EX
	$C_4H_8O_2$	Ethyl acetate	343.55	0.2990	101.33	EX
	$C_4H_8O_2$	Butanoic acid	372.95	0.9559	101.33	OX
	$C_4H_8O_2$	1,4-Dioxane	360.65	0.5280	101.33	OX
	$C_4H_8O_2$	Propyl formate	344.85	0.3090	101.33	EX
	$C_4H_8O_2$	Methyl propanoate	344.75	0.3050	101.33	EX
	C_4H_9Br	1-Bromobutane	353.95	0.4950	101.33	EX

Molecular formula						
Comp. 1	Comp. 2	Name	T_{Az}/K	$y_{1,Az}$	P_{Az}/kPa	Type
	C_4H_9Br	1-Bromo-2-methylpropane	348.45	0.3730	101.33	EX
	C_4H_9Cl	1-Chloro-2-methylpropane	333.95	0.1970	101.33	LX
	$C_4H_{10}O$	1-Butanol	365.45	0.7540	101.33	EX
	$C_4H_{10}O$	2-Butanol	360.50	0.6200	101.33	LX
	$C_4H_{10}O$	2-Methyl-2-propanol	353.00	0.4011	101.33	OX
	$C_4H_{11}N$	Butylamine	349.85	0.0700	101.33	OX
	C_5H_5N	Pyridine	367.30	0.7500	101.33	OX
	C_5H_8	2-Methyl-1,3-butadiene	305.85	0.0520	101.33	EX
	C_5H_8	Methylenecyclobutane	313.15	0.0212	101.30	EX
	C_5H_8O	Cyclopropyl methyl ketone	361.65	0.7060	101.19	EX
	$C_5H_8O_2$	Methyl methacrylate	354.45	0.4996	101.33	EX
	C_5H_{10}	2-Methyl-2-butene	309.75	0.0650	101.33	EX
	$C_5H_{10}O$	3-Methyl-2-buten-1-ol	369.55	0.9141	101.33	EX
	$C_5H_{10}O$	3-Methyl-3-buten-1-ol	333.15	0.8680	101.33	EX
	$C_5H_{10}O$	2-Methyl-3-buten-2-ol	359.25	0.5770	101.33	LX
	$C_5H_{10}O$	3-Pentanone	356.05	0.4750	101.33	EX
	$C_5H_{10}O_2$	Isopropyl acetate	349.75	0.3960	101.33	EX
	$C_5H_{10}O_2$	Propyl acetate	355.91	0.5228	101.33	EX
	$C_5H_{10}O_2$	Butyl formate	356.95	0.5360	101.33	EX
	$C_5H_{10}O_2$	Isobutyl formate	352.75	0.4460	101.33	EX
	$C_5H_{12}O$	3-Methyl-1-butanol	367.97	0.8265	101.33	EX
	$C_5H_{12}O$	2-Methyl-2-butanol	360.85	0.6355	101.75	EX
	$C_5H_{12}O$	1-Pentanol	369.08	0.8633	101.33	EX
	$C_5H_{12}O$	2-Pentanol	363.15	0.7550	92.49	EX
	C_6H_6	Benzene	342.35	0.2980	101.33	EX
	C_6H_7N	Aniline	372.55	0.9580	101.33	EX
	C_6H_7N	4-Methylpyridine	370.50	0.8972	101.33	OX
	C_6H_{10}	Cyclohexene	343.95	0.3090	101.33	EX
	$C_6H_{10}O$	Cyclohexanone	369.45	0.8694	101.33	EX
	$C_6H_{10}O$	Methyldihydropyran (unspecified isomer)	360.75	0.5841	100.93	EX
	$C_6H_{10}O_2$	4-Vinyl-1,3-dioxane	367.65	0.8955	101.33	EX
	C_6H_{12}	1-Hexene	318.15	0.1510	63.35	EX
	$C_6H_{12}O_2$	Butyl acetate	363.35	0.7013	101.33	EX
	$C_6H_{12}O_2$	Isobutyl acetate	361.05	0.6440	101.33	EX
	$C_6H_{12}O_2$	4,4-Dimethyl-1,3-dioxane	366.00	0.7779	101.33	EX
	$C_6H_{12}O_2$	4,5-Dimethyl-1,3-dioxane (unspecified isomer)	365.05	0.7966	101.50	EX
	$C_6H_{12}O_2$	4-Ethyl-1,3-dioxane	365.75	0.7257	101.30	EX
	$C_6H_{12}O_2$	Diacetone alcohol	370.00	0.9900	90.79	OX
	$C_6H_{12}O_2$	Propyl propanoate	362.05	0.6600	101.33	EX
	$C_6H_{13}N$	Cyclohexylamine	369.55	0.8692	101.33	OX
	C_6H_{14}	Hexane	334.75	0.2110	101.33	EX
	$C_6H_{14}O$	Butyl ethyl ether	349.85	0.4070	101.33	EX
	$C_6H_{14}O$	1-Hexanol	367.89	0.9432	101.33	EX
	$C_6H_{14}O_3$	Di(ethylene glycol) dimethyl ether	372.70	0.9679	101.33	OX
	$C_6H_{15}N$	Diisopropylamine	347.25	0.3654	101.33	EX
	$C_6H_{15}N$	Dipropylamine	359.00	0.6046	101.33	EX
	C_7H_8	Toluene	357.25	0.5230	101.33	EX
	C_7H_8O	Benzyl alcohol	373.05	0.9840	101.33	EX
	C_7H_9N	2,6-Dimethylpyridine	369.17	0.8647	101.33	EX
	$C_7H_{12}O_4$	1,2-Propanediol diacetate	358.15	0.9740	59.41	EX
	C_7H_{14}	1-Heptene	350.20	0.4100	101.33	EX
	$C_7H_{14}O_2$	Isopentyl acetate	367.05	0.7990	101.46	EX
	$C_7H_{14}O_2$	Butyl propanoate	367.95	0.8340	101.33	EX
	C_7H_{16}	Heptane	352.35	0.4510	101.33	EX
	$C_7H_{16}O$	1-Heptanol	371.99	0.9703	101.33	EX
	C_8H_8	Styrene	367.15	0.8000	101.33	EX
	C_8H_8O	Acetophenone	371.15	0.9675	101.19	EX
	C_8H_{10}	*m*-Xylene	365.15	0.7667	101.33	EX

Molecular formula						
Comp. 1	Comp. 2	Name	T_{Az}/K	$y_{1,Az}$	P_{Az}/kPa	Type
	C_8H_{10}	*p*-Xylene	365.15	0.7450	101.33	EX
	C_8H_{10}	Ethylbenzene	364.15	0.7221	101.33	EX
	$C_8H_{16}O_2$	Butyl butanoate	369.85	0.9110	101.33	EX
	C_8H_{18}	Octane	362.75	0.6850	101.33	EX
	C_8H_{18}	2,2,4-Trimethylpentane	351.95	0.4420	101.33	EX
	$C_8H_{18}O$	Dibutyl ether	368.65	0.7628	101.33	EX
	$C_8H_{18}O$	1-Octanol	372.75	0.9820	101.33	EX
	$C_8H_{19}N$	Dibutylamine	370.05	0.8850	101.33	EX
	C_9H_{10}	Isopropenylbenzene	369.95	0.8880	101.33	EX
	C_9H_{12}	Isopropylbenzene	368.15	0.8340	101.33	EX
	$C_9H_{12}O$	2-Phenyl-2-propanol	371.25	0.9718	101.33	EX
	C_9H_{20}	Nonane	367.95	0.8280	101.33	EX
	$C_9H_{20}O$	1-Nonanol	373.00	0.9846	101.33	EX
	$C_{10}H_{22}$	Decane	370.75	0.9180	101.33	EX
	$C_{10}H_{22}O$	1-Decanol	373.13	0.9865	101.33	EX
	$C_{12}H_{27}N$	Tributylamine	372.80	0.9762	101.46	EX
CCl_4		**Tetrachloromethane**				
	C_2H_6O	Ethanol	338.19	0.6140	101.33	OX
	C_3H_6O	Acetone	341.25	0.0337	149.93	OX
	C_3H_8O	1-Propanol	346.28	0.8032	101.33	OX
	C_3H_8O	2-Propanol	341.83	0.6686	101.33	OX
	C_4H_6O	2-Butenal	348.15	0.6500	97.86	OX
	C_4H_6O	2-Methylpropenal	339.15	0.6000	97.86	OX
	C_4H_8O	2-Butanone	346.99	0.6630	101.33	OX
	$C_4H_8O_2$	Ethyl acetate	347.95	0.5700	101.33	OX
	$C_4H_{10}O$	1-Butanol	349.71	0.9500	101.33	OX
	$C_4H_{10}O$	2-Methyl-1-propanol	348.95	0.9080	101.33	OX
	$C_5H_{10}O$	2-Methyl-3-buten-2-ol	348.45	0.9009	101.06	OX
CS_2		**Carbon disulfide**				
	CH_4O	Methanol	310.65	0.7000	101.33	LX
$CHCl_3$		**Trichloromethane**				
	CH_4O	Methanol	328.15	0.6480	107.99	OX
	C_2H_6O	Ethanol	332.45	0.8410	101.33	OX
	C_3H_6O	Acetone	337.58	0.6398	101.33	ON
	$C_3H_6O_2$	Methyl acetate	337.51	0.6760	101.33	ON
	C_3H_8O	2-Propanol	334.15	0.9500	101.33	OX
	C_4H_6O	2-Butenal	329.15	0.9950	97.86	OX
	C_6H_{12}	2-Methyl-1-pentene	333.95	0.6235	101.19	OX
	C_6H_{14}	Hexane	333.45	0.7840	101.33	OX
CHN		**Hydrogen cyanide**				
	C_3H_5Cl	3-Chloropropene	296.45	0.8016	101.33	OX
CH_2Cl_2		**Dichloromethane**				
	C_2H_6O	Ethanol	312.05	0.9600	101.33	OX
CH_2O_2		**Formic acid**				
	$C_2H_4Cl_2$	1,2-Dichloroethane	350.17	0.4275	101.33	OX
	$C_5H_{10}O_2$	Butyl formate	372.15	0.8700	101.33	OX
	C_8H_{10}	*m*-Xylene	365.95	0.8545	101.33	EX
CH_3NO_2		**Nitromethane**				
	C_2H_6O	Ethanol	333.15	0.2850	53.61	OX
	C_3H_7Br	1-Bromopropane	343.25	0.1020	99.82	OX
	$C_4H_8O_2$	1,4-Dioxane	373.25	0.4101	101.48	OX
	C_5H_{10}	2-Methyl-2-butene	311.15	0.0570	101.33	LX
	C_7H_{14}	Methylcyclohexane	354.85	0.5123	101.33	EX
	C_7H_{16}	Heptane	353.25	0.4790	101.33	EX
	C_8H_{18}	Octane	363.38	0.6964	99.73	EX
	C_9H_{20}	Nonane	369.29	0.8403	99.73	EX
	$C_{10}H_{22}$	Decane	371.96	0.9239	99.73	EX
	$C_{11}H_{24}$	Undecane	373.16	0.9619	99.73	EX

Molecular formula						
Comp. 1	Comp. 2	Name	T_{Az}/K	$y_{1,Az}$	P_{Az}/kPa	Type
	$C_{12}H_{26}$	Dodecane	373.75	0.9846	99.73	EX
CH_4O		**Methanol**				
	$C_2HBrClF_3$	2-Bromo-2-chloro-1,1,1-trifluoroethane	317.25	0.1890	93.33	OX
	C_2H_5Br	Bromoethane	308.05	0.1610	101.33	OX
	C_3H_5Cl	3-Chloropropene	312.15	0.2570	100.39	OX
	C_3H_6O	Acetone	328.29	0.2400	101.33	OX
	$C_3H_6O_2$	Methyl acetate	328.15	0.3480	107.19	OX
	$C_3H_6O_2$	1,3-Dioxolane	334.66	0.6910	101.30	OX
	$C_3H_6O_2$	Ethyl formate	318.15	0.3000	81.34	OX
	$C_3H_6O_3$	Dimethyl carbonate	337.25	0.8504	102.52	OX
	C_3H_7Cl	1-Chloropropane	313.35	0.2500	101.59	OX
	$C_4H_4F_6O$	Bis(2,2,2-trifluoroethyl) ether	326.28	0.4450	101.30	OX
	$C_4H_6O_2$	Vinyl acetate	332.05	0.6182	101.33	OX
	C_4H_8O	2-Butanone	323.15	0.8020	58.80	OX
	C_4H_8O	Tetrahydrofuran	332.75	0.5040	101.33	OX
	$C_4H_8O_2$	Ethyl acetate	335.66	0.7120	101.33	OX
	$C_4H_{10}O$	Diethyl ether	305.15	0.0500	93.33	OX
	$C_4H_{10}O_2$	Dimethylacetal	330.35	0.4700	101.33	OX
	$C_5H_3F_9O$	1,1,1,2,3,3-Hexafluoro-3-(2,2,2-trifluoroethoxy)propane	330.67	0.5600	101.30	OX
	C_5H_6	1,3-Cyclopentadiene	309.05	0.2120	101.33	OX
	C_5H_8	2-Methyl-1,3-butadiene	303.55	0.1670	101.33	OX
	C_5H_8	Methylenecyclobutane	309.05	0.2190	101.33	OX
	C_5H_8	1-Methylcyclobutene	304.85	0.1900	101.33	OX
	C_5H_8	*cis*-1,3-Pentadiene	311.10	0.2300	101.33	OX
	C_5H_8	*trans*-1,3-Pentadiene	309.65	0.2110	101.33	OX
	C_5H_{10}	2-Methyl-1-butene	300.55	0.1720	101.33	OX
	C_5H_{10}	3-Methyl-1-butene	291.05	0.0890	101.33	OX
	C_5H_{10}	2-Methyl-2-butene	306.25	0.2160	101.33	OX
	C_5H_{10}	1-Pentene	300.05	0.1469	102.47	OX
	$C_5H_{10}O$	2,3-Epoxy-2-methylbutane	334.95	0.6590	101.33	OX
	C_5H_{12}	Isopentane	297.05	0.0930	101.33	OX
	C_5H_{12}	Pentane	303.20	0.1930	101.30	OX
	$C_5H_{12}O$	Butyl methyl ether	330.00	0.5515	100.08	OX
	$C_5H_{12}O$	Methyl *tert*-butyl ether	325.00	0.3140	103.15	OX
	$C_5H_{12}O$	Ethyl propyl ether	330.00	0.4050	112.25	OX
	$C_5H_{12}O_2$	Diethoxymethane	336.03	0.8127	101.52	OX
	$C_5H_{12}O_2$	2,2-Dimethoxypropane	334.15	0.7250	100.00	OX
	$C_5H_{14}N_2$	*N,N,N',N'*-Tetramethylmethanediamine	335.15	0.7670	101.33	OX
	C_6F_6	Hexafluorobenzene	318.15	0.6100	61.73	OX
	C_6H_5F	Fluorobenzene	333.35	0.6625	101.62	OX
	C_6H_6	Benzene	331.56	0.6090	101.33	OX
	C_6H_{12}	Cyclohexane	328.75	0.6090	106.66	OX
	C_6H_{12}	2-Methyl-1-pentene	330.00	0.4517	141.80	OX
	C_6H_{14}	2,3-Dimethylbutane	313.15	0.3620	85.50	OX
	C_6H_{14}	Hexane	333.15	0.5160	149.64	OX
	$C_6H_{14}O$	*tert*-Butyl ethyl ether	330.95	0.6002	101.54	OX
	$C_6H_{14}O$	Diisopropyl ether	330.00	0.5390	101.61	OX
	$C_6H_{14}O$	Butyl ethyl ether	335.00	0.8010	98.84	OX
	$C_6H_{14}O$	2-Methoxy-2-methylbutane	335.55	0.7735	101.69	OX
	C_7H_8	Toluene	336.65	0.8820	101.33	OX
	C_7H_{14}	Methylcyclohexane	333.15	0.7520	102.87	EX
	C_7H_{16}	Heptane	331.95	0.7279	101.33	OX
	$C_7H_{16}O$	2-Ethoxy-2-methylbutane	335.15	0.8736	97.28	OX
	C_8H_{18}	Octane	335.55	0.8830	101.33	LX
	C_9H_{20}	Nonane	337.25	0.9526	101.33	OX
$C_2Cl_3F_3$		**1,1,2-Trichloro-1,2,2-trifluoroethane**				
	$C_2H_3F_3O$	2,2,2-Trifluoroethanol	316.58	0.7770	101.33	EX

Molecular formula						
Comp. 1	Comp. 2	Name	T_{Az}/K	$y_{1,Az}$	P_{Az}/kPa	Type
	C_2H_6O	Ethanol	317.75	0.8456	101.42	OX
	C_3H_8O	2-Propanol	319.35	0.9159	100.95	OX
	$C_4H_{10}O$	2-Methyl-2-propanol	319.95	0.9426	101.09	OX
C_2Cl_4		**Tetrachloroethene**				
	$C_2H_3Cl_3$	1,1,2-Trichloroethane	385.95	0.2115	101.33	OX
	C_8H_{16}	1-Octene	393.15	0.5900	101.33	OX
	C_8H_{16}	*cis*-4-Octene	393.65	0.7100	101.33	OX
	C_8H_{16}	*trans*-4-Octene	393.45	0.6700	101.33	OX
	C_8H_{18}	Octane	371.90	0.8781	53.44	OX
$C_2Cl_4F_2$		**1,1,2,2-Tetrachloro-1,2-difluoroethane**				
	$C_2H_4Cl_2$	1,2-Dichloroethane	353.80	0.2700	101.33	OX
$C_2HBrClF_3$		**2-Bromo-2-chloro-1,1,1-trifluoroethane**				
	$C_4H_{10}O$	Diethyl ether	323.65	0.7200	93.33	ON
C_2HCl_3		**Trichloroethene**				
	$C_2H_4Cl_2$	1,2-Dichloroethane	355.35	0.3324	101.36	OX
	C_2H_6O	Ethanol	343.85	0.4741	101.33	OX
	C_4H_6O	2-Butenal	360.15	0.9000	97.86	OX
	C_6H_{12}	Cyclohexane	353.40	0.0975	101.32	OX
$C_2H_2Cl_2$		***trans*-1,2-Dichloroethene**				
	$C_5H_3F_9O$	1,1,1,2,3,3-Hexafluoro-3-(2,2,2-trifluoroethoxy)propane	318.50	0.8390	101.30	OX
C_2H_3N		**Acetonitrile**				
	C_3H_8O	2-Propanol	348.15	0.5287	100.81	OX
	$C_4H_6O_2$	Vinyl acetate	344.65	0.1948	98.33	OX
	C_4H_8O	2-Butanone	352.15	0.3195	101.15	OX
	C_4H_8O	Tetrahydrofuran	338.95	0.0784	101.13	OX
	$C_4H_{10}O$	2-Methyl-2-propanol	333.15	0.6200	56.93	OX
	C_5H_8	2-Methyl-1,3-butadiene	306.75	0.0410	101.33	OX
	C_5H_8	Methylenecyclobutane	312.45	0.1450	101.33	OX
	$C_5H_8O_2$	Methyl methacrylate	355.25	0.9866	102.07	OX
	C_5H_{10}	2-Methyl-2-butene	308.95	0.1320	101.33	OX
	C_5H_{10}	1-Pentene	301.85	0.0830	101.33	OX
	C_5H_{12}	Isopentane	298.45	0.1040	101.33	EX
	C_6H_6	Benzene	328.15	0.4560	54.65	OX
	$C_6H_{14}O$	2-Methoxy-2-methylbutane	346.13	0.5835	100.56	OX
	$C_7H_{16}O$	2-Ethoxy-2-methylbutane	348.85	0.7219	98.99	OX
	$C_{10}H_{20}$	1-Decene	354.55	0.9924	100.51	OX
$C_2H_4Cl_2$		**1,1-Dichloroethane**				
	C_3H_8O	2-Propanol	329.55	0.8928	101.60	OX
	C_6H_{14}	Hexane	329.30	0.8025	101.21	OX
$C_2H_4Cl_2$		**1,2-Dichloroethane**				
	C_3H_8O	2-Propanol	347.25	0.5258	100.32	OX
	$C_4H_{10}O$	2-Methyl-1-propanol	356.05	0.9173	101.26	OX
	$C_4H_{10}O$	2-Methyl-2-propanol	349.45	0.5336	101.43	OX
	C_7H_{14}	Methylcyclohexane	354.65	0.8036	101.21	OX
	C_8H_{18}	2,2,4-Trimethylpentane	343.15	0.7600	73.13	OX
C_2H_4O		**Acetaldehyde**				
	C_4H_6	1,3-Butadiene	268.15	0.0520	101.33	OX
	C_5H_8	2-Methyl-1,3-butadiene	292.23	0.8140	101.33	OX
$C_2H_4O_2$		**Acetic acid**				
	C_5H_5N	Pyridine	411.25	0.5780	101.33	ON
	$C_5H_{12}O$	3-Methyl-2-butanol	392.65	0.7210	101.33	ON
	C_6H_7N	2-Methylpyridine	417.27	0.5120	101.33	ON
	$C_6H_{10}O_2$	Vinyl butanoate	386.45	0.5750	101.33	OX
	C_6H_{14}	Hexane	341.40	0.0839	101.33	OX
	C_7H_9N	2,4-Dimethylpyridine	435.45	0.3022	101.33	ON
	C_7H_{16}	Heptane	364.95	0.4490	101.33	OX
	C_8H_{10}	*o*-Xylene	389.75	0.8640	101.33	OX

Molecular formula						
Comp. 1	Comp. 2	Name	T_{Az}/K	$y_{1,Az}$	P_{Az}/kPa	Type
	C_8H_{10}	*p*-Xylene	388.40	0.8200	101.33	OX
	C_8H_{18}	Octane	378.85	0.6870	101.33	OX
	C_9H_{20}	Nonane	386.05	0.8250	101.33	OX
	$C_{10}H_{22}$	Decane	390.05	0.9250	101.33	OX
	$C_{11}H_{24}$	Undecane	391.15	0.9720	101.33	OX
$C_2H_4O_2$		**Methyl formate**				
	C_2H_5Br	Bromoethane	303.05	0.7360	101.33	OX
	$C_4H_{10}O$	Diethyl ether	301.55	0.6030	101.33	OX
	C_5H_8	2-Methyl-1,3-butadiene	298.90	0.5150	101.33	OX
	C_5H_{10}	2-Methyl-2-butene	297.75	0.5760	101.33	OX
	C_5H_{12}	Isopentane	291.55	0.4920	101.33	OX
	C_5H_{12}	Pentane	294.85	0.5740	101.33	OX
	C_6H_{14}	Hexane	302.65	0.8490	101.33	OX
C_2H_5Br		**Bromoethane**				
	C_5H_{10}	2-Methyl-2-butene	308.55	0.5110	101.33	OX
	C_5H_{12}	Isopentane	300.55	0.2180	101.33	OX
$C_2H_5NO_2$		**Nitroethane**				
	$C_4H_{10}O$	2-Methyl-1-propanol	375.81	0.4080	101.33	OX
	C_7H_{16}	Heptane	362.95	0.3520	101.33	OX
C_2H_6O		**Ethanol**				
	C_3H_3N	Acrylonitrile	343.95	0.4440	101.33	OX
	$C_3H_6O_2$	Methyl acetate	329.79	0.0362	101.33	OX
	$C_4H_3F_7O$	1,1,2,2-Tetrafluoroethyl 1,1,1-trifluoroethyl ether	326.67	0.2000	101.30	OX
	$C_4H_4F_6O$	Bis(2,2,2-trifluoroethyl) ether	331.90	0.2840	101.30	OX
	C_4H_8O	Butanal	345.45	0.3690	101.33	OX
	C_4H_8O	2-Butanone	347.15	0.5080	101.33	OX
	C_4H_8O	Tetrahydrofuran	344.95	0.1290	125.00	OX
	$C_4H_8O_2$	Ethyl acetate	344.85	0.4590	101.33	OX
	$C_4H_8O_2$	1,4-Dioxane	351.33	0.9480	101.33	OX
	$C_4H_8O_2$	Methyl propanoate	346.30	0.5140	103.91	OX
	$C_4H_{11}N$	Butylamine	354.99	0.5900	101.33	ON
	$C_5H_3F_9O$	1,1,1,2,3,3-Hexafluoro-3-(2,2,2-trifluoroethoxy)propane	337.88	0.3980	101.30	OX
	C_5H_8	2-Methyl-1,3-butadiene	305.95	0.1500	101.33	OX
	C_5H_8	Cyclopentene	323.40	0.1440	134.00	OX
	C_5H_{10}	2-Methyl-2-butene	309.79	0.0795	101.33	OX
	C_5H_{10}	Cyclopentane	323.44	0.1800	121.00	OX
	$C_5H_{10}O$	2,3-Epoxy-2-methylbutane	343.45	0.2930	101.33	OX
	$C_5H_{10}O$	3-Methyl-2-butanone	350.85	0.8250	101.33	OX
	$C_5H_{10}O$	2-Pentanone	351.15	0.9779	100.50	OX
	$C_5H_{10}O$	3-Pentanone	351.33	0.9590	101.33	OX
	$C_5H_{10}O_2$	Isopropyl acetate	349.85	0.7010	101.33	OX
	$C_5H_{10}O_2$	Methyl butanoate	346.30	0.8800	83.88	OX
	C_5H_{12}	Isopentane	299.95	0.0540	101.33	OX
	C_5H_{12}	Pentane	307.15	0.0537	101.33	OX
	$C_5H_{12}O$	Methyl *tert*-butyl ether	327.75	0.0380	101.33	OX
	$C_5H_{12}O_2$	Diethoxymethane	348.30	0.6497	102.35	OX
	C_6H_5F	Fluorobenzene	343.85	0.4752	101.54	OX
	C_6H_6	Benzene	341.25	0.4600	101.33	OX
	C_6H_{12}	Cyclohexane	337.95	0.4540	102.26	OX
	C_6H_{14}	Hexane	331.65	0.3410	101.33	OX
	$C_6H_{14}O$	*tert*-Butyl ethyl ether	339.95	0.3728	101.72	OX
	$C_6H_{14}O$	2-Methoxy-2-methylbutane	346.81	0.5820	101.32	OX
	C_7H_8	Toluene	349.75	0.8152	101.33	OX
	$C_7H_{16}O$	2-Ethoxy-2-methylbutane	349.35	0.7644	101.54	OX
	C_8H_{18}	Octane	349.85	0.8250	101.33	OX
	C_8H_{18}	2,2,4-Trimethylpentane	344.42	0.6450	101.33	OX

Molecular formula						
Comp. 1	Comp. 2	Name	T_{Az}/K	$y_{1,Az}$	P_{Az}/kPa	Type
	C_9H_{20}	Nonane	351.35	0.9400	101.33	OX
$C_2H_6O_2$		**1,2-Ethanediol**				
	$C_5H_{12}O_3$	Di(ethylene glycol) monomethyl ether	463.95	0.4388	101.33	OX
	$C_6H_{14}O_3$	Di(ethylene glycol) monoethyl ether	467.15	0.6480	101.33	OX
	C_7H_8O	*o*-Cresol	462.67	0.3797	101.33	OX
	$C_7H_{16}O_3$	Di(ethylene glycol) monoisopropyl ether	466.35	0.6964	101.33	OX
	$C_7H_{16}O_3$	Di(ethylene glycol) monopropyl ether	468.55	0.8448	101.33	OX
	$C_7H_{16}O_3$	Di(propylene glycol) monomethyl ether (unspecified isomer)	457.65	0.3500	101.33	OX
	$C_8H_{11}N$	2,4,6-Trimethylpyridine	443.65	0.1734	101.33	OX
	$C_8H_{18}O_3$	Di(ethylene glycol) monobutyl ether	469.15	0.9102	101.33	OX
	$C_8H_{18}O_3$	Di(ethylene glycol) monoisobutyl ether	467.55	0.8355	101.33	OX
	$C_8H_{18}O_3$	Di(propylene glycol) monoethyl ether (unspecified isomer)	458.65	0.4800	101.33	OX
	$C_9H_{20}O_3$	Di(propylene glycol) monopropyl ether (unspecified isomer)	463.15	0.6590	101.33	OX
	$C_{10}H_{22}O_3$	Di(propylene glycol) monobutyl ether (unspecified isomer)	465.75	0.8130	101.33	OX
C_3H_3N		**Acrylonitrile**				
	C_5H_8	Methylenecyclobutane	313.80	0.1275	101.33	OX
	C_6H_6	Benzene	347.45	0.5575	101.46	OX
	C_6H_{12}	Cyclohexane	337.75	0.4836	101.94	OX
	C_6H_{14}	Hexane	330.90	0.4048	101.05	OX
C_3H_4O		**Acrolein**				
	C_5H_8	2-Methyl-1,3-butadiene	306.15	0.1980	101.33	OX
C_3H_6O		**Propanal**				
	C_5H_8	2-Methyl-1,3-butadiene	306.35	0.1700	101.33	OX
	C_5H_8	Methylenecyclobutane	311.30	0.3600	101.33	OX
C_3H_6O		**Acetone**				
	$C_3H_6O_2$	Methyl acetate	328.85	0.6470	101.33	OX
	C_3H_7Br	1-Bromopropane	328.75	0.9915	99.75	OX
	C_4H_8O	Tetrahydrofuran	328.85	0.9603	100.35	OX
	C_4H_9Cl	2-Chloro-2-methylpropane	322.05	0.1944	102.11	OX
	C_5H_8	2-Methyl-1,3-butadiene	306.95	0.0610	101.33	OX
	C_5H_8	Methylenecyclobutane	311.25	0.2800	101.33	OX
	C_5H_8	1-Methylcyclobutene	307.75	0.2220	101.33	OX
	C_5H_{10}	2-Methyl-1-butene	303.25	0.1400	101.33	OX
	C_5H_{10}	2-Methyl-2-butene	308.75	0.2440	101.33	OX
	C_5H_{12}	Isopentane	298.75	0.1730	101.33	OX
	$C_5H_{12}O$	Methyl *tert*-butyl ether	324.35	0.4824	102.19	OX
	C_6H_{12}	Cyclohexane	330.05	0.7590	109.32	OX
	C_6H_{12}	1-Hexene	323.35	0.5973	101.40	OX
	C_6H_{12}	2-Methyl-1-pentene	333.40	0.5793	140.60	OX
	C_6H_{14}	Hexane	322.95	0.6480	101.33	OX
	$C_6H_{14}O$	Diisopropyl ether	327.10	0.7424	100.17	OX
	$C_6H_{15}N$	Triethylamine	318.15	0.9800	68.13	OX
	C_7H_{14}	Methylcyclohexane	318.15	0.9500	68.66	OX
C_3H_6O		**Allyl alcohol**				
	$C_5H_{10}O_2$	Ethyl propanoate	367.65	0.5597	99.79	OX
	C_6H_6	Benzene	349.90	0.2203	101.33	OX
	C_6H_{12}	Cyclohexane	333.15	0.2790	63.98	OX
$C_3H_6O_2$		**Methyl acetate**				
	C_3H_7Br	1-Bromopropane	329.60	0.9727	99.56	OX
	C_6H_{10}	Cyclohexene	330.35	0.9121	102.87	OX
	C_6H_{12}	Cyclohexane	328.65	0.8000	101.33	OX
	C_6H_{12}	Methylcyclopentane	325.85	0.6917	99.50	OX
	C_6H_{12}	1-Hexene	323.15	0.6340	92.08	OX
	C_6H_{12}	2-Methyl-1-pentene	325.15	0.5931	100.38	OX

Molecular formula						
Comp. 1	**Comp. 2**	**Name**	T_{Az}/K	$y_{1,Az}$	P_{Az}/kPa	**Type**
	C_6H_{14}	Hexane	326.65	0.6590	106.66	OX
	C_7H_{16}	Heptane	323.15	0.9570	79.48	OX
$C_3H_6O_2$		**Ethyl formate**				
	C_3H_7Br	2-Bromopropane	326.15	0.7090	101.33	OX
	C_6H_{12}	Cyclohexane	323.15	0.8210	91.46	OX
$C_3H_6O_2$		**Propanoic acid**				
	C_5H_5N	Pyridine	421.75	0.6860	101.33	ON
$C_3H_6O_3$		**Dimethyl carbonate**				
	$C_5H_{12}O_2$	Diethoxymethane	358.71	0.4437	100.42	OX
	C_6H_6	Benzene	353.50	0.1366	100.48	OX
	C_6H_{12}	Cyclohexane	346.95	0.3780	101.49	OX
	C_6H_{12}	Methylcyclopentane	342.35	0.2680	103.46	OX
	C_6H_{14}	Hexane	338.15	0.2540	98.46	OX
	$C_6H_{14}O$	Dipropyl ether	356.45	0.5044	100.73	OX
	C_7H_{16}	Heptane	355.15	0.5930	99.67	OX
C_3H_7Br		**1-Bromopropane**				
	C_3H_8O	2-Propanol	339.15	0.7349	99.97	OX
	C_6H_{12}	Cyclohexane	343.35	0.9219	98.84	OX
C_3H_7NO		***N,N*-Dimethylformamide**				
	C_7H_{16}	Heptane	370.15	0.0800	101.33	OX
	$C_{10}H_{16}$	1,4-Dimethyl-4-vinylcyclohexene	415.65	0.5880	101.33	OX
	$C_{10}H_{16}$	1-Methyl-3-(1-methylethylidene)cyclohexene	419.05	0.7250	101.33	OX
$C_3H_7NO_2$		**1-Nitropropane**				
	C_7H_{16}	Heptane	369.25	0.1630	101.33	OX
$C_3H_7NO_2$		**2-Nitropropane**				
	C_7H_{16}	Heptane	367.55	0.2920	101.33	OX
C_3H_8O		**1-Propanol**				
	$C_4H_3F_7O$	1,1,2,2-Tetrafluoroethyl 1,1,1-trifluoroethyl ether	329.23	0.0350	101.30	OX
	$C_4H_4F_6O$	Bis(2,2,2-trifluoroethyl) ether	336.22	0.1100	101.30	OX
	$C_4H_6O_2$	2,3-Butanedione	359.30	0.3600	100.67	OX
	$C_4H_8O_2$	1,4-Dioxane	365.30	0.6418	101.30	OX
	$C_5H_{10}O_2$	Propyl acetate	367.88	0.6190	101.33	OX
	$C_5H_{12}O_2$	Diethoxymethane	359.01	0.2320	99.43	OX
	C_6H_6	Benzene	350.20	0.2060	101.33	OX
	C_6H_{12}	Cyclohexane	347.68	0.2490	101.33	OX
	C_6H_{12}	Methylcyclopentane	340.85	0.1729	101.19	OX
	$C_6H_{12}O_2$	4,4-Dimethyl-1,3-dioxane	368.20	0.9597	101.30	OX
	C_6H_{14}	Hexane	348.15	0.1900	137.23	OX
	C_7H_8	Toluene	365.35	0.6770	101.33	OX
	C_7H_{16}	Heptane	357.65	0.4830	101.33	OX
	C_8H_8	Styrene	369.08	0.9884	98.13	OX
	C_8H_{10}	*o*-Xylene	369.85	0.9886	98.66	OX
	C_8H_{10}	*m*-Xylene	369.90	0.9531	99.06	OX
	C_8H_{10}	*p*-Xylene	369.60	0.9531	99.99	OX
	C_8H_{14}	1-Octyne	369.00	0.8600	101.33	OX
	C_8H_{18}	Octane	366.85	0.7483	101.33	OX
	C_8H_{18}	2,2,4-Trimethylpentane	357.89	0.4580	101.30	OX
	C_9H_{20}	Nonane	369.95	0.9225	101.33	OX
C_3H_8O		**2-Propanol**				
	$C_4H_4F_6O$	Bis(2,2,2-trifluoroethyl) ether	334.16	0.2230	101.30	OX
	$C_4H_6O_2$	2,3-Butanedione	350.85	0.6454	100.95	OX
	C_4H_8O	2-Butanone	350.55	0.3830	101.33	OX
	$C_4H_{10}O$	2-Methyl-2-propanol	343.05	0.5551	60.27	ON
	$C_5H_3F_9O$	1,1,1,2,3,3-Hexafluoro-3-(2,2,2-trifluoroethoxy)propane	341.23	0.3420	101.30	OX
	C_5H_8	2-Methyl-1,3-butadiene	307.05	0.0150	101.33	OX

Molecular formula						
Comp. 1	Comp. 2	Name	T_{Az}/K	$y_{1,Az}$	P_{Az}/kPa	Type
	C_5H_{10}	2-Methyl-2-butene	310.95	0.0460	101.33	OX
	$C_5H_{10}O$	2,3-Epoxy-2-methylbutane	346.10	0.1400	101.33	OX
	$C_5H_{10}O$	3-Methyl-2-butanone	354.75	0.8500	101.33	OX
	C_5H_{12}	Isopentane	298.15	0.1370	101.33	OX
	$C_5H_{12}O_2$	Diethoxymethane	351.45	0.6107	98.61	OX
	C_6H_5F	Fluorobenzene	347.75	0.4666	101.25	OX
	C_6H_6	Benzene	345.03	0.3960	101.33	OX
	C_6H_{10}	Cyclohexene	344.65	0.4271	101.40	OX
	C_6H_{12}	Cyclohexane	342.75	0.4050	101.33	OX
	C_6H_{12}	Methylcyclopentane	336.45	0.2900	98.14	OX
	C_6H_{14}	Hexane	338.15	0.2900	112.66	OX
	$C_6H_{14}O$	Diisopropyl ether	340.00	0.2050	103.36	OX
	$C_6H_{15}N$	Diisopropylamine	352.94	0.4890	101.33	OX
	C_7H_8	Toluene	354.65	0.8370	101.33	OX
	C_7H_{14}	Methylcyclohexane	350.85	0.6530	101.33	OX
	C_7H_{16}	Heptane	349.55	0.6023	101.33	OX
	$C_7H_{16}O$	*tert*-Butyl isopropyl ether	349.95	0.5306	102.70	OX
	C_8H_{18}	Octane	354.63	0.8990	101.33	OX
	C_8H_{18}	2,2,4-Trimethylpentane	349.58	0.6350	101.30	OX
$C_3H_8O_2$		**2-Methoxyethanol**				
	C_8H_8	Styrene	393.95	0.7787	98.93	OX
	C_8H_{10}	*o*-Xylene	392.65	0.7127	98.79	OX
	C_8H_{10}	*m*-Xylene	392.15	0.6397	99.73	OX
	C_8H_{10}	*p*-Xylene	392.65	0.6303	99.99	OX
	C_8H_{16}	1-Octene	380.75	0.4700	101.33	OX
	C_8H_{16}	*cis*-4-Octene	381.25	0.4900	101.33	OX
	C_8H_{16}	*trans*-4-Octene	381.05	0.4900	101.33	OX
$C_3H_8O_2$		**Dimethoxymethane**				
	C_5H_6	1,3-Cyclopentadiene	313.65	0.3350	101.33	OX
	C_5H_8	2-Methyl-1,3-butadiene	306.80	0.0160	101.33	OX
	C_5H_8	Methylenecyclobutane	310.35	0.4630	101.33	OX
	C_5H_8	1-Methylcyclobutene	309.05	0.2900	101.33	OX
$C_3H_8O_2$		**1,2-Propanediol**				
	$C_7H_{16}O_3$	Di(propylene glycol) monomethyl ether (unspecified isomer)	456.85	0.5691	101.33	OX
	$C_8H_{18}O_3$	Di(propylene glycol) monoethyl ether (unspecified isomer)	458.75	0.7778	101.33	OX
	$C_9H_{20}O_3$	Di(propylene glycol) monoisopropyl ether (unspecified isomer)	458.95	0.8130	101.33	OX
	$C_9H_{20}O_3$	Di(propylene glycol) monopropyl ether (unspecified isomer)	458.95	0.9010	101.33	OX
	$C_{10}H_{22}O_3$	Di(propylene glycol) monobutyl ether (unspecified isomer)	459.65	0.9721	101.33	OX
	$C_{10}H_{22}O_3$	Di(propylene glycol) monoisobutyl ether (unspecified isomer)	459.05	0.9255	101.33	OX
$C_3H_8O_2$		**1,3-Propanediol**				
	$C_5H_{12}O_3$	Di(ethylene glycol) monomethyl ether	455.25	0.6300	101.33	OX
	$C_6H_{14}O_3$	Di(ethylene glycol) monoethyl ether	459.25	0.9350	101.33	OX
C_4H_6		**1,3-Butadiene**				
	C_4H_8	2-Butene (unspecified isomer)	267.59	0.7650	101.33	OX
C_4H_6O		**2-Butenal**				
	C_7H_8	Toluene	374.15	0.5950	97.86	OX
	C_8H_{18}	Octane	353.15	0.4950	97.86	OX
$C_4H_6O_2$		**Vinyl acetate**				
	C_6H_{12}	Cyclohexane	340.45	0.6200	101.33	OX
	C_6H_{14}	Hexane	335.25	0.4450	101.33	OX
$C_4H_6O_2$		**2,3-Butanedione**				
	C_7H_8	Toluene	362.70	0.9513	101.34	OX

Molecular formula						
Comp. 1	Comp. 2	Name	T_{Az}/K	$y_{1,Az}$	P_{Az}/kPa	Type
$C_4H_6O_3$		**Acetic anhydride**				
	C_8H_{16}	1-Octene	367.53	0.2840	53.88	OX
	C_8H_{18}	Octane	397.65	0.3500	129.80	OX
C_4H_8O		**Butanal**				
	C_6H_{12}	2-Methyl-1-pentene	334.15	0.2293	101.48	OX
C_4H_8O		**2-Butanone**				
	$C_4H_8O_2$	Ethyl acetate	349.55	0.1700	101.33	OX
	C_6H_6	Benzene	351.53	0.4790	101.33	OX
	C_6H_{10}	Cyclohexene	343.29	0.5110	89.35	OX
	C_6H_{12}	1-Hexene	334.75	0.1760	100.58	OX
	C_6H_{14}	Hexane	337.15	0.3280	101.33	OX
	$C_6H_{14}O$	Diisopropyl ether	340.55	0.1938	101.56	OX
	$C_6H_{14}O$	Dipropyl ether	351.40	0.7785	100.88	OX
	C_7H_{14}	Methylcyclohexane	350.50	0.7984	98.93	OX
	C_7H_{16}	Heptane	350.15	0.7670	101.33	OX
C_4H_8O		**Tetrahydrofuran**				
	C_6H_{12}	2-Methyl-1-pentene	334.65	0.2867	101.29	OX
	C_6H_{14}	Hexane	323.15	0.5900	65.83	OX
$C_4H_8O_2$		**Ethyl acetate**				
	$C_4H_{10}O$	2-Methyl-2-propanol	349.75	0.7778	101.28	OX
	C_6H_6	Benzene	350.55	0.9453	102.45	OX
	C_6H_{10}	Cyclohexene	347.45	0.6183	100.87	OX
	C_6H_{12}	Cyclohexane	345.00	0.5390	102.45	OX
	C_6H_{12}	1-Hexene	333.15	0.1230	91.47	OX
	C_6H_{14}	Hexane	338.00	0.3430	101.32	OX
	C_7H_{14}	Methylcyclohexane	349.90	0.9001	101.83	OX
$C_4H_8O_2$		**Butanoic acid**				
	C_5H_5N	Pyridine	436.35	0.9117	101.33	ON
	$C_8H_{16}O_2$	Butyl butanoate	434.60	0.6532	93.33	OXD
	$C_8H_{16}O_2$	Butyl butanoate	434.78	0.8639	93.33	OND
	$C_{11}H_{24}$	Undecane	435.55	0.9060	101.33	OX
$C_4H_8O_2$		**1,4-Dioxane**				
	$C_4H_{10}O$	2-Butanol	371.75	0.4732	100.77	OX
	$C_5H_{10}O_2$	Propyl acetate	373.35	0.6334	101.13	OX
	$C_5H_{12}O$	2-Methyl-2-butanol	373.75	0.8119	99.62	OX
	C_6H_{10}	Cyclohexene	355.75	0.1065	101.44	OX
	C_6H_{12}	Methylcyclopentane	343.85	0.0538	99.79	OX
	$C_6H_{15}N$	Triethylamine	343.15	0.2500	56.80	OX
	C_7H_{16}	Heptane	364.30	0.4868	101.06	OX
	$C_7H_{16}O$	2-Ethoxy-2-methylbutane	369.15	0.5452	100.27	OX
$C_4H_8O_2$		**Propyl formate**				
	C_6H_6	Benzene	343.15	0.3770	76.08	OX
$C_4H_8O_2$		**Methyl propanoate**				
	C_7H_{14}	Methylcyclohexane	352.45	0.8956	101.33	OX
C_4H_9Cl		**1-Chlorobutane**				
	C_6H_{12}	Cyclohexane	348.31	0.5800	95.85	OX
C_4H_9NO		***N,N*-Dimethylacetamide**				
	C_8H_{10}	*o*-Xylene	416.95	0.0591	103.40	OX
	C_8H_{10}	Ethylbenzene	408.95	0.0037	101.70	OX
$C_4H_{10}O$		**1-Butanol**				
	C_5H_5N	Pyridine	392.00	0.7050	101.33	ON
	$C_5H_{10}O_3$	Diethyl carbonate	370.85	0.6346	53.20	OX
	C_6H_5Cl	Chlorobenzene	388.25	0.6950	101.33	OX
	C_6H_{12}	Cyclohexane	352.68	0.0787	101.33	OX
	$C_6H_{12}O_2$	Butyl acetate	389.97	0.7700	101.33	OX
	$C_6H_{12}O_2$	Isobutyl acetate	387.15	0.5980	101.33	OX
	C_6H_{14}	Hexane	341.35	0.0370	101.33	OX
	C_7H_8	Toluene	378.85	0.3320	101.33	OX

Molecular formula						
Comp. 1	Comp. 2	Name	T_{Az}/K	$y_{1,Az}$	P_{Az}/kPa	Type
	C_7H_{12}	3-Ethylcyclopentene	367.65	0.1900	101.33	OX
	C_7H_{16}	Heptane	366.55	0.2272	101.38	OX
	C_8H_8	Styrene	388.71	0.8923	98.39	OX
	C_8H_{10}	*o*-Xylene	388.05	0.8671	100.13	OX
	C_8H_{10}	*m*-Xylene	387.75	0.7865	101.46	OX
	C_8H_{10}	*p*-Xylene	387.85	0.7823	99.73	OX
	C_8H_{14}	1-Octyne	386.50	0.6200	101.33	OX
	C_8H_{14}	2-Octyne	398.30	0.7910	101.33	OX
	C_8H_{16}	1-Octene	363.45	0.4530	53.33	OX
	C_8H_{16}	*cis*-4-Octene	382.35	0.5300	101.33	OX
	C_8H_{16}	*trans*-4-Octene	382.15	0.5310	101.33	OX
	C_8H_{18}	Octane	383.15	0.5500	102.79	OX
	$C_8H_{18}O$	Dibutyl ether	390.59	0.8754	101.33	OX
	C_9H_{16}	1-Butylcyclopentene	356.70	0.8450	79.99	OX
	C_9H_{16}	1-Nonyne	390.60	0.9400	101.33	OX
	C_9H_{20}	Nonane	389.05	0.8128	101.33	OX
$C_4H_{10}O$		**2-Butanol**				
	$C_5H_{10}O$	3-Pentanone	370.50	0.6075	99.98	OX
	C_6H_{10}	Cyclohexene	352.75	0.2046	101.25	OX
	C_6H_{12}	Cyclohexane	349.90	0.1892	101.02	OX
	C_6H_{14}	Hexane	348.15	0.1010	128.66	OX
	$C_6H_{14}O$	2-Methoxy-2-methylbutane	359.15	0.0991	102.12	OX
	C_7H_8	Toluene	353.44	0.5550	56.67	OX
	C_7H_{16}	Heptane	361.95	0.4116	102.70	OX
	$C_7H_{16}O$	2-Ethoxy-2-methylbutane	367.75	0.4931	102.89	OX
	C_8H_{10}	*m*-Xylene	369.85	0.9717	101.06	OX
	C_8H_{10}	*p*-Xylene	369.55	0.9646	101.46	OX
	C_8H_{18}	Octane	371.05	0.8001	101.30	OX
$C_4H_{10}O$		**Diethyl ether**				
	C_5H_{12}	Pentane	306.85	0.5500	101.33	OX
$C_4H_{10}O$		**2-Methyl-1-propanol**				
	$C_5H_{10}O_2$	Isobutyl formate	370.90	0.1930	101.33	OX
	C_6H_6	Benzene	352.45	0.0780	101.33	OX
	C_6H_{10}	Cyclohexene	353.75	0.1363	100.31	OX
	C_6H_{12}	Cyclohexane	351.35	0.1325	101.45	OX
	C_6H_{12}	Methylcyclopentane	343.15	0.0567	100.35	OX
	C_7H_8	Toluene	374.35	0.4941	101.33	OX
	C_8H_{10}	*m*-Xylene	380.35	0.9300	101.33	OX
	C_8H_{10}	*p*-Xylene	380.30	0.9200	101.33	OX
	C_8H_{18}	Octane	376.58	0.6700	101.30	OX
$C_4H_{10}O$		**2-Methyl-2-propanol**				
	C_5H_8	Methylenecyclobutane	314.65	0.0150	101.33	OX
	C_6H_{10}	Cyclohexene	346.00	0.4172	99.61	OX
	C_6H_{12}	Methylcyclopentane	339.35	0.2559	99.93	OX
	C_6H_{12}	1-Hexene	333.25	0.2650	101.30	OX
	C_6H_{14}	Hexane	337.70	0.2502	101.30	OX
	$C_6H_{14}O$	*tert*-Butyl ethyl ether	342.85	0.2512	101.44	OX
	$C_6H_{14}O$	Diisopropyl ether	340.45	0.1058	101.72	OX
	$C_6H_{14}O$	2-Methoxy-2-methylbutane	353.20	0.5617	101.80	OX
	C_7H_8	Toluene	353.44	0.9200	93.61	OX
	$C_7H_{16}O$	*tert*-Butyl isopropyl ether	350.90	0.5390	102.94	OX
	C_8H_{18}	Octane	343.15	0.9680	61.18	OX
	C_8H_{18}	2,2,4-Trimethylpentane	339.28	0.6040	59.49	OX
$C_4H_{10}O_2$		**1,4-Butanediol**				
	$C_{15}H_{32}O$	1-Pentadecanol	502.75	0.9980	101.33	OX
$C_4H_{10}O_2$		**1,2-Dimethoxyethane**				
	C_7H_{14}	Methylcyclohexane	350.00	0.8190	79.42	OX
$C_4H_{10}O_2$		**2-Ethoxyethanol**				

Molecular formula						
Comp. 1	Comp. 2	Name	T_{Az}/K	$y_{1,Az}$	P_{Az}/kPa	Type
	C_8H_8	Styrene	405.75	0.6438	101.33	OX
	C_8H_{10}	*o*-Xylene	404.95	0.5965	101.36	OX
	C_8H_{10}	*m*-Xylene	401.75	0.5159	101.33	OX
	C_8H_{10}	*p*-Xylene	402.55	0.5042	102.19	OX
	C_8H_{10}	Ethylbenzene	401.05	0.4632	100.94	OX
$C_4H_{10}O_3$		**Di(ethylene glycol)**				
	$C_9H_{20}O$	1-Nonanol	486.65	0.0095	101.33	OX
$C_4H_{11}N$		**Butylamine**				
	C_6H_6	Benzene	343.15	0.7000	80.89	OX
C_5H_5N		**Pyridine**				
	C_7H_8	Toluene	383.19	0.2250	101.33	OX
	C_7H_{16}	Heptane	368.61	0.3002	101.33	OX
	C_9H_{20}	Nonane	388.15	0.9350	101.33	OX
C_5H_6		**2-Methyl-1-buten-3-yne**				
	C_5H_8	2-Methyl-1,3-butadiene	305.88	0.7210	101.33	OX
	C_5H_{10}	2-Methyl-1-butene	303.15	0.3450	101.33	OX
	C_5H_{12}	Isopentane	299.35	0.3620	101.33	OX
C_5H_6		**1,3-Cyclopentadiene**				
	C_5H_{10}	2-Methyl-2-butene	310.85	0.3000	101.33	OX
	C_5H_{12}	Pentane	307.75	0.1959	101.30	OX
C_5H_8		**2-Methyl-1,3-butadiene**				
	C_5H_{12}	Pentane	310.55	0.7421	114.66	OX
	$C_6F_{15}N$	Tris(perfluoroethyl)amine	303.35	0.8200	101.33	EX
C_5H_8		**3-Methyl-1-butyne**				
	C_5H_{12}	Isopentane	297.15	0.5650	101.33	OX
C_5H_8		**1-Pentyne**				
	C_5H_{10}	2-Methyl-2-butene	310.95	0.3300	101.33	OX
	C_5H_{12}	Pentane	307.55	0.3050	101.33	OX
C_5H_8O		**Cyclopentanone**				
	$C_5H_{12}O$	3-Methyl-1-butanol	402.02	0.5944	101.33	OX
	$C_5H_{12}O$	1-Pentanol	403.84	0.9196	101.33	OX
$C_5H_8O_2$		**Methyl methacrylate**				
	C_7H_{16}	Heptane	366.35	0.4597	99.94	OX
	C_8H_{18}	Octane	373.70	0.9651	100.16	OX
C_5H_{10}		**2-Methyl-1-butene**				
	$C_6F_{15}N$	Tris(perfluoroethyl)amine	301.95	0.8450	101.33	OX
C_5H_{10}		**2-Methyl-2-butene**				
	$C_6F_{15}N$	Tris(perfluoroethyl)amine	307.65	0.8170	101.33	OX
$C_5H_{10}O$		**3-Methyl-3-buten-1-ol**				
	C_6H_{12}	Cyclohexane	352.65	0.0215	101.10	OX
	$C_6H_{12}O_2$	4,4-Dimethyl-1,3-dioxane	403.05	0.7590	102.26	OX
	C_7H_8	Toluene	381.55	0.2391	101.60	OX
	C_7H_{16}	Heptane	370.00	0.2100	101.30	OX
$C_5H_{10}O$		**2-Methyl-3-buten-2-ol**				
	C_6H_{12}	Cyclohexane	350.15	0.1904	101.20	OX
	C_6H_{12}	1-Hexene	336.55	0.0479	101.30	OX
	C_7H_8	Toluene	366.55	0.7788	101.20	OX
$C_5H_{10}O$		**3-Pentanone**				
	C_7H_{14}	Methylcyclohexane	366.95	0.4441	99.82	OX
	$C_7H_{16}O$	2-Ethoxy-2-methylbutane	371.15	0.4764	100.21	OX
$C_5H_{10}O_2$		**Propyl acetate**				
	C_6H_{12}	Cyclohexane	353.15	0.0598	100.43	OX
	C_7H_{14}	Methylcyclohexane	368.40	0.4746	100.90	OX
	C_7H_{16}	Heptane	366.75	0.4215	101.38	OX
	$C_7H_{16}O$	2-Ethoxy-2-methylbutane	370.95	0.6529	100.03	OX
C_5H_{12}		**Isopentane**				
	$C_6F_{15}N$	Tris(perfluoroethyl)amine	299.65	0.9020	101.33	OX
$C_5H_{12}O$		**2-Methyl-1-butanol**				

Molecular formula						
Comp. 1	Comp. 2	Name	T_{Az}/K	$y_{1,Az}$	P_{Az}/kPa	Type
	C_8H_{10}	*o*-Xylene	402.05	0.7417	101.87	OX
	C_8H_{10}	*m*-Xylene	400.65	0.6316	101.85	OX
	C_8H_{10}	*p*-Xylene	400.15	0.6273	101.07	OX
	C_8H_{10}	Ethylbenzene	398.75	0.5657	99.46	OX
$C_5H_{12}O$		**3-Methyl-1-butanol**				
	$C_6H_{10}O$	Cyclohexanone	404.87	0.9094	101.33	OX
	C_7H_8	Toluene	383.15	0.1250	101.33	OX
	$C_7H_{14}O_2$	Isopentyl acetate	403.95	0.9900	101.33	OX
	C_7H_{16}	Heptane	368.15	0.1016	95.06	OX
$C_5H_{12}O$		**2-Methyl-2-butanol**				
	C_6H_6	Benzene	352.35	0.1500	101.33	OX
	C_6H_{12}	Cyclohexane	351.95	0.1100	101.33	OX
	C_6H_{12}	Methylcyclopentane	344.75	0.0551	101.80	OX
	C_6H_{14}	Hexane	339.06	0.0436	93.55	OX
	C_7H_{14}	Methylcyclohexane	366.60	0.3965	99.87	OX
	C_7H_{16}	Heptane	348.15	0.3140	56.83	OX
	$C_7H_{16}O$	2-Ethoxy-2-methylbutane	369.85	0.3904	100.52	OX
$C_5H_{12}O$		**1-Pentanol**				
	$C_6H_{10}O$	Cyclohexanone	392.37	0.9748	53.32	OX
	$C_7H_{14}O_2$	Isopentyl acetate	407.45	0.6000	101.33	OX
	C_7H_{16}	Heptane	371.45	0.0576	101.33	OX
	C_8H_{18}	Octane	393.15	0.2847	101.33	OX
	C_9H_{20}	Nonane	404.45	0.6242	101.33	OX
	$C_{10}H_{22}$	Decane	410.65	0.9221	101.33	OX
$C_5H_{12}O$		**3-Pentanol**				
	C_7H_{16}	Heptane	368.15	0.2001	98.62	OX
$C_5H_{12}O_2$		**Diethoxymethane**				
	C_6H_{12}	Cyclohexane	353.21	0.1774	101.39	OX
	C_6H_{14}	Hexane	361.27	0.9101	102.30	OX
C_6F_6		**Hexafluorobenzene**				
	C_6H_6	Benzene	353.60	0.7600	101.33	OND
	C_6H_6	Benzene	352.50	0.1832	101.33	OXD
$C_6F_{15}N$		**Tris(perfluoroethyl)amine**				
	C_6H_6	Benzene	329.95	0.5900	101.33	EX
	C_6H_{12}	Cyclohexane	329.35	0.5690	101.33	EX
	C_6H_{14}	Hexane	327.65	0.4840	101.33	OX
C_6H_5Br		**Bromobenzene**				
	$C_6H_{12}O$	Cyclohexanol	403.15	0.7390	52.45	OX
C_6H_6		**Benzene**				
	C_6H_{12}	Cyclohexane	353.15	0.5460	109.18	OX
	C_6H_{12}	Methylcyclopentane	333.15	0.1390	69.93	OX
	C_6H_{14}	Hexane	341.45	0.0500	101.33	OX
	C_7H_{16}	Heptane	353.25	0.9922	101.32	OX
	C_8H_{18}	2,2,4-Trimethylpentane	353.25	0.9751	101.32	OX
C_6H_6O		**Phenol**				
	C_6H_7N	Aniline	459.09	0.3884	101.33	ON
	C_6H_7N	2-Methylpyridine	458.33	0.7852	101.32	ON
	C_6H_7N	3-Methylpyridine	462.93	0.6918	101.32	ON
	C_7H_5N	Benzonitrile	465.11	0.2345	101.33	ON
	C_7H_6O	Benzaldehyde	447.00	0.6001	73.00	ON
	C_7H_9N	2,6-Dimethylpyridine	459.32	0.7539	101.32	ON
	C_8H_{18}	Octane	398.17	0.0690	101.32	OX
	C_9H_{12}	Propylbenzene	428.15	0.1150	91.85	OX
	C_9H_{12}	1,2,3-Trimethylbenzene	443.45	0.3936	101.33	OX
	C_9H_{12}	1,2,4-Trimethylbenzene	440.65	0.2409	101.33	OX
	C_9H_{12}	1,3,5-Trimethylbenzene	436.95	0.1828	101.33	OX
	C_9H_{18}	1-Nonene	413.15	0.1297	86.57	OX
	C_9H_{20}	Nonane	419.18	0.2180	101.32	OX

Molecular formula						
Comp. 1	Comp. 2	Name	T_{Az}/K	$y_{1,Az}$	P_{Az}/kPa	Type
	$C_{10}H_{12}$	1,2,3,4-Tetrahydronaphthalene	448.15	0.9031	84.25	OX
	$C_{10}H_{14}$	Butylbenzene	447.05	0.5535	101.33	OX
	$C_{10}H_{14}$	*sec*-Butylbenzene	441.15	0.3129	101.33	OX
	$C_{10}H_{14}$	*tert*-Butylbenzene	439.95	0.2773	101.33	OX
	$C_{10}H_{14}$	Diethylbenzene (unspecified isomer)	446.45	0.4705	101.33	OX
	$C_{10}H_{14}$	*o*-Diethylbenzene	447.15	0.5565	101.33	OX
	$C_{10}H_{14}$	*m*-Diethylbenzene	445.95	0.5152	101.33	OX
	$C_{10}H_{14}$	1-Ethyl-3,5-dimethylbenzene	447.95	0.5840	101.33	OX
	$C_{10}H_{14}$	2-Ethyl-1,4-dimethylbenzene	447.95	0.5840	101.33	OX
	$C_{10}H_{14}$	Isobutylbenzene	441.75	0.3522	101.33	OX
	$C_{10}H_{14}$	1-Isopropyl-2-methylbenzene	443.75	0.4643	101.33	OX
	$C_{10}H_{14}$	1-Isopropyl-3-methylbenzene	443.05	0.4027	101.33	OX
	$C_{10}H_{14}$	1-Isopropyl-4-methylbenzene	443.65	0.4430	101.33	OX
	$C_{10}H_{14}$	1-Methyl-2-propylbenzene	447.75	0.5801	101.33	OX
	$C_{10}H_{14}$	1-Methyl-3-propylbenzene	446.35	0.5264	101.33	OX
	$C_{10}H_{14}$	1-Methyl-4-propylbenzene	447.15	0.5575	101.33	OX
	$C_{10}H_{14}$	1,2,3,5-Tetramethylbenzene	454.25	0.7957	101.33	OX
	$C_{10}H_{14}$	1,2,4,5-Tetramethylbenzene	453.36	0.7857	101.33	OX
	$C_{10}H_{18}$	*trans*-Decahydronaphthalene	443.15	0.5419	99.85	OX
	$C_{10}H_{22}$	Decane	434.15	0.4450	101.32	OX
	$C_{11}H_{16}$	1-Butyl-2-methylbenzene	455.55	0.8504	101.33	OX
	$C_{11}H_{16}$	1-Butyl-3-methylbenzene	454.15	0.8099	101.33	OX
	$C_{11}H_{16}$	1-Butyl-4-methylbenzene	454.55	0.8230	101.33	OX
	$C_{11}H_{22}$	1-Undecene	443.15	0.6426	92.31	OX
	$C_{12}H_{26}$	Dodecane	450.73	0.7900	101.32	OX
	$C_{14}H_{30}$	Tetradecane	452.48	0.9650	101.32	OX
C_6H_7N		**Aniline**				
	C_7H_8O	*o*-Cresol	464.29	0.0953	101.33	ON
	C_9H_{12}	1,2,3-Trimethylbenzene	444.65	0.3331	101.33	OX
	C_9H_{12}	1,2,4-Trimethylbenzene	441.80	0.1850	101.33	OX
	C_9H_{12}	1,3,5-Trimethylbenzene	437.68	0.1071	101.33	OX
	C_9H_{20}	Nonane	422.35	0.1770	101.33	OX
	$C_{10}H_{14}$	Butylbenzene	448.65	0.4993	101.33	OX
	$C_{10}H_{14}$	*sec*-Butylbenzene	443.15	0.3021	101.33	OX
	$C_{10}H_{14}$	*tert*-Butylbenzene	438.25	0.2104	101.33	OX
	$C_{10}H_{14}$	*o*-Diethylbenzene	448.75	0.5024	101.33	OX
	$C_{10}H_{14}$	*m*-Diethylbenzene	447.45	0.4584	101.33	OX
	$C_{10}H_{14}$	*p*-Diethylbenzene	448.85	0.5086	101.33	OX
	$C_{10}H_{14}$	1-Ethyl-3,5-dimethylbenzene	449.65	0.5310	101.33	OX
	$C_{10}H_{14}$	2-Ethyl-1,4-dimethylbenzene	449.65	0.5310	101.33	OX
	$C_{10}H_{14}$	Isobutylbenzene	442.75	0.2890	101.33	OX
	$C_{10}H_{14}$	1-Isopropyl-2-methylbenzene	445.95	0.4052	101.33	OX
	$C_{10}H_{14}$	1-Isopropyl-3-methylbenzene	444.15	0.3419	101.33	OX
	$C_{10}H_{14}$	1-Isopropyl-4-methylbenzene	445.35	0.3829	101.33	OX
	$C_{10}H_{14}$	1-Methyl-2-propylbenzene	449.45	0.5270	101.33	OX
	$C_{10}H_{14}$	1-Methyl-3-propylbenzene	447.85	0.4711	101.33	OX
	$C_{10}H_{14}$	1-Methyl-4-propylbenzene	448.75	0.5035	101.33	OX
	$C_{10}H_{14}$	1,2,3,5-Tetramethylbenzene	456.55	0.7504	101.33	OX
	$C_{10}H_{14}$	1,2,4,5-Tetramethylbenzene	455.36	0.7349	101.33	OX
	$C_{10}H_{22}$	Decane	440.43	0.4660	101.33	OX
	$C_{11}H_{16}$	1-Butyl-2-methylbenzene	458.15	0.8091	101.33	OX
	$C_{11}H_{16}$	1-Butyl-3-methylbenzene	456.55	0.7661	101.33	OX
	$C_{11}H_{16}$	1-Butyl-4-methylbenzene	457.05	0.7807	101.33	OX
	$C_{11}H_{24}$	Undecane	449.05	0.6970	101.33	OX
	$C_{12}H_{26}$	Dodecane	453.52	0.8220	101.33	OX
	$C_{13}H_{28}$	Tridecane	456.22	0.9300	101.33	OX
	$C_{14}H_{30}$	Tetradecane	457.05	0.9770	101.33	OX
C_6H_7N		**2-Methylpyridine**				

Molecular formula						
Comp. 1	Comp. 2	Name	T_{Az}/K	$y_{1,Az}$	P_{Az}/kPa	Type
	C_8H_{18}	Octane	394.27	0.4610	101.33	OX
	C_9H_{20}	Nonane	402.35	0.8790	101.33	OX
C_6H_7N		**3-Methylpyridine**				
	C_7H_8O	*m*-Cresol	477.01	0.1556	101.32	ON
	C_7H_9N	2,6-Dimethylpyridine	416.64	0.2940	101.33	OX
C_6H_7N		**4-Methylpyridine**				
	C_7H_8O	*m*-Cresol	477.74	0.1822	101.32	ON
	C_7H_9N	2,6-Dimethylpyridine	417.08	0.2000	101.32	OX
$C_6H_{10}O$		**Methyldihydropyran (unspecified isomer)**				
	C_7H_8	Toluene	381.85	0.0207	101.30	OX
$C_6H_{10}O$		**4-Methylenetetrahydropyran**				
	C_7H_8	Toluene	381.15	0.5253	101.20	OX
$C_6H_{12}O$		**Cyclohexanol**				
	C_8H_{10}	*o*-Xylene	415.95	0.1426	101.33	OX
	C_8H_{10}	*m*-Xylene	411.85	0.0503	101.33	OX
	C_8H_{10}	*p*-Xylene	410.95	0.0505	101.33	OX
	C_9H_{20}	Nonane	410.20	0.3350	79.99	OX
$C_6H_{12}O_2$		**Butyl acetate**				
	C_8H_{16}	1-Octene	393.00	0.3030	101.33	OX
$C_6H_{12}O_2$		**4,4-Dimethyl-1,3-dioxane**				
	C_8H_{10}	*o*-Xylene	404.65	0.9662	101.30	OX
	C_8H_{18}	Octane	393.95	0.3343	101.20	OX
	C_9H_{20}	Nonane	402.15	0.8864	101.30	OX
	$C_{10}H_{22}$	Decane	405.35	0.9999	100.60	OX
$C_6H_{14}O$		**1-Hexanol**				
	C_8H_{18}	Octane	398.55	0.0886	101.33	OX
	C_9H_{20}	Nonane	416.95	0.3649	101.33	OX
	$C_{10}H_{22}$	Decane	427.05	0.7123	101.33	OX
$C_6H_{14}O_2$		**2,2-Dimethoxybutane**				
	C_7H_8	Toluene	380.15	0.9180	101.44	OX
C_7F_{16}		**Perfluoroheptane**				
	C_7H_{16}	Heptane	328.16	0.6100	53.60	OX
C_7H_5N		**Benzonitrile**				
	C_7H_8O	*o*-Cresol	468.91	0.5100	101.33	ON
	C_7H_8O	*m*-Cresol	476.10	0.1441	101.33	ON
	C_7H_8O	*p*-Cresol	476.95	0.0898	101.33	ON
	$C_8H_{10}O$	2,6-Xylenol	477.15	0.0807	101.33	ON
C_7H_8O		***o*-Cresol**				
	$C_8H_{11}N$	2,4,6-Trimethylpyridine	470.35	0.6561	101.33	ON
	$C_{10}H_{14}$	*sec*-Butylbenzene	444.65	0.0938	101.33	OX
	$C_{10}H_{14}$	Diethylbenzene (unspecified isomer)	453.10	0.2694	101.33	OX
	$C_{10}H_{14}$	1,2,4,5-Tetramethylbenzene	462.37	0.6273	101.33	OX
	$C_{10}H_{22}$	Decane	433.15	0.3100	78.71	OX
	$C_{11}H_{22}$	1-Undecene	448.15	0.5516	83.07	OX
	$C_{11}H_{24}$	Undecane	433.15	0.5800	56.40	OX
	$C_{12}H_{26}$	Dodecane	458.15	0.8466	93.55	OX
C_7H_8O		***m*-Cresol**				
	C_7H_9N	2,6-Dimethylpyridine	475.66	0.9869	101.32	ON
	C_9H_7N	Quinoline	511.20	0.0356	101.33	ON
	C_9H_{20}	Nonane	413.15	0.0400	76.54	OX
	$C_{10}H_8$	Naphthalene	474.65	0.9680	101.33	OX
	$C_{10}H_{12}$	1,2,3,4-Tetrahydronaphthalene	468.45	0.5900	93.10	OX
	$C_{10}H_{14}$	*sec*-Butylbenzene	445.85	0.0136	101.33	OX
	$C_{10}H_{14}$	Diethylbenzene (unspecified isomer)	454.10	0.1010	101.33	OX
	$C_{10}H_{14}$	1,2,4,5-Tetramethylbenzene	466.87	0.3591	101.33	OX
	$C_{10}H_{22}$	Decane	433.15	0.2170	75.85	OX
C_7H_8O		***p*-Cresol**				
	$C_{10}H_8$	Naphthalene	474.55	0.9414	101.33	OX

Molecular formula						
Comp. 1	Comp. 2	Name	T_{Az}/K	$y_{1,Az}$	P_{Az}/kPa	Type
	$C_{10}H_{14}$	*sec*-Butylbenzene	446.05	0.0186	101.33	OX
	$C_{10}H_{14}$	Diethylbenzene (unspecified isomer)	454.50	0.1105	101.33	OX
C_7H_8O		**Benzyl alcohol**				
	$C_{10}H_{22}$	Decane	445.75	0.2490	101.33	OX
C_7H_9N		**2-Methylaniline**				
	$C_{10}H_{22}$	Decane	446.91	0.1770	101.33	OX
	$C_{11}H_{24}$	Undecane	461.40	0.4930	101.33	OX
	$C_{12}H_{26}$	Dodecane	468.90	0.7650	101.33	OX
	$C_{13}H_{28}$	Tridecane	472.55	0.9070	101.33	OX
$C_7H_{14}O_2$		**Pentyl acetate**				
	C_9H_{20}	Nonane	419.20	0.5380	101.32	OX
$C_7H_{16}O$		**1-Heptanol**				
	C_9H_{20}	Nonane	423.45	0.1071	101.33	OX
	$C_{10}H_{22}$	Decane	438.75	0.4308	101.33	OX
	$C_{11}H_{24}$	Undecane	447.85	0.8014	101.33	OX
C_8H_{10}		***o*-Xylene**				
	C_9H_{20}	Nonane	417.40	0.8498	101.33	OX
$C_8H_{10}O$		**2,6-Xylenol**				
	C_9H_7N	Quinoline	511.00	0.0890	101.33	ON
	$C_{10}H_8$	Naphthalene	475.70	0.9381	101.33	OX
	$C_{10}H_{14}$	1,2,4,5-Tetramethylbenzene	468.85	0.3480	101.33	OX
$C_8H_{10}O$		**2,3-Xylenol**				
	C_9H_7N	Quinoline	513.30	0.2684	101.33	ON
	$C_{10}H_8$	Naphthalene	485.45	0.4123	101.33	OX
$C_8H_{10}O$		**2,4-Xylenol**				
	C_9H_7N	Quinoline	512.30	0.1717	101.33	ON
	$C_{10}H_8$	Naphthalene	481.25	0.6435	101.33	OX
	$C_{10}H_{14}$	1,2,4,5-Tetramethylbenzene	474.05	0.1869	101.33	OX
$C_8H_{10}O$		**2,5-Xylenol**				
	C_9H_7N	Quinoline	512.30	0.1717	101.33	ON
	$C_{10}H_8$	Naphthalene	481.25	0.6435	101.33	OX
	$C_{10}H_{14}$	1,2,4,5-Tetramethylbenzene	474.35	0.1763	101.33	OX
$C_8H_{10}O$		**3,4-Xylenol**				
	C_9H_7N	Isoquinoline	519.75	0.2955	101.33	ON
	C_9H_7N	Quinoline	514.77	0.3907	101.33	ON
	$C_{10}H_8$	Naphthalene	490.95	0.1158	101.33	OX
	$C_{10}H_9N$	3-Methylisoquinoline	524.35	0.0811	101.33	ON
	$C_{10}H_9N$	2-Methylquinoline	521.17	0.1647	101.33	ON
	$C_{10}H_9N$	3-Methylquinoline	523.60	0.1152	101.33	ON
	$C_{10}H_9N$	7-Methylquinoline	525.85	0.0466	101.33	ON
	$C_{11}H_{11}N$	2,3-Dimethylquinoline	521.60	0.2113	101.33	ON
$C_8H_{10}O$		**3,5-Xylenol**				
	C_9H_7N	Isoquinoline	518.05	0.1915	101.33	ON
	C_9H_7N	Quinoline	513.58	0.3287	101.33	ON
	$C_{10}H_8$	Naphthalene	489.33	0.2601	101.33	OX
	$C_{10}H_9N$	2-Methylquinoline	520.65	0.0094	101.33	ON
	$C_{11}H_{11}N$	2,3-Dimethylquinoline	520.70	0.0530	101.33	ON
$C_8H_{10}O$		**2-Ethylphenol**				
	C_9H_7N	Quinoline	511.75	0.1041	101.33	ON
	$C_{10}H_8$	Naphthalene	478.35	0.8005	101.33	OX
	$C_{10}H_{14}$	1,2,4,5-Tetramethylbenzene	471.45	0.3707	101.33	OX
$C_8H_{10}O$		**3-Ethylphenol**				
	C_9H_7N	Quinoline	512.70	0.2089	101.33	ON
	$C_{10}H_8$	Naphthalene	483.45	0.5551	101.33	OX
	$C_{10}H_{14}$	1,2,4,5-Tetramethylbenzene	475.95	0.1249	101.33	OX
$C_8H_{10}O$		**4-Ethylphenol**				
	C_9H_7N	Quinoline	513.45	0.2832	101.33	ON
	$C_{10}H_8$	Naphthalene	486.10	0.3762	101.33	OX

Molecular formula						
Comp. 1	**Comp. 2**	**Name**	**T_{Az}/K**	**$y_{1,Az}$**	**P_{Az}/kPa**	**Type**
$C_8H_{11}N$		**2,4-Dimethylaniline**				
	$C_{11}H_{24}$	Undecane	468.13	0.1490	101.33	OX
	$C_{12}H_{26}$	Dodecane	482.95	0.4520	101.33	OX
	$C_{13}H_{28}$	Tridecane	488.43	0.7880	101.33	OX
	$C_{14}H_{30}$	Tetradecane	490.53	0.9840	101.33	OX
$C_8H_{18}O$		**1-Octanol**				
	$C_{10}H_{22}$	Decane	446.45	0.1029	101.33	OX
	$C_{11}H_{24}$	Undecane	460.05	0.4772	101.33	OX
	$C_{12}H_{26}$	Dodecane	466.95	0.8836	101.33	OX
C_9H_7N		**Isoquinoline**				
	$C_{11}H_{10}$	2-Methylnaphthalene	513.90	0.2074	101.33	OX
C_9H_7N		**Quinoline**				
	$C_9H_{12}O$	3-Isopropylphenol	514.70	0.6109	101.33	ON
	$C_9H_{12}O$	2-Isopropylphenol	512.75	0.8015	101.33	ON
	$C_9H_{12}O$	2-Propylphenol	513.60	0.7243	101.33	ON
	$C_9H_{12}O$	3-Propylphenol	514.70	0.6109	101.33	ON
	$C_9H_{12}O$	4-Propylphenol	515.35	0.5451	101.33	ON
	$C_{10}H_{14}O$	2-Butylphenol	515.70	0.5350	101.33	ON
	$C_{10}H_{14}O$	2-*tert*-Butylphenol	513.70	0.7299	101.33	ON
	$C_{10}H_{14}O$	3-*tert*-Butylphenol	517.05	0.4315	101.33	ON
	$C_{10}H_{14}O$	4-Isobutylphenol	515.95	0.5061	101.33	ON
	$C_{10}H_{14}O$	2-*sec*-Butylphenol	514.70	0.6339	101.33	ON
	$C_{10}H_{14}O$	4-*sec*-Butylphenol	516.45	0.4551	101.33	ON
	$C_{11}H_{10}$	2-Methylnaphthalene	511.05	0.9213	101.33	OX
	$C_{11}H_{16}O$	2-*tert*-Butyl-5-methylphenol	515.45	0.5854	101.33	ON
	$C_{11}H_{16}O$	2-*sec*-Butyl-4-methylphenol	516.10	0.5139	101.33	ON
C_9H_{12}		**1,2,3-Trimethylbenzene**				
	$C_{10}H_{22}$	Decane	433.35	0.4010	72.54	OX
C_9H_{12}		**1,2,4-Trimethylbenzene**				
	$C_{10}H_{22}$	Decane	433.35	0.8600	80.25	OX
$C_9H_{12}O$		**2-Ethyl-4-methylphenol**				
	$C_{10}H_8$	Naphthalene	488.20	0.2218	101.33	OX
$C_9H_{12}O$		**2-Ethyl-5-methylphenol**				
	$C_{10}H_8$	Naphthalene	489.45	0.1710	101.33	OX
$C_9H_{12}O$		**2-Isopropylphenol**				
	$C_{10}H_8$	Naphthalene	483.15	0.5102	101.33	OX
	$C_{10}H_{14}$	1,2,4,5-Tetramethylbenzene	476.25	0.1036	101.33	OX
$C_9H_{12}O$		**2,4,6-Trimethylphenol**				
	$C_{10}H_8$	Naphthalene	486.70	0.3161	101.33	OX
$C_9H_{20}O$		**1-Nonanol**				
	$C_{11}H_{24}$	Undecane	468.45	0.0925	101.33	OX
	$C_{12}H_{26}$	Dodecane	480.65	0.5235	101.33	OX
$C_{10}H_{22}O$		**1-Decanol**				
	$C_{12}H_{26}$	Dodecane	489.25	0.1068	101.33	OX

VISCOSITY OF GASES

The following table gives the viscosity of some common gases as a function of temperature. Unless otherwise noted, the viscosity values refer to a pressure of 100 kPa (1 bar). The notation $P = 0$ indicates the low pressure limiting value is given. The difference between the viscosity at 100 kPa and the limiting value is generally less than 1%. Viscosity is given in units of μPa s; note that 1 μPa s = 10^{-5} poise. Substances are listed in the modified Hill order (see Introduction).

		Viscosity in micropascal seconds (μPa s)						
		100 K	200 K	300 K	400 K	500 K	600 K	Ref.
	Air	7.1	13.3	18.6	23.1	27.1	30.8	1
Ar	Argon	8.0	15.9	22.9	28.8	34.2	39.0	2,8
BF_3	Boron trifluoride		12.3	17.1	21.7	26.1	30.2	13
ClH	Hydrogen chloride			14.6	19.7	24.3		13
F_6S	Sulfur hexafluoride ($P = 0$)			15.3	19.8	23.9	27.7	10
H_2	Hydrogen ($P = 0$)	4.2	6.8	9.0	10.9	12.7	14.4	4
D_2	Deuterium ($P = 0$)	5.9	9.6	12.6	15.4	17.9	20.3	11
H_2O	Water			10.0	13.3	17.3	21.4	6
D_2O	Deuterium oxide			11.1	13.7	17.7	22.0	7
He	Helium ($P = 0$)	9.7	15.3	20.0	24.4	28.4	32.3	8
Kr	Krypton ($P = 0$)	8.8	17.1	25.6	33.1	39.8	45.9	8
NO	Nitric oxide		13.8	19.2	23.8	28.0	31.9	13
N_2	Nitrogen ($P = 0$)		12.9	17.9	22.2	26.1	29.6	12
N_2O	Nitrous oxide		10.0	15.0	19.4	23.6	27.4	13
Ne	Neon ($P = 0$)	14.4	24.3	32.1	38.9	45.0	50.8	8
O_2	Oxygen ($P = 0$)	7.5	14.6	20.8	26.1	30.8	35.1	12
O_2S	Sulfur dioxide		8.6	12.9	17.5	21.7		13
Xe	Xenon ($P = 0$)	8.3	15.4	23.2	30.7	37.6	44.0	8
CO	Carbon monoxide	6.7	12.9	17.8	22.1	25.8	29.1	13
CO_2	Carbon dioxide		10.0	15.0	19.7	24.0	28.0	9,10
$CHCl_3$	Chloroform			10.2	13.7	16.9	20.1	13
CH_4	Methane		7.7	11.2	14.3	17.0	19.4	10
CH_4O	Methanol				13.2	16.5	19.6	13
C_2H_2	Acetylene			10.4	13.5	16.5		13
C_2H_4	Ethylene		7.0	10.4	13.6	16.5	19.1	3
C_2H_6	Ethane		6.4	9.5	12.3	14.9	17.3	5
C_2H_6O	Ethanol				11.6	14.5	17.0	13
C_3H_8	Propane			8.3	10.9	13.4	15.8	5
C_4H_{10}	Butane			7.5	10.0	12.3	14.6	5
C_4H_{10}	Isobutane			7.6	10.0	12.3	14.6	5
$C_4H_{10}O$	Diethyl ether			7.6	10.1	12.4		13
C_5H_{12}	Pentane			6.7	9.2	11.4	13.4	13
C_6H_{14}	Hexane				8.6	10.8	12.8	13

References

1. K. Kadoya, N. Matsunaga, and A. Nagashima, Viscosity and thermal conductivity of dry air in the gaseous phase, *J. Phys. Chem. Ref. Data*, 14, 947, 1985.
2. B. A. Younglove and H. J. M. Hanley, The viscosity and thermal conductivity coefficients of gaseous and liquid argon, *J. Phys. Chem. Ref. Data*, 15, 1323, 1986.
3. P. M. Holland, B. E. Eaton, and H. J. M. Hanley, A correlation of the viscosity and thermal conductivity data of gaseous and liquid ethylene, *J. Phys. Chem. Ref. Data*, 12, 917, 1983.
4. M. J. Assael, S. Mixafendi, and W. A. Wakeham, The viscosity and thermal conductivity of normal hydrogen in the limit zero density, *J. Phys. Chem. Ref. Data*, 15, 1315, 1986.
5. B. A. Younglove and J. F. Ely, Thermophysical properties of fluids. II. Methane, ethane, propane, isobutane, and normal butane, *J. Phys. Chem. Ref. Data*, 16, 577, 1987.
6. J. V. Sengers and J. T. R. Watson, Improved international formulations for the viscosity and thermal conductivity of water substance, *J. Phys. Chem. Ref. Data*, 15, 1291, 1986.
7. N. Matsunaga and A. Nagashima, Transport properties of liquid and gaseous D_2O over a wide range of temperature and pressure, *J. Phys. Chem. Ref. Data*, 12, 933, 1983.
8. J. Kestin, et al., Equilibrium and transport properties of the noble gases and their mixtures at low density, *J. Phys. Chem. Ref. Data*, 13, 299, 1984.
9. V. Vescovic, et al., The transport properties of carbon dioxide, *J. Phys. Chem. Ref. Data*, 19, 1990.
10. R. D. Trengove and W. A. Wakeham, The viscosity of carbon dioxide, methane, and sulfur hexafluoride in the limit of zero density, *J. Phys. Chem. Ref. Data*, 16, 175, 1987.
11. M. J. Assael, S. Mixafendi, and W. A. Wakeham, The viscosity of normal deuterium in the limit of zero density, *J. Phys. Chem. Ref. Data*, 16, 189, 1987.
12. W. A. Cole and W. A. Wakeham, The viscosity of nitrogen, oxygen, and their binary mixtures in the limit of zero density, *J. Phys. Chem. Ref. Data*, 14, 209, 1985.
13. C. Y. Ho, Ed., *Properties of Inorganic and Organic Fluids, CINDAS Data Series on Materials Properties*, Vol. V-1, Hemisphere Publishing Corp., New York, 1988.

VISCOSITY OF LIQUIDS

The absolute viscosity of some common liquids at temperatures between −25 and 100 °C is given in this table. Values were derived by fitting experimental data to suitable expressions for the temperature dependence. The substances are arranged by molecular formula in the modified Hill order (see Preface). All values are given in units of millipascal seconds (mPa s); this unit is identical to centipoise (cp).

Viscosity values correspond to a nominal pressure of 1 atmosphere. If a value is given at a temperature above the normal boiling point, the applicable pressure is understood to be the vapor pressure of the liquid at that temperature. A few values are given at a temperature slightly below the normal freezing point; these refer to the supercooled liquid.

The accuracy ranges from 1% in the best cases to 5 to 10% in the worst cases. Additional significant figures are included in the table to facilitate interpolation.

References

1. Viswanath, D. S. and Natarajan, G., *Data Book on the Viscosity of Liquids*, Hemisphere Publishing Corp., New York, 1989.
2. Daubert, T. E., Danner, R. P., Sibul, H. M., and Stebbins, C. C., *Physical and Thermodynamic Properties of Pure Compounds: Data Compilation*, extant 1994 (core with 4 supplements), Taylor & Francis, Bristol, PA (also available as database).
3. Ho, C. Y., Ed., *CINDAS Data Series on Material Properties*, Vol. V-1, *Properties of Inorganic and Organic Fluids*, Hemisphere Publishing Corp., New York, 1988.
4. Stephan, K. and Lucas, K., *Viscosity of Dense Fluids*, Plenum Press, New York, 1979.
5. Vargaftik, N. B., *Tables of Thermophysical Properties of Liquids and Gases*, 2nd ed., John Wiley, New York, 1975.

Molecular formula	Name	Viscosity in mPa s					
		−25 °C	0 °C	25 °C	50 °C	75 °C	100 °C
Compounds not containing carbon							
Br_2	Bromine		1.252	0.944	0.746		
Cl_3HSi	Trichlorosilane		0.415	0.326			
Cl_3P	Phosphorous trichloride	0.870	0.662	0.529	0.439		
Cl_4Si	Tetrachlorosilane			99.4	96.2		
H_2O	Water		1.793	0.890	0.547	0.378	0.282
H_4N_2	Hydrazine			0.876	0.628	0.480	0.384
Hg	Mercury			1.526	1.402	1.312	1.245
NO_2	Nitrogen dioxide		0.532	0.402			
Compounds containing carbon							
CCl_3F	Trichlorofluoromethane	0.740	0.539	0.421			
CCl_4	Tetrachloromethane		1.321	0.908	0.656	0.494	
CS_2	Carbon disulfide		0.429	0.352			
$CHBr_3$	Tribromomethane			1.857	1.367	1.029	
$CHCl_3$	Trichloromethane	0.988	0.706	0.537	0.427		
CHN	Hydrogen cyanide		0.235	0.183			
CH_2Br_2	Dibromomethane	1.948	1.320	0.980	0.779	0.652	
CH_2Cl_2	Dichloromethane	0.727	0.533	0.413			
CH_2O_2	Formic acid			1.607	1.030	0.724	0.545
CH_3I	Iodomethane		0.594	0.469			
CH_3NO	Formamide		7.114	3.343	1.833		
CH_3NO_2	Nitromethane	1.311	0.875	0.630	0.481	0.383	0.317
CH_4O	Methanol	1.258	0.793	0.544			
CH_5N	Methylamine	0.319	0.231				
$C_2Cl_3F_3$	1,1,2-Trichlorotrifluoro-ethane	1.465	0.945	0.656	0.481		
C_2Cl_4	Tetrachloroethylene		1.114	0.844	0.663	0.535	0.442
C_2HCl_3	Trichloroethylene		0.703	0.545	0.444	0.376	
C_2HCl_5	Pentachloroethane		3.761	2.254	1.491	1.061	
$C_2HF_3O_2$	Trifluoroacetic acid			0.808	0.571		
$C_2H_2Cl_2$	*cis*-1,2-Dichloroethylene	0.786	0.575	0.445			
$C_2H_2Cl_2$	*trans*-1,2-Dichloroethylene	0.522	0.398	0.317	0.261		
$C_2H_2Cl_4$	1,1,1,2-Tetrachloroethane	3.660	2.200	1.437	1.006	0.741	0.570
$C_2H_3ClF_2$	1-Chloro-1,1-difluoro-ethane	0.477	0.376				
C_2H_3ClO	Acetyl chloride			0.368	0.294		
$C_2H_3Cl_3$	1,1,1-Trichloroethane	1.847	1.161	0.793	0.578	0.428	
C_2H_3N	Acetonitrile		0.400	0.369	0.284	0.234	
$C_2H_4Br_2$	1,2-Dibromoethane			1.595	1.116	0.837	0.661

Molecular formula	Name	Viscosity in mPa s					
		–25 °C	0 °C	25 °C	50 °C	75 °C	100 °C
$C_2H_4Cl_2$	1,1-Dichloroethane			0.464	0.362		
$C_2H_4Cl_2$	1,2-Dichloroethane		1.125	0.779	0.576	0.447	
$C_2H_4O_2$	Acetic acid			1.056	0.786	0.599	0.464
$C_2H_4O_2$	Methyl formate		0.424	0.325			
C_2H_5Br	Bromoethane	0.635	0.477	0.374			
C_2H_5Cl	Chloroethane	0.416	0.319				
C_2H_5I	Iodoethane		0.723	0.556	0.444	0.365	
C_2H_5NO	*N*-Methylformamide		2.549	1.678	1.155	0.824	0.606
$C_2H_5NO_2$	Nitroethane	1.354	0.940	0.688	0.526	0.415	0.337
C_2H_6O	Ethanol	3.262	1.786	1.074	0.694	0.476	
C_2H_6OS	Dimethyl sulfoxide			1.987	1.290		
$C_2H_6O_2$	Ethylene glycol			16.1	6.554	3.340	1.975
C_2H_6S	Dimethyl sulfide		0.356	0.284			
C_2H_6S	Ethanethiol		0.364	0.287			
C_2H_7N	Dimethylamine	0.300	0.232				
C_2H_7NO	Ethanolamine			21.1	8.560	3.935	1.998
C_3H_5Br	3-Bromopropene		0.620	0.471	0.373		
C_3H_5Cl	3-Chloropropene		0.408	0.314			
C_3H_5ClO	Epichlorohydrin	2.492	1.570	1.073	0.781	0.597	0.474
C_3H_5N	Propanenitrile			0.294	0.240	0.202	
C_3H_6O	Acetone	0.540	0.395	0.306	0.247		
C_3H_6O	Allyl alcohol			1.218	0.759	0.505	
C_3H_6O	Propanal			0.321	0.249		
$C_3H_6O_2$	Ethyl formate		0.506	0.380	0.300		
$C_3H_6O_2$	Methyl acetate		0.477	0.364	0.284		
$C_3H_6O_2$	Propanoic acid		1.499	1.030	0.749	0.569	0.449
C_3H_7Br	1-Bromopropane		0.645	0.489	0.387		
C_3H_7Br	2-Bromopropane		0.612	0.458	0.359		
C_3H_7Cl	1-Chloropropane		0.436	0.334			
C_3H_7Cl	2-Chloropropane		0.401	0.303			
C_3H_7I	1-Iodopropane		0.970	0.703	0.541	0.436	0.363
C_3H_7I	2-Iodopropane		0.883	0.653	0.506	0.407	
C_3H_7NO	*N,N*-Dimethylformamide		1.176	0.794	0.624		
$C_3H_7NO_2$	1-Nitropropane	1.851	1.160	0.798	0.589	0.460	0.374
C_3H_8O	1-Propanol	8.645	3.815	1.945	1.107	0.685	
C_3H_8O	2-Propanol		4.619	2.038	1.028	0.576	
$C_3H_8O_2$	1,2-Propylene glycol		248	40.4	11.3	4.770	2.750
$C_3H_8O_3$	Glycerol			934	152	39.8	14.8
C_3H_8S	1-Propanethiol		0.503	0.385			
C_3H_8S	2-Propanethiol		0.477	0.357	0.280		
C_3H_9N	Propylamine			0.376			
C_3H_9N	Isopropylamine		0.454	0.325			
C_4H_4O	Furan	0.661	0.475	0.361			
C_4H_5N	Pyrrole		2.085	1.225	0.828	0.612	
$C_4H_6O_3$	Acetic anhydride		1.241	0.843	0.614	0.472	0.377
C_4H_7N	Butanenitrile			0.553	0.418	0.330	0.268
C_4H_8O	2-Butanone	0.720	0.533	0.405	0.315	0.249	
C_4H_8O	Tetrahydrofuran	0.849	0.605	0.456	0.359		
$C_4H_8O_2$	1,4-Dioxane			1.177	0.787	0.569	
$C_4H_8O_2$	Ethyl acetate		0.578	0.423	0.325	0.259	
$C_4H_8O_2$	Methyl propionate		0.581	0.431	0.333	0.266	
$C_4H_8O_2$	Propyl formate		0.669	0.485	0.370	0.293	
$C_4H_8O_2$	Butanoic acid		2.215	1.426	0.982	0.714	0.542
$C_4H_8O_2$	2-Methylpropanoic acid		1.857	1.226	0.863	0.639	0.492
$C_4H_8O_2S$	Sulfolane				6.280	3.818	2.559
C_4H_8S	Tetrahydrothiophene			0.973	0.912		
C_4H_9Br	1-Bromobutane		0.815	0.606	0.471	0.379	
C_4H_9Cl	1-Chlorobutane		0.556	0.422	0.329	0.261	
C_4H_9N	Pyrrolidine	1.914	1.071	0.704	0.512		
C_4H_9NO	*N,N*-Dimethylacetamide			1.927			

Molecular formula	Name	Viscosity in mPa s					
		–25 °C	0 °C	25 °C	50 °C	75 °C	100 °C
C_4H_9NO	Morpholine			2.021	1.247	0.850	0.627
$C_4H_{10}O$	1-Butanol	12.19	5.185	2.544	1.394	0.833	0.533
$C_4H_{10}O$	2-Butanol			3.096	1.332	0.698	0.419
$C_4H_{10}O$	2-Methyl-2-propanol			4.312	1.421	0.678	
$C_4H_{10}O$	Diethyl ether		0.283	0.224			
$C_4H_{10}O_3$	Diethylene glycol			30.200	11.130	4.917	2.505
$C_4H_{10}S$	Diethyl sulfide		0.558	0.422	0.331	0.267	
$C_4H_{11}N$	Butylamine		0.830	0.574	0.409	0.298	
$C_4H_{11}N$	Isobutylamine		0.770	0.571	0.367		
$C_4H_{11}N$	Diethylamine			0.319	0.239		
$C_4H_{11}NO_2$	Diethanolamine				109.5	28.7	9.100
$C_5H_4O_2$	Furfural		2.501	1.587	1.143	0.906	0.772
C_5H_5N	Pyridine		1.361	0.879	0.637	0.497	0.409
C_5H_{10}	1-Pentene	0.313	0.241	0.195			
C_5H_{10}	2-Methyl-2-butene		0.255	0.203			
C_5H_{10}	Cyclopentane		0.555	0.413	0.321		
$C_5H_{10}O$	Mesityl oxide	1.291	0.838	0.602	0.465	0.381	0.326
$C_5H_{10}O$	2-Pentanone		0.641	0.470	0.362	0.289	0.238
$C_5H_{10}O$	3-Pentanone		0.592	0.444	0.345	0.276	0.227
$C_5H_{10}O_2$	Butyl formate		0.937	0.644	0.472	0.362	0.289
$C_5H_{10}O_2$	Propyl acetate		0.768	0.544	0.406	0.316	0.255
$C_5H_{10}O_2$	Ethyl propanoate		0.691	0.501	0.380	0.299	0.242
$C_5H_{10}O_2$	Methyl butanoate		0.759	0.541	0.406	0.318	0.257
$C_5H_{10}O_2$	Methyl isobutanoate		0.672	0.488	0.373	0.296	
$C_5H_{11}N$	Piperidine			1.573	0.958	0.649	0.474
C_5H_{12}	Pentane	0.351	0.274	0.224			
C_5H_{12}	Isopentane	0.376	0.277	0.214			
$C_5H_{12}O$	1-Pentanol	25.4	8.512	3.619	1.820	1.035	0.646
$C_5H_{12}O$	2-Pentanol			3.470	1.447	0.761	0.465
$C_5H_{12}O$	3-Pentanol			4.149	1.473	0.727	0.436
$C_5H_{12}O$	2-Methyl-1-butanol			4.453	1.963	1.031	0.612
$C_5H_{12}O$	3-Methyl-1-butanol		8.627	3.692	1.842	1.031	0.631
$C_5H_{13}N$	Pentylamine		1.030	0.702	0.493	0.356	
C_6F_6	Hexafluorobenzene			2.789	1.730	1.151	
$C_6H_4Cl_2$	*o*-Dichlorobenzene		1.958	1.324	0.962	0.739	0.593
$C_6H_4Cl_2$	*m*-Dichlorobenzene		1.492	1.044	0.787	0.628	0.525
C_6H_5Br	Bromobenzene		1.560	1.074	0.798	0.627	0.512
C_6H_5Cl	Chlorobenzene	1.703	1.058	0.753	0.575	0.456	0.369
C_6H_5ClO	*o*-Chlorophenol			3.589	1.835	1.131	0.786
C_6H_5ClO	*m*-Chlorophenol				4.041		
C_6H_5F	Fluorobenzene		0.749	0.550	0.423	0.338	
C_6H_5I	Iodobenzene		2.354	1.554	1.117	0.854	0.683
$C_6H_5NO_2$	Nitrobenzene		3.036	1.863	1.262	0.918	0.704
C_6H_6	Benzene			0.604	0.436	0.335	
C_6H_6ClN	*o*-Chloroaniline			3.316	1.913	1.248	0.887
C_6H_6O	Phenol				3.437	1.784	1.099
C_6H_7N	Aniline			3.847	2.029	1.247	0.850
$C_6H_8N_2$	Phenylhydrazine			13.0	4.553	1.850	0.848
C_6H_{10}	Cyclohexene		0.882	0.625	0.467	0.364	
$C_6H_{10}O$	Cyclohexanone			2.017	1.321	0.919	0.671
$C_6H_{11}N$	Hexanenitrile			0.912	0.650	0.488	0.382
C_6H_{12}	Cyclohexane			0.894	0.615	0.447	
C_6H_{12}	Methylcyclopentane	0.927	0.653	0.479	0.364		
C_6H_{12}	1-Hexene	0.441	0.326	0.252	0.202		
$C_6H_{12}O$	Cyclohexanol			57.5	12.3	4.274	1.982
$C_6H_{12}O$	2-Hexanone	1.300	0.840	0.583	0.429	0.329	0.262
$C_6H_{12}O$	4-Methyl-2-pentanone			0.545	0.406		
$C_6H_{12}O_2$	Butyl acetate		1.002	0.685	0.500	0.383	0.305
$C_6H_{12}O_2$	Isobutyl acetate			0.676	0.493	0.370	0.286
$C_6H_{12}O_2$	Ethyl butanoate			0.639	0.453		

Molecular formula	Name	Viscosity in mPa s					
		–25 °C	0 °C	25 °C	50 °C	75 °C	100 °C
$C_6H_{12}O_2$	Diacetone alcohol	28.7	6.621	2.798	1.829	1.648	
$C_6H_{12}O_3$	Paraldehyde			1.079	0.692	0.485	0.362
$C_6H_{13}N$	Cyclohexylamine			1.944	1.169	0.782	0.565
C_6H_{14}	Hexane		0.405	0.300	0.240		
C_6H_{14}	2-Methylpentane		0.372	0.286	0.226		
C_6H_{14}	3-Methylpentane		0.395	0.306			
$C_6H_{14}O$	Dipropyl ether		0.542	0.396	0.304	0.242	
$C_6H_{14}O$	1-Hexanol			4.578	2.271	1.270	0.781
$C_6H_{15}N$	Triethylamine		0.455	0.347	0.273	0.221	
$C_6H_{15}N$	Dipropylamine		0.751	0.517	0.377	0.288	0.228
$C_6H_{15}N$	Diisopropylamine			0.393	0.300	0.237	
$C_6H_{15}NO_3$	Triethanolamine			609	114	31.5	11.7
C_7H_5N	Benzonitrile			1.267	0.883	0.662	0.524
C_7H_7Cl	*o*-Chlorotoluene		1.390	0.964	0.710	0.547	0.437
C_7H_7Cl	*m*-Chlorotoluene		1.165	0.823	0.616	0.482	0.391
C_7H_7Cl	*p*-Chlorotoluene			0.837	0.621	0.483	0.390
C_7H_8	Toluene	1.165	0.778	0.560	0.424	0.333	0.270
C_7H_8O	*o*-Cresol				3.035	1.562	0.961
C_7H_8O	*m*-Cresol			12.9	4.417	2.093	1.207
C_7H_8O	Benzyl alcohol			5.474	2.760	1.618	1.055
C_7H_8O	Anisole			1.056	0.747	0.554	0.427
C_7H_9N	*N*-Methylaniline		4.120	2.042	1.222	0.825	0.606
C_7H_9N	*o*-Methyl aniline		10.3	3.823	1.936	1.198	0.839
C_7H_9N	*m*-Methyl aniline		8.180	3.306	1.679	1.014	0.699
C_7H_9N	Benzylamine			1.624	1.080	0.769	0.577
C_7H_{14}	Methylcyclohexane		0.991	0.679	0.501	0.390	0.316
C_7H_{14}	1-Heptene		0.441	0.340	0.273	0.226	
$C_7H_{14}O$	2-Heptanone			0.714	0.407	0.297	
$C_7H_{14}O_2$	Heptanoic acid			3.840	2.282	1.488	1.041
C_7H_{16}	Heptane	0.757	0.523	0.387	0.301	0.243	
C_7H_{16}	3-Methylhexane			0.350			
$C_7H_{16}O$	1-Heptanol			5.810	2.603	1.389	0.849
$C_7H_{16}O$	2-Heptanol			3.955	1.799	0.987	0.615
$C_7H_{16}O$	3-Heptanol				1.957	0.976	0.584
$C_7H_{16}O$	4-Heptanol			4.207	1.695	0.882	0.539
$C_7H_{17}N$	Heptylamine			1.314	0.865	0.600	0.434
C_8H_8	Styrene		1.050	0.695	0.507	0.390	0.310
C_8H_8O	Acetophenone			1.681			0.634
$C_8H_8O_2$	Methyl benzoate			1.857			
$C_8H_8O_3$	Methyl salicylate					1.102	0.815
C_8H_{10}	Ethylbenzene		0.872	0.631	0.482	0.380	0.304
C_8H_{10}	*o*-Xylene		1.084	0.760	0.561	0.432	0.345
C_8H_{10}	*m*-Xylene		0.795	0.581	0.445	0.353	0.289
C_8H_{10}	*p*-Xylene			0.603	0.457	0.359	0.290
$C_8H_{10}O$	Phenetole			1.197	0.817	0.594	0.453
$C_8H_{11}N$	*N,N*-Dimethylaniline		1.996	1.300	0.911	0.675	0.523
$C_8H_{11}N$	*N*-Ethylaniline		3.981	2.047	1.231	0.825	0.596
C_8H_{16}	Ethylcyclohexane		1.139	0.784	0.579		
$C_8H_{16}O_2$	Octanoic acid			5.020	2.656	1.654	1.147
C_8H_{18}	Octane		0.700	0.508	0.385	0.302	0.243
$C_8H_{18}O$	1-Octanol			7.288	3.232	1.681	0.991
$C_8H_{18}O$	4-Methyl-3-heptanol		1.904	1.085	0.702	0.497	0.375
$C_8H_{18}O$	5-Methyl-3-heptanol		2.052	1.178	0.762	0.536	0.401
$C_8H_{18}O$	2-Ethyl-1-hexanol		20.7	6.271	2.631	1.360	0.810
$C_8H_{18}O$	Dibutyl ether	1.417	0.918	0.637	0.466	0.356	0.281
$C_8H_{19}N$	Dibutylamine		1.509	0.918	0.619	0.449	0.345
$C_8H_{19}N$	Diisobutylamine		1.115	0.723	0.511	0.384	0.303
C_9H_7N	Quinoline			3.337	1.892	1.201	0.833
C_9H_{10}	Indane		2.230	1.357	0.931	0.692	0.545
C_9H_{12}	Cumene		1.075	0.737	0.547		

Molecular formula	Name	Viscosity in mPa s					
		−25 °C	0 °C	25 °C	50 °C	75 °C	100 °C
$C_9H_{14}O$	Isophorone		4.201	2.329	1.415	0.923	0.638
$C_9H_{18}O$	5-Nonanone			1.199	0.834	0.619	0.484
$C_9H_{18}O_2$	Nonanoic acid			7.011	3.712	2.234	1.475
C_9H_{20}	Nonane		0.964	0.665	0.488	0.375	0.300
$C_9H_{20}O$	1-Nonanol			9.123	4.032		
$C_{10}H_{10}O_4$	Dimethyl phthalate		63.2	14.4	5.309	2.824	1.980
$C_{10}H_{14}$	Butylbenzene			0.950	0.683	0.515	
$C_{10}H_{18}$	*cis*-Decahydronaphthalene	12.8	5.645	3.042	1.875	1.271	0.924
$C_{10}H_{18}$	*trans*-Decahydronaphthalene	6.192	3.243	1.948	1.289	0.917	0.689
$C_{10}H_{20}O_2$	Decanoic acid				4.327	2.651	
$C_{10}H_{22}$	Decane	2.188	1.277	0.838	0.598	0.453	0.359
$C_{10}H_{22}O$	1-Decanol			10.9	4.590		
$C_{11}H_{24}$	Undecane		1.707	1.098	0.763	0.562	0.433
$C_{12}H_{10}O$	Diphenyl ether				2.130	1.407	1.023
$C_{12}H_{26}$	Dodecane		2.277	1.383	0.930	0.673	0.514
$C_{13}H_{12}$	Diphenylmethane					1.265	0.929
$C_{13}H_{28}$	Tridecane		2.909	1.724	1.129	0.796	0.594
$C_{14}H_{30}$	Tetradecane			2.128	1.376	0.953	0.697
$C_{16}H_{22}O_4$	Dibutyl phthalate	483	66.4	16.6	6.470	3.495	2.425
$C_{16}H_{34}$	Hexadecane			3.032	1.879	1.260	0.899
$C_{18}H_{38}$	Octadecane				2.487	1.609	1.132

VISCOSITY OF CARBON DIOXIDE ALONG THE SATURATION LINE

The table below gives the viscosity of gas and liquid CO_2 along the liquid–vapor saturation line.

T/K	*P*/kPa	Gas η/μPa s	Liquid η/μPa s
205	227	10.33	
210	327	10.60	
215	465	10.87	
220	600	11.13	241.68
225	735	11.41	221.72
230	894	11.69	203.75
235	1075	11.98	187.48
240	1283	12.27	172.67
245	1519	12.58	159.13
250	1786	12.90	146.69
255	2085	13.24	135.20
260	2419	13.61	124.30
265	2790	14.02	114.63
270	3203	14.47	105.21
275	3658	14.99	96.44
280	4160	15.61	87.89
285	4712	16.37	79.64
290	5315	17.36	71.47
295	5984	18.79	63.01
300	6710	21.29	53.33
302	6997	23.52	48.30

References

1. Fenghour, A., Wakeham, W. A., and Vesovic, V., *J. Phys. Chem. Ref. Data*, 27, 31, 1998.
2. Angus, S., et al., *International Tables for the Fluid State: Carbon Dioxide*, Pergamon Press, Oxford, 1976.

VISCOSITY AND DENSITY OF AQUEOUS HYDROXIDE SOLUTIONS

The viscosity and density of aqueous hydroxide solutions at 25 °C are tabulated here as a function of concentration. Viscosity is given in millipascal second, which is equal to the c.g.s. unit centipoise (cP). The last entry in each column refers to the saturated solution.

Reference

Sipos, P. M., Hefter, G., and May, P. M., *J. Chem. Eng. Data*, 45, 613, 2000.

Viscosity in mPa s

c/mol L^{-1}	LiOH	NaOH	KOH	CsOH	$(CH_3)_4NOH$
0.5	1.017	0.997	0.937	0.91	1.017
1.0	1.169	1.116	0.990	0.94	1.186
1.5	1.340	1.248	1.050	0.97	1.430
2.0	1.537	1.396	1.116	1.03	1.762
3.0	2.050	1.754	1.269	1.19	3.031
4.0	2.734	2.228	1.448	1.41	7.238
5.0		2.867	1.657	1.67	
6.0		3.727	1.902	1.98	
7.0		4.869	2.196	2.40	
8.0		6.351	2.554	3.09	
9.0		8.230	3.005	4.31	
10.0		10.554	3.581	6.46	
11.0		13.362	4.328		
12.0		16.677	5.303		
13.0		20.503	6.577		
14.0		24.826	8.235		
15.0		29.604			
16.0		34.767			
17.0		40.212			
18.0		45.800			
19.0		51.354			
Sat.	3.311	51.911	8.526		8.850

Density in g/cm^3

c/mol L^{-1}	LiOH	NaOH	KOH	CsOH	$(CH_3)_4NOH$
0.5	1.012	1.019	1.022	1.063	0.999
1.0	1.025	1.040	1.045	1.128	1.002
1.5	1.038	1.059	1.068	1.193	1.005
2.0	1.050	1.078	1.090	1.257	1.009
3.0	1.072	1.115	1.133	1.383	1.019
4.0	1.093	1.149	1.174	1.508	1.030
5.0		1.182	1.214	1.632	
6.0		1.213	1.253	1.755	
7.0		1.243	1.290	1.876	
8.0		1.271	1.326	1.997	
9.0		1.299	1.362	2.117	
10.0		1.325	1.396	2.236	
11.0		1.350	1.429	2.354	
12.0		1.374	1.462	2.471	
13.0		1.397	1.494	2.587	
14.0		1.419	1.524	2.703	
15.0		1.441			
16.0		1.461			
17.0		1.481			
18.0		1.499			
19.0		1.517			
Sat.	1.109	1.519	1.529	2.800	1.032

VISCOSITY OF LIQUID METALS

This table gives the viscosity of several liquid metals as a function of temperature. Experimental data from some of the references were smoothed to produce the table. Viscosity is given in millipascal second (mPa s), which equals the c.g.s. unit centipoise (cP).

References

1. Shpil'rain, E. E., Yakimovich, K. A., Fomin, V. A., Skovorodjko, S. N., and Mozgovoi, A. G., in *Handbook of Thermodynamic and Transport Properties of the Alkali Metals*, Ohse, R. H., Ed., Blackwell Scientific Publishers, Oxford, 1985. [Li, Na, K, Rb, Cs]
2. Rothwell, E., *J. Inst. Metals*, 90, 389, 1961. [Al]
3. Culpin, M. F., *Proc. Phys. Soc.*, 70, 1079, 1957. [Ca]
4. *Landolt-Börnstein, Numerical Data and Functional Relationships in Science and Technology, Sixth Edition,* II/5a, *Transport Phenomena I (Viscosity and Diffusion)*, Springer-Verlag, Heidelberg, 1961. [Co, Au, Mg, Ni, Ag]
5. Spells, K. E., *Proc. Phys. Soc.*, 48, 299, 1936. [Ga]
6. Walsdorfer, H., Arpshofen, I., and Predel, B., *Z. Met.*, 79, 503, 1988. [In]

	Viscosity in mPa s					
t/°C	Lithium	Sodium	Potassium	Rubidium	Cesium	Gallium
50				0.542	0.598	1.921
100		0.687	0.441	0.435	0.469	1.608
150		0.542	0.358	0.365	0.389	1.397
200	0.566	0.451	0.303	0.316	0.334	1.245
250	0.503	0.387	0.263	0.280	0.294	1.130
300	0.453	0.341	0.234	0.252	0.264	1.040
350	0.412	0.306	0.211	0.230	0.240	0.968
400	0.379	0.278	0.193	0.212	0.221	0.909
450	0.352	0.255	0.178	0.197	0.206	0.859
500	0.328	0.237	0.166	0.185	0.192	0.817
550	0.308	0.221	0.155	0.174	0.181	0.781
600	0.290	0.208	0.146	0.165	0.171	0.750
650	0.275	0.196	0.138	0.157	0.163	0.722
700	0.261	0.186	0.132	0.150	0.156	0.698
750	0.249	0.177	0.126	0.143	0.149	0.677
800	0.238	0.170	0.120	0.138	0.143	0.657
850	0.228	0.163	0.115	0.133	0.138	0.640
900	0.219	0.156	0.111	0.128	0.134	0.624
950	0.211	0.151	0.107	0.124	0.129	0.610
1000	0.204	0.146	0.104	0.120	0.125	0.597
1050	0.197	0.141	0.101	0.117	0.122	0.585
1100	0.191	0.137	0.098	0.114	0.119	0.574
1150	0.185	0.133	0.095	0.111	0.116	
1200	0.180	0.129	0.092	0.108	0.113	
1250	0.175	0.126	0.090	0.105	0.110	
1300	0.170	0.123	0.088	0.103	0.108	
1350	0.166	0.120	0.086	0.101	0.106	
1400	0.162	0.117	0.084	0.099	0.104	
1450	0.158	0.115	0.082	0.097	0.102	
1500	0.155	0.113	0.081	0.095	0.100	
1550	0.151	0.110	0.079	0.093	0.098	
1600	0.148	0.108	0.078	0.092	0.097	
1650	0.145	0.106	0.076	0.090	0.095	
1700	0.142	0.105	0.075		0.094	
1750	0.139	0.103	0.074		0.092	
1800	0.137	0.101			0.091	
1850	0.135	0.100			0.090	
1900	0.132	0.098			0.089	
1950	0.130	0.097			0.088	
2000	0.128	0.096			0.086	

t/°C	Viscosity in mPa s							
	Aluminum	Calcium	Cobalt	Gold	Indium	Magnesium	Nickel	Silver
250					1.35			
300					1.22			
350					1.12			
400					1.04			
450					0.98			
700	1.289					1.10		
750	1.200					0.96		
800	1.115					0.84		
850	1.028	1.107				0.74		
900		0.959				0.67		
1000								3.80
1050								3.56
1100				5.130				3.31
1150				4.874				3.06
1200				4.640				2.82
1250				4.429				2.61
1300				4.240				2.42
1350								2.28
1400								2.20
1450								2.19
1500			4.15				4.35	
1550			3.89				4.09	
1600			3.64				3.87	
1650			3.41				3.67	
1700			3.20				3.49	
1750			2.99				3.32	

THERMAL CONDUCTIVITY OF GASES

This table gives the thermal conductivity of several gases as a function of temperature. Unless otherwise noted, the values refer to a pressure of 100 kPa (1 bar) or to the saturation vapor pressure if that is less than 100 kPa. The notation $P = 0$ indicates the low pressure limiting value is given. In general, the $P = 0$ and $P = 100$ kPa values differ by less than 1%. Units are milliwatts per meter kelvin. Substances are listed in the modified Hill order.

MF	Name	Thermal conductivity in mW/m K						Ref.
		100 K	200 K	300 K	400 K	500 K	600 K	
	Air	9.4	18.4	26.2	33.3	39.7	45.7	1
Ar	Argon	6.2	12.4	17.9	22.6	26.8	30.6	2,8
BF_3	Boron trifluoride			19.0	24.6			11
ClH	Hydrogen chloride		9.2	14.5	19.5	24.0	28.1	11
F_6S	Sulfur hexafluoride ($P = 0$)			13.0	20.6	27.5	33.8	16
H_2	Hydrogen ($P = 0$)	68.6	131.7	186.9	230.4			4
H_2O	Water			18.7	27.1	35.7	47.1	6
	Deuterium oxide				27.0	36.5	47.6	7
H_2S	Hydrogen sulfide			14.6	20.5	26.4	32.4	11
H_3N	Ammonia			24.4	37.4	51.6	66.8	11
He	Helium ($P = 0$)	75.5	119.3	156.7	190.6	222.3	252.4	8
Kr	Krypton ($P = 0$)	3.3	6.4	9.5	12.3	14.8	17.1	8
NO	Nitric oxide		17.8	25.9	33.1	39.6	46.2	11
N_2	Nitrogen	9.8	18.7	26.0	32.3	38.3	44.0	12
N_2O	Nitrous oxide		9.8	17.4	26.0	34.1	41.8	11
Ne	Neon ($P = 0$)	22.3	37.6	49.8	60.3	69.9	78.7	8
O_2	Oxygen	9.3	18.4	26.3	33.7	41.0	48.1	10
O_2S	Sulfur dioxide			9.6	14.3	20.0	25.6	11
Xe	Xenon ($P = 0$)	2.0	3.6	5.5	7.3	8.9	10.4	8
CCl_2F_2	Dichlorodifluoromethane			9.9	15.0	20.1	25.2	13
CF_4	Tetrafluoromethane ($P = 0$)			16.0	24.1	32.2	39.9	16
CO	Carbon monoxide ($P = 0$)			25.0	32.3	39.2	45.7	14
CO_2	Carbon dioxide		9.6	16.8	25.1	33.5	41.6	9
$CHCl_3$	Trichloromethane			7.5	11.1	15.1		11
CH_4	Methane		22.5	34.1	49.1	66.5	84.1	5,15
CH_4O	Methanol				26.2	38.6	53.0	11
$C_2Cl_2F_4$	1,2-Dichlorotetrafluoroethane			10.25	15.7	21.1		13
$C_2Cl_3F_3$	1,1,2-Trichlorotrifluoroethane			9.0	13.6	18.3		13
C_2H_2	Acetylene			21.4	33.3	45.4	56.8	11
C_2H_4	Ethylene		11.1	20.5	34.6	49.9	68.6	3
C_2H_6	Ethane		11.0	21.3	35.4	52.2	70.5	5
C_2H_6O	Ethanol			14.4	25.8	38.4	53.2	11
C_3H_6O	Acetone			11.5	20.2	30.6	42.7	11
C_3H_8	Propane			18.0	30.6	45.5	61.9	5
C_4F_8	Perfluorocyclobutane			12.5	19.5			13
C_4H_{10}	Butane			16.4	28.4	43.0	59.1	5
C_4H_{10}	Isobutane			16.1	27.9	42.1	57.6	5
$C_4H_{10}O$	Diethyl ether			15.1	25.0	37.1		11
C_5H_{12}	Pentane			14.4	24.9	37.8	52.7	11
C_6H_{14}	Hexane				23.4	35.4	48.7	11

References

1. Kadoya, K. Matsunaga, N., and Nagashima, A., Viscosity and thermal conductivity of dry air in the gaseous phase, *J. Phys. Chem. Ref. Data*, 14, 947, 1985.
2. Younglove, B. A. and Hanley, H. J. M., The viscosity and thermal conductivity coefficients of gaseous and liquid argon, *J. Phys. Chem. Ref. Data*, 15, 1323, 1986.
3. Holland, P. M., Eaton, B. E., and Hanley, H. J. M., A correlation of the viscosity and thermal conductivity data of gaseous and liquid ethylene, *J. Phys. Chem. Ref. Data*, 12, 917, 1983.
4. Assael, M. J., Mixafendi, S., and Wakeham, W. A., The viscosity and thermal conductivity of normal hydrogen in the limit of zero density, *J. Phys. Chem. Ref. Data*, 15, 1315, 1986.
5. Younglove, B. A. and Ely, J. F., Thermophysical properties of fluids. II. Methane, ethane, propane, isobutane, and normal butane, *J. Phys. Chem. Ref. Data*, 16, 577, 1987.

6. Sengers, J. V. and Watson, J. T. R., Improved international formulations for the viscosity and thermal conductivity of water substance, *J. Phys. Chem. Ref. Data*, 15, 1291, 1986.
7. Matsunaga, N. and Nagashima, A., Transport properties of liquid and gaseous D_2O over a wide range of temperature and pressure, *J. Phys. Chem. Ref. Data*, 12, 933, 1983.
8. Kestin, J. et al., Equilibrium and transport properties of the noble gases and their mixtures at low density, *J. Phys. Chem. Ref. Data*, 13, 229, 1984.
9. Vescovic, V. et al., The transport properties of carbon dioxide, *J. Phys. Chem. Ref. Data*, 19, 1990.
10. Younglove, B. A., Thermophysical properties of fluids. I. Argon, ethylene, parahydrogen, nitrogen, nitrogen trifluoride, and oxygen, *J. Phys. Chem. Ref. Data*, 11, Suppl. 1, 1982.
11. Ho, C. Y., Ed., *Properties of Inorganic and Organic Fluids, CINDAS Data Series on Materials Properties*, Volume V-1, Hemisphere Publishing Corp., New York, 1988.
12. Stephen, K., Krauss, R., and Laesecke, A., Viscosity and thermal conductivity of nitrogen for a wide range of fluid states, *J. Phys. Chem. Ref. Data*, 16, 993, 1987.
13. Krauss, R. and Stephan, K., Thermal conductivity of refrigerants in a wide range of temperature and pressure, *J. Phys. Chem. Ref. Data*, 18, 43, 1989.
14. Millat, J. and Wakeham, W. A., The thermal conductivity of nitrogen and carbon monoxide in the limit of zero density, *J. Phys. Chem. Ref. Data*, 18, 565, 1989.
15. Friend, D. G., Ely, J. F., and Ingham, H., Thermophysical properties of methane, *J. Phys. Chem. Ref. Data*, 18, 583, 1989.
16. Uribe, F. J., Mason, E. A., and Kestin, J., Thermal conductivity of nine polyatomic gases at low density, *J. Phys. Chem. Ref. Data*, 19, 1123, 1990.

THERMAL CONDUCTIVITY OF LIQUIDS

This table gives the thermal conductivity of about 275 liquids at temperatures between −25 and 100 °C. Values refer to nominal atmospheric pressure; when an entry is given for a temperature above the normal boiling point of the liquid, the pressure is understood to be the saturation vapor pressure at that temperature. Reference 1 contains data on many of these liquids at high pressures. Data on halocarbon refrigerants over a wide range of temperature and pressure may be found in Reference 6.

Values given to three decimal places (i.e., to 0.001 W/m K) have an uncertainty of 2% to 5%. Values given to 0.0001 W/m K should be accurate to 1% or better.

Substances are arranged by molecular formula in Hill order, except that compounds not containing carbon precede those that do contain carbon.

References

1. Vargaftik, N. B., Filippov, L. P., Tarzimanov, A. A., and Totskii, E. E., *Handbook of Thermal Conductivity of Liquids and Gases*, CRC Press, Boca Raton FL, 1994.
2. Daubert, T. E., Danner, R. P., Sibul, H. M., and Stebbins, C. C., *Physical and Thermodynamic Properties of Pure Compounds: Data Compilation*, extant 1994 (core with four supplements), Taylor and Francis, Bristol, PA (also available as a database).
3. Watanabe, H., *J. Chem. Eng. Data* 48, 124, 2003.
4. Watanabe, H., and Seong, D. J., *Int. J. Thermophys*. 23, 337, 2002.
5. Nieto de Castro, C. A., Li, S. F. Y., Nagashima, A., Trengove, R. D., and Wakeham, W. A., *J. Phys. Chem. Ref. Data* 15, 1073, 1986.
6. Krauss, R., and Stephan, K., *J. Phys. Chem. Ref. Data* 18, 43, 1989.
7. Assael, M. J., Ramires, M. L. V., Nieto de Castro, C. A., and Wakeham, W. A., *J. Phys. Chem. Ref. Data* 19, 113, 1990.
8. Ramires, M. L. V., Nieto de Castro, C. A., Nagasaka, Y., Nagashima, A., Assael, M. J., and Wakeham, W. A., *J. Phys. Chem. Ref. Data* 24, 1377, 1995.
9. Ramires, M. L. V., Nieto de Castro, C. A., Perkins, R. A., Nagasaka, Y., Nagashima, A., Assael, M. J., and Wakeham, W. A., *J. Phys. Chem. Ref. Data* 29, 133, 2000.
10. Marsh, K. N., Ed., *Recommended Reference Materials for the Realization of Physicochemical Properties*, Blackwell Scientific Publications, Oxford, 1987.
11. Beaton, C. F., and Hewitt, G. F., *Physical Property Data for the Design Engineer*, Hemisphere Publishing Corp., New York, 1989.

Molecular formula	Name	Thermal conductivity in W/m K						
		−25 °C	0° C	25 °C	50 °C	75 °C	100 °C	Ref.
Cl_4Ge	Germanium(IV) chloride	0.111	0.105	0.100	0.095	0.090	0.084	1
Cl_4Si	Tetrachlorosilane			0.099	0.096			2
Cl_4Sn	Tin(IV) chloride	0.123	0.117	0.112	0.106	0.101	0.095	1
Cl_4Ti	Titanium(IV) chloride		0.143	0.138	0.134	0.129	0.124	1
H_2O	Water		0.5562	0.6062	0.6423	0.6643	0.6729	8
Hg	Mercury	7.85	8.175	8.514	8.842	9.161	9.475	11
CCl_3F	Trichlorofluoromethane	0.102	0.096	0.089	0.083	0.076	0.070	1
CCl_4	Tetrachloromethane		0.109	0.103	0.098	0.092	0.087	1
$CHCl_3$	Trichloromethane	0.127	0.122	0.117	0.112	0.107	0.102	2
CH_2Br_2	Dibromomethane	0.120	0.114	0.108	0.103	0.097		2
CH_2Cl_2	Dichloromethane	0.158	0.149	0.140	0.133	0.128	0.127	1
CH_2I_2	Diiodomethane			0.098	0.093	0.088	0.083	1
CH_2O_2	Formic acid			0.267	0.265	0.263	0.261	1
CH_3NO_2	Nitromethane	0.226	0.215	0.204	0.193	0.182	0.171	1
CH_4O	Methanol	0.218	0.210	0.202	0.195	0.189	0.182	1
CS_2	Carbon disulfide		0.154	0.149				2
$C_2Br_2F_4$	1,2-Dibromotetrafluoroethane	0.071	0.066	0.061	0.057	0.053	0.049	1
$C_2Cl_3F_3$	1,1,2-Trichloro-1,2,2-trifluoroethane	0.0847	0.0790	0.0736	0.0683			6
C_2Cl_4	Tetrachloroethene		0.117	0.110	0.104	0.098	0.093	1
$C_2Cl_4F_2$	1,1,2,2-Tetrachloro-1,2-difluoroethane			0.082	0.078	0.074	0.069	1
C_2HCl_3	Trichloroethene	0.128	0.121	0.114	0.106	0.098	0.090	1
$C_2H_2Cl_4$	1,1,2,2-Tetrachloroethane	0.124	0.118	0.111	0.104	0.098	0.091	1
$C_2H_3Cl_3$	1,1,1-Trichloroethane		0.106	0.101	0.096			2
C_2H_3N	Acetonitrile	0.208	0.198	0.188	0.178	0.168		2
$C_2H_4Br_2$	1,2-Dibromoethane			0.100	0.096	0.092	0.088	1
$C_2H_4Cl_2$	1,2-Dichloroethane	0.144	0.139	0.133	0.128	0.122	0.117	1
$C_2H_4O_2$	Acetic acid			0.158	0.153	0.149	0.144	2
$C_2H_4O_2$	Methyl formate		0.194	0.187				1
C_2H_5Br	Bromoethane	0.107	0.104	0.101				1
C_2H_5Cl	Chloroethane	0.145	0.132	0.119	0.106	0.093		2
C_2H_5I	Iodoethane		0.091	0.087	0.083	0.079		1
C_2H_5NO	*N*-Methylformamide			0.203	0.201	0.199	0.196	2
$C_2H_5NO_2$	Nitroethane			0.173	0.161	0.149		1
C_2H_6O	Ethanol	0.181	0.174	0.167	0.160	0.153	0.148	1
$C_2H_6O_2$	1,2-Ethanediol		0.248	0.254	0.258	0.261	0.261	1
C_2H_7NO	Ethanolamine			0.240	0.238	0.236		1

Molecular formula	Name	Thermal conductivity in W/m K						Ref.
		–25 °C	0° C	25 °C	50 °C	75 °C	100 °C	
C_3F_8	Perfluoropropane	0.062	0.056	0.051	0.046	0.041	0.035	1
C_3H_3N	Acrylonitrile	0.186	0.176	0.166	0.156	0.146	0.136	1
C_3H_5ClO	Epichlorohydrin	0.142	0.137	0.131	0.125	0.119	0.114	2
C_3H_6O	Allyl alcohol			0.162				1
C_3H_6O	Acetone		0.169	0.161				2
C_3H_6O	Methyloxirane		0.181	0.171				1
$C_3H_6O_2$	Propanoic acid		0.147	0.144	0.141	0.139	0.136	1
$C_3H_6O_2$	Ethyl formate	0.181	0.171	0.160	0.149	0.138		1
$C_3H_6O_2$	Methyl acetate	0.174	0.164	0.153	0.143	0.133	0.122	2
C_3H_7Br	1-Bromopropane	0.108	0.104	0.099	0.094			1
C_3H_7Cl	1-Chloropropane	0.129	0.123	0.116	0.110	0.104	0.098	1
C_3H_7I	1-Iodopropane	0.096	0.092	0.087	0.083	0.078	0.074	1
C_3H_7I	2-Iodopropane	0.089	0.085	0.082	0.078	0.074	0.071	1
C_3H_7NO	*N*,*N*-Dimethylformamide			0.183	0.175	0.167	0.159	1
$C_3H_7NO_2$	1-Nitropropane			0.152	0.144	0.137		1
C_3H_8O	1-Propanol	0.162	0.158	0.154	0.149	0.145	0.141	2
C_3H_8O	2-Propanol	0.146	0.141	0.135	0.129	0.124	0.118	2
$C_3H_8O_2$	1,2-Propanediol	0.199	0.200	0.200	0.200	0.199	0.197	1
$C_3H_8O_2$	2-Methoxyethanol			0.190	0.180	0.170		1
$C_3H_8O_3$	Glycerol			0.285	0.288	0.292	0.296	1
C_3H_9N	Trimethylamine	0.143	0.133					2
C_4F_8	Perfluorocyclobutane	0.082	0.072	0.063	0.053	0.044	0.034	1
C_4H_4O	Furan	0.142	0.134	0.126				2
C_4H_4S	Thiophene			0.199	0.195	0.191	0.186	2
C_4H_6	1,2-Butadiene	0.147	0.134					1
C_4H_6	2-Butyne	0.137	0.129	0.121				2
$C_4H_6O_2$	Vinyl acetate			0.151	0.141	0.131	0.120	1
$C_4H_6O_3$	Acetic anhydride		0.170	0.164	0.158	0.152	0.146	1
C_4H_8O	Butanal		0.155	0.147	0.140	0.132		1
C_4H_8O	2-Butanone	0.158	0.151	0.145	0.139	0.133		2
C_4H_8O	Tetrahydrofuran	0.132	0.126	0.120	0.114			2
$C_4H_8O_2$	Propyl formate		0.151	0.144	0.137	0.130		1
$C_4H_8O_2$	Ethyl acetate		0.151	0.144	0.136			1
$C_4H_8O_2$	Methyl propanoate			0.141	0.137			1
$C_4H_8O_2$	1,4-Dioxane			0.159	0.147	0.135	0.123	2
C_4H_9Br	1-Bromobutane	0.112	0.107	0.103	0.098	0.093	0.088	1
C_4H_9I	1-Iodobutane		0.094	0.090	0.085	0.081	0.077	1
C_4H_9NO	*N*,*N*-Dimethylacetamide			0.175	0.172	0.168		1
$C_4H_{10}O$	1-Butanol		0.158	0.153	0.147	0.142	0.137	1
$C_4H_{10}O$	2-Methyl-2-propanol			0.112	0.110	0.109	0.108	1
$C_4H_{10}O$	Diethyl ether	0.150	0.140	0.130	0.120	0.110	0.100	2
$C_4H_{10}O_2$	2-Ethoxyethanol			0.190	0.182	0.174	0.165	1
C_5H_5N	Pyridine		0.171	0.166	0.162	0.157	0.153	1
$C_5H_6O_2$	Furfuryl alcohol			0.179				1
C_5H_8	2-Methyl-1,3-butadiene	0.141	0.130	0.119				1
C_5H_8	1-Pentyne	0.144	0.136	0.127	0.119			1
C_5H_8	Cyclopentene	0.143	0.136	0.129				2
$C_5H_8O_2$	Methyl methacrylate		0.156	0.147	0.137	0.127	0.117	1
$C_5H_8O_2$	2,4-Pentanedione			0.154	0.150	0.146	0.143	1
C_5H_9NO	*N*-Methyl-2-pyrrolidone			0.167	0.162	0.157		1
C_5H_{10}	1-Pentene	0.131	0.124	0.116				2
C_5H_{10}	Cyclopentane	0.140	0.133	0.126				2
$C_5H_{10}O$	Pentanal		0.146	0.139	0.133	0.127	0.121	1
$C_5H_{10}O$	2-Pentanone		0.149	0.142	0.135	0.128	0.121	1
$C_5H_{10}O$	3-Pentanone		0.151	0.144	0.137	0.129	0.122	1
$C_5H_{10}O_2$	Pentanoic acid			0.140	0.137	0.133	0.130	1
$C_5H_{10}O_2$	Butyl formate			0.136	0.130	0.123	0.117	1
$C_5H_{10}O_2$	Propyl acetate		0.146	0.140	0.135	0.130	0.124	1
$C_5H_{10}O_2$	Ethyl propanoate				0.133	0.121		1
$C_5H_{10}O_2$	Methyl butanoate			0.140				1
$C_5H_{11}Br$	1-Bromopentane	0.113	0.109	0.105	0.101	0.097	0.093	1
$C_5H_{11}Cl$	1-Chloropentane		0.125	0.120	0.115	0.109		1

Molecular formula	Name	Thermal conductivity in W/m K –25 °C	0° C	25 °C	50 °C	75 °C	100 °C	Ref.
$C_5H_{11}I$	1-Iodopentane		0.096	0.092	0.088	0.084	0.081	1
C_5H_{12}	Pentane	0.130	0.1207	0.1113	0.1018	0.0923	0.083	4
C_5H_{12}	Isopentane			0.111				1
$C_5H_{12}O$	1-Pentanol	0.159	0.155	0.150	0.145	0.141	0.136	1
$C_5H_{12}O$	2-Methyl-2-butanol		0.119	0.116	0.113	0.109	0.106	1
$C_5H_{12}O_2$	1,5-Pentanediol		0.221	0.222				1
$C_5H_{12}O_3$	Diethylene glycol monomethyl ether			0.190	0.185	0.180	0.175	1
C_6F_6	Hexafluorobenzene				0.083			1
C_6F_{14}	Perfluorohexane		0.067	0.065	0.064			1
$C_6H_3Cl_3$	1,2,3-Trichlorobenzene				0.110	0.108	0.106	1
$C_6H_3Cl_3$	1,2,4-Trichlorobenzene			0.112	0.109	0.106		1
$C_6H_4Cl_2$	*o*-Dichlorobenzene		0.125	0.121	0.117	0.113	0.109	1
$C_6H_4Cl_2$	*m*-Dichlorobenzene		0.120	0.116	0.113	0.109		1
$C_6H_4Cl_2$	*p*-Dichlorobenzene				0.112	0.108	0.105	1
C_6H_5Br	Bromobenzene	0.119	0.115	0.111	0.107	0.103	0.099	1
C_6H_5Cl	Chlorobenzene	0.137	0.132	0.127	0.123	0.118	0.113	1
C_6H_5F	Fluorobenzene			0.136	0.131	0.126		1
C_6H_5I	Iodobenzene	0.106	0.103	0.101	0.098	0.095	0.092	1
$C_6H_5NO_2$	Nitrobenzene			0.149	0.145	0.142	0.139	1
C_6H_6	Benzene			0.1411	0.1329	0.1247		7
C_6H_6ClN	2-Chloroaniline			0.148				1
C_6H_6O	Phenol				0.153	0.149	0.147	1
C_6H_7N	Aniline		0.175					1
$C_6H_8N_2$	Hexanedinitrile			0.174	0.168			1
C_6H_{10}	Cyclohexene	0.142	0.136	0.130	0.124	0.118		2
$C_6H_{10}O$	Cyclohexanone			0.138	0.134	0.130	0.126	1
$C_6H_{10}O$	Mesityl oxide	0.170	0.163	0.156	0.149	0.142	0.134	2
$C_6H_{10}O_3$	Ethyl acetoacetate			0.155	0.152	0.148	0.144	1
$C_6H_{10}O_4$	Diethyl oxalate			0.157				1
C_6H_{12}	1-Hexene	0.138	0.129	0.121	0.113			1
C_6H_{12}	Cyclohexane			0.123	0.117	0.111		2
$C_6H_{12}O$	2-Hexanone		0.156	0.145	0.134	0.124	0.115	1
$C_6H_{12}O$	Cyclohexanol			0.138	0.134	0.130	0.126	1
$C_6H_{12}O_2$	Hexanoic acid		0.148	0.142	0.137	0.131		1
$C_6H_{12}O_2$	Butyl acetate		0.143	0.136	0.130	0.123	0.116	1
$C_6H_{12}O_2$	Propyl propanoate				0.133			1
$C_6H_{12}O_2$	Ethyl butanoate		0.143	0.137	0.131	0.126		1
$C_6H_{12}O_2$	Methyl pentanoate		0.143	0.138	0.132	0.127		1
$C_6H_{12}O_3$	Paraldehyde			0.130				1
$C_6H_{13}Br$	1-Bromohexane	0.115	0.111	0.108	0.104	0.101	0.097	1
$C_6H_{13}I$	1-Iodohexane		0.098	0.095	0.091	0.088	0.084	1
C_6H_{14}	Hexane	0.133	0.1250	0.1167	0.1083	0.0999	0.092	4
C_6H_{14}	2-Methylpentane	0.120	0.1127	0.1050	0.0972	0.0894	0.082	3
C_6H_{14}	3-Methylpentane	0.122	0.1142	0.1064	0.0986	0.0909	0.083	3
C_6H_{14}	2,2-Dimethylbutane	0.108	0.1006	0.0934	0.0861	0.0788	0.072	3
C_6H_{14}	2,3-Dimethylbutane	0.115	0.1076	0.1003	0.0930	0.0857	0.078	3
$C_6H_{14}O$	1-Hexanol	0.161	0.157	0.152	0.147	0.142	0.137	1
$C_6H_{14}O$	Dipropyl ether		0.137	0.130	0.123	0.117		1
$C_6H_{14}O_2$	1,2-Diethoxyethane				0.140	0.133	0.125	1
$C_6H_{14}O_3$	Diethylene glycol monoethyl ether				0.188	0.184	0.180	1
$C_6H_{14}O_4$	Triethylene glycol		0.193	0.195	0.196	0.196	0.196	1
$C_6H_{15}N$	Triethylamine	0.146	0.139	0.132	0.125	0.118	0.111	1
C_7F_{16}	Perfluoroheptane	0.068	0.064	0.060	0.056	0.053		1
C_7H_5N	Benzonitrile			0.148	0.142	0.136	0.130	1
C_7H_6O	Benzaldehyde			0.153	0.148	0.143	0.139	1
C_7H_8	Toluene	0.1455	0.1385	0.1310	0.1235	0.1162	0.1095	9
C_7H_8O	*o*-Cresol							1
C_7H_8O	*m*-Cresol			0.149	0.147	0.145		1
C_7H_8O	Benzyl alcohol			0.159	0.158	0.156	0.154	1
C_7H_8O	Anisole			0.145	0.142	0.139	0.136	1
C_7H_9N	2-Methylaniline			0.162				1
C_7H_9N	3-Methylaniline			0.161				1

Molecular formula	Name	Thermal conductivity in W/m K						
		–25 °C	0° C	25 °C	50 °C	75 °C	100 °C	Ref.
C_7H_{14}	1-Heptene	0.139	0.132	0.125	0.118	0.111		1
C_7H_{14}	Cycloheptane			0.123	0.118	0.112	0.108	1
$C_7H_{14}O$	Heptanal			0.140				1
$C_7H_{14}O$	3-Heptanone		0.143	0.137	0.131	0.125	0.119	1
$C_7H_{14}O$	4-Heptanone			0.136	0.131	0.125	0.120	1
$C_7H_{14}O_2$	Hexyl formate			0.141	0.133	0.126	0.119	1
$C_7H_{14}O_2$	Heptanoic acid			0.140	0.137	0.133		1
$C_7H_{14}O_2$	Pentyl acetate		0.141	0.134	0.126	0.120	0.113	1
$C_7H_{14}O_2$	Butyl propanoate			0.139	0.133	0.126	0.121	1
$C_7H_{14}O_2$	Ethyl pentanoate			0.132				1
$C_7H_{14}O_2$	Methyl hexanoate			0.136	0.131	0.126	0.121	1
C_7H_{16}	Heptane	0.1378	0.1303	0.1228	0.1152	0.1077		5
C_7H_{16}	2-Methylhexane	0.125	0.1177	0.1105	0.1033	0.0961	0.089	3
C_7H_{16}	3-Methylhexane	0.126	0.1184	0.1112	0.1040	0.0968	0.090	3
C_7H_{16}	3-Ethylpentane	0.128	0.1203	0.1128	0.1053	0.0978	0.090	3
C_7H_{16}	2,2-Dimethylpentane	0.111	0.1046	0.0980	0.0913	0.0847	0.078	3
C_7H_{16}	2,3-Dimethylpentane	0.120	0.1127	0.1059	0.0990	0.0922	0.085	3
C_7H_{16}	2,4-Dimethylpentane	0.116	0.1089	0.1020	0.0951	0.0882	0.081	3
C_7H_{16}	3,3-Dimethylpentane	0.113	0.1068	0.1001	0.0934	0.0867	0.080	3
C_7H_{16}	2,2,3-Trimethylbutane	0.107	0.1011	0.0950	0.0889	0.0828	0.077	3
$C_7H_{16}O$	1-Heptanol	0.160	0.158	0.153	0.149	0.144	0.139	1
C_8F_{18}	Perfluorooctane		0.066	0.062	0.059	0.055	0.052	1
C_8H_8	Styrene	0.148	0.142	0.137	0.131	0.126	0.120	2
C_8H_8O	Acetophenone			0.147	0.146	0.144	0.142	1
$C_8H_8O_2$	Methyl benzoate			0.147				1
C_8H_{10}	Ethylbenzene	0.143	0.137	0.130	0.123	0.116	0.110	1
C_8H_{10}	*o*-Xylene			0.131	0.126	0.120	0.114	2
C_8H_{10}	*m*-Xylene			0.130	0.124	0.118	0.113	2
C_8H_{10}	*p*-Xylene			0.130	0.124	0.118	0.112	2
$C_8H_{10}O$	Ethoxybenzene	0.151	0.145	0.140	0.135	0.130		1
$C_8H_{10}O_2$	2-Phenoxyethanol			0.169	0.168	0.166	0.165	1
$C_8H_{11}N$	*N*-Ethylaniline			0.150				1
$C_8H_{11}N$	*N,N*-Dimethylaniline			0.122	0.119	0.115		1
C_8H_{16}	1-Octene	0.139	0.133	0.126	0.120	0.114	0.107	1
$C_8H_{16}O$	2-Octanone		0.141	0.135	0.129	0.124	0.118	1
$C_8H_{16}O_2$	Heptyl formate		0.141	0.137	0.132	0.128	0.123	1
$C_8H_{16}O_2$	Octanoic acid			0.146	0.143	0.139	0.135	1
$C_8H_{16}O_2$	Hexyl acetate			0.135	0.129	0.123	0.118	1
$C_8H_{16}O_2$	Pentyl propanoate			0.138	0.132			1
$C_8H_{16}O_2$	Ethyl hexanoate		0.142	0.137	0.133	0.128	0.123	1
$C_8H_{17}Cl$	1-Chlorooctane		0.130	0.127	0.124	0.121	0.119	1
C_8H_{18}	Octane	0.139	0.1317	0.1244	0.1171	0.1097	0.102	4
C_8H_{18}	2-Methylheptane	0.127	0.1206	0.1139	0.1072	0.1005	0.094	3
C_8H_{18}	3-Methylheptane	0.128	0.1216	0.1149	0.1081	0.1014	0.095	3
C_8H_{18}	2,2,4-Trimethylpentane	0.107	0.1007	0.0948	0.0888	0.0829	0.077	3
C_8H_{18}	2,3,4-Trimethylpentane	0.115	0.1093	0.1035	0.0976	0.0918	0.086	3
$C_8H_{18}O$	Ethyl hexyl ether		0.131	0.126	0.120	0.114	0.109	1
$C_8H_{18}O$	1-Octanol		0.162	0.158	0.153	0.148	0.143	1
$C_8H_{18}O$	Dibutyl ether		0.139	0.132	0.125	0.118	0.112	1
$C_8H_{18}O_3$	Diethylene glycol monobutyl ether			0.163	0.158	0.153	0.148	1
$C_8H_{18}O_4$	Triethylene glycol dimethyl ether			0.169	0.158	0.147		1
$C_8H_{18}O_5$	Tetraethylene glycol			0.191	0.192			1
C_9H_7N	Quinoline			0.147	0.144	0.141	0.138	1
C_9H_{10}	Indan			0.135				1
$C_9H_{10}O_2$	Ethyl benzoate			0.141				1
C_9H_{12}	Propylbenzene	0.134	0.130	0.125	0.120	0.115	0.109	1
C_9H_{12}	Isopropylbenzene	0.132	0.128	0.123	0.118	0.112	0.107	1
C_9H_{12}	1,2,4-Trimethylbenzene			0.129	0.124	0.118	0.114	1
C_9H_{12}	1,3,5-Trimethylbenzene	0.143	0.139	0.134	0.129	0.123	0.117	1
C_9H_{18}	1-Nonene	0.136	0.130	0.123	0.116	0.110	0.104	1
$C_9H_{18}O_2$	Nonanoic acid			0.150	0.146	0.142	0.138	1
$C_9H_{18}O_2$	Heptyl acetate			0.135	0.128	0.122	0.116	1

Molecular formula	Name	Thermal conductivity in W/m K						
		–25 °C	0° C	25 °C	50 °C	75 °C	100 °C	Ref.
$C_9H_{19}Br$	1-Bromononane		0.116	0.112	0.109	0.106	0.103	1
$C_9H_{19}Cl$	1-Chlorononane		0.132	0.128	0.124	0.120	0.115	1
$C_9H_{19}I$	1-Iodononane		0.105	0.102	0.099	0.095	0.092	1
C_9H_{20}	Nonane	0.141	0.1337	0.1269	0.1201	0.1133	0.106	4
$C_9H_{20}O$	1-Nonanol		0.164	0.159	0.155	0.150	0.145	1
$C_{10}H_7Br$	1-Bromonaphthalene			0.110	0.109	0.108	0.106	1
$C_{10}H_7Cl$	1-Chloronaphthalene			0.126				1
$C_{10}H_{10}O_4$	Dimethyl phthalate			0.1473	0.1443	0.1409	0.1373	10
$C_{10}H_{12}$	1,2,3,4-Tetrahydronaphthalene			0.131	0.129	0.128	0.126	1
$C_{10}H_{14}$	Butylbenzene			0.126	0.121	0.116	0.111	1
$C_{10}H_{14}$	*sec*-Butylbenzene, (±)-		0.129	0.124	0.119	0.114	0.108	1
$C_{10}H_{14}$	*tert*-Butylbenzene			0.117	0.114	0.110	0.106	1
$C_{10}H_{14}$	1-Isopropyl-4-methylbenzene	0.132	0.127	0.122	0.117	0.112	0.107	2
$C_{10}H_{14}$	*o*-Diethylbenzene		0.133	0.127	0.122	0.116	0.111	1
$C_{10}H_{18}$	*trans*-Decahydronaphthalene			0.113				1
$C_{10}H_{20}$	1-Decene	0.138	0.132	0.126	0.120	0.114	0.109	1
$C_{10}H_{20}O$	Decanal		0.149	0.144	0.139	0.134	0.129	1
$C_{10}H_{20}O_2$	Heptyl propanoate			0.137	0.132	0.127	0.122	1
$C_{10}H_{20}O_2$	Hexyl butanoate			0.137	0.132	0.127	0.121	1
$C_{10}H_{20}O_2$	Decanoic acid				0.148	0.144	0.140	1
$C_{10}H_{22}$	Decane	0.142	0.1360	0.1296	0.1232	0.1167	0.110	4
$C_{10}H_{22}O$	1-Decanol			0.162	0.159	0.155	0.151	1
$C_{10}H_{22}O$	Dipentyl ether			0.131	0.125	0.121	0.116	1
$C_{10}H_{22}O_2$	1,2-Dibutoxyethane			0.140	0.134	0.127	0.120	1
$C_{11}H_{16}$	Pentylbenzene		0.135	0.130	0.125	0.120	0.115	1
$C_{11}H_{22}$	1-Undecene			0.126	0.118	0.114	0.108	1
$C_{11}H_{22}O$	6-Undecanone			0.137	0.132	0.127		1
$C_{11}H_{22}O_2$	Undecanoic acid				0.153	0.149		1
$C_{11}H_{22}O_2$	Octyl propanoate			0.135	0.130	0.125	0.120	1
$C_{11}H_{22}O_2$	Heptyl butanoate			0.139	0.134	0.129	0.123	1
$C_{11}H_{24}$	Undecane			0.136	0.128	0.122	0.116	1
$C_{11}H_{24}O$	1-Undecanol			0.169	0.165	0.161	0.158	1
$C_{12}H_{10}O$	Diphenyl ether				0.139	0.135	0.131	2
$C_{12}H_{14}O_4$	Diethyl phthalate			0.172	0.169	0.166		1
$C_{12}H_{16}$	Cyclohexylbenzene			0.121	0.119	0.117		1
$C_{12}H_{18}$	Hexylbenzene		0.141	0.137	0.132	0.128	0.124	1
$C_{12}H_{24}O_2$	Decyl acetate			0.146	0.136	0.126		1
$C_{12}H_{24}O_2$	Octyl butanoate			0.139	0.134	0.129	0.125	1
$C_{12}H_{26}$	Dodecane			0.135	0.130	0.124	0.119	1
$C_{12}H_{26}O$	1-Dodecanol				0.167	0.163	0.159	1
$C_{12}H_{26}O_3$	Diethylene glycol dibutyl ether		0.150	0.146	0.143	0.139	0.135	1
$C_{12}H_{27}N$	Tributylamine			0.129				1
$C_{13}H_{26}$	1-Tridecene			0.130	0.125	0.120	0.115	1
$C_{13}H_{28}$	Tridecane			0.130	0.125	0.120	0.115	1
$C_{14}H_{28}$	1-Tetradecene			0.136	0.131	0.126	0.121	1
$C_{14}H_{30}$	Tetradecane			0.139	0.134	0.129	0.124	1
$C_{14}H_{30}O$	1-Tetradecanol				0.167	0.162	0.157	2
$C_{16}H_{22}O_4$	Dibutyl phthalate		0.139	0.136	0.134	0.131	0.129	1
$C_{16}H_{34}$	Hexadecane			0.140	0.135	0.130	0.125	2
$C_{18}H_{38}$	Octadecane				0.146	0.142	0.137	2
$C_{20}H_{40}O_2$	Butyl palmitate			0.151	0.148	0.144	0.140	1
$C_{22}H_{42}O_2$	Butyl oleate			0.157	0.153	0.149	0.145	1
$C_{22}H_{42}O_4$	Dioctyl hexanedioate			0.157	0.153	0.149	0.145	1

DIFFUSION IN GASES

This table gives binary diffusion coefficients D_{12} for a number of common gases as a function of temperature. Values refer to atmospheric pressure. The diffusion coefficient is inversely proportional to pressure as long as the gas is in a regime where binary collisions dominate. See Reference 1 for a discussion of the dependence of D_{12} on temperature and composition.

The first part of the table gives data for several gases in the presence of a large excess of air. The remainder applies to equimolar mixtures of gases. Each gas pair is ordered alphabetically according to the most common way of writing the formula. The listing of pairs then follows alphabetical order by the first constituent.

References

1. Marrero, T. R., and Mason, E. A., *J. Phys. Chem. Ref. Data*, 1, 1, 1972.
2. Kestin, J., et al., *J. Phys. Chem. Ref. Data*, 13, 229, 1984.

D_{12}/cm² s⁻¹ for p = 101.325 kPa and the Specified T/K

System	200	273.15	293.15	373.15	473.15	573.15	673.15
Large Excess of Air							
Ar-air		0.167	0.189	0.289	0.437	0.612	0.810
CH_4-air			0.210	0.321	0.485	0.678	0.899
CO-air			0.208	0.315	0.475	0.662	0.875
CO_2-air			0.160	0.252	0.390	0.549	0.728
H_2-air		0.668	0.756	1.153	1.747	2.444	3.238
H_2O-air			0.242	0.399	0.638	0.873	1.135
He-air		0.617	0.697	1.057	1.594	2.221	2.933
SF_6-air				0.150	0.233	0.329	0.438
Equimolar Mixture							
Ar-CH_4				0.306	0.467	0.657	0.876
Ar-CO		0.168	0.190	0.290	0.439	0.615	0.815
Ar-CO_2		0.129	0.148	0.235	0.365	0.517	0.689
Ar-H_2		0.698	0.794	1.228	1.876	2.634	3.496
Ar-He	0.381	0.645	0.726	1.088	1.617	2.226	2.911
Ar-Kr	0.064	0.117	0.134	0.210	0.323	0.456	0.605
Ar-N_2		0.168	0.190	0.290	0.439	0.615	0.815
Ar-Ne	0.160	0.277	0.313	0.475	0.710	0.979	1.283
Ar-O_2		0.166	0.187	0.285	0.430	0.600	0.793
Ar-SF_6				0.128	0.202	0.290	0.389
Ar-Xe	0.052	0.095	0.108	0.171	0.264	0.374	0.498
CH_4-H_2			0.708	1.084	1.648	2.311	3.070
CH_4-He			0.650	0.992	1.502	2.101	2.784
CH_4-N_2			0.208	0.317	0.480	0.671	0.890
CH_4-O_2			0.220	0.341	0.523	0.736	0.978
CH_4-SF_6				0.167	0.257	0.363	0.482
CO-CO_2			0.162	0.250	0.384		
CO-H_2	0.408	0.686	0.772	1.162	1.743	2.423	3.196
CO-He	0.365	0.619	0.698	1.052	1.577	2.188	2.882
CO-Kr		0.131	0.149	0.227	0.346	0.485	0.645
CO-N_2	0.133	0.208	0.231	0.336	0.491	0.673	0.878
CO-O_2			0.202	0.307	0.462	0.643	0.849
CO-SF_6				0.144	0.226	0.323	0.432
CO_2-C_3H_8			0.084	0.133	0.209		
CO_2-H_2	0.315	0.552	0.627	0.964	1.470	2.066	2.745
CO_2-H_2O			0.162	0.292	0.496	0.741	1.021
CO_2-He	0.300	0.513	0.580	0.878	1.321		
CO_2-N_2			0.160	0.253	0.392	0.553	0.733
CO_2-N_2O	0.055	0.099	0.113	0.177	0.276		
CO_2-Ne	0.131	0.227	0.258	0.395	0.603	0.847	
CO_2-O_2			0.159	0.248	0.380	0.535	0.710
CO_2-SF_6				0.099	0.155		
D_2-H_2	0.631	1.079	1.219	1.846	2.778	3.866	5.103
H_2-He	0.775	1.320	1.490	2.255	3.394	4.726	6.242
H_2-Kr	0.340	0.601	0.682	1.053	1.607	2.258	2.999

System	200	273.15	293.15	373.15	473.15	573.15	673.15
H_2-N_2	0.408	0.686	0.772	1.162	1.743	2.423	3.196
H_2-Ne	0.572	0.982	1.109	1.684	2.541	3.541	4.677
H_2-O_2		0.692	0.782	1.188	1.792	2.497	3.299
H_2-SF_6			0.412	0.649	0.998	1.400	1.851
H_2-Xe		0.513	0.581	0.890	1.349	1.885	2.493
H_2O-N_2			0.242	0.399			
H_2O-O_2			0.244	0.403	0.645	0.882	1.147
He-Kr	0.330	0.559	0.629	0.942	1.404	1.942	2.550
He-N_2	0.365	0.619	0.698	1.052	1.577	2.188	2.882
He-Ne	0.563	0.948	1.066	1.592	2.362	3.254	4.262
He-O_2		0.641	0.723	1.092	1.640	2.276	2.996
He-SF_6			0.400	0.592	0.871	1.190	1.545
He-Xe	0.282	0.478	0.538	0.807	1.201	1.655	2.168
Kr-N_2		0.131	0.149	0.227	0.346	0.485	0.645
Kr-Ne	0.131	0.228	0.258	0.392	0.587	0.812	1.063
Kr-Xe	0.035	0.064	0.073	0.116	0.181	0.257	0.344
N_2-Ne			0.317	0.483	0.731	1.021	1.351
N_2-O_2			0.202	0.307	0.462	0.643	0.849
N_2-SF_6				0.148	0.231	0.328	0.436
N_2-Xe		0.107	0.122	0.188	0.287	0.404	0.539
Ne-Xe	0.111	0.193	0.219	0.332	0.498	0.688	0.901
O_2-SF_6			0.097	0.154	0.238	0.334	0.441

DIFFUSION OF GASES IN WATER

This table gives values of the diffusion coefficient, *D*, for diffusion of several common gases in water at various temperatures. For simple one-dimensional transport, the diffusion coefficient describes the time–rate of change of concentration, d*c*/d*t*, through the equation

$$dc/dt = D\, d^2c/dx^2$$

where *x* is, for example, the perpendicular distance from a gas–liquid interface. The values below have been selected from the references indicated; in some cases data have been refitted to permit interpolation in temperature.

Gas–liquid diffusion coefficients are difficult to measure, and large differences are found between values obtained by different authors and through different experimental methods. See References 1 and 2 for a discussion of measurement techniques.

References

1. Jähne, B., Heinz, G., and Dietrich, W., *J. Geophys. Res.*, 92, 10767, 1987.
2. Himmelblau, D. M., *Chem. Rev.*, 64, 527, 1964.
3. Boerboom, A. J. H., and Kleyn, G., *J. Chem. Phys.*, 50, 1086, 1969.
4. O'Brien, R. N., and Hyslop, W. F., *Can. J. Chem.*, 55, 1415, 1977.
5. Maharajh, D. M., and Walkley, J., *Can. J. Chem.*, 51, 944, 1973.
6. *Landolt-Börnstein, Numerical Data and Functional Relationships in Science and Technology, Sixth Edition*, II/5a, *Transport Phenomena I (Viscosity and Diffusion)*, Springer-Verlag, Heidelberg, 1969.

***D*/10⁻⁵ cm² s⁻¹**

	10°C	15°C	20°C	25°C	30°C	35°C	Ref.
Ar				2.5			3,4
$CHCl_2F$				1.80			5
CH_3Br				1.35			5
CH_3Cl				1.40			5
CH_4	1.24	1.43	1.62	1.84	2.08	2.35	1
CO_2	1.26	1.45	1.67	1.91	2.17	2.47	1
C_2H_2	1.43	1.59	1.78	1.99	2.23		2
Cl_2		1.13	1.5	1.89			2,6
HBr				3.15			6
HCl				3.07			6
H_2	3.62	4.08	4.58	5.11	5.69	6.31	1
H_2S				1.36			2,6
He	5.67	6.18	6.71	7.28	7.87	8.48	1,3
Kr	1.20	1.39	1.60	1.84	2.11	2.40	1,3
NH_3		1.3	1.5				2
NO_2			1.23	1.4	1.59		2,6
N_2				2.0			2
N_2O		1.62	2.11	2.57			2,6
Ne	2.93	3.27	3.64	4.03	4.45	4.89	1,3
O_2		1.67	2.01	2.42			2,6
Rn	0.81	0.96	1.13	1.33	1.55	1.80	1
SO_2			1.62	1.83	2.07	2.32	2
Xe	0.93	1.08	1.27	1.47	1.70	1.95	1,3

DIFFUSION COEFFICIENTS IN LIQUIDS AT INFINITE DILUTION

This table lists diffusion coefficients D_{AB} at infinite dilution for some binary liquid mixtures. Values are given at 25 °C when available; it should be noted that the diffusion coefficient generally increases by 10% to 20% for a 10 °C increase above ambient temperature.

Solvents are listed in alphabetical order, as are the solutes within each solvent group.

References

1. Safi, A., Nicolas, C., Neau, E., and Chevalier, J.-L., *J. Chem. Eng. Data* 52, 977, 2007.
2. Safi, A., Nicolas, C., Neau, E., and Chevalier, J.-L., *J. Chem. Eng. Data* 52, 126, 2007.
3. Sanni, S. A., Fell, C. J. D., and Hutchison, H. P., *J. Chem. Eng. Data* 16, 424, 1971.
4. Fan, Y., Qian, R., Shi, M., and Shi, J., *J. Chem. Eng. Data* 40, 1053, 1995.
5. *Landolt-Börnstein Numerical Data and Functional Relationships in Science and Technology,* Sixth Edition, Vol. II/5a, Springer-Verlag, Heidelberg, 1969.

Solute	Solvent	t/°C	D_{AB}/ 10^{-5} cm^2s^{-1}	Ref.
Acetic acid	Acetone	25	3.31	5
Benzene	Acetone	25	4.25	2
Benzoic acid	Acetone	25	2.62	5
Formic acid	Acetone	25	3.77	5
Nitrobenzene	Acetone	20	2.94	5
Tetrachloromethane	Acetone	25	3.29	5
Trichloromethane	Acetone	25	3.64	5
Water	Acetone	25	4.56	5
Acetic acid	Benzene	25	2.09	5
Aniline	Benzene	25	1.96	5
Benzoic acid	Benzene	25	1.38	5
Bromobenzene	Benzene	8	1.45	5
2-Butanone	Benzene	30	2.09	5
Chloroethene	Benzene	8	1.77	5
Cyclohexane	Benzene	25	2.09	3
Ethanol	Benzene	25	3.02	5
Formic acid	Benzene	25	2.28	5
Heptane	Benzene	25	1.79	3
Methanol	Benzene	25	3.80	5
Toluene	Benzene	25	1.85	3
1,2,4-Trichlorobenzene	Benzene	8	1.34	5
Trichloromethane	Benzene	25	2.26	5
Benzene	1-Butanol	25	1.00	5
Biphenyl	1-Butanol	25	0.63	5
Butanoic acid	1-Butanol	30	0.51	5
p-Dichlorobenzene	1-Butanol	25	0.82	5
1,6-Hexanedioic acid	1-Butanol	30	0.40	5
Methanol	1-Butanol	30	0.59	5
cis-9-Octadecenoic acid	1-Butanol	30	0.25	5
Propane	1-Butanol	25	1.57	5
Water	1-Butanol	25	0.56	5
Benzene	Cyclohexane	25	1.92	1
Chlorobenzene	Cyclohexane	25	1.34	1
p-Chlorotoluene	Cyclohexane	25	1.28	1
Ethylbenzene	Cyclohexane	25	1.36	1
Naphthalene	Cyclohexane	25	1.18	1
Perylene	Cyclohexane	25	0.79	1
Pyrene	Cyclohexane	25	0.95	1
Tetrachloromethane	Cyclohexane	25	1.49	3
Toluene	Cyclohexane	25	1.66	1
Benzene	Decane	25	2.16	1
Chlorobenzene	Decane	25	1.98	1
p-Chlorotoluene	Decane	25	1.80	1
Ethylbenzene	Decane	25	1.79	1
Naphthalene	Decane	25	1.65	1
Perylene	Decane	25	1.08	1
Pyrene	Decane	25	1.23	1
Toluene	Decane	25	1.93	1
Trichloromethane	Diethyl ether	25	4.48	3
Allyl alcohol	Ethanol	20	0.98	5
Benzene	Ethanol	25	1.88	2
Iodine	Ethanol	25	1.32	5
Iodobenzene	Ethanol	20	1.00	5
3-Methyl-1-butanol	Ethanol	20	0.81	5
Pyridine	Ethanol	20	1.10	5
Tetrachloromethane	Ethanol	25	1.50	5
Water	Ethanol	25	1.24	5
Acetic acid	Ethyl acetate	20	2.18	5
Acetone	Ethyl acetate	20	3.18	5
2-Butanone	Ethyl acetate	30	2.93	5
Ethyl benzoate	Ethyl acetate	20	1.85	5
Nitrobenzene	Ethyl acetate	20	2.25	5
Water	Ethyl acetate	25	3.20	5
Benzene	Heptane	25	3.75	1
Chlorobenzene	Heptane	25	3.42	1
p-Chlorotoluene	Heptane	25	3.11	1
Ethylbenzene	Heptane	25	3.15	1
Naphthalene	Heptane	25	2.81	1
Perylene	Heptane	25	1.89	1
Pyrene	Heptane	25	2.16	1
Toluene	Heptane	25	3.42	1
Benzene	Hexane	25	4.70	1
Bromobenzene	Hexane	8	2.60	5

Solute	Solvent	t/°C	D_{AB}/ 10^{-5} cm^{2}s^{-1}	Ref.
2-Butanone	Hexane	30	3.74	5
Chlorobenzene	Hexane	25	4.16	1
p-Chlorotoluene	Hexane	25	3.74	1
Dodecane	Hexane	25	2.73	5
Ethylbenzene	Hexane	25	3.73	1
Iodine	Hexane	25	4.45	5
Methane	Hexane	25	0.09	5
Naphthalene	Hexane	25	3.55	1
Perylene	Hexane	25	2.30	1
Propane	Hexane	25	4.87	5
Pyrene	Hexane	25	2.58	1
Tetrachloromethane	Hexane	25	3.70	5
Toluene	Hexane	25	4.12	1
Benzene	Octane	25	3.19	1
Chlorobenzene	Octane	25	2.89	1
p-Chlorotoluene	Octane	25	2.62	1
Ethylbenzene	Octane	25	2.58	1
Naphthalene	Octane	25	2.35	1
Perylene	Octane	25	1.58	1
Pyrene	Octane	25	1.81	1
Toluene	Octane	25	2.83	1
1,3,5-Trimethylbenzene	Octane	30	2.21	4
o-Xylene	Octane	30	2.65	4
p-Xylene	Octane	30	2.82	4
Acetone	Tetrachloromethane	25	1.75	5
Benzene	Tetrachloromethane	25	1.42	5
Cyclohexane	Tetrachloromethane	25	1.30	5
Ethanol	Tetrachloromethane	25	1.90	5
Iodine	Tetrachloromethane	30	1.63	5
Trichloromethane	Tetrachloromethane	25	1.66	5
Acetic acid	Toluene	25	2.26	5
Benzene	Toluene	25	2.54	3
Benzoic acid	Toluene	25	1.49	5
Cyclohexane	Toluene	25	2.42	3
Formic acid	Toluene	25	2.65	5
Water	Toluene	25	6.19	5
Acetone	Trichloromethane	25	2.55	5
Benzene	Trichloromethane	25	2.89	5
2-Butanone	Trichloromethane	25	2.13	5
Butyl acetate	Trichloromethane	25	1.71	5
Cyclohexane	Trichloromethane	25	1.28	3
Diethyl ether	Trichloromethane	25	2.13	3
Ethanol	Trichloromethane	15	2.20	5
Ethyl acetate	Trichloromethane	25	2.02	5
Benzene	2,2,4-Trimethylpentane	30	3.46	4
Ethylbenzene	2,2,4-Trimethylpentane	30	2.63	4
Toluene	2,2,4-Trimethylpentane	30	3.15	4
1,3,5-Trimethylbenzene	2,2,4-Trimethylpentane	30	2.26	4
o-Xylene	2,2,4-Trimethylpentane	30	2.62	4
p-Xylene	2,2,4-Trimethylpentane	30	3.03	4
Acetic acid	Water	25	1.29	5
Acetone	Water	25	1.28	5
Acetonitrile	Water	15	1.26	5
L-Alanine	Water	25	0.91	5
Allyl alcohol	Water	15	0.90	5
Aniline	Water	20	0.92	5
DL-Arabinose	Water	20	0.69	5
Benzene	Water	20	1.02	5
1-Butanol	Water	25	0.56	5
Caprolactam	Water	25	0.87	5
Chloroethene	Water	25	1.34	5
Cyclohexane	Water	20	0.84	5
Diethylamine	Water	20	0.97	5
1,2-Ethanediol	Water	25	1.16	5
Ethanol	Water	25	1.24	5
Ethanolamine	Water	25	1.08	5
Ethyl acetate	Water	20	1.00	5
Ethylbenzene	Water	20	0.81	5
Ethyl carbamate	Water	15	0.80	5
α-*D*-Glucose	Water	25	0.67	5
Glycerol	Water	25	1.06	5
Glycine	Water	25	1.05	5
β-*D*-Lactose	Water	15	0.38	5
α-Maltose	Water	15	0.38	5
D-Mannitol	Water	15	0.50	5
Methane	Water	25	1.49	5
Methanol	Water	15	1.28	5
3-Methyl-1-butanol	Water	10	0.69	5
Methylcyclopentane	Water	20	0.85	5
Phenol	Water	20	0.89	5
1-Propanol	Water	15	0.87	5
Propene	Water	25	1.44	5
Pyridine	Water	25	0.58	5
Raffinose	Water	15	0.33	5
Sucrose	Water	25	0.52	5
Toluene	Water	20	0.85	5
Urea	Water	25	1.38	5

Section 7
Biochemistry

Properties of Amino Acids....7-1
Structures of Common Amino Acids....7-3
Properties of Purine and Pyrimidine Bases....7-5
The Genetic Code....7-6
Properties of Fatty Acids and Their Methyl Esters....7-7
Composition and Properties of Common Oils And Fats....7-9
Carbohydrate Names and Symbols....7-14
Standard Transformed Gibbs Energies of Formation for Biochemical Reactants....7-16
Apparent Equilibrium Constants for Enzyme-Catalyzed Reactions....7-19
Thermodynamic Quantities for the Ionization Reactions of Buffers in Water....7-23
Biological Buffers....7-26
Typical pH Values of Biological Materials and Foods....7-27
Structure and Functions of Some Common Drugs....7-28
Chemical Constituents of Human Blood....7-45
Chemical Composition of the Human Body....7-48
Nutrient Values of Foods....7-49

Section 7
Biochemistry

Properties of Amino Acids
Structures of Common Amino Acids
Properties of Purine and Pyrimidine Bases
The Genetic Code
Properties of Fatty Acids and Their Methyl Esters
Composition and Properties of Common Oils and Fats
Carbohydrate Names and Symbols
Standard Transformed Gibbs Energies of Formation for Biochemical Reactants
Apparent Equilibrium Constants for Enzyme-Catalyzed Reactions
Thermodynamic Quantities for the Ionization Reactions of Buffers in Water
Biological Buffers
Typical pH Values of Biological Materials and Foods
Structure and Functions of Some Common Drugs
Chemical Constituents of Human Blood
Typical Composition of the Human Body
Nutritive Value of Foods

PROPERTIES OF AMINO ACIDS

This table gives selected properties of some important amino acids and closely related compounds. The first part of the table lists the 20 "standard" amino acids that are the basic constituents of proteins. The second part includes other amino acids and related compounds of biochemical importance. Within each part of the table the compounds are listed by name in alphabetical order. Structures are given in the following table.

Symbol: Three-letter symbol for the standard amino acids
M_r: Molecular weight
t_m: Melting point
pK_a, pK_b, pK_c, pK_d: Negative of the logarithm of the acid dissociation constants for the COOH and NH_2 groups (and, in some cases, other groups) in the molecule (at 25 °C)
pI: pH at the isoelectric point
S: Solubility in water in units of grams of compound per kilogram of water; a temperature of 25 °C is understood unless otherwise stated in a superscript. When quantitative data are not available, the notations sl.s. (for slightly soluble), s. (for soluble), and v.s. (for very soluble) are used.
V_2^0: Partial molar volume in aqueous solution at infinite dilution (at 25 °C)

Data on the enthalpy of formation of many of these compounds are included in the table "Standard Thermodynamic Properties of Chemical Substances" in Section 5 of this *Handbook*. Absorption spectra and optical rotation data can be found in Reference 3. Partial molar volume is taken from Reference 5; other thermodynamic properties, including solubility as a function of temperature, are given in References 3 and 5. Most of the p*K* values come from References 1, 6, and 7.

References

1. Dawson, R. M. C., Elliott, D. C., Elliott, W. H., and Jones, K. M., *Data for Biochemical Research*, Third Edition, Clarendon Press, Oxford, 1986.
2. O'Neil, Maryadele J., Ed., *The Merck Index, Fourteenth Edition*, Merck & Co., Rahway, NJ, 2006.
3. Sober, H. A., Ed., *CRC Handbook of Biochemistry. Selected Data for Molecular Biology*, CRC Press, Boca Raton, FL, 1968.
4. Voet, D., and Voet, J. G., *Biochemistry, Second Edition*, John Wiley & Sons, New York, 1995.
5. Hinz, H. J., Ed., *Thermodynamic Data for Biochemistry and Biotechnology*, Springer-Verlag, Heidelberg, 1986.
6. Fasman, G. D., Ed. *Practical Handbook of Biochemistry and Molecular Biology*, CRC Press, Boca Raton, FL, 1989.
7. Smith, R. M., and Martell, A. E., *NIST Standard Reference Database 46: Critically Selected Stability Constants of Metal Complexes Database*, Version 3.0, National Institute of Standards and Technology, Gaithersburg, MD, 1997.
8. Ramasami, P., *J. Chem. Eng. Data*, 47, 1164, 2002.

Symbol	Name	Mol. form.	M_r	t_m/°C	pK_a	pK_b	pK_c	pK_d	pI	S/g kg^{-1}	V_2^0/ cm^3 mol^{-1}
Ala	L-Alanine	$C_3H_7NO_2$	89.09	297	2.33	9.71			6.00	166.9	60.54
Arg	L-Arginine	$C_6H_{14}N_4O_2$	174.20	244	2.03	9.00	12.10		10.76	182.6	127.42
Asn	L-Asparagine	$C_4H_8N_2O_3$	132.12	235	2.16	8.73			5.41	25.1	78.0
Asp	L-Aspartic acid	$C_4H_7NO_4$	133.10	270	1.95	9.66	3.71		2.77	5.04	74.8
Cys	L-Cysteine	$C_3H_7NO_2S$	121.16	240	1.91	10.28	8.14		5.07	v.s.	73.45
Gln	L-Glutamine	$C_5H_{10}N_2O_3$	146.14	185	2.18	9.00			5.65	42	
Glu	L-Glutamic acid	$C_5H_9NO_4$	147.13	160	2.16	9.58	4.15		3.22	8.6	89.85
Gly	Glycine	$C_2H_5NO_2$	75.07	290	2.34	9.58			5.97	239	43.26
His	L-Histidine	$C_6H_9N_3O_2$	155.15	287	1.70	9.09	6.04		7.59	43.5	98.3
Ile	L-Isoleucine	$C_6H_{13}NO_2$	131.17	284	2.26	9.60			6.02	34.2	105.80
Leu	L-Leucine	$C_6H_{13}NO_2$	131.17	293	2.32	9.58			5.98	23.8	107.77
Lys	L-Lysine	$C_6H_{14}N_2O_2$	146.19	224	2.15	9.16	10.67		9.74	5.8	108.5
Met	L-Methionine	$C_5H_{11}NO_2S$	149.21	281	2.16	9.08			5.74	56	105.57
Phe	L-Phenylalanine	$C_9H_{11}NO_2$	165.19	283	2.18	9.09			5.48	27.9	121.5
Pro	L-Proline	$C_5H_9NO_2$	115.13	221	1.95	10.47			6.30	1625	82.76
Ser	L-Serine	$C_3H_7NO_3$	105.09	228	2.13	9.05			5.68	250	60.62
Thr	L-Threonine	$C_4H_9NO_3$	119.12	256	2.20	8.96			5.60	90.6	76.90
Trp	L-Tryptophan	$C_{11}H_{12}N_2O_2$	204.23	289	2.38	9.34			5.89	13.2	143.8
Tyr	L-Tyrosine	$C_9H_{11}NO_3$	181.19	343	2.24	9.04	10.10		5.66	0.51	
Val	L-Valine	$C_5H_{11}NO_2$	117.15	315	2.27	9.52			5.96	88	90.75

Name	Mol. form.	M_r	t_m/°C	pK_a	pK_b	pK_c	pK_d	pI	S/g kg^{-1}	V_2^0/ cm^3 mol^{-1}
N-Acetylglutamic acid	$C_7H_{11}NO_5$	189.17	199						s.	
*N*6-Acetyl-L-lysine	$C_8H_{16}N_2O_3$	188.22	265	2.12	9.51					
β-Alanine	$C_3H_7NO_2$	89.09	200	3.51	10.08				890	58.28
2-Aminoadipic acid	$C_6H_{11}NO_4$	161.16	207	2.14	4.21	9.77		3.18	2.2^{40}	
DL-2-Aminobutanoic acid	$C_4H_9NO_2$	103.12	304	2.30	9.63			6.06	210	75.6
DL-3-Aminobutanoic acid	$C_4H_9NO_2$	103.12	194.3	3.43	10.05			7.30	1250	76.3
4-Aminobutanoic acid	$C_4H_9NO_2$	103.12	203	4.02	10.35				971	73.2
10-Aminodecanoic acid	$C_{10}H_{21}NO_2$	187.28	188.5							167.3

Name	Mol. form.	M_r	t_m/°C	pK_a	pK_b	pK_c	pK_d	pI	S/g kg^{-1}	V_2^0/ cm^3 mol^{-1}
7-Aminoheptanoic acid	$C_7H_{15}NO_2$	145.20	195						v.s.	120.0
6-Aminohexanoic acid	$C_6H_{13}NO_2$	131.17	205					7.29	850	104.2
L-3-Amino-2-methylpropanoic acid	$C_4H_9NO_2$	103.12	185						s.	
2-Amino-2-methylpropanoic acid	$C_4H_9NO_2$	103.12	335	2.36	10.21			5.72	137	77.55
9-Aminononanoic acid	$C_9H_{19}NO_2$	173.26	191							151.3
8-Aminooctanoic acid	$C_8H_{17}NO_2$	159.23	192							136.1
5-Amino-4-oxopentanoic acid	$C_5H_9NO_3$	131.13	118	4.05	8.90					
5-Aminopentanoic acid	$C_5H_{11}NO_2$	117.15	157 dec						s.	87.6
o-Anthranilic acid	$C_7H_7NO_2$	137.14	146	2.05	4.95				3.5^{14}	
Azaserine	$C_5H_7N_3O_4$	173.13	150		8.55				v.s.	
Canavanine	$C_5H_{12}N_4O_3$	176.17	172	2.50	6.60	9.25		7.93	v.s.	
L-γ-Carboxyglutamic acid	$C_6H_9NO_6$	191.14	167	1.70	9.90	4.75	3.20			
Carnosine	$C_9H_{14}N_4O_3$	226.23	260	2.51	9.35	6.76			322	
Citrulline	$C_6H_{13}N_3O_3$	175.19	222	2.32	9.30			5.92	s.	
Creatine	$C_4H_9N_3O_2$	131.13	303	2.63	14.30				16	
L-Cysteic acid	$C_3H_7NO_5S$	169.16	260	1.89	8.70	1.30			v.s.	
L-Cystine	$C_6H_{12}N_2O_4S_2$	240.30	260	1.50	8.80	2.05	8.03		0.17	
2,4-Diaminobutanoic acid	$C_4H_{10}N_2O_2$	118.13	118.1	1.85	8.24	10.44		9.27	s.	
3,5-Dibromo-L-tyrosine	$C_9H_9Br_2NO_3$	338.98	245						2.72	
3,5-Dichloro-L-tyrosine	$C_9H_9Cl_2NO_3$	250.08	247						1.97	
3,5-Diiodo-L-tyrosine	$C_9H_9I_2NO_3$	432.98	213	2.12	9.10	6.16			0.62	
Dopamine	$C_8H_{11}NO_2$	153.18			10.36	8.88			s.	
L-Ethionine	$C_6H_{13}NO_2S$	163.24	273	2.18	9.05	13.10				
N-Glycylglycine	$C_4H_8N_2O_3$	132.12	263	3.13	8.10				231	
Guanidinoacetic acid	$C_3H_7N_3O_2$	117.11	282	2.82					5	
Histamine	$C_5H_9N_3$	111.15	83		9.83	6.11			v.s.	
L-Homocysteine	$C_4H_9NO_2S$	135.19	232	2.15	8.57	10.38		5.55	s.	
Homocystine	$C_8H_{16}N_2O_4S_2$	268.35	264	1.59	9.44	2.54	8.52		0.2	
L-Homoserine	$C_4H_9NO_3$	119.12	203	2.27	9.28			6.17	1100	
3-Hydroxy-DL-glutamic acid	$C_5H_9NO_5$	163.13	209					3.28		
5-Hydroxylysine	$C_6H_{14}N_2O_3$	162.19		2.13	8.85	9.83		9.15		
trans-4-Hydroxy-L-proline	$C_5H_9NO_3$	131.13	274	1.82	9.47			5.74	361	84.49
L-3-Iodotyrosine	$C_9H_{10}INO_3$	307.08	205	2.20	9.10	8.70			sl.s.	
L-Kynurenine	$C_{10}H_{12}N_2O_3$	208.21	194						sl.s.	
L-Lanthionine	$C_6H_{12}N_2O_4S$	208.24	294						1.5	
Levodopa	$C_9H_{11}NO_4$	197.19	277	2.32	8.72	9.96	11.79		1.65^{20}	
L-1-Methylhistidine	$C_7H_{11}N_3O_2$	169.18	249	1.69	8.85	6.48			200	
L-Norleucine	$C_6H_{13}NO_2$	131.17	301	2.31	9.68			6.09	15	107.7
L-Norvaline	$C_5H_{11}NO_2$	117.15	307	2.31	9.65				107	91.8
L-Ornithine	$C_5H_{12}N_2O_2$	132.16	140	1.94	8.78	10.52		9.73	v.s.	
O-Phosphoserine	$C_3H_8NO_6P$	185.07	166	2.14	9.80	5.70				
L-Pyroglutamic acid	$C_5H_7NO_3$	129.12	162	3.32						
Sarcosine	$C_3H_7NO_2$	89.09	212	2.18	9.97				428	
Taurine	$C_2H_7NO_3S$	125.15	328	-0.3	9.06				105	
L-Thyroxine	$C_{15}H_{11}I_4NO_4$	776.87	235	2.20	10.01	6.45			sl.s.	

STRUCTURES OF COMMON AMINO ACIDS

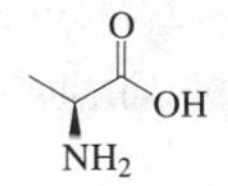

L-Alanine (Ala)

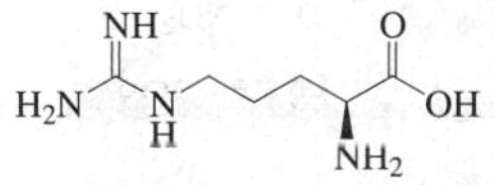

L-Arginine (Arg)

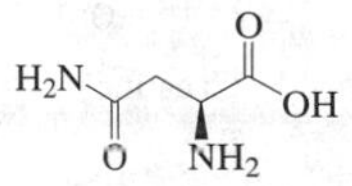

L-Asparagine (Asn)

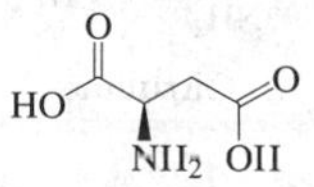

L-Aspartic acid (Asp)

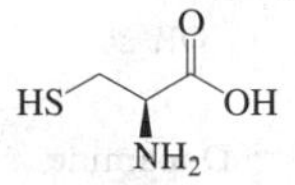

L-Cysteine (Cys)

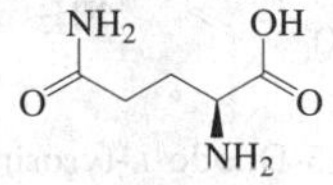

L-Glutamine (Gln)

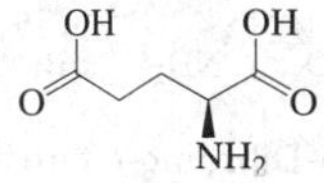

L-Glutamic acid (Glu)

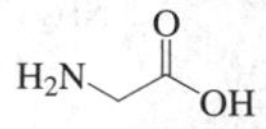

Glycine (Gly)

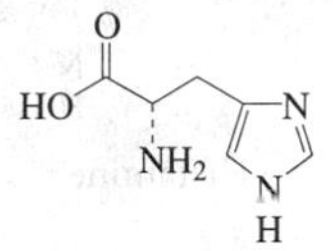

L-Histidine (His)

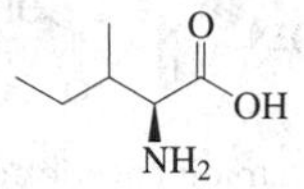

L-Isoleucine (Ile)

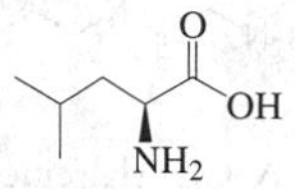

L-Leucine (Leu)

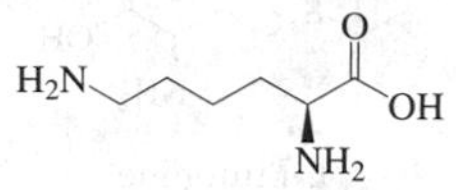

L-Lysine (Lys)

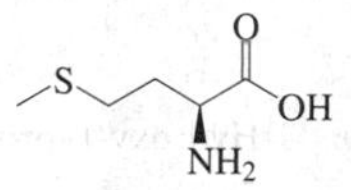

L-Methionine (Met)

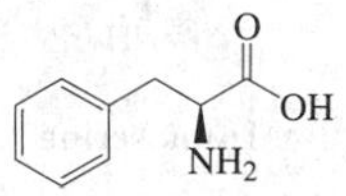

L-Phenylalanine (Phe)

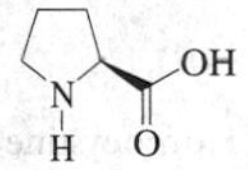

L-Proline (Pro)

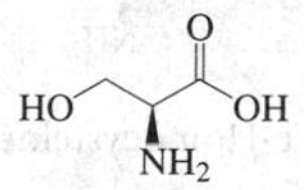

L-Serine (Ser)

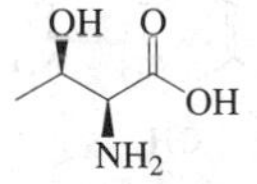

L-Threonine (Thr)

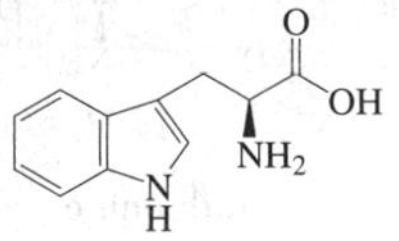

L-Tryptophan (Trp)

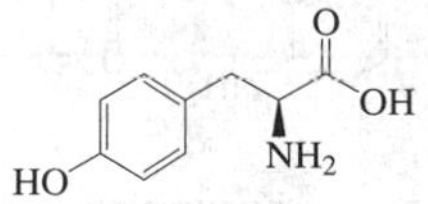

L-Tyrosine (Tyr)

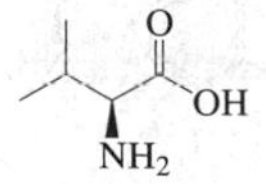

L-Valine (Val)

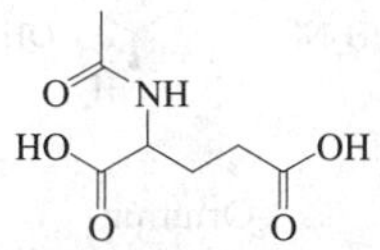

N-Acetylglutamic acid

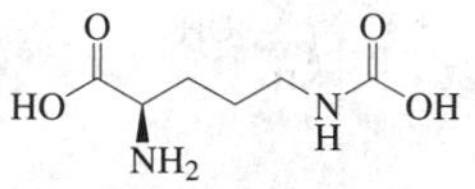

*N*6-Acetyl-L-lysine

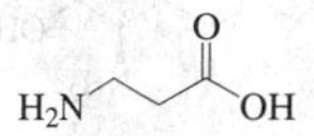

β-Alanine

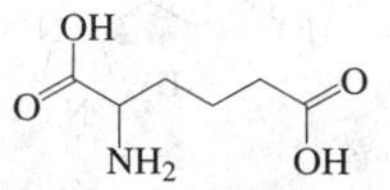

2-Aminoadipic acid

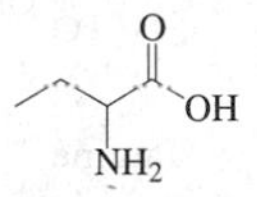

DL-2-Aminobutanoic acid

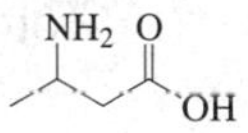

DL-3-Aminobutanoic acid

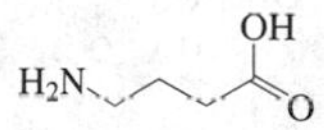

4-Aminobutanoic acid

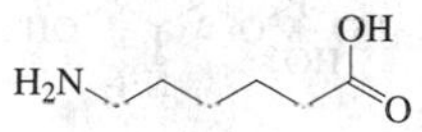

6-Aminohexanoic acid

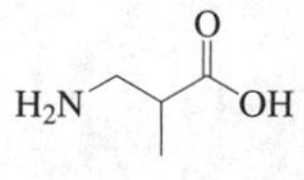

L-3-Amino-2-methylpropanoic acid

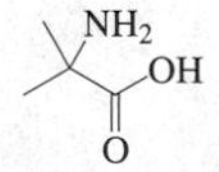

2-Amino-2-methylpropanoic acid

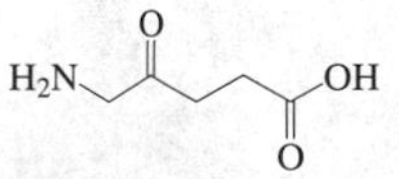

5-Amino-4-oxopentanoic acid

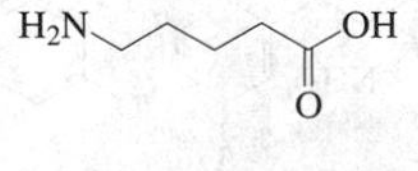

5-Aminopentanoic acid

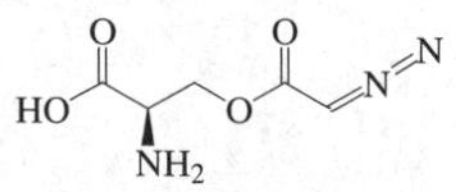

Azaserine

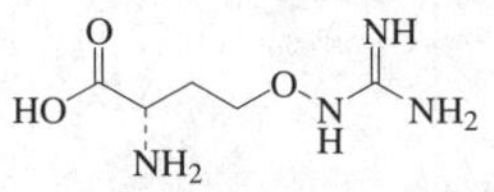

Canavanine

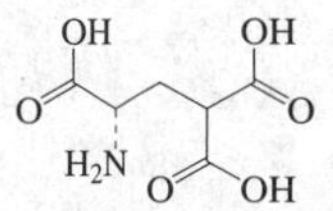

L-γ-Carboxyglutamic acid

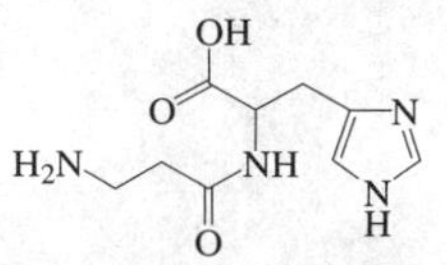

Carnosine

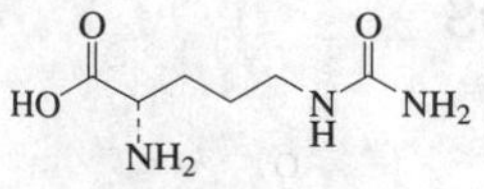

Citrulline

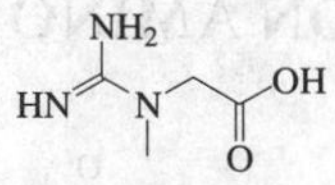

Creatine

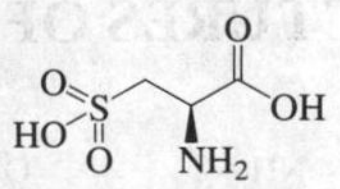

L-Cysteic acid

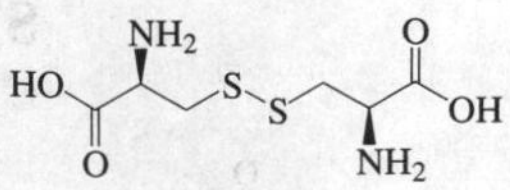

L-Cystine

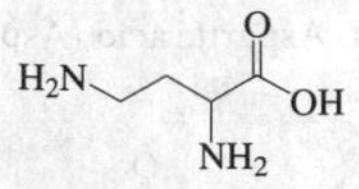

2,4-Diaminobutanoic acid

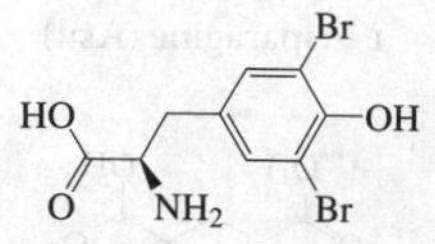

3,5-Dibromo-L-tyrosine

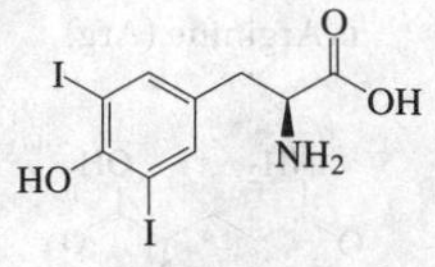

3,5-Diiodo-L-tyrosine

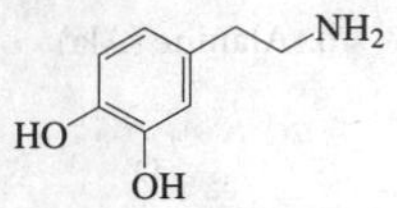

Dopamine

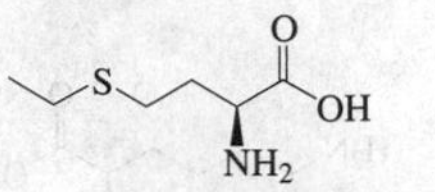

L-Ethionine

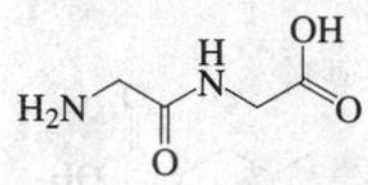

N-Glycylglycine

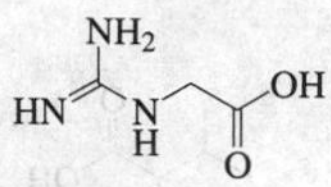

Guanidinoacetic acid

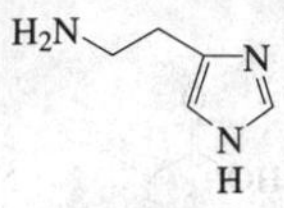

Histamine

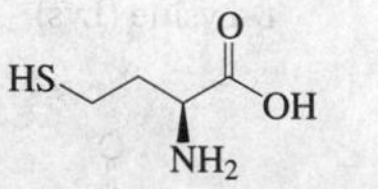

L-Homocysteine

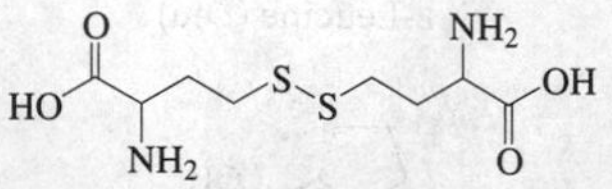

Homocystine

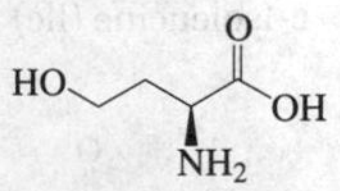

L-Homoserine

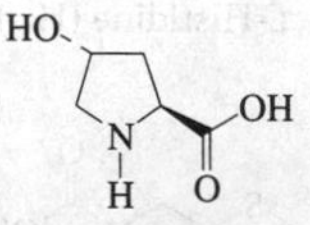

trans-4-Hydroxy-L-proline

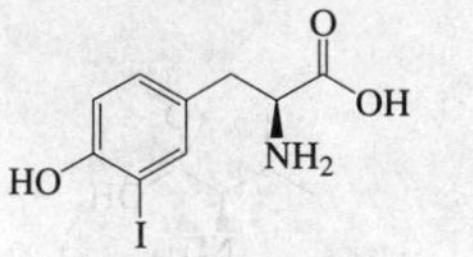

L-3-Iodotyrosine

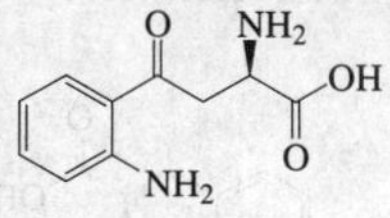

L-Kynurenine

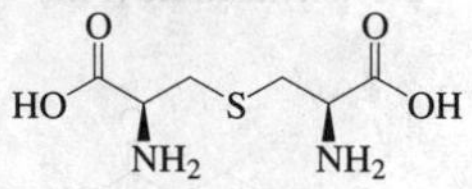

L-Lanthionine

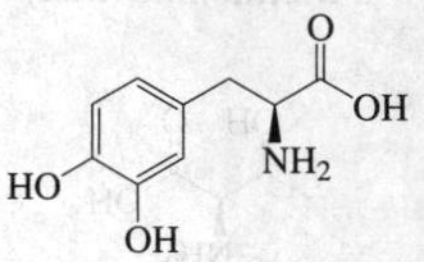

Levodopa

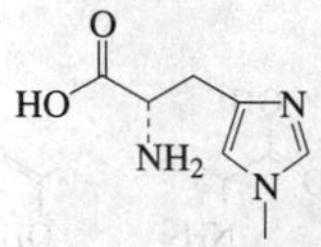

L-1-Methylhistidine

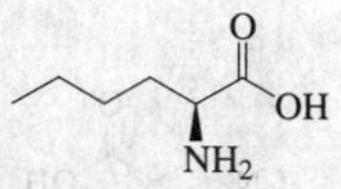

L-Norleucine

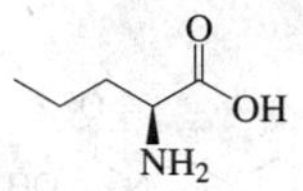

L-Norvaline

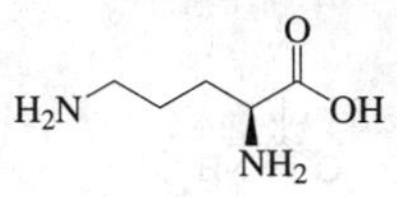

L-Ornithine

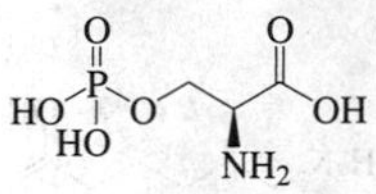

O-Phosphoserine

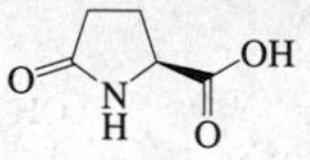

L-Pyroglutamic acid

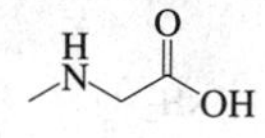

Sarcosine

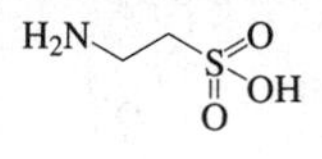

Taurine

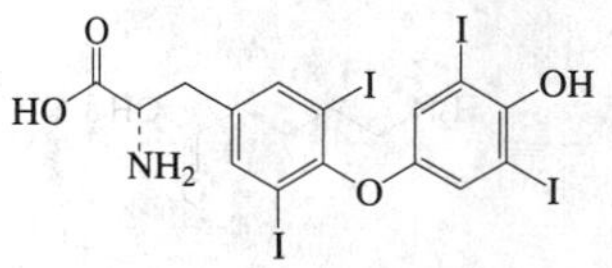

L-Thyroxine

PROPERTIES OF PURINE AND PYRIMIDINE BASES

This table lists some of the important purine and pyrimidine bases that occur in nucleic acids. The pK_a values (negative logarithm of the acid dissociation constant) are given for each ionization stage. The last column gives the aqueous solubility S at the indicated temperature in units of grams per 100 grams of solution.

The numbering system in the rings is:

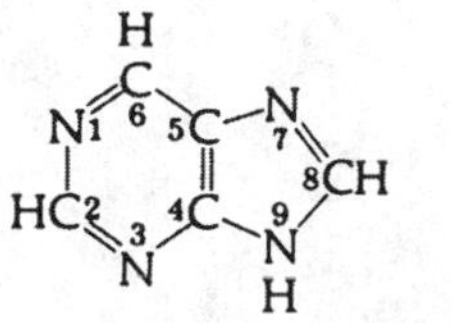

Purine

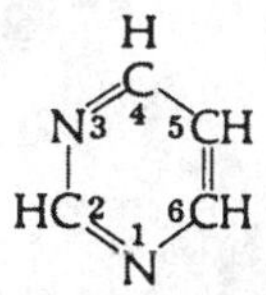

Pyrimidine

References

1. Dawson, R. M. C., et al., *Data for Biochemical Research*, 3rd ed., Clarendon Press, Oxford, 1986
2. O'Neil, M. J., Ed., *The Merck Index*, 13th ed., Merck and Co., Rahway, NJ, 2001.

Common name	Systematic name	Mol. form.	Mol. wt.	pK_a values			S/mass % (temp.)
Pyrimidines							
Cytosine	4-Amino-2-hydroxypyrimidine	$C_4H_5N_3O$	111.10	4.60	12.16		0.73 (25 °C)
5-Methylcytosine	4-Amino-2-hydroxy-5-methylpyrimidine	$C_5H_7N_3O$	125.13	4.6	12.4		0.45 (25 °C)
5-Hydroxymethylcytosine	4-Amino-2-hydroxy-5-hydroxymethylpyrimidine	$C_5H_7N_3O_2$	141.13	4.3	13		
Uracil	2,4-Dihydroxypyrimidine	$C_4H_4N_2O_2$	112.09	0.5	9.5	>13	0.27 (25 °C)
Thymine	5-Methyluracil	$C_5H_6N_2O_2$	126.11	9.94	>13		0.35 (25 °C)
Orotic acid	Uracil-6-carboxylic acid	$C_5H_4N_2O_4$	156.10	2.4	9.5	>13	0.18 (18 °C)
Purines							
Adenine	6-Aminopurine	$C_5H_5N_5$	135.14	<1	4.3	9.83	0.104 (25 °C)
Guanine	2-Amino-6-hydroxypurine	$C_5H_5N_5O$	151.13	3.3	9.2	12.3	0.0068 (40 °C)
7-Methylguanine	7-Methyl-2-amino-6-hydroxypurine	$C_6H_7N_5O$	165.16	3.5	9.9		
Isoguanine	6-Amino-2-hydroxypurine	$C_5H_5N_5O$	151.13	4.5	9.0		0.006 (25 °C)
Xanthine	2,6-Dioxopurine	$C_5H_4N_4O_2$	152.11	0.8	7.4	11.1	0.05 (20 °C)
Hypoxanthine	6-Hydroxypurine	$C_5H_4N_4O$	136.11	2.0	8.9	12.1	0.07 (19 °C)
Uric acid	2,6,8-Trihydroxypurine	$C_5H_4N_4O_3$	168.11	5.4	11.3		0.002 (20 °C)

THE GENETIC CODE

This table gives the correspondence between a messenger RNA codon and the amino acid that it specifies. The symbols for bases in the codon are:

U: uracil
C: cytosine
A: adenine
G: guanine

The amino acid symbols are given in the table entitled "Structures of Common Amino Acids." A chain-initiating codon is indicated by **init** and a chain-terminating codon by **term**.

Example: UCA codes for **Ser**, UAC codes for **Tyr**, etc.

First position	Second position: U	Second position: C	Second position: A	Second position: G	**Third position**
U	Phe	Ser	Tyr	Cys	U
	Phe	Ser	Tyr	Cys	C
	Leu	Ser	**term**	**term**	A
	Leu	Ser	**term**	Trp	G
C	Leu	Pro	His	Arg	U
	Leu	Pro	His	Arg	C
	Leu	Pro	Gln	Arg	A
	Leu	Pro	Gln	Arg	G
A	Ile	Thr	Asn	Ser	U
	Ile	Thr	Asn	Ser	C
	Ile	Thr	Lys	Arg	A
	Met (**init**)	Thr	Lys	Arg	G
G	Val	Ala	Asp	Gly	U
	Val	Ala	Asp	Gly	C
	Val	Ala	Glu	Gly	A
	Val (**init**)	Ala	Glu	Gly	G

PROPERTIES OF FATTY ACIDS AND THEIR METHYL ESTERS

This table gives the names and selected properties of some important fatty acids and their methyl esters. It includes most of the acids that are significant constituents of naturally occurring oils and fats. Compounds are listed first by number of carbon atoms and, secondly, by the degree of unsaturation. Both the systematic name and the common or trivial name are given, as well as the Chemical Abstracts Service Registry Number and the shorthand acid code that is frequently used. The first number in this code gives the number of carbon atoms; the number following the colon is the number of unsaturated centers (mainly double bonds). The location and orientation of the unsaturated centers follow. The symbols used are: c = *cis*; t = *trans*; a = acetylenic center; e = ethylenic center at end of chain; ep = *epoxy*. Thus 9c,11t indicates a double bond with *cis* orientation at the No. 9 carbon and another with *trans* orientation at the No. 11 carbon. More details on the codes can be found in Reference 1.

The table gives the molecular weight and melting point of the acid and the melting and boiling points of the methyl ester of the acid when available. A superscript on the boiling point indicates the pressure in mmHg (torr); if there is no superscript, the value refers to one atmosphere (760 mmHg). The references cover many other fatty acids beyond those listed here and give additional properties.

We are indebted to Frank D. Gunstone for advice on the content of the table.

References

1. Gunstone, F. D., Harwood, J. L., and Dijkstra, A. J., eds., *The Lipid Handbook, Third Edition,* CRC Press, Boca Raton, FL, 2006.
2. Gunstone, F. D., and Adlof, R. O., *Common (non-systematic) Names for Fatty Acids,* www.aocs.org/member/division/analytic/fanames.asp, 2003.
3. Firestone, D., *Physical and Chemical Characteristics of Oils, Fats, and Waxes,* 2nd Edition, AOCS Press, Urbana, IL, 2006.
4. Dawson, R. M. C., Elliott, D. C., Elliott, W. H., and Jones, K. M., *Data for Biochemical Research,* Third Edition, Clarendon Press, Oxford, 1986.
5. Altman, P. L., and Dittmer, D. S., eds., *Biology Data Book,* Second Edition, Vol. 1, Federation of American Societies for Experimental Biology, Bethesda, MD, 1972.
6. Fasman, G. D., Ed., *Practical Handbook of Biochemistry and Molecular Biology,* CRC Press, Boca Raton, FL, 1989.

Systematic name	Common name	Mol. form.	Acid code	CAS RN	Mol. weight	mp/°C	Methyl ester mp/°C	Methyl ester bp/°C
Butanoic acid	Butyric acid	$C_4H_8O_2$	4:0	107-92-6	88.106	–5.1	–85.8	102.8
Pentanoic acid	Valeric acid	$C_5H_{10}O_2$	5:0	109-52-4	102.132	–33.6		127.4
3-Methylbutanoic acid	Isovaleric acid	$C_5H_{10}O_2$	4:0 3-Me	503-74-2	102.132	–29.3		116.5
Hexanoic acid	Caproic acid	$C_6H_{12}O_2$	6:0	142-62-1	116.158	–3	–71	149.5
Heptanoic acid	Enanthic acid	$C_7H_{14}O_2$	7:0	111-14-8	130.185	–7.2	–56	174
Octanoic acid	Caprylic acid	$C_8H_{16}O_2$	8:0	124-07-2	144.212	16.5	–40	192.9
Nonanoic acid	Pelargonic acid	$C_9H_{18}O_2$	9:0	112-05-0	158.238	12.4		213.5
Decanoic acid	Capric acid	$C_{10}H_{20}O_2$	10:0	334-48-5	172.265	31.4	–18	224
9-Decenoic acid	Caproleic acid	$C_{10}H_{18}O_2$	10:1 9e	14436-32-9	170.249	26.5		120^{20}
Undecanoic acid		$C_{11}H_{22}O_2$	11:0	112-37-8	186.292	28.6		123^{10}
Dodecanoic acid	Lauric acid	$C_{12}H_{24}O_2$	12:0	143-07-7	200.318	43.8	5.2	267
cis-9-Dodecenoic acid	Lauroleic acid	$C_{12}H_{22}O_2$	12:1 9c	2382-40-3	198.302			
Tridecanoic acid		$C_{13}H_{26}O_2$	13:0	638-53-9	214.344	41.5	6.5	92^{1}
Tetradecanoic acid	Myristic acid	$C_{14}H_{28}O_2$	14:0	544-63-8	228.371	54.2	19	295
cis-9-Tetradecenoic acid	Myristoleic acid	$C_{14}H_{26}O_2$	14:1 9c	13147-06-3	226.355	–4		
Pentadecanoic acid		$C_{15}H_{30}O_2$	15:0	1002-84-2	242.398	52.3	18.5	153.5
Hexadecanoic acid	Palmitic acid	$C_{16}H_{32}O_2$	16:0	57-10-3	256.424	62.5	30	417
cis-9-Hexadecenoic acid	Palmitoleic acid	$C_{16}H_{30}O_2$	16:1 9c	373-49-9	254.408	0.5		140^{5}
Heptadecanoic acid	Margaric acid	$C_{17}H_{34}O_2$	17:0	506-12-7	270.451	61.3	30	185^{9}
Octadecanoic acid	Stearic acid	$C_{18}H_{36}O_2$	18:0	57-11-4	284.478	69.3	39.1	443
cis-6-Octadecenoic acid	Petroselinic acid	$C_{18}H_{34}O_2$	18:1 6c	593-39-5	282.462	29.8		
cis-9-Octadecenoic acid	Oleic acid	$C_{18}H_{34}O_2$	18:1 9c	112-80-1	282.462	13.4	–19.9	218.5^{20}
trans-9-Octadecenoic acid	Elaidic acid	$C_{18}H_{34}O_2$	18:1 9t	112-79-8	282.462	45	13.5	218^{24}
cis-11-Octadecenoic acid	*cis*-Vaccenic acid	$C_{18}H_{34}O_2$	18:1 11c	506-17-2	282.462	15		$163^{0.1}$
trans-11-Octadecenoic acid	Vaccenic acid	$C_{18}H_{34}O_2$	18:1 11t	693-72-1	282.462	44		172^{3}
cis-12,13-Epoxy-*cis*-9-octadecenoic acid	Vernolic acid	$C_{18}H_{32}O_3$	18:1 12,13-ep,9c	503-07-1	296.445	32.5		
12-Hydroxy-*cis*-9-octadecenoic acid	Ricinoleic acid	$C_{18}H_{34}O_3$	18:1 12-OH,9c	141-22-0	298.461	5.5		226^{15}
cis,trans-9,11-Octadecadienoic acid	Rumenic (CLA)	$C_{18}H_{32}O_2$	18:2 9c,11t	1839-11-8	280.446	20		
cis,cis-9,12-Octadecadienoic acid	Linoleic acid	$C_{18}H_{32}O_2$	18:2 9c,12c	60-33-3	280.446	–7	–35	215^{20}

Systematic name	Common name	Mol. form.	Acid code	CAS RN	Mol. weight	mp/°C	Methyl ester mp/°C	Methyl ester bp/°C
trans,cis-10,12-Octadecadienoic acid	(CLA)	$C_{18}H_{32}O_2$	18:2 10t,12c	22880-03-1	280.446	23	−12	
cis-9-Octadecen-12-ynoic acid	Crepenynic acid	$C_{18}H_{30}O_2$	18:2 9c,12a	2277-31-8	278.430			
cis,cis,cis-5,9,12-Octadecatrienoic acid	Pinolenic acid	$C_{18}H_{30}O_2$	18:3 5c,9c,12c	27213-43-0	278.430			
trans,cis,cis-5,9,12-Octadecatrienoic acid	Columbinic acid	$C_{18}H_{30}O_2$	18:3 5t,9c,12c	2441-53-4	278.430			
cis,cis,cis-6,9,12-Octadecatrienoic acid	γ-Linolenic acid	$C_{18}H_{30}O_2$	18:3 6c,9c,12c	506-26-3	278.430			$162^{0.5}$
trans,trans,cis-8,10,12-Octadecatrienoic acid	Calendic acid	$C_{18}H_{30}O_2$	18:3 8t,10t,12c	28872-28-8	278.430	40		
cis,trans,cis-9,11,13-Octadecatrienoic acid	Punicic acid	$C_{18}H_{30}O_2$	18:3 9c,11t,13c	544-72-9	278.430	45		
cis,trans,trans-9,11,13-Octadecatrienoic acid	α-Eleostearic acid	$C_{18}H_{30}O_2$	18:3 9c,11t,13t	506-23-0	278.430	49		148^{1}
trans,trans,cis-9,11,13-Octadecatrienoic acid	Catalpic acid	$C_{18}H_{30}O_2$	18:3 9t,11t,13c	4337-71-7	278.430	32		
trans,trans,trans-9,11,13-Octadecatrienoic acid	β-Eleostearic acid	$C_{18}H_{30}O_2$	18:3 9t,11t,13t	544-73-0	278.430	71.5	13	162^{1}
cis,cis,cis-9,12,15-Octadecatrienoic acid	α-Linolenic acid	$C_{18}H_{30}O_2$	18:3 9c,12c,15c	463-40-1	278.430	−11.3	−52	$109^{0.018}$
6,9,12,15-Octadecatetraenoic acid, all *cis*	Stearidonic acid	$C_{18}H_{28}O_2$	18:4 6c,9c,12c,15c	20290-75-9	276.414	−57		
cis,trans,trans,cis-9,11,13,15-Octadecatetraenoic acid	Parinaric acid	$C_{18}H_{28}O_2$	18:4 9c,11t,13t,15c	593-38-4	276.414	86		
Nonadecanoic acid		$C_{19}H_{38}O_2$	19:0	646-30-0	298.504	69.4	41.3	190^{4}
Eicosanoic acid	Arachidic acid	$C_{20}H_{40}O_2$	20:0	506-30-9	312.531	76.5	54.5	215^{10}
3,7,11,15-Tetramethylhexadecanoic acid	Phytanic acid	$C_{20}H_{40}O_2$	16:0 3,7,11,15-tetramethyl	14721-66-5	312.531	−65		
cis-5-Eicosenoic acid		$C_{20}H_{38}O_2$	20:1 5c	7050-07-9	310.515	27		
cis-9-Eicosenoic acid	Gadoleic acid	$C_{20}H_{38}O_2$	20:1 9c	29204-02-2	310.515	24.5		
cis-11-Eicosenoic acid	Gondoic acid	$C_{20}H_{38}O_2$	20:1 11c	2462-94-4	310.515	24		
cis,cis,cis-8,11,14-Eicosatrienoic acid	Dihomo-γ-linolenic acid	$C_{20}H_{34}O_2$	20:3 8c,11c,14c	1783-84-2				
5,8,11,14-Eicosatetraenoic acid, all *cis*	Arachidonic acid	$C_{20}H_{32}O_2$	20:4 5c,8c,11c,14c	506-32-1	304.467	−49.5		$195^{0.7}$
5,8,11,14,17-Eicosapentaenoic acid, all *cis*	Timnodonic acid, EPA	$C_{20}H_{30}O_2$	20:5 5c,8c,11c,14c,17c	10417-94-4	302.451	−54		
Heneicosanoic acid		$C_{21}H_{42}O_2$	21:0	2363-71-5	326.557	82	49	207^{4}
Docosanoic acid	Behenic acid	$C_{22}H_{44}O_2$	22:0	112-85-6	340.583	81.5	54	
cis-11-Docosenoic acid	Cetolic acid	$C_{22}H_{42}O_2$	22:1 11c	506-36-5	338.567	33		
cis-13-Docosenoic acid	Erucic acid	$C_{22}H_{42}O_2$	22:1 13c	112-86-7	338.567	34.7		221^{5}
trans-13-Docosenoic acid	Brassidic acid	$C_{22}H_{42}O_2$	22:1 13t	506-33-2	338.567	61.9	35	
cis,cis-5,13-Docosadienoic acid		$C_{22}H_{40}O_2$	22:2 5c,13c	676-39-1	336.552	−4		
7,10,13,16,19-Docosapentaenoic acid, all *cis*		$C_{22}H_{34}O_2$	22:5 7c,10c,13c,16c,19c					
4,7,10,13,16,19-Docosahexaenoic acid, all *cis*	Cervonic acid, DHA	$C_{22}H_{32}O_2$	22:6 4c,7c,10c,13c,16c,19c	2091-24-9		−45		
Tricosanoic acid		$C_{23}H_{46}O_2$	23:0	2433-96-7		79.6	53.4	
Tetracosanoic acid	Lignoceric acid	$C_{24}H_{48}O_2$	24:0	557-59-5	368.637	87.5	60	
cis-15-Tetracosenoic acid	Nervonic acid	$C_{24}H_{46}O_2$	24:1 15c	506-37-6	366.621	43	15	$165^{0.02}$
Pentacosanoic acid		$C_{25}H_{50}O_2$	25:0	506-38-7	382.664	77.5	62	
Hexacosanoic acid	Cerotic acid	$C_{26}H_{52}O_2$	26:0	506-46-7	396.690	88.5	63.8	286^{15}
Heptacosanoic acid		$C_{27}H_{54}O_2$	27:0	7138-40-1		87.6	64	
Octacosanoic acid	Montanic acid	$C_{28}H_{56}O_2$	28:0	506-48-9	424.744	90.9	67	
Nonacosanoic acid		$C_{29}H_{58}O_2$	29:0	4250-38-8	438.770	90.3	69	
Triacontanoic acid	Melissic acid	$C_{30}H_{60}O_2$	30:0	506-50-3	452.796	93.6	72	
Hentriacontanoic acid		$C_{31}H_{62}O_2$	31:0	38232-01-8	466.823	93.1		
Dotriacontanoic acid	Lacceric acid	$C_{32}H_{64}O_2$	32:0	3625-52-3	480.849	96.2		$192^{0.01}$

COMPOSITION AND PROPERTIES OF COMMON OILS AND FATS

This table lists some of the most common naturally occurring oils and fats. The list is separated into those of plant origin, fish and other marine life origin, and land animal origin. The oils and fats consist mainly of esters of glycerol (i.e., triglycerides) with fatty acids of 10 to 22 carbon atoms. The four fatty acids with the highest concentration are given for each oil; concentrations are given in weight percent. Because there is often a wide variation in composition depending on the source of the oil sample, a range (or sometimes an average) is generally given. More complete data on composition, including minor fatty acids, sterols, and tocopherols, can be found in the references.

The acids are labeled by the codes described in the previous table, "Properties of Fatty Acids and Their Methyl Esters," which gives the systematic and common names of the acids. Thus 18:2 9c,12c indicates a C_{18} acid with two double bonds in the 9 and 12 positions, both with a *cis* configuration (*cis,cis*-9,12-octadecadienoic acid, or linoleic acid).

The density and refractive index of the oils are typical values; superscripts indicate the temperature in °C.

Notes:

- The composition figure given for oleic acid (18:1 9c) often includes low levels of other 18:1 isomers.
- In some oils where a concentration is given for 18:2 9c,12c (linoleic acid), other isomers of 18:2 may be included.
- Likewise, where a concentration is given for 18:3 9c,12c,15c (α-linolenic acid), other isomers of 18:3 may be included.
- The acid 20:5 6c,9c,12c,15c,17c, which is prevalent in many fish oils, is often abbreviated as 20:5 ω-3 or 20:5 n-3.

The assistance of Frank D. Gunstone in preparing this table is gratefully acknowledged.

References

1. Firestone, D., *Physical and Chemical Characteristics of Oils, Fats, and Waxes,* 2nd Edition, AOCS Press, Urbana, IL, 2006.
2. Gunstone, F. D., Harwood, J. L., and Dijkstra, A. J., eds., *The Lipid Handbook,* Third Edition, CRC Press, Boca Raton, FL, 2006.
3. Dawson, R. M. C., Elliott, D. C., Elliott, W. H., and Jones, K. M., *Data for Biochemical Research,* Third Edition, Clarendon Press, Oxford, 1986.
4. Altman, P. L., and Dittmer, D. S., eds., *Biology Data Book,* Second Edition, Vol. 1, Federation of American Societies for Experimental Biology, Bethesda, MD, 1972.

Type of oil	Principal fatty acid components in weight %				mp/ °C	Density/ g cm⁻³	Refractive index	Iodine value	Saponification value
Plants									
Almond kernel oil	18:1 9c	43–70%	18:2 9c,12c	24–30%		0.910^{25}	1.467^{26}	89–101	188–200
	16:0	4–13%	18:0	1–10%					
Apricot kernel oil	18:1 9c	58–66%	18:2 9c,12c	29–33%		0.910^{25}	1.469^{25}	97–110	185–199
	16:0	4.6–6%	18:0	1%					
Argan seed oil	18:1 9c	42–55%	18:2 9c,12c	30–34%		0.912^{20}	1.467^{20}	92–102	189–195
	16:0	12–16%	18:0	2–7%					
Avocado pulp oil	18:1 9c	56–74%	18:2 9c,12c	10–17%		0.912^{25}	1.466^{25}	85–90	177–198
	16:0	9–18%	16:1 9c	3–9%					
Babassu palm oil	12:0	40–55%	14:0	11–27%	24	0.914^{25}	1.450^{40}	10–18	245–256
	18:1 9c	9–20%	16:0	5.2–11%					
Blackcurrant oil	18:2 9c,12c	45–50%	18:3 6c,9c,12c	14–20%		0.923^{20}	1.480^{20}	173–182	185–195
	18:3 9c,12c,15c	12–15%	18:1 9c	9–13%					
Borage (star-flower) oil	18:2 9c,12c	36–40%	18:3 6c,9c,12c	17–25%				141–160	189–192
	18:1 9c	14–21%	16:0	9.4–12%					
Borneo tallow	18:0	39–43%	18:1 9c	34–37%	38	0.855^{100}	1.456^{40}	29–38	189–200
	16:0	18–21%	20:0	1.0%					
Cameline oil	18:3 9c,12c,15c	33–38%	18:2 9c,12c	15–16%		0.924^{15}	1.477^{20}	127–155	180–190
	20:1 total	14–16%	18:1 9c	12–24%					
Canola (rapeseed)	18:1 9c	59–66%	18:2 9c,12c	24–29%	−10			91	
oil (low linolenic)	16:0	4–5%	18:3 9c,12c,15c	2–3%					
Canola (rapeseed)	18:1 9c	52–67%	18:2 9c,12c	16–25%	−10	0.915^{20}	1.466^{40}	110–126	182–193
oil (low erucic)	18:3 9c,12c,15c	6–14%	16:0	3.3–6.0%					
Caraway seed oil	18:1 9c	40%	18:2 9c,12c	30%			1.471^{35}	128	178
	18:1 6c	26%	16:0	3%					
Cashew nut oil	18:1 9c	57–80%	18:2 9c,12c	16–22%		0.914^{15}	1.463^{40}	79–89	180–196
	16:0	4–17%	18:0	2–12%					

Type of oil	Principal fatty acid components in weight %				mp/ °C	Density/ g cm⁻³	Refractive index	Iodine value	Saponification value
Castor oil	18:1 12-OH,9c	88%	18:2 9c,12c	3–5%	−18	0.952^{25}	1.475^{25}	81–91	176–187
	18:1 9c	2.9–6%	22:0	2.1%					
Cherry kernel oil	18:2 9c,12c	42–45%	18:1 9c	35–49%		0.918^{25}	1.468^{40}	110–118	190–198
	16:0	4–9%	18:3 9c,11t,13t	3–10%					
Chinese vegetable tallow	16:0	58–72%	18:1 9c	20–35%	44	0.887^{25}	1.456^{40}	16–29	200–218
	18:0	1–8%	14:0	0.5–3.7%					
Cocoa butter	18:0	31–37%	18:1 9c	31–35%	34	0.974^{25}	1.457^{40}	32–40	192–200
	16:0	25–27%	18:2 9c,12c	2.8–4.0%					
Coconut oil	12:0	45–51%	14:0	17–21%	25	0.913^{40}	1.449^{40}	5–13	248–265
	16:0	7.7–10.2%	18:1 9c	5.4–9.9%					
Cohune nut oil	12:0	44–48%	14:0	16–17%		0.914^{25}	1.450^{40}	9–14	251–260
	18:1 9c	8–10%	16:0	7–10%					
Coriander seed oil	18:1 6c	53%	18:1 9c	32%		0.908^{25}	1.464^{25}	86–100	182–191
	18:2 9c,12c	7–14%	16:0	3–8%					
Corn oil	18:2 9c,12c	40–66%	18:1 9c	20–42%	−20	0.919^{20}	1.472^{25}	107–135	187–195
	16:0	9–16%	18:0	0–3%					
Cottonseed oil	18:2 9c,12c	47–58%	16:0	18–26%	−1	0.920^{20}	1.462^{40}	96–115	189–198
	18:1 9c	14–22%	18:0	2.1–3.3%					
Crambe oil	22:1 13c	55–60%	18:1 9c	12–15%		0.906^{25}	1.470^{25}	87–113	
	18:2 9c,12c	8–10%	18:3 9c,12c,15c	6–7%					
Cuphea seed oil	8:0	65–78%	10:0	19–24%					
(caprylic acid rich)	18:2 9c,12c	1–4%	16:0	0.6–3%					
Euphorbia lagascae	18:1 12,13-ep,9c	64%	18:1 other	19%		0.952^{25}	1.473^{25}	102	
seed oil	18:2 9c,12c	9%	16:0	4%					
Evening primrose oil	18:2 9c,12c	65–80%	18:3 6c,9c,12c	8–14%			1.479^{20}	147–155	193–198
	16:0	6–10%	18:1 9c	5–12%					
Grape seed oil	18:2 9c,12c	58–78%	18:1 9c	12–28%		0.923^{20}	1.475^{40}	130–138	188–194
	16:0	5.5–11%	18:0	3–6%					
Hazelnut oil (Chilean)	18:1 9c	39%	16:1 11c	22.7%					
	20:1 total	9.7%	22:1 total	9.5%					
Hazelnut oil (Filbert)	18:1 9c	72–84%	18:2 9c,12c	5.7–22%		0.909^{25}	1.473^{25}	83–90	188–197
	16:0	4.1–7.2%	18:0	1.5–2.4%					
Hempseed oil	18:2 9c,12c	45–60%	18:3 9c,12c,15c	15–30%		0.921^{25}	1.472^{40}	145–166	190–195
	18:1 9c	11–16%	16:0	6–12%					
Illipe (mowrah) butter	18:1 9c	34%	16:0	23%	27	0.862^{100}	1.460^{40}	53–70	188–207
	18:0	23%	18:2 9c,12c	14%					
Jojoba oil[a]	20:1 total	66–74%	22:1 undefined	9–19%					
	18:1 9c	5–12%	24:1 15c	1–5%					
Kapok seed oil[b]	18:1 9c	45–65%	16:0	10–28%	30	0.926^{15}	1.469^{25}	86–110	189–197
	18:2 9c,12c	7–35%	18:0	2–9%					
Kokum butter	18:0	49–56%	18:1 9c	39–49%	41		1.456^{40}	33–37	192
	16:0	2–5%	18:2 9c,12c	1–2%					
Kusum oil	18:1 9c	57–62%	20:0	20–25%			1.461^{40}	48–58	220–230
	16:0	5–8%	18:0	2–6%					
Linola oil	18:2 9c,12c	72%	18:1 9c	16%				142	
	16:0	5.6%	18:0	4.0%					
Linseed oil	18:3 9c,12c,15c	52–58%	18:1 9c	18–20%	−24	0.924^{25}	1.480^{25}	170–203	188–196
	18:2 9c,12c	17%	18:2 9c,12c	16%					
Macadamia nut oil	18:1 9c	56–59%	16:1 9c	21–22%					
	16:0	8–9%	18:0	2–4%					

Type of oil	Principal fatty acid components in weight %				mp/ °C	Density/ g cm^{-3}	Refractive index	Iodine value	Saponification value
Mango seed oil	18:1 9c	38–50%	18:0	31–49%		0.912^{15}	1.461^{25}	39–48	188–195
	18:2 9c,12c	3–6%	20:0	2–6%					
Meadowfoam seed oil	20:1 5c	58–77%	22:1 total	8–24%			1.464^{40}	86–91	168
	22:2 5c,13c	7–15%	18:1 9c	1–3%					
Melon oil	18:2 9c,12c	67% (av.)	18:1 9c	12% (av.)					
	16:0	11% (av.)	18:0	9% (av.)					
Moringa peregrina seed oil	18:1 9c	70%	16:0	9%		0.903^{24}	1.460^{40}	70	185
	18:0	3.8%	22:0	2.4%					
Mustard seed oil	22:1 13c	43%	22:1 13c	22–50%		0.913^{20}	1.465^{40}	92–125	170–184
	18:3 9c,12c,15c	12%	18:2 9c,12c	10–24%					
Neem oil	18:1 9c	49–62%	18:0	14–24%	−3	0.912^{30}	1.462^{40}	68–71	195–205
	16:0	13–18%	18:2 9c,12c	7–15%					
Niger seed oil	18:2 9c,12c	52–78%	16:0	5–12%		0.924^{15}	1.468^{40}	126–135	188–193
	18:1 9c	4–10%	18:0	2–12%					
Nutmeg butter	14:0	76–83%	18:1 9c	5–11%	45		1.468^{40}	48–85	170–190
	16:0	4–10%	12:0	3–6%					
Oat oil	18:2 9c,12c	24–48%	18:1 9c	18–53%		0.917^{25}	1.467^{40}	105–116	190–199
	16:0	13–39%	18:0	0.5–4%					
Oiticica oil	18:3 9c,11t,13t, 4-oxo	70–80%	16:0	7%		0.972^{20}	1.514^{25}	140–150	188–193
	18:0	5%	18:1 9c	4–7%					
Olive oil	18:1 9c	55–83%	18:2 9c,12c	9%	−6	0.911^{20}	1.469^{20}	75–94	184–196
	16:0	7.5–20%	18:2 9c,12c	3.5–21%					
Palm kernel oil	12:0	40–55%	14:0	14–18%	24	0.922^{15}	1.450^{40}	14–21	230–250
	18:1 9c	12–21%	16:0	6.5–10%					
Palm oil	16:0	40–48%	18:1 9c	36–44%	35	0.914^{15}	1.455^{40}	49–55	190–209
	18:2 9c,12c	6.5–12%	18:0	3.5–6.5%					
Palm olein	18:1 9c	40–44%	16:0	38–43%		0.91^{40}	1.459^{40}	>56	194–202
	18:2 9c,12c	10–13%	18:0	3.7–4.8%					
Palm stearin	16:0	48–74%	18:1 9c	16–36%		0.884^{60}	1.449^{40}	<48	193–205
	18:0	3.9–5.6%	18:2 9c,12c	3.2–9.8%					
Parsley seed oil	18:1 6c	69–76%	18:1 9c	12–15%			1.4800^{40}	110–120	
	18:2 9c,12c	6–14%	16:0	2%					
Peanut oil	18:1 9c	36–67%	18:2 9c,12c	14–43%	3	0.914^{20}	1.463^{40}	86–107	187–196
	16:0	8.3–14%	22:0	2.1–4.4%					
Perilla oil	18:3 9c,12c,15c	59%	18:2 9c,12c	14–18%		0.924^{25}	1.477^{25}	192–208	188–197
	18:1 9c	11–13%	16:0	6–9%					
Phulwara butter	16:0	57–61%	18:1 9c	30–36%	43	0.862^{100}	1.458^{40}	40–51	188–200
	18:2 9c,12c	3–4%	18:0	3–4%					
Pine nut oil	18:2 9c,12c	47–51%	18:1 9c	36–39%		0.919^{15}		118–121	193–197
	16:0	6–8%	18:0	2–3%					
Poppy seed oil	18:2 9c,12c	62–73%	18:1 9c	16–30%	−15	0.916^{25}	1.469^{40}	132–146	188–196
	16:0	7–11%	18:0	1–4%					
Rice bran oil	18:1 9c	38–48%	18:2 9c,12c	16–36%		0.916^{25}	1.472^{25}	92–108	181–189
	16:0	16–28%	18:0	2–4%					
Safflower seed oil	18:2 9c,12c	68–83%	18:1 9c	8.4–30%		0.924^{15}	1.474^{25}	136–148	186–198
	16:0	5.3–8.0%	18:0	1.9–2.9%					
Safflower seed oil (high oleic)	18:1 9c	74–80%	18:2 9c,12c	13–18%		0.921^{20}	1.470^{25}	91–95	
	16:0	5–6%	18:0	1.5–2.0%					
Sal fat	18:0	33–57%	18:1 9c	31–52%	33		1.456^{40}	31–45	175–192
	16:0	6–23%	20:0	1–8%					

Type of oil	Principal fatty acid components in weight %				mp/ °C	Density/ g cm^{-3}	Refractive index	Iodine value	Saponification value
Sesame seed oil	18:2 9c,12c	40–51%	18:1 9c	33–44%	−6	0.917^{20}	1.467^{40}	104–120	187–195
	16:0	7.9–10.2%	18:0	4.4–6.7%					
Sheanut butter	18:1 9c	45–50%	18:0	36–41%	38	0.863^{100}	1.465^{40}	52–66	178–198
	16:0	4–8%	18:2 9c,12c	4–8%					
Soybean oil	18:2 9c,12c	50–57%	18:1 9c	18–28%	−16	0.920^{20}	1.468^{40}	118–139	189–195
	16:0	9–13%	18:3 9c,12c,15c	5.5–9.5%					
Stillingia seed kernel oil[c]	18:3 total	41–54%	18:2 9c,12c	24–30%		0.937^{25}	1.483^{25}	169–191	202–212
	18:1 9c	7–10%	16:0	6–9%					
Sunflower seed oil	18:2 9c,12c	48–74%	18:1 9c	13–40%	−17	0.919^{20}	1.474^{25}	118–145	188–194
	16:0	5–8%	18:0	2.5–7.0%					
Sunflower oil, high-oleic	18:1 9c	80%	18:2 9c,12c	10%					
(HO)	18:0	4.4%	16:0	3.5%		0.911^{25}	1.468^{25}	81	
Sunflower oil, mid-Oleic	18:1 9c	65%	18:2, 18:3	25%					
(NuSun oil)	16:0, 18:0	10%							
Tall oil	18:2 9c,12c	41–52%	18:1 9c	41–48%		0.969^{25}	1.494^{25}	140–180	154–180
	16:0	5–6%	18:0	2–3%					
Tung oil	18:3 9c,11t,13t	71–82%	18:2 9c,12c	8–15%	−2	0.912^{25}	1.517^{25}	160–175	189–195
	18:1 9c	4–10%	18:0	3%					
Ucuhuba butter oil	14:0	64–73%	12:0	13–15%		0.870^{100}	1.451^{50}	11–17	221–229
	18:1 9c	6–8%	16:0	3–9%					
Vernonia seed oil	18:1 12,13-ep,9c	62–72%	18:2 9c,12c	9–17%		0.901^{30}	1.486^{32}	55	176
	16:0	3–7%	18:0	2–6%					
Walnut oil	18:2 9c,12c	56–60%	18:1 9c	17–19%		0.921^{25}	1.474^{25}	138–162	189–197
	18:3 9c,12c,15c	13–14%	16:0	6–8%					
Wheatgerm oil	18:2 9c,12c	50–59%	18:1 9c	13–23%		0.926^{25}	1.479^{25}	100–128	179–217
	16:0	12–20%	18:3 9c,12c,15c	2–9%					
Marine animals									
Anchovy oil	20:5 6c,9c,12c,15c,17c	22%	16:0	17%				163–169	191–194
	16:1 undefined	13%	18:1 undefined	10%					
Capelin oil[d]	20:1 undefined	17%	22:1 undefined	15%			1.463^{50}	94–164	185–202
	18:1 undefined	14%	16:0	10%					
Cod liver oil	18:1 undefined	24%	20:1 undefined	13%		0.924^{15}	1.482^{25}	142–176	180–192
	22:6 4c,7c,10c,13c, 16c,19c	11%	16:0	10%					
Herring oil	22:1 undefined	19%	16:0	17%		0.914^{20}	1.474^{25}	115–160	161–192
	20:1 undefined	15%	18:1 undefined	14%					
Mackerel oil	22:1 undefined	15%	16:0	14%		0.929^{15}	1.481^{20}	136–167	
	18:1 undefined	13%	20:1 undefined	12%					
Menhaden oil	16:0	19%	20:5 6c,9c,12c,15c, 17c	14%		0.920^{15}		150–200	192–199
	16:1 undefined	12%	18:1 undefined	11%					
Salmon oil	22:6 4c,7c,10c,13c, 16c,19c	18%	20:5 6c,9c,12c,15c, 17c	13%		0.924^{15}	1.475^{25}	130–160	183–186
	16:0	9.8%	16:1	4.8%					
Sardine oil	16:0	18%	20:5 6c,9c,12c,15c, 17c	16%		0.915^{25}	1.464^{65}	159–192	188–199
	18:1 undefined	13%	16:1 undefined	10%					
Seal blubber oil, harp	18:1 9c	21%	20:1	12%					
	22:6 4c,7c,10c,13c, 16c,19c	7.6%	20:5 6c,9c,12c,15c, 17c	6.4%					

Type of oil	Principal fatty acid components in weight %				mp/ °C	Density/ g cm⁻³	Refractive index	Iodine value	Saponification value
Shark liver oil	18:1 undefined	45%	16:0	21%		0.917^{25}	1.476^{25}	150–300	170–190
	20:1	12%	22:1	9%					
Tuna oil	22:6 4c,7c,10c,13c, 16c,19c	22%	16:0	22%					
	18:1 undefined	21%	20:5 6c,9c,12c,15c, 17c	6%					
Trout lipids	16:0	21–24%	18:1 undefined	18–31%					
	18:2	7–16%	16:1	4–10%					
Whale oil, minke	18:1 undefined	18%	20:1	17%					
	22:1	11%	16:1	9%					
Land animals									
Beef tallow	18:1 undefined	31–50%	18:0	25–40%	47	0.902^{25}	1.454^{40}	33–47	190–200
	16:0	20–37%	14:0	1–6%					
Butterfat	16:0	28.1% (av.)	18:1 9c	20.8% (av.)	32	0.934^{15}	1.455^{40}	26–40	210–232
	14:0	10.8% (av.)	18:0	10.6% (av.)					
Chicken egg lipids, yolk	16:0	28%	18:1 9c	25%					
	18:0	17%	18:2 9c,12c	16%					
Chicken fat	18:1 undefined	37%	16:0	22%		0.918^{15}	1.456^{40}	76–80	
	18:2	20%	18:0	6%					
Milk fats, cow	16:0	28.2% (av.)	18:1 9c	21.4% (av.)					
	18:0	12.6% (av.)	14:0	10.6% (av.)					
Milk fats, human	18:1 9c	31.1% (av.)	16:0	21.6% (av.)					
	18:2 9c,12c	11.7% (av.)	14:0	6.6% (av.)					
Mutton tallow	18:1 undefined	30–42%	18:0	22–34%	48	0.946^{15}	1.455^{40}	35–46	
	16:0	20–27%	14:0	2–4%					
Pork lard	18:1 undefined	35–62%	16:0	20–32%	30	0.898^{20}			
	18:0	5–24%	18:2	3–16%					

[a] Jojoba oil consists primarily of wax esters of the acids listed here and long-chain alcohols.
[b] Kapok oil also contains up to 15% cyclopropene acids.
[c] Stillingia oil also contains 5–10% *trans,cis*-2,4-decadienoic acid (stillingic acid, 10:2 2t,4c).
[d] Capelin oil also contains about 10% 16:1.

CARBOHYDRATE NAMES AND SYMBOLS

The following table lists the systematic names and symbols for selected carbohydrates and some of their derivatives. The symbols for monosaccharide residues and derivatives are recommended by IUPAC for use in describing the structures of oligosaccharide chains. A more complete list can be found in the reference.

Reference

McNaught, A. D., *Pure Appl. Chem.*, 68, 1919–2008, 1996.

Common name	Symbol	Systematic name
Abequose	Abe	3,6-Dideoxy-D-*xylo*-hexose
N-Acetyl-2-deoxyneur-2-enaminic acid	Neu2en5Ac	
N-Acetylgalactosamine	GalNAc	
N-Acetylglucosamine	GlcNAc	
N-Acetylneuraminic acid	Neu5Ac	
Allose	All	*allo*-Hexose
Altrose	Alt	*altro*-Hexose
Apiose	Api	3-*C*-(Hydroxymethyl)-*glycero*-tetrose
Arabinitol	Ara-ol	Arabinitol
Arabinose	Ara	*arabino*-Pentose
Arcanose		2,6-Dideoxy-3-*C*-methyl-3-*O*-methyl-*xylo*-hexose
Ascarylose		3,6-Dideoxy-L-*arabino*-hexose
Boivinose		2,6-Dideoxy-D-gulose
Chalcose		4,6-Dideoxy-3-*O*-methyl-D-*xylo*-hexose
Cladinose		2,6-Dideoxy-3-*C*-methyl-3-*O*-methyl-L-*ribo*-hexose
Colitose		3,6-Dideoxy-L-*xylo*-hexose
Cymarose		6-Deoxy-3-*O*-methyl-*ribo*-hexose
3-Deoxy-D-*manno*-oct-2-ulosonic acid	Kdo	
2-Deoxyribose	dRib	2-Deoxy-*erythro*-pentose
2,3-Diamino-2,3-dideoxy-D-glucose	GlcN3N	
Diginose		2,6-Dideoxy-3-*O*-methyl-*lyxo*-hexose
Digitalose		6-Deoxy-3-*O*-methyl-D-galactose
Digitoxose		2,6-Dideoxy-D-*ribo*-hexose
3,4-Di-*O*-methylrhamnose	Rha3,4Me_2	
Ethyl glucopyranuronate	Glc*p*A6Et	
Evalose		6-Deoxy-3-*C*-methyl-D-mannose
Fructose	Fru	*arabino*-Hex-2-ulose
Fucitol	Fuc-ol	6-Deoxy-D-galactitol
Fucose	Fuc	6-Deoxygalactose
β-D-Galactopyranose 4-sulfate	β-D-Gal*p*4*S*	
Galactosamine	GalN	2-Amino-2-deoxygalactose
Galactose	Gal	*galacto*-Hexose
Glucitol	Glc-ol	
Glucosamine	GlcN	2-Amino-2-deoxyglucose
Glucose	Glc	*gluco*-Hexose
Glucuronic acid	GlcA	
N-Glycoloylneuraminic acid	Neu5Gc	
Gulose	Gul	*gulo*-Hexose
Hamamelose		2-*C*-(Hydroxymethyl)-D-ribose
Idose	Ido	*ido*-Hexose
Iduronic acid	IdoA	
Lactose	Lac	β-D-Galactopyranosyl-(1→4)-D-glucose
Lyxose	Lyx	*lyxo*-Pentose
Maltose		α-D-Glucopyranosyl-(1→4)-D-glucose
Mannose	Man	*manno*-Hexose
2-*C*-Methylxylose	Xyl2*C*Me	
Muramic acid	Mur	2-Amino-3-*O*-[(R)-1-carboxyethyl]-2-deoxy-D-glucose
Mycarose		2,6-Dideoxy-3-*C*-methyl-L-*ribo*-hexose
Mycinose		6-Deoxy-2,3-di-*O*-methyl-D-allose

Common name	Symbol	Systematic name
Neuraminic acid	Neu	5-Amino-3,5-dideoxy-D-*glycero*-D-*galacto*-non-2-ulosonic acid
Panose		α-D-Glucopyranosyl-(1→6)-α-D-glucopyranosyl-(1→4)-D-glucose
Paratose		3,6-Dideoxy-D-*ribo*-hexose
Primeverose		β-D-Xylopyranosyl-(1→6)-D-glucose
Psicose	Psi	*ribo*-Hex-2-ulose
Quinovose	Qui	6-Deoxyglucose
Raffinose		β-D-Fructofuranosyl-α-D-galactopyranosyl-(1→6)-α-D-glucopyranoside
Rhamnose	Rha	6-Deoxymannose
Rhodinose		2,3,6-Trideoxy-L-*threo*-hexose
Ribose	Rib	*ribo*-Pentose
Ribose 5-phosphate	Rib5*P*	
Ribulose	Ribulo (Rul)	*erythro*-Pent-2-ulose
Rutinose		α-L-Rhamnopyranosyl-(1→6)-D-glucose
Sarmentose		2,6-Dideoxy-3-*O*-methyl-D-*xylo*-hexose
Sedoheptulose		D-*altro*-Hept-2-ulose
Sorbose	Sor	*xylo*-Hex-2-ulose
Streptose		5-Deoxy-3-*C*-formyl-L-lyxose
Sucrose		β-D-Fructofuranosyl-α-D-glucopyranoside
Tagatose	Tag	*lyxo*-Hex-2-ulose
Talose	Tal	*talo*-Hexose
Turanose		α-D-Glucopyranosyl-(1→3)-D-fructose
Tyvelose	Tyv	3,6-Dideoxy-D-*arabino*-hexose
Xylose	Xyl	*xylo*-Pentose
Xylulose	Xylulo (Xul)	*threo*-Pent-2-ulose

STANDARD TRANSFORMED GIBBS ENERGIES OF FORMATION FOR BIOCHEMICAL REACTANTS

Robert N. Goldberg and Robert A. Alberty

This table contains values of the standard transformed Gibbs energies of formation $\Delta_f G'^\circ$ for 130 biochemical reactants. Values of $\Delta_f G'^\circ$ are given at pH 7.0, the temperature 298.15 K, and the pressure 100 kPa for three ionic strengths: $I = 0$, $I = 0.1$ mol/L and $I = 0.25$ mol/L. The table can be used for calculating apparent equilibrium constants K' and standard apparent reduction potentials E'° for biochemical reactions. Such a listing is more compact than tabulating the actual apparent equilibrium constants or standard apparent reduction potentials, which would require a very large number of reactant–product combinations. In the table, all reactants are in aqueous solution unless indicated otherwise.

A biochemical reactant is a sum of species. For example, ATP consists of an equilibrium mixture of the aqueous species ATP^{4-}, $HATP^{3-}$, H_2ATP^{2-}, $MgATP^{2-}$, etc. Similarly, phosphate refers to the equilibrium mixture of the aqueous species PO_4^{3-}, HPO_4^{2-}, $H_2PO_4^-$, H_3PO_4, $MgHPO_4$, etc. Biochemical reactions are written using biochemical reactants in terms of an apparent equilibrium constant K', which is distinct from the standard equilibrium constant K. This subject is discussed in an IUPAC report (see Reference 1 below).

The apparent equilibrium constant K' and the standard transformed Gibbs energy change $\Delta_r G'^\circ$ for a biochemical reaction can be calculated from the $\Delta_f G'^\circ$ values by using the relationship

$$-RT \ln K' = \Delta_r G'^\circ = \Sigma \nu'_i \Delta_f G'^\circ,$$

where the summation is over all of the biochemical reactants. The quantity ν'_i is the stoichiometric number of reactant i (ν'_i is positive for reactants on the right side of the equation and negative for reactants on the left side); R is the gas constant. As an example, the hydrolysis reaction of ATP is

$$\text{ATP} + H_2O(l) = \text{ADP} + \text{phosphate}.$$

At pH 7.00 and $I = 0.25$ M, $\Delta_r G'^\circ$ and K' are calculated as follows:

$$\Delta_r G'^\circ = \{-1424.70 - 1059.49 - (-2292.50 - 155.66)\}\cdot(\text{kJ mol}^{-1}) = -36.03 \text{ kJ mol}^{-1}$$

$$K' = \exp[-(-36030 \text{ J mol}^{-1})/\{(8.3145 \text{ J mol}^{-1} \text{ K}^{-1})\cdot(298.15 \text{ K})\} = 2.05\cdot 10^6$$

An example involving a biochemical half-cell reaction is

$$\text{acetaldehyde(aq)} + 2\,e^- = \text{ethanol(aq)}.$$

At 298.15 K, pH 7.00, and $I = 0$, the standard apparent reduction potential E'° can be calculated as follows

$$E'^\circ = -(1/nF)\cdot\{\Delta_f G'^\circ(\text{ethanol}) - \Delta_f G'^\circ(\text{acetaldehyde})\},$$

where n is the number of electrons in the half-cell reaction and F is the Faraday constant. Then,

$$E'^\circ = [-1/(2\cdot 9.6485\cdot 10^4 \text{ C mol}^{-1})]\cdot(58.10\cdot 10^3 \text{ J mol}^{-1} - 20.83\cdot 10^3 \text{ J mol}^{-1}) = -0.193 \text{ V}$$

References

1. Alberty, R.A., Cornish-Bowden, A., Gibson, Q.H., Goldberg, R.N., Hammes, G., Jencks, W., Tipton, K.F, Veech, R., Westerhoff, H.V., and Webb, E.C. *Pure Appl. Chem.* 66, 1641-1666, 1994.
2. Alberty, R.A., *Arch. Biochem. Biophys.*, 353, 116-130, 1998; 358, 25-39, 1998.
3. Alberty, R.A., *Thermodynamics of Biochemical Reactions,* Wiley-Interscience, New York, 2003.
4. Alberty, R.A., *BasicBiochemData2: Data and Programs for Biochemical Thermodynamics,* <http://library.wolfram.com/infocenter/MathSource/797>.

Reactant	$\Delta_f G'^\circ(I = 0)$ kJ mol^{-1}	$\Delta_f G'^\circ(I = 0.1$ M$)$ kJ mol^{-1}	$\Delta_f G'^\circ(I = 0.25$ M$)$ kJ mol^{-1}
Acetaldehyde	20.83	23.27	24.06
Acetate	-249.46	-248.23	-247.83
Acetone	80.04	83.71	84.90
Acetyl Coenzyme A	-60.49	-58.65	-58.06
Acetylphosphate	-1109.34	-1107.57	-1107.02
cis-Aconitate	-797.26	-800.93	-802.12
Adenine	510.45	513.51	514.50
Adenosine	324.93	332.89	335.46
Adenosine 5'-diphosphate (ADP)	-1428.93	-1425.55	-1424.70
Adenosine 5'-monophosphate (AMP)	-562.04	-556.53	-554.83
Adenosine 5'-triphosphate (ATP)	-2292.61	-2292.16	-2292.50
D-Alanine	-91.31	-87.02	-85.64
Ammonia	80.50	82.34	82.93
D-Arabinose	-342.67	-336.55	-334.57
L-Asparagine	-206.28	-201.38	-199.80
L-Aspartate	-456.14	-453.08	-452.09
1,3-Biphosphoglycerate	-2202.06	-2205.69	-2207.30
Butanoate	-72.94	-69.26	-68.08
1-Butanol	227.72	233.84	235.82

Reactant	$\Delta_f G'^\circ(I = 0)$ kJ mol^{-1}	$\Delta_f G'^\circ(I = 0.1\ M)$ kJ mol^{-1}	$\Delta_f G'^\circ(I = 0.25\ M)$ kJ mol^{-1}
Citrate	-963.46	-965.49	-966.23
Isocitrate	-956.82	-958.84	-959.58
Coenzyme A (CoA)	-7.98	-7.43	-7.26
CO(aq)	-119.90	-119.90	-119.90
CO(g)	-137.17	-137.17	-137.17
CO_2(aq)[total]	-547.33	-547.15	-547.10
CO_2(g)	-394.36	-394.36	-394.36
Creatine	100.41	105.92	107.69
Creatinine	256.55	260.84	262.22
L-Cysteine	-59.23	-55.01	-53.65
L-Cystine	-187.03	-179.69	-177.32
Cytochrome c [oxidized]	0.00	-5.51	-7.29
Cytochrome c [reduced]	-24.51	-26.96	-27.75
Dihydroxyacetone phosphate	-1096.60	-1095.91	-1095.70
Ethanol	58.10	61.77	62.96
Ethyl acetate	-18.00	-13.10	11.52
Ferredoxin [oxidized]	0.00	-0.61	-0.81
Ferredoxin [reduced]	38.07	38.07	38.07
Flavine adenine dinucleotide (FAD) [oxidized]	1238.65	1255.17	1260.51
Flavine adenine dinucleotide (FAD) [reduced]	1279.68	1297.43	1303.16
Flavin adenine dinucleotide-enzyme (FADenz) [oxidized]	1238.65	1255.17	1260.51
Flavin adenine dinucleotide-enzyme (FADenz) [reduced]	1229.96	1247.71	1253.44
Flavin mononucleotide (FMN) [oxidized]	759.17	768.35	771.32
Flavin mononucleotide (FMN) [reduced]	800.20	810.61	813.97
Formate	-311.04	-311.04	-311.04
D-Fructose	-436.03	-428.69	-426.32
D-Fructose 1,6-diphosphate	-2202.84	-2205.66	-2206.78
D-Fructose 6-phosphate	-1321.71	-1317.16	-1315.74
Fumarate	-521.97	-523.19	-523.58
D-Galactose	-429.45	-422.11	-419.74
α-*D*-Galactose 1-phosphate	-1317.50	-1313.01	-1311.60
D-Glucose	-436.42	-429.08	-426.71
α-*D*-Glucose 1-phosphate	-1318.03	-1313.34	-1311.89
D-Glucose 6-phosphate	-1325.00	-1320.37	-1318.92
Glutamate	-377.82	-373.54	-372.16
D-Glutamine	-128.46	-122.34	-120.36
Glutathione [oxidized]	1198.69	1214.60	1219.74
Glutathione [reduced]	625.75	634.76	637.62
Glutathione-coenzyme A	563.49	572.06	574.83
D-Glyceraldehyde 3-phosphate	-1088.94	-1088.25	-1088.04
Glycerol	-177.83	-172.93	-171.35
sn-Glycerol 3-phosphate	-1080.22	-1077.83	-1077.13
Glycine	-180.13	-177.07	-176.08
Glycolate	-411.08	-409.86	-409.46
Glycylglycine	-200.55	-195.65	-194.07
Glyoxylate	-428.64	-428.64	-428.64
H_2(aq)	97.51	98.74	99.13
H_2(g)	79.91	81.14	81.53
H_2O(l)	-157.28	-156.05	-155.66
H_2O_2(aq)	-54.12	-52.89	-52.50
3-Hydroxypropanoate	-318.62	-316.17	-315.38
Hypoxanthine	249.33	251.77	252.56
Indole	503.49	507.78	509.16
Lactate	-316.94	-314.49	-313.70
Lactose	-688.29	-674.83	-670.48
L-Leucine	167.18	175.14	177.71
L-Isoleucine	175.53	183.49	186.06
D-Lyxose	-349.58	-343.46	-341.48
Malate	-682.88	-682.85	-682.85
Maltose	-695.65	-682.19	-677.84

Reactant	$\Delta_f G'^\circ(I = 0)$ kJ mol^{-1}	$\Delta_f G'^\circ(I = 0.1$ M$)$ kJ mol^{-1}	$\Delta_f G'^\circ(I = 0.25$ M$)$ kJ mol^{-1}
D-Mannitol	-383.22	-374.65	-371.89
Mannose	-430.52	-423.18	-420.81
Methane(aq)	125.50	127.94	128.73
Methane(g)	109.11	111.55	112.34
Methanol	-15.48	-13.04	-12.25
L-Methionine	-63.40	-56.67	-54.49
N_2(aq)	18.70	18.70	18.70
N_2(g)	0.00	0.00	0.00
Nicotinamide Adenine Dinucleotide (NAD) [oxidized]	1038.86	1054.17	1059.11
Nicotinamide Adenine Dinucleotide (NAD) [reduced]	1101.47	1115.55	1120.09
Nicotinamide Adenine Dinucleotide Phosphate (NADP) [oxidized]	163.73	173.52	176.68
Nicotinamide Adenine Dinucleotide Phosphate (NADP) [reduced]	229.67	235.79	237.77
O_2(aq)	16.40	16.40	16.40
O_2(g)	0.00	0.00	0.00
Oxalate	-673.90	-676.35	-677.14
Oxaloacetate	-713.38	-714.60	-715.00
Oxalosuccinate	-979.05	-979.05	-979.05
2-Oxoglutarate	-633.58	-633.58	-633.58
Palmitate	979.25	997.61	1003.54
L-Phenylalanine	232.42	239.15	241.33
Phosphate	-1058.56	-1059.17	-1059.49
2-Phospho-*D*-glycerate	-1340.72	-1341.32	-1341.79
3-Phospho-*D*-glycerate	-1346.38	-1347.19	-1347.73
Phospho*enol*pyruvate	-1185.46	-1188.53	-1189.73
1-Propanol	143.84	148.74	150.32
2-Propanol	134.42	139.32	140.90
Pyrophosphate	-1934.95	-1939.13	-1940.66
Pyruvate	-352.40	-351.18	-350.78
Retinal	1118.78	1135.91	1141.45
Retinol	1170.78	1189.14	1195.07
Ribose	-339.23	-333.11	-331.13
Ribose 1-phosphate	-1215.87	-1212.24	-1211.14
Ribose 5-phosphate	-1223.95	-1220.32	-1219.22
Ribulose	-336.38	-330.26	-328.28
L-Serine	-231.18	-226.89	-225.51
Sorbose	-432.47	-425.13	-422.76
Succinate	-530.72	-530.65	-530.64
Succinyl Coenzyme A	-349.90	-348.06	-347.47
Sucrose	-685.66	-672.20	-667.85
Thioredoxin [oxidized]	0.00	0.00	0.00
Thioredoxin [reduced]	54.32	55.41	55.74
L-Tryptophan	364.78	372.12	374.49
L-Tyrosine	68.82	75.55	77.73
Ubiquinone [oxidized]	3596.07	3651.15	3668.94
Ubiquinone [reduced]	3586.06	3642.37	3660.55
Urate	-206.03	-204.81	-204.41
Urea	-42.97	-40.53	-39.74
Uric acid	-197.07	-194.63	-193.84
L-Valine	80.87	87.60	89.78
D-Xylose	-350.93	-344.81	-342.83
D-Xylulose	-346.59	-340.47	-338.49

APPARENT EQUILIBRIUM CONSTANTS FOR ENZYME-CATALYZED REACTIONS

Robert N. Goldberg

This table contains values of apparent equilibrium constants K' for selected enzyme-catalyzed reactions at specified temperatures T and pHs. In those cases where the ionic strength I and/or the pMg (pMg = $-\log_{10}[Mg^{2+}]$) have been reported, the values of these quantities are given. The Enzyme Commission numbers [Webb (1992)] of the enzymes that were used to catalyze the reactions are also given.

There are two fundamentally different types of equilibrium constants. This is illustrated by the following example for the hydrolysis of adenosine 5′-triphosphate (ATP) to adenosine 5′-diphosphate (ADP) and phosphate:

$$\text{ATP} + H_2O = \text{ADP} + \text{phosphate} \qquad (1)$$

The apparent equilibrium constant for the overall biochemical reaction (1) is

$$K' = [\text{ADP}][\text{phosphate}]/([\text{ATP}]c^{o}) \qquad (2)$$

The biochemical reactants ATP, ADP, and phosphate each exist in several different ionized and metal bound forms. For example, ATP is an equilibrium mixture of the species ATP^{4-}, $HATP^{3-}$, H_2ATP^{2-}, $MgATP^{2-}$, $MgHATP^{-}$, Mg_2ATP^{0}. Additional species would also have to be considered if Ca^{2+} were present. Thus, ATP has often been denoted in the literature as ΣATP or as $(\text{ATP})_{tot}$. When it is clear that one is dealing with total amounts of substances, it is not necessary to use either the Σ or "tot." Thus, these designations are not used in this table. In the above equation, c^{o} = 1 mol dm^{-3}; it is included to make K′ dimensionless. The standard transformed Gibbs energy of reaction $\Delta_r G'^{o}$ at specified conditions of temperature T, pressure P, ionic strength I, pH, and pMg can be calculated from K':

$$\Delta_r G'^{o} = -RT \ln K' \qquad (3)$$

The molar gas constant, R, is equal to 8.314 472 J K^{-1} mol^{-1}. $\Delta_r G'^{o}$ and the apparent equilibrium constant, K', can be used to calculate the position of equilibrium of overall biochemical reactions.

It is also possible to choose a chemical reference reaction that involves selected solute species:

$$ATP^{4-} + H_2O = ADP^{3-} + HPO_4^{2-} + H^{+} \qquad (4)$$

The equilibrium constant for this reference reaction is

$$K = [ADP^{3-}][HPO_4^{2-}][H^{+}]/\{[ATP^{4-}](c^{o})^2\} \qquad (5)$$

Equations and algorithms that relate these two different types of equilibrium constants have been published [Alberty (1969), Akers and Goldberg (2001), Alberty (2003)]. To calculate the equilibrium constant K for the reference reaction from the apparent equilibrium constant K', or vice versa, one needs the equilibrium constants for the binding of H^{+} and for the relevant metal ions to ATP^{4-}, ADP^{3-}, and HPO_4^{2-}.

To avoid confusion between the two different types of equilibrium constants (K' and K) and to avoid ambiguity about whether specific species or sums of species are intended, the word "ammonia," for example, rather than NH_3 or NH_4^{+}, is used for total ammonia, and chemical formulas are used for specific chemical species. Other substances such as carbon dioxide (CO_2, HCO_3^{-}, and CO_3^{2-}), and phosphate ($H_2PO_4^{-}$, HPO_4^{2-}, and PO_4^{3-}) are treated in the same manner. Exceptions are made for water, which is always written as H_2O, and for gaseous hydrogen and oxygen, which are written as $H_2(g)$ and $O_2(g)$, respectively.

For symmetrical reactions, there is no concern about the units used to calculate the value of an equilibrium constant. However, care must be exercised for reactions that are not symmetrical. In such cases, the units "mol dm^{-3}" have been used for all concentrations. As stated above, a c^{o} (1 mol dm^{-3}) is then used to make all equilibrium constants dimensionless.

All substances are assumed to be in aqueous solutions unless specified otherwise.

Values of $\Delta_r G'^{o}$ and K' can also be calculated for many biochemical reactions by using the table "Standard Transformed Gibbs Energies of Formation for Biochemical Reactants" in Section 7 of this Handbook.

Abbreviations

ADP	adenosine 5′-diphosphate
AMP	adenosine 5′-monophosphate
ATP	adenosine 5′-triphosphate
CoA	coenzyme A
GDP	guanosine 5′-diphosphate
GMP	guanosine 5′-monophosphate
GTP	guanosine 5′-triphosphate
IDP	inosine 5′-diphosphate
IMP	inosine 5′-monophosphate
ITP	inosine 5′-triphosphate
NAD_{ox}	β-nicotinamide-adenine dinucleotide, oxidized form
NAD_{red}	β-nicotinamide-adenine dinucleotide, reduced form
$NADP_{ox}$	β-nicotinamide-adenine dinucleotide phosphate, oxidized form
$NADP_{red}$	β-nicotinamide-adenine dinucleotide phosphate, reduced form
UDP	uridine 5′-diphosphate
UTP	uridine 5′-triphosphate

References

Akers, D.L., and Goldberg, R.N., *Mathematica J.*, 8, 86–113 (2001).

Alberty, R.A., *J. Biol. Chem.*, 243, 1337–1343, 1969.

Alberty, R.A., *Thermodynamics of Biochemical Reactions*, Wiley-Interscience, Hoboken, NJ, 2003.

Goldberg, R.N., Tewari, Y.B., Bell, D., Fazio, K., and Anderson, E., *J. Phys. Chem. Ref. Data*, 22, 515-582 1993.

Goldberg, R.N., and Tewari, Y.B., *J. Phys. Chem. Ref. Data*, 23, 547-617, 1994.

Goldberg, R.N., and Tewari, Y.B., *J. Phys. Chem. Ref. Data*, 23, 1035-1103, 1994.

Goldberg, R.N., and Tewari, Y.B., *J. Phys. Chem. Ref. Data*, 24, 1669-1698, 1995.

Goldberg, R.N., and Tewari, Y.B., *J. Phys. Chem. Ref. Data*, 24, 1765-1801, 1995.

Goldberg, R.N., *J. Phys. Chem. Ref. Data*, 28, 931–965, 1999.

Goldberg, R.N., Tewari, Y.B., and Bhat, T.N., *Bioinformatics*, 20, 2874–2877, 2004; <xpdb.nist.gov/enzyme_thermodynamics/>.

Goldberg, R.N., Tewari, Y.B., and Bhat, T.N., *J. Phys. Chem. Ref. Data*, 36, 1347–1397, 2007.

Webb, E.C., *Enzyme Nomenclature 1992*, Academic Press, New York, 1992. See also <www.chem.qmul.ac.uk/iubmb/enzyme/>.

Reaction	K'	Enzyme Commission Number	T K	pH	I mol dm^{-3}	pMg
benzyl alcohol + NAD$_{ox}$ = benzaldehyde + NAD$_{red}$	$9.8 \cdot 10^{-4}$	1.1.1.1	298.15	7.5		
1-butanol + NAD$_{ox}$ = butanal + NAD$_{red}$	$1.8 \cdot 10^{-3}$	1.1.1.1	298.15	8.3		
cyclohexanol + NAD$_{ox}$ = cyclohexanone + NADH$_{red}$	0.090	1.1.1.1	298.15	7.2		
1-hexanol + NAD$_{ox}$ = hexanal + NAD$_{red}$	$2.87 \cdot 10^{-3}$	1.1.1.1	298.15	8.3		
1-octanol + NAD$_{ox}$ = octanal + NAD$_{red}$	$1.1 \cdot 10^{-3}$	1.1.1.1	298.15	8.3		
L-homoserine + NADP$_{ox}$ = L-aspartate 4-semialdehyde + NADP$_{red}$	$6.3 \cdot 10^{-4}$	1.1.1.3	298.15	7.9		
xylitol + NAD$_{ox}$ = L-xylulose + NAD$_{red}$	$2.97 \cdot 10^{-4}$	1.1.1.10	298.15	7.00		
D-sorbitol + NAD$_{ox}$ = D-fructose + NAD$_{red}$	0.032	1.1.1.14	298.15	7.0		
quinate + NAD$_{ox}$ = 5-dehydroquinate + NAD$_{red}$	$4.61 \cdot 10^{-3}$	1.1.1.24	305.15	7.2		
shikimate + NADP$_{ox}$ = 5-dehydroshikimate + NADP$_{red}$	0.036	1.1.1.25	303.15	7.0		
2-hydroxybutanoate + NAD$_{ox}$ = 2-oxobutanoate + NAD$_{red}$	$3.0 \cdot 10^{-3}$	1.1.1.27	298.65	8.0		
(R)-3-hydroxybutanoate + NAD$_{ox}$ = 3-oxobutanoate + NAD$_{red}$	$1.9 \cdot 10^{-3}$	1.1.1.30	298.15	7.0		
D-glucose 6-phosphate + NADP$_{ox}$ = D-glucono-1,5-lactone 6-phosphate + NADP$_{red}$	1.50	1.1.1.49	301.15	6.40		
5α-androstane-3α-ol-17-one + NAD$_{ox}$ = 5α-androstane-3,17-dione + NAD$_{red}$	0.058	1.1.1.50	298.15	7.0		
5α-pregnane-3α,17α,21-triol-20-one + NAD$_{ox}$ = 5α-pregnane-17α,21-diol-3,20-dione + NAD$_{red}$	0.0113	1.1.1.50	298.15	7.0		
5α-androstane-3β,17α-diol + NAD$_{ox}$ = 5α-androstane-17α-ol-3-one + NAD$_{red}$	0.0211	1.1.1.51	298.15	7.0		
4-androstene-17β-ol-3-one + NAD$_{ox}$ = 4-androstene-3,17-dione + NAD$_{red}$	0.378	1.1.1.51	298.15	7.0		
1,2-propanediol + NADP$_{ox}$ = L-lactaldehyde + NADP$_{red}$	$6.0 \cdot 10^{-5}$	1.1.1.55	298.15	8.4		
ribitol + NAD$_{ox}$ = D-ribulose + NAD$_{red}$	$3.1 \cdot 10^{-3}$	1.1.1.56	310.15	7.4		
3-hydroxypropanoate + NAD$_{ox}$ = 3-oxopropanoate + NAD$_{red}$	$9.0 \cdot 10^{-3}$	1.1.1.59	298.15	9.0		
estradiol-17β + NAD$_{ox}$ = estrone + NAD$_{red}$	0.18	1.1.1.62	298.15	7.00		
benzyl alcohol + NAD$_{ox}$ = benzaldehyde + NAD$_{red}$	0.097	1.1.1.90	300.15	9.5		
L-carnitine + NAD$_{ox}$ = 3-dehydrocarnitine + NAD$_{red}$	$1.3 \cdot 10^{-4}$	1.1.1.108	303.15	7.0		
L-threonate + NAD$_{ox}$ = 3-oxo- L-threonate + NAD$_{red}$	$3.42 \cdot 10^{-4}$	1.1.1.129	298.15	7.0		
prostaglandin E$_1$ + NAD$_{ox}$ = 15-oxo-prostaglandin E$_1$ + NAD$_{red}$	0.65	1.1.1.141	298.15	7.0		
7,8-dihydrobiopterin + NADP$_{ox}$ = sepiapterin + NADP$_{red}$	0.045	1.1.1.153	298.15	8.0		
glycine + acetaldehyde = L-threonine	56	2.1.2.1	310.15	7.6		
sedoheptulose 7-phosphate + D-glyceraldehyde 3-phosphate = D-ribose 5-phosphate + D-xylulose 5-phosphate	0.48	2.2.1.1	311.15	7.0	0.25	3.0
acetyl-CoA + choline = CoA + O-acetylcholine	1.60	2.3.1.7	298.15	7.0	0.25	
acetyl-CoA + acyl-carrier protein = CoA + acetyl-[acyl-carrier protein]	2.09	2.3.1.38	311.15	6.5		
UDPglucose + D-fructose = UDP + sucrose	6.7	2.4.1.13	298.15	7.5		
cellobiose + orthophosphate = D-glucose + α-D-glucose 1-phosphate	0.23	2.4.1.20	310.15	7.0		
laminaritriose + orthophosphate = laminaribiose + α-D-glucose 1-phosphate	0.26	2.4.1.31	310.15	6.5		
α,α-trehalose + orthophosphate = D-glucose + β-D-glucose 1-phosphate	0.24	2.4.1.64	310.15	7.0		
UDPglucose + sinapate = UDP + 1-sinapoyl-D-glucose	0.21	2.4.1.120	303.15	6.0		
inosine + orthophosphate = hypoxanthine + α-D-ribose 1-phosphate	0.0164	2.4.2.1	311.15	7.0	0.25	3.0
xanthosine + orthophosphate = xanthine + α-D-ribose 1-phosphate	0.0156	2.4.2.1	311.15	7.0	0.25	3.0
uridine + orthophosphate = uracil + α-D-ribose 1-phosphate	0.44	2.4.2.2	310.15	7.0		
adenine + 5-phospho-α-D-ribose 1-diphosphate = AMP + pyrophosphate	$2 \cdot 10^{3}$	2.4.2.7	311.15	7.4	0.25	3.0
GMP + hypoxanthine = IMP + guanine	0.38	2.4.2.8	310.15	7.4		
guanine + 5-phospho-α-D-ribose 1-diphosphate = GMP + pyrophosphate	$1 \cdot 10^{5}$	2.4.2.8	311.15	7.4	0.25	3.0
hypoxanthine + 5-phospho-α-D-ribose 1-diphosphate = IMP + pyrophosphate	$1 \cdot 10^{5}$	2.4.2.8	311.15	7.4	0.25	3.0
ATP + ammonium carbamate = ADP + carbamoyl phosphate	0.042	2.7.2.2	283.15	9.4		
ATP + creatine = ADP + phosphocreatine	$5.78 \cdot 10^{-3}$	2.7.3.2	310.15	7.11	0.25	2.47
ATP + L-arginine = ADP + N$^{\omega}$-phospho-L-arginine	0.10	2.7.3.3	285.15	7.25		
ATP + sulfate = adenosine 5′-phosphosulfate + pyrophosphate	$4 \cdot 10^{-8}$	2.7.7.4	303.15	7.5		
UTP + α-D-glucose 1-phosphate = pyrophosphate + UDPglucose	0.48	2.7.7.9	310.15	8.0		
succinyl-CoA + acetoacetate = succinate + acetoacetyl-CoA	$2.8 \cdot 10^{-3}$	2.8.3.5	303.15	7.0		

Reaction	K'	Enzyme Commission Number	T K	pH	I mol dm^{-3}	pMg
acetylcholine + H_2O = acetate + choline	$5.38 \cdot 10^2$	3.1.1.7	296.15	5.1		
IMP + H_2O = inosine + orthophosphate	$1.58 \cdot 10^2$	3.1.3.1	298.15	8.55	1.53	4.44
phosphorylcholine + H_2O = choline + orthophosphate	49.9	3.1.3.1	311.15	6.90		
L-*O*-phosphoserine + H_2O = L-serine + orthophosphate	56	3.1.3.1	308.15	7.0		
cytidine 2′:3′-(cyclic)phosphate + H_2O = cytidine 3′-monophosphate	$1.06 \cdot 10^3$	3.1.27.5	298.15	6.0		
isomaltose + H_2O = 2 D-glucose	17.2	3.2.1.3	298.15	5.65		
β-gentiobiose + H_2O = 2 D-glucose	17.7	3.2.1.21	298.15	5.65		
3-O-β-D-galactopyranosyl-D-arabinose + H_2O = D-galactose + D-arabinose	$1.04 \cdot 10^2$	3.2.1.23	298.15	5.65		
lactulose + H_2O = D-galactose + D-fructose	$1.28 \cdot 10^2$	3.2.1.23	298.15	5.65		
4′,5′-anhydroadenosine + H_2O = adenosine	0.48	3.3.1.1	310.15	7.0		
pteroylglutamate + H_2O = pteroate + L-glutamate	15.6	3.4.19.9	310.15	7.3		
N-acetyl-L-phenylalanine methyl ester + H_2O = *N*-acetyl-L-phenylalanine + methanol	$5.88 \cdot 10^2$	3.4.21.1	293.15	5.5		
hippurylanilide + H_2O = hippuric acid + aniline	11	3.4.22.2	312.15	5.0		
ammonium carbamate + H_2O = 2 ammonia + carbon dioxide	$1.92 \cdot 10^3$	3.5.1.5	293.15	6.5		
ampicillin + H_2O = 6-aminopenicillanic acid + D(−)-α-aminophenylacetic acid	0.013	3.5.1.11	298.15	5.0		
cephalexin + H_2O = 7-aminodeacetoxycephalosporanic acid + D(−)-α-aminophenylacetic acid	0.044	3.5.1.11	298.15	5.8		
cephaloridine + H_2O = 2-thienylacetic acid + 7-amino-3-(1-pyridyl-methyl)-3-cephem-4-carboxylic acid	0.015	3.5.1.11	298.15	5.0		
penicillin G + H_2O = 6-aminopenicillanic acid + phenylacetic acid	0.445	3.5.1.11	298.15	6.71		
N-acetyl-L-alanine + H_2O = acetate + L-alanine	7	3.5.1.14	298.15	6.0		
ampicillin + H_2O = ampicillinoic acid	95	3.5.2.6	282.35	5.55		
penicillin G + H_2O = penicillinoic acid	2.9	3.5.2.6	298.15	6.01		
cytidine + H_2O = uridine + ammonia	$1.03 \cdot 10^4$	3.5.4.5	298.15	7.00		
N^4-methylcytidine + H_2O = uridine + methylamine	$4.88 \cdot 10^2$	3.5.4.5	298.15	7.50		
5,10-methenyltetrahydrofolate + H_2O = 10-formyltetrahydrofolate	50	3.5.4.9	298.15	7.0		
ITP + oxaloacetate + H_2O = IDP + phospho*enol*pyruvate + carbon dioxide	12	4.1.1.32	303.15	7.6		
2-deoxy-D-ribose 5-phosphate = D-glyceraldehyde 3-phosphate + acetaldehyde	$2.5 \cdot 10^{-4}$	4.1.2.4	295.15	7.5		
6-phospho-2-dehydro-3-deoxy-D-gluconate = pyruvate + D-glyceraldehyde 3-phosphate	$1.2 \cdot 10^{-3}$	4.1.2.14	298.15	8.0	0.37	
L-fuculose 1-phosphate = glycerone phosphate + (*S*)-lactaldehyde	$4.6 \cdot 10^{-4}$	4.1.2.17	310.15	7.2		
L-rhamnulose 1-phosphate = glycerone phosphate + (*S*)-lactaldehyde	0.083	4.1.2.19	310.15	7.5		
isocitrate = succinate + glyoxylate	$2.3 \cdot 10^{-3}$	4.1.3.1	303.15	7.7		
(*S*)-2-methylmalate = acetate + pyruvate	0.151	4.1.3.22	298.15	7.4	0.845	
isocitrate = citrate	14.7	4.2.1.3	298.15	7.4		
3-dehydroquinate = 3-dehydroshikimate + H_2O	15	4.2.1.10	302.15	7.4		
(3*R*)-3-hydroxybutanoyl-CoA = *cis*-but-2-enoyl-CoA + H_2O	0.18	4.2.1.17	298.15	7.5		
indole + D-glyceraldehyde 3-phosphate = 1-(indol-3-yl)glycerol 3-phosphate	$1.2 \cdot 10^4$	4.2.1.20	298.15	7.54		
(*R*)-malate = maleate + H_2O	$4.88 \cdot 10^{-4}$	4.2.1.31	298.15	7.00	0.10	
(*R*)-2-methylmalate = 2-methylmaleate + H_2O	0.0962	4.2.1.35	298.15	7.0	0.10	
D-glutamate = 5-oxo-D-proline + H_2O	24.3	4.2.1.48	293.4	7.9		
L-*threo*-3-methylaspartate = 2-methylfumarate + ammonia	0.238	4.3.1.2	298.15	7.9		
L-histidine = urocanate + ammonia	3.01	4.3.1.3	298.25	8.41	0.167	
L-phenylalanine = *trans*-cinnamate + ammonia	2.47	4.3.1.5	298.05	7.69		
ATP = adenosine 3′:5′-(cyclic)phosphate + diphosphate	0.065	4.6.1.1	298.15	7.0		
L,L-2,6-diaminoheptanedioate = *meso*-diaminoheptanedioate	1.9	5.1.1.7	310.15	7.0		
D-ribulose 5-phosphate = D-xylulose 5-phosphate	1.82	5.1.3.1	311.15	7.0	0.25	3.0
UDPglucose = UDPgalactose	0.33	5.1.3.2	298.15	8.7		
GDPmannose = GDP-L-galactose	0.52	5.1.3.18	310.15	8.0		
all-*trans*-retinal = 11-*cis*-retinal	0.05	5.2.1.3	309.15	7.0		

Reaction	K'	Enzyme Commission Number	T K	pH	I mol dm^{-3}	pMg
9-*cis*,12-*cis*-octadecadienoate = 9-*cis*,11-*trans*-octadecadienoate	61	5.2.1.5	308.15	7.0		
D-erythrose = D-erythrulose	2.3	5.3.1.2	308.15	5.8		
D-arabinose = D-ribulose	0.146	5.3.1.3	320.25	7.4		
L-fucose = L-fuculose	0.12	5.3.1.3	310.15	8.0		
L-arabinose = L-ribulose	0.11	5.3.1.4	298.15	7.0		
D-psicose = β-D-allose	2.15	5.3.1.4	317.25	7.4		
D-ribose 5-phosphate = D-ribulose 5-phosphate	0.83	5.3.1.6	311.15	7.0	0.25	3.0
D-rhamnose = D-rhamnulose	0.58	5.3.1.7	303.15	7.4		
D-mannose 6-phosphate = D-fructose 6-phosphate	0.99	5.3.1.8	298.15	8.50		
6-amino-D-glucose 6-phosphate = 6-amino-D-fructose 6-phosphate	0.202	5.3.1.9	278.85	8.7		
D-glucosamine 6-phosphate + H_2O = D-fructose 6-phosphate + ammonia	0.15	5.3.1.10	310.15	8.4		
D-lyxose = D-xylulose	0.23	5.3.1.15	298.15	7.0		
D-ribose = D-ribulose	0.391	5.3.1.20	313.15	7.4		
keto-phenylpyruvate = *enol*-phenylpyruvate	0.1	5.3.2.1	298.15	7.8		
L-lysine = (3*S*)-3,6-diaminohexanoate	5.3	5.4.3.2	303.15	7.7		
(*R*)-methylmalonyl-CoA = succinyl-CoA	23.1	5.4.99.2	298.15	7.4		
(−)-4-carboxymethyl-Δ^{α}-but-2-en-4-olide = *cis,trans*-hexadienedioate	4.0	5.5.1.1	303.15	8.0		
ATP + heptanoate + CoA = AMP + diphosphate + *n*-heptanoyl-CoA	1.11	6.2.1.2	311.15	8.0		
GTP + succinate + CoA = GDP + phosphate + succinyl-CoA	1.68	6.2.1.4	298.15	7.15	0.25	2.91
GTP + IMP + L-aspartate = GDP + phosphate + adenylosuccinate	2.9	6.3.4.4	310.15	8.0		
ATP + L-citrulline + L-aspartate = AMP + diphosphate + L-arginosuccinate	2.14	6.3.4.5	311.15	6.91		
ATP + propanoyl-CoA + carbon dioxide = ADP + phosphate + (*S*)-methylmalonyl-CoA	$8.1 \cdot 10^{-3}$	6.4.1.3	310.15	8.15		

THERMODYNAMIC QUANTITIES FOR THE IONIZATION REACTIONS OF BUFFERS IN WATER

Robert N. Goldberg, Nand Kishore, and Rebecca M. Lennen

This table contains selected values for the pK, standard molar enthalpy of reaction $\Delta_r H^\circ$, and standard molar heat-capacity change $\Delta_r C_p^\circ$ for the ionization reactions of 64 buffers many of which are relevant to biochemistry and to biology.[1] The values pertain to the temperature $T = 298.15$ K and the pressure $p = 0.1$ MPa. The standard state is the hypothetical ideal solution of unit molality. These data permit one to calculate values of the pK and of $\Delta_r H^\circ$ at temperatures in the vicinity $\{T \approx (274 \text{ K to } 350 \text{ K})\}$ of the reference temperature $\theta = 298.15$ K by using the following equations[2]

$$\Delta_r G_T^\circ = -RT\ln K_T = \ln(10) \cdot RT \cdot \mathrm{p}K_T, \tag{1}$$

$$R\ln K_T = -(\Delta_r G_\theta^\circ/\theta) + \Delta_r H_\theta^\circ\{(1/\theta) - (1/T)\} + \Delta_r C_{p\theta}^\circ\{(\theta/T) - 1 + \ln(T/\theta)\}, \tag{2}$$

$$\Delta_r H_T^\circ = \Delta_r H_\theta^\circ + \Delta_r C_{p\theta}^\circ(T - \theta). \tag{3}$$

Here, $\Delta_r G^\circ$ is the standard molar Gibbs energy change and K is the equilibrium constant for a reaction; R is the gas constant (8.314 472 J K^{-1} mol^{-1}). The subscripts T and θ denote the temperature to which a quantity pertains, the subscript p denotes constant pressure, and the subscript r denotes that the quantity refers to a reaction. Combination of equations (1) and (2) yields the following equation that gives pK as a function of temperature:

$$\mathrm{p}K_T = -\{R \cdot \ln(10)\}^{-1}[-\{\ln(10) \cdot RT \cdot \mathrm{p}K_\theta/\theta\} + \Delta_r H_\theta^\circ\{(1/\theta) - (1/T)\} + \Delta_r C_{p\theta}^\circ\{(\theta/T) - 1 + \ln(T/\theta)\}]. \tag{4}$$

The above equations neglect higher order terms that involve temperature derivatives of $\Delta_r C_p^\circ$. Also, it is important to recognize that the values of pK and $\Delta_r H^\circ$ effectively pertain to ionic strength $I = 0$. However, the values of pK and $\Delta_r H^\circ$ are almost always dependent on the ionic strength and the actual composition of the solution. These issues are discussed in Reference 1, which also gives an approximate method for making appropriate corrections.

References

1. Goldberg, R. N., Kishore, N., and Lennen, R. M., "Thermodynamic Quantities for the Ionization Reactions of Buffers," *J. Phys. Chem. Ref. Data*, 31, 231, 2002.
2. Clarke, E. C. W., and Glew, D. N., *Trans. Faraday Soc.*, 62, 539-547, 1966.

Selected Values of Thermodynamic Quantities for the Ionization Reactions of Buffers in Water at T = 298.15 K and p = 0.1 MPa

Buffer	Reaction	pK	$\Delta_r H^\circ$ kJ mol^{-1}	$\Delta_r C_p^\circ$ J mol^{-1} K^{-1}
ACES	$HL^\pm = H^+ + L^-$, ($HL = C_4H_{10}N_2O_4S$)	6.847	30.43	−49
Acetate	$HL = H^+ + L^-$, ($HL = C_2H_4O_2$)	4.756	−0.41	−142
ADA	$H_3L^+ = H^+ + H_2L^\pm$, ($H_2L = C_6H_{10}N_2O_5$)	1.59		
	$H_2L^\pm = H^+ + HL^-$	2.48	16.7	
	$HL^- = H^+ + L^{2-}$	6.844	12.23	−144
2-Amino-2-methyl-1,3-propanediol	$HL^+ = H^+ + L$, ($L = C_4H_{11}NO_2$)	8.801	49.85	−44
2-Amino-2-methyl-1-propanol	$HL^+ = H^+ + L$, ($L = C_4H_{11}NO$)	9.694	54.05	≈−21
3-Amino-1-propanesulfonic acid	$HL = H^+ + L^-$, ($HL = C_3H_9NO_3S$)	10.2		
Ammonia	$NH_4^+ = H^+ + NH_3$	9.245	51.95	8
AMPSO	$HL^\pm = H^+ + L^-$, ($HL = C_7H_{17}NO_5S$)	9.138	43.19	−61
Arsenate	$H_3AsO_4 = H^+ + H_2AsO_4^-$	2.31	−7.8	
	$H_2AsO_4^- = H^+ + HAsO_4^{2-}$	7.05	1.7	
	$HAsO_4^{2-} = H^+ + AsO_4^{3-}$	11.9	15.9	
Barbital	$H_2L = H^+ + HL^-$, ($H_2L = C_8H_{12}N_2O_3$)	7.980	24.27	−135
	$HL^- = H^+ + L^{2-}$	12.8		
BES	$HL^\pm = H^+ + L^-$, ($HL = C_6H_{15}NO_5S$)	7.187	24.25	−2
Bicine	$H_2L^+ = H^+ + HL^\pm$, ($HL = C_6H_{13}NO_4$)	2.0		
	$HL^\pm = H^+ + L^-$	8.334	26.34	0
Bis-tris	$H_3L^+ = H^+ + H_2L^\pm$, ($H_2L = C_8H_{19}NO_5$)	6.484	28.4	27
Bis-tris propane	$H_2L^{2+} = H^+ + HL^+$, ($L = C_{11}H_{26}N_2O_6$)	6.65		
	$HL^+ = H^+ + L$	9.10		
Borate	$H_3BO_3 = H^+ + H_2BO_3^-$	9.237	13.8	≈−240
Cacodylate	$H_2L^+ = H^+ + HL$, ($HL = C_2H_6AsO_2$)	1.78	−3.5	
	$HL = H^+ + L^-$	6.28	−3.0	−86
CAPS	$HL^\pm = H^+ + L^-$, ($HL = C_9H_{19}NO_3S$)	10.499	48.1	57
CAPSO	$HL^\pm = H^+ + L^-$, ($HL = C_9H_{19}NO_4S$)	9.825	46.67	21
Carbonate	$H_2CO_3 = H^+ + HCO_3^-$	6.351	9.15	−371
	$HCO_3^- = H^+ + CO_3^{2-}$	10.329	14.70	−249
CHES	$HL^\pm = H^+ + L^-$, ($HL = C_8H_{17}NO_3S$)	9.394	39.55	9

Buffer	Reaction	p*K*	$\Delta_r H°$ kJ mol^{-1}	$\Delta_r C_p°$ J mol^{-1} K^{-1}
Citrate	$H_3L = H^+ + H_2L^-$, ($H_3L = C_6H_8O_7$)	3.128	4.07	−131
	$H_2L^- = H^+ + HL^{2-}$	4.761	2.23	−178
	$HL^{2-} = H^+ + L^{3-}$	6.396	−3.38	−254
L-Cysteine	$H_3L^+ = H^+ + H_2L$, ($H_2L = C_3H_7NO_2S$)	1.71	≈−0.6	
	$H_2L = H^+ + HL^-$	8.36	36.1	≈−66
	$HL^- = H^+ + L^{2-}$	10.75	34.1	≈−204
Diethanolamine	$HL^+ = H^+ + L$, ($L = C_4H_{11}NO_2$)	8.883	42.08	36
Diglycolate	$H_2L = H^+ + HL^-$, ($H_2L = C_4H_6O_5$)	3.05	−0.1	≈−142
	$HL^- = H^+ + L^{2-}$	4.37	−7.2	≈−138
3,3-Dimethylglutarate	$H_2L = H^+ + HL^-$, ($H_2L = C_7H_{12}O_4$)	3.70		
	$HL^- = H^+ + L^{2-}$	6.34		
DIPSO	$HL^{\pm} = H^+ + L^-$, ($HL = C_7H_{17}NO_6S$)	7.576	30.18	42
Ethanolamine	$HL^+ = H^+ + L$, ($L = C_2H_7NO$)	9.498	50.52	26
N-Ethylmorpholine	$HL^+ = H^+ + L$, ($L = C_6H_{13}NO$)	7.77	27.4	
Glycerol 2-phosphate	$H_2L = H^+ + HL^-$, ($H_2L = C_3H_9NO_6P$)	1.329	−12.2	−330
	$HL^- = H^+ + L^{2-}$	6.650	−1.85	−212
Glycine	$H_2L^+ = H^+ + HL^{\pm}$, ($HL = C_2H_5NO_2$)	2.351	4.00	−139
	$HL^{\pm} = H^+ + L^-$	9.780	44.2	−57
Glycine amide	$HL^+ = H^+ + L$, ($L = C_2H_6N_2O$)	8.04	42.9	
Glycylglycine	$H_2L^+ = H^+ + HL^{\pm}$, ($HL = C_4H_8N_2O_3$)	3.140	0.11	−128
	$HL^{\pm} = H^+ + L^-$	8.265	43.4	−16
Glycylglycylglycine	$H_2L^+ = H^+ + HL^{\pm}$, ($HL = C_6H_{11}N_3O_4$)	3.224	0.84	
	$HL^{\pm} = H^+ + L^-$	8.090	41.7	
HEPES	$H_2L^+ = H^+ + HL^{\pm}$, ($HL = C_8H_{18}N_2O_4S$)	≈3.0		
	$HL^{\pm} = H^+ + L^-$	7.564	20.4	47
HEPPS	$HL^{\pm} = H^+ + L^-$, ($HL = C_6H_{20}N_2O_4S$)	7.957	21.3	48
HEPPSO	$HL^{\pm} = H^+ + L^-$, ($HL = C_9H_{20}N_2O_5S$)	8.042	23.70	47
L-Histidine	$H_3L^{2+} = H^+ + H_2L^+$, ($HL = C_6H_9N_3O_2$)	1.5_4	3.6	
	$H_2L^+ = H^+ + HL$	6.07	29.5	176
	$HL = H^+ + L^-$	9.34	43.8	−233
Hydrazine	$H_2L^{2+} = H^+ + HL^+$, ($L = H_4N_2$)	−0.99	38.1	
	$HL^+ = H^+ + L$	8.02	41.7	
Imidazole	$HL^+ = H^+ + L$, ($L = C_3H_4N_2$)	6.993	36.64	−9
Maleate	$H_2L = H^+ + HL^-$, ($H_2L = C_4H_4O_4$)	1.92	1.1	≈−21
	$HL^- = H^+ + L^{2-}$	6.27	−3.6	≈−31
2-Mercaptoethanol	$HL = H^+ + L^-$, ($HL = C_2H_6OS$)	9.7_5	26.2	
MES	$HL^{\pm} = H^+ + L^-$, ($HL = C_6H_{13}NO_4S$)	6.270	14.8	5
Methylamine	$HL^+ = H^+ + L$, ($L = CH_5N$)	10.645	55.34	33
2-Methylimidazole	$HL^+ = H^+ + L$, ($L = C_4H_6N_2$)	8.0_1	36.8	
MOPS	$HL^{\pm} = H^+ + L^-$, ($HL = C_7H_{15}NO_4S$)	7.184	21.1	25
MOPSO	$H_2L^+ = H^+ + HL^{\pm}$, ($HL = C_7H_{15}NO_5S$)	0.060		
	$HL^{\pm} = H^+ + L^-$	6.90	25.0	≈38
Oxalate	$H_2L = H^+ + HL^-$, ($H_2L = C_2H_2O_4$)	1.27	−3.9	≈−231
	$HL^- = H^+ + L^{2-}$	4.266	7.00	−231
Phosphate	$H_3PO_4 = H^+ + H_2PO_4^-$	2.148	−8.0	−141
	$H_2PO^-_4 = H^+ + HPO_4^{2-}$	7.198	3.6	−230
	$HPO_4^{2-} = H^+ + PO_4^{3-}$	12.35	16.0	−242
Phthalate	$H_2L = H^+ + HL^-$, ($H_2L = C_8H_6O_4$)	2.950	−2.70	−91
	$HL^- = H^+ + L^{2-}$	5.408	−2.17	−295
Piperazine	$H_2L^{2+} = H^+ + HL^+$, ($L = C_4H_{10}N_2$)	5.333	31.11	86
	$HL^+ = H^+ + L$	9.731	42.89	75
PIPES	$HL^{\pm} = H^+ + L^-$, ($HL = C_8H_{18}N_2O_6S_2$)	7.141	11.2	22
POPSO	$HL^{\pm} = H^+ + L^-$, ($HL = C_{10}H_{22}N_2O_8S_2$)	≈8.0		
Pyrophosphate	$H_4P_2O_7 = H^+ + H_3P_2O_7^-$	0.83	−9.2	≈−90
	$H_3P_2O^-_7 = H^+ + H_2P_2O_7^{2-}$	2.26	−5.0	≈−130
	$H_2P_2O_7^{2-} = H^+ + HP_2O_7^{3-}$	6.72	0.5	−136
	$HP_2O_7^{3-} = H^+ + P_2O_7^{4-}$	9.46	1.4	−141
Succinate	$H_2L = H^+ + HL^-$, ($H_2L = C_4H_6O_4$)	4.207	3.0	−121
	$HL^- = H^+ + L^{2-}$	5.636	−0.5	−217
Sulfate	$HSO^-_4 = H^+ + SO_4^{2-}$	1.987	−22.4	−258

Thermodynamic Quantities for the Ionization Reactions of Buffers in Water

Buffer	Reaction	p*K*	$\Delta_r H°$ kJ mol^{-1}	$\Delta_r C_p°$ J mol^{-1} K^{-1}
Sulfite	$H_2SO_3 = H^+ + HSO_3^-$	1.857	−17.80	−272
	$HSO^-_3 = H^+ + SO_3^{2-}$	7.172	−3.65	−262
TAPS	$HL^{\pm} = H^+ + L^-$, ($HL = C_7H_{17}NO_6S$)	8.44	40.4	15
TAPSO	$HL^{\pm} = H^+ + L^-$, ($HL = C_7H_{17}NO_7S$)	7.635	39.09	16
L(+)-Tartaric acid	$H_2L = H^+ + HL^-$, ($H_2L = C_4H_6O_6$)	3.036	3.19	−147
	$HL^- = H^+ + L^{2-}$	4.366	0.93	−218
TES	$HL^{\pm} = H^+ + L^-$, ($HL = C_6H_{15}NO_6S$)	7.550	32.13	0
Tricine	$H_2L^+ = H^+ + HL^{\pm}$, ($HL = C_6H_{13}NO_5$)	2.023	5.85	−196
	$HL^{\pm} = H^+ + L^-$	8.135	31.37	−53
Triethanolamine	$HL^+ = H^+ + L$, ($L = C_6H_{15}NO_3$)	7.762	33.6	50
Triethylamine	$HL^+ = H^+ + L$, ($L = C_6H_{15}N$)	10.72	43.13	151
Tris	$HL^+ = H^+ + L$, ($L = C_4H_{11}NO_3$)	8.072	47.45	−59

BIOLOGICAL BUFFERS

This table of frequently used buffers gives the pK_a value at 25 °C and the useful pH range of each buffer. The buffers are listed in order of increasing pH.

The table is reprinted with permission of Sigma Chemical Company, St. Louis, MO.

Acronym	Name	Mol. wt.	pK_a	Useful pH range
MES	2-(*N*-Morpholino)ethanesulfonic acid	195.2	6.1	5.5–6.7
BIS TRIS	Bis(2-hydroxyethyl)iminotris(hydroxymethyl)methane	209.2	6.5	5.8–7.2
ADA	*N*-(2-Acetamido)-2-iminodiacetic acid	190.2	6.6	6.0–7.2
ACES	2-[(2-Amino-2-oxoethyl)amino]ethanesulfonic acid	182.2	6.8	6.1–7.5
PIPES	Piperazine-*N,N′*-bis(2-ethanesulfonic acid)	302.4	6.8	6.1–7.5
MOPSO	3-(*N*-Morpholino)-2-hydroxypropanesulfonic acid	225.3	6.9	6.2–7.6
BIS TRISPROPANE	1,3-Bis[tris(hydroxymethyl)methylamino]propane	282.3	6.8[a]	6.3–9.5
BES	*N,N*-Bis(2-hydroxyethyl)-2-aminoethanesulfonic acid	213.2	7.1	6.4–7.8
MOPS	3-(*N*-Morpholino)propanesulfonic acid	209.3	7.2	6.5–7.9
HEPES	*N*-(2-Hydroxyethyl)piperazine-*N′*-(2-ethanesulfonic acid)	238.3	7.5	6.8–8.2
TES	*N*-Tris(hydroxymethyl)methyl-2-aminoethanesulfonic acid	229.2	7.5	6.8–8.2
DIPSO	3-[*N,N*-Bis(2-hydroxyethyl)amino]-2-hydroxypropanesulfonic acid	243.3	7.6	7.0–8.2
TAPSO	3-[*N*-Tris(hydroxymethyl)methylamino)-2-hydroxypropanesulfonic acid	259.3	7.6	7.0–8.2
TRIZMA	Tris(hydroxymethyl)aminomethane	121.1	8.1	7.0–9.1
HEPPSO	*N*-(2-hydroxyethyl)piperazine-*N′*-(2-hydroxypropanesulfonic acid)	268.3	7.8	7.1–8.5
POPSO	Piperazine-*N,N′*-bis(2-hydroxypropanesulfonic acid)	362.4	7.8	7.2–8.5
EPPS	*N*-(2-Hydroxyethyl)piperazine-*N′*-(3-propanesulfonic acid)	252.3	8.0	7.3–8.7
TEA	Triethanolamine	149.2	7.8	7.3–8.3
TRICINE	*N*-Tris(hydroxymethyl)methylglycine	179.2	8.1	7.4–8.8
BICINE	*N,N*-Bis(2-hydroxyethyl)glycine	163.2	8.3	7.6–9.0
TAPS	*N*-Tris(hydroxymethyl)methyl-3-aminopropanesulfonic acid	243.3	8.4	7.7–9.1
AMPSO	3-[(1,1-Dimethyl-2-hydroxyethyl)amino]-2-hydroxypropanesulfonic acid	227.3	9.0	8.3–9.7
CHES	2-(*N*-Cyclohexylamino)ethanesulfonic acid	207.3	9.3	8.6–10.0
CAPSO	3-(Cyclohexylamino)-2-hydroxy-1-propanesulfonic acid	237.3	9.6	8.9–10.3
AMP	2-Amino-2-methyl-1-propanol	89.1	9.7	9.0–10.5
CAPS	3-(Cyclohexylamino)-1-propanesulfonic acid	221.3	10.4	9.7–11.1

[a] pK_a = 9.0 for the second dissociation stage.

TYPICAL pH VALUES OF BIOLOGICAL MATERIALS AND FOODS

This table gives typical pH ranges for various biological fluids and common foods. All values refer to 25 °C.

Biological Materials	
Blood, human	7.35–7.45
Blood, dog	6.9–7.2
Spinal fluid, human	7.3–7.5
Saliva, human	6.5–7.5
Gastric contents, human	1.0–3.0
Duodenal contents, human	4.8–8.2
Feces, human	4.6–8.4
Urine, human	4.8–8.4
Milk, human	6.6–7.6
Bile, human	6.8–7.0

Foods	
Apples	2.9–3.3
Apricots	3.6–4.0
Asparagus	5.4–5.8
Bananas	4.5–4.7
Beans	5.0–6.0
Beers	4.0–5.0
Beets	4.9–5.5
Blackberries	3.2–3.6
Bread, white	5.0–6.0
Butter	6.1–6.4
Cabbage	5.2–5.4
Carrots	4.9–5.3
Cheese	4.8–6.4
Cherries	3.2–4.0
Cider	2.9–3.3
Corn	6.0–6.5
Crackers	6.5–8.5
Dates	6.2–6.4
Eggs, fresh white	7.6–8.0
Flour, wheat	5.5–6.5
Gooseberries	2.8–3.0
Grapefruit	3.0–3.3
Grapes	3.5–4.5
Hominy (lye)	6.8–8.0
Jams, fruit	3.5–4.0
Jellies, fruit	2.8–3.4
Lemons	2.2–2.4
Limes	1.8–2.0
Maple syrup	6.5–7.0
Milk, cows	6.3–6.6
Olives	3.6–3.8
Oranges	3.0–4.0
Oysters	6.1–6.6
Peaches	3.4–3.6
Pears	3.6–4.0
Peas	5.8–6.4
Pickles, dill	3.2–3.6
Pickles, sour	3.0–3.4
Pimento	4.6–5.2
Plums	2.8–3.0
Potatoes	5.6–6.0
Pumpkin	4.8–5.2
Raspberries	3.2–3.6
Rhubarb	3.1–3.2
Salmon	6.1–6.3
Sauerkraut	3.4–3.6
Shrimp	6.8–7.0
Soft drinks	2.0–4.0
Spinach	5.1–5.7
Squash	5.0–5.4
Strawberries	3.0–3.5
Sweet potatoes	5.3–5.6
Tomatoes	4.0–4.4
Tuna	5.9–6.1
Turnips	5.2–5.6
Vinegar	2.4–3.4
Water, drinking	6.5–8.0
Wines	2.8–3.8

STRUCTURE AND FUNCTIONS OF SOME COMMON DRUGS

This table lists the names, categories, therapeutic uses, and chemical structures of selected drugs. The generic (chemical) name of each drug is given, along with some of the trade names under which it is sold. When available, physical properties are given in italics in the fourth column. The structure given refers to the active drug, but many of these are packaged as salts or other derivatives. The drugs have been selected to represent a variety of categories; most are widely used throughout the world.

The list is divided into therapeutic categories; within each category the listing is alphabetical by generic name. The index that follows the table can be used to locate a drug by either generic or trade name.

References

1. *The Combined Chemical Dictionary on DVD, Version 12:1*, CRC Press, Boca Raton, FL, June 2008; also available on the Internet at www.chemnetbase.com.
2. Milne, G. W. A., *Drugs: Synonyms and Properties*, Ashgate Publishing, Aldershot, Hampshire, UK, 2000.
3. Corey, E. J., Czakó, B., and Kürti, L., *Molecules and Medicine*, John Wiley & Sons, Hoboken, NJ, 2007.
4. *Physicians' Desk Reference, 61st Edition*, Thomson PDR, Montvale, NJ, 2007.
5. O'Neil, M. J., Editor, *The Merck Index, 14th Edition*, Merck & Co., Whitehouse Station, NJ, 2006.

	Generic Name	Trade Names	Category and Properties	Applications	Structure
				Antiallergic Agents	
1	Albuterol	Proventil; Ventolin; Volmax	β_2-Andrenergic receptor agonist	Treatment of troubled breathing caused by asthma, emphysema, and other lung diseases	
2	Budesonide	Budeson; Budamax; Rhinocort; Pulmicort; Inflammide	Glucocorticoid *mp 226 °C*	Management of asthma and treatment of inflammatory bowel disease	
3	Cetirizine	Zyrtec	Histamine H_1-receptor antagonist	Treatment of seasonal allergies and hives	
4	Fexofenadine	Carboxyterfenadine; Allegra; Telfast	Histamine H_1-receptor antagonist	Treatment of allergic rhinitis	
5	Fluticasone	Flovent (as propanoate); Flonase (as propanoate); Advair (with salmeterol)	Anti-inflammatory glucocorticoid	Treatment of asthma & rhinitis	
6	Loratadine	Claritin; Claratyne; Alavert	Long-acting antihistamine *mp 132 °C*	Relief of allergy symptoms	

	Generic Name	Trade Names	Category and Properties	Applications	Structure
7	Montelukast	Singulair	Leukotriene LTD_4 receptor antagonist	Control of asthma and relief of seasonal allergies	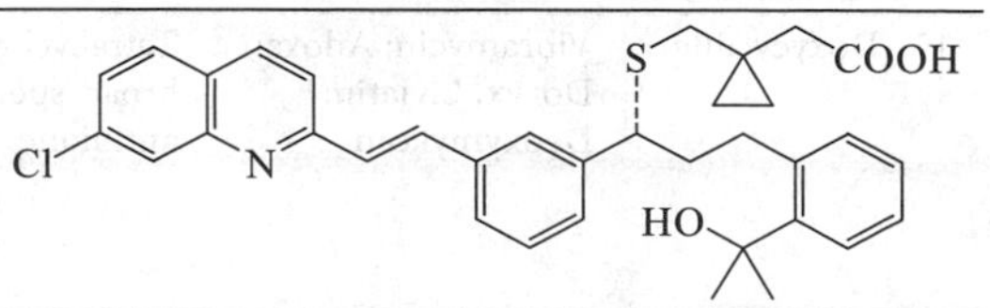
8	Salmeterol	Serevent	β_2-Adrenergic receptor agonist	Treatment of asthma & chronic obstructive pulmonary disease	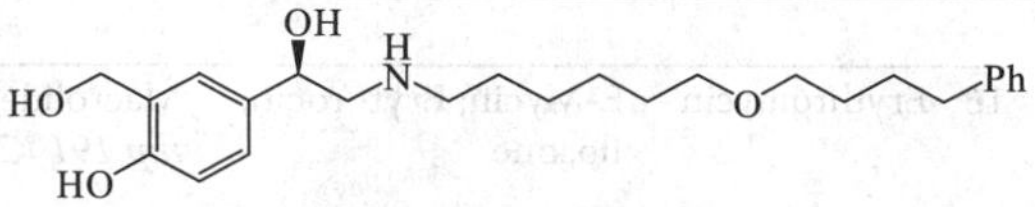
9	Tiotropium bromide	Spiriva	Long-acting antimuscarinic bronchodilator	Treatment of chronic obstructive pulmonary disease	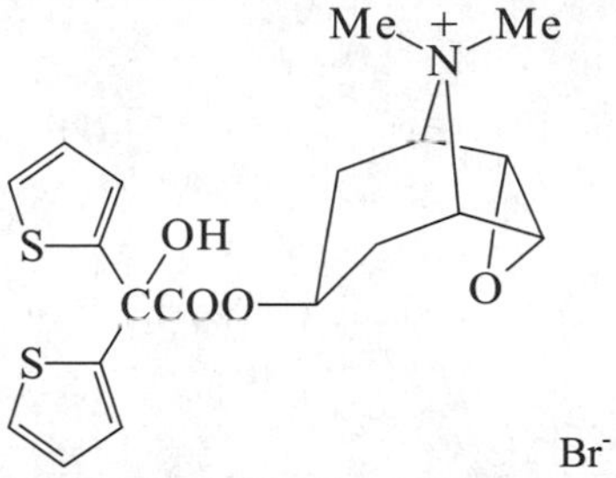
				Antibiotics	
10	Amikacin	Amikin; Biclin; Chemacin; Flexilite; Negasin	Aminoglycoside antibiotic	Treatment of serious infections resistant to other antibiotics	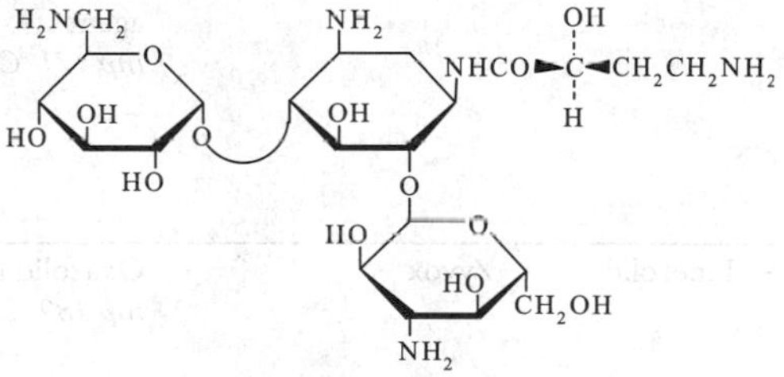
11	Amoxicillin	Amoxil; Isimoxin; Ospamox	β-Lactam/penicillin	Treatment of a broad spectrum of bacterial infections	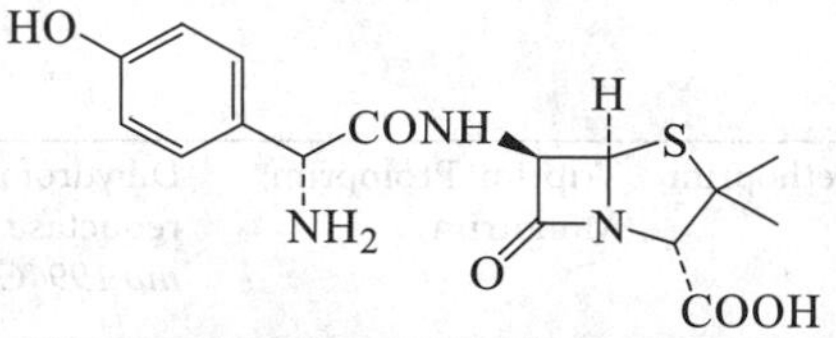
12	Azithromycin	Zithromax; Vinzam; Zmax; Azitrocin	Azalide/macrolide antibiotic *mp 155 °C*	Treatment of bacterial skin, ear, and respiratory infections	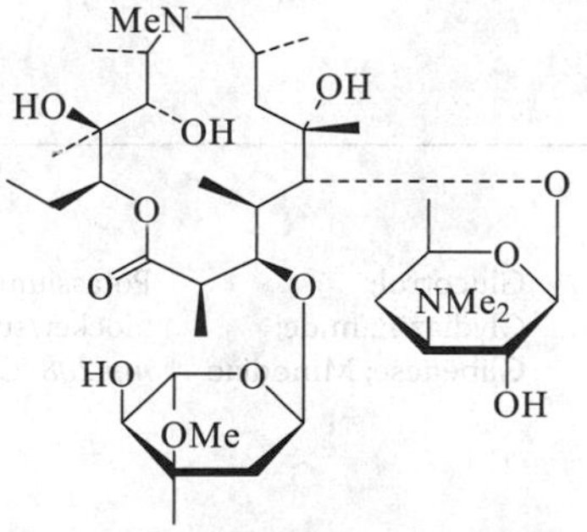
13	Cefaclor	Ceclor; Kefolar; Panacef; Panoral	β-Lactam/ cephalosporin	Treatment of bacterial infections, pneumonia, and urinary tract infections	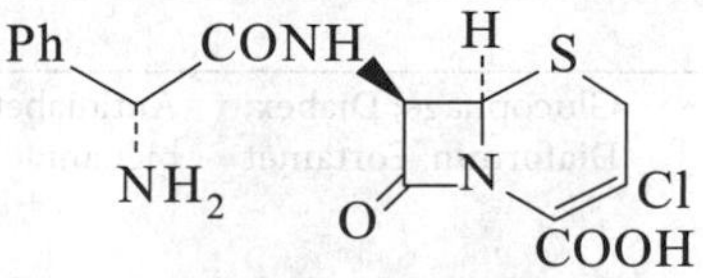
14	Ciprofloxacin	Cipro; Ciproxin; Ciprobay; Flociprin; Uniflox	Fluoroquinoline/ broad-spectrum antibiotic *mp 256 °C*	Treatment of urinary & respiratory tract infections, anthrax, and sexually-transmitted diseases	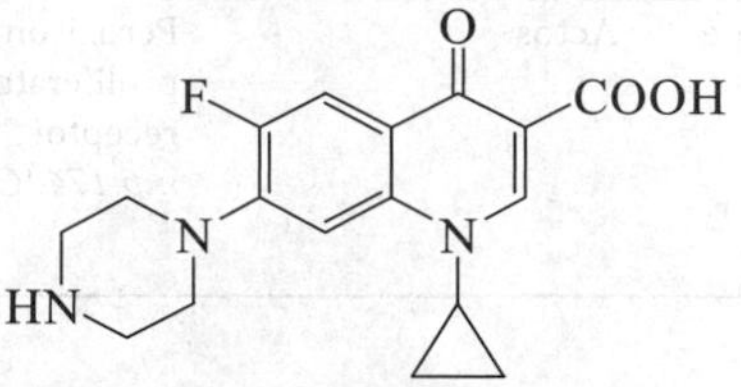

	Generic Name	Trade Names	Category and Properties	Applications	Structure
15	Doxycycline	Vibramycin; Adoxa; Doryx; Liviatin; Deoxymykoin	Tetracycline/ broad-spectrum antibiotic	Treatment of urinary tract, respiratory tract, and eye infections; anthrax, syphilis, cholera, etc.	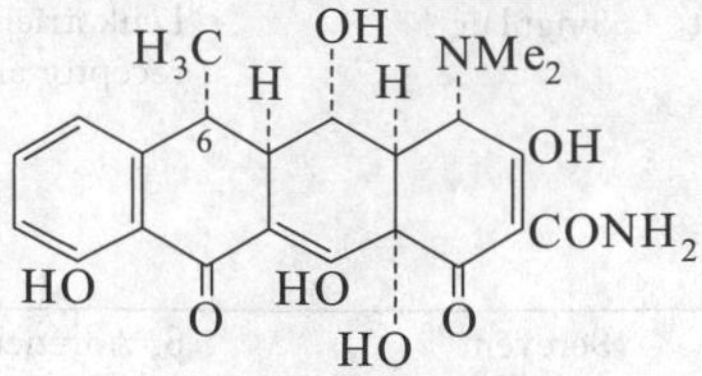
16	Erythromycin	E-Mycin; Erythrocin; Ilosone	Macrolide antibiotic *mp 191 °C*	Treatment of bacterial infections, including diphtheria, pertussis, rheumatic fever, venereal disease, etc.	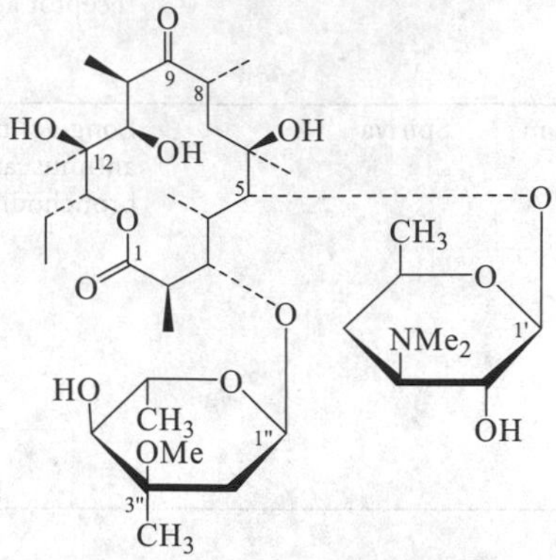
17	Isoniazid	Laniazid	Antimycobacterial agent *mp 171 °C*	Treatment of tuberculosis; reduction of tremors from multiple sclerosis	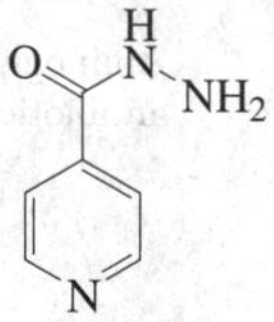
18	Linezolid	Zyvox	Oxazolidinone *mp 182 °C*	Treatment of serious Gram-positive infections resistant to other antibiotics	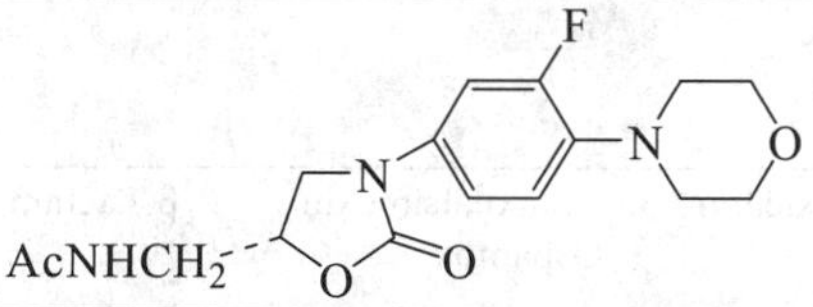
19	Trimethoprim	Triprim; Proloprim; Monotrim	Dihydrofolate reductase inhibitor *mp 199 °C*	Treatment of urinary tract infections, diarrhea, and ear infections	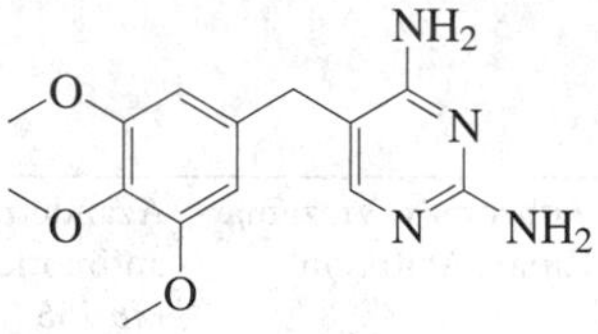

Antidiabetic Drugs

	Generic Name	Trade Names	Category and Properties	Applications	Structure
20	Glipizide	Glucotrol; Glydiazinamide; Glibenese; Minodiab	Potassium channel blocker/sulfonylurea *mp 208 °C*	Treatment of type 2 diabetes by stimulating insulin secretion in pancreas β-cells	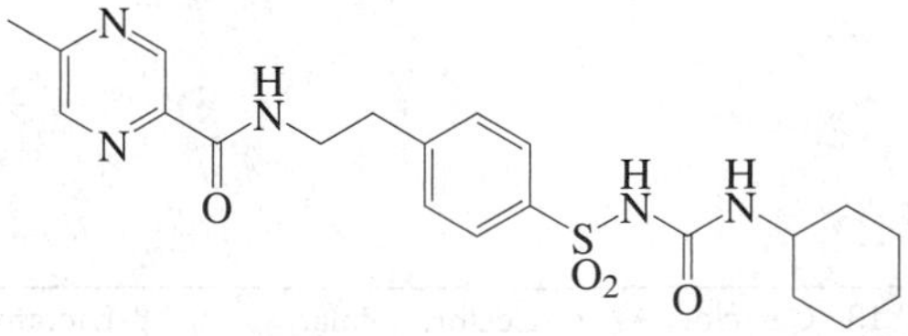
21	Metformin	Glucophage; Diabex; Diaformin; Fortamet	Antidiabetic biguanide	Treatment of type 2 diabetes by enhancing transport of glucose into muscle cells	 $Me_2NC(=NH)NHC(=NH)NH_2$
22	Pioglitazone	Actos	Peroxisome proliferator-activated receptor *mp 174 °C*	Treatment of type 2 diabetes by increasing glucose metabolism and insulin sensitivity	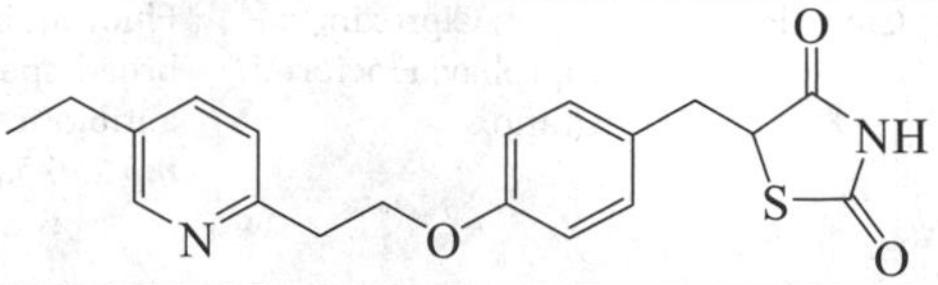

	Generic Name	Trade Names	Category and Properties	Applications	Structure
23	Rosiglitazone	Avandia	Thiazolidinedione	Treatment of type 2 diabetes by increasing insulin sensitivity. Possible adverse effects on patients with heart problems.	
24	Sitagliptin	Januvia	Dipeptidyl peptidase IV inhibitor	Treatment of type 2 diabetes by enhancing the body's ability to lower elevated glucose levels	
			Anti-Inflammatory Agents		
25	Acetylsalicylic acid	Aspirin	NSAID	Pain and fever relief; anticlotting agent	
26	Celecoxib	Celebrex; Onsenal	NSAID (COX-2 inhibitor) *mp158 °C*	Treatment of osteoarthritis and rheumatoid arthritis	
27	Ibuprofen	Advil	NSAID (cyclooxygenase inhibitor) *mp 76 °C*	Relief of inflammation and pain	
28	Meloxicam	Mobic; Metacam; Metacain	NSAID (cyclooxygenase inhibitor) *mp 254 °C*	Treatment of osteoarthritis and rheumatoid arthritis	
29	Naproxen	Aleve; Naprelan; Anaprox; Naprogesic	NSAID *mp 155 °C*	Relief of inflammation and pain	
30	Prednisone	Meticorten; Deltasone	Adrenocortical steroid; anti-inflammatory agent; immunosuppressant	Treatment of asthma and other inflammatory diseases	
			Anti-Ulcer Drugs		
31	Cimetidine	Tagamet; Cimetimax; Gastromet; Peptimax	Histamine H_2-receptor antagonist *mp 142 °C*	Treatment of peptic ulcer, gastrointestinal bleeding, and gastroesophageal reflux disease	
32	Lansoprazole	Prevacid; SoluTab; Prevpac; Zoton; Prezal	Proton pump inhibitor *mp 180 °C*	Treatment of duodenal ulcers & gastroesophageal reflux disease	

	Generic Name	Trade Names	Category and Properties	Applications	Structure
33	Omeprazole	Prilosec; Nexium (Mg salt); Losec; Mepral; Mopral; Zoltum	Proton pump inhibitor *mp 156 °C*	Treatment of peptic ulcer, dyspepsia, and gastroesophageal reflux disease	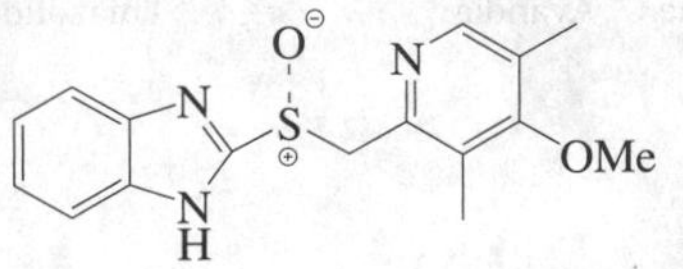
34	Pantoprazole	Protonix; Pantozol; Rifun	Proton pump inhibitor	Treatment of gastric acid-related conditions	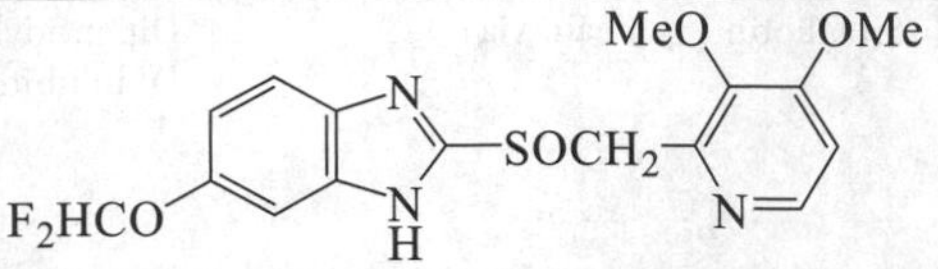
35	Ranitidine	Zantac; Azantac; Melfax; Rantec; Sostril; Taural	Histamine H_2-receptor antagonist *mp 70 °C*	Treatment of peptic ulcer, gastrointestinal bleeding, and gastroesophageal reflux disease	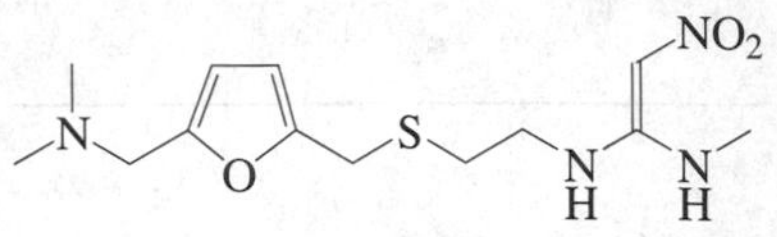
			Antiviral and Antifungal Agents		
36	Acyclovir	Zovirax; Zovir; Avirax; Mirolex	Viral DNA synthesis inhibitor *mp 225 °C*	Treatment of cold sores, genital herpes, chicken pox, etc.	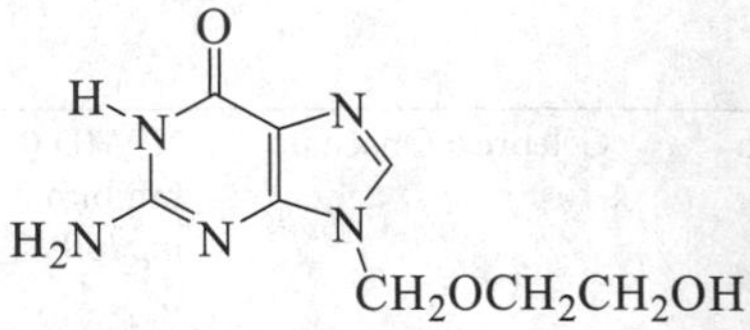
37	Amphotericin	Fungizone; Amfostet; Amphozone	Polyene macrolide antifungal agent *mp 170 °C*	Intravenous treatment of systemic fungal infections	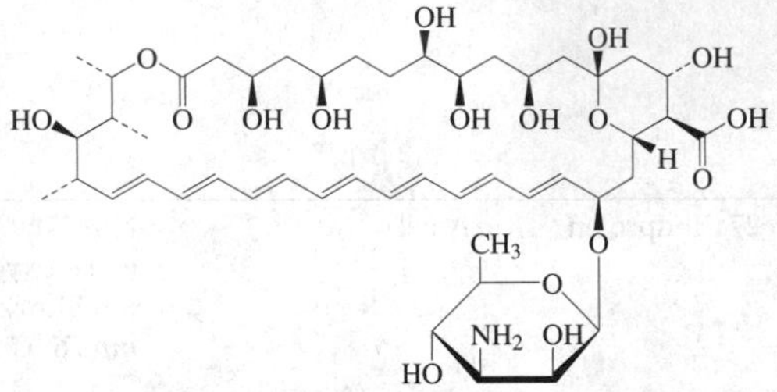
38	Efavirenz	Sustiva	Non-nucleoside HIV reverse transcriptase inhibitor *mp 131 °C*	Treatment of HIV-1 infections (as part of combination therapy)	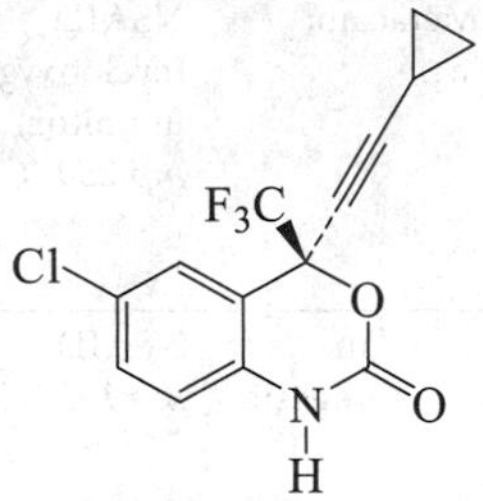
39	Fluconazole	Diflucan; Biozolene; Elazor; Triflucan	Cytochrome P450 14α-demethylase inhibitor *mp 139 °C*	Treatment and prevention of superficial & systemic fungal infections	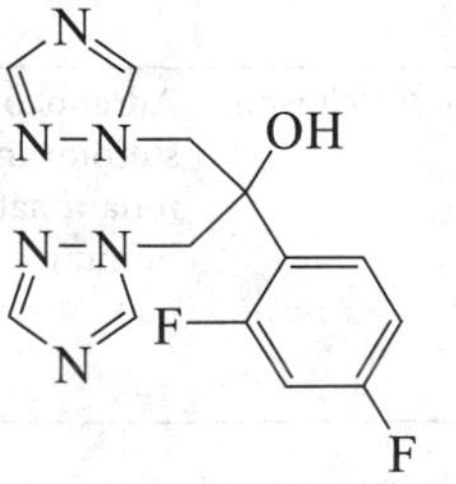
40	Lamivudine	Epivir; Heptodin; Zeffix	HIV reverse transcriptase inhibitor	Treatment of hepatitis B and human immunodeficiency virus	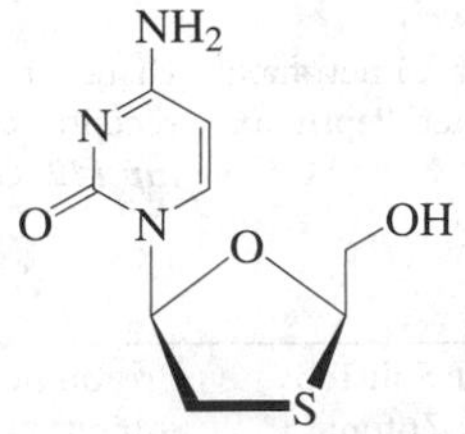

	Generic Name	Trade Names	Category and Properties	Applications	Structure
41	Nevirapine	Viramune	Non-nucleoside HIV reverse transcriptase inhibitor *mp 250 °C*	Treatment of HIV infections (as part of combination therapy)	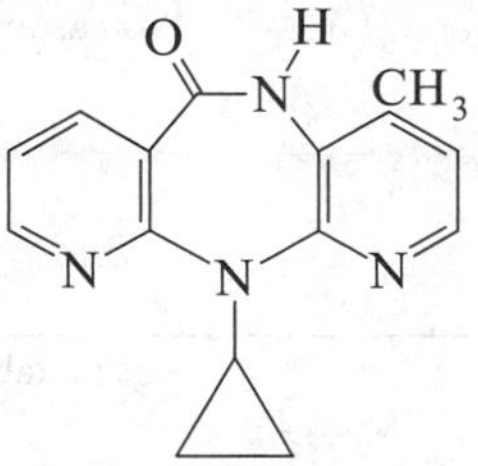
42	Oseltamivir	Tamiflu	Neuraminidase inhibitor *Foam*	Prevention & treatment of influenza A and B	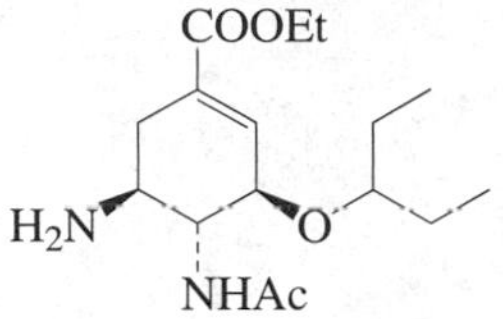
43	Ribavirin	Virazole; Rebetol; Copegus; Ribasphere; Viratek; Cotronak	Nucleoside antimetabolite *mp 175 °C*	Treatment of hepatitis C	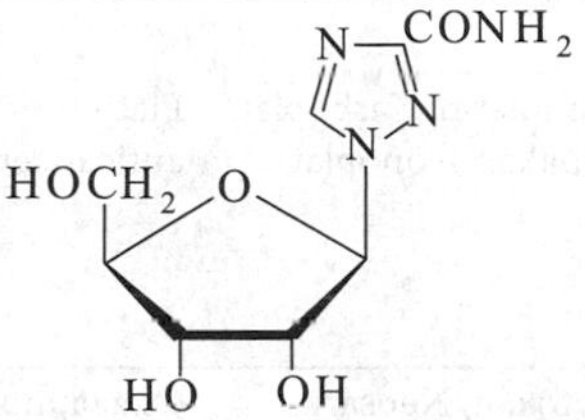
44	Terbinafine	Lamisil	Squalene epoxidase inhibitor	Treatment of fungal infections of the skin & nails	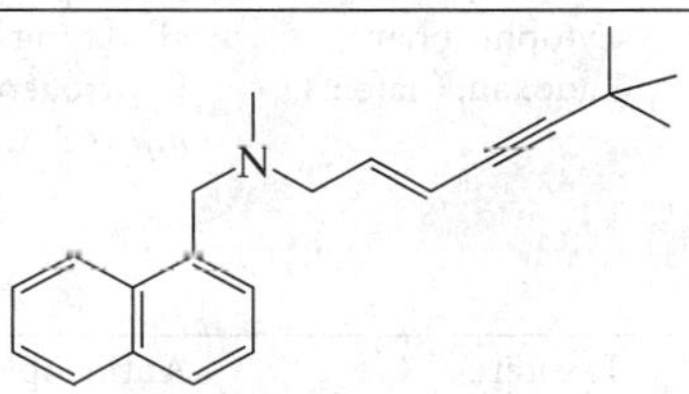
45	Zalcitabine	Hivid	Pyrimidine nucleoside reverse transcriptase inhibitor	Treatment of HIV infection & AIDS	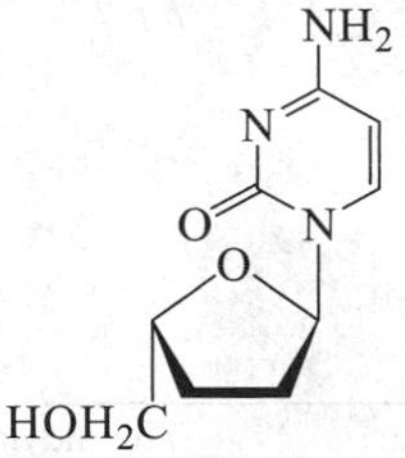
46	Zidovudine	Retrovir; AZT; Azidothymidine; Zidovir	Pyrimidine nucleoside reverse transcriptase inhibitor *mp 121 °C*	Treatment of HIV infection & prevention of mother-to-child transmission	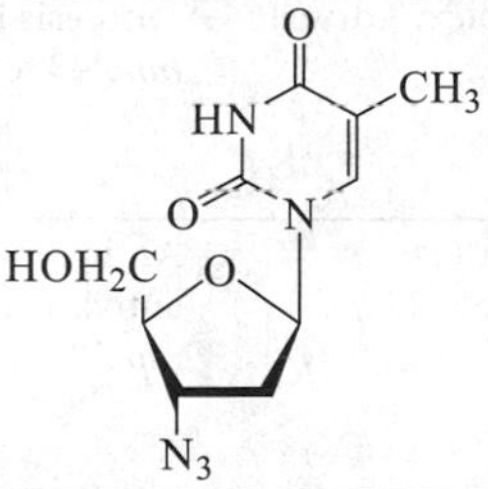
			Cancer Chemotherapy Drugs		
47	Anastrozole	Arimidex	Aromatase inhibitor; antineoplastic agent *mp 81 °C*	Treatment of breast cancer in postmenopausal women	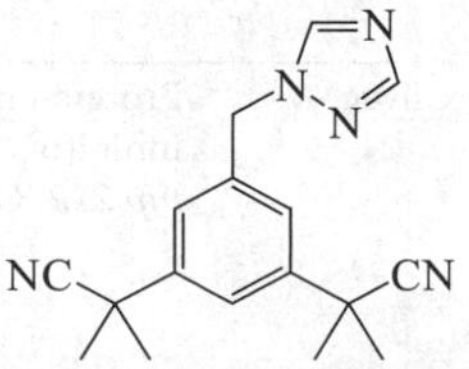
48	Bicalutamide	Casodex	Antiandrogen *mp 180 °C*	Treatment of advanced prostate cancer	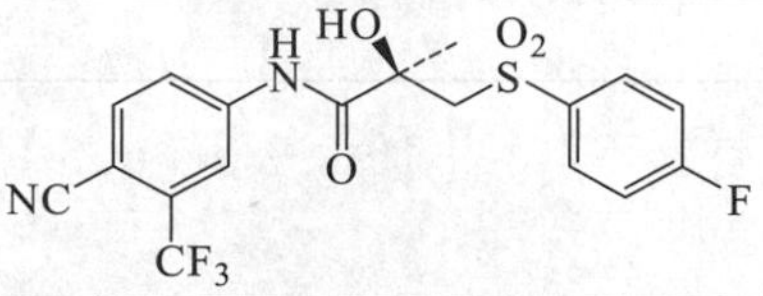

	Generic Name	Trade Names	Category and Properties	Applications	Structure
49	Bortezomib	Velcade	Proteasome inhibitor	Treatment of lymphomas and multiple myeloma	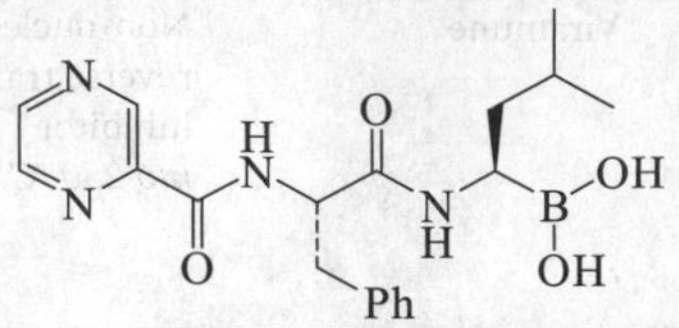
50	Capecitabine	Xeloda	Antimetabolite	Treatment of breast & colon cancer	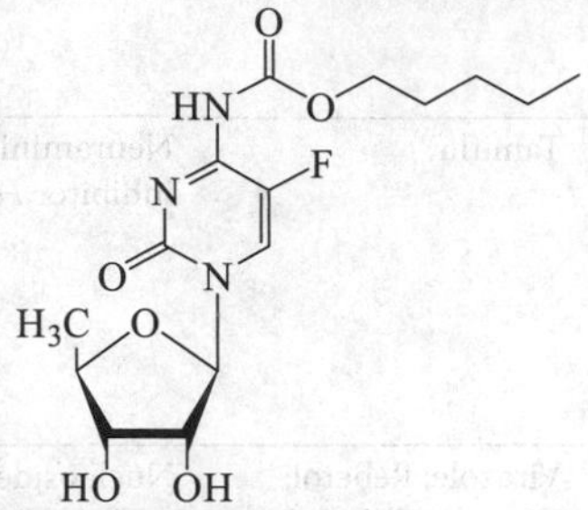
51	Carboplatin	Paraplatin; Carboplat; Erbakar; Nonoplat	Platinum-based anticancer agent	Treatment of ovarian cancer	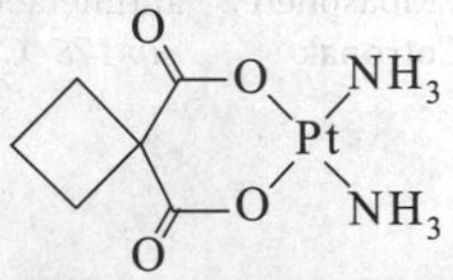
52	Cyclo-phosphamide	Cytoxan; Neosar; Cytophosphan; Endoxan; Clafen	Oxazaphosphorine alkylating agent ("Nitrogen mustard") *mp 143 °C*	Treatment of lymphomas & leukemias, multiple myeloma, and other cancers	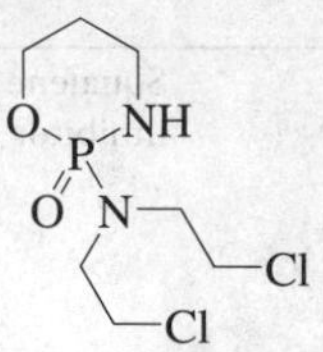
53	Docetaxel	Taxotere	Antineoplastic agent *mp 232 °C*	Treatment of ovarian, breast and bronchial carcinomas	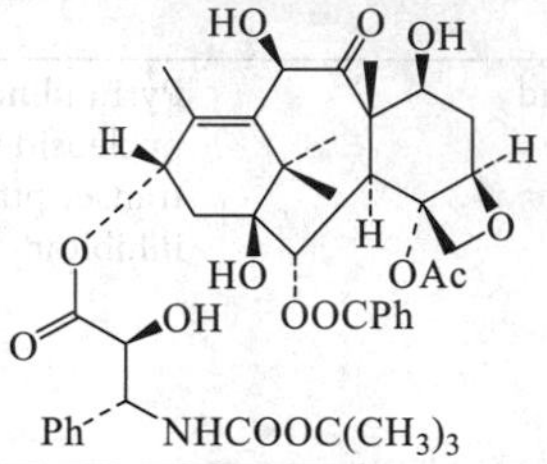
54	Fluorouracil	Efudex; Carac; Fluoroplex; Adrucil	Thymidylate synthesis inhibitor *mp 243 °C*	Constituent of several antineoplastic combinations	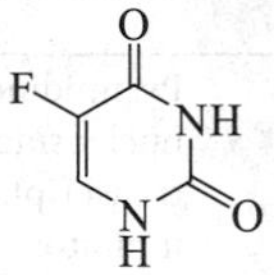
55	Gemcitabine	Gemzar (as hydrochloride)	Antineoplastic and antiviral agent *mp 290 °C*	Treatment of lung and pancreatic tumors	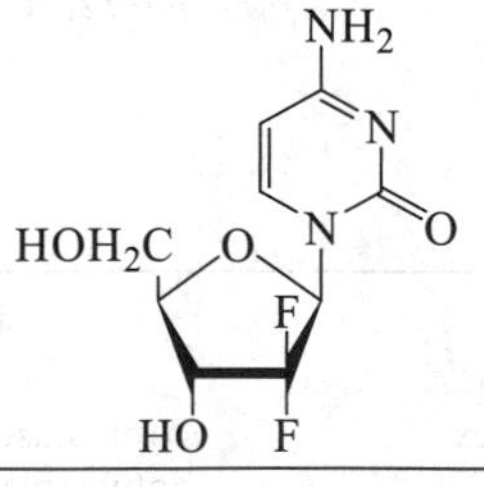
56	Imatinib	Gleevec; Glivec	Protein kinase inhibitor *mp 212 °C*	Treatment of myelogenous leukemia and gastrointestinal tumors	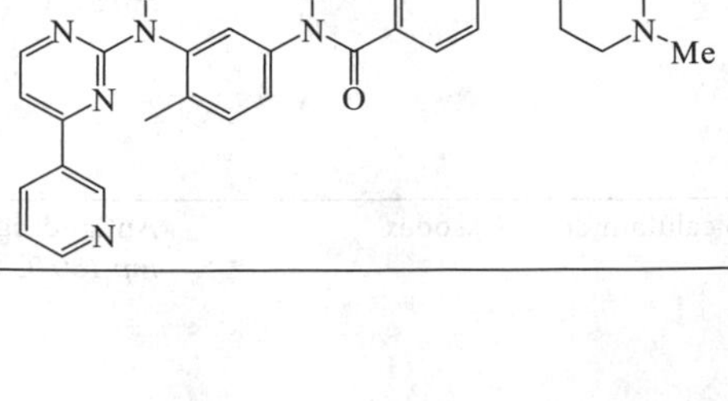

	Generic Name	Trade Names	Category and Properties	Applications	Structure
57	Irinotecan	Camptosar; Camptetin; Topotecin	Topoisomerase I inhibitor	Treatment of colorectal cancer	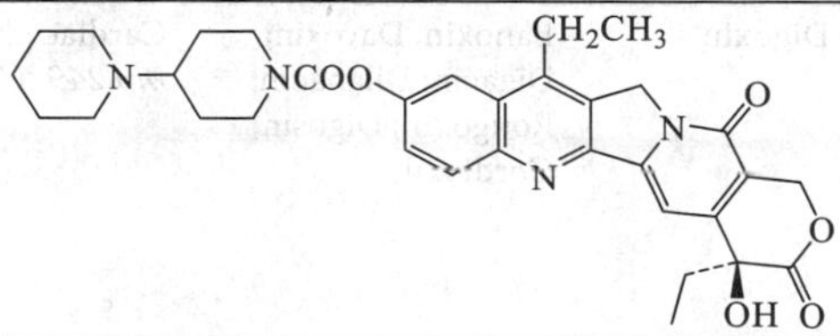
58	Paclitaxel	Taxol; Abraxane; Yewtaxan	Microtubule-stabilizing agent *mp 214 °C*	Treatment of ovarian, breast, and lung cancer	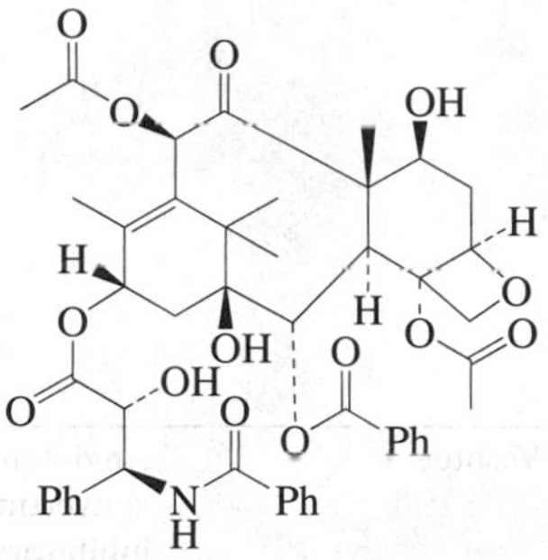
59	Sunitinib	Sutent	Tyrosine kineases inhibitor	Treatment of gastrointestinal & kidney tumors	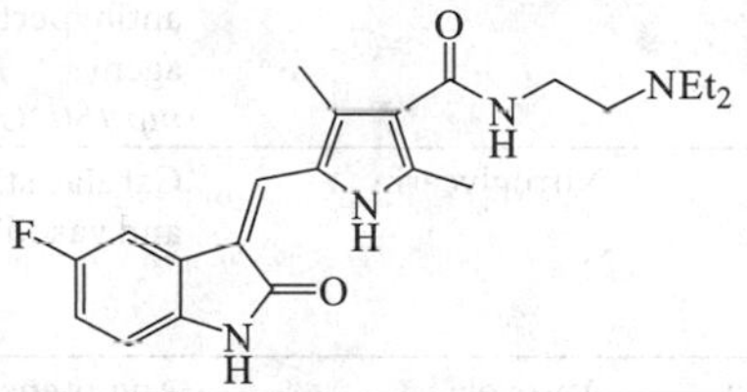
60	Tamoxifen	Nolvadex; Soltamox; Tamaxin; Tamoplex; Valodex	Selective estrogen receptor modulator *mp 97 °C*	Treatment of breast cancer	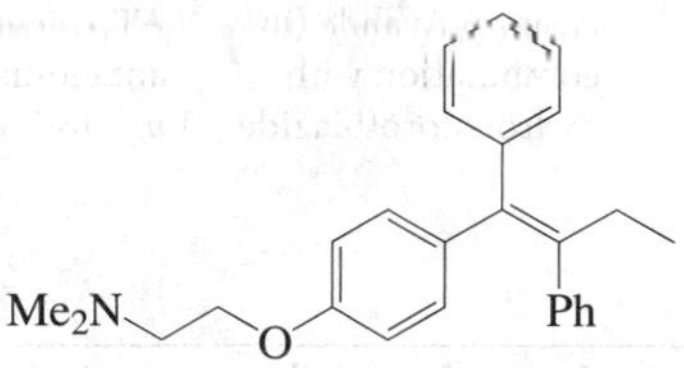
			Cardiovascular Agents		
61	Aliskiren	Tekturna; Rasilex	Renin inhibitor	Treatment of hypertension	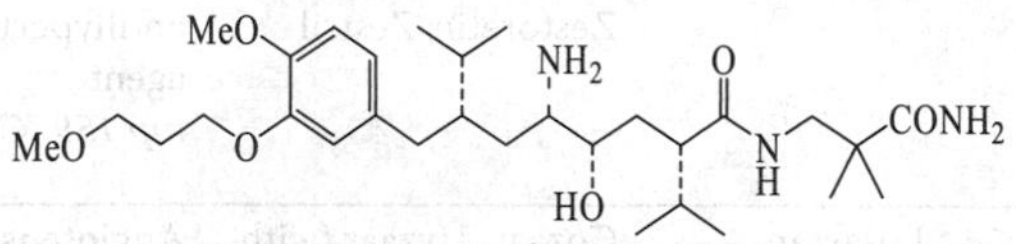
62	Amlodipine	Norvasc	Calcium channel blocker	Treatment of hypertension, atrial fibrillation, and angina	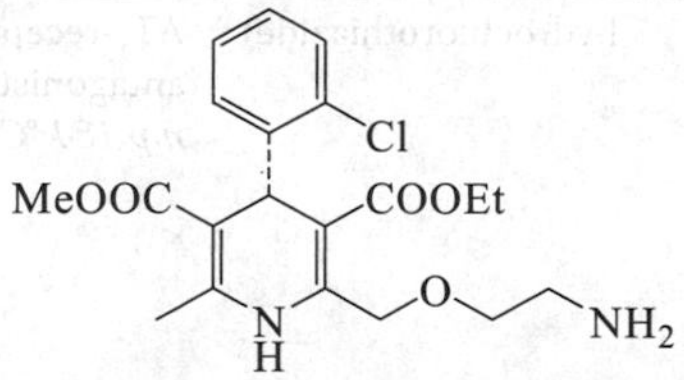
63	Atenolol	Tenormin	β-blocker *mp 147 °C*	Treatment of hypertension and excessive heart rate	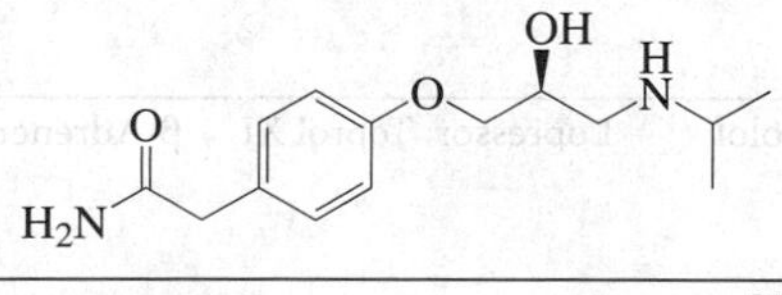
64	Carvedilol	Coreg; Dilatrend; Eucardic	Calcium channel blocker; β-adrenoceptor blocker	Treatment of heart failure and hypertension	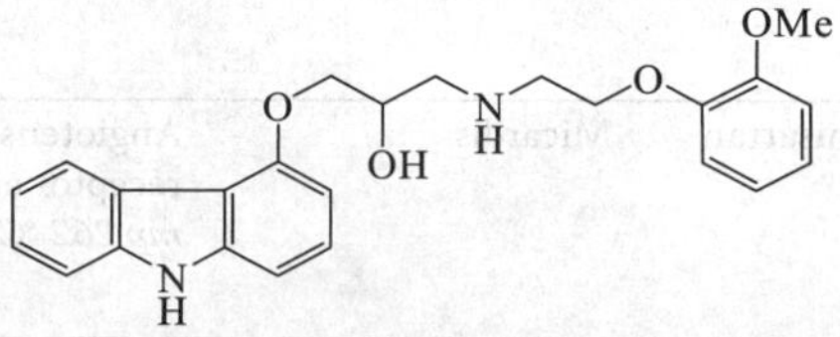
65	Clopidogrel	Plavix; Isocover; Meilax; Tipidyl	Antiplatelet agent	Prevention of blood clots after stroke or myocardial infarction	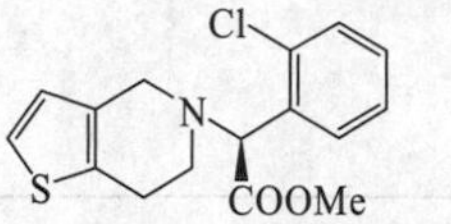

	Generic Name	Trade Names	Category and Properties	Applications	Structure
66	Digoxin	Lanoxin; Davoxin; Digacin; Dilanacin; Rougoxin; Digosin; Cordioxil	Cardiac glycoside *mp 249 °C*	Treatment of congestive heart failure	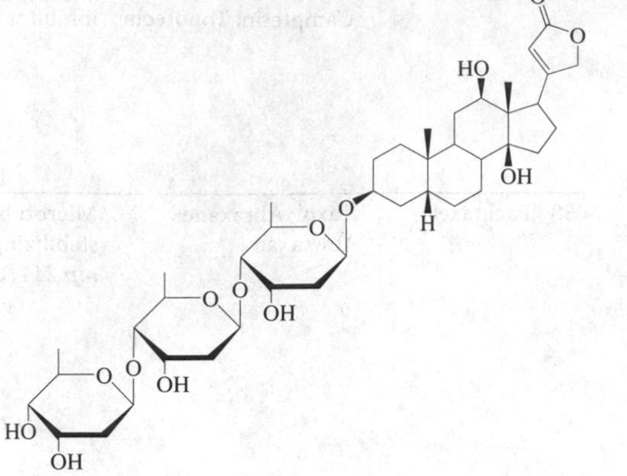
67	Enalaprilat	Vasotec	Angiotensin-converting enzyme inhibitor; antihypertensive agent *mp 150 °C*	Treatment of hypertension, atherosclerosis, and congestive heart failure	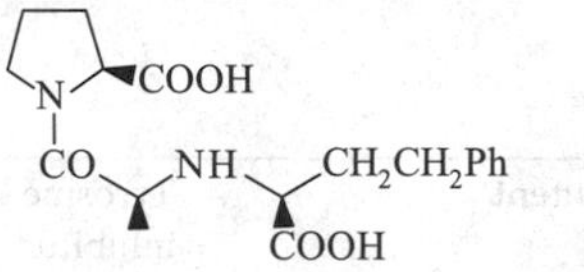
68	Glycerol trinitrate	Nitroglycerin	Cardiac stimulant and vasodilator	Treatment of angina and congestive heart failure	CH_2ONO_2 / $CHONO_2$ / CH_2ONO_2
69	Irbesartan	Aprovel; Avapro; Avalide (in combination with hydrochlorothiazide	Angiotensin II AT_1-receptor antagonist *mp 180 °C*	Treatment of hypertension and diabetes-related kidney disease	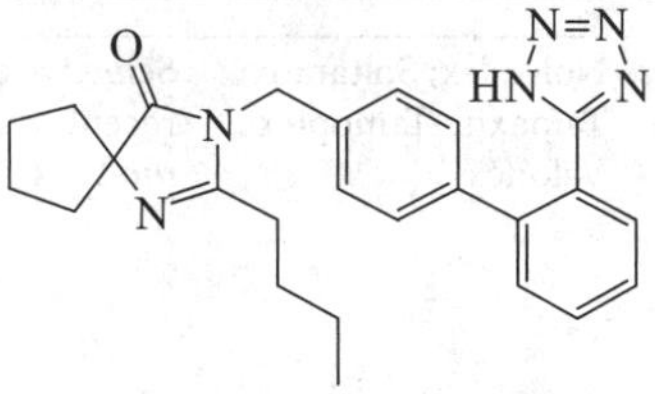
70	Lisinopril	Acecomb; Alapril; Carace; Novatec; Novazyd; Vivatec; Zestoretic; Zestril	Angiotensin-converting enzyme inhibitor; antihypertensive agent *mp 159 °C*	Treatment of hypertension	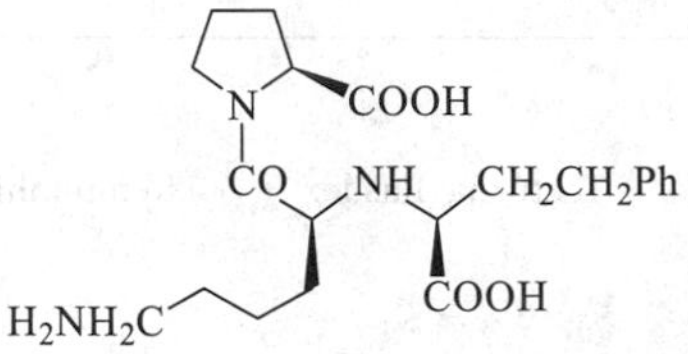
71	Losartan	Cozaar; Hyzaar (with hydrochlorothiazide)	Angiotensin II AT_1-receptor antagonist *mp 184 °C*	Treatment of congestive heart failure and hypertension	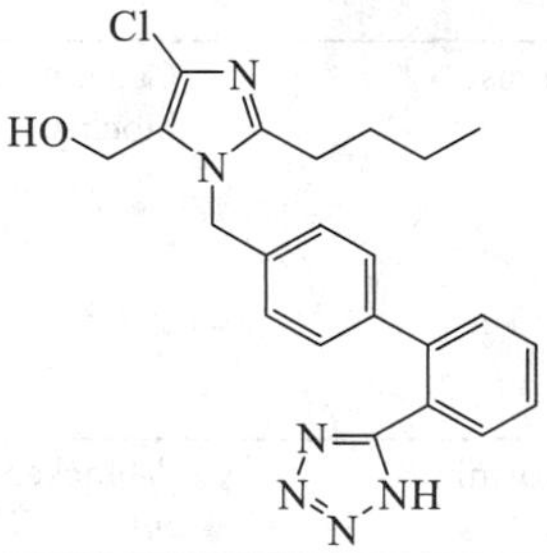
72	Metoprolol	Lopressor; Toprol XL	β-Adrenergic blocker	Treatment of angina and hypertension	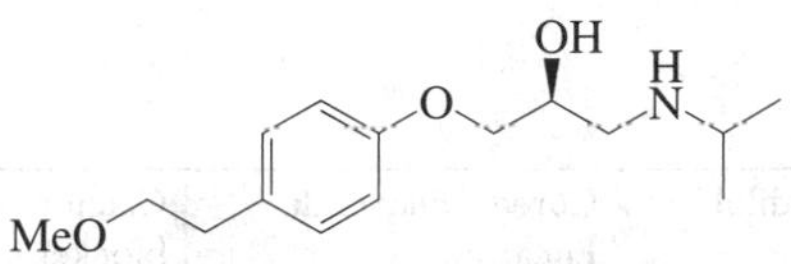
73	Telmisartan	Micardis	Angiotensin II (AT_1) receptor antagonist *mp 262 °C*	Treatment of hypertension	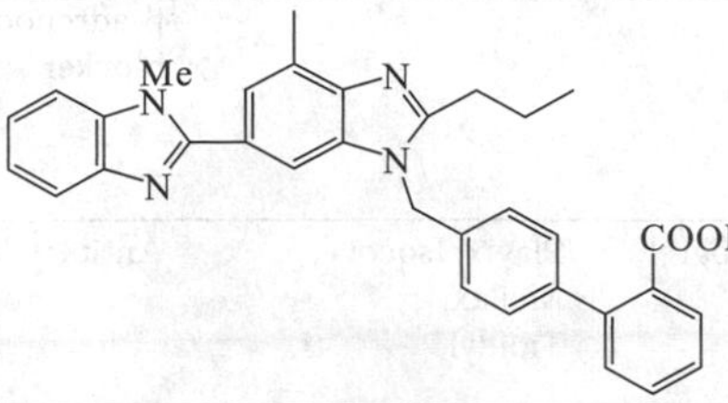

	Generic Name	Trade Names	Category and Properties	Applications	Structure
74	Terazosin	Hytrin; Itrin; Hytrinex; Magnurol; Teraprost; Vasocard; Uroflo; etc.	Antihypertensive agent (α_1-adrenoceptor antagonist)	Treatment of hypertension and benign prostatic hyperplasia	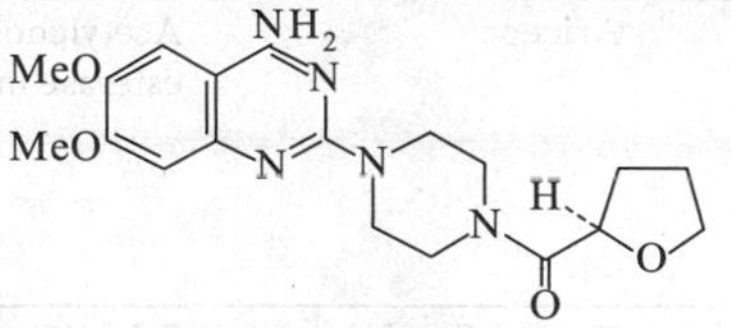
75	Warfarin	Coumadin	Anticlotting agent *mp 161 °C*	Reduction of possibility of stroke or coronary	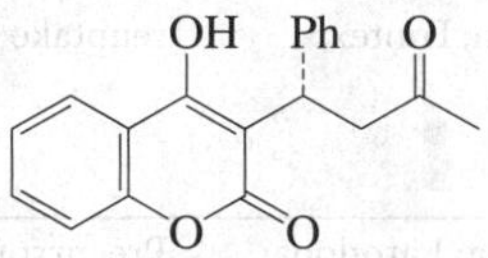
			Cholesterol-Lowering Drugs		
76	Atorvastatin	Lipitor; Caduet (in combination with amlodipine)	HMG-CoA reductase inhibitor (statin)	Reduction of LDL cholesterol levels by inhibiting cholesterol biosynthesis	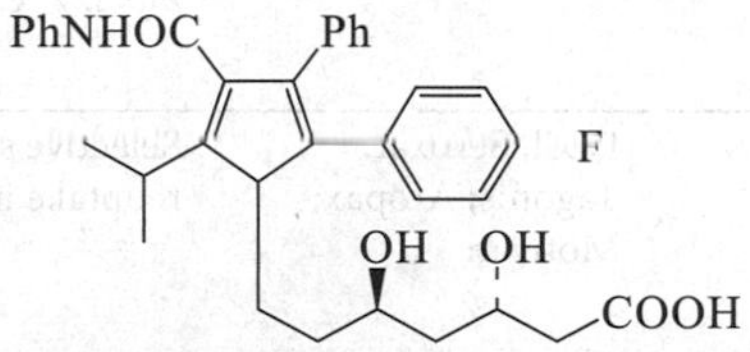
77	Ezetimibe	Zetia	Selective cholesterol absorption inhibitor *mp 165 °C*	Reduction of LDL cholesterol levels by inhibiting dietary cholesterol absorption	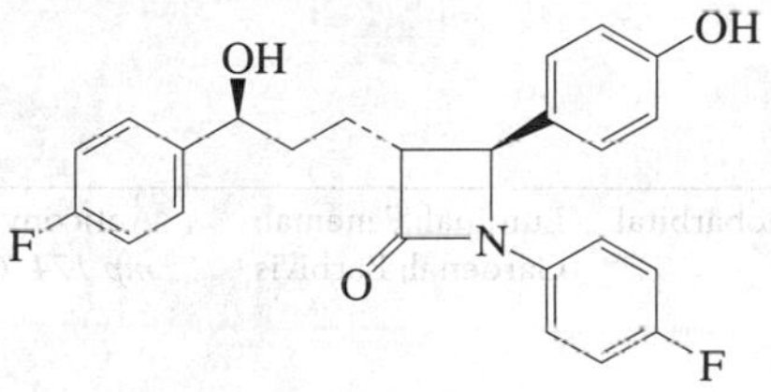
78	Nicotinic acid	Niacin	Lipoprotein synthesis inhibitor	Reduction of LDL cholesterol levels	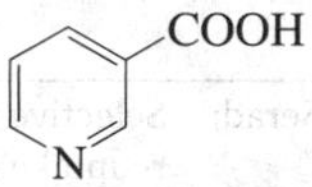
79	Pravastatin	Pravochol	HMG-CoA reductase inhibitor (statin)	Reduction of LDL cholesterol levels by inhibiting cholesterol biosynthesis	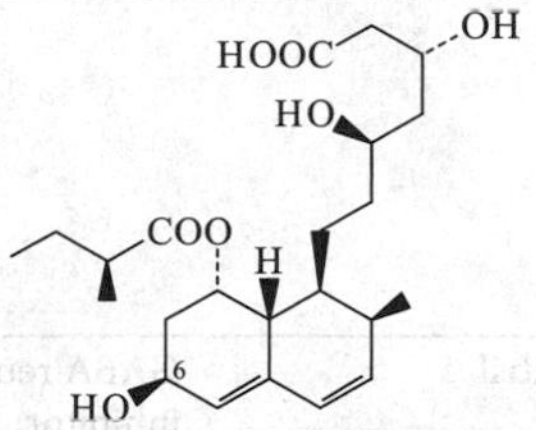
80	Simvastatin	Zocor; Vytorin (combination with Ezetimibe)	HMG-CoA reductase inhibitor (statin) *mp 136 °C*	Reduction of LDL cholesterol levels by inhibiting cholesterol biosynthesis	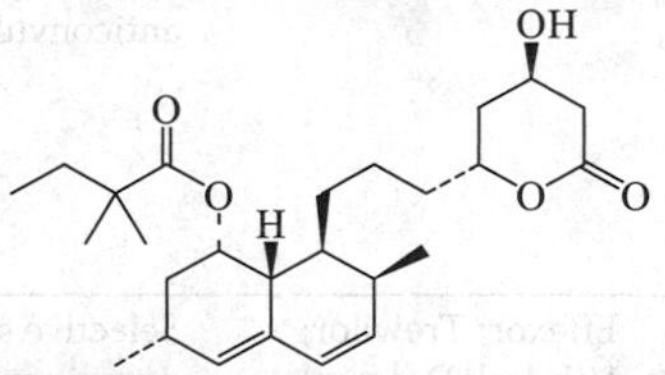
			Depression and Anxiety Drugs		
81	Bupropion	Wellbutrin; Amfebutamone; Zyban	Dopamine reuptake inhibitor *Pale yellow oil*	Antidepressant; smoking cessation aid	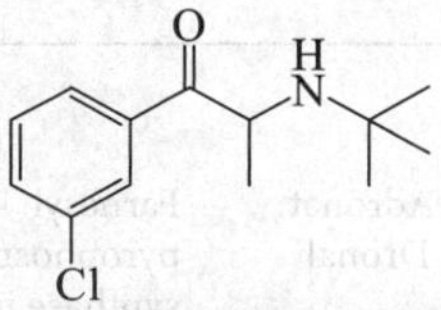
82	Diazepam	Valium; Stesolid; Seduxen; Antenex; Calmpose; Livotensin	Benzodiazepine central nervous system depressant *mp 125 °C*	Treatment of anxiety, seizures, muscular spasms, and insomnia	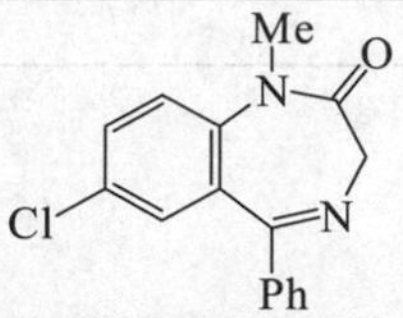

	Generic Name	Trade Names	Category and Properties	Applications	Structure
83	Donepezil	Aricept	Acetylcholine esterase inhibitor	Treatment of Alzheimer's disease	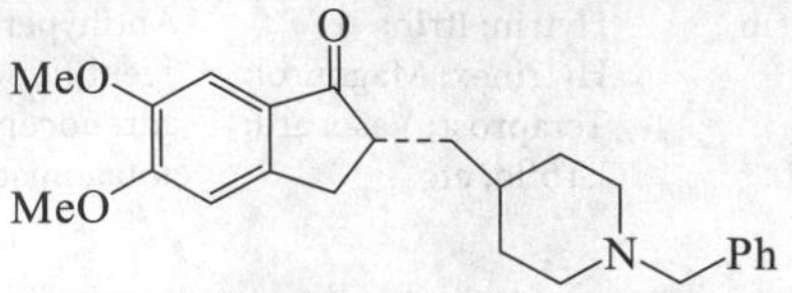
84	Fluoxetine	Prozac; Sarafem; Adofen; Fontex; Lorien	Selective serotonin reuptake inhibitor	Treatment of depression, panic attacks, and obsessive-compulsive disorder	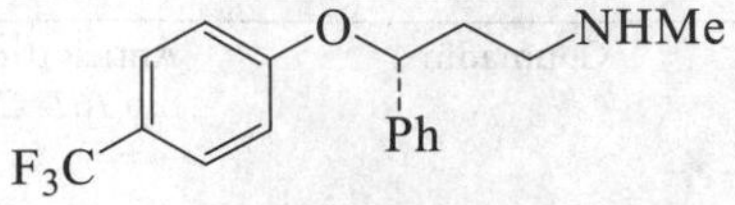
85	Levodopa	L-Dopa; Larodopa; Bendopa; Veldopa	Precursor to the neurotransmitter dopamine *mp 277 °C*	Treatment of Parkinson's disease	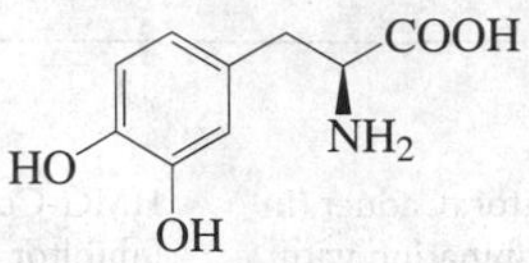
86	Paroxetine	Paxil; Seroxat; Tagonis; Aropax; Motivan	Selective serotonin reuptake inhibitor	Treatment of depression, panic attacks, and obsessive-compulsive disorder	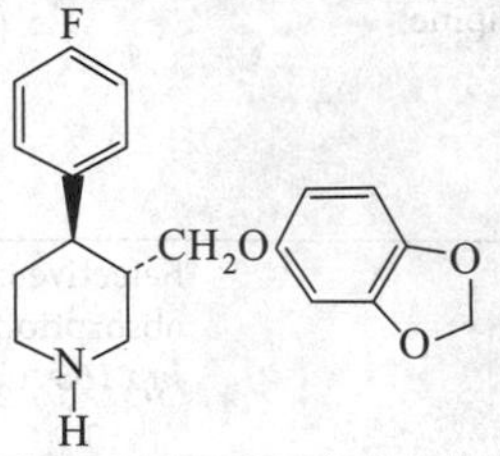
87	Phenobarbital	Luminal; Fenemal; Gardenal; Barbivis	Anticonvulsant *mp 174 °C*	Epilepsy control; also used as a sedative	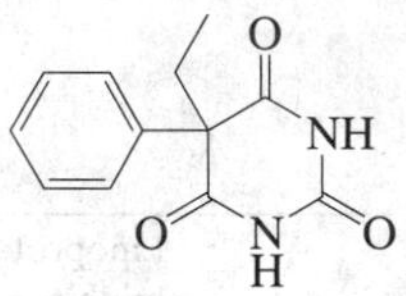
88	Sertraline	Zoloft; Lustral; Serad; Serlain; Tatig	Selective serotonin reuptake inhibitor	Treatment of depression, panic attacks, and obsessive-compulsive disorder	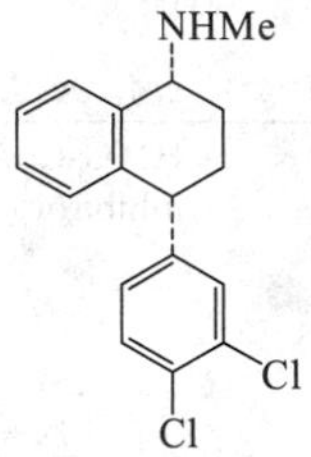
89	Tiagabine	Gabitril	GABA reuptake inhibitor; anticonvulsant	Treatment of epileptic seizures	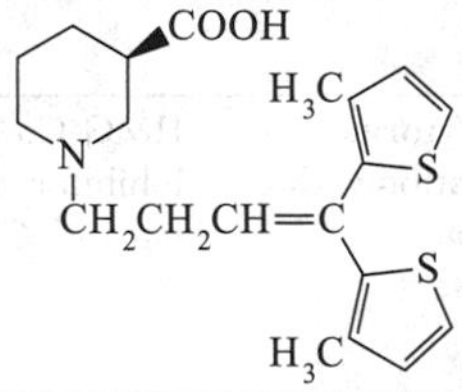
90	Venlafaxine	Effexor; Trewilor; Vandral; Dobupal	Selective serotonin & norepinephrine reuptake inhibitor	Treatment of anxiety and panic disorders	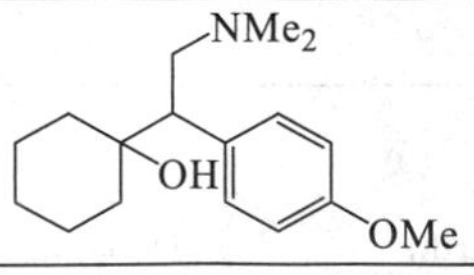

Osteoporosis Drugs

	Generic Name	Trade Names	Category and Properties	Applications	Structure
91	Alendronic acid	Fosamax; Adronat; Alendros; Dronal	Farnesyl pyrophosphate synthase inhibitor	Prevention and treatment of osteoporosis and Paget's disease	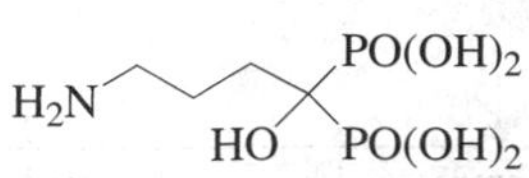

	Generic Name	Trade Names	Category and Properties	Applications	Structure
92	Calcitriol	Rocaltrol; Calcijex	Calcium and phosphate metabolism regulator *mp 170 °C*	Treatment of rickets and osteoporosis	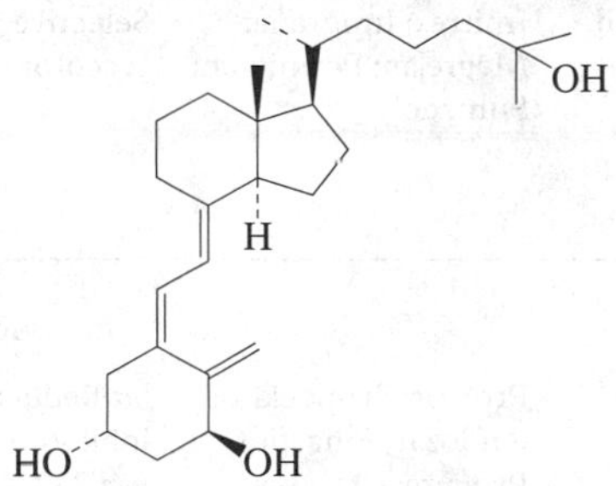
93	Raloxifene	Evista (as hydrochloride)	Selective estrogen receptor modulator	Prevention and treatment of osteoporosis	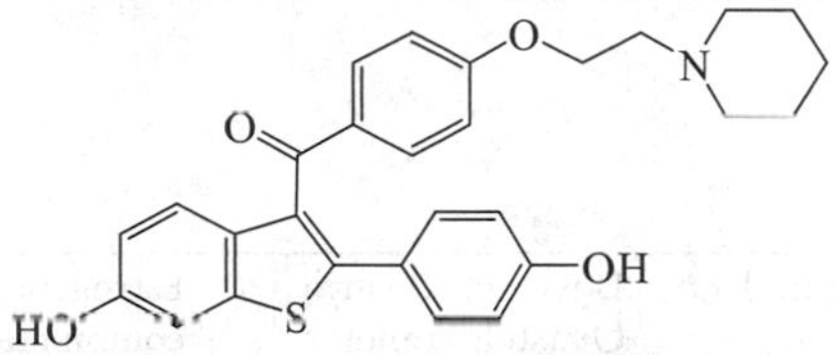
94	Risedronic acid	Actonel (as Na salt); Optinate	Calcium regulator	Treatment of osteoporosis & Paget's disease	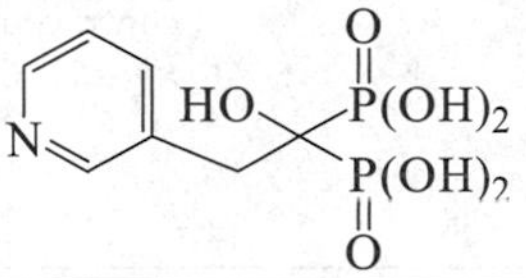
			Pain Relief Drugs		
95	Acetaminophen	Tylenol; APAP; Paracetamol; Hedrex; Tramil	Analgesic/antipyretic *mp 170 °C*	Relief of musculoskeletal, neuralgic, and other types of pain	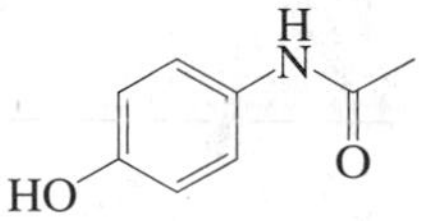
96	Fentanyl	Duragesic; Actiq; Fentora; Sublimaze	Opioid μ-receptor agonist *mp 87.5 °C*	Treatment of severe pain; spinal and epidural anesthesia	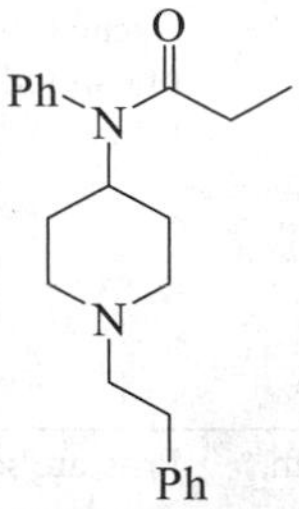
97	Gabapentin	Neurotin; Aclonium	Anticonvulsant/ analgesic; CNS depressant *mp 164 °C*	Treatment of neuralgia, pain from shingles, migraine, and epilepsy	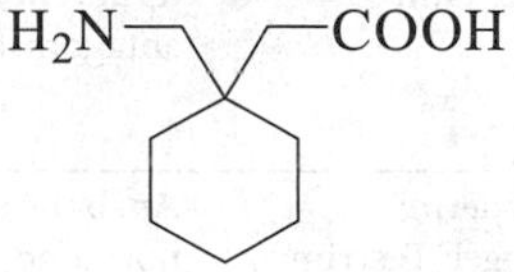
98	Lidocaine	Xylocaine; Xylocard; Lidamantle	Aminoamide anesthetic/ antiarrhythmic agent	Local anesthetic for dental procedures	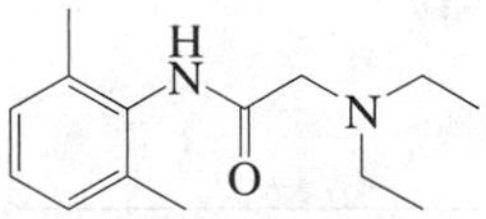
99	Morphine	Avinza; Contin; Kadian; Roxanol; Meconium; Morfine	Opioid analgesic *mp 255 °C*	Pain management, especially in malignant diseases	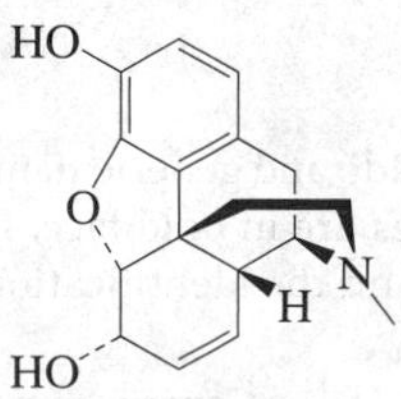
100	Thiopental	Pentothal; Penthiobarbital; Thiopentone	Short-acting barbiturate/ anesthetic	Induction of presurgical anesthesia	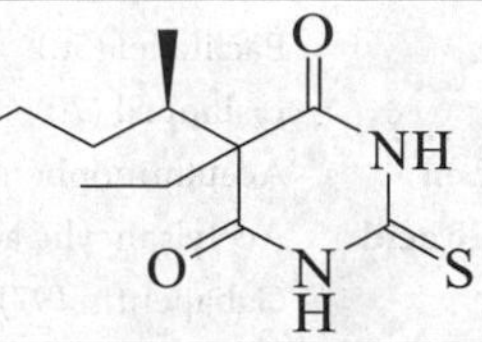

	Generic Name	Trade Names	Category and Properties	Applications	Structure
101	Sumatriptan	Imitrex; Imigrane; Megrelan; Permicran; Sumadol	Selective 5-HT_{1D} receptor agonist	Treatment of severe migraine headaches	
			Reproductive and Urinary System Drugs		
102	Finasteride	Proscar; Propecia; Andozac; Finastid; Procure; Urprosan	5α Reductase inhibitor *mp 252 °C*	Treatment of benign prostatic hyperplasia and male hair loss	$CONHC(CH_3)_3$
103	Mestranol	Devocin; Norinyl; Ovastol; Tranel	Estrogen, used in combination as oral contraceptive *mp 151 °C*	Prevention of unplanned pregnancy	
104	Mifepristone	Mifeprex; Mifegyne; Corlux; RU 486	Progesterone receptor modulator *mp 150 °C*	Termination of pregnancy	
105	Sildenafil	Viagra; Revatio	Cyclic GMP phosphodiesterase inhibitor	Treatment of erectile dysfunction and pulmonary hypertension	
106	Tamsulosin	Flomax; Amsulosin; Harnal; Omix	Prostate selective α_1-adrenoceptor antagonist	Treatment of benign prostatic hyperplasia	
107	Testosterone	Androderm; Androgel; Testrim; Striant	Anabolic steroid hormone	Treatment of male hypogonadism	

Index

The index below lists the trade and generic names for the drugs in this table; the generic names are in bold face. Each entry is referred to by its generic name and the identification number in the table. An asterisk* beside the trade name indicates a product that is a combination of two drugs.

Name	Generic Name
Abraxane	Paclitaxel (58)
Acecomb	Lisinopril (70)
Acetaminophen	Acetaminophen (95)
Acetylsalicylic acid	Acetylsalicylic acid (25)
Aclonium	Gabapentin (97)

Name	Generic Name
Actiq	Fentanyl (96)
Actonel	Risedronic acid (94)
Actos	Pioglitazone (22)
Acyclovir	Acyclovir (36)
Adofen	Fluoxetine (84)

Name	Generic Name
Adoxa	Doxycycline (15)
Adronat	Alendronic acid (91)
Adrucil	Fluorouracil (54)
Advair*	Fluticasone (5)
Advil	Ibuprofen (27)
Alapril	Lisinopril (70)
Alavert	Loratadine (6)
Albuterol	Albuterol (1)
Alendronic acid	Alendronic acid (91)
Alendros	Alendronic acid (91)
Aleve	Naproxen (29)
Aliskiren	Aliskiren (61)
Allegra	Fexofenadine (4)
Amfebutamone	Bupropion (81)
Amfostet	Amphotericin (37)
Amikacin	Amikacin (10)
Amikin	Amikacin (10)
Amlodipine	Amlodipine (62)
Amoxicillin	Amoxicillin (11)
Amoxil	Amoxicillin (11)
Amphotericin	Amphotericin (37)
Amphozone	Amphotericin (37)
Amsulosin	Tamsulosin (106)
Anaprox	Naproxen (29)
Anastrozole	Anastrozole (47)
Andozac	Finasteride (102)
Androderm	Testosterone (107)
Androgel	Testosterone (107)
Antenex	Diazepam (82)
APAP	Acetaminophen (95)
Aprovel	Irbesartan (69)
Aricept	Donepezil (83)
Arimidex	Anastrozole (47)
Aropax	Paroxetine (86)
Aspirin	Acetylsalicylic acid (25)
Atenolol	Atenolol (63)
Atorvastatin	Atorvastatin (76)
Avalide*	Irbesartan (69)
Avandia	Rosiglitazone (23)
Avapro	Irbesartan (69)
Avinza	Morphine (99)
Avirax	Acyclovir (36)
Azantac	Ranitidine (35)
Azidothymidine	Zidovudine (46)
Azithromycin	Azithromycin (12)
Azitrocin	Azithromycin (12)
AZT	Zidovudine (46)
Barbivis	Phenobarbital (87)
Bendopa	Levodopa (85)
Bicalutamide	Bicalutamide (48)
Biclin	Amikacin (10)
Biozolene	Fluconazole (39)
Bortezomib	Bortezomib (49)
Budamax	Budesonide (2)
Budeson	Budesonide (2)
Budesonide	Budesonide (2)
Bupropion	Bupropion (81)
Caduet*	Atorvastatin (76)
Calcijex	Calcitriol (92)
Calcitriol	Calcitriol (92)
Calmpose	Diazepam (82)
Camptetin	Irinotecan (57)
Camptosar	Irinotecan (57)
Capecitabine	Capecitabine (50)
Carac	Fluorouracil (54)
Carace	Lisinopril (70)
Carboplat	Carboplatin (51)
Carboplatin	Carboplatin (51)
Carboxyterfenadine	Fexofenadine (4)
Carvedilol	Carvedilol (64)
Casodex	Bicalutamide (48)
Ceclor	Cefaclor (13)
Cefaclor	Cefaclor (13)
Celebrex	Celecoxib (26)
Celecoxib	Celecoxib (26)
Cetirizine	Cetirizine (3)
Chemacin	Amikacin (10)
Cimetidine	Cimetidine (31)
Cimetimax	Cimetidine (31)
Cipro	Ciprofloxacin (14)
Ciprobay	Ciprofloxacin (14)
Ciprofloxacin	Ciprofloxacin (14)
Ciproxin	Ciprofloxacin (14)
Clafen	Cyclophosphamide (52)
Claratyne	Loratadine (6)
Claritin	Loratadine (6)
Clopidogrel	Clopidogrel (65)
Contin	Morphine (99)
Copegus	Ribavirin (43)
Cordioxil	Digoxin (66)
Coreg	Carvedilol (64)
Corlux	Mifepristone (104)
Cotronak	Ribavirin (43)
Coumadin	Warfarin (75)
Cozaar	Losartan (71)
Cyclophosphamide	Cyclophosphamide (52)
Cytophosphan	Cyclophosphamide (52)
Cytoxan	Cyclophosphamide (52)
Davoxin	Digoxin (66)
Deltasone	Prednisone (30)
Deoxymykoin	Doxycycline (15)
Devocin	Mestranol (103)
Diabex	Metformin (21)
Diaformin	Metformin (21)
Diazepam	Diazepam (82)
Diflucan	Fluconazole (39)
Digacin	Digoxin (66)
Digosin	Digoxin (66)
Digoxin	Digoxin (66)
Dilanacin	Digoxin (66)

Name	Generic Name
Dilatrend	Carvedilol (64)
Dobupal	Venlafaxine (90)
Docetaxel	Docetaxel (53)
Donepezil	Donepezil (83)
Doryx	Doxycycline (15)
Doxycycline	Doxycycline (15)
Dronal	Alendronic acid (91)
Duragesic	Fentanyl (96)
E-Mycin	Erythromycin (16)
Efavirenz	Efavirenz (38)
Effexor	Venlafaxine (90)
Efudex	Fluorouracil (54)
Elazor	Fluconazole (39)
Enalaprilat	Enalaprilat (67)
Endoxan	Cyclophosphamide (52)
Epivir	Lamivudine (40)
Erbakar	Carboplatin (51)
Erythrocin	Erythromycin (16)
Erythromycin	Erythromycin (16)
Eucardic	Carvedilol (64)
Evista	Raloxifene (93)
Ezetimibe	Ezetimibe (77)
Fenemal	Phenobarbital (87)
Fentanyl	Fentanyl (96)
Fentora	Fentanyl (96)
Fexofenadine	Fexofenadine (4)
Finasteride	Finasteride (102)
Finastid	Finasteride (102)
Flexilite	Amikacin (10)
Flociprin	Ciprofloxacin (14)
Flomax	Tamsulosin (106)
Flonase	Fluticasone (5)
Flovent	Fluticasone (5)
Fluconazole	Fluconazole (39)
Fluoroplex	Fluorouracil (54)
Fluorouracil	Fluorouracil (54)
Fluoxetine	Fluoxetine (84)
Fluticasone	Fluticasone (5)
Fontex	Fluoxetine (84)
Fortamet	Metformin (21)
Fosamax	Alendronic acid (91)
Fungizone	Amphotericin (37)
Gabapentin	Gabapentin (97)
Gabitril	Tiagabine (89)
Gardenal	Phenobarbital (87)
Gastromet	Cimetidine (31)
Gemcitabine	Gemcitabine (55)
Gemzar	Gemcitabine (55)
Gleevec	Imatinib (56)
Glibenese	Glipizide (20)
Glipizide	Glipizide (20)
Glivec	Imatinib (56)
Glucophage	Metformin (21)
Glucotrol	Glipizide (20)
Glycerol trinitrate	Glycerol trinitrate (68)

Name	Generic Name
Glydiazinamide	Glipizide (20)
Harnal	Tamsulosin (106)
Hedrex	Acetaminophen (95)
Heptodin	Lamivudine (40)
Hivid	Zalcitabine (45)
Hytrin	Terazosin (74)
Hytrinex	Terazosin (74)
Hyzaar*	Losartan (71)
Ibuprofen	Ibuprofen (27)
Ilosone	Erythromycin (16)
Imatinib	Imatinib (56)
Imigrane	Sumatriptan (101)
Imitrex	Sumatriptan (101)
Inflammide	Budesonide (2)
Irbesartan	Irbesartan (69)
Irinotecan	Irinotecan (57)
Isimoxin	Amoxicillin (11)
Isocover	Clopidogrel (65)
Isoniazid	Isoniazid (17)
Itrin	Terazosin (74)
Januvia	Sitagliptin (24)
Kadian	Morphine (99)
Kefolar	Cefaclor (13)
L-Dopa	Levodopa (85)
Lamisil	Terbinafine (44)
Lamivudine	Lamivudine (40)
Laniazid	Isoniazid (17)
Lanoxin	Digoxin (66)
Lansoprazole	Lansoprazole (32)
Larodopa	Levodopa (85)
Levodopa	Levodopa (85)
Lidamantle	Lidocaine (98)
Lidocaine	Lidocaine (98)
Linezolid	Linezolid (18)
Lipitor	Atorvastatin (76)
Lisinopril	Lisinopril (70)
Liviatin	Doxycycline (15)
Livotensin	Diazepam (82)
Lopressor	Metoprolol (72)
Loratadine	Loratadine (6)
Lorien	Fluoxetine (84)
Losartan	Losartan (71)
Losec	Omeprazole (33)
Luminal	Phenobarbital (87)
Lustral	Sertraline (88)
Magnurol	Terazosin (74)
Meconium	Morphine (99)
Megrelan	Sumatriptan (101)
Meilax	Clopidogrel (65)
Melfax	Ranitidine (35)
Meloxicam	Meloxicam (28)
Mepral	Omeprazole (33)
Mestranol	Mestranol (103)
Metacain	Meloxicam (28)
Metacam	Meloxicam (28)

Name	Generic Name
Metformin	Metformin (21)
Meticorten	Prednisone (30)
Metoprolol	Metoprolol (72)
Micardis	Telmisartan (73)
Mifegyne	Mifepristone (104)
Mifeprex	Mifepristone (104)
Mifepristone	Mifepristone (104)
Minodiab	Glipizide (20)
Mirolex	Acyclovir (36)
Mobic	Meloxicam (28)
Monotrim	Trimethoprim (19)
Montelukast	Montelukast (7)
Mopral	Omeprazole (33)
Morfine	Morphine (99)
Morphine	Morphine (99)
Motivan	Paroxetine (86)
Naprelan	Naproxen (29)
Naprogesic	Naproxen (29)
Naproxen	Naproxen (29)
Negasin	Amikacin (10)
Neosar	Cyclophosphamide (52)
Neurotin	Gabapentin (97)
Nevirapine	Nevirapine (41)
Nexium	Omeprazole (33)
Niacin	Nicotinic acid (78)
Nicotinic acid	Nicotinic acid (78)
Nitroglycerin	Glycerol trinitrate (68)
Nolvadex	Tamoxifen (60)
Nonoplat	Carboplatin (51)
Norinyl	Mestranol (103)
Norvasc	Amlodipine (62)
Novatec	Lisinopril (70)
Novazyd	Lisinopril (70)
Omeprazole	Omeprazole (33)
Omix	Tamsulosin (106)
Onsenal	Celecoxib (26)
Optinate	Risedronic acid (94)
Oseltamivir	Oseltamivir (42)
Ospamox	Amoxicillin (11)
Ovastol	Mestranol (103)
Paclitaxel	Paclitaxel (58)
Panacef	Cefaclor (13)
Panoral	Cefaclor (13)
Pantoprazole	Pantoprazole (34)
Pantozol	Pantoprazole (34)
Paracetamol	Acetaminophen (95)
Paraplatin	Carboplatin (51)
Paroxetine	Paroxetine (86)
Paxil	Paroxetine (86)
Penthiobarbital	Thiopental (100)
Pentothal	Thiopental (100)
Peptimax	Cimetidine (31)
Permicran	Sumatriptan (101)
Phenobarbital	Phenobarbital (87)
Pioglitazone	Pioglitazone (22)

Name	Generic Name
Plavix	Clopidogrel (65)
Pravastatin	Pravastatin (79)
Pravochol	Pravastatin (79)
Prednisone	Prednisone (30)
Prevacid	Lansoprazole (32)
Prevpac	Lansoprazole (32)
Prezal	Lansoprazole (32)
Prilosec	Omeprazole (33)
Procure	Finasteride (102)
Proloprim	Trimethoprim (19)
Propecia	Finasteride (102)
Proscar	Finasteride (102)
Protonix	Pantoprazole (34)
Proventil	Albuterol (1)
Prozac	Fluoxetine (84)
Pulmicort	Budesonide (2)
Raloxifene	Raloxifene (93)
Ranitidine	Ranitidine (35)
Rantec	Ranitidine (35)
Rasilex	Aliskiren (61)
Rebetol	Ribavirin (43)
Retrovir	Zidovudine (46)
Revatio	Sildenafil (105)
Rhinocort	Budesonide (2)
Ribasphere	Ribavirin (43)
Ribavirin	Ribavirin (43)
Rifun	Pantoprazole (34)
Risedronic acid	Risedronic acid (94)
Rocaltrol	Calcitriol (92)
Rosiglitazone	Rosiglitazone (23)
Rougoxin	Digoxin (66)
Roxanol	Morphine (99)
RU 486	Mifepristone (104)
Salmeterol	Salmeterol (8)
Sarafem	Fluoxetine (84)
Seduxen	Diazepam (82)
Serad	Sertraline (88)
Serevent	Salmeterol (8)
Serlain	Sertraline (88)
Seroxat	Paroxetine (86)
Sertraline	Sertraline (88)
Sildenafil	Sildenafil (105)
Simvastatin	Simvastatin (80)
Singulair	Montelukast (7)
Sitagliptin	Sitagliptin (24)
Soltamox	Tamoxifen (60)
SoluTab	Lansoprazole (32)
Sostril	Ranitidine (35)
Spiriva	Tiotropium bromide (9)
Stesolid	Diazepam (82)
Striant	Testosterone (107)
Sublimaze	Fentanyl (96)
Sumadol	Sumatriptan (101)
Sumatriptan	Sumatriptan (101)
Sunitinib	Sunitinib (59)

Name	Generic Name
Sustiva	Efavirenz (38)
Sutent	Sunitinib (59)
Tagamet	Cimetidine (31)
Tagonis	Paroxetine (86)
Tamaxin	Tamoxifen (60)
Tamiflu	Oseltamivir (42)
Tamoplex	Tamoxifen (60)
Tamoxifen	Tamoxifen (60)
Tamsulosin	Tamsulosin (106)
Tatig	Sertraline (88)
Taural	Ranitidine (35)
Taxol	Paclitaxel (58)
Taxotere	Docetaxel (53)
Tekturna	Aliskiren (61)
Telfast	Fexofenadine (4)
Telmisartan	Telmisartan (73)
Tenormin	Atenolol (63)
Teraprost	Terazosin (74)
Terazosin	Terazosin (74)
Terbinafine	Terbinafine (44)
Testosterone	Testosterone (107)
Testrim	Testosterone (107)
Thiopental	Thiopental (100)
Thiopentone	Thiopental (100)
Tiagabine	Tiagabine (89)
Tiotropium bromide	Tiotropium bromide (9)
Tipidyl	Clopidogrel (65)
Topotecin	Irinotecan (57)
Toprol XL	Metoprolol (72)
Tramil	Acetaminophen (95)
Tranel	Mestranol (103)
Trewilor	Venlafaxine (90)
Triflucan	Fluconazole (39)
Trimethoprim	Trimethoprim (19)
Triprim	Trimethoprim (19)
Tylenol	Acetaminophen (95)
Uniflox	Ciprofloxacin (14)
Uroflo	Terazosin (74)
Urprosan	Finasteride (102)
Valium	Diazepam (82)
Valodex	Tamoxifen (60)

Name	Generic Name
Vandral	Venlafaxine (90)
Vasocard	Terazosin (74)
Vasotec	Enalaprilat (67)
Velcade	Bortezomib (49)
Veldopa	Levodopa (85)
Venlafaxine	Venlafaxine (90)
Ventolin	Albuterol (1)
Viagra	Sildenafil (105)
Vibramycin	Doxycycline (15)
Vinzam	Azithromycin (12)
Viramune	Nevirapine (41)
Viratek	Ribavirin (43)
Virazole	Ribavirin (43)
Vivatec	Lisinopril (70)
Volmax	Albuterol (1)
Vytorin*	Simvastatin (80)
Warfarin	Warfarin (75)
Wellbutrin	Bupropion (81)
Xeloda	Capecitabine (50)
Xylocaine	Lidocaine (98)
Xylocard	Lidocaine (98)
Yewtaxan	Paclitaxel (58)
Zalcitabine	Zalcitabine (45)
Zantac	Ranitidine (35)
Zeffix	Lamivudine (40)
Zestoretic	Lisinopril (70)
Zestril	Lisinopril (70)
Zetia	Ezetimibe (77)
Zidovir	Zidovudine (46)
Zidovudine	Zidovudine (46)
Zithromax	Azithromycin (12)
Zmax	Azithromycin (12)
Zocor	Simvastatin (80)
Zoloft	Sertraline (88)
Zoltum	Omeprazole (33)
Zoton	Lansoprazole (32)
Zovir	Acyclovir (36)
Zovirax	Acyclovir (36)
Zyban	Bupropion (81)
Zyrtec	Cetirizine (3)
Zyvox	Linezolid (18)

CHEMICAL CONSTITUENTS OF HUMAN BLOOD

This table lists typical concentrations of some of the chemical constituents of human blood. The table covers elements and compounds of relatively low molecular weight. References 1 and 4 give extensive information on enzymes, hormones, vitamins, and other blood constituents.

The values given for the normal range refer to healthy adults who have not been exposed to unusual environmental agents. In keeping with IUPAC practice, all values refer to a volume of one liter, and thus are stated in units of g/L, mg/L, μg/L or mmol/L. Many clinical test results, especially in the United States, are reported on a deciliter (dL) rather than a liter basis; thus the values in this table should be divided by 10 to place them on a dL basis. The symbols S (for serum), P (plasma), and WB (whole blood) in the second column indicate the nature of the blood sample to which the values apply. In some cases only a single mean value has been reported, rather than a range; these are given in italics.

The total volume of blood in a 100 kg (220 lb) adult is 7.5 L for a male and 6.7 L for a female. The corresponding volume of plasma is 4.4 L and 4.3 L, respectively (Reference 1).

Values from Reference 1 are so-called "reference values" against which clinical tests of blood chemistry are compared. In these cases the "normal range" is understood to include about 95% of the population. The remaining 5% may show values outside the normal range without necessarily implying a medical problem. Note that these reference values may vary slightly from one testing laboratory to another, depending on the detailed test procedure.

Accurate measurements on trace elements are very difficult to make, and wide variations can be found in the literature. Preferred measurement methods are discussed in References 2 and 6. Values for the trace elements can also vary from one country to another, depending on dietary or environmental factors. Thus cadmium levels tend to be higher in Japan because of the prevalence of seafood in the diet, and lead levels are higher in regions where lead additives are still used in gasoline. Variations with gender, age, geography, and occurrence of diseases are reviewed in Reference 6.

The Critical Values column gives levels that deviate far enough from the normal range to suggest a probable medical issue. Such values from Reference 3 are the Biological Exposure Indexes (BEI) that are specified by the American Council of Government Industrial Hygienists (ACGIH) as danger signals for the levels of pollutants in the workplace.

References

1. Wallach, J., *Interpretation of Diagnostic Tests, Eight Edition*, Wolters Kluwer, Philadelphia, 2007.
2. IUPAC Commission on Toxicology, "Sample Collection Guidelines for Trace Elements in Blood and Urine," *Pure & Appl. Chem.*, 67, 1575, 1995.
3. *2008 TLV's and BEI's*, American Conference of Governmental Industrial Hygienists, 1330 Kemper Meadow Drive, Cincinnati, OH 45240–1634, 2008 (www.acgih.org).
4. Altman, P. L., and Dittmer, D. S., Eds., *Biology Data Book, Second Edition, Vol. III*, Federation of American Societies for Experimental Biology, Bethesda, MD, 1974.
5. Bowen, H. J. M., *Trace Elements in Biochemistry*, Academic Press, New York, 1966.
6. Versieck, J., and Cornelis, R., *Trace Elements in Human Plasma or Serum*, CRC Press, Boca Raton, FL, 1989.

Component		Unit	Normal Range Low	Normal Range High	Critical Values	Ref.
		Inorganic				
Aluminum	S	μg/L	1	10	>60	6,2
Ammonia	P	μg/L	190	600	>700	1
Antimony	S,P	μg/L		*1*		6
Arsenic	S	μg/L	0.5	5		6,2
Barium	S,P	μg/L		*79*		4,5
Beryllium	S,P	μg/L		*<4*		4,5
Bicarbonate (HCO_3^-)	WB	mmol/L	22	28	<10 or >40	1
Bromine	S,P	mg/L	2	11		6,4
Cadmium	S	μg/L	0.1	1	>5	6,2,3
Calcium, total	S	mg/L	90	105	<65 or >140	1
Calcium ion (Ca^{++})	WB	mg/L	30	45		1
Carbon dioxide	P	mmol/L	21	30	<11 or >40	1
Carbon monoxide*	WB	%CO-Hb	0	5%	30%	1
Cesium	S,P	μg/L	0.5	2.0		6
Chloride (Cl^-)	S	mmol/L	98	106	<80 or >115	1
Chromium	S	μg/L	0.1	0.4		6,2
Cobalt	S	μg/L	0.05	0.35	>1	6,2,3
Copper	S	mg/ L	0.7	1.4		1,2,6
Fluorine	S,P	μg/L	33	236		6
Hydrogen ion (H^+)	WB	pH	7.38	7.44	<7.10 or >7.59	1
Iodine (total)	S,P	μg/L	59	76		4
Iron	S	mg/L	0.5	1.7		1
Lead	S	μg/L	5	100	>300	1,3,6

Component		Unit	Normal Range Low	High	Critical Values	Ref.
Lithium	S,P	μg/L		*8*		6
Magnesium	S	mg/L	18	30	<10 or >47	1
Manganese	S	μg/L	0.3	1.0		6,2
Mercury	S	μg/L	0.5	3	>15	2,3
Molybdenum	S,P	μg/L	0.3	1.3		6
Nickel	S	μg/L	0.1	1.3		6,2
Oxygen (arterial)	WB	% saturation	96%	100%		1
Oxygen (venous)	WB	% saturation	60%	85%		1
Phosphorus (inorganic)	S	mg/L	30	45	<11	1
Potassium	S	mmol/L	3.5	5.0	<2.8 or >6.2	1
		mg/L	137	196		
Rubidium	S,P	μg/L	100	300		6
Selenium	S,P	μg/L	40	160		2,6
Silver	S,P	μg/L		*1*		6
Sodium	S	mmol/L	135	145	<120 or >160	1
		g/L	3.11	3.34		
Strontium	S,P	μg/L		*57*		4,5
Sulfur (total)	S,P	mg/L		*780*		4
Tellurium	S,P	μg/L		*30*		4,5
Titanium	S,P	μg/L		*33*		4,5
Tin	S,P	μg/L		*1*		4,5
Vanadium	S,P	μg/L	0.02	1.0		6
Zinc	S,P	mg/L	0.5	1.2		6,2,4
Zirconium	S,P	μg/L		*400*		4,5
Organic						
Acetoacetate ion	P	mg/L		*<10*		1
Acetone	S,P	mg/L	3	20		1
Alanine	S,P	mg/L	30	37		4
Arginine	S,P	mg/L	12	19		4
Asparagine	S,P	mg/L	5.4	6.5		4
Cholesterol, total	P	mg/L	1000	2000**	>2400	1,4
HDL Cholesterol	P	mg/L	400	600		1
LDL Cholesterol	P	mg/L	0	1000	>1900	1
Citrulline	S,P	mg/L	2.1	9.7		4
Creatine	S,P	mg/L	2.8	6.2		4
Creatinine	S	mg/L	5	15	>50	1
Fructose	WB	mg/L	5	50		4
Glucosamine	S,P	mg/L	760	1110		4
Glucose (fasting)	S	mg/L	600	1000	<450 or >1300	1
Glutamic acid	S,P	mg/L	4.3	11.5		4
Glutamine	S,P	mg/L	61	102		4
Glycine	S,P	mg/L	13.4	17.3		4
Histidine	S,P	mg/L	7.9	14.8		4
Homocysteine	P	mg/L	0.54	1.62		1
Isoleucine	S,P	mg/L	6.9	12.8		4
Lactate (venous)	P	mg/L	50	150		1
Leucine	S,P	mg/L	14	23		4
Lysine	S,P	mg/L	25	30		4
Methionine	S,P	mg/L	3.3	4.3		4
Ornithine	S,P	mg/L	6.2	8.0		4
Phenylalanine	S,P	mg/L	5.8	14.0		1
Proline	S,P	mg/L	20	33		4
Serine	S,P	mg/L	10.1	12.5		4

Component		Unit	Normal Range Low	High	Critical Values	Ref.
Taurine	S,P	mg/L	4.1	8.2		4
Threonine	S,P	mg/L	12	17		4
Triglyceride	S	mg/L	250	1750		1
Tyrosine	S,P	mg/L	8.1	14.5		4
Urea	S	mmol/L	3.5	7.0	<0.7 or >28	1
Urea nitrogen (BUN)	S	mg/L (of N)	100	200	<20 or >800	1
Uric acid (males)	S	mg/L	25	80		1
Uric acid (females)	S	mg/L	13	60		1
Valine	S,P	mg/L	24	37		4

* Measured as the percent of hemoglobin bound to CO. Typical value for heavy smokers is 5%–10%. Major symptoms begin around 30%, and respiratory failure sets in at >60%.

** This is the desirable upper limit. Values between 2000 and 2400 mg/L are considered borderline high.

CHEMICAL COMPOSITION OF THE HUMAN BODY

The elemental composition of the "standard man" of mass 70 kg is given below.

References

1. Padikal, T. N., and Fivozinsky, S. P., *Medical Physics Data Book, National Bureau of Standards Handbook 138*, U. S. Government Printing Office, Washington, DC, 1981.
2. Snyde, W. S., et al., *Reference Man: Anatomical, Physiological, and Metabolic Characteristics*, Pergamon, New York, 1975.

Element	**Amount (g)**	**Percent of total body mass**
Oxygen	43,000	61
Carbon	16,000	23
Hydrogen	7000	10
Nitrogen	1800	2.6
Calcium	1000	1.4
Phosphorus	780	1.1
Sulfur	140	0.20
Potassium	140	0.20
Sodium	100	0.14
Chlorine	95	0.12
Magnesium	19	0.027
Silicon	18	0.026
Iron	4.2	0.006
Fluorine	2.6	0.0037
Zinc	2.3	0.0033
Rubidium	0.32	0.00046
Strontium	0.32	0.00046
Bromine	0.20	0.00029
Lead	0.12	0.00017
Copper	0.072	0.00010
Aluminum	0.061	0.00009
Cadmium	0.050	0.00007
Boron	<0.048	0.00007
Barium	0.022	0.00003
Tin	<0.017	0.00002
Manganese	0.012	0.00002
Iodine	0.013	0.00002
Nickel	0.010	0.00001
Gold	<0.010	0.00001
Molybdenum	<0.0093	0.00001
Chromium	<0.0018	0.000003
Cesium	0.0015	0.000002
Cobalt	0.0015	0.000002
Uranium	0.00009	0.0000001
Beryllium	0.000036	
Radium	$3.1 \cdot 10^{-11}$	

NUTRIENT VALUES OF FOODS

The U. S. Department of Agriculture maintains the USDA National Nutrient Database for Standard Reference, which contains over 7000 food items with data on the energy content, minerals, vitamins, and other properties of nutritional interest. The table here includes about 600 common foods extracted from that database. The properties listed are the energy content (in effect, the enthalpy of combustion); the content of carbohydrates, proteins, and lipids (fats); the cholesterol content; and the amount of sodium, potassium, calcium, magnesium, iron, copper, zinc, manganese, phosphorus, and selenium. All values are given for a 100 gram sample of the food.

To conform with common practice in nutritional science, the energy content is given in kilocalories. For conversion to kilojoules, this number should be multiplied by 4.184. For conversion to the avoirdupois units frequently used in the United States, note that:

1 oz = 28.35 g (thus the 100 g basis in this table is approximately 3.5 oz)

1 lb = 453.6 g

The full USDA database covers specific proprietary brands of many processed foods, and it includes vitamin content as well as the minerals given in this table.

Reference

U.S. Department of Agriculture, Agricultural Research Service. 2005. USDA National Nutrient Database for Standard Reference, Release 18. Nutrient Data Laboratory Home Page, http://www.ars.usda.gov/ba/bhnrc/ndl

Food description	Energy kcal/ 100 g	Carb. g/ 100 g	Protein g/ 100 g	Fat g/ 100 g	Chol. mg/ 100 g	Na mg/ 100 g	K mg/ 100 g	Ca mg/ 100g	Mg mg/ 100 g	Fe mg/ 100 g	Cu mg/ 100 g	Zn mg/ 100 g	Mn mg/ 100 g	P mg/ 100 g	Se µg/ 100 g
Alcoholic beverage, beer, light	29	1.6	0.2	0.0	0	4	21	4	5	0.03	0.006	0.01	0.006	12	0.4
Alcoholic beverage, beer, regular	43	3.6	0.5	0.0	0	4	27	4	6	0.02	0.005	0.01	0.008	14	0.6
Alcoholic beverage, dessert wine, sweet	160	13.7	0.2	0.0	0	9	92	8	9	0.24	0.045	0.07	0.119	9	0.5
Alcoholic beverage, distilled (gin, rum, vodka, whiskey), 80 proof	231	0.0	0.0	0.0	0	1	2	0	0	0.04	0.021	0.04	0.018	4	0.0
Alcoholic beverage, distilled (gin, rum, vodka, whiskey), 90 proof	263	0.0	0.0	0.0	0	1	2	0	0	0.04	0.021	0.04	0.018	4	0.0
Alcoholic beverage, distilled, all, 100 proof	295	0.0	0.0	0.0	0	1	2	0	0	0.04	0.021	0.04	0.018	4	0.0
Alcoholic beverage, liqueur, coffee, 63 proof	308	32.2	0.1	0.3	0	8	30	1	3	0.06	0.040	0.03	0.017	6	0.3
Alcoholic beverage, table wine, red	85	2.6	0.1	0.0	0	4	127	8	12	0.46	0.011	0.14	0.132	23	0.2
Alcoholic beverage, table wine, white	83	2.6	0.1	0.0	0	5	71	9	10	0.27	0.004	0.12	0.117	18	0.1
Almonds, dry roasted, w/o salt	597	19.3	22.1	52.8	0	1	746	266	286	4.51	1.170	3.54	2.620	489	2.8
Almonds, oil roasted, w/o salt	607	17.7	21.2	55.2	0	1	699	291	274	3.68	0.955	3.07	2.460	466	2.8
Apple juice, canned or bottled, unsweetened, w/o vitamin C	47	11.7	0.1	0.1	0	3	119	7	3	0.37	0.022	0.03	0.113	7	0.1
Apples, raw, w/o skin	48	12.8	0.3	0.1	0	0	90	5	4	0.07	0.031	0.05	0.038	11	0.0
Apples, raw, with skin	52	13.8	0.3	0.2	0	1	107	6	5	0.12	0.027	0.04	0.035	11	0.0
Apricots, dried, sulfured, stewed, w. sugar	113	29.3	1.2	0.2	0	3	443	15	15	1.52	0.138	0.24	0.088	38	
Apricots, dried, sulfured, stewed, w/o sugar	85	22.2	1.2	0.2	0	4	411	19	11	0.94	0.121	0.14	0.083	25	0.8
Artichokes, (globe or French), boiled, w/o salt	50	11.2	3.5	0.2	0	95	354	45	60	1.29	0.233	0.49	0.259	86	0.2
Arugula, raw	25	3.7	2.6	0.7	0	27	369	160	47	1.46	0.076	0.47	0.321	52	0.3
Asparagus, boiled	22	4.1	2.4	0.2	0	14	224	23	14	0.91	0.165	0.60	0.154	54	6.1
Asparagus, raw	20	3.9	2.2	0.1	0	2	202	24	14	2.14	0.189	0.54	0.158	52	2.3
Avocados, raw, all commercial varieties	160	8.5	2.0	14.7	0	7	485	12	29	0.55	0.190	0.64	0.142	52	0.4
Bagel, plain, toasted, enriched w. calcium propanoate	257	50.5	10.0	1.6	0	448	75	89	22	6.05	0.130	1.90	0.515	87	22.8
Bagels, cinnamon-raisin, toasted	294	59.3	10.6	1.8	0	346	163	20	23	4.09	0.160	0.81	0.328	83	33.3
Baked beans, canned, no salt	105	20.6	4.8	0.4	0	1	296	50	32	0.29	0.206	1.40		104	4.5
Bamboo shoots, boiled, w/o salt	12	1.9	1.5	0.2	0	4	533	12	3	0.24	0.082	0.47	0.113	20	0.4
Bananas, raw	89	22.8	1.1	0.3	0	1	358	5	27	0.26	0.078	0.15	0.270	22	1.0
Bass, freshwater, mixed species, cooked, dry heat	146	0.0	24.2	4.7	87	90	456	103	38	1.91	0.119	0.83	1.140	256	16.2
Bass, striped, cooked, dry heat	124	0.0	22.7	3.0	103	88	328	19	51	1.08	0.040	0.51	0.019	254	46.8
Beans, French, mature seeds, boiled, w/o salt	129	24.0	7.1	0.8	0	6	370	63	56	1.08	0.115	0.64	0.382	102	1.2
Beans, kidney, all types, mature seeds, boiled, w/o salt	127	22.8	8.7	0.5	0	1	405	35	42	2.22	0.216	1.00	0.430	138	1.1
Beans, lima, immature seeds, canned	71	13.3	4.1	0.3	0	252	285	28	34	1.61	0.162	0.64	0.700	71	1.1
Beans, navy, mature seeds, canned	113	20.5	7.5	0.4	0	448	288	47	47	1.85	0.208	0.77	0.375	134	5.8
Beans, pinto, mature seeds, boiled, w/o salt	143	26.2	9.0	0.7	0	1	436	46	50	2.09	0.219	0.98	0.453	147	6.2

Food description	Energy kcal/ 100 g	Carb. g/ 100 g	Protein g/ 100 g	Fat g/ 100 g	Chol. mg/ 100 g	Na mg/ 100 g	K mg/ 100 g	Ca mg/ 100g	Mg mg/ 100 g	Fe mg/ 100 g	Cu mg/ 100 g	Zn mg/ 100 g	Mn mg/ 100 g	P mg/ 100 g	Se µg/ 100 g
Beans, snap, green, boiled, w/o salt	35	7.9	1.9	0.3	0	1	146	44	18	0.65	0.057	0.25	0.285	29	0.2
Beef, bottom sirloin, tri-tip roast, lean & fat, 0" fat, choice, roasted	218	0.0	25.7	12.4	94	50	308	17	20	1.70	0.083	4.52	0.009	189	27.3
Beef, brisket, flat half, lean & fat, 1/8" fat, select, braised	280	0.0	29.0	17.4	71	49	237	17	19	2.40	0.090	6.92	0.010	180	28.1
Beef, chuck, arm pot roast, lean, 1/8" fat, choice, braised	224	0.0	34.7	8.4	81	56	275	15	23	3.04	0.129	8.20	0.012	213	34.7
Beef, ground, 75% lean meat / 25% fat, patty, broiled	278	0.0	25.6	18.7	89	78	289	30	20	2.37	0.075	6.19	0.010	189	21.4
Beef, ground, 95% lean meat / 5% fat, patty, broiled	171	0.0	26.3	6.6	76	65	348	7	22	2.83	0.096	6.43	0.014	206	21.7
Beef, loin, porterhouse steak, lean & fat, 1/4" fat, all grades, broiled	329	0.0	22.5	25.8	72	62	255	8	20	2.68	0.118	4.13	0.014	177	19.3
Beef, loin, T-bone steak, lean & fat, 1/4" fat, all grades, broiled	306	0.0	23.5	22.8	65	67	282	7	22	3.09	0.122	4.30	0.014	185	11.6
Beef, rib, eye, small end (ribs 10-12), lean, 0" fat, choice, broiled	205	0.0	28.9	9.0	91	60	363	16	25	1.98	0.092	5.49	0.011	227	33.6
Beef, rib, large end (ribs 6-9), lean & fat, 1/4" fat, choice, roasted	383	0.0	22.3	32.0	85	63	283	10	19	2.27	0.086	5.58	0.013	168	21.9
Beef, round, bottom round roast, lean, 0" fat, select, roasted	169	0.0	28.3	5.3	72	38	238	6	19	2.35	0.079	4.92	0.010	183	35.7
Beef, round, bottom round, lean & fat, 1/8" fat, all grades, roasted	218	0.0	26.4	11.6	75	35	214	6	17	2.16	0.063	4.43	0.009	164	27.0
Beef, round, eye of round, lean & fat, 1/8" fat, all grades, roasted	208	0.0	28.3	9.7	62	37	227	7	18	2.29	0.067	4.70	0.010	174	28.7
Beef, tenderloin, lean & fat, 1/8" fat, all grades, broiled	267	0.0	26.5	17.1	90	54	329	19	22	1.69	0.079	4.76	0.009	205	28.6
Beef, top sirloin, lean & fat, 1/8" fat, select, broiled	230	0.0	27.1	12.7	67	57	345	22	23	1.66	0.074	4.93	0.009	217	29.3
Beets, boiled	44	10.0	1.7	0.2	0	77	305	16	23	0.79	0.074	0.35	0.326	38	0.7
Blackberries, raw	43	9.6	1.4	0.5	0	1	162	29	20	0.62	0.165	0.53	0.646	22	0.4
Blueberries, raw	57	14.5	0.7	0.3	0	1	77	6	6	0.28	0.057	0.16	0.336	12	0.1
Bluefish, cooked, dry heat	159	0.0	25.7	5.4	76	77	477	9	42	0.62	0.068	1.04	0.027	291	46.8
Bologna, beef	314	4.0	10.3	28.2	56	1080	172	31	14	1.10	0.068	9.10	0.044	172	0.0
Bologna, pork	247	0.7	15.3	19.9	59	1184	281	11	14	0.77	0.080	2.03	0.036	139	12.7
Bratwurst, beef & pork, smoked	297	2.0	12.2	26.3	78	848	283	7	15	1.00	0.080	2.47	0.041	130	14.1
Bratwurst, pork, cooked	333	2.9	13.7	29.2	74	846	259	28	21	0.53	0.079	2.49	0.012	225	26.9
Brazil nuts, dried, unblanched	656	12.3	14.3	66.4	0	3	659	160	376	2.43	1.743	4.06	1.223	725	1917.0
Bread, cracked-wheat	260	49.5	8.7	3.9	0	538	177	43	52	2.81	0.222	1.24	1.371	153	25.3
Bread, Italian	271	50.0	8.8	3.5	0	584	110	78	27	2.94	0.191	0.86	0.464	103	27.2
Bread, mixed-grain, toasted (includes whole-grain, 7-grain)	272	50.4	10.9	4.1	0	530	222	99	58	3.77	0.276	1.38	1.615	191	32.1
Bread, oatmeal	269	48.5	8.4	4.4	0	599	142	66	37	2.70	0.209	1.02	0.940	126	24.6
Bread, pita, white, enriched	275	55.7	9.1	1.2	0	536	120	86	26	2.62	0.168	0.84	0.481	97	27.1
Bread, pumpernickel	250	47.5	8.7	3.1	0	671	208	68	54	2.87	0.287	1.48	1.305	178	24.5
Bread, rye	259	48.3	8.5	3.3	0	660	166	73	40	2.83	0.186	1.14	0.824	125	30.9
Bread, white, commercially prepared	266	50.6	7.6	3.3	0	681	100	151	23	3.74	0.253	0.74	0.478	99	17.3
Bread, whole-wheat, commercially prepared	246	46.1	9.7	4.2	0	527	252	72	86	3.30	0.284	1.94	2.324	229	36.6
Broccoli, boiled, w/o salt	35	7.2	2.4	0.4	0	41	293	40	21	0.67	0.061	0.45	0.194	67	1.6
Broccoli, raw	34	6.6	2.8	0.4	0	33	316	47	21	0.73	0.049	0.41	0.210	66	2.5
Brussels sprouts, boiled, w/o salt	36	7.1	2.6	0.5	0	21	317	36	20	1.20	0.083	0.33	0.227	56	1.5
Butter, w. salt	717	0.1	0.9	81.1	215	576	24	24	2	0.02	0.000	0.09	0.000	24	1.0
Butter, w/o salt	717	0.1	0.9	81.1	215	11	24	24	2	0.02	0.016	0.09	0.004	24	1.0
Cabbage, boiled, w/o salt	22	4.5	1.0	0.4	0	8	97	31	8	0.17	0.012	0.09	0.117	15	0.6
Cabbage, raw	24	5.6	1.4	0.1	0	18	246	47	15	0.59	0.023	0.18	0.159	23	0.9
Cake, angel food, commercially prepared	258	57.8	5.9	0.8	0	749	93	140	12	0.52	0.078	0.07	0.085	32	7.3
Cake, carrot, dry mix, pudding-type	415	79.2	5.1	9.8	0	567	169	172	8	1.80	0.050	0.20	0.528	247	14.9
Cake, fruitcake, commercially prepared	324	61.6	2.9	9.1	5	270	153	33	16	2.07	0.050	0.27	0.220	52	2.0
Cake, pound, commercially prepared, butter	388	48.8	5.5	19.9	221	398	119	35	11	1.38	0.035	0.46	0.090	137	8.8
Cake, white, dry mix	426	78.0	4.5	10.9	0	664	117	192	11	1.39	0.081	0.46	0.205	337	8.6
Cake, yellow, dry mix, enriched	432	78.1	4.4	11.6	2	657	82	135	10	1.50	0.072	0.27	0.191	310	3.0
Candies, fudge, chocolate, prepared-from-recipe	411	76.5	2.4	10.4	14	47	131	45	36	1.77	0.333	1.10	0.423	69	2.5

Food description	Energy kcal/ 100 g	Carb. g/ 100 g	Protein g/ 100 g	Fat g/ 100 g	Chol. mg/ 100 g	Na mg/ 100 g	K mg/ 100 g	Ca mg/ 100g	Mg mg/ 100 g	Fe mg/ 100 g	Cu mg/ 100 g	Zn mg/ 100 g	Mn mg/ 100 g	P mg/ 100 g	Se µg/ 100 g
Candies, gumdrops, starch jelly pieces	396	98.9	0.0	0.0	0	44	5	3	1	0.40	0.012	0.00	0.010	1	0.8
Candies, hard	394	98.0	0.0	0.2	0	38	5	3	3	0.30	0.029	0.01	0.010	3	0.6
Candies, marshmallows	318	81.3	1.8	0.2	0	80	5	3	2	0.23	0.097	0.04	0.008	8	1.7
Candies, milk chocolate	535	59.4	7.7	29.7	23	79	372	189	63	2.35	0.491	2.01	0.471	208	4.5
Candies, milk chocolate w. almond bites	550	51.0	9.8	35.7	19	74	471	220	59	1.50	0.200	1.34	0.010	227	3.4
Candies, peanut brittle, prepared-from-recipe	486	71.2	7.6	19.0	12	445	168	27	42	1.22	0.254	0.87	0.593	106	2.6
Candies, semisweet chocolate	479	63.1	4.2	30.0	0	11	365	32	115	3.13	0.700	1.62	0.800	132	4.2
Candies, sweet chocolate	505	59.6	3.9	34.2	0	16	290	24	113	2.76	0.574	1.50	0.494	147	2.8
Candies, white chocolate	539	59.2	5.9	32.1	14	90	286	199	12	0.24	0.060	0.74	0.008	176	4.5
Capers, canned	23	4.9	2.4	0.9	0	2964	40	40	33	1.67	0.374	0.32	0.078	10	1.2
Carbonated beverage, cola, contains caffeine	37	9.6	0.1	0.0	0	4	2	2	0	0.11	0.001	0.02	0.002	10	0.1
Carbonated beverage, ginger ale	34	8.8	0.0	0.0	0	7	1	3	1	0.18	0.018	0.05	0.013	0	0.1
Carbonated beverage, orange	48	12.3	0.0	0.0	0	12	2	5	1	0.06	0.015	0.10	0.013	1	0.0
Carbonated beverage, root beer	41	10.6	0.0	0.0	0	13	1	5	1	0.05	0.007	0.07	0.013	0	0.1
Carbonated beverage, tonic water	34	8.8	0.0	0.0	0	12	0	1	0	0.01	0.006	0.10	0.001	0	0.0
Carrots, boiled, w/o salt	35	8.2	0.8	0.2	0	58	235	30	10	0.34	0.017	0.20	0.155	30	0.7
Carrots, raw	41	9.6	0.9	0.2	0	69	320	33	12	0.30	0.045	0.24	0.143	35	0.1
Cashew nuts, dry roasted, w. salt	574	32.7	15.3	46.4	0	640	565	45	260	6.00	2.220	5.60	0.826	490	11.7
Cashew nuts, oil roasted, w. salt	581	30.2	16.8	47.8	0	308	632	43	273	6.05	2.043	5.35	1.668	531	20.3
Cashew nuts, raw	553	30.2	18.2	43.9	0	12	660	37	292	6.68	2.195	5.78	1.655	593	19.9
Catfish, channel, farmed, cooked, dry heat	152	0.0	18.7	8.0	64	80	321	9	26	0.82	0.122	1.05	0.020	245	14.5
Catsup	97	25.1	1.7	0.4	0	1114	382	18	19	0.51	0.181	0.26	0.128	33	0.3
Cauliflower, boiled, w/o salt	23	4.1	1.8	0.5	0	15	142	16	9	0.33	0.027	0.18	0.138	32	0.5
Cauliflower, raw	25	5.3	2.0	0.1	0	30	303	22	15	0.44	0.042	0.28	0.156	44	0.6
Celery, raw	14	3.0	0.7	0.2	0	80	260	40	11	0.20	0.035	0.13	0.103	24	0.4
Cereals, corn grits, white, enriched, cooked w. water, w/o salt	59	12.9	1.4	0.2	0	2	21	3	5	0.60	0.018	0.07	0.018	11	3.1
Chard, Swiss, boiled, w/o salt	20	4.1	1.9	0.1	0	179	549	58	86	2.26	0.163	0.33	0.334	33	0.9
Chard, Swiss, raw	19	3.7	1.8	0.2	0	213	379	51	81	1.80	0.179	0.36	0.366	46	0.9
Cheese, American cheddar, imitation	239	11.6	16.7	14.0	36	1345	242	562	29	0.33	0.033	2.59		712	15.2
Cheese, blue	353	2.3	21.4	28.7	75	1395	256	528	23	0.31	0.040	2.66	0.009	387	14.5
Cheese, brie	334	0.5	20.8	27.7	100	629	152	184	20	0.50	0.019	2.38	0.034	188	14.5
Cheese, camembert	300	0.5	19.8	24.3	72	842	187	388	20	0.33	0.021	2.38	0.038	347	14.5
Cheese, cheddar	403	1.3	24.9	33.1	105	621	98	721	28	0.68	0.031	3.11	0.010	512	13.9
Cheese, cheshire	387	4.8	23.4	30.6	103	700	95	643	21	0.21	0.042	2.79	0.012	464	14.5
Cheese, colby	394	2.6	23.8	32.1	95	604	127	685	26	0.76	0.042	3.07	0.012	457	14.5
Cheese, cottage, creamed, large or small curd	103	2.7	12.5	4.5	15	405	84	60	5	0.14	0.028	0.37	0.003	132	9.0
Cheese, cottage, low fat, 1% milkfat	72	2.7	12.4	1.0	4	406	86	61	5	0.14	0.028	0.38	0.003	134	9.0
Cheese, cream	349	2.7	7.6	34.9	110	296	119	80	6	1.20	0.016	0.54	0.004	104	2.4
Cheese, cream, low fat	231	7.0	10.6	17.6	56	296	167	112	8	1.68	0.022	0.76	0.004	146	4.0
Cheese, edam	357	1.4	25.0	27.8	89	965	188	731	30	0.44	0.036	3.75	0.011	536	14.5
Cheese, feta	264	4.1	14.2	21.3	89	1116	62	493	19	0.65	0.032	2.88	0.028	337	15.0
Cheese, gouda	356	2.2	24.9	27.4	114	819	121	700	29	0.24	0.036	3.90	0.011	546	14.5
Cheese, gruyere	413	0.4	29.8	32.3	110	336	81	1011	36	0.17	0.032	3.90	0.017	605	14.5
Cheese, mozzarella, whole milk	300	2.2	22.2	22.4	79	627	76	505	20	0.44	0.011	2.92	0.030	354	17.0
Cheese, muenster	368	1.1	23.4	30.0	96	628	134	717	27	0.41	0.031	2.81	0.008	468	14.5
Cheese, parmesan, grated	431	4.1	38.5	28.6	88	1529	125	1109	38	0.90	0.238	3.87	0.085	729	17.7
Cheese, pimento	375	1.7	22.1	31.2	94	1428	162	614	22	0.42	0.033	2.98	0.016	744	14.5
Cheese, provolone	351	2.1	25.6	26.6	69	876	138	756	28	0.52	0.026	3.23	0.010	496	14.5
Cheese, Swiss	380	5.4	26.9	27.8	92	192	77	791	38	0.20	0.043	4.36	0.005	567	18.2
Cheese, tilsit	340	1.9	24.4	26.0	102	753	65	700	13	0.23	0.026	3.50	0.013	500	14.5
Cheesecake, commercially prepared	321	25.5	5.5	22.5	55	207	90	51	11	0.63	0.020	0.51	0.140	93	5.2
Cherries, sour, red, raw	50	12.2	1.0	0.3	0	3	173	16	9	0.32	0.104	0.10	0.112	15	0.0
Cherries, sweet, raw	63	16.0	1.1	0.2	0	0	222	13	11	0.36	0.060	0.07	0.070	21	0.0

Food description	Energy kcal/ 100 g	Carb. g/ 100 g	Protein g/ 100 g	Fat g/ 100 g	Chol. mg/ 100 g	Na mg/ 100 g	K mg/ 100 g	Ca mg/ 100g	Mg mg/ 100 g	Fe mg/ 100 g	Cu mg/ 100 g	Zn mg/ 100 g	Mn mg/ 100 g	P mg/ 100 g	Se µg/ 100 g
Chestnuts, European, boiled & steamed	131	27.8	2.0	1.4	0	27	715	46	54	1.73	0.472	0.25	0.854	99	
Chestnuts, European, roasted	245	53.0	3.2	2.2	0	2	592	29	33	0.91	0.507	0.57	1.180	107	1.2
Chicken, broilers or fryers, breast, meat & skin, fried	222	1.6	31.8	8.9	89	76	259	16	30	1.19	0.057	1.10	0.026	233	23.9
Chicken, broilers or fryers, breast, meat & skin, roasted	197	0.0	29.8	7.8	84	71	245	14	27	1.07	0.050	1.02	0.018	214	24.7
Chicken, broilers or fryers, breast, meat only, fried	187	0.5	33.4	4.7	91	79	276	16	31	1.14	0.054	1.08	0.021	246	26.2
Chicken, broilers or fryers, breast, meat only, roasted	165	0.0	31.0	3.6	85	74	256	15	29	1.04	0.049	1.00	0.017	228	27.6
Chicken, broilers or fryers, dark meat, meat only, roasted	205	0.0	27.4	9.7	93	93	240	15	23	1.33	0.080	2.80	0.021	179	18.0
Chicken, broilers or fryers, drumstick, meat & skin, fried	245	1.6	27.0	13.7	90	89	229	12	23	1.34	0.080	2.89	0.028	176	18.4
Chicken, broilers or fryers, thigh, meat & skin, roasted	247	0.0	25.1	15.5	93	84	222	12	22	1.34	0.078	2.36	0.021	174	19.5
Chicken, broilers or fryers, thigh, meat only, roasted	209	0.0	25.9	10.9	95	88	238	12	24	1.31	0.081	2.57	0.021	183	29.0
Chicken, broilers or fryers, wing, meat & skin, fried	321	2.4	26.1	22.2	81	77	177	15	19	1.25	0.061	1.76	0.028	150	21.3
Chicken, Cornish game hens, meat & skin, roasted	260	0.0	22.3	18.2	131	64	245	13	18	0.91	0.061	1.49	0.015	146	15.5
Chicken, Cornish game hens, meat only, roasted	134	0.0	23.3	3.9	106	63	250	13	19	0.77	0.059	1.53	0.015	149	20.8
Chickpeas, mature seeds, boiled, w/o salt	164	27.4	8.9	2.6	0	7	291	49	48	2.89	0.352	1.53	1.030	168	3.7
Chicory greens, raw	23	4.7	1.7	0.3	0	45	420	100	30	0.90	0.295	0.42	0.429	47	0.3
Chocolate syrup	279	65.1	2.1	1.1	0	72	224	14	65	2.11	0.512	0.73	0.382	129	2.7
Clam, mixed species, breaded & fried	202	10.3	14.2	11.2	61	364	326	63	14	13.91	0.356	1.46	0.540	188	28.9
Clam, mixed species, raw	74	2.6	12.8	1.0	34	56	314	46	9	13.98	0.344	1.37	0.500	169	24.3
Cocoa, dry powder, unsweetened	229	54.3	19.6	13.7	0	21	1524	128	499	13.86	3.788	6.81	3.837	734	14.3
Coconut meat, raw	354	15.2	3.3	33.5	0	20	356	14	32	2.43	0.435	1.10	1.500	113	10.1
Cod, Atlantic, cooked, dry heat	105	0.0	22.8	0.9	55	78	244	14	42	0.49	0.036	0.58	0.020	138	37.6
Cod, Atlantic, dried & salted	290	0.0	62.8	2.4	152	7027	1458	160	133	2.50	0.176	1.59	0.050	950	147.8
Coffee, brewed from grounds, prep w. tap h2o	1	0.0	0.1	0.0	0	2	49	2	3	0.01	0.002	0.02	0.023	3	0.0
Coffee, brewed, espresso, rest-prep	2	0.0	0.1	0.2	0	14	115	2	80	0.13	0.050	0.05	0.050	7	0.0
Coleslaw, home-prepared	69	12.4	1.3	2.6	8	23	181	45	10	0.59	0.023	0.20	0.097	32	0.7
Collards, boiled, w/o salt	26	4.9	2.1	0.4	0	16	116	140	20	1.16	0.038	0.23	0.436	30	0.5
Cookies, brownies, commercially prepared	405	63.9	4.8	16.3	17	312	149	29	31	2.25	0.224	0.72	0.128	101	6.3
Cookies, chocolate chip, commercially prepared, higher fat, unenriched	481	66.8	5.4	22.6	0	315	135	25	31	1.00	0.212	0.64	0.449	108	
Cookies, chocolate sandwich, w/creme filling	466	71.6	5.3	19.1	0	483	187	21	48	3.93	0.431	0.98	0.629	92	8.1
Cookies, fig bars	348	70.9	3.7	7.3	0	350	207	64	27	2.90	0.147	0.39	0.343	62	3.3
Cookies, gingersnaps	416	76.9	5.6	9.8	0	654	346	77	49	6.40	0.305	0.55	1.555	83	5.1
Cookies, graham crackers, plain or honey	423	76.8	6.9	10.1	0	605	135	24	30	3.73	0.202	0.81	0.804	104	10.2
Cookies, oatmeal, commercially prepared	450	68.7	6.2	18.1	0	383	142	37	33	2.58	0.134	0.79	0.839	138	9.8
Cookies, peanut butter sandwich, regular	478	65.6	8.8	21.1	0	368	192	53	49	2.60	0.237	1.06	0.912	188	7.7
Cookies, shortbread, commercially prepared, plain	502	64.5	6.1	24.1	20	455	100	35	17	2.74	0.144	0.53	0.428	108	7.3
Cookies, vanilla wafers, higher fat	473	71.1	4.3	19.4	0	306	107	25	12	2.21	0.124	0.33	0.384	64	11.3
Cookies, vanilla wafers, lower fat	441	73.6	5.0	15.2	51	312	97	48	14	2.38	0.100	0.36	0.262	104	11.3
Corn, sweet, white, boiled, w. salt	108	25.1	3.3	1.3	0	253	249	2	32	0.61	0.053	0.48	0.194	103	0.8
Corn, sweet, yellow, boiled, w. salt	108	25.1	3.3	1.3	0	253	249	2	32	0.61	0.053	0.48	0.194	103	0.2
Couscous, cooked	112	23.2	3.8	0.2	0	5	58	8	8	0.38	0.041	0.26	0.084	22	27.5
Cowpeas (black-eyed), immature seeds, boiled, w. salt	97	20.3	3.2	0.4	0	240	418	128	52	1.12	0.133	1.03	0.572	51	2.5
Crab, Alaska king, cooked, moist heat	97	0.0	19.4	1.5	53	1072	262	59	63	0.76	1.182	7.62	0.040	280	40.0
Crab, Alaska king, raw	84	0.0	18.3	0.6	42	836	204	46	49	0.59	0.922	5.95	0.035	219	36.4
Crab, blue, cooked, moist heat	102	0.0	20.2	1.8	100	279	324	104	33	0.91	0.645	4.22	0.190	206	40.2
Crab, blue, raw	87	0.0	18.1	1.1	78	293	329	89	34	0.74	0.669	3.54	0.150	229	37.4
Crab, Dungeness, cooked, moist heat	110	1.0	22.3	1.2	76	378	408	59	58	0.43	0.734	5.47	0.097	175	47.6
Crackers, cheese, regular	503	58.2	10.1	25.3	13	995	145	151	36	4.77	0.210	1.13	0.629	218	8.6
Crackers, matzo, plain	395	83.7	10.0	1.4	0	2	112	13	25	3.16	0.060	0.68	0.650	89	36.9
Crackers, saltines (includes oyster, soda, soup)	428	70.9	9.2	11.4	0	1072	154	68	22	5.64	0.298	0.83	0.653	101	10.3

Food description	Energy kcal/ 100 g	Carb. g/ 100 g	Protein g/ 100 g	Fat g/ 100 g	Chol. mg/ 100 g	Na mg/ 100 g	K mg/ 100 g	Ca mg/ 100g	Mg mg/ 100 g	Fe mg/ 100 g	Cu mg/ 100 g	Zn mg/ 100 g	Mn mg/ 100 g	P mg/ 100 g	Se µg/ 100 g
Crackers, wheat, regular	473	64.9	8.6	20.6	0	795	183	49	62	4.40	0.318	1.60	1.781	220	6.3
Cranberries, raw	46	12.2	0.4	0.1	0	2	85	8	6	0.25	0.061	0.10	0.360	13	0.1
Cranberry juice, unsweetened	46	12.2	0.4	0.1	0	2	77	8	6	0.25	0.055	0.10		13	0.1
Crayfish, mixed species, farmed, cooked, moist heat	87	0.0	17.5	1.3	137	97	238	51	33	1.11	0.580	1.48	0.217	241	34.2
Cream, fluid, half and half	130	4.3	3.0	11.5	37	41	130	105	10	0.07	0.010	0.51	0.001	95	1.8
Cream, half & half, fat free	59	9.0	2.6	1.4	5	144	206	96	16	0.00	0.016	0.81	0.002	151	2.9
Cream, sour, cultured	214	4.3	3.2	21.0	44	53	144	116	11	0.06	0.019	0.27	0.003	85	2.2
Cream, whipped, cream topping, pressurized	257	12.5	3.2	22.2	76	130	147	101	11	0.05	0.010	0.37	0.001	89	1.4
Croissants, butter	406	45.8	8.2	21.0	67	744	118	37	16	2.03	0.080	0.75	0.330	105	22.7
Croutons, plain	407	73.5	11.9	6.6	0	698	124	76	31	4.08	0.163	0.89	0.500	115	37.5
Cucumber, peeled, raw	12	2.2	0.6	0.2	0	2	136	14	12	0.22	0.071	0.17	0.073	21	0.1
Cucumber, with peel, raw	15	3.6	0.7	0.1	0	2	147	16	13	0.28	0.041	0.20	0.079	24	0.3
Curry powder	325	58.2	12.7	13.8	0	52	1543	478	254	29.59	0.815	4.05	4.289	349	17.1
Dandelion greens, boiled, w/o salt	33	6.4	2.0	0.6	0	44	232	140	24	1.80	0.115	0.28	0.230	42	0.3
Dandelion greens, raw	45	9.2	2.7	0.7	0	76	397	187	36	3.10	0.171	0.41	0.342	66	0.5
Duck, domesticated, meat & skin, roasted	337	0.0	19.0	28.4	84	59	204	11	16	2.70	0.227	1.86	0.019	156	20.0
Duck, domesticated, meat only, roasted	201	0.0	23.5	11.2	89	65	252	12	20	2.70	0.231	2.60	0.019	203	22.4
Egg, white, raw, fresh	52	0.7	10.9	0.2	0	166	163	7	11	0.08	0.023	0.03	0.011	15	20.0
Egg, whole, fried	201	0.9	13.6	15.3	457	204	147	59	13	1.98	0.111	1.20	0.041	208	34.2
Egg, whole, hard-boiled	155	1.1	12.6	10.6	424	124	126	50	10	1.19	0.013	1.05	0.026	172	30.8
Egg, whole, poached	147	0.8	12.5	9.9	422	294	133	53	12	1.83	0.102	1.10	0.039	190	31.6
Egg, whole, scrambled	166	2.2	11.1	12.2	352	280	138	71	12	1.20	0.014	1.00	0.022	170	22.5
Egg, yolk, raw, fresh	322	3.6	15.9	26.5	1234	48	109	129	5	2.73	0.077	2.30	0.055	390	50.0
Eggplant, boiled, w/o salt	35	8.7	0.8	0.2	0	1	123	6	11	0.25	0.059	0.12	0.113	15	0.1
Eggplant, raw	24	5.7	1.0	0.2	0	2	230	9	14	0.24	0.082	0.16	0.250	25	0.3
Endive, raw	17	3.4	1.3	0.2	0	22	314	52	15	0.83	0.099	0.79	0.420	28	0.2
English muffins, whole-wheat, toasted	221	44.1	9.6	2.3	0	692	228	288	77	2.66	0.225	1.74	1.946	307	43.8
English, muffins, plain, toasted, enriched, w. calcium propanoate (includes sourdough)	270	52.7	10.3	2.0	0	477	129	197	28	4.65	0.160	1.40	0.630	107	26.3
Fat, beef tallow	902	0.0	0.0	100.0	109	0	0	0	0	0.00	0.000	0.00		0	0.2
Fat, chicken	900	0.0	0.0	99.8	85	0	0	0	0	0.00	0.000	0.00		0	0.2
Fat, duck	900	0.0	0.0	99.8	100	0	0	0	0	0.00	0.000	0.00		0	0.2
Fat, goose	900	0.0	0.0	99.8	100	0	0	0	0	0.00		0.00		0	0.2
Fat, turkey	900	0.0	0.0	99.8	102	0	0	0	0	0.00	0.000	0.00		0	0.2
Fennel, bulb, raw	31	7.3	1.2	0.2	0	52	414	49	17	0.73	0.066	0.20	0.191	50	0.7
Figs, dried, stewed	107	27.6	1.4	0.4	0	4	294	70	29	0.88	0.124	0.24	0.220	29	0.2
Figs, raw	74	19.2	0.8	0.3	0	1	232	35	17	0.37	0.070	0.15	0.128	14	0.2
Fish oil, cod liver	902	0.0	0.0	100.0	570	0	0	0	0	0.00	0.000	0.00	0.000	0	0.0
Frankfurter beef	330	4.1	11.2	29.6	53	1140	156	14	14	1.51	0.184	2.46	0.082	160	8.2
Frankfurter, chick	257	6.8	12.9	19.5	101	1370	84	95	10	2.00	0.050	1.04	0.015	107	18.4
Frankfurter, pork	269	0.3	12.8	23.7	66	816	264	267	15	3.70	0.074	2.09	0.016	171	27.8
Frankfurter, turkey	226	1.5	14.3	17.7	107	1426	179	106	14	1.84	0.100	3.11	0.016	134	15.4
Game meat, beaver, roasted	212	0.0	34.9	7.0	117	59	403	22	29	10.00	0.189	2.27		292	43.1
Game meat, beefalo, roasted	188	0.0	30.7	6.3	58	82	459	24		3.05		6.40		250	13.1
Game meat, bison, lean, roasted	143	0.0	28.4	2.4	82	57	361	8	26	3.42	0.107	3.68	0.008	209	35.5
Game meat, boar, wild, roasted	160	0.0	28.3	4.4	77	60	396	16	27	1.12	0.056	3.01		134	13.0
Game meat, deer, roasted	158	0.0	30.2	3.2	112	54	335	7	24	4.47	0.300	2.75	0.046	226	12.9
Game meat, elk, roasted	146	0.0	30.2	1.9	73	61	328	5	24	3.63	0.142	3.16	0.013	180	13.0
Game meat, moose, roasted	134	0.0	29.3	1.0	78	69	334	6	24	4.22	0.079	3.68	0.009	176	12.8
Game meat, rabbit, domesticated, roasted	197	0.0	29.1	8.1	82	47	383	19	21	2.27	0.189	2.27	0.032	263	38.5
Game meat, raccoon, roasted	255	0.0	29.2	14.5	97	79	398	14	30	7.10	0.189	2.27		261	18.0
Game meat, squirrel, roasted	173	0.0	30.8	4.7	121	119	352	3	28	6.81	0.148	1.78	0.032	211	15.1
Goose, domesticated, meat & skin, roasted	305	0.0	25.2	21.9	91	70	329	13	22	2.83	0.264	2.62	0.023	270	21.8
Goose, domesticated, meat only, roasted	238	0.0	29.0	12.7	96	76	388	14	25	2.87	0.276	3.17	0.024	309	25.5

Food description	Energy kcal/ 100 g	Carb. g/ 100 g	Protein g/ 100 g	Fat g/ 100 g	Chol. mg/ 100 g	Na mg/ 100 g	K mg/ 100 g	Ca mg/ 100g	Mg mg/ 100 g	Fe mg/ 100 g	Cu mg/ 100 g	Zn mg/ 100 g	Mn mg/ 100 g	P mg/ 100 g	Se µg/ 100 g
Goose, liver, raw	133	6.3	16.4	4.3	515	140	230	43	24	30.53	7.522	3.07	0.000	261	68.1
Gooseberries, raw	44	10.2	0.9	0.6	0	1	198	25	10	0.31	0.070	0.12	0.144	27	0.6
Grape juice, canned or bottled, unsweetened, w/o vitamin C	61	15.0	0.6	0.1	0	3	132	9	10	0.24	0.028	0.05	0.360	11	0.1
Grapefruit juice, white, canned, unsweetened	38	9.0	0.5	0.1	0	1	153	7	10	0.20	0.038	0.09	0.020	11	0.1
Grapefruit, raw, pink & red, all areas	42	10.7	0.8	0.1	0	0	135	22	9	0.08	0.032	0.07	0.022	18	0.1
Grapefruit, raw, white, all areas	33	8.4	0.7	0.1	0	0	148	12	9	0.06	0.050	0.07	0.013	8	1.4
Grapes, American type (slip skin), raw	67	17.2	0.6	0.4	0	2	191	14	5	0.29	0.040	0.04	0.718	10	0.1
Grouper, mixed species, cooked, dry heat	118	0.0	24.8	1.3	47	53	475	21	37	1.14	0.045	0.51	0.012	143	46.8
Haddock, cooked, dry heat	112	0.0	24.2	0.9	74	87	399	42	50	1.35	0.033	0.48	0.030	241	40.5
Haddock, smoked	116	0.0	25.2	1.0	77	763	415	49	54	1.40	0.042	0.50	0.030	251	42.9
Halibut, Atlantic & Pacific, cooked, dry heat	140	0.0	26.7	2.9	41	69	576	60	107	1.07	0.035	0.53	0.020	285	46.8
Ham, sliced, regular (approx 11% fat)	163	3.8	16.6	8.6	57	1304	287	24	22	1.02	0.089	1.35	0.557	153	20.7
Herring, Atlantic, kippered	217	0.0	24.6	12.4	82	918	447	84	46	1.51	0.135	1.36	0.050	325	52.6
Hominy, canned, white or yellow	72	14.3	1.5	0.9	0	210	9	10	16	0.62	0.030	1.05	0.070	35	3.0
Hummus, commercial	166	14.3	7.9	9.6	0	379	228	38	71	2.44	0.527	1.83	0.773	176	2.6
Ice creams, chocolate	216	28.2	3.8	11.0	34	76	249	109	29	0.93	0.135	0.58	0.140	107	2.5
Ice creams, French vanilla, soft-serve	222	22.2	4.1	13.0	91	61	177	131	12	0.21	0.030	0.52	0.005	116	3.0
Ice creams, strawberry	192	27.6	3.2	8.4	29	60	188	120	14	0.21	0.037	0.34	0.078	100	1.9
Ice creams, vanilla	201	23.6	3.5	11.0	44	80	199	128	14	0.09	0.023	0.69	0.008	105	1.8
Jams and preserves	278	68.9	0.4	0.1	0	32	77	20	4	0.49	0.100	0.06	0.040	19	2.0
Jellies	266	70.0	0.2	0.0	0	30	54	7	6	0.19	0.011	0.03	0.132	6	0.4
Kale, boiled, w/o salt	28	5.6	1.9	0.4	0	23	228	72	18	0.90	0.156	0.24	0.416	28	0.9
Kale, raw	50	10.0	3.3	0.7	0	43	447	135	34	1.70	0.290	0.44	0.774	56	0.9
Kiwi fruit, (Chinese gooseberries), fresh, raw	61	14.7	1.1	0.5	0	3	312	34	17	0.31	0.130	0.14	0.098	34	0.2
Knockwurst, pork or beef	307	3.2	11.1	27.7	60	930	199	11	11	0.66	0.060	1.66	0.021	98	13.5
Kumquats, raw	71	15.9	1.9	0.9	0	10	186	62	20	0.86	0.095	0.17	0.135	19	0.0
Lamb, domestic, composite of retail cuts, lean & fat, 1/4" fat, choice, cooked	294	0.0	24.5	20.9	97	72	310	17	23	1.88	0.119	4.46	0.022	188	26.4
Lamb, domestic, composite of retail cuts, lean & fat, 1/8" fat, choice, cooked	271	0.0	25.5	18.0	96	72	318	16	24	1.93	0.121	4.74	0.024	193	27.2
Lamb, domestic, composite of retail cuts, lean, 1/4" fat, choice, cooked	206	0.0	28.2	9.5	92	76	344	15	26	2.05	0.128	5.27	0.028	210	26.1
Lamb, domestic, leg, shank half, lean & fat, 1/8" fat, choice, roasted	217	0.0	26.7	11.4	90	65	329	9	25	1.99	0.118	4.72	0.026	200	29.7
Lamb, domestic, loin, lean & fat, 1/4" fat, choice, roasted	309	0.0	22.6	23.6	95	64	246	18	23	2.12	0.119	3.41	0.020	180	24.6
Lamb, domestic, rib, lean & fat, 1/8" fat, choice, roasted	341	0.0	21.8	27.5	96	74	277	22	20	1.62	0.117	3.62	0.021	170	22.3
Lamb, ground, broiled	283	0.0	24.8	19.7	97	81	339	22	24	1.79	0.128	4.67	0.024	201	27.7
Lard	902	0.0	0.0	100.0	95	0	0	0	0	0.00	0.000	0.11	0.000	0	0.2
Leeks, (bulb & lower leaf-portion), boiled, w/o salt	31	7.6	0.8	0.2	0	10	87	30	14	1.10	0.062	0.06	0.247	17	0.5
Lemon juice, canned or bottled	21	6.5	0.4	0.3	0	21	102	11	8	0.13	0.037	0.06	0.020	9	0.1
Lemons, raw, with peel	20	10.7	1.2	0.3	0	3	145	61	12	0.70	0.260	0.10		15	
Lemons, raw, without peel	29	9.3	1.1	0.3	0	2	138	26	8	0.60	0.037	0.06	0.030	16	0.4
Lentils, mature seeds, boiled, w. salt	116	20.1	9.0	0.4	0	238	369	19	36	3.33	0.251	1.27	0.494	180	2.8
Lentils, sprouted, stir-fried, w. salt	101	21.3	8.8	0.5	0	246	284	14	35	3.10	0.337	1.60	0.502	153	0.6
Lettuce, iceberg (includes crisp head types), raw	14	3.0	0.9	0.1	0	10	141	18	7	0.41	0.025	0.15	0.125	20	0.1
Lettuce, romaine, raw	17	3.3	1.2	0.3	0	8	247	33	14	0.97	0.048	0.23	0.155	30	0.4
Lima beans, large, mature seeds, boiled, w/o salt	115	20.9	7.8	0.4	0	2	508	17	43	2.39	0.235	0.95	0.516	111	4.5
Lima beans, thin seeded (baby), mature seeds, boiled, w/o salt	126	23.3	8.0	0.4	0	3	401	29	53	2.40	0.215	1.03	0.585	127	4.9
Limes, raw	30	10.5	0.7	0.2	0	2	102	33	6	0.60	0.065	0.11	0.008	18	0.4
Liverwurst spread	305	5.9	12.4	25.5	118	700	170	22	12	8.85	0.240	2.30	0.155	230	58.0
Lobster, northern, cooked, moist heat	98	1.3	20.5	0.6	72	380	352	61	35	0.39	1.940	2.92	0.061	185	42.7
Macadamia nuts, dry roasted, w/o salt	718	13.4	7.8	76.1	0	4	363	70	118	2.65	0.570	1.29	3.036	198	3.6
Macaroni, cooked, enriched	158	30.9	5.8	0.9	0	1	45	7	18	1.33	0.103	0.50	0.317	58	26.4
Mackerel, Atlantic, cooked, dry heat	262	0.0	23.9	17.8	75	83	401	15	97	1.57	0.094	0.94	0.020	278	51.6

Food description	Energy kcal/ 100 g	Carb. g/ 100 g	Protein g/ 100 g	Fat g/ 100 g	Chol. mg/ 100 g	Na mg/ 100 g	K mg/ 100 g	Ca mg/ 100g	Mg mg/ 100 g	Fe mg/ 100 g	Cu mg/ 100 g	Zn mg/ 100 g	Mn mg/ 100 g	P mg/ 100 g	Se µg/ 100 g
Mackerel, king, cooked, dry heat	134	0.0	26.0	2.6	68	203	558	40	41	2.28	0.033	0.72	0.006	318	46.8
Mackerel, Pacific & jack, mixed species, cooked, dry heat	201	0.0	25.7	10.1	60	110	521	29	36	1.49	0.119	0.86	0.019	160	46.8
Mackerel, salted	305	0.0	18.5	25.1	95	4450	520	66	60	1.40	0.100	1.10		254	73.4
Mackerel, Spanish, cooked, dry heat	158	0.0	23.6	6.3	73	66	554	13	38	0.74	0.065	0.62	0.012	271	40.6
Mangos, raw	65	17.0	0.5	0.3	0	2	156	10	9	0.13	0.110	0.04	0.027	11	0.6
Margarine, regular, hard, corn (hydrogenated)	719	0.9	0.9	80.5	0	943	42	30	3	0.00				23	0.0
Margarine, regular, hard, soybean (hydrogenated) & palm (hydrogenated)	719	0.9	0.9	80.5	0	943	42	30	3	0.00		0.00		23	0.0
Margarine, regular, stick, unsalted, 80% fat	719	0.9	0.9	80.5	0	2	25	17	2	0.00	0.000	0.00		13	0.0
Margarine, regular, tub, unsalted, 80% fat	716	0.5	0.8	80.4	0	28	38	26	2	0.00	0.000	0.00		20	0.0
Marmalade, orange	246	66.3	0.3	0.0	0	56	37	38	2	0.15	0.090	0.04	0.020	4	0.6
Mayonnaise dressing, no cholesterol	688	0.3	0.0	77.8	0	486	14	7	1	0.23	0.000	0.13		25	1.6
Mayonnaise, low sodium, diet	231	16.0	0.3	19.2	24	110	10	0	0	0.00	0.000	0.11	0.000	0	1.6
Melons, honeydew, raw	36	9.1	0.5	0.1	0	18	228	6	10	0.17	0.024	0.09	0.027	11	0.7
Milk, buttermilk, fluid, cultured, reduced fat	56	5.3	4.1	2.0	8	86	180	143	13	0.06	0.008	0.24		82	2.3
Milk, goat, fluid	69	4.5	3.6	4.1	11	50	204	134	14	0.05	0.046	0.30	0.018	111	1.4
Milk, low fat, fluid, 1% milkfat, w. vitamin A	42	5.0	3.4	1.0	5	44	150	119	11	0.03	0.010	0.42	0.003	95	3.3
Milk, low sodium, fluid	61	4.5	3.1	3.5	14	3	253	101	5	0.05	0.010	0.38	0.004	86	2.0
Milk, reduced fat, fluid, 2% milkfat, w. added vitamin A	50	4.7	3.3	2.0	8	41	150	117	11	0.03	0.012	0.43	0.003	94	2.5
Muffins, blueberry, commercially prepared	277	48.0	5.5	6.5	30	447	123	57	16	1.61	0.074	0.49	0.440	197	11.2
Muffins, corn, commercially prepared	305	50.9	5.9	8.4	26	521	69	74	32	2.81	0.299	0.54	0.355	284	15.2
Muffins, oat bran	270	48.3	7.0	7.4	0	393	507	63	157	4.20	0.330	1.84	2.630	376	11.0
Mullet, striped, cooked, dry heat	150	0.0	24.8	4.9	63	71	458	31	33	1.41	0.141	0.88	0.022	244	46.8
Mushrooms, boiled, w/o salt	28	5.3	2.2	0.5	0	2	356	6	12	1.74	0.504	0.87	0.115	87	11.9
Mushrooms, oyster, raw	35	6.4	3.3	0.4	0	18	420	3	18	1.33	0.244	0.77	0.113	120	2.6
Mushrooms, portabella, grilled	35	5.1	4.1	0.8	0	10	521	4	15	0.56	0.499	0.73	0.075	150	17.7
Mushrooms, raw	22	3.3	3.1	0.3	0	5	318	3	9	0.50	0.318	0.52	0.047	86	9.3
Mushrooms, shiitake, cooked, w/o salt	56	14.4	1.6	0.2	0	4	117	3	14	0.44	0.896	1.33	0.204	29	24.8
Mussels, blue, cooked, moist heat	172	7.4	23.8	4.5	56	369	268	33	37	6.72	0.149	2.67	6.800	285	89.6
Mussels, blue, raw	86	3.7	11.9	2.2	28	286	320	26	34	3.95	0.094	1.60	3.400	197	44.8
Mustard greens, boiled, w/o salt	15	2.1	2.3	0.2	0	16	202	74	15	0.70	0.084	0.11	0.274	41	0.6
Nectarines, raw	44	10.6	1.1	0.3	0	0	201	6	9	0.28	0.086	0.17	0.054	26	0.0
Noodles, egg, cooked, enriched, w. salt	138	25.2	4.5	2.1	29	165	38	12	21	1.47	0.098	0.65	0.315	76	23.9
Noodles, Japanese, soba, cooked	99	21.4	5.1	0.1	0	60	35	4	9	0.48	0.008	0.12	0.374	25	
Nutmeg, ground	525	49.3	5.8	36.3	0	16	350	184	183	3.04	1.027	2.15	2.900	213	1.6
Ocean perch, Atlantic, cooked, dry heat	121	0.0	23.9	2.1	54	96	350	137	39	1.18	0.033	0.61	0.020	277	55.5
Oil, avocado	884	0.0	0.0	100.0		0	0	0	0	0.00	0.000	0.00	0.000	0	0.0
Oil, canola	884	0.0	0.0	100.0	0	0	0	0	0	0.00	0.000	0.00	0.000	0	0.0
Oil, canola and soybean	884	0.0	0.0	100.0	0	0	0	0	0	0.00	0.000	0.00		0	0.0
Oil, coconut	862	0.0	0.0	100.0	0	0	0	0	0	0.04	0.000	0.00	0.000	0	0.0
Oil, corn and canola	884	0.0	0.0	100.0	0	0	0	0	0	0.00	0.000	0.00		0	0.0
Oil, mustard	884	0.0	0.0	100.0		0	0	0	0	0.00	0.000	0.00	0.000	0	0.0
Oil, olive, salad or cooking	884	0.0	0.0	100.0	0	2	1	1	0	0.56	0.000	0.00	0.000	0	0.0
Oil, palm	884	0.0	0.0	100.0	0	0	0	0	0	0.01	0.000	0.00		0	0.0
Oil, palm kernel	862	0.0	0.0	100.0	0	0	0	0	0	0.00	0.000	0.00	0.000	0	0.0
Oil, peanut, salad or cooking	884	0.0	0.0	100.0	0	0	0	0	0	0.03	0.000	0.01		0	0.0
Oil, safflower, salad or cooking, linoleic >70%	884	0.0	0.0	100.0	0	0	0	0	0	0.00	0.000	0.00		0	0.0
Oil, safflower, salad or cooking, oleic >70%	884	0.0	0.0	100.0	0	0	0	0	0	0.00	0.000	0.00	0.000	0	0.0
Oil, sesame, salad or cooking	884	0.0	0.0	100.0	0	0	0	0	0	0.00	0.000	0.00		0	0.0
Oil, soybean lecithin	763	0.0	0.0	100.0	0	0	0	0	0	0.00	0.000	0.00	0.000	0	0.0
Oil, soybean, salad or cooking	884	0.0	0.0	100.0	0	0	0	0	0	0.02	0.000	0.00		0	0.0
Oil, sunflower, linoleic, approx. 65%	884	0.0	0.0	100.0	0	0	0	0	0	0.00	0.000	0.00		0	0.0
Okra, boiled, w/o salt	22	4.6	1.9	0.2	0	6	135	77	36	0.28	0.085	0.43	0.294	32	0.4

Food description	Energy kcal/ 100 g	Carb. g/ 100 g	Protein g/ 100 g	Fat g/ 100 g	Chol. mg/ 100 g	Na mg/ 100 g	K mg/ 100 g	Ca mg/ 100g	Mg mg/ 100 g	Fe mg/ 100 g	Cu mg/ 100 g	Zn mg/ 100 g	Mn mg/ 100 g	P mg/ 100 g	Se µg/ 100 g
Olives, pickled, canned or bottled, green	145	3.8	1.0	15.3	0	1556	42	52	11	0.49	0.120	0.04		4	0.9
Olives, ripe, canned (small-extra large)	115	6.3	0.8	10.7	0	872	8	88	4	3.30	0.251	0.22	0.020	3	0.9
Onions, boiled, w/o salt	44	10.2	1.4	0.2	0	3	166	22	11	0.24	0.067	0.21	0.153	35	0.6
Onions, raw	42	10.1	0.9	0.1	0	3	144	22	10	0.19	0.038	0.16	0.132	27	0.5
Orange juice, includes from concentrate	44	10.1	0.8	0.3	0	1	190	10	11	0.17	0.040	0.04	0.023	11	0.1
Oranges, raw, all commercial varieties	47	11.8	0.9	0.1	0	0	181	40	10	0.10	0.045	0.07	0.025	14	0.5
Oranges, raw, California, Valencias	49	11.9	1.0	0.3	0	0	179	40	10	0.09	0.037	0.06	0.023	17	
Oranges, raw, Florida	46	11.5	0.7	0.2	0	0	169	43	10	0.09	0.039	0.08	0.024	12	0.5
Oranges, raw, navels	49	12.5	0.9	0.2	0	1	166	43	11	0.13	0.039	0.08	0.029	23	0.0
Ostrich, ground, cooked, pan-broiled	175	0.0	26.2	7.1	83	80	323	8	23	3.43	0.136	4.33	0.017	224	33.5
Ostrich, inside leg, cooked	141	0.0	29.0	1.9	73	83	352	6	25	3.12	0.148	4.71	0.018	244	36.5
Ostrich, top loin, cooked	155	0.0	28.1	3.9	93	77	353	6	25	3.31	0.148	4.72	0.018	245	36.6
Oyster, eastern, breaded & fried	197	11.6	8.8	12.6	81	417	244	62	58	6.95	4.294	87.13	0.490	159	66.5
Oyster, eastern, farmed, raw	59	5.5	5.2	1.6	25	178	124	44	33	5.78	0.738	37.92	0.394	93	63.7
Oyster, eastern, wild, raw	68	3.9	7.1	2.5	53	211	156	45	47	6.66	4.452	90.81	0.367	135	63.7
Oyster, Pacific, cooked, moist heat	163	9.9	18.9	4.6	100	212	302	16	44	9.20	2.679	33.24	1.222	243	154.0
Oyster, Pacific, raw	81	5.0	9.5	2.3	50	106	168	8	22	5.11	1.576	16.62	0.643	162	77.0
Palm hearts, raw	115	25.6	2.7	0.2	0	14	1806	18	10	1.69	0.644	3.73		140	0.7
Papayas, raw	39	9.8	0.6	0.1	0	3	257	24	10	0.10	0.016	0.07	0.011	5	0.6
Parsley, raw	36	6.3	3.0	0.8	0	56	554	138	50	6.20	0.149	1.07	0.160	58	0.1
Pastrami beef 98% fat-free	95	1.5	19.6	1.2	47	1010	228	9	18	2.78	0.079	4.26	0.013	150	10.4
Pate de foie gras, canned, smoked	462	4.7	11.4	43.8	150	697	138	70	13	5.50	0.400	0.92	0.120	200	44.0
Pate, liver, not specified, canned	319	1.5	14.2	28.0	255	697	138	70	13	5.50	0.400	2.85	0.120	200	41.6
Peaches, raw	39	9.5	0.9	0.3	0	0	190	6	9	0.25	0.068	0.17	0.061	20	0.1
Peanut butter, chunk style, w/o salt	589	21.6	24.1	49.9	0	17	745	45	160	1.90	0.578	2.79	1.800	319	8.2
Peanut butter, smooth style, w/o salt	588	19.6	25.1	50.4	0	17	649	43	154	1.87	0.473	2.91	1.466	358	5.6
Peanuts, all types, dry-roasted, w. salt	585	21.5	23.7	49.7	0	813	658	54	176	2.26	0.671	3.31	2.083	358	7.5
Peanuts, all types, oil-roasted, w. salt	599	15.3	28.0	52.5	0	320	726	61	176	1.52	0.533	3.28	1.845	397	3.3
Pears, raw	58	15.5	0.4	0.1	0	1	119	9	7	0.17	0.082	0.10	0.049	11	0.1
Peas, edible-podded, boiled, w. salt	42	7.1	3.3	0.2	0	240	240	42	26	1.97	0.077	0.37	0.168	55	0.7
Peas, edible-podded, raw	42	7.6	2.8	0.2	0	4	200	43	24	2.08	0.079	0.27	0.244	53	0.7
Peas, green, boiled, w/o salt	84	15.6	5.4	0.2	0	3	271	27	39	1.54	0.173	1.19	0.525	117	1.9
Pecans, dry roasted, w. salt	710	13.6	9.5	74.3	0	383	424	72	132	2.80	1.167	5.07	3.933	293	4.0
Pecans, oil roasted, w. salt	715	13.0	9.2	75.2	0	393	392	67	121	2.47	1.200	4.47	3.700	263	6.0
Peppers, hot chili, green, raw	40	9.5	2.0	0.2	0	7	340	18	25	1.20	0.174	0.30	0.237	46	0.5
Peppers, hot chili, red, raw	40	8.8	1.9	0.4	0	9	322	14	23	1.03	0.129	0.26	0.187	43	0.5
Peppers, jalapeno, raw	30	5.9	1.4	0.6	0	1	215	10	19	0.70	0.133	0.23	0.250	31	0.3
Peppers, sweet, green, boiled, w/o salt	28	6.7	0.9	0.2	0	2	166	9	10	0.46	0.065	0.12	0.115	18	0.3
Peppers, sweet, green, raw	20	4.6	0.9	0.2	0	3	175	10	10	0.34	0.066	0.13	0.122	20	0.0
Peppers, sweet, red, boiled, w/o salt	28	6.7	0.9	0.2	0	2	166	9	10	0.46	0.065	0.12	0.115	18	0.3
Peppers, sweet, red, raw	26	6.0	1.0	0.3	0	2	211	7	12	0.43	0.017	0.25	0.112	26	0.1
Peppers, sweet, yellow, raw	27	6.3	1.0	0.2	0	2	212	11	12	0.46	0.107	0.17	0.117	24	0.3
Persimmons, native, raw	127	33.5	0.8	0.4	0	1	310	27		2.50				26	
Pheasant, cooked, total edible	247	0.0	32.4	12.1	89	43	271	16	22	1.43	0.084	1.37		242	20.7
Pickle, cucumber, sour	11	2.3	0.3	0.2	0	1208	23	0	4	0.40	0.085	0.02	0.011	14	0.0
Pickle, cucumber, sweet	117	31.8	0.4	0.3	0	939	32	4	4	0.59	0.105	0.08	0.015	12	0.0
Pickles, cucumber, dill	18	4.1	0.6	0.2	0	1282	116	9	11	0.53	0.079	0.14	0.015	21	0.0
Pie, apple, commercially prepared, enriched flour	237	34.0	1.9	11.0	0	266	65	11	7	0.45	0.046	0.16	0.182	24	1.0
Pie, blueberry, commercially prepared	232	34.9	1.8	10.0	0	325	50	8	5	0.30	0.046	0.16	0.176	23	1.4
Pie, cherry, commercially prepared	260	39.8	2.0	11.0	0	246	81	12	8	0.48	0.040	0.18	0.140	29	1.2
Pie, chocolate creme, commercially prepared	304	33.6	2.6	19.4	5	136	127	36	21	1.07	0.050	0.23	0.200	68	7.5
Pie, coconut creme, commercially prepared	298	37.2	2.1	16.6	0	255	65	29	20	0.80	0.068	0.47	0.438	85	5.3
Pie, egg custard, commercially prepared	210	20.8	5.5	11.6	33	240	106	80	11	0.58	0.024	0.52	0.060	112	7.1
Pie, lemon meringue, commercially prepared	268	47.2	1.5	8.7	45	146	89	56	15	0.61	0.001	0.49	0.060	105	3.0

Food description	Energy kcal/ 100 g	Carb. g/ 100 g	Protein g/ 100 g	Fat g/ 100 g	Chol. mg/ 100 g	Na mg/ 100 g	K mg/ 100 g	Ca mg/ 100g	Mg mg/ 100 g	Fe mg/ 100 g	Cu mg/ 100 g	Zn mg/ 100 g	Mn mg/ 100 g	P mg/ 100 g	Se µg/ 100 g
Pie, mince, prepared from recipe	289	48.0	2.6	10.8	0	254	203	22	14	1.49	0.113	0.22	0.263	42	6.6
Pie, peach	223	32.9	1.9	10.0	0	270	125	8	6	0.50	0.053	0.09	0.152	22	1.3
Pie, pecan, commercially prepared	400	57.2	4.0	18.5	32	424	74	17	18	1.04	0.195	0.57	0.789	77	5.0
Pie, pumpkin, commercially prepared	210	27.3	3.9	9.5	20	282	154	60	15	0.79	0.048	0.45	0.240	71	2.6
Pike, northern, cooked, dry heat	113	0.0	24.7	0.9	50	49	331	73	40	0.71	0.065	0.86	0.310	282	16.2
Pike, walleye, raw	93	0.0	19.1	1.2	86	51	389	110	30	1.30	0.178	0.62	0.800	210	12.6
Pimento, canned	23	5.1	1.1	0.3	0	14	158	6	6	1.68	0.049	0.19	0.092	17	0.2
Pine nuts, pinyon, dried	629	19.3	11.6	61.0	0	72	628	8	234	3.06	1.035	4.28	4.333	35	
Pineapple juice, canned, unsweetened, w/o vitamin C	53	12.9	0.4	0.1	0	2	130	13	12	0.31	0.069	0.11	0.504	8	0.1
Pineapple, raw, all variety	48	12.6	0.5	0.1	0	1	115	13	12	0.28	0.099	0.10	1.177	8	0.1
Pineapple, raw, traditional variety	45	11.8	0.6	0.1		1	125	13	12	0.25	0.081	0.08	1.593	9	0.0
Pistachio nuts, dry roasted, w/o salt	571	27.7	21.4	46.0	0	10	1042	110	120	4.20	1.325	2.30	1.275	485	9.3
Plantains, cooked	116	31.2	0.8	0.2	0	5	465	2	32	0.58	0.066	0.13		28	1.4
Plums, dried (prunes), uncooked	240	63.9	2.2	0.4	0	2	732	43	41	0.93	0.281	0.44	0.299	69	0.3
Plums, raw	46	11.4	0.7	0.3	0	0	157	6	7	0.17	0.057	0.10	0.052	16	0.0
Pollock, Atlantic, cooked, dry heat	118	0.0	24.9	1.3	91	110	456	77	86	0.59	0.064	0.60	0.019	283	46.8
Pollock, walleye, cooked, dry heat	113	0.0	23.5	1.1	96	116	387	6	73	0.28	0.055	0.60	0.020	482	43.4
Pomegranates, raw	68	17.2	1.0	0.3	0	3	259	3	3	0.30	0.070	0.12		8	0.6
Pompano, Florida, cooked, dry heat	211	0.0	23.7	12.1	64	76	636	43	31	0.67	0.078	0.69	0.025	341	46.8
Popcorn, air-popped	387	77.8	12.9	4.5	0	8	329	7	144	3.19	0.262	3.08	1.113	358	0.0
Pork sausage, fresh, cooked	339	0.0	19.4	28.4	84	749	294	13	17	1.36	0.086	2.08	0.005	163	0.0
Pork, cured, bacon, broiled, pan-fried or roasted	541	1.4	37.0	41.8	110	2310	565	11	33	1.44	0.164	3.50	0.022	533	62.0
Pork, cured, breakfast strips, cooked	459	1.1	29.0	30.7	105	2099	400	14	20	1.97	0.150	[illegible]	0.044	[illegible]	[illegible]
Pork, cured, Canadian-style bacon, grilled	185	1.4	24.2	8.4	58	1546	390	10	21	0.82	0.054	1.70	0.027	296	24.7
Pork, cured, ham, boneless, extra lean (approx 5% fat), roasted	145	1.5	20.9	5.5	53	1203	287	8	14	1.48	0.079	2.88	0.054	196	19.5
Pork, cured, ham, boneless, regular (approx 11% fat), roasted	178	0.0	22.6	9.0	59	1500	409	8	22	1.34	0.145	2.47	0.041	281	19.8
Pork, cured, ham, whole, lean & fat, roasted	243	0.0	21.6	16.8	62	1187	286	7	19	0.87	0.083	2.32	0.014	214	22.7
Pork, cured, salt pork, raw	748	0.0	5.1	80.5	86	1424	66	6	7	0.44	0.050	0.90	0.005	52	5.8
Pork, fresh, composite of retail cuts (leg, loin, & shoulder), lean, cooked	212	0.0	29.3	9.7	86	59	375	21	26	1.10	0.061	2.97	0.018	237	45.0
Pork, fresh, leg (ham), rump half, lean & fat, roasted	252	0.0	28.9	14.3	96	62	374	12	27	1.05	0.103	2.82	0.023	272	46.8
Pork, fresh, leg (ham), shank half, lean & fat, roasted	289	0.0	25.3	20.1	92	59	338	15	22	0.98	0.098	3.06	0.028	257	43.3
Pork, fresh, loin, blade (chops), bone-in, lean & fat, broiled	320	0.0	22.5	24.9	86	70	344	29	22	0.93	0.083	3.37	0.008	212	37.1
Pork, fresh, loin, center loin (chops), bone-in, lean & fat, broiled	240	0.0	28.7	13.1	82	58	358	33	25	0.80	0.046	2.26	0.003	232	44.3
Pork, fresh, loin, center rib (chops), bone-in, lean, broiled	219	0.0	30.8	9.7	81	65	420	31	28	0.82	0.070	2.38	0.020	245	47.3
Pork, fresh, loin, center rib (chops), boneless, lean & fat, broiled	260	0.0	27.6	15.8	82	62	401	28	26	0.77	0.068	2.26	0.018	237	44.0
Pork, fresh, loin, center rib (roasts), boneless, lean & fat, roasted	252	0.0	27.0	15.2	81	48	346	6	22	0.93	0.016	2.64	0.010	214	40.3
Pork, fresh, loin, country-style ribs, lean & fat, braised	296	0.0	23.9	21.5	87	59	328	29	17	1.22	0.093	3.56	0.012	167	39.7
Pork, fresh, loin, sirloin (chops), bone-in, lean & fat, broiled	259	0.0	26.7	16.1	86	68	383	17	29	0.99	0.058	2.57	0.010	246	47.7
Pork, fresh, loin, sirloin (chops), boneless, lean & fat, broiled	208	0.0	30.5	8.6	91	56	372	18	27	1.21	0.053	2.62	0.003	243	50.5
Pork, fresh, loin, tenderloin, lean & fat, broiled	201	0.0	29.9	8.1	94	64	444	5	35	1.39	0.067	2.89	0.012	290	47.7
Pork, fresh, spareribs, lean & fat, braised	397	0.0	29.1	30.3	121	93	320	47	24	1.85	0.142	4.60	0.014	261	37.4
Potato chips, plain, salted	547	49.7	6.6	37.5	0	525	1642	24	70	1.61	0.398	2.39	0.664	155	8.1
Potatoes, baked, flesh & skin, w. salt	93	21.2	2.5	0.1	0	244	535	15	28	1.08	0.118	0.36	0.219	70	0.4
Potatoes, boiled in skin, flesh, w. salt	87	20.1	1.9	0.1	0	240	379	5	22	0.31	0.188	0.30	0.138	44	0.3
Potatoes, boiled w/o skin, flesh, w. salt	86	20.0	1.7	0.1	0	241	328	8	20	0.31	0.167	0.27	0.140	40	0.3
Potatoes, French fried, all types, frozen, oven-heated	172	28.7	2.7	5.2	0	32	451	12	26	0.74	0.135	0.38	0.210	97	0.2

Food description	Energy kcal/ 100 g	Carb. g/ 100 g	Protein g/ 100 g	Fat g/ 100 g	Chol. mg/ 100 g	Na mg/ 100 g	K mg/ 100 g	Ca mg/ 100g	Mg mg/ 100 g	Fe mg/ 100 g	Cu mg/ 100 g	Zn mg/ 100 g	Mn mg/ 100 g	P mg/ 100 g	Se µg/ 100 g
Potatoes, hashed brown, home-prepared	265	35.1	3.0	12.5	0	342	576	14	35	0.55	0.293	0.47	0.247	70	0.5
Potatoes, mashed, home-prepared, whole milk & margarine added	113	16.9	2.0	4.2	1	333	328	22	19	0.26	0.153	0.30	0.112	49	0.8
Potatoes, red, flesh & skin, baked	89	19.6	2.3	0.2	0	8	545	9	28	0.70	0.174	0.40	0.173	72	0.5
Potatoes, white, flesh & skin, baked	94	21.1	2.1	0.2	0	7	544	10	27	0.64	0.127	0.35	0.189	75	0.5
Pretzels, hard, plain, salted	380	79.8	10.3	2.6	0	1357	146	18	29	5.20	0.264	1.43	1.789	113	6.0
Prunes, dehydrated (low-moisture), stewed	113	29.7	1.2	0.2	0	2	353	24	21	1.17	0.204	0.25	0.104	37	
Pumpkin, boiled, w/o salt	20	4.9	0.7	0.1	0	1	230	15	9	0.57	0.091	0.23	0.089	30	0.2
Pumpkin, raw	26	6.5	1.0	0.1	0	1	340	21	12	0.80	0.127	0.32	0.125	44	0.3
Quail, cooked, total edible	234	0.0	25.1	14.1	86	52	216	15	22	4.43	0.592	3.10		279	21.8
Quinces, raw	57	15.3	0.4	0.1	0	4	197	11	8	0.70	0.130	0.04		17	0.6
Radicchio, raw	23	4.5	1.4	0.3	0	22	302	19	13	0.57	0.341	0.62	0.138	40	0.9
Radishes, raw	16	3.4	0.7	0.1	0	39	233	25	10	0.34	0.050	0.28	0.069	20	0.6
Raisins, golden seedless	302	79.5	3.4	0.5	0	12	746	53	35	1.79	0.363	0.32	0.308	115	0.7
Raisins, seeded	296	78.5	2.5	0.5	0	28	825	28	30	2.59	0.302	0.18	0.267	75	0.6
Raisins, seedless	299	79.2	3.1	0.5	0	11	749	50	32	1.88	0.318	0.22	0.299	101	0.6
Raspberries, raw	52	11.9	1.2	0.7	0	1	151	25	22	0.69	0.090	0.42	0.670	29	0.2
Rhubarb, frozen, cooked, w/sugar	116	31.2	0.4	0.1	0	1	96	145	12	0.21	0.027	0.08	0.073	8	0.9
Rhubarb, raw	21	4.5	0.9	0.2	0	4	288	86	12	0.22	0.021	0.10	0.196	14	1.1
Rice, brown, long-grain, cooked	111	23.0	2.6	0.9	0	5	43	10	43	0.42	0.100	0.63	0.905	83	9.8
Rice, brown, medium-grain, cooked	112	23.5	2.3	0.8	0	1	79	10	44	0.53	0.081	0.62	1.097	77	
Rice, white, glutinous, cooked	97	21.1	2.0	0.2	0	5	10	2	5	0.14	0.049	0.41	0.262	8	5.6
Rice, white, long-grain, regular, cooked	130	28.2	2.7	0.3	0	1	35	10	12	1.20	0.069	0.49	0.472	43	7.5
Rice, white, medium-grain, cooked	130	28.6	2.4	0.2	0	0	29	3	13	1.49	0.038	0.42	0.377	37	7.5
Rice, white, short-grain, cooked	130	28.7	2.4	0.2	0	0	26	1	8	1.46	0.072	0.40	0.357	33	7.5
Rockfish, Pacific, mixed species, cooked, dry heat	121	0.0	24.0	2.0	44	77	520	12	34	0.53	0.037	0.53	0.020	228	46.8
Rolls, hamburger or hotdog, plain	279	49.5	9.5	4.3	0	479	94	138	21	3.32	0.220	0.66	0.272	62	19.5
Rutabagas, boiled, w/o salt	39	8.7	1.3	0.2	0	20	326	48	23	0.53	0.041	0.35	0.174	56	0.7
Sablefish, cooked, dry heat	250	0.0	17.2	19.6	63	72	459	45	71	1.64	0.028	0.41	0.019	215	46.8
Sablefish, smoked	257	0.0	17.7	20.1	64	737	471	50	74	1.69	0.036	0.43	0.020	222	50.2
Salami, cooked, turkey	152	0.4	15.3	9.4	76	1004	216	40	22	1.25	0.190	2.32	0.020	266	26.4
Salami, dry or hard, pork	407	1.6	22.6	33.7	79	2260	378	13	22	1.30	0.160	4.20	0.070	229	25.4
Salami, dry or hard, pork, beef	385	3.8	23.2	30.1	100	2010	378	8	17	1.51	0.080	3.23		142	26.1
Salami, Italian, pork	425	1.2	21.7	37.0	80	1890	340	10	22	1.52	0.160	4.20	0.070	229	25.4
Salmon, Atlantic, farmed, cooked, dry heat	206	0.0	22.1	12.4	63	61	384	15	30	0.34	0.049	0.43	0.016	252	41.4
Salmon, Atlantic, farmed, raw	183	0.0	19.9	10.9	59	59	362	12	28	0.36	0.049	0.40	0.015	233	36.5
Salmon, Atlantic, wild, cooked, dry heat	182	0.0	25.4	8.1	71	56	628	15	37	1.03	0.321	0.82	0.021	256	46.8
Salmon, Atlantic, wild, raw	142	0.0	19.8	6.3	55	44	490	12	29	0.80	0.250	0.64	0.016	200	36.5
Salmon, Chinook, cooked, dry heat	231	0.0	25.7	13.4	85	60	505	28	122	0.91	0.053	0.56	0.019	371	46.8
Salmon, Chinook, raw	179	0.0	19.9	10.4	50	47	394	26	95	0.25	0.041	0.44	0.015	289	36.5
Salmon, Chinook, smoked	117	0.0	18.3	4.3	23	784	175	11	18	0.85	0.230	0.31	0.017	164	32.4
Salmon, Chinook, smoked (lox)	117	0.0	18.3	4.3	23	2000	175	11	18	0.85	0.230	0.31	0.017	164	38.1
Salmon, chum, cooked, dry heat	154	0.0	25.8	4.8	95	64	550	14	28	0.71	0.071	0.60	0.019	363	46.8
Salmon, coho, farmed, cooked, dry heat	178	0.0	24.3	8.2	63	52	460	12	34	0.39	0.089	0.47	0.021	332	14.1
Salmon, coho, farmed, raw	160	0.0	21.3	7.7	51	47	450	12	31	0.34	0.048	0.43	0.012	292	12.6
Salmon, coho, wild, cooked, dry heat	139	0.0	23.5	4.3	55	58	434	45	33	0.61	0.071	0.56	0.019	322	38.0
Salmon, coho, wild, raw	146	0.0	21.6	5.9	45	46	423	36	31	0.56	0.051	0.41	0.014	262	36.5
Salmon, pink, cooked, dry heat	149	0.0	25.6	4.4	67	86	414	17	33	0.99	0.099	0.71	0.019	295	57.2
Salmon, pink, raw	116	0.0	19.9	3.5	52	67	323	13	26	0.77	0.077	0.55	0.015	230	44.6
Salmon, sockeye, cooked, dry heat	216	0.0	27.3	11.0	87	66	375	7	31	0.55	0.067	0.51	0.020	276	37.8
Salmon, sockeye, raw	168	0.0	21.3	8.6	62	47	391	6	24	0.47	0.052	0.54	0.014	215	33.7
Sausage, chicken, beef, pork, skinless, smoked	216	8.1	13.6	14.3	120	1034	246	100	14	4.80	0.063	2.68	0.015	132	20.2
Sausage, Italian, pork, cooked	344	4.3	19.1	27.3	57	1207	304	21	18	1.43	0.080	2.39		170	22.0
Sausage, Italian, sweet, links	149	2.1	16.1	8.4	30	570	194	25	12	1.19	0.040	1.52	0.007	103	10.8

Food description	Energy kcal/ 100 g	Carb. g/ 100 g	Protein g/ 100 g	Fat g/ 100 g	Chol. mg/ 100 g	Na mg/ 100 g	K mg/ 100 g	Ca mg/ 100g	Mg mg/ 100 g	Fe mg/ 100 g	Cu mg/ 100 g	Zn mg/ 100 g	Mn mg/ 100 g	P mg/ 100 g	Se µg/ 100 g
Sausage, Polish, beef w. chicken, hot	259	3.6	17.6	19.4	66	1540	237	12	14	0.88	0.090	1.93	0.049	136	17.7
Sausage, smoked link sausage, pork & beef	320	2.4	12.0	28.7	58	911	179	12	13	0.75	0.077	1.26	0.048	121	0.0
Sausage, turkey, breakfast links, mild	235	1.6	15.4	18.1	60	585	197	32	25	1.07	0.111	2.13	0.066	185	22.2
Scallops, (bay & sea), steamed	112	0.0	23.2	1.4	53	265	476	115	55	3.00	0.300	3.00		338	27.9
Scrapple, pork	213	14.1	8.1	13.9	49	659	158	7	13	1.89	0.212	1.06		76	17.4
Sea bass, mixed species, cooked, dry heat	124	0.0	23.6	2.6	53	87	328	13	53	0.37	0.024	0.52	0.020	248	46.8
Shad, American, cooked, dry heat	252	0.0	21.7	17.7	96	65	492	60	38	1.24	0.082	0.47	0.054	349	46.8
Shark, mixed species, batter-dipped & fried	228	6.4	18.6	13.8	59	122	155	50	43	1.11	0.042	0.48	0.050	194	34.0
Shortening, frying (heavy duty), beef tallow & cottonseed	900	0.0	0.0	100.0	100	0	0	0	0	0.00		0.00		0	0.0
Shortening, household, soybean (hydrogenated) & palm	884	0.0	0.0	100.0	0	0	0	0	0	0.00	0.000	0.00	0.000	0	0.0
Shrimp, mixed species, cooked, moist heat	99	0.0	20.9	1.1	195	224	182	39	34	3.09	0.193	1.56	0.034	137	39.6
Shrimp, mixed species, imitation, made from surimi	101	9.1	12.4	1.5	36	705	89	19	43	0.60	0.032	0.33	0.011	282	22.9
Shrimp, mixed species, raw	106	0.9	20.3	1.7	152	148	185	52	37	2.41	0.264	1.11	0.050	205	38.0
Smelt, rainbow, cooked, dry heat	124	0.0	22.6	3.1	90	77	372	77	38	1.15	0.178	2.12	0.900	295	46.8
Snails, raw	90	2.0	16.1	1.4	50	70	382	10	250	3.50	0.400	1.00		272	27.4
Snapper, mixed species, cooked, dry heat	128	0.0	26.3	1.7	47	57	522	40	37	0.24	0.046	0.44	0.017	201	49.0
Sour cream, light	136	7.1	3.5	10.6	35	71	212	141	10	0.07	0.016	0.50		71	3.1
Sour cream, reduced fat	181	7.0	7.0	14.1	35	70	211	141	11	0.06	0.010	0.27		85	4.1
Soybeans, green, boiled, w/o salt	141	11.1	12.4	6.4	0	14	539	145	60	2.50	0.117	0.91	0.502	158	1.4
Soybeans, green, raw	147	11.1	13.0	6.8	0	15	620	197	65	3.55	0.128	0.99	0.547	194	1.5
Soybeans, mature cooked, boiled, w/o salt	173	9.9	16.6	9.0	0	1	515	102	86	5.14	0.407	1.15	0.824	245	7.3
Spaghetti, cooked, enriched, w. salt	157	30.6	5.8	0.9	0	128	45	7	18	1.33	0.103	0.50	0.317	58	26.4
Spaghetti, whole-wheat, cooked	124	26.5	5.3	0.5	0	3	44	15	30	1.06	0.167	0.81	1.379	89	25.9
Spinach, boiled, w/o salt	23	3.8	3.0	0.3	0	70	466	136	87	3.57	0.174	0.76	0.935	56	1.5
Spinach, raw	23	3.6	2.9	0.4	0	79	558	99	79	2.71	0.130	0.53	0.897	49	1.0
Squash, summer, all varieties, boiled, w/o salt	20	4.3	0.9	0.3	0	1	192	27	24	0.36	0.103	0.39	0.213	39	0.2
Squash, summer, all varieties, raw	16	3.4	1.2	0.2	0	2	262	15	17	0.35	0.051	0.29	0.175	38	0.2
Squash, summer, zucchini, includes skin, boiled, w/o salt	16	3.9	0.6	0.1		3	253	13	22	0.35	0.086	0.18	0.178	40	0.2
Squash, summer, zucchini, includes skin, raw	16	3.4	1.2	0.2	0	10	262	15	17	0.35	0.051	0.29	0.175	38	0.2
Squash, winter, acorn, baked, w/o salt	56	14.6	1.1	0.1	0	4	437	44	43	0.93	0.086	0.17	0.242	45	0.7
Squash, winter, butternut, baked, w/o salt	40	10.5	0.9	0.1	0	4	284	41	29	0.60	0.065	0.13	0.172	27	0.5
Squash, winter, Hubbard, baked, w/o salt	50	10.8	2.5	0.6	0	8	358	17	22	0.47	0.045	0.15	0.170	23	0.6
Squash, zucchini, baby, raw	21	3.1	2.7	0.4	0	3	459	21	33	0.79	0.097	0.83	0.196	93	0.3
Strawberries, raw	32	7.7	0.7	0.3	0	1	153	16	13	0.42	0.048	0.14	0.386	24	0.4
Sturgeon, mixed species, cooked, dry heat	135	0.0	20.7	5.2	77	69	364	17	45	0.90	0.053	0.54	0.030	271	16.2
Sturgeon, mixed species, raw	105	0.0	16.1	4.0	60	54	284	13	35	0.70	0.041	0.42	0.025	211	12.6
Sturgeon, mixed species, smoked	173	0.0	31.2	4.4	80	739	379	17	47	0.93	0.050	0.56	0.030	281	20.1
Sugars, brown	377	97.3	0.0	0.0	0	39	346	85	29	1.91	0.298	0.18	0.320	22	1.2
Sugars, granulated	387	100.0	0.0	0.0	0	0	2	1	0	0.01	0.000	0.00	0.000	0	0.6
Sugars, maple	354	90.9	0.1	0.2	0	11	274	90	19	1.61	0.099	6.06	4.422	3	0.8
Sweet potato, baked in skin, w/o salt	90	20.7	2.0	0.2	0	36	475	38	27	0.69	0.161	0.32	0.497	54	0.2
Sweet potato, boiled, w/o skin	76	17.7	1.4	0.1	0	27	230	27	18	0.72	0.094	0.20	0.266	32	0.2
Swordfish, cooked, dry heat	155	0.0	25.4	5.1	50	115	369	6	34	1.04	0.162	1.47	0.020	337	61.7
Swordfish, raw	121	0.0	19.8	4.0	39	90	288	4	27	0.81	0.126	1.15	0.019	263	48.1
Syrups, corn, dark	286	77.6	0.0	0.0	0	155	44	18	8	0.37	0.053	0.04	0.100	11	2.9
Syrups, corn, high-fructose	281	76.0	0.0	0.0	0	2	0	0	0	0.03	0.029	0.02	0.094	0	0.7
Syrups, maple	261	67.1	0.0	0.2	0	9	204	67	14	1.20	0.074	4.16	3.298	2	0.6
Syrups, table blends, cane & 15% maple	278	69.5	0.0	0.0	0	104	53	12	2	0.19	0.017	0.63	0.495	0	0.5
Tangerines, (mandarin oranges), raw	53	13.3	0.8	0.3	0	2	166	37	12	0.15	0.042	0.07	0.039	20	0.1
Tempeh, cooked	196	9.4	18.2	11.4		14	401	96	77	2.13	0.540	1.57	1.285	253	0.0
Tilefish, cooked, dry heat	147	0.0	24.5	4.7	64	59	512	26	33	0.31	0.052	0.53	0.015	236	51.5
Tilefish, raw	96	0.0	17.5	2.3	50	53	433	26	28	0.25	0.041	0.37	0.010	187	36.5

Food description	Energy kcal/ 100 g	Carb. g/ 100 g	Protein g/ 100 g	Fat g/ 100 g	Chol. mg/ 100 g	Na mg/ 100 g	K mg/ 100 g	Ca mg/ 100g	Mg mg/ 100 g	Fe mg/ 100 g	Cu mg/ 100 g	Zn mg/ 100 g	Mn mg/ 100 g	P mg/ 100 g	Se µg/ 100 g
Tofu, fried	271	10.5	17.2	20.2	0	16	146	372	60	4.87	0.398	1.99	1.495	287	28.5
Tomato juice, canned, w. salt	17	4.2	0.8	0.1	0	269	229	10	11	0.43	0.061	0.15	0.070	18	0.3
Tomatoes, green, raw	23	5.1	1.2	0.2	0	13	204	13	10	0.51	0.090	0.07	0.100	28	0.4
Tomatoes, orange, raw	16	3.2	1.2	0.2	0	42	212	5	8	0.47	0.062	0.14	0.088	29	0.4
Tomatoes, red, ripe, cooked	18	4.0	1.0	0.1	0	11	218	11	9	0.68	0.075	0.14	0.105	28	0.5
Tomatoes, red, ripe, raw, year-round average	18	3.9	0.9	0.2	0	5	237	10	11	0.27	0.059	0.17	0.114	24	0.0
Tomatoes, sun-dried	258	55.8	14.1	3.0	0	2095	3427	110	194	9.09	1.423	1.99	1.846	356	5.5
Tomatoes, yellow, raw	15	3.0	1.0	0.3	0	23	258	11	12	0.49	0.101	0.28	0.120	36	0.4
Trout, mixed species, cooked, dry heat	190	0.0	26.6	8.5	74	67	463	55	28	1.92	0.241	0.85	1.091	314	16.2
Trout, rainbow, farmed, cooked, dry heat	169	0.0	24.3	7.2	68	42	441	86	32	0.33	0.061	0.49	0.020	266	15.0
Trout, rainbow, wild, cooked, dry heat	150	0.0	22.9	5.8	69	56	448	86	31	0.38	0.058	0.51	0.021	269	13.2
Tuna, fresh, bluefin, cooked, dry heat	184	0.0	29.9	6.3	49	50	323	10	64	1.31	0.110	0.77	0.020	326	46.8
Tuna, fresh, bluefin, raw	144	0.0	23.3	4.9	38	39	252	8	50	1.02	0.086	0.60	0.015	254	36.5
Tuna, fresh, skipjack, raw	103	0.0	22.0	1.0	47	37	407	29	34	1.25	0.086	0.82	0.015	222	36.5
Tuna, fresh, yellowfin, raw	108	0.0	23.4	1.0	45	37	444	16	50	0.73	0.064	0.52	0.015	191	36.5
Tuna, skipjack, fresh, cooked, dry heat	132	0.0	28.2	1.3	60	47	522	37	44	1.60	0.110	1.05	0.019	285	46.8
Tuna, white, canned in oil	186	0.0	26.5	8.1	31	396	333	4	34	0.65	0.130	0.47	0.016	267	60.1
Tuna, white, canned in water	128	0.0	23.6	3.0	42	377	237	14	33	0.97	0.039	0.48	0.019	217	65.7
Tuna, yellowfin, fresh, cooked, dry heat	139	0.0	30.0	1.2	58	47	569	21	64	0.94	0.082	0.67	0.019	245	46.8
Turbot, European, cooked, dry heat	122	0.0	20.6	3.8	62	192	305	23	65	0.46	0.047	0.28	0.022	165	46.8
Turkey breast meat	104	4.2	17.1	1.7	43	1015	302	8	21	1.44	0.057	1.33	0.018	162	22.8
Turkey, all classes, breast, meat & skin, roasted	189	0.0	28.7	7.4	74	63	288	21	27	1.40	0.047	2.03	0.020	210	29.1
Turkey, all classes, dark meat, meat & skin, raw	160	0.0	18.9	8.8	72	71	261	17	20	1.69	0.137	2.95	0.021	170	26.4
Turkey, all classes, leg, meat & skin, roasted	208	0.0	27.9	9.8	85	77	280	32	23	2.30	0.154	4.27	0.023	199	37.8
Turkey, all classes, meat only, roasted	170	0.0	29.3	5.0	76	70	298	25	26	1.78	0.094	3.10	0.021	213	36.8
Turkey, all classes, wing, meat & skin, roasted	229	0.0	27.4	12.4	81	61	266	24	25	1.46	0.056	2.10	0.020	197	29.9
Turnip greens, boiled, w/o salt	20	4.4	1.1	0.2	0	29	203	137	22	0.80	0.253	0.14	0.337	29	0.9
Turnips, boiled, w/o salt	22	5.1	0.7	0.1	0	16	177	33	9	0.18	0.002	0.12	0.071	26	0.2
Turnips, raw	28	6.4	0.9	0.1	0	67	191	30	11	0.30	0.085	0.27	0.134	27	0.7
Veal, composite of retail cuts, fat, cooked	642	0.0	9.4	66.7	73	57	173	4	10	1.00	0.044	0.87	0.012	116	5.5
Veal, composite of retail cuts, lean & fat, cooked	231	0.0	30.1	11.4	114	87	325	22	26	1.15	0.114	4.76	0.036	239	12.3
Veal, composite of retail cuts, lean, cooked	196	0.0	31.9	6.6	118	89	338	24	28	1.16	0.120	5.10	0.038	250	13.0
Veal, ground, broiled	172	0.0	24.4	7.6	103	83	337	17	24	0.99	0.103	3.87	0.035	217	13.7
Veal, leg (top round), lean & fat, roasted	160	0.0	27.7	4.7	103	68	389	6	28	0.91	0.129	3.04	0.030	234	11.2
Veal, loin, lean & fat, roasted	217	0.0	24.8	12.3	103	93	325	19	25	0.87	0.110	3.03	0.029	212	11.0
Veal, rib, lean & fat, roasted	228	0.0	24.0	14.0	110	92	295	11	22	0.97	0.099	4.09	0.030	197	10.5
Veal, shoulder, arm, lean & fat, roasted	183	0.0	25.5	8.3	108	90	348	26	26	1.15	0.141	4.18	0.030	221	11.1
Veal, sirloin, lean & fat, roasted	202	0.0	25.1	10.5	102	83	351	13	26	0.92	0.129	3.35	0.029	223	11.1
Veal, sirloin, lean, roasted	168	0.0	26.3	6.2	104	85	365	14	27	0.91	0.136	3.54	0.030	231	11.5
Vinegar, distilled	18	0.0	0.0	0.0	0	2	2	6	1	0.03	0.006	0.01	0.055	4	0.5
Watercress, raw	11	1.3	2.3	0.1	0	41	330	120	21	0.20	0.077	0.11	0.244	60	0.9
Watermelon, raw	30	7.6	0.6	0.2	0	1	112	7	10	0.24	0.042	0.10	0.038	11	0.4
Wheat flour, whole-grain	339	72.6	13.7	1.9	0	5	405	34	138	3.88	0.382	2.93	3.799	346	70.7
Wheat flours, bread, unenriched	361	72.5	12.0	1.7	0	2	100	15	25	0.90	0.182	0.85	0.792	97	39.7
Wolffish, Atlantic, cooked, dry heat	123	0.0	22.4	3.1	59	109	385	8	38	0.12	0.037	1.00	0.019	250	46.8
Wolffish, Atlantic, raw	96	0.0	17.5	2.4	46	85	300	6	30	0.09	0.029	0.78	0.015	200	36.5
Yam, boiled, or baked, w/o salt	116	27.6	1.5	0.1	0	8	670	14	18	0.52	0.152	0.20	0.371	49	0.7
Yellowtail, mixed species, cooked, dry heat	187	0.0	29.7	6.7	71	50	538	29	38	0.63	0.058	0.67	0.019	201	46.8
Yogurt, fruit varieties, non-fat	94	19.0	4.4	0.2	2	58	194	152	15	0.07	0.011	0.74	0.035	119	6.0
Yogurt, plain, low fat, 12 grams protein per 8 oz	63	7.0	5.3	1.6	6	70	234	183	17	0.08	0.013	0.89	0.004	144	3.3
Yogurt, plain, skim milk, 13 grams protein per 8 oz	56	7.7	5.7	0.2	2	77	255	199	19	0.09	0.015	0.97	0.005	157	3.6
Yogurt, plain, whole milk, 8 grams protein per 8 oz	61	4.7	3.5	3.3	13	46	155	121	12	0.05	0.009	0.59	0.004	95	2.2

Section 8
Analytical Chemistry

Preparation of Special Analytical Reagents 8-1
Standard Solutions of Acids, Bases, and Salts 8-5
Standard Solutions of Oxidation and Reduction Reagents 8-7
Organic Analytical Reagents for the Determination of Inorganic Substances 8-8
Flame and Bead Tests 8-13
Acid-Base Indicators 8-15
Fluorescent Indicators 8-18
Conversion Formulas for Concentration of Solutions 8-19
Electrochemical Series 8-20
Reduction and Oxidation Potentials for Certain Ion Radicals 8-30
Practical pH Measurements on Natural Waters 8-37
Buffer Solutions Giving Round Values of pH at 25 °C 8-39
Dissociation Constants of Inorganic Acids and Bases 8-40
Dissociation Constants of Organic Acids and Bases 8-42
Concentrative Properties of Aqueous Solutions: Density, Refractive Index, Freezing Point Depression, and Viscosity 8-52
Ionization Constant of Water 8-78
Ionization Constant of Normal and Heavy Water 8-79
Solubility of Selected Gases in Water 8-80
Solubility of Carbon Dioxide in Water at Various Temperatures and Pressures 8-84
Aqueous Solubility and Henry's Law Constants of Organic Compounds 8-85
Aqueous Solubility of Inorganic Compounds at Various Temperatures 8-121
Solubility Product Constants 8-127
Solubility of Common Salts at Ambient Temperatures 8-130
Solubility of Hydrocarbons in Seawater 8-131
Solubility of Organic Compounds in Pressurized Hot Water 8-133
Solubility Chart 8-136
Reduction of Weighings in Air to Vacuo 8-138
Volume of One Gram of Water 8-139
Properties of Carrier Gases for Gas Chromatography 8-140
Solvents for Ultraviolet Spectrophotometry 8-141
^{13}C Chemical Shifts of Useful Nmr Solvents 8-142
Mass Spectral Peaks of Common Organic Solvents 8-143
Proton NMR Shifts of Common Organic Solvents 8-150

PREPARATION OF SPECIAL ANALYTICAL REAGENTS

Aluminon (qualitative test for aluminum). Aluminon is a trade name for the ammonium salt of aurintricarboxylic acid. Dissolve 1 g of the salt in 1 L of distilled water. Shake the solution well to ensure thorough mixing.

Bang's reagent (for glucose estimation). Dissolve 100 g of K_2CO_3, 66 g of KCl and 160 g of $KHCO_3$ in the order given in about 700 mL of water at 30 °C. Add 4.4 g of $CuSO_4$ and dilute to 1 L after the CO_2 is evolved. This solution should be shaken only in such a manner as not to allow entry of air. After 24 hours 300 mL are diluted to 1 L with saturated KCl solution, shaken gently and used after 24 hours; 50 mL is equivalent to 10 mg glucose.

Barfoed's reagent (test for glucose). See Cupric acetate.

Baudisch's reagent. See Cupferron.

Benedict's solution (qualitative reagent for glucose). With the aid of heat, dissolve 173 g of sodium citrate and 100 g of Na_2CO_3 in 800 mL of water. Filter, if necessary, and dilute to 850 mL. Dissolve 17.3 g of $CuSO_4{\cdot}5H_2O$ in 100 mL of water. Pour the latter solution, with constant stirring, into the carbonate-citrate solution, and dilute to 1 L.

Benzidine hydrochloride solution (for sulfite determination). Make a paste of 8 g of benzidine hydrochloride ($C_{12}H_8(NH_3)_2{\cdot}2HCl$) and 20 mL of water, add 20 mL of HCl (sp. gr. 1.12) and dilute to 1 L with water. Each mL of this solution is equivalent to 0.00357 g of H_2SO_4.

Bertrand's reagent (glucose estimation). Consists of the following solutions:

1. Dissolve 200 g of Rochelle salt and 150 g of NaOH in sufficient water to make 1 L of solution.
2. Dissolve 40 g of $CuSO_4$ in enough water to make 1 L of solution.
3. Dissolve 50 g of $Fe_2(SO_4)_3$ and 200 g of H_2SO_4 (sp. gr. 1.84) in sufficient water to make 1 L of solution.
4. Dissolve 5 g of $KMnO_4$ in sufficient water to make 1 L of solution.

Bial's reagent (for pentose). Dissolve 1 g of orcinol (5-methyl-1,3-benzenediol) in 500 mL of 30% HCl to which 30 drops of a 10% solution of $FeCl_3$ have been added.

Boutron — Boudet soap solution:

1. Dissolve 100 g of pure castile soap in about 2.5 L of 56% ethanol.
2. Dissolve 0.59 g of $Ba(NO_3)_2$ in 1 L of water.

Adjust the castile soap solution so that 2.4 mL of it will give a permanent lather with 40 mL of solution (b). When adjusted, 2.4 mL of soap solution is equivalent to 220 parts per million of hardness (as $CaCO_3$) for a 40 mL sample. See also Soap solution.

Brucke's reagent (protein precipitation). See Potassium iodide-mercuric iodide.

Clarke's soap solution (estimation of hardness in water).

1. Dissolve 100 g of pure powdered castile soap in 1 L of 80% ethanol and allow to stand overnight.
2. Prepare a solution of $CaCl_2$ by dissolving 0.5 g of $CaCO_3$ in HCl (sp. gr. 1.19), neutralize with NH_4OH and make slightly alkaline to litmus, and dilute to 500 mL. One mL is equivalent to 1 mg of $CaCO_3$.

Titrate (1) against (2) and dilute (1) with 80% ethanol until 1 mL of the resulting solution is equivalent to 1 mL of (2) after making allowance for the lather factor (the amount of standard soap solution required to produce a permanent lather in 50 mL of distilled water). One mL of the adjusted solution after subtracting the lather factor is equivalent to 1 mg of $CaCO_3$. See also Soap solution.

Cobalticyanide paper (Rinnmann's test for Zn). Dissolve 4 g of $K_3Co(CN)_6$ and 1 g of $KClO_3$ in 100 mL of water. Soak filter paper in solution and dry at 100 °C. Apply drop of zinc solution and burn in an evaporating dish. A green disk is obtained if zinc is present.

Cochineal. Extract 1 g of cochineal for 4 days with 20 mL of alcohol and 60 mL of distilled water. Filter.

Congo red. Dissolve 0.5 g of Congo red in 90 mL of distilled water and 10 mL of alcohol.

Cupferron (Baudisch's reagent for iron analysis). Dissolve 6 g of the ammonium salt of *N*-hydroxy-*N*-nitrosoaniline (cupferron) in 100 mL of H_2O. Reagent good for 1 week only and must be kept in the dark.

Cupric acetate (Barfoed's reagent for reducing monosaccharides). Dissolve 66 g of cupric acetate and 10 mL of glacial acetic acid in water and dilute to 1 L.

Cupric oxide, ammoniacal; Schweitzer's reagent (dissolves cotton, linen, and silk, but not wool).

1. Dissolve 5 g of cupric sulfate in 100 mL of boiling water, and add sodium hydroxide until precipitation is complete. Wash the precipitate well, and dissolve it in a minimum quantity of ammonium hydroxide.
2. Bubble a slow stream of air through 300 mL of strong ammonium hydroxide containing 50 g of fine copper turnings. Continue for 1 hour.

Cupric sulfate in glycerin-potassium hydroxide (reagent for silk). Dissolve 10 g of cupric sulfate, $CuSO_4{\cdot}5H_2O$, in 100 mL of water and add 5 g of glycerol. Add KOH solution slowly until a deep blue solution is obtained.

Cupron (precipitates copper). Dissolve 5 g of benzoinoxime in 100 mL of 95% ethanol.

Cuprous chloride, acidic (reagent for CO in gas analysis).

1. Cover the bottom of a 2-L flask with a layer of cupric oxide about 0.5 inch deep, suspend a coil of copper wire so as to reach from the bottom to the top of the solution, and fill the flask with hydrochloric acid (sp. gr. 1.10). Shake occasionally. When the solution becomes nearly colorless, transfer to reagent bottles, which should also contain copper wire. The stock bottle may be refilled with dilute hydrochloric acid until either the cupric oxide or the copper wire is used up. Copper sulfate may be substituted for copper oxide in the above procedure.
2. Dissolve 340 g of $CuCl_2{\cdot}2H_2O$ in 600 mL of conc. HCl and reduce the cupric chloride by adding 190 mL of a saturated solution of stannous chloride or until the solution is colorless. The stannous chloride is prepared by treating 300 g of metallic tin in a 500 mL flask with conc. HCl until no more tin goes into solution.
3. (Winkler method). Add a mixture of 86 g of CuO and 17 g of finely divided metallic Cu, made by the reduction of

CuO with hydrogen, to a solution of HCl, made by diluting 650 mL of conc. HCl with 325 mL of water. After the mixture has been added slowly and with frequent stirring, a spiral of copper wire is suspended in the bottle, reaching all the way to the bottom. Shake occasionally, and when the solution becomes colorless, it is ready for use.

Cuprous chloride, ammoniacal (reagent for CO in gas analysis).

1. The acid solution of cuprous chloride as prepared above is neutralized with ammonium hydroxide until an ammonia odor persists. An excess of metallic copper must be kept in the solution.
2. Pour 800 mL of acidic cuprous chloride, prepared by the Winkler method, into about 4 L of water. Transfer the precipitate to a 250 mL graduate. After several hours, siphon off the liquid above the 50 mL mark and refill with 7.5% NH_4OH solution which may be prepared by diluting 50 mL of conc. NH_4OH with 150 mL of water. The solution is well shaken and allowed to stand for several hours. It should have a faint odor of ammonia.

Dichlorofluorescein indicator. Dissolve 1 g in 1 L of 70% alcohol or 1 g of the sodium salt in 1 L of water.

Dimethylglyoxime, 0.01 N. Dissolve 0.6 g of dimethylglyoxime (2,3-butanedione oxime) in 500 mL of 95% ethanol. This is an especially sensitive test for nickel, a very definite crimson color being produced.

Diphenylamine (reagent for rayon). Dissolve 0.2 g in 100 mL of concentrated sulfuric acid.

Diphenylamine sulfonate (for titration of iron with $K_2Cr_2O_7$). Dissolve 0.32 g of the barium salt of diphenylamine sulfonic acid in 100 mL of water, add 0.5 g of sodium sulfate and filter off the precipitate of $BaSO_4$.

Diphenylcarbazide. Dissolve 0.2 g of diphenylcarbazide in 10 mL of glacial acetic acid and dilute to 100 mL with 95% ethanol.

Esbach's reagent (estimation of protein). To a water solution of 10 g of picric acid and 20 g of citric acid, add sufficient water to make 1 L of solution.

Eschka's compound. Two parts of calcined ("light") magnesia are thoroughly mixed with 1 part of anhydrous sodium carbonate.

Fehling's solution (reagent for reducing sugars.)

1. Copper sulfate solution. Dissolve 34.66 g of $CuSO_4·5H_2O$ in water and dilute to 500 mL.
2. Alkaline tartrate solution. Dissolve 173 g of potassium sodium tartrate (Rochelle salt, $KNaC_4H_4O_6·4H_2O$) and 50 g of NaOH in water and dilute when cold to 500 mL.

Mix equal volumes of the two solutions at the time of using.

Ferric-alum indicator. Dissolve 140 g of ferric ammonium sulfate crystals in 400 mL of hot water. When cool, filter, and make up to a volume of 500 mL with dilute nitric acid.

Folin's mixture (for uric acid). To 650 mL of water add 500 g of $(NH_4)_2SO_4$, 5 g of uranium acetate, and 6 g of glacial acetic acid. Dilute to 1 L.

Formaldehyde — sulfuric acid (Marquis' reagent for alkaloids). Add 10 mL of formaldehyde solution to 50 mL of sulfuric acid.

Froehde's reagent. See Sulfomolybdic acid.

Fuchsin (reagent for linen). Dissolve 1 g of fuchsin in 100 mL of alcohol.

Fuchsin — sulfurous acid (Schiff's reagent for aldehydes). Dissolve 0.5 g of fuchsin and 9 g of sodium bisulfite in 500 mL of water, and add 10 mL of HCl. Keep in well-stoppered bottles and protect from light.

Gunzberg's reagent (detection of HCl in gastric juice). Prepare as needed a solution containing 4 g of phloroglucinol (1,3,5-benzenetriol) and 2 g of vanillin in 100 mL of absolute ethanol.

Hager's reagent. See Picric acid.

Hanus solution (for iodine number). Dissolve 13.2 g of resublimed iodine in 1 L of glacial acetic acid which will pass the dichromate test for reducible matter. Add sufficient bromine to double the halogen content, determined by titration (3 mL is about the proper amount). The iodine may be dissolved by the aid of heat, but the solution should be cold when the bromine is added.

Iodine, tincture of. To 50 mL of water add 70 g of I_2 and 50 g of KI. Dilute to 1 L with alcohol.

Iodo-potassium iodide (Wagner's reagent for alkaloids). Dissolve 2 g of iodine and 6 g of KI in 100 mL of water.

Litmus (indicator). Extract litmus powder three times with boiling alcohol, each treatment consuming an hour. Reject the alcoholic extract. Treat residue with an equal weight of cold water and filter; then exhaust with five times its weight of boiling water, cool and filter. Combine the aqueous extracts.

Magnesia mixture (reagent for phosphates and arsenates). Dissolve 55 g of magnesium chloride and 105 g of ammonium chloride in water, barely acidify with hydrochloric acid, and dilute to 1 L. The ammonium hydroxide may be omitted until just previous to use. The reagent, if completely mixed and stored for any period of time, becomes turbid.

Magnesium uranyl acetate. Dissolve 100 g of $UO_2(C_2H_3O_2)_2·2H_2O$ in 60 mL of glacial acetic acid and dilute to 500 mL. Dissolve 330 g of $Mg(C_2H_3O_2)_2·4H_2O$ in 60 mL of glacial acetic acid and dilute to 200 mL. Heat solutions to the boiling point until clear, pour the magnesium solution into the uranyl solution, cool and dilute to 1 L. Let stand overnight and filter if necessary.

Marme's reagent. See Potassium-cadmium iodide.

Marquis' reagent. See Formaldehyde-sulfuric acid.

Mayer's reagent (white precipitate with most alkaloids in slightly acid solutions). Dissolve 1.358 g of $HgCl_2$ in 60 mL of water and pour into a solution of 5 g of KI in 10 mL of H_2O. Add sufficient water to make 100 mL.

Methyl orange indicator. Dissolve 1 g of methyl orange in 1 L of water. Filter, if necessary.

Methyl orange, modified. Dissolve 2 g of methyl orange and 2.8 g of xylene cyanole FF in 1 L of 50% alcohol.

Methyl red indicator. Dissolve 1 g of methyl red in 600 mL of alcohol and dilute with 400 mL of water.

Methyl red, modified. Dissolve 0.50 g of methyl red and 1.25 g of xylene cyanole FF in 1 L of 90% alcohol. Or, dissolve 1.25 g of methyl red and 0.825 g of methylene blue in 1 L of 90% alcohol.

Millon's reagent (for albumins and phenols). Dissolve 1 part of mercury in 1 part of cold fuming nitric acid. Dilute with twice the volume of water and decant the clear solution after several hours.

Molisch's reagent. See 1-Naphthol.

1-Naphthol (Molisch's reagent for wool). Dissolve 15 g of 1-naphthol in 100 mL of alcohol or chloroform.

Nessler's reagent (for ammonia). Dissolve 50 g of KI in the smallest possible quantity of cold water (50 mL). Add a saturated solution of mercuric chloride (about 22 g in 350 mL of water will be needed) until an excess is indicated by the formation of a precipitate. Then add 200 mL of 5 N NaOH and dilute to 1 L. Let settle, and draw off the clear liquid.

Nickel oxide, ammoniacal (reagent for silk). Dissolve 5 g of nickel sulfate in 100 mL of water, and add sodium hydroxide solution until nickel hydroxide is completely precipitated. Wash the precipitate well and dissolve in 25 mL of concentrated ammonium hydroxide and 25 mL of water.

Nitron (detection of nitrate radical). Dissolve 10 g of nitron (1,4-diphenyl-3-(phenylamino)-1,2,4-triazolium hydroxide) in 5 mL of glacial acetic acid and 95 mL of water. The solution may be filtered with slight suction through an alundum crucible and kept in a dark bottle.

1-Nitroso-2-naphthol. Make a saturated solution in 50% acetic acid (1 part of glacial acetic acid with 1 part of water). Does not keep well.

Nylander's solution (carbohydrates). Dissolve 20 g of bismuth subnitrate and 40 g of Rochelle salt in 1 L of 8% NaOH solution. Cool and filter.

Obermayer's reagent (for indoxyl in urine). Dissolve 4 g of $FeCl_3$ in 1 L of HCl (sp. gr. 1.19).

Oxine. Dissolve 14 g of 8-hydroxyquinoline in 30 mL of glacial acetic acid. Warm slightly, if necessary. Dilute to 1 L.

Oxygen absorbent. Dissolve 300 g of ammonium chloride in 1 L of water and add 1 L of concentrated ammonium hydroxide solution. Shake the solution thoroughly. For use as an oxygen absorbent, a bottle half full of copper turnings is filled nearly full with the NH_4Cl-NH_4OH solution and the gas passed through.

Pasteur's salt solution. To 1 L of distilled water add 2.5 g of potassium phosphate, 0.25 g of calcium phosphate, 0.25 g of magnesium sulfate, and 12.00 g of ammonium tartrate.

Pavy's solution (glucose reagent). To 120 mL of Fehling's solution, add 300 mL of NH_4OH (sp. gr. 0.88) and dilute to 1 L with water.

Phenanthroline ferrous ion indicator. Dissolve 1.485 g of 1,10-phenanthroline monohydrate in 100 mL of 0.025 M ferrous sulfate solution.

Phenolphthalein. Dissolve 1 g of phenolphthalein in 50 mL of alcohol and add 50 mL of water.

Phenolsulfonic acid (determination of nitrogen as nitrate). Dissolve 25 g of phenol in 150 mL of conc. H_2SO_4, add 75 mL of fuming H_2SO_4 (15% SO_3), stir well and heat for 2 hours at 100 °C.

Phloroglucinol solution (pentosans). Make a 3% phloroglucinol (1,3,5-benzenetriol) solution in alcohol. Keep in a dark bottle.

Phosphomolybdic acid (Sonnenschein's reagent for alkaloids).

1. Prepare ammonium phosphomolybdate and after washing with water, boil with nitric acid and expel NH_3; evaporate to dryness and dissolve in 2 M nitric acid.
2. Dissolve ammonium molybdate in HNO_3 and treat with phosphoric acid. Filter, wash the precipitate, and boil with aqua regia until the ammonium salt is decomposed. Evaporate to dryness. The residue dissolved in 10% HNO_3 constitutes Sonnenschein's reagent.

Phosphoric acid — sulfuric acid mixture. Dilute 150 mL of conc. H_2SO_4 and 100 mL of conc. H_3PO_4 (85%) with water to a volume of 1 L.

Phosphotungstic acid (Scheibler's reagent for alkaloids).

1. Dissolve 20 g of sodium tungstate and 15 g of sodium phosphate in 100 mL of water containing a little nitric acid.
2. The reagent is a 10% solution of phosphotungstic acid in water. The phosphotungstic acid is prepared by evaporating a mixture of 10 g of sodium tungstate dissolved in 5 g of phosphoric acid (sp. gr. 1.13) and enough boiling water to effect solution. Crystals of phosphotungstic acid separate.

Picric acid (Hager's reagent for alkaloids, wool and silk). Dissolve 1 g of picric acid in 100 mL of water.

Potassium antimonate (reagent for sodium). Boil 22 g of potassium antimonate with 1 L of water until nearly all of the salt has dissolved, cool quickly, and add 35 mL of 10% potassium hydroxide. Filter after standing overnight.

Potassium-cadmium iodide (Marme's reagent for alkaloids). Add 2 g of CdI_2 to a boiling solution of 4 g of KI in 12 mL of water, and then mix with 12 mL of saturated KI solution.

Potassium hydroxide (for CO_2 absorption). Dissolve 360 g of KOH in water and dilute to 1 L.

Potassium iodide — mercuric iodide (Brucke's reagent for proteins). Dissolve 50 g of KI in 500 mL of water, and saturate with mercuric iodide (about 120 g). Dilute to 1 L.

Potassium pyrogallate (for oxygen absorption). For mixtures of gases containing less than 28% oxygen, add 100 mL of KOH solution (50 g of KOH to 100 mL of water) to 5 g of pyrogallol. For mixtures containing more than 28% oxygen the KOH solution should contain 120 g of KOH to 100 mL of water.

Pyrogallol, alkaline.

1. Dissolve 75 g of pyrogallic acid in 75 mL of water.
2. Dissolve 500 g of KOH in 250 mL of water. When cool, adjust until sp. gr. is 1.55.

For use, add 270 mL of solution (2) to 30 mL of solution (1).

Rosolic acid (indicator). Dissolve 1 g of rosolic acid in 10 mL of alcohol and add 100 mL of water.

Scheibler's reagent. See Phosphotungstic acid.

Schiff's reagent. See Fuchsin-sulfurous acid.

Schweitzer's reagent. See Cupric oxide, ammoniacal.

Soap solution (reagent for hardness in water). Dissolve 100 g of dry castile soap in 1 L of 80% alcohol (5 parts alcohol to 1 part water). Allow to stand several days and dilute with 70% to 80% alcohol until 6.4 mL produces a permanent lather with 20 mL of standard calcium solution. The latter solution is made by dissolving 0.2 g of $CaCO_3$ in a small amount of dilute HCl, evaporating to dryness and making up to 1 L.

Sodium bismuthate (oxidation of manganese). Heat 20 parts of NaOH nearly to redness in an iron or nickel crucible and add slowly 10 parts of basic bismuth nitrate which has been previously dried. Add 2 parts of sodium peroxide, and pour the brownish-yellow fused mass onto an iron plate to cool. When cold, break up in a mortar, extract with water, and collect on an asbestos filter.

Sodium hydroxide (for CO_2 absorption). Dissolve 330 g of NaOH in water and dilute to 1 L.

Sodium nitroprusside (reagent for hydrogen sulfide and wool). Use a freshly prepared solution of 1 g of sodium nitroferricyanide in 10 mL of water.

Sodium oxalate (primary standard). Dissolve 30 g of the commercial salt in 1 L of water, make slightly alkaline with sodium hydroxide, and let stand until perfectly clear. Filter and evaporate the filtrate to 100 mL. Cool and filter. Pulverize the residue and wash it several times with small volumes of water. The procedure is repeated until the mother liquor is free from sulfate and is neutral to phenolphthalein.

Sodium plumbite (reagent for wool). Dissolve 5 g of sodium hydroxide in 100 mL of water. Add 5 g of litharge (PbO) and boil until dissolved.

Sodium polysulfide. Dissolve 480 g of $Na_2S{\cdot}9H_2O$ in 500 mL of water, add 40 g of NaOH and 18 g of sulfur. Stir thoroughly and dilute to 1 L with water.

Sonnenschein's reagent. See Phosphomolybdic acid.

Starch solution.

1. Make a paste with 2 g of soluble starch and 0.01 g of HgI_2 with a small amount of water. Add the mixture slowly to 1 L of boiling water and boil for a few minutes. Keep in a glass stoppered bottle. If other than soluble starch is used, the solution will not clear on boiling; it should be allowed to stand and the clear liquid decanted.
2. A solution of starch that keeps indefinitely is made as follows: Mix 500 mL of saturated NaCl solution (filtered), 80 mL of glacial acetic acid, 20 mL of water and 3 g of starch. Bring slowly to a boil and boil for 2 minutes.
3. Make a paste with 1 g of soluble starch and 5 mg of HgI_2, using as little cold water as possible. Then pour about 200 mL of boiling water on the paste and stir immediately. This will give a clear solution if the paste is prepared correctly and the water actually boiling. Cool and add 4 g of KI. Starch solution decomposes on standing due to bacterial action, but this solution will keep well if stored under a layer of toluene.

Stoke's reagent. Dissolve 30 g of $FeSO_4$ and 20 g of tartaric acid in water and dilute to 1 L. Just before using, add concentrated NH_4OH until the precipitate first formed is redissolved.

Sulfanilic acid (reagent for nitrites). Dissolve 0.5 g of sulfanilic acid in a mixture of 15 mL of glacial acetic acid and 135 mL of recently boiled water.

Sulfomolybdic acid (Froehde's reagent for alkaloids and glucosides). Dissolve 10 g of molybdic acid or sodium molybdate in 100 mL of conc. H_2SO_4.

Tannic acid (reagent for albumin, alkaloids, and gelatin). Dissolve 10 g of tannic acid in 10 mL of alcohol and dilute with water to 100 mL.

Titration mixture (residual chlorine in water analyasis). Prepare 1 L of dilute HCl (100 mL of HCl (sp. gr. 1.19) in sufficient water to make 1 L). Dissolve 1 g of *o*-tolidine in 100 mL of the dilute HCl and dilute to 1 L with dilute HCl solution.

Trinitrophenol solution. See Picric acid.

Turmeric tincture (reagent for borates). Digest ground turmeric root with several quantities of water which are discarded. Dry the residue and digest it several days with six times its weight of alcohol. Filter.

Uffelmann's reagent (turns yellow in presence of lactic acid). To a 2% solution of pure phenol in water, add a water solution of $FeCl_3$ until the phenol solution becomes violet in color.

Wagner's reagent. See Iodo-potassium iodide.

Wagner's solution (used in phosphate rock analysis to prevent precipitation of iron and aluminum). Dissolve 25 g of citric acid and 1 g of salicylic acid in water and dilute to 1 L. Use 50 mL of the reagent.

Wij's iodine monochloride solution (for iodine number). Dissolve 13 g of resublimed iodine in 1 L of glacial acetic acid which will pass the dichromate test for reducible matter. Set aside 25 mL of this solution. Pass into the remainder of the solution dry chlorine gas (dried and washed by passing through H_2SO_4 (sp. gr. 1.84)) until the characteristic color of free iodine has been discharged. Now add the iodine solution which was reserved, until all free chlorine has been destroyed. A slight excess of iodine does little or no harm, but an excess of chlorine must be avoided. Preserve in well stoppered, amber colored bottles. Avoid use of solutions which have been prepared for more than 30 days.

Wij's special solution (for iodine number). To 200 mL of glacial acetic acid that will pass the dichromate test for reducible matter, add 12 g of dichloramine T (*N,N*-dichloro-4-methylbenzenesulfonamide), and 16.6 g of dry KI (in small quantities with continual shaking until all the KI has dissolved). Make up to 1 L with the same quality of acetic acid used above and preserve in a dark colored bottle.

Zimmermann-Reinhardt reagent (determination of iron). Dissolve 70 g of $MnSO_4{\cdot}4H_2O$ in 500 mL of water, add 125 mL of conc. H_2SO_4 and 125 mL of 85% H_3P0_4, and dilute to 1 L.

Zinc chloride solution, basic (reagent for silk). Dissolve 1000 g of zinc chloride in 850 mL of water, and add 40 g of zinc oxide. Heat until solution is complete.

Zinc uranyl acetate (reagent for sodium). Dissolve 10 g of $UO_2(C_2H_3O_2)_2{\cdot}2H_2O$ in 6 g of 30% acetic acid with heat, if necessary, and dilute to 50 mL. Dissolve 30 g of $Zn(C_2H_3O_2)_2{\cdot}H_2O$ in 3 g of 30% acetic acid and dilute to 50 mL. Mix the two solutions, add 50 mg of NaCl, allow to stand overnight and filter.

STANDARD SOLUTIONS OF ACIDS, BASES, AND SALTS

For each compound listed, the last column of this table gives the mass in grams that is contained in 1 liter of a solution whose amount-of-substance concentration divided by the equivalence factor of the compound equals 0.1 mol/L. In the older literature such a solution is often referred to as a "decinormal solution" (0.1 N).

Reference

Compendium of Analytical Nomenclature (IUPAC), Pergamon Press, Oxford, 1978.

Name	Formula	Atomic or molecular weight	Equivalence factor	Mass in grams
Acetic acid	$HC_2H_3O_2$	60.0530	1	6.0053
Ammonia	NH_3	17.0306	1	1.7031
Ammonium ion	NH_4^+	18.0386	1	1.8039
Ammonium chloride	NH_4Cl	53.4916	1	5.3492
Ammonium sulfate	$(NH_4)_2SO_4$	132.1388	1/2	6.6069
Ammonium thiocyanate	NH_4CNS	76.1204	1	7.6120
Barium	Ba	137.34	1/2	6.867
Barium carbonate	$BaCO_3$	197.3494	1/2	9.8675
Barium chloride hydrate	$BaCl_2 \cdot 2H_2O$	244.2767	1/2	12.2138
Barium hydroxide	$Ba(OH)_2$	171.3547	1/2	8.5677
Barium oxide	BaO	153.3394	1/2	7.6670
Bromine	Br	79.909	1	7.9909
Calcium	Ca	40.08	1/2	2.004
Calcium carbonate	$CaCO_3$	100.0894	1/2	5.0045
Calcium chloride	$CaCl_2$	110.9860	1/2	5.5493
Calcium chloride hydrate	$CaCl_2 \cdot 6H_2O$	219.0150	1/2	10.9508
Calcium hydroxide	$Ca(OH)_2$	74.0947	1/2	3.7047
Calcium oxide	CaO	56.0794	1/2	2.8040
Chlorine	Cl	35.453	1	3.5453
Citric acid	$C_6H_8O_7 \cdot H_2O$	210.1418	1/3	7.0047
Cobalt	Co	58.9332	1/2	2.9466
Copper	Cu	63.54	1/2	3.177
Copper oxide (cupric)	CuO	79.5394	1/2	3.9770
Copper sulfate hydrate	$CuSO_4 \cdot 5H_2O$	249.6783	1/2	12.4839
Hydrochloric acid	HCl	36.4610	1	3.6461
Hydrocyanic acid	HCN	27.0258	1	2.7026
Iodine	I	126.9044	1	12.6904
Lactic acid	$C_3H_6O_3$	90.0795	1	9.0080
Malic acid	$C_4H_6O_5$	134.0894	1/2	6.7045
Magnesium	Mg	24.312	1/2	1.2156
Magnesium carbonate	$MgCO_3$	84.3214	1/2	4.2161
Magnesium chloride	$MgCl_2$	95.2180	1/2	4.7609
Magnesium chloride hydrate	$MgCl_2 \cdot 6H_2O$	203.2370	1/2	10.1623
Magnesium oxide	MgO	40.3114	1/2	2.0156
Manganese	Mn	54.938	1/2	2.7469
Manganese sulfate	$MnSO_4$	150.9996	1/2	7.5500
Mercuric chloride	$HgCl_2$	271.4960	1/2	13.5748
Nickel	Ni	58.71	1/2	2.9356
Nitric acid	HNO_3	63.0129	1	6.3013
Oxalic acid	$H_2C_2O_4$	90.0358	1/2	4.5018
Oxalic acid hydrate	$H_2C_2O_4 \cdot 2H_2O$	126.0665	1/2	6.3033
Oxalic acid anhydride	C_2O_3	72.0205	1/2	3.6010
Phosphoric acid	H_3PO_4	97.9953	1/3	3.2665
Potassium	K	39.102	1	3.9102
Potassium bicarbonate	$KHCO_3$	100.1193	1	10.0119
Potassium carbonate	K_2CO_3	138.2134	1/2	6.9106

Name	Formula	Atomic or molecular weight	Equivalence factor	Mass in grams
Potassium chloride	KCl	74.5550	1	7.4555
Potassium cyanide	KCN	65.1199	1	6.5120
Potassium hydroxide	KOH	56.1094	1	5.6109
Potassium oxide	K_2O	94.2034	1/2	4.7102
Potassium tartrate	$K_2H_4C_4O_6$	226.2769	1/2	11.3139
Silver	Ag	107.87	1	10.787
Silver nitrate	$AgNO_3$	169.8749	1	16.9875
Sodium	Na	22.9898	1	2.2990
Sodium bicarbonate	$NaHCO_3$	84.0071	1	8.4007
Sodium carbonate	Na_2CO_3	105.9890	1/2	5.2995
Sodium chloride	NaCl	58.4428	1	5.8443
Sodium hydroxide	NaOH	39.9972	1	3.9997
Sodium oxide	Na_2O	61.9790	1/2	3.0990
Sodium sulfide	Na_2S	78.0436	1/2	3.9022
Succinic acid	$H_2C_4H_4O_4$	118.0900	1/2	5.9045
Sulfuric acid	H_2SO_4	98.0775	1/2	4.9039
Tartaric acid	$C_4H_6O_6$	150.0888	1/2	7.5044
Zinc	Zn	65.37	1/2	3.269
Zinc sulfate hydrate	$ZnSO_4 \cdot 7H_2O$	287.5390	1/2	14.3769

STANDARD SOLUTIONS OF OXIDATION AND REDUCTION REAGENTS

For each reagent listed, the last column of this table gives the mass in grams that is contained in a solution whose amount-of-substance concentration divided by the equivalence factor of the compound equals 0.1 mol/L. The equivalence factor given refers to the most common reactions of the reagent. In the older literature such a solution is often called a "decinormal solution" (0.1 N).

Reference

Compendium of Analytical Nomenclature (IUPAC), Pergamon Press, Oxford, 1978.

Name	Formula	Atomic or molecular weight	Equivalence factor	Mass in grams
Antimony	Sb	121.75	1/2	6.0875
Arsenic	As	74.9216	1/2	3.7461
Arsenic trisulfide	As_2S_3	246.0352	1/4	6.1509
Arsenous oxide	As_2O_3	197.8414	1/4	4.9460
Barium peroxide	BaO_2	169.3388	1/2	8.4669
Barium peroxide hydrate	$BaO_2 \cdot 8H_2O$	313.4615	1/2	15.6730
Calcium	Ca	40.08	1/2	2.004
Calcium carbonate	$CaCO_3$	100.0894	1/2	5.0045
Calcium hypochlorite	$Ca(OCl)_2$	142.9848	1/4	3.5746
Calcium oxide	CaO	56.0794	1/2	2.8040
Chlorine	Cl	35.453	1	3.5453
Chromium trioxide	CrO_3	99.9942	1/3	3.3331
Ferrous ammonium sulfate	$FeSO_4(NH_4)SO_4 \cdot 6H_2O$	392.0764	1	39.2076
Hydroferrocyanic acid	$H_4Fe(CN)_6$	215.9860	1	21.5986
Hydrogen peroxide	H_2O_2	34.0147	1/2	1.7007
Hydrogen sulfide	H_2S	34.0799	1/2	1.7040
Iodine	I	126.9044	1	12.6904
Iron	Fe	55.847	1	5.5847
Iron oxide (ferrous)	FeO	71.8464	1	7.1846
Iron oxide (ferric)	Fe_2O_3	159.6922	1/2	7.9846
Lead peroxide	PbO_2	239.1888	1/2	11.9594
Manganese dioxide	MnO_2	86.9368	1/2	4.3468
Nitric acid	HNO_3	63.0129	1/3	2.1004
Nitrogen trioxide	N_2O_3	76.0116	1/4	1.9002
Nitrogen pentoxide	N_2O_5	108.0104	1/6	1.8001
Oxalic acid	$C_2H_2O_4$	90.0358	1/2	4.5018
Oxalic acid hydrate	$C_2H_2O_4 \cdot 2H_2O$	126.0665	1/2	6.3033
Oxygen	O	15.9994	1/2	0.8000
Potassium dichromate	$K_2Cr_2O_7$	294.1918	1/6	4.9032
Potassium chlorate	$KClO_3$	122.5532	1/6	2.0425
Potassium chromate	K_2CrO_4	194.1076	1/3	6.4733
Potassium ferrocyanide	$K_4Fe(CN)_6$	368.3621	1	36.8362
Potassium ferrocyanide hydrate	$K_4Fe(CN)_6 \cdot 3H_2O$	422.4081	1	42.2408
Potassium iodide	KI	166.0064	1	16.6006
Potassium nitrate	KNO_3	101.1069	1/3	3.3702
Potassium perchlorate	$KClO_4$	138.5526	1/8	1.7319
Potassium permanganate	$KMnO_4$	158.0376	1/5	3.1608
Sodium chlorate	$NaClO_3$	106.4410	1/6	1.7740
Sodium nitrate	$NaNO_3$	84.9947	1/3	2.8332
Sodium thiosulfate hydrate	$Na_2S_2O_3 \cdot 5H_2O$	248.1825	1	24.8183
Stannous chloride	$SnCl_2$	189.5960	1/2	9.4798
Stannous oxide	SnO	134.6894	1/2	6.7345
Sulfur dioxide	SO_2	64.0628	1/2	3.2031
Tin	Sn	118.69	1/2	5.935

ORGANIC ANALYTICAL REAGENTS FOR THE DETERMINATION OF INORGANIC SUBSTANCES

G. Ackermann, L. Sommer, and D. Thorburn Burns

Determination	Reagents	Ref.
Aluminium	Alizarin Red S	Onishi, Part II a, p 28. (5), Snell, *Metals I*, p 587. (7)
	Aluminon	Fries/Getrost, p 16. (2), Onishi, IIa, p 21. (5), Snell, *Metals I*, p 590. (7)
	Aluminon + Cetyltrimethylammonium bromide	Huaxue Shiji, *8*, 85, (1986)
	Chrome Azurol S	Onishi, Part IIa, p 26. (5), Snell, *Metals I*, p 605. (7)
	Chrome Azurol S + Cetyltrimethylammonium bromide	Marczenko, p 133. (3), Snell, *Metals I*, p 606. (7)
	Chromazol KS + Cetylpyridinium bromide	*Analyst, 107*, 428, (1982).
	Eriochrome Cyanine R	Fries/Getrost, p 19. (2), Onishi, Part IIa, p 25. (5), Snell, *Metals I*, p 611. (7)
	Eriochrome Cyanine R + Cetyltrimethylammonium bromide	Snell, *Metals I*, p 613. (7), *Analyst, 107*, 1431, (1982).
	8-Hydroxyquinoline	Fries/Getrost, p 22. (2), Marczenko, p 131. (3), Onishi, Part IIa, p 31. (5), Snell, *Metals I* , p 622. (7)
Ammonia	Phenol + Sodium hypochlorite	Boltz, p 210 (1), Marczenko, p 413. (3), Snell, *Nonmetals*, p 604. (9)
Antimony	Brilliant Green	Onishi, Part IIa, p 102. (5), Snell, *Metals I*, p 384. (7)
	Bromopyrogallol Red	*Talanta, 13*, 507, (1966).
	Rhodamine B	Fries/Getrost, p 32. (2), Marczenko, p 141. (3), Onishi, Part IIa, p 93. (5), Snell, *Metals I*, p 404. (7)
	Silver diethyldithiocarbamate	Fries/Getrost, p 36. (2)
Arsenic	Silver diethyldithiocarbamate	Fries/Getrost, p 41. (2), Marczenko, p 153. (3), Onishi, Part IIa, p 153. (5), Snell, *Metals I*, p 370. (7)
Barium	Sulfonazo III	Fries/Getrost, p 46. (2), Snell, *Metals II*, p 1782. (8), Onishi, Part IIa, p 202. (5)
Beryllium	Beryllon II	Snell, *Metals I*, p 667. (7)
	Chrome Azurol S	Marczenko, p 163. (3), Snell, *Metals I*, p 672. (7)
	Chrome Azurol S + Cetyltrimethylammonium bromide	Marczenko, p 164. (3), Snell, *Metals I*, p 673. (7)
	Eriochrome Cyanine R	Snell, *Metals I*, p 675. (7), *Talanta, 31*, 249, (1984).
	Eriochrome Cyanine R + Cetyltrimethylammonium bromide	Zh. *Anal. Khim.*, *33*, 1298, (1978).
Bismuth	Dithizone	Onishi, Part IIa, p 262. (5), Snell, *Metals I*, p 303. (7)
	Pyrocatechol Violet	Fres. Z. *Anal. Chem.*, *186*, 418, (1962).
	Pyrocatechol Violet + Cetyltrimethylammonium bromide	Zh. *Anal. Khim.*, *38*, 216, (1983).
	Thiourea	Onishi, Part IIa, p 260. (5), Snell, *Metals I*, p 317. (7)
	Xylenol Orange	Friez/Getrost, p 57. (2), Marczenko, p 172. (3), Snell, *Metals I*, p 320. (7)
Boron	Azomethine H	Snell, Nonmetals, p 165. (9)
	Carminic acid	Boltz, p 14. (1), Fries/Getrost, p 65. (2), Snell, *Nonmetals*, p 170. (9), Williams, p 35. (11)
	Curcumin	Boltz, p 8. (1), Fries/Getrost, p 68. (2), Marczenko, p 180. (3), Snell, *Nonmetals*, p 180. (9), Fres. Z. *Anal. Chem.*, *323*, 266, (1986).
	Methylene Blue	Boltz, p 21. (1), Marczenko, p 183. (3), Snell, *Nonmetals*, p 205. (9), *Talanta*, 31, 547, (1984).
Bromide	Fluorescein	Boltz, p 48. (1), Snell, *Nonmetals*, p 276., Fres. Z. *Anal. Chem.*, *301*, 28 (1980).
	Phenol Red	Boltz, p 44. (1), Marczenko, p 190. (3), Snell, Nonmetals, p 28. (9)
Cadmium	2-(5-Bromo-2-pyridylazo)-5-diethylaminophenol	Marczenko, p 197. (3)
	Cadion	Onishi, Part IIa, p 323. (5)
	Dithizone	Fries/Getrost, p 78. (2), Onishi, Part IIa, p 315. (5), Snell, *Metals I*, p 279. (7), West, p 25. (10).
	4–(2-Pyridylazo)resorcinol	Fres. Z. *Anal. Chem.*, *310*, 51, (1982).
Calcium	Chlorophosphonazo III	Marczenko, p 207. (3), Snell, *Metals II*, p 1744. (8)
	Glyoxal-bis(2-hydroxyanil)	Fries/Getrost, p 86. (2), Onishi, Part IIa, p 352. (5), Snell, *Metals I*, p 1762. (8)
	Murexide	Onishi, Part IIa, p 357. (5), Snell, *Metals II*, p 1769. (8)
	Phthalein Purple	*Anal. Chim. Acta*, *34*, 71 (1966).
Cerium	*N*-benzoyl-*N*-phenylhydroxylamine	*Anal. Chim. Acta*, *48*, 155, (1969).

Determination	Reagents	Ref.
	8-Hydroxyquinoline	Fries/Getrost, p 93. (2), Marczenko, p 220. (3), Onishi, Part IIa, p 383. (7)
Chlorine	*N,N*-Diethyl-1,4-phenylenediamine	Boltz, p 92. (1), Fries/Getrost, p 101. (2), Snell, Nonmetals, p 225. (9), *Analyst, 90,* 187, (1965).
Chromium	1,5-Diphenylcarbazide	Fries/Getrost, p 105. (2), Onishi, Part IIa, p 412. (5), Snell, *Metals I,* p 714. (7), West, p 12. (10)
	4-(2-Pyridylazo)resorcinol	Snell, *Metals I,* p 736. (7), West, p 17. (10)
	4-(2-Pyridylazo)resorcinol + Tetradecyldimethyl-benzylammonium chloride	West, p 17. (10), *Anal. Chim. Acta, 67,* 297, (1973).
	4-(2-Pyridylazo)resorcinol + Hydrogen peroxide	Fres. Z. *Anal. Chem., 304,* 382, (1980).
Cobalt	Nitroso-R salt	Fries/Getrost, p 118. (2), Onishi, Part IIa, p 454. (5), Snell, *Metals I,* p 953. (7)
	1-Nitroso-2-naphthol	Fries/Getrost, p 111. (2), Marczenko, p 246. (3), Snell, *Metals I,* p 947. (5)
	2-Nitroso-1-naphthol	Fries/Getrost, p 113. (2), Onishi, Part IIa, p 459. (5), Snell, *Metals I,* p 949. (7), West, p 45. (10)
	4-(2-Pyridylazo)resorcinol	Snell, *Metals I,* p 969. (7), West, p 44. (10)
	4-(2-Pyridylazo)resorcinol + Diphenylguanidine	*Zh. Anal. Khim., 35,* 1306, (1980).
Copper	Bathocuproine	Fries/Getrost, p 135. (2), Snell, *Metals I,* p 148. (7)
	Bathocuproine disulfonic acid	Fries/Getrost, p 137. (2), West, p 52. (10)
	Dithizone	Marczenko, p 258. (3), Onishi, Part IIa, p 529. (5), Snell, *Metals I,* p 199. (7)
	Neocuproine	Snell, *Metals I,* p 217. (5), West, p 51. (10)
	Cuprizone	Onishi, Part IIa, p 534. (5), Snell, *Metals I,* p 157. (7), West, p 53. (10)
	4-(2-pyridylazo)resorcinol + Tetradecyldimethyl-benzylammonium chloride	*Anal. Chim. Acta, 138,* 321, (1982).
Cyanide	Barbituric Acid + Pyridine	Fries/Getrost, p 153. (2), Snell, *Nonmetals,* p 653. (9)
	Barbituric Acid + Pyridine-4-carboxylic acid	*Anal. Chim. Acta, 99,* 197, (1970).
Fluoride	Alizarin Fluorine blue + Lanthanum(III) ion	Boltz, p 129. (1), Fries/Getrost, p 158. (2), Snell, *Nonmetals,* p 333. (9), Williams, p 354. (11)
	Eriochrome Cyanine R + Zirconium(IV) ion	Boltz, p 119. (1), Snell, *Nonmetals,* p 359. (2), Williams, p 357. (11)
Gallium	Pyrocatechol violet + Diphenylguanidine	Snell, *Metals I,* p 500. (7)
	8-Hydroxyquinoline	Onishi Pt IIa, p 582. (5), Snell, *Metals I,* p 505. (7)
	1-(2-Pyridylazo)-2-naphthol	Snell, *Metals I,* p 512. (7)
	4-(2-Pyridylazo)resorcinol	Snell, *Metals I,* p 513. (7)
	Rhodamine B	Marczenko, p 284. (3), Onishi, Part IIa, p 578. (5), Snell, *Metals I,* p 515. (7)
	Xylenol Orange	Fries/Getrost, p 166. (2), Snell, *Metals I,* p 523. (7)
	Xylenol Orange + 8-Hydroxyquinoline	*Zh. Anal. Khim., 26,* 75, (1971).
Germanium	Brilliant Green + Molybdate	Snell, *Metals I,* p 562. (7)
	Phenylfluorone	Fries/Getrost, p 168. (2), Marczenko, p 292. (3), Onishi, Part IIa, p 607. (5), Snell, *Metals I,* p 570. (7)
Gold	5-(4-Diethylaminobenzylidene) rhodanine	Fries/Getrost, p 173. (2), Onishi, Part IIa, p 631. (5), Snell, *Metals II,* p 1516. (8)
	Rhodamine B	Fries/Getrost, p 175. (2), Marczenko, p 301. (3), Onishi, Part IIa, p 637. (5), Snell, *Metals II,* p 513. (8)
Hafnium	Arsenazo III	Snell, *Metals II,* p 1184. (8), *Talanta, 19,* 807, (1972).
Indium	Bromopyrogallol Red	Snell, *Metals I,* p 469. (7)
	Chrome Azurol S	Snell, *Metals I,* p 474. (7)
	Chrome Azurol S + Cetyltrimethylammonium bromide	*Anal. Chim. Acta, 67,* 107, (1973).
	Dithizone	Fries/Getrost, p 179. (2), Onishi, Part IIa, p 672. (5), Snell, *Metals I,* p 474. (7)
	8-Hydroxyquinoline	Onishi, Part IIa, p 670. (5), Snell, *Metals I,* p 475. (7)
	1-(2-Pyridylazo)-2-naphthol	Snell, *Metals I,* p 480. (7)
	4-(2-Pyridylazo)resorcinol	Marczenko, p 309. (3), Snell, *Metals* I, p 480. (7)
Iodide	Neocuproine + Copper(II)	*Anal. Chim. Acta, 69,* 321, (1974).
Iodine	Starch	Boltz, p 162. (1), Marczenko, p 316. (3), Snell, *Nonmetals,* p 307. (9)
Iridium	Rhodamine 6G + Tin(II)	Marczenko, p 323. (3)
	N,N-Dimethyl-4-nitrosoaniline	*Anal. Chem., 27,* 1776, (1955).
Iron	Bathophenanthroline	Fries/Getrost, p 189. (2), Onishi, Part IIa, p 729. (5), Snell, *Metals I,* p 763. (7)
	Bathophenanthroline disulfonic acid	Fries/Getrost, p 191. (2), Snell, *Metals I,* p 772. (7)

Determination	Reagents	Ref.
	2,2´-Bipyridyl	Snell, *Metals I,* p 750. (7)
	Chrome Azurol S + Cetyltrimethylammonium bromide	Snell, *Metals I,* p 757. (7), *Coll. Czech. Chem. Comm., 45,* 2656, (1980).
	1,10-Phenanthroline	Fries/Getrost, p 199. (2), Marczenko, p 331. (3), Onishi, Part IIa, p 725. (5), Snell, *Metals I,* p 795. (7)
	1,10-Phenanthroline + Bromothymol Blue	*Zh. Anal. Khim., 25,* 1348, (1970).
	Ferrozine	Onishi, Part IIa, p 730. (5), Snell, *Metals I,* p 783. (7)
Lanthanum	Arsenazo III	Marczenko, p 468. (3), Snell, *Metals II,* p 1910. (8)
Lead	Dithizone	Fries/Getrost, p 207. (2), Onishi, Part IIa, p 824. (5), Snell, *Metals I,* p 2. (7), West, p 34. (10)
	Sodium diethyldithiocarbamate	Fries/Getrost, p 214. (2), Snell, *Metals I,* p 27. (7)
	4-(2-Pyridylazo)resorcinol	Fries/Getrost, p 220. (2), Marczenko, p 347. (3), Snell, *Metals I,* p 34. (7)
Lithium	Thoron	Onishi, Part IIa, p 863. (5), Snell, *Metals II,* p 1726. (8), *Talanta,* 30, 587, (1983).
Magnesium	Eriochrome Black T	Fries/Getrost, p 226. (2), Marczenko, p 355. (3), Onishi, Part IIb, p 13. (6), Snell, *Metals II,* p 1932. (8)
	8-Hydroxyquinoline	Onishi, Part IIb, p 11. (6), Snell, *Metals II,* p 1938. (8)
	8-Hydroxyquinoline + Butylamine	Fries/Getrost, p 228. (2), Snell, *Metals II,* p 1938. (8)
	Titan Yellow	Fries/Getrost, p 234. (2), Marczenko, p 352. (3), Snell, *Metals II,* p 1945. (8)
	Xylidyl Blue	Fries/Getrost, p 231. (2), Onishi, Part IIb, p 14. (6), Snell, *Metals II,* p 1950. (8)
Manganese	Formaldoxime	Fries/Getrost, p 236. (2), Marczenko, p 364. (3), Onishi Part IIb, p 38. (6), Snell, *Metals II,* 1010. (8)
Mercury	Dithizone	Fries/Getrost, p 243. (2), Marczenko, p 373. (3), Onishi, Part IIb, p 66. (6), Snell, *Metals I,* p 107. (7), West, p 29. (10)
	Michler's thioketone	Marczenko, p 375. (3), Snell, *Metals I,* p 126. (7)
	Xylenol Orange	*Talanta, 16,* 1023, (1969)
Molybdenum	Bromopyrogallol Red + Cetylpyridium chloride	West, p 58. (10)
	Phenylfluorone	Snell, *Metals II,* p 1311. (8), *Microchem. J., 31,* 56, (1985).
	Toluene-3,4-dithiol	Fries/Getrost, p 251. (2), Marczenko, p 384. (3), Onishi, Part IIb, p 96. (6), Snell, *Metals II,* p 1301. (8)
Nickel	2-(5-Bromo-2-pyridylazo)-5-diethylaminophenol	Marczenko, p 397. (3), *Talanta 28,* 189, (1981).
	Dimethylglyoxime	Fries/Getrost, p 263. (2), Marczenko, p 393. (3), Onishi, Part IIb, p 125. (6), Snell, *Metals I,* p 887. (7)
	Dimethylglyoxime + Oxidant	Fries/Getrost, p 263. (2), Onishi, Part IIb, p 125. (6), Snell, *Metals I,* p 887. (7)
	2,2´-Furildioxime	Marczenko, p 396. (3), Snell, *Metals I,* p 904. (7)
	2-(2-Pyridylazo)-2-naphthol	Snell, *Metals I,* p 910. (7)
	4-(2-Pyridylazo)resorcinol	Snell, *Metals I,* p 911. (7), West, p 39. (10), *Anal. Chim. Acta, 82,* 431, (1976).
Niobium	*N*-Benzoyl-*N*-phenylhydroxylamine	Snell, *Metals II,* p 1425. (8)
	Pyrocatechol + EDTA or 2,2´Bipyridyl or 1-(2-thenoyl)-3,3,3,-trifluoroacetone	Snell, *Metals II,* p 1427. (8)
	Bromopyrogallol red	Marczenko, p 407. (3), Snell, *Metals II,* p 1426. (8)
	Bromopyrogallol red + Cetylpyridinium chloride	*Talanta, 32,* 189, (1985).
	4-(2-Pyridylazo)resorcinol	Fries/Getrost, p 274. (2), Marczenko, p 406. (3), Onishi, Part IIb, p 160. (7), Snell, *Metals II,* p 1447. (8)
	Sulfochlorophenol S	Onishi, Part IIb, p 161. (7), Snell, *Metals II,* p 1430. (8)
	Xylenol Orange	Onishi, Part IIb, p 164. (7)
Nitrate	Brucine	Boltz, p 227. (1), Fries/Getrost, p 280. (2), Snell, *Nonmetals,* p 546. (9)
	Chromotropic acid	Boltz, p 229. (1), Fries/Getrost, p 281. (2), Snell, *Nonmetals,* p 548. (9), Williams, p 132. (11), *Fres. Z. Anal. Chem., 320,* 490, (1985).
	Sulfanilamide + *N*-(1-Naphthyl)ethylenediamine dihydrochloride	Fries/Getrost, p 279. (2), Snell, *Nonmetals,* p 559. (9)
Nitrite	Sulfanilamide + *N*-(1-Naphthyl)ethylenediamine dihydrochlorine	Boltz, p 241. (1), Snell, *Nonmetals,* p 585. (8), *Analyst, 109,* 1281, (1984).
	Sulfanilic acid + 1-Naphthylamine	Boltz, p 237. (1), Fries/Getrost, p 285. (2), Marczenko, p 419. (3), Snell, *Nonmetals,* p 586. (9)
Osmium	1,5-Diphenylcarbazide	Marczenko, p 428. (3)
Palladium	2-(5 Bromo-2-pyridylazo)-5-diethylaminophenol	*Talanta, 33,* 939, (1986).

Determination	Reagents	Ref.
	Dithizone	Marczenko, p 440. (3), Onishi, Part IIb, p 227. (6), Snell, *Metals II,* p 1577. (8)
	2-Nitroso-1-naphthol	Fries/Getrost, p 294. (2), Onishi, Part IIb, p 226. (6), Snell, *Metals II,* p 1581. (8)
	4-(2-Pyridylazo)resorcinol	Snell, *Metals II,* p 1583. (8) *Analyst, 107,* 708, (1982).
Phosphate	Rhodamine B + Molybdate	Snell, *Nonmetals,* p 103. (9)
	Malachite Green + Molybdate	Snell, *Nonmetals,* p 12. (9), *Analyst, 108,* 361, (1983).
Platinum	Sulfochlorophenolazorhodamine	Onishi, Part IIb, p 253. (6), *Talanta, 34,* 87, (1987).
	Dithizone	Fries/Getrost, p 300. (2), Onishi, Part IIb, p 253. (6), Snell, *Metals II,* p 1534. (8)
	2-Mercaptobenzothiazole	Fries/Getrost, p 302. (2), *Zh. Anal. Khim., 24,* 1172, (1969).
Rare Earths	Arsenazo I	Marczenko, p 470. (3), Onishi, Part IIa, p 785. (5), Snell, *Metals II,* p 1857. (8)
	Arsenazo III	Fries/Getrost, p 309. (2), Marczenko, p 468. (3), Onishi, Part IIa, p 786. (5), Snell, *Metals II,* p 1862. (8)
	Xylenol Orange	Onishi, Part IIa, p 787. (5), Snell, *Metals II,* p 1874. (8)
Rhenium	2,2′-Furildioxime	Fries/Getrost, p 310. (2), Marczenko, p 481. (3), Onishi, Part IIb, p 288. (6), Snell, *Metals II,* p 1659. (8)
Rhodium	1-(2-Pyridylazo)-2-naphthol	Fries/Getrost, p 311. (2), Snell, *Metals II,* p 1553. (8)
Ruthenium	1,10-Phenanthroline	Onishi, Part IIb, p 331. (6), Snell, *Metals II,* p 1623. (8)
	Thiourea	Fries/Getrost, p 318. (2), Onishi, Part IIb, 329. (6), Snell, *Metals II,* p 1626. (8)
	1,4-Diphenylthiosemicarbazide	Marczenko, p 493. (3), Onishi, Part IIb, p 330. (8)
Scandium	Alizarin red S	Fries/Getrost, p 319. (2), Onishi, Part IIb, p 360. (6), Snell, *Metals I,* p 536. (7)
	Arsenazo III	Onishi, Part IIb, p 359. (6), Snell, *Metals I,* p 539. (7)
	Chrome Azurol S	Snell, *Metals* I, p 551. (7), *Anal. Chim. Acta, 159,* 309, (1984).
	Xylenol Orange	Marczenko, p 501. (3), Onishi, Part IIb, p 357. (6), Snell, *Metals I,* p 547. (7)
Selenium	3,3′-Diaminobenzidine	Boltz, p 391. (1), Fries/Getrost, p 323. (2), Marczenko, p 508. (3), Snell, *Nonmetals,* p 490. (9), West, p 4. (10).
	2,3-Diaminonaphthaline	Snell, *Nonmetals,* p 501. (9)
Silver	Dithizone	Fries/Getrost, p 328. (2), Marczenko, p 524. (3), Onishi, Part IIb, p 379. (6), Snell, *Metals I,* p 82. (7)
	Eosin + 1,10-Phenanthroline	Snell, *Metals I,* p 93. (7)
Sulfate	Methylthymol blue + Barium (II)	Snell, *Nonmetals,* p 457. (9)
Sulfide	*N,N,*-Dimethyl-1,4-phenylenediamine	Boltz, p 483. (1), Fries/Getrost, p 344. (2), Snell, *Nonmetals,* p 400. (9), Williams, p 578. (11)
Sulfite	Pararosaniline + Formaldehyde	Boltz, p 478. (1), Marczenko, p 540. (3), Snell, *Nonmetals,* p 430. (9), Williams, p 591. (11)
Tantalum	Methyl Violet	Marczenko, p 551. (3), Snell, *Metals II,* p 1485. (8)
	4-(2-Pyridylazo)resorcinol	Snell, *Metals II,* p 1488. (8)
	Phenylfluorone	Onishi, Part IIb, p 166. (6), Snell, *Metals II,* p 1486. (8)
Tellurium	Diethyldithiocarbamate	Boltz, p 402. (1), Fries/Getrost, p 348. (2), Snell, *Nonmetals,* p 533. (9), Williams, p 220. (10)
	Bismuthiol II	Boltz, p 401. (1), Marczenko, p 557. (3), Snell, *Nonmetals,* p 524. (9)
Thallium	Brilliant green	Fries/Getrost, p 352. (2), Marczenko, p 567. (3), Onishi, Part IIb, p 426. (6), Snell, *Metals I,* p 45. (7)
	Dithizone	Fries/Getrost, p 355. (2), Onishi, Part IIb, p 426. (6), Snell, *Metals I,* p 54. (7)
	Rhodamine B	Fries/Getrost, p 354. (2), Marczenko, p 566. (3), Onishi, Part IIb, p 424. (6), Snell, *Metals I,* p 63. (7)
Thorium	Arsenazo III	Fries/Getrost, p 360. (2), Marczenko, p 575. (3), Onishi, Part IIb, p 460. (6), Snell, *Metals II,* p 1820. (8)
	Thoron	Marczenko, p 574. (3), Onishi, Part IIb, p 463. (6), Snell, *Metals I,* p 1835. (7)
	Xylenol Orange	Snell, *Metals I,* p 1852. (7)
	Xylenol Orange + Cetyltrimethylammonium bromide	*Talanta, 26,* 499, (1979).
Tin	Pyrocatechol violet (and + Cetyltrimethylammonium bromide)	Marczenko, p 585. (3), Onishi, Part IIb, p 501. (6), Snell, *Metals I,* p 422. (7)
	Gallein	Onishi, Part IIb, p 507, 510. (6), Snell, *Metals I,* p 432. (7)
	Phenylfluorone	Fries/Getrost, p 368. (2), Marczenko, p 582. (3), Onishi, Part IIb, p 497. (6), Snell, *Metals I,* p 444. (7)

Determination	Reagents	Ref.
	Toluene-3,4-dithiol + Dispersant	Fries/Getrost, p 366. (2), Onishi, Part IIb, p 502. (6), Snell, *Metals I*, p 427. (7)
Titanium	Chromotropic acid	Marczenko, p 593. (3), Onishi, Part IIb, p 551. (6), Snell, *Metals II*, p 1080. (8)
	Diantipyrinylmethane	Onishi, Part IIb, p 545. (6), Snell, *Metals II*, 1085. (8)
	Tiron	Fries/Getrost, p 376. (2), Onishi, Part IIb, p 549. (6), Snell, *Metals II*, p 1114. (8)
Tungsten	Pyrocatechol Violet	Snell, *Metals II*, p 1265. (8)
	Tetraphenylarsonium chloride + Thiocyanate	Onishi, Part IIb, p 596. (6), Snell, *Metals II*, p 1278. (8)
	Toluene-3,5-dithiol	Marczenko, p 605. (3), Onishi, Part IIb, p 590. (6), Snell, *Metals II*, p 1267. (8)
Uranium	Arsenazo III	Marczenko, p 611. (3), Onishi, Part IIb, p 627. (6), Snell, *Metals II*, p 1356. (8)
	2-(5-Bromo-2-pyridylazo)diethylaminophenol	Fries/Getrost, p 388. (2), Onishi, Part IIb, p 625. (6)
	Chlorophosphonazo III	Snell, *Metals II*, p 1367. (8), *Fres. Z. Anal. Chem.*, *306*, 110, (1981).
	1-(2-Pyridylazo)-2-naphthol	Fries/Getrost, p 386. (2), Onishi, Part IIb, p 625. (6), Snell, *Metals II*, p 1387. (8)
Vanadium	*N*-Benzoyl-*N*-phenylhydroxylamine	Fries/Getrost, p 395. (2), Marczenko, p 625. (3), Snell, *Metals II*, p 1196. (8)
	8-Hydroxyquinoline	Marczenko, p 623. (3), Snell, *Metals II*, p 1209. (8)
	4-(2-pyridylazo)resorcinol	Fries/Getrost, p 404. (2), Marczenko, p 628. (3), Onishi, Part IIb, p 625. (6), Snell, *Metals II*, p 1226. (8)
Yttrium	Alizarin Red S	Fries/Getrost, p 406. (2), Onishi, Part IIa, p 784. (5), Snell, *Metals II*, p 1919. (8)
	Arsenazo III	Marczenko, p 468. (3), Onishi, Part IIa, p 786. (5), Snell, *Metals II*, p 1921. (8)
	Xylenol Orange	Fries/Getrost, p 406. (2), Onishi, Part IIa, p 787. (5), Snell, *Metals II*, p 1923. (8)
Zinc	Dithizone	Fries/Getrost, p 408. (2), Marczenko, p 637. (3), Onishi, Part IIb, p 708. (6), Snell, *Metals II*, p 1042. (8)
	1-(2-Pyridylazo)-2-naphthol	Marczenko, p 639. (3), Onishi, Part IIb, p 719. (6), Snell, *Metals II*, p 1056. (8)
	Xylenol Orange	Fries/Getrost, p 417. (2), Snell, *Metals II*, p 1062. (8), Talanta, *26*, 693, (1979).
	Zircon	Fries/Getrost, p 412. (2), Onishi, Part IIb, p 719. (6), Snell, *Metals II*, p 1063. (8), West, p 23. (10)
Zirconium	Alizarin Red S	Fries/Getrost, p 421. (2), Marczenko, p 647. (3), Onishi, Part IIb, p 763. (6), Snell, *Metals II*, p 1136. (8)
	Arsenazo III	Fries/Getrost, p 421. (2), Onishi, Part IIb, p 770. (6), Snell, *Metals II*, p 1143. (8)
	Pyrocatechol Violet	Onishi, Part IIb, p 771. (6), Snell, *Metals II*, p 1149. (8)
	Morin	Fries/Getrost, p 424. (2), Onishi, Part IIb, p 765. (6), Snell, *Metals II*, p 1158. (8)
	Xylenol Orange	Fries/Getrost, p 419. (2), Marczenko, p 648. (3), Onishi, Part IIb, p 767. (6), Snell, *Metals II*, p 1167. (8)

Reviews

Savvin, S. B., *Crit. Rev. Anal. Chem.*, 8, 55, 1979.

Sommer, L, Ackermann, G., Thorburn Burns, D., and Savvin, S. B., *Pure and Applied Chem.*, 62, 2147, 1990.

Sommer, L., Ackermann, G., and Thorburn Burns, D., *Pure and Applied Chem.*, 62, 2323, 1990)

Sommer, L., Komarek, J., and Thorburn Burns, D., *Pure and Applied Chem.*, *64*, 213, 1992.

Monographs

1. Boltz, D. F., and Howell, J. A., *Colorimetric Determination of Nonmetals*, 2nd ed, Wiley, New York, 1978.
2. Fries, J. and Getrost, H., *Organic Reagents for Trace Analysis*, E Merck, Darmstadt, 1977.
3. Marczenko, Z., *Separation and Spectrophotometric Determination of Elements*, Ellis Horwood, Chichester, 1986.
4. Sandell, E. B. and Onishi, H., *Photometric Determination of Traces of Metals. General Aspects, Part I*, 4th ed, J. Wiley, New York, 1978.
5. Onishi, H., *Photometric Determination of Traces of Metals. Part IIa: Individual Metals, Aluminium to Lithium*, 4th ed, J. Wiley, New York, 1986.
6. Onishi, H., *Photometric Determination of Traces of Metals. Part IIb: Individual Metals, Magnesium to Zinc*, 4th ed, J. Wiley, New York, 1989.
7. Snell, F. D., *Photometric and Fluorimetric Methods of Analysis, Metals Part 1*, J. Wiley, New York, 1978.
8. Snell, F. D., *Photometric and Fluorimetric Methods of Analysis, Metals Part 2*, J. Wiley, New York, 1978.
9. Snell, F. D., *Photometric and Fluorimetric Methods of Analysis, Nonmetals*, J. Wiley, New York, 1981.
10. West, T. S. and Nürnberg, H. W., Eds., *The Determination of Trace Metals in Natural Waters*, Blackwell, Oxford, 1988.
11. Williams, W. J., *Handbook of Anion Determination*, Butterworth, London, 1979.
12. Townshend, A., Burns, D. T., Guilbault, G. G., Lobinski, R., Marczenko, Z., Newman, E., and Onishi, H., *Dictionary of Analytical Reagents*, Chapman & Hall, London, 1993.

FLAME AND BEAD TESTS

Flame Colorations

Violet

Potassium compounds. Purple red through blue glass. Easily obscured by sodium flame. Bluish-green through green glass. Rubidium and cesium compounds impart same flame as potassium compounds.

Blues

Azure — Copper chloride. Copper bromide gives azure blue followed by green. Other copper compounds give same coloration when moistened with hydrochloric acid.

Light blue — Lead, arsenic, selenium.

Greens

Emerald — Copper compounds except the halides, and when not moistened with hydrochloric acid.

Pure green — Compounds of thallium and tellurium.

Yellowish — Barium compounds. Some molybdenum compounds. Borates, especially when treated with sulfuric acid or when burned with alcohol.

Bluish — Phosphates with sulfuric acid.

Feeble — Antimony compounds. Ammonium compounds.

Whitish — Zinc.

Reds

Carmine — Lithium compounds. Violet through blue glass. Invisible through green glass. Masked by barium flame.

Scarlet — Strontium compounds. Violet through blue glass. Yellowish through green glass. Masked by barium flame.

Yellowish — Calcium compounds. Greenish through blue glass. Green through green glass. Masked by barium flame.

Yellow

Yellow — All sodium compounds. Invisible with blue glass.

Bead Tests

Abbreviations employed: s = saturated; ss = supersaturated; ns = not saturated; h = hot; c = cold

Borax Beads

Substance	Oxidizing flame	Reducing flame
Aluminum	Colorless (h, c, ns); opaque (ss)	Colorless; opaque (s)
Antimony	Colorless; yellow or brownish (h, ss)	Gray and opaque
Barium	Colorless (ns)	
Bismuth	Colorless; yellow or brownish (h, ss)	Gray and opaque
Cadmium	Colorless	Gray and opaque
Calcium	Colorless (ns)	
Cerium	Red (h)	Colorless (h, c)
Chromium	Green (c)	Green
Cobalt	Blue (h, c)	Blue (h, c)
Copper	Green (h); blue (c)	Red (c); opaque (ss); colorless (h)
Iron	Yellow or brownish red (h, ns)	Green (ss)
Lead	Colorless; yellow or brownish (h, ss)	Gray and opaque
Magnesium	Colorless (ns)	
Manganese	Violet (h, c)	Colorless (h, c)
Molybdenum	Colorless	Yellow or brown (h)
Nickel	Brown; red (c)	Gray and opaque
Silicon	Colorless (h, c); opaque (ss)	Colorless; opaque (s)
Silver	Colorless (ns)	Gray and opaque
Strontium	Colorless (ns)	
Tin	Colorless (h, c); opaque (ss)	Colorless; opaque (s)
Titanium	Colorless	Yellow (h); violet (c)
Tungsten	Colorless	Brown
Uranium	Yellow or brownish (h, ns)	Green
Vanadium	Colorless	Green

Beads of Microcosmic Salt
$NaNH_4HPO_4$

Substance	Oxidizing flame	Reducing flame
Aluminum	Colorless; opaque (s)	Colorless; not clear (ss)
Antimony	Colorless (ns)	Gray and opaque
Barium	Colorless; opaque (s)	Colorless; not clear (ss)
Bismuth	Colorless (ns)	Gray and opaque
Cadmium	Colorless (ns)	Gray and opaque
Calcium	Colorless; opaque (s)	Colorless; not clear (ss)

Beads of Microcosmic Salt
$NaNH_4HPO_4$

Substance	Oxidizing flame	Reducing flame
Cerium	Yellow or brownish red (h, s)	Colorless
Chromium	Red (h, s); green (c)	Green (c)
Cobalt	Blue (h, c)	Blue (h, c)
Copper	Blue (c); green (h)	Red and opaque (c)
Iron	Yellow or brown (h, s)	Colorless; yellow or brownish (h)
Lead	Colorless (ns)	Gray and opaque
Magnesium	Colorless; opaque (s)	Colorless; not clear (ss)
Manganese	Violet (h, c)	Colorless
Molybdenum	Colorless; green (h)	Green (h)
Nickel	Yellow (c); red (h, s)	Yellow (c); red (h); gray and opaque
Silver		Gray and opaque
Strontium	Colorless; opaque (s)	Colorless; not clear (ss)
Tin	Colorless; opaque (s)	Colorless
Titanium	Colorless (ns)	Violet (c); yellow or brownish (h)
Uranium	Green; yellow or brownish	Green (h) (h, s)
Vanadium	Yellow	Green
Zinc	Colorless (ns)	Gray and opaque

Sodium Carbonate Bead

Substance	Oxidizing flame	Reducing flame
Manganese	Green	Colorless

ACID-BASE INDICATORS

A. K. Covington

The first part of this table lists some common acid-base indicators in alphabetical order along with the approximate pH range(s) at which a color change occurs. Following this is a table of the same indicators ordered by pH range, which includes the nature of the color change, instructions on preparation of the indicator solution, and the acid dissociation constant p*K*, when available.

The color code is:

C = colorless	A = amber	B/G = blue-green
Pk = pink	Y = yellow	V = violet
R = red	B = blue	P = purple
O = orange		

Reference

Bishop, E., Ed., *Indicators*, Pergamon, Oxford, 1972.

Indicator	pH Range
Alizarin	5.6-7.2; 11.0-12.4
Alizarin Red S	4.6-6.0
Alizarin Yellow R	10.1-12.0
Benzopurpurine 4B	2.2-4.2
4,4'-Bis(2-amino-1-naphthylazo)-2,2'-stilbenedisulfonic acid	3.0-4.0
4,4'-Bis(4-amino-1-naphthylazo)-2,2'-stilbenedisulfonic acid	8.0-9.0
Brilliant Yellow	6.6-7.8
Bromocresol Green	3.8-5.4
Bromocresol Purple	5.2-6.8
Bromophenol Blue	3.0-4.6
Bromothymol Blue	6.0-7.6
Chlorophenol Red	5.2-6.8
Clayton Yellow	12.2-13.2
Congo Red	3.0-5.0
o-Cresolphthalein	8.2-9.8
Cresol Red	0.0-1.0; 7.0-8.8
Crystal Violet	0.0-1.8
Curcumin (Turmaric)	7.4-8.6
p-(2,4-Dihydroxyphenylazo) benzenesulfonic acid, sodium salt	11.4-12.6
p-Dimethylaminoazobenzene	2.8-4.4
4-(4-Dimethylamino-1-naphylazo)-3-methoxybenzenesulfonic acid	3.5-4.8
2-(*p*-Dimethylamino-phenylazo)pyridine	0.2-1.8; 4.4-5.6
N,N-Dimethyl-*p*-(*m*-tolylazo)aniline	2.6-4.8
2,4-Dinitrophenol	2.0-4.7
2-(2,4 Dinitrophenylazo)-1-naphthol-3,6-disulfonic acid, disodium salt	6.0-7.0
6,8-Dinitro-2,4-(1*H*)quinazolinedione	6.4-8.0
Erythrosin, disodium salt	2.2-3.6
4-(*p*-Ethoxyphenylazo)-*m*-phenylene-diamine monohydrochloride	4.4-5.8
Ethyl bis(2,4-dimethylphenyl)ethanoate	8.4-9.6
Ethyl Orange	3.4-4.8
Ethyl Red	4.0-5.8
Ethyl Violet	0.0-2.4
5,5'-Indigodisulfonic acid, disodium salt	11.4-13.0
Malachite Green	0.2-1.8
Metacresol Purple	1.2-2.8; 7.4-9.0
Metanil Yellow	1.2-2.4
Methyl Green	0.2-1.8
Methyl Orange	3.2-4.4
Methyl Red	4.8-6.0
Methyl Violet	0.0-1.6
p-Naphtholbenzein	8.2-10.0
Neutral Red	6.8-8.0
p-Nitrophenol	5.4-6.6
m-Nitrophenol	6.8-8.6
Orange IV	1.4-2.8
Paramethyl Red	1.0-3.0
Phenolphthalein	8.2-10.0
Phenol Red	6.6-8.0
4-Phenylazodiphenylamine	1.2-2.6
4-Phenylazo-1-naphthylamine	4.0-5.6
Propyl Red	4.8-6.6
Quinaldine Red	1.4-3.2
Resazurin	3.8-6.4
Resorcin Blue	4.4-6.2
Tetrabromophenolphthalein ethyl ester, potassium salt	3.0-4.2
Thymol Blue	1.2-2.8; 8.0-9.6
Thymolphthalein	9.4-10.6
4-*o*-Tolylazo-*o*-toluidine	1.4-2.8
1,3,5-Trinitrobenzene	12.0-14.0
2,4,6-Trinitrotoluene	11.5-13.0
Turmaric	7.4-8.6

pH range	Color change	Indicator	pK	Preparation
0.0-1.0	R-Y	Cresol Red		0.1 g in 26.2 mL 0.01 M NaOH + 223.8 mL water
0.0-1.6	Y-B	Methyl Violet		0.01-0.05% in water
0.0-1.8	Y-B	Crystal Violet		0.02% in water
0.0-2.4	Y-B	Ethyl Violet		0.1 g in 50 mL 50% v/v methanol-water
0.2-1.8	Y-B/G	Malachite Green	1.3	water
0.2-1.8	Y-B	Methyl Green		0.1% in water
0.2-1.8	Y-R	2-(*p*-Dimethylaminophenylazo)pyridine		0.1% in ethanol
1.0-3.0	R-Y	Paramethyl Red		ethanol
1.2-2.4	R-Y	Metanil Yellow		0.01% in water
1.2-2.6	R-Y	4-Phenylazodiphenylamine		0.01 g in 1 mL 1 M HCl + 50 mL ethanol + 49 mL water
1.2-2.8	R-Y	Thymol Blue	1.65	0.1 g in 21.5 mL 0.01 M NaOH + 228.5 mL water
1.2-2.8	R-Y	Metacresol Purple	1.51	0.1 g in 26.2 mL 0.01 M NaOH + 223.8 mL water
1.4-2.8	R-Y	Orange IV		0.01% in water
1.4-2.8	O-Y	4-*o*-Tolylazo-*o*-toluidine		water
1.4-3.2	C-R	Quinaldine Red	2.63	1% in ethanol
2.0-4.7	C-Y	2,4-Dinitrophenol	3.96	sat. solution in water
2.2-3.6	O-R	Erythrosin, disodium salt		0.1% in water
2.2-4.2	V-R	Benzopurpurine 4B		0.1% in water
2.6-4.8	R-Y	*N,N*-Dimethyl-*p*-(*m*-tolylazo)aniline		0.1% in water
2.8-4.4	R-Y	*p*-Dimethylaminoazobenzene		0.1 g in 100 mL 90% v/v ethanol-water
3.0-4.0	P-R	4,4'-Bis(2-amino-1-naphthylazo)-2,2'-stilbenedisulfonic acid		0.1 g in 5.9 mL 0.05 M NaOH + 94.1 mL water
3.0-4.2	Y-B	Tetrabromophenolphthalein ethyl ester, potassium salt		0.1% in ethanol
3.0-4.6	Y-B	Bromophenol Blue	4.10	0.1 g in 14.9 mL 0.01 M NaOH + 235.1 mL water
3.0-5.0	B-R	Congo Red		0.1% in water
3.2-4.4	R-Y	Methyl Orange	3.46	0.1% in water
3.4-4.8	R-Y	Ethyl Orange	4.34	0.05-0.2% in water or aqueous ethanol
3.5-4.8	V-Y	4-(4-Dimethylamino-1-naphylazo)-3-methoxybenzenesulfonic acid		0.1% in 60% ethanol-water
3.8-5.4	Y-B	Bromocresol Green	4.90	0.1 g in 14.3 mL 0.01 M NaOH + 235.7 mL water
3.8-6.4	O-V	Resazurin		water
4.0-5.6	R-Y	4-Phenylazo-1-naphthylamine		0.1% in ethanol
4.0-5.8	C-R	Ethyl Red	5.42	0.1 g in 100 mL 50% v/v methanol-water 0.1% in ethanol
4.4-5.6	R-Y	2-(*p*-Dimethylaminophenylazo)pyridine		0.1% in ethanol
4.4-5.8	O-Y	4-(*p*-Ethoxyphenylazo)-*m*-phenylene-diamine monohydrochloride		0.1% in water
4.4-6.2	R-B	Resorcin Blue		0.2% in ethanol
4.6-6.0	Y-R	Alizarin Red S		water
4.8-6.0	R-Y	Methyl Red	5.00	0.02 g in 100 mL 60% v/v ethanol-water
4.8-6.6	R-Y	Propyl Red	5.48	ethanol
5.2-6.8	Y-P	Bromocresol Purple	6.40	0.1 g in 18.5 mL 0.01 M NaOH + 231.5 mL water
5.2-6.8	Y-R	Chlorophenol Red	6.25	0.1 g in 23.6 mL 0.01 M NaOH + 226.4 mL water
5.4-6.6	C-Y	*p*-Nitrophenol	7.15	0.1% in water
5.6-7.2	Y-R	Alizarin		0.1% in methanol
6.0-7.0	Y-B	2-(2,4-Dinitrophenylazo)-1-naphthol-3,6-disulfonic acid, disodium salt		0.1% in water
6.0-7.6	Y-B	Bromothymol Blue	7.30	0.1 g in 16 mL 0.01 M NaOH + 234 mL water
6.4-8.0	C-Y	6,8-Dinitro-2,4-(1*H*)quinazolinedione		25 g in 115 mL 1 M NaOH + 50 mL water at 100 °C
6.6-7.8	Y-R	Brilliant Yellow		1% in water
6.6-8.0	Y-R	Phenol Red	8.00	0.1 g in 28.2 mL 0.01 M NaOH + 221.8 mL water
6.8-8.0	R-A	Neutral Red		0.01 g in 100 mL 50% v/v ethanol-water
6.8-8.6	C-Y	*m*-Nitrophenol	8.28	0.3% in water
7.0-8.8	Y-R	Cresol Red	8.46	0.1 g in 26.2 mL 0.01 M NaOH + 223.8 mL water
7.4-8.6	Y-R	Turmaric (Curcumin)		ethanol
7.4-9.0	Y-P	Metacresol Purple	8.3	0.1 g in 26.2 mL 0.01 M NaOH + 223.8 mL water
8.0-9.0	B-R	4,4'-Bis(4-amino-1-naphthylazo)-2,2'-stilbenedisulfonic acid		0.1 g in 5.9 mL 0.05 M NaOH + 94.1 mL water
8.0-9.6	Y-B	Thymol Blue	9.20	0.1 g in 21.5 mL 0.01 M NaOH + 228.5 mL water
8.2-10.0	O-B	*p*-Naphtholbenzein		1% in dil. alkali
8.2-10.0	C-Pk	Phenolphthalein	9.5	0.5 g in 100 mL 50% v/v ethanol-water

pH range	Color change	Indicator	pK	Preparation
8.2-9.8	C-R	*o*-Cresolphthalein		0.04% in ethanol
8.4-9.6	C-B	Ethyl bis(2,4-dimethylphenyl)ethanoate		sat. solution in 50% acetone-ethanol
9.4-10.6	C-B	Thymolphthalein		0.04 g in 100 mL 50% v/v ethanol-water
10.1-12.0	Y-R	Alizarin Yellow R		0.01% in water
11.0-12.4	R-P	Alizarin		0.1% in methanol
11.4-12.6	Y-O	*p*-(2,4-Dihydroxyphenylazo) benzenesulfonic acid, sodium salt		0.1% in water
11.4-13.0	B-Y	5,5'-Indigodisulfonic acid, disodium salt		water
11.5-13.0	C-O	2,4,6-Trinitrotoluene		0.1-0.5% in ethanol
12.0-14.0	C-O	1,3,5-Trinitrobenzene		0.1-0.5% in ethanol
12.2-13.2	Y-A	Clayton Yellow		0.1% in water

FLUORESCENT INDICATORS

Jack DeMent

Fluorescent indicators are substances that show definite changes in fluorescence with change in pH. Some fluorescent materials are not suitable for indicators since their change in fluorescence is too gradual. Fluorescent indicators find greatest utility in the titration of opaque, highly turbid or deeply colored solutions. A long wavelength ultraviolet ("black light") lamp in a dimly lighted room provides the best environment for titrations involving fluorescent indicators, although bright daylight is sometimes sufficient to evoke a response in the bright green, yellow and orange fluorescent indicators. Titrations are carried out in non-fluorescent glassware. One should check the glassware prior to use to make certain that it does not fluoresce due to the wavelengths of light involved in the titration. The meniscus of the liquid in the burette can be followed when a few particles of an insoluble fluorescent solid are dropped onto its surface.

In this table the indicators are arranged by approximate pH range covered. In the case of some of the dyestuffs the end point may vary slightly with the source or manufacturer.

Indicator	C.I.	From pH	To pH
		pH 0 to 2	
Benzoflavine	—	0.3, yellow fl.	1.7, green fl.
3,6-Dioxyphthalimide	—	0, blue fl.	2.4, green fl.
Eosine YS	768	0,yellow colored	3.0, yellow fl.
Erythrosine	772	0, yellow colored	3.6, yellow fl.
Esculin	—	1.5, colorless	2, blue fl.
4-Ethoxyacridone	—	1.2, green fl.	3.2, blue fl.
3,6-Tetramethyldiaminooxanthone	—	1.2, green fl.	3.4, blue fl.
		pH 2 to 4	
Chromotropic acid	—	3.5, colorless	4.5, blue fl
Fluorescein	766	4, colorless	4.5, green fl.
Magdala Red	—	3.0, purple colored	4.0, fl
α-Naphthylamine	—	3.4, colorless	4.8, blue fl.
β-Naphthylamine	—	2.8, colorless	4.4, violet fl.
Phloxine	774	3.4, colorless	5.0, bright yellow fl.
Salicylic acid	—	2.5, colorless	3.5, blue fl.
		pH 4 to 6	
Acridine	788	4.9, green fl.	5.1, violet colored
Dichlorofluorescein	—	4.0, colorless	5.0, green fl.
3,6-Dioxyxanthone	—	5.4, colorless	7.6, blue-violet fl.
Erythrosine	772	4.0, colorless	4.5, yellow-green fl.
β-Methylesculetin	—	4.0, colorless	6.2, blue fl
Neville-Winther acid	—	6.0, colorless	6.5, blue fl.
Resorufin	—	4.4, yellow fl	6.4, weak orange fl.
Quininic acid	—	4.4, yellow colored	5.0, blue fl.
Quinine [first end point]	—	5.0, blue fl.	6.1, violet fl.
		pH 6 to 8	
Acid R Phosphine	—	(claimed for range pH 6.0–7.0)	
Brilliant Diazol Yellow	—	6.5, colorless	7.5, violet, fl.
Cleves acid	—	6.5, colorless	7.5, green fl.
Coumaric acid	—	7.2, colorless	9.0, green fl.
3,6-Dioxyphthalic dinitrile	—	5.8, blue fl.	8.2, green fl.
Magnesium 8-hydrozyquinolinate	—	6.5, colorless	7.5, golden fl.
β-Methylumbelliferone	—	7.0, colorless	7.5, blue fl.
1-Naphthol-4-sulfonic acid	—	6.0, colorless	6.5, blue fl.
Orcinaurine	—	6.5, colorless	8.0, green fl.
Patent Phosphine	789	(for the range pH 6.0–7.0, green-yellow fl.)	
Thioflavine	816	(for the region pH 6.5–7.0, yellow fl.)	
Umbelliferone	—	6.5, colorless	7.6, blue fl.
		pH 8 to 10	
Acridine Orange	788	8.4, orange colored	10.4, green fl.
Ethoxyphenylnaphthostilbazonium chloride	—	9, green fl.	11, non-fl.

G Salt	—	9.0, dull blue fl.	9.5, bright blue fl.
Naphthazol derivatives	—	8.2, colorless	10.0, yellow or green fl.
α-Naphthionic acid	—	9, blue fl.	11, green fl.
2-Naphthol-3,6-disulfonic acid	—	9.5, dark blue fl	Light blue fl. at higher pH
β-Naphthol	—	8.6, colorless	Blue fl. at higher pH
α-Naphtholsulfonic acid	—	8.0, dark blue fl.	9.0, bright violet fl.
1,4-Naphtholsulfonic acid	—	8.2, dark blue fl.	Light blue fl. at higher pH
Orcinsulfonphthalein	—	8.6, yellow colored	10.0 fl.
Quinine [second end point]	—	9.5, violet fl.	10.0, colorless
R-Salt	—	9.0, dull blue fl.	9.5, bright blue fl.
Sodium 1-naphthol-2-sulfonate	—	9.0, dark blue fl.	10.0, bright violet fl.
		pH 10 to 12	
Courmarin	—	9.8, deep green fl.	12, light green fl.
Eosine BN	771	10.5, colorless	14.0, yellow fl.
Papaverine (permanganate oxidized)	—	9.5, yellow fl.	11.0, blue fl.
Schaffers Salt	—	5.0, violet fl.	11.0, green-blue fl.
SS-Acid (sodium salt)	—	10.0, violet fl.	12.0, yellow-colored
		pH 12 to 14	
Cotarnine	—	12.0, yellow fl.	13.0, white fl.
α-Naphthionic acid	—	12, blue fl.	13, green fl.
β-Naphthionic acid	—	12, blue fl.	13, violet fl.

CONVERSION FORMULAS FOR CONCENTRATION OF SOLUTIONS

A = Weight percent of solute
B = Molecular weight of solvent
E = Molecular weight of solute
F = Grams of solute per liter of solution
G = Molality
M = Molarity
N = Mole fraction
R = Density of solution in grams per milliliter

Concentration of solute—SOUGHT	Concentration of solute—GIVEN				
	A	N	G	M	F
A	—	$\frac{100N \times E}{N \times E + (1-N)B}$	$\frac{100G \times E}{1000 + G \times E}$	$\frac{M \times E}{10R}$	$\frac{F}{10R}$
N	$\frac{\frac{A}{E}}{\frac{A}{E} + \frac{100-A}{B}}$	—	$\frac{B \times G}{B \times G + 1000}$	$\frac{B \times M}{M(B-E) + 1000R}$	$\frac{B \times F}{F(B-E) + 1000R \times E}$
G	$\frac{1000A}{E(100-A)}$	$\frac{1000N}{B - N \times B}$	—	$\frac{1000M}{1000R - (M \times E)}$	$\frac{1000F}{E(1000R - F)}$
M	$\frac{10R \times A}{E}$	$\frac{1000R \times N}{N \times E + (1-N)B}$	$\frac{1000R \times G}{1000 + E \times G}$	—	$\frac{F}{E}$
F	$10AR$	$\frac{1000R \times N \times E}{N \times E + (1-N)B}$	$\frac{1000R \times G \times E}{1000 + G \times E}$	$M \times E$	—

ELECTROCHEMICAL SERIES

Petr Vanýsek

There are three tables for this electrochemical series. Each table lists standard reduction potentials, E° values, at 298.15 K (25 °C), and at a pressure of 101.325 kPa (1 atm). Table 1 is an alphabetical listing of the elements, according to the symbol of the elements. Thus, data for silver (Ag) precede those for aluminum (Al). Table 2 lists only those reduction reactions that have E° values positive in respect to the standard hydrogen electrode. In Table 2, the reactions are listed in the order of increasing positive potential, and they range from 0.0000 V to + 3.4 V. Table 3 lists only those reduction potentials which have E° negative with respect to the standard hydrogen electrode. In Table 3, the reactions are listed in the order of decreasing potential and range from 0.0000 V to −4.10 V. The reliability of the potentials is not the same for all the data. Typically, the values with fewer significant figures have lower reliability. The values of reduction potentials, in particular those of less common reactions, are not definite; they are subject to occasional revisions.

Abbreviations: ac = acetate; bipy = 2,2′-dipyridine, or bipyridine; en = ethylenediamine; phen = 1,10-phenanthroline.

References

1. Milazzo, G., Caroli, S., and Sharma, V. K. *Tables of Standard Electrode Potentials*, Wiley, Chichester, 1978.
2. Bard, A. J., Parsons, R., and Jordan, J. *Standard Potentials in Aqueous Solutions*, Marcel Dekker, New York, 1985.
3. Bratsch, S. G. *J. Phys. Chem. Ref. Data*, 18, 1–21, 1989.

TABLE 1. Alphabetical Listing

Reaction	E°/V
$Ac^{3+} + 3\,e \rightleftharpoons Ac$	−2.20
$Ag^{+} + e \rightleftharpoons Ag$	0.7996
$Ag^{2+} + e \rightleftharpoons Ag^{+}$	1.980
$Ag(ac) + e \rightleftharpoons Ag + (ac)^{-}$	0.643
$AgBr + e \rightleftharpoons Ag + Br^{-}$	0.07133
$AgBrO_3 + e \rightleftharpoons Ag + BrO_3^{-}$	0.546
$Ag_2C_2O_4 + 2\,e \rightleftharpoons 2\,Ag + C_2O_4^{2-}$	0.4647
$AgCl + e \rightleftharpoons Ag + Cl^{-}$	0.22233
$AgCN + e \rightleftharpoons Ag + CN^{-}$	−0.017
$Ag_2CO_3 + 2\,e \rightleftharpoons 2\,Ag + CO_3^{2-}$	0.47
$Ag_2CrO_4 + 2\,e \rightleftharpoons 2\,Ag + CrO_4^{2-}$	0.4470
$AgF + e \rightleftharpoons Ag + F^{-}$	0.779
$Ag_4[Fe(CN)_6] + 4\,e \rightleftharpoons 4\,Ag + [Fe(CN)_6]^{4-}$	0.1478
$AgI + e \rightleftharpoons Ag + I^{-}$	−0.15224
$AgIO_3 + e \rightleftharpoons Ag + IO_3^{-}$	0.354
$Ag_2MoO_4 + 2\,e \rightleftharpoons 2\,Ag + MoO_4^{2-}$	0.4573
$AgNO_2 + e \rightleftharpoons Ag + 2\,NO_2^{-}$	0.564
$Ag_2O + H_2O + 2\,e \rightleftharpoons 2\,Ag + 2\,OH^{-}$	0.342
$Ag_2O_3 + H_2O + 2\,e \rightleftharpoons 2\,AgO + 2\,OH^{-}$	0.739
$Ag^{3+} + 2\,e \rightleftharpoons Ag^{+}$	1.9
$Ag^{3+} + e \rightleftharpoons Ag^{2+}$	1.8
$Ag_2O_2 + 4\,H^{+} + e \rightleftharpoons 2\,Ag + 2\,H_2O$	1.802
$2\,AgO + H_2O + 2\,e \rightleftharpoons Ag_2O + 2\,OH^{-}$	0.607
$AgOCN + e \rightleftharpoons Ag + OCN^{-}$	0.41
$Ag_2S + 2\,e \rightleftharpoons 2\,Ag + S^{2-}$	−0.691
$Ag_2S + 2\,H^{+} + 2\,e \rightleftharpoons 2\,Ag + H_2S$	−0.0366
$AgSCN + e \rightleftharpoons Ag + SCN^{-}$	0.08951
$Ag_2SeO_3 + 2\,e \rightleftharpoons 2\,Ag + SeO_4^{2-}$	0.3629
$Ag_2SO_4 + 2\,e \rightleftharpoons 2\,Ag + SO_4^{2-}$	0.654
$Ag_2WO_4 + 2\,e \rightleftharpoons 2\,Ag + WO_4^{2-}$	0.4660
$Al^{3+} + 3\,e \rightleftharpoons Al$	−1.662
$Al(OH)_3 + 3\,e \rightleftharpoons Al + 3\,OH^{-}$	−2.31
$Al(OH)_4^{-} + 3\,e \rightleftharpoons Al + 4\,OH^{-}$	−2.328
$H_2AlO_3^{-} + H_2O + 3\,e \rightleftharpoons Al + 4\,OH^{-}$	−2.33
$AlF_6^{3-} + 3\,e \rightleftharpoons Al + 6\,F^{-}$	−2.069
$Am^{4+} + e \rightleftharpoons Am^{3+}$	2.60
$Am^{2+} + 2\,e \rightleftharpoons Am$	−1.9
$Am^{3+} + 3\,e \rightleftharpoons Am$	−2.048
$Am^{3+} + e \rightleftharpoons Am^{2+}$	−2.3
$As + 3\,H^{+} + 3\,e \rightleftharpoons AsH_3$	−0.608
$As_2O_3 + 6\,H^{+} + 6\,e \rightleftharpoons 2\,As + 3\,H_2O$	0.234
$HAsO_2 + 3\,H^{+} + 3\,e \rightleftharpoons As + 2\,H_2O$	0.248
$AsO_2^{-} + 2\,H_2O + 3\,e \rightleftharpoons As + 4\,OH^{-}$	−0.68
$H_3AsO_4 + 2\,H^{+} + 2\,e^{-} \rightleftharpoons HAsO_2 + 2\,H_2O$	0.560
$AsO_4^{3-} + 2\,H_2O + 2\,e \rightleftharpoons AsO_2^{-} + 4\,OH^{-}$	−0.71
$At_2 + 2\,e \rightleftharpoons 2\,At^{-}$	0.3
$Au^{+} + e \rightleftharpoons Au$	1.692
$Au^{3+} + 2\,e \rightleftharpoons Au^{+}$	1.401
$Au^{3+} + 3\,e \rightleftharpoons Au$	1.498
$Au^{2+} + e^{-} \rightleftharpoons Au^{+}$	1.8
$AuOH^{2+} + H^{+} + 2\,e \rightleftharpoons Au^{+} + H_2O$	1.32
$AuBr_2^{-} + e \rightleftharpoons Au + 2\,Br^{-}$	0.959
$AuBr_4^{-} + 3\,e \rightleftharpoons Au + 4\,Br^{-}$	0.854
$AuCl_4^{-} + 3\,e \rightleftharpoons Au + 4\,Cl^{-}$	1.002
$Au(OH)_3 + 3\,H^{+} + 3\,e \rightleftharpoons Au + 3\,H_2O$	1.45
$H_2BO_3^{-} + 5\,H_2O + 8\,e \rightleftharpoons BH_4^{-} + 8\,OH^{-}$	−1.24
$H_2BO_3^{-} + H_2O + 3\,e \rightleftharpoons B + 4\,OH^{-}$	−1.79
$H_3BO_3 + 3\,H^{+} + 3\,e \rightleftharpoons B + 3\,H_2O$	−0.8698
$B(OH)_3 + 7\,H^{+} + 8\,e \rightleftharpoons BH_4^{-} + 3\,H_2O$	−0.481
$Ba^{2+} + 2\,e \rightleftharpoons Ba$	−2.912
$Ba^{2+} + 2\,e \rightleftharpoons Ba(Hg)$	−1.570
$Ba(OH)_2 + 2\,e \rightleftharpoons Ba + 2\,OH^{-}$	−2.99
$Be^{2+} + 2\,e \rightleftharpoons Be$	−1.847
$Be_2O_3^{2-} + 3\,H_2O + 4\,e \rightleftharpoons 2\,Be + 6\,OH^{-}$	−2.63
p–benzoquinone + $2\,H^{+} + 2\,e \rightleftharpoons$ hydroquinone	0.6992
$Bi^{+} + e \rightleftharpoons Bi$	0.5
$Bi^{3+} + 3\,e \rightleftharpoons Bi$	0.308
$Bi^{3+} + 2\,e \rightleftharpoons Bi^{+}$	0.2
$Bi + 3\,H^{+} + 3\,e \rightleftharpoons BiH_3$	−0.8
$BiCl_4^{-} + 3\,e \rightleftharpoons Bi + 4\,Cl^{-}$	0.16
$Bi_2O_3 + 3\,H_2O + 6\,e \rightleftharpoons 2\,Bi + 6\,OH^{-}$	−0.46
$Bi_2O_4 + 4\,H^{+} + 2\,e \rightleftharpoons 2\,BiO^{+} + 2\,H_2O$	1.593
$BiO^{+} + 2\,H^{+} + 3\,e \rightleftharpoons Bi + H_2O$	0.320
$BiOCl + 2\,H^{+} + 3\,e \rightleftharpoons Bi + Cl^{-} + H_2O$	0.1583
$Bk^{4+} + e \rightleftharpoons Bk^{3+}$	1.67
$Bk^{2+} + 2\,e \rightleftharpoons Bk$	−1.6
$Bk^{3+} + e \rightleftharpoons Bk^{2+}$	−2.8

Reaction	$E°$/V
$Br_2(aq) + 2\,e \rightleftharpoons 2\,Br^-$	1.0873
$Br_2(l) + 2\,e \rightleftharpoons 2\,Br^-$	1.066
$HBrO + H^+ + 2\,e \rightleftharpoons Br^- + H_2O$	1.331
$HBrO + H^+ + e \rightleftharpoons 1/2\,Br_2(aq) + H_2O$	1.574
$HBrO + H^+ + e \rightleftharpoons 1/2\,Br_2(l) + H_2O$	1.596
$BrO^- + H_2O + 2\,e \rightleftharpoons Br^- + 2\,OH^-$	0.761
$BrO_3^- + 6\,H^+ + 5\,e \rightleftharpoons 1/2\,Br_2 + 3\,H_2O$	1.482
$BrO_3^- + 6\,H^+ + 6\,e \rightleftharpoons Br^- + 3\,H_2O$	1.423
$BrO_3^- + 3\,H_2O + 6\,e \rightleftharpoons Br^- + 6\,OH^-$	0.61
$(CN)_2 + 2\,H^+ + 2\,e \rightleftharpoons 2\,HCN$	0.373
$2\,HCNO + 2\,H^+ + 2\,e \rightleftharpoons (CN)_2 + 2\,H_2O$	0.330
$(CNS)_2 + 2\,e \rightleftharpoons 2\,CNS^-$	0.77
$CO_2 + 2\,H^+ + 2\,e \rightleftharpoons HCOOH$	–0.199
$Ca^+ + e \rightleftharpoons Ca$	–3.80
$Ca^{2+} + 2\,e \rightleftharpoons Ca$	–2.868
$Ca(OH)_2 + 2\,e \rightleftharpoons Ca + 2\,OH^-$	–3.02
Calomel electrode, 1 molal KCl	0.2800
Calomel electrode, 1 molar KCl (NCE)	0.2801
Calomel electrode, 0.1 molar KCl	0.3337
Calomel electrode, saturated KCl (SCE)	0.2412
Calomel electrode, saturated NaCl (SSCE)	0.2360
$Cd^{2+} + 2\,e \rightleftharpoons Cd$	–0.4030
$Cd^{2+} + 2\,e \rightleftharpoons Cd(Hg)$	–0.3521
$Cd(OH)_2 + 2\,e \rightleftharpoons Cd(Hg) + 2\,OH^-$	–0.809
$CdSO_4 + 2\,e \rightleftharpoons Cd + SO_4^{2-}$	–0.246
$Cd(OH)_4^{2-} + 2\,e \rightleftharpoons Cd + 4\,OH^-$	–0.658
$CdO + H_2O + 2\,e \rightleftharpoons Cd + 2\,OH^-$	0.783
$Ce^{3+} + 3\,e \rightleftharpoons Ce$	–2.336
$Ce^{3+} + 3\,e \rightleftharpoons Ce(Hg)$	–1.4373
$Ce^{4+} + e \rightleftharpoons Ce^{3+}$	1.72
$CeOH^{3+} + H^+ + e \rightleftharpoons Ce^{3+} + H_2O$	1.715
$Cf^{4+} + e \rightleftharpoons Cf^{3+}$	3.3
$Cf^{3+} + e \rightleftharpoons Cf^{2+}$	–1.6
$Cf^{3+} + 3\,e \rightleftharpoons Cf$	–1.94
$Cf^{2+} + 2\,e \rightleftharpoons Cf$	–2.12
$Cl_2(g) + 2\,e \rightleftharpoons 2\,Cl^-$	1.35827
$HClO + H^+ + e \rightleftharpoons 1/2\,Cl_2 + H_2O$	1.611
$HClO + H^+ + 2\,e \rightleftharpoons Cl^- + H_2O$	1.482
$ClO^- + H_2O + 2\,e \rightleftharpoons Cl^- + 2\,OH^-$	0.81
$ClO_2 + H^+ + e \rightleftharpoons HClO_2$	1.277
$HClO_2 + 2\,H^+ + 2\,e \rightleftharpoons HClO + H_2O$	1.645
$HClO_2 + 3\,H^+ + 3\,e \rightleftharpoons 1/2\,Cl_2 + 2\,H_2O$	1.628
$HClO_2 + 3\,H^+ + 4\,e \rightleftharpoons Cl^- + 2\,H_2O$	1.570
$ClO_2^- + H_2O + 2\,e \rightleftharpoons ClO^- + 2\,OH^-$	0.66
$ClO_2^- + 2\,H_2O + 4\,e \rightleftharpoons Cl^- + 4\,OH^-$	0.76
$ClO_2(aq) + e \rightleftharpoons ClO_2^-$	0.954
$ClO_3^- + 2\,H^+ + e \rightleftharpoons ClO_2 + H_2O$	1.152
$ClO_3^- + 3\,H^+ + 2\,e \rightleftharpoons HClO_2 + H_2O$	1.214
$ClO_3^- + 6\,H^+ + 5\,e \rightleftharpoons 1/2\,Cl_2 + 3\,H_2O$	1.47
$ClO_3^- + 6\,H^+ + 6\,e \rightleftharpoons Cl^- + 3\,H_2O$	1.451
$ClO_3^- + H_2O + 2\,e \rightleftharpoons ClO_2^- + 2\,OH^-$	0.33
$ClO_3^- + 3\,H_2O + 6\,e \rightleftharpoons Cl^- + 6\,OH^-$	0.62
$ClO_4^- + 2\,H^+ + 2\,e \rightleftharpoons ClO_3^-\,H_2O$	1.189
$ClO_4^- + 8\,H^+ + 7\,e \rightleftharpoons 1/2\,Cl_2 + 4\,H_2O$	1.39
$ClO_4^- + 8\,H^+ + 8\,e \rightleftharpoons Cl^- + 4\,H_2O$	1.389
$ClO_4^- + H_2O + 2\,e \rightleftharpoons ClO_3^- + 2\,OH^-$	0.36
$Cm^{4+} + e \rightleftharpoons Cm^{3+}$	3.0
$Cm^{3+} + 3\,e \rightleftharpoons Cm$	–2.04
$Co^{2+} + 2\,e \rightleftharpoons Co$	–0.28
$Co^{3+} + e \rightleftharpoons Co^{2+}$	1.92
$[Co(NH_3)_6]^{3+} + e \rightleftharpoons [Co(NH_3)_6]^{2+}$	0.108
$Co(OH)_2 + 2\,e \rightleftharpoons Co + 2\,OH^-$	–0.73
$Co(OH)_3 + e \rightleftharpoons Co(OH)_2 + OH^-$	0.17
$Cr^{2+} + 2\,e \rightleftharpoons Cr$	–0.913
$Cr^{3+} + e \rightleftharpoons Cr^{2+}$	–0.407
$Cr^{3+} + 3\,e \rightleftharpoons Cr$	–0.744
$Cr_2O_7^{2-} + 14\,H^+ + 6\,e \rightleftharpoons 2\,Cr^{3+} + 7\,H_2O$	1.36
$CrO_2^- + 2\,H_2O + 3\,e \rightleftharpoons Cr + 4\,OH^-$	–1.2
$HCrO_4^- + 7\,H^+ + 3\,e \rightleftharpoons Cr^{3+} + 4\,H_2O$	1.350
$CrO_2 + 4\,H^+ + e \rightleftharpoons Cr^{3+} + 2H_2O$	1.48
$Cr(V) + e \rightleftharpoons Cr(IV)$	1.34
$CrO_4^{2-} + 4\,H_2O + 3\,e \rightleftharpoons Cr(OH)_3 + 5\,OH^-$	–0.13
$Cr(OH)_3 + 3\,e \rightleftharpoons Cr + 3\,OH^-$	–1.48
$Cs^+ + e \rightleftharpoons Cs$	–3.026
$Cu^+ + e \rightleftharpoons Cu$	0.521
$Cu^{2+} + e \rightleftharpoons Cu^+$	0.153
$Cu^{2+} + 2\,e \rightleftharpoons Cu$	0.3419
$Cu^{2+} + 2\,e \rightleftharpoons Cu(Hg)$	0.345
$Cu^{3+} + e \rightleftharpoons Cu^{2+}$	2.4
$Cu_2O_3 + 6\,H^+ + 2e \rightleftharpoons 2Cu^{2+} + 3\,H_2O$	2.0
$Cu^{2+} + 2\,CN^- + e \rightleftharpoons [Cu(CN)_2]^-$	1.103
$CuI_2^- + e \rightleftharpoons Cu + 2\,I^-$	0.00
$Cu_2O + H_2O + 2\,e \rightleftharpoons 2\,Cu + 2\,OH^-$	–0.360
$Cu(OH)_2 + 2\,e \rightleftharpoons Cu + 2\,OH^-$	–0.222
$2\,Cu(OH)_2 + 2\,e \rightleftharpoons Cu_2O + 2\,OH^- + H_2O$	–0.080
$2\,D^+ + 2\,e \rightleftharpoons D_2$	–0.013
$Dy^{2+} + 2\,e \rightleftharpoons Dy$	2.2
$Dy^{3+} + 3\,e \rightleftharpoons Dy$	–2.295
$Dy^{3+} + e \rightleftharpoons Dy^{2+}$	–2.6
$Er^{2+} + 2\,e \rightleftharpoons Er$	–2.0
$Er^{3+} + 3\,e \rightleftharpoons Er$	–2.331
$Er^{3+} + e \rightleftharpoons Er^{2+}$	–3.0
$Es^{3+} + e \rightleftharpoons Es^{2+}$	–1.3
$Es^{3+} + 3\,e \rightleftharpoons Es$	–1.91
$Es^{2+} + 2\,e \rightleftharpoons Es$	–2.23
$Eu^{2+} + 2\,e \rightleftharpoons Eu$	–2.812
$Eu^{3+} + 3\,e \rightleftharpoons Eu$	–1.991
$Eu^{3+} + e \rightleftharpoons Eu^{2+}$	–0.36
$F_2 + 2\,H^+ + 2\,e \rightleftharpoons 2\,HF$	3.053
$F_2 + 2\,e \rightleftharpoons 2\,F^-$	2.866
$F_2O + 2\,H^+ + 4\,e \rightleftharpoons H_2O + 2\,F^-$	2.153
$Fe^{2+} + 2\,e \rightleftharpoons Fe$	–0.447
$Fe^{3+} + 3\,e \rightleftharpoons Fe$	–0.037
$Fe^{3+} + e \rightleftharpoons Fe^{2+}$	0.771
$2\,HFeO_4^- + 8\,H^+ + 6\,e \rightleftharpoons Fe_2O_3 + 5\,H_2O$	2.09
$HFeO_4^- + 4\,H^+ + 3\,e \rightleftharpoons FeOOH + 2\,H_2O$	2.08
$HFeO_4^- + 7\,H^+ + 3\,e \rightleftharpoons Fe^{3+} + 4\,H_2O$	2.07
$Fe_2O_3 + 4\,H^+ + 2\,e \rightleftharpoons 2\,FeOH^+ + H_2O$	0.16
$[Fe(CN)_6]^{3-} + e \rightleftharpoons [Fe(CN)_6]^{4-}$	0.358
$FeO_4^{2-} + 8\,H^+ + 3\,e\quad Fe^{3+} + 4\,H_2O$	2.20
$[Fe(bipy)_2]^{3+} + e \rightleftharpoons Fe(bipy)_2]^{2+}$	0.78
$[Fe(bipy)_3]^{3+} + e \rightleftharpoons Fe(bipy)_3]^{2+}$	1.03
$Fe(OH)_3 + e \rightleftharpoons Fe(OH)_2 + OH^-$	–0.56
$[Fe(phen)_3]^{3+} + e \rightleftharpoons [Fe(phen)_3]^{2+}$	1.147
$[Fe(phen)_3]^{3+} + e \rightleftharpoons [Fe(phen)_3]^{2+}$ (1 molar H_2SO_4)	1.06
$[Ferricinium]^+ + e \rightleftharpoons$ ferrocene	0.400
$Fm^{3+} + e \rightleftharpoons Fm^{2+}$	–1.1
$Fm^{3+} + 3\,e \rightleftharpoons Fm$	–1.89
$Fm^{2+} + 2\,e \rightleftharpoons Fm$	–2.30

Reaction	E°/V
$Fr^+ + e \rightleftharpoons Fr$	–2.9
$Ga^{3+} + 3 e \rightleftharpoons Ga$	–0.549
$Ga^+ + e \rightleftharpoons Ga$	–0.2
$GaOH^{2+} + H^+ + 3 e \rightleftharpoons Ga + H_2O$	–0.498
$H_2GaO^-_3 + H_2O + 3 e \rightleftharpoons Ga + 4 OH^-$	–1.219
$Gd^{3+} + 3 e \rightleftharpoons Gd$	–2.279
$Ge^{2+} + 2 e \rightleftharpoons Ge$	0.24
$Ge^{4+} + 4 e \rightleftharpoons Ge$	0.124
$Ge^{4+} + 2 e \rightleftharpoons Ge^{2+}$	0.00
$GeO_2 + 2 H^+ + 2 e \rightleftharpoons GeO + H_2O$	–0.118
$H_2GeO_3 + 4 H^+ + 4 e \rightleftharpoons Ge + 3 H_2O$	–0.182
$2 H^+ + 2 e \rightleftharpoons H_2$	0.00000
$H_2 + 2 e \rightleftharpoons 2 H^-$	–2.23
$HO_2 + H^+ + e \rightleftharpoons H_2O_2$	1.495
$2 H_2O + 2 e \rightleftharpoons H_2 + 2 OH^-$	–0.8277
$H_2O_2 + 2 H^+ + 2 e \rightleftharpoons 2 H_2O$	1.776
$Hf^{4+} + 4 e \rightleftharpoons Hf$	–1.55
$HfO^{2+} + 2 H^+ + 4 e \rightleftharpoons Hf + H_2O$	–1.724
$HfO_2 + 4 H^+ + 4 e \rightleftharpoons Hf + 2 H_2O$	–1.505
$HfO(OH)_2 + H_2O + 4 e \rightleftharpoons Hf + 4 OH^-$	–2.50
$Hg^{2+} + 2 e \rightleftharpoons Hg$	0.851
$2 Hg^{2+} + 2 e \rightleftharpoons Hg_2^{2+}$	0.920
$Hg_2^{2+} + 2 e \rightleftharpoons 2 Hg$	0.7973
$Hg_2(ac)_2 + 2 e \rightleftharpoons 2 Hg + 2(ac)^-$	0.51163
$Hg_2Br_2 + 2 e \rightleftharpoons 2 Hg + 2 Br^-$	0.13923
$Hg_2Cl_2 + 2 e \rightleftharpoons 2 Hg + 2 Cl^-$	0.26808
$Hg_2HPO_4 + 2 e \rightleftharpoons 2 Hg + HPO_4^{2-}$	0.6359
$Hg_2I_2 + 2 e \rightleftharpoons 2 Hg + 2 I^-$	–0.0405
$Hg_2O + H_2O + 2 e \rightleftharpoons 2 Hg + 2 OH^-$	0.123
$HgO + H_2O + 2 e \rightleftharpoons Hg + 2 OH^-$	0.0977
$Hg(OH)_2 + 2 H^+ + 2 e \rightleftharpoons Hg + 2 H_2O$	1.034
$Hg_2SO_4 + 2 e \rightleftharpoons 2 Hg + SO_4^{2-}$	0.6125
$Ho^{2+} + 2 e \rightleftharpoons Ho$	–2.1
$Ho^{3+} + 3 e \rightleftharpoons Ho$	–2.33
$Ho^{3+} + e \rightleftharpoons Ho^{2+}$	–2.8
$I_2 + 2 e \rightleftharpoons 2 I^-$	0.5355
$I_3^- + 2 e \rightleftharpoons 3 I^-$	0.536
$H_3IO_6^{2-} + 2 e \rightleftharpoons IO^-_3 + 3 OH^-$	0.7
$H_5IO_6 + H^+ + 2 e \rightleftharpoons IO_3^- + 3 H_2O$	1.601
$2 HIO + 2 H^+ + 2 e \rightleftharpoons I_2 + 2 H_2O$	1.439
$HIO + H^+ + 2 e \rightleftharpoons I^- + H_2O$	0.987
$IO^- + H_2O + 2 e \rightleftharpoons I^- + 2 OH^-$	0.485
$2 IO_3^- + 12 H^+ + 10 e \rightleftharpoons I_2 + 6 H_2O$	1.195
$IO_3^- + 6 H^+ + 6 e \rightleftharpoons I^- + 3 H_2O$	1.085
$IO_3^- + 2 H_2O + 4 e \rightleftharpoons IO^- + 4 OH^-$	0.15
$IO_3^- + 3 H_2O + 6 e \rightleftharpoons IO^- + 6 OH^-$	0.26
$In^+ + e \rightleftharpoons In$	–0.14
$In^{2+} + e \rightleftharpoons In^+$	–0.40
$In^{3+} + e \rightleftharpoons In^{2+}$	–0.49
$In^{3+} + 2 e \rightleftharpoons In^+$	–0.443
$In^{3+} + 3 e \rightleftharpoons In$	–0.3382
$In(OH)_3 + 3 e \rightleftharpoons In + 3 OH^-$	–0.99
$In(OH)_4^- + 3 e \rightleftharpoons In + 4 OH^-$	–1.007
$In_2O_3 + 3 H_2O + 6 e \rightleftharpoons 2 In + 6 OH^-$	–1.034
$Ir^{3+} + 3 e \rightleftharpoons Ir$	1.156
$[IrCl_6]^{2-} + e \rightleftharpoons [IrCl_6]^{3-}$	0.8665
$[IrCl_6]^{3-} + 3 e \rightleftharpoons Ir + 6 Cl^-$	0.77
$Ir_2O_3 + 3 H_2O + 6 e \rightleftharpoons 2 Ir + 6 OH^-$	0.098
$K^+ + e \rightleftharpoons K$	–2.931
$La^{3+} + 3 e \rightleftharpoons La$	–2.379
$La(OH)_3 + 3 e \rightleftharpoons La + 3 OH^-$	–2.90
$Li^+ + e \rightleftharpoons Li$	–3.0401
$Lr^{3+} + 3 e \rightleftharpoons Lr$	–1.96
$Lu^{3+} + 3 e \rightleftharpoons Lu$	–2.28
$Md^{3+} + e \rightleftharpoons Md^{2+}$	–0.1
$Md^{3+} + 3 e \rightleftharpoons Md$	–1.65
$Md^{2+} + 2 e \rightleftharpoons Md$	–2.40
$Mg^+ + e \rightleftharpoons Mg$	–2.70
$Mg^{2+} + 2 e \rightleftharpoons Mg$	–2.372
$Mg(OH)_2 + 2 e \rightleftharpoons Mg + 2 OH^-$	–2.690
$Mn^{2+} + 2 e \rightleftharpoons Mn$	–1.185
$Mn^{3+} + e \rightleftharpoons Mn^{2+}$	1.5415
$MnO_2 + 4 H^+ + 2 e \rightleftharpoons Mn^{2+} + 2 H_2O$	1.224
$MnO_4^- + e \rightleftharpoons MnO_4^{2-}$	0.558
$MnO_4^- + 4 H^+ + 3 e \rightleftharpoons MnO_2 + 2 H_2O$	1.679
$MnO_4^- + 8 H^+ + 5 e \rightleftharpoons Mn^{2+} + 4 H_2O$	1.507
$MnO_4^- + 2 H_2O + 3 e \rightleftharpoons MnO_2 + 4 OH^-$	0.595
$MnO_4^{2-} + 2 H_2O + 2 e \rightleftharpoons MnO_2 + 4 OH^-$	0.60
$Mn(OH)_2 + 2 e \rightleftharpoons Mn + 2 OH^-$	–1.56
$Mn(OH)_3 + e \rightleftharpoons Mn(OH)_2 + OH^-$	0.15
$Mn_2O_3 + 6 H^+ + e \rightleftharpoons 2 Mn^{2+} + 3 H_2O$	1.485
$Mo^{3+} + 3 e \rightleftharpoons Mo$	–0.200
$MoO_2 + 4 H^+ + 4 e \rightleftharpoons Mo + 4 H_2O$	–0.152
$H_3Mo_7O_{24}^{3-} + 45 H^+ + 42 e \rightleftharpoons 7 Mo + 24 H_2O$	0.082
$MoO_3 + 6 H^+ + 6 e \rightleftharpoons Mo + 3 H_2O$	0.075
$N_2 + 2 H_2O + 6 H^+ + 6 e \rightleftharpoons 2 NH_4OH$	0.092
$3 N_2 + 2 H^+ + 2 e \rightleftharpoons 2 HN_3$	–3.09
$N_5^+ + 3 H^+ + 2 e \rightleftharpoons 2 NH_4^+$	1.275
$N_2O + 2 H^+ + 2 e \rightleftharpoons N_2 + H_2O$	1.766
$H_2N_2O_2 + 2 H^+ + 2 e \rightleftharpoons N_2 + 2 H_2O$	2.65
$N_2O_4 + 2 e \rightleftharpoons 2 NO_2^-$	0.867
$N_2O_4 + 2 H^+ + 2 e \rightleftharpoons 2 NHO_2$	1.065
$N_2O_4 + 4 H^+ + 4 e \rightleftharpoons 2 NO + 2 H_2O$	1.035
$2 NH_3OH^+ + H^+ + 2 e \rightleftharpoons N_2H_5^+ + 2 H_2O$	1.42
$2 NO + 2 H^+ + 2 e \rightleftharpoons N_2O + H_2O$	1.591
$2 NO + H_2O + 2 e \rightleftharpoons N_2O + 2 OH^-$	0.76
$HNO_2 + H^+ + e \rightleftharpoons NO + H_2O$	0.983
$2 HNO_2 + 4 H^+ + 4 e \rightleftharpoons H_2N_2O_2 + 2 H_2O$	0.86
$2 HNO_2 + 4 H^+ + 4 e \rightleftharpoons N_2O + 3 H_2O$	1.297
$NO_2^- + H_2O + e \rightleftharpoons NO + 2 OH^-$	–0.46
$2 NO_2^- + 2 H_2O + 4 e \rightleftharpoons N_2O_2^{2-} + 4 OH^-$	–0.18
$2 NO_2^- + 3 H_2O + 4 e \rightleftharpoons N_2O + 6 OH^-$	0.15
$NO_3^- + 3 H^+ + 2 e \rightleftharpoons HNO_2 + H_2O$	0.934
$NO_3^- + 4 H^+ + 3 e \rightleftharpoons NO + 2 H_2O$	0.957
$2 NO_3^- + 4 H^+ + 2 e \rightleftharpoons N_2O_4 + 2 H_2O$	0.803
$NO_3^- + H_2O + 2 e \rightleftharpoons NO_2^- + 2 OH^-$	0.01
$2 NO_3^- + 2 H_2O + 2 e \rightleftharpoons N_2O_4 + 4 OH^-$	–0.85
$Na^+ + e \rightleftharpoons Na$	–2.71
$Nb^{3+} + 3 e \rightleftharpoons Nb$	–1.099
$NbO_2 + 2 H^+ + 2 e \rightleftharpoons NbO + H_2O$	–0.646
$NbO_2 + 4 H^+ + 4 e \rightleftharpoons Nb + 2 H_2O$	–0.690
$NbO + 2 H^+ + 2 e \rightleftharpoons Nb + H_2O$	–0.733
$Nb_2O_5 + 10 H^+ + 10 e \rightleftharpoons 2 Nb + 5 H_2O$	–0.644
$Nd^{3+} + 3 e \rightleftharpoons Nd$	–2.323
$Nd^{2+} + 2 e \rightleftharpoons Nd$	–2.1
$Nd^{3+} + e \rightleftharpoons Nd^{2+}$	–2.7
$Ni^{2+} + 2 e \rightleftharpoons Ni$	–0.257
$Ni(OH)_2 + 2 e \rightleftharpoons Ni + 2 OH^-$	–0.72
$NiO_2 + 4 H^+ + 2 e \rightleftharpoons Ni^{2+} + 2 H_2O$	1.678
$NiO_2 + 2 H_2O + 2 e \rightleftharpoons Ni(OH)_2 + 2 OH^-$	–0.490

Reaction	$E°$/V
$\mathrm{No^{3+} + e \rightleftharpoons No^{2+}}$	1.4
$\mathrm{No^{3+} + 3\,e \rightleftharpoons No}$	–1.20
$\mathrm{No^{2+} + 2\,e \rightleftharpoons No}$	–2.50
$\mathrm{Np^{3+} + 3\,e \rightleftharpoons Np}$	–1.856
$\mathrm{Np^{4+} + e \rightleftharpoons Np^{3+}}$	0.147
$\mathrm{NpO_2 + H_2O + H^+ + e \rightleftharpoons Np(OH)_3}$	–0.962
$\mathrm{O_2 + 2\,H^+ + 2\,e \rightleftharpoons H_2O_2}$	0.695
$\mathrm{O_2 + 4\,H^+ + 4\,e \rightleftharpoons 2\,H_2O}$	1.229
$\mathrm{O_2 + H_2O + 2\,e \rightleftharpoons HO_2^- + OH^-}$	–0.076
$\mathrm{O_2 + 2\,H_2O + 2\,e \rightleftharpoons H_2O_2 + 2\,OH^-}$	–0.146
$\mathrm{O_2 + 2\,H_2O + 4\,e \rightleftharpoons 4\,OH^-}$	0.401
$\mathrm{O_3 + 2\,H^+ + 2\,e \rightleftharpoons O_2 + H_2O}$	2.076
$\mathrm{O_3 + H_2O + 2\,e \rightleftharpoons O_2 + 2\,OH^-}$	1.24
$\mathrm{O(g) + 2\,H^+ + 2\,e \rightleftharpoons H_2O}$	2.421
$\mathrm{OH + e \rightleftharpoons OH^-}$	2.02
$\mathrm{HO_2^- + H_2O + 2\,e \rightleftharpoons 3\,OH^-}$	0.878
$\mathrm{OsO_4 + 8\,H^+ + 8\,e \rightleftharpoons Os + 4\,H_2O}$	0.838
$\mathrm{OsO_4 + 4\,H^+ + 4\,e \rightleftharpoons OsO_2 + 2\,H_2O}$	1.02
$\mathrm{[Os(bipy)_2]^{3+} + e \rightleftharpoons [Os(bipy)_2]^{2+}}$	0.81
$\mathrm{[Os(bipy)_3]^{3+} + e \rightleftharpoons [Os(bipy)_3]^{2+}}$	0.80
$\mathrm{P(red) + 3\,H^+ + 3\,e \rightleftharpoons PH_3(g)}$	–0.111
$\mathrm{P(white) + 3\,H^+ + 3\,e \rightleftharpoons PH_3(g)}$	–0.063
$\mathrm{P + 3\,H_2O + 3\,e \rightleftharpoons PH_3(g) + 3\,OH^-}$	–0.87
$\mathrm{H_2P_2^- + e \rightleftharpoons P + 2\,OH^-}$	–1.82
$\mathrm{H_3PO_2 + H^+ + e \rightleftharpoons P + 2\,H_2O}$	–0.508
$\mathrm{H_3PO_3 + 2\,H^+ + 2\,e \rightleftharpoons H_3PO_2 + H_2O}$	–0.499
$\mathrm{H_3PO_3 + 3\,H^+ + 3\,e \rightleftharpoons P + 3\,H_2O}$	–0.454
$\mathrm{HPO_3^{2-} + 2\,H_2O + 2\,e \rightleftharpoons H_2PO_2^- + 3\,OH^-}$	–1.65
$\mathrm{HPO_3^{2-} + 2\,H_2O + 3\,e \rightleftharpoons P + 5\,OH^-}$	–1.71
$\mathrm{H_3PO_4 + 2\,H^+ + 2\,e \rightleftharpoons H_3PO_3 + H_2O}$	–0.276
$\mathrm{PO_4^{3-} + 2\,H_2O + 2\,e \rightleftharpoons HPO_3^{2-} + 3\,OH^-}$	–1.05
$\mathrm{Pa^{3+} + 3\,e \rightleftharpoons Pa}$	–1.34
$\mathrm{Pa^{4+} + 4\,e \rightleftharpoons Pa}$	–1.49
$\mathrm{Pa^{4+} + e \rightleftharpoons Pa^{3+}}$	–1.9
$\mathrm{Pb^{2+} + 2\,e \rightleftharpoons Pb}$	–0.1262
$\mathrm{Pb^{2+} + 2\,e \rightleftharpoons Pb(Hg)}$	–0.1205
$\mathrm{PbBr_2 + 2\,e \rightleftharpoons Pb + 2\,Br^-}$	–0.284
$\mathrm{PbCl_2 + 2\,e \rightleftharpoons Pb + 2\,Cl^-}$	–0.2675
$\mathrm{PbF_2 + 2\,e \rightleftharpoons Pb + 2\,F^-}$	–0.3444
$\mathrm{PbHPO_4 + 2\,e \rightleftharpoons Pb + HPO_4^{2-}}$	–0.465
$\mathrm{PbI_2 + 2\,e \rightleftharpoons Pb + 2\,I^-}$	–0.365
$\mathrm{PbO + H_2O + 2\,e \rightleftharpoons Pb + 2\,OH^-}$	–0.580
$\mathrm{PbO_2 + 4\,H^+ + 2\,e \rightleftharpoons Pb^{2+} + 2\,H_2O}$	1.455
$\mathrm{HPbO_2^- + H_2O + 2\,e \rightleftharpoons Pb + 3\,OH^-}$	–0.537
$\mathrm{PbO_2 + H_2O + 2\,e \rightleftharpoons PbO + 2\,OH^-}$	0.247
$\mathrm{PbO_2 + SO_4^{2-} + 4\,H^+ + 2\,e \rightleftharpoons PbSO_4 + 2\,H_2O}$	1.6913
$\mathrm{PbSO_4 + 2\,e \rightleftharpoons Pb + SO_4^{2-}}$	–0.3588
$\mathrm{PbSO_4 + 2\,e \rightleftharpoons Pb(Hg) + SO_4^{2-}}$	–0.3505
$\mathrm{Pd^{2+} + 2\,e \rightleftharpoons Pd}$	0.951
$\mathrm{[PdCl_4]^{2-} + 2\,e \rightleftharpoons Pd + 4\,Cl^-}$	0.591
$\mathrm{[PdCl_6]^{2-} + 2\,e \rightleftharpoons [PdCl_4]^{2-} + 2\,Cl^-}$	1.288
$\mathrm{Pd(OH)_2 + 2\,e \rightleftharpoons Pd + 2\,OH^-}$	0.07
$\mathrm{Pm^{2+} + 2\,e \rightleftharpoons Pm}$	–2.2
$\mathrm{Pm^{3+} + 3\,e \rightleftharpoons Pm}$	–2.30
$\mathrm{Pm^{3+} + e \rightleftharpoons Pm^{2+}}$	–2.6
$\mathrm{Po^{4+} + 2\,e \rightleftharpoons Po^{2+}}$	0.9
$\mathrm{Po^{4+} + 4\,e \rightleftharpoons Po}$	0.76
$\mathrm{Pr^{4+} + e \rightleftharpoons Pr^{3+}}$	3.2
$\mathrm{Pr^{2+} + 2\,e \rightleftharpoons Pr}$	–2.0
$\mathrm{Pr^{3+} + 3\,e \rightleftharpoons Pr}$	–2.353

Reaction	$E°$/V
$\mathrm{Pr^{3+} + e \rightleftharpoons Pr^{2+}}$	–3.1
$\mathrm{Pt^{2+} + 2\,e \rightleftharpoons Pt}$	1.18
$\mathrm{[PtCl_4]^{2-} + 2\,e \rightleftharpoons Pt + 4\,Cl^-}$	0.755
$\mathrm{[PtCl_6]^{2-} + 2\,e \rightleftharpoons [PtCl_4]^{2-} + 2\,Cl^-}$	0.68
$\mathrm{Pt(OH)_2 + 2\,e \rightleftharpoons Pt + 2\,OH^-}$	0.14
$\mathrm{PtO_3 + 2\,H^+ + 2\,e \rightleftharpoons PtO_2 + H_2O}$	1.7
$\mathrm{PtO_3 + 4\,H^+ + 2\,e \rightleftharpoons Pt(OH)_2^{2+} + H_2O}$	1.5
$\mathrm{PtOH^+ + H^+ + 2\,e \rightleftharpoons Pt + H_2O}$	1.2
$\mathrm{PtO_2 + 2\,H^+ + 2\,e \rightleftharpoons PtO + H_2O}$	1.01
$\mathrm{PtO_2 + 4\,H^+ + 4\,e \rightleftharpoons Pt + 2\,H_2O}$	1.00
$\mathrm{Pu^{3+} + 3\,e \rightleftharpoons Pu}$	–2.031
$\mathrm{Pu^{4+} + e \rightleftharpoons Pu^{3+}}$	1.006
$\mathrm{Pu^{5+} + e \rightleftharpoons Pu^{4+}}$	1.099
$\mathrm{PuO_2(OH)_2 + 2\,H^+ + 2\,e \rightleftharpoons Pu(OH)_4}$	1.325
$\mathrm{PuO_2(OH)_2 + H^+ + e \rightleftharpoons PuO_2OH + H_2O}$	1.062
$\mathrm{Ra^{2+} + 2\,e \rightleftharpoons Ra}$	–2.8
$\mathrm{Rb^+ + e \rightleftharpoons Rb}$	–2.98
$\mathrm{Re^{3+} + 3\,e \rightleftharpoons Re}$	0.300
$\mathrm{ReO_4^- + 4\,H^+ + 3\,e \rightleftharpoons ReO_2 + 2\,H_2O}$	0.510
$\mathrm{ReO_2 + 4\,H^+ + 4\,e \rightleftharpoons Re + 2\,H_2O}$	0.2513
$\mathrm{ReO_4^- + 2\,H^+ + e \rightleftharpoons ReO_3 + H_2O}$	0.768
$\mathrm{ReO_4^- + 4\,H_2O + 7\,e \rightleftharpoons Re + 8\,OH^-}$	–0.584
$\mathrm{ReO_4^- + 8\,H^+ + 7\,e \rightharpoonup Re + 4\,H_2O}$	0.368
$\mathrm{Rh^+ + e \rightleftharpoons Rh}$	0.600
$\mathrm{Rh^{3+} + 3\,e \rightleftharpoons Rh}$	0.758
$\mathrm{[RhCl_6]^{3-} + 3\,e \rightleftharpoons Rh + 6\,Cl^-}$	0.431
$\mathrm{RhOH^{2+} + H^+ + 3\,e \rightleftharpoons Rh + H_2O}$	0.83
$\mathrm{Ru^{2+} + 2\,e \rightleftharpoons Ru}$	0.455
$\mathrm{Ru^{3+} + e \rightleftharpoons Ru^{2+}}$	0.2487
$\mathrm{RuO_2 + 4\,H^+ + 2\,e \rightleftharpoons Ru^{2+} + 2\,H_2O}$	1.120
$\mathrm{RuO_4^- + e \rightleftharpoons RuO_4^{2-}}$	0.59
$\mathrm{RuO_4 + e \rightleftharpoons RuO_4^-}$	1.00
$\mathrm{RuO_4 + 6\,H^+ + 4\,e \rightleftharpoons Ru(OH)_2^{2+} + 2\,H_2O}$	1.40
$\mathrm{RuO_4 + 8\,H^+ + 8\,e \rightleftharpoons Ru + 4\,H_2O}$	1.038
$\mathrm{[Ru(bipy)_3]^{3+} + e^- \rightleftharpoons [Ru(bipy)_3]^{2+}}$	1.24
$\mathrm{[Ru(H_2O)_6]^{3+} + e^- \rightleftharpoons [Ru(H_2O)_6]^{2+}}$	0.23
$\mathrm{[Ru(NH_3)_6]^{3+} + e^- \rightleftharpoons [Ru(NH_3)_6]^{2+}}$	0.10
$\mathrm{[Ru(en)_3]^{3+} + e^- \rightleftharpoons [Ru(en)_3]^{2+}}$	0.210
$\mathrm{[Ru(CN)_6]^{3-} + e^- \rightleftharpoons [Ru(CN)_6]^{4-}}$	0.86
$\mathrm{S + 2\,e \rightleftharpoons S^{2-}}$	–0.47627
$\mathrm{S + 2H^+ + 2\,e \rightleftharpoons H_2S(aq)}$	0.142
$\mathrm{S + H_2O + 2\,e \rightleftharpoons SH^- + OH^-}$	–0.478
$\mathrm{2\,S + 2\,e \rightleftharpoons S_2^{2-}}$	–0.42836
$\mathrm{S_2O_6^{2-} + 4\,H^+ + 2\,e \rightleftharpoons 2\,H_2SO_3}$	0.564
$\mathrm{S_2O_8^{2-} + 2\,e \rightleftharpoons 2\,SO_4^{2-}}$	2.010
$\mathrm{S_2O_8^{2-} + 2\,H^+ + 2\,e \rightleftharpoons 2\,HSO_4^-}$	2.123
$\mathrm{S_4O_6^{2-} + 2\,e \rightleftharpoons 2\,S_2O_3^{2-}}$	0.08
$\mathrm{2\,H_2SO_3 + H^+ + 2\,e \rightleftharpoons HS_2O_4^- + 2\,H_2O}$	–0.056
$\mathrm{H_2SO_3 + 4\,H^+ + 4\,e \rightleftharpoons S + 3\,H_2O}$	0.449
$\mathrm{2\,SO_3^{2-} + 2\,H_2O + 2\,e \rightleftharpoons S_2O_4^{2-} + 4\,OH^-}$	–1.12
$\mathrm{2\,SO_3^{2-} + 3\,H_2O + 4\,e \rightleftharpoons S_2O_3^{2-} + 6\,OH^-}$	–0.571
$\mathrm{SO_4^{2-} + 4\,H^+ + 2\,e \rightleftharpoons H_2SO_3 + H_2O}$	0.172
$\mathrm{2\,SO_4^{2-} + 4\,H^+ + 2\,e \rightleftharpoons S_2O_6^{2-} + H_2O}$	–0.22
$\mathrm{SO_4^{2-} + H_2O + 2\,e \rightleftharpoons SO_3^{2-} + 2\,OH^-}$	–0.93
$\mathrm{Sb + 3\,H^+ + 3\,e \rightleftharpoons SbH_3}$	–0.510
$\mathrm{Sb_2O_3 + 6\,H^+ + 6\,e \rightleftharpoons 2\,Sb + 3\,H_2O}$	0.152
$\mathrm{Sb_2O_5}$ (senarmontite) $\mathrm{+ 4\,H^+ + 4\,e \rightleftharpoons Sb_2O_3 + 2\,H_2O}$	0.671
$\mathrm{Sb_2O_5}$ (valentinite) $\mathrm{+ 4\,H^+ + 4\,e \rightleftharpoons Sb_2O_3 + 2\,H_2O}$	0.649

Reaction	$E°$/V
$Sb_2O_5 + 6\,H^+ + 4\,e \rightleftharpoons 2\,SbO^+ + 3\,H_2O$	0.581
$SbO^+ + 2\,H^+ + 3\,e \rightleftharpoons Sb + 2\,H_2O$	0.212
$SbO_2^- + 2\,H_2O + 3\,e \rightleftharpoons Sb + 4\,OH^-$	–0.66
$SbO_3^- + H_2O + 2\,e \rightleftharpoons SbO_2^- + 2\,OH^-$	–0.59
$Sc^{3+} + 3\,e \rightleftharpoons Sc$	–2.077
$Se + 2\,e \rightleftharpoons Se^{2-}$	–0.924
$Se + 2\,H^+ + 2\,e \rightleftharpoons H_2Se(aq)$	–0.399
$H_2SeO_3 + 4\,H^+ + 4\,e \rightleftharpoons Se + 3\,H_2O$	0.74
$Se + 2\,H^+ + 2\,e \rightleftharpoons H_2Se$	–0.082
$SeO_3^{2-} + 3\,H_2O + 4\,e \rightleftharpoons Se + 6\,OH^-$	–0.366
$SeO_4^{2-} + 4\,H^+ + 2\,e \rightleftharpoons H_2SeO_3 + H_2O$	1.151
$SeO_4^{2-} + H_2O + 2\,e \rightleftharpoons SeO_3^{2-} + 2\,OH^-$	0.05
$SiF_6^{2-} + 4\,e \rightleftharpoons Si + 6\,F^-$	–1.24
$SiO + 2\,H^+ + 2\,e \rightleftharpoons Si + H_2O$	–0.8
SiO_2 (quartz) $+ 4\,H^+ + 4\,e \rightleftharpoons Si + 2\,H_2O$	0.857
$SiO_3^{2-} + 3\,H_2O + 4\,e \rightleftharpoons Si + 6\,OH^-$	–1.697
$Sm^{3+} + e \rightleftharpoons Sm^{2+}$	–1.55
$Sm^{3+} + 3\,e \rightleftharpoons Sm$	–2.304
$Sm^{2+} + 2\,e \rightleftharpoons Sm$	–2.68
$Sn^{2+} + 2\,e \rightleftharpoons Sn$	–0.1375
$Sn^{4+} + 2\,e \rightleftharpoons Sn^{2+}$	0.151
$Sn(OH)_3^+ + 3\,H^+ + 2\,e \rightleftharpoons Sn^{2+} + 3\,H_2O$	0.142
$SnO_2 + 4\,H^+ + 2\,e^- \rightleftharpoons Sn^{2+} + 2\,H_2O$	–0.094
$SnO_2 + 4\,H^+ + 4\,e \rightleftharpoons Sn + 2\,H_2O$	–0.117
$SnO_2 + 3\,H^+ + 2\,e \rightleftharpoons SnOH^+ + H_2O$	–0.194
$SnO_2 + 2\,H_2O + 4\,e \rightleftharpoons Sn + 4\,OH^-$	–0.945
$HSnO_2^- + H_2O + 2\,e \rightleftharpoons Sn + 3\,OH^-$	–0.909
$Sn(OH)_6^{2-} + 2\,e \rightleftharpoons HSnO_2^- + 3\,OH^- + H_2O$	–0.93
$Sr^+ + e \rightleftharpoons Sr$	–4.10
$Sr^{2+} + 2\,e \rightleftharpoons Sr$	–2.899
$Sr^{2+} + 2\,e \rightleftharpoons Sr(Hg)$	–1.793
$Sr(OH)_2 + 2\,e \rightleftharpoons Sr + 2\,OH^-$	–2.88
$Ta_2O_5 + 10\,H^+ + 10\,e \rightleftharpoons 2\,Ta + 5\,H_2O$	–0.750
$Ta^{3+} + 3\,e \rightleftharpoons Ta$	–0.6
$Tc^{2+} + 2\,e \rightleftharpoons Tc$	0.400
$TcO_4^- + 4\,H^+ + 3\,e \rightleftharpoons TcO_2 + 2\,H_2O$	0.782
$Tc^{3+} + e \rightleftharpoons Tc^{2+}$	0.3
$TcO_4^- + 8\,H^+ + 7\,e \rightleftharpoons Tc + 4\,H_2O$	0.472
$Tb^{4+} + e \rightleftharpoons Tb^{3+}$	3.1
$Tb^{3+} + 3\,e \rightleftharpoons Tb$	–2.28
$Te + 2\,e \rightleftharpoons Te^{2-}$	–1.143
$Te + 2\,H^+ + 2\,e \rightleftharpoons H_2Te$	–0.793
$Te^{4+} + 4\,e \rightleftharpoons Te$	0.568
$TeO_2 + 4\,H^+ + 4\,e \rightleftharpoons Te + 2\,H_2O$	0.593
$TeO_3^{2-} + 3\,H_2O + 4\,e \rightleftharpoons Te + 6\,OH^-$	–0.57
$TeO_4^- + 8\,H^+ + 7\,e \rightleftharpoons Te + 4\,H_2O$	0.472
$H_6TeO_6 + 2\,H^+ + 2\,e \rightleftharpoons TeO_2 + 4\,H_2O$	1.02
$Th^{4+} + 4\,e \rightleftharpoons Th$	–1.899
$ThO_2 + 4\,H^+ + 4\,e \rightleftharpoons Th + 2\,H_2O$	–1.789
$Th(OH)_4 + 4\,e \rightleftharpoons Th + 4\,OH^-$	–2.48
$Ti^{2+} + 2\,e \rightleftharpoons Ti$	–1.630
$Ti^{3+} + e \rightleftharpoons Ti^{2+}$	–0.9
$TiO_2 + 4\,H^+ + 2\,e \rightleftharpoons Ti^{2+} + 2\,H_2O$	–0.502
$Ti^{3+} + 3\,e \rightleftharpoons Ti$	–1.37
$TiOH^{3+} + H^+ + e \rightleftharpoons Ti^{3+} + H_2O$	–0.055

Reaction	$E°$/V
$Tl^+ + e \rightleftharpoons Tl$	–0.336
$Tl^+ + e \rightleftharpoons Tl(Hg)$	–0.3338
$Tl^{3+} + 2\,e \rightleftharpoons Tl^+$	1.252
$Tl^{3+} + 3\,e \rightleftharpoons Tl$	0.741
$TlBr + e \rightleftharpoons Tl + Br^-$	–0.658
$TlCl + e \rightleftharpoons Tl + Cl^-$	–0.5568
$TlI + e \rightleftharpoons Tl + I^-$	–0.752
$Tl_2O_3 + 3\,H_2O + 4\,e \rightleftharpoons 2\,Tl^+ + 6\,OH^-$	0.02
$TlOH + e \rightleftharpoons Tl + OH^-$	–0.34
$Tl(OH)_3 + 2\,e \rightleftharpoons TlOH + 2\,OH^-$	–0.05
$Tl_2SO_4 + 2\,e \rightleftharpoons Tl + SO_4^{2-}$	–0.4360
$Tm^{3+} + e \rightleftharpoons Tm^{2+}$	–2.2
$Tm^{3+} + 3\,e \rightleftharpoons Tm$	–2.319
$Tm^{2+} + 2\,e \rightleftharpoons Tm$	–2.4
$U^{3+} + 3\,e \rightleftharpoons U$	–1.798
$U^{4+} + e \rightleftharpoons U^{3+}$	–0.607
$UO_2^+ + 4\,H^+ + e \rightleftharpoons U^{4+} + 2\,H_2O$	0.612
$UO_2^{2+} + e \rightleftharpoons UO^+_2$	0.062
$UO_2^{2+} + 4\,H^+ + 2\,e \rightleftharpoons U^{4+} + 2\,H_2O$	0.327
$UO_2^{2+} + 4\,H^+ + 6\,e \rightleftharpoons U + 2\,H_2O$	–1.444
$V^{2+} + 2\,e \rightleftharpoons V$	–1.175
$V^{3+} + e \rightleftharpoons V^{2+}$	–0.255
$VO^{2+} + 2\,H^+ + e \rightleftharpoons V^{3+} + H_2O$	0.337
$VO_2^+ + 2\,H^+ + e \rightleftharpoons VO^{2+} + H_2O$	0.991
$V_2O_5 + 6\,H^+ + 2\,e \rightleftharpoons 2\,VO^{2+} + 3\,H_2O$	0.957
$V_2O_5 + 10\,H^+ + 10\,e \rightleftharpoons 2\,V + 5\,H_2O$	–0.242
$V(OH)_4^+ + 2\,H^+ + e \rightleftharpoons VO^{2+} + 3\,H_2O$	1.00
$V(OH)_4^+ + 4\,H^+ + 5\,e \rightleftharpoons V + 4\,H_2O$	–0.254
$[V(phen)_3]^{3+} + e \rightleftharpoons [V(phen)_3]^{2+}$	0.14
$W^{3+} + 3\,e \rightleftharpoons W$	0.1
$W_2O_5 + 2\,H^+ + 2\,e \rightleftharpoons 2\,WO_2 + H_2O$	–0.031
$WO_2 + 4\,H^+ + 4\,e \rightleftharpoons W + 2\,H_2O$	–0.119
$WO_3 + 6\,H^+ + 6\,e \rightleftharpoons W + 3\,H_2O$	–0.090
$WO_3 + 2\,H^+ + 2\,e \rightleftharpoons WO_2 + H_2O$	0.036
$2\,WO_3 + 2\,H^+ + 2\,e \rightleftharpoons W_2O_5 + H_2O$	–0.029
$H_4XeO_6 + 2\,H^+ + 2\,e \rightleftharpoons XeO_3 + 3\,H_2O$	2.42
$XeO_3 + 6\,H^+ + 6\,e \rightleftharpoons Xe + 3\,H_2O$	2.10
$XeF + e \rightleftharpoons Xe + F^-$	3.4
$Y^{3+} + 3\,e \rightleftharpoons Y$	–2.372
$Yb^{3+} + e \rightleftharpoons Yb^{2+}$	–1.05
$Yb^{3+} + 3\,e \rightleftharpoons Yb$	–2.19
$Yb^{2+} + 2\,e \rightleftharpoons Yb$	–2.76
$Zn^{2+} + 2\,e \rightleftharpoons Zn$	–0.7618
$Zn^{2+} + 2\,e \rightleftharpoons Zn(Hg)$	–0.7628
$ZnO_2^{2-} + 2\,H_2O + 2\,e \rightleftharpoons Zn + 4\,OH^-$	–1.215
$ZnSO_4 \cdot 7\,H_2O + 2\,e = Zn(Hg) + SO_4^{2-} + 7\,H_2O$ (Saturated $ZnSO_4$)	–0.7993
$ZnOH^+ + H^+ + 2\,e \rightleftharpoons Zn + H_2O$	–0.497
$Zn(OH)_4^{2-} + 2\,e \rightleftharpoons Zn + 4\,OH^-$	–1.199
$Zn(OH)_2 + 2\,e \rightleftharpoons Zn + 2\,OH^-$	–1.249
$ZnO + H_2O + 2\,e \rightleftharpoons Zn + 2\,OH^-$	–1.260
$ZrO_2 + 4\,H^+ + 4\,e \rightleftharpoons Zr + 2\,H_2O$	–1.553
$ZrO(OH)_2 + H_2O + 4\,e \rightleftharpoons Zr + 4\,OH^-$	–2.36
$Zr^{4+} + 4\,e \rightleftharpoons Zr$	–1.45

TABLE 2. Reduction Reactions Having $E°$ Values More Positive than That of the Standard Hydrogen Electrode

Reaction	$E°$/V
$2\,H^+ + 2\,e \rightleftharpoons H_2$	0.00000
$CuI_2^- + e \rightleftharpoons Cu + 2\,I^-$	0.00
$Ge^{4+} + 2\,e \rightleftharpoons Ge^{2+}$	0.00
$NO_3^- + H_2O + 2\,e \rightleftharpoons NO_2^- + 2\,OH^-$	0.01
$Tl_2O_3 + 3\,H_2O + 4\,e \rightleftharpoons 2\,Tl^+ + 6\,OH^-$	0.02
$SeO_4^{2-} + H_2O + 2\,e \rightleftharpoons SeO_3^{2-} + 2\,OH^-$	0.05
$WO_3 + 2\,H^+ + 2\,e \rightleftharpoons WO_2 + H_2O$	0.036
$UO_2^{2+} + e = UO_2^+$	0.062
$Pd(OH)_2 + 2\,e \rightleftharpoons Pd + 2\,OH^-$	0.07
$AgBr + e \rightleftharpoons Ag + Br^-$	0.07133
$MoO_3 + 6\,H^+ + 6\,e \rightleftharpoons Mo + 3\,H_2O$	0.075
$S_4O_6^{2-} + 2\,e \rightleftharpoons 2\,S_2O_3^{2-}$	0.08
$H_3Mo_7O_{24}^{3-} + 45\,H^+ + 42\,e \rightleftharpoons 7\,Mo + 24\,H_2O$	0.082
$AgSCN + e \rightleftharpoons Ag + SCN^-$	0.08951
$N_2 + 2\,H_2O + 6\,H^+ + 6\,e \rightleftharpoons 2\,NH_4OH$	0.092
$HgO + H_2O + 2\,e \rightleftharpoons Hg + 2\,OH^-$	0.0977
$Ir_2O_3 + 3\,H_2O + 6\,e \rightleftharpoons 2\,Ir + 6\,OH^-$	0.098
$2\,NO + 2\,e \rightleftharpoons N_2O_2^{2-}$	0.10
$[Ru(NH_3)_6]^{3+} + e \rightleftharpoons [Ru(NH_3)_6]^{2+}$	0.10
$W^{3+} + 3\,e \rightleftharpoons W$	0.1
$[Co(NH_3)_6]^{3+} + e \rightleftharpoons [Co(NH_3)_6]^{2+}$	0.108
$Hg_2O + H_2O + 2\,e \rightleftharpoons 2\,Hg + 2\,OH^-$	0.123
$Ge^{4+} + 4\,e \rightleftharpoons Ge$	0.124
$Hg_2Br_2 + 2\,e \rightleftharpoons 2\,Hg + 2\,Br^-$	0.13923
$Pt(OH)_2 + 2\,e \rightleftharpoons Pt + 2\,OH^-$	0.14
$[V(phen)_3]^{3+} + e \rightleftharpoons [V(phen)_3]^{2+}$	0.14
$S + 2H^+ + 2\,e \rightleftharpoons H_2S(aq)$	0.142
$Sn(OH)_3^+ + 3\,H^+ + 2\,e \rightleftharpoons Sn^{2+} + 3\,H_2O$	0.142
$Np^{4+} + e \rightleftharpoons Np^{3+}$	0.147
$Ag_4[Fe(CN)_6] + 4\,e \rightleftharpoons 4\,Ag + [Fe(CN)_6]^{4-}$	0.1478
$IO_3^- + 2\,H_2O + 4\,e \rightleftharpoons IO^- + 4\,OH^-$	0.15
$Mn(OH)_3 + e \rightleftharpoons Mn(OH)_2 + OH^-$	0.15
$2\,NO_2^- + 3\,H_2O + 4\,e \rightleftharpoons N_2O + 6\,OH^-$	0.15
$Sn^{4+} + 2\,e \rightleftharpoons Sn^{2+}$	0.151
$Sb_2O_3 + 6\,H^+ + 6\,e \rightleftharpoons 2\,Sb + 3\,H_2O$	0.152
$Cu^{2+} + e \rightleftharpoons Cu^+$	0.153
$BiOCl + 2\,H^+ + 3\,e \rightleftharpoons Bi + Cl^- + H_2O$	0.1583
$BiCl_4^- + 3\,e \rightleftharpoons Bi + 4\,Cl^-$	0.16
$Fe_2O_3 + 4\,H^+ + 2\,e \rightleftharpoons 2\,FeOH^+ + H_2O$	0.16
$Co(OH)_3 + e \rightleftharpoons Co(OH)_2 + OH^-$	0.17
$SO_4^{2-} + 4\,H^+ + 2\,e \rightleftharpoons H_2SO_3 + H_2O$	0.172
$Bi^{3+} + 2\,e \rightleftharpoons Bi^+$	0.2
$[Ru(en)_3]^{3+} + e \rightleftharpoons [Ru(en)_3]^{2+}$	0.210
$SbO^+ + 2\,H^+ + 3\,e \rightleftharpoons Sb + 2\,H_2O$	0.212
$AgCl + e \rightleftharpoons Ag + Cl^-$	0.22233
$[Ru(H_2O)_6]^{3+} + e \rightleftharpoons [Ru(H_2O)_6]^{2+}$	0.23
$As_2O_3 + 6\,H^+ + 6\,e \rightleftharpoons 2\,As + 3\,H_2O$	0.234
Calomel electrode, saturated NaCl (SSCE)	0.2360
$Ge^{2+} + 2\,e \rightleftharpoons Ge$	0.24
$Ru^{3+} + e \rightleftharpoons Ru^{2+}$	0.24
Calomel electrode, saturated KCl	0.2412
$PbO_2 + H_2O + 2\,e \rightleftharpoons PbO + 2\,OH^-$	0.247
$HAsO_2 + 3\,H^+ + 3\,e \rightleftharpoons As + 2\,H_2O$	0.248
$Ru^{3+} + e \rightleftharpoons Ru^{2+}$	0.2487
$ReO_2 + 4\,H^+ + 4\,e \rightleftharpoons Re + 2\,H_2O$	0.2513
$IO_3^- + 3\,H_2O + 6\,e \rightleftharpoons I^- + OH^-$	0.26
$Hg_2Cl_2 + 2\,e \rightleftharpoons 2\,Hg + 2\,Cl^-$	0.26808
Calomel electrode, 1 molal KCl	0.2800
Calomel electrode, 1 molar KCl (NCE)	0.2801
$At_2 + 2\,e \rightleftharpoons 2\,At^-$	0.3
$Re^{3+} + 3\,e \rightleftharpoons Re$	0.300
$Tc^{3+} + e \rightleftharpoons Tc^{2+}$	0.3
$Bi^{3+} + 3\,e \rightleftharpoons Bi$	0.308
$BiO^+ + 2\,H^+ + 3\,e \rightleftharpoons Bi + H_2O$	0.320
$UO_2^{2+} + 4\,H^+ + 2\,e \rightleftharpoons U^{4+} + 2\,H_2O$	0.327
$ClO_3^- + H_2O + 2\,e \rightleftharpoons ClO_2^- + 2\,OH^-$	0.33
$2\,HCNO + 2\,H^+ + 2\,e \rightleftharpoons (CN)_2 + 2\,H_2O$	0.330
Calomel electrode, 0.1 molar KCl	0.3337
$VO^{2+} + 2\,H^+ + e \rightleftharpoons V^{3+} + H_2O$	0.337
$Cu^{2+} + 2\,e \rightleftharpoons Cu$	0.3419
$Ag_2O + H_2O + 2\,e \rightleftharpoons 2\,Ag + 2\,OH^-$	0.342
$Cu^{2+} + 2\,e \rightleftharpoons Cu(Hg)$	0.345
$AgIO_3 + e \rightleftharpoons Ag + IO_3^-$	0.354
$[Fe(CN)_6]^{3-} + e \rightleftharpoons [Fe(CN)_6]^{4-}$	0.358
$ClO_4^- + H_2O + 2\,e \rightleftharpoons ClO^-_3 + 2\,OH^-$	0.36
$Ag_2SeO_3 + 2\,e \rightleftharpoons 2\,Ag + SeO_3^{2-}$	0.3629
$ReO_4^- + 8\,H^+ + 7\,e \rightleftharpoons Re + 4\,H_2O$	0.368
$(CN)_2 + 2\,H^+ + 2\,e \rightleftharpoons 2\,HCN$	0.373
$[Ferricinium]^+ + e \rightleftharpoons ferrocene$	0.400
$Tc^{2+} + 2\,e \rightleftharpoons Tc$	0.400
$O_2 + 2\,H_2O + 4\,e \rightleftharpoons 4\,OH^-$	0.401
$AgOCN + e \rightleftharpoons Ag + OCN^-$	0.41
$[RhCl_6]^{3-} + 3\,e \rightleftharpoons Rh + 6\,Cl^-$	0.431
$Ag_2CrO_4 + 2\,e \rightleftharpoons 2\,Ag + CrO_4^{2-}$	0.4470
$H_2SO_3 + 4\,H^+ + 4\,e \rightleftharpoons S + 3\,H_2O$	0.449
$Ru^{2+} + 2\,e \rightleftharpoons Ru$	0.455
$Ag_2MoO_4 + 2\,e \rightleftharpoons 2\,Ag + MoO_4^{2-}$	0.4573
$Ag_2C_2O_4 + 2\,e \rightleftharpoons 2\,Ag + C_2O_4^{2-}$	0.4647
$Ag_2WO_4 + 2\,e \rightleftharpoons 2\,Ag + WO_4^{2-}$	0.4660
$Ag_2CO_3 + 2\,e \rightleftharpoons 2\,Ag + CO_3^{2-}$	0.47
$TcO_4^- + 8\,H^+ + 7\,e \rightleftharpoons Tc + 4\,H_2O$	0.472
$TeO_4^- + 8\,H^+ + 7\,e \rightleftharpoons Te + 4\,H_2O$	0.472
$IO^- + H_2O + 2\,e \rightleftharpoons I^- + 2\,OH^-$	0.485
$NiO_2 + 2\,H_2O + 2\,e \rightleftharpoons Ni(OH)_2 + 2\,OH^-$	0.490
$Bi^+ + e \rightleftharpoons Bi$	0.5
$ReO_4^- + 4\,H^+ + 3\,e \rightleftharpoons ReO_2 + 2\,H_2O$	0.510
$Hg_2(ac)_2 + 2\,e \rightleftharpoons 2\,Hg + 2(ac)^-$	0.51163
$Cu^+ + e \rightleftharpoons Cu$	0.521
$I_2 + 2\,e \rightleftharpoons 2\,I^-$	0.5355
$I_3^- + 2\,e \rightleftharpoons 3\,I^-$	0.536
$AgBrO_3 + e \rightleftharpoons Ag + BrO_3^-$	0.546
$MnO_4^- + e \rightleftharpoons MnO_4^-$	0.558
$H_3AsO_4 + 2\,H^+ + 2\,e \rightleftharpoons HAsO_2 + 2\,H_2O$	0.560
$S_2O_6^{2-} + 4\,H^+ + 2\,e \rightleftharpoons 2\,H_2SO_3$	0.564
$AgNO_2 + e \rightleftharpoons Ag + NO_2^-$	0.564
$Te^{4+} + 4\,e \rightleftharpoons Te$	0.568
$Sb_2O_5 + 6\,H^+ + 4\,e \rightleftharpoons 2\,SbO^+ + 3\,H_2O$	0.581
$RuO_4^- + e \rightleftharpoons RuO_4^{2-}$	0.59
$[PdCl_4]^{2-} + 2\,e \rightleftharpoons Pd + 4\,Cl^-$	0.591
$TeO_2 + 4\,H^+ + 4\,e \rightleftharpoons Te + 2\,H_2O$	0.593
$MnO_4^- + 2\,H_2O + 3\,e \rightleftharpoons MnO_2 + 4\,OH^-$	0.595
$Rh^{2+} + 2\,e \rightleftharpoons Rh$	0.600

Reaction	$E°/V$
$Rh^+ + e \rightleftharpoons Rh$	0.600
$MnO_4^{2-} + 2 H_2O + 2 e \rightleftharpoons MnO_2 + 4 OH^-$	0.60
$2 AgO + H_2O + 2 e \rightleftharpoons Ag_2O + 2 OH^-$	0.607
$BrO_3^- + 3 H_2O + 6 e \rightleftharpoons Br^- + 6 OH^-$	0.61
$UO_2^+ + 4 H^+ + e \rightleftharpoons U^{4+} + 2 H_2O$	0.612
$Hg_2SO_4 + 2 e \rightleftharpoons 2 Hg + SO_4^{2-}$	0.6125
$ClO_3^- + 3 H_2O + 6 e \rightleftharpoons Cl^- + 6 OH^-$	0.62
$Hg_2HPO_4 + 2 e \rightleftharpoons 2 Hg + HPO_4^{2-}$	0.6359
$Ag(ac) + e \rightleftharpoons Ag + (ac)^-$	0.643
Sb_2O_5(valentinite) $+ 4 H^+ + 4 e \rightleftharpoons Sb_2O_3 + 2 H_2O$	0.649
$Ag_2SO_4 + 2 e \rightleftharpoons 2 Ag + SO_4^{2-}$	0.654
$ClO_2^- + H_2O + 2 e \rightleftharpoons ClO^- + 2 OH^-$	0.66
Sb_2O_5(senarmontite) $+ 4 H^+ + 4 e \rightleftharpoons Sb_2O_5 + 2 H_2O$	0.671
$[PtCl_6]^{2-} + 2 e \rightleftharpoons [PtCl_4]^{2-} + 2 Cl^-$	0.68
$O_2 + 2 H^+ + 2 e \rightleftharpoons H_2O_2$	0.695
p–benzoquinone $+ 2 H^+ + 2 e \rightleftharpoons$ hydroquinone	0.6992
$H_3IO_6^{2-} + 2 e \rightleftharpoons IO_3^- + 3 OH^-$	0.7
$Ag_2O_3 + H_2O + 2 e \rightleftharpoons 2 AgO + 2 OH^-$	0.739
$Tl^{3+} + 3 e \rightleftharpoons Tl$	0.741
$[PtCl_4]^{2-} + 2 e \rightleftharpoons Pt + 4 Cl^-$	0.755
$Rh^{3+} + 3 e \rightleftharpoons Rh$	0.758
$ClO_2^- + 2 H_2O + 4 e \rightleftharpoons Cl^- + 4 OH^-$	0.76
$2 NO + H_2O + 2 e \rightleftharpoons N_2O + 2 OH^-$	0.76
$Po^{4+} + 4 e \rightleftharpoons Po$	0.76
$BrO^- + H_2O + 2 e \rightleftharpoons Br^- + 2 OH^-$	0.761
$ReO_4^- + 2 H^+ + e \rightleftharpoons ReO_3 + H_2O$	0.768
$(CNS)_2 + 2 e \rightleftharpoons 2 CNS^-$	0.77
$[IrCl_6]^{3-} + 3 e \rightleftharpoons Ir + 6 Cl^-$	0.77
$Fe^{3+} + e \rightleftharpoons Fe^{2+}$	0.771
$AgF + e \rightleftharpoons Ag + F^-$	0.779
$[Fe(bipy)_2]^{3+} + e \rightleftharpoons [Fe(bipy)_2]^{2+}$	0.78
$TcO_4^- + 4 H^+ + 3 e \rightleftharpoons TcO_2 + 2 H_2O$	0.782
$Hg_2^{2+} + 2 e \rightleftharpoons 2 Hg$	0.7973
$Ag^+ + e \rightleftharpoons Ag$	0.7996
$[Os(bipy)_3]^{3+} + e \rightleftharpoons [Os(bipy)_3]^{2+}$	0.80
$2 NO_3^- + 4 H^+ + 2 e \rightleftharpoons N_2O_4 + 2 H_2O$	0.803
$[Os(bipy)_2]^{3+} + e \rightleftharpoons [Os(bipy)_2]^{2+}$	0.81
$RhOH^{2+} + H + 3 e \rightleftharpoons Rh + H_2O$	0.83
$OsO_4 + 8 H^+ + 8 e \rightleftharpoons Os + 4 H_2O$	0.838
$ClO^- + H_2O + 2 e \rightleftharpoons Cl^- + 2 OH^-$	0.841
$Hg^{2+} + 2 e \rightleftharpoons Hg$	0.851
$AuBr_4^- + 3 e \rightleftharpoons Au + 4 Br^-$	0.854
SiO_2(quartz) $+ 4 H^+ + 4 e \rightleftharpoons Si + 2 H_2O$	0.857
$2 HNO_2 + 4 H^+ + 4 e \rightleftharpoons H_2N_2O_2 + H_2O$	0.86
$[Ru(CN)_6]^{3-} + e^- \rightleftharpoons [Ru(CN)_6]^{4-}$	0.86
$[IrCl_6]^{2-} + e \rightleftharpoons [IrCl_6]^{3-}$	0.8665
$N_2O_4 + 2 e \rightleftharpoons 2 NO_2^-$	0.867
$HO_2^- + H_2O + 2 e \rightleftharpoons 3 OH^-$	0.878
$Po^{4+} + 2 e \rightleftharpoons Po^{2+}$	0.9
$2 Hg^{2+} + 2 e \rightleftharpoons Hg_2^{2+}$	0.920
$NO_3^- + 3 H^+ + 2 e \rightleftharpoons HNO_2 + H_2O$	0.934
$Pd^{2+} + 2 e \rightleftharpoons Pd$	0.951
$ClO_2(aq) + e \rightleftharpoons ClO_2^-$	0.954
$NO_3^- + 4 H^+ + 3 e \rightleftharpoons NO + 2 H_2O$	0.957
$V_2O_5 + 6 H^+ + 2 e \rightleftharpoons 2 VO^{2+} + 3 H_2O$	0.957
$AuBr_2^- + e \rightleftharpoons Au + 2 Br^-$	0.959
$HNO_2 + H^+ + e \rightleftharpoons NO + H_2O$	0.983

Reaction	$E°/V$
$HIO + H^+ + 2 e \rightleftharpoons I^- + H_2O$	0.987
$VO_2^+ + 2 H^+ + e \rightleftharpoons VO^{2+} + H_2O$	0.991
$PtO_2 + 4 H^+ + 4 e \rightleftharpoons Pt + 2 H_2O$	1.00
$RuO_4 + e \rightleftharpoons RuO_4^-$	1.00
$V(OH)_4^+ + 2 H^+ + e \rightleftharpoons VO^{2+} + 3 H_2O$	1.00
$AuCl_4^- + 3 e \rightleftharpoons Au + 4 Cl^-$	1.002
$Pu^{4+} + e \rightleftharpoons Pu^{3+}$	1.006
$PtO_2 + 2 H^+ + 2 e \rightleftharpoons PtO + H_2O$	1.01
$OsO_4 + 4 H + 4 e \rightleftharpoons OsO_2 + 2 H_2O$	1.02
$H_6TeO_6 + 2 H^+ + 2 e \rightleftharpoons TeO_2 + 4 H_2O$	1.02
$[Fe(bipy)_3]^{3+} + e \rightleftharpoons [Fe(bipy)_3]^{2+}$	1.03
$Hg(OH)_2 + 2 H^+ + 2 e \rightleftharpoons Hg + 2 H_2O$	1.034
$N_2O_4 + 4 H^+ + 4 e \rightleftharpoons 2 NO + 2 H_2O$	1.035
$RuO_4 + 8 H^+ + 8 e \rightleftharpoons Ru + 4H_2O$	1.038
$[Fe(phen)_3]^{3+} + e \rightleftharpoons [Fe(phen)_3]^{2+}$ (1 molar H_2SO_4)	1.06
$PuO_2(OH)_2 + H^+ + e \rightleftharpoons PuO_2OH + H_2O$	1.062
$N_2O_4 + 2 H^+ + 2 e \rightleftharpoons 2 HNO_2$	1.065
$Br_2(l) + 2 e \rightleftharpoons 2Br^-$	1.066
$IO_3^- + 6 H^+ + 6 e \rightleftharpoons I^- + 3 H_2O$	1.085
$Br_2(aq) + 2 e \rightleftharpoons 2Br^-$	1.0873
$Pu^{5+} + e \rightleftharpoons Pu^{4+}$	1.099
$Cu^{2+} + 2 CN^- + e \rightleftharpoons [Cu(CN)_2]^-$	1.103
$RuO_2 + 4 H^+ + 2 e \rightleftharpoons Ru^{2+} + 2 H_2O$	1.120
$[Fe(phen)_3]^{3+} + e \rightleftharpoons [Fe(phen)_3]^{2+}$	1.147
$SeO_4^{2-} + 4 H^+ + 2 e \rightleftharpoons H_2SeO_3 + H_2O$	1.151
$ClO_3^- + 2 H^+ + e \rightleftharpoons ClO_2 + H_2O$	1.152
$Ir^{3+} + 3 e \rightharpoonup Ir$	1.156
$Pt^{2+} + 2 e \rightleftharpoons Pt$	1.18
$ClO_4^- + 2 H^+ + 2 e \rightleftharpoons ClO_3^- + H_2O$	1.189
$2 IO_3^- + 12 H^+ + 10 e \rightleftharpoons I_2 + 6 H_2O$	1.195
$PtOH^+ + H^+ + 2 e \rightleftharpoons Pt + H_2O$	1.2
$ClO_3^- + 3 H^+ + 2 e \rightleftharpoons HClO_2 + H_2O$	1.214
$MnO_2 + 4 H^+ + 2 e \rightleftharpoons Mn^{2+} + 2 H_2O$	1.224
$O_2 + 4 H^+ + 4 e \rightleftharpoons 2 H_2O$	1.229
$O_3 + H_2O + 2 e \rightleftharpoons O_2 + 2 OH^-$	1.24
$[Ru(bipy)_3]^{3+} + e \rightleftharpoons [Ru(bipy)_3]^{2+}$	1.24
$Tl^{3+} + 2 e \rightleftharpoons Tl^+$	1.252
$N_2H_5^+ + 3 H^+ + 2 e \rightleftharpoons 2 NH_4^+$	1.275
$ClO_2 + H^+ + e \rightleftharpoons HClO_2$	1.277
$[PdCl_6]^{2-} + 2 e \rightleftharpoons [PdCl_4]^{2-} + 2 Cl^-$	1.288
$2 HNO_2 + 4 H^+ + 4 e \rightleftharpoons N_2O + 3 H_2O$	1.297
$AuOH^{2+} + H^+ + 2 e \rightleftharpoons Au^+ + H_2O$	1.32
$PuO_2(OH)_2 + 2 H^- + 2 e \rightleftharpoons Pu(OH)_4$	1.325
$HBrO + H^+ + 2 e \rightleftharpoons Br^- + H_2O$	1.331
$Cr(V) + e \rightleftharpoons Cr(IV)$	1.34
$HCrO_4^- + 7 H^+ + 3 e \rightleftharpoons Cr^{3+} + 4 H_2O$	1.350
$Cl_2(g) + 2 e \rightleftharpoons 2Cl^-$	1.35827
$Cr_2O_7^{2-} + 14 H^+ + 6 e \rightleftharpoons 2 Cr^{3+} + 7 H_2O$	1.36
$ClO_4^- + 8 H^+ + 8 e \rightleftharpoons Cl^- + 4 H_2O$	1.389
$ClO_4^- + 8 H^+ + 7 e \rightleftharpoons 1/2 Cl_2 + 4 H_2O$	1.39
$No^{3+} + e \rightleftharpoons No^{2+}$	1.4
$RuO_4 + 6 H^+ + 4 e \rightleftharpoons Ru(OH)_2^{2+} + 2 H_2O$	1.40
$Au^{3+} + 2 e \rightleftharpoons Au^+$	1.401
$2 NH_3OH^+ + H^+ + 2 e \rightleftharpoons N_2H_5^+ + 2 H_2O$	1.42
$BrO_3^- + 6 H^+ + 6 e \rightleftharpoons Br^- + 3 H_2O$	1.423
$2 HIO + 2 H^+ + 2 e \rightleftharpoons I_2 + 2 H_2O$	1.439
$Au(OH)_3 + 3 H^+ + 3 e \rightleftharpoons Au^- + 3 H_2O$	1.45

Reaction	$E°/V$
$3IO_3^- + 6 H^+ + 6 e \rightleftharpoons Cl^- + 3 H_2O$	1.451
$PbO_2 + 4 H^+ + 2 e \rightleftharpoons Pb^{2+} + 2 H_2O$	1.455
$ClO_3^- + 6 H^+ + 5 e \rightleftharpoons 1/2 Cl_2 + 3 H_2O$	1.47
$CrO_2 + 4 H^+ + e \rightleftharpoons Cr^{3+} + 2 H_2O$	1.48
$BrO_3^- + 6 H^+ + 5 e \rightleftharpoons 1/2 Br_2 + 3 H_2O$	1.482
$HClO + H^+ + 2 e \rightleftharpoons Cl^- + H_2O$	1.482
$Mn_2O_3 + 6 H^+ + e \rightleftharpoons 2 Mn^{2+} + 3 H_2O$	1.485
$HO_2 + H^+ + e \rightleftharpoons H_2O_2$	1.495
$Au^{3+} + 3 e \rightleftharpoons Au$	1.498
$PtO_3 + 4 H^+ + 2 e \rightleftharpoons Pt(OH)_2^{2+} + H_2O$	1.5
$MnO_4^- + 8 H^+ + 5 e \rightleftharpoons Mn^{2+} + 4 H_2O$	1.507
$Mn^{3+} + e \rightleftharpoons Mn^{2+}$	1.5415
$HClO_2 + 3 H^+ + 4 e \rightleftharpoons Cl^- + 2 H_2O$	1.570
$HBrO + H^+ + e \rightleftharpoons 1/2 Br_2(aq) + H_2O$	1.574
$2 NO + 2 H^+ + 2 e \rightleftharpoons N_2O + H_2O$	1.591
$Bi_2O_4 + 4 H^+ + 2 e \rightleftharpoons 2 BiO^+ + 2 H_2O$	1.593
$HBrO + H^+ + e \rightleftharpoons 1/2 Br_2(l) + H_2O$	1.596
$H_5IO_6 + H^+ + 2 e \rightleftharpoons IO_3^- + 3 H_2O$	1.601
$HClO + H^+ + e \rightleftharpoons 1/2 Cl_2 + H_2O$	1.611
$HClO_2 + 3 H^+ + 3 e \rightleftharpoons 1/2 Cl_2 + 2 H_2O$	1.628
$HClO_2 + 2 H^+ + 2 e \rightleftharpoons HClO + H_2O$	1.645
$Bk^{4+} + e \rightleftharpoons Bk^{3+}$	1.67
$NiO_2 + 4 H^+ + 2 e \rightleftharpoons Ni^{2+} + 2 H_2O$	1.678
$MnO_4^- + 4 H^+ + 3 e \rightleftharpoons MnO_2 + 2 H_2O$	1.679
$PbO_2 + SO_4^{2-} + 4 H^+ + 2 e \rightleftharpoons PbSO_4 + 2 H_2O$	1.6913
$Au^+ + e \rightleftharpoons Au$	1.692
$PtO_3 + 2 H^+ + 2 e \rightleftharpoons PtO_2 + H_2O$	1.7
$CeOH^{3+} + H^+ + e \rightleftharpoons Ce^{3+} + H_2O$	1.715
$Ce^{4+} + e \rightleftharpoons Ce^{3+}$	1.72
$N_2O + 2 H^+ + 2 e \rightleftharpoons N_2 + H_2O$	1.766
$H_2O_2 + 2 H^+ + 2 e \rightleftharpoons 2 H_2O$	1.776
$Ag^{3+} + e \rightleftharpoons Ag^{2+}$	1.8
$Au^{2+} + e^- \rightleftharpoons Au^+$	1.8
$Ag_2O_2 + 4 H^+ + e \rightleftharpoons 2 Ag + 2 H_2O$	1.802
$Co^{3+} + e \rightleftharpoons Co^{2+}$(2 molar H_2SO_4)	1.83
$Ag^{3+} + 2 e \rightleftharpoons Ag^+$	1.9
$Co^{3+} + e \rightleftharpoons Co^{2+}$	1.92
$Ag^{2+} + e \rightleftharpoons Ag^+$	1.980
$Cu_2O_3 + 6 H^+ + 2 e \rightleftharpoons 2 Cu^{2+} + 3 H_2O$	2.0
$S_2O_8^{2-} + 2 e \rightleftharpoons 2 SO_4^{2-}$	2.010
$OH + e \rightleftharpoons OH^-$	2.02
$HFeO_4^- + 7 H^+ + 3 e \rightleftharpoons Fe^{3+} + 4 H_2O$	2.07
$O_3 + 2 H^+ + 2 e \rightleftharpoons O_2 + H_2O$	2.076
$HFeO_4^- + 4 H^+ + 3 e \rightleftharpoons FeOOH + 2 H_2O$	2.08
$2 HFeO_4^- + 8 H^+ + 6 e \rightleftharpoons Fe_2O_3 + 5 H_2O$	2.09
$XeO_3 + 6 H^+ + 6 e \rightleftharpoons Xe + 3 H_2O$	2.10
$S_2O_8^{2-} + 2 H^+ + 2 e \rightleftharpoons 2 HSO_4^-$	2.123
$F_2O + 2 H^+ + 4 e \rightleftharpoons H_2O + 2 F^-$	2.153
$FeO_4^{2-} + 8 H^+ + 3 e \rightleftharpoons Fe^{3+} + 4 H_2O$	2.20
$Cu^{3+} + e \rightleftharpoons Cu^{2+}$	2.4
$H_4XeO_6 + 2 H^+ + 2 e \rightleftharpoons XeO_3 + 3 H_2O$	2.42
$O(g) + 2 H^+ + 2 e \rightleftharpoons H_2O$	2.421
$Am^{4+} + e \rightleftharpoons Am^{3+}$	2.60
$H_2N_2O_2 + 2 H^+ + 2 e \rightleftharpoons N_2 + 2 H_2O$	2.65
$F_2 + 2 e \rightleftharpoons 2 F^-$	2.866
$Cm^{4+} + e \rightleftharpoons Cm^{3+}$	3.0
$F_2 + 2 H^+ + 2 e \rightleftharpoons 2 HF$	3.053
$Tb^{4+} + e \rightleftharpoons Tb^{3+}$	3.1
$Pr^{4+} + e \rightleftharpoons Pr^{3+}$	3.2
$Cf^{4+} + e \rightleftharpoons Cf^{3+}$	3.3
$XeF + e \rightleftharpoons Xe + F^-$	3.4

TABLE 3. Reduction Reactions Having $E°$ Values More Negative than That of the Standard Hydrogen Electrode

Reaction	$E°/V$
$2 H^+ + 2 e \rightleftharpoons H_2$	0.00000
$2 D^+ + 2 e \rightleftharpoons D_2$	–0.013
$AgCN + e \rightleftharpoons Ag + CN^-$	–0.017
$2 WO_3 + 2 H^+ + 2 e \rightleftharpoons W_2O_5 + H_2O$	–0.029
$W_2O_5 + 2 H^+ + 2 e \rightleftharpoons 2 WO_2 + H_2O$	–0.031
$Ag_2S + 2 H^+ + 2 e \rightleftharpoons 2 Ag + H_2S$	–0.0366
$Fe^{3+} + 3 e \rightleftharpoons Fe$	–0.037
$Hg_2I_2 + 2 e \rightleftharpoons 2 Hg + 2 I^-$	–0.0405
$Tl(OH)_3 + 2 e \rightleftharpoons TlOH + 2 OH^-$	–0.05
$TiOH^{3+} + H^+ + e \rightleftharpoons Ti^{3+} + H_2O$	–0.055
$2 H_2SO_3 + H^+ + 2 e \rightleftharpoons HS_2O_4^- + 2 H_2O$	–0.056
$P(white) + 3 H^+ + 3 e \rightleftharpoons PH_3(g)$	–0.063
$O_2 + H_2O + 2 e \rightleftharpoons HO_2^- + OH^-$	–0.076
$2 Cu(OH)_2 + 2 e \rightleftharpoons Cu_2O + 2 OH^- + H_2O$	–0.080
$Se + 2 H^+ + 2 e \rightleftharpoons H_2Se$	–0.082
$WO_3 + 6 H^+ + 6 e \rightleftharpoons W + 3 H_2O$	–0.090
$SnO_2 + 4 H^+ + 2 e \rightleftharpoons Sn^{2+} + 2 H_2O$	–0.094
$Md^{3+} + e \rightleftharpoons Md^{2+}$	–0.1
$P(red) + 3 H^+ + 3 e \rightleftharpoons PH_3(g)$	–0.111
$SnO_2 + 4 H^+ + 4 e \rightleftharpoons Sn + 2 H_2O$	–0.117
$GeO_2 + 2 H^+ + 2 e \rightleftharpoons GeO + H_2O$	–0.118
$WO_2 + 4 H^+ + 4 e \rightleftharpoons W + 2 H_2O$	–0.119
$Pb^{2+} + 2 e \rightleftharpoons Pb(Hg)$	–0.1205
$Pb^{2+} + 2 e \rightleftharpoons Pb$	–0.1262
$CrO_4^{2-} + 4 H_2O + 3 e \rightleftharpoons Cr(OH)_3 + 5 OH^-$	–0.13
$Sn^{2-} + 2 e \rightleftharpoons Sn$	–0.1375
$In^+ + e \rightleftharpoons In$	–0.14
$O_2 + 2 H_2O + 2 e \rightleftharpoons H_2O_2 + 2 OH^-$	–0.146
$MoO_2 + 4 H^+ + 4 e \rightleftharpoons Mo + 4 H_2O$	–0.152
$AgI + e \rightleftharpoons Ag + I^-$	–0.15224
$2 NO_2^- + 2 H_2O + 4 e \rightleftharpoons N_2O_2^{2-} + 4 OH^-$	–0.18
$H_2GeO_3 + 4 H^+ + 4 e \rightleftharpoons Ge + 3 H_2O$	–0.182
$SnO_2 + 3 H^+ + 2 e \rightleftharpoons SnOH^+ + H_2O$	–0.194
$CO_2 + 2 H^+ + 2 e \rightleftharpoons HCOOH$	–0.199
$Mo^{3+} + 3 e \rightleftharpoons Mo$	–0.200
$Ga^+ + e \rightleftharpoons Ga$	–0.2
$2 SO_4^{2-} + 4 H^+ + 2 e \rightleftharpoons S_2O_6^{2-} + H_2O$	–0.22
$Cu(OH)_2 + 2 e \rightleftharpoons Cu + 2 OH^-$	–0.222
$V_2O_5 + 10 H^+ + 10 e \rightleftharpoons 2 V + 5 H_2O$	–0.242
$CdSO_4 + 2 e \rightleftharpoons Cd + SO_4^{2-}$	–0.246
$V(OH)_4^+ + 4 H^+ + 5 e \rightleftharpoons V + 4 H_2O$	–0.254
$V^{3+} + e \rightleftharpoons V^{2+}$	–0.255

Reaction	$E°$/V
$Ni^{2+} + 2\,e \rightleftharpoons Ni$	–0.257
$PbCl_2 + 2\,e \rightleftharpoons Pb + 2\,Cl^-$	–0.2675
$H_3PO_4 + 2\,H^+ + 2\,e \rightleftharpoons H_3PO_3 + H_2O$	–0.276
$Co^{2+} + 2\,e \rightleftharpoons Co$	–0.28
$PbBr_2 + 2\,e \rightleftharpoons Pb + 2\,Br^-$	–0.284
$Tl^+ + e \rightleftharpoons Tl(Hg)$	–0.3338
$Tl^+ + e \rightleftharpoons Tl$	–0.336
$In^{3+} + 3\,e \rightleftharpoons In$	–0.3382
$TlOH + e \rightleftharpoons Tl + OH^-$	–0.34
$PbF_2 + 2\,e \rightleftharpoons Pb + 2\,F^-$	–0.3444
$PbSO_4 + 2\,e \rightleftharpoons Pb(Hg) + SO_4^{2-}$	–0.3505
$Cd^{2+} + 2\,e \rightleftharpoons Cd(Hg)$	–0.3521
$PbSO_4 + 2\,e \rightleftharpoons Pb + SO_4^{2-}$	–0.3588
$Cu_2O + H_2O + 2\,e \rightleftharpoons 2\,Cu + 2\,OH^-$	–0.360
$Eu^{3+} + e \rightleftharpoons Eu^{2+}$	–0.36
$PbI_2 + 2\,e \rightleftharpoons Pb + 2\,I^-$	–0.365
$SeO_3^{2-} + 3\,H_2O + 4\,e \rightleftharpoons Se + 6\,OH^-$	–0.366
$Se + 2\,H^+ + 2\,e \rightleftharpoons H_2Se(aq)$	–0.399
$In^{2+} + e \rightleftharpoons In^+$	–0.40
$Cd^{2+} + 2\,e \rightleftharpoons Cd$	–0.4030
$Cr^{3+} + e \rightleftharpoons Cr^{2+}$	–0.407
$2\,S + 2\,e \rightleftharpoons S_2^{2-}$	–0.42836
$Tl_2SO_4 + 2\,e \rightleftharpoons Tl + SO_4^{2-}$	–0.4360
$In^{3+} + 2\,e \rightleftharpoons In^+$	–0.443
$Fe^{2+} + 2\,e \rightleftharpoons Fe$	–0.447
$H_3PO_3 + 3\,H^+ + 3\,e \rightleftharpoons P + 3\,H_2O$	–0.454
$Bi_2O_3 + 3\,H_2O + 6\,e \rightleftharpoons 2\,Bi + 6\,OH^-$	–0.46
$NO_2^- + H_2O + e \rightleftharpoons NO + 2\,OH$	–0.46
$PbHPO_4 + 2\,e \rightleftharpoons Pb + HPO_4^{2-}$	–0.465
$S + 2\,e \rightleftharpoons S^{2-}$	–0.47627
$S + H_2O + 2\,e \rightleftharpoons HS^- + OH^-$	–0.478
$B(OH)_3 + 7\,H^+ + 8\,e \rightleftharpoons BH_4^- + 3\,H_2O$	–0.481
$In^{3+} + e \rightleftharpoons In^{2+}$	–0.49
$ZnOH^+ + H^+ + 2\,e \rightleftharpoons Zn + H_2O$	–0.497
$GaOH^{2+} + H^+ + 3\,e \rightleftharpoons Ga + H_2O$	–0.498
$H_3PO_3 + 2\,H^+ + 2\,e \rightleftharpoons H_3PO_2 + H_2O$	–0.499
$TiO_2 + 4\,H^+ + 2\,e \rightleftharpoons Ti^{2+} + 2\,H_2O$	–0.502
$H_3PO_2 + H^+ + e \rightleftharpoons P + 2\,H_2O$	–0.508
$Sb + 3\,H^+ + 3\,e \rightleftharpoons SbH_3$	–0.510
$HPbO_2^- + H_2O + 2\,e \rightleftharpoons Pb + 3\,OH^-$	–0.537
$Ga^{3+} + 3\,e \rightleftharpoons Ga$	–0.549
$TlCl + e \rightleftharpoons Tl + Cl^-$	–0.5568
$Fe(OH)_3 + e \rightleftharpoons Fe(OH)_2 + OH^-$	–0.56
$TeO_3^{2-} + 3\,H_2O + 4\,e \rightleftharpoons Te + 6\,OH^-$	–0.57
$2\,SO_3^{2-} + 3\,H_2O + 4\,e \rightleftharpoons S_2O_3^{2-} + 6\,OH^-$	–0.571
$PbO + H_2O + 2\,e \rightleftharpoons Pb + 2\,OH^-$	–0.580
$ReO_2^- + 4\,H_2O + 7\,e \rightleftharpoons Re + 8\,OH^-$	–0.584
$SbO_3^- + H_2O + 2\,e \rightleftharpoons SbO_2^- + 2\,OH^-$	–0.59
$Ta^{3+} + 3\,e \rightleftharpoons Ta$	–0.6
$U^{4+} + e \rightleftharpoons U^{3+}$	–0.607
$As + 3\,H^+ + 3\,e \rightleftharpoons AsH_3$	–0.608
$Nb_2O_5 + 10\,H^+ + 10\,e \rightleftharpoons 2\,Nb + 5\,H_2O$	–0.644
$NbO_2 + 2\,H^+ + 2\,e \rightleftharpoons NbO + H_2O$	–0.646
$Cd(OH)_4^{2-} + 2\,e \rightleftharpoons Cd + 4\,OH^-$	–0.658
$TlBr + e \rightleftharpoons Tl + Br^-$	–0.658
$SbO_2^- + 2\,H_2O + 3\,e \rightleftharpoons Sb + 4\,OH^-$	–0.66
$AsO_2^- + 2\,H_2O + 3\,e \rightleftharpoons As + 4\,OH^-$	–0.68

Reaction	$E°$/V
$NbO_2 + 4\,H^+ + 4\,e \rightleftharpoons Nb + 2\,H_2O$	–0.690
$Ag_2S + 2\,e \rightleftharpoons 2\,Ag + S^{2-}$	–0.691
$AsO_4^{3-} + 2\,H_2O + 2\,e \rightleftharpoons AsO_2^- + 4\,OH^-$	–0.71
$Ni(OH)_2 + 2\,e \rightleftharpoons Ni + 2\,OH^-$	–0.72
$Co(OH)_2 + 2\,e \rightleftharpoons Co + 2\,OH^-$	–0.73
$NbO + 2\,H^+ + 2\,e \rightleftharpoons Nb + H_2O$	–0.733
$H_2SeO_3 + 4\,H^+ + 4\,e \rightleftharpoons Se + 3\,H_2O$	–0.74
$Cr^{3+} + 3\,e \rightleftharpoons Cr$	–0.744
$Ta_2O_5 + 10\,H^+ + 10\,e \rightleftharpoons 2\,Ta + 5\,H_2O$	–0.750
$TlI + e \rightleftharpoons Tl + I^-$	–0.752
$Zn^{2+} + 2\,e \rightleftharpoons Zn$	–0.7618
$Zn^{2+} + 2\,e \rightleftharpoons Zn(Hg)$	–0.7628
$CdO + H_2O + 2\,e \rightleftharpoons Cd + 2\,OH^-$	–0.783
$Te + 2\,H^+ + 2\,e \rightleftharpoons H_2Te$	–0.793
$ZnSO_4 \cdot 7H_2O + 2\,e \rightleftharpoons Zn(Hg) + SO_4^{2-} + 7\,H_2O$ (Saturated $ZnSO_4$)	–0.7993
$Bi + 3\,H^+ + 3\,e \rightleftharpoons BiH_3$	–0.8
$SiO + 2\,H^+ + 2\,e \rightleftharpoons Si + H_2O$	–0.8
$Cd(OH)_2 + 2\,e \rightleftharpoons Cd(Hg) + 2\,OH^-$	–0.809
$2\,H_2O + 2\,e \rightleftharpoons H_2 + 2\,OH^-$	–0.8277
$2\,NO_3^- + 2\,H_2O + 2\,e \rightleftharpoons N_2O_4 + 4\,OH^-$	–0.85
$H_3BO_3 + 3\,H^+ + 3\,e \rightleftharpoons B + 3\,H_2O$	–0.8698
$P + 3\,H_2O + 3\,e \rightleftharpoons PH_3(g) + 3\,OH^-$	–0.87
$Ti^{3+} + e \rightleftharpoons Ti^{2+}$	–0.9
$HSnO_2^- + H_2O + 2\,e \rightleftharpoons Sn + 3\,OH^-$	–0.909
$Cr^{2+} + 2\,e \rightleftharpoons Cr$	–0.913
$Se + 2\,e \rightleftharpoons Se^{2-}$	–0.924
$SO_4^{2-} + H_2O + 2\,e \rightleftharpoons SO_3^{2-} + 2\,OH^-$	–0.93
$Sn(OH)_6^{2-} + 2\,e \rightleftharpoons HSnO_2^- + 3\,OH^- + H_2O$	–0.93
$SnO_2 + 2\,H_2O + 4\,e \rightleftharpoons Sn + 4\,OH^-$	–0.945
$In(OH)_3 + 3\,e \rightleftharpoons In + 3\,OH^-$	–0.99
$NpO_2 + H_2O + H^+ + e \rightleftharpoons Np(OH)_3$	–0.962
$In(OH)_4^- + 3\,e \rightleftharpoons In + 4\,OH^-$	–1.007
$In_2O_3 + 3\,H_2O + 6\,e \rightleftharpoons 2\,In + 6\,OH^-$	–1.034
$PO_4^{3-} + 2\,H_2O + 2\,e \rightleftharpoons HPO_3^{2-} + 3\,OH^-$	–1.05
$Yb^{3+} + e \rightleftharpoons Yb^{2+}$	–1.05
$Nb^{3+} + 3\,e \rightleftharpoons Nb$	–1.099
$Fm^{3+} + e \rightleftharpoons Fm^{2+}$	–1.1
$2\,SO_3^{2-} + 2\,H_2O + 2\,e \rightleftharpoons S_2O_4^{2-} + 4\,OH^-$	–1.12
$Te + 2\,e \rightleftharpoons Te^{2-}$	–1.143
$V^{2+} + 2\,e \rightleftharpoons V$	–1.175
$Mn^{2+} + 2\,e \rightleftharpoons Mn$	–1.185
$Zn(OH)_4^{2-} + 2\,e \rightleftharpoons Zn + 4\,OH^-$	–1.199
$CrO_2 + 2\,H_2O + 3\,e \rightleftharpoons Cr + 4\,OH^-$	–1.2
$No^{3+} + 3\,e \rightleftharpoons No$	–1.20
$ZnO_2^- + 2\,H_2O + 2\,e \rightleftharpoons Zn + 4\,OH^-$	–1.215
$H_2GaO_3^- + H_2O + 3\,e \rightleftharpoons Ga + 4\,OH^-$	–1.219
$H_2BO_3^- + 5\,H_2O + 8\,e \rightleftharpoons BH_4^- + 8\,OH^-$	–1.24
$SiF_6^{2-} + 4\,e \rightleftharpoons Si + 6\,F^-$	–1.24
$Zn(OH)_2 + 2\,e \rightleftharpoons Zn + 2\,OH^-$	–1.249
$ZnO + H_2O + 2\,e \rightleftharpoons Zn + 2\,OH^-$	–1.260
$Es^{3+} + e \rightleftharpoons Es^{2+}$	–1.3
$Pa^{3+} + 3\,e \rightleftharpoons Pa$	–1.34
$Ti^{3+} + 3\,e \rightleftharpoons Ti$	–1.37
$Ce^{3+} + 3\,e \rightleftharpoons Ce(Hg)$	–1.4373
$UO_2^{2+} + 4\,H^+ + 6\,e \rightleftharpoons U + 2\,H_2O$	–1.444
$Zr^{4+} + 4\,e \rightleftharpoons Zr$	–1.45

Reaction	$E°/V$
$Cr(OH)_3 + 3\,e \rightleftharpoons Cr + 3\,OH^-$	–1.48
$Pa^{4+} + 4\,e \rightleftharpoons Pa$	–1.49
$HfO_2 + 4\,H^+ + 4\,e \rightleftharpoons Hf + 2\,H_2O$	–1.505
$Hf^{4+} + 4\,e \rightleftharpoons Hf$	–1.55
$Sm^{3+} + e \rightleftharpoons Sm^{2+}$	–1.55
$ZrO_2 + 4\,H^+ + 4\,e \rightleftharpoons Zr + 2\,H_2O$	–1.553
$Mn(OH)_2 + 2\,e \rightleftharpoons Mn + 2\,OH^-$	–1.56
$Ba^{2+} + 2\,e \rightleftharpoons Ba(Hg)$	–1.570
$Bk^{2+} + 2\,e \rightleftharpoons Bk$	–1.6
$Cf^{3+} + e \rightleftharpoons Cf^{2+}$	–1.6
$Ti^{2+} + 2\,e \rightleftharpoons Ti$	–1.630
$Md^{3+} + 3\,e \rightleftharpoons Md$	–1.65
$HPO_3^{2-} + 2\,H_2O + 2\,e \rightleftharpoons H_2PO_2^- + 3\,OH^-$	–1.65
$Al^{3+} + 3\,e \rightleftharpoons Al$	–1.662
$SiO_3^{2-} + H_2O + 4\,e \rightleftharpoons Si + 6\,OH^-$	–1.697
$HPO_3^{2-} + 2\,H_2O + 3\,e \rightleftharpoons P + 5\,OH^-$	–1.71
$HfO^{2+} + 2\,H^+ + 4\,e \rightleftharpoons Hf + H_2O$	–1.724
$ThO_2 + 4\,H^+ + 4\,e \rightleftharpoons Th + 2\,H_2O$	–1.789
$H_2BO_3^- + H_2O + 3\,e \rightleftharpoons B + 4\,OH^-$	–1.79
$Sr^{2+} + 2\,e \rightleftharpoons Sr(Hg)$	–1.793
$U^{3+} + 3\,e \rightleftharpoons U$	–1.798
$H_2PO^-_2 + e \rightleftharpoons P + 2\,OH^-$	–1.82
$Be^{2+} + 2\,e \rightleftharpoons Be$	–1.847
$Np^{3+} + 3\,e \rightleftharpoons Np$	–1.856
$Fm^{3+} + 3\,e \rightleftharpoons Fm$	–1.89
$Th^{4+} + 4\,e \rightleftharpoons Th$	–1.899
$Am^{2+} + 2\,e \rightleftharpoons Am$	–1.9
$Pa^{4+} + e \rightleftharpoons Pa^{3+}$	–1.9
$Es^{3+} + 3\,e \rightleftharpoons Es$	–1.91
$Cf^{3+} + 3\,e \rightleftharpoons Cf$	–1.94
$Lr^{3+} + 3\,e \rightleftharpoons Lr$	–1.96
$Eu^{3+} + 3\,e \rightleftharpoons Eu$	–1.991
$Er^{2+} + 2\,e \rightleftharpoons Er$	–2.0
$Pr^{2+} + 2\,e \rightleftharpoons Pr$	–2.0
$Pu^{3+} + 3\,e \rightleftharpoons Pu$	–2.031
$Cm^{3+} + 3\,e \rightleftharpoons Cm$	–2.04
$Am^{3+} + 3\,e \rightleftharpoons Am$	–2.048
$AlF_6^{3-} + 3\,e \rightleftharpoons Al + 6\,F^-$	–2.069
$Sc^{3+} + 3\,e \rightleftharpoons Sc$	–2.077
$Ho^{2+} + 2\,e \rightleftharpoons Ho$	–2.1
$Nd^{2+} + 2\,e \rightleftharpoons Nd$	–2.1
$Cf^{2+} + 2\,e \rightleftharpoons Cf$	–2.12
$Yb^{3+} + 3\,e \rightleftharpoons Yb$	–2.19
$Ac^{3+} + 3\,e \rightleftharpoons Ac$	–2.20
$Dy^{2+} + 2\,e \rightleftharpoons Dy$	–2.2
$Tm^{3+} + e \rightleftharpoons Tm^{2+}$	–2.2
$Pm^{2+} + 2\,e \rightleftharpoons Pm$	–2.2
$Es^{2+} + 2\,e \rightleftharpoons Es$	–2.23
$H_2 + 2\,e \rightleftharpoons 2\,H^-$	–2.23
$Gd^{3+} + 3\,e \rightleftharpoons Gd$	–2.279
$Tb^{3+} + 3\,e \rightleftharpoons Tb$	–2.28
$Lu^{3+} + 3\,e \rightleftharpoons Lu$	–2.28
$Dy^{3+} + 3\,e \rightleftharpoons Dy$	–2.295
$Am^{3+} + e \rightleftharpoons Am^{2+}$	–2.3
$Fm^{2+} + 2\,e \rightleftharpoons Fm$	–2.30
$Pm^{3+} + 3\,e \rightleftharpoons Pm$	–2.30
$Sm^{3+} + 3\,e \rightleftharpoons Sm$	–2.304
$Al(OH)_3 + 3\,e \rightleftharpoons Al + 3\,OH^-$	–2.31
$Tm^{3+} + 3\,e \rightleftharpoons Tm$	–2.319
$Nd^{3+} + 3\,e \rightleftharpoons Nd$	–2.323
$Al(OH)^- + 3\,e \rightleftharpoons Al + 4\,OH^-$	–2.328
$H_2AlO_3^- + H_2O + 3\,e \rightleftharpoons Al + 4\,OH^-$	–2.33
$Ho^{3+} + 3\,e \rightleftharpoons Ho$	–2.33
$Er^{3+} + 3\,e \rightleftharpoons Er$	–2.331
$Ce^{3+} + 3\,e \rightleftharpoons Ce$	–2.336
$Pr^{3+} + 3\,e \rightleftharpoons Pr$	–2.353
$ZrO(OH)_2 + H_2O + 4\,e \rightleftharpoons Zr + 4\,OH^-$	–2.36
$Mg^{2+} + 2\,e \rightleftharpoons Mg$	–2.372
$Y^{3+} + 3\,e \rightleftharpoons Y$	–2.372
$La^{3+} + 3\,e \rightleftharpoons La$	–2.379
$Tm^{2+} + 2\,e \rightleftharpoons Tm$	–2.4
$Md^{2+} + 2\,e \rightleftharpoons Md$	–2.40
$Th(OH)_4 + 4\,e \rightleftharpoons Th + 4\,OH$	–2.48
$HfO(OH)_2 + H_2O + 4\,e \rightleftharpoons Hf + 4\,OH^-$	–2.50
$No^{2+} + 2\,e \rightleftharpoons No$	–2.50
$Dy^{3+} + e \rightleftharpoons Dy^{2+}$	–2.6
$Pm^{3+} + e \rightleftharpoons Pm^{2+}$	–2.6
$Be_2O_3^{2-} + 3\,H_2O + 4\,e \rightleftharpoons 2\,Be + 6\,OH^-$	–2.63
$Sm^{2+} + 2\,e \rightleftharpoons Sm$	–2.68
$Mg(OH)_2 + 2\,e \rightleftharpoons Mg + 2\,OH^-$	–2.690
$Nd^{3+} + e \rightleftharpoons Nd^{2+}$	–2.7
$Mg^+ + e \rightleftharpoons Mg$	–2.70
$Na^+ + e \rightleftharpoons Na$	–2.71
$Yb^{2+} + 2\,e \rightleftharpoons Yb$	–2.76
$Bk^{3+} + e \rightleftharpoons Bk^{2+}$	–2.8
$Ho^{3+} + e \rightleftharpoons Ho^{2+}$	–2.8
$Ra^{2+} + 2\,e \rightleftharpoons Ra$	–2.8
$Eu^{2+} + 2\,e \rightleftharpoons Eu$	–2.812
$Ca^{2+} + 2\,e \rightleftharpoons Ca$	–2.868
$Sr(OH)_2 + 2\,e \rightleftharpoons Sr + 2\,OH^-$	–2.88
$Sr^{2+} + 2\,e \rightleftharpoons Sr$	–2.899
$Fr^+ + e \rightleftharpoons Fr$	–2.9
$La(OH)_3 + 3\,e \rightleftharpoons La + 3\,OH^-$	–2.90
$Ba^{2+} + 2\,e \rightleftharpoons Ba$	–2.912
$K^+ + e \rightleftharpoons K$	–2.931
$Rb^+ + e \rightleftharpoons Rb$	–2.98
$Ba(OH)_2 + 2\,e \rightleftharpoons Ba + 2\,OH^-$	–2.99
$Er^{3+} + e \rightleftharpoons Er^{2+}$	–3.0
$Ca(OH)_2 + 2\,e \rightleftharpoons Ca + 2\,OH^-$	–3.02
$Cs^+ + e \rightleftharpoons Cs$	–3.026
$Li^+ + e \rightleftharpoons Li$	–3.0401
$3\,N_2 + 2\,H^+ + 2\,e \rightleftharpoons 2\,HN_3$	–3.09
$Pr^{3+} + e \rightleftharpoons Pr^{2+}$	–3.1
$Ca^+ + e \rightleftharpoons Ca$	–3.80
$Sr^+ + e \rightleftharpoons Sr$	–4.10

REDUCTION AND OXIDATION POTENTIALS FOR CERTAIN ION RADICALS

Petr Vanýsek

There are two tables for ion radicals. The first table lists reduction potentials for organic compounds that produce anion radicals during reduction, a process described as $A + e^- \rightleftharpoons A^{-\cdot}$. The second table lists oxidation potentials for organic compounds that produce cation radicals during oxidation, a process described as $A \rightleftharpoons A^{+\cdot} + e^-$. To obtain reduction potential for a reverse reaction, the sign for the potential is changed.

Unlike the table of the Electrochemical Series, which lists *standard* potentials, values for radicals are experimental values with experimental conditions given in the second column. Since the measurements leading to potentials for ion radicals are very dependent on conditions, an attempt to report standard potentials for radicals would serve no useful purpose. For the same reason, the potentials are also reported as experimental values, usually a half-wave potential ($E_{1/2}$ in polarography) or a peak potential (E_p in cyclic voltammetry). Unless otherwise stated, the values are reported vs. SCE (saturated calomel electrode). To obtain a value vs. normal hydrogen electrode, 0.241 V has to be added to the SCE values. All the ion radicals chosen for inclusion in the tables result from electrochemically reversible reactions. More detailed data on ion radicals can be found in the *Encyclopedia of Electrochemistry of Elements*, (A. J. Bard, Ed.), Vols. XI and XII in particular, Marcel Dekker, New York, 1978.

Abbreviations are: CV — cyclic voltammetry; DMF — *N,N*-Dimethylformamide; *E* swp — potential sweep; $E°$ — standard potential; E_p — peak potential; $E_{p/2}$ — half-peak potential; $E_{1/2}$ — half wave potential; *M* — mol/L; MeCN — acetonitrile; pol — polarography; rot Pt dsk — rotated Pt disk; SCE — saturated calomel electrode; $TBABF_4$ — tetrabutylammonium tetrafluoroborate; TBAI — tetrabutylammonium iodide; TBAP — tetrabutylammonium perchlorate; TEABr — tetraethylammonium bromide; TEAP — tetraethylammonium perchlorate; THF — tetrahydrofuran; $TPACF_3SO_3$ — tetrapropylammonium trifluoromethanesulfite; TPAP — tetrapropylammonium perchlorate; and wr — wire.

Reduction Potentials (Products Are Anion Radicals)

Substance	Conditions/electrode/technique	Potential V (vs. SCE)
Acetone	DMF, 0.1 *M* TEABr/Hg/pol	$E_{1/2} = -2.84$
1-Naphthyphenylacetylene	DMF, 0.03 *M* TBAI/Hg/pol	$E_{1/2} = -1.91$
1-Naphthalenecarboxyaldehyde	-/Hg/pol	$E_{1/2} = -0.91$
2-Naphthalenecarboxyaldehyde	-/Hg/pol	$E_{1/2} = -0.96$
2-Phenanthrenecarboxaldehyde	-/Hg/pol	$E_{1/2} = -1.00$
3-Phenanthrenecarboxaldehyde	-/Hg/pol	$E_{1/2} = -0.94$
9-Phenanthrenecarboxaldehyde	-/Hg/pol	$E_{1/2} = -0.83$
1-Anthracenecarboxaldehyde	-/Hg/pol	$E_{1/2} = -0.75$
1-Pyrenecarboxaldehyde	-/Hg/pol	$E_{1/2} = -0.76$
2-Pyrenecarboxaldehyde	-/Hg/pol	$E_{1/2} = -1.00$
Anthracene	DMF, 0.1 *M* TBAP/Pt dsk/CV	$E_p = -2.00$
	DMF, 0.5 *M* $TBABF_4$/Hg/CV	$E_{1/2} = -1.93$
	MeCN, 0.1 *M* TEAP/Hg/CV	$E_{1/2} = -2.07$
	DMF, 0.1 *M* TBAI/Hg/pol	$E_{1/2} = -1.92$
9,10-Dimethylanthracene	DMF, 0.1 *M* TBAP/Pt/CV	$E_p = -2.08$
	MeCN, 0.1 *M* TBAP/Pt/CV	$E_p = -2.10$
1-Phenylanthracene	DMF, 0.5 *M* $TBABF_2$/Hg/CV	$E_{1/2} = -1.91$
	DMF, 0.1 *M* TBAI/Hg/pol	$E_{1/2} = -1.878$
2-Phenylanthracene	DMF, 0.1 *M* TBAI/Hg/pol	$E_{1/2} = -1.875$
8-Phenylanthracene	DMF, 0.5 *M* $TBABF_4$/Hg/CV	$E_{1/2} = -1.91$
9-Phenylanthracene	DMF, 0.5 *M* $TBABF_4$/Hg/CV	$E_{1/2} = -1.93$
	DMF, 0.1 *M* TBAI/Hg/pol	$E_{1/2} = -1.863$
1,8-Diphenylanthracene	DMF, 0.5 *M* $TBABF_4$/Hg/CV	$E_{1/2} = -1.88$
1,9-Diphenylanthracene	DMF, 0.1 *M* TBAI/Hg/pol	$E_{1/2} = -1.846$
1,10-Diphenylanthracene	DMF, 0.1 *M* TBAI/Hg/pol	$E_{1/2} = -1.786$
8,9-Diphenylanthracene	DMF, 0.5 *M* $TBABF_4$/Hg/CV	$E_{1/2} = -1.90$
9,10-Diphenylanthracene	MeCN, 0.1 *M* TBAP/rot Pt/E swp	$E_{1/2} = -1.83$
	DMF, 0.1 *M* TBAI/Hg/pol	$E_{1/2} = -1.835$
1,8,9-Triphenylanthracene	DMF, 0.5 *M* $TBABF_4$/Hg/CV	$E_{1/2} = -1.85$
1,8,10-Triphenylanthracene	DMF, 0.5 *M* $TBABF_4$/Hg/CV	$E_{1/2} = -1.81$
9,10-Dibiphenylanthracene	MeCN, 0.1 *M* TBAP/rot Pt/E swp	$E_{1/2} = -1.94$
Benz(a)anthracene	MeCN, 0.1 *M* TEAP/Hg/CV	$E_{1/2} = -2.11$
	MeCN, 0.1 *M* TEAP/Hg/pol	$E_{1/2} = -2.40$[a]
Azulene	DMF, 0.1 *M* TBAI/Hg/pol	$E_{1/2} = -1.10$[c]
Annulene	DMF, 0.5 *M* TBAP 0 °C/Hg/pol	$E_{1/2} = -1.23$
Benzaldehyde	DMF, 0.1 *M* TBAP/Hg/pol	$E_{1/2} = -1.67$
Benzil	DMSO, 0.1 *M* TBAP/Hg/pol	$E_{1/2} = -1.04$
Benzophenone	-/Hg/pol	$E_{1/2} = -1.80$
	DMF/Pt dsk/CV	$E° = -1.72$
Chrysene	MeCN, 0.1 *M* TEAP/Hg/pol	$E_{1/2} = -2.73$[a]
Fluoranthrene	DMF, 0.1 *M* TBAP/Pt dsk/CV	$E_p = -1.76$
Cyclohexanone	DMF, 0.1 *M* TEABr/Hg/pol	$E_{1/2} = -2.79$

Substance	Conditions/electrode/technique	Potential V (vs. SCE)
5,5-Dimethyl-3-phenyl-2-cyclohexen-1-one	DMF, 0.5 *M*/Hg/pol	$E_{1/2} = -1.71$
1,2,3-Indanetrione hydrate (ninhydrin)	DMF, 0.2 *M* $NaNO_3$/Hg/pol	$E_{1/2} = -0.039$
Naphthacene	DMF, 0.1 *M* TBAI/Hg/pol	$E_{1/2} = -1.53$
Naphthalene	DMF, 0.1 *M* TBAP/Pt dsk/CV	$E_p = -2.55$
	DMF, 0.5 *M* $TBABF_4$/Hg/CV	$E_{1/2} = -2.56$
	DMF, MeCN, 0.1 *M* TEAP/Hg/CV	$E_{1/2} = -2.63$
	DMF, 0.1 *M* TBAI/Hg/pol	$E_{1/2} = -2.50$
1-Phenylnaphthalene	DMF, 0.5 *M* $TBABF_4$/Hg/CV	$E_{1/2} = -2.36$
1,2-Diphenylnaphthalene	DMF, 0.5 *M* $TBABF_4$/Hg/CV	$E_{1/2} = -2.25$
Cyclopentanone	DMF, 0.1 *M* TEABr/Hg/pol	$E_{1/2} = -2.82$
Phenanthrene	MeCN, 0.1 *M* TBAP/Pt wr/CV	$E_{1/2} = -2.47$
	MeCN, 0.1 *M* TEAP/Hg/pol	$E_{1/2} = -2.88$[a]
Pentacene	THF, 0.1 *M* TBAP/rot Pt dsk/E swp	$E_{1/2} = -1.40$
Perylene	MeCN, 0.1 *M* TEAP/Hg/CV	$E_{1/2} = -1.73$
1,3-Diphenyl-1,3-propanedione	DMSO, 0.2 *M* TBAP/Hg/CV	$E_{1/2} = -1.42$
2,2-Dimethyl-1,3-diphenyl-1,3 propanedione	DMSO, TBAP/Hg/CV	$E_{1/2} = -1.80$
Pyrene	DMF, 0.1 *M* TBAP/Pt/CV	$E_p = -2.14$
	MeCN, 0.1 *M* TEAP/Hg/pol	$E_{1/2} = -2.49$[a]
Diphenylsulfone	DMF, TEABr	$E_{1/2} = -2.16$
Triphenylene	MeCN, 0.1 *M* TEAP/Hg/pol	$E_{1/2} = -2.87$[a]
9,10-Anthraquinone	DMF, 0.5 *M* TBAP, 20°/Pt dsk/CV	$E_{1/2} = -1.01$
1,4-Benzoquinone	MeCN, 0.1 *M* TEAP/Pt/CV	$E_p = -0.54$
1,4-Naphthohydroquinone, dipotassium salt	DMF, 0.5 *M* TBAP, 20°/Pt dsk/CV	$E_{1/2} = -1.55$
Rubrene	DMF, 0.1 *M* TBAP/Pt dsk/CV	$E_p = -1.48$
	DMF, 0.1 *M* TBAI/Hg/pol	$E_{1/2} = -1.410$
Benzocyclooctatetraene	THF, 0.1 *M* TBAP/Hg/pol	$E_{1/2} = -2.13$
sym-Dibenzocyclooctatetraene	THF, 0.1 *M* TBAP/Hg/pol	$E_{1/2} = -2.29$
Ubiquinone-6	MeCN, 0.1 *M* TEAP/Pt/CV	$E_p = -1.05$[e]
(9-Phenyl-fluorenyl)$^+$	10.2 *M* H_2SO_4/Hg/CV	$E_p = -0.01$[b]
(Triphenylcyclopropenyl)$^+$	MeCN, 0.1 *M* TEAP/Hg/CV	$E_p = -1.87$
(Triphenylmethyl)$^+$	MeCN, 0.1 *M* TBAP/Hg/pol	$E_{1/2} = 0.27$
	H_2SO_4, 10.2 *M*/Hg/CV	$E_p = -0.58$[b]
(Tribiphenylmethyl)$^+$	MeCN, 0.1 *M* TBAP/Hg/pol	$E_{1/2} = 0.19$
(Tri-4-*t*-butyl-5-phenylmethyl)$^+$	MeCN, 0.1 *M* TBAP/Hg/pol	$E_{1/2} = 0.13$
(Tri-4-isopropylphenylmethyl)$^+$	MeCN, 0.1 *M* TBAP/Hg/pol	$E_{1/2} = 0.07$
(Tri-4-methylphenylmethyl)$^+$	MeCN, 0.1 *M* TBAP/Hg/pol	$E_{1/2} = 0.05$
(Tri-4-cyclopropylphenylmethyl)$^+$	MeCN, 0.1 *M* TBAP/Hg/pol	$E_{1/2} = 0.01$
(Tropylium)$^+$	MeCN, 0.1 *M* TBAP/Hg/pol	$E_{1/2} = -0.17$
	DMF, 0.15 *M* TBAI/Hg/pol	$E_{1/2} = -1.55$
	DMF, 0.15 *M* TBAI/Hg/pol	$E_{1/2} = -1.55$
	DMF, 0.15 *M* TBAI/Hg/pol	$E_{1/2} = -1.57$
	DMF, 0.15 *M* TBAI/Hg/pol	$E_{1/2} = -1.60$
	DMF, 0.15 *M* TBAI/Hg/pol	$E_{1/2} = -1.87$
	DMF, 0.15 *M* TBAI/Hg/pol	$E_{1/2} = -1.96$
	DMF, 0.15 *M* TBAI/Hg/pol	$E_{1/2} = -2.05$

Oxidation Potentials (Products Are Cation Radicals)

Substance	Conditions/electrode/technique	Potential V (vs. SCE)
Anthracene	CH_2Cl_2, 0.2 *M* $TBABF_4$, −70 °C/Pt dsk/CV	$E_p = +0.73$[d]
9,10-Dimethylanthracene	MeCN, 0.1 *M* $LiClO_4$/Pt wr/CV	$E_p = +1.0$
9,10-Dipropylanthracene	MeCN, 0.1 *M* TEAP/Pt/CV	$E_p = +1.08$
1,8-Diphenylanthracene	CH_2Cl_2, 0.2 *M* $TPrACF_3SO_3$/rot Pt wr/E swp	$E_{1/2} = +1.34$
8,9-Diphenylanthracene	CH_2Cl_2, 0.2 *M* $TPrACF_3SO_3$/rot Pt wr/E swp	$E_{1/2} = +1.30$
9,10-Diphenylanthracene	MeCN/Pt/CV	$E_p = +1.22$
Perylene	MeCN, 0.1 *M* TBAP/Pt/CV	$E_p = +1.34$
Pyrene	DMF, 0.1 *M* TBAP/Pt dsk/CV	$E_p = +1.25$
Rubrene	DMF, 0.1 *M* TBAP/Pt dsk/CV	$E_p = +1.10$
Tetracene	CH_2Cl_2, 0.2 *M* $TBABF_4$, −70 °C/Pt wr/CV	$E_p = +0.35$[d]
1,4-Dithiabenzene	MeCN, 0.1 *M* TEAP/Pt dsk/rot	$E_{1/2} = +0.69$
1,4-Dithianaphthalene	MeCN, 0.1 *M* TEAP/Pt dsk/rot	$E_{1/2} = +0.80$
Thianthrene	0.1 *M* TPAP/Pt/CV	$E_{1/2} = +1.28$

[a] vs. 0.01 M Ag/$AgClO_4$
[b] vs. Hg/Hg_2SO_4, 17 M H_2SO_4
[c] vs. Hg pool
[d] vs. Ag/saturated $AgNO_3$
[e] vs. Ag/0.01 M Ag+

pH SCALE FOR AQUEOUS SOLUTIONS

A.K. Covington

A Working Party of IUPAC, after extensive considerations over five years, has produced a report (1) which sets pH firmly within the International System of Units (SI). A summary of these important developments is given below.

The concept of pH is unique amongst the commonly encountered physicochemical quantities in that, in terms of its definition,

$$\mathrm{pH} = -\lg a_H \quad (1)$$

it involves a single ion quantity, the activity of the hydrogen ion, which is immeasurable by any thermodynamically valid method and requires a convention for its evaluation.

pH was originally defined by Sørensen (2) in terms of the concentration of hydrogen ions (in modern nomenclature) as pH = $-\lg (c_H/c^o)$ where c_H is the hydrogen ion concentration in mol dm^{-3}, and c^o = 1 mol dm^{-3} is the standard amount concentration. Subsequently (3), it was accepted as more satisfactory to define pH in terms of the relative activity of hydrogen ions in solution

$$\mathrm{pH} = -\lg a_H = -\lg (m_H\gamma_H/m^o) \quad (2)$$

where a_H is the relative (molality basis) activity and γ_H is the molal activity coefficient of the hydrogen ion H^+ at the molality m_H, and m^o the standard molality. The quantity pH is intended to be a measure of the activity of hydrogen ions in solution. However, since it is defined in terms of a quantity that cannot be measured by a thermodynamically valid method, eqn. (2) can only be considered a *notional definition* of pH.

pH being a single ion quantity, it is not determinable in terms of a fundamental (or base) unit of any measurement system, and there is difficulty providing a proper basis for the traceability of pH measurements. A satisfactory approach is now available in that pH determinations can be incorporated into the International System (SI) if they can be traced to measurements made using a method that fulfils the definition of a 'primary method of measurement' (4).

The essential feature of a primary method is that it must operate according to a well-defined measurement equation in which all of the variables can be determined experimentally in terms of SI units. Any limitation in the determination of the experimental variables, or in the theory, must be included within the estimated uncertainty of the method if traceability to the SI is to be established. If a convention were used without an estimate of its uncertainty, true traceability to SI would not be established. The electrochemical cell without liquid junction, known as the Harned cell (5), fulfills the definition of a primary method for the measurement of the acidity function, $p(a_H\gamma_{Cl})$, and subsequently of the pH of buffer solutions.

The Harned cell is written as

$$\mathrm{Pt} \mid H_2 \mid \text{buffer S}, Cl^- \mid AgCl \mid Ag \quad \text{(Cell I)}$$

and contains a standard buffer, S, with chloride ions, as potassium or sodium chloride, added in order to use the silver–silver chloride electrode as reference electrode. The application of the Nernst equation to the spontaneous cell reaction of Cell I:

$$\tfrac{1}{2} H_2 + AgCl \rightarrow Ag(s) + H^+ + Cl^-$$

yields the potential difference E_I of the cell (corrected to 1 atm (101.325 kPa), the partial pressure of hydrogen gas used in electrochemistry in preference to 100 kPa) as

$$E_I = E^o - (RT/F)\ln 10 \lg [(m_H\gamma_H/m^o)(m_{Cl}\gamma_{Cl}/m^o)] \quad (3)$$

which can be rearranged, since $a_H = m_H\gamma_H/m^o$, to give the acidity function

$$p(a_H\gamma_{Cl}) = -\lg(a_H\gamma_{Cl}) = (E_I - E^o)/[(RT/F)\ln 10] + \lg(m_{Cl}/m^o) \quad (4)$$

where E^o is the standard potential difference of the cell, and hence of the silver–silver chloride electrode, and γ_{Cl} is the activity coefficient of the chloride ion.

The standard potential difference of the silver–silver chloride electrode, E^o, is determined from a Harned cell in which only HCl is present at a fixed molality (e.g., m = 0.01 mol kg^{-1})

$$\mathrm{Pt} \mid H_2 \mid HCl\,(m) \mid AgCl \mid Ag \quad \text{(Cell Ia)}$$

The application of the Nernst equation to the HCl cell (Ia) gives

$$E_{Ia} = E^o - (2RT/F)\ln 10 \lg[(m_{HCl}/m^o)(\gamma_{\pm HCl})] \quad (5)$$

where E_{Ia} has been corrected to 1 atmosphere partial pressure of hydrogen gas (101.325 kPa) and $\gamma_{\pm HCl}$ is the mean ionic activity coefficient of HCl.

Values of the activity coefficient ($\gamma_{\pm HCl}$) at molality 0.01 mol kg^{-1} and various temperatures were given by Bates and Robinson (6). The standard potential difference depends on the method of preparation of the electrodes, but individual determinations of the activity coefficient of HCl at 0.01 mol kg^{-1} are more uniform than values of E^o. Hence the practical determination of the potential difference of the cell with HCl at 0.01 mol kg^{-1} is recommended at 298.15 K at which the mean ionic activity coefficient is 0.904. (It is unnecessary to repeat the measurement of E^o at other temperatures but simply to correct published smoothed values by the observed difference in E^0 at 298.15 K.)

In national metrology institutes (NMIs), measurements of Cells I and Ia are often done simultaneously in a thermostat bath. Subtracting eqn. (5) from (3) gives

$$\Delta E = E_I - E_{Ia} = -(RT/F)\ln 10\{\lg[(m_H\gamma_H/m^o)(m_{Cl}\gamma_{Cl}/m^o)] - \lg[(m_{HCl}/m^o)^2\gamma^2_{\pm HCl}]\} \quad (6)$$

which is independent of the standard potential difference. Therefore, the subsequently calculated pH does not depend on the standard potential difference and hence does not depend on the assumption that the standard potential of the hydrogen electrode is zero at all temperatures. Therefore, the Harned cell gives an exact comparison between hydrogen ion activities at different temperatures.

The quantity $p(a_H\gamma_{Cl}) = -\lg (a_H\gamma_{Cl})$, on the left-hand side of (4), is called the acidity function (5). To obtain the quantity pH according to eqn. (2) from the acidity function, it is necessary to evaluate $\lg \gamma_{Cl}$ independently. This is done in two steps: (i) the value of lg $(a_H\gamma_{Cl})$ at zero chloride molality, lg $(a_H\gamma_{Cl})^o$, is evaluated and (ii) a value for the activity of the chloride ion γ^o_{Cl}, at zero chloride molality (sometimes referred to as the limiting or 'trace' activity coefficient) is calculated using the Bates-Guggenheim convention (7). The value of lg $(a_H\gamma_{Cl})^o$ corresponding to zero chloride molality is determined by linear extrapolation of measurements using Harned cells with at least three added molalities of sodium or potassium chloride ($I < 0.1$ mol kg^{-1}).

The value of lg $(a_H\gamma_{Cl})^o$ corresponding to zero chloride molality is determined by linear extrapolation of measurements using Harned cells with at least three added molalities of sodium or potassium chloride ($I < 0.1$ mol kg^{-1}) in accord with eqn. (7):

$$-\lg (a_H\gamma_{Cl}) = -\lg (a_H\gamma_{Cl})^o + Sm_{Cl} \quad (7)$$

where S is an empirical, temperature dependent, constant.

The Bates-Guggenheim convention (7) assumes that the trace activity coefficient of the chloride ion γ^o_{Cl} is given by

$$\lg \gamma^o_{Cl} = -A\, I^{1/2} / (1 + Ba\, I^{1/2}) \quad (8)$$

where A is the Debye-Hückel temperature dependent constant (limiting slope), a is the *mean* distance of closest approach of the ions (ion size parameter), Ba is set equal to 1.5 (mol kg^{-1})$^{-1/2}$ at all temperatures in the range 5-50 °C, and I is the ionic strength of the buffer (which for its evaluation requires knowledge of appropriate acid dissociation constants).

The various stages in the assignment of primary standard pH values are combined in eqn. (9), which is derived from eqns. (4), (5) and (8)

$$\text{pH(PS)} = \lim m_{Cl \to 0} \{(E_I - E^o)/[(RT/F)\ln 10] + \lg (m_{Cl}/m^o)\} - AI^{1/2}/[1 + 1.5\,(I/m^o)^{1/2}] \quad (9)$$

In order for a particular buffer solution to be considered a primary buffer solution, it must be of the "highest metrological" quality (4) in accordance with the definition of a primary standard. It is recommended that it have the following attributes (9):

1. High buffer value in the range 0.016-0.07 (mol OH$^-$)/pH.
2. Small dilution value at half concentration (change in pH with change in buffer concentration) in the range 0.01-0.20.
3. Small dependence of pH on temperature less than ± 0.01 K^{-1}.
4. Low residual liquid junction potential < 0.01 in pH.
5. Ionic strength ≤0.1 mol kg^{-1} to permit applicability of Bates-Guggenheim convention.
6. NMI certificate for specific batch.
7. Reproducible purity of preparation (lot to lot differences of |ΔpH(PS)| < 0.003).
8. Long-term stability of stored solid material.

Values for the above and other important parameters for the primary and secondary buffer materials are given in Table 1.

Primary Standard Buffers

As there can be significant variations in the purity of samples of a buffer of the same nominal chemical composition, it is essential that the primary buffer material used has been certified with values that have been measured with Cell I. The Harned cell is used by many national metrological institutes for accurate measurements of pH of buffer solutions.

Typical values of the pH(PS) of the seven solutions from the six accepted primary standard reference buffers, which meet the conditions stated above, are listed in Table 2. Batch-to-batch variations in purity can result in changes in the pH value of samples of at most 0.003. The typical values in Table 2 should not be used in place of the certified value (from a Harned cell measurement) for a specific batch of buffer material.

The required attributes listed above effectively limit the range of primary buffers available to between pH 3 and 10 (at 25°C). Calcium hydroxide and potassium tetraoxalate are excluded because the contribution of hydroxide or hydrogen ions to the ionic strength is significant. Also excluded are the nitrogen bases of the type BH$^+$ (such as tris(hydroxymethyl)aminomethane and piperazine phosphate) and the zwitterionic buffers (e.g., HEPES and MOPS (10)). These do not comply because either the Bates-Guggenheim convention is not applicable, or the liquid junction potentials are high. This means the choice of primary standards is restricted to buffers derived from oxy-carbon, -phosphorus, -boron and mono, di- and tri-protic carboxylic acids. The uncertainties (11) associated with Harned cell measurements are calculated (1) to be 0.004 in pH at NMIs, with typical variation between batches of primary standard buffers of 0.003.

Secondary Standards

Substances that do not fulfill all the criteria for primary standards, but to which pH values can be assigned using Cell I are considered to be secondary standards (Table 3). Reasons for their exclusion as primary standards include difficulties in achieving consistent and suitable chemical quality (e.g. acetic acid is a liquid), suspected high liquid junction potential, or inappropriateness of the Bates-Guggenheim convention (e.g., other charge-type buffers). The uncertainty is higher (e.g., 0.01) for biological buffers. Certain other substances, which cannot be used in cells containing hydrogen gas electrodes, are also classed as secondary standards.

Calibration Procedures

(a) One-point calibration

A single point calibration is insufficient to determine both slope and one-point parameters. The theoretical value for the slope can be assumed but the practical slope may be up to 5% lower. Alternatively, a value for the practical slope can be assumed from the manufacturer's prior calibration. The one-point calibration therefore yields only an estimate of pH(X). Since both parameters may change with age of the electrodes, this is not a reliable procedure.

(b) Two-point calibration [target uncertainty: 0.02–0.03 at 25 °C]

In the majority of practical applications, glass electrodes cells are calibrated by a two-point calibration, or bracketing, procedure using two standard buffer solutions, with pH values, pH(S_1) and pH(S_2), bracketing the unknown pH(X). Bracketing is often taken to mean that the pH(S_1) and pH(S_2) buffers selected from Table 2 should be those that are immediately above and below pH(X). This may not be appropriate in all situations and choice of a wider range may be better.

(c) Multi-point calibration [target uncertainty: 0.01–0.03 at 25 °C].
Multi-point calibration is carried out using up to five standard buffers. The use of more than five points yields no significant improvement in the statistical information obtainable.
Details of uncertainty computations (11) have been given (1).

Measurement of pH and choice of pH Standard Solutions

1a) If pH is not required to better than ±0.05 any pH standard solution may be selected.
1b) If pH is required to ±0.002 and interpretation in terms of hydrogen ion concentration or activity is desired, choose a standard solution, pH(PS), to match X as closely as possible in terms of pH, composition and ionic strength.
2) Alternatively, a bracketing procedure may be adopted whereby two standard solutions are chosen whose pH values, pH(S1), pH(S2) are on either side of pH(X). Then if the corresponding potential difference measurements are E(S1), E(S2), E(X), then pH(X) is obtained from

$$\text{pH(X)} = \text{pH(S1)} + [E(\text{X}) - E(\text{S1})]/\ \%k$$

where $\%k = 100[E(\text{S2}) - E(\text{S1})]/[\text{pH(S2)} - \text{pH(S1)}]$ is the apparent percentage slope. This procedure is very easily done on some pH meters simply by adjusting downwards the slope factor control with the electrodes in S2. The purpose of the bracketing procedure is to compensate for deficiencies in the electrodes and measuring system.

Information to be given about the measurement of pH(X)

The standard solutions selected for calibration of the pH meter system should be reported with the measurement as follows:

System calibrated with pH(S) = ... at ...K.

System calibrated with two primary standards, pH(PS1) =.... and pH(PS2) =....at ....K.

System calibrated with n standards, pH(S1) =.... , pH(S2) =.... etc. at....K.

Interpretation of pH(X) in terms of hydrogen ion concentration

The defined pH has no simple interpretation in terms of hydrogen ion concentration but the mean ionic activity coefficient of a typical 1:1 electrolyte can be used to obtain hydrogen ion concentration subject to an uncertainty of 3.9% in concentration, corresponding to 0.02 in pH.

References

1. Buck, R.P., Rondinini, S., Covington, A.K., Baucke, F.G.K., Brett, C.M.A., Camoes, M.F.C., Milton, M.J.T., Mussini, T., Naumann, R., Pratt, K.W., Spitzer, P., and Wilson, G.S. *Pure Appl. Chem.*, 74, 2105, 2002.
2. Sørensen, S.P.L. *Comp. Rend. Trav. Lab. Carlsberg*, 8, 1, 1909.
3. Sørensen, S.P.L., and Linderstrøm-Lang, K.L., *Comp. Rend. Trav. Lab. Carlsberg*, 15, 1924.
4. BIPM, *Com. Cons. Quantité de Matière* 4, 1998. See also M.J.T. Milton and T.J. Quinn, *Metrologia* 38, 289, 2001.
5. Harned H.S., and Owen, B.B., *The Physical Chemistry of Electrolytic Solutions*, Ch 14, Reinhold, New York, 1958.
6. Bates R.G., and Robinson, R.A., *J. Soln. Chem.* 9, 455, 1980.
7. Bates R.G., and Guggenheim, E.A., *Pure Appl. Chem.* 1, 163, 1960.
8. *International Vocabulary of Basic and General Terms in Metrology* (VIM), Beuth, Berlin, 2nd Edn. 1994.
9. Bates, R.G. *Determination of pH*, Wiley, New York, 1973.
10. Good, N.E. et al., *Biochem. J.* 5, 467, 1966.
11. *Guide to the Expression of Uncertainty* (GUM), BIPM, IEC, IFCC, ISO, IUPAC, IUPAP, OIML, 1993.

TABLE 1. Summary of Useful Properties of Some Primary and Secondary Standard Buffer Substances and Solutions

Salt or solid substance	Formula	Molality/ mol kg^{-1}	Molar mass/ g mol^{-1}	Density/ g/mL	Amount conc. at 20°C/ mol dm^{-3}	Mass/g to make 1 dm^3	Dilution value $\Delta pH_{1/2}$	Buffer value (β)/ mol OH$^-$ dm^{-3}	pH Temperature coefficient/ K^{-1}
Potassium tetroxalate dihydrate	$KH_3C_4O_8{\cdot}2H_2O$	0.1	254.191	1.0091	0.09875	25.101			
Potassium tetraoxalate dihydrate	$KH_3C_4O_8{\cdot}2H_2O$	0.05	254.191	1.0032	0.04965	12.620	0.186	0.070	0.001
Potassium hydrogen tartrate (sat at 25°C)	$KHC_4H_4O_6$	0.0341	188.18	1.0036	0.034	6.4	0.049	0.027	−0.0014
Potassium dihydrogen citrate	$KH_2C_6H_5O_7$	0.05	230.22	1.0029	0.04958	11.41	0.024	0.034	−0.022
Potassium hydrogen phthalate	$KHC_8H_4O_4$	0.05	204.44	1.0017	0.04958	10.12	0.052	0.016	0.00012
Disodium hydrogen orthophosphate +	Na_2HPO_4	0.025	141.958	1.0038	0.02492	3.5379	0.080	0.029	−0.0028

Salt or solid substance	Formula	Molality/ mol kg^{-1}	Molar mass/ g mol^{-1}	Density/ g/mL	Amount conc. at 20°C/ mol dm^{-3}	Mass/g to make 1 dm^3	Dilution value $\Delta pH_{1/2}$	Buffer value (β)/ mol OH$^-$ dm^{-3}	pH Temperature coefficient/ K^{-1}
potassium dihydrogen orthophosphate	KH_2PO_4	0.025	136.085			3.3912			
Disodium hydrogen orthophosphate +	Na_2HPO_4	0.03043	141.959	1.0020	0.08665	4.302	0.07	0.016	−0.0028
potassium dihydrogen orthophosphate	KH_2PO_4	0.00869	136.085		0.03032	1.179			
Disodium tetraborate decahydrate	$Na_2B_4O_7 \cdot 10H_2O$	0.05	381.367	1.0075	0.04985	19.012			
Disodium tetraborate decahydrate	$Na_2B_4O_7 \cdot 10H_2O$	0.01	381.367	1.0001	0.00998	3.806	0.01	0.020	−0.0082
Sodium hydrogen carbonate +	$NaHCO_3$	0.025	84.01	1.0013	0.02492	2.092	0.079	0.029	−0.0096
sodium carbonate	Na_2CO_3	0.025	105.99			2.640			
Calcium hydroxide (sat. at 25°C)	$Ca(OH)_2$	0.0203	74.09	0.9991	0.02025	1.5	-0.28	0.09	−0.033

TABLE 2. Typical Values of pH(PS) for Primary Standards at 0–50°C

	Temperature in °C										
Primary standards (PS)	**0**	**5**	**10**	**15**	**20**	**25**	**30**	**35**	**37**	**40**	**50**
Sat. potassium hydrogen tartrate (at 25°C)						3.557	3.552	3.549	3.548	3.547	3.549
0.05 mol kg^{-1} potassium dihydrogen citrate	3.863	3.840	3.820	3.802	3.788	3.776	3.766	3.759	3.756	3.754	3.749
0.05 mol kg^{-1} potassium hydrogen phthalate	4.000	3.998	3.997	3.998	4.000	4.005	4.011	4.018	4.022	4.027	4.050
0.025 mol kg^{-1} disodium hydrogen phosphate + 0.025 mol kg^{-1} potassium dihydrogen phosphate	6.984	6.951	6.923	6.900	6.881	6.865	6.853	6.844	6.841	6.838	6.833
0.03043 mol kg^{-1} disodium hydrogen phosphate + 0.008695 mol kg^{-1} potassium dihydrogen phosphate	7.534	7.500	7.472	7.448	7.429	7.413	7.400	7.389	7.386	7.380	7.367
0.01 mol kg^{-1} disodium tetraborate	9.464	9.395	9.332	9.276	9.225	9.180	9.139	9.102	9.088	9.068	9.011
0.025 mol kg^{-1} sodium hydrogen carbonate + 0.025 mol kg^{-1} sodium carbonate	10.317	10.245	10.179	10.118	10.062	10.012	9.966	9.926	9.910	9.889	9.828

TABLE 3. Values of pH(SS) of Some Secondary Standards from Harned Cell I Measurements

	Temperature in °C								
Secondary standards	**0**	**5**	**10**	**15**	**20**	**25**	**30**	**37**	**40**	**50**
0.05 mol kg^{-1} potassium tetroxalate[a]	1.67	1.67	1.67	1.67	1.68	1.68	1.68	1.69	1.69	1.71
0.05 mol kg^{-1} sodium hydrogen diglycolate[b]		3.47	3.47	3.48	3.48	3.49	3.50	3.52	3.53	3.56
0.1 mol dm^{-3} acetic acid + 0.1 mol dm^{-3} sodium acetate	4.68	4.67	4.67	4.66	4.66	4.65	4.65	4.66	4.66	4.68
mol dm^{-3} acetic acid + 0.1 mol dm^{-3} sodium acetate	4.74	4.73	4.73	4.72	4.72	4.72	4.72	4.73	4.73	4.75
0.02 mol kg^{-1} piperazine phosphate[c]	6.58	6.51	6.45	6.39	6.34	6.29	6.24	6.16	6.14	6.06
0.05 mol kg^{-1} tris hydrochloride + 0.01667 mol kg^{-1} tris[c]	8.47	8.30	8.14	7.99	7.84	7.70	7.56	7.38	7.31	7.07
0.05 mol kg^{-1} disodium tetraborate	9.51	9.43	9.36	9.30	9.25	9.19	9.15	9.09	9.07	9.01
Saturated (at 25 °C) calcium hydroxide	13.42	13.21	13.00	12.81	12.63	12.45	12.29	12.07	11.98	11.71

[a] Potassium trihydrogen dioxalate ($KH_3C_4O_8$)
[b] Sodium hydrogen 2,2'-oxydiacetate
[c] 2-Amino-2-(hydroxymethyl)-1,3 propanediol or tris(hydroxymethyl)aminomethane

PRACTICAL pH MEASUREMENTS ON NATURAL WATERS

A. K. Covington and W. Davison

(1) Dilute solutions and freshwater including 'acid-rain' samples ($I < 0.02$ mol kg^{-1})

Major problems could be encountered due to errors associated with the liquid junction. It is recommended that either a free diffusion junction is used or it is verified that the junction is working correctly using dilute solutions as follows. For commercial electrodes calibrated with IUPAC aqueous RVS or PS standards, the pH(X) of dilute solutions should be within ±0.02 of those given in Table 1. The difference in determined pH(X) between a stirred and unstirred dilute solution should be < 0.02. The characteristics of glass electrodes are such that below pH 5 the readings should be stable within 2 min, but for pH 5 to 8.8 or so minutes may be necessary to attain stability. Interpretation of pH(X) measured in this way in terms of activity of hydrogen ion, a_{H^+} is subject[1] to an uncertainty of ±0.02 in pH.

(2) Seawater

Measurements made by calibration of electrodes with IUPAC aqueous RVS or PS standards to obtain pH(X) are perfectly valid. However, the interpretation of pH(X) in terms of the activity of hydrogen ion is complicated by the nonzero residual liquid junction potential as well as by systematic differences between electrode pairs, principally attributable to the reference electrode. For 35‰ salinity seawater (S = 0.035) a_{H+} calculated from pH(X) is typically 12% too low. Special seawater pH scales have been devised to overcome this problem:

(i) The total hydrogen ion scale, pH_T, is defined in terms of the sum of free and complexed (total) hydrogen ion concentrations, where

$$^{T}C_H = [H^+] + [HSO_4^-] + [HF].$$

$$\text{So, } pH_T = -\log {^{T}C_H}$$

Calibration of the electrodes with a buffer having a composition similar to that of seawater, to which pH_T has been assigned, results in values of pHT(X) (Tables 2, 3) which are accurately interpretable in terms of $^{T}C_H$.

(ii) The free hydrogen ion scale, pH_F, is defined, and fully interpretable, in terms of the concentration of free hydrogen ions.

$$pH_F = -\log [H^+]$$

Values of pH_F as a function of temperature have been assigned to the same set of pH_T seawater buffers, and so alternatively can be used for calibration (Tables 2, 3).[2,3]

(3) Estuarine water

Prescriptions for seawater scale buffers are available for a range of salinities. Reliable estuarine pH measurements can be made by calibrating with a buffer of the same salinity as the sample. However, these buffers are difficult to prepare and their use presumes prior knowledge of salinity of the sample. Interpretable measurements of estuarine pH can be made by calibration with IUPAC aqueous RVS or PS standards if the electrode pair is additionally calibrated using a 20‰ salinity seawater buffer.[4] The difference between the assigned pH_{SWS} of the seawater buffer and its measured pH(X) value using RVS or PS standards is

$$\Delta pH = pH_{SWS} - pH(X)$$

Values of ΔpH should be in the range of 0.08 to 0.18. It empirically corrects for differences between the two pH scales and for measurement errors associated with the electrode pair. The pH(X) of samples measured using IUPAC aqueous buffers, can be converted to pH_T or pH_F using the appropriate measured ΔpH:

$$pH_T = pH(X) - \Delta pH$$

$$\text{or } pH_F = pH(X) - \Delta pH$$

This simple procedure is appropriate to pH measurement at salinities from 2‰ to 35‰. For salinities lower than 2‰ the procedures for freshwaters should be adopted.

References

1. Davison, W. and Harbinson, T. R., *Analyst*, 113, 709, 1988.
2. Culberson, C. H., in *Marine Electrochemistry*, Whitfield, M. and Jagner, D., Eds., Wiley, 1981.
3. Millero, F. J., *Limnol. Oceanogr.*, 31, 839, 1986.
4. Covington, A. K., Whalley, P. D., Davison, W., and Whitfield, M., in *The Determination of Trace Metals in Natural Waters*, West, T. S. and Nurnberg, H. W., Eds., Blackwell, Oxford, 1988.
5. Koch, W. F., Marinenko, G., and Paule, R. C., *J. Res. NBS*, 91, 33, 1986.

TABLE 1. pH of Dilute Solutions at 25 °C, Degassed and Equilibrated with Air, Suitable as Quality Control Standards

	Ionic strength mmol kg^{-1}	Concentration(x) mmol kg^{-1}	pH $p_{CO_2} = 0$	pH p_{CO_2} = air
Potassium hydrogen phthalate	10.7	10	4.12	4.12
	1.1	1	4.33	4.33
$x KH_2PO_4 + x Na_2HPO_4$	9.9	2.5	7.07	7.05
$x KH_2PO_4 + 3.5x Na_2HPO_4$	10	0.87	7.61	7.58
$Na_2B_4O_7 \cdot 10H_2O$	10	5	9.20	—
HCl	0.1	0.1	4.03	4.03
SRM2694-I[a]	—	—	4.30	—
SRM2694-II[a]	—	—	3.59	—

Note: The pH of solutions near to pH 4 is virtually independent of temperature over the range of 5 to 30 °C.

[a] Simulated rainwater samples are available (Reference 5) from NIST containing sulfate, nitrate, chloride, fluoride, sodium, potassium, calcium and magnesium.

TABLE 2. Composition of Seawater Buffer of Salinity S = 35‰ at 25 °C (Reference 3)

Solute	mol dm^{-3}	mol kg^{-1}	g kg^{-1}	g dm^{-3}
NaCl	0.3666	0.3493	20.416	20.946
Na_2SO_4	0.02926	0.02788	3.96	4.063
KCl	0.01058	0.01008	0.752	0.772
$CaCl_2$	0.01077	0.01026	1.139	1.169
$MgCl_2$	0.05518	0.05258	5.006	5.139
Tris	0.06	0.05717	6.926	7.106
Tris · HCl	0.06	0.05717	9.010	9.244

Tris = tris(hydroxymethyl)aminomethane $(HOCH_2)_3CNH_2$.
A 20‰ buffer is made by diluting the 35‰ in the ratio 20:35.

TABLE 3. Assigned Values of 20‰ and 35‰ Buffers on Free and Total Hydrogen Ion Scales. Calculated from Equations Provided by Millero (Reference 3)

Temp (°C)	pH_T S = 20‰	pH_T S = 35‰	pH_F S = 20‰	pH_F S = 35‰
5	8.683	8.718	8.759	8.81
10	8.513	8.542	8.597	8.647
15	8.351	8.374	8.442	8.491
20	8.195	8.212	8.292	8.341
25	8.045	8.057	8.149	8.197
30	7.901	7.908	8.011	8.059
35	7.762	7.764	7.879	7.926

BUFFER SOLUTIONS GIVING ROUND VALUES OF pH AT 25 °C

A pH	A x	B pH	B x	C pH	C x	D pH	D x	E pH	E x
1.00	67.0	2.20	49.5	4.10	1.3	5.80	3.6	7.00	46.6
1.10	52.8	2.30	45.8	4.20	3.0	5.90	4.6	7.10	45.7
1.20	42.5	2.40	42.2	4.30	4.7	6.00	5.6	7.20	44.7
1.30	33.6	2.50	38.8	4.40	6.6	6.10	6.8	7.30	43.4
1.40	26.6	2.60	35.4	4.50	8.7	6.20	8.1	7.40	42.0
1.50	20.7	2.70	32.1	4.60	11.1	6.30	9.7	7.50	40.3
1.60	16.2	2.80	28.9	4.70	13.6	6.40	11.6	7.60	38.5
1.70	13.0	2.90	25.7	4.80	16.5	6.50	13.9	7.70	36.6
1.80	10.2	3.00	22.3	4.90	19.4	6.60	16.4	7.80	34.5
1.90	8.1	3.10	18.8	5.00	22.6	6.70	19.3	7.90	32.0
2.00	6.5	3.20	15.7	5.10	25.5	6.80	22.4	8.00	29.2
2.10	5.10	3.30	12.9	5.20	28.8	6.90	25.9	8.10	26.2
2.20	3.9	3.40	10.4	5.30	31.6	7.00	29.1	8.20	22.9
		3.50	8.2	5.40	34.1	7.10	32.1	8.30	19.9
		3.60	6.3	5.50	36.6	7.20	34.7	8.40	17.2
		3.70	4.5	5.60	38.8	7.30	37.0	8.50	14.7
		3.80	2.9	5.70	40.6	7.40	39.1	8.60	12.2
		3.90	1.4	5.80	42.3	7.50	40.9	8.70	10.3
		4.00	0.1	5.90	43.7	7.60	42.4	8.80	8.5
						7.70	43.5	8.90	7.0
						7.80	44.5	9.00	5.7
						7.90	45.3		
						8.00	46.1		

F pH	F x	G pH	G x	H pH	H x	I pH	I x	J pH	J x
8.00	20.5	9.20	0.9	9.60	5.0	10.90	3.3	12.00	6.0
8.10	19.7	9.30	3.6	9.70	6.2	11.00	4.1	12.10	8.0
8.20	18.8	9.40	6.2	9.80	7.6	11.10	5.1	12.20	10.2
8.30	17.7	9.50	8.8	9.90	9.1	11.20	6.3	12.30	12.8
8.40	16.6	9.60	11.1	10.00	10.7	11.30	7.6	12.40	16.2
8.50	15.2	9.70	13.1	10.10	12.2	11.40	9.1	12.50	20.4
8.60	13.5	9.80	15.0	10.20	13.8	11.50	11.1	12.60	25.6
8.70	11.6	9.90	16.7	10.30	15.2	11.60	13.5	12.70	32.2
8.80	9.6	10.00	18.3	10.40	16.5	11.70	16.2	12.80	41.2
8.90	7.1	10.10	19.5	10.50	17.8	11.80	19.4	12.90	53.0
9.00	4.6	10.20	20.5	10.60	19.1	11.90	23.0	13.00	66.0
9.10	2.0	10.30	21.3	10.70	20.2	12.00	26.9		
		10.40	22.1	10.80	21.2				
		10.50	22.7	10.90	22.0				
		10.60	23.3	11.00	22.7				
		10.70	23.8						
		10.80	24.25						

A. 25 ml of 0.2 molar KCl + x ml of 0.2 molar HCl.
B. 50 ml of 0.1 molar potassium hydrogen phthalate + x ml of 0.1 molar HCl.
C. 50 ml of 0.1 molar potassium hydrogen phthalate + x ml of 0.1 molar NaOH.
D. 50 ml of 0.1 molar potassium dihydrogen phosphate + x ml of 0.1 molar NaOH.
E. 50 ml of 0.1 molar tris(hydroxymethyl)aminomethane + x ml of 0.1 M HCl.
F. 50 ml of 0.025 molar borax + x ml of 0.1 molar HCl.
G. 50 ml of 0.025 molar borax + x ml of 0.1 molar NaOH.
H. 50 ml of 0.05 molar sodium bicarbonate + x ml of 0.1 molar NaOH.
I. 50 ml of 0.05 molar disodium hydrogen phosphate + x ml of 0.1 molar NaOH.
J. 25 ml of 0.2 molar KCl + x ml of 0.2 molar NaOH.

Final volume of mixtures = 100 ml.

References

1. Bower, V. E., and Bates, R. G., *J. Res. Natl. Bur. Stand.*, 55, 197, 1955 (A–D).
2. Bates, R. G., and Bower, V. E., *Anal. Chem.*, 28, 1322, 1956 (E–J).

DISSOCIATION CONSTANTS OF INORGANIC ACIDS AND BASES

The data in this table are presented as values of pK_a, defined as the negative logarithm of the acid dissociation constant K_a for the reaction

$$BH \rightleftharpoons B^- + H^+$$

Thus $pK_a = -\log K_a$, and the hydrogen ion concentration $[H^+]$ can be calculated from

$$K_a = \frac{[H^+][B^-]}{[BH]}$$

In the case of bases, the entry in the table is for the conjugate acid; e.g., ammonium ion for ammonia. The OH^- concentration in the system

$$NH_3 + H_2O \rightleftharpoons NH_4^+ + OH^-$$

can be calculated from the equation

$$K_b = K_{water}/K_a = \frac{[OH^-][NH_4^+]}{[NH_3]}$$

where $K_{water} = 1.01 \times 10^{-14}$ at 25 °C. Note that $pK_a + pK_b = pK_{water}$.

All values refer to dilute aqueous solutions at zero ionic strength at the temperature indicated. The table is arranged alphabetically by compound name.

Reference

1. Perrin, D. D., *Ionization Constants of Inorganic Acids and Bases in Aqueous Solution, Second Edition,* Pergamon, Oxford, 1982.

Name	Formula	Step	t/°C	pK_a
Aluminum(III) ion	Al^{+3}		25	5.0
Ammonia	NH_3		25	9.25
Arsenic acid	H_3AsO_4	1	25	2.26
		2	25	6.76
		3	25	11.29
Arsenious acid	H_2AsO_3		25	9.29
Barium(II) ion	Ba^{+2}		25	13.4
Boric acid	H_3BO_3	1	20	9.27
		2	20	>14
Calcium(II) ion	Ca^{+2}		25	12.6
Carbonic acid	H_2CO_3	1	25	6.35
		2	25	10.33
Chlorous acid	$HClO_2$		25	1.94
Chromic acid	H_2CrO_4	1	25	0.74
		2	25	6.49
Cyanic acid	HCNO		25	3.46
Germanic acid	H_2GeO_3	1	25	9.01
		2	25	12.3
Hydrazine	N_2H_4		25	8.1
Hydrazoic acid	HN_3		25	4.6
Hydrocyanic acid	HCN		25	9.21
Hydrofluoric acid	HF		25	3.20
Hydrogen peroxide	H_2O_2		25	11.62
Hydrogen selenide	H_2Se	1	25	3.89
		2	25	11.0
Hydrogen sulfide	H_2S	1	25	7.05
		2	25	19
Hydrogen telluride	H_2Te	1	18	2.6
		2	25	11
Hydroxylamine	NH_2OH		25	5.94
Hypobromous acid	HBrO		25	8.55
Hypochlorous acid	HClO		25	7.40
Hypoiodous acid	HIO		25	10.5
Iodic acid	HIO_3		25	0.78
Lithium ion	Li^+		25	13.8
Magnesium(II) ion	Mg^{+2}		25	11.4
Nitrous acid	HNO_2		25	3.25
Perchloric acid	$HClO_4$		20	-1.6
Periodic acid	HIO_4		25	1.64
Phosphoric acid	H_3PO_4	1	25	2.16

Name	Formula	Step	t/°C	pK_a
		2	25	7.21
		3	25	12.32
Phosphorous acid	H_3PO_3	1	20	1.3
		2	20	6.70
Pyrophosphoric acid	$H_4P_2O_7$	1	25	0.91
		2	25	2.10
		3	25	6.70
		4	25	9.32
Selenic acid	H_2SeO_4	2	25	1.7
Selenious acid	H_2SeO_3	1	25	2.62
		2	25	8.32
Silicic acid	H_4SiO_4	1	30	9.9
		2	30	11.8
		3	30	12
		4	30	12
Sodium ion	Na^+		25	14.8
Strontium(II) ion	Sr^{+2}		25	13.2
Sulfamic acid	NH_2SO_3H		25	1.05
Sulfuric acid	H_2SO_4	2	25	1.99
Sulfurous acid	H_2SO_3	1	25	1.85
		2	25	7.2
Telluric acid	H_2TeO_4	1	18	7.68
		2	18	11.0
Tellurous acid	H_2TeO_3	1	25	6.27
		2	25	8.43
Tetrafluoroboric acid	HBF_4		25	0.5
Thiocyanic acid	HSCN		25	−1.8
Water	H_2O		25	13.995

DISSOCIATION CONSTANTS OF ORGANIC ACIDS AND BASES

This table lists the dissociation (ionization) constants of over 1070 organic acids, bases, and amphoteric compounds. All data apply to dilute aqueous solutions and are presented as values of pK_a, which is defined as the negative of the logarithm of the equilibrium constant K_a for the reaction

$$HA \rightleftharpoons H^+ + A^-$$

i.e.,

$$K_a = [H^+][A^-]/[HA]$$

where $[H^+]$, etc. represent the concentrations of the respective species in mol/L. It follows that pK_a = pH + log[HA] – log[A$^-$], so that a solution with 50% dissociation has pH equal to the pK_a of the acid.

Data for bases are presented as pK_a values for the conjugate acid, i.e., for the reaction

$$BH^+ \rightleftharpoons H^+ + B$$

In older literature, an ionization constant K_b was used for the reaction $B + H_2O \rightleftharpoons BH^+ + OH^-$. This is related to K_a by

$$pK_a + pK_b = pK_{water} = 14.00 \quad (\text{at } 25\ °C)$$

Compounds are listed by molecular formula in Hill order.

References

1. Perrin, D. D., *Dissociation Constants of Organic Bases in Aqueous Solution*, Butterworths, London, 1965; Supplement, 1972.
2. Serjeant, E. P., and Dempsey, B., *Ionization Constants of Organic Acids in Aqueous Solution*, Pergamon, Oxford, 1979.
3. Albert, A., "Ionization Constants of Heterocyclic Substances", in Katritzky, A. R., Ed., *Physical Methods in Heterocyclic Chemistry*, Academic Press, New York, 1963.
4. Sober, H.A., Ed., *CRC Handbook of Biochemistry*, CRC Press, Boca Raton, FL, 1968.
5. Perrin, D. D., Dempsey, B., and Serjeant, E. P., pK_a *Prediction for Organic Acids and Bases*, Chapman and Hall, London, 1981.
6. Albert, A., and Serjeant, E. P., *The Determination of Ionization Constants, Third Edition*, Chapman and Hall, London, 1984.
7. O'Neil, M.J., Ed., *The Merck Index, 14th Edition*, Merck & Co., Whitehouse Station, NJ, 2006.

Mol. form.	Name	Step	*t*/°C	pK_a
$CHNO$	Cyanic acid		25	3.7
CH_2N_2	Cyanamide		29	1.1
CH_2O	Formaldehyde		25	13.27
CH_2O_2	Formic acid		25	3.75
CH_3NO_2	Nitromethane		25	10.21
CH_3NS_2	Carbamodithioic acid		25	2.95
CH_4N_2O	Urea		25	0.10
CH_4N_2S	Thiourea		25	-1
CH_4O	Methanol		25	15.5
CH_4S	Methanethiol		25	10.33
CH_5N	Methylamine		25	10.66
CH_5NO	*O*-Methylhydroxylamine			12.5
CH_5N_3	Guanidine		25	13.6
C_2HCl_3O	Trichloroacetaldehyde		25	10.04
$C_2HCl_3O_2$	Trichloroacetic acid		20	0.66
$C_2HF_3O_2$	Trifluoroacetic acid		25	0.52
$C_2H_2Cl_2O_2$	Dichloroacetic acid		25	1.35
$C_2H_2O_3$	Glyoxylic acid		25	3.18
$C_2H_2O_4$	Oxalic acid	1	25	1.25
		2	25	3.81
$C_2H_3BrO_2$	Bromoacetic acid		25	2.90
$C_2H_3ClO_2$	Chloroacetic acid		25	2.87
$C_2H_3Cl_3O$	2,2,2-Trichloroethanol		25	12.24
$C_2H_3FO_2$	Fluoroacetic acid		25	2.59
$C_2H_3F_3O$	2,2,2-Trifluoroethanol		25	12.37
$C_2H_3IO_2$	Iodoacetic acid		25	3.18
$C_2H_3NO_4$	Nitroacetic acid		24	1.48
$C_2H_3N_3$	1H-1,2,3-Triazole		20	1.17
$C_2H_3N_3$	1H-1,2,4-Triazole		20	2.27
$C_2H_4N_2$	Aminoacetonitrile		25	5.34
C_2H_4O	Acetaldehyde		25	13.57
C_2H_4OS	Thioacetic acid		25	3.33
$C_2H_4O_2$	Acetic acid		25	4.756
$C_2H_4O_2S$	Thioglycolic acid		25	3.68
$C_2H_4O_3$	Glycolic acid		25	3.83
C_2H_5N	Ethyleneimine		25	8.04
C_2H_5NO	Acetamide		25	15.1
$C_2H_5NO_2$	Acetohydroxamic acid			8.70
$C_2H_5NO_2$	Nitroethane		25	8.46
$C_2H_5NO_2$	Glycine	1	25	2.35
		2	25	9.78
$C_2H_6N_2$	Ethanimidamide		25	12.1
C_2H_6O	Ethanol		25	15.5
C_2H_6OS	2-Mercaptoethanol		25	9.72
$C_2H_6O_2$	Ethyleneglycol		25	15.1
$C_2H_7AsO_2$	Dimethylarsinic acid	1	25	1.57
		2	25	6.27
C_2H_7N	Ethylamine		25	10.65
C_2H_7N	Dimethylamine		25	10.73
C_2H_7NO	Ethanolamine		25	9.50
$C_2H_7NO_3S$	2-Aminoethanesulfonic acid	1	25	1.5
		2	25	9.06
C_2H_7NS	Cysteamine	1	25	8.27
		2	25	10.53
$C_2H_7N_5$	Biguanide	1		11.52
		2		2.93
$C_2H_8N_2$	1,2-Ethanediamine	1	25	9.92
		2	25	6.86
$C_2H_8O_7P_2$	1-Hydroxy-1,1-diphosphonoethane	1		1.35
		2		2.87
		3		7.03
		4		11.3
$C_3H_2O_2$	2-Propynoic acid		25	1.84
C_3H_3NO	Oxazole		33	0.8
C_3H_3NO	Isoxazole		25	-2.0
$C_3H_3NO_2$	Cyanoacetic acid		25	2.47
C_3H_3NS	Thiazole		25	2.52
$C_3H_3N_3O_3$	Cyanuric acid	1		6.88
		2		11.40
		3		13.5
$C_3H_4N_2$	1*H*-Pyrazole		25	2.49
$C_3H_4N_2$	Imidazole		25	6.99

Mol. form.	Name	Step	t/°C	pK_a
$C_3H_4N_2S$	2-Thiazolamine		20	5.36
C_3H_4O	Propargyl alcohol		25	13.6
$C_3H_4O_2$	Acrylic acid		25	4.25
$C_3H_4O_3$	Pyruvic acid		25	2.39
$C_3H_4O_4$	Malonic acid	1	25	2.85
		2	25	5.70
$C_3H_4O_5$	Hydroxypropanedioic acid	1		2.42
		2		4.54
$C_3H_5BrO_2$	3-Bromopropanoic acid		25	4.00
$C_3H_5ClO_2$	2-Chloropropanoic acid		25	2.83
$C_3H_5ClO_2$	3-Chloropropanoic acid		25	3.98
$C_3H_6N_2$	3-Aminopropanenitrile		20	7.80
$C_3H_6N_6$	1,3,5-Triazine-2,4,6-triamine		25	5.00
C_3H_6O	Allyl alcohol		25	15.5
$C_3H_6O_2$	Propanoic acid		25	4.87
$C_3H_6O_2S$	(Methylthio)acetic acid		25	3.66
$C_3H_6O_3$	Lactic acid		25	3.86
$C_3H_6O_3$	3-Hydroxypropanoic acid		25	4.51
$C_3H_6O_4$	Glyceric acid		25	3.52
C_3H_7N	Allylamine		25	9.49
C_3H_7N	Azetidine		25	11.29
C_3H_7NO	2-Propanone oxime		25	12.42
$C_3H_7NO_2$	*L*-Alanine	1	25	2.34
		2	25	9.87
$C_3H_7NO_2$	β-Alanine	1	25	3.55
		2	25	10.24
$C_3H_7NO_2$	Sarcosine	1	25	2.21
		2	25	10.1
$C_3H_7NO_2S$	*L*-Cysteine	1	25	1.5
		2	25	8.7
		3	25	10.2
$C_3H_7NO_3$	*L*-Serine	1	25	2.19
		2	25	9.21
$C_3H_7NO_5S$	*DL*-Cysteic acid	1	25	1.3
		2	25	1.9
		3	25	8.70
$C_3H_7N_3O_2$	Glycocyamine		25	2.82
$C_3H_8O_2$	Ethylene glycol monomethyl ether		25	14.8
$C_3H_8O_3$	Glycerol		25	14.15
C_3H_9N	Propylamine		25	10.54
C_3H_9N	Isopropylamine		25	10.63
C_3H_9N	Trimethylamine		25	9.80
C_3H_9NO	2-Methoxyethylamine		25	9.40
C_3H_9NO	Trimethylamine oxide		20	4.65
$C_3H_{10}N_2$	1,2-Propanediamine, (±)	1	25	9.82
		2	25	6.61
$C_3H_{10}N_2$	1,3-Propanediamine	1	25	10.55
		2	25	8.88
$C_3H_{10}N_2O$	1,3-Diamino-2-propanol	1	20	9.69
		2	20	7.93
$C_3H_{11}N_3$	1,2,3-Triaminopropane	1	20	9.59
		2	20	7.95
$C_4H_4FN_3O$	Flucytosine			3.26
$C_4H_4N_2$	Pyrazine		20	0.65
$C_4H_4N_2$	Pyrimidine		20	1.23
$C_4H_4N_2$	Pyridazine		20	2.24
$C_4H_4N_2O_2$	Uracil		25	9.45
$C_4H_4N_2O_3$	Barbituric acid		25	4.01
$C_4H_4N_2O_5$	Alloxanic acid		25	6.64

Mol. form.	Name	Step	t/°C	pK_a
$C_4H_4N_4O_2$	5-Nitropyrimidinamine		20	0.35
$C_4H_4O_2$	2-Butynoic acid		25	2.62
$C_4H_4O_4$	Maleic acid	1	25	1.92
		2	25	6.23
$C_4H_4O_4$	Fumaric acid	1	25	3.02
		2	25	4.38
$C_4H_4O_5$	Oxaloacetic acid	1	25	2.55
		2	25	4.37
		3	25	13.03
C_4H_5N	Pyrrole		25	-3.8
$C_4H_5NO_2$	Succinimide		25	9.62
$C_4H_5N_3$	2-Pyrimidinamine		20	3.45
$C_4H_5N_3$	4-Pyrimidinamine		20	5.71
$C_4H_5N_3O$	Cytosine	1		4.60
		2		12.16
$C_4H_5N_3O_2$	6-Methyl-1,2,4-triazine-3,5(2H,4H) dione			7.6
$C_4H_6N_2$	1-Methylimidazol		25	6.95
$C_4H_6N_4O_3$	Allantoin		25	8.96
$C_4H_6N_4O_3S_2$	Acetazolamide			7.2
$C_4H_6O_2$	*trans*-Crotonic acid		25	4.69
$C_4H_6O_2$	3-Butenoic acid		25	4.34
$C_4H_6O_2$	Cyclopropanecarboxylic acid		25	4.83
$C_4H_6O_3$	2-Oxobutanoic acid		25	2.50
$C_4H_6O_3$	Acetoacetic acid		25	3.6
$C_4H_6O_4$	Succinic acid	1	25	4.21
		2	25	5.64
$C_4H_6O_4$	Methylmalonic acid	1	25	3.07
		2	25	5.76
$C_4H_6O_5$	Malic acid	1	25	3.40
		2	25	5.11
$C_4H_6O_6$	*DL*-Tartaric acid	1	25	3.03
		2	25	4.37
$C_4H_6O_6$	*meso*-Tartaric acid	1	25	3.17
		2	25	4.91
$C_4H_6O_6$	*L*-Tartaric acid	1	25	2.98
		2	25	4.34
$C_4H_6O_8$	Dihydroxytartaric acid		25	1.92
$C_4H_7ClO_2$	2-Chlorobutanoic acid			2.86
$C_4H_7ClO_2$	3-Chlorobutanoic acid			4.05
$C_4H_7ClO_2$	4-Chlorobutanoic acid			4.52
$C_4H_7NO_2$	4-Cyanobutanoic acid		25	2.42
$C_4H_7NO_3$	*N*-Acetylglycine		25	3.67
$C_4H_7NO_4$	Iminodiacetic acid	1		2.98
		2		9.89
$C_4H_7NO_4$	*L*-Aspartic acid	1	25	1.99
		2	25	3.90
		3	25	9.90
$C_4H_7N_3O$	Creatinine	1	25	4.8
		2		9.2
$C_4H_7N_5$	2,4,6-Pyrimidinetriamine		20	6.84
$C_4H_8N_2O_3$	*L*-Asparagine	1	20	2.1
		2	20	8.80
$C_4H_8N_2O_3$	*N*-Glycylglycine	1	25	3.14
		2		8.17
$C_4H_8O_2$	Butanoic acid		25	4.83
$C_4H_8O_2$	2-Methylpropanoic acid		20	4.84
$C_4H_8O_3$	3-Hydroxybutanoic acid, (±)		25	4.70
$C_4H_8O_3$	4-Hydroxybutanoic acid		25	4.72
$C_4H_8O_3$	Ethoxyacetic acid		18	3.65
C_4H_9N	Pyrrolidine		25	11.31

Mol. form.	Name	Step	t/°C	pK_a
C_4H_9NO	Morpholine		25	8.50
$C_4H_9NO_2$	2-Methylalanine	1	25	2.36
		2	25	10.21
$C_4H_9NO_2$	*N,N*-Dimethylglycine		25	9.89
$C_4H_9NO_2$	*DL*-2-Aminobutanoic acid	1	25	2.29
		2	25	9.83
$C_4H_9NO_2$	4-Aminobutanoic acid	1	25	4.031
		2	25	10.556
$C_4H_9NO_2S$	*DL*-Homocysteine	1	25	2.22
		2	25	8.87
		3	25	10.86
$C_4H_9NO_3$	*L*-Threonine	1	25	2.09
		2	25	9.10
$C_4H_9NO_3$	*L*-Homoserine	1	25	2.71
		2	25	9.62
$C_4H_9N_3O_2$	Creatine	1	25	2.63
		2	25	14.3
$C_4H_{10}N_2$	Piperazine	1	25	9.73
		2	25	5.33
$C_4H_{10}N_2O_2$	2,4-Diaminobutanoic acid	1	25	1.85
		2	25	8.24
		3	25	10.44
$C_4H_{10}O_4$	1,2,3,4-Butanetetrol			13.9
$C_4H_{11}N$	Butylamine		25	10.60
$C_4H_{11}N$	*sec*-Butylamine		25	10.56
$C_4H_{11}N$	*tert*-Butylamine		25	10.68
$C_4H_{11}N$	Diethylamine		25	10.84
$C_4H_{11}NO_3$	Tris(hydroxymethyl) methylamine		20	8.3
$C_4H_{12}N_2$	1,4-Butanediamine	1	25	10.80
		2	25	9.63
C_5H_4BrN	3-Bromopyridine		25	2.84
C_5H_4ClN	2-Chloropyridine		25	0.49
C_5H_4ClN	3-Chloropyridine		25	2.81
C_5H_4ClN	4-Chloropyridine		25	3.83
C_5H_4FN	2-Fluoropyridine		25	-0.44
$C_5H_4N_2O_2$	4-Nitropyridine		25	1.61
$C_5H_4N_4$	1*H*-Purine	1	20	2.30
		2	20	8.96
$C_5H_4N_4O$	Hypoxanthine		25	8.7
$C_5H_4N_4O$	Allopurinol			10.2
$C_5H_4N_4O_3$	Uric acid		12	3.89
$C_5H_4N_4S$	1,7-Dihydro-6H-purine-6-thione	1		7.77
		2		11.17
$C_5H_4O_2S$	2-Thiophenecarboxylic acid		25	3.49
$C_5H_4O_2S$	3-Thiophenecarboxylic acid		25	4.1
$C_5H_4O_3$	2-Furancarboxylic acid		25	3.16
$C_5H_4O_3$	3-Furancarboxylic acid		25	3.9
C_5H_5N	Pyridine		25	5.23
C_5H_5NO	2-Pyridinol	1	20	0.75
		2	20	11.65
C_5H_5NO	3-Pyridinol	1	20	4.79
		2	20	8.75
C_5H_5NO	4-Pyridinol	1	20	3.20
		2	20	11.12
C_5H_5NO	2(1H)-Pyridinone	1	20	0.75
		2	20	11.65
C_5H_5NO	Pyridine-1-oxide		24	0.79
$C_5H_5NO_2$	1*H*-Pyrrole-2-carboxylic acid		20	4.45
$C_5H_5NO_2$	1*H*-Pyrrole-3-carboxylic acid		20	5.00
$C_5H_5N_3O$	Pyrazinecarboxamide			0.5
$C_5H_5N_5$	Adenine	1		4.3
		2		9.83
$C_5H_5N_5O$	Guanine		40	9.92
$C_5H_6N_2$	2-Pyridinamine		20	6.82
$C_5H_6N_2$	3-Pyridinamine		25	6.04
$C_5H_6N_2$	4-Pyridinamine		25	9.11
$C_5H_6N_2$	2-Methylpyrazine		27	1.45
$C_5H_6N_2O_2$	Thymine		25	9.94
$C_5H_6O_4$	1,1-Cyclopropanedicarboxylic acid	1	25	1.82
		2	25	7.43
$C_5H_6O_4$	*trans*-1-Propene-1,2-dicarboxylic acid	1	25	3.09
		2	25	4.75
$C_5H_6O_4$	1-Propene-2,3-dicarboxylic acid	1	25	3.85
		2	25	5.45
$C_5H_6O_5$	2-Oxoglutaric acid	1	25	2.47
		2	25	4.68
$C_5H_7NO_3$	5,5-Dimethyl-2,4-oxazolidinedione		37	6.13
$C_5H_7NO_3$	*L*-Pyroglutamic acid		25	3.32
$C_5H_7N_3$	2,5-Pyridinediamine		20	6.48
$C_5H_7N_3$	Methylaminopyrazine		25	3.39
$C_5H_7N_3O_4$	Azaserine			8.55
$C_5H_8N_2$	2,4-Dimethylimidazole		25	8.36
$C_5H_8N_4O_3S_2$	Methazolamide			7.30
$C_5H_8O_2$	*trans*-3-Pentenoic acid		25	4.51
$C_5H_8O_4$	Dimethylmalonic acid		25	3.15
$C_5H_8O_4$	Glutaric acid	1	18	4.32
		2	25	5.42
$C_5H_8O_4$	Methylsuccinic acid	1	25	4.13
		2	25	5.64
$C_5H_9NO_2$	*L*-Proline	1	25	1.95
		2	25	10.64
$C_5H_9NO_3$	5-Amino-4-oxopentanoic acid	1	25	4.05
		2	25	8.90
$C_5H_9NO_3$	*trans*-4-Hydroxyproline	1	25	1.82
		2	25	9.66
$C_5H_9NO_4$	*L*-Glutamic acid	1	25	2.13
		2	25	4.31
		3		9.67
$C_5H_9N_3$	Histamine	1	25	6.04
		2	25	9.75
$C_5H_{10}N_2O_3$	Glycylalanine		25	3.15
$C_5H_{10}N_2O_3$	*L*-Glutamine	1	25	2.17
		2	25	9.13
$C_5H_{10}N_2O_4$	Glycylserine	1	25	2.98
		2	25	8.38
$C_5H_{10}O_2$	Pentanoic acid		20	4.83
$C_5H_{10}O_2$	2-Methylbutanoic acid		25	4.80
$C_5H_{10}O_2$	3-Methylbutanoic acid		25	4.77
$C_5H_{10}O_2$	2,2-Dimethylpropanoic acid		20	5.03
$C_5H_{10}O_4$	*D*-2-Deoxyribose		25	12.61
$C_5H_{10}O_5$	*L*-Ribose		25	12.22
$C_5H_{10}O_5$	*D*-Xylose		18	12.14
$C_5H_{11}N$	Piperidine		25	11.123
$C_5H_{11}N$	*N*-Methylpyrrolidine		25	10.46
$C_5H_{11}NO$	4-Methylmorpholine		25	7.38
$C_5H_{11}NO_2$	*L*-Valine	1	25	2.29
		2	25	9.74

Mol. form.	Name	Step	t/°C	pK_a
$C_5H_{11}NO_2$	*DL*-Norvaline	1		2.36
		2		9.72
$C_5H_{11}NO_2$	*L*-Norvaline	1	25	2.32
		2	25	9.81
$C_5H_{11}NO_2$	*N*-Propylglycine	1	25	2.35
		2	25	10.19
$C_5H_{11}NO_2$	5-Aminopentanoic acid	1	25	4.27
		2	25	10.77
$C_5H_{11}NO_2$	Betaine		0	1.83
$C_5H_{11}NO_2S$	*L*-Methionine	1	25	2.13
		2	25	9.27
$C_5H_{12}N_2O$	Tetramethylurea			2
$C_5H_{12}N_2O_2$	*L*-Ornithine	1	25	1.71
		2	25	8.69
		3	25	10.76
$C_5H_{13}N$	Pentylamine		25	10.63
$C_5H_{13}N$	3-Pentanamine		17	10.59
$C_5H_{13}N$	3-Methyl-1-butanamine		25	10.60
$C_5H_{13}N$	2-Methyl-2-butanamine		19	10.85
$C_5H_{13}N$	2,2-Dimethylpropylamine		25	10.15
$C_5H_{13}N$	Diethylmethylamine		25	10.35
$C_5H_{14}NO$	Choline		25	13.9
$C_5H_{14}N_2$	1,5-Pentanediamine	1	25	10.05
		2	25	10.93
$C_6H_3Cl_3N_2O_2$	4-Amino-3,5,6-trichloro-2-pyridinecarboxlic acid			3.6
$C_6H_3N_3O_7$	2,4,6-Trinitrophenol		24	0.42
$C_6H_4Cl_2O$	2,3-Dichlorophenol		25	7.44
$C_6H_4N_2O_5$	2,4-Dinitrophenol		25	4.07
$C_6H_4N_2O_5$	2,5-Dinitrophenol		15	5.15
$C_6H_4N_4$	Pteridine		20	4.05
C_6H_5BrO	2-Bromophenol		25	8.45
C_6H_5BrO	3-Bromophenol		25	9.03
C_6H_5BrO	4-Bromophenol		25	9.37
$C_6H_5Br_2N$	3,5-Dibromoaniline		25	2.34
C_6H_5ClO	2-Chlorophenol		25	8.56
C_6H_5ClO	3-Chlorophenol		25	9.12
C_6H_5ClO	4-Chlorophenol		25	9.41
$C_6H_5Cl_2N$	2,4-Dichloroaniline		22	2.05
C_6H_5FO	2-Fluorophenol		25	8.73
C_6H_5FO	3-Fluorophenol		25	9.29
C_6H_5FO	4-Fluorophenol		25	9.89
C_6H_5IO	2-Iodophenol		25	8.51
C_6H_5IO	3-Iodophenol		25	9.03
C_6H_5IO	4-Iodophenol		25	9.33
C_6H_5NO	2-Pyridinecarboxaldehyde		25	12.68
C_6H_5NO	4-Pyridinecarboxaldehyde		30	12.05
$C_6H_5NO_2$	Nitrobenzene		0	3.98
$C_6H_5NO_2$	2-Pyridinecarboxylic acid	1	20	0.99
		2	20	5.39
$C_6H_5NO_2$	3-Pyridinecarboxylic acid	1	25	2.00
		2	25	4.82
$C_6H_5NO_2$	4-Pyridinecarboxylic acid	1	25	1.77
		2	25	4.84
$C_6H_5NO_3$	2-Nitrophenol		25	7.23
$C_6H_5NO_3$	3-Nitrophenol		25	8.36
$C_6H_5NO_3$	4-Nitrophenol		25	7.15
$C_6H_5N_3$	1*H*-Benzotriazole		20	1.6
$C_6H_5N_5O$	2-Amino-4-hydroxypteridine	1	20	2.27
		2	20	7.96
$C_6H_5N_5O_2$	Xanthopterin	2	20	6.59

Mol. form.	Name	Step	t/°C	pK_a
		3	20	9.31
C_6H_6BrN	2-Bromoaniline		25	2.53
C_6H_6BrN	3-Bromoaniline		25	3.53
C_6H_6BrN	4-Bromoaniline		25	3.89
C_6H_6ClN	2-Chloroaniline		25	2.66
C_6H_6ClN	3-Chloroaniline		25	3.52
C_6H_6ClN	4-Chloroaniline		25	3.98
C_6H_6FN	2-Fluoroaniline		25	3.20
C_6H_6FN	3-Fluoroaniline		25	3.59
C_6H_6FN	4-Fluoroaniline		25	4.65
C_6H_6IN	2-Iodoaniline		25	2.54
C_6H_6IN	3-Iodoaniline		25	3.58
C_6H_6IN	4-Iodoaniline		25	3.81
$C_6H_6N_2O$	3-Pyridinecarboxamide		20	3.3
$C_6H_6N_2O$	2-Pyridinecarbox-aldehyde oxime	1	20	3.59
		2	20	10.18
$C_6H_6N_2O_2$	2-Nitroaniline		25	-0.25
$C_6H_6N_2O_2$	3-Nitroaniline		25	2.46
$C_6H_6N_2O_2$	4-Nitroaniline		25	1.02
C_6H_6O	Phenol		25	9.99
$C_6H_6O_2$	*p*-Hydroquinone	1	25	9.85
		2	25	11.4
$C_6H_6O_2$	Pyrocatechol	1	25	9.34
		2	25	12.6
$C_6H_6O_2$	Resorcinol	1	25	9.32
		2	25	11.1
$C_6H_6O_2S$	Benzenesulfinic acid		20	1.3
$C_6H_6O_3S$	Benzenesulfonic acid		25	0.70
$C_6H_6O_4$	5-Hydroxy-2-(hydroxy-methyl)-4H-pyran-4-one			7.9
$C_6H_6O_4S$	3-Hydroxybenzene-sulfonic acid		25	9.07
$C_6H_6O_4S$	4-Hydroxybenzene-sulfonic acid		25	9.11
$C_6H_6O_6$	*cis*-1-Propene-1,2,3-tricarboxylic acid		25	1.95
$C_6H_6O_6$	*trans*-1-Propene-1,2,3-tricarboxylic acid	1	25	2.80
		2	25	4.46
C_6H_6S	Benzenethiol		25	6.62
$C_6H_7BO_2$	Benzeneboronic acid			8.83
C_6H_7N	Aniline		25	4.87
C_6H_7N	2-Methylpyridine		25	6.00
C_6H_7N	3-Methylpyridine		25	5.70
C_6H_7N	4-Methylpyridine		25	5.99
C_6H_7NO	2-Aminophenol	1	20	4.78
		2	20	9.97
C_6H_7NO	3-Aminophenol	1	20	4.37
		2	20	9.82
C_6H_7NO	4-Aminophenol	1	25	5.48
		2	25	10.30
C_6H_7NO	2-Methoxypyridine		20	3.28
C_6H_7NO	3-Methoxypyridine		25	4.78
C_6H_7NO	4-Methoxypyridine		25	6.58
$C_6H_7NO_3S$	2-Aminobenzenesulfonic acid		25	2.46
$C_6H_7NO_3S$	3-Aminobenzenesulfonic acid		25	3.74
$C_6H_7NO_3S$	4-Aminobenzenesulfonic acid		25	3.23
$C_6H_8N_2$	*N*-Methylpyridinamine		20	9.65
$C_6H_8N_2$	*o*-Phenylenediamine	1	20	4.57

Mol. form.	Name	Step	t/°C	pK_a
		2	20	0.80
$C_6H_8N_2$	*m*-Phenylenediamine	1	20	5.11
		2	20	2.50
$C_6H_8N_2$	*p*-Phenylenediamine	1	20	6.31
		2	20	2.97
$C_6H_8N_2$	Phenylhydrazine		15	8.79
$C_6H_8O_2$	2,4-Hexadienoic acid		25	4.76
$C_6H_8O_2$	1,3-Cyclohexanedione		25	5.26
$C_6H_8O_4$	2,2-Dimethyl-1,3-dioxane-4,6-dione			5.1
$C_6H_8O_6$	*L*-Ascorbic acid	1	25	4.04
		2	16	11.7
$C_6H_8O_7$	Citric acid	1	25	3.13
		2	25	4.76
		3	25	6.40
$C_6H_8O_7$	Isocitric acid	1	25	3.29
		2	25	4.71
		3	25	6.40
$C_6H_9NO_6$	Nitrilotriacetic acid	1	20	3.03
		2	20	3.07
		3	20	10.70
$C_6H_9NO_6$	*L*-γ-Carboxyglutamic acid	1	25	1.7
		2	25	3.2
		3	25	4.75
		4	25	9.9
$C_6H_9N_3$	4,6-Dimethylpyrimidinamine		20	4.82
$C_6H_9N_3O_2$	*L*-Histidine	1	25	1.80
		2	25	6.04
		3	25	9.33
$C_6H_{10}O_2$	Cyclopentanecarboxylic acid		25	4.99
$C_6H_{10}O_3$	Ethyl acetoacetate		25	10.68
$C_6H_{10}O_4$	3-Methylglutaric acid		25	4.24
$C_6H_{10}O_4$	Adipic acid	1	18	4.41
		2	18	5.41
$C_6H_{11}NO_2$	2-Piperidinecarboxylic acid	1	25	2.28
		2	25	10.72
$C_6H_{11}NO_3$	Adipamic acid		25	4.63
$C_6H_{11}NO_4$	2-Aminoadipic acid	1	25	2.14
		2	25	4.21
		3	25	9.77
$C_6H_{11}N_3O_4$	*N*-(*N*-Glycylglycyl)glycine	1	25	3.225
		2	25	8.09
$C_6H_{11}N_3O_4$	Glycylasparagine	1	25	2.942
		2	18	8.44
$C_6H_{12}N_2$	Triethylenediamine	1		3.0
		2		8.7
$C_6H_{12}N_2O_4S_2$	*L*-Cystine	1		1
		2		2.1
		3		8.02
		4		8.71
$C_6H_{12}O_2$	Hexanoic acid		25	4.85
$C_6H_{12}O_2$	4-Methylpentanoic acid		18	4.84
$C_6H_{12}O_6$	β-*D*-Fructose		25	12.27
$C_6H_{12}O_6$	α-*D*-Glucose		25	12.46
$C_6H_{12}O_6$	*D*-Mannose		25	12.08
$C_6H_{13}N$	Cyclohexylamine		25	10.64
$C_6H_{13}N$	1-Methylpiperidine		25	10.38
$C_6H_{13}N$	1,2-Dimethylpyrrolidine		26	10.20
$C_6H_{13}NO$	*N*-Ethylmorpholine		25	7.67
$C_6H_{13}NO_2$	*L*-Leucine	1	25	2.33
		2	25	9.74
$C_6H_{13}NO_2$	*L*-Isoleucine	1	25	2.32
		2	25	9.76
$C_6H_{13}NO_2$	*L*-Norleucine	1	25	2.34
		2	25	9.83
$C_6H_{13}NO_2$	6-Aminohexanoic acid	1	25	4.37
		2	25	10.80
$C_6H_{13}NO_4$	*N,N*-Bis(2-hydroxyethyl)glycine	2	20	8.35
$C_6H_{13}N_3O_3$	Citrulline	1	25	2.43
		2	25	9.69
$C_6H_{14}N_2$	*cis*-1,2-Cyclohexanediamine	1	20	9.93
		2	20	6.13
$C_6H_{14}N_2$	*trans*-1,2-Cyclohexanediamine	1	20	9.94
		2	20	6.47
$C_6H_{14}N_2$	*cis*-2,5-Dimethylpiperazine	1	25	9.66
		2	25	5.20
$C_6H_{14}N_2O_2$	*L*-Lysine	1	25	2.16
		2	25	9.06
		3	25	10.54
$C_6H_{14}N_4O_2$	*L*-Arginine	1	25	1.82
		2	25	8.99
		3	25	12.5
$C_6H_{14}O_6$	*D*-Mannitol		18	13.5
$C_6H_{15}N$	Hexylamine		25	10.56
$C_6H_{15}N$	Diisopropylamine		25	11.05
$C_6H_{15}N$	Triethylamine		25	10.75
$C_6H_{15}NO_3$	Triethanolamine		25	7.76
$C_6H_{16}N_2$	1,6-Hexanediamine	1	0	11.86
		2	0	10.76
$C_6H_{16}N_2$	*N,N,N',N'*-Tetramethyl-1,2-ethanediamine	1	25	10.40
		2	25	8.26
$C_6H_{19}NSi_2$	Hexamethyldisilazane			7.55
$C_7HF_5O_2$	Pentafluorobenzoic acid		25	1.75
$C_7H_3Br_2NO$	3,5-Dibromo-4-hydroxybenzonitrile			4.06
$C_7H_3N_3O_8$	2,4,6-Trinitrobenzoic acid		25	0.65
$C_7H_4Cl_3NO_3$	Triclopyr			2.68
$C_7H_4N_2O_6$	2,4-Dinitrobenzoic acid		25	1.43
$C_7H_5BrO_2$	2-Bromobenzoic acid		25	2.85
$C_7H_5BrO_2$	3-Bromobenzoic acid		25	3.81
$C_7H_5BrO_2$	4-Bromobenzoic acid		25	3.96
$C_7H_5ClO_2$	2-Chlorobenzoic acid		25	2.90
$C_7H_5ClO_2$	3-Chlorobenzoic acid		25	3.84
$C_7H_5ClO_2$	4-Chlorobenzoic acid		25	4.00
$C_7H_5FO_2$	2-Fluorobenzoic acid		25	3.27
$C_7H_5FO_2$	3-Fluorobenzoic acid		25	3.86
$C_7H_5FO_2$	4-Fluorobenzoic acid		25	4.15
$C_7H_5F_3O$	2-(Trifluoromethyl)phenol		25	8.95
$C_7H_5F_3O$	3-(Trifluoromethyl)phenol		25	8.68
$C_7H_5IO_2$	2-Iodobenzoic acid		25	2.86
$C_7H_5IO_2$	3-Iodobenzoic acid		25	3.87
$C_7H_5IO_2$	4-Iodobenzoic acid		25	4.00
C_7H_5NO	2-Hydroxybenzonitrile		25	6.86
C_7H_5NO	3-Hydroxybenzonitrile		25	8.61
C_7H_5NO	4-Hydroxybenzonitrile		25	7.97
$C_7H_5NO_3S$	Saccharin		18	11.68
$C_7H_5NO_4$	2-Nitrobenzoic acid		25	2.17
$C_7H_5NO_4$	3-Nitrobenzoic acid		25	3.46
$C_7H_5NO_4$	4-Nitrobenzoic acid		25	3.43

Mol. form.	Name	Step	t/°C	pK_a
$C_7H_5NO_4$	2,3-Pyridinedicarboxylic acid	1	25	2.43
		2	25	4.78
$C_7H_5NO_4$	2,4-Pyridinedicarboxylic acid	1	25	2.15
$C_7H_5NO_4$	2,6-Pyridinedicarboxylic acid	1	25	2.16
		2	25	4.76
$C_7H_5NO_4$	3,5-Pyridinedicarboxylic acid	1	25	2.80
$C_7H_6ClN_3O_4S_2$	Chlorothiazide	1		6.85
		2		9.45
$C_7H_6F_3N$	3-(Trifluoromethyl)aniline		25	3.49
$C_7H_6F_3N$	4-(Trifluoromethyl)aniline		25	2.45
$C_7H_6N_2$	1*H*-Benzimidazole		25	5.53
$C_7H_6N_2$	2-Aminobenzonitrile		25	0.77
$C_7H_6N_2$	3-Aminobenzonitrile		25	2.75
$C_7H_6N_2$	4-Aminobenzonitrile		25	1.74
C_7H_6O	Benzaldehyde		25	14.90
$C_7H_6O_2$	Benzoic acid		25	4.204
$C_7H_6O_2$	Salicylaldehyde		25	8.37
$C_7H_6O_2$	3-Hydroxybenzaldehyde		25	8.98
$C_7H_6O_2$	4-Hydroxybenzaldehyde		25	7.61
$C_7H_6O_3$	2-Hydroxybenzoic acid	1	20	2.98
		2	20	13.6
$C_7H_6O_3$	3-Hydroxybenzoic acid	1	25	4.08
		2	19	9.92
$C_7H_6O_3$	4-Hydroxybenzoic acid	1	25	4.57
		2	25	9.46
$C_7H_6O_4$	2,4-Dihydroxybenzoic acid	1	25	3.11
		2	25	8.55
		3	25	14.0
$C_7H_6O_4$	2,5-Dihydroxybenzoic acid	1	25	2.97
$C_7H_6O_4$	3,4-Dihydroxybenzoic acid	1	25	4.48
		2	25	8.83
		3	25	12.6
$C_7H_6O_4$	3,5-Dihydroxybenzoic acid	1	25	4.04
$C_7H_6O_5$	2,4,6-Trihydroxybenzoic acid		25	1.68
$C_7H_6O_5$	3,4,5-Trihydroxybenzoic acid		25	4.41
C_7H_7NO	Benzamide		25	~13
$C_7H_7NO_2$	Aniline-2-carboxylic acid	1	25	2.17
		2	25	4.85
$C_7H_7NO_2$	Aniline-3-carboxylic acid	1	25	3.07
		2	25	4.79
$C_7H_7NO_2$	Aniline-4-carboxylic acid	1	25	2.50
		2	25	4.87
$C_7H_7NO_3$	4-Amino-2-hydroxybenzoic acid			3.25
$C_7H_8ClN_3O_4S_2$	Hydrochlorothiazide	1		7.9
		2		9.2
$C_7H_8N_4O_2$	Theobromine		18	7.89
$C_7H_8N_4O_2$	Theophylline		25	8.77
C_7H_8O	*o*-Cresol		25	10.29
C_7H_8O	*m*-Cresol		25	10.09
C_7H_8O	*p*-Cresol		25	10.26
C_7H_8OS	4-(Methylthio)phenol		25	9.53
$C_7H_8O_2$	2-Methoxyphenol		25	9.98
$C_7H_8O_2$	3-Methoxyphenol		25	9.65
$C_7H_8O_2$	4-Methoxyphenol		25	10.21
C_7H_8S	Benzenemethanethiol		25	9.43
C_7H_9N	Benzylamine		25	9.34

Mol. form.	Name	Step	t/°C	pK_a
C_7H_9N	2-Methylaniline		25	4.45
C_7H_9N	3-Methylaniline		25	4.71
C_7H_9N	4-Methylaniline		25	5.08
C_7H_9N	*N*-Methylaniline		25	4.85
C_7H_9N	2-Ethylpyridine		25	5.89
C_7H_9N	2,3-Dimethylpyridine		25	6.57
C_7H_9N	2,4-Dimethylpyridine		25	6.99
C_7H_9N	2,5-Dimethylpyridine		25	6.40
C_7H_9N	2,6-Dimethylpyridine		25	6.65
C_7H_9N	3,4-Dimethylpyridine		25	6.46
C_7H_9N	3,5-Dimethylpyridine		25	6.15
C_7H_9NO	2-Methoxyaniline		25	4.53
C_7H_9NO	3-Methoxyaniline		25	4.20
C_7H_9NO	4-Methoxyaniline		25	5.36
C_7H_9NS	2-(Methylthio)aniline		25	3.45
C_7H_9NS	4-(Methylthio)aniline		25	4.35
$C_7H_9N_5$	2-Dimethylaminopurine	1	20	4.00
		2	20	10.24
$C_7H_{11}N_3O_2$	*L*-1-Methylhistidine	1	25	1.69
		2	25	6.48
		3	25	8.85
$C_7H_{11}N_3O_2$	*L*-3-Methylhistidine	1	25	1.92
		2	25	6.56
		3	25	8.73
$C_7H_{12}O_2$	Cyclohexanecarboxylic acid		25	4.91
$C_7H_{12}O_4$	Heptanedioic acid	1	25	4.71
		2	25	5.58
$C_7H_{12}O_4$	Butylpropanedioic acid	1	5	2.96
$C_7H_{13}NO_4$	α-Ethylglutamic acid	1	25	3.846
		2	25	7.838
$C_7H_{14}O_2$	Heptanoic acid		25	4.89
$C_7H_{14}O_6$	α-Methylglucoside		25	13.71
$C_7H_{15}N$	1-Ethylpiperidine		23	10.45
$C_7H_{15}N$	1,2-Dimethylpiperidine,(±)		25	10.22
$C_7H_{15}NO_3$	Carnitine		25	3.80
$C_7H_{17}N$	Heptylamine		25	10.67
$C_7H_{17}N$	2-Heptanamine		19	10.7
$C_8H_5NO_2$	3-Cyanobenzoic acid		25	3.60
$C_8H_5NO_2$	4-Cyanobenzoic acid		25	3.55
$C_8H_6N_2$	Cinnoline		20	2.37
$C_8H_6N_2$	Quinazoline		29	3.43
$C_8H_6N_2$	Quinoxaline		20	0.56
$C_8H_6N_2$	Phthalazine		20	3.47
$C_8H_6N_4O_5$	Nitrofurantoin			7.2
$C_8H_6O_3$	3-Formylbenzoic acid		25	3.84
$C_8H_6O_3$	4-Formylbenzoic acid		25	3.77
$C_8H_6O_4$	Phthalic acid	1	25	2.943
		2	25	5.432
$C_8H_6O_4$	Isophthalic acid	1	25	3.70
		2	25	4.60
$C_8H_6O_4$	Terephthalic acid	1	25	3.54
		2	25	4.34
$C_8H_7ClO_2$	2-Chlorobenzeneacetic acid		25	4.07
$C_8H_7ClO_2$	3-Chlorobenzeneacetic acid		25	4.14
$C_8H_7ClO_2$	4-Chlorobenzeneacetic acid		25	4.19
$C_8H_7ClO_3$	2-Chlorophenoxyacetic acid		25	3.05
$C_8H_7ClO_3$	3-Chlorophenoxyacetic acid		25	3.10
$C_8H_7NO_4$	2-Nitrobenzeneacetic acid		25	4.00
$C_8H_7NO_4$	3-Nitrobenzeneacetic acid		25	3.97
$C_8H_7NO_4$	4-Nitrobenzeneacetic acid		25	3.85
$C_8H_8F_3N_3O_4S_2$	Hydroflumethiazide	1		8.9

Mol. form.	Name	Step	t/°C	pK$_a$
		2		9.7
$C_8H_8N_2$	2-Methyl-1*H*-benzimidazole	1	25	6.19
$C_8H_8O_2$	*o*-Toluic acid		25	3.91
$C_8H_8O_2$	*m*-Toluic acid		25	4.25
$C_8H_8O_2$	*p*-Toluic acid		25	4.37
$C_8H_8O_2$	Benzeneacetic acid		25	4.31
$C_8H_8O_2$	1-(2-Hydroxyphenyl)ethanone		25	10.06
$C_8H_8O_2$	1-(3-Hydroxyphenyl)ethanone		25	9.19
$C_8H_8O_2$	1-(4-Hydroxyphenyl)ethanone		25	8.05
$C_8H_8O_3$	2-Methoxybenzoic acid		25	4.08
$C_8H_8O_3$	3-Methoxybenzoic acid		25	4.10
$C_8H_8O_3$	4-Methoxybenzoic acid		25	4.50
$C_8H_8O_3$	Phenoxyacetic acid		25	3.17
$C_8H_8O_3$	Mandelic acid		25	3.37
$C_8H_8O_4$	2,5-Hydroxybenzeneacetic acid		25	4.40
C_8H_9NO	Acetanilide		25	0.5
$C_8H_9NO_2$	2-(Methylamino)benzoic acid		25	5.34
$C_8H_9NO_2$	3-(Methylamino)benzoic acid		25	5.10
$C_8H_9NO_2$	4-(Methylamino)benzoic acid		25	5.04
$C_8H_9NO_2$	*N*-Phenylglycine	1	25	1.83
		2		4.39
$C_8H_{10}BrN$	4-Bromo-*N,N*-dimethylaniline		25	4.23
$C_8H_{10}ClN$	3-Chloro-*N,N*-dimethylaniline		20	3.83
$C_8H_{10}ClN$	4-Chloro-*N,N*-dimethylaniline		20	4.39
$C_8H_{10}N_2O_2$	*N,N*-Dimethyl-3-nitroaniline		25	2.62
$C_8H_{11}N$	*N*-Ethylaniline		25	5.12
$C_8H_{11}N$	*N,N*-Dimethylaniline		25	5.07
$C_8H_{11}N$	2,6-Dimethylaniline		25	3.89
$C_8H_{11}N$	Benzeneethanamine		25	9.83
$C_8H_{11}N$	2,4,6-Trimethylpyridine		25	7.43
$C_8H_{11}NO$	2-Ethoxyaniline		28	4.43
$C_8H_{11}NO$	3-Ethoxyaniline		25	4.18
$C_8H_{11}NO$	4-Ethoxyaniline		28	5.20
$C_8H_{11}NO$	4-(2-Aminoethyl)phenol	1	25	9.74
		2	25	10.52
$C_8H_{11}NO$	2-(2-Methoxyethyl)pyridine			5.5
$C_8H_{11}NO_2$	Dopamine	1	25	8.9
		2	25	10.6
$C_8H_{11}NO_3$	Norepinephrine	1	25	8.64
		2	25	9.70
$C_8H_{11}N_3O_6$	6-Azauridine			6.70
$C_8H_{11}N_5$	Phenylbiguanide	1		10.76
		2		2.13
$C_8H_{12}N_2O_3$	Barbital		25	7.43
$C_8H_{12}O_2$	5,5-Dimethyl-1,3-cyclohexanedione		25	5.15
$C_8H_{13}NO_2$	Arecoline			6.84
$C_8H_{14}O_2S_2$	Thioctic acid			5.4
$C_8H_{14}O_4$	Octanedioic acid	1	25	4.52
$C_8H_{15}NO$	Tropine		15	3.80
$C_8H_{15}NO$	Pseudotropine		15	3.80
$C_8H_{16}N_2O_3$	*N*-Glycylleucine		25	3.18
$C_8H_{16}N_2O_3$	*N*-Leucylglycine	1	25	3.25
		2	25	8.2
$C_8H_{16}N_2O_4S_2$	Homocystine	1	25	1.59
		2	25	2.54
		3	25	8.52
		4	25	9.44
$C_8H_{16}O_2$	Octanoic acid		25	4.89
$C_8H_{16}O_2$	2-Propylpentanoic acid			4.6
$C_8H_{17}N$	2-Propylpiperidine,(*S*)			10.9
$C_8H_{17}N$	2,2,4-Trimethylpiperidine		30	11.04
$C_8H_{17}NO$	*trans*-6-Propyl-3-piperidinol,(3*S*)			10.3
$C_8H_{19}N$	Octylamine		25	10.65
$C_8H_{19}N$	*N*-Methyl-2-heptanamine		17	10.99
$C_8H_{19}N$	Dibutylamine		21	11.25
$C_8H_{20}N_2$	1,8-Octanediamine	1	20	11.00
		2	20	10.1
C_9H_6BrN	3-Bromoquinoline		25	2.69
$C_9H_7ClO_2$	*trans-o*-Chlorocinnamic acid		25	4.23
$C_9H_7ClO_2$	*trans-m*-Chlorocinnamic acid		25	4.29
$C_9H_7ClO_2$	*trans-p*-Chlorocinnamic acid		25	4.41
C_9H_7N	Quinoline		20	4.90
C_9H_7N	Isoquinoline		20	5.40
C_9H_7NO	2-Quinolinol	1	20	-0.31
		2	20	11.76
C_9H_7NO	3-Quinolinol	1	20	4.28
		2	20	8.08
C_9H_7NO	4-Quinolinol	1	20	2.23
		2	20	11.28
C_9H_7NO	6-Quinolinol	1	20	5.15
		2	20	8.90
C_9H_7NO	8-Quinolinol	1	25	4.91
		2	25	9.81
C_9H_7NO	7-Isoquinolinol	1	20	5.68
		2	20	8.90
$C_9H_7NO_3$	2-Cyanophenoxyacetic acid		25	2.98
$C_9H_7NO_3$	3-Cyanophenoxyacetic acid		25	3.03
$C_9H_7NO_3$	4-Cyanophenoxyacetic acid		25	2.93
$C_9H_7N_7O_2S$	Azathioprine			8.2
$C_9H_8N_2$	2-Quinolinamine		20	7.34
$C_9H_8N_2$	3-Quinolinamine		20	4.91
$C_9H_8N_2$	4-Quinolinamine		20	9.17
$C_9H_8N_2$	1-Isoquinolinamine		20	7.62
$C_9H_8N_2$	3-Isoquinolinamine		20	5.05
$C_9H_8O_2$	*cis*-Cinnamic acid		25	3.88
$C_9H_8O_2$	*trans*-Cinnamic acid		25	4.44
$C_9H_8O_2$	α-Methylenebenezene-acetic acid			4.35
$C_9H_8O_4$	2-(Acetyloxy)benzoic acid		25	3.48
$C_9H_9Br_2NO_3$	3,5-Dibromo-*L*-tyrosine	1		2.17
		2		6.45
		3		7.60
$C_9H_9ClO_2$	3-(2-Chlorophenyl)-propanoic acid		25	4.58
$C_9H_9ClO_2$	3-(3-Chlorophenyl)-propanoic acid		25	4.59
$C_9H_9ClO_2$	3-(4-Chlorophenyl)-propanoic acid		25	4.61
$C_9H_9I_2NO_3$	*L*-3,5-Diiodotyrosine	1	25	2.12

Mol. form.	Name	Step	t/°C	pK_a
		2	25	5.32
		3	25	9.48
$C_9H_9NO_3$	*N*-Benzoylglycine		25	3.62
$C_9H_9NO_4$	3-(2-Nitrophenyl)-propanoic acid		25	4.50
$C_9H_9NO_4$	3-(4-Nitrophenyl)-propanoic acid		25	4.47
$C_9H_9N_3O_2$	Carbendazim			4.48
$C_9H_9N_3O_2S_2$	Sulfathiazole			7.2
$C_9H_{10}INO_3$	*L*-3-Iodotyrosine	1	25	2.2
		2	25	8.7
		3	25	9.1
$C_9H_{10}N_2$	2-Ethylbenzimidazole		25	6.18
$C_9H_{10}O_2$	3,5-Dimethylbenzoic acid		25	4.32
$C_9H_{10}O_2$	Benzenepropanoic acid		25	4.66
$C_9H_{10}O_2$	α-Methylbenzeneacetic acid		25	4.64
$C_9H_{10}O_3$	α-Hydroxy-α-methyl-benezeneacetic acid		25	3.47
$C_9H_{11}Cl_2N_3O_4S_2$	Methylclothiazide			9.4
$C_9H_{11}N$	*N*-Allylaniline		25	4.17
$C_9H_{11}N$	1-Indanamine		22	9.21
$C_9H_{11}NO_2$	4-(Dimethylamino)-benzoic acid	1		6.03
		2		11.49
$C_9H_{11}NO_2$	Ethyl 4-aminobenzoate			2.5
$C_9H_{11}NO_2$	*L*-Phenylalanine	1	25	2.20
		2	25	9.31
$C_9H_{11}NO_3$	*L*-Tyrosine	1	25	2.20
		2	25	9.11
		3	25	10.1
$C_9H_{11}NO_4$	Levodopa	1	25	2.32
		2	25	8.72
		3	25	9.96
		4	25	11.79
$C_9H_{12}N_2O_2$	Tyrosineamide		25	7.33
$C_9H_{13}N$	*N*-Isopropylaniline		25	5.77
$C_9H_{13}NO_3$	Epinephrine	1	25	8.66
		2	25	9.95
$C_9H_{13}N_2O_9P$	5'-Uridylic acid	1		6.4
		2		9.5
$C_9H_{13}N_3O_5$	Cytidine	1		4.22
		2		12.5
$C_9H_{14}ClNO$	Phenylpropanolamine hydrochloride			9.44
$C_9H_{14}N_2O_3$	Metharbital			8.45
$C_9H_{14}N_3O_8P$	3'-Cytidylic acid	1		0.8
		2		4.28
		3		6.0
$C_9H_{14}N_4O_3$	Carnosine	1	20	2.73
		2	20	6.87
		3	20	9.73
$C_9H_{15}NO_3S$	Captopril	1		3.7
		2		9.8
$C_9H_{15}N_5O$	Minoxidil			4.61
$C_9H_{16}O_4$	Nonanedioic acid	1	25	4.53
		2	25	5.33
$C_9H_{18}O_2$	Nonanoic acid		25	4.96
$C_9H_{19}N$	*N*-Butylpiperidine		23	10.47
$C_9H_{19}N$	2,2,6,6-Tetramethyl-piperidine		25	11.07
$C_9H_{21}N$	Nonylamine		25	10.64
$C_{10}H_7NO_2$	8-Quinolinecarboxylic acid		25	1.82
$C_{10}H_8O$	1-Naphthol		25	9.39
$C_{10}H_8O$	2-Naphthol		25	9.63
$C_{10}H_9N$	1-Naphthylamine		25	3.92
$C_{10}H_9N$	2-Naphthylamine		25	4.16
$C_{10}H_9N$	2-Methylquinoline		20	5.83
$C_{10}H_9N$	4-Methylquinoline		20	5.67
$C_{10}H_9N$	5-Methylquinoline		20	5.20
$C_{10}H_9NO$	5-Amino-1-naphthol		25	3.97
$C_{10}H_9NO$	6-Methoxyquinoline		20	5.03
$C_{10}H_9NO_2$	1H-Indole-3-acetic acid			4.75
$C_{10}H_{10}O_2$	*o*-Methylcinnamic acid		25	4.50
$C_{10}H_{10}O_2$	*m*-Methylcinnamic acid		25	4.44
$C_{10}H_{10}O_2$	*p*-Methylcinnamic acid		25	4.56
$C_{10}H_{12}N_2$	Tryptamine		25	10.2
$C_{10}H_{12}N_2O$	5-Hydroxytryptamine	1	25	9.8
		2	25	11.1
$C_{10}H_{12}N_2O_5$	Dinoseb			4.62
$C_{10}H_{12}N_4O_3$	Dideoxyinosine			9.12
$C_{10}H_{12}O$	5,6,7,8-Tetrahydro-2-naphthalenol		25	10.48
$C_{10}H_{12}O_2$	Benzenebutanoic acid		25	4.76
$C_{10}H_{12}O_5$	Propyl 3,4,5-trihydroxy-benzoate			8.11
$C_{10}H_{13}N_5O_4$	Adenosine	1	25	3.6
		2	25	12.4
$C_{10}H_{14}N_2$	*L*-Nicotine	1		8.02
		2		3.12
$C_{10}H_{14}N_5O_7P$	5'-Adenylic acid	1		3.8
		2		6.2
$C_{10}H_{14}O$	2-*tert*-Butylphenol		25	10.62
$C_{10}H_{14}O$	3-*tert*-Butylphenol		25	10.12
$C_{10}H_{14}O$	4-*tert*-Butylphenol		25	10.23
$C_{10}H_{15}N$	*N tert* Butylaniline		25	7.00
$C_{10}H_{15}N$	*N,N*-Diethylaniline		25	6.57
$C_{10}H_{15}NO$	*d*-Ephedrine		10	10.139
$C_{10}H_{15}NO$	*l*-Ephedrine		10	9.958
$C_{10}H_{17}N_3O_6S$	*l*-Glutathione	1	25	2.12
		2	25	3.59
		3	25	8.75
		4	25	9.65
$C_{10}H_{18}N_4O_5$	*L*-Argininosuccinic acid	1	25	1.62
		2	25	2.70
		3	25	4.26
		4	25	9.58
$C_{10}H_{18}O_4$	Sebacic acid	1		4.59
		2		5.59
$C_{10}H_{19}N$	Bornylamine		25	10.17
$C_{10}H_{19}N$	Neobornylamine		25	10.01
$C_{10}H_{21}N$	Butylcyclohexylamine		25	11.23
$C_{10}H_{21}N$	1,2,2,6,6-Pentamethyl-piperidine		30	11.25
$C_{10}H_{23}N$	Decylamine		25	10.64
$C_{11}H_8N_2$	1*H*-Perimidine		20	6.35
$C_{11}H_8O_2$	1-Naphthalenecarboxylic acid		25	3.69
$C_{11}H_8O_2$	2-Naphthalenecarboxylic acid		25	4.16
$C_{11}H_{11}N$	Methyl-1-naphthylamine		27	3.67
$C_{11}H_{12}I_3NO_2$	Iopanoic acid			4.8
$C_{11}H_{12}N_2O_2$	*L*-Tryptophan	1	25	2.46
		2	25	9.41

Mol. form.	Name	Step	t/°C	pK_a
$C_{11}H_{12}N_4O_3S$	Sulfamethoxypyridazine			6.7
$C_{11}H_{13}F_3N_2O_3S$	Mefluidide			4.6
$C_{11}H_{13}NO_3$	Hydrastinine			11.38
$C_{11}H_{13}N_3O_3S$	Sulfisoxazole			5
$C_{11}H_{14}N_2O$	Cytisine	1		6.11
		2		13.08
$C_{11}H_{14}O_2$	2-*tert*-Butylbenzoic acid		25	3.54
$C_{11}H_{14}O_2$	3-*tert*-Butylbenzoic acid		25	4.20
$C_{11}H_{14}O_2$	4-*tert*-Butylbenzoic acid		25	4.38
$C_{11}H_{16}N_2O_2$	Pilocarpine	1	25	1.6
		2	25	6.9
$C_{11}H_{16}N_4O_4$	Pentostatin			5.2
$C_{11}H_{17}N$	*N,N*-Diethyl-2-methylaniline		25	7.24
$C_{11}H_{17}NO_3$	Isoproterenol			8.64
$C_{11}H_{17}N_3O_8$	Tetrodotoxin			8.76
$C_{11}H_{18}ClNO_3$	Methoxamine hydrochloride		25	9.2
$C_{11}H_{18}N_2O_3$	Amobarbital		25	8.0
$C_{11}H_{25}N$	Undecylamine		25	10.63
$C_{11}H_{26}NO_2PS$	Methylphosphonothioic acid *S*[2-[bis(1-isopropyl)amino]-ethyl], *O*-ethylester			7.9
$C_{12}H_6Cl_4O_2S$	Bithionol	1		4.82
		2		10.50
$C_{12}H_8N_2$	1,10-Phenanthroline		25	4.84
$C_{12}H_8N_2$	Phenazine		20	1.20
$C_{12}H_{10}O$	2-Hydroxybiphenyl		25	10.01
$C_{12}H_{10}O$	3-Hydroxybiphenyl		25	9.64
$C_{12}H_{10}O$	4-Hydroxybiphenyl		25	9.55
$C_{12}H_{11}N$	Diphenylamine		25	0.79
$C_{12}H_{11}N$	2-Aminobiphenyl		25	3.83
$C_{12}H_{11}N$	3-Aminobiphenyl		18	4.25
$C_{12}H_{11}N$	4-Aminobiphenyl		18	4.35
$C_{12}H_{11}N$	2-Benzylpyridine		25	5.13
$C_{12}H_{11}N_3$	4-Aminoazobenzene		25	2.82
$C_{12}H_{12}N_2$	*p*-Benzidine	1	20	4.65
		2	20	3.43
$C_{12}H_{12}N_2O_3$	Phenobarbital	1		7.3
		2		11.8
$C_{12}H_{13}I_3N_2O_3$	Iocetamic acid			4
$C_{12}H_{13}N$	*N,N*-Dimethyl-1-naphthylamine		25	4.83
$C_{12}H_{13}N$	*N,N*-Dimethyl-2-naphthylamine		25	4.566
$C_{12}H_{14}N_4O_2S$	Sulfamethazine	1		7.4
		2		2.65
$C_{12}H_{14}N_4O_3S$	Sulfacytine			6.9
$C_{12}H_{17}N_3O_4$	Agaritine	1		3.4
		2		8.86
$C_{12}H_{20}N_2O_2$	Aspergillic acid			5.5
$C_{12}H_{21}N_5O_2S_2$	Nizatidine	1		2.1
		2		6.8
$C_{12}H_{22}O_{11}$	Sucrose		25	12.7
$C_{12}H_{22}O_{11}$	α-Maltose		21	12.05
$C_{12}H_{23}N$	Dicyclohexylamine			10.4
$C_{12}H_{27}N$	Dodecylamine		25	10.63
$C_{13}H_9N$	Acridine		20	5.58
$C_{13}H_9N$	Phenanthridine		20	5.58
$C_{13}H_{10}N_2$	9-Acridinamine		20	9.99
$C_{13}H_{10}N_2$	2-Phenylbenzimidazole	1	25	5.23
		2	25	11.91
$C_{13}H_{10}O_2$	2-Phenylbenzoic acid		25	3.46
$C_{13}H_{10}O_3$	2-Phenoxybenzoic acid		25	3.53
$C_{13}H_{10}O_3$	3-Phenoxybenzoic acid		25	3.95
$C_{13}H_{10}O_3$	4-Phenoxybenzoic acid		25	4.57
$C_{13}H_{11}N_3$	3,6-Acridinediamine		20	9.65
$C_{13}H_{12}Cl_2O_4$	Ethacrynic acid			3.50
$C_{13}H_{12}N_2O$	Harmine			7.70
$C_{13}H_{12}N_2O_3S$	Sulfabenzamide		25	4.57
$C_{13}H_{13}N$	4-Benzylaniline		25	2.17
$C_{13}H_{14}N_2O_{13}$	Harmaline			4.2
$C_{13}H_{15}N_3O_3$	Imazapyr	1		1.9
		2		3.6
$C_{13}H_{16}ClNO$	Ketamine			7.5
$C_{13}H_{19}NO_4S$	4-[(Dipropylamino)-sulfonyl]benzoic acid			5.8
$C_{13}H_{21}N$	2,6-Di-*tert*-butylpyridine			3.58
$C_{13}H_{29}N$	(Tridecyl)amine		25	10.63
$C_{14}H_{12}F_3NO_4S_2$	Perfluidone			2.5
$C_{14}H_{12}O_2$	α-Phenylbenzeneacetic acid		25	3.94
$C_{14}H_{12}O_3$	α-Hydroxy-α-phenyl-benezeneacetic acid		25	3.04
$C_{14}H_{18}N_4O_3$	Trimethoprim			6.6
$C_{14}H_{19}NO_2$	Methylphenidate			8.9
$C_{14}H_{21}N_3O_3S$	Tolazamide		25	3.6
$C_{14}H_{22}N_2O_3$	Atenolol			9.6
$C_{14}H_{31}N$	Tetradecylamine		25	10.62
$C_{15}H_{10}ClN_3O_3$	Clonazepam	1		1.5
		2		10.5
$C_{15}H_{11}I_4NO_4$	*L*-Thyroxine	1	25	2.2
		2	25	6.45
		3	25	10.1
$C_{15}H_{14}O_3$	Fenoprofen			4.5
$C_{15}H_{15}NO_2$	Mefenamic acid			4.2
$C_{15}H_{15}N_3O_2$	Methyl Red	1		2.5
		2		9.5
$C_{15}H_{17}ClN_4$	NeutralRed			6.7
$C_{15}H_{19}NO_2$	Tropacocaine		15	4.32
$C_{15}H_{19}N_3O_3$	Imazethapyr	1		2.1
		2		3.9
$C_{15}H_{21}N_3O_2$	Physostigmine	1		6.12
		2		12.24
$C_{15}H_{26}N_2$	Sparteine	1	20	2.24
		2	20	9.46
$C_{15}H_{33}N$	Pentadecylamine		25	10.61
$C_{16}H_{13}ClN_2O$	Valium			3.4
$C_{16}H_{14}ClN_3O$	Chlorodiazepoxide			4.8
$C_{16}H_{16}N_2O_2$	Lysergic acid	1		3.44
		2		7.68
$C_{16}H_{17}N_3O_4S$	Cephalexin	1		5.2
		2		7.3
$C_{16}H_{19}N_3O_4S$	Cephradine	1		2.63
		2		7.27
$C_{16}H_{22}N_2$	Lycodine	1		3.97
		2		8.08
$C_{16}H_{35}N$	Hexadecylamine		25	10.61
$C_{17}H_{17}NO_2$	Apomorphine	1		7.0
		2		8.92
$C_{17}H_{19}NO_3$	Piperine		18	12.22
$C_{17}H_{19}NO_3$	Morphine	1	25	8.21
		2	20	9.85
$C_{17}H_{20}N_4O_6$	Riboflavin	1		1.7

Mol. form.	Name	Step	t/°C	pK_a
		2	25	9.69
$C_{17}H_{20}O_6$	Mycophenolic acid			4.5
$C_{17}H_{23}NO_3$	Hyoscyamine		21	9.7
$C_{17}H_{27}NO_4$	Nadolol			9.67
$C_{18}H_{19}ClN_4$	Clozapine	1		3.70
		2		7.60
$C_{18}H_{21}NO_3$	Codeine			8.21
$C_{18}H_{21}N_3O$	Dibenzepin			8.25
$C_{18}H_{32}O_2$	Linoleic acid		25	4.77
$C_{18}H_{33}ClN_2O_5S$	Clindamycin			7.6
$C_{18}H_{39}N$	Octadecylamine		25	10.60
$C_{19}H_{10}Br_4O_5S$	Bromophenol Blue			4.0
$C_{19}H_{14}O_5S$	Phenol Red			7.9
$C_{19}H_{16}ClNO_4$	Indomethacin			4.5
$C_{19}H_{17}N_3O_4S_2$	Cephaloridine			3.2
$C_{19}H_{20}N_2O_2$	Phenylbutazone			4.5
$C_{19}H_{21}N$	Protriptyline			8.2
$C_{19}H_{21}NO_3$	Thebaine		15	6.05
$C_{19}H_{22}N_2O$	Cinchonine	1		5.85
		2		9.92
$C_{19}H_{22}N_2O$	Cinchonidine	1		5.80
		2		10.03
$C_{19}H_{22}N_2O_2$	Cupreine			6.57
$C_{19}H_{22}O_6$	Gibberellic acid			4.0
$C_{19}H_{23}N_3O_2$	Ergometrinine			7.3
$C_{19}H_{23}N_3O_2$	Ergonovine			6.8
$C_{20}H_{14}O_4$	Phenolphthalein		25	9.7
$C_{20}H_{21}NO_4$	Papaverine			6.1
$C_{20}H_{23}N$	Amitriptyline			9.4
$C_{20}H_{23}N_7O_7$	Folinic acid	1		3.1
		2		4.8
		3		10.4
$C_{20}H_{24}N_2O_2$	Quinine	1	25	8.52
		2	25	4.13
$C_{20}H_{24}N_2O_2$	Quinidine	1	20	5.4
		2	20	10.0
$C_{20}H_{26}N_2O_2$	Hydroquinine			5.33
$C_{21}H_{14}Br_4O_5S$	Bromocresol Green			4.7
$C_{21}H_{16}Br_2O_5S$	Bromocresol Purple			6.3
$C_{21}H_{18}O_5S$	CresolRed			8.3
$C_{21}H_{21}NO_6$	Hydrastine			7.8
$C_{21}H_{22}N_2O_2$	Strychnine		25	8.26
$C_{21}H_{23}ClFNO_2$	Haloperidol			8.3
$C_{21}H_{31}NO_4$	Furethidine			7.48
$C_{21}H_{35}N_3O_7$	Lisinopril	1		2.5
		2		4.0
		3		6.7
		4		10.1
$C_{22}H_{18}O_4$	*o*-Cresolphthalein			9.4
$C_{22}H_{22}FN_3O_2$	Droperidol			7.64
$C_{22}H_{23}NO_7$	Noscapine			7.8
$C_{22}H_{25}NO_6$	Colchicine		20	12.36
$C_{22}H_{25}N_3O$	Benzpiperylon	1		6.73
		2		9.13
$C_{22}H_{33}NO_2$	Atisine			12.2
$C_{23}H_{26}N_2O_4$	Brucine	1		6.04
		2		11.07
$C_{24}H_{40}O_4$	Deoxycholic acid		20	5.15
$C_{24}H_{40}O_5$	Cholic acid		20	4.98
$C_{25}H_{29}I_2NO_3$	Amiodarone		25	6.56
$C_{25}H_{41}NO_9$	Aconine			9.52
$C_{26}H_{43}NO_6$	Glycocholic acid			4.4
$C_{26}H_{45}NO_7S$	Taurocholic acid			1.4
$C_{27}H_{28}Br_2O_5S$	Bromothymol Blue			7.0
$C_{27}H_{38}N_2O_4$	Verapamil			8.6
$C_{29}H_{32}O_{13}$	Etoposide			9.8
$C_{29}H_{40}N_2O_4$	Emetine	1		5.77
		2		6.64
$C_{30}H_{23}BrO_4$	Bromadiolone		21	4.04
$C_{30}H_{48}O_3$	Oleanolic acid			2.52
$C_{31}H_{36}N_2O_{11}$	Novobiocin	1		4.3
		2		9.1
$C_{32}H_{32}O_{13}S$	Teniposide			10.13
$C_{33}H_{40}N_2O_9$	Reserpine			6.6
$C_{34}H_{47}NO_{11}$	Aconitine			5.88
$C_{36}H_{51}NO_{11}$	Veratridine			9.54
$C_{37}H_{67}NO_{13}$	Erythromycin			8.8
$C_{43}H_{58}N_4O_{12}$	Rifampin	1		1.7
		2		7.9
$C_{45}H_{73}NO_{15}$	Solanine		15	6.66
$C_{46}H_{56}N_4O_{10}$	Vincristine			5.4
$C_{46}H_{58}N_4O_9$	Vinblastine	1		5.4
		2		7.4

CONCENTRATIVE PROPERTIES OF AQUEOUS SOLUTIONS: DENSITY, REFRACTIVE INDEX, FREEZING POINT DEPRESSION, AND VISCOSITY

This table gives properties of aqueous solutions of 66 substances as a function of concentration. All data refer to a temperature of 20 °C. The properties are:

Mass %: Mass of solute divided by total mass of solution, expressed as percent.
m Molality (moles of solute per kg of water).
c Molarity (moles of solute per liter of solution).
ρ Density of solution in g/cm³.
n Index of refraction, relative to air, at a wavelength of 589 nm (sodium D line); the index of pure water at 20 °C is 1.3330.
Δ Freezing point depression in °C relative to pure water.
η Absolute (dynamic) viscosity in mPa s (equal to centipoise, cP); the viscosity of pure water at 20 °C is 1.002 mPa s.

Density data for aqueous solutions over a wider range of temperatures and pressures (and for other compounds) may be found in Reference 2. Solutes are listed in the following order:

Acetic acid
Acetone
Ammonia
Ammonium chloride
Ammonium sulfate
Barium chloride
Calcium chloride
Cesium chloride
Citric acid
Copper sulfate
Disodium ethylenediamine tetraacetate (EDTA sodium)
Ethanol
Ethylene glycol
Ferric chloride
Formic acid
D-Fructose
D-Glucose
Glycerol
Hydrochloric acid
Lactic acid
Lactose
Lithium chloride
Magnesium chloride
Magnesium sulfate
Maltose
Manganese(II) sulfate
D-Mannitol
Methanol
Nitric acid
Oxalic acid
Phosphoric acid
Potassium bicarbonate
Potassium bromide
Potassium carbonate
Potassium chloride
Potassium hydroxide
Potassium iodide
Potassium nitrate
Potassium permanganate
Potassium hydrogen phosphate
Potassium dihydrogen phosphate
Potassium sulfate
1-Propanol
2-Propanol
Silver nitrate
Sodium acetate
Sodium bicarbonate
Sodium bromide
Sodium carbonate
Sodium chloride
Sodium citrate
Sodium hydroxide
Sodium nitrate
Sodium phosphate
Sodium hydrogen phosphate
Sodium dihydrogen phosphate
Sodium sulfate
Sodium thiosulfate
Strontium chloride
Sucrose
Sulfuric acid
Trichloroacetic acid
Tris(hydroxymethyl)methylamine
Urea
Zinc sulfate

References

1. Wolf, A. V., *Aqueous Solutions and Body Fluids*, Hoeber, 1966.
2. Söhnel, O., and Novotny, P., *Densities of Aqueous Solutions of Inorganic Substances*, Elsevier, Amsterdam, 1985.

Solute	Mass %	*m*/mol kg⁻¹	*c*/mol L⁻¹	ρ/g cm⁻³	*n*	Δ/°C	η/mPa s
Acetic acid	0.5	0.084	0.083	0.9989	1.3334	0.16	1.012
CH_3COOH	1.0	0.168	0.166	0.9996	1.3337	0.32	1.022
	2.0	0.340	0.333	1.0011	1.3345	0.63	1.042
	3.0	0.515	0.501	1.0025	1.3352	0.94	1.063
	4.0	0.694	0.669	1.0038	1.3359	1.26	1.084
	5.0	0.876	0.837	1.0052	1.3366	1.58	1.105
	6.0	1.063	1.006	1.0066	1.3373	1.90	1.125
	7.0	1.253	1.175	1.0080	1.3381	2.23	1.143
	8.0	1.448	1.345	1.0093	1.3388	2.56	1.162
	9.0	1.647	1.515	1.0107	1.3395	2.89	1.186
	10.0	1.850	1.685	1.0121	1.3402	3.23	1.210
	12.0	2.271	2.028	1.0147	1.3416	3.91	1.253
	14.0	2.711	2.372	1.0174	1.3430	4.61	1.298
	16.0	3.172	2.718	1.0200	1.3444	5.33	1.341
	18.0	3.655	3.065	1.0225	1.3458	6.06	1.380
	20.0	4.163	3.414	1.0250	1.3472	6.81	1.431
	22.0	4.697	3.764	1.0275	1.3485	7.57	1.478
	24.0	5.259	4.116	1.0299	1.3498	8.36	1.525

Solute	Mass %	*m*/mol kg⁻¹	*c*/mol L⁻¹	ρ/g cm⁻³	*n*	Δ/°C	η/mPa s
	26.0	5.851	4.470	1.0323	1.3512	9.17	1.572
	28.0	6.476	4.824	1.0346	1.3525	10.00	1.613
	30.0	7.137	5.180	1.0369	1.3537	10.84	1.669
	32.0	7.837	5.537	1.0391	1.3550	11.70	1.715
	34.0	8.579	5.896	1.0413	1.3562	12.55	1.762
	36.0	9.367	6.255	1.0434	1.3574	13.38	1.812
	38.0	10.207	6.615	1.0454	1.3586		1.852
	40.0	11.102	6.977	1.0474	1.3598		1.912
	50.0	16.653	8.794	1.0562	1.3653		2.158
	60.0	24.979	10.620	1.0629	1.3700		2.409
	70.0	38.857	12.441	1.0673	1.3738		2.629
	80.0	66.611	14.228	1.0680	1.3767		2.720
	90.0	149.875	15.953	1.0644	1.3771		2.386
	92.0	191.507	16.284	1.0629	1.3766		2.240
	94.0	260.894	16.602	1.0606	1.3759		2.036
	96.0	399.667	16.911	1.0578	1.3748		1.813
	98.0	815.987	17.198	1.0538	1.3734		1.535
	100.0		17.447	1.0477	1.3716		1.223
Acetone	0.5	0.087	0.086	0.9975	1.3334	0.16	1.013
$(CH_3)_2CO$	1.0	0.174	0.172	0.9968	1.3337	0.32	1.024
	2.0	0.351	0.343	0.9954	1.3344	0.65	1.047
	3.0	0.533	0.513	0.9940	1.3352	0.97	1.072
	4.0	0.717	0.684	0.9926	1.3359	1.30	1.099
	5.0	0.906	0.853	0.9912	1.3366	1.63	1.125
	6.0	1.099	1.023	0.9899	1.3373	1.96	1.150
	7.0	1.296	1.191	0.9886	1.3381	2.29	1.174
	8.0	1.497	1.360	0.9874	1.3388	2.62	1.198
	9.0	1.703	1.528	0.9861	1.3395	2.95	1.221
	10.0	1.913	1.696	0.9849	1.3402	3.29	1.244
Ammonia	0.5	0.295	0.292	0.9960	1.3332	0.55	1.009
NH_3	1.0	0.593	0.584	0.9938	1.3335	1.14	1.015
	2.0	1.198	1.162	0.9895	1.3339	2.32	1.029
	3.0	1.816	1.736	0.9853	1.3344	3.53	1.043
	4.0	2.447	2.304	0.9811	1.3349	4.78	1.057
	5.0	3.090	2.868	0.9770	1.3354	6.08	1.071
	6.0	3.748	3.428	0.9730	1.3359	7.43	1.085
	7.0	4.420	3.983	0.9690	1.3365	8.95	1.099
	8.0	5.106	4.533	0.9651	1.3370	10.34	1.113
	9.0	5.807	5.080	0.9613	1.3376	11.90	1.127
	10.0	6.524	5.622	0.9575	1.3381	13.55	1.141
	12.0	8.007	6.695	0.9502	1.3393	17.13	1.169
	14.0	9.558	7.753	0.9431	1.3404	21.13	1.195
	16.0	11.184	8.794	0.9361	1.3416	25.63	1.218
	18.0	12.889	9.823	0.9294	1.3428	30.70	1.237
	20.0	14.679	10.837	0.9228	1.3440	36.42	1.254
	22.0	16.561	11.838	0.9164	1.3453	43.36	1.268
	24.0	18.542	12.826	0.9102	1.3465	51.38	1.280
	26.0	20.630	13.801	0.9040	1.3477	60.77	1.288
	28.0	22.834	14.764	0.8980	1.3490	71.66	
	30.0	25.164	15.713	0.8920	1.3502	84.06	
Ammonium	0.5	0.094	0.093	0.9998	1.3340	0.32	0.999
chloride	1.0	0.189	0.187	1.0014	1.3349	0.64	0.996
NH_4Cl	2.0	0.382	0.376	1.0045	1.3369	1.27	0.992
	3.0	0.578	0.565	1.0076	1.3388	1.91	0.988
	4.0	0.779	0.756	1.0107	1.3407	2.57	0.985
	5.0	0.984	0.948	1.0138	1.3426	3.25	0.982
	6.0	1.193	1.141	1.0168	1.3445	3.94	0.979
	7.0	1.407	1.335	1.0198	1.3464	4.66	0.976

Solute	Mass %	m/mol kg^{-1}	c/mol L^{-1}	ρ/g cm^{-3}	n	Δ/°C	η/mPa s
	8.0	1.626	1.529	1.0227	1.3483	5.40	0.974
	9.0	1.849	1.726	1.0257	1.3502	6.16	0.972
	10.0	2.077	1.923	1.0286	1.3521	6.95	0.970
	12.0	2.549	2.320	1.0344	1.3559	8.60	0.969
	14.0	3.043	2.722	1.0401	1.3596		0.969
	16.0	3.561	3.128	1.0457	1.3634		0.971
	18.0	4.104	3.537	1.0512	1.3671		0.973
	20.0	4.674	3.951	1.0567	1.3708		0.978
	22.0	5.273	4.368	1.0621	1.3745		0.986
	24.0	5.903	4.789	1.0674	1.3782		0.996
Ammonium	0.5	0.038	0.038	1.0012	1.3338	0.17	1.008
sulfate	1.0	0.076	0.076	1.0042	1.3346	0.33	1.014
$(NH_4)_2SO_4$	2.0	0.154	0.153	1.0101	1.3363	0.63	1.027
	3.0	0.234	0.231	1.0160	1.3379	0.92	1.041
	4.0	0.315	0.309	1.0220	1.3395	1.21	1.057
	5.0	0.398	0.389	1.0279	1.3411	1.49	1.073
	6.0	0.483	0.469	1.0338	1.3428	1.77	1.090
	7.0	0.570	0.551	1.0397	1.3444	2.05	1.108
	8.0	0.658	0.633	1.0456	1.3460	2.33	1.127
	9.0	0.748	0.716	1.0515	1.3476	2.61	1.147
	10.0	0.841	0.800	1.0574	1.3492	2.89	1.168
	12.0	1.032	0.971	1.0691	1.3523	3.47	1.210
	14.0	1.232	1.145	1.0808	1.3555	4.07	1.256
	16.0	1.441	1.323	1.0924	1.3586	4.69	1.305
	18.0	1.661	1.504	1.1039	1.3616		1.359
	20.0	1.892	1.688	1.1154	1.3647		1.421
	22.0	2.134	1.876	1.1269	1.3677		1.490
	24.0	2.390	2.067	1.1383	1.3707		1.566
	26.0	2.659	2.262	1.1496	1.3737		1.650
	28.0	2.943	2.460	1.1609	1.3766		1.743
	30.0	3.243	2.661	1.1721	1.3795		1.847
	32.0	3.561	2.866	1.1833	1.3824		1.961
	34.0	3.898	3.073	1.1945	1.3853		2.086
	36.0	4.257	3.284	1.2056	1.3881		2.222
	38.0	4.638	3.499	1.2166	1.3909		2.371
	40.0	5.045	3.716	1.2277	1.3938		2.530
Barium	0.5	0.024	0.024	1.0026	1.3337	0.12	1.009
chloride	1.0	0.049	0.048	1.0070	1.3345	0.23	1.016
$BaCl_2$	2.0	0.098	0.098	1.0159	1.3360	0.46	1.026
	3.0	0.149	0.148	1.0249	1.3375	0.69	1.037
	4.0	0.200	0.199	1.0341	1.3391	0.93	1.049
	5.0	0.253	0.251	1.0434	1.3406	1.18	1.062
	6.0	0.307	0.303	1.0528	1.3422	1.44	1.075
	7.0	0.361	0.357	1.0624	1.3438	1.70	1.087
	8.0	0.418	0.412	1.0721	1.3454	1.98	1.101
	9.0	0.475	0.468	1.0820	1.3470	2.27	1.114
	10.0	0.534	0.524	1.0921	1.3487	2.58	1.129
	12.0	0.655	0.641	1.1128	1.3520	3.22	1.161
	14.0	0.782	0.763	1.1342	1.3555	3.92	1.195
	16.0	0.915	0.889	1.1564	1.3591	4.69	1.234
	18.0	1.054	1.019	1.1793	1.3627		1.277
	20.0	1.201	1.156	1.2031	1.3664		1.325
	22.0	1.355	1.297	1.2277	1.3703		1.378
	24.0	1.517	1.444	1.2531	1.3741		1.437
	26.0	1.687	1.597	1.2793	1.3781		1.503
Calcium	0.5	0.045	0.045	1.0024	1.3342	0.22	1.015
chloride	1.0	0.091	0.091	1.0065	1.3354	0.44	1.028
$CaCl_2$	2.0	0.184	0.183	1.0148	1.3378	0.88	1.050

Solute	Mass %	m/mol kg^{-1}	c/mol L^{-1}	ρ/g cm^{-3}	n	Δ/°C	η/mPa s
	3.0	0.279	0.277	1.0232	1.3402	1.33	1.078
	4.0	0.375	0.372	1.0316	1.3426	1.82	1.110
	5.0	0.474	0.469	1.0401	1.3451	2.35	1.143
	6.0	0.575	0.567	1.0486	1.3475	2.93	1.175
	7.0	0.678	0.667	1.0572	1.3500	3.57	1.208
	8.0	0.784	0.768	1.0659	1.3525	4.28	1.242
	9.0	0.891	0.872	1.0747	1.3549	5.04	1.279
	10.0	1.001	0.976	1.0835	1.3575	5.86	1.319
	12.0	1.229	1.191	1.1014	1.3625	7.70	1.408
	14.0	1.467	1.413	1.1198	1.3677	9.83	1.508
	16.0	1.716	1.641	1.1386	1.3730	12.28	1.625
	18.0	1.978	1.878	1.1579	1.3784	15.11	1.764
	20.0	2.253	2.122	1.1775	1.3839	18.30	1.930
	22.0	2.541	2.374	1.1976	1.3895	21.70	2.127
	24.0	2.845	2.634	1.2180	1.3951	25.30	2.356
	26.0	3.166	2.902	1.2388	1.4008	29.70	2.645
	28.0	3.504	3.179	1.2600	1.4066	34.70	3.000
	30.0	3.862	3.464	1.2816	1.4124	41.00	3.467
	32.0	4.240	3.759	1.3036	1.4183	49.70	4.035
	34.0	4.642	4.062	1.3260	1.4242		4.820
	36.0	5.068	4.375	1.3488	1.4301		5.807
	38.0	5.522	4.698	1.3720	1.4361		7.321
	40.0	6.007	5.030	1.3957	1.4420		8.997
Cesium	0.5	0.030	0.030	1.0020	1.3334	0.10	1.000
chloride	1.0	0.060	0.060	1.0058	1.3337	0.20	0.997
CsCl	2.0	0.121	0.120	1.0135	1.3345	0.40	0.992
	3.0	0.184	0.182	1.0214	1.3353	0.61	0.988
	4.0	0.247	0.245	1.0293	1.3361	0.81	0.984
	5.0	0.313	0.308	1.0374	1.3369	1.02	0.980
	6.0	0.379	0.373	1.0456	1.3377	1.22	0.977
	7.0	0.447	0.438	1.0540	1.3386	1.43	0.974
	8.0	0.516	0.505	1.0625	1.3394	1.64	0.971
	9.0	0.587	0.573	1.0711	1.3403	1.85	0.969
	10.0	0.660	0.641	1.0798	1.3412	2.06	0.966
	12.0	0.810	0.782	1.0978	1.3430	2.51	0.961
	14.0	0.967	0.928	1.1163	1.3448	2.97	0.955
	16.0	1.131	1.079	1.1355	1.3468	3.46	0.950
	18.0	1.304	1.235	1.1552	1.3487	3.96	0.945
	20.0	1.485	1.397	1.1756	1.3507	4.49	0.939
	22.0	1.675	1.564	1.1967	1.3528		0.934
	24.0	1.876	1.737	1.2185	1.3550		0.930
	26.0	2.087	1.917	1.2411	1.3572		0.926
	28.0	2.310	2.103	1.2644	1.3594		0.924
	30.0	2.546	2.296	1.2885	1.3617		0.922
	32.0	2.795	2.497	1.3135	1.3641		0.922
	34.0	3.060	2.705	1.3393	1.3666		0.924
	36.0	3.341	2.921	1.3661	1.3691		0.926
	38.0	3.640	3.146	1.3938	1.3717		0.930
	40.0	3.960	3.380	1.4226	1.3744		0.934
	42.0	4.301	3.624	1.4525	1.3771		0.940
	44.0	4.667	3.877	1.4835	1.3800		0.947
	46.0	5.060	4.142	1.5158	1.3829		0.956
	48.0	5.483	4.418	1.5495	1.3860		0.967
	50.0	5.940	4.706	1.5846	1.3892		0.981
	60.0	8.910	6.368	1.7868	1.4076		1.120
	64.0	10.560	7.163	1.8842	1.4167		1.238
Citric acid	0.5	0.026	0.026	1.0002	1.3336	0.05	1.013
$(HO)C(COOH)_3$	1.0	0.053	0.052	1.0022	1.3343	0.11	1.024
	2.0	0.106	0.105	1.0063	1.3356	0.21	1.048

Solute	Mass %	m/mol kg^{-1}	c/mol L^{-1}	ρ/g cm^{-3}	n	Δ/°C	η/mPa s
	3.0	0.161	0.158	1.0105	1.3368	0.32	1.073
	4.0	0.217	0.211	1.0147	1.3381	0.43	1.098
	5.0	0.274	0.265	1.0189	1.3394	0.54	1.125
	6.0	0.332	0.320	1.0232	1.3407	0.65	1.153
	7.0	0.392	0.374	1.0274	1.3420	0.76	1.183
	8.0	0.453	0.430	1.0316	1.3433	0.88	1.214
	9.0	0.515	0.485	1.0359	1.3446	1.00	1.247
	10.0	0.578	0.541	1.0402	1.3459	1.12	1.283
	12.0	0.710	0.655	1.0490	1.3486	1.38	1.357
	14.0	0.847	0.771	1.0580	1.3514	1.66	1.436
	16.0	0.991	0.889	1.0672	1.3541	1.95	1.525
	18.0	1.143	1.008	1.0764	1.3569	2.26	1.625
	20.0	1.301	1.130	1.0858	1.3598	2.57	1.740
	22.0	1.468	1.254	1.0953	1.3626	2.88	1.872
	24.0	1.644	1.380	1.1049	1.3655	3.21	2.017
	26.0	1.829	1.508	1.1147	1.3684	3.55	2.178
	28.0	2.024	1.639	1.1246	1.3714	3.89	2.356
	30.0	2.231	1.772	1.1346	1.3744	4.25	2.549
Copper sulfate $CuSO_4$	0.5	0.031	0.031	1.0033	1.3339	0.08	1.017
	1.0	0.063	0.063	1.0085	1.3348	0.14	1.036
	2.0	0.128	0.128	1.0190	1.3367	0.26	1.084
	3.0	0.194	0.194	1.0296	1.3386	0.37	1.129
	4.0	0.261	0.261	1.0403	1.3405	0.48	1.173
	5.0	0.330	0.329	1.0511	1.3424	0.59	1.221
	6.0	0.400	0.399	1.0620	1.3443	0.70	1.276
	7.0	0.472	0.471	1.0730	1.3462	0.82	1.336
	8.0	0.545	0.543	1.0842	1.3481	0.93	1.400
	9.0	0.620	0.618	1.0955	1.3501	1.05	1.469
	10.0	0.696	0.694	1.1070	1.3520	1.18	1.543
	12.0	0.854	0.850	1.1304	1.3560	1.45	1.701
	14.0	1.020	1.013	1.1545	1.3601	1.75	1.889
	16.0	1.193	1.182	1.1796	1.3644		2.136
	18.0	1.375	1.360	1.2059	1.3689		2.449
Disodium ethylenediamine tetraacetate (EDTA sodium) $Na_2C_{10}H_{14}N_2O_8$	0.5	0.015	0.015	1.0009	1.3339	0.07	1.017
	1.0	0.030	0.030	1.0036	1.3348	0.14	1.032
	1.5	0.045	0.045	1.0062	1.3356	0.21	1.046
	2.0	0.061	0.060	1.0089	1.3365	0.27	1.062
	2.5	0.076	0.075	1.0115	1.3374	0.33	1.077
	3.0	0.092	0.090	1.0142	1.3383	0.40	1.093
	3.5	0.108	0.106	1.0169	1.3392	0.46	1.109
	4.0	0.124	0.121	1.0196	1.3400	0.52	1.125
	4.5	0.140	0.137	1.0223	1.3409	0.58	1.142
	5.0	0.157	0.152	1.0250	1.3418	0.65	1.160
	5.5	0.173	0.168	1.0277	1.3427	0.71	1.178
	6.0	0.190	0.184	1.0305	1.3436	0.77	1.197
Ethanol CH_3CH_2OH	0.5	0.109	0.108	0.9973	1.3333	0.20	1.023
	1.0	0.219	0.216	0.9963	1.3336	0.40	1.046
	2.0	0.443	0.432	0.9945	1.3342	0.81	1.095
	3.0	0.671	0.646	0.9927	1.3348	1.23	1.140
	4.0	0.904	0.860	0.9910	1.3354	1.65	1.183
	5.0	1.142	1.074	0.9893	1.3360	2.09	1.228
	6.0	1.385	1.286	0.9878	1.3367	2.54	1.279
	7.0	1.634	1.498	0.9862	1.3374	2.99	1.331
	8.0	1.887	1.710	0.9847	1.3381	3.47	1.385
	9.0	2.147	1.921	0.9833	1.3388	3.96	1.442
	10.0	2.412	2.131	0.9819	1.3395	4.47	1.501
	12.0	2.960	2.551	0.9792	1.3410	5.56	1.627
	14.0	3.534	2.967	0.9765	1.3425	6.73	1.761

Solute	Mass %	*m*/mol kg⁻¹	*c*/mol L⁻¹	ρ/g cm⁻³	*n*	Δ/°C	η/mPa s
	16.0	4.134	3.382	0.9739	1.3440	8.01	1.890
	18.0	4.765	3.795	0.9713	1.3455	9.40	2.019
	20.0	5.427	4.205	0.9687	1.3469	10.92	2.142
	22.0	6.122	4.613	0.9660	1.3484	12.60	2.259
	24.0	6.855	5.018	0.9632	1.3498	14.47	2.370
	26.0	7.626	5.419	0.9602	1.3511	16.41	2.476
	28.0	8.441	5.817	0.9571	1.3524	18.43	2.581
	30.0	9.303	6.212	0.9539	1.3535	20.47	2.667
	32.0	10.215	6.601	0.9504	1.3546	22.44	2.726
	34.0	11.182	6.987	0.9468	1.3557	24.27	2.768
	36.0	12.210	7.370	0.9431	1.3566	25.98	2.803
	38.0	13.304	7.747	0.9392	1.3575	27.62	2.829
	40.0	14.471	8.120	0.9352	1.3583	29.26	2.846
	42.0	15.718	8.488	0.9311	1.3590	30.98	2.852
	44.0	17.055	8.853	0.9269	1.3598	32.68	2.850
	46.0	18.490	9.213	0.9227	1.3604	34.36	2.843
	48.0	20.036	9.568	0.9183	1.3610	36.04	2.832
	50.0	21.706	9.919	0.9139	1.3616	37.67	2.813
	60.0	32.559	11.605	0.8911	1.3638	44.93	2.547
	70.0	50.648	13.183	0.8676	1.3652		2.214
	80.0	86.824	14.649	0.8436	1.3658		1.881
	90.0	195.355	15.980	0.8180	1.3650		1.542
	92.0	249.620	16.225	0.8125	1.3646		1.475
	94.0	340.062	16.466	0.8070	1.3642		1.407
	96.0	520.946	16.697	0.8013	1.3636		1.342
	98.0		16.920	0.7954	1.3630		1.273
	100.0		17.133	0.7893	1.3614		1.203
Ethylene glycol $(CH_2OH)_2$	0.5	0.081	0.080	0.9988	1.3335	0.15	1.010
	1.0	0.163	0.161	0.9995	1.3339	0.30	1.020
	2.0	0.329	0.322	1.0007	1.3348	0.61	1.048
	3.0	0.498	0.484	1.0019	1.3358	0.92	1.074
	4.0	0.671	0.646	1.0032	1.3367	1.24	1.099
	5.0	0.848	0.809	1.0044	1.3377	1.58	1.125
	6.0	1.028	0.972	1.0057	1.3386	1.91	1.153
	7.0	1.213	1.136	1.0070	1.3396	2.26	1.182
	8.0	1.401	1.299	1.0082	1.3405	2.62	1.212
	9.0	1.593	1.464	1.0095	1.3415	2.99	1.243
	10.0	1.790	1.628	1.0108	1.3425	3.37	1.277
	12.0	2.197	1.959	1.0134	1.3444	4.16	1.348
	14.0	2.623	2.292	1.0161	1.3464	5.01	1.424
	16.0	3.069	2.626	1.0188	1.3484	5.91	1.500
	18.0	3.537	2.962	1.0214	1.3503	6.89	1.578
	20.0	4.028	3.300	1.0241	1.3523	7.93	1.661
	24.0	5.088	3.981	1.0296	1.3564	10.28	1.843
	28.0	6.265	4.669	1.0350	1.3605	13.03	2.047
	32.0	7.582	5.364	1.0405	1.3646	16.23	2.280
	36.0	9.062	6.067	1.0460	1.3687	19.82	2.537
	40.0	10.741	6.776	1.0514	1.3728	23.84	2.832
	44.0	12.659	7.491	1.0567	1.3769	28.32	3.166
	48.0	14.872	8.212	1.0619	1.3811	33.30	3.544
	52.0	17.453	8.939	1.0670	1.3851	38.81	3.981
	56.0	20.505	9.671	1.0719	1.3892	44.83	4.475
	60.0	24.166	10.406	1.0765	1.3931	51.23	5.026
Ferric chloride $FeCl_3$	0.5	0.031	0.031	1.0025	1.3344	0.21	1.024
	1.0	0.062	0.062	1.0068	1.3358	0.39	1.047
	2.0	0.126	0.125	1.0153	1.3386	0.75	1.093
	3.0	0.191	0.189	1.0238	1.3413	1.15	1.139
	4.0	0.257	0.255	1.0323	1.3441	1.56	1.187
	5.0	0.324	0.321	1.0408	1.3468	2.00	1.238

Solute	Mass %	m/mol kg^{-1}	c/mol L^{-1}	ρ/g cm^{-3}	n	Δ/°C	η/mPa s
	6.0	0.394	0.388	1.0493	1.3496	2.48	1.292
	7.0	0.464	0.457	1.0580	1.3524	2.99	1.350
	8.0	0.536	0.526	1.0668	1.3552	3.57	1.412
	9.0	0.610	0.597	1.0760	1.3581	4.19	1.480
	10.0	0.685	0.669	1.0853	1.3611	4.85	1.553
	12.0	0.841	0.817	1.1040	1.3670	6.38	1.707
	14.0	1.004	0.969	1.1228	1.3730	8.22	1.879
	16.0	1.174	1.126	1.1420		10.45	2.080
	18.0	1.353	1.289	1.1615		13.08	2.311
	20.0	1.541	1.457	1.1816		16.14	2.570
	24.0	1.947	1.810	1.2234		23.79	3.178
	28.0	2.398	2.189	1.2679		33.61	4.038
	32.0	2.901	2.595	1.3153		49.16	5.274
	36.0	3.468	3.030	1.3654			7.130
	40.0	4.110	3.496	1.4176			9.674
Formic acid	0.5	0.109	0.109	0.9994	1.3333	0.21	1.006
HCOOH	1.0	0.219	0.217	1.0006	1.3336	0.42	1.011
	2.0	0.443	0.436	1.0029	1.3342	0.82	1.017
	3.0	0.672	0.655	1.0053	1.3348	1.24	1.195
	4.0	0.905	0.876	1.0077	1.3354	1.67	1.032
	5.0	1.143	1.097	1.0102	1.3359	2.10	1.039
	6.0	1.387	1.320	1.0126	1.3365	2.53	1.046
	7.0	1.635	1.544	1.0150	1.3371	2.97	1.052
	8.0	1.889	1.768	1.0175	1.3376	3.40	1.058
	9.0	2.149	1.994	1.0199	1.3382	3.84	1.064
	10.0	2.414	2.221	1.0224	1.3387	4.27	1.070
	12.0	2.962	2.678	1.0273	1.3397	5.19	1.082
	14.0	3.537	3.139	1.0322	1.3408	6.11	1.094
	16.0	4.138	3.605	1.0371	1.3418	7.06	1.106
	18.0	4.769	4.074	1.0419	1.3428	8.08	1.119
	20.0	5.431	4.548	1.0467	1.3437	9.11	1.132
	28.0	8.449	6.481	1.0654	1.3475	13.10	1.179
	36.0	12.220	8.477	1.0839	1.3511	17.65	1.227
	44.0	17.070	10.529	1.1015	1.3547	22.93	1.281
	52.0	23.535	12.633	1.1183	1.3581	29.69	1.340
	60.0	32.587	14.813	1.1364	1.3612	38.26	1.410
	68.0	46.166	17.054	1.1544	1.3641		1.490
D-Fructose	0.5	0.028	0.028	1.0002	1.3337	0.05	1.015
$C_6H_{12}O_6$	1.0	0.056	0.056	1.0021	1.3344	0.10	1.028
	2.0	0.113	0.112	1.0061	1.3358	0.21	1.054
	3.0	0.172	0.168	1.0101	1.3373	0.32	1.080
	4.0	0.231	0.225	1.0140	1.3387	0.43	1.106
	5.0	0.292	0.283	1.0181	1.3402	0.54	1.134
	6.0	0.354	0.340	1.0221	1.3417	0.66	1.165
	7.0	0.418	0.399	1.0262	1.3431	0.78	1.198
	8.0	0.483	0.458	1.0303	1.3446	0.90	1.232
	9.0	0.549	0.517	1.0344	1.3461	1.03	1.270
	10.0	0.617	0.576	1.0385	1.3476	1.16	1.309
	12.0	0.757	0.697	1.0469	1.3507	1.43	1.391
	14.0	0.904	0.820	1.0554	1.3538	1.71	1.483
	16.0	1.057	0.945	1.0640	1.3569	2.01	1.587
	18.0	1.218	1.072	1.0728	1.3601	2.32	1.703
	20.0	1.388	1.201	1.0816	1.3634	2.64	1.837
	22.0	1.566	1.332	1.0906	1.3667	3.05	1.986
	24.0	1.753	1.465	1.0996	1.3700	3.43	2.154
	26.0	1.950	1.600	1.1089	1.3734	3.82	2.348
	28.0	2.159	1.738	1.1182	1.3768	4.20	2.562
	30.0	2.379	1.878	1.1276	1.3803		2.817
	32.0	2.612	2.020	1.1372	1.3839		3.112

Solute	Mass %	m/mol kg⁻¹	c/mol L⁻¹	ρ/g cm⁻³	n	Δ/°C	η/mPa s
	34.0	2.859	2.164	1.1469	1.3874		3.462
	36.0	3.122	2.312	1.1568	1.3911		3.899
	38.0	3.402	2.461	1.1668	1.3948		4.418
	40.0	3.700	2.613	1.1769	1.3985		5.046
	42.0	4.019	2.767	1.1871	1.4023		5.773
	44.0	4.361	2.925	1.1975	1.4062		6.644
	46.0	4.728	3.084	1.2080	1.4101		7.753
	48.0	5.124	3.247	1.2187	1.4141		9.060
D-Glucose	0.5	0.028	0.028	1.0001	1.3337	0.05	1.010
$C_6H_{12}O_6$	1.0	0.056	0.056	1.0020	1.3344	0.11	1.021
	2.0	0.113	0.112	1.0058	1.3358	0.21	1.052
	3.0	0.172	0.168	1.0097	1.3373	0.32	1.083
	4.0	0.231	0.225	1.0136	1.3387	0.43	1.113
	5.0	0.292	0.282	1.0175	1.3402	0.55	1.145
	6.0	0.354	0.340	1.0214	1.3417	0.67	1.179
	7.0	0.418	0.398	1.0254	1.3432	0.79	1.214
	8.0	0.483	0.457	1.0294	1.3447	0.91	1.250
	9.0	0.549	0.516	1.0334	1.3462	1.04	1.289
	10.0	0.617	0.576	1.0375	1.3477	1.17	1.330
	12.0	0.757	0.697	1.0457	1.3508	1.44	1.416
	14.0	0.904	0.819	1.0540	1.3539	1.73	1.512
	16.0	1.057	0.944	1.0624	1.3571	2.03	1.625
	18.0	1.218	1.070	1.0710	1.3603	2.35	1.757
	20.0	1.388	1.199	1.0797	1.3635	2.70	1.904
	22.0	1.566	1.329	1.0884	1.3668	3.07	2.063
	24.0	1.753	1.462	1.0973	1.3702	3.48	2.242
	26.0	1.950	1.597	1.1063	1.3736	3.90	2.458
	28.0	2.159	1.734	1.1154	1.3770	4.34	2.707
	30.0	2.379	1.873	1.1246	1.3805	4.79	2.998
	32.0	2.612	2.014	1.1340	1.3840		3.324
	34.0	2.859	2.158	1.1434	1.3876		3.704
	36.0	3.122	2.304	1.1529	1.3912		4.193
	38.0	3.402	2.452	1.1626	1.3949		4.786
	40.0	3.700	2.603	1.1724	1.3986		5.493
	42.0	4.019	2.756	1.1823	1.4024		6.288
	44.0	4.361	2.912	1.1924	1.4062		7.235
	46.0	4.728	3.071	1.2026	1.4101		8.454
	48.0	5.124	3.232	1.2130	1.4141		9.883
	50.0	5.551	3.396	1.2235	1.4181		11.884
	52.0	6.013	3.562	1.2342	1.4222		14.489
	54.0	6.516	3.732	1.2451	1.4263		17.916
	56.0	7.064	3.905	1.2562	1.4306		22.886
	58.0	7.665	4.081	1.2676	1.4349		29.389
	60.0	8.326	4.261	1.2793	1.4394		37.445
Glycerol	0.5	0.055	0.054	0.9994	1.3336	0.07	1.022
$CH_2OHCHOHCH_2OH$	1.0	0.110	0.109	1.0005	1.3342	0.18	1.034
	2.0	0.222	0.218	1.0028	1.3353	0.41	1.060
	3.0	0.336	0.327	1.0051	1.3365	0.63	1.088
	4.0	0.452	0.438	1.0074	1.3376	0.85	1.116
	5.0	0.572	0.548	1.0097	1.3388	1.08	1.145
	6.0	0.693	0.659	1.0120	1.3400	1.32	1.176
	7.0	0.817	0.771	1.0144	1.3412	1.56	1.207
	8.0	0.944	0.883	1.0167	1.3424	1.81	1.240
	9.0	1.074	0.996	1.0191	1.3436	2.06	1.275
	10.0	1.207	1.109	1.0215	1.3448	2.32	1.310
	12.0	1.481	1.337	1.0262	1.3472	2.88	1.386
	14.0	1.768	1.568	1.0311	1.3496	3.47	1.469
	16.0	2.068	1.800	1.0360	1.3521	4.09	1.560
	18.0	2.384	2.035	1.0409	1.3547	4.76	1.658

Solute	Mass %	m/mol kg^{-1}	c/mol L^{-1}	ρ/g cm^{-3}	n	Δ/°C	η/mPa s
	20.0	2.715	2.271	1.0459	1.3572	5.46	1.766
	24.0	3.429	2.752	1.0561	1.3624	7.01	2.01
	28.0	4.223	3.242	1.0664	1.3676	8.77	2.32
	32.0	5.110	3.742	1.0770	1.3730	10.74	2.69
	36.0	6.108	4.252	1.0876	1.3785	12.96	3.15
	40.0	7.239	4.771	1.0984	1.3841	15.50	3.73
	44.0	8.532	5.300	1.1092	1.3897		4.48
	48.0	10.024	5.838	1.1200	1.3954		5.45
	52.0	11.764	6.385	1.1308	1.4011		6.73
	56.0	13.820	6.944	1.1419	1.4069		8.47
	60.0	16.288	7.512	1.1530	1.4129		10.9
	64.0	19.305	8.092	1.1643	1.4189		14.3
	68.0	23.075	8.680	1.1755	1.4249		19.4
	72.0	27.923	9.277	1.1866	1.4310		27.2
	76.0	34.387	9.884	1.1976	1.4370		39.6
	80.0	43.436	10.498	1.2085	1.4431		60.6
	84.0	57.009	11.121	1.2192	1.4492		98
	88.0	79.632	11.753	1.2299	1.4553		170
	92.0	124.878	12.392	1.2404	1.4613		319
	96.0	260.615	13.039	1.2508	1.4674		648
	100.0		13.694	1.2611	1.4735		1460
Hydrochloric acid HCl	0.5	0.138	0.137	1.0007	1.3341	0.49	1.008
	1.0	0.277	0.275	1.0031	1.3353	0.99	1.015
	2.0	0.560	0.553	1.0081	1.3376	2.08	1.029
	3.0	0.848	0.833	1.0130	1.3399	3.28	1.044
	4.0	1.143	1.117	1.0179	1.3422	4.58	1.059
	5.0	1.444	1.403	1.0228	1.3445	5.98	1.075
	6.0	1.751	1.691	1.0278	1.3468	7.52	1.091
	7.0	2.064	1.983	1.0327	1.3491	9.22	1.108
	8.0	2.385	2.277	1.0377	1.3515	11.10	1.125
	9.0	2.713	2.574	1.0426	1.3538	13.15	1.143
	10.0	3.047	2.873	1.0476	1.3561	15.40	1.161
	12.0	3.740	3.481	1.0576	1.3607	20.51	1.199
	14.0	4.465	4.099	1.0676	1.3653		1.239
	16.0	5.224	4.729	1.0777	1.3700		1.282
	18.0	6.020	5.370	1.0878	1.3746		1.326
	20.0	6.857	6.023	1.0980	1.3792		1.374
	22.0	7.736	6.687	1.1083	1.3838		1.426
	24.0	8.661	7.362	1.1185	1.3884		1.483
	26.0	9.636	8.049	1.1288	1.3930		1.547
	28.0	10.666	8.748	1.1391	1.3976		1.620
	30.0	11.754	9.456	1.1492	1.4020		1.705
	32.0	12.907	10.175	1.1594	1.4066		1.799
	34.0	14.129	10.904	1.1693	1.4112		1.900
	36.0	15.427	11.642	1.1791	1.4158		2.002
	38.0	16.810	12.388	1.1886	1.4204		2.105
	40.0	18.284	13.140	1.1977	1.4250		
Lactic acid $CH_3CHOHCOOH$	0.5	0.056	0.055	0.9992	1.3335	0.10	1.014
	1.0	0.112	0.111	1.0002	1.3340	0.19	1.027
	2.0	0.227	0.223	1.0023	1.3350	0.38	1.056
	3.0	0.343	0.334	1.0043	1.3360	0.57	1.084
	4.0	0.463	0.447	1.0065	1.3370	0.76	1.110
	5.0	0.584	0.560	1.0086	1.3380	0.95	1.138
	6.0	0.709	0.673	1.0108	1.3390	1.16	1.167
	7.0	0.836	0.787	1.0131	1.3400	1.36	1.198
	8.0	0.965	0.902	1.0153	1.3410	1.57	1.229
	9.0	1.098	1.017	1.0176	1.3420	1.79	1.262
	10.0	1.233	1.132	1.0199	1.3430	2.02	1.296
	12.0	1.514	1.365	1.0246	1.3450	2.49	1.366

Solute	Mass %	m/mol kg^{-1}	c/mol L^{-1}	ρ/g cm^{-3}	n	Δ/°C	η/mPa s
	14.0	1.807	1.600	1.0294	1.3470	2.99	1.441
	16.0	2.115	1.837	1.0342	1.3491	3.48	1.522
	18.0	2.437	2.076	1.0390	1.3511	3.96	1.607
	20.0	2.775	2.318	1.0439	1.3532	4.44	1.699
	24.0	3.506	2.807	1.0536	1.3573		1.902
	28.0	4.317	3.305	1.0632	1.3615		2.136
	32.0	5.224	3.811	1.0728	1.3657		2.414
	36.0	6.244	4.325	1.0822	1.3700		2.730
	40.0	7.401	4.847	1.0915	1.3743		3.114
	44.0	8.722	5.377	1.1008	1.3786		3.566
	48.0	10.247	5.917	1.1105	1.3828		4.106
	52.0	12.026	6.466	1.1201	1.3871		4.789
	56.0	14.129	7.023	1.1297	1.3914		5.579
	60.0	16.652	7.588	1.1392	1.3958		6.679
	64.0	19.736	8.161	1.1486	1.4001		8.024
	68.0	23.590	8.741	1.1579	1.4045		9.863
	72.0	28.546	9.328	1.1670	1.4088		12.866
	76.0	35.154	9.922	1.1760	1.4131		16.974
	80.0	44.405	10.522	1.1848	1.4173		22.164
Lactose	0.5	0.015	0.015	1.0002	1.3337	0.03	1.013
$C_{12}H_{22}O_{11}$	1.0	0.030	0.029	1.0021	1.3345	0.06	1.026
	2.0	0.060	0.059	1.0061	1.3359	0.11	1.058
	3.0	0.090	0.089	1.0102	1.3375	0.17	1.089
	4.0	0.122	0.119	1.0143	1.3390	0.23	1.120
	5.0	0.154	0.149	1.0184	1.3406	0.29	1.154
	6.0	0.186	0.179	1.0225	1.3421	0.35	1.191
	7.0	0.220	0.210	1.0267	1.3437	0.42	1.232
	8.0	0.254	0.241	1.0308	1.3453	0.50	1.276
	9.0	0.289	0.272	1.0349	1.3468		1.321
	10.0	0.325	0.304	1.0390	1.3484		1.370
	12.0	0.398	0.367	1.0473	1.3515		1.476
	14.0	0.476	0.432	1.0558	1.3548		1.593
	16.0	0.556	0.498	1.0648	1.3582		1.724
	18.0	0.641	0.565	1.0746	1.3619		1.869
Lithium	0.5	0.119	0.118	1.0012	1.3341	0.42	1.019
chloride	1.0	0.238	0.237	1.0041	1.3351	0.84	1.037
LiCl	2.0	0.481	0.476	1.0099	1.3373	1.72	1.072
	3.0	0.730	0.719	1.0157	1.3394	2.68	1.108
	4.0	0.983	0.964	1.0215	1.3415	3.73	1.146
	5.0	1.241	1.211	1.0272	1.3436	4.86	1.185
	6.0	1.506	1.462	1.0330	1.3457	6.14	1.226
	7.0	1.775	1.715	1.0387	1.3478	7.56	1.269
	8.0	2.051	1.971	1.0444	1.3499	9.11	1.313
	9.0	2.333	2.230	1.0502	1.3520	10.79	1.360
	10.0	2.621	2.491	1.0560	1.3541	12.61	1.411
	12.0	3.217	3.022	1.0675	1.3583	16.59	1.522
	14.0	3.840	3.564	1.0792	1.3625	21.04	1.647
	16.0	4.493	4.118	1.0910	1.3668		1.787
	18.0	5.178	4.683	1.1029	1.3711		1.942
	20.0	5.897	5.260	1.1150	1.3755		2.128
	22.0	6.653	5.851	1.1274	1.3799		2.341
	24.0	7.449	6.453	1.1399	1.3844		2.600
	26.0	8.288	7.069	1.1527	1.3890		2.925
	28.0	9.173	7.700	1.1658	1.3936		3.318
	30.0	10.109	8.344	1.1791	1.3983		3.785
Magnesium	0.5	0.053	0.053	1.0022	1.3343	0.26	1.024
chloride	1.0	0.106	0.106	1.0062	1.3356	0.52	1.046
$MgCl_2$	2.0	0.214	0.213	1.0144	1.3381	1.06	1.091

Solute	Mass %	m/mol kg^{-1}	c/mol L^{-1}	ρ/g cm^{-3}	n	Δ/°C	η/mPa s
	3.0	0.325	0.322	1.0226	1.3406	1.65	1.139
	4.0	0.438	0.433	1.0309	1.3432	2.30	1.188
	5.0	0.553	0.546	1.0394	1.3457	3.01	1.241
	6.0	0.670	0.660	1.0479	1.3483		1.298
	7.0	0.791	0.777	1.0564	1.3508		1.358
	8.0	0.913	0.895	1.0651	1.3534		1.423
	9.0	1.039	1.015	1.0738	1.3560		1.493
	10.0	1.167	1.137	1.0826	1.3587		1.570
	12.0	1.432	1.387	1.1005	1.3641		1.745
	14.0	1.710	1.645	1.1189	1.3695		1.956
	16.0	2.001	1.911	1.1372	1.3749		2.207
	18.0	2.306	2.184	1.1553	1.3804		2.507
	20.0	2.626	2.467	1.1742	1.3859		2.867
	22.0	2.962	2.758	1.1938	1.3915		3.323
	24.0	3.317	3.060	1.2140	1.3972		3.917
	26.0	3.690	3.371	1.2346	1.4030		4.694
	28.0	4.085	3.692	1.2555	1.4089		5.709
	30.0	4.501	4.022	1.2763	1.4148		7.017
Magnesium	0.5	0.042	0.042	1.0033	1.3340	0.10	1.027
sulfate	1.0	0.084	0.084	1.0084	1.3350	0.19	1.054
$MgSO_4$	2.0	0.170	0.169	1.0186	1.3371	0.36	1.112
	3.0	0.257	0.256	1.0289	1.3391	0.52	1.177
	4.0	0.346	0.345	1.0392	1.3411	0.69	1.249
	5.0	0.437	0.436	1.0497	1.3431	0.87	1.328
	6.0	0.530	0.528	1.0602	1.3451	1.05	1.411
	7.0	0.625	0.623	1.0708	1.3471	1.24	1.498
	8.0	0.722	0.719	1.0816	1.3492	1.43	1.593
	9.0	0.822	0.817	1.0924	1.3512	1.64	1.702
	10.0	0.923	0.917	1.1034	1.3532	1.85	1.829
	12.0	1.133	1.122	1.1257	1.3572	2.31	2.104
	14.0	1.352	1.336	1.1484	1.3613	2.86	2.412
	16.0	1.582	1.557	1.1717	1.3654	3.67	2.809
	18.0	1.824	1.788	1.1955	1.3694		3.360
	20.0	2.077	2.027	1.2198	1.3735		4.147
	22.0	2.343	2.275	1.2447	1.3776		5.199
	24.0	2.624	2.532	1.2701	1.3817		6.498
	26.0	2.919	2.800	1.2961	1.3858		8.066
Maltose	0.5	0.015	0.015	1.0003	1.3337	0.03	1.016
$C_{12}H_{22}O_{11}$	1.0	0.030	0.029	1.0023	1.3345	0.06	1.030
	2.0	0.060	0.059	1.0063	1.3359	0.11	1.060
	3.0	0.090	0.089	1.0104	1.3374	0.17	1.092
	4.0	0.122	0.119	1.0144	1.3389	0.23	1.126
	5.0	0.154	0.149	1.0184	1.3404	0.29	1.162
	6.0	0.186	0.179	1.0224	1.3420	0.35	1.200
	7.0	0.220	0.210	1.0265	1.3435	0.42	1.239
	8.0	0.254	0.241	1.0305	1.3450	0.48	1.281
	9.0	0.289	0.272	1.0345	1.3466	0.55	1.325
	10.0	0.325	0.303	1.0385	1.3482	0.62	1.372
	12.0	0.398	0.367	1.0465	1.3513	0.77	1.474
	14.0	0.476	0.431	1.0545	1.3546	0.92	1.588
	16.0	0.556	0.497	1.0629	1.3578	1.08	1.715
	18.0	0.641	0.564	1.0716	1.3612	1.25	1.859
	20.0	0.730	0.631	1.0801	1.3644	1.43	2.021
	22.0	0.824	0.700	1.0894	1.3678	1.64	2.216
	24.0	0.923	0.770	1.0984	1.3714	1.85	2.463
	26.0	1.026	0.842	1.1080	1.3749	2.08	2.753
	28.0	1.136	0.914	1.1171	1.3785	2.34	3.066
	30.0	1.252	0.988	1.1269	1.3821	2.62	3.427
	40.0	1.948	1.375	1.1769	1.4013	4.41	6.926

Solute	Mass %	m/mol kg^{-1}	c/mol L^{-1}	ρ/g cm^{-3}	n	Δ/°C	η/mPa s
	50.0	2.921	1.797	1.2304	1.4217		17.786
	52.0	3.165	1.886	1.2416	1.4260		22.034
	54.0	3.429	1.976	1.2528	1.4308		28.757
	56.0	3.718	2.068	1.2638	1.4350		38.226
	58.0	4.034	2.159	1.2740	1.4394		49.298
	60.0	4.382	2.253	1.2855	1.4440		
Manganese(II)	1.0	0.067	0.067	1.0080	1.3348	0.16	1.046
sulfate	2.0	0.135	0.135	1.0178	1.3366	0.31	1.090
$MnSO_4$	3.0	0.205	0.204	1.0277	1.3384	0.44	1.137
	4.0	0.276	0.275	1.0378	1.3402	0.57	1.187
	5.0	0.349	0.347	1.0480	1.3420	0.70	1.242
	6.0	0.423	0.421	1.0583	1.3438	0.84	1.301
	7.0	0.498	0.495	1.0688	1.3457	0.98	1.363
	8.0	0.576	0.572	1.0794	1.3475	1.12	1.431
	9.0	0.655	0.650	1.0902	1.3494	1.28	1.505
	10.0	0.736	0.729	1.1012	1.3513	1.44	1.587
	12.0	0.903	0.893	1.1236	1.3551	1.80	1.779
	14.0	1.078	1.063	1.1467	1.3589	2.21	2.005
	16.0	1.261	1.240	1.1705	1.3629	2.67	2.272
	18.0	1.454	1.424	1.1950	1.3668	3.19	2.580
	20.0	1.656	1.616	1.2203	1.3708	3.80	2.938
D-Mannitol	0.5	0.028	0.027	1.0000	1.3337	0.05	1.019
$CH_2(CHOH)_4CH_2OH$	1.0	0.055	0.055	1.0017	1.3345	0.10	1.032
	2.0	0.112	0.110	1.0053	1.3359	0.21	1.057
	3.0	0.170	0.166	1.0088	1.3374	0.32	1.081
	4.0	0.229	0.222	1.0124	1.3389	0.43	1.107
	5.0	0.289	0.279	1.0159	1.3403	0.54	1.135
	6.0	0.350	0.336	1.0195	1.3418	0.66	1.166
	7.0	0.413	0.393	1.0230	1.3433	0.77	1.200
	8.0	0.477	0.451	1.0266	1.3447	0.90	1.236
	9.0	0.543	0.509	1.0302	1.3462	1.02	1.275
	10.0	0.610	0.567	1.0338	1.3477	1.15	1.314
	11.0	0.678	0.626	1.0375	1.3491	1.28	1.355
	12.0	0.749	0.686	1.0412	1.3506	1.41	1.398
	13.0	0.820	0.746	1.0450	1.3521	1.55	1.443
	14.0	0.894	0.806	1.0489	1.3536	1.69	1.489
	15.0	0.969	0.867	1.0529	1.3552	1.84	1.537
Methanol	0.5	0.157	0.156	0.9973	1.3331	0.28	1.022
CH_3OH	1.0	0.315	0.311	0.9964	1.3332	0.56	1.040
	2.0	0.637	0.621	0.9947	1.3334	1.14	1.070
	3.0	0.965	0.930	0.9930	1.3336	1.75	1.100
	4.0	1.300	1.238	0.9913	1.3339	2.37	1.131
	5.0	1.643	1.544	0.9896	1.3341	3.02	1.163
	6.0	1.992	1.850	0.9880	1.3343	3.71	1.196
	7.0	2.349	2.155	0.9864	1.3346	4.41	1.229
	8.0	2.714	2.459	0.9848	1.3348	5.13	1.264
	9.0	3.087	2.762	0.9832	1.3351	5.85	1.297
	10.0	3.468	3.064	0.9816	1.3354	6.60	1.329
	12.0	4.256	3.665	0.9785	1.3359	8.14	1.389
	14.0	5.081	4.262	0.9755	1.3365	9.72	1.446
	16.0	5.945	4.856	0.9725	1.3370	11.36	1.501
	18.0	6.851	5.447	0.9695	1.3376	13.13	1.554
	20.0	7.803	6.034	0.9666	1.3381	15.02	1.604
	22.0	8.803	6.616	0.9636	1.3387	16.98	1.652
	24.0	9.856	7.196	0.9606	1.3392	19.04	1.697
	26.0	10.966	7.771	0.9576	1.3397	21.23	1.735
	28.0	12.138	8.341	0.9545	1.3402	23.59	1.769
	30.0	13.376	8.908	0.9514	1.3407	25.91	1.795

Solute	Mass %	m/mol kg^{-1}	c/mol L^{-1}	ρ/g cm^{-3}	n	Δ/°C	η/mPa s
	32.0	14.688	9.470	0.9482	1.3411	28.15	1.814
	34.0	16.078	10.028	0.9450	1.3415	30.48	1.827
	36.0	17.556	10.580	0.9416	1.3419	32.97	1.835
	38.0	19.129	11.127	0.9382	1.3422	35.60	1.839
	40.0	20.807	11.669	0.9347	1.3425	38.60	1.837
	50.0	31.211	14.288	0.9156	1.3431	54.50	1.761
	60.0	46.816	16.749	0.8944	1.3426	74.50	1.600
	70.0	72.826	19.040	0.8715	1.3411		1.368
	80.0	124.844	21.144	0.8468	1.3385		1.128
	90.0	280.899	23.045	0.8204	1.3348		0.861
	100.0		24.710	0.7917	1.3290		0.586
Nitric acid	0.5	0.080	0.079	1.0009	1.3336	0.28	1.004
HNO_3	1.0	0.160	0.159	1.0037	1.3343	0.56	1.005
	2.0	0.324	0.320	1.0091	1.3356	1.12	1.007
	3.0	0.491	0.483	1.0146	1.3368	1.70	1.010
	4.0	0.661	0.648	1.0202	1.3381	2.32	1.014
	5.0	0.835	0.814	1.0257	1.3394	2.96	1.018
	6.0	1.013	0.982	1.0314	1.3407	3.63	1.022
	7.0	1.194	1.152	1.0370	1.3421	4.33	1.027
	8.0	1.380	1.324	1.0427	1.3434	5.05	1.032
	9.0	1.570	1.498	1.0485	1.3447	5.81	1.038
	10.0	1.763	1.673	1.0543	1.3460	6.60	1.044
	12.0	2.164	2.030	1.0660	1.3487	8.27	1.058
	14.0	2.583	2.395	1.0780	1.3514	10.08	1.075
	16.0	3.023	2.768	1.0901	1.3541	12.04	1.094
	18.0	3.484	3.149	1.1025	1.3569	14.16	1.116
	20.0	3.967	3.539	1.1150	1.3596		1.141
	22.0	4.476	3.937	1.1277	1.3624		1.169
	24.0	5.011	4.344	1.1406	1.3652		1.199
	26.0	5.576	4.760	1.1536	1.3680		1.233
	28.0	6.172	5.185	1.1668	1.3708		1.271
	30.0	6.801	5.618	1.1801	1.3736		1.311
	32.0	7.468	6.060	1.1934	1.3763		1.354
	34.0	8.175	6.512	1.2068	1.3790		1.400
	36.0	8.927	6.971	1.2202	1.3817		1.450
	38.0	9.727	7.439	1.2335	1.3842		1.504
	40.0	10.580	7.913	1.2466	1.3867		1.561
Oxalic acid	0.5	0.056	0.056	1.0006	1.3336	0.16	1.013
$(COOH)_2$	1.0	0.112	0.111	1.0030	1.3342	0.30	1.023
	1.5	0.169	0.167	1.0054	1.3347	0.44	1.033
	2.0	0.227	0.224	1.0079	1.3353	0.57	1.044
	2.5	0.285	0.281	1.0103	1.3359	0.71	1.055
	3.0	0.343	0.337	1.0126	1.3364	0.84	1.065
	3.5	0.403	0.395	1.0150	1.3370	0.97	1.076
	4.0	0.463	0.452	1.0174	1.3375	1.09	1.086
	4.5	0.523	0.510	1.0197	1.3381		1.097
	5.0	0.585	0.568	1.0220	1.3386		1.108
	6.0	0.709	0.684	1.0265	1.3397		1.129
	7.0	0.836	0.802	1.0310	1.3407		1.150
	8.0	0.966	0.920	1.0355	1.3418		1.172
Phosphoric	0.5	0.051	0.051	1.0010	1.3335	0.12	1.010
acid	1.0	0.103	0.102	1.0038	1.3340	0.24	1.020
H_3PO_4	2.0	0.208	0.206	1.0092	1.3349	0.46	1.050
	3.0	0.316	0.311	1.0146	1.3358	0.69	1.079
	4.0	0.425	0.416	1.0200	1.3367	0.93	1.108
	5.0	0.537	0.523	1.0254	1.3376	1.16	1.138
	6.0	0.651	0.631	1.0309	1.3385	1.38	1.169
	7.0	0.768	0.740	1.0363	1.3394	1.62	1.200

Solute	Mass %	m/mol kg^{-1}	c/mol L^{-1}	ρ/g cm^{-3}	n	Δ/°C	η/mPa s
	8.0	0.887	0.850	1.0418	1.3403	1.88	1.232
	9.0	1.009	0.962	1.0474	1.3413	2.16	1.267
	10.0	1.134	1.075	1.0531	1.3422	2.45	1.303
	12.0	1.392	1.304	1.0647	1.3441	3.01	1.382
	14.0	1.661	1.538	1.0765	1.3460	3.76	1.469
	16.0	1.944	1.777	1.0885	1.3480	4.45	1.565
	18.0	2.240	2.022	1.1009	1.3500	5.25	1.671
	20.0	2.551	2.273	1.1135	1.3520	6.23	1.788
	22.0	2.878	2.529	1.1263	1.3540	7.38	1.914
	24.0	3.223	2.791	1.1395	1.3561	8.69	2.049
	26.0	3.585	3.059	1.1528	1.3582	10.12	2.198
	28.0	3.968	3.333	1.1665	1.3604	11.64	2.365
	30.0	4.373	3.614	1.1804	1.3625	13.23	2.553
	32.0	4.802	3.901	1.1945	1.3647	14.94	2.766
	34.0	5.257	4.194	1.2089	1.3669	16.81	3.001
	36.0	5.740	4.495	1.2236	1.3691	18.85	3.260
	38.0	6.254	4.803	1.2385	1.3713	21.09	3.544
	40.0	6.803	5.117	1.2536	1.3735	23.58	3.856
Potassium bicarbonate $KHCO_3$	0.5	0.050	0.050	1.0014	1.3335	0.18	1.009
	1.0	0.101	0.100	1.0046	1.3341	0.34	1.015
	2.0	0.204	0.202	1.0114	1.3353	0.67	1.027
	3.0	0.309	0.305	1.0181	1.3365	0.98	1.040
	4.0	0.416	0.409	1.0247	1.3376	1.29	1.053
	5.0	0.526	0.515	1.0310	1.3386	1.60	1.067
	6.0	0.638	0.622	1.0379	1.3397	1.91	1.081
	7.0	0.752	0.730	1.0446	1.3409	2.22	1.096
	8.0	0.869	0.840	1.0514	1.3419	2.53	1.112
	9.0	0.988	0.951	1.0581	1.3430	2.84	1.128
	10.0	1.110	1.064	1.0650	1.3441	3.16	1.145
	12.0	1.362	1.293	1.0788	1.3462	3.79	1.183
	14.0	1.626	1.528	1.0929	1.3484	4.41	1.224
	16.0	1.903	1.770	1.1073	1.3506		1.270
	18.0	2.193	2.017	1.1221	1.3528		1.319
	20.0	2.497	2.272	1.1372	1.3550		1.373
	22.0	2.817	2.533	1.1527	1.3572		1.432
	24.0	3.154	2.801	1.1685	1.3595		1.497
Potassium bromide KBr	0.5	0.042	0.042	1.0018	1.3336	0.15	1.000
	1.0	0.085	0.084	1.0054	1.3342	0.29	0.998
	2.0	0.171	0.170	1.0127	1.3354	0.59	0.994
	3.0	0.260	0.257	1.0200	1.3366	0.88	0.990
	4.0	0.350	0.345	1.0275	1.3379	1.18	0.985
	5.0	0.442	0.435	1.0350	1.3391	1.48	0.981
	6.0	0.536	0.526	1.0426	1.3403	1.78	0.977
	7.0	0.633	0.618	1.0503	1.3416	2.10	0.974
	8.0	0.731	0.711	1.0581	1.3429	2.42	0.970
	9.0	0.831	0.806	1.0660	1.3441	2.74	0.967
	10.0	0.934	0.903	1.0740	1.3454	3.07	0.964
	12.0	1.146	1.099	1.0903	1.3481	3.76	0.958
	14.0	1.368	1.302	1.1070	1.3507	4.49	0.953
	16.0	1.601	1.512	1.1242	1.3535	5.25	0.949
	18.0	1.845	1.727	1.1419	1.3562	6.04	0.946
	20.0	2.101	1.950	1.1601	1.3591	6.88	0.944
	22.0	2.370	2.179	1.1788	1.3620	7.76	0.943
	24.0	2.654	2.416	1.1980	1.3650	8.70	0.943
	26.0	2.952	2.661	1.2179	1.3680	9.68	0.944
	28.0	3.268	2.914	1.2383	1.3711	10.72	0.947
	30.0	3.601	3.175	1.2593	1.3743	11.82	0.952
	32.0	3.954	3.445	1.2810	1.3776	12.98	0.959
	34.0	4.329	3.724	1.3033	1.3809		0.968

Solute	Mass %	m/mol kg⁻¹	c/mol L⁻¹	ρ/g cm⁻³	n	Δ/°C	η/mPa s
	36.0	4.727	4.012	1.3263	1.3843		0.979
	38.0	5.150	4.311	1.3501	1.3878		0.993
	40.0	5.602	4.620	1.3746	1.3914		1.010
Potassium carbonate K_2CO_3	0.5	0.036	0.036	1.0027	1.3339	0.18	1.013
	1.0	0.073	0.073	1.0072	1.3347	0.34	1.025
	2.0	0.148	0.147	1.0163	1.3365	0.66	1.048
	3.0	0.224	0.223	1.0254	1.3382	0.99	1.071
	4.0	0.301	0.299	1.0345	1.3399	1.32	1.094
	5.0	0.381	0.378	1.0437	1.3416	1.67	1.119
	6.0	0.462	0.457	1.0529	1.3433	2.03	1.146
	7.0	0.545	0.538	1.0622	1.3450	2.40	1.174
	8.0	0.629	0.620	1.0715	1.3467	2.77	1.204
	9.0	0.716	0.704	1.0809	1.3484	3.17	1.235
	10.0	0.804	0.789	1.0904	1.3501	3.57	1.269
	12.0	0.987	0.963	1.1095	1.3535	4.45	1.339
	14.0	1.178	1.144	1.1291	1.3569	5.39	1.414
	16.0	1.378	1.330	1.1490	1.3603	6.42	1.497
	18.0	1.588	1.523	1.1692	1.3637	7.55	1.594
	20.0	1.809	1.722	1.1898	1.3671	8.82	1.707
	24.0	2.285	2.139	1.2320	1.3739	11.96	1.978
	28.0	2.814	2.584	1.2755	1.3807	16.01	2.331
	32.0	3.405	3.057	1.3204	1.3874	21.46	2.834
	36.0	4.070	3.559	1.3665	1.3940	28.58	3.503
	40.0	4.824	4.093	1.4142	1.4006	37.55	4.360
	50.0	7.236	5.573	1.5404	1.4168		9.369
Potassium chloride KCl	0.5	0.067	0.067	1.0014	1.3337	0.23	1.000
	1.0	0.135	0.135	1.0046	1.3343	0.46	0.999
	2.0	0.274	0.271	1.0110	1.3357	0.92	0.999
	3.0	0.415	0.409	1.0174	1.3371	1.38	0.998
	4.0	0.559	0.549	1.0239	1.3384	1.85	0.997
	5.0	0.706	0.691	1.0304	1.3398	2.32	0.996
	6.0	0.856	0.835	1.0369	1.3411	2.80	0.994
	7.0	1.010	0.980	1.0434	1.3425	3.29	0.992
	8.0	1.166	1.127	1.0500	1.3438	3.80	0.990
	9.0	1.327	1.276	1.0566	1.3452	4.30	0.989
	10.0	1.490	1.426	1.0633	1.3466	4.81	0.988
	12.0	1.829	1.733	1.0768	1.3493	5.88	0.990
	14.0	2.184	2.048	1.0905	1.3521		0.994
	16.0	2.555	2.370	1.1043	1.3549		0.999
	18.0	2.944	2.701	1.1185	1.3577		1.004
	20.0	3.353	3.039	1.1328	1.3606		1.012
	22.0	3.783	3.386	1.1474	1.3635		1.024
	24.0	4.236	3.742	1.1623	1.3665		1.040
Potassium hydroxide KOH	0.5	0.090	0.089	1.0025	1.3340	0.30	1.010
	1.0	0.180	0.179	1.0068	1.3350	0.61	1.019
	2.0	0.364	0.362	1.0155	1.3369	1.24	1.038
	3.0	0.551	0.548	1.0242	1.3388	1.89	1.058
	4.0	0.743	0.736	1.0330	1.3408	2.57	1.079
	5.0	0.938	0.929	1.0419	1.3427	3.36	1.102
	6.0	1.138	1.124	1.0509	1.3445	4.14	1.126
	7.0	1.342	1.322	1.0599	1.3464	4.92	1.151
	8.0	1.550	1.524	1.0690	1.3483		1.177
	9.0	1.763	1.729	1.0781	1.3502		1.205
	10.0	1.980	1.938	1.0873	1.3520		1.233
	12.0	2.431	2.365	1.1059	1.3558		1.294
	14.0	2.902	2.806	1.1246	1.3595		1.361
	16.0	3.395	3.261	1.1435	1.3632		1.436
	18.0	3.913	3.730	1.1626	1.3670		1.521

Solute	Mass %	m/mol kg^{-1}	c/mol L^{-1}	ρ/g cm^{-3}	n	Δ/°C	η/mPa s
	20.0	4.456	4.213	1.1818	1.3707		1.619
	22.0	5.027	4.711	1.2014	1.3744		1.732
	24.0	5.629	5.223	1.2210	1.3781		1.861
	26.0	6.262	5.750	1.2408	1.3818		2.006
	28.0	6.931	6.293	1.2609	1.3854		2.170
	30.0	7.639	6.851	1.2813	1.3889		2.357
	40.0	11.882	9.896	1.3881	1.4068		3.879
	50.0	17.824	13.389	1.5024	1.4247		7.892
Potassium	0.5	0.030	0.030	1.0019	1.3337	0.11	0.999
iodide	1.0	0.061	0.061	1.0056	1.3343	0.22	0.997
KI	2.0	0.123	0.122	1.0131	1.3357	0.43	0.991
	3.0	0.186	0.184	1.0206	1.3370	0.64	0.986
	4.0	0.251	0.248	1.0282	1.3384	0.86	0.981
	5.0	0.317	0.312	1.0360	1.3397	1.08	0.976
	6.0	0.385	0.377	1.0438	1.3411	1.30	0.969
	7.0	0.453	0.443	1.0517	1.3425	1.53	0.963
	8.0	0.524	0.511	1.0598	1.3440	1.77	0.957
	9.0	0.596	0.579	1.0679	1.3454	2.01	0.951
	10.0	0.669	0.648	1.0762	1.3469	2.26	0.946
	12.0	0.821	0.790	1.0931	1.3498	2.77	0.937
	14.0	0.981	0.937	1.1105	1.3529	3.30	0.929
	16.0	1.147	1.088	1.1284	1.3560	3.87	0.921
	18.0	1.322	1.244	1.1469	1.3593	4.46	0.915
	20.0	1.506	1.405	1.1659	1.3626	5.09	0.910
	22.0	1.699	1.571	1.1856	1.3661	5.76	0.905
	24.0	1.902	1.744	1.2060	1.3696	6.46	0.901
	26.0	2.117	1.922	1.2270	1.3733	7.21	0.898
	28.0	2.343	2.106	1.2487	1.3771	8.01	0.895
	30.0	2.582	2.297	1.2712	1.3810	8.86	0.892
	32.0	2.835	2.495	1.2944	1.3851	9.76	0.891
	34.0	3.103	2.700	1.3185	1.3893	10.72	0.890
	36.0	3.388	2.913	1.3434	1.3936	11.73	0.890
	38.0	3.692	3.134	1.3692	1.3981	12.81	0.893
	40.0	4.016	3.364	1.3959	1.4027	13.97	0.897
Potassium	0.5	0.050	0.050	1.0014	1.3335	0.17	0.999
nitrate	1.0	0.100	0.099	1.0045	1.3339	0.33	0.996
KNO_3	2.0	0.202	0.200	1.0108	1.3349	0.64	0.990
	3.0	0.306	0.302	1.0171	1.3358	0.94	0.986
	4.0	0.412	0.405	1.0234	1.3368	1.22	0.983
	5.0	0.521	0.509	1.0298	1.3377	1.50	0.980
	6.0	0.631	0.615	1.0363	1.3386	1.76	0.977
	7.0	0.744	0.722	1.0428	1.3396	2.02	0.975
	8.0	0.860	0.830	1.0494	1.3405	2.27	0.973
	9.0	0.978	0.940	1.0560	1.3415	2.52	0.971
	10.0	1.099	1.051	1.0627	1.3425	2.75	0.970
	12.0	1.349	1.277	1.0762	1.3444		0.970
	14.0	1.610	1.509	1.0899	1.3463		0.972
	16.0	1.884	1.747	1.1039	1.3482		0.976
	18.0	2.171	1.991	1.1181	1.3502		0.982
	20.0	2.473	2.240	1.1326	1.3521		0.990
	22.0	2.790	2.497	1.1473	1.3541		0.999
	24.0	3.123	2.759	1.1623	1.3561		1.010
Potassium	0.5	0.032	0.032	1.0017		0.11	1.001
permanganate	1.0	0.064	0.064	1.0051		0.22	1.000
$KMnO_4$	1.5	0.096	0.096	1.0085		0.32	0.999
	2.0	0.129	0.128	1.0118		0.43	0.998
	3.0	0.196	0.193	1.0186			0.995
	4.0	0.264	0.260	1.0254			0.992

Solute	Mass %	m/mol kg^{-1}	c/mol L^{-1}	ρ/g cm^{-3}	n	Δ/°C	η/mPa s
	5.0	0.333	0.327	1.0322			0.989
	6.0	0.404	0.394	1.0390			0.985
Potassium	0.5	0.029	0.029	1.0025	1.3338	0.13	1.013
hydrogen	1.0	0.058	0.058	1.0068	1.3345	0.25	1.023
phosphate	1.5	0.087	0.087	1.0110	1.3353	0.37	1.034
K_2HPO_4	2.0	0.117	0.117	1.0153	1.3361	0.49	1.046
	2.5	0.147	0.146	1.0195	1.3368	0.61	1.057
	3.0	0.178	0.176	1.0238	1.3376	0.73	1.069
	3.5	0.208	0.207	1.0281	1.3384	0.86	1.081
	4.0	0.239	0.237	1.0324	1.3392	0.97	1.094
	4.5	0.271	0.268	1.0368	1.3399	1.10	1.107
	5.0	0.302	0.299	1.0412	1.3407	1.22	1.120
	6.0	0.366	0.362	1.0500	1.3422	1.46	1.147
	7.0	0.432	0.426	1.0590	1.3438	1.70	1.177
	8.0	0.499	0.491	1.0680	1.3453	1.95	1.209
Potassium	0.5	0.037	0.037	1.0018	1.3336	0.13	1.010
dihydrogen	1.0	0.074	0.074	1.0053	1.3342	0.25	1.019
phosphate	1.5	0.112	0.111	1.0089	1.3348	0.37	1.028
KH_2PO_4	2.0	0.150	0.149	1.0125	1.3354	0.49	1.038
	3.0	0.227	0.225	1.0197	1.3365	0.72	1.060
	4.0	0.306	0.302	1.0269	1.3377	0.96	1.083
	5.0	0.387	0.380	1.0342	1.3388	1.19	1.108
	6.0	0.469	0.459	1.0414	1.3400	1.41	1.133
	7.0	0.553	0.539	1.0486	1.3411	1.63	1.160
	8.0	0.639	0.621	1.0558	1.3422	1.84	1.187
	9.0	0.727	0.703	1.0630	1.3434	2.04	1.215
	10.0	0.816	0.786	1.0703	1.3445	2.23	1.245
Potassium	0.5	0.029	0.029	1.0022	1.3336	0.14	1.006
sulfate	1.0	0.058	0.058	1.0062	1.3343	0.26	1.011
K_2SO_4	2.0	0.117	0.116	1.0143	1.3355	0.50	1.021
	3.0	0.177	0.176	1.0224	1.3368	0.73	1.033
	4.0	0.239	0.237	1.0306	1.3380	0.95	1.045
	5.0	0.302	0.298	1.0388	1.3393	1.17	1.058
	6.0	0.366	0.360	1.0470	1.3405		1.072
	7.0	0.432	0.424	1.0553	1.3417		1.087
	8.0	0.499	0.488	1.0637	1.3428		1.102
	9.0	0.568	0.554	1.0721	1.3440		1.117
	10.0	0.638	0.620	1.0806	1.3452		1.132
1-Propanol	1.0	0.168	0.166	0.9963	1.3339	0.31	1.051
$CH_3CH_2CH_2OH$	2.0	0.340	0.331	0.9946	1.3348	0.61	1.100
	3.0	0.515	0.496	0.9928	1.3357	0.93	1.152
	4.0	0.693	0.660	0.9911	1.3366	1.24	1.208
	5.0	0.876	0.823	0.9896	1.3376	1.57	1.267
	6.0	1.062	0.987	0.9882	1.3385	1.91	1.325
	7.0	1.252	1.149	0.9868	1.3394	2.26	1.387
	8.0	1.447	1.312	0.9855	1.3404	2.61	1.449
	9.0	1.646	1.474	0.9842	1.3414	2.99	1.514
	10.0	1.849	1.635	0.9829	1.3423	3.36	1.577
	12.0	2.269	1.958	0.9804	1.3442	4.09	1.710
	14.0	2.709	2.278	0.9779	1.3460	4.91	1.849
	16.0	3.169	2.595	0.9749	1.3477	5.78	1.986
	18.0	3.652	2.911	0.9719	1.3494	6.67	2.106
	20.0	4.160	3.223	0.9686	1.3510	7.76	2.218
	24.0	5.254	3.838	0.9612	1.3539	9.12	2.432
	28.0	6.471	4.441	0.9533	1.3566	10.17	2.612
	32.0	7.830	5.033	0.9452	1.3592	10.66	2.765
	36.0	9.359	5.613	0.9370	1.3614		2.900

Solute	Mass %	m/mol kg^{-1}	c/mol L^{-1}	ρ/g cm^{-3}	n	Δ/°C	η/mPa s
	40.0	11.093	6.182	0.9288	1.3635		3.010
	60.0	24.958	8.860	0.8875	1.3734		3.186
	80.0	66.556	11.275	0.8470	1.3812		2.822
	100.0		13.368	0.8034	1.3852		2.227
2-Propanol	1.0	0.168	0.166	0.9960	1.3338	0.30	1.056
$CH_3CHOHCH_3$	2.0	0.340	0.331	0.9939	1.3346	0.60	1.112
	3.0	0.515	0.495	0.9920	1.3355	0.93	1.166
	4.0	0.693	0.659	0.9902	1.3364	1.26	1.225
	5.0	0.876	0.822	0.9884	1.3373	1.61	1.287
	6.0	1.062	0.985	0.9871	1.3382	1.96	1.352
	7.0	1.252	1.148	0.9855	1.3392	2.32	1.417
	8.0	1.447	1.310	0.9843	1.3400	2.68	1.485
	9.0	1.646	1.472	0.9831	1.3410	3.06	1.553
	10.0	1.849	1.633	0.9816	1.3420	3.48	1.629
	12.0	2.269	1.955	0.9793	1.3439	4.43	1.794
	14.0	2.709	2.276	0.9772	1.3459	5.29	1.970
	16.0	3.169	2.596	0.9751	1.3478	6.36	2.160
	18.0	3.652	2.913	0.9725	1.3496	7.40	2.352
	20.0	4.160	3.227	0.9696	1.3514	8.52	2.550
	40.0	11.093	6.191	0.9302	1.3642		
	60.0	24.958	8.809	0.8824	1.3717		
	80.0	66.556	11.103	0.8341	1.3742		
	100.0		13.058	0.7848	1.3776		
Silver	0.5	0.030	0.030	1.0027	1.3336	0.10	1.003
nitrate	1.0	0.059	0.059	1.0070	1.3342	0.20	1.005
$AgNO_3$	2.0	0.120	0.120	1.0154	1.3352	0.40	1.009
	3.0	0.182	0.181	1.0239	1.3363	0.59	1.013
	4.0	0.245	0.243	1.0327	1.3374	0.78	1.016
	5.0	0.310	0.307	1.0417	1.3385	0.96	1.020
	6.0	0.376	0.371	1.0506	1.3396	1.15	1.024
	7.0	0.443	0.437	1.0597	1.3407	1.33	1.027
	8.0	0.512	0.503	1.0690	1.3419	1.51	1.031
	9.0	0.582	0.571	1.0785	1.3431	1.69	1.035
	10.0	0.654	0.641	1.0882	1.3443	1.87	1.039
	12.0	0.803	0.783	1.1079	1.3467	2.21	1.049
	14.0	0.958	0.930	1.1284	1.3493	2.55	1.060
	16.0	1.121	1.083	1.1496	1.3519	2.86	1.072
	18.0	1.292	1.241	1.1715	1.3546		1.086
	20.0	1.472	1.406	1.1942	1.3574		1.101
	22.0	1.660	1.577	1.2177	1.3602		1.117
	24.0	1.859	1.755	1.2420	1.3632		1.135
	26.0	2.068	1.940	1.2672	1.3662		1.154
	28.0	2.289	2.132	1.2933	1.3694		1.176
	30.0	2.523	2.332	1.3204	1.3726		1.200
	32.0	2.770	2.541	1.3487	1.3760		1.227
	34.0	3.033	2.758	1.3780	1.3795		1.257
	36.0	3.311	2.985	1.4087	1.3832		1.290
	38.0	3.608	3.223	1.4407	1.3871		1.326
	40.0	3.925	3.472	1.4743	1.3911		1.366
Sodium	0.5	0.061	0.061	1.0008	1.3337	0.22	1.021
acetate	1.0	0.123	0.122	1.0034	1.3344	0.43	1.040
CH_3COONa	2.0	0.249	0.246	1.0085	1.3358	0.88	1.080
	3.0	0.377	0.371	1.0135	1.3372	1.34	1.124
	4.0	0.508	0.497	1.0184	1.3386	1.82	1.171
	5.0	0.642	0.624	1.0234	1.3400	2.32	1.222
	6.0	0.778	0.752	1.0283	1.3414	2.85	1.278
	7.0	0.918	0.882	1.0334	1.3428	3.40	1.337
	8.0	1.060	1.013	1.0386	1.3442	3.98	1.401

Solute	Mass %	m/mol kg^{-1}	c/mol L^{-1}	ρ/g cm^{-3}	n	Δ/°C	η/mPa s
	9.0	1.206	1.145	1.0440	1.3456	4.57	1.468
	10.0	1.354	1.279	1.0495	1.3470		1.539
	12.0	1.662	1.552	1.0607	1.3498		1.688
	14.0	1.984	1.829	1.0718	1.3526		1.855
	16.0	2.322	2.112	1.0830	1.3554		2.054
	18.0	2.676	2.400	1.0940	1.3583		2.284
	20.0	3.047	2.694	1.1050	1.3611		2.567
	22.0	3.438	2.993	1.1159	1.3639		2.948
	24.0	3.849	3.297	1.1268	1.3666		3.400
	26.0	4.283	3.606	1.1377	1.3693		3.877
	28.0	4.741	3.921	1.1488	1.3720		4.388
	30.0	5.224	4.243	1.1602	1.3748		4.940
Sodium	0.5	0.060	0.060	1.0018	1.3337	0.20	1.015
bicarbonate	1.0	0.120	0.120	1.0054	1.3344	0.40	1.028
$NaHCO_3$	1.5	0.181	0.180	1.0089	1.3351	0.59	1.042
	2.0	0.243	0.241	1.0125	1.3357	0.78	1.057
	2.5	0.305	0.302	1.0160	1.3364	0.98	1.071
	3.0	0.368	0.364	1.0196	1.3370	1.16	1.086
	3.5	0.432	0.426	1.0231	1.3377	1.35	1.102
	4.0	0.496	0.489	1.0266	1.3383	1.54	1.118
	4.5	0.561	0.552	1.0301	1.3390	1.72	1.134
	5.0	0.627	0.615	1.0337	1.3396	1.90	1.151
	5.5	0.693	0.679	1.0372	1.3403	2.08	1.168
	6.0	0.760	0.743	1.0408	1.3409	2.26	1.185
Sodium	0.5	0.049	0.049	1.0021	1.3337	0.17	1.004
bromide	1.0	0.098	0.098	1.0060	1.3344	0.34	1.007
NaBr	2.0	0.198	0.197	1.0139	1.3358	0.69	1.012
	3.0	0.301	0.298	1.0218	1.3372	1.04	1.017
	4.0	0.405	0.400	1.0298	1.3386	1.39	1.022
	5.0	0.512	0.504	1.0380	1.3401	1.76	1.028
	6.0	0.620	0.610	1.0462	1.3415	2.14	1.034
	7.0	0.732	0.717	1.0546	1.3430	2.53	1.040
	8.0	0.845	0.826	1.0630	1.3445	2.93	1.046
	9.0	0.961	0.937	1.0716	1.3460	3.34	1.053
	10.0	1.080	1.050	1.0803	1.3475	3.77	1.060
	12.0	1.325	1.281	1.0981	1.3506	4.67	1.077
	14.0	1.582	1.519	1.1164	1.3538	5.65	1.096
	16.0	1.851	1.765	1.1352	1.3570	6.74	1.119
	18.0	2.133	2.020	1.1546	1.3604		1.144
	20.0	2.430	2.283	1.1745	1.3638		1.174
	22.0	2.741	2.555	1.1951	1.3673		1.207
	24.0	3.069	2.837	1.2163	1.3708		1.244
	26.0	3.415	3.129	1.2382	1.3745		1.287
	28.0	3.780	3.431	1.2608	1.3783		1.336
	30.0	4.165	3.744	1.2842	1.3822		1.395
	32.0	4.574	4.069	1.3083	1.3862		1.465
	34.0	5.007	4.406	1.3333	1.3903		1.546
	36.0	5.467	4.755	1.3592	1.3946		1.639
	38.0	5.957	5.119	1.3860	1.3990		1.745
	40.0	6.479	5.496	1.4138	1.4035		1.866
Sodium	0.5	0.047	0.047	1.0034	1.3341	0.22	1.025
carbonate	1.0	0.095	0.095	1.0086	1.3352	0.43	1.049
Na_2CO_3	2.0	0.193	0.192	1.0190	1.3375	0.75	1.102
	3.0	0.292	0.291	1.0294	1.3397	1.08	1.159
	4.0	0.393	0.392	1.0398	1.3419	1.42	1.222
	5.0	0.497	0.495	1.0502	1.3440	1.77	1.292
	6.0	0.602	0.600	1.0606	1.3462	2.13	1.367
	7.0	0.710	0.707	1.0711	1.3483		1.448

Solute	Mass %	m/mol kg^{-1}	c/mol L^{-1}	ρ/g cm^{-3}	n	Δ/°C	η/mPa s
	8.0	0.820	0.816	1.0816	1.3504		1.538
	9.0	0.933	0.927	1.0922	1.3525		1.638
	10.0	1.048	1.041	1.1029	1.3547		1.754
	11.0	1.166	1.156	1.1136	1.3568		1.884
	12.0	1.287	1.273	1.1244	1.3589		2.028
	13.0	1.410	1.392	1.1353	1.3610		2.186
	14.0	1.536	1.514	1.1463	1.3631		2.361
	15.0	1.665	1.638	1.1574	1.3652		2.551
Sodium chloride NaCl	0.5	0.086	0.086	1.0018	1.3339	0.30	1.011
	1.0	0.173	0.172	1.0053	1.3347	0.59	1.020
	2.0	0.349	0.346	1.0125	1.3365	1.19	1.036
	3.0	0.529	0.523	1.0196	1.3383	1.79	1.052
	4.0	0.713	0.703	1.0268	1.3400	2.41	1.068
	5.0	0.901	0.885	1.0340	1.3418	3.05	1.085
	6.0	1.092	1.069	1.0413	1.3435	3.70	1.104
	7.0	1.288	1.256	1.0486	1.3453	4.38	1.124
	8.0	1.488	1.445	1.0559	1.3470	5.08	1.145
	9.0	1.692	1.637	1.0633	1.3488	5.81	1.168
	10.0	1.901	1.832	1.0707	1.3505	6.56	1.193
	12.0	2.333	2.229	1.0857	1.3541	8.18	1.250
	14.0	2.785	2.637	1.1008	1.3576	9.94	1.317
	16.0	3.259	3.056	1.1162	1.3612	11.89	1.388
	18.0	3.756	3.486	1.1319	1.3648	14.04	1.463
	20.0	4.278	3.928	1.1478	1.3684	16.46	1.557
	22.0	4.826	4.382	1.1640	1.3721	19.18	1.676
	24.0	5.403	4.847	1.1804	1.3757		1.821
	26.0	6.012	5.326	1.1972	1.3795		1.990
Sodium citrate (HO)C(COONa)$_3$	1.0	0.039	0.039	1.0049	1.3348	0.20	1.043
	2.0	0.079	0.078	1.0120	1.3366	0.39	1.081
	3.0	0.120	0.118	1.0186	1.3383	0.59	1.122
	4.0	0.161	0.159	1.0260	1.3401	0.79	1.166
	5.0	0.204	0.200	1.0331	1.3419	0.97	1.210
	6.0	0.247	0.242	1.0405	1.3437	1.17	1.263
	7.0	0.292	0.284	1.0482	1.3455	1.36	1.314
	8.0	0.337	0.327	1.0557	1.3473	1.57	1.371
	9.0	0.383	0.371	1.0632	1.3491	1.77	1.427
	10.0	0.431	0.415	1.0708	1.3509	1.96	1.499
	12.0	0.528	0.505	1.0861	1.3546	2.38	1.649
	14.0	0.631	0.598	1.1019	1.3583	2.82	1.832
	16.0	0.738	0.693	1.1173	1.3618	3.27	2.045
	18.0	0.851	0.790	1.1327	1.3656	3.82	2.290
	20.0	0.969	0.891	1.1492	1.3693	4.39	2.596
	24.0	1.224	1.099	1.1813	1.3767		3.409
	28.0	1.507	1.318	1.2151	1.3845		4.586
	32.0	1.823	1.548	1.2487	1.3923		6.541
	36.0	2.180	1.792	1.2843	1.4001		9.788
Sodium hydroxide NaOH	0.5	0.126	0.125	1.0039	1.3344	0.43	1.027
	1.0	0.253	0.252	1.0095	1.3358	0.86	1.054
	2.0	0.510	0.510	1.0207	1.3386	1.74	1.112
	3.0	0.773	0.774	1.0318	1.3414	2.64	1.176
	4.0	1.042	1.043	1.0428	1.3441	3.59	1.248
	5.0	1.316	1.317	1.0538	1.3467	4.57	1.329
	6.0	1.596	1.597	1.0648	1.3494	5.60	1.416
	7.0	1.882	1.883	1.0758	1.3520	6.69	1.510
	8.0	2.174	2.174	1.0869	1.3546	7.87	1.616
	9.0	2.473	2.470	1.0979	1.3572	9.12	1.737
	10.0	2.778	2.772	1.1089	1.3597	10.47	1.882
	12.0	3.409	3.393	1.1309	1.3648	13.42	2.201

Solute	Mass %	m/mol kg⁻¹	c/mol L⁻¹	ρ/g cm⁻³	n	Δ/°C	η/mPa s
	14.0	4.070	4.036	1.1530	1.3697	16.76	2.568
	15.0	4.412	4.365	1.1640	1.3722		2.789
	16.0	4.762	4.701	1.1751	1.3746		3.043
	18.0	5.488	5.387	1.1971	1.3793		3.698
	20.0	6.250	6.096	1.2192	1.3840		4.619
	22.0	7.052	6.827	1.2412	1.3885		5.765
	24.0	7.895	7.579	1.2631	1.3929		7.100
	26.0	8.784	8.352	1.2848	1.3971		8.744
	28.0	9.723	9.145	1.3064	1.4012		10.832
	30.0	10.715	9.958	1.3277	1.4051		13.517
	32.0	11.766	10.791	1.3488	1.4088		16.844
	34.0	12.880	11.643	1.3697	1.4123		20.751
	36.0	14.064	12.512	1.3901	1.4156		25.290
	38.0	15.324	13.398	1.4102	1.4186		30.461
	40.0	16.668	14.300	1.4299	1.4215		36.312
Sodium nitrate $NaNO_3$	0.5	0.059	0.059	1.0016	1.3336	0.20	1.004
	1.0	0.119	0.118	1.0050	1.3341	0.40	1.007
	2.0	0.240	0.238	1.0117	1.3353	0.79	1.012
	3.0	0.364	0.359	1.0185	1.3364	1.18	1.018
	4.0	0.490	0.483	1.0254	1.3375	1.56	1.025
	5.0	0.619	0.607	1.0322	1.3387	1.94	1.032
	6.0	0.751	0.734	1.0392	1.3398	2.32	1.040
	7.0	0.886	0.862	1.0462	1.3409	2.70	1.049
	8.0	1.023	0.991	1.0532	1.3421	3.08	1.059
	9.0	1.164	1.123	1.0603	1.3432	3.46	1.069
	10.0	1.307	1.256	1.0674	1.3443	3.84	1.081
	12.0	1.604	1.527	1.0819	1.3466	4.60	1.107
	14.0	1.915	1.806	1.0967	1.3489	5.37	1.138
	18.0	2.583	2.387	1.1272	1.3536	6.98	1.215
	20.0	2.941	2.689	1.1429	1.3559	7.81	1.263
	30.0	5.042	4.326	1.2256	1.3678		1.609
	40.0	7.844	6.200	1.3175	1.3802		2.226
Sodium phosphate Na_3PO_4	0.5	0.031	0.031	1.0042	1.3343	0.19	1.033
	1.0	0.062	0.062	1.0100	1.3356	0.37	1.064
	1.5	0.093	0.093	1.0158	1.3369	0.53	1.094
	2.0	0.124	0.125	1.0216	1.3381	0.67	1.126
	2.5	0.156	0.157	1.0275	1.3394	0.79	1.161
	3.0	0.189	0.189	1.0335	1.3406		1.198
	3.5	0.221	0.222	1.0395	1.3419		1.238
	4.0	0.254	0.255	1.0456	1.3432		1.281
	4.5	0.287	0.289	1.0517	1.3444		1.327
	5.0	0.321	0.323	1.0579	1.3457		1.375
	5.5	0.355	0.357	1.0642	1.3470		1.426
	6.0	0.389	0.392	1.0705	1.3482		1.480
	6.5	0.424	0.427	1.0768	1.3495		1.538
	7.0	0.459	0.462	1.0832	1.3507		1.598
	7.5	0.495	0.498	1.0896	1.3519		1.662
	8.0	0.530	0.535	1.0961	1.3532		1.729
Sodium hydrogen phosphate Na_2HPO_4	0.5	0.035	0.035	1.0032	1.3340	0.17	1.021
	1.0	0.071	0.071	1.0082	1.3349	0.32	1.042
	1.5	0.107	0.107	1.0131	1.3358	0.46	1.064
	2.0	0.144	0.143	1.0180	1.3368		1.088
	2.5	0.181	0.180	1.0229	1.3377		1.113
	3.0	0.218	0.217	1.0279	1.3386		1.138
	3.5	0.255	0.255	1.0328	1.3396		1.165
	4.0	0.293	0.292	1.0378	1.3405		1.193
	4.5	0.332	0.331	1.0428	1.3414		1.223
	5.0	0.371	0.369	1.0478	1.3424		1.254

Solute	Mass %	*m*/mol kg⁻¹	*c*/mol L⁻¹	ρ/g cm⁻³	*n*	Δ/°C	η/mPa s
	5.5	0.410	0.408	1.0528	1.3433		1.286
Sodium dihydrogen phosphate NaH_2PO_4	0.5	0.042	0.042	1.0019	1.3336	0.14	1.018
	1.0	0.084	0.084	1.0056	1.3343	0.28	1.035
	1.5	0.127	0.126	1.0094	1.3349	0.42	1.051
	2.0	0.170	0.169	1.0131	1.3356	0.56	1.068
	2.5	0.214	0.212	1.0168	1.3362	0.70	1.085
	3.0	0.258	0.255	1.0206	1.3369	0.84	1.103
	3.5	0.302	0.299	1.0244	1.3375	0.98	1.121
	4.0	0.347	0.343	1.0281	1.3382	1.12	1.140
	4.5	0.393	0.387	1.0319	1.3388	1.25	1.160
	5.0	0.439	0.432	1.0358	1.3395	1.39	1.180
	6.0	0.532	0.522	1.0434	1.3408	1.65	1.223
	7.0	0.627	0.613	1.0511	1.3421	1.89	1.270
	8.0	0.725	0.706	1.0589	1.3434	2.12	1.319
	9.0	0.824	0.800	1.0668	1.3447	2.35	1.371
	10.0	0.926	0.896	1.0747	1.3460	2.58	1.428
	12.0	1.137	1.091	1.0907	1.3486	3.06	1.552
	14.0	1.357	1.292	1.1070	1.3512	3.53	1.694
	16.0	1.588	1.499	1.1236	1.3538	4.03	1.861
	18.0	1.830	1.711	1.1404	1.3565	4.55	2.050
	20.0	2.084	1.930	1.1576	1.3592	5.10	2.283
	22.0	2.351	2.155	1.1752	1.3618		2.550
	24.0	2.632	2.387	1.1931	1.3646		2.850
	26.0	2.929	2.625	1.2113	1.3673		3.214
	28.0	3.242	2.870	1.2299	1.3700		3.682
	30.0	3.572	3.123	1.2488	1.3728		4.300
	32.0	3.923	3.383	1.2682	1.3756		5.079
	34.0	4.294	3.650	1.2879	1.3784		6.008
	36.0	4.689	3.925	1.3080	1.3812		7.098
	38.0	5.109	4.208	1.3285	1.3840		8.363
	40.0	5.557	4.499	1.3493	1.3869		9.814
Sodium sulfate Na_2SO_4	0.5	0.035	0.035	1.0027	1.3338	0.17	1.013
	1.0	0.071	0.071	1.0071	1.3345	0.32	1.026
	2.0	0.144	0.143	1.0161	1.3360	0.61	1.058
	3.0	0.218	0.217	1.0252	1.3376	0.87	1.091
	4.0	0.293	0.291	1.0343	1.3391	1.13	1.126
	5.0	0.371	0.367	1.0436	1.3406	1.36	1.163
	6.0	0.449	0.445	1.0526	1.3420	1.56	1.202
	7.0	0.530	0.523	1.0619	1.3435		1.244
	8.0	0.612	0.603	1.0713	1.3449		1.289
	9.0	0.696	0.685	1.0808	1.3464		1.337
	10.0	0.782	0.768	1.0905	1.3479		1.390
	12.0	0.960	0.938	1.1101	1.3509		1.508
	14.0	1.146	1.114	1.1301	1.3539		1.646
	16.0	1.341	1.296	1.1503	1.3567		1.812
	18.0	1.545	1.483	1.1705	1.3595		2.005
	20.0	1.760	1.677	1.1907	1.3620		2.227
	22.0	1.986	1.875	1.2106	1.3643		2.481
Sodium thiosulfate $Na_2S_2O_3$	0.5	0.032	0.032	1.0024	1.3340	0.14	1.012
	1.0	0.064	0.064	1.0065	1.3351	0.28	1.023
	2.0	0.129	0.128	1.0148	1.3371	0.57	1.044
	3.0	0.196	0.194	1.0231	1.3392	0.84	1.066
	4.0	0.264	0.261	1.0315	1.3413	1.09	1.090
	5.0	0.333	0.329	1.0399	1.3434	1.34	1.115
	6.0	0.404	0.398	1.0483	1.3454	1.59	1.141
	7.0	0.476	0.468	1.0568	1.3475	1.83	1.169
	8.0	0.550	0.539	1.0654	1.3496	2.06	1.199
	9.0	0.626	0.611	1.0740	1.3517	2.30	1.231

Solute	Mass %	m/mol kg⁻¹	c/mol L⁻¹	ρ/g cm⁻³	n	Δ/°C	η/mPa s
	10.0	0.703	0.685	1.0827	1.3538	2.55	1.267
	12.0	0.862	0.835	1.1003	1.3581	3.06	1.345
	14.0	1.030	0.990	1.1182	1.3624	3.60	1.435
	16.0	1.205	1.150	1.1365	1.3667	4.17	1.537
	18.0	1.388	1.315	1.1551	1.3711	4.76	1.657
	20.0	1.581	1.485	1.1740	1.3756	5.37	1.798
	30.0	2.711	2.417	1.2739	1.3987		2.903
	40.0	4.216	3.498	1.3827	1.4229		5.758
Strontium chloride $SrCl_2$	0.5	0.032	0.032	1.0027	1.3339	0.16	1.012
	1.0	0.064	0.064	1.0071	1.3348	0.31	1.021
	2.0	0.129	0.128	1.0161	1.3366	0.62	1.039
	3.0	0.195	0.194	1.0252	1.3384	0.93	1.057
	4.0	0.263	0.261	1.0344	1.3402	1.26	1.076
	5.0	0.332	0.329	1.0437	1.3421	1.61	1.096
	6.0	0.403	0.399	1.0532	1.3440	1.98	1.116
	7.0	0.475	0.469	1.0628	1.3459	2.38	1.136
	8.0	0.549	0.541	1.0726	1.3478	2.80	1.157
	9.0	0.624	0.615	1.0825	1.3498	3.25	1.180
	10.0	0.701	0.689	1.0925	1.3518	3.74	1.204
	12.0	0.860	0.843	1.1131	1.3558	4.81	1.258
	14.0	1.027	1.002	1.1342	1.3599	6.03	1.317
	16.0	1.202	1.167	1.1558	1.3641	7.41	1.383
	18.0	1.385	1.338	1.1780	1.3684	8.98	1.460
	20.0	1.577	1.515	1.2008	1.3728	10.74	1.549
	22.0	1.779	1.699	1.2241	1.3772	12.74	1.650
	24.0	1.992	1.890	1.2481	1.3817	14.99	1.765
	26.0	2.216	2.087	1.2728	1.3864		1.897
	28.0	2.453	2.293	1.2983	1.3911		2.056
	30.0	2.703	2.507	1.3248	1.3961		2.245
	32.0	2.968	2.730	1.3523	1.4013		2.527
	34.0	3.250	2.962	1.3811	1.4067		2.846
	36.0	3.548	3.205	1.4114	1.4124		3.206
Sucrose $C_{12}H_{22}O_{11}$	0.5	0.015	0.015	1.0002	1.3337	0.03	1.015
	1.0	0.030	0.029	1.0021	1.3344	0.06	1.028
	2.0	0.060	0.059	1.0060	1.3359	0.11	1.055
	3.0	0.090	0.089	1.0099	1.3373	0.17	1.084
	4.0	0.122	0.118	1.0139	1.3388	0.23	1.114
	5.0	0.154	0.149	1.0178	1.3403	0.29	1.146
	6.0	0.186	0.179	1.0218	1.3418	0.35	1.179
	7.0	0.220	0.210	1.0259	1.3433	0.42	1.215
	8.0	0.254	0.241	1.0299	1.3448	0.49	1.254
	9.0	0.289	0.272	1.0340	1.3463	0.55	1.294
	10.0	0.325	0.303	1.0381	1.3478	0.63	1.336
	12.0	0.398	0.367	1.0465	1.3509	0.77	1.429
	14.0	0.476	0.431	1.0549	1.3541	0.93	1.534
	16.0	0.556	0.497	1.0635	1.3573	1.10	1.653
	18.0	0.641	0.564	1.0722	1.3606	1.27	1.790
	20.0	0.730	0.632	1.0810	1.3639	1.47	1.945
	22.0	0.824	0.700	1.0899	1.3672	1.67	2.124
	24.0	0.923	0.771	1.0990	1.3706	1.89	2.331
	26.0	1.026	0.842	1.1082	1.3741	2.12	2.573
	28.0	1.136	0.914	1.1175	1.3776	2.37	2.855
	30.0	1.252	0.988	1.1270	1.3812	2.64	3.187
	32.0	1.375	1.063	1.1366	1.3848	2.94	3.762
	34.0	1.505	1.139	1.1464	1.3885	3.27	4.052
	36.0	1.643	1.216	1.1562	1.3922	3.63	4.621
	38.0	1.791	1.295	1.1663	1.3960	4.02	5.315
	40.0	1.948	1.375	1.1765	1.3999	4.45	6.162
	42.0	2.116	1.456	1.1868	1.4038	4.93	7.234

Solute	Mass %	*m*/mol kg^{-1}	*c*/mol L^{-1}	ρ/g cm^{-3}	*n*	Δ/°C	η/mPa s
	44.0	2.295	1.539	1.1972	1.4078		8.596
	46.0	2.489	1.623	1.2079	1.4118		10.301
	48.0	2.697	1.709	1.2186	1.4159		12.515
	50.0	2.921	1.796	1.2295	1.4201		15.431
	60.0	4.382	2.255	1.2864	1.4419		58.487
	70.0	6.817	2.755	1.3472	1.4654		481.561
	80.0	11.686	3.299	1.4117	1.4906		
Sulfuric acid	0.5	0.051	0.051	1.0016	1.3336	0.21	1.010
H_2SO_4	1.0	0.103	0.102	1.0049	1.3342	0.42	1.019
	2.0	0.208	0.206	1.0116	1.3355	0.80	1.036
	3.0	0.315	0.311	1.0183	1.3367	1.17	1.059
	4.0	0.425	0.418	1.0250	1.3379	1.60	1.085
	5.0	0.537	0.526	1.0318	1.3391	2.05	1.112
	6.0	0.651	0.635	1.0385	1.3403	2.50	1.136
	7.0	0.767	0.746	1.0453	1.3415	2.95	1.159
	8.0	0.887	0.858	1.0522	1.3427	3.49	1.182
	9.0	1.008	0.972	1.0591	1.3439	4.08	1.206
	10.0	1.133	1.087	1.0661	1.3451	4.64	1.230
	12.0	1.390	1.322	1.0802	1.3475	5.93	1.282
	14.0	1.660	1.563	1.0947	1.3500	7.49	1.337
	16.0	1.942	1.810	1.1094	1.3525	9.26	1.399
	18.0	2.238	2.064	1.1245	1.3551	11.29	1.470
	20.0	2.549	2.324	1.1398	1.3576	13.64	1.546
	22.0	2.876	2.592	1.1554	1.3602	16.48	1.624
	24.0	3.220	2.866	1.1714	1.3628	19.85	1.706
	26.0	3.582	3.147	1.1872	1.3653	24.29	1.797
	28.0	3.965	3.435	1.2031	1.3677	29.65	1.894
	30.0	4.370	3.729	1.2191	1.3701	36.21	2.001
	32.0	4.798	4.030	1.2353	1.3725	44.76	2.122
	34.0	5.252	4.339	1.2518	1.3749	55.28	2.255
	36.0	5.735	4.656	1.2685	1.3773		2.392
	38.0	6.249	4.981	1.2855	1.3797		2.533
	40.0	6.797	5.313	1.3028	1.3821		2.690
	42.0	7.383	5.655	1.3205	1.3846		2.872
	44.0	8.011	6.005	1.3386	1.3870		3.073
	46.0	8.685	6.364	1.3570	1.3895		3.299
	48.0	9.411	6.734	1.3759	1.3920		3.546
	50.0	10.196	7.113	1.3952	1.3945		3.826
	52.0	11.045	7.502	1.4149	1.3971		4.142
	54.0	11.969	7.901	1.4351	1.3997		4.499
	56.0	12.976	8.312	1.4558	1.4024		4.906
	58.0	14.080	8.734	1.4770	1.4050		5.354
	60.0	15.294	9.168	1.4987	1.4077		5.917
	70.0	23.790	11.494	1.6105			
	80.0	40.783	14.088	1.7272			
	90.0	91.762	16.649	1.8144			
	92.0	117.251	17.109	1.8240			
	94.0	159.734	17.550	1.8312			
	96.0	244.698	17.966	1.8355			
	98.0	499.592	18.346	1.8361			
	100.0		18.663	1.8305			
Trichloroacetic acid	0.5	0.031	0.031	1.0008	1.3337	0.11	1.011
CCl_3COOH	1.0	0.062	0.061	1.0034	1.3343	0.21	1.021
	2.0	0.125	0.123	1.0083	1.3356	0.42	1.044
	3.0	0.189	0.186	1.0133	1.3369	0.64	1.069
	4.0	0.255	0.249	1.0182	1.3381	0.86	1.096
	5.0	0.322	0.313	1.0230	1.3394	1.08	1.123
	6.0	0.391	0.377	1.0279	1.3406	1.30	1.150
	7.0	0.461	0.442	1.0328	1.3418	1.53	1.177

Solute	Mass %	m/mol kg^{-1}	c/mol L^{-1}	ρ/g cm^{-3}	n	Δ/°C	η/mPa s
	8.0	0.532	0.508	1.0378	1.3431	1.76	1.204
	9.0	0.605	0.574	1.0428	1.3444	1.99	1.233
	10.0	0.680	0.641	1.0479	1.3456	2.23	1.263
	12.0	0.835	0.777	1.0583	1.3483	2.73	1.326
	14.0	0.996	0.916	1.0692	1.3510	3.26	1.393
	16.0	1.166	1.058	1.0806	1.3539	3.82	1.462
	18.0	1.343	1.203	1.0921	1.3568		1.533
	20.0	1.530	1.351	1.1035	1.3597		1.608
	24.0	1.933	1.654	1.1260	1.3652		1.768
	28.0	2.380	1.968	1.1485	1.3705		1.935
	32.0	2.880	2.294	1.1713	1.3759		2.118
	36.0	3.443	2.632	1.1947	1.3813		2.320
	40.0	4.080	2.984	1.2188	1.3868		1.543
	44.0	4.809	3.349	1.2435	1.3923		2.797
	48.0	5.650	3.726	1.2682	1.3977		3.076
Tris (hydroxymethyl)-methylamine $H_2NC(CH_2OH)_3$	0.5	0.041	0.041	0.9994	1.3337	0.08	1.014
	1.0	0.083	0.083	1.0006	1.3344	0.16	1.027
	2.0	0.168	0.166	1.0030	1.3359	0.31	1.054
	3.0	0.255	0.249	1.0054	1.3374	0.47	1.083
	4.0	0.344	0.333	1.0078	1.3388	0.64	1.115
	5.0	0.434	0.417	1.0103	1.3403	0.80	1.148
	6.0	0.527	0.502	1.0128	1.3418	0.97	1.182
	7.0	0.621	0.587	1.0153	1.3433	1.15	1.218
	8.0	0.718	0.672	1.0179	1.3448	1.33	1.256
	9.0	0.816	0.758	1.0204	1.3463	1.51	1.295
	10.0	0.917	0.844	1.0230	1.3478	1.70	1.337
	12.0	1.126	1.019	1.0282	1.3508	2.08	1.427
	14.0	1.344	1.194	1.0335	1.3539	2.47	1.527
	16.0	1.572	1.372	1.0389	1.3570	2.90	1.642
	18.0	1.812	1.552	1.0443	1.3601	3.36	1.772
	20.0	2.064	1.733	1.0498	1.3633	3.85	1.920
	30.0	3.538	2.670	1.0781	1.3797		2.998
	40.0	5.503	3.657	1.1076	1.3970		5.208
Urea $(NH_2)_2CO$	0.5	0.084	0.083	0.9995	1.3337	0.16	1.007
	1.0	0.168	0.167	1.0007	1.3344	0.31	1.010
	2.0	0.340	0.334	1.0033	1.3358	0.62	1.012
	3.0	0.515	0.502	1.0058	1.3372	0.93	1.017
	4.0	0.694	0.672	1.0085	1.3387	1.24	1.025
	5.0	0.876	0.842	1.0111	1.3401	1.55	1.033
	6.0	1.063	1.013	1.0138	1.3416	1.88	1.041
	7.0	1.253	1.185	1.0165	1.3431	2.22	1.049
	8.0	1.448	1.358	1.0192	1.3446	2.56	1.057
	9.0	1.647	1.531	1.0220	1.3461	2.91	1.065
	10.0	1.850	1.706	1.0248	1.3476	3.26	1.074
	12.0	2.270	2.059	1.0304	1.3506	3.95	1.091
	14.0	2.710	2.415	1.0360	1.3537	4.66	1.109
	16.0	3.171	2.775	1.0417	1.3568	5.40	1.130
	18.0	3.655	3.139	1.0473	1.3599	6.19	1.153
	20.0	4.163	3.506	1.0530	1.3629	7.00	1.178
	22.0	4.696	3.878	1.0586	1.3661	7.81	1.205
	24.0	5.258	4.253	1.0643	1.3692	8.64	1.235
	26.0	5.850	4.632	1.0699	1.3723	9.52	1.266
	28.0	6.475	5.014	1.0756	1.3754	10.45	1.298
	30.0	7.136	5.401	1.0812	1.3785	11.40	1.332
	32.0	7.835	5.791	1.0869	1.3817	12.34	1.371
	34.0	8.577	6.185	1.0926	1.3848	13.27	1.413
	36.0	9.366	6.584	1.0984	1.3881	14.20	1.459
	38.0	10.205	6.988	1.1044	1.3913	15.11	1.509
	40.0	11.100	7.397	1.1106	1.3947	15.99	1.565

Solute	Mass %	m/mol kg^{-1}	c/mol L^{-1}	ρ/g cm^{-3}	n	Δ/°C	η/mPa s
	42.0	12.057	7.812	1.1171	1.3982	16.83	1.629
	44.0	13.082	8.234	1.1239	1.4018	17.62	1.700
	46.0	14.183	8.665	1.1313	1.4056		1.780
Zinc sulfate	0.5	0.031	0.031	1.0034	1.3339	0.08	1.021
$ZnSO_4$	1.0	0.063	0.062	1.0085	1.3348	0.15	1.040
	2.0	0.126	0.126	1.0190	1.3366	0.28	1.081
	3.0	0.192	0.191	1.0296	1.3384	0.41	1.126
	4.0	0.258	0.258	1.0403	1.3403	0.53	1.175
	5.0	0.326	0.326	1.0511	1.3421	0.65	1.227
	6.0	0.395	0.395	1.0620	1.3439	0.77	1.283
	7.0	0.466	0.465	1.0730	1.3457	0.89	1.341
	8.0	0.539	0.537	1.0842	1.3475	1.01	1.403
	9.0	0.613	0.611	1.0956	1.3494	1.14	1.470
	10.0	0.688	0.686	1.1071	1.3513	1.27	1.545
	12.0	0.845	0.840	1.1308	1.3551	1.55	1.716
	14.0	1.008	1.002	1.1553	1.3590	1.89	1.918
	16.0	1.180	1.170	1.1806	1.3630	2.31	2.152

IONIZATION CONSTANT OF WATER

Serguei N. Lvov and Allan H. Harvey

This table gives values of pK_w = $-\log_{10}(K_w)$, where K_w is the equilibrium constant of the reaction $2H_2O = H_3O^+(aq) + OH^-(aq)$. K_w is defined as $K_w = a_{H_3O^+}a_{OH^-} / a^2_{H_2O}$, where a_i is the dimensionless activity of species i. The activities are on the molality basis for ions and mole fraction basis for water molecules. It is assumed that the activity of H_3O^+(aq) is the same as the activity of H^+(aq), so that K_w is numerically equal to $a_{H^+}a_{OH^-} / a^2_{H_2O}$, the equilibrium constant for the ionization reaction of water, $H_2O = H^+(aq) + OH^-(aq)$, that is most commonly used in the literature. Values in the table are calculated using an analytical equation given in Refs. 1 and 2 where K_w is presented as a function of temperature and density from 0 °C to 800 °C and 0 g cm^{-3} to 1.25 g cm^{-3}.

References

1. International Association for the Properties of Water and Steam, *Release on the Ionization Constant of H_2O* (2007), available from www.iapws.org.
2. Bandura, A.V., and Lvov, S.N., *J. Phys. Chem. Ref. Data* 35, 15, 2006.

Pressure, MPa	Temperature, °C								
	0	25	50	75	100	150	200	250	300
0.1 or p_s	14.946	13.995	13.264	12.696	12.252	11.641	11.310	11.205	11.339
25	14.848	13.908	13.181	12.613	12.165	11.543	11.189	11.050	11.125
50	14.754	13.824	13.102	12.533	12.084	11.450	11.076	10.898	10.893
75	14.665	13.745	13.026	12.458	12.006	11.364	10.974	10.769	10.715
100	14.580	13.668	12.953	12.385	11.933	11.283	10.880	10.655	10.568
150	14.422	13.524	12.815	12.249	11.795	11.135	10.713	10.458	10.327
200	14.278	13.390	12.687	12.123	11.668	11.000	10.564	10.289	10.131
250	14.145	13.265	12.567	12.004	11.549	10.876	10.430	10.140	9.963
300	14.021	13.148	12.453	11.892	11.437	10.760	10.306	10.005	9.814
350	13.906	13.037	12.346	11.786	11.331	10.651	10.191	9.881	9.679
400	13.797	12.932	12.243	11.685	11.230	10.548	10.083	9.766	9.555
500	13.595	12.736	12.052	11.496	11.042	10.356	9.884	9.557	9.332
600	13.411	12.556	11.875	11.322	10.868	10.181	9.703	9.369	9.135
700	13.240	12.389	11.710	11.159	10.705	10.018	9.537	9.197	8.956
800	13.080	12.233	11.556	11.006	10.553	9.865	9.381	9.037	8.791
900	12.930	12.085	11.410	10.861	10.410	9.721	9.236	8.888	8.638
1000	12.788	11.946	11.272	10.725	10.273	9.585	9.098	8.748	8.495

Pressure, MPa	Temperature, °C						
	350	400	450	500	600	700	800
0.1 or p_s	11.920	47.961	47.873	47.638	46.384	43.925	40.785
25	11.551	16.566	18.135	18.758	19.425	19.829	20.113
50	11.076	11.557	12.710	14.195	15.621	16.279	16.693
75	10.802	11.045	11.491	12.162	13.507	14.301	14.791
100	10.600	10.744	11.005	11.381	12.296	13.040	13.544
150	10.295	10.345	10.464	10.642	11.117	11.613	12.032
200	10.062	10.063	10.119	10.220	10.513	10.853	11.171
250	9.869	9.839	9.859	9.917	10.112	10.360	10.609
300	9.702	9.651	9.646	9.677	9.810	9.998	10.199
350	9.554	9.487	9.465	9.476	9.567	9.712	9.877
400	9.420	9.341	9.305	9.302	9.361	9.475	9.613
500	9.182	9.086	9.031	9.007	9.024	9.094	9.191
600	8.974	8.866	8.798	8.761	8.749	8.790	8.861
700	8.787	8.670	8.593	8.546	8.514	8.536	8.587
800	8.616	8.493	8.409	8.354	8.308	8.314	8.352
900	8.458	8.330	8.240	8.180	8.122	8.117	8.144
1000	8.311	8.178	8.084	8.019	7.952	7.939	7.957

Note: Pressure for first row is 0.1 MPa at $t < 100$ °C and $t \geq 400$ °C, or p_s (saturated liquid) for 100 °C $\leq t \leq$ 350 °C

IONIZATION CONSTANT OF NORMAL AND HEAVY WATER

Serguei N. Lvov and Allan H. Harvey

This table gives the ionization constant for liquid H_2O and D_2O at temperatures from 0 °C to 100 °C at the saturated vapor pressure. The quantity tabulated is $-\log_{10}(K_w)$, where $K_w = a_{H_3O^+} a_{OH^-} / a^2_{H_2O}$ for H_2O and $K_w = a_{D_3O^+} a_{OD^-} / a^2_{D_2O}$ for D_2O. Values in the table are calculated using analytical equations given in Refs. 1 and 2 for H_2O and Ref. 3 for D_2O.

References

1. International Association for the Properties of Water and Steam, *Release on the Ionization Constant of H₂O* (2007), available from www.iapws.org.
2. Bandura, A. V., and Lvov, S. N., *J. Phys. Chem. Ref. Data* 35, 15, 2006.
3. Mesmer, R. E., and Herting, D. L., *J. Solution Chem.* 7, 901, 1978.

	$-\log_{10} K_w$	
t/°C	H_2O	D_2O
0	14.947	15.972
5	14.734	15.743
10	14.534	15.527
15	14.344	15.324
20	14.165	15.132
25	13.995	14.951
30	13.833	14.779
35	13.680	14.616
40	13.535	14.462
45	13.396	14.316
50	13.265	14.176
55	13.140	14.044
60	13.020	13.918
65	12.907	13.798
70	12.799	13.683
75	12.696	13.574
80	12.598	13.470
85	12.505	13.371
90	12.417	13.276
95	12.332	13.186
100	12.252	13.099

SOLUBILITY OF SELECTED GASES IN WATER

L. H. Gevantman

The values in this table are taken almost exclusively from the International Union of Pure and Applied Chemistry "Solubility Data Series." Unless noted, they comprise evaluated data fitted to a smoothing equation. The data at each temperature are then derived from the smoothing equation which expresses the mole fraction solubility X_1 of the gas in solution as:

$$\ln X_1 = A + B/T^* + C \ln T^*$$

where

$$T^* = T/100\ \mathrm{K}$$

All values refer to a partial pressure of the gas of 101.325 kPa (one atmosphere).

The equation constants, the standard deviation for $\ln X_1$ (except where noted), and the temperature range over which the equation applies are given in the column headed Equation constants. There are two exceptions. The equation for methane has an added term, DT^*. The equation for H_2Se and H_2S takes the form,

$$\ln X_1 = A + B/T + C \ln T + DT$$

where T is the temperature in kelvin.

Solubilities given for those gases that react with water, namely ozone, nitrogen oxides, chlorine and its oxides, carbon dioxide, hydrogen sulfide, hydrogen selenide and sulfur dioxide, are recorded as bulk solubilities; i.e., all chemical species of the gas and its reaction products with water are included.

Gas	T/K	Solubility (X_1)	Equation constants	Ref.
Hydrogen (H_2)	288.15	1.510×10^{-5}	$A = -48.1611$	1
$M_r = 2.01588$	293.15	1.455×10^{-5}	$B = 55.2845$	
	298.15	1.411×10^{-5}	$C = 16.8893$	
	303.15	1.377×10^{-5}	Std. dev. = ± 0.54%	
	308.15	1.350×10^{-5}	Temp. range = 273.15—353.15	
Deuterium (D_2)	283.15	$1.675 \times 10^{-5} \pm 0.57\%$	Averaged experimental values	1
$M_r = 4.0282$	288.15	$1.595 \times 10^{-5} \pm 0.57\%$		
	293.15	$1.512 \times 10^{-5} \pm 0.78\%$	Temp. range = 278.15—303.15	
	298.15	$1.460 \times 10^{-5} \pm 0.52\%$		
	303.15	$1.395 \times 10^{-5} \pm 0.37\%$		
Helium (He)	288.15	7.123×10^{-6}	$A = -41.4611$	2
$A_r = 4.0026$	293.15	7.044×10^{-6}	$B = 42.5962$	
	298.15	6.997×10^{-6}	$C = 14.0094$	
	303.15	6.978×10^{-6}	Std. dev. = ±0.54%	
	308.15	6.987×10^{-6}	Temp. range = 273.15—348.15	
Neon (Ne)	288.15	8.702×10^{-6}	$A = -52.8573$	2
$A_r = 20.1797$	293.15	8.395×10^{-6}	$B = 61.0494$	
	298.15	8.152×10^{-6}	$C = 18.9157$	
	303.15	7.966×10^{-6}	Std. dev. = ±0.47%	
	308.15	7.829×10^{-6}	Temp. range = 273.15—348.15	
Argon (Ar)	288.15	3.025×10^{-5}	$A = -57.6661$	3
$A_r = 39.948$	293.15	2.748×10^{-5}	$B = 74.7627$	
	298.15	2.519×10^{-5}	$C = 20.1398$	
	303.15	2.328×10^{-5}	Std. dev. = ±0.26%	
	308.15	2.169×10^{-5}	Temp. range = 273.15—348.15	
Krypton (Kr)	288.15	5.696×10^{-5}	$A = -66.9928$	4
$A_r = 83.80$	293.15	5.041×10^{-5}	$B = 91.0166$	
	298.15	4.512×10^{-5}	$C = 24.2207$	
	303.15	4.079×10^{-5}	Std. dev. = ±0.32%	
	308.15	3.725×10^{-5}	Temp. range = 273.15—353.15	
Xenon (Xe)	288.15	10.519×10^{-5}	$A = -74.7398$	4
$A_r = 131.29$	293.15	9.051×10^{-5}	$B = 105.210$	
	298.15	7.890×10^{-5}	$C = 27.4664$	
	303.15	6.961×10^{-5}	Std. dev. = ±0.35%	
	308.15	6.212×10^{-5}	Temp. range = 273.15—348.15	

Gas	T/K	Solubility (X_1)	Equation constants	Ref.
Radon-222(^{222}Rn)	288.15	2.299×10^{-4}	$A = -90.5481$	
A_r = 222	293.15	1.945×10^{-4}	$B = 130.026$	
	298.15	1.671×10^{-4}	$C = 35.0047$	
	303.15	1.457×10^{-4}	Std. dev. = ±1.02%	
	308.15	1.288×10^{-4}	Temp. range = 273.15—373.15	
Oxygen (O_2)	288.15	2.756×10^{-5}	$A = -66.7354$	5
M_r = 31.9988	293.15	2.501×10^{-5}	$B = 87.4755$	
	298.15	2.293×10^{-5}	$C = 24.4526$	
	303.15	2.122×10^{-5}	Std. dev. = ±0.36%	
	308.15	1.982×10^{-5}	Temp. range = 273.15—348.15	
Ozone (O_3)	293.15	1.885×10^{-6} ± 10%	Experimental value derived	5
M_r = 47.9982		pH = 7.0	from Henry's Law Constant Equation	
Nitrogen (N_2)	288.15	1.386×10^{-5}	$A = -67.3877$	6
M_r = 28.0134	293.15	1.274×10^{-5}	$B = 86.3213$	
	298.15	1.183×10^{-5}	$C = 24.7981$	
	303.15	1.108×10^{-5}	Std. dev. = ±0.72%	
	308.15	1.047×10^{-5}	Temp. range = 273.15—348.15	
Nitrous oxide (N_2O)	288.15	5.948×10^{-4}	$A = -60.7467$	7
M_r = 44.0129	293.15	5.068×10^{-4}	$B = 88.8280$	
	298.15	4.367×10^{-4}	$C = 21.2531$	
	303.15	3.805×10^{-4}	Std. dev. = ±1.2%	
	308.15	3.348×10^{-4}	Temp. range = 273.15—313.15	
Nitric oxide (NO)	288.15	4.163×10^{-5}	$A = -62.8086$	7
M_r = 30.0061	293.15	3.786×10^{-5}	$B = 82.3420$	
	298.15	3.477×10^{-5}	$C = 22.8155$	
	303.15	3.222×10^{-5}	Std. dev. = ±0.76%	
	308.15	3.012×10^{-5}	Temp. range = 273.15—358.15	
Carbon monoxide (CO)	288.15	2.095×10^{-5}	Derived from Henry's	8
M_r = 28.0104	293.15	1.918×10^{-5}	Law Constant Equation	
	298.15	1.774×10^{-5}	Std. dev. = ±0.043%	
	303.15	1.657×10^{-5}	Temp. range = 273.15—328.15	
	308.15	1.562×10^{-5}		
Carbon dioxide (CO_2)	288.15	8.21×10^{-4}	Derived from Henry's	9
M_r = 44.0098	293.15	7.07×10^{-4}	Law Constant Equation	
	298.15	6.15×10^{-4}	Std. dev. = ±1.1%	
	303.15	5.41×10^{-4}	Temp. range = 273.15—353.15	
	308.15	4.80×10^{-4}		
Hydrogen selenide (H_2Se)	288.15	1.80×10^{-3}	$A = 9.15$	10
M_r = 80.976	298.15	1.49×10^{-3}	$B = 974$	
	308.15	1.24×10^{-3}	$C = -3.542$	
			$D = 0.0042$	
			Std. dev. = $\pm 2.3 \times 10^{-5}$	
			Temp. range = 288.15—343.15	
Hydrogen sulfide (H_2S)	288.15	2.335×10^{-3}	$A = -24.912$	10
M_r = 34.082	293.15	2.075×10^{-3}	$B = 3477$	
	298.15	1.85×10^{-3}	$C = 0.3993$	
	303.15	1.66×10^{-3}	$D = 0.0157$	
	308.15	1.51×10^{-3}	Std. dev. = $\pm 6.5 \times 10^{-5}$	
			Temp. range = 283.15—603.15	
Sulfur dioxide (SO_2)	288.15	3.45×10^{-2}	$A = -25.2629$	11
M_r = 64.0648	293.15	2.90×10^{-2}	$B = 45.7552$	

Gas	T/K	Solubility (X_1)	Equation constants	Ref.
	298.15	2.46×10^{-2}	C = 5.6855	
	303.15	2.10×10^{-2}	Std. dev. = ±1.8%	
	308.15	1.80×10^{-2}	Temp. range = 278.15—328.15	
Chlorine (Cl_2)	283.15	2.48×10^{-3} ± 2%	Experimental data	11
M_r = 70.9054	293.15	1.88×10^{-3} ± 2%	Temp. range = 283.15—333.15	
	303.15	1.50×10^{-3} ± 2%		
	313.15	1.23×10^{-3} ± 2%		
Chlorine monoxide (Cl_2O)	273.15	5.25×10^{-1} ± 1%	Experimental data	11
M_r = 86.9048	276.61	4.54×10^{-1} ± 1%	Temp. range = 273.15—293.15	
	283.15	4.273×10^{-1} ± 1%		
	293.15	3.353×10^{-1} ± 1%		
Chlorine dioxide (ClO_2)	288.15	2.67×10^{-2}	A = 7.9163	11
M_r = 67.4515	293.15	2.20×10^{-2}	B = 0.4791	
	298.15	1.823×10^{-2}	C = 11.0593	
	303.15	1.513×10^{-2}	Std. dev. = ±4.6%	
	308.15	1.259×10^{-2}	Temp. range = 283.15—333.15	
Methane (CH_4)	288.15	3.122×10^{-5}	A = −115.6477	12
M_r = 16.0428	293.15	2.806×10^{-5}	B = 155.5756	
	298.15	2.552×10^{-5}	C = 65.2553	
	303.15	2.346×10^{-5}	D = −6.1698	
	308.15	2.180×10^{-5}	Std. dev. = ±0.056%	
			Temp. range = 273.15—328.15	
Ethane (C_2H_6)	288.15	4.556×10^{-5}	A = −90.8225	13
M_r = 30.0696	293.15	3.907×10^{-5}	B = 126.9559	
	298.15	3.401×10^{-5}	C = 34.7413	
	303.15	3.002×10^{-5}	Std. dev. = ±0.13%	
	308.15	2.686×10^{-5}	Temp. range = 273.15—323.15	
Propane (C_3H_8)	288.15	3.813×10^{-5}	A = −102.044	14
M_r = 44.097	293.15	3.200×10^{-5}	B = 144.345	
	298.15	2.732×10^{-5}	C = 39.4740	
	303.15	2.370×10^{-5}	Std. dev. = ±0.012%	
	308.15	2.088×10^{-5}	Temp. range = 273.15—347.15	
Butane (C_4H_{10})	288.15	3.274×10^{-5}	A = −102.029	14
M_r = 58.123	293.15	2.687×10^{-5}	B = 146.040	
	298.15	2.244×10^{-5}	C = 38.7599	
	303.15	1.906×10^{-5}	Std. dev. = ±0.026%	
	308.15	1.645×10^{-5}	Temp. range = 273.15—349.15	
2-Methyl propane (Isobutane)	288.15	2.333×10^{-5}	A = −129.714	14
(C_4H_{10})	293.15	1.947×10^{-5}	B = 183.044	
M_r = 58.123	298.15	1.659×10^{-5}	C = 53.4651	
	303.15	1.443×10^{-5}	Std. dev. = ±0.034%	
	308.15	1.278×10^{-5}	Temp. range = 278.15—318.15	

References

1. C. L. Young, Ed., *IUPAC Solubility Data Series*, Vol. 5/6, Hydrogen and Deuterium, Pergamon Press, Oxford, England, 1981.
2. H. L. Clever, Ed., *IUPAC Solubility Data Series*, Vol. 1, Helium and Neon, Pergamon Press, Oxford, England, 1979.
3. H. L. Clever, Ed., *IUPAC Solubility Data Series*, Vol. 4, Argon, Pergamon Press, Oxford, England, 1980.
4. H. L. Clever, Ed., *IUPAC Solubility Data Series*, Vol. 2, Krypton, Xenon and Radon, Pergamon Press, Oxford, England, 1979.
5. R. Battino, Ed., *IUPAC Solubility Data Series*, Vol. 7, Oxygen and Ozone, Pergamon Press, Oxford, England, 1981.
6. R. Battino, Ed., *IUPAC Solubility Data Series*, Vol. 10, Nitrogen and Air, Pergamon Press, Oxford, England, 1982.
7. C. L. Young, Ed., *IUPAC Solubility Data Series*, Vol. 8, Oxides of Nitrogen, Pergamon Press, Oxford, England, 1981.
8. R. W. Cargill, Ed., *IUPAC Solubility Data Series*, Vol. 43, Carbon Monoxide, Pergamon Press, Oxford, England, 1990.
9. R. Crovetto, Evaluation of Solubility Data for the System CO2-H2O, *J. Phys. Chem. Ref. Data*, 20, 575, 1991.

10. P. G. T. Fogg and C. L. Young, Eds., *IUPAC Solubility Data Series*, Vol. 32, Hydrogen Sulfide, Deuterium Sulfide, and Hydrogen Selenide, Pergamon Press, Oxford, England, 1988.
11. C. L. Young, Ed., *IUPAC Solubility Data Series*, Vol. 12, Sulfur Dioxide, Chlorine, Fluorine and Chlorine Oxides, Pergamon Press, Oxford, England, 1983.
12. H. L. Clever and C. L. Young, Eds., *IUPAC Solubility Data Series*, Vol. 27/28, Methane, Pergamon Press, Oxford, England, 1987.
13. W. Hayduk, Ed., *IUPAC Solubility Data Series*, Vol. 9, Ethane, Pergamon Press, Oxford, England, 1982.
14. W. Hayduk, Ed., *IUPAC Solubility Data Series*, Vol. 24, Propane, Butane and 2-Methylpropane, Pergamon Press, Oxford, England, 1986.

SOLUBILITY OF CARBON DIOXIDE IN WATER AT VARIOUS TEMPERATURES AND PRESSURES

The solubility of CO_2 in water, expressed as mole fraction of CO_2 in the liquid phase, is given for pressures up to atmospheric and temperatures of 0 to 100 °C. Note that 1 standard atmosphere equals 101.325 kPa. The references give data over a wider range of temperature and pressure. The estimated uncertainty is about 2%.

References

1. Carroll, J. J., Slupsky, J. D., and Mather, A. E., *J. Phys. Chem. Ref. Data*, 20, 1201, 1991.
2. Fernandez-Prini, R. and Crovetto, R., *J. Phys. Chem. Ref. Data*, 18, 1231, 1989.
3. Crovetto, R., *J. Phys. Chem. Ref. Data*, 20, 575,1991.

	1000 × mole fraction of CO_2 in liquid phase						
	Partial pressure of CO_2 in kPa						
t/°C	5	10	20	30	40	50	100
0	0.067	0.135	0.269	0.404	0.538	0.671	1.337
5	0.056	0.113	0.226	0.338	0.451	0.564	1.123
10	0.048	0.096	0.191	0.287	0.382	0.477	0.950
15	0.041	0.082	0.164	0.245	0.327	0.409	0.814
20	0.035	0.071	0.141	0.212	0.283	0.353	0.704
25	0.031	0.062	0.123	0.185	0.247	0.308	0.614
30	0.027	0.054	0.109	0.163	0.218	0.271	0.541
35	0.024	0.048	0.097	0.145	0.193	0.242	0.481
40	0.022	0.043	0.087	0.130	0.173	0.216	0.431
45	0.020	0.039	0.078	0.117	0.156	0.196	0.389
50	0.018	0.036	0.071	0.107	0.142	0.178	0.354
55	0.016	0.033	0.065	0.098	0.131	0.163	0.325
60	0.015	0.030	0.060	0.090	0.121	0.150	0.300
65	0.014	0.028	0.056	0.084	0.112	0.140	0.279
70	0.013	0.026	0.052	0.079	0.105	0.131	0.261
75	0.012	0.025	0.049	0.074	0.099	0.123	0.245
80	0.012	0.023	0.047	0.070	0.093	0.116	0.232
85	0.011	0.022	0.044	0.067	0.089	0.111	0.221
90	0.011	0.021	0.042	0.064	0.085	0.106	0.211
95	0.010	0.020	0.041	0.061	0.082	0.102	0.203
100	0.010	0.020	0.039	0.059	0.079	0.098	0.196

AQUEOUS SOLUBILITY AND HENRY'S LAW CONSTANTS OF ORGANIC COMPOUNDS

The solubility in water of about 1300 organic compounds, including many compounds of environmental interest, is tabulated here. When data are available, values are given at several temperatures between 0 °C and 100 °C. Solids, liquids, and gases are included; additional data on gases can be found in the table "Solubility of Selected Gases in Water" in Section 8.

Solubility of solids is defined as the concentration of the compound in a solution that is in equilibrium with the solid phase at the specified temperature and one atmosphere pressure. For liquids whose water mixtures separate into two phases, the solubility given here is the concentration of the specified compound in the water-rich phase at equilibrium. In the case of gases (i.e., compounds whose vapor pressure at the specified temperature exceeds one atmosphere) the solubility is defined here as the concentration in the water phase when the partial pressure of the compound above the solution is 101.325 kPa (1 atm). Values for gases are marked with an asterisk.

The solubility values in this table are expressed as mass percent of solute, $s = 100w_2$, where the mass fraction w_2 is defined as

$$w_2 = m_2/(m_1 + m_2)$$

where m_2 is the mass of solute and m_1 the mass of water. For convenience, the solubility expressed in grams of solute that will dissolve in 1 kilogram of water is tabulated in the adjacent column to mass percent. For compounds with low solubility (e.g., $s < 1\%$), that column is, to a high approximation, numerically identical to the solubility expressed in grams of solute per liter of solution.

The mass fraction w_2 is related to other common measures of solubility as follows:

Molality:	$m_2 = 1000\, w_2/M_2(1 - w_2)$
Molarity:	$c_2 = 1000\rho w_2/M_2$
Mole fraction:	$x_2 = (w_2/M_2)/\{(w_2/M_2) + (1 - w_2)/M_1\}$
Mass of solute per 100 g of H_2O:	$100w_2/(1 - w_2)$
Mass of solute per liter of solution:	$1000\rho w_2$

Here, M_2 is the molar mass of the solute, $M_1 = 18.015$ g/mol is the molar mass of water, and ρ is the density of the solution in g/mL.

Data have been selected from evaluated sources wherever possible, in particular the *IUPAC Solubility Data Series*. Many values come from experimental measurements reported in the *Journal of Chemical and Engineering Data* and the *Journal of Chemical Thermodynamics*, as well as critical review papers in the *Journal of Physical and Chemical Reference Data*. The primary source for each value is listed in the column following the solubility values; additional references of interest are sometimes given. Many of the references contain solubility data at other temperatures and pH values and in the presence of other compounds. The user is cautioned that wide variations of data are found in the literature for the lower solubility compounds. The references should be consulted for more information on these compounds.

The table also contains values of the Henry's Law constant k_H, which provides a measure of the partition of a substance between the atmosphere and the aqueous phase. Here, k_H is defined as the limit of p_2/c_2 as the concentration approaches zero, where p_2 is the partial pressure of the solute above the solution and c_2 is the concentration in the solution at equilibrium (other formulations of Henry's Law are often used; see Reference 5). The values of k_H listed here are based on direct experimental measurement whenever available, but many of them are simply calculated as the ratio of the pure compound vapor pressure to the solubility. This approximation is reliable only for compounds of very low solubility. In fact, values of k_H found in the literature frequently differ by a factor of two or three, and variations over an order of magnitude are not unusual (Reference 5). Therefore, the data given here should be taken only as a rough indication of the true Henry's Law constant, which is difficult to measure precisely.

All values of k_H refer to 25 °C. If the vapor pressure of the compound at 25 °C is greater than one atmosphere, it can be assumed that the k_H value has been calculated as $101.325/c_2$. The source of the Henry's Law data is given in the last column. The air-water partition coefficient (i.e., ratio of air concentration to water concentration when both are expressed in the same units) is equal to k_H/RT or $k_H/2.48$ in the units used here.

Compounds are listed by systematic name. To locate a compound by molecular formula or CAS Registry Number, use the indexes to the table "Physical Constants of Organic Compounds" in Section 3, which point to the entry in that table from which the name can be determined.

References

1. *Solubility Data Series, International Union of Pure and Applied Chemistry, Vol. 15*, Pergamon Press, Oxford, 1982.
2. *Solubility Data Series, International Union of Pure and Applied Chemistry, Vol. 20*, Pergamon Press, Oxford, 1985.
3. *Solubility Data Series, International Union of Pure and Applied Chemistry, Vol. 37*, Pergamon Press, Oxford, 1988.
4. *Solubility Data Series, International Union of Pure and Applied Chemistry, Vol. 38*, Pergamon Press, Oxford, 1988.
5. Mackay, D., and Shiu, W. Y., *J. Phys. Chem. Ref. Data*, 10, 1175, 1981.
6. Pearlman, R. S., and Yalkowsky, S. H., *J. Phys. Chem. Ref. Data*, 13, 975, 1984.
7. Shiu, W. Y., and Mackay, D., *J. Phys. Chem. Ref. Data*, 15, 911, 1986.
8. Varhanickova, D., Lee, S. C., Shiu, W. Y., and Mackay, D., *J. Chem. Eng. Data*, 40, 620, 1995.
9. Miller, M. M., Ghodbane, S., Wasik, S. P., Tewari, Y. B., and Martire, D. E., *J. Chem. Eng. Data*, 29, 184, 1984.
10. Riddick, J. A., Bunger, W. B., and Sakano, T. K., *Organic Solvents, Fourth Edition*, John Wiley & Sons, New York, 1986.
11. Mackay, D., Shiu, W. Y., and Ma, K. C., *Illustrated Handbook of Physical-Chemical Properties and Environmental Fate for Organic Chemicals, Vol. I*, Lewis Publishers/CRC Press, Boca Raton, FL, 1992.
12. Mackay, D., Shiu, W. Y., and Ma, K. C., *Illustrated Handbook of Physical-Chemical Properties and Environmental Fate for Organic Chemicals, Vol. II*, Lewis Publishers/CRC Press, Boca Raton, FL, 1992.
13. Mackay, D., Shiu, W. Y., and Ma, K. C., *Illustrated Handbook of Physical-Chemical Properties and Environmental Fate for Organic Chemicals, Vol. III*, Lewis Publishers/CRC Press, Boca Raton, FL, 1993.
14. Horvath, A. L., *Halogenated Hydrocarbons*, Marcel Dekker, New York, 1982.
15. Howard, P. H., *Handbook of Environmental Fate and Exposure Data for Organic Chemicals, Vol. I*, Lewis Publishers/CRC Press, Boca Raton, FL, 1989.

* Indicates a value of *s* for a gas at a partial pressure of 101.325 kPa (1 atm) in equilibrium with the solution.

16. Howard, P. H., *Handbook of Environmental Fate and Exposure Data for Organic Chemicals, Vol. II*, Lewis Publishers/CRC Press, Boca Raton, FL, 1990.
17. Banergee, S., Yalkowsky, S. H., and Valvani, S. C., *Environ. Sci. Technol.*, 14, 1227, 1980.
18. Gevantman, L. H., in *CRC Handbook of Chemistry and Physics, 90th Edition*, p. 8–80, CRC Press, Boca Raton, FL, 2009.
19. Wilhelm, E., Battino, R., and Wilcock, R. J., *Chem. Rev.* 77, 219, 1977.
20. Stephenson, R. M., *J. Chem. Eng. Data*, 37, 80, 1992.
21. Stephenson, R. M., Stuart, J., and Tabak, M., *J. Chem. Eng. Data*, 29, 287, 1984.
22. Shiu, W.-Y., and Ma, K.-C, *J. Phys. Chem. Ref. Data*, 29, 41, 2000.
23. Lun, R., Varhanickova, D., Shiu, W.-Y., and Mackay, D., *J. Chem. Eng. Data*, 42, 951 (1997).
24. Huang, G.-L., Xiao, H., Chi, J., Shiu, W.-Y., and Mackay, D., *J. Chem. Eng. Data*, 45, 411, 2000.
25. Horvath, A. L., Getzen, F. W., and Maczynska, Z., *J. Phys. Chem. Ref. Data*, 28, 395, 2000 [IUPAC No. 67].
26. Dawson, R. M. C., Elliott, D. C., Elliott, W. H., and Jones, K. M., *Data for Biochemical Research*, Third Edition, Clarendon Press, Oxford, 1986.
27. Stephen, H., and Stephen, T., *Solubilities of Organic and Inorganic Compounds*, MacMillan, New York, 1963.
28. Shiu, W.-Y., and Mackay, D., *J. Chem. Eng. Data* 42, 27, 1997.
29. Hinz, H.-J., ed., *Thermodynamic Data for Biochemistry and Biotechnology*, Springer-Verlag, Berlin, 1986.
30. Budavari, S., ed., *The Merck Index, Twelfth Edition*, Merck & Co., Rahway, NJ, 1996.
31. Bamford, H. A., Poster, D. L., and Baker, J. E., *J. Chem. Eng. Data*, 45, 1069, 2000.
32. Lide, D. R., and Milne, G. W. A., *Handbook of Data on Organic Compounds, Third Edition*, CRC Press, Boca Raton, FL, 1994.
33. Apelblat, A., and Manzurola, E., *J. Chem. Thermodynamics* 21, 1005, 1989.
34. Apelblat, A., and Manzurola, E., *J. Chem. Thermodynamics* 22, 289, 1990.
35. Horvath, A. L., and Getzen, F. W., *J. Phys. Chem. Ref. Data* 28, 649, 1999 [IUPAC No. 68].
36. Sazonov, V. P., Marsh, K. N., and Hefter, G. T., *J. Phys. Chem. Ref. Data* 29, 1165, 2000 [IUPAC No. 71].
37. Verbruggen, E. M. J., Hermens, J. L. M., and Tolls, J., *J. Phys. Chem. Ref. Data* 29, 1435, 2000.
38. Sazonov, V. P., Shaw, D. G., and Marsh, K. N., *J. Phys. Chem. Ref. Data* 31, 1, 2002 [IUPAC No. 77].
39. Sazonov, V. P., and Shaw, D. G., *J. Phys. Chem. Ref. Data* 31, 989, 2002 [IUPAC No. 78].
40. Yalkowsky, S. H., and He, Y., *Handbook of Aqueous Solubility Data*, CRC Press, Boca Raton, FL, 2003.
41. Shiu, W.-Y., and Ma, K.-C., *J. Phys. Chem. Ref. Data* 29, 387, 2000.
42. Shaw, D. G., and Maczynski, A., *J. Phys. Chem. Ref. Data* 35, 687, 2006 [IUPAC No. 81, Part 11].
43. Nordstrom, F. L., and Rasmuson, A. C., *J. Chem. Eng. Data* 51, 1668, 2006.
44. Nordstrom, F. L., and Rasmuson, A. C., *J. Chem. Eng. Data* 51, 1775, 2006.
45. Sapoundjiev, D., Lorenz, H,. and Seidel-Morgenstern, A., *J. Chem. Eng. Data* 51, 1562, 2006.
46. Marche, C., Ferronato, C., and Jose, J., *J. Chem. Eng. Data* 48, 967, 2003.
47. Lu, J., Wang, X., Yang, X., and Ching, C., *J. Chem. Eng. Data* 51, 1593, 2006.
48. Achard, C., Jaoui, M., Schwing, M., and Rogalski, M., *J. Chem. Eng. Data* 41, 504, 1996.
49. Shareef, A., et al., *J. Chem. Eng. Data* 51, 879, 2006.
50. Clever, H. L., et al., *J. Phys. Chem. Ref. Data* 34, 201, 2005 [IUPAC No. 80].
51. Jaoui, M., Achard, C., and Rogalski, M., *J. Chem. Eng. Data* 47, 297, 2002.
52. Fichan, I., Larroche, C., and Gros, J. B., *J. Chem. Eng. Data* 44, 56, 1999.
53. Freire, M. G., et al., *J. Chem. Eng. Data* 50, 237, 2005.
54. Domanska, U., and Kozlowska, M. K., *J. Chem. Eng. Data* 47, 456, 2002.
55. Phelan, J. M., and Barnett, J. L., *J. Chem. Eng. Data* 46, 375, 2001.
56. Long, B-W., Wang, L-S., and Wu, J-S., *J. Chem. Eng. Data* 50, 136, 2005.
57. Marche, C., Ferronato, C., and Jose, J., *J. Chem. Eng. Data* 49, 937, 2004.
58. Oleszek-Kudlak, S., Shibata, E., and Nakamura, T., *J. Chem. Eng. Data* 49, 570, 2004.
59. Lynch, J. C., et al., *J. Chem. Eng. Data* 46, 1549, 2001.
60. Xiao, H., Li, N., and Wania, F., *J. Chem. Eng. Data* 49, 173, 2004.
61. Ma, J. H. Y., Hung, H., Shiu, W-Y., and Mackay, D., *J. Chem. Eng. Data* 46, 619, 2001.
62. Carta, R., and Tola, G., *J. Chem. Eng. Data* 41, 414, 1996; 44, 563, 1999.
63. Kao, H. D., et al., *Pharm. Res.* 17, 978, 2000.
64. Heric, E. L., and Langford, R. E., *J. Chem. Eng. Data* 17, 471, 1972.
65. Marche, C., Delépine, H., Ferronato, C., and Jose, J., *J. Chem. Eng. Data* 48, 398, 2003.
66. Wang, L-C, and Wang, F-A, *J. Chem. Eng. Data* 49, 155, 2004.
67. Shen, L, and Wania, F., *J. Chem. Eng. Data* 50, 742, 2005.
68. Oleszek-Kudlak, S., Shibata, E., and Nakamura, T., *J. Chem. Eng. Data* 52, 1824, 2007.
69. Zhao, H-K., Li, R-R, Ji, H-Z, Zhang, D-S., Tang, C., and Yang, L-Q., *J. Chem. Eng. Data* 52, 2072, 2007.
70. Yang, X., Wang, X., and Ching, C. B., *J. Chem. Eng. Data* 53, 1133, 2008.
71. Liu, L., and Chen, J., *J. Chem. Eng. Data* 53, 1649, 2008.
72. Szterner, P., *J. Chem. Eng. Data* 53, 1738, 2008.
73. Kong, M-Z., Shi, X-H, Cao, Y-C., and Zhou, C-R., *J. Chem. Eng. Data* 53, 615, 2008.
74. Daneshfar, A., Ghaziaskar, H. S., and Homayoun, N., *J. Chem. Eng. Data* 53, 776, 2008.
75. Manzurola, E., and Apelblat, A., *J. Chem. Thermodynamics* 34, 1127, 2002.
76. Apelblat, A., Manzurola, E., and Balal, N. A., *J. Chem. Thermodynamics* 38, 565, 2006.
77. Apelblat, A., and Mishelevich, A., *J. Chem. Thermodynamics* 40, 897, 2008.
78. Góral, M., Wiśniewska-Goclowska, B., and Mączyński, A., *J. Phys. Chem. Ref. Data* 35, 1391, 2006.
79. Mączyński, A., Shaw, D. G., Góral, M., and Wiśniewska-Goclowska, B., *J. Phys. Chem. Ref. Data* 37, 1119, 2008.
80. Mączyński, A., Shaw, D. G., Góral, M., and Wiśniewska-Goclowska, B., *J. Phys. Chem. Ref. Data* 37, 1147, 2008.
81. Mączyński, A., Shaw, D. G., Góral, M., and Wiśniewska-Goclowska, B., *J. Phys. Chem. Ref. Data* 37, 1169, 2008.
82. Mączyński, A., Shaw, D. G., Góral, M., and Wiśniewska-Goclowska, B., *J. Phys. Chem. Ref. Data* 37, 1517, 2008.
83. Mączyński, A., Shaw, D. G., Góral, M., and Wiśniewska-Goclowska, B., *J. Phys. Chem. Ref. Data* 37, 1575, 2008.
84. Mączyński, A., Shaw, D. G., Góral, M., and Wiśniewska-Goclowska, B., *J. Phys. Chem. Ref. Data* 37, 1611, 2008.
85. Luning Prak, D. J., and O'Sullivam, D. W., *J. Chem. Eng. Data* 51, 448, 2006.
86. Shiu, W.-Y., Wania, F., Hung, H., and Mackay, D., *J. Chem. Eng. Data* 42, 293, 1997.
87. Mączyński, A., and Shaw, D. G., *J. Phys. Chem. Ref. Data* 36, 59, 2007.
88. Mączyński, A., and Shaw, D. G., *J. Phys. Chem. Ref. Data* 36, 133, 2007.
89. Góral, M., Wiśniewska-Goclowska, B., and Mączyński, A., *J. Phys. Chem. Ref. Data* 33, 1159, 2004.
90. Heric, E. L., and Langford, R. E., *J. Chem. Eng. Data* 17, 209, 1972.

Name	Mol. Form.	Mol. Wt.	t/°C	Solubility, s: 100 w_2 (mass%)	Solubility, s: g per kg H_2O	Ref.	Henry Const., k_H: kPa m³mol⁻¹	Ref.
Acenaphthene	$C_{12}H_{10}$	154.207	0	0.00015	0.0015	4		
			25	0.000380	0.00380	22	0.01217	22
			50	0.00092	0.0092	4		
Acenaphthylene	$C_{12}H_8$	152.192	20	0.0016	0.016	28	0.012	28
Acephate	$C_4H_{10}NO_3PS$	183.166	20	≈28	≈390	40		
Acetamide	C_2H_5NO	59.067	20	40.8	689	10		
Acetanilide	C_8H_9NO	135.163	20	0.52	5.2	27		
			70	2.7	28	27		
Acetazolamide	$C_4H_6N_4O_3S_2$	222.246	30	0.10	1.0	40		
Acetohexamide	$C_{15}H_{20}N_2O_4S$	324.396	37	0.0013	0.013	40		
Acetonitrile	C_2H_3N	41.052	−3	40.5	681	39		
			−10	31.7	464	39		
Acetophenone	C_8H_8O	120.149	20	0.67	6.7	84	0.00108	28
			50	0.81	8.2	84	0.00108	28
			80	1.16	11.7	84	0.00108	28
Acetylene	C_2H_2	26.037	25	0.108*	1.08*	19		
2-(Acetyloxy)benzoic acid	$C_9H_8O_4$	180.158		0.25	2.5	27		
2-(Acetyloxy)-5-bromobenzoic acid	$C_9H_7BrO_4$	259.054		0.07	0.7	30		
Acridine	$C_{13}H_9N$	179.217	25	0.00466	0.0466	6		
Acrolein	C_3H_4O	56.063	20	20.8	263	10		
Acrylamide	C_3H_5NO	71.078	20	≈27	≈370	40		
Acrylonitrile	C_3H_3N	53.063	20	7.35	79.3	10		
Adenine	$C_5H_5N_5$	135.128	25	0.104	1.04	29		
Adenosine	$C_{10}H_{13}N_5O_4$	267.242	25	0.51	5.1	29		
Alachlor	$C_{14}H_{20}ClNO_2$	269.768	23	0.024	0.24	40		
L-Alanine	$C_3H_7NO_2$	89.094	25	14.30	167	26		
β-Alanine	$C_3H_7NO_2$	89.094	25	47.1	890	26		
Aldicarb	$C_7H_{14}N_2O_2S$	190.263	20	0.60	6.0	40		
Aldrin	$C_{12}H_8Cl_6$	364.910	25	0.00002	0.0002	67		
Allopurinol	$C_5H_4N_4O$	136.112	25	0.057	0.57	40		
Ametryn	$C_9H_{17}N_5S$	227.330	20	0.0190	0.190	40		
2-Amino-9,10-anthracenedione	$C_{14}H_9NO_2$	223.227	25	0.000016	0.00016	40		
4-Aminoazobenzene	$C_{12}H_{11}N_3$	197.235	25	0.0030	0.030	40		
			97	0.068	0.68	40		
4-Aminobenzenesulfonamide	$C_6H_8N_2O_2S$	172.205	20	0.71	7.2	40		
4-Aminobenzenesulfonic acid	$C_6H_7NO_3S$	173.190	7	0.59	5.9	27		
DL-2-Aminobutanoic acid	$C_4H_9NO_2$	103.120	25	17.4	211	26		
DL-3-Aminobutanoic acid	$C_4H_9NO_2$	103.120	25	55.6	1250	26		
4-Amino-*N*-[(butylamino)carbonyl] benzenesulfonamide	$C_{11}H_{17}N_3O_3S$	271.336	37	0.053	0.53	40		
3-Amino-2,5-dichlorobenzoic acid	$C_7H_5Cl_2NO_2$	206.027	25	0.070	0.70	40		
6-Amino-1,3-dihydro-2*H*-purin-2-one	$C_5H_5N_5O$	151.127	25	0.006	0.06	26		
4-(2-Aminoethyl)phenol	$C_8H_{11}NO$	137.179	15	1.03	10.4	40		
6-Aminohexanoic acid	$C_6H_{13}NO_2$	131.173	25	46	852	29		
4-Amino-2-hydroxybenzoic acid	$C_7H_7NO_3$	153.136	20	0.20	2.0	40		
2-Amino-2-methylpropanoic acid	$C_4H_9NO_2$	103.120	25	12.1	138	26		
4-Amino-5-methyl-2(1*H*)-pyrimidinone	$C_5H_7N_3O$	125.129	25	0.45	4.5	26		
2-Aminophenol	C_6H_7NO	109.126	20	1.92	19.6	40		
3-Aminophenol	C_6H_7NO	109.126	20	2.56	26.3	40		

Name	Mol. Form.	Mol. Wt.	t/°C	Solubility, s: 100 w_2 (mass%)	Solubility, s: g per kg H_2O	Ref.	Henry Const., k_H: kPa m^3mol^{-1}	Ref.
			70	≈24	≈320	40		
4-Aminophenol	C_6H_7NO	109.126	20	1.55	15.7	40		
Aminopyrine	$C_{13}H_{17}N_3O$	231.293	25	4.8	50	40		
Amitriptyline	$C_{20}H_{23}N$	277.404	24	0.00097	0.0097	40		
Amobarbital	$C_{11}H_{18}N_2O_3$	226.272	25	0.06	0.6	40		
Anilazine	$C_9H_5Cl_3N_4$	275.522	20	0.001	0.01	40		
Aniline	C_6H_7N	93.127	25	3.38	35.0	10	14	15
Aniline-2-carboxylic acid	$C_7H_7NO_2$	137.137	20	0.349	3.49	40		
Aniline-4-carboxylic acid	$C_7H_7NO_2$	137.137	25	0.54	5.4	40		
Aniline hydrochloride	C_6H_8ClN	129.588	15	15.1	178	27		
Anisole	C_7H_8O	108.138	20	0.203	2.03	20	0.025	13
			40	0.184	1.84	20	0.025	13
			81	0.294	2.95	20	0.025	13
Anthracene	$C_{14}H_{10}$	178.229	0	0.0000022	0.000022	42,4		
			25	0.0000044	0.000044	42,22	0.00396	22
			50	0.000029	0.00029	42		
9,10-Anthracenedione	$C_{14}H_8O_2$	208.213	25	0.00014	0.0014	40		
Apomorphine	$C_{17}H_{17}NO_2$	267.323	25	2.0	20	40		
L-Arginine	$C_6H_{14}N_4O_2$	174.201	25	15.44	183	26		
L-Ascorbic acid	$C_6H_8O_6$	176.124	25	25.2	337	33		
			50	41.0	695	33		
L-Asparagine	$C_4H_8N_2O_3$	132.118	25	2.45	25.1	26		
L-Aspartic acid	$C_4H_7NO_4$	133.104	10	0.29	2.9	77		
			25	0.49	4.9	77		
			50	1.31	13.3	77		
Atrazine	$C_8H_{14}ClN_5$	215.684	25	0.007	0.07	26		
Atropine	$C_{17}H_{23}NO_3$	289.370	20	0.3	3	40		
Azinphos-methyl	$C_{10}H_{12}N_3O_3PS_2$	317.324	20	0.00209	0.0209	40		
trans-Azobenzene	$C_{12}H_{10}N_2$	182.220	20	0.03	0.3	27		
Bayleton	$C_{14}H_{16}ClN_3O_2$	293.749	20	0.026	0.26	40		
Bendiocarb	$C_{11}H_{13}NO_4$	223.226	25	0.004	0.04	40		
Bentazon	$C_{10}H_{12}N_2O_3S$	240.278	20	0.050	0.50	40		
Benzaldehyde	C_7H_6O	106.122	20	0.3	3	10		
Benzamide	C_7H_7NO	121.137	12	0.577	5.77	27		
Benz[a]anthracene	$C_{18}H_{12}$	228.288	10	0.00000038	0.0000038	42		
			25	0.00000093	0.0000093	42,22	0.00058	22
Benzene	C_6H_6	78.112	10	0.174	1.74	22		
			20	0.177	1.77	22		
			30	0.183	1.83	22		
			40	0.192	1.92	22		
			50	0.206	2.06	22		
			70	0.249	2.50	65		
			101	0.398	4.00	65		
Benzeneacetic acid	$C_8H_8O_2$	136.149	25	1.71	17.4	27		
1,2-Benzenediamine	$C_6H_8N_2$	108.141	20	3.02	31.1	40		
1,3-Benzenediamine	$C_6H_8N_2$	108.141	20	3.48	36.1	40		
1,4-Benzenediamine	$C_6H_8N_2$	108.141	24	3.45	35.7	40		
1,2-Benzenedicarboxamide	$C_8H_8N_2O_2$	164.162	30	0.59	5.9	40		
Benzeneethanol	$C_8H_{10}O$	122.164	25	1.72	17.5	40		

Name	Mol. Form.	Mol. Wt.	t/°C	Solubility, s: 100 w_2 (mass%)	Solubility, s: g per kg H_2O	Ref.	Henry Const., k_H: kPa m³mol⁻¹	Ref.
Benzenehexacarboxylic acid	$C_{12}H_6O_{12}$	342.169	25	49.3	972	76		
Benzenepentacarboxylic acid	$C_{11}H_6O_{10}$	298.160	10	11.9	135	76		
			25	21.1	267	76		
			50	36.2	567	76		
1,2,3,4-Benzenetetracarboxylic acid	$C_{10}H_6O_8$	254.150	10	11.0	124	76		
			25	20.9	264	76		
			50	39.5	653	76		
1,2,3,5-Benzenetetracarboxylic acid	$C_{10}H_6O_8$	254.150	10	7.50	81.1	76		
			25	10.1	112	76		
			50	15.8	188	76		
1,2,4,5-Benzenetetracarboxylic acid	$C_{10}H_6O_8$	254.150	10	0.51	5.1	76		
			25	1.06	10.7	76		
			50	3.82	39.7	76		
1,2,3-Benzenetricarboxylic acid	$C_9H_6O_6$	210.140	10	2.39	24.5	76		
			25	4.78	50.2	76		
			50	17.4	211	76		
1,2,4-Benzenetricarboxylic acid	$C_9H_6O_6$	210.140	10	1.02	10.3	76		
			25	1.92	19.6	76		
			50	5.45	57.6	76		
1,3,5-Benzenetricarboxylic acid	$C_9H_6O_6$	210.140	10	0.110	1.10	76		
			25	0.207	2.07	76		
			50	0.598	6.02	76		
1,2,3-Benzenetriol	$C_6H_6O_3$	126.110	25	38.5	626	27		
1,3,5-Benzenetriol	$C_6H_6O_3$	126.110	20	1.12	11.3	27		
p-Benzidine	$C_{12}H_{12}N_2$	184.236	24	0.0360	0.360	40		
1*H*-Benzimidazole	$C_7H_6N_2$	118.136	15	0.33	3.3	54		
			20	0.201	2.01	6		
1,3-Benzodioxole-5-carboxaldehyde	$C_8H_6O_3$	150.132	20	0.35	3.5	40		
Benzo[b]fluoranthene	$C_{20}H_{12}$	252.309	20	0.0000002	0.000002	40		
Benzo[k]fluoranthene	$C_{20}H_{12}$	252.309	25	0.00000008	0.0000008	40		
11*H*-Benzo[a]fluorene	$C_{17}H_{12}$	216.277	25	0.0000045	0.000045	42,4		
11*H*-Benzo[b]fluorene	$C_{17}H_{12}$	216.277	25	0.0000002	0.000002	42,4		
Benzoic acid	$C_7H_6O_2$	122.122	10	0.209	2.09	76		
			25	0.343	3.44	76		
			50	0.842	8.49	76		
Benzoin	$C_{14}H_{12}O_2$	212.244	25	0.03	0.3	40		
Benzonitrile	C_7H_5N	103.122	25	0.2	2	10		
Benzo[ghi]perylene	$C_{22}H_{12}$	276.330	25	0.000000026	0.00000026	42,4	0.000075	12
Benzophenone	$C_{13}H_{10}O$	182.217	20	0.0075	0.075	40		
2*H*-1-Benzopyran-2-one	$C_9H_6O_2$	146.143	20	0.190	1.90	40		
			60	0.69	6.9	40		
Benzo[a]pyrene	$C_{20}H_{12}$	252.309	25	0.00000043	0.0000043	42,22	0.0000465	22
Benzo[e]pyrene	$C_{20}H_{12}$	252.309	8	0.00000032	0.0000032	42		
			17	0.00000044	0.0000044	42,22	0.0000467	22
			25	0.00000048	0.0000048	42		
Benzo[f]quinoline	$C_{13}H_9N$	179.217	25	0.0079	0.079	6		
p-Benzoquinone	$C_6H_4O_2$	108.095	25	1.36	13.8	27		
Benzo[b]thiophene	C_8H_6S	134.199	20	0.0130	0.130	6		
Benzo[b]triphenylene	$C_{22}H_{14}$	278.346	25	0.0000027	0.000027	4		

Name	Mol. Form.	Mol. Wt.	t/°C	Solubility, s: 100 w_2 (mass%)	Solubility, s: g per kg H_2O	Ref.	Henry Const., k_H: kPa m³mol⁻¹	Ref.
Benzoxazole	C_7H_5NO	119.121	20	0.834	8.34	6		
N-Benzoylglycine	$C_9H_9NO_3$	179.172	25	0.37	3.7	29		
Benzoyl peroxide	$C_{14}H_{10}O_4$	242.227	20	0.000016	0.00016	40		
N-Benzoyl-*L*-phenylalanine	$C_{16}H_{15}NO_3$	269.295	25	0.085	0.85	29		
Benzyl acetate	$C_9H_{10}O_2$	150.174	25	0.150	1.50	40		
Benzyl alcohol	C_7H_8O	108.138	20	0.08	0.8	10		
Benzyl formate	$C_8H_8O_2$	136.149	20	1.07	10.8	20		
			80	1.43	14.5	20		
Bifenthrin	$C_{23}H_{22}ClF_3O_2$	422.868	25	0.00001	0.0001	32		
Biotin	$C_{10}H_{16}N_2O_3S$	244.310	25	0.035	0.35	40		
Biphenyl	$C_{12}H_{10}$	154.207	0	0.000272	0.00272	4		
			25	0.00054	0.0054	58,22	0.0280	22
			50	0.0022	0.022	4		
2,2'-Bipyridine	$C_{10}H_8N_2$	156.184	25	0.61	6.1	40		
2,2'-Biquinoline	$C_{18}H_{12}N_2$	256.301	24	0.000102	0.00102	6		
Bis(4-aminophenyl) sulfone	$C_{12}H_{12}N_2O_2S$	248.300	25	0.016	0.16	40		
Bis(2-chloroethyl) ether	$C_4H_8Cl_2O$	143.012	20	1.04	10.5	20	0.003	13
			81	1.26	12.8	20		
1,1-Bis(4-chlorophenyl)-2,2,2-trichloroethanol	$C_{14}H_9Cl_5O$	370.485	25	0.00013	0.0013	40		
Bis(2-ethylhexyl) phthalate	$C_{24}H_{38}O_4$	390.557	25	0.000027	0.00027	40		
2,2-Bis(4-hydroxyphenyl)propane	$C_{15}H_{16}O_2$	228.287	25	0.0300	0.30	49		
1,3-Bis(trifluoromethyl)benzene	$C_8H_4F_6$	214.108	25	0.0041	0.041	2		
Borneol	$C_{10}H_{18}O$	154.249	25	0.046	0.46	52		
Bromacil	$C_9H_{13}BrN_2O_2$	261.115	25	0.082	0.82	40		
Bromobenzene	C_6H_5Br	157.008	10	0.0387	0.387	2		
			25	0.0445	0.445	2	0.250	28
			40	0.0516	0.516	2		
2-Bromobenzoic acid	$C_7H_5BrO_2$	201.018	25	0.185	1.85	27		
3-Bromobenzoic acid	$C_7H_5BrO_2$	201.018	25	0.040	0.40	27		
4-Bromobenzoic acid	$C_7H_5BrO_2$	201.018	25	0.0056	0.056	27		
1-Bromobutane	C_4H_9Br	137.018	25	0.087	0.87	35	1.2	13
4-Bromo-1-butene	C_4H_7Br	135.003	25	0.076	0.76	35		
1-Bromo-2-chlorobenzene	C_6H_4BrCl	191.453	25	0.0124	0.124	2		
1-Bromo-3-chlorobenzene	C_6H_4BrCl	191.453	25	0.0118	0.118	2		
1-Bromo-4-chlorobenzene	C_6H_4BrCl	191.453	25	0.00442	0.0442	2		
1-Bromo-2-chloroethane	C_2H_4BrCl	143.410	30	0.683	6.83	25		
Bromochloromethane	CH_2BrCl	129.384	25	1.7	17	10	0.18	13
1-Bromo-3-chloropropane	C_3H_6BrCl	157.437	25	0.223	2.23	35		
2-Bromo-2-chloro-1,1,1-trifluoroethane	$C_2HBrClF_3$	197.381	10	0.52	5.2	25		
			25	0.41	4.1	25		
			40	0.40	4.0	25		
Bromodichloromethane	$CHBrCl_2$	163.829	30	0.300	3.00	40		
Bromoethane	C_2H_5Br	108.965	0	1.05	10.6	25		
			25	0.90	9.0	25	1.23	13
1-Bromoheptane	$C_7H_{15}Br$	179.098	25	0.00067	0.0067	35		
1-Bromohexane	$C_6H_{13}Br$	165.071	25	0.00258	0.0258	35		
1-Bromo-4-iodobenzene	C_6H_4BrI	282.904	25	0.000794	0.00794	2		
Bromomethane	CH_3Br	94.939	20	1.80*	18.3*	5	0.63	13
1-Bromo-3-methylbutane	$C_5H_{11}Br$	151.045	16	0.020	0.20	35		

Name	Mol. Form.	Mol. Wt.	t/°C	Solubility, s: 100 w_2 (mass%)	Solubility, s: g per kg H_2O	Ref.	Henry Const., k_H: kPa $m^3 mol^{-1}$	Ref.
1-Bromo-2-methylpropane	C_4H_9Br	137.018	18	0.051	0.51	35		
1-Bromooctane	$C_8H_{17}Br$	193.125	25	0.000167	0.00167	35		
1-Bromopentane	$C_5H_{11}Br$	151.045	25	0.0127	0.127	35		
4-Bromophenol	C_6H_5BrO	173.007	25	1.86	19.0	2		
1-Bromopropane	C_3H_7Br	122.992	0	0.298	2.98	35		
			25	0.234	2.34	35	3.8	13
2-Bromopropane	C_3H_7Br	122.992	20	0.32	3.2	35	1.27	13
3-Bromopropene	C_3H_5Br	120.976	25	0.38	3.8	35		
4-Bromotoluene	C_7H_7Br	171.035	25	0.011	0.11	2		
Bromotrifluoromethane	$CBrF_3$	148.910	25	0.032*	0.32*	14		
5-Bromouracil	$C_4H_3BrN_2O_2$	190.983	25	0.288	2.89	72		
Brucine	$C_{23}H_{26}N_2O_4$	394.463	20	0.012	0.12	27		
1,3-Butadiene	C_4H_6	54.091	25	0.0735*	0.735*	5	20.7	13
Butanal	C_4H_8O	72.106	25	7.1	76	10		
Butanamide	C_4H_9NO	87.120	25	≈19	≈230	40		
Butane	C_4H_{10}	58.122	25	0.00724*	0.0724*	18	95.9	5
2,3-Butanedione	$C_4H_6O_2$	86.090	20	31.7	464	20		
			80	21.8	279	20		
Butanenitrile	C_4H_7N	69.106	20	3.3	34	10		
1,2,3,4-Butanetetrol	$C_4H_{10}O_4$	122.120	20	38.0	613	27		
1-Butanethiol	$C_4H_{10}S$	90.187	20	0.0597	0.597	10		
1-Butanol	$C_4H_{10}O$	74.121	0	10.5	117	78,1		
			25	7.3	79	78,1		
			50	6.4	68	78,1		
			100	8.8	96	78		
2-Butanol	$C_4H_{10}O$	74.121	10	23.9	314	1,87		
			25	18.1	221	1,87		
			50	14.0	163	1,87		
2-Butanone	C_4H_8O	72.106	0	35.9	560	82		
			25	25.6	344	82		
			40	21.5	274	82		
			70	18.1	221	20		
			100	19.3	239	82		
trans-2-Butenal	C_4H_6O	70.090	20	15.6	185	10		
1-Butene	C_4H_8	56.107	25	0.0222*	0.222*	5	25.6	13
trans-2-Butenoic acid	$C_4H_6O_2$	86.090	20	7.1	76	26		
cis-2-Buten-1-ol	C_4H_8O	72.106	20	16.6	199	10		
3-Buten-2-one	C_4H_6O	70.090	28	54.3	1190	82		
			50	35.6	553	82		
			80	37.6	603	82		
Butyl acetate	$C_6H_{12}O_2$	116.158	20	0.68	6.8	10		
sec-Butyl acetate	$C_6H_{12}O_2$	116.158	20	0.62	6.2	10		
Butyl 4-aminobenzoate	$C_{11}H_{15}NO_2$	193.243	25	0.018	0.18	40		
Butylbenzene	$C_{10}H_{14}$	134.218	25	0.00138	0.0138	22,89	1.33	22
sec-Butylbenzene	$C_{10}H_{14}$	134.218	25	0.0014	0.014	4,89	1.89	11
tert-Butylbenzene	$C_{10}H_{14}$	134.218	25	0.0032	0.032	4	1.28	11
Butyl ethyl ether	$C_6H_{14}O$	102.174	20	0.65	6.5	20		
			70	0.39	3.9	20		
Butyl 4-hydroxybenzoate	$C_{11}H_{14}O_3$	194.227	25	0.020	0.20	40		

Name	Mol. Form.	Mol. Wt.	t/°C	Solubility, s 100 w_2 (mass%)	g per kg H_2O	Ref.	Henry Const., k_H kPa m³mol⁻¹	Ref.
Butyl methyl ether	$C_5H_{12}O$	88.148	0	2.51	25.7	79		
			25	0.89	9.0	79		
4-*tert*-Butylphenol	$C_{10}H_{14}O$	150.217	25	0.058	0.58	40		
Butyl propanoate	$C_7H_{14}O_2$	130.185	22	0.572	5.72	27		
Butyl stearate	$C_{22}H_{44}O_2$	340.583	25	0.2	2	10		
Butyl vinyl ether	$C_6H_{12}O$	100.158	20	0.3	3	10		
1-Butyne	C_4H_6	54.091	25	0.287*	2.87*	5	1.91	5
Caffeine	$C_8H_{10}N_4O_2$	194.191	25	2.12	21.7	29		
Camphor, (+)	$C_{10}H_{16}O$	152.233	20	0.01	0.1	10		
trans-Camphoric acid	$C_{10}H_{16}O_4$	200.232	25	0.8	8	27		
Cantharidin	$C_{10}H_{12}O_4$	196.200	20	0.003	0.03	40		
Caprolactam	$C_6H_{11}NO$	113.157	25	84.0	5250	10		
Captafol	$C_{10}H_9Cl_4NO_2S$	349.061	20	0.000142	0.00142	40		
Captan	$C_9H_8Cl_3NO_2S$	300.590	20	0.00005	0.0005	40		
Carbaryl	$C_{12}H_{11}NO_2$	201.221	20	0.0102	0.102	40		
Carbazole	$C_{12}H_9N$	167.206	22	0.000120	0.00120	6		
Carbofuran	$C_{12}H_{15}NO_3$	221.252	20	0.032	0.32	40		
Carbon dioxide	CO_2	44.010	25	0.150*	1.50*	18		
Carbon disulfide	CS_2	76.141	20	0.210	2.10	10		
Carbon monoxide	CO	28.010	25	0.00276*	0.0276*	18		
Carboxin	$C_{12}H_{13}NO_2S$	235.302	25	0.017	0.17	40		
Carminic acid	$C_{22}H_{20}O_{13}$	492.386	20	0.13	1.3	40		
Carnosine	$C_9H_{14}N_4O_3$	226.232	25	24.4	323	26		
Carvenol	$C_{10}H_{16}O$	152.233	25	0.29	2.9	52		
Carvenone, (*S*)-	$C_{10}H_{16}O$	152.233	15	0.22	2.2	27		
Carvone	$C_{10}H_{14}O$	150.217	15	0.13	1.3	27		
(*S*)-Carvone	$C_{10}H_{14}O$	150.217	25	0.13	1.3	52		
Cephalexin	$C_{16}H_{17}N_3O_4S$	347.389	25	1.2	12	40		
Chloramphenicol	$C_{11}H_{12}Cl_2N_2O_5$	323.129	25	0.38	3.8	40		
Chlordane	$C_{10}H_6Cl_8$	409.779	25	0.00006	0.0006	67		
2-Chloroaniline	C_6H_6ClN	127.572	25	0.876	8.76	10		
3-Chloroaniline	C_6H_6ClN	127.572	20	0.54	5.4	40		
4-Chloroaniline	C_6H_6ClN	127.572	20	0.275	2.75	40		
Chlorobenzene	C_6H_5Cl	112.557	5	0.050	0.50	61		
			25	0.050	0.50	61		
			45	0.055	0.55	61,2		
Chlorobenzilate	$C_{16}H_{14}Cl_2O_3$	325.186	20	0.001	0.01	32		
2-Chlorobenzoic acid	$C_7H_5ClO_2$	156.567	25	0.209	2.09	27		
3-Chlorobenzoic acid	$C_7H_5ClO_2$	156.567	25	0.040	0.40	27		
4-Chlorobenzoic acid	$C_7H_5ClO_2$	156.567	25	0.072	0.72	27		
2-Chlorobiphenyl	$C_{12}H_9Cl$	188.652	25	0.00055	0.0055	7	0.0701	7
1-Chlorobutane	C_4H_9Cl	92.567	1	0.062	0.62	35		
			25	0.087	0.87	35	1.54	13
2-Chlorobutane	C_4H_9Cl	92.567	0	0.107	1.07	35		
			25	0.092	0.92	35		
3-Chloro-2-butanone	C_4H_7ClO	106.551	19	2.80	28.8	20		
			92	3.38	35.0	20		
Chlorodiazepoxide	$C_{16}H_{14}ClN_3O$	299.754	20	0.2	2	40		
Chlorodibromomethane	$CHBr_2Cl$	208.280	30	0.251	2.51	40		

Name	Mol. Form.	Mol. Wt.	t/°C	Solubility, s: 100 w_2 (mass%)	Solubility, s: g per kg H_2O	Ref.	Henry Const., k_H: kPa m³mol⁻¹	Ref.
Chlorodifluoromethane	$CHClF_2$	86.469	25	0.30*	3.0*	10	3.0	13
4-Chloro-2,5-dimethylphenol	C_8H_9ClO	156.609	25	0.89	8.9	2		
4-Chloro-2,6-dimethylphenol	C_8H_9ClO	156.609	25	0.52	5.2	2		
4-Chloro-3,5-dimethylphenol	C_8H_9ClO	156.609	25	0.34	3.4	2		
1-Chloro-2,4-dinitrobenzene	$C_6H_3ClN_2O_4$	202.552	25	0.00092	0.0092	40		
Chloroethane	C_2H_5Cl	64.514	0	0.45	4.5	25		
			25	0.67*	6.7*	25	1.02	13
Chloroethene	C_2H_3Cl	62.498	25	0.27*	2.7*	5	2.68	13
1-Chloro-2-fluorobenzene	C_6H_4ClF	130.547	25	0.0502	0.502	40		
Chlorofluoromethane	CH_2ClF	68.478	25	1.05*	10.6*	14		
1-Chloroheptane	$C_7H_{15}Cl$	134.647	25	0.00136	0.0136	35		
1-Chlorohexane	$C_6H_{13}Cl$	120.620	5	0.0047	0.047	35		
			25	0.0064	0.064	35		
2-Chloro-4-hydroxy-5-methoxybenzaldehyde	$C_8H_7ClO_3$	186.593	25	0.013	0.13	8		
3-Chloro-4-hydroxy-5-methoxybenzaldehyde	$C_8H_7ClO_3$	186.593	25	0.093	0.93	8		
1-Chloro-2-iodobenzene	C_6H_4ClI	238.453	25	0.00689	0.0689	2		
1-Chloro-3-iodobenzene	C_6H_4ClI	238.453	25	0.00674	0.0674	2		
1-Chloro-4-iodobenzene	C_6H_4ClI	238.453	25	0.00311	0.0311	2		
Chloromethane	CH_3Cl	50.488	25	0.535*	5.35*	5	0.98	13
1-Chloro-2-methoxyethane	C_3H_7ClO	94.540	20	7.79	84.5	20		
			70	6.31	67.3	20		
(Chloromethyl)benzene	C_7H_7Cl	126.584	20	0.0493	0.493	10		
3-(Chloromethyl)heptane	$C_8H_{17}Cl$	148.674	20	0.01	0.1	10		
2-Chloro-6-methylphenol	C_7H_7ClO	142.583	25	0.36	3.6	2		
4-Chloro-2-methylphenol	C_7H_7ClO	142.583	25	0.68	6.8	2		
4-Chloro-3-methylphenol	C_7H_7ClO	142.583	25	0.40	4.0	2		
(4-Chloro-2-methylphenoxy)acetic acid	$C_9H_9ClO_3$	200.618	25	0.117	1.17	40		
1-Chloro-2-methylpropane	C_4H_9Cl	92.567	25	0.92	9.2	35		
2-Chloro-2-methylpropane	C_4H_9Cl	92.567	15	0.29	2.9	35		
1-Chloro-2-methylpropene	C_4H_7Cl	90.552	25	0.916	9.16	5	0.12	5
1-Chloronaphthalene	$C_{10}H_7Cl$	162.616	25	0.00224	0.0224	5	0.0363	28
2-Chloronaphthalene	$C_{10}H_7Cl$	162.616	25	0.00117	0.0117	5	0.0335	28
1-Chloro-2-nitrobenzene	$C_6H_4ClNO_2$	157.555	20	0.0441	0.441	40		
1-Chloro-3-nitrobenzene	$C_6H_4ClNO_2$	157.555	20	0.0273	0.273	40		
1-Chloro-4-nitrobenzene	$C_6H_4ClNO_2$	157.555	20	0.0453	0.453	40		
3-Chloro-2-nitrobenzoic acid	$C_7H_4ClNO_4$	201.565	25	0.047	0.47	27		
5-Chloro-2-nitrobenzoic acid	$C_7H_4ClNO_4$	201.565	25	0.96	9.6	27		
1-Chlorooctane	$C_8H_{17}Cl$	148.674	25	0.0345	0.345	35		
Chloropentafluoroethane	C_2ClF_5	154.466	25	0.006*	0.06*	10	260	13
1-Chloropentane	$C_5H_{11}Cl$	106.594	5	0.020	0.20	35		
			25	0.0201	0.201	35	2.37	13
3-Chloropentane	$C_5H_{11}Cl$	106.594	25	0.025	0.25	35		
5-Chloro-2-pentanone	C_5H_9ClO	120.577	22	4.7	49	20		
			71	13.5	156	20		
2-Chlorophenol	C_6H_5ClO	128.556	25	2.27	23.2	48,51,2		
3-Chlorophenol	C_6H_5ClO	128.556	25	2.2	22	2		
4-Chlorophenol	C_6H_5ClO	128.556	25	2.55	26.2	48,51,2		
N'-(4-Chlorophenyl)-*N,N*-dimethylurea	$C_9H_{11}ClN_2O$	198.648	25	0.023	0.23	26		
1-Chloropropane	C_3H_7Cl	78.541	25	0.250	2.50	35	1.41	13

Name	Mol. Form.	Mol. Wt.	t/°C	Solubility, s: 100 w_2 (mass%)	Solubility, s: g per kg H_2O	Ref.	Henry Const., k_H: kPa m^3mol^{-1}	Ref.
2-Chloropropane	C_3H_7Cl	78.541	0	0.44	4.4	35		
			20	0.30	3.0	35		
3-Chloropropene	C_3H_5Cl	76.525	25	0.40	4.0	35	1.10	5
			50	0.13	1.3	35		
Chloropropham	$C_{10}H_{12}ClNO_2$	213.661	25	0.0080	0.080	40		
1-Chlorotetradecane	$C_{14}H_{29}Cl$	232.833	25	0.0232	0.232	35		
Chlorothalonil	$C_8Cl_4N_2$	265.911	25	0.00006	0.0006	40		
Chlorothiazide	$C_7H_6ClN_3O_4S_2$	295.724	25	0.0283	0.283	40		
2-Chlorotoluene	C_7H_7Cl	126.584	25	0.0117	0.117	61		
3-Chlorotoluene	C_7H_7Cl	126.584	25	0.0117	0.117	61		
4-Chlorotoluene	C_7H_7Cl	126.584	25	0.0123	0.123	61		
Chlorotrifluoromethane	$CClF_3$	104.459	25	0.009*	0.09*	10	6.9	13
3-Chloro-1,1,1-trifluoropropane	$C_3H_4ClF_3$	132.512	20	0.133	1.33	35		
2-Chloro-1,3,5-trinitrobenzene	$C_6H_2ClN_3O_6$	247.549	15	0.018	0.18	40		
5-Chlorouracil	$C_4H_3ClN_2O_2$	146.532	25	0.250	2.51	72		
Chlorpyrifos	$C_9H_{11}Cl_3NO_3PS$	350.586	20	0.000073	0.00073	40		
Chlorsulfuron	$C_{12}H_{12}ClN_5O_4S$	357.773	25	2.71	27.9	32		
Cholic acid	$C_{24}H_{40}O_5$	408.572	20	0.028	0.28	26		
Chrysene	$C_{18}H_{12}$	228.288	7	0.00000007	0.0000007	42		
			25	0.00000019	0.0000019	42,22	0.000065	22
trans-Cinnamaldehyde	C_9H_8O	132.159	25	0.135	1.35	40		
trans-Cinnamic acid	$C_9H_8O_2$	148.159	20	0.1	1	26		
			98	0.59	5.9	26		
Citric acid	$C_6H_8O_7$	192.124	20	59	1440	26		
Clopyralid	$C_6H_3Cl_2NO_2$	192.000	20	0.1	1	40		
Clorophene	$C_{13}H_{11}ClO$	218.678	20	0.42	4.2	40		
Cocaine	$C_{17}H_{21}NO_4$	303.354	25	0.17	1.7	27		
Codeine	$C_{18}H_{21}NO_3$	299.365	25	0.79	7.9	27		
Colchicine	$C_{22}H_{25}NO_6$	399.437	20	4	42	26		
Coronene	$C_{24}H_{12}$	300.352	25	0.000000014	0.00000014	42,4		
Creatine	$C_4H_9N_3O_2$	131.133	25	1.6	16	26		
o-Cresol	C_7H_8O	108.138	40	3.08	31.8	10		
m-Cresol	C_7H_8O	108.138	40	2.51	25.7	10		
p-Cresol	C_7H_8O	108.138	40	2.26	23.1	10		
Crufomate	$C_{12}H_{19}ClNO_3P$	291.711	20	0.50	5.0	40		
Cyanazine	$C_9H_{13}ClN_6$	240.692	25	0.0171	0.171	40		
2-Cyanoacetamide	$C_3H_4N_2O$	84.076	20	11.5	130	40		
Cyanogen	C_2N_2	52.034	25	0.8*	8*	30		
Cyanogen chloride	CClN	61.471	0	5.7	60	40		
Cyanoguanidine	$C_2H_4N_4$	84.080	25	3.8	40	40		
Cyanuric acid	$C_3H_3N_3O_3$	129.074	25	0.259	2.59	40		
Cycloheptane	C_7H_{14}	98.186	25	0.0030	0.030	3	9.59	13
Cycloheptanone	$C_7H_{12}O$	112.169	20	3.61	37.5	20		
			92	2.82	29.0	20		
1,3,5-Cycloheptatriene	C_7H_8	92.139	25	0.064	0.64	3	0.47	13
Cycloheptene	C_7H_{12}	96.170	25	0.0066	0.066	3	4.9	13
1,4-Cyclohexadiene	C_6H_8	80.128	25	0.08	0.8	3	1.03	13
Cyclohexane	C_6H_{12}	84.159	25	0.0058	0.058	3	19.4	13
			70	0.0092	0.092	65		

Name	Mol. Form.	Mol. Wt.	t/°C	Solubility, s: 100 w_2 (mass%)	Solubility, s: g per kg H_2O	Ref.	Henry Const., k_H: kPa m^3mol^{-1}	Ref.
			100	0.0163	0.163	65		
Cyclohexanecarboxylic acid	$C_7H_{12}O_2$	128.169	15	0.201	2.01	27		
Cyclohexanol	$C_6H_{12}O$	100.158	10	4.62	48.4	1		
			25	3.8	40	1		
			40	3.30	34.1	1		
Cyclohexanone	$C_6H_{10}O$	98.142	10	12.2	139	83		
			25	9.5	105	83		
			50	7.6	82	83		
			80	6.8	73	20		
Cyclohexanone oxime	$C_6H_{11}NO$	113.157	25	1.57	16.0	40		
Cyclohexene	C_6H_{10}	82.143	25	0.016	0.16	3	4.57	13
Cyclohexyl butanoate	$C_{10}H_{18}O_2$	170.249	20	0.11	1.1	20		
			90	0.09	0.90	20		
Cyclooctane	C_8H_{16}	112.213	25	0.00079	0.0079	4	10.7	13
1,3-Cyclopentadiene	C_5H_6	66.102	25	0.068	0.68	3		
Cyclopentane	C_5H_{10}	70.133	25	0.0157	0.157	3	19.1	13
Cyclopentanol	$C_5H_{10}O$	86.132	19	10.6	119	88		
			50	8.3	91	88		
			90	9.2	101	88		
Cyclopentanone	C_5H_8O	84.117	0	37.7	605	20		
			20	31.0	449	20		
			80	24.8	330	20		
Cyclopentene	C_5H_8	68.118	25	0.054	0.54	3	6.56	13
Cyclopropane	C_3H_6	42.080	25	0.0484*	0.484*	19		
Cyfluthrin	$C_{22}H_{18}Cl_2FNO_3$	434.287	20	0.0000002	0.000002	32		
Cygon	$C_5H_{12}NO_3PS_2$	229.258	20	2.6	27	40		
Cyhalothrin	$C_{23}H_{19}ClF_3NO_3$	449.850	20	0.0000005	0.000005	32		
Cypermethrin	$C_{22}H_{19}Cl_2NO_3$	416.297	20	0.000001	0.00001	32		
L-Cystine	$C_6H_{12}N_2O_4S_2$	240.300	25	0.0166	0.166	62		
Cytisine	$C_{11}H_{14}N_2O$	190.241	16	≈30	≈430	40		
Cytosine	$C_4H_5N_3O$	111.102	25	0.73	7.3	29		
Daminozide	$C_6H_{12}N_2O_3$	160.170	25	9.1	100	40		
Dazomet	$C_5H_{10}N_2S_2$	162.276	25	0.12	1.2	40		
Decabromobiphenyl ether	$C_{12}Br_{10}O$	959.167	25	0.0000025	0.000025	40		
Decachlorobiphenyl	$C_{12}Cl_{10}$	498.658	25	0.00000000012	0.0000000012	7	0.0208	7
cis-Decahydronaphthalene	$C_{10}H_{18}$	138.250	25	0.000089	0.00089	37		
trans-Decahydronaphthalene	$C_{10}H_{18}$	138.250	25	0.000089	0.00089	4	3	13
Decane	$C_{10}H_{22}$	142.282	0	0.0000015	0.000015	4	479	13
Decanedioic acid	$C_{10}H_{18}O_4$	202.248	20	0.10	1.0	40		
Decanoic acid	$C_{10}H_{20}O_2$	172.265	20	0.015	0.15	26		
1-Decanol	$C_{10}H_{22}O$	158.281	25	0.0037	0.037	1		
2-Decanone	$C_{10}H_{20}O$	156.265	25	0.0079	0.079	84		
4-Decanone	$C_{10}H_{20}O$	156.265	20	0.0238	0.238	20		
			80	0.0064	0.064	20		
1-Decene	$C_{10}H_{20}$	140.266	25	0.00057	0.0057	4		
2'-Deoxyadenosine	$C_{10}H_{13}N_5O_3$	251.242	25	0.67	6.7	29		
Dexamethasone	$C_{22}H_{29}FO_5$	392.460	25	0.009	0.09	40		
Dibenz[a,j]acridine	$C_{21}H_{13}N$	279.335	25	0.000016	0.00016	6		
Dibenz[a,h]anthracene	$C_{22}H_{14}$	278.346	25	0.00000005	0.0000005	42,4		

Name	Mol. Form.	Mol. Wt.	t/°C	Solubility, s 100 w_2 (mass%)	g per kg H_2O	Ref.	Henry Const., k_H kPa m^3mol^{-1}	Ref.
Dibenz[a,j]anthracene	$C_{22}H_{14}$	278.346	27	0.0000012	0.000012	42,4		
13*H*-Dibenzo[a,i]carbazole	$C_{20}H_{13}N$	267.324	24	0.00000104	0.0000104	6		
Dibenzo[b,e][1,4]dioxin	$C_{12}H_8O_2$	184.191	25	0.000126	0.00126	68		
Dibenzofuran	$C_{12}H_8O$	168.191	25	0.000475	0.00475	41	0.011	12
Dibenzothiophene	$C_{12}H_8S$	184.257	25	0.000103	0.00103	6		
Dibenzyl ether	$C_{14}H_{14}O$	198.260	35	0.0040	0.040	10		
o-Dibromobenzene	$C_6H_4Br_2$	235.904	25	0.00748	0.0748	2		
m-Dibromobenzene	$C_6H_4Br_2$	235.904	25	0.0064	0.064	2		
p-Dibromobenzene	$C_6H_4Br_2$	235.904	25	0.0020	0.020	2		
1,4-Dibromobutane	$C_4H_8Br_2$	215.915	25	0.035	0.35	35		
1,2-Dibromo-1-chloroethane	$C_2H_3Br_2Cl$	222.306	20	0.060	0.60	25		
1,2-Dibromo-3-chloropropane	$C_3H_5Br_2Cl$	236.333	20	0.123	1.23	35		
1,2-Dibromo-1,2-dichloroethane	$C_2H_2Br_2Cl_2$	256.751	20	0.070	0.70	25		
1,2-Dibromoethane	$C_2H_4Br_2$	187.861	20	0.412	4.14	20		
			50	0.493	4.95	20	0.066	13
			80	0.572	5.75	20		
1,2-Dibromo-1,1,2,3,3,3-hexafluoropropane	$C_3Br_2F_6$	309.830	21	0.0068	0.068	35		
3,5-Dibromo-4-hydroxybenzonitrile	$C_7H_3Br_2NO$	276.913	25	0.013	0.13	40		
Dibromomethane	CH_2Br_2	173.835	20	1.28	13.0	20	0.086	13
			90	1.51	15.3	20		
2,4-Dibromophenol	$C_6H_4Br_2O$	251.903	25	0.2	2	2		
1,2-Dibromopropane	$C_3H_6Br_2$	201.888	25	0.143	1.43	10		
1,3-Dibromopropane	$C_3H_6Br_2$	201.888	25	0.169	1.69	35		
1,2-Dibromotetrafluoroethane	$C_2Br_2F_4$	259.823	25	0.00030	0.0030	25		
Dibutylamine	$C_8H_{19}N$	129.244	20	0.47	4.7	10		
Dibutyl ether	$C_8H_{18}O$	130.228	0	0.040	0.40	20	0.48	13
			20	0.023	0.23	20	0.48	13
			90	0.010	0.10	20		
Dibutyl phthalate	$C_{16}H_{22}O_4$	278.344	25	0.00112	0.0112	15		
Dibutyl sebacate	$C_{18}H_{34}O_4$	314.461	20	0.004	0.04	10		
o-Dichlorobenzene	$C_6H_4Cl_2$	147.002	5	0.012	0.12	61,58,2		
			25	0.015	0.15	61,58,2		
			45	0.020	0.20	61,58,2		
m-Dichlorobenzene	$C_6H_4Cl_2$	147.002	10	0.0103	0.103	41,2		
			25	0.0120	0.120	41,2	0.376	11
			45	0.0141	0.141	61,2		
p-Dichlorobenzene	$C_6H_4Cl_2$	147.002	10	0.00512	0.0512	2		
			25	0.0080	0.080	41	0.244	28
			50	0.0167	0.167	2		
3,5-Dichloro-1,2-benzenediol	$C_6H_4Cl_2O_2$	179.001	25	0.78	7.8	8		
4,5-Dichloro-1,2-benzenediol	$C_6H_4Cl_2O_2$	179.001	25	1.19	12.0	8		
3,3'-Dichloro-*p*-benzidine	$C_{12}H_{10}Cl_2N_2$	253.126	25	0.00031	0.0031	40		
2,5-Dichlorobiphenyl	$C_{12}H_8Cl_2$	223.098	25	0.0002	0.002	7	0.0201	7
2,6-Dichlorobiphenyl	$C_{12}H_8Cl_2$	223.098	25	0.00014	0.0014	7		
1,1-Dichloro-2,2-bis(*p*-chlorophenyl)ethane	$C_{14}H_{10}Cl_4$	320.041	25	0.000009	0.00009	40		
			45	0.000024	0.00024	40		
1,1-Dichlorobutane	$C_4H_8Cl_2$	127.013	25	0.050	0.50	35		
1,4-Dichlorobutane	$C_4H_8Cl_2$	127.013	25	0.16	1.6	35		
2,3-Dichlorobutane	$C_4H_8Cl_2$	127.013	20	0.056	0.56	35		

Name	Mol. Form.	Mol. Wt.	t/°C	Solubility, s: 100 w_2 (mass%)	Solubility, s: g per kg H_2O	Ref.	Henry Const., k_H: kPa m³mol⁻¹	Ref.
2,7-Dichlorodibenzo-*p*-dioxin	$C_{12}H_6Cl_2O_2$	253.081	25	0.00000041	0.0000041	68		
1,2-Dichloro-1,1-difluoroethane	$C_2H_2Cl_2F_2$	134.940	24	0.49	4.9	25		
Dichlorodifluoromethane	CCl_2F_2	120.914	20	0.028*	0.28*	5	41	13
1,3-Dichloro-5,5-dimethyl hydantoin	$C_5H_6Cl_2N_2O_2$	197.019	20	0.050	0.50	40		
1,1-Dichloroethane	$C_2H_4Cl_2$	98.959	0	0.62	6.2	25		
			25	0.50	5.0	25	0.63	13
			50	0.50	5.0	25		
1,2-Dichloroethane	$C_2H_4Cl_2$	98.959	0	0.92	9.2	25		
			25	0.86	8.6	25	0.14	13
			50	1.05	10.6	25		
			100	2.17	22.2	25		
1,1-Dichloroethene	$C_2H_2Cl_2$	96.943	5	0.310	3.10	25		
			25	0.242	2.42	25	2.62	13
			50	0.225	2.25	25		
			90	0.355	3.55	25		
cis-1,2-Dichloroethene	$C_2H_2Cl_2$	96.943	10	0.76	7.6	25		
			25	0.64	6.4	25	0.46	13
			40	0.66	6.6	25		
trans-1,2-Dichloroethene	$C_2H_2Cl_2$	96.943	10	0.53	5.3	25		
			25	0.45	4.5	25	0.96	13
			40	0.41	4.1	25		
1,1-Dichloro-1-fluoroethane	$C_2H_3Cl_2F$	116.949	25	0.042	0.42	25		
Dichlorofluoromethane	$CHCl_2F$	102.923	25	0.95*	9.5*	10		
1,2-Dichloro-1,1,2,3,3,3-hexafluoropropane	$C_3Cl_2F_6$	220.928	21	0.0096	0.096	35		
1,4-Dichloro-5-isopropyl-2-methylbenzene	$C_{10}H_{12}Cl_2$	203.108	25	0.00049	0.0049	23		
Dichloromethane	CH_2Cl_2	84.933	25	1.73	17.6	20	0.30	13
3,6-Dichloro-2-methoxybenzoic acid	$C_8H_6Cl_2O_3$	221.038	25	0.45	4.5	40		
(Dichloromethyl)benzene	$C_7H_6Cl_2$	161.029	30	0.025	0.25	10		
2,3-Dichloro-2-methylbutane	$C_5H_{10}Cl_2$	141.038	25	0.029	0.29	35		
2,4-Dichloro-6-methylphenol	$C_7H_6Cl_2O$	177.028	25	0.0283	0.283	2		
2,6-Dichloro-4-methylphenol	$C_7H_6Cl_2O$	177.028	25	0.0673	0.673	2		
2,3-Dichloro-1,4-naphthalenedione	$C_{10}H_4Cl_2O_2$	227.044	25	0.00001	0.0001	40		
1,2-Dichloro-4-nitrobenzene	$C_6H_3Cl_2NO_2$	192.000	20	0.0121	0.121	40		
1,2-Dichloropentane	$C_5H_{10}Cl_2$	141.038	25	0.029	0.29	35		
1,5-Dichloropentane	$C_5H_{10}Cl_2$	141.038	19	0.02	0.2	35		
2,3-Dichloropentane	$C_5H_{10}Cl_2$	141.038	25	0.029	0.29	35		
Dichlorophene	$C_{13}H_{10}Cl_2O_2$	269.123	25	0.003	0.03	40		
2,3-Dichlorophenol	$C_6H_4Cl_2O$	163.001	25	0.82	8.3	40		
2,4-Dichlorophenol	$C_6H_4Cl_2O$	163.001	25	0.55	5.5	48,51,24		
2,6-Dichlorophenol	$C_6H_4Cl_2O$	163.001	25	0.262	2.62	40		
(2,4-Dichlorophenoxy)acetic acid	$C_8H_6Cl_2O_3$	221.038	25	0.07	0.7	40		
4-(2,4-Dichlorophenoxy)butanoic acid	$C_{10}H_{10}Cl_2O_3$	249.090	25	0.0046	0.046	40		
2-(2,4-Dichlorophenoxy)propanoic acid	$C_9H_8Cl_2O_3$	235.064	25	0.083	0.83	40		
1,2-Dichloropropane	$C_3H_6Cl_2$	112.986	5	0.270	2.70	35		
			25	0.274	2.74	35	0.29	13
			40	0.297	2.97	35		
1,3-Dichloropropane	$C_3H_6Cl_2$	112.986	5	0.218	2.18	35		
			25	0.280	2.80	35		
cis-1,3-Dichloropropene	$C_3H_4Cl_2$	110.970	20	0.27	2.7	5	0.24	5

Name	Mol. Form.	Mol. Wt.	t/°C	Solubility, s 100 w_2 (mass%)	g per kg H_2O	Ref.	Henry Const., k_H kPa m³mol⁻¹	Ref.
trans-1,3-Dichloropropene	$C_3H_4Cl_2$	110.970	20	0.28	2.8	5	0.18	5
2,3-Dichloropropene	$C_3H_4Cl_2$	110.970	25	0.215	2.15	5	0.36	5
1,2-Dichloro-1,1,2,2-tetrafluoroethane	$C_2Cl_2F_4$	170.921	25	0.013*	0.13*	10	127	13
2,4-Dichlorotoluene	$C_7H_6Cl_2$	161.029	25	0.00260	0.0260	61		
2,6-Dichlorotoluene	$C_7H_6Cl_2$	161.029	25	0.00233	0.0233	61		
2,2-Dichloro-1,1,1-trifluoroethane	$C_2HCl_2F_3$	152.930	25	0.46	4.6	25		
Diclofop-methyl	$C_{16}H_{14}Cl_2O_4$	341.186	20	0.0003	0.003	32		
Dieldrin	$C_{12}H_8Cl_6O$	380.909	25	0.000020	0.00020	67		
Diethanolamine	$C_4H_{11}NO_2$	105.136	20	95.4	20700	10		
1,1-Diethoxyethane	$C_6H_{14}O_2$	118.174	25	5	53	10		
1,2-Diethoxyethane	$C_6H_{14}O_2$	118.174	20	21.0	266	10		
2-(Diethylamino)-*N*-(2,6-dimethylphenyl) acetamide	$C_{14}H_{22}N_2O$	234.337	25	0.38	3.8	40		
o-Diethylbenzene	$C_{10}H_{14}$	134.218	20	0.0071	0.071	40		
p-Diethylbenzene	$C_{10}H_{14}$	134.218	20	0.0025	0.025	40		
Diethyl carbonate	$C_5H_{10}O_3$	118.131	20	1.8	18	40		
Diethyl ether	$C_4H_{10}O$	74.121	0	12.5	143	79	0.088	13
			25	5.9	63	79	0.088	13
			38	4.6	48	79	0.088	13
			82	3.1	32	79	0.088	13
Diethyl glutarate	$C_9H_{16}O_4$	188.221	30	1.20	12.1	20		
			91	0.91	9.2	20		
Diethyl maleate	$C_8H_{12}O_4$	172.179	20	1.56	15.8	20		
			91	1.75	17.8	20		
Diethyl malonate	$C_7H_{12}O_4$	160.168	20	2.26	23.1	20		
			91	2.47	25.3	20		
Diethyl phthalate	$C_{12}H_{14}O_4$	222.237	25	0.12	1.2	40		
trans-Diethylstilbestrol	$C_{18}H_{20}O_2$	268.351	20	0.01	0.1	40		
Diethyl succinate	$C_8H_{14}O_4$	174.195	20	0.19	1.9	40		
Diethyl sulfide	$C_4H_{10}S$	90.187	25	0.307	3.07	40		
Diflubenzuron	$C_{14}H_9ClF_2N_2O_2$	310.683	20	0.00002	0.0002	40		
o-Difluorobenzene	$C_6H_4F_2$	114.093	25	0.114	1.14	2		
m-Difluorobenzene	$C_6H_4F_2$	114.093	25	0.114	1.14	2		
p-Difluorobenzene	$C_6H_4F_2$	114.093	25	0.122	1.22	2		
1,1-Difluoroethane	$C_2H_4F_2$	66.050	20	0.29*	2.9*	50		
Digitoxin	$C_{41}H_{64}O_{13}$	764.939	25	0.0004	0.004	40		
Diglycolic acid	$C_4H_6O_5$	134.088	24	40.0	667	34		
			50	59.9	1490	34		
Digoxin	$C_{41}H_{64}O_{14}$	780.939	25	0.0059	0.059	40		
Dihexyl ether	$C_{12}H_{26}O$	186.333	20	0.019	0.19	20		
			90	0.019	0.19	20		
1,2-Dihydrobenz[j]aceanthrylene	$C_{20}H_{14}$	254.325	27	0.00000035	0.0000035	42,6		
1,3-Dihydro-2*H*-benzimidazol-2-one	$C_7H_6N_2O$	134.135	24	0.37	3.7	54		
1,2-Dihydro-3-methylbenz[j]aceanthrylene	$C_{21}H_{16}$	268.352	25	0.00000022	0.0000022	42,6		
			27	0.00000028	0.0000028	42		
2,3-Dihydro-6-propyl-2-thioxo-4(1*H*)-pyrimidinone	$C_7H_{10}N_2OS$	170.231	25	0.120	1.20	40		
1,7-Dihydro-6*H*-purine-6-thione	$C_5H_4N_4S$	152.178	25	0.0124	0.124	40		
3,4-Dihydro-2*H*-pyran	C_5H_8O	84.117	20	1.04	10.5	20		
			82	2.26	23.1	20		

Name	Mol. Form.	Mol. Wt.	t/°C	Solubility, s: 100 w_2 (mass%)	Solubility, s: g per kg H_2O	Ref.	Henry Const., k_H: kPa m^3mol^{-1}	Ref.
1,4-Dihydroxy-9,10-anthracenedione	$C_{14}H_8O_4$	240.212	25	0.0000096	0.000096	40		
3,4-Dihydroxybenzoic acid	$C_7H_6O_4$	154.121	14	1.8	18	26		
			80	21.3	271	26		
3,12-Dihydroxycholan-24-oic acid, (3α,5β,12α)	$C_{24}H_{40}O_4$	392.573	20	0.001	0.01	40		
17,21-Dihydroxypregna-1,4-diene-3,11,20-trione	$C_{21}H_{26}O_5$	358.428	25	0.012	0.12	40		
17,21-Dihydroxypregn-4-ene-3,11,20-trione	$C_{21}H_{28}O_5$	360.444	25	0.028	0.28	30		
o-Diiodobenzene	$C_6H_4I_2$	329.905	25	0.00192	0.0192	2		
m-Diiodobenzene	$C_6H_4I_2$	329.905	25	0.000185	0.00185	2		
p-Diiodobenzene	$C_6H_4I_2$	329.905	25	0.000893	0.00893	2		
cis-1,2-Diiodoethene	$C_2H_2I_2$	279.846	25	0.046	0.46	25		
trans-1,2-Diiodoethene	$C_2H_2I_2$	279.846	25	0.015	0.15	25		
Diiodomethane	CH_2I_2	267.836	30	0.124	1.24	10	0.032	13
3,5-Diiodo-*L*-tyrosine	$C_9H_9I_2NO_3$	432.981	25	0.062	0.62	26		
Diisopentyl ether	$C_{10}H_{22}O$	158.281	20	0.02	0.2	10		
Diisopropyl ether	$C_6H_{14}O$	102.174	20	0.79	8.0	20	0.26	13
			61	0.22	2.2	20		
1,2-Dimethoxybenzene	$C_8H_{10}O_2$	138.164	20	0.716	7.21	20		
			92	1.073	10.85*	20		
3,3'-Dimethoxybenzidine	$C_{14}H_{16}N_2O_2$	244.289	25	0.006	0.06	40		
Dimethoxymethane	$C_3H_8O_2$	76.095	16	24.4	323	10		
4-(Dimethylamino)azobenzene	$C_{14}H_{15}N_3$	225.289	20	0.00014	0.0014	40		
2',3-Dimethyl-4-aminoazobenzene	$C_{14}H_{15}N_3$	225.289	37	0.0007	0.007	40		
2,5-Dimethylaniline	$C_8H_{11}N$	121.180	20	0.66	6.6	27		
N,N-Dimethylaniline	$C_8H_{11}N$	121.180	25	0.111	1.11	40		
9,10-Dimethylanthracene	$C_{16}H_{14}$	206.282	25	0.0000056	0.000056	42,4		
Dimethylarsinic acid	$C_2H_7AsO_2$	137.998	25	≈41	≈700	40		
7,12-Dimethylbenz[a]anthracene	$C_{20}H_{16}$	256.341	25	0.0000061	0.000061	42		
2,2-Dimethylbutane	C_6H_{14}	86.175	25	0.0021	0.021	3	199	13
2,3-Dimethylbutane	C_6H_{14}	86.175	25	0.0021	0.021	3	144	13
2,2-Dimethyl-1-butanol	$C_6H_{14}O$	102.174	25	0.78	7.9	78,1		
2,3-Dimethyl-2-butanol	$C_6H_{14}O$	102.174	25	4.2	44	1		
3,3-Dimethyl-2-butanol	$C_6H_{14}O$	102.174	25	2.4	25	1		
3,3-Dimethyl-2-butanone	$C_6H_{12}O$	100.158	0	2.92	30.1	83		
			19	1.97	20.1	20		
			25	1.85	18.8	83		
			50	1.46	14.8	83		
			90	1.14	11.5	20		
2,3-Dimethyl-1-butene	C_6H_{12}	84.159	30	0.046	0.46	3		
cis-1,2-Dimethylcyclohexane	C_8H_{16}	112.213	25	0.00060	0.0060	4	36	5
trans-1,2-Dimethylcyclohexane	C_8H_{16}	112.213	30	0.00050	0.0050	57,4	88.2	5
			100	0.00293	0.0293	57,4		
Dimethyl ether	C_2H_6O	46.068	25	35.3*	546	79		
			50	29.2*	412	79		
Dimethylglyoxime	$C_4H_8N_2O_2$	116.119	20	0.06	0.6	40		
3,5-Dimethyl-4-heptanol	$C_9H_{20}O$	144.254	15	0.072	0.72	1		
2,6-Dimethyl-4-heptanone	$C_9H_{18}O$	142.238	21	0.045	0.45	20		
			91	0.037	0.37	20		
1,2-Dimethyl-1*H*-imidazole	$C_5H_8N_2$	96.131	19	94.3	16500	54		
Dimethyl maleate	$C_6H_8O_4$	144.126	25	8.0	87	10		

Name	Mol. Form.	Mol. Wt.	t/°C	Solubility, s 100 w_2 (mass%)	g per kg H_2O	Ref.	Henry Const., k_H kPa m³mol⁻¹	Ref.
Dimethyl malonate	$C_5H_8O_4$	132.116	19	14.9	175	20		
			90	29.8	425	20		
1,3-Dimethylnaphthalene	$C_{12}H_{12}$	156.223	25	0.0008	0.008	4		
1,4-Dimethylnaphthalene	$C_{12}H_{12}$	156.223	25	0.00114	0.0114	4		
1,5-Dimethylnaphthalene	$C_{12}H_{12}$	156.223	25	0.00031	0.0031	4	0.036	28
2,3-Dimethylnaphthalene	$C_{12}H_{12}$	156.223	25	0.00025	0.0025	4		
2,6-Dimethylnaphthalene	$C_{12}H_{12}$	156.223	25	0.00017	0.0017	4		
Dimethyl oxalate	$C_4H_6O_4$	118.089	20	5.82	61.8	27		
2,2-Dimethylpentane	C_7H_{16}	100.202	25	0.00044	0.0044	3	318	5
2,3-Dimethylpentane	C_7H_{16}	100.202	25	0.00052	0.0052	3	175	5
2,4-Dimethylpentane	C_7H_{16}	100.202	25	0.00042	0.0042	3	323	13
3,3-Dimethylpentane	C_7H_{16}	100.202	25	0.00059	0.0059	3	186	5
2,3-Dimethyl-2-pentanol	$C_7H_{16}O$	116.201	25	1.5	15	1		
2,4-Dimethyl-2-pentanol	$C_7H_{16}O$	116.201	25	1.3	13	1		
2,2-Dimethyl-3-pentanol	$C_7H_{16}O$	116.201	25	0.82	8.2	1		
2,3-Dimethyl-3-pentanol	$C_7H_{16}O$	116.201	25	1.6	16	1		
2,4-Dimethyl-3-pentanol	$C_7H_{16}O$	116.201	25	0.70	7.0	1		
2,4-Dimethyl-3-pentanone	$C_7H_{14}O$	114.185	20	0.52	5.2	20		
			90	0.30	3.0	20		
N,N-Dimethyl-*N'*-phenylurea	$C_9H_{12}N_2O$	164.203	25	0.32	3.2	40		
Dimethyl phthalate	$C_{10}H_{10}O_4$	194.184	25	0.40	4.0	15		
2,2-Dimethyl-1-propanol	$C_5H_{12}O$	88.148	12	3.87	40.3	78,1		
			25	3.26	33.7	78,1		
			80	2.84	29.2	78,1		
4-(1,1-Dimethylpropyl)phenol	$C_{11}H_{16}O$	164.244	25	0.017	0.17	40		
Dimethyl succinate	$C_6H_{10}O_4$	146.141	21	12.4	142	20		
			92	17.1	206	20		
Dimethyl sulfate	$C_2H_6O_4S$	126.132	18	2.7	28	27		
Dimethyl sulfide	C_2H_6S	62.134	25	2	20	10		
Dimethyl sulfoxide	C_2H_6OS	78.133	25	25.3	339	10		
Dimethyl terephthalate	$C_{10}H_{10}O_4$	194.184	25	0.00328	0.0328	40		
Dimethyl tetrachloroterephthalate	$C_{10}H_6Cl_4O_4$	331.965	25	0.00005	0.0005	40		
N,N-Dimethyl-*N'*-[3-(trifluoromethyl)phenyl]urea	$C_{10}H_{11}F_3N_2O$	232.201	20	0.0105	0.105	40		
2,4-Dinitroaniline	$C_6H_5N_3O_4$	183.122	25	0.0078	0.078	40		
1,2-Dinitrobenzene	$C_6H_4N_2O_4$	168.107	20	0.21	2.1	27		
1,3-Dinitrobenzene	$C_6H_4N_2O_4$	168.107	20	2.09	21.3	27		
1,4-Dinitrobenzene	$C_6H_4N_2O_4$	168.107	20	1.30	13.2	27		
3,5-Dinitrobenzoic acid	$C_7H_4N_2O_6$	212.116	25	0.134	1.34	27		
2,4-Dinitrophenol	$C_6H_4N_2O_5$	184.106	25	0.069	0.69	48,51		
			35	0.098	0.98	48,51		
Dipentyl ether	$C_{10}H_{22}O$	158.281	25	0.11	1.1	81		
Diphenamid	$C_{16}H_{17}NO$	239.312	27	0.026	0.26	32		
Diphenylamine	$C_{12}H_{11}N$	169.222	20	0.0055	0.055	40		
			50	0.0058	0.058	40		
1,2-Diphenylethane	$C_{14}H_{14}$	182.261	25	0.00044	0.0044	6	0.017	12
Diphenyl ether	$C_{12}H_{10}O$	170.206	25	0.0018	0.0180	6	0.027	13
Diphenylmethane	$C_{13}H_{12}$	168.234	25	0.00014	0.0014	42,4	0.001	12
Diphenyl phthalate	$C_{20}H_{14}O_4$	318.323	24	0.000008	0.00008	40		
1,3-Diphenyl-1-triazene	$C_{12}H_{11}N_3$	197.235	20	0.050	0.50	40		

Name	Mol. Form.	Mol. Wt.	t/°C	Solubility, s: 100 w_2 (mass%)	Solubility, s: g per kg H_2O	Ref.	Henry Const., k_H: kPa m^3mol^{-1}	Ref.
N,N'-Diphenylurea	$C_{13}H_{12}N_2O$	212.246	20	0.015	0.15	40		
Dipropylamine	$C_6H_{15}N$	101.190	20	2.5	26	10		
Dipropyl ether	$C_6H_{14}O$	102.174	0	2.67	27.4	80	0.26	13
			25	0.91	9.2	80	0.26	13
Diuron	$C_9H_{10}Cl_2N_2O$	233.093	25	0.0042	0.042	40		
Docosane	$C_{22}H_{46}$	310.600	22	0.0000006	0.000006	37		
Dodecane	$C_{12}H_{26}$	170.334	25	0.00000037	0.0000037	4	750	5
Dodecanedioic acid	$C_{12}H_{22}O_4$	230.301	20	0.004	0.04	40		
Dodecanoic acid	$C_{12}H_{24}O_2$	200.318	20	0.0055	0.055	26		
1-Dodecanol	$C_{12}H_{26}O$	186.333	25	0.0004	0.004	1		
Droperidol	$C_{22}H_{22}FN_3O_2$	379.427	30	0.00041	0.0041	40		
Eicosane	$C_{20}H_{42}$	282.547	25	0.00000019	0.0000019	42,4		
Emetine	$C_{29}H_{40}N_2O_4$	480.639	15	0.096	0.96	40		
Endrin	$C_{12}H_8Cl_6O$	380.909	25	0.000025	0.00025	67		
l-Ephedrine	$C_{10}H_{15}NO$	165.232	25	0.57	5.7	40		
Epichlorohydrin	C_3H_5ClO	92.524	20	6.58	70.4	10	0.003	13
			65	7.2	78	40		
Epinephrine	$C_9H_{13}NO_3$	183.204	20	0.018	0.18	40		
1,2-Epoxy-4-(epoxyethyl)cyclohexane	$C_8H_{12}O_2$	140.180	20	13.4	155	40		
2,3-Epoxy-α-pinane	$C_{10}H_{16}O$	152.233	25	0.039	0.39	52		
Erythromycin	$C_{37}H_{67}NO_{13}$	733.927	30	0.12	1.2	40		
			80	0.04	0.4	40		
Estra-1,3,5(10)-triene-3,17-diol (17β)	$C_{18}H_{24}O_2$	272.383	25	0.000151	0.00151	49		
Estrone	$C_{18}H_{22}O_2$	270.367	25	0.000130	0.00130	49		
Ethane	C_2H_6	30.069	25	0.00568*	0.0568*	18	50.6	5
1,2-Ethanediol, diacetate	$C_6H_{10}O_4$	146.141	25	13.3	153	40		
Ethinylestradiol	$C_{20}H_{24}O_2$	296.404	25	0.000921	0.00921	49		
Ethoxybenzene	$C_8H_{10}O$	122.164	25	0.12	1.2	10		
2-Ethoxyethyl acetate	$C_6H_{12}O_3$	132.157		14	163	30		
N-(4-Ethoxyphenyl)acetamide	$C_{10}H_{13}NO_2$	179.216	25	0.0502	0.502	40		
Ethyl acetate	$C_4H_8O_2$	88.106	25	8.08	87.9	10		
Ethyl acetoacetate	$C_6H_{10}O_3$	130.141	25	12	136	10		
Ethyl acrylate	$C_5H_8O_2$	100.117	25	1.50	15.2	10		
Ethylbenzene	C_8H_{10}	106.165	0	0.020	0.20	4,89		
			25	0.0161	0.161	22,89	0.843	22
			40	0.0200	0.200	4,89		
Ethyl benzoate	$C_9H_{10}O_2$	150.174	25	0.083	0.83	20		
Ethyl butanoate	$C_6H_{12}O_2$	116.158	20	0.49	4.9	10		
2-Ethyl-1-butanol	$C_6H_{14}O$	102.174	20	0.92	9.3	78		
			50	0.80	8.1	78		
Ethyl carbamate	$C_3H_7NO_2$	89.094	15	48	920	27		
Ethyl cyanoacetate	$C_5H_7NO_2$	113.116	20	25.9	350	10		
Ethylcyclohexane	C_8H_{16}	112.213	30	0.00061	0.0061	57,4		
			100	0.00212	0.0212	57,4		
Ethylcyclopentane	C_7H_{14}	98.186	20	0.012	0.12	3		
Ethyl decanoate	$C_{12}H_{24}O_2$	200.318	20	0.0015	0.015	27		
Ethylene	C_2H_4	28.053	25	0.01336*	0.1336*	19	21.7	5
Ethyleneimine	C_2H_5N	43.068	20	0.90	9.1	40		
Ethyl formate	$C_3H_6O_2$	74.079	25	11.8	134	10		

Name	Mol. Form.	Mol. Wt.	t/°C	Solubility, s: 100 w_2 (mass%)	Solubility, s: g per kg H_2O	Ref.	Henry Const., k_H: kPa m^3mol^{-1}	Ref.
Ethyl heptanoate	$C_9H_{18}O_2$	158.238	20	0.029	0.29	27		
Ethyl hexanoate	$C_8H_{16}O_2$	144.212	20	0.063	0.63	27		
2-Ethyl-1-hexanol	$C_8H_{18}O$	130.228	25	0.071	0.71	78		
			50	0.074	0.74	78		
2-Ethylhexylamine	$C_8H_{19}N$	129.244	20	0.25	2.5	10		
Ethyl 4-hydroxybenzoate	$C_9H_{10}O_3$	166.173	25	0.0080	0.080	40		
Ethyl isopropyl ether	$C_5H_{12}O$	88.148	25	0.52	5.2	79		
Ethyl 2-methylbutanoate, (+)	$C_7H_{14}O_2$	130.185	19	0.257	2.58	20		
			91	0.151	1.51	20		
Ethyl 3-methylbutanoate	$C_7H_{14}O_2$	130.185	20	0.2	2	10		
Ethyl *N*-methylcarbamate	$C_4H_9NO_2$	103.120	15	69	2230	27		
1-Ethylnaphthalene	$C_{12}H_{12}$	156.223	25	0.00101	0.0101	4	0.039	12
2-Ethylnaphthalene	$C_{12}H_{12}$	156.223	25	0.00080	0.0080	4	0.078	12
O-Ethyl *O*-*p*-nitrophenyl benzenethiophosphonate	$C_{14}H_{14}NO_4PS$	323.304	22	0.00031	0.0031	40		
N-Ethyl-*N*-nitrosourea	$C_3H_7N_3O_2$	117.107	20	1.3	13	40		
Ethyl nonanoate	$C_{11}H_{22}O_2$	186.292	20	0.003	0.03	27		
Ethyl octanoate	$C_{10}H_{20}O_2$	172.265	20	0.007	0.07	27		
Ethyl pentanoate	$C_7H_{14}O_2$	130.185	25	0.3	3	27		
3-Ethyl-3-pentanol	$C_7H_{16}O$	116.201	25	1.7	17	1		
4-Ethylphenol	$C_8H_{10}O$	122.164	20	0.59	5.9	40		
Ethyl propanoate	$C_5H_{10}O_2$	102.132	20	1.92	19.6	10		
Ethyl *N*-propylcarbamate	$C_6H_{13}NO_2$	131.173	15	7.70	83.4	27		
Ethyl propyl ether	$C_5H_{12}O$	88.148	25	1.87	19.2	79		
2-Ethyltoluene	C_9H_{12}	120.191	25	0.0075	0.075	89,5	0.529	13
4-Ethyltoluene	C_9H_{12}	120.191	25	0.0094	0.094	5	0.500	13
Ethyl vinyl ether	C_4H_8O	72.106	20	0.9	9	10		
Etoposide	$C_{29}H_{32}O_{13}$	588.556	20	0.02	0.2	40		
Eucalyptol	$C_{10}H_{18}O$	154.249	21	0.379	3.79	40		
			50	0.170	1.70	40		
Fenamiphos	$C_{13}H_{22}NO_3PS$	303.358	20	0.0329	0.329	40		
Fenbutatin oxide	$C_{60}H_{78}OSn_2$	1052.68	23	0.0000005	0.000005	32		
α-Fenchol, (+)-	$C_{10}H_{18}O$	154.249	25	0.083	0.83	52		
Fenchone	$C_{10}H_{16}O$	152.233	20	0.2	2	84		
Fenoxycarb	$C_{17}H_{19}NO_4$	301.338	20	0.0006	0.006	32		
Ferbam	$C_9H_{18}FeN_3S_6$	416.494	20	0.013	0.13	40		
Fluoranthene	$C_{16}H_{10}$	202.250	20	0.000017	0.00017	42		
			25	0.000021	0.00021	42,22	0.00096	22
9*H*-Fluorene	$C_{13}H_{10}$	166.218	0	0.00007	0.0007	42,4		
			25	0.00019	0.0019	42,22	0.00787	22
			50	0.00063	0.0063	42,4		
Fluorescein	$C_{20}H_{12}O_5$	332.306	20	0.005	0.05	27		
Fluorobenzene	C_6H_5F	96.102	19	0.170	1.70	20	0.70	11
			80	0.188	1.88	20	0.70	11
2-Fluorobenzoic acid	$C_7H_5FO_2$	140.112	25	0.72	7.2	27		
3-Fluorobenzoic acid	$C_7H_5FO_2$	140.112	25	0.15	1.5	27		
4-Fluorobenzoic acid	$C_7H_5FO_2$	140.112	25	0.12	1.2	27		
Fluoroethane	C_2H_5F	48.059	25	0.216*	2.16*	14		
Fluoromethane	CH_3F	34.033	0	0.420*	4.20*	50		
			25	0.201*	2.01*	50		

Name	Mol. Form.	Mol. Wt.	t/°C	Solubility, s: 100 w_2 (mass%)	Solubility, s: g per kg H_2O	Ref.	Henry Const., k_H: kPa m³mol⁻¹	Ref.
			80	0.082*	0.82*	50		
1-Fluoropropane	C_3H_7F	62.086	14	0.386*	3.86*	14		
2-Fluoropropane	C_3H_7F	62.086	15	0.366*	3.66*	14		
5-Fluorouracil	$C_4H_3FN_2O_2$	130.077	25	1.77	18.0	72		
Folic acid	$C_{19}H_{19}N_7O_6$	441.397	0	0.001	0.01	26		
			100	0.05	0.5	26		
Folpet	$C_9H_4Cl_3NO_2S$	296.558	20	0.00010	0.0010	40		
β-*D*-Fructose	$C_6H_{12}O_6$	180.155	20	≈31	≈450	40		
Furan	C_4H_4O	68.074	25	1	10	10	0.54	13
2-Furancarboxylic acid	$C_5H_4O_3$	112.084	25	4.76	50.0	33		
			50	25.2	337	33		
Furfural	$C_5H_4O_2$	96.085	20	8.2	89	10		
Galactaric acid	$C_6H_{10}O_8$	210.138	14	0.33	3.3	40		
D-Galactose	$C_6H_{12}O_6$	180.155	20	40.6	684	27		
D-Glucitol	$C_6H_{14}O_6$	182.171	20	≈41	≈700	40		
α-*D*-Glucose	$C_6H_{12}O_6$	180.155	15	45.0	818	27		
			30	54.6	1200	27		
			80	81.5	4400	27		
DL-Glutamic acid	$C_5H_9NO_4$	147.130	25	2.30	23.5	29		
L-Glutamic acid	$C_5H_9NO_4$	147.130	10	0.444	4.46	75		
			25	0.874	8.81	75		
			50	2.13	21.8	75		
L-Glutamine	$C_5H_{10}N_2O_3$	146.144	25	4.0	42	26		
Glycerol triacetate	$C_9H_{14}O_6$	218.203	25	5.8	62	10		
Glycine	$C_2H_5NO_2$	75.067	25	18.5	227	70		
			36	22.1	284	70		
			50	26.1	353	70		
Glycolic acid	$C_2H_4O_3$	76.051	25	71.2	2470	34		
			55	77.9	3520	34		
N-Glycylglycine	$C_4H_8N_2O_3$	132.118	25	18.8	232	47,29		
Glyphosate	$C_3H_8NO_5P$	169.074	25	1.2	12	32		
Guanidinoacetic acid	$C_3H_7N_3O_2$	117.107	25	0.5	5	26		
Guanine	$C_5H_5N_5O$	151.127	25	0.0068	0.068	29		
Guanosine	$C_{10}H_{13}N_5O_5$	283.241	25	0.0500	0.500	29		
Haloperidol	$C_{21}H_{23}ClFNO_2$	375.865	30	0.0003	0.003	40		
Heptachlor	$C_{10}H_5Cl_7$	373.318	25	0.000018	0.00018	67		
2,2′,3,3′,4,4′,6-Heptachlorobiphenyl	$C_{12}H_3Cl_7$	395.323	25	0.0000002	0.000002	7	0.0054	7
Heptadecanoic acid	$C_{17}H_{34}O_2$	270.451	20	0.00042	0.0042	26		
1,6-Heptadiyne	C_7H_8	92.139	25	0.125	1.25	3		
Heptanal	$C_7H_{14}O$	114.185	11	0.124	1.24	27		
Heptane	C_7H_{16}	100.202	25	0.000242	0.00242	46		
			50	0.000341	0.00341	46	209	13
			75	0.000570	0.00570	46		
			100	0.00108	0.0108	46		
Heptanedioic acid	$C_7H_{12}O_4$	160.168	25	6.347	67.77	33		
			50	42.80	748	33		
Heptanoic acid	$C_7H_{14}O_2$	130.185	15	0.24	2.4	27		
1-Heptanol	$C_7H_{16}O$	116.201	0	0.236	2.37	78		
			25	0.164	1.64	78,1	0.00562	28

Name	Mol. Form.	Mol. Wt.	t/°C	Solubility, s: 100 w_2 (mass%)	Solubility, s: g per kg H_2O	Ref.	Henry Const., k_H: kPa m³mol⁻¹	Ref.
			50	0.164	1.64	78,1		
			90	0.245	2.46	78		
2-Heptanol	$C_7H_{16}O$	116.201	30	0.33	3.3	1		
3-Heptanol, (*S*)-	$C_7H_{16}O$	116.201	25	0.43	4.3	1		
4-Heptanol	$C_7H_{16}O$	116.201	25	0.47	4.7	1		
2-Heptanone	$C_7H_{14}O$	114.185	25	0.435	4.37	20	0.0171	28
			90	0.353	3.53	20	0.0171	28
3-Heptanone	$C_7H_{14}O$	114.185	20	0.479	4.81	20		
			90	0.309	3.10	20		
4-Heptanone	$C_7H_{14}O$	114.185	20	0.457	4.57	20		
			90	0.316	3.16	20		
1-Heptene	C_7H_{14}	98.186	25	0.032	0.32	3	40.3	13
trans-2-Heptene	C_7H_{14}	98.186	25	0.015	0.15	3	42.2	13
Heptyl butanoate	$C_{11}H_{22}O_2$	186.292	20	0.028	0.28	20		
			80	0.020	0.20	20		
1-Heptyne	C_7H_{12}	96.170	25	0.0094	0.094	3	4.47	13
Hesperetin	$C_{16}H_{14}O_6$	302.278	15	0.00004	0.0004	71		
			25	0.00014	0.0014	71		
			35	0.00052	0.0052	71		
Hexachlorobenzene	C_6Cl_6	284.782	25	0.00000096	0.0000096	58	0.131	11
			35	0.0000018	0.000018	58		
			55	0.0000038	0.000038	58		
2,2',3,3',4,4'-Hexachlorobiphenyl	$C_{12}H_4Cl_6$	360.878	25	0.00000006	0.0000006	7	0.0354	31
2,2',4,4',6,6'-Hexachlorobiphenyl	$C_{12}H_4Cl_6$	360.878	25	0.0000003	0.000003	41	0.818	7
2,2',3,3',6,6'-Hexachlorobiphenyl	$C_{12}H_4Cl_6$	360.878	25	0.0000004	0.000004	41		
Hexachloro-1,3-butadiene	C_4Cl_6	260.761	25	0.41	4.1	35		
1,2,3,4,5,6-Hexachlorocyclohexane, (1α,2α,3β,4α,5α,6β)	$C_6H_6Cl_6$	290.830	25	0.00078	0.0078	60		
			45	0.0015	0.015	60		
1,2,3,4,5,6-Hexachlorocyclohexane, (1α,2α,3β,4α,5β,6β)	$C_6H_6Cl_6$	290.830	25	0.00018	0.0018	60		
1,2,3,4,5,6-Hexachlorocyclohexane, (1α,2β,3α,4β,5α,6β)	$C_6H_6Cl_6$	290.830	25	0.00002	0.0002	60		
Hexachloroethane	C_2Cl_6	236.739	25	0.005	0.05	25	0.85	13
Hexachloropropene	C_3Cl_6	248.750	20	0.00118	0.0118	35		
Hexacosafluorododecane	$C_{12}F_{26}$	638.086	20	0.00000096	0.0000096	35		
Hexacosane	$C_{26}H_{54}$	366.707	25	0.00000017	0.0000017	42,37		
Hexadecane	$C_{16}H_{34}$	226.441	25	0.0000004	0.000004	42,37		
Hexadecanoic acid	$C_{16}H_{32}O_2$	256.424	20	0.00072	0.0072	26		
1-Hexadecanol	$C_{16}H_{34}O$	242.440	25	0.000003	0.00003	1		
1,5-Hexadiene	C_6H_{10}	82.143	25	0.017	0.17	3		
Hexafluorobenzene	C_6F_6	186.054	8	0.0778	0.778	53		
			28	0.0616	0.616	53		
			67	0.0636	0.636	53		
Hexahydro-1,3,5-trinitro-1,3,5-triazine	$C_3H_6N_6O_6$	222.116	3	0.0014	0.014	59		
			20	0.0037	0.037	59		
			25	0.0060	0.060	17		
			34	0.0086	0.086	59		
Hexamethylenetetramine	$C_6H_{12}N_4$	140.186	12	44.8	812	27		
Hexane	C_6H_{14}	86.175	25	0.00098	0.0098	46		

				Solubility, s			Henry Const., k_H	
Name	Mol. Form.	Mol. Wt.	t/°C	100 w_2 (mass%)	g per kg H_2O	Ref.	kPa m^3mol^{-1}	Ref.
			50	0.00114	0.0114	46		
			75	0.00167	0.0167	46	183	13
			100	0.00291	0.0291	46		
1,6-Hexanediamine	$C_6H_{16}N_2$	116.204	5	≈42	≈720	40		
Hexanedinitrile	$C_6H_8N_2$	108.141	20	0.80	8.0	16		
1,6-Hexanedioic acid	$C_6H_{10}O_4$	146.141	15	1.48	15.0	26		
			100	61.5	1600	26		
Hexanoic acid	$C_6H_{12}O_2$	116.158	25	1.01	10.2	64		
			35	1.09	11.0	64		
			60	1.16	11.7	26		
1-Hexanol	$C_6H_{14}O$	102.174	0	0.79	7.9	1		
			10	0.70	7.0	78		
			25	0.59	5.9	78,1		
			50	0.55	5.5	78,1		
2-Hexanol	$C_6H_{14}O$	102.174	25	1.4	14	1		
3-Hexanol	$C_6H_{14}O$	102.174	25	1.6	16	1		
2-Hexanone	$C_6H_{12}O$	100.158	10	1.91	19.5	83		
			25	1.49	15.1	83		
			50	1.17	11.8	83		
3-Hexanone	$C_6H_{12}O$	100.158	25	1.47	14.9	83		
Hexatriacontane	$C_{36}H_{74}$	506.973	25	0.00000017	0.0000017	42,37		
Hexazinone	$C_{12}H_{20}N_4O_2$	252.313	25	3.2	33	40		
1-Hexene	C_6H_{12}	84.159	25	0.0053	0.053	3	41.8	5
trans-2-Hexene	C_6H_{12}	84.159	25	0.0067	0.067	3		
1-Hexen-3-ol	$C_6H_{12}O$	100.158	25	2.52	25.9	1		
4-Hexen-2-ol	$C_6H_{12}O$	100.158	25	3.81	39.6	1		
Hexyl acetate	$C_8H_{16}O_2$	144.212	20	0.02	0.2	10		
sec-Hexyl acetate	$C_8H_{16}O_2$	144.212	20	0.13	1.3	10		
Hexylbenzene	$C_{12}H_{18}$	162.271	25	0.00021	0.0021	4		
4-Hexyl-1,3-benzenediol	$C_{12}H_{18}O_2$	194.270	18	0.05	0.5	40		
Hexyl butanoate	$C_{10}H_{20}O_2$	172.265	29	0.021	0.21	20		
1-Hexyne	C_6H_{10}	82.143	25	0.036	0.36	3	4.14	13
L-Histidine	$C_6H_9N_3O_2$	155.154	25	4.17	43.5	26		
Homocystine	$C_8H_{16}N_2O_4S_2$	268.354	25	0.02	0.2	26		
L-Homoserine	$C_4H_9NO_3$	119.119	25	52.4	1100	26		
Hydramethylnon	$C_{25}H_{24}F_6N_4$	494.476	20	0.0000006	0.000006	32		
Hydrochlorothiazide	$C_7H_8ClN_3O_4S_2$	297.740	25	0.007	0.07	40		
Hydrocortisone	$C_{21}H_{30}O_5$	362.460	25	0.029	0.29	40		
Hydroflumethiazide	$C_8H_8F_3N_3O_4S_2$	331.293	37	0.068	0.68	40		
p-Hydroquinone	$C_6H_6O_2$	110.111	25	7.42	80.1	27		
17-Hydroxyandrost-4-en-3-one, (17β)	$C_{19}H_{28}O_2$	288.424	25	0.0024	0.024	40		
4-Hydroxybenzaldehyde	$C_7H_6O_2$	122.122	30	1.27	12.9	40		
2-Hydroxybenzamide	$C_7H_7NO_2$	137.137	10	0.122	1.22	44		
			25	0.241	2.42	44		
			50	0.737	7.42	44		
α-Hydroxybenzeneacetic acid	$C_8H_8O_3$	152.148	25	11.3	127	27		
2-Hydroxybenzoic acid	$C_7H_6O_3$	138.121	10	0.119	1.19	43,33		
			25	0.189	1.89	43,33		
			50	0.521	5.24	43,33		

Name	Mol. Form.	Mol. Wt.	t/°C	Solubility, s: 100 w_2 (mass%)	Solubility, s: g per kg H_2O	Ref.	Henry Const., k_H: kPa m^3mol^{-1}	Ref.
4-Hydroxybenzoic acid	$C_7H_6O_3$	138.121	15	0.8	8	26		
			75	2.5	26	27		
2-Hydroxybiphenyl	$C_{12}H_{10}O$	170.206	25	0.07	0.7	40		
4-Hydroxybiphenyl	$C_{12}H_{10}O$	170.206	25	0.0056	0.056	40		
4-Hydroxy-3-methoxybenzaldehyde	$C_8H_8O_3$	152.148	25	0.247	2.47	8		
3-Hydroxy-4-oxo-4*H*-pyran-2,6-dicarboxylic acid	$C_7H_4O_7$	200.103	25	0.84	8.4	27		
N-(4-Hydroxyphenyl)acetamide	$C_8H_9NO_2$	151.163	25	1.3	13	40		
trans-4-Hydroxy-*L*-proline	$C_5H_9NO_3$	131.130	25	26.5	361	26		
Hyoscyamine	$C_{17}H_{23}NO_3$	289.370	20	0.36	3.6	40		
Hypoxanthine	$C_5H_4N_4O$	136.112	25	0.070	0.70	29		
Ibuprofen	$C_{13}H_{18}O_2$	206.281	25	0.0011	0.011	40		
			60	0.0048	0.048	40		
Imazaquin	$C_{17}H_{17}N_3O_3$	311.335	20	0.009	0.09	32		
Imidacloprid	$C_9H_{10}ClN_5O_2$	255.66	30	0.038	0.38	73		
			51	0.117	1.17	73		
Imidazole	$C_3H_4N_2$	68.077	19	67.3	2060	54		
2,4-Imidazolidinedione	$C_3H_4N_2O_2$	100.076	25	3.93	40.9	29		
Imidodicarbonic diamide	$C_2H_5N_3O_2$	103.080	15	1.5	15	40		
Iminodiacetic acid	$C_4H_7NO_4$	133.104	5	2.32	23.8	40		
Indan	C_9H_{10}	118.175	25	0.010	0.10	4		
1*H*-Indazole	$C_7H_6N_2$	118.136	20	0.0827	0.827	6		
Indeno[1,2,3-cd]pyrene	$C_{22}H_{12}$	276.330	20	0.00000002	0.0000002	40		
1*H*-Indole	C_8H_7N	117.149	20	0.187	1.87	6		
Indomethacin	$C_{19}H_{16}ClNO_4$	357.788	25	0.001	0.01	40		
Inosine	$C_{10}H_{12}N_4O_5$	268.226	20	1.6	16	29		
Iodobenzene	C_6H_5I	204.008	10	0.0193	0.193	2		
			25	0.0226	0.226	2	0.078	11
			45	0.0279	0.279	2		
2-Iodobenzoic acid	$C_7H_5IO_2$	248.018	25	0.095	0.95	27		
3-Iodobenzoic acid	$C_7H_5IO_2$	248.018	25	0.016	0.16	27		
4-Iodobenzoic acid	$C_7H_5IO_2$	248.018	25	0.0027	0.027	27		
1-Iodobutane	C_4H_9I	184.018	17	0.021	0.21	10	1.87	13
Iodoethane	C_2H_5I	155.965	0	0.44	4.4	25		
			25	0.40	4.0	25	0.52	13
1-Iodoheptane	$C_7H_{15}I$	226.098	25	0.00035	0.0035	35		
Iodomethane	CH_3I	141.939	20	1.4	14	10	0.54	13
1-Iodopropane	C_3H_7I	169.992	0	0.114	1.14	35		
			20	0.100	1.00	35	0.93	13
2-Iodopropane	C_3H_7I	169.992	0	0.167	1.67	35		
			20	0.140	1.40	35		
5-Iodouracil	$C_4H_3IN_2O_2$	237.983	25	0.49	4.9	72		
trans-β-Ionone	$C_{13}H_{20}O$	192.297	25	0.017	0.17	52		
Iopanoic acid	$C_{11}H_{12}I_3NO_2$	570.932	37	0.034	0.34	40		
Iprodione	$C_{13}H_{13}Cl_2N_3O_3$	330.166	20	0.0013	0.013	40		
Isobutanal	C_4H_8O	72.106	20	9.1	100	10		
Isobutane	C_4H_{10}	58.122	25	0.00535*	0.0535*	18	120	5
Isobutene	C_4H_8	56.107	25	0.0263*	0.263*	5	21.6	13
Isobutyl acetate	$C_6H_{12}O_2$	116.158	20	0.63	6.3	10		
Isobutylbenzene	$C_{10}H_{14}$	134.218	25	0.0010	0.010	4	3.32	11

Name	Mol. Form.	Mol. Wt.	t/°C	Solubility, s: 100 w_2 (mass%)	Solubility, s: g per kg H_2O	Ref.	Henry Const., k_H: kPa m³mol⁻¹	Ref.
Isobutyl formate	$C_5H_{10}O_2$	102.132	22	1.0	10	10		
Isobutyl isobutanoate	$C_8H_{16}O_2$	144.212	20	0.5	5	10		
Isobutyl propanoate	$C_7H_{14}O_2$	130.185	19	0.225	2.26	20		
			91	0.142	1.42	20		
1*H*-Isoindole-1,3(2*H*)-dione	$C_8H_5NO_2$	147.132	25	0.036	0.36	40		
L-Isoleucine	$C_6H_{13}NO_2$	131.173	25	3.31	34.2	26		
Isoniazid	$C_6H_7N_3O$	137.139	25	11.0	124	40		
Isopentane	C_5H_{12}	72.149	25	0.00485	0.0485	3	479	13
Isopentyl acetate	$C_7H_{14}O_2$	130.185	20	0.2	2	10		
Isopentyl formate	$C_6H_{12}O_2$	116.158	22	0.3	3	27		
Isophorone	$C_9H_{14}O$	138.206	20	1.57	16.0	20		
			80	1.27	12.9	20		
Isophthalic acid	$C_8H_6O_4$	166.132	10	0.0062	0.062	76		
			25	0.0154	0.154	56		
			50	0.0395	0.395	56		
			80	0.123	1.23	56		
Isopropenylbenzene	C_9H_{10}	118.175	20	0.0116	0.116	40		
Isopropyl acetate	$C_5H_{10}O_2$	102.132	20	2.9	30	10		
Isopropylbenzene	C_9H_{12}	120.191	25	0.0050	0.050	22	1.466	22
1-Isopropyl-2-methylbenzene	$C_{10}H_{14}$	134.218	25	0.00482	0.0482	23		
1-Isopropyl-3-methylbenzene	$C_{10}H_{14}$	134.218	25	0.00425	0.0425	23		
1-Isopropyl-4-methylbenzene	$C_{10}H_{14}$	134.218	25	0.0051	0.051	23	0.80	5
Isopropyl phenylcarbamate	$C_{10}H_{13}NO_2$	179.216	20	0.01	0.1	40		
Isoquinoline	C_9H_7N	129.159	20	0.452	4.52	6		
Isosorbide dinitrate	$C_6H_8N_2O_8$	236.136	25	0.055	0.55	40		
Kepone	$C_{10}Cl_{10}O$	490.636	100	0.4	4	40		
L-Lanthionine	$C_6H_{12}N_2O_4S$	208.235	25	0.15	1.5	26		
Lasiocarpine	$C_{21}H_{33}NO_7$	411.490	20	0.67	6.7	40		
L-Leucine	$C_6H_{13}NO_2$	131.173	25	2.32	23.8	62		
Levodopa	$C_9H_{11}NO_4$	197.188	20	0.165	1.65	63		
d-Limonene	$C_{10}H_{16}$	136.234	0	0.001	0.01	4		
			25	0.0020	0.020	52		
Linalol	$C_{10}H_{18}O$	154.249	25	0.156	1.56	52		
Linuron	$C_9H_{10}Cl_2N_2O_2$	249.093	25	0.0075	0.075	40		
L-Lysine	$C_6H_{14}N_2O_2$	146.187	25	0.58	5.8	26		
Maleic acid	$C_4H_4O_4$	116.073	25	44.1	789	26		
Malic acid	$C_4H_6O_5$	134.088	26	59	1440	26		
Malonic acid	$C_3H_4O_4$	104.062	0	37.9	610	26		
			20	42.4	736	26		
			50	48.1	927	26		
Malononitrile	$C_3H_2N_2$	66.061	20	10.6	119	40		
α-Maltose	$C_{12}H_{22}O_{11}$	342.296	20	51.9	1080	27		
D-Mannitol	$C_6H_{14}O_6$	182.171	25	17.7	215	27		
Mefenamic acid	$C_{15}H_{15}NO_2$	241.286	20	0.0026	0.026	40		
Melphalan	$C_{13}H_{18}Cl_2N_2O_2$	305.200	30	0.44	4.4	40		
Mercury(II) phenyl acetate	$C_8H_8HgO_2$	336.74	20	0.2	2	30		
Mesityl oxide	$C_6H_{10}O$	98.142	20	2.8	29	83		
Methacrylic acid	$C_4H_6O_2$	86.090	20	8.9	98	10		
Methane	CH_4	16.043	25	0.00227*	0.0227*	18	67.4	5

Name	Mol. Form.	Mol. Wt.	t/°C	Solubility, s: 100 w_2 (mass%)	Solubility, s: g per kg H_2O	Ref.	Henry Const., k_H: kPa m^{3}mol^{-1}	Ref.
Methazolamide	$C_5H_8N_4O_3S_2$	236.273	15	0.0472	0.472	40		
Methazole	$C_9H_6Cl_2N_2O_3$	261.061	24	0.00015	0.0015	40		
Methidathion	$C_6H_{11}N_2O_4PS_3$	302.330	20	0.0187	0.187	40		
L-Methionine	$C_5H_{11}NO_2S$	149.212	25	5.3	56	26		
Methomyl	$C_5H_{10}N_2O_2S$	162.210	25	5.5	58	40		
Methoxsalen	$C_{12}H_8O_4$	216.190	30	0.0048	0.048	40		
2-Methoxyaniline	C_7H_9NO	123.152	25	1.24	12.6	40		
4-Methoxyaniline	C_7H_9NO	123.152	20	1.14	11.5	40		
4-Methoxybenzaldehyde	$C_8H_8O_2$	136.149	25	0.429	4.29	40		
4-Methoxybenzoic acid	$C_8H_8O_3$	152.148	25	0.023	0.23	27		
Methoxychlor	$C_{16}H_{15}Cl_3O_2$	345.648	25	0.000005	0.00005	40		
2-Methoxy-2-methylbutane	$C_6H_{14}O$	102.174	20	1.10	11.1	20		
			79	0.36	3.6	20		
4-Methoxyphenol	$C_7H_8O_2$	124.138	20	2.51	25.7	40		
Methyclothiazide	$C_9H_{11}Cl_2N_3O_4S_2$	360.237	20	0.005	0.05	40		
Methyl acetate	$C_3H_6O_2$	74.079	20	24.5	325	10		
Methyl acrylate	$C_4H_6O_2$	86.090	25	4.94	52.0	10		
2-Methylacrylonitrile	C_4H_5N	67.090	20	2.57	26.4	10		
2-Methylaniline	C_7H_9N	107.153	20	1.66	16.9	10		
4-Methylaniline	C_7H_9N	107.153	21	7.35	79.3	10		
N-Methylaniline	C_7H_9N	107.153	25	0.56	5.6	40		
2-Methylanthracene	$C_{15}H_{12}$	192.256	6	0.0000007	0.000007	42		
			25	0.0000021	0.000021	42,22		
9-Methylanthracene	$C_{15}H_{12}$	192.256	25	0.000026	0.00026	42,4		
9-Methylbenz[a]anthracene	$C_{19}H_{14}$	242.314	27	0.0000066	0.000066	42,4		
10-Methylbenz[a]anthracene	$C_{19}H_{14}$	242.314	25	0.0000055	0.000055	42,4		
2-Methylbenzenesulfonamide	$C_7H_9NO_2S$	171.217	25	0.162	1.62	27		
3-Methylbenzenesulfonamide	$C_7H_9NO_2S$	171.217	25	0.78	7.8	27		
4-Methylbenzenesulfonamide	$C_7H_9NO_2S$	171.217	25	0.316	3.16	27		
2-Methyl-1*H*-benzimidazole	$C_8H_8N_2$	132.163	20	0.145	1.45	6		
Methyl benzoate	$C_8H_8O_2$	136.149	20	0.21	2.1	10		
2-Methyl-1,3-butadiene	C_5H_8	68.118	25	0.061	0.61	3	7.78	5
			50	0.076*	0.76*	3		
Methyl butanoate	$C_5H_{10}O_2$	102.132		1.6	16	30		
3-Methylbutanoic acid	$C_5H_{10}O_2$	102.132	20	4.0	42	26		
2-Methyl-1-butanol	$C_5H_{12}O$	88.148	10	3.38	35.0	78		
			25	2.75	28.3	78		
			50	2.35	24.1	78		
3-Methyl-1-butanol	$C_5H_{12}O$	88.148	10	3.17	32.7	78,1		
			25	2.59	26.6	78,1		
			70	2.24	22.9	78,1		
2-Methyl-2-butanol	$C_5H_{12}O$	88.148	25	11.0	124	88,1		
			60	6.6	71	88,1		
3-Methyl-2-butanol	$C_5H_{12}O$	88.148	25	5.6	59	1		
3-Methyl-2-butanone	$C_5H_{10}O$	86.132	0	9.4	104	82		
			25	6.1	65	82		
			40	5.2	55	82		
3-Methyl-1-butene	C_5H_{10}	70.133	25	0.013*	0.13*	3	54.7	5

Name	Mol. Form.	Mol. Wt.	t/°C	Solubility, s: 100 w_2 (mass%)	Solubility, s: g per kg H_2O	Ref.	Henry Const., k_H: kPa m^3mol^{-1}	Ref.
2-Methyl-2-butene	C_5H_{10}	70.133	25	0.041	0.41	3		
2-Methyl-3-buten-2-ol	$C_5H_{10}O$	86.132	18	27.4	377	88		
			29	18.4	225	88		
Methyl *tert*-butyl ether	$C_5H_{12}O$	88.148	0	7.72	83.7	79		
			25	3.25	33.6	79		
			35	2.56	26.3	79		
			70	1.64	16.7	79		
Methyl carbamate	$C_2H_5NO_2$	75.067	15	69	2230	27		
5-Methylchrysene	$C_{19}H_{14}$	242.314	27	0.0000062	0.000062	42,4		
Methylcyclohexane	C_7H_{14}	98.186	26	0.00161	0.0161	3	43.3	13
			100	0.00548	0.0548	3		
2-Methylcyclohexanone	$C_7H_{12}O$	112.169	0	2.93	30.2	84		
			20	1.98	20.2	20		
			31	1.72	17.5	84		
			60	1.44	14.6	84		
			90	1.54	15.6	20		
4-Methylcyclohexanone	$C_7H_{12}O$	112.169	20	2.43	24.9	20		
			80	1.95	19.9	20		
1-Methylcyclohexene	C_7H_{12}	96.170	25	0.0052	0.052	3		
Methylcyclopentane	C_6H_{12}	84.159	25	0.0043	0.043	3	36.7	5
1-Methyl-2,4-dinitrobenzene	$C_7H_6N_2O_4$	182.134	12	0.0130	0.130	55		
			32	0.0270	0.270	85		
			62	0.098	0.98	85		
2-Methyl-4,6-dinitrophenol	$C_7H_6N_2O_5$	198.133		0.0130	0.130	40		
Methyl formate	$C_2H_4O_2$	60.052	25	23	300	10		
3-Methylheptane	C_8H_{18}	114.229	25	0.000079	0.00079	4	376	5
2-Methyl-2-heptanol	$C_8H_{18}O$	130.228	30	0.25	2.5	1		
5-Methyl-3-heptanone	$C_8H_{16}O$	128.212	20	0.192	1.92	20		
			90	0.131	1.31	20		
2-Methylhexane	C_7H_{16}	100.202	25	0.00025	0.0025	3	346	5
3-Methylhexane	C_7H_{16}	100.202	25	0.00026	0.0026	3	249	13
2-Methyl-2-hexanol	$C_7H_{16}O$	116.201	25	1.0	10	1		
5-Methyl-2-hexanol	$C_7H_{16}O$	116.201	25	0.49	4.9	1		
3-Methyl-3-hexanol	$C_7H_{16}O$	116.201	25	1.2	12	1		
5-Methyl-2-hexanone	$C_7H_{14}O$	114.185	19	0.537	5.40	20		
			90	0.417	4.19	20		
5-Methyl-3-hexanone	$C_7H_{14}O$	114.185	20	0.47	4.7	20		
			81	0.32	3.2	20		
Methyl 4-hydroxybenzoate	$C_8H_8O_3$	152.148	25	0.24	2.4	40		
2-Methyl-1*H*-imidazole	$C_4H_6N_2$	82.104	18	23.2	302	54		
3-Methyl-1*H*-indole	C_9H_9N	131.174	20	0.050	0.50	6		
3-Methylisoquinoline	$C_{10}H_9N$	143.185	20	0.092	0.92	6		
Methyl isothiocyanate	C_2H_3NS	73.117	20	0.75	7.6	40		
Methylmalonic acid	$C_4H_6O_4$	118.089	0	30.1	431	26		
			20	40	670	26		
Methyl methacrylate	$C_5H_8O_2$	100.117	20	1.56	15.8	10		
2-Methyl-3-(2-methylphenyl)-4(3*H*)-quinazolinone	$C_{16}H_{14}N_2O$	250.294	23	0.03	0.3	40		
1-Methylnaphthalene	$C_{11}H_{10}$	142.197	25	0.00281	0.0281	22	0.045	22

Name	Mol. Form.	Mol. Wt.	t/°C	Solubility, s: 100 w_2 (mass%)	Solubility, s: g per kg H_2O	Ref.	Henry Const., k_H: kPa m^3mol^{-1}	Ref.
2-Methylnaphthalene	$C_{11}H_{10}$	142.197	25	0.0025	0.025	4	0.051	12
2-Methyl-1,4-naphthalenedione	$C_{11}H_8O_2$	172.181	25	0.016	0.16	40		
N-Methyl-*N*-nitrosourea	$C_2H_5N_3O_2$	103.080	14	2.3	24	40		
4-Methyloctane	C_9H_{20}	128.255	25	0.0000115	0.000115	4	1000	5
Methyloxirane	C_3H_6O	58.079	20	40.5	681	10	0.0087	13
Methyl parathion	$C_8H_{10}NO_5PS$	263.208	10	0.00218	0.0218	40		
			20	0.00380	0.0380	40		
			30	0.0059	0.059	40		
2-Methylpentane	C_6H_{14}	86.175	25	0.00137	0.0137	3	176	13
3-Methylpentane	C_6H_{14}	86.175	25	0.00129	0.0129	3	170	13
2-Methyl-1-pentanol	$C_6H_{14}O$	102.174	25	0.76	7.7	78,1		
			50	0.70	7.0	78		
4-Methyl-1-pentanol	$C_6H_{14}O$	102.174	25	0.76	7.6	1		
2-Methyl-2-pentanol	$C_6H_{14}O$	102.174	25	3.2	33	1		
3-Methyl-2-pentanol	$C_6H_{14}O$	102.174	25	1.9	19	1		
4-Methyl-2-pentanol	$C_6H_{14}O$	102.174	27	1.5	15	1		
2-Methyl-3-pentanol	$C_6H_{14}O$	102.174	25	2.0	20	1		
3-Methyl-3-pentanol	$C_6H_{14}O$	102.174	25	4.3	45	1		
4-Methyl-2-pentanone	$C_6H_{12}O$	100.158	0	2.92	30.1	83		
			25	1.85	18.8	83		
			50	1.46	14.8	83		
2-Methyl-3-pentanone	$C_6H_{12}O$	100.158	25	1.5	15	83		
2-Methyl-1-pentene	C_6H_{12}	84.159	25	0.0078	0.078	3	28.1	5
4-Methyl-1-pentene	C_6H_{12}	84.159	25	0.0048	0.048	3	63.2	5
1-Methylphenanthrene	$C_{15}H_{12}$	192.256	7	0.0000095	0.000095	42		
			25	0.0000269	0.000269	42,4		
Methylprednisolone	$C_{22}H_{30}O_5$	374.470	25	0.012	0.12	40		
Methyl propanoate	$C_4H_8O_2$	88.106		6	60	30		
2-Methylpropanoic acid	$C_4H_8O_2$	88.106	20	22.8	295	10		
2-Methyl-1-propanol	$C_4H_{10}O$	74.121	0	12.2	139	78,1		
			25	8.1	88	78,1	0.00273	28
			50	7.0	70	78,1		
Methyl propyl ether	$C_4H_{10}O$	74.121	0	5.4	57	79		
			25	3.0	31	79		
2-Methyl-2-propyl-1,3-propanediol dicarbamate	$C_9H_{18}N_2O_4$	218.250	25	0.33	3.3	40		
Methyl salicylate	$C_8H_8O_3$	152.148	30	0.74	7.4	10		
17-Methyltestosterone	$C_{20}H_{30}O_2$	302.451	25	0.0033	0.033	40		
2-Methyltetrahydrofuran	$C_5H_{10}O$	86.132	19	14.4	168	20	0.67	13
			71	6.0	64	20		
N-Methyl-*N*,2,4,6-tetranitroaniline	$C_7H_5N_5O_8$	287.144	20	0.0074	0.074	40		
Methylthiouracil	$C_5H_6N_2OS$	142.179	25	0.0533	0.533	40		
1-Methyl-2,3,4-trinitrobenzene	$C_7H_5N_3O_6$	227.131	14	0.0091	0.091	85,59		
			23	0.0116	0.116	85,59		
			61	0.0643	0.643	85,59		
Metronidazole	$C_6H_9N_3O_3$	171.153	20	0.93	9.4	40		
Mirex	$C_{10}Cl_{12}$	545.543	25	0.0000085	0.000085	40		
Morphine	$C_{17}H_{19}NO_3$	285.338	20	0.015	0.15	27		
β-Myrcene	$C_{10}H_{16}$	136.234	25	0.030	0.30	52		
Naphthacene	$C_{18}H_{12}$	228.288	25	0.00000007	0.0000007	42,4	0.000004	12

Name	Mol. Form.	Mol. Wt.	t/°C	Solubility, s: 100 w_2 (mass%)	Solubility, s: g per kg H_2O	Ref.	Henry Const., k_H: kPa m^3mol^{-1}	Ref.
Naphthalene	$C_{10}H_8$	128.171	10	0.0019	0.019	4		
			25	0.00316	0.0316	22	0.043	22
			50	0.0082	0.082	4		
1-Naphthaleneacetic acid	$C_{12}H_{10}O_2$	186.206	25	0.0415	0.415	40		
1-Naphthalenecarboxylic acid	$C_{11}H_8O_2$	172.181	25	0.0058	0.058	27		
1-Naphthalenylthiourea	$C_{11}H_{10}N_2S$	202.275	20	0.06	0.6	40		
1-Naphthol	$C_{10}H_8O$	144.170	20	0.111	1.11	40		
2-Naphthol	$C_{10}H_8O$	144.170	20	0.064	0.64	40		
			80	0.67	6.7	40		
1-Naphthylamine	$C_{10}H_9N$	143.185	20	0.17	1.7	40		
2-Naphthylamine	$C_{10}H_9N$	143.185	20	0.0189	0.189	40		
Narceine	$C_{23}H_{27}NO_8$	445.462	13	0.078	0.78	27		
Neopentane	C_5H_{12}	72.149	25	0.00332*	0.0332*	3	220	13
Nitrapyrin	$C_6H_3Cl_4N$	230.907	20	0.0040	0.040	40		
2-Nitroaniline	$C_6H_6N_2O_2$	138.124	30	1.47	14.9	27		
3-Nitroaniline	$C_6H_6N_2O_2$	138.124	30	0.121	1.21	27		
4-Nitroaniline	$C_6H_6N_2O_2$	138.124	30	0.073	0.73	27		
2-Nitroanisole	$C_7H_7NO_3$	153.136	30	0.169	1.69	10		
4-Nitroanisole	$C_7H_7NO_3$	153.136	30	0.059	0.59	27		
3-Nitrobenzaldehyde	$C_7H_5NO_3$	151.120	25	0.16	1.6	27		
4-Nitrobenzaldehyde	$C_7H_5NO_3$	151.120	25	0.23	2.3	27		
Nitrobenzene	$C_6H_5NO_2$	123.110	25	0.21	2.1	17		
3-Nitro-1,2-benzenedicarboxylic acid	$C_8H_5NO_6$	211.129	25	1.63	16.6	69		
			40	2.97	30.6	69		
4-Nitro-1,2-benzenedicarboxylic acid	$C_8H_5NO_6$	211.129	25	60.9	1560	69		
			40	68.0	2125	69		
2-Nitrobenzoic acid	$C_7H_5NO_4$	167.120	25	0.55	5.5	40		
3-Nitrobenzoic acid	$C_7H_5NO_4$	167.120	10	0.197	1.97	75		
			25	0.313	3.14	75		
			50	0.90	9.1	75		
4-Nitrobenzoic acid	$C_7H_5NO_4$	167.120	25	0.0422	0.422	40		
Nitroethane	$C_2H_5NO_2$	75.067	25	4.4	46	38		
			50	5.3	56	38		
Nitrofen	$C_{12}H_7Cl_2NO_3$	284.095	22	0.00095	0.0095	40		
Nitrofurantoin	$C_8H_6N_4O_5$	238.158	30	0.011	0.11	40		
Nitrofurazone	$C_6H_6N_4O_4$	198.137	20	0.0238	0.238	40		
Nitroguanidine	$CH_4N_4O_2$	104.069	25	1.2	12	40		
Nitromethane	CH_3NO_2	61.041	0	9.2	101	36		
			25	11.0	124	36		
			50	14.8	174	36		
1-Nitronaphthalene	$C_{10}H_7NO_2$	173.169	18	0.005	0.05	40		
2-Nitrophenol	$C_6H_5NO_3$	139.109	25	0.170	1.70	48,51		
3-Nitrophenol	$C_6H_5NO_3$	139.109	20	2.14	21.9	27		
4-Nitrophenol	$C_6H_5NO_3$	139.109	20	1.56	15.8	48,51		
1-Nitropropane	$C_3H_7NO_2$	89.094	25	1.54	15.6	38		
			90	2.29	23.4	20		
2-Nitropropane	$C_3H_7NO_2$	89.094	25	1.75	17.8	38		
			90	2.36	24.2	20		
N-Nitrosodiethylamine	$C_4H_{10}N_2O$	102.134	24	9.6	106	40		

Name	Mol. Form.	Mol. Wt.	t/°C	Solubility, s: 100 w_2 (mass%)	Solubility, s: g per kg H_2O	Ref.	Henry Const., k_H: kPa m³mol⁻¹	Ref.
N-Nitrosodiphenylamine	$C_{12}H_{10}N_2O$	198.219	25	0.0035	0.035	17		
2-Nitrotoluene	$C_7H_7NO_2$	137.137	30	0.065	0.65	27		
3-Nitrotoluene	$C_7H_7NO_2$	137.137	30	0.050	0.50	27		
4-Nitrotoluene	$C_7H_7NO_2$	137.137	30	0.044	0.44	27		
2,2',3,3',4,5,5',6,6'-Nonachlorobiphenyl	$C_{12}HCl_9$	464.213	25	0.0000000018	0.000000018	7		
1,8-Nonadiyne	C_9H_{12}	120.191	25	0.0125	0.125	4		
Nonane	C_9H_{20}	128.255	25	0.000017	0.00017	4	333	13
			50	0.000022	0.00022	4		
Nonanedioic acid	$C_9H_{16}O_4$	188.221	25	0.1780	1.780	34		
			65	1.322	13.40	34		
Nonanoic acid	$C_9H_{18}O_2$	158.238	20	0.0284	0.284	26		
1-Nonanol	$C_9H_{20}O$	144.254	25	0.0129	0.129	78,1		
			90	0.0291	0.291	78		
2-Nonanol	$C_9H_{20}O$	144.254	15	0.026	0.26	1		
3-Nonanol	$C_9H_{20}O$	144.254	15	0.032	0.32	1		
4-Nonanol	$C_9H_{20}O$	144.254	15	0.0026	0.026	1		
5-Nonanol	$C_9H_{20}O$	144.254	15	0.0032	0.032	1		
2-Nonanone	$C_9H_{18}O$	142.238	20	0.038	0.38	20		
			70	0.034	0.34	20		
3-Nonanone	$C_9H_{18}O$	142.238	30	0.056	0.56	20		
			80	0.046	0.46	20		
5-Nonanone	$C_9H_{18}O$	142.238	20	0.054	0.54	20		
			80	0.029	0.29	20		
1-Nonene	C_9H_{18}	126.239	25	0.000112	0.00112	40		
Nonyl formate	$C_{10}H_{20}O_2$	172.265	10	0.012	0.12	20		
			90	0.039	0.39	20		
4-Nonylphenol	$C_{15}H_{24}O$	220.351	25	0.000636	0.00636	40		
1-Nonyne	C_9H_{16}	124.223	25	0.00072	0.0072	4		
Norethisterone	$C_{20}H_{26}O_2$	298.419	25	0.00063	0.0063	40		
Norflurazon	$C_{12}H_9ClF_3N_3O$	303.666	25	0.0028	0.028	40		
L-Norleucine	$C_6H_{13}NO_2$	131.173	25	1.5	15	26		
L-Norvaline	$C_5H_{11}NO_2$	117.147	25	9.7	107	26		
Noscapine	$C_{22}H_{23}NO_7$	413.421	25	0.03	0.3	40		
2,2',3,3',5,5',6,6'-Octachlorobiphenyl	$C_{12}H_2Cl_8$	429.768	25	0.00000015	0.0000015	41	0.0381	7
Octachlorodibenzo-*p*-dioxin	$C_{12}Cl_8O_2$	459.751	25	$2.3{\cdot}10^{-11}$	$2.3{\cdot}10^{-10}$	68		
Octachloro-1,3-pentadiene	C_5Cl_8	343.678	20	0.000020	0.00020	35		
Octacosane	$C_{28}H_{58}$	394.761	22	0.0000006	0.000006	37		
Octadecane	$C_{18}H_{38}$	254.495	25	0.00000021	0.0000021	42,37		
1-Octadecanol	$C_{18}H_{38}O$	270.494	34	0.000011	0.00011	1		
Octane	C_8H_{18}	114.229	25	0.000073	0.00073	46	311	13
			50	0.000102	0.00102	47		
			75	0.000179	0.00179	46		
			100	0.000377	0.00377	46		
Octanedioic acid	$C_8H_{14}O_4$	174.195	25	0.242	2.43	34		
			50	0.557	5.570	34		
Octanoic acid	$C_8H_{16}O_2$	144.212	25	0.080	0.80	26		
1-Octanol	$C_8H_{18}O$	130.228	25	0.0460	0.460	78		
			60	0.0536	0.536	78		
2-Octanol	$C_8H_{18}O$	130.228	25	0.4	4	1		

Name	Mol. Form.	Mol. Wt.	t/°C	Solubility, s 100 w_2 (mass%)	g per kg H_2O	Ref.	Henry Const., k_H kPa m³mol⁻¹	Ref.
2-Octanone	$C_8H_{16}O$	128.212	20	0.134	1.34	84		
			50	0.098	0.98	84		
			80	0.091	0.91	84		
3-Octanone	$C_8H_{16}O$	128.212	20	0.137	1.37	20		
			91	0.106	1.06	20		
1-Octene	C_8H_{16}	112.213	25	0.00027	0.0027	4	96.3	13
Octyl acetate	$C_{10}H_{20}O_2$	172.265	19	0.020	0.20	20		
			92	0.012	0.12	20		
1-Octyne	C_8H_{14}	110.197	25	0.0024	0.024	4	7.87	13
Orotic acid	$C_5H_4N_2O_4$	156.097	18	0.18	1.8	26		
Oryzalin	$C_{12}H_{18}N_4O_6S$	346.359	25	0.00024	0.0024	40		
Ouabain	$C_{29}H_{44}O_{12}$	584.652	25	1.3	13	40		
Oxalic acid	$C_2H_2O_4$	90.035	20	8.69	95.2	27		
			80	45.8	845	27		
Oxamyl	$C_7H_{13}N_3O_3S$	219.261	25	≈21	≈270	40		
4-Oxopentanoic acid	$C_5H_8O_3$	116.116	10	63.6	1750	34		
			25	84.0	5250	34		
4-Oxo-4*H*-pyran-2,6-dicarboxylic acid	$C_7H_4O_6$	184.103	25	1.45	14.7	27		
Papaverine	$C_{20}H_{21}NO_4$	339.386	37	0.0037	0.037	40		
Paraldehyde	$C_6H_{12}O_3$	132.157	25	11	124	30		
Parathion	$C_{10}H_{14}NO_5PS$	291.261	20	0.00129	0.0129	40		
Pendimethalin	$C_{13}H_{19}N_3O_4$	281.308	20	0.00003	0.0003	40		
Pentachlorobenzene	C_6HCl_5	250.337	25	0.000050	0.00050	41	0.085	11
2,3,4,5,6-Pentachlorobiphenyl	$C_{12}H_5Cl_5$	326.433	25	0.0000008	0.000008	7		
2,2',4,5,5'-Pentachlorobiphenyl	$C_{12}H_5Cl_5$	326.433	25	0.000001	0.00001	7	0.0421	31
Pentachloroethane	C_2HCl_5	202.294	25	0.049	0.49	25	0.25	13
Pentachloronitrobenzene	$C_6Cl_5NO_2$	295.335	20	0.000044	0.00044	40		
Pentachlorophenol	C_6HCl_5O	266.336	25	0.0021	0.021	48,51,24		
2,3,4,5,6-Pentachlorotoluene	$C_7H_3Cl_5$	264.364	25	0.0000028	0.000028	61		
Pentadecanoic acid	$C_{15}H_{30}O_2$	242.398	20	0.0012	0.012	26		
1-Pentadecanol	$C_{15}H_{32}O$	228.414	25	0.000010	0.00010	1		
1,4-Pentadiene	C_5H_8	68.118	25	0.056	0.56	3	12	5
Pentaerythritol	$C_5H_{12}O_4$	136.147	15	5.3	56	30		
Pentaerythritol tetranitrate	$C_5H_8N_4O_{12}$	316.138	20	0.0002	0.002	40		
Pentanal	$C_5H_{10}O$	86.132	25	1.2	12	40		
Pentane	C_5H_{12}	72.149	25	0.0041	0.041	3	128	13
Pentanedioic acid	$C_5H_8O_4$	132.116	25	58.3	1400	33		
			50	78.1	3570	33		
2,4-Pentanedione	$C_5H_8O_2$	100.117	20	16.1	192	20		
			80	32.2	475	20		
Pentanoic acid	$C_5H_{10}O_2$	102.132	16	3.6	37	26		
			25	4.32	45.2	90		
			35	5.26	55.5	90		
1-Pentanol	$C_5H_{12}O$	88.148	0	3.23	33.4	78,1		
			25	2.14	21.9	78,1		
			50	1.83	18.6	78,1		
			90	2.12	21.7	78		
2-Pentanol	$C_5H_{12}O$	88.148	25	4.3	45	21		
3-Pentanol	$C_5H_{12}O$	88.148	25	5.6	59	21		

Name	Mol. Form.	Mol. Wt.	t/°C	Solubility, s: 100 w_2 (mass%)	Solubility, s: g per kg H_2O	Ref.	Henry Const., k_H: kPa m^3mol^{-1}	Ref.
2-Pentanone	$C_5H_{10}O$	86.132	0	8.7	95	20	0.00847	28
			25	5.5	58	20	0.00847	28
			80	3.8	40	20	0.00847	28
3-Pentanone	$C_5H_{10}O$	86.132	0	7.6	82	82		
			25	4.9	52	82		
			80	3.6	37	82		
1-Pentene	C_5H_{10}	70.133	25	0.0148	0.148	3	40.3	5
cis-2-Pentene	C_5H_{10}	70.133	25	0.0203	0.203	3	22.8	5
Pentyl acetate	$C_7H_{14}O_2$	130.185	20	0.17	1.7	10		
sec-Pentyl acetate (*S*)-	$C_7H_{14}O_2$	130.185	25	0.2	2	27		
Pentylbenzene	$C_{11}H_{16}$	148.245	25	0.00043	0.0043	89,5	1.69	11
Pentylcyclopentane	$C_{10}H_{20}$	140.266	25	0.0000115	0.000115	4	185	5
Pentyl propanoate	$C_8H_{16}O_2$	144.212	20	0.1	1	27		
1-Pentyne	C_5H_8	68.118	25	0.157	1.57	3	2.5	5
Perfluorocyclobutane	C_4F_8	200.030	5	0.00638*	0.0638*	50		
			25	0.00247*	0.0247*	50		
			45	0.00158*	0.0158*	50		
Perfluorodecane	$C_{10}F_{22}$	538.072	20	0.000031	0.00031	35		
Perfluoroheptane	C_7F_{16}	388.049	25	0.0000013	0.000013	35		
Perfluorohexane	C_6F_{14}	338.042	25	0.0000098	0.000098	35		
Perfluoro-2-methylpentane	C_6F_{14}	338.042	25	0.000017	0.00017	35		
Perfluorooctane	C_8F_{18}	438.057	25	0.00000017	0.0000017	35		
Perfluoropentane	C_5F_{12}	288.035	25	0.00012	0.0012	35		
Perfluoropropane	C_3F_8	188.019	15	0.0015*	0.015*	14		
Perfluoropropene	C_3F_6	150.022	25	0.0194*	0.194*	14		
Permethrin	$C_{21}H_{20}Cl_2O_3$	391.288	20	0.00002	0.0002	32		
Perylene	$C_{20}H_{12}$	252.309	25	0.00000004	0.0000004	42,4	0.000003	12
Phenanthrene	$C_{14}H_{10}$	178.229	0	0.000039	0.00039	42		
			10	0.000047	0.00047	42,4		
			25	0.00012	0.0012	42,22	0.00324	22
			50	0.00042	0.0042	42,4		
Phenmedipham	$C_{16}H_{16}N_2O_4$	300.309	25	0.00047	0.0047	32		
Phenobarbital	$C_{12}H_{12}N_2O_3$	232.234	25	0.12	1.2	40		
			45	0.26	2.6	40		
Phenol	C_6H_6O	94.111	15	7.60	82.3	48,51		
			25	8.40	91.7	48,51		
			35	9.31	102.7	48,51		
Phenolphthalein	$C_{20}H_{14}O_4$	318.323	20	0.018	0.18	27		
10*H*-Phenothiazine	$C_{12}H_9NS$	199.271	25	0.00016	0.0016	40		
2-Phenoxyethanol	$C_8H_{10}O_2$	138.164	20	2.53	26.0	40		
Phenyl acetate	$C_8H_8O_2$	136.149	20	0.59	5.9	20		
			91	0.91	9.2	20		
DL-Phenylalanine	$C_9H_{11}NO_2$	165.189	25	1.40	14.2	29		
L-Phenylalanine	$C_9H_{11}NO_2$	165.189	25	2.71	27.9	26		
Phenylbutazone	$C_{19}H_{20}N_2O_2$	308.374	25	0.0034	0.034	40		
1-Phenyl-1-propanone	$C_9H_{10}O$	134.174	19	0.32	3.2	20		
			80	0.24	2.4	20		
Phenylthiourea	$C_7H_8N_2S$	152.217	25	2.55	26.2	27		
Phenytoin	$C_{15}H_{12}N_2O_2$	252.268	37	0.0038	0.038	40		

Name	Mol. Form.	Mol. Wt.	t/°C	Solubility, s: 100 w_2 (mass%)	Solubility, s: g per kg H_2O	Ref.	Henry Const., k_H: kPa m^3mol^{-1}	Ref.
Phosalone	$C_{12}H_{15}ClNO_4PS_2$	367.808	20	0.00026	0.0026	40		
Phosmet	$C_{11}H_{12}NO_4PS_2$	317.321	25	0.0025	0.025	40		
Phthalic acid	$C_8H_6O_4$	166.132	10	0.464	4.66	76		
			25	0.719	7.24	76		
			50	1.76	17.9	76		
			65	3.57	37.0	33		
Phthalic anhydride	$C_8H_4O_3$	148.116	27	0.62	6.20	40		
Picene	$C_{22}H_{14}$	278.346	27	0.00000025	0.0000025	42,4		
α-Pinene, (-)	$C_{10}H_{16}$	136.234	25	0.00050	0.0050	52		
β-Pinene, (1*S*)-	$C_{10}H_{16}$	136.234	25	0.00110	0.0110	52		
2,5-Piperazinedione	$C_4H_6N_2O_2$	114.103	25	1.64	16.7	29		
2-Pivaloyl-1,3-indandione	$C_{14}H_{14}O_3$	230.259	25	0.0018	0.018	40		
Prednisolone	$C_{21}H_{28}O_5$	360.444	25	0.03	0.3	40		
Progesterone	$C_{21}H_{30}O_2$	314.462	25	0.00088	0.0088	40		
			41	0.00206	0.0206	40		
L-Proline	$C_5H_9NO_2$	115.131	25	61.9	1625	26		
Prometone	$C_{10}H_{19}N_5O$	225.291	20	0.075	0.75	40		
Prometryn	$C_{10}H_{19}N_5S$	241.357	20	0.0048	0.048	32		
Propachlor	$C_{11}H_{14}ClNO$	211.688	20	0.07	0.7	40		
Propanal	C_3H_6O	58.079	25	30.6	441	10		
Propane	C_3H_8	44.096	25	0.00669*	0.0669*	10	71.6	5
Propanenitrile	C_3H_5N	55.079	25	10.3	115	10		
Propanil	$C_9H_9Cl_2NO$	218.079	20	0.013	0.13	40		
Propazine	$C_9H_{16}ClN_5$	229.710	20	0.00086	0.0086	40		
Propene	C_3H_6	42.080	25	0.0200*	0.200*	5	21.3	5
1-Propene-2,3-dicarboxylic acid	$C_5H_6O_4$	130.100	20	7.7	83	26		
trans-1-Propene-1,2,3-tricarboxylic acid	$C_6H_6O_6$	174.108	25	20.9	264	26		
			90	52.5	1105	26		
Propoxur	$C_{11}H_{15}NO_3$	209.242	20	0.193	1.93	40		
Propyl acetate	$C_5H_{10}O_2$	102.132	20	2.3	24	10		
Propylbenzene	C_9H_{12}	120.191	25	0.0052	0.052	22	1.041	22
Propyl butanoate	$C_7H_{14}O_2$	130.185	17	0.162	1.62	27		
Propylcyclopentane	C_8H_{16}	112.213	25	0.00020	0.0020	4	90.2	5
Propyl formate	$C_4H_8O_2$	88.106	22	2.05	20.9	10		
Propyl 4-hydroxybenzoate	$C_{10}H_{12}O_3$	180.200	25	0.04	0.4	40		
Propyl propanoate	$C_6H_{12}O_2$	116.158	25	0.6	6	27		
Propyne	C_3H_4	40.064	25	0.364*	3.64*	5	1.11	5
Propyzamide	$C_{12}H_{11}Cl_2NO$	256.127	25	0.0015	0.015	32		
Pyrene	$C_{16}H_{10}$	202.250	0	0.0000049	0.000049	42		
			15	0.0000069	0.000069	42		
			25	0.0000139	0.000139	42,22	0.00092	22
			50	0.000053	0.00053	42,4		
			75	0.000231	0.00231	42		
3-Pyridinecarboxamide	$C_6H_6N_2O$	122.124	20	≈33	≈490	40		
3-Pyridinecarboxylic acid	$C_6H_5NO_2$	123.110	24	1.63	16.6	66		
			52	3.40	35.2	66		
			72	5.20	54.9	66		
Pyrocatechol	$C_6H_6O_2$	110.111	20	31.1	451	27		
Pyrrole	C_4H_5N	67.090	25	4.5	47	10		

Name	Mol. Form.	Mol. Wt.	t/°C	Solubility, s: 100 w_2 (mass%)	Solubility, s: g per kg H_2O	Ref.	Henry Const., k_H: kPa m³mol⁻¹	Ref.
Quinic acid	$C_7H_{12}O_6$	192.166	9	29	410	26		
Quinidine	$C_{20}H_{24}N_2O_2$	324.417	20	0.020	0.20	27		
Quinine	$C_{20}H_{24}N_2O_2$	324.417	25	0.057	0.57	27		
Quinoline	C_9H_7N	129.159	20	0.633	6.33	6		
8-Quinolinol	C_9H_7NO	145.158	25	0.065	0.65	40		
Quinoxaline	$C_8H_6N_2$	130.147	50	54	1170	6		
Raffinose	$C_{18}H_{32}O_{16}$	504.437	20	12.5	143	27		
Reserpine	$C_{33}H_{40}N_2O_9$	608.679	30	0.0073	0.073	40		
Resorcinol	$C_6H_6O_2$	110.111	20	63.7	1750	27		
Riboflavin	$C_{17}H_{20}N_4O_6$	376.364	25	0.0075	0.075	40		
Ronnel	$C_8H_8Cl_3O_3PS$	321.546	20	0.00011	0.0011	40		
Rotenone	$C_{23}H_{22}O_6$	394.417	25	0.000017	0.00017	40		
Saccharin	$C_7H_5NO_3S$	183.185	25	0.40	4.0	27		
			100	4.0	42	27		
Salicylaldehyde	$C_7H_6O_2$	122.122	86	1.68	17.1	10		
Sarcosine	$C_3H_7NO_2$	89.094	25	30.0	429	26		
L-Serine	$C_3H_7NO_3$	105.093	25	20	250	26		
Shikimic acid	$C_7H_{10}O_5$	174.151		15	176	26		
Silvex	$C_9H_7Cl_3O_3$	269.509	25	0.014	0.14	40		
Solanine	$C_{45}H_{73}NO_{15}$	868.060	15	0.0026	0.026	40		
L-Sorbose	$C_6H_{12}O_6$	180.155	17	≈26	≈350	40		
Stearic acid	$C_{18}H_{36}O_2$	284.478	20	0.00029	0.0029	26		
trans-Stilbene	$C_{14}H_{12}$	180.245	25	0.000029	0.00029	42,4	0.040	12
Streptozotocin	$C_8H_{15}N_3O_7$	265.221	25	0.50	5.0	40		
Strychnine	$C_{21}H_{22}N_2O_2$	334.412	20	0.013	0.13	27		
Styrene	C_8H_8	104.150	25	0.032	0.32	22	0.286	22
			50	0.046	0.46	4,89	0.30	13
Succinamide	$C_4H_8N_2O_2$	116.119	50	18.4	225	27		
Succinic acid	$C_4H_6O_4$	118.089	25	7.71	83.5	27		
			100	55	1220	27		
Succinonitrile	$C_4H_4N_2$	80.088	25	11.5	130	10		
Sucrose	$C_{12}H_{22}O_{11}$	342.296	20	67.1	2040	27		
			50	72.3	2610	27		
			100	83.0	4880	27		
Sulfamethazine	$C_{12}H_{14}N_4O_2S$	278.330	20	0.053	0.53	40		
Sulfamethoxazole	$C_{10}H_{11}N_3O_3S$	253.277	25	0.0281	0.281	40		
Sulfathiazole	$C_9H_9N_3O_2S_2$	255.316	20	0.048	0.48	40		
Sulfisoxazole	$C_{11}H_{13}N_3O_3S$	267.304	37	0.03	0.3	40		
DL-Tartaric acid	$C_4H_6O_6$	150.087	0	8.95	98.3	26		
			20	17.1	206	26		
			100	65	1860	26		
L-Tartaric acid	$C_4H_6O_6$	150.087	20	58	1380	26		
			100	77	3350	26		
Tebuthiuron	$C_9H_{16}N_4OS$	228.314	20	0.23	2.3	40		
Terbacil	$C_9H_{13}ClN_2O_2$	216.664	25	0.071	0.71	40		
Terephthalic acid	$C_8H_6O_4$	166.132	10	0.0082	0.082	76		
			25	0.0065	0.065	76		
			50	0.0074	0.074	76		
o-Terphenyl	$C_{18}H_{14}$	230.304	25	0.000124	0.00124	42,40		

Name	Mol. Form.	Mol. Wt.	t/°C	Solubility, s: 100 w_2 (mass%)	Solubility, s: g per kg H_2O	Ref.	Henry Const., k_H: kPa m³mol⁻¹	Ref.
m-Terphenyl	$C_{18}H_{14}$	230.304	25	0.000152	0.00152	42,40		
p-Terphenyl	$C_{18}H_{14}$	230.304	25	0.00000180	0.000018	42,40		
α-Terpineol	$C_{10}H_{18}O$	154.249	25	0.189	1.89	52		
1,2,4,5-Tetrabromobenzene	$C_6H_2Br_4$	393.696	25	0.00000434	0.0000434	2		
1,1,2,2-Tetrabromoethane	$C_2H_2Br_4$	345.653	0	0.052	0.52	25		
			25	0.068	0.68	25		
			50	0.106	1.06	25		
			100	0.307	3.07	25		
Tetrabromomethane	CBr_4	331.627	30	0.024	0.24	14		
1,2,3,4-Tetrachlorobenzene	$C_6H_2Cl_4$	215.892	25	0.0007	0.007	41	0.144	11
1,2,3,5-Tetrachlorobenzene	$C_6H_2Cl_4$	215.892	25	0.00035	0.0035	41	0.59	11
1,2,4,5-Tetrachlorobenzene	$C_6H_2Cl_4$	215.892	25	0.000060	0.00060	41	0.122	11
3,4,5,6-Tetrachloro-1,2-benzenediol	$C_6H_2Cl_4O_2$	247.891	25	0.071	0.71	8		
2,2′,4′,5-Tetrachlorobiphenyl	$C_{12}H_6Cl_4$	291.988	25	0.0000016	0.000016	9		
2,3,4,5-Tetrachlorobiphenyl	$C_{12}H_6Cl_4$	291.988	25	0.000002	0.00002	7		
2,3,5,6-Tetrachloro-2,5-cyclohexadiene-1,4-dione	$C_6Cl_4O_2$	245.875	20	0.025	0.25	40		
2,3,7,8-Tetrachlorodibenzo-*p*-dioxin	$C_{12}H_4Cl_4O_2$	321.971	22	0.0000000019	0.000000019	40		
1,1,2,2-Tetrachloro-1,2-difluoroethane	$C_2Cl_4F_2$	203.830	27	0.016	0.16	25		
1,1,1,2-Tetrachloroethane	$C_2H_2Cl_4$	167.849	0	0.120	1.20	25		
			25	0.107	1.07	25	0.24	13
			50	0.123	1.23	25		
1,1,2,2-Tetrachloroethane	$C_2H_2Cl_4$	167.849	5	0.302	3.02	25		
			25	0.283	2.83	25	0.026	13
			50	0.318	3.18	25		
Tetrachloroethene	C_2Cl_4	165.833	0	0.0273	0.273	20		
			20	0.0286	0.286	20	1.73	13
			80	0.0380	0.380	20		
Tetrachloromethane	CCl_4	153.823	25	0.065	0.65	20	2.99	13
			75	0.115	1.15	20	2.99	13
2,3,4,6-Tetrachloro-5-methylphenol	$C_7H_4Cl_4O$	245.918	25	0.00061	0.0061	2		
2,3,4,6-Tetrachlorophenol	$C_6H_2Cl_4O$	231.891	25	0.017	0.17	24		
1,1,1,3-Tetrachloro-2,2,3,3-tetrafluoropropane	$C_3Cl_4F_4$	253.838	21	0.0052	0.052	35		
Tetracosane	$C_{24}H_{50}$	338.654	22	0.0000004	0.000004	37		
Tetradecane	$C_{14}H_{30}$	198.388	25	0.00000023	0.0000023	42,5		
Tetradecanoic acid	$C_{14}H_{28}O_2$	228.371	20	0.0020	0.020	26		
1-Tetradecanol	$C_{14}H_{30}O$	214.387	25	0.000031	0.00031	1		
Tetraethylsilane	$C_8H_{20}Si$	144.331	25	0.0000325	0.000325	10		
Tetrafluoroethene	C_2F_4	100.015	25	0.0158*	0.158*	19,50		
			70	0.0090*	0.090*	50		
Tetrafluoromethane	CF_4	88.005	0	0.00390*	0.0390*	50		
			25	0.00185*	0.0185*	50,19		
			50	0.00134*	0.0134*	50		
Tetrahydro-2,5-dimethoxyfuran	$C_6H_{12}O_3$	132.157	21	32	470	20		
			90	19	235	20		
1,2,3,4-Tetrahydronaphthalene	$C_{10}H_{12}$	132.202	20	0.0045	0.045	40		
Tetrahydropyran	$C_5H_{10}O$	86.132	20	8.57	93.7	20		
			81	4.29	44.8	20		
1,2,4,5-Tetramethylbenzene	$C_{10}H_{14}$	134.218	25	0.000348	0.00348	4	2.55	11
N,N,N′,N′-Tetramethyl-4,4′-diaminobenzophenone	$C_{17}H_{20}N_2O$	268.353	20	0.04	0.4	40		

Name	Mol. Form.	Mol. Wt.	t/°C	Solubility, s: 100 w_2 (mass%)	Solubility, s: g per kg H_2O	Ref.	Henry Const., k_H: kPa m^3mol^{-1}	Ref.
Tetramethylsilane	$C_4H_{12}Si$	88.224	25	0.00196	0.0196	10		
Theophylline	$C_7H_8N_4O_2$	180.165	20	0.52	5.2	29		
Thioacetamide	C_2H_5NS	75.133	25	12.3	140	40		
Thiourea	CH_4N_2S	76.121	20	10.6	119	40		
			80	≈37	≈590	40		
2-Thioxo-4-thiazolidinone	$C_3H_3NOS_2$	133.192	25	0.225	2.25	40		
Thiram	$C_6H_{12}N_2S_4$	240.432	20	0.003	0.03	40		
DL-Threonine	$C_4H_9NO_3$	119.119	10	14.34	167	45		
			20	15.69	186	45		
			40	19.84	248	45		
L-Threonine	$C_4H_9NO_3$	119.119	10	7.34	79.2	45		
			20	8.31	90.6	45		
			40	10.78	121	45		
Thymidine	$C_{10}H_{14}N_2O_5$	242.228	25	5.1	54	29		
Thymine	$C_5H_6N_2O_2$	126.114	25	0.35	3.5	29		
Thymol	$C_{10}H_{14}O$	150.217		0.1	1	30		
Tolazamide	$C_{14}H_{21}N_3O_3S$	311.400	30	0.0065	0.065	40		
Tolbutamide	$C_{12}H_{18}N_2O_3S$	270.347	25	0.011	0.11	40		
o-Tolidine	$C_{14}H_{16}N_2$	212.290	25	0.13	1.3	40		
Toluene	C_7H_8	92.139	5	0.054	0.54	61		
			25	0.0519	0.519	61,22	0.660	22
			45	0.063	0.63	61		
			90	0.12	1.2	22		
p-Toluenesulfonic acid	$C_7H_8O_3S$	172.202	40	≈33	≈490	40		
o-Toluic acid	$C_8H_8O_2$	136.149	25	0.118	1.18	27		
m-Toluic acid	$C_8H_8O_2$	136.149	25	0.098	0.98	27		
p-Toluic acid	$C_8H_8O_2$	136.149	10	0.030	0.30	75		
			25	0.036	0.36	75		
			50	0.089	0.89	75		
1,3,5-Triazine-2,4,6-triamine	$C_3H_6N_6$	126.120	20	0.323	3.23	40		
			95	4.2	44	40		
1*H*-1,2,4-Triazol-3-amine	$C_2H_4N_4$	84.080	23	22	280	26		
1,2,4-Tribromobenzene	$C_6H_3Br_3$	314.800	25	0.0010	0.010	2		
1,3,5-Tribromobenzene	$C_6H_3Br_3$	314.800	25	0.0000789	0.000789	2		
1,1,2-Tribromoethane	$C_2H_3Br_3$	266.757	20	0.050	0.50	25		
Tribromofluoromethane	CBr_3F	270.721	25	0.040	0.40	14		
Tribromomethane	$CHBr_3$	252.731	25	0.30	3.0	5	0.047	13
2,4,6-Tribromophenol	$C_6H_3Br_3O$	330.799	15	0.0007	0.007	2		
Tributylamine	$C_{12}H_{27}N$	185.349	25	0.0142	0.142	40		
Tributyl phosphate	$C_{12}H_{27}O_4P$	266.313	25	0.039	0.39	10		
Tributyrin	$C_{15}H_{26}O_6$	302.363	20	0.010	0.10	40		
Trichloroacetaldehyde	C_2HCl_3O	147.387	25	≈39	≈640	40		
Trichloroacetic acid	$C_2HCl_3O_2$	163.387	25	92.3	11990	27		
1,2,3-Trichlorobenzene	$C_6H_3Cl_3$	181.447	25	0.0021	0.021	41	0.242	11
1,2,4-Trichlorobenzene	$C_6H_3Cl_3$	181.447	15	0.0029	0.029	61		
			25	0.0037	0.037	61,41	0.277	11
			45	0.0047	0.047	61		
1,3,5-Trichlorobenzene	$C_6H_3Cl_3$	181.447	25	0.0008	0.008	41	1.1	11
3,4,5-Trichloro-1,2-benzenediol	$C_6H_3Cl_3O_2$	213.446	25	0.051	0.51	8		

Name	Mol. Form.	Mol. Wt.	t/°C	Solubility, s: 100 w_2 (mass%)	Solubility, s: g per kg H_2O	Ref.	Henry Const., k_H: kPa m^3mol^{-1}	Ref.
2,4,5-Trichlorobiphenyl	$C_{12}H_7Cl_3$	257.543	25	0.000014	0.00014	7	0.0379	31
2,4,6-Trichlorobiphenyl	$C_{12}H_7Cl_3$	257.543	25	0.00002	0.0002	7	0.0495	7
1,1,1-Trichloro-2,2-bis(4-chlorophenyl)ethane	$C_{14}H_9Cl_5$	354.486	25	0.0000004	0.000004	67		
2,4,6-Trichloro-3,5-dimethylphenol	$C_8H_7Cl_3O$	225.500	25	0.00050	0.0050	2		
1,1,1-Trichloroethane	$C_2H_3Cl_3$	133.404	0	0.134	1.34	25		
			25	0.129	1.29	25	1.76	13
			50	0.138	1.38	25		
1,1,2-Trichloroethane	$C_2H_3Cl_3$	133.404	0	0.425	4.25	25		
			25	0.459	4.59	25	0.092	13
			50	0.536	5.36	25		
Trichloroethene	C_2HCl_3	131.388	0	0.145	1.45	25		
			25	0.128	1.28	25	1.03	13
			60	0.133	1.33	25		
Trichlorofluoromethane	CCl_3F	137.368	20	0.11	1.1	5	10.2	13
Trichloromethane	$CHCl_3$	119.378	25	0.80	8.0	20	0.43	13
			59	0.79	7.9	20	0.43	13
1,2,4-Trichloro-5-methylbenzene	$C_7H_5Cl_3$	195.474	25	0.00023	0.0023	61		
(Trichloromethyl)benzene	$C_7H_5Cl_3$	195.474	5	0.0053	0.053	10		
2,4,6-Trichloro-3-methylphenol	$C_7H_5Cl_3O$	211.473	25	0.0112	0.112	2		
Trichloronitromethane	CCl_3NO_2	164.376	0	0.227	2.27	40		
			25	0.162	1.62	40		
1,1,1-Trichloro-2,2,3,3,3-pentafluoropropane	$C_3Cl_3F_5$	237.383	21	0.0058	0.058	35		
2,4,5-Trichlorophenol	$C_6H_3Cl_3O$	197.446	25	0.1	1	2		
2,4,6-Trichlorophenol	$C_6H_3Cl_3O$	197.446	25	0.069	0.69	48,51,24		
2,4,5-Trichlorophenoxyacetic acid	$C_8H_5Cl_3O_3$	255.483	25	0.028	0.28	40		
1,2,3-Trichloropropane	$C_3H_5Cl_3$	147.431	10	0.14	1.4	35		
			25	0.20	2.0	35	0.038	13
1,1,2-Trichloro-1,2,2-trifluoroethane	$C_2Cl_3F_3$	187.375	25	0.017	0.17	25	32	13
Tri-*p*-cresyl phosphate	$C_{21}H_{21}O_4P$	368.363	25	0.00004	0.0004	40		
Tridecane	$C_{13}H_{28}$	184.361	25	0.000000033	0.00000033	37		
Tridecanoic acid	$C_{13}H_{26}O_2$	214.344	20	0.0033	0.033	26		
Triethylamine	$C_6H_{15}N$	101.190	20	5.5	58	10		
Triethylamine hydrochloride	$C_6H_{16}ClN$	137.651	25	57.8	1370	27		
Trifluoromethane	CHF_3	70.014	25	0.15*	1.5*	50,14		
3,4,5-Trihydroxybenzoic acid	$C_7H_6O_5$	170.120	25	1.52	15.4	74		
			50	3.82	39.7	74		
			100	25.0	333	27		
Triiodomethane	CHI_3	393.732	25	0.012	0.12	14		
Trimethoprim	$C_{14}H_{18}N_4O_3$	290.318	25	0.04	0.4	40		
1,2,3-Trimethylbenzene	C_9H_{12}	120.191	25	0.0070	0.070	22	0.343	22
1,2,4-Trimethylbenzene	C_9H_{12}	120.191	25	0.0057	0.057	22	0.569	22
1,3,5-Trimethylbenzene	C_9H_{12}	120.191	25	0.0050	0.050	22	0.781	22
2,3,3-Trimethyl-2-butanol	$C_7H_{16}O$	116.201	40	2.2	22	1		
1,1,3-Trimethylcyclohexane	C_9H_{18}	126.239	25	0.000177	0.00177	4	105	13
1,1,3-Trimethylcyclopentane	C_8H_{16}	112.213	25	0.00037	0.0037	4	159	5
2,2,5-Trimethylhexane	C_9H_{20}	128.255	25	0.00008	0.0008	4	246	13
1,4,5-Trimethylnaphthalene	$C_{13}H_{14}$	170.250	25	0.00021	0.0021	42,4		
2,6,8-Trimethyl-4-nonanone	$C_{12}H_{24}O$	184.318	10	0.012	0.12	20		
			80	0.014	0.14	20		

Name	Mol. Form.	Mol. Wt.	t/°C	Solubility, s 100 w_2 (mass%)	g per kg H_2O	Ref.	Henry Const., k_H kPa m^3mol^{-1}	Ref.
2,2,4-Trimethylpentane	C_8H_{18}	114.229	25	0.00022	0.0022	4	307	13
2,3,4-Trimethylpentane	C_8H_{18}	114.229	25	0.00018	0.0018	4	206	13
Trimethyl phosphate	$C_3H_9O_4P$	140.074	25	≈33	≈490	40		
1,3,5-Trinitrobenzene	$C_6H_3N_3O_6$	213.104	15	0.028	0.28	40		
2,4,6-Trinitrobenzoic acid	$C_7H_3N_3O_8$	257.114	23	1.97	20.1	40		
Trinitroglycerol	$C_3H_5N_3O_9$	227.087	25	0.13	1.3	40		
			80	0.34	3.4	40		
2,4,6-Trinitrophenol	$C_6H_3N_3O_7$	229.104	25	1.25	12.7	40		
			90	4.9	52	40		
2,4,6-Trinitrotoluene	$C_7H_5N_3O_6$	227.131	20	0.012	0.12	40		
			100	0.15	1.5	40		
2,4,6-Trinitro-*N*-(2,4,6-trinitrophenyl)aniline	$C_{12}H_5N_7O_{12}$	439.208	17	0.0060	0.060	40		
1,3,5-Trioxane	$C_3H_6O_3$	90.078	25	17.4	211	30		
Triphenylene	$C_{18}H_{12}$	228.288	25	0.0000043	0.000043	42,4	0.00001	12
Triphenyl phosphate	$C_{18}H_{15}O_4P$	326.283	24	0.000073	0.00073	40		
Triphenyltin hydroxide	$C_{18}H_{16}OSn$	367.029	20	0.0001	0.001	32		
Tris(hydroxymethyl)methylamine	$C_4H_{11}NO_3$	121.135	25	≈41	≈700	40		
L-Tryptophan	$C_{11}H_{12}N_2O_2$	204.225	25	1.30	13.2	26		
DL-Tyrosine	$C_9H_{11}NO_3$	181.188	25	0.35	3.5	30		
L-Tyrosine	$C_9H_{11}NO_3$	181.188	25	0.0507	0.507	62		
Undecane	$C_{11}H_{24}$	156.309	25	0.0000004	0.000004	37		
Uracil	$C_4H_4N_2O_2$	112.087	25	0.460	4.62	72		
Urea	CH_4N_2O	60.055	5	44	790	26		
			25	54.4	1200	26		
Uric acid	$C_5H_4N_4O_3$	168.111	20	0.002	0.02	26		
L-Valine	$C_5H_{11}NO_2$	117.147	25	8.13	88.5	26		
Valium	$C_{16}H_{13}ClN_2O$	284.739	25	0.005	0.05	40		
Vidarabine	$C_{10}H_{15}N_5O_5$	285.257	20	0.051	0.51	40		
Vinclozolin	$C_{12}H_9Cl_2NO_3$	286.110	20	0.1	1	32		
Vinyl acetate	$C_4H_6O_2$	86.090	20	2.0	20	10		
4-Vinylcyclohexene	C_8H_{12}	108.181	25	0.005	0.05	4		
Warfarin	$C_{19}H_{16}O_4$	308.328	20	0.004	0.04	40		
Xanthine	$C_5H_4N_4O_2$	152.112	20	0.05	0.5	26		
o-Xylene	C_8H_{10}	106.165	25	0.0171	0.171	22	0.551	22
			45	0.021	0.21	4		
m-Xylene	C_8H_{10}	106.165	0	0.0203	0.203	4		
			25	0.0161	0.161	22	0.730	22
			40	0.022	0.22	4		
p-Xylene	C_8H_{10}	106.165	0	0.0160	0.160	4		
			25	0.0181	0.181	22	0.690	22
			40	0.022	0.22	4		
2,3-Xylenol	$C_8H_{10}O$	122.164	25	0.457	4.57	40		
2,4-Xylenol	$C_8H_{10}O$	122.164	25	0.787	7.87	10		
2,5-Xylenol	$C_8H_{10}O$	122.164	25	0.354	3.54	40		
2,6-Xylenol	$C_8H_{10}O$	122.164	25	0.60	6.0	40		
3,4-Xylenol	$C_8H_{10}O$	122.164	25	0.477	4.77	40		
3,5-Xylenol	$C_8H_{10}O$	122.164	29	0.62	6.2	10		
D-Xylose	$C_5H_{10}O_5$	150.130	25	≈30	≈430	40		
Ziram	$C_6H_{12}N_2S_4Zn$	305.841	20	0.0065	0.065	40		

AQUEOUS SOLUBILITY OF INORGANIC COMPOUNDS AT VARIOUS TEMPERATURES

The solubility of over 300 common inorganic compounds in water is tabulated here as a function of temperature. Solubility is defined as the concentration of the compound in a solution that is in equilibrium with a solid phase at the specified temperature. In this table the solid phase is generally the most stable crystalline phase at the temperature in question. An asterisk * on solubility values in adjacent columns indicates that the solid phase changes between those two temperatures (usually from one hydrated phase to another or from a hydrate to the anhydrous solid). In such cases the slope of the solubility vs. temperature curve may show a discontinuity.

All solubility values are expressed as mass percent of solute, $100 \cdot w_2$, where

$$w_2 = m_2/(m_1 + m_2)$$

and m_2 is the mass of solute and m_1 the mass of water. This quantity is related to other common measures of solubility as follows:

Molarity: $c_2 = 1000\, \rho w_2/M_2$
Molality: $m_2 = 1000 w_2/M_2(1-w_2)$
Mole fraction: $x_2 = (w_2/M_2)/\{(w_2/M_2) + (1-w_2)/M_1\}$
Mass of solute per 100 g of H_2O: $r_2 = 100 w_2/(1-w_2)$

Here M_2 is the molar mass of the solute and M_1 = 18.015 g/mol is the molar mass of water. ρ is the density of the solution in g cm^{-3}.

The data in the table have been derived from the references indicated; in many cases the data have been refitted or interpolated in order to present solubility at rounded values of temperature. Where available, values were taken from the IUPAC *Solubility Data Series* (Reference 1) or the related papers in the *Journal of Physical and Chemical Reference Data* (References 2 to 5), which present carefully evaluated data.

The solubility of sparingly soluble compounds that do not appear in this table may be calculated from the data in the table "Solubility Product Constants." Solubility of inorganic gases may be found in the table "Solubility of Selected Gases in Water."

Compounds are listed alphabetically by chemical formula in the most commonly used form (e.g., NaCl, NH_4NO_3, etc.).

References

1. *Solubility Data Series*, International Union of Pure and Applied Chemistry. Volumes 1 to 53 were published by Pergamon Press, Oxford, from 1979 to 1994; subsequent volumes were published by Oxford University Press, Oxford. The number following the colon is the volume number in the series.
2. Clever, H. L., and Johnston, F. J., *J. Phys. Chem. Ref. Data*, 9, 751, 1980.
3. Marcus, Y., *J. Phys. Chem. Ref. Data*, 9, 1307, 1980.
4. Clever, H. L., Johnson, S. A., and Derrick, M. E., *J. Phys. Chem. Ref. Data*, 14, 631, 1985.
5. Clever, H. L., Johnson, S. A., and Derrick, M. E., *J. Phys. Chem. Ref. Data*, 21, 941, 1992.
6. Söhnel, O., and Novotny, P., *Densities of Aqueous Solutions of Inorganic Substances*, Elsevier, Amsterdam, 1985.
7. Krumgalz, B.S., *Mineral Solubility in Water at Various Temperatures*, Israel Oceanographic and Limnological Research Ltd., Haifa, 1994.
8. Potter, R. W., and Clynne, M. A., *J. Research U.S. Geological Survey*, 6, 701, 1978; Clynne, M. A., and Potter, R. W., *J. Chem. Eng. Data*, 24, 338, 1979.
9. Marshal, W. L., and Slusher, R., *J. Phys. Chem.*, 70, 4015, 1966; Knacke, O., and Gans, W., *Zeit. Phys. Chem.*, NF, 104, 41, 1977.
10. Stephen, H., and Stephen, T., *Solubilities of Inorganic and Organic Compounds, Vol. 1*, Macmillan, New York, 1963.

Compound	0 °C	10 °C	20 °C	25 °C	30 °C	40 °C	50 °C	60 °C	70 °C	80 °C	90 °C	100 °C	Ref.
$AgBrO_3$				0.193							1.32		7
$AgClO_2$	0.17	0.31	0.47	0.55	0.64	0.82	1.02	1.22	1.44	1.66	1.88	2.11	7
$AgClO_3$				15									7
$AgClO_4$	81.6	83.0	84.2	84.8	85.3	86.3	86.9	87.5	87.9	88.3	88.6	88.8	6
$AgNO_2$	0.155			0.413									7
$AgNO_3$	55.9	62.3	67.8	70.1	72.3	76.1	79.2	81.7	83.8	85.4	86.7	87.8	6
Ag_2SO_4	0.56	0.67	0.78	0.83	0.88	0.97	1.05	1.13	1.20	1.26	1.32	1.39	7
$AlCl_3$	30.84	30.91	31.03	31.10	31.18	31.37	31.60	31.87	32.17	32.51	32.90	33.32	7
$Al(ClO_4)_3$	54.9										64.4		7
AlF_3	0.25	0.34	0.44	0.50	0.56	0.68	0.81	0.96	1.11	1.28	1.45	1.64	7
$Al(NO_3)_3$	37.0	38.2	39.9	40.8	42.0	44.5	47.3	50.4	53.8*			61.5*	6
$Al_2(SO_4)_3$	27.5			27.8	28.2	29.2	30.7	32.6	34.9	37.6	40.7	44.2	7
As_2O_3	1.19	1.48	1.80	2.01	2.27	2.86	3.43	4.11	4.89	5.77	6.72	7.71	10
$BaBr_2$	47.6	48.5	49.5	50.0	50.4	51.4	52.5	53.5	54.5	55.5	56.6	57.6	6
$Ba(BrO_3)_2$	0.285	0.442	0.656	0.788	0.935	1.30	1.74	2.27	2.90	3.61	4.40	5.25	1:14
$Ba(C_2H_3O_2)_2$	37.0			44.2									7
$BaCl_2$	23.30	24.88	26.33	27.03	27.70	29.00	30.27	31.53	32.81	34.14	35.54	37.05	8
$Ba(ClO_2)_2$	30.5			31.3								44.7	7
$Ba(ClO_3)_2$	16.90	21.23	23.66	27.50	29.43	33.16	36.69	40.05	43.04	45.90	48.70	51.17	1:14
$Ba(ClO_4)_2$	67.30	70.96	74.30	75.75	77.05	79.23	80.92	82.21	83.16	83.88	84.43	84.90	7
BaF_2		0.158		0.161									7
BaI_2	62.5	64.7	67.3	68.8	69.1	69.5	70.1	70.7	71.3	72.0	72.7	73.4	6

Compound	0 °C	10 °C	20 °C	25 °C	30 °C	40 °C	50 °C	60 °C	70 °C	80 °C	90 °C	100 °C	Ref.
$Ba(IO_3)_2$	0.0182	0.0262	0.0342	0.0396	0.045*	0.058*	0.073	0.090	0.109	0.131	0.156	0.182	1:14
$Ba(NO_2)_2$	31.1	36.6	41.8	44.3	46.8	51.6	56.2	60.5	64.6	68.5	72.1	75.6	10
$Ba(NO_3)_2$	4.7	6.3	8.2	9.3	10.2	12.4	14.7	17.0	19.3	21.5	23.5	25.5	6
$Ba(OH)_2$	1.67			4.68	8.4	19	33	52	74	100			7
BaS	2.79	4.78	6.97	8.21	9.58	12.67	16.18	20.05	24.19	28.55	33.04	37.61	7
$Ba(SCN)_2$				62.6									7
$BaSO_3$				0.0011									1:26
$BeCl_2$	40.5			41.7									7
$Be(ClO_4)_2$				59.5									7
$BeSO_4$	26.69	27.58	28.61	29.22	29.90	31.51	33.39	35.50	37.78	40.21	42.72	45.28	7
$CaBr_2$	55	56	59	61	63	68	71	73					10
$CaCl_2$	36.70	39.19	42.13	44.83*	49.12*	52.85*	56.05*	56.73	57.44	58.21	59.04	59.94	8
$Ca(ClO_3)_2$	63.2	64.2	65.5	66.3	67.2	69.0	71.0	73.2	75.5*	77.4*	77.7	78.0	1:14
$Ca(ClO_4)_2$				65.3									7
CaF_2	0.0013			0.0016									10
CaI_2	64.6	66.0	67.6	68.3	69.0	70.8	72.4	74.0	76.0	78.0	79.6	81.0	7
$Ca(IO_3)_2$	0.082	0.155	0.243	0.305	0.384*	0.517*	0.590	0.652	0.811*	0.665*	0.668		1:14
$Ca(NO_2)_2$	38.6	39.5	44.5	48.6									7
$Ca(NO_3)_2$	50.1	53.1	56.7	59.0	60.9	65.4	77.8	78.1	78.2	78.3	78.4	78.5	6
$CaSO_3$			0.0059	0.0054	0.0049	0.0041	0.0035	0.0030	0.0026	0.0023	0.0020	0.0019	1:26
$CaSO_4$	0.174	0.191	0.202	0.205	0.208	0.210	0.207	0.201	0.193	0.184	0.173	0.163	9
$CdBr_2$	36.0	43.0	49.9	53.4	56.4	60.3*	60.3*	60.5	60.7	60.9	61.3	61.6	6
CdC_2O_4				0.0060									5
$CdCl_2$	47.2	50.1	53.2	54.6	56.3*	57.3*	57.5	57.8	58.1	58.51	58.98	59.5	6
$Cd(ClO_4)_2$				58.7								66.9	7
CdF_2		5.82	4.65	4.18	3.76								5
CdI_2	44.1	44.9	45.8	46.3	46.8	47.9	49.0	50.2	51.5	52.7	54.1	55.4	6
$Cd(IO_3)_2$				0.091									5
$Cd(NO_3)_2$	55.4	57.1	59.6	61.0	62.8	66.5	70.6	86.1	86.5	86.8	87.1	87.4	6
$CdSO_4$	43.1	43.1	43.2	43.4	43.6	44.1	43.5	42.5	41.4	40.2	38.5	36.7	6
$CdSeO_4$	42.04	40.59	39.02	38.18	37.29	35.35	33.15	30.65	27.84	24.69	21.24	17.49	5
$Ce(NO_3)_3$	57.99	59.80	61.89	63.05	64.31*	67.0*	68.6	71.1*	74.9*	79.2	80.9	83.1	1:13
$CoCl_2$	30.30	32.60	34.87	35.99	37.10	39.27	41.38	43.46	45.50	47.51	49.51	51.50	7
$Co(ClO_4)_2$	50.0			53.0									7
CoF_2				1.4									7
CoI_2	58.00	61.78	65.35	66.99	68.51	71.17	73.41	75.29	76.89	78.28	79.52	80.70	7
$Co(NO_2)_2$	0.076			0.49									7
$Co(NO_3)_2$	45.5	47.0	49.4	50.8	52.4	56.0	60.1	62.6	64.9	67.7			6
$CoSO_4$	19.9	23.0	26.1	27.7	29.2	32.3	34.4	35.9	35.5	33.2	30.6	27.8	6
$Co(SCN)_2$				50.7									7
CrO_3	62.2	62.3	62.6	62.8	63.0	63.5	64.1	64.7	65.5	66.2	67.1	67.9	6
$CsBr$				55.2									7
$CsBrO_3$	1.16	1.93	3.01	3.69	4.46	6.32	8.60	11.32	14.45	17.96	21.83	25.98	1:30
$CsCl$	61.83	63.48	64.96	65.64	66.29	67.50	68.60	69.61	70.54	71.40	72.21	72.96	1:47
$CsClO_3$	2.40	3.87	5.94	7.22	8.69	12.15	16.33	21.14	26.45	32.10	37.89	43.42	1:30
$CsClO_4$	0.79	1.01	1.51	1.96	2.57	4.28	6.55	9.29	12.41	15.80	19.39	23.07	7
CsI	30.9	37.2	43.2	45.9	48.6	53.3	57.3	60.7	63.6	65.9	67.7	69.2	6
$CsIO_3$	1.08	1.58	2.21	2.59	3.02	3.96	5.06	6.29	7.70	9.20	10.79	12.45	1:30
$CsNO_3$	8.46	13.0	18.6	21.8	25.1	32.0	39.0	45.7	51.9	57.3	62.1	66.2	6
$CsOH$					75								7
Cs_2SO_4	62.6	63.4	64.1	64.5	64.8	65.5	66.1	66.7	67.3	67.8	68.3	68.8	6
$CuBr_2$				55.8									7
$CuCl_2$	40.8	41.7	42.6	43.1	43.7	44.8	46.0	47.2	48.5	49.9	51.3	52.7	6
$Cu(ClO_4)_2$	54.3				59.3								7
CuF_2				0.075									7
$Cu(NO_3)_2$	45.2	49.8	56.3	59.2	61.1	62.0	63.1	64.5	65.9	67.5	69.2	71.0	6
$CuSO_4$	12.4	14.4	16.7	18.0	19.3	22.2	25.4	28.8	32.4	36.3	40.3	43.5	6
$CuSeO_4$	10.6			16.0									7
$Dy(NO_3)_3$	58.79	59.99	61.49	62.35	63.29	65.43	68.04	71.58					1:13
$Er(NO_3)_3$	61.58	63.15	64.84	65.75	66.69	68.70	70.96	73.64	77.75				1:13
$Eu(NO_3)_3$	55.2	56.7	58.5	59.4	60.4	62.5	64.6						1:13

Compound	0 °C	10 °C	20 °C	25 °C	30 °C	40 °C	50 °C	60 °C	70 °C	80 °C	90 °C	100 °C	Ref.
$FeBr_2$				54.6								64.8*	7
$FeCl_2$	33.2*			39.4*								48.7*	7
$FeCl_3$	42.7	44.9	47.9	47.7	51.6	74.8	76.7	84.6	84.3	84.3	84.4	84.7	6
$Fe(ClO_4)_2$	63.39			67.76									7
FeF_3				5.59									7
$Fe(NO_3)_3$	40.15			46.57									7
$Fe(NO_3)_2$	41.44			46.67									7
$FeSO_4$	13.5	17.0	20.8	22.8	24.8	28.8	32.8	35.5	33.6	30.4	27.1	24.0	6
$Gd(NO_3)_3$	56.3	57.7	59.2	60.1	61.0	62.9	65.2	67.9	71.5				1:13
HIO_3	73.45	74.10	74.98	75.48	76.03	77.20	78.46	79.78	81.13	82.48	83.82	85.14	1:30
H_3BO_3	2.61	3.57	4.77	5.48	6.27	8.10	10.3	12.9	15.9	19.3	23.1	27.3	6
$HgBr_2$	0.26	0.37	0.52	0.61	0.72	0.96	1.26	1.63	2.08	2.61	3.23	3.95	4
$Hg(CN)_2$	6.57	7.83	9.33	10.2	11.1	13.1	15.5	18.2	21.2	24.6	28.3	32.3	6
$HgCl_2$	4.24	5.05	6.17	6.81	7.62	9.53	12.02	15.18	19.16	24.06	29.90	36.62	4
HgI_2			0.0041	0.0055	0.0072	0.0122	0.0199						4
$Hg(SCN)_2$				0.070									4
Hg_2Cl_2				0.0004									3
$Hg_2(ClO_4)_2$	73.8			79.8*								85.3*	7
Hg_2SO_4	0.038	0.043	0.048	0.051	0.054	0.059	0.065	0.070	0.076	0.082	0.088	0.093	4
$Ho(NO_3)_3$				63.8									1:13
KBF_4	0.28	0.34	0.45	0.55	0.75	1.38	2.09	2.82	3.58	4.34	5.12	5.90	10
KBr	35.0	37.3	39.4	40.4	41.4	43.2	44.8	46.2	47.6	48.8	49.8	50.8	6
$KBrO_3$	2.97	4.48	6.42	7.55	8.79	11.57	14.71	18.14	21.79	25.57	29.42	33.28	1:30
$KC_2H_3O_2$	68.40	70.29	72.09	72.92	73.70	75.08	76.27	77.31	78.22	79.04	79.80	80.55	7
KCl	21.74	23.61	25.39	26.22	27.04	28.59	30.04	31.40	32.66	33.86	34.99	36.05	1:47
$KClO_3$	3.03	4.67	6.74	7.93	9.21	12.06	15.26	18.78	22.65	26.88	31.53	36.65	1:30
$KClO_4$	0.70	1.10	1.67	2.04	2.47	3.54	4.94	6.74	8.99	11.71	14.94	18.67	6
KF	30.90	39.8	47.3	50.41	53.2					60.0			7
$KHCO_3$	18.62	21.73	24.92	26.6	28.13	31.32	34.46	37.51	40.45				6
$KHSO_4$	27.1	29.7	32.3	33.6	35.0	37.8	40.5	43.4	46.2	49.02	51.82	54.6	6
KH_2PO_4	11.74	14.91	18.25	19.97	21.77	25.28	28.95	32.76	36.75	40.96	45.41	50.12	1:31
KI	56.0	57.6	59.0	59.7	60.4	61.6	62.8	63.8	64.8	65.7	66.6	67.4	6
KIO_3	4.53	5.96	7.57	8.44	9.34	11.09	13.22	15.29	17.41	19.58	21.78	24.03	1:30
KIO_4	0.16	0.22	0.37	0.51	0.70	1.24	1.96	2.83	3.82	4.89	6.02	7.17	7
$KMnO_4$	2.74	4.12	5.96	7.06	8.28	11.11	14.42	18.16					6
KNO_2	73.7	74.6	75.3	75.7	76.0	76.7	77.4	78.0	78.5	79.1	79.6	80.1	6
KNO_3	12.0	17.6	24.2	27.7	31.3	38.6	45.7	52.2	58.0	63.0	67.3	70.8	6
KOH	48.7	50.8	53.2	54.7	56.1	57.9	58.6	59.5	60.6	61.8	63.1	64.6	6
KSCN	63.8	66.4	69.1	70.4	71.6	74.1	76.5	78.9	81.1	83.3	85.3	87.3	6
K_2CO_3	51.3	51.7	52.3	52.7	53.1	54.0	54.9	56.0	57.2	58.4	59.6	61.0	6
K_2CrO_4	37.1	38.1	38.9	39.4	39.8	40.5	41.3	41.9	42.6	43.2	43.8	44.3	6
$K_2Cr_2O_7$	4.30	7.12	10.9	13.1	15.5	20.8	26.3	31.7	36.9	41.5	45.5	48.9	6
K_2HAsO_4	48.5*			63.6*								79.8*	7
K_2HPO_4	57.0	59.1	61.5	62.7	64.1	67.7*		72.7*					1:31
K_2MoO_4				64.7							66.5		7
K_2SO_3	51.30	51.39	51.49	51.55	51.62	51.76	51.93	52.11	52.32	52.54	52.79	53.06	1:26
K_2SO_4	7.11	8.46	9.95	10.7	11.4	12.9	14.2	15.5	16.7	17.7	18.6	19.3	6
$K_2S_2O_3$	49.0*			62.3*							75.7*		7
$K_2S_2O_5$	22.1	26.7	31.1	33.1	35.2	39.0	42.6	46.0	49.1	52.0	54.6		1:26
K_2SeO_3	68.4*			68.5*								68.5*	7
K_2SeO_4	52.70	52.93	53.17	53.30	53.43	53.70	53.99	54.30	54.61	54.94	55.26	55.60	7
K_3AsO_4	51.5*			55.6*								73*	7
$K_3Fe(CN)_6$	23.9	27.6	31.1	32.8	34.3	37.2	39.6	41.7	43.5	45.0	46.1	47.0	6
K_3PO_4	44.3			51.4									7
$K_4Fe(CN)_6$	12.5	17.3	22.0	23.9	25.6	29.2	32.5	35.5	38.2	40.6	41.4	43.1	6
$LaCl_3$	49.0	48.5	48.6	48.9	49.3	50.5	52.1	54.0	56.3	58.9	61.7		6
$La(NO_3)_3$	55.0	56.9	58.9	60.0	61.1	63.6	66.3	69.9*	74.1*				1:13
LiBr	58.4	60.1	62.7	64.4	65.9	67.8	68.3	69.0	69.8	70.7	71.7	72.8	6
$LiBrO_3$	61.03	62.62	64.44	65.44	66.51	68.90	71.68*	73.24*	74.43	75.66	76.93	78.32	1:30
$LiC_2H_3O_2$	23.76	26.49	29.42	31.02	32.72	36.48	40.65	45.15	49.93	54.91	60.04	65.26	7
LiCl	40.45	42.46*	45.29*	45.81	46.25	47.30	48.47	49.78	51.27	52.98	54.98*	56.34*	1:47

Compound	0 °C	10 °C	20 °C	25 °C	30 °C	40 °C	50 °C	60 °C	70 °C	80 °C	90 °C	100 °C	Ref.
$LiClO_3$	73.2	75.6*	80.8*	82.1	83.4	85.9*	87.1*	88.2	89.6	91.3	93.4	95.7	1:30
$LiClO_4$	30.1	32.6	35.5	37.0	38.6	41.9	45.5	49.2	53.2	57.2	61.3	71.4	6
LiF	0.120	0.126	0.131	0.134									7
LiH_2PO_4	55.8												7
LiI	59.4	60.5	61.7	62.3	63.0	64.3	65.8	67.3	68.8	81.3	81.7	82.6	6
$LiIO_3$				43.8									1:30
$LiNO_2$	41	45	49	51	53	56	60	63	66	68			10
$LiNO_3$	34.8	37.6	42.7	50.5	57.9	60.1	62.2	64.0	65.7	67.2	68.5	69.7	6
LiOH	10.8	10.8	11.0	11.1	11.3	11.7	12.2	12.7	13.4	14.2	15.1	16.1	6
LiSCN				54.5									7
Li_2CO_3	1.54	1.43	1.33	1.28	1.24	1.15	1.07	0.99	0.92	0.85	0.78	0.72	7
$Li_2C_2O_4$				5.87									7
Li_2HPO_3	9.07	8.40	7.77	7.47	7.18	6.64	6.16	5.71	5.30	4.91	4.53	4.16	7
Li_2SO_4	26.3	25.9	25.6	25.5	25.3	25.0	24.8	24.5	24.3	24.0	23.8	23.6	6
Li_3PO_4				0.027									1:31
$Lu(NO_3)_3$				71.1									1:13
$MgBr_2$	49.3	49.8	50.3	50.6	50.9	51.5	52.1	52.8	53.5	54.2	55.0	55.7	6
$Mg(BrO_3)_2$	43.0	45.2	48.0	49.4	51.0	54.3	57.9	61.6	65.3	69.0*	70.9*	71.7	1:14
$Mg(C_2H_3O_2)_2$	36.18	37.55	38.92	39.61									7
MgC_2O_4				0.038									7
$MgCl_2$	33.96	34.85	35.58	35.90	36.20	36.77	37.34	37.97	38.71	39.62	40.75	42.15	8
$Mg(ClO_3)_2$	53.35	54.40	56.81	58.66	60.91*	65.46*	67.33	69.27	71.01	72.44	73.48		1:14
$Mg(ClO_4)_2$	47.8	48.7	49.6	50.1	50.5	51.3	52.1						6
$MgCrO_4$	32.06*			35.39*									7
$MgCr_2O_7$				58.9						67.0			7
MgF_2				0.013									7
MgI_2	54.7	56.1	58.2	59.4	60.8	63.9	65.0	65.0	65.0	65.0	65.1	65.2	6
$Mg(IO_3)_2$	3.19*	6.70*	7.92	8.52	9.11	10.45	11.99	13.7	15.6	17.6	19.6		1:14
$Mg(NO_2)_2$				47									7
$Mg(NO_3)_2$	38.4	39.5	40.8	41.6	42.4	44.1	45.9	47.9	50.0	52.2	70.6	72.0	6
$MgSO_3$	0.32	0.37	0.46	0.52	0.61	0.87*	0.85*	0.76	0.69	0.64	0.62	0.60	1:26
$MgSO_4$	18.2	21.7	25.1	26.3	28.2	30.9	33.4	35.6	36.9	35.9	34.7	33.3	6
MgS_2O_3	30.7			34.1									7
$MgSeO_4$	31.4*			35.7*								47*	7
$MnBr_2$	56.00	57.72	59.39	60.19	60.96	62.41	63.75	65.01	66.19	67.32	68.42	69.50	7
$MnCl_2$	38.7	40.6	42.5	43.6	44.7	47.0	49.4	54.1	54.7	55.2	55.7	56.1	6
MnF_2	0.80*			1.01*								0.48	7
$Mn(IO_3)_2$				0.27							0.34		7
$Mn(NO_3)_2$	50.5			61.7									7
$MnSO_4$	34.6	37.3	38.6	38.9	38.9	37.7	36.3	34.6	32.8	30.8	28.8	26.7	6
NH_4Br	37.5	40.2	42.7	43.9	45.1	47.3	49.4	51.3	53.0	54.6	56.1	57.4	7
NH_4Cl	22.92	25.12	27.27	28.34	29.39	31.46	33.50	35.49	37.46	39.40	41.33	43.24	1:47
NH_4ClO_4	10.8	14.1	17.8	19.7	21.7	25.8	29.8	33.6	37.3	40.7	43.8	46.6	6
NH_4F	41.7	43.2	44.7	45.5	46.3	47.8	49.3	50.9	52.5	54.1			7
NH_4HCO_3	10.6	13.7	17.6	19.9	22.4	27.9	34.2	41.4	49.3	58.1	67.6	78.0	7
$NH_4H_2AsO_4$	25.2	29.0	32.7	34.5	36.3	39.7	43.1	46.2	49.3	52.2	55.0		7
$NH_4H_2PO_4$	17.8	22.0	26.4	28.8	31.2	36.2	41.6	47.2	53.0	59.2	65.7	72.4	7
NH_4I	60.7	62.1	63.4	64.0	64.6	65.8	66.8	67.8	68.7	69.6	70.4	71.1	6
NH_4IO_3				3.70	4.20	5.64	7.63						1:30
NH_4NO_2	55.7	59.0	64.9	68.8									7
NH_4NO_3	54.0	60.1	65.5	68.0	70.3	74.3	77.7	80.8	83.4	85.8	88.2	90.3	6
NH_4SCN				64.4					81.1				7
$(NH_4)_2C_2O_4$	2.31	3.11	4.25	4.94	5.73	7.56	9.73	12.2	15.1	18.3	21.8	25.7	7
$(NH_4)_2HPO_4$	36.4	38.2	40.0	41.0	42.0	44.1	46.2	48.5	50.9	53.3	55.9	58.6	7
$(NH_4)_2S_2O_5$	65.5	67.9	69.8	70.5	71.3	72.3	72.9	73.1					1:26
$(NH_4)_2S_2O_8$	37.00	40.45	43.84	45.49	47.11	50.25	53.28	56.23	59.13	62.00			7
$(NH_4)_2SO_3$	32.2	34.9	37.7	39.1	40.6	43.7	47.0	50.6	54.5	58.9			1:26
$(NH_4)_2SO_4$	41.3	42.1	42.9	43.3	43.8	44.7	45.6	46.6	47.5	48.5	49.5	50.5	6
$(NH_4)_2SeO_3$	49.0	51.1	53.4	54.7	56.0	58.9	62.0	65.4	69.1				7
$(NH_4)_2SeO_4$				54.02									7
$(NH_4)_3PO_4$				15.5									7

Compound	0 °C	10 °C	20 °C	25 °C	30 °C	40 °C	50 °C	60 °C	70 °C	80 °C	90 °C	100 °C	Ref.
NaBr	44.4	45.9	47.7	48.6	49.6	51.6	53.7	54.1	54.3	54.5	54.7	54.9	6
$NaBrO_3$	20.0	23.22	26.65	28.28	29.86	32.83	35.55	38.05	40.37	42.52			1:30
$NaCHO_2$	30.8	37.9	45.7	48.7	50.6	52.0	53.5	55.0					6
$NaC_2H_3O_2$	26.5	28.8	31.8	33.5	35.5	39.9	45.1	58.3	59.3	60.5	61.7	62.9	6
NaCl	26.28	26.32	26.41	26.45	26.52	26.67	26.84	27.03	27.25	27.50	27.78	28.05	1:47
NaClO	22.7			44.4									7
$NaClO_2$				97.0*				95.3*					7
$NaClO_3$	44.27	46.67	49.3	50.1	51.2	53.6	55.5	57.0	58.5	60.5	63.3	67.1	1:30
$NaClO_4$	61.9	64.1	66.2	67.2	68.3	70.4	72.5	74.1	74.7	75.4	76.1	76.7	6
NaF	3.52	3.72	3.89	3.97	4.05	4.20	4.34	4.46	4.57	4.66	4.75	4.82	6
$NaHCO_3$	6.48	7.59	8.73	9.32	9.91	11.13	12.40	13.70	15.02	16.37	17.73	19.10	7
$NaHSO_4$				22.2								33.3	10
NaH_2PO_4	36.54	41.07	46.00	48.68	51.54	57.89*	61.7*	62.3*	65.9	68.7			1:31
NaI	61.2	62.4	63.9	64.8	65.7	67.7	69.8	72.0	74.7	74.8	74.9	75.1	6
$NaIO_3$	2.43	4.40	7.78*	8.65*	9.60	11.67	13.99	16.52	19.25*	21.1*	22.9	24.7	1:30
$NaIO_4$				12.62									7
$NaNO_2$	41.9	43.4	45.1	45.9	46.8	48.7	50.7	52.8	55.0	57.2	59.5	61.8	6
$NaNO_3$	42.2	44.4	46.6	47.7	48.8	51.0	53.2	55.3	57.5	59.6	61.7	63.8	6
NaOH	30	39	46	50	53	58	63	67	71	74	76	79	10
NaSCN		52.9	57.1	60.2	62.7	63.5	64.2	65.0	65.9	66.9	67.9	69.0	6
$Na_2B_4O_7$	1.23	1.71	2.50	3.07	3.82	6.02	9.7	14.9	17.1	19.9	23.5	28.0	6
Na_2CO_3	6.44	10.8	17.9	23.5	28.7	32.8	32.2	31.7	31.3	31.1	30.9	30.9	6
$Na_2C_2O_4$	2.62	2.95	3.30	3.48	3.65	4.00	4.36	4.71	5.06	5.41	5.75	6.08	6
Na_2CrO_4	22.6	32.3	44.6	46.7	46.9	48.9	51.0	53.4	55.3	55.5	55.8	56.1	6
$Na_2Cr_2O_7$	62.1	63.1	64.4	65.2	66.1	68.0	70.1	72.3	74.6	77.0	79.6	80.7	6
Na_2HAsO_4	5.6*			29.3*								67*	7
Na_2HPO_4	1.66	4.19	7.51	10.55	16.34*	35.17*	44.64*	45.20	46.81	48.78	50.52	51.53	1:31
Na_2MoO_4	30.6	38.8	39.4	39.4	39.8	40.3	41.0	41.7	42.6	43.5	44.5	45.5	6
Na_2S	11.1	13.2	15.7	17.1	18.6	22.1	26.7	28.1	30.2	33.0	36.4	41.0	6
Na_2SO_3	12.0	16.1	20.9	23.5	26.3*	27.3*	25.9	24.8	23.7	22.8	22.1	21.5	1:26
Na_2SO_4			16.13	21.94	29.22*	32.35*	31.55	30.90	30.39	30.02	29.79	29.67	8
$Na_2S_2O_3$	33.1	36.3	40.6	43.3	45.9	52.0	62.3	65.7	68.8	69.4	70.1	71.0	6
$Na_2S_2O_5$		38.4	39.5	40.0	40.6	41.8	43.0	44.2	45.5	46.8	48.1	49.5	1:26
Na_2SeO_3				47.3*								45*	7
Na_2SeO_4	11.7			36.9*								42.1*	7
Na_2WO_4	41.6	41.9	42.3	42.6	42.9	43.6	44.4	45.3	46.2	47.3	48.4	49.5	6
Na_3PO_4	4.28	7.30	10.8	12.6	14.1	16.6	22.9	28.4	32.4	37.6	40.4	43.5	6
$Na_4P_2O_7$	2.23	3.28	4.81	6.62	7.00	10.10	14.38	20.07	27.31	36.03	32.37	30.67	6
$NdCl_3$	49.0	49.3	49.7	50.0	50.4	51.2	52.2	53.3	54.5	55.8	57.1	58.5	6
$Nd(NO_3)_3$	55.76	57.49	59.37	60.38	61.43	63.69	66.27	69.47					1:13
$NiCl_2$	34.7	36.1	38.5	40.3	41.7	42.1	43.2	45.0	46.1	46.2	46.4	46.6	6
$Ni(ClO_4)_2$	51.1			52.8									7
NiF_2				2.50							2.52		7
NiI_2	55.40	57.68	59.78	60.69	61.50	62.80	63.73	64.38	64.80	65.09	65.30		7
$Ni(NO_3)_2$	44.1	46.0	48.4	49.8	51.3	54.6	58.3	61.0	63.1	65.6	67.9	69.0	6
$NiSO_4$	21.4	24.4	27.4	28.8	30.3*	32.0*	34.1	35.8	37.7	39.9	42.3	44.8	6
$Ni(SCN)_2$				35.48									7
$NiSeO_4$	21.6		26.2*									45.6*	7
$PbBr_2$	0.449	0.620	0.841	0.966	1.118	1.46	1.89						2
$PbCl_2$	0.66	0.81	0.98	1.07	1.17	1.39	1.64	1.93	2.24	2.60	2.99	3.42	2
$Pb(ClO_4)_2$				81.5									7
PbF_2		0.0603	0.0649	0.0670	0.0693								2
PbI_2	0.041	0.052	0.067	0.076	0.086	0.112	0.144	0.187	0.243	0.315			2
$Pb(IO_3)_2$				0.0025									7
$Pb(NO_3)_2$	28.46	32.13	35.67	37.38	39.05	42.22	45.17	47.90	50.42	52.72	54.82	56.75	2
$PbSO_4$	0.0033	0.0038	0.0042	0.0044	0.0047	0.0052	0.0058						2
$PrCl_3$	48.0	48.1	48.6	49.0	49.5	50.8	52.3	54.1	56.1	58.3			6
$Pr(NO_3)_3$	57.50	59.20	61.16	62.24	63.40*	65.7*	67.8	70.2	73.4				1:13
RbBr	47.4	50.1	52.6	53.8	54.9	57.0	58.8	60.6	62.1	63.5	64.8	65.9	6
$RbBrO_3$	0.97	1.55	2.36	2.87	3.45	4.87	6.64	8.78	11.29	14.15	17.32	20.76	1:30
RbCl	43.58	45.65	47.53	48.42	49.27	50.86	52.34	53.67	54.92	56.08	57.16	58.15	1:47

Compound	0 °C	10 °C	20 °C	25 °C	30 °C	40 °C	50 °C	60 °C	70 °C	80 °C	90 °C	100 °C	Ref.
$RbClO_3$	2.10	3.38	5.14	6.22	7.45	10.35	13.85	17.93	22.53	27.57	32.96	38.60	1:30
$RbClO_4$	1			1.5								17	7
RbF			75										7
$RbHCO_3$			53.7										7
RbI	55.8	58.6	61.1	62.3	63.4	65.4	67.2	68.8	70.3	71.6	72.7	73.8	6
$RbIO_3$	1.09	1.53	2.07	2.38	2.74	3.52	4.41	5.42	6.52	7.74	9.00	10.36	1:30
$RbNO_3$	16.4	25.0	34.6	39.4	44.2	53.1	60.8	67.2	72.2	76.1	79.0	81.2	6
RbOH					63.4								7
Rb_2CrO_4	38.27			43.26									7
Rb_2SO_4	27.3	30.0	32.5	33.7	34.8	36.9	38.7	40.3	41.8	43.0	44.1	44.9	6
$SbCl_3$	85.7			90.8									7
SbF_3	79.4			83.1									7
$Sc(NO_3)_3$	57.0	59.3	61.6	62.8	63.9	66.2	68.5						1:13
$Sm(NO_3)_3$	54.83	56.33	58.08	59.05	60.08	62.38	65.05*	68.1*	70.8	74.2			1:13
$SmCl_3$		48.0	48.2	48.4	48.6	49.2	50.0						6
$SnCl_2$	46	64											7
SnI_2			0.97									3.87	7
$SrBr_2$	46.0	48.3	50.6	51.7	52.9	55.2	57.6	59.9	62.3	64.6	66.8	69.0	6
$Sr(BrO_3)_2$	18.53	22.00	25.39	27.02	28.59	31.55	34.21	36.57	38.64*	40.2*	40.8	41.0	1:14
$SrCl_2$	31.94	32.93	34.43	35.37	36.43	38.93	41.94	45.44*	46.81*	47.69	48.70	49.87	8
$Sr(ClO_2)_2$	13.0	13.6	14.1	14.3	14.5	14.9	15.3	15.6	15.9				7
$Sr(ClO_3)_2$	63.29	63.42	63.64	63.77	63.93	64.29	64.70	65.16	65.65	66.18	66.74	67.31	1:14
$Sr(ClO_4)_2$	70.04*			75.35*		78.44*							7
SrF_2	0.011			0.021									7
SrI_2	62.5	62.8	63.5	63.9	64.5	65.8	67.3	69.0	70.8	72.7	74.7	79.2	6
$Sr(IO_3)_2$	0.102	0.126	0.152	0.165	0.179	0.206	0.233	0.259	0.284	0.307	0.328	0.346	1:14
$Sr(MnO_4)_2$	2.5												7
$Sr(NO_2)_2$					41.9	44.3						58.6	7
$Sr(NO_3)_2$	28.2	34.6	41.0	44.5	47.0	47.4	47.9	48.4	48.9	49.5	50.1	50.7	6
$Sr(OH)_2$	0.9			2.2									7
$SrSO_3$				0.0015									1:26
$SrSO_4$				0.0135									7
SrS_2O_3	8.8	13.2	17.7	20.0	22.2	26.8							7
$Tb(NO_3)_3$			60.6	61.02									1:13
Tl_2SO_4	2.65	3.56	4.61	5.19	5.80	7.09	8.46	9.89	11.33	12.77	14.18	15.53	6
$Tm(NO_3)_3$				67.9									1:13
$UO_2(NO_3)_2$	49.52	51.82	54.42	55.85	57.55	61.59	67.07						1:55
$Y(NO_3)_3$	55.57	56.93	58.75	59.86	61.11*	63.3*	64.9	67.9	72.5				1:13
$Yb(NO_3)_3$				70.5									1:13
$ZnBr_2$	79.3	80.1	81.8	83.0	84.1	85.6	85.8	86.1	86.3	86.6	86.8	87.1	6
ZnC_2O_4		0.0010	0.0019	0.0026									5
$ZnCl_2$		76.6	79.0	80.3	81.4	81.8	82.4	83.0	83.7	84.4	85.2	86.0	6
$Zn(ClO_4)_2$	44.29*			46.27*			48.70						7
ZnF_2				1.53									5
ZnI_2	81.1	81.2	81.3	81.4	81.5	81.7	82.0	82.3	82.6	83.0	83.3	83.7	6
$Zn(IO_3)_2$			0.58	0.64	0.69	0.77	0.82						5
$Zn(NO_3)_2$	47.8	50.8	54.4	54.6	58.5	79.1	80.1	87.5	89.9				6
$ZnSO_3$			0.1786	0.1790	0.1794	0.1803	0.1812						5
$ZnSO_4$	29.1	32.0	35.0	36.6	38.2	41.3	43.0	42.1	41.0	39.9	38.8	37.6	6
$ZnSeO_4$	33.06	34.98	37.38	38.79	40.34								5

SOLUBILITY PRODUCT CONSTANTS

The solubility product constant K_{sp} is a useful parameter for calculating the aqueous solubility of sparingly soluble compounds under various conditions. It may be determined by direct measurement or calculated from the standard Gibbs energies of formation $\Delta_f G°$ of the species involved at their standard states. Thus if $K_{sp} = [M^+]^m, [A^-]^n$ is the equilibrium constant for the reaction

$$M_mA_n(s) \rightleftharpoons mM^+(aq) + nA^-(aq),$$

where M_mA_n is the slightly soluble substance and M^+ and A^- are the ions produced in solution by the dissociation of M_mA_n, then the Gibbs energy change is

$$\Delta G° = m\,\Delta_f G°(M^+,aq) + n\,\Delta_f G°(A^-,aq) - \Delta_f G°(M_mA_n, s)$$

The solubility product constant is calculated from the equation

$$\ln K_{sp} = -\Delta G°/RT$$

The first table below gives selected values of K_{sp} at 25 °C. Many of these have been calculated from standard state thermodynamic data in References 1 and 2; other values are taken from publications of the IUPAC Solubility Data Project (References 3 to 7).

The above formulation is not convenient for treating sulfides because the S^{-2} ion is usually not present in significant concentrations (see Reference 8). This is due to the hydrolysis reaction

$$S^{-2} + H_2O \rightleftharpoons HS^- + OH^-$$

which is strongly shifted to the right except in very basic solutions. Furthermore, the equilibrium constant for this reaction, which depends on the second ionization constant of H_2S, is poorly known. Therefore it is more useful in the case of sulfides to define a different solubility product K_{spa} based on the reaction

$$M_mS_n(s) + 2H^+ \rightleftharpoons mM^+ + nH_2S\ (aq)$$

Values of K_{spa}, taken from Reference 8, are given for several sulfides in the auxiliary table following the main table. Additional discussion of sulfide equilibria may be found in References 7 and 9.

References

1. Wagman, D. D., Evans, W. H., Parker, V. B., Schumm, R. H., Halow, I., Bailey, S. M., Churney, K. L., and Nuttall, R. L., *The NBS Tables of Chemical Thermodynamic Properties, J. Phys. Chem. Ref. Data*, Vol. 11, Suppl. 2, 1982.
2. Garvin, D., Parker, V. B., and White, H. J., *CODATA Thermodynamic Tables*, Hemisphere, New York, 1987.
3. *Solubility Data Series* (53 Volumes), International Union of Pure and Applied Chemistry, Pergamon Press, Oxford, 1979–1992.
4. Clever, H. L., and Johnston, F. J., *J. Phys. Chem. Ref. Data*, 9, 751, 1980.
5. Marcus, Y., *J. Phys. Chem. Ref. Data*, 9, 1307, 1980.
6. Clever, H. L., Johnson, S. A., and Derrick, M. E., *J. Phys. Chem. Ref. Data*, 14, 631, 1985.
7. Clever, H. L., Johnson, S. A., and Derrick, M. E., *J. Phys. Chem. Ref. Data*, 21, 941, 1992.
8. Myers, R. J., *J. Chem. Educ.*, 63, 687, 1986.
9. Licht, S., *J. Electrochem. Soc.*, 135, 2971, 1988.

Compound	Formula	K_{sp}
Aluminum phosphate	$AlPO_4$	$9.84 \cdot 10^{-21}$
Barium bromate	$Ba(BrO_3)_2$	$2.43 \cdot 10^{-4}$
Barium carbonate	$BaCO_3$	$2.58 \cdot 10^{-9}$
Barium chromate	$BaCrO_4$	$1.17 \cdot 10^{-10}$
Barium fluoride	BaF_2	$1.84 \cdot 10^{-7}$
Barium hydroxide octahydrate	$Ba(OH)_2 \cdot 8H_2O$	$2.55 \cdot 10^{-4}$
Barium iodate	$Ba(IO_3)_2$	$4.01 \cdot 10^{-9}$
Barium iodate monohydrate	$Ba(IO_3)_2 \cdot H_2O$	$1.67 \cdot 10^{-9}$
Barium molybdate	$BaMoO_4$	$3.54 \cdot 10^{-8}$
Barium selenate	$BaSeO_4$	$3.40 \cdot 10^{-8}$
Barium sulfate	$BaSO_4$	$1.08 \cdot 10^{-10}$
Barium sulfite	$BaSO_3$	$5.0 \cdot 10^{-10}$
Beryllium hydroxide	$Be(OH)_2$	$6.92 \cdot 10^{-22}$
Bismuth arsenate	$BiAsO_4$	$4.43 \cdot 10^{-10}$
Bismuth iodide	BiI_3	$7.71 \cdot 10^{-19}$
Cadmium arsenate	$Cd_3(AsO_4)_2$	$2.2 \cdot 10^{-33}$
Cadmium carbonate	$CdCO_3$	$1.0 \cdot 10^{-12}$
Cadmium fluoride	CdF_2	$6.44 \cdot 10^{-3}$
Cadmium hydroxide	$Cd(OH)_2$	$7.2 \cdot 10^{-15}$
Cadmium iodate	$Cd(IO_3)_2$	$2.5 \cdot 10^{-8}$
Cadmium oxalate trihydrate	$CdC_2O_4 \cdot 3H_2O$	$1.42 \cdot 10^{-8}$
Cadmium phosphate	$Cd_3(PO_4)_2$	$2.53 \cdot 10^{-33}$
Calcium carbonate (calcite)	$CaCO_3$	$3.36 \cdot 10^{-9}$
Calcium fluoride	CaF_2	$3.45 \cdot 10^{-11}$
Calcium hydroxide	$Ca(OH)_2$	$5.02 \cdot 10^{-6}$
Calcium iodate	$Ca(IO_3)_2$	$6.47 \cdot 10^{-6}$
Calcium iodate hexahydrate	$Ca(IO_3)_2 \cdot 6H_2O$	$7.10 \cdot 10^{-7}$
Calcium molybdate	$CaMoO_4$	$1.46 \cdot 10^{-8}$
Calcium oxalate monohydrate	$CaC_2O_4 \cdot H_2O$	$2.32 \cdot 10^{-9}$
Calcium phosphate	$Ca_3(PO_4)_2$	$2.07 \cdot 10^{-33}$
Calcium sulfate	$CaSO_4$	$4.93 \cdot 10^{-5}$
Calcium sulfate dihydrate	$CaSO_4 \cdot 2H_2O$	$3.14 \cdot 10^{-5}$
Calcium sulfite hemihydrate	$CaSO_3 \cdot 0.5H_2O$	$3.1 \cdot 10^{-7}$
Cesium perchlorate	$CsClO_4$	$3.95 \cdot 10^{-3}$
Cesium periodate	$CsIO_4$	$5.16 \cdot 10^{-6}$
Cobalt(II) arsenate	$Co_3(AsO_4)_2$	$6.80 \cdot 10^{-29}$
Cobalt(II) hydroxide (blue)	$Co(OH)_2$	$5.92 \cdot 10^{-15}$
Cobalt(II) iodate dihydrate	$Co(IO_3)_2 \cdot 2H_2O$	$1.21 \cdot 10^{-2}$
Cobalt(II) phosphate	$Co_3(PO_4)_2$	$2.05 \cdot 10^{-35}$
Copper(I) bromide	$CuBr$	$6.27 \cdot 10^{-9}$
Copper(I) chloride	$CuCl$	$1.72 \cdot 10^{-7}$
Copper(I) cyanide	$CuCN$	$3.47 \cdot 10^{-20}$
Copper(I) iodide	CuI	$1.27 \cdot 10^{-12}$
Copper(I) thiocyanate	$CuSCN$	$1.77 \cdot 10^{-13}$
Copper(II) arsenate	$Cu_3(AsO_4)_2$	$7.95 \cdot 10^{-36}$
Copper(II) iodate monohydrate	$Cu(IO_3)_2 \cdot H_2O$	$6.94 \cdot 10^{-8}$
Copper(II) oxalate	CuC_2O_4	$4.43 \cdot 10^{-10}$
Copper(II) phosphate	$Cu_3(PO_4)_2$	$1.40 \cdot 10^{-37}$
Europium(III) hydroxide	$Eu(OH)_3$	$9.38 \cdot 10^{-27}$
Gallium(III) hydroxide	$Ga(OH)_3$	$7.28 \cdot 10^{-36}$
Iron(II) carbonate	$FeCO_3$	$3.13 \cdot 10^{-11}$
Iron(II) fluoride	FeF_2	$2.36 \cdot 10^{-6}$

Compound	Formula	K_{sp}
Iron(II) hydroxide	$Fe(OH)_2$	$4.87 \cdot 10^{-17}$
Iron(III) hydroxide	$Fe(OH)_3$	$2.79 \cdot 10^{-39}$
Iron(III) phosphate dihydrate	$FePO_4 \cdot 2H_2O$	$9.91 \cdot 10^{-16}$
Lanthanum iodate	$La(IO_3)_3$	$7.50 \cdot 10^{-12}$
Lead(II) bromide	$PbBr_2$	$6.60 \cdot 10^{-6}$
Lead(II) carbonate	$PbCO_3$	$7.40 \cdot 10^{-14}$
Lead(II) chloride	$PbCl_2$	$1.70 \cdot 10^{-5}$
Lead(II) fluoride	PbF_2	$3.3 \cdot 10^{-8}$
Lead(II) hydroxide	$Pb(OH)_2$	$1.43 \cdot 10^{-20}$
Lead(II) iodate	$Pb(IO_3)_2$	$3.69 \cdot 10^{-13}$
Lead(II) iodide	PbI_2	$9.8 \cdot 10^{-9}$
Lead(II) selenate	$PbSeO_4$	$1.37 \cdot 10^{-7}$
Lead(II) sulfate	$PbSO_4$	$2.53 \cdot 10^{-8}$
Lithium carbonate	Li_2CO_3	$8.15 \cdot 10^{-4}$
Lithium fluoride	LiF	$1.84 \cdot 10^{-3}$
Lithium phosphate	Li_3PO_4	$2.37 \cdot 10^{-11}$
Magnesium carbonate	$MgCO_3$	$6.82 \cdot 10^{-6}$
Magnesium carbonate trihydrate	$MgCO_3 \cdot 3H_2O$	$2.38 \cdot 10^{-6}$
Magnesium carbonate pentahydrate	$MgCO_3 \cdot 5H_2O$	$3.79 \cdot 10^{-6}$
Magnesium fluoride	MgF_2	$5.16 \cdot 10^{-11}$
Magnesium hydroxide	$Mg(OH)_2$	$5.61 \cdot 10^{-12}$
Magnesium oxalate dihydrate	$MgC_2O_4 \cdot 2H_2O$	$4.83 \cdot 10^{-6}$
Magnesium phosphate	$Mg_3(PO_4)_2$	$1.04 \cdot 10^{-24}$
Manganese(II) carbonate	$MnCO_3$	$2.24 \cdot 10^{-11}$
Manganese(II) iodate	$Mn(IO_3)_2$	$4.37 \cdot 10^{-7}$
Manganese(II) oxalate dihydrate	$MnC_2O_4 \cdot 2H_2O$	$1.70 \cdot 10^{-7}$
Mercury(I) bromide	Hg_2Br_2	$6.40 \cdot 10^{-23}$
Mercury(I) carbonate	Hg_2CO_3	$3.6 \cdot 10^{-17}$
Mercury(I) chloride	Hg_2Cl_2	$1.43 \cdot 10^{-18}$
Mercury(I) fluoride	Hg_2F_2	$3.10 \cdot 10^{-6}$
Mercury(I) iodide	Hg_2I_2	$5.2 \cdot 10^{-29}$
Mercury(I) oxalate	$Hg_2C_2O_4$	$1.75 \cdot 10^{-13}$
Mercury(I) sulfate	Hg_2SO_4	$6.5 \cdot 10^{-7}$
Mercury(I) thiocyanate	$Hg_2(SCN)_2$	$3.2 \cdot 10^{-20}$
Mercury(II) bromide	$HgBr_2$	$6.2 \cdot 10^{-20}$
Mercury(II) iodide	HgI_2	$2.9 \cdot 10^{-29}$
Neodymium carbonate	$Nd_2(CO_3)_3$	$1.08 \cdot 10^{-33}$
Nickel(II) carbonate	$NiCO_3$	$1.42 \cdot 10^{-7}$
Nickel(II) hydroxide	$Ni(OH)_2$	$5.48 \cdot 10^{-16}$
Nickel(II) iodate	$Ni(IO_3)_2$	$4.71 \cdot 10^{-5}$
Nickel(II) phosphate	$Ni_3(PO_4)_2$	$4.74 \cdot 10^{-32}$
Palladium(II) thiocyanate	$Pd(SCN)_2$	$4.39 \cdot 10^{-23}$
Potassium hexachloroplatinate	K_2PtCl_6	$7.48 \cdot 10^{-6}$
Potassium perchlorate	$KClO_4$	$1.05 \cdot 10^{-2}$
Potassium periodate	KIO_4	$3.71 \cdot 10^{-4}$
Praseodymium hydroxide	$Pr(OH)_3$	$3.39 \cdot 10^{-24}$
Radium iodate	$Ra(IO_3)_2$	$1.16 \cdot 10^{-9}$
Radium sulfate	$RaSO_4$	$3.66 \cdot 10^{-11}$
Rubidium perchlorate	$RbClO_4$	$3.00 \cdot 10^{-3}$
Scandium fluoride	ScF_3	$5.81 \cdot 10^{-24}$
Scandium hydroxide	$Sc(OH)_3$	$2.22 \cdot 10^{-31}$
Silver(I) acetate	$AgCH_3COO$	$1.94 \cdot 10^{-3}$
Silver(I) arsenate	Ag_3AsO_4	$1.03 \cdot 10^{-22}$
Silver(I) bromate	$AgBrO_3$	$5.38 \cdot 10^{-5}$
Silver(I) bromide	$AgBr$	$5.35 \cdot 10^{-13}$
Silver(I) carbonate	Ag_2CO_3	$8.46 \cdot 10^{-12}$
Silver(I) chloride	$AgCl$	$1.77 \cdot 10^{-10}$
Silver(I) chromate	Ag_2CrO_4	$1.12 \cdot 10^{-12}$
Silver(I) cyanide	$AgCN$	$5.97 \cdot 10^{-17}$
Silver(I) iodate	$AgIO_3$	$3.17 \cdot 10^{-8}$
Silver(I) iodide	AgI	$8.52 \cdot 10^{-17}$
Silver(I) oxalate	$Ag_2C_2O_4$	$5.40 \cdot 10^{-12}$
Silver(I) phosphate	Ag_3PO_4	$8.89 \cdot 10^{-17}$
Silver(I) sulfate	Ag_2SO_4	$1.20 \cdot 10^{-5}$
Silver(I) sulfite	Ag_2SO_3	$1.50 \cdot 10^{-14}$
Silver(I) thiocyanate	$AgSCN$	$1.03 \cdot 10^{-12}$
Strontium arsenate	$Sr_3(AsO_4)_2$	$4.29 \cdot 10^{-19}$
Strontium carbonate	$SrCO_3$	$5.60 \cdot 10^{-10}$
Strontium fluoride	SrF_2	$4.33 \cdot 10^{-9}$
Strontium iodate	$Sr(IO_3)_2$	$1.14 \cdot 10^{-7}$
Strontium iodate monohydrate	$Sr(IO_3)_2 \cdot H_2O$	$3.77 \cdot 10^{-7}$
Strontium iodate hexahydrate	$Sr(IO_3)_2 \cdot 6H_2O$	$4.55 \cdot 10^{-7}$
Strontium sulfate	$SrSO_4$	$3.44 \cdot 10^{-7}$
Thallium(I) bromate	$TlBrO_3$	$1.10 \cdot 10^{-4}$
Thallium(I) bromide	$TlBr$	$3.71 \cdot 10^{-6}$
Thallium(I) chloride	$TlCl$	$1.86 \cdot 10^{-4}$
Thallium(I) chromate	Tl_2CrO_4	$8.67 \cdot 10^{-13}$
Thallium(I) iodate	$TlIO_3$	$3.12 \cdot 10^{-6}$
Thallium(I) iodide	TlI	$5.54 \cdot 10^{-8}$
Thallium(I) thiocyanate	$TlSCN$	$1.57 \cdot 10^{-4}$
Thallium(III) hydroxide	$Tl(OH)_3$	$1.68 \cdot 10^{-44}$
Tin(II) hydroxide	$Sn(OH)_2$	$5.45 \cdot 10^{-27}$
Yttrium carbonate	$Y_2(CO_3)_3$	$1.03 \cdot 10^{-31}$
Yttrium fluoride	YF_3	$8.62 \cdot 10^{-21}$
Yttrium hydroxide	$Y(OH)_3$	$1.00 \cdot 10^{-22}$
Yttrium iodate	$Y(IO_3)_3$	$1.12 \cdot 10^{-10}$
Zinc arsenate	$Zn_3(AsO_4)_2$	$2.8 \cdot 10^{-28}$
Zinc carbonate	$ZnCO_3$	$1.46 \cdot 10^{-10}$
Zinc carbonate monohydrate	$ZnCO_3 \cdot H_2O$	$5.42 \cdot 10^{-11}$
Zinc fluoride	ZnF_2	$3.04 \cdot 10^{-2}$
Zinc hydroxide	$Zn(OH)_2$	$3 \cdot 10^{-17}$
Zinc iodate dihydrate	$Zn(IO_3)_2 \cdot 2H_2O$	$4.1 \cdot 10^{-6}$
Zinc oxalate dihydrate	$ZnC_2O_4 \cdot 2H_2O$	$1.38 \cdot 10^{-9}$
Zinc selenide	$ZnSe$	$3.6 \cdot 10^{-26}$
Zinc selenite monohydrate	$ZnSeO_3 \cdot H_2O$	$1.59 \cdot 10^{-7}$

Sulfides

Compound	Formula	K_{spa}
Cadmium sulfide	CdS	$8 \cdot 10^{-7}$
Copper(II) sulfide	CuS	$6 \cdot 10^{-16}$
Iron(II) sulfide	FeS	$6 \cdot 10^{2}$
Lead(II) sulfide	PbS	$3 \cdot 10^{-7}$
Manganese(II) sulfide (green)	MnS	$3 \cdot 10^{7}$
Mercury(II) sulfide (red)	HgS	$4 \cdot 10^{-33}$
Mercury(II) sulfide (black)	HgS	$2 \cdot 10^{-32}$
Silver(I) sulfide	Ag_2S	$6 \cdot 10^{-30}$
Tin(II) sulfide	SnS	$1 \cdot 10^{-5}$
Zinc sulfide (sphalerite)	ZnS	$2 \cdot 10^{-4}$
Zinc sulfide (wurtzite)	ZnS	$3 \cdot 10^{-2}$

SOLUBILITY OF COMMON SALTS AT AMBIENT TEMPERATURES

This table gives the aqueous solubility of selected salts at temperatures from 10 °C to 40 °C. Values are given in molality terms.

References

1. Apelblat, A., *J. Chem. Thermodynamics*, 24, 619, 1992.
2. Apelblat, A., *J. Chem. Thermodynamics*, 25, 63, 1993.
3. Apelblat, A., *J. Chem. Thermodynamics*, 25, 1513, 1993.
4. Apelblat, A. and Korin, E., *J. Chem. Thermodynamics*, 30, 59, 1998.

Salt	10 °C	15 °C	20 °C	25 °C	30 °C	35 °C	40 °C	Ref.
$BaCl_2$	1.603	1.659	1.716	1.774	1.834	1.895	1.958	1
$Ca(NO_3)_2$	6.896	7.398	7.986	8.675	9.480	10.421		1
$CuSO_4$	1.055	1.153	1.260	1.376	1.502	1.639		3
$FeSO_4$	1.352	1.533	1.729	1.940	2.165	2.405		3
KBr	5.002	5.237	5.471	5.703	5.932	6.157		3
KIO_3	0.291	0.333	0.378	0.426	0.478	0.534	0.593	4
K_2CO_3	7.756	7.846	7.948	8.063	8.191	8.331	8.483	1
LiCl	19.296	19.456	19.670	19.935				2
$Mg(NO_3)_2$	4.403	4.523	4.656	4.800	4.958	5.130	5.314	1
$MnCl_2$	5.421	5.644	5.884	6.143	6.422	6.721		3
NH_4Cl	6.199	6.566	6.943	7.331				2
NH_4NO_3	18.809	21.163	23.721	26.496				2
$(NH_4)_2SO_4$	5.494	5.589	5.688	5.790	5.896	6.005		3
NaBr	8.258	8.546	8.856	9.191	9.550	9.937	10.351	4
NaCl	6.110	6.121	6.136	6.153	6.174	6.197	6.222	4
$NaNO_2$	11.111	11.484	11.883	12.310	12.766	13.253	13.772	4
$NaNO_3$	9.395	9.819	10.261	10.723	11.204	11.706	12.230	4
RbCl	6.911	7.180	7.449	7.717	7.986	8.253	8.520	4
$ZnSO_4$	2.911	3.116	3.336	3.573	3.827	4.099	4.194	1

SOLUBILITY OF HYDROCARBONS IN SEAWATER

Concern about pollution of the oceans has stimulated measurements of the solubility of organic compounds in seawater. This table gives the solubility of several hydrocarbons in seawater. The data are derived from a review in the IUPAC Solubility Data Series (Reference 1).

Solubility is expressed in this table as parts per million by mass, i.e.,

$$S/\text{ppm(mass)} = 10^6 \times w_2 = 10^6 \times m_2/(m_1 + m_2)$$

where m_1 and m_2 are the masses of solvent (seawater) and solute, respectively, under saturation conditions, and w_2 is the mass fraction. Since the solubilities in this table are very low, the value of S is effectively the mass of hydrocarbon in grams per 1000 kg of seawater.

The temperature and salinity of each measurement are given in the table. Salinity is a standardized measure of the concentration of dissolved salts, as explained in the table "Properties of Seawater" in Section 14. Salinity values in the open oceans at mid-latitude typically fall between 34 and 36.

Reference 1 gives details of the method of measurement and an indication of the reliability of the measurements.

Reference

1. Shaw, David G., and Maczynski, A., IUPAC-NIST Solubility Data Series 81. Hydrocarbons with Water and Seawater — Revised and Updated. Part 12. C_5-C_{26} Hydrocarbons with Seawater, *J. Phys. Chem. Ref. Data* 35, 785, 2006.

Name	Mol. Form.	Salinity	t/°C	S/ppm (mass)
Acenaphthene	$C_{12}H_{10}$	35	15	0.21
Acenaphthene	$C_{12}H_{10}$	35	25	1.8
Anthracene	$C_{14}H_{10}$	35	25	0.031
Benz[a]anthracene	$C_{18}H_{12}$	35	25	0.0056
Benzene	C_6H_6	34.4	0	1320
Benzene	C_6H_6	35	25	1360
Benzo[ghi]perylene	$C_{22}H_{12}$	6	25	0.00021
Benzo[a]pyrene	$C_{20}H_{12}$	6	25	0.00017
Benzo[e]pyrene	$C_{20}H_{12}$	30	25	0.0033
Benzo[b]triphenylene	$C_{22}H_{14}$	6	25	0.027
Biphenyl	$C_{12}H_{10}$	35	25	4.76
Butylbenzene	$C_{10}H_{14}$	34.5	25	7.1
sec-Butylbenzene	$C_{10}H_{14}$	34.5	25	12
tert-Butylbenzene	$C_{10}H_{14}$	34.5	25	21
Chrysene	$C_{18}H_{12}$	35	25	0.0011
Dibenz[a,h]anthracene	$C_{22}H_{14}$	6	25	0.021
Dibenz[a,j]anthracene	$C_{22}H_{14}$	6	25	0.010
Dodecane	$C_{12}H_{26}$	35	25	0.0029
Eicosane	$C_{20}H_{42}$	35	25	0.0008
Ethylbenzene	C_8H_{10}	34.4	0	140
Ethylbenzene	C_8H_{10}	34.4	10	129
Ethylbenzene	C_8H_{10}	34.4	25	111
Fluoranthene	$C_{16}H_{10}$	35	25	0.124
9*H*-Fluorene	$C_{13}H_{10}$	35	25	1.2
Heptane	C_7H_{16}	6	25	10.3
Hexacosane	$C_{26}H_{54}$	35	25	0.0001
Hexadecane	$C_{16}H_{34}$	35	25	0.0004
Hexane	C_6H_{14}	35.3	25	7.9
Isopropylbenzene	C_9H_{12}	34.5	25	43
2-Methylanthracene	$C_{15}H_{12}$	35	25	0.013
Methylcyclopentane	C_6H_{12}	34.5	25	29
1-Methylnaphthalene	$C_{11}H_{10}$	30	25	23
1-Methylphenanthrene	$C_{15}H_{12}$	35	25	0.20
Naphthalene	$C_{10}H_8$	35	25	22.8
Nonane	C_9H_{20}	6	25	0.43
Octadecane	$C_{18}H_{38}$	35	25	0.0008
Pentane	C_5H_{12}	34.5	25	28
Phenanthrene	$C_{14}H_{10}$	34	25	0.69

Name	Mol. Form.	Salinity	*t*/°C	*S*/ppm (mass)
Pyrene	$C_{16}H_{10}$	35	25	0.086
Tetradecane	$C_{14}H_{30}$	35	25	0.0017
Toluene	C_7H_8	34.4	0	450
Toluene	C_7H_8	35	25	387
1,2,3-Trimethylbenzene	C_9H_{12}	34.5	25	49
1,2,4-Trimethylbenzene	C_9H_{12}	34.5	25	40
1,3,5-Trimethylbenzene	C_9H_{12}	34.5	25	31
Undecane	$C_{11}H_{24}$	6	25	0.01
o-Xylene	C_8H_{10}	34.5	25	130
m-Xylene	C_8H_{10}	34.5	25	106
p-Xylene	C_8H_{10}	34.5	25	111

SOLUBILITY OF ORGANIC COMPOUNDS IN PRESSURIZED HOT WATER

Liquid water at elevated temperatures and pressures, but still in the subcritical region, is of interest as a solvent in various laboratory and industrial processes. In effect, this means water at a temperature between about 100 °C and 373 °C, the critical temperature, and at pressures up to 400 bar or greater. Since the dielectric constant of water decreases with increasing temperature, the solubility of many compounds, especially non-polar compounds, increases dramatically at higher temperature. The fact that solubility can be fine-tuned by controlling temperature and pressure makes pressurized hot water a useful tool in various extraction and reaction processes.

This table gives a sample of the variations of solubility with temperature and pressure for several compounds, mostly hydrocarbons. The solubility is expressed in both mole fraction of solute, x_2, and mass percent, $100w_2$, where w_2 is the mass fraction. More information is available in the references.

References

1. *Solubility Data Series, International Union of Pure and Applied Chemistry, Vol. 38*, Pergamon Press, Oxford, 1988.
2. Shaw, D. G., and Maczynski, A., *J. Phys. Chem. Ref. Data* 35, 687, 2006.
3. Stephenson, R. M., *J. Chem. Eng. Data* 37, 80, 1992.
4. Lun, R., Varhanickova, D., Shiu, W.-Y., and Mackay, D., *J. Chem. Eng. Data* 42, 951, 1997.
5. Miller, D. J., et al., *J. Chem. Eng. Data* 43, 1043, 1998.
6. Miller, D. J., and Hawthorne, S. B., *J. Chem. Eng. Data* 45, 78, 2000.
7. Ma, J. H. Y., Hung, H., Shiu, W-Y., and Mackay, D., *J. Chem. Eng. Data* 46, 619, 2001.
8. Marche, C., Ferronato, C., and Jose, J., *J. Chem. Eng. Data* 48, 967, 2003.
9. Oleszek-Kudlak, S., Shibata, E., and Nakamura, T., *J. Chem. Eng. Data* 49, 570, 2004.
10. Marche, C., Ferronato, C., and Jose, J., *J. Chem. Eng. Data* 49, 937, 2004.
11. Andersson, T. A., Hartonen, K. M., and Riekkola, M-L., *J. Chem. Eng. Data* 50, 1177, 2005.
12. Karasek, P., Planeta, J., and Roth, M., *J. Chem. Eng. Data* 51, 616, 2006.
13. Shiu, W.-Y., and Ma, K.-C, *J. Phys. Chem. Ref. Data* 29, 41, 2000.

Name	Mol. Form.	t/°C	p/bar	Solubility 10^3x_2	Solubility Mass%	Ref.
Acenaphthene	$C_{12}H_{10}$	25	1	0.000444	0.000380	13
		250	50	1.25	1.06	11
Anthracene	$C_{14}H_{10}$	25	1	0.0000074	0.0000044	2
		50	50	0.000017	0.000017	5
		100	45	0.00032	0.00032	5
		100	39	0.000457	0.00045	12
		150	50	0.0102	0.0101	11
		200	77	0.13	0.13	12
		250	50	0.497	0.49	11
		300	100	3.78	3.62	11
Benz[a]anthracene	$C_{18}H_{12}$	25	1	0.00000073	0.00000093	2
		60	50	0.00000846	0.0000107	12
		100	50	0.000113	0.000143	12
		120	52	0.000418	0.00053	12
		150	49	0.00296	0.00375	12
Benzene	C_6H_6	25	1	0.40	0.178	13
		25	65	0.40	0.173	6
		25	400	0.33	0.143	6
		50	65	0.47	0.203	6
		100	65	0.89	0.38	6
		150	65	2.2	0.95	6
		200	65	5.0	2.13	6
		200	400	4.1	1.75	6
Carbazole	$C_{12}H_9N$	25	1	0.00013	0.00012	5
		25	54	0.00011	0.000102	5
		50	56	0.00045	0.00042	5
		100	54	0.0099	0.0092	5
		150	54	0.162	0.150	5
		200	52	1.9	1.74	5
Chrysene	$C_{18}H_{12}$	25	1	0.00000016	0.00000019	2

Name	Mol. Form.	t/°C	p/bar	Solubility 10^3x_2	Solubility Mass%	Ref.
		25	32	0.00000063	0.0000008	5
		50	36	0.000001	0.0000013	5
		100	38	0.000013	0.000016	5
		150	43	0.00060	0.00076	5
		200	45	0.0158	0.020	5
		225	62	0.0758	0.096	5
o-Dichlorobenzene	$C_6H_4Cl_2$	25	1	0.018	0.0094	9
		50	65	0.023	0.019	6
		100	65	0.055	0.045	6
		150	65	0.18	0.15	6
		200	65	0.57	0.46	6
trans-1,2-Dimethylcyclohexane	C_8H_{16}	25	1	0.008	0.00050	10
		101	7	0.0047	0.0029	10
		131	7	0.0108	0.0067	10
		151	7	0.0223	0.0139	10
		170	7	0.0356	0.0222	10
Ethylcyclohexane	C_8H_{16}	25	1	0.00098	0.00061	10
		100	7	0.00340	0.00212	10
		131	7	0.0085	0.0053	10
		151	7	0.01665	0.0104	10
		171	7	0.0334	0.0208	10
Heptane	C_7H_{16}	25	1	0.0004352	0.000242	8
		50	7	0.000613	0.00034096	8
		100	7	0.001938	0.00108	8
		125	7	0.00400	0.00222	8
		150	7	0.00878	0.00488	8
		170	7	0.01701	0.00946	8
Hexane	C_6H_{14}	25	1	0.002045	0.00098	8
		100	7	0.006074	0.0029	8
		125	7	0.01192	0.0057	8
		150	7	0.02555	0.0122	8
		170	7	0.04935	0.0236	8
1-Isopropyl-4-methylbenzene	$C_{10}H_{14}$	25	1	0.0030	0.0051	4
		50	60	0.0040	0.0030	6
		100	60	0.011	0.0082	6
		150	60	0.043	0.032	6
		200	60	0.20	0.15	6
Methylcyclohexane	C_7H_{14}	25	1	0.00293	0.00151	10
		100	7	0.01006	0.0055	10
		131	7	0.0244	0.0133	10
		151	7	0.0423	0.0231	10
		171	7	0.0708	0.0386	10
Naphthalene	$C_{10}H_8$	25	1	0.00444	0.00316	13
		40	50	0.00692	0.0049	12
		50	50	0.0114	0.0081	12
		65	50	0.0264	0.0188	12
		75	50	0.0435	0.0309	12
Octane	C_8H_{18}	25	1	0.0001158	0.000073	8
		100	7	0.0005943	0.000377	8
		125	7	0.0014163	0.000898	8
		150	7	0.0036957	0.00234	8
		170	7	0.0083483	0.00529	8
		200	65	0.029	0.018	6
Perylene	$C_{20}H_{12}$	25	1	0.00000003	0.0000004	2

Name	Mol. Form.	t/°C	p/bar	Solubility 10^3x_2	Solubility Mass%	Ref.
		50	50	0.00000029	0.0000004	5
		100	45	0.00000210	0.00000294	5
		150	47	0.000120	0.000168	5
		200	48	0.0050	0.0070	5
Pyrene	$C_{16}H_{10}$	25	1	0.000012	0.0000139	2
		100	50	0.000637	0.00072	11
		100	200	0.00078	0.00087	5
		140	50	0.0054	0.0061	11
		200	50	0.0492	0.055	11
		250	50	0.205	0.23	11
		300	50	1.41	1.56	11
p-Terphenyl	$C_{18}H_{14}$	25	1	0.00000141	0.00000180	2
		100	49	0.0000219	0.000028	12
		140	51	0.000372	0.000476	12
		180	55	0.00626	0.0080	12
		200	53	0.0241	0.0308	12
		210	54	0.0393	0.0502	12
Tetrachloroethene	C_2Cl_4	25	1	0.0285	0.0286	3
		50	65	0.027	0.025	6
		100	65	0.059	0.054	6
		150	65	0.18	0.17	6
		200	65	0.59	0.54	6
Toluene	C_7H_8	25	1	0.107	0.0519	7,13
		50	50	0.125	0.064	6
		100	50	0.27	0.138	6
		150	50	0.66	0.337	6
		200	50	1.9	0.96	6
2,2,4-Trimethylpentane	C_8H_{18}	25	1	0.00035	0.00022	1
		50	65	0.00052	0.00033	6
		100	65	0.0020	0.00127	6
		150	65	0.0102	0.0065	6
		200	65	0.061	0.0387	6
Triphenylene	$C_{18}H_{12}$	25	1	0.0000034	0.0000043	2
		100	51	0.0000899	0.000114	12
		140	50	0.00126	0.00160	12
		180	64	0.0123	0.0156	12
		195	60	0.0283	0.0359	12
m-Xylene	C_8H_{10}	25	1	0.028	0.0161	13
		50	60	0.036	0.021	6
		100	60	0.085	0.050	6
		150	60	0.27	0.159	6
		200	60	0.88	0.516	6

SOLUBILITY CHART

Abbreviations: **W**, soluble in water; **A**, insoluble in water but soluble in acids; **w**, sparingly soluble in water but soluble in acids; **a**, insoluble in water and only sparingly soluble in acids; **I**, insoluble in water and acids; **d**, decomposes in water. * Indicates two modifications of the salt.

No.		Al	NH_4	Sb	Ba	Bi	Cd	Ca	Cr	Co	Cu	Au (I)	Au (II)	H	Fe (II)	Fe (III)
1	Acetate	W	W		W	W	W	W	W	W	W	W	W	W		
	—($C_2H_3O_2$)	$Al(—)_3$	$NH_4(—)$		$Ba(—)_2$	$Bi(—)_3$	$Cd(—)_2$	$Ca(—)_2$	$Cr(—)_3$	$Co(—)_2$	$Cu(—)_2$			$C_2H_4O_2$	$Fe(—)_2$	$Fe_2(—)_6$
2	Arsenate	a	W	A	w	A	A	w		A	A			W	A	A
	—(AsO_4)	Al(—)	$(NH_4)_3(—)$	Sb(—)	$Ba_3(—)_2$	Bi(—)	$Cd_3(—)_2$	$Ca_3(—)_2$		$Co_3(—)_2$	$Cu_3(—)_2$			H_3AsO_4	$Fe_3(—)_2$	Fe(—)
3	Arsenite		W	A				w		A	A					
	—(AsO_3)		NH_4AsO_2	Sb(—)				$Ca_3(—)_2$		$Co_3H_6(—)_4$	CuH(—)					
4	Benzoate		W		W	A	W	W		W	w			W	W	A
	—($C_7H_5O_2$)		$NH_4(—)$		$Ba(—)_2$	$Bi(—)_3$	$Cd(—)_2$	$Ca(—)_2$		$Co(—)_2$	$Cu(—)_2$			$C_7H_6O_2$	$Fe(—)_2$	$Fe_2(—)_6$
5	Bromide	W	W	d	W	d	W	W	W(I)*	W	W	w	W	W	W	W
		$AlBr_2$	NH_4Br	$SbBr_3$	$BaBr_2$	$BiBr_3$	$CdBr_2$	$CaBr_2$	$CrBr_3$	$CoBr_2$	$CuBr_2$	AuBr	$AuBr_3$	HBr	$FeBr_2$	$FeBr_3$
6	Carbonate		W		w		A	w	W	A					w	
			$(NH_4)_2CO_3$		$BaCO_3$		$CdCO_3$	$CaCO_3$	$CrCO_3$	$CoCO_3$					$FeCO_3$	
7	Chlorate	W	W		W	W	W	W		W	W			W	W	W
	—(ClO_3)	$Al(—)_3$	$NH_4(—)$		$Ba(—)_2$	$Bi(—)_3$	$Cd(—)_2$	$Ca(—)_2$		$Co(—)_2$	$Cu(—)_2$			$HClO_3$	$Fe(—)_2$	$Fe(—)_3$
8	Chloride	W	W	W	W	d	W	W	I	W	W	w	W	W	W	W
		$AlCl_3$	NH_4Cl	$SbCl_3$	$BaCl_2$	$BiCl_3$	$CdCl_2$	$CaCl_2$	$CrCl_3$	$CoCl_2$	$CuCl_2$	AuCl	$AuCl_3$	HCl	$FeCl_2$	$FeCl_3$
9	Chromate		W		A		A	W		A						A
	—(CrO_4)		$(NH_4)_2(—)$		Ba(—)		Cd(—)	Ca(—)		Co(—)						$Fe_2(—)_3$
10	Citrate	W	W		w	A	A	w		w				W		W
	—($C_6H_5O_7$)	Al(—)	$(NH_4)_3(—)$		$Ba_3(—)_2$	Bi(—)	$Cd_3(—)_2$	$Ca_3(—)_2$		$Co_3(—)_2$				$C_6H_8O_7$		Fe(—)
11	Cyanide		W		W	w	W	W	A	A	A	w	W	W	a	
			NH_4CN		$Ba(CN)_2$	$Bi(CN)_3$	$Cd(CN)_2$	$Ca(CN)_2$	$Cr(CN)_3$	$Co(CN)_2$	$Cu(CN)_2$	AuCN	$Au(CN)_3$	HCN	$Fe(CN)_2$	
12	Ferricy'de		W		w		A	W		I	I			W	I	
	—($Fe(CN)_6$)		$(NH_4)_3(—)$		$Ba_3(—)_2$		$Cd_3(—)_2$	$Ca_3(—)_2$		$Co_3(—)_2$	$Cu_3(—)_2$			$H_3(—)$	$Fe_3(—)_2$	
13	Ferrocy'de	w	W		W		A	W		I	I			W	I	a
	—($Fe(CN)_6$)	$Al_4(—)_3$	$(NH_4)_4(—)$		$Ba_2(—)$		$Cd_2(—)$	$Ca_2(—)$		$Co_2(—)$	$Cu_2(—)$			$H_4(—)$	$Fe_2(—)$	$Fe_4(—)_3$
14	Fluoride	W	W	W	w	W	W	w	W(a)*	W	w			W	w	w
		AlF_3	NH_4F	SbF_3	BaF_2	BiF_3	CdF_2	CaF_2	CrF_3	CoF_2	CuF_2			HF	FeF_2	FeF_3
15	Formate	W	W		W	W	W	W		W	W			W	W	W
	—(CHO_2)	$Al(—)_3$	$NH_4(—)$		$Ba(—)_2$	$Bi(—)_3$	$Cd(—)_2$	$Ca(—)_2$		$Co(—)_2$	$Cu(—)_2$			CH_2O_2	$Fe(—)_2$	$Fe(—)_3$
16	Hydroxide	A	W		W	A	A	W	A	A	A	W	A		A	A
		$Al(OH)_3$	NH_4OH		$Ba(OH)_2$	$Bi(OH)_3$	$Cd(OH)_2$	$Ca(OH)_2$	$Cr(OH)_3$	$Co(OH)_2$	$Cu(OH)_2$	AuOH	$Au(OH)_3$		$Fe(OH)_2$	$Fe(OH)_3$
17	Iodide	W	W	d	W	A	W	W	W	W	a	a	a	W	W	W
		AlI_3	NH_4I	SbI_3	BaI_2	BiI_3	CdI_2	CaI_2	CrI_3	CoI_2	CuI	AuI	AuI_3	HI	FeI_2	FeI_3
18	Nitrate	W	W		W	d	W	W	W	W	W			W	W	W
		$Al(NO_3)_3$	NH_4NO_3		$Ba(NO_3)_2$	$Bi(NO_3)_3$	$Cd(NO_3)_2$	$Ca(NO_3)_2$	$Cr(NO_3)_3$	$Co(NO_3)_2$	$Cu(NO_3)_2$			HNO_3	$Fe(NO_3)_2$	$Fe(NO)_3$
19	Oxalate	A	W		w	A	w	A	W	A	A			W	A	W
	—(C_2O_4)	$Al_2(—)_3$	$(NH_4)_2(—)$		Ba(—)	$Bi_2(—)_3$	Cd(—)	Ca(—)	Cr(—)	Co(—)	Cu(—)			$C_2H_2O_4$	Fe(—)	$Fe_2(—)_3$
20	Oxide	a		w	W	A	A	w	a	A	A		A	W	A	A
		Al_2O_3		Sb_2O_3	BaO	Bi_2O_3	CdO	CaO	Cr_2O_3	CoO	CuO	Au_2O	Au_2O_3	H_2O_2	FeO	Fe_2O_3
21	Phosphate	A	W		A	A	A	w	w	A	A			W	A	w
		$AlPO_4$	$NH_4H_2PO_4$		$Ba_3(PO_4)_2$	$BiPO_4$	$Cd_3(PO_4)_2$	$Ca_3(PO_4)_2$	$Cr_2(PO_4)_2$	$Co_3(PO_4)_2$	$Cu_3(PO_4)_2$		H_3PO_4	$Fe_3(PO_4)_2$	$FePO_4$	
22	Silicate,	I			W		A	w		A	A			I		
	—(SiO_3)	$Al_2(—)_3$			Ba(—)		Cd(—)	Ca(—)		Co_2SiO_4	Cu(—)			H_2SiO_3		
23	Sulfate	W	W	A	a	d	W	w	W(I)*	W	W			W	W	w
		$Al_2(SO_4)_3$	$(NH_4)_2SO_4$	$Sb_2(SO_4)_3$	$BaSO_4$	$Bi_2(SO_4)_3$	$CdSO_4$	$CaSO_4$	$Cr_2(SO_4)_3$	$CoSO_4$	$CuSO_4$			H_2SO_4	$FeSO_4$	$Fe(SO_4)_3$
24	Sulfide	d	W	A	d	A	A	w	d	A	A	I	I	W	A	d
		Al_2S_3	$(NH_4)_2S$	Sb_2S_3	BaS	Bi_2S_3	CdS	CaS	Cr_2S_3	CoS	CuS	Au_2S	Au_2S_3	H_2S	FeS	Fe_2S_3
25	Tartrate	w	W	W	w	A	A	w	d	A	A	I	I	W	A	d
	—($C_4H_4O_6$)	$Al_2(—)_3$	$(NH_4)_2(—)$	$Sb_2(—)_3$	Ba(—)	$Bi_2(—)_3$	Cd(—)	Ca(—)		Co(—)	Cu(—)			$C_4H_6O_6$	Fe(—)	$Fe_2(—)_3$
26	Thiocy'te		W		W			W		W	d			W	W	W
			NH_4CNS		$Ba(CNS)_2$			Ca(CNS)		$Co(CNS)_2$	CuCNS			CNSH	$Fe(CNS)_2$	$Fe(CNS)_3$

No.		Pb	Mg	Mn	Hg (I)	Hg (II)	Ni	K	Pt	Ag	Na	Sn (IV)	Sn (II)	Sr	Zn
1	Acetate	W	W	W	w	W	W	W		w	W	W	d	W	W
	—($C_2H_3O_2$)	$Pb(—)_2$	$Mg(—)_2$	$Mn(—)_2$	Hg(—)	$Hg(—)_2$	$Ni(—)_2$	K(—)		Ag(—)	Na(—)	$Sn(—)_4$	$Sn(—)_2$	$Sr(—)_2$	$Zn(—)_2$
2	Arsenate	A	A	w	A	w	A	W		A	W			w	A
	—(AsO_4)	PbH(—)	$Mg_3(—)$	MnH(—)	$Hg_3(—)$	$Hg_3(—)_2$	$Ni_3(—)_2$	$K_3(—)$		$Ag_3(—)$	$Na_3(—)$			SrH(—)	$Zn_3(—)_2$
3	Arsenite		W	A	A	A	A	W		A	W		A	w	
	—(AsO_3)		$Mg_3(—)_2$	$Mn_3H_6(—)_4$	$Hg_3(—)$	$Hg_3(—)$	$Ni_3H_6(—)_4$	K_3AsO_3		$Ag_3(—)$	$Na_2H(—)$		$Sn_3(—)_2$	$Sr_3(—)_2$	
4	Benzoate	w	W	W	A	w	w	W		w	W				W
	—($C_7H_5O_2$)	$Pb(—)_2$	$Mg(—)_2$	$Mn(—)_2$	$Hg_2(—)_2$	$Hg(—)_2$	$Ni(—)_2$	K(—)		Ag(—)	Na(—)				$Zn(—)_2$
5	Bromide	W	W	W	A	W	W	W	w	a	W	W	W	W	W
		$PbBr_2$	$MgBr_2$	$MnBr_2$	HgBr	$HgBr_2$	$NiBr_2$	KBr	$PtBr_4$	AgBr	NaBr	$SnBr_4$	$SnBr_2$	$SrBr_2$	$ZnBr_2$
6	Carbonate	A	w	w	A		w	W		A	W			w	w
		$PbCO_3$	$MgCO_3$	$MnCO_3$	Hg_2CO_3		$NiCO_3$	K_2CO_3		Ag_2CO_3	Na_2CO_3			$SrCO_3$	$ZnCO_3$
7	Chlorate	W	W	W	W	W	W	W		W	W		W	W	W
	—(ClO_3)	$Pb(—)_2$	$Mg(—)_2$	$Mn(—)_2$	Hg(—)	$Hg(—)_2$	$Ni(—)_2$	K(—)		Ag(—)	Na(—)		$Sn(—)_2$	$Sr(—)_2$	$Zn(—)_2$
8	Chloride	W	W	W	a	W	W	W	W	a	W	W	W	W	W
		$PbCl_2$	$MgCl_2$	$MnCl_2$	HgCl	$HgCl_2$	$NiCl_2$	KCl	$PtCl_4$	AgCl	NaCl	$SnCl_4$	$SnCl_2$	$SrCl_2$	$ZnCl_2$
9	Chromate	A	W		w	w	A	W		w	W	W	A	w	w
	—(CrO_4)	Pb(—)	Mg(—)		$Hg_2(—)$	Hg(—)	Ni(—)	$K_2(—)$		$Ag_2(—)$	$Ma_2(—)$	$Sn(—)_2$	Sn(—)	Sr(—)	Zn(—)
10	Citrate	W	W	w	w		W	W		w	W			A	w
	—($C_6H_5O_7$)	$Pb_3(—)_2$	$Mg_3(—)_2$	MnH(—)	$Hg_3(—)$		$Ni_3(—)_2$	$K_3(—)$		$Ag_3(—)$	$Na_3(—)$			SrH(—)	$Zn_3(—)_2$
11	Cyanide	w	W		A	W	a	W	I	a	W			W	A
		$Pb(CN)_2$	$Mg(CN)_2$		HgCN	$Hg(CN)_2$	$Ni(CN)_2$	KCN	$Pt(CN)_2$	AgCN	NaCN			$Sr(CN)_2$	$Zn(CN)_2$
12	Ferricy'de	w	W			A	I	W		I	W		A	W	A
	—$Fe(CN)_6$	$Pb_3(—)_2$	$Mg_3(—)_2$			$Hg_3(—)_2$	$Ni_3(—)_2$	$K_3(—)$		$Ag_3(—)$	$Na_3(—)$		$Sn_3(—)_2$	$Sr_3(—)_2$	$Zn_3(—)_2$
13	Ferrocy'de	a	W	A		I	I	W		I	W		a	W	I
	—$Fe(CN)_6$	$Pb_2(—)$	$Mg_2(—)$	$Mn_2(—)$		$Hg_2(—)$	$Ni_2(—)$	$K_4(—)$		$Ag_4(—)$	$Na_4(—)$		$Sn_2(—)$	$Sr_2(—)$	$Zn_2(—)$
14	Fluoride	w	w	A	d	d	w	W	W	W	W	W	W	w	w
		PbF_2	MgF_2	MnF_2	HgF	HgF_2	NiF_2	KF	PtF_4	AgF	NaF	SnF_4	SnF_2	SrF_2	ZnF_2
15	Formate	W	W	W	w	W	W	W		W	W			W	W
	—(CHO_2)	$Pb(—)_2$	$Mg(—)_2$	$Mn(—)_2$	Hg(—)	$Hg(—)_2$	$Ni(—)_2$	K(—)		Ag(—)	Na(—)			$Sr(—)_2$	$Zn(—)_2$
16	Hydroxide	w	A	A		A	w	W	A		W	w	A	W	A
		$Pb(OH)_2$	$Mg(OH)_2$	$Mn(OH)_2$		$Hg(OH)_2$	$Ni(OH)_2$	KOH	$Pt(OH)_4$		NaOH	$Sn(OH)_4$	$Sn(OH)_2$	$Sr(OH)_2$	$Zn(OH)_2$
17	Iodide	w	W	W	A	w	W	W	I	I	W	d	W	W	W
		PbI_2	MgI_2	MnI_2	HgI	HgI_2	NiI_2	KI	PtI_2	AgI	NaI	SnI_4	SnI_2	SrI_2	ZnI_2
18	Nitrate	W	W	W	W	W	W	W	W	W	W		d	W	W
		$Pb(NO_3)_2$	$Mg(NO_3)_2$	$Mn(NO_3)_2$	$HgNO_3$	$Hg(NO_3)_2$	$Ni(NO_3)_2$	KNO_3	$Pt(NO_3)_4$	$AgNO_3$	$NaNO_3$		$Sn(NO_3)_2$	$Sr(NO_3)_2$	$Zn(NO_3)_2$
19	Oxalate	A	w	w	a	A	A	W		a	W		A	w	A
	—(C_2O_4)	Pb(—)	Mg(—)	Mn(—)	$Hg_2(—)$	Hg(—)	Ni(—)	$K_2(—)$		$Ag_2(—)$	$Na_2(—)$		Sn(—)	Sr(—)	Zn(—)
20	Oxide	w	A	A	A	w	A	W	A	w	d	A	A	W	w
		PbO	MgO	MnO	Hg_2O	HgO	NiO	K_2O	PtO	Ag_2O	Na_2O	SnO_2	SnO	SrO	ZnO
21	Phosphate	A	w	w	A	A	A	W		A	W		A	A	A
		$Pb_3(PO_4)_2$	$Mg_3(PO_4)_2$	$Mn_3(PO_4)_2$	Hg_3PO_4	$Hg_3(PO_4)_2$	$Ni_3(PO_4)_2$	K_3PO_4		Ag_3PO_4	Na_3PO_4		$Sn_3(PO_4)$	$Sr_3(PO_4)_2$	$Zn_3(PO_4)_2$
22	Silicate	A	A	I				W			W			A	A
	—(SiO_3)	Pb(—)	Mg(—)	Mn(—)				$K_2(—)$			$Na_2(—)$			Sr(—)	Zn(—)
23	Sulfate	w	W	W	w	d	W	W	W	w	W	W	W	w	W
		$PbSO_4$	$MgSO_4$	$MnSO_4$	Hg_2SO_4	$HgSO_4$	$NiSO_4$	K_2SO_4	$Pt(SO_4)_2$	Ag_2SO_4	Na_2SO_4	$Sn(SO_4)_2$	$SnSO_4$	$SrSO_4$	$ZnSO_4$
24	Sulfide	A	d	A	I	I	A	W	I	A	W	A	A	W	A
		PbS	MgS	MnS	Hg_2S	HgS	NiS	K_2S	PtS	Ag_2S	Na_2S	SnS_2	SnS	SrS	ZnS
25	Tartrate	A	w	w	I	I	A	W	I	A	W	A	A	W	A
	—($C_4H_4O_6$)	Pb(—)	Mg(—)	Mn(—)	$Hg_2(—)$		Ni(—)	$K_2(—)$		$Ag_2(—)$	$Na_2(—)$		Sn(—)	Sr(—)	Zn(—)
26	Thiocy'te	w	W	W	A	w		W		I	W			W	W
		$Pb(CNS)_2$	$Mg(CNS)_2$	$Mn(CNS)_2$	HgCNS	$Hg(CNS)_2$		KCNS		AgCNS	NaCNS			$Sr(CNS)_2$	$Zn(CNS)_2$

REDUCTION OF WEIGHINGS IN AIR TO VACUO

When the mass M of a body is determined in air, a correction is necessary for the buoyancy of the air. The corrected mass is given by $M + kM/1000$, where k is a function of the material used for the weights, given by

$$k = 1000\rho_{air}(1/\rho_{body} - 1/\rho_{weight})$$

and ρ is density. The table below is computed for an air density of 0.0012 g/cm^3 and for densities of three common weights: platinum-iridium (21.6 g/cm^3), brass (8.5 g/cm^3), and aluminum or quartz (2.65 g/cm^3).

References

1. Kaye, G. W. C., and Laby, T. H., *Tables of Physical and Chemical Constants, 16th Edition*, pp. 25–28, Longman, London, 1995.
2. Giacomo, P., *Metrologia*, 18, 33, 1982.
3. Davis, R. S., *Metrologia*, 29, 67, 1992.

Density of body (g/cm^3)	Value of k for weights of: Pt-Ir	Brass	Quartz or Al
0.5	2.34	2.26	1.95
0.6	1.94	1.86	1.55
0.7	1.66	1.57	1.26
0.8	1.44	1.36	1.05
0.9	1.28	1.19	0.88
1.0	1.14	1.06	0.75
1.1	1.04	0.95	0.64
1.2	0.94	0.86	0.55
1.3	0.87	0.78	0.47
1.4	0.80	0.72	0.40
1.5	0.74	0.66	0.35
1.6	0.69	0.61	0.30
1.7	0.65	0.56	0.25
1.8	0.61	0.53	0.21
1.9	0.58	0.49	0.18
2.0	0.54	0.46	0.15
2.5	0.42	0.34	0.03
3.0	0.34	0.26	-0.05
4.0	0.24	0.16	-0.15
6.0	0.14	0.06	-0.25
8.0	0.09	0.01	-0.30
10.0	0.06	-0.02	-0.33
15.0	0.02	-0.06	-0.37
20.0	0.00	-0.08	-0.39
22.0	0.00	-0.09	-0.40

For a more accurate calculation, use the following values of the density of air (assuming 50% relative humidity and 0.04% CO_2).

P/kPa	Air temperature 10 °C	20 °C	30 °C
85	0.001043	0.001005	0.000968
90	0.001105	0.001065	0.001025
95	0.001166	0.001124	0.001083
100	0.001228	0.001184	0.001140
105	0.001290	0.001243	0.001198

Formulas for calculating the density of air over more extended ranges of temperature, pressure, and humidity may be found in the references.

VOLUME OF ONE GRAM OF WATER

The following table, which is designed for gravimetric calibration of volumetric apparatus, gives the specific volume of water at standard atmospheric pressure as a function of temperature.

Reference

Marsh, K. N., Ed., *Recommended Reference Materials for the Realization of Physicochemical Properties*, pp. 25–27, Blackwell Scientific Publications, Oxford, 1987.

t/°C	Volume of 1 g H_2O in cm^3
10	1.0002980
11	1.0003928
12	1.0005007
13	1.0006212
14	1.0007542
15	1.0008992
16	1.0010561
17	1.0012246
18	1.0014044
19	1.0015952
20	1.0017969
21	1.0020092
22	1.0022320
23	1.0024649
24	1.0027079
25	1.0029607
26	1.0032234
27	1.0034956
28	1.0037771
29	1.0040679
30	1.0043679

PROPERTIES OF CARRIER GASES FOR GAS CHROMATOGRAPHY

The following is a list of carrier gases sometimes used in gas chromatography, with properties relevant to the design of chromatographic systems. All data refer to normal atmospheric pressure (101.325 kPa).

M_r: Molecular weight (relative molar mass)
ρ_{25}: Density at 25 °C in g/L
λ: Thermal conductivity in mW/m °C
η: Viscosity in μPa s (equal to 10^{-3} cP)
c_p: Specific heat at 25 °C in J/g °C

References

1. Lide, D. R., and Kehiaian, H. V., *CRC Handbook of Thermophysical and Thermochemical Data*, CRC Press, Boca Raton, FL, 1994.
2. Bruno, T. J., and Svoronos, P. D. N., *CRC Handbook of Basic Tables for Chemical Analysis*, CRC Press, Boca Raton, FL, 1989.

			At 25 °C		At 250 °C		
Gas	M_r	ρ_{25} g L^{-1}	λ mW/m °C	η μPa s	λ mW/m °C	η μPa s	c_p(25 °C) J/g °C
Hydrogen	2.016	0.0824	185.9	8.9	280	13.1	14.3
Helium	4.003	0.1636	154.6	19.9	230	29.5	5.20
Argon	39.95	1.6329	17.8	22.7	27.7	35.3	0.521
Nitrogen	28.01	1.1449	25.9	17.9	39.6	26.8	1.039
Oxygen	32.00	1.3080	26.2	20.7	42.6	31.8	0.919
Carbon monoxide	28.01	1.1449	24.8	17.8	40.7	26.5	1.039
Carbon dioxide	44.01	1.7989	16.7	14.9	35.5	24.9	0.843
Sulfur hexafluoride	146.05	5.9696	13.1	28.1	15.3	24.8	0.664
Methane	16.04	0.6556	34.5	11.1	75.0	17.6	2.23
Ethane	30.07	1.2291	20.9	9.4	57.7	15.5	1.75
Ethylene	28.05	1.1465	20.5	10.3	53.8	17.2	1.53
Propane	44.10	1.8025	17.9	8.3	49.2	14.0	1.67

SOLVENTS FOR ULTRAVIOLET SPECTROPHOTOMETRY

This table lists some solvents commonly used for sample preparation for ultraviolet spectrophotometry. The properties given are:

- λ_c: cutoff wavelength, below which the solvent absorption becomes excessive.
- ε: dielectric constant (relative permittivity); the temperature in °C is given as a superscript.
- t_b: normal boiling point.

References

1. Bruno, T. J., and Svoronos, P. D. N., *CRC Handbook of Basic Tables for Chemical Analysis*, CRC Press, Boca Raton, FL, 1989.
2. *Landolt-Börnstein, Numerical Data and Functional Relationships in Science and Technology, New Series, IV/6, Static Dielectric Constants of Pure Liquids and Binary Liquid Mixtures,* Springer–Verlag, Heidelberg, 1991.

Name	λ_c/nm	ε	t_b/°C
Acetic acid	260	6.20^{20}	117.9
Acetone	330	21.01^{20}	56.0
Acetonitrile	190	36.64^{20}	81.6
Benzene	280	2.28^{20}	80.0
2-Butanol	260	17.26^{20}	99.5
Butyl acetate	254	5.07^{20}	126.1
Carbon disulfide	380	2.63^{20}	46
Carbon tetrachloride	265	2.24^{20}	76.8
1-Chlorobutane	220	7.28^{20}	78.6
Chloroform	245	4.81^{20}	61.1
Cyclohexane	210	2.02^{20}	80.7
1,2-Dichloroethane	226	10.42^{20}	83.5
Dichloromethane	235	8.93^{25}	40
Diethyl ether	218	4.27^{20}	34.5
N,N-Dimethylacetamide	268	38.85^{21}	165
N,N-Dimethylformamide	270	38.25^{20}	153
Dimethyl sulfoxide	265	47.24^{20}	189
1,4-Dioxane	215	2.22^{20}	101.5
Ethanol	210	25.3^{20}	78.2
Ethyl acetate	255	6.08^{20}	77.1
Ethylene glycol dimethyl ether	240	7.30^{24}	85
Ethylene glycol monoethyl ether	210	13.38^{25}	135
Ethylene glycol monomethyl ether	210	17.2^{25}	124.1
Glycerol	207	46.53^{20}	290
Heptane	197	1.92^{20}	98.5
Hexadecane	200	2.05^{20}	286.8
Hexane	210	1.89^{20}	68.7
Methanol	210	33.0^{20}	64.6
Methylcyclohexane	210	2.02^{20}	100.9
Methyl ethyl ketone	330	18.56^{20}	79.5
Methyl isobutyl ketone	335	13.11^{20}	116.5
2-Methyl-1-propanol	230	17.93^{20}	107.8
N-Methyl-2-pyrrolidone	285	32.55^{20}	202
Nitromethane	380	37.27^{20}	101.1
Pentane	210	1.84^{20}	36.0
Pentyl acetate	212	4.79^{20}	149.2
1-Propanol	210	20.8^{20}	97.2
2-Propanol	210	20.18^{20}	82.3
Pyridine	330	13.26^{20}	115.2
Tetrachloroethylene	290	2.27^{30}	121.3
Tetrahydrofuran	220	7.52^{22}	65
Toluene	286	2.38^{23}	110.6
1,1,2-Trichloro-1,2,2-trifluoroethane	231	2.41^{25}	47.7
2,2,4-Trimethylpentane	215	1.94^{20}	99.2
Water	191	80.10^{20}	100.0
o-Xylene	290	2.56^{20}	144.5
m-Xylene	290	2.36^{20}	139.1
p-Xylene	290	2.27^{20}	138.3

^{13}C CHEMICAL SHIFTS OF USEFUL NMR SOLVENTS

The following table gives the expected carbon-13 chemical shifts, relative to tetramethylsilane, for various useful NMR solvents. In some solvents, slight changes can occur with change of concentration.[2,3]

References

1. Bruno, T. J., and Svoronos, P. D. N., *CRC Handbook of Basic Tables for Chemical Analysis*, CRC Press, Boca Raton, FL, 1989.
2. Silverstein, R. M., Bassler, G. C., and Morrill, T. C., *Spectrometric Identification of Organic Compounds*, John Wiley & Sons, Now York, 1981.
3. Rahman, A. U., *Nuclear Magnetic Resonance. Basic Principles*, Springer-Verlag, New York, 1986.
4. Pretsch, E., Clerc, T., Seibl, J., and Simon, W., *Spectral Data for Structure Determination of Organic Compounds, Second Edition*, Springer-Verlag, Heidelberg, 1989.

Solvent	Formula	Chemical shift (ppm)
Acetic acid-d_4	CD_3COOD	20.0 (CD_3) 205.8 (C=O)
Acetone	$(CH_3)_2C{=}O$	30.7 (CH_3) 206.7 (C=O)
Acetone-d_6	$(CD_3)_2C{=}O$	29.2 (CD_3) 204.1 (C=O)
Acetonitrile-d_3	$CD_3C{\equiv}N$	1.3 (CD_3) 117.1 (C≡N)
Benzene	C_6H_6	128.5
Benzene-d_6	C_6D_6	128.4
Carbon disulfide	CS_2	192.3
Carbon tetrachloride	CCl_4	96.0
Chloroform	$CHCl_3$	77.2
Chloroform-d_3	$CDC1_3$	77.05
Cyclohexane-d_{12}	C_6D_{12}	27.5
Dichloromethane-d_2	CD_2C1_2	53.6
Dimethylformamide-d_7	$(CD_3)_2NCDO$	31 (CD_3) 36 (CD_3) 162.4 (C=O)
Dimethylsulfoxide-d_6	$(CD_3)_2S{=}O$	39.6
Dioxane-d_8	$C_4D_3O_2$	67.4
Formic acid-d_2	DCOOD	165.5
Methanol-d_4	CD_3OD	49.3
Nitromethane-d_3	CD_3NO_2	57.3
Pyridine	C_5H_5N	123.6 (C_3) 135.7 (C_4) 149.8 (C_2)
Pyridine-d_5	C_5D_5N	123.9 (C_3) 135.9 (C_4) 150.2 (C_2)
1,1,2,2-Tetrachloroethane-d_2	$CDC1_2CDC1_2$	75.5
Tetrahydrofuran-d_8	C_4D_8O	25.8 (C_2) 67.9 (C_1)
Trichlorofluoromethane	$CFC1_3$	117.6

MASS SPECTRAL PEAKS OF COMMON ORGANIC SOLVENTS

The strongest peaks in the mass spectra of 375 important organic solvents and other liquid reagents are listed in this table. The *e*/*m* value for each peak is followed by the relative intensity in parentheses, with the strongest peak assigned an intensity of 100. The peaks for each compound are listed in order of decreasing intensity. Compounds are listed by the name used in this *Handbook*, with other common names given in parentheses.

Data on the physical properties of the same compounds may be found in Section 15 in the table *Laboratory Solvents and Other Liquid Reagents*.

References

1. NIST/EPA/NIH Mass Spectral Database, National Institute of Standards and Technology, Gaithersburg, MD, 20899.
2. Lide, D. R., and Milne, G. W. A., Editors, *Handbook of Data on Organic Compounds, Third Edition*, CRC Press, Boca Raton, FL, 1994. (Also available as a CD-ROM database.)
3. Lide, D. R., Editor, *Properties of Organic Compounds*, <www.chemnetbase.com/scripts/pocweb.exe>.

Compound	*e*/*m* (intensity)									
Acetic acid	43(100)	45(87)	60(57)	15(42)	42(14)	29(13)	14(13)	28(7)	18(6)	16(6)
Acetic anhydride	43(100)	42(35)	45(29)	60(18)	29(9)	41(8)	40(2)	26(2)	87(1)	61(1)
Acetone	43(100)	15(34)	58(23)	27(9)	14(9)	42(8)	26(7)	29(5)	28(5)	39(4)
Acetonitrile	41(100)	40(46)	39(13)	14(9)	38(6)	28(4)	26(4)	25(3)	42(2)	27(2)
Acrolein (2-Propenal)	27(100)	56(74)	28(65)	26(54)	55(52)	29(37)	25(8)	53(5)	38(5)	57(4)
Acrylonitrile	53(100)	26(85)	52(79)	51(34)	27(13)	50(8)	25(7)	38(5)	54(3)	37(3)
Allyl alcohol	57(100)	31(34)	29(32)	28(31)	58(25)	39(22)	27(20)	30(16)	32(14)	26(11)
Allylamine	30(100)	56(80)	28(76)	57(33)	39(21)	29(20)	27(18)	26(13)	41(8)	18(8)
2-Amino-2-methyl-1-propanol (2-Aminoisobutanol)	58(100)	41(18)	18(17)	42(13)	28(11)	56(10)	30(10)	29(8)	43(6)	59(5)
Aniline (Benzenamine)	93(100)	66(32)	65(16)	39(13)	92(10)	94(7)	41(5)	40(5)	67(4)	64(3)
Anisole (Methoxybenzene)	108(100)	65(76)	78(60)	39(44)	51(21)	77(20)	93(16)	79(14)	50(13)	63(12)
Benzaldehyde	51(100)	77(81)	50(55)	106(44)	105(43)	52(26)	78(16)	39(13)	27(10)	74(8)
Benzene	78(100)	77(20)	52(19)	51(17)	50(15)	39(12)	79(6)	76(5)	74(4)	38(4)
Benzeneacetonitrile (Benzyl cyanide)	117(100)	90(43)	116(35)	89(22)	51(13)	39(11)	63(10)	118(9)	91(8)	50(8)
Benzenethiol (Phenyl mercaptan)	110(100)	66(28)	109(23)	39(15)	77(14)	51(14)	84(13)	65(11)	50(11)	45(11)
Benzonitrile	103(100)	76(34)	50(13)	104(9)	75(7)	51(7)	77(5)	52(4)	39(4)	74(3)
Benzyl acetate	108(100)	43(76)	91(60)	90(48)	79(27)	107(17)	77(17)	65(15)	51(15)	89(14)
Benzyl alcohol	79(100)	108(83)	77(74)	107(66)	51(35)	105(23)	106(22)	50(17)	78(16)	39(16)
Bis(2-aminoethyl)amine (Diethylenetriamine)	44(100)	73(59)	30(35)	19(18)	56(16)	28(16)	27(16)	42(11)	99(8)	43(8)
Bis(2-chloroethyl) ether	93(100)	63(74)	27(38)	95(32)	65(24)	31(9)	49(4)	28(4)	94(3)	62(3)
Bis(2-ethylhexyl) phthalate	149(100)	57(32)	167(29)	71(21)	43(21)	70(18)	150(11)	113(10)	55(10)	41(9)
Bis(2-hydroxyethyl) sulfide	61(100)	45(68)	31(38)	104(36)	91(34)	47(26)	44(26)	27(24)	60(18)	43(17)
Bromobenzene	77(100)	158(64)	156(64)	51(39)	50(17)	78(8)	76(6)	75(6)	28(5)	159(4)
1-Bromobutane (Butyl bromide)	57(100)	41(56)	29(45)	27(29)	56(13)	39(12)	28(12)	55(7)	43(7)	138(6)
2-Bromobutane (*sec*-Butyl bromide)	57(100)	41(60)	29(57)	27(34)	39(20)	28(9)	26(9)	136(1)		
1-Bromo-2-chloroethane	63(100)	27(85)	65(31)	26(23)	144(8)	81(8)	79(8)	28(7)	142(6)	93(6)
Bromochloromethane	49(100)	130(67)	128(52)	51(31)	93(23)	81(20)	79(20)	95(17)	132(16)	47(8)
1-Bromodecane (Decyl bromide)	43(100)	135(94)	137(91)	57(81)	41(58)	55(56)	71(38)	69(36)	85(33)	29(27)
Bromoethane (Ethyl bromide)	108(100)	110(97)	29(62)	27(51)	28(35)	26(14)	93(6)	32(6)	95(5)	81(5)
2-Bromo-2-methylpropane (*tert*-Butyl bromide)	57(100)	41(67)	29(45)	39(30)	27(18)	28(8)	40(5)	38(5)	58(4)	55(4)
1-Bromonaphthalene	44(100)	206(39)	127(39)	208(37)	36(31)	69(29)	131(13)	29(13)	126(12)	63(12)
1-Bromopentane (Pentyl bromide)	43(100)	71(80)	41(56)	27(51)	42(37)	29(34)	55(33)	39(30)	28(12)	26(9)
1-Bromopropane (Propyl bromide)	43(100)	41(77)	28(69)	27(60)	39(49)	124(42)	122(41)	42(34)	32(20)	29(15)
2-Bromopropane (Isopropyl bromide)	43(100)	27(47)	41(43)	39(22)	124(8)	122(8)	26(7)	81(6)	79(6)	38(6)
2-Bromopropene	41(100)	39(58)	122(37)	120(36)	38(12)	37(8)	40(6)	81(5)	79(5)	42(4)
Butanal	44(100)	43(74)	72(57)	41(56)	27(55)	29(48)	57(23)	39(22)	28(15)	42(11)
Butanenitrile	41(100)	29(62)	27(37)	28(10)	39(9)	26(7)	40(5)	42(4)	38(4)	15(4)
1-Butanethiol (Butyl mercaptan)	56(100)	41(74)	90(66)	47(43)	27(43)	28(36)	29(33)	57(17)	39(16)	61(15)
Butanoic acid	60(100)	27(50)	73(27)	42(25)	41(24)	43(22)	29(21)	45(19)	39(15)	28(11)
Butanoic anhydride	71(100)	43(59)	27(26)	41(19)	42(10)	39(10)	73(9)	28(7)	72(5)	55(5)
1-Butanol (Butyl alcohol)	31(100)	56(81)	41(62)	43(60)	27(50)	42(31)	29(31)	28(17)	39(16)	55(12)
2-Butanol (*sec*-Butyl alcohol)	45(100)	31(22)	27(22)	59(20)	29(18)	43(13)	41(12)	44(8)	18(8)	28(5)
2-Butanone (Methyl ethyl ketone)	43(100)	72(24)	29(19)	27(12)	57(7)	42(5)	26(4)	28(3)	44(2)	39(2)
trans-2-Butenal (*trans*-Crotonaldehyde)	41(100)	39(97)	70(82)	69(65)	27(49)	29(39)	42(30)	38(29)	40(27)	37(18)
2-Butoxyethanol (Ethylene glycol monobutyl ether)	57(100)	45(38)	29(35)	41(31)	87(16)	27(12)	56(11)	31(9)	75(7)	28(7)
2-Butoxyethyl acetate (Ethylene glycol monobutyl ether acetate)	57(100)	43(86)	56(50)	87(33)	41(26)	29(22)	85(18)	88(11)	44(11)	27(7)
Butyl acetate	43(100)	56(34)	41(17)	27(16)	29(15)	73(11)	61(10)	28(7)	55(6)	39(6)

Compound	*e*/*m* (intensity)									
sec-Butyl acetate	43(100)	56(21)	87(15)	41(14)	29(8)	57(6)	73(4)	61(4)	55(4)	27(4)
Butylamine	30(100)	73(10)	28(5)	41(3)	27(3)	18(3)	44(2)	42(2)	31(2)	29(2)
tert-Butylamine	58(100)	41(21)	42(15)	18(9)	30(8)	15(8)	39(7)	57(6)	28(6)	59(4)
Butylbenzene	91(100)	92(55)	134(20)	65(13)	27(10)	39(9)	105(8)	51(7)	78(6)	41(6)
sec-Butylbenzene	105(100)	134(18)	91(14)	77(10)	27(9)	106(9)	51(7)	79(7)		
tert-Butylbenzene	119(100)	91(65)	41(40)	134(24)	39(15)	79(14)	77(13)	51(13)	120(11)	65(7)
Butyl butanoate	43(100)	71(90)	56(80)	89(69)	41(67)	27(52)	29(47)	57(29)	39(22)	60(19)
Butyl formate	56(100)	41(60)	31(58)	29(53)	27(45)	43(34)	28(21)	39(19)	55(11)	42(11)
1-*tert*-Butyl-4-methylbenzene	133(100)	105(38)	41(23)	148(18)	93(16)	91(14)	115(13)	134(11)	39(11)	116(10)
Butyl vinyl ether	29(100)	41(74)	56(56)	57(43)	27(42)	44(26)	15(16)	85(14)	39(14)	43(13)
γ-Butyrolactone	28(100)	42(74)	29(48)	27(33)	41(27)	56(25)	86(24)	26(18)	85(10)	39(10)
Caprolactam	55(100)	113(87)	30(81)	56(66)	84(60)	85(57)	42(51)	41(33)	28(26)	43(17)
Carbon disulfide	76(100)	32(22)	44(17)	78(9)	38(6)	28(5)	77(3)	64(1)	46(1)	39(1)
2-Chloroaniline	127(100)	129(32)	92(17)	65(16)	128(10)	91(9)	64(9)	39(8)	63(7)	99(6)
Chlorobenzene	112(100)	77(63)	114(33)	51(29)	50(14)	75(8)	113(7)	78(5)	76(5)	28(4)
1-Chlorobutane (Butyl chloride)	56(100)	41(65)	27(50)	43(35)	29(24)	39(18)	28(16)	26(11)	55(8)	15(8)
2-Chlorobutane (*sec*-Butyl chloride)	56(100)	57(100)	41(90)	27(77)	29(57)	63(46)	39(34)	92(1)		
1-Chloro-1,1-difluoroethane	65(100)	45(31)	85(14)	31(10)	64(8)	44(7)	35(6)	26(6)	87(5)	81(4)
Chloroethane (Ethyl chloride)	64(100)	28(91)	29(84)	27(75)	66(32)	26(28)	49(25)	51(8)	63(6)	65(4)
2-Chloroethanol (Ethylene chlorohydrin)	31(100)	15(13)	29(10)	28(10)	27(9)	43(8)	44(7)	26(5)	18(5)	14(5)
(Chloromethyl)benzene (Benzyl chloride)	91(100)	126(20)	65(14)	92(9)	39(9)	63(8)	128(6)	45(6)	89(5)	125(3)
1-Chloro-3-methylbutane (Isopentyl chloride)	43(100)	55(56)	41(55)	27(51)	70(49)	42(37)	29(34)	57(30)	39(30)	56(15)
1-Chloro-2-methylpropane (Isobutyl chloride)	43(100)	41(67)	42(50)	27(33)	39(26)	15(11)	29(10)	56(7)	49(6)	38(5)
2-Chloro-2-methylpropane (*tert*-Butyl chloride)	57(100)	41(80)	77(44)	29(26)	39(24)	79(14)	27(14)	59(7)	56(7)	38(6)
1-Chloronaphthalene	162(100)	127(36)	164(30)	126(20)	163(10)	77(8)	101(6)	75(6)	128(5)	28(4)
1-Chlorooctane (Octyl chloride)	91(100)	41(83)	43(76)	27(62)	29(56)	55(55)	39(34)	93(32)	57(32)	69(29)
Chloropentafluoroethane	85(100)	69(61)	31(38)	87(32)	50(17)	35(8)	119(6)	66(4)	100(3)	47(3)
1-Chloropentane (Pentyl chloride)	42(100)	41(90)	70(89)	55(87)	27(73)	29(55)	43(40)	39(40)	28(19)	57(18)
1-Chloropropane (Propyl chloride)	42(100)	29(46)	27(37)	41(23)	28(15)	43(14)	39(12)	78(6)	63(6)	49(5)
3-Chloropropene (Allyl chloride)	41(100)	39(73)	76(28)	38(16)	37(13)	40(12)	27(12)	26(11)	78(9)	49(5)
2-Chlorotoluene	91(100)	126(68)	89(23)	128(22)	90(18)	65(17)	125(15)	92(13)	127(10)	63(6)
3-Chlorotoluene	91(100)	126(27)	63(15)	65(12)	89(11)	39(11)	128(9)	125(8)	92(8)	62(6)
Cyclohexane	56(100)	84(71)	41(70)	27(37)	55(36)	39(35)	42(30)	69(23)	28(18)	43(14)
Cyclohexanol	57(100)	44(68)	41(68)	39(51)	32(40)	43(38)	31(32)	42(22)	67(18)	82(16)
Cyclohexanone	55(100)	42(85)	41(34)	27(33)	98(31)	39(27)	69(26)	70(20)	43(14)	28(14)
Cyclohexene	67(100)	54(72)	82(37)	41(35)	39(33)	27(15)	53(12)	81(9)	51(8)	79(6)
Cyclohexylamine	56(100)	43(23)	28(17)	99(10)	70(8)	57(6)	30(6)	93(5)	54(4)	41(4)
Cyclopentane	42(100)	70(30)	55(29)	41(29)	39(22)	27(15)	40(7)	29(5)	28(4)	43(3)
Cyclopentanone	55(100)	28(50)	84(42)	41(38)	56(29)	27(24)	39(19)	42(15)	26(9)	29(7)
cis-Decahydronaphthalene (*cis*-Decalin)	67(100)	81(87)	41(81)	138(67)	96(62)	82(62)	39(50)	55(45)	27(44)	95(42)
trans-Decahydronaphthalene (*trans*-Decalin)	41(100)	68(91)	67(88)	82(67)	27(65)	96(61)	95(55)	138(51)	81(51)	29(51)
Decane	43(100)	57(90)	41(41)	71(33)	29(30)	85(24)	27(20)	56(17)	55(14)	42(14)
Diacetone alcohol	43(100)	59(41)	58(17)	101(10)	41(9)	31(9)	83(6)	56(6)	55(6)	29(6)
1,2-Dibromoethane	27(100)	107(77)	109(72)	26(24)	28(10)	81(5)	79(5)	25(5)	95(4)	93(4)
Dibromofluoromethane	111(100)	113(98)	192(29)	43(16)	41(16)	190(15)	194(14)	81(9)	79(9)	122(7)
Dibromomethane	174(100)	93(96)	95(84)	172(53)	176(50)	91(11)	81(9)	79(9)	94(5)	65(5)
1,2-Dibromopropane	41(100)	121(66)	123(65)	39(48)	27(28)	107(11)	38(10)	26(10)	109(9)	42(9)
1,2-Dibromotetrafluoroethane	179(100)	181(97)	129(34)	131(33)	100(17)	31(13)	260(12)	50(8)	69(7)	262(6)
Dibutylamine	86(100)	72(52)	30(48)	44(40)	29(31)	57(24)	41(21)	73(15)	28(15)	43(13)
Dibutyl ether	57(100)	41(34)	29(30)	56(25)	87(21)	27(9)	58(8)	55(6)	39(5)	28(5)
Dibutyl oxalate	57(100)	41(61)	56(24)	44(9)	55(8)	43(7)	58(5)	42(4)	103(2)	73(2)
Dibutyl phthalate	149(100)	86(18)	57(18)	223(17)	205(17)	150(17)	104(17)	56(17)	41(17)	65(16)
Dibutyl sebacate	241(100)	185(71)	41(37)	56(35)	55(32)	57(31)	242(23)	143(21)	98(21)	125(20)
o-Dichlorobenzene	146(100)	148(64)	111(38)	75(23)	113(12)	74(12)	50(11)	150(10)	73(9)	147(7)
m-Dichlorobenzene	146(100)	148(65)	111(36)	75(25)	50(19)	74(16)	150(11)	113(11)	73(11)	147(8)
1,1-Dichloroethane	63(100)	27(71)	65(31)	26(19)	83(11)	85(7)	61(7)	35(6)	98(5)	62(5)
1,2-Dichloroethane	62(100)	27(91)	49(40)	64(32)	26(31)	63(19)	98(14)	51(13)	61(12)	100(9)
1,1-Dichloroethene	61(100)	96(61)	98(38)	63(32)	26(16)	60(15)	62(7)	25(7)	100(6)	35(6)
cis-1,2-Dichloroethene	61(100)	96(73)	98(47)	63(32)	26(30)	60(21)	25(13)	35(12)	62(9)	100(8)
trans-1,2-Dichloroethene	61(100)	96(67)	98(43)	26(34)	63(32)	60(24)	25(15)	62(10)	100(7)	47(7)
Dichlorofluoromethane	67(100)	69(32)	47(13)	35(13)	31(9)	32(8)	48(7)	83(5)	102(4)	49(4)

Compound	e/m (intensity)									
Dichloromethane (Methylene chloride)	49(100)	84(64)	86(39)	51(31)	47(14)	48(8)	88(6)	50(3)	85(2)	83(2)
(Dichloromethyl)benzene (Benzal chloride)	125(100)	127(32)	160(14)	89(13)	162(9)	63(9)	126(8)	62(7)	105(5)	39(5)
1,2-Dichloropropane	63(100)	62(71)	27(57)	41(49)	39(32)	65(31)	76(27)	64(25)	49(13)	77(12)
1,2-Dichloro-1,1,2,2-tetrafluoroethane	85(100)	135(52)	87(33)	137(17)	101(9)	31(9)	103(6)	100(6)	50(5)	69(4)
2,4-Dichlorotoluene	125(100)	160(61)	162(40)	127(32)	89(23)	159(16)	161(14)	63(13)	62(11)	126(10)
3,4-Dichlorotoluene	125(100)	160(47)	127(32)	162(31)	89(15)	159(11)	161(10)	63(10)	126(8)	62(8)
Diethanolamine	30(100)	74(82)	28(77)	56(69)	18(50)	42(46)	29(36)	27(34)	45(30)	43(19)
1,1-Diethoxyethane (Acetal)	44(100)	43(92)	29(77)	31(76)	45(74)	27(52)	72(48)	73(23)	28(17)	46(15)
1,2-Diethoxyethane (Ethylene glycol diethyl ether)	31(100)	59(71)	29(58)	45(43)	27(33)	74(27)	43(15)	15(14)	28(12)	44(10)
Diethylamine	30(100)	58(81)	44(28)	73(18)	29(18)	28(17)	72(12)	42(11)	27(11)	59(4)
Diethyl carbonate	29(100)	45(70)	31(53)	27(39)	91(24)	28(15)	63(11)	26(10)	30(6)	43(5)
Diethylene glycol	45(100)	75(23)	31(20)	44(16)	27(14)	76(12)	29(12)	43(11)	42(9)	41(4)
Diethylene glycol dimethyl ether (Diglyme)	59(100)	58(43)	31(34)	29(32)	45(28)	28(19)	89(15)	43(9)	27(5)	60(4)
Diethylene glycol monobutyl ether acetate	43(100)	87(93)	57(82)	41(28)	45(25)	56(18)	29(18)	72(14)	85(12)	101(10)
Diethylene glycol monoethyl ether (Carbitol)	45(100)	59(56)	72(37)	73(22)	60(14)	31(13)	75(11)	44(9)	104(8)	103(7)
Diethylene glycol monoethyl ether acetate	43(100)	29(51)	31(42)	45(40)	59(24)	72(18)	44(10)	73(9)	42(9)	30(6)
Diethylene glycol monomethyl ether	45(100)	31(42)	59(41)	29(38)	28(32)	58(21)	43(14)	27(13)	44(11)	32(10)
Diethyl ether	31(100)	29(63)	59(40)	27(35)	45(33)	74(23)	15(17)	43(9)	28(9)	26(9)
Diethyl oxalate	29(100)	31(16)	45(14)	27(14)	74(11)	28(9)	43(4)	30(4)	73(3)	75(2)
Diethyl sulfide	75(100)	47(81)	90(73)	62(60)	29(55)	61(54)	27(48)	28(23)	46(17)	45(17)
Diisopropylamine	44(100)	86(30)	58(14)	42(13)	28(13)	41(12)	43(11)	27(11)	15(11)	39(6)
Diisopropyl ether	45(100)	43(39)	87(15)	41(12)	59(10)	27(8)	39(4)	69(3)	42(3)	31(3)
1,2-Dimethoxybenzene (Veratrole)	138(100)	95(65)	77(48)	123(44)	52(42)	41(33)	65(30)	51(29)	39(19)	63(17)
1,2-Dimethoxyethane (Ethylene glycol dimethyl ether)	45(100)	60(13)	29(13)	90(7)	58(6)	31(5)	28(5)	43(4)	59(3)	46(2)
Dimethoxymethane (Methylal)	45(100)	75(61)	29(59)	31(13)	30(6)	15(6)	47(5)	76(2)	46(2)	44(2)
N,N-Dimethylacetamide	44(100)	87(69)	43(46)	45(23)	42(19)	72(15)	15(11)	30(8)	28(5)	88(4)
Dimethylamine	44(100)	45(81)	18(32)	28(30)	43(19)	42(15)	15(9)	46(5)	41(5)	27(5)
2,4-Dimethylaniline (2,4-Xylidine)	121(100)	120(70)	100(57)	77(16)	[illegible]	[illegible]	[illegible]	[illegible]	[illegible]	[illegible]
2,2-Dimethylbutane (Neohexane)	43(100)	57(98)	71(73)	41(61)	29(51)	27(36)	56(29)	39(25)	55(15)	15(12)
2,3-Dimethylbutane	43(100)	42(97)	41(27)	71(25)	27(11)	86(10)	39(9)	29(6)	55(5)	44(4)
Dimethyl disulfide	94(100)	45(63)	79(59)	46(38)	47(26)	15(18)	48(14)	61(12)	64(11)	96(9)
N,N-Dimethylformamide	73(100)	44(86)	42(36)	30(22)	28(20)	29(8)	43(7)	72(6)	58(5)	74(4)
Dimethyl glutarate	59(100)	100(51)	55(49)	42(33)	129(32)	101(32)	41(26)	43(22)	128(19)	87(15)
2,6-Dimethyl-4-heptanone (Isovalerone)	57(100)	85(82)	41(46)	43(39)	58(33)	28(30)	26(30)	39(22)	42(12)	142(11)
2,5-Dimethylhexane	57(100)	43(93)	42(31)	41(26)	99(19)	71(19)	29(11)	70(10)	55(9)	27(9)
Dimethyl maleate	113(100)	59(83)	26(72)	29(40)	85(37)	54(31)	114(25)	53(24)	55(13)	82(9)
2,2-Dimethylpentane	57(100)	43(73)	41(46)	56(40)	85(34)	29(31)	27(23)	39(15)	15(7)	55(6)
2,4-Dimethylpentane	43(100)	57(73)	56(41)	41(35)	42(24)	85(17)	29(16)	27(12)	39(9)	58(3)
2,4-Dimethyl-3-pentanone (Diisopropyl ketone)	43(100)	71(31)	41(13)	27(9)	70(6)	39(6)	114(5)	42(5)	44(3)	72(2)
2,4-Dimethylpyridine (2,4-Lutidine)	107(100)	106(63)	79(44)	92(22)	65(22)	51(19)	77(17)	80(12)	52(11)	50(11)
2,6-Dimethylpyridine (2,6-Lutidine)	107(100)	39(39)	106(29)	66(22)	92(18)	65(18)	38(12)	27(11)	79(9)	63(9)
Dimethyl sulfoxide	63(100)	78(70)	15(40)	45(35)	29(16)	61(13)	46(12)	31(11)	48(10)	47(10)
1,4-Dioxane	28(100)	29(37)	88(31)	58(24)	31(17)	15(17)	27(15)	30(13)	43(11)	26(9)
1,3-Dioxolane	73(100)	29(56)	44(53)	45(28)	28(21)	43(20)	27(13)	31(7)	74(5)	42(3)
Dipentene	68(100)	93(50)	67(44)	94(22)	39(22)	107(18)	92(18)	53(18)	136(16)	79(16)
Dipentyl ether (Amyl ether)	71(100)	43(92)	29(43)	70(40)	41(36)	27(34)	42(23)	69(17)	55(16)	39(16)
Dipropylamine	30(100)	72(79)	44(40)	43(32)	27(25)	28(24)	41(22)	86(11)	58(10)	42(10)
Dodecane	57(100)	43(91)	71(53)	41(45)	85(31)	29(27)	55(19)	56(18)	28(16)	42(14)
1-Dodecene	43(100)	56(83)	55(83)	41(75)	69(62)	70(58)	57(57)	83(47)	29(38)	84(33)
Epichlorohydrin	57(100)	27(39)	29(32)	49(25)	31(22)	62(18)	28(16)	92(1)		
1,2-Epoxybutane (Ethyloxirane)	43(100)	44(58)	27(52)	29(50)	45(46)	26(27)	28(27)	72(5)		
1,2-Ethanediamine	30(100)	18(13)	42(6)	43(5)	27(5)	44(4)	29(4)	17(4)	15(4)	41(3)
1,2-Ethanediol (Ethylene glycol)	31(100)	33(35)	29(13)	32(11)	43(6)	27(5)	28(4)	62(3)	30(3)	44(2)
1,2-Ethanediol, diacetate (Ethylene glycol diacetate)	43(100)	86(11)	42(7)	15(7)	116(4)	73(4)	44(3)	29(3)	103(2)	45(2)
Ethanol	31(100)	45(44)	46(18)	27(18)	29(15)	43(14)	30(6)	42(3)	19(3)	14(3)
Ethanolamine	30(100)	18(30)	28(15)	42(7)	31(6)	17(6)	61(5)	15(5)	43(3)	29(3)
Ethoxybenzene (Phenetole)	94(100)	122(39)	28(12)	66(11)	39(9)	77(8)	95(7)	65(7)	51(7)	29(6)
2-Ethoxyethanol (Ethylene glycol monoethyl ether; Cellosolve)	31(100)	29(52)	59(50)	27(27)	45(26)	72(14)	43(14)	15(14)	28(8)	26(6)
2-Ethoxyethyl acetate (Ethylene glycol monoethyl ether acetate)	43(100)	31(34)	59(31)	72(28)	44(25)	29(24)	45(12)	27(11)	15(11)	87(7)
Ethyl acetate	43(100)	29(46)	27(33)	45(32)	61(28)	28(25)	42(18)	73(11)	88(10)	70(10)
Ethyl acetoacetate	43(100)	29(24)	88(18)	28(16)	85(14)	27(12)	42(11)	60(9)	130(6)	45(6)

Compound	e/m (intensity)									
Ethyl acrylate (Ethyl propenoate)	55(100)	27(32)	29(15)	56(12)	45(9)	73(8)	28(8)	26(6)	99(5)	85(5)
Ethylamine	30(100)	28(32)	44(20)	45(19)	27(13)	15(10)	42(9)	29(8)	41(5)	40(5)
Ethylbenzene	91(100)	106(31)	51(14)	39(10)	77(8)	65(8)	105(7)	92(7)	78(7)	27(6)
Ethyl benzoate	105(100)	77(65)	122(34)	51(34)	27(17)	150(16)	29(14)	50(13)	106(12)	78(6)
Ethyl butanoate	43(100)	71(88)	29(83)	27(43)	88(40)	41(28)	60(22)	45(20)	73(17)	42(17)
Ethyl cyanoacetate	29(100)	68(59)	27(34)	40(21)	28(16)	41(14)	15(13)	45(10)	43(9)	26(9)
Ethylcyclohexane	83(100)	55(65)	82(42)	41(36)	112(23)	56(11)	67(10)	39(10)	42(9)	84(8)
Ethylene carbonate	29(100)	44(62)	43(54)	88(40)	30(16)	28(11)	45(7)	58(6)	42(6)	73(4)
Ethyl formate	31(100)	28(73)	27(51)	29(38)	45(34)	26(17)	74(11)	43(9)	47(8)	56(4)
2-Ethyl-1,3-hexanediol	56(100)	55(71)	41(60)	43(55)	29(46)	27(44)	31(34)	57(33)	73(32)	39(20)
2-Ethyl-1-hexanol	57(100)	43(41)	41(40)	29(29)	55(28)	83(27)	56(23)	70(20)	27(17)	31(13)
N-Ethyl-*N*-isopropyl-2-propanamine	72(100)	114(77)	44(41)	129(22)	42(21)	30(20)	43(18)	41(17)	27(15)	70(11)
Ethyl 3-methylbutanoate	29(100)	41(52)	27(51)	57(43)	43(42)	60(39)	88(38)	85(38)	61(26)	45(24)
3-Ethyl-2-methylpentane	43(100)	70(50)	41(27)	71(25)	29(19)	27(19)	85(18)	55(18)	57(15)	42(14)
Ethyl propanoate	57(100)	29(84)	102(17)	27(17)	75(15)	28(14)	45(13)	74(12)	73(7)	43(6)
Fluorobenzene	96(100)	70(17)	97(7)	95(6)	75(6)	50(6)	51(4)	39(4)	69(3)	57(3)
2-Fluorotoluene	109(100)	110(55)	83(18)	57(11)	63(10)	39(9)	51(6)	50(6)	107(5)	62(5)
3-Fluorotoluene	109(100)	110(54)	83(11)	57(5)	111(4)	107(4)	63(3)	39(3)	108(2)	89(2)
4-Fluorotoluene	109(100)	110(60)	83(13)	57(10)	108(8)	39(8)	63(7)	107(5)	51(5)	50(5)
Furan	68(100)	39(64)	40(9)	38(9)	42(6)	29(6)	37(5)	69(4)	34(2)	67(1)
Furfural	39(100)	96(55)	95(52)	38(38)	29(35)	37(29)	40(11)	97(9)	50(7)	42(7)
Furfuryl alcohol	98(100)	41(65)	39(59)	81(55)	53(53)	97(51)	42(49)	69(39)	70(36)	29(28)
Glycerol	61(100)	43(90)	31(57)	44(54)	29(38)	18(32)	27(12)	42(11)	60(10)	45(10)
Glycerol triacetate (Triacetin)	43(100)	103(44)	145(34)	116(17)	115(13)	44(10)	86(9)	28(8)	73(7)	42(7)
Heptane	43(100)	41(56)	29(49)	57(47)	27(46)	71(45)	56(27)	42(26)	39(23)	70(18)
1-Heptanol	41(100)	70(87)	56(86)	31(78)	43(72)	29(70)	55(67)	27(65)	42(54)	69(41)
3-Heptanol	59(100)	69(73)	87(37)	41(37)	29(33)	55(30)	31(24)	116(0)		
2-Heptanone (Methyl pentyl ketone)	43(100)	58(60)	71(14)	41(11)	27(11)	59(9)	39(8)	29(8)	42(5)	114(4)
3-Heptanone (Ethyl butyl ketone)	57(100)	29(76)	41(32)	85(29)	72(22)	43(15)	39(11)	114(10)	27(7)	55(5)
1-Heptene	41(100)	56(88)	29(70)	55(59)	42(52)	27(45)	39(41)	70(37)	69(27)	57(27)
Hexane	57(100)	43(78)	41(77)	29(61)	27(57)	56(45)	42(39)	39(27)	28(16)	86(14)
Hexanedinitrile (Adiponitrile)	41(100)	68(50)	54(42)	40(21)	55(20)	27(17)	39(16)	28(13)	52(7)	42(6)
Hexanenitrile	41(100)	54(68)	27(59)	55(55)	29(44)	39(35)	57(30)	43(30)	68(24)	28(22)
Hexanoic acid (Caproic acid)	60(100)	73(42)	27(36)	41(33)	43(27)	29(26)	45(20)	39(16)	42(15)	55(14)
1-Hexanol	56(100)	43(78)	31(74)	41(71)	27(64)	29(59)	55(58)	42(53)	39(37)	69(27)
2-Hexanone (Butyl methyl ketone)	43(100)	58(60)	57(17)	100(16)	29(15)	41(13)	85(8)	27(8)	71(7)	59(5)
Hexyl acetate	43(100)	56(66)	41(38)	55(37)	61(35)	42(33)	84(31)	69(19)	73(17)	57(7)
3-Hydroxypropanenitrile (Hydracrylonitrile)	41(100)	31(97)	29(20)	42(17)	52(15)	40(13)	53(10)	51(9)	39(7)	26(7)
Iodobenzene	204(100)	77(82)	51(32)	50(20)	127(8)	205(6)	78(6)	74(6)	102(5)	76(5)
1-Iodobutane (Butyl iodide)	57(100)	29(78)	41(56)	27(39)	184(36)	39(17)	28(16)	26(7)	127(6)	55(6)
2-Iodobutane (*sec*-Butyl iodide)	57(100)	29(36)	184(36)	41(36)	27(10)	39(9)	127(6)	58(4)		
Iodoethane (Ethyl iodide)	156(100)	29(75)	27(63)	127(31)	26(14)	28(9)	128(8)	141(2)	25(2)	140(1)
Iodomethane (Methyl iodide)	142(100)	127(38)	141(14)	15(13)	139(5)	140(4)	128(3)	14(1)	13(1)	71(0)
1-Iodopropane (Propyl iodide)	43(100)	170(68)	41(35)	27(26)	127(14)	39(11)	44(4)	141(3)	128(3)	42(3)
2-Iodopropane (Isopropyl iodide)	43(100)	170(46)	41(45)	27(44)	39(19)	127(17)	42(4)	38(4)	128(3)	44(3)
Isobutanal (2-Methyl-1-propanal)	43(100)	41(84)	27(47)	72(46)	39(30)	29(25)	42(10)	70(5)	38(5)	28(5)
Isobutyl acetate	43(100)	56(26)	73(15)	41(10)	29(5)	71(3)	57(3)	39(3)	27(3)	86(2)
Isobutylamine	30(100)	28(9)	41(6)	73(5)	27(5)	39(4)	29(3)	15(3)	58(2)	56(2)
Isobutylbenzene	91(100)	92(58)	43(22)	134(20)	65(12)	41(12)	39(11)	27(8)	51(7)	93(5)
Isobutyl formate	43(100)	56(82)	41(78)	31(65)	29(64)	27(53)	60(39)	39(36)	42(26)	15(14)
Isobutyl isobutanoate	43(100)	41(94)	56(76)	71(70)	57(68)	27(62)	29(59)	89(37)	39(32)	42(17)
Isopentyl acetate	43(100)	70(49)	55(38)	61(15)	42(15)	41(14)	27(12)	87(11)	29(10)	73(9)
Isophorone	82(100)	39(20)	138(17)	54(13)	27(12)	41(10)	53(8)	83(7)	29(7)	55(6)
Isopropyl acetate	43(100)	61(17)	41(14)	87(9)	59(8)	27(8)	42(7)	39(4)	45(3)	44(2)
Isopropylbenzene (Cumene)	105(100)	120(25)	77(13)	51(12)	79(10)	106(9)	39(9)	27(8)	103(6)	91(5)
1-Isopropyl-4-methylbenzene (*p*-Cymene)	119(100)	91(42)	134(33)	39(27)	41(20)	117(18)	65(18)	77(17)	27(16)	120(15)
Isoquinoline	129(100)	102(26)	51(20)	128(18)	50(11)	130(10)	75(10)	76(9)	103(8)	74(7)
d-Limonene (Citrene)	68(100)	93(50)	67(49)	41(22)	94(21)	79(21)	39(21)	136(20)	53(19)	121(16)
Mesityl oxide	55(100)	83(89)	43(73)	29(42)	98(36)	39(32)	27(28)	53(11)	41(10)	56(5)
Methanol	31(100)	29(72)	32(67)	15(42)	28(12)	14(10)	30(9)	13(6)	12(3)	16(2)
2-Methoxyethanol (Ethylene glycol monomethyl ether)	45(100)	31(15)	29(14)	28(11)	47(9)	76(6)	43(6)	58(4)	46(4)	27(4)

Compound	e/m (intensity)									
2-Methoxyethyl acetate (Ethylene glycol monomethyl ether acetate)	43(100)	45(48)	58(42)	29(10)	42(4)	31(4)	73(3)	27(3)	59(2)	26(2)
Methyl acetate	43(100)	74(52)	28(38)	42(19)	59(17)	44(8)	32(8)	29(6)	31(4)	75(2)
2-Methylacrylonitrile	41(100)	39(54)	67(44)	40(26)	38(24)	52(23)	27(23)	37(20)	66(17)	51(12)
Methylamine	30(100)	31(87)	28(56)	29(19)	32(15)	15(12)	27(9)			
2-Methylaniline (*o*-Toluidine)	106(100)	107(83)	77(17)	79(13)	39(12)	53(10)	52(10)	54(9)	51(9)	28(9)
3-Methylaniline (*m*-Toluidine)	106(100)	107(84)	79(17)	77(17)	108(7)	78(6)	80(5)	89(4)	65(4)	53(4)
N-Methylaniline	106(100)	107(79)	77(23)	51(12)	79(11)	65(9)	39(9)	78(8)	108(7)	50(6)
Methyl benzoate	105(100)	77(81)	51(45)	136(24)	50(18)	106(8)	78(6)	28(6)	39(5)	27(5)
Methyl butanoate	43(100)	74(90)	71(66)	41(32)	27(31)	59(28)	87(19)	42(15)	28(15)	39(14)
3-Methylbutanoic acid (Isovaleric acid)	60(100)	43(61)	41(54)	27(33)	45(31)	29(27)	74(24)	39(24)	87(21)	57(20)
3-Methyl-1-butanol (Isopentyl alcohol)	55(100)	42(90)	43(82)	41(81)	70(71)	31(61)	29(59)	27(59)	39(44)	57(31)
Methyl *tert*-butyl ether	41(100)	73(78)	57(71)	29(60)	43(48)	39(34)	28(28)	56(23)	55(18)	45(16)
Methyl cyanoacetate	59(100)	15(65)	68(60)	40(38)	28(17)	29(16)	55(11)	54(10)	39(10)	67(8)
Methylcyclohexane	83(100)	55(82)	41(60)	98(44)	42(35)	56(30)	27(29)	39(27)	69(23)	70(22)
cis-4-Methylcyclohexanol	57(100)	58(54)	70(38)	81(36)	96(33)	55(25)	41(23)	114(17)	71(15)	56(12)
N-Methylformamide	59(100)	30(54)	28(34)	29(13)	58(8)	15(7)	60(3)	41(3)	27(3)	31(2)
Methyl formate	31(100)	29(63)	32(34)	60(28)	30(7)	28(7)	44(2)	18(2)	61(1)	59(1)
2-Methylheptane	43(100)	57(91)	42(39)	41(27)	29(16)	70(15)	27(13)	71(12)	99(10)	55(9)
4-Methylheptane	43(100)	71(53)	70(46)	41(27)	29(23)	27(23)	55(15)	57(14)	42(14)	39(10)
6-Methyl-1-heptanol (Isooctyl alcohol)	41(100)	55(94)	43(93)	69(84)	57(81)	56(73)	29(50)	84(48)	70(47)	27(46)
2-Methylhexane	43(100)	42(38)	41(35)	85(32)	57(26)	29(22)	27(22)	56(20)	39(11)	55(5)
5-Methyl-2-hexanone (Methyl isopentyl ketone)	43(100)	58(34)	27(14)	41(13)	15(13)	57(11)	39(9)	71(8)	59(8)	29(8)
Methyl methacrylate (Methyl 2-methyl-2-propenoate)	41(100)	69(66)	39(40)	100(34)	15(20)	40(10)	59(8)	99(6)	38(6)	55(5)
2-Methyloctane	43(100)	57(51)	41(35)	71(28)	42(28)	29(23)	27(21)	85(17)	56(15)	84(13)
2-Methylpentane	43(100)	42(53)	41(35)	27(31)	71(29)	39(20)	29(18)	57(11)	15(10)	70(7)
3-Methylpentane	57(100)	56(76)	41(68)	29(60)	27(40)	43(29)	39(22)	55(9)	15(9)	28(8)
2-Methyl-2,4-pentanediol (Hexylene glycol)	59(100)	43(61)	56(25)	45(17)	41(16)	57(13)	42(13)	85(11)	61(10)	31(10)
4-Methylpentanenitrile	55(100)	41(52)	43(46)	27(39)	39(29)	57(27)	54(26)	29(22)	82(13)	28(12)
4-Methyl-2-pentanol	45(100)	43(47)	69(30)	41(27)	27(19)	39(13)	29(12)	87(11)	84(10)	57(10)
3-Methyl-3-pentanol	73(100)	55(38)	43(35)	45(28)	27(25)	29(21)	41(12)	87(11)	31(11)	15(9)
4-Methyl-2-pentanone (Methyl isobutyl ketone)	43(100)	58(84)	29(65)	41(56)	57(44)	27(42)	39(31)	85(19)	100(14)	42(14)
2-Methylpropanenitrile (Isobutyronitrile)	42(100)	68(45)	28(45)	54(26)	41(26)	27(26)	29(25)	26(15)	39(13)	15(13)
Methyl propanoate	57(100)	29(72)	59(31)	88(26)	27(18)	28(9)	31(5)	44(4)	26(4)	58(3)
2-Methylpropanoic acid (Isobutyric acid)	43(100)	41(42)	27(40)	73(22)	39(15)	45(14)	42(11)	29(9)	88(7)	28(6)
2-Methyl-1-propanol (Isobutyl alcohol)	43(100)	33(73)	31(72)	41(66)	42(60)	27(43)	29(18)	39(17)	28(8)	74(6)
2-Methyl-2-propanol (*tert*-Butyl alcohol)	59(100)	31(33)	41(22)	43(18)	29(13)	27(11)	57(10)	42(4)	60(3)	28(3)
2-Methylpyridine (2-Picoline)	93(100)	66(41)	39(31)	92(20)	78(19)	51(19)	65(16)	38(13)	50(12)	52(11)
3-Methylpyridine (3-Picoline)	93(100)	39(51)	66(46)	92(31)	65(29)	40(19)	38(18)	67(13)	63(11)	51(11)
N-Methyl-2-pyrrolidinone	99(100)	44(89)	98(80)	42(60)	41(38)	43(17)	28(17)	71(13)	39(11)	70(10)
Methyl salicylate	120(100)	92(59)	152(47)	121(32)	65(22)	39(22)	93(15)	64(14)	18(14)	63(13)
2-Methylthiophene	97(100)	98(57)	45(22)	39(14)	53(9)	99(8)	27(8)	69(6)	58(6)	59(5)
Morpholine	57(100)	29(100)	87(69)	28(69)	30(38)	56(33)	86(28)	31(28)	27(12)	15(7)
Nitrobenzene	77(100)	51(59)	123(42)	50(25)	30(15)	65(14)	39(10)	93(9)	74(7)	78(6)
Nitroethane	29(100)	30(12)	28(11)	26(9)	27(8)	43(5)	41(5)	14(5)	15(3)	46(2)
Nitromethane	30(100)	61(64)	46(39)	28(30)	45(8)	27(8)	44(7)	29(7)	60(5)	43(4)
1-Nitropropane	43(100)	27(93)	41(90)	39(34)	30(25)	44(20)	42(20)	26(20)	28(13)	54(12)
2-Nitropropane	43(100)	41(73)	27(71)	39(30)	30(18)	15(11)	42(9)	28(8)	26(8)	38(6)
Nonane	43(100)	57(75)	41(29)	84(26)	85(22)	29(22)	71(18)	56(16)	27(13)	42(12)
Octane	43(100)	57(30)	85(25)	41(25)	71(19)	29(17)	56(14)	70(10)	42(10)	27(10)
1-Octanol	41(100)	56(85)	43(82)	55(81)	31(69)	27(69)	29(68)	42(62)	70(53)	69(48)
2-Octanone (Hexyl methyl ketone)	43(100)	58(79)	41(56)	59(52)	71(49)	27(46)	29(36)	39(27)	57(18)	55(17)
1-Octene	43(100)	41(82)	55(80)	56(67)	42(67)	70(54)	29(44)	27(31)	69(30)	39(29)
Pentachloroethane	167(100)	165(91)	117(90)	119(89)	83(58)	169(54)	130(43)	132(42)	60(40)	85(37)
Pentane	43(100)	42(55)	41(45)	27(42)	29(26)	39(19)	57(13)	28(9)	15(9)	72(8)
1,5-Pentanediol (Pentamethylene glycol)	31(100)	56(85)	41(67)	57(59)	55(51)	44(45)	29(37)	43(31)	68(29)	27(26)
2,4-Pentanedione (Acetylacetone)	43(100)	85(31)	100(20)	27(12)	42(10)	29(10)	41(7)	39(7)	31(5)	26(5)
Pentanenitrile (Valeronitrile)	41(100)	43(97)	54(54)	27(34)	55(21)	28(19)	29(16)	39(15)	42(5)	26(5)
Pentanoic acid (Valeric acid)	60(100)	73(34)	27(33)	29(28)	41(21)	43(17)	45(16)	28(14)	42(12)	39(12)
1-Pentanol (Amyl alcohol)	42(100)	70(72)	55(65)	41(56)	31(47)	29(41)	27(26)	57(22)	28(22)	43(21)
2-Pentanol (sec-Amyl alcohol)	45(100)	43(20)	55(18)	27(17)	29(11)	41(9)	31(9)	15(9)	44(8)	39(8)
3-Pentanol (Diethyl carbinol)	59(100)	31(83)	41(42)	27(35)	29(34)	58(15)	43(15)	57(14)	39(12)	15(9)

Compound	e/m (intensity)									
2-Pentanone (Methyl propyl ketone)	43(100)	41(17)	86(12)	42(12)	27(11)	39(8)	71(7)	58(7)	45(7)	44(3)
3-Pentanone (Diethyl ketone)	57(100)	29(50)	86(26)	27(13)	58(4)	56(4)	28(4)	26(3)	43(2)	42(2)
Pentyl acetate (Amyl acetate)	43(100)	70(90)	42(52)	28(51)	61(50)	55(41)	73(21)	41(20)	29(14)	69(11)
Pentylamine (Amylamine)	30(100)	87(8)	41(4)	28(4)	45(3)	42(3)	27(3)	56(2)	44(2)	43(2)
Pentyl lactate	43(100)	70(44)	71(32)	55(28)	41(22)	29(20)	27(15)	45(12)	42(9)	57(7)
α-Pinene	93(100)	92(30)	39(24)	41(23)	77(22)	91(21)	27(21)	79(18)	121(13)	53(10)
β-Pinene	93(100)	41(64)	69(47)	39(33)	27(31)	79(20)	77(18)	53(14)	94(13)	91(13)
Piperidine	84(100)	85(53)	56(46)	57(43)	28(41)	29(37)	44(34)	42(30)	30(30)	43(25)
Propanal	29(100)	58(59)	28(58)	27(39)	57(20)	31(4)	30(4)	42(3)	39(3)	59(2)
1,2-Propanediol (1,2-Propylene glycol)	45(100)	18(46)	29(21)	43(19)	31(18)	27(17)	28(11)	19(8)	44(6)	61(5)
1,3-Propanediol (Trimethylene glycol)	28(100)	58(93)	31(76)	57(70)	29(40)	27(26)	45(24)	43(23)	19(18)	30(17)
Propanenitrile	28(100)	54(63)	26(20)	27(17)	52(11)	55(10)	51(9)	15(9)	53(7)	25(7)
Propanoic acid	28(100)	29(84)	74(79)	27(62)	45(56)	73(48)	57(30)	26(21)	55(17)	56(16)
Propanoic anhydride	57(100)	29(55)	28(20)	27(20)	74(19)	73(12)	45(11)	26(5)	30(4)	58(3)
1-Propanol (Propyl alcohol)	31(100)	27(19)	29(18)	59(11)	42(9)	60(7)	41(7)	28(7)	43(3)	32(3)
2-Propanol (Isopropyl alcohol)	45(100)	43(19)	27(17)	29(12)	41(7)	31(6)	19(6)	42(5)	44(4)	59(3)
Propargyl alcohol (3-Hydroxy-1-propyne)	55(100)	39(25)	28(20)	27(19)	29(16)	38(14)	26(11)	37(8)	53(6)	56(4)
Propyl acetate	43(100)	61(19)	31(18)	27(15)	42(11)	73(9)	41(9)	29(9)	59(5)	39(5)
Propylamine	30(100)	28(13)	59(8)	27(7)	41(5)	42(3)	39(3)	29(3)	26(3)	18(3)
Propylbenzene	91(100)	120(21)	92(10)	38(10)	65(9)	78(6)	51(6)	27(5)	63(4)	105(3)
Propylene carbonate	28(100)	57(69)	43(66)	29(51)	27(45)	30(42)	26(19)	42(17)	31(16)	58(13)
Propyl formate	27(100)	29(92)	31(81)	42(60)	41(53)	43(29)	39(29)	26(28)	47(22)	30(16)
Pyridine	79(100)	52(62)	51(31)	50(19)	78(11)	53(7)	39(7)	80(6)	27(3)	77(2)
Pyrrole	67(100)	41(58)	39(58)	40(51)	28(42)	38(20)	37(12)	66(7)	68(5)	27(3)
Pyrrolidine	43(100)	28(52)	70(33)	71(26)	42(22)	41(20)	27(16)	39(15)	29(10)	30(9)
2-Pyrrolidone	85(100)	42(43)	41(36)	28(33)	30(29)	56(16)	84(14)	40(12)	27(12)	29(9)
Quinoline	129(100)	51(28)	76(25)	128(24)	44(24)	50(20)	32(19)	75(18)	74(12)	103(11)
Salicylaldehyde (2-Hydroxybenzaldehyde)	28(100)	122(98)	121(92)	39(79)	65(52)	76(31)	32(31)	44(30)	93(26)	38(23)
Styrene	104(100)	103(41)	78(32)	51(28)	77(23)	105(12)	50(12)	52(11)	39(11)	102(10)
Sulfolane	41(100)	28(94)	56(82)	55(72)	120(37)	27(32)	39(19)	29(17)	26(11)	48(5)
α-Terpinene	121(100)	93(85)	136(43)	91(40)	77(34)	39(33)	27(33)	79(27)	41(26)	43(18)
1,1,1,2-Tetrachloro-2,2-difluoroethane	167(100)	169(96)	117(85)	119(82)	171(31)	85(29)	121(26)	82(14)	47(14)	101(13)
1,1,2,2-Tetrachloro-1,2-difluoroethane	101(100)	103(64)	167(54)	169(52)	117(19)	119(18)	171(17)	105(11)	31(11)	132(9)
1,1,1,2-Tetrachloroethane	131(100)	133(96)	117(76)	119(73)	95(34)	135(31)	121(23)	97(23)	61(19)	60(18)
1,1,2,2-Tetrachloroethane	83(100)	85(63)	95(11)	87(10)	168(8)	133(8)	131(8)	96(8)	61(8)	60(8)
Tetrachloroethene	166(100)	164(82)	131(71)	129(71)	168(45)	94(38)	47(31)	96(24)	133(20)	59(17)
Tetrachloromethane (Carbon tetrachloride)	117(100)	119(98)	121(31)	82(24)	47(23)	84(16)	35(14)	49(8)	28(8)	36(6)
Tetraethylene glycol	45(100)	89(10)	44(8)	43(6)	31(6)	29(6)	27(6)	101(5)	75(5)	28(5)
Tetrahydrofuran	42(100)	41(52)	27(33)	72(29)	71(27)	39(24)	43(22)	29(22)	40(13)	15(10)
1,2,3,4-Tetrahydronaphthalene	104(100)	132(53)	91(43)	51(17)	39(17)	131(15)	117(15)	115(14)	78(13)	77(13)
Tetrahydropyran	41(100)	28(64)	56(57)	45(57)	29(51)	27(49)	85(47)	86(42)	39(28)	55(23)
Tetrahydrothiophene	60(100)	88(54)	45(37)	46(32)	47(26)	27(24)	59(18)	87(16)	39(14)	54(13)
Tetramethylsilane	73(100)	43(14)	45(12)	74(8)	29(7)	15(5)	75(4)	44(4)	42(4)	31(4)
Tetramethylurea	72(100)	44(27)	116(24)	42(13)	15(13)	17(7)	28(5)	73(4)	56(4)	18(4)
1*H*-Tetrazole	42(100)	28(60)	27(25)	29(24)	41(13)	43(11)	70(8)	26(7)	40(2)	38(2)
Thiophene	84(100)	58(65)	45(58)	39(29)	57(13)	38(8)	69(7)	37(7)	83(6)	50(6)
Toluene	91(100)	92(73)	39(20)	65(14)	63(11)	51(11)	50(7)	27(6)	93(5)	90(5)
Tribromomethane (Bromoform)	173(100)	171(50)	175(49)	93(22)	91(22)	79(18)	81(17)	94(13)	92(13)	254(11)
Tributylamine	142(100)	100(19)	143(11)	29(8)	185(7)	57(6)	44(6)	41(6)	30(5)	86(4)
1,2,4-Trichlorobenzene	180(100)	182(96)	184(30)	145(30)	109(26)	147(19)	75(11)	74(9)	111(8)	181(7)
1,1,1-Trichloroethane	97(100)	99(64)	61(58)	26(31)	27(24)	117(19)	63(19)	119(18)	35(17)	62(11)
1,1,2-Trichloroethane	97(100)	83(95)	99(62)	85(60)	61(58)	26(23)	96(21)	63(19)	27(17)	98(15)
Trichloroethene	95(100)	130(90)	132(85)	60(65)	97(64)	35(40)	134(27)	47(26)	62(21)	59(13)
Trichloroethylsilane	135(100)	133(100)	126(67)	128(46)	137(37)	98(22)	63(21)	35(16)	100(14)	127(11)
Trichlorofluoromethane	101(100)	103(66)	66(13)	105(11)	35(11)	47(9)	31(8)	82(4)	68(4)	37(4)
Trichloromethane (Chloroform)	83(100)	85(64)	47(35)	35(19)	48(16)	49(12)	87(10)	37(6)	50(5)	84(4)
(Trichloromethyl)benzene (Benzotrichloride)	159(100)	161(64)	89(14)	163(11)	28(10)	63(9)	160(8)	123(8)	62(8)	124(6)
1,2,3-Trichloropropane	75(100)	39(58)	49(42)	110(38)	61(34)	77(33)	112(22)	27(16)	97(15)	38(15)
1,1,2-Trichloro-1,2,2-trifluoroethane	101(100)	151(68)	103(64)	85(45)	31(45)	153(44)	35(20)	66(19)	47(18)	87(14)
Tridecane	57(100)	43(91)	71(51)	41(34)	85(24)	29(23)	56(14)	55(12)	27(11)	42(10)
Triethanolamine	118(100)	56(69)	45(60)	42(56)	44(27)	43(25)	41(14)	116(8)	57(8)	86(7)

Compound	*e/m* (intensity)									
Triethylamine	86(100)	30(68)	58(37)	28(24)	29(23)	27(19)	44(18)	101(17)	42(16)	56(8)
Triethylene glycol	45(100)	58(11)	89(9)	31(8)	29(8)	75(7)	44(7)	43(7)	27(7)	28(5)
Triethyl phosphate	99(100)	81(71)	155(56)	82(45)	45(45)	109(44)	127(41)	43(24)	125(16)	111(14)
Trifluoroacetic acid	45(100)	69(70)	51(36)	28(28)	50(15)	44(11)	43(7)	97(5)	31(5)	29(5)
2,2,2-Trifluoroethanol	31(100)	33(24)	61(19)	29(19)	51(16)	69(9)	32(6)	49(5)	83(4)	81(4)
(Trifluoromethyl)benzene (Benzotrifluoride)	146(100)	145(40)	127(34)	96(28)	77(10)	51(10)	147(8)	75(6)	50(6)	128(3)
Trimethylamine	58(100)	59(47)	30(29)	42(26)	44(17)	15(14)	28(10)	18(10)	43(8)	57(7)
1,2,3-Trimethylbenzene (Hemimellitene)	105(100)	120(47)	39(22)	77(17)	91(14)	51(14)	27(14)	79(12)	119(11)	106(9)
1,2,4-Trimethylbenzene (Pseudocumene)	105(100)	120(56)	119(17)	77(15)	39(15)	51(11)	91(10)	27(10)	106(9)	79(7)
1,3,5-Trimethylbenzene (Mesitylene)	105(100)	120(64)	119(15)	77(13)	39(11)	106(9)	91(9)	51(8)	27(7)	121(6)
2,2,3-Trimethylbutane (Triptane)	57(100)	43(71)	56(63)	41(53)	85(30)	29(26)	27(18)	39(14)	15(6)	55(5)
2,2,5-Trimethylhexane	57(100)	56(35)	71(18)	41(17)	43(14)	29(8)	70(4)	58(4)	113(3)	55(3)
2,2,4-Trimethylpentane (Isooctane)	57(100)	41(31)	56(28)	43(24)	29(16)	27(9)	39(7)	58(4)	55(4)	99(2)
2,3,3-Trimethylpentane	43(100)	71(45)	70(36)	57(36)	41(29)	85(25)	27(18)	55(16)	29(16)	39(11)
2,3,4-Trimethylpentane	43(100)	71(62)	70(41)	41(25)	27(18)	55(17)	57(16)	29(14)	39(10)	42(7)
Trimethyl phosphate	110(100)	109(35)	79(34)	95(25)	80(23)	15(20)	140(18)	47(10)	31(7)	139(5)
2,4,6-Trimethylpyridine (2,4,6-Collidine)	121(100)	39(27)	79(26)	120(24)	106(17)	27(16)	77(13)	51(11)	42(11)	122(10)
1-Undecene	41(100)	43(87)	55(80)	70(67)	56(67)	69(55)	29(55)	83(51)	57(50)	27(46)
Vinyl acetate	43(100)	28(45)	42(26)	44(24)	86(20)	31(10)	32(7)	29(7)	45(2)	41(2)
o-Xylene	91(100)	106(40)	39(21)	105(17)	51(17)	77(15)	27(12)	65(10)	92(8)	79(8)
m-Xylene	91(100)	106(65)	105(29)	39(18)	51(15)	77(14)	27(10)	92(8)	79(8)	78(8)
p-Xylene	91(100)	106(62)	105(30)	51(16)	39(16)	77(13)	27(11)	92(7)	78(7)	65(7)

PROTON NMR SHIFTS OF COMMON ORGANIC SOLVENTS

The table below lists the ^{1}H chemical shifts for over 300 organic solvents and liquid reagents. The solvent in which the shift was measured is given in the second column. Shifts are given in parts per million relative to tetramethylsilane (TMS) and are listed in order of smallest to largest shift. In many cases the peaks show additional small splittings. Compounds are listed by the name used in this *Handbook*, with other common names given in parentheses.

References

1. Lide, D. R., Editor, *Properties of Organic Compounds*, <www.chemnetbase.com/scripts/pocweb.exe>; Lide, D. R., and Milne, G. W. A., Editors, *Handbook of Data on Organic Compounds, Third Edition*, CRC Press, Boca Raton, FL, 1993.
2. Spectral Database for Organic Compounds, SDBS, National Institute of Advanced Industrial Science and Technology (AIST), Japan, <riodb01.ibase.aist.go.jp>.
3. NMRShiftDB, <nmrshiftdb.ice.mpg.de>.

Compound	Solvent	HNMR shifts (ppm relative to TMS)					Ref.
Acetic acid	$CDCl_3$	2.10	11.4				2
Acetic anhydride	CCl_4	2.2					1
Acetone	$CDCl_3$	2.1					1
Acetonitrile	CCl_4	1.9					1
Acrolein (2-Propenal)	$CDCl_3$	6.4	9.5				1
Acrylonitrile	$CDCl_3$	6.3					1
Allyl alcohol	$CDCl_3$	3.6	4.1	5.1	5.3	6.0	1
Allylamine	$CDCl_3$	1.5	3.3	5.0	5.1	5.9	1
2-Amino-2-methyl-1-propanol (2-Aminoisobutanol)	CCl_4	1.1	2.8	3.2			1
Aniline (Benzenamine)	CCl_4	3.3	6.4	6.6	7.0		1
Anisole (Methoxybenzene)	$CDCl_3$	3.8	7.1				1
Benzaldehyde	$CDCl_3$	7.7	10.0				1
Benzene	$CDCl_3$	7.34					2
Benzeneacetonitrile (Benzyl cyanide)	CCl_4	0.8	0.9	1.6	2.3		1
Benzenethiol (Phenyl mercaptan)	CCl_4	3.2	6.9				1
Benzonitrile	CCl_4	7.5					1
Benzyl acetate	$CDCl_3$	2.1	5.1	7.3			1
Benzyl alcohol	$CDCl_3$	2.4	4.6	7.3			1
Bis(2-aminoethyl)amine (Diethylenetriamine)	$CDCl_3$	1.23	2.69	2.79			2
Bis(2-chloroethyl) ether	$CDCl_3$	3.66	3.77				2
Bis(2-ethylhexyl) phthalate	CCl_4	0.9	1.5	4.2	7.4	7.7	1
Bis(2-hydroxyethyl) sulfide	$CDCl_3$	2.8	3.8	4.2			1
Bromobenzene	CCl_4	7.1	7.4				1
1-Bromobutane (Butyl bromide)	CCl_4	1.0	1.4	1.8	3.4		1
2-Bromobutane (*sec*-Butyl bromide)	$CDCl_3$	1.1	1.7	1.8	4.1		1
1-Bromo-2-chloroethane	$CDCl_3$	3.3	4.0				1
Bromochloromethane	CCl_4	5.2					1
1-Bromodecane (Decyl bromide)	CCl_4	0.9	1.8	3.3			1
Bromoethane (Ethyl bromide)	$CDCl_3$	1.7	3.4				1
2-Bromo-2-methylpropane (*tert*-Butyl bromide)	CCl_4	1.8					1
1-Bromonaphthalene	CCl_4	7.4	8.1				1
1-Bromopentane (Pentyl bromide)	CCl_4	0.9	1.4	1.9	3.3		1
1-Bromopropane (Propyl bromide)	CCl_4	1.0	1.9	3.4			1
2-Bromopropane (Isopropyl bromide)	$CDCl_3$	1.7	4.3				1
2-Bromopropene	$CDCl_3$	2.3	5.3	5.5			1
Butanal	$CDCl_3$	1.0	1.7	2.4	9.7		1
Butanenitrile	CCl_4	1.1	1.7	2.3			1
1-Butanethiol (Butyl mercaptan)	$CDCl_3$	0.9	1.2	1.5	2.5		1
Butanoic acid	CCl_4	0.9	1.7	2.3	12.0		1
Butanoic anhydride	CCl_4	1.0	1.7	2.4			1

Compound	Solvent	HNMR shifts (ppm relative to TMS)						Ref.
1-Butanol (Butyl alcohol)	$CDCl_3$	0.94	1.39	1.53	2.24	3.63		2
2-Butanol (*sec*-Butyl alcohol)	$CDCl_3$	0.93	1.17	1.46	2.37	3.71		2
2-Butanone (Methyl ethyl ketone)	$CDCl_3$	1.06	2.14	2.45				2
trans-2-Butenal (*trans*-Crotonaldehyde)	$CDCl_3$	2.0	6.1	6.9	9.5			1
2-Butoxyethanol (Ethylene glycol monobutyl ether)	CCl_4	0.9	1.3	3.3	3.7			1
Butyl acetate	$CDCl_3$	0.9	1.4	2.0	4.1			1
Butylamine	$CDCl_3$	0.92	1.33	1.43	1.77	2.68		2
tert-Butylamine	$CDCl_3$	1.1	1.2					1
Butylbenzene	CCl_4	0.9	1.4	2.6	7.1			1
sec-Butylbenzene	CCl_4	0.8	1.2	1.6	2.5	7.1		1
tert-Butylbenzene	CCl_4	1.3	7.2					1
Butyl formate	$CDCl_3$	0.9	1.5	4.2	8.1			1
1-*tert*-Butyl-4-methylbenzene	$CDCl_3$	1.30	2.31	7.11	7.26			2
Butyl vinyl ether	CCl_4	0.9	1.4	3.6	3.8	4.0	6.3	1
γ-Butyrolactone	$CDCl_3$	4.4						1
Caprolactam	$CDCl_3$	1.7	2.4	3.2	7.8			1
2-Chloroaniline	CCl_4	3.8	6.8					1
Chlorobenzene	$CDCl_3$	7.3						2
2-Chlorobutane (*sec*-Butyl chloride)	CCl_4	1.1	1.5	1.7	3.9			1
Chloroethane (Ethyl chloride)	$CDCl_3$	1.5	3.6					1
2-Chloroethanol (Ethylene chlorohydrin)	$CDCl_3$	2.8	3.7	3.8				1
(Chloromethyl)benzene (Benzyl chloride)	CCl_4	4.5	7.3					1
1-Chloro-3-methylbutane (Isopentyl chloride)	$CDCl_3$	0.9	1.7	3.6				1
1-Chloro-2-methylpropane (Isobutyl chloride)	CCl_4	1.0	1.9	3.3				1
2-Chloro-2-methylpropane (*tert*-Butyl chloride)	CCl_4	1.6						1
1-Chloronaphthalene	CCl_4	7.1	7.5	8.2				1
1-Chlorooctane (Octyl chloride)	CCl_4	0.9	1.3	1.8	3.5			1
1-Chloropentane (Pentyl chloride)	CCl_4	0.9	1.6	3.4				1
1-Chloropropane (Propyl chloride)	CCl_4	1.0	1.8	3.4				1
3-Chloropropene (Allyl chloride)	CCl_4	4.0	5.2	5.3	5.9			1
2-Chlorotoluene	$CDCl_3$	2.4	7.2					1
3-Chlorotoluene	CCl_4	2.3	7.1					1
Cyclohexane	$CDCl_3$	1.43						2
Cyclohexanol	CCl_4	1.6	3.5	4.2				1
Cyclohexanone	CCl_4	1.8	2.3					1
Cyclohexene	CCl_4	1.6	2.0	5.6				1
Cyclohexylamine	CCl_4	1.4	1.5	2.6				1
Cyclopentane	CCl_4	1.5						1
Cyclopentanone	CCl_4	2.0						1
cis-Decahydronaphthalene (*cis*-Decalin)	$CDCl_3$	1.42	1.62					2
trans-Decahydronaphthalene (*trans*-Decalin)	$CDCl_3$	0.87	0.93	1.23	1.54	1.67		2
Decane	CCl_4	0.9	1.3					1
Diacetone alcohol	$CDCl_3$	1.3	2.2	2.6	3.7			1
1,2-Dibromoethane	$CDCl_3$	3.65						2
Dibromomethane	CCl_4	4.9						1
1,2-Dibromopropane	CCl_4	1.8	3.5	3.8	4.2			1
Dibutylamine	CCl_4	0.5	0.9	1.4	2.5			1
Dibutyl ether	CCl_4	0.9	1.4	3.3				1
Dibutyl sebacate	CCl_4	1.0	1.5	2.2	4.0			1
o-Dichlorobenzene	CCl_4	7.2						1
m-Dichlorobenzene	CCl_4	7.2	7.4					1
1,1-Dichloroethane	$CDCl_3$	2.06	5.90					2

Compound	Solvent	HNMR shifts (ppm relative to TMS)					Ref.
1,2-Dichloroethane	CCl_4	3.7					1
1,1-Dichloroethene	CCl_4	5.5					1
cis-1,2-Dichloroethene	$CDCl_3$	6.28					2
trans-1,2-Dichloroethene	$(CH_3)_4Si$	6.24					2
Dichloromethane (Methylene chloride)	CCl_4	5.3					1
(Dichloromethyl)benzene (Benzal chloride)	CCl_4	6.6	7.4				1
1,2-Dichloropropane	$CDCl_3$	1.61	3.59	3.74	4.14		2
2,4-Dichlorotoluene	CCl_4	2.3	7.0	7.3			1
3,4-Dichlorotoluene	CCl_4	2.3	7.0				1
Diethanolamine	D_2O	2.7	3.7				1
1,1-Diethoxyethane (Acetal)	$CDCl_3$	1.2	1.3	3.5	3.7	4.7	2
1,2-Diethoxyethane (Ethylene glycol diethyl ether)	$CDCl_3$	1.22	3.54	3.58			1
Diethylamine	CCl_4	0.9	1.0	2.6			1
Diethyl carbonate	$CDCl_3$	1.3	4.2				1
Diethylene glycol	$CDCl_3$	3.7	4.2				1
Diethylene glycol dimethyl ether (Diglyme)	$CDCl_3$	3.3	3.5				1
Diethylene glycol monoethyl ether (Carbitol)	CCl_4	1.2	3.1	3.5	3.6		1
Diethylene glycol monoethyl ether acetate	$CDCl_3$	1.22	2.08	3.54	3.71	4.23	2
Diethylene glycol monomethyl ether	$CDCl_3$	3.3	3.4	3.6			1
Diethyl ether	$CDCl_3$	1.21	3.47				2
Diethyl sulfide	CCl_4	1.2	2.5				1
Diisopropylamine	CCl_4	0.7	1.0	2.9			1
Diisopropyl ether	CCl_4	1.0	3.5				1
1,2-Dimethoxybenzene (Veratrole)	CCl_4	3.7	6.8				1
1,2-Dimethoxyethane (Ethylene glycol dimethyl ether)	CCl_4	3.3	3.4				1
Dimethoxymethane (Methylal)	CCl_4	3.2	4.4				1
N,N-Dimethylacetamide	$CDCl_3$	2.1	2.9	3.0			1
2,4-Dimethylaniline (2,4-Xylidine)	CCl_4	2.0	2.2	3.4	6.4	6.7	1
2,2-Dimethylbutane (Neohexane)	CCl_4	0.9	1.1	1.3	1.1	1.3	1
2,3-Dimethylbutane	CCl_4	0.9	1.5				1
Dimethyl disulfide		2.45					3
N,N-Dimethylformamide	$CDCl_3$	2.9	3.0	8.0			1
Dimethyl glutarate	$CDCl_3$	2.0	2.4	3.7			1
2,6-Dimethyl-4-heptanone (Isovalerone)	CCl_4	0.9	2.1				1
2,5-Dimethylhexane	CCl_4	0.9	1.4				1
Dimethyl maleate	CCl_4	3.7	6.2				1
2,2-Dimethylpentane	$CDCl_3$	0.9	0.9	1.2			1
2,4-Dimethylpentane	CCl_4	0.9	1.1	1.6			1
2,4-Dimethyl-3-pentanone (Diisopropyl ketone)	CCl_4	1.0	2.6				1
2,4-Dimethylpyridine (2,4-Lutidine)	$CDCl_3$	2.3	2.5	7.0	7.0	8.4	1
2,6-Dimethylpyridine (2,6-Lutidine)	$CDCl_3$	2.51	6.93	7.42			2
Dimethyl sulfoxide	$CDCl_3$	2.62					2
1,4-Dioxane	$CDCl_3$	3.69					2
1,3-Dioxolane	$CDCl_3$	3.88	4.90				2
Dipentyl ether (Amyl ether)	$CDCl_3$	0.9	1.4	3.4			1
Dipropylamine	$CDCl_3$	1.5	2.6				1
Dodecane	CCl_4	0.9	1.3				1
1-Dodecene	CCl_4	0.9	1.3	2.0	5.4		1
Epichlorohydrin	CCl_4	2.6	2.8	3.2	3.5	3.6	1
1,2-Epoxybutane (Ethyloxirane)	CCl_4	1.0	1.5	2.3	2.6	2.7	1
1,2-Ethanediamine	CCl_4	1.2	2.6				1
1,2-Ethanediol (Ethylene glycol)	D_2O	3.7					1

Compound	Solvent	HNMR shifts (ppm relative to TMS)						Ref.
1,2-Ethanediol, diacetate (Ethylene glycol diacetate)	CCl_4	2.0	4.2					1
Ethanol	$CDCl_3$	1.23	2.61	3.69				2
Ethanolamine	$CDCl_3$	2.7	2.8	3.5				1
Ethoxybenzene (Phenetole)	CCl_4	1.3	3.9	6.9				1
2-Ethoxyethanol (Ethylene glycol monoethyl ether (Cellosolve)	$CDCl_3$	1.22	2.70	3.55	3.72			2
2-Ethoxyethyl acetate (Ethylene glycol monoethyl ether acetate)	CCl_4	1.2	2.0	3.4	3.5	4.1		1
Ethyl acetate	$CDCl_3$	1.26	2.04	4.12				2
Ethyl acetoacetate	$CDCl_3$	1.3	1.9	2.2	3.3	4.1	4.9	1
Ethyl acrylate (Ethyl propenoate)	CCl_4	1.3	4.1	5.7	6.1	6.3		1
Ethylamine	D_2O	1.1	2.6					1
Ethylbenzene	$CDCl_3$	1.3	2.7	7.2				1
Ethyl benzoate	CCl_4	1.3	4.3	7.4	8.0			1
Ethyl butanoate	CCl_4	0.9	1.2	1.7	2.2	4.1		1
Ethyl cyanoacetate	CCl_4	1.3	3.4	4.3				1
Ethylcyclohexane	$CDCl_3$	0.9	1.9	1.4				1
Ethylene carbonate	$CDCl_3$	4.5						1
Ethyl formate	CCl_4	1.3	4.2	7.9				1
2-Ethyl-1,3-hexanediol	$CDCl_3$	1.0	1.4	3.8				1
2-Ethyl-1-hexanol	$CDCl_3$	0.9	1.3	1.8	3.5			1
Ethyl 3-methylbutanoate	$CDCl_3$	1.0	1.3	1.9	4.1			1
3-Ethyl-2-methylpentane	CCl_4	0.9	0.9	1.5				1
Fluorobenzene	CCl_4	7.0						1
2-Fluorotoluene	CCl_4	2.2	6.9					1
3-Fluorotoluene	CCl_4	2.3	6.9					1
4-Fluorotoluene	CCl_4	2.2	6.8	7.0				1
Furan	$CDCl_3$	6.38	7.44					2
Furfural	$CDCl_3$	6.6	7.3	7.7	9.7			1
Furfuryl alcohol	$CDCl_3$	2.8	4.6	6.3	7.4			1
Glycerol	D_2O	3.6						1
Glycerol triacetate (Triacetin)	$CDCl_3$	2.1	4.2	4.3	5.2			1
Heptane	$CDCl_3$	0.88	1.27	1.30				2
1-Heptanol	CCl_4	0.9	1.4	3.4	3.5			1
3-Heptanol	CCl_4	0.9	1.4	2.3	3.4			1
2-Heptanone (Methyl pentyl ketone)	CCl_4	0.9	1.3	2.0	2.3			1
3-Heptanone (Ethyl butyl ketone)	CCl_4	1.0	1.4	2.3				1
1-Heptene	CCl_4	0.9	1.4	2.0	4.9	5.7		1
Hexane	$CDCl_3$	0.89	1.27	1.29				2
Hexanedinitrile (Adiponitrile)	$CDCl_3$	1.8	2.5					1
Hexanenitrile	$CDCl_3$	0.9	1.5	2.3				1
Hexanoic acid (Caproic acid)	$CDCl_3$	0.9	1.4	2.4	11.4			1
1-Hexanol	$CDCl_3$	0.90	1.32	1.56	1.79	3.62		2
Hexyl acetate	CCl_4	0.9	1.4	2.0	4.0			1
3-Hydroxypropanenitrile (Hydracrylonitrile)	$CDCl_3$	2.6	3.4	3.9				1
Iodobenzene	CCl_4	6.8	7.5	7.7				1
1-Iodobutane (Butyl iodide)	$CDCl_3$	1.0	1.7	1.9	4.2			1
2-Iodobutane (*sec*-Butyl iodide)	$CDCl_3$	1.0	1.7	1.9	4.2			1
Iodoethane (Ethyl iodide)	$CDCl_3$	1.2	2.6	3.7				1
Iodomethane (Methyl iodide)	$CDCl_3$	2.2						1
1-Iodopropane (Propyl iodide)	CCl_4	1.0	1.8	3.2				1
2-Iodopropane (Isopropyl iodide)	$CDCl_3$	1.9	4.3					1
Isobutanal (2-Methyl-1-propanal)	CCl_4	1.1	2.4	9.6				1
Isobutyl acetate	CCl_4	0.9	1.9	2.0	3.8			1

Compound	Solvent	HNMR shifts (ppm relative to TMS)								Ref.
Isobutylbenzene	CCl_4	0.9	1.9	2.4	7.1					1
Isobutyl formate	$CDCl_3$	1.0	2.0	3.9	8.0					1
Isobutyl isobutanoate	CCl_4	0.9	1.2	1.9	2.5	3.8				1
Isopentyl acetate	$CDCl_3$	0.9	1.5	2.0	4.0					1
Isophorone	$CDCl_3$	1.04	1.95	2.19	5.88					2
Isopropyl acetate	$CDCl_3$	0.9	1.4	1.6	2.4					1
Isopropylbenzene (Cumene)	$CDCl_3$	1.3	2.4	2.9	7.3					1
1-Isopropyl-4-methylbenzene (*p*-Cymene)	$CDCl_3$	1.2	2.3	2.9	7.1					1
Isoquinoline	$CDCl_3$	8.5	9.3							1
d-Limonene (Citrene)	CCl_4	1.4	1.7	1.9	4.6	5.3				1
Mesityl oxide	$CDCl_3$	1.89	2.14	2.16	6.09					2
Methanol	$CDCl_3$	3.43	3.66							2
2-Methoxyethanol (Ethylene glycol monomethyl ether)	$CDCl_3$	2.5	3.4	3.5	3.7	3.5	3.7			1
2-Methoxyethyl acetate (Ethylene glycol monomethyl ether acetate)	$CDCl_3$	2.09	3.39	3.59	4.22					2
Methyl acetate	CCl_4	2.0	3.7							1
2-Methylacrylonitrile	$CDCl_3$	2.0	5.7	5.8						1
2-Methylaniline (*o*-Toluidine)	CCl_4	2.0	3.2	6.7						1
3-Methylaniline (*m*-Toluidine)	CCl_4	2.2	3.3	6.4	6.9					1
N-Methylaniline	CCl_4	2.7	3.3	6.4	6.6	7.1				1
Methyl benzoate	CCl_4	3.8	7.4	8.0						1
3-Methylbutanoic acid (Isovaleric acid)	CCl_4	1.0	2.2	11.0						1
3-Methyl-1-butanol (Isopentyl alcohol)	CCl_4	0.9	1.5	3.5	4.1					1
Methyl cyanoacetate	$CDCl_3$	3.5	3.8							1
Methylcyclohexane	CCl_4	0.9	1.4							1
cis-4-Methylcyclohexanol	$CDCl_3$	0.9	1.5	2.9	3.9					1
N-Methylformamide	$CDCl_3$	2.82	7.4	8.16						2
Methyl formate	$CDCl_3$	3.76	8.07							2
2-Methylheptane	CCl_4	0.9	1.3							1
4-Methylheptane	$CDCl_3$	0.8	0.9	1.4						1
2-Methylhexane	$CDCl_3$	0.9	0.9	1.4						1
5-Methyl-2-hexanone (Methyl isopentyl ketone)	CCl_4	0.9	1.4	2.0	2.3					1
Methyl methacrylate (Methyl 2-methyl-2-propenoate)	$CDCl_3$	2.0	3.8	5.6	6.1					1
2-Methyloctane	CCl_4	0.9	1.0	1.3						1
2-Methylpentane	CCl_4	0.8	0.9	1.5						1
3-Methylpentane	CCl_4	0.8	1.5							1
2-Methyl-2,4-pentanediol (Hexylene glycol)	CCl_4	1.1	1.2	1.3	1.4	1.6	4.2	4.7	4.9	1
4-Methylpentanenitrile	CCl_4	1.0	1.6	2.3						1
4-Methyl-2-pentanol	CCl_4	0.9	1.1	1.3	1.8	3.5	3.7			1
3-Methyl-3-pentanol	CCl_4	0.9	1.1	1.4	1.8					1
4-Methyl-2-pentanone (Methyl isobutyl ketone)	$CDCl_3$	0.9	2.1	2.3						1
2-Methylpropanenitrile (Isobutyronitrile)	$CDCl_3$	1.3	2.7							1
2-Methylpropanoic acid (Isobutyric acid)	CCl_4	1.2	2.6							1
2-Methyl-1-propanol (Isobutyl alcohol)	CCl_4	0.9	1.7	3.3	4.0					1
2-Methyl-2-propanol (*tert*-Butyl alcohol)	$CDCl_3$	1.3	1.4							1
2-Methylpyridine (2-Picoline)	CCl_4	2.5	6.9	7.4	8.4					1
3-Methylpyridine (3-Picoline)	CCl_4	2.3	7.0	7.4	8.4					1
N-Methyl-2-pyrrolidinone	$CDCl_3$	2.1	2.4	2.8	3.4					1
Methyl salicylate	CCl_4	3.9	6.7	6.9	7.3	7.7	10.6			1
2-Methylthiophene	$CDCl_3$	2.5	6.7	6.9	7.0					1
Morpholine	$CDCl_3$	2.59	2.86	3.67						2
Nitrobenzene	CCl_4	7.52	7.65	8.19						2
Nitroethane	$CDCl_3$	1.6	4.3							1

Compound	Solvent	HNMR shifts (ppm relative to TMS)								Ref.
Nitromethane	CCl_4	4.2								1
1-Nitropropane	$CDCl_3$	1.0	2.1	4.4						1
2-Nitropropane		1.55	4.70							3
Nonane	$CDCl_3$	0.9	1.3							1
Octane	$CDCl_3$	0.88	1.26							1
1-Octanol	$CDCl_3$	0.88	1.29	1.5	2.40	3.60				2
2-Octanone (Hexyl methyl ketone)	CCl_4	0.9	1.4	2.1	2.4					1
1-Octene	CCl_4	0.9	1.3	2.1	4.8	4.9	5.7			1
Pentane	$CDCl_3$	0.88	1.26	1.30						2
1,5-Pentanediol (Pentamethylene glycol)	D_2O	1.5	3.6							1
2,4-Pentanedione (Acetylacetone)	CCl_4	2.0	2.2	3.5	5.4	14.7				1
Pentanenitrile (Valeronitrile)	CCl_4	1.0	1.6	2.3						1
Pentanoic acid (Valeric acid)	CCl_4	0.9	1.5	2.3	11.7					1
1-Pentanol (Amyl alcohol)	CCl_4	0.9	1.4	3.5	4.4					1
2-Pentanol (*sec*-Amyl alcohol)	$CDCl_3$	0.9	1.2	1.3	2.2	3.7				1
3-Pentanol (Diethyl carbinol)	CCl_4	0.9	1.4	3.3	3.4					1
2-Pentanone (Methyl propyl ketone)	CCl_4	0.9	1.6	2.0	2.3					1
3-Pentanone (Diethyl ketone)	CCl_4	1.0	2.4							1
Pentyl acetate (Amyl acetate)	CCl_4	0.9	1.4	2.0	4.0					1
Pentylamine (Amylamine)	$CDCl_3$	0.9	1.4	2.7						1
α-Pinene	$CDCl_3$	0.84	1.16	1.27	1.66	1.93	2.19	2.33	5.19	2
Piperidine	$CDCl_3$	1.53	2.18	2.79						2
Propanal	$CDCl_3$	1.1	2.4	9.8						1
1,2-Propanediol (1,2-Propylene glycol)	$CDCl_3$	1.1	3.4	3.9	4.3					1
1,3-Propanediol (Trimethylene glycol)	D_2O	1.8	3.7							1
Propanenitrile	CCl_4	1.3	2.3							1
Propanoic acid	$CDCl_3$	1.1	2.4	10.5						1
Propanoic anhydride	CCl_4	1.2	2.4							1
1-Propanol (Propyl alcohol)	$CDCl_3$	0.9	1.6	2.3	3.6					1
2-Propanol (Isopropyl alcohol)	$CDCl_3$	1.2	1.6	4.0						1
Propargyl alcohol (3-Hydroxy-1-propyne)	$CDCl_3$	2.5	2.8	4.3						1
Propyl acetate	CCl_4	0.9	1.6	2.0	4.0					1
Propylamine	CCl_4	0.9	1.5	2.1	2.7					1
Propylbenzene	$CDCl_3$	0.9	1.6	2.6						1
Propyl formate	CCl_4	1.0	1.7	4.1	7.9					1
Pyridine	$CDCl_3$	7.23	7.62	8.59						2
Pyrrole	$CDCl_3$	6.2	6.7	8.0						1
Pyrrolidine	C_6H_6	1.5	2.4	2.7						1
2-Pyrrolidone	$CDCl_3$	2.2	3.4	7.7						1
Quinoline	CCl_4	7.1	7.8	8.8						1
Salicylaldehyde (2-Hydroxybenzaldehyde)	$CDCl_3$	7.0	7.4	9.8	11.0					1
Styrene	CCl_4	5.1	5.6	6.6	7.2					1
Sulfolane	$CDCl_3$	2.2	3.0							1
α-Terpinene	$CDCl_3$	1.02	1.77	2.09	2.28	5.59	5.62			2
1,1,1,2-Tetrachloroethane	$CDCl_3$	4.29								2
1,1,2,2-Tetrachloroethane	$CDCl_3$	5.91								2
Tetraethylene glycol	CCl_4	3.5	3.6							1
Tetrahydrofuran	$CDCl_3$	1.84	3.73							2
1,2,3,4-Tetrahydronaphthalene	$CDCl_3$	1.8	2.8	7.1						1
Tetrahydropyran	CCl_4	1.6	3.6							1
Tetrahydrothiophene	$CDCl_3$	1.9	2.8							1
Tetramethylsilane	CCl_4	0.0								1

Compound	Solvent	HNMR shifts (ppm relative to TMS)						Ref.
Tetramethylurea	CCl_4	2.8						1
Thiophene	$CDCl_3$	7.1	7.3					1
Toluene	$CDCl_3$	2.34	7.18					2
Tribromomethane (Bromoform)	CCl_4	6.8						1
Tributylamine	CCl_4	0.9	1.3	2.3				1
1,1,1-Trichloroethane	CCl_4	2.7						1
1,1,2-Trichloroethane	$CDCl_3$	4.0	5.8					1
Trichloroethene	CCl_4	6.5						1
Trichloroethylsilane	CCl_4	1.3						1
Trichloromethane (Chloroform)	CCl_4	7.2						1
(Trichloromethyl)benzene (Benzotrichloride)	CCl_4	7.3	7.8					1
Tridecane	CCl_4	0.9	1.3					1
Triethanolamine	D_2O	2.7	3.6					1
Triethylamine	CCl_4	1.0	2.4					1
Triethylene glycol	$CDCl_3$	3.5	3.7					1
Triethyl phosphate	$CDCl_3$	1.4	4.1					1
2,2,2-Trifluoroethanol	$CDCl_3$	3.4	3.9					1
(Trifluoromethyl)benzene (Benzotrifluoride)	CCl_4	7.5						1
Trimethylamine	CCl_4	2.12						2
1,2,3-Trimethylbenzene (Hemimellitene)	$CDCl_3$	2.2	2.3	7.0				1
1,2,4-Trimethylbenzene (Pseudocumene)	CCl_4	2.2	6.8					1
1,3,5-Trimethylbenzene (Mesitylene)	$CDCl_3$	2.3	6.8					1
2,2,3-Trimethylbutane (Triptane)	CCl_4	0.8	1.3					1
2,2,5-Trimethylhexane	CCl_4	0.9	1.2					1
2,3,3-Trimethylpentane	CCl_4	0.8	0.8	1.4				1
2,3,4-Trimethylpentane	CCl_4	0.8	1.9					1
Trimethyl phosphate	$CDCl_3$	3.78						2
2,4,6-Trimethylpyridine (2,4,6-Collidine)	CCl_4	2.2	2.4	6.6				1
1-Undecene	CCl_4	0.9	1.3	2.0	4.8	4.9	5.6	1
Vinyl acetate	$CDCl_3$	2.1	4.6	4.9	7.3			1
o-Xylene	$CDCl_3$	2.22	7.07					2
m-Xylene	$CDCl_3$	2.28	6.95	7.11				2
p-Xylene	$CDCl_3$	2.30	7.05					2

Section 9
Molecular Structure and Spectroscopy

Bond Lengths in Crystalline Organic Compounds 9-1
Bond Lengths in Organometallic Compounds 9-17
Structure of Free Molecules in the Gas Phase 9-19
Characteristic Bond Lengths in Free Molecules 9-48
Atomic Radii of the Elements 9-49
Dipole Moments 9-51
Hindered Internal Rotation 9-60
Bond Dissociation Energies 9-65
Electronegativity 9-99
Force Constants for Bond Stretching 9-100
Fundamental Vibrational Frequencies of Small Molecules 9-101
Spectroscopic Constants of Diatomic Molecules 9-104
Infrared Correlation Charts 9-110
Nuclear Spins, Moments, and Other Data Related to NMR Spectroscopy 9-115
Proton NMR Chemical Shifts for Characteristic Organic Structures 9-118
^{13}C-NMR Absorptions of Major Functional Groups 9-119

BOND LENGTHS IN CRYSTALLINE ORGANIC COMPOUNDS

The following table gives average interatomic distances for bonds between the elements H, B, C, N, O, F, Si, P, S, Cl, As, Se, Br, Te, and I as determined from X-ray and neutron diffraction measurements on organic crystals. The table has been derived from an analysis of high-precision structure data on about 10,000 crystals contained in the 1985 version of the Cambridge Structural Database, which is maintained by the Cambridge Crystallographic Data Center. The explanation of the columns is:

Column 1:	Specification of elements in the bond, with coordination number given in parentheses, and bond type (single, double, etc.). For carbon, the hybridization state is given.
Column 2:	Substructure in which the bond is found. The target bond is set in boldface. Where X is not specified, it denotes any element type. C# indicates any sp^3 carbon atom, and C* denotes an sp^3 carbon whose bonds, in addition to those specified in the linear formulation, are to C and H atoms only.
Column 3:	*d* is the unweighted mean in Å units of all the values for that bond length found in the sample.
Column 4:	*m* is the median in Å units of all values.
Column 5:	σ is the standard deviation in the sample.
Column 6:	q_1 is the lower quartile for the sample (i.e., 25% of values are less than q_1 and 75% exceed it).
Column 7:	q_u is the upper quartile for the sample.
Column 8:	*n* is number of observations in the sample.
Column 9:	Notes refer to the footnotes in Appendix 1.

References to special cases are given in a shorthand form and listed in Appendix 2. Further information on the method of analysis of the data may be found in the reference cited below.

The table is reprinted with permission of the authors, the Royal Society of Chemistry, and the International Union of Crystallography.

Reference

Frank H. Allen, Olga Kennard, David G. Watson, Lee Brammer, A. Guy Orpen, and Robin Taylor, *J. Chem. Soc. Perkin Trans.* II, S1–S19, 1987.

Bond	Substructure	*d*	*m*	σ	q_1	q_u	*n*	Note
As(3)–As(3)	X_2–**As–As**–X_2	2.459	2.457	0.011	2.456	2.466	8	
As–B	see CUDLOC (2.065), CUDLUI (2.041)							
As–BR	see CUDDEE, CUDDII (2.340–3.203)							
As(4)–C	**X_3–As–CH_3**	1.903	1.907	0.016	1.893	1.916	12	
	$(X)_2$(C,O,S=)**As–C***sp*^{3}	1.927	1.929	0.017	1.921	1.937	16	
	As–C*ar* in Ph_4As^+	1.905	1.909	0.012	1.897	1.912	108	
	$(X)_2$(C,O,S=)**As–C***ar*	1.922	1.927	0.016	1.908	1.934	36	
As(3)–C	X_2–**As–C***sp*^{3}	1.963	1.965	0.017	1.948	1.978	6	
	X_2–**As–C***ar*	1.956	1.956	0.015	1.944	1.964	41	
As(3)–Cl	X_2–**As–Cl**	2.268	2.256	0.039	2.247	2.281	10	
As(6)–F	in **AsF_6^-**	1.678	1.676	0.020	1.659	1.695	36	
As(3)–I	see OPIMAS (2.579, 2.590)							
As(3)–N(3)	X_2–**As–N**–X_2	1.858	1.858	0.029	1.839	1.873	19	
As(4)=N(2)	see TPASSN (1.837)							
As(4)–O	$(X)_2$(O=)**As–OH**	1.710	1.712	0.017	1.695	1.726	6	
As(3)–O	see ASAZOC, PHASOC01 (1.787–1.845)							
As(4)=O	X_3–**As=O**	1.661	1.661	0.016	1.652	1.667	9	
As(3)=P(3)	see BELNIP (2.350, 2.362)							†
As(3)–P(3)	see BUTHAZ10 (2.124)							†
As(3)–S	X_2–**As–S**	2.275	2.266	0.032	2.247	2.298	14	
As(4)=S	X_3–**As=S**	2.083	2.082	0.004	2.080	2.086	9	
As(3)–Se(2)	see COSDIX, ESEARS (2.355–2.401)							†
As(3)–Si(4)	see BICGEZ, MESIAD (2.351–2.365)							†
As(3)–Te(2)	see ETEARS (2.571, 2.576)							†
B(n)–B(n)	*n* = 5–7 in boron cages	1.775	1.773	0.031	1.763	1.786	688	
B(4)–B(4)	see CETTAW (2.041)							
B(4)–B(3)	see COFVOI (1.698)							
B(3)–B(3)	X_2–**B–B**–X_2	1.701	1.700	0.014	1.691	1.712	8	
B(6)–BR		1.967	1.971	0.014	1.954	1.979	7	†
B(4)–BR		2.017	2.008	0.031	1.990	2.044	15	†
B(*n*)–C	*n* = 5–7: **B–C** in cages	1.716	1.717	0.020	1.707	1.728	96	
	n = 3–4: **B–C***sp*^{3} not cages	1.597	1.599	0.022	1.585	1.611	29	1
	n = 4: **B–C***ar*	1.606	1.607	0.012	1.596	1.615	41	
	n = 4: **B–C***ar* in Ph_4B^-	1.643	1.643	0.006	1.641	1.645	16	
B(*n*)–C	*n* = 3: **B–C***ar*	1.556	1.552	0.015	1.546	1.566	24	
B(*n*)–Cl	**B**(5)–**Cl** and **B**(3)–**Cl**	1.751	1.751	0.011	1.743	1.761	14	

Bond	Substructure	d	m	σ	q_l	q_u	n	Note
	B(4)–Cl	1.833	1.833	0.013	1.821	1.843	22	
B(4)–F	**B–F** (B neutral)	1.366	1.368	0.017	1.356	1.375	25	
	$\mathbf{B^-}$**–F** in BF_4^-	1.365	1.372	0.029	1.352	1.390	84	
B(4)–I	see TMPBTI (2.220, 2.253)							
B(4)–N(3)	X_3–**B–N**(=C)(X)	1.611	1.617	0.013	1.601	1.625	8	
	in pyrazaboles	1.549	1.552	0.015	1.536	1.560	10	
B(3)–N(3)	X_2–**B–N**–C_2: all coplanar	1.404	1.404	0.014	1.389	1.408	40	2
	for τ(BN) > 30° see BOGSUL, BUSHAY, CILRUK (1.434–1.530)							
	S_2–**B–N**–X_2	1.447	1.443	0.013	1.435	1.470	14	
B(4)–O	$\mathbf{B^-}$**–O** in BO_4^-	1.468	1.468	0.022	1.453	1.479	24	
	for neutral B–O see Note 3							3
B(3)–O(2)	X_2–**B–O**–X	1.367	1.367	0.024	1.349	1.382	35	
B(n)–P	n = 4: **B–P**	1.922	1.927	0.027	1.900	1.954	10	
	n = 3: see BUPSIB10 (1.892, 1.893)							
B(4)–S	**B(4)–S(3)**	1.930	1.927	0.009	1.925	1.934	10	
	B(4)–S(2)	1.896	1.896	0.004	1.893	1.899	6	
B(3)–S	N–**B**–S_2	1.806	1.806	0.010	1.799	1.816	28	
	(=X–)(N–)**B–S**	1.851	1.854	0.013	1.842	1.859	10	
Br–Br	see BEPZEB, TPASTB	2.542	2.548	0.015	2.526	2.551	4	
Br–C	**Br–C***	1.966	1.967	0.029	1.951	1.983	100	4
	Br–C*sp*³ (cyclopropane)	1.910	1.910	0.010	1.900	1.914	8	
	Br–C*sp*²	1.883	1.881	0.015	1.874	1.894	31	4
	Br–C*ar* (mono-Br + *m.p*-Br_2)	1.899	1.899	0.012	1.892	1.906	119	4
	Br–C*ar* (o-Br_2)	1.875	1.872	0.011	1.864	1.884	8	4
⁻Br(2)–Cl	see TEACBR (2.362–2.402)							†
Br–I	see DTHIBR10 (2.646), TPHOSI (2.695)							
Br–N	see NBBZAM (1.843)							
Br–O	see CIYFOF	1.581	1.581	0.007	1.574	1.587	4	
Br–P	see CISTED (2.366)							
Br–S(2)	see BEMLIO (2.206)							†
Br–S(3)	see CIWYIQ (2.435, 2.453)							†
Br–S(3)⁺	see THINBR (2.321)							†
Br–SE	see CIFZUM (2.508, 2.619)							
Br–Si	see BIZJAV (2.284)							
Br–Te	In $\mathbf{Br_6Te^{2-}}$ see CUGBAH (2.692–2.716)							
	Br–Te(4) see BETUTE10 (3.079, 3.015)							
	Br–Te(3) see BTUPTE (2.835)							
C*sp*³–C*sp*³	C#–$\mathbf{CH_2}$–$\mathbf{CH_3}$	1.513	1.514	0.014	1.507	1.523	192	
	$(C\#)_2$–**CH**–$\mathbf{CH_3}$	1.524	1.526	0.015	1.518	1.534	226	
	$(C\#)_3$–**C**–$\mathbf{CH_3}$	1.534	1.534	0.011	1.527	1.541	825	
	C#–$\mathbf{CH_2}$–$\mathbf{CH_2}$–C#	1.524	1.524	0.014	1.516	1.532	2459	
	$(C\#)_2$–**CH**–$\mathbf{CH_2}$–C#	1.531	1.531	0.012	1.524	1.538	1217	
	$(C\#)_3$–**C**–$\mathbf{CH_2}$–C#	1.538	1.539	0.010	1.533	1.544	330	
	$(C\#)_2$–**CH–CH**–$(C\#)_2$	1.542	1.542	0.011	1.536	1.549	321	
	$(C\#)_3$–**C–CH**–$(C\#)_2$	1.556	1.556	0.011	1.549	1.562	215	
	$(C\#)_3$–**C–C**–$(C\#)_3$	1.588	1.580	0.025	1.566	1.610	21	
	C*–C* (overall)	1.530	1.530	0.015	1.521	1.539	5777	5,6
	in cyclopropane (any subst.)	1.510	1.509	0.026	1.497	1.523	888	7
	in cyclobutane (any subst.)	1.554	1.553	0.021	1.540	1.567	679	8
	in cyclopentane (C,H-subst.)	1.543	1.543	0.018	1.532	1.554	1641	
	in cyclohexane (C,H-subst.)	1.535	1.535	0.016	1.525	1.545	2814	
	cyclopropyl-C* (exocyclic)	1.518	1.518	0.019	1.505	1.531	366	7
	cyclobutyl-C* (exocyclic)	1.529	1.529	0.016	1.519	1.539	376	8
	cyclopentyl-C* (exocyclic)	1.540	1.541	0.017	1.527	1.549	956	
	cyclohexyl-C* (exocyclic)	1.539	1.538	0.016	1.529	1.549	2682	
	in cyclobutene (any subst.)	1.573	1.574	0.017	1.566	1.586	25	8
	in cyclopentene (C,H-subst.)	1.541	1.539	0.015	1.532	1.549	208	
	in cyclohexene (C,H-subst.)	1.541	1.541	0.020	1.528	1.554	586	
	in oxirane (epoxide)	1.466	1.466	0.015	1.458	1.474	249	9
	in aziridine	1.480	1.481	0.021	1.465	1.496	67	9

Bond	Substructure	*d*	*m*	σ	q_1	q_u	*n*	Note
	in oxetane	1.541	1.541	0.019	1.527	1.557	16	
	in azetidine	1.548	1.543	0.018	1.536	1.558	22	
	oxiranyl-C* (exocyclic)	1.509	1.507	0.018	1.497	1.519	333	9
	aziridinyl-C* (exocyclic)	1.512	1.512	0.018	1.496	1.526	13	9
C*sp*^{3}–C*sp*^{2}	**CH$_3$**–C=C	1.503	1.504	0.011	1.497	1.509	215	
	C#–**CH$_2$**–C=C	1.502	1.502	0.013	1.494	1.510	483	
	(C#)$_2$–**CH**–C=C	1.510	1.510	0.014	1.501	1.518	564	
	(C#)$_3$–**C**–C=C	1.522	1.522	0.016	1.511	1.533	193	
C*sp*^{3}–C*sp*^{2}	C*–C=C (overall)	1.507	1.507	0.015	1.499	1.517	1456	5
	C*–C=C (endocyclic)							
	in cyclopropene	1.509	1.508	0.016	1.500	1.516	20	10
	in cyclobutene	1.513	1.512	0.018	1.500	1.525	50	8
	in cyclopentene	1.512	1.512	0.014	1.502	1.521	208	
	in cyclohexene	1.506	1.505	0.016	1.495	1.516	391	
	in cyclopentadiene	1.502	1.503	0.019	1.490	1.515	18	
	in cyclohexa-1,3-diene	1.504	1.504	0.017	1.491	1.517	56	
	C*–C=C (exocyclic):							
	cyclopropenyl-C*	1.478	1.475	0.012	1.470	1.485	7	10
	cyclobutenyl-C*	1.489	1.483	0.015	1.479	1.496	11	8
	cyclopentenyl-C*	1.504	1.506	0.012	1.495	1.512	115	
	cyclohexenyl-C*	1.511	1.511	0.013	1.502	1.519	292	
	C*CH=O in aldehydes	1.510	1.510	0.008	1.501	1.518	7	
	(C*)$_2$–C=O							
	in ketones	1.511	1.511	0.015	1.501	1.521	952	11
	in cyclobutanone	1.529	1.530	0.016	1.514	1.545	18	
	in cyclopentanone	1.514	1.514	0.016	1.505	1.523	312	
	acyclic and 6 + rings	1.509	1.509	0.016	1.499	1.519	626	
	C*–COOH in carboxylic acids	1.502	1.502	0.014	1.495	1.510	176	
	C*–COO$^-$ in carboxylate anions	1.520	1.521	0.011	1.516	1.528	57	
	C*–C(=O)(–OC*)							
	in acyclic esters	1.497	1.496	0.018	1.484	1.509	553	12
	in β-lactones	1.519	1.519	0.020	1.500	1.538	4	13
	in γ-lactones	1.512	1.512	0.015	1.501	1.521	110	12
	in δ-lactones	1.504	1.502	0.013	1.495	1.517	27	12
	cyclopropyl **(C)–C=O** in ketones, acids and esters	1.486	1.485	0.018	1.474	1.497	105	7
	C*–C(=O)(–NH$_2$) in acyclic amides	1.514	1.512	0.016	1.506	1.526	32	14
	C*–C(=O)(–NHC*) in acyclic amides	1.506	1.505	0.012	1.498	1.515	78	14
	C*–C(=O)[–N(C*)$_2$] in acyclic amides	1.505	1.505	0.011	1.496	1.517	15	14
C*sp*^{3}–C*ar*	**CH$_3$**–C*ar*	1.506	1.507	0.011	1.501	1.513	454	
	C#–**CH$_2$**–C*ar*	1.510	1.510	0.009	1.505	1.516	674	
	(C#)$_2$–**CH**–C*ar*	1.515	1.515	0.011	1.508	1.522	363	
	(C#)$_3$–**C**–C*ar*	1.527	1.530	0.016	1.517	1.539	308	
	C*–C*ar* (overall)	1.513	1.513	0.014	1.505	1.521	1813	
	cyclopropyl **(C)–C*ar***	1.490	1.490	0.015	1.479	1.503	90	7
C*sp*^{3}–C*sp*^{1}	**C*–C≡C**	1.466	1.465	0.010	1.460	1.469	21	15
	C#–C≡C	1.472	1.472	0.012	1.464	1.481	88	15
	C*–C≡N	1.470	1.469	0.013	1.463	1.479	106	7b
	cyclopropyl **(C)–C≡N**	1.444	1.447	0.010	1.436	1.451	38	7
C*sp*^{2}–C*sp*^{2}	C=C–C=C							
	(conjugated)	1.455	1.455	0.011	1.447	1.463	30	16,18
	(unconjugated)	1.478	1.476	0.012	1.470	1.479	8	17,18
	(overall)	1.460	1.460	0.015	1.450	1.470	38	
	C=C–C=C–C=C	1.443	1.445	0.013	1.431	1.454	29	18
	C=C–C=C (endocyclic in TCNQ)	1.432	1.433	0.012	1.424	1.441	280	19
	C=C–C(=O)(–C*)							
	(conjugated)	1.464	1.462	0.018	1.453	1.476	211	16,18
	(unconjugated)	1.484	1.486	0.017	1.475	1.497	14	17,18
	(overall)	1.465	1.462	0.018	1.453	1.478	226	
	C=C–C(=O)–C=C							
	in benzoquinone (C,H-subst. only)	1.478	1.476	0.011	1.469	1.488	28	
	in benzoquinone (any subst.)	1.478	1.478	0.031	1.464	1.498	172	

Bond	Substructure	*d*	*m*	*σ*	q_1	q_u	*n*	Note
	non-quinonoid	1.456	1.455	0.012	1.447	1.464	28	
	C=C–COOH	1.475	1.476	0.015	1.461	1.488	22	
	C=C–COOC*	1.488	1.489	0.014	1.478	1.497	113	
	C=C–COO$^-$	1.502	1.499	0.017	1.488	1.510	11	
	HOOC–COOH	1.538	1.537	0.007	1.535	1.541	9	
	HOOC–COO$^-$	1.549	1.552	0.009	1.546	1.553	13	
	$^-$OOC–COO$^-$	1.564	1.559	0.022	1.554	1.568	9	
	formal Csp^2–Csp^2 single bond in selected non-fused heterocycles:							
	in 1*H*-pyrrole (C3–C4)	1.412	1.410	0.016	1.401	1.427	29	
	in furan (C3–C4)	1.423	1.423	0.016	1.412	1.433	62	
	in thiophene (C3–C4)	1.424	1.425	0.015	1.415	1.433	40	
	in pyrazole (C3–C4)	1.410	1.412	0.016	1.400	1.418	20	
	in isoxazole (C3–C4)	1.425	1.425	0.016	1.413	1.438	9	
	in furazan (C3–C4)	1.428	1.427	0.007	1.422	1.435	6	
	in furoxan (C3–C4)	1.417	1.417	0.006	1.412	1.422	14	
Csp^2–*Car*	C=C–*Car*							
	(conjugated)	1.470	1.470	0.015	1.463	1.480	37	16,18
Csp^2–*Car*		1.488	1.490	0.012	1.480	1.496	87	17,18
	(overall)	1.483	1.483	0.015	1.472	1.494	124	
	cyclopropenyl (C=C)–*Car*	1.447	1.448	0.006	1.441	1.452	8	10
	Car–C(=O)–C*	1.488	1.489	0.016	1.478	1.500	84	
	Car–C(=O)–*Car*	1.480	1.481	0.017	1.468	1.494	58	
	Car–COOH	1.484	1.485	0.014	1.474	1.491	75	
	Car–C(=O)(–OC*)	1.487	1.487	0.012	1.480	1.494	218	
	Car–COO$^-$	1.504	1.509	0.014	1.495	1.512	26	
	Car–C(–O)–NH_2	1.500	1.503	0.020	1.498	1.510	19	
	Car–C=N–C#							
	(conjugated)	1.476	1.478	0.014	1.466	1.486	27	16
	(unconjugated)	1.491	1.490	0.008	1.485	1.496	48	17
	(overall)	1.485	1.487	0.013	1.481	1.493	75	
	in indole (C3–C3a)	1.434	1.434	0.011	1.428	1.439	40	
Csp^2–Csp^1	C=C–C≡C	1.431	1.427	0.014	1.425	1.441	11	7b
	C=C–C≡N in TCNQ	1.427	1.427	0.010	1.420	1.433	280	19
Car–*Car*	in biphenyls (*ortho* subst. all **H**)	1.487	1.488	0.007	1.484	1.493	30	
	(≥1 non-**H** *ortho*-subst.)	1.490	1.491	0.010	1.486	1.495	212	
Car–Csp^1	*Car*–C≡C	1.434	1.436	0.006	1.430	1.437	37	
	Car–C≡N	1.443	1.444	0.008	1.436	1.448	31	
Csp^1–Csp^1	C≡C–C=C	1.377	1.378	0.012	1.374	1.384	21	
Csp^2=Csp^2	C*–CH=CH_2	1.299	1.300	0.027	1.280	1.311	42	
	$(C^*)_2$–C=CH_2	1.321	1.321	0.013	1.313	1.328	77	
	C*–CH=CH–C*							
	(*cis*)	1.317	1.318	0.013	1.310	1.323	106	
	(*trans*)	1.312	1.311	0.011	1.304	1.320	19	
	(overall)	1.316	1.317	0.015	1.309	1.323	127	
	$(C^*)_2$–C=CH–C*	1.326	1.328	0.011	1.319	1.334	168	
	$(C^*_2$–C=C–$(C^*)_2$	1.331	1.330	0.009	1.326	1.334	89	
	$(C^*,H)_2$–C=C–$(C^*,H)_2$ (overall)	1.322	1.323	0.014	1.315	1.331	493	5
	in cyclopropene (any subst.)	1.294	1.288	0.017	1.284	1.302	10	10
	in cyclobutene (any subst.)	1.335	1.335	0.019	1.324	1.347	25	8
	in cyclopentene (C,H-subst.)	1.323	1.324	0.013	1.314	1.331	104	
	in cyclohexene (C,H-subst.)	1.326	1.325	0.012	1.318	1.334	196	
	C=C=C (allenes, any subst.)	1.307	1.307	0.005	1.303	1.310	18	
	C=C–C=C (C,H subst., conjugated)	1.330	1.330	0.014	1.322	1.338	76	16
	C=C–C=C–C=C (C,H subst., conjugated)	1.345	1.345	0.012	1.337	1.350	58	16
	C=C–*Car* (C,H subst., conjugated)	1.339	1.340	0.011	1.334	1.346	124	16
	C=C in cyclopenta-1,3-diene (any subst.)	1.341	1.341	0.017	1.328	1.356	18	
	C=C in cyclohexa-1,3-diene (any subst.)	1.332	1.332	0.013	1.323	1.341	56	
	in C=C–C=O							
	(C,H subst., conjugated)	1.340	1.340	0.013	1.332	1.348	211	16,18
	(C,H subst., unconjugated)	1.331	1.330	0.008	1.326	1.339	14	17,18

Bond	Substructure	*d*	*m*	*σ*	q_1	q_u	*n*	Note
	(C,**H** subst., overall)	1.340	1.339	0.013	1.332	1.348	226	
	in cyclohexa-2,5-dien-1-ones	1.329	1.327	0.011	1.321	1.335	28	
	in *p*-benzoquinones							
	(C*,**H** subst.)	1.333	1.337	0.011	1.325	1.338	14	
	(any subst.)	1.349	1.339	0.030	1.330	1.364	86	
	in TCNQ							
	(endocyclic)	1.352	1.353	0.010	1.345	1.358	142	19
	(exocyclic)	1.392	1.391	0.017	1.379	1.405	139	19
	C=**C**–OH in enol tautomers	1.362	1.360	0.020	1.349	1.370	54	
	in heterocycles (any subst.):							
	1*H*-pyrrole (C2–C3, C4–C5)	1.375	1.377	0.018	1.361	1.388	58	
	furan (C2–C3, C4–C5)	1.341	1.342	0.021	1.329	1.351	125	
	thiophene (C2–C3, C4–C5)	1.362	1.359	0.025	1.346	1.377	60	
	pyrazole (C4–C5)	1.369	1.372	0.019	1.362	1.383	20	
	imidazole (C4–C5)	1.360	1.361	0.014	1.352	1.367	44	
	isoxazole (C4–C5)	1.341	1.336	0.012	1.331	1.355	9	
	indole (C2–C3)	1.361	1.363	0.012	1.355	1.371	40	
Car ≃ *Car*	in phenyl rings with C*, **H** subst. only							
	H–**C** ≃ **C**–H	1.380	1.381	0.013	1.372	1.388	2191	
	C*–**C** ≃ **C**–H	1.387	1.388	0.010	1.382	1.393	891	
	C*–**C** ≃ **C**–C*	1.397	1.397	0.009	1.392	1.403	182	
	C ≃ **C** (overall)	1.384	1.384	0.013	1.375	1.391	3264	
	F–**C** ≃ **C**–F	1.372	1.374	0.011	1.366	1.380	84	4
	Cl–**C** ≃ **C**–Cl	1.388	1.389	0.014	1.380	1.398	152	4
	in naphthalene (D_{2h}, any subst.)							
	C1–C2	1.364	1.364	0.014	1.356	1.373	440	
	C2–C3	1.406	1.406	0.014	1.397	1.415	218	
	C1–C8a	1.420	1.419	0.012	1.412	1.426	440	
	C4a–C8a	1.422	1.424	0.011	1.417	1.429	109	
Car ≃ *Car*	in anthracene (D_{2h}, any subst.)							
	C1–C2	1.356	1.356	0.009	1.350	1.360	56	
	C2–C3	1.410	1.410	0.010	1.401	1.416	34	
	C1–C9a	1.430	1.430	0.006	1.426	1.434	56	
	C4a–C9a	1.435	1.436	0.007	1.429	1.440	34	
	C9–C9a	1.400	1.402	0.009	1.395	1.406	68	
	in pyridine (C,H subst.)	1.379	1.381	0.012	1.371	1.387	276	20
	(any subst.)	1.380	1.380	0.015	1.371	1.389	537	20
	in pyridinium cation							
	(N^+ –H; C,H subst. on C)							
	C2–C3	1.373	1.375	0.012	1.368	1.380	30	
	C3–C4	1.379	1.380	0.011	1.371	1.388	30	
	(N^+ –X; C,H subst. on C)							
	C2–C3	1.373	1.372	0.019	1.362	1.382	151	
	C3–C4	1.383	1.385	0.019	1.372	1.394	151	
	in pyrazine (H subst. on C)	1.379	1.377	0.010	1.370	1.388	10	
	(any subst. on C)	1.405	1.405	0.024	1.388	1.420	60	
	in pyrimidine (C,H subst. on C)	1.387	1.389	0.018	1.379	1.400	28	
Csp^1≡Csp^1	X–**C**≡**C**–X	1.183	1.183	0.014	1.174	1.193	119	15
	C,H–**C**≡**C**–C,H	1.181	1.181	0.014	1.173	1.192	104	15
	in **C**≡**C**–C(sp^2,*ar*)	1.189	1.193	0.010	1.181	1.195	38	15
	in **C**≡**C**–C≡C	1.192	1.192	0.010	1.187	1.197	42	15
	in **CH**≡**C**–C#	1.174	1.174	0.011	1.167	1.180	42	15
Csp^3–Cl	Omitting 1,2-dichlorides:							
	C–**CH$_2$**–**Cl**	1.790	1.790	0.007	1.783	1.795	13	4
	C_2–**CH**–**Cl**	1.803	1.802	0.003	1.800	1.807	8	4
	C_3–**C**–**Cl**	1.849	1.856	0.011	1.837	1.858	5	4
	X–CH_2–Cl (X = C,H,N,O)	1.790	1.791	0.011	1.783	1.797	37	4
	X_2–CH–Cl (X = C,H,N,O)	1.805	1.803	0.014	1.800	1.812	26	4
	X_3–**C**–**Cl** (X = C,H,N,O)	1.843	1.838	0.014	1.835	1.858	7	4
	X_2–**C**–**Cl$_2$** (X = C,H,N,O)	1.779	1.776	0.015	1.769	1.790	18	4

Bond	Substructure	d	m	σ	q_l	q_u	n	Note
	X–C–Cl_3 (X = C,H,N,O)	1.768	1.765	0.011	1.761	1.776	33	4
	Cl–CH(–C)–CH(–C)–Cl	1.793	1.793	0.013	1.786	1.800	66	4
	Cl–C(–C_2)–C(–C_2)–Cl	1.762	1.760	0.010	1.757	1.765	54	4
	cyclopropyl–Cl	1.755	1.756	0.011	1.749	1.763	64	
*Csp*²–Cl	C=C–Cl (C,H,N,O subst. on C)	1.734	1.729	0.019	1.719	1.748	63	4
	C=C–Cl_2 (C,H,N,O subst. on C)	1.720	1.716	0.013	1.708	1.729	20	4
	Cl–C=C–Cl	1.713	1.711	0.011	1.705	1.720	80	4
Car–Cl	*Car*–Cl (mono–Cl + m,p-Cl_2)	1.739	1.741	0.010	1.734	1.745	340	4
	Car–Cl (*o*–Cl_2)	1.720	1.720	0.010	1.713	1.717	364	4
*Csp*¹Cl	see HCLENE10 (1.634, 1.646)							
*Csp*³–F	Omitting 1,2-difluorides							
	C–CH_2–F and C_2–CH–F	1.399	1.399	0.017	1.389	1.408	25	4
	C_3–C–F	1.428	1.431	0.009	1.421	1.435	11	4
	(C*,H)$_2$–C–F_2	1.349	1.347	0.012	1.342	1.356	58	4
	C*–C–F_3	1.336	1.334	0.007	1.330	1.344	12	4
	F–C*–C*–F	1.371	1.374	0.007	1.362	1.375	26	4
	X_3–C–F (X = C,H,N,O)	1.386	1.389	0.033	1.373	1.408	70	4
	X_2–C–F_2 (X = C,H,N,O)	1.351	1.349	0.013	1.342	1.356	58	4
	X–C–F_3 (X = C,H,N,O)	1.322	1.323	0.015	1.314	1.332	309	4
	F–C(–X)$_2$–C(–X)$_2$–F (X = C,H,N,O)	1.373	1.374	0.009	1.362	1.377	30	4
	F–C(–X)$_2$–NO_2 (X = any subst.)	1.320	1.319	0.009	1.312	1.327	18	
*Csp*²–F	C=C–F (C,H,N,O subst. on C)	1.340	1.340	0.013	1.334	1.346	34	4
Car–F	*Car*–F (mono-F + *m,p*-F_2)	1.363	1.362	0.008	1.357	1.368	38	4
	Car–F (*o*-F_2)	1.340	1.340	0.009	1.336	1.344	167	4
*Csp*³–H	C–C–H_3 (methyl)	1.059	1.061	0.030	1.039	1.083	83	21
	C_2–C–H_2 (primary)	1.092	1.095	0.013	1.088	1.099	100	21
	C_3–C–H (secondary)	1.099	1.097	0.004	1.095	1.103	14	21
	$C_{2,3}$–C–H (primary and secondary)	1.093	1.095	0.012	1.089	1.100	118	21
	X–C–H_3 (methyl)	1.066	1.074	0.028	1.049	1.087	160	21
	X_2–C–H_2 (primary)	1.092	1.095	0.012	1.088	1.099	230	21
	X_3–C–H (secondary)	1.099	1.099	0.007	1.095	1.103	117	21
	$X_{2,3}$–C–H (primary and secondary)	1.094	1.096	0.011	1.091	1.100	348	21
*Csp*²–H	C–C=C–H	1.077	1.079	0.012	1.074	1.085	14	21
Car–H	*Car*–H	1.083	1.083	0.011	1.080	1.087	218	21
*Csp*³–I	C*–I	2.162	2.159	0.015	2.149	2.179	15	4
Car–I	*Car*–I	2.095	2.095	0.015	2.089	2.104	51	4
*Csp*³–N(4)	C*–NH_3^+	1.488	1.488	0.013	1.482	1.495	298	
	(C*)$_2$–NH_2^+	1.494	1.493	0.016	1.484	1.503	249	
	(C*)$_3$–NH^+	1.502	1.502	0.015	1.491	1.512	509	
	(C*)$_4$–N^+	1.510	1.509	0.020	1.496	1.523	319	
	C*–N^+ (overall)	1.499	1.498	0.018	1.488	1.510	1370	
*Csp*³–N(3)	C*–N^+ in N-subst. pyridinium	1.485	1.484	0.009	1.477	1.490	32	
	C*–NH_2 (N*sp*³: pyramidal)	1.469	1.470	0.010	1.462	1.474	19	22
	(C*)$_2$–NH (N*sp*³: pyramidal)	1.469	1.467	0.012	1.461	1.477	152	5,22
	(C*)$_3$–N (N*sp*³: pyramidal)	1.469	1.468	0.014	1.460	1.476	1042	5,22
	C*–N*sp*³ (overall)	1.469	1.468	0.014	1.460	1.476	1201	
	*Csp*³–N*sp*³							
	in aziridine	1.472	1.471	0.016	1.464	1.482	134	
	in azetidine	1.484	1.481	0.018	1.472	1.495	21	
	in tetrahydropyrrole	1.475	1.473	0.016	1.464	1.483	66	
	in piperidine	1.473	1.473	0.013	1.460	1.479	240	
	*Csp*³–N*sp*² (N planar) in:							23
	acyclic amides C*–NH–C=O	1.454	1.451	0.011	1.446	1.461	78	14
	β-lactams C*–N(–X)–C=O (endo)	1.464	1.465	0.012	1.458	1.475	23	13
	γ-lactams							
	C*–NH–C=O (endo)	1.457	1.458	0.011	1.449	1.465	20	13
	C*–N(–C*)–C=O (endo)	1.462	1.461	0.010	1.453	1.466	15	13
	C*–N(–C*)–C=O (exo)	1.458	1.456	0.014	1.448	1.465	15	13
	δ-lactams							
	C*–NH–C=O (endo)	1.478	1.472	0.016	1.467	1.491	6	14
	C*–N(–C*)–C=O (endo)	1.479	1.476	0.007	1.475	1.482	15	14

Bond	Substructure	*d*	*m*	*σ*	q_1	q_u	*n*	Note
	C*–N(–C*)–C=O (exo)	1.468	1.471	0.009	1.462	1.477	15	14
	nitro compounds (1,2-dinitro omitted):							
	C–**CH**$_2$–**NO**$_2$	1.485	1.483	0.020	1.478	1.502	8	
	C$_2$–**CH–NO**$_2$	1.509	1.509	0.011	1.502	1.511	12	
	C$_3$–**C–NO**$_2$	1.533	1.533	0.013	1.530	1.539	17	
	C$_2$–**C**–(**NO**$_2$)$_2$	1.537	1.536	0.016	1.525	1.550	19	
	1,2-dinitro: NO$_2$–C*–C*–NO$_2$	1.552	1.550	0.023	1.536	1.572	32	
Csp^3–N(2)	**C#–N**=N	1.493	1.493	0.020	1.477	1.506	54	
	C*–N=C–*Car*	1.465	1.468	0.011	1.461	1.472	75	
Csp^2–N(3)	C=**C–NH**$_2$ Nsp^2 planar	1.336	1.344	0.017	1.317	1.348	10	23
	C=**C–NH**–C# Nsp^2 planar	1.339	1.340	0.016	1.327	1.351	17	23
	C=**C–N**–(C#)$_2$							
	Nsp^2 planar	1.355	1.358	0.014	1.341	1.363	22	23
	Nsp^3 pyramidal	1.416	1.418	0.018	1.397	1.432	18	22
	Csp^2–Nsp^2 (N planar) in:							23
	acyclic amides							
	NH$_2$–**C**=O	1.325	1.323	0.009	1.318	1.331	32	14
	C*–**NH–C**=O	1.334	1.333	0.011	1.326	1.343	78	14
	(C*)$_2$–**N–C**=O	1.346	1.342	0.011	1.339	1.356	5	14
	β-lactams C*–**NH–C**=O	1.385	1.388	0.019	1.374	1.396	23	13
	γ-lactams							
	C*–**NH–C**=O	1.331	1.331	0.011	1.326	1.337	20	13
	C*–**N**(–C*)–**C**=O	1.347	1.344	0.014	1.335	1.359	15	13
	δ-lactams							
	C*–**NH–C**=O	1.334	1.334	0.006	1.330	1.339	6	14
	(C*)–**N**(–C*)–**C**=O	1.352	1.353	0.010	1.344	1.356	15	14
	peptides C#–**N**(–X)–**C**(–C#)(=O)	1.333	1.334	0.013	1.326	1.340	380	24
	ureas							
	(**NH**$_2$)$_2$–**C**=O	1.334	1.334	0.008	1.329	1.339	48	25,26
	(C#–**NH**)$_2$–**C**=O	1.347	1.345	0.010	1.341	1.354	26	25
	[(C#)$_n$–**N**]$_2$–**C**=O	1.363	1.359	0.014	1.354	1.370	40	25,27
	thioureas	1.346	1.343	0.023	1.328	1.361	192	
	(X$_2$**N**)$_2$–**C**=S							
	imides							
	[C#–**C**(=O)]$_2$–**NH**	1.376	1.377	0.012	1.369	1.383	64	
	[C#–**C**(=O)]$_2$–**N**–C#	1.389	1.383	0.017	1.376	1.404	38	
	[Csp^2–**C**(=O)]$_2$–**N**–C#	1.396	1.396	0.010	1.389	1.403	46	
	[Csp^2–**C**(=O)]$_2$–**N**–Csp^2	1.409	1.406	0.020	1.391	1.419	28	
	guanidinium [**C**–(**NH**$_2$)$_3$]$^+$ (unsubst.)	1.321	1.320	0.008	1.314	1.327	39	
	(any subst.)	1.328	1.325	0.015	1.317	1.333	140	
	in heterocyclic systems (any subst.)							
	1*H*-pyrrole (N1–C2, N1–C5)	1.372	1.374	0.016	1.363	1.384	58	
	indole (N1–C2)	1.370	1.370	0.012	1.364	1.377	40	
	pyrazole (N1–C5)	1.357	1.359	0.012	1.347	1.365	20	
	imidazole (N1–C2)	1.349	1.349	0.018	1.338	1.358	44	
	imidazole (N1–C5)	1.370	1.370	0.010	1.365	1.377	44	
Csp^2–N(2)	in imidazole (N3–C4)	1.376	1.377	0.011	1.369	1.384	44	
Car–N(4)	**Car–N**$^+$–(C,H)$_3$	1.465	1.466	0.007	1.461	1.470	23	
Car–N(3)	**Car–NH**$_2$							
	(Nsp^2: planar)	1.355	1.360	0.020	1.340	1.372	33	23
	(Nsp^3: pyramidal)	1.394	1.396	0.011	1.385	1.403	25	22
	(overall)	1.375	1.377	0.025	1.363	1.394	98	28
Car–N(3)	**Car–NH**–C#							
	(Nsp^2: planar)	1.353	1.353	0.007	1.347	1.359	16	23
	(Nsp^3: pyramidal)	1.419	1.423	0.017	1.412	1.432	8	22
	(overall)	1.380	1.364	0.032	1.353	1.412	31	28
	Car–N–(C#)$_2$							
	(Nsp^2: planar)	1.371	1.370	0.016	1.363	1.382	41	23
	(Nsp^3: pyramidal)	1.426	1.425	0.011	1.421	1.431	22	22
	(overall)	1.390	1.385	0.030	1.366	1.420	69	28
	in indole (N1–C7a)	1.372	1.372	0.007	1.367	1.376	40	

Bond	Substructure	*d*	*m*	*σ*	q_l	q_u	*n*	Note
	Car–NO_2	1.468	1.469	0.014	1.460	1.476	556	
Car–N(2)	*Car*–N=N	1.431	1.435	0.020	1.422	1.442	26	
Csp^2=N(3)	in furoxan ($^+$N2=C3)	1.316	1.316	0.009	1.311	1.324	14	
Csp^2=N(2)	*Car*–C=N–C#	1.279	1.279	0.008	1.275	1.285	75	
	$(C,H)_2$–C=N–OH in oximes	1.281	1.280	0.013	1.273	1.288	67	
	S–C=N–X	1.302	1.302	0.021	1.285	1.319	36	
	in pyrazole (N2=C3)	1.329	1.331	0.014	1.315	1.339	20	
	in imidazole (C2=N3)	1.313	1.314	0.011	1.307	1.319	44	
	in isoxazole (N2=C3)	1.314	1.315	0.009	1.305	1.320	9	
	in furazan (N2=C3, C4=N5)	1.298	1.299	0.006	1.294	1.303	12	
	in furoxan (C4=N5)	1.304	1.306	0.008	1.300	1.308	14	
Car ≃ N(3)	C ≃ N^+–H (pyrimidinium)	1.335	1.334	0.015	1.325	1.342	30	
	C ≃ N^+–C* (pyrimidinium)	1.346	1.346	0.010	1.340	1.352	64	
	C ≃ N^+–O^- (pyrimidinium)	1.362	1.359	0.013	1.353	1.369	56	
Car ≃ N(2)	C ≃ N (pyridine)	1.337	1.338	0.012	1.330	1.344	269	
	C ≃ N (pyrazine)	1.336	1.335	0.022	1.319	1.347	120	
	C ≃ N ≃ C (pyrimidine)	1.339	1.338	0.015	1.333	1.342	28	
	N ≃ C ≃ N (pyrimidine)	1.333	1.335	0.013	1.326	1.337	28	
	C ≃ N (pyrimidine) (overall)	1.336	1.337	0.014	1.331	1.339	56	
	in any 6-membered N-containing aromatic ring:							
	H–C ≃ N ≃ C–H	1.334	1.334	0.014	1.327	1.341	146	
	H–C ≃ N ≃ C–C*	1.339	1.341	0.013	1.336	1.345	38	
	C*–C ≃ N ≃ C–C*	1.345	1.345	0.008	1.342	1.348	24	
	C ≃ N ≃ C (overall)	1.336	1.337	0.014	1.329	1.344	204	
Csp^1≡N(2)	X–S–N≡C^- (isothiocyanide)	1.144	1.147	0.006	1.140	1.148	6	
Csp^1≡N(1)	C*–**C≡N**	1.136	1.137	0.010	1.131	1.142	140	
	C=C–**C≡N** in TCNQ	1.144	1.144	0.008	1.139	1.149	284	19
	Car–**C≡N**	1.138	1.138	0.007	1.133	1.143	31	
	X–**C≡N**	1.144	1.141	0.012	1.138	1.151	10	
	(S–**C≡N**)$^-$	1.155	1.156	0.012	1.147	1.165	14	
Csp^3–O(2)	in alcohols							
	CH_3–OH	1.413	1.414	0.018	1.395	1.425	17	
	C–**CH_2–OH**	1.426	1.426	0.011	1.420	1.431	75	
	C_2–**CH–OH**	1.432	1.431	0.011	1.425	1.439	266	
	C_3–**C–OH**	1.440	1.440	0.012	1.432	1.449	106	
	C*–OH (overall)	1.432	1.431	0.013	1.424	1.441	464	
	in dialkyl ethers							29
	CH_3–O–C*	1.416	1.418	0.016	1.405	1.426	110	
	C–**CH_2–O**–C*	1.426	1.424	0.011	1.418	1.435	34	
	C_2–**CH–O**–C*	1.429	1.430	0.010	1.420	1.437	53	
	C_3–**C–O**–C*	1.452	1.450	0.011	1.445	1.458	39	
	C*–O–C* (overall)	1.426	1.425	0.019	1.414	1.437	236	5
	in aryl alkyl ethers							29
	CH_3–O–*Car*	1.424	1.424	0.012	1.417	1.431	616	
	C–**CH_2–O**–*Car*	1.431	1.430	0.013	1.422	1.438	188	
	C_2–**CH–O**–*Car*	1.447	1.446	0.020	1.435	1.466	58	
	C_3–**C–O**–*Car*	1.470	1.469	0.018	1.456	1.483	55	
	C*–O–*Car* (overall)	1.429	1.427	0.018	1.419	1.436	917	
	in alkyl esters of carboxylic acids							12,29
	CH_3–O–C(=O)–C*	1.448	1.449	0.010	1.442	1.455	200	
	C–CH_2–O–C(=O)–C*	1.452	1.453	0.009	1.445	1.458	32	
	C_2–**CH–O**–C(=O)–C*	1.460	1.460	0.010	1.454	1.465	78	
	C_3–**C–O**–C(=O)–C*	1.477	1.475	0.008	1.472	1.484	6	
	C*–O–C(=O)–C* (overall)	1.450	1.451	0.014	1.442	1.459	314	
	in alkyl esters of α,β-unsaturated acids:							
	C*–O–C(=O)–C=C (overall)	1.453	1.452	0.013	1.444	1.459	112	
	in alkyl esters of benzoic acid							
	C*–O–C(=O)–C(phenyl) (overall)	1.454	1.454	0.012	1.446	1.463	219	
	in ring systems							
	oxirane (epoxides) (any subst.)	1.446	1.446	0.014	1.438	1.456	498	9

Bond	Substructure	*d*	*m*	*σ*	q_1	q_u	*n*	Note
	oxetane (any subst.)	1.463	1.460	0.015	1.451	1.474	16	
	tetrahydrofuran (C,H subst.)	1.442	1.441	0.017	1.430	1.451	154	
Csp^3–O(2)	tetrahydropyran (C,H subst.)	1.441	1.442	0.015	1.431	1.451	22	
	β-lactones: **C*–O**–C(=O)	1.492	1.494	0.010	1.481	1.501	4	16
	γ-lactones: **C*–O**–C(=O)	1.464	1.464	0.012	1.455	1.473	110	12
	δ-lactones: **C*–O**–C(=O)	1.461	1.464	0.017	1.452	1.473	27	12
	O–C–O system in *gem*-diols, and pyranose and furanose sugars:							30,31
	HO–C*–OH	1.397	1.401	0.012	1.388	1.405	18	
	C_5–O_5–C_1–O_1H in pyranoses							
	O_1 axial (α):							
	C_5–O_5	1.439	1.440	0.008	1.432	1.445	29	
	O_5–C_1	1.427	1.426	0.012	1.421	1.432	29	
	C_1–O_1	1.403	1.400	0.012	1.391	1.412	29	
	O_1 equatorial (β):							
	C_5–O_5	1.435	1.436	0.008	1.429	1.440	17	
	O_5–C_1	1.430	1.431	0.010	1.424	1.436	17	
	C_1–O_1	1.393	1.393	0.007	1.386	1.399	17	
	α + β (overall):							
	C_5–O_5	1.439	1.440	0.008	1.432	1.446	60	
	O_5–C_1	1.430	1.429	0.012	1.421	1.436	60	
	C_1–O_1	1.401	1.399	0.011	1.392	1.407	60	
	C_4–O_4–C_1–O_1H in furanoses (overall values)							
	C_4–O_4	1.442	1.446	0.012	1.436	1.449	18	
	O_4–C_1	1.432	1.432	0.012	1.421	1.443	18	
	C_1–O_1	1.404	1.405	0.013	1.397	1.409	18	
	C_5–O_5–C_1–O_1–C* in pyranoses							
	O_1 axial (α):							
	C_5–O_5	1.439	1.438	0.010	1.433	1.446	67	
	O_5–C_1	1.417	1.417	0.009	1.410	1.424	67	
	C_1–O_1	1.409	1.409	0.014	1.401	1.417	67	
	O_1–C*	1.435	1.435	0.013	1.427	1.443	67	
	O_1 equatorial (β):							
	C_5–O_5	1.434	1.435	0.006	1.429	1.439	39	
	O_5–C_1	1.424	1.424	0.008	1.418	1.431	39	
	C_1–O_1	1.390	1.390	0.011	1.381	1.400	39	
	O_1–C*	1.437	1.438	0.013	1.428	1.445	39	
	α + β (overall):							
	C_5–O_5	1.436	1.436	0.009	1.431	1.442	126	
	O_5–C_1	1.419	1.419	0.011	1.412	1.426	126	
	C_1–O_1	1.402	1.403	0.016	1.391	1.413	126	
	O_1–C*	1.436	1.436	0.013	1.428	1.445	126	
	C_4–O_4–C_1–O_1–C* in furanoses (overall values)							
	C_4–O_4	1.443	1.445	0.013	1.429	1.453	23	
	O_4–C_1	1.421	1.418	0.012	1.413	1.431	23	
	C_1–O_1	1.410	1.409	0.014	1.401	1.420	23	
	O_1–C*	1.439	1.437	0.014	1.429	1.449	23	
	Miscellaneous:							
	C#–O–SiX_3	1.416	1.416	0.017	1.405	1.428	29	
	C*–O–SO_2–C	1.465	1.461	0.014	1.454	1.475	33	
Csp^2–O(2)	in enols: C=**C–OH**	1.333	1.331	0.017	1.324	1.342	53	
	in enol esters: C=**C–O**–**C***	1.354	1.353	0.016	1.341	1.363	40	
	in acids:							
	C*–**C**(=O)–**OH**	1.308	1.311	0.019	1.298	1.320	174	
	C=C–**C**(=O)–**OH**	1.293	1.295	0.019	1.279	1.307	22	
	Car–**C**(=O)–**OH**	1.305	1.311	0.020	1.291	1.317	75	
	in esters:							
	C*–**C**(=O)–**O**–C*	1.336	1.337	0.014	1.328	1.346	551	12,29
	C=C–**C**(=O)–**O**–C*	1.332	1.331	0.011	1.324	1.339	112	

Bond	Substructure	*d*	*m*	*σ*	q_l	q_u	*n*	Note
	Car–C(=O)–**O**–C*	1.337	1.335	0.013	1.329	1.344	219	12
	C*–**C**(=O)–**O**–C=C	1.362	1.359	0.018	1.351	1.374	26	
	C*–C(=O)–**O**–**C**=C	1.407	1.405	0.017	1.394	1.420	26	
	C*–C(=O)–**O**–*Car*	1.360	1.359	0.011	1.355	1.367	40	12
	in anhydrides: O=**C**–**O**–C=O	1.386	1.386	0.011	1.379	1.393	70	
	in ring systems:							
	furan (O1–C2, O1–C5)	1.368	1.369	0.015	1.359	1.377	125	
	isoxazole (O1–C5)	1.354	1.354	0.010	1.345	1.360	9	
	β-lactones: C*–**C**(=O)–**O**–C*	1.359	1.359	0.013	1.348	1.371	4	13
	γ-lactones: C*–**C**(=O)–**O**–C*	1.350	1.349	0.012	1.342	1.359	110	12
	δ-lactones: C*–**C**(=O)–**O**–C*	1.339	1.339	0.016	1.332	1.347	27	12
Car–O(2)	in phenols: ***Car*–OH**	1.362	1.364	0.015	1.353	1.373	551	
	in aryl alkyl ethers: ***Car*–O**–C*	1.370	1.370	0.011	1.363	1.377	920	29,32
Car–O(2)	in diaryl ethers: ***Car*–O**–*Car*	1.384	1.381	0.014	1.375	1.391	132	
	in esters: ***Car*–O**–C(=O)–C*	1.401	1.401	0.010	1.394	1.408	40	12
*Csp*²=O(1)	in aldehydes and ketones:							
	C*–**CH=O**	1.192	1.192	0.005	1.188	1.197	7	
	$(C^*)_2$–**C=O**	1.210	1.210	0.008	1.206	1.215	474	5
	(C#)$_2$–**C=O**							
	in cyclobutanones	1.198	1.198	0.007	1.194	1.204	12	
	in cyclopentanones	1.208	1.208	0.007	1.203	1.212	155	
	in cyclohexanones	1.211	1.211	0.009	1.207	1.216	312	
	C=C–**C=O**	1.222	1.222	0.010	1.216	1.229	225	
	(C=C)$_2$–**C=O**	1.233	1.229	0.010	1.226	1.242	28	
	Car–**C=O**	1.221	1.218	0.014	1.212	1.229	85	
	(*Car*)$_2$–**C=O**	1.230	1.226	0.015	1.220	1.238	66	
	C=O in benzoquinones	1.222	1.220	0.013	1.211	1.231	86	
	delocalized double bonds in carboxylate anions:							
	H–**C ≃ O**$_2^-$ (formate)	1.242	1.243	0.012	1.234	1.252	24	
	C*–**C ≃ O**$_2^-$	1.254	1.253	0.010	1.247	1.261	114	
	C=C–**C ≃ O**$_2^-$	1.250	1.248	0.017	1.238	1.261	52	
	Car–**C ≃ O**$_2^-$	1.255	1.253	0.010	1.249	1.262	22	
	HOOC–**C ≃ O**$_2^-$ (hydrogen oxalate)	1.243	1.247	0.015	1.232	1.256	26	
	$^-$**O**$_2$ ≃ **C**–**C ≃ O**$_2^-$ (oxalate)	1.251	1.251	0.007	1.248	1.254	18	
	in carboxylic acids (X–COOH)							
	C*–**C(=O)**–OH	1.214	1.214	0.019	1.203	1.224	175	
	C=C–**C(=O)**–OH	1.229	1.226	0.017	1.218	1.237	22	
	Car–**C(=O)**–OH	1.226	1.223	0.020	1.211	1.241	75	
	in esters:							
	C*–**C(=O)**–O–C*	1.196	1.196	0.010	1.190	1.202	551	12
	C=C–**C(=O)**–O–C*	1.199	1.198	0.009	1.193	1.203	113	
	Car–**C(=O)**–O–C*	1.202	1.201	0.009	1.196	1.207	218	12
	C*–**C(=O)**–O–C=C	1.190	1.190	0.014	1.184	1.198	26	
	C*–**C(=O)**–O–*Car*	1.187	1.188	0.011	1.181	1.195	40	12
	in anhydrides: **O=C**–O–**C=O**	1.187	1.187	0.010	1.184	1.193	70	
	in β-lactones: C*–**C(=O)**–O–C*	1.193	1.193	0.006	1.187	1.198	4	13
	γ-lactones: C*–**C(=O)**–O–C*	1.201	1.202	0.009	1.196	1.206	109	12
	δ-lactones: C*–**C(=O)**–O–C*	1.205	1.207	0.008	1.201	1.209	27	12
	in amides:							
	NH$_2$–**C**(–C*)=**O**	1.234	1.233	0.012	1.225	1.243	32	14
	(C*–)(C*,H–)N–**C**(–C*)=**O**	1.231	1.231	0.012	1.224	1.238	378	14
	β-lactams: C*–NH–**C=O**	1.198	1.200	0.012	1.193	1.204	23	13
	γ-lactams:							
	C*–NH–**C=O**	1.235	1.235	0.008	1.232	1.240	20	13
	C*–N(–C*)–C=O	1.225	1.226	0.011	1.217	1.233	15	13
	δ-lactams:							
	C*–NH–**C=O**	1.240	1.241	0.003	1.237	1.243	6	14
	C*–N(–C*)–**C=O**	1.233	1.233	0.007	1.229	1.239	15	14
	in ureas:							
	(NH)$_2$)$_2$–**C=O**	1.256	1.256	0.007	1.249	1.261	24	25,26
	(C#–NH)$_2$–**C=O**	1.241	1.237	0.011	1.235	1.245	13	25

Bond	Substructure	*d*	*m*	*σ*	q_l	q_u	*n*	Note
	[(C#)$_n$–N]$_2$–**C=O**	1.230	1.230	0.007	1.224	1.234	20	25,27
Csp^3–P(4)	C$_3$–**P$^+$–C***	1.800	1.802	0.015	1.790	1.812	35	33
	C$_2$–**P**(=O)–**C**H$_3$	1.791	1.790	0.006	1.786	1.795	10	
	C$_2$–**P**(=O)–**C**H$_2$–C	1.806	1.806	0.009	1.801	1.813	45	
	C$_2$–**P**(=O)–**C**H–C$_2$	1.821	1.821	0.009	1.815	1.828	15	
	C$_2$–**P**(=O)–**C**–C$_3$	1.841	1.842	0.008	1.835	1.847	14	
	C$_2$–**P**(=O)–**C*** (overall)	1.813	1.811	0.017	1.800	1.822	84	
Csp^3–P(3)	C$_2$–**P–C***	1.855	1.857	0.019	1.840	1.870	23	
Car–P(4)	C$_3$–**P$^+$–*Car***	1.793	1.792	0.011	1.786	1.800	276	
	C$_2$–**P**(=O)–***Car***	1.801	1.802	0.011	1.796	1.807	98	
	Ph$_3$–P=N$^+$=**P–Ph$_3$**	1.795	1.795	0.008	1.789	1.800	197	
Car–P(3)	C$_2$–**P–*Car***	1.836	1.837	0.010	1.830	1.844	102	
	(N≃)$_2$P–*Car*(P ≃ N aromatic)	1.795	1.793	0.011	1.788	1.803	43	
Csp^3–S(4)	**C*–SO$_2$**–C (C* = CH$_3$ excluded)	1.786	1.782	0.018	1.774	1.797	75	
	C*–SO$_2$–C (overall)	1.779	1.778	0.020	1.764	1.790	94	
	C*–SO$_2$–O–X	1.745	1.744	0.009	1.738	1.754	7	34
	C*–SO$_2$–N–X$_2$	1.758	1.756	0.018	1.746	1.773	17	34
Csp^3–S(3)	**C*–S**(=O)–C (C* = CH$_3$ excluded)	1.818	1.814	0.024	1.802	1.829	69	
	C*–S(=O)–C (overall)	1.809	1.806	0.025	1.793	1.820	88	
	CH$_3$–S$^+$–X$_2$	1.786	1.787	0.007	1.779	1.792	21	
	C*–S$^+$–X$_2$ (C* = CH$_3$ excluded)	1.823	1.820	0.016	1.812	1.834	18	
	C*–S$^+$–X$_2$ (overall)	1.804	1.794	0.025	1.788	1.820	41	
Csp^3–S(2)	**C*–SH**	1.808	1.805	0.010	1.800	1.819	6	
	CH$_3$–S–C*	1.789	1.787	0.008	1.784	1.794	9	
Csp^3–S(2)	C–**CH$_2$–S**–C*	1.817	1.816	0.013	1.808	1.824	92	
	C$_2$–**CH–S**–C*	1.819	1.819	0.011	1.811	1.825	32	
	C$_3$–**C–S**–C*	1.856	1.860	0.011	1.854	1.863	26	
	C*–S–C* (overall)	1.819	1.817	0.019	1.809	1.827	242	
	in thiirane	1.834	1.835	0.025	1.810	1.858	4	9
	in thiirane: see ZCMXSP (1.817, 1.844)							
	in tetrahydrothiophene	1.827	1.826	0.018	1.811	1.837	20	
	in tetrahydrothiopyran	1.823	1.821	0.014	1.812	1.832	24	
	C–**CH$_2$–S**–S–X	1.823	1.820	0.014	1.813	1.832	41	
	C$_3$–**C–S**–S–X	1.863	1.865	0.015	1.848	1.878	11	
	C*–S–S–X (overall)	1.833	1.828	0.022	1.818	1.848	59	
Csp^2–S(2)	C=**C–S**–C*	1.751	1.755	0.017	1.740	1.764	61	
	C=**C–S**–C=C (in tetrathiafulvalene)	1.741	1.741	0.011	1.733	1.750	88	
	C=**C–S**–C=C (in thiophene)	1.712	1.712	0.013	1.703	1.722	60	
	O=**C–S**–C#	1.762	1.759	0.018	1.747	1.778	20	
Car–S(4)	***Car*–SO$_2$**–C	1.763	1.764	0.009	1.756	1.769	96	
	***Car*–SO$_2$**–O–X	1.752	1.750	0.008	1.749	1.756	27	
	***Car*–SO$_2$**–N–X$_2$	1.758	1.759	0.013	1.749	1.765	106	35
Car–S(3)	***Car*–S**(=O)–C	1.790	1.790	0.010	1.783	1.798	41	
	***Car*–S$^+$**–X$_2$	1.778	1.779	0.010	1.771	1.787	10	
Car–S(2)	***Car*–S**–C*	1.773	1.774	0.009	1.765	1.779	44	
	***Car*–S**–*Car*	1.768	1.767	0.010	1.762	1.774	158	
	***Car*–S**–*Car* (in phenothiazine)	1.764	1.764	0.008	1.760	1.769	48	
	***Car*–S**–S–X	1.777	1.777	0.012	1.767	1.785	47	
Csp^1–S(2)	N≡**C–S**–X	1.679	1.683	0.026	1.645	1.698	10	
Csp^1–S(1)	(N≡**C–S**)$^-$	1.630	1.630	0.014	1.619	1.641	14	
Csp^2=S(1)	(C*)$_2$–**C=S**: see IPMUDS (1.599)							
	(*Car*)$_2$–**C=S**: see CELDOM (1.611)							
	(X)$_2$–**C=S** (X = C,N,O,S)	1.671	1.675	0.024	1.656	1.689	245	
	X$_2$N–**C**(=**S**)–S–X	1.660	1.660	0.016	1.648	1.674	38	
	(X$_2$N)$_2$–**C=S** (thioureas)	1.681	1.684	0.020	1.669	1.693	96	
	N–**C**(=**S**)$_2$	1.720	1.721	0.012	1.709	1.731	20	
Csp^3–Se	**C#–Se**	1.970	1.967	0.032	1.948	1.998	21	
Csp^2–Se(2)	C=**C–Se**–C=C (in tetraselenafulvalene)	1.893	1.895	0.013	1.882	1.902	32	
Car–Se(3)	**Ph$_3$–Se$^+$**	1.930	1.929	0.006	1.924	1.936	13	
Csp^3–Si(5)	**C#–Si$^-$**–X$_4$	1.874	1.876	0.015	1.859	1.884	9	
Csp^3–Si(4)	**CH$_3$–Si**–X$_3$	1.857	1.857	0.018	1.848	1.869	552	

Bond	Substructure	d	m	σ	q_1	q_u	n	Note
	C*–Si–X_3 (C* = CH_3 excluded)	1.888	1.887	0.023	1.872	1.905	124	
	C*–Si–X_3 (overall)	1.863	1.861	0.024	1.850	1.875	681	
Car–Si(4)	***Car*–Si**–X_3	1.868	1.868	0.014	1.857	1.878	178	
Csp^1–Si(4)	C≡**C–Si**–X_3	1.837	1.840	0.012	1.824	1.849	8	
Csp^3–Te	**C#–Te**	2.158	2.159	0.030	2.128	2.177	13	
Car–Te	***Car*–Te**	2.116	2.115	0.020	2.104	2.130	72	
Csp^2=Te	see CEDCUJ (2.044)							
Cl–Cl	see PHASCL (2.306, 2.227)							
Cl–I	see CMBIDZ (2.563), HXPASC (2.541, 2.513), METAMM (2.552), BQUINI (2.416, 2.718)							
Cl–N	see BECTAE (1.743–1.757), BOGPOC (1.705)							
Cl–O(1)	in ClO_4^-	1.414	1.419	0.026	1.403	1.431	252	
Cl–P	$(N{=})_2$**P–Cl** (N ≃ P aromatic)	1.997	1.994	0.015	1.989	2.004	46	
	Cl–P (overall)	2.008	2.001	0.035	1.986	2.028	111	
Cl–S	**Cl–S** (overall)	2.072	2.079	0.023	2.047	2.091	6	
	see also longer bonds in CILSAR (2.283), BIHXIZ (2.357), CANLUY (2.749)							
Cl–Se	see BIRGUE10, BIRHAL10, CTCNSE (2.234–2.851)							
Cl–Si(4)	**Cl–Si**–X_3 (monochloro)	2.072	2.075	0.009	2.066	2.078	5	
	$\mathbf{Cl_2}$**–Si**–X_2 and $\mathbf{Cl_3}$**–Si**–X	2.020	2.012	0.015	2.007	2.036	5	
Cl–Te	Cl–Te in range 2.34–2.60	2.520	2.515	0.034	2.493	2.537	22	36
	see also longer bonds in BARRIV, BOJPUL, CETUTE, EPHTEA, OPNTEC10 (2.73–2.94)							
F–N(3)	**F–N**–C_2 and $\mathbf{F_2}$**–N**–C	1.406	1.404	0.016	1.395	1.416	9	
F–P(6)	in hexafluorophosphate, PF_6^-	1.579	1.587	0.025	1.563	1.598	72	
F–P(3)	$(N{=})_2$**P–F**(N ≃ P aromatic)	1.495	1.497	0.016	1.481	1.510	10	
F–S	43 observations in range 1.409–1.770 in a wide variety of environments; F–S(6) in F_2–SO_2–C_2 (see FPSULF10, BETJOZ)	1.640	1.646	0.011	1.626	1.649	6	
	F–S(4) in $\mathbf{F_2}$**–S**(=O)–N (see BUDTEZ)	1.527	1.528	0.004	1.524	1.530	24	37
F–Si(6)	in SiF_6^{2-}	1.694	1.701	0.013	1.677	1.703	6	
F–Si(5)	**F–Si**$^-$–X_4	1.636	1.639	0.035	1.602	1.657	10	
F–Si(4)	**F–Si**–X_3	1.588	1.587	0.014	1.581	1.599	24	
F–Te	see CUCPlZ (F–Te(6) = 1.942, 1.937), FPHTEL(F–Te(4) = 2.006)							
H–N(4)	X_3–**N**$^+$**–H**	1.033	1.036	0.022	1.026	1.045	87	21
H–N(3)	X_3–**N–H**	1.009	1.010	0.019	0.997	1.023	95	21
H–O(2)	in alcohols C*–**O–H**	0.967	0.969	0.010	0.959	0.974	63	21
	C#–**O–H**	0.967	0.970	0.010	0.959	0.974	73	21
	in acids O=C–**O–H**	1.015	1.017	0.017	1.001	1.031	16	21,38
I–I	in I_3^-	2.917	2.918	0.011	2.907	2.927	6	
I–N	see BZPRIB, CMBIDZ, HMTITI, HMTNTI, IFORAM, IODMAM (2.042–2.475)							
I–O	X–**I–O**(see BZPRIB, CAJMAB, IBZDAC11)	2.144	2.144	0.028	2.127	2.164	6	
	for IO_6^- see BOVMEE (1.829–1.912)							
I–P(3)	see CEHKAB (2.490–2.493)							†
I–S	sec DTHIBR10 (2.687), ISUREA10 (2.629), BZTPPI (3.251)							
I–Te(4)	**I–Te**–X_3	2.926	2.928	0.026	2.902	2.944	8	
N(4)–N(3)	X_3–**N**$^+$**–N**^{0}–X_2 (N^0 planar)	1.414	1.414	0.005	1.412	1.418	13	
N(3)–N(3)	(C)(C,H)–$\mathbf{N_a}$–$\mathbf{N_b}$(C)(C,H)							5,39
	$\mathbf{N_a}$, $\mathbf{N_b}$ pyramidal	1.454	1.452	0.021	1.444	1.457	44	40
	$\mathbf{N_a}$ pyramidal, N_b planar	1.420	1.420	0.015	1.407	1.433	68	40
	$\mathbf{N_a}$, N_b planar	1.401	1.401	0.018	1.384	1.418	40	40
	overall	1.425	1.425	0.027	1.407	1.443	139	
N(3)–N(2)	in pyrazole (N1–N2)	1.366	1.366	0.019	1.350	1.375	20	
	in pyridaznium (Nl$^+$≃N2)	1.350	1.349	0.010	1.345	1.361	7	
N(2) ≃ N(2)	**N ≃ N** (aromatic) in pyridazine with C,H as *ortho* substituents	1.304	1.300	0.019	1.287	1.326	6	

Bond	Substructure	*d*	*m*	*σ*	q_l	q_u	*n*	Note
	with N,Cl as *ortho* substituents	1.368	1.373	0.011	1.362	1.375	9	
N(2)=N(2)	C#–**N=N**–C#							
	cis	1.245	1.244	0.009	1.239	1.252	21	
	trans	1.222	1.222	0.006	1.218	1.227	6	
	(overall)	1.240	1.241	0.012	1.230	1.251	27	
	*C*ar–**N=N**–*C*ar	1.255	1.253	0.016	1.247	1.262	13	
	X–**N=N**=N (azides)	1.216	1.226	0.028	1.202	1.237	19	
N(2)=N(1)	X–N=**N=N** (azides)	1.124	1.128	0.015	1.114	1.137	19	
N(3)–O(2)	$(C,H)_2$–**N–OH** (N*sp*^{2}: planar)	1.396	1.394	0.012	1.390	1.401	28	
	C_2–**N–O**–C							
	(N*sp*^{3}: pyramidal)	1.463	1.465	0.012	1.457	1.468	22	
	(N*sp*^{2}: planar)	1.397	1.394	0.011	1.388	1.409	12	
	in furoxan (N2–O1)	1.438	1.436	0.009	1.430	1.447	14	
N(3)–O(1)	$(C\approx)_2$**N$^+$–O$^-$** in pyridine *N*-oxides	1.304	1.299	0.015	1.291	1.316	11	
	in furoxan ($^+$N2–O6$^-$)	1.234	1.234	0.008	1.228	1.240	14	
N(2)–O(2)	in oximes							
	$(C\#)_2$–C=**N–OH**	1.416	1.418	0.006	1.416	1.420	7	
	(H)(C*sp*^{2})–C=**N–OH**	1.390	1.390	0.011	1.380	1.401	20	
	(C#)(C*sp*^{2})–C=**N–OH**	1.402	1.403	0.010	1.393	1.410	18	
	$(Csp^2)_2$–C=**N–OH**	1.378	1.377	0.017	1.365	1.393	16	
	$(C,H)_2$–C=**N–OH** (overall)	1.394	1.395	0.018	1.379	1.408	67	
	in furazan (O1–N2, O1–N5)	1.385	1.383	0.013	1.378	1.392	12	
	in furoxan (O1–N5)	1.380	1.380	0.011	1.370	1.388	14	
	in isoxazole (O1–N2)	1.425	1.425	0.010	1.417	1.434	9	
N(3)=O(1)	in nitrate ions **NO_3^-**	1.239	1.240	0.020	1.227	1.251	105	
	in nitro groups							
	C*–**NO_2**	1.212	1.214	0.012	1.206	1.221	84	
	C#–**NO_2**	1.210	1.210	0.011	1.205	1.218	231	
	*C*ar–**NO_2**	1.217	1.218	0.011	1.211	1.215	1116	
	C–NO_2 (overall)	1.218	1.219	0.013	1.210	1.226	1733	
N(3)–P(4)	X_2–**P**(=X)–**N**X_2							
	N*sp*^{2}: planar	1.652	1.651	0.024	1.634	1.670	205	
	N*sp*^{3}: pyramidal	1.683	1.683	0.005	1.680	1.686	6	
	(overall)	1.662	1.662	0.029	1.639	1.682	358	
	subsets of this group are:							
	O_2–**P**(=S)–**N**X_2	1.628	1.624	0.015	1.615	1.634	9	
	C–**P**(=S)–$(\mathbf{N}X_2)_2$	1.691	1.694	0.018	1.678	1.703	28	
	O–**P**(=S)–$(\mathbf{N}X_2)_2$	1.652	1.654	0.014	1.642	1.664	28	
	P(=O)–$(\mathbf{N}X_2)_3$	1.663	1.668	0.026	1.640	1.679	78	
N(3)–P(3)	–**N**X–**P**(–X)–NX–P(–X)–(P_2N_2 ring)	1.730	1.721	0.017	1.716	1.748	20	
	–**N**X–**P**(=S)–NX–**P**(=S)–(P_2N_2 ring)	1.697	1.697	0.015	1.690	1.703	44	
	in P-substituted phosphazenes:							
	$(N\approx)_2$**P–N** (amino)	1.637	1.638	0.014	1.625	1.651	16	
	(aziridinyl)	1.672	1.674	0.010	1.665	1.676	15	
N(2)=P(4)	Ph_3–**P=N$^+$**=P–Ph_3	1.571	1.573	0.013	1.563	1.580	66	
N(2)=P(3)	Ph_3–**P=N**–C,S	1.599	1.597	0.018	1.580	1.615	7	
N(2) ≈ P(3)	**N ≈ P** aromatic							
	in phosphazenes	1.582	1.582	0.019	1.571	1.594	126	
	in P ≈ N ≈ S	1.604	1.606	0.009	1.594	1.612	36	
N(3)–S(4)	C–**SO_2–N**H_2	1.600	1.601	0.012	1.591	1.610	14	35
	C–**SO_2–N**H–C#	1.633	1.633	0.019	1.615	1.652	47	35
	C–**SO_2–N**–$C(\#)_2$	1.642	1.641	0.024	1.623	1.659	38	35
N(3)–S(2)	C–**S–N**X_2 N*sp*^{2}: planar	1.710	1.707	0.019	1.698	1.722	22	23
	(for N*sp*^{3} pyramidal see MODIAZ: 1.765)							
	X–S–NX_2 N*sp*^{2}: planar	1.707	1.705	0.012	1.699	1.715	30	23
N(2)–S(2)	C=**N–S**–X	1.656	1.663	0.027	1.632	1.677	36	
N(2) ≈ S(2)	**N ≈ S** aromatic in **P ≈ N ≈ S**	1.560	1.558	0.011	1.554	1.563	37	
N(2)=S(2)	**N=S** in **N=S=N** and **N=S=S**	1.541	1.546	0.022	1.521	1.558	37	
N(3)–SE	see COJCUZ (1.830), DSEMOR10 (1.846, 1.852), MORTRS10 (1.841)							
N(2)–Se	see SEBZQI (1.805), NAPSEZ10 (1.809, 1.820)							

Bond	Substructure	*d*	*m*	*σ*	q_1	q_u	*n*	Note
N(2)=Se	see CISMUM (1.790, 1.791)							
N(3)–Si(5)	see DMESIP01, BOJLER, CASSAQ, CASYOK, CECXEN, CINTEY, CIPBUY, FMESIB, MNPSIL, PNPOSI (1.973–2.344)							
N(3)–Si(4)	X_3–**Si–N**X_2 (overall)	1.748	1.746	0.022	1.735	1.757	170	
	subsets of this group are:							
	X_3–**Si–N**HX	1.714	1.719	0.014	1.702	1.727	16	
	X_3–**Si–N**X–Si–X_3 acyclic	1.743	1.744	0.016	1.731	1.755	45	
	N–Si–N in 4-membered rings	1.742	1.742	0.009	1.735	1.748	53	
	N–Si–N in 5-membered rings	1.741	1.742	0.019	1.726	1.749	33	
N(2)–Si(4)	X_3–**Si–N**$^-$–Si–X_3	1.711	1.712	0.019	1.693	1.729	15	
N–Te	see ACLTEP (2.402), BIBLAZ (1.980), CESSAU (2.023)							
O(2)–O(2)	C*–**O–O**–C*,H							
	τ(OO) = 70–85°	1.464	1.464	0.009	1.458	1.472	12	
	τ(OO) *ca.* 180°	1.482	1.480	0.005	1.478	1.486	5	
	overall	1.469	1.471	0.012	1.461	1.478	17	
	O=C–**O–O**–C=O see ACBZPO01 (1.446), CEYLUN (1.452), CIMHIP (1.454)							
	Si–**O–O**–Si	1.496	1.499	0.005	1.490	1.499	10	
O(2)–P(5)	X–**P**–(**O**X)$_4$							41
	trigonal bipyramidal:							
	axial	1.689	1.685	0.024	1.675	1.712	20	
	equatorial	1.619	1.622	0.024	1.604	1.628	20	
	square pyramidal	1.662	1.661	0.020	1.649	1.673	28	
O(2)–P(4)	C–**O–P**(≃ O)$_3^{2-}$	1.621	1.622	0.007	1.615	1.628	12	
	(H–**O**)$_2$–**P**(≃ O)$_2^-$	1.560	1.561	0.009	1.555	1.566	16	
	(C–**O**)$_2$–**P**(≃ O)$_2^-$	1.608	1.607	0.013	1.599	1.615	16	
	(C#–**O**)$_3$–**P**=O	1.558	1.554	0.011	1.550	1.564	30	
	(*Car*–**O**)$_3$–**P**=O	1.587	1.588	0.014	1.572	1.599	19	
	X–**O–P**(=O)–(C,N)$_2$	1.590	1.585	0.016	1.577	1.601	33	
	(X–**O**)$_2$–**P**(=O)–(C,N)	1.571	1.572	0.013	1.563	1.579	70	
O(2)–P(3)	(N≃)$_2$**P–O**–C (N ≃ P aromatic)	1.573	1.573	0.011	1.563	1.584	16	
O(1)=P(4)	C–O–**P**(≃ **O**)$_3^{2-}$ (delocalized)	1.513	1.512	0.008	1.508	1.518	42	
	(H–O)$_2$–**P**(≃ **O**)$_2^-$ (delocalized)	1.503	1.503	0.005	1.499	1.508	16	
	(C–O)$_2$–**P**(≃ **O**)$_2^-$ (delocalized)	1.483	1.485	0.008	1.474	1.490	16	
	(C–O)$_3$–**P=O**	1.449	1.448	0.007	1.446	1.452	18	
	C_3–**P=O**	1.489	1.486	0.010	1.481	1.496	72	
	N_3–**P=O**	1.461	1.462	0.014	1.449	1.470	26	
	(C)$_2$(N)–**P=O**	1.487	1.489	0.007	1.479	1.493	5	
	(C,N)$_2$(O)–**P=O**	1.467	1.462	0.007	1.462	1.472	33	
	(C,N)(O)$_2$–**P=O**	1.457	1.458	0.009	1.454	1.462	35	
O(2)–S(4)	C–**O–SO**$_2$–C	1.577	1.576	0.015	1.566	1.584	41	
	C–**O–SO**$_2$–CH_3	1.569	1.569	0.013	1.556	1.582	7	
	C–**O–SO**$_2$–*Car*	1.580	1.578	0.015	1.571	1.588	27	
O(1)=S(4)	C–**SO**$_2$–C	1.436	1.437	0.010	1.431	1.442	316	42
	X–**SO**$_2$–NX_2	1.428	1.428	0.010	1.422	1.434	326	
	C–**SO**$_2$–N–(C,H)$_2$	1.430	1.430	0.009	1.425	1.435	206	
	C–**SO**$_2$–O–C	1.423	1.423	0.008	1.418	1.428	82	
	in **SO**$_4^{2-}$	1.472	1.473	0.013	1.463	1.481	104	
O(1)=S(3)	C–**S**(=**O**)–C	1.497	1.498	0.013	1.489	1.505	90	5
O–Se	see BAPPAJ, BIRGUE10, BIRHAL10, CXMSEO, DGLYSE, SPSEBU (1.597 for **O=Se** to 1.974 for **O–Se**)							
O(2)–Si(5)	(X–**O**)$_3$–**Si**–(N)(C)	1.663	1.658	0.023	1.650	1.665	21	
O(2)–Si(4)	X_3–**Si–O–X** (overall)	1.631	1.630	0.022	1.617	1.646	191	
O(2)–Si(4)	subsets of this group are:							
	X_3–**Si–O**–C#	1.645	1.647	0.012	1.634	1.652	29	
	X_3–**Si–O**–Si–X_3	1.622	1.625	0.014	1.614	1.631	70	
	X_3–**Si–O–O–Si**–X_3	1.680	1.676	0.008	1.673	1.688	10	
O(2)–Te(6)	(X–**O**)$_6$–**Te**	1.927	1.927	0.020	1.908	1.942	16	

Bond	Substructure	*d*	*m*	*σ*	q_l	q_u	*n*	Note
O(2)–Te(4)	$(X–\mathbf{O})_2–\mathbf{Te}–X_2$	2.133	2.136	0.054	2.078	2.177	12	
P(4)–P(4)	$X_3–\mathbf{P}–\mathbf{P}–X_3$	2.256	2.259	0.025	2.243	2.277	6	
P(4)–P(3)	see CECHEX (2.197), COZPIQ (2.249)							
P(3)–P(3)	$X_2–\mathbf{P}–\mathbf{P}–X_2$	2.214	2.210	0.022	2.200	2.224	41	
P(4)=P(4)	see BUTSUE (2.054)							
P(3)=P(3)	see BALXOB (2.034)							
P(4)=S(1)	$C_3–\mathbf{P}=\mathbf{S}$	1.954	1.952	0.005	1.950	1.957	13	
	$(N,O)_2(C)–\mathbf{P}=\mathbf{S}$	1.922	1.924	0.014	1.913	1.927	26	
	$(N,O)_3–\mathbf{P}=\mathbf{S}$	1.913	1.914	0.014	1.906	1.921	50	
P(4)=Se(1)	$X_3–\mathbf{P}=\mathbf{Se}$	2.093	2.099	0.019	2.075	2.108	12	
P(3)–Si(4)	$X_2–\mathbf{P}–\mathbf{Si}–X_3$: 3- and 4-rings excluded (see BOPFER, BOPFIV, CASTOF10, COZVIW: 2.201–2.317)	2.264	2.260	0.019	2.249	2.283	22	
P(4)=Te(1)	see MOPHTE (2.356), TTEBPZ (2.327)							
S(2)–S(2)	C–**S**–**S**–C							
	τ(SS) = 75–105°	2.031	2.029	0.015	2.021	2.038	46	
	τ(SS) = 0–20°	2.070	2.068	0.022	2.057	2.077	28	
	(overall)	2.048	2.045	0.026	2.028	2.068	99	
	In polysulphide chain–**S**–**S**–S–	2.051	2.050	0.022	2.037	2.065	126	
S(2)–S(1)	X–N=S–S	1.897	1.896	0.012	1.887	1.908	5	
S–Se(4)	see BUWZUO (2.264, 2.269)							
S–Se(2)	X–**Se**–**S** (any)	2.193	2.195	0.015	2.174	2.207	9	
S(2)–Si(4)	$X_3–\mathbf{Si}–S–X$	2.145	2.138	0.020	2.130	2.158	19	
S(2)–Te	X–**S**–**Te** (any)	2.405	2.406	0.022	2.383	2.424	10	
	X=**S**–**Te** (any)	2.682	2.686	0.035	2.673	2.694	28	
Se(2)–Se(2)	X–**Se**–**Se**–X	2.340	2.340	0.024	2.315	2.361	15	
Se(2)–Te(2)	see BAWFUA, BAWGAH (2.524–2.561)							†
Si(4)–Si(4)	X_3 **Si** **Si** X_3 3–membered rings excluded: see CIHRAM (2.511)	2.359	2.359	0.012	2.349	2.366	42	
Te–Te	see CAHJOK (2.751, 2.704)							

† The standard deviation in the sample for the bond type is greater than for the other entries.

Appendix 1. (Footnotes to Table)

1. Sample dominated by $B–CH_3$. For longer bonds in $B^-–CH_3$ see LITMEB10 [$B(4)–CH_3$ = 1.621–1.644Å].
2. p(π)–p(π) Bonding with Bsp^2 and Nsp^2 coplanar (τBN = 0 ± 15°) predominates. See G. Schmidt, R. Boese, and D. Bläser, *Z. Naturforsch.*, 1982, **37b**, 1230.
3. 84 observations range from 1.38 to 1.61 Å and individual values depend on substituents on B and O. For a discussion of borinic acid adducts see S. J. Rettig and J. Trotter, *Can. J. Chem.*, 1982, **60**, 2957.
4. See M. Kaftory in *The Chemistry of Functional Groups. Supplement D: The Chemistry of Halides, Pseudohalides, and Azides*, S. Patai and Z. Rappoport, Eds., Wiley: New York, 1983, Part 2, ch. 24.
5. Bonds which are endocyclic or exocyclic to any 3- or 4-membered rings have been omitted from all averages in this section.
6. The overall average given here is for $Csp^3–Csp^3$ bonds which carry only C or H substituents. The value cited reflects the relative abundance of each 'substitution' group. The 'mean of means' for the 9 subgroups is 1.538 (σ = 0.022) Å.
7. See F. H. Allen, (a) *Acta Crystallogr.*, 1980, **B36**, 81; (b) 1981, **B37**, 890.
8. See F. H. Allen, *Acta Crystallogr.*, 1984, **B40**, 64.
9. See F. H. Allen, *Tetrahedron*, 1982, **38**, 2843.
10. See F. H. Allen, *Tetrahedron*, 1982, **38**, 645.
11. Cyclopropanones and cyclobutanones excluded.
12. See W. B. Schweizer and J. D. Dunitz, *Helv. Chim. Acta*, 1982, **65**, 1547.
13. See L. Norskov-Lauritsen, H.-B. Bürgi, P. Hoffmann, and H. R. Schmidt, *Helv. Chim. Acta*, 1985, **68**, 76.
14. See P. Chakrabarti and J. D. Dunitz, *Helv. Chim. Acta*, 1982, **65**, 1555.
15. See J. L. Hencher in *The Chemistry of the C≡C Triple Bond*, S. Patai, Ed., Wiley, New York, 1978, ch. 2.
16. Conjugated: torsion angle about central C–C single bond is 0 ± 20° (*cis*) or 180 ± 20° (*trans*).
17. Unconjugated: torsion angle about central C–C single bond is 20–160°.
18. Other conjugative substituents excluded.
19. TCNQ is tetracyanoquinodimethane.
20. No difference detected between C2 ≃ C3 and C3 ≃ C4 bonds.
21. Derived from neutron diffraction results only.
22. Nsp^3: pyramidal; mean valence angle at N is in range 108–114°.
23. Nsp^2: planar; mean valence angle at N is ≥ 117.5°.
24. Cyclic and acyclic peptides.
25. See R. H. Blessing, *J. Am. Chem. Soc.*, 1983, **105**, 2776.
26. See L. Lebioda, *Acta Crystallogr.*, 1980, **B36**, 271.
27. *n* = 3 or 4, i.e. tri- or tetra-substituted ureas.
28. Overall value also includes structures with mean valence angle at N in the range 115–118°.
29. See F. H. Allen and A. J. Kirby, *J. Am. Chem. Soc.*, 1984, **106**, 6197.
30. See A. J. Kirby, 'The Anomeric Effect and Related Stereoelectronic Effects at Oxygen,' Springer, Berlin, 1983.
31. See B. Fuchs, L. Schleifer, and E. Tartakovsky, *Nouv. J. Chim.*, 1984, **8**, 275.
32. See S. C. Nyburg and C. H. Faerman, *J. Mol. Struct.*, 1986, **140**, 347.
33. Sample dominated by $P–CH_3$ and $P–CH_2–C$.
34. Sample dominated by C* = methyl.
35. See A. Kalman, M. Czugler, and G. Argay, *Acta Crystallogr.*, 1981, **B37**, 868.
36. Bimodal distribution resolved into 22 'short' bonds and 5 longer outliers.
37. All 24 observations come from BUDTEZ.
38. 'Long' O–H bonds in centrosymmetric O---H---O H–bonded dimers are excluded.
39. N–N bond length also dependent on torsion angle about N–N bond and on nature of substituent C atoms; these effects are ignored here.
40. N pyramidal has average angle at N in range 100–113.5°; N planar has average angle of ≥ 117.5°.
41. See R. R. Holmes and J. A. Deiters, *J. Amer. Chem. Soc.*, 1977, **99**, 3318.
42. No detectable variation in S=O bond length with type of C-substituent.

Appendix 2

Short-form references to individual CSD entries cited by reference code in the Table. A full list of CSD bibliographic entries is given in SUP 56701.

Refcode	Reference
ACBZPO01	*J. Am. Chem. Soc.*, 1975, **97**, 6729.
ACLTEP	*J. Organomet. Chem.*, 1980, **184**, 417.
ASAZOC	*Dokl. Akad. Nauk SSSR*, 1979, **249**, 120.
BALXOB	*J. Am. Chem. Soc.*, 1981, **103**, 4587.
BAPPAJ	*Inorg. Chem.*, 1981, **20**, 3071.
BARRIV	*Acta Chem. Scand., Ser. A*, 1981, **35**, 443.
BAWFUA	*Cryst. Struct. Commun.*, 1981, **10**, 1345.
BAWGAH	*Cryst. Struct. Commun.*, 1981, **10**, 1353.
BECTAE	*J. Org. Chem.*, 1981, **46**, 5048, 1981.
BELNIP	*Z. Naturforsch., Teil B*, 1982, **37**, 299.
BEMLIO	*Chem. Ber.*, 1982, **115**, 1126.
BEPZEB	*Cryst. Struct. Commun.*, 1982, **11**, 175.
BETJOZ	*J. Am. Chem. Soc.*, 1982, **104**, 1683.
BETUTE10	*Acta Chem. Scand., Ser. A*, 1976, **30**, 719.
BIBLAZ	*Zh. Strukt. Khim.*, 1981, **22**, 118.
BICGEZ	*Z. Anorg. Allg. Chem.*, 1982, **486**, 90.
BIHXIZ	*J. Chem. Soc., Chem. Commun.*, 1982, 982.
BIRGUE10	*Z. Naturforsch., Teil B*, 1983, **38**, 20.
BIRHAL10	*Z. Naturforsch., Teil B*, 1982, **37**, 1410.
BIZJAV	*J. Organomet. Chem.*, 1982, **238**, C1.
BOGPOC	*Z. Naturforsch., Teil B*, 1982, **37**, 1402.
BOGSUL	*Z. Naturforsch., Teil* B, 1982, **37**, 1230.
BOJLER	*Z. Anorg. Allg. Chem.*, 1982, **493**, 53.
BOJPUL	*Acta Chem. Scand., Ser. A*, 1982, **36**, 829.
BOPFER	*Chem. Ber.*, 1983, **116**, 146.
BOPFIV	*Chem. Ber.*, 1983, **116**, 146.
BOVMEE	*Acta Crystallogr., Sect. B*, 1982, **38**, 1048.
BQUINI	*Acta Crystallogr., Sect. B*, 1979, **35**, 1930.
BTUPTE	*Acta Chem. Scand., Ser. A*, 1975, **29**, 738.
BUDTEZ	*Z. Naturforsch., Teil B*, 1983, **38**, 454.
BUPSIB10	*Z. Anorg. Allg. Chem.*, 1981, **474**, 31.
BUSHAY	*Z. Naturforsch., Teil. B*, 1983, **38**, 692.
BUTHAZ10	*Inorg. Chem.*, 1984, **23**, 2582.
BUTSUE	*J. Chem. Soc., Chem. Commun.*, 1983, 862.
BUWZUO	*Acta Chem. Scand., Ser A*, 1983, **37**, 219.
BZPRIB	*Z. Naturforsch., Teil B*, 1981, **36**, 922.
BZTPPI	*Inorg. Chem.*, 1978, **17**, 894.
CAHJOK	*Inorg. Chem.*, 1983, **22**, 1809.
CAJMAB	*Chem. Z*, 1983, **107**, 169.
CANLUY	*Tetrahedron Lett.*, 1983, **24**, 4337.
CASSAQ	*J. Struct. Chem.*, 1983, **2**, 101.
CASTOF10	*Acta Crystallogr., Sect. C*, 1984, **40**, 1879.
CASYOK	*J. Struct. Chem.*, 1983, **2**, 107.
CECHEX	*Z. Anorg. Allg. Chem.*, 1984, **508**, 61.
CECXEN	*J. Struct. Chem.*, 1983, **2**, 207.
CEDCUJ	*J. Org. Chem.*, 1983, **48**, 5149.
CEHKAB	*Z. Naturforsch., Teil B*, 1984, **39**, 139.
CELDOM	*Acta Crystallogr., Sect. C*, 1984, **40**, 556.
CESSAU	*Acta Crystallogr., Sect. C*, 1984, **40**, 653.
CETTAW	*Chem. Ber.*, 1984, **117**, 1089.
CETUTE	*Acta Chem. Scand., Ser A*, 1975, **29**, 763.
CEYLUN	*Izv. Akad. Nauk SSSR, Ser. Khim.*, 1983, 2744.
CIFZUM	*Acta Chem. Scand., Ser A*, 1984, **38**, 289.
CIHRAM	*Angew. Chem., Int. Ed. Engl.*, 1984, **23**, 302.
CILRUK	*J. Chem. Soc., Chem. Commun.*, 1984, 1023.
CILSAR	*J. Chem. Soc., Chem. Commun.*, 1984, 1021.
CIMHIP	*Acta Crystallogr., C*, 1984, **40**, 1458.
CINTEY	*Dokl. Akad. Nauk SSSR*, 1984, **274**, 615.
CIPBUY	*J. Struct. Chem.*, 1983, **2**, 281.
CISMUM	*Z. Naturforsch., Teil B*, 1984, **39**, 485.
CISTED	*Z. Anorg. Allg. Chem.*, 1984, **511**, 95.
CIWYIQ	*Inorg. Chem.*, 1984, **23**, 1946.
CIYFOF	*Inorg. Chem.*, 1984, **23**, 1790.
CMBIDZ	*J. Org. Chem.*, 1979, **44**, 1447.
CODDEE	*Z. Naturforsch., Teil B*, 1984, **39**, 1257.
CODDII	*Z. Naturforsch., Teil B*, 1984, **39**, 1257.
COFVOI	*Z. Naturforsch., Teil B*, 1984, **39**, 1027.
COJCUZ	*Chem. Ber.*, 1984, **117**, 2686.
COSDIX	*Z. Naturforsch., Teil B*, 1984, **39**, 1344.
COZPIQ	*Chem. Ber.*, 1984, **117**, 2063.
COZVIW	*Z. Anorg. Allg. Chem.*, 1984, **515**, 7.
CTCNSE	*J. Am. Chem. Soc.*, 1980, **102**, 5430.
CUCPIZ	*J. Am. Chem. Soc.*, 1984, **106**, 7529.
CUDLOC	*J. Cryst. Spectrosc.*, 1985, **15**, 53.
CUDLUI	*J. Cryst. Spectrosc.*, 1985, **15**, 53.
CUGBAH	*Acta Crystallogr., Sect. C*, 1985, **41**, 476.
CXMSEO	*Acta Crystallogr., Sect. B*, 1973, **29**, 595.
DGLYSE	*Acta Crystallogr., Sect. B*, 1975, **31**, 1785.
DMESIP01	*Acta Crystallogr., Sect. C*, 1984, **40**, 895.
DSEMOR10	*J. Chem. Soc., Dalton Trans.*, 1980, 628.
DTHIBR10	*Inorg. Chem.*, 1971, **10**, 697.
EPHTEA	*Inorg. Chem.*, 1980, **19**, 2487.
ESEARS	*J. Chem. Soc. C*, 1971, 1511.
ETEARS	*J. Chem. Soc. C*, 1971, 1511.
FMESIB	*J. Organomet. Chem.*, 1980, **197**, 275.
FPHTEL	*J. Chem. Soc., Dalton Trans.*, 1980, 2306.
FPSULF10	*J. Am. Chem. Soc.*, 1982, **104**, 1683.
HCLENE10	*Acta Crystallogr., Sect. B*, 1982, **38**, 3139.
HMTITI	*Acta Crystallogr., Sect. B*, 1975, **31**, 1505.
HMTNTI	*Z. Anorg. Allg. Chem.*, 1974, **409**, 237.
HXPASC	*J. Chem. Soc., Dalton Trans.*, 1975, 1381.
IBZDAC11	*J. Chem. Soc., Dalton Trans.*, 1979, 854.
IFORAM	*Monatsh. Chem.*, 1974, **105**, 621.
IODMAM	*Acta Crystallogr., Sect. B*, 1977, **33**, 3209.
IPMUDS	*Acta Crystallogr., Sect. B*, 1973, **29**, 2128.
ISUREA10	*Acta Crystallogr., Sect. B*, 1972, **28**, 643.
LITMEB10	*J. Am. Chem. Soc.*, 1975, **97**, 6401.
MESIAD	*Z. Naturforsch., Teil B*, 1980, **35**, 789.
METAMM	*Acta Crystallogr.*, 1964, **17**, 1336.
MNPSIL	*J. Am. Chem. Soc.*, 1969, **91**, 4134.
MODIAZ	*J. Heterocycl. Chem.*, 1980, **17**, 1217.
MOPHTE	*Acta Chem. Scand., Ser. A*, 1980, **34**, 333.
MORTRS10	*J. Chem. Soc., Dalton Trans.*, 1980, 628.
NAPSEZ10	*J. Am. Chem. Soc.*, 1980, **102**, 5070.
NBBZAM	*Z. Naturforsch., Teil B*, 1977, **32**, 1416.
OPIMAS	*Aust. J. Chem.*, 1977, **30**, 2417.
OPNTEC10	*J. Chem. Soc., Dalton Trans.*, 1982, 251.
PHASCL	*Acta Crystallogr., Sect. B*, 1981, **37**, 1357.
PHASOC01	*Aust. J. Chem.*, 1975, **28**, 15.
PNPOSI	*J. Am. Chem. Soc.*, 1968, **90**, 5102.
SEBZQI	*J. Chem. Soc., Chem. Commun.*, 1977, 325.
SPSEBU	*Acta Chem. Scand., Ser. A*, 1979, **33**, 403.
TEACBR	*Cryst. Struct. Commun.*, 1974, **3**, 753.
THINBR	*J. Am. Chem. Soc.*, 1970, **92**, 4002.
TMPBTI	*Acta Crystallogr., Sect. B*, 1975, **31**, 1116.
TPASSN	*J. Chem. Soc., Dalton Trans.*, 1977, 514.
TPASTB	*Cryst. Struct. Commun.*, 1976, **5**, 39.
TPHOSI	*Z. Naturforsch., Teil B*, 1979, **34**, 1064.
TTEBPZ	*Z. Naturforsch., Teil B*, 1979, **34**, 256.
ZCMXSP	*Cryst. Struct. Commun.*, 1977, **6**, 93.

BOND LENGTHS IN ORGANOMETALLIC COMPOUNDS

This table summarizes the average values of interatomic distances of representative metal–ligand bonds. Sigma bonds between *d*- and *f*-block metals and the elements C, N, O, P, S, and As are included. The values are extracted from a much larger list in Reference 1. The tabulated values are the unweighted means of reported measurements on compounds in each category. If four or more measurements are available, the standard deviation is given in parentheses. All values are in Ångstrom units (10^{-10} m).

The first part of the table covers metal–carbon bonds in different ligand categories, while the second part covers metal bonds to other elements. R stands for any alkyl group; Me for a CH_3 group; C_6R_5 indicates an aryl group; and C(=O)R an acyl group. Metals are listed in atomic number order.

Reference

1. Orpen, A. G., Brammer, L., Allen, F. H., Kennard, O., Watson, D. G., and Taylor, R., *J. Chem. Soc. Dalton Trans.*, 1989, S1-S83.

M	M–CH_3	M–CH_2R	M–CR=CR_2	M–C_6R_5	M–C(=O)R
Ti		2.167	2.215(0.042)	2.148	
V				2.114(0.012)	
Cr	2.168		2.035(0.009)	2.075(0.019)	
Mn	2.095(0.030)	2.176(0.024)	2.007	2.064(0.021)	2.044
Fe	2.074	2.091(0.030)	1.991(0.039)	2.031(0.062)	1.997(0.033)
Co	2.014(0.023)	2.039(0.032)	1.934(0.019)	1.974	1.990
Ni	2.029	1.964	1.892(0.017)	1.917(0.038)	1.850(0.059)
Cu				2.020	
Zn		1.964			
Zr	2.292(0.049)		2.257		
Nb	2.336	1.319			
Mo	2.254(0.065)	2.250(0.061)	2.204(0.049)	2.193(0.054)	2.109
Ru	2.179(0.045)	2.036(0.010)	2.063	2.092(0.057)	2.091
Rh	2.092(0.027)	2.100	2.040(0.054)	2.011(0.026)	1.995(0.031)
Pd		2.028	2.000(0.024)	1.981(0.032)	1.982(0.029)
Hf	2.275(0.049)		2.205		
Ta	2.217(0.035)	2.225(0.056)		2.199(0.073)	
W	2.189(0.039)	2.175	2.224		
Re	2.173(0.051)	2.290		2.027	2.190(0.027)
Os		2.221	2.052	2.090(0.032)	2.161
Ir	2.175		2.071(0.044)	2.070(0.038)	2.019
Pt	2.083(0.045)	2.062(0.031)	2.024(0.037)	2.049(0.046)	1.991(0.025)
Au	2.066(0.045)		2.042	2.059(0.024)	
Hg	2.072(0.026)	2.125		2.086(0.040)	
Th	2.567				

M	M–NH_3	M–OH_2	M–PMe_3	M–SR	M–AsR_3
Ti		2.066(0.052)		2.369	2.686
V		2.129(0.131)	2.510(0.010)	2.378(0.007)	
Cr	2.069(0.008)	1.997(0.070)	2.389(0.069)	2.362	2.460(0.040)
Mn		2.189(0.040)	2.455(0.164)	2.366(0.054)	2.400(0.013)
Fe		2.085(0.066)	2.246(0.042)	2.271(0.028)	2.352(0.043)
Co	1.965(0.021)	2.085(0.064)	2.217(0.043)	2.254(0.025)	2.323(0.021)
Ni	2.074(0.093)	2.079(0.038)	2.204(0.031)	2.187(0.007)	2.333(0.035)
Cu	1.987(0.017)	2.186(0.215)			2.367(0.016)
Zn	2.044	2.090(0.061)		2.295	
Y		2.398(0.068)			
Zr			2.692		
Nb		2.248(0.137)			2.741(0.008)
Mo	2.217	2.201(0.094)	2.462(0.046)	2.401(0.050)	2.582(0.036)
Ru	2.126(0.024)	2.074(0.051)	2.307(0.050)		2.446(0.031)
Rh	2.114(0.018)	2.190(0.096)	2.266(0.036)		2.416(0.039)
Pd	2.032	2.200	2.287(0.018)		2.386(0.052)
Ag		2.350			
Cd		2.318(0.065)		2.444	
La		2.556(0.062)			
Ce		2.565(0.063)			
Pr		2.518(0.038)			
Nd		2.533(0.058)			
Sm		2.459(0.050)			
Eu		2.441(0.055)			
Gd		2.443(0.074)			
Tb		2.455			
Dy		2.409(0.074)			
Ho		2.407(0.069)			
Er		2.404(0.083)			
Yb		2.353(0.066)			
Lu		2.404(0.116)			
Ta			2.589(0.044)		
W		2.115(0.065)	2.485(0.039)		
Re	2.253	2.199(0.091)	2.369(0.065)		2.575(0.006)
Os	2.136	2.166	2.328(0.029)		
Ir	2.050(0.021)		2.323(0.028)	2.461	
Pt			2.295(0.036)	2.320(0.015)	2.366(0.058)
Au		2.157		2.293	
Hg		2.690(0.083)		2.402(0.065)	
Th		2.483(0.032)			
U		2.455(0.047)			

STRUCTURE OF FREE MOLECULES IN THE GAS PHASE

This table gives information on the geometric structure of selected molecules in the gas phase, including the overall geometry, interatomic distances, and bond angles. The molecules have been chosen to provide data on a wide variety of chemical bonds and to illustrate the influence of molecular environment on bond distances and angles. The table is restricted to molecules with conventional covalent or ionic bonds, but it should be pointed out that structure data on many loosely bonded complexes of the van der Waals type have recently become available. The references below contain data on many molecules that are not included here and give additional information such as uncertainties and isotopic variations.

The two techniques for gas phase structure determination are spectroscopy and electron diffraction. The following codes are used to indicate the method used for each set of data:

ED – Gas phase electron diffraction
MW – Microwave spectroscopy, including both measurements in bulk gases and molecular beams
IR – Infrared spectroscopy
R – Raman spectroscopy
UV – Electronic spectroscopy in the ultraviolet and visible regions, including fluorescence measurements
ESR – Electron spin resonance.

In some cases data from two sources have been combined to derive the structure; these are labeled by "ED, MW," for example.

Because of the internal vibrations that are present in all molecules, even in their lowest energy state, the definition of interatomic distance is not a simple matter. The ideal measure is the equilibrium distance in the hypothetical non-vibrating state, designated by r_e. This is the value of the separation of the atoms at the minimum of the potential function that describes the forces between the two atoms. All other measures represent some form of average, generally complex, over the vibrational motions. Since the potential function is asymmetric and less steep at distances beyond the potential minimum, the average distance is normally greater than r_e. Distances determined by electron diffraction (ED) represent an average over all vibrational states that are populated at the temperature of the measurement; the most common measure is designated r_g. Distances determined by spectroscopy (MW, IR, R, or UV) through measurements on the ground vibrational state of the molecule, designated by r_0, describe some form of average, not easily defined, over the zero-point vibrations. Another measure that is frequently used in microwave spectroscopy is the "substitution" distance r_s, which is operationally defined through a series of measurements on different isotopic species. In simple cases, r_s often lies between r_0 and r_e and is therefore a closer approximation to r_e. Several other types of averages have been used; good discussions can be found in Volumes II/25 and II/28 of the *Landolt-Börnstein* series (Reference 1) and in References 4 and 5.

Unless otherwise specified, distances and angles given in this table are r_0 values if the method is spectroscopic and r_g values if the method is electron diffraction. When given, equilibrium and substitution distances are designated by r_e and r_s, respectively.

Many interatomic distances and angles calculated by *ab initio* techniques have been reported in the recent literature. However, it should be emphasized that all data in this table are obtained from direct experimental measurements. In a few cases, *ab initio* calculations of vibration-rotation interaction constants have been combined with the primary experimental measurements to derive r_e values in the table.

The number of significant figures in the values is an indication of the precision of the measurement; thus a distance quoted to three decimal places is probably reliable to about 0.005 Å or better. However, discrepancies between r_e, r_0, and r_g values for the same bond are often the order of 0.01 Å because of vibrational averaging considerations, so care must be taken in comparing bond distances in different molecules. Some distances in simple molecules are given here to four or five decimal places, but little chemical significance can be attached to differences beyond the third decimal place.

The table is presented in two parts: Part A covers molecules that do not contain carbon while Part B lists carbon-containing molecules. Because many of the entries in Part A are free radicals or other transient species whose systematic chemical names are unfamiliar, the listing in Part A is in order of chemical formula. Part B is ordered by name. In both parts the second column gives information on the overall configuration of the molecule, often indicated by the point group of the equilibrium geometry. Columns 3 through 8 give the values of the bond distances and angles, and the last column indicates the experimental method. Distances are given in Å units, where 1 Å = 10^{-10} m or 0.1 nm. Angles are given in degrees.

The efforts of Kozo Kuchitsu in preparing an earlier version of this table and in giving advice on the new version are gratefully acknowledged.

References

1. *Landolt-Börnstein Numerical Data and Functional Relationships in Science and Technology*, Springer-Verlag, Berlin. The following volumes are in the series *Structure Data of Free Polyatomic Molecules*:
   II/7, 1976
   II/15,1987
   II/21, *Supplement to II/7 and II/15*, 1992
   II/23, *Supplement to II/7, II/15, and II/21*, 1995
   II/25A, *Inorganic Molecules*, 1998
   II/25B, *Molecules Containing One or Two Carbon Atoms*, 1999
   II/25C, *Molecules Containing Three or Four Carbon Atoms*, 2000
   II/25D, *Molecules Containing Five or More Carbon Atoms*, 2003
   II/28A, *Inorganic Molecules*, 2006
   II/28B, *Molecules Containing One or Two Carbon Atoms*, 2006
   II/28C, *Molecules Containing Three or Four Carbon Atoms*, 2007
   II/28D, *Molecules Containing Five or More Carbon Atoms*, 2007.
2. Harmony, M. D., Laurie, V. W., Kuczkowski, R. L., Schwendeman, R. H., Ramsay, D. A., Lovas, F. J., Lafferty, W. J., and Maki, A. G., "Molecular Structure of Gas-Phase Polyatomic Molecules Determined by Spectroscopic Methods", *J. Phys. Chem. Ref. Data*, 8, 619, 1979.
3. Huber, K. P., and Herzberg, G., *Molecular Spectra and Molecular Structure IV. Constants of Diatomic Molecules*, Van Nostrand Reinhold, London, 1979.
4. Hargittai, M., "Molecular Structure of Metal Halides," *Chem. Rev.* 100, 2233-2301, 2000.
5. Harmony, M. D., and Berry, R. J., *Struct. Chem.* 1, 49, 1989.

Part 1 Molecules Not Containing Carbon

Formula	Structure	Bond distances in Å and angles in degrees						Method
AgBr		Ag—Br (r_e)	2.3931					MW
AgCl		Ag—Cl (r_e)	2.2808					MW
AgF		Ag—F (r_e)	1.9832					MW
AgH		Ag—H (r_e)	1.617					UV
AgI		Ag—I (r_e)	2.5446					MW
AgLi		Ag—Li	2.41					UV
AgO		Ag—O (r_e)	2.0030					UV
AgOH	bent	Ag—O	2.016	O—H	0.952	∠HOAg	108.3 (ass.)	MW
AlBr		Al—Br (r_e)	2.295					UV
$AlBr_3$	D_{3h}	Al—Br	2.221					ED
AlCa		Al—Ca	3.148					UV
AlCl		Al—Cl (r_e)	2.1301					MW
$AlCl_3$	D_{3h}	Al—Cl	2.063					ED
AlCo		Al—Co	2.283					UV
AlCu		Al—Cu	2.339					UV
AlF		Al—F (r_e)	1.6544					MW
AlF_3	D_{3h}	Al—F	1.633					ED
AlH		Al—H (r_e)	1.6482					UV
AlI		Al—I (r_e)	2.5371					MW
AlI_3	D_{3h}	Al—I	2.461					ED
AlK		Al—K	3.88					UV
AlMn		Al—Mn	2.638					UV
AlNi		Al—Ni	2.321					UV
AlO		Al—O (r_e)	1.6176					UV
AlS		Al—S (r_e)	2.029					UV
AlV		Al—V	2.620					UV
AlZn		Al—Zn	2.696					UV
Al_2		Al—Al (r_e)	2.701					UV
Al_2Br_6	Br_a Br_b Br_a / Al Al / Br_a Br_b Br_a D_{2h}	Al—Br_a	2.234	Al—Br_b	2.433			ED
		∠Br_bAlBr_b	91.6	∠Br_aAlBr_a	122			
Al_2Cl_6	See Al_2Br_6	Al—Cl_a	2.061	Al—Cl_b	2.250			ED
	D_{2h}	∠Cl_bAlCl_b	90.0	∠Cl_aAlCl_a	122			
$AsBr_3$	C_{3v}	As—Br	2.324	∠BrAsBr	99.6			ED
$AsCl_3$	C_{3v}	As—Cl	2.165	∠ClAsCl	98.6			ED, MW
AsF_3	C_{3v}	As—F	1.710	∠FAsF	95.9			ED, MW
AsF_5	F_a, F_b, F_b—As, F_b, F_a D_{3h}	As—F_a	1.711	As—F_b	1.656			ED
AsH		As—H (r_e)	1.5232					UV
AsH_3	C_{3v}	As—H (r_e)	1.511	∠HAsH (θ_e)	92.1			MW, IR
AsI_3	C_{3v}	As—I	2.557	∠IAsI	100.2			ED
AsN		As—N (r_e)	1.6184					UV
AsO		As—O (r_e)	1.6236					UV
AsP		As—P (r_e)	1.99954					MW
As_2		As—As (r_e)	2.1026					UV
AuH		Au—H (r_e)	1.5237					UV
Au_2		Au—Au (r_e)	2.4719					UV
BBr		B—Br (r_e)	1.888					UV
BBr_3	D_{3h}	B—Br	1.893					ED
BCl		B—Cl (r_e)	1.7153					UV
$BClF_2$	C_{2v}	B—Cl (r_s)	1.728	B—F	1.315	∠FBF	118.1	MW

Formula	Structure	Bond distances in Å and angles in degrees						Method
BCl_3	D_{3h}	B—Cl	1.742					ED
BF		B—F (r_e)	1.2626					UV
BF_2H		B—H	1.189	B—F	1.311	∠FBF	118.3	MW
BF_2OH	F_aF_bBOH	B—F_a (r_e)	1.3229	B—F_b (r_e)	1.3129	B—O (r_e)	1.3448	MW
	planar	∠FBF (θ_e)	118.36	∠F_aBO (θ_e)	122.25	∠BOH (θ_e)	113.14	
	F_a *cis* to OH	O—H (r_e)	0.9574					
BF_3	D_{3h}	B—F	1.313					ED, IR
BH		B—H (r_e)	1.2325					UV
BH_2NH_2	planar	B—N	1.391	B—H	1.195	N—H	1.004	MW
		∠HBH	122.2	∠HNH	114.2			
BH_3	planar	B—H	1.1900					IR
BH_3PH_3	staggered form	B—P	1.937	B—H	1.212	P—H	1.399	MW
		∠PBH	103.6	∠BPH	116.9	∠HBH	114.6	
		∠HPH	101.3					
BI_3	D_{3h}	B—I	2.118					ED
BN		B—N (r_e)	1.281					UV
BO		B—O (r_e)	1.2045					EPR
BO_2	linear	B—O	1.265					UV
BS		B—S	1.6091					UV
B_2		B—B (r_e)	1.590					UV
B_2H_6	H_a H_b H_a / B B / H_a H_b H_a	B—H_a	1.19	B—H_b	1.33	B···B	1.77	IR, ED
		∠H_aBH_a	122	∠H_bBH_b	97			
$B_3H_3O_3$		B—O	1.376	∠BOB	120	∠OBO	120	ED
$B_3H_6N_3$	C_2	B—N	1.435	B—H	1.26	N—H	1.05	ED
		∠BNB	121	∠NBN	118			
BaBr		Ba—Br (r_e)	2.8445					UV
$BaBr_2$		Ba—Br	2.912	∠BrBaBr	137.0			ED
BaCl		Ba—Cl (r_e)	2.6828					UV
BaF		Ba—F (r_e)	2.163					UV
BaH		Ba—H (r_e)	2.2318					UV
BaI		Ba—I (r_e)	3.0848					UV
BaI_2		Ba—I	3.150	∠IBaI	137.6			ED
BaO		Ba—O (r_e)	1.9397					MW
BaOH	linear	Ba—O	2.200	O—H	0.927			UV
BaS		Ba—S (r_e)	2.5074					MBE
$BeCl_2$	linear	Be—Cl (r_e)	1.791					ED,IR
BeF		Be—F (r_e)	1.3609					UV
BeF_2	linear	Be—F (r_e)	1.3730					IR
BeH		Be—H (r_e)	1.3431					UV
BeH_2	linear	Be—H(r_e)	1.3264					IR
BeO		Be—O (r_e)	1.3308					UV
BeS		Be—S (r_e)	1.7415					UV
BiBr		Bi—Br (r_e)	2.6095					MW
$BiBr_3$	C_{3v}	Bi—Br	2.577	∠BrBiBr	98.6			ED
BiCl		Bi—Cl (r_e)	2.4716					MW
$BiCl_3$	C_{3v}	Bi—Cl	2.424	∠ClBiCl	97.5			ED
BiF		Bi—F (r_e)	2.0516					MW
BiF_3	C_{3v}	Bi—F	1.987	∠FBiF	96.1			ED
BiH		Bi—H (r_e)	1.805					UV
BiI		Bi—I (r_e)	2.8005					MW
BiI_3	C_{3v}	Bi—I	2.807	∠IBiI	99.5			ED
BiO		Bi—O (r_e)	1.934					UV
BiP		Bi—P (r_e)	2.29345					IR
Bi_2		Bi—Bi (r_e)	2.6596					UV
BrCl		Br—Cl (r_e)	2.1361					MW
BrF		Br—F (r_e)	1.7590					MW
BrF_3	F_a—Br—F_a with F_b on Br; C_{2v}	Br—F_a	1.810	∠F_{ax}BrF_{eq}	85.1	∠F_aBrF_b	86.2	MW
		Br—F_b	1.721					

Formula	Structure	Bond distances in Å and angles in degrees						Method
BrF_5	C_{4v}	Br—F (av.)	1.753	(Br—F_{eq}) – (Br—F_{ax})	0.069	∠F_{ax}BrF_{eq}	85.1	ED, MW
BrN_3	Br$N_aN_bN_c$ planar	N_a—N_b ∠NNN	1.113 (ass.) 170.7	N_b—N_c ∠BrNN	1.247 109.7	N_a—Br	1.899	ED
BrO		Br—O (r_e)	1.7172					MW
BrO_2	C_{2v}	Br—O (r_e)	1.644	∠OBrO (θ_e)	114.3			MW
Br_2		Br—Br (r_e)	2.2811					R
$CaBr_2$	linear	Ca—Br	2.62					ED
CaCl		Ca—Cl (r_e)	2.43676					UV
$CaCl_2$	linear	Ca—Cl	2.483					ED
CaF		Ca—F (r_e)	1.967					UV
CaH		Ca—H (r_e)	2.002					UV
CaI		Ca—I (r_e)	2.8286					UV
CaI_2	linear	Ca—I	2.840					ED
CaO		Ca—O (r_e)	1.8221					UV
CaOH	linear	Ca—O	1.985	O—H	0.921			UV
CaS		Ca—S (r_e)	2.3178					UV
CdH		Cd—H (r_e)	1.781					EPR
CdH_2	linear	Cd—H	1.6792					IR
$CdBr_2$	linear	Cd—Br	2.394					ED
$CdCl_2$	linear	Cd—Cl	2.284					ED
CdI_2	linear	Cd—I	2.582					ED
CeF_4	T_d	Ce—F	2.036					ED
CeI_3	quasiplanar	Ce—I	2.948					ED
ClBS	linear	B—Cl	1.681	B—S	1.606			MW
ClF		Cl—F (r_e)	1.6283					MW
ClF_3	F_a—Cl—F_a F_b	Cl—F_a	1.698	Cl—F_b	1.598	∠F_aClF_b	87.5	MW
ClN_3	Cl$N_aN_bN_c$ planar	N_a—N_b ∠NNN	1.253 171.0	N_b—N_c ∠ClNN	1.113 108.7	N_a—Cl	1.746	MW
ClO		Cl—O (r_e)	1.5696					MW, UV
ClO_2	C_{2v}	Cl—O	1.470	∠OClO	117.38			MW
Cl_2		Cl—Cl (r_e)	1.9878					UV
Cl_2O	C_{2v}	Cl—O	1.6959	∠ClOCl	110.89			MW
$CoBr_2$	linear	Co—Br	2.241					ED
$CoCl_2$	linear	Co—Cl	2.113					ED
CoF_2	linear	Co—F	1.754	[Co—F (r_e)]	1.738			ED
CoF_3	D_{3h}	Co—F	1.732					ED
CoH		Co—H (r_e)	1.542					UV
CrF_2	linear	Cr—F	1.795					ED
CrF_3	D_{3h}	Cr—F	1.732					ED
CrF_4	T_d	Cr—F	1.706					ED
CrH		Cr—H (r_e)	1.656					UV
CrO		Cr—O (r_e)	1.615					UV
CsBr		Cs—Br (r_e)	3.0723					MW
CsCl		Cs—Cl (r_e)	2.9063					MW
CsF		Cs—F (r_e)	2.3454					MW
CsH		Cs—H (r_e)	2.4938					UV
CsI		Cs—I (r_e)	3.3152					MW
CsO		Cs—O (r_e)	2.3007					MW
CsOH	linear; large amplitude bending mode	Cs—O (r_e)	2.395	O—H (r_e)	0.97			MW
Cs_2		Cs—Cs (r_e)	4.47					UV
CuBr		Cu—Br (r_e)	2.1734					MW
CuCl		Cu—Cl (r_e)	2.0512					MW
CuF		Cu—F (r_e)	1.7449					MW
CuF_2	linear	Cu—F	1.713					ED
CuH		Cu—H (r_e)	1.4626					UV
CuI		Cu—I (r_e)	2.3383					MW
CuLi		Cu—Li	2.26					UV
CuO		Cu—O (r_e)	1.7244					UV

Formula	Structure	Bond distances in Å and angles in degrees						Method
CuOH	bent	Cu—O (r_s)	1.769	O—H	0.952	∠HOCu	110.24 (θ_s)	MW
CuS		Cu—S	2.051					UV
Cu_2		Cu—Cu (r_e)	2.2197					UV
$DyBr_3$	quasiplanar	Dy—Br	2.609					ED
$DyCl_3$	quasiplanar	Dy—Cl	2.461					ED
FN_3	$FN_aN_bN_c$ planar	N_a—N_b ∠NNN	1.253 170.3	N_b—N_c ∠FNN	1.132 103.8	N_a—F	1.439	MW
F_2		F—F (r_e)	1.4119					R
$FeBr_2$	linear	Fe—Br	2.294					ED
$FeCl_2$	linear	Fe—Cl	2.132					UV,ED
FeF_2	linear	Fe—F	1.769	[Fe—F (r_e)]	1.755			ED
FeF_3	D_{3h}	Fe—F	1.763					ED
FeH		Fe—H	1.620					IR
FeO		Fe—O	1.444					UV
FeS		Fe—S	2.017					MW
GaBr		Ga—Br (r_e)	2.3525					MW
$GaBr_3$	D_{3h}	Ga—Br	2.249					ED
GaCl		Ga—Cl (r_e)	2.2017					MW
$GaCl_3$	D_{3h}	Ga—Cl	2.110					ED
GaF		Ga—F (r_e)	1.7744					MW
GaF_3	D_{3h}	Ga—F	1.725					ED
GaH		Ga—H (r_e)	1.663					UV
GaI		Ga—I (r_e)	2.5747					MW
GaI_3	D_{3h}	Ga—I	2.458					ED
GaO		Ga—O	1.744					UV
Ga_2Br_6	See Al_2Br_6 D_{2h}	Ga—Br_a ∠Br_aGaBr_a	2.250 92.7	Ga—Br_b ∠Br_bGaBr_b	2.453 123			ED
Ga_2Cl_6	See Al_2Br_6 D_{2h}	Ga—Cl_a ∠Cl_aGaCl_a	2.116 90	Ga—Cl_b ∠Cl_bGaCl_b	2.305 124.5			ED
$GdBr_3$	C_{3v}	Gd—Br	2.641					ED
$GdCl_3$	C_{3v}	Gd—Cl	2.488					ED
GdF_3	C_{3v}	Gd—F	2.053					ED
GdI_3	C_{3v}	Gd—I	2.840	∠IGdI	108			ED
$GeBrH_3$	C_{3v}	Ge—H	1.526	Ge—Br	2.299	∠HGeH	106.2	MW, IR
$GeBr_2$		Ge—Br (r_e)	2.359	∠BrGeBr	101.0			ED
$GeBr_4$	T_d	Ge—Br	2.272					ED
$GeClH_3$	C_{3v}	Ge—H	1.537	Ge—Cl	2.150	∠HGeH	111.0	IR, MW
$GeCl_2$		Ge—Cl (r_e)	2.186	∠ClGeCl	100.3			ED
$GeCl_4$	T_d	Ge—Cl	2.113					ED
$GeFH_3$	C_{3v}	Ge—H	1.522	Ge—F	1.732	∠HGeH	113.0	MW, IR
GeF_2		Ge—F (r_e)	1.7321	∠FGeF (θ_e)	97.15			MW
GeH		Ge—H (r_e)	1.5880					UV
GeHI		Ge—I	2.525	Ge—H	1.593	∠HGeI	93.5	UV
GeH_4	T_d	Ge—H	1.5251					IR, R
GeI_2		Ge—I	2.540	∠IGeI	102.1			ED
GeI_4	T_d	Ge—I	2.515					ED
GeO		Ge—O (r_e)	1.6246					MW
GeS		Ge—S (r_e)	2.0121					MW
GeSe		Ge—Se (r_e)	2.1346					MW
GeTe		Ge—Te (r_e)	2.3402					MW
Ge_2H_6		Ge—Ge ∠HGeH	2.403 106.4	Ge—H ∠GeGeH	1.541 112.5			ED
HBr		H—Br (r_e)	1.4145					MW
HCl		H—Cl (r_e)	1.2746					MW
HClO	ClOH (bent)	Cl—O	1.690	O—H	0.975	∠HOCl	102.5	MW, IR
$HClO_4$	H / O_b / Cl / O_a O_a O_a	Cl—O_a ∠O_aClO_a	1.407 114.3	Cl—O_b ∠O_aClO_b	1.639 104.1			ED

Formula	Structure	Bond distances in Å and angles in degrees						Method
HF		H—F (r_e)	0.9169					MW
HFO	FOH (bent)	F—O	1.442	O—H	0.96	∠HOF	97.2	MW
HI		H—I (r_e)	1.6090					MW
HIO	IOH (bent)	I—O	1.9941	O—H	0.967	∠HOI	103.9	MW
HNO	bent	N—O	1.212	N—H	1.063	∠HNO	108.6	UV
HNO_2	O_a=N—O_bH	*s-trans* conformer		*s-cis* conformer				MW
		O_b—H	0.958	O_b—H	0.98			
		N—O_b	1.432	N—O_b	1.39			
		N—O_a	1.170	N—O_a	1.19			
		∠O_aNO_b	110.7	∠O_aNO_b	114			
		∠NO_bH	102.1	∠NO_bH	104			
HNO_3	H—O_c—N(=O_a)(=O_b)	N—O_a	1.20	N—O_b	1.21	N—O_c	1.41	MW
	planar	O_c—H	0.96	∠O_cNO_b	115.9	∠HO_cN	102.2	
		∠O_cNO_a	113.9					
HNSO	planar *cis*	N—S	1.512	S—O	1.451	N—H	1.029	MW
		∠NSO	120.4	∠HNS	115.8			
HN_3	$HN_aN_bN_c$ planar	N_a—N_b	1.245	N_b—N_c	1.134	N_a—H	1.015	MW
		∠NNN	171.8	∠HNN	109.2			
HPO		P—O	1.4843	P—H	1.473	∠HPO	104.57	MW
H_2		H—H (r_e)	0.74144					UV
H_2O	C_{2v}	O—H (r_e)	0.9575	∠HOH (θ_e)	104.51			MW, IR
H_2O_2	C_2	O—O	1.475	∠OOH	94.8	dihedral angle	119.8	IR
H_2S	C_{2v}	H—S (r_e)	1.3356	∠HSH (θ_e)	92.12			MW, IR
H_2SO_4	H_b–O_b, H_a–O_a, S, O_d, O_c	O—H	0.97	S—O_a	1.574	S—O_c	1.422	MW
		∠O_aSO_b	101.3	∠O_cSO_d	123.3	∠O_aSO_c	108.6	
		∠O_aSO_d	106.4	∠H_aO_aS	108.5	dihedral angle between the H_aO_aS and O_aSO_c planes	20.8	
	C_2	dihedral angle between the H_aO_aS and O_aSO_b planes	90.9	dihedral angle between the H_aSO_b and O_cSO_d planes	88.4			
H_2S_2	C_2	S—S	2.055	S—H	1.327	∠SSH	91.3	ED, MW
		dihedral angle	90.6					
$HfBr_4$	T_d	Hf—Br	2.450					ED
$HfCl_4$	T_d	Hf—Cl	2.316					ED
HfF		Hf—F	1.8596					UV
HfF_4	T_d	Hf—F	1.909					ED
HfI_4	T_d	Hf—I	2.662					ED
$HgBr_2$	linear	Hg—Br	2.384					ED
$HgCl_2$	linear	Hg—Cl	2.252					ED
HgH		Hg—H (r_e)	1.7404					UV
HgI_2	linear	Hg—I	2.568					ED
$HoCl_3$		Ho—Cl	2.462					ED
HoF_3		Ho—F	2.007					ED
HoO		Ho—O	1.797					UV
IBr		I—Br (r_e)	2.4691					MW
ICl		I—Cl (r_e)	2.3210					MW
IF		I—F (r_e)	1.9098					UV
IF_5	C_{4v}	I—F (av.)	1.860	(I—F_{eq}) – (I—F_{ax})	0.03	∠$F_{ax}IF_{eq}$	82.1	ED, MW
IO		I—O (r_e)	1.8676					MW
I_2		I—I (r_e)	2.6663					R
InBr		In—Br (r_e)	2.5432					MW
InCl		In—Cl (r_e)	2.4012					MW
$InCl_3$		In—Cl	2.291					ED
InF		In—F (r_e)	1.9854					MW
InH		In—H (r_e)	1.8376					UV

Formula	Structure	Bond distances in Å and angles in degrees						Method
InI		In—I (r_e)	2.7537					MW
IrF_6	O_h	Ir—F	1.831					ED
KBH_4	$H_a(BH_3)K$ (C_{3v})	B—H (BH_3)	1.272	B—H_a	1.233	K—B	2.656	MW
KBr		K—Br (r_e)	2.8208					MW
KCl		K—Cl (r_e)	2.6667					MW
KF		K—F (r_e)	2.1716					MW
KH		K—H (r_e)	2.244					UV
KI		K—I (r_e)	3.0478					MW
KOH	linear; large amplitude bending mode	K—O	2.212	O—H	0.91			MW
K_2		K—K (r_e)	3.9051					UV
KrF_2	linear	Kr—F	1.89					ED
LaBr		La—Br (r_e)	2.65208					MW
$LaBr_3$	C_{3v}	La—Br	2.742					ED
LaCl		La—Cl (r_e)	2.49804					MW
$LaCl_3$	C_{3v}	La—Cl	2.589					ED
LaF		La—F (r_e)	2.02338					MW
LaI		La—I (r_e)	2.87885					MW
LaO		La—O (r_e)	1.82591					UV
$LiBH_4$	$H_a(BH_3)Li$ (C_{3v})	B—H (H_3)	1.257	B—H_a	1.218	Li—B	1.939	MW
LiBr		Li—Br (r_e)	2.1704					MW
LiCl		Li—Cl (r_e)	2.0207					MW
LiF		Li—F (r_e)	1.5639					MW
LiH		Li—H (r_e)	1.5949					MW
LiI		Li—I (r_e)	2.3919					MW
LiO		Li—O (r_e)	1.68822					UV
LiOH	linear	Li—O (r_e)	1.5776	O—H (r_e)	0.949			MW
Li_2		Li—Li (r_e)	2.0729					UV
Li_2Cl_2	Li Cl Cl Li	Li—Cl	2.23	Cl—Cl	3.61	∠ClLiCl	108	ED
Li_2O	linear	Li—O	1.606					UV
$LuBr_3$	C_{3v}	Lu—Br	2.557					ED
$LuCl_3$	C_{3v}	Lu—Cl	2.417	∠ClLuCl	112			ED
LuI_3	C_{3v}	Lu—I	2.768					ED
MgBr		Mg—Br (r_e)	2.34742					MW
MgCl		Mg—Cl (r_e)	2.1964					UV
$MgCl_2$	linear	Mg—Cl	2.179					ED
MgF		Mg—F (r_e)	1.7500					UV
MgF_2	linear	Mg—F	1.771					ED
MgH		Mg—H (r_e)	1.7297					UV
MgO		Mg—O (r_e)	1.749					UV
MgOH	linear	Mg—O	1.770	O—H	0.912			UV
Mg_2		Mg—Mg (r_e)	3.891					UV
$MnBr_2$	linear	Mn—Br	2.344					ED
$MnCl_2$	linear	Mn—Cl	2.202					ED
MnF_2	linear	Mn—F	1.811	[Mn—F (r_e)]	1.797			ED
MnH		Mn—H (r_e)	1.7308					UV
MnI_2	linear	Mn—I	2.538					ED
$MoCl_4O$	C_{4v}	Mo—Cl	2.279	Mo—O	1.658			ED
		∠ClMoCl	87.2					
MoF_4		Mo—F	1.851					ED
MoF_6	O_h	Mo—F	1.821					ED
NBr		N—Br (r_e)	1.79					UV
NCl		N—Cl (r_e)	1.6107					UV
$NClH_2$		N—H	1.017	N—Cl	1.748			MW, IR
		∠HNCl	103.7	∠HNH	107			
NCl_3		N—Cl	1.759	∠ClNCl	107.1			ED

Formula	Structure	Bond distances in Å and angles in degrees						Method
NF		N—F (r_e)	1.3170					UV
NF_2		N—F	1.3528	∠FNF	103.18			MW
NH_2		N—H	1.024	∠HNH	103.3			UV
NH_2NO_2		N—N	1.427	N—H	1.005			MW
		dihedral angle between NH_2 and NNO_2 planes	128.2	∠HNH	115.2	∠ONO	130.1	
NH_3	C_{3v}	N—H (r_e)	1.012	∠HNH (θ_e)	106.7			IR
NH_4Cl	$H_3N{\cdots}HCl$ (C_{3v})	N—Cl	3.136					MW
NH		N—H (r_e)	1.0362					LMR
NH_2OH	bisector of HNH angle is *trans* to OH bond	N—O	1.453	N—H	1.02	O—H	0.962	MW
		∠HNO	103.3	∠HNH	107	∠NOH	101.4	
NO		N—O (r_e)	1.1506					IR
NOCl		N—O	1.14	N—Cl	1.975	∠ONCl	113	MW
NOF		N—O	1.136	N—F	1.512	∠FNO	110.1	MW
NO_2		N—O	1.193	∠ONO	134.1			MW
NO_2Cl	C_{2v}	N—O	1.202	N—Cl	1.840	∠ONO	130.6	MW
NO_2F	C_{2v}	N—O	1.1798	N—F	1.467	∠ONO	136	MW
NS		N—S (r_e)	1.4940					IR
N_2		N—N (r_e)	1.0977					UV
N_2H_4	H_a atom is closer to the C_2 axis, H_b is farther from the C_2 axis	N—N	1.449	N—H	1.021	$\angle NNH_b$	106	ED, MW
		∠HNH	106.6 (ass.)	$\angle NNH_a$	112			
		dihedral angle of internal rotation	91					
N_2O		N—N (r_e)	1.1284	N—O (r_e)	1.1841			MW, IR
N_2O_3	O_a O_b / N_a—N_b / O_c	N_a—N_b	1.864	N_a—O_a	1.142			MW
		N_b—O_b	1.202	N_b—O_c	1.217			
		$\angle O_aN_aN_b$	105.05	$\angle N_aN_bO_b$	112.72	$\angle N_aN_bO_c$	117.47	
N_2O_4	O O / N—N / O O; D_{2h}	N—N	1.782	N—O	1.190	∠ONO	135.4	ED
$NaBH_4$	$H_a(BH_3)Na$ (C_{3v})	B—H (BH_3)	1.278	B—H_a	1.238	Na—B	2.308	MW
NaBr		Na—Br (r_e)	2.5020					MW
NaCl		Na—Cl (r_e)	2.3609					MW
NaF		Na—F (r_e)	1.9260					MW
NaH		Na—H (r_e)	1.8873					UV
NaI		Na—I (r_e)	2.7115					MW
NaO		Na—O (r_e)	2.05155					UV
Na_2		Na—Na (r_e)	3.0789					UV
$NbCl_4$	T_d	Nb—Cl	2.279					ED
$NbCl_5$	D_{3h}	Nb—Cl_{ax}	2.307	Nb—Cl_{eq}	2.276			ED
NbO		Nb—O (r_e)	1.691					UV
NdI_3	C_{3v}	Nd—I	2.879					ED
NiBr		Ni—Br	2.1963					UV
$NiBr_2$	linear	Ni—Br	2.201					ED
$NiCl_2$	linear	Ni—Cl	2.076					ED
NiF_2	linear	Ni—F	1.729	[Ni—F (r_e)]	1.715			ED
NiH		Ni—H (r_e)	1.476					UV
NiI		Ni—I	2.348					UV
NpF_6	O_h	Np—F	1.982					ED
OF		O—F (r_e)	1.3579					LMR
OF_2	C_{2v}	O—F (r_e)	1.4053	∠FOF (θ_e)	103.07			MW
OH		O—H (r_e)	0.96966					UV
$O(SiH_3)_2$		Si—H	1.486	Si—O	1.634	∠SiOSi	144.1	ED
O_2		O—O (r_e)	1.2074					MW

Formula	Structure	Bond distances in Å and angles in degrees						Method
O_2F_2	C_2	O—O	1.217	F—O	1.575	∠OOF	109.5	MW
		dihedral angle of internal rotation	87.5					
O_3	C_{2v}	O—O (r_e)	1.2716	∠OOO (θ_e)	117.47			MW
OsF_6	O_h	Os—F	1.832					ED
OsO_4	T_d	Os—O	1.712					ED
PBr_3	C_{3v}	P—Br	2.220	∠BrPBr	101.0			ED
PCl		P—Cl (r_e)	2.01461					UV
PCl_3	C_{3v}	P—Cl	2.039	∠ClPCl	100.27			ED
PCl_5	Cl_a, Cl_b, Cl_b—P, Cl_b, Cl_a D_{3h}	P—Cl_a	2.124	P—Cl_b	2.020			ED
PF		P—F (r_e)	1.5896					UV
PF_3	C_{3v}	P—F	1.570	∠FPF	97.8			ED, MW
PF_5	D_{3h}	P—F_{eq}	1.534	P—F_{ax}	1.577			ED
PH		P—H (r_e)	1.4223					LMR
PH_2		P—H	1.418	∠HPH	91.70			UV
PH_3	C_{3v}	P—H	1.4200	∠HPH	93.345			MW
PN		N—P (r_e)	1.49087					MW
PO		O—P (r_e)	1.4759					UV
$POCl_3$	C_{3v}	P—O	1.449	P—Cl	1.993	∠ClPCl	103.3	ED
POF_3	C_{3v}	P—O	1.436	P—F	1.524	∠FPF	101.3	ED, MW
P_2		P—P (r_e)	1.8931					UV
P_2F_4	*trans conformer*	P—F	1.587	P—P	2.281	∠FPF	99.1	P_2F_4
		∠PPF	95.4					
P_4	T_d	P—P	2.21					ED
P_4O_6	T_d	P—O	1.638	∠POP	126.4			ED
$PbBr_2$	bent	Pb—Br (r_e)	2.598					ED
$PbCl_2$	bent	Pb—Cl (r_e)	2.444					ED
$PbCl_4$	T_d	Pb—Cl	2.369					ED
PbF		Pb—F (r_e)	2.0575					UV
PbF_2	bent	Pb—F (r_e)	2.041					ED
PbH		Pb—H (r_e)	1.839					UV
PbI_2	bent	Pb—I (r_e)	2.807					ED
PbO		Pb—O (r_e)	1.9218					MW
PbS		Pb—S (r_e)	2.2869					MW
PbSe		Pb—Se (r_e)	2.4022					MW
PbTe		Pb—Te (r_e)	2.5950					MW
$PrCl_3$	C_{3v}	Pr—Cl	2.554					ED
PrF_3	C_{3v}	Pr—F	2.091					ED
PrI_3	C_{3v}	Pr—I	2.901	∠IPrI	113			ED
PtC		Pt—C (r_e)	1.6767					UV
PtH		Pt—H (r_e)	1.52852					UV
PtN		Pt—N (r_e)	1.682					MW
PtO		Pt—O (r_e)	1.7273					UV
PtS		Pt—S (r_e)	2.03983					MW
PtSi		Pt—Si (r_e)	2.0612					MW
PuF_6	O_h	Pu—F	1.972					ED
RbBr		Rb—Br (r_e)	2.9447					MW
RbCl		Rb—Cl (r_e)	2.7869					MW
RbF		Rb—F (r_e)	2.2703					MW
RbH		Rb—H (r_e)	2.367					UV
RbI		Rb—I (r_e)	3.1768					MW
RbO		Rb—O (r_e)	2.25420					UV
RbOH	linear; large amplitude bending mode	Rb—O	2.301	O—H	0.957			MW
$ReClO_3$	C_{3v}	Re—O	1.702	Re—Cl	2.229	∠ClReO	109.4	MW

Formula	Structure	Bond distances in Å and angles in degrees						Method
$ReClO_4$	C_{4v}	Re—O	1.663	Re—Cl	2.270	∠ClReO	105.5	ED
$ReCl_5$	D_{3h}	Re—Cl_{eq}	2.238	Re—Cl_{ax}	2.263			ED
ReF_6	O_h	Re—F	1.832					ED
ReF_7	pseudorotation	Re—F	1.835					ED
RhB		Rh—B	1.691					UV
RhC		Rh—C	1.614					UV
RhS		Rh—S	2.059					UV
RuO_4	T_d	Ru—O	1.706					ED
SCl_2	C_{2v}	S—Cl	2.006	∠ClSCl	103.0			ED
SF		S—F (r_e)	1.6006					MW
SF_2		S—F	1.5921	∠FSF	98.20			MW
SF_6	O_h	S—F	1.561					ED
SH		S—H (r_e)	1.34066					UV
SO		S—O (r_e)	1.4811					MW
$SOCl_2$		S—O	1.44	S—Cl	2.072			MW
		∠ClSCl	97.2	∠OSCl	108.0			
SOF_2		S—O	1.420	S—F	1.583			ED
		∠FSF	92.2	∠OSF	106.2			
SOF_4	F_b F_b / F_a—S—F_a / O	S—O	1.403	S—F_a	1.575	S—F_b	1.552	ED
		∠OSF_a	90.7	∠OSF_b	124.9			
		∠F_aSF_b	89.6	∠F_bSF_b	110.2			
	C_{2v}							
SO_2		S—O (r_e)	1.4308	∠OSO (θ_e)	119.329			MW
SO_2Cl_2	C_{2v}	S—Cl	2.011	S—O	1.404			ED
		∠ClSCl	100.0	∠OSO	123.5			
SO_2F_2	C_{2v}	S—F	1.530	S—O	1.397			ED
		∠FSF	97	∠OSO	123			
SO_3	D_{3h}	S—O	1.4198					IR
$S(SiH_3)_2$		Si—S	2.136	Si—H	1.494	∠SiSSi	97.4	ED
S_2		S—S (r_e)	1.8892					R
S_2Br_2	C_2	S—Br	2.24	S—S	1.98	∠SSBr	105	ED
		dihedral angle of internal rotation	83.5					
S_2Cl_2	C_2	S—Cl	2.057	S—S	1.931	∠SSCl	108.2	ED
		dihedral angle of internal rotation	84.1					
S_2O_2	planar *cis* form	S—S	2.025	S—O	1.458	∠OSS	112.8	MW
S_8	S S S S S S S S (ring)	S—S	2.07	∠SSS	105		(D_{4d})	ED
$SbBr_3$	C_{3v}	Sb—Br	2.490	∠BrSbBr	98.2			ED
$SbCl_3$	C_{3v}	Sb—Cl	2.334	∠ClSbCl	97.1			ED
$SbCl_5$	D_{3h}	Sb—Cl_{eq}	2.277	Sb—Cl_{ax}	2.338			ED
SbF		Sb—F (r_e)	1.918					UV
SbF_3	C_{3v}	Sb—F	1.880	∠FSbF	94.9			ED
SbH		Sb—H	1.723					UV
SbH_3	C_{3v}	Sb—H	1.704	∠HSbH	91.6			MW
SbI_3	C_{3v}	Sb—I	2.721	∠ISbI	99.0			ED
SbO		Sb—O (r_e)	1.826					UV
SbP		Sb—P (r_e)	2.20544					MW
$ScCl_3$	D_{3h}	Sc—Cl	2.291					ED
ScF		Sc—F (r_e)	1.788					UV
ScF_3	D_{3h}	Sc—F	1.847					ED
SeF		Se—F	1.742					MW
SeF_6	O_h	Se—F	1.69					ED
SeH		Se—H (r_e)	1.48					UV
SeO		Se—O (r_e)	1.6393					MW

Formula	Structure	Bond distances in Å and angles in degrees						Method
$SeOF_2$		Se—O	1.576	Se—F	1.730			MW
		∠OSeF	104.82	∠FSeF	92.22			
SeO_2		Se—O (r_e)	1.6076	∠OSeO (θ_e)	113.83			MW
SeO_3	D_{3h}	Se—O	1.69					ED
Se_2		Se—Se (r_e)	2.1660					UV
Se_6	six-membered ring with chair conformation	Se—Se	2.34	∠SeSeSe	102			ED
$SiBrF_3$	C_{3v}	Si—F	1.559	Si—Br	2.156	∠FSiBr	108.5	MW
$SiBrH_3$	C_{3v}	Si—Br	2.210	Si—H	1.486	∠HSiBr	107.8	MW
SiCl		Si—Cl (r_e)	2.058					UV
$SiClH_3$	C_{3v}	Si—Cl	2.049	Si—H	1.486	∠HSiCl	107.9	MW
$SiCl_4$	T_4	Si—Cl	2.019					ED
SiF		Si—F	1.6008					UV
$SiFH_3$	C_{3v}	Si—F	1.593	Si—H	1.486	∠HSiH	110.63	MW, IR
SiF_2		Si—F (r_e)	1.590	∠FSiF (θ_e)	100.8			MW
SiF_3H	C_{3v}	Si—H ((r_e)	1.4468	Si—F (r_e)	1.5624	∠HSiF (θ_e)	110.64	MW
SiF_4	T_d	Si—F	1.553					ED
SiH		Si—H (r_e)	1.5201					UV
SiH_3I	C_{3v}	Si—I	2.437	Si—H	1.486	∠HSH	107.8	MW
SiH_4	T_d	Si—H	1.4798					IR
SiN		Si—N (r_e)	1.572					UV
SiO		Si—O (r_e)	1.5097					MW
SiS		Si—S (r_e)	1.9293					MW
SiSe		Si—Se (r_e)	2.0583					MW
Si_2		Si—Si (r_e)	2.246					UV
Si_2Cl_6		Si—Si	2.32	Si—Cl	2.009	∠ClSiCl	109.7	ED
Si_2F_6		Si—Si	2.317	Si—F	1.564	∠FSiF	108.6	ED
Si_2H_6		Si—Si	2.331	Si—H	1.492			ED
		∠SiSiH	110.3	∠HSiH	108.6			
$SnBr_2$		Sn—Br (r_e)	2.501	∠BrSnBr	100.0			ED
SnCl		Sn—Cl (r_e)	2.361					UV
$SnCl_2$		Sn—Cl (r_e)	2.335	∠ClSnCl	99.1			ED
$SnCl_4$	T_d	Sn—Cl	2.281					ED
SnF		Sn—F (r_e)	1.944					UV
SnH		Sn—H (r_e)	1.7815					UV
SnH_4	T_d	Sn—H	1.711					R, IR
SnI_2		Sn—I (r_e)	2.688					ED
SnO		Sn—O (r_e)	1.8325					MW,UV
SnS		Sn—S (r_e)	2.2090					MW
SnSe		Sn—Se (r_e)	2.3256					MW
SnTe		Sn—Te (r_e)	2.5228					MW
SrBr		Sr—Br (r_e)	2.7352					UV
$SrBr_2$	quasilinear	Sr—Br	2.783					ED
$SrCl_2$		Sr—Cl	2.630	∠ClSrCl	155			ED
SrF		Sr—F (r_e)	2.0754					UV
SrH		Sr—H (r_e)	2.1456					UV
SrI		Sr—I (r_e)	2.9436					UV
SrI_2	linear	Sr—I	3.01					ED
SrO		Sr—O (r_e)	1.9198					MW
SrOH		Sr—O	2.111	O—H	0.922			UV
SrS		Sr—S (r_e)	2.4405					UV
$TaBr_5$	D_{3h}	Ta—Br_{eq}	2.412	Ta—Br_{ax}	2.473			ED
$TaCl_5$	D_{3h}	Ta—Cl_{eq}	2.268	Ta—Cl_{ax}	2.315			ED
TaO		Ta—O (r_e)	1.6875					UV
$TbCl_3$	C_{3v}	Tb—Cl	2.476					ED
TeF_6	O_h	Te—F	1.815					ED
TeH		Te—H	1.74					UV
TeO		Te—O (r_e)	1.825					UV
Te_2		Te—Te (r_e)	2.5574					UV
$ThCl_4$	T_d	Th—Cl	2.567					ED

Formula	Structure	Bond distances in Å and angles in degrees						Method
ThF_4	T_d	Th—F	2.124					ED
ThO		Th—O (r_e)	1.84032					UV
$TiBr_4$	T_d	Ti—Br	2.339					ED
$TiCl_3$	D_{3h}	Ti—Cl	2.208					ED
$TiCl_4$	T_d	Ti—Cl	2.170					ED
TiF		Ti—F	1.8342					MW
TiF_4	T_d	Ti—F	1.756					ED
TiI_3	D_{3h}	Ti—I	2.568					ED
TiI_4	T_d	Ti—I	2.546					ED
TiO		Ti—O (r_e)	1.620					UV
TiS		Ti—S (r_e)	2.0825					UV
TlBr		Tl—Br (r_e)	2.6182					MW
TlCl		Tl—Cl (r_e)	2.4848					MW
TlF		Tl—F (r_e)	2.0844					MW
TlH		Tl—H (r_e)	1.870					UV
TlI		Tl—I (r_e)	2.8137					MW
UCl_4	T_d	U—Cl	2.506					ED
UCl_6	O_h	U—F	2.46					ED
UF_4	T_d	U—F	2.059					ED
UF_6	O_h	U—F	2.000					ED
UI_3	C_{3v}	U—I	2.88					ED
VCl_3O	C_{3v}	V—O	1.570	V—Cl	2.142	∠ClVCl	111.3	ED, MW
VBr_4	T_d (Jahn-Teller effect)	V—Br	2.276					ED
VCl_4	T_d (Jahn-Teller effect)	V—Cl	2.138					ED
VF_3	D_{3h}	V—F	1.751					ED
VF_5		V—F_{eq}	1.709	V—F_{ax}	1.736			ED
VMo		V—Mo	1.876					UV
VO		V—O (r_e)	1.5893					UV
$WClF_5$	Cl F_b F_b—W—F_b F_b F_a	W—F (av.)	1.836	W—Cl	2.251	∠F_aWF_b	88.7	MW
WCl_5	D_{3h}	W—Cl_{eq}	2.243	W—Cl_{ax}	2.293			ED
WCl_6	O_h	W—Cl	2.290					ED
WF_4O	C_{4v}	W—O	1.666	W—F	1.847	∠FWF	86.2	ED
WF_6	O_h	W—F	1.833					ED
XeF_2	linear	Xe—F	1.977					IR
XeF_4	D_{4h}	Xe—F	1.94					ED
XeF_6	O_h	Xe—F	1.890					ED
XeO_4	T_d	Xe—O	1.736					ED
YCl		Y—Cl	2.385					UV
YCl_3		Y—Cl	2.437					ED
YF		Y—F (r_e)	1.9257					UV
YI_3		Y—I	2.817					ED
YO		Y—O (r_e)	1.790					UV
YbBr		Yb—Br (r_e)	2.6454					UV
YbH		Yb—H (r_e)	2.0526					UV
$ZnBr_2$	linear	Zn—Br	2.204					ED
$ZnCl_2$	linear	Zn—Cl	2.072					ED
ZnF		Zn—F (r_e)	1.7677					MW
ZnF_2	linear	Zn—F	1.742	[Zn—F (r_e)]	1.729			ED
ZnH		Zn—H (r_e)	1.5949					UV
ZnI_2	linear	Zn—I	2.401					ED
$ZrBr_4$	T_d	Zr—Br	2.465					ED
$ZrCl_4$	T_d	Zr—Cl	2.328					ED
ZrF_4	T_d	Zr—F	1.902					ED
ZrI_4	T_d	Zr—I	2.660					ED
ZrO		Zr—O (r_e)	1.7116					UV

Part 2. Molecules containing carbon

Compound	Structure	Bond distances in Å and angles in degrees						Method
Acetaldehyde	$C_bH_3-C_a(=O)-H$	C_a—O	1.210	C_a—C_b	1.515			ED, MW
		C_a—H	1.128	C_b—H	1.107			
		∠C_bC_aO	124.1	∠C_bC_aH	115.3	∠HC_bH	109.8	
Acetamide	CH_3CONH_2	C—O	1.220	C—N	1.380			ED
		C—C	1.519	N—H	1.022	C—H	1.124	
		∠CCN	115.1	∠NCO	122.0			
Acetic acid	$CH_3-C(=O_a)-O_b-H$	C—C	1.520	C—O_a	1.214	C—O_b	1.364	ED
		C—H	1.10	∠CCO_a	126.6	∠CCO_b	110.6	
Acetone	$(CH_3)_2CO$ Symmetry axis of each CH_3 is tilted 2° from the C—C bond	C—C	1.520	C—O	1.213	C—H	1.103	ED, MW
		∠CCC	116.0	∠HCH	108.5			
Acetonitrile	CH_3CN (C_{3v})	C—N	1.159	C—C	1.468	C—H	1.107	ED, MW
		∠CCH	109.7					
Acetonitrile-*N*-oxide	CH_3CNO (C_{3v})	C—C	1.442	C—N	1.169	N—O	1.217	MW
Acetyl chloride	CH_3COCl	C—C	1.506	C—O	1.187	C—H	1.105	ED, MW
		C—Cl	1.798	∠HCH	108.6	∠OCCl	121.2	
		∠CCCl	111.6					
Acetylene	HC≡CH	C—C (r_e)	1.203	C—H (r_e)	1.060			IR
Acrolein	$H_2C_a=C_bH-C_cH=O$ (planar *s-trans* form)	C_a—C_b	1.345	C_b—C_c	1.484	C_c—O	1.217	ED, MW
		C_a—H	1.10	C_c—H	1.13	∠HC_cC_b	114	
		∠$C_aC_bC_c$	120.3	∠C_bC_cO	123.3	Other CCH (av.)	122	
Acrylonitrile	$H_2C_a=C_bH-C_c\equiv N$	C_a—C_b	1.343	C_b—C_c	1.438	C_c—N	1.167	ED, MW
		C_a—H	1.114	∠C_bC_cN	178	∠$C_aC_bC_c$	121.7	
		∠HCC	120					
Allene	CH_2=C=CH_2	C—C	1.3084	C—H	1.087	∠HCH	118.2	IR
Aniline	$C_6H_5NH_2$	C—C	1.392	C—N	1.431	N—H	0.998	MW
		∠HNH	113.9	dihedral angle between NH_2 plane and N—C bond	140.6			
Azetidine	CH_2—CH_2 | | CH_2—NH	C—N	1.482	C—C	1.553			ED
		C—H	1.107	N—H	1.03			
		∠CCC	86.9	∠CCN	85.8	∠CNC	92.2	
		dihedral angle between CCC and CNC planes	147					
Benzamide	C_6H_5—C_aONH_2	C—C (ring)	1.401	C (ring)—C_a	1.511	C_a—O	1.225	ED
		C—H	1.112	C—N	1.380			
		∠CCN	117.8	∠CCC (ring)	120(ass.)	∠CCO	121.2	
Benzene	C_6H_6	C—C	1.399	C—H	1.101			ED, IR
p-Benzoquinone	$O=C_a(C_bH=C_bH)_2C_a=O$ (ring)	C_a—O	1.225	C_a—C_b	1.481	C_b—C_b	1.344	ED
		∠$C_bC_aC_b$	118.1					

Compound	Structure	Bond distances in Å and angles in degrees						Method
Bicyclo[1.1.0]butane	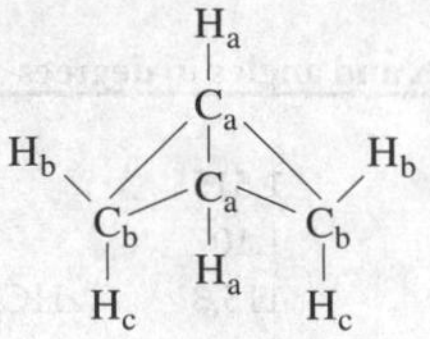	$C_a—C_a$	1.497	$C_a—C_b$	1.498	$C_a—H_a$	1.071	MW
		$C_b—H_b$	1.093	$C_b—H_c$	1.093	$\angle H_bC_bH_c$	115.6	
		$\angle C_bC_aH_a$	130.4	$\angle C_aC_aH_a$	128.4	$\angle C_aC_bC_a$	60.0	
		dihedral angle between the two $C_aC_aC_b$ planes	121.7					
Bicyclo[2.2.1]heptane	See preceeding structure C_7H_{12}	$C_a—C_b$	1.54	$C_b—C_b$	1.56	$C_a—C_c$	1.56	ED
		C—C (av.)	1.549	$\angle C_aC_cC_a$	93.1			
		dihedral angle between the two $C_aC_bC_bC_a$ planes	113.1					
Bicyclo[2.2.0]hexa-2,5-diene	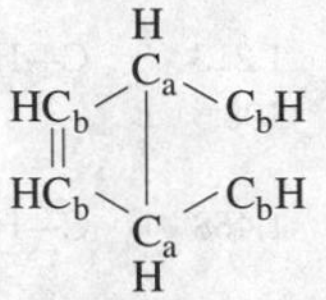	$C_b—C_b$	1.345	$C_a—C_a$	1.574	$C_a—C_b$	1.524	ED
		dihedral angle between the two $C_aC_bC_bC_a$ planes	117.3					
Bicyclo[2.2.2]octane	$HC_a(C_bH_2C_bH_2)_3C_aH$ large-amplitude torsional motion about D_{3h} symmetry axis	$C_a—C_b$	1.54	$C_b—C_b$	1.55	C—C (av.)	1.542	ED
		$\angle C_aC_bC_b$	109.7					
Bicyclo[1.1.1]pentane	C_5H_8	C—C	1.557	∠CCC	74.2			ED
Bicyclo[2.1.0]pentane	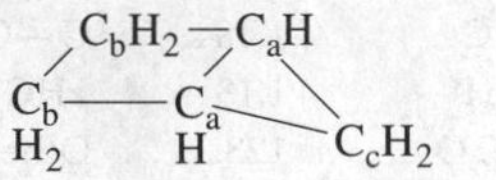	$C_a—C_a$	1.536	$C_b—C_b$	1.565	$C_a—C_c$	1.507	MW
		$C_a—C_b$	1.528	dihedral angle between the $C_aC_aC_bC_b$ and $C_aC_aC_c$ planes	112.7			
Biphenyl	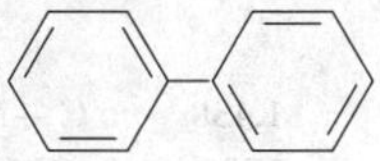	C—C (intra-ring)	1.396	C—C (inter-ring)	1.49			ED
		torsional dihedral angle between the two rings	≈40					
4,4′-Bipyridyl	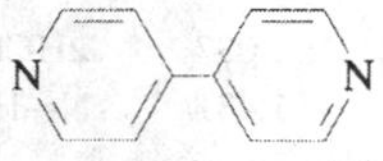	C—C (inter-ring)	1.465	C—C (intra-ring)	1.375	C—N (intra-ring)	1.375	ED
		torsional dihedral angle between the two rings	≈37					
Bis(cyclopentadienyl) beryllium	$(C_5H_5)_2Be$ (C_{5v})	Be—(cyclopentadienyl plane)	1.470, 1.92	C—C	1.423			ED
Bis(cyclopentadienyl) iron	$(C_5H_5)_2Fe$ (D_{5h})	Fe—C	2.064	C—C	1.440	C—H	1.104	ED
Bis(cyclopentadienyl) lead	$(C_5H_5)_2Pb$ (D_{5h})	Pb—C	2.79	C—C	1.430			ED
		dihedral angle between the two C_5H_5 planes	40~50 (The two rings are not parallel)					
Bis(cyclopentadienyl) manganese	$(C_5H_5)_2Mn$ (D_{5h})	Mn—C	2.383	C—C	1.429			ED
Bis(cyclopentadienyl) nickel	$(C_5H_5)_2Ni$ (D_{5h})	Ni—C	2.196	C—C	1.430			ED
Bis(cyclopentadienyl) ruthenium	$(C_5H_5)_2Ru$ (D_{5h})	Ru—C	2.196	C—C	1.439			ED
Bis(cyclopentadienyl) tin	$(C_5H_5)_2Sn$ (D_{5h})	Sn—C	2.71	C—C	1.431	C—H	1.14	ED
Borane carbonyl	BH_3CO (C_{3v})	C—O	1.131	B—C	1.540	B—H	1.194	MW
		∠BCO	180	∠HBH	113.9			

Compound	Structure	Bond distances in Å and angles in degrees						Method
Bromobenzene	Br–C_a; HC_b=C_bH; HC_c–C_cH; C_d–H (ring)	C_a—C_b	1.42	C_b—C_c	1.375	C_c—C_d	1.401	MW
		C—Br	1.85	C—H	1.072	$\angle C_bC_aC_b$	117.4	
Bromochloroacetylene	ClC≡CBr	C—Cl	1.636	C—Br	1.784	C—C	1.206	ED
Bromoiodoacetylene	IC≡CBr	C—I	1.972	C—Br	1.795	C—C	1.206	ED
Bromomethane	CH_3Br	C—Br (r_e)	1.933	C—H (r_e)	1.086	$\angle$HCH (θ_e)	111.2	MW, IR
Bromomethyl	CH_2Br (planar)	C—Br	1.848	C—H	1.084	$\angle$HCH (ass.)	124.5	MW
Bromomethylene	CHBr (bent)	C—Br	1.857	C—H	1.110	$\angle$HCH	101.0	UV
Bromomethylmercury	CH_3HgBr (C_{3v})	C—Hg	2.07	Hg—Br	2.406			MW
1,3-Butadiene	C_aH_2=C_bH—C_bH=C_aH_2 (C_{2h})	C_a—C_b	1.349	C_b—C_b	1.467	C—H (av.)	1.108	ED
		$\angle$CCC	124.4	$\angle C_bC_aH$	120.9			
1,3-Butadiyne	$HC_a≡C_bC_b≡C_aH$ (linear)	C_a—C_b	1.218	C_b—C_b	1.384	C—H	1.09	ED
Butane	$CH_3CH_2CH_2CH_3$	C—C	1.531	C—H	1.117	$\angle$CCC	113.8	ED
		$\angle$CCH	111.0	dihedral angle for the *gauche* conformer	65			
2,3-Butanedione	$CH_3COCOCH_3$ *trans* conformer	C—O	1.215	C—C (av.)	1.524	C—H	1.108	ED
		$\angle$CCC	116.2	$\angle$CCO	119.5			
2-Butanone	C_aH_3—C_bH_2—C_c(=O)—C_dH_3 *trans* conformer	C—C (av.)	1.518	C_c—O	1.219	C—H (av.)	1.102	ED
		$\angle C_aC_bC_c$	113.5	$\angle C_bC_cO$	121.9	$\angle C_dC_cO$	121.9	
1,2,3-Butatriene	$H_2C_a=C_b=C_b=C_aH_2$ (D_{2h})	C_a—C_b	1.32	C_b—C_b	1.28	C—H	1.08	ED
cis-2-Butene	$C_aH_3C_bH=C_bHC_aH_3$	C_a—C_b	1.506	C_b—C_b	1.346	$\angle C_aC_bC_b$	125.4	ED
trans-2-Butene	$C_aH_3C_bH=C_bHC_aH_3$	C_a—C_b	1.508	C_b—C_b	1.347	$\angle C_aC_bC_b$	123.8	ED
1-Buten-3-yne	H_a, H_b on C_a=C_b(H_c)—C_c≡C_d—H_d	C_a—C_b	1.344	C_b—C_c	1.434	C_c—C_d	1.215	ED, MW
		C_a—H_a	1.11	C_d—H_d	1.09			
		$\angle C_aC_bC_c$	123.1	$\angle C_bC_cC_d$	178	$\angle H_aC_aC_b$	119	
		$\angle H_bC_aC_b$	122	$\angle H_cC_bC_a$	122	$\angle C_cC_dH_d$	182	
tert-Butyl chloride	$(CH_3)_3CCl$	C—C	1.528	C—Cl	1.828	C—H	1.102	ED, MW
		$\angle$CCCl	107.3	$\angle$CCH	110.8	$\angle$CCC	111.6	
2-Butyne	C_aH_3—$C_b≡C_b$—C_aH_3	C_b—C_b	1.214	C_a—C_b	1.468	C—H	1.116	ED
		$\angle C_bC_aH$	110.7					
Carbon dimer	C_2	C—C (r_e)	1.2425					UV
Carbon trimer	C_3 (linear)	C—C	1.277					UV
Carbon dioxide	CO_2 (linear)	C—O (r_e)	1.1600					IR
Carbon disulfide	CS_2 (linear)	C—S (r_e)	1.5526					IR
Carbon monobromide	CBr	C—Br	1.8209					UV
Carbon monoselenide	CSe	C—Se (r_e)	1.67609					UV
Carbon monosulfide	CS	C—S (r_e)	1.5349					MW
Carbon monoxide	CO	C—O (r_e)	1.1283					MW
Carbon oxyselenide	OCSe (linear)	C—O	1.159	C—Se	1.709			MW
Carbon oxysulfide	OCS (linear)	C—O (r_e)	1.1578	C—S (r_e)	1.5601			MW
Carbon phosphide	CP	C—P (r_e)	1.562					UV
Carbon sulfide selenide	SCSe (linear)	C—S	1.553	C—Se	1.693			MW
Carbon sulfide telluride	SCTe (linear)	C—S	1.557	C—Te	1.904			MW
Carbon suboxide	OCCCO (linear)	C—C	1.289	C—O	1.163			ED

Compound	Structure	Bond distances in Å and angles in degrees						Method
Carbonyl bromide	$COBr_2$	C—O	1.178	C—Br	1.923	∠BrCBr	112.3	ED, MW
Carbonyl chloride	$COCl_2$	C—O	1.179	C—Cl	1.742	∠ClCCl	111.8	ED, MW
Carbonyl chloride fluoride	COClF	C—O	1.173	C—F	1.334	C—Cl	1.725	ED, MW
		∠ClCO	127.5	∠FCCl	108.8			
Carbonyl dicyanide	$CO(CN)_2$	C—O	1.209	C—C	1.466	C—N	1.153	ED, MW
		∠CCC	115	∠CCN	180			
Carbonyl fluoride	COF_2	C—O	1.172	C—F	1.3157	∠FCF	107.71	ED, MW
Chloroacetylene	HC≡CCl	C—Cl	1.6368	C—C	1.2033	C—H	1.0550	MW
Chlorobenzene	C_6H_5Cl	C—C	1.400	C—Cl	1.737	C—H	1.083	ED
Chlorocyanoacetylene	ClC≡C—CN	C—Cl	1.624	C—N	1.160	C—C	1.205	ED
		C—CN	1.362					
Chloroethane	H_b Cl H_b—C_b—C_a—H_a H_b H_a	C—C	1.528	C—Cl	1.802	C—H	1.103	ED, MW
		∠CCCl	110.7	∠$H_bC_bH_b$	109.8	∠$H_aC_aH_a$	109.2	
		∠$C_bC_aH_a$	110.6	C_a—H_a = C_b—H_b (ass.)				
2-Chloroethanol	$ClCH_2CH_2OH$ (*gauche*)	C—O	1.413	C—C	1.519	C—Cl	1.801	ED
		O—H	1.033	C—H	1.093			
		∠CCCl	110.7	∠CCO	113.8	dihedral angle of internal rotation	62.4	
Chloroiodoacetylene	ClC≡CI	C—Cl	1.63	C—I	1.99	C—C	1.209 (ass)	MW
Chloromethane	CH_3Cl	C—Cl	1.785	C—H	1.090	∠HCH	110.8	MW, IR
Chloromethylidyne	CCl	C—Cl	1.6512					UV
Chloromethylmercury	CH_3HgCl (C_{3v})	C—Hg	2.06	Hg—Cl	2.282			MW
trans-1-Chloropropene	$CH_3CH=CHCl$	C—Cl	1.728	∠CCCl	121.9			MW
3-Chloropropene	$CH_2ClCH=CH_2$ *cis* conformer	C—Cl	1.811	∠CCCl	115.2			MW
	skew conformer	C—Cl	1.809	∠CCCl	109.6	dihedral angle of internal rotation	122.4	
Chlorotrifluoromethane	$CClF_3$ (C_{3v})	C—Cl	1.752	C—F	1.325	∠FCF	108.6	ED, MW
Chromium carbonyl	$Cr(CO)_6$	Cr—C	1.92	C—O	1.16	∠CrCO	180	ED
Cobalt cyanide	CoC≡N	Co—C	1.883	C—N	1.131			MW
Copper cyanide	CuC≡N	Cu—C	1.832	C—N	1.158			MW
Cyanamide	$H_2N_aCN_b$	N_a—C	1.346	C—N_b	1.160	N—H	1.00	MW
		∠HNH	114	dihedral angle between NH_2 plane and N—C bond	142			
Cyanide	CN	C—N (r_e)	1.1718					MW
Cyanoacetylene	$HC_a≡C_b—C_cN$	C_a—C_b	1.205	C_b—C_c	1.378	C—H	1.058	MW
		C_c—N	1.159					
Cyanocyclopropane	$C_3H_5C_aN$	C—C (ring)	1.513	C—C_a	1.472	C_a—N	1.157	MW
		C—H	1.107	∠C_aCH	119.6	∠HCH	114.6	
Cyanogen	N≡C—C≡N (linear)	C—N	1.163	C—C	1.393			ED
Cyanogen azide	N≡C—N=N≡N (planar)	C—N	1.312	N=N	1.252	N≡N	1.133	MW
		C≡N	1.164	∠CNN	120.2	∠NCN	176.0	
Cyanogen bromide	BrCN (linear)	C—N (r_e)	1.157	C—Br (r_e)	1.790			MW
Cyanogen chloride	ClCN (linear)	C—Cl (r_e)	1.629	C—N (r_e)	1.160			MW
Cyanogen fluoride	FCN (linear)	C—F	1.262	C—N	1.159			MW
Cyanogen iodide	ICN (linear)	C—I	1.995	C—N	1.159			MW
1-Cyano-2-propyne	$HC_a≡C_bC_cH_2C_d≡N$	C_a—C_b	1.207 (ass.)	C_b—C_c(ass.)	1.465	C_c—C_d	1.454	MW
		C_d—N	1.159 (ass.)	C_a—H(ass.)	1.057	C_c—H(ass.)	1.090	
		∠$C_bC_cC_d$	113.4	∠HC_cH	109.4 (ass.)	∠C_bC_cH	111.3	

Compound	Structure	Bond distances in Å and angles in degrees						Method
Cyclobutane	$(CH_2)_4$	C—C	1.555	C—H	1.113			ED
		dihedral angle between the two CCC planes	145					
Cyclobutanone	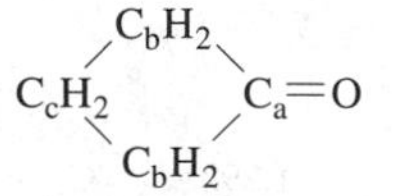 C_bH_2 C_cH_2 C_a=O C_bH_2	C_a—C_b	1.527	C_b—C_c	1.556			MW
		$\angle C_bC_aC_b$	93.1	$\angle C_aC_bC_c$	88.0			
Cyclobutene	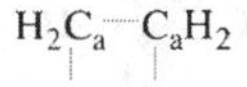 H_2C_a—C_aH_2 HC_b=C_bH	C_a—C_a	1.566	C_b—C_b	1.342	C_a—C_b	1.517	MW
		C_a—H	1.094	C_b—H	1.083			
		$\angle C_aC_aC_b$	94.2	$\angle C_aC_bH$	133.5	$\angle HC_aH$	109.2	
		$\angle C_aC_aH$	114.5	$\angle C_aC_aC_b$	85.8	dihedral angle between CH_2 plane and C_a—C_a bond	135.8	
2,4,6-Cycloheptatrien-1-one	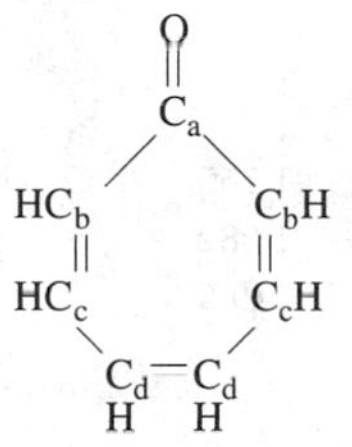 O ‖ C_a HC_b C_bH ‖ ‖ HC_c C_cH C_d=C_d H H (C_{2v})	C_a—C_b	1.45	C_b—C_c	1.36	C_c—C_d	1.46	ED
		C_d—C_d	1.34	C_a—O	1.23	$\angle C_bC_aC_b$	122	
		$\angle C_aC_bC_c$	133	$\angle C_bC_cC_d$	126	$\angle C_cC_dC_d$	130	
Cyclohexane	C_6H_{12} (chair form)	C—C	1.536	C—H	1.119	∠CCC	111.3	ED
Cyclohexene	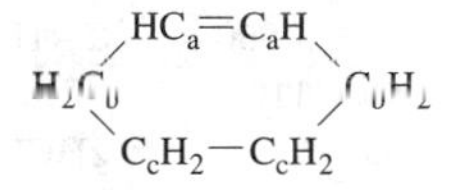 HC_a=C_aH H_2C_b C_bH_2 C_cH_2—C_cH_2 half-chair form (C_2)	C_a—C_a	1.334	C_a—C_b	1.50	C_b—C_c	1.52	ED
		C_c—C_c	1.54	$\angle C_aC_aC_b$	123.4	$\angle C_aC_bC_c$	112.0	
		$\angle C_bC_cC_c$	110.9					
Cyclooctatetraene	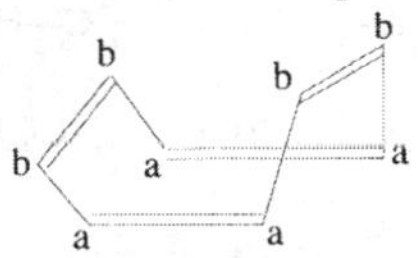 tub form (D_{2d})	C_a—C_b	1.476	C_a—C_a	1.340	C_b—C_b	1.340	ED
		C—H	1.100	$\angle C_bC_aC_a$	126.1	$\angle C_aC_bC_b$	126.1	
		dihedral angle between $C_aC_aC_aC_a$ and $C_aC_bC_bC_a$ planes	136.9					
1,3-Cyclopentadiene	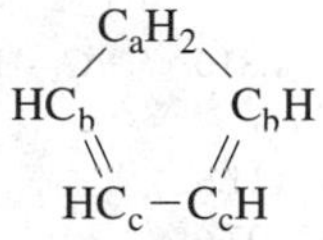 C_aH_2 HC_b C_bH HC_c—C_cH	C_a—C_b	1.509	C_b—C_c	1.342	C_c—C_c	1.469	MW
		$\angle C_aC_bC_c$	109.3	$\angle C_bC_cC_c$	109.4	$\angle C_bC_aC_b$	102.8	
Cyclopentadienylindium	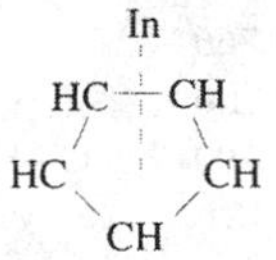 In HC—CH HC CH CH	C—In	2.621	C—C	1.426	(C_{5v})		ED
Cyclopentane	$(CH_2)_5$	C—C	1.546	C—H	1.114	∠CCH	111.7	ED
Cyclopentene	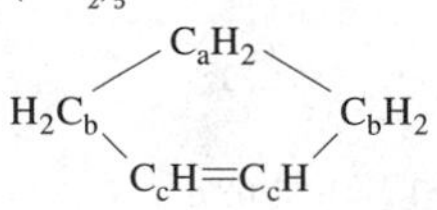 C_aH_2 H_2C_b C_bH_2 C_cH=C_cH	C_a—C_b	1.546	C_b—C_c	1.519	C_c—C_c	1.342	ED
		$\angle C_aC_bC_c$	103.0	$\angle C_bC_cC_c$	110.0	$\angle C_bC_aC_b$	104.0	
		dihedral angle between $C_bC_aC_b$ and $C_bC_cC_cC_b$ planes	151.2					
Cyclopropane	$(CH_2)_3$	C—C	1.512	C—H	1.083	∠HCH	114.0	R

Compound	Structure	Bond distances in Å and angles in degrees						Method
Cyclopropanone	H_2C_b | $C_a=O$ | H_2C_b	C_a—C_b	1.475	C_b—C_b	1.575	C_a—O	1.191	MW
		C—H	1.086	∠$C_aC_bC_b$	57.7	∠HC_bH	114	
		dihedral angle between CH_2 plane and C_b—C_b bond	151					
Cyclopropene	C_aH_2 | HC_b=C_bH	C_a—C_b	1.505	C_b—C_b	1.293	C_a—H	1.085	MW
		C_b—H	1.072	∠C_aC_bH	150	∠HC_aH	114.3	
Cyclopropenone	H, H | $C_b=C_c$ | C_a | O | C_{2v}	C_a—C_b (r_s)	1.423	C_b—C_c (r_s)	1.349	C_a—O (r_s)	1.212	MW
		C—H (r_s)	1.079	∠HC_bC_c (θ_s)	144.3	$C_bC_aC_c$ (θ_s)	56.6	
Decalin	$C_{10}H_{18}$	C—C (av.)	1.530	C—H (av.)	1.113	∠CCC (av.)	111.4	ED
Diazirine	CH_2 | N=N	C—N	1.482	N—N	1.228	C—H	1.09	MW
		∠HCH	117					
Diazoacetonitrile	H | C_b | C_a N_b | N_a N_c	C_a—C_b	1.424	C_a—N_a	1.165	C_b—N_b	1.280	MW
		N_b—N_c	1.132	C—H	1.082			
		∠C_aC_bH	117	∠$C_aC_bN_b$	119.5			
Diazomethane	CH_2N_2	C—N	1.32	N—N	1.12	C—H	1.075	MW, IR
		∠HCH	126.0					
1,2-Dibromoethane	CH_2BrCH_2Br	C—C	1.506	C—Br	1.950	C—H	1.108	ED
		∠CCBr	109.5	∠CCH	110			
Dibromomethane	CH_2Br_2	C—Br	1.924	C—H	1.08	∠HCBr	109	ED
		∠BrCBr	113.2					
2,2'-Dichlorobiphenyl	C_6H_4Cl—C_6H_4Cl	C—C (rings)	1.398	C—C (inter-ring)	1.495	C—H	1.10	ED
		C—Cl	1.732	∠CCCl	121.4	∠CCH	126	
		dihedral angle between the two rings (defined as 0 for *cis* conformer)	74					
trans-1,4-Dichlorocyclohexane	$C_6H_{10}Cl_2$	C—C	1.530	C—Cl	1.810	C—H	1.102	ED
		∠CCC	111.5					
	equatorial:	∠CCCl	108.6	∠HCCl	111.5			
	axial:	∠CCCl	110.6	∠HCCl	107.6			
1,1-Dichloroethane	$CHCl_2CH_3$	C—C	1.540	C—Cl	1.766			MW
		∠ClCCl	112.0	∠CCCl	111.0			
1,2-Dichloroethane	CH_2ClCH_2Cl	C—C	1.531	C—Cl	1.790	C—H	1.11	ED
		∠CCCl	109.0	∠CCH	113			
1,1-Dichloroethene	CH_2=CCl_2 (C_{2v})	C—C	1.32 (ass.)	C—Cl	1.73			MW
		∠ClCC	123					
cis-1,2-Dichloroethene	CHCl=CHCl	C—C	1.354	C—Cl	1.718			ED
		∠ClCC	123.8					
Dichloromethane	CH_2Cl_2	C—Cl (r_e)	1.765	C—H (r_e)	1.087			MW, IR
		∠ClCCl (θ_e)	112.0	∠HCH (θ_e)	111.5			

Compound	Structure	Bond distances in Å and angles in degrees						Method
1,2-Dicyanocyclobutene	$C_c{\equiv}N$ / $H_2C_b - C_a$ | || $H_2C_{b'} - C_{a'}$ \ $C_{c'}{\equiv}N$ C_{2v}	$C_a—C_{a'}$	1.361	$C_a—C_b$	1.515	$C_b—C_{b'}$	1.567	MW
		$C_a—C_c$	1.420	$C_c—N$	1.157	$C_b—H$	1.088	
		$\angle C_{a'}C_aC_b$	93.9	$\angle C_aC_bC_{b'}$	86.1	$\angle C_aC_cN$	178.2	
		$\angle C_bC_aC_c$	133.3	$\angle C_aC_bH$	114.7	$\angle C_{a'}C_aC_bH$	115.8	
Difluorocyanamide	$F_2N_b—C{\equiv}N_a$	$C—N_a$	1.158	$C—N_b$	1.386	$N_b—F$	1.399	MW
		$\angle N_aCN_b$	174	$\angle CN_bF$	105.4	$\angle FN_bF$	102.8	
Difluorocyclopropenone	F \ $C_b{=}C_c$ / F; C_a (ring); $C_a{=}O$ C_{2v}	$C_a—C_b$	1.453	$C_b—C_c$	1.324	$C_a—O$	1.192	MW
		C—F	1.314	$\angle FC_bC_c$	145.7			
Difluorodimethylsilane	$(CH_3)_2SiF_2$	C—Si	1.844	Si—F	1.585	C—H (ass.)	1.093	MW
		∠CSiC	115.2	∠FSiF	106.1	∠SiCH (ass.)	110.8	
1,1-Difluoroethane	CH_3CHF_2	C—C	1.498	C—F	1.364	C—H (av.)	1.081	ED
		∠CCF	110.7	∠CCH (av.)	111.0	dihedral angle between CCF planes	118.9	
1,2-Difluoroethane	CH_2FCH_2F	C—C	1.503	C—F	1.389	C—H	1.103	ED
		∠CCF	110.3	∠CCH	111	dihedral angle of internal rotation	109	
1,1-Difluoroethene	$CH_2{=}CF_2$	C—C	1.340	C—F	1.315	C—H	1.091	ED, MW
		∠CCF	124.7	∠CCH	119.0			
cis-1,2-Difluoroethene	CHF=CHF	C—C	1.33	C—F	1.342	C—H	1.099	ED, MW
		∠CCF	122.0	∠CCH	124.1			
Difluoromethane	CH_2F_2	C—F	1.357	C—H	1.093			MW
		∠FCF	108.3	∠HCH	113.7			
Dimethoxymethane	$H_3C_a—O—C_bH_2—O—C_aH_3$	$C_a—O$	1.432	$C_b—O$	1.382	C—H (av.)	1.108	ED
		∠COC	114.6	∠OCO	114.3	∠OCH	110.3	
Dimethylamine	$(CH)_2NH$	C—N	1.455	N—H	1.00	C—H	1.106	ED
		∠CNC	111.8	∠CNH	107	∠NCH	112	
		∠HCH	107					
Dimethylberyllium	$(CH_3)_2Be$ (CBeC linear)	C—Be	1.698	C—H	1.127	∠BeCH	113.9	ED
Dimethyl cadmium	$(CH_3)_2Cd$	C—Cd	2.112	∠HCH	108.4			R
Dimethyl carbonate	$(C_aH_3O_a)_2C_b{=}O_b$	$C_b—O_b$	1.209	$C_b—O_a$	1.34	$C_a—O_a$	1.42	ED
		$\angle O_aC_bO_a$	107	$\angle C_bO_aC_a$	114.5			
Dimethylcyanamide	$(C_aH_3)_2N_a—C_b{\equiv}N_b$	$C_b—N_b$	1.161	$C_a—N_a$	1.463	$C_b—N_a$	1.338	ED
trans-Dimethyldiazene	$CH_3N{=}NCH_3$	C—N	1.482	N—N	1.247	∠CNN	112.3	ED
		$\angle C_aNC_a$	115.5	$\angle C_aNC_b$	116.0			
1,2-Dimethyldiborane	CH_3, H_b, CH_3 / B, B / H_t, H_b, H_t	B—B	1.799	B—C	1.580			ED
		$B—H_b$ (*cis*)	1.358	$B—H_b$ (*trans*)	1.365	$B—H_t$	1.24	
		∠BBC (*cis*)	122.6	∠BBC (*trans*)	121.8			
Dimethyl diselenide	$(CH_3)_2Se_2$	C—Se	1.95	Se—Se	2.326	C—H	1.13	ED
		∠CSeSe	98.9	∠HCSe	108	CSeSeC dihedral angle	88	
Dimethyl disulfide	$(CH_3)_2S_2$	C—S	1.816	S—S	2.029	C—H	1.105	ED
		∠SSC	103.2	∠SCH	111.3	CSSC dihedral angle	85	

Compound	Structure	Bond distances in Å and angles in degrees						Method
S,S′-Dimethyl dithiocarbonate	$C_aH_3SC_b(=O)SC_aH_3$	C_a—S	1.802	C_b—S	1.777	C_b—O	1.206	ED
	syn-syn conformer	∠OCS	124.9	∠CSC	99.3			
Dimethyl ether	$(CH_3)_2O$	C—O	1.416	C—H	1.121			ED
		∠COC	112	∠HCH	108			
N,N′-Dimethylhydrazine	$CH_3NH—NHCH_3$	C—N	1.46	N—N	1.42	N—H	1.03	ED
		C—H	1.12	∠NNC	112	CNNC dihedral angle	90	
Dimethyl mercury	$(CH_3)_2Hg$	C—Hg	2.083	C—H	1.160 (ass.)	Hg···H	2.71	ED
Dimethylphosphine	$(CH_3)_2PH$	C—P	1.848	P—H	1.419			MW
		∠CPC	99.7	∠CPH	97.0			
2,2-Dimethylpropanenitrile	$(C_cH_3)_3C_b—C_a≡N$	$C_a—C_b$	1.495	$C_b—C_c$	1.536	C_a—N	1.159	MW
		$∠C_cC_bC_c$	110.5					
Dimethyl selenide	$(CH_3)_2Se$	C—Se	1.943	C—H	1.093			MW
		∠CSeC	96.2	∠SeCH	108.7	∠HCH	110.3	
Dimethyl silane	$(CH_3)_2SiH_2$	C—Si	1.868	C—H	1.089	Si—H	1.482	MW
		∠CSiC	110.9	∠CSiH	109.5	∠SiCH	110.9	
		∠HSiH	107.8					
Dimethyl sulfide	$(CH_3)_2S$	C—S	1.802	C—H	1.090			ED, MW
		∠CSC	98.80	∠HCH	109.3			
Dimethyl sulfone	$(CH_3)_2SO_2$	C—S	1.771	S—O	1.435	C—H	1.114	ED
		∠CSC	102	∠OSO	121			
Dimethyl sulfoxide	$(CH_3)_2SO$	C—S	1.799	S—O	1.485	C—H	1.081	MW
		∠CSC	96.6	∠CSO	106.7	∠HCH	110.3	
		dihedral angle between SCC plane and S—O bond	115.5					
Dimethyl zinc	$(CH_3)_2Zn$	C—Zn	1.929	∠HCH	107.7			R
1,4-Dioxane	O(CH_2CH_2)$_2$O ring, chair form	C—C	1.523	C—O	1.423	C—H	1.112	ED
		∠CCO	109.2	∠COC	112.45			
Ethane	C_2H_6 staggered conformation	C—C	1.5351	C—H	1.0940	∠CCH	111.17	MW
		C—C (r_e)	1.522					
1,2-Ethanediamine	$H_2NCH_2CH_2NH_2$ *gauche* conformer	C—C	1.545	C—N	1.469	C—H	1.11	ED
		∠CCN	110.2	dihedral angle between NCC and CCN planes	64			
Ethanethiol	$C_bH_3—C_aH_2—SH$	$C_a—C_b$	1.530	C_a—S	1.829	S—H	1.350	MW
		C_a—H	1.090	C_b—H	1.093	$∠C_aSH$	96.4	
		$∠C_bC_aS$	108.3	$∠C_bC_aH$	109.6	$∠C_aC_bH$	109.7	
Ethanol	$C_bH_3C_aH_2OH$ staggered conformation	C—C	1.512	C—O	1.431	O—H	0.971	MW
		C_a—H	1.10	C_b—H	1.09	∠COH	105	
		∠CCO	107.8	$∠C_bC_aH$	111	$∠C_aC_bH$	110	
Ethylene	$CH_2{=}CH_2$	C—C (r_s)	1.329	C—H (r_s)	1.082	∠HCH ($θ_s$)	117.2	MW, IR
Ethyleneimine	aziridine ring: N bearing H_a; each C bearing H_b and H_c	C—C	1.481	N—C	1.475			MW
		C—H	1.084	N—H	1.016			
		∠CNC	60.3	$∠H_aNC$	109.3	$∠H_bCH_c$	115.7	
		$∠H_bCC$	117.8	$∠H_bCN$	118.3	$∠H_cCC$	119.3	
		$∠H_cCN$	114.3					
Ethyl methyl ether	$C_2H_5OCH_3$	C—C	1.520	C—O (av.)	1.418	C—H (av.)	1.118	ED
		∠COC	111.9	∠OCC	109.4	∠HCH	109.0	

Compound	Structure	Bond distances in Å and angles in degrees						Method
Ethyl methyl sulfide	$C_2H_5SCH_3$	C—C	1.536	C—S (av.)	1.813	C—H	1.111	ED
	gauche conformer	∠CSC	97	∠SCC	114.0	∠HCH	110	
Fluoroketene	HFC=C=O	C—C	1.317	C—O	1.167	C—F	1.360	MW
		C—H	1.102	∠CCO	178.0	∠CCF	119.5	
		∠CCH	122.3					
Fluoromethane	CH_3F	C—F (r_e)	1.382	C—H (r_e)	1.095	∠HCH (θ_e)	110.45	MW, IR
Fluoromethylidyne	CF	C—F (r_e)	1.2718					UV
(Fluoromethylidyne) phosphine	FC≡P	C—F	1.285	C—P	1.541			MW
2-Fluoropropane	CH_3CHFCH_3	C—C	1.522	C—F	1.398			MW
		∠CCC	113.4	∠CCF	108.2			
Formaldehyde	H_2CO	C—O	1.208	C—H	1.116	∠HCH	116.5	MW
Formaldehyde azine	H_2C=N—N=CH_2	C—N	1.277	N—N	1.418	C—H	1.094	ED
	trans conformer	∠CNN	111.4	∠HCN	120.7			
Formaldehyde oxime	H_a OH_c C=N H_b	C—N	1.276	N—O	1.408	O—H_c	0.956	MW
		C—H_a	1.085	C—H_b	1.086	∠CNO	110.2	
		∠H_aCN	121.8	∠H_bCN	115.6	∠NOH_c	102.7	
Formamide	H_c O N—C H_b H_a	C—N	1.368	C—O	1.212	C—H_a	1.125	ED, MW
		N—H	1.027	∠CNH (av.)	119.2	∠NCO	125.0	
Formic acid	O_a H—C O_b—H	C—O_a	1.202	C—O_b	1.343	O_b—H	0.972	MW
		C—H	1.097					
		∠O_aCO_b	124.9	∠HCO_a	124.1	∠CO_bH	106.3	
	(planar)							
Formic acid dimer	O_a····H—O_b HC CH O_b—H····O_a	C—O_a	1.220	C—O_b	1.323	O_a···O_b	2.703	ED
		∠O_aCO_b	126.2	∠C$O_a$$O_b$	108.5			
Formyl radical	HC=O	C—O	1.1712	C—H	1.110	∠HCO	127.43	MW
Fulvene	C_dH_2 C_a HC_b C_bH HC_c—C_cH	C_a—C_d	1.349	C_a—C_b	1.470	C_b—C_c	1.355	MW
		C_c—C_c	1.476	C_b—H	1.078	C_c—H	1.080	
		C_d—H	1.13	∠$C_bC_aC_b$	106.6	∠$C_bC_cC_c$	109	
		∠$C_aC_bC_c$	107.7	∠C_aC_bH	124.7	∠C_bC_cH	126.4	
		∠HC_dH	117					
Furan	H_a O H_a C_a C_a C_b—C_b H_b H_b	C_a—C_b	1.361	C_b—C_b	1.431	C_a—O	1.362	MW
		C_a—H_a	1.075	C_b—H_b	1.077			
		∠$C_aC_bC_b$	106.1	∠C_bC_aO	110.7	∠C_aOC_a	106.6	
		∠$C_aC_bH_b$	128.0	∠OC_aH_a	115.9			
Furfural	H H C_c—C_b O_b H—C_d C_a—C_e O_a H	C_a—C_e	1.458	C_e—O_b	1.250	C_e—H	1.088	MW
		∠C_aC_eO	121.6	∠$C_eC_aC_b$	133.9	∠C_aC_eH	116.9	
		trans conformer (with respect to O_a and O_b atoms)						
Glycolaldehyde	H_c O_b C_b H_b—C_a H_a H_b O_a	C_a—C_b	1.499	C_a—O_a	1.437	C_b—O_b	1.209	MW
		C_a—H_b	1.093	C_b—H_c	1.102	O_a—H_a	1.051	
		∠$C_aC_bO_b$	122.7	∠$C_bC_aO_a$	111.5			
		∠$C_aC_bH_c$	115.3	∠$C_bC_aH_b$	109.2	∠$H_bC_aH_b$	107.6	
		∠$C_aO_aH_a$	101.6	∠$H_bC_aO_a$	109.7			
Glyoxal	CHOCHO	C—C	1.526	C—O	1.212	C—H	1.132	ED, UV
	trans conformer	∠CCO	121.2	∠HCO	112			
Hexachloroethane	Cl_3CCCl_3	C—C	1.56	C—Cl	1.769	∠CCCl	110.0	ED
2,4-Hexadiyne	$C_aH_3C_b$≡C_cC_c≡$C_bC_aH_3$	C_a—C_b	1.450	C_b—C_c	1.208	C_c—C_c	1.377	ED
		C_a—H	1.09					
Hexafluoroethane	F_3CCF_3	C—C	1.545	CF	1.326	∠CCF	109.8	ED

Compound	Structure	Bond distances in Å and angles in degrees						Method
Hexafluoropropene	CF_2=$CFCF_3$	C—C	1.513	C=C	1.329 (ass.)	C—F	1.329 (ass.)	ED
		∠CCC	127.8	∠FCC (CF)	120	∠FCC(CF_2)	124	
		∠FCC(CF_3)	110					
trans-1,3,5-Hexatriene	H_2C_a=C_bHC_cH=C_cHC_bH=C_aH_2	C_a—C_b	1.337	C_b—C_c	1.458	C_c—C_c	1.368	ED
		∠$C_aC_bC_c$	121.7	∠$C_bC_cC_c$	124.4			
Hydrogen cyanide	HCN (linear)	C—H (r_e)	1.0655	C—N (r_e)	1.1532			MW, IR
Iminocyanide radical	HNCN	N—H	1.034	N···N	2.470			UV
		∠HNC	116.5	∠NCN	~180			
Iodoacetylene	IC≡CH	C—C	1.218	C—I	1.980	C—H	1.059	IR
Iodocyanoacetylene	IC_a≡C_bC_c≡N (linear)	C_a—C_b	1.207	C_b—C_c	1.370	C_c—N	1.160	MW
		C_a—I	1.985					
Iodomethane	CH_3I	C—I (r_e)	2.132	C—H (r_e)	1.084	∠HCH (θ_e)	111.2	MW, IR
Iron pentacarbonyl	$Fe(CO)_5$ (D_{3h})	Fe—C (av.)	1.821	$(Fe—C)_{eq}$ – $(Fe—C)_{ax}$	0.020	C—O (av.)	1.153	ED
Isobutane	$(C_bH_3)_3C_aH$	C_a—C_b	1.535	C_a—H	1.122	C_b—H	1.113	ED, MW
		∠$C_bC_aC_b$	110.8	∠C_aC_bH	111.4			
Isobutene	$(C_aH_3)_2C_b$=C_cH_2 (C_aH_3, C_aH_3, C_b=C_c, H_c, H)	C_a—C_b	1.508	C_b—C_c	1.342	C_a—H	1.119	ED, MW
		C_c—H_c	1.10					
		∠$C_aC_bC_a$	115.6	∠$C_aC_bC_c$	122.2	∠C_bC_cH	121	
		∠HC_aC_b (av.)	111.4	∠HC_aH	107.9	∠$H_cC_cH_c$	118.5	
Isocyanic acid	HNCO (bent)	N—C	1.209	C—O	1.166	N—H	0.986	MW
		∠NCO	180	∠HNC	128.0			
Isocyanomethane	C_aH_3—N≡C_b	C_a—N	1.424	N— C_b	1.166	C_a—H	1.102	MW
		∠NC_aH	109.12					
		∠HCH	123.0					
Isofulminic acid	HCNO (linear)	C—N	1.161	N—O	1.207	H—C	1.027	MW
Isothiocyanic acid	HNCS	N—C	1.216	C—S	1.561	N—H	0.989	MW
		∠NCS	180	∠HNC	135.0			
Ketene	H_2C=C=O	C—C	1.315	C—O	1.163			MW
		C—H	1.090	∠HCH	123.5			
Malononitrile	$CH_2(CN)_2$	C—C	1.480	C—N	1.147	C—H	1.091	MW
		∠CCC	110.4	∠CCN	176.6	∠HCH	108.4	
Methane	CH_4	C—H (r_e)	1.0870					IR
Methanethioamide	S=C(H_c)—N(H_a)(H_b)	C—S	1.626	C—N	1.358	C—H_c	1.10	MW
		N—H_a	1.002	N—H_b	1.007			
		∠NCS	125.3	∠H_aNC	117.9	∠H_bNC	120.4	
		∠SCH_c	127	∠H_aNH_b	121.7	∠NCH_c	108	
Methanethiol	CH_3SH	C—S	1.819	S—H	1.34	C—H	1.09	MW
		∠HSC	96.5	∠HCH	109.8	angle between CH_3 symmetry axis and C—S bond	2.2	
Methanol	CH_3OH	C—O	1.4246	C—H	1.0936	O—H	0.9451	MW
		∠COH	108.53	∠HCH	108.63	angle between CH_3 symmetry axis and C—O bond	3.27	
Methyl	·CH_3 planar (D_{3h})	C—H	1.076					R
N-Methylacetamide	H_3C_a—C_b(=O)—N(H)—C_cH_3	C_a—C_b	1.520	C_b—N	1.386	C_c—N	1.469	ED
		C_b—O	1.225	C—H	1.107			
		∠C_bNC_c	119.7	∠NC_bO	121.8	∠C_aC_bN	114.1	

Compound	Structure	Bond distances in Å and angles in degrees						Method
Methylamine	CH_3NH_2	C—N	1.471	N—H	1.019	C—H	1.095	MW
		∠HNC	110.3	∠HNH	106.6	∠HCH	108.1	
		angle between CH_3 symmetry axis and C—N bond	4.3					
Methyl azide	CH_3—N_a—N_b—N_c NNN linear	C—N_a	1.468	N_a—N_b	1.216	N_b—N_c	1.113	ED
		C—H	1.09	∠CN_aN_b	116.8			
3-Methyl-3*H*-diazirine	CH_3CH<(N=N)	C—C	1.501	C—N	1.481	N—N	1.235	MW
		∠NCN	49.3	dihedral angle between CNN plane and C—C bond	122.3			
Methylene	:CH_2	C—H (r_e)	1.0748	∠HCH (θ_e)	133.84			IR,MW
Methylenecyclopropane	(C_cH_2)$_2$$C_b$=$C_aH_2$	C_a—C_b	1.332	C_b—C_c	1.457	C_c—C_c	1.542	MW
		C_c—H	1.09	∠$C_cC_bC_c$	63.9	∠HC_aH	114.3	
		∠HC_cH	113.5	dihedral angle between C_cH_2 plane and C_c—C_c bond	150.8			
3-Methyleneoxetane	O(C_cH_2)$_2$$C_b$=$C_aH_2$	C_a—C_b	1.33	C_b—C_c	1.52	C_c—O	1.45	MW
		C—H	1.09 (ass)	∠HC_cH	114 (ass)	∠HC_aH	120 (ass)	
		∠$C_cC_bC_c$	87					
Methylenephosphine	CH_cH_t=PH planar	C—P	1.673	C—H_c	1.09	C—H_t	1.09	MW
		P—H	1.420	∠CPH	97.4			
		∠HCH	117.2	∠PCH_c	124.4	∠PCH_t	118.4	
Methyl formate	C_aH_3—O_a—C_b(=O_b)—H_b	C_b—O_b	1.206	C—O (av.)	1.393	C_a—H	1.08	ED
		C_b—H	1.101 (ass.)					
		∠COC	114	∠$O_aC_bO_b$	127	∠O_aC_aH	110	
Methylgermane	CH_3GeH_3	C—Ge	1.945	Ge—H	1.529	C—H	1.083	MW
		∠HGeH	109.3	∠HCH	108.4			
Methyl hypochlorite	CH_3OCl	C—O	1.389	O—Cl	1.674	C—H	1.103	MW
		∠COCl	112.8	∠HCH	109.6			
Methylidyne	:CH	C—H (r_e)	1.1198					UV
Methylidynephosphine	HCP	C—P (r_e)	1.5398	C—H (r_e)	1.0692			MW
Methylketene	C_cH_3(H)C_b=C_a=O	C_a—C_b	1.306	C_b—C_c	1.518	C_a —O	1.171	MW
		C_b—H	1.083	C_c—H	1.10			
		∠OC_aC_b	180.5	∠$C_aC_bC_c$	122.6	∠C_aC_bH	113.7	
		∠C_cC_bH	123.7	∠HCH	109.2			
Methyl nitrate	H_a,H_a,H_b—C—O—N(=O_a)—O_b	C—O	1.437	C—H_a	1.10	C—H_b	1.09	MW
		O—N	1.402	N—O_a	1.205	N—O_b	1.208	
		∠CON	112.7	∠ONO_a	118.1	∠ONO_b	112.4	
		∠OCH_a	110	∠OCH_b	103			
Methyloxirane	$C_aH_3C_bH$—C_cH_2 (O bridge)	C_a—C_b	1.51	∠$C_aC_bC_c$	121.0	dihedral angle between C_bC_cO plane and C_aC_b bond	123.8	MW
Methylphosphine	CH_3PH_2	C—P	1.858	C—H	1.094			ED
Methylphosphonic difluoride	CH_3POF_2	C—P	1.770	P—O	1.444	P—F	1.545	ED,MW
		∠OPC	117.8	∠FPC	103.7	∠FPF	99.2	
Methylsilane	CH_3SiH_3	C—Si	1.867	Si—H	1.485	C—H	1.093	MW
		∠HCH	107.7	∠HSiH	108.3			
Methylstannane	CH_3SnH_3	C—Sn	2.143	Sn—H	1.700			MW

Compound	Structure	Bond distances in Å and angles in degrees								Method
Methyl thiocyanate	C_aH_3–S–C_b–N	S–C_a	1.824	S–C_b	1.684	C_b–N	1.170			MW
		C–H	1.081							
		∠C_aSC_b	99.0	∠HCH	110.6	∠HCS	108.3			
Methyltrioxorhenium	CH_3ReO_3	Re–C	2.074	Re–O	1.703	C–H	1.088			MW
		∠ReCH	108.9	∠CReO	106.4					
Molybdenum carbide	MoC	Mo–C	1.676							UV
Molybdenum carbonyl	$Mo(CO)_6$ (O_h)	Mo–C	2.063	C–O	1.145					ED
Naphthalene		C_a–C_b	1.37	C_b–C_b	1.41	C_a–C_c	1.42			ED
		C_c–C_c	1.42	C–C (av.)	1.40	∠$C_aC_cC_c$	119.4			
Neopentane	$C(CH_3)_4$	C–C	1.537	C–H	1.114	∠CCH	112			ED
Nickel carbonyl	$Ni(CO)_4$ (T_d)	Ni–C	1.839	C–O	1.121					IR
Nickel monocarbonyl	NiCO (linear)	Ni–C	1.64	C–O	1.19					IR
Nickel cyanide	NiC≡N (linear)	Ni–C	1.828	C–N	1.158					MW
Nitromethane	CH_3NO_2	C–N	1.489	N–O	1.224	C–H	1.088 (ass.)			MW
		∠ONO	125.3	∠NCH	107					
N-Nitrosodimethylamine	$(CH_3)_2NNO$	C–N	1.461	N–O	1.235	N–N	1.344			ED
		∠CNC	123.2	∠CNN	116.4	∠ONN	113.6			
Nitrosomethane	CH_3NO	C–N	1.49	N–O	1.22	C–H	1.084			MW
		∠CNO	112.6	∠NCH	109.0					
2,5-Norbornadiene	(C_{2v})	C_a–C_b	1.535	C_b–C_b	1.343	C_a–C_c	1.573			ED
		C–H	1.12	∠$C_aC_cC_a$	94					
		dihedral angle between the two $C_aC_bC_bC_a$ planes	115.6							
1,2,5-Oxadiazole	(planar)	C–C	1.421	C–N	1.300	O–N	1.380			MW
		C–H	1.076	∠CCH	130.2	∠NCH	120.9			
		∠CCN	109.0	∠NON	110.4	∠ONC	105.8			
1,3,4-Oxadiazole	(planar)	C–O	1.348	C–N	1.297	N–N	1.399			MW
		C–H	1.075	∠OCH	118.1	∠NCH	128.5			
		∠CNN	105.6	∠COC	102.0	∠OCN	113.4			
Oxalic acid		C–C	1.544	C–O_a	1.205	C–O_b	1.336			ED
		O_b–H	1.05							
		∠CCO_a	123.1	∠O_aCO_b	125.0	∠CO_bH	104			
Oxalyl chloride		C–C	1.534	C–O	1.182	C–Cl	1.744			ED
		∠CCO	124.2	∠CCCl	111.7	68% trans, 32% gauche at 0 °C				
Oxetane		C–C	1.546	C–O	1.448	C–H (av.)	1.090			MW
		∠CCC	85	∠COC	92	∠OCC	92			
		∠HCH (av.)	109.9							

Compound	Structure	Bond distances in Å and angles in degrees						Method
Oxirane	CH_2—O—CH_2 (ring)	C—C	1.466	C—O	1.431	C—H	1.085	MW
		∠HCH	116.6	dihedral angle between NH_2 plane and N—C bond	158.0			
Phenol	H—O—C_a; HC_b, C_bH; HC_c, C_cH; C_dH (ring)	C—C (av.)	1.397	C_a—O	1.364	O—H	0.956	MW
		C_b—H	1.084	C_c—H	1.076	C_d—H	1.082	
		∠COH	109.0					
Phosphirane	CH_2—PH—CH_2 (ring)	C—C	1.502	C—P	1.867	P—H	1.43	MW
		C—H	1.09	∠CPC	47.4	∠HPC	95.2	
		∠HCH	114.4	∠CCH	118	dihedral angle between PCC plane and PH bond	95.7	
Piperazine	NH(CH_2—CH_2)₂NH (ring) (C_{2h})	C—C	1.540	C—N	1.467	C—H	1.110	ED
		∠CNC	109.0	∠CCN	110.4			
Palladium carbide	PdC	Pd—C	1.712					UV
Platinum carbide	PtC	Pt—C (r_e)	1.6767					UV
Potassium carbide	KC	K—C	2.528					MW
Propane	C_3H_8	C—C	1.532	C—H	1.107			ED
		∠CCC	112	∠HCH	107			
Propene	H_d—C_c(H_d)(H_d)—C_b(H_c)=C_a(H_b)(H_a)	C_a—C_b	1.341	C_b—C_c	1.506			ED, MW
		C_a—H_a	1.104	C_c—H_d	1.117			
		∠$C_aC_bC_c$	124.3	∠$C_bC_aH_{a,b,c}$	121.3	∠$C_bC_cH_d$	110.7	
2-Propenoyl chloride	H_2C_a=C_bH—C_c(Cl)=O	C_a—C_b	1.35	C_b—C_c	1.48	C_c—Cl	1.82	MW
		C_c—O	1.19	C—H	1.086 (ass.)			
		∠$C_aC_bC_c$	123	∠C_bC_cCl	116	∠C_bC_cO	127	
		∠C_aC_bH	120 (ass.)	∠C_bC_aH	121.5 (ass.)			
2-Propynal	H_aC_a≡C_b—C_cH_cO (planar)	C_a—C_b	1.211	C_b—C_c	1.453	C_c—O	1.214	ED, MW
		C_a—H_a	1.085	C_c—H_c	1.130			
		∠$C_aC_bC_c$	178.6	∠C_bC_cO	124.2	∠$C_bC_cH_c$	113.7	
Propyne	H_3C_c—C_b≡C_aH	C_c—C_b	1.459	C_b—C_a	1.206			MW
		C_a—H	1.056	C_c—H	1.105	∠HC_cC_b	110.2	
Propynal isocyanide	H_3C_c—C_b≡C_a—N≡C	C_c—C_b (r_s)	1.456	C_b—C_a (r_s)	1.206	C_a—N (r_s)	1.316	MW
		N—C (r_s)	1.175	C_c—H (r_s)	1.090	∠HC_cC_b (θ_s)	110.7	
Pyrazine	(ring with N, N)	C—C	1.339	C—N	1.403	C—H	1.115	ED
		∠CCH	123.9	∠CCN	115.6			
Pyridazine	HC_b=C_bH; HC_a, C_aH; N—N (ring)	C_a—C_b	1.393	C_b—C_b	1.375	C_a—N	1.341	ED, MW
		N—N	1.330	∠NCC	123.7	∠NNC	119.3	

Compound	Structure	Bond distances in Å and angles in degrees						Method
Pyridine		C_a—C_b	1.395	C_b—C_c	1.394	C_a—N	1.340	MW
		C_a—H_a	1.084	C_b—H_b	1.081	C_c—H_c	1.077	
		∠$C_aC_bC_c$	118.5	∠$C_bC_cC_b$	118.3	∠$C_cC_bH_b$	121.3	
		∠C_aNC_a	116.8	∠NC_aC_b	123.9	∠NC_aH_a	115.9	
Pyrimidine	(C_{2v} assumed)	C—C	1.393	C—N	1.340			ED
		∠NCN	127.6	∠CNC	115.5			
Pyrrole		C_a—C_b	1.382	C_b—C_b	1.417	C_a—N	1.370	MW
		C_a—H_a	1.076	C_b—H_b	1.077	N—H	0.996	
		∠$C_aC_bC_b$	107.4	∠C_aNC_a	109.8	∠NC_aC_b	107.7	
		∠C_bC_aH	127.1	∠NC_aH_a	121.5			
Pyruvonitrile		C_a—C_b	1.518	C_b—C_c	1.477	C—H	1.12	ED, MW
		C—N	1.17	C—O	1.208	∠HCH	109.2	
		∠$C_aC_bC_c$	114.2	∠C_aC_bO	124.5	∠CCN	179	
Ruthenium carbide	RuC	Ru—C	1.607					UV
Silacyclobutane	CH_2—CH_2 CH_2—SiH_2	C—C	1.571	C—Si	1.885	C—H	1.100	ED
		Si—H	1.47	∠CCC	99.8	∠CSiC	77.2	
		∠SiCC	84.8	dihedral angle between CCC and CSiC planes	146			
Silaethene	$H_2Si{=}CH_2$	Si—C (r_e)	1.704	Si—H (r_e)	1.467	C—H (r_e)	1.082	MW
		∠HCSi	122.0	∠HSiC	122.4			
Silicon dicarbide	CSiC (ring)	C—C(r_s)	1.269	Si—C (r_s)	1.832	∠CSiC (θ_s)	40.5	MW
Silylchloroacetylene	$SiH_3C{\equiv}CCl$	C—C	1.234	Si—C	1.812	C—Cl	1.620	ED
		Si—H	1.488	∠HSiC	109.4			
Silyl cyanide	$SiH_3C{\equiv}N$	Si—C	1.850	C—N	1.156	Si—H	1.487	ED,MW
		∠HSiC	107.25					
Sodium carbide	NaC	Na—C	2.232					MW
Spiro[2.2]pentane	(D_{2d})	C_b—C_b	1.52	C_a—C_b	1.47	C—H	1.09	ED
		∠$C_bC_aC_b$	62	∠HCH	118			
Strontium methyl	$SrCH_3$	Sr—C	2.487	C—H (ass.)	1.104	∠HCH	105.8	UV
Succinonitrile	CH_2CN | CH_2CN	C—C	1.561	C—C(N)	1.465	C—N	1.161	ED
		C—H	1.09	∠CCC	110.4	dihedral angle of CCCC for *gauche* conformer	75	
Tetrabromomethane	CBr_4 (T_d)	C—Br	1.935					ED
Tetrachloroethene	$CCl_2{=}CCl_2$	C—C	1.354	C—Cl	1.718	∠ClCCl	115.7	ED
Tetrachloromethane	CCl_4 (T_d)	C—Cl	1.767					ED
Tetracyanoethene	$(CN)_2C{=}C(CN)_2$	C—C	1.435	C=C	1.357	C—N	1.162	ED
		∠CC=C	121.1					
2,2,4,4-Tetrafluoro-1,3-dithietane	(D_{2h} assumed)	C—S	1.785	C—F	1.314	∠CSC	83.2	ED
		∠FCS	113.7					
Tetrafluoroethene	$CF_2{=}CF_2$	C—C	1.31	C—F	1.319	∠CCF	123.8	ED
Tetrafluoromethane	CF_4 (T_d)	C—F	1.323					ED

Compound	Structure	Bond distances in Å and angles in degrees						Method
Tetrahydrofuran	CH_2CH_2—O—CH_2CH_2 (ring)	C—C	1.536	C—O	1.428	C—H	1.115	ED
Tetrahydropyran	H_2C ring with O (chair form)	C—C	1.531	C—O	1.420	C—H	1.116	ED
		∠COC	111.5	∠OCC	111.8	∠CCC (C)	108	
		∠CCC (O)	111					
Tetrahydrothiophene	CH_2CH_2—S—CH_2CH_2 (ring)	C—C	1.536	C—S	1.839	C—H	1.120	ED
		∠CCC	105.0	∠CSC	93.4	∠SCC	106.1	
Tetraiodomethane	CI_4 (T_d)	C—I	2.15					ED
Tetramethylgermane	$(CH_3)_4Ge$	C—Ge	1.945	C—H	1.12	∠GeCH	108	ED
Tetramethyl lead	$(CH_3)_4Pb$	C—Pb	2.238					ED
Tetramethylsilane	$(CH_3)_4Si$	C—Si	1.875	C—H	1.115	∠HCH	109.8	ED
Tetramethylstannane	$(CH_3)_4Sn$	C—Sn	2.144	C—H	1.12			ED
1,2,5-Thiadiazole	S, N, N, HC—CH ring (planar)	C—C	1.420	C—N	1.328	S—N	1.631	MW
		C—H	1.079					
		∠CCN	113.8	∠NSN	99.6	∠CCH	126.2	
1,3,4-Thiadiazole	S, HC, CH, N—N ring (planar)	C—S	1.721	C—N	1.302	N—N	1.371	MW
		C—H	1.08	∠CSC	86.4	∠SCN	114.6	
		∠CCN	112.2	∠NCH	123.5	∠SCH	121.9	
Thietane	CH_2—CH_2, CH_2—S (ring)	C—C	1.549	C—S	1.847	C—H (av.)	1.100	ED, MW
		∠CSC	76.8	∠HCH (av.)	112	dihedral angle between CCC and CSC planes	154	
Thiirane	H_2C—S—H_2C (ring)	C—C	1.484	C—S	1.815	C—H	1.083	MW
		∠CSC	48.3	∠CCS	65.9	∠HCH	116	
		dihedral angle between CH_2 plane and C—C bond	152					
Thioacetaldehyde	H_3C_b—C_a(=S)H	C_a—S (r_s)	1.610	C_a—C_b (r_s)	1.506			MW
		C_a—H (r_s)	1.089	C_b—H (r_s)	1.094 (av.)			
		∠C_bC_aS (θ_s)	125.3	∠C_bC_aH (θ_s)	119.4	∠HC_bC_a (θ_s)	110.6 (av.)	
Thiocarbonyl fluoride	F_2CS	C—S	1.589	C—F	1.315	∠FCF	107.1	MW
Thioformaldehyde	CH_2S	C—S	1.611	C—H	1.093	∠HCH	116.9	MW
Thioketene	H_2C=C=S	C—C (r_s)	1.314	C—S (r_s)	1.554	C—H (r_s)	1.080	IR
	C_{2v}	∠HCH (θ_s)	119.8					
Thiophene	H_b, C_b—C_b, H_aC_a, C_aH_a, S (ring)	C_a—C_b	1.370	C_b—C_b	1.423	C_a—S	1.714	MW
		C_a—H_a	1.078	C_b—H_b	1.081			
		∠$C_aC_bC_b$	112.5	∠C_aSC_a	92.2	∠SC_aC_b	115.5	
		∠SC_aH_a	119.9	∠$C_aC_bH_b$	124.3			
Toluene	C_6H_5—CH_3	C—C (ring)	1.399	C—CH_3	1.524	C—H (av.)	1.11	ED
1,1,1-Tribromoethane	CH_3CBr_3	C—C	1.51 (ass.)	C—Br	1.93	C—H	1.095 (ass.)	MW
		∠BrCBr	111	∠CCBr	108	∠CCH	109.0 (ass.)	

Compound	Structure	Bond distances in Å and angles in degrees						Method
Tribromomethane	$CHBr_3$ (C_{3v})	C—Br	1.924	C—H	1.11	∠BrCBr	111.7	ED, MW
Tri-*tert*-butyl methane	$HC_a[C_b(C_cH_3)_3]_3$	C_a—C_b	1.611	C_b—C_c	1.548	C—H	1.111	ED
		∠$C_aC_bC_c$	113.0					
Trichloroacetonitrile	CCl_3CN	C—C	1.460	C—N	1.165	C—Cl	1.763	ED
		∠ClCCl	110.0					
1,1,1-Trichloroethane	CH_3CCl_3	C—C	1.541	C—Cl	1.771	C—H	1.090	MW
		∠CCCl	109.6	∠ClCCl	109.4	∠HCH	110.0	
		∠CCH	108.9					
Trichlorofluoromethane	CCl_3F	C—Cl	1.754	C—F	1.362	∠ClCCl	111	MW
Trichloromethane	$CHCl_3$	C—Cl	1.758	C—H	1.100	∠ClCCl	111.3	MW
Trichloromethylgermane	CH_3GeCl_3	C—Ge	1.89	Ge—Cl	2.132	C—H	1.103 (ass.)	ED, MW
		∠ClGeCl	106.4	∠GeCH	110.5 (ass.)			
Trichloromethylsilane	CH_3SiCl_3	C—Si	1.876	Si—Cl	2.021			MW
Trichloromethylstannane	CH_3SnCl_3	C—Sn	2.10	Sn—Cl	2.304	C—H	1.100	ED
1,1,1-Trichloro-2,2,2-trifluoroethane	CF_3CCl_3 (staggered configuration)	C—C	1.54	C—F	1.33	C—Cl	1.77	MW
		∠CCF	110	∠CCCl	109.6			
		∠CSnCl	113.9	∠ClSnCl	104.7	∠SnCH	108	
Triethylenediamine	$N(CH_2CH_2)_3N$ (D_{3h})	C—C	1.562	C—N	1.472	∠CNC	108.7	ED
		∠NCC	110.2					
Trifluoroacetic acid	$CF_3C(=O_a)O_bH$	C—C	1.546	C—O_a	1.192	C—O_b	1.35	ED
		C—F	1.325	O—H	0.96 (ass.)			
		∠CCO_a	126.8	∠CCO_b	111.1	∠CCF	109.5	
1,1,1-Trifluoroethane	CH_3CF_3	C—C	1.494	C—F	1.340	C—H	1.081	ED
Trifluoroiodomethane	CF_3I (C_{3v})	C—F	1.330	C—I	2.138	∠FCF	108.1	ED, MW
Trifluoromethane	CHF_3 (C_{3v})	C—F	1.332	C—H	1.098	∠FCF	108.8	MW
Trifluoromethanesulfonyl fluoride	$CF_3SO_2F_a$	C—S	1.835	C—F (av.)	1.325	S—O	1.410	ED
		S—F_a	1.543	∠CSF_a	95.4	∠CSO	108.5	
		∠OSO	124.1	∠FCF	109.8			
Trifluoromethylimino-sulfurdifluoride	$CF_3N{=}SF_2$	C—N	1.409	S—N	1.477	S—F	1.594	ED,MW
		C—F	1.331	∠CNS	127.2	∠NSF	112.7	
		∠FSF	92.8	∠FCF	108.1			
Trifluoromethyl peroxide	CF_3OOCF_3	O—O	1.42	C—O	1.399	C—F	1.320	ED
		∠COO	107	∠FCF	109.0	COOC dihedral angle of internal rotation	123	
		∠CCF	119.2	∠CCH	112			
Trimethyl aluminium	$(CH_3)_3Al$	C—Al	1.957	C—H	1.113			ED
		∠CAlC	120	∠AlCH	111.7			
Trimethylamine	$(CH_3)_3N$	C—N	1.458	C—H	1.100			ED
		∠CNC	110.9	∠HCH	110			
Trimethylarsine	$(CH_3)_3As$	C—As	1.979	∠CAsC	98.8	∠AsCH	111.4	ED
Trimethyl bismuth	$(CH_3)_3Bi$	C—Bi	2.263	C—H	1.07	∠CBiC	97.1	ED
Trimethylborane	$(CH_3)_3B$	C—B	1.578	C—H	1.114			ED
		∠CBC	120	∠BCH	112.5			
Trimethylphosphine	$(CH_3)_3P$	C—P	1.847	C—H	1.091			ED
		∠CPC	98.6	∠PCH	110.7			
1,3,5-Trioxane	cyclic $(-O-CH_2-)_3$ ring (H_2C, CH_2, $C H_2$, O, O, O)	C—O	1.422	∠OCO	112.2	∠COC	110.3	MW

Compound	Structure	Bond distances in Å and angles in degrees						Method
Triphenylamine	$(C_6H_5)_3N$ (C_3)	C—C	1.392	C—N	1.42	∠CNC	116	ED
		torsional dihedral angle of phenyl rings	47					
Tungsten carbide	WC	W—C	1.7135					UV
Tungsten carbonyl	$W(CO)_6$ (O_h)	W—C	2.059	C—O	1.149			ED
Vanadium carbonyl	$V(CO)_6$ (O_h, involving dynamic Jahn-Teller effect)	V—C	2.015	C—O	1.138			ED
Vinyl bromide	See Vinyl chloride	C—C	1.3256	C—Br	1.8835	C—H_a	1.0780	MW
		C—H_b	1.0804	C—H_c	1.0794	∠CCBr	122.62	
		∠CCH_a	124.34	∠CCH_b	119.28	∠CCH_c	122.03	
Vinyl chloride	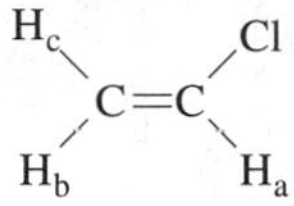	C—C	1.3262	C—Cl	1.7263	C—H_a	1.0783	MW
		C—H_b	1.0796	C—H_c	1.0796	∠CCCl	122.75	
		∠CCH_a	123.91	∠CCH_b	119.28	∠CCH_c	121.77	
Vinyl fluoride	See Vinyl chloride	C—C	1.3210	C—F	1.3428	C—H_a	1.0796	MW
		C—H_b	1.0774	C—H_c	1.0789	∠CCF	121.70	
		∠CCH_a	125.95	∠CCH_b	118.97	∠CCH_c	121.34	
Vinyl iodide	See Vinyl chloride	C—C	1.3276	C—I	2.0830	C—H_a	1.0787	MW
		C—H_b	1.0823	C—H_c	1.0799	∠CCI	122.97	
		∠CCH_a	123.54	∠CCH_b	119.36	∠CCH_c	122.30	
Zinc cyanide	ZnC≡N (linear)	Zn—C	1.955	C—N	1.146			MW

CHARACTERISTIC BOND LENGTHS IN FREE MOLECULES

This is a summary of typical bond lengths in gas-phase molecules. The value given for each bond is near the mid-range of values found in simple molecules. Bond lengths usually vary by 1 or 2%, and often by more, depending on the nature of the other bonds attached to the two atoms in question. References 1 and 2 give bond lengths in individual gas-phase molecules, as determined by spectroscopic and electron diffraction methods.

All bond distances are given in Å (1 Å = 10^{-10} m).

References

1. "Bond Lengths and Angles in Gas-Phase Molecules", *CRC Handbook of Chemistry and Physics*, 86th Edition, 2005, p. **9**-19.
2. Harmony, M. D., Laurie, V. W., Kuczkowski, R. L., Schwendeman, R. H., Ramsay, D. A., Lovas, F. J., Lafferty, W. J., and Maki, A. G., Molecular structure of gas-phase polyatomic molecules determined by spectroscopic methods, *J. Phys. Chem. Ref. Data* 8, 619, 1979.
3. Lide, D. R., "A survey of carbon-carbon bond lengths", *Tetrahedron* 17, 125, 1962.

A. Characteristic lengths of single bonds.

	As	Br	C	Cl	F	Ge	H	I	N	O	P	S	Sb	Se	Si
As	2.10														
Br	2.32	2.28													
C	1.96	1.94	1.53												
Cl	2.17	2.14	1.79	1.99											
F	1.71	1.76	1.39	1.63	1.41										
Ge		2.30	1.95	2.15	1.73	2.40									
H	1.51	1.41	1.09	1.28	0.92	1.53	0.74								
I		2.47	2.13	2.32	1.91	2.51	1.61	2.67							
N			1.46	1.90	1.37		1.02		1.45						
O			1.42	1.70	1.42		0.96		1.43	1.48					
P		2.22	1.85	2.04	1.57		1.42		1.65		2.25				
S		2.24	1.82	2.05	1.56		1.34					2.00			
Sb				2.33			1.70								
Se			1.95		1.71		1.47							2.33	
Si		2.21	1.87	2.05	1.58		1.48	2.44		1.63		2.14			2.33
Sn			2.14	2.28			1.71	2.67							
Te					1.82		1.66								

B. Lengths of multiple bonds (non-ring molecules).

C=C	1.34
C≡C	1.20
C=N	1.21
C≡N	1.16
C=O	1.21
C=S	1.61
N=N	1.24
N≡N	1.13
N=O	1.18
O=O	1.21

C. Effect of environment on carbon-carbon single bonds (other single bonds not shown). From Reference 3.

Configuration	C–C length	Examples of molecules
C–C	1.526	$H_3C–CH_3$
C–C=	1.501	$H_3C–CH=CH_2$
C–C≡	1.459	$H_3C–C≡CH$
=C–C=	1.467	$H_2C=CH–CH=CH_2$
≡C–C=	1.445	$HC≡C–CH=CH_2$
≡C–C≡	1.378	HC≡C–C≡CH

D. Some metal-carbon bond lengths in gas-phase molecules.

Al–C	1.96	Bi–C	2.26	Pb–C	2.24
B–C	1.58	Cd–C	2.11	Sn–C	2.14
Be–C	1.70	Hg–C	2.08	Zn–C	1.93

ATOMIC RADII OF THE ELEMENTS

Manjeera Mantina, Rosendo Valero, Christopher J. Cramer, and Donald G. Truhlar

Atomic radii are not precisely defined, but are nevertheless very widely used parameters in modeling and understanding molecular structure and interactions. Three main classes of radii may be identified: van der Waals radii (which include radii used to characterize steric interactions), covalent radii, and ionic radii. This section is concerned with the first two; ionic radii are covered in another section of the *Handbook* called "Ionic Radii in Crystals."

There are many scales of van der Waals radii, but they are not fully consistent with one another. The van der Waals radii determined by Bondi (Ref. 1) from x-ray diffraction data, crystal densities, gas kinetic collision cross sections, critical densities, and liquid-state properties are the most widely used values, but Bondi recommended radius values for only 28 of the 44 main-group elements in the periodic table plus 9 transition metals and one actinide. Rowland and Taylor (Ref. 2) redetermined nine of the main-group radii from crystal structure data and recommended that Bondi's values be accepted except for H, for which they recommended a new value; we accepted their recommendation for H and adopted Bondi's 27 other values for main-group elements. Radii for the 16 remaining main-group elements were determined from electronic structure calculations on selected van der Waals molecules by analyzing the results in a way designed to yield radii compatible with Bondi's scale for main-group elements (Ref. 3). Bondi's values for the transition metals and actinide are smaller than expected based on ionization potentials and covalent radii, so we do not adopt them, but defer to later recommendations based on analysis of a more extensive set of crystal data. Van der Waals radii for the remaining elements through atomic number $Z = 93$ were taken from Hu et al. (Ref. 4), who determined them from average atomic volumes in crystals as obtained by statistical analysis of the Cambridge Structural Database. Van der Waals radii for the elements with $Z = 94–103$ were based on the work of Guzei and Wendt (Ref. 5), who modeled the zero-potential distance for the interaction potentials involved in nonbonded steric interactions; we increased their values by 6 %, which is the value they found is needed to make their steric radii scale most consistent with Bondi's van der Waals radii.

Covalent radii are more straightforward, especially for elements that tend to form only single bonds, although some researchers distinguish metallic bonds from covalent bonds. The covalent radii tabulated here are recommendations for single covalent bonds, and they are based on a comprehensive evaluation of experimental data by Cordero et al. (Ref. 6), who recommended covalent radii for all elements up to $Z = 96$, and on an analysis combining experimental data and theoretical calculations by Pyykkö and Atsumi (Ref. 7), who recommended single-bond covalent radii for all elements up to $Z = 118$. If one is interested in a specific coordination number, oxidation state, or type of ligand, one might find a more appropriate radius in the specialized literature since the values in the table below are generic average values. In particular, we give the Pyykkö-Atsumi values for carbon and for $Z = 97-118$ and an average of the values from the two sources for all other elements. For Mn, Fe, and Co, Cordero et al. give two values, and we used the lower of these in the average because it corresponds to a smoother periodic trend.

All values are rounded to the nearest 0.01 Å, but in most cases the uncertainty in the value is of the order of 0.1 Å.

References

1. Bondi, A., *J. Phys. Chem.* 68, 441, 1964.
2. Rowland, R. S., and Taylor, R., *J. Phys. Chem.* 100, 7384, 1996.
3. Mantina, M., Chamberlin, A. C., Valero, R., Cramer, C. J., and Truhlar, D. G., *J. Phys. Chem. A* 113, 5806, 2009.
4. Hu, S.-Z., Zhou, Z.-H., and Tsai, K.-R., *Acta Physico-Chimica Sinica* 19, 1073, 2003.
5. Guzei, I. A., and Wendt, M., *Dalton Trans.* 2006, 3991, 2006.
6. Cordero, B., Gómez, V., Platero-Prats, A. E., Revés, M., Echeverría, J., Cremades, E., Barragán, F., and Alvarez, S., *Dalton Trans.* 2008, 2832, 2008.
7. Pyykkö, P., and Atsumi, M., *Chem. Eur. J.* 15, 186, 2009.

Element	Symbol	r_{vdW}/Å	r_{cov}/Å
Actinium	Ac	2.60	2.01
Aluminum	Al	1.84	1.24
Americium	Am	2.44	1.73
Antimony	Sb	2.06	1.40
Argon	Ar	1.88	1.01
Arsenic	As	1.85	1.20
Astatine	At	2.02	1.48
Barium	Ba	2.68	2.06
Berkelium	Bk	2.44	1.68
Beryllium	Be	1.53	0.99
Bismuth	Bi	2.07	1.50
Bohrium	Bh		1.41
Boron	B	1.92	0.84
Bromine	Br	1.83	1.17
Cadmium	Cd	2.30	1.40
Calcium	Ca	2.31	1.74
Californium	Cf	2.45	1.68
Carbon	C	1.70	0.75
Cerium	Ce	2.35	1.84
Cesium	Cs	3.43	2.38

Element	Symbol	r_{vdW}/Å	r_{cov}/Å
Chlorine	Cl	1.75	1.00
Chromium	Cr	1.89	1.30
Cobalt	Co	1.92	1.18
Copernicum	Cp		1.22
Copper	Cu	1.86	1.22
Curium	Cm	2.45	1.68
Darmstadtium	Ds		1.28
Dubnium	Db		1.49
Dysprosium	Dy	2.29	1.80
Einsteinium	Es	2.45	1.65
Erbium	Er	2.35	1.77
Europium	Eu	2.33	1.83
Fermium	Fm	2.45	1.67
Fluorine	F	1.47	0.60
Francium	Fr	3.48	2.42
Gadolinium	Gd	2.37	1.82
Gallium	Ga	1.87	1.23
Germanium	Ge	2.11	1.20
Gold	Au	2.17	1.30
Hafnium	Hf	2.12	1.64

Element	Symbol	r_{vdW}/Å	r_{cov}/Å
Hassium	Hs		1.34
Helium	He	1.40	0.37
Holmium	Ho	2.16	1.79
Hydrogen	H	1.10	0.32
Indium	In	1.93	1.42
Iodine	I	1.98	1.36
Iridium	Ir	2.02	1.32
Iron	Fe	1.94	1.24
Krypton	Kr	2.02	1.16
Lanthanum	La	2.40	1.94
Lawrencium	Lr	2.46	1.61
Lead	Pb	2.02	1.45
Lithium	Li	1.81	1.30
Lutetium	Lu	2.21	1.74
Magnesium	Mg	1.73	1.40
Manganese	Mn	1.97	1.29
Meitnerium	Mt		1.29
Mendelevium	Md	2.46	1.73
Mercury	Hg	2.09	1.32
Molybdenum	Mo	2.09	1.46
Neodymium	Nd	2.29	1.88
Neon	Ne	1.54	0.62
Neptunium	Np	2.21	1.80
Nickel	Ni	1.84	1.17
Niobium	Nb	2.07	1.56
Nitrogen	N	1.55	0.71
Nobelium	No	2.46	1.76
Osmium	Os	2.16	1.36
Oxygen	O	1.52	0.64
Palladium	Pd	2.02	1.30
Phosphorus	P	1.80	1.09
Platinum	Pt	2.09	1.30
Plutonium	Pu	2.43	1.80
Polonium	Po	1.97	1.42
Potassium	K	2.75	2.00
Praseodymium	Pr	2.39	1.90
Promethium	Pm	2.36	1.86
Protactinium	Pa	2.43	1.84
Radium	Ra	2.83	2.11

Element	Symbol	r_{vdW}/Å	r_{cov}/Å
Radon	Rn	2.20	1.46
Rhenium	Re	2.17	1.41
Rhodium	Rh	1.95	1.34
Roentgenium	Rg		1.21
Rubidium	Rb	3.03	2.15
Ruthenium	Ru	2.07	1.36
Rutherfordium	Rf		1.57
Samarium	Sm	2.29	1.85
Scandium	Sc	2.16	1.59
Seaborgium	Sg		1.43
Selenium	Se	1.90	1.18
Silicon	Si	2.10	1.14
Silver	Ag	2.03	1.36
Sodium	Na	2.27	1.60
Strontium	Sr	2.49	1.90
Sulfur	S	1.80	1.04
Tantalum	Ta	2.17	1.58
Technetium	Tc	2.09	1.38
Tellurium	Te	2.06	1.37
Terbium	Tb	2.21	1.81
Thallium	Tl	1.96	1.44
Thorium	Th	2.37	1.90
Thulium	Tm	2.27	1.77
Tin	Sn	2.17	1.40
Titanium	Ti	1.87	1.48
Tungsten	W	2.10	1.50
Ununhexium	Uuh		1.75
Ununoctium	Uuo		1.57
Ununpentium	Uup		1.62
Ununquadium	Uuq		1.43
Ununseptium	Uus		1.65
Ununtrium	Uut		1.36
Uranium	U	2.40	1.83
Vanadium	V	1.79	1.44
Xenon	Xe	2.16	1.36
Ytterbium	Yb	2.42	1.78
Yttrium	Y	2.19	1.76
Zinc	Zn	2.10	1.20
Zirconium	Zr	1.86	1.64

DIPOLE MOMENTS

This table gives selected values of the electric dipole moment for over 800 molecules. When available, values determined by microwave spectroscopy, molecular beam electric resonance, and other high-resolution spectroscopic techniques were selected. Otherwise, the values come from measurements of the dielectric constant in the gas phase or, if these do not exist, in the liquid phase. Entries are listed alphabetically; compounds not containing carbon are listed first, followed by compounds containing carbon.

The dipole moment is given in debye units (D). The conversion factor to SI units is 1 D = 3.33564×10^{-30} C m.

Dipole moments of individual conformers (rotational isomers) are given when they have been measured. The conformers are designated as *gauche, trans, axial,* etc. The meaning of these terms can be found in the references. In some cases an average value, obtained from measurements on the bulk gas, is also given. Other information on molecules that have been studied by spectroscopy, such as the components of the dipole moment in the molecular framework and the variation with vibrational state and isotopic species, is given in the references.

When the accuracy of a value is explicitly stated (i.e., 1.234 ± 0.005), the stated uncertainty generally indicates two or three standard deviations. When no uncertainty is given, the value may be assumed to be precise to a few units in the last decimal place. However, if more than three decimal places are given, the exact interpretation of the final digits may require analysis of the vibrational averaging.

Values measured in the gas phase that are questionable because of undetermined error sources are indicated as approximate (≈). Values obtained by liquid phase measurements, which sometimes have large errors because of association effects, are enclosed in brackets, e.g., [1.8].

References

1. Nelson, R. D., Lide, D. R., and Maryott, A. A., *Selected Values of Electric Dipole Moments for Molecules in the Gas Phase,* Natl. Stand. Ref. Data Ser. — Nat. Bur. Stnds. 10, 1967.
2. *Landolt-Börnstein, Numerical Data and Functional Relationships in Science and Technology, New Series,* II/6 (1974), Springer-Verlag, Heidelberg.
3. *Landolt-Börnstein, Numerical Data and Functional Relationships in Science and Technology, New Series,* II/14a (1982), Springer-Verlag, Heidelberg.
4. *Landolt-Börnstein, Numerical Data and Functional Relationships in Science and Technology, New Series,* II/14b (1983), Springer-Verlag, Heidelberg.
5. *Landolt-Börnstein, Numerical Data and Functional Relationships in Science and Technology, New Series,* II/19c (1992), Springer-Verlag, Heidelberg.
6. *Landolt-Börnstein, Numerical Data and Functional Relationships in Science and Technology, New Series,* II/24c (2002), Springer-Verlag, Heidelberg.
7. Riddick, J. A., Bunger, W. B., and Sakano, T. K., *Organic Solvents, Fourth Edition,* John Wiley & Sons, New York, 1986.
8. Kasuya, T., Lafferty, W. J., and Lide, D. R., *J. Chem. Phys.* 48, 1, 1968.
9. Kirchhoff, W. H., and Lide, D. R., *J. Chem. Phys.* 51, 467, 1969.
10. Durig, J. R., Li, Y. S., and Rizzolo, J. J., *J. Chem. Phys.* 77, 5885, 1982.
11. Ogata, T., Mochizuki, A. and Yamashita, E., *J. Chem. Phys.* 87, 2531, 1987.
12. Rego, A., and Cox, A. P., *J. Chem. Phys.* 89, 124, 1988.
13. Tyblewski, M., et al., *J. Chem. Phys.* 97, 6168, 1992.
14. Kawashima, Y., et al., *J. Chem. Phys.* 99, 820, 1993.
15. Caminati, W., Melandri, S., and Favero, L., *J. Chem. Phys.* 100, 8569, 1994.
16. Cederberg, J., et al., *J. Chem. Phys.* 105, 3361, 1996.
17. Bauder, A., et al., *J. Chem. Phys.* 106, 7558, 1997.
18. Muller, H. S. P., Miller, C. E., and Cohen, E. A., *J. Chem. Phys.* 107, 8292, 1997.
19. Burgh, D. J., Suenram, R. D., and Stevens, W. J., *J. Chem. Phys.* 111, 3526, 1999.
20. Blake, T. A., et al., *J. Chem. Phys.* 98, 6031, 1993.
21. Ruoff, R. S., et al., *J. Chem. Phys.* 89, 138, 1988.
22. Muenter, J. S., *J. Chem. Phys.* 90, 4048, 1989.
23. Peterson, J. I., Suenram, R. D., and Lovas, F. J., *J. Chem. Phys.* 90, 5964, 1989.
24. Suenram, R. D., Lovas, F. J., and Matsumura, K., *Astrophys. J. Lett.* 342, 103, 1989.
25. Groner, P., et al., *J. Chem. Phys.* 91, 1434, 1989.
26. Suenram, R. D., Lovas, F. J., Fraser, G. T., and Matsumura, K., *J. Chem. Phys.* 92, 4724, 1990.
27. Andrews, A. M., et al., *J. Chem. Phys.* 93, 7030, 1990.
28. Peterson, K. I., Suenram, R. D., and Lovas, F. J., *J. Chem. Phys.* 94, 106, 1991.
29. Iida, M., Ohshima, Y., and Endo, Y., *J. Chem. Phys.* 95, 4772, 1991.
30. Andrews, A. M., Hillig, K. W., and Kuczkowski, R. L., *J. Chem. Phys.* 96, 1784, 1992.
31. Ruoff, R. S., et al., *J. Chem. Phys.* 96, 3441, 1992.
32. Germann, T. C., Tschopp, S. L., and Gutowsky, H. S., *J. Chem. Phys.* 97, 1619, 1992.
33. Taleb-Bendiab, A., Hillig, K. W., and Kuczkowski, R. L., *J. Chem. Phys.* 97, 2996, 1992.
34. Taleb-Bendiab, A., Hillig, K. W., and Kuczkowski, R. L., *J. Chem. Phys.* 98, 3627, 1993.
35. Xu, L-W., and Kuczkowski, R. L., *J. Chem. Phys.* 100, 15, 1994.
36. Peterson, K. I., Suenram, R. D., and Lovas, F. J., *J. Chem. Phys.* 102, 7807, 1995.
37. Tatamitani, Y., and Ogata, T., *J. Chem. Phys.* 121, 9885, 2004.
38. Medvedev, I., et al., *Astrophys. J. Suppl.* 148, 593, 2003.
39. Lesarri, A., Suenram, R. D., and Brugh, D., *J. Chem. Phys.* 117, 9651, 2002.
40. Arunan, E., et al., *J. Chem. Phys.* 117, 9766, 2002.
41. Smith, T. C., Clouthier, D. J., and Steimle, T. C., *J. Chem. Phys.* 115, 817, 2001.
42. Peebles, S. A., Sun, L., and Kuczkowski, R. L., *J. Chem. Phys.* 110, 6804, 1999.
43. Namiki, K. C., Robinson, J. S., and Steimle, T. C., *J. Chem. Phys.* 109, 5283, 1998.
44. Peebles, S. A., and Kuczkowski, R. L., *J. Chem. Phys.* 109, 5276, 1998.
45. Sauer, B. E., Wang, J., and Hinds, E. A., *J. Chem. Phys.* 105, 7412, 1996.
46. Fry, J. L., Drouin, B. J., and Miller, C. E., *J. Chem. Phys.* 124, 084304, 2006.
47. Christiansen, J. J., *J. Mol. Spectrosc.* 231, 131, 2005.
48. Kisiel, Z., et al., *Chem. Phys. Lett.* 325, 523, 2000.
49. Lovas, F. J., et al., *Astrophys. J. Lett.* 455, 201, 1995.
50. Suenram, R. D., and Lovas, F. J., *Astrophys. J. Lett.* 429, 89, 1994.
51. Biermann, S., et al., *J. Chem. Phys.* 105, 9754, 1996.
52. McGlone, S., and Bauder, A., *J. Chem. Phys.* 109, 5383, 1998.
53. Peebles, S. A., and Kuczkowski, R. L., *J. Chem. Phys.* 111, 10511, 1999.
54. Plusquellic, D. F., et al., *J. Chem. Phys.* 115, 3057, 2001.
55. Muller, H. S. P., and Cohen, E. A., *J. Chem. Phys.* 116, 2407, 2002.
56. Andrews, A. M., and Kuczkowski, R. L., *J. Chem. Phys.* 98, 791, 1993.
57. Lovas, F. J., et al., *J. Chem. Phys.* 92, 891, 1990.
58. Klots, T. D., Emilsson, T., and Gutowsky, H. S., *J. Chem. Phys.* 97, 5335, 1992.
59. Careless, A. J., Kroto, H. W., and Landsberg, B. M., *Chem. Phys.* 1, 371, 1973.
60. Costain, C. C., and Kroto, H. W., *Can. J. Phys.* 50, 1453, 1972.
61. Kroto, H. W., Nixon, J. F., and Ohno, K., *J. Mol. Spectrosc.* 90, 367, 1981.
62. Kroto, H. W., Nixon, J. F., and Simmons, N. P. C., *J. Mol. Spectrosc.* 82, 185, 1980.
63. Kroto, H. W., Nixon, J. F., and Ohno, K., *J. Mol. Spectrosc.* 77, 270, 1979.
64. Cox, A. P., Ewart, I. C., and Gayton, T. R., *J. Mol. Spectrosc.* 125, 76, 1987.
65. Cohen, E. A., and Pickett, H. M., *J. Mol. Spectrosc.* 87, 582, 1981.
66. Peebles, S. A., and Peebles, R. A., *J. Mol. Struct.* 607, 19, 2002.
67. Suenram, R. D., Lovas, F. J., and Pickett, H. M., *J. Mol. Spectrosc.* 116, 406, 1986.
68. Kroto, H. W., and Landsberg, B. M., *J. Mol. Spectrosc.* 62, 346, 1976.

Name	Mol. Form.	μ/D	Ref.
Compounds not containing carbon			
Aluminum monofluoride	AlF	1.53 ± 0.15	1
Ammonia	H_3N	1.4718 ± 0.0002	5
Arsenic(III) chloride	$AsCl_3$	1.59 ± 0.08	1
Arsenic(III) fluoride	AsF_3	2.59 ± 0.05	1
Arsine	AsH_3	0.217 ± 0.003	5
Barium oxide	BaO	7.954 ± 0.003	5
Barium sulfide	BaS	10.86 ± 0.02	3
Bromine chloride	BrCl	0.519 ± 0.004	3
Bromine dioxide	BrO_2	2.8 ± 0.1	18
Bromine fluoride	BrF	1.422 ± 0.016	3
Bromine oxide	BrO	1.76 ± 0.04	2
Bromine pentafluoride	BrF_5	1.51 ± 0.15	1
Bromosilane	BrH_3Si	1.319	3
Bromotrifluorosilane	BrF_3Si	0.835 ± 0.007	64
Calcium monochloride	CaCl	≈3.6	4
Cesium chloride	ClCs	10.387 ± 0.004	2
Cesium fluoride	CsF	7.884 ± 0.001	2
Cesium sodium	CsNa	4.75 ± 0.20	2
Chlorine fluoride	ClF	0.888061	5
Chlorine oxide	ClO	1.297 ± 0.001	5
Chlorine trifluoride	ClF_3	0.6 ± 0.1	1
Chloroborane	$BClH_2$	0.75 ± 0.05	14
Chlorogermane	$ClGeH_3$	2.13 ± 0.02	1
Chlorosilane	ClH_3Si	1.31 ± 0.01	1
Chlorosyl fluoride	ClFO	1.93 ± 0.02	55
Chlorotrifluorosilane	ClF_3Si	0.636 ± 0.004	5
Chromium monoxide	CrO	3.88 ± 0.13	5
Copper(I) fluoride	CuF	5.77 ± 0.29	2
Copper(II) oxide	CuO	4.5 ± 0.5	5
Dichlorosilane	Cl_2H_2Si	1.17 ± 0.02	1
Difluoramine	F_2HN	1.92 ± 0.02	1
Difluorine dioxide	F_2O_2	1.44 ± 0.07	1
Difluoroborane	BF_2H	0.971 ± 0.010	8
cis-Difluorodiazine	F_2N_2	0.16 ± 0.01	1
Difluorosilane	F_2H_2Si	1.55 ± 0.02	1
Difluorosilylene	F_2Si	1.23 ± 0.02	2
Disiloxane	H_6OSi_2	0.24 ± 0.02	1
Fluoramine	FH_2N	2.27 ± 0.18	5
Fluorine azide	FN_3	≈1.3	5
Fluorine monoxide	F_2O	0.308180	5
Fluorine oxide	FO	0.0043 ± 0.0004	5
Fluoroborane	BF	≈0.5	2
Fluorogermane	$FGeH_3$	2.33 ± 0.12	2
Fluorosilane	FH_3Si	1.2969 ± 0.0006	5
Gallium monofluoride	FGa	2.45 ± 0.05	2
Germanium(II) fluoride	F_2Ge	2.61 ± 0.02	2
Germanium(II) oxide	GeO	3.2823 ± 0.0001	2
Germanium(II) selenide	GeSe	1.65 ± 0.05	2
Germanium(II) sulfide	GeS	2.00 ± 0.06	2
Germanium(II) telluride	GeTe	1.06 ± 0.07	2
Germylazide	GeH_3N_3	2.58 ± 0.02	25
Hafnium monoxide	HfO	3.431 ± 0.005	26
Hafnium(IV) oxide	HfO_2	7.92 ± 0.01	39
Hexaborane(10)	B_6H_{10}	2.50 ± 0.05	3
Hydrazine	H_4N_2	1.75 ± 0.09	1
Hydrazoic acid	HN_3	1.70 ± 0.09	3
Hydrogen bromide	BrH	0.8272 ± 0.0003	3
Hydrogen chloride	ClH	1.1086 ± 0.0003	3
Hydrogen fluoride	FH	1.826178	2
Hydrogen iodide	HI	0.448 ± 0.001	2
Hydrogen peroxide	H_2O_2	1.573 ± 0.001	65
Hydrogen sulfide	H_2S	0.97833	5
Hydroxyl	HO	1.655 ± 0.001	5
Hydroxylamine	H_3NO	0.59 ± 0.05	2
Hypochlorous acid	ClHO	≈1.3	2
Hypofluorous acid	FHO	2.23 ± 0.11	3
Imidogen	HN	1.39 ± 0.07	3
Indium(I) chloride	ClIn	3.79 ± 0.19	2
Indium(I) fluoride	FIn	3.40 ± 0.07	2
Iodine bromide	BrI	0.726 ± 0.003	5
Iodine chloride	ClI	1.24 ± 0.02	2
Iodine fluoride	FI	1.948 ± 0.020	3
Iodine monoxide	IO	2.45 ± 0.05	2
Iodine pentafluoride	F_5I	2.18 ± 0.11	1
Lanthanum monoxide	LaO	3.207 ± 0.011	26
Lead(II) oxide	OPb	4.64 ± 0.50	2
Lead(II) sulfide	PbS	3.59 ± 0.18	2
Lithium bromide	BrLi	7.268 ± 0.001	2
Lithium chloride	ClLi	7.12887	2
Lithium fluoride	FLi	6.3274 ± 0.0002	3
Lithium fluoride–sodium fluoride complex	FLi•FNa	2.62 ± 0.02	51
Lithium hydride	HLi	5.884 ± 0.001	2
Lithium hydroxide	HLiO	4.754 ± 0.002	3
Lithium iodide	ILi	7.428 ± 0.001	2
Lithium monoxide	LiO	6.84 ± 0.03	2
Lithium potassium	KLi	3.45 ± 0.20	2
Lithium rubidium	LiRb	4.0 ± 0.1	2
Lithium sodium	LiNa	0.463 ± 0.002	2
Magnesium oxide	MgO	6.2 ± 0.6	5
Mercapto	HS	0.7580 ± 0.0001	3
Nitric acid	HNO_3	2.17 ± 0.02	1
Nitric oxide	NO	0.15872	2
Nitrogen dioxide	NO_2	0.316 ± 0.010	1
Nitrogen sulfide	NS	1.81 ± 0.02	2
Nitrogen trichloride	Cl_3N	0.39 ± 0.01	3
Nitrogen trifluoride	F_3N	0.235 ± 0.004	1
Nitrogen trioxide	N_2O_3	2.122 ± 0.010	2
Nitrosyl bromide	BrNO	≈1.8	1
Nitrosyl fluoride	FNO	1.730 ± 0.003	3
Nitrosyl hydride	HNO	1.62 ± 0.03	3
Nitrous acid (*cis*)	HNO_2	1.423 ± 0.005	2
Nitrous acid (*trans*)	HNO_2	1.855 ± 0.016	2
Nitrous oxide	N_2O	0.16083	3
Nitryl chloride	$ClNO_2$	0.53	1
Nitryl fluoride	FNO_2	0.466 ± 0.005	2
Ozone	O_3	0.53373	3
Pentaborane(9)	B_5H_9	2.13 ± 0.04	1
Perchloryl fluoride	$ClFO_3$	0.023 ± 0.001	3
Peroxynitrous acid	HNO_3	1.07 ± 0.002	46

Name	Mol. Form.	μ/D	Ref.
Peroxynitric acid	HNO_4	1.99 ± 0.02	67
Phosphine	H_3P	0.5740 ± 0.0003	3
Phosphorothioc trifluoride	F_3PS	0.64 ± 0.02	1
Phosphorus(III) chloride	Cl_3P	0.56 ± 0.02	2
Phosphorus(III) fluoride	F_3P	1.03 ± 0.01	1
Phosphorus monoxide	OP	1.88 ± 0.07	5
Phosphorus nitride	NP	2.7470 ± 0.0001	2
Phosphoryl chloride	Cl_3OP	2.54 ± 0.05	2
Phosphoryl fluoride	F_3OP	1.8685 ± 0.0001	3
Potassium bromide	BrK	10.628 ± 0.001	2
Potassium chloride	ClK	10.269 ± 0.001	2
Potassium fluoride	FK	8.585 ± 0.003	2
Potassium hydroxide	HKO	7.415 ± 0.002	16
Potassium iodide	IK	≈10.8	2
Potassium sodium	KNa	2.693 ± 0.014	3
Rubidium bromide	BrRb	≈10.9	2
Rubidium chloride	ClRb	10.510 ± 0.005	2
Rubidium fluoride	FRb	8.5465 ± 0.0005	2
Rubidium iodide	IRb	≈11.5	2
Rubidium sodium	NaRb	3.1 ± 0.3	2
Selenium dioxide	O_2Se	2.62 ± 0.05	2
Selenium tetrafluoride	F_4Se	1.78 ± 0.09	2
Silicon monosulfide	SSi	1.73 ± 0.09	2
Silicon monoxide	OSi	3.0982	2
Silver(I) bromide	AgBr	5.62 ± 0.03	5
Silver(I) chloride	AgCl	6.08 ± 0.06	5
Silver(I) fluoride	AgF	6.22 ± 0.30	2
Silver(I) iodide	AgI	4.55 ± 0.05	5
Sodium bromide	BrNa	9.1183 ± 0.0006	2
Sodium chloride	ClNa	9.00117	2
Sodium fluoride	FNa	8.156 ± 0.001	2
Sodium iodide	INa	9.236 ± 0.003	2
Stibine	H_3Sb	0.12 ± 0.05	1
Strontium oxide	OSr	8.900 ± 0.003	2
Sulfur dichloride	Cl_2S	0.36 ± 0.01	3
Sulfur difluoride	F_2S	1.05 ± 0.05	2
Sulfur dioxide	O_2S	1.63305	3
Sulfur monofluoride	FS	0.794 ± 0.02	3
Sulfur monoxide	OS	1.55 ± 0.02	1
Sulfur oxide (SSO)	OS_2	1.47 ± 0.03	1
Sulfur tetrafluoride	F_4S	0.632 ± 0.003	1
Sulfuryl chloride	Cl_2O_2S	1.81 ± 0.04	1
Sulfuryl fluoride	F_2O_2S	1.12 ± 0.02	1
Tetraborane(10)	B_4H_{10}	0.486 ± 0.002	3
Tetrafluorohydrazine (*gauche*)	F_4N_2	0.257 ± 0.002	5
Tetrafluorosilane–ammonia complex	$F_4Si{\bullet}H_3N$	5.61 ± 0.02	31
Thallium(I) bromide	BrTl	4.49 ± 0.05	2
Thallium(I) chloride	ClTl	4.54299	2
Thallium(I) fluoride	FTl	4.2282 ± 0.0008	2
Thallium(I) iodide	ITl	4.61 ± 0.07	2
Thionitrosyl chloride (NSCl)	ClNS	1.87 ± 0.02	2
Thionitrosyl fluoride (NSF)	FNS	1.902 ± 0.012	2
Thionyl chloride	Cl_2OS	1.45 ± 0.03	1
Thionyl fluoride	F_2OS	1.63 ± 0.01	1
Tin(II) oxide	OSn	4.32 ± 0.22	2
Tin(II) sulfide	SSn	3.18 ± 0.16	2
Titanium(II) oxide	OTi	2.96 ± 0.05	5
Trichlorofluorosilane	Cl_3FSi	0.49 ± 0.01	2
Trichlorosilane	Cl_3HSi	0.86 ± 0.01	2
Trifluoramine oxide	F_3NO	0.0390 ± 0.0004	9
1,1,1-Trifluorodisilane	$F_3H_3Si_2$	2.03 ± 0.10	3
Trifluoroiodosilane	F_3ISi	1.11 ± 0.03	5
Trifluorosilane	F_3HSi	1.27 ± 0.03	1
Water	H_2O	1.8546 ± 0.0040	3
Water dimer–hydrogen bromide complex	$H_4O_2{\bullet}BrH$	2.281 ± 0.003	48
Water dimer–hydrogen chloride complex	$H_4O_2{\bullet}ClH$	2.328 ± 0.003	48
Ytterbium monofluoride	FYb	3.91 ± 0.04	45
Yttrium monoxide	OY	4.524 ± 0.007	26
Zirconium(II) oxide	OZr	2.55 ± 0.01	26
Zirconium(IV) oxide	O_2Zr	7.80 ± 0.02	19
Compounds containing carbon			
Acenaphthene	$C_{12}H_{10}$	≈0.85	1
Acetaldehyde	C_2H_4O	2.750 ± 0.006	3
Acetamide	C_2H_5NO	3.68 ± 0.03	5
Acetic acid	$C_2H_4O_2$	1.70 ± 0.03	2
Acetic anhydride	$C_4H_6O_3$	≈2.8	1
Acetone	C_3H_6O	2.88 ± 0.03	1
Acetonitrile	C_2H_3N	3.92519	5
Acetophenone	C_8H_8O	3.02 ± 0.06	1
Acetyl chloride	C_2H_3ClO	2.72 ± 0.14	1
Acetylene–carbon dioxide complex	$C_2H_2{\bullet}CO_2$	0.161 ± 0.001	22
Acetylene–carbon monoxide complex	$C_2H_2{\bullet}CO$	0.311 ± 0.001	32
Acetylene–carbon oxysulfide trimer complex	$C_2H_2{\bullet}C_3O_3S_3$	1.23 ± 0.02	53
Acetylene–hydrogen cyanide complex	$C_2H_2{\bullet}CHN$	3.29 ± 0.03	32
Acetyl fluoride	C_2H_3FO	2.96 ± 0.03	1
Acrolein (*cis*)	C_3H_4O	2.552 ± 0.003	5
Acrolein (*trans*)	C_3H_4O	3.117 ± 0.004	5
Acrylonitrile	C_3H_3N	3.92 ± 0.07	5
Allyl alcohol (*gauche*)	C_3H_6O	1.55 ± 0.08	3
Allyl alcohol (*average*)	C_3H_6O	1.60 ± 0.08	1
Allylamine	C_3H_7N	≈1.2	1
Aniline	C_6H_7N	1.13 ± 0.02	3
Anisole	C_7H_8O	1.38 ± 0.07	1
Azulene	$C_{10}H_8$	0.80 ± 0.02	1
Benzaldehyde	C_7H_6O	[3.0]	7
Benzeneacetonitrile	C_8H_7N	[3.5]	7
Benzene–hydrogen sulfide complex	$C_6H_6{\bullet}H_2S$	1.14 ± 0.02	40
Benzene–krypton complex	$C_6H_6{\bullet}Kr$	0.136 ± 0.002	58
Benzene–sulfur dioxide complex	$C_6H_6{\bullet}O_2S$	2.061 ± 0.002	33
Benzenethiol	C_6H_6S	[1.23]	7
Benzonitrile	C_7H_5N	4.18 ± 0.08	1
Benzyl acetate	$C_9H_{10}O_2$	[1.22]	7
Benzyl alcohol	C_7H_8O	1.71 ± 0.09	1

Name	Mol. Form.	μ/D	Ref.
Benzyl benzoate	$C_{14}H_{12}O_2$	[2.06]	7
Bis(2-aminoethyl)amine	$C_4H_{13}N_3$	[1.89]	7
Bis(2-chloroethyl) ether	$C_4H_8Cl_2O$	[2.58]	7
Bis(2-ethylhexyl) phthalate	$C_{24}H_{38}O_4$	[2.84]	7
Borane carbonyl	CH_3BO	1.698 ± 0.020	3
Bromoacetylene	C_2HBr	0.22962	5
Bromobenzene	C_6H_5Br	1.70 ± 0.03	1
1-Bromobutane	C_4H_9Br	2.08 ± 0.10	1
2-Bromobutane	C_4H_9Br	2.23 ± 0.11	1
1-Bromo-2-chloroethane	C_2H_4BrCl	[1.2]	7
Bromochlorofluoromethane	CHBrClF	1.5 ± 0.3	17
Bromochloromethane	CH_2BrCl	[1.66]	7
1-Bromodecane	$C_{10}H_{21}Br$	[1.93]	7
Bromoethane	C_2H_5Br	2.04 ± 0.02	5
Bromoethene	C_2H_3Br	1.42 ± 0.03	1
Bromofluoroacetylene	C_2BrF	0.448 ± 0.002	5
1-Bromoheptane	$C_7H_{15}Br$	2.16 ± 0.11	1
Bromomethane	CH_3Br	1.8203 ± 0.0004	5
2-Bromo-2-methylpropane	C_4H_9Br	[2.17]	7
1-Bromonaphthalene	$C_{10}H_7Br$	[1.55]	7
1-Bromopentane	$C_5H_{11}Br$	2.20 ± 0.11	1
1-Bromopropane	C_3H_7Br	2.18 ± 0.11	1
2-Bromopropane	C_3H_7Br	2.21 ± 0.11	1
2-Bromopropene	C_3H_5Br	[1.51]	7
3-Bromopropene	C_3H_5Br	≈1.9	1
Bromotrifluoromethane	$CBrF_3$	0.65 ± 0.05	1
1,2-Butadiene	C_4H_6	0.403 ± 0.002	1
Butanal	C_4H_8O	2.72 ± 0.05	1
1,4-Butanediol	$C_4H_{10}O_2$	[2.58]	7
Butanenitrile (*gauche*)	C_4H_7N	3.91 ± 0.04	5
Butanenitrile (*anti*)	C_4H_7N	3.73 ± 0.06	5
1-Butanethiol	$C_4H_{10}S$	[1.53]	7
Butanoic acid	$C_4H_8O_2$	[1.65]	7
1-Butanol	$C_4H_{10}O$	1.66 ± 0.03	1
2-Butanone	C_4H_8O	2.779 ± 0.015	2
trans-2-Butenal	C_4H_6O	3.67 ± 0.07	1
1-Butene (*cis*)	C_4H_8	0.438 ± 0.007	2
1-Butene (*skew*)	C_4H_8	0.359 ± 0.011	2
cis-2-Butene	C_4H_8	0.253 ± 0.005	2
cis-2-Butene-1,4-diol	$C_4H_8O_2$	[2.48]	7
trans-2-Butene-1,4-diol	$C_4H_8O_2$	[2.45]	7
trans-2-Butenoic acid	$C_4H_6O_2$	[2.13]	7
cis-2-Buten-1-ol	C_4H_8O	1.96 ± 0.03	5
trans-2-Buten-1-ol	C_4H_8O	1.90 ± 0.02	5
1-Buten-3-yne	C_4H_4	0.22 ± 0.02	3
2-Butoxyethanol	$C_6H_{14}O_2$	[2.08]	7
Butyl acetate	$C_6H_{12}O_2$	[1.87]	7
sec-Butyl acetate	$C_6H_{12}O_2$	[1.87]	7
Butylamine	$C_4H_{11}N$	≈1.0	1
sec-Butylamine	$C_4H_{11}N$	[1.28]	7
tert-Butylamine	$C_4H_{11}N$	[1.29]	7
tert-Butylbenzene	$C_{10}H_{14}$	≈0.83	1
Butyl ethyl ether	$C_6H_{14}O$	[1.24]	7
Butyl formate	$C_5H_{10}O_2$	[2.03]	7
Butyl stearate	$C_{22}H_{44}O_2$	[1.88]	7
Butyl vinyl ether	$C_6H_{12}O$	[1.25]	7
1-Butyne	C_4H_6	0.782 ± 0.004	5
γ-Butyrolactone	$C_4H_6O_2$	4.27 ± 0.03	3
Calcium methoxide	CH_3CaO	1.58 ± 0.08	43
Camphor, (+)	$C_{10}H_{16}O$	[3.1]	7
Caprolactam	$C_6H_{11}NO$	[3.9]	7
Carboimidic difluoride	CHF_2N	1.393 ± 0.001	5
Carbon dioxide dimer–water complex	$C_2O_4{\bullet}H_2O$	1.989 ± 0.002	23
Carbon dioxide–mercury complex	$CO_2{\bullet}Hg$	0.107 ± 0.003	29
Carbon dioxide–water dimer complex	$CO_2{\bullet}H_4O_2$	1.746 ± 0.010	28
Carbon disulfide–sulfur dioxide complex	$CO_2{\bullet}O_2S$	1.096 ± 0.001	42
Carbon monoselenide	CSe	1.99 ± 0.04	3
Carbon monosulfide	CS	1.958 ± 0.005	2
Carbon monoxide	CO	0.10980	3
Carbon monoxide dimer–water complex	$C_2O_2{\bullet}H_2O$	1.57 ± 0.05	36
Carbon oxyselenide	COSe	0.73 ± 0.02	1
Carbon oxysulfide	COS	0.715189	5
Carbon oxysulfide–carbon dioxide dimer complex	$COS{\bullet}C_2O_4$	0.69 ± 0.05	44
Carbon oxysulfide–water complex	$COS{\bullet}H_2O$	2.668 ± 0.003	37
Carbonyl chloride	CCl_2O	1.17 ± 0.01	1
Carbonyl fluoride	CF_2O	0.95 ± 0.01	1
Chloroacetyl chloride	$C_2H_2Cl_2O$	2.23 ± 0.11	1
Chloroacetylene	C_2HCl	0.44408	5
2-Chloroaniline	C_6H_6ClN	[1.77]	7
Chlorobenzene	C_6H_5Cl	1.69 ± 0.03	1
1-Chlorobutane	C_4H_9Cl	2.05 ± 0.04	1
2-Chlorobutane	C_4H_9Cl	2.04 ± 0.10	1
Chlorocyclohexane (*axial*)	$C_6H_{11}Cl$	1.91 ± 0.02	5
Chlorocyclohexane (*equitorial*)	$C_6H_{11}Cl$	2.44 ± 0.07	5
1-Chloro-1,1-difluoroethane	$C_2H_3ClF_2$	2.14 ± 0.04	1
Chlorodifluoromethane	$CHClF_2$	1.42 ± 0.03	1
Chloroethane	C_2H_5Cl	2.05 ± 0.02	1
2-Chloroethanol	C_2H_5ClO	1.78 ± 0.09	1
Chloroethene	C_2H_3Cl	1.45 ± 0.03	1
1-Chloro-4-fluorobenzene	C_6H_4ClF	0.12 ± 0.01	66
1-Chloro-1-fluoroethane	C_2H_4ClF	2.068 ± 0.014	3
Chlorofluoromethane	CH_2ClF	1.82 ± 0.04	1
Chloromethane	CH_3Cl	1.8963 ± 0.0002	5
(Chloromethyl)benzene	C_7H_7Cl	[1.82]	7
1-Chloro-3-methylbutane	$C_5H_{11}Cl$	[1.92]	7
1-Chloro-2-methylpropane	C_4H_9Cl	2.00 ± 0.10	1
2-Chloro-2-methylpropane	C_4H_9Cl	2.13 ± 0.04	1
1-Chloronaphthalene	$C_{10}H_7Cl$	[1.57]	7
1-Chloro-2-nitrobenzene	$C_6H_4ClNO_2$	4.64 ± 0.09	1
1-Chloro-3-nitrobenzene	$C_6H_4ClNO_2$	3.73 ± 0.07	1
1-Chloro-4-nitrobenzene	$C_6H_4ClNO_2$	2.83 ± 0.06	1
1-Chlorooctane	$C_8H_{17}Cl$	[2.00]	7
Chloropentafluoroethane	C_2ClF_5	0.52 ± 0.05	1
1-Chloropentane	$C_5H_{11}Cl$	2.16 ± 0.11	1
4-Chlorophenol	C_6H_5ClO	2.11 ± 0.11	1

Name	Mol. Form.	μ/D	Ref.
1-Chloropropane (*gauche*)	C_3H_7Cl	2.02 ± 0.03	5
1-Chloropropane (*trans*)	C_3H_7Cl	1.95 ± 0.02	5
1-Chloropropane (*average*)	C_3H_7Cl	2.05 ± 0.04	1
2-Chloropropane	C_3H_7Cl	2.17 ± 0.11	1
cis-1-Chloropropene	C_3H_5Cl	1.67 ± 0.08	1
trans-1-Chloropropene	C_3H_5Cl	1.97 ± 0.10	1
2-Chloropropene	C_3H_5Cl	1.647 ± 0.010	3
3-Chloropropene	C_3H_5Cl	1.94 ± 0.10	1
4-Chloropyridine	C_5H_4ClN	0.756 ± 0.005	3
2-Chlorotoluene	C_7H_7Cl	1.56 ± 0.08	1
3-Chlorotoluene	C_7H_7Cl	[1.82]	7
4-Chlorotoluene	C_7H_7Cl	2.21 ± 0.04	1
Chlorotrifluoroethene	C_2ClF_3	0.40 ± 0.10	1
Chlorotrifluoromethane	$CClF_3$	0.50 ± 0.01	1
o-Cresol	C_7H_8O	[1.45]	7
m-Cresol	C_7H_8O	[1.48]	7
p-Cresol	C_7H_8O	[1.48]	7
Cyanamide	CH_2N_2	4.28 ± 0.10	5
Cyanoacetylene	C_3HN	3.73172	5
Cyanoformamide	$C_2H_2N_2O$	4.10 ± 0.12	47
Cyanogen azide (NCN_3)	CN_4	2.96 ± 0.07	60
Cyanogen chloride	CClN	2.8331 ± 0.0002	3
Cyanogen fluoride	CFN	2.120 ± 0.001	3
Cyanogen iodide	CIN	3.67 ± 0.02	5
Cyanomethylmercury	C_2H_3HgN	4.7 ± 0.1	12
Cyclobutanecarbonitrile	C_5H_7N	4.04 ± 0.04	5
Cyclobutanone	C_4H_6O	2.89 ± 0.03	2
Cyclobutene	C_4H_6	0.132 ± 0.001	1
1,3-Cycloheptadiene	C_7H_{10}	0.740	3
2,4,6-Cycloheptatrien-1-one	C_7H_6O	4.1 ± 0.3	3
3,5-Cyclohexadiene-1,2-dione	$C_6H_4O_2$	4.23 ± 0.02	3
Cyclohexanone	$C_6H_{10}O$	3.246 ± 0.006	5
Cyclohexene (*half-chair*)	C_6H_{10}	0.332 ± 0.012	2
Cyclohexylamine	$C_6H_{13}N$	[1.26]	7
1,3-Cyclopentadiene	C_5H_6	0.419 ± 0.004	1
2,4-Cyclopentadien-1-one	C_5H_4O	3.132 ± 0.007	3
Cyclopentanone	C_5H_8O	≈3.3	1
Cyclopentene	C_5H_8	0.20 ± 0.02	1
3-Cyclopenten-1-one	C_5H_6O	2.79 ± 0.03	3
Cyclopropane-sulfur dioxide complex	$C_3H_6{\bullet}O_2S$	1.681 ± 0.001	30
Cyclopropanone	C_3H_4O	2.67 ± 0.13	2
Cyclopropene	C_3H_4	0.454 ± 0.010	1
Cyclopropylamine	C_3H_7N	1.19 ± 0.01	2
Cyclopropyl methyl ketone	C_5H_8O	2.62 ± 0.25	2
Diacetone alcohol	$C_6H_{12}O_2$	[3.24]	7
Diazomethane	CH_2N_2	1.50 ± 0.01	1
Dibromodifluoromethane	CBr_2F_2	0.66 ± 0.05	1
1,2-Dibromoethane	$C_2H_4Br_2$	[1.19]	7
Dibromomethane	CH_2Br_2	1.43 ± 0.03	1
1,2-Dibromopropane	$C_3H_6Br_2$	[1.2]	7
Dibutylamine	$C_8H_{19}N$	[0.98]	7
Dibutyl ether	$C_8H_{18}O$	1.17 ± 0.06	1
Dibutyl phthalate	$C_{16}H_{22}O_4$	[2.82]	7
Dibutyl sebacate	$C_{18}H_{34}O_4$	[2.48]	7

Name	Mol. Form.	μ/D	Ref.
Dibutyl sulfide	$C_8H_{18}S$	[1.61]	7
o-Dichlorobenzene	$C_6H_4Cl_2$	2.50 ± 0.05	1
m-Dichlorobenzene	$C_6H_4Cl_2$	1.72 ± 0.09	1
1,4-Dichlorobutane	$C_4H_8Cl_2$	2.22 ± 0.11	1
1,1-Dichloro-2,2-difluoroethene	$C_2Cl_2F_2$	0.50	7
Dichlorodifluoromethane	CCl_2F_2	0.51 ± 0.05	1
1,1-Dichloroethane	$C_2H_4Cl_2$	2.06 ± 0.04	1
1,2-Dichloroethane	$C_2H_4Cl_2$	[1.83]	7
1,1-Dichloroethene	$C_2H_2Cl_2$	1.34 ± 0.01	1
cis-1,2-Dichloroethene	$C_2H_2Cl_2$	1.90 ± 0.04	1
Dichlorofluoromethane	$CHCl_2F$	1.29 ± 0.03	1
1,1-Dichloro-2-fluoropropene	$C_3H_3Cl_2F$	2.43 ± 0.02	3
Dichloromethane	CH_2Cl_2	1.60 ± 0.03	1
(Dichloromethyl)benzene	$C_7H_6Cl_2$	[2.07]	7
Dichloromethylborane	CH_3BCl_2	1.419 ± 0.013	5
1,2-Dichloropropane	$C_3H_6Cl_2$	[1.85]	7
1,3-Dichloropropane	$C_3H_6Cl_2$	2.08 ± 0.04	1
1,2-Dichloro-1,1,2,2-tetrafluoroethane	$C_2Cl_2F_4$	≈0.5	1
2,4-Dichlorotoluene	$C_7H_6Cl_2$	[1.70]	7
3,4-Dichlorotoluene	$C_7H_6Cl_2$	[2.95]	7
Diethanolamine	$C_4H_{11}NO_2$	[2.8]	7
1,1-Diethoxyethane	$C_6H_{14}O_2$	[1.38]	7
Diethylamine	$C_4H_{11}N$	0.92 ± 0.06	1
Diethyl carbonate	$C_5H_{10}O_3$	1.10 ± 0.06	1
Diethylene glycol	$C_4H_{10}O_3$	[2.31]	7
Diethylene glycol dimethyl ether	$C_6H_{14}O_3$	[1.97]	7
Diethylene glycol monoethyl ether	$C_6H_{14}O_3$	[1.6]	7
Diethylene glycol monoethyl ether acetate	$C_8H_{16}O_4$	[1.8]	7
Diethylene glycol monomethyl ether	$C_5H_{12}O_3$	[1.6]	7
Diethyl ether	$C_4H_{10}O$	1.098 ± 0.001	38
Diethyl malonate	$C_7H_{12}O_4$	[2.54]	7
Diethyl oxalate	$C_6H_{10}O_4$	[2.49]	7
Diethyl sulfide (*trans-trans*)	$C_4H_{10}S$	1.556 ± 0.004	54
Diethyl sulfide (*trans-gauche*)	$C_4H_{10}S$	1.591 ± 0.009	54
Diethyl sulfide (*gauche-gauche*)	$C_4H_{10}S$	1.645 ± 0.001	54
o-Difluorobenzene	$C_6H_4F_2$	2.46 ± 0.05	2
m-Difluorobenzene	$C_6H_4F_2$	1.51 ± 0.02	2
1,1-Difluorocyclohexane	$C_6H_{10}F_2$	2.556 ± 0.010	3
3,3-Difluorocyclopropene	$C_3H_2F_2$	2.98 ± 0.02	3
1,1-Difluoroethane	$C_2H_4F_2$	2.27 ± 0.05	1
1,2-Difluoroethane (*gauche*)	$C_2H_4F_2$	2.67 ± 0.13	2
1,1-Difluoroethene	$C_2H_2F_2$	1.3893 ± 0.0002	5
cis-1,2-Difluoroethene	$C_2H_2F_2$	2.42 ± 0.02	1
Difluoromethane	CH_2F_2	1.9785 ± 0.02	3
Difluoromethylborane	CH_3BF_2	1.668 ± 0.003	3
Difluoromethylene	CF_2	0.47 ± 0.02	3
1,1-Difluoro-1-propene	$C_3H_4F_2$	0.889 ± 0.007	2
2,3-Dihydro-1,4-dioxin	$C_4H_6O_2$	0.939 ± 0.008	3
3,6-Dihydro-1,2-dioxin	$C_4H_6O_2$	2.329 ± 0.001	3

Name	Mol. Form.	μ/D	Ref.
2,3-Dihydrofuran	C_4H_6O	1.32 ± 0.03	2
2,5-Dihydrofuran	C_4H_6O	1.63 ± 0.01	5
Dihydro-3-methyl-2(3*H*)-furanone	$C_5H_8O_2$	4.56 ± 0.02	5
Dihydro-5-methyl-2(3*H*)-furanone	$C_5H_8O_2$	4.71 ± 0.05	5
3,4-Dihydro-2*H*-pyran	C_5H_8O	1.400 ± 0.008	5
3,6-Dihydro-2*H*-pyran	C_5H_8O	1.283 ± 0.005	3
2,3-Dihydrothiophene	C_4H_6S	1.61 ± 0.20	5
2,5-Dihydrothiophene	C_4H_6S	1.75 ± 0.01	3
Diiodomethane	CH_2I_2	[1.08]	7
Diisopentyl ether	$C_{10}H_{22}O$	[1.23]	7
Diisopropylamine	$C_6H_{15}N$	[1.15]	7
Diisopropyl ether	$C_6H_{14}O$	1.13 ± 0.10	1
Diketene	$C_4H_4O_2$	3.53 ± 0.07	1
1,2-Dimethoxybenzene	$C_8H_{10}O_2$	[1.29]	7
Dimethoxymethane	$C_3H_8O_2$	[0.74]	7
N,N-Dimethylacetamide	C_4H_9NO	[3.7]	7
Dimethylamine	C_2H_7N	1.01 ± 0.02	2
N,N-Dimethylaniline	$C_8H_{11}N$	1.68 ± 0.17	1
2,4-Dimethylaniline	$C_8H_{11}N$	[1.40]	7
2,6-Dimethylaniline	$C_8H_{11}N$	[1.63]	7
3,3-Dimethyl-1-butyne	C_6H_{10}	0.661 ± 0.004	1
1,1-Dimethylcyclopropane	C_5H_{10}	0.142 ± 0.001	3
3,3-Dimethylcyclopropene	C_5H_8	0.287 ± 0.003	3
Dimethyl disulfide	$C_2H_6S_2$	[1.85]	7
Dimethyl ether	C_2H_6O	1.30 ± 0.01	1
N,N-Dimethylformamide	C_3H_7NO	3.82 ± 0.08	1
2,6-Dimethyl-4-heptanone	$C_9H_{18}O$	[2.66]	7
Dimethyl maleate	$C_6H_8O_4$	[2.48]	7
2,4-Dimethyl-3-pentanone	$C_7H_{14}O$	[2.74]	7
2,2-Dimethylpropanal	$C_5H_{10}O$	2.66 ± 0.05	2
2,2-Dimethylpropanenitrile	C_5H_9N	3.95 ± 0.04	1
2,4-Dimethylpyridine	C_7H_9N	[2.30]	7
2,6-Dimethylpyridine	C_7H_9N	[1.66]	7
Dimethyl sulfide	C_2H_6S	1.554 ± 0.004	3
Dimethyl sulfoxide	C_2H_6OS	3.96 ± 0.04	1
1,3-Dioxane	$C_4H_8O_2$	2.06 ± 0.04	2
1,3-Dioxolane	$C_3H_6O_2$	1.19 ± 0.06	3
Dipentyl ether	$C_{10}H_{22}O$	[1.20]	7
Diphenyl ether	$C_{12}H_{10}O$	≈1.3	1
Dipropylamine	$C_6H_{15}N$	[1.03]	7
Dipropyl ether	$C_6H_{14}O$	1.21 ± 0.06	1
1,3-Dithiane	$C_4H_8S_2$	2.14 ± 0.04	5
Divinyl ether	C_4H_6O	0.78 ± 0.05	2
Epichlorohydrin	C_3H_5ClO	[1.8]	7
1,2-Epoxybutane	C_4H_8O	1.891 ± 0.011	3
1,2-Ethanediamine	$C_2H_8N_2$	1.99 ± 0.10	1
1,2-Ethanediol, diacetate	$C_6H_{10}O_4$	[2.34]	7
1,2-Ethanedithiol	$C_2H_6S_2$	2.03 ± 0.08	5
Ethanethiol (*gauche*)	C_2H_6S	1.61 ± 0.08	3
Ethanethiol (*trans*)	C_2H_6S	1.58 ± 0.08	3
Ethanol (*gauche*)	C_2H_6O	1.68 ± 0.03	3
Ethanol (*trans*)	C_2H_6O	1.44 ± 0.03	2
Ethanol (*average*)	C_2H_6O	1.69 ± 0.03	1
Ethanolamine	C_2H_7NO	[2.27]	7
Ethoxybenzene	$C_8H_{10}O$	1.45 ± 0.15	1
2-Ethoxyethanol	$C_4H_{10}O_2$	[2.08]	7
2-Ethoxyethyl acetate	$C_6H_{12}O_3$	[2.25]	7
Ethyl acetate	$C_4H_8O_2$	1.78 ± 0.09	1
Ethyl acrylate	$C_5H_8O_2$	[1.96]	7
Ethylamine (*gauche*)	C_2H_7N	1.210 ± 0.015	5
Ethylamine (*trans*)	C_2H_7N	1.304 ± 0.011	5
Ethylamine (*average*)	C_2H_7N	1.22 ± 0.10	1
Ethylbenzene	C_8H_{10}	0.59 ± 0.05	1
Ethyl benzoate	$C_9H_{10}O_2$	2.00 ± 0.10	1
Ethyl butanoate	$C_6H_{12}O_2$	[1.74]	7
Ethyl *trans*-cinnamate	$C_{11}H_{12}O_2$	[1.84]	7
Ethyl cyanate	C_3H_5NO	4.72 ± 0.09	5
Ethyl cyanoacetate	$C_5H_7NO_2$	[2.17]	7
Ethylene carbonate	$C_3H_4O_3$	[4.9]	7
Ethylene glycol (*average*)	$C_2H_6O_2$	2.36 ± 0.10	5
Ethyleneimine	C_2H_5N	1.90 ± 0.01	1
Ethylene–sulfur dioxide complex	$C_2H_4{\bullet}O_2S$	1.650 ± 0.003	27
Ethylene–water complex	$C_2H_4{\bullet}H_2O$	1.10 ± 0.01	56
Ethyl formate (*gauche*)	$C_3H_6O_2$	1.81 ± 0.02	2
Ethyl formate (*trans*)	$C_3H_6O_2$	1.98 ± 0.02	2
Ethyl formate (*average*)	$C_3H_6O_2$	1.93	1
2-Ethyl-1-hexanol	$C_8H_{18}O$	[1.74]	7
2-Ethylhexyl acetate	$C_{10}H_{20}O_2$	[1.8]	7
Ethyl lactate	$C_5H_{10}O_3$	[2.4]	7
Ethyl methyl ether (*trans*)	C_3H_8O	1.17 ± 0.02	3
Ethyl methyl sulfide (*gauche*)	C_3H_8S	1.593 ± 0.004	5
Ethyl methyl sulfide (*trans*)	C_3H_8S	1.56 ± 0.03	3
Ethyl propanoate	$C_5H_{10}O_2$	[1.74]	7
Ethyl vinyl ether	C_4H_8O	[1.26]	7
Fluoroacetylene	C_2HF	0.7207 ± 0.0003	3
Fluorobenzene	C_6H_5F	1.60 ± 0.08	1
Fluorocyclohexane (*equitorial*)	$C_6H_{11}F$	2.11 ± 0.04	2
Fluorocyclohexane (*axial*)	$C_6H_{11}F$	1.81 ± 0.04	2
1-Fluorocyclohexene	C_6H_9F	1.942 ± 0.010	5
Fluoroethane	C_2H_5F	1.937 ± 0.007	5
Fluoroethene	C_2H_3F	1.468 ± 0.003	5
Fluoromethane	CH_3F	1.858 ± 0.002	3
Fluoromethylidyne	CF	0.645 ± 0.005	3
(Fluoromethylidyne) phosphine (FCP)	CFP	0.279 ± 0.001	62
Fluoromethylsilane	CH_5FSi	1.700 ± 0.008	5
1-Fluoro-4-nitrobenzene	$C_6H_4FNO_2$	2.87 ± 0.06	1
1-Fluoropropane (*gauche*)	C_3H_7F	1.90 ± 0.10	1
1-Fluoropropane (*trans*)	C_3H_7F	2.05 ± 0.04	1
2-Fluoropropane	C_3H_7F	1.958 ± 0.001	5
cis-1-Fluoropropene	C_3H_5F	1.46 ± 0.03	1
trans-1-Fluoropropene	C_3H_5F	≈1.9	1
2-Fluoropropene	C_3H_5F	1.61 ± 0.03	1
3-Fluoropropene (*gauche*)	C_3H_5F	1.939 ± 0.015	1
3-Fluoropropene (*cis*)	C_3H_5F	1.765 ± 0.014	1
3-Fluoropropyne	C_3H_3F	1.73 ± 0.02	5
3-Fluoropyridine	C_5H_4FN	2.09 ± 0.26	3
2-Fluorotoluene	C_7H_7F	1.37 ± 0.07	1

Name	Mol. Form.	μ/D	Ref.
3-Fluorotoluene	C_7H_7F	1.82 ± 0.04	2
4-Fluorotoluene	C_7H_7F	2.00 ± 0.10	1
Formaldehyde	CH_2O	2.332 ± 0.002	3
Formaldehyde dimer	$C_2H_4O_2$	0.858 ± 0.005	57
Formamide	CH_3NO	3.73 ± 0.07	1
Formic acid	CH_2O_2	1.425 ± 0.002	5
Formyl fluoride	CHFO	2.081 ± 0.001	5
Fulminic acid	CHNO	3.09934	5
Fulvene	C_6H_6	0.4236 ± 0.013	2
Furan	C_4H_4O	0.66 ± 0.01	1
Furfural	$C_5H_4O_2$	[3.54]	7
Furfuryl alcohol	$C_5H_6O_2$	[1.92]	7
Glycerol	$C_3H_8O_3$	[2.56]	7
Glycine (Conformer I)	$C_2H_5NO_2$	1.147 ± 0.005	49
Glycine (Conformer II)	$C_2H_5NO_2$	5.45 ± 0.05	49
Glycolaldehyde	$C_2H_4O_2$	2.73 ± 0.05	2
Glyoxal (*cis*)	$C_2H_2O_2$	4.8 ± 0.2	2
2-Heptanol	$C_7H_{16}O$	[1.71]	7
3-Heptanol	$C_7H_{16}O$	[1.71]	7
2-Heptanone	$C_7H_{14}O$	[2.59]	7
3-Heptanone	$C_7H_{14}O$	[2.78]	7
Hexamethylphosphoric triamide	$C_6H_{18}N_3OP$	[5.5]	7
Hexanoic acid	$C_6H_{12}O_2$	[1.13]	7
2-Hexanone	$C_6H_{12}O$	[2.66]	7
sec-Hexyl acetate	$C_8H_{16}O_2$	[1.9]	7
1-Hexyne	C_6H_{10}	0.83 ± 0.05	1
Hydrogen cyanide	CHN	2.985188	5
Hydrogen cyanide trimer	$C_3H_3N_3$	10.6	21
Hydrogen isocyanide	CHN	3.05 ± 0.15	3
p-Hydroquinone	$C_6H_6O_2$	2.38 ± 0.05	15
3-Hydroxypropanenitrile (*gauche*)	C_3H_5NO	3.17 ± 0.02	5
Imidazole	$C_3H_4N_2$	3.8 ± 0.4	2
Iodoacetylene	C_2HI	0.02525	5
Iodobenzene	C_6H_5I	1.70 ± 0.09	1
1-Iodobutane	C_4H_9I	[1.93]	7
2-Iodobutane	C_4H_9I	2.12 ± 0.11	1
Iodoethane	C_2H_5I	1.976 ± 0.002	5
Iodoethene	C_2H_3I	1.311 ± 0.005	5
Iodomethane	CH_3I	1.6406 ± 0.0004	5
1-Iodo-2-methylpropane	C_4H_9I	[1.87]	7
Iodomethylsilane	CH_5ISi	1.862 ± 0.005	5
1-Iodopropane	C_3H_7I	2.04 ± 0.10	1
2-Iodopropane	C_3H_7I	[1.95]	7
Isobutanal (*gauche*)	C_4H_8O	2.69 ± 0.01	5
Isobutanal (*trans*)	C_4H_8O	2.86 ± 0.01	5
Isobutane	C_4H_{10}	0.132 ± 0.002	1
Isobutene	C_4H_8	0.503 ± 0.010	1
Isobutyl acetate	$C_6H_{12}O_2$	[1.86]	7
Isobutylamine	$C_4H_{11}N$	[1.27]	7
Isobutyl formate	$C_5H_{10}O_2$	[1.88]	7
Isobutyl isobutanoate	$C_8H_{16}O_2$	[1.9]	7
Isocyanic acid (HNCO)	CHNO	≈1.6	2
Isocyanobenzene	C_7H_5N	4.018 ± 0.003	5
Isocyanocyclopropane	C_4H_5N	4.03 ± 0.10	3
2-Isocyanopropane	C_4H_7N	4.055 ± 0.001	5
Isopentane	C_5H_{12}	0.13 ± 0.05	1
Isopentyl acetate	$C_7H_{14}O_2$	[1.86]	7
Isopropylamine	C_3H_9N	1.19 ± 0.06	3
Isopropylbenzene	C_9H_{12}	≈0.79	1
Isopropyl methyl ether	$C_4H_{10}O$	1.247 ± 0.003	5
Isoquinoline	C_9H_7N	2.73 ± 0.14	1
Isoxazole	C_3H_3NO	2.95 ± 0.04	3
Isoxazole–carbon monoxide complex	$C_3H_3NO{\bullet}CO$	2.873 ± 0.004	52
Ketene	C_2H_2O	1.42215	3
Mesityl oxide	$C_6H_{10}O$	[2.79]	7
Methacrylic acid	$C_4H_6O_2$	[1.65]	7
Methanethiol	CH_4S	1.52 ± 0.08	1
Methanol	CH_4O	1.70 ± 0.02	1
2-Methoxyethanol (*gauche*)	$C_3H_8O_2$	2.36 ± 0.05	2
2-Methoxyethyl acetate	$C_5H_{10}O_3$	[2.13]	7
1-Methoxy-1,2-propadiene	C_4H_6O	0.963 ± 0.020	5
N-Methylacetamide	C_3H_7NO	[4.3]	7
Methyl acetate	$C_3H_6O_2$	1.72 ± 0.09	1
Methyl acrylate	$C_4H_6O_2$	[1.77]	7
2-Methylacrylonitrile	C_4H_5N	3.69 ± 0.18	1
Methylamine	CH_5N	1.31 ± 0.03	1
2-Methylaniline	C_7H_9N	[1.60]	7
3-Methylaniline	C_7H_9N	[1.45]	7
4-Methylaniline	C_7H_9N	[1.52]	7
Methyl azide	CH_3N_3	2.17 ± 0.04	2
Methyl benzoate	$C_8H_8O_2$	[1.94]	7
2-Methyl-1,3-butadiene	C_5H_8	0.25 ± 0.01	1
3-Methylbutanoic acid	$C_5H_{10}O_2$	[0.63]	7
2-Methyl-1-butanol	$C_5H_{12}O$	[1.88]	7
2-Methyl-2-butanol	$C_5H_{12}O$	[1.82]	7
3-Methyl-1-butene (*gauche*)	C_5H_{10}	0.398 ± 0.004	3
3-Methyl-1-butene (*trans*)	C_5H_{10}	0.320 ± 0.010	3
3-Methyl-2-butenenitrile	C_5H_7N	4.61 ± 0.13	10
2-Methyl-1-buten-3-yne	C_5H_6	0.513 ± 0.02	2
Methyl cyanate	C_2H_3NO	4.26 ± 0.18	5
cis-3-Methylcyclohexanol	$C_7H_{14}O$	[1.91]	7
trans-3-Methylcyclohexanol	$C_7H_{14}O$	[1.75]	7
3-Methylcyclopentanone	$C_6H_{10}O$	3.14 ± 0.03	5
3-Methyl-2-cyclopenten-1-one	C_6H_8O	4.33 ± 0.002	5
Methylcyclopropane	C_4H_8	0.139 ± 0.004	2
Methyldiborane(6)	CH_8B_2	0.566 ± 0.006	3
Methyldifluorophosphine	CH_3F_2P	2.056 ± 0.006	3
Methylenecyclohexane	C_7H_{12}	0.62 ± 0.01	5
Methylenecyclopropene	C_4H_4	1.90 ± 0.01	5
Methylenephosphine (CH_2 = PH)	CH_3P	0.869 ± 0.003	61
N-Methylformamide	C_2H_5NO	3.83 ± 0.08	1
Methyl formate	$C_2H_4O_2$	1.77 ± 0.04	1
2-Methylfuran	C_5H_6O	0.65 ± 0.05	2
3-Methylfuran	C_5H_6O	1.03 ± 0.02	2
5-Methyl-2(3*H*)-furanone	$C_5H_6O_2$	4.08 ± 0.02	5
Methyl hydroperoxide	CH_4O_2	≈0.65	13
Methylidyne	CH	≈1.46	2

Name	Mol. Form.	μ/D	Ref.
Methyl isocyanate	C_2H_3NO	≈2.8	1
Methyl isothiocyanate	C_2H_3NS	3.453 ± 0.003	5
4-Methylisoxazole	C_4H_5NO	3.583 ± 0.005	5
Methyl methacrylate	$C_5H_8O_2$	[1.67]	7
2-Methyloxazole	C_4H_5NO	1.37 ± 0.07	5
4-Methyloxazole	C_4H_5NO	1.08 ± 0.05	5
5-Methyloxazole	C_4H_5NO	2.16 ± 0.04	5
Methyloxirane	C_3H_6O	2.01 ± 0.02	1
2-Methyl-2,4-pentanediol	$C_6H_{14}O_2$	[2.9]	7
4-Methylpentanenitrile	$C_6H_{11}N$	[3.5]	7
Methylphosphonic difluoride	CH_3F_2OP	3.69 ± 0.26	3
N-Methylpropanamide	C_4H_9NO	3.61	7
2-Methylpropanenitrile	C_4H_7N	4.29 ± 0.09	3
2-Methyl-2-propanethiol	$C_4H_{10}S$	1.66 ± 0.03	3
2-Methylpropanoic acid	$C_4H_8O_2$	[1.08]	7
2-Methyl-1-propanol	$C_4H_{10}O$	1.64 ± 0.08	1
2-Methyl-2-propanol	$C_4H_{10}O$	[1.66]	7
2-Methylpropenal	C_4H_6O	2.68 ± 0.13	1
2-Methyl-2-propenol (*skew*)	C_4H_8O	1.295 ± 0.022	5
Methyl propyl ether (*trans-trans*)	$C_4H_{10}O$	1.107 ± 0.013	3
2-Methylpyridine	C_6H_7N	1.85 ± 0.04	2
3-Methylpyridine	C_6H_7N	[2.40]	7
4-Methylpyridine	C_6H_7N	2.70 ± 0.02	2
2-Methylpyrimidine	$C_5H_6N_2$	1.676 ± 0.010	3
5-Methylpyrimidine	$C_5H_6N_2$	2.881 ± 0.006	3
N-Methylpyrrolidine	$C_5H_{11}N$	0.572 ± 0.003	5
N-Methyl-2-pyrrolidinone	C_5H_9NO	[4.1]	7
Methyl salicylate	$C_8H_8O_3$	[2.47]	7
Methylsilane	CH_6Si	0.73456	5
Methyl silyl ether	CH_6OSi	1.15 ± 0.02	2
3-Methylthietane	C_4H_8S	2.046 ± 0.009	5
2-Methylthiophene	C_5H_6S	0.674 ± 0.005	2
3-Methylthiophene	C_5H_6S	0.914 ± 0.015	3
Methyl vinyl ether	C_3H_6O	0.965 ± 0.002	5
Morpholine	C_4H_9NO	1.55 ± 0.03	3
2-Nitroanisole	$C_7H_7NO_3$	[5.0]	7
Nitrobenzene	$C_6H_5NO_2$	4.22 ± 0.08	1
Nitroethane	$C_2H_5NO_2$	3.23 ± 0.03	2
Nitromethane	CH_3NO_2	3.46 ± 0.02	1
1-Nitropropane	$C_3H_7NO_2$	3.66 ± 0.07	1
2-Nitropropane	$C_3H_7NO_2$	3.73 ± 0.07	1
Nonanoic acid	$C_9H_{18}O_2$	[0.79]	7
2,5-Norbornadiene	C_7H_8	0.0587 ± 0.0001	5
cis-9-Octadecenoic acid	$C_{18}H_{34}O_2$	[1.18]	7
Octanoic acid	$C_8H_{16}O_2$	[1.15]	7
1-Octanol	$C_8H_{18}O$	[1.76]	7
2-Octanol	$C_8H_{18}O$	[1.71]	7
2-Octanone	$C_8H_{16}O$	[2.70]	7
1,4-Oxathiane	C_4H_8OS	0.295 ± 0.003	3
Oxazole	C_3H_3NO	1.503 ± 0.030	3
Oxetane	C_3H_6O	1.94 ± 0.01	1
2-Oxetanone	$C_3H_4O_2$	4.18 ± 0.03	1
3-Oxetanone	$C_3H_4O_2$	0.887 ± 0.005	2
Oxirane	C_2H_4O	1.89 ± 0.01	1
Paraldehyde	$C_6H_{12}O_3$	1.43 ± 0.07	1
Pentachloroethane	C_2HCl_5	0.92 ± 0.05	1
cis-1,3-Pentadiene	C_5H_8	0.500 ± 0.015	2
trans-1,3-Pentadiene	C_5H_8	0.585 ± 0.010	2
1,3-Pentadiyne	C_5H_4	1.207 ± 0.001	5
1,5-Pentanediol	$C_5H_{12}O_2$	[2.5]	7
2,4-Pentanedione	$C_5H_8O_2$	[2.78]	7
Pentanenitrile	C_5H_9N	4.12 ± 0.08	1
Pentanoic acid	$C_5H_{10}O_2$	[1.61]	7
1-Pentanol	$C_5H_{12}O$	[1.7]	7
2-Pentanol	$C_5H_{12}O$	[1.66]	7
3-Pentanol	$C_5H_{12}O$	[1.64]	7
2-Pentanone	$C_5H_{10}O$	[2.70]	7
3-Pentanone	$C_5H_{10}O$	[2.82]	7
1,2,3-Pentatriene	C_5H_6	0.51 ± 0.05	11
1-Pentene	C_5H_{10}	≈0.5	1
1-Penten-3-yne	C_5H_6	0.66 ± 0.02	2
cis-3-Penten-1-yne	C_5H_6	0.78 ± 0.02	2
trans-3-Penten-1-yne	C_5H_6	1.06 ± 0.05	2
Pentyl acetate	$C_7H_{14}O_2$	1.75 ± 0.10	1
Pentyl formate	$C_6H_{12}O_2$	1.90 ± 0.10	1
1-Pentyne (*gauche*)	C_5H_8	0.769 ± 0.028	2
1-Pentyne (*trans*)	C_5H_8	0.842 ± 0.010	2
Perfluoropyridine	C_5F_5N	0.98 ± 0.08	3
Phenol	C_6H_6O	1.224 ± 0.008	3
Phenylacetylene	C_8H_6	0.656 ± 0.005	3
Phenylsilane	C_6H_8Si	0.845 ± 0.012	3
1-Phosphapropyne (CH_3CP)	C_2H_3P	1.499 ± 0.001	63
Piperidine (*equitorial*)	$C_5H_{11}N$	0.82 ± 0.02	3
Piperidine (*axial*)	$C_5H_{11}N$	1.19 ± 0.02	3
Piperidine (*average*)	$C_5H_{11}N$	[1.19]	3
Propanal (*gauche*)	C_3H_6O	2.86 ± 0.01	5
Propanal (*cis*)	C_3H_6O	2.52 ± 0.05	1
Propanal (*average*)	C_3H_6O	2.72	1
Propane	C_3H_8	0.084 ± 0.001	1
1,2-Propanediol	$C_3H_8O_2$	[2.25]	7
1,3-Propanediol	$C_3H_8O_2$	[2.55]	7
Propanenitrile	C_3H_5N	4.05 ± 0.03	3
1-Propanethiol (*gauche*)	C_3H_8S	1.683 ± 0.010	3
1-Propanethiol (*trans*)	C_3H_8S	1.60 ± 0.08	3
2-Propanethiol (*gauche*)	C_3H_8S	1.53 ± 0.03	3
2-Propanethiol (*trans*)	C_3H_8S	1.61 ± 0.03	3
Propanoic acid (*cis*)	$C_3H_6O_2$	1.46 ± 0.07	2
Propanoic acid (*average*)	$C_3H_6O_2$	1.75 ± 0.09	1
1-Propanol (*gauche*)	C_3H_8O	1.58 ± 0.03	2
1-Propanol (*trans*)	C_3H_8O	1.55 ± 0.03	2
2-Propanol (*trans*)	C_3H_8O	1.58 ± 0.03	2
Propargyl alcohol	C_3H_4O	1.13 ± 0.06	2
Propene	C_3H_6	0.366 ± 0.001	1
Propene–sulfur dioxide complex	$C_3H_6{\bullet}O_2S$	1.34 ± 0.003	35
Propyl acetate	$C_5H_{10}O_2$	[1.78]	7
Propylamine	C_3H_9N	1.17 ± 0.06	1
Propylene carbonate	$C_4H_6O_3$	[4.9]	7
Propyleneimine (*cis*)	C_3H_7N	1.77 ± 0.09	2
Propyleneimine (*trans*)	C_3H_7N	1.57 ± 0.03	2
Propyl formate	$C_4H_8O_2$	[1.89]	7

Name	Mol. Form.	μ/D	Ref.
2-Propynal	C_3H_2O	2.78 ± 0.02	5
Propyne	C_3H_4	0.784 ± 0.001	3
Propyne-argon complex	$C_3H_4{\cdot}Ar$	0.730 ± 0.005	20
4*H*-Pyran-4-one	$C_5H_4O_2$	3.79 ± 0.02	5
4*H*-Pyran-4-thione	C_5H_4OS	3.95 ± 0.05	5
1*H*-Pyrazole	$C_3H_4N_2$	2.20 ± 0.01	3
Pyridazine	$C_4H_4N_2$	4.22 ± 0.02	2
Pyridine	C_5H_5N	2.215 ± 0.010	3
2-Pyridinecarbonitrile	$C_6H_4N_2$	5.78 ± 0.11	3
3-Pyridinecarbonitrile	$C_6H_4N_2$	3.66 ± 0.11	3
4-Pyridinecarbonitrile	$C_6H_4N_2$	1.96 ± 0.03	3
3-Pyridinecarboxaldehyde	C_6H_5NO	1.44	3
4-Pyridinecarboxaldehyde	C_6H_5NO	1.66	3
2-Pyridinecarboxaldehyde	C_6H_5NO	3.56 ± 0.07	3
Pyrimidine	$C_4H_4N_2$	2.334 ± 0.010	2
Pyrrole	C_4H_5N	1.767 ± 0.001	5
Pyrrolidine	C_4H_9N	[1.57]	7
2-Pyrrolidone	C_4H_7NO	[3.5]	7
Quinoline	C_9H_7N	2.29 ± 0.11	1
Salicylaldehyde	$C_7H_6O_2$	[2.86]	7
Selenoformaldehyde	CH_2Se	1.41 ± 0.01	5
Silicon dicarbide	C_2Si	2.393 ± 0.006	24
Silicon methylidyne	$CHSi$	0.066 ± 0.002	41
Styrene	C_8H_8	0.123 ± 0.003	5
Succinonitrile	$C_4H_4N_2$	[3.7]	7
Sulfolane	$C_4H_8O_2S$	[4.8]	7
1,1,2,2-Tetrabromoethane	$C_2H_2Br_4$	[1.38]	7
1,1,2,2-Tetrachloroethane	$C_2H_2Cl_4$	1.32 ± 0.07	1
1,2,3,4-Tetrafluorobenzene	$C_6H_2F_4$	2.42 ± 0.05	3
1,2,3,5-Tetrafluorobenzene	$C_6H_2F_4$	1.46 ± 0.06	3
1,1,1,2-Tetrafluoroethane	$C_2H_2F_4$	1.80 ± 0.22	5
Tetrahydrofuran	C_4H_8O	1.75 ± 0.04	2
Tetrahydrofurfuryl alcohol	$C_5H_{10}O_2$	[2.1]	7
Tetrahydropyran (*chair*)	$C_5H_{10}O$	1.58 ± 0.03	3
Tetrahydro-4*H*-pyran-4-one	$C_5H_8O_2$	1.720 ± 0.003	3
1,2,5,6-Tetrahydropyridine	C_5H_9N	1.007 ± 0.003	3
Tetrahydrothiophene	C_4H_8S	[1.90]	7
Tetramethylurea	$C_5H_{12}N_2O$	[3.5]	7
1*H*-Tetrazole	CH_2N_4	2.19 ± 0.05	3
Thiacyclohexane	$C_5H_{10}S$	1.781 ± 0.010	3
1,2,5-Thiadiazole	$C_2H_2N_2S$	1.579 ± 0.007	3
Thietane	C_3H_6S	1.85 ± 0.09	1
Thietane 1,1-dioxide	$C_3H_6O_2S$	4.8 ± 0.1	5
Thioacetaldehyde	C_2H_4S	2.33 ± 0.02	68
Thiocarbonyl fluoride	CF_2S	0.080	59
Thioformaldehyde	CH_2S	1.6491 ± 0.0004	3
Thiophene	C_4H_4S	0.55 ± 0.01	2
2-Thiophenecarbonitrile	C_5H_3NS	4.59 ± 0.02	3
3-Thiophenecarbonitrile	C_5H_3NS	4.13 ± 0.02	3
4*H*-Thiopyran-4-thione	$C_5H_4S_2$	3.9 ± 0.2	5
Toluene	C_7H_8	0.375 ± 0.010	3
Toluene-sulfur dioxide complex	$C_7H_8{\cdot}O_2S$	1.87 ± 0.03	34
1*H*-1,2,4-Triazole	$C_2H_3N_3$	2.7 ± 0.1	3
Tribromomethane	$CHBr_3$	0.99 ± 0.02	1
Tributylamine	$C_{12}H_{27}N$	[0.78]	7
Tributyl borate	$C_{12}H_{27}BO_3$	[0.77]	7
Tributyl phosphate	$C_{12}H_{27}O_4P$	[3.07]	7
Tricarbon monosulfide	C_3S	3.704 ± 0.009	50
1,1,1-Trichloroethane	$C_2H_3Cl_3$	1.755 ± 0.015	2
1,1,2-Trichloroethane	$C_2H_3Cl_3$	[1.4]	7
Trichloroethene	C_2HCl_3	[0.8]	7
Trichloroethylsilane	$C_2H_5Cl_3Si$	[2.04]	7
Trichlorofluoromethane	CCl_3F	0.46 ± 0.02	2
Trichloromethane	$CHCl_3$	1.04 ± 0.02	2
(Trichloromethyl)benzene	$C_7H_5Cl_3$	[2.03]	7
Trichloromethylsilane	CH_3Cl_3Si	1.91 ± 0.01	2
Tri-*o*-cresyl phosphate	$C_{21}H_{21}O_4P$	[2.87]	7
Tri-*m*-cresyl phosphate	$C_{21}H_{21}O_4P$	[3.05]	7
Tri-*p*-cresyl phosphate	$C_{21}H_{21}O_4P$	[3.18]	7
Triethanolamine	$C_6H_{15}NO_3$	[3.57]	7
Triethylamine	$C_6H_{15}N$	0.66 ± 0.05	1
Triethyl phosphate	$C_6H_{15}O_4P$	[3.12]	7
Trifluoroacetic acid	$C_2HF_3O_2$	2.28 ± 0.25	1
Trifluoroacetonitrile	C_2F_3N	1.262 ± 0.010	3
1,2,4-Trifluorobenzene	$C_6H_3F_3$	1.402 ± 0.009	5
1,1,1-Trifluoroethane	$C_2H_3F_3$	2.347 ± 0.005	3
Trifluoroethene	C_2HF_3	1.32 ± 0.03	2
Trifluoroiodomethane	CF_3I	1.048 ± 0.003	3
Trifluoroisocyanomethane	C_2F_3N	1.153 ± 0.010	5
Trifluoromethane	CHF_3	1.65150	3
(Trifluoromethyl)benzene	$C_7H_5F_3$	2.86 ± 0.06	1
Trifluoromethylsilane	CH_3F_3Si	2.3394 ± 0.0002	5
(Trifluoromethyl)silane	CH_3F_3Si	2.32 ± 0.02	5
3,3,3-Trifluoropropene	$C_3H_3F_3$	2.45 ± 0.05	1
3,3,3-Trifluoro-1-propyne	C_3HF_3	2.317 ± 0.013	5
Trimethylamine	C_3H_9N	0.612 ± 0.003	1
Trimethyl phosphate	$C_3H_9O_4P$	[3.18]	7
2,4,6-Trimethylpyridine	$C_8H_{11}N$	[2.05]	7
1,3,5-Trioxane	$C_3H_6O_3$	2.08 ± 0.02	1
Vinyl acetate	$C_4H_6O_2$	[1.79]	7
Vinyl formate	$C_3H_4O_2$	1.49 ± 0.01	1
2-Vinylfuran	C_6H_6O	0.69 ± 0.07	5
Vinylsilane	C_2H_6Si	0.657 ± 0.002	5
o-Xylene	C_8H_{10}	0.640 ± 0.005	2
2,4-Xylenol	$C_8H_{10}O$	[1.4]	7
2,5-Xylenol	$C_8H_{10}O$	[1.45]	7
2,6-Xylenol	$C_8H_{10}O$	[1.40]	7
3,4-Xylenol	$C_8H_{10}O$	[1.56]	7
3,5-Xylenol	$C_8H_{10}O$	[1.55]	7

HINDERED INTERNAL ROTATION

I. Ozier and N. Moazzen-Ahmadi

In asymmetric tops like methyl alcohol, CH_3OH, and symmetric rotors like CH_3SiH_3, the methyl group can undergo internal rotation relative to the rest of the molecule, traditionally called the frame (LS59, OM07). Although various different tops are considered here, all have three-fold symmetry. In such cases, the potential V hindering the internal rotation can be written:

$$V(\alpha)= V_3(\tfrac{1}{2})(1-\cos 3\alpha) + V_6(\tfrac{1}{2})(1-\cos 6\alpha) + V_9(\tfrac{1}{2})(1-\cos 9\alpha)+\ldots ,$$

where α is the deviation from equilibrium of the angle between the top and frame that measures the torsional motion. If only the first two terms are retained, then V_3 is the height of the hindering potential and V_6 is the shape parameter. For symmetric tops like CH_3CH_3 where the top and frame are identical, α is replaced by 2γ and the origin for γ is often taken as the eclipsed configuration. In the expansion, $-\cos 6n\gamma$ is then replaced by $(-1)^{n+1}\cos 6n\gamma$, where n = 1,2,... In cases where different forms of the expansion have been used in the original works, the values of the parameters published there have been converted to the conventions defined here.

In Tables 1 and 2, values are given for V_3 for a selection of asymmetric and symmetric tops, respectively. In cases where the higher order parameters have been determined, these are given in the Comments column. Where appropriate, this column also indicates the specific top, isomer, state, and/or isotopomer that has been studied. For ethane, three symmetric top isotopomer are listed to illustrate the isotopic dependence of V_3 and V_6. In all other cases, only one isotopomer is listed, even if several have been studied. In all but one of these cases, the isotopomer reported is the one with the highest natural abundance. However, CH_3OCDO is listed because the results obtained are more precise than for CH_3OCHO. The molecules are listed alphabetically in Hill order according to the molecular formula.

The determinations listed for the potential parameters are effective values that incorporate to varying degrees effects from other molecular parameters. For example, the apparent value of V_3 can be changed significantly if the reduced rotational constant F is calculated from the structure, rather than being determined independently (LS59). Other examples include such mechanisms as coupling to excited skeletal vibrations (OM07) and redundancies connecting some of the torsional parameters (LB68, MO87). The experimental uncertainties quoted are taken from the original works; no attempt has been made to standardize the definitions. All the potential parameters are given in cm^{-1}. Where the original work has reported these values in other units, the conversion to cm^{-1} has been carried out using standard factors (LB02):

1 calorie = 4.1868 joules;
1 calorie/mole = 0.34998915 cm^{-1}.

A variety of different methods have been used to measure V_3, V_6, and V_9 (LS59, OM07); only a few of the more important will be discussed here. *For asymmetric rotors,* both the pure rotational spectrum and its torsion-rotation counterpart are electric dipole allowed and are affected in lowest order by the leading terms in the torsional Hamiltonian. Both types of spectra have been used extensively to determine V_3 (LS59). *For symmetric tops* with a single torsional degree of freedom, either the permanent electric dipole moment vanishes, as in CH_3CH_3, or the normal rotational spectrum is independent of V_3 in lowest order, as in CH_3SiH_3. In the latter case, the molecular beam avoided crossing method can often be used (OM07). The torsion-rotation spectrum is forbidden in lowest order, but becomes weakly allowed through interactions with the infrared active skeletal vibrations (OM07). By employing long absorption path lengths, this spectrum has been used to determine V_3 in a number of molecules. For both asymmetric and symmetric tops, the most precise determinations of the molecular parameters have been made in cases where both rotational and torsion-rotation spectra have been investigated.

References

ALA97 Antolínez, S., López, J. C., and Alonso, J. L., *J. Chem. Soc., Faraday Trans.* 93, 1291, 1997.

ALB97 Alonso, J. L., López, J. C., Blanco, S., and Guarnieri, A., *J. Mol. Spectrosc.* 182, 148, 1997.

ALL97 Alonso, J. L., Lesarri, A., López, J. C., Blanco, S., Kleiner, I., and Demaison, J., *Molec. Phys.* 91, 731, 1997.

BL85 Bestmann, G., Lalowski, W., and Dreizler, H., *Z. Naturforsch.* A 40, 271, 1985.

BM07 Borvayeh, L., Moazzen-Ahmadi, N., and Horneman, V.-M., *J. Mol. Spectrosc.* 242, 77, 2007.

BW64 Butcher, S. S., and Wilson, E. B., *J. Chem. Phys.* 40, 1671, 1964.

CA96 Charro, M. E., and Alonso, J. L., *J. Mol. Spectrosc.* 176, 251, 1996.

CB61 Cahill, P., and Butcher, S., *J. Chem. Phys.* 35, 2255, 1961.

DG87 Durig, J. R., Guirgis, G. A., and Van Der Veken, B. J., *J. Raman Spectrosc.* 18, 549, 1987.

DL73 Durig, J. R., Li, Y. S., Carreira, L. A., and Odom, J. D., *J. Amer. Chem. Soc.* 95, 2491, 1973.

DM87 Demaison, J., Maes, H., van Eijck, B. P., Wlodarczak, G., and Lasne, M. C., *J. Mol. Spectrosc.* 125, 214, 1987.

DS81 Dreizler, H., and Scappini, F., *Z. Naturforsch.* A 36, 1187, 1981.

EG96 Eltayeb, S., Guirgis, G. A., Fanning, A. R., and Durig, J. R., *J. Raman Spectrosc.* 27, 111, 1996.

FD83 Fliege, E., Dreizler, H., Demaison, J., Boucher, D., Burie, J., and Dubrulle A., *J. Chem. Phys.* 78, 3541, 1983.

G00 Groner, P., *J. Mol. Structure* 550–551, 473, 2000.

GA89 Groner, P., Attia, G. M., Mohamad, A. B., Sullivan, J. F., Li, Y. S., and Durig, J. R., *J. Chem. Phys.* 91, 1434, 1989.

GG04 Groner, P., Gillies, C. W., Gillies, J. Z., Zhang, Y., and Block, E., *J. Mol. Spectrosc.* 226, 169, 2004.

GH02 Grabow, J.-U., Hartwig, H., Heineking, N., Jäger, W., Mäder, H., Nicolaisen, H. W., and Stahl, W., *J. Mol. Structure* 612, 349, 2002.

IA03 Ilyushin, V. V., Alekseev, E. A., Dyubko, S. F., and Kleiner, I., *J. Mol. Spectrosc.* 220, 170, 2003.

K60 Krisher, L. C., *J. Chem. Phys.* 33, 1237, 1960.

KD86 Kasten, W., and Dreizler, H., *Z. Naturforsch.* A 41, 944, 1986.

KH96 Kleiner, I., Hougen, J. T., Grabow, J.-U., Belov, S. P., Tretyakov, M. Yu., and Cosléou, J., *J. Mol. Spectrosc.* 179, 41, 1996.

KL67 Kuczkowski, R. L., and Lide, D. R., *J. Chem. Phys.* 46, 357, 1967.

KW74 Krisher, L. C., Watson, W. A., and Morrison, J. A., *J. Chem. Phys.* 61, 3429, 1974.

L59 Laurie, V. W., *J. Chem. Phys.* 30, 1210, 1959.

LB68 Lees, R. M., and Baker, J. G., *J. Chem. Phys.* 48, 5299, 1968.

LB97 di Lauro, C., Bunker, P. R., Johns, J. W. C., and McKellar, A. R. W., *J. Mol. Spectrosc.* 184, 177, 1997.

LB02 Demaison, J., and Wlodarczak, G., Hindered rotation-Asymmetric top molecules, in: Hüttner, W. (Ed), *Landolt-Börnstein Numerical Data and Functional Relationships in Science and Technology, New Series: Group II: Molecules and Radicals, volume 24, subvolume C, Molecular Constants Mostly from Microwave, Molecular Beam, and Sub-Doppler Laser Spectroscopy*, Springer-Verlag, Heidelberg, 2002.

LE97 de Luis, A., Eugenia Sanz, M., Lorenzo, F. J., López, J. C., and Alonso, J. L., *J. Mol. Spectrosc.* 184, 60, 1997.

LJ72 Lide, D. R., Johnson, D. R., Sharp, K. G., and Coyle, T. D., *J. Chem. Phys.* 57, 3699, 1972.
LS59 Lin, C. C., and Swalen, J. D., *Rev. Mod. Phys.* 31, 841 1959.
MK66 Moloney, M. J., and Krisher, L. C., *J. Chem. Phys.* 45, 3277, 1966.
MM96 Marstokk, K.-M., Møllendal, H., and Samdal, S., *J. Mol. Structure* 376, 11, 1996.
MM01 Marstokk, K.-M., Mollendal, H., Samdal, S., and Steinborn, D., *J. Mol. Structure* 567, 41, 2001.
MO87 Moazzen-Ahmadi, N., and Ozier, I., *J. Mol. Spectrosc.* 126, 99, 1987.
MS02 Merke, I., Stahl, W., Kassi, S., Petitprez, D., and Wlodarczak, G., *J. Mol. Spectrosc.* 216, 437, 2002.
ND06 Nair, K. P. R., Demaison, J., Wlodarczak, G., and Merke, I., *J. Mol. Spectrosc.* 237, 137, 2006.
NH02 Niide, Y., and Hayashi, M., *J. Mol. Spectrosc.* 216, 61, 2002.
NH04 Niide, Y., and Hayashi, M., *J. Mol. Spectrosc.* 223, 152, 2004.
NY85 Nakagawa, J., Yamada, K., Bester, M., and Winnewisser, G., *J. Mol. Spectrosc.* 110, 74, 1985.
OH07 Ohashi, N., and Hougen, J. T., *J. Mol. Spectrosc.* 243, 162, 2007.
OK75 Odom, J. D., Kalasinsky, V. F., and Durig, J. R., *Inorg. Chem.* 14, 2837, 1975.
OM07 Ozier, I., and Moazzen-Ahmadi, N., Internal rotation in symmetric tops, in: Arimondo, E., Berman, P. R., and Lin, C. C., (Eds.), *Advances in Atomic, Molecular and Optical Physics*, vol. 54, p. 423, Elsevier, Amsterdam, 2007.
PK59 Pierce, L., and Krisher, L. C., *J. Chem. Phys.* 31, 875, 1959.
S75 Stiefvater, O. L., *J. Chem. Phys.* 62, 233, 1975.
SD04 Suenram, R. D., DaBell, R. S., Hight Walker, A. R., Lavrich, R. J., Plusquellic, D. F., Ellzy, M. W., Lochner, J. M., Cash, L., Jensen, J. O., and Samuels, A. C., *J. Mol. Spectrosc.* 224, 176, 2004.
SG05 Schnell, M., Grabow, J.-U., Hartwig, H, Heineking, N., Meyer, M., Stahl, W., and Caminati, W., *J. Mol. Spectrosc.* 229, 1, 2005.
SH82 Shiki, Y., Hasegawa, A., and Hayashi, M., *J. Mol. Structure* 78, 185, 1982.
SH86 Sastry, K. V. L. N., Herbst, E., Booker, R. A., and De Lucia, F. C., *J. Mol. Spectrosc.* 116, 120, 1986.
SL02 Suenram, R. D., Lovas, F. J., Plusquellic, D. F., Lesarri, A., Kawashima, Y., Jensen, J. O., and Samuels, A. C., *J. Mol. Spectrosc.* 211, 110, 2002.
SL06 Suenram, R. D., Lovas, F. J., Plusquellic, D. F., Ellzy, M. W., Lochner, J. M., Jensen, J. O., and Samuels, A. C., *J. Mol. Spectrosc.* 235, 18, 2006.
ST06 Styger, C., Ozier, I., Wang, S.-X., and Bauder, A., *J. Mol. Spectrosc.* 239, 115, 2006.
TB86 Typke, V., Botskor, I., and Wiedenmann, K.-H., *J. Mol. Spectrosc.* 120, 435, 1986.
TH86 Tyblewksi, M., Ha, T.-K., and Bauder, A., *J. Mol. Spectrosc.* 115, 353, 1986.
VD88 Vormann, K., and Dreizler, H., *Z. Naturforsch.* A 43, 338, 1988.
VG96 Voges, K., Gripp, J., Hartwig, H., and Dreizler, H., *Z. Naturforsch.* A 51, 299, 1996.
VR75 Varma, R., Ramaprasad, K. R., and Nelson, J. F., *J. Chem. Phys.* 63, 915, 1975.
WA02 Wang, S.-X., Schroderus, J., Ozier, I., Moazzen-Ahmadi, N., Horneman, V.-M., Ilyushyn, V. V., Alekseev, E. A., Katrich, A. A., and Dyubko, S. F., *J. Mol. Spectrosc.* 214, 69, 2002.

TABLE 1. Asymmetric Top Potential Parameters

	Name	Molecular Formula	Line Formula	Ref.	V_3/cm^{-1}	Comments
1	Trifluoromethanethiol	CHF_3S	CF_3SH	LB02	500.83 ± 0.03	
2	Methylphosphonic difluoride	CH_3F_2OP	$CH_3P(=O)F_2$	SL06	676 ± 25	
3	Methanol	CH_4O	CH_3OH	LB02	373.594 ± 0.007	V_6 = −1.597 ± 0.051 V_9 = 1.04 ± 0.20
4	Methanethiol	CH_4S	CH_3SH	SH86	443.029 ± 0.070	V_6 = −1.6451 ± 0.0144
5	Methyldisulfane	CH_4S_2	CH_3SSH	TH86	609.0 ± 14.0	
6	Trifluoromethyl isocyanate	C_2F_3NO	$CF_3N{=}C{=}O$	LB02	47.8769 ± 0.0051	
7	Trifluoroacetaldehyde	C_2HF_3O	$CF_3C(H){=}O$	DG87	298 ± 10	
8	Pentafluoroethane	C_2HF_5	CF_3CHF_2	EG96	1190 ± 4	
9	Acetyl bromide	C_2H_3BrO	$CH_3C(Br){=}O$	K60	456.7 ± 10.5	
10	1 Chloro-1,1-difluoroethane	$C_2H_3ClF_2$	CH_3CClF_2	ALB97	1311.8 ± 1.4	
11	Acetyl chloride	C_2H_3ClO	$CH_3C(Cl){=}O$	LB02	442.74 ± 1.05	^{35}Cl
12	Acetyl fluoride	C_2H_3FO	$CH_3C(F){=}O$	PK59	364.3 ± 2.1	
13	Methyl fluoroformate	$C_2H_3FO_2$	$CH_3OC(F){=}O$	LB02	374.1 ± 0.2	
14	Methyl trifluoromethyl ether	$C_2H_3F_3O$	CH_3OCF_3	LB02	382 ± 10	CH_3
15	Acetyl iodide	C_2H_3IO	$CH_3C(=O)I$	MK66	455.3 ± 10.5	
16	Methyl cyanate	C_2H_3NO	$CH_3OC{\equiv}N$	LB02	399.0 ± 17.5	
17	1-Chloro-1-fluoroethane	C_2H_4ClF	CH_3CHClF	LB02	1334.9 ± 3.8	
18	1,1-Difluoroethane	$C_2H_4F_2$	CH_3CHF_2	LB02	1163.0 ± 2.5	
19	Acetaldehyde	C_2H_4O	$CH_3C(H){=}O$	KH96	407.716 ± 0.010	V_6 =−12.068 ± 0.037
20	Thioacetaldehyde *S*-oxide	C_2H_4OS	$CH_3C(H){=}S{=}O$	LB02	285.6 ± 0.3	*Z* isomer
21	Acetic acid	$C_2H_4O_2$	CH_3COOH	IA03	170.1742 ± 0.0002	V_6 = −6.4725 ± 0.0001
22	Methyl formate	$C_2H_3DO_2$	$CH_3OC(D){=}O$	LB02	400.60 ± 0.03	deuterated
23	Fluoroethane	C_2H_5F	CH_3CH_2F	FD83	1172.1 ± 1.4	
24	Nitrosoethane	C_2H_5NO	$CH_3CH_2N{=}O$	LB02	903 ± 25	*gauche* conformer
		C_2H_5NO	$CH_3CH_2N{=}O$	LB02	911 ± 25	*cis* conformer
25	Acetamide	C_2H_5NO	$CH_3C(NH_2){=}O$	LB02	24.949 ± 0.008	
26	Difluorodimethylsilane	$C_2H_6F_2Si$	$(CH_3)_2SiF_2$	SG05	439.4 ± 2.5	

Name	Molecular Formula	Line Formula	Ref.	V_3/cm^{-1}	Comments
27 *N*-Nitrosodimethylamine	$C_2H_6N_2O$	$(CH_3)_2NN=O$	LB02	145.8 ± 0.25	*cis* CH_3
	$C_2H_6N_2O$	$(CH_3)_2NN=O$	LB02	737.4 ± 13.3	*trans* CH_3
28 Ethanol	C_2H_6O	CH_3CH_2OH	LB02	1173.76 ± 2.20	*trans* isomer
29 Dimethyl ether	C_2H_6O	$(CH_3)_2O$	NH04	926.0 ± 3.5	
30 Dimethyl sulfide	C_2H_6S	$(CH_3)_2S$	NH04	751.1 ± 4.8	
31 Vinylsilane	C_2H_6Si	$SiH_3C(H)=CH_2$	SH82	520.1 ± 1.8	
32 Dimethyl disulfide	$C_2H_6S_2$	CH_3SSCH_3	LB02	535.1 ± 1.8	
33 Dimethyl diselenide	$C_2H_6Se_2$	$CH_3SeSeCH_3$	GG04	395 ± 2	
34 Dimethylsilane	C_2H_8Si	$(CH_3)_2SiH_2$	NH04	578.0 ± 3.5	
35 3,3,3-Trifluoropropene	$C_3H_3F_3$	$CF_3C(H)=CH_2$	ALL97	653.06 ± 0.83	
36 Methyl cyanoformate	$C_3H_3NO_2$	$CH_3OC(C{\equiv}N)=O$	LB02	406.6 ± 1.1	*s-trans* conformer
37 (Methylthio)acetylene	C_3H_4S	$CH_3SC{\equiv}CH$	DM87	592.0 ± 3.3	
38 1,1,1-Trifluoropropane	$C_3H_5F_3$	$CH_3CH_2CF_3$	ALA97	922.2 ± 1.4	
39 2-Iodopropene	C_3H_5I	$CH_3C(I)=CH_2$	LB02	905.8 ± 4.2	
40 Ethyl isocyanide	C_3H_5N	$CH_3CH_2N{\equiv}C$	LB02	1167.6 ± 18.2	
41 Propene	C_3H_6	$CH_3C(H)=CH_2$	LB02	697.499 ± 0.048	$V_6 = -13.0$ (fixed)
42 Propanal	C_3H_6O	$CH_3CH_2C(H)=O$	BW64	798 ± 39	*cis* conformer
43 Acetone	C_3H_6O	$(CH_3)_2C=O$	G00	251.4 ± 2.6	$V_6 = -6.92 \pm 0.65$
44 (Methylthio)ethene	C_3H_6S	$CH_3SC(H)=CH_2$	MM01	1138 ± 13	
45 Propanoic acid	$C_3H_6O_2$	CH_3CH_2COOH	S75	819.0 ± 10.5	*cis* conformer
46 Methyl mercaptoacetate	$C_3H_6O_2S$	$CH_3OC(=O)C(H_2)SH$	LB02	411 ± 8	state 0$^+$
	$C_3H_6O_2S$	$CH_3OC(=O)C(H_2)SH$	LB02	412 ± 9	state 0$^-$
47 2-Bromopropane	C_3H_7Br	$(CH_3)_2CHBr$	LB02	1437.0 ± 2.5	^{79}Br
48 1-Chloropropane	C_3H_7Cl	$CH_3C(H_2)C(H_2)Cl$	LE97	1017.8 ± 1.4	*gauche* conformer
	C_3H_7Cl	$CH_3C(H_2)C(H_2)Cl$	LE97	966.0 ± 7.0	*trans* conformer
49 2-Chloropropane	C_3H_7Cl	$(CH_3)_2CHCl$	LB02	1374.03 ± 1.00	^{35}Cl
50 1-Fluoropropane	C_3H_7F	$CH_3C(H_2)C(H_2)F$	KD86	965.3 ± 12.2	*gauche* conformer
	C_3H_7F	$CH_3C(H_2)C(H_2)F$	KD86	948.5 ± 2.8	*trans* conformer
51 2-Fluoropropane	C_3H_7F	$(CH_3)_2CHF$	LB02	1162.79 ± 0.84	
52 Butanenitrile	C_4H_7N	$CH_3C(H_2)C(H_2)C{\equiv}N$	VD88	1087.4 ± 8.4	*gauche* conformer
	C_4H_7N	$CH_3C(H_2)C(H_2)C{\equiv}N$	VD88	1088.5 ± 13.3	*trans* conformer
53 Propanamide	C_3H_7NO	$CH_3CH_2C(=O)NH_2$	MM96	761 ± 42	*syn* conformer
54 *N,N*-Dimethylformamide	C_3H_7NO	$(CH_3)_2NC(H)=O$	LB02	366.04 ± 0.26	*cis* CH_3
	C_3H_7NO	$(CH_3)_2NC(H)=O$	LB02	772.4 ± 7.4	*trans* CH_3
55 Propane	C_3H_8	$(CH_3)_2CH_2$	BL85	1108.1 ± 9.5	
56 Cyclopropylgermane	C_3H_8Ge	$C(H_2)C(H_2)C(H)(GeH_3)$	LB02	466.6 ± 16.7	GeH_3
57 *N*-Nitrosoethylmethylamine	$C_3H_8N_2O$	$CH_3CH_2N(CH_3)N=O$	LB02	310 ± 30	*N*-methyl top, OGM conformer
58 1-Propanol	C_3H_8O	$CH_3C(H_2)C(H_2)OH$	DS81	956 ± 21	*trans* conformer
59 Cyclopropylsilane	C_3H_8Si	$C(H_2)C(H_2)C(H)(SiH_3)$	TB86	670.9 ± 1.5	
60 Dimethyl(methylene)silane	C_3H_8Si	$(CH_3)_2Si=CH_2$	LB02	351.4 ± 5.9	
61 Dimethyl methylphosphonate	$C_3H_9O_3P$	$(OCH_3)_2P(=O)CH_3$	SL02	662 ± 6	*P*-methyl top
	$C_3H_9O_3P$	$(OCH_3)_2P(=O)CH_3$	OH07	278.82 ± 0.06	*O*-methyl top #1
	$C_3H_9O_3P$	$(OCH_3)_2P(=O)CH_3$	OH07	181.82 ± 0.01	*O*-methyl top #2
62 But-2-ynoyl fluoride	C_4H_3FO	$CH_3C{\equiv}CC(F)=O$	LB02	2.20 ± 0.12	
63 *cis*-2-Butenenitrile	C_4H_5N	$CH_3C(H)=C(H)C{\equiv}N$	LB02	485.50 ± 0.25	
64 2-Methylacrylonitrile	C_4H_5N	$CH_2=C(CH_3)C{\equiv}N$	LB02	695.2 ± 2.1	
65 2-Methyloxazole	C_4H_5NO	$N=C(CH_3)OC(H)=C(H)$	LB02	251.70 ± 1.17	
66 4-Methyloxazole	C_4H_5NO	$N=C(H)OC(H)=C(CH_3)$	LB02	429.44 ± 0.33	

	Name	Molecular Formula	Line Formula	Ref.	V_3/cm^{-1}	Comments
67	5-Methyloxazole	C_4H_5NO	N=C(H)OC(CH$_3$)=C(H)	LB02	477.90 ± 1.34	
68	5-Methylisoxazole	C_4H_5NO	C(H)=NOC(CH$_3$)=C(H)	LB02	272.05 ± 1.00	
69	2-Methylthiazole	C_4H_5NS	N=C(CH$_3$)SC(H)=C(H)	GH02	34.938 ± 0.020	
70	4-Methylisothiazole	C_4H_5NS	N=C(H)C(CH$_3$)=C(H)S	LB02	105.767 ± 0.043	
71	4-Methyl-2-oxetanone	$C_4H_6O_2$	OC(=O)C(H$_2$)C(H)(CH$_3$)	LB02	1256.5 ± 10.5	
72	*trans*-1-Fluoro-2-butene	C_4H_7F	CH$_3$C(H)=C(H)CH$_2$F	LB02	596 ± 7	anticlinal conformer
73	1-Isocyanopropane	C_4H_7N	CH$_3$C(H$_2$)C(H$_2$)N≡C	LB02	1012.3 ± 8.4	*gauche* conformer
		C_4H_7N	CH$_3$C(H$_2$)C(H$_2$)N≡C	LB02	1033.8 ± 7.7	*trans* conformer
74	Isobutene	C_4H_8	(CH$_3$)$_2$C=CH$_2$	LB02	761.58 ± 1.05	
75	*cis*-2-Butene	C_4H_8	CH$_3$CH=CHCH$_3$	LB02	259.89 ± 0.42	
76	3-Methoxy-1-propene	C_4H_8O	CH$_3$OC(H$_2$)C(H)=CH$_2$	LB02	728.0 ± 10.5	*skew gauche* conformer
		C_4H_8O	CH$_3$OC(H$_2$)C(H)=CH$_2$	LB02	829.5 ± 10.5	*syn-trans* conformer
77	2,2-Dimethyloxirane	C_4H_8O	OC(CH$_3$)(CH$_3$)C(H$_2$)	LB02	945.61 ± 0.75	
78	*cis*-2,3-Dimethyloxirane	C_4H_8O	OC(H)(CH$_3$)C(H)(CH$_3$)	LB02	577.80 ± 1.84	*cis* conformer
		C_4H_8O	OC(H)(CH$_3$)C(H)(CH$_3$)	LB02	862.52 ± 1.84	*trans* conformer
79	2-Methyloxetane	C_4H_8O	OC(H$_2$)C(H$_2$)C(H)(CH$_3$)	LB02	1166.5 ± 4.9	
80	3-Methyloxetane	C_4H_8O	OC(H$_2$)C(H)(CH$_3$)C(H$_2$)	LB02	1149.4 ± 4.2	
81	3-Methoxythietane	C_4H_8OS	SC(H$_2$)C(H)(OCH$_3$)C(H$_2$)	LB02	1071.0 ± 10.5	
82	3-(Methylthio)-1-propene	C_4H_8S	CH$_3$SC(H$_2$)C(H)=CH$_2$	LB02	619 ± 28	
83	2,2-Dimethylthiirane	C_4H_8S	SC(CH$_3$)(CH$_3$)C(H$_2$)	LB02	1268.3 ± 3.0	
84	Butane	C_4H_{10}	CH$_3$C(H$_2$)C(H$_2$)CH$_3$	LB02	948 ± 24	
85	*N*-Methyl-*N*-nitrosopropylamine	$C_4H_{10}N_2O$	CH$_3$C(H$_2$)C(H$_2$)N(CH$_3$)N=O	LB02	320 ± 30	*N*-methyl top, conformer OMGA
86	Dihydro-3-methyl-2(3*H*)-furanone	$C_5H_8O_2$	OC(=O)C(H)(CH$_3$)C(H$_2$)C(H$_2$)	LB02	913.8 ± 2.5	
87	Dihydro-4-methyl-2(3*H*)-furanone	$C_5H_8O_2$	OC(=O)C(H$_2$)C(H)(CH$_3$)C(H$_2$)	CA96	1437.8 ± 8.4	
88	Dihydro-5-methyl-2(3*H*)-furanone	$C_5H_8O_2$	OC(=O)C(H$_2$)C(H$_2$)C(H)(CH$_3$)	CA96	1233.0 ± 4.2	
89	*tert*-Butyl isocyanate	C_5H_9NO	(CH$_3$)$_3$C≡N=C=O	LB02	41.510 ± 0.015	(CH$_3$)$_3$ C group
90	Methyl *tert*-butyl ether	$C_5H_{12}O$	(CH$_3$)$_3$COCH$_3$	LB02	498.6 ± 1.5	*O*-methyl top
91	2-Methylcyclopentanone	$C_6H_{10}O$	C(=O)C(H)(CH$_3$)C(H$_2$)C(H$_2$)C(H$_2$)	LB02	844.2 ± 2.4	
92	3-Methylcyclopentanone	$C_6H_{10}O$	C(=O)C(H$_2$)C(H)(CH$_3$)C(H$_2$)C(H$_2$)	LB02	1233.8 ± 1.7	
93	*tert*-Butyl ethyl ether	$C_6H_{14}O$	(CH$_3$)$_3$COC(H$_2$)CH$_3$	LB02	1025 ± 3	ethyl CH$_3$
94	2,4-Difluorotoluene	$C_7H_6F_2$	C(H)=C(CH$_3$)C(F)=C(H)C(F)=C(H)	LB02	204.04 ± 0.23	
95	2-Chlorotoluene	C_7H_7Cl	C(H)=C(H)C(Cl)=C(CH$_3$)C(H)=C(H)	ND06	513.8 ± 2.7	^{35}Cl
96	2,6-Dimethylpyridine	C_7H_9N	C(H)=C(H)C(CH$_3$)=NC(CH$_3$)=C(H)	LB02	98.24 ± 0.27	
97	1,2,2-Trimethylpropyl methylphosphonofluoridate	$C_7H_{16}FO_2P$	(CH$_3$)$_3$CC(H)(CH$_3$)OP(O)(F)CH$_3$	SD04	821 ± 5	*P*-methyl top, conformer GD-I
		$C_7H_{16}FO_2P$	(CH$_3$)$_3$CC(H)(CH$_3$)OP(O)(F)CH$_3$	SD04	738 ± 5	*P*-methyl top, conformer GD-II
98	Germyl azide	GeH_3N_3	GeH$_3$-N=N≡N	GA89	86.598 ± 0.062	
99	Silylphospine	H_5PSi	SiH$_3$PH$_2$	VR75	537.2 ± 14.0	

TABLE 2. Symmetric Top Potential Parameters

	Name	Molecular Formula	Line Formula	Ref.	V_3/cm^{-1}	Comments
1	Phosphine-trifluoroborane	BF_3H_3P	H_3PBF_3	OK75	1169 ± 123	
2	Trihydro(phosphorus trifluoride)boron	BF_3H_3P	F_3PBH_3	KL67	1134 ± 53	
3	Trihydro(phosphine)boron	BH_6P	H_3PBH_3	DL73	864.5 ± 17.5	
4	Trifluoro(trifluoromethyl)silane	CF_6Si	CF_3SiF_3	LJ72	489 ± 50	
5	Trifluoromethylgermane	CH_3F_3Ge	CF_3GeH_3	KW74	448 ± 53	
6	Trifluoromethylsilane	CH_3F_3Si	CH_3SiF_3	ST06	414.147 ± 0.030	
7	Methylgermane	CH_6Ge	CH_3GeH_3	L59	433.6 ± 8.8	
8	Methylsilane	CH_6Si	CH_3SiH_3	OM07	603.3878 ± 0.0037	
9	Methylstannane	CH_6Sn	CH_3SnH_3	CB61	227 ± 10	
10	1,1,1-Trifluoroethane	$C_2H_3F_3$	CH_3CF_3	WA02	1112.24 ± 0.16	
11	Ethane	C_2H_6	CH_3CH_3	OM07	1013.28 ± 0.10	V_6 = 8.798 ± 0.041
12	Ethane-1,1,1-d_3	$C_2H_3D_3$	CH_3CD_3	OM07	1001.876 ± 0.023	V_6 = 9.328 ± 0.018
13	Ethane-d_6	C_2D_6	CD_3CD_3	OM07	989.946 ± 0.090	V_6 = 9.51 ± 0.10
14	1-Silylpropyne	C_3H_6Si	$CH_3C{\equiv}CSiH_3$	NY85	3.77 ± 0.70	
15	Trimethylchlorosilane	C_3H_9ClSi	$(CH_3)_3SiCl$	MS02	576.9 ± 0.9	
16	2-Butyne	C_4H_6	$CH_3C{\equiv}CCH_3$	LB97	6.067 ± 0.040	V_6 = 0.1240 ± 0.0144 V_9 = −0.0916 ± 0.0180
17	Ethynyltrimethylgermane	$C_5H_{10}Ge$	$(CH_3)_3GeC{\equiv}CH$	VG96	376.2 ± 16.7	
18	Disilane	H_6Si_2	SiH_3SiH_3	BM07	412.033 ± 0.010	

BOND DISSOCIATION ENERGIES

Yu-Ran Luo

The bond dissociation energy (enthalpy) is also referred to as bond disruption energy, bond energy, bond strength, or binding energy (abbreviation: BDE, BE, or D). It is defined as the standard enthalpy change of the following fission: R–X → R + X. The BDE, denoted by $D^o(\text{R–X})$, is usually derived by the thermochemical equation, $D^o(\text{R–X}) = \Delta_f H^o(\text{R}) + \Delta_f H^o(\text{X}) - \Delta_f H^o(\text{RX})$. The enthalpy of formation $\Delta_f H^o$ of a large number of atoms, free radicals, ions, clusters and compounds is available from the websites of NIST, NASA, CODATA, and IUPAC. Most authors prefer to use the BDE values at 298.15 K.

The following seven tables provide essential information of experimental BDE values of R–X and R^+–X bonds.

(1) Table 1: Bond Dissociation Energies in Diatomic Molecules
(2) Table 2: Enthalpy of Formation of Gaseous Atoms
(3) Table 3: Bond Dissociation Energies in Polyatomic Molecules
(4) Table 4: Enthalpies of Formation of Free Radicals and Other Transient Species
(5) Table 5: Bond Dissociation Energies of Common Organic Molecules
(6) Table 6: Bond Dissociation Energies in Diatomic Cations
(7) Table 7: Bond Dissociation Energies in Polyatomic Cations

The data in these tables have been revised through September 2009.

TABLE 1. Bond Dissociation Energies in Diatomic Molecules

The BDEs in diatomic species have usually been measured by spectroscopy or mass spectrometry. In the absence of data on the enthalpy function, the values at 0 K, $D^o(\text{A–B})$, are converted to D^o_{298} by the approximate equation:

$$D^o_{298}(\text{A–B}) \approx D^o(\text{A–B}) + (3/2)RT = D^o(\text{A–B}) + 3.7181 \text{ kJ mol}^{-1}$$

This table has been arranged in an alphabetical order of the atoms A in the diatomics A–B.

A–B	D^o_{298}/kJ mol^{-1}	Ref.
Ac–O	794	1
Ag–Ag	162.9 ± 2.9	1
Ag–Al	183.7 ± 9.2	1
Ag–Au	202.5 ± 9.6	1
Ag–Bi	192 ± 42	1
Ag–Br	280.3 ± 1.3	1
Ag–Cl	279.1 ± 8.4	1
Ag–Cu	171.5 ± 9.6	1
Ag–D	226.8	1
Ag–Dy	130 ± 19	1
Ag–Eu	127 ± 13	1
Ag–F	356.9 ± 5.8	1
Ag–Ga	159 ± 17	1
Ag–Ge	174.5 ± 21	1
Ag–H	202.4 ± 9.1	1
Ag–Ho	124 ± 19	1
Ag–I	234 ± 29	1
Ag–In	166.5 ± 4.9	1
Ag–Li	186.1	1
Ag–Mn	99.2 ± 21	1
Ag–Na	133.1 ± 12.6	1
Ag–Nd	<213	1
Ag–O	221 ± 21	1
Ag–S	216.7 ± 14.6	1
Ag–Se	210.0 ± 14.6	1
Ag–Si	185.1 ± 9.6	1
Ag–Sn	136 ± 21	1
Ag–Te	195.8 ± 14.6	1
Al–Al	264.3 ± 0.5	1
Al–Ar	5.69	1
Al–As	202.7 ± 7.1	1
Al–Au	325.9 ± 6.3	1
Al–Br	429.2 ± 5.8	1
Al–C	267.7	1
Al–Ca	52.7	1
Al–Cl	502	1
Al–Co	181.6 ± 0.2	1
Al–Cr	222.9 ± 0.9	1
Al–Cu	227.1 ± 1.2	1
Al–D	290.4	1
Al–F	675	1
Al–H	288 ± 13	1
Al–I	369.9 ± 2.1	1
Al–Kr	6.05	1
Al–Li	76.1	1
Al–N	≤368 ± 15	1
Al–Ne	3.9	1
Al–Ni	224.7 ± 4.8	1
Al–O	501.9 ± 10.6	1
Al–P	216.7 ± 12.6	1
Al–Pd	254.4 ± 12.1	1
Al–S	332 ± 10	1
Al–Sb	216.3 ± 6	1
Al–Se	318 ± 13	1
Al–Si	246.9 ± 12.6	1
Al–Te	268 ± 13	1
Al–Ti	263.4	1
Al–U	326 ± 29	1
Al–V	147.4 ± 1.0	1
Al–Xe	7.39	1
Am–O	553 ± 36	1
Ar–Ar	4.91	1
Ar–B	4.62	1
Ar–Br	~5.0	1
Ar–C	5.158	1
Ar–Ca	4.44 ± 0.60	1
Ar–Cd	5.57 ± 0.05	1
Ar–Ga	3.96	1
Ar–Ge	<5.4	1
Ar–He	3.96	1
Ar–Hg	5.32	1
Ar–I	~5.3	1
Ar–In	4.18	1
Ar–Kr	5.11	1
Ar–Li	~7.82	1
Ar–Mg	~3.7	1
Ar–Na	~4.2	1
Ar–Ne	4.27	1
Ar–Si	5.86	1
Ar–Sn	<5.1	1
Ar–Tl	4.09	1
Ar–Xe	5.28	1
Ar–Zn	5.0	1
As–As	385.8 ± 10.5	1
As–Cl	448	1
As–D	270.3	1
As–F	410	1
As–Ga	202.5 ± 4.8	1
As–H	274.0 ± 2.9	1
As–I	296.6 ± 24	1
As–In	201 ± 10	1
As–N	489 ± 2.1	1
As–O	484 ± 8	1
As–P	433.5 ± 12.6	1
As–S	379.5 ± 6.3	1
As–Sb	330.5 ± 5.4	1
As–Se	96	1
As–Tl	198.3 ± 14.6	1
Au–Au	226.2 ± 0.5	1
Au–B	367.8 ± 10.5	1
Au–Ba	254.8 ± 10.0	1
Au–Be	237.7 ± 4.0	1
Au–Bi	293 ± 8.4	1
Au–Br	213 ± 21	1

A–B	D^o_{298}/kJ mol^{-1}	Ref.	A–B	D^o_{298}/kJ mol^{-1}	Ref.	A–B	D^o_{298}/kJ mol^{-1}	Ref.	A–B	D^o_{298}/kJ mol^{-1}	Ref.
Au–Ca	250.4 ± 4.0	1	B–H	345.2 ± 2.5	1	Bi–O	337.2 ± 12.6	1	Br–Sb	314 ± 59	1
Au–Ce	322 ± 18	1	B–I	361	1	Bi–P	281.7 ± 13	1	Br–Sc	444 ± 63	1
Au–Cl	280 ± 13	1	B–Ir	512.2 ± 17	1	Bi–Pb	142.4 ± 3.0	1	Br–Se	297 ± 84	1
Au–Co	218.0 ± 16.4	1	B–La	335 ± 63	1	Bi–S	315.5 ± 4.6	1	Br–Si	358.2 ± 8.4	1
Au–Cr	223.7 ± 28.9	1	B–N	377.9 ± 8.7	1	Bi–Sb	252.7 ± 3.9	1	Br–Sm	331.4	1
Au–Cs	253 ± 3.5	1	B–Ne	3.97	1	Bi–Se	280.3 ± 5.9	1	Br–Sn	337 ± 13	1
Au–Cu	227.1 ± 1.2	1	B–O	809	1	Bi–Sn	193 ± 13	1	Br–Sr	365	1
Au–D	322.2	1	B–P	347 ± 16.7	1	Bi–Te	232.2 ± 11.3	1	Br–T	372.77	1
Au–Dy	259 ± 24	1	B–Pd	351.5 ± 16.7	1	Bi–Tl	120.9 ± 12.6	1	Br–Tb	382.8	1
Au–Eu	245 ± 12	1	B–Pt	477.8 ± 16.7	1	Bk–O	598	1	Br–Th	364	1
Au–F	294.1	1	B–Rh	475.8 ± 21	1	Br–Br	193.859 ± 0.120	1	Br–Ti	373	1
Au–Fe	187.0 ± 19.3	1	B–Ru	446.9 ± 21	1	Br–C	318.0 ± 8.4	1	Br–Tl	331 ± 21	1
Au–Ga	290 ± 15	1	B–S	577 ± 9.2	1	Br–Ca	339	1	Br–Tm	299.1	1
Au–Ge	273.2 ± 14.6	1	B–Sc	272 ± 63	1	Br–Cd	159 ± 96	1	Br–U	377 ± 15	1
Au–H	300.5 ± 2.6	4	B–Se	462 ± 14.6	1	Br–Ce	373.2	1	Br–V	439 ± 42	1
Au–Ho	267 ± 35	1	B–Si	317 ± 12	1	Br–Cl	219.32 ± 0.05	1	Br–W	329.3	1
Au–I	276	1	B–Te	354 ± 20	1	Br–Co	326 ± 42	1	Br–Xe	5.94 ± 0.02	1
Au–In	286.0 ± 5.7	1	B–Th	297 ± 33	1	Br–Cr	328.0 ± 24.3	1	Br–Y	481 ± 84	1
Au–La	457 ± 28	1	B–Ti	272 ± 63	1	Br–Cs	389.1 ± 4.2	1	Br–Yb	295.4	1
Au–Li	284.5 ± 6.7	1	B–U	322 ± 33	1	Br–Cu	331 ± 25	1	Br–Zn	138 ± 29	1
Au–Lu	332 ± 19	1	B–Y	289 ± 63	1	Br–D	370.74	1	Br–Zr	420	1
Au–Mg	179.1 ± 2.7	1	Ba–Br	402	1	Br–Dy	339.3 ± 10.5	1	C–C	618.3 ± 15.4	1
Au–Mn	197.7 ± 21	1	Ba–Cl	443	1	Br–Er	361.3	1	C–Ce	443 ± 30	1
Au–Na	215.1 ± 12.6	1	Ba–D	≤193.7	1	Br–Eu	548	1	C–Cl	394.9 ± 13.4	1
Au–Nd	294 ± 29	1	Ba–F	580.6	1	Br–F	280 ± 12	1	C–D	341.4	1
Au–Ni	247 ± 16.4	1	Ba–H	192.0	1	Br–Fe	243 ± 84	1	C–F	513.8 ± 10.0	1
Au–O	223 ± 21	1	Ba–I	322.6 ± 6.3	1	Br–Ga	402 ± 13	1	C–Fe	376.3 ± 28.9	1
Au–Pb	133 ± 42	1	Ba–O	562 ± 13.4	1	Br–Gd	372.0	1	C–Ge	455.7 ± 11	1
Au–Pd	142.7 ± 21	1	Ba–Pd	221.8 ± 5.0	1	Br–Ge	347 ± 8	1	C–H	338.4 ± 1.2	1
Au–Pr	311 ± 25	1	Ba–Rh	259.4 ± 25	1	Br–H	366.16 ± 0.20	1	C–Hf	540 ± 25	1
Au–Rb	243 ± 3.5	1	Ba–S	418 ± 21	1	Br–Hg	74.9	1	C–I	253.1 ± 35.6	1
Au–Rh	232.6 ± 29	1	Be–Be	59	1	Br–Ho	321.8	1	C–Ir	631 ± 5	1
Au–S	253.6 ± 14.6	1	Be–Br	316	1	Br–I	179.1 ± 0.4	1	C–La	463 ± 20	1
Au–Sc	280 ± 40	1	Be–Cl	434	1	Br–In	409 ± 10	1	C–Mo	482 ± 16	1
Au–Se	251.0 ± 14.6	1	Be–D	203.1	1	Br–K	379.1 ± 4.2	1	C–N	750.0 ± 2.9	1
Au–Si	304.6 ± 6.0	1	Be–F	573	1	Br–La	446.2	1	C–Nb	523.8 ± 14.5	1
Au–Sn	256.5 ± 7.2	1	Be–H	221	1	Br–Li	418.8 ± 4.2	1	C–Ni	337.0	1
Au–Sr	264 ± 42	1	Be–I	261	1	Br–Lu	301.5	1	C–O	1076.38 ± 0.67	1
Au–Tb	285 ± 33	1	Be–O	437	1	Br–Mg	317.96	1	C–Os	608 ± 25	1
Au–Te	237.2 ± 14.6	1	Be–S	372 ± 59	1	Br–Mn	314.2 ± 9.6	1	C–P	507.5 ± 8.8	1
Au–U	318 ± 29	1	Be–T	204.4	1	Br–Mo	313.4	1	C–Pd	436 ± 20	1
Au–V	246.0 ± 8.7	1	Bi–Bi	204.4	1	Br–N	280.8 ± 21	1	C–Pt	577.8 ± 6.8	13
Au–Y	310 ± 12	1	Bi–Br	240.2	1	Br–Na	363.1 ± 4.2	1	C–Rh	580 ± 4	1
B–B	290	1	Bi–Cl	300.4 ± 4.2	1	Br–Nd	339.7	1	C–Ru	648 ± 13	1
B–Br	390.9 ± 0.5	1	Bi–D	283.7	1	Br–Ni	360 ± 13	1	C–S	713.3 ± 1.2	1
B–C	448 ± 29	1	Bi–F	366.5 ± 12.5	1	Br–O	237.6 ± 0.4	1	C–Sc	444 ± 21	1
B–Cd	301.0	1	Bi–Ga	158.6 ± 16.7	1	Br–P	≤329	1	C–Se	590.4 ± 5.9	1
B–Ce	305 ± 21	1	Bi–H	≤283.3	1	Br–Pb	248.5 ± 14.6	1	C–Si	447	1
B–Cl	427	1	Bi–I	186.1 ± 5.8	1	Br–Pr	344.5	1	C–Tc	564 ± 29	1
B–D	341.0 ± 6.3	1	Bi–In	153.6 ± 1.7	1	Br–Rb	380.7 ± 4.2	1	C–Th	453 ± 17	1
B–F	732	1	Bi–Li	149.4	1	Br–S	218 ± 17	1	C–Ti	423 ± 30	1

A–B	D^o_{298}/kJ mol^{-1}	Ref.
C–U	455 ± 15	1
C–V	423 ± 24	1
C–Y	418 ± 14	1
C–Zr	495.8 ± 38.6	1
Ca–Ca	16.52 ± 0.11	1
Ca–Cl	409 ± 8.7	1
Ca–D	≤169.9	1
Ca–F	529	1
Ca–H	223.8	1
Ca–I	284.7 ± 8.4	1
Ca–Kr	5.15 ± 0.72	1
Ca–Li	84.9 ± 8.4	1
Ca–O	383.3 ± 5.0	1
Ca–Pd	347 - 360	1
Ca–S	335 ± 21	1
Ca–Xe	7.31 ± 0.96	1
Cd–Cd	7.36	1
Cd–Cl	208.4	1
Cd–F	305 ± 21	1
Cd–H	69.0 ± 0.4	1
Cd–I	97.2 ± 2.1	1
Cd–In	134	1
Cd–K	7.8	1
Cd–Kr	5.17	1
Cd–Na	10.2	1
Cd–Ne	3.97	1
Cd–O	236 ± 84	1
Cd–S	208.5 ± 20.9	1
Cd–Se	127.6 ± 25.1	1
Cd–Te	100.0 ± 15.1	1
Cd–Xe	6.54	1
Ce–Ce	251.7	1
Ce–Cl	457.0	1
Ce–F	582 ± 42	1
Ce–I	333.8	1
Ce–Ir	575 ± 9	1
Ce–N	519 ± 21	1
Ce–O	790	1
Ce–Os	524 ± 20	1
Ce–Pd	319 ± 21	1
Ce–Pt	550 ± 5	1
Ce–Rh	545 ± 7	1
Ce–Ru	494 ± 12	1
Ce–S	569	1
Ce–Se	494.5 ± 14.6	1
Ce–Te	189.4 ± 12.6	1
Cf–O	498	1
Cl–Cl	436.303 ± 0.011	8
Cl–Co	343.9	11
Cl–Cr	380.3	11
Cl–Cs	445.7 ± 7.7	1

A–B	D^o_{298}/kJ mol^{-1}	Ref.
Cl–Cu	377.8 ± 7.5	1
Cl–D	436.303 ± 0.011	1
Cl–Dy	392.4	1
Cl–Er	448.6	1
Cl–Eu	405.5	1
Cl–F	260.83	1
Cl–Fe	335.5	11
Cl–Ga	463 ± 13	1
Cl–Gd	451.0	1
Cl–Ge	390.8 ± 9.6	1
Cl–H	431.361 ± 0.013	1
Cl–Hg	92.0 ± 9.2	1
Cl–Ho	409.1	1
Cl–I	211.3 ± 0.4	1
Cl–In	436 ± 8	1
Cl–K	433.0 ± 8.4	1
Cl–La	521.6	1
Cl–Li	469 ± 13	1
Cl–Lu	325.7 ± 2	1
Cl–Mg	312	1
Cl–Mn	337.6	11
Cl–N	333.9 ± 9.6	1
Cl–Na	412.1 ± 8.4	1
Cl–Nd	418.7	1
Cl–Ni	372.3	11
Cl–O	267.47 ± 0.08	1
Cl–P	≤376	1
Cl–Pb	301 ± 50	1
Cl–Pr	423.5	1
Cl–Ra	343 ± 75	1
Cl–Rb	427.6 ± 8.4	1
Cl–S	241.8	1
Cl–Sb	360 ± 50	1
Cl–Sc	331	1
Cl–Se	322	1
Cl–Si	416.7 ± 6.3	1
Cl–Sm	418.7	1
Cl–Sn	350 ± 8	1
Cl–Sr	409	1
Cl–T	438.64	1
Cl–Ta	544	1
Cl–Tb	470.1	1
Cl–Th	489	1
Cl–Ti	405.4 ± 10.5	1
Cl–Tl	372.8 ± 2.1	1
Cl–Tm	378.0	1
Cl–U	439	1
Cl–V	477 ± 63	1
Cl–W	419	1
Cl–Xe	7.08	1
Cl–Y	523 ± 84	1

A–B	D^o_{298}/kJ mol^{-1}	Ref.
Cl–Yb	374.5	1
Cl–Zn	229 ± 8	1
Cl–Zr	530	1
Cm–O	710 ± 45	15
Co–Co	<127	1
Co–Cu	161.1 ± 16.4	1
Co–D	270.2 ± 5.8	1
Co–F	431 ± 63	1
Co–Ge	230 ± 21	1
Co–H	244.9 ± 4.8	1
Co–I	280 ± 21	1
Co–Mn	50 ± 8	1
Co–Nb	267.02 ± 0.10	1
Co–O	397.4 ± 8.7	1
Co–S	331	1
Co–Sc	240.1	7
Co–Si	274.4 ± 17	1
Co–Ti	235.37 ± 0.10	1
Co–Y	253.71 ± 0.10	1
Co–Zr	306.39 ± 0.10	1
Cr–Cr	152.0 ± 6	1
Cr–Cu	154.4 ± 14.5	1
Cr–F	523 ± 19	1
Cr–Fe	~75	1
Cr–Ge	154 ± 7	1
Cr–H	189.9 ± 6.7	1
Cr–I	287.0 ± 24.3	1
Cr–N	377.8 ± 18.8	1
Cr–Nb	295.72 ± 0.06	1
Cr–O	461 ± 8.7	1
Cr–Pb	105 ± 2	1
Cr–S	331	1
Cr–Sn	141 ± 3	1
Cs–Cs	43.919 ± 0.010	1
Cs–F	517.1 ± 7.7	1
Cs–H	175.364	1
Cs–Hg	8	1
Cs–I	338.5 ± 2.1	1
Cs–Li	72.9 ± 1.2	5
Cs–Na	63.2 ± 1.3	1
Cs–O	293 ± 25	1
Cs–Rb	49.57 ± 0.01	1
Cu–Cu	201	1
Cu–D	270.3	1
Cu–Dy	144 ± 18	1
Cu–F	414	1
Cu–Ga	215.9 ± 15	1
Cu–Ge	208.8 ± 21	1
Cu–H	254.8 ± 6	1
Cu–Ho	144 ± 19	1
Cu–I	289 ± 63	1

A–B	D^o_{298}/kJ mol^{-1}	Ref.
Cu–In	187.4 ± 7.9	1
Cu–Li	191.9	1
Cu–Na	176.1 ± 16.7	1
Cu–Ni	201.7 ± 9.6	1
Cu–O	287.4 ± 11.6	1
Cu–S	274.5 ± 14.6	1
Cu–Se	255.2 ± 14.6	1
Cu–Si	221.3 ± 6.3	1
Cu–Sn	170 ± 10	1
Cu–Tb	191 ± 18	1
Cu–Te	230.5 ± 14.6	1
D–D	443.3197 ± 0.0003	1
D–F	576.236 ± 0.011	1
D–Ga	≤276.5	1
D–Ge	≤322	1
D–H	439.2223 ± 0.0002	1
D–Hg	42.05	1
D–I	302.33	1
D–In	246	1
D–K	182.4	1
D–Li	240.24	1
D–Lu	302	1
D–Mg	161.33 ± 0.33	1
D–Mn	312 ± 6	1
D–N	341.6	1
D–Ni	≤302.9	1
D–O	429.64	1
D–P	299.0	1
D–Pt	≤350.2	1
D–S	350.62 ± 1.20	1
D–Si	302.5	1
D–Sr	167.7	1
D–T	444.91	1
D–Tl	193.0	1
D–Zn	88.7	1
Dy–Dy	70.3	1
Dy–F	531	1
Dy–I	269.0 ± 8.4	1
Dy–O	615	1
Dy–S	414 ± 42	1
Dy–Se	322 ± 20	1
Dy–Te	234 ± 20	1
Er–Er	75 ± 29	1
Er–F	565 ± 17	1
Er–I	315.8	1
Er–O	606	1
Er–S	418 ± 21	1
Er–Se	326 ± 20	1
Er–Te	238 ± 20	1
Es–O	460	1
Eu–Eu	45.2	1

A–B	D^o_{298}/kJ mol^{-1}	Ref.
Eu–F	544	1
Eu–I	288.3	1
Eu–Li	268.1 ± 12.6	1
Eu–O	473	1
Eu–Rh	238 ± 34	1
Eu–S	365.7 ± 13.4	1
Eu–Se	302.9 ± 14.6	1
Eu–Te	251.0 ± 14.6	1
F–F	158.670 ± 0.096	1
F–Fe	447	1
F–Ga	584 ± 13	1
F–Gd	590 ± 17	1
F–Ge	523 ± 13	1
F–H	569.680 ± 0.011	1
F–Hf	650 ± 15	1
F–Hg	~180	1
F–Ho	540	1
F–I	≤271.5	1
F–In	516 ± 13	1
F–K	489.2	1
F–Kr	6.6	1
F–La	659.0 ± 17.2	1
F–Li	577 ± 21	1
F–Lu	405 ± 19	1
F–Mg	463	1
F–Mn	445.2 ± 7.5	1
F–Mo	464	1
F–N	≤349	1
F–Na	477.3	1
F–Nd	545.2 ± 12.6	1
F–Ni	439.7 ± 5.9	2
F–Np	430 ± 50	1
F–O	220	1
F–P	≤405	1
F–Pb	355 ± 13	1
F–Pr	582 ± 46	1
F–Pu	538 ± 29	1
F–Rb	494 ± 21	1
F–Ru	402	1
F–S	343.5 ± 6.7	1
F–Sb	439 ± 96	1
F–Sc	599.1 ± 13.4	1
F–Se	339 ± 42	1
F–Si	576.4 ± 17	1
F–Sm	565	1
F–Sn	476 ± 8	1
F–Sr	538	1
F–T	579.009 ± 0.108	1
F–Ta	573 ± 13	1
F–Tb	561 ± 42	1
F–Th	652	1
F–Ti	569 ± 33	1
F–Tl	439 ± 21	1
F–Tm	510	1
F–U	648	1
F–V	590 ± 63	1
F–W	≤544	1
F–Xe	14.18	1
F–Y	685.3 ± 13.4	1
F–Yb	≥517.6 ± 9.6	1
F–Zn	364 ± 63	1
F–Zr	627.2 ± 10.5	1
Fe–Fe	118	1
Fe–Ge	210.9 ± 29	1
Fe–H	148 ± 3	1
Fe–I	123	1
Fe–O	407.0 ± 1.0	1
Fe–S	328.9 ± 14.6	1
Fe–Si	297 ± 25	1
Fm–O	443	1
Ga–Ga	<106.4	1
Ga–H	265.9 ± 5.9	4
Ga–I	334 ± 13	1
Ga–In	94.0 ± 3	1
Ga–Kr	4.08	1
Ga–Li	133.1 ± 14.6	1
Ga–O	374 ± 21	1
Ga–P	229.7 ± 12.6	1
Ga–Sb	192.0 ± 12.6	1
Ga–Te	265 ± 21	1
Ga–Xe	5.27	1
Gd–Gd	206.3 ± 67.5	1
Gd–I	333.8	1
Gd–O	715	1
Gd–S	526.8 ± 10.5	1
Gd–Se	430 ± 15	1
Gd–Te	341 ± 15	1
Ge–Ge	264.4 ± 6.8	1
Ge–H	263.2 ± 4.8	1
Ge–I	268 ± 25	1
Ge–Ni	290.3 ± 10.9	1
Ge–O	657.5 ± 4.6	4
Ge–Pb	145.3 ± 6.9	6
Ge–Pd	254.7 ± 10.5	1
Ge–S	534 ± 3	1
Ge–Sc	270 ± 11	1
Ge–Se	484.7 ± 1.7	1
Ge–Si	297	1
Ge–Sn	230.1 ± 13	1
Ge–Te	396.7 ± 3.3	1
Ge–Y	279 ± 11	1
H–H	435.7799 ± 0.0001	1
H–Hg	39.844	1
H–I	298.26 ± 0.10	1
H–In	243.1	1
H–K	174.576	1
H–Li	238.039 ± 0.006	1
H–Mg	127.18 ± 0.006	10
H–Mn	251 ± 5	1
H–Mo	202.5 ± 18.3	9
H–N	≤338.9	1
H–Na	185.69 ± 0.29	1
H–Nb	>221.9 ± 9.6	1
H–Ni	240 ± 8	1
H–O	429.91 ± 0.29	1
H–P	297.0 ± 2.1	1
H–Pb	≤157	1
H–Pd	234 ± 25	1
H–Pt	330	1
H–Rb	172.6	1
H–Rh	241.0 ± 5.9	1
H–Ru	223 ± 15	1
H–S	353.57 ± 0.30	1
H–Sb	239.7 ± 4.2	1
H–Sc	205 ± 17	1
H–Se	312.5	1
H–Si	293.3 ± 1.9	1
H–Sn	264 ± 17	1
H–Sr	164 ± 8	1
H–T	440.49	1
H–Te	270.7 ± 1.7	1
H–Ti	204.6 ± 8.8	1
H–Tl	195.4 ± 4	1
H–V	209.3 ± 6.8	1
H–Yb	183.1 ± 2.0	1
H–Zn	85.8 ± 2	1
He–He	3.809	1
He–Hg	3.8	1
He–Xe	3.8	1
Hf–Hf	328 ± 58	1
Hf–N	535 ± 30	1
Hf–O	801 ± 13	1
Hg–Hg	8.10 ± 0.18	1
Hg–I	34.69 ± 0.96	1
Hg–K	8.8	1
Hg–Kr	5.75	1
Hg–Li	13.16 ± 0.38	1
Hg–Na	10.8	1
Hg–Ne	4.14	1
Hg–O	269	1
Hg–Rb	8.4	1
Hg–S	217.3 ± 22.2	1
Hg–Se	144.3 ± 30.1	1
Hg–T	43.14	1
Hg–Te	<142	1
Hg–Tl	2.9	1
Hg–Xe	6.65	1
Hg–Zn	7.3	1
Ho–Ho	70.3	1
Ho–I	275.1	1
Ho–O	606	1
Ho–S	428.4 ± 14.6	1
Ho–Se	333 ± 15	1
Ho–Te	≤259 ± 15	1
I–I	152.25 ± 0.57	1
I–In	306.9 ± 1.1	1
I–K	322.5 ± 2.1	1
I–Kr	5.67	1
I–La	411.7	1
I–Li	345.2 ± 4.2	1
I–Lu	263.2	1
I–Mg	229	1
I–Mn	282.8 ± 9.6	1
I–Mo	266.9	1
I–N	159 ± 17	1
I–Na	304.2 ± 2.1	1
I–Nd	301.5	1
I–Ni	293 ± 21	1
I–O	233.4 ± 1.3	12
I–Pb	194 ± 38	1
I–Pr	306.2	1
I–Rb	318.8 ± 2.1	1
I–Si	243.1 ± 8.4	1
I–Sm	293.1	1
I–Sn	235 ± 3	1
I–Sr	301	1
I–Tb	336.2	1
I–Te	192 ± 42	1
I–Th	361 ± 25	1
I–Ti	306	1
I–Tl	285 ± 21	1
I–Tm	260.8	1
I–U	299 ± 27	1
I–Xe	~6.9	1
I–Y	422.6 ± 12.5	1
I–Yb	257.3	1
I–Zn	153.1 ± 6.3	1
I–Zr	127	1
In–In	82.0 ± 5.7	1
In–Kr	4.85	1
In–Li	92.5 ± 14.6	1
In–O	346 ± 30	1
In–P	197.9 ± 8.4	1
In–S	287.9 ± 14.6	1

A–B	D^o_{298}/kJ mol^{-1}	Ref.
In–Sb	151.9 ± 10.5	1
In–Se	245.2 ± 14.6	1
In–Te	215.5 ± 14.6	1
In–Xe	6.48	1
In–Zn	32.2	1
Ir–Ir	361 ± 68	1
Ir–La	577 ± 12	1
Ir–Nb	465 ± 25	1
Ir–O	414 ± 42	1
Ir–Si	462.8 ± 21	1
Ir–Th	574 ± 42	1
Ir–Ti	422 ± 13	1
Ir–Y	457 ± 15	1
K–K	56.96	1
K–Kr	4.6	1
K–Li	82.0 ± 4.2	1
K–Na	65.994 ± 0.008	1
K–Zn	6.5	1
K–O	271.5 ± 12.6	1
K–Rb	53.723 ± 0.005	1
K–Xe	5.0	1
Kr–Kr	5.39	1
Kr–Li	~12.1	1
Kr–Mg	6.71 ± 0.96	1
Kr–Na	~4.53	1
Kr–Ne	4.31	1
Kr–O	<8	1
Kr–Tl	4.14	1
Kr–Xe	5.66	1
Kr–Zn	5.0	1
La–La	244.9	1
La–N	519 ± 42	1
La–O	798	1
La–Pt	505 ± 12	1
La–Rh	550 ± 12	1
La–S	573.4 ± 1.7	1
La–Se	485.7 ± 14.6	1
La–Te	385.6 ± 15	1
La–Y	197 ± 21	1
Li–Li	105.0	1
Li–Mg	67.4 ± 6.3	1
Li–Na	87.181 ± 0.001	1
Li–O	340.5 ± 6.3	1
Li–Pb	78.7 ± 8	1
Li–S	312.5 ± 7.5	1
Li–Sb	169.0 ± 10.0	1
Li–Si	149	1
Li–Sm	193.3 ± 18.8	1
Li–Tm	276.1 ± 14.6	1
Li–Xe	~12.1	1
Li–Yb	143.5 ± 12.6	1
Lr–O	665	1
Lu–Lu	142 ± 33	1
Lu–O	669	1
Lu–Pt	402 ± 34	1
Lu–S	508.4 ± 14.4	1
Lu–Se	418 ± 15	1
Lu–Te	325 ± 15	1
Md–O	418	1
Mg–Mg	11.3	1
Mg–Ne	~4.1	1
Mg–O	358.2 ± 7.2	1
Mg–S	234	1
Mg–Xe	9.70 ± 1.79	1
Mn–Mn	61.6 ± 9.6	1
Mn–O	362 ± 25	1
Mn–S	301 ± 17	1
Mn–Se	239.3 ± 9.2	1
Mo–Mo	435.5 ± 1.0	1
Mo–Nb	452 ± 25	1
Mo–O	502	1
N–N	944.84 ± 0.10	1
N–O	631.62 ± 0.18	1
N–P	617.1 ± 20.9	1
N–Pt	374.2 ± 9.6	1
N–Pu	469 ± 63	1
N–S	467 ± 24	1
N–Sb	460 ± 84	1
N–Sc	464 ± 84	1
N–Si	437.1 ± 9.9	1
N–Ta	607 ± 84	1
N–Th	577 ± 33	1
N–Ti	476 ± 33	1
N–U	531 ± 21	1
N–V	523 ± 38	1
N–Xe	26.9	1
N–Y	477 ± 63	1
N–Zr	565 ± 25	1
Na–Na	74.805 ± 0.586	1
Na–Ne	~3.8	1
Na–O	270 ± 4	1
Na–Rb	63.887 ± 0.024	1
Na–Xe	~5.12	1
Nb–Nb	513	1
Nb–Ni	271.9 ± 0.1	1
Nb–O	726.5 ± 10.6	1
Nb–Ti	302.0 ± 0.1	1
Nb–V	369.3 ± 0.1	1
Nd–Nd	82.8	1
Nd–O	703	1
Nd–S	471.5 ± 14.6	1
Nd–Se	393.9	1
Nd–Te	305 ± 15	1
Ne–Ne	4.070	1
Ne–Xe	4.31	1
Ne–Zn	3.92	1
Ni–Ni	204	1
Ni–O	366 ± 30	1
Ni–Pd	140.9	1
Ni–Pt	273.7 ± 0.3	1
Ni–S	356 ± 21	1
Ni–Si	318 ± 17	1
Ni–V	206.3 ± 0.2	1
Ni–Y	283.92 ± 0.10	1
Ni–Zr	279.8 ± 0.1	1
No–O	268	1
Np–O	731	1
O–O	498.36 ± 0.17	1
O–Os	575	1
O–P	589	1
O–Pa	792	1
O–Pb	382.4 ± 3.3	4
O–Pd	238.1 ± 12.6	1
O–Pr	740	1
O–Pt	418.6 ± 11.6	10
O–Pu	656.1	1
O–Rb	276 ± 12.6	1
O–Re	627 ± 84	1
O–Rh	405 ± 42	1
O–Ru	528 ± 42	1
O–S	517.90 ± 0.05	1
O–Sb	434 ± 42	1
O–Sc	671.4 ± 1.0	1
O–Se	429.7 ± 6.3	1
O–Si	799.6 ± 13.4	1
O–Sm	573	1
O–Sn	528	1
O–Sr	426.3 ± 6.3	1
O–Ta	839	1
O–Tb	694	1
O–Tc	548	1
O–Te	377 ± 21	1
O–Th	877	1
O–Ti	666.5 ± 5.6	1
O–Tl	213 ± 84	1
O–Tm	514	1
O–U	755	1
O–V	637	1
O–W	720 ± 71	1
O–Xe	36.4	1
O–Y	714.1 ± 10.2	1
O–Yb	387.7 ± 10	1
O–Zn	≤250	1
O–Zr	766.1 ± 10.6	1
Os–Os	415 ± 77	1
P–P	489.1	1
P–Pt	≤416.7 ± 16.7	1
P–Rh	353.1 ± 16.7	1
P–S	442 ± 10	1
P–Sb	356.9 ± 4.2	1
P–Se	363.7 ± 10.0	1
P–Si	363.6	1
P–Te	297.9 ± 10.0	1
P–Th	372 ± 29	1
P–Tl	209 ± 13	1
P–U	293 ± 21	1
P–W	305 ± 4	1
Pb–Pb	86.6 ± 0.8	1
Pb–S	398	1
Pb–Sb	161.5 ± 10.5	1
Pb–Se	302.9 ± 4.2	1
Pb–Si	168.8 ± 7.3	6
Pb–Te	249.8 ± 10.5	1
Pd–Pd	>136	1
Pd–Pt	191.0	1
Pd–Si	261 ± 12	1
Pd–Y	241 ± 15	1
Po–Po	187	1
Pr–Pr	129.1	1
Pr–S	492.5 ± 4.6	1
Pr–Se	446.4 ± 23.0	1
Pr–Te	326 ± 20	1
Pt–Pt	306.7 ± 1.9	1
Pt–Si	501 ± 18	1
Pt–Th	551 ± 42	1
Pt–Ti	397.5 ± 10.6	1
Pt–Y	474 ± 12	1
Rb–Rb	48.898 ± 0.005	1
Re–Re	432 ± 30	1
Rh–Rh	235.85 ± 0.05	1
Rh–Sc	444 ± 11	1
Rh–Si	395.0 ± 18.0	1
Rh–Th	513 ± 21	1
Rh–Ti	390.8 ± 14.6	1
Rh–U	519 ± 17	1
Rh–V	364 ± 29	1
Rh–Y	446 ± 11	1
Ru–Ru	193.0 ± 19.3	1
Ru–Si	397.1 ± 21	1
Ru–Th	592 ± 42	1
Ru–V	414 ± 29	1
S–S	425.30	1
S–Sb	378.7	1
S–Sc	478.2 ± 12.6	1

A–B	D^o_{298}/kJ mol^{-1}	Ref.	A–B	D^o_{298}/kJ mol^{-1}	Ref.	A–B	D^o_{298}/kJ mol^{-1}	Ref.	A–B	D^o_{298}/kJ mol^{-1}	Ref.
S–Se	371.1 ± 6.7	1	Sb–Tl	126.7 ± 10.5	1	Si–Te	429.2	3	Ti–Ti	117.6	1
S–Si	617 ± 5	1	Sc–Sc	163 ± 21	1	Si–Y	258 ± 17	1	Ti–V	203.2 ± 0.1	1
S–Sm	389	1	Sc–Se	385 ± 17	1	Sm–Sm	54 ± 21	1	Ti–Zr	214.3 ± 0.1	1
S–Sn	467	1	Sc–Si	227.2 ± 14	1	Sm–Te	272.4 ± 14.6	1	Tl–Tl	59.4	1
S–Sr	338.5 ± 16.7	1	Sc–Te	289 ± 17	1	Sn–Sn	187.1 ± 0.3	1	Tl–Xe	4.18	1
S–Ta	669.5 ± 13.5	1	Se–Se	330.5	1	Sn–Te	338.1 ± 6.3	1	Tm–Tm	54 ± 17	1
S–Tb	515 ± 42	1	Se–Si	538 ± 13	1	Sr–Sr	16.64 ± 1.12	1	U–U	222 ± 21	1
S–Te	335 ± 42	1	Se–Sm	331.0 ± 14.6	1	T–T	446.67	1	V–V	269.3 ± 0.1	1
S–Ti	418 ± 3	1	Se–Sn	401.2 ± 5.9	1	Ta–Ta	390 ± 96	1	V–Zr	260.6 ± 0.3	1
S–Tm	368 ± 21	1	Se–Sr	251.0 ± 12.6	1	Tb–Tb	138.8	1	W–W	666	1
S–U	528.4 ± 10.5	1	Se–Tb	423 ± 20	1	Tb–Te	339 ± 42	1	Xe–Xe	6.023	1
S–V	449.4 ± 14.6	1	Se–Te	293.3	1	Tc–Tc	330	1	Y–Y	~270 ± 39	1
S–Y	528.4 ± 10.5	1	Se–Ti	381 ± 42	1	Te–Te	257.6 ± 4.1	1	Yb–Yb	16.3	1
S–Yb	167	1	Se–Tm	274 ± 40	1	Te–Ti	289 ± 17	1	Zn–Zn	22.2 ± 6.3	1
S–Zn	224.8 ± 12.6	1	Se–V	347 ± 21	1	Te–Tm	182 ± 40	1	Zr–Zr	298.2 ± 0.1	1
S–Zr	572.0 ± 11.6	1	Se–Y	435 ± 13	1	Te–Y	339 ± 13	1			
Sb–Sb	301.7 ± 6.3	1	Se–Zn	170.7 ± 25.9	1	Te–Zn	117.6 ± 18.0	1			
Sb–Te	277.4 ± 3.8	1	Si–Si	310	1	Th–Th	≤289 ± 33	1			

References

1. Luo, Y. R. *Comprehensive Handbook of Chemical Bond Energies*, CRC Press, Boca Raton, FL, 2007.
2. Hildenbrand, D. L., and Lau, K.H., *J. Phys. Chem. A* 110, 11886, 2006.
3. Chattopadhyaya, S., Pramanik, A., Banerjee, A., and Das, K. K., *J. Phys. Chem. A* 110, 12303, 2006.
4. Brutti, S., Balducci, G., and Gigli, G., *Rapid Commun. Mass Spectrom.* 21, 89, 2007.
5. Staanum, P., Pashov, A., Knöckel, H., and Tiemann, E., *Phys. Rev. A* 75, 042513, 2007.
6. Ciccioli, A., Gigli, G., Meloni, G., and Testani, E., *J. Chem. Phys.* 127, 054303/1, 2007.
7. Nagarajan, R., and Morse, M. D., *J. Chem. Phys.* 127, 074304/1, 2007.
8. Li, J., Hao, Y., Yang, J., Zhou, C., and Mo, Y., *J. Chem. Phys.* 127, 104307/1, 2007.
9. Armentrout, P. B., *Organometallics* 26, 5473, 2007.
10. Shayesteh, A., Henderson, R. D. E., Le Roy, R. J., and Bernath, P. F., *J. Phys. Chem. A* 111, 12495, 2007.
11. Hildenbrand, D. H., *J. Phys. Chem. A* 112, 3813, 2008.
12. Dooley, K. S., Geidosch, J. N., and North, S. W., *Chem. Phys. Lett.* 457, 303, 2008.
13. Citir, M., Metz, R. B., Belau, L., and Ahmed, M., *J. Phys. Chem. A* 112, 9584, 2008.
14. Hildenbrand, D. L., Lau, K. H., Perez-Mariano, J., and Sanjurjo, A., *J. Phys. Chem. A* 112, 9978, 2008.
15. Gibson, J. K., Haire, R. G., Santos, M., Pires de Matos, A., and Marçalo, J., *J. Phys. Chem. A* 112, 11373, 2008.

TABLE 2. Enthalpy of Formation of Gaseous Atoms

Atom	$\Delta_f H^o_{298}$/kJ mol^{-1}	Ref.	Atom	$\Delta_f H^o_{298}$/kJ mol^{-1}	Ref.	Atom	$\Delta_f H^o_{298}$/kJ mol^{-1}	Ref.	Atom	$\Delta_f H^o_{298}$/kJ mol^{-1}	Ref.
Ac	406	5	Cr	397.48 ± 4.2	3	La	431.0 ± 2.1	4	Pu	345	6
Ag	284.9 ± 0.8	2	Cs	76.5 ± 1.0	2	Li	159.3 ± 1.0	2	Ra	159	5
Al	330.9 ± 4.0	2	Cu	337.4 ± 1.2	2	Lu	427.6 ± 2.1	4	Rb	80.9 ± 0.8	2
Am	284	6	Dy	290.4 ± 2.1	4	Mg	147.1 ± 0.8	2	Re	774 ± 6.3	1
As	302.5 ± 13	1	Er	316.4 ± 2.1	4	Mn	283.3 ± 4.2	3	Rh	556 ± 4	1
Au	368.2 ± 2.1	1	Es	133	6	Mo	658.98 ± 3.8	3	Ru	650.6 ± 6.3	1
B	565 ± 5	2	Eu	177.4 ± 2.1	4	N	472.68 ± 0.40	2	S	277.17 ± 0.15	2
Ba	179.1 ± 5.0	3	F	79.38 ± 0.30	2	Na	107.5 ± 0.7	3	Sb	264.4 ± 2.5	1
Be	324 ± 5	2	Fe	415.5 ± 1.3	3	Nb	733.0 ± 8	3	Sc	377.8 ± 4	1
Bi	209.6 ± 2.1	1	Ga	271.96 ± 2.1	3	Nd	326.9 ± 2.1	4	Se	227.2 ± 4	1
Bk	310	6	Gd	397.5 ± 2.1	4	Ni	430.1 ± 8.4	3	Si	450.0 ± 8	2
Br	111.87 ± 0.12	3	Ge	372 ± 3	2	Np	464.8	6	Sm	206.7 ± 2.1	4
C	716.68 ± 0.45	2	H	217.998 ± 0.006	2	O	249.229 ± 0.002	7	Sn	301.2 ± 1.5	2
Ca	177.8 ± 0.8	2	Hf	618.4 ± 6.3	3	Os	787 ± 6.3	1	Sr	164.0 ± 1.7	3
Cd	111.80 ± 0.20	2	Hg	61.38 ± 0.04	2	P	316.5 ± 1.0	2	Ta	782.0 ± 2.5	1
Ce	420.1 ± 2.1	4	Ho	300.6 ± 2.1	4	Pa	563	5	Tb	388.7 ± 2.1	4
Cf	196	6	I	106.76 ± 0.04	2	Pb	195.2 ± 0.8	2	Tc	678	5
Cl	121.301 ± 0.008	2	In	243 ± 4	1	Pd	376.6 ± 2.1	1	Te	196.6 ± 2.1	1
Cm	386	6	Ir	669 ± 4	1	Pr	356.9 ± 2.1	4	Th	602 ± 6	2
Co	426.7	3	K	89.0 ± 0.8	2	Pt	565.7 ± 1.3	1	Ti	473 ± 3	2

Atom	$\Delta_f H^o_{298}$/kJ mol^{-1}	Ref.	Atom	$\Delta_f H^o_{298}$/kJ mol^{-1}	Ref.	Atom	$\Delta_f H^o_{298}$/kJ mol^{-1}	Ref.	Atom	$\Delta_f H^o_{298}$/kJ mol^{-1}	Ref.
Tl	182.2 ± 0.4	1	U	533 ± 8	2	W	851.0 ± 6.3	3	Yb	155.6 ± 2.1	4
Tm	232.2 ± 2.1	4	V	515.5 ± 8	3	Y	424.7 ± 2.1	4	Zn	130.40 ± 0.40	2
									Zr	610.0 ± 8.4	3

References

1. Brewer, L., and Rosenblatt, G. M., *Adv. High Temp. Chem.* 2, 1, 1969.
2. Cox, J. D., Wagman, D. D., and Medvedev, V. A., Eds., *CODATA Key Values for Thermodynamics*, Hemisphere Publishing Corporation, New York, 1989; updated e-version: http://www.codata.org/codata.
3. NIST Chemistry WebBook, http://webbook.nist.gov, *NIST-JANAF Thermochemical Table*, 4th Edn., Chase, Jr., M. W., Ed., ACS, AIP, New York, 1998.
4. Chandrasekharaiah, M.S., and Gingerich, K.A., Thermodynamic properties of gaseous species, in *Handbook on the Chemistry and Physics of Rare Earths*, Gschneidner, Jr., K.A., and Ering, L., Eds., Elsevier, Amsterdam, 1989, Vol. 12, Chap. 86, pp. 409–431.
5. Lias, S.G., Bartmess, J. E., Liebman, J. F., Holmes, J. L., Levin, R. D., and Mallard, W. G., *J. Phys. Chem. Ref. Data* 17, Suppl. 1, 1988.
6. Kleinschmidt, P.D., Ward, J. W., Matlack, G. M., and Haire, R. G., *High Temp. Sci.* 19, 267, 1985.
7. Ruscic, B., Pinzon, R.E., Morton, M. E., Srinivasan, N. K., Su, M.-C., Sutherland, J. W., and Michael, J. V., *J. Phys. Chem. A* 110, 6592, 2006.

TABLE 3. Bond Dissociation Energies in Polyatomic Molecules

The D^o_{298} values in polyatomic molecules are notoriously difficult to measure accurately since the mechanism of the kinetic systems involved in many of the measurements are seldom straightforward. Thus, much lively controversy has taken place in the literature and is likely to continue for some time to come. We will continue updating and presenting our assessment of the most reliable BDE data every year.

The references relating to each of the D^o_{298} values listed in Table 3 are contained in the *Comprehensive Handbook of Chemical Bond Energies*, by Yu-Ran Luo, CRC Press, 2007. Many D^o_{298} in Table 3 are derived from the equation

$$D^o_{298}(\text{R-X}) = \Delta_f H^o(\text{R}) + \Delta_f H^o(\text{X}) - \Delta_f H^o(\text{RX})$$

Here, the enthalpies of formation of the atoms and radicals are taken from Tables 2 and 4, respectively, and the enthalpies of formation of the molecules are from reference sources listed in the above *Comprehensive Handbook of Chemical Bond Energies.*

Table 3 presents **H-C**, **C-C**, **C-halogen**, **O-**, **N-**, **S-**, **Si-**, **Ge-**, **Sn-**, **Pb-**, **P-**, **As-**, **Sb-**, **Bi-**, **Se-**, **Te-**, and **metal**-X BDEs. The **boldface** in the species indicates the dissociated fragment. The **metal**-X BDEs are arranged on the basis of the Periodic Table with the new IUPAC notation for Groups 1 to 18, see inside front cover of this *Handbook*.

Bond	D^o_{298}/kJ mol^{-1}	Ref.
(1) C–H BDEs		
CH_3–**H**	439.3 ± 0.4	1
CH_3CH_2–**H**	420.5 ± 1.3	1
$CH_3CH_2CH_2$–**H**	422.2 ± 2.1	1
$CH_3\mathbf{CH_2}CH_3$	410.5 ± 2.9	1
$CH_3CH_2CH_2CH_2$–**H**	421.3	1
$CH_3\mathbf{CH_2}CH_2CH_3$	411.1 ± 2.2	1
$(CH_3)_2CHCH_2$–**H**	419.2 ± 4.2	1
$(CH_3)_3C$–**H**	400.4 ± 2.9	1
$(CH_3)_3CCH_2$–**H**	419.7 ± 4.2	1
$(CH_3CH_2)\mathbf{CH}(CH_3)_2$	400.8	1
$CH_3\mathbf{CH_2}(CH_2)_2CH_3$	415.1	1
$(C_3H_7)\mathbf{CH}(CH_3)_2$	396.2 ± 8.4	1
$CH_3CH(CH_3)\mathbf{CH}(CH_3)_2$	399.2 ± 13.0	1
$CH_3\mathbf{CH_2}(CH_2)_3CH_3$	410	1
$CH_3\mathbf{CH_2}(CH_2)_4CH_3$	410	1
HCC–**H**	557.81 ± 0.30	1
HCCCC–**H**	539 ± 12	1
$CHCCH_2$–**H**	384.1 ± 4.2	1
CH_3CCCH_2–**H**	379.5	1
$HCC\mathbf{CH_2}CH_3$	373.0	1
CH_2=$CHCCCH_2$–**H**	363.3	1
$CH_3CC\mathbf{CH_2}CH_3$	365.3 ± 9.6	1
$HCC\mathbf{CH_2}CH_2CH_3$	349.8 ± 8.4	1
$HCC\mathbf{CH}(CH_3)_2$	345.2 ± 8.4	1
$CH_3CC\mathbf{CH}(CH_3)_2$	344.3 ± 11.3	1
HCCCCCC–**H**	~543 ± 13	1
H_2C=CH–**H**	464.2 ± 2.5	1
CH_2=C=CH–**H**	371.1 ± 12.6	1
CH_3CH=CH–**H**	464.8	1
CH_2=$CHCH_2$–**H**	369 ± 3	1
CH_2=CH-CH_2CH_2–**H**	410.5	1
CH_2=$CH\mathbf{CH_2}CH_3$	350.6	1
CH_2=$C(CH_3)CH_2$–**H**	372.8	1
CH_2=CHCH=$CHCH_2$–**H**	347.3 ± 12.6	1
$(CH_2$=$CH)_2CH$–**H**	320.5 ± 4.2	1
CH_2=$CH\mathbf{CH_2}CH_2CH_3$	348.8	1
CH_2=$CH\mathbf{CH}(CH_3)_2$	332.6 ± 7.1	1
CH_2=$C(CH_3CH_2)CH_2$–**H**	356.1 ± 8.4	1
$(CH_2$=$CH)_2C(CH_3)$ –**H**	322.2	1
H-*cyclo*-C_3H_5	444.8 ± 1.0	1
H-CH_2-*cyclo*-C_3H_5	407.5 ± 6.7	1
H-*cyclo*-C_4H_7	409.2 ± 1.3	1
H-*cyclo*-C_5H_9	400.0 ± 4.2	1
H-*cyclo*-C_6H_{11}	416.3	1
H–C_6H_5	472.2 ± 2.2	1
H–$CH_2C_6H_5$	375.5 ± 5.0	1
H–$CH(CH_3)C_6H_5$	357.3 ± 6.3	1
H–$CH(C_6H_5)_2$	353.5 ± 2.1	1
H–$CH(C_6H_4$-p-$OH)_2$	375.8 ± 4.7	1
H–$C(CH_3)_2C_6H_5$	348.1 ± 4.2	1
H–$C(C_6H_5)_3$	338.9 ± 8.4	1
1-**H**-$C_{10}H_7$	469.4 ± 5.4	1
2-**H**-$C_{10}H_7$	468.2 ± 5.9	1
H–CF_3	445.2 ± 2.9	1
H–CHF_2	431.8 ± 4.2	1
H–CH_2F	423.8 ± 4.2	1
H–$CClF_2$	421.3 ± 8.4	1
H–CCl_2F	410.9 ± 8.4	1
H–$CBrF_2$	415.5 ± 12.6	1
H–CHClF	421.7 ± 10.0	1
H–CCl_3	392.5 ± 2.5	1
H–$CHCl_2$	400.6 ± 2.0	1
H–CH_2Cl	419.0 ± 2.3	1
H–CFClBr	413 ± 21	1

Bond	D^o_{298}/kJ mol^{-1}	Ref.
$\mathbf{H}\text{–}CHClBr$	406.0 ± 2.4	1
$\mathbf{H}\text{–}CCl_2Br$	387 ± 21	1
$\mathbf{H}\text{–}CClBr_2$	371 ± 21	1
$\mathbf{H}\text{–}CBr_3$	399.2 ± 8.4	1
$\mathbf{H}\text{–}CHBr_2$	412.6 ± 2.7	3
$\mathbf{H}\text{–}CH_2Br$	427.2 ± 2.4	1
$\mathbf{H}\text{–}CI_3$	423 ± 29	1
$\mathbf{H}\text{–}CHI_2$	431.0 ± 8.4	1
$\mathbf{H}\text{–}CH_2I$	431.6 ± 2.8	1
$CF_3CF_2\text{–}\mathbf{H}$	429.7 ± 2.1	1
$CHF_2CF_2\text{–}\mathbf{H}$	431.0 ± 18.8	1
$CH_2FCF_2\text{–}\mathbf{H}$	433.0 ± 14.6	1
$CHF_2CFH\text{–}\mathbf{H}$	426.8 ± 14.6	1
$CF_3CH_2\text{–}\mathbf{H}$	446.4 ± 4.5	1
$CH_3CF_2\text{–}\mathbf{H}$	416.3 ± 4.2	1
$CH_2FCHF\text{–}\mathbf{H}$	413.4 ± 12.6	1
$CHF_2CH_2\text{–}\mathbf{H}$	433.0 ± 14.6	1
$CH_2FCH_2\text{–}\mathbf{H}$	433.5 ± 8.4	1
$CH_3CHF\text{–}\mathbf{H}$	410.9 ± 8.4	1
$CF_3CHCl\text{–}\mathbf{H}$	425.9 ± 6.3	1
$CF_3CClBr\text{–}\mathbf{H}$	404.2 ± 6.3	1
$CClF_2CHF\text{–}\mathbf{H}$	412.1 ± 2.1	1
$CCl_3CCl_2\text{–}\mathbf{H}$	397.5 ± 8.4	1
$CHCl_2CCl_2\text{–}\mathbf{H}$	393.3 ± 8.4	1
$CH_3CCl_2\text{–}\mathbf{H}$	397.9 ± 5.0	1
$CH_3CHCl\text{–}\mathbf{H}$	406.6 ± 1.5	1
$CH_2ClCH_2\text{–}\mathbf{H}$	423.1 ± 2.4	1
$CH_3CBr_2\text{–}\mathbf{H}$	397.1 ± 5.0	1
$CH_2BrCH_2\text{–}\mathbf{H}$	415.1 ± 8.4	1
$CH_3CHBr\text{–}\mathbf{H}$	415.0 ± 2.7	3
$CF_2{=}CF\text{–}\mathbf{H}$	464.4 ± 8.4	1
$CF_3CF_2CF_2\text{–}\mathbf{H}$	432.2	1
$CH_3CH_2CHCl\text{–}\mathbf{H}$	407.0 ± 3.5	1
$CH_2{=}CH\text{-}CHF\text{–}\mathbf{H}$	370.7 ± 4.6	1
$CH_2{=}CHCHCl\text{–}\mathbf{H}$	370.7 ± 4.6	1
$CH_2{=}CHCHBr\text{–}\mathbf{H}$	374.0 ± 4.6	1
$\mathbf{H}\text{–}C_6F_5$	487.4	1
$\mathbf{H}\text{–}CH_2OH$	401.92 ± 0.63	1
$\mathbf{CH_2}CHOH$	467 ± 11	1
$CH_3\mathbf{CH_2}OH$	401.2 ± 4.2	1
$\mathbf{CH_3}CH_2OH$	421.7 ± 8	1
$CH_3CH_2\mathbf{CH_2}OH$	392	1
$CH_3\mathbf{CH_2}CH_2OH$	394.6 ± 8.4	1
$\mathbf{CH_3}CH_2CH_2OH$	406.3 ± 8.4	1
$(CH_3)_2\mathbf{CH}OH$	383.7 ± 8.4	1
$(\mathbf{CH_3})_2CHOH$	394.6 ± 8.4	1
$CH_2{=}CH\mathbf{CH_2}OH$	341.4 ± 7.5	1
$(\mathbf{CH_3})_3COH$	418.4 ± 8.4	1
$(CH_2{=}CH)_2\mathbf{CH}OH$	288.7	1
$Ph_2\mathbf{CH}OH$	326	1
$CH_3\mathbf{CH}(OH)_2$	~385	1
$(\mathbf{CH_2}OH)_2$	385.3	1
$HOCH_2(CH_2)_2(OH)CH\text{–}\mathbf{H}$	399.2	1
$\mathbf{CH_3}OCH_3$	402.1	1
$\mathbf{CHF_2}OCF_3$	443.5 ± 4.2	1
$\mathbf{CHF_2}OCHF_2$	435.1 ± 4.2	1
$\mathbf{CH_3}OCF_3$	426.8 ± 4.2	1
$\mathbf{CH_3}OCH_2CH_3$	389.1	1
$(\mathbf{CH_3})_3COC(CH_3)_3$	402.1	1
$CH_3\mathbf{CH_2}OCH_2CH_3$	389.1	1
$CH_3\mathbf{CH_2}Ot\text{-}C(CH_3)_3$	405.4	1
$\mathbf{CH_3}OPh$	385.0	1
H-2-oxiran-2-yl	420.5 ± 6.5	1
H-tetrahydrofuran-2-yl	385.3 ± 6.7	1
$HC(O)\text{–}\mathbf{H}$	368.40 ± 0.67	1
$FC(O)\text{–}\mathbf{H}$	423.0	1
$CH_3C(O)\text{–}\mathbf{H}$	374.0 ± 1.3	1
$CF_3C(O)\text{–}\mathbf{H}$	390.4	1
$C_2H_5C(O)\text{–}\mathbf{H}$	374.5	1
$CH_2{=}CHC(O)\text{–}\mathbf{H}$	372.8	1
$C_3H_7C(O)\text{–}\mathbf{H}$	371.2	1
iso-$C_3H_7C(O)\text{–}\mathbf{H}$	364.5	1
$C_4H_9C(O)\text{–}\mathbf{H}$	372.0	1
$(CH_3)_2CHCH_2C(O)\text{–}\mathbf{H}$	362.5	1
$C_2H_5CH(CH_3)C(O)\text{–}\mathbf{H}$	360.8	1
tert-$BuC(O)\text{–}\mathbf{H}$	375.1	1
$Et_2CHC(O)\text{–}\mathbf{H}$	367.2	1
$CH_3(CH_2)_8C(O)\text{–}\mathbf{H}$	373.3	1
$C_6H_5C(O)\text{–}\mathbf{H}$	371.1 ± 10.9	1
$PhCH_2C(O)\text{–}\mathbf{H}$	362.0	1
$PhC(CH_3)_2C(O)\text{–}\mathbf{H}$	362.9	1
$\mathbf{H}\text{–}CH{=}C{=}O$	448.1	1
$\mathbf{CH_3}C(O)H$	394.5 ± 9.2	1
$\mathbf{CH_3}C(O)Cl$	≤423.4	1
$CH_3\mathbf{CH_2}C(O)H$	383.7	1
$\mathbf{CH_3}COCH_3$	401.2 ± 2.9	1
$CF_3C(O)\mathbf{CH_3}$	465.6	1
$\mathbf{CH_3}COCH_2CH_3$	403.8	1
$MeCO\mathbf{CH_2}Me$	386.2 ± 7.1	1
$EtCO\mathbf{CH_2}Me$	396.5 ± 2.8	1
$CH_3\mathbf{CH_2}COC_6H_5$	402.8 ± 3.6	1
$Me\mathbf{CH_2}COPh$	388.7	1
$\mathbf{H}\text{–}C(O)OH$	404.2	1
$\mathbf{CH_3}C(O)OH$	398.7 ± 12.1	1
$Cl\mathbf{CH_2}C(O)OH$	398.9	1
$\mathbf{H}\text{–}C(O)OCH_3$	399.2 ± 8.4	1
$\mathbf{CH_3}C(O)OCH_3$	406.3 ± 10.5	1
$CH_3C(O)O\mathbf{CH_3}$	404.6	1
$\mathbf{CH_3}C(O)OCH_2CH_3$	401.7	1
$\mathbf{CH_3}C(O)OPh$	419.2 ± 5.4	1
$CH_3\mathbf{CH_2}C(O)OEt$	400	1
$Ph\mathbf{CH_2}C(O)OEt$	370.7	1
$Me_2\mathbf{CH}C(O)OEt$	387.4	1
$Ph\mathbf{CH}Me(C(O)OEt)$	358.2	1
H-furaylmethyl	361.9 ± 8.4	1
$\mathbf{CH_3}NH_2$	392.9 ± 8.4	1
$\mathbf{CH_3}N{=}CH_2$	407.9 ± 14.6	1
$CH_3\mathbf{CH_2}NH_2$	377.0 ± 8.4	1
$C_2H_5\mathbf{CH_2}NH_2$	380.7 ± 8.4	1
$C_3H_7\mathbf{CH_2}NH_2$	393.3 ± 8.4	1
$C_4H_9\mathbf{CH_2}NH_2$	387.7 ± 8.4	1
$HOCH_2\mathbf{CH_2}NH_2$	379.5 ± 8.4	1
$(CH_3\mathbf{CH_2})_2NH$	370.7 ± 8.4	1
$(C_3H_7\mathbf{CH_2})_2NH$	379.9 ± 8.4	1
$(C_4H_9\mathbf{CH_2})_2NH$	384.5 ± 8.4	1
$(C_2H_5)_2N\mathbf{CH_2}CH_3$	379.5 ± 1.7	1
$(C_2H_5\mathbf{CH_2})_3N$	376.6 ± 8.4	1
$((CH_3)_2C\mathbf{CH_2})_3N$	388.3 ± 8.4	1
$(Bu)_2N\mathbf{CH_2}(nPr)$	381 ± 10.0	1
$((CH_3)_2\mathbf{CH})_3N$	387.0 ± 8.4	1
$(CH_3)_2\mathbf{CH}NH_2$	372.0 ± 8.4	1
$\mathbf{CH_3}NHCH_3$	364.0 ± 8.4	1
$(\mathbf{CH_3})_3N$	380.7 ± 8.4	1
tert-$BuN(\mathbf{CH_3})_2$	376.6 ± 8.4	1
$((HOCH_2CH_2)_2(\mathbf{CH_3}))N$	364.4 ± 8.4	1
$(HOCH_2\mathbf{CH_2})_3N$	379.9 ± 8.4	1
$((HOCH_2)\mathbf{CH}(CH_3))_3N$	379.9 ± 8.4	1
$Ph\mathbf{CH_2}NH_2$	368.2	1
$PhN(\mathbf{CH_2}CH_3)_2$	383.3 ± 4.2	1
$Ph_2N\mathbf{CH_3}$	379.5 ± 1.7	1
$PhN(\mathbf{CH_2}Ph)_2$	357.3 ± 8.8	1
$N(\mathbf{CH_2}Ph)_3$	372.8 ± 2.5	1
$PhN(\mathbf{CH_2}CH{=}CH_2)_2$	339.3 ± 2.9	1
$N(\mathbf{CH_2}CH{=}CH_2)_3$	345.6 ± 3.3	1
$H_2NNH(\mathbf{CH_3})$	410	1
$HNN(\mathbf{CH_3})_2$	410	1
$(\mathbf{CH_3})_2NC_6H_5$	383.7 ± 5.4	1
$\mathbf{H}\text{–}CN$	528.5 ± 0.8	1
$\mathbf{CH_3}CN$	405.8 ± 4.2	1
$CH_3\mathbf{CH_2}CN$	393.3 ± 12.6	1
$Ph\mathbf{CH_2}CN$	344.3	1
$C_6F_5\mathbf{CH_2}CN$	350.6	1
$\mathbf{CH_2}(CN)_2$	366.5	1
$\mathbf{CH_2}(CN)(NH_2)$	355.2	1
$(CH_3)_2\mathbf{CH}CN$	384.5	1
$\mathbf{CH_3}NC$	389.1 ± 12.6	1
$\mathbf{H}\text{–}HCNN$	405.8 ± 8.4	1
$\mathbf{H}\text{–}CNN$	331 ± 17	1
$\mathbf{CH_3}NO_2$	415.4	1
$CH_3\mathbf{CH_2}NO_2$	410.5	1
$C_2H_5\mathbf{CH_2}NO_2$	410.5	1
$Me_2\mathbf{CH}NO_2$	394.9	1
$C_6H_5C(NO_2)\mathbf{CH}CH_3$	357.3	1

Bond	D^o_{298}/kJ mol^{-1}	Ref.
H–C(S)H	399.6 ± 5.0	1
CH_3SH	392.9 ± 8.4	1
CH_3SCH_3	392.0 ± 5.9	1
$PhSCH_3$	389.1	1
$PhCH_2SPh$	352.3	1
$(PhS)_2CHPh$	341.0	1
$PhSCHPh_2$	344.8	1
CH_3SOCH_3	393.3	1
$CH_3SO_2CH_3$	414.2	1
$CH_3SO_2CF_3$	431.0	1
CH_3SO_2Ph	414.2	1
$PhCH_2SO_2Me$	380.7	1
$PhCH_2SO_2CF_3$	372.4	1
$PhCH_2SO_2tBu$	376.6	1
Ph_2CHSO_2Ph	365.3	1
$CH_2(SPh)_2$	372.4	1
H–CH_2SiMe_3	418 ± 6.3	1
H–$CH_2C(CH_3)_2SiMe_3$	409 ± 5	1
H–CH_2SiMe_2Ph	410.1	1
H–$CH((CH_3)_3Si)_2$	397 ± 13	1
H–$CH_2B(RO)_2$	412.5	1
H–$CH((CH_3)_2P)_2$	385 ± 13	1
(2) C–C BDEs		
CH_3–CH_3	377.4 ± 0.8	1
CH_3–C_2H_5	370.3 ± 2.1	1
CH_3–C_3H_7	372.0 ± 2.9	1
CH_3–*iso*-C_3H_7	369.0 ± 3.8	1
CH_3–C_4H_9	371.5 ± 2.9	1
CH_3–*iso*-C_4H_9	370.3 ± 4.6	1
CH_3–*sec*-C_4H_9	368.2 ± 2.9	1
CH_3–*tert*-C_4H_9	363.6 ± 2.9	1
CH_3–C_5H_{11}	368.4 ± 6.3	1
CH_3–$CH(C_2H_5)_2$	365.7 ± 4.2	1
CH_3–$C(CH_3)_2(CH_2CH_3)$	360.9 ± 6.3	1
CH_3–C_6H_{13}	368.2 ± 6.3	1
C_2H_5–C_2H_5	363.2 ± 2.5	1
C_3H_7–C_3H_7	366.1 ± 3.3	1
iso-C_3H_7–*iso*-C_3H_7	353.5 ± 4.6	1
C_4H_9–C_4H_9	364.0 ± 3.8	1
iso-C_4H_9–*iso*-C_4H_9	362.3 ± 6.3	1
sec-C_4H_9–*sec*-C_4H_9	348.5 ± 3.3	1
tert-C_4H_9–*tert*-C_4H_9	322.6 ± 4.2	1
CH_3–*cyclo*-C_5H_9	358.2 ± 5.0	1
CH_3–*cyclo*-C_6H_{11}	377.0 ± 7.5	1
cyclo-C_6H_{11}–*cyclo*-C_6H_{11}	369.0 ± 8.4	1
CH_3–$CH_2C{\equiv}CH$	320.5 ± 5.0	1
CH_3–$CH_2C{\equiv}CCH_3$	308.4 ± 6.3	1
CH_3–$CH(CH_3)C{\equiv}CH$	305.4 ± 8.4	1
CH_3–$CH(CH_3)C{\equiv}CCH_3$	320.9 ± 6.3	1
CH_3–$C(CH_3)_2C{\equiv}CH$	295.8 ± 6.3	1
CH_3–$C(CH_3)_2C{\equiv}CCH_3$	303.3 ± 6.3	1

Bond	D^o_{298}/kJ mol^{-1}	Ref.
CH_3–$CHCH_2$	426.3 ± 6.3	1
CH_3–$CH{=}CCH_2$	359.8 ± 5.9	1
CH_3–cyclopro-en-1-yl	340.6 ± 20.9	1
CH_3–$CH_2CH{=}CH_2$	317.6 ± 3.8	1
CH_3–$CH_2C(CH_3){=}CH_2$	310.0 ± 4.2	1
CH_3–$CH(CH_3)CH{=}CH_2$	302.5 ± 6.3	1
CH_3–$C(CH_3)_2CH{=}CH_2$	282.4 ± 6.3	1
CH_3–*cyclo*-C_5H_7	299.2 ± 8.4	1
CH_3–C_6H_5	426.8 ± 4.2	1
HCC–C_6H_5	590.8 ± 5.9	1
C_2H_3–C_6H_5	482.0 ± 5.4	1
CH_3-$CH_2C_6H_5$	325.1 ± 4.2	1
CH_3–$CH(CH_3)C_6H_5$	318.8 ± 8.4	1
CH_3–$C(CH_3)_2C_6H_5$	303.3 ± 8.4	1
CH_3–$CH_2CHCHPh$	295.4	1
CH_3–$CH(C_6H_5)_2$	315.9 ± 6.3	1
CH_3–$C(CH_3)(C_6H_5)_2$	290.8 ± 8.4	1
C_6H_5–C_6H_5	478.6 ± 6.3	1
C_6H_5–$CH_2C_6H_5$	383.7 ± 8.4	1
$C_6H_5CH_2$–$CH_2C_6H_5$	272.8 ± 9.2	1
C_6H_5–$CH(C_6H_5)_2$	361.1 ± 8.4	1
C_6H_5–$C(C_6H_5)_3$	324.3 ± 12.6	1
Ph_2CH–$CHPh_2$	247.3 ± 8.4	1
$PhCH_2$–CPh_3	234.7 ± 14.6	1
R-R, π-dimer, R = phenalenyl	42	1
R-R, σ-dimer, R = phenalenyl	42.7	1
R-R, R = 9-phenylfluorenyl	63.6	1
CF_3–CF_3	413.0 ± 5.0	1
CF_3–CHF_2	399.6 ± 8.4	1
CF_3–$CClF_2$	373.6 ± 12.5	1
CF_3–CH_2F	397.5 ± 8.4	1
CF_3–CCl_3	332.2 ± 5.4	1
CF_3–$CHBrCl$	377.0 ± 10.5	1
CF_3–CH_2Br	399.6 ± 8.4	1
CF_3–CH_2I	408.4 ± 10.5	1
CF_3–CH_3	429.3 ± 5.0	1
CHF_2–CHF_2	382.4 ± 15.5	1
$CClF_2$–$CClF_2$	378.7 ± 12.6	1
CF_2Cl–$CFCl_2$	358.6 ± 12.6	1
CHF_2–CH_2F	394.1 ± 16.7	1
CH_2F–CH_2F	368.2 ± 8.4	1
CHF_2–CH_3	405.0 ± 8.4	1
CH_2F–CH_3	388.3 ± 8.4	1
$CHClF$–CH_3	399.6 ± 12.6	1
CF_2Br–$CHClF$	369.4	1
CF_2Br–CH_3	396.6 ± 15.1	1
CCl_3–CCl_3	285.8 ± 6.3	1
CCl_3–$CClF_2$	282.0 ± 12.6	1
CCl_3–$CHCl_2$	303.3 ± 6.3	1

Bond	D^o_{298}/kJ mol^{-1}	Ref.
CCl_3–CH_2Cl	323.8 ± 8.4	1
CCl_3–CH_3	362.3 ± 6.3	1
$CHCl_2$–$CHCl_2$	326.9 ± 4.1	1
$CHCl_2$–CH_2Cl	352.2 ± 5.9	1
$CHCl_2$–CH_3	361.3 ± 2.5	1
$CHBrCl$–CH_3	384.5	1
$CHClBr$–$CHClBr$	317.1 ± 12.6	1
CH_2Cl–CH_2Cl	360.7 ± 8.4	1
CH_2Cl–CH_3	375.7 ± 9.2	1
Br_3C–CH_3	356.9 ± 12.6	1
Br_3C–CBr_3	278.7 ± 16.7	1
$CHBr_2$–CH_3	372.8	1
CH_2Br–CH_2Cl	378.2	1
CH_2Br–CH_2Br	379.9 ± 8.4	1
CH_2I–CH_2I	387.0 ± 10.5	1
CH_3–CH_2Br	381.6 ± 8.4	1
CH_3–CH_2I	384.5 ± 8.4	1
CF_3–CF_2CF_3	424.3 ± 13.6	1
CF_3–$CF{=}CF_2$	420.5	1
CH_3–CH_2CH_2Cl	371.4 ± 2.8	1
CH_3–$CHClCH_3$	367.5 ± 2.0	1
CH_2Cl–$CHClCH_3$	356.5 ± 8.4	1
CH_2Cl–CH_2CClH_2	369.0 ± 8.4	1
CH_3–CCl_2CH_3	362.8 ± 8.4	1
CH_2Br–$CHBrCH_3$	369.4 ± 8.4	1
CH_2ClCH_2–$CHClCH_3$	364.4 ± 8.4	1
CH_2ClCH_2–CH_2CClH_2	369.0 ± 8.4	1
CH_3CHBr–$CHBrCH_3$	355.6 ± 8.4	1
CF_3–C_6H_5	463.2 ± 12.6	1
CCl_3–C_6H_5	388.7 ± 8.4	1
CH_3–C_6F_5	439.3	1
CF_3–C_6F_5	435.1	1
CF_3–$CH_2C_6H_5$	365.7 ± 12.6	1
C_6F_5–C_6F_5	488.3	1
CF_3–$CHPh_2$	352.3 ± 16.7	1
CF_3–CPh_3	290.8 ± 16.7	1
CF_2CF–$CFCF_2$	558.1 ± 12.6	1
CH_2FCH_2–CPh_3	274.9 ± 16.7	1
CHF_2CH_2–CPh_3	264.0 ± 16.7	1
CH_3–CH_2OH	364.8 ± 4.2	1
CF_3–CH_2OH	405.4 ± 6.3	1
C_2H_5–CH_2OH	356.9 ± 5.0	1
C_3H_7–CH_2OH	357.3 ± 3.3	1
iso-C_3H_7–CH_2OH	354.8 ± 4.2	1
C_4H_9–CH_2OH	355.6 ± 4.2	1
sec-C_4H_9–CH_2OH	352.7 ± 4.2	1
iso-C_4H_9–CH_2OH	354.0 ± 5.4	1
C_6H_5–CH_2OH	413.4 ± 5.4	1
HOH_2C–CH_2OH	358.2 ± 6.3	1
NH_2CH_2–CH_2OH	335.6 ± 10.5	1
CH_3–CH_2OCH_3	363.2 ± 5.0	1

Bond	D^o_{298}/kJ mol^{-1}	Ref.
CH_3OCH_2–CH_2OCH_3	338.9 ± 10.5	1
CH_3–C(O)H	354.8 ± 1.7	1
CCl_3–C(O)H	309.2 ± 5.0	1
CH_3–C(O)F	417.6 ± 6.3	1
CH_3–C(O)Cl	367.8 ± 6.3	1
CCl_3–C(O)Cl	289.1 ± 6.3	1
$CHCl_2$–C(O)Cl	312.5 ± 8.4	1
$CClH_2$–C(O)Cl	340.2 ± 8.4	1
C_6H_5–C(O)H	408.4 ± 4.2	1
C_6H_5–C(O)Cl	417.6 ± 6.3	1
CH_3–C(O)CH_3	351.9 ± 2.1	1
C_2H_5–C(O)CH_3	347.3 ± 2.9	1
C_3H_7–C(O)CH_3	348.5 ± 2.9	1
iso-C_3H_7–C(O)CH_3	340.2 ± 3.8	1
C_4H_7–C(O)CH_3	346.9 ± 5.4	1
tert-C_4H_9–C(O)CH_3	329.3 ± 4.2	1
C_6H_5–C(O)CH_3	406.7 ± 4.6	1
$C_6H_5CH_2$–C(O)CH_3	299.7 ± 8.4	1
HC(O)–C(O)H	295.8 ± 6.3	1
ClC(O)–C(O)Cl	292.5 ± 8.4	1
CH_3C(O)–C(O)H	302.5 ± 8.4	1
CH_3C(O)–C(O)CH_3	307.1 ± 4.2	1
C_6H_5C(O)–C(O)C_6H_5	288.3 ± 16.7	1
CH_3–C(O)OH	384.9 ± 8.4	1
CF_3–C(O)OH	370.7 ± 8.4	1
CCl_3–C(O)OH	310.5 ± 12.6	1
$CClH_2$–C(O)OH	357.7 ± 8.4	1
CH_2Br–C(O)OH	358.2 ± 8.4	1
NH_2CH_2–C(O)OH	349.4 ± 8.4	1
CH_3NHCH_2–C(O)OH	300.4 ± 8.4	1
C_6H_5–C(O)OH	429.7 ± 8.4	1
C_6F_5–C(O)OH	470.0 ± 10.5	1
$HOCH_2$–C(O)OH	371.5 ± 5.4	1
HOC(O)–C(O)OH	334.7 ± 6.3	1
CH_3NHCH_2–C(O)OH	301.2 ± 16.7	1
$CH_3CH(NH_2)$–C(O)OH	331.4 ± 16.7	1
NH_2CH_2–CH_2C(O)OH	325.5 ± 16.7	1
CN–CN	571.9 ± 6.7	1
HC(O)–CN	455.2 ± 8.4	1
HC(S)–CN	530.1 ± 8.4	1
CF_3–CN	469.0 ± 4.2	1
CH_3–CN	521.7 ± 9.2	1
NCC–CN	462.3	1
C_2H_5–CN	506.7 ± 7.5	1
CH_3–CH_2CN	348.1 ± 12.6	1
C_6H_5–CH_2CN	386.6 ± 8.4	1
CH_3–CH(CH_3)CN	332.6 ± 8.4	1
CH_3–C$(CH_3)_2$CN	340.6 ± 16.7	1
CH_3–C(CH_3)(CN)C_6H_5	250.6	
$(Ph)_2$(CN)C–C(CN)$(Ph)_2$	109.6	
$(NO_2)_3$C–C$(NO_2)_3$	308.8	
C_{58}–C_2	955.2 ± 14.5	1
(3) C–halogen BDEs		
F–CN	482.8	1
F–CF_3	546.8 ± 2.1	1
F–CHF_2	533.9 ± 5.9	1
F–CH_2F	496.2 ± 8.8	1
F–CF_2Cl	511.7	1
F–$CFCl_2$	482.0 ± 10.5	1
F–CHFCl	462.3 ± 10.0	1
F–CCl_3	439.3 ± 4	1
F–CH_2Cl	465.3 ± 9.6	1
F–CH_3	460.2 ± 8.4	1
F–C≡CH	521.3	1
F–C≡CF	519 ± 21	1
F–CF=CF_2	546.4 ± 12.6	1
F–CF_2CF_3	532.2 ± 6.3	1
F–CH_2CF_3	457.7	1
F–CF_2CH_3	522.2 ± 8.4	1
F–C_2H_3	517.6 ± 12.6	1
F–C_2H_5	467.4 ± 8.4	1
F–C_3H_7	474.9 ± 8.4	1
F–*iso*-C_3H_7	483.8 ± 8.4	1
F–*tert*-C_4H_9	495.8 ± 8.4	1
F–C_6H_5	525.5 ± 8.4	1
F–C_6F_5	485 ± 25	1
F–$CH_2C_6H_5$	412.8 ± 4.2	1
F–COH	497.9 ± 10.5	1
F–COF	510.3	1
F–COCl	484.5	1
F–C(O)CH_3	511.7 ± 12.6	1
Cl–CN	422.6 ± 8.4	1
Cl–CF_3	365.3 ± 3.8	1
Cl–CHF_2	364 ± 8	1
Cl–CH_2F	354.4 ± 11.7	1
Cl–CF_2Cl	333.9 ± 10.5	1
Cl–$CFCl_2$	320.9 ± 8.4	1
Cl–CHFCl	346.0 ± 13.4	1
Cl–CCl_3	296.6	1
Cl–$CHCl_2$	311.1 ± 2.0	1
Cl–CH_2Cl	338.0 ± 3.3	1
Cl–$CBrCl_2$	287 ± 10.5	1
Cl–CH_2Br	332.8 ± 4.6	1
Cl–CH_2I	328.2 ± 6.9	1
Cl–CH_3	350.2 ± 1.7	1
Cl–C≡CCl	443 ± 50	1
Cl–C≡CH	435.6 ± 8.4	1
Cl–CH_2CN	267.4	1
Cl–CCl=CCl_2	383.7	1
Cl–CH=CH_2	394.1 ± 3.1	2
Cl–CF=CF_2	434.7 ± 8.4	1
Cl–CF_2CF_3	346.0 ± 7.1	1
Cl–CF_2CF_2Cl	331.4 ± 20.9	1
Cl–CCl_2CF_3	307.9	1
Cl–CCl_2CCl_3	303.8	1
Cl–$CHClCCl_3$	330.5 ± 4.2	1
Cl–CCl_2CHCl_2	311.7	1
Cl–$CHClCH_3$	327.9 ± 1.8	1
Cl–CH_2CH_2Cl	345.1 ± 5.0	1
Cl–$CHBrCH_3$	331.8 ± 8.4	1
Cl–CH_2CH_3	352.3 ± 3.3	1
Cl–CH_2CH=CH_2	298.3 ± 5.0	1
Cl–C_3H_7	352.7 ± 4.2	1
Cl–$CH_2CH_2CH_2Cl$	348.9	1
Cl–*iso*-C_3H_7	354.0 ± 6.3	1
Cl–CH_2CHCH=CH_2	342.7	1
Cl–C_4H_9	350.6 ± 6.3	1
Cl–*sec*-C_4H_9	350.2 ± 6.3	1
Cl–*tert*-C_4H_9	351.9 ± 6.3	1
CH_2CHCH**Cl**(CH_3)	300.0 ± 6.3	1
Cl–C_5H_{11}	350.6 ± 6.3	1
Cl–C$(CH_3)_2$$(C_2H_5)$	352.7 ± 6.3	1
Cl–*cyclo*-C_6H_{11}	360.2 ± 6.5	1
Cl–C_6H_5	399.6 ± 6.3	1
Cl–C_6F_5	383.3 ± 8.4	1
Cl–$CH_2C_6H_5$	299.9 ± 4.3	1
Cl–C(O)Cl	318.8 ± 8.4	1
Cl–COF	376.6	1
Cl–C(O)CH_3	354.0 ± 8.4	1
Cl–C(O)CH_2CH_3	353.3 ± 6.3	1
Cl–C(O)C_6H_5	341.0 ± 8.4	1
Cl–CH_2C(O)C_6H_5	309	1
Cl–CH_2C(O)OH	310.9 ± 2.2	1
Cl–C(O)OC_6H_5	364	1
Cl–C$(NO_2)_3$	302.1	1
Br–CN	364.8 ± 4.2	1
Br–CF_3	296.2 ± 1.3	1
Br–CHF_2	288.7 ± 8.4	1
Br–CF_2Cl	269.9 ± 6.3	1
Br–CCl_3	231.4 ± 4.2	1
Br–CH_2Cl	277.3 ± 3.6	1
Br–CBr_3	242.3 ± 8.4	1
Br–$CHBr_2$	274.9 ± 13.0	1
Br–CH_2Br	276.1 ± 5.3	1
Br–CH_2I	274.5 ± 7.5	1
Br–CH_3	294.1 ± 2.1	1
Br–C≡CH	410.5	1
Br–CH=CH_2	338.3 ± 3.1	1
Br–CF_2CF_3	283.3 ± 6.3	1
Br–$CClBrCF_3$	251.0 ± 6.3	1
Br–CF_2CF_2Br	282.8 ± 6.7	1
Br–$CHClCF_3$	274.9 ± 6.3	1
Br–CF_2CH_3	287.0 ± 5.4	1

Bond	D^o_{298}/kJ mol^{-1}	Ref.
Br–CH_2CH_2Cl	292.5 ± 8.4	1
Br–$CHClCH_3$	272.0 ± 8.4	1
Br–C_2H_5	292.9 ± 4.2	1
Br–$CH_2CH{=}CH_2$	237.2 ± 5.0	1
Br–C_3H_7	298.3 ± 4.2	1
Br–*iso*-C_3H_7	299.2 ± 6.3	1
Br–$CH_2CH_2CH_2Br$	324.7	1
Br–$CF_2CF_2CF_3$	278.2 ± 10.5	1
CF_3CF**Br**CF_3	274.2 ± 4.6	1
Br–C_4H_9	296.6 ± 4.2	1
Br–*sec*-C_4H_9	300.0 ± 4.2	1
Br–*tert*-C_4H_9	292.9 ± 6.3	1
Br–C_6H_5	336.4 ± 6.3	1
Br–C_6F_5	~328	1
Br–$CH_2C_6H_5$	239.3 ± 6.3	1
Br–$CH_2C_6F_5$	225.1 ± 6.3	1
Br–1–$C_{10}H_7$	339.7	1
Br–2–$C_{10}H_7$	341.8	1
Br–anthracenyl	322.6	1
Br–$C(O)CH_3$	292.0 ± 8.4	1
Br–$C(O)C_6H_5$	276.6 ± 8.4	1
Br–$CH_2C(O)CH_3$	257.9 ± 10.5	1
Br–$CH_2C(O)C_6H_5$	271	1
Br–$CH_2C(O)OH$	257.4 ± 3.7	1
Br–$C(NO_2)_3$	218.4	1
I–CN	320.1	1
I–CF_3	227.2 ± 1.3	1
I–CCl_3	168 ± 42	1
I–CH_2Cl	221.8 ± 4.2	1
I–CH_2Br	219.2 ± 5.4	1
I–CH_2I	216.9 ± 7.9	1
I–CH_3	238.9 ± 2.1	1
I–CH_2CN	187.0 ± 6.3	1
I–CF_2CF_3	219.2 ± 2.1	1
I–CF_2CF_2I	217.6 ± 6.7	1
I–CH_2CF_3	235.6 ± 4.2	1
I–$CHFCClF_2$	202 ± 2	1
I–CF_2CH_3	217.6 ± 4.2	1
I-CFICH3	218.0 ± 4.2	1
CF_3CFICF_3	215.1	1
I–$CH{=}CH_2$	259.0 ± 4.2	1
I–C_2H_5	233.5 ± 6.3	1
I–$CH_2CH{=}CH_2$	185.8 ± 6.3	1
I–C_3H_7	236.8 ± 4.2	1
I–*iso*-C_3H_7	234.7 ± 6.3	1
I–C_4F_9	205.8	1
I–*tert*-C_4H_9	227.2 ± 6.3	1
I–C_6H_5	272.0 ± 4.2	1
I–C_6F_5	<301.7	1
I–$CH_2C_6H_5$	187.8 ± 4.8	1
I–1-naphthyl	274.5 ± 10.5	1

Bond	D^o_{298}/kJ mol^{-1}	Ref.
I–2-naphthyl	272.0 ± 10.5	1
I–CH_2CN	187.0 ± 8.4	1
I–CH_2OCH_3	229.4 ± 8.4	1
I–CH_2SCH_3	216.8 ± 6.3	1
I–$C(O)CH_3$	223.0 ± 8.4	1
I–$C(O)C_6H_5$	212.1 ± 8.4	1
I–$CH_2C(O)OH$	197.5 ± 2.7	1
I–$C(NO_2)_3$	144.8	1
(4) O–X BDEs		
HO–H	497.10 ± 0.29	1
FO–H	425.1	1
ClO–H	393.7	1
BrO–H	405	1
IO–H	403.3	1
CH_3O–H	440.2 ± 3	1
CF_3O–H	497.1	1
HC≡CO–H	443.1	1
C_2H_5O–H	441.0 ± 5.9	1
$CH_2{=}CHO$–H	355.6	1
CF_3CH_2O–H	447.7 ± 10.5	1
C_3H_7O–H	≤433 ± 2	1
iso-C_3H_7O–H	442.3 ± 2.8	1
C_4H_9O–H	432.3	1
sec-C_4H_9O–H	441.4 ± 4.2	1
tert-C_4H_9O–H	444.9 ± 2.8	1
tert-BuCH_2O–H	436.1	1
$C_6H_5CH_2O$–H	442.7 ± 8.8	1
$CH_3C(OH)O$–H	446.9 ± 6.3	1
$(CH_3)_2C(OH)O$–H	450.6 ± 6.3	1
HC(O)O–H	468.6 ± 12.6	1
$CH_3C(O)O$–H	468.6 ± 12.6	1
$C_2H_5C(O)O$–H	472.8	1
iso-$C_3H_7C(O)O$–H	472.8	1
$C_6H_5C(O)O$–H	464.4 ± 16.7	1
HOO–H	366.06 ± 0.29	1
CH_3OO–H	370.3 ± 2.1	1
CF_3OO–H	383	1
CH_2FOO–H	379	1
CCl_3OO–H	386	1
$CHCl_2OO$–H	383	1
CH_2ClOO–H	379	1
CBr_3OO–H	383	1
CH_2BrOO–H	379	1
C_2H_5OO–H	354.8 ± 9.2	1
$CH_3CHClOO$–H	377	1
CH_3CCl_2OO–H	383	1
$CF_3CHClOO$–H	384	1
C_2Cl_5OO–H	383	1
iso-C_3H_7OO–H	356	1
$CH_2{=}CHCH_2OO$–H	372.4	1
tert-C_4H_9OO–H	352.3 ± 8.8	1

Bond	D^o_{298}/kJ mol^{-1}	Ref.
C_6H_5OO–**H**	384	1
$C_6H_5CH_2OO$–**H**	363	1
$(C_6H_5)_2CHOO$–**H**	370	1
$CH_3C(O)OO$–**H**	386	1
$CCl_2(CN)OO$–**H**	384	1
$OHCH_2OO$–**H**	368	1
H–ONO	330.7	1
H–OONO	299.2	1
H–ONH_2	318	1
H–ONO_2	426.8	1
H–ONNOH	189	1
H–OPO_2	465.7 ± 12.6	1
H–OSO_2OH	441.4 ± 14.6	1
H–OSiMe3	495	1
$(CH_3)CHNO$-**H**	354.4	1
$(CH_3)_2CNO$-**H**	354.0	1
$(C_6H_5)CHNO$-**H**	368.6	1
PhO–**H**	362.8 ± 2.9	1
α-tocopherol RO-**H**	323.4	1
β-tocopherol RO-**H**	335.6	1
γ-tocopherol RO-**H**	335.1	1
δ-tocopherol RO-**H**	342.8	1
p-$C_6H_5CH_2$-C_6H_4O-**H**	356.2	1
O–O_2	106.6	1
HO–OH	210.66 ± 0.42	1
HO–OF	199.7 ± 8.4	1
HO–OCl	~146	1
HO–OBr	138.5 ± 8.4	1
FO–OF	199.6	1
ClO–OCl	72.4 ± 2.8	1
IO–OI	74.9 ± 17	1
***trans-perp*-HO**–ONO	≤67.8 ± 0.4	1
***cis-cis*-HO**–ONO	83.3 ± 2.1	1
HO–ONO_2	163.2 ± 8.4	1
HO–OCH_3	189.1 ± 4.2	1
HO–OCF_3	201.3 ± 20.9	1
HO–OC_2H_5	178.7 ± 6.3	1
HO–O-*iso*-C_3H_7	185.8 ± 6.3	1
HO–O-*tert*-C_4H_9	186.2 ± 4.2	1
HO–$OC(O)CH_3$	169.9 ± 2.1	1
HO–$OC(O)C_2H_5$	169.9 ± 2.1	1
CH_3O–OCH_3	167.4 ± 6.3	1
CF_3O–OCF_3	198.7 ± 2.1	1
C_2H_5O–OC_2H_5	166.1	1
C_3H_7O–OC_3H_7	155.2 ± 4.2	1
***iso*-C_3H_7O**–O-*iso*-C_3H_7	157.7	1
***sec*-C_4H_9O**–O-*sec*-C_4H_9	152.3 ± 4.2	1
***tert*-BuO**–O-*tert*-Bu	162.8 ± 2.1	1
***tert*-BuCH_2O**–OCH_2-*tert*-Bu	152.3	1
EtC(Me)$_2$O–OC(Me)$_2$Et	164.4 ± 4.2	1
$(CF_3)_3CO$–$OC(CF_3)_3$	148.5 ± 4.6	1

Bond	D^o_{298}/kJ mol^{-1}	Ref.
Ph_3CO–$OCPh_3$	131.4	1
SF_5O–OSF_5	155.6	1
SF_5O–$OOSF_5$	126.8	1
$(CH_3)_3CO$–$OSi(CH_3)_3$	196.6	1
***tert*-BuO**–$OGeEt_3$	192.5	1
***tert*-BuO**–$OSnEt_3$	192.5	1
CF_3OO–OCF_3	126.8 ± 8.4	1
HC(O)O–**OH**	199.2 ± 8.4	1
FC(O)O–OC(O)F	96.2	1
$CH_3C(O)O$–ONO_2	131.4 ± 8.4	1
$CH_3C(O)O$–$OC(O)CH_3$	140.2 ± 21	1
$CF_3C(O)O$–$OC(O)CF_3$	125.5	1
$CF_3OC(O)O$–OC(O)F	121.3 ± 4.2	1
$CF_3OC(O)O$–OCF_3	142.3 ± 2.9	1
$CF_3OC(O)O$–$OC(O)OCF_3$	119.2	1
$C_2H_5C(O)O$–$OC(O)C_2H_5$	150.6	1
$C_3H_7C(O)O$–$OC(O)C_3H_7$	150.6	1
$FS(O)_2O$–$OS(O)_2F$	92–100	1
HO–CF_3	≤482.0 ± 1.3	1
FO–CF_3	408 ± 17	1
HO–CH_3	384.93 ± 0.71	1
HO–C_2H_5	391.2 ± 2.9	1
HO–CH_2CF_3	408.4 ± 8.4	1
HO–CH_2CH=CH_2	332.6 ± 4.2	1
HO–C_3H_7	392.0 ± 2.9	1
HO–*iso*-C_3H_7	397.9 ± 4.2	1
HO–C_4H_9	389.9 ± 4.2	1
HO–*sec*-C_4H_9	396.1 ± 4.2	1
HO–*iso*-C_4H_9	394.1 ± 4.2	1
HO–*tert*-C_4H_9	398.3 ± 4.2	1
HO–$CH(CH_3)(nC_3H_7)$	398.3 ± 4.2	1
HO–$CH(C_2H_5)_2$	399.2 ± 4.2	1
HO–$C(CH_3)_2(C_2H_5)$	395.8 ± 6.3	1
HO–C_6H_5	463.6 ± 4.2	1
HO–C_6F_5	446.9 ± 9.2	1
HO–$CH_2C_6H_5$	334.1 ± 2.6	1
HO–$C(CH_3)_2C_6H_5$	339.3 ± 6.3	1
cyclo-C_5H_9–**OH**	385.8 ± 6.3	1
1-$C_{10}H_7$–**OH**	468.6 ± 6.3	1
2-$C_{10}H_7$–**OH**	467.8 ± 6.3	1
$(CH_3)_2(NH_2)C$–**OH**	310.4 ± 6.3	1
$CH_3C(O)$-**OH**	459.4 ± 4.2	1
$HOCH_2$-**OH**	411.3	1
CH_3–OCH_3	351.9 ± 4.2	1
ICH_2–OCH_3	373.2 ± 12.6	1
CH_3O–**C_2H_5**	355.2 ± 5.4	1
CH_3O–**$CHClCH_3$**	370.3 ± 8.4	1
CH_3O–**C_3H_7**	358.6 ± 6.3	1
CH_3O–***iso*-C_3H_7**	360.7 ± 4.2	1
CH_3O-**C_4H_9**	346.0 ± 6.3	1
CH_3O-***tert*-C_4H_9**	353.1 ± 6.3	1
C_6H_5–OCH_3	418.8 ± 5.9	1
$C_6H_5CH(CH_3)$–OCH_3	313.4 ± 9.6	1
C_6H_5–OC_6H_5	326.8 ± 4.2	1
CH_3–OC(O)H	383.7 ± 12.6	1
HC(O)–OH	457.7 ± 2.1	1
$CH_3C(O)$–OH	459.4 ± 4.2	1
$C_6H_5C(O)$–OH	447.7 ± 10.5	1
HO–$CH_2C(O)OH$	368.2 ± 10.5	1
CH_3-$OC(O)CH_3$	380.3 ± 12.6	1
HC(O)–OCH_3	423.8 ± 4.2	1
$CH_3C(O)$–OCH_3	424.3 ± 6.3	1
$C_6H_5C(O)$–OCH_3	421.3 ± 12.6	1
$C_6H_5C(O)$–OC_6H_5	307.5 ± 8.4	1
CH_3OCH_2–OCH_3	367.5 ± 8.4	1
$CH_3C(O)$–$OC(O)CH_3$	382.4 ± 12.6	1
$C_6H_5C(O)$–$OC(O)C_6H_5$	384.9 ± 16.7	1
CH_3–OOH	300.4 ± 12.6	1
C_2H_5–OOH	332.2 ± 20.9	1
C_3H_7–OOH	364.4	1
***iso*-C_3H_7**–OOH	298.3	1
***tert*-C_4H_9**–OOH	309.2 ± 4.2	1
CH_3–$OOCH_3$	292.5 ± 8.4	1
CF_3–$OOCF_3$	361.5 ± 8.4	1
CH_3–**OO**	137.0 ± 3.8	1
CF_3–**OO**	169.0	1
$CClF_2$–**OO**	127.6	1
CCl_2F-**OO**	124.7	1
CH_2Cl–**OO**	122.4 ± 10.5	1
$CHCl_2$–**OO**	108.2 ± 8.2	1
CCl_3–**OO**	92.0 ± 6.4	1
HC(O)–**OOH**	290.0	1
$CH_3C(O)$–$OOC(O)CH_3$	315.1	1
ClO–CF_3	≤369.9 ± 1.3	1
CH_3–**ONO**	245.2	1
C_2H_5–**ONO**	260.2	1
C_3H_7–**ONO**	249.4 ± 6.3	1
iso-C_3H_7–**ONO**	254.4 ± 6.3	1
C_4H_9–**ONO**	256.5 ± 6.3	1
iso-C_4H_9–**ONO**	254.0 ± 6.3	1
sec-C_4H_9–**ONO**	253.6 ± 6.3	1
tert-C_4H_9–**ONO**	252.7 ± 6.3	1
$(C_2H_5)(CH_3)_2C$–**ONO**	254.0 ± 8.4	1
CH_3–**ONO_2**	340.2	1
C_2H_5–**ONO_2**	344.8	1
CH_3O–CH_2CN	393.3	1
O–**N_2**	167.4 ± 0.4	1
O–**NO**	306.21 ± 0.13	1
O–**NO_2**	206.3	1
NO–**NO**	40.6 ± 2.1	1
O_2N–**ONO_2**	95.4 ± 1.5	1
cis-HO-**NO**	207.0	1
trans-HO-**NO**	200.64 ± 0.19	1
FO–**NO**	132.5 ± 17	1
cis-ClO–**NO**	127.6 ± 8.4	1
trans-ClO–**NO**	116.6 ± 8.4	1
cis-BrO–**NO**	138.1 ± 8.4	1
trans-BrO–**NO**	121.6 ± 8.4	1
trans-perp-HOO–**NO**	114.2 ± 4	1
CH_3O–**NO**	176.6 ± 3.3	1
C_2H_5O–**NO**	185.4 ± 4.2	1
C_3H_7O–**NO**	179.1 ± 6.3	1
iso-C_3H_7O–**NO**	175.3 ± 4.2	1
C_4H_9O–**NO**	177.8 ± 6.5	1
iso-C_4H_9O–**NO**	175.7 ± 6.5	1
sec-C_4H_9O–**NO**	173.6 ± 3.3	1
tert-C_4H_9O–**NO**	176.1 ± 5.9	1
tert-AmO–**NO**	171.1 ± 0.4	1
C_6H_5O–**NO**	87.0	1
HO–**NO_2**	205.4	1
FO–**NO_2**	131.8 ± 12.6	1
ClO–**NO_2**	110.9	4
BrO–**NO_2**	118.0 ± 6.3	1
IO–**NO_2**	~100	1
CH_3O–**NO_2**	176.1 ± 4.2	1
C_2H_5O–**NO_2**	174.5 ± 4.2	1
C_3H_7O–**NO_2**	177.0 ± 4.2	1
iso-C_3H_7O–**NO_2**	175.7 ± 4.2	1
HOO–**NO_2**	99.2 ± 4.6	1
CH_3OO–**NO_2**	86.6 ± 8.4	1
CF_3OO–**NO_2**	105	1
CF_2ClOO–**NO_2**	106.7	1
$CFCl_2OO$–**NO_2**	106.7	1
CCl_3OO–**NO_2**	95.8	1
$CH_3N(O)$-**O**	305.3 ± 4.4	1
$C_6H_5N(O)$-**O**	392 ± 8	1
C_5H_5N-**O**	264.9 ± 2.0	1
C_6H_5N=$N(O)(C_6H_5)$-**O**	309.4 ± 3.5	1
$C_6H_5(O)N$=$N(O)(C_6H_5)$-**O**	309.4 ± 3.6	1
O–SO	551.1	1
O–SOF_2	513.3	1
O–$SOCl_2$	398.5	1
O–$S(OH)_2$	493.7 ± 25	1
HO–SH	293.3 ± 16.7	1
HO–SOH	313.4 ± 12.6	1
HO–$S(OH)O_2$	384.9 ± 8.4	1
HO–SCH_3	303.8 ± 12.6	1
HO–SO_2CH_3	360.2 ± 12.6	1
F–OH	215.1	1
F–OF	164.1	1

Bond	D^o_{298}/kJ mol^{-1}	Ref.
F–OCF_3	200.8 ± 4.2	1
F–OCH_3	>196.6	1
F–ONO_2	143.1	1
Cl–OH	233.5	1
Cl–OCl	142	1
Cl–OCF_3	≤220.9 ± 8.4	1
Cl–OCH_3	200.8	1
Cl–O-*tert*-C_4H_9	198.3	1
Cl–OOCl	91.2	1
Cl–ONO_2	172.0	1
Br–OH	209.6 ± 4.2	1
Br–OBr	125	1
Br–O-*tert*-C_4H_9	183.3	1
Br–ONO_2	143.1 ± 6.3	1
I–OH	213.4	1
I–OI	130.1	1
I-ONO_2	>140.6	1

(5) N–X BDEs

Bond	D^o_{298}/kJ mol^{-1}	Ref.
H–NH_2	450.08 ± 0.24	1
H–NF_2	316.7 ± 10.5	1
H–NNH	254.4	1
H–N_3	≤389	1
H–N=CH_2	364 ± 25	1
H–NO	199.5	1
H–NHOH	341	1
H–NCO	460.7 ± 2.1	1
H–NCS	≤396.6 ± 4.6	1
H–NCS	347.3 ± 8.4	1
CH_3**NH_2**	425.1 ± 8.4	1
tert-Bu**NH_2**	397.5 ± 8.4	1
$C_6H_5CH_2$**NH_2**	418.4	1
$(CH_3)_2$**NH**	395.8 ± 8.4	1
H–NHNH(CH_3)	276 ± 21	1
H–NHN$(CH_3)_2$	356 ± 21	1
NH_2CN	414.2	1
$(NH_2)_2$C=O	464.4	1
$(NH_2)_2$C=S	389.1	1
CH_3CS**NH_2**	380.7	1
PhCS**NH_2**	380.7	1
(Ph**NH**)$_2$C=S	364.0	1
$(NH_2)_2$C=NH	435.1	1
Ph_2C=**NH**	489.5	1
H–N$(SiMe_3)_2$	464	1
H–NHPh	375.3	1
C_6H_5**NHOH**	292	1
C_6H_5**NH**(CONMe2)	387.9	1
H–NPh_2	364.8	1
HN–N_2	63	1
ON–N	480.7 ± 0.4	1
ON–NO	8.49 ± 0.12	1
ON–NO_2	42.5	1
O_2N–NO_2	57.3 ± 1	1
H_2N–NH_2	277.0 ± 1.3	1
F_2N–NF_2	92.9 ± 12.6	1
H_2N–$NHCH_3$	275.8 ± 8.4	1
H_2N–N$(CH_3)_2$	259.8 ± 8.4	1
H_2N–NHC_6H_5	227.6 ± 8.4	1
H_2N–NO_2	230	1
$H_2NN(CH_3)$–**NO**	179.6	1
$(C_6H_5)_2$N-**NO**	94.6	1
N_3–CH_3	335.1 ± 20.5	1
N_3–C_6H_5	375.7 ± 20.9	1
N_3–$CH_2C_6H_5$	211.3 ± 14.2	1
CH_3–**NC**	413.0 ± 3.3	1
C_2H_5–**NC**	413.4 ± 8.4	1
iso-C_3H_7–**NC**	423.0 ± 8.4	1
tert-C_4H_9–**NC**	399.6 ± 5.4	1
NC **NO**	204.4	1
CH_3–**NO**	172	1
CF_3–**NO**	167	1
CCl_3–**NO**	125	1
C_2H_5–**NO**	171.5	1
CH_2CHCH_2–**NO**	110	1
iso-C_3H_7–**NO**	152.7 ± 12.6	1
tert-C_4H_9–**NO**	167	1
C_6H_5–**NO**	226.8 ± 2.1	1
C_6F_5–**NO**	211.3 ± 4.2	1
$C_6H_5CH_2$–**NO**	123	1
CH_3–**NO_2**	260.7 ± 2.1	1
C_2H_5–**NO_2**	254.4	1
C_3H_7–**NO_2**	256.5	1
iso-C_3H_7–**NO_2**	259.8	1
C_4H_9–**NO_2**	254.8	1
sec-C_4H_9–**NO_2**	263.2	1
tert-C_4H_9–**NO_2**	258.6	1
C_6H_5–**NO_2**	295.8 ± 4.2	1
$C_6H_5CH_2$–**NO_2**	210.3 ± 6.3	1
$(NO_2)CH_2$–**NO_2**	207.1	1
$(NO_2)_3C$–**NO_2**	176.1	1
CF_3–**NF_2**	280.7	1
$C_6H_5CH_2$–**NF_2**	237.2 ± 14.6	1
CH_3–**NH_2**	356.1 ± 2.1	1
C_2H_5–**NH_2**	352.3 ± 6.3	1
C_3H_7–**NH_2**	356.1 ± 2.9	1
iso-C_3H_7–**NH_2**	357.7 ± 3.8	1
C_4H_9–**NH_2**	356.1 ± 2.9	1
sec-C_4H_9–**NH_2**	359.0 ± 2.9	1
iso-C_4H_9–**NH_2**	254.8 ± 5.0	1
tert-C_4H_9–**NH_2**	355.6 ± 6.3	1
pyridin-2-yl–**NH_2**	431	1
C_6H_5–**NH_2**	429.3 ± 4.2	1
$C_6H_5CH_2$–**NH_2**	306.7 ± 6.3	1
$C_6H_5CH(CH)_3$–**NH_2**	307.5 ± 9.6	1
HC(O)–**NH_2**	421.7 ± 8.4	1
$CH_3C(O)$–**NH_2**	414.6 ± 8.4	1
HS–**NO**	138.9	1
CH_3S–**NO**	104.6 ± 4.2	1
tert-BuS–**NO**	115.1	1
$PhCH_2S$–**NO**	120.5	1
C_6H_5S-**NO**	81.2 ± 5.4	1
SCN–SCN	255.6	1
FSO_2–NF_2	163	1
F–NO	235.26	1
F–NO_2	221.3	1
F–NF_2	254.0	1
F–NH_2	286.6	1
Cl–NO	158.8 ± 0.8	1
Cl–NO_2	141.8 ± 1.3	1
Cl–NF_2	~134	1
Cl–NH_2	253.1	1
Br–NO	120.1 ± 0.8	1
Br–NO_2	82.0 ± 7.1	1
Br–NF_2	<227.2	1
I–NO	75.6 ± 4	1
I–NO_2	79.6 ± 4	1

(6) S–X BDEs

Bond	D^o_{298}/kJ mol^{-1}	Ref.
H–SH	381.18 ± 0.05	1
H–SCH_3	365.7 ± 2.1	1
H–$SCHCH_2$	351.5 ± 8.4	1
H–SC_2H_5	365.3	1
H–SC_3H_7	365.7	1
H–S-*iso*-C_3H_7	369.9 ± 8.4	1
H–S-*tert*-C_4H_9	362.3 ± 9.2	1
H–SOH	330.5 ± 14.6	1
H–$SCOCH_3$	370.7	1
H–SCOPh	364	1
H–SO_2CH_3	≤397	1
H–$SSCH_3$	330.5 ± 14.6	1
H–SPh	349.4 ± 4.5	1
H–SSH	318.0 ± 14.6	1
H–SSSH	292.9 ± 6.5	1
HS–SH	270.7 ± 8.4	1
FS–SF	362.3	1
ClS–SCl	329.7	1
HS–SCH_3	272.0	1
HS–SPh	255.2 ± 6.3	1
CH_3S–SCH_3	272.8 ± 3.8	1
C_2H_5S–SC_2H_5	276.6	1
MeS–SPh	272.0 ± 6.3	1
C_6H_5S–SC_6H_5	214.2 ± 12.6	1
F_5S–SF_5	305 ± 21	1

Bond	D^o_{298}/kJ mol^{-1}	Ref.
HS–CH_3	312.5 ± 4.2	1
HS–C_2H_5	307.9 ± 2.1	1
HS–C_3H_7	310.5 ± 2.9	1
HS–*iso*-C_3H_7	307.1 ± 3.8	1
HS–C_4H_9	309.2 ± 2.9	1
HS–*sec*-C_4H_9	307.5 ± 2.9	1
HS–*iso*-C_4H_9	310.0 ± 4.6	1
HS–*tert*-C_4H_9	301.2 ± 3.8	1
HS–C_6H_5	360.7 ± 6.3	1
HS–$CH_2C_6H_5$	258.2 ± 6.3	1
HS–C(O)H	309.6 ± 8.4	1
HS–C(O)CH_3	307.9 ± 6.3	1
CH_3S–CH_3	307.9 ± 3.3	1
HOS–CH_3	284.9 ± 12.6	1
CH_3SO–CH_3	221.8 ± 8.4	1
$HOSO_2$–CH_3	324.3 ± 12.6	1
CH_3SO_2–CH_3	279.5	1
F_5S–CF_3	392 ± 43	1
F–SF_5	391.6	1
F–SO_2(F)	379	1
Cl–SF_5	<272	1
Cl–SO_2CH_3	293	1
Cl-SO_2Ph	297	1
Br-SBr	259 ± 17	1
Br–SF_5	<230	1
I-SH	206.7 ± 8.4	1
I-SCH_3	206.3 ± 7.1	1
(7) Si-, Ge-, Sn-, and Pb–X BDEs		
SiH_3–H	383.7 ± 2.1	1
Me_3Si–H	396 ± 7	1
H_5Si_2–H	373 ± 8	1
$(C_2H_5)_3Si$–H	396 ± 4	1
$C_6H_5SiH_2$–H	382 ± 5	1
$(CH_3S)_3Si$–H	364.0	1
$(iPrS)_3Si$–H	376.6	1
$PhMe_2Si$–H	377 ± 7	1
Ph_2SiH–H	379 ± 7	1
Ph_2MeSi–H	361 ± 10	1
SiF_3–H	432 ± 5	1
$SiCl_3$–H	391	5
$SiBr_3$–H	334 ± 8	1
SiH_3–SiH_3	321 ± 4	1
SiH_3–Si_2H_5	313 ± 8	1
Ph_3Si–$SiPh_3$	368.2	1
F_3Si–SiF_3	453.1 ± 25	1
SiH_3–CH_3	375 ± 5	1
SiF_3–CH_3	355.6	1
H_3Si–NO	158.2 ± 5.7	1
H_3Si–PH_2	331.4	1
SiH_3–F	638 ± 5	1
SiH_3–Cl	458 ± 7	1
SiH_3–Br	376 ± 9	1
SiH_3–I	299 ± 8	1
GeH_3–H	348.9 ± 8.4	1
Me_3Ge–H	364.0	1
Ph_3Ge–H	359.8	1
$(CH_3)_3Ge$-$Ge(CH_3)_3$	280.3	1
$(CH_3)_3Ge$-CH_3	288.7	1
Me_3Sn–H	326.4	1
Ph_3Sn–H	294.6	1
$(CH_3)_3Sn$-$Sn(CH_3)_3$	257.7	1
$(CH_3)_3Sn$-Cl	425 ± 17	1
$(CH_3)_3Pb$-$Pb(CH_3)_3$	228.4	1
Cl_3Pb-Cl	271 ± 84	1
$(CH_3)_3Pb$-CH_3	238 ± 21	1
(8) P-, As-, Sb-, Bi–X BDEs		
H_2P–H	351.0 ± 2.1	1
CH_3PH–H	322.2 ± 12.6	1
H_2P–PH_2	256.1	1
$(C_2H_5)_2P$–$P(C_2H_5)_2$	359.8	1
F_2P–F	549	1
Cl_2P–Cl	356 ± 8	1
Br_2P–Br	<259	1
I_2P–I	217	1
H_2P–SiH_3	331.4	1
H_2As–H	319.2 ± 0.8	1
H_2Sb-H	288.3 ± 2.1	1
F_2Bi–F	435 ± 19	1
Br_2Bi–Br	>297.1	1
(9) Se- and Te–X BDEs		
H–SeH	334.93 ± 0.75	1
H–SeC_6H_5	326.4 ± 16.7	1
PhSe–SePh	280 ± 19	1
H–TeH	277.0 ± 5.0	1
H–TeC_6H_5	≤264	1
PhTe–TePh	138.1 ± 12.6	1
(10) Metal-Centered BDEs		
Arranged by the Periodic Table		
(10.1) Group 1		
Li–OH	431.0	1
Li–C_2H_5	214.6 ± 8.4	1
Li–nC_4H_9	197.9 ± 16.3	1
Na–OH	342.3	1
Na–O_2	<200	1
K–OH	359	1
Rb–OH	356.2 ± 4.2	1
Cs–OH	373	1
(10.2) Group 2		
BeO–H	469	1
Be(OH)–OH	476	1
MgO–H	441	1
Mg(OH)–OH	349	1
BrMg–CH_3	253	1
BrMg–CH_2CH_3	205	1
BrMg–i-C_3H_7	184	1
BrMg–t-C_4H_9	174	1
BrMg–C_6H_5	289	1
BrMg–$CH_2C_6H_5$	201	1
BrMg–$C(C_6H_5)_3$	180	1
Ca(OH)–OH	409	1
Sr(OH)–OH	407	1
Ba(OH)–OH	443	1
(10.3) Group 3		
Sc–CH_3	116 ± 29	1
Sc–C_6H_6	60.8	1
La(η^5-C_5Me_5)$_2$–CH($SiMe_3$)$_2$	278.7 ± 10.5	1
Nd(η^5-C_5Me_5)$_2$–CH($SiMe_3$)$_2$	236.8 ± 10.5	1
(η^5-C_5Me_5)$_2$Sm–H	226.8 ± 12.6	1
(η^5-C_5Me_5)$_2$Sm–OCH_3	343.1	1
(η^5-C_5Me_5)$_2$Sm–(η^3-C_3H_5)	188.3 ± 6.3	1
(η^5-C_5Me_5)$_2$Sm–S-nC_3H_7	295.4 ± 10.0	1
(η^5-C_5Me_5)$_2$Sm–N(CH_3)$_2$	201.7 ± 7.5	1
(η^5-C_5Me_5)$_2$Sm–SiH($SiMe_3$)$_2$	179.9 ± 21	1
(η^5-C_5Me_5)$_2$Sm–P(Et)$_2$	136.4 ± 8.4	1
(η^5-C_5Me_5)$_2$Eu–I	238.9 ± 8.4	1
(η^5-C_5Me_5)$_2$Yb–I	256.1 ± 6.3	1
Lu(η^5-C_5Me_5)$_2$–CH($SiMe_3$)$_2$	279.1 ± 10.5	1
(η^5-$C_5H_4SiMe_3$)$_3$Th–H	277 ± 6	1
(η^5-$C_5H_4SiMe_3$)$_3$Th–O	371 ± 24	1
(η^5-C_5H_5)$_3$Th–CH_3	375 ± 9	1
(η^5-C_5H_5)$_3$Th–$CH_2Si(CH_3)_3$	369 ± 12	1
(C_9H_7)$_3$Th–$CH_2C_6H_5$	342 ± 9	1
(η^5-C_5H_4tBu)$_3$U–H	249.7 ± 5.7	1
(η^5-$C_5H_4SiMe_3$)$_3$U–H	253.7 ± 5.1	1
[HB(3,5-Me_2Pz)$_3$]U(Cl)$_2$–Cl	422.6	1
(η^5-$C_5H_4SiMe_3$)$_3$U–I	265.6 ± 4.3	1
(η^5-C_5H_4tBu)$_3$U–O	307 ± 9	1
(η^5-$C_5H_4SiMe_3$)$_3$U–CO	43.1 ± 0.8	1
(C_9H_7)$_3$U–CH_3	196.3 ± 6.6	1
(η^5-C_5Me_5)$_2$U(Cl)–C_6H_5	358 ± 11	1
(η^5-$C_5H_4SiMe_3$)$_3$U–THF	41.0 ± 0.8	1
(10.4) Group 4		
Ti(η^5-C_5H_5)$_2$–Cl	471	1
Ti(Cl)(η^5-C_5H_5)$_2$–Cl	390	1
Ti(η^5-C_5Me_5)$_2$–I	219	1

Bond	D^o_{298}/kJ mol^{-1}	Ref.
$Ti(\eta^5\text{-}C_5H_5)_2\text{–}CO$	174	1
$Ti(CO)(\eta^5\text{-}C_5H_5)_2\text{–}CO$	170	1
$Ti\text{–}CH_3$	174 ± 29	1
$Ti(Cl)(\eta^5\text{-}C_5H_5)_2\text{–}CH_3$	276	1
$Ti(Cl)((\eta^5\text{-}C_5H_5)_2\text{–}C_6H_5$	292	1
$Ti(C_6H_6)\text{–}C_6H_5$	308.7	1
$Zr(\eta^5\text{-}C_5Me_5)_2\text{–}H$	351.0 ± 7.5	1
$Zr(H)(\eta^5\text{-}C_5Me_5)_2\text{–}H$	326.4 ± 4	1
$Zr(\eta^5\text{-}C_5Me_5)_2\text{–}Cl$	481.2	1
$Zr(\eta^5\text{-}C_5Me_5)_2\text{–}Br$	410.0	1
$Zr(I)(\eta^5\text{-}C_5Me_5)_2\text{–}I$	336.4 ± 2.1	1
$Zr(\eta^5\text{-}C_5Me_5)_2(Ph)\text{–}OH$	482.4 ± 6.3	1
$Zr(\eta^5\text{-}C_5Me_5)_2(Ph)(OH)\text{–}OH$	482.8 ± 10.5	1
$Zr(\eta^5\text{-}C_5Me_5)_2(NH_2)H\text{–}NH_2$	421.3 ± 15.1	1
$Zr(\eta^5\text{-}C_5Me_5)_3\text{–}CH_3$	276 ± 10	1
$Zr(\eta^5\text{-}C_5H_5)_2(C_6H_5)\text{–}C_6H_5$	300 ± 10	1
$Zr(\eta^5\text{-}C_5H_5)_2(Si(SiMe_3)_3)\text{–}SiMe_3$	188 ± 30	1
$Hf(H)(\eta^5\text{-}C_5Me_5)_2\text{–}H$	346.0 ± 7.9	1
$Hf(\eta^5\text{-}C_5Me_5)(C_4H_9)\text{–}C_4H_9$	274 ± 10	1
(10.5) Group 5		
$(\eta^5\text{-}C_5H_5)(CO)_3V\text{–}\eta^2H_2$	90 ± 20	1
$(\eta^5\text{-}C_5H_5)(CO)_3V\text{–}CO$	146 ± 21	1
$V\text{–}CH_3$	169 ± 18	1
$V\text{–}C_6H_6$	76.2	1
$V(C_6H_6)\text{–}C_6H_6$	307.8	1
$Nb(\eta^5\text{-}C_5H_5)_2H_3\text{–}TFE$	18.8 ± 1.3	1
$Ta(CH_3)_5\text{–}CH_3$	261 ± 5	1
$(Me_3SiCH_2)_4Ta\text{–}(CH_2SiMe_3)$	184.1 ± 8.4	1
(10.6) Group 6		
$[Cr(CO)_3(\eta^5\text{-}C_5Me_5)]_2\text{–}Hg$	61.5	1
$[Cr(CO)_3(\eta^5\text{-}C_5Me_5)]\text{–}Hg$	111.3	1
$Cr(CO)_5\text{–}Xe$	37.7 ± 3.8	1
$(CO)_2(PPh_3)(\eta^5\text{-}C_5H_5)Cr\text{–}H$	250.2 ± 4.2	1
$(\eta^5\text{-}C_5H_5)Cr(CO)_3\text{–}H,$	257	1
$Cr(CO)_5\text{–}H_2$	78 ± 4	1
$(P(C_6H_{11})_3)_2(CO)_3Cr\text{–}H_2$	30.5 ± 0.4	1
$(\eta^6\text{-}C_6H_6)(CO)_3Cr\text{–}H_2$	251 ± 17	1
$Cr(CO)_5\text{–}N_2$	81 ± 4	1
$(P(C_6H_{11})_3)_2(CO)_3Cr\text{–}N_2$	38.9 ± 0.8	1
$(\eta^5\text{-}C_5Me_5)(CO)_3Cr\text{–}SH$	193	1
$Cr(CO)_5\text{–}CO$	154.0 ± 8.4	1
$Cr(CO)_5\text{–}CH_4$	~33.5 ± 8	1
$Cr\text{–}C_6H_6$	9.6 ± 5.8	1
$Cr(C_6H_6)\text{–}C_6H_6$	268.2 ± 15.4	1
$Cr(CO)_5\text{–}C_6H_6$	57.3 ± 3.3	1
$(P(C_6H_{11})_3)_2(CO)_3Cr\text{–}P(OMe_3)_3$	68.6 ± 2.5	1
$(\eta^5\text{-}C_5H_5))Mo(CO)_3\text{–}H$	290	1
$Mo(\eta^5\text{-}C_5H_5)_2\text{–}H$	246	1
$Mo(H)(\eta^5\text{-}C_5H_5)_2\text{–}H$	256.9 ± 8.4	1
$Mo(CO)_3(\eta^5\text{-}C_5H_5)\text{–}I$	216.7 ± 4.2	1
$(\eta^5\text{-}C_5Me_5)_2Mo\text{–}O$	272	1
$(P(C_6H_{11})_3)_2(CO)_3Mo\text{–}H_2$	27.2 ± 0.8	1
$(P(C_6H_{11})_3)_2(CO)_3Mo\text{–}N_2$	37.7 ± 2.5	1
$Mo(CO)_5\text{–}CO$	169.5 ± 8.4	1
$Mo(CO)_3(\eta^5\text{-}C_5H_5)\text{–}CH_3$	203 ± 8	1
$W(CO)_5\text{–}Xe$	35.1 ± 0.8	1
$W(CO)_3(\eta^5\text{-}C_5H_5)\text{–}H$	303	1
$W(H)(\eta^5\text{-}C_5H_5)_2\text{–}H$	310.9 ± 4.2	1
$W(I)(\eta^5\text{-}C_5H_5)_2\text{–}H$	273 ± 14	1
$(CO)_5W\text{–}H_2$	≥67	1
$(P(C_6H_{11})_3)(CO)_3W\text{–}(\eta^2\text{-}H_2)$	28.5 ± 2.1	1
$W(CO)_5\text{–}CO$	192.5 ± 8.48.4	1
$W(CH_3)(\eta^5\text{-}C_5H_5)_2\text{–}CH_3$	220.9 ± 4	1
(10.7) Group 7		
$F_3Mn\text{–}MnF_3$	210.9 ± 2.5	1
$(CO)_5Mn\text{–}Mn(CO)_5$	185 ± 8	1
$(CO)_5Mn\text{–}H$	284.5	1
$(PPh_3)Mn(CO)_4\text{–}H$	286.2	1
$MnBr(CO)_4\text{–}CO$	184	1
$(\eta^5\text{-}C_5H_5)(CO)_2Mn\text{–}CO$	195.8 ± 9.2	1
$Mn\text{–}CH_3$	>35 ± 12	1
$Mn(CO)_5\text{–}CH_3$	187.0 ± 3.8	1
$Mn(CO)_5\text{–}C_6H_5$	207 ± 11	1
$(CO)_5Mn\text{–}Re(CO)_5$	149 ± 11	1
$(\eta^5\text{-}C_5H_5)Mn(CO)_2\text{–}PhMe$	59.4 ± 3.3	1
$(CO)_5Tc\text{–}Tc(CO)_5$	177.5 ± 1.9	1
$(CO)_5Re\text{–}Re(CO)_5$	187 ± 4.8	1
$(CO)_5Re\text{–}H$	313	1
$(CO)_5Re\text{–}CH_3$	220 ± 8	1
(10.8) Group 8		
$(CO)_4Fe\text{–}Fe(CO)_5$	171.5	1
$(CO)_4Fe(H)_x\text{–}H$	259.4 ± 8.4	1
$(\eta^5\text{-}C_5H_5)(CO)_2Fe\text{–}H$	239	1
$Fe(CO)_3(N_2)\text{–}N_2$	37.7 ± 19.2	1
$Fe(C_2H_2)(CO)_4\text{–}CO$	88 ± 2.3	1
$Fe(CO)_2(PMe_3)\text{–}CO$	>125	1
$Fe(CO)_3(PPh_3)\text{–}CO$	<177.8 ± 5	1
$Fe\text{–}NH_3$	31.4 ± 4.2	1
$Fe\text{–}CH_2$	364 ± 29	1
$Fe\text{–}CH_3$	135 ± 29	1
$Fe(C_2H_4)(CO)_3\text{–}C_2H_4$	89.1 ± 8	1
$Fe\text{–}C_3H_5$	218	1
$Fe\text{–}C_3H_6$	79	1
$Fe(CO)_5\text{–}Ni(CO)_4$	37.7	1
$Fe(CO)_5\text{–}(\eta^3\text{-}C_3H_5)$	176	1
$Fe(C_3H_6)(CO)_3\text{–}C_3H_6$	~79.5	1
$(CO_2)(\eta^5\text{-}C_5H_5)Ru\text{–}H$	272	1
$(PMe_3)_2(\eta^5\text{-}C_5Me_5)Ru\text{–}H$	167.4	1
$(CO)_2(\eta^5\text{-}C_5Me_5)Ru\text{–}Cl$	337.6	1
$(\eta^5\text{-}C_5Me_5)(PMe_3)_2Ru\text{–}Cl$	<138	1
$(\eta^5\text{-}C_5Me_5)(PMe_3)_2Ru\text{–}OH$	204.6	1
$(CO)_4Ru\text{–}CO$	115 ± 1.7	1
$(\eta^5\text{-}C_5Me_5)(PMe_3)_2Ru\text{–}CH_3$	142.3	1
$Os(H)(CO)_4\text{–}H$	326.4	1
$(CO)_4Os\text{–}CO$	133 ± 2.6	1
$Os(C_2H_2)(CO)_4\text{–}CO$	99.5 ± 0.8	1
(10.9) Group 9		
$(CO)_4Co\text{–}Co(CO)_4$	83 ± 29	1
$(CO)_4Co\text{–}Mn(CO)_5$	96 ± 12	1
$(CO)_4Co\text{–}Re(CO)_5$	113 ± 15	1
$Co(CO)_4\text{–}H$	278	1
$Co(CO)_3(PPh_3)\text{–}H$	272	1
$(CO)_3HCo\text{–}CO$	~54	1
$(\eta^5\text{-}C_5H_5)Co(CO)\text{–}CO$	184.3 ± 4.8	1
$Co\text{–}CH_2$	331 ± 38	1
$Co\text{–}CH_3$	178 ± 8	1
cobalamin–CH_3	150.6	1
cobinamide–iC_4H_9	104	1
Co–C bonds in B_{12}	123.8 ± 6.3	1
$Cl(CO)_2Rh\ Rh(CO)_2Cl$	94.6	1
HRh(m-xylyl)Rh–H	255.6 ± 1.7	1
$(PiPr_3)_2(Cl)Rh\text{–}H_2$	136.0	1
$(PiPr_3)_2(Cl)Rh\text{–}N_2$	69.0	1
$(PiPr_3)_2(Cl)Rh\text{–}CO$	201.7	1
HRh(m-xylyl)Rh–CH_2OH	195.4 ± 7.5	1
$Ir(Cl)(CO)(PMe_3)_2\text{–}H$	251	1
$Ir(H)(\eta^5\text{-}C_5Me_5)(PMe_3)\text{–}H$	310.5 ± 21	1
$Ir(Cl)(H)(CO)(PEt_3)_2\text{–}H$	243.1	1
$Ir(Cl)(H)(CO)(PPh_3)_2\text{–}H$	246.9	1
$(Cl)(CO)(PPh_3)_2Ir\text{–}H_2$	62.8	1
$(Cl)(CO)(PPh_3)_2Ir\text{–}CO$	45.2	1
$Ir(H)(\eta^5\text{-}C_5Me_5)(PMe_3)\text{–}C_6H_5$	321	1
(10.10) Group 10		
$Ni\text{–}H_2O$	~29	1
$Ni(CO)_3\text{–}N_2$	~42	1

Bond	D^o_{298}/kJ mol^{-1}	Ref.
$Ni(CO)_3$–CO	104.6 ± 8.4	1
Ni–CH_3	208 ± 8	1
Ni–C_2H_2	193 ± 25	1
Ni–C_2H_4	147.3 ± 17.6	1
Ni–propyne	155 ± 21	1
Ni–2-butyne	121 ± 21	1
Pd–OH	213	1
trans-Pt$(PPh_3)_2$(Cl)–H	307 ± 37	1
$[Ph_2PCH_2]_2$MePt–H	104.6	1
$[Ph_2PCH_2]_2$MePt –OH	167.4	1
$[Ph_2PCH_2]_2$MePt –SH	90.0	1
Pt(η^5-C_5H_5)$(CH_3)_2$–CH_3	163 ± 21	1
cis-Pt$(PEt_3)_2(CH_3)$–CH_3	269 ± 13	1
(10.11) Group 11		
Cu–OH	>406	1
Cu–CO	25 ± 5	1
Cu–CH_3	223 ± 5	1
Cu–NH_3	47 ± 15	1
Cu(NH_3)–NH_3	83.7 ± 4.2	1
Cu–C_6H_6	16.4 ± 12.5	1
Cu(C_6H_6)–C_6H_6	27.0 ± 19.3	1
Ag–CH_3	134.1 ± 6.8	1
Ag–NH_3	8 ± 13	1
Ag(NH_3)–NH_3	62.8 ± 4.2	1
Au–OH	>262	1
Au–NH_3	76 ± 6	1
Au–CH_3	≥191.6	1
Au–C_6H_6	8.4	1
(10.12) Group 12		
Zn–CH_3	70 ± 10	1
Zn(CH_3)–CH_3	266.5 ± 6.3	1
Zn–C_2H_5	92.0 ± 17.6	1
Zn(C_2H_5)–C_2H_5	219.2 ± 8.4	1
Cd–CH_3	63.6 ± 10.0	1
Cd(CH_3)–CH_3	234.3 ± 6.3	1
Hg–CH_3	22.6 ± 12.6	1
Hg(CH_3)–CH_3	239.3 ± 6.3	1
ClHg–CH_3	280.0 ± 12.6	1
BrHg–CH_3	270 ± 38	1
IHg–CH_3	258.6 ± 12.6	1
(10.13) Group 13		
H_3B–BH_3	172	1
H_3B–NH_3	130.1 ± 4.2	1
$(CH_3)_3B$–NH_3	57.7 ± 1.3	1
F_3B–$N(CH_3)_3$	130 ± 4.6	1
Cl_3B–$N(CH_3)_3$	127.6	1
F_2B–CH_3	397 - 418	1
Al–OH	547 ± 13	1
Al–C_2H_2	>54	1
Cl_3Al–$N(CH_3)_3$	198.7 ± 8.4	1
$(CH_3)_3Al$–$N(CH_3)_3$	130	1
$(CH_3)_3Al$–$O(CH_3)_2$	92	1
$(CH_3)_3Ga$–$O(C_2H_5)_2$	50.6 ± 0.8	1
Cl_3Ga-$S(C_2H_5)_2$	235.1	1
In–CH_3	216.3	1
In$(CH_3)_1$–CH_3	318.8	1
In$(CH_3)_2$–CH_3	587.4	1
$(CH_3)_3In$–$N(CH_3)_3$	83.3 ± 2.1	1
Tl–OH	330 ± 30	1

References

1. Luo, Y. R. *Comprehensive Handbook of Chemical Bond Energies*, CRC Press, Boca Raton, FL, 2007.
2. Shuman, N. S., Ochieng, M. A., Sztáray, B., and Baer, T., *J. Phys. Chem. A* 112, 5647, 2008.
3. Seetula, J. A., and Eskola, A. J., *Chem. Phys.* 351, 141, 2008.
4. Golden, D. M., *Int. J. Chem. Kinet.* 41, 573, 2009.
5. Shuman, N. S., Spencer, A. P., and Baer, T., *J. Phys. Chem. A* 113, 9458, 2009.

TABLE 4. Enthalpies of Formation of Free Radicals and Other Transient Species

References: Yu-Ran Luo, *Comprehensive Handbook of Chemical Bond Energies*, CRC Press, 2007.

Radical	$\Delta_f H^o_{298}$/kJ mol^{-1}	Ref.
(1) Carbon-Centered Species		
CH	595.8 ± 0.6	1
CH_2 (triplet)	391.2 ± 1.6	1
CH_2 (singlet)	428.8 ± 1.6	1
$^{\bullet}CH_3$, methyl	146.7 ± 0.3	1
$^{\bullet}C_2H$, acetenyl, $CH{\equiv}C^{\bullet}$	567.4 ± 2.1	1
$^{\bullet}C_2H_2$, vinylidene $CH_2{=}C^{\bullet\bullet}$	419.7 ± 16.7	1
$^{\bullet}C_2H_3$, vinyl, $CH_2{=}C^{\bullet}H$	299.6 ± 3.3	1
$^{\bullet}C_2H_5$, ethyl, $CH_3C^{\bullet}H_2$	118.8 ± 1.3	1
$^{\bullet}C_3H_3$, propargyl, $CH{\equiv}CC^{\bullet}H_2$	351.9	2
$^{\bullet}C_3H_3$, $CH_3C{\equiv}C^{\bullet}$	515 ± 13	1
$^{\bullet}C_3H_3$, $CH_2{=}C{=}CH^{\bullet} \leftrightarrow CH{\equiv}CC^{\bullet}H_2$	351.9	2
$^{\bullet}C_3H_3$, cyclopro-2-en-1-yl	439.7 ± 17.2	1
$^{\bullet}C_3H_5$, allyl, $CH_2{=}CHC^{\bullet}H_2$	171.0 ± 3.0	1
$^{\bullet}C_3H_5$, $CH_3CH{=}C^{\bullet}H$	267 ± 6	1
$^{\bullet}C_3H_5$, $CH_3C^{\bullet}{=}CH_2$	231.4	1
$^{\bullet}C_3H_5$, cyclopropyl	279.9 ± 10.5	1
n-$C_3H_7^{\bullet}$, n-propyl, $CH_3CH_2C^{\bullet}H_2$	100 ± 2	1
i-$C_3H_7^{\bullet}$, i-propyl, $CH_3C^{\bullet}HCH_3$	88 ± 3	1
$^{\bullet}$n-C_4H_3, $CH{\equiv}CCH{=}C^{\bullet}H$	547.3	1
$^{\bullet}$i-C_4H_3, $CH_2{=}C^{\bullet}C{\equiv}CH$	499.2	1
$^{\bullet}C_4H_5$, $CH_3C{\equiv}CC^{\bullet}H_2$	304.5	1
$^{\bullet}C_4H_5$, $CH{\equiv}CC^{\bullet}HCH_3$	316.5	1
$^{\bullet}C_4H_5$, $^{\bullet}CH{=}CHCHCH_2$	364.4	1
$^{\bullet}C_4H_5$, $CH_2{=}CHC^{\bullet}CH_2$	313.3	1
$^{\bullet}C_4H_7$, $CH_3CH{=}CHC^{\bullet}H_2$	146 ± 8	1
$^{\bullet}C_4H_7$, $CH_2{=}CHCH_2C^{\bullet}H_2$	192.5	1
$^{\bullet}C_4H_7$, $CH_2{=}C(CH_3)C^{\bullet}H_2$	137.9	1
$^{\bullet}C_4H_7$, $CH_2{=}CHC^{\bullet}HCH_3$	136.2	1
$^{\bullet}C_4H_7$, cyclopropylmethyl	213.8 ± 6.7	1
$^{\bullet}C_4H_7$, cyclobutyl	219.2 ± 4.2	1
n-$C_4H_9^{\bullet}$, n-butyl, $CH_3CH_2CH_2C^{\bullet}H_2$	77.8 ± 2.1	1
i-$C_4H_9^{\bullet}$, i-butyl, $(CH_3)_2CHC^{\bullet}H_2$	70 ± 4	1
s-$C_4H_9^{\bullet}$, s-butyl, $CH_3C^{\bullet}HCH_2CH_3$	67.8 ± 2.1	1
t-$C_4H_9^{\bullet}$, t-butyl, $(CH_3)_3C^{\bullet}$	48 ± 3	1

Radical	$\Delta_f H^\circ_{298}$/kJ mol^{-1}	Ref.
$^{\bullet}C_5H_3$, $CH\equiv C-C\equiv CC^{\bullet}H_2$	579.1	1
$^{\bullet}C_5H_3$, $(CH\equiv C)_2C^{\bullet}H$	573.2	1
$^{\bullet}C_5H_5$, $CH_2=CHC\equiv CC^{\bullet}H_2$	351.5	1
$^{\bullet}C_5H_5$, $CH_2=CH-C^{\bullet}H-C\equiv CH$	372.4	1
$^{\bullet}C_5H_5$, cyclopenta-1,3-dien-5-yl	274.1 ± 7.3	1
$^{\bullet}C_5H_7$, $CH_3C\equiv CC^{\bullet}HCH_3$	272.8 ± 9.2	1
$^{\bullet}C_5H_7$, $CH\equiv CC^{\bullet}HC_2H_5$	277.0 ± 8.4	1
$^{\bullet}C_5H_7$, $CH\equiv CC^{\bullet}(CH_3)_2$	257.3 ± 9.2	1
$^{\bullet}C_5H_7$, $CH_2=CHCH=CHC^{\bullet}H_2$	205.0 ± 12.6	1
$^{\bullet}C_5H_7$, $(CH_2=CH)_2C^{\bullet}H$	208.0 ± 4.2	1
$^{\bullet}C_5H_7$, $CH_3CH=C=CHC^{\bullet}H_2$	278.0	1
$^{\bullet}C_5H_7$, spiropentyl	380.7 ± 4.2	1
$^{\bullet}C_5H_7$, cyclopent-1-en-3-yl	160.7 ± 4.2	1
$^{\bullet}C_5H_9$, cyclopentyl	105.9 ± 4.2	1
$^{\bullet}C_5H_9$, $CH_2=CHC^{\bullet}HCH_2CH_3$	109.6 ± 8.4	1
$^{\bullet}C_5H_9$, $CH_3CH=CHC^{\bullet}H(CH_3)$	92	1
$^{\bullet}C_5H_9$, $CH_3CH=C(CH_3)C^{\bullet}H_2$	92.0	1
$^{\bullet}C_5H_9$, $CH_2=CHC^{\bullet}(CH_3)_2$	87.0 ± 8.4	1
$^{\bullet}C_5H_9$, $CH_2=C(CH_3)C^{\bullet}H(CH_3)$	93.7	1
$^{\bullet}C_5H_9$, $CH_2=C(C^{\bullet}H_2)CH_2CH_3$	114.2	1
$^{\bullet}C_5H_9$, $CH_2=CH(CH_2)_2C^{\bullet}H_2$	179.5	1
$nC_5H_{11}^{\bullet}$, $CH_3CH_2CH_2CH_2C^{\bullet}H_2$	54.4	1
$^{\bullet}C_5H_{11}$, $(C_2H_5)_2C^{\bullet}H$	47.0	1
$^{\bullet}C_5H_{11}$, $(nC_3H_7)(CH_3)C^{\bullet}H$	50.2	1
$^{\bullet}C_5H_{11}$, $(CH_3)_3C^{\bullet}CH_2$	36.4 ± 8.4	1
$^{\bullet}C_5H_{11}$, $(C_2H_5)(CH_3)_2C^{\bullet}$	29	1
$^{\bullet}C_6H_5$, phenyl	330.1 ± 3.3	1
$^{\bullet}C_6H_7$, cyclohexa-1,3-dien-5-yl	199.2	1
$^{\bullet}C_6H_7$, cyclohexa-1,4-dien-3-yl	208.0 ± 3.9	5
$^{\bullet}C_6H_9$, $CH_3C\equiv CC^{\bullet}(CH_3)_2$	221.8 ± 9.2	1
$^{\bullet}C_6H_9$, $(CH_2=CH)_2C^{\bullet}(CH_3)$	193.7	1
$^{\bullet}C_6H_9$, cyclohexa-1-en-3-yl	119.7	1
$^{\bullet}C_6H_{11}$, $CH_2=CH(CH_2)_3C^{\bullet}H_2$	158.6	1
$^{\bullet}C_6H_{11}$, $CH_2=CHC^{\bullet}H(CH_2)_2CH_3$	89.0	1
$^{\bullet}C_6H_{11}$, $CH_2=C(CH_3)C^{\bullet}(CH_3)_2$	37.7 ± 6.3	1
$^{\bullet}C_6H_{11}$, $(CH_3)_2C=C(CH_3)C^{\bullet}H_2$	39.7 ± 6.3	1
$^{\bullet}C_6H_{11}$, $(CH_3)_2C=CHC^{\bullet}H(CH_3)$	47.3	1
$^{\bullet}C_6H_{11}$, (Z)-$CH_3CH=CHC^{\bullet}(CH_3)_2$	54.4	1
$^{\bullet}C_6H_{11}$, cyclohexyl	75.3 ± 6.3	1
$nC_6H_{13}^{\bullet}$, $CH_3CH_2CH_2CH_2CH_2C^{\bullet}H_2$	33.5	1
$^{\bullet}C_6H_{13}$, $(nC_4H_9)(CH_3)C^{\bullet}H$	29.3	1
$^{\bullet}C_6H_{13}$, 2-methyl-2-pentyl	3.3 ± 8.4	1
$^{\bullet}C_6H_{13}$, 3-methyl-3-pentyl	14.2	1
$^{\bullet}C_6H_{13}$, 2,3-dimethyl-2-butyl	3.1 ± 10	1
$^{\bullet}C_7H_3$, $(CH\equiv C)_3C^{\bullet}$	784.5	1
$^{\bullet}C_7H_7$, benzyl, $C_6H_5C^{\bullet}H_2$	208.0 ± 1.7	1
$^{\bullet}C_7H_7$, quadricyclolan-5-yl	578.6 ± 5.4	1
$^{\bullet}C_7H_7$, quadricyclolan-4-yl	587.4 ± 5.4	1
$^{\bullet}C_7H_7$, norborna-2,5-dien-7-yl	511.7 ± 7.9	1
$^{\bullet}C_7H_7$, cyclohepta-1,3,5-trien-7-yl	285.3 ± 12.6	1
$^{\bullet}C_7H_9$, $CH_2=CH(CH=CH)_2CC^{\bullet}H_2$	251.0	1

Radical	$\Delta_f H^\circ_{298}$/kJ mol^{-1}	Ref.
$^{\bullet}C_7H_9$, $(CH_2=CH)_3C^{\bullet}$	274.0	1
$^{\bullet}C_7H_{11}$, norborn-1-yl	136.4 ± 10.5	1
$^{\bullet}C_7H_{11}$, cycloheptenyl	119.2	1
$^{\bullet}C_7H_{13}$, cycloheptyl	50.6 ± 4.2	1
$^{\bullet}C_7H_{13}$, cyclo-$[C^{\bullet}(CH_3)(CH_2)_5]$	22.6	1
$^{\bullet}C_7H_{13}$, cyclo-$[C^{\bullet}(CH_2CH_3)(CH_2)_4]$	47.0	1
$^{\bullet}C_7H_{15}$, $(nC_5H_{11})(CH_3)CH^{\bullet}$	8.4	1
$^{\bullet}C_7H_{15}$, $(CH_3)_2CHCHC^{\bullet}(CH_3)_2$	−21.8 ± 5.2	1
$^{\bullet}C_8H_7$, cubyl	831.0 ± 16.7	1
$^{\bullet}C_8H_7$, $C_6H_5C^{\bullet}=CH_2$	309.6	1
$^{\bullet}C_8H_7$, $C_6H_5CH=CH^{\bullet}$	387.0	1
$^{\bullet}C_8H_9$, $C_6H_5C^{\bullet}H(CH_3)$	175.7 ± 7.5	1
$^{\bullet}C_8H_9$, $C_6H_5CH_2C^{\bullet}H_2$	236.0 ± 7.5	1
$^{\bullet}C_8H_9$, p-$CH_3C_6H_4C^{\bullet}H_2$	167.4	1
$^{\bullet}C_8H_9$, m-$CH_3C_6H_4C^{\bullet}H_2$	167.4	1
$^{\bullet}C_8H_9$, o-$CH_3C_6H_4C^{\bullet}H_2$	167.4	1
$^{\bullet}C_8H_9$, 1-vinyl-cyclohexa-2,4-dienyl	247.7 ± 14.2	1
$^{\bullet}C_8H_9$, 2-vinyl-cyclohexa-2,4-dienyl	249.8 ± 14.2	1
$^{\bullet}C_8H_9$, 3-vinyl-cyclohexa-2,4-dienyl	269.4 ± 14.2	1
$^{\bullet}C_8H_9$, 6-vinyl-cyclohexa-2,4-dienyl	284.5 ± 14.2	1
$^{\bullet}C_8H_{13}$, $CH_2=CHCH=CHC^{\bullet}H(CH_2)_2CH_3$	130.5	1
$^{\bullet}C_8H_{13}$, $CH_2=CHC^{\bullet}H(CH_2)_3CH=CH_2$	130.5	1
$^{\bullet}C_8H_{13}$, bicyclooct-1-yl	92.0	1
$^{\bullet}C_8H_{15}$, $CH_2=CHC^{\bullet}H(CH_2)_4CH_3$	49.8	1
$^{\bullet}C_8H_{15}$, (E)-$CH_3CH=C^{\bullet}(CH_2)_4CH_3$	29.7	1
$^{\bullet}C_8H_{15}$, (Z)-$(CH_3)_2C^{\bullet}CH=CHCH(CH_3)_2$	9.2	1
$^{\bullet}C_8H_{15}$, cyclooctanyl	59.4	1
$^{\bullet}C_8H_{15}$, cyclo-$[C^{\bullet}(CH_2CH_3)(CH_2)_5]$	10.0	1
$^{\bullet}C_9H_7$, indenyl	297.1	1
$^{\bullet}C_9H_9$, indanyl-1	204.2 ± 8.4	1
$^{\bullet}C_9H_{11}$, 2,6-dimethylbenzyl	124.7	1
$^{\bullet}C_9H_{11}$, 3,6-dimethylbenzyl	124.7	1
$^{\bullet}C_9H_{11}$, 3,5-dimethylbenzyl	124.7	1
$^{\bullet}C_9H_{11}$, $C_6H_5C^{\bullet}(CH_3)_2$	133.9 ± 4.2	1
$^{\bullet}C_9H_{11}$, o-$^{\bullet}C_6H_4C_2H_5$	279.5 ± 7.5	1
$^{\bullet}C_9H_{17}$, cyclononanyl	52.3	1
$^{\bullet}C_{10}H_7$, naphth-1-yl	401.7 ± 5.4	1
$^{\bullet}C_{10}H_7$, naphth-2-yl	400.4 ± 5.9	1
$^{\bullet}C_{10}H_{11}$, tetralin-1-yl	154.8 ± 5.0	1
$^{\bullet}C_{10}H_{13}$, 1-phenyl-but-4-yl	192.0	1
$^{\bullet}C_{10}H_{13}$, $(C_6H_5CH_2)(C_2H_5)C^{\bullet}H$	184.5	1
$^{\bullet}C_{10}H_{13}$, $(C_6H_5CH_2CH_2)(CH_3)C^{\bullet}H$	184.5	1
$^{\bullet}C_{10}H_{13}$, $(C_6H_5C^{\bullet}HCH_2CH_2CH_3$	134.7	1
$^{\bullet}C_{10}H_{15}$, 1-adamantyl	51.5	1
$^{\bullet}C_{10}H_{15}$, 2-adamantyl	61.9	1
$^{\bullet}C_{10}H_{19}$, cyclodecanyl	32.2	1
$^{\bullet}C_{11}H_9$, 1-naphthylmethyl	252.7	1
$^{\bullet}C_{11}H_{21}$, cycloundecanyl	7.5	1
$^{\bullet}C_{12}H_{23}$, cyclododecanyl	−38.5	1
$^{\bullet}C_{13}H_9$, 9-fluorenyl	297.5	1
$^{\bullet}C_{13}H_{11}$, $(C_6H_5)_2C^{\bullet}H$	302.1 ± 4.2	1

Radical	$\Delta_f H^\circ_{298}$/kJ mol^{-1}	Ref.
${}^{\bullet}C_{13}H_{11}$, 9-methyl-9-fluorenyl	268.2	1
${}^{\bullet}C_{14}H_{11}$, 9,10-dihydroanthracen-9-yl	261.0	1
${}^{\bullet}C_{15}H_{11}$, 9-anthracenylmethyl	337.6	1
${}^{\bullet}C_{15}H_{11}$, 9-phenanthrenylmethyl	311.3	1
${}^{\bullet}C_{16}H_{31}$, $CH_2{=}CHC^{\bullet}H(CH_2)_{12}CH_3$	−118.8	1
${}^{\bullet}C_{19}H_{15}$, trityl, $(C_6H_5)_3C^{\bullet}$	392.0 ± 8.4	1
${}^{\bullet}C_{35}H_{25}$, pentamethylcyclopentadienyl	67.4	1
CF	255.2 ± 8	1
CF_2	−182.0 ± 6.3	1
$FC^{\bullet}(O)$	−161.2 ± 8.4	1
CHF	143.0 ± 12.6	1
CClF	31.0 ± 13.4	1
CCl	443.1 ± 13.0	1
CCl_2	226	1
$ClC^{\bullet}(O)$	−21.8 ± 2.5	1
CHCl	326.4 ± 8.4	1
CClBr	267	1
CBr	510 ± 63	1
CHBr	373 ± 18	1
CBr_2	343.5	1
CI	570 ± 35	1
CI_2	468 ± 60	1
${}^{\bullet}CF_3$	−465.7 ± 2.1	1
${}^{\bullet}CHF_2$	−238.9 ± 4.2	1
${}^{\bullet}CH_2F$	−31.8 ± 4.2	1
${}^{\bullet}CClF_2$	−279.0 ± 8.4	1
${}^{\bullet}CCl_2F$	−89.0 ± 8.4	1
${}^{\bullet}CBrClF$	−35.5 ± 6.3	1
${}^{\bullet}CHClF$	−60.7 ± 10.0	1
${}^{\bullet}CBrF_2$	−224.7 ± 12.6	1
${}^{\bullet}CCl_3$	71.1 ± 2.5	1
${}^{\bullet}CHCl_2$	87.1 ± 1.6	1
${}^{\bullet}CH_2Cl$	117.2 ± 2.9	1
${}^{\bullet}CHBrCl$	140 ± 4	1
${}^{\bullet}CHBr_2$	199.1 ± 2.7	3
${}^{\bullet}CBr_2Cl$	163 ± 8	1
${}^{\bullet}CBrCl_2$	124 ± 8	1
${}^{\bullet}CBr_3$	214.8	1
${}^{\bullet}CH_2Br$	171.1 ± 2.7	1
${}^{\bullet}CI_3$	424.9 ± 2.8	1
${}^{\bullet}CHI_2$	314.4 ± 3.3	1
${}^{\bullet}CH_2I$	229.7 ± 8.4	1
${}^{\bullet}C_2F$, $FC{\equiv}C^{\bullet}$	460.0 ± 21.0	1
${}^{\bullet}C_2Cl$, $ClC{\equiv}C^{\bullet}$	568 ± 26	1
${}^{\bullet}C_2F_3$, $CF_2{=}C^{\bullet}F$	−192.0 ± 8.4	1
${}^{\bullet}C_2F_2H$, $CF_2{=}C^{\bullet}H$	−92.9 ± 8.4	1
${}^{\bullet}C_2F_2H$, $CHF{=}C^{\bullet}F$	−50.6 ± 8.4	1
${}^{\bullet}CCl_2H$, $CHCl{=}C^{\bullet}Cl$	234.7 ± 8.4	1
${}^{\bullet}CClH_2$, $CH_2{=}C^{\bullet}Cl$	>251	1
${}^{\bullet}C_2F_5$, $CF_3C^{\bullet}F_2$	−892.9 ± 4.2	1
${}^{\bullet}C_2HF_4$, $CF_3C^{\bullet}HF$	−680.8 ± 9.6	1
${}^{\bullet}C_2HF_4$, $CHF_2C^{\bullet}F_2$	−664.8	1
${}^{\bullet}C_2H_2F_3$, $CF_3C^{\bullet}H_2$	−517.1 ± 8.4	1
${}^{\bullet}C_2H_2F_3$, $CHF_2C^{\bullet}HF$	−456.0	1
${}^{\bullet}C_2H_2F_3$, $CH_2FC^{\bullet}F_2$	−449.8	1
${}^{\bullet}C_2H_2F_2Cl$, $CF_2ClC^{\bullet}H_2$	−310.9 ± 7.0	1
${}^{\bullet}C_2H_3F_2$, $CH_3C^{\bullet}F_2$	−302.5 ± 8.4	1
${}^{\bullet}C_2H_3F_2$, $CHF_2C^{\bullet}H_2$	−285.8	1
${}^{\bullet}C_2H_3F_2$, $CH_2FC^{\bullet}HF$	−238.5	1
${}^{\bullet}C_2H_4F$, $CH_3C^{\bullet}HF$	−70.3 ± 8.4	1
${}^{\bullet}C_2H_4F$, $CH_2FC^{\bullet}H_2$	−59.4 ± 8.4	1
${}^{\bullet}C_2H_2F_2Cl$, $CF_2ClC^{\bullet}H_2$	−315.2 ± 6	1
${}^{\bullet}C_2F_4Cl$, $CF_2ClC^{\bullet}F_2$	−686.0	1
${}^{\bullet}C_2HF_3Cl$, $CClF_2C^{\bullet}HF$	−450.6 ± 12.6	1
${}^{\bullet}C_2F_4Cl$, $CF_3C^{\bullet}FCl$	−728.0	1
${}^{\bullet}C_2F_3Cl_2$, $CF_3C^{\bullet}Cl_2$	−564.0	1
${}^{\bullet}C_2F_3ClBr$, $CF_3C^{\bullet}ClBr$	−504.2 ± 8.4	1
${}^{\bullet}C_2Cl$, $ClC{\equiv}C^{\bullet}$	534 ± 50	1
${}^{\bullet}C_2Cl_3$, $CCl_2{=}C^{\bullet}Cl$	190 ± 50	1
${}^{\bullet}C_2Cl_5$, $CCl_3C^{\bullet}Cl_2$	35.1 ± 5.4	1
${}^{\bullet}C_2HCl_4$, $CHCl_2C^{\bullet}Cl_2$	23.4 ± 8.4	1
${}^{\bullet}C_2HCl_4$, $CCl_3C^{\bullet}HCl$	51.0	1
${}^{\bullet}C_2H_2Cl_3$, $CH_2ClC^{\bullet}Cl_2$	26.4	1
${}^{\bullet}C_2H_2Cl_3$, $CHCl_2C^{\bullet}HCl$	46.4	1
${}^{\bullet}C_2H_2Cl_3$, $CCl_3C^{\bullet}H_2$	71.5 ± 8	1
${}^{\bullet}C_2H_3Cl_2$, $CH_3C^{\bullet}Cl_2$	42.5 ± 1.7	1
${}^{\bullet}C_2H_3Cl_2$, $CH_2ClC^{\bullet}ClH$	65.3	1
${}^{\bullet}C_2H_3Cl_2$, $CHCl_2C^{\bullet}H_2$	90.1 ± 0.8	1
${}^{\bullet}C_2H_4Cl$, $CH_3C^{\bullet}HCl$	76.5 ± 1.6	1
${}^{\bullet}C_2H_4Cl$, $CH_2ClC^{\bullet}H_2$	93.0 ± 2.4	1
${}^{\bullet}C_2H_3Br_2$, $CH_3C^{\bullet}Br_2$	140.2 ± 5.4	1
${}^{\bullet}C_2H_4Br$, $BrCH_2C^{\bullet}H_2$	135.1	1
${}^{\bullet}C_2H_4Br$, $CH_3C^{\bullet}HBr$	133.4 ± 3.4	3
${}^{\bullet}C_2Br$, $CBrC^{\bullet}$	623.8	1
${}^{\bullet}C_2Br_3$, $CBr_2C^{\bullet}Br$	385.3	1
${}^{\bullet}C_2Br_5$, $CBr_3C^{\bullet}Br_2$	283.3	1
${}^{\bullet}C_3H_6Cl$, $CH_3CH_2C^{\bullet}HCl$	56.6	1
${}^{\bullet}C_3H_6Cl$, $CH_3C^{\bullet}ClCH_3$	29.9 ± 0.6	1
${}^{\bullet}C_3H_6Br$, $C^{\bullet}H_2CH_2CH_2Br$	120.1 ± 1.3	1
${}^{\bullet}C_3H_6Br$, $CH_3C^{\bullet}HCH_2Br$	96.7 ± 5.9	1
${}^{\bullet}C_3H_6Br$, $CH_3CH_2C^{\bullet}HBr$	107.5 ± 2.5	1
${}^{\bullet}C_6F_5$	−547.7 ± 8.4	1
${}^{\bullet}CH_3O$, $HOC^{\bullet}H_2$	−17.0 ± 0.7	1
${}^{\bullet}CH_2ClO$, $HOC^{\bullet}ClH$	−60.7 ± 7.5	1
${}^{\bullet}CHCl_2O$, $HOC^{\bullet}Cl_2$	−94.1 ± 7.5	1
${}^{\bullet}CH_2ClO$, $ClOC^{\bullet}H_2$	135.6 ± 9.2	1
${}^{\bullet}CH_2BrO$, $BrOC^{\bullet}H_2$	151 ± 16	1
${}^{\bullet}C_2H_3O$, $C^{\bullet}H{=}CHOH$	121 ± 11	1
${}^{\bullet}C_2H_3O$, $C^{\bullet}H_2CHO$	13.0 ± 2	1
${}^{\bullet}C_2H_5O$, $CH_3C^{\bullet}HOH$	−54.0	1
${}^{\bullet}C_2H_4ClO$, $CH_3C^{\bullet}ClOH$	−108.4 ± 8.8	1
${}^{\bullet}C_2H_4ClO$, $C^{\bullet}H_2CHClOH$	−73.2 ± 8.8	1

Radical	$\Delta_f H^\circ_{298}$/kJ mol^{-1}	Ref.
$^\bullet C_2H_3Cl_2O$, $C^\bullet H_2CCl_2OH$	-99.6 ± 8.8	1
$^\bullet C_2H_5O$, $C^\bullet H_2CH_2OH$	-31 ± 7	1
$^\bullet C_2H_3O$, oxiran-2-yl	149.8 ± 6.3	1
$^\bullet C_3H_5O$,CH_2=$CHC^\bullet HOH$	0 ± 8.4	1
$^\bullet C_3H_7O$, $CH_3CH_2C^\bullet HOH$	-81 ± 4	1
$^\bullet C_3H_7O$, $(CH_3)C^\bullet HCH_2OH$	-78.7 ± 8.4	1
$^\bullet C_3H_7O$, $HOCH_2CH_2C^\bullet H_2$	-66.9 ± 8.4	1
$^\bullet C_3H_7O$, $(CH_3)_2C^\bullet OH$	-96.4	1
$^\bullet C_3H_7O$, $^\bullet CH_2CH(OH)CH_3$	-62.8 ± 11.7	1
$^\bullet C_4H_9O$, $^\bullet CH_2C(OH)(CH_3)_2$	-147.3 ± 8.4	1
$^\bullet C_2H_5O_3$, $C^\bullet H_2OCH_2OOH$	109.6 ± 4.2	1
$PhCH^\bullet OH$	29.3 ± 8.4	1
$Ph_2C^\bullet OH$	152.3 ± 6.3	1
$^\bullet C_2H_5O$, $CH_3OC^\bullet H_2$	0 ± 4.2	1
$^\bullet C_3H_7O$, $CH_3OC^\bullet HCH_3$	-57.7 ± 8.4	1
$^\bullet C_3H_7O$, $CH_3CH_2OC^\bullet H_2$	-45.2 ± 8.4	1
$^\bullet C_3H_7O$, $C^\bullet H_2CH_2OCH_3$	-7.1 ± 4.2	1
$^\bullet C_4H_9O$, $(CH_3)_2CHOC^\bullet H_2$	-70.3 ± 7.1	1
$^\bullet C_4H_9O$, $CH_3CH_2OC^\bullet HCH_3$	-81.2 ± 4.2	1
$^\bullet C_4H_9O$, $C^\bullet H_2CH(CH_3)OCH_3$	-42.3 ± 3.8	1
$^\bullet C_4H_9O$, $(CH_3)_2C^\bullet OCH_3$	-72.4 ± 10	1
$^\bullet C_5H_{11}O$, $(CH_3)_3COC^\bullet H_2$	-102.5 ± 8.4	1
$^\bullet C_2H_5O_2$, $HOCH_2C^\bullet HOH$	-220.1 ± 8.4	1
$C^\bullet H$=C=O, ketenyl	177.5 ± 8.8	1
$HC^\bullet(O)$	42.5 ± 0.5	1
$C^\bullet CO$	381.2 ± 2.1	1
$CH_3C^\bullet(O)$	-10.3 ± 1.8	1
$CF_3C^\bullet(O)$	-608.7	1
$CH_2ClC^\bullet(O)$	-21 ± 12.6	1
$CHCl_2C^\bullet(O)$	-17.6 ± 23	1
$CCl_3C^\bullet(O)$	-19.7	1
$CH_3CH_2C^\bullet(O)$	-31.7 ± 3.4	1
$CH_2CHC^\bullet(O)$	88.5	1
$CH_2C(CH_3)C^\bullet(O)$	58.6 ± 16.7	1
$CH_3CH_2CH_2C^\bullet(O)$	54.4 ± 4.2	1
$(CH_3)_2CHC^\bullet(O)$	-64.0 ± 3.8	1
$(CH_3)_3CC^\bullet(O)$	-102.9 ± 6.3	1
$C_6H_5C^\bullet(O)$	116.3 ± 10.9	1
$HC(O)CH_2^\bullet$	10.5 ± 9.2	1
$ClC(O)CH_2^\bullet$	-52.7 ± 13	1
E-$C^\bullet HClC(O)H$	-27.2 ± 10.5	1
Z-$C^\bullet HClC(O)H$	-23.4 ± 10.5	1
$C^\bullet Cl_2C(O)H$	-55.6 ± 14.2	1
E-$C^\bullet HClC(O)Cl$	-88.7 ± 15.1	1
$C^\bullet H_2C(O)F$	-273.0 ± 5.8	1
Z-$C^\bullet HClC(O)Cl$	-84.9 ± 13.8	1
$C^\bullet Cl_2C(O)Cl$	-101.7 ± 15.5	1
$CH_3C(O)CH_2^\bullet$	-34 ± 3	1
$CH_3C(O)C^\bullet HCH_3$	-70.3 ± 7.1	1
$CH_3C(O)C^\bullet$=CH_2	113.4	1
$C_2H_5C(O)C^\bullet HCH_3$	-107.5 ± 20.9	1
iPrC(O)$C^\bullet(CH_3)_2$	-173.6 ± 20.9	1
$tC_4H_9C(O)C^\bullet H_2$	-115.5 ± 12.6	1
$PhC(O)C^\bullet H_2$	84.5 ± 12.6	1
$PhC(O)C^\bullet HCH_3$	41.4 ± 20.9	1
$PhC^\bullet HC(O)CH_2Ph$	134.3 ± 20.9	1
$PhC(O)OC^\bullet H_2$	-69.9	1
$^\bullet C(O)OH$-trans	$\geq -194.6 \pm 2.9$	1
$^\bullet C(O)OH$-cis	-219.7	1
$^\bullet C(O)OCH_3$	-161.5	1
$C^\bullet H_2C(O)OH$	-248.9 ± 12.0	1
$C^\bullet H(CH_3)C(O)OH$	-293 ± 3	1
$C^\bullet H_2C(O)OCH_3$	-236.8 ± 8.4	1
$C^\bullet H_2C(O)OCH_2CH_3$	-260.2 ± 12.6	1
$C^\bullet H_2C(O)OPh$	-28.0	1
$^\bullet C_4H_7O$, tetrahydrofuran-2-yl	-18.0 ± 6.3	1
$^\bullet C_4H_8O$, cyclopentanon-2-yl	-41.8 ± 12.6	1
$^\bullet C_4H_7O_2$, 1,4-dioxan-2-yl	-131.8 ± 12.6	1
$^\bullet C_7H_5O_2$, 2-C(O)OH-$^\bullet C_6H_4$	-33.0	1
$^\bullet C_7H_5O_2$, 3-C(O)OH-$^\bullet C_6H_4$	-35.0	1
$^\bullet C_7H_5O_2$, 4-C(O)OH-$^\bullet C_6H_4$	-36.0	1
$^\bullet CH_3O_2$, $C^\bullet H_2OOH$	66.1	1
$^\bullet C_2H_5O_2$, $C^\bullet H_2CH_2OOH$	46.0 ± 4.6	1
$^\bullet C_2H_5O_2$, $CH_3CH^\bullet OOH$	26.9	1
$^\bullet C_3H_7O_2$, $CH_3CH^\bullet CH_2OOH$	10.9 ± 5.4	1
$^\bullet C_3H_7O_2$, $C^\bullet H_2CH(OOH)CH_3$	2.9 ± 6.3	1
$^\bullet C_4H_9O_2$, $(CH_3)_2C^\bullet CH_2OOH$	-30.1 ± 5.4	1
$^\bullet C_4H_9O_2$, $C^\bullet H_2C(CH_3)_2OOH$	-26.8 ± 5.4	1
$^\bullet C_2H_3O_3$, $C^\bullet H_2C(O)OOH$	-137.9	1
$^\bullet CHN_2$	494.5	1
$^\bullet CH_2N$=CH_2	263.6 ± 12.6	1
$^\bullet CH_2NH_2$	151.9 ± 8.4	1
$CH_3C^\bullet HNH_2$	111.7 ± 8.4	1
$(CH_3)_2C^\bullet NH_2$	69.9 ± 8.4	1
$^\bullet CH_2NHCH_3$	156.6	1
$^\bullet CH_2N(CH_3)_2$	148.0	1
$(C_2H_5)_2NC^\bullet HCH_3$	68.6 ± 2.1	1
$^\bullet CH_2N(CH_3)Ph$	266.0 ± 12.6	1
$^\bullet CN$	439.3 ± 2.9	1
$^\bullet CH_2CN$	252.6 ± 4	1
$CH_3C^\bullet HCN$	226.7 ± 12.6	1
$^\bullet CH_2CH_2CN$	245.4 ± 12.6	1
$(CH_3)_2C^\bullet CN$	190.4 ± 12.6	1
$Ph(CH_3)C^\bullet CN$	248.5 ± 8.4	1
$NCC^\bullet HCH_2CN$	381.8 ± 12.6	1
$^\bullet CH_2NC$	334.7 ± 16.7	1
$^\bullet C(O)NC$	210.0 ± 10	1
$^\bullet C(O)NH_2$	-15.1 ± 4	1
$C^\bullet NN$	569 ± 21	1
$HC^\bullet NN$	460 ± 8	1
$H_2C^\bullet NN$	292.5 ± 2.1	1
$^\bullet CH_2NO$	157 ± 4	1

Radical	$\Delta_f H^o_{298}$/kJ mol^{-1}	Ref.
•CH_2NO_2	115.1 ± 12.6	1
CH_3C•HNO_2	61.9 ± 12.6	1
$(CH_3)_2C$•NO_2	6.3 ± 12.6	1
PhC•HNO_2	169.0 ± 12.6	1
•C_6H_6N, 3-NH_2-C_6H_4	320.1	1
•C_6H_6N, 4-NH_2-C_6H_4	327.8	1
•$C_6H_4NO_2$, 3-NO_2-C_6H_4	340.6 ± 10.0	1
•$C_6H_4NO_2$, 4-NO_2-C_6H_4	302.7	1
•$C_6H_4CH_3$, 2-Me-C_6H_4	315.1 ± 10.5	1
•$C_6H_4CH_3$, 4-Me-C_6H_4	296.6 ± 9.6	1
•$C_6H_3N_2O_4$, 3,5-$(NO_2)_2$-C_6H_3	305.4	1
•$C_7H_6NO_2$, 2-Me-4-NO_2-C_6H_3	295.4 ± 8.4	1
•C_4H_3N, pyrrol-2-yl	385.8	1
•C_4H_3N, pyrrol-3-yl	385.8	1
•C_4H_8N, pyrrolidin-2-yl	142.7 ± 12.6	1
•C_5H_4N, pyrid-2-yl	362.0	1
•C_5H_4N, pyrid-3-yl	391.0	1
•C_5H_4N, pyrid-4-yl	391.0	1
•$C_4H_7N_2$, piperad-2-yl	119.7	1
•$C_4H_3N_2$, pyrazin-2-yl	409.2 ± 12.6	1
•$C_4H_3N_2$, pyrimid-2-yl	388.0 ± 12.6	1
•$C_4H_3N_2$, pyrimid-4-yl	409.0 ± 12.6	1
•$C_4H_3N_2$, pyrimid-5-yl	446.4 ± 12.6	1
•$CH(NO_2)_2$	139.1	1
•$C(NO_2)_3$	201.2	1
•$CH_2C(NO_2)_3$	150.6	1
•$CH_2CH(NO_2)_2$	103.3	1
•$CH_2CH_2C(NO_2)_3$	133.9	1
•$CH_2N(NO_2)CH_2C(NO_2)_3$	173.6	1
•$CH_2N(NO_2)CH_2CH(NO_2)_2$	126.4	1
•$CH_2CH_2N(NO_2)CH_2C(NO_2)_3$	168.6	1
•$CH_2CH_2ONO_2$	37.7	1
•$CH_2(ONO_2)CHCH_2ONO_2$	–25.5	1
•$CH(CH_2ONO_2)_2$	–57.3	1
•$CH_2C(CH_2ONO_2)_3$	–158.2	1
•CH_2NHNO_2	164.8	1
•$CH_2N(NO_2)CH_3$	149.4	1
•$CH_2N(NO_2)_2$	210.5	1
•$CH_2CH_2N(NO_2)CH_3$	144.3	1
•$CH_2N(NO_2)CH_2N(NO_2)CH_3$	202.1	1
•$CH_2N(NO_2)(CH_2)N(NO_2)CH_3$	173.2	1
C•(S)H	300.4 ± 8.4	1
•CH_2SH	151.9 ± 8.4	1
•CH_2SCH_3	136.8 ± 5.9	1
•CH_2SPh	268.6 ± 12.6	1
•CH_2SOCH_3	23.8 ± 12.6	1
HOC•(S)S	110.5	1
•$CH_2SO_2CH_3$	–177.0 ± 12.6	1
•CH_2SO_2Ph	–57.3 ± 12.6	1
PhC•HSO_2CH_3	–109.2 ± 12.6	1
PhC•HSO_2Ph	7 ± 12.6	1

Radical	$\Delta_f H^o_{298}$/kJ mol^{-1}	Ref.
Ph_2C•SO_2Ph	102 ± 12.6	1
Ph_2C•SPh	435.6 ± 12.6	1
NC•(O)	127.2	1
•CNH	207.9 ± 12.1	1
•CNO	323 ± 30	1
•CH_2SiMe_3	–32 ± 6	1
•$CH_2C(CH_3)_2SiMe_3$	–125	1
•CP	450 ± 9	1
(2) Oxygen-Centered Species		
HO•	37.36 ± 0.13	1
FO•	109 ± 10	1
ClO•	101.63 ± 0.1	1
BrO•	126.2 ± 1.7	1
IO•	115.9 ± 5.0	1
HOO•	12.30 ± 0.25	1
FOO•	25.4 ± 2	1
ClOO•	98.0 ± 4	1
BrOO•	108 ± 40	1
IOO•	96.6 ± 15	1
OFO•	378.6 ± 20	1
OClO•	95.4	1
ClOOClO•	142 ± 12	1
ClClO•	90 ± 30	1
NCO•	184.1	1
CNO•	386.6	1
HONNO•	172	1
sym-ClO_3	217.2 ± 21	1
HSO•	–21.8 ± 2.1	1
HSOO•	112	1
CH_3SOO•	76	1
CF_3SO_2O•	–912	1
NCO•	184.0	1
O_2NO•	73.7 ± 1.4	1
ONOO•	82.8	1
$HOS(O)_2O$•	–511.7	1
CH_3O•	21.0 ± 2.1	1
CF_3O•	–635.1 ± 7.1	1
CCl_3O•	–38.1 ± 9.2	1
CH_2ClO•	–21.3 ± 9.2	1
$CHCl_2O$•	–32.2 ± 9.2	1
CH_2=CH-O•	18.4 ± 1.3	1
CF_3CHFO•	–851.0	1
C_2H_5O•	–13.6 ± 3.3	1
CH_3CHClO•	–61.9 ± 12.1	1
CH_3CCl_2O•	–91.6 ± 11.7	1
nC_3H_7O•	–30.1 ± 8.4	1
iC_3H_7O•	–48.5 ± 3.3	1
$(CH_3)_2CClO$•	–108.4 ± 8.4	1
nC_4H_9O•	–62.8	1
sC_4H_9O•	–69.5	1
tC_4H_9O•	–85.8 ± 3.8	1

Radical	$\Delta_f H^\circ_{298}$/kJ mol^{-1}	Ref.
$CH_2=CHCH_2O^\bullet$	87.0	1
$C_6H_5O^\bullet$	48.5 ± 2.9	1
o-Cl-$C_6H_4O^\bullet$	30.6	1
$C_6Cl_5O^\bullet$	~63	1
p-Cl-$C_6H_4O^\bullet$	~9	1
o-OH-$C_6H_4O^\bullet$	−186.3	1
p-OH-$C_6H_4O^\bullet$	−143.6	1
o-CH_3O-$C_6H_4O^\bullet$	−125.5	1
p-CH_3O-$C_6H_4O^\bullet$	−81.1	1
$C_6H_5CH_2O^\bullet$	136.0 ± 12.6	1
$C_{10}H_7O^\bullet$, naphthoxy-1	165.3	1
$C_{10}H_7O^\bullet$, naphthoxy-2	174.1	1
$HC(O)O^\bullet$	−129.7 ± 12.6	1
$FC(O)O^\bullet$	368.0	1
$CH_3C(O)O^\bullet$	−179.9 ± 12.6	1
$CF_3C(O)O^\bullet$	−797.0	1
$CF_3OC(O)O^\bullet$	−958.1 ± 16.7	1
$C_6H_5C(O)O^\bullet$	−50.2 ± 16.7	1
$CH_3OO^\bullet$	20.1 ± 5.1	1
$C_2H_3OO^\bullet$, $CH_2=CHOO^\bullet$	101.7 ± 1.7	1
$C_2H_5OO^\bullet$	−28.5 ± 9.6	1
$C_3H_5OO^\bullet$, $CH_2=CHCH_2OO^\bullet$	88.7	1
$iC_3H_7OO^\bullet$	−65.4 ± 11.3	1
$C_4H_7OO^\bullet$, $CH_3CH=CHCH_2OO^\bullet$	82.6 ± 5.3	1
$tC_4H_9OO^\bullet$	−101.5 ± 9.2	1
neo-$C_5H_{11}OO^\bullet$	−115.5	1
$HOCH_2OO^\bullet$	−162.1	1
$HOOCH_2CH_2OO^\bullet$	100	1
$C_6H_5CH_2OO^\bullet$	114.6 ± 4.2	1
c-$C_6H_{11}OO^\bullet$	−25.0 ± 10.5	1
$(C_2H_5)N(CH3)CHOO^\bullet$	−36.0 ± 12.6	1
$CF_3OO^\bullet$	−635.0	1
$CF_2ClOO^\bullet$	−406.7 ± 14.6	1
$CFCl_2OO^\bullet$	−213.7	1
$CH_2ClOO^\bullet$	−5.1 ± 13.6	1
$CHCl_2OO^\bullet$	−19.2 ± 11.2	1
$CCl_3OO^\bullet$	−20.9 ± 8.9	1
$CH_3CHClOO^\bullet$	−54.7 ± 3.4	1
$CH_3CCl_2OO^\bullet$	−63.8 ± 9.8	1
$CH_3OCH_2OO^\bullet$	−142.2 ± 4.2	1
$CH_3C(O)CH_2OO^\bullet$	−142.1 ± 4	1
$CH_3C(O)OO^\bullet$	−154.4 ± 5.8	1
$HOOO^\bullet$	>12.84	4
$CH_3OOO^\bullet$	33.4 ± 12.6	1
$C_2H_5OOO^\bullet$	5.4 ± 12.6	1

(3) **Nitrogen-Centered Species**

Radical	$\Delta_f H^\circ_{298}$/kJ mol^{-1}	Ref.
ON	91.04 ± 0.08	1
NO_2	33.97 ± 0.08	1
N_2O	82.05 ± 0.4	1
NH	357 ± 1	1
$^\bullet NH_2$	186.2 ± 1.0	1
$^\bullet NNH$	249.5	1
$^\bullet NCO$	131.8	1
$^\bullet N_3$	414.2 ± 20.9	1
$^\bullet N_2H_3$	243.5	1
(Z)-N_2H_2	213.0 ± 10.9	1
NF	209.2	1
$^\bullet NF_2$	42.3 ± 8	1
$^\bullet NHF$	112 ± 15	1
NBr	301 ± 21	1
HNO	107.1 ± 2.5	1
FNO	−65.7 ± 1.7	1
ClNO	51.71 ± 0.42	1
BrNO	82.13 ± 0.8	1
INO	112.1 ± 20.9	1
NCO	120.9	1
NCN	464.8 ± 2.9	1
NSi	372 ± 63	1
$NH_2C(O)N^\bullet H$	0.8 ± 12.6	1
$CH_3C(O)N^\bullet H$	−6.7 ± 12.6	1
$NH_2C(S)N^\bullet H$	194 ± 12.6	1
$CH_3C(S)N^\bullet H$	173 ± 12.6	1
$PhC(S)N^\bullet H$	307 ± 12.6	1
$HCON^\bullet H$	49.8 ± 12.6	1
$NH_2C(NH)N^\bullet H$	250.6 ± 12.6	1
$^\bullet NHCN$	319.2 ± 2.9	1
$CH_2N^\bullet H$	104.6 ± 12.6	1
$CH_3N^\bullet H$	184.1 ± 8.4	1
$tBuN^\bullet H$	95.4 ± 12.6	1
$C_6H_5CH_2N^\bullet H$	288.3 ± 12.6	1
$C_6H_5N^\bullet H$	244.3 ± 4.2	1
$(CH_3)_2N^\bullet$	158.2 ± 4.2	1
$(C_6H_5)(CH_3)N^\bullet$	241.0 ± 6.3	1
$(C_6H_5)_2N^\bullet$	366.0 ± 6.3	1
1-pyrrolyl	269.2 ± 12.6	1
1 pyrazolyl	413.0 ± 2.1	1
carbazol-9-yl	383.3 ± 8.4	1
$CH_3N_2^\bullet$	215.5 ± 7.5	1
$C_2H_5N_2^\bullet$	187.4 ± 10.5	1
$iC_3H_7N_2^\bullet$	146.0 ± 8.4	1
$nC_4H_9N_2^\bullet$	140.6 ± 8.4	1
$tC_4H_9N_2^\bullet$	97.5 ± 4.2	1
$(NO_2)HN^\bullet$	162.3	1
$(CH_3)(NO_2)N^\bullet$	139.0	1
$(NO_2)_2N^\bullet$	200.0	1
$CH_3N^\bullet CH_2N(NO_2)CH_3$	185.4	1

(4) **Sulfur-Centered Species**

Radical	$\Delta_f H^\circ_{298}$/kJ mol^{-1}	Ref.
$HOS^\bullet$	−6.7 ± 2.1	1
$HC(O)S^\bullet$	56.5	1
$HS^\bullet O_2$	−221.8	1
$HOS^\bullet O_2$	−384.9	1
$NCS^\bullet$	300 ± 8	1

Radical	$\Delta_f H^\circ_{298}$/kJ mol^{-1}	Ref.
$HS^\bullet$	143.0 ± 0.8	1
$CH_3S^\bullet$	124.7 ± 1.7	1
$C_2H_5S^\bullet$	101	1
$nC_3H_7S^\bullet$	80	1
$iC_3H_7S^\bullet$	74.9 ± 8.4	1
$tC_4H_9S^\bullet$	43.9 ± 8.4	1
$C_6H_5S^\bullet$	242.7 ± 4.6	1
$C_6Cl_5S^\bullet$	~184	1
$C_6H_5CH_2S^\bullet$	246	1
$CH_3S^\bullet O$	−67 ± 10	1
$CH_3S^\bullet O_2$	−239.3	1
$HSS^\bullet$	115.5 ± 14.6	1
$CH_3SS^\bullet$	68.6 ± 8.4	1
$C_2H_5SS^\bullet$	43.5 ± 8.4	1
$iC_3H_7SS^\bullet$	13.8 ± 8.4	1
$tC_4H_9SS^\bullet$	−19.2 ± 8.4	1
$HOC(S)S^\bullet$	110.5 ± 4.6	1
$HC(O)S^\bullet$	56.5	1
SF	13.0 ± 6.3	1
SF_2	−296.7 ± 16.7	1
SF_3	−503.0 ± 33.5	1
SF_4	−763.2 ± 20.9	1
SF_5	−879.9 ± 15.1	1
$ClS^\bullet$	156.5 ± 16.7	1
SN	263.6 ± 105	1
SCl	156.5 ± 16.7	1
(5) Si-, Ge-, Sn-, Pb-Centered Species		
SiF	−20.1 ± 12.6	1
SiF_2	−638 ± 6	1
$^\bullet SiF_3$	−987 ± 20	1
SiCl	198.3 ± 6.7	1
$SiCl_2$	−169 ± 3	1
$^\bullet SiCl_3$	321 ± 8	6
SiBr	235 ± 46	1
$SiBr_2$	46 ± 8	1
$^\bullet SiBr_3$	−201.7 ± 63	1
SiI	313.8 ± 42	1
SiI_2	92.5 ± 8.4	1
$^\bullet SiI_3$	35.3 ± 63	1
SiH	376.6 ± 8.4	1
$SiH_2(^1A_1)$	273 ± 2	1
$SiH_2(^3B_1)$	360.7	1
$^\bullet SiH_3$	200.4 ± 2.5	1
$MeSi^\bullet H_2$	141 ± 6	1
$Me_2Si^\bullet H$	78 ± 6	1
$Me_3Si^\bullet$	15 ± 7	1
$^\bullet Si_2H_3$	~402	1
$H_3SiSi^\bullet H_2$	234 ± 6	1
$C_6H_5Si^\bullet H_2$	274	1
$H_3SiSi^\bullet H$	312 ± 8	1
$MeSi^\bullet$	302.2	1
$MeSi^\bullet H$	202 ± 6	1
$Me_2Si^{\bullet\bullet}$	135 ± 8	1
SiN	313.8 ± 42	1
$^\bullet GeH_3$	221.8 ± 8.4	1
GeF	−71 ± 10	1
GeF_2	−574 ± 20	1
$^\bullet GeF_3$	−807 ± 50	1
GeCl	69 ± 18	1
$GeCl_2$	−171 ± 5	1
$^\bullet GeCl_3$	−268 ± 50	1
GeBr	137 ± 5	1
$GeBr_2$	−61 ± 5	1
$^\bullet GeBr_3$	−119 ± 50	1
GeI	211 ± 25	1
GeI_2	50.2 ± 4	1
$^\bullet GeI_3$	42 ± 50	1
SnF	−95 ± 7.2	1
SnF_2	−511 ± 9.2	1
$^\bullet SnF_3$	−647 ± 50	1
SnCl	35 ± 12	1
$SnCl_2$	−202.6 ± 7.1	1
$^\bullet SnCl_3$	−292 ± 50	1
SnBr	76 ± 12	1
$SnBr_2$	−119 ± 2.8	1
$^\bullet SnBr_3$	−159 ± 50	1
SnI	173 ± 12	1
SnI_2	−8.1 ± 4.2	1
$^\bullet SnI_3$	−8 ± 50	1
$^\bullet Sn(CH_3)_3$	132.2	1
$^\bullet Sn(C_6H_5)_3$	518.8 ± 21	1
PbH	236.2 ± 19.2	1
PbF	−80.3 ± 10.5	1
PbF_2	−435.1 ± 8.4	1
$^\bullet PbF_3$	−490 ± 60	1
PbCl	15.1 ± 50	1
$PbCl_2$	−174.1 ± 1.3	1
$^\bullet PbCl_3$	−178 ± 80	1
PbBr	70.9 ± 42	1
$PbBr_2$	−104.4 ± 6.3	1
$^\bullet PbBr_3$	−104 ± 80	1
PbI	107.4 ± 37.7	1
PbI_2	−3.2 ± 4.2	1
$^\bullet PbI_3$	22 ± 80	1

References

1. Luo, Y. R. *Comprehensive Handbook of Chemical Bond Energies*, CRC Press, Boca, Raton, FL, 2007.
2. Wheeler, S. E., Robertson, K. A., Allen, W. D., Schaefer III, H. F., Bomble, Y. J., and Stanton, J. F., *J. Phys. Chem. A* 111, 3819, 2007.
3. Seetula, J. A., and Eskola, A. J., *Chem. Phys.* 351, 141, 2008.
4. Denis, P. A., and Ornellas, F. R., *J. Phys. Chem. A* 113, 499, 2009.
5. Gao, Y., DeYonker, N. J., Garrett III, E. C., Wilson, A. K., Cundari, T. R., and Marshall, P., *J. Phys. Chem. A* 113, 6955, 2009.
6. Shuman, N. S., Spencer, A. P., and Baer, T., *J. Phys. Chem. A* 113, 9458, 2009.

TABLE 5. Bond Dissociation Energies of Some Organic Molecules

D^o_{298}(R-X)/ kJ mol^{-1} of some organic compounds are listed below. All data are from Tables 1 and 3.

	X=H	F	Cl	Br	I	OH	OCH_3	NH_2	NO	CH_3	$COCH_3$	CF_3	CCl_3
R=H	435.7799	569.658	431.361	366.16	298.26	497.10	440.2	450.08	199.5	439.3	374.0	445.2	392.5
CH_3	439.3	460.2	350.2	294.1	238.9	384.93	351.9	356.1	172.0	377.4	351.9	429.3	362.3
C_2H_5	420.5	447.4	352.3	292.9	233.5	391.2	355.2	352.3	171.5	370.3	347.3	—	—
i-C_3H_7	410.5	483.8	354.0	299.2	234.7	397.9	360.7	357.7	152.7	369.0	340.2	—	—
t-C_4H_9	400.4	495.8	351.9	292.9	227.2	398.3	353.1	355.6	167	363.6	329.3	—	—
C_6H_5	472.2	525.5	399.6	336.4	272.0	463.6	418.8	429.3	226.8	426.8	406.7	463.2	388.7
$C_6H_5CH_2$	375.5	412.8	299.9	239.3	187.8	334.1	—	306.7	123	325.1	299.7	365.7	—
CCl_3	392.5	439.3	296.6	231.4	168	—	—	—	125	362.3	—	332.2	285.8
CF_3	445.2	546.8	365.3	296.2	227.2	≤482.0	—	—	167	429.3	—	413.0	332.2
C_2F_5	429.7	532.2	346.0	283.3	219.2	—	—	—	—	—	—	424.3	—
CH_3CO	374.0	511.7	354.0	292.0	223.0	459.4	424.3	414.6	—	351.9	307.1	—	—
CN	528.5	482.8	422.6	364.8	320.1	—	—	—	204.4	521.7	—	469.0	—
C_6F_5	487.4	485	383.3	~328	<301.7	446.9	—	—	211.3	439.3	—	435.1	—

TABLE 6. Bond Dissociation Energies in Diatomic Cations

From thermochemistry, we have

$$D^o_{298}(A^+{-}B) \equiv \Delta_f H^o(A^+) + \Delta_f H^o(B) - \Delta_f H^o(AB^+) = D^o_{298}(A{-}B) + IP(A) - IP(AB)$$

Thus, $D^o_{298}(A^+{-}B)$ may be derived using the Table 1 and the ionization potentials of species A and AB. The following Table has been arranged in an alphabetical order of the atoms. The **boldface** in the species indicates the dissociated fragment.

A^+–B	D^o_{298} kJ/mol^{-1}	Ref.	A^+–B	D^o_{298} kJ/mol^{-1}	Ref.	A^+–B	D^o_{298} kJ/mol^{-1}	Ref.
$\mathbf{Ag^+}$–Ag	167.9 ± 8.7	1	$\mathbf{Au^+}$–Be	401 ± 29	1	$\mathbf{Be^+}$–Ar	49.0 ± 2.4	1
$\mathbf{Ag^+}$–Cl	32 ± 30	1	$\mathbf{Au^+}$–C	311.5 ± 7.7	4	$\mathbf{Be^+}$–Au	410 ± 29	1
$\mathbf{Ag^+}$–F	24 ± 27	1	$\mathbf{Au^+}$–F	79	1	$\mathbf{Be^+}$–Be	196 ± 0.5	8
$\mathbf{Ag^+}$–H	43.5 ± 5.9	1	$\mathbf{Au^+}$–Ge	292 ± 24	1	$\mathbf{Be^+}$–Cl	417 ± 50	1
$\mathbf{Ag^+}$–O	123 ± 5	1	$\mathbf{Au^+}$–H	213.1 ± 7.7	4	$\mathbf{Be^+}$–F	575 ± 98	1
$\mathbf{Ag^+}$–S	123 ± 13	1	$\mathbf{Au^+}$–I	230~280	1	$\mathbf{Be^+}$–H	307.3 ± 5.0	1
$\mathbf{Al^+}$–Al	121	1	$\mathbf{Au^+}$–Xe	130 ± 13	1	$\mathbf{Be^+}$–O	362.0 ± 6.2	1
$\mathbf{Al^+}$–Ar	15.47	1	$\mathbf{B^+}$–Ar	32.7	1	$\mathbf{Bi^+}$–Bi	199 ± 10	1
$\mathbf{Al^+}$–Ca	148.5	1	$\mathbf{B^+}$–B	187	1	$\mathbf{Bi^+}$–O	174	1
$\mathbf{Al^+}$–Cl	173 ± 42	1	$\mathbf{B^+}$–Br	164 ± 21	1	$\mathbf{Bi^+}$–S	179 ± 50	1
$\mathbf{Al^+}$–F	314 ± 21	1	$\mathbf{B^+}$–C	284 ± 58	1	$\mathbf{Bi^+}$–Se	184 ± 29	1
$\mathbf{Al^+}$–Kr	5.54	1	$\mathbf{B^+}$–Cl	308 ± 21	1	$\mathbf{Bi^+}$–Te	125 ± 50	1
$\mathbf{Al^+}$–O	166.7 ± 12.0	1	$\mathbf{B^+}$–F	460 ± 10	1	$\mathbf{Bi^+}$–Tl	100 ± 42	1
$\mathbf{Al^+}$–Se	114 ± 49	1	$\mathbf{B^+}$–H	198 ± 5	1	$\mathbf{Bk^+}$–O	610	1
$\mathbf{Ar^+}$–Ar	130.323 ± 0.087	1	$\mathbf{B^+}$–O	326 ± 48	1	$\mathbf{Br^+}$–Br	318.858 ± 0.024	1
$\mathbf{Ar^+}$–He	2.9 ± 0.8	1	$\mathbf{B^+}$–Pt	314 ± 98	1	$\mathbf{Br^+}$–C	451.5 ± 8.6	1
$\mathbf{Ar^+}$–Ne	7.5 ± 0.8	1	$\mathbf{B^+}$–Se	298 ± 98	1	$\mathbf{Br^+}$–Cl	303.000 ± 0.048	1
$\mathbf{As^+}$–As	364 ± 22	1	$\mathbf{B^+}$–Si	365 ± 15	1	$\mathbf{Br^+}$–F	251.5 ± 12.6	1
$\mathbf{As^+}$–H	290.8 ± 3.0	1	$\mathbf{Ba^+}$–Ar	11.85	1	$\mathbf{Br^+}$–H	379.26 ± 2.89	1
$\mathbf{As^+}$–O	495	1	$\mathbf{Ba^+}$–Br	418 ± 10	1	$\mathbf{Br^+}$–O	365.7 ± 3.1	1
$\mathbf{As^+}$–P	367 ± 59	1	$\mathbf{Ba^+}$–Cl	468.2 ± 10	1	$\mathbf{C^+}$–Ar	72.3	1
$\mathbf{As^+}$–S	433.2 ± 12.5	1	$\mathbf{Ba^+}$–D	245.2 ± 9.6	1	$\mathbf{C^+}$–Br	398 ± 8.6	1
$\mathbf{Au^+}$–Al	170 ± 30	1	$\mathbf{Ba^+}$–F	640 ± 29	1	$\mathbf{C^+}$–C	601.9 ± 19.3	1
$\mathbf{Au^+}$–Au	234.5	1	$\mathbf{Ba^+}$–I	335 ± 10	1	$\mathbf{C^+}$–Cl	614	1
$\mathbf{Au^+}$–B	329 ± 50	1	$\mathbf{Ba^+}$–O	441.4 ± 15	1	$\mathbf{C^+}$–F	721 ± 40	1

A^+–B	D^o_{298} kJ/mol^{-1}	Ref.
C^+–H	397.848 ± 0.013	1
C^+–N	524.5 ± 4.2	1
C^+–O	810.7 ± 0.8	1
C^+–P	587 ± 50	1
C^+–S	706.6 ± 2.1	1
C^+–Se	587 ± 50	1
Ca^+–Al	144.7	1
Ca^+–Ar	12.99 ± 0.60	1
Ca^+–Au	306 ± 29	1
Ca^+–Br	417.6 ± 10	1
Ca^+–Ca	104.1	1
Ca^+–Cl	433.4 ± 12	1
Ca^+–F	556.5 ± 8.4	1
Ca^+–H	284.2 ± 10	1
Ca^+–I	293.7 ± 10.8	1
Ca^+–Kr	18.60 ± 0.72	1
Ca^+–Ne	4.95 ± 0.06	1
Ca^+–O	348 ± 5	1
Ca^+–Xe	25.38 ± 0.96	1
Cd^+–Cd	122.5 ± 10	1
Cd^+–H	179.5	1
Ce^+–Au	278 ± 34	1
Ce^+–Br	341.0	1
Ce^+–C	254 ± 96	1
Ce^+–Ce	207 ± 42	1
Ce^+–Cl	429.5	1
Ce^+–F	586 ± 63	1
Ce^+–I	295.5	1
Ce^+–Ir	530 ± 96	1
Ce^+–N	494 ± 63	1
Ce^+–O	852 ± 15	1
Ce^+–Pd	255 ± 53	1
Ce^+–Pt	467 ± 96	1
Ce^+–Rh	423 ± 96	1
Ce^+–S	524 ± 59	1
Cl^+–Ar	169	1
Cl^+–Cl	385.746 ± 0.096	6
Cl^+–D	457.284 ± 0.017	1
Cl^+–F	291 ± 10	1
Cl^+–H	452.714 ± 0.018	1
Cl^+–N	650 ± 10	1
Cl^+–O	468.0 ± 2.1	1
Cm^+–O	670 ± 40	7
Co^+–Ar	52.89 ± 0.06	1
Co^+–Br	>289	1
Co^+–C	351 ± 29	1
Co^+–Cl	285 ± 12	1
Co^+–Co	269	1
Co^+–D	199.6 ± 5.8	1
Co^+–H	195 ± 6	1
Co^+–He	16.4 ± 0.4	1
Co^+–I	211.7 ± 8.4	1
Co^+–Kr	68.37 ± 0.18	1
Co^+–Ne	12.8 ± 0.4	1
Co^+–O	317.3 ± 4.8	1
Co^+–S	288.3 ± 8.7	1
Co^+–Si	317.1 ± 6.7	1
Co^+–Xe	85.7 ± 6.8	1
Cr^+–Ar	31.7 ± 3.9	1
Cr^+–C	277 ± 24	1
Cr^+–Cl	>211	1
Cr^+–Cr	129	1
Cr^+–D	135 ± 9	1
Cr^+–F	279 ± 42	1
Cr^+–H	136 ± 9	1
Cr^+–He	7.8 ± 0.4	1
Cr^+–Ne	9.5 ± 0.4	1
Cr^+–O	359	1
Cr^+–S	258.6 ± 16.4	1
Cr^+–Si	203 ± 15	1
Cr^+–Xe	71.9 ± 10.0	1
Cs^+–Ar	8.2	1
Cs^+–Br	60.5 ± 10	1
Cs^+–Cl	107.4 ± 10	1
Cs^+–Cs	62.6 ± 9.6	1
Cs^+–F	43.7 ± 10	1
Cs^+–He	5.1	1
Cs^+–I	29.3 ± 10	1
Cs^+–Kr	15.1	1
Cs^+–Na	48.1 ± 4.2	1
Cs^+–Ne	6.11	1
Cs^+–O	59	1
Cs^+–Rb	68.3 ± 10	1
Cs^+–Xe	14.7	1
Cu^+–Ar	51.9 ± 6.8	1
Cu^+–Cl	91 ± 10	1
Cu^+–Cu	155.2 ± 7.7	1
Cu^+–F	117 ± 21	1
Cu^+–Ge	231 ± 23	1
Cu^+–H	93 ± 13	1
Cu^+–Kr	24.3 ± 0.8	1
Cu^+–O	133.9 ± 11.6	1
Cu^+–S	203.3 ± 14.5	1
Cu^+–Si	260 ± 8	1
Cu^+–Xe	102.1 ± 5.8	1
D^+–D	263.4405 ± 0.0003	1
Dy^+–Br	324.2	1
Dy^+–Cl	407.9	1
Dy^+–Cu	196 ± 42	1
Dy^+–F	535 ± 24	1
Dy^+–I	279.9	1
Dy^+–O	597 ± 15	1
Er^+–Br	315.8	1
Er^+–Cl	406.7	1
Er^+–F	546 ± 34	1
Er^+–I	271.6	1
Er^+–O	583 ± 15	1
Es^+–O	470 ± 60	1
Eu^+–Ag	85 ± 50	1
Eu^+–Au	252 ± 97	1
Eu^+–Br	333.8	1
Eu^+–Cl	430.7	1
Eu^+–F	543 ± 29	1
Eu^+–I	290.7	1
Eu^+–O	393 ± 15	1
Eu^+–S	257 ± 32	1
F^+–Ar	161.1	1
F^+–F	325.393 ± 0.096	1
F^+–He	181.62 ± 0.08	1
F^+–Kr	152.4	1
F^+–Xe	188	1
Fe^+–Ar	14.2 ± 7.7	1
Fe^+–Br	>293	1
Fe^+–C	356.1 ± 17.2	1
Fe^+–Cl	>343	1
Fe^+–Co	259 ± 21	1
Fe^+–Cr	209 ± 29	1
Fe^+–Cu	222 ± 29	1
Fe^+–D	227	1
Fe^+–F	360 – 423	1
Fe^+–Fe	272	1
Fe^+–H	211.2 ± 9.6	1
Fe^+–I	>239	1
Fe^+–Kr	33.5 ± 6.7	1
Fe^+–N	485	1
Fe^+–Nb	285 ± 21	1
Fe^+–Ni	268 ± 21	1
Fe^+–O	334 ± 6	9
Fe^+–S	295.2 ± 5.8	1
Fe^+–Sc	200 ± 21	1
Fe^+–Si	277 ± 9	1
Fe^+–Ta	301 ± 21	1
Fe^+–Ti	251 ± 25	1
Fe^+–V	314 ± 21	1
Fe^+–Xe	46.0 ± 5.8	1
Ga^+–Bi	62 ± 98	1
Ga^+–Br	56.5 ± 16	1
Ga^+–Cl	86 ± 21	1
Ga^+–F	136 ± 15	1
Ga^+–Ga	126.3	1
Ga^+–I	41.6 ± 15	1
Ga^+–O	46 ± 50	1
Ga^+–Sb	38 ± 96	1

A$^+$–B	D^o_{298} kJ/mol^{-1}	Ref.
Ga$^+$–Te	19 ± 29	1
Gd$^+$–Cd	122.5 ± 10	1
Gd$^+$–H	179.5	1
Ge$^+$–Br	398 ± 42	1
Ge$^+$–C	223 ± 31	1
Ge$^+$–Cl	473 ± 50	1
Ge$^+$–F	565 ± 21	1
Ge$^+$–Ge	274 ± 10	1
Ge$^+$–H	377 ± 84	1
Ge$^+$–O	344 ± 21	1
Ge$^+$–S	283 ± 21	1
Ge$^+$–Se	234 ± 10	1
Ge$^+$–Si	268 ± 21	1
Ge$^+$–Te	233 ± 19	1
H$^+$–D	261.1021 ± 0.0002	1
H$^+$–H	259.4659 ± 0.0002	1
He$^+$–H	123.9	1
He$^+$–He	229.687 ± 0.019	1
Hf$^+$–C	311.5 ± 2.9	10
Hf$^+$–H	193.8 ± 10.6	2
Hf$^+$–O	670.4 ± 10.6	10
Hg$^+$–Ar	22.2 ± 1.2	1
Hg$^+$–H	207	1
Hg$^+$–Hg	134	1
Hg$^+$–Kr	37.9 ± 1.3	1
Hg$^+$–Xe	72.2 ± 1.3	1
Ho$^+$–Ag	155 ± 61	1
Ho$^+$–Au	250 ± 60	1
Ho$^+$–Br	320.6	1
Ho$^+$–Cl	410.3	1
Ho$^+$–Cu	214 ± 35	1
Ho$^+$–F	542 ± 50	1
Ho$^+$–Ho	88 ± 96	1
Ho$^+$–I	270.4	1
Ho$^+$–O	561 ± 25	1
I$^+$–Br	184.90 ± 0.02	1
I$^+$–Cl	247.5 ± 0.4	1
I$^+$–F	262.9 ± 2.1	1
I$^+$–H	304.70 ± 0.10	1
I$^+$–I	262.90 ± 0.04	1
I$^+$–O	316.3 ± 10.5	1
In$^+$–Br	65.2 ± 12.6	1
In$^+$–Cl	193 ± 21	1
In$^+$–F	148 ± 50	1
In$^+$–I	51.5 ± 21	1
In$^+$–In	81 ± 30	1
In$^+$–S	171 ± 50	1
In$^+$–Sb	73 ± 50	1
In$^+$–Se	118 ± 50	1
In$^+$–Te	41 ± 50	1
Ir$^+$–C	635.8 ± 4.8	3
Ir$^+$–D	302.8 ± 5.8	1
Ir$^+$–H	305.7 ± 5.8	1
Ir$^+$–O	247	1
K$^+$–Ar	14 ± 7	1
K$^+$–Br	35.7 ± 10.5	1
K$^+$–Cl	51 ± 19	1
K$^+$–He	6.00	1
K$^+$–I	18 ± 45	1
K$^+$–K	83.86 ± 0.15	1
K$^+$–Kr	15.8	1
K$^+$–Li	59.9 ± 5.9	1
K$^+$–Na	58.69 ± 0.08	1
K$^+$–Ne	7.79	1
K$^+$–O	13	1
K$^+$–Xe	19.5	1
Kr$^+$–Ar	55.31 ± 0.14	1
Kr$^+$–H	464	1
Kr$^+$–He	2.1 ± 0.8	1
Kr$^+$–Kr	110.967 ± 0.033	1
Kr$^+$–N	136.9 ± 13	1
Kr$^+$–Ne	3.8 ± 0.8	1
La$^+$–Au	436 ± 97	1
La$^+$–Br	425.9	1
La$^+$–C	427 ± 33	1
La$^+$–Cl	503.6	1
La$^+$–F	589 ± 34	1
La$^+$–H	243 ± 9	1
La$^+$–I	392.4	1
La$^+$–Ir	356 ± 97	1
La$^+$–O	875 ± 25	1
La$^+$–Pt	522 ± 78	1
La$^+$–Rh	345 ± 97	1
La$^+$–S	629 ± 96	1
La$^+$–Si	277.0 ± 9.6	1
Li$^+$–Ar	33 ± 14	1
Li$^+$–Bi	91 ± 50	1
Li$^+$–Br	41.8 ± 10.6	1
Li$^+$–Cl	66 ± 15	1
Li$^+$–F	7 ± 21	1
Li$^+$–He	10.66	1
Li$^+$–I	51.1 ± 6.3	1
Li$^+$–Kr	48.1	1
Li$^+$–Li	137.3 ± 6.3	1
Li$^+$–Ne	15.32	1
Li$^+$–O	38.9 ± 9.6	1
Li$^+$–Sb	129.6 ± 13.9	1
Li$^+$–Xe	56.4	1
Lu$^+$–Br	86.1	1
Lu$^+$–Cl	180.6	1
Lu$^+$–F	376.8	1
Lu$^+$–H	204 ± 15	1
Lu$^+$–I	40.7	1
Lu$^+$–O	524 ± 15	1
Lu$^+$–Si	107 ± 13	1
Mg$^+$–Ar	19.20	1
Mg$^+$–Au	267 ± 29	1
Mg$^+$–Cl	327 ± 6.5	1
Mg$^+$–D	203.6 ± 0.8	1
Mg$^+$–F	477 ± 50	1
Mg$^+$–H	190.8 ± 5.8	1
Mg$^+$–Kr	25.39	1
Mg$^+$–Mg	125	1
Mg$^+$–Ne	4.9 ± 0.6	1
Mg$^+$–O	245.2 ± 10	1
Mg$^+$–Xe	53.74	1
Mn$^+$–Cl	>211	1
Mn$^+$–F	321 ± 24	1
Mn$^+$–H	202.5 ± 5.9	1
Mn$^+$–I	>211	1
Mn$^+$–Mn	129	1
Mn$^+$–O	285 ± 13	1
Mn$^+$–S	247 ± 23	1
Mn$^+$–Se	165 ± 50	1
Mo$^+$–C	442.7 ± 13.5	1
Mo$^+$–F	376 ± 29	1
Mo$^+$–H	170 ± 6	1
Mo$^+$–Mo	449.4 ± 1.0	1
Mo$^+$–O	488.2 ± 1.9	1
Mo$^+$–S	355.1 ± 5.8	1
Mo$^+$–Xe	>53.1 ± 6.8	1
N$^+$–Ar	208.4 ± 9.6	1
N$^+$–F	584 ± 42	1
N$^+$–H	≥435.67 ± 0.77	1
N$^+$–N	843.85 ± 0.10	1
N$^+$–O	115	1
Na$^+$–Ar	19 ± 8	1
Na$^+$–Br	58.2 ± 10.6	1
Na$^+$–Cl	20.3 ± 10	1
Na$^+$–He	7.55	1
Na$^+$–I	64.9 ± 3.0	1
Na$^+$–Kr	~24.9	1
Na$^+$–Li	95.8 ± 3.9	1
Na$^+$–Na	98.64 ± 0.29	1
Na$^+$–Na	6.4	1
Na$^+$–Ne	~9.04	1
Na$^+$–O	37 ± 19	1
Na$^+$–Xe	~28.6	1
Nb$^+$–Ar	40.87 ± 0.13	1
Nb$^+$–C	509 ± 15	1
Nb$^+$–Fe	>251	1
Nb$^+$–H	220 ± 7	1
Nb$^+$–Nb	576.8 ± 9.6	1

A^+–B	D^o_{298} kJ/mol^{-1}	Ref.
Nb$^+$–O	688 ± 11	1
Nb$^+$–S	501.7 ± 20.3	1
Nb$^+$–V	404.7 ± 0.2	1
Nb$^+$–Xe	73.28 ± 0.12	1
Nd$^+$–Au	267 ± 84	1
Nd$^+$–Br	352.9	1
Nd$^+$–Cl	441.4	1
Nd$^+$–F	309.6	1
Nd$^+$–I	596 ± 32	1
Nd$^+$–O	753 ± 15	1
Ne$^+$–H	1239	1
Ne$^+$–He	13.0 ± 0.8	1
Ne$^+$–Ne	125.29 ± 1.93	1
Ni$^+$–Ar	53.9	1
Ni$^+$–Br	>289	1
Ni$^+$–C	418	1
Ni$^+$–Cl	192 ± 4	1
Ni$^+$–D	166.0 ± 7.7	1
Ni$^+$–F	≥456	1
Ni$^+$–H	158.1 ± 7.7	1
Ni$^+$–He	12.4 ± 0.4	1
Ni$^+$–I	>297	1
Ni$^+$–Ne	9.9 ± 0.4	1
Ni$^+$–Ni	208	1
Ni$^+$–O	275.9 ± 7.7	1
Ni$^+$–S	241.0 ± 3.9	1
Ni$^+$–Si	326 ± 6.7	1
Np$^+$–F	730 ± 100	1
Np$^+$–O	≥752	1
O$^+$–Ar	33.8	1
O$^+$–F	301.8 ± 8.4	1
O$^+$–H	487.9 ± 0.34	1
O$^+$–N	1050.64 ± 0.13	1
O$^+$–O	647.75 ± 0.17	1
Os$^+$–H	238.9	1
Os$^+$–O	418 ± 50	1
P$^+$–C	512 ± 42	1
P$^+$–Cl	289	1
P$^+$–F	490.6 ± 8.4	1
P$^+$–H	329.6 ± 2.1	1
P$^+$–N	483 ± 21	1
P$^+$–O	791.3 ± 8.4	1
P$^+$–P	481 ± 50	1
P$^+$–S	606 ± 34	1
Pa$^+$–O	~800	1
Pb$^+$–Br	260 ± 63	1
Pb$^+$–Cl	285 ± 63	1
Pb$^+$–F	347 ± 32	1
Pb$^+$–O	247 ± 8.4	1
Pb$^+$–Pb	214 ± 29	1
Pb$^+$–S	293 ± 50	1
Pb$^+$–Se	169.4 ± 6.3	1
Pb$^+$–Te	163 ± 63	1
Pd$^+$–C	528 ± 5	1
Pd$^+$–H	208.4 ± 8.7	1
Pd$^+$–O	145 ± 11	1
Pd$^+$–Pd	197 ± 29	1
Pd$^+$–S	197 ± 6	1
Pd$^+$–Si	289 ± 50	1
Pr$^+$–Au	317 ± 81	1
Pr$^+$–Br	357.7	1
Pr$^+$–Cl	445.0	1
Pr$^+$–F	557 ± 63	1
Pr$^+$–I	317.0	1
Pr$^+$–O	796 ± 15	1
Pt$^+$–Ar	36.4 ± 8.7	1
Pt$^+$–B	398 ± 105	1
Pt$^+$–C	530.5 ± 4.8	1
Pt$^+$–Cl	249.8 ± 14.5	1
Pt$^+$–H	275 ± 5	1
Pt$^+$–N	326.9 ± 9.6	1
Pt$^+$–O	318.4 ± 6.7	1
Pt$^+$–Pt	318 ± 23	1
Pt$^+$–Si	515 ± 50	1
Pt$^+$–Xe	86.6 ± 28.9	1
Pu$^+$–F	562 ± 50	1
Pu$^+$–O	655	1
Rb$^+$–Ar	12.0	1
Rb$^+$–Br	17.6v5.1	1
Rb$^+$–Cl	10.5 ± 10.5	1
Rb$^+$–I	27 ± 42	1
Rb$^+$–Kr	14.9	1
Rb$^+$–Na	50.1 ± 3.9	1
Rb$^+$–Ne	6.95	1
Rb$^+$–O	29	1
Rb$^+$–Rb	75.6 ± 9.6	1
Rb$^+$–Xe	21.5	1
Re$^+$–C	497.7 ± 3.9	1
Re$^+$–H	224.7 ± 6.7	1
Re$^+$–O	435 ± 59	1
Rh$^+$–C	414 ± 17	1
Rh$^+$–H	164.8 ± 3.8	1
Rh$^+$–O	295.0 ± 5.8	1
Rh$^+$–S	226 ± 13	1
Ru$^+$–C	453.5 ± 10.6	1
Ru$^+$–H	160.2 ± 5.0	1
Ru$^+$–O	372 ± 5	1
Ru$^+$–S	288 ± 6	1
S$^+$–C	620.8 ± 1.3	1
S$^+$–F	343.5 ± 4.8	1
S$^+$–H	348.2 ± 1.7	1
S$^+$–N	516 ± 34	1
S$^+$–O	524.3 ± 0.4	1
S$^+$–P	573 ± 21	1
S$^+$–S	522.4 ± 0.5	1
Sc$^+$–C	326 ± 6	1
Sc$^+$–Cl	410 ± 42	1
Sc$^+$–F	605 ± 32	1
Sc$^+$–Fe	201 ± 21	1
Sc$^+$–H	235 ± 8	1
Sc$^+$–O	689 ± 5	1
Sc$^+$–S	529.7 ± 17.4	1
Sc$^+$–Se	475.8 ± 8.4	1
Sc$^+$–Si	242.3 ± 10.5	1
Se$^+$–F	364 ± 42	1
Se$^+$–H	304	1
Se$^+$–P	514 ± 25	1
Se$^+$–S	392 ± 19	1
Se$^+$–Se	413 ± 19	1
Si$^+$–Au	175 ± 50	1
Si$^+$–B	351 ± 15	1
Si$^+$–Br	276 ± 96	1
Si$^+$–C	365 ± 50	1
Si$^+$–Cl	591.0 ± 0.6	1
Si$^+$–F	684.1 ± 5.4	1
Si$^+$–H	316.6 ± 2.1	1
Si$^+$–O	478 ± 13.4	1
Si$^+$–P	272 ± 50	1
Si$^+$–Pd	237 ± 50	1
Si$^+$–Pt	525 ± 50	1
Si$^+$–S	387.5 ± 6.0	1
Si$^+$–Si	334 ± 19	1
Si$^+$–Te	347 ± 50	1
Sm$^+$–Br	343.3	1
Sm$^+$–Cl	435.4	1
Sm$^+$–F	620.9	1
Sm$^+$–I	299.1	1
Sm$^+$–O	569 ± 15	1
Sn$^+$–Br	335 ± 50	1
Sn$^+$–Cu	184 ± 96	1
Sn$^+$–F	364 ± 29	1
Sn$^+$–O	281 ± 10	1
Sn$^+$–S	240 ± 19	1
Sn$^+$–Se	174 ± 6.3	1
Sn$^+$–Sn	193	1
Sn$^+$–Te	168.7 ± 8.4	1
Sr$^+$–Ar	13.32 ± 2.92	1
Sr$^+$–Br	378.1 ± 8.4	1
Sr$^+$–Cl	427 ± 8.4	1
Sr$^+$–F	615 ± 50	1
Sr$^+$–H	209 ± 5	1
Sr$^+$–I	308.2	1
Sr$^+$–Kr	18.13 ± 6.94	1

A^+–B	D°_{298} kJ/mol^{-1}	Ref.
Sr^+–Ne	4.52 ± 9.6	1
Sr^+–O	298.7	1
Sr^+–Sr	108.5 ± 1.6	1
Ta^+–C	369.4 ± 3.9	10
Ta^+–H	230 ± 6	1
Ta^+–O	688.7 ± 11.6	10
Ta^+–Ta	666	1
Tb^+–Cu	245 ± 34	1
Tb^+–O	722 ± 15	1
Tc^+–H	197.5	1
Tc^+–O	>167	1
Te^+–H	305 ± 12	1
Te^+–O	339 ± 50	1
Te^+–P	415 ± 97	1
Te^+–Se	342 ± 19	1
Te^+–Si	339.6	5
Te^+–Te	278 ± 29	1
Th^+–Cl	499 ± 29	1
Th^+–F	682 ± 29	1
Th^+–O	875 ± 16	1
Th^+–Pt	388 ± 193	1
Th^+–Rh	504 ± 67	1
Ti^+–C	395 ± 23	1
Ti^+–Cl	426.8	1
Ti^+–F	≥456	1
Ti^+–H	226.6 ± 10.6	1
Ti^+–N	501 ± 13	1
Ti^+–O	667 ± 7	1
Ti^+–Pt	82 ± 96	1
Ti^+–S	461.1 ± 6.8	1
Ti^+–Si	249 ± 16	1
Ti^+–Ti	229	1
Tl^+–Br	52 ± 50	1
Tl^+–Cl	26 ± 4	1
Tl^+–F	13 ± 21	1
Tl^+–I	133 ± 21	1
Tl^+–Tl	22 ± 50	1
Tm^+–Br	312.2	1
Tm^+–Cl	407.9	1
Tm^+–F	537 ± 16	1
Tm^+–I	266.8	1
Tm^+–O	482 ± 15	1
U^+–Br	345 ± 29	1
U^+–C	300 ± 96	1
U^+–Cl	431 ± 34	1
U^+–D	283.4 ± 9.6	1
U^+–F	668 ± 29	1
U^+–H	284 ± 8	1
U^+–N	~485	1
U^+–O	757 ± 42	1
U^+–P	186	1
U^+–S	518 ± 29	1
V^+–Ar	39.39 ± 0.12	1
V^+–C	373 ± 13.5	1
V^+–D	202 ± 6	1
V^+–Fe	314 ± 21	1
V^+–H	202 ± 6	1
V^+–Kr	49.46 ± 0.18	1
V^+–N	448.6 ± 5.8	1
V^+–Nb	403.5 ± 0.2	1
V^+–O	581.6 ± 9.6	1
V^+–S	358.9 ± 8.7	1
V^+–Si	229 ± 15	1
V^+–V	302	1
V^+–Xe	66.4 ± 0.6	1
W^+–C	463.0 ± 8.7	10
W^+–F	444 ± 96	1
W^+–H	222.5 ± 5	1
W^+–O	656.9 ± 6.8	10
Xe^+–Ar	13.4	1
Xe^+–H	355	1
Xe^+–Kr	41.65 ± 0.08	1
Xe^+–N	66.4 ± 9.6	1
Xe^+–Ne	2.1 ± 0.8	1
Xe^+–Xe	99.6	1
Y^+–C	281 ± 12	1
Y^+–F	677 ± 21	1
Y^+–H	260.5 ± 5.8	1
Y^+–O	718 ± 25	1
Y^+–Pt	466 ± 192	1
Y^+–S	533.9 ± 8	1
Y^+–Si	243 ± 13	1
Y^+–Te	360 ± 96	1
Y^+–Y	281 ± 21	1
Yb^+–Br	307.4	1
Yb^+–Cl	399.6	1
Yb^+–F	557.5 ± 14.4	1
Yb^+–I	262.0	1
Yb^+–O	376 ± 15	1
Yb^+–Yb	238 ± 96	1
Zn^+–Ar	28.7 ± 1.2	1
Zn^+–H	216 ± 15	1
Zn^+–O	161.1 ± 4.8	1
Zn^+–S	198 ± 12	1
Zn^+–Si	274.1 ± 9.6	1
Zn^+–Zn	60 ± 19	1
Zr^+–Ar	36.09 ± 0.24	1
Zr^+–C	445.8 ± 15.4	1
Zr^+–H	218.8 ± 9.6	1
Zr^+–N	443 ± 46	1
Zr^+–O	753 ± 11	1
Zr^+–S	549.0 ± 9.6	1
Zr^+–Zr	407.0 ± 9.6	1

References

1. Luo, Y. R., *Comprehensive Handbook of Chemical Bond Energies*, CRC Press, Boca, Raton, 2007.
2. Parke, L. G., Hinton, C. S., and Armentrout, P. B., *Int. J. Mass Spectrom.* 254, 168, 2006.
3. Li, F.-X., Zhang, X.-G., and Armentrout, P. B., *Int. J. Mass Spectrom.* 255/256, 279, 2006.
4. Li, F.-X., and Armentrout, P. B., *J. Chem. Phys.* 125, 133114/1, 2006.
5. Chattopadhyaya, S., Pramanik, A., Banerjee, A., and Das, K. K., *J. Phys. Chem. A* 110, 12303, 2006.
6. Li, J., Hao, Y., Yang, J., Zhou, C., and Mo, Y., *J. Chem. Phys.* 127, 104307/1, 2007.
7. Gibson, J. K., Haire, R. G., Santos, M., Pires de Matos, A., and Marçalo, J., *J. Phys. Chem. A* 112, 11373, 2008.
8. Merritt, J. M., Kaledin, A. L., Bondybey, V. E., and Heaven, M. C., *Phys. Chem. Chem. Phys.* 10, 4006, 2008.
9. Schröder, D., *J. Phys. Chem. A* 112, 13215, 2008.
10. Hinton, C. S., Li, F.-X., and Armentrout, P. B., *Int. J. Mass Spectrom.* 280, 226, 2009.

TABLE 7. Bond Dissociation Energies in Polyatomic Cations

This Table has been arranged on the basis of the Periodic Table with the IUPAC notation for Groups 1 to 18, see inside front cover of this *Handbook*. The **boldface** in the species indicates the dissociated fragment.

Bond	Do^2_{98}/kJ mol^{-1}	Ref.
(1) Group 1		
Li$^+$–H_2	27.2	1
Li$^+$–CO	57 ± 13	1
Li$^+$–H_2O	139 ± 8	1
Li$^+$–NH_3	156 ± 8	1
Li$^+$–CH_4	130	1
Li$^+$–CH_3OH	156 ± 8	1
Li$^+$–CH_3OCH_3	167 ± 10	1
Li$^+$–pyridine	183.0 ± 14.5	1
Li$^+$–Gly (glycine)	220 ± 9	1
Na$^+$–H_2	10.4 ± 0.8	1
Na$^+$–N_2	33.5	1
Na$^+$–CO	31 ± 8	1
Na$^+$–CO_2	66.5	1
Na$^+$–SO_2	79.1	1
Na$^+$–O_3	52.3	1
Na$^+$–H_2O	91.2 ± 6.3	1
Na$^+$(H_2O)–H_2O	82.0 ± 5.8	1
Na$^+$(H_2O)$_2$–H_2O	66.1	1
Na$^+$(H_2O)$_3$–H_2O	52.7 ± 0.8	1
Na$^+$(glycine)–H_2O	75.1 ± 5.3	1
Na$^+$(glutamine)–H_2O	52 ± 1	1
Na$^+$–NH_3	106.2 ± 5.4	1
Na$^+$–HNO_3	86.2	1
Na$^+$–CH_4	30.1	1
Na$^+$–CH_3OH	98.8 ± 5.7	1
Na$^+$–CH_3CN	125.5 ± 9.6	1
Na$^+$–C_2H_4	44.6 ± 4.4	1
Na$^+$–CH_3OCH_3	101.4 ± 5.7	1
Na$^+$–$CH_3C(O)H$	114.4 ± 3.4	1
Na$^+$– MeCOMe	131.3 ± 4.1	1
Na$^+$–C_6H_6	97.0 ± 5.9	1
Na$^+$–pyrrole	103.7 ± 4.8	1
Na$^+$–Gly (glycine)	166.7 ± 5.1	1
Na$^+$–Ala (alanine)	167 ± 4	1
Na$^+$–GlyGly (glycylglycine)	203 ± 8	1
K$^+$–H_2	6.1 ± 0.8	1
K$^+$–CO_2	35.6	1
K$^+$–H_2O	74.9	1
K$^+$(H_2O)$_2$–H_2O	67.4	1
K$^+$(H_2O)$_3$–H_2O	55.2	1
K$^+$(H_2O)$_4$–H_2O	11.8	1
K$^+$(H_2O)$_5$–H_2O	44.8	1
K$^+$(H_2O)$_6$–H_2O	41.8	1
K$^+$–NH_3	79 ± 7	1
K$^+$–C_6H_6	80.3	1
K$^+$–adenine	95.1 ± 3.2	1
K$^+$–indole	104.6 ± 12.6	1
K$^+$–Phe (phenylalanine)	150.5 ± 5.8	1
K$^+$–Tyr (tyrosine)	165.0 ± 5.8	1
Rb$^+$–H_2O	66.9 ± 12.6	1
Rb$^+$–NH_3	78.2	1
Rb$^+$–CH_3CN	86.6 ± 1.3	1
Rb$^+$–C_6H_5OH	70.2 ± 3.7	1
Cs$^+$–H_2O	57.3	1
Cs$^+$–$C_6H_5NH_2$	70.8 ± 4.5	1
(2) Group 2		
CH_3Be^+–CH_3	192.9 ± 13.4	1
***tert*-C(CH_3)$_3$Be$^+$**–*tert*-C(CH_3)$_3$	121.8 ± 13.4	1
Mg$^+$–OH	314 ± 33	1
Mg$^+$–CO	43.1 ± 5.8	1
Mg$^+$–CO_2	58.4 ± 5.8	1
Mg$^+$–H_2O	122.5 ± 12.5	1
Mg$^+$–NH_3	158.9 ± 11.6	1
Mg$^+$–CH_4	29.8 ± 6.8	1
Mg$^+$–MeOH	147.6 ± 6.8	1
Mg$^+$–C_6H_6	155.2	1
Mg$^+$–pyridine	200.0 ± 6.4	1
Mg$^+$–imidazole	243.9 ± 10.4	1
Mg^{2+}(H_2O)$_5$–H_2O	101.3	1
Mg^{2+}(Me_2CO)$_5$–Me_2CO	93.3	1
Ca$^+$–OH	435.1 ± 14.5	1
Ca$^+$–H_2O	117.2	1
Ca$^+$–C_6H_6	134	1
Ca$^+$–imidazole	186.3 ± 3.9	1
Ca^{2+}(H_2O)$_4$–H_2O	110.0 ± 5.9	1
Ca^{2+}(Me_2CO)$_5$–Me_2CO	101.3	1
Sr$^+$–CO	20.3	1
Sr$^+$–CO_2	41.9	1
Sr$^+$–H_2O	144.3	1
Sr$^+$–C_6H_6	117	1
Sr^{2+}(H_2O)$_5$–H_2O	87.4	1
Ba$^+$–OH	530.7 ± 19.3	1
Ba^{2+}(H_2O)$_4$–H_2O	90.8	1
(3) Group 3		
Sc$^+$–H_2	23.0 ± 1.3	1
Sc$^+$–CH_2	412 ± 22	1
Sc$^+$–CH_3	233 ± 10	1
Sc$^+$–C_2H_2	240 ± 20	1
Sc$^+$–C_2H_4	≥131	1
Sc$^+$–C_6H_6	222 ± 21	1
Sc$^+$–H_2O	131	1

Bond	Do^2_{98}/kJ mol^{-1}	Ref.
Sc^+–NH	483 ± 10	1
Sc^+–NH_2	347 ± 5	1
Sc^+–pyridine	231.5 ± 10.3	1
Y^+–CH_2	398 ± 13	1
Y^+–CH_3	249 ± 5.0	1
Y^+–C_2H_2	218 ± 13	1
Y^+–C_2H_4	>138	1
Y^+–CO	29.9 ± 10.6	1
Y^+–CS	137.0 ± 7.7	1
Y^+(O)–CO_2	86 ± 5	1
La^+–CH	523 ± 33	1
La^+–CH_2	401 ± 7	1
La^+–CH_3	217 ± 15	1
La^+–C_2H_2	262 ± 30	1
La^+–C_2H_4	192.5	1
Lu^+–CH_2	>230 ± 6	1
Lu^+–CH_3	176 ± 20	1
U^+(F)–F	552 ± 44	1
$U^+(F)_2$–F	523 ± 38	1
$U^+(F)_3$–F	381 ± 19	1
$U^+(F)_4$–F	243 ± 17	1
$U^+(F)_5$–F	26 ± 11	1
(4) Group 4		
Ti^+–CH	478 ± 5	1
Ti^+–CH_2	391 ± 15	1
Ti^+–CH_3	213.8 ± 3	1
Ti^+–CH_4	70.3 ± 2.5	1
Ti^+–C_2H_2	213 ± 13	1
Ti^+–C_2H_4	146 ± 11	1
Ti^+–C_6H_6	259 ± 9	1
Ti^+–CO	117.7 ± 5.8	1
Ti^+–H_2O	157.7 ± 5.9	1
Ti^+–NH	466 ± 12	1
Ti^+–NH_2	356 ± 13	1
Ti^+–NH_3	197 ± 7	1
Ti^+–pyridine	217.2 ± 9.3	1
Ti^+–imidazole	≤232.4 ± 8.2	1
Zr^+–CH	568 ± 13	1
Zr^+–CH_2	444.8 ± 5	1
Zr^+–CH_3	227.7 ± 9.6	1
Zr^+–C_2H_2	273 ± 14	1
Zr^+–CO	77 ± 10	1
Zr^+–CS	257.6 ± 10.6	1
Hf^+–CH	492.1 ± 14.5	2
Hf^+–CH_2	421.6 ± 6.8	2
Hf^+–CH_2	204.5 ± 25.1	2
Hf^+–C_2H_2	150.6	1
(5) Group 5		
$(CO)_6V^+$–H	220 ± 14	1
V^+–H_2	42.7 ± 2.1	1

Bond	Do^2_{98}/kJ mol^{-1}	Ref.
V^+–CH	470 ± 5	1
V^+–CH_2	326 ± 6	1
V^+–CH_3	193 ± 7	1
V^+–C_2H_2	172 ± 8	1
V^+–C_2H_4	124 ± 8	1
V^+–(η^5-C_5H_5)	530.7	1
V^+–C_6H_6	234 ± 10	1
V^+–CO	114.8 ± 2.9	1
V^+–CO_2	72.4 ± 3.8	1
V^+–H_2O	149.8 ± 5.0	1
V^+–NH	423 ± 29	1
V^+–NH_2	293 ± 6	1
V^+–NH_3	192 ± 11	1
V^+–pyridine	218.7 ± 13.5	1
V^+–imidazole	≤243.4 ± 8.0	1
Nb^+–H_2	61.9	1
Nb^+–CH	581 ± 19	1
Nb^+–CH_2	428.4 ± 8.7	1
Nb^+–CH_3	198.8 ± 10.6	1
Nb^+–CH_3NH_2	134	1
Nb^+–C_3H_6	117.7	1
$(NbFe)^+$–C_3H_4	>163	1
Nb^+–CO	95.5 ± 4.8	1
Nb^+–CS	242.2 ± 10.6	1
Nb_7^+–N_2	<215	1
Ta^+–CH	561.5 ± 15.4	6
Ta^+–CH_2	464.1 ± 2.9	6
Ta^+–CH_3	259.5 ± 13.5	6
Ta^+–C_6H_6	251~301	1
(6) Group 6		
$(CO)_6Cr^+$–H	230 ± 10	1
(η^5-C_5H_5)(NO)$(CO)_2Cr^+$–H	207.1 ± 14	1
Cr^+–H_2	31.8 ± 2.1	1
Cr^+–CH	294 ± 29	1
Cr^+–CH_2	216 ± 4	1
Cr^+–CH_3	110 ± 4	1
Cr^+–C_6H_6	170 ± 10	1
Cr^+–indole	196.6 ± 16.7	1
Cr^+–CO	89.7 ± 5.8	1
Cr^+–OH	298 ± 14	1
Cr^+–H_2O	132.6 ± 8.8	1
Cr^+–N_2	59 ± 4	1
Cr^+–NH_3	183 ± 10	1
$(CO)_6Mo^+$–H	260 ± 9	1
Mo^+–CH	513.3 ± 13.5	1
Mo^+–CH_2	344.4 ± 10	1
Mo^+–CH_3	151.5 ± 8.7	1
Mo^+–CO	193.9 ± 9.6	1
Mo^+–CO_2	49.2 ± 7	1
Mo^+–CS	162 ± 18	1
Mo^+–CS_2	67.5 ± 12.5	1

Bond	D^{o}_{298}/kJ mol^{-1}	Ref.
Mo^+–NH	<385	1
Mo^+–pyrrole	>289	1
$(CO)_6W^+$–H	257 ± 9	1
W^+–CH	580 ± 27	1
W^+–CH_2	456.4 ± 5.8	1
W^+–CH_3	~222.9 ± 9.6	1
$(PMe_3)_3(CO)_3W^+$–H	259.4	1
W^+–pyrrole	>209	1
(7) Group 7		
$(CO)_5Mn^+$–H	172 ± 10	1
Mn^+–H_2	7.9 ± 1.7	1
Mn^+–CH_2	295 ± 13	1
Mn^+–CH_3	215 ± 10	1
$Mn^+(CO)_5$–CH_3	132 ± 15	1
$Mn^+(CO)_5$–CH_4	>30	1
Mn^+–(η^5-C_5H_5)	326.1 ± 9.6	1
Mn^+–C_6H_6	145 ± 10	1
Mn^+–OH	332 ± 24	1
Mn^+–CO	25 ± 10	1
Mn^+–H_2O	121.8 ± 5.9	1
Mn^+–CH_3OH	134 ± 29	1
Mn^+–$OC(CH_3)_2$	159 ± 14	1
Mn^+–CS	80.0 ± 21	1
Mn^+–NH_2	254 ± 20	1
Mn^+–NH_3	147 ± 8	1
Tc^+–CH_2	<464	1
Tc^+–C_2H_2	<320	1
$Re^+(CH_3)(CO)_5$–H	294 ± 13	1
$(PMe_3)(CO)_2Re^+$–H	300.4	1
(8) Group 8		
$Fe^+(O)$–H	444 ± 17	1
$Fe^+(CO)$–H	120 ± 23	1
$Fe^+(H_2O)$–H	215 ± 14	1
$Fe^+(\eta^5$-$C_5H_5)$–H	193 ± 21	1
$(CO_2)_5Fe^+$–H	299 ± 15	1
Fe^+–H_2	45.2 ± 2.5	1
Fe^+–CH	423 ± 29	1
Fe^+–CH_2	≤342 ± 2	1
Fe^+–CH_3	229 ± 5	1
Fe^+–CH_4	73.2	1
Fe^+–C_2H_2	159.0 ± 2.1	1
Fe^+–C_2H_3	238 ± 10	1
Fe^+–C_2H_4	145 ± 11	1
Fe^+–C_2H_5	233 ± 9	1
Fe^+–C_2H_6	64 ± 6	1
Fe^+–OH	366 ± 12	1
Fe^+–CO	129.3 ± 3.9	1
Fe^+D–CO	53 ± 13	1
Fe^+–CO_2	74.3 ± 7.7	1
Fe^+–H_2O	128.9 ± 0.8	1

Bond	D^{o}_{298}/kJ mol^{-1}	Ref.
Fe^+–N_2	53 ± 4	1
Fe^+–NH_3	184 ± 12	1
Fe^+–CS_2	166.1 ± 4.6	1
Fe^+–imidazole	246.1 ± 13.8	1
Fe^+–SiH	254 ± 13	1
Fe^+–SiH_2	181 ± 9	1
Fe^+–SiH_3	183 ± 9	1
$Ru^+(\eta^5$-$C_5H_5)_2$–H	292 ± 16	1
$(\eta^5$-$C_5Me_5)_2Ru^+$–H	284.5	1
Ru^+–CH	501.7 ± 11.6	1
Ru^+–CH_2	344.4 ± 4.8	1
Ru^+–CH_3	160.2 ± 5.8	1
Ru^+–CS	253 ± 20	1
$OsO_4{}^+$–H	552 ± 13	1
(9) Group 9		
$(\eta^5$-$C_5H_5)(CO)_2Co^+$–H	245 ± 12	1
$(CH_3OD)Co^+$–H	147.6 ± 7.7	1
Co^+–H_2	76.1 ± 4.2	1
$(\eta^5$-$C_5H_5)Co^+$–H_2	67.8	1
Co^+–CH	420 ± 37	1
Co^+–CH_2	317 ± 5	1
Co^+–CH_3	203 ± 4	1
Co^+–CH_4	96.7	1
Co^+–C_{60}	243 ± 67	1
Co^+–CO	173.7 ± 6.7	1
Co^+–H_2O	164.4 ± 5.9	1
Co^+–CS	259 ± 33	1
Co^+–N_2	96.2 ± 7.1	1
Co^+–NH_2	247 ± 7	1
Co^+–NH_3	219 ± 16	1
Co^+–CH_3CN	>255 ± 17	1
Co^+–$P(CH_3)_3$	278 ± 11	1
Co^+–$P(C_2H_5)_3$	339 ± 16	1
$(CH)Rh^+$–H	372 ± 21	1
$(\eta^5$-$C_5H_5)(CO)_2Rh^+$–H	287 ± 12	1
Rh^+–CH	444 ± 12	1
Rh^+–CH_2	356 ± 8	1
Rh^+–CH_3	142 ± 6	1
Rh^+–NO	167 ± 21	1
Rh^+–CS	234 ± 19	1
$(CO)(\eta^5$-$C_5H_5)(PPh_3)Ir^+$–H	313.4	1
$(CO)_2(\eta^5$-$C_5Me_5)Ir^+$–H	298.3	1
Ir^+–CH	666.7 ± 22.2	3
Ir^+–CH_2	474.7 ± 2.9	3
Ir^+–CH_3	313.6 ± 17.4	3
Ir^+–C_2H_4	234.3	1
(10) Group 10		
$(CO)_4Ni^+$–H	248 ± 9	1
$(\eta^5$-$C_5H_5)(NO)Ni^+$–H	315 ± 14	1
$(\eta^5$-$C_5H_5)(\eta^5$-$C_5H_5)Ni^+$–H	215 ± 13	1

Bond	Do^2_{98}/kJ mol^{-1}	Ref.
Ni$^+$–H_2	72.4 ± 1.3	1
Ni$^+$–CH	301.0 ± 11.6	1
Ni$^+$–CH_2	306 ± 4	1
Ni$^+$–CH_3	169.8 ± 6.8	1
Ni$^+$–CH_4	96.5 ± 4	1
Ni$^+$–OH	235 ± 19	1
Ni$^+$–CO	175 ± 11	1
Ni$^+$–CO_2	104 ± 1	1
Ni$^+$–H_2O	183.7 ± 3.3	1
Ni$^+$–CS	234.5 ± 9.6	1
Ni$^+$–N_2	110.9 ± 10.5	1
Ni$^+$–NO	227.6 ± 7.5	1
Ni$^+$–NH_2	232.5 ± 7.7	1
Ni$^+$–NH_3	238 ± 19	1
Pd$^+$–CH	536 ± 10	1
Pd$^+$–CH_2	463 ± 3	1
Pd$^+$–CH_3	258 ± 8	1
Pd$^+$–CH_4	170.8 ± 7.7	1
Pd$^+$–CS	200 ± 14	1
Pd$^+$–C_2H_2	>28.9 ± 4.8	1
Pt$^+$–H_2	146.7 ± 11.6	1
Pt$^+$–CH	536.4 ± 9.6	1
Pt$^+$–CH_2	471	1
Pt$^+$–CH_3	257.6 ± 7.7	1
Pt$^+$–CH_4	170.8 ± 7.7	1
Pt$^+$–O_2	64.6 ± 4.8	1
Pt$^+$–CO	218.1 ± 8.7	1
Pt$^+$–CO_2	59.8 ± 4.8	1
Pt$^+$–NH_3	274 ± 12	1
Pt$^+$–C_2H_4	229.7	1
(11) Group 11		
Cu$^+$–H_2	51.9 ± 0.4	1
Cu$^+$–CH_2	267.3 ± 6.8	1
Cu$^+$–CH_3	111 ± 7	1
Cu$^+$–C_2H_2	>21.2 ± 9.6	1
Cu$^+$–C_2H_4	176 ± 14	1
Cu$^+$–C_6H_6	218.0 ± 9.6	1
Cu$^+$–CO	149 ± 7	1
Cu$^+$–N_2	89 ± 30	1
Cu$^+$–NO	109.0 ± 4.8	1
Cu$^+$–H_2O	160.7 ± 7.5	1
Cu$^+$–NH_2	192 ± 13	1
Cu$^+$–NH_3	237 ± 15	1
Cu$^+$–CS	238.3 ± 11.6	1
Cu$^+$–SiH	246 ± 27	1
Cu$^+$–SiH_2	≥231 ± 7	1
Cu$^+$–SiH_3	97 ± 25	1
Ag$^+$–CH_2	≥107 ± 4	1
Ag$^+$–CH_3	66.6 ± 4.8	1
Ag$^+$–C_2H_5	65.7 ± 7.5	1
Ag$^+$–C_6H_6	167 ± 19	1
Ag$^+$–O_2	29.7 ± 0.8	1
Ag$^+$–CO	89 ± 5	1
Ag$^+$–H_2O	134 ± 8	1
Ag$^+$–CS	152 ± 20	1
Ag$^+$–NH_3	170 ± 13	1
Au$^+$–CH_2	357.0 ± 6.8	5
Au$^+$–CH_3	209.4 ± 23.2	5
Au$^+$–C_2H_4	344.5	1
Au$^+$–C_6H_6	289 ± 29	1
Au$^+$–CO	201 ± 8	1
Au$^+$–H_2O	164.0 ± 9.6	1
Au$^+$–H_2S	230 ± 25	1
Au$^+$–NH_3	297 ± 29	1
Au$^+$–PH_3	402 ± 33	1
(12) Group 12		
Zn$^+$–H_2	15.7 ± 1.7	1
Zn$^+$–CH_3	280 ± 7	1
Zn$^+$–OH	127.2	1
Zn$^+$–H_2O	163	1
Zn$^+$–NO	76.2 ± 9.6	1
Zn$^+$–pyrimidine	209.6 ± 7.7	1
Zn$^+$–CS	149 ± 23	1
Cd$^+$–CH_3	220 ± 0	1
Cd$^+$**(**CH_3**)**–CH_3	109 ± 3	1
Cd$^+$–C_6H_6	136 ± 19	1
Hg$^+$–CH_3	285 ± 3	1
Hg$^+$**(**CH_3**)**–CH_3	96 ± 3	1
(13) Group 13		
B$^+$–H_2	15.9 ± 0.8	1
HB$^+$–H_2	61.5 ± 2.1	1
(CH_3**)**$_2$**B**$^+$–CH_3	32.6 ± 4.2	1
Al$^+$–H_2	5.6 ± 0.6	1
Al$^+$–N_2	5.6	1
Al$^+$–CO_2	≥29.3	1
Al$^+$–H_2O	104 ± 15	1
Al$^+$–MeOH	139.7	1
Al$^+$–EtC(O)Et	191.2	1
Al$^+$–C_6H_6	147.3 ± 8.4	1
Al$^+$–pyridine	190.3 ± 10.3	1
Al$^+$–phenol	154.8 ± 16.7	1
Al$^+$–imidazole	232.4 ± 8.2	1
Ga$^+$–NH_3	122.5	1
In$^+$–NH_3	111.0	1
(14) Group 14		
C_{58}^+–C_2	955 ± 15	1
C_{60}^+–C_2	822.0 ± 12.5	1
C_{62}^+–C_2	846.2 ± 10.6	1
C_{78}^+–C_2	938.8 ± 10.6	1
HC_2^+–H	574.749	1
$C_6H_5^+$–H	376.3 ± 4.8	1

Bond	D^{o}_{298}/kJ mol^{-1}	Ref.
$C_2H_3^+$–Cl	249 ± 1.0	7
$C_2H_5^+$–Br	206.3 ± 1.0	7
$C_6H_5^+$–Br	266.3	1
$C_2H_3^+$–I	196.2 ± 1.4	7
CH_3^+–H_2	186	1
CH_5^+–H_2	7.9 ± 0.4	1
$C_2H_5^+$–H_2	17	1
CH_3^+–O_2	80 ± 7	4
CO^+–N_2	67.5 ± 19.3	1
H_2CH^+–N_2	31.8	1
CO^+–CO	173.7 ± 14.6	1
$CO^+(CO)$–CO	52.3	1
$CO^+(CO)_2$–CO	30.2	1
$CO^+(CO)_3$–CO	18.4	1
$(CO_2)^+$–CO_2	70.3	1
$(CO_2)^+(CO_2)$–CO_2	34.7	1
$(CO_2)^+(CO_2)_2$–CO_2	21.3	1
$(CO_2)^+(CO_2)_3$–CO_2	20.1 ± 1.3	1
CH_3^+–N_2O	221.3	1
CH_3^+–SO_2	253.6	1
CH_3^+–OCS	239.3	1
CH_3^+–CS_2	251.9	1
CH_3^+–H_2O	279	1
$CH_3^+(H_2O)$–H_2O	106.3	1
$CH_3^+(H_2O)_2$–H_2O	87.9	1
$CH_3^+(H_2O)_3$–H_2O	61.9	1
$CH_3^+(H_2O)_4$–H_2O	48.5	1
CH_3^+–H_2S	344.8	1
CH_2^+–CH_2O	303.0 ± 2.9	1
CH_3^+–NH_3	431.4	1
$(CH_3)^+$–CH_3	209.2 ± 4.2	1
CH_3^+–CH_4	166.5	1
CF_3^+–CH_4	19.0	1
$(CH_5)^+$–CH_4	28.7 ± 1.3	1
$C_6H_6^+$–CH_4	12.0	1
CH_3^+–CH_3F	230	1
CH_3^+–CF_3Cl	221	1
CH_3^+–CH_3Cl	259	1
tert-$C_4H_9^+$–CH_3OH	63	1
tert-$C_4H_9^+$–CH_3CN	85	1
tert-$C_4H_9^+$–SO_2F_2	43.5	1
CH_3^+–C_2H_3O	338.7 ± 2.9	1
CH_3^+–CF_3ClOCl	252	1
tert--$C_4H_9^+$–$(CH_3)_2S$	185	1
tert-$C_4H_9^+$–C_2H_5OH	85	1
tert-$C_4H_9^+$–C_3H_8	27.6	1
tert-$C_4H_9^+$–t-C_4H_9Cl	339	1
tert-$C_4H_9^+$–$(CH_3)_3CH$	30.1	1
tert-$C_4H_9^+$–C_6H_6	92	1
$(C_6H_6)^+$–C_6H_6	73.6	1
$(C_6H_6)^+$–indole	54.8	1
$C_6F_6^+$–C_6F_6	30.1 ± 4	1
C_{60}^+–C_{60}	35.89 ± 7.72	1
$PhSiH_2^+$–H	159	1
$Si^+(CH_3)_3$–Cl	178.5 ± 1.9	1
SiH_3^+–CO	≥151	1
SiF_3^+–CO	174.1 ± 1.3	1
$(CH_3)_3Si^+$–H_2O	125.9 ± 7.9	1
$(CH_3)_3Si^+$–NH_3	194.6	1
$Si^+(CH_3)(Cl)_2$–CH_3	60.8 ± 2.9	1
$Si^+(CH_3)_2(Cl)$–CH_3	41.5 ± 1.9	1
Si^+–CH_3	413.9 ± 5.8	1
$Si^+(CH_3)$–CH_3	123 ± 48	1
$Si^+(CH_3)_2$–CH_3	513 ± 27	1
$Si^+(CH_3)_3$–CH_3	66.6 ± 5.8	1
$(CH_3)_3Si^+$–CH_3OH	164.0	1
$(CH_3)_3Si^+$–$(C_2H_5)_2O$	184.9	1
$(CH_3)_3Si^+$–C_6H_6	100.0	1
$(CH_3)_3Si^+$–CH_3NH_2	231.8	1
$(CH_3)_3Ge^+$–H_2O	119.7 ± 2.1	1
$(C_2H_5)_3Ge^+$–H_2O	104.2 ± 2.1	1
$(CH_3)_3Sn^+$–NH_3	154	1
$(CH_3)_3Sn^+$–H_2O	108	1
$(CH_3)_3Sn^+$–$(CH_3)_2CO$	157	1
$(CH_3)_3Sn^+$–C_3H_7SH	143	1
Pb^+–H_2O	93.7	1
Pb^+–NH_3	118.4 ± 0.8	1
Pb^+–CH_3OH	97.5 ± 0.8	1
Pb^+–CH_3NH_2	148.1 ± 1.3	1
Pb^+–C_6H_6	110 ± 2	1
(15) Group 15		
H_2N^+–H	544.43 ± 0.10	1
H_3N^+–H	515.1	1
Me_3N^+–H	376	1
Et_3N^+–H	362	1
(imidazole)$^+$ –Zn	216.1 ± 3.9	1
N_2H^+–H_2	24.7 ± 0.8	1
ON^+–O_2	14.2	1
N^+–N_2	303.8	1
ON^+–N_2	21.3	1
N_2^+–N_2	102.3 ± 14.6	1
HN_2^+–N_2	60.7	1
N_3^+–N_2	18.8 ± 1.3	1
O_2N^+–N_2	19.2 ± 1.3	1
H_4N^+–N_2	54 ± 21	1
ON^+–NO	59.4 ± 0.8	1
ON^+–CO	27.2 ± 1.3	1
ON^+–O_3	<58	1
ON^+–CO_2	32.2	1
N_2O^+–ON_2	72.8 ± 6.3	1
NO^+–ON_2	36.4 ± 0.8	1
$(HON_2)^+$–ON_2	69.9 ± 4	1

Bond	D°_{298}/kJ mol^{-1}	Ref.
ON^+–H_2O	95	1
$ON^+(H_2O)$–H_2O	67.4	1
$ON^+(H_2O)_2$–H_2O	56.5	1
H_4N^+–H_2O	86.2 ± 4.2	1
$H_4N^+(H_2O)$–H_2O	72.8 ± 4.2	1
$H_4N^+(H_2O)_2$–H_2O	57.3 ± 4.2	1
$H_4N^+(H_2O)_3$–H_2O	51.0	1
$H_4N^+(H_2O)_4$–H_2O	44.4	1
(glycine)H^+–H_2O	77.2 ± 11.0	1
(tryptophan)H^+–H_2O	31.2 ± 2.5	1
(tryptophanylglicine)H^+–H_2O	56.0 ± 5.3	1
H_4N^+–H_2S	47.7	1
$H^+(NH_3)$–NH_3	108.8	1
$H^+(NH_3)_2$–NH_3	69.5	1
$H^+(NH_3)_3$–NH_3	57.3	1
$H^+(NH_3)_4$–NH_3	49.0	1
$H^+(NH_3)_5$–NH_3	29.3	1
$H^+(NH_3)_6$–NH_3	27.2	1
NH_4^+–CH_4	15.0	1
ON^+–CH_3OH	97.6	1
O_2N^+–CH_3OH	80.3 ± 9.6	1
$(CH_3CNH)^+$–CH_3CN	130.1 ± 9.6	1
(pyridineH)$^+$–pyridine	105.4 ± 4	1
(valine H)$^+$–valine	86.6 ± 8.4	1
(betainH)$^+$–betaine	139.9 ± 4.8	1
H_4P^+–H_2O	54.4	1
$(H_4P)^+$–PH_3	48.1	1
AsH_2^+–H	257	1
I_2As^+–acetone	106 ± 17	1
I_2As^+–benzene	77 ± 17	1
Bi^+–H_2O	95.4	1
Bi^+–NH_3	149	1
Bi^+–C_6H_6	≤149	1
(16) Group 16		
$(H_3O)^+$–H_2	14.6 ± 2.1	1
O^+–O_2	179.5	1
$O^+(O_2)_1$–O_2	28.9	1
$O^+(O_2)_2$–O_2	3.9	1
O_2^+–O_2	38.3 ± 2.1	1
$O_2^+(O_2)$–O_2	24.6 ± 1.3	1
$O_2^+(O_2)_2$–O_2	10.4 ± 0.8	1
$O_2^+(O_2)_3$–O_2	9.0 ± 0.8	1
$O_2^+(O_2)_4$–O_2	8.0 ± 0.8	1
$O_2^+(O_2)_5$–O_2	7.9 ± 1.3	1
O^+–N_2	231.4	1
O_2^+–N_2	22.6	1
$(H_3O)^+$–N_2	22.2 ± 2.1	1
O_4^+–N_2	12.3	1
O_2^+–CO	31.8	1
O_2^+–CO_2	41.0 ± 2.1	1
CO_2^+–CO_2	65.3 ± 4	1

Bond	D°_{298}/kJ mol^{-1}	Ref.
$(H_3O)^+$–CO_2	64.0	1
$(H_3O)^+(CO_2)$–CO_2	51.9	1
$(H_3O)^+(CO_2)_2$–CO_2	43.9	1
$(H_3O)^+(CO_2)_3$–CO_2	18.0	1
O_2^+–ON_2	56.1 ± 4	1
$(H_3O)^+$–ON_2	70.7 ± 6.5	1
$(H_3O)^+(H_2O)$–ON_2	50.6 ± 2.1	1
$(H_3O)^+(H_2O)_2$–ON_2	42.7 ± 2.1	1
O_3^+–O_3	67.5 ± 39	1
$OClO^+$–OClO	246 ± 48	1
O_2^+–H_2O	>67	1
$(OH)^+(H_2O)_2$–H_2O	87.4	1
$(OH)^+(H_2SO_4)(H_2O)_4$–H_2O	56.9	1
$(OH)^+(H_2SO_4)(H_2O)_5$–H_2O	49.8	1
$(OH)^+(H_2SO_4)(H_2O)_6$–H_2O	44.8	1
$(H_2O)^+$–H_2O	164.0	1
$(H_3O)^+$–H_2O	140.2	1
$(H_3O)^+(H_2O)$–H_2O	93.3	1
$(H_3O)^+(H_2O)_2$–H_2O	71.1	1
$(H_3O)^+(H_2O)_3$–H_2O	64.0	1
$(H_3O)^+(H_2O)_4$–H_2O	54.4	1
$(H_3O)^+(H_2O)_5$–H_2O	49.0	1
$(H_3O)^+(H_2O)_6$–H_2O	43.1	1
(HCOOH)H^+–H_2O	100.8	1
$CH_3OH_2^+$–H_2O	115.6	1
CH_3CHOH^+–H_2O	104.6	1
$(CH_3)_2OH^+$–H_2O	100.4	1
(tetrahydrofuranH)$^+$–H_2O	82.8	1
(furanH)$^+$–H_2O	43.5	1
furane$^+$–H_2O	41.0	1
(phenol)$^+$–H_2O	78.0	1
(1-naphthol)$^+$–H_2O	66.4	1
H_3O^+–HC(O)H	137.7	1
H_3O^+–NH_3	229.3	1
$H_3O^+(NH_3)$–NH_3	77.0	1
$H_3O^+(NH_3)_2$–NH_3	71.5	1
$H_3O^+(NH_3)_3$–NH_3	62.8	1
H_3O^+–PH_3	144	1
H_3O^+–SO_3	74	1
$(HCOOH)^+$–HCOOH	96.5 ± 9.6	1
H_3O^+–CH_4	33.5	1
$(CH_3OH)^+$–CH_3OH	115.8 ± 19.3	1
$CH_3OH_2^+$–CH_3OH	136.4	1
H_3O^+–CH_3CN	195.4	1
furan$^+$–furan	94.1	1
BH^+–B, B = tetrahydofuran	125.1	1
S^+–CS_2	166	1
CS^+–CS_2	150.6	1
CS_2^+–CS_2	104.2	1
HCS_2^+–CS_2	46.4	1
OS^+–SO_2	57.7	1

Bond	Do^2_{98}/kJ mol^{-1}	Ref.
O_2S^+–SO_2	63.6	1
OCS^+–OCS	100.0	1
OCS^+–CO_2	72.0	1
SO_2^+–CO_2	42.7	1
H_3S^+–H_2O	91.6	1
thiopheneH$^+$–H_2O	42.7	1
H_3S^+–H_2S	53.6 ± 6.3	1
H_3S^+–CH_4	16.3	1
$(CH_3)_2Se^{\bullet+}$–$Se(CH_3)_2$	~95 ± 3	1
$(CH_3)_2Te^{\bullet+}$–$Te(CH_3)_2$	97 ± 2	1
(17) Group 17		
HF^+–HF	≥138	1
$(H_2Cl)^+$ –Cl	39.6	1
HCl^+–HCl	83.9	1
Cl^+–CCl_3	446.7 ± 9.6	1
Cl^+–C_2H_3	685.0 ± 4.8	1
HBr^+–HBr	96	1
I^+–CH_3	330.0	1
$I^+(CH_3I)$–CH_3	51.1	1
$I^+(CH_3I)_2$–CH_3	112.9	1
(18) Group 18		
$He^+(He)_1$–He	17.6	1
$He^+(He)_2$–He	2.7 ± 0.6	1
$Ne^+(Ne)$–Ne	10.3 ± 0.6	1
$Ne^+(Ne)_2$–Ne	3.3 ± 0.6	1
$Ar^+(Ar)$–Ar	20.4 ± 0.6	1
$Ar^+(Ar)_2$–Ar	7.0 ± 0.6	1
$Ar^+(N_2)$–Ar	25.1	1
$Ar^+(N_2)(Ar)$–Ar	7.1	1
$Ar^+(N_2)(Ar)_2$–Ar	7.1	1
$Kr^+(Kr)$–Kr	23.3 ± 0.6	1
$Kr^+(Kr)_2$–Kr	9.0 ± 0.6	1
$Xe^+(Xe)$–Xe	25.2 ± 0.6	1
$Xe^+(Xe)_2$–Xe	11.0 ± 0.6	1
Ar^+–H_2	93.7	1
Ar^+–N_2	127.6	1
$Ar^+(N_2)$–N_2	31.0	1
$Ar^+(N_2)_2$–N_2	10.9	1
Ar^+–CO	75 ± 17	1
$Ar^+(CO)$–CO	13	1
Kr^+–CO	103.3 ± 7.5	1
Kr^+–CO_2	79.1 ± 2.9	1

References

1. Luo, Y. R., *Comprehensive Handbook of Chemical Bond Energies*, CRC Press, Boca Raton, FL, 2007.
2. Parke, L. G., Hinton, C. S., and Armentrout, P. B., *Int. J. Mass Spectrom.* 254, 168, 2006.
3. Li, F.-X., Zhang, *X*.-G., and Armentrout, P. B., *Int. J. Mass Spectrom.* 255/256, 279, 2006.
4. Meloni, G., Zou, P., Klippenstein, S. J., Ahmed, M., Leone, S. R., Taatjes, C. A., and Osborn, D. L., *J. Am. Chem. Soc.* 128, 13559, 2006.
5. Li, F.-X., and Armentrout, P. B., *J. Chem. Phys.* 125, 133114/1, 2006.
6. Parke, L. G., Hinton, C. S., and Armentrout, P. B., *J. Phys. Chem. C* 111, 17773, 2007.
7. Shuman, N. S., Ochieng, M. A., Sztáray, B., and Baer, T., *J. Phys. Chem. A* 112, 5647, 2008.

ELECTRONEGATIVITY

Electronegativity is a parameter originally introduced by Pauling which describes, on a relative basis, the tendency of an atom in a molecule to attract bonding electrons. While electronegativity is not a precisely defined molecular property, the electronegativity difference between two atoms provides a useful measure of the polarity and ionic character of the bond between them. This table gives the electronegativity *X*, on the Pauling scale, for the most common oxidation state. Other scales are described in the references.

References

1. Pauling, L., *The Nature of the Chemical Bond, Third Edition*, Cornell University Press, Ithaca, NY, 1960.
2. Allen, L. C., *J. Am. Chem. Soc.*, 111, 9003, 1989.
3. Allred, A. L., *J. Inorg. Nucl. Chem.*, 17, 215, 1961.

Z	Symbol	*X*	*Z*	Symbol	*X*	*Z*	Symbol	*X*
1	H	2.20	33	As	2.18	65	Tb	—
2	He	—	34	Se	2.55	66	Dy	1.22
3	Li	0.98	35	Br	2.96	67	Ho	1.23
4	Be	1.57	36	Kr	—	68	Er	1.24
5	B	2.04	37	Rb	0.82	69	Tm	1.25
6	C	2.55	38	Sr	0.95	70	Yb	—
7	N	3.04	39	Y	1.22	71	Lu	1.0
8	O	3.44	40	Zr	1.33	72	Hf	1.3
9	F	3.98	41	Nb	1.6	73	Ta	1.5
10	Ne	—	42	Mo	2.16	74	W	1.7
11	Na	0.93	43	Tc	2.10	75	Re	1.9
12	Mg	1.31	44	Ru	2.2	76	Os	2.2
13	Al	1.61	45	Rh	2.28	77	Ir	2.2
14	Si	1.90	46	Pd	2.20	78	Pt	2.2
15	P	2.19	47	Ag	1.93	79	Au	2.4
16	S	2.58	48	Cd	1.69	80	Hg	1.9
17	Cl	3.16	49	In	1.78	81	Tl	1.8
18	Ar	—	50	Sn	1.96	82	Pb	1.8
19	K	0.82	51	Sb	2.05	83	Bi	1.9
20	Ca	1.00	52	Te	2.1	84	Po	2.0
21	Sc	1.36	53	I	2.66	85	At	2.2
22	Ti	1.54	54	Xe	2.60	86	Rn	—
23	V	1.63	55	Cs	0.79	87	Fr	0.7
24	Cr	1.66	56	Ba	0.89	88	Ra	0.9
25	Mn	1.55	57	La	1.10	89	Ac	1.1
26	Fe	1.83	58	Ce	1.12	90	Th	1.3
27	Co	1.88	59	Pr	1.13	91	Pa	1.5
28	Ni	1.91	60	Nd	1.14	92	U	1.7
29	Cu	1.90	61	Pm	—	93	Np	1.3
30	Zn	1.65	62	Sm	1.17	94	Pu	1.3
31	Ga	1.81	63	Eu	—			
32	Ge	2.01	64	Gd	1.20			

FORCE CONSTANTS FOR BOND STRETCHING

Representative force constants (*f*) for stretching of chemical bonds are listed in this table. Except where noted, all force constants are derived from values of the harmonic vibrational frequencies ω_e. Values derived from the observed vibrational fundamentals ν, which are noted by a, are lower than the harmonic force constants, typically by 2 to 3% in the case of heavy atoms (often by 5 to 10% if one of the atoms is hydrogen). Values are given in the SI unit newton per centimeter (N/cm), which is identical to the commonly used cgs unit mdyn/Å.

References

1. Huber, K. P., and Herzberg, G., *Molecular Spectra and Molecular Structure. IV. Constants of Diatomic Molecules*, Van Nostrand Reinhold, New York, 1979.
2. Shimanouchi, T., The Molecular Force Field, in Eyring, H., Henderson, D., and Yost, W., Eds., *Physical Chemistry: An Advanced Treatise*, Vol. IV, Academic Press, New York, 1970.
3. Tasumi, M., and Nakata, M., *Pure and Appl. Chem.*, 57, 121–147, 1985.

Bond	Molecule	*f*/(N/cm)	Note	Bond	Molecule	*f*/(N/cm)	Note
H-H	H_2	5.75			OCS	7.44	
Be-H	BeH	2.27		C-N	CN	16.29	
B-H	BH	3.05			HCN	18.78	
C-H	CH	4.48			CH_3CN	18.33	
	CH_4	5.44	b		CH_3NH_2	5.12	a,c
	C_2H_6	4.83	a,b,c	C-P	CP	7.83	
	CH_3CN	5.33	b	Si-Si	Si_2	2.15	
	CH_3Cl	5.02	a,b,c	Si-O	SiO	9.24	
	$CCl_2{=}CH_2$	5.57	b	Si-F	SiF	4.90	
	HCN	6.22		Si-Cl	SiCl	2.63	
N-H	NH	5.97		N-N	N_2	22.95	
O-H	OH	7.80			N_2O	18.72	
	H_2O	8.45		N-O	NO	15.95	
P-H	PH	3.22			N_2O	11.70	
S-H	SH	4.23		P-P	P_2	5.56	
	H_2S	4.28		P-O	PO	9.45	
F-H	HF	9.66		O-O	O_2	11.77	
Cl-H	HCl	5.16			O_3	5.74	a
Br-H	HBr	4.12		S-O	SO	8.30	
I-H	HI	3.14			SO_2	10.33	a
Li-H	LiH	1.03		S-S	S_2	4.96	
Na-H	NaH	0.78		F-F	F_2	4.70	
K-H	KH	0.56		Cl-F	ClF	4.48	
Rb-H	RbH	0.52		Br-F	BrF	4.06	
Cs-H	CsH	0.47		Cl-Cl	Cl_2	3.23	
C-C	C_2	12.16		Br-Cl	BrCl	2.82	
	$CCl_2{=}CH_2$	8.43		Br-Br	Br_2	2.46	
	C_2H_6	4.50	a,c	I-I	I_2	1.72	
	CH_3CN	5.16		Li-Li	Li_2	0.26	
C-F	CF	7.42		Li-Na	LiNa	0.21	
	CH_3F	5.71	a,c	Na-Na	Na_2	0.17	
C-Cl	CCl	3.95		Li-F	LiF	2.50	
	CH_3Cl	3.44	a,c	Li-Cl	LiCl	1.43	
	$CCl_2{=}CH_2$	4.02	b	Li-Br	LiBr	1.20	
C-Br	CH_3Br	2.89	a,c	Li-I	LiI	0.97	
C-I	CH_3I	2.34	a,c	Na-F	NaF	1.76	
C-O	CO	19.02		Na-Cl	NaCl	1.09	
	CO_2	16.00		Na-Br	NaBr	0.94	
	OCS	16.14		Na-I	NaI	0.76	
	CH_3OH	5.42	a,c	Be-O	BeO	7.51	
C-S	CS	8.49		Mg-O	MgO	3.48	
	CS_2	7.88		Ca-O	CaO	3.61	

[a] Derived from fundamental frequency, without anharmonicity correction.
[b] Average of symmetric and antisymmetric (or degenerate) modes.
[c] Calculated from Local Symmetry Force Field (see Reference 2).

FUNDAMENTAL VIBRATIONAL FREQUENCIES OF SMALL MOLECULES

This table lists the fundamental vibrational frequencies of selected three-, four-, and five-atom molecules. Both stable molecules and transient free radicals are included. The data have been taken from evaluated sources. In general, the selected values are based on gas-phase infrared, Raman, or ultraviolet spectra; when these were not available, liquid-phase or matrix-isolation spectra were used.

Molecules are grouped by structural type. Within each group, related molecules appear together for convenient comparison.

The vibrational modes are described by their approximate character in terms of stretching, bending, deformation, etc. However, it should be emphasized that most such descriptions are only approximate, and that the true normal mode usually involves a mixture of motions. Abbreviations are:

sym.	symmetric
antisym.	antisymmetric
str.	stretch
deform.	deformation
scis.	scissors
rock.	rocking
deg.	degenerate

In the case of free radicals, strong interactions may exist between the electronic and bending vibrational motions. Details can be found in References 3 and 4. The references should be consulted for information on the accuracy of the data and for data on other molecules not listed here.

All fundamental frequencies (more precisely, wavenumbers) are given in units of cm^{-1}.

XY_2 Molecules

Point groups $D_{\infty h}$(linear) and C_{2v}(bent)

Molecule	Structure	Sym. str.	Bend	Antisym. str.
CO_2	Linear	1333	667	2349
CS_2	Linear	658	397	1535
C_3	Linear	1224	63	2040
CNC	Linear		321	1453
NCN	Linear	1197	423	1476
BO_2	Linear	1056	447	1278
BS_2	Linear	510	120	1015
KrF_2	Linear	449	233	590
XeF_2	Linear	515	213	555
$XeCl_2$	Linear	316		481
H_2O	Bent	3657	1595	3756
D_2O	Bent	2671	1178	2788
F_2O	Bent	928	461	831
Cl_2O	Bent	639	296	686
O_3	Bent	1103	701	1042
H_2S	Bent	2615	1183	2626
D_2S	Bent	1896	855	1999
SF_2	Bent	838	357	813
SCl_2	Bent	525	208	535
SO_2	Bent	1151	518	1362
H_2Se	Bent	2345	1034	2358
D_2Se	Bent	1630	745	1696
NH_2	Bent	3219	1497	3301
NO_2	Bent	1318	750	1618
NF_2	Bent	1075	573	942
ClO_2	Bent	945	445	1111
CH_2	Bent		963	
CD_2	Bent		752	
CF_2	Bent	1225	667	1114
CCl_2	Bent	721	333	748
CBr_2	Bent	595	196	641
SiH_2	Bent	2032	990	2022
SiD_2	Bent	1472	729	1468
SiF_2	Bent	855	345	870
$SiCl_2$	Bent	515		505
$SiBr_2$	Bent	403		400
GeH_2	Bent	1887	920	1864
$GeCl_2$	Bent	399	159	374
SnF_2	Bent	593	197	571
$SnCl_2$	Bent	352	120	334
$SnBr_2$	Bent	244	80	231
PbF_2	Bent	531	165	507
$PbCl_2$	Bent	314	99	299
ClF_2	Bent	500		576

XYZ Molecules

Point Groups $C_{\infty v}$ (linear) and C_s(bent)

Molecule	Structure	XY str.	Bend	YZ str.
HCN	Linear	3311	712	2097
DCN	Linear	2630	569	1925
FCN	Linear	1077	451	2323
ClCN	Linear	744	378	2216
BrCN	Linear	575	342	2198
ICN	Linear	486	305	2188
CCN	Linear	1060	230	1917
CCO	Linear	1063	379	1967
HCO	Bent	2485	1081	1868
HCC	Linear	3612		1848
OCS	Linear	2062	520	859
NCO	Linear	1270	535	1921
NNO	Linear	2224	589	1285
HNB	Linear	3675		2035
HNC	Linear	3653		2032
HNSi	Linear	3583	523	1198
HBO	Linear		754	1817
FBO	Linear		500	2075
ClBO	Linear	676	404	1958
BrBO	Linear	535	374	1937
FNO	Bent	766	520	1844
ClNO	Bent	596	332	1800
BrNO	Bent	542	266	1799
HNF	Bent		1419	1000

XYZ Molecules

Point Groups $C_{\infty v}$ (linear) and C_s (bent)

Molecule	Structure	XY str.	Bend	YZ str.
HNO	Bent	2684	1501	1565
HPO	Bent	2095	983	1179
HOF	Bent	3537	886	1393
HOCl	Bent	3609	1242	725
HOO	Bent	3436	1392	1098
FOO	Bent	579	376	1490
ClOO	Bent	407	373	1443
BrOO	Bent			1487
HSO	Bent		1063	1009
NSF	Bent	1372	366	640
NSCl	Bent	1325	273	414
HCF	Bent		1407	1181
HCCl	Bent		1201	815
HSiF	Bent	1913	860	834
HSiCl	Bent		808	522
HSiBr	Bent	1548	774	408

Symmetric XY_3 Molecules

Point Groups D_{3h} (planar) and C_{3v} (pyramidal)

Molecule	Structure	Sym. str.	Sym. deform.	Deg. str.	Deg. deform.
NH_3	Pyram.	3337	950	3444	1627
ND_3	Pyram.	2420	748	2564	1191
PH_3	Pyram.	2323	992	2328	1118
AsH_3	Pyram.	2116	906	2123	1003
SbH_3	Pyram.	1891	782	1894	831
NF_3	Pyram.	1032	647	907	492
PF_3	Pyram.	892	487	860	344
AsF_3	Pyram.	741	337	702	262
PCl_3	Pyram.	504	252	482	198
PI_3	Pyram.	303	111	325	79
AsI_3	Pyram.	219	94	224	71
$AlCl_3$	Pyram.	375	183	595	150
SO_3	Planar	1065	498	1391	530
BF_3	Planar	888	691	1449	480
BH_3	Planar		1125	2808	1640
CH_3	Planar		606	3161	1396
CD_3	Planar		453	2369	1029
CF_3	Pyram.	1090	701	1260	510
SiF_3	Pyram.	830	427	937	290

Linear XYYX Molecules

Point Group $D_{\infty h}$

Molecule	Sym. XY str.	Antisym. XY str.	YY str.	Bend	Bend
C_2H_2	3374	3289	1974	612	730
C_2D_2	2701	2439	1762	505	537
C_2N_2	2330	2158	851	507	233

Planar X_2YZ Molecules

Point Group C_{2v}

Molecule	Sym.XY str.	YZ str.	YX_2 scis.	Antisym. XY str.	YX_2 rock	YX_2 wag
H_2CO	2783	1746	1500	2843	1249	1167
D_2CO	2056	1700	1106	2160	990	938
F_2CO	965	1928	584	1249	626	774
Cl_2CO	567	1827	285	849	440	580
O_2NF	1310	822	568	1792	560	742
O_2NCl	1286	793	370	1685	408	652

Tetrahedral XY_4 Molecules

Point Group T_d

Molecule	Sym. str.	Deg. deform.(e)	Deg. str.(f)	Deg. deform.(f)
CH_4	2917	1534	3019	1306
CD_4	2109	1092	2259	996
CF_4	909	435	1281	632
CCl_4	459	217	776	314
CBr_4	267	122	672	182
CI_4	178	90	555	125
SiH_4	2187	975	2191	914
SiD_4	1558	700	1597	681
SiF_4	800	268	1032	389
$SiCl_4$	424	150	621	221
GeH_4	2106	931	2114	819
GeD_4	1504	665	1522	596
$GeCl_4$	396	134	453	172
$SnCl_4$	366	104	403	134
$TiCl_4$	389	114	498	136
$ZrCl_4$	377	98	418	113
$HfCl_4$	382	102	390	112
RuO_4	885	322	921	336
OsO_4	965	333	960	329

References

1. T. Shimanouchi, Tables of Molecular Vibrational Frequencies, Consolidated Volume I, Natl. Stand. Ref. Data Ser., Natl. Bur. Stand. (U.S.), 39, 1972.
2. T. Shimanouchi, Tables of Molecular Vibrational Frequencies, Consolidated Volume II, *J. Phys. Chem. Ref. Data*, 6, 993, 1977.
3. G. Herzberg, *Electronic Spectra and Electronic Structure of Polyatomic Molecules*, D. Van Nostrand Co., Princeton, NJ, 1966.
4. M. E. Jacox, Ground state vibrational energy levels of polyatomic transient molecules, *J. Phys. Chem. Ref. Data*, 13, 945, 1984.

SPECTROSCOPIC CONSTANTS OF DIATOMIC MOLECULES

David R. Lide

This table lists the leading spectroscopic constants and equilibrium internuclear distance r_e in the ground electronic state for selected diatomic molecules. The constants are those describing the vibrational and rotational energy through the expressions:

$$E_{vib}/hc = \omega_e(v + ½) - \omega_e x_e(v + ½)^2 + \cdots$$

$$E_{rot}/hc = B_v J(J + 1) - D_v[J(J + 1)]^2 + \cdots$$

where

$$B_v = B_e - \alpha_e(v + ½) + \cdots$$

$$D_v = D_e + \cdots$$

Here v and J are the vibrational and rotational quantum numbers, respectively, h is Planck's constant, and c is the speed of light. In this customary formulation the constants ω_e, B_e, etc. have dimensions of inverse length; in this table they are given in units of cm^{-1}.

Users should note that higher order terms in the above energy expressions are required for very precise calculations. The references contain constants for many of these higher terms, as well as more precise values of the lower constants. Also, if the ground electronic state is not $^1\Sigma$, additional terms are needed to account for the interaction between electronic and pure rotational angular momentum. For some molecules in the table the data have been analyzed in terms of the Dunham series expansion:

$$E/hc = \Sigma_{lm} Y_{lm}(v + ½)^l J^m(J + 1)^m$$

In such cases it has been assumed that $Y_{10} = \omega_e$, $Y_{01} = B_e$, etc., although in the highest approximations these identities are not precisely correct. Some of the values of r_e in the table have been corrected for breakdown of the Born-Oppenheimer approximation, which can affect the last decimal place. Because of differences in the method of data analysis and limitations in the model, care should be taken in comparing r_e values for different molecules to a precision beyond 0.001 Å.

Entries in the ω_e column that are marked by * give the interval between $v = 0$ and $v = 1$ states instead of a value of ω_e.

Molecules are listed in alphabetical order by formula as most commonly written. In most cases this form places the more electropositive element first, but there are exceptions such as OH, NH, CH, etc. References 1–5 are evaluated compilations covering many molecules and giving references to the original literature.

References

1. Huber, K. P., and Herzberg, G., *Molecular Spectra and Molecular Structure IV. Constants of Diatomic Molecules*, Van Nostrand Reinhold, New York, 1979.
2. *Landolt-Börnstein, Numerical Data and Functional Relationships in Science and Technology, New Series*, II/6 (1974), II/14a (1982), II/14b (1983), II/19a (1992), II/19d-1 (1995), II/24a (1998), *Molecular Constants*, Springer-Verlag, Heidelberg.
3. Lovas, F. J., and Tiemann, E., *J. Phys. Chem. Ref. Data* 3, 609, 1974.
4. Irikura, K. K., *J. Phys. Chem. Ref. Data* 35, 389, 2007.
5. Lovas, F. J., Tiemann, E., Coursey, J. S., Kotochigova, S. A., Chang, J., Olsen, K., and Dragoset, R. A., *Diatomic Spectral Database* (version 2.0). Available: http://physics.nist.gov/Diatomic, National Institute of Standards and Technology, Gaithersburg, MD, November 2009.
6. Wormsbecher, R. F., Hessel, M. M., and Lovas, F. J., *J. Chem. Phys.* 74, 6893, 1981.
7. Hedderich, H. G., Dulick, M., and Bernath, P. F., *J. Chem. Phys.* 99, 8363, 1993.
8. Ram, R. S., and Bernath, P. F., *J. Mol. Spectrosc.* 176, 320, 1996.
9. Miller, C. E., and Drouin, B. J., *J. Mol. Spectrosc.*. 205, 312, 2001.
10. Tellinghuisen, P. C., Tellinghuisen, J., Coxon, J. A., Velazco, J. E., and Setser, D. W., *J. Chem. Phys.* 68, 5187, 1978.
11. Muntianu, A., Guo, B., and Bernath, P. F., *J. Mol. Spectrosc.* 176, 274, 1996.
12. Yamada, C., Chang, M. C., and Hirota, E., *J. Chem. Phys.* 86, 3804, 1987.
13. Wang, X., Magnes, J., Marjatta Lyyra, A., Ross, A. J., Martin, F., Dove, P. M., and Le Roy, R. J., *J. Chem. Phys.* 117, 9339, 2002.
14. Yamada, C., and Hirota, E., *J. Chem. Phys.* 99, 8489, 1993.
15. James, A. M., Kowalczyk, P., Fournier, R., and Simard, B., *J. Chem. Phys.* 99, 8504, 1993.
16. Campbell, J. M., Dulick, M., Klapstein, D., White, J. B., and Bernath, P. F., *J. Chem. Phys.* 99, 8379, 1993.
17. White, J. B., Dulick, M., and Bernath, P. F., *J. Chem. Phys.* 99, 8371, 1993.
18. Shayesteh, A., Appadoo, D. R. T., Gordon, I., Le Roy, R. J., and Bernath, P. F., *J. Chem. Phys.* 120, 10002, 2004.
19. Sanz, M. E., McCarthy, M. C., and Thaddeus, P., *J. Chem. Phys.* 119, 11715, 2003.
20. Müller, W., and Meyer, W., *J. Chem. Phys.* 80, 3311, 1984.
21. Staanum, P., Pashov, A., Knöckel, H., and Tiemann, E., *Phys. Rev. A* 75, 042513, 2007.
22. Lovas, F. J., Maki, A. G., and Olson, W. B., *J. Mol. Spectrosc.* 87, 449, 1981.
23. Bogey, M., Demuynck, C., and Destombes, J. L., *Chem. Phys.* 66, 99, 1982.
24. Babou, Y., Rivière, Ph., Perrin, M. Y., and Soufiani, A., *Int. J. Thermophys.* 30, 416, 2009.
25. Skatrud, D. D., DeLucia, F. C., Blake, G. A., and Sastry, K. V. L. N., *J. Mol. Spectrosc.* 99, 35, 1983.
26. Le Floch, A., *Mol. Phys.* 72, 133, 1991.
27. George, T., Urban, W., and Le Floch, A., *J. Mol. Spectrosc.* 165, 500, 1994.
28. Mürtz, P., Thümmel, H., Pfelzer, C., and Urban, W., *Mol. Phys.* 86, 1362, 1995.
29. Maki, A. G., Lovas, F. J., and Suenram, R. D., *J. Mol. Spectrosc.* 91, 424, 1982.
30. Engelke, F., Ennen, G., and Meiwes, K. H., *Chem. Phys.* 66, 391, 1982.
31. Tanaka, T., Tamura, M., and Tanaka, K., *J. Mol. Struct.* 413, 153, 1997.

Molecule	State	ω_e cm^{-1}	$\omega_e x_e$ cm^{-1}	B_e cm^{-1}	α_e cm^{-1}	D_e 10^{-6} cm^{-1}	r_e Å	Ref.
$^{107}Ag^{79}Br$	$^1\Sigma^+$	249.57	0.63	0.064833	0.0002361	0.0175	2.39311	1,2,3
$^{107}Ag^{35}Cl$	$^1\Sigma^+$	343.49	1.17	0.12298388	0.00059541	0.06305	2.28079	1,2,3
$^{107}Ag^{19}F$	$^1\Sigma^+$	513.45	2.59	0.2657020	0.0019206	0.284	1.98318	1,2,3
$^{107}Ag^{1}H$	$^1\Sigma^+$	1759.9	34.06	6.449	0.201	344	1.618	1,2
$^{107}Ag^{2}H$	$^1\Sigma^+$	1250.70	17.17	3.2572	0.0722	85.9	1.6180	1,2

Molecule	State	ω_e cm^{-1}	$\omega_e x_e$ cm^{-1}	B_e cm^{-1}	α_e cm^{-1}	D_e 10^{-6} cm^{-1}	r_e Å	Ref.
$^{107}Ag^{127}I$	$^1\Sigma^+$	206.50	0.46	0.04486821	0.0001414	0.00847	2.54463	1,2,3
$^{107}Ag^{16}O$	$^2\Pi_{1/2}$	490.2	3.1	0.3020	0.0025	0.45	2.003	1,2
$^{27}Al_2$	$^3\Pi_u$	285.8	0.9	0.17127	0.0008		2.701	1,2
$^{27}Al^{79}Br$	$^1\Sigma^+$	378.0	1.28	0.15919713	0.00086045	0.11285	2.29481	1,2,3
$^{27}Al^{35}Cl$	$^1\Sigma^+$	481.77	2.10	0.24393007	0.00161108	0.25017	2.13014	7
$^{27}Al^{19}F$	$^1\Sigma^+$	802.32	4.85	0.55248021	0.00498426	1.0464	1.65437	4,5
$^{27}Al^{1}H$	$^1\Sigma^+$	1682.37	29.05	6.3937842	0.1870527	368.53	1.64736	17
$^{27}Al^{2}H$	$^1\Sigma^+$	1211.77	15.06	3.3183929	0.0698773	99.42	1.64637	17
$^{27}Al^{127}I$	$^1\Sigma^+$	316.1	1.0	0.11769985	0.00055859		2.53710	1,2,3
$^{27}Al^{16}O$	$^2\Sigma^+$	979.49	7.01	0.6413856	0.0057796	1.08	1.61782	4
$^{27}Al^{32}S$	$^2\Sigma^+$	617.11	3.33	0.2800368	0.0017823	0.22	2.02828	4
$^{75}As_2$	$^1\Sigma_g^+$	429.55	1.12	0.10179	0.000333		2.1026	1,2
$^{75}As^{1}H$	$^3\Sigma^-$	2130*		7.3067	0.2117	327	1.52315	1,2
$^{75}As^{2}H$	$^3\Sigma^-$	1484*		3.6688		90	1.5306	1,2
$^{75}As^{14}N$	$^1\Sigma^+$	1068.54	5.41	0.54551	0.003366	0.53	1.6184	1,2
$^{75}As^{16}O$	$^2\Pi_{1/2}$	967.08	4.85	0.48482	0.003299	0.49	1.6236	1,2
$^{197}Au_2$	$^1\Sigma_g^+$	190.9	0.42	0.028013	0.0000723	0.00250	2.4719	1,2
$^{197}Au^{1}H$	$^1\Sigma^+$	2305.01	43.12	7.2401	0.2136	279	1.5239	1,2
$^{197}Au^{2}H$	$^1\Sigma^+$	1634.98	21.65	3.6415	0.07614	70.9	1.5238	1,2
$^{11}B_2$	$^3\Sigma_g$	1051.3	9.35	1.212	0.014		1.590	1,2
$^{11}B^{79}Br$	$^1\Sigma^+$	684.31	3.52	0.4894	0.0035	1.00	1.888	1,2
$^{11}B^{35}Cl$	$^1\Sigma^+$	840.29	5.49	0.684282	0.006812	1.80	1.71528	1,29
$^{11}B^{19}F$	$^1\Sigma^+$	1402.16	11.82	1.51674399	0.01904848	7.11	1.26267	4
$^{11}B^{1}H$	$^1\Sigma^+$	2366.73	49.34	12.025755	0.421565	1242	1.23217	4
$^{11}B^{2}H$	$^1\Sigma^+$	1703.3	28	6.54	0.17	400	1.2324	1,2
$^{11}B^{14}N$	$^3\Pi$	1514.6	12.3	1.666	0.025	8.1	1.281	1,2
$^{11}B^{16}O$	$^2\Sigma^+$	1885.29	11.69	1.781110	0.016516	6.32	1.20475	4
$^{11}B^{32}S$	$^2\Sigma^+$	1179.91	6.25	0.79478	0.00578	1.4	1.60935	4
$^{138}Ba^{79}Br$	$^2\Sigma^+$	193.77	0.41	0.0415082	0.0001219	0.00762	2.84449	1,2
$^{138}Ba^{35}Cl$	$^2\Sigma^+$	279.92	0.82	0.08396717	0.00033429	0.03022	2.68276	1,2
$^{138}Ba^{19}F$	$^2\Sigma^+$	468.9	1.79	0.2159	0.0012	0.175	2.163	1,2,3
$^{138}Ba^{1}H$	$^2\Sigma^+$	1168.31	14.50	3.38285	0.06599	112.67	2.23175	1,2
$^{138}Ba^{2}H$	$^2\Sigma^+$	829.77	7.32	1.7071	0.02363	28.77	2.2304	1,2
$^{138}Ba^{127}I$	$^2\Sigma^+$	152.14	0.27	0.02680587	0.00006634	0.00333	3.08476	1,2
$^{138}Ba^{16}O$	$^1\Sigma^+$	669.76	2.03	0.3126140	0.0013921	0.2724	1.93969	1,2,3
$^{138}Ba^{32}S$	$^1\Sigma^+$	379.42	0.88	0.10331	0.0003188	0.0306	2.5074	1,2
$^{9}Be^{19}F$	$^2\Sigma^+$	1247.36	9.12	1.4889	0.0176	8.28	1.3610	1,2
$^{9}Be^{1}H$	$^2\Sigma^+$	2061.24	37.33	10.31992	0.3084	1022	1.34241	4
$^{9}Be^{2}H$	$^2\Sigma^+$	1530.32	20.71	5.6872	0.1225	313.8	1.3419	1,2
$^{9}Be^{16}O$	$^1\Sigma^+$	1487.32	11.83	1.6510	0.0190	8.20	1.3309	1,2
$^{9}Be^{32}S$	$^1\Sigma^+$	997.94	6.14	0.79059	0.00664	2.00	1.7415	1
$^{209}Bi_2$	$^1\Sigma_g^+$	172.71	0.34	0.022781	0.000055	0.00150	2.6596	1,2
$^{209}Bi^{79}Br$	O^+	209.62	0.52	0.04321526	0.00013269	0.007347	2.60950	5
$^{209}Bi^{35}Cl$	O^+	308.18	1.09	0.9212553	0.0004020	0.0329	2.47152	5
$^{209}Bi^{19}F$	O^+	513.0	2.35	0.22998897	0.00150262	0.185	2.05154	5
$^{209}Bi^{1}H$	$^3\Sigma^-$	1635.73	31.6	5.137	0.148	183	1.805	1,2
$^{209}Bi^{2}H$	$^3\Sigma^-$	1173.32	16.1	2.592	0.054	50.6	1.804	1,2
$^{209}Bi^{127}I$	O^+	164.12	0.32	0.02722281	0.00006979	0.00300	2.80050	5
$^{79}Br_2$	$^1\Sigma_g^+$	325.32	1.08	0.082107	0.0003187	0.02092	2.2811	1,2
$^{79}Br^{35}Cl$	$^1\Sigma^+$	444.28	1.84	0.152470	0.000770	0.07183	2.13607	1,2,3
$^{79}Br^{19}F$	$^1\Sigma^+$	670.75	4.05	0.35584	0.00261	0.401	1.75894	1,2,3
$^{79}Br^{16}O$	$^2\Pi_{3/2}$	779	6.8	0.429598	0.003639	0.523	1.717	1,2,3
$^{12}C_2$	$^1\Sigma_g^+$	1855.01	13.56	1.82010	0.01801	6.96	1.24244	24
$^{12}C^{35}Cl$	$^2\Pi_{1/2}$	876.90	5.45	0.697137	0.006853	1.9	1.64518	4
$^{12}C^{19}F$	$^2\Pi_{1/2}$	1307.93	11.08	1.41626	0.01844	6.6	1.27218	4
$^{12}C^{1}H$	$^2\Pi_{1/2}$	2860.75	64.44	14.45988	0.53654	1450	1.1199	4
$^{12}C^{2}H$	$^2\Pi_{1/2}$	2101.05	34.73	7.8079823	0.212240	420	1.11887	4
$^{12}C^{14}N$	$^2\Sigma^+$	2068.65	13.10	1.8997830	0.0173717	6.4034	1.17181	24,25
$^{12}C^{16}O$	$^1\Sigma^+$	2169.81	13.29	1.931280985	0.01750439	6.1216	1.12832	24,26,27
$^{12}C^{31}P$	$^2\Sigma^+$	1239.79	6.83	0.79886775	0.00596933	1.33	1.56198	4

Molecule	State	ω_e cm^{-1}	$\omega_e x_e$ cm^{-1}	B_e cm^{-1}	α_e cm^{-1}	D_e 10^{-6} cm^{-1}	r_e Å	Ref.
$^{12}C^{32}S$	$^1\Sigma^+$	1285.15	6.50	0.82004356	0.00591835	1.336	1.53482	1,4
$^{12}C^{80}Se$	$^1\Sigma^+$	1035.36	4.86	0.5750	0.00379	0.71	1.67609	1,2,3
$^{40}Ca^{79}Br$	$^2\Sigma^+$	285.3	0.86	0.09446622	0.00040360	0.0413	2.59358	1,5
$^{40}Ca^{35}Cl$	$^2\Sigma^+$	367.53	1.31	0.1522302	0.0007990	0.1029	2.43676	1,2
$^{40}Ca^{19}F$	$^2\Sigma^+$	581.1	2.74	0.339	0.0026	0.45	1.967	1,2
$^{40}Ca^{1}H$	$^2\Sigma^+$	1298.34	19.10	4.2766	0.0970	183.7	2.0025	1,2
$^{40}Ca^{2}H$	$^2\Sigma^+$	910*		2.1769	0.035	47.9	2.002	1,2
$^{40}Ca^{127}I$	$^2\Sigma^+$	238.70	0.63	0.0693263	0.0002634	0.0234	2.82859	1,2
$^{40}Ca^{16}O$	$^1\Sigma^+$	732.03	4.83	0.444441	0.003282	0.6541	1.8221	1,2
$^{40}Ca^{32}S$	$^1\Sigma^+$	462.23	1.78	0.1766757	0.0008270	0.1032	2.31775	1,2
$^{114}Cd^{1}H$	$^2\Sigma^+$	1337.1*		5.323		314	1.781	1,2
$^{114}Cd^{2}H$	$^2\Sigma^+$			2.704		76	1.775	1,2
$^{35}Cl_2$	$^1\Sigma_g^+$	559.75	2.69	0.24415	0.00152	0.186	1.9872	4
$^{35}Cl^{19}F$	$^1\Sigma^+$	783.45	4.95	0.5164805	0.0043585	0.88	1.62831	4
$^{35}Cl^{16}O$	$^2\Pi_{3/2}$	853.64	5.52	0.62345797	0.0059357	1.33	1.56962	4
$^{52}Cr^{1}H$	$^6\Sigma^+$	1581*		6.220	0.179	347	1.656	1,2
$^{52}Cr^{2}H$	$^6\Sigma^+$	1182*		3.14		88.8	1.664	1,2
$^{52}Cr^{16}O$	$^5\Pi$	898.4	6.8	0.5231	0.0070		1.615	1,2
$^{133}Cs_2$	$^1\Sigma_g^+$	42.02	0.08	0.0127	0.0000264	0.00464	4.47	1,2
$^{133}Cs^{79}Br$	$^1\Sigma^+$	149.66	0.37	0.03606925	0.00012401	0.00838	3.07225	1,2,3
$^{133}Cs^{35}Cl$	$^1\Sigma^+$	214.17	0.73	0.07209149	0.00033756	0.03268	2.90627	1,2,3
$^{133}Cs^{19}F$	$^1\Sigma^+$	352.56	1.62	0.18436969	0.0011756	0.20168	2.34535	1,2,3
$^{133}Cs^{1}H$	$^1\Sigma^+$	891.0	12.9	2.7099	0.0579	113	2.4938	1,2
$^{133}Cs^{2}H$	$^1\Sigma^+$	619.1*		1.354		20	2.505	1,2
$^{133}Cs^{127}I$	$^1\Sigma^+$	119.18	0.25	0.02362736	0.00006826	0.00371	3.31519	1,2,3
$^{133}Cs^{16}O$	$^2\Sigma^+$	357.5*		0.223073	0.001303	0.348	2.3007	1,2
$^{63}Cu_2$	$^1\Sigma_g^+$	264.55	1.02	0.10874	0.000614	0.0716	2.2197	1,2
$^{63}Cu^{79}Br$	$^1\Sigma^+$	314.8	0.96	0.10192625	0.00045214	0.04274	2.17344	1,2
$^{65}Cu^{35}Cl$	$^1\Sigma^+$	415.29	1.58	0.17628802	0.00099647	0.12706	2.05118	1,2
$^{63}Cu^{19}F$	$^1\Sigma^+$	622.7	3.95	0.3794029	0.0032298	0.563	1.74493	1,2,3
$^{63}Cu^{1}H$	$^1\Sigma^+$	1941.26	37.51	7.9441	0.2563	520	1.46263	1,2
$^{63}Cu^{2}H$	$^1\Sigma^+$	1384.14	18.97	4.0381	0.0917	136.2	1.4626	1,2
$^{63}Cu^{127}I$	$^1\Sigma^+$	264.5	0.60	0.07328742	0.00028390	0.02244	2.33832	1,2
$^{63}Cu^{16}O$	$^2\Pi_{3/2}$	640.17	4.43	0.44454	0.00456	0.85	1.7244	1,2
$^{63}Cu^{32}S$	$^2\Pi_{3/2}$	415.0	1.75	0.1891		0.18	2.051	1,2
$^{19}F_2$	$^1\Sigma_g^+$	916.93	11.32	0.889294	0.0125952	3.3	1.41264	4
$^{19}F^{16}O$	$^2\Pi_{3/2}$	1053.01	9.92	1.0587076	0.013295	4.2823	1.35411	9
$^{56}Fe^{16}O$	$^5\Delta$	965*		0.650		0.72	1.444	1,2
$^{69}Ga^{81}Br$	$^1\Sigma^+$	263.0	0.81	0.081839	0.0003207	0.032	2.35248	1,2,3
$^{69}Ga^{35}Cl$	$^1\Sigma^+$	365.67	1.25	0.1499046	0.0007936	0.1008	2.20169	1,2,3
$^{69}Ga^{19}F$	$^1\Sigma^+$	622.2	3.2	0.3595161	0.0028642	0.50	1.77437	1,2,3
$^{69}Ga^{1}H$	$^1\Sigma^+$	1603.94	28.41	6.1434095	0.1906376	359.70	1.66208	16
$^{69}Ga^{2}H$	$^1\Sigma^+$	1143.23	14.43	3.1218854	0.0689978	93.021	1.66113	16
$^{69}Ga^{127}I$	$^1\Sigma^+$	216.38	0.47	0.0569359	0.0001897	0.015770	2.57464	1,2,3
$^{69}Ga^{16}O$	$^2\Sigma$	767.5	6.24	0.4271		0.37	1.744	1,2
$^{74}Ge^{79}Br$	$^2\Pi_{1/2}$	295	0.7					1,2
$^{74}Ge^{35}Cl$	$^2\Pi_{1/2}$	407.6	1.36					1,2
$^{72}Ge^{1}H$	$^2\Pi_{1/2}$	1833.77	37	6.726	0.192	326	1.5880	1,2
$^{72}Ge^{2}H$	$^2\Pi_{1/2}$	1320.09	19	3.415	0.070	83.2	1.5874	1,2
$^{74}Ge^{16}O$	$^1\Sigma^+$	986.49	4.47	0.4856981	0.0030787	0.4709	1.62464	1,2
$^{74}Ge^{32}S$	$^1\Sigma^+$	575.8	1.80	0.18656576	0.00074910	0.07883	2.01209	1,2
$^{74}Ge^{80}Se$	$^1\Sigma^+$	408.7	1.36	0.09634051	0.00028904	0.02207	2.13463	1,2
$^{74}Ge^{130}Te$	$^1\Sigma^+$	323.9	0.75	0.06533821	0.00017246	0.012	2.34017	1,2
$^{1}H_2$	$^1\Sigma_g^+$	4401.21	121.34	60.853	3.062	47100	0.74144	1,2
$^{2}H_2$	$^1\Sigma_g^+$	3115.50	61.82	30.444	1.0786	11410	0.74152	1,2
$^{3}H_2$	$^1\Sigma_g^+$	2546.5	41.23	20.335	0.5887		0.74142	1,2
$^{1}H^{81}Br$	$^1\Sigma^+$	2648.97	45.22	8.46488	0.23328	345.8	1.41444	1,2,3
$^{2}H^{81}Br$	$^1\Sigma^+$	1884.75	22.72	4.245596	0.084	88.32	1.4145	1,2
$^{1}H^{35}Cl$	$^1\Sigma^+$	2990.92	52.80	10.5933002	0.3069985	531.94	1.27456	4
$^{2}H^{35}Cl$	$^1\Sigma^+$	2145.16	27.18	5.448796	0.113292	140	1.27458	1,2

Molecule	State	ω_e cm^{-1}	$\omega_e x_e$ cm^{-1}	B_e cm^{-1}	α_e cm^{-1}	D_e 10^{-6} cm^{-1}	r_e Å	Ref.
$^{1}H^{19}F$	$^{1}\Sigma^{+}$	4138.39	89.94	20.953712	0.7933704	2150	0.91685	4
$^{2}H^{19}F$	$^{1}\Sigma^{+}$	2998.19	45.76	11.0102	0.3017	594	0.91694	1,2
$^{1}H^{127}I$	$^{1}\Sigma^{+}$	2309.01	39.64	6.4263650	0.1689	206.9	1.60916	1,2,3
$^{202}Hg^{1}H$	$^{2}\Sigma^{+}$	1203.24*		5.3888		395.3	1.7662	1,2
$^{202}Hg^{2}H$	$^{2}\Sigma^{+}$	896.12*		2.739		91	1.757	1,2
$^{127}I_2$	$^{1}\Sigma_g^{+}$	214.50	0.61	0.03737	0.000114	0.0043	2.666	1,2
$^{127}I^{79}Br$	$^{1}\Sigma^{+}$	268.64	0.81	0.0568325	0.0001969	0.0102	2.46899	1,2,3
$^{127}I^{35}Cl$	$^{1}\Sigma^{+}$	384.29	1.50	0.1141587	0.0005354	0.0403	2.32088	1,2,3
$^{127}I^{19}F$	$^{1}\Sigma^{+}$	610.24	3.12	0.2797111	0.0018738	0.2356	1.90976	1,2,3
$^{127}I^{16}O$	$^{2}\Pi_{3/2}$	681.5	4.3	0.34026	0.00270	0.36	1.8676	1,2
$^{115}In^{81}Br$	$^{1}\Sigma^{+}$	221.0	0.65	0.05489468	0.00018672	0.01350	2.54315	1,2,3
$^{115}In^{35}Cl$	$^{1}\Sigma^{+}$	317.39	1.03	0.1090583	0.0005177	0.0515	2.40117	1,2,3
$^{115}In^{19}F$	$^{1}\Sigma^{+}$	535.4	2.6	0.2623241	0.0018798	0.252	1.98540	1,2,3
$^{115}In^{1}H$	$^{1}\Sigma^{+}$	1476.0	25.61	4.995	0.143	223	1.8380	1,2
$^{115}In^{2}H$	$^{1}\Sigma^{+}$	1048.2	12.4	2.523	0.051	58	1.837	1,2
$^{115}In^{127}I$	$^{1}\Sigma^{+}$	177.08	0.34	0.03686702	0.00010411	0.00639	2.75364	1,2,3
$^{39}K_2$	$^{1}\Sigma_g^{+}$	92.02	0.28	0.056743	0.000165	0.0863	3.9051	1,2
$^{39}K^{79}Br$	$^{1}\Sigma^{+}$	213	0.80	0.08122109	0.00040481	0.04462	2.82078	1,2,3
$^{39}K^{35}Cl$	$^{1}\Sigma^{+}$	281	1.30	0.1286348	0.0007899	0.1087	2.66665	1,2,3
$^{39}K^{19}F$	$^{1}\Sigma^{+}$	426.26	2.45	0.27993741	0.00233492	0.4829	2.17146	1,2,3
$^{39}K^{1}H$	$^{1}\Sigma^{+}$	983.6	14.3	3.416400	0.085313	163.55	2.243	1,2
$^{39}K^{2}H$	$^{1}\Sigma^{+}$	707	7.7	1.754	0.0318	50	2.240	1,2
$^{39}K^{127}I$	$^{1}\Sigma^{+}$	186.53	0.57	0.06087473	0.00026776	0.02593	3.04784	1,2,3
$^{39}K^{23}Na$	$^{1}\Sigma^{+}$	124.03	0.50	0.09519989	0.00044966	0.2206	3.49958	6
$^{139}La^{16}O$	$^{2}\Sigma^{+}$	812.8	2.22	0.35252001	0.00142365	0.2626	1.82591	1,2
$^{7}Li_2$	$^{1}\Sigma_g^{+}$	351.41	2.58	0.672530	0.007046	9.79	2.6733	13
$^{7}Li^{79}Br$	$^{1}\Sigma^{+}$	563.2	3.5	0.555399	0.005644	2.159	2.17043	1,2,3
$^{7}Li^{35}Cl$	$^{1}\Sigma^{+}$	642.95	4.47	0.7065225	0.0080102	3.409	2.02067	1,5
$^{7}Li^{133}Cs$	$^{1}\Sigma^{+}$	184.70	1.00	0.188003	0.001248	0.7784	3.6681	21
$^{7}Li^{19}F$	$^{1}\Sigma^{+}$	910.57	8.21	1.34525715	0.02028749	11.75	1.56386	4,5
$^{7}Li^{1}H$	$^{1}\Sigma^{+}$	1405.50	21.17	7.5137315	0.2163911	859	1.59490	4,5
$^{7}Li^{2}H$	$^{1}\Sigma^{+}$	1054.94	13.06	4.23308131	0.09149428	272	1.59526	4,5
$^{7}Li^{127}I$	$^{1}\Sigma^{+}$	496.85	2.85	0.4431766	0.0040862	1.4104	2.39192	1,2,3
$^{7}Li^{39}K$	$^{1}\Sigma^{+}$	207		0.265			3.27	1,20
$^{7}Li^{23}Na$	$^{1}\Sigma^{+}$	256.99	1.66	0.376833	0.003810	3.340	2.88851	4,30
$^{7}Li^{16}O$	$^{2}\Pi$	814.62	7.78	1.212830	0.017899	0.1079	1.68822	3,14
$^{24}Mg_2$	$^{1}\Sigma_g^{+}$	51.12	1.64	0.09287	0.00378	1.22	3.891	1
$^{24}Mg^{35}Cl$	$^{2}\Sigma^{+}$	462.12*		0.2456154	0.0016204	0.2723	2.19639	1,2
$^{24}Mg^{19}F$	$^{2}\Sigma^{+}$	711.69*		0.51922	0.00470	1.080	1.7500	1,2
$^{24}Mg^{1}H$	$^{2}\Sigma^{+}$	1492.78	29.85	5.825523	0.177298	354.56	1.72972	18
$^{24}Mg^{2}H$	$^{2}\Sigma^{+}$	1077.30	15.52	3.034344	0.066607	96.25	1.72916	18
$^{24}Mg^{16}O$	$^{1}\Sigma^{+}$	785.21	5.13	0.5748414	0.0053223	1.233	1.74817	28,4
$^{24}Mg^{32}S$	$^{1}\Sigma^{+}$	528.74	2.70	0.26797	0.00176	0.276	2.1425	1
$^{55}Mn^{1}H$	$^{7}\Sigma$	1548.0	28.8	5.6841	0.1570	303.9	1.7311	1,2
$^{55}Mn^{2}H$	$^{7}\Sigma$	1103	13.9	2.8957	0.051	79.5	1.7310	1,2
$^{14}N_2$	$^{1}\Sigma_g^{+}$	2358.56	14.32	1.998236	0.017310	5.737	1.09769	24
$^{14}N^{79}Br$	$^{3}\Sigma^{-}$	691.75	4.72	0.444	0.0040		1.79	1,2
$^{14}N^{35}Cl$	$^{3}\Sigma^{-}$	827.96	5.30	0.64976739	0.00641432	1.596	1.61071	1,5
$^{14}N^{19}F$	$^{3}\Sigma^{-}$	1141.37	8.99	1.205679	0.014889	5.4	1.31698	4
$^{14}N^{1}H$	$^{3}\Sigma^{-}$	3282.72	79.04	16.66792	0.65038	1710	1.03719	4
$^{14}N^{2}H$	$^{3}\Sigma^{-}$	2399.13	42.11	8.9087	0.2546	491.7	1.03665	8
$^{14}N^{16}O$	$^{2}\Pi_{1/2}$	1904.20	14.07	1.67195	0.0171	0.5	1.15077	1,24
$^{14}N^{32}S$	$^{2}\Pi_{1/2}$	1218.7	7.28	0.769602	0.0064	1.2	1.4940	1,2,3
$^{23}Na_2$	$^{1}\Sigma_g^{+}$	159.09	0.71	0.15473537	0.0086375	0.58	3.07858	1,4,20
$^{23}Na^{79}Br$	$^{1}\Sigma^{+}$	302	1.5	0.1512533	0.0009410	0.1554	2.50204	1,2,3
$^{23}Na^{35}Cl$	$^{1}\Sigma^{+}$	364.68	1.78	0.21806302	0.00162479	0.31202	2.36080	4,5
$^{23}Na^{19}F$	$^{1}\Sigma^{+}$	535.66	3.58	0.43690153	0.0045592	1.16296	1.92595	11
$^{23}Na^{1}H$	$^{1}\Sigma^{+}$	1171.97	19.70	4.90327	0.1370	343.8	1.8870	4,5
$^{23}Na^{2}H$	$^{1}\Sigma^{+}$	826.1*		2.557089	0.051600	93.46	1.88654	1,2
$^{23}Na^{127}I$	$^{1}\Sigma^{+}$	258	1.1	0.1178056	0.0006478	0.0973	2.71145	1,2,3

Molecule	State	ω_e cm^{-1}	$\omega_e x_e$ cm^{-1}	B_e cm^{-1}	α_e cm^{-1}	D_e 10^{-6} cm^{-1}	r_e Å	Ref.
$^{23}Na^{16}O$	$^2\Pi$	492.3		0.424630	0.004506	1.2638	2.05155	1,2
$^{93}Nb_2$	$^3\Sigma_g^-$	424.89	0.94	0.084054	0.000242	0.016	2.0778	15
$^{93}Nb^{16}O$	$^4\Sigma^-$	989.0	3.8	0.4321	0.0021	0.22	1.691	1,2
$^{58}Ni^1H$	$^2\Delta_{5/2}$	1926.6	38	7.700	0.23	481	1.476	1,2
$^{58}Ni^2H$	$^2\Delta_{5/2}$	1390.1	19	3.992	0.092	130	1.465	1,2
$^{16}O_2$	$^3\Sigma_g^-$	1580.19	11.98	1.445622	0.015933	4.839	1.20752	24
$^{16}O^1H$	$^2\Pi_{3/2}$	3737.76	84.88	18.911	0.7242	1938	0.96966	1,2,3
$^{16}O^2H$	$^2\Pi_{3/2}$	2720.24	44.05	10.021	0.276	537.4	0.9698	1,2
$^{31}P_2$	$^1\Sigma_g^+$	780.77	2.84	0.30362	0.00149	0.188	1.8934	1
$^{31}P^{35}Cl$	$^3\Sigma^-$	551.38	2.23	0.2528748	0.0015119	0.2124	2.01461	1,2
$^{31}P^{19}F$	$^3\Sigma^-$	846.73	4.49	0.5667427	0.004639	1.0156	1.58933	12
$^{31}P^1H$	$^3\Sigma^-$	2363.77	43.91	8.53904	0.2534	4.462	1.42218	8
$^{31}P^2H$	$^3\Sigma^-$	1699.2	23.0	4.4081	0.0928	116	1.4220	1,2
$^{31}P^{14}N$	$^1\Sigma^+$	1336.95	6.90	0.7864844	0.0055337	1.091	1.49087	4
$^{31}P^{16}O$	$^2\Pi_{1/2}$	1233.34	6.56	0.733223657	0.005466162	1.3	1.47637	1,4
$^{208}Pb_2$		110.5	0.35					1,2
$^{208}Pb^{79}Br$	$^2\Pi_{1/2}$	207.5	0.50					1,2
$^{208}Pb^{35}Cl$	$^2\Pi_{1/2}$	303.9	0.88					1,2
$^{208}Pb^{19}F$	$^2\Pi_{1/2}$	502.73	2.28	0.22875	0.001473	0.183	2.0575	1,2
$^{208}Pb^1H$	$^2\Pi_{1/2}$	1564.1	29.75	4.971	0.144	201	1.839	1,2
$^{208}Pb^{16}O$	$^1\Sigma^+$	720.96	3.52	0.30730373	0.00190977	0.2138	1.92181	1,2,3
$^{208}Pb^{32}S$	$^1\Sigma^+$	429.17	1.26	0.11632307	0.00043510	0.03418	2.28678	1,2,3
$^{208}Pb^{80}Se$	$^1\Sigma^+$	277.6	0.51	0.05059953	0.00012993	0.0070	2.40218	1,2,3
$^{208}Pb^{130}Te$	$^1\Sigma^+$	212.0	0.43	0.03130774	0.00006743	0.0027	2.59492	1,2,3
$^{195}Pt^{12}C$	$^1\Sigma^+$	1051.13	4.86	0.53044	0.003273	0.546	1.6767	1,2
$^{195}Pt^1H$	$^2\Delta_{5/2}$	2294.68*		7.1963	0.1996	261	1.52852	1,2
$^{195}Pt^2H$	$^2\Delta_{5/2}$	1644.3*		3.640	0.071	66	1.524	1,2
$^{85}Rb^{79}Br$	$^1\Sigma^+$	169.46	0.46	0.04752798	0.00018596	0.01496	2.94474	1,2,3
$^{85}Rb^{35}Cl$	$^1\Sigma^+$	228	0.92	0.0876404	0.0004537	0.04947	2.78673	1,2,3
$^{85}Rb^{19}F$	$^1\Sigma^+$	376	1.9	0.2106640	0.0015228	0.2684	2.27033	1,2,3
$^{85}Rb^1H$	$^1\Sigma^+$	936.9	14.21	3.020	0.072	123	2.367	1,2
$^{85}Rb^{127}I$	$^1\Sigma^+$	138.51	0.33	0.03283293	0.00010946	0.00738	3.17688	1,2,3
$^{85}Rb^{16}O$	$^2\Sigma^+$	388.4*		0.246481	0.002174	0.397	2.25420	1,2
$^{32}S_2$	$^3\Sigma_g^-$	725.71	2.86	0.29539516	0.00159754	0.19	1.88941	4
$^{32}S^{19}F$	$^2\Pi_{3/2}$	837.64	4.47	0.555173	0.004459	0.975	1.59624	4,5
$^{32}S^1H$	$^2\Pi_{3/2}$	2696.25	48.74	9.60025	0.27990	480	1.34061	4
$^{32}S^2H$	$^2\Pi_{3/2}$	1885	31	4.95130	0.10308	130	1.34049	1,2
$^{32}S^{16}O$	$^3\Sigma^-$	1149.2	5.6	0.7208171	0.005737	1.134	1.48109	1,19,23
$^{121}Sb^{35}Cl$	$^3\Sigma^-$	374.7	0.6					1,2
$^{121}Sb^{19}F$	$^3\Sigma^-$	605.0	2.6	0.2792	0.0020	0.23	1.918	1,2
$^{121}Sb^1H$	$^3\Sigma^-$			5.684		240	1.723	1,2
$^{121}Sb^2H$	$^3\Sigma^-$			2.8782		45	1.7194	1,2
$^{121}Sb^{14}N$	$^1\Sigma^+$	942.0	5.6					1,2
$^{121}Sb^{16}O$	$^2\Pi_{1/2}$	816	4.2	0.3580	0.0022	0.270	1.826	1,2
$^{45}Sc^{19}F$	$^1\Sigma^+$	735.6	3.8	0.3950	0.00266		1.788	1,2
$^{80}Se_2$	$^3\Sigma_g^-$	385.30	0.96	0.08992	0.000288	0.024	2.166	1,2
$^{80}Se^1H$	$^2\Pi_{3/2}$	2400*		8.02	0.23	330	1.48	1,2
$^{80}Se^2H$	$^2\Pi_{3/2}$	1708*		3.94			1.48	1,2
$^{80}Se^{16}O$	$^3\Sigma^-$	914.69	4.52	0.4655	0.00323	0.5	1.648	1,2
$^{28}Si_2$	$^3\Sigma_g^-$	510.98	2.02	0.2390	0.0014	0.21	2.246	1
$^{28}Si^{35}Cl$	$^2\Pi_{1/2}$	535.59	2.18	0.256103	0.001582	0.25	2.05794	4
$^{28}Si^{19}F$	$^2\Pi_{1/2}$	857.33	4.83	0.58125735	0.00503859	1.065	1.60100	31
$^{28}Si^1H$	$^2\Pi_{1/2}$	2042.52	36.06	7.503898	0.21814	400	1.51966	4
$^{28}Si^2H$	$^2\Pi_{1/2}$	1469.32	18.23	3.8840	0.0781	105.4	1.5199	1,2
$^{28}Si^{14}N$	$^2\Sigma^+$	1151.28	6.46	0.730927	0.005685	1.2	1.57207	4
$^{28}Si^{16}O$	$^1\Sigma^+$	1241.54	5.97	0.7267521	0.0050379	0.9923	1.50975	1,19,22
$^{28}Si^{32}S$	$^1\Sigma^+$	749.64	2.58	0.30352788	0.00147308	0.201	1.92926	1,19
$^{28}Si^{80}Se$	$^1\Sigma^+$	580.0	1.78	0.1920117	0.0007767	0.0842	2.05832	1,2,3
$^{120}Sn^{79}Br$	$^2\Pi_{1/2}$	247.2	0.6					1,2
$^{120}Sn^{35}Cl$	$^2\Pi_{1/2}$	351.1	1.06	0.1117	0.0004		2.361	1,2

Molecule	State	ω_e cm^{-1}	$\omega_e x_e$ cm^{-1}	B_e cm^{-1}	α_e cm^{-1}	D_e 10^{-6} cm^{-1}	r_e Å	Ref.
$^{118}Sn^{19}F$	$^2\Pi_{1/2}$	577.6	2.69	0.2727	0.0014	0.26	1.944	1,2
$^{120}Sn^1H$	$^2\Pi_{1/2}$			5.31488		207.5	1.78146	1,2
$^{120}Sn^2H$	$^2\Pi_{1/2}$	1188.0*		2.6950	0.049	53.4	1.7770	1,2
$^{120}Sn^{127}I$	$^2\Pi_{1/2}$	199.0	0.6					1,2
$^{120}Sn^{16}O$	$^1\Sigma^+$	822.13	3.72	0.35571998	0.00214432	0.26638	1.83251	1,2,3
$^{120}Sn^{32}S$	$^1\Sigma^+$	487.26	1.36	0.13686139	0.00050563	0.0424	2.20898	1,2,3
$^{120}Sn^{80}Se$	$^1\Sigma^+$	331.2	0.74	0.0649978	0.0001705	0.011	2.32557	1,2,3
$^{120}Sn^{130}Te$	$^1\Sigma^+$	259.5	0.50	0.04247917	0.00009543	0.0055	2.52280	1,2,3
$^{88}Sr^{79}Br$	$^2\Sigma^+$	216.60	0.52	0.0541847	0.0001827	0.01356	2.73522	1,2
$^{88}Sr^{35}Cl$	$^2\Sigma^+$	302.3	0.95					1,2
$^{88}Sr^{19}F$	$^2\Sigma^+$	502.4	2.3	0.2505346	0.0015513	0.2498	2.07537	1,2
$^{88}Sr^1H$	$^2\Sigma^+$	1206.2	17.0	3.6751	0.0814	135	2.1456	1,2
$^{88}Sr^2H$	$^2\Sigma^+$	841	8.6	1.8609	0.0292	34.7	2.1449	1,2
$^{88}Sr^{127}I$	$^2\Sigma^+$	173.77	0.35	0.0367097	0.0001060	0.00655	2.94364	1,2
$^{88}Sr^{16}O$	$^1\Sigma^+$	653.5	3.96	0.33798	0.00219	0.36	1.91983	1,2
$^{181}Ta^{16}O$	$^2\Delta_{3/2}$	1028.69	3.51	0.40284	0.00182	0.2450	1.68746	1,2
$^{130}Te_2$	$^3\Sigma_g^-$	247.07	0.51	0.039681	0.000106	0.0044	2.5574	1,2
$^{(130)}Te^1H$	$^2\Pi_{3/2}$			5.56			1.74	1,2
$^{130}Te^{16}O$	0^+	797.11	4.00	0.3554	0.00237	0.27	1.825	1,2
$^{232}Th^{16}O$	$^1\Sigma^+$	895.77	2.39	0.332644	0.001302	0.1833	1.84032	1,2
$^{48}Ti^{16}O$	$^3\Delta_1$	1009.02	4.50	0.53541	0.00301	0.603	1.6202	1,2
$^{205}Tl^{81}Br$	$^1\Sigma^+$	192.10	0.39	0.0423899	0.0001276	0.0083	1.61817	1,2,3
$^{205}Tl^{35}Cl$	$^1\Sigma^+$	284.71	0.86	0.09139702	0.00039784	0.0377	2.48483	1,2,3
$^{205}Tl^{19}F$	$^1\Sigma^+$	476.86	2.24	0.22315014	0.00150380	0.1955	2.08439	1,2,3
$^{205}Tl^1H$	$^1\Sigma^+$	1390.7	22.7	4.806	0.154	254	1.870	1,2
$^{205}Tl^2H$	$^1\Sigma^+$	987.7	12.04	2.419	0.057	60	1.869	1,2
$^{205}Tl^{127}I$	$^1\Sigma^+$	150*		0.0271676	0.0000664	0.0036	2.81361	1,2,3
$^{51}V^{16}O$	$^4\Sigma^-$	1011.3	4.86	0.54825	0.00352	0.6	1.5893	1,2
$^{(132)}Xe^{19}F$	$^2\Sigma$	225.4	10.9	0.19326	0.00699	0.536	2.293	10
$^{89}Y^{35}Cl$	$^1\Sigma$	380.7	1.3	0.1160	0.0003	0.09	2.41	1,2
$^{89}Y^{19}F$	$^1\Sigma^+$	631.29	2.50	0.29042	0.00163	0.237	1.9257	1,2
$^{89}Y^{16}O$	$^2\Sigma^+$	861.0	2.9	0.3881	0.0018	0.32	1.790	1,2
$^{174}Yb^1H$	$^2\Sigma^+$	1249.54	21.06	3.9931	0.0957	161.8	2.0526	1,2
$^{174}Yb^2H$	$^2\Sigma^+$	886.6	10.57	2.01162	0.03425	41.60	2.0516	1,2
$^{64}Zn^{35}Cl$	$^2\Sigma$	390.5	1.6					1,2
$^{64}Zn^{19}F$	$^2\Sigma$	628	3.5					1,2
$^{64}Zn^1H$	$^2\Sigma^+$	1607.6	55.14	6.6794	0.2500	466	1.5949	1,2
$^{64}Zn^2H$	$^2\Sigma^+$	1072	28	3.350		124	1.6054	1,2
$^{64}Zn^{127}I$	$^2\Sigma$	223.4	0.6					1,2
$^{90}Zr^{16}O$	$^1\Sigma^+$	969.8	4.9	0.42263	0.0023	0.319	1.7116	1,2

* Indicates a value for the interval between $v = 0$ and $v = 1$ states instead of a value of ω_e.

INFRARED CORRELATION CHARTS

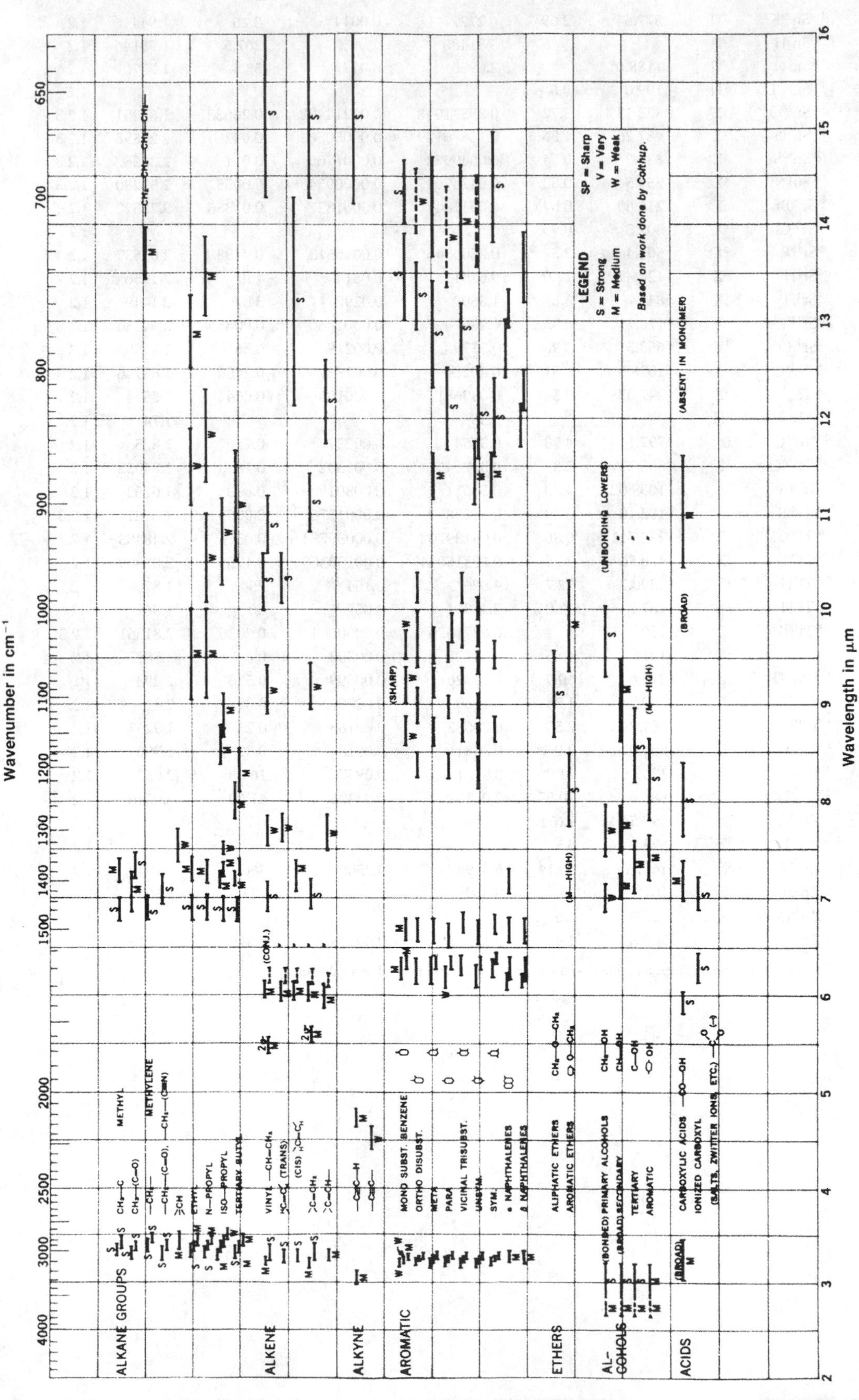

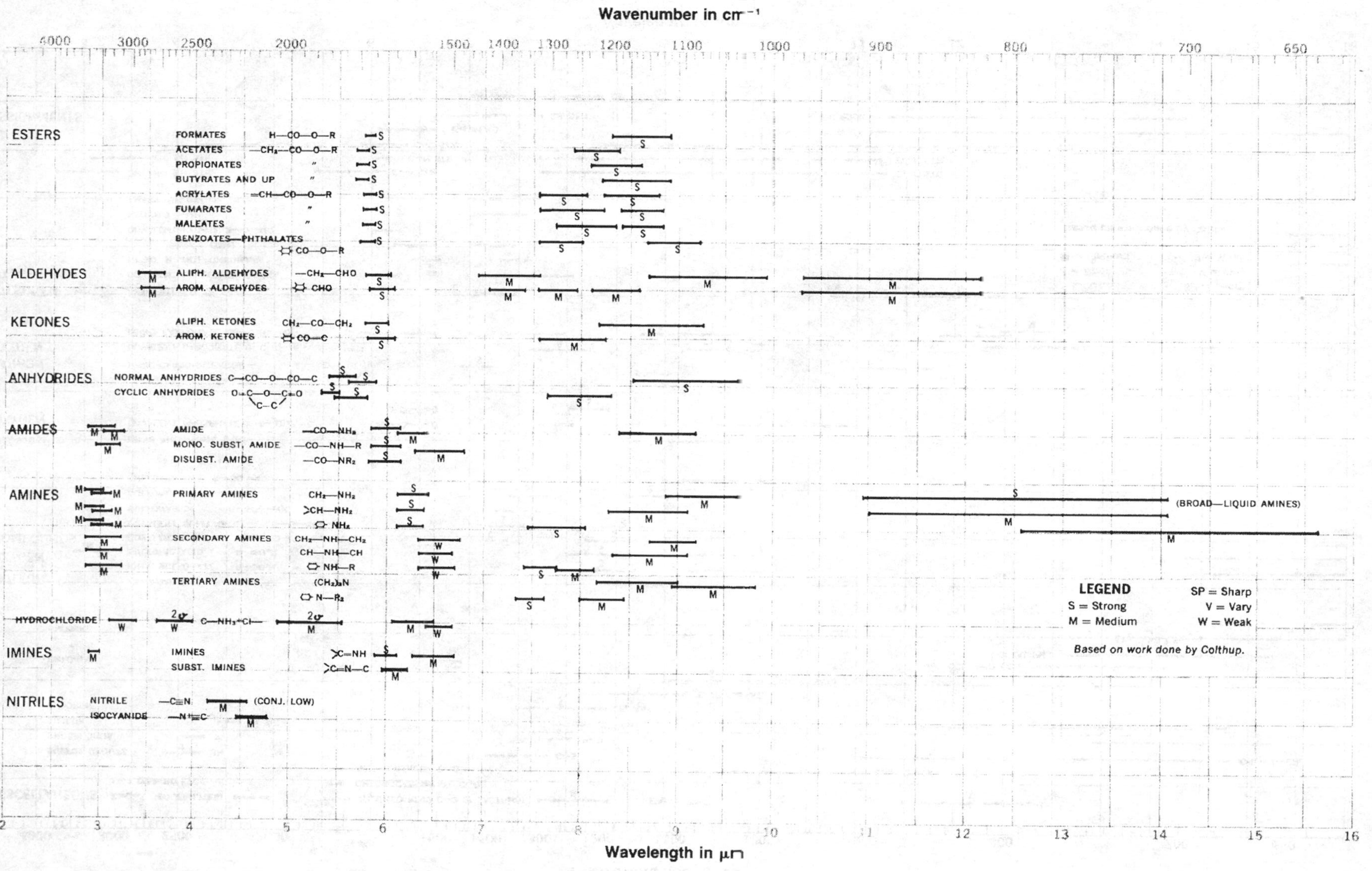
Wavenumber in cm⁻¹
Wavelength in μm
ESTERS
FORMATES H—CO—O—R
ACETATES —CH₂—CO—O—R
PROPIONATES
BUTYRATES AND UP
ACRYLATES =CH—CO—O—R
FUMARATES
MALEATES
BENZOATES—PHTHALATES
CO—O—R
ALDEHYDES
ALIPH. ALDEHYDES —CH₂—CHO
AROM. ALDEHYDES CHO
KETONES
ALIPH. KETONES CH₂—CO—CH₂
AROM. KETONES CO—C
ANHYDRIDES
NORMAL ANHYDRIDES C—CO—O—CO—C
CYCLIC ANHYDRIDES
AMIDES
AMIDE —CO—NH₂
MONO. SUBST. AMIDE —CO—NH—R
DISUBST. AMIDE —CO—NR₂
AMINES
PRIMARY AMINES CH₂—NH₂
>CH—NH₂
NH₂
SECONDARY AMINES CH₂—NH—CH₂
CH—NH—CH
NH—R
TERTIARY AMINES (CH₂)₃N
N—R₂
HYDROCHLORIDE
C—NH₃⁺Cl⁻
IMINES
IMINES >C=NH
SUBST. IMINES >C=N—C
NITRILES
NITRILE —C≡N
(CONJ. LOW)
ISOCYANIDE —N≡C
(BROAD—LIQUID AMINES)
LEGEND
S = Strong
M = Medium
SP = Sharp
V = Vary
W = Weak
Based on work done by Colthup.

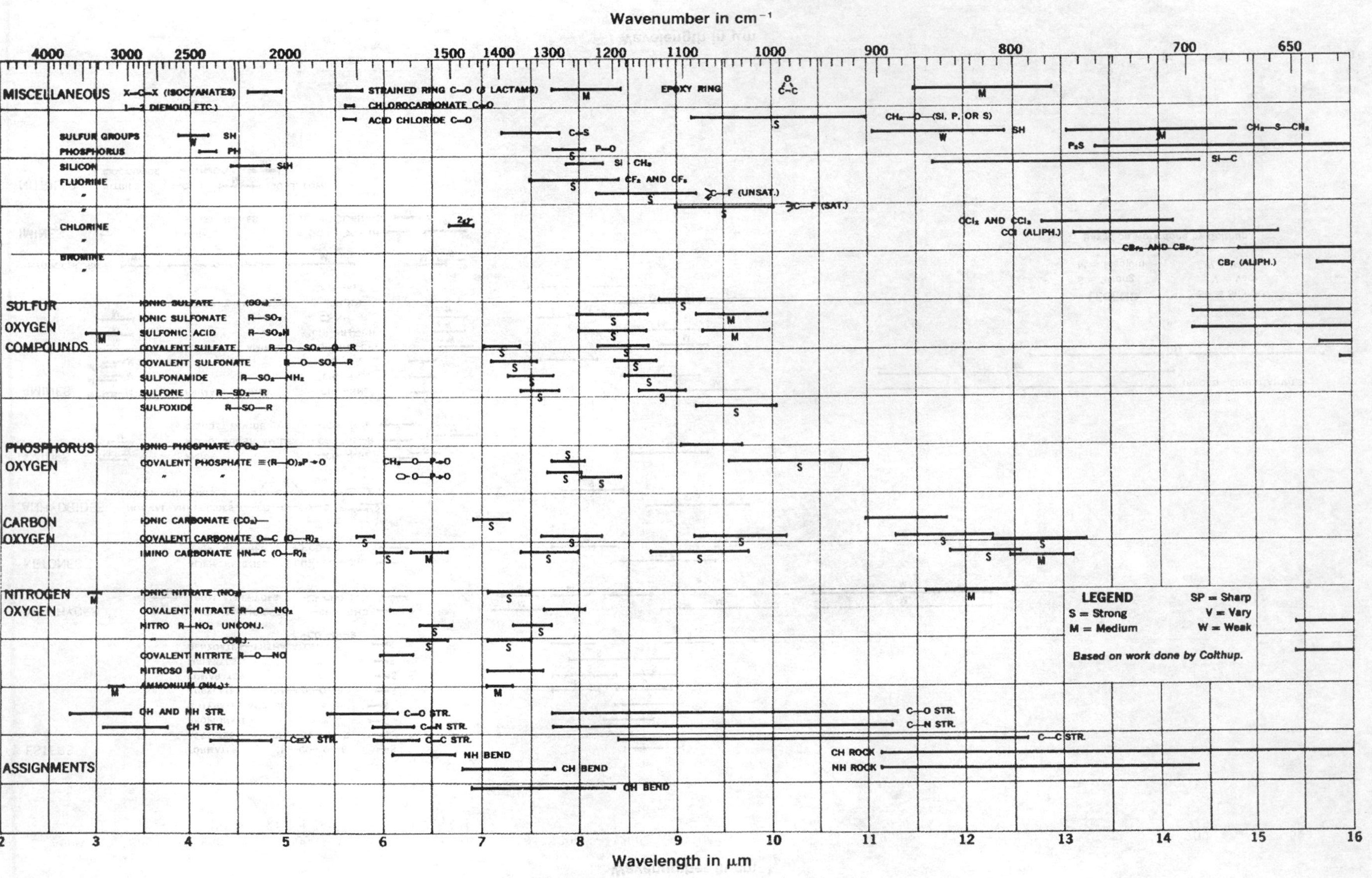
Wavenumber in cm−1
4000 3000 2500 2000 1500 1400 1300 1200 1100 1000 900 800 700 650
MISCELLANEOUS
X—C—X (ISOCYANATES)
STRAINED RING C—O (β LACTAMS)
CHLOROCARBONATE C=O
ACID CHLORIDE C—O
EPOXY RING
CH₂—O—(SI, P, OR S)
SULFUR GROUPS
PHOSPHORUS
SILICON
FLUORINE
CF₂ AND CF₃
C—F (UNSAT.)
C—F (SAT.)
CH₂—S—CH₃
SI—C
CHLORINE
CCl₂ AND CCl₃
CCl (ALIPH.)
BROMINE
CBr₂ AND CBr₃
CBr (ALIPH.)
SULFUR
OXYGEN
COMPOUNDS
IONIC SULFATE
IONIC SULFONATE
SULFONIC ACID
COVALENT SULFATE
COVALENT SULFONATE
SULFONAMIDE
SULFONE
SULFOXIDE
PHOSPHORUS
OXYGEN
IONIC PHOSPHATE
COVALENT PHOSPHATE
CARBON
OXYGEN
IONIC CARBONATE
COVALENT CARBONATE
IMINO CARBONATE
NITROGEN
OXYGEN
IONIC NITRATE
COVALENT NITRATE
NITRO R—NO₂ UNCONJ.
CONJ.
COVALENT NITRITE
NITROSO
AMMONIUM
LEGEND
S = Strong
M = Medium
SP = Sharp
V = Vary
W = Weak
Based on work done by Colthup.
CH AND NH STR.
CH STR.
C—O STR.
C—N STR.
C—C STR.
NH BEND
CH BEND
CH ROCK
NH ROCK
ASSIGNMENTS
2 3 4 5 6 7 8 9 10 11 12 13 14 15 16
Wavelength in μm

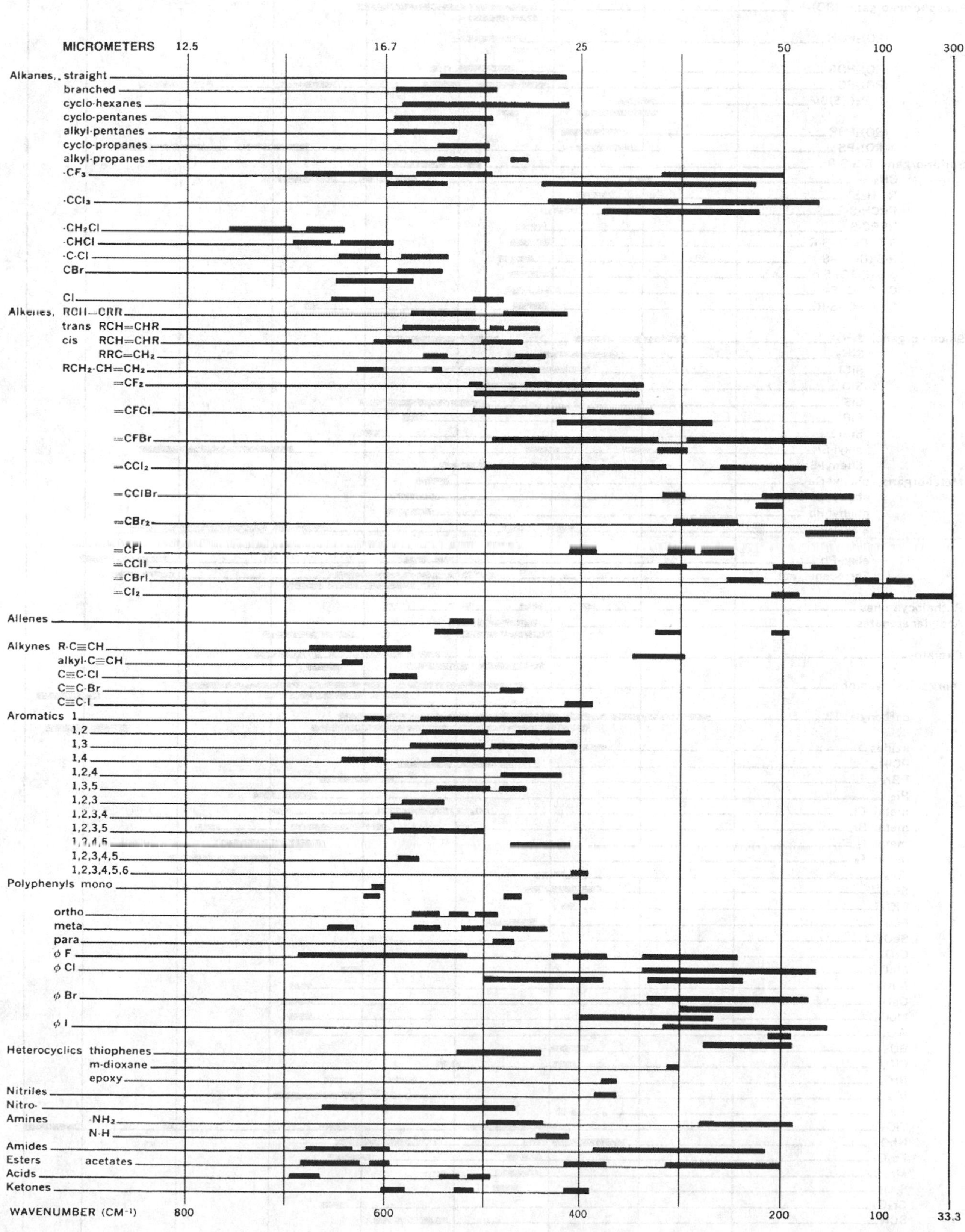
Far Infrared Region
MICROMETERS 12.5 16.7 25 50 100 300
Alkanes, straight
branched
cyclo-hexanes
cyclo-pentanes
alkyl-pentanes
cyclo-propanes
alkyl-propanes
-CF₃
-CCl₃
-CH₂Cl
-CHCl
-C-Cl
CBr
CI
Alkenes, RCH=CRR
trans RCH=CHR
cis RCH=CHR
RRC=CH₂
RCH₂-CH=CH₂
=CF₂
=CFCl
=CFBr
=CCl₂
=CClBr
=CBr₂
=CFI
=CCII
=CBrI
=CI₂
Allenes
Alkynes R-C≡CH
alkyl-C≡CH
C≡C-Cl
C≡C-Br
C≡C-I
Aromatics 1
1,2
1,3
1,4
1,2,4
1,3,5
1,2,3
1,2,3,4
1,2,3,5
1,2,4,5
1,2,3,4,5
1,2,3,4,5,6
Polyphenyls mono
ortho
meta
para
φ F
φ Cl
φ Br
φ I
Heterocyclics thiophenes
m-dioxane
epoxy
Nitriles
Nitro-
Amines -NH₂
N-H
Amides
Esters acetates
Acids
Ketones
WAVENUMBER (CM⁻¹) 800 600 400 200 100 33.3

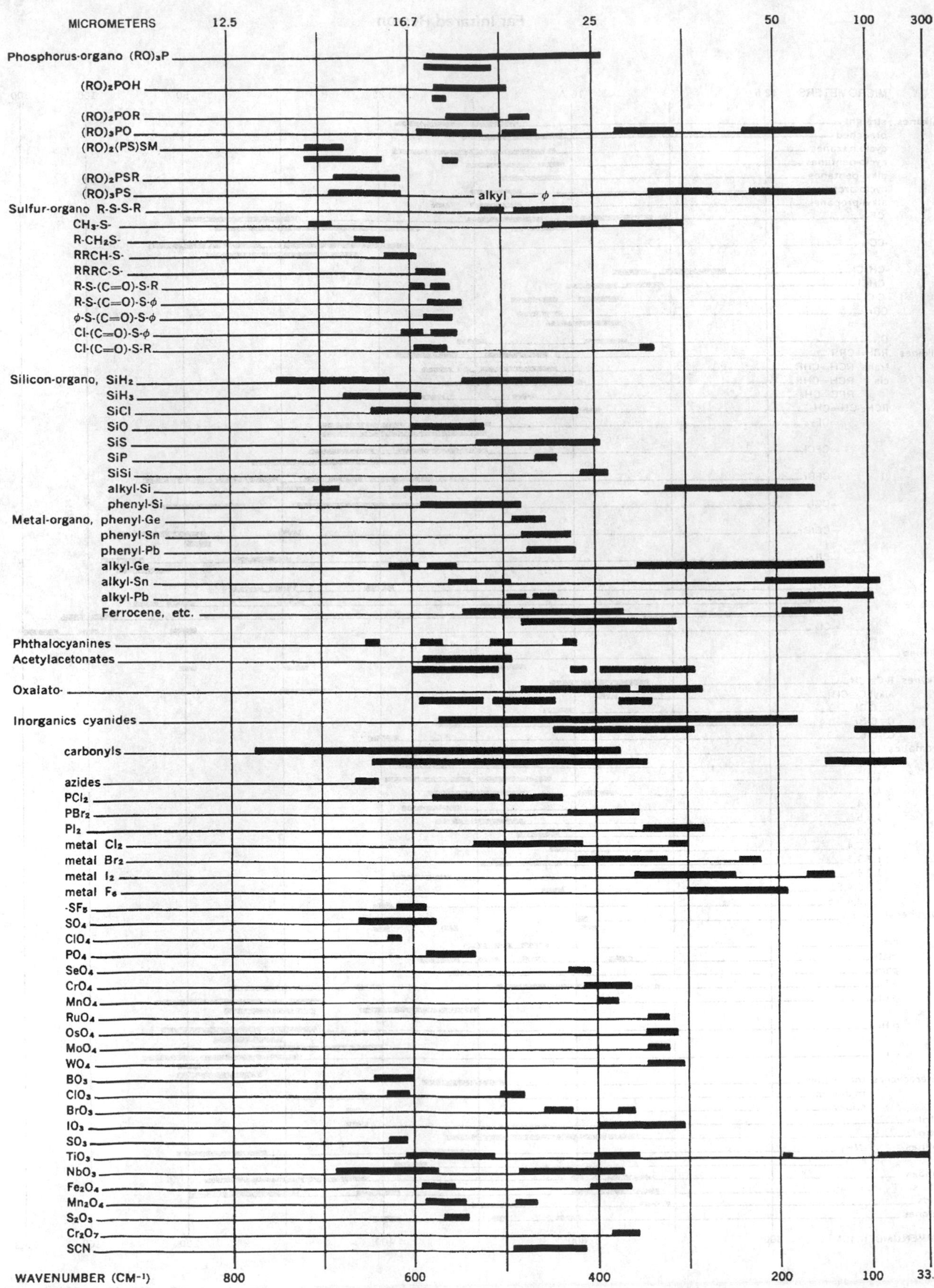
MICROMETERS
12.5
16.7
25
50
100
300
Phosphorus-organo (RO)$_3$P
(RO)$_2$POH
(RO)$_2$POR
(RO)$_3$PO
(RO)$_2$(PS)SM
(RO)$_2$PSR
(RO)$_3$PS
alkyl — ϕ
Sulfur-organo R-S-S-R
CH$_3$-S-
R-CH$_2$S-
RRCH-S-
RRRC-S-
R-S-(C=O)-S-R
R-S-(C=O)-S-ϕ
ϕ-S-(C=O)-S-ϕ
Cl-(C=O)-S-ϕ
Cl-(C=O)-S-R
Silicon-organo, SiH$_2$
SiH$_3$
SiCl
SiO
SiS
SiP
SiSi
alkyl-Si
phenyl-Si
Metal-organo, phenyl-Ge
phenyl-Sn
phenyl-Pb
alkyl-Ge
alkyl-Sn
alkyl-Pb
Ferrocene, etc.
Phthalocyanines
Acetylacetonates
Oxalato-
Inorganics cyanides
carbonyls
azides
PCl$_2$
PBr$_2$
PI$_2$
metal Cl$_2$
metal Br$_2$
metal I$_2$
metal F$_6$
-SF$_5$
SO$_4$
ClO$_4$
PO$_4$
SeO$_4$
CrO$_4$
MnO$_4$
RuO$_4$
OsO$_4$
MoO$_4$
WO$_4$
BO$_3$
ClO$_3$
BrO$_3$
IO$_3$
SO$_3$
TiO$_3$
NbO$_3$
Fe$_2$O$_4$
Mn$_2$O$_4$
S$_2$O$_3$
Cr$_2$O$_7$
SCN
WAVENUMBER (CM^{-1})
800
600
400
200
100
33.

NUCLEAR SPINS, MOMENTS, AND OTHER DATA RELATED TO NMR SPECTROSCOPY

This table presents the following data relevant to nuclear magnetic resonance spectroscopy:

Z: Atomic number

Isotope: Element symbol and mass number

Abundance: Natural abundance of the isotope in percent. An * indicates a radioactive nuclide; if no value is given, the nuclide is not present in nature or its abundance is highly variable.

I: Nuclear spin

ν: Resonant frequency in megahertz for an applied field H_0 of 1 tesla (in cgs units, 10 kilogauss). The resonant frequency scales with H_0.

Relative sensitivity: Sensitivity relative to ^{1}H (=1) assuming an equal number of nuclei and constant temperature. Values were calculated from the expressions:

For constant H_0: $0.0076508(\mu/\mu_N)^3(I+1)/I^2$
For constant ν: $0.23871(\mu/\mu_N)(I+1)$

μ/μ_N: Nuclear magnetic moment in units of the nuclear magneton μ_N

Q: Nuclear quadrupole moment in units of femtometers squared (1 fm^2 = 10^{-2} barn). Because the determination of quadrupole moments requires knowledge of the electron configuration near the nucleus, values of *Q* in the literature tend to scatter considerably. The values quoted here come mainly from the review of Pyykkö (Ref. 3), otherwise from Ref. 1.

The table includes all stable nuclides of non-zero spin for which spin and magnetic moment values have been measured, as well as selected radioactive nuclides of current or potential interest. At least one isotope is included for each element through *Z* = 95 for which data are available. See Reference 1 for a complete listing of spins and moments.

The assistance of P. Pyykkö in providing data on nuclear quadrupole moments is gratefully acknowledged.

References

1. Holden, N. E., "Table of the Isotopes", in Lide, D. R., Ed., *CRC Handbook of Chemistry and Physics*, 90th Ed., CRC Press, Boca Raton, FL, 2009.
2. Raghavan, P., *At. Data Nuc. Data Tables*, 42, 189, 1989.
3. Pyykkö, P., *Mol. Phys.* 106, 1965, 2008.
4. Stone, N. J., *At. Data Nucl. Data Tables*, 90, 75, 2005.
5. IUPAC Commission on Physiochemical Symbols, Terminology and Units, *Quantities, Units, and Symbols in Physical Chemistry, Third Edition*, Royal Society of Chemistry, Cambridge, 2007.

Z	Isotope	Abundance %	I	ν/MHz for H_0 = 1 T	Relative Sensitivity Const. H_0	Const. ν	μ/μ_N	*Q*/fm^2
1	^{1}n		1/2	29.1647	0.32139	0.6850	−1.91304272	
1	^{1}H	99.9885	1/2	42.5775	1.00000	1.0000	+2.792847337	
1	^{2}H	0.0115	1	6.5359	0.00965	0.4094	+0.857438228	+0.2860
1	^{3}H	*	1/2	45.4148	1.21354	1.0667	+2.9789625	
2	^{3}He	0.000134	1/2	32.4380	0.44220	0.7619	−2.127750	
3	^{6}Li	7.59	1	6.2661	0.00850	0.3925	+0.8220467	−0.0808
3	^{7}Li	92.41	3/2	16.5483	0.29356	1.9434	+3.25644	−4.01
4	^{9}Be	100	3/2	5.9842	0.01388	0.7028	−1.1776	+5.288
5	^{10}B	19.9	3	4.5752	0.01985	1.7193	+1.800645	+8.459
5	^{11}B	80.1	3/2	13.6630	0.16522	1.6045	+2.688649	+4.059
6	^{13}C	1.07	1/2	10.7084	0.01591	0.2515	+0.7024118	
7	^{14}N	99.636	1	3.0777	0.00101	0.1928	+0.4037610	+2.044
7	^{15}N	0.364	1/2	4.3173	0.00104	0.1014	−0.2831888	
8	^{17}O	0.038	5/2	5.7742	0.02910	1.5822	−1.89379	−2.558
9	^{19}F	100	1/2	40.0776	0.83400	0.9413	+2.628868	
10	^{21}Ne	0.27	3/2	3.3631	0.00246	0.3949	−0.661797	+10.155
11	^{23}Na	100	3/2	11.2688	0.09270	1.3234	+2.217522	+10.4
12	^{25}Mg	10.00	5/2	2.6083	0.00268	0.7147	−0.85545	+19.94
13	^{27}Al	100	5/2	11.1031	0.20689	3.0424	+3.641507	+14.66
14	^{29}Si	4.685	1/2	8.4655	0.00786	0.1988	−0.55529	
15	^{31}P	100	1/2	17.2515	0.06652	0.4052	+1.13160	
16	^{33}S	0.75	3/2	3.2717	0.00227	0.3842	+0.6438212	−6.78
17	^{35}Cl	75.76	3/2	4.1765	0.00472	0.4905	+0.8218743	−8.165
17	^{37}Cl	24.24	3/2	3.4765	0.00272	0.4083	+0.6841236	−6.435
18	^{37}Ar	*	3/2	5.819	0.01276	0.6833	+1.145	+7.6
18	^{39}Ar	*	7/2	3.46	0.01130	1.7080	−1.59	−12
19	^{39}K	93.2581	3/2	1.9893	0.00051	0.2336	+0.3914662	+5.85
19	^{40}K	0.0117	4	2.4737	0.00523	1.5493	−1.298100	−7.3
19	^{41}K	6.7302	3/2	1.0919	0.00008	0.1282	+0.2148701	+7.11

Z	Isotope	Abundance %	I	ν/MHz for H_0 = 1 T	Relative Sensitivity Const. H_0	Relative Sensitivity Const. ν	μ/μ_N	Q/fm^2
20	^{43}Ca	0.135	7/2	2.8697	0.00643	1.4154	−1.317643	−4.08
21	^{45}Sc	100	7/2	10.3591	0.30244	5.1094	+4.756487	−22.0
22	^{47}Ti	7.44	5/2	2.4041	0.00210	0.6588	−0.78848	+30.2
22	^{49}Ti	5.41	7/2	2.4048	0.00378	1.1861	−1.10417	+24.7
23	^{50}V	0.250	6	4.2505	0.05571	5.5905	+3.345689	+21
23	^{51}V	99.750	7/2	11.2133	0.38360	5.5307	+5.1487057	−5.2
24	^{53}Cr	9.501	3/2	2.4115	0.00091	0.2832	−0.47454	−15
25	^{55}Mn	100	5/2	10.5763	0.17881	2.8981	+3.46872	+33
26	^{57}Fe	2.119	1/2	1.3816	0.00003	0.0324	+0.0906230	+16
27	^{59}Co	100	7/2	10.077	0.27841	4.9703	+4.627	+42
28	^{61}Ni	1.1399	3/2	3.8114	0.00359	0.4476	−0.75002	+16.2
29	^{63}Cu	69.15	3/2	11.3188	0.09393	1.3292	+2.2273456	−22.0
29	^{65}Cu	30.85	3/2	12.1027	0.11484	1.4213	+2.38161	−20.4
30	^{67}Zn	4.102	5/2	2.6685	0.00287	0.7312	+0.875205	+15.0
31	^{69}Ga	60.108	3/2	10.2478	0.06971	1.2035	+2.01659	+17.1
31	^{71}Ga	39.892	3/2	13.0208	0.14300	1.5291	+2.56227	+10.7
32	^{73}Ge	7.76	9/2	1.4897	0.00141	1.1547	−0.8794677	−19.6
33	^{75}As	100	3/2	7.3150	0.02536	0.8590	+1.439475	+31.4
34	^{77}Se	7.63	1/2	8.1568	0.00703	0.1916	+0.5350422	
35	^{79}Br	50.69	3/2	10.7042	0.07945	1.2570	+2.106400	+31.3
35	^{81}Br	49.31	3/2	11.5384	0.09951	1.3550	+2.270562	+26.2
36	^{83}Kr	11.500	9/2	1.6442	0.00190	1.2744	−0.970669	+25.9
37	^{85}Rb	72.17	5/2	4.1253	0.01061	1.1304	+1.35298	+27.6
37	^{87}Rb	27.83	3/2	13.9814	0.17704	1.6419	+2.75131	+13.35
38	^{87}Sr	7.00	9/2	1.8525	0.00272	1.4358	−1.093603	+30.5
39	^{89}Y	100	1/2	2.0949	0.00012	0.0492	−0.1374154	
40	^{91}Zr	11.22	5/2	3.9748	0.00949	1.0892	−1.30362	−17.6
41	^{93}Nb	100	9/2	10.4523	0.48821	8.1013	+6.1705	−32
42	^{95}Mo	15.90	5/2	2.7874	0.00327	0.7638	−0.9142	−2.2
42	^{97}Mo	9.56	5/2	2.8463	0.00349	0.7799	−0.9335	+25.5
43	^{99}Tc	*	9/2	9.6294	0.38174	7.4635	+5.6847	−12.9
44	^{99}Ru	12.76	5/2	1.9553	0.00113	0.5358	−0.6413	+7.9
44	^{101}Ru	17.06	5/2	2.1916	0.00159	0.6005	−0.7188	+45.7
45	^{103}Rh	100	1/2	1.3477	0.00003	0.0317	−0.08840	
46	^{105}Pd	22.33	5/2	1.957	0.00113	0.5364	−0.642	+66.0
47	^{107}Ag	51.839	1/2	1.7331	0.00007	0.0407	−0.1136796	
47	^{109}Ag	48.161	1/2	1.9924	0.00010	0.0468	−0.1306906	
48	^{111}Cd	12.80	1/2	9.0692	0.00966	0.2130	−0.5948861	
48	^{113}Cd	12.22	1/2	9.4871	0.01106	0.2228	−0.6223009	
49	^{113}In	4.29	9/2	9.3655	0.35121	7.2589	+5.5289	+75.9
49	^{115}In	95.71	9/2	9.3856	0.35348	7.2745	+5.5408	+77.0
50	^{115}Sn	0.34	1/2	14.0077	0.03561	0.3290	−0.91883	
50	^{117}Sn	7.68	1/2	15.2610	0.04605	0.3584	−1.00104	
50	^{119}Sn	8.59	1/2	15.9660	0.05273	0.3750	−1.04728	
51	^{121}Sb	57.21	5/2	10.2551	0.16302	2.8101	+3.3634	−54.3
51	^{123}Sb	42.79	7/2	5.5532	0.04659	2.7390	+2.5498	−69.2
52	^{123}Te	0.89	1/2	11.2349	0.01837	0.2639	−0.7369478	
52	^{125}Te	7.07	1/2	13.5446	0.03219	0.3181	−0.8884509	
53	^{127}I	100	5/2	8.5778	0.09540	2.3504	+2.813273	−69.6
54	^{129}Xe	26.4006	1/2	11.8604	0.02162	0.2786	−0.7779763	
54	^{131}Xe	21.2324	3/2	3.5159	0.00282	0.4129	+0.6918619	−11.4
55	^{133}Cs	100	7/2	5.6234	0.04838	2.7736	+2.582025	−0.343
56	^{135}Ba	6.592	3/2	4.2617	0.00501	0.5005	+0.838627	+16.0
56	^{137}Ba	11.232	3/2	4.7634	0.00700	0.5594	+0.937365	+24.5
57	^{138}La	0.090	5	5.6615	0.09404	5.3189	+3.713646	+45

Z	Isotope	Abundance %	I	ν/MHz for H_0 = 1 T	Relative Sensitivity Const. H_0	Relative Sensitivity Const. ν	μ/μ_N	Q/fm²
57	^{139}La	99.910	7/2	6.0612	0.06058	2.9895	+2.7830455	+20.0
58	^{137}Ce	*	3/2	4.88	0.00752	0.5729	0.96	
58	^{139}Ce	*	3/2	5.39	0.01012	0.6326	1.06	
58	^{141}Ce	*	7/2	2.37	0.00364	1.1709	1.09	
59	^{141}Pr	100	5/2	13.0359	0.33483	3.5720	+4.2754	−5.9
60	^{143}Nd	12.2	7/2	2.319	0.00339	1.1440	−1.065	−63
60	^{145}Nd	8.3	7/2	1.429	0.00079	0.7047	−0.656	−33
61	^{143}Pm	*	5/2	11.59	0.23510	3.1748	+3.80	
61	^{147}Pm	*	7/2	5.62	0.04827	2.7714	+2.58	+74
62	^{147}Sm	14.99	7/2	1.7748	0.00152	0.8754	−0.8149	−26
62	^{149}Sm	13.82	7/2	1.4631	0.00085	0.7216	−0.6718	+7.4
63	^{151}Eu	47.81	5/2	10.5856	0.17929	2.9006	+3.4718	+90.3
63	^{153}Eu	52.19	5/2	4.6745	0.01544	1.2809	+1.5331	+241
64	^{155}Gd	14.80	3/2	1.312	0.00015	0.1541	−0.2582	+127
64	^{157}Gd	15.65	3/2	1.720	0.00033	0.2020	−0.3385	+135
65	^{159}Tb	100	3/2	10.23	0.06945	1.2019	+2.014	+143.2
66	^{161}Dy	18.889	5/2	1.4654	0.00048	0.4015	−0.4806	+250.7
66	^{163}Dy	24.896	5/2	2.0508	0.00130	0.5619	+0.6726	+265
67	^{165}Ho	100	7/2	9.0883	0.20423	4.4826	+4.173	+358
68	^{167}Er	22.869	7/2	1.2281	0.00050	0.6057	−0.5639	+356.5
69	^{169}Tm	100	1/2	3.531	0.00057	0.0829	−0.2316	−120
70	^{171}Yb	14.28	1/2	7.5261	0.00552	0.1768	+0.49367	
70	^{173}Yb	16.13	5/2	2.0730	0.00135	0.5680	−0.67989	+280
71	^{175}Lu	97.41	7/2	4.8626	0.03128	2.2084	+2.2327	+349
71	^{176}Lu	2.59	7	3.451	0.03975	6.0518	+3.169	+497
72	^{177}Hf	18.60	7/2	1.7282	0.00140	0.8524	+0.7935	+336.5
72	^{179}Hf	13.62	9/2	1.0856	0.00055	0.8414	−0.6409	+379.3
73	^{181}Ta	99.988	7/2	5.1627	0.03744	2.5464	+2.3705	+317
74	^{183}W	14.31	1/2	1.7957	0.00008	0.0422	+0.1177848	
75	^{185}Re	37.40	5/2	9.7176	0.13870	2.6628	+3.1871	+218
75	^{187}Re	62.60	5/2	9.8170	0.14300	2.6900	+3.2197	+207
76	^{187}Os	1.96	1/2	0.9856	0.00001	0.0231	+0.06465189	
76	^{189}Os	16.15	3/2	3.3536	0.00244	0.3938	+0.659933	+85.6
77	^{191}Ir	37.3	3/2	0.7658	0.00003	0.0899	+0.1507	+81.6
77	^{193}Ir	62.7	3/2	0.8319	0.00004	0.0977	+0.1637	+75.1
78	^{195}Pt	33.832	1/2	9.2922	0.01039	0.2182	+0.60952	
79	^{197}Au	100	3/2	0.7406	0.00003	0.0870	+0.145746	+54.7
80	^{199}Hg	16.87	1/2	7.7123	0.00594	0.1811	+0.5058855	
80	^{201}Hg	13.18	3/2	2.8469	0.00149	0.3343	−0.5602257	+38.7
81	^{203}Tl	29.52	1/2	24.7316	0.19598	0.5809	+1.6222579	
81	^{205}Tl	70.48	1/2	24.9749	0.20182	0.5866	+1.6382146	
82	^{207}Pb	22.1	1/2	9.0340	0.00955	0.2122	+0.59258	
83	^{209}Bi	100	9/2	6.9630	0.14433	5.3968	+4.1106	−51.6
84	^{209}Po	*	1/2	11.7	0.02096	0.2757	+0.77	
86	^{211}Rn	*	1/2	9.16	0.00997	0.2152	+0.601	
87	^{223}Fr	*	3/2	5.95	0.01362	0.6982	+1.17	+117
88	^{223}Ra	*	3/2	1.3746	0.00017	0.1614	+0.2705	+121
88	^{225}Ra	*	1/2	11.187	0.01814	0.2627	−0.7338	
89	^{227}Ac	*	3/2	5.6	0.01131	0.6565	+1.1	+170
90	^{229}Th	*	5/2	1.40	0.00042	0.3843	+0.46	+430
91	^{231}Pa	100	3/2	10.2	0.06903	1.1995	2.01	−172
92	^{235}U	0.7204	7/2	0.83	0.00015	0.4082	−0.38	+493.6
93	^{237}Np	*	5/2	9.57	0.13264	2.6234	+3.14	+388.6
94	^{239}Pu	*	1/2	3.09	0.00038	0.0727	+0.203	
95	^{243}Am	*	5/2	4.6	0.01446	1.2532	+1.5	+421

PROTON NMR CHEMICAL SHIFTS FOR CHARACTERISTIC ORGANIC STRUCTURES

The chart below summarizes the range of chemical shifts for protons in several classes or organic compounds and substituent groups. The chemical shifts δ are given in parts per million relative to tetramethylsilane.

Reference

Mohacsi, E., *J. Chem. Edu.*, 41, 38, 1964 (with permission).

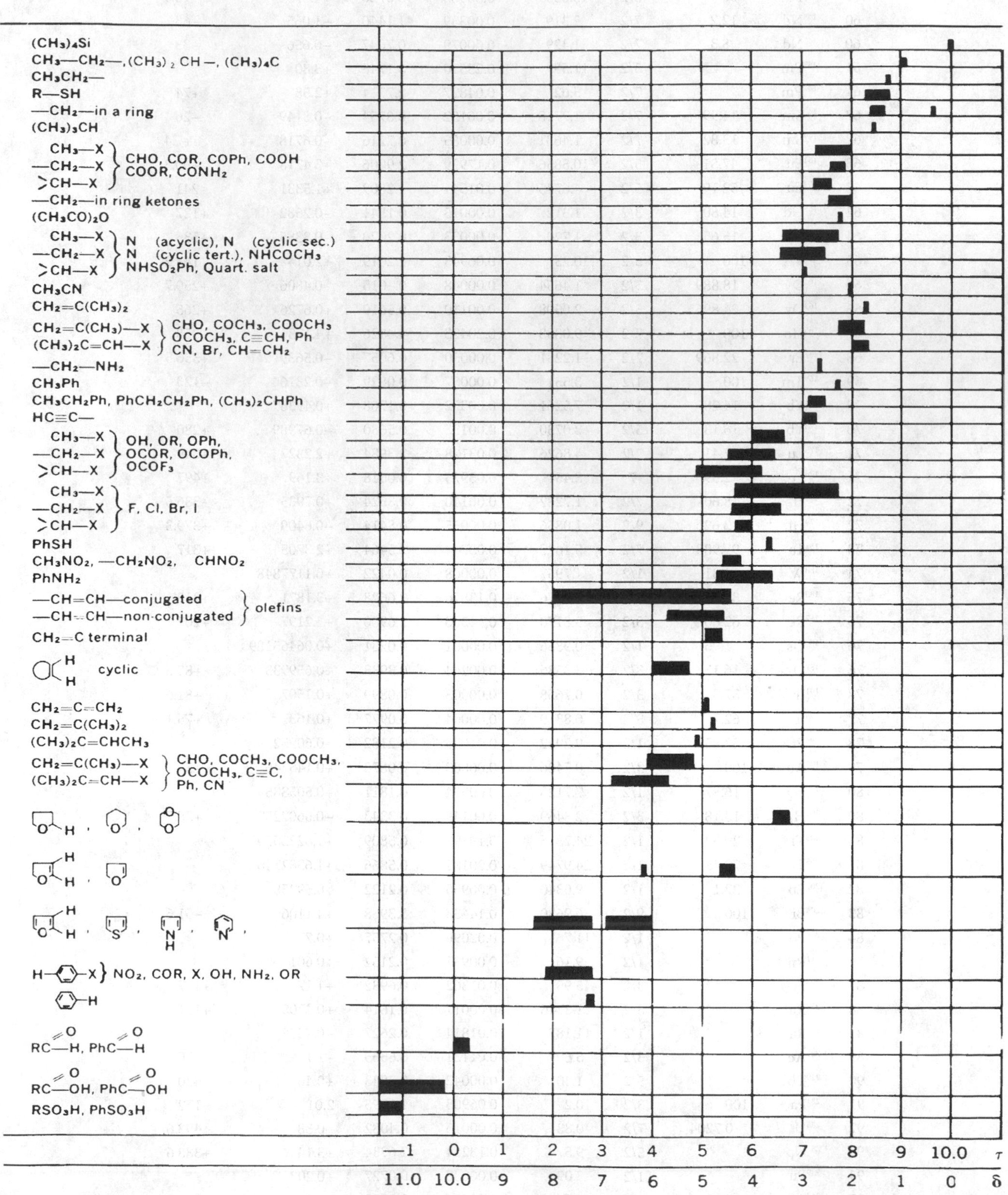

^{13}C-NMR ABSORPTIONS OF MAJOR FUNCTIONAL GROUPS

The table below lists the range of ^{13}C chemical shifts δ in parts per million relative to tetramethylsilane, in descending order, for various functional groups. Examples of simple compounds for each family are given to illustrate the correlations. The shifts for the carbons of interest, which are italicized, are given in parentheses; when two or more values appear, they refer to the sequence of italicized carbon atoms from left to right in the formula.

References

1. Yoder, C. H. and Schaeffer, C. D., Jr., *Introduction to Multinuclear NMR: Theory and Application*, Benjamin/Cummings, Menlo Park, CA, 1987.
2. Silverstein, R. M., Bassler, G. C., and Morrill, T. C., *Spectrometric Identification of Organic Compounds*, John Wiley & Sons, New York, 1981.
3. Brown, D. W., A Short Set of ^{13}C NMR Correlation Tables, *J. Chem. Educ.*, 62, 209, 1985.

δ (ppm)	Group	Family	Example (δ of italicized carbon)	
220-165	>C=O	Ketones	$\mathrm{(CH_3)_2}\mathit{C}\mathrm{O}$	(206.0)
			$\mathrm{(CH_3)_2CH}\mathit{C}\mathrm{OCH_3}$	(212.1)
		Aldehydes	$\mathrm{CH_3}\mathit{C}\mathrm{HO}$	(199.7)
		α,β-Unsaturated	$\mathrm{CH_3CH{=}CH}\mathit{C}\mathrm{HO}$	(192.4)
		carbonyls	$\mathrm{CH_2{=}CH}\mathit{C}\mathrm{OCH_3}$	(169.9)
		Carboxylic acids	$\mathrm{H}\mathit{C}\mathrm{O_2H}$	(166.0)
			$\mathrm{CH_3}\mathit{C}\mathrm{O_2H}$	(178.1)
		Amides	$\mathrm{H}\mathit{C}\mathrm{ONH_2}$	(165.0)
			$\mathrm{CH_3}\mathit{C}\mathrm{ONH_2}$	(172.7)
		Esters	$\mathrm{CH_3}\mathit{C}\mathrm{O_2CH_2CH_3}$	(170.3)
			$\mathrm{CH_2{=}CH}\mathit{C}\mathrm{O_2CH_3}$	(165.5)
140-120	>C=C<	Aromatic	$\mathit{C}_6\mathrm{H_6}$	(128.5)
		Alkenes	$\mathit{C}\mathrm{H_2{=}CH_2}$	(123.2)
			$\mathit{C}\mathrm{H_2{=}}\mathit{C}\mathrm{HCH_3}$	(115.9, 136.2)
			$\mathit{C}\mathrm{H_2{=}}\mathit{C}\mathrm{HCH_2Cl}$	(117.5, 133.7)
			$\mathrm{CH_3CH{=}}\mathit{C}\mathrm{HCH_2CH_3}$	(132.7)
125-115	-CN	Nitriles	$\mathrm{CH_3\text{-}}\mathit{C}\mathrm{N}$	(117.7)
80-70	-CC-	Alkynes	$\mathrm{H}\mathit{C}\mathrm{CH}$	(71.9)
			$\mathrm{CH_3}\mathit{C}\mathrm{CH_3}$	(73.9)
70-45	-C-O	Esters	$\mathit{C}\mathrm{H_3OO}\mathit{C}\mathrm{H_2CH_3}$	(57.6, 67.9)
		Alcohols	$\mathrm{HO}\mathit{C}\mathrm{H_3}$	(49.0)
			$\mathrm{HO}\mathit{C}\mathrm{H_2CH_3}$	(57.0)
40-20	$\mathrm{-C\text{-}NH_2}$	Amines	$\mathit{C}\mathrm{H_3NH_2}$	(26.9)
			$\mathrm{CH_3}\mathit{C}\mathrm{H_2NH_2}$	(35.9)
30-15	$\mathrm{-S\text{-}CH_3}$	Sulfides (thioethers)	$\mathrm{C_6H_5\text{-}S\text{-}}\mathit{C}\mathrm{H_3}$	15.6
30-(-2.3)	-C-H	Alkanes, cycloalkanes	$\mathit{C}\mathrm{H_4}$	(-2.3)
			$\mathit{C}\mathrm{H_3CH_3}$	(5.7)
			$\mathit{C}\mathrm{H_3}\mathit{C}\mathrm{H_2CH_3}$	(15.8, 16.3)
			$\mathit{C}\mathrm{H_3}\mathit{C}\mathrm{H_2CH_2CH_3}$	(13.4, 25.2)
			$\mathit{C}\mathrm{H_3}\mathit{C}\mathrm{H_2}\mathit{C}\mathrm{H_2CH_2CH_3}$	(13.9, 22.8, 34.7)
		Cyclohexane		(26.9)

Section 10
Atomic, Molecular, and Optical Physics

Line Spectra of the Elements........ 10-1
Atomic Transition Probabilities........ 10-93
Electron Affinities........ 10-147
Proton Affinities........ 10-167
Atomic and Molecular Polarizabilities........ 10-186
Ionization Energies of Atoms and Atomic Ions........ 10-196
Ionization Energies of Gas-Phase Molecules........ 10-199
X-Ray Atomic Energy Levels........ 10-217
Electron Binding Energies of the Elements........ 10-221
Natural Width of X-Ray Lines........ 10-227
Photon Attenuation Coefficients........ 10-228
Classification of Electromagnetic Radiation........ 10-233
Sensitivity of the Human Eye to Light of Different Wavelengths........ 10-235
Black Body Radiation........ 10-236
Characteristics of Infrared Detectors........ 10-238
Index of Refraction of Inorganic Crystals........ 10-239
Refractive Index and Transmittance of Representative Glasses........ 10-243
Index of Refraction of Water........ 10-244
Index of Refraction of Liquids for Calibration Purposes........ 10-245
Index of Refraction of Air........ 10-246
Index of Refraction of Gases........ 10-247
Characteristics of Laser Sources........ 10-248
Infrared Laser Frequencies........ 10-254
Infrared and Far-Infrared Absorption Frequency Standards........ 10-261

LINE SPECTRA OF THE ELEMENTS

Joseph Reader and Charles H. Corliss

The original tables from which this table was derived were prepared under the auspices of the Committee on Line Spectra of the Elements of the National Academy of Sciences National Research Council. The table contains the outstanding spectral lines of neutral (I) and singly ionized (II) atoms of the elements from hydrogen through plutonium (Z = 1–94); selected strong lines from doubly ionized (III), triply ionized (IV), and quadruply ionized (V) atoms are also included. Listed are lines that appear in emission from the vacuum ultraviolet to the far infrared. These lines were selected from much larger lists in such a way as to include the stronger observed lines in each spectral region. A more extensive list may be found in Reference 1.

The data were compiled by the following contributors.

J. G. Conway — Lawrence Berkeley Laboratory
C. H. Corliss — National Bureau of Standards
R. D. Cowan — Los Alamos Scientific Laboratory
C. R. Cowley — University of Michigan
Henry M. and Hannah Crosswhite — Argonne National Laboratory
S. P. Davis — University of California, Berkeley
V. Kaufman — National Bureau of Standards
R. L. Kelly — Naval Postgraduate School
J. F. Kielkopf — University of Louisville
W. C. Martin — National Bureau of Standards
T. K. McCubbin — Pennsylvania State University
L. J. Radziemski — Los Alamos Scientific Laboratory
J. Reader — National Bureau of Standards
C. J. Sansonetti — National Bureau of Standards
G. V. Shalimoff — Lawrence Berkeley Laboratory
R. W. Stanley — Purdue University
J. O. Stoner, Jr. — University of Arizona
H. H. Stroke — New York University
D. R. Wood — Wright State University
E. F. Worden — Lawrence Livermore Laboratory
J. J. Wynne — International Business Machines Corporation
R. Zalubas — National Bureau of Standards

All wavelengths are given in Ångstrom units (10^{-10} m). Below 2000 Å the wavelengths are in vacuum (except for the Cu II line at 1999.698 Å, which is in air); above 2000 Å the wavelengths are in air. Wavelengths given to three decimal places have an uncertainty of less than 0.001 Å and are therefore suitable for calibration purposes. In the air region, the elements used most commonly for calibration are Ne, Ar, Kr, Fe, Th, and Hg; in the vacuum region, the most common are C, N, O, Si, Cu.

All data refer to natural isotopic abundance of the elements except that Kr I and Kr II lines below 11,000 Å given to three decimal places are for ^{86}Kr. A separate table for ^{198}Hg contains accurately known wavelengths that are frequently used for calibration.

A large number of the lines for neutral and singly ionized atoms were extracted from the National Bureau of Standards (NBS) *Tables of Spectral Line Intensities* (Reference 2). The intensities of these lines represent quantitative estimates of relative line strengths that take into account varying detection sensitivity at different wavelengths. They are on a linear scale. For nearly all of the other lines the intensities represent qualitative estimates of the relative strengths of lines not greatly separated in wavelength. Because different observers frequently use different scales for their intensity estimates, these intensities are useful only as a rough indication of the appearance of a spectrum. In some cases the intensity scale is not intended to be linear. In the first and second spectra the intensities of the lines of the singly ionized atom (II) relative to those of the neutral atom (I) should be used with caution, inasmuch as the concentration of ions in a light source depends greatly on the excitation conditions.

Descriptive symbols that follow the wavelength have the following meanings:

c — complex
d — line consists of two unresolved lines
h — hazy
l — shaded to longer wavelengths
s — shaded to shorter wavelengths
p — perturbed by a close line
r — easily reversed
w — wide

The table is arranged alphabetically by element name (not symbol); for each element the lines are listed by wavelength. References to the sources of data for each element are given at the end of the table, starting on page 10-89.

General References

1. Reader, J., Corliss, C. H., Wiese, W. L., and Martin, G. A., *Tables of Line Spectra of the Elements, Part 1. Wavelengths and Intensities*, Nat. Stand. Ref. Data Sys.- Nat. Bur. Standards (U.S.), No. 68, 1980.
2. Meggers, W. F., Corliss, C. H., and Scribner, B. F., *Tables of Spectral Line Intensities, Part 1. Arranged by Elements*, Nat. Bur. Stand. (U.S.), Monograph 145, 1975.
3. Fuhr, J. R., Martin, W. C., Musgrove, A., Sugar, J., and Wiese, W. L., "NIST Atomic Spectroscopic Database" ver. 1.1, January 1996. *NIST Physical Reference Data*, National Institute of Standards and Technology, Gaithersburg, MD. Available at the WWW address: http://physics.nist.gov/PhysRefData/contents.html

Intensity	Wavelength/Å	
Actinium Ac Z = 89		
2000 h	2952.55	III
2000 h	3392.78	III
3000	3487.59	III
2000 s	3863.12	II
3000 s	4088.44	II
3000 s	4168.40	II
100	4179.98	I
20	4183.12	I
20	4194.40	I
20 l	4384.53	I
20	4396.71	I
2000 h	4413.09	III
20	4462.73	I
3000 h	4569.87	III
1000	5910.85	II
20	6359.86	I
20 l	6691.27	I
Aluminum Al Z = 13		
900	125.53	V
800	126.07	V
800	130.41	V
1000	130.85	V
900	131.00	V
900	131.44	V
800	160.07	IV
1000	278.69	V
900	281.39	V
70	486.884	III
30	486.912	III
250	511.138	III
150	511.191	III
500	560.317	III
200	560.433	III
100	670.068	III
200	671.118	III
500	695.829	III
400	696.217	III
200	725.683	III
300	726.915	III
400	855.034	III
500	856.746	III
400	892.024	III
50	893.887	III
450	893.897	III
800	1042.17	IV
50	1191.812	II
900	1237.19	IV
900	1257.62	IV
800	1264.18	IV
1000	1272.76	IV
150	1350.18	II
800	1384.13	III
800	1447.51	IV
800	1494.79	IV
1000	1526.14	V
800	1537.54	IV
800	1539.830	II
1000	1557.25	IV
100	1569.385	II
900	1582.04	IV
800	1584.46	IV
125	1596.059	II
700	1605.766	III
100	1611.814	III
800	1611.874	III
150	1625.627	II
800	1639.06	IV
100	1644.235	II
100	1644.809	II
1000	1670.787	II
100	1686.250	II
800	1719.440	II
500	1721.244	II
900	1721.271	II
500	1724.952	II
900	1724.984	II
350	1760.104	II
300	1761.975	II
290	1763.00	I
500	1763.869	II
700	1763.952	II
450	1765.64	I
300	1765.815	II
450	1766.38	I
400	1767.731	II
450	1769.14	I
1000	1818.56	IV
600	1828.588	II
400	1832.837	II
250	1834.808	II
1000	1854.716	III
300	1855.929	II
700	1858.026	II
120	1859.980	II
1000	1862.311	II
600	1862.790	III
200	1929.978	II
150	1931.048	II
200	1932.377	II
400	1934.503	II
150	1934.713	II
300	1935.840	III
200	1935.949	III
150	1936.907	II
220	1939.261	II
700	1990.531	II
150	2016.052	II
150	2016.234	II
100	2016.368	II
200	2074.008	II
700	2094.264	II
150	2094.744	II
300	2094.791	II
100	2095.104	II
200	2095.141	II
400	2269.10	I
120	2269.22	I
140	2321.56	I
460	2367.05	I
110	2367.61	I
110	2368.11	I
180	2369.30	I
140	2370.22	I
160	2372.07	I
850	2373.12	I
170	2373.35	I
110	2373.57	I
240	2567.98	I
480	2575.10	I
110	2637.70	II
150	2652.48	I
200	2660.39	I
160	2669.17	II
650	2816.19	II
150	3041.28	II
360	3050.07	I
450	3057.14	I
150	3074.64	II
4500 r	3082.153	I
7200 r	3092.710	I
1800 r	3092.839	I
150	3428.92	II
150	3443.64	I
900	3492.23	IV
800	3508.46	IV
450	3586.56	II
360	3587.07	II
290	3587.45	II
870	3601.63	III
220	3651.06	II
110	3651.10	II
150	3654.98	II
290	3655.00	II
450	3900.68	II
4500 r	3944.006	I
9000 r	3961.520	I
110	3995.86	II
290	4226.81	II
870	4529.19	III
150	4585.82	II
110	4588.19	II
550	4666.80	II
110	4898.76	II
110	4902.77	II
150	5280.21	II
290	5283.77	II
150	5285.85	II
110	5312.32	II
220	5316.07	II
150	5371.84	II
180	5557.06	I
110	5557.95	I
450	5593.23	II
1200	5696.60	III
1000	5722.73	III
110	5853.62	II
220	5971.94	II
290	6001.76	II
220	6001.88	II
450	6006.42	II
150	6061.11	II
290	6068.43	II
110	6068.53	II
450	6073.23	II
110	6181.57	II
150	6181.68	II
290	6182.28	II
220	6182.45	II
450 h	6183.42	II
450	6201.52	II
360	6201.70	II
290	6226.18	II
360	6231.78	II
450	6243.36	II
450	6335.74	II
360	6696.02	I
230	6698.67	I
110	7361.57	I
140	7362.30	I
230	7835.31	I
290	7836.13	I
110	8075.35	I
290	8640.70	II
360	8772.87	I
450	8773.90	I
110	8828.91	I
180	8841.28	I
140	8923.56	I
150	9290.65	II
110	9290.75	II
150	10076.29	II
110	10768.36	I
140	10782.04	I
110	10872.98	I
230	10891.73	I
450	11253.19	I
570	11254.88	I
570	13123.41	I
450	13150.76	I
230	16718.96	I
300	16750.56	I
140	16763.36	I
300	21093.04	I
360	21163.75	I
Antimony Sb Z = 51		
15	722.86	III
15	732.33	III
	861.5	IV
4	876.84	II
4	921.07	II
6	983.57	II
15	999.62	III
6	1001.13	II
6	1009.43	II
40	1011.94	III
6	1052.21	II
8	1056.27	II
8	1057.32	II
40	1065.90	III
6	1073.81	II
30	1075.82	III
	1087.6	IV

Intensity	Wavelength/Å	
8	1104.32	V
30	1151.49	III
40	1157.74	III
	1199.1	IV
50	1205.20	III
50	1210.64	III
12	1226.00	V
6	1230.30	II
8	1274.98	II
20	1306.69	III
8	1327.40	II
6	1358.04	II
8	1384.70	II
20	1404.18	III
6	1407.83	II
8	1436.49	II
20 r	1486.57	I
40 h	1491.36	I
	1499.2	IV
12	1505.70	V
50 r	1512.57	I
12	1524.47	V
120 r	1532.74	I
80 r	1535.06	I
6	1565.51	II
8	1576.11	II
7	1581.36	II
80 r	1599.96	I
10	1606.98	II
200 w	1612.8	I
100 w	1623.3	I
20	1657.04	II
100 w	1662.6	I
15	1673.89	III
15	1711.84	III
80 r	1716.93	I
150 r	1717.45	I
150 r	1723.43	I
15	1725.33	III
100 r	1736.19	I
100 h	1765.76	I
100 r	1780.87	I
100 r	1788.24	I
150	1800.18	I
50 r	1810.50	I
80 r	1814.20	I
100	1829.50	I
50 r	1868.17	I
300 r	1871.15	I
150 r	1882.56	I
100	1927.08	I
200 r	1950.39	I
60 r	2029.49	I
70 r	2039.77	I
150 r	2049.57	I
1000 r	2068.33	I
100	2079.56	I
50 r	2098.41	I
80 r	2118.48	I
100 r	2127.39	I
50 r	2137.05	I

Intensity	Wavelength/Å	
100 r	2139.69	I
10	2141.80	II
50 r	2141.83	I
100 r	2144.86	I
1500 r	2175.81	I
250 r	2179.19	I
200 r	2201.32	I
300 r	2208.45	I
150 r	2220.73	I
100	2221.98	I
120 r	2224.93	I
300 r	2262.51	I
120	2288.98	I
150 r	2293.44	I
300 r	2306.46	I
2500 r	2311.47	I
150	2315.89	I
400 h	2373.67	I
300 h	2383.64	I
100	2395.22	I
150	2422.13	I
250	2426.35	I
400 r	2445.51	I
400	2478.32	I
150	2480.44	I
100	2510.54	I
2000 r	2528.52	I
15	2528.54	II
10	2567.75	II
150	2574.06	I
15	2590.13	III
1500 r	2598.05	I
500 r	2598.09	I
300 r	2612.31	I
12	2617.17	III
200 r	2652.60	I
20	2669.39	III
300 r	2670.64	I
200 r	2682.76	I
120	2692.25	I
150 r	2718.90	I
400 r	2769.95	I
1000 r	2877.92	I
15	2980.96	II
500 r	3029.83	I
600 r	3232.52	I
20	3241.28	II
700 r	3267.51	I
15	3498.46	II
25	3637.80	II
250	3637.83	I
20	3722.78	II
200 r	3722.79	I
20	3850.22	II
200	4033.55	I
20	4033.56	II
20	4133.63	II
15	4140.54	II
15	4195.17	II
20	4219.07	II
20	4314.32	II

Intensity	Wavelength/Å	
15	4514.50	II
30	4596.90	II
20	4599.09	II
15	4604.77	II
30	4647.32	II
20	4675.74	II
40	4711.26	II
20	4757.81	II
20	4765.36	II
30	4784.03	II
20	4802.01	II
20	4832.82	II
20	4877.24	II
15	4947.40	II
15	5044.56	II
20	5238.94	II
20	5354.24	II
40 h	5556.10	I
100 l	5632.02	I
30	5639.75	II
60 h	5830.34	I
100	6005.21	II
20	6053.41	II
30	6079.80	II
50	6130.04	II
20	6154.94	II
20	6611.49	I
30	6647.44	II
30 h	7648.28	I
80	7844.44	I
200	7924.65	I
60	8411.69	I
150	8572.64	I
100	8619.55	I
400	9518.68	I
400	9949.14	I
200	10078.49	I
300	10261.01	I
200	10585.60	I
1000	10677.41	I
800	10741.94	I
80	10794.11	I
600	10839.73	I
200	10868.58	I
400	10879.55	I
300	11012.79	I
150	11266.23	I
5	12116.06	I

Argon Ar Z = 18

Intensity	Wavelength/Å	
3	336.56	V
3	337.56	V
6	338.00	V
2	338.43	V
2	339.01	V
3	339.89	V
3	350.88	V
4	396.87	IV
4	398.55	IV
2	436.67	V
5	446.00	V
8	446.95	V
4	447.53	V
18	449.06	V
4	449.49	V
3	458.12	V
2	458.98	V
6 p	461.23	V
3	462.42	V
7	463.94	V
30	487.227	II
50	490.650	II
30	490.701	II
30	519.327	II
3	522.09	V
5	524.19	V
6	527.69	V
30	542.912	II
200	543.203	II
70	547.461	II
2	554.50	V
70	556.817	II
5	558.48	V
70	573.362	II
30	576.736	II
70	580.263	II
30	583.437	II
70	597.700	II
30	602.858	II
30	612.372	II
6	623.77	IV
3	635.12	V
500	661.867	II
30	664.562	II
200	666.011	II
1000	670.946	II
3000	671.851	II
70	676.242	II
30	677.952	II
30	679.218	II
200	679.401	II
10	683.28	IV
7	688.39	IV
12 p	689.01	IV
6	699.41	IV
8	700.28	IV
3	705.35	V
5	709.20	V
4	715.60	V
3	715.65	V
200	718.090	II
3000	723.361	II
2	725.11	V
500	725.548	II
70	730.930	II
200	740.269	II
200	744.925	II
70	745.322	II
4	754.20	IV
5	761.47	IV
12	769.15	III
5	800.57	IV
10	801.09	IV

Intensity	Wavelength/Å	
10	801.41	IV
5	801.91	IV
20	802.859	I
100	806.471	I
60	806.869	I
30	807.218	I
40	807.653	I
50	809.927	I
120	816.232	I
70	816.464	I
80	820.124	I
4	822.16	V
120	825.346	I
120	826.365	I
5	827.05	V
3	827.35	V
150	834.392	I
4 p	834.88	V
100	835.002	I
2	836.13	V
15	840.03	IV
100	842.805	I
20	843.77	IV
25	850.60	IV
180	866.800	I
150	869.754	I
10	871.10	III
9	875.53	III
180 r	876.058	I
12	878.73	III
8	879.62	III
180 r	879.947	I
9	883.18	III
10	887.40	III
150	894.310	I
5	900.36	IV
9	901.17	IV
1000	919.781	II
1000	932.054	II
1000 r	1048.220	I
500 r	1066.660	I
7	1669.67	III
7	1673.42	III
7	1675.48	III
9	1914.40	III
7	1915.56	III
10	2125.16	III
15	2133.87	III
10	2138.59	III
10	2148.73	III
15	2166.19	III
10	2168.26	III
20	2170.23	III
25	2177.22	III
8	2184.06	III
10	2188.22	III
15	2192.06	III
7	2248.73	III
10	2279.10	III
7	2281.22	III
7	2282.21	III

Intensity	Wavelength/Å	
12	2293.03	III
4	2299.72	IV
10	2300.85	III
15	2302.17	III
9	2317.00	III
15	2317.47	III
12	2318.04	III
10	2319.13	III
10	2319.37	III
9	2345.17	III
7	2351.67	III
9	2360.26	III
10	2395.63	III
12	2399.15	III
10	2413.20	III
7	2415.61	III
10	2418.82	III
5	2420.456	II
12	2423.52	III
12	2423.93	III
7	2443.69	III
8	2447.71	IV
8	2472.95	III
7	2476.10	III
12	2488.86	III
12	2513.28	IV
10	2516.789	II
6	2518.40	IV
9	2525.69	IV
10	2534.709	II
15	2562.087	II
12	2562.17	IV
10	2568.07	IV
7	2569.53	IV
12	2599.47	IV
10	2608.06	IV
7	2608.44	IV
12	2615.68	IV
6	2619.98	IV
12	2621.36	IV
12	2624.92	IV
7	2631.90	III
15	2640.34	IV
10	2654.63	III
8	2674.02	III
9	2678.38	III
9	2682.63	IV
10	2724.84	III
14	2757.92	IV
7	2762.23	III
10	2776.26	IV
12	2784.47	IV
14	2788.96	IV
7	2797.11	IV
16	2809.44	IV
10	2830.25	IV
7	2842.88	III
8	2855.29	III
6	2874.40	IV
9	2884.12	III
25	2891.612	II

Intensity	Wavelength/Å	
12	2913.00	IV
11	2926.33	IV
200	2942.893	II
100	2979.050	II
10	3010.02	III
12	3024.05	III
50	3033.508	II
6	3037.98	IV
12	3054.82	III
10	3064.77	III
8	3077.40	IV
10	3078.15	III
50	3093.402	II
7	3110.41	III
7	3127.90	III
8	3200.37	I
20	3243.689	II
25	3285.85	III
25	3293.640	II
20	3301.88	III
20	3307.228	II
15	3311.25	III
7	3319.34	I
7	3323.59	III
25	3336.13	III
20	3344.72	III
25	3350.924	II
15	3358.49	III
7	3361.28	III
7	3373.47	I
25	3376.436	II
25	3388.531	II
15	3391.85	III
7	3393.73	I
7	3417.49	III
9	3424.25	III
8	3438.04	III
7	3461.07	I
9	3471.32	III
70	3476.747	II
20	3478.232	II
20	3480.55	III
50	3491.244	II
100	3491.536	II
12	3499.67	III
15	3503.58	III
70	3509.778	II
8	3511.12	III
70	3514.388	II
70	3545.596	II
70	3545.845	II
7	3554.306	I
100	3559.508	II
100	3561.030	II
70	3576.616	II
25	3581.608	II
50	3582.355	II
70	3588.441	II
7	3606.522	I
25	3622.138	II
20	3639.833	II

Intensity	Wavelength/Å	
35	3718.206	II
70	3729.309	II
50	3737.889	II
150	3765.270	II
50	3766.119	II
20	3770.369	I
20	3770.520	II
25	3780.840	II
20	3795.37	III
25	3803.172	II
50	3809.456	II
7	3834.679	I
70	3850.581	II
10	3858.32	III
35	3868.528	II
7	3907.84	III
35	3925.719	II
50	3928.623	II
25	3932.547	II
70	3946.097	II
7	3947.505	I
35	3948.979	I
8	3960.53	III
20	3979.356	II
35	3994.792	II
50	4013.857	II
6	4023.60	III
50	4033.809	II
20	4035.460	II
150	4042.894	II
50	4044.418	I
100	4052.921	II
200	4072.005	II
70	4072.385	II
25	4076.628	II
35	4079.574	II
25	4082.387	II
150	4103.912	II
300	4131.724	II
5	4146.70	III
35	4156.086	II
400	4158.590	I
50	4164.180	I
35	4179.297	II
50	4181.884	I
100	4190.713	I
50	4191.029	I
200	4198.317	I
400	4200.674	I
25	4218.665	II
25	4222.637	II
25	4226.988	II
100	4228.158	II
100	4237.220	II
25	4251.185	I
200	4259.362	I
100	4266.286	I
70	4266.527	II
150	4272.169	I
550	4277.528	II
20	4282.898	II

Intensity	Wavelength/Å	
100	4300.101	I
25	4300.650	II
70	4309.239	II
200	4331.200	II
50	4332.030	II
100	4333.561	I
50	4335.338	I
25	4345.168	I
800	4348.064	II
50	4352.205	II
25	4362.066	II
50	4367.832	II
200	4370.753	II
70	4371.329	II
50	4375.954	II
150	4379.667	II
50	4385.057	II
70	4400.097	II
200	4400.986	II
400	4426.001	II
150	4430.189	II
50	4430.996	II
50	4433.838	II
20	4439.461	II
35	4448.879	II
100	4474.759	II
200	4481.811	II
100	4510.733	I
20	4522.323	I
20	4530.552	II
400	4545.052	II
20	4564.405	II
400	4579.350	II
400	4589.898	II
15	4596.097	I
550	4609.567	II
7	4628.441	I
35	4637.233	II
400	4657.901	II
15	4702.316	I
20	4721.591	II
550	4726.868	II
50	4732.053	II
300	4735.906	II
800	4764.865	II
550	4806.020	II
150	4847.810	II
50	4865.910	II
800	4879.864	II
70	4889.042	II
20	4904.752	II
35	4933.209	II
200	4965.080	II
50	5009.334	II
70	5017.163	II
70	5062.037	II
20	5090.495	II
100	5141.783	II
70	5145.308	II
5	5151.391	I
15	5162.285	I

Intensity	Wavelength/Å	
25	5165.773	II
20	5187.746	I
20	5216.814	II
7	5221.271	I
5	5421.352	I
10	5451.652	I
25	5495.874	I
5	5506.113	I
25	5558.702	I
10	5572.541	I
35	5606.733	I
20	5650.704	I
10	5739.520	I
5	5834.263	I
10	5860.310	I
15	5882.624	I
25	5888.584	I
50	5912.085	I
15	5928.813	I
5	5942.669	I
7	5987.302	I
5	5998.999	I
5	6025.150	I
70	6032.127	I
35	6043.223	I
10	6052.723	I
20	6059.372	I
7	6098.803	I
10	6105.635	I
100	6114.923	II
10	6145.441	I
7	6170.174	I
150	6172.278	II
10	6173.096	I
10	6212.503	I
5	6215.938	I
25	6243.120	II
7	6296.872	I
15	6307.657	I
7	6369.575	I
20	6384.717	I
70	6416.307	I
25	6483.082	II
15	6538.112	I
15	6604.853	I
25	6638.221	II
20	6639.740	II
50	6643.698	II
5	6660.676	I
5	6664.051	I
25	6666.359	II
100	6677.282	I
35	6684.293	II
150	6752.834	I
5	6756.163	I
15	6766.612	I
20	6861.269	II
150	6871.289	I
5	6879.582	I
10	6888.174	I
50	6937.664	I

Intensity	Wavelength/Å	
7	6951.478	I
7	6960.250	I
10000	6965.431	I
150	7030.251	I
10000	7067.218	I
100	7068.736	I
25	7107.478	I
25	7125.820	I
1000	7147.042	I
15	7158.839	I
70	7206.980	I
15	7265.172	I
7	7270.664	I
2000	7272.936	I
35	7311.716	I
25	7316.005	I
5	7350.814	I
70	7353.293	I
200	7372.118	I
20	7380.426	II
10000	7383.980	I
20	7392.980	I
15	7412.337	I
10	7425.294	I
25	7435.368	I
10	7436.297	I
20000	7503.869	I
15000	7514.652	I
25000	7635.106	I
15000	7723.761	I
10000	7724.207	I
10	7891.075	I
20000	7948.176	I
20000	8006.157	I
25000	8014.786	I
7	8053.308	I
20000	8103.693	I
35000	8115.311	I
10000	8264.522	I
20	8392.27	I
15000	8408.210	I
20000	8424.648	I
15000	8521.442	I
7	8605.776	I
4500	8667.944	I
20	8771.860	II
180	8849.91	I
20	9075.394	I
35000	9122.967	I
550	9194.638	I
15000	9224.499	I
400	9291.531	I
1600	9354.220	I
25000	9657.786	I
4500	9784.503	I
180	10052.06	I
30	10332.72	I
100	10467.177	II
1600	10470.054	I
13	10478.034	I
180	10506.50	I

Intensity	Wavelength/Å	
200	10673.565	I
11	10681.773	I
7	10683.034	II
30	10733.87	I
30	10759.16	I
7	10812.896	II
11	11078.869	I
30	11106.46	I
12	11441.832	I
400	11488.109	I
200	11668.710	I
12	11719.488	I
200	12112.326	I
50	12139.738	I
50	12343.393	I
200	12402.827	I
200	12439.321	I
100	12456.12	I
200	12487.663	I
150	12702.281	I
30	12733.418	I
12	12746.232	I
200	12802.739	I
50	12933.195	I
500	12956.659	I
200	13008.264	I
200	13213.99	I
200	13228.107	I
100	13230.90	I
500	13272.64	I
1000	13313.210	I
1000	13367.111	I
30	13499.41	I
1000	13504.191	I
11	13573.617	I
30	13599.333	I
400	13622.659	I
200	13678.550	I
1000	13718.577	I
10	13825.715	I
10	13907.478	I
200	14093.640	I
100	15046.50	I
25	15172.69	I
10	15329.34	I
30	15989.49	I
30	16519.86	I
500	16940.58	I
12	18427.76	I
50	20616.23	I
30	20986.11	I
20	23133.20	I
20	23966.52	I

Arsenic As $Z = 33$

Intensity	Wavelength/Å	
510	871.7	III
325	889.0	III
325	927.5	III
325	937.2	III
325	953.6	III
325	963.8	III
250	987.7	V

Intensity	Wavelength/Å	
340	1021.96	II
250	1029.5	V
340	1082.35	II
500	1139.40	II
615	1149.31	II
555	1181.51	II
555	1189.87	II
615	1196.38	II
615	1196.56	II
340	1207.44	II
800	1211.17	II
800	1218.10	II
340	1223.15	II
760	1241.31	II
965	1243.08	II
870	1245.67	II
800	1258.58	II
965	1263.77	II
800	1266.34	II
800	1267.59	II
715	1280.99	II
715	1287.54	II
715	1305.70	II
340	1307.74	II
760	1333.15	II
965	1341.55	II
760	1355.93	II
965	1369.77	II
800	1373.65	II
1000	1375.07	II
760	1375.78	II
800	1394.64	II
800	1400.31	II
500	1448.59	II
500	1558.88	II
500	1570.99	II
100 r	1593.60	I
500	1660.55	II
340	1860.34	II
1000 r	1890.42	I
500	1912.94	II
800 r	1937.59	I
585 r	1972.62	I
170 r	1990.35	I
100 r	1991.13	I
100 r	1995.43	I
230 r	2003.34	I
100 r	2009.19	I
200	2263.2	IV
350 r	2288.12	I
200	2301.0	IV
350 r	2349.84	I
100 r	2370.77	I
135 r	2381.18	I
250	2417.5	IV
250	2454.0	IV
170 r	2456.53	I
200	2461.4	IV
340	2602.00	II
170 r	2780.22	I
300	2830.359	II
300	2831.164	II
100 r	2860.44	I
300	2884.406	II
80	2926.3	III
615	2959.572	II
300	3003.819	II
300	3116.516	II
340	3842.60	II
325	3922.6	III
715	4190.082	II
615	4197.40	II
615	4242.982	II
500	4315.657	II
500	4323.867	II
500	4336.64	II
500	4352.145	II
425	4352.864	II
375	4371.17	II
615	4427.106	II
615	4431.562	II
715	4458.469	II
340	4461.075	II
715	4466.348	II
500	4474.46	II
800	4494.230	II
850	4507.659	II
615	4539.74	II
715	4543.483	II
615	4602.427	II
340	4629.787	II
340	4707.586	II
340	4730.67	II
340	4888.557	II
340	5105.58	II
500	5107.55	II
425	5231.38	II
500	5331.23	II
340	5497.727	II
425	5558.09	II
425	5651.32	II
425	6110.07	II
500	6170.27	II
300	6511.74	II
300	7092.27	II
300	7102.72	II
340	7990.53	II
300	8174.51	II
200	9300.61	I
230	9597.95	I
290	9626.70	I
230	9833.76	I
170	9915.71	I
290	9923.05	I
290	10024.04	I
170	10614.07	I
Astatine At	***Z* = 85**	
8	2162.25	I
10	2244.01	I
Barium Ba	***Z* = 56**	
14	555.48	III
14	587.57	III
18	647.27	III
300	719.86	V
150	721.85	V
1000	766.87	V
40000	794.89	IV
300	877.41	V
50000	923.74	IV
200	946.26	V
200	1486.72	II
400	1504.01	II
300	1554.38	II
200	1572.73	II
	1573.92	II
	1630.40	II
100	1674.51	II
400	1694.37	II
	1697.16	II
	1761.75	II
	1771.03	II
	1786.93	II
100	1904.15	II
500	1924.70	II
	1985.60	II
300	1999.54	II
10	2001.30	III
	2009.20	II
400	2023.95	II
	2052.68	II
	2054.57	II
500	2214.7	II
800	2245.61	II
1000	2254.73	II
1400	2304.24	II
60	2331.10	III
2000	2335.27	II
190	2347.58	II
40	2512.28	III
40	2523.83	III
60	2528.51	II
50	2559.54	III
8 h	2596.64	I
100	2634.78	II
40	2681.89	III
8	2702.63	I
18	2771.36	II
15	2785.28	I
100 r	3071.58	I
40	3079.14	III
10 h	3108.21	I
8	3132.60	I
8 h	3135.72	I
10	3137.70	I
10	3155.34	I
10	3155.67	I
12	3158.05	I
12 h	3158.54	I
25	3165.60	I
15 h	3173.69	I
30	3183.16	I
15	3183.96	I
10	3193.91	I
25 h	3203.70	I
30	3221.63	I
40	3222.19	I
50	3261.96	I
60 r	3262.34	I
40	3281.50	I
15	3281.77	I
50	3322.80	I
80 h	3356.80	I
50	3368.18	III
60 r	3377.08	I
20	3377.39	I
70 r	3420.32	I
25	3421.01	I
30 h	3421.48	I
40	3463.74	I
200 r	3501.11	I
80 h	3524.97	I
30 h	3531.35	I
80 h	3544.66	I
20 h	3547.68	I
100	3552.45	II
200	3567.73	II
100	3576.28	II
30	3577.62	I
80 h	3579.67	I
200	3596.57	II
40	3630.64	I
40 h	3636.83	I
20 h	3688.47	I
400	3735.75	II
200	3816.69	II
200	3842.80	II
100	3854.76	II
20	3889.33	I
1400 l	3891.78	II
20	3892.65	I
40	3909.91	I
500	3914.73	II
25	3926.85	III
50	3935.72	I
20	3937.87	I
200	3939.67	II
500	3949.51	II
25	3993.06	III
80	3993.40	I
30	3995.66	I
300	4036.26	II
200	4083.77	II
30 h	4084.86	I
1500 h	4130.66	II
20	4132.43	I
200	4166.00	II
500	4216.04	II
800	4267.95	II
100	4283.10	I
300	4287.80	II
200	4297.60	II
800	4309.32	II
20 h	4323.00	I

Intensity	Wavelength/Å	
600	4325.73	II
200	4326.74	II
300	4329.62	II
80	4350.33	I
60	4402.54	I
400	4405.23	II
40	4431.89	I
60 h	4488.98	I
50 h	4493.64	I
40	4505.92	I
200	4509.63	II
60 h	4523.17	I
130	4524.93	II
65000	4554.03	II
40	4573.85	I
80	4579.64	I
30	4599.75	I
20 h	4619.92	I
25 h	4628.33	I
300	4644.10	II
30	4673.62	I
35	4691.62	I
20	4700.43	I
800	4708.94	II
40	4726.44	I
800	4843.46	II
300	4847.14	II
200	4850.84	II
30 h	4877.65	I
400	4899.97	II
15	4902.90	I
20000	4934.09	II
8	4947.35	I
1000	4957.15	II
300	4997.81	II
1000	5013.00	II
20 h	5159.94	I
20	5267.03	I
800	5361.35	II
1000	5391.60	II
200	5421.05	II
100	5424.55	I
200	5428.79	II
300	5480.30	II
200	5519.05	I
1000 r	5535.48	I
20 h	5620.40	I
10	5680.18	I
400	5777.62	I
800	5784.18	II
100	5800.23	I
20	5805.69	I
150	5826.28	I
2800	5853.68	II
15	5907.64	I
100	5971.70	I
800	5981.25	II
100	5997.09	I
300	5999.85	II
100	6019.47	I
200	6063.12	I

Intensity	Wavelength/Å	
300	6110.78	I
400	6135.83	II
20000	6141.72	II
150	6341.68	I
500	6378.91	II
10	6383.76	III
90	6450.85	I
150	6482.91	I
12000	6496.90	II
300	6498.76	I
150	6527.31	I
3000	6595.33	I
150	6654.10	I
1500	6675.27	I
1800	6693.84	I
1000	6769.62	II
600	6865.69	I
300 h	6867.85	I
1000	6874.09	II
6000	7059.94	I
2400 hs	7120.33	I
600	7195.24	I
600 hl	7228.84	I
3000	7280.30	I
1200	7392.41	I
300	7417.53	I
900 hl	7459.78	I
600	7488.08	I
450 hl	7636.90	I
600 hl	7642.91	I
1800	7672.09	I
1200	7780.48	I
180 h	7839.57	I
1500	7905.75	I
600	7911.34	I
900 h	8210.24	I
8	8308.69	III
1800 h	8559.97	I
100	8710.74	II
100	8737.71	II
300 h	8799.76	I
300	8860.98	I
450	8914.99	I
300	9219.69	I
300	9308.08	I
300 h	9324.58	I
1500	9370.06	I
300	9455.92	I
8	9521.76	III
450	9589.37	I
900	9608.88	I
300 h	9645.72	I
1500 hl	9830.37	I
900	10001.08	I
600	10032.10	I
1200 h	10233.23	I
300	10471.26	I
120 hl	10791.25	I
180 hl	11012.69	I
150 h	11114.42	I
240	11303.04	I

Intensity	Wavelength/Å	
120 h	11697.45	I
120	13207.30	I
120	13810.50	I
120	14077.90	I
120	15000.40	I
120	20712.00	I
150	25515.70	I
150	29223.90	I

Beryllium Be $Z = 4$

Intensity	Wavelength/Å	
	58.13	IV
	58.57	IV
	59.32	IV
	60.74	IV
	64.06	IV
	75.93	IV
1 h	76.10	III
2	76.48	III
3	78.53	III
4	78.66	III
1 h	78.92	III
5	81.89	III
10	82.38	III
	82.58	II
20	83.20	III
	83.66	II
30	84.76	III
50	88.31	III
	89.16	I
	89.80	II
	90.04	II
	90.21	I
	90.67	I
	91.06	II
	91.36	II
	91.74	II
	92.19	I
	92.61	II
	93.14	II
	93.42	II
	93.93	II
	94.78	II
	95.76	II
	96.29	I
	97.24	I
	97.44	I
	97.86	I
	97.97	I
	98.12	I
	98.37	I
	98.66	I
	98.94	I
	99.19	I
100	100.25	III
	100.86	I
	101.20	I
	102.13	I
	102.49	II
	104.40	II
	104.67	I
	105.80	I
	107.26	I

Intensity	Wavelength/Å	
1	107.38	I
3	509.99	III
2	549.31	III
6	582.08	III
4	661.32	III
8	675.59	III
	714.0	II
4	725.59	III
5	725.71	II
5	743.58	II
7	746.23	III
2	767.75	III
8	775.37	II
20	842.06	II
	865.3	II
2	925.25	II
10	943.56	II
10	973.27	II
	981.4	II
	1020.1	II
8	1026.93	II
5	1036.32	II
15	1048.23	II
1	1114.69	III
20	1143.03	II
	1155.9	II
60	1197.19	II
2	1213.12	III
1	1214.32	III
2	1362.25	III
1	1401.52	III
10	1421.26	III
5	1422.86	III
	1426.12	I
1	1435.17	III
2	1440.77	III
	1491.76	I
20	1512.30	II
60	1512.43	II
100	1661.49	I
2 h	1754.69	III
15	1776.12	II
20	1776.34	II
	1907.	I
	1909.0	II
	1912.	I
3	1917.03	III
	1919.	I
5	1929.67	I
10	1943.68	I
60 h	1954.97	III
	1956.	I
50	1964.59	I
5	1985.13	I
	1997.95	I
	1997.98	I
60	1998.01	I
	2033.25	I
	2033.28	I
	2033.38	I
50	2055.90	I

Intensity	Wavelength/Å	
100	2056.01	I
75 h	2076.94	III
60 h	2080.38	III
25	2118.56	III
15 h	2122.27	III
10	2125.57	I
20	2125.68	I
15 h	2127.20	III
5	2137.25	III
25	2145.	I
55	2174.99	I
55	2175.10	I
5	2191.57	III
	2273.5	II
	2324.6	II
	2337.0	I
950	2348.61	I
20	2350.66	I
60	2350.71	I
200	2350.83	I
2	2413.34	II
16	2413.46	II
20	2453.84	II
	2480.6	I
35	2494.54	I
35	2494.58	I
100	2494.73	I
16	2507.43	II
5	2617.99	II
20	2618.13	II
100	2650.45	I
60	2650.55	I
200	2650.62	I
60	2650.69	I
100	2650.76	I
5	2697.46	II
20	2697.58	II
20	2728.88	II
30	2738.05	I
	2764.2	II
20	2898.13	I
10	2898.19	I
20	2898.25	I
30	2986.06	I
10	2986.42	I
60	3019.33	I
30	3019.49	I
30	3019.53	I
20	3019.60	I
10	3046.52	II
30	3046.69	II
	3090.3	I
10	3110.81	I
10	3110.92	I
20	3110.99	I
	3120.	I
480	3130.42	II
320	3131.07	II
	3136.	I
	3150.	I
	3160.6	I
	3163.	I
	3168.	I
	3180.7	II
	3187.	I
20	3193.81	I
20	3197.10	II
30	3197.15	II
20	3208.60	I
	3220.	I
60	3229.63	I
2	3233.52	II
10	3241.62	II
30	3241.83	II
15	3269.02	I
100	3274.58	II
30	3274.67	II
30	3282.91	I
30	3321.01	I
30	3321.09	I
220	3321.34	I
20	3345.43	I
60	3367.63	I
	3405.6	II
5	3451.37	I
300	3455.18	I
20	3476.56	I
300	3515.54	I
10	3555.	I
100	3720.36	III
	3720.92	III
	3722.98	III
100	3736.30	I
700	3813.45	I
40	3865.13	I
80	3865.42	I
1	3865.51	I
6	3865.72	I
100	3866.03	I
90 h	4249.14	III
100	4253.05	I
60	4253.76	I
300	4360.66	II
500	4360.99	II
400	4407.94	I
2	4485.52	III
100 h	4487.30	III
1	4495.09	III
140 h	4497.8	III
	4526.6	I
	4548.	I
12	4572.66	I
700	4673.33	II
1000	4673.42	II
6	4709.37	I
200	4828.16	II
40	4849.16	I
2 h	4858.22	II
80	5087.75	I
8	5218.12	II
20	5218.33	II
3	5255.86	II
64	5270.28	II
500	5270.81	II
20	5403.04	II
20	5410.21	II
	5558.	I
140 h	6142.01	III
10	6229.11	I
16	6279.43	II
30	6279.73	II
30	6473.54	I
60	6547.89	II
60	6558.36	II
30	6564.52	I
2 h	6636.44	II
1	6756.72	II
2	6757.13	II
30	6786.56	I
1 h	6884.22	I
6 h	6884.44	I
100	6982.75	I
6 h	7154.40	I
40 h	7154.65	I
100	7209.13	I
3	7401.20	II
2	7401.43	II
10	7551.90	I
10 h	7618.68	I
20 h	7618.88	I
60	8090.06	I
5 h	8158.99	I
10 h	8159.24	I
4	8254.07	I
10 h	8287.07	I
30	8547.36	I
60	8547.67	I
300	8801.37	I
6	8882.18	I
40	9190.45	I
20 h	9243.92	I
1 h	9343.89	II
40	9392.74	I
2	9476.43	II
16	9477.03	II
20	9847.32	I
10 h	9895.63	I
20 h	9895.96	I
80	9939.78	I
16	10095.52	II
20	10095.73	II
60	10119.92	II
80	10331.03	I
30	11066.46	I
	11173.	II
1	11173.73	II
120	11496.39	I
2 h	11625.16	II
	11659.	II
2	11660.25	II
100	12095.36	II
30	12098.18	II
100	14643.92	I
60	14644.75	I
200	16157.72	I
80	17855.38	I
120	17856.63	I
100	18143.54	I
160	31775.05	I
200	31778.70	I

Bismuth Bi *Z* = 83

Intensity	Wavelength/Å	
6	420.7	IV
6	431.2	IV
2	488.39	V
3	563.62	V
5	670.76	III
6	686.88	V
5	730.71	V
10	738.17	V
4	775.16	III
6	790.5	IV
6	790.6	IV
8	792.5	IV
10	820.3	IV
9	822.9	IV
12	824.9	IV
15 d	864.45	V
15	872.6	IV
12	923.9	IV
15	943.3	IV
25	1039.99	III
50 h	1045.76	III
30	1051.81	III
15	1058.88	II
20	1085.47	II
10	1099.20	II
24	1103.4	IV
20	1139.01	III
50	1224.64	III
10	1225.43	II
15	1232.78	II
10	1241.05	II
10	1265.35	II
15	1283.73	II
10	1306.18	II
60	1317.0	IV
20	1325.46	II
40	1326.84	III
20	1329.47	II
60	1346.12	III
20	1350.07	II
25	1372.61	II
15	1376.02	II
20	1393.92	II
35	1423.33	III
35	1423.52	III
45	1436.83	II
25	1447.94	II
50	1455.11	II
60 h	1461.00	III
25	1462.14	II
35	1486.93	II
20	1502.50	II
40	1520.57	II

Intensity	Wavelength/Å	
40	1533.17	II
30	1536.77	II
35	1538.06	II
20	1563.67	II
40	1573.70	II
60	1591.79	II
25	1601.58	II
60 h	1606.40	III
40	1609.70	II
40	1611.38	II
20	1652.81	II
20	1749.29	II
80	1777.11	II
60	1787.47	II
70	1791.93	II
70	1823.80	II
100	1902.41	II
9000	1954.53	I
7000	1960.13	I
25	1989.35	II
7000	2021.21	I
9000	2061.70	I
45 h	2068.9	II
4600	2110.26	I
2500	2133.63	I
15	2143.40	II
15	2143.46	II
60	2186.9	II
40 h	2214.0	II
360	2228.25	I
1700	2230.61	I
340	2276.58	I
100	2311.	IV
100	2326.	IV
16	2368.12	II
12	2368.25	II
100	2376.	IV
190	2400.88	I
75 h	2414.6	III
10	2501.0	II
25	2515.69	I
70	2524.49	I
20 h	2544.5	II
700	2627.91	I
100	2629.	IV
100	2677.	IV
12	2693.0	II
280 c	2696.76	I
20	2713.3	II
140 d	2730.50	I
100	2767.	IV
100	2772.	IV
360	2780.52	I
100	2786.	IV
15	2803.42	II
11	2803.70	II
12	2805.3	II
140 c	2809.62	I
100	2842.	IV
80 h	2855.6	III
4000	2897.98	I

Intensity	Wavelength/Å	
100	2924.	IV
100	2933.	IV
100	2936.	IV
15	2936.7	II
3200	2938.30	I
20	2950.4	II
12	2963.4	II
2800	2989.03	I
700	2993.34	I
100	3012.	IV
2400	3024.64	I
60	3034.87	I
100	3042.	IV
9000 c	3067.72	I
140	3076.66	I
35	3115.0	III
100	3239.	IV
550 c	3397.21	I
10	3430.83	II
12	3431.23	II
40 h	3451.0	III
40	3473.8	III
35	3485.5	III
500 c	3510.85	I
380 c	3596.11	I
45	3613.4	III
100	3643.	IV
12	3654.2	II
100	3682.	IV
50	3695.32	III
50	3695.68	III
100	3734.	IV
70 h	3792.5	II
12	3811.1	II
20	3815.8	II
10	3845.8	II
30	3863.9	II
100	3868.	IV
40 h	4079.1	II
10	4097.2	II
140	4121.53	I
140	4121.86	I
75 h	4259.4	II
25	4272.0	II
70 h	4301.7	II
12 h	4339.8	II
25 h	4340.5	II
12 h	4379.4	II
25 h	4476.8	II
60 h	4705.3	II
600 c	4722.52	I
30	4730.3	II
20	4749.7	II
40 h	4797.4	III
12	4908.2	II
10	4916.6	II
12	4969.7	II
20	4993.6	II
45 h	5079.3	III
10	5091.6	II
50 h	5124.3	II

Intensity	Wavelength/Å	
60 h	5144.3	II
20	5201.5	II
75 h	5209.2	II
40 h	5270.3	II
10	5397.8	II
10 c	5552.35	I
3	5599.41	I
20	5655.2	II
40 h	5719.2	II
6	5742.55	I
12	5818.3	II
20	5860.2	II
20	5973.0	II
15	6059.1	II
15	6128.0	II
6	6134.82	I
3	6475.73	I
3	6476.24	I
15	6497.7	II
10	6577.2	II
40 h	6600.2	II
50 h	6808.6	II
4 h	6991.12	I
12	7033.	II
2	7036.15	I
10 h	7381.	II
2	7502.33	I
10 h	7637.	II
10	7750.	II
3	7838.70	I
2	7840.33	I
20	7965.	II
40	8008.	III
12 h	8050.	II
50	8070.	III
15	8328.	II
15	8388.	II
30	8532.	II
2	8544.54	I
1	8579.74	I
25	8653.	II
2	8754.88	I
3	8761.54	I
25	8863.	II
2	8907.81	I
2000 d	9657.04	I
40	9827.78	I
20	10104.5	I
15	10138.8	I
20	10300.6	I
20	10536.19	I
50	11072.44	I
1500 d	11710.37	I
40	11999.49	I
200	12165.08	I
200	12690.04	I
100	12817.8	I
200	14330.5	I
50	16001.5	I
60	22551.6	I

Intensity	Wavelength/Å	
Boron B Z = 5		
	41.00	V
30	48.59	V
10	52.68	IV
30	60.31	IV
	194.37	V
	262.37	V
160	344.0	IV
450	385.0	IV
40	411.80	III
285	418.7	IV
20	510.77	III
40	510.85	III
	512.53	V
150	518.24	III
75	518.27	III
110	677.00	III
160	677.14	III
40	693.95	II
40	731.36	II
40	731.44	II
	749.74	V
40	758.48	III
70	758.67	III
110	882.54	II
110	882.68	II
40	984.67	II
110	1081.88	II
110	1082.07	II
70	1112.2	IV
450	1168.9	IV
70	1170.9	IV
110	1230.16	II
220	1362.46	II
70	1600.46	I
120	1600.73	I
160	1623.58	II
110	1623.77	II
220	1624.02	II
70	1624.16	II
160	1624.34	II
100	1663.04	I
150	1666.87	I
200	1667.29	I
150	1817.86	I
200	1818.37	I
300	1825.91	I
300	1826.41	I
110	1842.81	II
20	1953.83	III
550	2065.78	III
250	2066.38	I
250	2066.65	I
100	2066.93	I
300	2067.19	I
450	2067.23	III
160	2077.09	III
500	2088.91	I
500	2089.57	I
70	2220.30	II
40	2234.09	III

Intensity	Wavelength/Å	
70	2234.59	III
40	2323.03	II
40	2328.67	II
40	2393.20	II
220	2395.05	II
40	2459.69	II
40	2459.90	II
1000	2496.77	I
1000	2497.73	I
70	2524.7	IV
160	2530.3	IV
450	2821.68	IV
70	2824.57	IV
285	2825.85	IV
160	2918.08	II
110	3032.26	II
70	3179.33	II
110	3323.18	II
110	3323.60	II
450	3451.29	II
285	4121.93	II
110	4194.79	II
40	4242.98	III
70	4243.61	III
110	4472.10	II
110	4472.85	II
220	4487.05	III
360	4497.73	III
70	4784.21	II
110	4940.38	II
110	6080.44	II
70	6285.47	II
70	7030.20	II
40	7031.90	II
110	7835.25	III
70	7841.41	III
20	8667.22	I
70	8668.57	I
800	11660.04	I
570	11662.47	I
125	15629.08	I
200	16240.38	I
250	16244.67	I
235	18994.33	I

***Bromine* Br *Z* = 35**

Intensity	Wavelength/Å	
700	379.73	IV
700	400.37	IV
800	482.11	V
900	531.97	V
1000	545.43	IV
1000	547.90	V
1000	559.76	IV
1000	569.19	IV
1000	576.59	IV
1000	585.10	IV
1000	586.71	IV
1000	597.51	IV
1000	600.09	IV
1000	601.27	IV
1000	607.03	IV
1000	617.85	IV
1000	619.87	IV
1000	630.14	IV
1000	642.23	IV
1000	661.53	IV
1000	683.51	IV
1000	697.72	IV
1000	715.39	IV
1000	731.00	IV
1000	800.12	IV
700	812.95	V
1000	813.66	IV
1000	850.81	V
1000	889.23	II
1000	948.97	II
1000	1015.54	II
1000	1049.00	II
1000	1069.15	V
900	1112.13	V
1000	1143.56	V
1000	1189.28	I
1000	1189.50	I
1000	1210.73	I
1000	1221.13	I
1000	1223.24	I
1200	1224.41	I
1200	1226.90	I
7500	1232.43	I
1200	1243.90	I
1500	1251.66	I
1000	1255.80	I
1500	1259.20	I
1200	1261.66	I
1200	1266.20	I
1000	1279.48	I
1000	1286.26	I
3000	1309.91	I
3000	1316.74	I
1000	1317.37	I
2000	1317.70	I
12000	1384.60	I
3000	1449.90	I
50000	1488.45	I
30000	1531.74	I
25000	1540.65	I
30000	1574.84	I
20000	1576.39	I
25000	1582.31	I
75000	1633.40	I
1000	2133.79	IV
1000	2145.02	IV
1000	2257.21	IV
1000	2272.73	IV
1000	2307.40	IV
1000	2408.16	IV
1000	2411.58	IV
700	2491.14	IV
1000	2581.19	IV
600	2661.40	IV
1000	2842.88	IV
1100 h	2907.71	IV
500 h	2972.26	II
500	3041.18	IV
500	3074.42	III
500	3349.64	III
500	3380.56	IV
500	3540.16	III
500	3562.43	III
1200	3815.65	I
1500	3992.36	I
1000	4223.89	II
2000	4365.14	I
1000	4365.60	II
1500	4425.14	I
10000	4441.74	I
10000	4472.61	I
20000	4477.72	I
1000	4490.42	I
3000	4513.44	I
15000	4525.59	I
3000	4575.74	I
2500	4614.58	I
2500	4752.28	I
4000	4780.31	I
1600	4785.19	I
4000	4979.76	I
1200	5395.48	I
1200	5466.22	I
1800	5852.08	I
1600	5940.48	I
2400	6122.14	I
40000	6148.60	I
2000	6177.39	I
1500	6335.48	I
60000	6350.73	I
2500	6410.32	I
1800	6483.56	I
1000	6514.62	I
20000	6544.57	I
1500	6548.09	I
50000 c	6559.80	I
1000	6571.31	I
1800	6579.14	I
20000	6582.17	I
1500	6620.47	I
50000 c	6631.62	I
20000	6682.28	I
10000	6692.13	I
8000	6728.28	I
2000	6760.06	I
2000	6779.48	I
2200	6786.74	I
6500	6790.04	I
1600 c	6791.48	I
1800	6861.15	I
10000	7005.19	I
2000	7260.45	I
10000	7348.51	I
40000	7512.96	I
1600	7591.61	I
1800	7595.07	I
2000	7616.41	I
30000	7803.02	I
1200	7827.23	I
2500 s	7881.45	I
2500	7881.57	I
2500	7925.81	I
30000 c	7938.68	I
3000	7947.94	I
3000	7950.18	I
8000	7978.44	I
10000	7978.57	I
30000	7989.94	I
2000	8026.35	I
2500	8026.54	I
30000	8131.52	I
1000 c	8152.65	I
10000	8153.75	I
25000	8154.00	I
5000	8246.86	I
15000	8264.96	I
75000 c	8272.44	I
20000	8334.70	I
10000	8343.70	I
1200	8384.04	I
40000	8446.55	I
4000	8477.45	I
1500	8513.38	I
1000	8557.73	I
1000	8566.28	I
20000	8638.66	I
4000	8698.53	I
10000 c	8793.47	I
15000	8819.96	I
25000	8825.22	I
4000	8888.98	I
30000	8897.62	I
6000	8932.40	I
1800	8949.39	I
9000	8964.00	I
30000	9166.06	I
15000	9173.63	I
20000	9178.16	I
40000	9265.42	I
15000	9320.86	I
6000	9793.48	I
10000	9896.40	I
3000	10140.08	I
6000	10237.74	I
1000	10299.62	I
1500	10377.65	I
30000	10457.96	I
1000	10742.14	I
3000	10755.92	I
1700	13217.17	I
1800	14354.57	I
1250	14888.70	I
1800	16731.19	I
1200	18568.31	I
3500	19733.62	I
1000	20281.73	I
1000	20624.67	I
1200	21787.24	I
4000	22865.65	I

Intensity	Wavelength/Å	
1000	23513.15	I
500	28346.50	I
500	30380.85	I
600	31630.13	I
150	38345.75	I
120	39964.36	I
Cadmium Cd Z = 48		
50	427.01	IV
50	447.85	IV
60	480.90	IV
70	493.00	IV
70	495.13	IV
70	498.14	IV
70	498.53	IV
80	504.09	IV
70	504.20	IV
70	504.50	IV
80	506.31	IV
60	508.01	IV
50	508.95	IV
70	509.55	IV
70	511.40	IV
80	513.00	IV
70	514.50	IV
60	519.42	IV
80	524.41	IV
70	524.47	IV
70	525.10	IV
60	525.19	IV
70	527.07	IV
80	531.09	IV
80	531.51	IV
70	534.29	IV
70	536.77	IV
60	540.90	IV
70	541.74	IV
80	542.60	IV
80	546.55	IV
60	553.06	IV
80	554.05	IV
60	567.01	IV
150	1118.16	IV
100	1164.65	IV
100	1183.40	IV
100	1256.00	II
150	1296.43	II
100	1326.50	II
60	1370.48	IV
150	1370.91	II
60	1418.89	IV
200	1514.26	II
50	1545.17	III
200	1571.58	II
100	1668.60	II
50	1702.47	II
40	1707.16	III
40	1722.95	III
50	1724.41	II
40	1747.67	III
40	1773.06	III
100	1785.84	II
75	1793.40	III
40	1823.41	III
100	1827.70	II
50	1844.66	III
40	1851.13	III
40	1855.85	III
200	1856.67	III
150	1874.08	III
300	1922.23	II
100	1943.54	II
40	1965.54	II
30	1986.89	II
200	1995.43	II
100	2007.49	II
50	2032.45	II
75	2036.23	II
40	2039.83	III
50	2045.61	III
75	2087.91	III
150	2096.00	II
50	2111.60	III
1000 r	2144.41	II
50	2155.06	II
100	2187.79	II
1000	2194.56	II
1000	2265.02	II
1500 r	2288.022	I
1000	2312.77	II
200	2321.07	II
40	2376.82	II
50	2418.69	II
50	2469.73	II
40	2487.93	II
40	2495.58	II
50	2509.11	II
30	2516.22	II
25 h	2525.196	I
50	2544.613	I
50	2551.98	II
25	2553.465	I
3	2565.789	I
500	2572.93	II
50	2580.106	I
30	2592.026	I
25 h	2602.048	I
50	2628.979	I
40	2632.190	I
75	2639.420	I
40	2659.23	II
50 h	2660.325	I
25	2668.20	II
50	2672.62	II
100	2677.540	I
25	2677.748	I
50	2707.00	II
75	2712.505	I
50	2733.820	I
1000	2748.54	II
100 h	2763.894	I
50 h	2764.230	I
50	2774.958	I
30	2823.19	II
200	2836.900	I
25	2856.46	II
100	2868.180	I
200 r	2880.767	I
50 r	2881.224	I
200	2914.67	II
50	2927.87	II
200	2929.27	II
1000 r	2980.620	I
200 r	2981.362	I
50	2981.845	I
50	3030.60	II
150	3080.822	I
25	3081.48	II
30	3082.593	I
100	3092.34	II
200	3133.167	I
50	3146.79	II
150	3250.33	II
300	3252.524	I
300	3261.055	I
50	3343.21	II
50	3385.49	II
30	3388.88	II
800	3403.652	I
50	3417.49	II
50	3442.42	II
100	3464.43	II
1000	3466.200	I
800	3467.655	I
25	3483.08	II
150	3495.44	II
25	3499.952	I
100	3524.11	II
100	3535.69	II
1000	3610.508	I
800	3612.873	I
60	3614.453	I
20	3649.558	I
10	3981.926	I
100	4029.12	II
200	4134.77	II
50	4141.49	II
100	4285.08	II
8	4306.672	I
100	4412.41	II
3	4412.989	I
1000	4415.63	II
30	4440.45	II
8	4662.352	I
200	4678.149	I
30	4744.69	II
300	4799.912	I
50	4881.72	II
50	5025.50	II
1000 h	5085.822	I
6	5154.660	I
100	5268.01	II
100	5271.60	II
1000	5337.48	II
1000	5378.13	II
200	5381.89	II
40	5843.30	II
50	5880.22	II
300	6099.142	I
100	6111.49	I
100	6325.166	I
30	6330.013	I
400	6354.72	II
500	6359.98	II
2000	6438.470	I
400	6464.94	II
25	6567.65	II
500	6725.78	II
100	6759.19	II
30	6778.116	I
50	7237.01	II
100	7284.38	II
1000	7345.670	I
50	8066.99	II
5	8200.309	I
20	9289.	I
15	11652.	I
35	14487.	I
80	15708.	I
55 d	19120.	I
25	24371.	I
35	25448.	I
Calcium Ca Z = 20		
250	190.46	V
250	196.97	V
300	199.55	V
250	200.51	V
265	257.98	V
400	267.77	V
300	270.31	V
400	280.99	V
300	284.98	V
450 c	286.96	V
500	322.17	V
800	323.22	V
300	330.94	V
300	334.55	V
250 c	342.45	IV
250	343.93	IV
450	352.92	V
250	377.18	V
200	387.08	V
750	425.00	V
600	434.57	IV
250	437.77	IV
750	443.82	IV
500	450.57	IV
500	558.60	V
400	637.93	V
300	643.12	V
400	646.57	V
750	656.00	IV
300	656.76	V
500	669.70	IV
24	1341.89	II

Intensity	Wavelength/Å	
12	1342.54	II
20	1433.75	II
20	1545.29	III
60	1649.86	II
20	1807.34	II
40	1814.50	II
40	1838.01	II
60	1840.06	II
20	1843.09	II
40	1850.69	II
17	2123.03	III
16	2152.43	III
16	2687.76	III
19	2881.78	III
21	2899.79	III
19	2924.33	III
20	2988.63	III
10	3006.86	I
15	3028.59	III
3	3055.32	I
19	3119.67	III
170	3158.87	II
180	3179.33	II
150	3181.28	II
20	3316.51	II
12	3361.92	I
19	3372.67	III
20	3461.87	II
13	3487.60	I
18	3537.77	III
20	3644.41	I
30	3683.70	II
40	3694.11	II
170	3706.03	II
180	3736.90	II
20	3755.67	II
30	3758.39	II
230	3933.66	II
220	3968.47	II
50	4097.10	II
60	4109.82	II
30	4110.28	II
40	4206.18	II
50	4220.07	II
50	4226.73	I
24	4283.01	I
22	4289.36	I
22	4298.99	I
25	4302.53	I
20	4302.81	III
23	4307.74	I
22	4318.65	I
20	4355.08	I
19	4399.59	III
25	4425.44	I
26	4434.96	I
25	4435.69	I
30	4454.78	I
28	4455.89	I
20	4456.61	I
20	4472.04	II

Intensity	Wavelength/Å	
20	4489.18	II
19	4499.88	III
23	4526.94	I
22	4578.55	I
23	4581.40	I
23	4581.47	I
24	4585.87	I
24	4585.96	I
20	4685.27	I
30	4716.74	II
40	4721.03	II
40	4799.97	II
25	4878.13	I
70	5001.48	II
80	5019.97	II
40	5021.14	II
23	5041.62	I
25	5188.85	I
22	5261.71	I
23	5262.24	I
22	5264.24	I
24	5265.56	I
25	5270.27	I
60	5285.27	II
70	5307.22	II
50	5339.19	II
27	5349.47	I
23	5512.98	I
25	5581.97	I
27	5588.76	I
24	5590.12	I
26	5594.47	I
25	5598.49	I
24	5601.29	I
24	5602.85	I
30	5857.45	I
27	6102.72	I
29	6122.22	I
22	6161.29	I
30	6162.17	I
22	6163.76	I
24	6166.44	I
26	6169.06	I
28	6169.56	I
35	6439.07	I
30	6449.81	I
22	6455.60	I
80	6456.87	II
34	6462.57	I
29	6471.66	I
32	6493.78	I
28	6499.65	I
23	6572.78	I
30	6717.69	I
33	7148.15	I
31	7202.19	I
33	7326.15	I
30	7575.81	II
60	7581.11	II
80	7601.30	II
20	7602.32	II

Intensity	Wavelength/Å	
40	7820.78	II
60	7843.38	II
20	8017.50	II
20	8020.50	II
70	8133.05	II
100	8201.72	II
110	8248.80	II
70	8254.73	II
130	8498.02	II
170	8542.09	II
160	8662.14	II
100	8912.07	II
110	8927.36	II
110	9213.90	II
90	9312.00	II
100	9319.56	II
110	9320.65	II
25	9416.97	I
100	9567.97	II
110	9599.24	II
80	9601.82	II
80	9854.74	II
110	9890.63	II
90	9931.39	II
100	10223.04	II
20	10343.81	I
20	11838.99	II
25	12816.04	I
24	12823.86	I
25	12909.10	I
30	13033.57	I
21	13086.44	I
24	13134.95	I
20	16150.77	I
22	16157.36	I
21	16197.04	I
20	18925.47	I
24	18970.14	I
30	19046.14	I
48	19309.20	I
49	19452.99	I
47	19505.72	I
50	19776.79	I
35	19853.10	I
34	19862.22	I
23	19917.19	I
24	19933.70	I
25	22624.93	I
30	22651.23	I

Carbon C $Z = 6$

Intensity	Wavelength/Å	
110	34.973	V
450	40.268	V
110	227.19	V
250	244.91	IV
160	248.66	V
160	248.74	V
200	289.14	IV
250	289.23	IV
570	312.42	IV
500	312.46	IV
250	371.69	III
250	371.75	III
150	371.78	III
650	384.03	IV
700	384.18	IV
500	386.203	III
400	419.52	IV
500	419.71	IV
200	450.734	III
400	459.46	III
500	459.52	III
570	459.63	III
250	511.522	III
250	535.288	III
300	538.080	III
350	538.149	III
400	538.312	III
350	574.281	III
9	595.022	II
30	687.053	II
50	687.345	II
10	858.092	II
20	858.559	II
30	903.624	II
60	903.962	II
150	904.142	II
30	904.480	II
800	977.03	III
9	1009.86	II
10	1010.08	II
10	1010.37	II
80	1036.337	II
150	1037.018	II
150	1157.910	I
150	1158.019	I
150	1158.035	I
370	1174.93	III
350	1175.26	III
330	1175.59	III
500	1175.71	III
350	1175.99	III
370	1176.37	III
150	1188.992	I
150	1189.447	I
200	1189.631	I
300	1193.009	I
300	1193.031	I
300	1193.240	I
300	1193.264	I
100	1193.393	I
150	1193.649	I
150	1193.679	I
100	1194.064	I
100	1194.488	I
100	1261.552	I
250	1277.245	I
250	1277.282	I
300	1277.513	I
300	1277.550	I
200	1280.333	I
100	1311.363	I
9	1323.951	II

Intensity	Wavelength/Å	
120	1329.578	I
120	1329.600	I
150	1334.532	II
300	1335.708	II
100	1354.288	I
150	1355.84	I
120	1364.164	I
100	1459.032	I
200	1463.336	I
120	1467.402	I
150	1481.764	I
1000	1548.202	IV
900	1550.774	IV
150	1560.310	I
400	1560.683	I
400	1560.708	I
100	1561.341	I
400	1561.438	I
150	1656.266	I
120	1656.928	I
300	1657.008	I
120	1657.380	I
120	1657.907	I
150	1658.122	I
500	1751.823	I
1000	1930.905	I
250	2162.94	III
40	2270.91	V
5	2277.25	V
20	2277.92	V
800	2296.87	III
800	2478.56	I
250	2509.12	II
350	2512.06	II
200 l	2524.41	IV
300 s	2529.98	IV
250 h	2574.83	II
150	2697.75	III
110 l	2724.85	III
150 l	2725.30	III
150 l	2725.90	III
350 l	2741.28	II
250	2746.49	II
1000	2836.71	II
800	2837.60	II
200	2982.11	III
800 h	2992.62	II
350	3876.19	II
350	3876.41	II
350	3876.66	II
570	3918.98	II
800	3920.69	II
150	4056.06	III
200	4067.94	III
250	4068.91	III
250	4070.26	III
250	4074.52	II
350 l	4075.85	II
150	4162.86	III
250 h	4186.90	III
800	4267.00	II
1000	4267.26	II
200	4325.56	III
600	4647.42	III
520	4650.25	III
375	4651.47	III
200 w	4658.30	IV
200	4665.86	III
200	4771.75	I
200	4932.05	I
5	4943.88	V
5	4944.56	V
200	5052.17	I
350	5132.94	II
350	5133.28	II
350	5143.49	II
570	5145.16	II
400	5151.09	II
300	5380.34	I
250	5648.07	II
350	5662.47	II
450	5695.92	III
250	5801.33	IV
200	5811.98	IV
150	5826.42	III
570	5889.77	II
350	5891.59	II
200	6001.13	I
250	6006.03	I
110	6007.18	I
150	6010.68	I
300	6013.22	I
250	6014.84	I
800	6578.05	II
570	6582.88	II
200	6587.61	I
150	6744.38	III
250	6783.90	II
150 h	7037.25	III
250	7113.18	I
250	7115.19	I
250	7115.63	II
200	7116.99	I
350	7119.90	II
800	7231.32	II
1000	7236.42	II
150	7612.65	III
90 w	7726.2	IV
200	7860.89	I
200	8058.62	I
300 h	8196.48	III
150	8332.99	III
520	8335.15	I
300	8500.32	III
250	9061.43	I
200	9062.47	I
200	9078.28	I
250	9088.51	I
450	9094.83	I
300	9111.80	I
800	9405.73	I
150	9603.03	I
250	9620.80	I
300	9658.44	I
200	10683.08	I
300	10691.25	I
12	11619.29	I
23	11628.83	I
13	11658.85	I
47	11659.68	I
24	11669.63	I
85	11748.22	I
142	11753.32	I
114	11754.76	I
11	11777.54	I
17	11892.91	I
30	11895.75	I
26	12614.10	I
20	13502.27	I
38	14399.65	I
16	14403.25	I
61	14420.12	I
12	14429.03	I
13	14442.24	I
12	16559.66	I
50	16890.38	I
10	17338.56	I
11	17448.60	I
13	18139.80	I
25	19721.99	I

Cerium Ce Z = 58

Intensity	Wavelength/Å	
300	399.36	V
200	482.96	V
40	741.79	IV
30	754.60	IV
75	1332.16	IV
75	1372.72	IV
100	2000.42	IV
100	2009.94	IV
10000	2318.64	III
10000	2372.34	III
10000	2380.12	III
10000	2431.45	III
15000	2439.80	III
10000	2454.32	III
10000	2469.95	III
10000	2483.82	III
10000	2497.50	III
20000	2531.99	III
10000	2603.59	III
340	2651.01	II
270	2830.90	II
250	2874.14	II
10000	2923.81	III
10000	2931.54	III
400	2976.91	II
10000	3022.75	III
50000	3031.58	III
95000	3055.59	III
20000	3056.56	III
40000	3057.23	III
20000	3057.58	III
680	3063.01	II
40000	3085.10	III
20000	3106.98	III
30000	3110.53	III
30000	3121.56	III
20000	3141.29	III
20000	3143.96	III
20000	3147.06	III
710	3194.83	II
990	3201.71	II
710	3218.94	II
880	3221.17	II
710	3227.11	II
20000	3228.57	III
710	3234.16	II
990	3272.25	II
20000	3353.29	III
10000	3395.77	III
30000	3427.36	III
40000	3443.63	III
30000	3454.39	III
40000	3459.39	III
60000	3470.92	III
710	3485.05	II
50000	3497.81	III
60000	3504.64	III
770	3539.08	II
50000	3544.07	III
1200	3560.80	II
1000	3577.45	II
1800	3655.85	II
880	3660.64	II
880	3667.98	II
1000	3709.29	II
1000	3709.93	II
1400	3716.37	II
800	3728.42	II
860	3786.63	II
2500	3801.52	II
800	3803.09	II
1000	3808.11	II
1100	3838.54	II
860	3848.59	II
860	3853.15	II
1200	3854.18	II
1200	3854.31	II
1100	3878.36	II
1500	3882.45	II
1000	3889.98	II
770	3907.29	II
980	3912.44	II
770	3918.28	II
770	3931.09	II
770	3940.34	II
2000	3942.15	II
2700	3942.75	II
770	3943.89	II
3100	3952.54	II
980	3956.28	II
770	3960.91	II
770	3967.05	II
770	3978.65	II

Intensity	Wavelength/Å	
770	3984.68	II
700	3992.39	II
910	3993.82	II
2800	3999.24	II
910	4003.77	II
2700	4012.39	II
910	4014.90	II
840	4024.49	II
840	4028.41	II
840	4031.34	II
2100	4040.76	II
910	4042.58	II
700	4053.51	II
1100	4071.81	II
1800	4073.48	II
1500	4075.71	II
1500	4075.85	II
910	4083.23	II
770	4118.14	II
980	4123.87	II
980	4127.37	II
2700	4133.80	II
2000	4137.65	II
770	4142.40	II
980	4149.94	II
1400	4151.97	II
1300	4165.61	II
3500	4186.60	II
840	4198.72	II
910	4202.94	II
1500	4222.60	II
770	4227.75	II
980	4239.92	II
1100	4248.68	II
2000	4289.94	II
1500	4296.67	II
770	4300.33	II
770	4306.72	II
980	4337.77	II
700	4349.79	II
910	4364.66	II
910	4382.17	II
700	4386.84	II
1700	4391.66	II
980	4418.78	II
770	4449.34	II
2400	4460.21	II
1400	4471.24	II
700	4479.36	II
700	4483.90	II
840	4486.91	II
770	4523.08	II
840	4527.35	II
840	4528.47	II
840	4539.75	II
2100	4562.36	II
1100	4572.28	II
840	4593.93	II
1700	4628.16	II
310	4737.28	II
470	5079.68	II
280	5159.69	I
280	5161.48	I
370	5187.46	II
260	5223.46	I
260	5245.92	I
340	5274.23	II
450	5353.53	II
300	5393.40	II
280	5409.23	II
260	5512.08	II
300	5696.99	I
370	5699.23	I
240	5719.03	I
230	5940.86	I
55	6001.90	I
55	6005.86	I
55	6006.82	I
75	6013.42	I
110	6024.20	I
10000	6032.54	III
110	6043.39	II
55	6047.40	I
10000	6060.91	III
45	6098.34	II
45	6123.67	I
35	6143.36	II
35	6186.17	I
35	6208.98	I
35	6228.94	I
23	6232.45	II
28	6237.45	I
45	6272.05	II
35	6295.58	I
28	6299.51	II
23	6300.21	I
35	6310.01	I
35	6343.95	II
35	6371.11	II
28	6386.84	I
23	6393.02	II
35	6430.07	I
23	6436.40	I
35	6458.03	I
28	6467.39	I
35	6473.72	I
23	6513.59	II
45	6555.65	I
23	6579.10	I
22	6612.06	I
30	6628.93	I
22	6652.72	II
26	6700.66	I
35	6704.27	I
30	6774.28	II
35	6775.59	I
30	6924.81	I
30	6986.02	I
35	7061.75	II
35	7086.35	II
22	7238.36	II
25	7252.75	I
25	7329.91	I
25	7397.77	I
25	7616.11	II
25	7689.17	II
22	7844.94	II
22	7857.54	II
30	8025.56	II
25	8772.14	II
30	8891.20	II

Cesium Ce Z = 55

Intensity	Wavelength/Å	
10000	614.01	III
2000	638.17	III
2500	666.25	III
5000	691.60	III
3500	703.89	III
15000	718.14	II
20000	721.79	III
20000	722.20	III
5000	731.56	III
12000	740.29	III
15000	808.76	II
15000	813.84	II
7500	830.39	III
35000	901.27	II
15000	920.35	III
40000	926.66	II
25000 c	1054.79	III
17 c	1673.99	III
12	1705.25	III
10	1801.83	III
20 c	1822.40	III
11	1823.93	III
12	1824.70	III
12	1841.80	III
25	1915.50	III
25 c	1923.29	III
12	1961.33	III
17	1996.56	III
710	2035.11	III
120	2056.43	III
330	2076.43	III
540	2077.30	III
410	2088.68	III
210	2101.63	III
200	2141.47	III
1000	2316.88	III
230	2325.95	III
390	2340.49	III
1600	2455.81	III
1600	2477.57	III
890	2485.45	III
410	2495.07	III
1400	2525.67	III
430	2573.05	III
16000	2596.86	III
390	2610.12	III
6200	2630.51	III
370	2700.32	III
710	2701.20	III
390	2776.44	III
270	2810.87	III
630	2845.70	III
3100	2859.32	III
200	2893.85	III
180	2921.13	III
3200	2976.86	III
210	3001.28	III
1700	3066.59	III
1100 c	3149.36	III
1400	3152.36	III
8400	3268.32	III
1300	3315.51	III
550	3340.60	III
430	3344.02	III
1200	3349.46	III
400	3463.45	III
580	3476.83	III
480	3559.82	III
7200	3597.45	III
1300	3608.31	III
2300	3618.19	III
300 c	3641.34	III
520	3651.08	III
4800	3661.40	III
640	3699.50	III
430	3837.46	III
2100 c	3876.15	I
2900	3888.37	III
600 c	3888.61	I
2700	3925.60	III
680 c	4001.70	III
3100	4006.55	III
420	4006.78	III
520	4043.42	III
14000	4264.70	II
18000 w	4277.13	II
370	4403.86	III
1200	4410.22	III
940	4425.68	III
530	4471.48	III
12000	4501.55	II
1200	4506.72	III
590	4522.86	III
20000	4526.74	II
1000 c	4555.28	I
460 c	4593.17	I
99900	4603.79	II
420 h	4620.61	III
210	4665.52	III
25000	4830.19	II
140	4851.59	III
19000	4870.04	II
37000	4952.85	II
370	5035.72	III
27000	5043.80	II
75000	5227.04	II
29000	5249.38	II
11000	5274.05	II
10000 c	5349.13	II
22000	5370.99	II
230	5380.79	III
60 c	5465.94	I

Intensity	Wavelength/Å	
37	5502.88	I
39000	5563.02	II
100	5635.21	I
210 c	5664.02	I
27	5745.72	I
24000	5831.14	II
59 c	5838.83	I
300	5845.14	I
51000	5925.63	II
140	5950.14	III
110	5979.97	III
640 c	6010.49	I
86	6034.09	I
150	6043.99	III
870	6079.86	III
9800	6128.61	II
330	6150.42	III
1000	6213.10	I
170	6217.60	I
450	6242.96	III
320 c	6354.55	I
510	6456.33	III
8300	6495.53	II
10000 w	6536.44	II
490	6586.51	I
97	6628.66	I
8800	6646.57	II
3300 c	6723.28	I
9600	6724.47	II
400	6753.12	III
200	6824.65	I
300	6870.45	I
37000	6955.50	II
4800	6973.30	I
16000	6979.67	II
980	6983.49	I
13000 w	7149.54	II
1900 c	7219.60	III
790	7228.53	I
130	7279.90	I
1100	7279.96	I
2600 c	7608.90	I
3300	7943.88	I
22000	7997.44	II
3500	8015.73	I
510	8078.94	I
4500	8079.04	I
59000 c	8521.13	I
15000 c	8761.41	I
61000 c	8943.47	I
18000	9172.32	I
5200	9208.53	I
19000	10024.36	I
4800	10123.41	I
26000	10123.60	I
2900	13424.31	I
38000 c	13588.29	I
8400	13602.56	I
5700	13758.81	I
55000 c	14694.91	I
820	16535.63	I

Intensity	Wavelength/Å	
1500	17012.32	I
760	20138.47	I
880	22811.86	I
1100	23037.98	I
3900	23344.47	I
4400	24251.21	I
850	24374.96	I
890 d	25763.51	I
500	25764.73	I
680 c	29310.06	I
2800	30103.27	I
610 c	30953.06	I
1100	34900.13	I
190	36131.00	I
2 c	39177.28	I
2 d	39421.25	I
1	39424.11	I

Chlorine Cl Z = 17

Intensity	Wavelength/Å	
500	392.43	V
800	486.17	IV
800	534.73	IV
700	535.67	IV
600	536.15	IV
900	537.61	IV
500	538.03	V
600	538.12	IV
800	542.23	V
600	542.30	V
1000	545.11	V
600	546.33	V
1000	547.63	V
500	549.22	IV
700	552.02	IV
600	553.30	IV
700	554.62	IV
600	556.23	III
700	556.61	III
700	557.12	III
350	559.305	II
700	561.53	III
700	561.68	III
700	561.74	III
400	571.904	II
800	574.406	II
500	601.50	IV
500	604.59	IV
500	606.35	III
700	618.057	II
600	619.982	II
800	620.298	II
700	626.735	II
800	635.881	II
1000	636.626	II
1000	650.894	II
1000	659.811	II
1300	661.841	II
2000	663.074	II
1500	682.053	II
1500	687.656	II
1500	693.594	II
2000	725.271	II
2500	728.951	II
2000	777.562	II
5000	787.580	II
5000	788.740	II
5000	793.342	II
500	834.84	IV
500	834.97	IV
6000	839.297	II
8000	839.599	II
600	840.93	IV
5000	851.691	II
2000	888.026	II
2000	893.549	II
2000	961.499	II
500	973.21	IV
600	977.56	IV
40	978.284	I
700	984.95	IV
25	998.372	I
25	998.432	I
75	1002.346	I
500	1005.28	III
600	1008.78	III
150	1013.664	I
700	1015.02	III
90	1025.553	I
6000	1063.831	II
3000	1067.945	II
9000	1071.036	II
6000	1071.767	II
5000	1075.230	II
5000	1079.080	II
200	1084.667	I
200	1085.171	I
250	1085.304	I
400	1088.06	I
350	1090.271	I
250	1090.982	I
250	1092.437	I
400	1094.769	I
350	1095.148	I
350	1095.662	I
400	1095.797	I
250	1096.810	I
300	1097.369	I
200	1098.068	I
200	1099.523	I
500	1107.528	I
800	1139.214	II
800	1167.148	I
3000	1179.293	I
1200	1188.774	I
900	1201.353	I
3000	1335.726	I
10000	1347.240	I
5000	1351.657	I
12000	1363.447	I
2500	1373.116	I
20000	1379.528	I
25000	1389.693	I
20000	1389.957	I
12000	1396.527	I
500	1441.470	II
500	1528.569	II
500	1542.942	II
500	1558.144	II
500	1565.050	II
600	1822.50	III
500	1828.40	III
500	1857.488	II
500	1901.61	III
500	1983.61	III
450 h	1997.370	II
450	2032.116	II
350 h	2088.583	II
350 h	2091.458	II
700	2253.07	III
500	2268.95	III
500	2278.34	III
700	2283.93	III
600	2323.50	III
500	2336.45	III
600	2340.64	III
600	2359.67	III
600	2370.37	III
700	2416.42	III
600	2447.14	III
600	2448.58	III
500	2486.91	III
500	2532.48	III
600	2580.67	III
500	2603.59	III
500	2632.67	III
500	2633.18	III
600	2665.54	III
700	2710.37	III
500	2724.03	IV
500	2751.23	IV
700	2782.47	IV
600	2965.56	III
500	3063.13	IV
600	3076.68	IV
600	3104.46	III
800	3139.34	III
900	3191.45	III
700	3289.80	III
700	3320.57	III
800	3329.06	III
900	3340.42	III
800	3392.89	III
800	3393.45	III
900	3530.03	III
800	3560.68	III
900	3602.10	III
800	3612.85	III
700	3622.69	III
700	3656.95	III
700	3670.28	III
700	3682.05	III
600	3705.45	III
600	3707.34	III
800	3720.45	III

Intensity	Wavelength/Å	
800	3748.81	III
500	3779.35	III
10000	3850.99	II
25000	3860.83	II
500	3925.87	III
700	3991.50	III
600	4018.50	III
600	4059.07	III
500	4104.23	III
500	4106.83	III
10000 h	4132.50	II
500	4608.21	III
40	4623.938	I
50	4654.040	I
80	4661.208	I
45	4691.523	I
40	4721.255	I
45	4740.729	I
13000	4781.32	II
99000	4794.55	II
29000	4810.06	II
16000	4819.47	II
81000	4896.77	II
47000	4904.78	II
26000	4917.73	II
10000	4995.48	II
26000	5078.26	II
30	5099.789	I
56000	5217.94	II
23000	5221.36	II
15000	5392.12	II
99000	5423.23	II
10000	5423.51	II
19000	5443.37	II
10000	5444.21	II
40	5532.162	I
50 d	5796.305	I
45	5799.914	I
30	5856.742	I
50	6019.812	I
200	6140.245	I
160	6194.757	I
150	6434.833	I
300	6932.903	I
300	6981.886	I
600	7086.814	I
7500	7256.62	I
5000	7414.11	I
550	7462.370	I
550	7489.47	I
700	7492.118	I
11000	7547.072	I
2300	7672.42	I
450	7702.828	I
7000	7717.581	I
10000	7744.97	I
2200	7769.16	I
650	7771.09	I
2200	7821.36	I
1700	7830.75	I
3000	7878.22	I

Intensity	Wavelength/Å	
2300	7899.31	I
1800	7915.08	I
3000	7924.645	I
2100	7933.89	I
1700	7935.012	I
650	7952.52	I
1500	7974.72	I
1300	7976.97	I
600	7980.60	I
2900	7997.85	I
2200	8015.61	I
1100	8023.33	I
400	8051.07	I
1700	8084.51	I
2200	8085.56	I
3000	8086.67	I
1300	8087.73	I
2500	8194.42	I
2200	8199.13	I
2200	8200.21	I
800	8203.78	I
18000	8212.04	I
3000	8220.45	I
20000	8221.74	I
18000	8333.31	I
99900	8375.94	I
400	8406.199	I
15000	8428.25	I
2200	8467.34	I
2200	8550.44	I
20000	8575.24	I
750	8578.02	I
75000	8585.97	I
450	8628.54	I
300	8641.71	I
3500	8686.26	I
2200	8912.92	I
3000	8948.06	I
2000	9038.982	I
2500	9045.43	I
1000	9069.656	I
2000	9073.17	I
7500	9121.15	I
3000	9191.731	I
500	9197.596	I
4000	9288.86	I
1500	9393.862	I
3500	9452.10	I
500	9486.964	I
1000	9584.801	I
3500	9592.22	I
250	9632.509	I
1000	9702.439	I
250	9744.426	I
200	9807.057	I
400	9875.970	I
331	10392.549	I
300	11123.05	I
269	11409.69	I
1000	11436.33	I
350	13243.8	I

Intensity	Wavelength/Å	
310	13296.0	I
550	13346.8	I
525	13821.7	I
294	14931.7	I
269	15108.0	I
381	15465.1	I
1094	15520.3	I
1487	15730.1	I
2780	15869.7	I
277	15883.3	I
342	15928.9	I
735	15960.0	I
283	15970.5	I
259	16198.5	I
717	19755.3	I
100	24470.0	I
	39716.0	I
	40085.5	I
	40089.5	I
	40532.2	I

Chromium Cr Z = 24

Intensity	Wavelength/Å	
100	438.62	V
100	464.02	V
100	620.66	IV
100	629.26	IV
80	630.30	IV
100	666.55	IV
100	693.92	IV
60	1030.47	III
100	1033.69	III
100	1036.03	III
80	1055.89	IV
80	1068.41	III
100	1116.48	V
150	1121.07	V
150	1127.63	V
100	1263.50	V
100	1417.42	IV
150	1465.86	V
150	1497.97	V
170	1519.03	V
220	1579.70	V
170	1591.72	V
150	1603.19	V
120	1672.66	IV
120	1758.51	IV
140	1802.72	IV
130	1812.41	IV
200	1837.44	V
140	1873.89	IV
140	1967.18	IV
120	1972.07	IV
19000	2055.52	II
14000	2061.49	II
8900	2065.42	II
200	2226.72	III
200	2235.91	III
150	2237.59	III
150	2244.10	III
150	2284.44	III
150	2324.88	III

Intensity	Wavelength/Å	
130	2383.33	I
140	2408.62	I
170	2496.31	I
110	2502.53	I
190	2504.31	I
110	2516.92	I
390	2519.52	I
190	2527.12	I
160	2549.54	I
130	2560.69	I
150	2571.74	I
100	2577.65	I
380	2591.85	I
250	2653.59	II
250	2658.59	II
320	2663.42	II
440	2666.02	II
280	2668.71	II
350	2671.81	II
280	2672.83	II
1800	2677.16	II
320	2678.79	II
230	2687.09	II
280	2691.04	II
180	2698.41	II
180	2698.69	II
110	2701.99	I
140	2712.31	II
170	2722.75	II
420 h	2726.51	I
280 h	2731.91	I
170 h	2736.47	I
250	2743.64	II
110 h	2748.29	I
330	2748.98	II
390	2750.73	II
280	2751.87	II
110 h	2752.88	I
150	2757.10	I
350	2757.72	II
750	2762.59	II
750	2766.54	II
250 h	2769.92	I
610	2780.70	I
180	2822.37	II
180	2830.47	II
2500	2835.63	II
110	2840.02	II
1700	2843.25	II
1200	2849.84	II
120	2851.36	II
880	2855.68	II
610	2858.91	II
440	2860.93	II
790	2862.57	II
750	2865.11	II
610	2866.74	II
480	2867.65	II
210	2870.44	II
110	2871.63	I
160	2873.48	II

Intensity	Wavelength/Å		Intensity	Wavelength/Å		Intensity	Wavelength/Å		Intensity	Wavelength/Å	
320	2875.99	II	220	3197.08	II	110	3807.93	I	110	4374.16	I
230	2876.24	II	170	3209.18	II	180	3815.43	I	530	4384.98	I
180	2877.98	II	140	3217.40	II	180	3819.56	I	110	4458.54	I
120	2879.27	I	120	3245.54	I	130	3826.42	I	660	4496.86	I
170	2887.00	I	130	3251.84	I	130	3830.03	I	380	4526.47	I
700	2889.29	I	130	3257.82	I	380	3841.28	I	380	4530.74	I
370	2893.25	I	130	3339.80	II	190	3848.98	I	240	4535.72	I
190	2894.17	I	110	3342.59	II	140	3849.36	I	240	4540.50	I
210	2896.75	I	170	3358.50	II	290	3850.04	I	240	4540.72	I
180	2905.49	I	160	3360.30	II	140	3852.22	I	140	4544.62	I
260	2909.05	I	430	3368.05	II	190	3854.22	I	600	4545.96	I
260	2910.90	I	140	3382.68	II	110	3855.29	I	120	4565.51	I
250	2911.14	I	170	3403.32	II	140	3855.57	I	120	4571.68	I
480	2967.64	I	360	3408.76	II	260	3857.63	I	360	4580.06	I
480	2971.11	I	210	3421.21	II	660	3883.29	I	360	4591.39	I
210	2971.91	II	270	3422.74	II	570	3885.22	I	480	4600.75	I
480	2975.48	I	140	3433.31	II	380	3886.79	I	240	4613.37	I
190	2979.74	II	270	3433.60	I	260	3894.04	I	600	4616.14	I
350	2980.79	I	160	3436.19	I	360	3902.92	I	550	4626.19	I
110	2985.32	II	140	3441.44	I	960	3908.76	I	1600	4646.17	I
480	2985.85	I	170	3445.62	I	120 hd	3911.82	I	570	4651.28	I
1500	2986.00	I	170	3447.43	I	120	3915.84	I	840	4652.16	I
2100	2986.47	I	190	3453.33	I	190	3916.24	I	240 d	4698.46	I
660	2988.65	I	130	3455.60	I	1900	3919.16	I	190	4708.04	I
160	2989.19	II	100	3460.43	I	600	3921.02	I	240	4718.43	I
480	2991.89	I	120	3550.64	I	600	3928.64	I	120	4730.71	I
230	2994.07	I	130	3566.16	I	410	3941.49	I	140	4737.35	I
300	2995.10	I	130	3573.64	I	1900	3963.69	I	340	4756.11	I
700	2996.58	I	330 h	3574.80	I	120	3969.06	I	190	4789.32	I
210	2998.79	I	19000	3578.69	I	1600	3969.75	I	120	4801.03	I
1100	3000.89	I	160 h	3584.33	I	1600	3976.66	I	110	4829.38	I
750	3005.06	I	130	3585.30	II	960	3983.91	I	140	4870.80	I
140	3013.03	I	17000	3593.49	I	190	3984.34	I	130	4887.01	I
710	3013.71	I	350	3601.67	I	160	3989.99	I	260	4922.27	I
710	3014.76	I	13000	3605.33	I	960	3991.12	I	110	4936.33	I
1400	3014.92	I	130	3632.84	I	160	3991.67	I	70	4942.50	I
710	3015.19	I	350	3636.59	I	190	3992.84	I	110	4954.81	I
2800	3017.57	I	630	3639.80	I	160	4001.44	I	60	5013.32	I
430	3018.50	I	220	3641.83	I	120	4012.47	II	70	5166.23	I
240	3018.82	I	220	3649.00	I	120	4026.17	I	70	5184.59	I
430	3020.67	I	170	3653.91	I	190	4039.10	I	70	5192.00	I
2800	3021.56	I	220	3656.26	I	160	4048.78	I	85	5196.44	I
1100	3024.35	I	130	3663.21	I	120	4058.77	I	5300	5204.52	I
170	3029.16	I	120	3685.55	I	140	4126.52	I	8400	5206.04	I
710	3030.24	I	130	3686.80	I	120	4153.82	I	11000	5208.44	I
140	3031.35	I	130	3687.25	I	140	4163.62	I	85	5224.94	I
390	3034.19	I	130	3730.81	I	170	4174.80	I	290	5247.56	I
550	3037.04	I	150	3732.03	I	170	4179.26	I	530	5264.15	I
550	3040.85	I	480	3743.58	I	110	4209.37	I	180	5265.72	I
110	3050.14	II	570	3743.88	I	20000	4254.35	I	95 h	5275.17	I
710	3053.88	I	340	3749.00	I	110	4263.14	I	70 h	5276.03	I
240	3118.65	II	230	3757.66	I	16000	4274.80	I	340	5296.69	I
430	3120.37	II	260	3768.24	I	10000	4289.72	I	70 h	5297.36	I
470	3124.94	II	130	3791.38	I	780	4337.57	I	660	5298.27	I
120	3128.70	II	130	3792.14	I	1100	4339.45	I	85	5300.75	I
590	3132.06	II	120	3793.29	I	380	4339.72	I	340 h	5328.34	I
140	3136.68	II	130	3793.88	I	1900	4344.51	I	70 h	5329.17	I
140	3147.23	II	140	3797.13	I	380	4351.05	I	780	5345.81	I
100	3155.15	I	200	3797.72	I	2300	4351.77	I	380	5348.32	I
100	3163.76	I	530	3804.80	I	570	4359.63	I	40	5400.61	I
240	3180.70	II	110	3806.83	I	530	4371.28	I	1400	5409.79	I

Intensity	Wavelength/Å	
24	5628.64	I
7	5642.36	I
24	5664.04	I
24	5694.73	I
40	5698.33	I
24	5702.31	I
24	5712.78	I
24 h	5783.11	I
30 h	5783.93	I
24 h	5785.00	I
19 h	5785.82	I
60 h	5787.99	I
180 h	5791.00	I
35	6330.10	I
22	6362.87	I
19	6661.08	I
21 h	6883.03	I
27 h	6924.13	I
30 h	6978.48	I
85	7355.90	I
130	7400.21	I
150	7462.31	I
40	8947.15	I
19	8976.83	I

***Cobalt Co** $Z = 27$*

Intensity	Wavelength/Å	
20	355.52	V
18	355.88	V
12	356.06	V
66	609.16	IV
70	609.21	IV
64	609.28	IV
10	1018.36	V
10	1021.14	V
15	1231.73	V
50	1277.01	V
80	1299.58	II
80	1306.95	II
50	1345.67	V
1000	1696.01	III
800	1697.99	III
1000	1707.35	III
5000	1760.35	III
5000	1773.57	III
2000	1780.05	III
3000	1782.97	III
1000	1787.08	III
1000	1789.07	III
1000	1823.08	III
2000	1830.09	III
2000	1831.44	III
5000	1835.00	III
1500	1842.34	I
1800	1847.89	I
1800	1852.71	I
2400	1855.05	I
2000	1863.83	III
1500	1878.28	I
1800	1936.58	I
1500	1946.79	I
1500	1951.90	I
1800	1954.22	I
1800	1955.17	I
1500	1958.55	I
1500	1961.59	I
1500 h	1968.69	I
1500 h	1968.93	I
3000	1970.71	I
1800 h	1971.16	I
1800 h	1972.52	I
1500	1973.85	I
1800	1976.97	I
2400 h	1980.89	I
1500	1989.80	I
1800	1990.34	I
1500 l	1998.49	I
1500	2002.32	I
900	2008.04	I
50	2011.51	II
1200 h	2014.58	I
900	2016.17	I
50	2022.35	II
50	2027.04	II
900	2031.96	I
1500	2039.95	I
1200	2041.11	I
50	2065.54	II
1500 h	2077.76	I
900	2085.67	I
900	2087.55	I
900	2089.35	I
900	2093.40	I
900	2094.86	I
900	2095.77	I
1200	2097.51	I
1500	2104.73	I
1500	2106.80	I
900	2108.98	I
900 s	2117.68	I
900	2137.78	I
900	2138.97	I
900	2163.03	I
1100	2174.60	I
200	2193.60	II
200	2256.73	II
150	2260.00	II
200	2283.52	II
1000	2286.15	II
200	2291.98	II
300 d	2293.38	II
300	2301.40	II
800 d	2307.85	II
2600	2309.02	I
500	2311.60	II
500	2314.05	II
300	2314.96	II
200 p	2317.06	II
2400	2323.14	I
300 p	2324.31	II
200 d	2326.11	II
500	2326.47	II
1400	2335.99	I
1600	2338.67	I
200	2347.39	II
1600	2352.85	I
200 d	2353.41	II
2000	2353.42	I
500	2363.80	II
400	2378.62	II
1400	2380.48	I
200	2381.76	II
300 p	2383.45	II
1400	2384.86	I
200	2386.36	II
500	2388.92	II
200	2397.38	II
1100 d	2402.06	I
200 p	2404.16	II
5300	2407.25	I
5300	2411.62	I
1600	2412.76	I
4800	2414.46	I
4800	2415.30	I
300	2417.65	II
4100	2424.93	I
3300	2432.21	I
2900	2436.66	I
2400	2439.05	I
200	2442.63	II
200 d	2446.03	II
200 p	2447.69	II
200	2450.00	II
200	2464.20	II
200	2486.44	II
200	2498.82	II
570	2504.52	I
500	2506.46	II
360	2506.88	I
200	2511.16	II
860	2517.87	I
500	2519.82	II
4300	2521.36	I
200 h	2524.65	II
300	2524.97	II
500	2528.62	II
2900	2528.97	I
200 p	2530.09	II
720	2530.13	I
860	2532.18	I
200 d	2533.82	II
2900	2535.96	I
860	2536.49	I
300	2541.94	II
1700	2544.25	I
200	2546.74	II
340	2548.34	I
310	2553.37	I
310	2555.07	I
300	2559.41	II
200	2560.03	II
960	2562.15	I
500	2564.04	II
1100	2567.35	I
960	2574.35	I
800	2580.32	II
300 d	2582.22	II
500	2587.22	II
500	2587.52	II
200	2588.91	II
100 p	2605.71	II
100	2612.50	II
100	2614.36	II
100 p	2628.77	II
100	2632.26	II
100	2636.07	II
310	2646.42	I
770	2648.64	I
100	2653.72	II
100	2663.53	II
200	2666.73	II
100	2675.85	II
100	2684.42	II
100	2702.02	II
200	2706.62	II
200	2707.35	II
190	2715.99	I
100	2727.78	II
80	2734.54	II
190	2745.10	I
100	2753.22	II
190	2764.19	I
100	2766.70	II
100	2774.97	II
100	2791.00	II
100	2793.73	II
150	2815.56	I
80	2835.63	II
80	2847.35	II
80	2871.22	II
190	2886.44	I
100	2918.38	II
100	2930.24	II
100	2954.73	II
690	2987.16	I
690	2989.59	I
60	3022.59	II
3100	3044.00	I
1700	3061.82	I
80	3387.70	II
1100	3388.17	I
2200	3395.38	I
11000	3405.12	I
4500	3409.18	I
6700	3412.34	I
2200	3412.63	I
2700	3417.16	I
50	3423.84	II
2500	3431.58	I
4500	3433.04	I
1600	3442.93	I
8800	3443.64	I
50	3446.39	II
4100	3449.17	I
2100	3449.44	I
21000	3453.50	I

Intensity	Wavelength/Å	
1000	3455.23	I
5100	3462.80	I
5100	3465.80	I
8000	3474.02	I
1900	3483.41	I
4800	3489.40	I
2400	3495.69	I
50	3501.72	II
9600	3502.28	I
7000	3506.32	I
50	3507.77	II
2900	3509.84	I
1400	3510.43	I
4800	3512.64	I
3800	3513.48	I
4800	3518.35	I
1300	3520.08	I
2700	3521.57	I
3800	3523.43	I
60	3523.51	II
6400	3526.85	I
2700	3529.03	I
7300	3529.81	I
1900	3533.36	I
50	3545.03	II
1100	3560.89	I
80	3561.07	II
8800	3569.38	I
50	3574.95	II
1600	3574.96	I
60	3575.32	II
2500	3575.36	I
60	3577.96	II
1000	3585.16	I
6700	3587.19	I
1900	3594.87	I
1600	3602.08	I
100	3621.21	II
1000	3627.81	I
80	3643.61	II
60	3681.35	II
1100	3745.50	I
1400	3842.05	I
6900	3845.47	I
5500	3873.12	I
2800	3873.96	I
7900	3894.08	I
1500	3935.97	I
80 h	3963.10	II
6000	3995.31	I
970	3997.91	I
350	4020.90	I
370	4045.39	I
350	4066.37	I
830	4092.39	I
550	4110.54	I
2800	4118.77	I
4400	4121.32	I
90	4190.71	I
90	4469.56	I
690	4530.96	I

Intensity	Wavelength/Å	
90	4549.66	I
140	4565.59	I
190	4581.60	I
120	4629.38	I
85	4663.41	I
110	4792.86	I
100	4840.27	I
150	4867.88	I
80 h	4964.18	II
50	5212.71	I
50	5230.22	I
50	5247.93	I
50	5342.71	I
50	5352.05	I

Copper Cu Z = 29

Intensity	Wavelength/Å	
80	685.141	II
100	709.313	II
100	718.179	II
150	724.489	II
200	735.520	II
250	736.032	II
80	779.295	II
100	797.455	II
150	810.998	II
200	813.883	II
300	826.996	II
150	848.808	II
250	851.303	II
250	858.487	II
400	861.994	II
400	865.390	II
250	869.336	II
150	873.263	II
200	876.723	II
250	877.012	II
200	877.555	II
500	878.699	II
100	884.133	II
250	885.847	II
600	886.943	II
600	890.567	II
500	892.414	II
800	893.678	II
400	894.227	II
600	896.759	II
400	896.976	II
600	901.073	II
400	906.113	II
800	914.213	II
600	922.019	II
500	924.239	II
400	935.232	II
600	935.898	II
600	943.335	II
600	945.525	II
500	945.965	II
200	954.383	II
250	956.290	II
400	958.154	II
200	960.414	II
250	968.042	II

Intensity	Wavelength/Å	
200	974.759	II
250	977.567	II
100	987.657	II
250	992.953	II
300	1004.055	II
300	1008.569	II
300	1008.728	II
300	1010.269	II
250	1012.597	II
500	1018.707	II
500	1027.831	II
250	1028.328	II
200	1030.263	II
600	1036.470	II
600	1039.348	II
600	1039.582	II
800	1044.519	II
800	1044.744	II
500	1049.755	II
600	1054.690	II
400	1055.797	II
600	1056.955	II
400	1058.799	II
600	1059.096	II
600	1060.634	II
600	1063.005	II
200	1065.782	II
200	1066.134	II
500	1069.195	II
300	1073.745	II
200	1088.395	II
300	1094.402	II
250	1097.053	II
150	1119.947	II
200	1142.640	II
300	1144.856	II
100	1250.048	II
150	1265.506	II
300	1275.572	II
150	1282.455	II
150	1287.468	II
150	1298.395	II
300	1308.297	II
300	1314.337	II
100	1320.686	II
100	1326.395	II
150	1350.594	II
250	1351.837	II
150	1355.305	II
300	1358.773	II
200	1359.009	II
200	1362.600	II
250	1367.951	II
200	1371.840	II
300 r	1376.79	III
200 r	1377.49	III
100	1393.128	II
100	1398.642	II
150	1402.777	II
150	1407.169	II
100	1414.898	II

Intensity	Wavelength/Å	
250	1418.426	II
250	1421.759	II
200	1427.829	II
400	1430.243	II
250	1434.904	II
150	1436.236	II
150	1442.139	II
200	1445.984	II
200	1449.058	II
250	1450.304	II
200	1452.294	II
300	1458.002	II
250	1459.412	II
200	1463.752	II
400	1463.838	II
200	1466.070	II
400	1470.697	II
200	1472.395	II
250	1473.978	II
200	1474.935	II
150	1476.059	II
300 r	1481.23	III
200	1481.544	II
200	1485.328	II
750	1488.831	II
300	1492.834	II
250	1493.366	II
250	1495.430	II
350	1496.687	II
150	1503.368	II
250	1504.757	II
200	1505.388	II
300	1508.632	II
350	1510.506	II
200	1512.465	II
200	1513.366	II
500	1514.492	II
200	1517.631	II
500	1519.492	II
600	1519.837	II
200	1520.540	II
200	1524.860	II
150	1525.764	II
500	1531.856	II
300	1532.131	II
250	1533.986	II
250	1535.002	II
500	1537.559	II
200	1540.239	II
300	1540.389	II
300	1540.588	II
750	1541.703	II
400	1544.677	II
100	1547.958	II
300	1550.653	II
300	1551.389	II
500	1552.646	II
250	1553.896	II
400	1555.134	II
500	1555.703	II
300	1558.345	II

Intensity	Wavelength/Å	
400	1565.924	II
400	1566.415	II
100	1569.416	II
300	1579.492	II
300	1580.626	II
400	1581.995	II
500	1583.682	II
400	1590.165	II
600	1593.556	II
500 r	1593.75	III
400	1598.402	II
400	1602.388	II
200	1604.848	II
300	1605.281	II
400	1606.834	II
250	1608.639	II
150	1610.296	II
200	1617.915	II
600	1621.426	II
400	1622.428	II
250	1630.268	II
100	1636.605	II
1000 r	1642.21	III
250	1649.458	II
30 r	1655.32	I
200	1656.322	II
200	1660.001	II
300	1663.002	II
100	1672.776	II
30	1688.09	I
30	1691.08	I
30 r	1703.84	I
50 r	1713.36	I
150	1717.721	II
50 r	1725.66	I
100	1736.551	II
50 r	1741.57	I
150	1753.281	II
200 r	1774.82	I
100 r	1825.35	I
250	1929.751	II
250	1944.597	II
100	1946.493	II
200	1957.518	II
150	1970.495	II
150	1977.027	II
500	1979.956	II
300	1989.855	II
250	1999.698	II
270	2035.854	II
250	2037.127	II
350	2043.802	II
300	2054.980	II
100	2078.663	II
110	2098.398	II
320	2104.797	II
300	2112.100	II
320	2117.310	II
350	2122.980	II
350	2126.044	II
420	2134.341	II
900	2135.981	II
400	2148.984	II
150	2161.320	II
1300 r	2165.09	I
250	2174.982	II
1600 r	2178.94	I
700	2179.410	II
1700 r	2181.72	I
700	2189.630	II
900	2192.268	II
400	2195.683	II
1700 r	2199.58	I
1300 r	2199.75	I
100	2200.509	II
200	2209.806	II
750	2210.268	II
1600 r	2214.58	I
250	2215.106	II
1000 r	2215.65	I
750	2218.108	II
2100 r	2225.70	I
150	2226.780	II
1600 r	2227.78	I
350	2228.868	II
2500 r	2230.08	I
1100 r	2238.45	I
900	2242.618	II
2300 r	2244.26	I
1000	2247.002	II
1300 r	2260.53	I
2200 r	2263.08	I
150	2263.786	II
200	2276.258	II
100	2286.645	II
2500 r	2293.84	I
170	2294.368	II
1000	2303.12	I
150	2369.890	II
2500 r	2392.63	I
120	2403.337	II
1500	2406.66	I
1000 r	2441.64	I
100	2485.792	II
2000 r	2492.15	I
150	2506.273	II
120	2526.593	II
300	2544.805	II
100	2571.756	II
150	2590.529	II
200	2600.270	II
2500 r	2618.37	I
200	2666.291	II
750	2689.300	II
700	2700.962	II
650	2703.184	II
700	2713.508	II
650	2718.778	II
300	2721.677	II
120	2737.342	II
270	2745.271	II
2500 r	2766.37	I
800	2769.669	II
200	2791.795	II
170	2799.528	II
100	2810.804	II
1250 r	2824.37	I
350	2837.368	II
100	2857.748	II
600	2877.100	II
270	2884.196	II
2500 r	2961.16	I
100	2986.335	II
2000	2997.36	I
2000	3010.84	I
2500	3036.10	I
2500	3063.41	I
1400	3073.80	I
1500	3093.99	I
1250	3099.93	I
2000	3108.60	I
1400 h	3126.11	I
1500	3194.10	I
1400	3208.23	I
1500 h	3243.16	I
10000 r	3247.54	I
10000 r	3273.96	I
1400 h	3282.72	I
400	3290.418	II
1500 h	3290.54	I
110	3300.881	II
250	3301.229	II
2500 h	3307.95	I
200	3316.276	II
1500	3337.84	I
150	3338.648	II
200	3365.648	II
450	3370.454	II
300	3374.952	II
200	3380.712	II
100	3384.945	II
1250 h	3483.76	I
1250	3524.23	I
2000	3530.38	I
1400	3599.13	I
1400	3602.03	I
1000	3686.555	II
150	3786.270	II
170	3797.849	II
100	3818.879	II
140	3826.921	II
160	3864.137	II
280	3884.131	II
150	3892.924	II
170	3903.177	II
140	3920.654	II
120	3933.268	II
120	3987.024	II
150	3993.302	II
140	4003.476	II
1250	4022.63	I
100	4032.647	II
600	4043.484	II
500	4043.751	II
2000	4062.64	I
120	4068.106	II
500	4131.363	II
200	4143.017	II
300	4153.623	II
500	4161.140	II
370	4164.284	II
400	4171.851	II
500	4179.512	II
500	4211.866	II
320	4230.449	II
200	4255.635	II
950	4275.11	I
300	4279.962	II
500	4292.470	II
400	4365.370	II
100	4444.831	II
400	4506.002	II
150	4516.049	II
150	4541.032	II
500	4555.920	II
100	4596.906	II
120	4649.271	II
2000	4651.12	I
120	4661.363	II
320	4671.702	II
300	4673.577	II
450	4681.994	II
100	4758.433	II
400	4812.948	II
120	4851.262	II
300	4854.988	II
100	4873.304	II
150	4901.427	II
1000	4909.734	II
500	4918.376	II
200	4926.424	II
900	4931.698	II
120	4943.026	II
700	4953.724	II
500	4985.506	II
400	5006.801	II
350	5009.851	II
400	5012.620	II
350	5021.279	II
200	5039.016	II
300	5047.348	II
900	5051.793	II
400	5058.910	II
500	5065.459	II
450	5067.094	II
350	5072.302	II
450	5088.277	II
420	5093.816	II
350	5100.067	II
1500	5105.54	I
250	5124.476	II
2000	5153.24	I
100	5158.093	II
100	5183.367	II

Intensity	Wavelength/Å	
2500	5218.20	I
100	5269.991	II
100	5276.525	II
1650	5292.52	I
100	5368.383	II
1500	5700.24	I
1500	5782.13	I
150	5805.989	II
100	5833.515	II
200	5897.971	II
120	5937.577	II
400	5941.196	II
100	5993.260	II
650	6000.120	II
100	6023.264	II
250	6072.218	II
150	6080.343	II
150	6099.990	II
160	6107.412	II
300	6114.493	II
600	6150.384	II
750	6154.222	II
500	6172.037	II
550	6186.884	II
400	6188.676	II
300	6198.092	II
470	6204.261	II
450	6208.457	II
750	6216.939	II
700	6219.844	II
500	6261.848	II
1000	6273.349	II
350	6288.696	II
900	6301.009	II
550	6305.972	II
400	6312.492	II
120	6326.466	II
400	6373.268	II
750	6377.840	II
400	6403.384	II
850	6423.884	II
200	6442.965	II
750	6448.559	II
170	6466.246	II
950	6470.168	II
750	6481.437	II
400	6484.421	II
220	6517.317	II
400	6530.083	II
120	6551.286	II
200	6577.080	II
750	6624.292	II
800	6641.396	II
450	6660.962	II
100	6770.362	II
300	6806.216	II
400	6809.647	II
320	6823.202	II
250	6844.157	II
320	6868.791	II
270	6872.231	II

Intensity	Wavelength/Å	
270	6879.404	II
220	6937.553	II
150	6952.871	II
150	6977.572	II
200	7022.860	II
300	7194.896	II
400	7326.008	II
300	7331.694	II
250	7382.277	II
1000	7404.354	II
270	7434.156	II
500	7562.015	II
700	7652.333	II
1000	7664.648	II
150	7681.788	II
450	7744.097	II
800	7778.738	II
750	7805.184	II
1500	7807.659	II
1000	7825.654	II
350	7860.577	II
300	7890.567	II
700	7902.553	II
1500	7933.13	I
400	7944.438	II
400	7972.033	II
1200	7988.163	II
2000	8092.63	I
500	8277.560	II
800	8283.160	II
250	8503.396	II
750	8511.061	II
200	8609.134	II
500	9813.213	II
250	9827.978	II
200	9830.798	II
600	9861.280	II
600	9864.137	II
200	9883.969	II
550	9916.419	II
500	9917.954	II
550	9925.594	II
450	9938.998	II
500	9960.354	II
450	10006.588	II
550	10022.969	II
550	10038.093	II
650	10054.938	II
450	10080.354	II

***Dysprosium* Dy $Z = 66$**

Intensity	Wavelength/Å	
260	2356.91	II
240	2410.01	II
260	2439.84	II
220	2585.30	I
440	2634.80	II
220	2755.75	II
300	2816.39	II
390	2913.95	II
610	3038.28	II
830	3135.38	II
500	3141.14	II

Intensity	Wavelength/Å	
1200	3156.52	II
670	3162.83	II
1000	3169.99	II
470	3215.19	II
830	3216.63	II
490	3235.89	II
490	3245.12	II
1200	3251.27	II
890	3280.09	II
490	3282.77	II
1100	3308.88	II
780	3316.32	II
1000	3319.88	II
780	3341.00	II
510	3353.58	II
510	3368.11	II
5300	3385.02	II
610	3388.85	II
3800	3393.57	II
1300	3396.16	II
5300	3407.80	II
1300	3413.78	II
530	3414.82	II
780	3419.63	II
530	3425.06	II
1900	3434.37	II
560	3440.93	II
1300	3441.45	II
3800	3445.57	II
830	3446.99	II
2700	3454.32	II
1300	3456.56	II
4400	3460.97	II
720	3468.43	II
560	3471.14	II
560 d	3471.53	II
1300	3477.07	II
4400	3494.49	II
560	3496.34	II
830	3498.71	II
830	3504.53	II
830	3505.45	II
1300	3506.81	II
560	3517.26	II
4400	3523.98	II
22000	3531.70	II
4400	3534.96	II
5500	3536.02	II
4400	3538.52	II
1700	3542.33	II
1400	3546.83	II
4400	3550.22	II
2200	3551.62	II
440 h	3558.23	II
440	3559.30	II
2200	3563.15	II
560	3563.69	II
780	3573.83	II
1400	3574.15	II
4400	3576.24	II
1700	3576.87	II

Intensity	Wavelength/Å	
830	3577.98	II
440	3580.04	II
3300	3585.06	II
1400	3585.78	II
560	3586.11	II
1100	3591.41	II
560	3591.81	II
560	3592.11	II
1800	3595.04	II
560	3600.38	II
1800	3606.12	II
440	3618.51	II
560	3620.16	II
470	3624.27	II
1100	3629.42	II
4000	3630.24	II
440	3632.78	II
1100	3640.25	II
11000	3645.40	II
1000	3648.78	II
700	3664.62	II
990	3672.30	II
420	3672.70	II
1400	3674.08	II
2200	3676.59	II
640	3678.51	I
820	3684.85	I
1300	3685.78	I
4700	3694.81	II
990	3698.21	II
540	3701.63	II
440	3707.57	II
440	3708.22	II
420	3710.07	II
1600	3724.45	II
930	3739.34	I
1200	3747.82	II
1400	3753.51	II
1400	3753.75	II
1200	3757.05	I
4700	3757.37	II
640	3767.63	I
640	3773.05	I
420	3781.47	I
3300	3786.18	II
1600	3788.44	II
700	3791.87	II
510	3804.14	II
580	3806.27	II
470	3812.27	I
470	3813.67	II
1400	3816.76	II
700	3825.68	II
2300	3836.50	II
1400	3841.31	II
420	3846.34	II
420	3847.02	I
1200	3853.03	II
420	3858.40	I
560	3868.45	II
1600	3868.81	I

Intensity	Wavelength/Å	
820	3869.86	II
7000	3872.11	II
1200	3873.99	II
470	3879.11	II
5800	3898.53	II
540	3914.87	II
540	3915.59	II
540 d	3917.29	I
420	3927.86	I
540	3930.14	I
2100	3931.52	II
10000	3944.68	II
800	3957.79	II
14000	3968.39	II
2700	3978.57	II
1400	3981.92	II
1600	3983.65	II
800	3984.21	II
540	3991.32	II
1600	3996.69	II
8000	4000.45	II
420	4005.84	I
540	4011.29	II
540	4013.82	I
540	4014.70	II
420	4027.78	II
520 d	4028.32	II
520	4032.47	II
420	4033.65	II
420	4036.32	II
12000	4045.97	I
1600	4050.56	II
520	4055.14	II
2500	4073.12	II
7400	4077.96	II
3900	4103.30	II
860	4103.87	I
1500	4111.34	II
490	4124.63	II
990	4129.42	II
1200	4143.10	II
990	4146.06	I
5700	4167.97	I
930	4183.72	I
12000	4186.82	I
2200	4191.64	I
6800	4194.84	I
800	4198.02	I
680	4201.30	I
680	4202.24	I
16000	4211.72	I
1800	4213.18	I
3700	4215.16	I
4400	4218.09	I
4400	4221.11	I
2700	4225.16	I
1000	4308.63	II
540	4409.38	II
740	4449.70	II
420	4577.78	I
2100	4589.36	I
990	4612.26	I
170	4731.84	II
120 h	4775.79	I
480	4957.34	II
70	5022.12	I
160	5042.63	I
95	5070.68	I
120	5077.67	I
80	5090.38	II
80	5110.32	I
130 h	5120.04	I
190	5139.60	II
110	5169.69	II
80	5185.30	I
290	5192.86	II
95	5197.66	II
70	5259.88	I
130	5260.56	I
65	5267.11	I
55	5282.07	I
160	5301.58	I
65	5340.30	I
85	5389.58	II
80	5419.13	I
70	5423.32	I
95	5451.11	I
65	5547.27	I
100	5639.50	I
55 h	5645.99	I
80	5652.01	I
70 h	5718.46	I
55	5745.53	I
55 h	5868.11	II
70	5945.80	I
120	5974.49	I
140	5988.56	I
140	6088.26	I
100	6168.43	I
270	6259.09	I
160	6579.37	I
75	6667.86	I
180	6835.42	I
80	6852.96	I
65	6899.32	II
55	7426.86	II
55	7543.73	I
80	7662.36	I
100	8201.57	II
45	8791.39	II

Erbium Er Z = 68

Intensity	Wavelength/Å	
600	2277.65	III
290	2586.73	II
490	2670.26	II
500	2739.27	III
610	2755.63	II
1000	2904.47	II
1500	2910.36	II
1500	2964.52	II
1200	3002.41	II
1000	3055.10	III
1000	3070.40	III
610	3073.34	II
720	3082.08	II
610	3084.02	II
770	3122.72	II
1500	3166.25	III
870	3181.92	II
870	3220.73	II
610	3223.31	II
2300	3230.58	II
2700	3264.78	II
720	3279.33	II
720	3280.22	II
2000	3301.23	III
2300	3312.42	II
770	3323.19	II
770	3332.70	II
1300	3346.04	II
1400	3364.08	II
1400 d	3368.02	II
7700	3372.71	II
970	3374.17	II
1700	3385.08	II
2300	3392.00	II
770	3441.13	II
970	3471.71	II
610	3479.41	II
970	3485.85	II
6700	3499.10	II
610	3502.78	I
610	3524.91	II
820	3549.84	II
1500	3558.02	I
1000	3559.90	II
920	3570.75	II
1000	3580.52	II
610	3590.76	I
610	3599.50	II
1000	3599.83	II
3100	3616.56	II
720	3628.04	I
1000	3633.54	II
1600	3638.68	I
900	3645.94	II
7900	3692.65	II
1300	3729.52	II
900	3742.64	II
900	3747.43	I
1800	3786.84	II
1600	3810.33	I
4000	3816.78	III
3600	3830.48	II
680	3855.90	I
7500	3862.85	I
1500	3880.61	II
1200	3882.89	II
4200	3892.68	I
5200	3896.23	II
11000	3906.31	II
3200	3937.01	I
2100	3938.63	II
3200	3944.42	I
2700	3973.04	I
3200	3973.58	I
1400	3974.72	II
810	3977.02	I
1100	3982.33	I
810	3987.66	I
14000	4007.96	I
1100	4012.58	I
3000	4020.51	I
1000	4046.96	I
940	4055.47	II
690	4059.78	II
3500	4087.63	I
1100	4098.10	I
6900	4151.11	I
1000	4190.70	I
1400	4218.43	I
690	4286.56	I
40000	4290.06	III
20000	4386.86	III
810	4409.34	I
1000	4606.61	I
570	4675.62	II
15000	4735.56	III
2000	4783.12	III
250	5007.25	I
200	5035.94	I
210	5042.05	II
120	5124.56	I
130	5127.41	II
120	5131.53	I
130	5133.83	II
170	5164.77	II
130	5172.78	I
160	5188.90	II
150	5206.52	I
140	5255.93	II
80	5272.91	I
90	5348.06	I
60	5414.63	II
180	5456.62	I
90	5468.32	I
80	5485.97	II
80	5593.46	I
60	5611.82	I
70	5622.01	I
80	5626.53	II
90	5640.36	I
70	5664.95	I
70	5719.55	I
100	5739.19	I
290	5762.80	I
70	5784.66	I
70	5800.79	I
430	5826.79	I
100	5850.07	I
120	5855.31	I
140	5872.35	I
120	5881.14	I
8000	5903.30	III

Intensity	Wavelength/Å	
70	6022.56	I
70	6061.25	I
60	6076.45	II
360	6221.02	I
55	6262.56	I
60	6268.87	I
130	6308.77	I
55	6326.13	I
55	6492.35	I
60	6583.48	I
70	6601.11	I
70	6759.87	I
35	6790.92	I
70	6848.10	I
55	6865.13	I
55	7459.55	I
120	7469.51	I
35	7680.01	I
35	7797.47	I
35	7921.85	I
30	7937.84	I
35	8312.82	I
55	8409.90	I
9	8866.84	II
Europium Eu Z = 63		
30	2124.69	III
200	2350.61	III
4000	2375.46	III
100 d	2435.14	III
1000	2444.38	III
4000	2445.99	III
2000	2513.76	III
200	2522.14	III
160	2564.17	II
110	2568.17	II
230	2577.14	II
1000	2638.77	II
380	2641.27	II
640	2668.34	II
110	2673.42	II
250	2678.29	II
250	2685.66	II
550	2692.03	II
700	2701.14	II
800	2701.90	II
240	2705.28	II
180	2709.99	I
700	2716.98	II
4200	2727.78	II
160	2740.62	II
120	2744.26	II
480	2781.89	II
1900	2802.84	II
220	2811.75	II
3400	2813.94	II
550	2816.18	II
2000	2820.78	II
400 cw	2828.72	II
260	2859.67	II
280	2862.57	II
200	2892.54	I
140	2893.03	I
360	2893.83	I
3200	2906.68	II
160	2908.99	I
850	2925.04	II
200 cw	2952.68	II
260	2960.21	II
300	2991.33	II
100 c	3023.93	III
200 c	3026.79	III
320 cw	3054.94	II
120	3058.98	I
220	3077.36	II
120	3097.45	II
320	3106.18	I
950	3111.43	I
120	3130.73	II
50 c	3171.00	III
50 c	3183.78	III
420	3210.57	I
1000	3212.81	I
420	3213.75	I
150	3272.77	II
210	3277.78	II
150	3301.95	II
140	3308.02	II
140	3313.33	II
950	3334.33	I
110	3350.40	I
140	3369.06	II
190	3391.99	II
280	3396.58	II
150	3425.02	II
150	3441.00	II
130	3461.38	II
470 cw	3521.09	II
150	3542.15	II
180	3552.52	II
150	3603.20	II
6400	3688.42	II
20000 cw	3724.94	II
350	3741.31	II
260	3761.12	II
39000 cw	3819.67	II
140	3844.23	II
190	3865.57	I
150	3884.75	I
28000 cw	3907.10	II
32000 cw	3930.48	II
30000 cw	3971.96	II
180	4011.69	II
150	4017.58	II
120	4039.19	I
120	4085.38	II
33000 cw	4129.70	II
60000 cw	4205.05	II
150	4298.73	I
240	4355.09	II
14000 cw	4435.56	II
3000	4522.57	II
11000	4594.03	I
9800	4627.22	I
8300	4661.88	I
110	4867.62	I
150	4907.18	I
180	4911.40	I
180	5013.17	I
170	5022.91	I
110	5029.54	I
170	5114.37	I
170	5129.10	I
210	5133.52	I
270	5160.07	I
210	5166.70	I
200	5199.85	I
110	5200.96	I
120	5206.44	I
750	5215.10	I
300	5223.49	I
120	5239.24	I
200	5266.40	I
390	5271.96	I
110	5272.48	I
150	5282.82	I
120	5291.26	I
120	5294.64	I
540	5357.61	I
120	5361.61	I
110	5376.94	I
120	5392.94	I
450	5402.77	I
380	5451.51	I
260	5452.94	I
120	5488.65	I
120	5510.52	I
200	5547.44	I
150	5570.33	I
200	5577.14	I
120	5580.03	I
210	5645.80	I
330	5765.20	I
180	5783.69	I
170	5818.74	II
600 cw	5830.98	I
330	5966.07	II
480 cw	5967.10	I
170	5972.75	I
240	5992.83	I
110	6012.56	I
420	6018.15	I
170	6029.00	I
420	6049.51	II
140	6057.36	I
240	6083.84	I
240	6099.35	I
120	6118.78	I
330	6173.05	II
110	6178.76	I
260 cw	6188.13	I
140	6195.07	I
240	6262.25	I
170	6299.77	I
230	6303.41	II
120 cw	6350.04	I
120 cw	6400.93	I
180	6410.04	I
140	6411.32	I
830	6437.64	II
120	6457.96	I
1400	6645.11	II
50	6666.35	III
140	6802.72	I
360	6864.54	I
120	7040.20	I
330	7077.10	II
570	7194.81	II
570	7217.55	II
540	7301.17	II
720	7370.22	II
300	7426.57	II
160	7583.91	I
60 cw	7742.57	I
70	7746.19	I
35	7887.99	I
24 cw	8209.80	I
21 cw	8642.67	I
18	8870.30	I
Fluorine F Z = 9		
50	148.00	V
50	163.56	V
90	165.98	V
100	166.18	V
50	186.84	V
60	190.57	V
70	190.84	V
50	196.39	IV
60	196.45	IV
70	200.09	IV
80	201.16	IV
90	208.25	IV
90	240.08	IV
100	251.03	IV
140	419.65	IV
150	420.05	IV
160	420.73	IV
100	429.51	III
110	430.15	III
150	430.76	IV
90	464.29	III
120	465.98	V
130	490.57	IV
160	491.00	IV
50	497.38	IV
60	497.83	IV
70	498.80	IV
90	506.16	V
100	508.08	V
120	508.39	III
60	514.08	V
90	525.29	V
100	526.30	V
120	567.69	III
110	567.75	III

Intensity	Wavelength/Å	
140	570.64	IV
140	571.30	IV
150	571.39	IV
160	572.66	IV
90	605.67	II
100	606.80	II
90	630.20	III
100	647.77	V
110	647.87	V
130	654.03	V
120	656.12	III
130	656.87	III
110	657.23	V
140	657.33	V
140	658.33	III
140	676.12	IV
130	677.15	IV
150	677.22	IV
130	678.99	IV
160	679.21	IV
60	757.04	V
150	806.96	I
125	809.60	I
500	951.87	I
1000	954.83	I
750	955.55	I
500	958.52	I
20	972.40	I
350	973.90	I
100	976.22	I
40	976.51	I
100	977.75	I
60	1082.31	V
70	1088.39	V
80	1219.03	III
80	1266.87	III
90	1267.71	III
70	1297.54	III
70	1359.92	III
110	1498.93	III
120	1502.01	III
110	1504.18	III
140	1504.79	III
130	1506.30	III
110	1506.77	III
100	1553.02	III
110	1557.59	III
100	1563.73	III
100	1565.54	III
100	1623.40	III
100	1650.76	III
130	1670.39	III
140	1677.40	III
100	1716.99	III
120	1770.09	III
150	1770.67	III
110	1772.93	III
140	1773.36	III
160	1791.65	III
110	1803.03	III
170	1805.90	III
110	1839.30	III
120	1839.97	III
110	1840.14	III
100	2027.44	III
120	2030.32	III
120	2217.17	III
50	2298.29	IV
40	2451.58	IV
120	2452.07	III
50	2456.92	IV
130	2464.85	III
130	2470.29	III
120	2478.73	III
150	2484.37	III
120	2542.77	III
120	2580.04	III
130	2583.81	III
120	2593.23	III
130	2595.53	III
140	2599.28	III
130	2625.01	III
140	2629.70	III
120	2656.44	III
130	2755.55	III
160	2759.63	III
120	2788.15	III
160	2811.45	III
40	2820.74	IV
50	2826.13	IV
140	2833.99	III
150	2835.63	III
150	2860.33	III
120	2862.86	III
140	2887.58	III
150	2889.45	III
120	2905.30	III
140	2913.29	III
160	2916.34	III
140	2932.49	III
140	2994.28	III
120 h	2997.21	III
130	2997.53	III
120	2999.47	III
130	3039.25	III
120	3039.75	III
160	3042.80	III
150	3049.14	III
140	3113.62	III
160	3115.70	III
180	3121.54	III
140	3124.79	III
140	3134.23	III
140	3146.99	III
180	3174.17	III
170	3174.76	III
120	3214.00	III
200	3501.45	II
200	3501.57	II
200	3502.96	II
6	3594.10	I
12	3668.17	I
270	3847.09	II
260	3849.99	II
250	3851.67	II
5	3898.48	I
8	3930.69	I
5	3934.26	I
5	3948.56	I
240	4024.73	II
220	4025.01	II
230	4025.49	II
200	4103.51	II
200	4246.23	II
200	4299.17	II
140 h	4420.30	III
120 h	4427.35	III
120 h	4432.32	III
140 h	4479.99	III
6	4960.65	I
150	5012.54	III
160	5110.99	III
15	5230.41	I
12	5279.01	I
18	5540.52	I
12	5552.43	I
10	5577.33	I
20	5624.06	I
12	5626.93	I
15	5659.15	I
40	5667.53	I
90	5671.67	I
18	5689.14	I
25	5700.82	I
25	5707.31	I
140	5753.17	III
120	5761.20	III
12	5950.15	I
25	5959.19	I
70	5965.28	I
50	5994.43	I
150	6015.83	I
80	6038.04	I
900	6047.54	I
100	6080.11	I
150	6091.82	III
140	6125.50	III
800	6149.76	I
400	6210.87	I
130	6233.57	III
13000	6239.65	I
10000	6348.51	I
140	6363.05	III
8000	6413.65	I
450	6569.69	I
300	6580.39	I
400	6650.41	I
1800	6690.48	I
400	6708.28	I
7000	6773.98	I
1500	6795.53	I
9000	6834.26	I
50000	6856.03	I
8000	6870.22	I
15000	6902.48	I
6000	6909.82	I
4000	6966.35	I
45000	7037.47	I
30000	7127.89	I
15000	7202.36	I
1000	7309.03	I
15000	7311.02	I
700	7314.30	I
5000	7331.96	I
120	7336.77	III
130	7354.94	III
10000	7398.69	I
4000	7425.65	I
2200	7482.72	I
2500	7489.16	I
900	7514.92	I
5000	7552.24	I
5000	7573.38	I
7000	7607.17	I
18000	7754.70	I
15000	7800.21	I
300	7879.18	I
500	7898.59	I
350	7936.31	I
300	7956.32	I
80	8016.01	II
1000	8040.93	I
900	8075.52	I
350	8077.52	I
350	8126.56	I
600	8129.26	I
300	8159.51	I
600	8179.34	I
300	8191.24	I
350	8208.63	I
2500	8214.73	I
3000	8230.77	I
500	8232.19	I
1500	8274.62	I
2000	8298.58	I
600	8302.40	I
900	8807.58	I
1000	8900.92	I
300	8912.78	I
350	9025.49	I
400	9042.10	I
350	9178.68	I
200	9433.67	I
25	9505.30	I
12	9662.04	I
25	9734.34	I
15	9822.11	I
12	9902.65	I
80 h	10047.98	II
15	10285.45	I
20	10862.31	I

***Francium* Fr *Z* = 87**

Intensity	Wavelength/Å	
	7177.	I

Gadolinium Gd Z = 64

Intensity	Wavelength/Å	
1200	1007.24	IV
1200	1063.84	IV
1600	1476.98	IV
1500	1705.03	IV
1600	1706.01	IV
2000	1736.24	IV
1500	1815.32	IV
2200	1975.24	III
3400	2018.07	III
2800	2359.31	III
1400	2397.87	IV
2800	2697.39	III
2800	2703.28	III
2700	2727.89	III
9000	2904.73	III
9500	2955.53	III
1200	2999.04	II
2100	3010.13	II
1900	3027.60	II
2100	3032.84	II
1600	3034.05	II
2100	3081.99	II
3500	3100.50	II
930	3145.00	II
980	3156.53	II
980	3161.37	II
4000	3176.66	III
1400	3331.38	II
1100	3336.18	II
5400	3350.47	II
4300	3358.62	II
5400	3362.23	II
1100	3392.53	II
1100 d	3407.56	II
6900	3422.47	II
1700	3439.21	II
2700	3439.99	II
1400	3450.38	II
1100	3451.23	II
2700	3463.98	II
1700	3467.27	II
1700	3468.99	II
1400	3473.22	II
2200	3481.28	II
1700	3481.80	II
1700	3494.40	II
1400	3505.51	II
1100	3512.50	II
4300	3545.80	II
3900	3549.36	II
1400	3557.05	II
5400	3584.96	II
1100	3592.71	II
1100	3604.87	I
6100	3646.19	II
3900	3654.62	II
3100	3656.15	II
1400	3662.26	II
2700	3664.60	II
2000	3671.20	II
2000	3684.13	I
3100	3687.74	II
2000	3697.73	II
1300	3699.73	II
2700	3712.70	II
2000	3713.57	I
1400	3716.36	II
2000	3717.48	I
1800 d	3719.45	II
1500	3730.84	II
4500	3743.47	II
1400	3758.31	II
8700	3768.39	II
1400	3770.69	II
2900	3783.05	I
5100	3796.37	II
3700	3813.97	II
3300	3850.69	II
5100	3850.97	II
4300	3852.45	II
1600	3866.99	I
1500	3894.70	II
2200	3916.51	II
1200	3934.79	I
1400	3945.54	I
1200	3957.67	II
1100	4023.14	I
1100	4028.15	I
1400	4037.33	II
1600	4045.01	I
1300	4049.43	II
2200	4049.86	II
2600	4053.64	I
2600	4058.22	I
1900	4063.39	II
1300	4078.44	II
2800	4078.70	I
1500	4085.56	II
1100	4092.71	I
2600	4098.61	II
2200	4130.37	II
1100	4132.28	II
2400	4175.54	I
2400	4184.25	II
2200	4190.78	I
1300	4212.00	II
4800	4225.85	I
1700	4251.73	II
1600	4262.09	I
1100	4306.34	I
1800	4313.84	I
2600 d	4325.57	II
1900	4327.12	I
1000	4344.30	II
2200	4346.46	I
1400	4401.86	I
1400	4422.41	I
1100	4430.63	I
1100	4519.66	I
910	4537.81	I
520	4614.50	I
700	4694.33	I
410	4743.65	I
470	4767.24	I
300	4784.62	I
320	4821.69	I
280	4934.12	I
750	5015.04	I
75	5039.09	I
5000	5091.70	III
130	5098.38	II
910	5103.45	I
180	5108.91	II
120	5125.56	II
860	5155.84	I
190	5176.28	II
410	5197.77	I
280	5219.40	I
130	5233.93	I
320	5251.18	I
120	5252.14	II
140	5255.80	I
280	5283.08	I
280	5301.67	I
220	5302.76	I
280	5307.30	I
130	5321.50	I
280	5321.78	I
110	5327.32	I
170	5333.30	I
300	5343.00	I
200	5348.67	I
300	5350.38	I
240	5353.26	I
3000	5365.96	III
150	5370.63	I
4000	5553.30	III
3000	5587.88	III
190	5617.91	I
110	5632.25	I
260	5643.24	I
3000	5658.98	III
390	5696.22	I
120	5733.86	II
240	5791.38	I
220	5851.63	I
280	5856.22	I
110	5904.56	I
170	5911.45	II
85	5930.29	I
85	5936.84	I
65	5937.71	I
110 h	5988.02	I
430	6114.07	I
75	6305.15	II
40	6331.35	I
40	6380.95	II
40 h	6538.15	I
55	6564.78	I
50	6634.36	II
35	6681.23	II
85	6730.73	I
50	6752.67	II
26	6786.33	II
100	6828.25	I
100	6916.57	I
50	6985.89	II
75	6991.92	I
60	6996.76	II
45	7006.16	II
35	7122.57	I
170	7168.37	I
28	7189.57	II
28	7262.66	I
35	7441.85	I
40	7464.36	I
55	7562.97	I
80	7733.50	I
35	7846.35	II
35	7856.93	I
25	7930.25	II
18	8146.15	I
21	8668.63	I
21 h	8832.06	II
14 h	8849.14	I
18 h	8867.31	I
5000	14332.88	III

Gallium Ga Z = 31

Intensity	Wavelength/Å	
14	294.50	IV
61	295.67	IV
30	298.44	V
30	300.01	V
30	302.86	V
41	304.99	IV
30	307.03	V
30	308.26	V
30	311.79	V
30	313.68	V
40	319.41	V
40	322.31	V
50	322.99	V
30	323.10	V
40	324.25	V
40	324.95	V
40	326.14	V
30	326.77	V
30	328.65	V
25	423.18	IV
16	439.92	IV
50	620.00	III
40	622.01	III
90	806.51	III
90	817.30	III
50	828.70	III
20	878.17	V
40	973.21	V
40	989.75	V
90	1014.47	V
90	1019.71	V
120	1050.48	V
80	1054.56	V
90	1058.12	V
80	1066.69	V

Intensity	Wavelength/Å	
60	1069.60	V
80	1073.77	V
90	1078.83	V
110	1079.60	V
60	1080.99	V
80	1085.00	III
250	1085.01	V
80	1087.37	V
90	1091.71	V
100	1094.36	V
80	1095.10	V
160	1102.83	V
140	1103.03	V
60	1105.61	III
75	1105.62	V
70	1106.17	V
80	1118.34	V
120	1126.40	V
130	1128.10	V
120	1128.53	V
100	1129.94	V
130	1136.07	V
67	1137.06	IV
130	1150.23	V
90	1150.27	III
70	1156.10	IV
120	1156.51	V
35	1157.74	V
70	1163.60	IV
25	1169.40	V
75	1170.58	IV
48	1171.71	IV
40	1178.95	V
68	1185.23	IV
40	1186.06	IV
73	1190.89	IV
73	1193.02	IV
75	1195.02	IV
69	1201.54	IV
72	1206.89	IV
80	1213.17	V
63	1216.15	IV
50	1228.03	IV
60	1236.38	IV
60	1238.59	IV
75	1245.53	IV
83	1258.77	IV
81	1264.66	IV
82	1267.15	IV
90	1267.16	III
81	1279.24	IV
15	1283.64	V
80	1285.33	IV
80	1293.46	III
60	1295.36	III
82	1295.86	IV
83	1299.46	IV
82	1303.53	IV
80	1309.68	IV
80	1314.82	IV
60	1323.15	III

Intensity	Wavelength/Å	
85	1338.09	IV
77	1347.03	IV
76	1351.06	IV
70	1353.92	III
74	1364.63	IV
60	1395.54	IV
77	1402.55	IV
70	1405.32	IV
73	1465.87	IV
90	1495.07	III
50	1534.46	III
10	1813.98	II
15	1845.30	II
20	2091.34	II
90	2417.70	III
90	2423.98	III
15	2424.36	III
10	2632.66	I
10	2665.05	I
20	2700.47	II
15	2780.15	II
50	3521.77	III
80	3581.19	III
100	3589.34	III
10	3731.10	III
10	3806.60	III
10	4032.99	I
10	4172.04	I
15	4251.16	II
10 h	4254.04	II
10	4255.77	II
40	4262.00	II
100	4380.69	III
150	4381.76	III
100	4863.00	III
150	4993.78	III
10	5808.28	III
20	5848.25	III
15	5993.51	III
10	6334.2	II
2000	6396.56	I
1000	6413.44	I
10 h	7251.4	I
20 h	7403.0	I
30 h	7464.0	I
10 h	7620.5	I
50 h	7734.77	I
100 h	7800.01	I
15 h	8002.55	I
20 h	8074.25	I
100 h	8311.86	I
200 h	8386.49	I
200 h	9492.92	I
200 h	9493.12	I
300 h	9589.36	I
100 h	10905.95	I
400	11949.12	I
200	12109.78	I
60	14996.64	I
60	22016.81	I
70	22568.71	I

Intensity	Wavelength/Å	
Germanium Ge Z = 32		
700	294.51	V
1000	295.64	V
200	304.98	V
20	621.52	V
50	724.21	V
60	746.88	V
60	760.05	V
10	862.234	II
15	875.493	II
15	905.977	II
20	920.554	II
300	971.35	V
300	990.66	V
50	999.101	II
300	1004.38	V
100	1016.638	II
900	1045.71	V
700	1072.66	V
100	1075.072	II
300	1085.51	II
40	1088.45	III
800	1089.49	V
200	1098.71	II
500	1106.74	II
1000	1116.94	V
500	1120.46	II
700	1163.39	V
200	1164.27	II
500	1181.19	II
500	1181.65	II
200	1188.73	II
20	1188.99	IV
300	1191.26	II
700	1222.30	V
20	1229.81	IV
500	1237.059	II
500	1261.905	II
100	1264.710	II
200	1401.24	II
200	1538.091	II
500	1576.855	II
75	1581.070	II
100	1602.486	II
3 r	1615.57	I
2 r	1624.130	I
2 r	1630.173	I
3 r	1636.31	I
4 r	1639.730	I
2	1647.531	I
200	1649.194	II
2	1651.528	I
4 r	1651.955	I
3	1661.345	I
4 r	1663.539	I
10 h	1665.275	I
4	1667.802	I
3 r	1670.608	I
100 r	1691.090	I
200 r	1716.784	I
100 h	1739.102	I

Intensity	Wavelength/Å	
100	1742.195	I
50	1746.065	I
200	1750.043	I
100	1758.279	I
100 h	1764.185	I
100 h	1765.284	I
50 h	1766.433	I
200	1774.176	I
200	1785.046	I
100 h	1793.071	I
75 h	1801.432	I
200 h	1841.328	I
200 h	1842.410	I
100 h	1844.410	I
100 h	1845.872	I
100 h	1846.958	I
200	1853.134	I
500 r	1860.086	I
100	1865.052	I
300 r	1874.256	I
100	1895.197	I
500 r	1904.702	I
50 h	1908.434	I
30	1912.409	I
300 r	1917.592	I
100 h	1923.467	I
500 r	1929.826	I
10 h	1934.048	I
100 r	1937.483	I
500	1938.008	II
100 r	1938.300	I
500	1938.891	II
30 s	1944.116	I
200	1944.731	I
200	1955.115	I
500	1962.013	I
30 h	1963.373	I
30	1965.383	I
200	1970.880	I
200	1979.274	II
300 h	1987.849	I
300	1988.267	I
500 r	1998.887	I
200	2011.29	I
1700	2019.068	I
2400 r	2041.712	I
1600 r	2043.770	I
420	2054.461	I
220 h	2057.238	I
750 r	2065.215	I
2600 r	2068.656	I
420	2086.021	I
2000 r	2094.258	I
25	2104.45	III
240	2105.824	I
95 h	2124.744	I
50 h	2186.451	I
340 r	2198.714	I
15	2220.375	I
18	2256.001	I
18	2314.201	I

Intensity	Wavelength/Å	
24	2327.918	I
15	2359.233	I
20	2379.144	I
10	2389.472	I
15	2397.885	I
130	2417.367	I
30	2436.412	I
30	2488.25	IV
90	2497.962	I
500	2500.54	II
70	2533.230	I
20	2542.44	IV
3	2556.298	I
28	2589.188	I
500	2592.534	I
8	2644.184	I
1200	2651.172	I
500	2651.568	I
500	2691.341	I
850	2709.624	I
400	2729.78	II
40	2740.426	I
650	2754.588	I
70	2793.925	I
80	2829.008	I
1000	2831.843	II
1000	2845.527	II
750	3039.067	I
600	3067.021	I
20	3124.816	I
35	3211.86	III
40	3255.05	III
110	3269.489	I
40	3434.03	III
300	3499.21	II
60	3554.19	IV
50	3676.65	IV
200	4178.96	III
70	4226.562	I
200	4260.85	III
150	4291.71	III
10	4685.829	I
1000	4741.806	II
1000	4814.608	II
50	4824.097	II
100	5131.752	II
200	5178.648	II
3	5194.583	I
6	5265.892	I
6	5513.263	I
8	5564.741	I
8	5607.010	I
6	5616.135	I
7	5621.426	I
6	5664.226	I
5	5664.842	I
9	5691.954	I
6	5701.776	I
5	5717.877	I
6	5801.029	I
9	5802.093	I
1000	5893.389	II
500	6021.041	II
75	6283.452	II
100	6336.377	II
100	6484.181	II
6	6557.488	I
50	7049.369	II
30	7145.390	II
5	7353.334	I
7	7384.208	I
10	7833.575	I
10	8031.039	I
6	8044.165	I
10	8256.013	I
10	8482.21	I
9	8700.60	I
5	9060.785	I
5	9095.957	I
6	9398.868	I
20	9474.993	II
20	9475.645	II
4	9492.559	I
7	9625.664	I
5	10039.436	I
4	10200.952	I
10	10382.427	I
10	10404.913	I
8	10734.068	I
8	10947.416	I
10	11125.130	I
230	11252.83	I
600	11714.76	I
1300	12069.20	I
1050	12391.58	I
235	13107.61	I
470	14822.38	I
150	16759.79	I
135	17214.34	I
70	18811.86	I
62	19279.24	I
28	20673.64	I

Gold Au $Z = 79$

Intensity	Wavelength/Å	
100	843.44	III
100	845.14	III
200	945.10	III
100	1040.63	III
80	1044.49	III
80	1046.81	III
100 h	1239.96	III
100	1278.51	III
100	1314.84	III
100 h	1328.37	I
200	1336.72	III
180	1341.68	III
100	1348.89	III
150	1350.32	III
150	1355.61	III
150	1356.13	III
50	1363.98	I
500	1365.40	III
200	1367.17	III
60	1368.62	I
70	1374.82	I
50	1375.76	I
180	1377.73	III
150	1378.69	III
150	1379.98	III
125	1380.53	III
200	1381.36	III
80	1382.75	I
50	1385.33	I
300	1385.79	III
100	1389.41	III
180	1391.46	III
60	1392.27	I
180	1396.00	III
50	1402.12	I
100	1402.91	III
70	1407.38	I
100	1408.45	I
225	1409.50	III
250	1413.80	III
100	1414.27	III
100	1417.09	III
125	1417.39	III
150	1427.42	III
300	1428.93	III
80	1429.19	I
250	1430.06	III
275	1433.37	III
50	1435.79	I
250	1435.81	III
300	1439.12	III
200	1441.21	III
150	1446.37	III
250	1448.42	III
250	1454.95	III
100	1464.72	III
150	1471.28	III
100	1474.73	III
150	1481.10	III
100	1481.76	I
300	1487.15	III
250	1487.91	III
200	1489.47	III
250	1500.37	III
200	1502.47	III
200	1503.74	III
100	1542.00	III
100	1548.50	III
200	1567.54	III
200	1574.85	III
200	1579.44	III
150	1584.10	III
200	1587.16	I
200	1589.56	III
150	1593.41	III
70	1598.24	I
200	1600.51	III
250	1617.16	III
100	1617.78	III
500	1621.93	III
100	1624.34	I
300 d	1629.13	III
250	1638.88	III
50	1639.90	I
100	1644.17	III
150	1646.67	I
250	1652.74	III
250	1664.77	III
100	1665.76	I
100	1668.11	III
125	1673.93	III
1000	1693.94	III
150	1697.09	III
200	1698.98	III
200	1699.34	I
200	1700.00	III
200	1702.25	III
100	1707.53	III
250	1710.16	III
200	1715.69	III
100	1716.71	III
300	1717.83	III
500	1727.31	III
100 d	1733.17	III
300	1738.48	III
150	1744.39	III
500	1746.10	III
500	1756.92	III
500	1761.95	III
300	1767.42	III
100	1774.42	III
800	1775.17	III
200	1776.40	III
100	1780.57	III
60	1783.22	II
300	1786.11	III
150	1792.65	III
500	1793.76	III
200	1801.98	III
400	1805.24	III
100	1809.81	III
400	1821.17	III
400	1844.89	III
150	1848.83	III
500	1861.80	III
150	1871.92	III
100	1879.83	I
150	1918.28	III
100	1932.04	III
100	1935.42	III
200	1948.79	III
100	1958.47	III
400	1989.63	III
150	1996.85	III
11000	2012.00	I
2600	2021.38	I
150	2082.09	II
300	2083.09	III
60	2110.68	II
100	2159.08	III
80	2167.33	III

Intensity	Wavelength/Å	
200	2172.20	III
100	2184.11	III
500	2188.97	III
70	2248.56	II
80	2263.62	II
300	2322.27	III
180	2352.65	I
100	2382.40	III
120	2387.75	I
150	2402.71	III
150	2405.12	III
2600	2427.95	I
60	2533.52	II
250	2641.48	I
3400	2675.95	I
1100	2748.25	I
100	2780.82	I
1000	2802.04	II
300	2819.79	II
100	2822.55	II
100 h	2825.44	II
300	2837.85	II
100	2846.92	II
100	2856.74	II
300	2883.45	I
300	2891.96	I
100	2893.25	II
100	2907.04	II
300	2913.52	II
300	2918.24	II
100	2954.22	II
100 h	2990.27	II
300	2994.80	II
320	3029.20	I
300	3065.42	I
100	3122.50	II
1600	3122.78	I
100	3194.72	I
100	3227.99	III
300	3230.63	I
300	3308.30	I
300	3309.64	I
100	3309.86	III
100	3320.12	I
100	3355.15	I
100 h	3391.31	I
100	3395.40	I
100	3467.21	I
300	3557.36	I
300	3586.73	I
100	3611.57	I
100 h	3631.31	I
300	3637.90	I
100 h	3645.02	I
100	3650.74	I
100	3709.62	I
100	3796.01	I
100	3874.73	I
100 h	3892.26	I
400	3897.86	I
300	3909.38	I
100	3927.69	I
400	4040.93	I
700	4065.07	I
100	4084.10	I
100	4241.80	I
200	4315.11	I
120 h	4437.27	I
250	4488.25	I
900 h	4607.51	I
100 h	4620.56	I
500	4792.58	I
100	4811.60	I
100	5147.44	I
300	5230.26	I
100 h	5261.76	I
100	5655.77	I
100 h	5721.36	I
300	5837.37	I
100 h	5862.93	I
300 h	5956.96	I
600	6278.17	I
100	6562.68	I
600	7510.73	I
10	8145.06	I
10	9254.28	I

***Hafnium Hf** Z = 72*

Intensity	Wavelength/Å	
220	545.41	V
180	600.00	V
200	618.27	IV
200	644.54	IV
400	647.39	IV
600	665.65	IV
200	673.49	IV
270	867.25	V
180	875.88	V
180	885.58	V
180	896.14	V
180	901.54	V
180	919.10	V
270	921.67	V
245	951.62	V
180	960.12	V
180	964.74	V
160	971.51	V
160	1092.76	V
160	1201.76	V
270	1232.03	V
200	1233.59	V
160	1237.42	V
160	1239.53	V
160	1244.46	V
440	1396.66	V
270	1400.09	V
160	1401.70	V
370	1407.17	V
370	1408.38	V
270	1412.28	V
270	1413.51	V
160	1421.96	V
220	1422.53	V
370	1433.43	V
370	1437.27	V
500	1437.73	V
370	1445.40	V
270	1457.91	V
100	1717.21	IV
270	1719.32	V
550	1729.08	V
750	1731.83	V
750	1733.96	V
440	1741.74	V
1000	1749.11	V
1000	1750.19	V
500	1760.89	V
370	1765.62	V
270	1774.02	V
6200	2012.78	II
8500	2028.18	II
300	2070.94	III
1200	2096.18	II
200 h	2099.30	III
200 h	2110.31	III
200	2155.66	III
200	2183.50	III
200	2195.44	III
540	2210.82	II
200	2234.59	III
320	2254.01	II
160	2255.15	II
250	2266.83	II
620	2277.16	II
200 h	2313.44	III
230	2321.14	II
580	2322.47	II
300	2323.25	II
300	2324.89	II
200	2332.97	II
300	2336.47	III
200	2337.33	II
230	2343.32	II
320	2347.44	II
540	2351.22	II
250	2380.30	II
250	2383.540	III
170	2393.18	II
450	2393.36	II
670	2393.83	II
540	2405.42	II
370	2410.14	II
320	2417.69	II
390	2447.25	II
450	2460.49	II
400	2461.74	III
430	2464.19	II
210	2469.18	II
2000	2495.16	III
290	2496.99	II
580	2512.69	II
580	2513.03	II
1000	2515.16	III
890	2516.88	II
340	2531.19	II
200	2537.33	II
320	2551.40	II
400 h	2560.74	III
250	2563.61	II
300 h	2567.46	III
890	2571.67	II
320	2573.90	II
320	2576.82	II
300	2578.14	II
320	2582.54	II
390	2606.37	II
450	2607.03	II
230	2613.60	II
450	2622.74	II
160	2637.00	I
1100	2638.71	II
1100	2641.41	II
160	2642.75	I
670	2647.29	II
160	2657.84	II
210	2661.88	II
290	2683.35	II
200	2687.22	III
670	2705.61	I
210	2712.42	II
250	2718.59	I
710	2738.76	II
200	2743.64	I
360	2751.81	II
500	2753.60	III
450	2761.63	I
160	2766.96	I
170	2773.02	I
980	2773.36	II
180	2774.02	II
390	2779.37	I
230	2808.00	II
230	2813.86	II
170	2814.48	II
230	2817.68	I
200	2819.74	I
1200	2820.22	II
490	2822.68	II
180	2833.28	I
410	2845.83	I
270	2849.21	II
270	2850.96	I
180	2851.21	II
180	2860.56	I
760	2861.01	II
760	2861.70	II
2100	2866.37	I
210	2887.14	I
800	2889.62	I
1800	2898.26	I
1200	2904.41	I
890	2904.75	I
2000	2916.48	I
580	2918.58	I
320	2919.59	II
180	2924.62	I

Intensity	Wavelength/Å	
490	2929.63	II
450	2929.90	I
710	2937.80	II
2000	2940.77	I
160	2944.71	I
1200	2950.68	I
1100	2954.20	I
540	2958.02	I
1400	2964.88	I
620	2966.93	I
710	2968.81	II
890	2975.88	II
1100	2980.81	I
210	2982.72	I
170	3000.10	II
800	3005.56	I
1100	3012.90	II
540	3016.78	I
1100	3016.94	II
980	3018.31	I
1200	3020.53	I
410	3031.16	II
710	3050.76	I
1100	3057.02	I
850	3067.41	I
2100	3072.88	I
170	3074.10	I
250	3074.79	I
430	3080.84	I
200	3096.76	I
340	3101.40	II
710	3109.12	II
710	3131.81	I
850	3134.72	II
170	3139.65	II
220	3145.32	II
220	3148.41	I
450	3156.63	I
270	3159.82	I
710	3162.61	II
450	3168.39	I
890	3172.94	I
450	3176.86	II
220	3181.01	I
360	3193.53	II
670	3194.19	II
200	3196.93	I
310	3206.11	I
180	3210.98	I
180	3217.30	II
180	3220.61	II
360	3247.66	I
220	3249.53	I
890	3253.70	II
270	3255.28	II
180	3273.66	II
200 h	3279.67	III
270	3279.98	II
160	3291.05	I
210	3306.12	I
340	3310.27	I

Intensity	Wavelength/Å	
670	3312.86	I
180	3317.99	II
890	3332.73	I
370	3352.06	II
230	3358.91	I
180	3360.06	I
180	3378.93	I
230	3384.70	II
170	3386.21	I
800	3389.83	II
230	3392.81	I
230	3394.59	II
230	3397.26	I
230	3397.60	I
2300	3399.80	II
170	3400.21	I
180	3402.51	I
230	3410.17	II
230	3417.34	I
410	3419.18	I
200	3428.37	II
250	3438.24	II
710	3472.40	II
200	3478.99	II
480	3479.28	II
250	3495.75	II
250	3497.16	I
980	3497.49	I
1200	3505.23	II
980	3523.02	I
980	3535.54	II
760	3536.62	I
180	3548.81	I
540	3552.70	II
1300	3561.66	II
270	3567.36	I
1100	3569.04	II
210	3597.42	II
540	3599.87	I
800	3616.89	I
320	3630.87	II
800	3644.36	II
320	3649.10	I
200	3651.84	I
220	3665.35	II
200	3672.27	I
480	3675.74	I
2200	3682.24	I
280	3696.51	I
240	3699.72	II
340	3701.15	II
1000	3717.80	I
650	3719.28	II
160	3729.10	I
460	3733.79	I
160	3737.88	II
400	3746.80	I
170	3766.92	II
200	3768.25	I
1400	3777.64	I
1400	3785.46	I

Intensity	Wavelength/Å	
650	3793.37	II
850 d	3800.38	I
320	3811.78	I
1300	3820.73	I
280	3830.02	I
800	3849.18	I
600	3858.31	I
230	3860.91	I
200	3872.55	II
160	3877.10	II
380	3880.82	II
200	3882.52	I
200	3889.23	I
200	3889.33	I
620	3899.94	I
620	3918.09	II
200	3923.90	II
320	3931.38	I
410	3951.83	I
160	3968.01	I
200	3973.48	I
180	4032.27	I
230	4062.84	I
180	4083.35	I
540	4093.16	II
1100	4174.34	I
160	4206.58	II
190	4209.70	I
170	4228.08	I
170	4232.44	II
170	4260.98	I
200	4263.39	I
170	4272.85	II
320	4294.79	I
160	4330.27	I
180	4336.66	II
250	4356.33	I
180	4370.97	II
160	4417.91	I
200	4438.04	I
250	4565.94	I
500 d	4598.80	I
230	4620.86	I
210	4655.19	I
120	4699.01	I
160	4782.74	I
310	4800.50	I
130	4859.24	I
120	4975.25	I
95	5018.20	I
95	5047.45	I
75	5170.18	I
230	5181.86	I
110	5243.99	I
120	5294.87	I
110	5354.73	I
110	5373.86	I
75	5452.92	I
230	5550.60	I
230	5552.12	I
95	5613.27	I

Intensity	Wavelength/Å	
160	5719.18	I
95	6098.67	I
95	6185.13	I
85	6789.27	I
160	6818.94	I
160	7063.83	I
570	7131.81	I
650	7237.10	I
410	7240.87	I
360	7624.40	I
110	7740.17	I
310	7845.35	I
130	7920.71	I
250	7994.73	I
130	8204.58	I
150	8546.48	I
160	8640.06	I
40	8711.24	I
65	9004.73	I

Helium He $Z = 2$

Intensity	Wavelength/Å	
15	231.454	II
20	232.584	II
30	234.347	II
50	237.331	II
100	243.027	II
300	256.317	II
1000	303.780	II
500	303.786	II
10	320.293	I
2	505.500	I
3	505.684	I
4	505.912	I
5	506.200	I
7	506.570	I
10	507.058	I
15	507.718	I
20	508.643	I
25	509.998	I
35	512.098	I
50	515.616	I
100	522.213	I
400	537.030	I
1000	584.334	I
50	591.412	I
5	958.70	II
6	972.11	II
8	992.36	II
15	1025.27	II
30	1084.94	II
35	1215.09	II
50	1215.17	II
120	1640.34	II
180	1640.47	II
7	2385.40	II
9	2511.20	II
50	2577.6	I
1	2723.19	I
12	2733.30	II
2	2763.80	I
10	2818.2	I
4	2829.08	I

Intensity	Wavelength/Å	
10	2945.11	I
40	3013.7	I
20	3187.74	I
3	3202.96	II
15	3203.10	II
1	3354.55	I
2	3447.59	I
1	3587.27	I
3	3613.64	I
2	3634.23	I
3	3705.00	I
1	3732.86	I
10	3819.607	I
1	3819.76	I
500	3888.65	I
20	3964.729	I
1	4009.27	I
50	4026.191	I
5	4026.36	I
12	4120.82	I
2	4120.99	I
3	4143.76	I
10	4387.929	I
3	4437.55	I
200	4471.479	I
25	4471.68	I
6	4685.4	II
30	4685.7	II
30	4713.146	I
4	4713.38	I
20	4921.931	I
100	5015.678	I
10	5047.74	I
5	5411.52	II
500	5875.62	I
100	5875.97	I
8	6560.10	II
100	6678.15	I
3	6867.48	I
200	7065.19	I
30	7065.71	I
50	7281.35	I
1	7816.15	I
2	8361.69	I
2	9063.27	I
2	9210.34	I
10	9463.61	I
4	9516.60	I
3	9526.17	I
1	9529.27	I
1	9603.42	I
3	9702.60	I
6	10027.73	I
2	10031.16	I
15	10123.6	II
1	10138.50	I
10	10311.23	I
2	10311.54	I
3	10667.65	I
300	10829.09	I
1000	10830.25	I

Intensity	Wavelength/Å	
2000	10830.34	I
9	10913.05	I
3	10917.10	I
4	11626.4	II
30	11969.12	I
20	12527.52	I
50	12784.99	I
20	12790.57	I
7	12845.96	I
10	12968.45	I
2	12984.89	I
12	15083.64	I
200	17002.47	I
1	18555.55	I
6	18636.8	II
500	18685.34	I
200	18697.23	I
100	19089.38	I
20	19543.08	I
1000	20581.30	I
80	21120.07	I
10	21121.43	I
20	21132.03	I
3	30908.5	II
4	40478.90	I

Holmium Ho $Z = 67$

Intensity	Wavelength/Å	
170	2502.91	II
170	2533.80	I
190	2605.86	II
270	2750.35	II
270	2769.89	II
300	2824.20	II
270 c	2831.69	II
270	2849.10	II
360	2880.26	II
460	2880.98	II
570 c	2909.41	II
410 c	2979.63	II
410	2987.64	II
480 c	3049.38	II
410 c	3054.00	II
500 c	3057.45	II
500 c	3082.34	II
910	3084.36	II
430 c	3086.54	II
760	3118.50	II
580 c	3166.62	II
810	3173.78	II
810 c	3181.50	II
980 c	3281.97	II
630 c	3337.23	II
980 c	3343.58	II
8100 c	3398.98	II
810 c	3410.26	II
1400 c	3414.90	II
5400	3416.46	II
1200	3421.63	II
2000 c	3425.34	II
2000 c	3428.13	II
3200	3453.14	II
16000 c	3456.00	II

Intensity	Wavelength/Å	
1600	3461.97	II
810 c	3473.91	II
5400 c	3474.26	II
6300	3484.84	II
2500 c	3494.76	II
810 c	3498.88	II
810	3510.73	I
4100 c	3515.59	II
410 c	3519.94	II
630	3540.76	II
1600	3546.05	II
1100 c	3556.78	II
410	3560.15	II
410 c	3573.24	II
630 c	3574.80	II
810	3579.12	I
410	3580.75	II
410	3581.83	II
630 c	3592.23	II
1100 cw	3598.77	II
540 c	3600.95	II
410	3618.43	I
430 c	3626.69	II
490	3627.25	II
430 c	3631.76	II
430 c	3638.30	II
1600 c	3662.29	I
1400	3667.97	I
720	3679.19	I
670	3679.70	I
720	3682.65	I
580	3690.65	I
410	3700.04	I
490 c	3702.35	II
430	3712.88	I
450	3720.72	I
1100	3731.40	I
810	3736.35	I
3200 cw	3748.17	II
8900 c	3796.75	II
8900 c	3810.73	II
410 c	3835.35	II
1300 cw	3837.51	II
410 c	3842.05	II
1100	3843.86	II
490 c	3846.73	II
1800 c	3854.07	II
2700 c	3861.68	II
3000 c	3888.96	II
13000 c	3891.02	II
1300 cw	3905.68	II
580	3955.73	I
490	3959.68	I
2700	4040.81	I
5400 c	4045.44	II
8100	4053.93	I
1700	4065.09	II
720	4068.05	I
8900	4103.84	I
2900	4108.62	I
1500	4120.20	I

Intensity	Wavelength/Å	
1300	4125.65	I
4300	4127.16	I
1500	4136.22	I
980 cw	4152.61	II
8100	4163.03	I
2500	4173.23	I
540	4194.35	I
2000	4227.04	I
1300 cw	4254.43	I
490	4264.05	I
1300	4350.73	I
300	4477.64	II
290	4629.10	II
80	4701.69	II
130	4709.84	II
130 c	4717.52	I
290	4742.04	II
100 c	4757.01	I
65	4782.92	I
290	4939.01	I
250 c	4967.21	II
220	4979.97	I
90	4995.05	I
130	5042.37	I
80	5093.07	I
140	5127.81	I
130	5142.59	II
110	5143.22	II
160	5149.59	II
90 c	5167.88	I
130 c	5182.11	I
90	5190.11	II
65	5251.82	I
90	5301.25	I
80	5330.11	I
90	5359.99	I
100	5407.08	I
70	5566.52	I
65	5627.60	I
140	5659.58	I
140 c	5691.47	I
140 c	5696.57	I
140 c	5860.28	I
70 c	5882.99	I
70	5921.76	I
70 cw	5948.03	I
70	5972.76	I
90	5973.52	I
230 c	5982.90	I
120	6081.79	I
70	6208.65	I
70 c	6305.36	I
70	6550.97	I
260	6604.94	I
120	6628.99	I
55 cw	6694.32	I
55 c	6785.43	I
40 cw	6939.49	I
45 cw	6950.39	I
140	7555.09	I
40 c	7628.42	I

Intensity	Wavelength/Å	
50 c	7693.15	I
60 cw	7815.48	I
60	7894.64	I
50	8512.94	I
40	8670.19	I
90	8915.98	II
Hydrogen H Z = 1		
15	926.226	I
20	930.748	I
30	937.803	I
50	949.743	I
100	972.537	I
300	1025.722	I
1000	1215.668	I
500	1215.674	I
5	3835.384	I
6	3889.049	I
8	3970.072	I
15	4101.74	I
30	4340.47	I
80	4861.33	I
120	6562.72	I
180	6562.852	I
5	9545.97	I
7	10049.4	I
12	10938.1	I
20	12818.1	I
40	18751.0	I
5	21655.3	I
8	26251.5	I
15	40511.6	I
4	46525.1	I
6	74578.	I
Indium In Z = 49		
17	378.61	V
17	386.21	V
14	388.91	V
25	393.89	V
25	400.57	V
25	402.39	V
622	472.71	IV
689	479.39	IV
709	498.62	IV
10	882.24	III
10	890.84	III
10	915.87	III
85	954.67	IV
87	973.50	IV
86	991.60	IV
89	1024.68	IV
85	1024.79	IV
88	1031.45	IV
82	1031.98	IV
80	1054.43	IV
84	1063.03	IV
83	1068.25	IV
82	1069.82	IV
86	1077.64	IV
90	1082.10	IV
83	1086.33	IV
82	1096.81	IV
84	1097.18	IV
85	1116.10	IV
80	1124.06	IV
90	1131.46	IV
85	1144.43	IV
80	1145.41	IV
89	1146.62	IV
83	1154.11	IV
84	1154.60	IV
90	1157.71	IV
90	1157.82	IV
85	1159.78	IV
88	1176.50	IV
85	1191.58	IV
83	1204.87	IV
90	1206.55	IV
88	1221.50	IV
85	1221.90	IV
85	1233.58	IV
87	1235.84	IV
90	1373.20	IV
88	1398.77	IV
81	1412.09	IV
100	1625.42	III
100	1748.83	III
30	1842.41	III
40	1850.30	III
30	2154.08	III
50	2205.28	II
40	2281.64	II
100 c	2306.05	II
90 d	2313.21	II
70 d	2327.95	II
80 h	2334.57	II
110 d	2382.63	II
70 h	2427.20	II
100	2447.90	II
110 d	2488.62	II
90	2488.95	II
80	2498.59	II
100	2499.60	II
90 d	2500.99	II
110 d	2512.31	II
100	2521.37	I
100	2527.41	III
160 d	2554.44	II
1100	2560.15	I
200	2601.76	I
90 d	2654.70	II
100 d	2662.63	II
140 d	2668.65	II
140 d	2674.56	II
80	2683.12	II
1600	2710.26	I
300	2713.94	I
80	2726.15	III
130 d	2749.75	II
700	2753.88	I
90 d	2818.97	II
180 c	2836.92	I
80	2865.68	II
120 d	2890.18	II
1100	2932.63	I
100	2941.05	II
100	2982.80	III
110 c	2999.40	II
100	3008.08	III
8000	3039.36	I
110 d	3099.80	II
180 c	3101.8	II
130 c	3138.60	II
80 c	3142.75	II
130 d	3146.70	II
150	3155.77	II
100 c	3158.40	II
90 c	3176.30	II
90 d	3198.11	II
13000	3256.09	I
3000	3258.56	I
90 c	3338.50	II
100 c	3404.28	II
110 d	3438.40	II
180 c	3693.91	II
95 c	3708.13	II
380 w	3716.14	II
120 c	3718.30	II
160 c	3718.72	II
160 c	3723.40	II
170 w	3795.21	II
230 c	3799.21	II
250 c	3834.65	II
200 c	3842.18	II
100	3852.82	III
100	3889.78	II
100 c	3902.07	II
250 w	3962.35	II
120 c	4004.66	II
140 d	4013.92	II
100	4023.77	III
150	4032.32	III
410 w	4056.94	II
100	4071.57	III
100	4072.93	III
17000	4101.76	I
140 c	4205.14	II
100 d	4213.04	II
110 d	4219.66	II
100	4252.68	III
150 d	4372.87	II
150 c	4500.78	II
18000	4511.31	I
110 c	4549.01	II
140 c	4570.85	II
180 w	4578.02	II
180 w	4578.40	II
140 c	4616.08	II
170 c	4617.17	II
250 c	4620.14	II
150 w	4620.70	II
170 c	4627.30	II
140 c	4637.04	II
380 c	4638.16	II
220 c	4644.58	II
360 c	4655.62	II
320 w	4656.74	II
190 c	4681.11	II
450 w	4684.8	II
90 d	4907.06	II
150 c	4973.77	II
80 h	5109.36	II
100 w	5115.14	II
140 c	5117.40	II
270 c	5120.80	II
200 w	5121.75	II
80 d	5129.85	II
240 c	5175.42	II
140 c	5184.44	II
200	5248.77	III
150 c	5309.45	II
80	5411.41	II
140 c	5418.45	II
220 w	5436.70	II
130 c	5497.50	II
140 c	5507.08	II
320 c	5513.00	II
250 w	5523.28	II
130 c	5536.50	II
190 w	5555.45	II
240 c	5576.90	II
200 w	5636.70	II
100	5645.15	III
160 c	5708.50	II
100 c	5721.80	II
100	5819.50	III
210 c	5853.15	II
490 w	5903.4	II
260 w	5915.4	II
120 c	5918.78	II
130 c	6062.9	II
250 c	6095.95	II
210 c	6108.66	II
180 w	6115.9	II
230 w	6128.7	II
240 w	6129.4	II
320 w	6132.1	II
150 c	6140.0	II
90	6143.23	II
140 c	6148.10	II
190 w	6149.5	II
80	6161.15	II
180 w	6162.45	II
200	6197.72	III
100 c	6224.28	II
280 w	6228.3	II
140 w	6231.1	II
270 w	6304.8	II
290 w	6362.3	II
300 w	6469.0	II
210 c	6541.20	II
190 c	6751.88	II
180 c	6765.9	II
100 c	6783.72	II

Intensity	Wavelength/Å	
320 w	6891.5	II
380 w	7182.9	II
180 c	7255.0	II
210 c	7276.5	II
180 c	7303.4	II
320 c	7350.6	II
100 c	7632.7	II
100 c	7682.9	II
210 c	7740.7	II
100 c	7776.96	II
180 c	7789.0	II
90 c	7840.9	II
240 c	8227.0	II
100 w	8813.5	II
80 c	8832.6	II
120 c	9197.7	II
120 c	9202.0	II
220 w	9213.0	II
160 d	9241.1	II
100	9977.86	I
200	10257.03	I
100 h	10744.31	I
20	11334.72	I
20	11731.48	I
10	12912.59	I
9	13429.96	I
7	14719.08	I
6	22291.06	I
7	23879.13	I

Iodine I $Z = 53$

Intensity	Wavelength/Å	
30	363.78	V
36	380.74	V
45	565.53	V
50	607.57	V
6	612.46	IV
6	666.81	III
8	705.11	III
7	784.64	III
7	784.80	III
8	795.52	III
7	919.28	IV
10000	1034.66	II
8	1094.20	III
10000	1139.80	II
10000	1160.56	II
20000	1166.48	II
10000	1178.65	II
15000	1187.34	II
10000	1190.85	II
200	1218.41	I
20000	1220.89	II
600	1224.05	I
600	1224.08	I
500	1228.89	I
20000	1234.06	II
600	1251.34	I
8	1252.35	III
2500	1259.15	I
3000	1259.51	I
800	1261.27	I
600	1267.57	I
600	1267.60	I
1500	1275.26	I
3000	1289.40	I
10000	1300.34	I
3000	1302.98	I
3000	1313.95	I
3000	1317.54	I
2000	1330.19	I
20000	1336.52	II
5000	1355.10	I
3000	1357.97	I
5000	1360.97	I
3000	1361.11	I
2500	1367.71	I
2500	1368.22	I
4000	1383.23	I
3000	1390.75	I
2000	1392.90	I
2000	1400.01	I
8000	1425.49	I
5000	1446.26	I
5000	1453.18	I
5000	1457.39	I
5000	1457.47	I
10000	1457.98	I
2500	1458.79	I
4000	1459.15	I
2500	1465.83	I
1000	1485.92	I
5000	1492.89	I
5000	1507.04	I
5000	1514.68	I
15000	1518.05	I
2500	1526.45	I
5000	1593.58	I
5000	1617.60	I
2500	1640.78	I
15000	1702.07	I
12000	1782.76	I
5000	1799.09	I
75000	1830.38	I
15000	1844.45	I
2000	2061.63	I
7	2361.13	IV
6	2372.45	IV
7	2376.46	IV
8	2387.11	IV
9	2426.10	IV
8	2475.35	IV
8	2519.74	IV
8	2545.67	IV
7	2545.71	III
8	2652.23	IV
5000	3078.75	II
200	4102.23	I
200	4129.21	I
100 d	4134.15	I
500 d	4321.84	I
250	4763.31	I
1000	4862.32	I
200	4916.94	I
10000	5119.29	I
3000 c	5161.20	II
1000	5234.57	I
3000 c	5245.71	II
10000	5338.22	II
5000 c	5345.15	II
600 c	5427.06	I
3000	5435.83	II
2000 c	5464.62	II
10000	5625.69	II
2000 c	5690.91	II
4000 c	5710.53	II
1000 d	5764.33	I
2000	5894.03	I
5000	5950.25	II
300	5984.86	I
2000 d	6024.08	I
2000 c	6074.98	II
1000	6082.43	I
2000 c	6127.49	II
800	6191.88	I
500	6213.10	I
800	6244.48	I
1000	6293.98	I
500	6313.13	I
800	6330.37	I
400	6333.50	I
2000	6337.85	I
1000	6339.44	I
500	6359.16	I
1000	6566.49	I
2000	6583.75	I
1000	6585.27	I
5000	6619.66	I
500	6661.11	I
500 c	6697.29	I
400	6732.03	I
4000	6812.57	II
500	6989.78	I
500	7120.05	I
1200	7122.05	I
2000	7142.06	I
1000	7164.79	I
400 d	7191.66	I
700	7227.30	I
1000	7236.78	I
500	7237.84	I
5000	7402.06	I
1000	7410.50	I
500	7416.48	I
5000	7468.99	I
500 c	7490.52	I
2000	7554.18	I
500 d	7556.65	I
2000 c	7700.20	I
600	7897.98	I
500	7969.48	I
1000	8003.63	I
99000	8043.74	I
300 d	8065.70	I
1000	8090.76	I
800 c	8169.38	I
500 d	8222.57	I
4000	8240.05	I
10000 c	8393.30	I
1000	8486.11	I
1500 c	8664.95	I
500 c	8700.80	I
250 d	8748.22	I
1000	8853.24	I
2000	8853.80	I
3000	8857.50	I
1000 d	8898.50	I
400	8964.69	I
400	8993.13	I
5000	9022.40	I
15000	9058.33	I
1000	9098.86	I
12000	9113.91	I
600	9128.03	I
600	9227.74	I
1000	9335.05	I
4000	9426.71	I
3000	9427.15	I
2000	9598.22	I
2000	9649.61	I
3000 d	9653.06	I
5000	9731.73	I
500	10003.05	I
750	10131.16	I
1000	10238.82	I
400	10375.20	I
400	10391.74	I
5000	10466.54	I
400	11236.56	I
350	11558.46	I
320	11778.34	I
450	11996.86	I
300	12033.69	I
150	12304.58	I
60	13149.16	I
140	13958.27	I
200	14287.02	I
100	14460.00	I
225	15032.57	I
105	15528.65	I
150	16037.33	I
15	18275.71	I
20	18348.52	I
15	18982.41	I
35	19070.17	I
110	19105.12	I
50	19370.02	I
10	20648.69	I
220	22183.03	I
150	22226.53	I
30	22309.21	I
32	24420.82	I
12	27365.42	I
9	27573.05	I
10	30361.93	I
8	30383.88	I

Intensity	Wavelength/Å	
10	34295.73	I
9	34513.11	I
3	40228.54	I
2	41633.80	I

Iridium Ir $Z = 77$

Intensity	Wavelength/Å	
9900	2010.65	I
8700	2022.35	I
15000	2033.57	I
6200	2052.22	I
5000	2060.64	I
3700	2083.22	I
3100	2085.74	I
17000	2088.82	I
14000	2092.63	I
2700	2112.68	I
1800	2119.54	I
2000	2125.44	I
4500	2126.81	II
2000	2127.52	I
4500	2127.94	I
3700	2148.22	I
2500	2150.54	I
3500	2152.68	II
2900	2155.81	I
7900	2158.05	I
2100	2162.88	I
5800	2169.42	II
4500	2175.24	I
2700	2178.17	I
1600	2187.43	II
1100	2190.38	II
740	2191.64	I
910	2208.09	II
1300	2220.37	I
790	2221.07	II
2500	2242.68	II
620	2245.76	II
2100	2253.38	I
2100	2255.10	I
1400	2255.81	I
350	2258.51	I
1400	2258.86	I
830	2264.61	I
1100	2266.33	I
1000	2268.90	I
660	2280.00	I
950	2281.02	II
660	2281.91	I
330	2284.60	I
330	2295.08	I
790	2298.05	I
460	2299.53	I
910	2300.50	I
2700	2304.22	I
410	2305.47	I
210	2307.27	I
910	2308.93	I
460	2315.38	I
410	2321.45	I
410	2321.58	I
210	2327.98	I
540	2333.30	I
740	2333.84	I
580	2334.50	I
1600	2343.18	I
740	2343.61	I
580	2355.00	I
230	2357.53	II
410	2358.16	I
500	2360.73	I
2500	2363.04	I
370	2368.04	II
3500	2372.77	I
290	2375.09	II
250	2377.28	I
250	2377.98	I
500	2379.38	I
540	2381.62	I
210	2383.17	I
1300	2386.89	I
2500	2390.62	I
2700	2391.18	I
230	2407.59	I
290	2409.37	I
290	2410.17	I
290	2410.73	I
540	2413.31	I
370	2415.86	I
620	2418.11	I
210	2424.89	I
370	2424.99	I
290	2425.66	I
540	2427.61	I
540	2431.24	I
1300	2431.94	I
270	2435.14	I
250	2445.34	I
250	2447.76	I
910	2452.81	I
1300	2455.61	I
230	2455.87	I
210	2457.03	I
210	2457.23	I
870	2467.30	I
3300	2475.12	I
210	2478.11	I
2100	2481.18	I
620	2493.08	I
210	2496.27	I
250	2502.63	I
4100	2502.98	I
210	2513.71	I
990	2533.13	I
1100	2534.46	I
580	2537.22	I
580	2542.02	I
7900	2543.97	I
790	2546.03	I
210	2551.40	I
210	2555.35	I
910	2564.18	I
210	2569.88	I
230	2572.70	I
740	2577.26	I
740	2592.06	I
740	2599.04	I
700	2608.25	I
1800	2611.30	I
210	2614.98	I
330	2617.78	I
210	2619.88	I
250	2625.32	I
700	2634.17	I
250	2639.42	I
3500	2639.71	I
210	2644.19	I
1800	2661.98	I
350	2662.63	I
2700	2664.79	I
520	2669.91	I
520	2671.84	I
330	2673.61	I
270	2692.34	I
3000	2694.23	I
330	2772.46	I
250	2775.55	I
520	2781.29	I
330	2785.22	I
540	2797.35	I
1600	2797.70	I
380	2798.18	I
410	2800.82	I
680	2823.18	I
1200	2824.45	I
820	2836.40	I
1100	2839.16	I
820	2840.22	I
3800	2849.72	I
380	2875.60	I
380	2875.98	I
270	2877.68	I
820	2882.64	I
650	2897.15	I
260	2901.95	I
260	2904.80	I
200	2907.24	I
440	2916.36	I
230	2918.57	I
4400	2924.79	I
1200	2934.64	I
880	2936.68	I
250	2938.47	I
2700	2943.15	I
230	2946.97	I
200	2949.76	I
1200	2951.22	I
200	2974.95	I
440	2980.65	I
300	2996.08	I
220	3002.25	I
600	3003.63	I
270	3017.31	I
380	3029.36	I
330	3039.26	I
300	3047.16	I
300	3049.44	I
300	3057.28	I
1600	3068.89	I
320	3083.22	I
240	3086.44	I
390	3088.04	I
510	3100.29	I
510	3100.45	I
340	3120.76	I
200	3121.78	I
3400	3133.32	I
490	3168.88	I
370	3177.58	I
370	3198.92	I
610	3212.12	I
370	3219.51	I
5100	3220.78	I
300	3229.28	I
470	3241.52	I
200	3262.01	I
390	3266.44	I
200	3322.60	I
560	3368.48	I
660	3437.02	I
410	3448.97	I
[illegible]	3513.64	I
220	3515.95	I
410	3522.03	I
320	3558.99	I
1200	3573.72	I
320	3594.39	I
220	3609.77	I
660	3628.67	I
220	3636.20	I
300	3661.71	I
300	3664.62	I
320	3674.98	I
200	3687.08	I
200	3731.36	II
530	3747.20	I
3100	3800.12	I
230	3817.24	I
480	3902.51	I
480	3915.38	I
400	3934.84	I
590	3976.31	I
460	3992.12	I
350	4033.76	I
370	4069.92	I
150	4070.68	I
100	4092.61	I
140	4115.78	I
90	4172.56	I
260	4268.10	I
220	4311.50	I
160	4399.47	I
65	4403.78	I
110	4426.27	I
55	4478.48	I

Intensity	Wavelength/Å	
55	4545.68	I
30	4548.48	I
35	4568.09	I
75	4616.39	I
26	4656.18	I
50	4728.86	I
26	4756.46	I
65	4778.16	I
30	4795.67	I
50	4938.09	I
26	4970.48	I
25	4999.74	I
25	5002.74	I
30	5014.98	I
30	5123.66	I
35	5364.32	I
75	5449.50	I
45	5625.55	I
35	5894.06	I
20	6110.67	I
12	6288.28	I
10	6686.08	I
6	7834.32	I

***Iron* Fe Z = 26**

Intensity	Wavelength/Å	
350	386.16	V
350	386.88	V
400	387.20	V
400	387.50	V
400	387.76	V
400	387.78	V
400	395.90	V
400	404.62	V
400	405.50	V
800	407.42	V
600	407.44	V
400	407.49	V
500	407.75	V
400	409.71	V
400	410.20	V
600	411.55	V
700	417.39	V
700	418.04	V
500	418.47	V
700	421.06	V
500	421.78	V
500	422.31	V
500	426.06	V
500	426.11	V
350	426.97	V
17	525.69	IV
15	526.29	IV
13	526.63	IV
14	536.61	IV
15	537.10	IV
13	537.26	IV
14	537.79	IV
13	537.94	IV
13	552.14	IV
14	607.53	IV
13	608.80	IV
10	813.38	III
10	844.28	III
10 p	861.83	III
10	891.17	III
10	950.33	III
10	981.37	III
10 w	983.88	III
12	1055.27	II
15	1068.36	II
15	1071.60	II
15	1096.89	II
12	1099.12	II
18	1112.09	II
12	1121.99	II
12	1122.86	II
12	1128.07	II
12	1130.43	II
15	1133.41	II
12	1133.68	II
12	1138.64	II
12	1142.33	II
12	1143.23	II
18	1144.95	II
12	1147.41	II
15	1148.29	II
12	1151.16	II
12	1267.44	II
12	1272.00	II
400	1317.86	V
400	1323.27	V
400	1330.40	V
400	1359.01	V
600	1361.82	V
700	1373.59	V
600	1373.67	V
500	1376.34	V
500	1378.56	V
800	1387.94	V
400	1397.97	V
600	1400.24	V
800	1402.39	V
400	1406.67	V
500	1406.82	V
400	1407.25	V
600	1409.45	V
400	1415.20	V
600	1420.46	V
800	1430.57	V
13	1431.43	IV
800	1440.53	V
400	1442.22	V
800	1446.62	V
700	1448.85	V
400	1449.93	V
700	1456.16	V
500	1459.83	V
400	1460.73	V
500	1462.63	V
700	1464.68	V
500	1465.38	V
400	1466.65	V
500	1469.00	V
500	1479.47	V
10 h	1505.17	III
13	1526.60	IV
13	1530.26	IV
14	1532.63	IV
13	1532.91	IV
15	1533.86	IV
13	1533.95	IV
14	1536.58	IV
10 h	1538.63	III
13	1542.16	IV
14	1542.70	IV
12 h	1550.20	III
13	1566.26	IV
14	1568.27	IV
13	1591.51	IV
13	1592.05	IV
13	1598.01	IV
13	1600.58	IV
10 h	1601.21	III
13	1601.67	IV
13	1603.18	IV
13	1603.73	IV
13	1604.88	IV
13	1605.68	IV
15	1605.97	IV
13	1606.98	IV
17	1609.10	IV
14	1609.83	IV
13	1610.47	IV
13	1611.20	IV
13	1613.64	IV
15	1614.02	IV
13	1614.64	IV
13	1615.00	IV
16	1616.68	IV
14	1617.68	IV
12	1618.47	II
14	1619.02	IV
13	1621.16	IV
14	1621.57	IV
13	1623.38	IV
13	1623.53	IV
15	1626.47	IV
14	1626.90	IV
13	1628.54	IV
13	1630.18	IV
17	1631.08	IV
15	1631.12	II
14	1632.40	IV
13	1634.01	IV
18	1635.40	II
15	1636.32	II
15	1639.40	II
14	1639.40	IV
16	1640.04	IV
14	1640.16	IV
12	1641.76	II
15	1641.87	IV
15	1647.09	IV
12	1647.16	II
15	1651.58	IV
15	1652.90	IV
13	1653.41	IV
13	1656.11	IV
15	1656.65	IV
14	1660.10	IV
13	1662.32	IV
13	1662.52	IV
13	1663.54	IV
13	1668.09	IV
12	1670.74	II
14	1671.04	IV
13	1673.68	IV
14	1675.66	IV
13	1681.36	IV
15	1687.69	IV
15	1698.88	IV
12	1702.04	II
13	1709.81	IV
15	1711.41	IV
14	1712.76	IV
14	1717.90	IV
14	1718.16	IV
14	1719.46	IV
14	1722.71	IV
14	1724.06	IV
16	1725.63	IV
13	1761.08	IV
12	1761.38	II
20	1785.26	II
20	1786.74	II
18	1788.07	II
13	1792.10	IV
13	1796.93	IV
13	1827.98	IV
10	1869.83	III
12	1877.99	III
10	1882.05	III
12	1886.76	III
13	1890.67	III
11	1893.98	III
20	1895.46	III
10 s	1907.58	III
19	1914.06	III
15	1915.08	III
15	1922.79	III
10 p	1926.01	III
18	1926.30	III
15	1930.39	III
14	1931.51	III
30	1934.538	I
25	1937.269	I
14	1937.34	III
10 l	1938.90	III
14 s	1943.48	III
12	1945.34	III
50	1946.988	I
10	1950.33	III
12	1951.01	III
25	1951.571	I
30	1952.59	I

Intensity	Wavelength/Å	
11	1952.65	III
30	1953.005	I
13	1953.32	III
10	1953.49	III
10 w	1954.22	III
60	1957.823	I
11	1958.58	III
60	1960.144	I
13	1960.32	III
30	1961.25	I
50	1962.111	I
12	1963.11	II
15	1987.50	III
14	1991.61	III
13	1994.07	III
12	1995.56	III
12	1996.42	III
10	2061.55	III
12	2068.24	III
14	2078.99	III
100	2084.122	I
10	2084.35	III
12	2090.14	III
15	2097.48	III
12	2097.69	III
12	2103.80	III
10	2107.32	III
15	2151.78	III
12	2157.71	III
50	2157.794	I
12	2158.47	III
10	2161.27	III
40	2166.773	I
12	2166.95	III
12	2171.04	III
15	2174.66	III
300	2178.118	I
12	2180.41	III
250	2186.486	I
60	2186.892	I
120	2187.195	I
250	2191.839	I
150	2196.043	I
80	2200.390	I
80	2200.724	I
15	2208.41	II
10 p	2208.85	III
20	2213.65	II
12	2218.26	II
20	2220.38	II
10	2221.83	III
10	2229.27	III
10	2232.43	III
10	2232.69	III
10	2235.91	III
10	2238.16	III
12 p	2241.54	III
50	2250.790	I
60	2251.874	I
300	2259.511	I
12	2261.59	III
60	2264.389	I
80	2267.085	I
10	2267.42	III
80	2267.469	I
50	2270.862	I
150	2272.070	I
150	2276.026	I
80	2279.937	I
150	2284.086	I
150	2287.250	I
300	2292.524	I
10	2293.06	III
15	2295.86	III
200	2297.787	I
600	2298.169	I
80	2299.220	I
300	2300.142	I
50	2301.684	I
100	2303.424	I
150	2303.581	I
120	2308.999	I
150	2313.104	I
10 p	2317.70	III
10	2319.22	III
200	2320.358	I
10 p	2321.71	III
10	2326.95	III
100	2327.40	II
100	2331.31	II
300	2332.80	II
10 p	2336.77	III
200	2338.01	II
10	2338.96	III
600	2343.49	II
80	2343.96	II
150	2344.28	II
200	2348.11	II
250	2348.30	II
200	2359.12	II
150	2360.00	II
120	2360.29	II
200	2364.83	II
80	2365.76	II
80	2368.59	II
80	2369.456	I
80	2369.95	II
120	2371.430	I
300	2373.624	I
150	2373.74	II
120	2374.518	I
120	2376.43	II
80	2379.27	II
120	2380.76	II
150	2381.835	I
1000	2382.04	II
300	2388.63	II
200	2389.973	I
1000	2395.62	II
300	2399.24	II
800	2404.88	II
250	2406.66	II
80	2406.97	II
300	2410.52	II
200	2411.07	II
150	2413.31	II
80	2417.87	II
60	2420.396	I
60	2423.089	I
150	2424.14	II
120	2428.36	II
120	2430.08	II
80	2432.26	II
60	2438.182	I
150	2439.30	II
150	2439.74	I
100	2442.57	I
250	2443.872	I
100	2444.51	II
50	2445.212	I
100	2445.57	II
60	2447.709	I
100	2453.476	I
1500	2457.598	I
150	2458.78	II
80	2461.28	II
100	2461.86	II
100	2462.181	I
1500	2462.647	I
50	2463.730	I
800	2465.149	I
60	2467.732	I
600	2468.879	I
80	2470.67	II
80	2470.965	I
800	2472.336	I
1000	2472.895	I
200	2473.16	I
600	2474.814	I
60	2476.657	I
120	2479.480	I
1200	2479.776	I
100	2480.16	II
80	2482.12	II
100	2482.66	II
10000	2483.271	I
300	2483.533	I
1000	2484.185	I
50	2485.990	I
800	2486.373	I
100	2486.691	I
100	2487.066	I
120	2487.370	I
4000	2488.143	I
100	2488.945	I
80	2489.48	II
1000	2489.750	I
50	2489.913	I
3000	2490.644	I
100	2490.71	II
2000	2491.155	I
100	2491.40	II
100	2493.18	II
500	2493.26	II
60	2494.000	I
50	2494.251	I
100	2495.87	I
600	2496.533	I
150	2498.90	I
1000	2501.132	I
50	2501.693	I
80	2506.09	II
500	2507.900	I
50	2508.753	I
1000	2510.835	I
120	2511.76	II
80	2512.275	I
400	2512.365	I
80	2516.570	I
300	2517.661	I
800	2518.102	I
150	2519.629	I
50	2522.480	I
4000	2522.849	I
200	2523.66	I
500	2524.293	I
100	2525.02	I
200	2525.39	II
300	2526.29	II
2000	2527.435	I
800	2529.135	I
250	2529.55	II
150	2529.836	I
200	2530.687	I
120	2533.63	II
100	2534.42	II
120	2535.49	II
400	2535.607	I
200	2536.792	I
200	2536.80	II
100	2538.80	II
100	2538.91	II
150	2538.99	II
50	2539.357	I
200	2540.66	II
600	2540.972	I
80	2541.10	II
300	2542.10	I
250	2543.92	I
150	2544.70	I
800	2545.978	I
80	2546.67	II
100	2548.74	II
80	2549.08	II
80	2549.39	II
600	2549.613	I
400	2562.53	II
200	2563.48	II
150	2574.36	II
300	2576.691	I
100	2582.58	II
1500	2584.54	I
650	2585.88	II
90	2591.54	II

Intensity	Wavelength/Å	
90	2593.73	II
650	2598.37	II
2000	2599.40	II
300	2599.57	I
60	2605.657	I
300	2606.51	II
800	2606.827	I
650	2607.09	II
600	2611.87	II
320	2613.82	II
320	2617.62	II
250	2618.018	I
90	2620.41	II
400	2623.53	I
200	2625.67	II
150	2628.29	II
250	2631.05	II
250	2631.32	II
100	2632.237	I
300	2635.809	I
50	2641.646	I
200	2643.998	I
300	2666.812	I
60	2666.965	I
600	2679.062	I
500	2684.75	II
400	2689.212	I
10 h	2695.13	III
200	2699.106	I
80	2706.012	I
400	2706.582	I
60	2708.571	I
200	2711.655	I
80	2714.41	II
50	2716.257	I
50	2717.786	I
250	2718.436	I
4000	2719.027	I
100	2719.420	I
50	2720.197	I
1500	2720.903	I
400	2723.578	I
150	2724.953	I
50	2726.235	I
80	2727.54	II
200	2728.020	I
50	2728.820	I
80	2728.90	II
1000	2733.581	I
60	2734.005	I
50	2734.268	I
500	2735.475	I
50	2735.612	I
500	2737.310	I
120	2737.83	I
400	2739.55	II
250	2742.254	I
800	2742.405	I
200	2743.20	II
150	2743.565	I
200	2744.068	I
80	2744.527	I
300	2746.48	II
100	2749.32	II
500	2749.48	II
1200	2750.140	I
80	2753.29	II
150	2754.032	I
100	2754.426	I
800	2755.73	II
250	2756.328	I
100	2757.316	I
120	2761.780	I
150	2761.81	II
150	2762.026	I
120	2762.772	I
120	2763.109	I
80	2766.910	I
250	2767.522	I
300	2772.07	I
600	2778.220	I
3000	2788.10	I
200	2797.78	I
400	2804.521	I
1500	2806.98	I
10 p	2813.24	III
2500	2813.287	I
300	2823.276	I
600	2825.56	I
50	2825.687	I
120	2828.808	I
1500	2832.436	I
120	2835.950	I
200	2838.119	I
200	2843.631	I
1000	2843.977	I
100	2845.594	I
800	2851.797	I
50	2869.307	I
50	2872.334	I
80	2874.172	I
50	2894.504	I
12	2904.43	III
10	2907.50	III
12	2907.70	III
120	2912.157	I
120	2929.007	I
1200	2936.903	I
60	2941.343	I
1000	2947.876	I
600	2953.940	I
250	2957.364	I
150	2965.254	I
1500	2966.898	I
120	2969.36	I
800	2970.099	I
1200	2973.132	I
500	2973.235	I
600	2981.445	I
1000	2983.570	I
1000	2994.427	I
250	2994.502	I
500	2999.512	I
120	3000.451	I
800	3000.948	I
12	3001.62	III
60	3001.655	I
12 h	3007.28	III
200	3007.282	I
500	3008.14	I
120	3009.569	I
15	3013.17	III
60	3017.627	I
60	3018.983	I
500	3020.491	I
1500	3020.639	I
600	3021.073	I
500	3024.032	I
150	3025.638	I
500	3025.842	I
80	3030.148	I
60	3031.214	I
60	3034.484	I
800	3037.389	I
80	3041.637	I
800	3047.604	I
600	3057.446	I
1000	3059.086	I
250	3067.244	I
120	3075.719	I
120	3091.577	I
80	3098.189	I
100	3099.895	I
100	3099.968	I
60	3100.303	I
100	3100.665	I
10 p	3136.43	III
10	3174.09	III
80	3175.445	I
10	3175.99	III
10	3178.01	III
150	3184.895	I
250	3191.659	I
500	3193.226	I
800	3193.299	I
200	3196.928	I
80	3199.500	I
50	3205.398	I
100	3211.88	I
200	3214.011	I
200	3214.396	I
60	3215.938	I
50	3217.377	I
80	3219.583	I
60	3219.766	I
300	3222.045	I
600	3225.78	I
80	3227.796	I
50	3233.967	I
120	3234.613	I
300	3236.222	I
100	3239.433	I
80	3244.187	I
80	3246.005	I
80	3265.046	I
50	3265.617	I
13	3266.88	III
50	3271.000	I
11	3276.08	III
150	3286.75	I
10	3288.81	III
120	3305.97	I
200	3306.343	I
400	3355.227	I
80	3355.517	I
60	3369.546	I
120	3370.783	I
50	3378.678	I
50	3380.110	I
60	3383.978	I
50	3392.304	I
150	3392.651	I
150	3399.333	I
80	3404.353	I
500	3407.458	I
250	3413.131	I
60	3424.284	I
500	3427.119	I
60	3428.748	I
6000	3440.606	I
2500	3440.989	I
1000	3443.876	I
200	3445.149	I
1200	3465.860	I
2000	3475.450	I
500	3476.702	I
2500	3490.574	I
500	3497.840	I
10	3501.76	III
250	3513.817	I
300	3521.261	I
400	3526.040	I
100	3526.166	I
60	3526.237	I
60	3526.381	I
60	3526.467	I
100	3533.199	I
200	3536.556	I
300	3541.083	I
250	3542.075	I
80	3553.739	I
400	3554.925	I
200	3556.878	I
400	3558.515	I
1000	3565.379	I
1200	3570.097	I
800	3570.25	I
120	3571.996	I
100	3573.393	I
60	3573.829	I
60	3573.888	I
4000	3581.19	I
150	3582.199	I
150	3584.660	I

Intensity	Wavelength/Å	
120	3584.929	I
300	3585.319	I
150	3585.705	I
10	3586.04	III
200	3586.103	I
400	3586.984	I
100	3594.633	I
11	3600.94	III
150	3603.204	I
11	3603.88	III
200	3605.454	I
500	3606.680	I
1500	3608.859	I
250	3610.16	I
60	3612.068	I
150	3617.788	I
1500	3618.768	I
200	3621.462	I
150	3622.004	I
150	3623.19	I
100	3631.096	I
1200	3631.463	I
60	3632.041	I
100	3638.298	I
200	3640.389	I
80	3643.717	I
1500	3647.842	I
250	3649.506	I
80	3650.279	I
200	3651.467	I
120	3670.024	I
150	3670.089	I
100	3676.311	I
150	3677.629	I
1500	3679.913	I
200	3682.242	I
120	3683.054	I
150	3684.107	I
120	3685.998	I
500	3687.456	I
120	3689.477	I
150	3694.008	I
120	3695.051	I
150	3701.086	I
80	3704.462	I
1200	3705.566	I
60	3707.041	I
150	3707.821	I
300	3707.919	I
600	3709.246	I
120	3716.442	I
8000	3719.935	I
1500	3722.563	I
120	3724.377	I
60	3725.491	I
60	3727.093	I
500	3727.619	I
150	3732.396	I
1200	3733.317	I
5000	3734.864	I
120	3735.324	I
6000	3737.131	I
100	3738.306	I
400	3743.362	I
6000	3745.561	I
1200	3745.899	I
3000	3748.262	I
80	3748.964	I
3000	3749.485	I
1500	3758.232	I
400	3760.05	I
1500	3763.788	I
400	3765.54	I
600	3767.191	I
60	3776.452	I
250	3785.95	I
100	3786.68	I
250	3787.880	I
250	3790.092	I
150	3794.34	I
400	3795.002	I
120	3797.518	I
250	3798.511	I
400	3799.547	I
200	3805.345	I
80	3806.696	I
600	3812.964	I
60	3813.059	I
1500	3815.840	I
2500	3820.425	I
150	3821.179	I
80	3824.306	I
2500	3824.444	I
1500	3825.880	I
1200	3827.823	I
1000	3834.222	I
120	3839.257	I
500	3840.437	I
800	3841.047	I
120	3843.256	I
80	3846.800	I
200	3849.96	I
120	3850.817	I
2500	3856.372	I
150	3859.212	I
10000	3859.911	I
150	3865.523	I
60	3867.215	I
250	3872.501	I
150	3873.761	I
250	3878.018	I
2000	3878.573	I
4000	3886.282	I
200	3887.048	I
300	3888.513	I
800	3895.656	I
1200	3899.707	I
400	3902.945	I
250	3906.479	I
80	3916.731	I
600	3920.258	I
1200	3922.911	I
1200	3927.920	I
2000	3930.296	I
60	3948.774	I
60	3949.953	I
50	3951.164	I
50	3952.601	I
16	3954.33	III
60	3956.454	I
250	3956.68	I
60	3966.614	I
11	3968.72	III
100	3969.257	I
80	3977.741	I
10 w	3979.42	III
40	3981.771	I
50	3983.956	I
60	3994.114	I
200	3997.392	I
40	3998.053	I
400	4005.241	I
60	4009.713	I
100	4021.867	I
10	4035.42	III
50	4040.638	I
4000	4045.813	I
11	4053.11	III
1500	4063.594	I
50	4066.975	I
50	4067.977	I
1200	4071.737	I
40	4076.629	I
12	4081.00	III
40	4100.737	I
40	4107.489	I
150	4118.544	I
10	4120.90	III
11	4122.02	III
11	4122.78	III
40	4127.608	I
400	4132.058	I
80	4134.676	I
40	4136.997	I
15	4137.76	III
13	4139.35	III
200	4143.415	I
800	4143.869	I
40	4153.898	I
50	4154.500	I
60	4156.799	I
18	4164.73	III
13	4166.84	III
50	4172.744	I
13	4174.26	III
60	4174.912	I
50	4175.635	I
50	4177.593	I
120	4181.754	I
50	4184.891	I
120	4187.038	I
120	4187.795	I
80	4191.430	I
40	4195.329	I
150	4198.304	I
40	4199.095	I
300	4202.029	I
40	4203.984	I
80	4206.696	I
80	4210.343	I
400	4216.183	I
100	4219.360	I
50	4222.212	I
11	4222.27	III
50	4225.956	I
200	4227.423	I
100	4233.602	I
13	4235.56	III
250	4235.936	I
50	4238.809	I
12	4243.75	III
50	4247.425	I
200	4250.118	I
300	4250.787	I
40	4258.315	I
800	4260.473	I
250	4271.153	I
1200	4271.759	I
12 h	4273.40	III
12	4279.72	III
1200	4282.402	I
14 h	4286.16	III
80	4291.462	I
16 h	4296.85	III
250	4299.234	I
18 h	4304.78	III
1200	4307.901	I
20 h	4310.36	III
150	4315.084	I
1500	4325.761	I
80	4352.734	I
80	4369.771	I
11 h	4372.31	III
14 h	4372.53	III
18 h	4372.81	III
800	4375.929	I
3000	4383.544	I
1200	4404.750	I
300	4415.122	I
12	4419.60	III
600	4427.299	I
400	4461.652	I
120	4466.551	I
80	4476.017	I
80	4482.169	I
200	4482.252	I
50	4489.739	I
50	4528.613	I
30	4647.433	I
30	4736.771	I
50	4859.741	I
120	4871.317	I
60	4872.136	I
30	4878.208	I

Intensity	Wavelength/Å	
100	4890.754	I
250	4891.492	I
30	4903.309	I
150	4918.992	I
500	4920.502	I
1500	4957.597	I
80	5001.862	I
30	5005.711	I
100	5006.117	I
60	5012.067	I
30	5014.941	I
150	5041.755	I
30	5049.819	I
30	5051.634	I
25	5074.748	I
150	5110.357	I
40	5139.251	I
100	5139.462	I
25	5151.910	I
12	5156.12	III
80	5166.281	I
2500	5167.487	I
80	5168.897	I
500	5171.595	I
50	5191.454	I
80	5192.343	I
200	5194.941	I
10	5199.08	III
30	5204.582	I
25	5215.179	I
150	5216.274	I
60	5226.862	I
1000	5227.150	I
250	5232.939	I
10	5235.66	III
18	5243.31	III
13 l	5260.34	III
100	5266.555	I
1200	5269.537	I
800	5270.357	I
14	5272.98	III
15	5276.48	III
30	5281.789	I
16	5282.30	III
60	5283.621	I
12	5284.83	III
11	5298.12	III
12	5299.93	III
25	5302.299	I
14 w	5302.60	III
10	5306.76	III
10	5322.74	III
150	5324.178	I
800	5328.038	I
300	5328.531	I
100	5332.899	I
80	5339.928	I
500	5341.023	I
11	5346.88	III
12	5353.77	III
12	5363.76	III

Intensity	Wavelength/Å	
10	5368.06	III
400	5371.489	I
11 l	5375.47	III
40	5393.167	I
300	5397.127	I
250	5405.774	I
250	5429.695	I
100	5434.523	I
200	5446.871	I
120	5455.609	I
25	5497.516	I
20	5501.464	I
30	5506.778	I
30	5569.618	I
60	5572.841	I
120	5586.755	I
200	5615.644	I
20	5624.541	I
50	5662.515	I
11	5719.88	III
10	5756.38	III
20	5762.990	I
18	5833.93	III
10	5854.62	III
30	5862.353	I
15	5891.91	III
30	5914.114	I
10 p	5920.13	III
18 p	5929.69	III
10	5952.31	III
14	5953.62	III
12	5979.32	III
30	5986.956	I
12 h	5989.08	III
18	5999.54	III
16	6032.59	III
13	6036.56	III
11	6048.72	III
11	6054.18	III
40	6065.482	I
30	6102.159	I
40	6136.614	I
40	6137.694	I
40	6191.558	I
30	6213.429	I
30	6219.279	I
40	6230.726	I
20	6246.317	I
80	6247.56	II
30	6252.554	I
20	6393.602	I
30	6399.999	I
20	6411.647	I
20	6421.349	I
30	6430.844	I
200	6456.38	II
60	6494.981	I
20	6546.239	I
20	6592.913	I
40	6677.989	I
25	7164.443	I

Intensity	Wavelength/Å	
80	7187.313	I
30	7207.381	I
30	7445.746	I
40	7495.059	I
60	7511.045	I
80	7937.131	I
60	7945.984	I
80	7998.939	I
60	8046.047	I
50	8085.176	I
150	8220.41	I
120	8327.053	I
20	8331.908	I
120	8387.770	I
30	8468.404	I
15	8514.069	I
60	8661.898	I
150	8688.621	I
52	11422.32	I
87	11439.12	I
91	11593.59	I
255	11607.57	I
160	11638.26	I
230	11689.98	I
160	11783.26	I
580	11882.84	I
225	11884.08	I
1030	11973.05	I
96	14400.56	I
72	14512.23	I
50	14555.06	I
40	14826.43	I
94	15294.58	I
41	15769.42	I
105	18856.65	I

Krypton Kr $Z = 36$

Intensity	Wavelength/Å	
30	467.35	III
150	472.16	V
100	484.39	V
250	496.25	V
120	500.77	V
200	507.20	V
30	540.86	III
60	548.04	V
30	565.64	III
30	569.16	III
30	571.98	III
30	579.83	III
30	585.14	III
30	585.96	III
30	593.70	III
30	594.10	III
30	596.41	III
40	600.17	III
30	603.67	III
50	605.86	III
35	606.47	III
50	611.12	III
35	616.72	III
40	621.45	III
45	622.80	III
50	625.02	III
30	625.76	III
45	628.59	III
50	630.04	III
35	633.09	III
120	637.87	V
50	639.98	III
60	646.41	III
50	651.20	III
50	659.72	III
30	664.86	III
40	672.34	III
35	672.85	III
35	676.57	III
35	680.13	III
35	683.68	III
45	686.25	III
45	687.98	III
	690.86	V
	691.75	V
45	691.93	III
50	695.61	III
30	698.05	III
50	708.36	III
600	708.85	V
50	714.00	III
100 p	722.04	III
60	729.40	II
30	746.70	III
200	761.18	II
100	763.98	II
60	766.20	II
200	771.03	II
60 p	773.69	II
200	782.10	II
100	783.72	II
60	785.97	III
	793.44	IV
	794.11	IV
7	805.76	IV
	810.70	V
18	816.82	IV
60	818.15	II
60	830.38	II
50	837.66	III
22	842.04	IV
100	844.06	II
50	854.73	III
60	862.58	III
60	864.82	II
60	868.87	II
40	870.84	III
50	876.08	III
200	884.14	II
1000	886.30	II
400	891.01	II
75	897.81	III
200	911.39	II
2000	917.43	II
50	945.44	I
50	946.54	I

Intensity	Wavelength/Å	
20	951.06	I
50	953.40	I
50	963.37	I
2000	964.97	II
50	987.29	III
100	1001.06	I
100	1003.55	I
100	1030.02	I
30	1158.74	III
200	1164.87	I
650	1235.84	I
6	1638.82	III
6	1914.09	III
3	2237.34	IV
6	2291.26	IV
3	2329.3	IV
4	2336.75	IV
4	2348.27	IV
3	2358.5	IV
3	2388.05	IV
40	2393.94	III
4	2416.9	IV
3	2428.04	IV
5	2442.68	IV
4	2451.7	IV
6	2459.74	IV
100 h	2464.77	II
5	2474.06	IV
60	2492.48	II
40	2494.01	III
4	2517.0	IV
5	2518.02	IV
6	2519.38	IV
5	2524.5	IV
5	2546.0	IV
6	2547.0	IV
4	2558.08	IV
30	2563.25	III
3	2586.9	IV
5	2606.17	IV
10	2609.5	IV
8	2615.3	IV
7	2621.11	IV
60	2639.76	III
30	2680.32	III
40	2681.19	III
80 h	2712.40	II
3	2730.55	IV
8	2748.18	IV
6	2774.70	IV
3	2829.60	IV
100	2833.00	II
3	2836.08	IV
30	2841.00	III
30	2851.16	III
5	2853.0	IV
3	2859.3	IV
50	2870.61	III
100	2892.18	III
30	2909.17	III
50	2952.56	III
60	2992.22	III
50	3022.30	III
80	3024.45	III
50	3046.93	III
30	3056.72	III
60	3063.13	III
40	3097.16	III
60	3112.25	III
30	3120.61	III
100	3124.39	III
60	3141.35	III
3	3142.01	IV
100	3189.11	III
80	3191.21	III
6	3224.99	IV
40	3239.52	III
40	3240.44	III
300	3245.69	III
3	3261.70	IV
150	3264.81	III
100	3268.48	III
30	3271.65	III
30	3285.89	III
30	3304.75	III
50	3311.47	III
200	3325.75	III
60	3330.76	III
50	3342.48	III
100	3351.93	III
40	3374.96	III
100	3439.46	III
70	3474.65	III
100	3488.59	III
200	3507.42	III
100	3564.23	III
100 h	3607.88	II
200	3631.889	II
30	3641.34	III
250	3653.928	II
80	3665.324	I
150	3669.01	II
100	3679.559	I
80	3686.182	II
30	3690.65	III
300 h	3718.02	II
200	3718.595	II
150	3721.350	II
200	3741.638	II
150	3744.80	II
80	3754.245	II
500	3778.089	II
500	3783.095	II
3	3809.30	IV
5	3860.58	IV
40 h	3868.70	III
150 h	3875.44	II
150	3906.177	II
200	3920.081	II
5	3934.29	IV
100	3994.840	II
100 h	3997.793	II
300	4057.037	II
300	4065.128	II
50	4067.37	III
500	4088.337	II
250	4098.729	II
100	4109.248	II
40	4131.33	III
250	4145.122	II
40	4154.46	III
150	4250.580	II
1000	4273.969	I
100	4282.967	I
600	4292.923	II
200	4300.49	II
500 h	4317.81	II
400	4318.551	I
1000	4319.579	I
150 h	4322.98	II
100	4351.359	I
3000	4355.477	II
500	4362.641	I
200	4369.69	II
800	4376.121	I
300 h	4386.54	II
200	4399.965	I
100	4425.189	I
500	4431.685	II
600	4436.812	II
600	4453.917	I
800	4463.689	I
800	4475.014	II
400 h	4489.88	II
600	4502.353	I
400 h	4523.14	II
200 h	4556.61	II
800	4577.209	II
300	4582.978	II
150 h	4592.80	II
500	4615.292	II
1000	4619.166	II
800	4633.885	II
2000	4658.876	II
500	4680.406	II
100	4691.301	II
200	4694.360	II
3000	4739.002	II
300	4762.435	II
1000	4765.744	II
300	4811.76	II
300	4825.18	II
800	4832.077	II
700	4846.612	II
150	4857.20	II
300	4945.59	II
20 h	5016.45	III
200	5022.40	II
250	5086.52	II
400 h	5125.73	II
500	5208.32	II
200	5308.66	II
500	5333.41	II
200	5468.17	II
10	5501.43	III
500	5562.224	I
2000	5570.288	I
80	5580.386	I
100	5649.561	I
400	5681.89	II
200 h	5690.35	II
100	5832.855	I
3000	5870.914	I
200	5992.22	II
60	5993.849	I
10 h	6037.17	III
60	6056.125	I
10 h	6078.38	III
10	6310.22	III
300	6420.18	II
100	6421.026	I
200	6456.288	I
150	6570.07	II
60	6699.228	I
100	6904.678	I
250	7213.13	II
100	7224.104	I
80	7287.258	I
400	7289.78	II
400	7407.02	II
60	7425.541	I
200	7435.78	II
100	7486.862	I
300	7524.46	II
1000	7587.411	I
2000	7601.544	I
150	7641.16	II
1000	7685.244	I
1200	7694.538	I
250	7735.69	II
150	7746.827	I
800	7854.821	I
200	7913.423	I
180	7928.597	I
200	7933.22	II
120	7973.62	II
100	7982.401	I
1500	8059.503	I
4000	8104.364	I
6000	8112.899	I
60	8132.967	I
3000	8190.054	I
200	8202.72	II
80	8218.365	I
3000	8263.240	I
100	8272.353	I
1500	8281.050	I
5000	8298.107	I
100	8412.430	I
3000	8508.870	I
150	8764.110	I
6000	8776.748	I
2000	8928.692	I
500	9238.48	II

Intensity	Wavelength/Å	
500 hl	9293.82	II
200 h	9320.99	II
300	9361.95	II
100	9362.082	I
200 h	9402.82	II
200 h	9470.93	II
500	9577.52	II
500 h	9605.80	II
400 h	9619.61	II
200	9663.34	II
200 h	9711.60	II
2000	9751.758	I
500	9803.14	II
500	9856.314	I
1000	10221.46	II
100	11187.108	I
200	11257.711	I
150	11259.126	I
500	11457.481	I
150	11792.425	I
1500	11819.377	I
600	11997.105	I
160	12077.224	I
100	12861.892	I
1100	13177.412	I
1000	13622.415	I
2400	13634.220	I
800	13658.394	I
200	13711.036	I
600	13738.851	I
150	13974.027	I
550	14045.657	I
140	14104.298	I
180	14402.22	I
2000	14426.793	I
100	14517.84	I
1600	14734.436	I
550	14762.672	I
450	14765.472	I
400	14961.894	I
120	15005.307	I
140	15209.526	I
1700	15239.615	I
130	15326.480	I
1500	15334.958	I
700	15372.037	I
200	15474.026	I
180	15681.02	I
120	15820.09	I
200	16726.513	I
2000	16785.128	I
1000	16853.488	I
2400	16890.441	I
1600	16896.753	I
1800	16935.806	I
600	17098.771	I
700	17367.606	I
120	17404.443	I
150	17616.854	I
650	17842.737	I
700	18002.229	I
2600	18167.315	I
100	18399.786	I
150	18580.896	I
300	18696.294	I
170	18785.460	I
200	18797.703	I
140	20209.878	I
300	20423.964	I
140	20446.971	I
600	21165.471	I
1800	21902.513	I
120	22485.775	I
180	23340.416	I
120	24260.506	I
180	24292.221	I
600	25233.820	I
180	28610.55	I
1000	28655.72	I
150	28769.71	I
140	28822.49	I
300	29236.69	I
300	30663.54	I
300	30979.16	I
500	39300.6	I
1100	39486.52	I
220	39557.25	I
100	39572.60	I
1400	39588.4	I
1100	39589.6	I
500	39954.8	I
300	39966.6	I
1300	40306.1	I
250	40685.16	I

***Lanthanum* La *Z* = 57**

Intensity	Wavelength/Å	
100	344.12	IV
400	390.72	V
1000	432.11	V
2500	435.28	V
10000	463.14	IV
5000	482.16	V
7000	498.08	V
15000	499.54	IV
10000	503.58	V
12000	526.76	V
10000	531.07	V
15000	533.23	V
8000	547.44	V
40000	552.02	IV
5000	600.24	V
30000	631.26	IV
400	796.99	III
2000	870.40	III
1000	882.34	III
400	942.86	III
50000	1081.61	III
95000	1099.73	III
2000	1255.63	III
10000	1349.18	III
25000	1368.04	IV
20000	1463.47	IV
15000	1507.87	IV
10000	1523.79	III
4000	1808.66	IV
5000	1902.97	IV
4000 c	2197.45	IV
770	2256.76	II
25000 w	2417.58	IV
50000	2532.75	IV
45000	2582.05	IV
95000 w	2597.50	IV
70000 w	2662.75	IV
420	2808.39	II
50000 w	2848.30	IV
30000 c	2962.58	IV
70000 w	3009.51	IV
90000 c	3056.68	IV
1000	3171.63	III
1500	3171.74	III
510	3245.13	II
550	3265.67	II
800	3303.11	II
1500	3337.49	II
870	3344.56	II
1500	3380.91	II
320	3628.83	II
1000	3645.42	II
550	3713.54	II
2400	3759.08	II
3700	3790.83	II
3900	3794.78	II
600	3840.72	II
1600	3849.02	II
3400	3871.64	II
1700	3886.37	II
1300	3916.05	II
1100	3921.54	II
2200	3929.22	II
9000	3949.10	II
4400	3988.52	II
3600	3995.75	II
2800	4031.69	II
3000	4042.91	II
850	4067.39	II
2800	4077.35	II
5500	4086.72	II
4400	4123.23	II
550	4141.74	II
1100	4151.97	II
1500	4196.55	II
1600	4238.38	II
480	4269.50	II
600	4286.97	II
600	4296.05	II
440	4322.51	II
4600	4333.74	II
550	4354.40	II
2000	4429.90	II
850	4522.37	II
420	4526.12	II
400	4558.46	II
400	4574.88	II
410	4613.39	II
410	4619.88	II
540	4655.50	II
360	4662.51	II
230	4692.50	II
230	4728.42	II
500	4740.28	II
390	4743.09	II
320	4748.73	II
320	4860.91	II
850	4899.92	II
1000	4920.98	II
1000	4921.79	II
370	4949.77	I
340	4970.39	II
370	4986.83	II
720	4999.47	II
210	5050.57	I
470	5114.56	II
470	5122.99	II
450	5145.42	I
290	5158.69	I
580	5177.31	I
850	5183.42	II
260	5188.22	II
720	5211.86	I
520	5234.27	I
340	5253.46	I
370	5271.19	I
370	5301.98	II
180	5303.55	II
500	5455.15	I
470	5501.34	I
240	5648.25	I
180	5740.66	I
370	5769.34	I
320	5789.24	I
450	5791.34	I
140	5821.99	I
320	5930.62	I
720	6249.93	I
260 d	6262.30	II
450	6394.23	I
250	6455.99	I
180	6709.50	I
110	7045.96	I
160	7066.23	II
50	7161.25	I
110 w	7282.34	II
110 w	7334.18	I
75 cw	7483.50	II
50	7498.83	I
85	7539.23	I
40	7964.83	I
75	8086.05	I
85	8324.69	I
95	8346.53	I
65	8545.44	I
300	8583.45	III
40	8674.43	I
35	8825.82	I
120	9184.38	III

Intensity	Wavelength/Å	
100	9212.63	III
140	10284.79	III

Lead Pb Z = 82

Intensity	Wavelength/Å	
10	496.38	IV
12	499.94	IV
14	529.78	IV
20	570.16	IV
10	648.50	IV
20	703.73	V
12	749.46	V
10	752.52	V
10	761.09	IV
18	767.45	V
18	769.49	V
14	771.42	V
14	782.79	V
15	797.02	V
18	802.07	IV
12	802.82	IV
18	809.63	V
10	812.59	IV
10	827.41	IV
12	832.60	IV
12	845.94	IV
18	857.64	IV
16	862.33	IV
20	863.97	V
14	870.44	IV
6	873.71	II
12	879.96	IV
18	883.90	V
14	884.96	IV
14	884.99	IV
14	888.37	V
8	889.68	II
16	890.72	IV
14	894.40	V
12	896.08	V
12	908.51	IV
14	915.71	V
12	917.90	IV
12	918.09	V
12	920.28	V
12	920.66	V
10	922.12	IV
12	922.49	IV
10	927.64	IV
14	932.20	IV
12	954.35	V
10	967.23	II
10	986.71	II
10	995.89	II
10	1016.61	II
14	1028.61	IV
20	1032.05	IV
16	1041.24	IV
18	1044.14	IV
12	1048.9	III
10	1049.82	II
10	1050.77	II
10	1051.26	V
15	1056.53	IV
10	1060.66	II
12	1072.09	IV
18	1080.81	IV
20	1084.17	IV
10	1088.86	V
10	1103.94	II
10	1108.43	II
10	1109.84	II
20	1116.08	IV
10	1119.57	II
10	1121.36	II
10	1133.14	II
18	1137.84	IV
14	1144.93	IV
12	1157.88	V
14	1185.43	V
20	1189.95	IV
10	1203.63	II
10	1231.20	II
11	1233.50	V
10	1291.10	IV
20	1313.05	IV
10	1331.65	II
10	1335.20	II
12	1343.06	IV
10	1348.37	II
16	1388.94	IV
18	1400.26	IV
10	1404.34	IV
10	1433.96	II
10	1512.42	II
14	1535.71	IV
20	1553.1	III
10	1671.53	II
10	1682.15	II
20	1726.75	II
10	1796.670	II
10	1822.050	II
10	1904.77	I
7	1921.471	II
12	1959.34	IV
16	1973.16	IV
10	1998.83	V
5 r	2022.02	I
10	2042.58	IV
12	2049.34	IV
8 r	2053.28	I
12	2079.22	IV
6	2111.758	I
10	2115.066	I
15	2154.01	IV
500 r	2170.00	I
7	2175.580	I
12	2177.46	IV
7	2187.888	I
8	2189.603	I
10	2203.534	II
20	2237.425	I
20	2246.86	I
25	2246.89	I
20	2259.01	V
150	2332.418	I
16	2359.53	IV
180	2388.797	I
550 r	2393.792	I
140	2399.597	I
320 r	2401.940	I
320 r	2411.734	I
16	2417.61	IV
15	2424.81	V
150 r	2443.829	I
160 r	2446.181	I
130 r	2476.378	I
80 r	2577.260	I
500 r	2613.655	I
900 r	2614.175	I
160	2628.262	I
4	2634.256	II
10	2657.094	I
700	2663.154	I
10	2697.541	I
25000 r	2801.995	I
100	2822.58	I
14000 r	2823.189	I
35000 r	2833.053	I
6	2840.557	II
14000 r	2873.311	I
3	2914.442	II
15	2966.460	I
15	2972.991	I
15	2980.157	I
4	2986.876	II
10	3043.85	III
150	3118.894	I
10	3137.81	III
10	3176.50	III
600	3220.528	I
100	3229.613	I
400	3240.186	I
200	3262.355	I
35000	3572.729	I
50000 r	3639.568	I
20000	3671.491	I
70000 r	3683.462	I
10	3713.982	II
25000	3739.935	I
12	3854.08	III
15000	4019.632	I
95000	4057.807	I
14000	4062.136	I
10	4157.814	I
10000	4168.033	I
8	4272.66	III
200	4340.413	I
10	4496.15	IV
6	4499.34	III
16	4534.60	IV
7	4571.21	III
10	4579.051	II
6	4761.12	III
1000	5005.416	I
100	5006.572	I
50	5089.484	I
10	5107.242	I
2000	5201.437	I
10	5372.099	II
40	5692.346	I
200	5895.624	I
2000	6001.862	I
500	6011.667	I
500	6059.356	I
40	6081.409	II
50	6110.520	I
100	6235.266	I
50 c	6660.20	II
20000	7228.965	I
10	7346.676	I
20	7809.259	I
5	7896.737	I
10	8168.001	I
6	8191.886	I
5	8217.711	I
40	8272.690	I
20	8409.384	I
10	8478.492	I
5	8722.810	I
10	8857.457	I
8	9293.476	I
15	9438.05	I
15	9604.297	I
15	9674.351	I
200	10290.458	I
100	10498.965	I
50	10649.249	I
15	10886.688	I
40	10969.53	I
	13512.6	I
	14743.0	I
	15349.6	I
	39039.4	I

Lithium Li Z = 3

Intensity	Wavelength/Å	
	102.9	III
	103.4	III
	104.1	III
	105.5	III
	108.0	III
	113.9	III
	125.5	II
	135.0	III
	136.5	II
	140.5	II
	167.21	II
	168.74	II
	171.58	II
	178.02	II
	199.28	II
	207.5	II
	456.	II
	483.	II
	540.	II
	540.0	III
	729.	II

Intensity	Wavelength/Å	
	729.1	III
	800.	II
	820.	II
	861.	II
	905.5	II
	917.5	II
	936.	II
	945.	II
	965.	II
	972.	II
	988.	II
	1018.	II
	1032.	II
	1036.	II
	1093.	II
	1103.	II
	1109.	II
	1116.	II
	1132.1	II
	1141.	II
	1166.4	II
	1198.09	II
	1215.	II
	1238.	II
	1253.8	II
	1420.89	II
	1424.	II
3	1492.93	II
5	1492.97	II
1	1493.04	II
	1555.	II
3	1653.08	II
5	1653.13	II
1	1653.21	II
	1681.66	II
	1755.33	II
	2009.	II
	2039.	I
	2068.	II
	2131.	II
	2164.	II
	2173.4	I
	2183.	II
	2214.	II
	2222.	II
	2237.	II
h	2249.21	II
	2286.82	II
	2302.57	II
	2303.33	II
	2304.59	I
	2304.92	I
	2305.36	I
	2305.83	I
	2306.29	I
	2306.82	I
	2307.44	I
	2308.97	I
	2309.88	I
	2310.94	I
	2312.11	I

Intensity	Wavelength/Å	
	2313.49	I
	2315.08	I
	2316.95	I
	2319.18	I
	2321.88	I
	2325.11	I
	2329.02	I
	2329.84	II
	2333.94	I
3	2336.88	II
5	2336.91	II
2	2337.00	II
	2340.15	I
	2348.22	I
	2358.93	I
	2373.54	I
	2381.54	II
	2383.20	II
1	2394.39	I
	2402.33	II
	2410.84	II
3	2425.43	I
	2429.81	II
	2460.2	I
10	2475.06	I
	2506.94	II
	2508.78	II
	2518.	I
	2539.49	II
24	2551.7	II
	2559.	II
15	2562.31	I
	2605.08	II
2	2657.29	II
3	2657.30	II
	2674.46	II
	2728.24	II
5	2728.29	II
2	2728.32	II
3	2730.47	II
1	2730.55	II
5	2741.20	I
	2766.99	II
	2790.31	II
	2801.	I
	2846.	I
	2868.	I
	2895.	I
2	2934.02	II
2	2934.07	II
5	2934.12	II
1	2934.25	II
	2968.	I
3	3029.12	II
3	3029.14	II
	3144.	I
3	3155.31	II
4	3155.33	II
1	3196.26	II
9	3196.33	II
4	3196.36	II

Intensity	Wavelength/Å	
5	3199.33	II
2	3199.43	II
17	3232.66	I
	3249.87	II
	3306.28	II
	3488.	I
	3579.8	I
	3618.	I
	3662.	I
	3684.32	II
1	3714.00	II
5	3714.16	II
6 d	3714.27	II
8	3714.29	II
7 d	3714.40	II
10	3714.41	II
1	3714.51	II
	3714.58	II
3	3718.7	I
6	3794.72	I
20	3915.30	I
20	3915.35	I
10	3985.48	I
10	3985.54	I
40	4132.56	I
40	4132.62	I
	4196.	I
20	4273.07	I
20	4273.13	I
5	4325.42	II
5	4325.47	II
1	4325.54	II
	4516.45	II
13	4602.83	I
13	4602.89	I
	4607.34	II
	4671.51	II
6	4671.65	II
2	4671.70	II
3	4678.06	II
1	4678.29	II
	4760.	I
	4763.	II
	4788.36	II
	4843.0	II
4	4881.32	II
4	4881.39	II
1	4881.49	II
8	4971.66	I
8	4971.75	I
	5037.92	II
	5271.	I
	5315.	I
	5395.	I
	5440.	I
600 c	5483.55	II
600 c	5485.65	II
320	6103.54	I
320	6103.65	I
3600	6707.76	I
3600	6707.91	I

Intensity	Wavelength/Å	
48	8126.23	I
48	8126.45	I
	8517.37	II
	9581.42	II
	10120.	II
	12232.	I
	12782.	I
	13566.	I
	17552.	I
	18697.	I
	19290.	I
	24467.	I
	40475.	I

Lutetium Lu ***Z* = 71**

Intensity	Wavelength/Å	
100	563.72	V
500	810.73	III
2000	832.28	III
100	861.92	V
400	876.80	IV
100	880.32	V
100	891.81	V
100	914.72	V
400	1001.18	III
100	1272.42	IV
800	1333.79	IV
400	1429.38	IV
200	1441.76	V
200	1453.35	V
200	1468.99	V
400	1472.12	V
200	1473.71	V
200	1485.58	V
400	1511.26	IV
600	1772.57	IV
100 c	1786.25	V
1000	1854.57	III
1500	2065.35	III
1500 c	2070.56	III
600 c	2086.47	IV
1000 c	2104.41	IV
1000 c	2108.31	IV
1700 h	2195.54	II
1000	2236.14	III
2000	2236.22	III
95	2276.94	II
190	2297.41	II
1300	2392.19	II
120	2399.14	II
80	2419.21	II
130	2459.64	II
370	2536.95	II
930	2571.23	II
1700	2578.79	II
4500 c	2603.35	III
1800	2613.40	II
18000	2615.42	II
1800	2619.26	II
2700	2657.80	II
570 h	2685.08	I
4200	2701.71	II
180 d	2719.09	I

Intensity	Wavelength/Å	
480 h	2728.95	I
3600	2754.17	II
750 h	2765.74	I
2000	2772.55	III
2700	2796.63	II
270 c	2834.35	II
330 h	2845.13	I
3000	2847.51	II
570 h	2885.14	I
6300	2894.84	II
4500	2900.30	II
300	2903.05	I
9000	2911.39	II
270 h	2949.73	I
1200	2951.69	II
4200	2963.32	II
2400	2969.82	II
1800	2989.27	I
3000	3020.54	II
2100	3056.72	II
1000	3057.86	III
7500	3077.60	II
390	3080.11	I
5100 h	3081.47	I
3000	3118.43	I
2400	3171.36	I
260	3191.80	II
1400	3198.12	II
4800	3254.31	II
3800	3278.97	I
7600	3281.74	I
6200	3312.11	I
7600	3359.56	I
6200	3376.50	I
950	3385.50	I
160 h	3391.55	I
1400	3396.82	I
4100	3397.07	II
4800	3472.48	II
8300 c	3507.39	II
1600	3508.42	I
4800	3554.43	II
4800	3567.84	I
340	3596.34	I
800	3623.99	II
680	3636.25	I
2600	3647.77	I
110	3756.70	I
110	3756.79	I
150	3800.67	I
2700	3841.18	I
530	3876.65	II
50	3918.86	I
480	3968.46	I
670	4054.45	I
310	4122.49	I
3100	4124.73	I
150 c	4131.79	I
460	4154.08	I
1600	4184.25	II
150	4277.50	I

Intensity	Wavelength/Å	
250	4281.03	I
330 d	4295.97	I
150	4309.57	I
190 c	4430.48	I
190	4450.81	I
3300	4518.57	I
100 h	4648.21	I
1000	4658.02	I
85 h	4659.03	I
150	4785.42	II
85	4815.05	I
460	4904.88	I
180	4942.34	I
800	4994.13	II
800	5001.14	I
140	5134.05	I
2700	5135.09	I
170	5196.61	I
500	5402.57	I
140 c	5421.90	I
100	5437.88	I
2100	5476.69	II
550	5736.55	I
80	5800.59	I
690 cw	5983.9	II
140	5997.13	I
1400	6004.52	I
440	6055.03	I
150	6159.94	II
600	6198.13	III
160	6199.66	II
2100	6221.87	II
80	6235.36	II
160	6242.34	II
70 h	6345.35	I
1100	6463.12	II
29	6477.67	I
55 c	6523.18	I
35 cw	6611.28	II
23 c	6677.14	I
30 c	6793.77	I
45	6917.31	I
23	7031.24	I
45	7125.84	II
14 ch	7237.98	I
11 c	7441.52	I
9 c	8178.16	I
17	8382.08	I
35	8459.19	II
10 d	8478.50	I
29 c	8508.08	I
35 c	8610.98	I

Magnesium Mg Z = 12

Intensity	Wavelength/Å	
400	146.95	IV
20	186.51	III
20	187.20	III
10	188.53	III
100	231.73	III
80	234.26	III
35	276.58	V
4000	320.99	IV

Intensity	Wavelength/Å	
3000	323.31	IV
30	353.09	V
150	857.29	IV
50	919.03	IV
250	1037.41	IV
300	1210.99	IV
300	1342.19	IV
800	1346.57	IV
300	1346.68	IV
600	1352.05	IV
900	1384.46	IV
500	1385.77	IV
800	1387.53	IV
300	1404.68	IV
1000	1409.36	IV
500	1437.53	IV
1000	1437.64	IV
300	1447.42	IV
300	1459.54	IV
400	1459.62	IV
400	1481.51	IV
350	1490.45	IV
300	1495.50	IV
300	1607.11	IV
5	1668.43	I
500	1683.02	IV
10	1683.41	I
400	1698.81	IV
15	1707.06	I
40	1734.84	II
50	1737.62	II
20	1747.80	I
40	1750.65	II
50	1753.46	II
30	1827.93	I
300	1844.17	IV
9	2025.82	I
25	2064.90	III
20	2091.96	III
20	2177.70	III
3	2329.58	II
20	2395.15	III
6	2449.57	II
1	2557.23	I
1	2560.94	I
1	2562.26	I
1	2564.94	I
1	2570.91	I
1	2572.25	I
2	2574.94	I
1	2577.89	I
1	2580.59	I
1	2584.22	I
2	2585.56	I
3	2588.28	I
1	2591.89	I
1	2593.23	I
2	2595.97	I
2	2602.50	I
4	2603.85	I
5	2606.62	I

Intensity	Wavelength/Å	
1	2613.36	I
2	2614.73	I
3	2617.51	I
3	2628.66	I
6	2630.05	I
8	2632.87	I
2	2644.80	I
3	2646.21	I
4	2649.06	I
8	2660.76	II
8	2660.82	II
6	2668.12	I
8	2669.55	I
10	2672.46	I
3	2693.72	I
5	2695.18	I
6	2698.14	I
8	2731.99	I
10	2733.49	I
12	2736.53	I
5	2765.22	I
7	2768.34	I
38	2776.69	I
32	2778.27	I
90	2779.83	I
8	2781.29	I
32	2781.42	I
36	2782.97	I
1000	2795.53	II
600	2802.70	II
3	2809.76	I
2	2811.11	I
1	2811.78	I
12	2846.72	I
12	2846.75	I
14	2848.34	I
14	2848.42	I
16	2851.65	I
16	2851.66	I
6000	2852.13	I
2	2902.92	I
4	2906.36	I
3	2915.45	I
10	2936.74	I
12	2938.47	I
2	2942.00	I
13	2942.00	I
20	3091.08	I
22	3092.99	I
14	3096.90	I
9	3104.71	II
8	3104.81	II
6	3168.98	II
6	3172.71	II
7	3175.78	II
2	3197.62	I
17	3329.93	I
6	3332.15	I
9	3336.68	I
7	3535.04	II
8	3538.86	II

Intensity	Wavelength/Å	
7	3549.52	II
8	3553.37	II
140	3829.30	I
300	3832.30	I
500	3838.29	I
8	3848.24	II
7	3850.40	II
3	3878.31	I
3	3895.57	I
4	3903.86	I
6	3938.40	I
8	3986.75	I
10	4057.50	I
15	4167.27	I
20	4351.91	I
9	4384.64	II
10	4390.59	II
8	4428.00	II
9	4433.99	II
14	4481.16	II
13	4481.33	II
28	4571.10	I
10	4730.03	I
7	4851.10	II
75	5167.33	I
220	5172.68	I
400	5183.61	I
8	5264.21	II
7	5264.37	II
9	5401.54	II
6	5528.41	I
30	5711.09	I
10	6318.72	I
9	6319.24	I
7	6319.49	I
10	6346.74	II
9	6346.96	II
11	6545.97	II
7	6781.45	II
8	6787.85	II
7	6812.86	II
8	6819.27	II
10	7193.17	I
10	7291.06	I
12	7387.69	I
20	7657.60	I
19	7659.15	I
17	7659.90	I
15	7691.55	I
12	7877.05	II
13	7896.37	II
10	8098.72	I
9	8115.22	II
8	8120.43	II
10	8209.84	I
20	8213.03	I
10	8213.99	II
11	8234.64	II
10	8310.26	I
15	8346.12	I
10	8710.18	I
12	8712.69	I
13	8717.83	I
10	8734.99	II
17	8736.02	I
11	8745.66	II
14	8806.76	I
10	8824.32	II
11	8835.08	II
20	8923.57	I
10	8997.16	I
14	9218.25	II
13	9244.27	II
12	9246.50	I
30	9255.78	I
10	9327.54	II
10	9340.54	II
25	9414.96	I
17	9429.81	I
19	9432.76	I
20	9438.78	I
12	9631.89	II
11	9632.43	II
15	9953.20	I
15	9983.20	I
17	9986.47	I
18	9993.21	I
14	10092.16	II
35	10811.08	I
11	10914.23	II
10	10951.78	II
25	10953.32	I
27	10957.30	I
28	10965.45	I
15	11032.10	I
14	11033.66	I
45	11828.18	I
30	12083.66	I
28	14877.62	I
35	15024.99	I
30	15040.24	I
25	15047.70	I
10	15765.84	I
30	17108.66	I
5	26392.90	I

Manganese Mn Z = 25

Intensity	Wavelength/Å	
600	410.30	V
600	410.60	V
600	415.62	V
650	415.98	V
600	428.59	V
600	435.67	V
1000	441.72	V
850	442.49	V
60	579.79	IV
60	581.44	IV
60	581.65	IV
60	585.21	IV
90	1242.25	IV
90	1244.50	IV
95	1251.93	IV
95	1257.28	IV
90	1264.41	IV
500	1283.58	III
400	1287.59	III
300	1291.62	III
1000	1360.72	III
800	1365.20	III
500 h	1609.17	III
1000	1614.14	III
2000	1620.60	III
500	1633.80	III
80	1667.00	IV
80	1698.30	IV
20	1726.47	II
30	1732.70	II
50	1733.55	II
40	1734.49	II
30	1737.93	II
20	1740.16	II
20	1742.00	II
85	1742.10	IV
85	1766.27	IV
80	1795.65	IV
80	1795.79	IV
30	1853.27	II
20	1857.92	II
50	1902.95	II
20	1907.84	II
75	1910.25	IV
30	1911.41	II
20 d	1914.68	II
100	1915.10	II
20	1918.64	II
30	1919.64	II
80	1921.25	II
20	1923.07	II
20	1923.34	II
30	1925.52	II
50	1926.59	II
30	1931.40	II
500	1941.28	III
800	1943.21	III
20	1945.15	II
20	1947.93	II
20	1950.14	II
500	1952.36	III
1000	1952.52	III
30	1953.23	II
20 d	1954.81	II
30	1959.25	II
20	1969.24	II
500	1978.95	III
30	1994.23	II
9700	1996.06	I
14000	1999.51	I
18000	2003.85	I
1000 w	2027.83	III
500 w	2028.14	III
50	2037.31	II
40	2037.64	II
40	2039.97	II
500	2049.68	III
500	2066.38	III
1000	2069.02	III
30	2076.21	II
900	2077.38	III
800	2084.23	III
600	2090.05	III
1500	2092.16	I
500	2094.78	III
20	2097.46	II
500	2097.93	III
500	2099.97	III
20	2102.50	II
1700	2109.58	I
30	2113.96	II
1000	2169.78	III
700	2174.15	III
900	2176.87	III
800	2181.86	III
800	2184.87	III
290	2208.81	I
540	2213.85	I
900	2220.55	III
770	2221.84	I
1000	2227.42	III
20	2373.36	II
20	2427.38	II
50	2427.72	II
30	2427.94	II
30	2437.37	II
20	2437.84	II
30	2452.49	II
50	2499.00	II
30	2507.60	II
20	2516.60	II
30	2516.74	II
20	2521.66	II
20	2530.72	II
20	2531.80	II
50	2532.78	II
50	2533.33	II
30	2534.10	II
80	2534.22	II
100	2535.66	II
30	2535.98	II
100	2537.92	II
50	2541.11	II
80	2542.92	II
50	2543.45	II
100	2548.75	II
50	2551.85	II
30	2553.27	II
75	2556.57	II
30	2556.89	II
95	2558.59	II
30	2559.41	II
150	2563.65	II
30	2565.22	II
580	2572.76	I
480	2575.51	I
12000	2576.10	II
550	2584.31	I

Intensity	Wavelength/Å		Intensity	Wavelength/Å		Intensity	Wavelength/Å		Intensity	Wavelength/Å	
30	2588.97	II	70	2886.68	II	290	3619.28	I	800	4451.59	I
45	2589.71	II	160	2889.58	II	220	3623.79	I	160	4453.00	I
250	2592.94	I	55	2892.39	II	140	3629.74	I	130	4455.01	I
6200	2593.73	II	50	2898.70	II	100	3660.40	I	160	4455.32	I
250	2595.76	I	80	2900.16	II	280	3693.67	I	110	4455.82	I
95	2598.90	II	140 h	2914.60	I	180	3696.57	I	210	4457.55	I
30	2602.72	II	190 h	2925.57	I	210	3706.08	I	270	4458.26	I
45	2603.72	II	1100	2933.06	II	130	3718.93	I	150	4461.08	I
4300	2605.69	II	1500	2939.30	II	130	3731.93	I	510	4462.02	I
190	2610.20	II	250 h	2940.39	I	260	3790.22	I	290	4464.68	I
500	2618.14	II	1900	2949.20	II	110	3800.55	I	200	4470.14	I
140	2622.90	I	30	3019.92	II	3200	3806.72	I	130	4472.79	I
150	2624.04	I	55	3031.06	II	700	3809.59	I	170	4490.08	I
40	2624.80	II	30	3035.35	II	2100	3823.51	I	240	4498.90	I
200	2625.58	II	330	3044.57	I	390	3823.89	I	240	4502.22	I
190	2632.35	II	120	3045.59	I	200	3829.68	I	160	4709.72	I
130	2638.17	II	200	3047.01	I	480	3833.86	I	180	4727.48	I
80	2639.84	II	30	3050.65	II	1300	3834.36	I	130	4739.11	I
27	2650.99	II	250	3054.36	I	350	3839.78	I	1000	4754.04	I
60	2655.91	II	140	3062.12	I	670	3841.08	I	180	4761.53	I
30	2666.77	II	170	3066.02	I	350	3843.98	I	750	4762.38	I
30	2667.03	II	170	3070.27	I	120	3926.47	I	300	4765.86	I
110	2672.59	II	160	3073.13	I	130	3982.58	I	500	4766.43	I
55	2673.37	II	140 h	3178.50	I	150	3985.24	I	940	4783.42	I
55	2674.43	II	220	3212.88	I	190	3986.83	I	1000	4823.52	I
45	2680.34	II	1000	3228.09	I	150	3987.10	I	19	5004.91	I
30	2680.68	II	300	3230.72	I	1500	4018.10	I	30	5074.79	I
30	2681.25	II	850	3236.78	I	150	4026.44	I	200	5079.20	III
55	2684.55	II	330	3243.78	I	27000	4030.76	I	150	5100.03	III
55	2685.94	II	650	3248.52	I	19000	4033.07	I	60	5117.94	I
110	2688.25	II	100	3251.14	I	11000	4034.49	I	50	5150.89	I
27	2693.19	II	310	3252.95	I	1500	4035.73	I	50	5196.59	I
55	2695.36	II	310	3256.14	I	5600	4041.36	I	85	5255.32	I
27	2698.97	II	220	3258.41	I	210 d	4045.13	I	160	5341.06	I
85	2701.00	II	180	3260.23	I	1100	4048.76	I	19	5349.88	I
50	2701.17	II	180	3264.71	I	150	4055.21	I	95	5377.63	I
160	2701.70	II	200	3330.78	II	1900	4055.54	I	95	5394.67	I
100	2703.98	II	720	3441.99	II	210	4057.95	I	50	5399.49	I
130	2705.74	II	50	3460.03	II	1100	4058.93	I	95	5407.42	I
80	2707.53	II	360	3460.33	II	150	4059.39	I	35	5413.69	I
110	2708.45	II	360 h	3474.04	II	730	4061.74	I	85	5420.36	I
45	2709.96	II		3474.13	II	730	4063.53	I	35	5432.55	I
80	2710.33	II	290	3482.91	II	290	4070.28	I	150	5454.07	III
110	2711.58	II	180	3488.68	II	730	4079.24	I	12	5457.47	I
30	2716.80	II	140	3495.84	II	730	4079.42	I	60	5470.64	I
30	2717.53	II	50	3496.81	II	1100	4082.94	I	200	5474.68	III
30	2719.01	II	100	3497.54	II	1100	4083.63	I	40	5481.40	I
50	2719.74	II	360	3531.85	I	200	4110.90	I	30	5505.87	I
30	2722.10	II	1100	3532.12	I	150	4131.12	I	50	5516.77	I
30	2724.46	II	1300	3547.80	I	120	4135.04	I	40	5537.76	I
55	2728.61	II	1100	3548.03	I	150	4176.60	I	21	5551.98	I
6200	2794.82	I	390	3548.20	I	120	4189.99	I	200	5946.65	III
5100	2798.27	I	2200	3569.49	I	370	4235.14	I	140	6013.50	I
220	2799.84	I	720	3569.80	I	510	4235.29	I	200	6016.64	I
3700	2801.06	I	1400	3577.88	I	190	4239.72	I	290	6021.80	I
110	2809.11	I	720	3586.54	I	290	4257.66	I	200	6231.21	III
60	2815.02	II	290	3595.12	I	290	4265.92	I	17	6440.97	I
30	2816.33	II	150	3601.72	III	270	4281.10	I	24	6491.71	I
60	2870.08	II	420	3607.54	I	50	4323.63	II	14 h	6942.52	I
30	2872.94	II	420	3608.49	I	350	4414.88	I	12	6989.96	I
80	2879.49	II	360	3610.30	I	210	4436.35	I	14	7069.84	I

Intensity	Wavelength/Å	
12	7184.25	I
24 h	7283.82	I
35 h	7302.89	I
50	7326.51	I
12	7680.20	I
12 h	8672.06	I
12 h	8701.05	I
17 h	8703.76	I
30 h	8740.93	I

Mercury 198 Hg Z = 80

Intensity	Wavelength/Å	
80	1250.564	I
8	1259.242	I
100	1268.825	I
5	1307.751	I
20	1402.619	I
10	1435.503	I
1000	1849.492	I
60	2262.210	II
20	2302.065	I
20	2345.440	I
100	2378.325	I
20	2380.004	I
40	2399.349	I
20	2399.729	I
20	2446.900	I
15	2464.064	I
40	2481.999	I
30	2482.713	I
40	2483.821	I
90	2534.769	I
15000	2536.506	I
25	2563.861	I
25	2576.290	I
250	2652.043	I
400	2653.683	I
100	2655.130	I
50	2698.831	I
80	2752.783	I
20	2759.710	I
40	2803.471	I
30	2804.438	I
750	2847.675	II
50	2856.939	I
150	2893.598	I
150	2916.227	II
60	2925.413	I
1200	2967.283	I
300	3021.500	I
120	3023.476	I
30	3025.608	I
50	3027.490	I
400	3125.670	I
320	3131.551	I
320	3131.842	I
80	3341.481	I
2800	3650.157	I
300	3654.839	I
80	3662.883	I
240	3663.281	I
30	3701.432	I
35	3704.170	I
30	3801.660	I
20	3901.867	I
60	3906.372	I
200	3983.839	II
1800	4046.572	I
150	4077.838	I
40	4108.057	I
250	4339.224	I
400	4347.496	I
4000	4358.337	I
80	4916.068	I
1100	5460.753	I
160	5675.922	I
240	5769.598	I
280	5790.663	I
20	6072.713	I
30	6234.402	I
160	6716.429	I
250	6907.461	I
240	11287.407	I

Mercury Hg Z = 80

Intensity	Wavelength/Å	
3	621.44	III
2	679.68	III
2	878.59	III
1	886.48	III
400	893.08	II
300	915.83	II
150	923.39	II
200	940.80	II
100	962.74	II
50	969.13	II
1	988.89	III
2	1009.29	III
5	1068.03	III
800	1099.26	II
2	1161.95	III
80	1250.58	I
8	1259.24	I
100	1268.82	I
5	1307.75	I
300	1307.93	II
400	1321.71	II
400	1331.74	II
80	1350.07	II
200	1361.27	II
20	1402.62	I
200	1414.43	II
10	1435.51	I
15	1619.46	II
120	1623.95	II
20	1628.25	II
150	1649.94	II
50	1653.64	II
200	1672.41	II
9	1681.40	III
100	1702.73	II
100	1707.40	II
120	1727.18	II
250	1732.14	II
15	1759.75	III
20	1775.68	I
40	1783.70	II
30	1796.22	II
200	1796.90	II
60	1798.74	II
30	1803.89	II
40	1808.29	II
400	1820.34	II
5	1832.74	I
1000	1849.50	I
160	1869.23	II
300	1870.55	II
200	1875.54	II
1	1894.77	III
20	1900.28	II
30	1927.60	II
300	1942.27	II
100	1972.94	II
200	1973.89	II
150	1987.98	II
90	2026.97	II
90	2052.93	II
70	2148.00	II
5	2247.55	I
60	2262.23	II
20	2302.06	I
7	2314.15	III
15	2323.20	I
5	2340.57	I
20	2345.43	I
20	2352.48	I
100	2378.32	I
20	2380.00	I
4	2380.55	III
40	2399.38	I
20	2399.73	I
10	2400.49	I
60	2407.35	II
50	2414.13	II
8	2431.65	III
5	2441.06	I
20	2446.90	I
15	2464.06	I
5	2480.56	III
40	2482.00	I
30	2482.72	I
40	2483.82	I
7	2484.50	III
90	2534.77	I
15000	2536.52	I
25	2563.86	I
25	2576.29	I
5	2578.91	I
2	2612.92	III
4	2617.97	III
15	2625.19	I
5	2639.78	I
250	2652.04	I
400	2653.69	I
100	2655.13	I
3	2670.49	III
5	2674.91	I
50	2698.83	I
50	2699.38	I
80	2705.36	II
70	2724.43	III
80	2752.78	I
20	2759.71	I
6	2769.22	III
40	2803.46	I
30	2804.43	I
2	2805.34	I
2	2806.77	I
150	2814.93	II
3	2844.76	III
750	2847.68	II
50	2856.94	I
150	2893.60	I
150	2916.27	II
60	2925.41	I
150	2935.94	II
400	2947.08	II
1200	2967.28	I
300	3021.50	I
120	3023.47	I
30	3025.61	I
50	3027.49	I
15	3090.05	III
400	3125.67	I
320	3131.55	I
320	3131.84	I
400	3208.20	II
400	3264.06	II
5	3283.02	III
12	3312.28	III
80	3341.48	I
100	3385.25	II
8	3389.01	III
5	3450.77	III
400	3451.69	II
3	3500.35	III
4	3538.88	III
200	3549.42	II
5	3557.24	III
2800	3650.15	I
300	3654.84	I
80	3662.88	I
240	3663.28	I
30	3701.44	I
35	3704.17	I
30	3801.66	I
15	3803.51	III
100	3806.38	II
20	3901.87	I
60	3906.37	I
100	3918.92	II
200	3983.96	II
1800	4046.56	I
150	4077.83	I
40	4108.05	I
70	4122.07	III
10	4140.34	III
100	4216.74	III

Intensity	Wavelength/Å	
250	4339.22	I
400	4347.49	I
4000	4358.33	I
100	4398.62	II
15	4470.58	III
12	4552.84	III
90	4660.28	II
50	4797.01	III
80	4855.72	II
10	4869.85	III
5	4883.00	I
5	4889.91	I
80	4916.07	I
5	4970.37	I
80	4973.57	III
5	4980.64	I
20	5102.70	I
40	5120.64	I
100	5128.45	II
20	5137.94	I
30	5210.82	III
20	5290.74	I
5	5316.78	I
60	5354.05	I
30	5384.63	I
1100	5460.74	I
30	5549.63	I
160	5675.86	I
6	5695.71	III
240	5769.60	I
100	5789.66	I
280	5790.66	I
140	5803.78	I
60	5859.25	I
60	5871.73	II
20	5871.98	I
20	6072.72	I
1000	6149.50	II
25	6220.35	III
30	6234.40	I
35	6418.98	III
40	6501.38	III
80	6521.13	II
10	6584.26	III
6	6610.12	III
30	6709.29	III
160	6716.43	I
250	6907.52	I
250	7081.90	I
200	7091.86	I
40	7346.37	II
100	7485.87	II
12	7517.46	III
20	7728.82	I
7	7808.10	III
100	7944.66	II
25	7946.75	III
50	7984.51	III
5	8151.64	III
2000	10139.75	I
240	11287.40	I

Intensity	Wavelength/Å	
120	13209.95	I
140	13426.57	I
60	13468.38	I
80	13505.58	I
500	13570.21	I
450	13673.51	I
200	13950.55	I
500	15295.82	I
100	16881.48	I
400	16920.16	I
300	16942.00	I
500	17072.79	I
400	17109.93	I
20	17116.75	I
20	17198.67	I
20	17213.20	I
70	17329.41	I
30	17436.18	I
50	18130.38	I
40	19700.17	I
	22493.28	I
250	23253.07	I
	32148.06	I
	36303.03	I

Molybdenum Mo Z = 42

Intensity	Wavelength/Å	
50	867.92	IV
100	884.19	IV
60	886.05	IV
50	891.74	IV
100	1169.33	III
100	1254.93	III
100	1258.52	III
100	1262.21	III
100	1263.74	III
100	1274.37	III
100	1276.40	III
200	1277.40	III
200	1277.58	III
200	1278.40	III
150	1281.90	III
150	1283.60	III
100	1854.73	III
80	1926.26	IV
100	1929.24	IV
80	1971.06	IV
70	2010.92	IV
19000	2015.11	II
40000	2020.30	II
21000	2038.44	II
17000	2045.98	II
50	2060.38	IV
4800	2081.68	II
2400	2089.52	II
2200	2092.50	II
4000	2093.11	II
2700	2100.84	II
1500	2104.29	II
1400	2108.02	II
100	2184.37	III
100	2211.02	III
400	2269.69	II

Intensity	Wavelength/Å	
150	2269.71	III
200	2294.97	III
160	2304.25	II
160	2306.97	II
150	2330.93	III
110	2332.12	II
190	2341.59	II
100	2359.76	III
110	2389.20	II
140	2403.61	II
120	2413.01	II
85	2498.28	II
200	2506.19	III
440	2538.46	II
330	2542.67	II
80	2558.88	II
85	2561.31	II
250	2593.70	II
250	2602.80	II
400	2616.78	I
440	2629.85	I
330	2636.67	II
720	2638.76	II
410	2640.99	I
600	2644.35	II
370	2646.49	II
640	2649.46	I
480	2653.35	II
560 h	2655.03	I
640	2660.58	II
720	2672.84	II
250	2673.27	II
1000	2679.85	I
95	2681.36	II
640	2683.23	II
880	2684.14	II
560	2687.99	II
480	2701.42	II
190	2713.51	II
290	2717.35	II
85	2726.97	II
140	2729.68	II
80	2730.20	II
330	2732.88	II
160	2736.96	II
80 h	2737.88	II
290	2746.30	II
110	2756.07	II
220	2763.62	II
240	2769.76	II
160	2773.78	II
190	2774.39	II
1700	2775.40	II
65	2777.86	II
880	2780.04	II
400	2784.99	II
100	2807.74	III
400	2807.76	II
1700	2816.15	II
220	2817.44	II
80	2827.74	II

Intensity	Wavelength/Å	
80	2834.39	II
80	2835.33	II
160	2842.15	II
1700	2848.23	II
370	2853.23	II
370	2863.81	II
220	2866.69	II
1700	2871.51	II
85	2872.88	II
220	2879.05	II
65	2888.15	II
1300	2890.99	II
95	2891.28	II
190	2892.81	II
950	2894.45	II
140	2897.63	II
70	2900.80	II
290	2903.07	II
80	2907.12	II
600	2909.12	II
1100	2911.92	II
120	2918.83	II
1300	2923.39	II
140	2924.32	II
1100	2930.50	II
800	2934.30	II
95	2940.10	II
110	2941.22	II
150	2944.82	II
140	2946.69	II
95	2947.28	II
125	2947.32	III
95	2955.84	II
240	2956.06	II
70	2956.90	II
95	2960.24	II
250	2963.79	II
210	2965.27	II
70	2971.91	II
250	2972.61	II
80	2975.40	II
95	2992.84	II
95	3027.77	II
100	3060.78	II
800	3064.28	I
250	3065.04	II
800	3074.37	I
85	3077.66	II
800	3085.62	I
270	3087.62	II
190	3092.07	II
560	3094.66	I
560	3101.34	I
1400	3112.12	I
290	3122.00	II
14000	3132.59	I
110	3138.72	II
220	3152.82	II
55	3155.64	II
6000	3158.16	I
8700	3170.35	I

Intensity	Wavelength/Å	
95	3172.03	II
160	3172.74	II
120 d	3187.59	II
7600	3193.97	I
880	3205.88	I
3000	3208.83	I
560	3215.07	I
880	3228.22	I
600	3229.79	I
1100	3233.14	I
950	3237.08	I
65	3240.71	II
950	3256.21	I
480	3264.40	I
800	3270.90	I
200	3271.69	III
1100	3289.02	I
950	3290.82	I
190	3292.31	II
100	3313.62	II
190	3320.90	II
640	3323.95	I
1300	3344.75	I
95	3346.40	II
1600	3358.12	I
950	3363.78	I
950	3379.97	I
1900	3384.62	I
130	3395.36	II
640	3404.34	I
1300	3405.94	I
640	3437.22	I
130	3446.08	II
3200	3447.12	I
640	3449.07	I
950	3456.39	I
640	3460.78	I
800	3504.41	I
560	3508.12	I
480	3521.41	I
640	3537.28	I
520	3558.10	I
400	3563.14	I
1400	3581.89	I
1400	3624.46	I
1000	3635.43	I
400	3657.35	I
540	3664.81	I
590	3672.82	I
1300	3680.60	I
65	3688.31	II
180	3692.64	II
1400	3694.94	I
500	3727.69	I
80	3744.37	II
29000	3798.25	I
520	3826.70	I
940	3828.87	I
1700	3833.75	I
29000	3864.11	I
580	3869.08	I
580	3886.82	I
19000	3902.96	I
65	3941.48	II
1400	4062.08	I
2300	4069.88	I
1300	4081.44	I
940	4084.38	I
730	4107.47	I
630	4120.10	I
2900	4143.55	I
480	4185.82	I
2500	4188.32	I
1500	4232.59	I
890	4276.91	I
1200	4277.24	I
1400	4288.64	I
680	4292.13	I
890	4293.21	I
840	4326.14	I
1900	4381.64	I
2500	4411.57	I
990	4434.95	I
480	4457.36	I
630	4474.56	I
400	4536.80	I
460	4626.47	I
640	4707.26	I
700	4731.44	I
770	4760.19	I
410	4819.25	I
410	4830.51	I
180	5014.60	I
80	5029.00	I
65	5030.78	I
100	5047.71	I
50	5055.00	I
200	5059.88	I
100	5080.02	I
100	5096.65	I
130	5097.52	I
130	5109.71	I
80	5114.97	I
150	5145.38	I
110	5147.39	I
80	5163.19	I
100	5167.76	I
160 d	5171.08	I
230 h	5172.94	I
160 h	5174.18	I
110	5200.17	I
50	5200.74	I
50	5211.86	I
80	5219.40	I
65	5231.06	I
100	5234.26	I
460 h	5238.20	I
230 h	5240.88	I
110 h	5242.81	I
100	5245.51	I
150	5259.04	I
65	5261.14	I
65	5279.65	I
210	5280.86	I
55	5292.08	I
55	5295.47	I
55	5313.89	I
80	5354.88	I
65	5356.48	I
560 hl	5360.56	I
110 hl	5364.28	I
65	5394.52	I
50	5400.47	I
55	5435.68	I
65	5437.75	I
50	5501.54	I
7800	5506.49	I
5200	5533.05	I
50	5543.12	I
55	5556.28	I
2500	5570.45	I
100	5610.93	I
330	5632.47	I
50	5634.86	I
230	5650.13	I
55	5674.47	I
460	5689.14	I
80	5705.72	I
210	5722.74	I
620	5751.40	I
520	5791.85	I
55 h	5849.73	I
50 h	5851.52	I
520	5858.27	I
50	5869.33	I
820	5888.33	I
50 h	5893.38	I
160 h	5928.88	I
35	6025.49	I
1300	6030.66	I
40	6101.87	I
40	6357.22	I
35	6401.07	I
100	6424.37	I
230	6619.13	I
50	6650.38	I
110	6733.98	I
50	6746.27	I
35	6753.97	I
40	6838.88	I
35	6914.01	I
110	7109.87	I
150	7242.50	I
40	7245.85	I
40	7391.36	I
140	7485.74	I
27	7720.77	I
40 h	8328.44	I
45 h	8389.32	I
45 h	8483.39	I

Neodymium Nd Z = 60

Intensity	Wavelength/Å	
75	2764.98	I
80	2993.20	II
95	3007.97	II
95	3014.19	II
95	3018.35	II
140	3056.71	II
130	3069.73	II
160	3075.38	II
240	3092.92	II
260	3115.18	II
290	3133.60	II
220	3134.90	II
170	3141.46	II
170	3142.44	II
150	3203.47	II
220	3259.24	II
220	3265.12	II
320	3275.22	II
290	3285.10	II
410	3328.28	II
320	3353.59	II
410	3560.75	II
470	3587.51	II
370	3615.82	II
410	3653.15	II
470	3662.26	II
540	3665.18	II
540	3672.36	II
580	3673.54	II
1200	3685.80	II
440	3687.30	II
410	3689.69	II
410	3697.56	II
470	3713.70	II
640 d	3714.73	II
470	3715.68	II
410	3718.54	II
410	3721.35	II
780	3723.50	II
410	3724.87	II
710	3728.13	II
470	3730.58	II
1000 d	3735.54	II
440	3737.10	II
1000	3738.06	II
580	3752.49	II
510	3757.82	II
930	3758.95	II
930	3763.47	II
510	3769.65	II
1400	3775.50	II
710	3779.47	II
580	3780.40	II
510	3781.32	II
2400	3784.25	II
370	3801.12	II
1200	3803.47	II
2500	3805.36	II
470	3807.23	II
540	3808.77	II
440	3809.06	II
580	3810.49	II
710	3814.73	II

Intensity	Wavelength/Å	
410	3822.47	II
1200	3826.42	II
540	3828.85	II
440	3829.16	II
510	3830.47	II
740	3836.54	II
1700	3838.98	II
410 d	3841.82	II
1700 d	3848.24	II
1500	3848.52	II
470	3850.22	II
2400 d	3851.66	II
3700 d	3863.33	II
850	3869.07	II
470	3875.87	II
1100	3878.58	II
1000	3879.55	II
780	3880.38	II
1200	3880.78	II
540	3887.87	II
1300	3889.93	II
1300	3890.58	II
1300	3890.94	II
580	3891.51	II
470	3892.06	II
810	3894.63	II
440	3897.63	II
2000	3900.21	II
1300	3901.84	II
1700	3905.89	II
510	3907.84	II
2000	3911.16	II
850	3912.23	II
440	3915.13	II
610	3915.95	II
1100	3920.96	II
510	3927.10	II
610	3934.82	II
410	3936.11	II
510	3938.86	II
2000	3941.51	II
2000	3951.16	II
810	3952.20	II
590	3958.00	II
510	3962.21	II
1400	3963.12	II
1100	3973.30	II
740	3973.69	II
740	3976.85	II
740	3979.49	II
470	3986.25	II
1400	3990.10	II
1000	3991.74	II
1100	3994.68	II
410	4000.50	II
540	4004.02	II
410	4007.43	II
3700	4012.25	II
540	4012.70	II
1000	4020.87	II
1000	4021.34	II
1000	4021.78	II
1200	4023.00	II
410	4030.47	II
1200	4031.82	II
3000	4040.80	II
410	4043.59	II
410	4048.81	II
850	4051.15	II
850	4059.96	II
4700	4061.09	II
1100	4069.28	II
710	4075.12	II
470	4075.28	II
470	4080.23	II
1400	4109.08	II
2500	4109.46	II
510	4110.48	II
410	4123.88	II
470	4133.36	II
510	4135.33	II
3000	4156.08	II
510	4156.26	II
410	4168.00	II
810	4175.61	II
2400	4177.32	II
640	4179.59	II
470	4205.60	II
470	4211.29	II
440	4227.73	II
1300	4232.38	II
2000	4247.38	II
850	4252.44	II
410	4261.84	II
470	4282.44	II
710	4284.52	II
5400	4303.58	II
470	4314.52	II
1100	4325.76	II
510	4327.93	II
540	4338.70	II
680	4351.29	II
850	4358.17	II
470 d	4374.93	II
710	4385.66	II
540	4400.83	II
510	4411.06	II
580	4446.39	II
1400	4451.57	II
740	4462.99	II
410	4501.82	II
250	4516.36	II
340	4541.27	II
340	4542.61	II
340	4563.22	II
300	4621.94	I
510	4634.24	I
340	4641.10	I
250	4645.77	II
300	4649.67	I
310	4683.45	I
470	4706.54	II
240	4719.02	I
240	4811.34	II
350	4825.48	II
280	4859.02	II
350	4883.81	I
220	4890.70	II
240	4891.07	I
280	4896.93	I
210	4901.84	I
330	4920.68	II
470	4924.53	I
260	4944.83	I
290	4954.78	I
290	4959.13	II
250	4989.94	II
360	5076.59	II
360	5092.80	II
360	5107.59	II
340	5123.79	II
680	5130.60	II
500	5191.45	II
630	5192.62	II
330	5200.12	II
310	5212.37	II
450	5234.20	II
250	5239.79	II
720	5249.59	II
360	5255.51	II
590	5273.43	II
680	5293.17	II
220	5311.46	II
500	5319.82	II
290	5361.47	II
160	5431.53	II
240	5594.43	II
220	5620.54	I
140 d	5675.97	I
220	5688.53	II
130	5702.24	II
160	5708.28	II
100	5729.29	I
160	5804.02	II
80	5811.57	II
70	5825.87	II
80	5842.39	II
55	5858.91	I
45	6007.67	I
45	6034.24	II
55	6066.03	I
45	6178.59	I
45	6223.39	I
55	6310.49	I
65	6385.20	II
45	6630.14	I
45	6650.57	II
40	6740.11	II
40	6900.43	II
35	7037.30	II
40	7066.89	II
29	7129.35	II
24	7189.42	II
20	7192.01	II
15	7236.54	II
12	7316.81	II
10	7406.62	II
10	7418.18	II
12	7511.16	II
17	7513.73	II
12	7528.99	II
10	7538.26	II
12	7696.56	II
10	7750.95	II
10	7808.47	II
12	7863.04	II
12	7917.01	II
12	7958.95	I
12	7965.73	II
15	7982.09	II
12	7982.68	II
12	8000.76	II
10	8120.93	II
12	8122.07	II
12	8141.75	II
12	8143.27	II
10	8231.52	II
10	8307.72	II
12	8346.36	II
17	8839.10	II

Neon Ne Z = 10

Intensity	Wavelength/Å	
66	119.01	V
200	122.52	V
66	125.12	V
45	131.99	V
50	132.04	V
150	140.76	V
150	140.79	V
100	142.44	V
100	142.50	V
150	142.72	V
100	143.27	V
150	143.34	V
150	147.13	V
66	151.23	V
120	151.42	V
15	151.82	IV
15	152.23	IV
45	154.50	V
15	158.65	IV
15	158.82	IV
100	164.02	V
100	164.14	V
80	172.62	IV
500	173.93	V
80	177.16	IV
150	186.58	IV
100	194.28	IV
100	208.48	IV
100	208.73	IV
80	208.90	IV
150	212.56	IV
140	223.24	IV
120	223.60	IV

Intensity	Wavelength/Å	
140	234.32	IV
120	234.70	IV
20	251.14	III
20	251.56	III
20	251.73	III
40	267.06	III
40	267.52	III
20	267.71	III
40	283.18	III
160	283.21	III
110	283.69	III
40	283.89	III
220	301.12	III
220	313.05	III
220	313.68	III
40	313.95	III
90	352.956	I
60	354.962	I
50	357.83	IV
400	357.96	V
500	358.47	V
200	358.72	IV
500	359.38	V
90	361.433	II
60	362.455	II
1000	365.59	V
220	379.31	III
125	387.14	IV
100	388.22	IV
150	405.854	II
120	407.138	II
800	416.20	V
150	421.61	IV
200	445.040	II
300	446.256	II
250	446.590	II
180	447.815	II
150	454.654	II
200	455.274	II
10	456.275	II
120	456.348	II
90	456.896	II
1000	460.728	II
500	462.391	II
140	469.77	IV
200	469.82	IV
180	469.87	IV
140	469.92	IV
250	480.41	V
150	481.28	V
250	481.36	V
500	482.99	V
285	488.10	III
220	488.87	III
450	489.50	III
70	489.64	III
220	490.31	III
360	491.05	III
120	521.74	IV
140	521.82	IV
80	541.13	IV
100	542.07	IV
150	543.89	IV
400	568.42	V
250	569.76	V
500	569.83	V
250	572.11	V
800	572.34	V
35	587.213	I
35	589.179	I
35	589.911	I
70	591.830	I
100	595.920	I
75	598.706	I
35	598.891	I
70	600.036	I
170	602.726	I
170	615.628	I
170	618.672	I
120	619.102	I
200	626.823	I
200	629.739	I
1000	735.896	I
400	743.720	I
60	993.88	I
70	1068.65	I
90	1131.72	I
100	1131.85	II
90	1229.83	I
20	1255.03	III
110	1255.68	III
160	1257.19	III
90	1418.38	I
90	1428.58	I
90	1436.09	I
120	1681.68	II
180	1688.36	II
100	1888.11	II
100	1889.71	II
200	1907.49	II
500	1916.08	II
300	1930.03	II
200	1938.83	II
100 c	1945.46	II
80	2007.01	II
65	2018.44	IV
110	2022.19	IV
80	2025.56	II
150	2085.47	II
200	2086.96	III
300	2089.43	III
240	2092.44	III
400	2095.54	III
180	2096.11	II
120	2096.25	II
200	2161.22	III
300	2163.77	III
200	2180.89	III
30	2203.88	IV
200	2209.35	III
200	2211.85	III
240	2213.76	III
300	2216.07	III
10	2220.81	IV
75	2227.42	V
110	2232.41	V
65	2245.48	V
250	2258.02	IV
65	2259.57	V
175	2262.08	IV
240	2263.21	III
65	2263.39	V
110	2264.54	IV
200	2264.91	III
250	2265.71	V
550	2285.79	IV
30	2293.14	IV
250	2293.49	IV
250	2350.84	IV
450	2352.52	IV
700	2357.96	IV
250	2362.68	IV
250	2363.28	IV
110	2365.49	IV
350	2372.16	IV
65	2384.20	IV
350	2384.95	IV
300	2412.73	III
240	2412.94	III
200	2413.78	III
200	2473.40	III
80 p	2562.12	II
90 w	2567.12	II
800	2590.04	III
600	2593.60	III
400	2595.68	III
300	2610.03	III
240	2613.41	III
200	2615.87	III
80	2623.11	II
80	2629.89	II
90 w	2636.07	II
80	2638.29	II
200	2638.70	III
200	2641.07	III
80	2644.10	II
600	2677.90	III
500	2678.64	III
80	2762.92	II
90	2792.02	II
80	2794.22	II
100	2809.48	II
80	2906.59	II
80	2906.82	II
90	2910.06	II
90	2910.41	II
80	2911.14	II
80	2915.12	II
80	2925.62	II
80 w	2932.10	II
80	2940.65	II
90	2946.04	II
150	2955.72	II
150	2963.24	II
150	2967.18	II
100	2973.10	II
15	2974.72	I
100	2979.46	II
12	2982.67	I
150	3001.67	II
120 p	3017.31	II
300	3027.02	II
300	3028.86	II
100	3030.79	II
120	3034.46	II
100	3035.92	II
100	3037.72	II
100	3039.59	II
100	3044.09	II
100	3045.56	II
120	3047.56	II
100	3054.34	II
100	3054.68	II
100	3059.11	II
100	3062.49	II
100	3063.30	II
100	3070.89	II
100	3071.53	II
100	3075.73	II
120	3088.17	II
100	3092.09	II
120	3092.90	II
100	3094.01	II
100	3095.10	II
100	3097.13	II
100	3117.98	II
120	3118.16	II
10	3126.199	I
300	3141.33	II
100	3143.72	II
100 p	3148.68	II
100	3164.43	II
100	3165.65	II
100	3188.74	II
120	3194.58	II
500	3198.59	II
60	3208.96	II
120	3209.36	II
120	3213.74	II
150	3214.33	II
150	3218.19	II
120	3224.82	II
120	3229.57	II
200	3230.07	II
120	3230.42	II
120	3232.02	II
150	3232.37	II
100	3243.40	II
100	3244.10	II
100	3248.34	II
100	3250.36	II
150	3297.73	II
150	3309.74	II
300	3319.72	II

Intensity	Wavelength/Å	
1000	3323.74	II
150	3327.15	II
100	3329.16	II
200	3334.84	II
150	3344.40	II
300	3345.45	II
150	3345.83	II
200	3355.02	II
120	3357.82	II
200	3360.60	II
120	3362.16	II
100	3362.71	II
120	3367.22	II
12	3369.808	I
40	3369.908	I
100	3371.80	II
500	3378.22	II
150	3388.42	II
120	3388.94	II
300	3392.80	II
100	3404.82	II
120	3406.95	II
100	3413.15	II
120	3416.91	II
120	3417.69	II
50	3417.904	I
15	3418.006	I
120	3428.69	II
60	3447.703	I
50	3454.195	I
100	3456.61	II
100	3459.32	II
25	3460.524	I
30	3464.339	I
30	3466.579	I
60	3472.571	I
150	3479.52	II
200	3480.72	II
200	3481.93	II
25	3498.064	I
30	3501.216	I
25	3515.191	I
150	3520.472	I
120	3542.85	II
120	3557.80	II
100	3561.20	II
250	3568.50	II
100	3574.18	II
200	3574.61	II
50	3593.526	I
30	3593.640	I
15	3600.169	I
20	3633.665	I
150	3643.93	II
200	3664.07	II
20	3682.243	I
12	3685.736	I
200	3694.21	II
10	3701.225	I
150	3709.62	II
250	3713.08	II
250	3727.11	II
800	3766.26	II
1000	3777.13	II
100	3818.43	II
120	3829.75	II
150	4219.74	II
100	4233.85	II
120	4250.65	II
120	4369.86	II
70	4379.40	II
150	4379.55	II
100	4385.06	II
200	4391.99	II
150	4397.99	II
150	4409.30	II
100	4413.22	II
100	4421.39	II
100 p	4428.52	II
100 p	4428.63	II
150 p	4430.90	II
150 p	4430.94	II
120	4457.05	II
100	4522.72	II
10	4537.754	I
10	4540.380	I
100	4569.06	II
15	4704.395	I
12	4708.862	I
10	4710.067	I
10	4712.066	I
15	4715.347	I
10	4752.732	I
12	4788.927	I
10	4790.22	I
10	4827.344	I
10	4884.917	I
4	5005.159	I
10	5037.751	I
10	5144.938	I
25	5330.778	I
20	5341.094	I
8	5343.283	I
60	5400.562	I
5	5562.766	I
10	5656.659	I
5	5719.225	I
12	5748.298	I
80	5764.419	I
12	5804.450	I
40	5820.156	I
500	5852.488	I
100	5872.828	I
100	5881.895	I
60	5902.462	I
60	5906.429	I
100	5944.834	I
100	5965.471	I
100	5974.627	I
120	5975.534	I
80	5987.907	I
100	6029.997	I
100	6074.338	I
80	6096.163	I
60	6128.450	I
100	6143.063	I
120	6163.594	I
250	6182.146	I
150	6217.281	I
150	6266.495	I
60	6304.789	I
100	6334.428	I
120	6382.992	I
200	6402.246	I
150	6506.528	I
60	6532.882	I
150	6598.953	I
70	6652.093	I
90	6678.276	I
20	6717.043	I
100	6929.467	I
90	7024.050	I
100	7032.413	I
50	7051.292	I
80	7059.107	I
100	7173.938	I
150	7213.20	II
150	7235.19	II
100	7245.167	I
150	7343.94	II
40	7472.439	I
90	7488.871	I
100	7492.10	II
150	7522.82	II
80	7535.774	I
60	7544.044	I
100	7724.628	I
120	7740.74	II
300	7839.055	I
120	7926.20	II
400	7927.118	I
700	7936.996	I
2000	7943.181	I
2000	8082.458	I
100	8084.34	II
1000	8118.549	I
600	8128.911	I
3000	8136.406	I
2500	8259.379	I
100	8264.81	II
2500	8266.077	I
800	8267.117	I
6000	8300.326	I
100	8315.00	II
1500	8365.749	I
100	8372.11	II
8000	8377.606	I
1000	8417.159	I
4000	8418.427	I
1500	8463.358	I
800	8484.444	I
5000	8495.360	I
600	8544.696	I
1000	8571.352	I
4000	8591.259	I
6000	8634.647	I
3000	8647.041	I
15000	8654.383	I
4000	8655.522	I
100	8668.26	II
5000	8679.492	I
5000	8681.921	I
2000	8704.112	I
4000	8771.656	I
12000	8780.621	I
10000	8783.753	I
500	8830.907	I
7000	8853.867	I
1000	8865.306	I
1000	8865.755	I
3000	8919.501	I
2000	8988.57	I
100	9079.46	II
6000	9148.67	I
6000	9201.76	I
4000	9220.06	I
2000	9221.58	I
2000	9226.69	I
1000	9275.52	I
200	9287.56	II
6000	9300.85	I
1500	9310.58	I
3000	9313.97	I
6000	9326.51	I
2000	9373.31	I
5000	9425.38	I
3000	9459.21	I
5000	9486.68	I
5000	9534.16	I
3000	9547.40	I
120	9577.01	II
1000	9665.42	I
100	9808.86	II
800	10295.42	I
2000	10562.41	I
1500	10798.07	I
2000	10844.48	I
3000	11143.020	I
3500	11177.528	I
1600	11390.434	I
1100	11409.134	I
3000	11522.746	I
1500	11525.020	I
950	11536.344	I
500	11601.537	I
1200	11614.081	I
300	11688.002	I
2000	11766.792	I
1500	11789.044	I
500	11789.889	I
1000	11984.912	I
3000	12066.334	I
800	12459.389	I
1000	12689.201	I

Intensity	Wavelength/Å	
1100	12912.014	I
700	13219.241	I
800	15230.714	I
400	17161.930	I
400	18035.80	I
1000	18083.21	I
350	18221.11	I
250	18227.02	I
2500	18276.68	I
2000	18282.62	I
1200	18303.97	I
250	18359.12	I
1200	18384.85	I
2000	18389.95	I
1000	18402.84	I
1200	18422.39	I
300	18458.65	I
400	18475.79	I
900	18591.55	I
1600	18597.70	I
350	18618.96	I
550	18625.16	I
1200	21041.295	I
750	21708.145	I
300	22247.35	I
350	22428.13	I
2250	22530.40	I
400	22661.81	I
600	23100.51	I
1000	23260.30	I
1050	23373.00	I
850	23565.36	I
3500	23636.52	I
300	23701.64	I
1100	23709.2	I
1800	23951.42	I
600	23956.46	I
1000	23978.12	I
200	24098.54	I
500	24161.42	I
600	24249.64	I
1500	24365.05	I
800	24371.60	I
400	24447.85	I
700	24459.4	I
300	24776.46	I
550	24928.88	I
250	25161.69	I
650	25524.37	I
125	28386.21	I
150	30200.	I
250	33173.09	I
450	33352.35	I
1300	33901.	I
2200	33912.10	I
600	34131.31	I
100	34471.44	I
120	35834.78	I

Neptunium Np Z = 93

Intensity	Wavelength/Å	
300	3481.93	I
300 h	3501.50	I
300 l	3986.89	I
300 s	5044.66	I
300 l	5601.70	I
300 l	5652.75	I
300 l	5784.39	I
300 l	5878.04	I
300 s	6011.22	I
300	6056.09	I
300 s	6073.90	I
300 s	6080.05	I
300 l	6120.49	I
300	6188.59	I
300 l	6200.00	I
300 s	6215.90	I
300 s	6317.84	I
300 l	6341.38	I
300 l	6566.11	I
300 l	6720.68	I
300 s	6751.32	I
300 s	6795.21	I
300 l	6802.62	I
300 l	6805.81	I
300 s	6816.44	I
300 l	6865.45	I
300 s	6907.13	I
300 h	6912.91	I
1000 s	6930.31	I
300 l	6963.63	I
3000 s	6972.09	I
300	7014.02	I
300 l	7018.91	I
300 s	7039.14	I
300 s	7080.01	I
300 l	7174.83	I
300 l	7184.93	I
300 l	7284.28	I
300 l	7292.29	I
300 l	7332.52	I
300 s	7370.60	I
300 l	7381.03	I
300 l	7381.65	I
300 l	7402.70	I
300 s	7512.22	I
300 l	7515.15	I
300 l	7546.05	I
300 l	7624.83	I
300	7626.85	I
300 s	7681.01	I
300 s	7685.25	I
1000 l	7735.14	I
300 l	7761.61	I
1000 l	7765.75	I
300 s	7776.07	I
300	7787.46	I
1000 l	7791.38	I
300 l	7851.44	I
300 l	7887.88	I
300 l	7901.71	I
300 l	7975.98	I
300 h	8080.32	I
300 s	8124.59	I
300	8155.11	I
300 l	8167.42	I
300 l	8183.06	I
300 l	8188.61	I
300 l	8247.82	I
300 l	8287.11	I
300 s	8287.75	I
300 l	8306.22	I
300 s	8313.66	I
1000 l	8339.12	I
300	8356.79	I
300 l	8367.11	I
3000	8372.88	I
3000	8529.96	I
1000 s	8696.23	I
1000 s	8906.02	I
1000	8942.70	I
1000 s	9004.75	I
1000 l	9006.31	I
10000 l	9016.18	I
3000 l	9141.30	I
3000 s	9379.33	I
3000 l	9468.66	I
3000 s	9679.13	I
3000 l	9930.55	I
10000 l	10091.99	I
10000 s	10817.45	I
10000 l	11695.15	I
10000 l	11776.64	I
10000 s	12148.18	I
10000 s	12377.42	I
10000 l	12407.99	I
10000 l	13834.33	I

Nickel Ni Z = 28

Intensity	Wavelength/Å	
55	315.24	V
56	315.71	V
72	354.18	V
76	354.42	V
68	354.49	V
500	630.71	III
500	676.94	III
300	713.33	III
300	713.38	III
500	718.48	III
300	722.09	III
500	729.82	III
400	731.70	III
300	732.16	III
300	747.99	III
300	750.05	III
300	757.80	III
400	770.22	III
500	778.81	III
300	788.04	III
500	811.57	III
500	826.14	III
500	842.14	III
400	845.24	III
300	847.43	III
300	860.64	III
300	862.88	III
300	863.22	III
300	867.51	III
300	973.79	III
400	979.59	III
500	1317.22	II
76	1398.19	IV
74	1411.45	IV
70	1438.82	IV
73	1449.01	IV
76	1452.22	IV
73	1482.25	IV
72	1489.83	IV
75	1525.31	IV
74	1527.68	IV
74	1527.80	IV
76	1534.71	IV
73	1537.25	IV
75	1543.41	IV
74	1546.23	IV
300	1604.54	III
300	1652.87	III
400	1687.90	III
1000	1692.51	III
800	1709.90	III
650	1715.30	III
500	1719.46	III
400	1722.28	III
500	1738.25	III
300	1739.78	III
1000	1741.55	II
300	1741.96	III
550	1747.01	III
300	1752.43	III
400	1753.01	III
800	1764.69	III
500	1767.94	III
2000	1769.64	III
400	1776.07	III
300	1807.24	III
300	1819.28	III
800	1823.06	III
400	1830.01	III
650	1847.28	III
800	1854.15	III
300	1858.75	III
1000	2165.55	II
2000	2169.10	II
2000	2174.67	II
1500	2175.15	II
2500	2185.50	II
3000	2192.09	II
5000	2205.55	II
4000	2206.72	II
6000	2216.48	II
1000	2264.46	II
2000	2270.21	II
1600	2289.98	I
630	2300.78	I
1000	2303.00	II
2000	2310.96	I
1700	2312.34	I

Intensity	Wavelength/Å	
1400	2313.66	I
1400	2313.98	I
1000	2316.04	II
1400	2317.16	I
2600	2320.03	I
1900	2321.38	I
1400	2325.79	I
940	2329.96	I
1200	2345.54	I
400	2347.52	I
1000	2375.42	II
240	2386.58	I
1000	2394.52	II
2000	2416.13	II
240	2419.31	I
160	2472.06	I
150	2798.65	I
250	2821.29	I
500	2943.91	I
570	2981.65	I
500	2992.60	I
1000	2994.46	I
4000	3002.49	I
2200	3003.63	I
3700	3012.00	I
1700	3037.94	I
3500	3050.82	I
1300	3054.32	I
1900	3057.64	I
500	3064.62	I
2600	3101.55	I
1300	3101.88	I
2900	3134.11	I
1100	3232.96	I
600	3243.06	I
660	3315.66	I
2000	3331.88	II
2900	3369.57	I
3300	3380.57	I
1300	3391.05	I
3300	3392.99	I
8200	3414.76	I
1600	3423.71	I
2600	3433.56	I
990	3437.28	I
4800	3446.26	I
1300	3452.89	I
5000	3458.47	I
5000	3461.65	I
1600	3472.54	I
550	3483.77	I
5500	3492.96	I
660	3500.85	I
2600	3510.34	I
6600	3515.05	I
660	3519.77	I
8200	3524.54	I
5000	3566.37	I
990	3571.87	I
1300	3597.70	I
1300	3610.46	I

Intensity	Wavelength/Å	
530	3612.74	I
6600	3619.39	I
200	3664.10	I
130	3669.24	I
180	3670.43	I
260	3674.15	I
160	3688.42	I
80	3693.93	I
120	3722.48	I
150	3736.81	I
60	3739.23	I
600	3775.57	I
700	3783.53	I
700	3807.14	I
110	3831.69	I
1200	3858.30	I
110	3973.56	I
110	4401.55	I
85	4459.04	I
55	4470.48	I
65	4605.00	I
75	4648.66	I
110	4714.42	I
45	4786.54	I
45	4855.41	I
40	4904.41	I
45	4980.16	I
45	4984.13	I
50	5017.59	I
100	5035.37	I
100	5080.52	I
65	5081.11	I
40 h	5146.48	I
40 h	5155.76	I
180	5476.91	I
23	5709.56	I
16	5754.68	I
10	5857.76	I
10	5892.88	I
10	6108.12	I
10	6176.81	I
10	6191.18	I
13	6256.36	I
16	6643.64	I
22	6767.77	I
10	6914.56	I
26	7122.20	I
16	7393.60	I
16	7409.35	I
23	7422.28	I
13	7522.76	I
19	7555.60	I
23	7617.00	I
16	7714.32	I
19	7727.61	I
19	7748.89	I
10	7788.94	I
13	7797.59	I
1000	8096.75	II
700	8121.48	II
9	8862.55	I

Intensity	Wavelength/Å	
500 w	9900.92	II

***Niobium* Nb *Z* = 41**

Intensity	Wavelength/Å	
80	464.55	V
80	468.32	V
80	763.77	V
80	774.02	V
60	993.54	IV
400	1005.72	IV
500	1007.05	IV
500	1010.19	IV
100	1116.08	IV
150	1120.02	IV
100	1258.87	V
60	1314.56	III
80	1445.43	III
80	1445.98	III
80	1447.09	III
100	1456.68	III
80	1484.73	III
100	1495.94	III
80	1498.02	III
80	1499.45	III
100	1501.99	III
60	1502.30	IV
80	1513.81	III
60	1524.36	IV
100	1524.91	III
100	1590.21	III
80	1598.86	III
80	1604.72	III
80	1639.51	III
100	1682.77	III
100	1705.44	III
100	1707.14	III
100	1758.33	V
100	1877.34	V
100	1892.92	III
60	1922.41	IV
100	1938.84	III
60	1978.22	IV
3300	2029.32	II
65	2032.53	IV
3000	2032.99	II
2000	2109.42	II
1700	2125.21	II
1100	2126.54	II
80 h	2130.24	III
1500	2131.18	II
80	2273.92	III
100	2275.23	III
80	2279.36	III
100	2281.51	III
80	2284.40	III
100	2290.36	III
370	2295.68	II
280	2302.08	II
100	2313.30	III
100	2338.09	III
80	2344.12	III
90	2349.21	III
80	2355.54	III

Intensity	Wavelength/Å	
100	2362.06	III
80	2362.50	III
80	2365.70	III
100	2372.73	III
170	2376.40	II
110	2387.09	II
100	2387.41	III
140	2387.52	II
80	2388.23	III
45	2388.27	II
160	2398.48	II
80	2404.89	III
55	2405.34	II
55	2405.85	II
140	2412.46	II
100	2413.94	III
160	2416.99	II
140	2418.69	II
100	2421.91	III
75	2433.80	II
40	2435.95	II
45	2437.42	II
40	2442.14	II
28	2442.68	II
65	2451.87	II
65	2453.95	II
100	2456.99	III
55	2458.09	II
65	2462.89	I
80	2468.72	III
80	2475.87	III
110	2477.38	II
65	2478.29	II
65	2479.94	II
35	2483.88	II
100	2499.73	III
110	2511.00	II
110	2521.40	II
390	2544.80	II
100	2545.64	III
110	2551.38	II
130	2556.94	II
80	2557.94	III
130	2562.41	II
110	2571.33	II
390	2583.99	II
390	2590.94	II
80	2598.86	III
80	2633.17	III
200	2642.24	II
320	2646.26	II
330	2647.50	I
330	2654.45	I
310	2656.08	II
80	2657.99	III
110	2665.25	II
110	2666.59	II
110	2667.30	II
400	2671.93	II
200	2673.57	II
200	2675.94	II

Intensity	Wavelength/Å	
160	2691.77	II
1000	2697.06	II
320	2698.86	II
320	2702.20	II
150	2702.52	II
470	2716.62	II
470	2721.98	II
310	2733.26	II
110	2737.09	II
240	2768.13	II
310	2773.20	I
270	2780.24	II
110	2793.05	II
190	2827.08	II
250	2841.15	II
280	2842.65	II
160	2846.28	II
240	2861.09	II
100	2865.61	II
500	2868.52	II
800	2875.39	II
270	2876.95	II
530	2877.03	II
100	2880.72	II
570	2883.18	II
280	2888.83	II
470	2897.81	II
400	2899.24	II
470	2908.24	II
670	2910.59	II
470	2911.74	II
1100	2927.81	II
110	2931.47	II
870	2941.54	II
110 h	2945.88	II
110	2946.12	II
110	2946.90	II
1100	2950.88	II
400	2972.57	II
320	2974.10	II
210	2977.68	II
200	2982.11	II
330	2990.26	II
470	2994.73	II
80	3001.84	III
140	3024.74	II
350	3028.44	II
300	3032.77	II
100	3044.76	II
100	3055.52	II
220	3064.53	II
110	3069.68	II
100	3070.90	II
110	3071.56	II
100	3073.24	II
400	3076.87	II
110	3080.35	II
1800	3094.18	II
140	3099.19	II
270	3127.53	II
1500	3130.79	II
80	3142.26	III
390	3145.40	II
1200	3163.40	II
150	3175.78	II
390	3180.29	II
300	3191.10	II
150	3191.43	II
1000	3194.98	II
120	3203.35	II
300	3206.34	II
390	3215.60	II
800	3225.48	II
140	3229.56	II
400	3236.40	II
200	3247.47	II
120	3248.94	II
320	3254.07	II
230	3260.56	II
160	3263.37	II
200	3283.46	II
160	3292.02	II
320	3296.01	I
400	3312.60	I
120	3319.58	II
130	3341.60	II
1300	3341.97	I
1300	3343.71	I
1700	3349.06	I
420	3349.52	I
340	3354.74	I
1700	3358.42	I
130	3365.58	II
340	3366.96	I
130	3369.16	II
350	3374.92	I
170	3386.24	II
350	3392.34	I
230	3408.68	II
180	3409.19	II
230	3412.94	II
230	3425.42	II
230	3426.57	II
180	3432.70	II
180	3440.59	II
200	3479.56	II
100	3484.05	II
500	3498.63	I
460	3507.96	I
200	3510.26	II
200	3515.42	II
200	3517.67	II
2000	3535.30	I
1300	3537.48	I
250	3540.96	II
500	3544.02	I
300	3550.45	I
1000	3554.66	I
630	3563.50	I
630	3563.62	I
1500	3575.85	I
5000	3580.27	I
500	3584.97	I
750	3589.11	I
500	3589.36	I
500	3593.97	I
500	3602.56	I
300	3619.51	II
420	3649.85	I
400	3651.19	II
200	3659.61	II
630	3660.37	I
900	3664.70	I
1500	3697.85	I
330	3711.34	I
3300	3713.01	I
480	3716.99	I
2700	3726.24	I
2700	3739.80	I
670	3740.73	II
1700	3742.39	I
530	3763.49	I
350	3765.08	I
530	3771.85	I
870	3781.01	I
1700	3787.06	I
1300	3790.15	I
3500	3791.21	I
2700	3798.12	I
2700	3802.92	I
670	3803.88	I
530	3804.74	I
670	3810.49	I
530	3811.03	I
530	3815.51	I
210	3818.86	II
670	3824.88	I
350	3835.18	I
350	3863.38	I
530	3877.56	I
870	3878.82	I
670	3883.14	I
1100	3885.44	I
670	3885.68	I
580	3891.30	I
670	3914.70	I
530	3920.20	I
670	3937.44	I
520	3943.67	I
910 d	3966.09	I
1100	4032.52	I
16000 c	4058.94	I
350	4060.79	I
12000	4079.73	I
440	4100.40	I
6700	4100.92	I
310	4116.90	I
5300	4123.81	I
670	4129.43	I
770	4129.93	I
2300	4137.10	I
440	4139.44	I
2700	4139.71	I
350	4143.21	I
870	4150.12	I
4400	4152.58	I
870	4163.47	I
4400	4163.66	I
4000	4164.66	I
3500	4168.13	I
310	4184.44	I
1200	4190.88	I
870	4192.07	I
870	4195.09	I
1300	4195.66	I
310	4198.51	I
350	4201.52	I
870	4205.31	I
350	4214.73	I
420	4217.94	I
420	4229.15	I
770	4262.05	I
420	4266.02	I
400	4286.99	I
580	4299.60	I
580	4300.99	I
390	4311.27	I
350	4326.33	I
390	4331.37	I
330	4410.21	I
150	4503.04	I
530	4523.41	I
480	4546.82	I
370	4564.53	I
720	4573.08	I
480	4581.62	I
1200	4606.77	I
170	4616.17	I
450	4630.11	I
450	4648.95	I
450	4663.83	I
340	4666.24	I
240	4667.22	I
580	4672.09	I
530	4675.37	I
320	4685.14	I
130 c	4706.14	I
260	4708.29	I
150	4713.50	I
220 c	4749.70	I
130 c	4967.78	I
190	4988.97	I
230	5017.75	I
150	5026.36	I
210	5039.04	I
170	5058.01	I
130	5065.25	I
750	5078.96	I
420	5095.30	I
170	5100.16	I
170	5120.30	I
210	5134.75	I
250	5160.33	I
250	5164.38	I

Intensity	Wavelength/Å	
230	5180.31	I
190	5189.20	I
170	5193.08	I
150	5195.84	I
150	5232.81	I
150 d	5251.62	I
270	5271.53	I
130 c	5276.20	I
250	5318.60	I
460	5344.17	I
340	5350.74	I
110	5437.27	I
85	5551.35	I
170	5642.11	I
130	5664.71	I
170	5665.63	I
130	5729.19	I
110	5760.34	I
110	5819.43	I
130 d	5838.64	I
190 cw	5900.62	I
150	5983.22	I
75	6221.96	I
85 c	6430.46	I
65	6544.61	I
210 cw	6660.84	I
150 cw	6677.33	I
190 c	6723.62	I
85	6828.11	I
85	6990.32	I
190 c	7046.81	I
130	7159.43	I
190 cw	7372.50	I
65	7515.93	I
170 c	7574.58	I
75 c	7726.68	I
35	7885.31	I
40	8135.20	I
29 cw	8320.93	I
29	8346.08	I
35	8905.78	I

Nitrogen N Z = 7

Intensity	Wavelength/Å	
400	181.75	IV
52	186.069	V
62	186.153	V
400	191.7	IV
400	192.9	IV
500	196.87	IV
500	197.23	IV
500	202.60	IV
500	205.94	IV
500	205.97	IV
500	206.03	IV
90	209.303	V
500	217.20	IV
500 d	217.90	IV
500 d	223.4	IV
800 w	225.12	IV
800	225.21	IV
600 w	234.12	IV
600 w	234.20	IV
600 w	234.25	IV
550	236.07	IV
500	237.99	IV
500 w	238.7	IV
600	238.80	IV
500 w	239.62	IV
900	247.20	IV
90	247.561	V
120	247.706	V
500 w	248.43	IV
500 w	248.46	IV
500 w	248.48	IV
500	257.95	III
650	258.50	III
700	259.19	III
800	260.09	III
600	260.45	IV
800	261.28	III
500	262.91	III
500	265.23	III
500	265.27	III
150	266.196	V
200	266.379	V
500	268.70	III
650	270.99	IV
250	283.42	IV
300	283.48	IV
350	285.50	IV
600	285.56	IV
600 w	297.7	IV
700	297.82	IV
650	300.32	IV
90	303.123	IV
500	303.28	IV
150	314.715	III
200	314.850	III
90	314.877	III
150	315.053	IV
120	322.503	IV
150	322.570	IV
200	322.724	IV
120	323.175	IV
600	323.26	III
300	335.050	IV
500	338.35	III
500	340.20	III
500 w	351.93	IV
500	351.98	III
700	353.06	IV
120	362.833	III
150	362.881	III
150	362.946	III
90	362.985	III
300	374.204	III
350	374.441	III
500	387.48	III
500	420.77	IV
250	451.869	III
300	452.226	III
650	463.74	IV
285	644.634	II
360	644.837	II
450	645.178	II
140	647.50	I
360	660.286	II
170	671.016	II
285	671.386	II
150	671.630	II
160	671.773	II
170	672.001	II
500	684.996	III
570	685.513	III
650	685.816	III
500	686.335	III
350	692.70	I
90	713.518	V
150	713.860	V
285	746.984	II
150	748.195	V
200	748.291	V
500	763.336	III
570	764.359	III
570	765.148	IV
250	771.544	III
300	771.901	III
350	772.385	III
200	772.891	III
150	772.975	III
650	775.965	II
90	885.67	I
90	909.697	I
80	910.278	I
40	910.645	I
450	915.612	II
450	915.962	II
550	916.012	II
650	916.701	II
520	921.992	IV
500	922.519	IV
480	923.057	IV
520	924.283	IV
90	953.415	I
100	953.655	I
130	953.970	I
1000	955.335	IV
130	963.990	I
115	964.626	I
70	965.041	I
650	979.842	III
700	979.919	III
900	989.790	III
700	991.514	III
1000	991.579	III
150 w	1036.16	IV
90	1067.614	I
60	1068.612	I
90	1078.71	IV
450	1083.990	II
600	1084.580	II
430	1085.546	II
650	1085.701	II
175	1097.237	I
115	1098.095	I
115	1098.260	I
105	1100.360	I
40	1100.465	I
90	1101.291	I
360	1134.165	I
385	1134.415	I
410	1134.980	I
105	1143.65	I
130	1163.884	I
60	1164.206	I
105	1164.325	I
270	1167.448	I
105	1168.334	I
60	1168.417	I
195	1168.536	I
230	1176.510	I
105	1176.630	I
195	1177.695	I
500	1183.031	III
570	1184.550	III
90	1188.01	IV
410	1199.550	I
385	1200.223	I
360	1200.710	I
175	1225.026	I
160	1225.37	I
190	1228.41	I
160	1228.79	I
1000	1238.821	V
900	1242.804	V
360	1243.179	I
315	1243.306	I
290	1310.540	I
250	1310.95	I
230	1319.00	I
315	1319.68	I
115	1326.57	I
115	1327.92	I
150	1387.371	III
360	1411.94	I
700	1492.625	I
490	1492.820	I
640	1494.675	I
90	1549.336	V
200 l	1616.33	V
350 l	1619.69	V
1000	1718.55	IV
250	1729.945	III
775	1742.729	I
700	1745.252	I
570	1747.848	III
350	1751.218	III
650	1751.657	III
150	1804.486	III
200	1805.669	III
150	1846.42	III
90 w	1860.37	V
350	1885.06	III
400	1885.22	III
200	1907.99	III

Intensity	Wavelength/Å	
150	1919.55	III
150	1919.77	III
300	1920.65	III
150	1920.84	III
200	1921.30	III
200	2064.01	III
250	2064.42	III
120	2068.68	III
90	2071.09	III
90	2080.34	IV
160	2095.53	II
70	2096.20	II
110	2096.86	II
90	2117.59	III
90	2121.50	III
110	2130.18	II
160	2142.78	II
90	2147.31	III
200	2188.20	III
150	2188.38	III
160	2206.09	II
160	2286.69	II
110	2288.44	II
220	2316.49	II
160	2316.69	II
285	2317.05	II
90 w	2318.09	IV
160	2461.27	II
150	2477.69	IV
110	2496.83	II
70	2496.97	II
110	2520.22	II
160	2520.79	II
220	2522.23	II
110	2590.94	II
250	2645.65	IV
300	2646.18	IV
350	2646.96	IV
250 w	2682.18	III
90	2689.20	III
160	2709.84	II
110	2799.22	II
110	2823.64	II
60 l	2859.16	V
160	2885.27	II
90 l	2974.52	V
150 w	2980.78	V
250 w	2981.31	V
60 w	2998.43	V
220	3006.83	II
90	3078.25	IV
120	3367.34	III
360	3437.15	II
90	3463.37	IV
570	3478.71	IV
500	3482.99	IV
400	3484.96	IV
90	3747.54	IV
90	3754.67	III
120	3771.05	III
285	3838.37	II
360	3919.00	II
90	3938.52	III
450	3955.85	II
1000	3995.00	II
150	3998.63	III
200	4003.58	III
360	4035.08	II
550	4041.31	II
360	4043.53	II
150	4057.76	IV
250	4097.33	III
140	4099.94	I
200	4103.43	III
185	4109.95	I
285	4176.16	II
120	4195.76	III
150	4200.10	III
285	4227.74	II
285	4236.91	II
220	4237.05	II
450	4241.78	II
90	4332.91	III
120	4345.68	III
300	4379.11	III
285	4432.74	II
650	4447.03	II
90	4510.91	III
120	4514.86	III
360	4530.41	II
550	4601.48	II
350	4603.73	V
90	4606.33	IV
450	4607.16	II
360	4613.87	II
250	4619.98	V
450	4621.39	II
870	4630.54	II
90	4634.14	III
120	4640.64	III
550	4643.08	II
285	4788.13	II
450	4803.29	II
180	4847.38	I
90	4858.82	III
150	4867.15	III
285	4895.11	II
160	4914.94	I
210	4935.12	I
200 w	4944.56	V
160	4950.23	I
350	4963.98	I
285	4987.37	II
450	4994.36	II
650	5001.48	II
360	5002.70	II
870	5005.15	II
550	5007.32	II
450	5010.62	II
360	5016.39	II
360	5025.66	II
550	5045.10	II
185	5281.20	I
140	5292.68	I
90	5314.35	III
200	5320.82	III
150	5327.18	III
450	5495.67	II
285	5535.36	II
650	5666.63	II
550	5676.02	II
870	5679.56	II
450	5686.21	II
450	5710.77	II
285	5747.30	II
700	5752.50	I
240	5764.75	I
265	5829.54	I
235	5854.04	I
360	5927.81	II
550	5931.78	II
285	5940.24	II
650	5941.65	II
285	5952.39	II
160	5999.43	I
210	6008.47	I
285	6167.76	II
360	6379.62	II
150	6380.77	IV
185	6411.65	I
210	6420.64	I
210	6423.02	I
210	6428.32	I
185	6437.68	I
235	6440.94	I
90	6454.11	III
185	6457.90	I
120	6467.02	III
300	6468.44	I
265	6481.71	I
750	6482.05	II
360	6482.70	I
300	6483.75	I
325	6484.80	I
160	6491.22	I
210	6499.54	I
185	6506.31	I
750	6610.56	II
185	6622.54	I
185	6636.94	I
235	6644.96	I
185	6646.50	I
235	6653.46	I
210	6656.51	I
185	6722.62	I
210	7398.64	I
160	7406.12	I
265	7406.24	I
685	7423.64	I
785	7442.29	I
900	7468.31	I
185	7608.80	I
60 w	7618.46	V
450	7762.24	II
400	8184.87	I
400	8188.02	I
250	8200.36	I
300	8210.72	I
570	8216.34	I
400	8223.14	I
400	8242.39	I
550	8438.74	II
500	8567.74	I
570	8594.00	I
650	8629.24	I
500	8655.89	I
220	8676.08	II
700	8680.28	I
650	8683.40	I
500	8686.15	I
110	8687.43	II
110 h	8699.00	II
500	8703.25	I
160 h	8710.54	II
570	8711.70	I
500	8718.83	I
250	8728.89	I
200	8747.36	I
500	9386.80	I
570	9392.79	I
250	9460.68	I
200	9863.33	I
160 h	9865.41	II
110 h	9868.21	II
160 h	9887.39	II
220 h	9891.09	II
160 h	9961.86	II
220 h	9969.34	II
285 h	10023.27	II
220 h	10035.45	II
220 h	10065.15	II
160 h	10070.12	II
250	10105.13	I
300	10108.89	I
350	10112.48	I
400	10114.64	I
110 h	10126.27	II
250	10539.57	I
200	12074.51	I
380	12186.82	I
225	12288.97	I
290	12328.76	I
310	12381.65	I
180	12438.40	I
510	12461.25	I
920	12469.62	I
500	13429.61	I
840	13581.33	I
180	13587.73	I
180	13602.27	I
290	13624.18	I
250	14757.07	I
100	14868.87	I
160	14966.60	I

Intensity	Wavelength/Å	
180	15582.27	I
120 s	17516.58	I
100 l	17584.86	I
100	17878.26	I

Osmium Os Z = 76

Intensity	Wavelength/Å	
9600	2001.45	I
13000	2003.73	I
17000	2010.15	I
29000	2018.14	I
14000	2022.76	I
14000	2028.23	I
18000	2034.44	I
26000	2045.36	I
8600	2058.69	I
13000	2061.69	I
7800	2067.21	II
4200	2070.67	II
7200	2076.95	I
14000	2079.97	I
2900	2082.54	I
2900	2089.03	I
2900	2089.21	I
6000	2097.60	I
5300	2100.63	I
2100	2117.66	I
4800	2117.96	I
5300	2137.11	I
2600	2154.59	I
1300	2157.84	I
1200	2158.53	I
3100	2166.90	I
1100	2167.75	I
2100	2171.65	I
1100	2234.61	I
1300	2252.15	I
2000	2255.85	II
1400	2264.60	I
1400	2282.26	II
500	2367.35	II
2600	2377.03	I
1700	2387.29	I
1100	2395.88	I
200	2423.07	II
1400	2424.97	I
110	2454.91	II
1800	2461.42	I
110	2468.90	II
530	2486.24	II
4500	2488.55	I
2600	2498.41	I
2400	2513.25	I
780	2538.00	II
1000	2542.51	I
1000	2590.76	I
1800	2613.06	I
3800	2637.13	I
1900	2644.11	I
1900	2658.60	I
2100	2689.82	I
3000	2714.64	I
1300	2720.04	I
960	2770.71	I
2800	2806.91	I
5100	2838.63	I
2300	2844.40	I
1500	2850.76	I
1500	2860.96	I
9600	2909.06	I
2100	2912.33	I
2100	2919.79	I
1100 h	2948.23	I
1400	2949.53	I
4400	3018.04	I
1100	3030.70	I
2900	3040.90	I
120	3042.74	II
8600	3058.66	I
1100	3077.72	I
3100	3156.25	I
180	3173.93	II
150	3213.31	II
1900	3232.06	I
3100	3262.29	I
3100	3267.94	I
1200	3290.26	I
7600	3301.56	I
960	3336.15	I
960	3370.59	I
620	3387.84	I
620	3401.86	I
620	3504.66	I
1200	3528.60	I
1200	3560.86	I
620	3598.11	I
95	3604.48	II
480	3670.89	I
3700	3752.52	I
2100	3782.20	I
730	3876.77	I
1000	3963.63	I
730	3977.23	I
960	4066.69	I
1200	4112.02	I
2500	4135.78	I
1200	4173.23	I
1200	4211.86	I
4900	4260.85	I
560	4293.95	I
560	4311.40	I
4900	4420.47	I
540	4550.41	I
670	4793.99	I
55	5031.83	I
45	5039.12	I
35	5072.88	I
35	5074.77	I
35	5079.09	I
90	5103.50	I
55	5110.81	I
140	5149.74	I
40	5193.52	I
270	5202.63	I
35	5203.23	I
45	5255.82	I
55	5265.15	I
40	5298.78	I
110	5376.79	I
120	5416.34	I
45	5416.69	I
28	5417.51	I
55	5443.31	I
22	5446.93	I
22	5457.30	I
28	5470.00	I
22	5509.33	I
270	5523.53	I
22	5546.82	I
80	5584.44	I
35	5620.08	I
22	5642.56	I
28	5645.25	I
28	5680.88	I
170	5721.93	I
22	5765.05	I
170	5780.82	I
40	5800.60	I
110	5857.76	I
28	5860.64	I
65	5996.00	I
35	6227.70	I
22	6269.41	I
22	6403.15	I
27	6729.56	I
22	7145.54	I
26	7602.95	I
7	8041.29	I

Oxygen O Z = 8

Intensity	Wavelength/Å	
80	124.616	V
110	135.523	V
80	138.109	V
110	139.029	V
80	151.447	V
110	151.477	V
150	151.546	V
80	164.574	V
110	164.657	V
80	164.709	V
80	166.235	V
150	167.99	V
110	170.219	V
450	172.169	V
250	185.745	V
375	192.751	V
450	192.799	V
520	192.906	V
80	193.003	V
200	194.593	V
150	195.86	IV
200	196.01	IV
80	202.161	V
80	202.224	V
80	202.283	V
80	202.334	V
150	202.393	V
110	203.78	V
150	203.82	V
100	203.85	V
200	203.89	V
100	203.94	V
110	207.18	IV
150	207.24	IV
300	207.794	V
150	215.040	V
200	215.103	V
250	215.245	V
250	216.018	V
520	220.352	V
80	227.372	V
80	227.469	V
150	227.511	V
80	227.549	V
80	227.634	V
80	227.689	V
150	231.823	V
140	233.46	IV
150	233.50	IV
110	233.52	IV
200	233.56	IV
110	233.60	IV
90	238.36	IV
180	238.57	IV
110	248.459	V
110	252.56	IV
110	252.95	IV
150	253.08	IV
300	260.39	IV
250	260.56	IV
80 d	264.34	III
110	264.48	III
110	266.97	III
150	266.98	III
150	267.03	III
150	277.38	III
300	279.63	IV
375	279.94	IV
110	285.71	IV
150	285.84	IV
110	286.448	V
80	295.62	III
110	295.66	III
120	295.72	III
150	303.41	III
150	303.46	III
140	303.52	III
160	303.62	III
160	303.69	III
250	303.80	III
200	305.60	III
250	305.66	III
190	305.70	III
300	305.77	III
190	305.84	III
200	306.62	IV
150	306.88	IV

Intensity	Wavelength/Å	
450	320.979	III
300	328.45	III
250	328.74	III
300	345.31	III
110	355.14	III
90	355.33	III
80	355.47	III
200	359.02	III
190	359.22	III
150	359.38	III
210	373.80	III
200	374.00	III
300	374.08	III
190	374.16	III
200	374.33	III
210	374.44	III
450	395.558	III
300	434.98	III
800	507.391	III
900	507.683	III
1000	508.182	III
1000	525.795	III
250	537.83	II
300	538.26	II
220	539.09	II
200	539.55	II
150	539.85	II
700	553.330	IV
775	554.075	IV
850	554.514	IV
700	555.261	IV
700	597.818	III
1000	599.598	III
580	608.398	IV
110	609.70	III
640	609.829	IV
160	610.04	III
200	610.75	III
100	610.85	III
270	616.952	IV
150	617.005	IV
200	617.036	IV
520	624.617	IV
580	625.130	IV
640	625.852	IV
1000	629.730	V
150	644.148	II
200	672.95	II
150	673.77	II
230	681.272	V
70	685.544	I
800	702.332	III
800	702.822	III
900	702.899	III
1000	703.850	III
900	718.484	II
600	718.562	II
70	744.794	I
700	758.678	V
640	759.441	V
580	760.228	V

Intensity	Wavelength/Å	
775	760.445	V
640	761.128	V
700	762.003	V
70	770.793	I
90	771.056	I
520	774.518	V
70	775.321	I
200	779.734	IV
315	779.821	IV
360	779.912	IV
200	779.997	IV
640	787.711	IV
520	790.109	IV
700	790.199	IV
70	791.973	I
300	796.66	II
200	802.200	IV
160	802.255	IV
90	804.267	I
70	804.848	I
70	805.295	I
80	805.810	I
240	832.762	II
600	832.927	III
450	833.332	II
780	833.742	III
600	834.467	II
600	835.096	III
800	835.292	III
40	877.879	I
130	921.296	IV
160	921.366	IV
80	922.008	I
200	923.367	IV
130	923.433	IV
90	935.193	I
40	948.686	I
90	971.738	I
40	976.448	I
160	988.773	I
40	990.204	I
250	1025.762	I
90	1027.431	I
160	1039.230	I
60	1040.942	I
40	1152.152	I
900	1302.168	I
600	1304.858	I
300	1306.029	I
200	1338.612	IV
130	1342.992	IV
230	1343.512	IV
640	1371.292	V
160	1476.89	III
160 w	1506.72	V
285	1590.01	III
160	1591.33	III
315 w	1643.68	V
160	1707.996	V
220	1760.12	III
110	1760.42	III

Intensity	Wavelength/Å	
220	1763.22	III
220	1764.48	III
750	1767.78	III
550	1768.24	III
360	1771.67	III
110	1773.00	III
110	1773.85	III
220	1779.16	III
160	1781.03	III
160	1784.85	III
220	1789.66	III
110	1848.26	III
110	1856.62	III
285	1872.78	III
285	1872.87	III
285	1874.94	III
160	1920.04	III
110	1920.75	III
110	1921.52	III
220	1923.49	III
110	1923.82	III
110	1926.94	III
360	2013.27	III
160	2026.96	III
220	2045.67	III
160	2052.74	III
30 d	2283.42	II
30 d	2284.89	II
110	2293.32	II
200	2300.35	II
30 d	2313.05	II
30 d	2316.12	II
30 d	2316.79	II
50 d	2319.68	II
30 d	2322.15	II
30 d	2339.31	II
200 d	2390.44	III
80	2394.33	III
110	2411.60	II
80	2422.84	III
80	2425.55	II
250	2433.56	II
80 d	2436.06	II
80 d	2438.83	III
80	2444.26	II
300	2445.55	II
200	2449.372	IV
200	2450.040	IV
200	2454.99	III
200	2493.44	IV
200	2493.77	IV
200	2507.73	IV
230	2509.19	IV
200	2517.2	IV
200	2558.06	III
80	2687.53	III
110	2695.49	III
300	2733.34	II
110	2747.46	II
1000	2781.01	V
920	2786.99	V

Intensity	Wavelength/Å	
775	2789.85	V
160	2836.26	IV
160	2921.45	IV
200	2941.33	V
210	2941.65	V
80	2959.68	III
265	2972.29	I
250	2983.78	III
80	3017.63	III
80	3023.45	III
80	3043.02	III
200	3047.13	III
110	3059.30	III
460	3063.42	IV
410	3071.61	IV
80	3121.71	III
160	3122.62	II
220	3129.44	II
110	3132.86	III
450	3134.82	II
285	3138.44	II
160	3144.66	V
160	3209.66	IV
80	3238.57	III
200	3260.98	III
300	3265.46	III
80	3267.31	III
220	3270.98	II
220	3273.52	II
220	3277.69	II
360	3287.59	II
160	3305.15	II
160	3306.60	II
80	3312.30	III
110	3340.74	III
230	3348.08	IV
270	3349.11	IV
160	3354.27	IV
200	3375.40	IV
220	3377.20	II
130	3378.06	IV
360	3381.20	IV
360	3385.52	IV
285	3390.25	II
270	3396.79	IV
360	3403.52	IV
220	3407.38	II
230	3409.66	IV
160	3409.84	II
410	3411.69	IV
230	3413.64	IV
80	3444.10	III
80	3455.12	III
285	3470.81	II
200	3489.83	IV
160	3492.24	IV
230	3560.39	IV
270	3563.33	IV
80	3698.70	III
80	3702.75	III
80	3703.37	III

Intensity	Wavelength/Å	
110	3707.24	III
220	3712.75	II
110	3715.08	III
315 w	3725.93	IV
285	3727.33	II
360	3729.03	IV
410	3736.85	IV
160	3739.92	II
110	3744.00	III
230	3744.89	IV
360	3749.49	II
150	3754.67	III
80	3757.21	III
250	3759.87	III
110	3791.26	III
160	3803.14	II
120	3823.41	I
450	3911.96	II
160	3919.29	II
185	3947.29	I
160	3947.48	I
140	3947.59	I
220	3954.37	II
100	3954.61	I
200	3961.59	III
450	3973.26	II
220	3982.20	II
160	4069.90	II
285	4072.16	II
450	4075.87	II
80 d	4083.91	II
50 d	4087.14	II
150 d	4089.27	II
110	4097.24	II
220	4105.00	II
285	4119.22	II
100	4123.99	V
160	4132.81	II
50	4146.06	II
220	4153.30	II
285	4185.46	II
450	4189.79	II
80	4233.27	I
50 d	4253.74	II
50 d	4253.98	II
50 d	4275.47	II
50 d	4303.78	II
285	4317.14	II
160	4336.86	II
220	4345.56	II
285	4349.43	II
220	4366.90	II
100	4368.25	I
220	4395.95	II
450	4414.91	II
285	4416.98	II
160	4448.21	II
160	4452.38	II
50	4465.45	II
50 d	4466.28	II
50	4467.83	II

Intensity	Wavelength/Å	
50	4469.41	II
360	4590.97	II
285	4596.17	II
80 d	4609.39	II
160	4638.85	II
360	4641.81	II
450	4649.14	II
160	4650.84	II
360	4661.64	II
285	4676.23	II
220	4699.21	II
285	4705.36	II
160	4924.60	II
230 w	4930.27	V
220	4943.06	II
135	5329.10	I
160	5329.68	I
190	5330.74	I
90	5435.18	I
110	5435.78	I
135	5436.86	I
120	5577.34	I
110	5592.37	III
130	5597.91	V
160	5958.39	I
190	5958.58	I
80	5995.28	I
160	6046.23	I
190	6046.44	I
110	6046.49	I
100	6106.27	I
400	6155.98	I
450	6156.77	I
490	6158.18	I
80	6256.83	I
100	6261.55	I
100	6366.34	I
100	6374.32	I
320	6453.60	I
360	6454.44	I
400	6455.98	I
130	6500.24	V
80	6604.91	I
100	6653.83	I
360	7001.92	I
450	7002.23	I
210	7156.70	I
400	7254.15	I
450	7254.45	I
320	7254.53	I
210	7476.44	I
100	7477.24	I
120	7479.08	I
120	7480.67	I
100	7706.75	I
870	7771.94	I
810	7774.17	I
750	7775.39	I
80	7886.27	I
100	7943.15	I
100	7947.17	I

Intensity	Wavelength/Å	
235	7947.55	I
210	7950.80	I
185	7952.16	I
110	7981.94	I
135	7982.40	I
190	7986.98	I
135	7987.33	I
250	7995.07	I
400	8221.82	I
265	8227.65	I
265	8230.02	I
325	8233.00	I
120	8235.35	I
120	8426.16	I
810	8446.25	I
1000	8446.36	I
935	8446.76	I
325	8820.43	I
160 d	9057.01	I
120	9118.29	I
80	9134.71	I
80	9150.14	I
80	9151.48	I
235	9156.01	I
450	9260.81	I
490	9260.84	I
450	9260.94	I
400	9262.58	I
540	9262.67	I
590	9262.77	I
490	9265.94	I
640	9266.01	I
185	9399.19	I
120	9481.16	I
120 d	9482.88	I
235	9487.43	I
140	9492.71	I
265	9497.97	I
160	9499.30	I
235	9505.59	I
210	9521.96	I
120	9523.36	I
120	9523.96	I
100	9528.28	I
100	9622.13	I
120	9625.29	I
160	9677.38	I
80	9694.66	I
65	9694.91	I
235	9741.50	I
235	9760.65	I
120	9909.05	I
140	9936.98	I
120	9940.41	I
160	9995.31	I
120 d	10421.18	I
590	11286.34	I
640	11286.91	I
490	11287.02	I
490	11287.32	I
490	11295.10	I

Intensity	Wavelength/Å	
540	11297.68	I
590	11302.38	I
265	11358.69	I
490	12464.02	I
450	12570.04	I
120	12990.77	I
160	13076.91	I
700	13163.89	I
750	13164.85	I
640	13165.11	I
160	16212.06	I
120	17966.70	I
590	18021.21	I
120	18041.48	I
120	18042.19	I
120	18046.23	I
140	18229.23	I
540	18243.63	I
140	26173.56	I

***Palladium* Pd *Z* = 46**

Intensity	Wavelength/Å	
200	705.49	III
200	727.72	III
500	763.06	III
500	766.42	III
2000	781.02	III
500	794.08	III
500	797.52	III
500	800.03	III
500	800.10	III
500	803.67	III
500	825.35	III
500	840.58	III
500	856.47	III
500	864.04	III
500	880.59	III
500	888.84	III
1000	889.29	III
300	1596.89	III
500	1741.62	III
4000	1782.55	III
400	1843.49	III
1500	1851.59	III
2000	1852.27	III
1000	1859.21	III
1500	1874.63	III
2000	1885.83	III
1000	1887.40	III
1500	1891.34	III
4000	1914.62	III
1000	1930.33	III
2000	1941.64	III
800	2002.16	III
1000	2004.47	III
500	2055.11	III
500	2149.82	III
500	2177.55	III
500	2177.63	III
100 r	2231.59	II
200 r	2296.53	II
100	2426.87	II
100	2430.94	II

Intensity	Wavelength/Å	
100	2433.11	II
100	2435.32	II
150	2446.17	II
1100	2447.91	I
100	2457.29	II
150	2469.29	II
100	2471.18	II
1700	2476.42	I
250	2486.52	II
300	2488.92	II
200	2498.81	II
150	2505.73	II
150	2551.84	II
150	2565.51	II
100	2569.56	II
150	2658.75	II
1900	2763.09	I
150 h	2776.85	II
100 h	2787.92	II
200	2854.59	II
100 h	2871.37	II
100 h	2878.01	II
520	2922.49	I
650	3002.65	I
1500	3027.91	I
1100	3065.31	I
2600	3114.04	I
11000	3242.70	I
2700	3251.64	I
3500	3258.78	I
3600	3302.13	I
5000	3373.00	I
24000	3404.58	I
13000	3421.24	I
5000	3433.45	I
6400	3441.40	I
7700	3460.77	I
10000	3481.15	I
2000	3489.77	I
12000	3516.94	I
12000	3553.08	I
4500	3571.16	I
20000	3609.55	I
20000	3634.70	I
5500	3690.34	I
1400	3718.91	I
1500	3799.19	I
1500	3832.29	I
2200	3894.20	I
1500	3958.64	I
290	4087.34	I
2500	4212.95	I
180	4473.59	I
160	5163.84	I
120	5295.63	I
55	5542.80	I
75	5670.07	I
55 h	5695.09	I
65	6784.52	I
75	7368.12	I
120	7764.03	I

Intensity	Wavelength/Å	
45	7915.80	I
55	8132.82	I
45	8300.83	I
65	8761.35	I

Phosphorus P Z = 15

Intensity	Wavelength/Å	
250	328.78	V
150	359.899	IV
500	388.318	IV
250	389.50	V
300	390.70	V
300	445.158	IV
375	475.60	V
120	498.180	III
520	542.57	V
600	544.92	V
200	569.853	III
200	581.831	III
350	629.008	IV
400	629.914	IV
500	631.779	IV
450	673.90	V
10	810.24	II
650	823.179	IV
700	824.730	IV
800	827.932	IV
300	847.669	III
350	855.624	III
500	859.652	III
10	865.44	II
450	865.45	V
600	871.39	V
700	877.476	IV
300	913.971	III
300	917.120	III
350	918.665	III
1000	950.655	IV
250	1003.598	III
570	1025.563	IV
500	1028.096	IV
570	1030.517	IV
500	1033.111	IV
500	1035.517	IV
900	1117.98	V
570	1118.551	IV
700	1128.01	V
20	1249.82	II
20	1301.87	II
20	1304.47	II
15	1304.68	II
35	1305.48	II
60	1310.70	II
500	1334.808	III
650	1344.327	III
300	1344.845	III
500	1366.695	IV
15	1372.033	I
400	1372.674	IV
15	1373.500	I
10	1374.732	I
15	1377.080	I
15	1377.937	I

Intensity	Wavelength/Å	
25	1379.429	I
25	1381.469	I
15	1381.637	I
500	1484.507	IV
400	1487.788	IV
350	1502.228	III
80	1532.51	II
120	1535.90	II
450	1610.50	V
150	1618.632	III
200	1618.907	III
140	1671.070	I
100	1671.510	I
180	1671.680	I
140	1672.035	I
140	1672.474	I
600	1674.591	I
600	1679.695	I
140	1685.976	I
100	1694.028	I
100	1694.486	I
100	1706.376	I
100	1707.553	I
600	1774.951	I
500	1782.838	I
400	1787.656	I
140	1834.801	I
140	1847.165	I
100	1849.820	I
140	1851.194	I
100	1852.069	I
500	1858.886	I
400	1859.393	I
140	1864.348	I
650	1888.523	IV
180	1905.481	I
140	1906.403	I
280	1907.665	I
280	2023.489	I
180	2024.516	I
400	2032.432	I
400	2033.477	I
400	2135.465	I
400	2136.182	I
400	2149.145	I
280	2152.940	I
500	2154.080	I
180	2235.732	I
450	2440.93	V
250	2478.256	IV
750	2533.976	I
950	2535.603	I
750	2553.262	I
500	2554.915	I
250	2605.506	IV
300	2632.713	III
400	2644.295	IV
400	2728.770	IV
500	2739.309	IV
250	2739.872	IV
450	2978.55	V

Intensity	Wavelength/Å	
700	3175.09	V
520	3204.04	V
300	3219.307	III
400	3233.602	III
650	3347.736	IV
570	3364.467	IV
400	3371.122	IV
300	3957.641	III
350	3978.307	III
400	4059.312	III
300	4080.084	III
500	4222.195	III
350	4246.720	III
400	4420.71	II
250	4479.776	III
250	4540.288	IV
250	4541.112	IV
500	4588.04	II
500	4589.86	II
600	4602.08	II
300	4626.70	II
300	4658.31	II
500	4943.53	II
300	4954.39	II
300	4969.71	II
100	5079.381	I
100	5098.221	I
100	5100.974	I
140	5109.628	I
140	5154.844	I
180	5162.290	I
300	5253.52	II
140	5293.539	I
400	5296.13	II
250	5316.07	II
300	5344.75	II
180	5345.851	I
100	5364.631	I
250	5378.20	II
300	5386.88	II
400	5425.91	II
100	5428.094	I
400	5450.74	II
140	5458.305	I
180	5477.672	I
140	5477.860	I
140	5478.267	I
100	5514.774	I
100	5516.997	I
250	5588.34	II
500	6024.18	II
400	6034.04	II
500	6043.12	II
250	6055.50	II
150	6083.409	III
350	6087.82	II
180	6097.690	I
350	6165.59	II
500	6199.024	I
180	6210.499	I
140	6375.681	I

Intensity	Wavelength/Å	
100	6388.579	I
250	6435.32	II
600	6459.99	II
600	6503.46	II
600	6507.97	II
100	6717.411	I
150	6992.690	III
100	7102.200	I
100	7158.367	I
180	7165.465	I
180	7175.102	I
180	7176.660	I
200	7443.657	IV
250	7845.63	II
100	8046.801	I
150	8113.528	III
140	8278.058	I
100	8367.856	I
140	8531.475	I
140	8613.835	I
180	8637.578	I
400	8741.529	I
100	8872.174	I
180	9175.819	I
950	9193.85	I
600	9278.88	I
1250	9304.94	I
500	9323.50	I
950	9435.069	I
950	9441.86	I
600	9452.83	I
1250	9493.56	I
1700	9525.73	I
1500	9545.18	I
280	9556.81	I
1700	9563.439	I
280	9593.50	I
750	9609.04	I
400	9638.939	I
500	9676.24	I
180	9706.533	I
1500	9734.750	I
280	9736.680	I
1500	9750.77	I
600	9790.21	I
1700	9796.85	I
280	9834.80	I
400	9903.68	I
280	9976.67	I
229	10084.27	I
458	10511.58	I
962	10529.52	I
1235	10581.57	I
415	10596.90	I
435	10681.40	I
265	10813.13	I
764	11183.23	I
402	11186.75	I
479	14241.64	I
256	14307.83	I
714	15711.52	I
228	15962.53	I
296	16254.77	I
203	16292.97	I
1627	16482.92	I
588	16590.07	I
225	16613.05	I
221	16738.68	I
419	16803.39	I
471	17112.48	I
289	17286.91	I
299	17423.67	I
287	23844.97	I
311	29097.16	I

Platinum Pt Z = 78

Intensity	Wavelength/Å	
30	1621.66	II
30	1723.13	II
30	1751.70	II
50 r	1777.09	II
30	1781.86	II
30	1879.09	II
40	1883.05	II
50	1889.52	II
50	1911.70	II
30	1929.25	II
30	1929.68	II
30	1939.80	II
30	1949.90	II
30	1983.74	II
40	2014.93	II
3200	2030.63	I
4400	2032.41	I
100	2036.46	II
40	2041.57	II
5500	2049.37	I
1500	2067.50	I
3000	2084.59	I
1000	2103.33	I
30	2115.57	II
950	2128.61	I
30	2130.69	II
1900	2144.23	I
100	2144.24	II
600	2165.17	I
1500	2174.67	I
30	2190.32	II
400	2202.22	I
50 h	2202.58	II
320	2222.61	I
50 h	2233.11	II
30 h	2240.99	II
100	2245.52	II
150	2249.30	I
30	2251.52	II
30 h	2251.92	II
190	2268.84	I
30 h	2271.72	II
280	2274.38	I
50 h	2287.50	II
30	2288.20	II
150	2289.27	I
150	2292.40	I
240	2308.04	I
50	2310.96	II
90	2315.50	I
220	2318.29	I
100	2326.10	I
170	2340.18	I
280	2357.10	I
180	2368.28	I
50	2377.28	II
130	2383.64	I
40	2386.81	I
120	2389.53	I
35	2396.17	I
70	2401.87	I
200	2403.09	I
100	2418.06	I
50	2424.87	II
80	2428.04	I
50	2428.20	I
25	2429.10	I
180	2436.69	I
650	2440.06	I
60	2450.97	I
440	2467.44	I
35	2471.01	I
1000	2487.17	I
25	2488.74	II
200	2490.12	I
160	2495.82	I
240	2498.50	I
50	2505.93	I
120	2508.50	I
50	2514.07	I
60	2515.03	I
240	2515.58	I
140	2524.30	I
40	2529.41	I
50	2536.49	I
160	2539.20	I
18	2549.46	I
50	2552.25	I
50	2596.00	I
70	2603.14	I
30	2616.76	II
50	2619.57	I
30	2625.34	II
1100	2628.03	I
130	2639.35	I
1000	2646.89	I
500	2650.86	I
20	2658.17	I
2800	2659.45	I
40	2674.57	I
440	2677.15	I
200	2698.43	I
2000	2702.40	I
1600	2705.89	I
60	2713.13	I
1300	2719.04	I
130	2729.92	I
1800	2733.96	I
70	2738.48	I
70	2747.61	I
80	2753.86	I
200	2754.92	I
30	2769.84	I
500	2771.67	I
40	2773.24	I
20	2774.00	I
50	2774.77	II
50	2793.27	I
100	2794.21	II
40 h	2799.98	II
140	2803.24	I
10	2808.51	I
50	2818.25	I
30 h	2822.27	II
1400	2830.30	I
70	2834.71	I
16	2853.11	I
80 h	2860.68	II
40 h	2865.05	II
40 h	2875.85	II
100 h	2877.52	II
25	2888.20	I
25	2893.22	I
600	2893.86	I
300	2897.87	I
60	2905.90	I
120	2912.26	I
120	2913.54	I
70	2919.34	I
30	2921.38	I
1700	2929.79	I
30	2942.76	I
30	2944.75	I
25	2959.10	I
60	2960.75	I
1800	2997.97	I
35	3001.17	II
220	3002.27	I
30	3017.88	I
30 h	3031.22	II
130	3036.45	I
800	3042.64	I
3200	3064.71	I
30	3071.94	I
130	3100.04	I
320	3139.39	I
140	3156.56	I
120	3200.71	I
320	3204.04	I
30	3230.29	I
20	3233.42	I
20	3250.36	I
40	3251.98	I
160	3255.92	I
25	3268.42	I
25	3281.97	I
120	3290.22	I
500	3301.86	I
60	3315.05	I

Intensity	Wavelength/Å	
35	3323.80	I
340	3408.13	I
35	3427.93	I
60	3483.43	I
160	3485.27	I
120	3628.11	I
70	3638.79	I
70	3643.17	I
50	3663.10	I
80	3671.99	I
80	3674.04	I
35	3699.91	I
18	3706.53	I
80	3818.69	I
40	3900.73	I
110	3922.96	I
35	3948.40	I
100	3966.36	I
20	3996.57	I
110	4118.69	I
80	4164.56	I
40	4192.43	I
18	4327.06	I
18	4391.83	I
80	4442.55	I
14	4445.55	I
25	4498.76	I
12	4520.90	I
35	4552.42	I
12	4879.53	I
14	5044.04	I
30	5059.48	I
35	5227.66	I
40	5301.02	I
12	5368.99	I
12	5390.79	I
14	5475.77	I
14	5478.50	I
6	5763.57	I
20	5840.12	I
8	5844.84	I
6	6026.04	I
7	6318.37	I
8	6326.58	I
9	6523.45	I
10	6710.42	I
20	6760.02	I
60	6842.60	I
20	7113.73	I
10	8224.74	I

Plutonium Pu Z = 94

Intensity	Wavelength/Å	
10000	2806.11	II
10000	2950.06	II
10000	3000.31	II
10000	3200.23	II
10000	3418.88	II
10000	3805.93	I
10000	4097.12	I
10000	4170.95	I
10000	4367.41	I
10000	5590.54	I

Intensity	Wavelength/Å	
10000	7068.90	I
10000	8691.94	I
3000	9533.07	I
3000	12144.46	I
3000	16897.38	I

Polonium Po Z = 84

Intensity	Wavelength/Å	
1500 w	2450.08	I
1500 w	2558.01	I
2500 w	3003.21	I
1200	4170.52	I
800	4493.21	I
500	8618.26	I

Potassium K Z = 19

Intensity	Wavelength/Å	
100	214.35	V
150	271.82	IV
100	273.06	IV
150	282.35	V
150	293.33	V
300	294.84	V
200	296.17	V
200	297.06	V
200	300.25	V
200	300.50	V
200	311.24	V
250	312.77	V
200	315.18	V
250	327.38	V
25	330.68	III
300	340.46	IV
150	340.74	IV
30	341.92	III
15	348.00	III
200	349.50	V
300	354.93	IV
150	356.26	IV
300	359.73	IV
200	359.91	IV
250	362.08	IV
150	362.15	IV
150	363.02	IV
500	372.15	V
200	372.46	V
200	372.77	V
300	375.96	IV
300	375.96	V
250	377.76	V
30	379.12	III
300	379.12	V
300	379.88	IV
25	380.48	III
250	380.48	IV
200	381.70	IV
30	382.23	III
300	382.23	IV
150	382.49	IV
200	382.65	IV
300	382.91	IV
250	384.10	IV
200	386.61	IV
300	387.80	V
250	388.92	IV
250	389.07	IV
250	389.07	V
250	390.11	V
250	390.42	IV
300	390.57	IV
200	391.46	IV
200	392.47	IV
500	393.14	IV
250	395.40	V
200	398.36	V
15	398.63	III
200	398.88	V
200	399.75	V
400	400.21	IV
20	402.10	III
300	402.91	IV
250	403.97	IV
150	404.41	IV
30	406.48	III
250	408.08	IV
40	408.96	III
50	413.79	III
30	414.87	III
250	415.05	V
200	415.79	V
30	416.00	III
150	417.28	IV
30	417.54	III
30	418.62	III
400	422.18	V
300	425.16	V
500	425.59	V
75	434.72	III
50	435.68	III
250	438.02	V
25	441.81	II
200	442.30	IV
300	443.57	IV
75	444.34	III
200	445.61	IV
250	446.83	IV
75	448.60	III
750	448.60	IV
200	449.71	V
200	452.90	V
250	455.67	V
400	456.33	IV
400	456.33	V
75	466.79	III
100	470.09	III
75	471.57	III
45	474.92	III
10	476.03	II
40	479.18	III
10	482.11	III
10	482.41	III
200	482.71	V
200	483.75	V
30	495.14	II
75	497.10	III
10	514.94	III
50	520.61	III
250	523.00	IV
25	523.79	III
200	526.45	IV
150	527.62	IV
40	529.80	III
15	539.71	III
15	546.12	III
750	580.32	V
250	585.51	V
500	586.32	V
30	600.77	II
250	602.27	V
400	603.43	V
25	607.93	II
30	612.62	II
250	638.67	V
750	646.19	IV
300	687.50	V
20	708.84	III
300	720.43	V
400	724.42	V
600	731.86	V
500	737.14	IV
500	741.95	IV
500	745.26	IV
400	746.35	IV
300	749.99	IV
150	754.19	IV
400	754.67	IV
20	765.31	III
30	765.64	III
150	770.29	V
150	771.46	V
35	778.53	III
20	872.31	III
10	873.86	III
15	874.04	III
6	2550.02	III
5	2635.11	III
5	2689.90	III
5	2938.45	III
5	2986.20	III
6	2992.42	III
6	3052.07	III
5	3056.84	III
5	3062.18	II
4	3101.79	I
3	3102.04	I
7	3217.16	I
6	3217.62	I
11	3446.37	I
10	3447.38	I
3	3648.84	I
4	3648.98	I
18	4044.14	I
17	4047.21	I
10	4641.88	I
11	4642.37	I

Intensity	Wavelength/Å	
4	4740.91	I
6	4744.35	I
5	4753.93	I
7	4757.39	I
5	4786.49	I
7	4791.05	I
6	4799.75	I
8	4804.35	I
7	4849.86	I
8	4856.09	I
8	4863.48	I
9	4869.76	I
8	4942.02	I
9	4950.82	I
9	4956.15	I
10	4965.03	I
10	5084.23	I
11	5097.17	I
11	5099.20	I
12	5112.25	I
12	5323.28	I
13	5339.69	I
12	5342.97	I
14	5359.57	I
16	5782.38	I
17	5801.75	I
15	5812.15	I
17	5831.89	I
8	6120.27	II
7	6307.29	II
19	6911.08	I
12	6936.28	I
20	6938.77	I
7	6964.18	I
12	6964.67	I
25	7664.90	I
24	7698.96	I
5	7955.37	I
4	7956.83	I
7	8078.11	I
6	8079.62	I
9	8250.18	I
8	8251.74	I
3	8390.22	I
11	8503.45	I
10	8505.11	I
4	8763.96	I
3	8767.05	I
13	8902.19	I
12	8904.02	I
5	8923.31	I
4	8925.44	I
7	9347.24	I
3	9349.25	I
6	9351.59	I
15	9595.70	I
14	9597.83	I
6	9949.67	I
5	9954.14	I
9	10479.63	I
5	10482.15	I
8	10487.11	I
17	11019.87	I
16	11022.67	I
17	11690.21	I
16	11769.62	I
17	11772.83	I
	12522.11	I
	13377.86	I
	13397.09	I
	15163.08	I
	15168.40	I
	40158.37	I

***Prascodymium* Pr $Z = 59$**

Intensity	Wavelength/Å	
7000	865.90	V
5000	869.17	V
2000	1228.59	IV
5000	1293.22	IV
5000	1295.28	IV
5000	1321.36	IV
5000	1333.57	IV
5000	1354.66	IV
2000	1360.64	IV
2000	1365.77	IV
5000	1374.41	IV
5000	1435.56	IV
2000	1520.98	IV
5000	1574.55	IV
5000	1575.10	IV
3000	1578.38	IV
2000	1622.30	IV
10000	1884.87	IV
2000	2083.23	IV
3300	2246.20	V
2000 c	2378.98	IV
40 h	2598.04	II
100 h	2707.37	II
60	2760.35	II
270	3168.24	II
200 d	3195.99	II
190	3219.48	II
200	3584.21	II
250	3645.66	II
250	3646.30	II
370	3668.83	II
290	3714.05	II
410	3739.18	II
680	3761.87	II
680	3800.30	II
390	3811.84	II
1300 h	3816.02	II
680	3818.28	II
310	3821.80	II
960	3830.72	II
480	3840.99	II
580	3846.59	II
1200	3850.79	II
720 c	3851.55	II
960	3852.80	II
480 c	3865.45	II
480	3876.19	II
1700 c	3877.18	II
680	3880.47	II
440 c	3885.19	II
440 c	3889.34	II
770 c	3908.05	II
630	3912.90	II
310	3913.55	II
1300 c	3918.85	II
420	3919.63	II
960	3925.47	II
480	3927.46	II
370	3929.29	II
370	3935.82	II
730 c	3947.63	II
900 c	3949.43	II
900 c	3953.51	II
380	3956.75	II
470	3962.45	II
560	3964.26	II
1600 c	3964.81	II
560 c	3966.57	II
500	3971.16	II
320	3971.67	II
620 c	3972.14	II
320	3974.85	II
1300 c	3989.68	II
340	3992.16	II
1600	3994.79	II
560 c	3997.04	II
320	3999.12	II
620 c	4000.17	II
730	4004.70	II
1900	4008.69	II
620	4010.60	II
730	4015.39	II
620	4020.96	II
470	4022.71	II
360	4025.54	II
360 c	4029.72	II
730 c	4031.75	II
960	4033.83	II
730	4038.45	II
470	4039.34	II
1300	4044.81	II
340	4047.08	II
450	4051.13	II
2200	4054.88	II
2200	4056.54	II
450	4058.80	II
3400	4062.81	II
500 c	4079.77	II
500 c	4080.98	II
790	4081.85	II
500	4083.34	II
560	4096.82	II
380	4098.40	II
2900 c	4100.72	II
1700 c	4118.46	II
340	4130.77	II
1500 c	4141.22	II
2700	4143.11	II
1700 c	4164.16	II
620	4171.82	II
730	4172.25	II
5200	4179.39	II
2500	4189.48	II
560 c	4191.60	II
2500 c	4206.72	II
500	4208.32	II
320	4211.86	II
320	4217.81	II
3800	4222.93	II
3800	4225.35	II
320	4233.11	II
320 c	4236.15	II
960	4241.01	II
340	4243.51	II
840 c	4247.63	II
500	4254.40	II
320	4269.09	II
790 c	4272.27	II
470 c	4280.07	II
790 c	4282.42	II
450 c	4298.98	II
1500	4305.76	II
1300	4333.97	II
360	4338.70	II
620 cw	4344.30	II
470 c	4347.49	II
340	4350.40	II
450	4354.91	II
410 c	4359.79	II
1200	4368.33	II
320	4371.62	II
430	4405.83	II
1700	4408.82	II
410	4413.77	II
1200 c	4429.13	II
730	4449.83	II
960	4468.66	II
1100	4496.46	II
790	4510.15	II
340 c	4534.15	II
340	4535.92	II
270 c	4628.74	II
270 c	4672.09	II
290	4695.77	I
250	4736.69	I
200	4924.60	I
320	4939.74	I
380	4951.37	I
270	5034.41	II
320	5045.52	I
360	5110.38	II
560	5110.76	II
410	5129.52	II
620	5173.90	II
360	5206.55	II
360	5219.05	II
560	5220.11	II
680	5259.73	II
340 c	5292.02	II
340	5292.62	II

Intensity	Wavelength/Å	
430	5322.76	II
65	5509.15	II
150	5535.17	II
110	5623.05	II
90	5624.45	II
90	5756.17	II
90	5779.28	I
160 d	5815.17	II
90	5823.72	II
90	5859.68	II
160	5939.90	II
7000 w	5956.05	III
90	5956.60	II
110	5967.82	II
90	6006.33	II
150	6017.80	II
150	6025.72	II
140	6055.13	I
65	6087.52	II
9000 w	6090.02	III
65	6114.38	II
65	6148.23	I
5000	6160.24	III
190	6161.18	II
270	6165.94	II
45	6244.35	II
110	6281.28	II
55 c	6359.03	I
55	6411.23	I
45	6429.63	II
45	6431.84	II
45	6486.55	I
45	6566.77	II
55	6616.67	I
75	6656.83	II
55	6673.41	II
75	6673.78	II
35 c	6747.09	I
55 cw	6798.60	I
35 cw	6827.60	II
7000	6910.14	III
40	7021.51	II
5000	7030.39	III
4500	7076.62	III
20	7114.55	I
24	7227.70	II
16	7407.56	II
20 c	7451.74	II
14	7541.02	II
20	7645.66	II
16	7721.84	I
14	7871.67	I
14	8067.44	I
10 cw	8122.78	II
11	8141.10	I
5000 w	8602.74	III
10	8714.59	II

Promethium Pm Z = 61

Intensity	Wavelength/Å	
1000	3892.15	II
1000	3910.26	II
1000	3919.10	II
1000	3957.74	II
1000 r	3998.96	II
1000	4417.96	II
900 r	4728.36	I
900	6100.21	I
1000 d	6520.45	I

Protactinium Pa Z = 91

Intensity	Wavelength/Å	
3000	2599.16	II
3000	2699.22	II
3000	2822.79	II
3000 h	2871.42	II
3000 h	2891.14	II
3000 l	3011.10	II
3000 s	3033.59	II
3000 l	3071.24	II
3000 l	3093.23	II
3000 l	3126.23	II
3000 l	3146.28	II
3000 l	3170.89	II
3000 l	3171.54	II
3000 l	3240.58	II
3000	3274.46	II
3000 l	3332.69	II
3000 s	3346.66	II
3000 l	3452.82	II
3000	3504.97	I
3000 s	3530.65	II
3000	3570.56	I
3000	3571.82	I
3000	3618.07	I
10000	3636.52	I
3000	3702.74	I
3000	3752.67	I
3000	3873.35	I
3000	3931.83	I
3000 s	3952.62	II
10000 l	3957.85	II
3000 s	3970.07	II
3000	3981.82	I
10000	3982.23	I
3000 l	4012.96	II
3000 s	4018.21	II
3000	4030.16	II
3000 s	4046.93	II
10000 s	4056.20	II
10000 s	4070.40	II
3000 l	4176.18	II
10000 l	4217.23	II
10000 s	4248.08	II
3000 s	4291.34	II
3000 s	4601.43	II
3000 l	6035.78	I
3000	6162.56	I
3000 l	6358.61	I
3000	6379.25	I
3000 l	6438.97	I
3000 h	6792.75	I
10000	6945.72	I
3000	6960.09	I
3000 h	6961.78	I
3000 s	6992.73	I
3000	7076.27	I
3000 h	7100.94	I
10000 s	7114.89	I
3000 h	7171.55	I
3000	7227.13	I
3000	7318.79	I
10000 l	7368.25	I
3000 h	7471.89	I
10000 h	7493.15	I
3000 h	7558.26	I
10000 h	7608.20	I
10000	7626.79	I
10000 s	7635.18	I
10000	7669.34	I
3000	7679.20	I
10000 h	7749.19	I
3000	7872.95	I
3000 l	7945.56	I
10000	8039.34	I
10000 h	8099.84	I
10000	8199.04	I
10000	8271.87	I
3000 s	8358.98	I
3000 s	8369.60	I
3000 h	8441.04	I
10000 h	8532.66	I
10000 s	8572.96	I
3000 h	8639.91	I
3000 h	8653.51	I
10000	8735.27	I
3000	10923.32	I
10000	11791.73	I
10000	14344.76	I
3000	18478.61	I

Radium Ra Z = 88

Intensity	Wavelength/Å	
100	3649.55	II
200	3814.42	II
100	4340.64	II
100	4682.28	II
100	4825.91	I
50	5660.81	I
50	7141.21	I
50	8019.70	II

Radon Rn Z = 86

Intensity	Wavelength/Å	
100	4349.60	I
200	7055.42	I
100	7268.11	I
300	7450.00	I
100	7809.82	I
100	8099.51	I
100	8270.96	I
100	8600.07	I

Rhenium Re Z = 75

Intensity	Wavelength/Å	
25000	2003.53	I
16000	2017.87	I
27000	2049.08	I
10000	2085.59	I
9800	2097.12	I
3400	2139.04	II
3700	2156.67	I
4900	2167.94	I
3400	2176.21	I
4200 c	2214.26	II
5200 c	2275.25	II
2900	2287.51	I
2700	2294.49	I
390	2298.09	II
610	2302.99	I
680	2306.54	I
800	2322.49	I
300	2328.66	I
860	2344.78	I
230	2349.39	I
680	2352.07	I
250	2356.50	I
1200	2365.90	I
570	2367.68	I
520	2369.27	I
220	2370.76	II
320	2375.07	I
370	2379.77	I
340	2388.57	I
230	2393.65	I
320	2394.37	I
320	2396.79	I
210 d	2400.72	I
210	2401.68	I
1500	2405.06	I
740	2405.60	I
320	2406.70	I
270	2410.37	I
1200	2419.81	I
300	2421.73	I
300	2421.88	I
2500	2428.58	I
490	2431.54	I
420	2432.18	I
340 c	2441.47	I
230	2442.51	I
250	2444.94	I
610	2446.98	I
610	2449.71	I
390	2461.20	I
800 c	2461.84	II
1200	2483.92	I
390	2485.81	I
980	2487.33	I
370	2496.04	I
370	2501.72	I
570	2502.35	II
230	2504.60	II
270	2505.94	I
1800 c	2508.99	I
570	2520.01	I
540	2521.50	I
370	2534.80	I
570	2540.51	I
740 d	2544.74	I
370	2545.48	I
300	2552.02	I
370	2554.63	II

Intensity	Wavelength/Å	
1000	2556.51	I
250	2559.08	I
340	2564.19	I
540	2568.64	II
370	2571.81	II
380	2586.79	I
290	2599.86	I
290	2603.89	I
660	2608.50	II
610 d	2611.54	I
310	2635.83	II
550	2636.64	I
270	2642.75	I
270	2649.05	I
660	2651.90	I
400	2654.12	I
220	2663.63	I
940	2674.34	I
220	2688.53	I
1300	2715.47	I
220	2732.21	I
610	2733.04	II
220	2758.00	I
210	2763.79	I
310	2767.74	I
220	2768.85	I
220	2769.32	I
350	2770.42	I
550	2783.57	I
220	2791.29	I
220	2814.68	I
880	2819.95	I
310	2834.08	I
220	2843.00	I
270	2850.98	I
240	2867.19	I
2900	2887.68	I
490	2896.01	I
830 c	2902.48	I
210	2905.58	I
550	2909.82	I
830 c	2927.42	I
270	2930.61	I
440	2943.14	I
270	2962.27	I
720	2965.11	I
1500	2965.76	I
310	2976.29	I
210	2978.15	I
220	2980.82	I
220	2982.19	I
220	2988.47	I
1800	2992.36	I
5500	2999.60	I
350	3001.14	I
220	3004.14	I
500	3016.02	I
300	3016.49	I
380	3030.45	I
240	3047.25	I
1600	3067.40	I
320	3069.94	I
260	3071.16	I
550	3082.43	I
340	3088.76	I
700	3100.67	I
700	3108.81	I
340	3110.86	I
340 c	3118.19	I
340	3121.36	I
420	3128.94	I
260	3134.02	I
250	3141.38	I
440	3151.64	I
330	3153.79	I
360 c	3158.31	I
220	3164.52	I
700	3168.37	I
220	3174.61	I
440	3177.71	I
260	3178.61	I
600	3182.87	I
1100	3184.76	I
1100	3185.57	I
260	3190.78	I
260	3192.36	I
220	3198.58	I
1100 c	3204.25	I
380	3235.94	I
600	3258.85	I
600	3259.55	I
300	3268.89	I
280	3296.70	I
280	3296.99	I
280	3301.60	I
240	3302.23	I
320	3303.21	II
280	3303.75	I
240	3313.95	I
600	3322.48	I
2000	3338.18	I
1600	3342.24	I
810	3344.32	I
320	3346.20	I
240 d	3356.33	I
240	3377.74	I
320	3379.06	II
320	3379.70	I
240	3389.43	I
4000	3399.30	I
650	3404.72	I
650	3405.89	I
240	3408.67	I
320	3409.83	I
320	3417.77	I
810	3419.41	I
8000	3424.62	I
400	3426.19	I
300	3427.61	I
320	3437.71	I
400	3449.37	I
16000 c	3451.88	I
240	3453.50	I
55000 c	3460.46	I
40000 c	3464.73	I
400	3467.96	I
240	3476.44	I
400	3480.38	I
320	3480.85	I
240	3482.23	I
560	3503.06	I
320	3516.65	I
320	3517.33	I
320	3537.46	I
240	3549.89	I
240	3570.26	I
360	3579.12	I
810 c	3580.15	II
650	3580.97	I
810	3583.02	I
320	3617.08	I
810	3637.84	I
440	3651.97	I
320	3670.53	I
860 c	3689.50	I
1500 c	3691.48	I
520	3703.24	I
240	3709.93	I
360 c	3717.28	I
4000	3725.76	I
240 c	3735.01	I
810	3735.31	I
910	3740.10	I
300 cw	3745.44	I
700	3787.52	I
240	3869.94	I
240	3875.26	I
240	3876.86	I
380 c	3917.27	I
550	3929.85	I
280	3961.04	I
350 c	3962.48	I
220	4033.31	I
240	4081.43	I
240 c	4110.89	I
240 cw	4133.42	I
1800	4136.45	I
700	4144.36	I
220	4182.90	I
220	4183.06	I
650	4221.08	I
3600 c	4227.46	I
260 c	4257.60	I
380	4358.69	I
360 cw	4394.38	I
2600	4513.31	I
260	4516.64	I
500	4522.73	I
2200 cw	4889.14	I
220	4923.90	I
1300	5270.95	I
1600 cw	5275.56	I
100	5667.88	I
110 c	5752.93	I
110 cw	5776.83	I
550	5834.31	I
200	6307.70	I
200	6321.90	I
100 cw	6605.19	I
180 c	6813.41	I
260	6829.90	I
50 cw	7640.94	I
65 cw	7912.94	I
Rhodium Rh Z = 45		
50	813.44	III
80	882.51	III
100	925.75	III
150	937.28	III
500	991.62	III
400	992.48	III
500 d	1009.60	III
200	1012.22	III
200	1015.17	III
200	1073.87	III
150	1784.24	III
200	1784.94	III
150	1796.50	III
200	1816.03	III
1000	1832.05	III
500	1859.85	III
800	1880.66	III
500	1884.91	III
500	1887.36	III
700	1888.62	III
800	1901.32	III
500	1910.16	III
600	1919.37	III
500	1927.07	III
700	1931.79	III
500	1954.25	III
500	1994.26	III
800	2013.71	III
500	2017.47	III
500	2028.53	III
800	2036.72	III
600	2037.61	III
1000	2040.18	III
3000	2048.67	III
2000	2064.11	III
800	2076.84	III
1000	2118.53	III
1000	2118.63	III
1000	2139.44	III
1000	2152.23	III
3000	2158.17	III
3000	2163.19	III
3000	2167.33	III
150	2276.21	II
140	2288.57	I
110	2309.82	I
350	2322.58	I
140	2326.47	I
190	2334.77	II
300	2361.92	I

Intensity	Wavelength/Å	
110	2368.34	I
270	2382.89	I
230	2383.40	I
270	2386.14	II
80	2415.84	II
130	2427.68	I
230	2429.52	I
110	2437.90	I
330	2440.34	I
90	2461.04	II
130	2473.09	I
150	2487.47	I
100	2490.77	II
130	2502.46	I
300	2504.29	II
150	2505.67	I
350	2509.70	I
300	2511.03	II
200	2515.75	I
130	2520.53	II
110	2537.04	II
350	2545.70	I
550	2555.36	I
150	2622.58	I
230	2625.88	I
100	2630.42	I
110	2647.28	I
400	2652.66	I
100	2680.63	I
400	2703.73	I
100	2715.31	II
180	2718.54	I
160	2728.94	I
100	2771.51	I
130	2783.03	I
150	2826.43	I
180	2826.68	I
280	2862.94	I
110	2878.66	I
140	2882.37	I
160	2907.21	I
65	2910.17	II
180	2924.02	I
130	2929.11	I
130	2931.94	I
230	2968.66	I
160	2977.68	I
450	2986.20	I
110	3004.46	I
50	3006.43	III
130	3023.91	I
50	3052.44	III
180	3083.96	I
140	3121.76	I
240	3123.70	I
130	3155.78	I
140	3189.05	I
470	3191.19	I
190	3197.13	I
520	3263.14	I
520	3271.61	I

Intensity	Wavelength/Å	
2300	3280.55	I
2300	3283.57	I
280	3289.14	I
210	3294.28	I
260	3300.46	I
50	3310.69	III
4200	3323.09	I
330	3338.54	I
280	3360.80	I
420	3368.38	I
1100	3372.25	I
110	3377.14	I
110	3385.78	I
5600	3396.82	I
820	3399.70	I
160	3406.55	I
820	3412.27	I
330	3421.22	I
120 d	3424.38	I
8200	3434.89	I
1400	3440.53	I
120	3447.74	I
120	3450.29	I
400	3455.22	I
180	3457.07	I
220	3457.93	I
5900	3462.04	I
180	3469.62	I
4700	3470.66	I
120	3472.25	I
4700	3474.78	I
2100	3478.91	I
110	3494.44	I
1200	3498.73	I
5900	3502.52	I
2800	3507.32	I
8800	3528.02	I
880 d	3538.14	I
280	3541.91	I
1200	3543.95	I
1800	3549.54	I
1200	3570.18	I
4700	3583.10	I
4700	3596.19	I
5900	3597.15	I
3100	3612.47	I
1800	3626.59	I
8200	3657.99	I
1300	3666.22	I
560	3681.04	I
1900	3690.70	I
9400	3692.36	I
940	3695.52	I
280	3698.26	I
380	3698.60	I
7600	3700.91	I
940	3713.02	I
650	3735.28	I
420	3737.27	I
420	3744.17	I
1200	3748.22	I

Intensity	Wavelength/Å	
240	3754.12	I
380	3754.27	I
490	3755.58	I
1000	3760.40	I
2300	3765.08	I
490	3769.97	I
380	3778.13	I
1000	3788.47	I
1300	3792.18	I
3800	3793.22	I
4900	3799.31	I
760	3805.92	I
1300	3806.76	I
470	3815.01	I
760	3816.47	I
1300	3818.19	I
3800	3822.26	I
2300	3828.48	I
2000	3833.89	I
5900	3856.52	I
490	3870.01	I
380	3877.34	I
120	3913.51	I
240	3922.19	I
2000	3934.23	I
590	3942.72	I
3800	3958.86	I
45	3964.54	II
380	3975.31	I
240	3984.40	I
240	3995.61	I
380	3996.15	I
120	4023.14	I
560	4082.78	I
140	4097.52	I
120	4119.68	I
1100	4121.68	I
1500	4128.87	I
2100	4135.27	I
240	4154.37	I
330	4196.50	I
3300	4211.14	I
820	4288.71	I
4200	4374.80	I
130	4569.00	I
150	4675.03	I
70	4745.11	I
70	5090.63	I
60	5155.54	I
60	5175.97	I
95	5193.14	I
130	5354.40	I
95	5390.44	I
160	5599.42	I
40	5686.38	I
29	5792.66	I
40	5806.91	I
35	5831.58	I
130	5983.60	I
35	6102.72	I
14	6199.99	I

Intensity	Wavelength/Å	
16	6253.72	I
29	6319.53	I
40	6752.35	I
13	6827.33	I
11	6857.68	I
20	6879.94	I
65	6965.67	I
16	6979.15	I
16	7001.58	I
18	7101.64	I
15	7104.45	I
18	7268.18	I
35	7270.82	I
18 h	7442.39	I
12	7475.74	I
12	7495.24	I
11	7557.67	I
29	7791.61	I
55	7824.91	I
21	8029.91	I
29	8045.36	I
15	8136.20	I
8	8425.59	I

***Rubidium* Rb $Z = 37$**

Intensity	Wavelength/Å	
30	465.85	III
40	481.118	II
500	482.83	III
500	489.66	III
600	493.48	III
90	497.430	II
20	508.434	II
150	513.266	II
300	530.173	II
75	533.801	II
1200	535.86	III
40	542.887	II
200	555.036	II
1200	556.19	III
1500	566.71	III
1000	572.82	III
1500	576.65	III
2500	579.63	III
1500	581.26	III
2500	589.419	II
1000	594.94	III
1300	595.88	III
1200	598.49	III
1500	643.878	II
25	663.76	IV
3000	697.049	II
6000	711.187	II
25	716.24	IV
50	740.85	IV
10000	741.456	II
5000	769.04	III
25	776.89	IV
2500	815.28	III
15	850.18	IV
1000	1604.12	II
5000	1760.50	II
2000	2068.92	II

Intensity	Wavelength/Å	
10000	2075.95	II
30000	2143.83	II
10000	2217.08	II
5000	2291.71	II
50000	2472.20	II
1000	2631.75	III
2000	2956.07	III
500	3086.84	III
500	3111.36	III
5000 c	3148.90	II
25	3157.54	I
50	3227.98	I
500	3286.41	III
60	3348.72	I
75	3350.82	I
100	3587.05	I
40	3591.57	I
5000	3600.60	II
10000	3600.64	II
25000	3940.51	II
1000	4201.80	I
500	4215.53	I
90000	4244.40	II
15000	4273.14	II
20000	4571.77	II
10000	4648.57	II
30000	4775.95	II
2	5087.987	I
2	5132.471	I
10	5150.134	I
10000	5152.08	II
1	5165.023	I
2	5165.142	I
15	5195.278	I
2	5233.968	I
20	5260.034	I
1	5260.228	I
3	5322.380	I
40	5362.601	I
4	5390.568	I
75	5431.532	I
3	5431.830	I
6	5578.788	I
40	5647.774	I
20	5653.750	I
60	5724.121	I
3	5724.614	I
75	6070.755	I
30 c	6159.626	I
75 c	6206.309	I
120 c	6298.325	I
5	6299.224	I
10000	6458.33	II
5000	6560.81	II
100 l	7279.997	I
150	7408.173	I
200 l	7618.933	I
300	7757.651	I
60	7759.436	I
90000 c	7800.27	I
45000 c	7947.60	I
40 l	8271.41	I
30	8271.71	I
40 l	8868.512	I
30	8868.852	I
30 l	9522.65	I
20 l	9540.18	I
2000 c	9689.05	II
35 l	10075.282	I
30 l	10075.708	I
100	13235.17	I
20	13442.81	I
30	13443.57	I
75	13665.01	I
1000	14752.41	I
800	15288.43	I
150	15289.48	I
20	22529.65	I
4	27314.31	I

Ruthenium Ru Z = 44

Intensity	Wavelength/Å	
250	850.09	III
200	850.30	III
250	919.74	III
500	940.09	III
500	966.54	III
750	974.14	III
900	979.43	III
500	981.35	III
900	986.84	III
900	994.56	III
300	1001.65	III
500	1009.13	III
900	1009.87	III
500	1014.68	III
800	1190.51	III
500	1200.07	III
500	1207.17	III
500	1209.77	III
300	1211.31	III
500	1941.35	III
500	2009.28	III
2400	2076.43	I
2600	2083.77	I
2400	2090.89	I
690	2255.52	I
780	2272.09	I
780	2279.57	I
480	2317.80	I
120	2334.96	II
190 h	2342.85	II
310	2351.33	I
170	2357.91	II
780	2402.72	II
150	2407.92	II
180	2455.53	II
150	2456.44	II
370	2456.57	II
280	2478.93	II
140	2498.42	II
140	2498.57	II
260	2507.01	II
110	2513.32	II
110	2517.32	II
150	2535.59	II
550	2549.58	I
370	2609.06	I
830	2612.07	I
460	2642.96	I
330	2651.84	I
400	2659.62	I
330	2661.61	II
690	2678.76	II
330	2692.06	II
200	2712.41	II
690	2719.52	I
140	2725.47	II
310	2734.35	II
1800	2735.72	I
100	2778.38	II
110	2787.83	II
350	2810.03	I
1700	2810.55	I
350	2818.36	I
400	2829.16	I
640	2854.07	I
420	2861.41	I
550	2866.64	I
1800	2874.98	I
740	2886.54	I
370	2908.88	I
1100	2916.26	I
180	2945.67	II
370	2949.50	I
550	2965.16	I
170	2965.55	II
140	2976.59	II
550	2976.92	I
1400	2988.95	I
460	2994.96	I
440	3006.59	I
330	3017.24	I
310	3020.88	I
390	3064.84	I
330	3096.57	I
830	3099.28	I
740	3100.84	I
490	3294.11	I
370	3301.59	I
930	3339.55	I
3100	3417.35	I
4900	3428.31	I
6400	3436.74	I
8300	3498.94	I
640	3514.49	I
790	3539.37	I
690	3570.59	I
6400	3589.22	I
6900	3593.02	I
6400	3596.18	I
1300	3599.76	I
3100	3634.93	I
6200	3661.35	I
830	3663.37	I
650	3669.49	I
550	3726.10	I
8700	3726.93	I
11000	3728.03	I
7100	3730.43	I
3500	3742.28	I
870	3742.78	I
2800	3745.59	I
760	3753.54	I
870	3755.93	I
1200	3759.84	I
600	3761.51	I
600	3767.35	I
1500	3777.59	I
600	3782.74	I
3900	3786.06	I
6000	3790.51	I
760	3798.05	I
7600	3798.90	I
7600	3799.35	I
600	3812.72	I
760	3817.27	I
760	3819.03	I
650	3822.09	I
550	3824.93	I
760	3831.80	I
930	3839.70	I
760	3850.43	I
1300	3857.55	I
650	3862.69	I
1300	3867.84	I
650	3892.21	I
760	3909.08	I
1500	3923.47	I
3300	3925.92	I
600	3931.76	I
760	3945.57	I
600	3978.44	I
600	3979.42	I
870	3984.86	I
1500	4022.16	I
600	4023.83	I
1400	4051.40	I
710	4054.05	I
760	4068.37	I
980	4076.73	I
6000	4080.60	I
930	4097.79	I
1900	4112.74	I
2000	4144.16	I
650	4145.74	I
870	4167.51	I
550	4197.58	I
550	4198.88	I
7600	4199.90	I
1500	4206.02	I
5400	4212.06	I
760	4214.44	I
930	4217.27	I
550	4230.31	I
760	4241.05	I

Intensity	Wavelength/Å	
760	4243.06	I
760	4284.33	I
550	4295.93	I
3700	4297.71	I
930	4307.60	I
550	4319.87	I
550	4342.07	I
710	4354.13	I
870	4361.21	I
2400	4372.21	I
870	4385.39	I
1300	4385.65	I
1700	4390.44	I
1600	4410.03	I
1100	4460.04	I
5400	4554.51	I
1700	4584.44	I
720	4647.61	I
1400	4709.48	I
500	4757.84	I
550	4869.15	I
160	5011.23	I
450	5057.33	I
120	5076.32	I
200	5093.83	I
530	5136.55	I
170	5142.76	I
250	5147.24	I
110	5151.07	I
500	5155.14	I
920	5171.03	I
180	5195.02	I
130	5284.08	I
260	5309.27	I
110	5335.93	I
130	5361.77	I
110 h	5401.04	I
80	5484.32	I
130	5510.71	I
90	5559.75	I
290	5636.24	I
180	5699.05	I
65	5814.98	I
55	5919.34	I
80	5921.45	I
21 h	5973.38	I
16	5988.67	I
35	5993.65	I
18	6116.77	I
26	6199.42	I
26	6225.20	I
18	6295.22	I
16	6390.23	I
26 h	6444.84	I
21	6663.14	I
55	6690.00	I
21	6766.95	I
30	6775.02	I
21	6824.17	I
26	6911.48	I
110	6923.23	I
26	6982.01	I
26	7027.98	I
35	7238.92	I
16	7393.93	I
18	7468.91	I
26	7485.79	I
70	7499.75	I
26	7559.61	I
18	7621.50	I
18	7722.87	I
22	7791.86	I
30	7847.80	I
80	7881.49	I
18	8264.96	I
22	8710.84	I

Samarium Sm Z = 62

Intensity	Wavelength/Å	
150	2789.38	II
410	3152.52	II
720	3183.92	II
600	3211.73	II
530	3216.85	II
600	3218.61	II
720	3230.56	II
720	3236.64	II
720	3239.66	II
720	3250.37	II
850	3254.38	II
1700	3306.39	II
1200	3321.18	II
1200	3365.86	II
1200	3382.40	II
4200	3568.27	II
4200	3592.60	II
1700	3604.28	II
3400	3609.49	II
1700	3621.23	II
3400	3634.29	II
2200	3661.36	II
2200	3670.84	II
1100	3693.99	II
1600	3728.47	II
2100	3731.26	II
1600	3735.98	II
2900	3739.12	II
1200	3743.87	II
930	3745.46	I
800	3756.41	I
1200	3757.53	II
1900	3760.69	II
1100	3764.37	II
370 d	3773.33	I
1100	3778.14	II
1500	3788.12	II
1600	3793.97	II
1600	3797.73	II
1600	3826.20	II
1100	3831.50	II
560	3834.48	I
1600	3843.50	II
530	3853.30	I
2700	3854.21	II
480	3854.56	I
400	3858.74	I
3700	3885.29	II
1600	3896.98	II
1300	3903.42	II
2500	3922.40	II
1900	3928.28	II
1300	3941.87	II
470	3951.89	I
1500	3963.00	II
1500	3971.40	II
620	3974.66	I
1000	3976.43	II
1500	3990.00	II
1400	4064.58	II
1000	4092.27	II
1900	4118.55	II
1200	4152.21	II
1000	4188.13	II
1100	4203.05	II
1000	4225.33	II
1200	4236.74	II
2100	4256.39	II
1300	4262.68	II
1200	4279.68	II
2200	4280.79	II
710	4282.21	I
470	4282.83	I
1600	4296.74	I
1900	4318.94	II
470	4319.53	I
1800	4329.02	II
440	4330.02	I
1300	4334.15	II
880	4336.14	I
1100	4347.80	II
440	4362.91	I
530	4380.42	I
1600	4390.86	II
410	4401.17	I
470	4419.33	I
1500	4420.53	II
2900	4424.34	II
470	4429.66	I
1600	4433.88	II
1800	4434.32	II
530	4441.81	I
440	4442.28	I
710	4445.15	I
1300	4452.73	II
1200	4454.63	II
1000	4458.52	II
2200	4467.34	II
810	4470.89	I
370	4499.11	I
440	4581.73	I
380	4649.49	I
470 d	4670.75	I
1100	4674.60	II
370	4688.73	I
530	4704.40	II
730	4716.10	I
770	4728.42	I
470	4745.68	II
730	4760.27	I
580	4783.10	I
350	4785.86	I
970	4841.70	I
730	4883.97	I
630	4910.40	I
350	4913.25	II
430	4918.99	I
400	5044.28	I
540	5071.20	I
510	5117.16	I
350	5122.14	I
360	5155.03	II
470	5175.42	I
250	5200.59	I
260	5251.92	I
400	5271.40	I
250	5282.91	I
220	5453.00	I
230	5493.72	I
230	5516.09	I
140	5550.40	I
140	5659.86	I
120	5696.73	I
85	5706.20	I
70	5773.77	I
60	5778.33	I
70 d	5786.98	II
60	5788.38	I
60	5800.52	I
65	5802.84	I
65	5867.79	I
50	5874.21	I
50	5898.96	I
65	5965.71	II
50	6045.00	I
50	6070.06	I
45	6084.12	I
45 h	6159.56	I
45	6256.54	I
100	6267.28	II
140	6569.31	II
110	6589.72	II
50	6671.51	I
120 d	6731.84	II
95	6794.20	II
120	6860.93	I
120	6955.29	II
90	7020.44	II
90	7039.22	II
90	7042.24	II
90	7051.52	II
90	7082.37	II
26	7088.30	I
30	7095.50	I
30	7104.54	I
26	7115.96	I
85 d	7149.60	II

Intensity	Wavelength/Å	
23	7213.82	I
60	7240.90	II
26	7347.30	I
30	7444.56	I
26	7445.41	I
13	7470.76	I
45	7645.09	II
12	7645.82	I
40 w	7835.08	II
16	7895.96	I
90	7928.14	II
40	8048.70	II
16	8065.16	I
45	8068.46	II
40 w	8305.79	II
19	8383.71	I
45 w	8485.99	II
45 w	8708.43	II
95	8913.66	II

Scandium Sc Z = 21

Intensity	Wavelength/Å	
350	180.14	V
500	243.87	V
500	252.85	V
500	253.73	V
900	283.91	V
800	284.45	V
600	286.29	V
900	289.59	V
15	289.85	IV
1000 d	291.93	V
800	293.25	V
15	296.31	IV
15	299.04	IV
700	300.00	V
1000	573.36	V
600	587.94	V
10	785.12	IV
25	1168.61	III
15	1550.80	IV
180	1603.06	III
150	1610.19	III
160	2010.42	III
12	2118.97	IV
11	2185.43	IV
11	2205.46	IV
14	2222.22	IV
11	2271.33	IV
110	2438.62	I
560	2545.22	II
2900	2552.37	II
560	2555.82	II
2300	2560.25	II
1100	2563.21	II
11	2586.93	IV
120	2692.78	I
350	2699.07	III
360	2706.77	I
210	2707.95	I
580	2711.35	I
230	2734.05	III
340	2965.86	I
1200	2974.01	I
1400	2980.75	I
340	2988.95	I
2200	3015.36	I
2700	3019.34	I
360	3030.76	I
120 h	3056.31	I
130	3065.11	II
990	3251.32	II
1500	3255.69	I
4400	3269.91	I
5500	3273.63	I
110 d	3343.28	II
270	3352.05	II
9900	3353.73	II
2000	3359.68	II
1700	3361.27	II
1700	3361.94	II
4000	3368.95	II
6600	3372.15	II
130	3418.51	I
200	3429.21	I
200	3429.48	I
270	3431.36	I
530	3435.56	I
270	3457.45	I
180	3462.19	I
130 d	3469.65	I
110	3471.13	I
200	3498.91	I
2700	3535.73	II
6600	3558.55	II
6100	3567.70	II
13000	3572.53	II
9900	3576.35	II
7700	3580.94	II
4000	3589.64	II
4000	3590.48	II
28000	3613.84	II
110	3617.43	I
20000	3630.75	II
13000	3642.79	II
6600	3645.31	II
110	3646.90	I
5300	3651.80	II
110	3664.25	II
290	3666.54	II
75 h	3717.10	I
270	3833.07	II
610	3843.03	II
20000	3907.49	I
23000	3911.81	I
4400	3933.38	I
5500	3996.61	I
530	4014.49	II
20000	4020.40	I
20000	4023.69	I
220	4030.67	I
140	4031.39	I
220	4043.80	I
200	4046.48	I
2700	4047.79	I
120	4049.95	I
5500	4054.55	I
220	4056.59	I
100	4068.66	III
160 h	4074.97	I
160	4078.57	I
6100	4082.40	I
200	4086.67	I
400	4087.16	I
440 h	4133.00	I
530 h	4140.30	I
720	4152.36	I
1100 h	4165.19	I
110 h	4218.26	I
110 h	4219.73	I
180	4231.93	I
200	4233.61	I
400	4238.05	I
15000	4246.83	II
290	4294.77	II
350	4305.71	II
4200	4314.09	II
3300	4320.74	II
2400	4325.01	II
180	4354.61	II
110	4358.64	I
2000	4374.46	II
130	4384.81	II
1100	4400.37	II
880	4415.56	II
120 h	4557.24	I
160 h	4573.99	I
350	4670.40	II
120	4706.97	I
120	4709.34	I
200	4728.77	I
490	4729.23	I
590	4734.10	I
690	4737.65	I
790	4741.02	I
1200	4743.81	I
200	4753.16	I
220	4779.35	I
170	4839.44	I
90	4909.76	I
90	4922.84	I
90	4934.25	I
170	4954.06	I
120	4973.66	I
150	4980.37	I
140	4991.92	I
530	5031.02	II
250	5064.32	I
530	5070.23	I
250	5075.81	I
2100	5081.56	I
1200	5083.72	I
1100	5085.55	I
750	5086.95	I
390	5087.14	I
270	5089.89	I
390	5096.73	I
620	5099.23	I
370	5101.12	I
180	5109.06	I
150	5112.86	I
320	5116.69	I
390	5210.52	I
280	5219.67	I
350	5239.82	II
280	5258.33	I
210	5285.76	I
120	5341.05	I
350	5349.30	I
120	5349.71	I
210	5355.75	I
530	5356.10	I
270	5375.35	I
370	5392.08	I
270	5446.20	I
120	5451.34	I
750	5481.99	I
530	5484.62	I
570	5514.22	I
660	5520.50	I
660	5526.82	II
70	5564.86	I
110	5591.33	I
80	5640.98	II
250	5657.88	II
1500	5671.81	I
1200	5686.84	I
1100	5700.21	I
10	5706.82	IV
190	5708.61	I
880	5711.75	I
230	5717.28	I
180	5724.08	I
14	5771.63	IV
620	6210.68	I
320	6239.78	I
120	6245.63	II
110	6249.96	I
80	6256.01	III
250	6258.96	I
750	6305.67	I
60	6378.82	I
90	6413.35	I
60	6604.60	II
65	6737.87	I
50	6819.52	I
50	6835.03	I
90	7449.16	III
55 h	7741.17	I
30	7800.44	I
19 h	8761.40	I
50	8829.78	III
30 h	8834.35	I
400	22051.86	I
150	22065.05	I

Selenium Se Z = 34

Intensity	Wavelength/Å	
360	613.0	V
360	652.7	IV
450	670.1	IV
360	724.3	III
450	746.4	IV
450	759.1	V
360	808.7	V
360	830.3	V
360	832.7	II
450	839.5	V
360	843.0	III
360	845.8	V
360	912.9	II
360	959.6	IV
360	974.8	III
450	996.7	IV
360	1013.4	II
360	1014.0	II
450	1033.6	II
450	1049.6	II
360	1057.4	II
360	1094.7	V
360	1099.1	III
450	1119.2	III
360	1141.9	II
450	1192.3	II
450	1227.6	V
285	1291.0	II
285	1308.9	II
285	1314.4	IV
120	1435.3	I
120	1435.8	I
150	1449.2	I
150	1500.9	I
250	1530.4	I
150	1531.3	I
200	1531.8	I
150	1575.3	I
150	1577.6	I
150	1577.9	I
150	1579.5	I
200	1580.0	I
150	1587.5	I
150	1593.2	I
250	1606.5	I
200	1617.4	I
150	1621.2	I
150	1643.4	I
250	1671.2	I
250	1675.3	I
250	1690.7	I
250	1793.3	I
300	1795.3	I
300	1855.2	I
250	1858.8	I
400	1898.6	I
350	1913.8	I
300	1919.2	I
500	1960.9	I
500	2039.8	I
285	2057.5	III
500	2074.8	I
285	2136.6	IV
500	2164.2	I
600	2413.5	I
300	2548.0	I
360	2665.5	IV
285	2724.3	IV
285	2767.2	III
220	2773.8	III
160	2951.6	IV
450	3387.2	III
450	3413.9	III
450	3457.8	III
450	3637.6	III
450	3738.7	III
450	3800.9	III
450	4169.1	III
360	4175.3	II
450	4180.9	II
285	4382.9	II
285	4446.0	II
220	4449.2	II
285	4467.6	II
500	4730.8	I
400	4739.0	I
300	4742.2	I
285	4840.6	II
360	4845.0	II
450	5227.5	II
360	5305.4	II
100	5365.5	I
120	5369.9	I
110	5374.1	I
285	5522.4	II
285	5566.9	II
285	5866.3	II
450	6056.0	II
285	6303.8	III
200	6325.6	I
360	6444.2	II
285	6490.5	II
285	6535.0	II
150	6831.3	I
120	6990.690	I
100	6991.792	I
200	7010.809	I
150	7013.875	I
300	7062.065	I
200	7575.1	I
250	7583.4	I
150	7592.2	I
300	8001.0	I
200	8036.4	I
150	8093.2	I
150	8094.7	I
180	8149.3	I
150	8152.0	I
200	8157.7	I
180	8163.1	I
150	8182.9	I
150	8440.47	I
150	8450.38	I
150	8742.33	I
300	8918.86	I
200	9001.97	I
200	9038.61	I
100	9432.50	I
200	10217.25	I
377	10307.45	I
900	10327.26	I
640	10386.36	I
275	11946.87	I
170	11952.64	I
205	11972.93	I
315	14817.93	I
410	14917.47	I
500	15151.44	I
320	15471.00	I
265	15520.97	I
395	15618.40	I
360	16659.44	I
505	16813.78	I
205	16866.54	I
235	21374.24	I
680	21442.56	I
415	21473.48	I
270	21716.36	I
240	21730.60	I
150	23388.85	I
265	24148.18	I
375	24385.99	I
255	25017.51	I
510	25127.43	I

Silicon Si Z = 14

Intensity	Wavelength/Å	
10	85.18	V
15	96.44	V
10	97.14	V
20	117.86	V
20	118.97	V
4	457.82	IV
8	566.61	III
8	653.33	III
7	815.05	IV
8	818.13	IV
9	823.41	III
40 h	845.77	II
100	889.72	II
200	892.00	II
9	967.95	III
100	989.87	II
200	992.68	II
10	993.52	III
13	994.79	III
16	997.39	III
50	1023.69	II
8	1066.63	IV
14	1108.37	III
16	1109.97	III
18	1113.23	III
8	1122.49	IV
10	1128.34	IV
100	1190.42	II
200	1193.28	II
250	1194.50	II
100	1197.39	II
30	1206.51	III
30	1206.53	III
9	1207.52	III
10	1210.46	III
50	1226.81	II
100	1227.60	II
150	1228.75	II
200	1229.39	II
100	1246.74	II
150	1248.43	II
100	1250.09	II
150	1250.43	II
200	1251.16	II
40	1256.49	I
50	1258.80	I
1000	1260.42	II
2000	1264.73	II
200	1265.02	II
17	1294.54	III
14	1296.73	III
15	1298.89	III
18	1298.96	III
14	1301.15	III
16	1303.32	III
100	1304.37	II
50 h	1305.59	II
200	1309.27	II
13	1312.59	III
100	1346.87	II
100	1348.54	II
150	1350.06	II
100	1352.64	II
100	1353.72	II
15	1393.76	IV
12	1402.77	IV
13	1417.24	III
90 h	1485.02	II
100 h	1485.51	II
12	1500.24	III
10	1501.19	III
9	1501.87	III
100 h	1509.10	II
50 h	1512.07	II
60 p	1516.91	II
500	1526.72	II
1000	1533.45	II
150	1594.55	I
100	1622.87	I
300	1629.43	I
200	1629.92	I
100	1667.62	I
100	1668.52	I
100	1672.59	I
200	1675.20	I
200	1696.20	I
200	1697.94	I
100 h	1770.92	I

Intensity	Wavelength/Å	
100 h	1776.83	I
100 h	1799.12	I
150	1808.00	II
500 h	1814.07	I
200	1816.92	II
200	1836.51	I
200	1841.44	I
9	1842.55	III
200	1843.77	I
300	1845.51	I
400	1847.47	I
200	1848.14	I
500	1850.67	I
200	1852.46	I
500 h	1874.84	I
200	1881.85	I
200	1887.70	I
200 h	1893.25	I
1000 h	1901.33	I
100 h	1902.46	II
50 h	1910.62	II
50	1941.67	II
100	1949.56	II
100	1954.97	I
50	2058.65	II
50	2059.01	II
200	2072.02	II
200	2072.70	II
100	2124.12	I
50 h	2136.56	II
110	2207.98	I
115	2210.89	I
110	2211.74	I
120	2216.67	I
120	2218.06	I
10	2296.87	III
10	2308.19	III
100 h	2356.30	II
30 h	2357.18	II
50 h	2357.97	II
300	2435.15	I
11	2449.48	III
425	2506.90	I
375	2514.32	I
500	2516.113	I
7	2517.51	IV
350	2519.202	I
425	2524.108	I
450	2528.509	I
110	2532.381	I
25	2541.82	III
10	2546.09	III
14	2559.21	III
30	2563.679	I
85	2568.641	I
45	2577.151	I
190	2631.282	I
11	2640.79	III
14	2655.51	III
9	2817.11	III
1000	2881.579	I
300	2904.28	II
500	2905.69	II
55	2970.355	I
150	2987.645	I
50	3006.739	I
75	3020.004	I
100 h	3030.00	II
9	3040.93	III
100 h	3043.69	II
50 h	3048.30	II
150 h	3053.18	II
25	3086.24	III
20	3093.42	III
16	3096.83	III
9	3165.71	IV
16	3185.13	III
13	3186.02	III
150	3188.97	II
150	3193.09	II
100	3195.41	II
14	3196.50	III
200	3199.51	II
100 h	3203.87	II
200 h	3210.03	II
15	3210.55	III
75	3214.66	II
12	3230.50	III
14	3233.95	III
15	3241.62	III
12	3258.66	III
10	3276.26	III
300	3333.14	II
500	3339.82	II
15	3486.91	III
9	3525.94	III
20	3590.47	III
8	3762.44	IV
20 c	3791.41	III
25	3796.11	III
30	3806.54	III
100 h	3853.66	II
500 h	3856.02	II
200 h	3862.60	II
300	3905.523	I
20	3924.47	III
10	4088.85	IV
70	4102.936	I
9	4116.10	IV
300 h	4128.07	II
500 h	4130.89	II
100 h	4190.72	II
50	4198.13	II
9	4338.50	III
30	4552.62	III
25	4567.82	III
20	4574.76	III
100	4621.42	II
150	4621.72	II
9 h	4631.24	IV
10 h	4654.32	IV
9	4683.02	III
16	4716.65	III
50	4782.991	I
35	4792.212	I
80	4792.324	I
15	4813.33	III
16	4819.72	III
18	4828.97	III
30	4947.607	I
40	5006.061	I
1000	5041.03	II
1000	5055.98	II
10 h	5091.42	III
100	5181.90	II
100 h	5185.25	II
200 h	5192.86	II
500 h	5202.41	II
100 h	5405.34	II
100 h	5438.62	II
100 h	5456.45	II
500 h	5466.43	II
500 h	5466.87	II
100 h	5469.21	II
200 h	5496.45	II
35	5517.535	I
100 h	5540.74	II
150 h	5576.66	II
30	5622.221	I
100 h	5632.97	II
200 h	5639.48	II
90	5645.611	I
150 h	5660.66	II
80	5665.554	I
1000 h	5669.56	II
120	5684.484	I
300 h	5688.81	II
100	5690.425	I
90	5701.105	I
200 h	5701.37	II
100 h	5706.37	II
160	5708.397	I
20	5739.73	III
45	5747.667	I
45	5753.625	I
45	5754.220	I
45	5762.977	I
70	5772.145	I
70	5780.384	I
90	5793.071	I
100	5797.859	I
150 h	5800.47	II
200	5806.74	II
50	5846.13	II
300 h	5868.40	II
40	5873.764	I
10 h	5898.79	III
150	5915.22	II
200	5948.545	I
500	5957.56	II
500	5978.93	II
90	6125.021	I
85	6131.574	I
90	6131.850	I
100	6142.487	I
100	6145.015	I
160	6155.134	I
160	6237.320	I
40	6238.287	I
125	6243.813	I
125	6244.468	I
180	6254.188	I
45	6331.954	I
1000	6347.10	II
1000	6371.36	II
45	6526.609	I
45	6527.199	I
45	6555.462	I
50 h	6660.52	II
100	6671.88	II
7	6701.21	IV
50 h	6717.04	II
100	6721.853	I
50	6829.82	II
30	6848.568	I
80	6976.523	I
180	7003.567	I
180	7005.883	I
90	7017.646	I
250	7034.903	I
6 h	7047.94	IV
200	7165.545	I
100	7226.206	I
100	7235.326	I
180	7250.625	I
160	7275.294	I
400	7289.173	I
375	7405.774	I
200	7409.082	I
275	7415.946	I
425	7423.497	I
9 h	7466.32	III
12 h	7612.36	III
100	7680.267	I
6 h	7723.82	IV
30	7800.008	I
400	7848.80	II
500	7849.72	II
30	7849.967	I
90	7918.386	I
120	7932.349	I
140	7944.001	I
35	7970.306	I
35	8035.619	I
70	8093.241	I
9 h	8102.86	III
11 h	8103.45	III
35	8230.642	I
9 h	8262.57	III
40	8443.982	I
40	8501.547	I
60	8502.221	I
40	8536.165	I
120	8556.780	I

Intensity	Wavelength/Å	
50	8648.462	I
40	8728.011	I
75	8742.451	I
100	8752.009	I
35	8790.389	I
100	9412.72	II
100	9413.506	I
30	10371.269	I
120	10585.141	I
120	10603.431	I
120	10660.975	I
30	10694.251	I
30	10727.408	I
60	10749.384	I
30	10784.550	I
80	10786.856	I
140	10827.091	I
60	10843.854	I
130	10869.541	I
30	10882.802	I
30	10885.336	I
80	10979.308	I
30	10982.061	I
80	11017.965	I
370	11984.19	I
220	11991.57	I
440	12031.51	I
190	15888.39	I
95	16060.03	I
110	19722.50	I

Silver Ag Z = 47

Intensity	Wavelength/Å	
25	730.83	II
30	752.80	II
400	799.41	III
15	1005.32	II
10	1065.49	II
12	1072.23	II
250	1074.22	II
150	1107.03	II
150	1112.46	II
60	1195.83	II
50	1223.33	II
50	1240.80	II
50	1246.87	II
55	1256.81	II
55	1257.55	II
50	1266.63	II
70	1273.67	II
65	1297.51	II
85	1311.20	II
55	1313.81	II
50	1314.61	II
60	1323.84	II
60	1342.09	II
50	1342.57	II
70	1346.62	II
50	1353.54	II
150	1364.50	II
100	1396.00	II
100	1410.93	II
90	1419.72	II
95	1432.60	II
100	1464.72	II
50	1466.23	II
50 r	1507.37	I
100 r	1515.63	I
50 r	1548.58	I
100	1555.16	II
100	1644.50	II
60	1651.52	I
50	1652.10	I
700	1656.18	III
120	1682.82	II
500	1693.51	III
10	1708.11	I
50	1709.27	II
125	1736.44	II
750	1751.03	III
10 h	1766.14	I
75	1790.37	II
600	1917.08	III
700	1957.62	III
100	1967.38	II
600	1975.92	III
500	1977.03	III
600	2000.24	III
150	2015.96	II
150	2033.98	II
200	2061.17	I
100	2069.85	I
80 r	2113.82	II
60	2145.60	II
600	2161.89	III
50	2186.76	II
60	2229.53	II
100 r	2246.43	II
500	2246.51	III
75 r	2248.74	II
75	2280.03	II
30 h	2309.56	I
700	2310.04	III
70 r	2317.05	II
80 r	2320.29	II
70 r	2324.68	II
80 r	2331.40	II
70	2357.92	II
50 h	2375.02	I
75	2411.41	II
90 r	2413.23	II
100 r	2437.81	II
80	2447.93	II
80	2473.84	II
60	2506.63	II
50 h	2575.63	I
60	2660.49	II
60	2721.77	I
75	2767.54	II
100 h	2824.39	I
30 h	3130.02	I
90	3180.70	II
100	3267.35	II
55000 r	3280.68	I
28000 r	3382.89	I
30	3469.16	I
70	3475.82	II
80	3495.28	II
50	3542.61	I
50 h	3624.68	I
75	3682.46	II
30	3682.50	I
80	3683.34	II
50 h	3709.20	I
200	3810.94	I
50	3811.78	I
100 h	3840.74	I
50 h	3907.41	I
50	3909.31	II
50 h	3914.40	I
70	3920.10	II
60	3949.43	II
100 h	3981.58	I
70	3985.19	II
100 h	4055.48	I
80	4085.91	II
100	4185.48	II
90 h	4210.96	I
100	4212.82	I
50	4311.07	I
50 h	4476.04	I
30 h	4615.69	I
80	4620.04	II
50	4620.46	II
60 h	4668.48	I
30 h	4677.60	I
100	4788.40	II
30 h	4847.82	I
100	4874.10	I
80	5027.35	II
1000	5209.08	I
1000	5465.50	I
100	5471.55	I
100	5667.34	I
10 h	6268.50	I
320	7687.78	I
25	8005.4	II
500	8273.52	I
25	8403.8	II
30 h	8645.70	I
10 h	8704.85	I
12	8747.6	II
15	9000.9	II
10	12551.0	I
60	16819.5	I
20	17416.7	I
15	18307.9	I
15	18382.3	I

Sodium Na Z = 11

Intensity	Wavelength/Å	
7	142.232	IV
8	146.064	IV
9	150.298	IV
8	150.687	IV
8	155.510	IV
8	156.537	IV
12	162.448	IV
10	163.190	IV
12	168.411	IV
10	168.546	IV
5	183.95	III
10	190.445	IV
10	199.772	IV
8	202.49	III
8	202.76	III
8 p	203.06	III
8	203.28	III
8	203.33	III
15	229.87	III
50 c	250.52	III
30	251.37	III
25	266.90	III
70	267.65	III
50	267.87	III
50	268.63	III
20 p	272.08	III
20	272.45	III
10	319.644	IV
300	372.08	II
350	376.38	II
100	378.14	III
70	380.10	III
12	408.684	IV
10	409.614	IV
15	410.372	IV
10	411.334	IV
13	412.242	IV
11	1582.18	IV
11 d	1583.98	IV
12	1584.14	IV
12 d	1587.05	IV
11	1615.92	IV
12	1618.57	IV
11	1655.47	IV
15 c	1701.97	IV
20 d	1887.47	III
12	1960.76	IV
11	1965.08	IV
12 d	2106.33	IV
30	2230.33	III
16	2232.19	III
20 h	2246.70	III
300	2315.65	II
18	2386.99	III
17	2394.03	III
300	2420.99	II
300	2424.73	II
25	2459.31	III
18	2468.85	III
20	2474.73	III
1000	2493.15	II
25	2497.03	III
17	2510.26	III
20	2543.84	I
10	2543.87	I
70	2593.87	I
35	2593.92	I

Intensity	Wavelength/Å	
850	2611.81	II
850	2661.00	II
1000	2671.83	II
200	2680.34	I
100	2680.43	I
1000	2841.72	II
400	2852.81	I
200	2853.01	I
2	2893.62	I
1100	2904.92	II
1100	2917.52	II
1100	2919.05	II
1200	2919.85	II
1300	2920.95	II
1000	2923.49	II
1200	2951.24	II
1100	2952.40	II
1000	2977.13	II
1100	2979.66	II
1100	2980.63	II
1300	2984.19	II
1700	3124.42	II
2500	3135.48	II
1700	3137.86	II
2000	3149.28	II
2000	3163.74	II
1000	3179.06	II
1700	3189.79	II
1600	3212.19	II
1500	3257.96	II
1700	3285.60	II
1700	3301.35	II
1200	3302.37	I
600	3302.98	I
1500	3304.96	II
1000	3318.04	II
50	3426.86	I
1500	3533.05	II
1200	3631.27	II
6	4238.99	I
10	4242.08	I
1	4249.41	I
2	4252.52	I
15	4273.64	I
20	4276.79	I
2	4287.84	I
3	4291.01	I
30	4321.40	I
40	4324.62	I
3	4341.49	I
5	4344.74	I
40	4390.03	I
60	4393.34	I
5	4419.88	I
8	4423.25	I
60	4494.18	I
100	4497.66	I
10	4541.63	I
15	4545.19	I
120	4664.811	I
200	4668.560	I

Intensity	Wavelength/Å	
20	4747.941	I
30	4751.822	I
200	4978.541	I
400	4982.813	I
40	5148.838	I
80	5153.402	I
280	5682.633	I
70	5688.193	I
560	5688.205	I
80000	5889.950	I
40000	5895.924	I
120	6154.225	I
240	6160.747	I
130	6530.70	II
130	6544.04	II
130	6545.75	II
20	7373.23	I
10	7373.49	I
50	7809.78	I
25	7810.24	I
4400	8183.256	I
800	8194.790	I
8800	8194.824	I
100	8649.92	I
60	8650.89	I
25	8942.96	I
40	9153.88	I
60	9465.94	I
80	9961.28	I
20	10566.00	I
60	10572.28	I
200	10746.44	I
80	10749.29	I
120	10834.87	I
35	11190.19	I
50	11197.21	I
400	11381.45	I
1000	11403.78	I
400	12679.17	I
60	14767.48	I
100	14779.73	I
60	16373.85	I
100	16388.85	I
400	18465.25	I
50	22056.44	I
25	22083.67	I
60	23348.41	I
100	23379.13	I

Strontium Sr Z = 38

Intensity	Wavelength/Å	
15	298.12	IV
15	300.12	IV
125	330.67	III
500	351.62	III
75	358.80	III
250	363.49	III
150	371.21	III
20	378.53	IV
75	392.44	IV
50	393.00	IV
50	396.22	IV
1000	437.24	III
1875	491.79	III
1250	507.04	III
3750	514.38	III
10	517.28	V
2500	562.75	III
25	578.01	V
30	624.93	V
25	642.23	V
50	649.21	V
25	660.94	V
200	664.43	IV
35	686.23	V
100	710.35	IV
12	747.82	V
50	1025.23	III
35	1125.49	III
50	1236.23	III
1400	2152.84	II
1400	2165.96	II
100	2273.71	III
100	2340.13	III
50	2346.97	IV
160	2428.10	I
100	2486.52	III
40	2555.60	IV
40	2571.04	IV
100	3002.61	III
200	3012.32	III
10	3019.29	IV
100	3021.73	III
50	3061.43	III
50	3182.61	III
100	3235.39	III
400	3351.25	I
650	3380.71	II
50	3430.76	III
950	3464.46	II
600	3969.26	I
1300	4030.38	I
46000	4077.71	II
32000	4215.52	II
9	4298.57	IV
340	4305.45	II
65000	4607.33	I
9	4685.08	IV
3200	4722.28	I
2200	4741.92	I
1400	4784.32	I
4800	4811.88	I
3600	4832.08	I
3000	4872.49	I
2000	4876.32	I
1000	4891.98	I
8000	4962.26	I
1300	4967.94	I
800 h	5156.07	I
1400	5222.20	I
2000	5225.11	I
2000	5229.27	I
2800	5238.55	I
4800	5256.90	I
40	5257.71	III
40	5443.48	III
1500	5450.84	I
7000	5480.84	I
3500	5504.17	I
2600	5521.83	I
2000	5534.81	I
2000	5540.05	I
1000	6380.75	I
900 h	6386.50	I
600 h	6388.24	I
9000	6408.47	I
5500	6504.00	I
1000	6546.79	I
1700	6550.26	I
3000	6617.26	I
1800	6791.05	I
4800	6878.38	I
1200	6892.59	I
5500	7070.10	I
2500	7309.41	I
500	7621.50	I
400 h	7673.06	I
200 h	8422.80	I
120	8505.69	II
200	8688.91	II
100	9294.10	I
400 h	9448.95	I
600	9596.00	I
300	9624.70	I
100	9638.10	I
100 h	9647.70	II
300	10036.66	II
1000	10327.31	II
200	10914.88	II
700	11241.25	I
100	12014.76	II
60	12445.90	II
40	12495.00	I
75	12974.70	II
100	13123.80	II
50	17447.40	I
230	20261.40	I
120	20700.70	I
30	26023.60	I

Sulfur S Z = 16

Intensity	Wavelength/Å	
5	437.4	V
5	438.2	V
5	439.6	V
20	519.3	IV
20	520.1	IV
40	520.8	IV
20	522.0	IV
20	522.5	IV
20	551.2	IV
40	652.5	IV
40	653.0	IV
70	653.6	IV
40	654.0	IV
70	655.6	IV
20	655.9	IV

Intensity	Wavelength/Å	
110	657.3	IV
40	658.3	V
70	659.8	V
40	660.9	IV
160	661.4	IV
110	663.2	V
40	663.7	IV
40	664.8	IV
70	666.1	IV
20	678.1	V
40	680.3	V
110	680.9	V
40	681.6	V
20	693.5	V
70	729.5	III
110	732.42	III
70	735.2	III
70	738.5	III
110	744.9	IV
110	748.4	IV
110	750.2	IV
110	753.8	IV
285	786.5	V
70	789.0	III
70	796.7	III
70	800.5	IV
70	804.0	IV
70	809.7	IV
110	816.0	IV
70	824.9	III
70	836.3	III
160	849.2	V
110	852.2	V
220	854.8	V
110	857.9	V
110	860.5	V
40	906.9	II
40	910.5	II
40	912.7	II
40	937.4	II
40	937.7	II
160	1062.7	IV
160	1073.0	IV
70	1073.5	IV
285	1077.1	III
40	1102.3	II
70	1194.0	III
70	1201.0	III
40	1234.1	II
40	1250.5	II
110	1253.8	II
110	1259.5	II
275	1270.782	I
250	1277.216	I
280	1295.653	I
275	1302.337	I
235	1302.863	I
235	1303.110	I
245	1303.430	I
260	1305.883	I
265	1310.194	I

Intensity	Wavelength/Å	
355	1316.542	I
290	1316.618	I
375	1323.515	I
355	1326.643	I
775	1381.552	I
710	1385.510	I
960	1388.435	I
640	1389.154	I
775	1392.588	I
1000	1396.112	I
300	1409.337	I
510	1425.030	I
425	1433.280	I
300	1436.968	I
300	1448.229	I
425	1472.972	I
550	1473.995	I
300	1474.380	I
355	1481.665	I
485	1483.039	I
300	1483.233	I
330	1485.622	I
390	1487.150	I
20	1624.0	IV
20	1629.2	IV
680	1666.688	I
640	1687.530	I
710	1807.311	I
680	1820.343	I
640	1826.245	I
710	1900.286	I
550	1914.698	I
20	2387.0	IV
40	2398.9	IV
110	2460.5	III
110	2489.6	III
160	2496.2	III
160	2499.1	III
220	2508.2	III
70	2636.9	III
220	2665.4	III
110	2691.8	III
110	2702.8	III
220	2718.9	III
110	2721.4	III
220	2726.8	III
220	2731.1	III
110	2741.0	III
285	2756.9	III
110	2775.2	III
160	2785.5	III
110	2863.5	III
160	2904.3	III
160	2986.0	III
110	3097.5	IV
110	3497.3	III
160	3632.0	III
110	3709.4	III
160	3717.8	III
160	3838.3	III
285	3867.6	I

Intensity	Wavelength/Å	
285	3902.0	I
160	3928.6	III
360	3933.3	II
450	4120.8	I
280	4142.3	II
360	4145.1	II
450	4153.1	II
450	4162.7	II
360	4253.6	III
450	4694.1	I
285	4695.4	I
160	4696.2	I
280	4716.2	II
450	4815.5	II
360	4924.1	II
450	4925.3	II
285	4993.5	I
360	5428.6	II
650	5432.8	II
1000	5453.8	II
1000	5473.6	II
1000	5509.7	II
280	5564.9	II
1000	5606.1	II
450	5640.0	II
450	5640.3	II
280	5647.0	II
650	5659.9	II
450	5664.7	II
160	5706.1	I
450	5819.2	II
450	6052.7	I
280	6286.4	II
450	6287.1	II
450	6305.5	II
450	6312.7	II
280	6384.9	II
280	6397.3	II
280	6398.0	II
360	6413.7	II
160	6743.6	I
285	6748.8	I
450	6757.2	I
450	7579.0	I
450	7629.8	I
285	7686.1	I
450	7696.7	I
1000	7924.0	I
160	7928.8	I
285	7930.3	I
450	7931.7	I
450	7967.4	I
450	7967.4	II
450	8314.7	I
450	8314.7	II
450	8585.6	I
285	8680.5	I
450	8694.7	I
360	8874.5	I
110	8882.5	I
220	8884.2	I

Intensity	Wavelength/Å	
160	9035.9	I
450	9212.9	I
450	9228.1	I
450	9237.5	I
285	9413.5	I
285	9421.9	I
285	9437.1	I
650	9649.9	I
450	9672.3	I
450	9680.8	I
450	9693.7	I
285	9697.3	I
285	9739.7	I
285	9932.3	I
285	9949.8	I
285	9958.9	I
285	10455.5	I
285	10459.5	I
Tantalum Ta Z = 73		
60	493.07	V
1000	890.87	V
500	947.30	V
67	999.34	IV
79	1116.10	IV
78	1136.17	IV
85	1175.51	IV
80	1189.28	IV
80	1192.67	IV
85	1213.09	IV
500	1213.42	V
85	1215.53	IV
90	1223.73	IV
88	1238.12	IV
95	1240.06	IV
87	1258.34	IV
94	1264.91	IV
98	1272.42	IV
94	1275.48	IV
86	1275.94	IV
92	1308.51	IV
87	1315.58	IV
92	1332.38	IV
86	1343.30	IV
92	1365.88	IV
5000	1392.56	V
91	1398.78	IV
93	1413.40	IV
91	1454.32	IV
92	1464.41	IV
93	1469.82	IV
90	1495.25	IV
95	1514.19	IV
85	1607.70	IV
7000	1709.10	V
85	1712.16	IV
85	1716.13	IV
85	2055.75	IV
1100	2140.13	II
1500	2146.87	II
1200	2182.71	II
1100	2193.88	II

Intensity	Wavelength/Å	
1500	2196.03	II
90	2199.58	IV
1500	2199.67	II
90	2207.64	IV
1400 d	2210.03	II
1400	2239.48	II
1200	2250.76	II
840	2261.42	II
990	2262.30	II
990	2272.59	II
790	2285.25	II
600	2286.59	II
990	2289.16	II
440	2302.24	II
440	2302.93	II
440	2312.60	II
420	2315.46	II
690	2331.98	II
550	2332.19	II
250	2357.30	I
260	2361.09	I
600	2364.24	II
320	2371.58	I
1400	2387.06	II
2400	2400.63	II
320	2416.89	II
360	2427.64	I
360	2429.71	II
480	2432.70	II
380	2470.90	II
600	2474.62	I
500	2484.95	I
600	2488.70	II
500	2490.46	I
600	2504.45	I
600	2507.45	I
1200 d	2526.35	I
600	2532.12	II
1200	2559.43	I
460	2562.10	I
600	2577.37	II
600	2603.49	II
1400	2608.63	I
1200	2635.58	II
860	2636.90	I
2400	2647.47	I
2600	2653.27	I
1900	2656.61	I
1500	2661.34	I
770	2675.90	II
1500	2685.17	II
470	2694.52	II
1000	2698.30	I
1200	2710.13	I
2600	2714.67	I
470	2727.44	II
1200	2748.78	I
860	2749.83	I
410	2752.49	II
1000	2758.31	I
430	2761.68	II
770	2775.88	I
680	2796.34	I
680	2797.76	II
510	2806.58	I
640	2844.25	I
560	2848.52	I
1500	2850.49	I
1900	2850.98	I
360	2861.98	I
470	2871.42	I
380	2880.02	I
770	2891.84	I
560	2902.05	I
310	2915.49	I
410	2925.19	I
310	2932.70	I
1700	2933.55	I
470	2940.06	I
1200	2940.22	I
510	2951.92	I
340	2953.56	I
1500	2963.32	I
770	2965.13	II
770	2965.54	I
340	2969.47	I
430	2975.56	I
1800	3012.54	II
290 d	3027.48	I
530	3049.56	I
530	3069.24	I
360	3077.24	I
560	3103.25	I
380	3124.97	I
380	3130.58	I
270	3132.64	I
320	3170.29	I
270	3173.59	I
600	3180.95	I
300	3223.83	I
1100	3311.16	I
680	3318.84	I
330 d	3330.99	II
640	3371.54	I
360	3385.05	I
450	3406.94	I
490	3480.52	I
380	3497.85	I
490	3511.04	I
750	3607.41	I
980	3626.62	I
500	3642.06	I
210	3918.51	I
210	3970.10	I
210	3996.17	I
410	4061.40	I
310	4067.91	I
300	4205.88	I
360 c	4510.98	I
340	4574.31	I
260	4619.51	I
450	4681.88	I
200	5037.37	I
100	5067.87	I
110	5115.84	I
100	5141.62	I
100	5143.69	I
330	5156.56	I
110	5212.74	I
110 d	5218.45	I
140	5341.05	I
200	5402.51	I
130	5419.19	I
90	5518.91	I
150	5645.91	I
130	5664.90	I
130	5776.77	I
90	5780.71	I
130	5811.10	I
240	5877.36	I
130	5882.30	I
90	5901.91	I
90	5918.95	I
130	5939.76	I
240	5944.02	I
190 c	5997.23	I
100	6020.72	I
250	6045.39	I
100	6047.25	I
100	6101.58	I
65	6144.56	I
130	6154.50	I
150	6256.68	I
150	6268.70	I
150	6309.58	I
75	6325.08	I
65	6341.17	I
75	6356.16	I
65	6360.84	I
90	6389.45	I
65	6428.60	I
250	6430.79	I
200	6450.36	I
380	6485.37	I
65	6505.52	I
100	6514.39	I
100	6516.10	I
100	6574.84	I
110	6611.95	I
75	6621.30	I
100	6673.73	I
180	6675.53	I
75 c	6740.73	I
75	6771.74	I
160 c	6813.25	I
210	6866.23	I
180	6875.27	I
150	6902.10	I
140	6927.38	I
140	6928.54	I
65	6951.26	I
180	6966.13	I
110 d	6995.22	I
75	7006.96	I
150	7148.63	I
110	7172.90	I
140	7301.74	I
160	7346.41	I
140 c	7352.86	I
100	7356.96	I
90 cw	7369.09	I
160	7407.89	I
100	7882.37	I
75	8026.50	I
75	8281.62	I

Technetium Tc Z = 43

Intensity	Wavelength/Å	
10000 c	3636.07	I
20000 c	4031.63	I
15000	4095.67	I
20000	4262.27	I
30000	4297.06	I
20000	4853.59	I

Tellurium Te Z = 52

Intensity	Wavelength/Å	
8	802.28	II
8	1059.51	II
8	1077.66	II
10	1161.42	II
10	1174.34	II
12	1175.79	II
9	1208.54	II
9	1220.98	II
9	1253.62	II
9	1270.52	II
10	1324.92	II
9	1363.24	II
8	1366.73	II
10	1374.80	II
10	1608.41	II
10	1613.15	II
5	1655.4	I
5	1688.5	I
6	1700.0	I
5	1708.0	I
10	1822.4	I
26000	2002.02	I
6500	2081.16	I
18000	2142.81	I
3200	2147.25	I
500	2259.02	I
1200	2383.26	I
1500	2385.78	I
50	2438.69	II
120	2530.72	I
100	2649.66	II
80	2661.10	II
110	2677.13	I
100	2858.29	II
150	2895.41	II
70	2967.29	II
70	3047.00	II
100	3175.14	I
60	3256.80	II
60	3329.22	II

Intensity	Wavelength/Å	
150	3406.79	II
50	3442.25	II
50	3521.11	II
50	3552.19	II
100	3611.78	II
50	3617.57	II
50	4006.52	II
70	4127.32	II
100	4169.77	II
80	4225.73	II
100	4261.11	II
60	4273.43	II
80	4285.85	II
150	4364.00	II
75	4385.10	II
170	4478.63	II
80	4537.07	II
100	4557.78	II
70	4630.62	II
100	4641.12	II
180	4654.37	II
200	4686.91	II
100	4696.38	II
100	4706.53	II
100	4766.05	II
100	4784.87	II
100	4827.14	II
150	4831.28	II
150	4842.90	II
130	4865.12	II
200	4866.24	II
8	5083.0	I
50	5449.84	II
50	5487.95	II
150	5576.35	II
150	5649.26	II
100	5666.20	II
200	5708.12	II
150	5755.85	II
100	5974.68	II
50	6367.13	II
10 h	6790.0	I
20 h	6837.6	I
20 h	6854.7	I
15 h	7191.1	I
20 h	7263.5	I
12	7460.98	II
15	7468.75	II
15	7921.69	II
15	7943.14	II
10	7950.34	II
30 h	8061.4	I
10	8122.44	II
20	8186.44	II
15	8273.53	II
15	8672.95	II
10	8733.81	II
205	8758.18	I
81	9004.37	I
5660	9722.74	I
532	9868.92	I
689	9956.30	I
325	9977.13	I
5950	10051.41	I
4097	10091.01	I
381	10118.08	I
397	10300.56	I
745	10493.57	I
1880	10918.34	I
10200	11089.56	I
508	11163.74	I
6620	11487.23	I
1580	13247.75	I
1050	14513.51	I
1480	15452.45	I
2430	15546.23	I
3760	16403.90	I
1960	17303.54	I
2780	18291.59	I
1020	21043.73	I
464	21602.50	I
74	22555.29	I
38	26539.17	I

Terbium Tb Z = 65

Intensity	Wavelength/Å	
1000	1259.40	IV
1000	1327.67	IV
1000	1373.86	IV
5000	1595.39	IV
2000	1633.19	IV
2000	2027.79	IV
1000	2089.98	IV
1000	2332.54	IV
110	2584.61	II
110	2608.57	II
130	2628.69	II
140	2669.29	II
190	2704.07	II
270	2769.53	II
320	2897.44	II
250	2956.21	II
230	3010.59	II
230	3016.18	II
460	3053.55	II
460	3070.05	II
670	3078.86	II
480	3082.36	II
480	3089.58	II
480	3102.96	II
440	3139.64	II
480	3187.26	II
480	3199.56	II
1100	3218.93	II
1200	3219.98	II
480	3252.32	II
760	3280.31	II
760	3281.40	II
1000	3285.04	II
1500	3293.07	II
3800	3324.40	II
760	3349.42	II
760	3364.93	II
810	3454.06	II
810 d	3472.79	II
810	3500.84	II
5700	3509.17	II
1300	3523.66	II
1100	3540.24	II
810	3543.89	II
3200	3561.74	II
810	3567.35	II
4200	3568.52	II
1600	3568.98	II
1100	3579.20	II
710	3585.03	II
810	3596.38	II
1600	3600.44	II
810	3625.54	II
2300	3650.40	II
810	3654.88	II
2000	3658.88	II
3800	3676.35	II
810	3682.26	II
450	3693.58	I
450	3700.12	I
4700	3702.86	II
2400	3703.92	II
1000 d	3711.76	II
650	3745.04	I
870	3747.17	II
870	3747.34	II
1100	3755.24	II
650	3759.35	I
1700	3765.14	I
2100	3776.49	II
600	3783.53	I
410	3789.92	I
760 d	3806.85	II
1500	3830.26	I
540	3833.42	I
920 d	3842.50	II
3700	3848.73	II
3500 w	3874.17	II
480	3888.22	I
490	3894.64	I
2400	3899.20	II
1600	3901.33	I
480	3908.06	I
650	3915.43	I
760	3925.45	II
810 d	3939.52	II
2200 d	3976.84	II
1800	3981.87	II
970	4002.59	II
1900	4005.47	II
760	4012.75	II
870	4032.28	II
2100	4033.03	II
430	4054.12	I
410	4060.37	I
1300	4061.58	I
650	4105.37	I
1100	4144.41	II
350	4158.53	I
390	4196.74	I
650	4203.74	I
600	4206.49	I
480	4215.09	I
480	4232.82	I
650	4266.34	I
760 cw	4278.52	II
450	4310.42	I
2200	4318.83	I
600	4322.23	I
600	4325.83	II
3000	4326.43	I
600	4332.12	I
870	4336.43	I
600	4337.64	I
1700	4338.41	I
700	4340.62	I
870	4356.81	I
330	4382.45	I
300	4388.23	I
260	4390.91	I
350	4423.10	I
240	4436.12	I
240	4448.04	I
430	4493.07	I
75	4514.31	II
110	4549.07	I
110	4550.45	I
110	4556.46	I
110	4563.69	II
210	4578.69	II
65	4584.84	II
65	4591.56	II
75 d	4626.32	II
95	4626.94	II
65	4632.07	I
65 h	4636.59	I
85	4641.00	II
210	4641.98	II
260 cw	4645.31	II
80	4647.23	I
80	4662.79	I
80	4676.90	I
70 c	4681.87	I
80	4688.63	II
80	4693.11	II
200	4702.41	II
110	4707.94	II
80	4739.93	I
70	4747.80	I
410 cw	4752.53	II
180	4786.78	I
100	4813.77	I
80	4875.57	II
80	4881.15	II
95	4915.90	I
65	4931.79	I
85	4993.82	II
110	5078.25	I
75	5089.12	II
85	5186.13	I

Intensity	Wavelength/Å	
120	5228.12	I
75	5248.71	I
75 w	5262.11	II
75	5281.05	I
65	5304.72	I
110	5319.23	I
65 w	5337.90	I
160	5354.88	I
75	5369.72	I
75	5375.98	I
50	5424.10	II
55	5459.81	I
55	5509.61	I
50	5514.54	I
65	5524.12	I
85 c	5747.58	I
75	5795.64	I
75	5803.13	II
65	5815.36	I
65	5851.07	I
65	5870.62	I
65 c	5920.78	I
75	5967.34	II
35	6331.68	II
35 cw	6518.68	I
35	6581.82	I
90	6677.94	II
40 cw	6702.61	I
130	6794.58	II
55	6896.37	II
45 h	6899.95	I
40	6901.98	I
65	7204.28	I
40	7257.73	I
45	7348.88	II
45	7496.12	I
27 h	7582.03	II
45	7590.24	I
65	7596.44	I
30	7627.81	I
30	7737.63	I
30	7855.79	II
27	7927.90	II
30	8025.42	II
30	8085.06	II
65	8194.82	II
95	8212.57	I
40	8450.06	II
30 h	8511.80	I
45	8583.45	II
30	8603.40	I
65	8765.74	II

Thallium Tl Z = 81

Intensity	Wavelength/Å	
10	570.49	IV
5 r	670.87	II
15 r	696.30	II
5 r	709.23	II
10 r	817.18	II
5 r	836.34	II
8 r	1018.85	II
30	1028.69	IV
20	1034.73	IV
20	1036.61	IV
10 r	1049.73	II
8 r	1050.30	II
5 r	1074.97	II
30	1079.68	IV
10 r	1130.17	II
15 r	1162.55	II
10 r	1167.43	II
10 r	1183.41	II
12 r	1194.84	II
5 r	1246.00	II
10	1266.33	III
15 r	1307.50	II
8 r	1310.20	II
25 r	1321.71	II
8 r	1330.40	II
10 r	1373.52	II
10	1477.14	III
8 r	1489.65	I
10 r	1499.30	II
10 r	1507.82	II
15 r	1561.58	II
10 r	1568.57	II
7 r	1593.26	II
5 h	1616.	I
5	1685.40	I
10 r	1792.76	II
12 r	1814.85	II
25 r	1908.64	II
100 r	2007.56	I
100 r	2210.71	I
30	2298.04	II
140	2315.98	I
900 h	2379.69	I
20	2530.86	II
700	2580.14	I
420	2709.23	I
4400 d	2767.87	I
10	2849.80	II
2800	2918.32	I
20	3091.56	II
15	3185.51	II
15	3186.56	II
15	3187.74	II
1200	3229.75	I
15	3291.01	II
15	3369.15	II
9	3456.34	III
20000	3519.24	I
5000	3529.43	I
8	3540.08	II
9	3560.68	II
12000 w	3775.72	I
10	3832.30	II
10	3887.15	II
7	4109.85	III
6	4269.81	III
20	4274.98	II
40	4306.80	II
20	4737.05	II
15	4981.35	II
25	5078.54	II
25	5152.14	II
18000	5350.46	I
15 d	5384.85	II
10	5410.97	II
25	5949.48	II
10	6179.98	II
10	6378.32	II
16 h	6549.84	I
10	6966.5	II
10	7815.80	I
20	8373.6	I
10	8474.27	I
10	8664.1	II
20	9130.	II
20	9130.5	I
40	9509.4	I
20	9930.4	I
30	10011.9	I
40	10488.80	I
1000	11512.82	I
150	12736.4	I
700	13013.2	I

Thorium Th Z = 90

Intensity	Wavelength/Å	
150	1707.37	IV
200	1950.02	IV
200	2002.34	IV
200	2413.50	III
200	2427.94	III
200	2431.68	III
200	2441.24	III
500	2565.593	II
480	2692.415	II
520	2747.156	II
410	2752.166	II
800	2832.315	II
1200	2837.295	II
100	2848.084	I
550	2870.406	II
100	2936.086	I
100	2943.729	I
420	3049.092	II
450	3067.729	II
670	3078.828	II
480	3080.217	II
510	3108.296	II
100	3116.263	I
510	3119.526	II
510	3122.963	II
480	3125.507	II
100	3136.216	I
420	3139.306	II
420	3142.835	II
420	3175.726	II
1100	3180.193	II
770	3188.233	II
560	3221.292	II
560	3229.009	II
480	3235.84	II
590	3238.116	II
910	3256.274	II
180	3257.366	I
910	3262.668	II
620	3287.789	II
910	3291.739	II
620	3292.520	II
240	3301.650	I
480	3304.238	I
510	3321.450	II
840	3325.120	II
250	3330.476	I
620	3334.604	II
620	3337.870	II
310	3348.768	I
980	3351.228	II
620	3358.602	II
250	3374.971	I
1300	3392.035	II
200	3396.727	I
250	3398.544	I
200	3405.558	I
250	3413.012	I
390	3421.210	I
270	3423.989	I
980	3433.998	II
770	3435.976	II
1300	3469.920	II
170	3471.218	I
200	3486.552	I
670	3539.587	II
180	3544.018	I
170	3549.595	I
200	3555.013	I
530	3559.451	II
200	3576.557	I
270	3592.780	I
270	3598.120	I
980	3609.445	II
200	3612.427	I
480	3615.133	II
270	3635.943	I
210	3642.248	I
170	3649.735	I
220	3663.202	I
280	3669.968	I
700	3675.567	II
150	3682.486	I
170	3692.566	I
180	3698.105	I
340	3706.767	I
590	3719.435	I
770	3721.825	II
1300	3741.183	II
310	3747.539	I
650	3752.569	II
180	3770.056	I
590	3803.075	I
450	3828.384	I
840	3839.746	II
450	3863.405	II
210	3875.374	I

Intensity	Wavelength/Å	
340	3895.419	I
590	3929.669	II
200	3932.911	I
390	3967.392	I
200	3972.155	I
150	3980.089	I
530	3994.549	II
220	4008.210	I
220	4009.056	I
280	4012.495	I
4200	4019.129	II
250	4030.842	I
250	4036.047	I
250	4063.407	I
700	4086.520	II
700	4094.747	II
150	4100.341	I
840	4108.421	II
240	4112.754	I
280	4115.758	I
1100	4116.713	II
200	4127.411	I
200	4134.067	I
450	4149.986	II
620	4178.060	II
620	4208.890	II
110	4253.538	I
110	4260.333	I
480	4277.313	II
700	4282.042	II
130	4337.277	I
1300	4381.860	II
1100	4391.110	II
110	4498.940	I
280	4510.527	II
90	4723.438	I
50	4840.843	I
280	4863.163	II
260	5017.255	II
110	5067.974	I
120	5148.211	II
95	5216.596	II
110	5231.160	I
95	5247.654	II
60	5343.581	I
60	5587.026	I
95	5707.103	II
70	5760.551	I
85	5989.044	II
60	6169.822	I
50	6182.622	I
50	6274.116	II
50	6274.117	II
50	6355.911	II
60	6457.283	I
50	6462.614	I
50 h	6531.342	I
55	6989.656	I
30	7045.795	II
30	7084.171	I
30	7168.896	I
40	7191.132	II
35	7208.006	I
50	7525.508	II
30	7647.380	I
30	8330.451	I
40	8967.641	I
20	9833.42	I
20	10726.93	I
20	10942.24	II
30	11230.259	I
20	11984.67	II
20	17208.22	II
15	18811.88	I
10	22264.35	II

***Thulium* Tm *Z* = 69**

Intensity	Wavelength/Å	
5000	2185.94	III
360	2284.79	II
20000	2296.21	III
5000	2305.03	III
20000	2311.16	III
5000	2312.72	III
5000	2326.19	III
6000	2328.50	III
6000	2329.29	III
3000	2331.80	III
3000	2357.05	III
4000	2406.63	III
450	2409.02	II
450	2426.17	II
770	2480.13	II
30000	2489.44	III
2000	2504.71	III
1300	2509.08	II
3000	2519.78	III
130	2527.02	I
10000	2552.46	III
360	2552.76	I
540	2561.65	II
430	2588.27	II
170 h	2596.49	I
810	2607.06	II
730	2624.33	II
5000	2682.32	III
2000	2707.03	III
3000	2719.47	III
540	2721.19	II
3000	2724.44	III
4000	2727.56	III
680	2794.60	II
730	2797.27	II
2000	2806.77	III
580	2827.92	II
200	2854.17	I
1600	2869.23	II
1000	2947.72	III
490	2973.22	I
1000	2998.28	III
1500	3015.30	II
360	3081.12	I
7400	3131.26	II
2300	3133.89	II
1900	3151.04	II
1500	3157.34	II
450	3172.65	I
2300	3172.83	II
1200	3236.81	II
1600	3240.23	II
2300	3241.54	II
320	3246.96	I
1900	3258.05	II
1600	3266.64	II
1200	3267.40	II
1100	3276.81	II
1200	3283.40	II
1200	3285.61	II
2300	3291.00	II
2000	3302.46	II
1200	3309.80	II
230	3349.99	I
4000	3362.61	II
1700	3397.50	II
850	3410.05	I
340	3412.59	I
340	3416.59	I
6400	3425.08	II
340	3429.33	I
4900	3441.50	II
4900	3453.66	II
8500	3462.20	II
210	3467.51	I
340	3476.69	I
340	3480.98	I
420	3487.38	I
340	3499.95	I
250	3517.60	I
1700	3535.52	II
420	3537.91	I
210	3555.82	I
340	3560.92	I
420	3563.88	I
1300	3566.47	II
420	3567.36	I
280	3586.07	I
2100	3608.77	II
1000	3629.09	III
380	3638.41	I
1100	3668.09	II
4800	3700.26	II
3800	3701.36	II
7700	3717.91	I
2400	3734.12	II
5000	3744.06	I
1700	3751.81	I
6000	3761.33	II
4800	3761.91	II
7100	3795.75	II
770	3798.54	I
600	3807.72	I
290	3826.39	I
1300	3838.20	II
290	3840.87	I
8900	3848.02	II
6800	3883.13	I
1800	3883.44	II
5400	3887.35	I
440	3896.62	I
3500	3916.48	I
1500	3949.27	I
1500	3958.10	II
1800	3996.52	II
220	4024.23	I
380	4044.47	I
10000	4094.19	I
9500	4105.84	I
1100	4138.33	I
8800	4187.62	I
6000	4203.73	I
380	4222.67	I
3000	4242.15	II
270	4271.71	I
150	4298.36	I
2700	4359.93	I
1400	4386.43	I
200	4394.42	I
140	4396.50	I
120	4454.03	I
540	4481.26	II
150	4519.60	I
260	4522.57	II
110	4548.60	I
270	4599.02	I
300	4615.94	II
80	4626.33	II
95	4626.56	II
110	4634.26	II
120	4655.09	I
160	4681.92	I
120	4691.11	I
110	4724.26	I
680	4733.34	I
70	4759.90	I
80	4831.20	II
140	4957.18	I
160	5009.77	II
160	5034.22	II
150	5060.90	I
95	5113.97	I
80	5213.38	I
650	5307.12	I
80	5346.49	II
270	5631.41	I
520	5675.84	I
40	5684.76	II
35	5709.97	II
190	5764.29	I
35	5838.76	II
240	5895.63	I
140	5971.26	I
200	6460.26	I
95	6604.96	I
110	6779.77	I
120	6844.26	I
80	6845.76	I

Intensity	Wavelength/Å	
10	6937.37	I
10	7017.90	I
12	7034.34	I
10	7106.14	I
17	7272.62	I
14	7310.51	I
14	7432.18	I
75	7481.08	I
75	7490.20	I
140	7558.33	I
80	7731.53	I
40	7856.08	I
55	7927.51	I
110	7930.84	I
95	8017.90	I
27	8472.01	II

Tin Sn ***Z* = 50**

Intensity	Wavelength/Å	
7	169.47	II
150	361.01	V
100	753.01	III
200	910.92	III
500	956.25	IV
7	985.13	II
500	1019.72	IV
1000	1044.49	IV
1000	1073.41	IV
200	1089.35	V
8	1108.19	II
1000	1119.34	IV
1000	1139.29	III
1000	1158.33	III
200	1160.74	V
10	1161.43	II
1000	1184.25	III
2000	1210.52	III
9	1219.07	II
13	1223.70	II
11	1243.00	II
2000	1251.38	V
1000	1259.92	III
20	1290.86	II
200	1294.36	V
1000	1305.97	III
1000	1314.55	IV
20	1316.59	II
1000	1327.34	III
1000	1347.65	III
1000	1386.74	III
25	1400.52	II
1000	1437.52	IV
20	1475.15	II
9	1489.22	II
1000	1570.36	III
10 r	1737.21	I
15 r	1751.46	I
20 r	1764.98	I
30 r	1790.75	I
80 r	1804.60	I
15	1811.34	II
500	1811.71	III
40 r	1815.74	I
120 r	1823.00	I
9	1831.89	II
50 r	1848.75	I
200 r	1860.32	I
80	1886.05	I
100	1891.40	I
12	1899.91	II
50	1909.30	I
80	1925.31	I
500	1941.86	III
150	1952.15	I
50 h	1977.6	I
80	1984.20	I
50	2040.66	I
50	2054.03	I
70	2058.31	I
80	2068.58	I
100	2072.89	I
100	2073.08	I
200	2096.39	I
100	2100.93	I
100 r	2113.93	I
50	2121.26	I
40 r	2148.73	I
20 r	2151.43	I
30	2151.54	II
80	2171.32	I
150 r	2194.49	I
300 r	2199.34	I
400 r	2209.65	I
80 r	2231.72	I
400 r	2246.05	I
60	2251.17	I
400 r	2268.91	I
200 r	2286.68	I
600 r	2317.23	I
300 r	2334.80	I
1000 r	2354.84	I
22	2368.33	II
100	2408.15	I
800 r	2421.70	I
1000 r	2429.49	I
15	2448.98	II
300	2483.39	I
13	2483.48	II
10	2486.99	II
200	2495.70	I
400	2546.55	I
500 r	2571.58	I
200	2594.42	I
200 r	2661.24	I
700 r	2706.51	I
150	2779.81	I
1400 r	2839.99	I
1000 r	2863.32	I
700 r	3009.14	I
850 r	3034.12	I
12	3047.50	II
550 r	3175.05	I
550 r	3262.34	I
50	3283.21	II
110	3330.62	I
60	3351.97	II
10	3472.46	II
11	3575.45	II
280 r	3801.02	I
10	5332.36	II
20	5561.95	II
25	5588.92	II
500	5631.71	I
15	5799.18	II
50	5925.44	I
100	5970.30	I
150	6037.70	I
250	6069.00	I
100	6073.46	I
400	6149.71	I
200	6154.60	I
150	6171.50	I
100	6310.78	I
70	6453.50	II
25	6844.05	II
20	7191.40	II
10	7387.79	II
13	7741.80	II
100	7754.97	I
100 h	8030.5	I
200	8114.09	I
80	8357.04	I
300	8422.72	I
400	8552.60	I
50 h	8681.7	I
50 h	9410.86	I
80 h	9415.37	I
150	9616.40	I
50	9741.1	I
100 h	9742.8	I
300 h	9805.38	I
500	9850.52	I
54	10894.00	I
70	11191.85	I
56	11277.66	I
200	11454.59	I
200	11616.26	I
258	11739.78	I
96	11825.18	I
106	11835.82	I
254	11932.99	I
48	12009.50	I
111	12313.24	I
42	12530.87	I
42	12536.5	I
89	12888.5	I
187	12981.7	I
187	13018.5	I
68	13081.5	I
378	13460.2	I
144	13608.2	I
40	20861.7	I
4	24738.2	I

Titanium Ti ***Z* = 22**

Intensity	Wavelength/Å	
17	252.96	V
15	498.26	V
14	502.08	V
13	526.57	V
18	779.07	IV
20	1298.66	III
20	1298.97	III
23	1455.19	III
20	1467.34	IV
11	1717.40	V
10	1841.49	V
20	2067.56	IV
18	2103.16	IV
180	2273.28	I
190	2279.96	I
190	2305.67	I
22	2413.99	III
25	2516.05	III
360	2525.60	II
24	2527.84	III
210	2529.85	I
190	2531.25	II
190	2534.62	II
130	2535.87	II
23	2540.06	III
24	2563.44	III
23	2565.42	III
22	2567.56	III
270	2599.92	I
340	2605.15	I
510	2611.28	I
300	2619.94	I
640	2641.10	I
800	2644.26	I
950	2646.64	I
250	2742.32	I
250	2802.50	I
190	2841.94	II
180	2877.44	II
280	2884.11	II
450	2912.08	I
340	2928.34	I
1100	2942.00	I
1300	2948.26	I
1600	2956.13	I
22	2984.75	III
1300 d	3066.22	II
1100	3072.97	II
1600	3075.22	II
2300	3078.64	II
3600	3088.02	II
720	3119.72	I
500	3161.20	II
780	3161.77	II
1000	3162.57	II
1600	3168.52	II
2400	3186.45	I
1000	3190.87	II
3100	3191.99	I
3800	3199.92	I
780	3202.54	II
1100	3217.06	II

Intensity	Wavelength/Å	
1300	3222.84	II
6600	3234.52	II
5200	3236.57	II
4100	3239.04	II
2600	3241.99	II
1200	3248.60	II
1200	3252.91	II
1200	3254.25	II
1200	3261.60	II
840	3314.42	I
2900	3322.94	II
2100	3329.46	II
1800	3335.20	II
1100	3340.34	II
5700	3341.88	I
4300	3349.04	II
12000	3349.41	II
4100	3354.64	I
7200	3361.21	II
1100	3370.44	I
4300	3371.45	I
5700	3372.80	II
2900 d	3377.48	I
1400	3380.28	II
5700	3383.76	II
1400	3385.95	I
1400	3387.84	II
1100	3394.58	II
890	3444.31	II
600	3461.50	II
600	3477.18	II
480	3491.05	II
890	3504.89	II
600	3510.84	II
17	3576.44	IV
600	3610.16	I
4800	3635.46	I
6600	3642.68	I
7200	3653.50	I
600	3671.67	I
3100	3685.20	II
600	3689.91	I
2900	3729.82	I
3300	3741.06	I
330	3741.64	II
5200	3752.86	I
3300	3759.30	II
2900	3761.32	II
840	3786.04	I
500	3882.89	I
530	3900.54	II
2600	3904.78	I
500	3913.46	II
500	3914.34	I
15	3915.47	III
1100	3924.53	I
890	3929.88	I
1100	3947.78	I
4500	3948.67	I
4500	3956.34	I
5200	3958.21	I
950	3962.85	I
950	3964.27	I
4800	3981.76	I
570	3982.48	I
5700	3989.76	I
7800	3998.64	I
950	4008.93	I
1200	4024.57	I
840	4078.47	I
890	4286.01	I
840	4287.40	I
950	4289.07	I
840	4290.94	I
840	4295.76	I
2000	4298.66	I
200	4300.05	II
2900	4300.56	I
4100	4301.09	I
6000	4305.92	I
1200	4314.80	II
330	4395.04	II
890	4427.10	I
230	4443.80	II
840	4449.15	I
550	4450.90	I
840	4453.32	I
950	4455.33	I
1100	4457.43	I
240	4468.50	II
530	4481.26	I
780	4512.74	I
1000	4518.03	I
1000	4522.80	I
780	4527.31	I
6000	4533.24	I
240	4533.97	II
3600	4534.78	I
2400	4535.58	I
1200	4535.92	I
1200	4536.05	I
720	4544.69	I
950	4548.77	I
240	4549.63	II
15	4549.84	III
950	4552.46	I
720	4555.49	I
240	4571.98	II
15 d	4572.20	III
950	4617.27	I
480	4623.09	I
720	4656.47	I
840	4667.59	I
950	4681.92	I
470	4840.87	I
400	4885.08	I
380	4899.91	I
5800	4981.73	I
4600	4991.07	I
4000	4999.51	I
3600	5007.21	I
3200 d	5014.19	I
840	5020.03	I
840	5022.87	I
1200	5035.91	I
840	5036.47	I
740	5038.40	I
1200	5039.95	I
1400	5064.66	I
1100	5173.75	I
1300	5192.98	I
1400	5210.39	I
17	5278.12	III
20	5398.93	IV
340	5512.53	I
270	5514.35	I
320	5514.54	I
250	5644.14	I
130	5675.44	I
95	5689.47	I
95	5715.13	I
85	5739.51	I
400	5866.46	I
230	5899.32	I
120	5918.55	I
150	5922.12	I
120	5941.76	I
300	5953.17	I
200	5965.84	I
270	5978.56	I
340	5999.04	I
110	6064.63	I
120	6085.23	I
120	6091.17	I
120	6126.22	I
17	6246.65	IV
380	6258.10	I
380	6258.70	I
300	6261.10	I
55	6546.28	I
65	6554.23	I
75	6556.07	I
18	6621.58	III
18	6667.99	III
80	6743.12	I
20	7072.64	III
18	7084.57	III
260	7209.44	I
130	7244.86	I
130	7251.72	I
120	7344.72	I
90	7357.74	I
60	7364.11	I
60	7978.88	I
55	8024.84	I
75	8364.24	I
100	8377.85	I
100	8382.54	I
75	8396.87	I
120	8412.36	I
170	8426.52	I
490	8434.94	I
240	8435.70	I
20	8466.87	III
90	8675.39	I

***Tungsten* W *Z* = 74**

Intensity	Wavelength/Å	
5800	2001.71	II
13000	2008.07	II
5100	2009.98	II
4100	2010.23	II
4100	2014.23	II
7300	2026.08	II
15000	2029.98	II
5300	2049.63	II
9700	2079.11	II
6100	2094.75	II
2100	2118.87	II
2400	2121.59	II
1500	2166.32	II
1300	2204.48	II
460	2249.80	I
510	2277.58	I
530 d	2294.49	I
340	2309.02	I
440	2313.17	I
460	2321.63	I
390 d	2326.56	I
320	2354.61	I
580	2360.44	I
850	2363.07	I
510	2374.47	I
670	2384.82	I
730	2397.09	II
560	2397.73	I
560	2397.98	I
1700 d	2405.58	I
610	2415.68	I
870	2424.21	I
1800	2435.96	I
580	2444.06	I
780	2451.48	II
870	2452.00	I
630	2454.98	I
780	2455.51	I
780	2456.53	I
1100	2459.30	I
1400	2466.85	I
480	2472.51	I
1200	2474.15	I
870	2480.13	I
1500	2481.44	I
480 d	2482.10	I
580	2484.74	I
390	2487.50	I
390	2489.23	II
630	2495.26	I
680	2504.70	I
75	2510.47	II
310	2520.46	I
780	2521.32	I
270	2522.04	II
780	2523.41	I
430	2527.76	I
780	2533.64	I

Intensity	Wavelength/Å	
1200	2547.14	I
780	2550.38	I
2700	2551.35	I
730	2561.97	I
870	2580.49	I
390	2584.39	I
390	2589.17	II
370	2601.96	I
680	2606.39	I
370	2608.32	I
970	2613.08	I
480	2613.82	I
400	2620.25	I
400	2622.21	I
400	2625.22	I
400	2632.48	I
400	2632.70	I
810	2633.13	I
400 d	2638.62	I
650	2646.18	I
400	2646.73	I
1600	2656.54	I
810	2662.84	I
810	2671.47	I
650	2677.28	I
2100	2681.42	I
650	2695.67	I
650	2699.59	I
400	2700.01	I
400	2706.58	I
400	2708.59	I
400 d	2708.80	I
400	2715.50	I
2100	2718.91	I
2600	2724.35	I
400	2725.03	I
650	2748.84	I
400	2762.34	I
400	2764.27	II
400	2769.74	I
810	2770.88	I
810	2774.00	I
810	2774.48	I
810	2792.70	I
400	2799.93	I
810	2818.06	I
1600	2831.38	I
810	2833.63	I
810	2848.02	I
1500	2896.44	I
690	2935.00	I
2400	2944.40	I
2400	2946.99	I
730 d	2979.71	I
360	3013.79	I
520	3016.47	I
770	3017.44	I
210	3024.93	I
310 d	3026.67	I
440 d	3041.73	I
270	3043.80	I

Intensity	Wavelength/Å	
440	3046.44	I
810	3049.69	I
180	3073.28	I
180 d	3084.83	I
370	3093.50	I
240	3107.23	I
240	3108.02	I
230	3117.57	I
260	3120.18	I
290	3163.42	I
320	3176.60	I
190	3181.82	I
390	3191.57	I
390	3198.84	I
520	3207.25	I
1000	3215.56	I
190	3232.49	I
210	3254.36	I
210	3259.66	I
210 d	3266.62	I
730	3300.82	I
440	3311.38	I
440	3326.20	I
440	3331.69	I
390	3373.75	I
230	3429.59	I
240	3443.00	I
400	3495.24	I
650	3545.22	I
240	3570.65	I
1900	3617.52	I
650	3682.08	I
400	3683.30	I
570	3688.06	I
810	3707.92	I
510	3757.92	I
680	3760.13	I
1000	3768.45	I
340	3773.71	I
1000	3780.77	I
290	3809.22	I
190	3810.38	I
260	3810.79	I
1400	3817.48	I
1100	3835.06	I
730	3846.22	I
1800	3867.99	I
730	3881.41	I
8600	4008.75	I
540	4015.22	I
910	4045.59	I
730	4069.95	I
5000	4074.36	I
1000	4102.70	I
540	4137.46	I
450	4171.17	I
220	4207.05	I
250	4219.37	I
540	4244.36	I
1400	4269.38	I
4100	4294.61	I

Intensity	Wavelength/Å	
2200	4302.11	I
200	4378.48	I
180	4384.85	I
200	4408.28	I
640	4484.19	I
170	4588.73	I
640	4659.87	I
640	4680.51	I
790	4843.81	I
380	4886.90	I
220	4982.59	I
820	5053.28	I
770	5224.66	I
220	5514.68	I
65	5648.37	I
55	5735.09	I
45	5801.85	I
40	5902.64	I
55	5947.57	I
55	5965.86	I
55	6012.78	I
40	6021.52	I
45	6292.02	I
35	6404.21	I
40	6445.12	I
17	6611.62	I
13	6678.42	I
15	6693.08	I
13	6984.27	I
15	7140.52	I
9	7162.64	I
11	7200.16	I
10	7278.24	I
15	7285.81	I
15	7296.55	I
10	7509.00	I
17	7569.92	I
17	7614.15	I
13	7688.97	I
11	7784.15	I
22	8017.19	I
22	8055.64	I
13	8123.82	I
10	8338.08	I
27	8585.11	I
10	8594.42	I
13	8865.53	I

***Uranium* U Z = 92**

Intensity	Wavelength/Å	
440	2565.41	II
610	2635.53	II
830	2793.94	II
870	2802.56	II
630	2807.05	II
630	2817.96	II
870	2821.12	II
680	2828.90	II
920	2832.06	II
970	2865.68	II
1200	2889.62	II
780	2906.80	II
780	2908.28	II

Intensity	Wavelength/Å	
580	2931.41	II
530 p	2940.37	II
1300	2941.92	II
830	2943.90	II
580	2956.06	II
580	2967.94	II
580	2971.06	II
530	2984.61	II
630	3022.21	II
630	3031.99	II
580	3050.20	II
630	3057.91	II
630	3062.54	II
580	3072.78	II
580	3093.01	II
580	3102.39	II
970	3111.62	II
530	3119.35	II
680	3124.95	II
530	3139.61	II
680	3149.24	II
530	3153.11	II
730	3229.50	II
680	3232.16	II
730	3291.33	II
1100	3305.89	II
730	3390.38	I
580	3424.56	II
580	3435.49	I
630	3466.30	I
680	3482.49	II
1600	3489.37	I
530	3496.41	II
630	3500.08	I
780	3507.34	I
1600	3514.61	I
630	3533.57	II
530	3540.47	II
1200	3550.82	II
680	3555.32	I
1200	3561.80	I
2300	3566.59	I
530	3569.08	I
630	3578.72	II
3200	3584.88	I
840	3638.20	I
2800	3670.07	II
1100	3701.52	II
600	3738.04	II
680	3746.42	II
950	3748.68	II
600	3751.17	I
1900	3782.84	II
570	3793.10	II
1900	3811.99	I
750	3826.51	II
2000	3831.46	II
1200	3839.63	I
2400	3854.64	II
4900	3859.57	II
1900	3865.92	II

Intensity	Wavelength/Å	
1500	3871.03	I
1000	3881.45	II
2200	3890.36	II
2000	3932.02	II
1200	3943.82	I
1200	3985.79	II
1000	4042.75	I
1600	4050.04	II
880	4062.54	II
2200	4090.13	II
810	4116.10	II
880	4153.97	I
1400	4171.59	II
1000	4241.67	II
600	4472.33	II
620	4543.63	II
170	4689.07	II
150	4756.81	I
110	5008.21	II
170	5027.38	I
80	5160.32	II
70	5280.38	I
80	5475.70	II
70	5480.26	II
70	5481.20	II
160	5492.95	II
70	5780.59	I
70	5798.53	II
230	5915.39	I
100	5976.32	I
90	6077.29	I
55	6372.46	I
90	6395.42	I
110	6449.16	I
90	6826.92	I
45	7533.93	I
50	7881.94	I
35	8445.39	I
75	8607.95	I
30	8757.76	I
100	10554.93	I
75	11167.84	I
100	11384.13	I
100	11859.42	I
100	11908.83	I
100	12250.46	I
100	13185.16	I
75	13306.23	I
100	13961.58	I
75	18634.43	I
75	21910.22	I

Vanadium V Z = 23

Intensity	Wavelength/Å	
20	225.46	V
20	251.66	V
20	286.84	V
35	483.01	V
50	633.94	III
200	677.34	IV
500	684.37	IV
400	737.85	IV
100	864.27	III
500	1006.46	III
500	1149.94	III
100	1426.65	IV
1000	1643.03	III
1000	1650.14	III
100	1680.20	V
1000	1694.78	III
1000	1760.07	III
1000	1788.26	III
1000	1794.60	III
1000	1812.19	III
300	1861.56	IV
500	1939.06	IV
400	1951.43	IV
500	1997.72	IV
2100	2092.44	I
500	2268.30	IV
1000	2292.86	III
2500	2330.42	III
2500	2371.06	III
1000	2382.46	III
240	2507.78	I
410	2526.22	I
210	2527.90	II
80 h	2570.72	IV
230	2574.02	I
250	2593.05	III
250	2595.10	III
80 h	2645.54	IV
180	2661.42	I
1100	2687.96	II
680	2700.94	II
530	2706.17	II
640	2715.69	II
180	2731.35	I
240	2864.36	I
900	2891.64	II
900	2892.66	II
1400	2893.32	II
900	2906.46	II
2400	2908.82	II
710	2923.62	I
2400	2924.02	II
1700	2924.64	II
900	2941.37	II
1100	2944.57	II
410	2962.77	I
600	2968.38	II
1200	3056.33	I
1400	3060.46	I
2400	3066.38	I
3800	3093.11	II
3000	3102.30	II
2600	3110.71	II
2000	3118.38	II
1500	3125.28	II
3200	3183.41	I
5300	3183.98	I
3800	3185.40	I
410	3187.71	II
530	3188.51	II
750	3190.68	II
1100	3267.70	II
900	3271.12	II
750	3276.12	II
80 h	3514.25	IV
560	3517.30	II
560	3533.68	I
560	3545.20	II
560	3556.80	II
560	3589.76	II
490	3592.02	II
560	3592.53	I
100	3679.86	III
1300	3688.07	I
1000	3690.28	I
1500	3692.22	I
1000	3695.86	I
3800	3703.58	I
1800	3704.70	I
320	3715.47	II
250	3727.34	II
280	3732.76	II
520	3790.32	I
1100	3794.96	I
570	3799.91	I
570	3803.47	I
1000	3813.49	I
1300	3818.24	I
1700	3828.56	I
2600	3840.75	I
1200	3855.37	I
3000	3855.84	I
1300	3864.86	I
1500	3875.08	I
700	3890.18	I
2400	3902.25	I
700	3909.89	I
540	3990.57	I
430	3998.73	I
170	4005.71	II
1100	4090.58	I
1800	4092.69	I
890	4095.49	I
2800	4099.80	I
590	4102.16	I
2800	4105.17	I
2300	4109.79	I
8900	4111.78	I
4300	4115.18	I
1800	4116.47	I
2000	4123.57	I
3100	4128.07	I
3100	4132.02	I
2300	4134.49	I
20	4200.32	V
360	4232.46	I
560	4268.64	I
460	4271.55	I
460	4276.96	I
430	4284.06	I
460	4330.02	I
510	4332.82	I
760	4341.01	I
1000	4352.87	I
12000	4379.24	I
7000	4384.72	I
4800	4389.97	I
3600	4395.23	I
1400	4400.58	I
2300	4406.64	I
2800	4407.64	I
3600	4408.20	I
4600	4408.51	I
640	4416.47	I
640	4421.57	I
640	4437.84	I
830	4441.68	I
640	4444.21	I
610	4452.01	I
1000	4459.76	I
2000	4460.29	I
610	4462.36	I
510	4577.17	I
640	4580.40	I
830	4586.36	I
1300	4594.11	I
230	4619.77	I
100	4635.18	I
130	4646.40	I
160	4670.49	I
130	4776.36	I
110	4786.51	I
130	4796.92	I
130	4807.53	I
130	4827.45	I
150	4831.64	I
120	4832.43	I
320	4851.48	I
480	4864.74	I
620	4875.48	I
740	4881.56	I
110	5128.53	I
110	5138.42	I
110	5192.99	I
110	5194.83	I
110	5234.07	I
110	5240.87	I
100	5401.93	I
140	5415.26	I
140	5584.50	I
100	5592.42	I
200	5624.60	I
400	5627.64	I
110	5657.44	I
110	5668.36	I
310	5670.85	I
1200	5698.52	I
920	5703.56	I
570	5706.98	I
850	5727.03	I
230	5731.25	I
230	5737.06	I

Intensity	Wavelength/Å	
450	6039.73	I
480	6081.44	I
1300	6090.22	I
600	6119.52	I
450	6199.19	I
450	6216.37	I
430	6230.74	I
710	6243.10	I
280	6251.82	I
130	6268.82	I
170	6274.65	I
200	6285.16	I
200	6292.83	I
170	6296.49	I
110	6531.43	I
65 c	6753.00	I
50 c	6766.49	I
40	6784.98	I
40	7338.92	I
35	7356.54	I
29 c	8027.39	I
120 w	8116.80	I
70 c	8161.07	I
60 c	8919.85	I

Xenon Xe Z = 54

Intensity	Wavelength/Å	
8	657.8	III
8	660.1	III
9	673.8	III
9	674.0	III
9	676.6	III
10	694.0	III
20	698.5	III
12	705.1	III
10	721.2	III
15	731.0	III
10	733.3	III
350	740.41	II
15	742.6	III
10	756.0	III
10	761.5	III
10	769.1	III
25	779.1	III
15	792.9	III
12	796.1	III
15	802.0	III
350	803.07	II
25	823.2	III
30	824.9	III
25	853.0	III
600	880.80	II
350	885.54	II
15	889.3	III
20	894.0	III
20	896.0	III
600	925.87	II
250	935.40	II
10	965.5	III
800	972.77	II
700	976.68	II
35	1003.4	III
35	1017.7	III
500	1032.44	II
700	1037.68	II
1100	1041.31	II
10	1047.8	III
1000	1048.27	II
1200	1051.92	II
12	1066.4	III
2000	1074.48	II
600	1083.86	II
1200	1100.43	II
30	1130.3	III
600	1158.47	II
250	1169.63	II
800 p	1183.05	II
250	1192.04	I
25	1232.1	III
600	1244.76	II
250	1250.20	I
1000	1295.59	I
600	1469.61	I
80	2668.98	III
100	2717.33	III
30	2814.45	III
40	2815.91	III
30	2827.45	III
40	2847.65	III
30	2862.40	III
200	2864.73	II
80 w	2871.10	III
60 w	2871.24	III
30	2871.7	III
150 h	2895.22	II
30	2896.62	III
50	2906.6	III
40	2911.89	III
80 w	2912.36	III
40	2940.2	III
60	2945.2	III
40	2947.5	III
40	2948.1	III
80 w	2970.47	III
400	2979.32	II
40	2992.87	III
30	3004.25	III
100 h	3017.43	II
100	3023.81	III
40	3083.5	III
50	3091.1	III
30	3106.46	III
300	3128.87	II
100 w	3138.3	III
80 c	3150.82	III
40	3185.2	III
100	3242.86	III
80	3268.98	III
30	3287.82	III
80 w	3301.55	III
40	3331.6	III
30	3358.0	III
200 h	3366.72	II
80	3384.12	III
2	3400.07	I
2	3418.37	I
2	3420.00	I
3	3442.66	I
60	3444.2	III
70	3454.2	III
100 w	3458.7	III
100 h	3461.26	II
40	3468.22	III
4	3469.81	I
4	3472.36	I
5	3506.74	I
80	3522.83	III
50	3542.3	III
10	3549.86	I
50	3552.1	III
10	3554.04	I
40	3561.4	III
100	3579.7	III
80	3583.6	III
100 w	3595.4	III
100	3606.06	III
40	3607.0	III
15	3610.32	I
8	3613.06	I
100 w	3615.9	III
40	3623.1	III
600	3624.08	III
6	3633.06	I
10	3669.91	I
50	3676.67	III
40	3685.90	I
40	3693.49	I
40	3776.3	III
300	3781.02	III
100	3841.5	III
200	3877.8	III
60	3880.5	III
100 l	3907.91	II
500	3922.55	III
300	3950.59	III
100	4037.59	II
200	4050.07	III
200 l	4057.46	II
60	4060.4	III
100 h	4098.89	II
100	4109.1	III
100	4145.7	III
200 l	4158.04	II
1000 h	4180.10	II
500 h	4193.15	II
300 h	4208.48	II
100 h	4209.47	II
300 h	4213.72	II
100	4215.60	II
300 h	4223.00	II
400 h	4238.25	II
500 h	4245.38	II
100 l	4251.57	II
30	4285.9	III
500 h	4296.40	II
500 h	4310.51	II
1000 l	4330.52	II
200 h	4369.20	II
100 l	4373.78	II
500 h	4393.20	II
500 l	4395.77	II
200 l	4406.88	II
150 l	4416.07	II
50	4434.2	III
500 h	4448.13	II
100 w	4462.1	III
1000 h	4462.19	II
500 l	4480.86	II
100 l	4521.86	II
100 w	4569.1	III
100 w	4570.1	III
100 w	4641.4	III
30	4673.7	III
60	4683.57	III
30	4723.60	III
600	4734.152	I
100 w	4757.3	III
150	4792.619	I
500	4807.02	I
400	4829.71	I
300	4843.29	I
40	4869.5	III
500	4916.51	I
500	4923.152	I
200 l	4971.71	II
400	4972.71	II
300	4988.77	II
100 l	4991.17	II
200	5028.280	I
200	5044.92	II
1000	5080.62	II
300	5122.42	II
100	5125.70	II
100	5178.82	II
300	5188.04	II
400	5191.37	II
100	5192.10	II
60	5239.0	III
500	5260.44	II
500	5261.95	II
2000	5292.22	II
300	5309.27	II
1000	5313.87	II
2000	5339.33	II
200	5363.20	II
30	5367.1	III
200	5368.07	II
500	5372.39	II
100	5392.80	I
50	5401.0	III
3000	5419.15	II
800	5438.96	II
300	5445.45	II
200	5450.45	II
400	5460.39	II
1000	5472.61	II

Intensity	Wavelength/Å	
100 l	5494.86	II
40	5524.4	III
200	5525.53	II
600	5531.07	II
100	5566.62	I
300	5616.67	II
300	5659.38	II
600	5667.56	II
150	5670.91	II
100	5695.75	I
200	5699.61	II
200	5716.10	II
500	5726.91	II
500	5751.03	II
300	5758.65	II
300	5776.39	II
100	5815.96	II
300	5823.89	I
150	5824.80	I
100	5875.02	I
300	5893.29	II
100	5894.99	I
200	5905.13	II
100	5934.17	I
500	5945.53	II
300	5971.13	II
2000	5976.46	II
200	6008.92	II
1000	6036.20	II
2000	6051.15	II
600	6093.50	II
1500	6097.59	II
400	6101.43	II
100	6115.08	II
100	6146.45	II
150	6178.30	I
120	6179.66	I
300	6182.42	I
500	6194.07	II
100	6198.26	I
60	6205.97	III
100	6220.02	II
25	6221.7	III
60	6238.2	III
60	6259.05	III
500	6270.82	II
400	6277.54	II
100	6284.41	II
100	6286.01	I
250	6300.86	II
500	6318.06	I
400	6343.96	II
600	6356.35	II
200	6375.28	II
100	6397.99	II
300	6469.70	I
150	6472.84	I
120	6487.76	I
100	6498.72	I
200 h	6504.18	I
300	6512.83	II

Intensity	Wavelength/Å	
200	6528.65	II
100	6533.16	I
1000	6595.01	II
100	6595.56	I
400	6597.25	II
100	6598.84	II
150	6668.92	I
300	6694.32	II
200	6728.01	I
150	6788.71	II
100	6790.37	II
1000	6805.74	II
200	6827.32	I
100	6872.11	I
300	6882.16	I
80	6910.22	II
100	6925.53	I
800 h	6942.11	II
100	6976.18	I
2000	6990.88	II
150	7082.15	II
500	7119.60	I
50 s	7147.50	II
200	7149.03	II
500	7164.83	II
100	7284.34	II
200	7301.80	II
200	7339.30	II
100	7386.00	I
150	7393.79	I
300	7548.45	II
200	7584.68	I
80	7618.57	II
500	7642.02	I
100	7643.91	I
200	7670.66	II
60	7787.04	II
100	7802.65	I
100	7881.32	I
300	7887.40	I
500	7967.34	I
100	8029.67	I
200	8057.26	I
150	8061.34	I
100	8101.98	I
150 h	8151.80	II
100	8171.02	I
700	8206.34	I
10000	8231.635	I
500	8266.52	I
7000	8280.116	I
2000	8346.82	I
100	8347.24	II
2000	8409.19	I
50 h	8515.19	II
200	8576.01	I
50 h	8604.23	II
250	8648.54	I
100	8692.20	I
200	8696.86	I
50 h	8716.19	II

Intensity	Wavelength/Å	
300	8739.39	I
100	8758.20	I
5000	8819.41	I
300	8862.32	I
200	8908.73	I
200	8930.83	I
1000	8952.25	I
100	8981.05	I
200	8987.57	I
400	9045.45	I
500	9162.65	I
100	9167.52	I
100	9374.76	I
200	9513.38	I
50 h	9591.35	II
150	9685.32	I
50 l	9698.68	II
100	9718.16	I
2000	9799.70	I
3000	9923.19	I
100	10838.37	I
90	11742.01	I
375	12235.24	I
100	12257.76	I
300	12590.20	I
2500	12623.391	I
250	13544.15	I
2000	13657.055	I
1250	14142.444	I
800	14240.96	I
375	14364.99	I
140	14660.81	I
3000	14732.806	I
100	15099.72	I
2500	15418.394	I
150	15557.13	I
250	15979.54	I
100	16039.90	I
1000	16053.28	I
125	16554.49	I
1500	16728.15	I
1500	17325.77	I
350	18788.13	I
150	20187.19	I
3000	20262.242	I
250	21470.09	I
1250	23193.33	I
110	23279.54	I
1800	24824.71	I
175	25145.84	I
2000	26269.08	I
2500	26510.86	I
250	28381.54	I
750	28582.25	I
300	29384.41	I
150	29448.06	I
100	29649.58	I
100	29813.62	I
600	30253.14	I
1500	30475.46	I
100	30504.12	I

Intensity	Wavelength/Å	
500	30794.18	I
6000	31069.23	I
125	31336.01	I
550	31607.91	I
100	32293.08	I
1800	32739.26	I
3500	33666.69	I
150	34014.67	I
450	34335.27	I
170	34744.00	I
5000	35070.25	I
110	35246.92	I
250	36209.21	I
150	36231.74	I
450	36508.36	I
850	36788.83	I
140	38685.98	I
175	38737.82	I
270	38939.60	I
120	39955.14	I

Ytterbium Yb Z = 70

Intensity	Wavelength/Å	
1000	1050.24	IV
1000	1054.46	IV
5000	1134.43	IV
900	1316.04	IV
800	1326.36	IV
900	1350.26	IV
80	1561.42	III
80 h	1765.21	III
800	1791.06	IV
100	1863.32	III
800	1873.91	III
500	1898.25	III
500	1998.82	III
900	2116.65	IV
2500	2116.67	II
800	2123.32	IV
3000	2126.74	II
800	2139.99	IV
20000	2144.77	IV
15000	2154.18	IV
370	2161.60	II
850	2185.71	II
640	2224.46	II
300	2240.11	III
300	2305.32	III
140	2320.81	II
170	2390.74	II
460	2464.50	I
140	2512.06	II
270	2538.67	II
2000	2567.61	III
1000	2579.57	III
800	2599.14	III
600	2621.11	III
1000	2642.56	III
1000	2651.74	III
700	2652.25	III
990	2653.75	II
200	2665.04	II
2000	2666.13	III

Intensity	Wavelength/Å	
2000	2666.99	III
390	2671.96	I
390	2672.66	II
170	2718.35	II
230	2748.66	II
1300	2750.48	II
170	2776.28	II
600	2795.60	III
1000	2803.43	III
600	2816.92	III
1000	2818.72	III
140	2821.15	II
190	2830.99	II
230 h	2847.18	II
360	2851.13	II
430	2859.80	II
140	2861.21	II
200	2867.06	II
45	2873.49	I
200	2888.04	II
3600	2891.38	II
600	2898.30	III
1000	2906.31	III
170	2914.21	II
140	2915.28	II
280	2919.35	II
35	2934.36	I
140	2945.91	II
2000	2970.56	II
200	2983.99	II
170	2994.80	II
800	2998.00	III
310	3005.77	II
160	3017.56	II
160	3026.67	II
2000	3029.49	III
920	3031.11	II
3000	3092.50	III
28	3100.74	I
170	3107.90	II
190	3117.81	II
4000	3126.01	III
1000	3138.58	III
230	3140.94	II
28	3162.29	I
800	3191.35	III
390	3192.88	II
240	3201.16	II
2000	3228.58	III
35	3239.58	I
18000	3289.37	II
130	3305.25	I
140	3305.73	II
80	3319.41	I
2000	3325.51	III
240	3337.17	II
280 d	3342.93	II
240	3375.48	II
2000	3384.01	III
140	3387.50	I
50	3412.45	I
140	3418.39	I
360	3426.04	I
240	3431.11	I
85	3452.40	I
500	3454.08	II
190 d	3458.29	II
360	3460.27	I
2400	3464.37	I
500	3476.30	II
500	3478.84	II
50	3517.00	I
230	3520.29	II
35	3559.03	I
200	3560.33	II
170	3560.70	II
360	3585.47	II
200	3619.80	II
240	3637.76	II
70	3648.15	I
90	3655.73	I
240	3669.69	II
140	3675.08	II
32000	3694.19	II
70	3700.58	I
400	3711.91	III
180	3734.69	I
550	3770.10	I
80	3774.32	I
60 h	3791.74	I
170	3839.91	I
340	3872.85	I
340	3900.85	I
140	3911.27	I
500	3931.23	III
32000	3987.99	I
930	3990.88	I
50	4007.36	I
2000	4028.14	III
70	4052.28	I
440	4089.68	I
170	4119.07	I
120	4174.56	I
340	4180.81	II
300	4213.64	III
150 d	4218.56	II
120	4231.97	I
70	4277.74	I
120	4305.97	I
60 h	4393.69	I
60 h	4430.21	I
440	4439.19	I
85 h	4482.42	I
100	4517.58	III
85 h	4563.95	I
640	4576.21	I
200	4582.36	I
70	4589.21	I
140	4590.83	I
40	4684.27	I
190	4726.08	II
170 h	4781.87	I
170	4786.61	II
35	4816.43	I
40	4837.46	I
40 h	4894.60	I
27	4912.36	I
710	4935.50	I
140	4966.90	I
30	5067.80	I
70	5069.14	I
220	5074.34	I
50	5076.74	I
60	5196.08	I
85	5211.60	I
100	5244.11	I
150 h	5277.04	I
170	5335.15	II
30 h	5351.29	I
150	5352.95	II
30	5363.66	I
40	5449.27	II
60	5481.92	I
40	5505.49	I
17	5524.54	I
85 h	5539.05	I
2400	5556.47	I
60	5651.98	II
220	5719.99	I
27	5771.66	II
35	5833.99	II
35	5837.14	II
27	5854.51	I
17	5989.33	I
40	5991.51	II
60	6152.57	II
60	6274.78	II
200	6328.52	III
35 h	6400.35	I
35 h	6417.91	I
340	6489.06	I
180	6667.82	I
25	6727.61	II
690	6799.60	I
9 h	7244.41	I
8 h	7305.22	I
10 h	7313.05	I
16 h	7350.04	I
25	7448.28	I
30 h	7527.46	I
750	7699.48	I
100	7971.46	III
70 h	8922.56	II
200	10110.60	III
100	10830.36	III

Yttrium Y Z = 39

Intensity	Wavelength/Å	
150	264.64	IV
150	273.03	IV
900	333.09	V
500	333.80	V
400	335.14	V
500	336.62	V
500	339.02	V
500	344.59	V
900	355.86	IV
300	370.42	IV
300	372.05	V
400	379.96	V
500	386.82	IV
300	403.45	V
300	420.74	V
600	425.03	IV
300	473.10	IV
4000	584.98	V
2000	630.97	V
5000	805.20	III
7000	809.92	III
15000	989.21	III
25000	996.37	III
5000	1314.51	III
4000	1334.04	III
4000	2068.98	III
10000	2127.98	III
16000	2191.16	III
350	2243.06	II
10000	2284.34	III
10000	2327.31	III
50	2354.20	I
50000	2367.23	III
40000	2414.64	III
560	2422.20	II
60	2694.21	I
95	2723.00	I
70	2742.53	I
140	2760.10	I
90000	2817.04	III
45	2822.56	I
70	2854.43	II
95	2886.48	I
160	2919.05	I
99000	2946.01	III
390	2948.40	I
350	2964.96	I
180	2971.59	I
750	2984.26	I
140	2996.94	I
130	3021.73	I
190	3045.37	I
95	3095.88	II
110	3173.06	II
220	3179.41	II
70	3191.31	I
2300	3195.62	II
2200	3200.27	II
2200	3203.32	II
3900	3216.69	II
6200	3242.28	II
4700	3327.89	II
85	3388.59	I
85	3412.47	I
170	3485.73	I
1700	3496.09	II
3900	3549.01	II
130	3551.80	I

Intensity	Wavelength/Å	
540	3552.69	I
170	3558.76	I
190	3571.43	I
260	3576.05	I
3300	3584.52	II
300	3587.75	I
100	3589.69	I
2800	3592.92	I
10000	3600.73	II
6200	3601.92	II
7800	3611.05	II
4300	3620.94	I
1900	3628.71	II
7800	3633.12	II
3000	3664.61	II
170	3692.53	I
13000	3710.30	II
1200	3747.55	II
10000	3774.33	II
1400	3776.56	II
7400	3788.70	II
1300	3818.35	II
4000	3832.88	II
80	3876.82	I
480	3878.28	II
4400	3950.36	II
3600	3982.60	II
940	4039.83	I
2400	4047.64	I
9400	4077.38	I
2000	4083.71	I
9900	4102.38	I
8900	4128.31	I
7500	4142.85	I
100 h	4157.63	I
2400	4167.52	I
2000	4174.14	I
8000	4177.54	II
160	4217.80	I
280 h	4220.63	I
600	4235.73	II
2200	4235.94	I
300	4251.20	I
360 h	4302.30	I
2800	4309.63	II
110	4330.78	I
440 h	4348.79	I
120	4357.73	I
800	4358.73	II
120	4366.03	I
12000	4374.94	II
150 h	4375.61	I
100	4387.74	I
1800	4398.02	II
890	4422.59	II
100	4443.66	I
130	4446.63	I
170	4475.72	I
180	4476.96	I
160	4477.45	I
110	4487.28	I
300	4487.47	I
500	4505.95	I
890	4527.25	I
440	4527.80	I
100	4544.32	I
100	4559.37	I
130	4596.55	I
95	4604.80	I
2000	4643.70	I
200 h	4658.32	I
2000	4674.84	I
180	4696.81	I
170	4728.53	I
160	4752.79	I
410	4760.98	I
120	4781.04	I
170	4786.89	I
180	4799.30	I
140	4819.64	I
120	4822.13	I
770	4839.87	I
550	4845.68	I
410	4852.69	I
120	4854.25	I
890	4854.87	II
330	4859.84	I
1900	4883.69	II
95	4893.44	I
1100	4900.12	II
100	4906.11	I
150	4921.87	I
120	4974.30	I
100	5006.97	I
75	5070.21	I
75	5072.19	I
1100	5087.42	II
180	5135.20	I
960	5200.41	II
1500	5205.72	II
10000	5238.10	III
180	5240.81	I
75	5380.62	I
220	5402.78	II
90	5424.37	I
190	5438.24	I
710	5466.46	I
100	5468.47	I
240	5497.41	II
300	5503.45	I
250	5509.90	II
120	5521.63	I
740	5527.54	I
120	5544.50	I
180	5577.42	I
620	5581.87	I
120	5606.33	I
560	5630.13	I
120	5644.69	I
120	5648.47	I
740	5662.94	II
90	5675.27	I
160	5706.73	I
90	5743.85	I
75	5765.64	I
100	5781.69	II
120	6009.19	I
120	6023.41	I
120	6135.04	I
150	6138.43	I
1200	6191.73	I
300	6222.59	I
1000	6435.00	I
90	6538.60	I
70	6557.39	I
95	6613.75	II
40	6650.61	I
150	6687.58	I
70	6700.71	I
190	6793.71	I
21	6815.16	I
45	6845.24	I
29	6887.22	I
24 h	6950.31	I
24	6979.88	I
29	7052.94	I
35	7191.66	I
35	7264.17	II
50	7346.46	I
29	7450.30	II
9000	7558.71	III
35	7563.13	I
29	7855.52	I
110	7881.90	II
10000	7991.43	III
24	8344.43	I
10000	8796.21	III
95	8800.62	I
19 h	8835.85	II

Zinc Zn Z = 30

Intensity	Wavelength/Å	
200	425.90	IV
200	428.54	IV
200	430.59	IV
1000	677.63	III
750	677.96	III
200	713.90	III
60	1193.23	II
50	1239.12	IV
50	1249.69	IV
500	1265.74	IV
500	1306.66	IV
200	1456.72	III
200	1459.98	IV
300	1499.42	III
300	1500.42	III
300	1505.92	III
300	1515.85	III
300	1552.30	III
90	1572.99	II
200	1629.19	III
200	1639.33	III
200	1673.05	III
80 d	1735.61	II
100	1767.69	III
100	1797.64	II
100 d	1811.05	II
100 d	1833.57	II
100	1864.12	II
100	1866.08	II
100	1872.13	II
100 d	1918.96	II
100 d	1929.67	II
100	1969.40	II
100	1982.11	II
100	1986.99	II
500	2025.48	II
500	2062.00	II
200	2064.23	II
120	2079.08	I
300	2099.94	II
200	2102.18	II
800 r	2138.56	I
1000	2501.99	II
150	2515.81	I
1000	2557.95	II
300	2582.49	I
200	2608.56	I
300	2608.64	I
200	2670.53	I
300	2684.16	I
300	2712.49	I
200	2756.45	I
300	2770.86	I
300	2770.98	I
400	2800.87	I
100	2801.06	I
200	3035.78	I
200	3072.06	I
300	3196.31	II
500 r	3282.33	I
800	3302.58	I
700 r	3302.94	I
800	3345.02	I
500	3345.57	I
50	3883.34	I
300	4680.14	I
400	4722.15	I
400	4810.53	I
800	4911.62	II
500	4924.03	II
200	5181.98	I
500	5894.33	II
500	6021.18	II
500	6102.49	II
500	6214.61	II
1000 h	6362.34	I
300	7588.5	II
300	7732.5	II
100	11054.25	I
100	13053.63	I
100	13150.59	I
100	14038.70	I
20	16483.45	I
20	16491.98	I

Intensity	Wavelength/Å	
20	16505.23	I
10	24375.02	I

***Zirconium* Zr *Z* = 40**

Intensity	Wavelength/Å	
500	304.01	V
60	480.66	IV
60	497.23	IV
60	500.22	IV
600	628.66	IV
500	633.56	IV
50	690.39	III
2000	740.61	V
10000	800.00	V
10000	806.89	V
10000	812.05	V
3000	841.40	V
300	863.65	IV
500	864.59	IV
9000	1183.97	IV
9000	1201.77	IV
10000	1219.86	IV
500	1303.93	V
500 p	1323.81	V
1000	1469.47	IV
10000	1546.17	IV
10000	1598.95	IV
5000	1607.95	IV
100	1612.00	III
700	1725.02	V
200	1790.19	III
150	1793.56	III
125	1798.13	III
600	1860.86	V
200	1940.25	III
600	2028.54	V
125	2070.43	III
200	2086.78	III
10000	2091.49	IV
10000	2092.36	IV
600	2132.42	V
10000	2163.68	IV
100	2175.80	III
100	2191.15	III
10000	2286.67	IV
100	2301.60	III
90	2539.65	I
570	2567.64	II
1600	2568.87	II
2100	2571.39	II
250	2620.56	III
200	2643.79	III
150	2664.26	III
1800	2678.63	II
90	2687.75	I
750	2700.13	II
1300	2722.61	II
800	2726.49	II
1400	2734.86	II
1100	2742.56	II
660	2745.86	II
660	2752.21	II
530	2758.81	II

Intensity	Wavelength/Å	
620	2814.90	I
390	2818.74	II
530	2825.56	II
710	2837.23	I
660	2844.58	II
350	2848.52	I
350	2851.97	II
340	2869.81	II
490	2875.98	I
300	2915.99	II
270	2918.24	II
320	2926.99	II
320	2948.94	II
320	2955.78	II
320	2960.87	I
320	2962.68	II
320	2968.96	II
320	2978.05	II
820	2985.39	I
320	3003.74	II
820	3011.75	II
350	3020.47	II
500	3028.04	II
880	3029.52	I
350 d	3036.39	II
690	3054.84	II
690	3106.58	II
350	3120.74	I
500	3129.18	II
500	3129.76	II
350	3132.07	I
690	3138.68	II
540	3164.31	II
880	3165.97	II
880	3182.86	II
540	3191.21	I
540	3212.01	I
760	3214.19	II
630	3231.69	II
630	3234.12	I
760	3241.05	II
1000	3273.05	II
1300	3279.26	II
880	3284.71	II
540	3305.15	II
880	3306.28	II
380	3322.99	II
380	3326.80	II
380	3334.25	II
760	3340.56	II
380	3344.79	II
760	3356.09	II
540	3357.26	II
380	3374.73	II
570	3387.87	II
760	3388.30	II
5700	3391.98	II
570	3393.12	II
570	3404.83	II
760	3410.25	II
380	3414.66	I

Intensity	Wavelength/Å	
1000	3430.53	II
4700	3438.23	II
600	3447.36	I
410	3457.56	II
820	3463.02	II
600	3471.19	I
1200	3479.39	II
1300	3481.15	II
4100	3496.21	II
820	3505.67	II
1000	3509.32	I
2000	3519.60	I
440	3525.81	II
440	3533.22	I
630	3542.62	II
1800	3547.68	I
630	3550.46	I
1800	3551.95	II
2100	3556.60	II
1100	3566.10	I
2100	3572.47	II
1100	3575.79	I
1300	3576.85	II
880	3586.29	I
3500	3601.19	I
690	3611.89	II
1100	3613.10	II
1100	3614.77	II
1100	3623.86	I
1100	3663.65	I
390	3671.27	II
800	3674.72	II
390	3697.46	II
960	3698.17	II
720	3709.26	II
560	3745.98	II
880	3751.60	II
480	3764.39	I
480	3766.72	I
340	3766.82	II
720	3780.54	I
560	3791.40	I
560	3822.41	I
2200	3835.96	I
1300	3836.76	II
550	3843.02	II
550	3847.01	I
550	3849.25	I
2900	3863.87	I
770	3864.34	I
990	3877.60	I
1500	3885.42	I
2900	3890.32	I
2000	3891.38	I
610	3921.79	I
1200	3929.53	I
940	3958.22	II
490	3966.66	I
990	3968.26	I
660	3973.50	I
770	3991.13	II

Intensity	Wavelength/Å	
770	3998.97	II
400	4023.98	I
770	4024.92	I
990	4027.20	I
400	4029.68	II
490	4030.04	I
400	4035.89	I
610	4043.58	I
490	4044.56	I
400	4045.61	II
610	4048.67	II
770	4055.03	I
600	4055.71	I
1500	4064.16	I
2000	4072.70	I
240	4078.31	I
2000	4081.22	I
400	4121.46	I
1200	4149.20	II
400	4161.21	II
400	4166.36	I
660	4187.56	I
400	4194.76	I
610	4199.09	I
610	4201.46	I
610	4208.98	II
400	4213.86	I
2000	4227.76	I
2000	4239.31	I
770	4240.34	I
770	4241.20	I
1200	4241.69	I
550	4282.20	I
550	4294.79	I
550	4341.13	I
1000	4347.89	I
290	4359.74	II
310	4360.81	I
350	4366.45	I
550	4507.12	I
610	4535.75	I
490	4542.22	I
490	4575.52	I
350	4602.57	I
700	4633.98	I
2300	4687.80	I
510	4688.45	I
1900	4710.08	I
1400	4739.48	I
870	4772.31	I
700	4815.63	I
250	5046.58	I
360	5064.91	I
470	5078.25	I
300	5155.45	I
200	5158.00	I
100	5191.60	II
270	5385.14	I
160	5664.51	I
160	5797.74	I
340	5879.80	I

Intensity	Wavelength/Å		Intensity	Wavelength/Å		Intensity	Wavelength/Å		Intensity	Wavelength/Å	
170	6045.85	I	300	6313.02	I	170	7103.72	I	790	8070.08	I
170	6121.91	I	150	6953.84	I	590	7169.09	I	390	8132.99	I
680	6127.44	I	150	6990.84	I	160	7944.61	I	280	8212.53	I
340	6134.55	I	540	7097.70	I	160	8005.27	I			
440	6143.20	I	280	7102.91	I	150	8063.09	I			

Sources of Data for Each Element

Numbers following the element name refer to the references on the following pages.

Actinium: 193
Aluminum: 6,8,81,89,127,144,146,227,228,282
Americium: 92
Antimony: 164,167,194,386,406
Argon: 190,203,204,219,367,368,372,373,374,375,414,421
Arsenic: 163,168,197,244,280
Astatine: 188
Barium: 1,78,111,252,259,277,279
Berkelium: 53,339
Beryllium: 15,44,73,102,115,134,135,171,175,198,335
Bismuth: 1,357,358,359,360,361
Boron: 66,69,74,94,104,171,221,222
Bromine: 42,122,124,139,142,240,243,246,248,249,250,316
Cadmium: 44,285,296,353,399
Calcium: 16,25,70,150,270
Californium: 52,331
Carbon: 22,66,211
Cerium: 1,136,166,261,305
Cesium: 78,82,154,155,200,201,259,263,325
Chlorine: 11,28,30,31,85,233,238,239
Chromium: 1,379,380,412
Cobalt: 1,100,125,159,236,276,291
Copper: 199,273,290,295,324
Curium: 51,332
Dysprosium: 1
Einsteinium: 333
Erbium: 1,301
Europium: 1,312
Fluorine: 68,169,224,225,226
Francium: 408
Gadolinium: 1,46,137,151,152
Gallium: 2,19,62,132,140,141,143,195,281
Germanium: 5,119,293,340,341,342
Gold: 38,72,234,393,395
Hafnium: 1,369,404,410,425
Helium: 16,94,173,183,317
Holmium: 1
Hydrogen: 214
Indium: 1,132,348,349,350,351,352,353,435,436
Iodine: 20,21,58,84,124,153,161,176,184
Iridium: 1
Iron: 56,63,71,101,105,138,174,278,381,382
Krypton: 61,121,123,147,208,232,366,390,409,417,421
Lanthanum: 1,78,79,220,309
Lead: 54,64,106,256,274,297,283,329,330
Lithium: 3,15,17,18,37,44,112,284,321,335
Lutetium: 1,148,310,401
Magnesium: 4,7,49,83,103,128,129,177,217,269,315,335
Manganese: 1,126,385,405,433
Mercury (198): 43,50,69,145,229,242
Mercury (Natural): 34,45,90,117,133,189,235,304,327,328,343
Molybdenum: 1,383,420
Neodymium: 1
Neon: 56,58,69,118,150,230,364,365,371,388,389,400,402,413,430
Neptunium: 93
Nickel: 1,294,415,416,422
Niobium: 1,392,407,431
Nitrogen: 66,107,108,212,213,318
Osmium: 1
Oxygen: 23,24,36,66,69,209,210,215
Palladium: 1,287,424
Phosphorus: 179,180,182,336
Platinum: 1,288
Plutonium: 91
Polonium: 47,48
Potassium: 32,59,60,75,76,86,150,160,172,268,314,322
Praseodymium: 1,149,306,308,337,338
Promethium: 196,260
Protactinium: 96
Radium: 253,254
Radon: 251
Rhenium: 1
Rhodium: 1,396
Rubidium: 12,109,130,241,257,258,262,264
Ruthenium: 1,423
Samarium: 1
Scandium: 1,88,150,298,323
Selenium: 9,80,181,216,245,247,275
Silicon: 87,170,237,292,319,320
Silver: 13,99,255,286,289,363,387,398
Sodium: 178,205,206,207,268,299,334
Strontium: 1,109,110,218,231,265,279,313
Sulfur: 29,144,202,209,210,266
Tantalum: 1,411,426
Technetium: 35
Tellurium: 1,344,345,346,347
Terbium: 1,302
Thallium: 1,195,348,354,355,356
Thorium: 1,97,98,156,157,165,434
Thulium: 1,307
Tin: 187,191,399,423
Titanium: 1,378,427,428
Tungsten: 1
Uranium: 1,303
Vanadium: 1,394,397,432
Xenon: 33,116,118,120,232,384,391,429
Ytterbium: 1,40,192,311
Yttrium: 1,77,265,419
Zinc: 39,55,113,131,185,186,370,376,377
Zirconium: 1,362,403,418

References

1. Meggers, W. F., Corliss, C. H., and Scribner, B. F., *Natl. Bur. Stand. (U.S.) Monogr.*, 145, Washington, D.C., 1975.
2. Aksenov, V. P. and Ryabtsev, A. N., *Opt. Spectrosc.*, 37, 860, 1970.
3. Andersen, N., Bickel, W. S., Carriveau, G. W., Jensen, K., and Veje, E., *Phys. Scr.*, 4, 113, 1971.
4. Andersson, E. and Johannesson, G. A., *Phys. Scr.*, 3, 203, 1971.
5. Andrew, K. L. and Meissner, K. W., *J. Opt. Soc. Am.*, 49, 146, 1959.
6. Artru, M. C. and Brillet, W. U. L., *J. Opt. Soc. Am.*, 64, 1063, 1974.
7. Artru, M. C. and Kaufman, V., *J. Opt. Soc. Am.*, 62, 949, 1972.
8. Artru, M. C. and Kaufman, V., *J. Opt. Soc. Am.*, 65, 594, 1975.
9. Badami, J. S. and Rao, K. R., *Proc. R. Soc. London*, 140(A), 387, 1933.
10. Baird, K. M. and Smith, D. S., *J. Opt. Soc. Am.*, 48, 300, 1958.
11. Bashkin, S. and Martinson, I., *J. Opt. Soc. Am.*, 61, 1686, 1971.
12. Beacham, J. R., Ph.D. thesis, Purdue University, 1970.
13. Benschop, H., Joshi, Y. N., and van Kleef, T. A. M., *Can. J. Phys.*, 53, 700, 1975.
14. Berry, H. G., Bromander, J., and Buchta, R., *Phys. Scr.*, 1, 181, 1970.
15. Berry, H. G., Bromander, J., Martinson, I., and Buchta, R., *Phys. Scr.*, 3, 63, 1971.
16. Berry, H. G., Desesquelles, J., and Dufay, M., *Phys. Rev. Sect A.*, 6, 600, 1972.
17. Berry, H. G., Desesquelles, J., and Dufay, M., *Nucl. Instrum. Methods*, 110, 43, 1973.
18. Berry, H. G., Pinnington, E. H., and Subtil, J. L., *J. Opt. Soc. Am.*, 62, 767, 1972.
19. Bidelman, W. P. and Corliss, C. H., *Astrophys. J.*, 135, 968, 1962.
20. Bloch, L. and Bloch, E., *Ann. Phys.* (Paris), 10(11), 141, 1929.
21. Bloch, L., Bloch, E., and Felici, N., *J. Phys. Radium*, 8, 355, 1937.
22. Bockasten, K., *Ark. Fys.*, 9, 457, 1955.
23. Bockasten, K., Hallin, R., Johansson, K. B., and Tsui, P., *Phys. Lett.* (Netherlands), 8, 181, 1964.
24. Bockasten, K. and Johansson, K. B., *Ark. Fys.*, 38, 563, 1969.
25. Borgstrom, A., *Ark. Fys.*, 38, 243, 1968.
26. Borgstrom, A., *Phys. Scr.*, 3, 157, 1971.
27. Bowen, I. S., *Phys. Rev.*, 29, 231, 1927.
28. Bowen, I. S., *Phys. Rev.*, 31, 34, 1928.
29. Bowen, I. S., *Phys. Rev.*, 39, 8, 1932.
30. Bowen, I. S., *Phys. Rev.*, 45, 401, 1934.
31. Bowen, I. S., *Phys. Rev.*, 46, 377, 1934.
32. Bowen, I. S., *Phys. Rev.*, 46, 791, 1934.
33. Boyce, J. C., *Phys. Rev.*, 49, 730, 1936.
34. Boyce, J. C. and Robinson, H. A., *J. Opt. Soc. Am.*, 26, 133, 1936.
35. Bozman, W. R., Meggers, W. F., and Corliss, C. H., *J. Res. Natl. Bur. Stand. Sect. A*, 71, 547, 1967.
36. Bromander, J., *Ark. Fys.*, 40, 257, 1969.
37. Bromander, J. and Buchta, R., *Phys. Scr.*, 1, 184, 1970.
38. Brown, C. M. and Ginter, M. L., *J. Opt. Soc. Am.*, 68, 243, 1978.
39. Brown, C. M., Tilford, S. G., and Ginter, M. L., *J. Opt. Soc. Am.*, 65, 1404, 1975.
40. Bryant, B. W., *Johns Hopkins Spectroscopic Report* No. 21, 1961.
41. Buchet, J. P., Buchet-Poulizac, M. C., Berry, H. G., and Drake, G. W. F., *Phys. Rev. Sect. A*, 7, 922, 1973.
42. Budhiraja, C. J. and Joshi, Y. N., *Can. J. Phys.*, 49, 391, 1971.
43. Burns, K. and Adams, K. B., *J. Opt. Soc. Am.*, 42, 56, 1952.
44. Burns, K. and Adams, K. B., *J. Opt. Soc. Am.*, 46, 94, 1956.
45. Burns, K., Adams, K. B., and Longwell, J., *J. Opt. Soc. Am.*, 40, 339, 1950.
46. Callahan, W. R., Ph.D. thesis, Johns Hopkins University, 1962.
47. Charles, G. W., *J. Opt. Soc. Am.*, 56, 1292, 1966.
48. Charles, G. W., Hunt, D. J., Pish, G., and Timma, D. L., *J. Opt. Soc. Am.*, 45, 869, 1955.
49. Codling, K., *Proc. Phys. Soc.*, 77, 797, 1961.
50. Comite Consulatif Pour La Definition du Metre, *J. Phys. Chem. Ref. Data*, 3, 852, 1974.
51. Conway, J. G., Blaise, J., and Verges, J., *Spectrochim. Acta Part B*, 31, 31, 1976.
52. Conway, J. G., Worden, E. F., Blaise, J., and Verges, J., *Spectrochim. Acta Part B*, 32, 97, 1977.
53. Conway, J. G., Worden, E. F., Blaise, J., Camus, P., and Verges, J., *Spectrochim. Acta Part B*, 32, 101, 1977.
54. Crooker, A. M., *Can. J. Res. Sect.* A, 14, 115, 1936.
55. Crooker, A. M. and Dick, K. A., *Can. J. Phys.*, 46, 1241, 1968.
56. Crosswhite, H. M., *J. Res. Natl. Bur. Stand. Sect. A*, 79, 17, 1975.
58. Crosswhite, H. M. and Dieke, G. H., *American Institute of Physics Handbook*, Section 7, 1972.
59. de Bruin, T. L., *Z. Phys.*, 38, 94, 1926.
60. de Bruin, T. L., *Z. Phys.*, 53, 658, 1929.
61. de Bruin, T. L., Humphreys, C. J., and Meggers, W. F., *J. Res. Natl. Bur. Stand.*, 11, 409, 1933.
62. Dick, K. A., *J. Opt. Soc. Am.*, 64, 702, 1973.
63. Dobbie, J. C., *Ann. Solar Phys. Observ.* (Cambridge), 5, 1, 1938.
64. Earls, L. T. and Sawyer, R. A., *Phys. Rev.*, 47, 115, 1935.
65. Edlen, B., *Z. Phys.*, 85, 85, 1933.
66. Edlen, B., *Nova Acta Reglae Soc. Sci. Ups.*, (IV) 9, No. 6, 1934.
67. Edlen, B., *Z. Phys.*, 93, 726, 1935.
68. Edlen, B., *Z. Phys.*, 94, 47, 1935.
69. Edlen, B., *Rep. Prog. Phys.*, 26, 181, 1963.
70. Edlen, B. and Risberg, P., *Ark. Fys.*, 10, 553, 1956.
71. Edlen, B. and Swings, P., *Astrophys. J.*, 95, 532, 1942.
72. Ehrhardt, J. C. and Davis, S. P., *J. Opt. Soc. Am.*, 61, 1342, 1971.
73. Eidelsberg, M., *J. Phys. B*, 5, 1031, 1972.
74. Eidelsberg, M., *J. Phys. B*, 7, 1476, 1974.
75. Ekberg, J. O. and Svensson, L. A., *Phys. Scr.*, 2, 283, 1970.
76. Ekefors, E., *Z. Phys.*, 71, 53, 1931.
77. Epstein, G. L. and Reader, J., *J. Opt. Soc. Am.*, 65, 310, 1975.
78. Epstein, G. L. and Reader, J., *J. Opt. Soc. Am.*, 66, 590, 1976.
79. Epstein, G. L. and Reader, J., unpublished.
80. Eriksson, K. B. S., *Phys. Lett. A.*, 41, 97, 1972.
81. Eriksson, K. B. S. and Isberg, H. B. S., *Ark. Fys.*, 23, 527, 1963.
82. Eriksson, K. B. S. and Wenaker, I., *Phys. Scr.*, 1, 21, 1970.
83. Esteva, J. M. and Mehlman, G., *Astrophys. J.*, 193, 747, 1974.
84. Even-Zohar, M. and Fraenkel, B. S., *J. Phys. B*, 5, 1596, 1972.
85. Fawcett, B. C., *J. Phys. B*, 3, 1732, 1970.
86. Fawcett, B. C., Culham Laboratory Report ARU-R4, 1971.
87. Ferner, E., *Ark. Mat. Astron. Fys.*, 28(A), 4, 1941.
88. Fischer, R. A., Knopf, W. C., and Kinney, F. E., *Astrophys. J.*, 130, 683, 1959.
89. Fowler, A., *Report on Series in Line Spectra*, Fleetway Press, London, 1922.
90. Fowles, G. R., *J. Opt. Soc. Am.*, 44, 760, 1954.
91. Fred, M., *Argonne Natl. Lab.*, unpublished, 1977.
92. Fred, M. and Tomkins, F. S., *J. Opt. Soc. Am.*, 47, 1076, 1957.
93. Fred, M., Tomkins, F. S., Blaise, J. E., Camus, P., and Verges, J., Argonne National Laboratory Report No. 76–68, 1976.
94. Garcia, J. D. and Mack, J. E., *J. Opt. Soc. Am.*, 55, 654, 1965.
96. Giacchetti, A., *Argonne Natl. Lab.*, unpublished, 1975.
97. Giacchetti, A., Blaise, J., Corliss, C. H., and Zalubas, R., *J. Res. Natl. Bur. Stand. Sect. A*, 78, 247, 1974.
98. Giacchetti, A., Stanley, R. W., and Zalubas, R., *J. Opt. Soc. Am.*, 69, 474, 1970.
99. Gilbert, W. P., *Phys. Rev.*, 47, 847, 1935.
100. Gilroy, H. T., *Phys. Rev.*, 38, 2217, 1931.
101. Glad, S., *Ark. Fys.*, 10, 291, 1956.
102. Goldsmith, S., *J. Phys. B*, 2, 1075, 1969.
103. Goorvitch, D., Mehlmam-Balloffet, G., and Valero, F. P. J., *J. Opt. Soc. Am.*, 60, 1458, 1970.
104. Goorvitch, D. and Valero, F. P. J., *Astrophys.* J., 171, 643, 1972.
105. Green, L. C., *Phys. Rev.*, 55, 1209, 1939.
106. Gutman, F., *Diss. Abstr. Int.* B, 31, 363, 1970.
107. Hallin, R., *Ark. Fys.*, 31, 511, 1966.
108. Hallin, R., *Ark. Fys.*, 32, 201, 1966.
109. Hansen, J. E. and Persson, W., *J. Opt. Soc. Am.*, 64, 696, 1974.
110. Hansen, J. E. and Persson, W., *Phys. Scr.*, 13, 166, 1976.
111. Hellintin, P., *Phys. Scr.*, 13, 155, 1976.
112. Herzberg, G. and Moore, H. R., *Can. J. Phys.*, 37, 1293, 1959.
113. Hetzler, C. W., Boreman, R. W., and Burns, K., *Phys. Rev.*, 48, 656, 1935.
114. Holmstrom, J. E. and Johansson, L., *Ark. Fys.*, 40, 133, 1969.

115. Hontzeas, S., Martinson, I., Erman, P., and Buchta, R., *Nucl. Instrum. Methods*, 110, 51, 1973.
116. Humphreys, C. J., *J. Res. Natl. Bur. Stand.*, 22, 19, 1939.
117. Humphreys, C. J., *J. Opt. Soc. Am.*, 43, 1027, 1953.
118. Humphreys, C. J., *J. Phys. Chem. Ref. Data*, 2, 519, 1973.
119. Humphreys, C. J. and Andrew, K. L., *J. Opt. Soc. Am.*, 54, 1134, 1964.
120. Humphreys, C. J. and Meggers, W. F., *J. Res. Natl. Bur. Stand.*, 10, 139, 1933.
121. Humphreys, C. J. and Paul, E., Jr., *J. Opt. Soc. Am.*, 60, 200, 1970.
122. Humphreys, C. J. and Paul, E., Jr., *J. Opt. Soc. Am.*, 62, 432, 1972.
123. Humphreys, C. J., Paul, E., Jr., Cowan, R. D., and Andrew, K. L., *J. Opt. Soc. Am.*, 57, 855, 1967.
124. Humphreys, C. J., Paul, E., Jr., and Minnhagen, L., *J. Opt. Soc. Am.*, 61, 110, 1971.
125. Iglesias, L., Inst. of Optics, Madrid, unpublished, 1977.
126. Iglesias, L. and Velasco, R., *Publ. Inst. Opt. Madrid*, No. 23, 1964.
127. Isberg, B., *Ark. Fys.*, 35, 551, 1967.
128. Johannesson, G. A., Lundstrom, T., and Minnhagen, L., *Phys. Scr.*, 6, 129, 1972.
129. Johannesson, G. A. and Lundstrom, T., *Phys. Scr.*, 8, 53, 1973.
130. Johansson, I. *Ark. Fys.*, 20, 135, 1961.
131. Johansson, I. and Contreras, R., *Ark. Fys.*, 37, 513, 1968.
132. Johansson, I. and Litzen, U., *Ark. Fys.*, 34, 573, 1967.
133. Johansson, I. and Svensson, K. F., *Ark. Fys.*, 16, 353, 1960.
134. Johansson, L., *Ark. Fys.*, 20, 489, 1961.
135. Johansson, L., *Ark. Fys.*, 23, 119, 1963.
136. Johansson, S. and Litzen, U., *Phys. Scr.*, 6, 139, 1972.
137. Johansson, S. and Litzen, U., *Phys. Scr.*, 8, 43, 1973.
138. Johansson, S. and Litzen, U., *Phys. Scr.*, 10, 121, 1974.
139. Joshi, Y. N., St. Francis Xavier Univ., Nova Scotia, unpublished.
140. Joshi, Y. N., Bhatia, K. S., and Jones, W. E., *Sci. Light Tokyo*, 21, 113, 1972.
141. Joshi, Y. N., Bhatia, K. S., and Jones, W. E., *Spectrochim. Acta Part B*, 28, 149, 1973.
142. Joshi, Y. N. and Budhiraja, C. J., *Can. J. Phys.*, 49, 670, 1971.
143. Joshi, Y. N. and van Kleef, T. A. M., *Can. J. Phys.*, 52, 1891, 1974.
144. Kaufman, V., *Natl. Bur. Stand.*, unpublished.
145. Kaufman, V., *J, Opt. Soc. Am.*, 52, 866, 1962.
146. Kaufman, V., Artru, M. C., and Brillet, W. U. L., *J. Opt. Soc. Am.*, 64, 197, 1974.
147. Kaufman, V. and Humphreys, C. J., *J. Opt. Soc. Am.*, 59, 1614, 1969.
148. Kaufman, V. and Sugar, J., *J. Opt. Soc. Am.*, 61, 1693, 1971.
149. Kaufman, V. and Sugar, J., *J. Res. Natl. Bur. Stand. Sect. A*, 71, 583, 1967.
150. Kelly, R. L. and Palumbo, L. J., *Naval Research Laboratory Report 7599*, Washington, DC., 1973.
151. Kielkopf, J. F., *Univ. of Louisville*, unpublished. 1975.
152. Kielkopf, J. F., *Univ. of Louisville*, unpublished, 1976.
153. Kiess, C. C. and Corliss, C. H., *J. Res. Natl. Bur. Stand. Sect. A*, 63, 1, 1959.
154. Kleiman, H., *J. Opt. Soc. Am.*, 52, 441, 1962.
155. Eriksson, K. B., Johansson, I., and Norlen, G., *Ark. Fys.*, 28, 233, 1964.
156. Klinkenberg, P. F. A., *Physica*, 15, 774, 1949.
157. Klinkenberg, P. F. A., *Physica*, 16, 618, 1950.
158. Krishnamurty, S. G., *Proc. Phys. Soc. London*, 48, 277, 1936.
159. Kruger, P. G. and Gilroy, H. T., *Phys. Rev.*, 48, 720, 1935.
160. Kruger, P. G. and Pattin, H. S., *Phys. Rev.*, 52, 621, 1937.
161. Lacroute, P., *Ann. Phys.* (Paris), 3, 5, 1935.
162. Lang, R. J., *Phys. Rev.*, 30, 762, 1927.
163. Lang, R. J., *Phys. Rev.*, 32, 737, 1928.
164. Lang, R. J., *Phys. Rev.*, 35, 445, 1930.
165. Lang, R. J., *Can. J. Res. Sect. A*, 14, 43, 1936.
166. Lang, R. J., *Can. J. Res. Sect. A*, 14, 127, 1936.
167. Lang, R. J. and Vestine, E. H., *Phys. Rev.*, 42, 233, 1932.
168. Li, H. and Andrew, K. L., *J. Opt. Soc. Am.*, 61, 96, 1971.
169. Liden, K., *Ark. Fys.*, 1, 229, 1949.
170. Litzen, U., *Ark. Fys.*, 28, 239, 1965.
171. Litzen, U., *Phys. Scr.*, 1, 251, 1970.
172. Litzen, U., *Phys. Scr.*, 1, 253, 1970.
173. Litzen, U., *Phys. Scr.*, 2, 103, 1970.
174. Litzen, U. and Verges, I., *Phys. Scr.*, 13, 240, 1976.
175. Lofstrand, B., *Phys. Scr.*, 8, 57, 1973.
176. Luc-Koenig, E., Morillon, C., and Verges, J., *Phys. Scr.*, 12, 199, 1975.
177. Lundstrom, T., *Phys. Scr.*, 7, 62, 1973.
178. Lundstrom, T. and Minnhagen, L., *Phys. Scr.*, 5, 243, 1972.
179. Magnusson, C. E. and Zetterberg, P. O., *Phys. Scr.*, 10, 177, 1974.
180. Magnusson, C. E., and Zetterberg, P. O., *Phys. Scr.*, 15, 237, 1977.
181. Martin, D. C., *Phys. Rev.*, 48. 938, 1935.
182. Svendenius, N., *Phys. Scr.*, 22, 240, 1980.
183. Martin, W. C., *J. Res. Natl. Bur. Stand. Sect. A*, 64, 19, 1960.
184. Martin, W. C. and Corliss, C. H., *J. Res. Natl. Bur. Stand. Sect. A*, 64, 443, 1960.
185. Martin, W. C. and Kaufman, V., *J. Res. Natl. Bur. Stand. Sect. A*, 74, 11, 1970.
186. Martin, W. C. and Kaufman, V., *J. Opt. Soc. Am.*, 60, 1096, 1970.
187. McCormick, W. W. and Sawyer, R. A., *Phys. Rev.*, 54, 71, 1938.
188. McLaughlin, R., *J. Opt. Soc. Am.*, 54, 965, 1964.
189. McLennan, J. C., McLay, A. B., and Crawford, M. F., *Proc. R. Soc. London Ser. A*, 134, 41, 1931.
190. Meissner, K. W., *Z. Phys.*, 39, 172, 1926.
191. Meggers, W. F., *J. Res. Natl. Bur. Stand.*, 24, 153, 1940.
192. Meggers, W. F. and Corliss, C. H., *J. Res. Natl. Bur. Stand. Sect. A*, 70, 63, 1966.
193. Meggers, W. F., Fred, M., and Tomkins, F. S., *J. Res. Natl. Bur. Stand.*, 58, 297, 1957.
194. Meggers, W. F. and Humphreys, C. J., *J. Res. Natl. Bur. Stand.*, 28, 463, 1942.
195. Meggers, W. F. and Murphy, R. J., *J. Res. Natl. Bur. Stand.*, 48, 334, 1952.
196. Meggers, W. F., Scribner, B. F., and Bozman, W. R., *J. Res. Natl. Bur. Stand.*, 46, 85, 1951.
197. Meggers, W. F., Shenstone, A. G., and Moore, C. E., *J. Res. Natl. Bur. Stand.*, 45, 346, 1950.
198. Mehlman, G. and Esteva, J. M., *Astrophys. J.*, 188, 191, 1974.
199. Meinders, E., *Physica*, 84(C), 117, 1976.
200. Sansonetti, C. J., Dissertation, Purdue University, 1981.
201. Sansonetti, C. J., *Natl. Bur. Stand. (U.S.)*, unpublished.
202. Millikan, R. A. and Bowen, I. S., *Phys. Rev.*, 25, 600, 1925.
203. Minnhagen, L., *J. Opt. Soc. Am.*, 61, 1257, 1925.
204. Minnhagen, L., *J. Opt. Soc. Am.*, 63, 1185, 1973.
205. Minnhagen, L., *Phys. Scr.*, 11, 38, 1975.
206. Minnhagen, L., *J. Opt. Soc. Am.*, 66, 659, 1976.
207. Minnhagen, L. and Nietsche, H., *Phys. Scr.*, 5, 237, 1972.
208. Minnhagen. L., Strihed, H., and Petersson, B., *Ark. Fys.*, 39, 471, 1969.
209. Moore, C. E., *Natl. Bur. Stand. (U.S.) Circ.*, 488, 1950.
210. Moore, C. E., *Revised Multiplet Table*, Princeton University Observatory No. 20, 1945.
211. Moore, C. E., National Standard Reference Data Series - National Bureau of Standards 3, Sect. 3, 1970.
212. Moore, C. E., National Standard Reference Data Series - National Bureau of Standards 3, Sect. 4, 1971.
213. Moore, C. E., National Standard Reference Data Series - National Bureau of Standards 3, Sect. 5, 1975.
214. Moore, C. E., National Standard Reference Data Series - National Bureau of Standards 3, Sect. 6, 1972.
215. Moore, C. E., National Standard Reference Data Series - National Bureau of Standards 3, Sect. 7, 1975.
216. Morillon, C. and Verges, J., *Phys. Scr.*, 10, 227, 1974.
217. Newsom, G. H., *Astrophys. J.*, 166, 243, 1971.
218. Newsom, G. H., O'Connor, S., and Learner, R. C. M., *J. Phys. B*, 6, 2162, 1973.
219. Norlen, G., *Phys. Scr.*, 8, 249, 1973.
220. Odabasi, H., *J. Opt. Soc. Am.*, 57, 1459, 1967.
221. Olme, A., *Ark. Fys.*, 40, 35, 1969.
222. Olme, A., *Phys. Scr.*, 1, 256, 1970.
223. Johansson, S., and Litzen, U., *J. Opt. Soc. Am.*, 61, 1427, 1971.
224. Palenius, H. P., *Ark. Fys.*, 39, 15, 1969.
225. Palenius, H. P., *Phys. Scr.*, 1, 113, 1970.
226. Palenius, H. P., *Univ. of Lund, Sweden*, unpublished.
227. Paschen, F., *Ann. Phys.*, Series 5, 12, 509, 1932.
228. Paschen, F. and Ritschl, R., *Ann. Phys.*, Series 5, 18, 867, 1933.

229. Peck, E. R., Khanna, B. N., and Anderholm, N. C., *J. Opt. Soc. Am.*, 52, 53, 1962.
230. Persson, W., *Phys. Scr.*, 3, 133, 1971.
231. Persson, W. and Valind, S., *Phys. Scr.*, 5, 187, 1972.
232. Petersson, B., *Ark. Fys.*, 27, 317, 1964.
233. Phillips, L. W. and Parker, W. L., *Phys. Rev.*, 60, 301, 1941.
234. Platt, J. R. and Sawyer, R. A., *Phys. Rev.*, 60, 866, 1941.
235. Plyer, E. K., Blaine, L. R., and Tidwell, E., *J. Res. Natl. Bur. Stand.*, 55, 279, 1955.
236. Poppe, R., van Kleef, T. A. M., and Raassen, A. J. J., *Physica*, 77, 165, 1974.
237. Radziemski, L. J., Jr. and Andrew, K. L., *J. Opt. Soc. Am.*, 55, 474, 1965.
238. Radziemski, L. J., Jr. and Kaufman, V., *J. Opt. Soc. Am.*, 59, 424, 1969.
239. Radziemski, L. J., Jr. and Kaufman, V., *J. Opt. Soc. Am.*, 64, 366, 1974.
240. Ramanadham, R. and Rao, K. R., *Indian J. Phys.*, 18, 317, 1944.
241. Ramb, R., *Ann. Phys.*, 10, 311, 1931.
242. Rank, D. H., Bennett, J. M., and Bennett, H. E., *J. Opt. Soc. Am.*, 40, 477, 1950.
243. Rao, A. S. and Krishnamurty, S. G., *Proc. Phys. Soc. London*, 46, 531, 1943.
244. Rao, K. R., *Proc. R. Soc. London, Ser. A*, 134, 604, 1932.
245. Rao, K. R. and Badami, *J. S., Proc. R. Soc. London Ser. A*, 131, 154, 1931.
246. Rao, K. R. and Krishnamurty, S. G., *Proc. R. Soc. London Ser. A*, 161, 38, 1937.
247. Rao, K. R. and Murti, S. G. K., *Proc. R. Soc. London* Ser. A, 145, 681, 1934.
248. Rao, Y. B., *Indian J. Phys.*, 32, 497, 1958.
249. Rao, Y. B., *Indian J. Phys.*, 33, 546, 1959.
250. Rao, Y. B., *Indian J. Phys.*, 35, 386, 1961.
251. Rasmussen, E., *Z. Phys.*, 80, 726, 1933.
252. Rasmussen, E., *Z. Phys.*, 83, 404, 1933.
253. Rasmussen, E., *Z. Phys.*, 86, 24, 1934.
254. Rasmussen, E., *Z. Phys.*, 87, 607, 1934.
255. Rasmussen, E., *Phys. Rev.*, 57, 840, 1940.
256. Rau, A. S. and Narayan, A. L., *Z. Phys.*, 59, 687, 1930.
257. Reader, J., *J. Opt. Soc. Am.*, 65, 286, 1975.
258. Reader, J., *J. Opt. Soc. Am.*, 65, 988, 1975.
259. Reader, J., *J. Opt Soc. Am.*, 73, 349, 1983.
260. Reader, J. and Davis, S., *J. Res. Natl. Bur. Stand. Sect. A*, 71, 587, 1967, and unpublished.
261. Reader, J. and Ekberg, J. O., *J. Opt. Soc. Am.*, 62, 464, 1972.
262. Reader, J. and Epstein, G. L., *J. Opt. Soc. Am.*, 62, 1467, 1972.
263. Reader, J. and Epstein, G. L., *J. Opt. Soc. Am.*, 65, 638, 1975.
264. Reader, J. and Epstein, G. L., *Natl. Bur. Stand.*, unpublished.
265. Reader, J., Epstein, G. L., and Ekberg, J. O., *J. Opt. Soc. Am.*, 62, 273, 1972.
266. Kaufman, V., *Phys. Scr.*, 26, 439, 1982.
267. Ricard, R., Givord, M., and George, F., *C. R. Acad. Sci. Paris*, 205, 1229, 1937.
268. Risberg, P., *Ark. Fys.*, 10, 583, 1956.
269. Risberg, G., *Ark. Fys.*, 28, 381, 1965.
270. Risberg, G., *Ark. Fys.*, 37, 231, 1968.
271. Robinson, H. A., *Phys. Rev.*, 49, 297, 1936.
272. Robinson, H. A., *Phys. Rev.*, 50, 99, 1936.
273. Ross, C. B., Jr., Doctoral dissertation, Purdue University, 1969.
274. Ross, C. B., Wood, D. R., and Scholl, P. S., *J. Opt. Soc. Am.*, 66, 36, 1976.
275. Ruedy, J. E. and Gibbs, R. C., *Phys. Rev.*, 46, 880, 1934.
276. Russell, H. N., King, R. B., and Moore, C. E., *Phys. Rev.*, 58, 407, 1940.
277. Russell, H. N. and Moore, C. E., *J. Res. Natl. Bur. Stand.*, 55, 299, 1955.
278. Russell, H. N., Moore, C. E., and Weeks, D. W., *Trans. Am. Philos. Soc.*, 34(2), 111, 1944.
279. Saunders, F., Schneider, E., and Buckingham, E., *Proc. Natl. Acad. Sci.*, 20, 291, 1934.
280. Sawyer, R. A. and Humphreys, C. J., *Phys. Rev.*, 32, 583, 1928.
281. Sawyer, R. A. and Lang, R. J., *Phys. Rev.*, 34, 712, 1929.
282. Sawyer, R. A. and Paschen, F., *Ann. Phys.*, 84(4),1, 1927.
283. Scholl, P. S., M.S. thesis, Wright State Univ., 1975.
284. Schurmann, D., *Z. Phys.*, 17, 4, 1975.
285. Seguier, J., *C. R. Acad. Sci. Paris*, 256, 1703, 1963.
286. Shenstone, A. G., *Phys. Rev.*, 31, 317, 1928.
287. Shenstone, A. G., *Phys. Rev.*, 32, 30, 1928.
288. Shenstone, A. G., *Trans. R. Soc. London*, 237(A), 57, 1938.
289. Shenstone, A. G., *Phys. Rev.*, 57, 894, 1940.
290. Shenstone, A. G. *Philos. Trans. R. Soc. London Ser. A*, 241, 297, 1948.
291. Shenstone, A. G., *Can. J. Phys.*, 38, 677, 1960.
292. Shenstone, A. G., *Proc. R. Soc. London*, 261(A), 153, 1961.
293. Shenstone, A. G., *Proc. R. Soc. London*, 276(A), 293, 1963.
294. Shenstone, A. G., *J. Res. Natl. Bur. Stand. Sect. A*, 74, 801, 1970.
295. Shenstone, A. G., *J. Res. Natl. Bur. Stand. Sect. A*, 79, 497, 1975.
296. Shenstone, A. G. and Pittenger, J. T., *J. Opt. Soc. Am.*, 39, 219, 1949.
297. Smith, S., *Phys. Rev.*, 36, 1, 1930.
298. Smitt, R., *Phys. Scr.*, 8, 292, 1973.
299. Soderqvist, J., *Ark. Mat. Astronom. Fys.*, 32(A), 1, 1946.
300. Sommer, L. A., *Ann. Phys.*, 75, 163, 1924.
301. Spector, N., *J. Opt. Soc. Am.*, 63, 358, 1973.
302. Spector, N. and Sugar, J., *J. Opt. Soc. Am.*, 66, 436, 1976.
303. Steinhaus, D. W., Radziemski, L. J., Jr., and Blaise, J., *Los Alamos Sci. Lab.*, unpublished, 1975.
304. Subbaraya, T. S., *Z. Phys.*, 78, 541, 1932.
305. Sugar, J., *J. Opt. Soc. Am.*, 55, 33, 1965.
306. Sugar, J., *J. Res. Natl. Bur. Stand. Sect. A*, 73, 333, 1969.
307. Sugar, J., *J. Opt. Soc. Am.*, 60, 454, 1970.
308. Sugar, J., *J. Res. Natl. Bur. Stand. Sect. A*, 78, 555, 1974.
309. Sugar, J. and Kaufman, V., *J. Opt. Soc. Am.*, 55, 1283, 1965.
310. Sugar, J. and Kaufman, V., *J. Opt. Soc. Am.*, 62, 562, 1972.
311. Sugar, J., Kaufman, V., and Spector, N., *J. Res. Natl. Bur. Stand., Sect. A*, 83, 233, 1978.
312. Sugar, J. and Spector, N., *J. Opt. Soc. Am.*, 64, 1484, 1974.
313. Sullivan, F. J. *Univ. Pittsburgh Bull.*, 35, 1, 1938.
314. Svensson, L. A. and Ekberg, J. O., *Ark. Fys.*, 37, 65, 1968.
315. Swensson, J. W. and Risberg, G., *Ark. Fys.*, 31, 237, 1966.
316. Tech, J. L., *J. Res. Natl. Bur. Stand. Sect. A*, 67, 505, 1963.
317. Tech, J. L. and Ward, J. F., *Phys. Rev. Lett.*, 27, 367, 1971.
318. Tilford, S. G., *J. Opt. Soc. Am.*, 53, 1051, 1963.
319. Toresson, Y. G., *Ark. Fys.*, 17, 179, 1960.
320. Toresson, Y. G., *Ark. Fys.*, 18, 389, 1960.
321. Toresson, Y. G. and Edlen, B., *Ark. Fys.*, 23, 117, 1963.
322. Tsien, W. Z., *Chin. J. Phys.*, Peiping, 3, 117, 1939.
323. van Deurzen, C. H. H., Conway, J., and Davis, S. P., *J. Opt. Soc. Am.*, 63, 158, 1973.
324. van Kleef, T. A. M., Raassen, A. J. J., and Joshi, Y. N., *Physica*, 84(C), 401, 1976.
325. Sansonetti, C. J., Andrew, K. L., and Verges, J., *J. Opt. Soc. Am.*, 71, 423, 1981.
326. Wheatley, M. A. and Sawyer, R. A., *Phys. Rev.*, 61, 591, 1942.
327. Wilkinson, P. G., *J. Opt. Soc. Am.*, 45, 862, 1955.
328. Wilkinson, P. G. and Andrew, K. L., *J. Opt. Soc. Am.*, 53, 710, 1963.
329. Wood, D. and Andrew, K. L., *J. Opt. Soc. Am.*, 58, 818, 1968.
330. Wood, D. R., Ron, C. B., Scholl, P. S., and Hoke, M., *J. Opt. Soc. Am.*, 64, 1159, 1974.
331. Worden, E. F. and Conway, J. G., *Lawrence Livermore Lab.*, unpublished, 1977.
332. Worden, E. F., Hulet, E. K., Gutmacher, R. G., Conway, J. G., *At. Data Nucl. Data Tables*, 18, 459, 1976.
333. Worden, E. F., Lougheed, R. W., Gutmacher, R. G., and Conway, J. G., *J. Opt. Soc. Am.*, 64, 77, 1974.
334. Wu, C. M., Ph.D. thesis, University of British Columbia, 1971.
335. Zaidel, A. N., Prokofev, V. K., Raiskii, S. M., Slavnyi, V. A., and Schreider, E. Y., *Tables of Spectral Lines*, 3rd ed., Plenum, New York, 1970.
336. Zetterberg, P. O. and Magnusson, C. E., *Phys. Scr.*, 15, 189, 1977.
337. Sugar, J., *J. Opt. Soc. Am.*, 55, 1058, 1965.
338. Sugar, J., *J. Opt. Soc. Am.*, 61, 727, 1971.
339. Worden, E. F., and Conway, J. G., *At. Data Nucl. Data Tables*, 22, 329, 1978.
340. Kaufman, V. and Edlen, B., *J. Phys. Chem. Ref. Data*, 3, 825, 1974.
341. Lang, R. J., *Phys. Rev.*, 34, 697, 1929.
342. Ryabtsev, A. N., *Opt. Spectros.*, 39, 455, 1975.

343. Foster, E. W., *Proc. R. Soc. London*, 200(A), 429, 1950.
344. Morillon, C. and Verges, J., *Phys. Scr.*, 12, 129, 1975.
345. Ruedy, J. E., *Phys. Rev.*, 41, 588, 1932.
346. McLennan, J. C., McLay, A. B., and McLeod, J. H., *Philos. Mag.*, 4, 486, 1927.
347. Handrup, M. B. and Mack, J. E., *Physica*, 30, 1245, 1964.
348. Clearman, H. E., *J. Opt. Soc. Am.*, 42, 373, 1952.
349. Paschen, F., *Ann. Physik*, 424, 148, 1938.
350. Paschen, F. and Campbell, J. S., *Ann. Phys.*, 31(5), 29, 1938.
351. Nodwell, R., *Univ. of British Columbia, Vancouver*, unpublished, 1955.
352. Gibbs, R. C. and White, H. E., *Phys. Rev.*, 31, 776, 1928.
353. Green, M., *Phys. Rev.*, 60, 117, 1941.
354. Ellis, C. B. and Sawyer, R. A., *Phys. Rev.*, 49, 145, 1936.
355. McLennan, J. C., McLay, A. B., and Crawford, M. F., *Proc. R. Soc. London Ser. A*, 125, 50, 1929.
356. Mack, J. E. and Fromer, M., *Phys. Rev.*, 48, 346, 1935.
357. Humphreys, C. J. and Paul, E., U.S. Nav. Ord. Lab., Navord Rep. 4589, 25, 1956.
358. Walters, F. M., *Sci. Pap. Bur. Stand.*, 17, 161, 1921.
359. Crawford, M. F. and McLay, A. B., *Proc. R. Soc. London Ser. A*, 143, 540, 1934.
360. McLay, A. D. and Crawford, M. F., *Phys. Rev.*, 44, 986, 1933.
361. Schoepfle, G. K., *Phys. Rev.*, 47, 232, 1935.
362. Acquista, N., and Reader, J., *J. Opt. Soc. Am.*, 70, 789, 1980.
363. Benschop, H., Joshi, Y. N., and van Kleef, T. A. M., *Can. J. Phys.*, 53, 498, 1975.
364. Bockasten, K., Hallin, R., and Hughes, T. P., *Proc. Phys. Soc.*, 81, 522, 1963.
365. Boyce, J. C., *Phys. Rev.*, 46, 378, 1934.
366. Boyce, J. C., *Phys. Rev.*, 47, 718, 1935.
367. Boyce, J. C., *Phys. Rev.*, 48, 396, 1935.
368. Boyce, J. C., *Phys. Rev.*, 49, 351, 1936.
369. Corliss, C. H. and Meggers, W. F., *J. Res. Natl. Bur. Stand.*, 61, 269, 1958.
370. Crooker, A. M. and Dick, K. A., *Can. J. Phys.*, 42, 766, 1964.
371. De Bruin, T. L., *Z. Physik*, 77, 505, 1932.
372. De Bruin, T. L., *Proc. Roy. Acad. Amsterdam*, 36, 727, 1933.
373. De Bruin, T. L., *Zeeman Verhandelingen*, (The Hague), 1935, p. 415.
374. De Bruin, T. L., *Physica*, 3, 809, 1936.
375. De Bruin, T. L., Proc. Roy. Acad. Amsterdam, 40, 339, 1937.
376. Dick, K. A., *Can. J. Phys.*, 46, 1291, 1968.
377. Dick, K. A., unpublished, 1978.
378. Edlen, B. and Swensson, J. W., *Phys. Scr.*, 12, 21, 1975.
379. Ekberg, J. O., *Phys. Scr.*, 7, 55, 1973.
380. Ekberg, J. O., *Phys. Scr.*, 7, 59, 1973.
381. Ekberg, J. O., *Phys. Scr.*, 12, 42, 1975.
382. Ekberg, J. O. and Edlen, B., *Phys. Scr.*, 18, 107, 1978.
383. Eliason, A. Y., *Phys. Rev.*, 43, 745, 1933.
384. Gallardo, M., Massone, C. A., Tagliaferri, A. A., Garavaglia, M., and Persson, W., *Phys. Scr.*, 19, 538, 1979.
385. Garcia-Riquelme, O., *Optica Pura Y Aplicada*, 1, 53, 1968.
386. Gibbs, R. C., Vieweg, A. M., and Gartlein, C. W., *Phys. Rev.*, 34, 406, 1929.
387. Gilbert, W. P., *Phys. Rev.*, 48, 338, 1935.
388. Goldsmith, S. and Kaufman, A. S., *Proc. Phys. Soc.*, 81, 544, 1963.
389. Hermansdorfer, H., *J. Opt. Soc. Am.*, 62, 1149, 1972.
390. Humphreys, C. J., *Phys. Rev.*, 47, 712, 1935.
391. Humphreys, C. J., *J. Res. Natl. Bur. Stand.*, 16, 639, 1936.
392. Iglesias, L., *J. Opt. Soc. Am.*, 45, 856, 1955.
393. Iglesias, L., *J. Res. Natl. Bur. Stand.*, 64A, 481, 1960.
394. Iglesias, L., Anales Fisica Y Quimica, 58A, 191, 1962.
395. Iglesias, L., *J. Res. Natl. Bur. Stand.*, 70A, 465, 1966.
396. Iglesias, L., *Can. J. Phys.*, 44, 895, 1966.
397. Iglesias, L., *J. Res. Natl. Bur. Stand.*, 72A, 295, 1968.
398. Joshi, Y. N., *Can. Spectrosc.*, 15, 96, 1970.
399. Joshi, Y. N. and van Kleef, T. A. M., *Can. J. Phys.*, 55, 714, 1977.
400. Kaufman, A. S., Hughes, T. P., and Williams, R. V., *Proc. Phys. Soc.*, 76, 17, 1960.
401. Kaufman, V. and Sugar, J., *J. Opt. Soc. Am.*, 68, 1529, 1978.
402. Keussler, V., Z. *Physik, 85*, 1, 1933.
403. Kiess, C. C., *J. Res. Natl. Bur. Stand.*, 56, 167, 1956.
404. Klinkenberg, P. F. A., van Kleef, T. A. M., and Noorman, P. E., *Physica*, 27, 1177, 1961.
405. Kovalev, V. I., Romanos, A. A., and Ryabtsev, A. N., *Opt. Spectrosc.*, 43, 10, 1977.
406. Lang, R. J., *Proc. Natl. Acad. Sci.*, 13, 341, 1927.
407. Lang, R. J., *Zeeman Verhandelingen*, (The Hague), 44, 1935.
408. Liberman, S., et al., *C. R. Acad. Sci.* (Paris), 286, 253, 1978.
409. Livingston, A. E., *J. Phys.*, B9, L215, 1976.
410. Meijer, F. G., *Physica*, 72, 431, 1974.
411. Meijer, F. G. and Metsch, B. C., *Physica*, 94C, 259, 1978.
412. Moore, F. L., thesis, Princeton, 1949.
413. Paul, F. W. and Polster, H. D., *Phys. Rev.*, 59, 424, 1941.
414. Phillips, L. W. and Parker, W. L., *Phys. Rev.*, 60, 301, 1941.
415. Poppe, R., *Physica*, 81C, 351, 1976.
416. Raassen, A. J. J., van Kleef, T. A. M., and Metsch, B. C., *Physica*, 84C, 133, 1976.
417. Rao, A. B. and Krishnamurty, S. G., *Proc. Phys. Soc. (London)*, 51, 772, 1939.
418. Reader, J. and Acquista, N., *J. Opt. Soc. Am.*, 69, 239, 1979.
419. Reader, J. and Epstein, G. L., *J. Opt. Soc. Am.*, 62, 619, 1972.
420. Rico, F. R., *Anales, Real Soc. Esp. Fis. Quim.*, 61, 103, 1965.
421. Schonheit, E., *Optik*, 23, 409, 1966.
422. Shenstone, A. G., *J. Opt. Soc. Am.*, 44, 749, 1954.
423. Shenstone, A. G., unpublished, 1958.
424. Shenstone, A. G., *J. Res. Natl. Bur. Stand.*, 67A, 87, 1963.
425. Sugar, J. and Kaufman, V., *J. Opt. Soc. Am.*, 64, 1656, 1974.
426. Sugar, J. and Kaufman, V., *Phys. Rev.*, C12, 1336, 1975.
427. Svensson, L. A., *Phys. Scr.*, 13, 235, 1976.
428. Swensson, J. W. and Edlen, B., *Phys. Scr.*, 9, 335, 1974.
429. Tagliaferri, A. A., Gallego Lluesma, E., Garavaglia, M., Gallardo, M., and Massone, C. A., *Optica Pura Y Aplica*, 7, 89.
430. Tilford, S. G. and Giddings, L. E., *Astrophys.* J., 141, 1222, 1965.
431. Trawick, M. W., *Phys. Rev.*, 46, 63, 1934.
432. Van Deurzen, C. H. H., *J. Opt. Soc. Am.*, 67, 476, 1977.
433. Yarosewick, S. L. and Moore, F. L., *J. Opt. Soc. Am.*, 57, 1381, 1967.
434. Zalubas, R., unpublished, 1979.
435. Bhatia, K. S., Jones, W. E., and Crooker, A. M., *Can. J. Phys.*, 50, 2421, 1972.
436. van Kleef, T. A. M. and Joshi, Y. N., *Phys. Scr.*, 24, 557, 1981.

ATOMIC TRANSITION PROBABILITIES

J. R. Fuhr, W. L. Wiese, L. I. Podobedova, and D. E. Kelleher

For the 91st edition of this *Handbook*, we include new, more accurate data for H, He, Li, Be, B, C I and C II, N I and N II, Na I – IV, Mg I – IV, Al I – III, and Si I – V. The new printed tables contain critically evaluated atomic transition probabilities for about 9000 selected lines of all elements for which reliable data are available on an absolute scale. The material is largely for neutral and singly ionized spectra, but also includes some prominent lines of doubly and more highly charged ions of important elements. A more extensive database can be found in the Internet and CD-ROM editions of the *CRC Handbook of Chemistry and Physics*.

Most of the data are obtained from comprehensive compilations of the Data Center on Atomic Transition Probabilities at the National Institute of Standards and Technology. Specifically, data have been taken from recent critical compilations on H, He, and Li (Ref. 1); on Be and B (Ref. 2); on neutral and singly-ionized C and N (Ref. 3); and on Na (Ref. 4), Mg (Ref. 4), Al (Ref. 5), and Si (Ref. 6). Material from earlier compilations for the elements H through Ne (Refs. 7 and 8) and Na through Ca (Ref. 9) was supplemented by some more recent material taken directly from the original literature. Most of the original literature is cited in the above tables and in recent bibliographies (Refs. 10 and 11); for lack of space, individual literature references are not cited here.

The wavelength range for the neutral species is normally the visible spectrum or shorter wavelengths; only the very prominent near infrared lines are included. For the higher ions, most of the strong lines are located in the far UV. The tabulation is limited to electric dipole - including intercombination - lines and comprises essentially the fairly strong transitions with estimated uncertainties in the 10 % to 50 % range. With the exception of hydrogen, helium, and the alkali metals, most transitions are between states with low principal quantum numbers.

The transition probability, A, is given in units of 10^8 s^{-1} and is listed with as many digits as is consistent with the indicated accuracy. Generally, the estimated uncertainties of the A-values are in the range from 25 % to 50 % for two-digit numbers, 10 % to 25 % for three-digit numbers and 1 % or better for four- and five-digit numbers.

Each transition is identified by the wavelength λ in angstroms and the statistical weights, g_i and g_k, of the lower (i) and upper (k) states [the product g_kA (or g_if) is needed for many applications]. Whenever the wavelengths of individual lines within a multiplet are extremely close, only an average wavelength for the multiplet as well as the multiplet A-value are given, and this is indicated by an asterisk (*) to the left of the wavelength. This also has been done when the transition probability for an entire multiplet has been taken from the literature and values for individual lines cannot be determined because of insufficient knowledge of the coupling of electrons. The wavelength data have been taken either from recent compilations or from the original literature cited in bibliographies published by the Atomic Energy Levels Data Center (Refs. 12 and 13) at the National Institute of Standards and Technology. Wavelength values are consistent with those given in the table "Line Spectra of the Elements," which appears elsewhere in this *Handbook*.

In addition to the transition probability A, the atomic oscillator strength f and the line strength S are often used in the literature. The conversion factors between these quantities are (for electric-dipole transitions):

$$g_i f = 1.499 \cdot 10^{-8}\, \lambda^2 g_k A = 303.8\, \lambda^{-1} S$$

where λ is in Å, A is in 10^8 s^{-1}, and S is in atomic units, which are $a_0^2 e^2 = 7.188 \cdot 10^{-59}$ m^{2}C^2.

The table for hydrogen is presented first, followed by the tables for other elements in alphabetical sequence by element name (not symbol). Within each element, the tables are ordered by increasing ionization stage (e.g., Al I, Al II, etc.).

The transition probabilities for hydrogen and hydrogen-like ions are known precisely. Because of the hydrogen degeneracy, a "transition" is actually the sum of all fine-structure transitions between the principal quantum numbers; therefore, the hydrogen table gives weighted average A-values. For hydrogen-like ions of nuclear charge Z, the following scaling laws hold:

$$A_Z = Z^4 A_{\text{Hydrogen}} \quad f_Z = f_{\text{Hydrogen}} \quad S_Z = Z^{-2} S_{\text{Hydrogen}} \quad \lambda_Z = Z^{-2} \lambda_{\text{Hydrogen}}$$

For very highly-charged hydrogen-like ions, starting at about $Z >$ 25, relativistic values must be used.

References

1. Wiese, W. L., and Fuhr, J. R., *J. Phys. Chem. Ref. Data* 38, 565, 2009.
2. Fuhr, J. R., and Wiese, W. L., *J. Phys. Chem. Ref. Data* 39, 013101, 2010.
3. Wiese, W. L., and Fuhr, J. R., *J. Phys. Chem. Ref. Data* 36, 1287, 2007.
4. Kelleher, D. E., and Podobedova, L. I., *J. Phys. Chem. Ref. Data* 37, 267, 2008.
5. Kelleher, D. E., and Podobedova, L. I., *J. Phys. Chem. Ref. Data* 37, 709, 2008.
6. Kelleher, D. E., and Podobedova, L. I., *J. Phys. Chem. Ref. Data* 37, 1285, 2008.
7. Wiese, W. L., Smith, M. W., and Glennon, B. M., *Atomic Transition Probabilities (H through Ne - A Critical Data Compilation)*, National Standard Reference Data Series, National Bureau of Standards 4, Vol. I, U.S. Government Printing Office, Washington, D.C., 1966.
8. Wiese, W. L., Fuhr, J. R., and Deters, T. M., *Atomic Transition Probabilities of Carbon, Nitrogen, and Oxygen, J. Phys. Chem. Ref. Data, Monograph 7*, 1996.
9. Wiese, W. L., Smith, M. W., and Miles, B. M., *Atomic Transition Probabilities (Na through Ca - A Critical Data Compilation)*, National Standard Reference Data Series, National Bureau of Standards 22, Vol. II, U. S. Government Printing Office, Washington, D.C., 1969.
10. Fuhr, J. R., Miller, B. J., and Martin, G. A., *Bibliography on Atomic Transition Probabilities (1914 through October 1997)*, National Bureau of Standards Special Publication 505, 1978; Miller, B. J., Fuhr, J. R., and Martin, G. A., *Bibliography on Atomic Transition Probabilities (November 1977 through February 1980)*, National Bureau of Standards Special Publication 505, Supplement 1, 1980.
11. Wiese, W. L., Reports on Astronomy, *Trans. Int. Astron. Union* 18A, 116, 1982; 19A, 122, 1985; 20A, 117, 1988, Reidel, D., Ed., Kluwer, Dordrecht, The Netherlands.
12. Moore, C. E., *Bibliography on the Analyses of Optical Atomic Spectra*, National Bureau of Standards Special Publication 306 - Section 1, 1968; Sections 2-4, 1969.
13. Hagan, L., and Martin, W. C., *Bibliography on Atomic Energy Levels and Spectra (July 1968 through June 1971)*, National Bureau of Standards Special Publication 363, 1972; Hagan, L., *Bibliography on Atomic Energy Levels and Spectra (July 1971 through June 1975)*, National Bureau of Standards Special Publication 363, Supplement 1, 1977; Zalubas, R., and Albright, A., *Bibliography on Atomic Energy Levels and Spectra (July 1975 through June 1979)*, National Bureau of Standards Special Publication 363, Supplement 2, 1980; Musgrove, A., and Zalubas, R., *Bibliography on Atomic Energy Levels and Spectra (July 1979 through December 1983)*, National Bureau of Standards Special Publication 363, Supplement 3, 1985.

λ Å	Weights g_i	Weights g_k	A 10^8 s^{-1}
Hydrogen			
H I			
912.765	2	1800	$5.1673\cdot10^{-6}$
912.837	2	1682	$6.1221\cdot10^{-6}$
912.916	2	1568	$7.2967\cdot10^{-6}$
913.004	2	1458	$8.7524\cdot10^{-6}$
913.102	2	1352	$1.0571\cdot10^{-5}$
913.212	2	1250	$1.2862\cdot10^{-5}$
913.337	2	1152	$1.5776\cdot10^{-5}$
913.478	2	1058	$1.9519\cdot10^{-5}$
913.639	2	968	$2.4380\cdot10^{-5}$
913.823	2	882	$3.0769\cdot10^{-5}$
914.036	2	800	$3.9276\cdot10^{-5}$
914.284	2	722	$5.0767\cdot10^{-5}$
914.574	2	648	$6.6540\cdot10^{-5}$
914.917	2	578	$8.8574\cdot10^{-5}$
915.327	2	512	$1.1997\cdot10^{-4}$
915.821	2	450	$1.6572\cdot10^{-4}$
916.427	2	392	$2.3409\cdot10^{-4}$
917.178	2	338	$3.3927\cdot10^{-4}$
918.127	2	288	$5.0659\cdot10^{-4}$
919.349	2	242	$7.8340\cdot10^{-4}$
920.961	2	200	$1.2631\cdot10^{-3}$
923.148	2	162	$2.1425\cdot10^{-3}$
926.223	2	128	$3.8694\cdot10^{-3}$
930.748	2	98	$7.5684\cdot10^{-3}$
937.803	2	72	$1.6440\cdot10^{-2}$
949.743	2	50	$4.1250\cdot10^{-2}$
972.537	2	32	$1.2785\cdot10^{-1}$
1025.72	2	18	$5.5751\cdot10^{-1}$
1215.67	2	8	4.6986
3662.23	8	1800	$2.8474\cdot10^{-6}$
3663.37	8	1682	$3.3742\cdot10^{-6}$
3664.65	8	1568	$4.0224\cdot10^{-6}$
3666.07	8	1458	$4.8261\cdot10^{-6}$
3667.65	8	1352	$5.8304\cdot10^{-6}$
3669.43	8	1250	$7.0963\cdot10^{-6}$
3671.45	8	1152	$8.7069\cdot10^{-6}$
3673.73	8	1058	$1.0777\cdot10^{-5}$
3676.33	8	968	$1.3467\cdot10^{-5}$
3679.32	8	882	$1.7005\cdot10^{-5}$
3682.78	8	800	$2.1719\cdot10^{-5}$
3686.80	8	722	$2.8093\cdot10^{-5}$
3691.52	8	648	$3.6851\cdot10^{-5}$
3697.12	8	578	$4.9101\cdot10^{-5}$
3703.82	8	512	$6.6583\cdot10^{-5}$
3711.94	8	450	$9.2102\cdot10^{-5}$
3721.91	8	392	$1.3032\cdot10^{-4}$
3734.34	8	338	$1.8927\cdot10^{-4}$
3750.12	8	288	$2.8337\cdot10^{-4}$
3770.60	8	242	$4.3972\cdot10^{-4}$
3797.87	8	200	$7.1225\cdot10^{-4}$
3835.35	8	162	$1.2156\cdot10^{-3}$
3889.02	8	128	$2.2148\cdot10^{-3}$
3970.08	8	98	$4.3889\cdot10^{-3}$
4101.74	8	72	$9.7320\cdot10^{-3}$
4340.47	8	50	$2.5304\cdot10^{-2}$
4861.34	8	32	$8.4193\cdot10^{-2}$
6562.83	8	18	$4.4101\cdot10^{-1}$
8392.19	18	800	$1.5167\cdot10^{-5}$
8413.11	18	722	$1.9643\cdot10^{-5}$
8437.75	18	648	$2.5804\cdot10^{-5}$
8467.04	18	578	$3.4442\cdot10^{-5}$
8502.27	18	512	$4.6801\cdot10^{-5}$
8545.17	18	450	$6.4901\cdot10^{-5}$
8598.18	18	392	$9.2117\cdot10^{-5}$
8664.80	18	338	$1.3431\cdot10^{-4}$
8750.25	18	288	$2.0207\cdot10^{-4}$
8862.55	18	242	$3.1558\cdot10^{-4}$
9014.67	18	200	$5.1558\cdot10^{-4}$
9228.77	18	162	$8.9050\cdot10^{-4}$
9545.70	18	128	$1.6506\cdot10^{-3}$
10049.4	18	98	$3.3585\cdot10^{-3}$
10938.1	18	72	$7.7829\cdot10^{-3}$
12818.1	18	50	$2.2008\cdot10^{-2}$
16406.4	32	288	$1.6205\cdot10^{-4}$
16805.7	32	242	$2.5565\cdot10^{-4}$
17361.2	32	200	$4.2347\cdot10^{-4}$
18173.2	32	162	$7.4593\cdot10^{-4}$
18751.0	18	32	$8.9860\cdot10^{-2}$
19444.5	32	128	$1.4242\cdot10^{-3}$
21655.2	32	98	$3.0415\cdot10^{-3}$
26251.4	32	72	$7.7110\cdot10^{-3}$
27573.0	50	288	$1.4024\cdot10^{-4}$
28719.8	50	242	$2.2460\cdot10^{-4}$
30381.1	50	200	$3.7999\cdot10^{-4}$
32957.8	50	162	$6.9078\cdot10^{-4}$
37391.4	50	128	$1.3877\cdot10^{-3}$
40511.4	32	50	$2.6993\cdot10^{-2}$
43747.2	72	288	$1.2884\cdot10^{-4}$
46524.9	50	98	$3.2528\cdot10^{-3}$
46706.2	72	242	$2.1096\cdot10^{-4}$
51279.2	72	200	$3.6881\cdot10^{-4}$
59072.4	72	162	$7.0652\cdot10^{-4}$
74598.3	50	72	$1.0254\cdot10^{-2}$
75009.1	72	128	$1.5609\cdot10^{-3}$
123718.6	72	98	$4.5608\cdot10^{-3}$
Aluminum			
Al I			
2118.312	2	4	$1.03\cdot10^{-1}$
2123.362	4	6	$1.22\cdot10^{-1}$
2129.663	2	4	$1.52\cdot10^{-1}$
2134.733	4	6	$1.81\cdot10^{-1}$
2145.555	2	4	$2.06\cdot10^{-1}$
2150.699	4	6	$2.46\cdot10^{-1}$
2168.805	2	4	$2.96\cdot10^{-1}$
2174.028	4	6	$3.53\cdot10^{-1}$
2199.150	2	2	$1.75\cdot10^{-2}$
2204.590	4	2	$3.49\cdot10^{-2}$
2204.660	2	4	$4.37\cdot10^{-1}$
2210.046	4	6	$5.20\cdot10^{-1}$
2257.999	2	2	$3.77\cdot10^{-2}$
2263.462	2	4	$6.83\cdot10^{-1}$
2263.731	4	2	$7.50\cdot10^{-2}$
2269.096	4	6	$7.58\cdot10^{-1}$
2269.220	4	4	$1.26\cdot10^{-1}$
2367.052	2	4	$7.61\cdot10^{-1}$
2372.070	2	2	$5.76\cdot10^{-2}$
2373.124	4	6	$9.07\cdot10^{-1}$
2373.349	4	4	$1.51\cdot10^{-1}$
2378.368	4	2	$1.14\cdot10^{-1}$
2567.984	2	4	$1.92\cdot10^{-1}$
2575.094	4	6	$3.60\cdot10^{-1}$
2575.393	4	4	$5.99\cdot10^{-2}$
2652.484	2	2	$1.42\cdot10^{-1}$
2660.393	4	2	$2.84\cdot10^{-1}$
3082.1529	2	4	$5.87\cdot10^{-1}$
3092.7099	4	6	$7.29\cdot10^{-1}$
3092.8386	4	4	$1.16\cdot10^{-1}$
3944.0058	2	2	$4.99\cdot10^{-1}$
3961.5200	4	2	$9.85\cdot10^{-1}$
5557.063	2	4	$2.30\cdot10^{-3}$
5557.948	2	2	$2.29\cdot10^{-3}$
6696.015	2	4	$1.00\cdot10^{-2}$
6698.673	2	2	$1.00\cdot10^{-2}$
7835.309	4	6	$3.71\cdot10^{-2}$
7836.134	6	8	$3.97\cdot10^{-2}$
8772.866	4	6	$6.47\cdot10^{-2}$
8773.896	6	8	$6.95\cdot10^{-2}$
8828.909	2	2	$6.72\cdot10^{-3}$
8841.277	4	2	$1.34\cdot10^{-2}$
8912.900	2	4	$2.28\cdot10^{-3}$
8923.555	4	6	$2.73\cdot10^{-3}$
8925.504	4	4	$4.54\cdot10^{-4}$
Al II			
1047.8893	1	3	$2.33\cdot10^{-1}$
1048.5588	3	5	$3.14\cdot10^{-1}$
1049.9233	5	7	$4.17\cdot10^{-1}$
1189.1854	3	3	$9.30\cdot10^{-1}$
1190.0518	3	5	1.12
1191.8111	5	7	1.48
1350.1782	3	5	4.80
1539.8303	3	5	6.70
1670.7867	1	3	$1.41\cdot10^{1}$
1719.4400	1	3	6.55
1721.2435	3	3	4.93
1721.2714	3	5	8.82
1724.9519	5	5	2.97
1724.9838	5	7	$1.18\cdot10^{1}$
1760.1044	3	5	3.13
1761.9751	1	3	4.12
1763.8692	3	3	3.04
1763.9521	5	5	9.21
1765.8150	3	1	$1.23\cdot10^{1}$
1767.7308	5	3	5.13
1772.802	1	3	9.41
1774.002	3	5	$1.26\cdot10^{1}$
1774.770	3	3	7.01
1776.975	5	7	$1.67\cdot10^{1}$
1777.825	5	5	4.18
1818.352	7	7	5.45
1820.124	3	3	4.57
1855.9286	1	3	$8.38\cdot10^{-1}$
1858.0262	3	3	2.49
1862.3111	5	3	4.08

λ Å	Weights g_i	g_k	A 10^8 s^{-1}
1904.326	1	3	2.70
1906.4082	3	5	2.00
1906.596	3	3	2.00
1906.674	3	1	8.20
1910.8252	5	5	5.80
1911.013	5	3	3.40
1931.0481	3	1	$1.04 \cdot 10^{1}$
1958.77	7	5	5.70
1990.5310	3	5	$1.38 \cdot 10^{1}$
2192.604	7	9	2.47
2194.189	7	7	$2.74 \cdot 10^{-1}$
2195.502	3	5	2.07
2816.185	3	1	3.57
2994.277	1	3	$4.04 \cdot 10^{-2}$
2995.525	3	5	$5.44 \cdot 10^{-2}$
2998.150	5	7	$7.26 \cdot 10^{-2}$
3088.516	3	5	$1.08 \cdot 10^{-1}$
3649.184	1	3	$1.26 \cdot 10^{-1}$
3649.232	1	3	$1.26 \cdot 10^{-1}$
3651.065	3	5	$1.70 \cdot 10^{-1}$
3651.096	3	5	$1.70 \cdot 10^{-1}$
3654.981	5	7	$2.26 \cdot 10^{-1}$
3654.998	5	7	$2.26 \cdot 10^{-1}$
3703.219	3	5	$3.80 \cdot 10^{-1}$
3731.952	1	3	$4.30 \cdot 10^{-2}$
3733.908	3	3	$1.30 \cdot 10^{-1}$
3738.015	5	3	$2.10 \cdot 10^{-1}$
3866.160	3	1	$3.70 \cdot 10^{-1}$
3900.675	3	5	$4.80 \cdot 10^{-3}$
4663.056	5	3	$5.81 \cdot 10^{-1}$
5388.48	1	3	$1.20 \cdot 10^{-2}$
5593.302	3	5	$9.26 \cdot 10^{-1}$
5613.291	5	7	$3.43 \cdot 10^{-2}$
5853.62	7	9	$1.28 \cdot 10^{-1}$
5861.53	5	7	$1.14 \cdot 10^{-1}$
5867.81	3	5	$2.00 \cdot 10^{-1}$
5971.980	3	5	$4.90 \cdot 10^{-2}$
5999.70	1	3	$2.07 \cdot 10^{-2}$
5999.83	1	3	$2.07 \cdot 10^{-2}$
6001.76	3	5	$2.79 \cdot 10^{-2}$
6001.88	3	3	$1.55 \cdot 10^{-2}$
6006.410	5	7	$3.40 \cdot 10^{-2}$
6061.124	3	1	$7.60 \cdot 10^{-2}$
6066.32	1	3	$8.88 \cdot 10^{-3}$
6066.44	1	3	$8.88 \cdot 10^{-3}$
6068.43	3	3	$2.66 \cdot 10^{-2}$
6068.53	3	3	$2.66 \cdot 10^{-2}$
6073.198	5	3	$4.20 \cdot 10^{-2}$
6226.18	1	3	$6.20 \cdot 10^{-1}$
6231.745	3	5	$8.40 \cdot 10^{-1}$
6243.36	5	7	1.11
6335.701	5	3	$1.40 \cdot 10^{-1}$
6816.69	1	3	$1.10 \cdot 10^{-1}$
6823.48	3	3	$3.40 \cdot 10^{-1}$
6837.14	5	3	$5.70 \cdot 10^{-1}$
6917.93	5	7	$1.60 \cdot 10^{-1}$
6919.96	3	1	$9.60 \cdot 10^{-1}$
7042.06	3	5	$5.78 \cdot 10^{-1}$
7056.60	3	3	$5.74 \cdot 10^{-1}$
7063.64	3	1	$5.73 \cdot 10^{-1}$

λ Å	Weights g_i	g_k	A 10^8 s^{-1}
7449.42	3	5	$1.20 \cdot 10^{-1}$
7471.41	5	7	$5.57 \cdot 10^{-1}$
7624.48	1	3	$4.59 \cdot 10^{-2}$
7627.85	3	5	$6.20 \cdot 10^{-2}$
7635.33	5	7	$9.00 \cdot 10^{-2}$
8354.318	7	9	$4.27 \cdot 10^{-1}$
8359.23	7	7	$4.75 \cdot 10^{-2}$
8359.492	5	7	$3.79 \cdot 10^{-1}$
8363.251	5	5	$6.64 \cdot 10^{-2}$
8363.469	3	5	$4.20 \cdot 10^{-1}$
8640.705	1	3	$3.00 \cdot 10^{-1}$

Al III

λ Å	Weights g_i	g_k	A 10^8 s^{-1}
560.390	2	4	$6.10 \cdot 10^{-1}$
560.390	2	2	$6.31 \cdot 10^{-1}$
695.817	2	4	$6.83 \cdot 10^{-1}$
696.212	2	2	$7.17 \cdot 10^{-1}$
1162.66	6	6	$1.39 \cdot 10^{-1}$
1162.66	6	8	2.09
1162.67	4	6	1.95
1352.857	6	8	4.26
1352.857	6	6	$2.84 \cdot 10^{-1}$
1352.857	4	6	3.98
1379.670	2	2	4.61
1384.140	4	2	9.22
1605.7661	2	4	$1.16 \cdot 10^{1}$
1611.8141	4	4	2.30
1611.8735	4	6	$1.38 \cdot 10^{1}$
1854.7164	2	4	5.44
1862.7895	2	2	5.36
1935.8404	6	8	$1.19 \cdot 10^{1}$
1935.8404	6	6	$7.92 \cdot 10^{-1}$
1935.9489	4	6	$1.11 \cdot 10^{1}$
3283.316	2	4	$1.99 \cdot 10^{-2}$
3287.302	4	6	$2.29 \cdot 10^{-2}$
3601.628	6	4	1.31
3601.926	4	4	$1.46 \cdot 10^{-1}$
3612.356	4	2	1.45
3702.106	2	2	1.13
3713.123	4	2	2.27
3980.14	6	8	$2.33 \cdot 10^{-1}$
3980.14	4	6	$2.17 \cdot 10^{-1}$
4149.915	6	8	2.05
4150.173	4	6	1.91
4357.562	2	4	$7.40 \cdot 10^{-2}$
4364.642	4	6	$8.67 \cdot 10^{-2}$
4512.564	2	4	2.09
4528.942	4	4	$4.15 \cdot 10^{-1}$
4529.194	4	6	2.49
4701.148	6	4	$7.67 \cdot 10^{-2}$
4701.412	8	6	$7.31 \cdot 10^{-2}$
4904.10	6	8	$3.51 \cdot 10^{-1}$
4904.10	4	6	$3.27 \cdot 10^{-1}$
5696.603	2	4	$8.77 \cdot 10^{-1}$
5722.728	2	2	$8.65 \cdot 10^{-1}$

Argon

Ar I

λ Å	Weights g_i	g_k	A 10^8 s^{-1}
1048.22	1	3	5.36

λ Å	Weights g_i	g_k	A 10^8 s^{-1}
1066.66	1	3	1.29
3948.98	5	3	$4.55 \cdot 10^{-3}$
4044.42	3	5	$3.33 \cdot 10^{-3}$
4158.59	5	5	$1.40 \cdot 10^{-2}$
4181.88	1	3	$5.61 \cdot 10^{-3}$
4190.71	5	5	$2.80 \cdot 10^{-3}$
4191.03	1	3	$5.39 \cdot 10^{-3}$
4198.32	3	1	$2.57 \cdot 10^{-2}$
4200.67	5	7	$9.67 \cdot 10^{-3}$
4259.36	3	1	$3.98 \cdot 10^{-2}$
4266.29	3	5	$3.12 \cdot 10^{-3}$
4272.17	3	3	$7.97 \cdot 10^{-3}$
4300.10	3	5	$3.77 \cdot 10^{-3}$
4333.56	3	5	$5.68 \cdot 10^{-3}$
4335.34	3	3	$3.87 \cdot 10^{-3}$
4510.73	3	1	$1.18 \cdot 10^{-2}$
4752.94	3	3	$4.5 \cdot 10^{-3}$
4768.68	3	5	$8.6 \cdot 10^{-3}$
4798.74	7	9	$8.8 \cdot 10^{-4}$
4835.97	7	9	$9.3 \cdot 10^{-4}$
4876.26	3	5	$7.8 \cdot 10^{-3}$
4886.29	7	9	$1.2 \cdot 10^{-3}$
4887.95	3	3	$1.3 \cdot 10^{-2}$
4894.69	3	1	$1.8 \cdot 10^{-2}$
4956.75	7	9	$1.8 \cdot 10^{-3}$
4989.95	5	7	$1.1 \cdot 10^{-3}$
5048.81	3	5	$4.6 \cdot 10^{-3}$
5054.18	3	3	$4.5 \cdot 10^{-3}$
5060.08	7	9	$3.7 \cdot 10^{-3}$
5070.99	5	3	$2.6 \cdot 10^{-3}$
5087.09	5	7	$1.6 \cdot 10^{-3}$
5118.21	5	7	$2.7 \cdot 10^{-3}$
5151.39	3	1	$2.39 \cdot 10^{-2}$
5162.29	3	3	$1.90 \cdot 10^{-2}$
5177.54	7	5	$2.4 \cdot 10^{-3}$
5194.02	3	1	$7.8 \cdot 10^{-3}$
5210.49	7	7	$1.1 \cdot 10^{-3}$
5214.77	5	3	$2.1 \cdot 10^{-3}$
5221.27	7	9	$8.8 \cdot 10^{-3}$
5241.09	5	5	$1.3 \cdot 10^{-3}$
5246.24	5	7	$1.2 \cdot 10^{-3}$
5252.79	5	7	$5.4 \cdot 10^{-3}$
5254.47	3	5	$3.6 \cdot 10^{-3}$
5286.07	5	7	$9.6 \cdot 10^{-4}$
5309.52	5	5	$1.2 \cdot 10^{-3}$
5317.73	5	7	$2.6 \cdot 10^{-3}$
5373.50	3	5	$2.7 \cdot 10^{-3}$
5410.48	5	7	$2.0 \cdot 10^{-3}$
5421.35	7	5	$6.0 \cdot 10^{-3}$
5439.99	3	3	$1.9 \cdot 10^{-3}$
5442.24	7	7	$9.3 \cdot 10^{-4}$
5451.65	3	5	$4.7 \cdot 10^{-3}$
5457.42	5	3	$3.6 \cdot 10^{-3}$
5473.46	5	3	$2.0 \cdot 10^{-3}$
5492.09	3	1	$5.6 \cdot 10^{-3}$
5495.87	7	9	$1.69 \cdot 10^{-2}$
5506.11	5	7	$3.6 \cdot 10^{-3}$
5524.96	7	7	$1.7 \cdot 10^{-3}$
5534.49	5	3	$2.7 \cdot 10^{-3}$
5558.70	3	5	$1.42 \cdot 10^{-2}$

λ Å	Weights g_i	g_k	A 10^8 s^{-1}
5559.66	3	5	$2.2 \cdot 10^{-3}$
5572.54	5	7	$6.6 \cdot 10^{-3}$
5588.72	5	5	$1.5 \cdot 10^{-3}$
5597.48	5	7	$4.2 \cdot 10^{-3}$
5606.73	3	3	$2.20 \cdot 10^{-2}$
5618.01	3	3	$2.1 \cdot 10^{-3}$
5623.78	5	5	$1.4 \cdot 10^{-3}$
5635.58	3	5	$9.6 \cdot 10^{-4}$
5639.12	1	3	$2.1 \cdot 10^{-3}$
5650.70	3	1	$3.20 \cdot 10^{-2}$
5659.13	5	5	$2.6 \cdot 10^{-3}$
5681.90	5	7	$2.0 \cdot 10^{-3}$
5683.73	5	5	$2.0 \cdot 10^{-3}$
5700.87	5	7	$5.9 \cdot 10^{-3}$
5739.52	3	5	$8.7 \cdot 10^{-3}$
5772.11	5	7	$2.0 \cdot 10^{-3}$
5773.99	5	5	$1.1 \cdot 10^{-3}$
5802.08	5	3	$4.2 \cdot 10^{-3}$
5882.62	3	1	$1.23 \cdot 10^{-2}$
5888.58	7	5	$1.29 \cdot 10^{-2}$
5928.81	5	3	$1.1 \cdot 10^{-2}$
5942.67	5	5	$1.8 \cdot 10^{-3}$
5949.26	3	3	$1.5 \cdot 10^{-3}$
5968.32	3	3	$1.8 \cdot 10^{-3}$
5971.60	3	1	$1.1 \cdot 10^{-2}$
5987.30	7	7	$1.2 \cdot 10^{-3}$
5999.00	5	5	$1.4 \cdot 10^{-3}$
6005.73	5	3	$1.4 \cdot 10^{-3}$
6013.68	7	5	$1.4 \cdot 10^{-3}$
6025.15	5	3	$9.0 \cdot 10^{-3}$
6043.22	5	7	$1.47 \cdot 10^{-2}$
6052.73	3	5	$1.9 \cdot 10^{-3}$
6064.76	5	7	$5.8 \cdot 10^{-4}$
6090.79	1	3	$3.0 \cdot 10^{-3}$
6098.81	3	3	$5.2 \cdot 10^{-3}$
6101.16	3	3	$3.3 \cdot 10^{-3}$
6105.64	3	5	$1.21 \cdot 10^{-2}$
6128.73	3	5	$8.6 \cdot 10^{-4}$
6145.44	5	7	$7.6 \cdot 10^{-3}$
6155.24	5	3	$5.1 \cdot 10^{-3}$
6165.12	5	5	$9.89 \cdot 10^{-4}$
6170.17	5	5	$5.0 \cdot 10^{-3}$
6173.10	3	5	$6.7 \cdot 10^{-3}$
6212.50	5	7	$3.9 \cdot 10^{-3}$
6215.94	5	5	$5.7 \cdot 10^{-3}$
6248.41	3	5	$6.8 \cdot 10^{-4}$
6296.87	3	5	$9.0 \cdot 10^{-3}$
6307.66	5	5	$6.0 \cdot 10^{-3}$
6364.89	3	1	$5.6 \cdot 10^{-3}$
6369.58	5	3	$4.2 \cdot 10^{-3}$
6384.72	3	3	$4.21 \cdot 10^{-3}$
6416.31	3	5	$1.16 \cdot 10^{-2}$
6466.55	1	3	$1.5 \cdot 10^{-3}$
6538.11	7	7	$1.1 \cdot 10^{-3}$
6604.02	7	5	$2.8 \cdot 10^{-3}$
6660.68	3	1	$7.8 \cdot 10^{-3}$
6664.05	5	5	$1.5 \cdot 10^{-3}$
6698.88	5	3	$1.6 \cdot 10^{-3}$
6719.22	1	3	$2.4 \cdot 10^{-3}$
6752.84	3	5	$1.93 \cdot 10^{-2}$
6754.37	3	3	$2.1 \cdot 10^{-3}$
6756.10	5	5	$3.6 \cdot 10^{-3}$
6766.61	5	3	$4.0 \cdot 10^{-3}$
6779.93	1	3	$1.21 \cdot 10^{-3}$
6827.25	5	3	$2.4 \cdot 10^{-3}$
6851.88	3	5	$6.7 \cdot 10^{-4}$
6871.29	3	3	$2.78 \cdot 10^{-2}$
6879.59	3	5	$1.8 \cdot 10^{-3}$
6887.10	5	7	$1.3 \cdot 10^{-3}$
6888.17	3	5	$2.5 \cdot 10^{-3}$
6925.01	3	3	$1.2 \cdot 10^{-3}$
6937.67	3	1	$3.08 \cdot 10^{-2}$
6951.46	5	5	$2.2 \cdot 10^{-3}$
6960.23	5	5	$2.4 \cdot 10^{-3}$
6965.43	5	3	$6.39 \cdot 10^{-2}$
6992.17	3	1	$7.5 \cdot 10^{-3}$
7030.25	7	5	$2.67 \cdot 10^{-2}$
7067.22	5	5	$3.80 \cdot 10^{-2}$
7068.73	5	3	$2.0 \cdot 10^{-2}$
7086.70	1	3	$1.5 \cdot 10^{-3}$
7107.48	5	5	$4.5 \cdot 10^{-3}$
7125.83	3	3	$6.0 \cdot 10^{-3}$
7147.04	5	3	$6.25 \cdot 10^{-3}$
7158.83	3	1	$2.1 \cdot 10^{-2}$
7206.98	5	3	$2.48 \cdot 10^{-2}$
7229.93	5	5	$6.6 \cdot 10^{-4}$
7265.17	3	3	$1.7 \cdot 10^{-3}$
7270.66	7	7	$1.1 \cdot 10^{-3}$
7272.93	3	3	$1.83 \cdot 10^{-2}$
7285.44	5	3	$1.2 \cdot 10^{-3}$
7311.72	3	3	$1.7 \cdot 10^{-2}$
7316.01	3	3	$9.6 \cdot 10^{-3}$
7350.78	3	1	$1.2 \cdot 10^{-2}$
7353.32	5	7	$9.6 \cdot 10^{-3}$
7372.12	7	9	$1.9 \cdot 10^{-2}$
7383.98	3	5	$8.47 \cdot 10^{-2}$
7392.97	5	3	$7.2 \cdot 10^{-3}$
7412.33	3	5	$3.9 \cdot 10^{-3}$
7422.26	3	5	$6.6 \cdot 10^{-4}$
7425.29	5	7	$3.1 \cdot 10^{-3}$
7435.33	5	5	$9.0 \cdot 10^{-3}$
7436.25	7	5	$2.7 \cdot 10^{-3}$
7484.24	3	5	$3.4 \cdot 10^{-3}$
7503.84	3	1	$4.45 \cdot 10^{-1}$
7510.42	5	5	$4.5 \cdot 10^{-3}$
7514.65	3	1	$4.02 \cdot 10^{-1}$
7618.33	3	5	$2.9 \cdot 10^{-3}$
7628.86	3	5	$2.9 \cdot 10^{-3}$
7635.11	5	5	$2.45 \cdot 10^{-1}$
7670.04	5	3	$2.8 \cdot 10^{-3}$
7704.81	5	7	$6.3 \cdot 10^{-4}$
7723.76	5	3	$5.18 \cdot 10^{-2}$
7724.21	1	3	$1.17 \cdot 10^{-1}$
7798.55	3	5	$8.7 \cdot 10^{-4}$
7868.20	1	3	$3.50 \cdot 10^{-3}$
7891.08	5	5	$9.5 \cdot 10^{-3}$
7916.45	3	3	$1.2 \cdot 10^{-3}$
7948.18	1	3	$1.86 \cdot 10^{-1}$
8006.16	3	5	$4.90 \cdot 10^{-2}$
8014.79	5	5	$9.28 \cdot 10^{-2}$
8037.23	1	3	$3.59 \cdot 10^{-3}$
8046.13	3	1	$1.12 \cdot 10^{-2}$
8053.31	5	3	$8.6 \cdot 10^{-3}$
8066.60	5	5	$1.4 \cdot 10^{-3}$
8103.69	3	3	$2.5 \cdot 10^{-1}$
8115.31	5	7	$3.31 \cdot 10^{-1}$
8264.52	3	3	$1.53 \cdot 10^{-1}$
8384.73	5	7	$2.4 \cdot 10^{-3}$
8408.21	3	5	$2.23 \cdot 10^{-1}$
8424.65	3	5	$2.15 \cdot 10^{-1}$
8490.30	3	5	$9.6 \cdot 10^{-4}$
8521.44	3	3	$1.39 \cdot 10^{-1}$
8605.78	5	5	$1.04 \cdot 10^{-2}$
8620.46	1	3	$9.2 \cdot 10^{-3}$
8667.94	1	3	$2.43 \cdot 10^{-2}$
8761.69	3	5	$9.5 \cdot 10^{-3}$
8784.61	3	1	$2.4 \cdot 10^{-3}$
8799.08	5	3	$4.6 \cdot 10^{-3}$
8962.19	3	3	$1.6 \cdot 10^{-3}$
9075.42	3	1	$1.2 \cdot 10^{-2}$
9122.97	5	3	$1.89 \cdot 10^{-1}$
9194.64	3	3	$1.76 \cdot 10^{-2}$
9224.50	3	5	$5.03 \cdot 10^{-2}$
9291.53	3	1	$3.26 \cdot 10^{-2}$
9354.22	3	3	$1.06 \cdot 10^{-2}$
9657.78	3	3	$5.43 \cdot 10^{-2}$
9784.50	3	5	$1.47 \cdot 10^{-2}$
10470.05	1	3	$9.8 \cdot 10^{-3}$
10478.0	3	3	$2.44 \cdot 10^{-2}$
10950.7	5	3	$3.96 \cdot 10^{-3}$
11078.9	5	5	$8.3 \cdot 10^{-3}$
11393.7	3	1	$2.22 \cdot 10^{-2}$
11441.8	5	3	$1.39 \cdot 10^{-2}$
11467.5	3	5	$3.69 \cdot 10^{-3}$
11488.11	3	3	$1.9 \cdot 10^{-3}$
11668.7	5	5	$3.76 \cdot 10^{-2}$
11719.5	5	3	$9.52 \cdot 10^{-3}$
12026.6	1	3	$4.2 \cdot 10^{-3}$
12112.2	7	7	$3.1 \cdot 10^{-2}$
12139.8	3	3	$4.5 \cdot 10^{-2}$
12343.7	5	7	$2.0 \cdot 10^{-2}$
12402.9	3	3	$1.1 \cdot 10^{-1}$
12439.2	3	5	$4.9 \cdot 10^{-2}$
12456.1	5	3	$8.9 \cdot 10^{-2}$
12487.6	7	5	$1.1 \cdot 10^{-1}$
12554.4	7	5	$1.2 \cdot 10^{-3}$
12702.4	3	3	$7.1 \cdot 10^{-2}$
12733.6	5	5	$1.1 \cdot 10^{-2}$
12746.3	3	3	$2.0 \cdot 10^{-2}$
12802.7	5	5	$5.7 \cdot 10^{-2}$
12933.3	3	1	$1.0 \cdot 10^{-1}$
12956.6	3	3	$7.4 \cdot 10^{-2}$
13008.5	5	3	$8.9 \cdot 10^{-2}$
13214.7	3	1	$8.1 \cdot 10^{-2}$
13273.1	5	7	$1.5 \cdot 10^{-1}$
13313.4	3	5	$1.3 \cdot 10^{-1}$
13504.0	5	7	$1.1 \cdot 10^{-1}$
13599.2	5	5	$2.2 \cdot 10^{-2}$
13622.4	3	5	$7.3 \cdot 10^{-2}$
13678.5	3	5	$6.2 \cdot 10^{-2}$

λ Å	Weights g_i	g_k	A $10^8\ s^{-1}$
14093.6	1	3	$4.3 \cdot 10^{-2}$
14739.1	5	7	$8.8 \cdot 10^{-4}$
15046.4	1	3	$5.2 \cdot 10^{-2}$
15172.3	1	3	$1.3 \cdot 10^{-2}$
15329.6	5	5	$1.2 \cdot 10^{-3}$
15555.5	5	7	$9.8 \cdot 10^{-5}$
15734.9	5	3	$2.9 \cdot 10^{-4}$
15816.8	5	3	$8.7 \cdot 10^{-4}$
15989.3	1	3	$1.9 \cdot 10^{-2}$
16122.7	5	3	$3.9 \cdot 10^{-4}$
16180.0	5	5	$1.2 \cdot 10^{-3}$
16264.1	3	3	$3.0 \cdot 10^{-4}$
16520.1	3	5	$2.6 \cdot 10^{-3}$
16739.8	3	5	$3.1 \cdot 10^{-3}$
16940.4	5	5	$2.5 \cdot 10^{-2}$
20317.0	1	3	$1.6 \cdot 10^{-4}$
20616.5	5	5	$3.9 \cdot 10^{-3}$
20812.0	5	7	$7.6 \cdot 10^{-4}$
21332.2	3	3	$3.2 \cdot 10^{-4}$
21534.9	3	5	$1.1 \cdot 10^{-3}$
22039.2	3	1	$1.2 \cdot 10^{-3}$
22077.4	5	3	$1.4 \cdot 10^{-3}$
23133.4	3	3	$1.7 \cdot 10^{-3}$
23844.8	9	7	$1.1 \cdot 10^{-2}$
23967.5	3	1	$3.6 \cdot 10^{-3}$

Ar II

λ Å	Weights g_i	g_k	A $10^8\ s^{-1}$
2942.9	4	4	$5.3 \cdot 10^{-1}$
2979.1	2	2	$4.16 \cdot 10^{-1}$
3139.0	6	6	$5.2 \cdot 10^{-1}$
3169.7	4	6	$4.9 \cdot 10^{-1}$
3181.0	6	4	$3.7 \cdot 10^{-1}$
3243.7	4	2	1.06
3249.8	2	4	$6.3 \cdot 10^{-1}$
3263.6	2	4	$1.55 \cdot 10^{-1}$
3281.7	2	2	$4.2 \cdot 10^{-1}$
3454.1	6	4	$3.14 \cdot 10^{-1}$
3476.7	6	6	1.25
3491.2	4	4	1.79
3491.5	6	8	2.31
3509.8	2	2	2.55
3514.4	4	6	1.36
3520.0	6	6	$5.2 \cdot 10^{-1}$
3521.3	8	8	$2.27 \cdot 10^{-1}$
3535.3	2	4	$5.7 \cdot 10^{-1}$
3548.5	4	4	$8.7 \cdot 10^{-1}$
3559.5	6	8	2.88
3565.0	2	4	$5.5 \cdot 10^{-1}$
3576.6	6	8	2.75
3581.6	2	4	1.76
3582.4	4	6	2.53
3588.4	8	10	3.03
3656.0	6	6	$7.6 \cdot 10^{-2}$
3717.2	6	8	$5.2 \cdot 10^{-2}$
3729.3	6	4	$4.80 \cdot 10^{-1}$
3763.5	8	6	$1.78 \cdot 10^{-1}$
3780.8	8	8	$7.7 \cdot 10^{-1}$
3799.4	6	4	$1.7 \cdot 10^{-1}$
3826.8	6	6	$2.81 \cdot 10^{-1}$
3841.5	4	2	$2.69 \cdot 10^{-1}$
3844.7	6	8	$4.8 \cdot 10^{-2}$
3850.6	4	4	$3.87 \cdot 10^{-1}$
3868.5	4	6	1.4
3872.1	4	4	$1.5 \cdot 10^{-1}$
3880.3	2	2	$2.32 \cdot 10^{-1}$
3900.6	4	6	$7.2 \cdot 10^{-2}$
3928.6	2	4	$2.44 \cdot 10^{-1}$
3932.5	4	4	$9.3 \cdot 10^{-1}$
3952.7	4	4	$2.08 \cdot 10^{-1}$
3979.4	4	2	$9.8 \cdot 10^{-1}$
4013.9	8	8	$1.05 \cdot 10^{-1}$
4042.9	4	4	$4.06 \cdot 10^{-1}$
4052.9	2	4	$6.7 \cdot 10^{-1}$
4072.0	6	6	$5.8 \cdot 10^{-1}$
4079.6	6	4	$1.19 \cdot 10^{-1}$
4131.7	4	2	$8.5 \cdot 10^{-1}$
4228.2	4	6	$1.31 \cdot 10^{-1}$
4237.2	4	4	$1.12 \cdot 10^{-1}$
4266.5	6	6	$1.64 \cdot 10^{-1}$
4277.5	6	4	$8.0 \cdot 10^{-1}$
4282.9	4	2	$1.32 \cdot 10^{-1}$
4300.6	6	6	$5.7 \cdot 10^{-2}$
4331.2	4	4	$5.74 \cdot 10^{-1}$
4332.0	4	2	$1.92 \cdot 10^{-1}$
4348.1	6	8	1.17
4352.2	2	2	$2.12 \cdot 10^{-1}$
4362.1	4	6	$5.5 \cdot 10^{-2}$
4370.8	4	4	$6.6 \cdot 10^{-1}$
4371.3	6	4	$2.21 \cdot 10^{-1}$
4376.0	4	2	$2.05 \cdot 10^{-1}$
4379.7	2	2	1.00
4400.1	4	4	$1.60 \cdot 10^{-1}$
4401.0	8	6	$3.04 \cdot 10^{-1}$
4412.9	6	8	$6.1 \cdot 10^{-2}$
4426.0	4	6	$8.17 \cdot 10^{-1}$
4430.2	2	4	$5.69 \cdot 10^{-1}$
4431.0	6	6	$1.09 \cdot 10^{-1}$
4474.8	4	2	$2.90 \cdot 10^{-1}$
4481.8	6	6	$4.55 \cdot 10^{-1}$
4545.1	4	4	$4.71 \cdot 10^{-1}$
4579.4	2	2	$8.0 \cdot 10^{-1}$
4589.9	4	6	$6.64 \cdot 10^{-1}$
4598.8	4	4	$6.7 \cdot 10^{-2}$
4609.6	6	8	$7.89 \cdot 10^{-1}$
4637.2	6	6	$7.1 \cdot 10^{-2}$
4657.9	4	2	$8.92 \cdot 10^{-1}$
4726.9	4	4	$5.88 \cdot 10^{-1}$
4732.1	6	4	$6.7 \cdot 10^{-2}$
4735.9	6	4	$5.80 \cdot 10^{-1}$
4764.9	2	4	$6.4 \cdot 10^{-1}$
4806.0	6	6	$7.80 \cdot 10^{-1}$
4847.8	4	2	$8.49 \cdot 10^{-1}$
4879.9	4	6	$8.23 \cdot 10^{-1}$
4889.0	2	2	$1.9 \cdot 10^{-1}$
4904.8	6	8	$3.7 \cdot 10^{-2}$
4933.2	4	4	$1.44 \cdot 10^{-1}$
4965.1	2	4	$3.94 \cdot 10^{-1}$
4972.2	2	2	$9.7 \cdot 10^{-2}$
5009.3	4	6	$1.51 \cdot 10^{-1}$
5017.2	4	6	$2.07 \cdot 10^{-1}$
5062.0	2	4	$2.23 \cdot 10^{-1}$
5141.8	6	8	$8.1 \cdot 10^{-2}$
5145.3	4	6	$1.06 \cdot 10^{-1}$
6114.9	10	8	$2.00 \cdot 10^{-1}$
6172.3	8	6	$2.00 \cdot 10^{-1}$
6243.1	8	6	$3.0 \cdot 10^{-2}$
6483.1	4	2	$1.06 \cdot 10^{-1}$
6638.2	6	4	$1.37 \cdot 10^{-1}$
6639.7	4	2	$1.69 \cdot 10^{-1}$
6643.7	10	8	$1.47 \cdot 10^{-1}$
6666.4	2	2	$8.8 \cdot 10^{-2}$
6684.3	8	6	$1.07 \cdot 10^{-1}$
6756.6	4	4	$2.0 \cdot 10^{-2}$
6863.5	6	6	$2.5 \cdot 10^{-2}$
7233.5	2	4	$3.7 \cdot 10^{-2}$
7380.4	4	4	$5.6 \cdot 10^{-2}$
7589.3	6	4	$1.07 \cdot 10^{-1}$

Ar III

λ Å	Weights g_i	g_k	A $10^8\ s^{-1}$
769.15	5	3	6.0
871.10	5	3	1.59
875.53	3	1	3.74
878.73	5	5	2.79
879.62	3	3	$9.2 \cdot 10^{-1}$
883.18	1	3	1.22
887.40	3	5	$9.0 \cdot 10^{-1}$
3024.1	5	7	2.6
3027.2	5	5	$6.4 \cdot 10^{-1}$
3054.8	3	5	1.9
3064.8	3	3	1.0
3078.2	1	3	1.4
3285.9	5	7	2.0
3301.9	5	5	2.0
3311.3	5	3	2.0
3336.1	7	9	2.0
3344.7	5	7	1.8
3352.1	7	7	$2.2 \cdot 10^{-1}$
3358.5	3	5	1.6
3361.3	5	5	$3.0 \cdot 10^{-1}$
3472.6	5	7	$2.0 \cdot 10^{-1}$
3480.6	7	7	1.6
3499.7	3	3	1.3
3500.6	3	5	$2.6 \cdot 10^{-1}$
3502.7	5	3	$4.3 \cdot 10^{-1}$
3503.6	5	5	1.2
3511.7	7	5	$2.6 \cdot 10^{-1}$

Ar IV

λ Å	Weights g_i	g_k	A $10^8\ s^{-1}$
840.03	4	2	2.73
843.77	4	4	2.70
850.60	4	6	2.63

Ar VI

λ Å	Weights g_i	g_k	A $10^8\ s^{-1}$
292.15	2	2	$6.9 \cdot 10^{1}$
294.05	4	2	$1.36 \cdot 10^{2}$

Ar VII

λ Å	Weights g_i	g_k	A $10^8\ s^{-1}$
*250.41	9	3	$2.78 \cdot 10^{2}$
*477.54	9	15	$9.92 \cdot 10^{1}$
585.75	1	3	$7.83 \cdot 10^{1}$

λ Å	Weights g_i	g_k	A 10^8 s^{-1}
*637.30	9	9	$6.7 \cdot 10^{1}$

Ar VIII

λ Å	Weights g_i	g_k	A 10^8 s^{-1}
158.92	2	4	$1.1 \cdot 10^{2}$
159.18	2	2	$1.11 \cdot 10^{2}$
229.44	2	2	$1.12 \cdot 10^{2}$
230.88	4	2	$2.21 \cdot 10^{2}$
337.09	4	4	$1.2 \cdot 10^{1}$
337.26	6	4	$1.0 \cdot 10^{2}$
338.22	4	2	$1.1 \cdot 10^{2}$
519.43	2	4	$6.3 \cdot 10^{1}$
526.46	4	6	$7.2 \cdot 10^{1}$
526.87	4	4	$1.2 \cdot 10^{1}$
700.24	2	4	$2.55 \cdot 10^{1}$
713.81	2	2	$2.4 \cdot 10^{1}$

Ar IX

λ Å	Weights g_i	g_k	A 10^8 s^{-1}
48.739	1	3	$1.69 \cdot 10^{3}$

Ar XIII

λ Å	Weights g_i	g_k	A 10^8 s^{-1}
162.96	5	3	$3.4 \cdot 10^{2}$
*163.08	9	3	$5.3 \cdot 10^{2}$
184.90	5	5	$1.66 \cdot 10^{2}$
186.38	1	3	$8.8 \cdot 10^{1}$
*207.89	9	9	$9.5 \cdot 10^{1}$
*245.10	9	15	$3.7 \cdot 10^{1}$

Ar XIV

λ Å	Weights g_i	g_k	A 10^8 s^{-1}
180.29	2	4	$4.5 \cdot 10^{1}$
183.41	2	2	$1.69 \cdot 10^{2}$
187.95	4	4	$1.97 \cdot 10^{2}$
191.35	4	2	$7.5 \cdot 10^{1}$
194.39	2	2	$4.6 \cdot 10^{1}$
203.35	4	2	$7.8 \cdot 10^{1}$

Ar XV

λ Å	Weights g_i	g_k	A 10^8 s^{-1}
25.05	1	3	$1.7 \cdot 10^{4}$
221.10	1	3	$9.55 \cdot 10^{1}$
*265.3	9	9	$8.1 \cdot 10^{1}$

Ar XVI

λ Å	Weights g_i	g_k	A 10^8 s^{-1}
*23.52	2	6	$1.43 \cdot 10^{4}$
*24.96	6	10	$4.4 \cdot 10^{4}$
353.88	2	4	$1.5 \cdot 10^{1}$
389.11	2	2	$1.1 \cdot 10^{1}$
1268	2	4	1.9
1401	2	2	1.4
2975	2	4	$9.0 \cdot 10^{-2}$
3514	4	6	$6.5 \cdot 10^{-2}$

Arsenic

As I

λ Å	Weights g_i	g_k	A 10^8 s^{-1}
1890.4	4	6	2.0
1937.6	4	4	2.0
1972.6	4	2	2.0
2288.1	6	4	2.8
2344.0	2	4	$3.5 \cdot 10^{-1}$
2349.8	4	2	3.1
2369.7	4	4	$6.0 \cdot 10^{-1}$
2370.8	4	6	$4.2 \cdot 10^{-1}$
2456.5	6	4	$7.2 \cdot 10^{-2}$
2492.9	4	2	$1.2 \cdot 10^{-1}$
2745.0	2	4	$2.6 \cdot 10^{-1}$
2780.2	4	4	$7.8 \cdot 10^{-1}$
2860.4	2	2	$5.5 \cdot 10^{-1}$
2898.7	4	2	$9.9 \cdot 10^{-2}$

Barium

Ba I

λ Å	Weights g_i	g_k	A 10^8 s^{-1}
2427.41	1	3	$5.60 \cdot 10^{-3}$
2472.74	1	3	$4.60 \cdot 10^{-3}$
3071.58	1	3	$4.20 \cdot 10^{-1}$
3501.11	1	3	$3.50 \cdot 10^{-1}$
3889.33	1	3	$1.10 \cdot 10^{-2}$
4132.43	1	3	$1.50 \cdot 10^{-2}$
4181.09	3	5	$4.99 \cdot 10^{-4}$
4181.66	3	5	$5.42 \cdot 10^{-4}$
4182.27	3	5	$6.11 \cdot 10^{-4}$
4182.94	3	5	$6.65 \cdot 10^{-4}$
4183.64	3	5	$6.70 \cdot 10^{-4}$
4184.40	3	5	$7.93 \cdot 10^{-4}$
4185.25	3	5	$8.43 \cdot 10^{-4}$
4186.16	3	5	$9.24 \cdot 10^{-4}$
4187.15	3	5	$9.90 \cdot 10^{-4}$
4188.25	3	5	$1.03 \cdot 10^{-3}$
4189.44	3	5	$1.13 \cdot 10^{-3}$
4190.76	3	5	$1.28 \cdot 10^{-3}$
4192.20	3	5	$1.36 \cdot 10^{-3}$
4193.81	3	5	$1.58 \cdot 10^{-3}$
4195.59	3	5	$1.78 \cdot 10^{-3}$
4323.00	3	5	$8.80 \cdot 10^{-2}$
4402.54	3	5	$2.70 \cdot 10^{-1}$
4488.98	5	7	$2.80 \cdot 10^{-1}$
4493.64	5	5	$1.95 \cdot 10^{-1}$
4573.85	3	1	1.21
4579.64	5	5	$7.00 \cdot 10^{-1}$
4599.72	3	1	$4.07 \cdot 10^{-1}$
4619.92	1	3	$2.70 \cdot 10^{-2}$
4700.42	3	3	$6.10 \cdot 10^{-2}$
4726.43	5	3	$3.30 \cdot 10^{-1}$
4801.30	9	3	$1.39 \cdot 10^{-1}$
4902.85	5	3	$5.40 \cdot 10^{-2}$
5169.53	5	3	$9.00 \cdot 10^{-4}$
5519.04	3	5	$5.70 \cdot 10^{-1}$
5535.48	1	3	1.19
5777.62	5	7	$8.00 \cdot 10^{-1}$
5784.04	3	5	$2.10 \cdot 10^{-1}$
5800.23	5	5	$2.39 \cdot 10^{-1}$
5826.27	5	3	$4.50 \cdot 10^{-1}$
5971.70	5	5	$1.62 \cdot 10^{-1}$
5997.09	3	3	$2.80 \cdot 10^{-1}$
6019.47	3	1	$8.10 \cdot 10^{-1}$
6063.11	5	3	$5.60 \cdot 10^{-1}$
6083.39	3	1	$1.10 \cdot 10^{-1}$
6129.23	3	1	$6.00 \cdot 10^{-2}$
6341.68	5	7	$1.16 \cdot 10^{-1}$
6450.85	3	5	$1.10 \cdot 10^{-1}$
6498.76	7	7	$5.40 \cdot 10^{-1}$
6527.31	5	5	$3.30 \cdot 10^{-1}$
6527.40	15	15	$6.15 \cdot 10^{-1}$
6595.33	3	3	$3.80 \cdot 10^{-1}$
6675.27	5	3	$1.89 \cdot 10^{-1}$
6693.84	7	5	$1.46 \cdot 10^{-1}$
6986.80	5	3	$5.20 \cdot 10^{-3}$
7059.94	7	9	$5.00 \cdot 10^{-1}$
7120.33	3	5	$1.10 \cdot 10^{-1}$
7195.23	1	3	$5.60 \cdot 10^{-2}$
7213.60	5	5	$6.50 \cdot 10^{-4}$
7280.30	5	7	$3.20 \cdot 10^{-1}$
7392.41	3	3	$1.81 \cdot 10^{-1}$
7417.54	7	5	$7.70 \cdot 10^{-3}$
7488.08	7	7	$7.30 \cdot 10^{-2}$
7528.18	5	5	$2.70 \cdot 10^{-2}$
7610.48	5	5	$1.10 \cdot 10^{-2}$
7644.90	9	3	$5.03 \cdot 10^{-1}$
7672.09	3	5	$1.52 \cdot 10^{-1}$
7780.48	5	5	$7.60 \cdot 10^{-2}$
7877.80	3	5	$1.60 \cdot 10^{-2}$
7905.75	5	3	$2.65 \cdot 10^{-1}$
8147.70	5	5	$6.30 \cdot 10^{-2}$
8560.00	5	5	$2.00 \cdot 10^{-1}$
8654.08	5	7	$3.10 \cdot 10^{-3}$
9370.12	5	5	$7.60 \cdot 10^{-2}$
9645.60	7	5	$1.10 \cdot 10^{-1}$
9704.31	3	1	$1.60 \cdot 10^{-1}$
9821.48	3	1	$5.50 \cdot 10^{-2}$
10370.30	3	5	$1.30 \cdot 10^{-2}$
10540.10	5	3	$1.80 \cdot 10^{-2}$
10649.10	5	5	$2.70 \cdot 10^{-2}$
11303.00	5	3	$1.10 \cdot 10^{-3}$
11373.70	3	1	$1.30 \cdot 10^{-1}$
12342.30	3	3	$9.00 \cdot 10^{-4}$
14723.10	3	5	$8.60 \cdot 10^{-3}$
14999.90	5	3	$2.50 \cdot 10^{-3}$
17186.90	3	1	$2.70 \cdot 10^{-2}$
18202.80	5	3	$1.20 \cdot 10^{-2}$
21567.70	5	3	$2.60 \cdot 10^{-3}$
30685.30	5	3	$6.50 \cdot 10^{-3}$

Ba II

λ Å	Weights g_i	g_k	A 10^8 s^{-1}
2528.41	2	4	$6.91 \cdot 10^{-1}$
2634.78	4	6	$7.33 \cdot 10^{-1}$
3891.78	2	4	2.17
4130.65	4	6	2.18
4166.00	4	4	$3.54 \cdot 10^{-1}$
4267.92	6	8	$3.10 \cdot 10^{-1}$
4309.26	8	10	$3.10 \cdot 10^{-1}$
4325.75	4	6	$5.65 \cdot 10^{-2}$
4524.93	2	2	$6.63 \cdot 10^{-1}$
4554.03	2	4	1.11
4708.90	2	4	$8.47 \cdot 10^{-2}$
4843.48	4	6	$9.34 \cdot 10^{-2}$
4899.93	4	2	1.04
4934.08	2	2	$9.53 \cdot 10^{-1}$
4957.09	6	8	$5.13 \cdot 10^{-1}$
5012.95	8	10	$5.15 \cdot 10^{-1}$
5361.35	4	6	$4.01 \cdot 10^{-2}$
5391.59	6	8	$4.22 \cdot 10^{-2}$

λ Å	Weights g_i	g_k	A $10^8\ s^{-1}$
5784.15	2	4	$1.59{\cdot}10^{-1}$
5853.67	4	4	$6.00{\cdot}10^{-2}$
5981.26	4	6	$1.73{\cdot}10^{-1}$
6135.60	2	2	$6.64{\cdot}10^{-2}$
6141.71	6	4	$4.12{\cdot}10^{-1}$
6378.92	4	2	$1.18{\cdot}10^{-1}$
6496.90	4	2	$3.10{\cdot}10^{-1}$
6769.48	6	8	$9.35{\cdot}10^{-1}$
6874.08	8	10	$9.26{\cdot}10^{-1}$
7115.03	8	10	$8.80{\cdot}10^{-3}$
8496.80	2	4	$3.31{\cdot}10^{-2}$
8661.90	6	4	$1.27{\cdot}10^{-2}$
8703.69	4	6	$3.69{\cdot}10^{-2}$
8710.77	6	8	$7.88{\cdot}10^{-1}$
8737.75	4	6	$7.29{\cdot}10^{-1}$
8760.61	8	6	$1.17{\cdot}10^{-2}$
8897.46	6	6	$4.93{\cdot}10^{-2}$
9603.12	2	4	$4.16{\cdot}10^{-1}$
10115.00	4	6	$4.27{\cdot}10^{-1}$
10212.80	4	4	$6.92{\cdot}10^{-2}$
10768.00	2	4	$5.56{\cdot}10^{-2}$
11088.50	4	6	$6.11{\cdot}10^{-2}$
11127.50	4	4	$1.01{\cdot}10^{-2}$
11519.50	2	2	$2.47{\cdot}10^{-2}$
11577.10	2	2	$1.75{\cdot}10^{-1}$
11931.90	4	2	$4.44{\cdot}10^{-2}$
12475.00	4	2	$2.80{\cdot}10^{-1}$
13057.80	2	4	$2.14{\cdot}10^{-1}$
14211.50	2	2	$1.66{\cdot}10^{-1}$
17738.90	6	8	$2.16{\cdot}10^{-1}$
18530.70	8	10	$1.96{\cdot}10^{-1}$
18729.70	2	4	$1.23{\cdot}10^{-1}$
19642.60	4	6	$1.28{\cdot}10^{-1}$
19845.10	4	4	$2.07{\cdot}10^{-2}$
22994.70	2	2	$6.18{\cdot}10^{-2}$
24612.50	4	4	$4.75{\cdot}10^{-3}$
24699.00	4	2	$9.98{\cdot}10^{-2}$
25923.20	6	4	$3.66{\cdot}10^{-2}$
27687.20	2	4	$6.10{\cdot}10^{-2}$
29058.90	4	2	$2.89{\cdot}10^{-2}$
30196.00	2	2	$4.70{\cdot}10^{-2}$
42934.70	6	8	$4.82{\cdot}10^{-3}$
43294.30	4	6	$4.39{\cdot}10^{-3}$
47520.80	6	6	$2.37{\cdot}10^{-4}$

Beryllium

Be I

λ Å	Weights g_i	g_k	A $10^8\ s^{-1}$
1661.48	1	3	$7.23{\cdot}10^{-2}$
1929.62	3	5	$5.12{\cdot}10^{-2}$
1929.71	5	7	$6.82{\cdot}10^{-2}$
1943.59	1	3	$5.43{\cdot}10^{-2}$
1943.62	3	5	$7.32{\cdot}10^{-2}$
1943.71	5	7	$9.77{\cdot}10^{-2}$
1964.54	1	3	$8.15{\cdot}10^{-2}$
1964.56	3	3	$6.12{\cdot}10^{-2}$
1964.56	3	5	$1.10{\cdot}10^{-1}$
1964.65	5	7	$1.47{\cdot}10^{-1}$
1964.65	5	5	$3.67{\cdot}10^{-2}$
1997.95	1	3	$1.31{\cdot}10^{-1}$
1997.98	3	5	$1.76{\cdot}10^{-1}$
1997.98	3	3	$9.80{\cdot}10^{-2}$
1998.07	5	5	$5.88{\cdot}10^{-2}$
1998.07	5	7	$2.35{\cdot}10^{-1}$
2032.72	5	3	$5.3{\cdot}10^{-2}$
2055.88	1	3	$2.28{\cdot}10^{-1}$
2055.90	3	3	$1.71{\cdot}10^{-1}$
2055.90	3	5	$3.08{\cdot}10^{-1}$
2056.00	5	5	$1.03{\cdot}10^{-1}$
2056.00	5	7	$4.11{\cdot}10^{-1}$
2125.57	3	3	$6.1{\cdot}10^{-2}$
2125.68	5	3	$1.01{\cdot}10^{-1}$
2174.96	1	3	$4.51{\cdot}10^{-1}$
2174.99	3	3	$3.39{\cdot}10^{-1}$
2174.99	3	5	$6.09{\cdot}10^{-1}$
2175.10	5	5	$2.03{\cdot}10^{-1}$
2175.10	5	7	$8.13{\cdot}10^{-1}$
2348.61	1	3	5.54
2350.66	1	3	$4.68{\cdot}10^{-2}$
2350.70	3	3	$1.41{\cdot}10^{-1}$
2350.83	5	3	$2.35{\cdot}10^{-1}$
2494.54	1	3	1.07
2494.58	3	5	1.44
2494.58	3	3	$8.02{\cdot}10^{-1}$
2494.73	5	3	$5.35{\cdot}10^{-2}$
2494.73	5	7	1.93
2494.73	5	5	$4.81{\cdot}10^{-1}$
2650.45	3	5	1.06
2650.55	1	3	1.41
2650.60	3	3	1.06
2650.62	5	5	3.17
2650.70	3	1	4.23
2650.76	5	3	1.76
3193.83	3	5	$5.08{\cdot}10^{-2}$
3229.62	3	5	$7.17{\cdot}10^{-2}$
3282.91	3	5	$1.06{\cdot}10^{-1}$
3321.01	1	3	$1.70{\cdot}10^{-1}$
3321.08	3	3	$5.10{\cdot}10^{-1}$
3321.34	5	3	$8.50{\cdot}10^{-1}$
3367.60	3	5	$1.64{\cdot}10^{-1}$
3476.56	3	1	$3.11{\cdot}10^{-2}$
3515.54	3	5	$2.73{\cdot}10^{-1}$
3736.30	3	1	$5.09{\cdot}10^{-2}$
3813.45	3	5	$4.87{\cdot}10^{-1}$
4407.94	3	1	$1.01{\cdot}10^{-1}$
4572.66	3	5	$7.61{\cdot}10^{-1}$
5252.81	1	3	$3.12{\cdot}10^{-3}$
5365.49	1	3	$3.65{\cdot}10^{-3}$
5546.45	1	3	$5.4{\cdot}10^{-3}$
5857.01	1	3	$7.0{\cdot}10^{-3}$
6085.75	5	3	$1.4{\cdot}10^{-3}$
6229.11	5	7	$2.26{\cdot}10^{-2}$
6319.62	5	3	$2.2{\cdot}10^{-3}$
6473.54	1	3	$7.7{\cdot}10^{-3}$
6564.52	5	7	$3.49{\cdot}10^{-2}$
6711.74	1	3	$3.62{\cdot}10^{-3}$
6711.74	3	5	$4.89{\cdot}10^{-3}$
6711.74	3	3	$2.71{\cdot}10^{-3}$
6711.91	5	7	$6.52{\cdot}10^{-3}$
6711.91	5	5	$1.63{\cdot}10^{-3}$
6725.96	5	3	$3.1{\cdot}10^{-3}$
6786.56	3	5	$1.7{\cdot}10^{-3}$
6786.56	3	3	$1.7{\cdot}10^{-3}$
6884.26	3	3	$3.81{\cdot}10^{-3}$
6884.26	3	5	$6.86{\cdot}10^{-3}$
6884.26	1	3	$5.08{\cdot}10^{-3}$
6884.44	5	7	$9.16{\cdot}10^{-3}$
6884.44	5	5	$2.29{\cdot}10^{-3}$
7049.72	1	3	$9.66{\cdot}10^{-4}$
7049.72	3	3	$2.90{\cdot}10^{-3}$
7049.91	5	3	$4.83{\cdot}10^{-3}$
7154.46	3	3	$5.55{\cdot}10^{-3}$
7154.46	1	3	$7.39{\cdot}10^{-3}$
7154.46	3	5	$9.99{\cdot}10^{-3}$
7154.65	5	7	$1.33{\cdot}10^{-2}$
7154.65	5	5	$3.33{\cdot}10^{-3}$
7209.13	5	7	$5.71{\cdot}10^{-2}$
7308.29	3	5	$3.33{\cdot}10^{-3}$
7434.42	1	3	$1.53{\cdot}10^{-3}$
7434.42	3	3	$4.59{\cdot}10^{-3}$
7434.63	5	3	$7.64{\cdot}10^{-3}$
7448.87	3	1	$5.63{\cdot}10^{-3}$
7498.42	3	5	$4.46{\cdot}10^{-3}$
7551.90	5	3	$4.3{\cdot}10^{-3}$
7618.66	3	5	$1.51{\cdot}10^{-2}$
7618.66	1	3	$1.12{\cdot}10^{-2}$
7618.66	3	3	$8.41{\cdot}10^{-3}$
7618.88	5	7	$2.02{\cdot}10^{-2}$
7618.88	5	5	$5.05{\cdot}10^{-3}$
7714.38	3	1	$8.21{\cdot}10^{-3}$
7792.05	3	5	$6.02{\cdot}10^{-3}$
8090.07	1	3	$3.29{\cdot}10^{-3}$
8153.74	3	1	$1.27{\cdot}10^{-2}$
8158.98	1	3	$2.65{\cdot}10^{-3}$
8158.98	3	3	$7.94{\cdot}10^{-3}$
8159.23	5	3	$1.32{\cdot}10^{-2}$
8254.07	3	1	$3.38{\cdot}10^{-1}$
8286.90	3	5	$8.00{\cdot}10^{-3}$
8547.36	3	3	$1.31{\cdot}10^{-2}$
8547.36	1	3	$1.74{\cdot}10^{-2}$
8547.36	3	5	$2.36{\cdot}10^{-2}$
8547.63	5	7	$3.14{\cdot}10^{-2}$
8547.63	5	5	$7.86{\cdot}10^{-3}$
8547.63	5	3	$8.73{\cdot}10^{-4}$
8801.37	5	7	$9.58{\cdot}10^{-2}$
8882.16	7	5	$1.34{\cdot}10^{-3}$
8882.16	5	3	$1.06{\cdot}10^{-3}$
8882.16	3	1	$1.59{\cdot}10^{-3}$
8979.19	3	1	$2.13{\cdot}10^{-2}$
9190.45	5	7	$1.43{\cdot}10^{-2}$
9190.45	3	5	$1.35{\cdot}10^{-2}$
9190.45	7	7	$1.79{\cdot}10^{-3}$
9190.45	5	5	$2.50{\cdot}10^{-3}$
9243.88	3	5	$9.47{\cdot}10^{-3}$
9392.74	5	3	$1.86{\cdot}10^{-3}$
9392.74	3	3	$6.19{\cdot}10^{-4}$
9392.74	5	5	$3.71{\cdot}10^{-4}$
9392.74	7	5	$2.08{\cdot}10^{-3}$
9392.74	3	1	$2.47{\cdot}10^{-3}$
9847.31	5	3	$5.2{\cdot}10^{-3}$

λ Å	Weights g_i	Weights g_k	A 10^8 s^{-1}
9895.58	3	3	$1.59 \cdot 10^{-2}$
9895.58	1	3	$5.30 \cdot 10^{-3}$
9895.95	5	3	$2.65 \cdot 10^{-2}$
9939.78	5	7	$2.51 \cdot 10^{-2}$
9939.78	5	5	$4.40 \cdot 10^{-3}$
9939.78	7	7	$3.14 \cdot 10^{-3}$
9939.78	3	5	$2.38 \cdot 10^{-2}$

Be II

λ Å	Weights g_i	Weights g_k	A 10^8 s^{-1}
973.276	4	6	$9.69 \cdot 10^{-1}$
1026.89	2	4	1.46
1026.96	4	6	1.76
1036.30	2	4	1.720
1142.96	2	4	3.14
1143.04	4	6	3.76
1512.27	2	4	9.217
1512.41	4	6	$1.106 \cdot 10^{1}$
1512.42	4	4	1.843
1776.10	2	2	1.361
1776.31	4	2	2.722
2296.81	2	4	$1.09 \cdot 10^{-1}$
2296.91	4	6	$1.30 \cdot 10^{-1}$
2302.96	4	6	$7.28 \cdot 10^{-2}$
2302.98	6	8	$7.80 \cdot 10^{-2}$
2381.95	4	6	$1.13 \cdot 10^{-1}$
2381.98	6	8	$1.22 \cdot 10^{-1}$
2413.34	2	4	$1.64 \cdot 10^{-1}$
2413.45	4	6	$1.97 \cdot 10^{-1}$
2453.84	2	4	$1.06 \cdot 10^{-1}$
2507.41	4	6	$1.82 \cdot 10^{-1}$
2507.45	6	8	$1.94 \cdot 10^{-1}$
2617.99	2	4	$2.65 \cdot 10^{-1}$
2618.13	4	6	$3.18 \cdot 10^{-1}$
2697.59	4	2	$1.48 \cdot 10^{-1}$
2728.85	4	6	$3.23 \cdot 10^{-1}$
2728.89	6	8	$3.46 \cdot 10^{-1}$
3046.52	2	4	$4.66 \cdot 10^{-1}$
3046.69	4	6	$5.60 \cdot 10^{-1}$
3046.70	4	4	$9.33 \cdot 10^{-2}$
3130.42	2	4	1.1292
3131.07	2	2	1.1285
3197.10	4	6	$6.82 \cdot 10^{-1}$
3197.15	6	8	$7.31 \cdot 10^{-1}$
3197.16	6	6	$4.87 \cdot 10^{-2}$
3233.54	6	4	$3.37 \cdot 10^{-2}$
3241.63	2	2	$1.39 \cdot 10^{-1}$
3241.83	4	2	$2.78 \cdot 10^{-1}$
3274.59	2	4	$1.41 \cdot 10^{-1}$
3274.67	2	2	$1.41 \cdot 10^{-1}$
4360.66	2	4	$9.12 \cdot 10^{-1}$
4360.99	4	6	1.09
4361.03	4	4	$1.82 \cdot 10^{-1}$
4673.33	4	6	2.06
4673.42	6	8	2.21
4673.45	6	6	$1.47 \cdot 10^{-1}$
4702.34	2	4	$4.73 \cdot 10^{-2}$
4702.52	4	6	$5.68 \cdot 10^{-2}$
4828.12	6	4	$7.89 \cdot 10^{-2}$
4828.18	4	2	$8.77 \cdot 10^{-2}$
5218.12	2	4	$7.08 \cdot 10^{-2}$
5218.34	4	6	$8.49 \cdot 10^{-2}$
5218.34	4	4	$1.41 \cdot 10^{-2}$
5270.27	2	2	$3.234 \cdot 10^{-1}$
5270.81	4	2	$6.466 \cdot 10^{-1}$
5416.12	2	2	$2.84 \cdot 10^{-2}$
5416.36	4	2	$5.68 \cdot 10^{-2}$
6279.42	2	4	$1.12 \cdot 10^{-1}$
6279.74	4	4	$2.23 \cdot 10^{-2}$
6279.74	4	6	$1.34 \cdot 10^{-1}$
6756.75	2	2	$4.90 \cdot 10^{-2}$
6757.12	4	2	$9.80 \cdot 10^{-2}$
7401.20	2	4	$2.54 \cdot 10^{-2}$
7401.43	2	2	$2.54 \cdot 10^{-2}$
9048.14	2	4	$2.47 \cdot 10^{-2}$
9048.49	4	6	$2.96 \cdot 10^{-2}$
9048.49	4	4	$4.94 \cdot 10^{-3}$
9476.41	2	4	$1.81 \cdot 10^{-1}$
9477.03	4	6	$2.17 \cdot 10^{-1}$
9477.14	4	4	$3.62 \cdot 10^{-2}$

Bismuth

Bi I

λ Å	Weights g_i	Weights g_k	A 10^8 s^{-1}
1954.5	4	6	1.2
2021.2	4	4	$6.0 \cdot 10^{-2}$
2061.7	4	6	$9.9 \cdot 10^{-1}$
2110.3	4	2	$9.1 \cdot 10^{-1}$
2177.3	4	2	$2.6 \cdot 10^{-2}$
2228.3	4	4	$8.9 \cdot 10^{-1}$
2230.6	4	6	2.6
2276.6	4	4	$2.5 \cdot 10^{-1}$
2515.7	4	6	$4.3 \cdot 10^{-2}$
2627.9	4	4	$4.7 \cdot 10^{-1}$
2696.8	4	6	$6.4 \cdot 10^{-2}$
2780.5	4	2	$3.09 \cdot 10^{-1}$
2798.7	6	6	$3.6 \cdot 10^{-2}$
2898.0	4	2	1.53
2938.3	6	4	1.23
2989.0	4	4	$5.5 \cdot 10^{-1}$
2993.3	4	6	$1.6 \cdot 10^{-1}$
3024.6	6	6	$8.8 \cdot 10^{-1}$
3067.7	4	2	2.07
3076.7	4	4	$3.5 \cdot 10^{-2}$
3397.2	6	4	$1.81 \cdot 10^{-1}$
3402.9	6	6	$1.6 \cdot 10^{-2}$
3510.9	6	4	$6.8 \cdot 10^{-2}$
3596.1	2	4	$1.98 \cdot 10^{-1}$
3888.2	2	2	$6.9 \cdot 10^{-2}$
4121.5	2	2	$1.64 \cdot 10^{-1}$
4308.5	2	4	$1.6 \cdot 10^{-2}$
4493.0	2	4	$1.5 \cdot 10^{-2}$
4722.5	4	2	$1.17 \cdot 10^{-1}$
6134.8	4	4	$1.8 \cdot 10^{-2}$

Boron

B I

λ Å	Weights g_i	Weights g_k	A 10^8 s^{-1}
1151.42	4	4	2.44
1378.65	2	4	3.42
1378.87	2	2	$1.37 \cdot 10^{1}$
1378.94	4	4	$1.71 \cdot 10^{1}$
1379.17	4	2	6.84
1465.56	2	4	3.3
1465.66	4	4	6.6
1465.79	6	4	9.9
1566.66	4	6	$3.36 \cdot 10^{-1}$
1587.38	2	4	$6.9 \cdot 10^{-1}$
1587.45	4	2	1.4
1587.59	4	6	$5.0 \cdot 10^{-1}$
1587.66	6	4	$7.5 \cdot 10^{-1}$
1587.75	6	6	1.2
1600.37	2	4	$4.67 \cdot 10^{-1}$
1600.76	4	6	$5.60 \cdot 10^{-1}$
1666.85	2	4	$8.66 \cdot 10^{-1}$
1667.27	4	6	1.04
1818.35	4	2	$6.22 \cdot 10^{-1}$
1825.89	2	4	1.70
1826.40	4	6	2.04
1826.40	4	4	$3.39 \cdot 10^{-1}$
2066.38	4	6	$6.8 \cdot 10^{-1}$
2066.65	6	6	1.6
2066.73	2	4	$9.4 \cdot 10^{-1}$
2066.93	4	4	$3.0 \cdot 10^{-1}$
2067.20	6	4	1.0
2067.20	4	2	1.9
2088.89	2	4	$3.61 \cdot 10^{-1}$
2089.57	4	6	$4.32 \cdot 10^{-1}$
2496.77	2	2	$8.37 \cdot 10^{-1}$
2497.72	4	2	1.67
5761.90	6	8	$5.7 \cdot 10^{-3}$
5942.62	6	8	$8.4 \cdot 10^{-3}$
5942.73	4	6	$7.8 \cdot 10^{-3}$
6244.56	6	8	$1.3 \cdot 10^{-2}$
6244.68	4	6	$1.2 \cdot 10^{-2}$
6819.52	6	8	$2.2 \cdot 10^{-2}$
6819.66	4	6	$2.0 \cdot 10^{-2}$
7208.59	4	2	$1.72 \cdot 10^{-2}$
8211.79	6	8	$3.42 \cdot 10^{-2}$
8211.79	6	6	$2.28 \cdot 10^{-3}$
8212.00	4	6	$3.19 \cdot 10^{-2}$
8667.23	2	2	$1.83 \cdot 10^{-2}$
8668.57	4	2	$3.67 \cdot 10^{-2}$
9576.21	4	6	$5.01 \cdot 10^{-3}$
9576.34	6	8	$5.37 \cdot 10^{-3}$

B II

λ Å	Weights g_i	Weights g_k	A 10^8 s^{-1}
882.676	5	7	$2.44 \cdot 10^{1}$
984.698	5	7	$2.62 \cdot 10^{1}$
1230.17	3	5	$1.36 \cdot 10^{1}$
1624.02	5	5	6.48
2005.87	7	7	2.22
2123.86	5	7	1.58
2220.30	3	3	3.49
2918.08	7	9	$6.57 \cdot 10^{-1}$
3323.21	3	5	$8.62 \cdot 10^{-1}$
3323.60	5	7	1.15
3451.30	3	5	$5.41 \cdot 10^{-1}$
4121.93	5	5	$3.68 \cdot 10^{-1}$
4121.93	5	7	2.10
4121.93	7	9	2.36
4121.93	7	7	$2.63 \cdot 10^{-1}$

λ Å	Weights g_i	g_k	A $10^8\ s^{-1}$
4121.93	3	5	1.98
4472.15	3	3	$5.44 \cdot 10^{-1}$
4472.86	5	3	$9.07 \cdot 10^{-1}$
4784.20	7	5	$1.94 \cdot 10^{-1}$
4940.37	5	7	1.88
6148.91	5	7	$1.96 \cdot 10^{-1}$
6148.91	3	5	$1.85 \cdot 10^{-1}$
6148.91	7	9	$2.20 \cdot 10^{-1}$
6285.51	5	3	$3.30 \cdot 10^{-1}$
6571.12	3	5	$1.14 \cdot 10^{-1}$
6717.65	5	7	$1.71 \cdot 10^{-1}$
7030.27	3	5	$3.97 \cdot 10^{-1}$
7032.03	3	3	$3.97 \cdot 10^{-1}$
7032.33	3	1	$3.97 \cdot 10^{-1}$
7159.55	3	5	$1.19 \cdot 10^{-1}$
7160.16	1	3	$1.59 \cdot 10^{-1}$
7165.11	3	3	$1.19 \cdot 10^{-1}$
7168.46	3	1	$4.75 \cdot 10^{-1}$
7170.45	5	5	$3.56 \cdot 10^{-1}$
7176.02	5	3	$1.97 \cdot 10^{-1}$
7228.46	1	3	$1.39 \cdot 10^{-1}$
7638.62	5	7	$1.71 \cdot 10^{-1}$
7638.62	3	5	$1.28 \cdot 10^{-1}$
7638.62	1	3	$9.5 \cdot 10^{-2}$
8655.83	5	3	$6.4 \cdot 10^{-2}$
9121.00	3	1	$3.23 \cdot 10^{-1}$
9226.43	7	9	$4.37 \cdot 10^{-1}$
9226.43	5	7	$3.88 \cdot 10^{-1}$
9226.43	3	5	$3.67 \cdot 10^{-1}$
9226.43	7	7	$4.87 \cdot 10^{-2}$
9226.43	5	5	$6.8 \cdot 10^{-2}$
9446.36	3	5	$2.08 \cdot 10^{-1}$
9659.49	3	5	$6.3 \cdot 10^{-2}$

Bromine

Br I

λ Å	Weights g_i	g_k	A $10^8\ s^{-1}$
1488.5	4	4	1.2
1540.7	4	4	1.4
1574.8	2	4	$2.0 \cdot 10^{-1}$
1576.4	4	6	$2.1 \cdot 10^{-2}$
1633.4	2	4	$8.1 \cdot 10^{-2}$
4365.1	2	4	$7.5 \cdot 10^{-3}$
4425.1	4	2	$4.2 \cdot 10^{-3}$
4441.7	6	4	$7.5 \cdot 10^{-3}$
4472.6	4	4	$9.3 \cdot 10^{-3}$
4477.7	6	8	$1.3 \cdot 10^{-2}$
4513.4	6	4	$2.8 \cdot 10^{-3}$
4525.6	6	6	$7.2 \cdot 10^{-3}$
4575.7	4	4	$1.6 \cdot 10^{-2}$
4614.6	4	6	$5.4 \cdot 10^{-3}$
4979.8	4	4	$2.6 \cdot 10^{-3}$
5245.1	2	4	$3.1 \cdot 10^{-3}$
5345.4	2	4	$7.6 \cdot 10^{-3}$
7348.5	4	6	$1.2 \cdot 10^{-1}$
7513.0	6	4	$1.2 \cdot 10^{-1}$
7803.0	2	4	$5.3 \cdot 10^{-2}$
7938.7	6	6	$1.9 \cdot 10^{-1}$
8131.5	2	4	$3.8 \cdot 10^{-2}$
8343.7	2	2	$2.2 \cdot 10^{-1}$
8446.6	4	4	$1.2 \cdot 10^{-1}$
8638.7	6	4	$9.7 \cdot 10^{-2}$

Br II

λ Å	Weights g_i	g_k	A $10^8\ s^{-1}$
4704.9	5	7	1.1
4785.5	5	5	$9.4 \cdot 10^{-1}$
4816.7	5	3	1.1

Cadmium

Cd I

λ Å	Weights g_i	g_k	A $10^8\ s^{-1}$
2288.0	1	3	5.3
2836.9	1	3	$2.8 \cdot 10^{-1}$
2880.8	3	5	$4.2 \cdot 10^{-1}$
2881.2	3	3	$2.4 \cdot 10^{-1}$
2980.6	5	7	$5.9 \cdot 10^{-1}$
2981.4	5	5	$1.5 \cdot 10^{-1}$
3261.1	1	3	$4.06 \cdot 10^{-3}$
3403.7	1	3	$7.7 \cdot 10^{-1}$
3466.2	3	5	1.2
3467.7	3	3	$6.7 \cdot 10^{-1}$
3610.5	5	7	1.3
3612.9	5	5	$3.5 \cdot 10^{-1}$
4140.5	3	5	$4.7 \cdot 10^{-2}$
4662.4	3	5	$5.5 \cdot 10^{-2}$
4678.1	1	3	$1.3 \cdot 10^{-1}$
4799.9	3	3	$4.1 \cdot 10^{-1}$
5085.8	5	3	$5.6 \cdot 10^{-1}$
6438.5	3	5	$5.9 \cdot 10^{-1}$

Cd II

λ Å	Weights g_i	g_k	A $10^8\ s^{-1}$
2144.4	2	4	2.8
2265.0	2	2	3.0
2572.9	2	2	1.7
2748.5	4	2	2.8
4415.6	4	6	$1.4 \cdot 10^{-2}$

Calcium

Ca I

λ Å	Weights g_i	g_k	A $10^8\ s^{-1}$
3006.9	5	5	$7.5 \cdot 10^{-1}$
3361.9	5	7	$2.23 \cdot 10^{-1}$
3630.8	3	5	$2.97 \cdot 10^{-1}$
3644.4	5	7	$3.55 \cdot 10^{-1}$
4098.5	7	9	$1.3 \cdot 10^{-1}$
4108.5	5	7	$9.0 \cdot 10^{-1}$
4226.7	1	3	2.18
4283.0	3	5	$4.34 \cdot 10^{-1}$
4289.4	1	3	$6.0 \cdot 10^{-1}$
4299.0	3	3	$4.66 \cdot 10^{-1}$
4302.5	5	5	1.36
4307.7	3	1	1.99
4318.7	5	3	$7.4 \cdot 10^{-1}$
4355.1	5	7	$1.9 \cdot 10^{-1}$
4425.4	1	3	$4.98 \cdot 10^{-1}$
4435.0	3	5	$6.7 \cdot 10^{-1}$
4435.7	3	3	$3.42 \cdot 10^{-1}$
4454.8	5	7	$8.7 \cdot 10^{-1}$
4455.9	5	5	$2.0 \cdot 10^{-1}$
4526.9	5	3	$4.1 \cdot 10^{-1}$
4578.6	3	5	$1.76 \cdot 10^{-1}$
4581.4	5	7	$2.09 \cdot 10^{-1}$
4585.9	7	9	$2.29 \cdot 10^{-1}$
4878.1	5	7	$1.88 \cdot 10^{-1}$
5041.6	5	3	$3.3 \cdot 10^{-1}$
5188.9	3	5	$4.0 \cdot 10^{-1}$
5261.7	3	3	$1.5 \cdot 10^{-1}$
5262.2	3	1	$6.0 \cdot 10^{-1}$
5264.2	5	5	$9.1 \cdot 10^{-2}$
5265.6	5	3	$4.4 \cdot 10^{-1}$
5270.3	7	5	$5.0 \cdot 10^{-1}$
5582.0	5	7	$6.0 \cdot 10^{-2}$
5588.8	7	7	$4.9 \cdot 10^{-1}$
5590.1	3	5	$8.3 \cdot 10^{-2}$
5594.5	5	5	$3.8 \cdot 10^{-1}$
5598.5	3	3	$4.3 \cdot 10^{-1}$
5601.3	7	5	$8.6 \cdot 10^{-2}$
5602.9	5	3	$1.4 \cdot 10^{-1}$
5857.5	3	5	$6.6 \cdot 10^{-1}$
6102.7	1	3	$9.6 \cdot 10^{-2}$
6122.2	3	3	$2.87 \cdot 10^{-1}$
6162.2	5	3	$3.54 \cdot 10^{-1}$
6169.1	5	3	$1.7 \cdot 10^{-1}$
6169.6	7	5	$1.9 \cdot 10^{-1}$
6439.1	7	9	$5.3 \cdot 10^{-1}$
6449.8	3	5	$9.0 \cdot 10^{-2}$
6462.6	5	7	$4.7 \cdot 10^{-1}$
6471.7	7	7	$5.9 \cdot 10^{-2}$
6493.8	3	5	$4.4 \cdot 10^{-1}$
6499.7	5	5	$8.1 \cdot 10^{-2}$

Ca II

λ Å	Weights g_i	g_k	A $10^8\ s^{-1}$
1341.9	2	4	$1.5 \cdot 10^{-2}$
1342.5	2	2	$1.5 \cdot 10^{-2}$
1649.9	2	4	$3.2 \cdot 10^{-3}$
1652.0	2	2	$3.1 \cdot 10^{-3}$
1673.9	2	4	$2.24 \cdot 10^{-1}$
1680.1	4	6	$2.65 \cdot 10^{-1}$
1680.1	4	4	$4.41 \cdot 10^{-2}$
1807.3	2	4	$3.54 \cdot 10^{-1}$
1814.5	4	6	$4.2 \cdot 10^{-1}$
1814.7	4	4	$7.0 \cdot 10^{-2}$
1843.1	2	2	$1.6 \cdot 10^{-1}$
1850.7	4	2	$3.08 \cdot 10^{-1}$
2103.2	2	4	$8.2 \cdot 10^{-1}$
2112.8	4	6	$9.7 \cdot 10^{-1}$
2113.2	4	4	$1.6 \cdot 10^{-1}$
2197.8	2	2	$3.1 \cdot 10^{-1}$
2208.6	4	2	$6.2 \cdot 10^{-1}$
3158.9	2	4	3.1
3179.3	4	6	3.6
3181.3	4	4	$5.8 \cdot 10^{-1}$
3706.0	2	2	$8.8 \cdot 10^{-1}$
3736.9	4	2	1.7
3933.7	2	4	1.47
3968.5	2	2	1.4

Ca III

λ Å	Weights g_i	g_k	A $10^8\ s^{-1}$
357.97	1	3	$8.8 \cdot 10^{2}$
439.69	1	3	$1.9 \cdot 10^{-1}$
490.55	1	3	$1.6 \cdot 10^{-2}$

Ca V

λ Å	Weights g_i	Weights g_k	A 10^8 s^{-1}
558.60	5	3	$2.2\cdot10^{1}$
637.93	5	3	3.9
643.12	3	1	9.1
646.57	5	5	6.9
647.88	3	3	2.3
651.55	1	3	2.9
656.76	3	5	2.1

Ca VII

λ Å	Weights g_i	Weights g_k	A 10^8 s^{-1}
550.20	5	5	$1.8\cdot10^{1}$
624.39	1	3	3.3
630.54	3	5	4.5
630.79	3	3	2.2
639.15	5	7	5.7
640.41	5	5	1.3

Ca VIII

λ Å	Weights g_i	Weights g_k	A 10^8 s^{-1}
182.71	2	2	$1.6\cdot10^{2}$
184.16	4	2	$3.2\cdot10^{2}$

Ca IX

λ Å	Weights g_i	Weights g_k	A 10^8 s^{-1}
163.23	5	3	$3.76\cdot10^{2}$
371.89	1	3	$8.8\cdot10^{1}$
373.81	3	5	$1.16\cdot10^{2}$
378.08	5	7	$1.5\cdot10^{2}$
395.03	3	5	$2.2\cdot10^{2}$
466.24	1	3	$1.12\cdot10^{2}$
498.01	3	5	$2.49\cdot10^{1}$
506.18	5	5	$7.2\cdot10^{1}$
515.57	5	3	$3.75\cdot10^{1}$

Ca X

λ Å	Weights g_i	Weights g_k	A 10^8 s^{-1}
110.96	2	4	$2.9\cdot10^{2}$
111.20	2	2	$2.92\cdot10^{2}$
151.84	2	2	$2.3\cdot10^{2}$
153.02	4	2	$4.5\cdot10^{2}$
206.57	4	4	$2.9\cdot10^{1}$
206.75	6	4	$2.6\cdot10^{2}$
207.39	4	2	$2.8\cdot10^{2}$
411.70	2	4	$8.3\cdot10^{1}$
419.75	4	6	$9.5\cdot10^{1}$
420.47	4	4	$1.6\cdot10^{1}$
557.76	2	4	$3.50\cdot10^{1}$
574.01	2	2	$3.2\cdot10^{1}$

Ca XI

λ Å	Weights g_i	Weights g_k	A 10^8 s^{-1}
30.448	1	3	$6.2\cdot10^{3}$
30.867	1	3	$4.9\cdot10^{4}$
35.212	1	3	$2.0\cdot10^{3}$

Ca XII

λ Å	Weights g_i	Weights g_k	A 10^8 s^{-1}
140.05	4	2	$3.7\cdot10^{2}$
147.27	2	2	$1.6\cdot10^{2}$

Ca XV

λ Å	Weights g_i	Weights g_k	A 10^8 s^{-1}
141.69	5	3	$4.08\cdot10^{2}$
*142.23	9	3	$6.3\cdot10^{2}$
161.00	5	5	$1.9\cdot10^{2}$

Ca XVII

λ Å	Weights g_i	Weights g_k	A 10^8 s^{-1}
19.558	1	3	$3.8\cdot10^{4}$
21.198	3	5	$4.9\cdot10^{4}$
192.82	1	3	$1.21\cdot10^{2}$
218.82	3	5	$2.76\cdot10^{1}$
223.02	1	3	$3.44\cdot10^{1}$
228.72	3	3	$2.37\cdot10^{1}$
232.83	5	5	$6.5\cdot10^{1}$
244.06	5	3	$3.28\cdot10^{1}$

Ca XVIII

λ Å	Weights g_i	Weights g_k	A 10^8 s^{-1}
*18.71	2	6	$2.31\cdot10^{4}$
*19.74	6	10	$7.0\cdot10^{4}$
302.19	2	4	$2.0\cdot10^{1}$
344.76	2	2	$1.3\cdot10^{1}$

Carbon

C I

λ Å	Weights g_i	Weights g_k	A 10^8 s^{-1}
1121.52	1	3	$1.74\cdot10^{-2}$
1121.64	3	1	$1.10\cdot10^{-1}$
1121.66	3	3	$3.50\cdot10^{-2}$
1121.92	3	3	$3.60\cdot10^{-2}$
1122.00	5	3	$3.96\cdot10^{-2}$
1122.00	3	5	$8.49\cdot10^{-2}$
1122.10	5	5	$6.71\cdot10^{-2}$
1122.18	5	7	$3.38\cdot10^{-2}$
1122.33	5	7	$1.06\cdot10^{-1}$
1122.45	1	3	$9.04\cdot10^{-2}$
1122.52	1	3	$2.03\cdot10^{-2}$
1122.65	3	3	$2.73\cdot10^{-2}$
1122.79	3	5	$7.25\cdot10^{-2}$
1123.11	5	7	$3.58\cdot10^{-2}$
1128.07	1	3	$2.37\cdot10^{-2}$
1128.17	1	3	$1.83\cdot10^{-2}$
1128.25	3	1	$1.55\cdot10^{-1}$
1128.28	3	3	$4.96\cdot10^{-2}$
1128.63	5	3	$5.80\cdot10^{-2}$
1128.69	3	3	$4.79\cdot10^{-2}$
1128.75	5	5	$1.02\cdot10^{-1}$
1128.82	3	5	$1.61\cdot10^{-1}$
1128.90	5	7	$3.98\cdot10^{-2}$
1129.03	5	3	$1.82\cdot10^{-2}$
1129.13	5	7	$1.70\cdot10^{-1}$
1129.20	1	3	$1.35\cdot10^{-1}$
1129.32	1	3	$3.46\cdot10^{-2}$
1129.40	3	3	$4.86\cdot10^{-2}$
1129.42	5	5	$1.19\cdot10^{-1}$
1129.59	3	1	$6.41\cdot10^{-2}$
1129.62	3	5	$7.29\cdot10^{-2}$
1129.87	5	3	$2.22\cdot10^{-2}$
1129.92	5	7	$4.84\cdot10^{-2}$
1138.38	1	3	$4.13\cdot10^{-2}$
1138.56	3	1	$2.31\cdot10^{-1}$
1138.56	1	3	$1.70\cdot10^{-2}$
1138.60	3	3	$7.31\cdot10^{-2}$
1138.95	5	3	$8.98\cdot10^{-2}$
1139.09	5	5	$1.60\cdot10^{-1}$
1139.30	3	3	$6.24\cdot10^{-2}$
1139.43	5	7	$4.31\cdot10^{-2}$
1139.51	3	5	$2.72\cdot10^{-1}$
1139.65	5	3	$2.26\cdot10^{-2}$
1139.77	3	5	$4.63\cdot10^{-2}$
1139.79	1	3	$2.11\cdot10^{-1}$
1139.81	5	7	$2.96\cdot10^{-1}$
1139.86	5	5	$1.30\cdot10^{-2}$
1140.01	3	3	$9.45\cdot10^{-2}$
1140.01	1	3	$6.10\cdot10^{-2}$
1140.12	5	5	$1.84\cdot10^{-1}$
1140.32	3	1	$1.15\cdot10^{-1}$
1140.36	3	5	$6.39\cdot10^{-2}$
1140.57	5	3	$4.17\cdot10^{-2}$
1140.64	5	7	$6.29\cdot10^{-2}$
1155.81	1	3	$8.21\cdot10^{-2}$
1155.98	3	1	$3.74\cdot10^{-1}$
1156.03	3	3	$1.14\cdot10^{-1}$
1156.20	3	5	$2.39\cdot10^{-2}$
1156.39	5	3	$1.51\cdot10^{-1}$
1156.56	5	5	$2.69\cdot10^{-1}$
1157.33	5	7	$3.99\cdot10^{-2}$
1157.41	3	3	$7.50\cdot10^{-2}$
1157.77	5	3	$2.50\cdot10^{-2}$
1157.77	3	5	$2.62\cdot10^{-1}$
1157.91	1	3	$3.52\cdot10^{-1}$
1158.02	5	7	$5.57\cdot10^{-1}$
1158.03	3	5	$3.10\cdot10^{-1}$
1158.13	3	3	$1.98\cdot10^{-1}$
1158.13	5	5	$2.58\cdot10^{-1}$
1158.32	1	3	$1.09\cdot10^{-1}$
1158.40	5	5	$8.20\cdot10^{-2}$
1158.54	3	3	$3.36\cdot10^{-2}$
1158.67	3	1	$2.19\cdot10^{-1}$
1158.73	3	5	$5.11\cdot10^{-2}$
1158.91	5	3	$8.37\cdot10^{-2}$
1158.97	5	7	$7.22\cdot10^{-2}$
1188.83	1	3	$1.95\cdot10^{-1}$
1188.99	3	1	$7.18\cdot10^{-1}$
1189.07	3	3	$2.01\cdot10^{-1}$
1189.25	3	5	$1.09\cdot10^{-1}$
1189.45	5	3	$2.96\cdot10^{-1}$
1189.63	5	5	$5.28\cdot10^{-1}$
1191.84	5	7	$2.91\cdot10^{-2}$
1192.22	1	3	$2.15\cdot10^{-2}$
1192.83	5	3	$2.19\cdot10^{-2}$
1193.01	3	5	$7.91\cdot10^{-1}$
1193.03	1	3	$6.39\cdot10^{-1}$
1193.24	5	7	1.11
1193.26	3	3	$4.29\cdot10^{-1}$
1193.39	5	5	$3.46\cdot10^{-1}$
1193.65	5	3	$2.25\cdot10^{-2}$
1193.68	3	5	$2.54\cdot10^{-1}$
1194.00	1	3	$1.94\cdot10^{-1}$
1194.06	5	5	$2.98\cdot10^{-1}$
1194.23	3	3	$8.34\cdot10^{-2}$
1194.30	3	5	$4.10\cdot10^{-2}$
1194.41	3	1	$4.40\cdot10^{-1}$
1194.49	5	7	$6.77\cdot10^{-2}$
1194.61	5	3	$1.77\cdot10^{-1}$
1260.74	1	3	$5.70\cdot10^{-1}$

λ Å	Weights g_i	g_k	A $10^8\ s^{-1}$
1260.93	3	1	1.81
1261.00	3	3	$4.68 \cdot 10^{-1}$
1261.12	3	5	$4.00 \cdot 10^{-1}$
1261.43	5	3	$7.51 \cdot 10^{-1}$
1261.55	5	5	1.34
1274.11	5	7	$1.20 \cdot 10^{-2}$
1276.48	1	3	$5.04 \cdot 10^{-2}$
1276.75	3	3	$9.38 \cdot 10^{-2}$
1277.19	5	3	$1.81 \cdot 10^{-2}$
1277.25	1	3	1.26
1277.28	3	5	1.70
1277.51	3	3	$9.24 \cdot 10^{-1}$
1277.55	5	7	2.28
1277.72	5	5	$6.30 \cdot 10^{-1}$
1277.95	5	3	$5.94 \cdot 10^{-2}$
1279.89	3	5	$3.43 \cdot 10^{-1}$
1280.14	1	3	$3.55 \cdot 10^{-1}$
1280.33	5	5	$6.38 \cdot 10^{-1}$
1280.40	3	3	$1.87 \cdot 10^{-1}$
1280.60	3	1	$8.82 \cdot 10^{-1}$
1280.85	5	3	$3.62 \cdot 10^{-1}$
1328.83	1	3	$7.31 \cdot 10^{-1}$
1329.09	3	1	2.22
1329.10	3	5	$5.39 \cdot 10^{-1}$
1329.12	3	3	$5.55 \cdot 10^{-1}$
1329.58	5	5	1.64
1329.60	5	3	$9.19 \cdot 10^{-1}$
1459.03	5	3	$5.45 \cdot 10^{-1}$
1463.34	5	7	1.79
1467.40	5	3	$5.49 \cdot 10^{-1}$
1470.09	5	7	$1.41 \cdot 10^{-2}$
1472.23	5	3	$7.79 \cdot 10^{-3}$
1481.76	5	5	$3.45 \cdot 10^{-1}$
1560.31	1	3	$6.54 \cdot 10^{-1}$
1560.68	3	5	$8.82 \cdot 10^{-1}$
1560.71	3	3	$4.89 \cdot 10^{-1}$
1561.34	5	5	$2.93 \cdot 10^{-1}$
1561.37	5	3	$3.25 \cdot 10^{-2}$
1561.44	5	7	1.17
1656.27	3	5	$8.72 \cdot 10^{-1}$
1656.93	1	3	1.16
1657.01	5	5	2.61
1657.38	3	3	$8.66 \cdot 10^{-1}$
1657.91	3	1	3.47
1658.12	5	3	1.44
1751.83	1	3	$8.38 \cdot 10^{-1}$
1763.91	1	3	$2.44 \cdot 10^{-2}$
1930.90	5	3	3.39
2478.56	1	3	$2.80 \cdot 10^{-1}$

C II

λ Å	Weights g_i	g_k	A $10^8\ s^{-1}$
687.053	2	4	$2.35 \cdot 10^{1}$
687.345	4	6	$2.82 \cdot 10^{1}$
687.352	4	4	4.70
806.384	4	6	2.53
806.533	2	4	3.51
806.568	6	6	5.89
806.677	4	4	1.12
806.687	2	2	1.40
806.830	4	2	7.01
806.861	6	4	3.79
858.092	2	2	1.49
858.559	4	2	2.93
903.623	2	4	6.78
903.962	2	2	$2.70 \cdot 10^{1}$
904.142	4	4	$3.39 \cdot 10^{1}$
904.480	4	2	$1.36 \cdot 10^{1}$
1009.86	2	4	5.65
1010.08	4	4	$1.13 \cdot 10^{1}$
1010.37	6	4	$1.69 \cdot 10^{1}$
1036.34	2	2	7.38
1037.02	4	2	$1.46 \cdot 10^{1}$
1065.89	6	4	$1.30 \cdot 10^{1}$
1065.92	4	4	1.44
1066.13	4	2	$1.45 \cdot 10^{1}$
1126.99	2	4	$1.01 \cdot 10^{-6}$
1127.13	2	2	$5.55 \cdot 10^{-7}$
1127.27	4	4	$1.91 \cdot 10^{-7}$
1127.41	4	2	$9.06 \cdot 10^{-8}$
1127.63	6	4	$1.78 \cdot 10^{-6}$
1323.86	6	4	$4.94 \cdot 10^{-1}$
1323.91	4	4	4.38
1323.95	6	6	4.55
1324.00	4	6	$3.27 \cdot 10^{-1}$
1334.53	2	4	2.41
1335.66	4	4	$4.76 \cdot 10^{-1}$
1335.71	4	6	2.88
1384.00	2	4	$4.07 \cdot 10^{-1}$
1384.36	2	2	$3.88 \cdot 10^{-1}$
1490.38	6	4	$2.02 \cdot 10^{-6}$
1720.46	2	4	$8.65 \cdot 10^{-1}$
1721.01	2	2	3.46
1721.68	4	4	4.33
1722.24	4	2	1.72
1760.40	6	4	$3.68 \cdot 10^{-1}$
1760.47	4	4	$4.08 \cdot 10^{-2}$
1760.82	4	2	$4.09 \cdot 10^{-1}$
1915.32	2	4	$9.36 \cdot 10^{-2}$
1916.01	2	2	$9.50 \cdot 10^{-2}$
2323.50	2	4	$1.40 \cdot 10^{-8}$
2324.69	2	2	$5.99 \cdot 10^{-7}$
2325.40	4	6	$4.43 \cdot 10^{-7}$
2326.93	4	4	$8.49 \cdot 10^{-8}$
2328.12	4	2	$6.78 \cdot 10^{-7}$
2509.13	2	4	$4.71 \cdot 10^{-1}$
2511.74	4	4	$9.27 \cdot 10^{-2}$
2512.06	4	6	$5.61 \cdot 10^{-1}$
2836.71	2	4	$3.30 \cdot 10^{-1}$
2837.60	2	2	$3.29 \cdot 10^{-1}$
3183.50	2	4	$2.40 \cdot 10^{-7}$
3187.70	4	4	$8.09 \cdot 10^{-7}$
3922.08	2	4	$1.44 \cdot 10^{-7}$
4309.31	4	4	$3.92 \cdot 10^{-3}$
4309.58	6	4	$3.52 \cdot 10^{-2}$
4312.80	4	2	$3.92 \cdot 10^{-2}$
4735.46	2	4	$9.05 \cdot 10^{-5}$
4737.97	2	2	$5.47 \cdot 10^{-4}$
4744.77	4	4	$5.98 \cdot 10^{-4}$
4747.28	4	2	$2.33 \cdot 10^{-4}$
6578.05	2	4	$3.67 \cdot 10^{-1}$
6582.88	2	2	$3.66 \cdot 10^{-1}$
7231.33	2	4	$3.49 \cdot 10^{-1}$
7236.42	4	6	$4.18 \cdot 10^{-1}$
7237.17	4	4	$6.96 \cdot 10^{-2}$

C III

λ Å	Weights g_i	g_k	A $10^8\ s^{-1}$
310.170	1	3	6.56
386.203	1	3	$3.46 \cdot 10^{1}$
459.466	1	3	$5.91 \cdot 10^{1}$
459.514	3	5	$7.97 \cdot 10^{1}$
459.627	5	7	$1.06 \cdot 10^{2}$
574.281	3	5	$6.24 \cdot 10^{1}$
977.020	1	3	$1.767 \cdot 10^{1}$
1174.93	3	5	3.293
1175.26	1	3	4.385
1175.59	3	3	3.287
1175.71	5	5	9.856
1175.99	3	1	$1.313 \cdot 10^{1}$
1176.37	5	3	5.468
1247.38	3	1	$2.082 \cdot 10^{1}$
2296.87	3	5	1.376
2849.05	3	1	$1.95 \cdot 10^{-1}$
3703.70	3	3	$5.90 \cdot 10^{-1}$
4325.56	3	5	$1.24 \cdot 10^{-1}$
4647.42	3	5	$7.26 \cdot 10^{-1}$
4650.25	3	3	$7.25 \cdot 10^{-1}$
4651.02	3	5	$2.28 \cdot 10^{-1}$
4651.47	3	1	$7.24 \cdot 10^{-1}$
4652.05	1	3	$3.04 \cdot 10^{-1}$
4659.06	3	3	$2.27 \cdot 10^{-1}$
4663.64	3	1	$9.05 \cdot 10^{-1}$
4665.86	5	5	$6.78 \cdot 10^{-1}$
4673.95	5	3	$3.75 \cdot 10^{-1}$
5244.66	1	3	$5.30 \cdot 10^{-2}$
5253.58	3	3	$1.58 \cdot 10^{-1}$
5272.52	5	3	$2.61 \cdot 10^{-1}$
5695.92	3	5	$4.27 \cdot 10^{-1}$
5858.34	3	1	$1.34 \cdot 10^{-1}$
5863.25	3	3	$3.35 \cdot 10^{-2}$
5871.68	5	3	$1.00 \cdot 10^{-1}$
5880.56	5	5	$1.99 \cdot 10^{-2}$
5894.07	7	5	$1.11 \cdot 10^{-1}$
6727.48	1	3	$1.12 \cdot 10^{-1}$
6731.04	3	5	$1.50 \cdot 10^{-1}$
6742.15	3	3	$8.32 \cdot 10^{-2}$
6744.39	5	7	$1.99 \cdot 10^{-1}$
6762.17	5	5	$4.95 \cdot 10^{-2}$
6773.39	5	3	$5.47 \cdot 10^{-3}$
6851.18	3	5	$7.60 \cdot 10^{-3}$
6853.68	5	7	$5.64 \cdot 10^{-3}$
6857.24	3	3	$3.79 \cdot 10^{-2}$
6862.69	5	5	$3.51 \cdot 10^{-2}$
6868.78	5	3	$1.26 \cdot 10^{-2}$
6872.04	7	7	$4.46 \cdot 10^{-2}$
6881.10	7	5	$7.80 \cdot 10^{-3}$
7353.88	5	3	$3.09 \cdot 10^{-2}$
7707.43	3	5	$1.30 \cdot 10^{-1}$
7771.76	3	1	$1.77 \cdot 10^{-1}$
7780.41	3	3	$1.76 \cdot 10^{-1}$
7796.00	3	5	$1.75 \cdot 10^{-1}$

λ Å	Weights g_i	g_k	A $10^8\ s^{-1}$
8500.32	1	3	$1.01\cdot10^{-1}$
9593.32	3	3	$5.32\cdot10^{-3}$
9651.47	5	5	$1.57\cdot10^{-2}$
9696.48	5	7	$7.53\cdot10^{-3}$
9696.54	3	5	$7.12\cdot10^{-3}$
9699.57	7	9	$8.47\cdot10^{-3}$
9701.10	1	3	$4.40\cdot10^{-2}$
9705.41	3	5	$5.93\cdot10^{-2}$
9706.44	3	3	$3.29\cdot10^{-2}$
9715.09	5	7	$7.88\cdot10^{-2}$
9717.75	5	5	$1.97\cdot10^{-2}$
9718.79	5	3	$2.19\cdot10^{-3}$

C IV

λ Å	Weights g_i	g_k	A $10^8\ s^{-1}$
*312.43	2	6	$4.63\cdot10^{1}$
*384.13	6	10	$1.76\cdot10^{2}$
1548.19	2	4	2.65
1550.77	2	2	2.64
5801.31	2	4	$3.17\cdot10^{-1}$
5811.97	2	2	$3.16\cdot10^{-1}$

C V

λ Å	Weights g_i	g_k	A $10^8\ s^{-1}$
34.9728	1	3	$2.554\cdot10^{3}$
40.2678	1	3	$8.873\cdot10^{3}$
*227.19	3	9	$1.363\cdot10^{2}$
247.315	1	3	$1.278\cdot10^{2}$
*248.71	9	15	$4.247\cdot10^{2}$
*260.19	9	3	$6.680\cdot10^{1}$
267.267	3	5	$3.947\cdot10^{2}$
*2273.9	3	9	$5.646\cdot10^{-1}$
3526.66	1	3	$1.663\cdot10^{-1}$
8420.72	3	5	$6.898\cdot10^{-2}$
*8433.2	3	9	$6.868\cdot10^{-2}$
8448.12	3	1	$6.832\cdot10^{-2}$
8449.19	3	3	$6.829\cdot10^{-2}$

Cesium

Cs I

λ Å	Weights g_i	g_k	A $10^8\ s^{-1}$
3203.5	2	4	$7.6\cdot10^{-6}$
3205.3	2	4	$7.9\cdot10^{-6}$
3207.5	2	4	$8.5\cdot10^{-6}$
3210.0	2	4	$9.4\cdot10^{-6}$
3212.8	2	4	$1.19\cdot10^{-5}$
3216.2	2	4	$1.49\cdot10^{-5}$
3220.1	2	4	$1.7\cdot10^{-5}$
3220.2	2	2	$1.07\cdot10^{-7}$
3224.8	2	4	$2.0\cdot10^{-5}$
3225.0	2	2	$1.43\cdot10^{-7}$
3230.5	2	4	$2.5\cdot10^{-5}$
3230.7	2	2	$1.97\cdot10^{-7}$
3237.4	2	4	$2.8\cdot10^{-5}$
3237.6	2	2	$2.63\cdot10^{-7}$
3245.9	2	4	$3.45\cdot10^{-5}$
3246.2	2	2	$3.7\cdot10^{-7}$
3256.7	2	4	$4.25\cdot10^{-5}$
3257.1	2	2	$7.0\cdot10^{-7}$
3270.5	2	4	$5.6\cdot10^{-5}$
3271.0	2	2	$9.8\cdot10^{-7}$
3288.6	2	4	$1.0\cdot10^{-4}$
3289.3	2	2	$2.7\cdot10^{-6}$
3313.1	2	4	$1.6\cdot10^{-4}$
3314.0	2	2	$5.2\cdot10^{-6}$
3347.5	2	4	$2.2\cdot10^{-4}$
3348.8	2	2	$1.1\cdot10^{-5}$
3397.9	2	4	$4.0\cdot10^{-4}$
3400.0	2	2	$2.4\cdot10^{-5}$
3476.8	2	4	$6.6\cdot10^{-4}$
3480.0	2	2	$6.6\cdot10^{-5}$
3611.4	2	4	$1.5\cdot10^{-3}$
3617.3	2	2	$2.5\cdot10^{-4}$
3876.1	2	4	$3.8\cdot10^{-3}$
3888.6	2	2	$9.7\cdot10^{-4}$
4555.3	2	4	$1.88\cdot10^{-2}$
4593.2	2	2	$8.0\cdot10^{-3}$
8521.1	2	4	$3.276\cdot10^{-1}$
8943.5	2	2	$2.87\cdot10^{-1}$

Chlorine

Cl I

λ Å	Weights g_i	g_k	A $10^8\ s^{-1}$
1188.8	4	6	2.33
1188.8	4	4	$2.71\cdot10^{-1}$
1201.4	2	4	2.39
1335.7	4	2	1.74
1347.2	4	4	4.19
1351.7	2	2	3.23
1363.4	2	4	$7.5\cdot10^{-1}$
4323.3	4	4	$1.1\cdot10^{-2}$
4363.3	4	6	$6.8\cdot10^{-3}$
4379.9	4	4	$1.4\cdot10^{-2}$
4389.8	6	8	$1.4\cdot10^{-2}$
4526.2	4	4	$5.1\cdot10^{-2}$
4601.0	2	2	$4.2\cdot10^{-2}$
4661.2	2	4	$1.2\cdot10^{-2}$
7256.6	6	4	$1.5\cdot10^{-1}$
7414.1	6	4	$4.7\cdot10^{-2}$
7547.1	4	4	$1.2\cdot10^{-1}$
7717.6	4	4	$3.0\cdot10^{-2}$
7745.0	2	4	$6.3\cdot10^{-2}$
7769.2	6	6	$6.0\cdot10^{-2}$
7821.4	6	8	$9.8\cdot10^{-2}$
7830.8	4	4	$9.7\cdot10^{-2}$
7878.2	6	6	$1.8\cdot10^{-2}$
7899.3	4	6	$5.1\cdot10^{-2}$
7924.6	2	4	$2.1\cdot10^{-2}$
7935.0	6	8	$3.9\cdot10^{-2}$
7997.9	4	4	$2.1\cdot10^{-2}$

Cl II

λ Å	Weights g_i	g_k	A $10^8\ s^{-1}$
3329.1	5	7	1.5
3522.1	7	7	1.4
3798.8	5	7	1.6
3805.2	7	9	1.8
3809.5	3	5	1.5
3851.0	5	7	1.8
3851.4	5	5	1.6
3854.7	3	5	2.2
3861.9	5	7	2.4
3868.6	7	9	2.7
3913.9	9	9	$8.2\cdot10^{-1}$
3990.2	5	7	$8.4\cdot10^{-1}$
4132.5	5	5	1.6
4276.5	9	7	$7.6\cdot10^{-1}$
4768.7	3	5	$7.7\cdot10^{-1}$
4781.3	5	7	1.0
4794.6	5	7	1.04
4810.1	5	5	$9.9\cdot10^{-1}$
4819.5	5	3	1.00
4904.8	5	7	$8.1\cdot10^{-1}$
4917.7	3	5	$7.5\cdot10^{-1}$
5078.3	7	7	$7.7\cdot10^{-1}$
5219.1	3	9	$8.6\cdot10^{-1}$
5392.1	5	7	1.0

Cl III

λ Å	Weights g_i	g_k	A $10^8\ s^{-1}$
2298.5	4	4	4.2
2340.6	6	6	4.2
2370.4	8	6	2.8
2531.8	2	4	4.4
2532.5	4	6	5.3
2577.1	4	6	4.3
2580.7	6	8	4.7
2601.2	2	4	4.6
2603.6	4	6	5.0
2609.5	6	8	5.7
2617.0	8	10	6.6
2661.6	4	6	3.4
2665.5	6	8	4.8
2691.5	4	4	3.5
2710.4	4	6	3.5
3340.4	6	6	1.5
3392.9	4	4	1.9
3393.5	6	6	1.9
3530.0	6	8	1.8
3560.7	4	6	1.7
3602.1	6	8	1.7
3612.9	4	6	1.2
3720.5	4	6	1.7

Chromium

Cr I

λ Å	Weights g_i	g_k	A $10^8\ s^{-1}$
1999.95	9	9	1.4
2726.50	5	7	$7.5\cdot10^{-1}$
2769.90	7	5	1.1
2780.70	9	7	1.4
2889.22	9	9	$6.6\cdot10^{-1}$
2893.25	7	7	$5.2\cdot10^{-1}$
2967.64	7	9	$3.9\cdot10^{-1}$
2971.10	5	7	$7.1\cdot10^{-1}$
2975.48	3	5	$8.9\cdot10^{-1}$
2988.64	5	7	$5.2\cdot10^{-1}$
2996.57	5	3	2.0
3000.88	7	5	1.6
3005.06	9	7	$9.2\cdot10^{-1}$
3013.72	3	5	$8.3\cdot10^{-1}$
3015.20	1	3	1.63
3020.67	3	3	1.5
3021.58	9	11	2.91

λ Å	Weights g_i	g_k	A $10^8\ s^{-1}$
3024.36	5	5	1.27
3030.25	7	7	1.1
3037.05	9	9	$5.4 \cdot 10^{-1}$
3040.84	7	5	$7.4 \cdot 10^{-1}$
3053.87	9	7	$7.97 \cdot 10^{-1}$
3148.44	9	11	$5.6 \cdot 10^{-1}$
3155.16	11	13	$5.7 \cdot 10^{-1}$
3163.76	13	15	$6.0 \cdot 10^{-1}$
3237.73	9	9	1.3
3578.68	7	9	1.48
3593.48	7	7	1.50
3605.32	7	5	1.62
3639.80	13	11	1.8
3743.89	13	13	$7.61 \cdot 10^{-1}$
3757.66	7	7	$4.13 \cdot 10^{-1}$
3768.24	5	5	$5.10 \cdot 10^{-1}$
3804.80	9	9	$6.9 \cdot 10^{-1}$
3963.69	13	15	1.3
3969.75	11	13	1.2
3983.90	7	9	1.05
3991.12	5	7	1.07
4001.44	9	11	$6.8 \cdot 10^{-1}$
4039.10	15	15	$6.7 \cdot 10^{-1}$
4048.78	13	13	$6.4 \cdot 10^{-1}$
4058.78	11	11	$6.7 \cdot 10^{-1}$
4065.71	9	11	$3.5 \cdot 10^{-1}$
4165.52	11	13	$7.5 \cdot 10^{-1}$
4204.48	13	11	$3.1 \cdot 10^{-1}$
4254.33	7	9	$3.15 \cdot 10^{-1}$
4263.15	15	17	$6.4 \cdot 10^{-1}$
4274.81	7	7	$3.07 \cdot 10^{-1}$
4275.98	11	11	$2.2 \cdot 10^{-1}$
4280.42	13	15	$4.7 \cdot 10^{-1}$
4289.73	7	5	$3.16 \cdot 10^{-1}$
4291.97	7	5	$2.4 \cdot 10^{-1}$
4297.75	11	13	$4.9 \cdot 10^{-1}$
4298.05	9	9	$2.6 \cdot 10^{-1}$
4300.52	9	7	$1.9 \cdot 10^{-1}$
4301.19	11	9	$2.6 \cdot 10^{-1}$
4302.78	11	11	$2.5 \cdot 10^{-1}$
4337.25	5	7	$2.0 \cdot 10^{-1}$
4373.65	9	9	$2.8 \cdot 10^{-1}$
4376.80	13	13	$3.2 \cdot 10^{-1}$
4413.86	7	5	$2.7 \cdot 10^{-1}$
4422.70	5	5	$2.7 \cdot 10^{-1}$
4424.29	9	7	$2.1 \cdot 10^{-1}$
4432.77	15	15	$4.9 \cdot 10^{-1}$
4490.55	9	7	$3.9 \cdot 10^{-1}$
4492.31	5	3	$4.47 \cdot 10^{-1}$
4495.28	9	7	$2.0 \cdot 10^{-1}$
4500.29	7	7	$2.1 \cdot 10^{-1}$
4506.84	13	11	$2.7 \cdot 10^{-1}$
4540.72	11	11	$3.14 \cdot 10^{-1}$
4564.17	11	13	$5.1 \cdot 10^{-1}$
4595.60	13	13	$4.7 \cdot 10^{-1}$
4622.47	7	7	$4.1 \cdot 10^{-1}$
4665.90	3	3	$3.0 \cdot 10^{-1}$
4689.38	7	5	$2.3 \cdot 10^{-1}$
4698.46	9	7	$2.2 \cdot 10^{-1}$
4708.02	11	9	$4.31 \cdot 10^{-1}$

λ Å	Weights g_i	g_k	A $10^8\ s^{-1}$
4718.43	13	11	$3.4 \cdot 10^{-1}$
4730.69	7	5	$3.83 \cdot 10^{-1}$
4737.33	9	7	$3.38 \cdot 10^{-1}$
4741.09	3	5	$2.2 \cdot 10^{-1}$
4752.07	13	13	$6.2 \cdot 10^{-1}$
4756.09	11	9	$4.0 \cdot 10^{-1}$
4792.49	7	5	$2.6 \cdot 10^{-1}$
4801.02	9	7	$3.06 \cdot 10^{-1}$
4816.13	9	9	$1.8 \cdot 10^{-1}$
4870.79	7	9	$3.5 \cdot 10^{-1}$
4887.01	9	11	$3.2 \cdot 10^{-1}$
4922.28	11	13	$4.0 \cdot 10^{-1}$
5204.51	5	3	$5.09 \cdot 10^{-1}$
5206.02	5	5	$5.14 \cdot 10^{-1}$
5208.42	5	7	$5.06 \cdot 10^{-1}$
5243.38	5	3	$2.19 \cdot 10^{-1}$
5297.37	7	9	$3.88 \cdot 10^{-1}$
5297.99	7	7	$3.0 \cdot 10^{-1}$
5328.36	9	11	$6.2 \cdot 10^{-1}$
5329.17	9	9	$2.25 \cdot 10^{-1}$
5783.11	3	3	$2.1 \cdot 10^{-1}$
5783.89	5	5	$2.02 \cdot 10^{-1}$
5787.97	5	7	$2.35 \cdot 10^{-1}$

Cr II

λ Å	Weights g_i	g_k	A $10^8\ s^{-1}$
2653.57	4	6	$3.5 \cdot 10^{-1}$
2658.59	2	4	$5.8 \cdot 10^{-1}$
2666.02	6	8	$5.9 \cdot 10^{-1}$
2668.71	4	2	1.4
2671.80	6	4	1.0
2672.83	8	6	$5.5 \cdot 10^{-1}$
2744.97	4	6	$8.5 \cdot 10^{-1}$
2787.61	6	6	1.5
2822.38	14	16	2.3
2835.63	10	12	2.0
2840.01	10	12	2.7
2843.24	8	10	$6.4 \cdot 10^{-1}$
2849.83	6	8	$9.2 \cdot 10^{-1}$
2851.35	8	10	2.2
2856.77	4	6	$4.3 \cdot 10^{-1}$
2857.40	6	8	$2.8 \cdot 10^{-1}$
2860.92	2	4	$6.9 \cdot 10^{-1}$
2862.57	8	8	$6.3 \cdot 10^{-1}$
2866.72	4	4	1.2
2867.09	4	4	1.1
2867.65	2	2	1.1
2870.43	6	6	1.3
2873.81	4	2	$8.8 \cdot 10^{-1}$
2880.86	6	4	$7.9 \cdot 10^{-1}$
2898.53	10	12	1.2
2921.81	8	10	$9.0 \cdot 10^{-1}$
2930.83	2	4	1.1
2935.12	6	8	1.8
2953.34	2	2	1.8
2966.03	10	8	$5.4 \cdot 10^{-1}$
2971.90	14	14	2.0
2979.73	12	12	1.8
2985.32	10	10	2.2
2989.18	8	8	2.2
3118.64	2	4	1.7

λ Å	Weights g_i	g_k	A $10^8\ s^{-1}$
3120.36	4	6	1.5
3122.59	12	12	$4.4 \cdot 10^{-1}$
3128.69	4	4	$8.1 \cdot 10^{-1}$
3136.68	6	6	$6.4 \cdot 10^{-1}$
4588.22	8	6	$1.2 \cdot 10^{-1}$

Cobalt

Co I

λ Å	Weights g_i	g_k	A $10^8\ s^{-1}$
2407.25	10	12	3.6
2414.46	6	8	3.4
2415.29	4	6	3.6
2424.93	10	10	3.2
2432.21	8	8	2.6
2436.66	6	6	2.6
2439.04	4	4	2.7
2511.02	10	10	$9.2 \cdot 10^{-1}$
2521.36	10	8	3.0
2528.97	8	6	2.8
2535.96	6	4	1.9
3405.12	10	10	1.0
3409.17	8	8	$4.2 \cdot 10^{-1}$
3412.34	8	10	$6.1 \cdot 10^{-1}$
3433.05	4	4	1.0
3443.64	8	8	$6.9 \cdot 10^{-1}$
3449.17	6	6	$7.6 \cdot 10^{-1}$
3453.51	10	12	1.1
3462.80	4	6	$7.9 \cdot 10^{-1}$
3474.02	6	8	$5.6 \cdot 10^{-1}$
3489.40	8	6	1.3
3495.68	4	6	$4.9 \cdot 10^{-1}$
3502.28	10	8	$8.0 \cdot 10^{-1}$
3506.32	8	6	$8.2 \cdot 10^{-1}$
3512.64	6	4	1.0
3518.34	6	4	1.6
3529.82	8	10	$4.6 \cdot 10^{-1}$
3569.37	8	8	1.6
3587.19	6	6	1.4
3845.47	8	10	$4.6 \cdot 10^{-1}$
3894.07	6	8	$6.9 \cdot 10^{-1}$
3995.31	8	10	$2.5 \cdot 10^{-1}$
4121.32	8	10	$1.9 \cdot 10^{-1}$
5146.75	8	8	$1.5 \cdot 10^{-1}$
5212.70	10	10	$1.9 \cdot 10^{-1}$
5280.63	10	8	$2.8 \cdot 10^{-1}$
5352.05	12	10	$2.7 \cdot 10^{-1}$
6082.43	10	10	$5.4 \cdot 10^{-2}$
6455.00	8	10	$9.0 \cdot 10^{-2}$
7838.12	8	10	$5.4 \cdot 10^{-2}$
8093.93	12	10	$2.0 \cdot 10^{-1}$
8372.79	10	10	$8.7 \cdot 10^{-2}$

Co II

λ Å	Weights g_i	g_k	A $10^8\ s^{-1}$
2286.15	11	13	3.3
2307.85	9	11	2.6
2311.61	7	9	2.8
2314.05	5	7	2.8
2314.97	3	5	2.7
2330.36	5	3	1.32
2344.28	3	3	1.5

λ Å	Weights g_i	g_k	A 10^8 s^{-1}
2353.41	7	7	1.9
2363.80	9	9	2.1
2378.62	11	9	1.9
2383.45	9	7	1.8
2388.92	11	11	2.8
2389.54	5	3	1.5
2404.17	3	3	1.5
2417.66	9	9	$8.5·10^{-1}$

Copper

Cu I

λ Å	Weights g_i	g_k	A 10^8 s^{-1}
*2024.3	2	6	$9.8·10^{-2}$
2165.1	2	4	$5.1·10^{-1}$
2178.9	2	4	$9.13·10^{-1}$
2181.7	2	2	1.0
2225.7	2	2	$4.6·10^{-1}$
2244.3	2	4	$1.19·10^{-2}$
2441.6	2	2	$2.0·10^{-2}$
2492.2	2	4	$3.11·10^{-2}$
2618.4	6	4	$3.07·10^{-1}$
2766.4	4	4	$9.6·10^{-2}$
2824.4	6	6	$7.8·10^{-2}$
2961.2	6	8	$3.76·10^{-2}$
3063.4	4	4	$1.55·10^{-2}$
3194.1	4	4	$1.55·10^{-2}$
3247.5	2	4	1.39
3274.0	2	2	1.37
3337.8	6	8	$3.8·10^{-3}$
4022.6	2	4	$1.90·10^{-1}$
4062.6	4	6	$2.10·10^{-1}$
4249.0	2	2	$1.95·10^{-1}$
4275.1	6	8	$3.45·10^{-1}$
4480.4	2	2	$3.0·10^{-2}$
4509.4	4	2	$2.75·10^{-1}$
4530.8	4	2	$8.4·10^{-2}$
4539.7	6	4	$2.12·10^{-1}$
4587.0	8	6	$3.20·10^{-1}$
4651.1	10	8	$3.80·10^{-1}$
4704.6	8	8	$5.5·10^{-2}$
5105.5	6	4	$2.0·10^{-2}$
5153.2	2	4	$6.0·10^{-1}$
5218.2	4	6	$7.5·10^{-1}$
5220.1	4	4	$1.50·10^{-1}$
5292.5	8	8	$1.09·10^{-1}$
5700.2	4	4	$2.4·10^{-3}$
5782.1	4	2	$1.65·10^{-2}$

Cu II

λ Å	Weights g_i	g_k	A 10^8 s^{-1}
2489.7	5	5	$1.5·10^{-2}$
2544.8	9	7	1.1
2689.3	7	7	$4.1·10^{-1}$
2701.0	5	5	$6.7·10^{-1}$
2703.2	3	3	1.2
2713.5	5	5	$6.8·10^{-1}$
2769.7	7	7	$6.1·10^{-1}$

Dysprosium

Dy I

λ Å	Weights g_i	g_k	A 10^8 s^{-1}
2862.7	17	15	$6.5·10^{-2}$
2964.6	17	17	$6.5·10^{-2}$
3147.7	15	17	$1.1·10^{-1}$
3263.2	15	13	$1.4·10^{-1}$
3511.0	15	13	$3.1·10^{-1}$
3571.4	15	13	$2.0·10^{-1}$
3757.1	17	19	3.0
3868.8	17	17	3.1
3967.5	17	19	$8.7·10^{-1}$
4046.0	17	15	1.5
4103.9	13	11	1.7
4186.8	17	17	1.32
4194.8	17	17	$7.2·10^{-1}$
4211.7	17	19	2.08
4218.1	15	15	1.85
4221.1	15	17	1.52
4225.2	13	15	4.5
4268.3	15	15	$3.6·10^{-2}$
4276.7	13	13	$7.3·10^{-1}$
4292.0	15	15	$5.8·10^{-2}$
4577.8	17	19	$2.2·10^{-2}$
4589.4	17	15	$1.3·10^{-1}$
4612.3	17	15	$8.2·10^{-2}$
5077.7	17	17	$5.7·10^{-3}$
5301.6	17	15	$1.1·10^{-2}$
5547.3	17	17	$2.7·10^{-3}$
5639.5	17	19	$4.7·10^{-3}$
5974.5	17	17	$4.0·10^{-3}$
5988.6	17	15	$5.3·10^{-3}$
6010.8	15	15	$2.6·10^{-2}$
6088.3	15	13	$3.5·10^{-2}$
6168.4	15	17	$2.5·10^{-2}$
6259.1	17	19	$8.5·10^{-3}$
6579.4	17	15	$7.5·10^{-3}$

Erbium

Er I

λ Å	Weights g_i	g_k	A 10^8 s^{-1}
3862.9	13	13	2.5
4008.0	13	15	2.6
4151.1	13	11	1.8

Europium

Eu I

λ Å	Weights g_i	g_k	A 10^8 s^{-1}
2372.9	8	6	$1.9·10^{-1}$
2375.3	8	8	$2.0·10^{-1}$
2379.7	8	10	$2.0·10^{-1}$
2710.0	8	10	$1.4·10^{-1}$
2724.0	8	8	$1.2·10^{-1}$
2892.5	8	8	$1.0·10^{-1}$
2893.0	8	6	$1.0·10^{-1}$
2909.0	8	10	$6.9·10^{-2}$
3106.2	8	10	$5.5·10^{-2}$
3111.4	8	10	$3.0·10^{-1}$
3168.3	8	10	$6.9·10^{-2}$
3210.6	8	8	$1.1·10^{-1}$
3212.8	8	8	$2.9·10^{-1}$
3213.8	8	6	$1.8·10^{-1}$
3334.3	8	6	$3.4·10^{-1}$
4594.0	8	10	1.4
4627.2	8	8	1.3
4661.9	8	6	1.3
5765.2	8	8	$1.1·10^{-2}$
6018.2	8	10	$8.5·10^{-3}$
6864.5	8	10	$5.8·10^{-3}$

Fluorine

F I

λ Å	Weights g_i	g_k	A 10^8 s^{-1}
806.96	4	6	3.3
809.60	2	4	2.8
951.87	4	2	2.6
954.83	4	4	5.77
955.55	2	2	5.1
958.52	2	4	1.3
6239.7	6	4	$2.5·10^{-1}$
6348.5	4	4	$1.8·10^{-1}$
6413.7	2	4	$1.1·10^{-1}$
6708.3	6	4	$1.4·10^{-2}$
6774.0	6	6	$1.0·10^{-1}$
6795.5	4	2	$5.2·10^{-2}$
6834.3	4	4	$2.1·10^{-1}$
6856.0	6	8	$4.94·10^{-1}$
6870.2	2	2	$3.8·10^{-1}$
6902.5	4	6	$3.2·10^{-1}$
6909.8	2	4	$2.2·10^{-1}$
6966.4	4	2	$1.1·10^{-1}$
7037.5	4	4	$3.0·10^{-1}$
7127.9	2	2	$3.8·10^{-1}$
7309.0	6	8	$4.7·10^{-1}$
7311.0	4	2	$3.9·10^{-1}$
7314.3	4	6	$4.8·10^{-1}$
7332.0	6	4	$3.1·10^{-1}$
7398.7	6	6	$2.85·10^{-1}$
7425.7	4	2	$3.4·10^{-1}$
7482.7	4	4	$5.6·10^{-2}$
7489.2	2	2	$1.1·10^{-1}$
7514.9	2	2	$5.2·10^{-2}$
7552.2	4	6	$7.8·10^{-2}$
7573.4	2	4	$1.0·10^{-1}$
7607.2	4	4	$7.0·10^{-2}$
7754.7	4	6	$3.82·10^{-1}$
7800.2	2	4	$2.1·10^{-1}$

Gallium

Ga I

λ Å	Weights g_i	g_k	A 10^8 s^{-1}
2195.4	2	2	$1.9·10^{-2}$
2199.7	4	2	$3.3·10^{-2}$
2214.4	4	6	$1.2·10^{-2}$
2235.9	4	2	$4.3·10^{-2}$
2255.0	2	2	$3.1·10^{-2}$
2259.2	4	6	$3.1·10^{-2}$
2294.2	2	4	$7.0·10^{-2}$
2297.9	4	2	$5.8·10^{-2}$
2338.2	4	6	$9.8·10^{-2}$
2371.3	2	2	$5.7·10^{-2}$
2418.7	4	2	$1.0·10^{-1}$
2450.1	2	4	$2.8·10^{-1}$
2500.2	4	6	$3.4·10^{-1}$
2659.9	2	2	$1.2·10^{-1}$
2719.7	4	2	$2.3·10^{-1}$

λ Å	Weights g_i	g_k	A 10^8 s^{-1}
2874.2	2	4	1.2
2943.6	4	6	1.4
2944.2	4	4	$2.7\cdot10^{-1}$
4033.0	2	2	$4.9\cdot10^{-1}$
4172.0	4	2	$9.2\cdot10^{-1}$

Ga II

λ Å	Weights g_i	g_k	A 10^8 s^{-1}
829.60	1	3	$2.2\cdot10^{-1}$
1414.4	1	3	$1.88\cdot10^{1}$

Germanium

Ge I

λ Å	Weights g_i	g_k	A 10^8 s^{-1}
1944.7	3	1	$7.0\cdot10^{-1}$
1955.1	3	3	$2.8\cdot10^{-1}$
1988.3	5	3	$2.5\cdot10^{-1}$
1998.9	5	5	$5.5\cdot10^{-1}$
2041.7	1	3	1.1
2065.2	3	3	$8.5\cdot10^{-1}$
2068.7	3	5	1.2
2086.0	3	5	$4.0\cdot10^{-1}$
2094.3	5	7	$9.7\cdot10^{-1}$
2105.8	5	5	$1.7\cdot10^{-1}$
2256.0	5	5	$3.2\cdot10^{-2}$
2417.4	5	5	$9.6\cdot10^{-1}$
2498.0	1	3	$1.3\cdot10^{-1}$
2533.2	3	3	$1.0\cdot10^{-1}$
2589.2	5	3	$5.1\cdot10^{-2}$
2592.5	3	5	$7.1\cdot10^{-1}$
2651.2	5	5	2.0
2651.6	1	3	$8.5\cdot10^{-1}$
2691.3	3	3	$6.1\cdot10^{-1}$
2709.6	3	1	2.8
2754.6	5	3	1.1
3039.1	5	3	2.8
3124.8	5	5	$3.1\cdot10^{-2}$
3269.5	5	3	$2.9\cdot10^{-1}$
4226.6	1	3	$2.1\cdot10^{-1}$
4685.8	1	3	$9.5\cdot10^{-2}$

Ge II

λ Å	Weights g_i	g_k	A 10^8 s^{-1}
999.10	2	4	1.9
1016.6	4	6	2.1
1017.1	4	4	$3.5\cdot10^{-1}$
1055.0	2	2	$6.9\cdot10^{-1}$
1075.1	4	2	1.3
1237.1	2	4	$1.9\cdot10^{1}$
1261.9	4	6	$2.2\cdot10^{1}$
1264.7	4	4	3.5
1602.5	2	2	3.4
1649.2	4	2	6.5
4741.8	2	4	$4.6\cdot10^{-1}$
4814.6	4	6	$5.1\cdot10^{-1}$
4824.1	4	4	$8.6\cdot10^{-2}$
5131.8	4	6	1.9
5178.5	6	6	$1.3\cdot10^{-1}$
5178.6	6	8	2.0
5893.4	2	4	$9.2\cdot10^{-1}$
6021.0	2	2	$8.4\cdot10^{-1}$
6336.4	2	2	$4.4\cdot10^{-1}$
6484.2	4	2	$8.5\cdot10^{-1}$

Gold

Au I

λ Å	Weights g_i	g_k	A 10^8 s^{-1}
2427.95	2	4	1.99
2675.95	2	2	1.64
3122.78	6	4	$1.90\cdot10^{-1}$
6278.30	4	2	$3.4\cdot10^{-2}$

Helium

He I

λ Å	Weights g_i	g_k	A 10^8 s^{-1}
507.058	1	3	$1.5929\cdot10^{-1}$
507.718	1	3	$2.1826\cdot10^{-1}$
508.643	1	3	$3.1031\cdot10^{-1}$
509.998	1	3	$4.6224\cdot10^{-1}$
512.099	1	3	$7.3174\cdot10^{-1}$
515.617	1	3	1.2582
522.213	1	3	2.4356
537.030	1	3	5.6634
584.334	1	3	$1.7989\cdot10^{1}$
*2677.13	3	9	$4.4174\cdot10^{-3}$
*2696.12	3	9	$6.0234\cdot10^{-3}$
*2723.19	3	9	$8.4996\cdot10^{-3}$
*2763.80	3	9	$1.2508\cdot10^{-2}$
*2829.08	3	9	$1.9389\cdot10^{-2}$
*2945.10	3	9	$3.2006\cdot10^{-2}$
*3187.74	3	9	$5.6361\cdot10^{-2}$
3231.270	1	3	$5.1015\cdot10^{-3}$
3258.273	1	3	$6.9627\cdot10^{-3}$
3296.773	1	3	$9.8432\cdot10^{-3}$
3354.555	1	3	$1.4537\cdot10^{-2}$
3447.589	1	3	$2.2691\cdot10^{-2}$
*3554.42	9	15	$7.5971\cdot10^{-3}$
*3562.99	9	3	$4.8363\cdot10^{-3}$
*3587.28	9	15	$1.8107\cdot10^{-2}$
*3599.32	9	3	$6.7245\cdot10^{-3}$
3613.642	1	3	$3.8022\cdot10^{-2}$
*3634.25	9	15	$2.6062\cdot10^{-2}$
*3652.00	9	3	$9.7111\cdot10^{-3}$
*3705.02	9	15	$3.9528\cdot10^{-2}$
*3732.88	9	3	$1.4895\cdot10^{-2}$
*3819.62	9	15	$6.4351\cdot10^{-2}$
3833.549	3	5	$9.6470\cdot10^{-3}$
3838.100	3	1	$3.7425\cdot10^{-3}$
*3867.49	9	3	$2.4466\cdot10^{-2}$
3871.786	3	5	$1.3386\cdot10^{-2}$
3878.177	3	1	$5.1753\cdot10^{-3}$
*3888.64	3	9	$9.4746\cdot10^{-2}$
3926.544	3	5	$1.9371\cdot10^{-2}$
3935.945	3	1	$7.4475\cdot10^{-3}$
3964.729	1	3	$6.9507\cdot10^{-2}$
4009.256	3	5	$2.9612\cdot10^{-2}$
4023.980	3	1	$1.1281\cdot10^{-2}$
*4026.21	9	15	$1.1600\cdot10^{-1}$
*4120.84	9	3	$4.4529\cdot10^{-2}$
4143.759	3	5	$4.8812\cdot10^{-2}$
4168.971	3	1	$1.8298\cdot10^{-2}$
4387.929	3	5	$8.9889\cdot10^{-2}$
4437.553	3	1	$3.2689\cdot10^{-2}$
*4471.50	9	15	$2.4578\cdot10^{-1}$
*4713.17	9	3	$9.5209\cdot10^{-2}$
4921.931	3	5	$1.9863\cdot10^{-1}$
5015.678	1	3	$1.3372\cdot10^{-1}$
5047.738	3	1	$6.7712\cdot10^{-2}$
*5875.66	9	15	$7.0703\cdot10^{-1}$
6678.152	3	5	$6.3705\cdot10^{-1}$
*7065.25	9	3	$2.7853\cdot10^{-1}$
7281.350	3	1	$1.8299\cdot10^{-1}$
*7298.04	3	9	$1.2913\cdot10^{-3}$
*7499.85	3	9	$1.7942\cdot10^{-3}$
*7816.14	3	9	$2.5748\cdot10^{-3}$
8094.115	1	3	$1.3791\cdot10^{-3}$
8265.701	1	3	$1.8722\cdot10^{-3}$
*8361.73	3	9	$3.8126\cdot10^{-3}$
8518.036	1	3	$2.6252\cdot10^{-3}$
*8582.64	9	15	$4.1927\cdot10^{-3}$
*8632.74	9	3	$2.7471\cdot10^{-3}$
*8776.74	9	15	$5.7758\cdot10^{-3}$
*8849.18	9	3	$3.8377\cdot10^{-3}$
8914.772	1	3	$3.8260\cdot10^{-3}$
8999.736	5	7	$2.8406\cdot10^{-3}$
*9063.32	9	15	$8.2702\cdot10^{-3}$
9085.421	3	5	$3.6807\cdot10^{-3}$
9111.026	3	1	$2.2000\cdot10^{-3}$
*9174.52	9	3	$5.5996\cdot10^{-3}$
*9210.34	15	21	$4.7381\cdot10^{-3}$
9303.163	3	5	$5.1030\cdot10^{-3}$
9340.143	3	1	$3.0562\cdot10^{-3}$
*9463.58	3	9	$5.6868\cdot10^{-3}$
*9516.60	9	15	$1.2439\cdot10^{-2}$
9603.441	1	3	$5.8286\cdot10^{-3}$
9625.697	3	5	$7.3744\cdot10^{-3}$
9682.388	3	1	$4.4271\cdot10^{-3}$
*9702.65	9	3	$8.6511\cdot10^{-3}$
10138.424	3	5	$1.1248\cdot10^{-2}$
10233.102	3	1	$6.7731\cdot10^{-3}$
*10311.27	9	15	$1.9945\cdot10^{-2}$
*10667.71	9	3	$1.4471\cdot10^{-2}$
*10830.17	3	9	$1.0216\cdot10^{-1}$
*10913.00	15	21	$1.9801\cdot10^{-2}$
*10996.65	15	9	$1.4253\cdot10^{-3}$
11013.072	1	3	$9.2496\cdot10^{-3}$
11044.983	3	5	$1.8457\cdot10^{-2}$
11225.937	3	1	$1.1168\cdot10^{-2}$
*10969.11	9	15	$3.4781\cdot10^{-2}$
*12527.48	3	9	$7.0932\cdot10^{-3}$
12755.688	5	3	$1.2754\cdot10^{-3}$
*12846.01	9	3	$2.7317\cdot10^{-2}$
12968.430	3	5	$3.3615\cdot10^{-2}$
*12984.88	15	9	$2.7292\cdot10^{-3}$
13411.683	3	1	$2.0572\cdot10^{-2}$

Indium

In I

λ Å	Weights g_i	g_k	A 10^8 s^{-1}
2560.2	2	4	$4.0\cdot10^{-1}$
2710.3	4	6	$4.0\cdot10^{-1}$
3039.4	2	4	1.3
3256.1	4	6	1.3

λ Å	Weights g_i	g_k	A 10^8 s^{-1}
4101.8	2	2	$5.6 \cdot 10^{-1}$
4511.3	4	2	1.02
In II			
2941.1	3	1	1.4
Iodine			
I I			
1782.8	4	4	2.71
1830.4	4	6	$1.6 \cdot 10^{-1}$
Iridium			
Ir I			
2475.12	10	10	$2.1 \cdot 10^{-1}$
2502.98	10	12	$3.2 \cdot 10^{-1}$
2639.71	10	10	$4.7 \cdot 10^{-1}$
2661.98	10	10	$2.5 \cdot 10^{-1}$
2664.79	10	8	$4.0 \cdot 10^{-1}$
2694.23	10	12	$4.8 \cdot 10^{-1}$
2849.72	10	10	$2.2 \cdot 10^{-1}$
2924.79	10	12	$1.42 \cdot 10^{-1}$
2934.64	8	10	$2.0 \cdot 10^{-1}$
3220.78	10	8	$2.4 \cdot 10^{-1}$
3573.72	8	10	$5.4 \cdot 10^{-2}$
3661.71	8	10	$4.0 \cdot 10^{-2}$
4033.76	8	10	$2.7 \cdot 10^{-2}$
4069.92	6	8	$3.6 \cdot 10^{-2}$
4913.35	12	12	$3.3 \cdot 10^{-2}$
Iron			
Fe I			
2166.77	9	7	2.7
2191.84	5	5	1.2
2196.04	3	3	1.2
2200.39	1	3	$8.9 \cdot 10^{-1}$
2298.17	9	9	$3.09 \cdot 10^{-1}$
2439.74	13	13	3.46
2442.57	11	11	3.12
2457.60	11	11	$4.81 \cdot 10^{-1}$
2462.65	9	9	$5.85 \cdot 10^{-1}$
2465.15	9	9	$4.35 \cdot 10^{-1}$
2468.88	11	11	$2.40 \cdot 10^{-1}$
2472.89	7	7	1.30
2474.81	7	7	$6.13 \cdot 10^{-1}$
2479.78	5	5	1.74
2483.27	9	11	4.80
2484.19	3	3	2.26
2487.07	3	3	$6.40 \cdot 10^{-1}$
2488.14	7	9	4.20
2489.75	1	3	2.31
2490.64	5	7	3.44
2491.16	3	5	2.91
2491.99	9	9	$3.25 \cdot 10^{-1}$
2496.53	9	11	$2.15 \cdot 10^{-1}$
2501.13	9	7	$6.75 \cdot 10^{-1}$
2505.01	9	11	$2.56 \cdot 10^{-1}$
2506.57	7	9	$2.04 \cdot 10^{-1}$
2507.90	7	9	$1.93 \cdot 10^{-1}$
2510.83	7	5	1.29
2518.10	5	3	1.93
2522.85	9	9	2.13
2524.29	3	1	3.23
2527.27	13	13	$3.46 \cdot 10^{-1}$
2527.44	7	7	1.93
2529.14	5	5	$9.91 \cdot 10^{-1}$
2529.31	5	7	4.86
2533.14	11	11	$2.07 \cdot 10^{-1}$
2535.61	1	3	$9.59 \cdot 10^{-1}$
2537.17	13	15	3.70
2540.97	3	5	$9.59 \cdot 10^{-1}$
2542.10	11	13	4.47
2543.92	9	11	4.70
2545.98	5	7	$7.16 \cdot 10^{-1}$
2549.61	7	9	$2.31 \cdot 10^{-1}$
2584.54	11	13	$3.15 \cdot 10^{-1}$
2606.83	9	11	$2.43 \cdot 10^{-1}$
2609.22	7	7	$4.60 \cdot 10^{-1}$
2623.53	7	9	$2.13 \cdot 10^{-1}$
2635.81	5	7	$2.11 \cdot 10^{-1}$
2656.15	13	15	$1.63 \cdot 10^{-1}$
2669.49	11	13	$1.34 \cdot 10^{-1}$
2679.06	11	11	$1.50 \cdot 10^{-1}$
2706.01	13	13	$2.28 \cdot 10^{-1}$
2706.58	7	5	$2.69 \cdot 10^{-1}$
2708.57	9	9	$6.49 \cdot 10^{-1}$
2719.03	9	7	1.42
2719.06	7	7	$7.40 \cdot 10^{-1}$
2719.42	11	11	$3.20 \cdot 10^{-1}$
2720.90	7	5	1.04
2723.58	5	3	$5.69 \cdot 10^{-1}$
2728.82	9	9	$2.98 \cdot 10^{-1}$
2733.58	11	9	$7.10 \cdot 10^{-1}$
2735.48	9	7	$5.03 \cdot 10^{-1}$
2737.31	3	3	$7.25 \cdot 10^{-1}$
2737.64	13	11	$1.14 \cdot 10^{-1}$
2742.25	7	5	$3.41 \cdot 10^{-1}$
2742.41	5	5	$4.70 \cdot 10^{-1}$
2750.14	7	7	$2.74 \cdot 10^{-1}$
2762.03	7	7	$1.76 \cdot 10^{-1}$
2767.52	9	9	$1.48 \cdot 10^{-1}$
2769.30	13	13	$1.80 \cdot 10^{-1}$
2788.10	11	13	$5.92 \cdot 10^{-1}$
2789.80	11	9	$2.36 \cdot 10^{-1}$
2804.86	9	7	$2.40 \cdot 10^{-1}$
2806.98	9	11	$1.15 \cdot 10^{-1}$
2813.29	9	11	$3.42 \cdot 10^{-1}$
2825.56	7	9	$1.32 \cdot 10^{-1}$
2832.44	7	9	$2.38 \cdot 10^{-1}$
2843.98	5	7	$3.17 \cdot 10^{-1}$
2851.80	3	5	$3.37 \cdot 10^{-1}$
2894.50	5	5	$4.83 \cdot 10^{-1}$
2899.41	5	3	$4.68 \cdot 10^{-1}$
2901.91	11	11	$1.78 \cdot 10^{-1}$
2907.52	9	11	$1.61 \cdot 10^{-1}$
2918.02	13	13	1.18
2923.29	11	11	1.39
2923.85	11	11	$2.97 \cdot 10^{-1}$
2925.36	7	9	$1.69 \cdot 10^{-1}$
2929.12	9	9	1.53
2936.90	9	9	$1.40 \cdot 10^{-1}$
2947.88	7	7	$1.83 \cdot 10^{-1}$
2948.43	9	9	$3.32 \cdot 10^{-1}$
2953.49	7	7	$3.64 \cdot 10^{-1}$
2959.99	11	13	$5.02 \cdot 10^{-1}$
2966.90	9	11	$2.72 \cdot 10^{-1}$
2973.24	7	9	$1.83 \cdot 10^{-1}$
2980.53	7	7	$1.66 \cdot 10^{-1}$
2981.85	7	9	$1.86 \cdot 10^{-1}$
2983.57	9	7	$2.79 \cdot 10^{-1}$
2990.39	9	11	$3.5 \cdot 10^{-1}$
2994.43	7	5	$4.39 \cdot 10^{-1}$
2999.51	11	11	$1.70 \cdot 10^{-1}$
3000.95	5	3	$6.42 \cdot 10^{-1}$
3008.14	3	1	1.07
3009.57	9	9	$1.43 \cdot 10^{-1}$
3011.48	7	9	$3.79 \cdot 10^{-1}$
3020.49	5	5	$1.94 \cdot 10^{-1}$
3020.64	9	9	$7.59 \cdot 10^{-1}$
3021.07	7	7	$4.55 \cdot 10^{-1}$
3025.64	13	13	$5.86 \cdot 10^{-1}$
3025.84	1	3	$3.48 \cdot 10^{-1}$
3030.15	11	11	$5.04 \cdot 10^{-1}$
3037.39	3	5	$2.91 \cdot 10^{-1}$
3047.60	5	7	$2.84 \cdot 10^{-1}$
3057.45	11	9	$3.13 \cdot 10^{-1}$
3059.09	7	9	$1.63 \cdot 10^{-1}$
3067.00	11	13	$1.71 \cdot 10^{-1}$
3067.24	9	7	$3.12 \cdot 10^{-1}$
3075.72	7	5	$3.14 \cdot 10^{-1}$
3079.99	9	11	$8.35 \cdot 10^{-2}$
3083.74	5	3	$3.08 \cdot 10^{-1}$
3100.30	5	5	$1.87 \cdot 10^{-1}$
3100.67	7	7	$1.35 \cdot 10^{-1}$
3125.68	13	11	$8.46 \cdot 10^{-2}$
3132.52	9	7	$3.39 \cdot 10^{-1}$
3143.99	9	9	$6.10 \cdot 10^{-1}$
3156.27	7	7	$6.36 \cdot 10^{-1}$
3157.04	9	11	$1.26 \cdot 10^{-1}$
3157.89	5	7	$1.61 \cdot 10^{-1}$
3160.66	9	9	$1.93 \cdot 10^{-1}$
3166.44	9	7	$1.14 \cdot 10^{-1}$
3171.35	9	7	$1.85 \cdot 10^{-1}$
3175.44	11	11	$1.44 \cdot 10^{-1}$
3178.01	11	9	$1.28 \cdot 10^{-1}$
3180.22	7	9	$4.42 \cdot 10^{-1}$
3181.52	7	5	$1.84 \cdot 10^{-1}$
3182.97	5	7	$1.42 \cdot 10^{-1}$
3188.82	3	5	$2.53 \cdot 10^{-1}$
3192.80	3	5	$5.01 \cdot 10^{-1}$
3193.30	5	7	$3.07 \cdot 10^{-1}$
3196.12	11	9	$1.40 \cdot 10^{-1}$
3196.93	9	11	$5.97 \cdot 10^{-1}$
3199.53	9	9	$2.23 \cdot 10^{-1}$
3205.40	3	3	$9.77 \cdot 10^{-1}$
3210.23	9	11	$1.15 \cdot 10^{-1}$
3210.83	5	3	$9.24 \cdot 10^{-1}$
3211.99	11	9	$4.64 \cdot 10^{-1}$
3214.01	7	7	$8.38 \cdot 10^{-1}$

λ Å	Weights g_i	g_k	A 10^8 s^{-1}
3214.06	7	5	1.18
3215.94	5	5	$6.19 \cdot 10^{-1}$
3217.38	11	9	$1.50 \cdot 10^{-1}$
3219.58	7	9	$4.64 \cdot 10^{-1}$
3219.80	9	7	$3.61 \cdot 10^{-1}$
3222.07	11	11	$8.65 \cdot 10^{-1}$
3225.79	11	13	1.18
3227.80	9	7	$4.96 \cdot 10^{-1}$
3228.25	5	3	$3.72 \cdot 10^{-1}$
3229.99	9	11	$1.06 \cdot 10^{-1}$
3230.21	5	5	$2.06 \cdot 10^{-1}$
3230.96	7	5	$3.7 \cdot 10^{-1}$
3233.05	13	15	$4.19 \cdot 10^{-1}$
3233.97	9	9	$2.08 \cdot 10^{-1}$
3239.43	9	9	$2.95 \cdot 10^{-1}$
3244.19	9	11	$3.06 \cdot 10^{-1}$
3248.20	7	7	$1.92 \cdot 10^{-1}$
3253.60	7	9	$1.62 \cdot 10^{-1}$
3254.36	11	13	$4.24 \cdot 10^{-1}$
3265.62	7	5	$3.06 \cdot 10^{-1}$
3271.00	5	3	$6.4 \cdot 10^{-1}$
3280.26	9	11	$4.21 \cdot 10^{-1}$
3282.89	3	5	$3.42 \cdot 10^{-1}$
3286.75	7	7	$5.99 \cdot 10^{-1}$
3292.02	7	9	$5.77 \cdot 10^{-1}$
3292.59	3	3	$3.0 \cdot 10^{-1}$
3305.97	5	7	$4.05 \cdot 10^{-1}$
3306.34	9	9	$5.74 \cdot 10^{-1}$
3306.35	3	5	$4.84 \cdot 10^{-1}$
3307.23	13	13	$1.97 \cdot 10^{-1}$
3314.74	5	7	$7.25 \cdot 10^{-1}$
3322.47	9	11	$8.21 \cdot 10^{-2}$
3323.74	5	5	$2.8 \cdot 10^{-1}$
3328.87	11	11	$2.21 \cdot 10^{-1}$
3355.23	9	9	$2.59 \cdot 10^{-1}$
3369.55	9	9	$2.15 \cdot 10^{-1}$
3370.78	11	11	$2.89 \cdot 10^{-1}$
3380.11	7	7	$1.66 \cdot 10^{-1}$
3392.65	7	7	$1.88 \cdot 10^{-1}$
3399.33	5	5	$2.76 \cdot 10^{-1}$
3402.26	13	13	$2.19 \cdot 10^{-1}$
3404.35	5	7	$1.09 \cdot 10^{-1}$
3406.44	3	5	$2.7 \cdot 10^{-1}$
3407.46	7	9	$6.09 \cdot 10^{-1}$
3410.17	3	5	$5.07 \cdot 10^{-1}$
3413.13	5	7	$3.23 \cdot 10^{-1}$
3417.84	3	3	$4.01 \cdot 10^{-1}$
3418.51	3	1	$9.88 \cdot 10^{-1}$
3422.66	3	5	$1.38 \cdot 10^{-1}$
3424.28	7	7	$1.61 \cdot 10^{-1}$
3425.01	9	7	$2.57 \cdot 10^{-1}$
3426.67	11	11	$1.07 \cdot 10^{-1}$
3427.12	7	9	$5.04 \cdot 10^{-1}$
3428.19	5	5	$1.71 \cdot 10^{-1}$
3440.61	9	7	$1.71 \cdot 10^{-1}$
3440.99	7	5	$1.24 \cdot 10^{-1}$
3445.15	5	7	$2.34 \cdot 10^{-1}$
3450.33	3	3	$2.34 \cdot 10^{-1}$
3459.91	5	3	$2.17 \cdot 10^{-1}$
3469.01	9	9	$8.58 \cdot 10^{-2}$
3476.34	7	7	$2.70 \cdot 10^{-1}$
3489.67	11	13	$7.47 \cdot 10^{-2}$
3495.29	9	7	$9.46 \cdot 10^{-2}$
3497.10	7	7	$9.02 \cdot 10^{-2}$
3508.47	9	11	$6.46 \cdot 10^{-2}$
3526.24	7	9	$1.70 \cdot 10^{-1}$
3526.38	7	7	$4.13 \cdot 10^{-1}$
3526.47	5	5	$1.29 \cdot 10^{-1}$
3526.67	5	5	$5.26 \cdot 10^{-1}$
3527.79	9	9	$2.17 \cdot 10^{-1}$
3529.82	3	3	$7.75 \cdot 10^{-1}$
3530.39	13	13	$4.65 \cdot 10^{-2}$
3533.01	1	3	$8.52 \cdot 10^{-1}$
3533.20	3	5	$8.25 \cdot 10^{-1}$
3536.56	5	7	$9.95 \cdot 10^{-1}$
3537.89	11	11	$8.0 \cdot 10^{-2}$
3540.12	7	9	$9.48 \cdot 10^{-2}$
3541.08	9	11	$8.65 \cdot 10^{-1}$
3542.08	7	9	$9.51 \cdot 10^{-1}$
3543.67	3	5	$1.6 \cdot 10^{-1}$
3545.64	9	9	$2.05 \cdot 10^{-1}$
3547.19	9	9	$7.13 \cdot 10^{-2}$
3552.83	5	5	$1.74 \cdot 10^{-1}$
3553.74	11	9	1.09
3554.92	11	13	1.40
3556.88	9	11	$4.1 \cdot 10^{-1}$
3558.52	5	7	$1.77 \cdot 10^{-1}$
3559.50	3	3	$2.2 \cdot 10^{-1}$
3560.70	7	9	$7.4 \cdot 10^{-2}$
3565.38	7	9	$4.29 \cdot 10^{-1}$
3567.03	5	7	$8.34 \cdot 10^{-2}$
3568.82	7	9	$6.72 \cdot 10^{-2}$
3570.10	9	11	$6.76 \cdot 10^{-1}$
3572.00	11	11	$2.89 \cdot 10^{-1}$
3573.39	5	7	$1.05 \cdot 10^{-1}$
3573.89	9	7	$5.73 \cdot 10^{-1}$
3575.25	11	9	$7.43 \cdot 10^{-2}$
3575.37	5	5	$3.06 \cdot 10^{-1}$
3576.76	11	9	$8.8 \cdot 10^{-2}$
3581.19	11	13	1.02
3582.20	13	11	$2.35 \cdot 10^{-1}$
3584.66	11	11	$3.29 \cdot 10^{-1}$
3584.79	7	5	$1.56 \cdot 10^{-1}$
3584.96	11	9	$6.74 \cdot 10^{-1}$
3585.32	7	7	$1.17 \cdot 10^{-1}$
3586.11	13	11	$7.02 \cdot 10^{-1}$
3586.98	5	5	$1.66 \cdot 10^{-1}$
3587.24	7	9	$7.73 \cdot 10^{-2}$
3588.61	11	11	$1.19 \cdot 10^{-1}$
3588.92	5	3	$2.15 \cdot 10^{-1}$
3589.45	9	7	$1.05 \cdot 10^{-1}$
3594.63	9	9	$3.14 \cdot 10^{-1}$
3597.02	5	3	$1.8 \cdot 10^{-1}$
3599.63	11	9	$2.33 \cdot 10^{-1}$
3602.46	7	7	$1.02 \cdot 10^{-1}$
3602.53	7	5	$2.12 \cdot 10^{-1}$
3603.20	11	11	$2.59 \cdot 10^{-1}$
3605.45	9	9	$4.66 \cdot 10^{-1}$
3605.50	13	11	$2.12 \cdot 10^{-1}$
3606.68	11	13	$8.29 \cdot 10^{-1}$
3608.14	9	11	$6.22 \cdot 10^{-2}$
3608.86	3	5	$8.13 \cdot 10^{-1}$
3610.16	13	13	$5.90 \cdot 10^{-1}$
3612.07	11	13	$1.11 \cdot 10^{-1}$
3617.79	5	7	$7.09 \cdot 10^{-1}$
3618.39	9	9	$8.88 \cdot 10^{-2}$
3618.77	5	7	$7.22 \cdot 10^{-1}$
3621.46	9	11	$4.45 \cdot 10^{-1}$
3621.72	11	9	$1.07 \cdot 10^{-1}$
3622.00	7	7	$5.14 \cdot 10^{-1}$
3623.19	13	13	$6.68 \cdot 10^{-2}$
3625.14	11	9	$8.15 \cdot 10^{-2}$
3630.35	9	7	$1.04 \cdot 10^{-1}$
3631.10	11	11	$2.15 \cdot 10^{-1}$
3631.46	7	9	$5.17 \cdot 10^{-1}$
3632.01	3	5	$6.74 \cdot 10^{-1}$
3634.33	9	7	$1.05 \cdot 10^{-1}$
3636.22	5	7	$2.20 \cdot 10^{-1}$
3637.87	9	9	$5.9 \cdot 10^{-2}$
3638.30	7	9	$2.36 \cdot 10^{-1}$
3640.39	9	11	$3.57 \cdot 10^{-1}$
3645.82	1	3	$4.87 \cdot 10^{-1}$
3647.42	3	3	$3.38 \cdot 10^{-1}$
3647.84	9	11	$2.91 \cdot 10^{-1}$
3649.51	11	9	$3.94 \cdot 10^{-1}$
3650.03	7	7	$2.26 \cdot 10^{-1}$
3650.28	11	11	$6.15 \cdot 10^{-2}$
3651.47	7	9	$5.83 \cdot 10^{-1}$
3655.46	5	5	$1.18 \cdot 10^{-1}$
3659.52	9	9	$6.31 \cdot 10^{-2}$
3667.25	9	7	$1.3 \cdot 10^{-1}$
3669.15	9	7	$8.03 \cdot 10^{-2}$
3669.52	9	7	$2.34 \cdot 10^{-1}$
3670.09	11	13	$7.20 \cdot 10^{-2}$
3676.31	9	11	$4.63 \cdot 10^{-2}$
3677.31	5	7	$2.28 \cdot 10^{-1}$
3677.63	7	5	$6.08 \cdot 10^{-1}$
3682.17	7	5	$1.04 \cdot 10^{-1}$
3682.24	5	5	1.5
3684.11	9	7	$2.97 \cdot 10^{-1}$
3684.14	9	7	$9.29 \cdot 10^{-2}$
3686.00	9	11	$3.34 \cdot 10^{-1}$
3687.46	11	9	$8.00 \cdot 10^{-2}$
3687.66	9	9	$7.38 \cdot 10^{-2}$
3688.46	7	9	$7.3 \cdot 10^{-2}$
3689.46	9	9	$3.70 \cdot 10^{-1}$
3690.73	11	11	$2.99 \cdot 10^{-1}$
3694.01	5	7	$8.35 \cdot 10^{-1}$
3695.05	7	9	$2.01 \cdot 10^{-1}$
3697.43	7	7	$1.94 \cdot 10^{-1}$
3701.09	7	9	$6.35 \cdot 10^{-1}$
3703.69	9	11	$6.31 \cdot 10^{-2}$
3704.46	11	9	$1.42 \cdot 10^{-1}$
3707.92	7	5	$3.32 \cdot 10^{-1}$
3709.25	9	7	$1.56 \cdot 10^{-1}$
3711.41	3	5	$1.28 \cdot 10^{-1}$
3716.44	9	7	$3.49 \cdot 10^{-1}$
3719.93	9	11	$1.62 \cdot 10^{-1}$
3721.50	5	5	$1.94 \cdot 10^{-1}$
3724.38	5	7	$1.04 \cdot 10^{-1}$

λ Å	Weights g_i	g_k	A 10^8 s^{-1}
3726.93	5	5	$4.57\cdot10^{-1}$
3727.09	9	7	$1.71\cdot10^{-1}$
3727.62	7	5	$2.24\cdot10^{-1}$
3727.81	7	5	$1.91\cdot10^{-1}$
3730.39	9	11	$9.73\cdot10^{-2}$
3732.40	5	5	$2.69\cdot10^{-1}$
3734.86	11	11	$9.01\cdot10^{-1}$
3735.32	9	9	$2.70\cdot10^{-1}$
3737.13	7	9	$1.41\cdot10^{-1}$
3738.31	11	13	$3.44\cdot10^{-1}$
3740.24	7	9	$1.3\cdot10^{-1}$
3742.62	9	9	$6.75\cdot10^{-2}$
3743.36	5	3	$2.60\cdot10^{-1}$
3743.47	11	11	$6.05\cdot10^{-1}$
3744.10	5	3	$3.17\cdot10^{-1}$
3745.56	5	7	$1.15\cdot10^{-1}$
3746.93	7	7	$2.33\cdot10^{-1}$
3748.96	9	11	$1.48\cdot10^{-1}$
3749.49	9	9	$7.63\cdot10^{-1}$
3753.61	7	5	$1.22\cdot10^{-1}$
3756.94	11	11	$2.2\cdot10^{-1}$
3758.23	7	7	$6.34\cdot10^{-1}$
3759.15	13	11	$4.55\cdot10^{-2}$
3760.05	13	15	$4.47\cdot10^{-2}$
3763.79	5	5	$5.44\cdot10^{-1}$
3765.54	13	15	$9.51\cdot10^{-1}$
3767.19	3	3	$6.39\cdot10^{-1}$
3778.51	7	5	$1.17\cdot10^{-1}$
3785.95	11	13	$4.14\cdot10^{-2}$
3786.19	5	5	$1.3\cdot10^{-1}$
3787.16	5	5	$9.9\cdot10^{-2}$
3787.88	3	5	$1.29\cdot10^{-1}$
3793.48	7	7	$7.92\cdot10^{-2}$
3795.00	5	7	$1.15\cdot10^{-1}$
3797.51	13	13	$4.57\cdot10^{-1}$
3799.55	7	9	$7.31\cdot10^{-2}$
3801.98	11	13	$3.7\cdot10^{-2}$
3805.34	9	11	$8.60\cdot10^{-1}$
3806.22	3	3	$2.5\cdot10^{-1}$
3806.70	11	11	$4.35\cdot10^{-1}$
3807.54	3	5	$9.37\cdot10^{-2}$
3810.76	5	3	$1.94\cdot10^{-1}$
3813.88	13	11	$6.62\cdot10^{-2}$
3815.84	9	7	1.12
3817.64	11	11	$7.7\cdot10^{-2}$
3820.43	11	9	$6.67\cdot10^{-1}$
3821.18	11	13	$5.54\cdot10^{-1}$
3825.88	9	7	$5.97\cdot10^{-1}$
3827.82	7	5	1.05
3834.22	7	5	$4.52\cdot10^{-1}$
3836.33	5	5	$3.29\cdot10^{-1}$
3839.26	9	9	$2.35\cdot10^{-1}$
3840.44	5	3	$4.70\cdot10^{-1}$
3841.05	5	3	1.36
3843.26	9	7	$3.70\cdot10^{-1}$
3846.41	11	9	$1.68\cdot10^{-1}$
3846.80	7	7	$6.20\cdot10^{-1}$
3849.97	3	1	$6.05\cdot10^{-1}$
3859.21	13	11	$7.25\cdot10^{-2}$
3859.91	9	9	$9.69\cdot10^{-2}$
3865.52	3	3	$1.55\cdot10^{-1}$
3867.22	5	5	$3.16\cdot10^{-1}$
3871.75	11	11	$5.83\cdot10^{-2}$
3872.50	5	5	$1.05\cdot10^{-1}$
3873.76	11	9	$6.57\cdot10^{-2}$
3878.02	7	7	$7.72\cdot10^{-2}$
3878.67	9	7	$7.02\cdot10^{-2}$
3878.73	3	3	$5.34\cdot10^{-1}$
3883.28	7	7	$1.28\cdot10^{-1}$
3888.51	5	5	$2.50\cdot10^{-1}$
3888.82	5	3	$1.95\cdot10^{-1}$
3891.93	3	3	$2.71\cdot10^{-1}$
3893.39	11	11	$1.00\cdot10^{-1}$
3894.01	5	5	$1.03\cdot10^{-1}$
3897.89	11	13	$6.20\cdot10^{-2}$
3900.52	7	7	$7.9\cdot10^{-2}$
3902.95	7	7	$2.14\cdot10^{-1}$
3903.90	9	9	$7.61\cdot10^{-2}$
3906.75	5	7	$7.05\cdot10^{-2}$
3916.73	13	11	$9.83\cdot10^{-2}$
3918.42	3	1	$4.22\cdot10^{-1}$
3918.64	7	7	$1.17\cdot10^{-1}$
3925.94	1	3	$1.67\cdot10^{-1}$
3926.01	7	7	$7.26\cdot10^{-2}$
3928.08	9	9	$5.64\cdot10^{-2}$
3935.81	5	5	$1.14\cdot10^{-1}$
3941.28	5	5	$9.1\cdot10^{-2}$
3942.44	3	5	$9.62\cdot10^{-2}$
3946.99	9	11	$3.91\cdot10^{-2}$
3948.10	7	9	$1.31\cdot10^{-1}$
3948.77	11	9	$2.08\cdot10^{-1}$
3951.16	3	5	$4.29\cdot10^{-1}$
3955.34	3	3	$1.5\cdot10^{-1}$
3956.46	13	11	$1.76\cdot10^{-1}$
3956.68	11	13	$1.22\cdot10^{-1}$
3957.02	5	7	$1.67\cdot10^{-1}$
3963.10	3	5	$1.5\cdot10^{-1}$
3967.42	9	7	$1.52\cdot10^{-1}$
3967.96	7	9	$6.09\cdot10^{-2}$
3969.26	9	7	$2.26\cdot10^{-1}$
3971.32	11	9	$4.97\cdot10^{-2}$
3973.65	5	7	$5.81\cdot10^{-2}$
3976.61	3	5	$1.20\cdot10^{-1}$
3983.96	9	7	$5.72\cdot10^{-2}$
3985.39	5	5	$8.53\cdot10^{-2}$
3996.96	9	9	$7.95\cdot10^{-2}$
3997.39	9	11	$1.26\cdot10^{-1}$
3998.05	11	9	$5.70\cdot10^{-2}$
4005.24	7	5	$2.04\cdot10^{-1}$
4006.31	11	9	$5.1\cdot10^{-2}$
4014.53	9	7	$1.53\cdot10^{-1}$
4021.87	7	9	$8.55\cdot10^{-2}$
4024.73	7	9	$8.09\cdot10^{-2}$
4030.49	9	11	$1.04\cdot10^{-1}$
4043.90	7	7	$8.69\cdot10^{-2}$
4045.59	9	7	$7.39\cdot10^{-2}$
4045.81	9	9	$8.62\cdot10^{-1}$
4062.44	3	3	$1.85\cdot10^{-1}$
4063.59	7	7	$6.65\cdot10^{-1}$
4067.98	9	9	$1.51\cdot10^{-1}$
4070.77	7	5	$1.1\cdot10^{-1}$
4071.74	5	5	$7.64\cdot10^{-1}$
4073.76	5	3	$1.68\cdot10^{-1}$
4076.63	9	9	$1.32\cdot10^{-1}$
4084.49	11	9	$8.66\cdot10^{-2}$
4085.30	7	7	$8.92\cdot10^{-2}$
4098.18	7	7	$7.49\cdot10^{-2}$
4107.49	5	3	$1.74\cdot10^{-1}$
4109.80	3	3	$1.51\cdot10^{-1}$
4112.96	11	13	$1.1\cdot10^{-1}$
4118.55	11	13	$4.96\cdot10^{-1}$
4125.62	9	11	$9.9\cdot10^{-2}$
4126.18	11	11	$4.2\cdot10^{-2}$
4127.61	1	3	$1.43\cdot10^{-1}$
4132.06	5	7	$1.18\cdot10^{-1}$
4132.90	3	5	$7.70\cdot10^{-2}$
4134.68	5	7	$1.25\cdot10^{-1}$
4137.00	3	5	$2.75\cdot10^{-1}$
4142.59	3	5	$7.5\cdot10^{-2}$
4143.41	9	9	$2.70\cdot10^{-1}$
4143.87	7	9	$1.33\cdot10^{-1}$
4149.37	11	13	$4.23\cdot10^{-2}$
4153.90	7	9	$2.05\cdot10^{-1}$
4154.50	5	3	$2.64\cdot10^{-1}$
4154.81	9	11	$1.40\cdot10^{-1}$
4156.80	5	5	$1.20\cdot10^{-1}$
4157.78	5	7	$2.18\cdot10^{-1}$
4158.79	3	5	$1.6\cdot10^{-1}$
4172.12	7	5	$9.80\cdot10^{-2}$
4175.64	3	5	$1.14\cdot10^{-1}$
4181.75	5	7	$2.32\cdot10^{-1}$
4184.89	5	5	$1.03\cdot10^{-1}$
4187.04	7	5	$2.15\cdot10^{-1}$
4187.80	9	7	$1.52\cdot10^{-1}$
4191.43	5	3	$2.73\cdot10^{-1}$
4195.33	11	11	$1.11\cdot10^{-1}$
4196.21	7	7	$1.09\cdot10^{-1}$
4198.25	9	9	$1.47\cdot10^{-1}$
4198.30	11	9	$8.03\cdot10^{-2}$
4198.63	5	5	$1.25\cdot10^{-1}$
4199.10	9	11	$4.92\cdot10^{-1}$
4200.92	7	9	$6.25\cdot10^{-2}$
4202.03	9	9	$8.22\cdot10^{-2}$
4203.94	13	13	$2.97\cdot10^{-2}$
4203.98	3	5	$7.37\cdot10^{-2}$
4210.34	3	3	$1.48\cdot10^{-1}$
4217.55	3	5	$2.46\cdot10^{-1}$
4219.36	11	13	$2.88\cdot10^{-1}$
4222.21	7	7	$5.76\cdot10^{-2}$
4224.17	9	11	$1.06\cdot10^{-1}$
4224.51	3	5	$6.81\cdot10^{-2}$
4225.45	5	7	$1.65\cdot10^{-1}$
4227.43	11	13	$5.29\cdot10^{-1}$
4233.60	3	5	$1.85\cdot10^{-1}$
4235.94	9	9	$1.88\cdot10^{-1}$
4238.81	7	9	$2.41\cdot10^{-1}$
4247.43	9	11	$1.94\cdot10^{-1}$
4250.12	5	7	$2.07\cdot10^{-1}$
4250.79	7	7	$1.02\cdot10^{-1}$
4260.47	11	11	$3.99\cdot10^{-1}$

λ	Weights		A
Å	g_i	g_k	$10^8\ s^{-1}$
4271.15	7	9	$1.82\cdot10^{-1}$
4271.76	9	11	$2.28\cdot10^{-1}$
4282.40	7	5	$1.21\cdot10^{-1}$
4299.23	9	11	$1.29\cdot10^{-1}$
4307.90	7	9	$3.38\cdot10^{-1}$
4315.08	5	5	$7.76\cdot10^{-2}$
4325.76	5	7	$5.16\cdot10^{-1}$
4327.10	5	5	$1.12\cdot10^{-1}$
4369.77	9	9	$6.09\cdot10^{-2}$
4383.55	9	11	$5.00\cdot10^{-1}$
4388.41	7	7	$1.03\cdot10^{-1}$
4401.29	7	7	$6.4\cdot10^{-2}$
4404.75	7	9	$2.75\cdot10^{-1}$
4415.12	5	7	$1.19\cdot10^{-1}$
4433.22	5	3	$2.1\cdot10^{-1}$
4443.19	1	3	$1.02\cdot10^{-1}$
4455.03	9	7	$4.1\cdot10^{-2}$
4466.55	5	7	$1.20\cdot10^{-1}$
4469.38	5	7	$1.59\cdot10^{-1}$
4476.02	3	5	$1.01\cdot10^{-1}$
4484.22	7	9	$5.04\cdot10^{-2}$
4485.68	3	3	$1.1\cdot10^{-1}$
4528.61	7	9	$5.44\cdot10^{-2}$
4547.85	5	7	$4.48\cdot10^{-2}$
4556.13	7	5	$1.05\cdot10^{-1}$
4619.29	7	5	$5.2\cdot10^{-2}$
4654.61	7	7	$3.68\cdot10^{-2}$
4667.45	7	9	$6.03\cdot10^{-2}$
4673.16	5	7	$3.81\cdot10^{-2}$
4678.85	7	9	$4.97\cdot10^{-2}$
4736.77	9	11	$4.78\cdot10^{-2}$
4789.65	5	7	$4.57\cdot10^{-2}$
4800.65	7	9	$3.01\cdot10^{-2}$
4859.74	5	3	$1.62\cdot10^{-1}$
4871.32	7	5	$2.44\cdot10^{-1}$
4872.14	3	3	$2.54\cdot10^{-1}$
4878.21	1	3	$1.21\cdot10^{-1}$
4890.76	5	5	$2.25\cdot10^{-1}$
4891.49	9	7	$3.08\cdot10^{-1}$
4903.31	3	5	$6.58\cdot10^{-2}$
4918.99	7	7	$1.79\cdot10^{-1}$
4920.50	11	9	$3.58\cdot10^{-1}$
4957.30	9	9	$1.18\cdot10^{-1}$
4957.60	13	11	$4.22\cdot10^{-1}$
4966.09	11	11	$3.31\cdot10^{-2}$
4973.10	3	3	$1.1\cdot10^{-1}$
4978.60	5	3	$1.19\cdot10^{-1}$
4985.25	5	5	$1.48\cdot10^{-1}$
4988.95	7	7	$5.2\cdot10^{-2}$
5001.86	9	7	$3.7\cdot10^{-1}$
5006.12	11	11	$5.87\cdot10^{-2}$
5014.94	7	5	$2.64\cdot10^{-1}$
5021.59	7	9	$6.18\cdot10^{-2}$
5022.24	5	3	$2.4\cdot10^{-1}$
5048.44	3	5	$4.88\cdot10^{-2}$
5068.77	9	7	$3.37\cdot10^{-2}$
5074.75	9	11	$1.4\cdot10^{-1}$
5090.77	7	5	$1.9\cdot10^{-1}$
5121.64	5	5	$7.9\cdot10^{-2}$
5137.38	11	9	$1.0\cdot10^{-1}$
5139.25	7	5	$9.16\cdot10^{-2}$
5139.46	9	9	$8.69\cdot10^{-2}$
5184.27	5	7	$3.8\cdot10^{-2}$
5191.46	5	3	$2.32\cdot10^{-1}$
5192.34	7	7	$1.34\cdot10^{-1}$
5208.59	7	5	$6.23\cdot10^{-2}$
5215.18	5	3	$1.10\cdot10^{-1}$
5226.86	5	5	$1.36\cdot10^{-1}$
5232.94	9	11	$1.94\cdot10^{-1}$
5235.39	9	7	$3.75\cdot10^{-2}$
5242.49	13	11	$2.38\cdot10^{-2}$
5263.31	5	5	$6.36\cdot10^{-2}$
5266.56	7	9	$1.10\cdot10^{-1}$
5273.16	1	3	$8.12\cdot10^{-2}$
5281.79	5	7	$5.00\cdot10^{-2}$
5283.62	7	7	$1.02\cdot10^{-1}$
5302.30	3	5	$9.04\cdot10^{-2}$
5324.18	9	9	$2.06\cdot10^{-1}$
5339.93	5	7	$6.36\cdot10^{-2}$
5364.87	5	7	$5.59\cdot10^{-1}$
5367.47	7	9	$7.13\cdot10^{-1}$
5369.96	9	11	$7.22\cdot10^{-1}$
5373.71	7	9	$3.7\cdot10^{-2}$
5383.37	11	13	$7.81\cdot10^{-1}$
5393.17	7	9	$4.91\cdot10^{-2}$
5398.28	5	5	$9.0\cdot10^{-2}$
5404.15	9	11	$6.92\cdot10^{-1}$
5410.91	7	9	$6.33\cdot10^{-1}$
5415.20	11	13	$7.67\cdot10^{-1}$
5463.28	9	9	$2.9\cdot10^{-1}$
5473.90	7	7	$5.2\cdot10^{-2}$
5476.29	7	9	$2.87\cdot10^{-2}$
5476.56	9	9	$8.70\cdot10^{-2}$
5563.60	5	7	$3.4\cdot10^{-2}$
5569.62	5	3	$2.34\cdot10^{-1}$
5572.84	7	5	$2.28\cdot10^{-1}$
5576.09	3	1	$2.5\cdot10^{-1}$
5586.76	9	7	$2.19\cdot10^{-1}$
5594.66	9	9	$5.20\cdot10^{-2}$
5602.95	3	3	$1.00\cdot10^{-1}$
5615.64	11	9	$2.64\cdot10^{-1}$
5624.54	5	5	$7.41\cdot10^{-2}$
5633.95	11	13	$7.7\cdot10^{-2}$
5638.26	9	7	$4.4\cdot10^{-2}$
5649.99	3	5	$5.1\cdot10^{-2}$
5655.18	7	9	$4.7\cdot10^{-2}$
5658.82	7	7	$4.34\cdot10^{-2}$
5662.52	11	9	$6.18\cdot10^{-2}$
5679.02	5	7	$3.7\cdot10^{-2}$
5686.53	9	11	$6.71\cdot10^{-2}$
5705.99	7	9	$6.1\cdot10^{-2}$
5753.12	3	5	$8.26\cdot10^{-2}$
5762.99	5	7	$9.6\cdot10^{-2}$
5816.37	9	11	$4.49\cdot10^{-2}$
5905.67	5	3	$1.1\cdot10^{-1}$
6301.50	5	5	$6.43\cdot10^{-2}$
6400.00	7	9	$9.27\cdot10^{-2}$
6411.65	5	7	$4.43\cdot10^{-2}$
6419.95	7	7	$1.2\cdot10^{-1}$
6469.19	3	3	$8.3\cdot10^{-2}$
6496.47	5	5	$7.8\cdot10^{-2}$
6569.22	7	9	$6.0\cdot10^{-2}$
6633.75	7	7	$3.44\cdot10^{-2}$
6841.34	5	7	$3.4\cdot10^{-2}$
6855.16	7	9	$2.86\cdot10^{-2}$
7187.32	9	11	$8.36\cdot10^{-2}$
7511.02	11	11	$1.35\cdot10^{-1}$
8220.38	13	11	$1.69\cdot10^{-1}$
8699.45	7	9	$4.08\cdot10^{-2}$
9012.07	11	9	$4.46\cdot10^{-2}$
9401.11	9	11	$2.64\cdot10^{-2}$
9414.04	7	9	$3.98\cdot10^{-2}$
9443.80	5	7	$6.39\cdot10^{-2}$
9569.91	11	11	$2.50\cdot10^{-2}$
9626.50	9	9	$4.51\cdot10^{-2}$
9738.57	11	13	$7.64\cdot10^{-7}$
9763.38	3	5	$5.42\cdot10^{-2}$
9861.74	7	9	$5.49\cdot10^{-2}$
9889.04	9	11	$2.22\cdot10^{-2}$
22473.28	11	11	$3.32\cdot10^{-2}$
23566.67	9	11	$2.21\cdot10^{-2}$
24547.95	11	9	$3.72\cdot10^{-2}$
24729.10	13	11	$5.08\cdot10^{-2}$

Fe II

λ (Å)	g_i	g_k	A ($10^8\ s^{-1}$)
1055.26	10	8	$4.6\cdot10^{-1}$
1063.97	10	8	$3.5\cdot10^{-1}$
1068.35	8	8	1.59
1071.58	6	8	1.14
1096.88	10	8	2.26
1112.05	10	12	$2.0\cdot10^{-1}$
1121.97	10	8	1.92
1122.84	8	6	1.81
1125.45	10	8	1.03
1128.05	2	4	1.40
1130.44	6	8	$3.1\cdot10^{-1}$
1133.40	8	10	$2.6\cdot10^{-1}$
1133.67	10	8	$3.1\cdot10^{-1}$
1138.63	8	8	$5.5\cdot10^{-1}$
1142.37	10	8	$2.6\cdot10^{-1}$
1143.23	10	10	$9.8\cdot10^{-1}$
1144.94	10	12	3.52
1147.41	8	8	1.24
1148.28	8	10	3.35
1151.15	6	8	2.23
1267.42	8	6	$9.3\cdot10^{-1}$
1272.61	6	4	$3.3\cdot10^{-1}$
1371.02	14	12	1.74
1563.79	8	8	1.33
1580.63	8	10	$5.8\cdot10^{-1}$
1608.45	10	8	1.91
1610.92	10	10	$1.94\cdot10^{-1}$
1618.47	8	8	$5.53\cdot10^{-1}$
1621.69	8	6	1.32
1623.09	8	8	$1.99\cdot10^{-1}$
1625.52	8	10	$4.04\cdot10^{-1}$
1625.91	6	8	$1.02\cdot10^{-1}$
1629.16	6	6	$8.66\cdot10^{-1}$
1631.13	6	4	$6.93\cdot10^{-1}$
1633.91	6	8	$3.85\cdot10^{-1}$

λ Å	Weights g_i	Weights g_k	A $10^8\ s^{-1}$
1634.35	4	6	$3.21\cdot10^{-1}$
1635.40	8	6	2.28
1636.33	4	4	$9.63\cdot10^{-1}$
1637.40	10	8	$3.57\cdot10^{-1}$
1639.40	2	4	$6.85\cdot10^{-1}$
1641.76	6	4	1.76
1647.16	6	6	$4.98\cdot10^{-1}$
1670.75	10	8	1.06
1676.86	8	8	$6.75\cdot10^{-2}$
1702.04	10	12	1.02
1761.37	8	8	1.42
1785.27	6	8	$1.2\cdot10^{1}$
1786.75	6	6	$1.2\cdot10^{1}$
1788.08	6	4	4.6
1818.52	8	8	$5.70\cdot10^{-2}$
2020.75	6	8	$1.83\cdot10^{-1}$
2078.16	10	10	$2.84\cdot10^{-2}$
2162.02	10	10	$2.54\cdot10^{-1}$
2182.36	10	8	$8.6\cdot10^{-2}$
2191.98	8	8	$7.54\cdot10^{-1}$
2201.59	6	8	$7.77\cdot10^{-1}$
2208.41	10	10	1.59
2209.03	10	8	1.27
2213.66	14	14	$3.26\cdot10^{-1}$
2218.26	8	10	1.57
2220.38	12	12	$4.19\cdot10^{-1}$
2228.73	6	8	1.59
2249.18	10	8	$3.00\cdot10^{-2}$
2253.13	8	8	$4.41\cdot10^{-2}$
2255.77	6	4	$4.75\cdot10^{-1}$
2260.08	10	10	$3.18\cdot10^{-2}$
2267.59	6	8	$3.69\cdot10^{-2}$
2279.92	8	10	$4.49\cdot10^{-2}$
2327.40	6	4	$6.55\cdot10^{-1}$
2327.88	10	12	1.08
2331.31	10	8	$3.17\cdot10^{-1}$
2332.80	8	6	1.31
2338.01	4	4	1.13
2338.54	10	12	$5.6\cdot10^{-2}$
2343.50	10	8	1.73
2343.96	8	6	$3.13\cdot10^{-1}$
2344.28	2	4	$9.27\cdot10^{-1}$
2345.34	14	12	$7.3\cdot10^{-1}$
2348.12	10	8	$6.50\cdot10^{-1}$
2348.30	6	6	1.15
2351.20	12	10	$7.19\cdot10^{-1}$
2351.67	6	6	1.80
2352.31	2	4	4.38
2353.47	12	14	4.98
2353.68	8	8	1.30
2354.48	10	8	$8.13\cdot10^{-1}$
2354.89	6	4	$2.67\cdot10^{-1}$
2359.11	4	6	$5.0\cdot10^{-1}$
2359.60	10	10	$2.25\cdot10^{-1}$
2360.00	10	10	$3.59\cdot10^{-1}$
2360.29	8	6	$6.23\cdot10^{-1}$
2360.53	6	8	$2.22\cdot10^{-1}$
2361.73	8	8	$2.40\cdot10^{-1}$
2362.02	8	8	$1.41\cdot10^{-1}$
2363.86	8	10	5.3
2364.83	8	8	$5.90\cdot10^{-1}$
2365.76	6	6	2.16
2366.59	6	6	$1.01\cdot10^{-1}$
2366.88	8	10	$3.51\cdot10^{-2}$
2368.60	6	4	$6.06\cdot10^{-1}$
2369.95	10	12	5.9
2370.50	4	4	$1.73\cdot10^{-1}$
2373.74	10	10	$4.25\cdot10^{-1}$
2375.19	4	2	$9.81\cdot10^{-1}$
2376.43	12	14	6.4
2378.55	8	8	$1.70\cdot10^{-1}$
2378.70	8	8	$1.49\cdot10^{-1}$
2379.28	8	8	$2.73\cdot10^{-1}$
2379.42	10	10	$3.68\cdot10^{-1}$
2380.76	6	8	$3.10\cdot10^{-1}$
2382.04	10	12	3.13
2382.36	4	6	$3.19\cdot10^{-2}$
2382.90	12	14	$1.62\cdot10^{-1}$
2383.06	8	6	$1.0\cdot10^{-1}$
2383.24	6	6	$3.59\cdot10^{-1}$
2384.39	4	4	$3.22\cdot10^{-1}$
2385.01	6	8	$3.60\cdot10^{-2}$
2388.39	10	12	$2.02\cdot10^{-1}$
2388.63	8	8	1.05
2390.10	14	16	5.5
2390.76	6	6	1.17
2391.48	8	10	$3.77\cdot10^{-2}$
2394.00	12	10	$9.4\cdot10^{-2}$
2395.42	6	4	$2.67\cdot10^{-1}$
2395.63	8	10	2.59
2396.72	10	12	$2.15\cdot10^{-1}$
2399.24	6	6	1.39
2400.05	12	14	4.57
2401.29	6	8	1.89
2402.45	10	10	$5.8\cdot10^{-1}$
2402.63	8	8	$8.19\cdot10^{-1}$
2404.43	4	2	$6.44\cdot10^{-1}$
2404.89	6	8	1.96
2406.66	4	4	1.61
2410.27	8	8	$7.65\cdot10^{-1}$
2410.52	4	6	1.55
2411.07	2	2	2.37
2411.81	10	12	4.33
2412.01	6	8	$1.66\cdot10^{-1}$
2413.31	2	4	1.02
2416.45	8	10	2.38
2417.87	12	12	$9.5\cdot10^{-1}$
2418.44	6	8	2.28
2419.89	10	10	$2.2\cdot10^{-2}$
2422.69	6	8	1.46
2422.93	10	8	$2.94\cdot10^{-2}$
2423.21	4	6	1.40
2424.15	10	12	2.21
2424.39	6	8	$1.61\cdot10^{-1}$
2424.59	6	6	1.24
2424.65	8	8	$6.55\cdot10^{-2}$
2428.36	8	10	2.68
2428.80	4	4	1.38
2429.04	2	4	1.23
2429.39	4	4	$6.9\cdot10^{-1}$
2429.86	8	8	1.51
2430.08	8	10	1.91
2432.26	6	8	1.57
2432.87	14	14	2.86
2433.50	10	12	$1.30\cdot10^{-1}$
2434.06	8	6	$7.2\cdot10^{-1}$
2434.24	8	10	2.01
2434.73	12	12	2.79
2434.95	4	6	1.39
2435.00	8	8	2.02
2436.62	6	8	2.70
2439.30	12	14	2.25
2440.42	6	8	1.18
2441.13	10	10	$8.95\cdot10^{-1}$
2442.38	10	12	2.75
2443.71	8	10	1.44
2444.52	6	8	2.78
2445.11	12	12	2.03
2445.57	4	6	2.07
2445.80	4	6	1.23
2446.11	8	8	1.06
2446.47	12	14	$2.99\cdot10^{-1}$
2447.21	6	6	1.15
2447.33	4	2	2.56
2447.76	12	10	1.97
2449.96	4	4	1.24
2450.21	2	4	1.26
2453.98	8	10	1.31
2454.58	14	12	1.16
2455.71	8	8	1.01
2455.90	4	6	1.73
2457.10	6	4	$4.71\cdot10^{-1}$
2458.78	10	12	2.31
2458.97	6	4	2.51
2460.44	10	12	5.39
2461.28	6	8	2.34
2461.86	8	10	2.43
2463.28	12	10	$7.1\cdot10^{-1}$
2464.01	10	8	1.32
2464.91	6	4	2.22
2465.91	8	6	1.62
2466.50	2	4	2.40
2466.67	4	2	2.64
2466.82	6	4	1.77
2468.30	10	10	$9.8\cdot10^{-2}$
2469.37	10	8	$2.23\cdot10^{-2}$
2469.52	8	6	2.58
2470.41	6	6	$6.0\cdot10^{-1}$
2470.67	8	6	1.54
2471.28	10	8	$4.15\cdot10^{-1}$
2472.61	8	10	3.22
2473.32	2	2	2.74
2475.12	4	6	3.72
2475.54	6	8	3.18
2476.27	8	10	$9.7\cdot10^{-2}$
2477.35	8	8	$1.70\cdot10^{-1}$
2478.57	6	6	$9.1\cdot10^{-1}$
2480.16	10	8	1.55
2481.05	12	12	$1.46\cdot10^{-1}$
2482.12	14	14	$6.5\cdot10^{-1}$

λ Å	Weights g_i	g_k	A $10^8\ s^{-1}$
2482.33	4	4	2.23
2482.66	12	10	1.25
2482.87	6	4	1.69
2483.72	8	10	$5.4 \cdot 10^{-1}$
2484.24	4	6	$8.3 \cdot 10^{-2}$
2484.44	8	8	2.16
2489.48	12	12	$5.1 \cdot 10^{-1}$
2489.83	12	12	1.94
2490.71	10	12	1.44
2490.86	8	8	$8.8 \cdot 10^{-1}$
2491.40	10	8	1.01
2492.34	10	12	$2.30 \cdot 10^{-1}$
2493.26	14	16	3.04
2493.88	6	6	1.74
2494.12	12	10	$2.97 \cdot 10^{-2}$
2497.82	6	6	1.68
2500.92	6	8	2.41
2501.35	2	2	1.48
2502.39	8	8	1.43
2503.33	12	12	$7.3 \cdot 10^{-1}$
2503.54	8	8	$3.32 \cdot 10^{-1}$
2503.57	10	10	$2.53 \cdot 10^{-1}$
2503.88	10	10	2.23
2506.09	10	10	$9.9 \cdot 10^{-1}$
2506.80	8	10	1.98
2508.34	8	10	$3.79 \cdot 10^{-1}$
2510.57	8	8	$1.54 \cdot 10^{-1}$
2511.76	8	10	2.30
2513.15	10	8	$2.49 \cdot 10^{-1}$
2514.38	8	8	2.11
2517.14	2	4	$9.2 \cdot 10^{-1}$
2519.05	8	6	2.10
2521.09	6	4	2.05
2521.82	8	8	2.36
2525.39	14	14	1.91
2525.92	8	8	$7.4 \cdot 10^{-1}$
2526.08	10	8	$3.52 \cdot 10^{-1}$
2526.29	6	6	2.47
2527.10	12	10	$3.67 \cdot 10^{-1}$
2527.71	10	8	$9.1 \cdot 10^{-1}$
2528.68	10	8	$2.3 \cdot 10^{-2}$
2529.08	4	6	1.80
2529.23	12	10	$3.27 \cdot 10^{-1}$
2529.55	10	10	2.20
2530.10	4	6	$6.6 \cdot 10^{-1}$
2533.63	12	12	1.92
2534.42	8	8	1.83
2535.36	6	4	2.46
2535.49	10	8	$7.47 \cdot 10^{-1}$
2536.67	12	12	$5.7 \cdot 10^{-1}$
2536.81	10	10	1.69
2536.84	12	14	$6.8 \cdot 10^{-1}$
2537.14	10	10	1.44
2538.21	14	12	1.26
2538.40	6	8	$3.7 \cdot 10^{-2}$
2538.50	8	6	$5.9 \cdot 10^{-1}$
2538.68	6	8	$7.4 \cdot 10^{-1}$
2538.91	10	8	1.28
2538.99	14	12	1.93
2539.81	8	8	$5.6 \cdot 10^{-2}$
2540.52	2	2	1.26
2540.66	6	8	1.70
2541.10	8	6	$9.6 \cdot 10^{-1}$
2541.84	8	6	$8.2 \cdot 10^{-1}$
2542.74	2	2	1.61
2543.38	10	12	$6.7 \cdot 10^{-1}$
2543.43	6	4	$8.3 \cdot 10^{-1}$
2544.97	4	6	$3.93 \cdot 10^{-1}$
2545.22	8	10	$5.3 \cdot 10^{-1}$
2545.44	8	10	$1.52 \cdot 10^{-1}$
2546.67	8	8	$7.98 \cdot 10^{-1}$
2547.34	8	8	$2.28 \cdot 10^{-1}$
2548.32	4	6	$2.69 \cdot 10^{-1}$
2548.59	10	10	$2.67 \cdot 10^{-1}$
2548.74	4	2	2.43
2548.92	12	10	$6.0 \cdot 10^{-1}$
2549.08	10	8	1.89
2549.40	4	4	1.65
2549.46	6	6	1.12
2549.77	8	6	$2.35 \cdot 10^{-1}$
2550.03	10	10	1.74
2550.15	8	10	$3.91 \cdot 10^{-1}$
2550.57	12	12	$1.6 \cdot 10^{-2}$
2550.68	12	12	1.07
2551.20	10	8	$2.48 \cdot 10^{-1}$
2554.94	8	8	$7.6 \cdot 10^{-2}$
2555.07	6	8	$1.96 \cdot 10^{-1}$
2555.45	4	6	$2.49 \cdot 10^{-1}$
2557.08	8	10	$2.8 \cdot 10^{-2}$
2557.51	10	8	$1.53 \cdot 10^{-1}$
2559.24	8	8	$6.4 \cdot 10^{-2}$
2559.77	6	8	$2.42 \cdot 10^{-1}$
2559.93	6	8	$2.47 \cdot 10^{-1}$
2560.28	4	4	1.77
2562.09	4	2	1.62
2562.54	8	6	1.79
2563.48	6	4	1.51
2566.22	8	10	2.61
2566.40	8	6	2.29
2566.62	10	12	$7.1 \cdot 10^{-2}$
2566.91	4	2	1.15
2568.41	2	4	$4.77 \cdot 10^{-1}$
2568.89	8	8	$2.8 \cdot 10^{-2}$
2569.78	2	4	1.11
2570.85	8	6	1.84
2571.55	10	10	$2.89 \cdot 10^{-2}$
2572.97	6	8	$7.89 \cdot 10^{-2}$
2573.21	8	10	$1.42 \cdot 10^{-1}$
2573.76	8	8	$2.3 \cdot 10^{-2}$
2574.37	6	4	2.43
2576.86	10	12	1.32
2577.92	2	2	1.24
2581.11	6	6	$7.61 \cdot 10^{-2}$
2582.41	6	8	$2.22 \cdot 10^{-1}$
2582.58	4	4	$8.80 \cdot 10^{-1}$
2583.05	8	10	$2.16 \cdot 10^{-2}$
2585.62	10	10	$3.09 \cdot 10^{-1}$
2585.88	10	8	$8.94 \cdot 10^{-1}$
2586.06	6	4	$5.8 \cdot 10^{-2}$
2587.95	8	10	1.69
2588.19	2	2	$1.5 \cdot 10^{-1}$
2588.80	8	8	$8.4 \cdot 10^{-2}$
2590.55	4	6	$7.9 \cdot 10^{-2}$
2591.54	6	6	$5.72 \cdot 10^{-1}$
2592.79	14	16	2.74
2593.73	2	4	$1.63 \cdot 10^{-1}$
2594.96	8	8	$1.0 \cdot 10^{-1}$
2598.37	8	6	1.43
2599.40	10	10	2.35
2604.05	8	8	$1.49 \cdot 10^{-1}$
2605.04	6	8	2.34
2605.31	4	4	1.99
2605.43	6	6	$3.40 \cdot 10^{-1}$
2605.90	4	2	1.27
2606.52	6	6	2.31
2607.09	6	4	1.73
2608.85	10	8	$5.0 \cdot 10^{-2}$
2609.13	8	10	$2.77 \cdot 10^{-1}$
2609.44	6	8	$6.0 \cdot 10^{-2}$
2609.87	8	8	$1.34 \cdot 10^{-1}$
2611.07	4	6	$7.28 \cdot 10^{-2}$
2611.87	8	8	1.20
2613.57	10	12	$2.0 \cdot 10^{-2}$
2613.82	4	2	2.12
2614.19	8	10	$3.3 \cdot 10^{-2}$
2614.59	10	8	$3.37 \cdot 10^{-2}$
2614.87	8	6	$3.5 \cdot 10^{-2}$
2617.62	6	6	$4.88 \cdot 10^{-1}$
2619.08	10	10	$2.48 \cdot 10^{-1}$
2620.17	6	6	$1.1 \cdot 10^{-1}$
2620.41	4	4	$4.30 \cdot 10^{-2}$
2620.70	8	8	$3.43 \cdot 10^{-1}$
2621.67	2	2	$5.60 \cdot 10^{-1}$
2623.13	14	14	$8.8 \cdot 10^{-2}$
2623.72	6	6	$1.92 \cdot 10^{-1}$
2625.49	12	14	2.55
2625.67	8	10	$3.52 \cdot 10^{-1}$
2626.50	4	6	$3.48 \cdot 10^{-1}$
2626.70	8	8	$1.94 \cdot 10^{-2}$
2628.29	2	4	$8.71 \cdot 10^{-1}$
2628.58	6	6	$3.4 \cdot 10^{-2}$
2629.59	6	8	$7.2 \cdot 10^{-1}$
2630.07	4	6	$5.1 \cdot 10^{-1}$
2631.05	4	6	$8.16 \cdot 10^{-1}$
2631.32	6	8	$6.29 \cdot 10^{-1}$
2631.61	10	12	$6.6 \cdot 10^{-1}$
2633.20	6	4	1.21
2636.70	4	4	$8.8 \cdot 10^{-2}$
2637.50	6	6	$6.2 \cdot 10^{-1}$
2637.64	2	4	$6.6 \cdot 10^{-1}$
2639.57	2	2	$8.0 \cdot 10^{-1}$
2641.12	4	4	$3.7 \cdot 10^{-2}$
2642.01	6	6	$2.29 \cdot 10^{-1}$
2646.21	12	10	$1.44 \cdot 10^{-2}$
2649.47	6	8	1.98
2650.48	6	8	1.60
2652.57	10	8	$4.45 \cdot 10^{-2}$
2654.63	4	4	$8.1 \cdot 10^{-1}$
2657.92	10	10	$3.2 \cdot 10^{-2}$
2658.25	8	8	$2.12 \cdot 10^{-1}$

λ Å	Weights g_i	g_k	A 10^8 s^{-1}
2662.56	2	2	1.33
2664.66	8	10	1.91
2666.64	6	8	1.87
2667.22	4	6	1.02
2669.93	2	4	$5.2 \cdot 10^{-1}$
2670.38	6	8	$6.0 \cdot 10^{-2}$
2671.39	2	4	$6.5 \cdot 10^{-1}$
2680.23	6	8	$1.10 \cdot 10^{-1}$
2682.51	8	10	$9.2 \cdot 10^{-1}$
2683.00	4	6	$7.3 \cdot 10^{-1}$
2684.75	8	10	1.57
2691.74	10	8	$5.04 \cdot 10^{-2}$
2692.60	10	12	1.40
2693.86	8	6	$4.2 \cdot 10^{-2}$
2697.33	4	4	$2.48 \cdot 10^{-1}$
2697.46	4	2	1.65
2697.73	10	8	$2.6 \cdot 10^{-2}$
2699.20	4	4	$6.2 \cdot 10^{-1}$
2703.99	8	8	1.38
2704.58	8	8	$1.66 \cdot 10^{-2}$
2707.13	4	6	$8.3 \cdot 10^{-1}$
2709.06	4	6	$3.88 \cdot 10^{-1}$
2711.84	12	14	$4.36 \cdot 10^{-1}$
2712.39	10	12	$1.29 \cdot 10^{-1}$
2714.41	8	6	$5.70 \cdot 10^{-1}$
2716.22	6	6	1.15
2716.44	6	6	$2.8 \cdot 10^{-2}$
2716.57	14	12	1.35
2717.88	16	14	1.51
2718.64	10	8	1.18
2719.30	6	8	$4.44 \cdot 10^{-1}$
2721.81	12	10	$5.1 \cdot 10^{-2}$
2722.06	8	8	$1.42 \cdot 10^{-1}$
2722.74	6	8	$8.2 \cdot 10^{-1}$
2724.88	6	6	$9.58 \cdot 10^{-2}$
2726.52	6	8	$5.0 \cdot 10^{-2}$
2727.38	12	10	$3.12 \cdot 10^{-1}$
2727.54	6	4	$9.38 \cdot 10^{-1}$
2728.91	8	10	$1.25 \cdot 10^{-1}$
2730.73	4	4	$2.79 \cdot 10^{-1}$
2732.01	10	8	$7.05 \cdot 10^{-2}$
2732.94	8	6	$9.5 \cdot 10^{-1}$
2736.97	4	2	1.22
2739.55	8	8	2.21
2741.39	6	6	$2.03 \cdot 10^{-1}$
2743.20	2	4	1.97
2744.90	6	8	$3.62 \cdot 10^{-2}$
2746.48	4	6	2.05
2746.98	6	6	1.69
2749.18	4	4	1.21
2749.32	6	8	2.16
2749.49	2	2	1.16
2750.01	10	10	$1.8 \cdot 10^{-2}$
2751.13	4	4	$2.92 \cdot 10^{-1}$
2752.15	4	4	$7.7 \cdot 10^{-1}$
2753.29	10	12	1.89
2754.89	8	6	1.21
2755.74	8	10	2.15
2756.51	6	8	$7.3 \cdot 10^{-2}$
2757.03	10	8	$8.07 \cdot 10^{-2}$
2761.81	2	4	$1.38 \cdot 10^{-1}$
2762.33	6	6	$6.0 \cdot 10^{-1}$
2762.45	6	4	$3.3 \cdot 10^{-2}$
2763.66	14	12	1.34
2763.91	8	6	$2.9 \cdot 10^{-2}$
2764.79	12	12	$1.1 \cdot 10^{-2}$
2765.13	10	8	1.47
2767.50	12	14	1.58
2768.93	4	6	$4.75 \cdot 10^{-2}$
2769.15	8	10	$6.6 \cdot 10^{-2}$
2769.35	12	14	$2.07 \cdot 10^{-1}$
2770.50	12	10	$4.08 \cdot 10^{-2}$
2771.19	10	12	$4.3 \cdot 10^{-2}$
2774.69	2	4	$2.73 \cdot 10^{-1}$
2776.18	6	8	$2.66 \cdot 10^{-2}$
2776.91	8	8	$4.08 \cdot 10^{-1}$
2779.30	10	8	1.00
2779.91	2	4	$2.56 \cdot 10^{-1}$
2780.05	2	2	$3.3 \cdot 10^{-1}$
2783.69	12	10	1.06
2784.28	2	4	$3.4 \cdot 10^{-2}$
2785.19	12	10	1.53
2787.24	8	6	$1.83 \cdot 10^{-1}$
2790.56	8	10	$2.1 \cdot 10^{-2}$
2793.89	10	12	$1.26 \cdot 10^{-1}$
2796.63	10	10	$2.0 \cdot 10^{-1}$
2797.92	10	10	$3.2 \cdot 10^{-2}$
2799.29	10	8	$1.55 \cdot 10^{-1}$
2805.32	4	6	$2.5 \cdot 10^{-2}$
2805.79	8	8	$3.22 \cdot 10^{-2}$
2809.78	8	8	$3.10 \cdot 10^{-1}$
2811.27	12	10	$1.2 \cdot 10^{-2}$
2812.49	4	4	$2.9 \cdot 10^{-2}$
2813.61	8	10	$3.40 \cdot 10^{-2}$
2817.09	6	4	$3.37 \cdot 10^{-1}$
2819.34	12	12	$9.7 \cdot 10^{-3}$
2826.03	8	6	$4.5 \cdot 10^{-2}$
2827.43	12	14	$2.4 \cdot 10^{-2}$
2828.63	12	10	$6.9 \cdot 10^{-2}$
2831.56	4	6	$7.6 \cdot 10^{-1}$
2833.09	6	6	$4.55 \cdot 10^{-1}$
2835.71	4	6	$5.1 \cdot 10^{-1}$
2836.19	4	4	$5.4 \cdot 10^{-2}$
2836.51	2	4	$9.8 \cdot 10^{-2}$
2837.30	10	12	$1.9 \cdot 10^{-2}$
2838.22	4	2	$8.6 \cdot 10^{-1}$
2839.51	10	8	1.47
2839.80	8	10	$5.8 \cdot 10^{-1}$
2840.34	12	12	$7.7 \cdot 10^{-2}$
2840.65	2	4	$7.6 \cdot 10^{-1}$
2840.76	10	12	$1.49 \cdot 10^{-1}$
2842.08	8	8	$1.5 \cdot 10^{-2}$
2843.32	10	10	$1.40 \cdot 10^{-2}$
2843.48	4	6	$9.6 \cdot 10^{-2}$
2844.96	2	2	$5.5 \cdot 10^{-1}$
2845.60	8	6	1.57
2847.77	4	4	$5.1 \cdot 10^{-1}$
2848.11	6	6	$9.9 \cdot 10^{-1}$
2848.32	6	4	1.59
2848.91	12	10	$5.3 \cdot 10^{-2}$
2849.61	10	12	$4.6 \cdot 10^{-2}$
2853.21	6	6	$2.3 \cdot 10^{-2}$
2855.67	8	10	$9.2 \cdot 10^{-2}$
2856.15	10	10	$5.0 \cdot 10^{-2}$
2856.38	6	8	$4.42 \cdot 10^{-1}$
2856.91	8	8	1.32
2857.17	6	8	$1.22 \cdot 10^{-1}$
2857.42	6	6	$2.0 \cdot 10^{-2}$
2858.34	10	12	$4.85 \cdot 10^{-1}$
2864.97	8	8	$4.3 \cdot 10^{-2}$
2869.16	8	10	$1.4 \cdot 10^{-2}$
2869.31	4	6	$4.04 \cdot 10^{-1}$
2871.06	10	12	$2.2 \cdot 10^{-2}$
2871.13	12	10	$3.0 \cdot 10^{-2}$
2872.38	10	8	$1.70 \cdot 10^{-1}$
2873.40	8	10	$4.56 \cdot 10^{-1}$
2875.35	8	10	$1.35 \cdot 10^{-1}$
2876.80	8	8	$9.56 \cdot 10^{-2}$
2879.25	10	8	$3.6 \cdot 10^{-2}$
2880.76	8	8	$2.21 \cdot 10^{-2}$
2883.71	12	14	$1.48 \cdot 10^{-1}$
2884.76	6	8	$2.46 \cdot 10^{-1}$
2885.93	14	12	$3.8 \cdot 10^{-2}$
2888.10	4	6	$6.1 \cdot 10^{-2}$
2894.78	10	12	$5.7 \cdot 10^{-2}$
2895.22	8	10	$1.09 \cdot 10^{-1}$
2897.27	6	4	$1.8 \cdot 10^{-1}$
2902.46	10	10	$3.2 \cdot 10^{-2}$
2906.12	2	4	$4.4 \cdot 10^{-2}$
2910.76	8	8	$1.5 \cdot 10^{-2}$
2917.08	6	8	$2.7 \cdot 10^{-2}$
2922.02	8	10	$3.8 \cdot 10^{-2}$
2926.59	8	10	$5.1 \cdot 10^{-2}$
2944.40	4	2	$3.5 \cdot 10^{-1}$
2947.65	6	4	$2.01 \cdot 10^{-1}$
2949.18	10	8	$2.45 \cdot 10^{-1}$
2953.77	6	8	$5.2 \cdot 10^{-2}$
2954.05	8	8	$1.2 \cdot 10^{-2}$
2959.60	8	6	$9.7 \cdot 10^{-2}$
2959.84	8	6	$1.36 \cdot 10^{-1}$
2964.13	8	6	$4.6 \cdot 10^{-2}$
2964.62	2	2	$6.5 \cdot 10^{-2}$
2965.03	4	4	$9.43 \cdot 10^{-2}$
2969.94	8	6	$2.28 \cdot 10^{-1}$
2970.52	4	6	$2.70 \cdot 10^{-2}$
2970.69	10	8	$4.15 \cdot 10^{-2}$
2982.06	4	6	$2.41 \cdot 10^{-1}$
2984.82	6	6	$4.29 \cdot 10^{-1}$
2985.54	2	4	$2.39 \cdot 10^{-1}$
2997.30	6	8	$8.6 \cdot 10^{-2}$
3000.06	8	6	$3.0 \cdot 10^{-2}$
3002.32	6	8	$2.0 \cdot 10^{-2}$
3002.65	4	6	$1.79 \cdot 10^{-1}$
3036.96	6	6	$2.22 \cdot 10^{-1}$
3044.84	8	10	$1.2 \cdot 10^{-2}$
3048.99	4	4	$3.84 \cdot 10^{-1}$
3056.80	14	12	$1.7 \cdot 10^{-2}$
3062.24	12	10	$1.36 \cdot 10^{-1}$
3065.32	6	6	$2.9 \cdot 10^{-2}$
3070.69	10	8	$1.28 \cdot 10^{-2}$

λ Å	Weights g_i	g_k	A 10^8 s^{-1}
3071.12	2	4	$2.59\cdot10^{-1}$
3076.44	4	6	$3.75\cdot10^{-1}$
3077.17	14	12	$1.35\cdot10^{-1}$
3078.68	6	8	$5.5\cdot10^{-1}$
3089.38	6	8	$2.2\cdot10^{-2}$
3096.29	8	8	$1.9\cdot10^{-2}$
3105.17	4	2	$7.5\cdot10^{-2}$
3105.55	2	2	$7.0\cdot10^{-2}$
3106.57	8	8	$1.88\cdot10^{-2}$
3114.30	4	4	$6.4\cdot10^{-2}$
3114.69	2	4	$2.5\cdot10^{-2}$
3116.58	6	4	$5.5\cdot10^{-2}$
3133.05	4	6	$1.5\cdot10^{-2}$
3135.36	6	6	$8.8\cdot10^{-2}$
3144.75	8	6	$2.7\cdot10^{-2}$
3154.20	10	10	$2.06\cdot10^{-1}$
3162.80	8	8	$5.5\cdot10^{-2}$
3167.86	8	8	$1.59\cdot10^{-1}$
3177.53	8	8	$1.74\cdot10^{-1}$
3179.50	6	8	$1.11\cdot10^{-1}$
3180.15	4	6	$7.7\cdot10^{-2}$
3186.74	4	4	$3.85\cdot10^{-2}$
3187.30	10	10	$5.0\cdot10^{-2}$
3192.91	6	6	$1.27\cdot10^{-2}$
3193.80	2	2	$5.4\cdot10^{-2}$
3193.86	8	8	$3.86\cdot10^{-2}$
3196.07	6	8	$1.61\cdot10^{-2}$
3210.44	2	4	$3.63\cdot10^{-2}$
3213.31	4	6	$6.12\cdot10^{-2}$
3227.74	6	8	$8.9\cdot10^{-2}$
3231.71	6	8	$1.4\cdot10^{-2}$
3232.78	8	6	$5.0\cdot10^{-2}$
3237.82	2	4	$6.8\cdot10^{-2}$
3243.72	10	8	$5.1\cdot10^{-2}$
3247.17	4	6	$7.1\cdot10^{-2}$
3258.77	6	8	$9.39\cdot10^{-2}$
3259.05	8	10	$6.7\cdot10^{-2}$
3276.60	6	8	$1.0\cdot10^{-2}$
3289.35	8	8	$2.1\cdot10^{-2}$
3323.06	8	10	$1.4\cdot10^{-2}$
3366.97	8	6	$2.2\cdot10^{-2}$
3381.01	6	4	$3.0\cdot10^{-2}$
3453.62	8	10	$8.5\cdot10^{-3}$
3468.68	8	8	$2.0\cdot10^{-2}$
3493.47	10	10	$3.2\cdot10^{-2}$
3621.27	2	4	$2.2\cdot10^{-2}$
3748.48	6	4	$3.4\cdot10^{-2}$
3759.46	4	2	$3.2\cdot10^{-2}$
3906.04	6	8	$1.1\cdot10^{-2}$
3935.96	8	10	$8.3\cdot10^{-3}$
3938.97	4	6	$8.4\cdot10^{-3}$
4233.17	6	8	$7.22\cdot10^{-3}$
4522.63	6	4	$8.4\cdot10^{-3}$
4549.19	4	6	$9.2\cdot10^{-3}$
4549.47	8	6	$1.00\cdot10^{-2}$
4583.84	10	8	$7.22\cdot10^{-3}$
4635.32	6	8	$1.0\cdot10^{-2}$
4923.93	6	4	$4.28\cdot10^{-2}$
4990.51	6	8	$5.2\cdot10^{-1}$
5001.96	12	14	1.57

λ Å	Weights g_i	g_k	A 10^8 s^{-1}
5018.44	6	6	$2.0\cdot10^{-2}$
5030.63	10	10	$7.1\cdot10^{-1}$
5035.71	10	12	$9.4\cdot10^{-1}$
5144.35	4	6	$8.5\cdot10^{-1}$
5149.47	8	10	$9.0\cdot10^{-1}$
5169.03	6	8	$4.22\cdot10^{-2}$
5197.58	6	4	$5.4\cdot10^{-3}$
5227.48	12	14	1.22
5247.95	4	6	1.43
5251.23	6	8	$8.0\cdot10^{-1}$
5264.18	8	10	$4.76\cdot10^{-1}$
5272.40	6	6	$3.9\cdot10^{-3}$
5276.00	10	8	$3.76\cdot10^{-3}$
5306.18	6	8	$3.28\cdot10^{-1}$
5316.22	14	14	$3.69\cdot10^{-1}$
5316.62	12	10	$3.89\cdot10^{-3}$
5387.06	12	14	$5.2\cdot10^{-1}$
5395.86	6	8	$5.5\cdot10^{-1}$
5402.06	10	12	$5.6\cdot10^{-1}$
5427.83	12	10	$5.9\cdot10^{-3}$
5429.99	8	10	$6.0\cdot10^{-1}$
5465.93	6	8	$6.2\cdot10^{-1}$
5482.31	10	12	$4.78\cdot10^{-1}$
5493.83	8	10	$4.01\cdot10^{-1}$
5506.19	12	14	1.14
5510.78	10	12	$2.28\cdot10^{-1}$
5529.05	6	6	$2.01\cdot10^{-1}$
5544.76	12	12	$2.49\cdot10^{-1}$
5783.63	8	10	$4.62\cdot10^{-1}$
5885.01	4	6	$6.4\cdot10^{-1}$
5902.83	8	10	$4.98\cdot10^{-1}$
5955.70	6	8	$4.19\cdot10^{-1}$
5961.71	10	12	$7.4\cdot10^{-1}$
5965.62	10	10	$2.19\cdot10^{-1}$
6175.15	8	8	$1.8\cdot10^{-3}$
6305.30	10	10	$1.4\cdot10^{-3}$
6331.95	6	8	$1.8\cdot10^{-3}$
6446.41	8	10	$1.3\cdot10^{-3}$
6456.38	8	6	$1.7\cdot10^{-3}$
Fe III			
1843.4	9	7	4.8
1844.3	7	5	4.9
1846.9	5	3	5.5
1854.38	3	1	5.7
1865.20	7	7	6.1
1893.98	11	9	5.5
1896.80	13	11	5.0
1904.3	5	5	5.7
1907.58	15	13	5.3
1915.08	13	15	6.0
1922.79	11	13	5.5
1930.39	9	11	5.1
1931.51	9	11	5.3
1937.35	7	9	5.1
1943.48	5	7	5.0
1950.33	13	15	5.5
1951.01	11	11	5.3
1952.65	9	9	4.9
1953.32	7	7	5.1

λ Å	Weights g_i	g_k	A 10^8 s^{-1}
1987.50	13	13	4.9
Fe VII			
150.807	5	7	$1.3\cdot10^{3}$
150.852	7	9	$1.3\cdot10^{3}$
151.023	9	11	$1.6\cdot10^{3}$
151.046	7	7	$2.2\cdot10^{2}$
151.145	9	9	$2.1\cdot10^{2}$
151.432	5	7	$2.2\cdot10^{2}$
151.512	5	5	$5.3\cdot10^{2}$
151.675	7	7	$3.9\cdot10^{2}$
151.782	9	9	$2.4\cdot10^{2}$
154.307	3	1	$8.9\cdot10^{2}$
154.335	5	7	$1.2\cdot10^{3}$
154.363	3	3	$4.2\cdot10^{2}$
154.565	5	3	$3.5\cdot10^{2}$
154.650	5	5	$8.8\cdot10^{2}$
154.848	1	3	$7.7\cdot10^{2}$
154.921	3	5	$9.7\cdot10^{2}$
154.941	3	3	$2.4\cdot10^{2}$
154.949	5	7	$1.0\cdot10^{3}$
155.994	9	11	$1.8\cdot10^{3}$
158.481	9	9	$2.3\cdot10^{2}$
165.087	1	3	$6.9\cdot10^{2}$
165.919	7	5	$2.8\cdot10^{3}$
166.365	9	7	$2.9\cdot10^{3}$
173.441	9	9	$3.6\cdot10^{3}$
176.744	9	9	$2.7\cdot10^{3}$
176.928	7	7	$2.4\cdot10^{3}$
177.172	5	5	$1.5\cdot10^{3}$
235.221	5	3	$1.7\cdot10^{2}$
240.053	3	1	$1.3\cdot10^{2}$
243.379	9	7	$2.1\cdot10^{2}$
Fe VIII			
112.472	4	4	$3.6\cdot10^{2}$
112.486	6	6	$4.3\cdot10^{2}$
116.196	4	6	$4.5\cdot10^{2}$
117.197	6	8	$3.8\cdot10^{2}$
167.486	4	4	$3.0\cdot10^{3}$
168.172	6	6	$3.1\cdot10^{3}$
168.545	6	4	$2.0\cdot10^{3}$
168.929	4	2	$2.1\cdot10^{3}$
185.213	6	8	$1.0\cdot10^{3}$
186.601	4	6	$9.4\cdot10^{2}$
Fe X			
76.822	2	2	$1.8\cdot10^{3}$
77.865	4	6	$1.6\cdot10^{3}$
100.026	8	10	$2.6\cdot10^{3}$
101.733	6	8	$1.8\cdot10^{3}$
101.846	4	6	$1.7\cdot10^{3}$
102.095	10	12	$2.9\cdot10^{3}$
102.192	10	12	$2.9\cdot10^{3}$
102.829	4	6	$2.1\cdot10^{3}$
103.319	6	8	$2.6\cdot10^{3}$
103.724	6	8	$1.7\cdot10^{3}$
104.638	8	10	$2.1\cdot10^{3}$
174.534	4	6	$1.8\cdot10^{3}$
175.266	2	4	$1.72\cdot10^{3}$

λ Å	Weights g_i	g_k	A 10^8 s^{-1}
Fe XI			
72.166	5	7	$2.9 \cdot 10^{3}$
72.310	5	5	$1.5 \cdot 10^{3}$
72.635	5	7	$1.6 \cdot 10^{3}$
91.394	5	7	$2.6 \cdot 10^{3}$
91.472	7	9	$2.5 \cdot 10^{3}$
91.63	3	5	$2.3 \cdot 10^{3}$
91.63	7	9	$3.4 \cdot 10^{3}$
91.63	5	7	$2.8 \cdot 10^{3}$
91.733	9	11	$4.1 \cdot 10^{3}$
92.81	9	11	$3.7 \cdot 10^{3}$
92.87	11	13	$3.9 \cdot 10^{3}$
93.433	9	11	$3.2 \cdot 10^{3}$
179.762	5	7	$1.67 \cdot 10^{3}$
Fe XII			
65.905	4	4	$2.0 \cdot 10^{3}$
66.526	6	8	$1.7 \cdot 10^{3}$
66.960	4	6	$1.6 \cdot 10^{3}$
67.164	4	2	$1.1 \cdot 10^{3}$
67.821	4	6	$1.4 \cdot 10^{3}$
68.382	2	4	$1.7 \cdot 10^{3}$
80.541	6	6	$8.7 \cdot 10^{2}$
81.943	6	4	$1.4 \cdot 10^{3}$
82.226	4	2	$1.9 \cdot 10^{3}$
84.48	4	6	$4.5 \cdot 10^{3}$
84.48	8	10	$4.9 \cdot 10^{3}$
84.52	10	12	$5.2 \cdot 10^{3}$
84.52	6	8	$4.0 \cdot 10^{3}$
84.85	6	8	$2.3 \cdot 10^{3}$
85.14	8	10	$3.4 \cdot 10^{3}$
85.477	10	12	$4.6 \cdot 10^{3}$
186.880	6	8	$1.0 \cdot 10^{3}$
192.394	4	2	$9.0 \cdot 10^{2}$
193.509	4	4	$9.1 \cdot 10^{2}$
195.119	4	6	$8.6 \cdot 10^{2}$
Fe XIII			
62.353	1	3	$2.0 \cdot 10^{3}$
62.46	5	7	$1.2 \cdot 10^{3}$
62.699	3	5	$2.3 \cdot 10^{3}$
63.188	5	7	$3.9 \cdot 10^{3}$
64.139	1	3	$2.1 \cdot 10^{3}$
74.845	5	5	$1.0 \cdot 10^{3}$
75.892	5	3	$7.7 \cdot 10^{2}$
76.117	5	3	$2.1 \cdot 10^{3}$
78.452	9	11	$6.3 \cdot 10^{3}$
84.270	7	9	$5.5 \cdot 10^{3}$
107.384	7	5	$1.8 \cdot 10^{3}$
Fe XIV			
72.80	10	12	$7.9 \cdot 10^{3}$
76.022	4	6	$6.6 \cdot 10^{3}$
76.152	6	8	$7.0 \cdot 10^{3}$
190	6	8	$2.8 \cdot 10^{2}$
211.316	2	4	$3.6 \cdot 10^{2}$
216	6	8	$1.7 \cdot 10^{2}$
217	6	8	$4.0 \cdot 10^{2}$
217	6	6	$2.6 \cdot 10^{2}$
219	2	4	$4.8 \cdot 10^{2}$
219	4	6	$2.4 \cdot 10^{2}$
219.123	4	6	$3.9 \cdot 10^{2}$
220	4	4	$3.2 \cdot 10^{2}$
221	4	6	$5.9 \cdot 10^{2}$
226	2	4	$3.9 \cdot 10^{2}$
264.787	4	4	$3.38 \cdot 10^{2}$
268	6	6	$2.1 \cdot 10^{2}$
280	4	6	$2.8 \cdot 10^{2}$
283	6	8	$2.7 \cdot 10^{2}$
288.45	6	4	$1.6 \cdot 10^{2}$
Fe XV			
38.95	1	3	$1.69 \cdot 10^{3}$
52.911	1	3	$2.94 \cdot 10^{3}$
59.404	3	5	$3.4 \cdot 10^{3}$
63.959	5	7	$1.6 \cdot 10^{3}$
65.370	1	3	$3.2 \cdot 10^{2}$
65.612	3	3	$9.8 \cdot 10^{2}$
66.238	5	3	$1.6 \cdot 10^{3}$
68.860	9	11	$9.2 \cdot 10^{3}$
69.7	3	1	$1.9 \cdot 10^{3}$
69.942	3	5	$7.4 \cdot 10^{3}$
69.989	5	7	$7.9 \cdot 10^{3}$
70.052	7	9	$8.8 \cdot 10^{3}$
70.224	1	3	$4.13 \cdot 10^{3}$
70.53	7	5	$2.6 \cdot 10^{2}$
70.59	7	7	$1.7 \cdot 10^{3}$
73.199	7	9	$8.8 \cdot 10^{3}$
73.473	5	7	$6.2 \cdot 10^{3}$
233.857	5	7	$2.2 \cdot 10^{2}$
235	1	3	$2.5 \cdot 10^{2}$
243	1	3	$2.4 \cdot 10^{2}$
243	5	7	$2.3 \cdot 10^{2}$
243.790	3	5	$4.2 \cdot 10^{2}$
248	3	1	$5.4 \cdot 10^{2}$
284.160	1	3	$2.28 \cdot 10^{2}$
Fe XVI			
39.827	2	4	$2.1 \cdot 10^{3}$
40.153	4	6	$2.5 \cdot 10^{3}$
40.199	4	6	$1.7 \cdot 10^{3}$
40.245	6	8	$1.8 \cdot 10^{3}$
46.661	4	6	$3.46 \cdot 10^{3}$
46.718	6	8	$3.7 \cdot 10^{3}$
50.350	2	4	$1.86 \cdot 10^{3}$
54.142	2	4	$3.41 \cdot 10^{3}$
54.728	4	6	$4.16 \cdot 10^{3}$
62.879	2	2	$1.05 \cdot 10^{3}$
63.719	4	2	$2.18 \cdot 10^{3}$
66.263	4	6	$9.39 \cdot 10^{3}$
66.368	6	8	$1.00 \cdot 10^{4}$
66.392	6	6	$6.69 \cdot 10^{2}$
76.502	6	4	$6.7 \cdot 10^{2}$
76.796	4	2	$7.72 \cdot 10^{2}$
80.192	4	6	$5.2 \cdot 10^{2}$
80.270	6	8	$5.4 \cdot 10^{2}$
85.587	2	4	$4.0 \cdot 10^{2}$
86.133	4	6	$4.8 \cdot 10^{2}$
96.256	4	6	$8.7 \cdot 10^{2}$
96.348	6	8	$9.3 \cdot 10^{2}$
117.2	2	4	$3.93 \cdot 10^{2}$
117.7	2	2	$3.9 \cdot 10^{2}$
123.4	2	4	$5.9 \cdot 10^{2}$
124.5	4	6	$7.0 \cdot 10^{2}$
144.06	4	6	$1.6 \cdot 10^{3}$
144.25	6	8	$1.6 \cdot 10^{3}$
148	4	2	$6.5 \cdot 10^{2}$
266.7	4	6	$3.9 \cdot 10^{2}$
267.0	6	8	$4.3 \cdot 10^{2}$
Fe XVII			
11.023	1	3	$2.1 \cdot 10^{4}$
12.123	1	3	$8.0 \cdot 10^{4}$
12.264	1	3	$5.9 \cdot 10^{4}$
12.526	1	3	$3.0 \cdot 10^{3}$
12.681	1	3	$3.5 \cdot 10^{3}$
13.823	1	3	$3.3 \cdot 10^{4}$
13.891	1	3	$3.4 \cdot 10^{3}$
15.015	1	3	$2.28 \cdot 10^{5}$
15.262	1	3	$6.0 \cdot 10^{4}$
16.777	1	3	$8.29 \cdot 10^{3}$
17.054	1	3	$9.33 \cdot 10^{3}$
41.37	9	11	$4.8 \cdot 10^{3}$
49.427	3	3	$4.0 \cdot 10^{3}$
50.26	7	9	$6.0 \cdot 10^{3}$
58.76	9	11	$1.2 \cdot 10^{4}$
Fe XIX			
13.413	5	3	$1.3 \cdot 10^{4}$
13.426	5	7	$4.8 \cdot 10^{4}$
13.47	3	1	$1.5 \cdot 10^{5}$
13.520	5	7	$2.0 \cdot 10^{5}$
13.56	3	5	$1.0 \cdot 10^{4}$
13.68	3	1	$8.0 \cdot 10^{4}$
13.69	5	7	$2.3 \cdot 10^{4}$
13.700	1	3	$2.7 \cdot 10^{5}$
13.71	5	5	$2.2 \cdot 10^{4}$
13.738	5	7	$1.0 \cdot 10^{4}$
13.796	5	7	$7.0 \cdot 10^{4}$
13.83	5	5	$1.4 \cdot 10^{4}$
13.934	1	3	$4.51 \cdot 10^{4}$
13.961	3	3	$2.0 \cdot 10^{4}$
14.668	5	7	$1.1 \cdot 10^{4}$
14.671	5	3	$1.1 \cdot 10^{4}$
14.929	3	3	$1.2 \cdot 10^{4}$
14.966	5	3	$2.5 \cdot 10^{4}$
14.995	5	5	$2.2 \cdot 10^{4}$
15.015	1	3	$1.4 \cdot 10^{4}$
16.668	3	1	$1.1 \cdot 10^{4}$
Fe XX			
12.77	4	4	$2.1 \cdot 10^{5}$
12.78	4	2	$6.9 \cdot 10^{4}$
12.78	2	4	$1.4 \cdot 10^{5}$
12.82	4	4	$1.1 \cdot 10^{5}$
12.88	6	4	$2.7 \cdot 10^{4}$
12.89	4	4	$4.4 \cdot 10^{4}$
12.90	4	6	$1.4 \cdot 10^{5}$
12.93	4	6	$1.6 \cdot 10^{5}$

λ Å	Weights g_i	g_k	A $10^8\ s^{-1}$
12.98	2	2	$6.7 \cdot 10^{4}$
12.99	6	6	$5.1 \cdot 10^{4}$
13.01	2	4	$3.0 \cdot 10^{4}$
13.03	4	2	$8.6 \cdot 10^{4}$
13.13	2	4	$8.9 \cdot 10^{4}$
13.79	6	6	$1.2 \cdot 10^{4}$
Fe XXI			
8.56	5	7	$2.0 \cdot 10^{4}$
8.64	5	7	$1.5 \cdot 10^{4}$
8.65	5	7	$3.9 \cdot 10^{4}$
9.42	3	3	$3.3 \cdot 10^{4}$
9.44	3	5	$1.7 \cdot 10^{4}$
9.45	1	3	$5.2 \cdot 10^{4}$
9.47	5	7	$4.9 \cdot 10^{4}$
9.67	1	3	$5.7 \cdot 10^{4}$
12.02	1	3	$1.3 \cdot 10^{4}$
12.13	3	3	$1.8 \cdot 10^{4}$
12.18	5	7	$2.2 \cdot 10^{4}$
12.21	3	1	$1.5 \cdot 10^{5}$
12.21	3	3	$1.2 \cdot 10^{5}$
12.25	1	3	$2.1 \cdot 10^{5}$
12.28	5	3	$5.2 \cdot 10^{4}$
12.30	5	7	$2.1 \cdot 10^{5}$
12.36	3	3	$3.6 \cdot 10^{4}$
12.37	5	7	$3.1 \cdot 10^{5}$
12.47	5	7	$5.8 \cdot 10^{4}$
12.47	5	3	$1.3 \cdot 10^{4}$
12.49	5	7	$1.3 \cdot 10^{4}$
12.53	5	5	$1.5 \cdot 10^{4}$
12.57	1	3	$7.2 \cdot 10^{4}$
12.73	5	5	$8.2 \cdot 10^{3}$
12.95	3	5	$6.2 \cdot 10^{3}$
13.03	5	5	$1.3 \cdot 10^{4}$
Fe XXII			
9.002	4	6	$5.5 \cdot 10^{4}$
9.006	6	8	$5.7 \cdot 10^{4}$
9.006	6	6	$5.3 \cdot 10^{4}$
9.163	4	6	$6.9 \cdot 10^{4}$
9.183	6	8	$8.3 \cdot 10^{4}$
9.241	4	6	$5.1 \cdot 10^{4}$
11.748	4	4	$1.2 \cdot 10^{5}$
11.748	4	6	$1.6 \cdot 10^{5}$
11.748	4	2	$1.8 \cdot 10^{5}$
11.763	2	4	$1.6 \cdot 10^{5}$
11.789	2	2	$2.6 \cdot 10^{5}$
11.789	6	8	$1.2 \cdot 10^{5}$
11.797	2	4	$1.7 \cdot 10^{5}$
11.823	6	4	$7.9 \cdot 10^{4}$
11.837	6	8	$2.3 \cdot 10^{5}$
11.837	6	6	$1.7 \cdot 10^{5}$
11.886	4	6	$1.3 \cdot 10^{5}$
11.898	2	4	$8.2 \cdot 10^{4}$
11.922	4	6	$1.8 \cdot 10^{5}$
11.976	6	8	$5.9 \cdot 10^{4}$
12.027	2	4	$6.9 \cdot 10^{4}$
12.045	6	8	$2.4 \cdot 10^{5}$
12.045	4	4	$9.7 \cdot 10^{4}$
12.053	4	6	$6.1 \cdot 10^{4}$
12.077	2	4	$1.0 \cdot 10^{5}$
12.077	4	6	$2.4 \cdot 10^{5}$
12.095	6	6	$7.8 \cdot 10^{4}$
12.193	2	4	$7.2 \cdot 10^{4}$
12.193	4	6	$9.9 \cdot 10^{4}$
12.325	2	2	$1.5 \cdot 10^{5}$
Fe XXIII			
8.614	5	7	$7.7 \cdot 10^{4}$
8.752	5	7	$1.2 \cdot 10^{5}$
10.927	5	7	$6.0 \cdot 10^{4}$
10.934	3	5	$5.4 \cdot 10^{4}$
11.165	3	5	$6.7 \cdot 10^{4}$
11.298	1	3	$1.3 \cdot 10^{5}$
11.325	3	5	$1.7 \cdot 10^{5}$
11.338	3	3	$9.3 \cdot 10^{4}$
11.433	3	3	$1.2 \cdot 10^{5}$
11.441	5	7	$2.2 \cdot 10^{5}$
11.445	5	5	$5.6 \cdot 10^{4}$
11.485	3	5	$1.40 \cdot 10^{5}$
11.519	5	5	$1.16 \cdot 10^{5}$
11.520	1	3	$2.16 \cdot 10^{5}$
11.524	5	7	$2.3 \cdot 10^{5}$
11.593	5	7	$3.58 \cdot 10^{5}$
11.613	3	5	$1.0 \cdot 10^{5}$
11.691	5	7	$7.7 \cdot 10^{4}$
11.698	5	5	$7.3 \cdot 10^{4}$
11.737	3	5	$1.8 \cdot 10^{5}$
11.898	1	3	$2.03 \cdot 10^{5}$
Fe XXIV			
1.8523	2	2	$1.0 \cdot 10^{5}$
1.8552	2	4	$4.82 \cdot 10^{6}$
1.8563	4	2	$2.43 \cdot 10^{6}$
1.8572	2	2	$3.06 \cdot 10^{6}$
1.858	2	4	$1.2 \cdot 10^{5}$
1.8614	4	4	$6.24 \cdot 10^{6}$
1.8626	2	4	$3.16 \cdot 10^{6}$
1.8627	2	2	$5.47 \cdot 10^{6}$
1.8637	2	2	$1.91 \cdot 10^{6}$
1.8655	4	6	$2.14 \cdot 10^{6}$
1.8672	4	2	$1.63 \cdot 10^{6}$
1.8678	4	4	$3.5 \cdot 10^{5}$
1.8721	4	6	$3.2 \cdot 10^{5}$
1.8721	2	2	$2.0 \cdot 10^{5}$
1.8730	2	4	$1.5 \cdot 10^{5}$
1.8739	4	4	$8.3 \cdot 10^{4}$
1.891	2	2	$9.7 \cdot 10^{4}$
1.897	4	2	$9.8 \cdot 10^{4}$
8.231	2	4	$6.10 \cdot 10^{4}$
8.316	4	6	$7.07 \cdot 10^{4}$
10.619	2	4	$7.28 \cdot 10^{4}$
10.663	2	2	$7.51 \cdot 10^{4}$
11.030	2	4	$1.84 \cdot 10^{5}$
11.171	4	6	$2.18 \cdot 10^{5}$
Fe XXV			
1.4607	1	3	$2.54 \cdot 10^{5}$
1.4945	1	3	$5.05 \cdot 10^{5}$
1.5730	1	3	$1.24 \cdot 10^{6}$
1.5749	1	3	$1.5 \cdot 10^{5}$
1.778	3	3	$8.7 \cdot 10^{4}$
1.782	3	1	$4.69 \cdot 10^{6}$
1.787	1	3	$2.57 \cdot 10^{6}$
1.787	5	5	$1.19 \cdot 10^{6}$
1.788	3	5	$2.68 \cdot 10^{6}$
1.788	3	5	$1.63 \cdot 10^{6}$
1.789	1	3	$1.78 \cdot 10^{6}$
1.790	3	3	$1.23 \cdot 10^{6}$
1.791	3	5	$4.10 \cdot 10^{6}$
1.791	3	3	$2.59 \cdot 10^{6}$
1.792	3	1	$4.92 \cdot 10^{6}$
1.792	5	5	$2.81 \cdot 10^{6}$
1.793	3	1	$2.67 \cdot 10^{6}$
1.794	5	3	$2.22 \cdot 10^{6}$
1.797	3	5	$8.8 \cdot 10^{5}$
1.798	3	3	$1.0 \cdot 10^{5}$
1.800	1	3	$8.6 \cdot 10^{4}$
1.802	3	1	$4.1 \cdot 10^{5}$
1.810	3	1	$5.9 \cdot 10^{5}$
1.8502	1	3	$4.57 \cdot 10^{6}$
1.8593	1	3	$4.42 \cdot 10^{5}$
10.038	3	3	$8.08 \cdot 10^{4}$
Krypton			
Kr I			
1164.9	1	3	3.16
1235.8	1	3	3.12
4274.0	5	5	$2.6 \cdot 10^{-2}$
4351.4	3	1	$3.2 \cdot 10^{-2}$
4362.6	5	3	$8.4 \cdot 10^{-3}$
4376.1	3	1	$5.6 \cdot 10^{-2}$
4400.0	3	5	$2.0 \cdot 10^{-2}$
4410.4	3	3	$4.4 \cdot 10^{-3}$
4425.2	3	3	$9.7 \cdot 10^{-3}$
4453.9	3	5	$7.8 \cdot 10^{-3}$
4463.7	3	3	$2.3 \cdot 10^{-2}$
4502.4	3	5	$9.2 \cdot 10^{-3}$
5562.2	5	5	$2.8 \cdot 10^{-3}$
5570.3	5	3	$2.1 \cdot 10^{-2}$
5649.6	1	3	$3.7 \cdot 10^{-3}$
5870.9	3	5	$1.8 \cdot 10^{-2}$
6904.7	3	5	$1.3 \cdot 10^{-2}$
7224.1	3	5	$1.4 \cdot 10^{-2}$
7587.4	3	1	$5.1 \cdot 10^{-1}$
7601.5	5	5	$3.1 \cdot 10^{-1}$
7685.2	3	1	$4.9 \cdot 10^{-1}$
7694.5	5	3	$5.6 \cdot 10^{-2}$
7854.8	1	3	$2.3 \cdot 10^{-1}$
8059.5	1	3	$1.9 \cdot 10^{-1}$
8104.4	5	5	$1.3 \cdot 10^{-1}$
8112.9	5	7	$3.6 \cdot 10^{-1}$
8190.1	3	5	$1.1 \cdot 10^{-1}$
8263.2	3	5	$3.5 \cdot 10^{-1}$
8281.1	3	3	$1.9 \cdot 10^{-1}$
8298.1	3	3	$3.2 \cdot 10^{-1}$
8508.9	3	3	$2.4 \cdot 10^{-1}$
8776.7	3	5	$2.7 \cdot 10^{-1}$
8928.7	5	3	$3.7 \cdot 10^{-1}$

λ Å	Weights g_i	g_k	A 10^8 s^{-1}
Kr II			
4250.6	4	4	$1.2 \cdot 10^{-1}$
4292.9	4	4	$9.6 \cdot 10^{-1}$
4355.5	6	8	1.0
4431.7	2	2	1.8
4436.8	2	4	$6.6 \cdot 10^{-1}$
4577.2	6	8	$9.6 \cdot 10^{-1}$
4583.0	6	4	$7.6 \cdot 10^{-1}$
4615.3	4	4	$5.4 \cdot 10^{-1}$
4619.2	4	6	$8.1 \cdot 10^{-1}$
4633.9	4	6	$7.1 \cdot 10^{-1}$
4658.9	6	4	$6.5 \cdot 10^{-1}$
4739.0	6	6	$7.6 \cdot 10^{-1}$
4762.4	2	4	$4.2 \cdot 10^{-1}$
4765.7	4	6	$6.7 \cdot 10^{-1}$
4811.8	2	4	$1.7 \cdot 10^{-1}$
4825.2	2	4	$1.9 \cdot 10^{-1}$
4832.1	4	2	$7.3 \cdot 10^{-1}$
5208.3	4	4	$1.4 \cdot 10^{-1}$
5308.7	4	6	$2.4 \cdot 10^{-2}$
7407.0	6	6	$7.0 \cdot 10^{-2}$
Lead			
Pb I			
2022.0	1	3	$5.2 \cdot 10^{-2}$
2053.3	1	3	$1.2 \cdot 10^{-1}$
2170.0	1	3	1.5
2401.9	3	3	$1.9 \cdot 10^{-1}$
2446.2	3	3	$2.5 \cdot 10^{-1}$
2476.4	3	5	$2.8 \cdot 10^{-1}$
2577.3	5	3	$5.0 \cdot 10^{-1}$
2613.7	3	3	$2.7 \cdot 10^{-1}$
2614.2	3	5	1.9
2628.3	5	3	$3.1 \cdot 10^{-2}$
2657.1	3	5	$9.8 \cdot 10^{-4}$
2663.2	5	5	$7.1 \cdot 10^{-1}$
2802.0	5	7	1.6
2823.2	5	5	$2.6 \cdot 10^{-1}$
2833.1	1	3	$5.8 \cdot 10^{-1}$
2873.3	5	5	$3.7 \cdot 10^{-1}$
3572.7	5	3	$9.9 \cdot 10^{-1}$
3639.6	3	3	$3.4 \cdot 10^{-1}$
3671.5	5	3	$4.4 \cdot 10^{-1}$
3683.5	3	1	1.5
3739.9	5	5	$7.3 \cdot 10^{-1}$
4019.6	5	7	$3.5 \cdot 10^{-2}$
4057.8	5	3	$8.9 \cdot 10^{-1}$
4062.1	5	3	$9.2 \cdot 10^{-1}$
4168.0	5	5	$1.2 \cdot 10^{-2}$
5005.4	1	3	$2.7 \cdot 10^{-1}$
5201.4	1	3	$1.9 \cdot 10^{-1}$
7229.0	5	3	$8.9 \cdot 10^{-3}$
Lithium			
Li I			
*2394.4	2	6	$2.664 \cdot 10^{-3}$
*2425.4	2	6	$3.823 \cdot 10^{-3}$
*2475.1	2	6	$5.735 \cdot 10^{-3}$

λ Å	Weights g_i	g_k	A 10^8 s^{-1}
*2562.3	2	6	$8.865 \cdot 10^{-3}$
*2741.2	2	6	$1.248 \cdot 10^{-2}$
*3232.7	2	6	$1.002 \cdot 10^{-2}$
*3671.7	6	10	$1.678 \cdot 10^{-2}$
*3720.9	6	10	$2.413 \cdot 10^{-2}$
*3746.6	6	2	$1.01 \cdot 10^{-2}$
*3795.1	6	10	$3.649 \cdot 10^{-2}$
*3835.6	6	2	$1.56 \cdot 10^{-2}$
*3915.3	6	10	$5.957 \cdot 10^{-2}$
*3985.5	6	2	$2.59 \cdot 10^{-2}$
*4132.6	6	10	$1.08 \cdot 10^{-1}$
*4273.1	6	2	$4.76 \cdot 10^{-2}$
*4602.9	6	10	$2.322 \cdot 10^{-1}$
*4971.7	6	2	$1.038 \cdot 10^{-1}$
*6103.6	6	10	$6.8563 \cdot 10^{-1}$
*6707.8	2	6	$3.6891 \cdot 10^{-1}$
*8126.4	6	2	$3.3466 \cdot 10^{-1}$
*10510	6	10	$1.97 \cdot 10^{-2}$
*11032	6	2	$1.46 \cdot 10^{-2}$
*12237	6	10	$3.49 \cdot 10^{-2}$
*12782	10	14	$4.578 \cdot 10^{-2}$
*12929	10	6	$2.28 \cdot 10^{-3}$
*13557	6	2	$2.84 \cdot 10^{-2}$
*17545	6	10	$6.791 \cdot 10^{-2}$
*18697	10	14	$1.383 \cdot 10^{-1}$
*19276	10	6	$5.375 \cdot 10^{-3}$
*24463	6	2	$7.453 \cdot 10^{-2}$
*26880	2	6	$3.738 \cdot 10^{-2}$
*38079	6	10	$1.37 \cdot 10^{-2}$
*41792	10	6	$2.77 \cdot 10^{-3}$
Li II			
199.279	1	3	$2.5569 \cdot 10^{2}$
*935.88	9	15	$6.1345 \cdot 10^{-1}$
*944.72	3	9	1.4329
*965.13	9	15	1.0002
*1017.9	9	15	1.8076
1093.43	1	3	1.3533
1102.46	3	5	1.4070
*1131.9	9	15	3.8492
*1198.1	3	9	2.8969
1237.28	3	5	3.1179
1420.89	1	3	2.8309
*1493.0	9	15	$1.1215 \cdot 10^{1}$
*1653.1	9	3	2.8585
1681.66	3	5	$1.0069 \cdot 10^{1}$
1755.33	3	1	2.0499
*2329.8	3	9	$1.1758 \cdot 10^{-1}$
*2674.4	3	9	$1.9081 \cdot 10^{-1}$
*3684.7	3	9	$3.0580 \cdot 10^{-1}$
*5484.5	3	9	$2.2727 \cdot 10^{-1}$
9581.43	1	3	$5.1423 \cdot 10^{-2}$
*21061	3	9	$2.5664 \cdot 10^{-2}$
Lutetium			
Lu I			
3376.5	4	4	2.23
3567.8	4	6	$5.9 \cdot 10^{-1}$
3620.3	6	4	$1.1 \cdot 10^{-2}$

λ Å	Weights g_i	g_k	A 10^8 s^{-1}
3841.2	6	6	$2.5 \cdot 10^{-1}$
4518.6	4	4	$2.1 \cdot 10^{-1}$
Magnesium			
Mg I			
1683.412	1	3	$1.88 \cdot 10^{-2}$
1707.061	1	3	$3.28 \cdot 10^{-2}$
1747.794	1	3	$6.62 \cdot 10^{-2}$
1827.934	1	3	$1.60 \cdot 10^{-1}$
2025.824	1	3	$6.12 \cdot 10^{-1}$
2731.993	1	3	$6.97 \cdot 10^{-2}$
2733.493	3	5	$9.37 \cdot 10^{-2}$
2736.542	5	7	$1.25 \cdot 10^{-1}$
2776.690	3	5	1.32
2778.270	1	3	1.82
2779.831	3	3	1.36
2779.831	5	5	4.09
2781.416	3	1	5.43
2782.972	5	3	2.14
2809.761	7	7	2.50
2811.112	5	5	1.96
2811.781	3	3	2.11
2846.716	1	3	$1.31 \cdot 10^{-1}$
2848.342	3	5	$1.77 \cdot 10^{-1}$
2851.660	5	7	$2.35 \cdot 10^{-1}$
2852.127	1	3	4.91
2915.453	5	5	4.09
2936.739	1	3	$1.37 \cdot 10^{-2}$
2938.473	3	3	$4.12 \cdot 10^{-2}$
2941.995	5	3	$6.83 \cdot 10^{-2}$
3091.065	1	3	$3.09 \cdot 10^{-1}$
3092.984	3	5	$3.74 \cdot 10^{-1}$
3096.890	5	7	$4.96 \cdot 10^{-1}$
3329.919	1	3	$3.09 \cdot 10^{-2}$
3332.146	3	3	$1.02 \cdot 10^{-1}$
3336.674	5	3	$1.70 \cdot 10^{-1}$
3829.3549	1	3	$8.99 \cdot 10^{-1}$
3832.2996	3	3	$6.74 \cdot 10^{-1}$
3832.3037	3	5	1.21
3838.2918	5	7	1.61
3838.2943	5	5	$4.03 \cdot 10^{-1}$
3890.241	1	3	1.31
3891.906	3	5	1.77
3893.304	3	3	$9.81 \cdot 10^{-1}$
3895.572	5	7	2.35
3898.120	5	5	$5.88 \cdot 10^{-1}$
3938.400	3	5	$5.47 \cdot 10^{-2}$
3986.7533	3	5	$7.30 \cdot 10^{-2}$
4057.5052	3	5	$1.02 \cdot 10^{-1}$
4167.2712	3	5	$1.38 \cdot 10^{-1}$
4351.9056	3	5	$1.84 \cdot 10^{-1}$
4571.0956	1	3	$2.54 \cdot 10^{-6}$
4702.9909	3	5	$2.19 \cdot 10^{-1}$
4730.0285	3	1	$1.34 \cdot 10^{-2}$
5167.3216	1	3	$1.13 \cdot 10^{-1}$
5172.6843	3	3	$3.37 \cdot 10^{-1}$
5183.6042	5	3	$5.61 \cdot 10^{-1}$
5528.4047	3	5	$1.39 \cdot 10^{-1}$
5711.0880	3	1	$3.86 \cdot 10^{-2}$

λ Å	Weights g_i	g_k	A 10^8 s^{-1}
6318.716	3	5	$2.63 \cdot 10^{-3}$
6319.236	3	3	$2.64 \cdot 10^{-3}$
6319.493	3	1	$2.63 \cdot 10^{-3}$
7291.060	1	3	$6.27 \cdot 10^{-4}$
7657.603	3	5	$1.23 \cdot 10^{-2}$
7659.152	3	3	$1.23 \cdot 10^{-2}$
7659.902	3	1	$1.23 \cdot 10^{-2}$
7881.667	5	7	$4.56 \cdot 10^{-3}$
7930.806	7	9	$5.38 \cdot 10^{-3}$
7930.806	3	5	$4.52 \cdot 10^{-3}$
7947.07	1	3	$2.91 \cdot 10^{-4}$
7953.39	5	3	$1.45 \cdot 10^{-3}$
8047.73	1	3	$3.53 \cdot 10^{-3}$
8049.854	3	5	$4.78 \cdot 10^{-3}$
8054.232	5	7	$6.36 \cdot 10^{-3}$
8098.724	7	9	$7.68 \cdot 10^{-3}$
8098.724	3	5	$6.46 \cdot 10^{-3}$
8209.839	5	3	$1.81 \cdot 10^{-3}$
8213.034	5	7	$4.38 \cdot 10^{-2}$
8303.313	1	3	$5.16 \cdot 10^{-3}$
8305.596	3	5	$9.28 \cdot 10^{-3}$
8310.264	5	7	$9.24 \cdot 10^{-3}$
8346.120	7	9	$1.15 \cdot 10^{-2}$
8346.120	3	5	$9.59 \cdot 10^{-3}$
8710.175	1	3	$7.97 \cdot 10^{-3}$
8712.689	3	5	$1.07 \cdot 10^{-2}$
8717.825	5	7	$1.43 \cdot 10^{-2}$
8736.021	7	9	$1.83 \cdot 10^{-2}$
8736.021	3	5	$1.54 \cdot 10^{-2}$
8806.757	3	5	$1.27 \cdot 10^{-1}$
8923.569	1	3	$5.86 \cdot 10^{-3}$
9246.499	5	3	$2.99 \cdot 10^{-3}$
9255.778	5	7	$7.95 \cdot 10^{-2}$
9414.964	5	7	$2.88 \cdot 10^{-2}$
9414.964	7	9	$3.24 \cdot 10^{-2}$
9414.964	3	5	$2.72 \cdot 10^{-2}$
9429.814	1	3	$1.34 \cdot 10^{-2}$
9432.764	3	5	$1.81 \cdot 10^{-2}$
9438.783	5	7	$2.41 \cdot 10^{-2}$
9665.54	3	5	$2.44 \cdot 10^{[illegible]}$
9983.20	1	3	$1.57 \cdot 10^{-3}$
9986.475	3	3	$4.70 \cdot 10^{-3}$
9993.209	5	3	$7.81 \cdot 10^{-3}$

Mg II

λ Å	Weights g_i	g_k	A 10^8 s^{-1}
870.2	2	4	$1.04 \cdot 10^{-2}$
870.2	2	2	$1.08 \cdot 10^{-2}$
884.7	2	4	$1.38 \cdot 10^{-2}$
884.7	2	2	$1.44 \cdot 10^{-2}$
907.4	2	4	$1.94 \cdot 10^{-2}$
907.4	2	2	$2.02 \cdot 10^{-2}$
946.703	2	4	$2.69 \cdot 10^{-2}$
946.769	2	2	$2.81 \cdot 10^{-2}$
1025.962	2	4	$3.43 \cdot 10^{-2}$
1026.108	2	2	$3.63 \cdot 10^{-2}$
1239.936	2	4	$1.35 \cdot 10^{-2}$
1240.399	2	2	$1.52 \cdot 10^{-2}$
1248.511	2	2	$3.83 \cdot 10^{-2}$
1249.932	4	2	$7.64 \cdot 10^{-2}$
1271.243	2	4	$1.23 \cdot 10^{-2}$
1271.943	2	2	$4.49 \cdot 10^{-2}$
1272.725	4	6	$1.44 \cdot 10^{-2}$
1273.427	4	2	$8.97 \cdot 10^{-2}$
1306.711	2	4	$2.19 \cdot 10^{-2}$
1307.877	2	2	$6.72 \cdot 10^{-2}$
1308.282	4	6	$2.58 \cdot 10^{-2}$
1309.439	4	2	$1.34 \cdot 10^{-1}$
1365.545	2	4	$4.53 \cdot 10^{-2}$
1367.260	4	6	$5.35 \cdot 10^{-2}$
1367.704	2	2	$1.08 \cdot 10^{-1}$
1369.425	4	2	$2.17 \cdot 10^{-1}$
1476.004	2	4	$1.10 \cdot 10^{-1}$
1478.013	4	6	$1.30 \cdot 10^{-1}$
1480.890	2	2	$1.93 \cdot 10^{-1}$
1482.902	4	2	$3.85 \cdot 10^{-1}$
1734.845	2	4	$4.29 \cdot 10^{-1}$
1737.618	4	6	$5.09 \cdot 10^{-1}$
1750.654	2	2	$4.00 \cdot 10^{-1}$
1753.456	4	2	$7.98 \cdot 10^{-1}$
2329.578	6	8	$1.36 \cdot 10^{-1}$
2329.578	4	6	$1.27 \cdot 10^{-1}$
2449.590	6	8	$2.16 \cdot 10^{-1}$
2449.590	4	6	$2.02 \cdot 10^{-1}$
2660.755	6	8	$3.81 \cdot 10^{-1}$
2660.817	4	6	$3.56 \cdot 10^{-1}$
2790.776	2	4	4.01
2795.528	2	4	2.60
2797.998	4	6	4.79
2802.704	2	2	2.57
2928.634	2	2	1.15
2936.509	4	2	2.30
2968.020	4	6	$1.98 \cdot 10^{-2}$
2969.145	2	2	$2.25 \cdot 10^{-2}$
2971.839	4	2	$4.49 \cdot 10^{-2}$
3104.722	6	8	$7.97 \cdot 10^{-1}$
3104.809	4	6	$7.44 \cdot 10^{-1}$
3165.878	2	4	$2.89 \cdot 10^{-2}$
3168.951	4	6	$3.43 \cdot 10^{-2}$
3172.706	2	2	$3.41 \cdot 10^{-2}$
3175.783	4	2	$6.81 \cdot 10^{[illegible]}$
3534.972	2	4	$5.80 \cdot 10^{-2}$
3538.813	4	6	$6.90 \cdot 10^{-2}$
3549.516	2	2	$5.64 \cdot 10^{-2}$
3553.366	4	2	$1.12 \cdot 10^{-1}$
3613.781	2	4	$1.79 \cdot 10^{-3}$
3615.583	2	2	$1.56 \cdot 10^{-3}$
3848.209	6	4	$2.96 \cdot 10^{-2}$
3848.335	4	4	$3.29 \cdot 10^{-3}$
3850.385	4	2	$3.24 \cdot 10^{-2}$
4384.637	2	4	$1.45 \cdot 10^{-1}$
4390.564	4	6	$1.73 \cdot 10^{-1}$
4427.994	2	2	$1.05 \cdot 10^{-1}$
4433.990	4	2	$2.10 \cdot 10^{-1}$
4436.486	6	8	$6.38 \cdot 10^{-2}$
4436.598	4	6	$5.95 \cdot 10^{-2}$
4481.130	6	8	2.33
4481.327	4	6	2.17
4630.878	6	4	$2.75 \cdot 10^{-3}$
4631.405	4	2	$3.01 \cdot 10^{-3}$
4739.588	6	8	$8.09 \cdot 10^{-2}$
4739.712	4	6	$7.55 \cdot 10^{-2}$
4868.845	6	4	$1.50 \cdot 10^{-3}$
4868.845	8	6	$1.41 \cdot 10^{-3}$
5068.937	6	4	$3.97 \cdot 10^{-3}$
5069.802	4	2	$4.35 \cdot 10^{-3}$
5264.215	6	8	$1.27 \cdot 10^{-1}$
5264.368	4	6	$1.19 \cdot 10^{-1}$
5434.039	6	4	$2.45 \cdot 10^{-3}$
5434.039	8	6	$2.33 \cdot 10^{-3}$
5451.259	2	4	$1.03 \cdot 10^{-2}$
5460.019	2	2	$1.11 \cdot 10^{-2}$
5464.136	4	2	$2.22 \cdot 10^{-2}$
5916.429	6	4	$6.42 \cdot 10^{-3}$
5918.158	4	2	$7.04 \cdot 10^{-3}$
5923.366	2	4	$1.40 \cdot 10^{-2}$
5928.233	4	6	$1.66 \cdot 10^{-2}$
5938.629	2	2	$1.36 \cdot 10^{-2}$
5943.499	4	2	$2.71 \cdot 10^{-2}$
6346.737	6	8	$2.20 \cdot 10^{-1}$
6346.962	4	6	$2.05 \cdot 10^{-1}$
6620.440	6	4	$4.61 \cdot 10^{-3}$
6620.569	8	6	$4.37 \cdot 10^{-3}$
6781.451	2	4	$2.49 \cdot 10^{-2}$
6787.851	4	6	$2.96 \cdot 10^{-2}$
6812.860	2	2	$2.11 \cdot 10^{-2}$
6819.270	4	2	$4.21 \cdot 10^{-2}$
7786.500	2	4	$1.97 \cdot 10^{-3}$
7790.978	2	2	$1.84 \cdot 10^{-3}$
7877.051	2	4	$6.58 \cdot 10^{-1}$
7896.368	4	6	$7.86 \cdot 10^{-1}$
8115.220	6	4	$1.22 \cdot 10^{-2}$
8120.434	4	2	$1.34 \cdot 10^{-2}$
8213.989	2	2	$2.65 \cdot 10^{-1}$
8234.639	4	2	$5.29 \cdot 10^{-1}$
8734.990	2	4	$5.34 \cdot 10^{-2}$
8745.657	4	6	$6.37 \cdot 10^{-2}$
8824.323	2	2	$3.69 \cdot 10^{-2}$
8835.082	4	2	$7.36 \cdot 10^{-2}$
9218.248	2	4	$3.64 \cdot 10^{-1}$
9244.266	2	2	$3.61 \cdot 10^{-1}$
9631.888	6	8	$4.21 \cdot 10^{-1}$
9632.435	4	6	$3.93 \cdot 10^{-1}$

Mg III

λ Å	Weights g_i	g_k	A 10^8 s^{-1}
186.5149	1	3	$1.86 \cdot 10^{2}$
187.1977	1	3	$1.26 \cdot 10^{2}$
188.5296	1	3	2.5
231.7333	1	3	$9.12 \cdot 10$
234.2631	1	3	4.98
1229.389	3	3	$5.72 \cdot 10^{-1}$
1239.827	3	1	1.82
1274.831	3	5	2.53
1350.156	5	3	$3.50 \cdot 10^{-1}$
1378.700	3	1	5.32
1393.391	7	5	6.38
1405.170	5	5	1.54
1422.118	3	5	$1.80 \cdot 10^{-1}$
1431.136	5	3	4.30
1435.550	3	1	3.04
1439.770	1	3	$3.54 \cdot 10^{-1}$

λ Å	Weights g_i	g_k	A 10^8 s^{-1}
1443.738	3	3	1.81
1446.254	3	1	$4.15 \cdot 10^{-1}$
1447.260	5	3	2.22
1458.172	3	1	3.21
1462.305	5	5	1.21
1467.188	3	3	$9.37 \cdot 10^{-1}$
1482.67	3	5	$1.55 \cdot 10^{-1}$
1493.097	5	5	1.02
1506.826	3	5	$3.91 \cdot 10^{-1}$
1550.82	3	1	$1.04 \cdot 10$
1572.712	3	5	6.92
1586.237	3	3	8.02
1592.360	3	1	8.53
1626.093	5	3	$1.77 \cdot 10^{-1}$
1635.946	7	5	$2.44 \cdot 10^{-1}$
1642.826	7	7	$9.17 \cdot 10^{-1}$
1648.822	3	3	$5.38 \cdot 10^{-1}$
1652.218	5	5	$9.10 \cdot 10^{-1}$
1659.244	5	7	$8.48 \cdot 10^{-1}$
1663.287	5	5	$1.68 \cdot 10^{-1}$
1675.710	3	5	$4.19 \cdot 10^{-1}$
1679.470	5	3	$5.88 \cdot 10^{-1}$
1687.091	3	5	1.41
1697.282	7	7	2.14
1703.108	5	3	$1.71 \cdot 10^{-1}$
1703.731	3	3	2.97
1704.368	7	5	$1.79 \cdot 10^{-1}$
1714.783	5	7	$4.76 \cdot 10^{-1}$
1722.041	5	5	3.23
1730.733	7	7	1.03
1730.778	3	3	1.44
1731.786	5	5	1.60
1738.835	7	9	$1.14 \cdot 10$
1739.475	5	7	$3.26 \cdot 10^{-4}$
1743.947	5	5	$3.76 \cdot 10^{-1}$
1745.009	5	3	$5.37 \cdot 10^{-2}$
1747.561	3	5	7.38
1748.932	5	7	8.71
1757.176	7	5	$2.04 \cdot 10^{-1}$
1757.888	1	3	2.10
1761.740	5	3	$3.16 \cdot 10^{-3}$
1763.805	3	3	3.17
1772.982	3	5	7.67
1775.942	5	5	$1.28 \cdot 10^{-1}$
1783.253	5	7	9.39
1787.927	5	5	1.46
1791.375	3	3	2.32
1793.207	5	3	$3.36 \cdot 10^{-1}$
1794.582	3	5	7.73
1800.662	5	7	8.21
1803.087	3	5	$1.03 \cdot 10^{-1}$
1820.421	1	3	2.98
1820.896	3	3	$3.05 \cdot 10^{-2}$
1826.750	3	3	$1.32 \cdot 10^{-1}$
1828.974	3	1	$4.97 \cdot 10^{-1}$
1838.336	5	7	1.09
1839.878	3	5	$4.55 \cdot 10^{-1}$
1847.561	5	7	$3.74 \cdot 10^{-2}$
1858.186	5	3	1.28
1868.225	5	5	1.94

λ Å	Weights g_i	g_k	A 10^8 s^{-1}
1879.492	5	5	1.76
1887.308	5	3	$7.56 \cdot 10^{-1}$
1896.304	5	3	$4.20 \cdot 10^{-1}$
1901.572	3	5	$3.78 \cdot 10^{-1}$
1901.572	3	3	$5.47 \cdot 10^{-1}$
1908.500	3	1	5.15
1918.777	5	5	1.98
1921.374	3	3	$2.26 \cdot 10^{-1}$
1923.896	3	5	1.37
1930.374	3	1	$8.30 \cdot 10^{-1}$
1930.672	5	5	1.70
1937.843	1	3	1.46
1938.936	5	3	$8.97 \cdot 10^{-1}$
1941.500	3	5	$1.23 \cdot 10^{-1}$
1941.500	3	3	$3.27 \cdot 10^{-1}$
1954.831	1	3	$7.97 \cdot 10^{-1}$
1962.145	3	3	$7.76 \cdot 10^{-1}$
1971.514	3	1	2.06
1977.554	3	5	$4.94 \cdot 10^{-1}$
1979.327	1	3	1.22
1979.43	1	3	$4.10 \cdot 10^{-1}$
2004.860	5	3	$3.33 \cdot 10^{-1}$
2039.553	5	5	1.54
2055.491	3	3	2.35
2064.902	5	7	4.21
2085.891	3	3	1.61
2091.963	3	5	2.57
2094.207	3	1	$1.21 \cdot 10^{-1}$
2097.936	1	3	1.50
2112.773	3	5	1.69
2134.054	3	3	2.36
2177.694	3	5	2.12
2273.414	3	3	$1.27 \cdot 10^{-2}$
2318.125	3	5	$4.44 \cdot 10^{-2}$
2395.149	5	3	1.67
2467.751	3	3	$6.91 \cdot 10^{-1}$
2490.534	1	3	1.61
2529.190	1	3	$1.89 \cdot 10^{-1}$
2618.011	1	3	$8.72 \cdot 10^{-1}$
2905.419	1	3	$6.84 \cdot 10^{-3}$
Mg IV			
129.857	4	6	$5.61 \cdot 10^{2}$
320.9943	4	2	$1.13 \cdot 10^{2}$
323.3076	2	2	$5.52 \cdot 10^{1}$
840.366	6	4	$1.24 \cdot 10^{1}$
842.087	8	6	$1.19 \cdot 10^{1}$
1346.543	6	8	$1.07 \cdot 10^{1}$
1352.020	4	6	6.59
1384.425	8	10	$1.45 \cdot 10^{1}$
1437.61	6	8	$1.21 \cdot 10^{1}$
Manganese			
Mn I			
2794.82	6	8	3.7
2798.27	6	6	3.6
2801.08	6	4	3.7
3016.45	10	12	$2.9 \cdot 10^{-1}$
3043.36	8	8	$5.9 \cdot 10^{-1}$

λ Å	Weights g_i	g_k	A 10^8 s^{-1}
3044.57	10	8	$5.7 \cdot 10^{-1}$
3045.59	10	10	$6.7 \cdot 10^{-1}$
3047.03	12	12	$6.1 \cdot 10^{-1}$
3082.71	14	14	$2.9 \cdot 10^{-1}$
3228.09	10	12	$6.4 \cdot 10^{-1}$
3230.72	8	8	$3.5 \cdot 10^{-1}$
3243.78	6	6	$5.3 \cdot 10^{-1}$
3256.14	4	6	$5.0 \cdot 10^{-1}$
3267.79	14	14	$3.5 \cdot 10^{-1}$
3268.72	6	8	$3.3 \cdot 10^{-1}$
3270.35	12	12	$2.6 \cdot 10^{-1}$
3273.02	10	10	$2.7 \cdot 10^{-1}$
3463.66	8	8	$3.2 \cdot 10^{-1}$
3511.83	12	12	$2.7 \cdot 10^{-1}$
3577.87	10	8	$9.4 \cdot 10^{-1}$
3601.27	12	10	$2.3 \cdot 10^{-1}$
3608.49	6	6	$3.6 \cdot 10^{-1}$
3660.40	12	14	$9.1 \cdot 10^{-1}$
3675.67	6	8	$2.2 \cdot 10^{-1}$
3676.96	10	12	$7.3 \cdot 10^{-1}$
3680.15	12	10	$1.9 \cdot 10^{-1}$
3682.09	8	10	$7.6 \cdot 10^{-1}$
3684.87	6	8	$2.6 \cdot 10^{-1}$
3706.08	12	14	1.4
3718.92	10	12	$9.6 \cdot 10^{-1}$
3731.94	8	10	1.0
3771.44	14	14	$1.9 \cdot 10^{-1}$
3773.86	12	12	$2.5 \cdot 10^{-1}$
3800.55	6	8	$2.7 \cdot 10^{-1}$
3806.72	10	12	$5.9 \cdot 10^{-1}$
3823.51	8	10	$5.21 \cdot 10^{-1}$
3834.37	6	8	$4.29 \cdot 10^{-1}$
3841.07	4	6	$3.3 \cdot 10^{-1}$
3889.46	12	14	$3.1 \cdot 10^{-1}$
3924.08	2	4	$9.4 \cdot 10^{-1}$
3926.48	6	8	$5.4 \cdot 10^{-1}$
3952.84	6	6	$4.1 \cdot 10^{-1}$
3982.90	6	4	$5.5 \cdot 10^{-1}$
4011.91	8	8	$2.3 \cdot 10^{-1}$
4018.11	10	8	$2.54 \cdot 10^{-1}$
4030.76	6	8	$1.7 \cdot 10^{-1}$
4041.36	10	10	$7.87 \cdot 10^{-1}$
4048.75	6	4	$7.5 \cdot 10^{-1}$
4052.48	6	8	$3.8 \cdot 10^{-1}$
4055.55	8	8	$4.31 \cdot 10^{-1}$
4058.94	4	2	$7.25 \cdot 10^{-1}$
4065.08	12	14	$2.5 \cdot 10^{-1}$
4066.24	10	8	$2.2 \cdot 10^{-1}$
4079.42	2	4	$3.8 \cdot 10^{-1}$
4082.95	4	6	$2.95 \cdot 10^{-1}$
4083.63	6	8	$2.8 \cdot 10^{-1}$
4089.94	8	10	$1.7 \cdot 10^{-1}$
4105.37	10	8	$1.7 \cdot 10^{-1}$
4135.03	12	12	$3.0 \cdot 10^{-1}$
4141.06	10	10	$2.6 \cdot 10^{-1}$
4148.80	8	8	$2.3 \cdot 10^{-1}$
4176.61	14	12	$2.4 \cdot 10^{-1}$
4189.99	12	10	$2.0 \cdot 10^{-1}$
4201.78	10	8	$2.3 \cdot 10^{-1}$
4235.30	8	6	$9.17 \cdot 10^{-1}$

λ Å	Weights g_i	g_k	A 10^8 s^{-1}
4265.93	4	4	$4.92·10^{-1}$
4281.10	6	6	$2.3·10^{-1}$
4411.87	12	10	$2.6·10^{-1}$
4414.89	8	6	$2.93·10^{-1}$
4419.77	10	8	$2.1·10^{-1}$
4436.36	6	4	$4.37·10^{-1}$
4451.58	8	8	$7.98·10^{-1}$
4453.01	4	2	$5.44·10^{-1}$
4455.82	4	6	$1.7·10^{-1}$
4457.55	6	6	$4.27·10^{-1}$
4458.26	6	8	$4.62·10^{-1}$
4461.09	8	8	$1.7·10^{-1}$
4462.03	8	10	$7.00·10^{-1}$
4464.68	6	6	$4.39·10^{-1}$
4470.14	4	4	$3.00·10^{-1}$
4479.40	8	10	$3.4·10^{-1}$
4490.08	2	4	$2.49·10^{-1}$
4498.90	4	6	$2.49·10^{-1}$
4502.22	6	8	$1.86·10^{-1}$
4605.37	10	12	$3.6·10^{-1}$
4626.54	12	14	$3.6·10^{-1}$
4709.71	8	8	$1.72·10^{-1}$
4727.46	6	6	$1.7·10^{-1}$
4739.11	4	4	$2.40·10^{-1}$
4754.05	6	8	$3.03·10^{-1}$
4761.53	2	4	$5.35·10^{-1}$
4762.38	8	10	$7.83·10^{-1}$
4765.86	4	6	$4.1·10^{-1}$
4766.43	6	8	$4.6·10^{-1}$
4783.43	8	8	$4.01·10^{-1}$
4823.53	10	8	$4.99·10^{-1}$
6013.48	4	6	$1.72·10^{-1}$
6021.79	8	6	$3.32·10^{-1}$

Mn II

λ Å	Weights g_i	g_k	A 10^8 s^{-1}
2593.72	7	7	2.6
2605.68	7	5	2.7
2933.05	5	3	2.0
2939.31	5	5	1.9
2949.20	5	7	1.9
3441.99	9	7	$4.3·10^{-1}$
3460.32	7	5	$3.2·10^{-1}$
3474.13	5	3	$1.5·10^{-1}$
3482.90	5	5	$2.0·10^{-1}$
3488.68	3	3	$2.5·10^{-1}$

Mn VI

λ Å	Weights g_i	g_k	A 10^8 s^{-1}
307.999	9	9	$3.7·10^{1}$
309.440	9	7	$5.7·10^{1}$
309.579	7	5	$4.4·10^{1}$
310.058	7	7	$3.4·10^{1}$
310.182	5	5	$2.8·10^{1}$
311.748	5	3	$5.7·10^{1}$
320.598	3	5	$1.5·10^{1}$
320.681	1	3	$2.2·10^{1}$
320.874	3	1	$7.8·10^{1}$
320.979	3	3	$2.2·10^{1}$
321.176	5	5	$6.0·10^{1}$
321.541	5	3	$2.7·10^{1}$
325.146	9	7	$1.3·10^{2}$
328.431	5	5	$4.4·10^{1}$
328.558	3	5	$1.2·10^{1}$
329.043	1	3	$1.1·10^{1}$
1236.23	5	3	$1.3·10^{1}$
1255.77	3	1	$1.2·10^{1}$
1285.10	5	7	$1.1·10^{1}$
1333.87	7	9	$1.0·10^{1}$

Mercury

Hg I

λ Å	Weights g_i	g_k	A 10^8 s^{-1}
2536.52	1	3	$8.00·10^{-2}$
2652.04	3	5	$3.88·10^{-1}$
2655.13	3	5	$1.1·10^{-1}$
2752.78	1	3	$6.10·10^{-2}$
2856.94	3	1	$1.1·10^{-2}$
2893.60	3	3	$1.6·10^{-1}$
2925.4	5	3	$7.7·10^{-2}$
2967.3	1	3	$4.5·10^{-1}$
3021.50	5	7	$5.09·10^{-1}$
3023.48	5	5	$9.4·10^{-2}$
3027.49	5	5	$2.0·10^{-2}$
3125.66	3	5	$6.56·10^{-1}$
3341.48	5	3	$1.68·10^{-1}$
3650.15	5	7	1.3
3654.83	5	5	$1.8·10^{-1}$
4046.56	1	3	$2.1·10^{-1}$
4077.81	3	1	$4.0·10^{-2}$
4108.1	3	1	$3.0·10^{-2}$
4339.22	3	5	$2.88·10^{-2}$
4347.50	3	5	$8.4·10^{-2}$
4358.34	3	3	$5.57·10^{-1}$
4916.07	3	1	$5.8·10^{-2}$
5025.64	3	3	$2.7·10^{-4}$
5460.75	5	3	$4.87·10^{-1}$
5769.59	3	5	$2.36·10^{-1}$
6234.4	1	3	$5.3·10^{-3}$
6716.4	1	3	$4.3·10^{-3}$
6907.5	3	5	$2.8·10^{-2}$
7728.8	1	3	$9.7·10^{-3}$
10139.79	3	1	$2.71·10^{-1}$

Molybdenum

Mo I

λ Å	Weights g_i	g_k	A 10^8 s^{-1}
2616.79	3	5	$7.34·10^{-1}$
2629.85	5	7	$7.75·10^{-1}$
2638.30	5	5	$7.57·10^{-1}$
2640.98	7	5	1.20
2649.46	7	9	$9.84·10^{-1}$
2655.02	9	7	$4.08·10^{-1}$
2658.11	7	7	$6.43·10^{-1}$
2679.85	9	11	1.31
2684.16	9	9	$4.18·10^{-1}$
2733.39	5	7	$2.95·10^{-1}$
2751.47	7	9	$2.54·10^{-1}$
2761.53	9	11	$2.06·10^{-1}$
2787.83	9	7	$2.85·10^{-1}$
2826.75	7	7	$4.23·10^{-1}$
2876.54	9	9	$2.84·10^{-1}$
2886.60	11	11	$4.74·10^{-1}$
2906.06	3	3	$8.04·10^{-1}$
2915.38	5	3	$7.31·10^{-1}$
2936.50	11	11	$2.33·10^{-1}$
2945.43	7	7	$3.66·10^{-1}$
2959.48	9	11	$1.75·10^{-1}$
2977.27	9	7	$3.28·10^{-1}$
2987.92	3	5	$8.43·10^{-1}$
2988.23	5	7	$4.28·10^{-1}$
2989.80	9	7	$9.27·10^{-1}$
3000.85	5	7	$2.58·10^{-1}$
3013.39	7	5	$6.06·10^{-1}$
3016.78	9	9	$2.75·10^{-1}$
3025.00	5	5	$8.49·10^{-1}$
3036.31	3	5	$5.81·10^{-1}$
3041.70	13	11	$5.94·10^{-1}$
3046.80	13	11	$1.63·10^{-1}$
3047.31	11	9	$5.01·10^{-1}$
3055.32	9	7	$4.29·10^{-1}$
3061.59	7	5	$4.41·10^{-1}$
3064.27	13	13	$8.46·10^{-1}$
3065.04	13	13	$3.08·10^{-1}$
3069.96	11	11	$2.72·10^{-1}$
3070.89	9	11	$1.87·10^{-1}$
3074.37	11	11	1.42
3079.88	9	11	$9.55·10^{-1}$
3080.10	7	9	$8.61·10^{-1}$
3085.62	9	9	1.63
3089.71	5	7	$2.34·10^{-1}$
3094.66	7	7	1.63
3100.88	7	9	1.20
3101.34	5	5	1.92
3117.54	13	13	$1.89·10^{-1}$
3132.59	7	9	1.79
3135.90	9	11	$3.68·10^{-1}$
3136.75	9	11	$1.57·10^{-1}$
3142.75	3	5	$4.10·10^{-1}$
3147.35	13	11	$2.41·10^{-1}$
3155.19	7	7	$2.75·10^{-1}$
3158.17	7	7	$4.63·10^{-1}$
3170.34	7	7	1.37
3171.38	5	7	$2.03·10^{-1}$
3175.59	13	11	$8.40·10^{-1}$
3179.78	11	13	$2.33·10^{-1}$
3183.03	11	9	$3.98·10^{-1}$
3184.58	7	5	$2.77·10^{-1}$
3185.10	7	7	$2.54·10^{-1}$
3185.71	5	3	$6.10·10^{-1}$
3188.10	7	9	$3.45·10^{-1}$
3188.41	5	7	$4.40·10^{-1}$
3192.79	9	11	$1.88·10^{-1}$
3193.98	7	5	1.53
3194.88	9	11	$1.75·10^{-1}$
3195.96	9	7	$4.10·10^{-1}$
3198.85	15	13	$7.22·10^{-1}$
3205.22	1	3	$4.27·10^{-1}$
3205.43	9	11	$2.55·10^{-1}$
3205.89	9	9	$5.35·10^{-1}$
3208.84	7	5	$2.77·10^{-1}$
3210.97	7	5	$6.94·10^{-1}$
3214.44	9	7	$2.01·10^{-1}$

λ	Weights		A	λ	Weights		A	λ	Weights		A
Å	g_i	g_k	10^8 s^{-1}	Å	g_i	g_k	10^8 s^{-1}	Å	g_i	g_k	10^8 s^{-1}
3215.07	3	5	4.20·10^{-1}	3475.03	3	3	4.68·10^{-1}	3720.25	7	9	2.86·10^{-1}
3216.78	15	13	2.10·10^{-1}	3479.42	7	5	2.26·10^{-1}	3725.55	7	7	1.60·10^{-1}
3221.73	3	1	1.41	3489.43	7	7	3.27·10^{-1}	3727.68	9	11	1.51·10^{-1}
3228.21	5	7	3.85·10^{-1}	3504.41	7	9	8.06·10^{-1}	3728.50	7	9	2.20·10^{-1}
3229.79	9	11	1.44·10^{-1}	3505.31	7	9	2.25·10^{-1}	3733.02	7	7	1.45·10^{-1}
3233.14	13	13	6.33·10^{-1}	3508.11	9	9	1.59·10^{-1}	3733.41	13	13	2.80·10^{-1}
3237.06	7	9	2.95·10^{-1}	3510.77	13	13	4.75·10^{-1}	3735.62	11	11	1.66·10^{-1}
3251.65	3	5	3.05·10^{-1}	3517.55	11	11	5.41·10^{-1}	3742.28	7	7	1.56·10^{-1}
3256.21	5	3	6.89·10^{-1}	3518.21	3	3	3.64·10^{-1}	3747.19	5	7	3.07·10^{-1}
3259.16	11	13	1.62·10^{-1}	3521.38	9	9	1.39·10^{-1}	3748.48	9	11	3.95·10^{-1}
3262.63	7	9	3.62·10^{-1}	3521.41	9	11	6.06·10^{-1}	3755.16	9	9	2.48·10^{-1}
3264.40	11	9	5.42·10^{-1}	3524.98	7	9	2.25·10^{-1}	3758.52	9	9	1.22·10^{-1}
3265.14	5	7	2.60·10^{-1}	3538.92	11	11	2.24·10^{-1}	3759.60	9	7	1.82·10^{-1}
3266.16	9	11	1.95·10^{-1}	3540.57	5	3	4.46·10^{-1}	3760.88	9	9	2.16·10^{-1}
3270.90	7	7	3.59·10^{-1}	3542.17	7	5	4.93·10^{-1}	3768.73	9	9	2.88·10^{-1}
3285.35	9	7	4.49·10^{-1}	3552.71	9	7	3.64·10^{-1}	3769.99	7	9	2.46·10^{-1}
3289.01	9	9	5.08·10^{-1}	3555.64	3	3	3.46·10^{-1}	3777.72	13	11	1.66·10^{-1}
3290.82	7	5	5.44·10^{-1}	3558.09	5	7	5.43·10^{-1}	3788.25	7	9	2.87·10^{-1}
3305.91	7	9	3.06·10^{-1}	3566.05	9	9	2.67·10^{-1}	3794.43	9	9	1.22·10^{-1}
3323.95	9	7	2.82·10^{-1}	3566.74	7	7	1.43·10^{-1}	3798.25	7	9	6.90·10^{-1}
3336.56	9	9	1.64·10^{-1}	3570.64	15	15	7.18·10^{-1}	3801.84	9	7	3.16·10^{-1}
3344.73	3	5	6.04·10^{-1}	3573.88	3	5	3.58·10^{-1}	3805.99	5	5	2.44·10^{-1}
3346.83	11	11	1.13·10^{-1}	3580.54	13	11	5.49·10^{-1}	3819.78	9	11	1.47·10^{-1}
3358.12	5	7	7.59·10^{-1}	3581.88	11	13	3.81·10^{-1}	3824.78	5	7	1.40·10^{-1}
3361.37	9	9	1.38·10^{-1}	3585.57	7	5	3.95·10^{-1}	3827.15	7	7	1.94·10^{-1}
3363.78	5	7	2.74·10^{-1}	3590.74	7	9	2.23·10^{-1}	3828.88	7	7	1.35·10^{-1}
3375.65	7	9	1.56·10^{-1}	3595.55	5	5	2.32·10^{-1}	3830.81	5	5	1.83·10^{-1}
3378.46	13	13	3.75·10^{-1}	3598.88	13	11	5.67·10^{-1}	3831.07	7	9	1.20·10^{-1}
3379.96	5	5	4.11·10^{-1}	3600.73	9	9	2.07·10^{-1}	3832.11	9	9	3.05·10^{-1}
3384.61	7	9	7.32·10^{-1}	3601.88	7	9	1.15·10^{-1}	3833.75	9	9	1.70·10^{-1}
3385.87	9	11	3.30·10^{-1}	3602.94	5	7	2.96·10^{-1}	3846.18	7	7	1.26·10^{-1}
3389.79	5	7	1.85·10^{-1}	3604.07	9	7	3.25·10^{-1}	3848.30	9	9	1.26·10^{-1}
3392.17	9	9	1.97·10^{-1}	3615.16	7	9	1.96·10^{-1}	3851.99	11	9	1.78·10^{-1}
3393.65	11	11	2.08·10^{-1}	3623.22	11	9	5.58·10^{-1}	3864.10	7	7	6.24·10^{-1}
3404.33	7	7	2.10·10^{-1}	3624.46	9	11	5.27·10^{-1}	3866.69	3	5	1.74·10^{-1}
3413.37	11	11	1.25·10^{-1}	3624.62	5	7	1.37·10^{-1}	3874.15	7	5	1.67·10^{-1}
3415.27	9	9	1.83·10^{-1}	3638.20	5	3	3.51·10^{-1}	3902.95	7	5	6.17·10^{-1}
3415.61	7	9	1.29·10^{-1}	3638.21	5	3	3.33·10^{-1}	3909.54	9	7	1.13·10^{-1}
3416.14	9	11	2.45·10^{-1}	3640.62	7	5	1.94·10^{-1}	3919.55	11	13	2.24·10^{-1}
3420.04	5	5	3.28·10^{-1}	3647.84	7	7	2.11·10^{-1}	3955.48	13	11	1.71·10^{-1}
3422.31	9	9	2.52·10^{-1}	3657.36	5	7	2.03·10^{-1}	3973.76	11	13	4.39·10^{-1}
3425.13	11	11	2.29·10^{-1}	3658.13	9	9	1.86·10^{-1}	3977.90	9	7	1.35·10^{-1}
3427.90	11	13	4.09·10^{-1}	3659.36	7	9	6.70·10^{-1}	3980.20	5	3	2.70·10^{-1}
3434.79	7	7	1.75·10^{-1}	3662.15	7	9	1.45·10^{-1}	3991.85	11	9	1.29·10^{-1}
3435.45	15	15	1.50	3662.99	11	11	3.48·10^{-1}	4010.13	5	3	4.38·10^{-1}
3437.21	11	9	8.06·10^{-1}	3663.27	7	5	2.30·10^{-1}	4021.01	9	11	2.65·10^{-1}
3445.03	7	9	1.53·10^{-1}	3664.81	11	13	9.54·10^{-1}	4051.18	13	11	1.36·10^{-1}
3445.26	7	5	2.96·10^{-1}	3669.34	9	7	2.16·10^{-1}	4062.08	11	9	1.96·10^{-1}
3447.12	9	11	8.75·10^{-1}	3672.81	9	11	1.95·10^{-1}	4069.88	13	11	3.25·10^{-1}
3449.07	7	9	1.52·10^{-1}	3672.82	9	9	1.13·10^{-1}	4076.19	9	9	1.16·10^{-1}
3449.85	5	7	1.65·10^{-1}	3680.68	11	11	2.96·10^{-1}	4084.37	9	7	1.94·10^{-1}
3452.60	7	7	2.48·10^{-1}	3681.72	9	7	1.68·10^{-1}	4107.46	7	5	2.02·10^{-1}
3456.15	5	5	3.60·10^{-1}	3687.96	5	7	2.12·10^{-1}	4120.09	13	15	6.05·10^{-1}
3460.78	9	7	6.03·10^{-1}	3688.97	11	9	3.26·10^{-1}	4131.92	9	11	1.56·10^{-1}
3466.19	9	7	2.11·10^{-1}	3690.59	11	9	2.07·10^{-1}	4148.98	9	11	1.56·10^{-1}
3466.96	7	7	1.52·10^{-1}	3694.94	5	7	6.36·10^{-1}	4157.40	13	11	2.17·10^{-1}
3467.85	5	7	2.63·10^{-1}	3696.04	11	11	3.59·10^{-1}	4157.90	9	11	1.60·10^{-1}
3469.22	5	3	6.96·10^{-1}	3708.55	7	9	1.28·10^{-1}	4185.82	11	13	3.82·10^{-1}
3469.63	13	15	1.51·10^{-1}	3715.75	9	7	2.38·10^{-1}	4188.32	11	13	3.32·10^{-1}
3470.92	3	5	2.91·10^{-1}	3718.48	5	7	1.34·10^{-1}	4194.56	11	11	2.70·10^{-1}

λ Å	Weights g_i	g_k	A 10^8 s^{-1}
4232.59	9	11	$3.17 \cdot 10^{-1}$
4240.83	5	5	$1.68 \cdot 10^{-1}$
4246.02	11	13	$2.00 \cdot 10^{-1}$
4251.88	13	11	$1.76 \cdot 10^{-1}$
4254.95	7	9	$2.01 \cdot 10^{-1}$
4269.28	11	11	$1.36 \cdot 10^{-1}$
4276.91	7	9	$2.85 \cdot 10^{-1}$
4277.24	9	11	$1.35 \cdot 10^{-1}$
4317.92	15	15	$1.28 \cdot 10^{-1}$
4325.80	3	3	$1.84 \cdot 10^{-1}$
4326.14	5	7	$2.56 \cdot 10^{-1}$
4340.74	5	7	$1.23 \cdot 10^{-1}$
4381.63	13	13	$2.93 \cdot 10^{-1}$
4382.41	11	13	$3.83 \cdot 10^{-1}$
4409.94	13	13	$1.38 \cdot 10^{-1}$
4411.69	11	11	$2.63 \cdot 10^{-1}$
4434.95	9	9	$2.51 \cdot 10^{-1}$
4446.42	11	11	$1.90 \cdot 10^{-1}$
4457.35	7	7	$1.28 \cdot 10^{-1}$
4474.57	5	5	$2.10 \cdot 10^{-1}$
4491.65	11	11	$2.09 \cdot 10^{-1}$
4536.80	13	15	$5.03 \cdot 10^{-1}$
4598.23	1	3	$1.47 \cdot 10^{-1}$
4624.23	9	9	$1.32 \cdot 10^{-1}$
4633.08	3	5	$2.35 \cdot 10^{-1}$
4652.24	5	7	$1.55 \cdot 10^{-1}$
4686.08	3	3	$1.72 \cdot 10^{-1}$
4688.21	13	15	$1.54 \cdot 10^{-1}$
4707.25	7	9	$3.63 \cdot 10^{-1}$
4718.86	5	5	$2.17 \cdot 10^{-1}$
4723.05	9	9	$1.23 \cdot 10^{-1}$
4731.44	9	11	$4.49 \cdot 10^{-1}$
4758.50	11	9	$3.01 \cdot 10^{-1}$
4760.18	11	13	$4.67 \cdot 10^{-1}$
4764.11	9	7	$2.16 \cdot 10^{-1}$
4811.05	13	11	$4.36 \cdot 10^{-1}$
4819.25	11	9	$2.71 \cdot 10^{-1}$
4830.51	9	7	$4.07 \cdot 10^{-1}$
4858.39	13	11	$1.24 \cdot 10^{-1}$
4868.02	7	5	$3.11 \cdot 10^{-1}$
5037.18	9	7	$1.14 \cdot 10^{-1}$
5044.36	7	5	$1.31 \cdot 10^{-1}$
5163.18	9	11	$2.03 \cdot 10^{-1}$
5171.06	5	7	$1.84 \cdot 10^{-1}$
5172.94	5	5	$4.11 \cdot 10^{-1}$
5174.18	5	3	$5.83 \cdot 10^{-1}$
5191.45	7	9	$1.62 \cdot 10^{-1}$
5238.21	7	9	$3.74 \cdot 10^{-1}$
5240.87	7	7	$3.89 \cdot 10^{-1}$
5242.80	7	5	$2.01 \cdot 10^{-1}$
5261.53	5	7	$1.13 \cdot 10^{-1}$
5280.85	5	5	$1.28 \cdot 10^{-1}$
5355.52	9	9	$1.21 \cdot 10^{-1}$
5356.46	11	11	$2.11 \cdot 10^{-1}$
5360.51	9	11	$6.19 \cdot 10^{-1}$
5364.28	9	9	$2.26 \cdot 10^{-1}$
5460.50	5	3	$3.46 \cdot 10^{-1}$
5493.76	7	5	$2.13 \cdot 10^{-1}$
5506.49	5	7	$3.61 \cdot 10^{-1}$
5533.03	5	5	$3.72 \cdot 10^{-1}$

λ Å	Weights g_i	g_k	A 10^8 s^{-1}
5570.44	5	3	$3.30 \cdot 10^{-1}$
5849.71	3	3	$3.02 \cdot 10^{-1}$
5851.50	3	5	$1.55 \cdot 10^{-1}$
5893.36	5	5	$2.60 \cdot 10^{-1}$
5895.93	5	7	$3.12 \cdot 10^{-1}$
5926.37	7	7	$1.63 \cdot 10^{-1}$
5928.88	7	9	$5.32 \cdot 10^{-1}$
7154.11	9	9	$3.45 \cdot 10^{-1}$

Neodymium

Nd II

λ Å	Weights g_i	g_k	A 10^8 s^{-1}
3780.4	16	18	$1.4 \cdot 10^{-1}$
3805.4	14	16	$6.9 \cdot 10^{-1}$
3807.2	10	12	$4.9 \cdot 10^{-2}$
3863.3	8	10	$1.5 \cdot 10^{-1}$
3941.5	10	10	$6.1 \cdot 10^{-1}$
3951.2	12	12	$6.0 \cdot 10^{-1}$
3973.3	18	18	$6.3 \cdot 10^{-1}$
3979.5	10	12	$2.7 \cdot 10^{-1}$
3990.1	16	16	$5.2 \cdot 10^{-1}$
4012.3	18	20	$5.5 \cdot 10^{-1}$
4061.1	16	18	$4.4 \cdot 10^{-1}$
4106.6	14	16	$6.8 \cdot 10^{-2}$
4109.5	14	16	$3.7 \cdot 10^{-1}$
4133.4	14	12	$1.5 \cdot 10^{-1}$
4156.1	12	14	$3.4 \cdot 10^{-1}$
4205.6	18	16	$1.8 \cdot 10^{-1}$
4284.5	18	18	$8.5 \cdot 10^{-2}$
4303.6	8	10	$4.7 \cdot 10^{-1}$
4325.8	16	16	$1.6 \cdot 10^{-1}$
4358.2	14	14	$1.5 \cdot 10^{-1}$
4382.7	12	10	$4.0 \cdot 10^{-2}$
4400.8	10	10	$6.8 \cdot 10^{-2}$
4451.6	12	14	$2.5 \cdot 10^{-1}$
4456.4	16	18	$6.4 \cdot 10^{-2}$
4463.0	14	16	$1.8 \cdot 10^{-1}$
4958.1	12	10	$1.2 \cdot 10^{-2}$
5130.6	22	20	$1.6 \cdot 10^{-1}$
5192.6	20	18	$1.7 \cdot 10^{-1}$
5249.6	18	16	$1.8 \cdot 10^{-1}$
5276.9	12	10	$1.2 \cdot 10^{-1}$
5293.2	16	14	$1.2 \cdot 10^{-1}$
5302.3	20	18	$1.1 \cdot 10^{-1}$
5311.5	14	12	$1.1 \cdot 10^{-1}$
5319.8	12	10	$1.6 \cdot 10^{-1}$
5357.0	18	16	$1.8 \cdot 10^{-1}$
5371.9	20	20	$5.1 \cdot 10^{-2}$
5485.7	18	18	$5.7 \cdot 10^{-2}$
5594.4	16	16	$7.0 \cdot 10^{-2}$
5620.6	18	18	$1.3 \cdot 10^{-1}$
5688.5	14	14	$5.9 \cdot 10^{-2}$
5718.1	16	16	$8.7 \cdot 10^{-2}$
5726.8	10	10	$5.6 \cdot 10^{-2}$
5740.9	12	12	$7.2 \cdot 10^{-2}$
5804.0	10	10	$4.6 \cdot 10^{-2}$
5865.1	16	18	$1.3 \cdot 10^{-2}$
6051.9	12	10	$1.1 \cdot 10^{-2}$

Neon

Ne I

λ Å	Weights g_i	g_k	A 10^8 s^{-1}
615.63	1	3	$3.8 \cdot 10^{-1}$
618.67	1	3	$9.3 \cdot 10^{-1}$
619.10	1	3	$3.3 \cdot 10^{-1}$
626.82	1	3	$7.4 \cdot 10^{-1}$
629.74	1	3	$4.8 \cdot 10^{-1}$
735.90	1	3	6.11
743.72	1	3	$4.86 \cdot 10^{-1}$
3369.8	5	5	$1.0 \cdot 10^{-3}$
3369.9	5	3	$7.6 \cdot 10^{-3}$
3375.6	5	3	$2.2 \cdot 10^{-3}$
3417.9	3	5	$9.2 \cdot 10^{-3}$
3418.0	3	3	$2.2 \cdot 10^{-3}$
3423.9	3	3	$1.0 \cdot 10^{-3}$
3447.7	5	5	$2.1 \cdot 10^{-2}$
3450.8	5	3	$4.9 \cdot 10^{-3}$
3454.2	3	1	$3.7 \cdot 10^{-2}$
3460.5	1	3	$7.0 \cdot 10^{-3}$
3464.3	5	5	$6.7 \cdot 10^{-3}$
3466.6	1	3	$1.3 \cdot 10^{-2}$
3472.6	5	7	$1.7 \cdot 10^{-2}$
3498.1	3	5	$5.1 \cdot 10^{-3}$
3501.2	3	3	$1.2 \cdot 10^{-2}$
3510.7	5	3	$2.2 \cdot 10^{-3}$
3515.2	3	5	$6.9 \cdot 10^{-3}$
3520.5	3	1	$9.3 \cdot 10^{-2}$
3593.5	3	5	$9.9 \cdot 10^{-3}$
3593.6	3	3	$6.6 \cdot 10^{-3}$
3600.2	3	3	$4.3 \cdot 10^{-3}$
3633.7	3	1	$1.1 \cdot 10^{-2}$
3682.2	3	5	$1.6 \cdot 10^{-3}$
3685.7	3	3	$3.9 \cdot 10^{-3}$
3701.2	3	5	$2.2 \cdot 10^{-3}$
4536.3	3	3	$5.0 \cdot 10^{-3}$
4702.5	3	3	$2.1 \cdot 10^{-3}$
4708.9	3	3	$4.2 \cdot 10^{-2}$
4955.4	3	3	$3.3 \cdot 10^{-3}$
5113.7	3	3	$1.0 \cdot 10^{-2}$
5120.5	3	3	$5.6 \cdot 10^{-3}$
5154.4	3	3	$1.9 \cdot 10^{-2}$
5191.3	3	3	$1.3 \cdot 10^{-2}$
5326.4	3	3	$6.8 \cdot 10^{-3}$
5333.3	3	3	$5.3 \cdot 10^{-3}$
5341.1	3	3	$1.1 \cdot 10^{-1}$
5400.6	3	1	$9.0 \cdot 10^{-3}$
5418.6	3	3	$5.2 \cdot 10^{-3}$
5433.7	3	3	$2.83 \cdot 10^{-3}$
5652.6	3	3	$8.9 \cdot 10^{-3}$
5662.5	3	3	$6.9 \cdot 10^{-3}$
5852.5	3	1	$6.82 \cdot 10^{-1}$
5868.4	3	3	$1.4 \cdot 10^{-2}$
5881.9	5	3	$1.15 \cdot 10^{-1}$
5913.6	3	3	$4.8 \cdot 10^{-2}$
5939.3	5	3	$2.00 \cdot 10^{-3}$
5944.8	5	5	$1.13 \cdot 10^{-1}$
5961.6	3	3	$3.3 \cdot 10^{-2}$
5975.5	5	3	$3.51 \cdot 10^{-2}$

λ Å	Weights g_i	g_k	A $10^8\ s^{-1}$
6030.0	3	3	$5.61 \cdot 10^{-2}$
6046.1	3	3	$2.26 \cdot 10^{-3}$
6074.3	3	1	$6.03 \cdot 10^{-1}$
6096.2	3	5	$1.81 \cdot 10^{-1}$
6118.0	5	3	$6.09 \cdot 10^{-3}$
6128.5	3	3	$6.7 \cdot 10^{-3}$
6143.1	5	5	$2.82 \cdot 10^{-1}$
6150.3	3	3	$1.5 \cdot 10^{-2}$
6163.6	1	3	$1.46 \cdot 10^{-1}$
6217.3	5	3	$6.37 \cdot 10^{-2}$
6266.5	1	3	$2.49 \cdot 10^{-1}$
6273.0	3	3	$9.7 \cdot 10^{-3}$
6293.7	3	3	$6.39 \cdot 10^{-3}$
6304.8	3	5	$4.16 \cdot 10^{-2}$
6328.2	5	3	$3.39 \cdot 10^{-2}$
6330.9	3	3	$2.3 \cdot 10^{-2}$
6334.4	5	5	$1.61 \cdot 10^{-1}$
6351.9	1	3	$3.45 \cdot 10^{-3}$
6383.0	3	3	$3.21 \cdot 10^{-1}$
6401.1	3	3	$1.39 \cdot 10^{-2}$
6402.2	5	7	$5.14 \cdot 10^{-1}$
6506.5	3	5	$3.00 \cdot 10^{-1}$
6532.9	1	3	$1.08 \cdot 10^{-1}$
6599.0	3	3	$2.32 \cdot 10^{-1}$
6602.9	3	3	$5.9 \cdot 10^{-3}$
6652.1	3	1	$2.9 \cdot 10^{-3}$
6678.3	3	5	$2.33 \cdot 10^{-1}$
6717.0	3	3	$2.17 \cdot 10^{-1}$
6721.1	3	3	$4.9 \cdot 10^{-4}$
6929.5	3	5	$1.74 \cdot 10^{-1}$
7024.1	3	3	$1.89 \cdot 10^{-2}$
7032.4	5	3	$2.53 \cdot 10^{-1}$
7051.3	3	3	$3.0 \cdot 10^{-2}$
7059.1	3	5	$6.8 \cdot 10^{-2}$
7173.9	3	5	$2.87 \cdot 10^{-2}$
7245.2	3	3	$9.35 \cdot 10^{-2}$
7304.8	1	3	$2.55 \cdot 10^{-3}$
7438.9	1	3	$2.31 \cdot 10^{-2}$
7472.4	3	3	$4.0 \cdot 10^{-2}$
7535.8	3	3	$4.3 \cdot 10^{-1}$
7937.0	5	5	$7.8 \cdot 10^{-3}$
8082.5	3	3	$1.2 \cdot 10^{-3}$
8118.5	3	3	$4.9 \cdot 10^{-2}$
8128.9	3	5	$7.2 \cdot 10^{-3}$
8259.4	5	5	$2.03 \cdot 10^{-2}$
8571.4	3	3	$5.5 \cdot 10^{-2}$
8582.9	3	5	$1.00 \cdot 10^{-2}$
8647.0	5	5	$3.91 \cdot 10^{-2}$
8681.9	3	3	$2.1 \cdot 10^{-1}$
8767.5	3	3	$1.1 \cdot 10^{-3}$
8771.7	3	3	$1.6 \cdot 10^{-1}$
8783.8	3	5	$3.13 \cdot 10^{-1}$
8865.3	3	3	$9.4 \cdot 10^{-3}$
9201.8	3	3	$9.1 \cdot 10^{-2}$
9433.0	3	3	$1.1 \cdot 10^{-3}$
9486.7	3	3	$2.5 \cdot 10^{-2}$
9534.2	3	3	$6.3 \cdot 10^{-2}$
10621	3	3	$2.4 \cdot 10^{-3}$
11409	3	3	$4.2 \cdot 10^{-2}$
11525	3	3	$8.4 \cdot 10^{-2}$
11767	3	3	$6.9 \cdot 10^{-2}$
12459	3	3	$1.5 \cdot 10^{-2}$
Ne II			
*357.03	6	10	$3.8 \cdot 10^{1}$
*361.77	6	2	$1.6 \cdot 10^{1}$
*406.28	6	10	$1.8 \cdot 10^{1}$
*446.37	6	6	$4.07 \cdot 10^{1}$
460.73	4	2	$4.7 \cdot 10^{1}$
462.39	2	2	$2.3 \cdot 10^{1}$
1907.5	4	2	$2.8 \cdot 10^{-1}$
1916.1	4	4	$6.9 \cdot 10^{-1}$
1930.0	2	2	$5.7 \cdot 10^{-1}$
1938.8	2	4	$1.3 \cdot 10^{-1}$
2858.0	6	6	$7.9 \cdot 10^{-1}$
2870.0	6	6	$1.7 \cdot 10^{-1}$
2873.0	6	4	$3.8 \cdot 10^{-1}$
2876.3	4	6	$7.8 \cdot 10^{-1}$
2876.5	6	4	$3.3 \cdot 10^{-1}$
2878.1	2	2	$6.9 \cdot 10^{-2}$
2888.4	4	6	$7.0 \cdot 10^{-2}$
2891.5	4	4	$6.1 \cdot 10^{-2}$
2897.0	6	8	$5.2 \cdot 10^{-2}$
2906.8	2	4	$5.5 \cdot 10^{-1}$
2910.1	4	2	1.7
2910.4	2	4	$5.9 \cdot 10^{-1}$
2916.2	6	4	$9.6 \cdot 10^{-2}$
2925.6	2	2	$5.6 \cdot 10^{-1}$
2933.7	6	6	$6.9 \cdot 10^{-2}$
2955.7	6	4	1.2
3001.7	4	4	$8.7 \cdot 10^{-1}$
3017.3	6	4	$3.5 \cdot 10^{-1}$
3027.0	6	6	1.4
3028.7	4	2	$8.5 \cdot 10^{-1}$
3028.9	2	4	$4.7 \cdot 10^{-1}$
3034.5	6	8	3.1
3037.7	4	4	2.1
3045.6	2	2	2.5
3047.6	4	6	1.8
3054.7	2	4	$9.4 \cdot 10^{-1}$
3092.9	6	6	1.3
3097.1	8	8	1.3
3118.0	8	6	$4.2 \cdot 10^{-2}$
3134.1	6	4	$2.6 \cdot 10^{-1}$
3140.4	8	6	$2.4 \cdot 10^{-1}$
3151.1	6	6	$4.8 \cdot 10^{-2}$
3154.8	8	6	$1.8 \cdot 10^{-2}$
3164.4	8	8	$1.6 \cdot 10^{-1}$
3165.7	6	6	$1.2 \cdot 10^{-1}$
3173.6	6	4	$4.5 \cdot 10^{-2}$
3176.1	4	6	$6.0 \cdot 10^{-2}$
3187.6	4	6	$1.4 \cdot 10^{-2}$
3188.7	6	6	$3.9 \cdot 10^{-1}$
3190.9	4	6	$1.5 \cdot 10^{-1}$
3194.6	4	4	$5.2 \cdot 10^{-1}$
3198.6	6	8	1.7
3198.9	4	4	$2.3 \cdot 10^{-1}$
3209.0	8	8	$1.6 \cdot 10^{-1}$
3209.4	2	4	$6.0 \cdot 10^{-1}$
3213.7	2	4	1.7
3214.3	4	6	2.2
3218.2	8	10	3.6
3224.8	6	8	3.5
3229.5	8	8	$1.3 \cdot 10^{-1}$
3229.6	8	10	3.6
3230.1	6	6	1.8
3230.4	4	6	$1.4 \cdot 10^{-1}$
3232.0	6	4	$2.7 \cdot 10^{-1}$
3232.4	4	4	1.6
3243.4	6	6	$2.3 \cdot 10^{-1}$
3244.1	6	8	1.5
3248.1	4	4	$2.4 \cdot 10^{-1}$
3255.4	6	4	$3.8 \cdot 10^{-2}$
3263.4	2	4	$3.9 \cdot 10^{-1}$
3269.9	4	6	$5.1 \cdot 10^{-1}$
3270.8	6	4	$5.7 \cdot 10^{-2}$
3297.7	6	6	$4.3 \cdot 10^{-1}$
3309.7	4	2	$3.1 \cdot 10^{-1}$
3310.5	4	4	$6.9 \cdot 10^{-2}$
3311.3	4	2	$2.6 \cdot 10^{-1}$
3314.7	6	6	$4.4 \cdot 10^{-2}$
3319.7	4	2	1.6
3320.2	8	6	$2.1 \cdot 10^{-1}$
3323.7	4	4	1.6
3327.2	4	4	$9.1 \cdot 10^{-1}$
3329.2	8	8	$8.8 \cdot 10^{-1}$
3330.7	6	6	$3.9 \cdot 10^{-2}$
3334.8	6	8	1.8
3336.1	4	6	1.1
3344.4	2	2	1.5
3345.5	6	4	1.4
3345.8	4	4	$2.2 \cdot 10^{-1}$
3353.6	4	2	$1.2 \cdot 10^{-1}$
3355.0	4	6	1.3
3356.3	6	6	$2.0 \cdot 10^{-1}$
3357.8	6	6	$5.0 \cdot 10^{-1}$
3360.3	2	4	$8.6 \cdot 10^{-1}$
3360.6	2	4	$8.2 \cdot 10^{-1}$
3362.9	4	2	$3.5 \cdot 10^{-1}$
3371.8	4	6	$2.2 \cdot 10^{-1}$
3374.1	4	4	$3.0 \cdot 10^{-1}$
3378.2	2	2	1.7
3379.3	2	2	$3.0 \cdot 10^{-1}$
3386.2	4	6	$5.5 \cdot 10^{-2}$
3388.4	4	6	2.2
3390.6	2	4	$7.7 \cdot 10^{-2}$
3392.8	2	4	$4.4 \cdot 10^{-1}$
3404.8	4	6	1.9
3407.0	6	8	2.3
3411.4	4	2	$6.1 \cdot 10^{-1}$
3413.2	4	4	1.8
3414.9	4	6	$1.8 \cdot 10^{-2}$
3416.9	6	6	$6.4 \cdot 10^{-1}$
3417.7	6	8	1.6
3438.9	2	2	1.4
3440.7	2	4	$3.5 \cdot 10^{-1}$
3453.1	4	4	$4.6 \cdot 10^{-1}$
3454.8	4	4	1.6
3456.6	2	4	$9.6 \cdot 10^{-1}$
3457.1	4	6	$9.9 \cdot 10^{-2}$

λ Å	Weights g_i	g_k	A $10^8\ s^{-1}$
3459.3	6	6	1.6
3475.2	4	4	$1.2\cdot10^{-2}$
3477.6	4	6	$4.3\cdot10^{-1}$
3481.9	4	2	1.4
3503.6	2	2	2.0
3522.7	4	2	$2.3\cdot10^{-2}$
3538.0	4	2	$7.6\cdot10^{-1}$
3539.9	4	4	$3.6\cdot10^{-2}$
3542.2	6	4	$6.0\cdot10^{-1}$
3542.9	4	6	1.2
3546.2	2	4	$6.3\cdot10^{-2}$
3551.6	2	4	$3.7\cdot10^{-2}$
3557.8	2	2	$1.9\cdot10^{-1}$
3561.2	4	6	$2.1\cdot10^{-1}$
3565.8	4	4	$6.2\cdot10^{-1}$
3568.5	6	8	1.4
3571.2	4	4	$6.3\cdot10^{-1}$
3574.2	6	6	$1.0\cdot10^{-1}$
3574.6	4	6	1.3
3590.4	4	6	$3.6\cdot10^{-2}$
3594.2	4	2	1.3
3612.3	2	4	$2.6\cdot10^{-1}$
3628.0	4	4	$6.0\cdot10^{-1}$
3632.7	4	4	$1.3\cdot10^{-1}$
3643.9	4	4	$3.2\cdot10^{-1}$
[illegible]	2	1	$9.9\cdot10^{-1}$
3659.9	4	6	$6.7\cdot10^{-2}$
3664.1	6	4	$7.0\cdot10^{-1}$
3679.8	4	2	$3.2\cdot10^{-1}$
3694.2	6	6	1.0
3697.1	2	2	$2.8\cdot10^{-1}$
3701.8	4	6	$2.7\cdot10^{-1}$
3709.6	4	2	1.1
3713.1	4	6	1.3
3721.8	4	6	$2.0\cdot10^{-1}$
3726.9	4	4	$1.2\cdot10^{-1}$
3727.1	2	4	$9.8\cdot10^{-1}$
3734.9	4	4	$1.9\cdot10^{-1}$
3744.6	2	4	$2.6\cdot10^{-1}$
3751.2	2	2	$1.8\cdot10^{-1}$
3753.8	4	6	$4.5\cdot10^{-1}$
3766.3	4	6	$2.9\cdot10^{-1}$
3777.1	2	4	$4.2\cdot10^{-1}$
3800.0	4	4	$3.7\cdot10^{-1}$
3818.4	2	4	$6.1\cdot10^{-1}$
3829.8	4	6	$8.4\cdot10^{-1}$
3942.3	4	6	$1.0\cdot10^{-2}$
Ne V			
*142.61	9	9	$6.7\cdot10^{2}$
*143.32	9	15	$1.2\cdot10^{3}$
147.13	5	7	$1.5\cdot10^{3}$
151.23	5	5	$3.38\cdot10^{2}$
154.50	1	3	$7.0\cdot10^{2}$
*167.69	9	9	$1.5\cdot10^{2}$
*358.93	9	3	$2.1\cdot10^{2}$
365.59	5	3	$1.35\cdot10^{2}$
*482.15	9	9	$3.01\cdot10^{1}$
*571.04	9	15	$1.0\cdot10^{1}$
2259.6	3	5	1.9

λ Å	Weights g_i	g_k	A $10^8\ s^{-1}$
2265.7	5	7	2.4
Ne VII			
97.502	1	3	$1.07\cdot10^{3}$
*115.46	9	3	$4.8\cdot10^{2}$
116.69	3	5	$1.6\cdot10^{3}$
127.66	3	1	$1.9\cdot10^{2}$
465.22	1	3	$4.09\cdot10^{1}$
558.61	3	5	8.11
559.95	1	3	$1.07\cdot10^{1}$
561.38	3	3	7.99
561.73	5	5	$2.39\cdot10^{1}$
562.99	3	1	$3.17\cdot10^{1}$
564.53	5	3	$1.31\cdot10^{1}$
Ne VIII			
*88.09	2	6	$8.4\cdot10^{2}$
*98.208	6	10	$2.77\cdot10^{3}$
770.41	2	4	5.90
780.32	2	2	5.69
2820.7	2	4	$7.20\cdot10^{-1}$
2860.1	2	2	$6.88\cdot10^{-1}$
Nickel			
Ni I			
1976.87	7	9	1.1
1990.25	5	7	$8.3\cdot10^{-1}$
2014.25	3	5	$9.3\cdot10^{-1}$
2085.57	5	5	2.6
2158.31	7	5	$6.9\cdot10^{-1}$
2289.98	9	7	2.1
2300.77	7	7	$7.5\cdot10^{-1}$
2312.34	7	7	5.5
2313.98	5	5	5.0
2317.16	7	5	3.8
2320.03	9	11	6.9
2321.38	5	7	5.6
2325.79	7	9	3.5
2329.96	5	3	5.3
2345.54	9	7	2.2
2346.63	7	5	$5.5\cdot10^{-1}$
3002.48	7	7	$8.0\cdot10^{-1}$
3003.62	5	5	$6.9\cdot10^{-1}$
3012.00	5	5	1.3
3037.93	7	7	$2.8\cdot10^{-1}$
3050.82	7	9	$6.0\cdot10^{-1}$
3054.31	5	5	$4.0\cdot10^{-1}$
3057.64	3	3	1.0
3101.56	5	7	$6.3\cdot10^{-1}$
3101.88	5	7	$4.9\cdot10^{-1}$
3134.11	3	5	$7.3\cdot10^{-1}$
3369.56	9	7	$1.8\cdot10^{-1}$
3380.57	5	3	1.3
3392.98	7	7	$2.4\cdot10^{-1}$
3414.76	7	9	$5.5\cdot10^{-1}$
3423.71	3	3	$3.3\cdot10^{-1}$
3433.56	7	7	$1.7\cdot10^{-1}$
3446.26	5	5	$4.4\cdot10^{-1}$
3452.88	5	7	$9.8\cdot10^{-2}$

λ Å	Weights g_i	g_k	A $10^8\ s^{-1}$
3458.46	3	5	$6.1\cdot10^{-1}$
3461.66	7	9	$2.7\cdot10^{-1}$
3472.55	5	7	$1.2\cdot10^{-1}$
3492.96	5	3	$9.8\cdot10^{-1}$
3510.33	3	1	1.2
3515.05	5	7	$4.2\cdot10^{-1}$
3524.54	7	5	1.0
3566.37	5	5	$5.6\cdot10^{-1}$
3619.39	5	7	$6.6\cdot10^{-1}$
4027.67	5	7	$1.3\cdot10^{-1}$
4295.88	9	7	$1.7\cdot10^{-1}$
4401.54	9	11	$3.8\cdot10^{-1}$
4462.46	3	5	$1.7\cdot10^{-1}$
4470.48	5	7	$1.9\cdot10^{-1}$
4600.37	5	3	$2.6\cdot10^{-1}$
4604.99	9	7	$2.3\cdot10^{-1}$
4606.23	5	3	$1.0\cdot10^{-1}$
4648.66	11	9	$2.4\cdot10^{-1}$
4686.22	5	5	$1.4\cdot10^{-1}$
4701.54	9	9	$1.4\cdot10^{-1}$
4714.42	13	11	$4.6\cdot10^{-1}$
4715.78	7	7	$2.0\cdot10^{-1}$
4732.47	7	9	$9.3\cdot10^{-2}$
4752.43	3	3	$2.0\cdot10^{-1}$
4756.52	9	9	$1.5\cdot10^{-1}$
4786.54	11	11	$1.8\cdot10^{-1}$
4829.03	5	7	$1.9\cdot10^{-1}$
4831.18	9	7	$1.6\cdot10^{-1}$
4838.64	9	7	$2.2\cdot10^{-1}$
4855.41	5	5	$5.7\cdot10^{-1}$
4904.41	5	3	$6.2\cdot10^{-1}$
4912.03	3	3	$1.5\cdot10^{-1}$
4913.97	5	3	$2.2\cdot10^{-1}$
4918.36	9	7	$2.3\cdot10^{-1}$
4935.83	7	5	$2.4\cdot10^{-1}$
4937.34	9	9	$1.2\cdot10^{-1}$
4953.20	5	5	$1.2\cdot10^{-1}$
4980.17	9	11	$1.9\cdot10^{-1}$
5000.34	7	7	$1.4\cdot10^{-1}$
5012.46	7	7	$1.1\cdot10^{-1}$
5017.58	11	11	$2.0\cdot10^{-1}$
5035.37	7	9	$5.7\cdot10^{-1}$
5042.20	3	5	$1.4\cdot10^{-1}$
5048.85	7	7	$1.6\cdot10^{-1}$
5080.53	9	11	$3.2\cdot10^{-1}$
5081.11	7	9	$5.7\cdot10^{-1}$
5082.35	3	3	$2.5\cdot10^{-1}$
5084.08	7	9	$3.1\cdot10^{-1}$
5099.95	7	7	$2.9\cdot10^{-1}$
5115.40	11	9	$2.2\cdot10^{-1}$
5129.37	7	5	$1.2\cdot10^{-1}$
5155.14	5	5	$1.1\cdot10^{-1}$
5155.76	5	7	$2.9\cdot10^{-1}$
5176.57	5	5	$1.8\cdot10^{-1}$
5371.33	7	7	$1.6\cdot10^{-1}$
5476.91	1	3	$9.5\cdot10^{-2}$
5637.12	3	3	$1.1\cdot10^{-1}$
5664.02	5	7	$1.1\cdot10^{-1}$
5695.00	3	3	$1.7\cdot10^{-1}$
6086.29	3	5	$1.1\cdot10^{-1}$

λ Å	Weights g_i	g_k	A 10^8 s^{-1}
6175.42	3	3	$1.7 \cdot 10^{-1}$
7122.24	5	7	$2.1 \cdot 10^{-1}$
7381.94	9	11	$9.7 \cdot 10^{-2}$
7422.30	7	5	$1.8 \cdot 10^{-1}$
7727.66	7	7	$1.1 \cdot 10^{-1}$

Ni II

λ Å	Weights g_i	g_k	A 10^8 s^{-1}
2165.55	10	10	2.4
2169.10	8	8	1.58
2174.67	8	10	1.43
2175.15	6	6	1.77
2184.61	4	4	2.90
2201.41	4	6	1.3
2206.72	6	8	1.66
2216.48	10	12	3.4
2220.40	6	8	2.3
2222.96	10	10	$9.8 \cdot 10^{-1}$
2224.86	8	8	1.55
2226.33	6	6	1.3
2253.85	4	6	1.98
2264.46	6	8	1.43
2270.21	8	10	1.56
2278.77	8	6	2.8
2287.09	6	4	2.8
2296.55	8	8	1.98
2297.14	6	4	2.70
2297.49	4	2	3.0
2298.27	6	6	2.8
2303.00	8	6	2.9
2316.04	10	8	2.88
2334.58	8	8	$8.0 \cdot 10^{-1}$
2375.42	6	8	$6.6 \cdot 10^{-1}$
2394.52	8	10	1.70
2416.13	6	8	2.1
2437.89	8	10	$5.4 \cdot 10^{-1}$
2510.87	8	10	$5.8 \cdot 10^{-1}$

Ni III

λ Å	Weights g_i	g_k	A 10^8 s^{-1}
1692.51	11	13	7.9
1709.90	9	11	6.3
1719.46	5	7	6.0
1722.28	3	5	5.9
1724.52	3	1	6.7
1741.96	9	7	5.7
1752.43	7	5	5.5
1760.56	5	3	6.5
1769.64	11	11	6.2
1823.06	9	9	5.6

Nitrogen

N I

λ Å	Weights g_i	g_k	A 10^8 s^{-1}
951.079	4	6	$8.29 \cdot 10^{-3}$
951.295	4	4	$1.71 \cdot 10^{-3}$
952.303	4	6	$1.12 \cdot 10^{-1}$
952.415	4	4	$1.45 \cdot 10^{-1}$
952.523	4	2	$7.62 \cdot 10^{-2}$
953.415	4	2	1.90
953.655	4	4	1.81
953.970	4	6	1.62
954.104	4	6	$1.95 \cdot 10^{-1}$
955.264	4	6	$3.37 \cdot 10^{-3}$
955.437	4	4	$1.40 \cdot 10^{-4}$
955.529	4	2	$2.63 \cdot 10^{-3}$
955.882	4	4	$4.29 \cdot 10^{-3}$
959.494	4	4	$3.75 \cdot 10^{-3}$
960.201	4	2	$1.69 \cdot 10^{-3}$
963.990	4	6	$5.94 \cdot 10^{-1}$
964.626	4	4	$5.66 \cdot 10^{-1}$
965.041	4	2	$5.52 \cdot 10^{-1}$
1003.37	4	4	$1.86 \cdot 10^{-6}$
1003.38	4	6	$8.40 \cdot 10^{-6}$
1134.17	4	2	1.51
1134.41	4	4	1.49
1134.98	4	6	1.44
1159.82	4	4	$4.94 \cdot 10^{-4}$
1160.94	4	2	$2.72 \cdot 10^{-4}$
1163.88	6	6	$3.25 \cdot 10^{-1}$
1164.00	4	6	$2.53 \cdot 10^{-3}$
1164.21	6	4	$2.98 \cdot 10^{-2}$
1164.32	4	4	$2.82 \cdot 10^{-1}$
1165.59	6	8	$1.69 \cdot 10^{-2}$
1165.72	6	6	$8.31 \cdot 10^{-4}$
1165.84	4	6	$1.99 \cdot 10^{-3}$
1165.88	6	4	$2.43 \cdot 10^{-4}$
1166.00	4	4	$6.79 \cdot 10^{-4}$
1167.45	6	8	1.10
1167.74	6	4	$1.17 \cdot 10^{-4}$
1167.86	4	4	$7.03 \cdot 10^{-4}$
1168.22	6	6	$1.30 \cdot 10^{-2}$
1168.33	4	6	$1.24 \cdot 10^{-1}$
1168.42	6	6	$3.38 \cdot 10^{-2}$
1168.54	4	6	$9.32 \cdot 10^{-1}$
1169.69	6	8	$2.79 \cdot 10^{-2}$
1170.16	6	6	$3.03 \cdot 10^{-3}$
1170.28	4	6	$3.06 \cdot 10^{-2}$
1170.42	6	4	$4.91 \cdot 10^{-4}$
1170.54	4	4	$1.11 \cdot 10^{-3}$
1170.67	4	2	$5.86 \cdot 10^{-2}$
1171.08	6	4	$1.25 \cdot 10^{-1}$
1171.20	4	4	$3.93 \cdot 10^{-3}$
1176.51	6	4	$8.52 \cdot 10^{-1}$
1176.63	4	4	$9.91 \cdot 10^{-2}$
1177.69	4	2	1.03
1183.28	6	6	$1.88 \cdot 10^{-5}$
1183.40	4	6	$2.16 \cdot 10^{-6}$
1184.24	6	4	$2.45 \cdot 10^{-3}$
1184.36	4	4	$3.34 \cdot 10^{-4}$
1184.98	4	2	$1.78 \cdot 10^{-3}$
1199.55	4	6	4.07
1200.22	4	4	4.03
1200.71	4	2	4.00
1243.17	6	4	$3.33 \cdot 10^{-1}$
1243.18	6	6	3.22
1243.31	4	4	3.10
1243.31	4	6	$2.36 \cdot 10^{-1}$
1310.54	4	6	$7.68 \cdot 10^{-1}$
1310.94	2	4	$6.05 \cdot 10^{-1}$
1310.95	4	4	$1.75 \cdot 10^{-1}$
1312.87	4	6	$5.53 \cdot 10^{-4}$
1313.07	2	4	$2.79 \cdot 10^{-3}$
1313.08	4	4	$7.66 \cdot 10^{-5}$
1313.28	2	2	$2.37 \cdot 10^{-3}$
1313.28	4	2	$1.15 \cdot 10^{-3}$
1314.97	2	2	$1.14 \cdot 10^{-3}$
1314.98	4	2	$6.83 \cdot 10^{-4}$
1315.43	2	4	$3.65 \cdot 10^{-3}$
1315.44	4	4	$7.40 \cdot 10^{-4}$
1316.04	4	6	$1.06 \cdot 10^{-3}$
1316.29	4	6	$1.12 \cdot 10^{-2}$
1318.50	4	6	$1.36 \cdot 10^{-4}$
1318.82	2	4	$1.45 \cdot 10^{-5}$
1318.83	4	4	$1.00 \cdot 10^{-2}$
1319.00	2	2	$4.59 \cdot 10^{-1}$
1319.00	4	2	$2.28 \cdot 10^{-1}$
1319.67	2	4	$1.70 \cdot 10^{-1}$
1319.68	4	4	$5.76 \cdot 10^{-1}$
1326.56	2	4	$9.72 \cdot 10^{-3}$
1326.57	4	4	$6.71 \cdot 10^{-2}$
1327.92	2	2	$9.47 \cdot 10^{-2}$
1327.92	4	2	$4.43 \cdot 10^{-2}$
1335.18	4	6	$3.25 \cdot 10^{-6}$
1336.39	2	4	$1.02 \cdot 10^{-4}$
1336.40	4	4	$5.93 \cdot 10^{-4}$
1337.19	2	2	$3.16 \cdot 10^{-4}$
1337.20	4	2	$1.22 \cdot 10^{-4}$
1411.93	2	4	$4.46 \cdot 10^{-1}$
1411.94	4	4	$1.01 \cdot 10^{-1}$
1411.95	4	6	$5.34 \cdot 10^{-1}$
1492.63	6	4	3.11
1492.82	4	4	$3.26 \cdot 10^{-1}$
1494.68	4	2	3.46
1742.72	2	4	$2.12 \cdot 10^{-1}$
1742.73	4	4	1.05
1745.25	2	2	$8.35 \cdot 10^{-1}$
1745.26	4	2	$4.01 \cdot 10^{-1}$
6884.31	2	4	$2.90 \cdot 10^{-6}$
6900.36	4	4	$1.27 \cdot 10^{-5}$
7380.10	4	6	$1.94 \cdot 10^{-5}$
7403.39	2	4	$1.14 \cdot 10^{-4}$
7405.68	6	6	$4.39 \cdot 10^{-5}$
7421.94	4	4	$1.63 \cdot 10^{-4}$
7423.64	2	4	$5.64 \cdot 10^{-2}$
7442.30	4	4	$1.19 \cdot 10^{-1}$
7447.81	6	4	$1.49 \cdot 10^{-4}$
7468.31	6	4	$1.96 \cdot 10^{-1}$
8184.86	4	6	$8.21 \cdot 10^{-2}$
8188.01	2	4	$1.25 \cdot 10^{-1}$
8200.36	2	2	$4.68 \cdot 10^{-2}$
8210.72	4	4	$5.23 \cdot 10^{-2}$
8216.34	6	6	$2.26 \cdot 10^{-1}$
8223.13	4	2	$2.62 \cdot 10^{-1}$
8242.39	6	4	$1.31 \cdot 10^{-1}$
8567.74	2	4	$4.86 \cdot 10^{-2}$
8594.00	2	2	$2.09 \cdot 10^{-1}$
8629.24	4	4	$2.67 \cdot 10^{-1}$
8655.88	4	2	$1.07 \cdot 10^{-1}$
8664.39	2	4	$1.63 \cdot 10^{-3}$
8680.28	6	8	$2.53 \cdot 10^{-1}$
8683.40	4	6	$1.88 \cdot 10^{-1}$

λ Å	Weights g_i	Weights g_k	A $10^8\ s^{-1}$
8686.15	2	4	$1.15\cdot10^{-1}$
8703.25	2	2	$2.16\cdot10^{-1}$
8711.70	4	4	$1.29\cdot10^{-1}$
8718.84	6	6	$6.54\cdot10^{-2}$
8728.90	4	2	$3.75\cdot10^{-2}$
8747.37	6	4	$9.65\cdot10^{-3}$
8758.21	2	4	$2.62\cdot10^{-4}$
8767.35	2	2	$1.22\cdot10^{-3}$
9020.69	2	4	$3.68\cdot10^{-3}$
9028.92	2	2	$3.20\cdot10^{-1}$
9060.48	2	4	$3.21\cdot10^{-1}$
9386.81	2	4	$2.13\cdot10^{-1}$
9392.79	4	6	$2.51\cdot10^{-1}$
9419.39	2	4	$2.93\cdot10^{-4}$
9460.68	4	4	$3.73\cdot10^{-2}$
9464.17	2	2	$3.50\cdot10^{-3}$
9493.77	4	4	$4.06\cdot10^{-6}$
9660.12	2	4	$2.49\cdot10^{-4}$
9708.45	2	2	$3.18\cdot10^{-5}$
9716.46	6	4	$1.34\cdot10^{-4}$
9740.39	4	2	$9.03\cdot10^{-5}$
9742.12	8	6	$4.91\cdot10^{-7}$
9776.90	2	4	$1.40\cdot10^{-2}$
9786.78	4	6	$1.18\cdot10^{-2}$
9788.29	2	2	$3.83\cdot10^{-2}$
9798.56	1	1	$3.33\cdot10^{-}$[illegible]
9810.01	4	2	$5.42\cdot10^{-2}$
9814.02	6	8	$7.00\cdot10^{-3}$
9822.75	6	6	$5.74\cdot10^{-2}$
9834.61	6	4	$4.50\cdot10^{-2}$
9863.33	8	8	$1.03\cdot10^{-1}$
9872.15	8	6	$2.97\cdot10^{-2}$
9883.38	2	2	$1.83\cdot10^{-2}$
9905.52	4	2	$1.83\cdot10^{-3}$
9909.22	2	4	$4.07\cdot10^{-3}$
9931.47	4	4	$2.02\cdot10^{-2}$
9965.75	4	6	$3.48\cdot10^{-3}$
9968.51	6	4	$2.23\cdot10^{-3}$
9980.42	4	6	$4.86\cdot10^{-3}$

N II

λ Å	Weights g_i	Weights g_k	A $10^8\ s^{-1}$
525.983	1	3	$3.94\cdot10^{-2}$
526.118	3	3	$8.74\cdot10^{-4}$
526.345	5	3	$5.91\cdot10^{-3}$
528.529	5	7	$1.86\cdot10^{-2}$
529.355	1	3	7.17
529.413	3	1	$2.44\cdot10^{1}$
529.491	3	3	6.86
529.637	3	5	4.83
529.722	5	3	$1.04\cdot10^{1}$
529.867	5	5	$1.96\cdot10^{1}$
533.511	1	3	$2.46\cdot10^{1}$
533.581	3	5	$3.29\cdot10^{1}$
533.650	3	3	$1.70\cdot10^{1}$
533.729	5	7	$4.24\cdot10^{1}$
533.815	5	5	9.35
533.884	5	3	$9.54\cdot10^{-1}$
534.637	3	5	$4.96\cdot10^{-2}$
534.872	5	5	$7.14\cdot10^{-4}$
536.300	3	5	$6.40\cdot10^{-2}$
536.365	5	7	$1.52\cdot10^{-1}$
536.536	5	5	$2.35\cdot10^{-2}$
572.069	5	3	$4.24\cdot10^{-3}$
574.650	5	7	$3.83\cdot10^{1}$
576.060	5	3	$4.22\cdot10^{-4}$
576.232	5	5	$3.23\cdot10^{-2}$
580.802	5	7	$1.39\cdot10^{-2}$
580.904	5	5	$1.38\cdot10^{-1}$
582.156	5	5	$2.74\cdot10^{1}$
583.925	5	7	$2.32\cdot10^{-2}$
584.128	5	5	$2.35\cdot10^{-1}$
599.644	1	3	$1.02\cdot10^{-4}$
599.819	3	3	$1.69\cdot10^{-3}$
600.115	5	3	$3.15\cdot10^{-4}$
635.197	1	3	$2.58\cdot10^{1}$
640.121	1	3	$2.30\cdot10^{-2}$
644.634	1	3	$1.19\cdot10^{1}$
644.837	3	3	$3.56\cdot10^{1}$
645.178	5	3	$5.94\cdot10^{1}$
646.209	1	3	$1.43\cdot10^{-2}$
660.286	5	3	$3.17\cdot10^{1}$
670.296	1	3	$3.80\cdot10^{-1}$
670.515	3	3	$3.02\cdot10^{-1}$
670.884	5	3	$3.68\cdot10^{-1}$
671.016	3	5	2.84
671.386	5	5	8.5[illegible]
671.411	1	3	3.40
671.630	3	3	2.53
671.773	3	1	$1.13\cdot10^{1}$
672.001	5	3	4.38
693.774	3	5	$5.66\cdot10^{-5}$
694.169	5	5	$4.14\cdot10^{-4}$
715.254	5	3	$1.06\cdot10^{-3}$
745.841	1	3	$1.35\cdot10^{1}$
746.984	5	3	$3.69\cdot10^{1}$
747.606	5	5	$7.17\cdot10^{-5}$
748.369	5	3	3.81
775.965	5	5	$3.14\cdot10^{1}$
816.740	1	3	$1.85\cdot10^{-4}$
834.070	5	7	$8.13\cdot10^{-7}$
834.740	5	5	$2.42\cdot10^{-7}$
835.163	5	3	$3.04\cdot10^{-8}$
858.376	1	3	$2.56\cdot10^{-1}$
860.205	1	3	$2.65\cdot10^{-2}$
915.612	1	3	4.23
915.962	3	1	$1.27\cdot10^{1}$
916.012	3	5	3.14
916.020	3	3	3.21
916.701	5	5	9.55
916.710	5	3	5.27
1064.95	5	5	$2.30\cdot10^{-5}$
1064.96	5	3	$5.16\cdot10^{-5}$
1083.99	1	3	2.10
1084.56	3	3	1.54
1084.58	3	5	2.82
1085.53	5	3	$9.96\cdot10^{-2}$
1085.55	5	5	$9.10\cdot10^{-1}$
1085.70	5	7	3.72
1162.50	3	1	$8.42\cdot10^{-5}$
1275.04	7	5	$5.10\cdot10^{-1}$
1275.25	5	5	$9.34\cdot10^{-2}$
1275.28	3	5	$6.32\cdot10^{-3}$
1276.20	5	3	$4.54\cdot10^{-1}$
1276.22	3	3	$1.54\cdot10^{-1}$
1276.80	3	1	$6.08\cdot10^{-1}$
1299.81	5	5	$7.33\cdot10^{-6}$
1300.04	5	7	$4.14\cdot10^{-5}$
1304.77	5	3	$6.74\cdot10^{-4}$
1304.79	3	3	$2.32\cdot10^{-4}$
1306.71	1	3	$2.06\cdot10^{-5}$
1343.34	7	7	$6.40\cdot10^{-2}$
1343.57	5	7	$7.55\cdot10^{-3}$
1345.08	7	5	$1.27\cdot10^{-2}$
1345.31	5	5	$4.90\cdot10^{-2}$
1345.34	3	5	$1.03\cdot10^{-2}$
1346.41	5	3	$1.94\cdot10^{-2}$
1346.44	3	3	$5.28\cdot10^{-2}$
1381.97	5	3	$1.30\cdot10^{-7}$
1382.00	3	3	$7.07\cdot10^{-5}$
1538.57	3	5	$1.59\cdot10^{-5}$
1627.35	3	5	$7.03\cdot10^{-3}$
1627.38	5	5	$2.39\cdot10^{-2}$
1628.90	3	3	$5.17\cdot10^{-3}$
1628.92	5	3	$1.98\cdot10^{-2}$
1629.08	1	3	$7.93\cdot10^{-3}$
1629.83	3	1	$3.31\cdot10^{-}$[illegible]
1675.73	3	3	$2.70\cdot10^{-1}$
1675.75	5	3	$4.39\cdot10^{-1}$
1675.92	1	3	$9.09\cdot10^{-2}$
1678.89	1	3	$3.62\cdot10^{-6}$
1740.31	5	7	$2.21\cdot10^{-1}$
1743.20	3	5	$1.67\cdot10^{-1}$
1743.23	5	5	$5.40\cdot10^{-2}$
1745.05	3	3	$9.07\cdot10^{-2}$
1745.08	5	3	$5.97\cdot10^{-3}$
1745.26	1	3	$1.23\cdot10^{-1}$
1805.24	3	3	$2.48\cdot10^{-4}$
1805.28	5	3	$3.05\cdot10^{-5}$
1805.47	1	3	$4.06\cdot10^{-5}$
2139.01	3	5	$5.15\cdot10^{-7}$
2142.77	5	5	$1.27\cdot10^{-6}$
3329.70	5	5	$2.01\cdot10^{-2}$
3408.13	3	1	$1.91\cdot10^{-1}$
3437.14	3	1	1.91
3775.61	5	5	$6.61\cdot10^{-6}$
3919.00	3	3	$7.56\cdot10^{-1}$
3955.85	3	5	$1.21\cdot10^{-1}$
3977.31	5	5	$7.79\cdot10^{-5}$
3995.00	3	5	1.22
4046.74	5	3	$1.44\cdot10^{-5}$
4109.59	3	1	$1.24\cdot10^{-6}$
4114.33	3	3	$1.65\cdot10^{-3}$
4123.12	3	5	$4.37\cdot10^{-4}$
4236.36	3	3	$1.93\cdot10^{-3}$
4319.06	3	1	$5.59\cdot10^{-5}$
4374.99	3	5	$6.31\cdot10^{-3}$
4379.58	3	3	$1.76\cdot10^{-3}$
4393.85	5	7	$5.21\cdot10^{-4}$
4412.50	7	7	$2.93\cdot10^{-5}$
4445.03	5	7	$1.20\cdot10^{-6}$

λ Å	Weights g_i	g_k	A 10^8 s^{-1}
4447.03	3	5	1.12
4459.94	3	1	$7.99 \cdot 10^{-2}$
4464.13	5	5	$5.85 \cdot 10^{-8}$
4465.53	3	3	$1.59 \cdot 10^{-2}$
4475.89	3	5	$3.96 \cdot 10^{-4}$
4476.27	5	3	$3.40 \cdot 10^{-5}$
4477.68	5	3	$6.44 \cdot 10^{-2}$
4488.09	5	5	$8.62 \cdot 10^{-3}$
4507.56	7	5	$7.39 \cdot 10^{-2}$
4564.76	3	5	$1.65 \cdot 10^{-2}$
4601.48	3	5	$2.22 \cdot 10^{-1}$
4607.15	1	3	$3.15 \cdot 10^{-1}$
4613.87	3	3	$2.12 \cdot 10^{-1}$
4621.39	3	1	$9.04 \cdot 10^{-1}$
4630.54	5	5	$7.48 \cdot 10^{-1}$
4643.09	5	3	$4.39 \cdot 10^{-1}$
4654.53	3	5	$1.92 \cdot 10^{-2}$
4667.21	3	3	$2.31 \cdot 10^{-2}$
4674.91	3	1	$8.54 \cdot 10^{-2}$
4709.44	3	3	$1.49 \cdot 10^{-3}$
4774.24	3	5	$3.07 \cdot 10^{-2}$
4779.72	3	3	$2.49 \cdot 10^{-1}$
4781.19	5	7	$1.92 \cdot 10^{-2}$
4788.14	5	5	$2.50 \cdot 10^{-1}$
4793.65	5	3	$7.73 \cdot 10^{-2}$
4803.29	7	7	$3.17 \cdot 10^{-1}$
4810.30	7	5	$4.75 \cdot 10^{-2}$
4860.17	3	5	$1.87 \cdot 10^{-2}$
4874.57	5	5	$2.17 \cdot 10^{-5}$
4895.12	5	3	$3.04 \cdot 10^{-2}$
4897.54	7	5	$5.61 \cdot 10^{-6}$
4987.38	3	1	$6.98 \cdot 10^{-1}$
4994.37	3	3	$7.11 \cdot 10^{-1}$
5001.13	3	5	$9.65 \cdot 10^{-1}$
5001.47	5	7	1.04
5002.70	1	3	$8.33 \cdot 10^{-2}$
5005.15	7	9	1.14
5007.33	3	5	$7.43 \cdot 10^{-1}$
5010.62	3	3	$2.10 \cdot 10^{-1}$
5016.38	5	5	$1.59 \cdot 10^{-1}$
5025.66	7	7	$1.04 \cdot 10^{-1}$
5040.71	7	5	$3.65 \cdot 10^{-3}$
5045.10	5	3	$3.37 \cdot 10^{-1}$
5073.59	3	3	$2.43 \cdot 10^{-2}$
5114.28	1	3	$3.88 \cdot 10^{-4}$
5123.53	3	3	$5.69 \cdot 10^{-5}$
5138.90	5	3	$3.34 \cdot 10^{-5}$
5238.18	3	5	$3.16 \cdot 10^{-5}$
5355.00	5	7	$7.02 \cdot 10^{-5}$
5383.72	3	5	$3.69 \cdot 10^{-3}$
5390.69	3	3	$2.04 \cdot 10^{-3}$
5452.07	1	3	$9.82 \cdot 10^{-2}$
5454.22	3	1	$3.70 \cdot 10^{-1}$
5462.58	3	3	$1.11 \cdot 10^{-1}$
5478.09	3	5	$5.22 \cdot 10^{-2}$
5480.05	5	3	$1.44 \cdot 10^{-1}$
5493.23	3	5	$3.11 \cdot 10^{-4}$
5495.65	5	5	$2.66 \cdot 10^{-1}$
5666.63	3	5	$3.45 \cdot 10^{-1}$
5674.00	3	5	$4.82 \cdot 10^{-7}$
5676.02	1	3	$2.80 \cdot 10^{-1}$
5679.56	5	7	$4.96 \cdot 10^{-1}$
5686.21	3	3	$1.78 \cdot 10^{-1}$
5710.77	5	5	$1.17 \cdot 10^{-1}$
5730.66	5	3	$1.26 \cdot 10^{-2}$
5747.30	3	5	$3.27 \cdot 10^{-2}$
5767.45	3	3	$2.39 \cdot 10^{-2}$
5927.81	1	3	$3.19 \cdot 10^{-1}$
5931.78	3	5	$4.23 \cdot 10^{-1}$
5940.24	3	3	$2.22 \cdot 10^{-1}$
5941.65	5	7	$5.47 \cdot 10^{-1}$
5952.39	5	5	$1.24 \cdot 10^{-1}$
5960.91	5	3	$1.29 \cdot 10^{-2}$
6065.00	3	5	$2.57 \cdot 10^{-3}$
6086.54	5	5	$5.87 \cdot 10^{-5}$
6284.32	5	3	$4.63 \cdot 10^{-2}$
6285.69	5	7	$1.10 \cdot 10^{-3}$
6286.11	3	5	$4.50 \cdot 10^{-4}$
6309.25	5	5	$1.71 \cdot 10^{-4}$
6366.79	1	3	$8.14 \cdot 10^{-5}$
6379.62	3	3	$3.52 \cdot 10^{-2}$
6433.44	3	5	$1.22 \cdot 10^{-4}$
6457.68	3	3	$1.17 \cdot 10^{-4}$
6472.43	3	1	$1.13 \cdot 10^{-4}$
6482.05	3	3	$2.58 \cdot 10^{-1}$
6610.56	5	7	$6.01 \cdot 10^{-1}$
6802.17	5	3	$1.43 \cdot 10^{-5}$
6826.23	5	5	$2.49 \cdot 10^{-4}$
7262.55	3	3	$1.42 \cdot 10^{-7}$
7528.12	5	7	$6.61 \cdot 10^{-5}$
7545.36	5	5	$4.53 \cdot 10^{-4}$
7559.05	5	3	$1.90 \cdot 10^{-5}$
7762.24	5	5	$8.49 \cdot 10^{-2}$
8089.08	5	7	$1.01 \cdot 10^{-4}$
8128.14	5	5	$5.55 \cdot 10^{-4}$
8687.43	3	1	$1.16 \cdot 10^{-2}$
9399.64	1	3	$1.04 \cdot 10^{-4}$
N III			
374.198	2	4	$9.89 \cdot 10^{1}$
451.871	2	2	$1.03 \cdot 10^{1}$
452.227	4	2	$2.05 \cdot 10^{1}$
684.998	2	4	9.63
685.515	2	2	$3.83 \cdot 10^{1}$
685.817	4	4	$4.54 \cdot 10^{1}$
686.336	4	2	$1.95 \cdot 10^{1}$
763.334	2	2	9.58
764.351	4	2	$1.85 \cdot 10^{1}$
771.545	2	4	8.19
771.901	4	4	$1.64 \cdot 10^{1}$
772.384	6	4	$2.45 \cdot 10^{1}$
772.889	6	4	$2.09 \cdot 10^{1}$
772.955	4	2	$2.34 \cdot 10^{1}$
979.832	4	4	8.84
979.905	6	6	9.21
989.799	2	4	4.18
991.511	4	4	$8.17 \cdot 10^{-1}$
991.577	4	6	4.97
1747.85	2	4	1.28
1751.22	4	4	$2.48 \cdot 10^{-1}$
1751.66	4	6	1.51
2972.55	2	2	$6.67 \cdot 10^{-1}$
2977.33	4	2	$3.32 \cdot 10^{-1}$
2978.84	2	4	$1.66 \cdot 10^{-1}$
2983.64	4	4	$8.24 \cdot 10^{-1}$
3342.76	2	2	$3.80 \cdot 10^{-1}$
3353.98	2	4	$7.66 \cdot 10^{-1}$
3354.32	4	6	$5.51 \cdot 10^{-1}$
3355.46	4	2	$7.51 \cdot 10^{-1}$
3358.78	2	2	$3.05 \cdot 10^{-1}$
3360.98	4	4	$2.44 \cdot 10^{-1}$
3365.80	4	2	1.52
3367.36	6	6	1.27
3374.07	6	4	$8.13 \cdot 10^{-1}$
3745.95	2	4	$1.90 \cdot 10^{-1}$
3752.63	2	2	$6.67 \cdot 10^{-2}$
3754.69	4	4	$3.78 \cdot 10^{-1}$
3762.60	4	4	$4.24 \cdot 10^{-2}$
3771.03	6	4	$5.59 \cdot 10^{-1}$
3771.36	6	4	$8.28 \cdot 10^{-2}$
3792.97	8	6	$1.03 \cdot 10^{-1}$
3934.50	2	4	$7.49 \cdot 10^{-1}$
3938.51	4	6	$8.96 \cdot 10^{-1}$
3942.88	4	4	$1.49 \cdot 10^{-1}$
4097.36	2	4	$8.70 \cdot 10^{-1}$
4103.39	2	2	$8.67 \cdot 10^{-1}$
4195.74	2	4	$9.37 \cdot 10^{-1}$
4200.07	4	6	1.12
4215.77	4	4	$1.85 \cdot 10^{-1}$
4318.78	2	4	$5.40 \cdot 10^{-2}$
4321.22	2	2	$1.08 \cdot 10^{-1}$
4321.39	4	6	$5.03 \cdot 10^{-2}$
4325.43	4	4	$8.60 \cdot 10^{-2}$
4327.69	6	8	$3.06 \cdot 10^{-2}$
4327.88	4	2	$1.07 \cdot 10^{-1}$
4332.95	6	6	$1.23 \cdot 10^{-1}$
4337.01	6	4	$7.47 \cdot 10^{-2}$
4345.81	8	8	$1.82 \cdot 10^{-1}$
4351.11	8	6	$4.01 \cdot 10^{-2}$
4510.88	2	4	$2.84 \cdot 10^{-1}$
4510.96	4	6	$4.77 \cdot 10^{-1}$
4514.85	6	8	$6.80 \cdot 10^{-1}$
4518.14	2	2	$5.65 \cdot 10^{-1}$
4523.56	4	4	$3.61 \cdot 10^{-1}$
4530.86	4	2	$1.12 \cdot 10^{-1}$
4534.58	6	6	$2.01 \cdot 10^{-1}$
4547.30	6	4	$3.33 \cdot 10^{-2}$
4634.13	2	4	$6.36 \cdot 10^{-1}$
4640.64	4	6	$7.60 \cdot 10^{-1}$
4641.85	4	4	$1.26 \cdot 10^{-1}$
4858.70	2	4	$4.35 \cdot 10^{-1}$
4858.98	4	6	$4.66 \cdot 10^{-1}$
4861.27	6	8	$5.32 \cdot 10^{-1}$
4867.12	4	4	$1.73 \cdot 10^{-1}$
4867.17	8	10	$6.18 \cdot 10^{-1}$
4873.60	6	6	$1.50 \cdot 10^{-1}$
4881.78	6	4	$1.22 \cdot 10^{-2}$
4884.14	8	8	$8.71 \cdot 10^{-2}$
4896.58	8	6	$5.86 \cdot 10^{-3}$
5260.86	2	2	$2.80 \cdot 10^{-2}$

λ Å	Weights g_i	g_k	A 10^8 s^{-1}
5270.57	2	4	$6.95 \cdot 10^{-2}$
5272.68	4	2	$1.39 \cdot 10^{-1}$
5282.43	4	4	$2.21 \cdot 10^{-2}$
5297.75	4	6	$4.93 \cdot 10^{-2}$
5298.95	6	4	$7.38 \cdot 10^{-2}$
5314.36	6	6	$1.14 \cdot 10^{-1}$
5320.87	6	8	$5.68 \cdot 10^{-1}$
5327.19	4	6	$5.29 \cdot 10^{-1}$
5352.46	6	6	$3.72 \cdot 10^{-2}$
6365.84	2	2	$2.18 \cdot 10^{-1}$
6394.75	2	4	$2.15 \cdot 10^{-1}$
6445.34	2	4	$8.89 \cdot 10^{-2}$
6450.79	2	2	$1.77 \cdot 10^{-1}$
6454.08	4	6	$1.49 \cdot 10^{-1}$
6463.09	4	4	$1.13 \cdot 10^{-1}$
6467.02	6	8	$2.11 \cdot 10^{-1}$
6468.57	4	2	$3.52 \cdot 10^{-2}$
6478.76	6	6	$6.31 \cdot 10^{-2}$
6487.84	6	4	$1.05 \cdot 10^{-2}$
7371.51	4	4	$3.53 \cdot 10^{-2}$
7404.54	6	6	$3.61 \cdot 10^{-2}$
8307.51	2	4	$1.65 \cdot 10^{-2}$
8344.95	2	2	$6.52 \cdot 10^{-2}$
8386.39	4	4	$8.03 \cdot 10^{-2}$
8424.56	4	2	$3.17 \cdot 10^{-2}$

N IV

λ Å	Weights g_i	g_k	A 10^8 s^{-1}
247.205	1	3	$1.19 \cdot 10^{2}$
*283.52	9	15	$3.05 \cdot 10^{2}$
*322.64	9	3	$8.99 \cdot 10^{1}$
335.047	3	5	$1.845 \cdot 10^{2}$
387.356	3	1	$2.55 \cdot 10^{1}$
765.147	1	3	$2.320 \cdot 10^{1}$
*923.16	9	9	$1.759 \cdot 10^{1}$
955.334	3	1	$2.919 \cdot 10^{1}$
1718.55	3	5	2.321
2649.88	3	3	1.07
3052.20	1	3	$1.33 \cdot 10^{-1}$
3059.60	3	3	$3.95 \cdot 10^{-1}$
3075.19	5	3	$6.18 \cdot 10^{-1}$
3443.61	3	5	$3.46 \cdot 10^{-1}$
3445.22	1	3	$4.60 \cdot 10^{-1}$
3454.65	3	3	$3.42 \cdot 10^{-1}$
3461.36	3	1	1.36
3463.36	5	5	1.02
3474.53	5	3	$5.61 \cdot 10^{-1}$
3478.72	3	5	1.06
*3480.8	3	9	1.06
3483.00	3	3	1.06
3484.93	3	1	1.06
3689.94	3	1	$9.10 \cdot 10^{-2}$
3694.14	3	3	$2.27 \cdot 10^{-2}$
3707.39	5	3	$6.73 \cdot 10^{-2}$
3714.43	5	5	$1.34 \cdot 10^{-2}$
3735.43	7	5	$7.37 \cdot 10^{-2}$
3747.54	3	5	$9.92 \cdot 10^{-1}$
4057.76	3	5	$6.62 \cdot 10^{-1}$
4740.26	3	5	$1.53 \cdot 10^{-2}$
4747.96	3	3	$7.60 \cdot 10^{-2}$
4752.49	5	7	$1.13 \cdot 10^{-2}$
4762.09	5	5	$6.99 \cdot 10^{-2}$
4769.86	5	3	$2.50 \cdot 10^{-2}$
4786.92	7	7	$8.79 \cdot 10^{-2}$
4796.66	7	5	$1.53 \cdot 10^{-2}$
5200.41	3	5	$2.67 \cdot 10^{-1}$
5204.28	5	7	$3.55 \cdot 10^{-1}$
5205.15	1	3	$1.97 \cdot 10^{-1}$
5226.70	3	3	$1.46 \cdot 10^{-1}$
5245.60	5	5	$8.66 \cdot 10^{-2}$
5272.35	5	3	$9.48 \cdot 10^{-3}$
5288.25	5	3	$3.22 \cdot 10^{-2}$
5736.93	3	5	$1.84 \cdot 10^{-1}$
5776.31	1	3	$1.85 \cdot 10^{-2}$
5784.76	3	1	$5.51 \cdot 10^{-2}$
5795.09	3	3	$1.37 \cdot 10^{-2}$
5812.31	3	5	$1.36 \cdot 10^{-2}$
5826.43	5	3	$2.25 \cdot 10^{-2}$
5843.84	5	5	$4.01 \cdot 10^{-2}$
6380.75	1	3	$1.42 \cdot 10^{-1}$
7103.24	1	3	$6.28 \cdot 10^{-2}$
7109.35	3	5	$8.46 \cdot 10^{-2}$
7111.28	3	3	$4.70 \cdot 10^{-2}$
*7116.8	9	15	$1.12 \cdot 10^{-1}$
7122.98	5	7	$1.12 \cdot 10^{-1}$
7127.25	5	5	$2.80 \cdot 10^{-2}$
7127.25	5	3	$3.11 \cdot 10^{-3}$
9165.07	3	5	$4.23 \cdot 10^{-2}$
9182.16	5	7	$4.45 \cdot 10^{-2}$
9222.99	7	9	$4.95 \cdot 10^{-2}$
9247.04	5	5	$7.66 \cdot 10^{-3}$
9311.55	7	7	$5.36 \cdot 10^{-3}$

N V

λ Å	Weights g_i	g_k	A 10^8 s^{-1}
*209.29	2	6	$1.21 \cdot 10^{2}$
*247.66	6	10	$4.26 \cdot 10^{2}$
1238.82	2	4	3.40
1242.80	2	2	3.37
4603.74	2	4	$4.14 \cdot 10^{-1}$
4619.97	2	2	$4.10 \cdot 10^{-1}$

N VI

λ Å	Weights g_i	g_k	A 10^8 s^{-1}
24.8980	1	3	$5.158 \cdot 10^{3}$
28.7870	1	3	$1.809 \cdot 10^{4}$
*161.220	3	9	$2.859 \cdot 10^{2}$
173.275	1	3	$2.697 \cdot 10^{2}$
*173.93	9	15	$8.756 \cdot 10^{2}$
185.192	3	5	$8.205 \cdot 10^{2}$
*1901	3	9	$6.780 \cdot 10^{-1}$
2896.4	1	3	$2.079 \cdot 10^{-1}$
*6991.1	3	9	$8.384 \cdot 10^{-2}$
9622.0	1	3	$3.276 \cdot 10^{-2}$

Oxygen

O I

λ Å	Weights g_i	g_k	A 10^8 s^{-1}
791.973	5	5	4.94
792.938	1	3	2.19
792.967	3	5	1.64
877.798	5	3	2.85
877.879	5	5	5.12
922.008	5	7	1.23
935.193	5	5	1.33
1028.16	1	3	$4.22 \cdot 10^{-1}$
1152.15	5	5	5.28
1217.65	1	3	2.06
1302.17	5	3	3.41
1304.86	3	3	2.03
1306.03	1	3	$6.76 \cdot 10^{-1}$
3823.41	7	7	$6.63 \cdot 10^{-3}$
3823.87	5	3	$1.87 \cdot 10^{-3}$
3824.35	5	5	$5.19 \cdot 10^{-3}$
3825.02	3	3	$5.59 \cdot 10^{-3}$
3825.19	5	7	$8.31 \cdot 10^{-4}$
3855.01	5	5	$1.63 \cdot 10^{-2}$
3947.29	5	7	$4.91 \cdot 10^{-3}$
3947.48	5	5	$4.88 \cdot 10^{-3}$
3947.59	5	3	$4.87 \cdot 10^{-3}$
3951.93	3	1	$3.10 \cdot 10^{-3}$
3952.98	5	3	$1.29 \cdot 10^{-3}$
3953.00	1	3	$1.03 \cdot 10^{-3}$
3954.52	3	5	$7.73 \cdot 10^{-4}$
3954.61	5	5	$2.32 \cdot 10^{-3}$
3997.95	5	3	$2.41 \cdot 10^{-2}$
4217.09	3	1	$5.44 \cdot 10^{-3}$
4222.77	5	3	$2.26 \cdot 10^{-3}$
4222.82	1	3	$1.81 \cdot 10^{-3}$
4233.27	5	5	$4.04 \cdot 10^{-3}$
4368.19	3	1	$7.56 \cdot 10^{-3}$
4368.24	3	5	$7.59 \cdot 10^{-3}$
4967.38	3	5	$4.43 \cdot 10^{-3}$
4967.88	5	7	$8.44 \cdot 10^{-3}$
4968.79	7	9	$1.27 \cdot 10^{-2}$
5019.29	5	5	$7.13 \cdot 10^{-3}$
5020.22	7	5	$9.98 \cdot 10^{-3}$
5329.11	3	5	$9.48 \cdot 10^{-3}$
5329.69	5	7	$1.81 \cdot 10^{-2}$
5330.74	7	9	$2.71 \cdot 10^{-2}$
5435.18	3	5	$7.74 \cdot 10^{-3}$
5435.77	5	5	$1.29 \cdot 10^{-2}$
5436.86	7	5	$1.80 \cdot 10^{-2}$
5512.60	3	5	$2.69 \cdot 10^{-3}$
5512.77	5	7	$3.58 \cdot 10^{-3}$
5554.83	3	3	$5.83 \cdot 10^{-3}$
5555.00	5	3	$9.71 \cdot 10^{-3}$
5958.39	3	5	$6.80 \cdot 10^{-3}$
5958.58	5	7	$9.06 \cdot 10^{-3}$
6046.23	3	3	$1.05 \cdot 10^{-2}$
6046.44	5	3	$1.75 \cdot 10^{-2}$
6046.49	1	3	$3.50 \cdot 10^{-3}$
6155.99	3	5	$2.67 \cdot 10^{-2}$
6156.78	5	7	$5.08 \cdot 10^{-2}$
6158.19	7	9	$7.62 \cdot 10^{-2}$
6324.84	7	5	$3.76 \cdot 10^{-5}$
6453.60	3	5	$1.65 \cdot 10^{-2}$
6454.44	5	5	$2.75 \cdot 10^{-2}$
6455.98	7	5	$3.85 \cdot 10^{-2}$
6726.28	5	5	$1.18 \cdot 10^{-5}$
6726.54	5	3	$6.44 \cdot 10^{-6}$
7001.92	3	5	$2.65 \cdot 10^{-2}$
7002.23	5	7	$3.53 \cdot 10^{-2}$

λ Å	Weights g_i	g_k	A 10^8 s^{-1}
7254.15	3	3	$2.24 \cdot 10^{-2}$
7254.45	5	3	$3.73 \cdot 10^{-2}$
7254.53	1	3	$7.45 \cdot 10^{-3}$
7771.94	5	7	$3.69 \cdot 10^{-1}$
7774.17	5	5	$3.69 \cdot 10^{-1}$
7775.39	5	3	$3.69 \cdot 10^{-1}$
7981.94	3	3	$2.33 \cdot 10^{-4}$
7982.40	1	3	$3.09 \cdot 10^{-4}$
7986.98	3	5	$4.19 \cdot 10^{-4}$
7987.33	5	5	$1.41 \cdot 10^{-4}$
7995.07	5	7	$5.63 \cdot 10^{-4}$
8221.82	7	7	$2.89 \cdot 10^{-1}$
8227.65	5	3	$8.13 \cdot 10^{-2}$
8230.00	5	5	$2.26 \cdot 10^{-1}$
8233.00	3	3	$2.43 \cdot 10^{-1}$
8235.35	3	5	$4.86 \cdot 10^{-2}$
8446.25	3	1	$3.22 \cdot 10^{-1}$
8446.36	3	5	$3.22 \cdot 10^{-1}$
8446.76	3	3	$3.22 \cdot 10^{-1}$
8820.42	5	7	$2.93 \cdot 10^{-1}$
9260.81	3	1	$4.46 \cdot 10^{-1}$
9260.85	3	3	$3.34 \cdot 10^{-1}$
9260.94	3	5	$1.56 \cdot 10^{-1}$
9262.58	5	3	$1.11 \cdot 10^{-1}$
9262.67	5	5	$2.60 \cdot 10^{-1}$
9262.78	5	7	$2.97 \cdot 10^{-1}$
9265.83	7	5	$2.97 \cdot 10^{-2}$
9265.93	7	7	$1.48 \cdot 10^{-1}$
9266.01	7	9	$4.45 \cdot 10^{-1}$
9482.89	5	3	$2.34 \cdot 10^{-1}$
9622.11	5	3	$5.22 \cdot 10^{-4}$
9622.16	3	3	$1.57 \cdot 10^{-3}$
9625.26	7	5	$3.25 \cdot 10^{-4}$
9625.30	7	7	$1.85 \cdot 10^{-3}$
9694.66	5	7	$4.54 \cdot 10^{-4}$
9694.91	5	5	$4.54 \cdot 10^{-4}$
9695.06	5	3	$4.54 \cdot 10^{-4}$

O II

λ Å	Weights g_i	g_k	A 10^8 s^{-1}
429.918	4	2	$4.25 \cdot 10^{1}$
430.041	4	4	$4.13 \cdot 10^{1}$
430.176	4	6	$4.36 \cdot 10^{1}$
483.760	4	2	$2.05 \cdot 10^{1}$
483.980	6	4	$1.80 \cdot 10^{1}$
484.027	4	4	3.22
485.087	6	8	$2.60 \cdot 10^{1}$
485.470	6	6	1.20
485.518	4	6	$1.93 \cdot 10^{1}$
2290.85	2	4	$7.41 \cdot 10^{-2}$
2293.30	2	2	$3.25 \cdot 10^{-1}$
2300.33	4	4	$4.17 \cdot 10^{-1}$
2302.81	4	2	$1.67 \cdot 10^{-1}$
2365.14	4	2	$1.52 \cdot 10^{-1}$
2375.72	6	4	$1.35 \cdot 10^{-1}$
2406.38	6	4	$1.85 \cdot 10^{-1}$
2407.48	4	4	$2.25 \cdot 10^{-1}$
2411.60	4	2	$2.05 \cdot 10^{-1}$
2411.64	2	2	$1.10 \cdot 10^{-1}$
2415.13	4	2	$2.20 \cdot 10^{-1}$
2418.46	6	6	$2.30 \cdot 10^{-1}$
2425.57	6	6	$1.77 \cdot 10^{-1}$
2433.54	2	4	$4.21 \cdot 10^{-1}$
2436.06	4	4	$1.69 \cdot 10^{-1}$
2444.25	4	4	$7.56 \cdot 10^{-2}$
2445.53	4	6	$4.98 \cdot 10^{-1}$
2517.96	4	6	$7.72 \cdot 10^{-2}$
2523.21	2	2	$9.63 \cdot 10^{-2}$
2526.87	4	4	$1.20 \cdot 10^{-1}$
2530.28	6	8	$8.16 \cdot 10^{-2}$
2571.46	2	4	$1.15 \cdot 10^{-1}$
2575.28	4	6	$1.37 \cdot 10^{-1}$
3134.73	8	6	1.23
3273.43	8	6	$9.99 \cdot 10^{-1}$
3377.15	2	2	1.27
3390.21	2	4	1.22
3407.28	6	6	1.02
3712.74	2	4	$2.84 \cdot 10^{-1}$
3727.32	4	4	$5.81 \cdot 10^{-1}$
3749.48	6	4	$9.31 \cdot 10^{-1}$
3833.07	6	8	$1.02 \cdot 10^{-2}$
3842.81	2	4	$7.45 \cdot 10^{-2}$
3843.58	4	6	$3.55 \cdot 10^{-2}$
3847.89	2	2	$1.95 \cdot 10^{-1}$
3850.80	4	6	$6.00 \cdot 10^{-3}$
3851.03	4	4	$1.59 \cdot 10^{-1}$
3851.47	8	8	$2.72 \cdot 10^{-2}$
3856.13	4	2	$2.28 \cdot 10^{-1}$
3857.16	6	6	$6.59 \cdot 10^{-2}$
3863.50	6	8	$6.49 \cdot 10^{-2}$
3864.13	2	2	$9.12 \cdot 10^{-2}$
3864.43	6	6	$2.15 \cdot 10^{-1}$
3864.67	6	4	$1.80 \cdot 10^{-1}$
3874.09	2	4	$3.26 \cdot 10^{-2}$
3875.80	8	6	$3.38 \cdot 10^{-2}$
3882.19	8	8	$5.50 \cdot 10^{-1}$
3882.45	4	4	$8.94 \cdot 10^{-2}$
3883.14	8	6	$1.13 \cdot 10^{-1}$
3893.52	4	6	$1.89 \cdot 10^{-2}$
3907.45	6	6	$8.64 \cdot 10^{-2}$
3911.96	6	4	1.09
3912.12	4	4	$1.41 \cdot 10^{-1}$
3919.27	4	2	1.22
3945.04	2	4	$2.05 \cdot 10^{-1}$
3954.36	2	2	$8.57 \cdot 10^{-1}$
3973.26	4	4	1.04
3982.71	4	2	$4.27 \cdot 10^{-1}$
4069.62	2	4	1.52
4069.88	4	6	1.53
4072.15	6	8	1.98
4075.86	8	10	2.11
4078.84	4	4	$5.52 \cdot 10^{-1}$
4084.65	6	8	$7.28 \cdot 10^{-2}$
4085.11	6	6	$4.55 \cdot 10^{-1}$
4092.93	8	8	$2.65 \cdot 10^{-1}$
4094.14	6	4	$4.70 \cdot 10^{-2}$
4096.53	4	6	$1.73 \cdot 10^{-1}$
4097.22	2	4	$3.62 \cdot 10^{-1}$
4103.00	2	2	$5.09 \cdot 10^{-1}$
4104.72	4	6	$3.14 \cdot 10^{-1}$
4104.99	4	4	$9.14 \cdot 10^{-1}$
4106.02	8	6	$1.70 \cdot 10^{-2}$
4109.84	6	6	$1.21 \cdot 10^{-2}$
4110.19	6	4	$2.54 \cdot 10^{-1}$
4110.79	4	2	$7.70 \cdot 10^{-1}$
4112.02	6	6	$1.81 \cdot 10^{-1}$
4113.83	8	6	$2.41 \cdot 10^{-1}$
4119.22	6	8	1.33
4120.28	6	6	$2.15 \cdot 10^{-1}$
4120.55	6	4	$2.60 \cdot 10^{-1}$
4121.46	2	2	$5.60 \cdot 10^{-1}$
4129.32	4	2	$1.79 \cdot 10^{-1}$
4132.80	2	4	$9.13 \cdot 10^{-1}$
4140.70	4	4	$4.09 \cdot 10^{-2}$
4153.30	4	6	$7.91 \cdot 10^{-1}$
4156.53	6	4	$2.11 \cdot 10^{-1}$
4169.22	6	6	$2.71 \cdot 10^{-1}$
4185.44	6	8	1.91
4189.58	8	8	$7.06 \cdot 10^{-2}$
4189.79	8	10	1.98
4192.51	6	4	$3.21 \cdot 10^{-1}$
4196.27	4	4	$3.56 \cdot 10^{-2}$
4196.70	4	2	$3.56 \cdot 10^{-1}$
4317.14	2	4	$3.70 \cdot 10^{-1}$
4319.63	4	6	$2.55 \cdot 10^{-1}$
4319.87	2	2	$5.62 \cdot 10^{-1}$
4325.76	2	2	$1.47 \cdot 10^{-1}$
4327.46	6	6	$6.76 \cdot 10^{-1}$
4327.85	6	4	$7.24 \cdot 10^{-2}$
4328.59	4	2	1.12
4331.47	4	6	$4.82 \cdot 10^{-2}$
4331.86	4	4	$6.50 \cdot 10^{-1}$
4336.86	4	4	$1.57 \cdot 10^{-1}$
4345.56	4	2	$8.31 \cdot 10^{-1}$
4347.22	6	4	$1.19 \cdot 10^{-1}$
4347.41	4	4	$9.32 \cdot 10^{-1}$
4349.43	6	6	$6.91 \cdot 10^{-1}$
4351.26	6	6	$9.89 \cdot 10^{-1}$
4351.46	4	6	$5.82 \cdot 10^{-2}$
4359.40	4	6	$1.44 \cdot 10^{-2}$
4366.89	6	4	$3.98 \cdot 10^{-1}$
4369.27	4	4	$3.57 \cdot 10^{-1}$
4395.93	6	6	$3.91 \cdot 10^{-1}$
4405.98	6	4	$4.30 \cdot 10^{-2}$
4414.90	4	6	$8.34 \cdot 10^{-1}$
4416.97	2	4	$7.13 \cdot 10^{-1}$
4443.01	6	6	$5.05 \cdot 10^{-1}$
4443.52	6	8	$1.89 \cdot 10^{-2}$
4447.68	8	6	$2.52 \cdot 10^{-2}$
4448.19	8	8	$5.10 \cdot 10^{-1}$
4452.38	4	4	$1.37 \cdot 10^{-1}$
4466.24	2	4	$9.00 \cdot 10^{-1}$
4467.46	2	2	$9.00 \cdot 10^{-1}$
4563.18	4	4	$7.18 \cdot 10^{-3}$
4590.97	6	8	$8.85 \cdot 10^{-1}$
4595.96	6	6	$4.87 \cdot 10^{-2}$
4596.18	4	6	$8.34 \cdot 10^{-1}$
4638.86	2	4	$3.71 \cdot 10^{-1}$
4641.81	4	6	$5.96 \cdot 10^{-1}$
4649.13	6	8	$7.81 \cdot 10^{-1}$
4650.84	2	2	$6.86 \cdot 10^{-1}$

λ Å	Weights g_i	g_k	A 10^8 s^{-1}
4661.63	4	4	$4.10 \cdot 10^{-1}$
4673.73	4	2	$1.35 \cdot 10^{-1}$
4676.23	6	6	$2.05 \cdot 10^{-1}$
4690.89	2	4	$1.86 \cdot 10^{-1}$
4691.42	2	2	$7.43 \cdot 10^{-1}$
4696.35	6	4	$3.25 \cdot 10^{-2}$
4698.44	6	6	$6.59 \cdot 10^{-2}$
4699.01	6	8	$9.88 \cdot 10^{-1}$
4699.22	4	6	$9.36 \cdot 10^{-1}$
4701.18	4	4	$9.23 \cdot 10^{-1}$
4701.71	4	2	$3.69 \cdot 10^{-1}$
4703.16	4	6	$9.20 \cdot 10^{-1}$
4705.35	6	8	1.10
4710.01	4	6	$2.98 \cdot 10^{-1}$
4741.70	6	6	$4.71 \cdot 10^{-2}$
4751.28	6	8	$6.39 \cdot 10^{-2}$
4752.69	6	6	$1.45 \cdot 10^{-2}$
4844.92	4	6	$1.02 \cdot 10^{-2}$
4856.39	4	6	$5.58 \cdot 10^{-2}$
4856.76	4	4	$1.00 \cdot 10^{-1}$
4860.97	2	4	$4.70 \cdot 10^{-1}$
4864.88	4	2	$8.07 \cdot 10^{-2}$
4871.52	4	6	$5.60 \cdot 10^{-1}$
4872.02	4	4	$9.34 \cdot 10^{-2}$
4890.86	4	2	$4.80 \cdot 10^{-1}$
4906.83	4	4	$4.54 \cdot 10^{-1}$
4924.53	4	6	$5.43 \cdot 10^{-1}$
4941.07	2	4	$5.87 \cdot 10^{-1}$
4943.01	4	6	$7.78 \cdot 10^{-1}$
4955.71	4	4	$1.82 \cdot 10^{-1}$
5159.94	2	2	$3.29 \cdot 10^{-1}$
5175.90	4	2	$1.49 \cdot 10^{-1}$
5190.50	2	4	$1.26 \cdot 10^{-1}$
5206.65	4	4	$3.58 \cdot 10^{-1}$
5583.22	2	4	$2.17 \cdot 10^{-2}$
5611.07	2	2	$2.14 \cdot 10^{-2}$
6627.37	4	4	$1.73 \cdot 10^{-1}$
6641.03	2	2	$9.88 \cdot 10^{-2}$
6666.66	4	2	$6.78 \cdot 10^{-2}$
6677.87	2	4	$3.37 \cdot 10^{-2}$
6717.75	2	2	$1.33 \cdot 10^{-1}$
6721.39	4	2	$1.81 \cdot 10^{-1}$
6810.48	6	8	$1.64 \cdot 10^{-3}$
6844.10	4	6	$2.97 \cdot 10^{-3}$
6846.80	8	8	$3.17 \cdot 10^{-2}$
6869.48	6	6	$5.35 \cdot 10^{-2}$
6884.88	4	4	$6.12 \cdot 10^{-2}$
6895.10	10	8	$2.72 \cdot 10^{-1}$
6906.44	8	6	$2.48 \cdot 10^{-1}$
6907.87	4	2	$3.03 \cdot 10^{-1}$
6910.56	6	4	$2.43 \cdot 10^{-1}$

O III

λ Å	Weights g_i	g_k	A 10^8 s^{-1}
305.656	3	5	$1.62 \cdot 10^{2}$
305.767	5	7	$2.16 \cdot 10^{2}$
320.978	5	7	$2.17 \cdot 10^{2}$
328.448	5	5	$1.04 \cdot 10^{2}$
345.312	1	3	$1.35 \cdot 10^{2}$
507.680	3	3	$4.82 \cdot 10^{1}$
508.178	5	3	$8.04 \cdot 10^{1}$
525.794	5	3	$9.60 \cdot 10^{1}$
599.590	5	5	$5.41 \cdot 10^{1}$
835.289	5	7	5.99
1760.41	3	5	$8.38 \cdot 10^{-1}$
1764.46	5	5	2.50
1766.63	1	3	1.11
1772.28	3	1	3.29
1772.97	5	3	1.37
2390.43	3	3	1.62
2454.97	3	1	3.43
2665.68	3	5	$6.75 \cdot 10^{-1}$
2674.58	5	5	1.11
2683.66	3	1	1.85
2686.15	7	5	1.54
2687.55	3	3	1.84
2695.48	3	5	1.82
2959.69	3	5	1.83
2983.78	3	5	2.15
2996.48	3	3	$4.64 \cdot 10^{-1}$
3004.34	5	5	$4.27 \cdot 10^{-1}$
3017.62	7	7	$5.38 \cdot 10^{-1}$
3023.43	3	5	$4.79 \cdot 10^{-1}$
3024.54	1	3	$6.16 \cdot 10^{-1}$
3035.41	3	3	$4.59 \cdot 10^{-1}$
3042.07	3	1	1.94
3047.10	5	5	1.40
3059.28	5	3	$8.72 \cdot 10^{-1}$
3064.98	1	3	$2.17 \cdot 10^{-1}$
3068.13	3	1	$6.49 \cdot 10^{-1}$
3068.67	3	5	$2.27 \cdot 10^{-1}$
3074.14	5	7	$1.84 \cdot 10^{-1}$
3074.72	5	3	$3.76 \cdot 10^{-1}$
3075.13	5	5	$1.61 \cdot 10^{-1}$
3075.95	7	9	$1.07 \cdot 10^{-1}$
3083.65	7	7	$3.20 \cdot 10^{-1}$
3084.64	7	5	$2.55 \cdot 10^{-1}$
3088.04	9	9	$5.30 \cdot 10^{-1}$
3095.79	9	7	$1.35 \cdot 10^{-1}$
3115.67	3	1	1.39
3121.63	3	3	1.38
3132.79	3	5	1.37
3201.14	3	3	$4.77 \cdot 10^{-1}$
3207.61	5	5	$4.40 \cdot 10^{-1}$
3216.07	7	7	$5.58 \cdot 10^{-1}$
3260.86	5	7	1.68
3265.33	7	9	1.88
3267.20	3	5	1.58
3281.83	5	5	$2.89 \cdot 10^{-1}$
3284.45	7	7	$2.06 \cdot 10^{-1}$
3299.39	1	3	$1.64 \cdot 10^{-1}$
3312.33	3	3	$4.60 \cdot 10^{-1}$
3326.06	3	3	$2.65 \cdot 10^{-1}$
3330.30	3	5	$6.81 \cdot 10^{-1}$
3330.32	3	5	$4.76 \cdot 10^{-1}$
3332.41	5	3	$7.92 \cdot 10^{-1}$
3332.93	5	7	$5.04 \cdot 10^{-1}$
3336.67	3	3	$3.76 \cdot 10^{-1}$
3340.76	5	3	$6.57 \cdot 10^{-1}$
3344.20	5	5	$1.25 \cdot 10^{-1}$
3344.51	5	7	$3.48 \cdot 10^{-1}$
3347.98	7	5	$4.86 \cdot 10^{-1}$
3350.62	5	3	1.12
3350.92	7	7	$9.91 \cdot 10^{-1}$
3355.86	7	7	$6.89 \cdot 10^{-1}$
3362.31	7	5	$6.87 \cdot 10^{-1}$
3376.61	3	1	1.49
3376.76	3	3	1.12
3377.26	3	5	$5.20 \cdot 10^{-1}$
3382.61	5	7	$9.86 \cdot 10^{-1}$
3383.31	5	3	$3.70 \cdot 10^{-1}$
3383.81	5	5	$8.62 \cdot 10^{-1}$
3384.90	7	9	1.48
3394.22	7	7	$4.88 \cdot 10^{-1}$
3395.43	7	5	$9.75 \cdot 10^{-2}$
3406.88	1	3	$1.93 \cdot 10^{-1}$
3408.13	3	1	$5.79 \cdot 10^{-1}$
3415.26	3	3	$1.44 \cdot 10^{-1}$
3428.63	3	5	$1.42 \cdot 10^{-1}$
3430.57	5	3	$2.37 \cdot 10^{-1}$
3444.05	5	5	$4.21 \cdot 10^{-1}$
3446.68	3	5	$9.71 \cdot 10^{-1}$
3447.15	1	3	$8.09 \cdot 10^{-1}$
3447.97	5	7	1.19
3450.91	7	9	1.44
3451.30	3	3	$8.06 \cdot 10^{-1}$
3451.81	5	5	$6.09 \cdot 10^{-1}$
3454.99	9	11	1.72
3459.94	7	7	$5.14 \cdot 10^{-1}$
3466.13	9	9	$2.84 \cdot 10^{-1}$
3520.94	1	3	$1.50 \cdot 10^{-1}$
3531.22	3	1	$4.45 \cdot 10^{-1}$
3534.90	3	5	$1.11 \cdot 10^{-1}$
3555.24	5	3	$1.82 \cdot 10^{-1}$
3556.78	5	5	$3.26 \cdot 10^{-1}$
3695.38	3	5	$4.01 \cdot 10^{-1}$
3698.72	5	7	$7.62 \cdot 10^{-1}$
3703.36	7	9	1.14
3704.75	3	3	$8.53 \cdot 10^{-1}$
3707.27	3	5	$7.34 \cdot 10^{-1}$
3709.54	3	1	1.13
3712.49	5	5	$6.59 \cdot 10^{-1}$
3714.03	3	3	$4.06 \cdot 10^{-1}$
3715.09	5	7	$9.73 \cdot 10^{-1}$
3720.89	7	7	$3.74 \cdot 10^{-1}$
3721.95	5	3	$2.80 \cdot 10^{-1}$
3725.31	5	5	$2.41 \cdot 10^{-1}$
3728.51	5	7	1.29
3728.84	7	9	1.45
3729.80	3	5	1.22
3734.83	7	5	$7.40 \cdot 10^{-2}$
3742.63	5	5	$2.24 \cdot 10^{-1}$
3746.90	7	7	$1.59 \cdot 10^{-1}$
3754.70	3	5	$7.53 \cdot 10^{-1}$
3757.23	1	3	$5.56 \cdot 10^{-1}$
3759.88	5	7	$9.79 \cdot 10^{-1}$
3774.03	3	3	$3.91 \cdot 10^{-1}$
3791.28	5	5	$2.24 \cdot 10^{-1}$
3961.57	5	7	1.25
4072.64	1	3	$3.37 \cdot 10^{-1}$
4073.98	3	5	$4.54 \cdot 10^{-1}$

λ Å	Weights g_i	g_k	A 10^8 s^{-1}
4081.02	5	7	$6.02 \cdot 10^{-1}$
4089.30	3	3	$2.49 \cdot 10^{-1}$
4103.07	5	5	$1.48 \cdot 10^{-1}$
4440.09	5	3	$4.42 \cdot 10^{-1}$
4447.69	5	5	$4.40 \cdot 10^{-1}$
4461.61	5	7	$4.36 \cdot 10^{-1}$
4524.22	3	1	$3.38 \cdot 10^{-1}$
4532.78	5	3	$1.40 \cdot 10^{-1}$
4535.29	3	3	$8.40 \cdot 10^{-2}$
4555.39	5	5	$2.49 \cdot 10^{-1}$
4557.91	3	5	$8.27 \cdot 10^{-2}$
5268.30	1	3	$3.50 \cdot 10^{-1}$
5508.24	5	5	$1.06 \cdot 10^{-1}$
5592.25	3	3	$3.27 \cdot 10^{-1}$
O IV			
238.570	4	6	$3.54 \cdot 10^{2}$
554.513	4	4	$6.06 \cdot 10^{1}$
625.853	6	4	$3.19 \cdot 10^{1}$
779.820	4	4	$1.31 \cdot 10^{1}$
779.912	6	6	$1.36 \cdot 10^{1}$
923.367	4	4	$1.10 \cdot 10^{1}$
1343.51	4	6	2.57
2132.64	4	4	1.29
2493.39	2	4	1.18
2493.75	4	6	$8.48 \cdot 10^{-1}$
2507.73	4	2	2.32
2509.22	6	6	1.94
2510.58	4	2	1.19
2517.37	6	4	1.24
2805.87	2	4	$2.90 \cdot 10^{-1}$
2816.53	4	4	$5.74 \cdot 10^{-1}$
2836.27	6	4	$8.43 \cdot 10^{-1}$
2916.31	2	4	1.06
2921.46	4	6	1.27
3063.43	2	4	1.30
3071.60	2	2	1.29
3194.78	6	6	$1.71 \cdot 10^{-1}$
3209.65	8	8	$2.53 \cdot 10^{-1}$
3348.06	2	4	$8.51 \cdot 10^{-1}$
3349.11	4	6	1.02
3354.27	4	2	$7.71 \cdot 10^{-1}$
3362.55	4	4	$7.65 \cdot 10^{-1}$
3375.40	4	6	$7.56 \cdot 10^{-1}$
3378.02	4	4	$1.66 \cdot 10^{-1}$
3381.21	4	6	$7.19 \cdot 10^{-1}$
3381.30	2	4	$4.28 \cdot 10^{-1}$
3385.52	6	8	1.02
3390.19	2	2	$8.49 \cdot 10^{-1}$
3396.80	4	4	$5.40 \cdot 10^{-1}$
3409.70	6	6	$3.00 \cdot 10^{-1}$
3411.30	4	4	$1.69 \cdot 10^{-1}$
3411.69	4	6	1.02
3489.89	4	6	$7.29 \cdot 10^{-1}$
3492.21	2	4	$6.06 \cdot 10^{-1}$
3560.39	4	6	1.03
3563.33	6	8	1.10
3725.89	2	4	$5.61 \cdot 10^{-1}$
3725.94	4	6	$6.01 \cdot 10^{-1}$
3729.03	6	8	$6.86 \cdot 10^{-1}$
3736.68	4	4	$2.23 \cdot 10^{-1}$
3736.85	8	10	$7.95 \cdot 10^{-1}$
3744.89	6	6	$1.92 \cdot 10^{-1}$
3758.39	8	8	$1.11 \cdot 10^{-1}$
3974.58	4	6	$6.62 \cdot 10^{-2}$
3977.09	6	4	$9.91 \cdot 10^{-2}$
3995.08	6	6	$1.52 \cdot 10^{-1}$
4687.03	2	4	$2.79 \cdot 10^{-1}$
4772.60	2	4	$1.23 \cdot 10^{-1}$
4779.10	2	2	$2.45 \cdot 10^{-1}$
4783.42	4	6	$2.06 \cdot 10^{-1}$
4794.18	4	4	$1.56 \cdot 10^{-1}$
4798.27	6	8	$2.91 \cdot 10^{-1}$
4813.15	6	6	$8.65 \cdot 10^{-2}$
5305.51	4	4	$6.10 \cdot 10^{-2}$
5362.51	6	6	$6.12 \cdot 10^{-2}$
6931.60	2	2	$7.35 \cdot 10^{-2}$
7004.11	4	4	$8.90 \cdot 10^{-2}$
7061.30	4	2	$3.48 \cdot 10^{-2}$
O V			
172.169	1	3	$2.94 \cdot 10^{2}$
*192.85	9	15	$6.90 \cdot 10^{2}$
*215.17	9	3	$1.83 \cdot 10^{2}$
220.353	3	5	$4.292 \cdot 10^{2}$
248.460	3	1	$5.59 \cdot 10^{1}$
629.732	1	3	$2.872 \cdot 10^{1}$
758.677	3	5	5.547
759.442	1	3	7.373
760.227	3	3	5.514
760.446	5	5	$1.652 \cdot 10^{1}$
761.128	3	1	$2.197 \cdot 10^{1}$
762.004	5	3	9.125
774.518	3	1	$3.804 \cdot 10^{1}$
1371.30	3	5	3.336
2729.31	3	5	$4.52 \cdot 10^{-1}$
2731.45	1	3	$5.90 \cdot 10^{-1}$
2743.61	3	3	$4.38 \cdot 10^{-1}$
2752.23	3	1	1.82
2755.13	5	5	1.37
2769.69	5	3	$7.88 \cdot 10^{-1}$
2781.01	3	5	1.40
*2784.0	3	9	1.40
2786.99	3	3	1.39
2789.85	3	1	1.38
3058.68	3	5	1.39
3144.66	3	5	$8.86 \cdot 10^{-1}$
3219.24	3	1	$1.54 \cdot 10^{-1}$
3222.29	1	3	$1.16 \cdot 10^{-1}$
3227.54	3	3	$3.38 \cdot 10^{-2}$
3239.21	3	3	$3.28 \cdot 10^{-1}$
3248.28	5	3	$1.18 \cdot 10^{-1}$
3263.54	5	5	$1.86 \cdot 10^{-2}$
3275.64	5	3	$4.76 \cdot 10^{-1}$
3297.62	7	5	$1.30 \cdot 10^{-1}$
3690.17	3	5	$1.97 \cdot 10^{-2}$
3698.36	3	3	$1.03 \cdot 10^{-1}$
3702.72	5	7	$1.41 \cdot 10^{-2}$
3717.31	5	5	$9.63 \cdot 10^{-2}$
3725.63	5	3	$2.91 \cdot 10^{-2}$
3746.64	7	7	$1.18 \cdot 10^{-1}$
3761.58	7	5	$1.61 \cdot 10^{-2}$
4119.37	3	5	$3.66 \cdot 10^{-1}$
4120.49	3	1	$3.33 \cdot 10^{-1}$
4123.96	5	7	$4.81 \cdot 10^{-1}$
4125.49	1	3	$2.70 \cdot 10^{-1}$
4134.11	3	3	$3.34 \cdot 10^{-1}$
4153.27	3	3	$1.92 \cdot 10^{-1}$
4158.86	3	5	$3.39 \cdot 10^{-1}$
4178.46	5	5	$1.12 \cdot 10^{-1}$
4213.35	5	3	$1.19 \cdot 10^{-2}$
4522.66	5	3	$1.02 \cdot 10^{-2}$
4554.53	3	5	$2.41 \cdot 10^{-1}$
5114.06	1	3	$1.80 \cdot 10^{-1}$
5339.94	1	3	$1.85 \cdot 10^{-2}$
5349.74	3	1	$7.04 \cdot 10^{-2}$
5372.71	3	3	$1.42 \cdot 10^{-2}$
5414.59	3	5	$9.29 \cdot 10^{-3}$
5428.38	5	3	$2.68 \cdot 10^{-2}$
5471.12	5	5	$4.86 \cdot 10^{-2}$
5571.81	1	3	$8.33 \cdot 10^{-2}$
5580.12	3	5	$1.11 \cdot 10^{-1}$
5583.23	3	3	$6.20 \cdot 10^{-2}$
*5589.9	9	15	$1.49 \cdot 10^{-1}$
5597.89	5	7	$1.48 \cdot 10^{-1}$
5604.27	5	5	$3.68 \cdot 10^{-2}$
5607.41	5	3	$4.08 \cdot 10^{-3}$
6330.05	5	7	$1.21 \cdot 10^{-1}$
6460.12	3	5	$9.37 \cdot 10^{-2}$
6466.14	5	7	$1.01 \cdot 10^{-1}$
6500.24	7	9	$1.11 \cdot 10^{-1}$
6543.77	5	5	$1.64 \cdot 10^{-2}$
6601.28	7	7	$1.14 \cdot 10^{-2}$
6764.72	1	3	$4.37 \cdot 10^{-2}$
6789.62	3	5	$5.79 \cdot 10^{-2}$
6817.40	3	3	$3.00 \cdot 10^{-2}$
6828.95	5	7	$7.35 \cdot 10^{-2}$
6878.76	5	5	$1.65 \cdot 10^{-2}$
O VI			
*150.10	2	6	$2.62 \cdot 10^{2}$
*173.03	6	10	$8.78 \cdot 10^{2}$
1031.91	2	4	4.16
1037.61	2	2	4.09
3811.35	2	4	$5.14 \cdot 10^{-1}$
3834.24	2	2	$5.05 \cdot 10^{-1}$
O VII			
18.6270	1	3	$9.365 \cdot 10^{3}$
21.6020	1	3	$3.309 \cdot 10^{4}$
*120.33	3	9	$5.334 \cdot 10^{2}$
128.411	1	3	$8.982 \cdot 10^{2}$
*128.46	9	15	$1.615 \cdot 10^{3}$
135.820	3	5	$1.523 \cdot 10^{3}$
*1630.3	3	9	$7.935 \cdot 10^{-1}$
2448.98	1	3	$2.514 \cdot 10^{-1}$
*5933.1	3	9	$1.002 \cdot 10^{-1}$
8241.76	1	3	$3.864 \cdot 10^{-2}$

λ Å	Weights g_i	g_k	A 10^8 s^{-1}
Phosphorus			
P I			
1671.7	4	2	$3.9\cdot10^{-1}$
1674.6	4	4	$4.0\cdot10^{-1}$
1679.7	4	6	$3.9\cdot10^{-1}$
1775.0	4	6	2.17
1782.9	4	4	2.14
1787.7	4	2	2.13
2135.5	4	4	$2.11\cdot10^{-1}$
2136.2	6	4	2.83
2149.1	4	2	3.18
2152.9	2	4	$4.85\cdot10^{-1}$
2154.1	4	4	$1.73\cdot10^{-1}$
2154.1	4	6	$5.8\cdot10^{-1}$
2534.0	2	4	$2.00\cdot10^{-1}$
2535.6	4	4	$9.5\cdot10^{-1}$
2553.3	2	2	$7.1\cdot10^{-1}$
2554.9	4	2	$3.00\cdot10^{-1}$
P II			
1301.9	1	3	$5.0\cdot10^{-1}$
1304.5	3	1	1.5
1304.7	3	3	$3.7\cdot10^{-1}$
1305.5	3	5	$3.8\cdot10^{-1}$
1309.9	5	3	$6.2\cdot10^{-1}$
1310.7	5	5	1.1
4475.3	5	7	1.3
4499.2	5	7	1.4
4530.8	3	5	1.0
4554.8	3	5	$9.6\cdot10^{-1}$
4588.0	5	7	1.7
4589.9	3	5	1.6
4602.1	7	9	1.9
4943.5	7	5	$6.3\cdot10^{-1}$
5253.5	3	5	1.0
5425.9	5	5	$6.9\cdot10^{-1}$
6024.2	3	5	$5.1\cdot10^{-1}$
6043.1	5	7	$6.8\cdot10^{-1}$
P III			
1334.8	2	4	$5.5\cdot10^{-1}$
1344.3	4	6	$6.4\cdot10^{-1}$
1344.8	4	4	$1.1\cdot10^{-1}$
4057.4	4	4	$1.0\cdot10^{-1}$
4059.3	6	4	$9.0\cdot10^{-1}$
4080.1	4	2	$9.9\cdot10^{-1}$
Potassium			
K I			
4044.1	2	4	$1.24\cdot10^{-2}$
4047.2	2	2	$1.24\cdot10^{-2}$
5084.2	2	2	$3.50\cdot10^{-3}$
5099.2	4	2	$7.0\cdot10^{-3}$
5323.3	2	2	$6.3\cdot10^{-3}$
5339.7	4	2	$1.26\cdot10^{-2}$
5343.0	2	4	$4.0\cdot10^{-3}$
5359.6	4	6	$4.6\cdot10^{-3}$
5782.4	2	2	$1.23\cdot10^{-2}$
5801.8	4	2	$2.46\cdot10^{-2}$
5812.2	2	4	$2.8\cdot10^{-3}$
5831.9	4	6	$3.2\cdot10^{-3}$
6911.1	2	2	$2.72\cdot10^{-2}$
6938.8	4	2	$5.4\cdot10^{-2}$
7664.9	2	4	$3.87\cdot10^{-1}$
7699.0	2	2	$3.82\cdot10^{-1}$
K II			
607.93	1	3	$1.3\cdot10^{-2}$
K III			
2550.0	6	4	2.0
2635.1	4	4	1.2
2992.4	6	8	2.5
3052.1	4	6	1.7
3202.0	4	4	1.8
3289.1	4	6	2.0
3322.4	6	6	1.3
3421.8	2	4	1.5
K XVI			
206.27	1	3	$9.4\cdot10^{1}$
K XVII			
22.020	2	4	$4.7\cdot10^{4}$
22.163	4	6	$5.6\cdot10^{4}$
22.18	4	4	$9.3\cdot10^{3}$
22.60	2	2	$2.5\cdot10^{3}$
22.76	4	2	$4.7\cdot10^{3}$
Praseodymium			
Pr II			
3997.0	15	15	$1.87\cdot10^{-1}$
4062.8	13	15	1.00
4100.7	17	19	$8.4\cdot10^{-1}$
4143.1	15	17	$5.8\cdot10^{-1}$
4179.4	13	15	$5.2\cdot10^{-1}$
4222.9	11	13	$3.91\cdot10^{-1}$
4241.0	17	15	$2.30\cdot10^{-1}$
4359.8	15	15	$1.1\cdot10^{-1}$
4405.8	17	17	$9.0\cdot10^{-2}$
4429.3	15	15	$2.28\cdot10^{-1}$
4449.8	13	13	$1.24\cdot10^{-1}$
4468.7	11	13	$1.54\cdot10^{-1}$
4510.2	13	15	$1.16\cdot10^{-1}$
4534.2	15	17	$4.9\cdot10^{-2}$
4734.2	15	13	$2.5\cdot10^{-2}$
4879.1	15	15	$1.8\cdot10^{-2}$
4886.0	15	15	$1.3\cdot10^{-2}$
4912.6	17	15	$5.7\cdot10^{-2}$
5034.4	19	19	$1.1\cdot10^{-1}$
5110.8	21	19	$2.78\cdot10^{-1}$
5135.1	17	17	$1.25\cdot10^{-1}$
5173.9	19	17	$3.18\cdot10^{-1}$
5219.1	15	15	$9.5\cdot10^{-2}$
5220.1	17	15	$2.35\cdot10^{-1}$
5251.7	15	13	$1.1\cdot10^{-2}$
5259.7	15	13	$2.24\cdot10^{-1}$
5292.6	13	13	$9.3\cdot10^{-2}$
5810.6	17	19	$2.3\cdot10^{-2}$
5879.3	15	15	$7.6\cdot10^{-2}$
6200.8	15	17	$1.8\cdot10^{-2}$
6278.7	13	15	$2.6\cdot10^{-2}$
6398.0	11	13	$1.9\cdot10^{-2}$
Rhodium			
Rh I			
3121.76	6	6	$1.1\cdot10^{-1}$
3189.05	6	6	$3.03\cdot10^{-1}$
3263.14	6	6	$1.3\cdot10^{-1}$
3271.61	6	4	$2.0\cdot10^{-1}$
3280.55	8	8	$2.36\cdot10^{-1}$
3283.57	6	8	$4.4\cdot10^{-1}$
3323.09	8	10	$6.3\cdot10^{-1}$
3396.82	10	10	$6.5\cdot10^{-1}$
3399.70	6	8	$1.2\cdot10^{-1}$
3462.04	6	6	$6.2\cdot10^{-1}$
3470.66	4	4	$8.5\cdot10^{-1}$
3478.91	6	6	$3.32\cdot10^{-1}$
3498.73	4	6	$2.12\cdot10^{-1}$
3502.52	10	10	$4.3\cdot10^{-1}$
3507.32	6	8	$3.4\cdot10^{-1}$
3528.02	8	8	$8.5\cdot10^{-1}$
3543.95	4	4	$4.65\cdot10^{-1}$
3549.54	6	6	$2.22\cdot10^{-1}$
3570.18	4	6	$1.82\cdot10^{-1}$
3583.10	8	10	$2.6\cdot10^{-1}$
3596.19	6	4	$5.5\cdot10^{-1}$
3597.15	6	8	$5.9\cdot10^{-1}$
3612.47	4	2	$8.90\cdot10^{-1}$
3654.87	8	8	$6.0\cdot10^{-2}$
3657.99	8	6	$8.8\cdot10^{-1}$
3666.22	6	8	$8.4\cdot10^{-2}$
3690.70	6	4	$3.23\cdot10^{-1}$
3692.36	10	8	$9.1\cdot10^{-1}$
3700.91	8	10	$3.9\cdot10^{-1}$
3788.47	4	6	$1.4\cdot10^{-1}$
3793.22	8	6	$4.2\cdot10^{-1}$
3799.31	8	8	$5.5\cdot10^{-1}$
3806.76	6	6	$6.2\cdot10^{-2}$
3818.19	6	4	$5.8\cdot10^{-1}$
3822.26	6	6	$8.5\cdot10^{-1}$
3828.48	6	6	$6.2\cdot10^{-1}$
3833.89	6	4	$5.8\cdot10^{-1}$
3856.52	8	10	$5.9\cdot10^{-1}$
3934.23	8	8	$1.58\cdot10^{-1}$
3942.72	4	2	$7.15\cdot10^{-1}$
3958.86	6	8	$5.5\cdot10^{-1}$
3984.40	4	4	$1.1\cdot10^{-1}$
4082.78	6	4	$1.4\cdot10^{-1}$
4121.68	6	6	$9.8\cdot10^{-2}$
4128.87	6	8	$1.73\cdot10^{-1}$
4135.27	8	8	$1.0\cdot10^{-1}$
4196.50	6	8	$3.9\cdot10^{-2}$
4211.14	8	10	$1.62\cdot10^{-1}$
4288.71	6	8	$6.1\cdot10^{-2}$
4374.80	8	10	$1.64\cdot10^{-1}$

λ Å	Weights g_i	g_k	A 10^8 s^{-1}
5983.60	10	10	$2.1\cdot10^{-2}$

Rubidium

Rb I

λ Å	Weights g_i	g_k	A 10^8 s^{-1}
3022.5	2	4	$4.13\cdot10^{-5}$
3032.0	2	4	$4.93\cdot10^{-5}$
3044.2	2	4	$8.2\cdot10^{-5}$
3060.2	2	4	$1.05\cdot10^{-4}$
3082.0	2	4	$1.49\cdot10^{-4}$
3112.6	2	4	$2.5\cdot10^{-4}$
3113.1	2	2	$1.3\cdot10^{-4}$
3157.5	2	4	$3.38\cdot10^{-4}$
3158.3	2	2	$2.0\cdot10^{-4}$
3228.0	2	4	$6.4\cdot10^{-4}$
3229.2	2	2	$3.8\cdot10^{-4}$
3348.7	2	4	$1.37\cdot10^{-3}$
3350.8	2	2	$8.9\cdot10^{-4}$
3587.1	2	4	$3.97\cdot10^{-3}$
3591.6	2	2	$2.9\cdot10^{-3}$
4201.8	2	4	$1.8\cdot10^{-2}$
4215.5	2	2	$1.5\cdot10^{-2}$
7800.3	2	4	$3.70\cdot10^{-1}$
7947.6	2	2	$3.40\cdot10^{-1}$

Scandium

Sc I

λ Å	Weights g_i	g_k	A 10^8 s^{-1}
3015.37	4	6	$7.8\cdot10^{-1}$
3019.35	6	8	$8.7\cdot10^{-1}$
3269.90	4	2	3.13
3273.63	6	4	2.81
3907.48	4	6	1.66
3911.81	6	8	1.79
4020.39	4	4	1.63
4023.68	6	6	1.65
4031.38	6	6	$2.9\cdot10^{-1}$
4043.80	8	8	$3.11\cdot10^{-1}$
4067.00	6	8	$1.91\cdot10^{-1}$
4074.96	4	6	$3.7\cdot10^{-1}$
4078.56	2	4	$4.3\cdot10^{-1}$
4086.66	6	8	$3.7\cdot10^{-1}$
4132.98	4	6	1.19
4140.27	6	8	1.17
4161.85	8	8	$1.77\cdot10^{-1}$
4233.59	6	6	$4.0\cdot10^{-1}$
4238.05	8	8	$7.1\cdot10^{-1}$
4706.94	4	6	$2.81\cdot10^{-1}$
4709.31	6	8	$4.0\cdot10^{-1}$
4728.77	8	8	$1.16\cdot10^{-1}$
4729.24	6	6	$1.93\cdot10^{-1}$
4734.11	4	2	1.10
4737.65	6	4	$8.8\cdot10^{-1}$
4741.02	8	6	$9.1\cdot10^{-1}$
4743.82	10	8	$9.8\cdot10^{-1}$
4973.67	4	2	$8.4\cdot10^{-1}$
4980.36	6	4	$5.6\cdot10^{-1}$
4983.43	4	4	$2.58\cdot10^{-1}$
4991.91	6	6	$3.8\cdot10^{-1}$
5018.41	6	4	$2.09\cdot10^{-1}$
5021.52	4	4	$2.30\cdot10^{-1}$
5070.17	6	8	$1.16\cdot10^{-1}$
5081.56	10	10	$7.6\cdot10^{-1}$
5083.72	8	8	$6.2\cdot10^{-1}$
5085.55	6	6	$5.7\cdot10^{-1}$
5086.94	4	4	$6.6\cdot10^{-1}$
5099.27	4	6	$1.50\cdot10^{-1}$
5339.43	6	6	$1.06\cdot10^{-1}$
5341.07	4	2	$3.8\cdot10^{-1}$
5349.34	6	4	$5.9\cdot10^{-1}$
5355.79	6	4	$3.0\cdot10^{-1}$
5356.10	8	6	$5.7\cdot10^{-1}$
5375.37	8	6	$3.4\cdot10^{-1}$
5392.06	10	8	$4.2\cdot10^{-1}$
5446.20	8	8	$2.8\cdot10^{-1}$
5451.37	6	6	$1.50\cdot10^{-1}$
5472.19	8	6	$9.7\cdot10^{-2}$
5482.01	8	8	$5.2\cdot10^{-1}$
5484.63	6	6	$5.2\cdot10^{-1}$
5514.23	6	8	$4.1\cdot10^{-1}$
5520.52	8	10	$4.3\cdot10^{-1}$
5671.83	10	12	$5.4\cdot10^{-1}$
5686.86	8	10	$4.9\cdot10^{-1}$
5700.19	6	8	$4.6\cdot10^{-1}$
5711.79	4	6	$4.5\cdot10^{-1}$
5717.31	8	8	$7.5\cdot10^{-2}$
6262.22	4	6	$8.4\cdot10^{-2}$
7741.16	10	10	$3.8\cdot10^{-2}$
7800.42	8	8	$5.1\cdot10^{-2}$

Sc II

λ Å	Weights g_i	g_k	A 10^8 s^{-1}
1880.6	5	3	5.0
2064.3	7	5	2.2
2068.0	5	3	2.0
2273.1	1	3	7.7
2545.20	5	5	$4.0\cdot10^{-1}$
2552.35	7	5	2.21
2555.79	3	3	$6.9\cdot10^{-1}$
2560.23	5	3	2.01
2563.19	3	1	2.70
2611.19	5	5	2.2
2667.70	3	5	1.5
2746.36	3	1	3.9
2782.31	5	5	1.3
2789.15	7	7	1.3
2801.31	9	9	1.3
2819.49	3	5	2.3
2822.12	5	7	2.5
2826.64	7	9	2.8
2870.85	5	3	1.1
2912.98	5	3	1.1
2979.68	3	5	1.2
2988.92	5	7	2.9
3039.92	7	9	3.5
3045.73	5	7	3.68
3052.92	7	9	3.92
3060.54	7	7	$3.0\cdot10^{-1}$
3065.12	9	11	4.00
3075.36	9	9	$2.5\cdot10^{-1}$
3128.27	3	3	1.9
3133.07	5	5	1.8
3139.72	7	7	2.1
3190.98	3	3	1.1
3199.33	5	3	1.9
3312.72	5	7	1.2
3320.40	5	3	1.2
3343.23	9	7	1.1
3353.72	5	7	1.51
3359.67	5	5	$2.16\cdot10^{-1}$
3361.26	3	3	$3.4\cdot10^{-1}$
3361.93	3	1	1.17
3368.94	5	3	$8.3\cdot10^{-1}$
3372.15	7	5	$9.9\cdot10^{-1}$
3379.16	3	3	2.5
3535.71	5	3	$6.1\cdot10^{-1}$
3558.53	5	7	$3.0\cdot10^{-1}$
3567.70	3	5	$3.5\cdot10^{-1}$
3572.53	7	7	1.38
3576.34	5	5	1.06
3580.93	3	3	1.23
3589.63	5	3	$4.6\cdot10^{-1}$
3590.47	7	5	$2.9\cdot10^{-1}$
3613.83	7	9	1.48
3630.74	5	7	1.20
3642.78	3	5	1.13
3645.31	7	7	$2.74\cdot10^{-1}$
3651.80	5	5	$3.0\cdot10^{-1}$
3859.59	7	5	1.1
4246.82	5	5	1.29
4314.08	9	7	$4.1\cdot10^{-1}$
4320.75	7	5	$4.0\cdot10^{-1}$
4325.00	5	3	$4.3\cdot10^{-1}$
4374.46	9	9	$1.48\cdot10^{-1}$
4400.39	7	7	$1.43\cdot10^{-1}$
4415.54	5	5	$1.47\cdot10^{-1}$
4670.41	5	7	$1.16\cdot10^{-1}$
5031.01	5	3	$3.5\cdot10^{-1}$
5239.81	1	3	$1.39\cdot10^{-1}$
5526.79	9	7	$3.3\cdot10^{-1}$
5657.91	5	5	$1.04\cdot10^{-1}$
5669.06	3	1	$1.31\cdot10^{-1}$

Silicon

Si I

λ Å	Weights g_i	g_k	A 10^8 s^{-1}
2207.98	1	3	$2.62\cdot10^{-1}$
2210.89	3	5	$3.46\cdot10^{-1}$
2211.74	3	3	$1.81\cdot10^{-1}$
2216.67	5	7	$4.54\cdot10^{-1}$
2218.06	5	5	$1.09\cdot10^{-1}$
2218.91	5	3	$1.05\cdot10^{-2}$
2438.77	1	3	$7.91\cdot10^{-3}$
2443.36	3	3	$1.32\cdot10^{-2}$
2452.12	5	3	$1.73\cdot10^{-3}$
2506.90	3	5	$5.47\cdot10^{-1}$
2514.32	1	3	$7.39\cdot10^{-1}$
2516.113	5	5	1.68
2519.202	3	3	$5.49\cdot10^{-1}$
2524.108	3	1	2.22
2528.509	5	3	$9.04\cdot10^{-1}$
2881.579	5	3	2.17

λ Å	Weights g_i	g_k	A 10^8 s^{-1}
2970.355	5	5	$6.00 \cdot 10^{-4}$
2987.645	5	3	$1.34 \cdot 10^{-2}$
3006.739	3	5	$1.1 \cdot 10^{-5}$
3020.004	5	5	$3.3 \cdot 10^{-5}$
3905.523	1	3	$1.33 \cdot 10^{-1}$
4102.936	1	3	$6.09 \cdot 10^{-4}$
4782.991	5	3	$1.7 \cdot 10^{-2}$
4792.212	3	1	$2.2 \cdot 10^{-2}$
4792.324	5	5	$1.7 \cdot 10^{-2}$
4947.607	3	1	$4.2 \cdot 10^{-2}$
5006.061	3	5	$2.8 \cdot 10^{-2}$
5622.221	3	3	$1.6 \cdot 10^{-2}$
5645.611	3	5	$9.7 \cdot 10^{-3}$
5665.554	1	3	$6.31 \cdot 10^{-3}$
5684.484	5	3	$2.6 \cdot 10^{-2}$
5690.425	3	3	$9.26 \cdot 10^{-3}$
5701.105	3	1	$1.83 \cdot 10^{-2}$
5708.397	5	5	$1.4 \cdot 10^{-2}$
5754.220	5	3	$1.23 \cdot 10^{-2}$
5772.145	3	1	$3.6 \cdot 10^{-2}$
5780.384	1	3	$9.8 \cdot 10^{-3}$
5793.071	3	5	$1.3 \cdot 10^{-2}$
5797.859	5	7	$2.53 \cdot 10^{-3}$
5948.545	3	5	$2.2 \cdot 10^{-2}$
6331.954	3	3	$1.0 \cdot 10^{-4}$
6555.102	7	9	6.9·10^[illegible]
6721.853	3	5	$3.4 \cdot 10^{-2}$
6976.523	3	5	$3.26 \cdot 10^{-2}$
7003.567	5	7	$3.42 \cdot 10^{-2}$
7005.883	7	9	$3.83 \cdot 10^{-2}$
7680.267	3	5	$4.62 \cdot 10^{-2}$
7918.386	3	5	$5.22 \cdot 10^{-2}$
7932.349	5	7	$5.13 \cdot 10^{-2}$
7944.001	7	9	$5.75 \cdot 10^{-2}$
7970.306	5	5	$7.11 \cdot 10^{-3}$
8035.619	7	7	$8.11 \cdot 10^{-3}$
8093.241	3	3	$1.5 \cdot 10^{-2}$
9413.506	3	1	$2.26 \cdot 10^{-1}$

Si II

λ Å	Weights g_i	g_k	A 10^8 s^{-1}
843.72	2	4	$7.05 \cdot 10^{-1}$
845.77	4	4	$1.42 \cdot 10^{-1}$
889.72	2	4	1.43
892.00	4	4	$2.88 \cdot 10^{-1}$
899.41	2	2	$4.63 \cdot 10^{-1}$
901.74	4	2	$9.23 \cdot 10^{-1}$
989.87	2	4	6.81
992.68	4	6	7.11
1020.70	2	2	$8.91 \cdot 10^{-1}$
1023.69	4	2	1.77
1190.42	2	4	6.53
1193.28	2	2	$2.69 \cdot 10^{1}$
1194.50	4	4	$3.45 \cdot 10^{1}$
1197.39	4	2	$1.40 \cdot 10^{1}$
1224.25	4	4	6.72
1224.97	4	6	4.39
1226.81	2	2	$2.77 \cdot 10^{1}$
1226.89	6	4	6.55
1226.99	2	4	$1.39 \cdot 10^{1}$
1227.60	6	6	$1.46 \cdot 10^{1}$
1228.44	4	2	5.53
1228.62	4	4	$1.77 \cdot 10^{1}$
1228.75	4	6	$2.32 \cdot 10^{1}$
1229.39	6	8	$2.25 \cdot 10^{1}$
1246.74	2	4	4.03
1248.43	4	4	8.30
1251.16	6	4	$1.30 \cdot 10^{1}$
1260.42	2	4	$2.57 \cdot 10^{1}$
1264.73	4	6	$3.04 \cdot 10^{1}$
1265.02	4	4	4.73
1304.37	2	2	3.64
1309.27	4	2	6.23
1348.54	2	4	3.36
1350.06	6	6	5.34
1350.52	4	4	1.61
1350.66	2	2	2.02
1352.64	4	2	6.12
1353.72	6	4	3.22
1410.22	6	4	3.47
1509.10	6	4	2.85
1512.07	4	2	3.15
1526.72	2	2	3.81
1533.45	4	2	7.52
1808.00	2	4	$2.54 \cdot 10^{-2}$
1816.92	4	6	$2.65 \cdot 10^{-2}$
1817.45	4	4	3.20·10^[illegible]
2072.02	4	6	$9.6 \cdot 10^{-1}$
2072.70	6	8	1.0
2334.40	2	2	$5.51 \cdot 10^{-5}$
2334.61	4	6	$2.44 \cdot 10^{-5}$
2344.20	4	4	$1.31 \cdot 10^{-5}$
2350.17	4	2	$4.70 \cdot 10^{-5}$
2682.21	4	4	$3.49 \cdot 10^{-2}$
2887.51	6	4	$6.39 \cdot 10^{-2}$
2904.28	4	6	$3.58 \cdot 10^{-1}$
2905.69	6	8	$3.83 \cdot 10^{-1}$
3203.87	2	4	$4.45 \cdot 10^{-1}$
3210.03	4	6	$5.29 \cdot 10^{-1}$
3333.14	2	2	$1.00 \cdot 10^{-1}$
3339.82	4	2	$2.00 \cdot 10^{-1}$
3853.66	4	4	$5.11 \cdot 10^{-2}$
3856.02	6	4	$4.40 \cdot 10^{-1}$
3862.60	4	2	$3.91 \cdot 10^{-1}$
4075.45	6	4	$4.00 \cdot 10^{-2}$
4076.78	4	2	$4.00 \cdot 10^{-2}$
4128.07	4	6	1.49
4130.89	6	8	1.74
4621.42	4	6	$1.28 \cdot 10^{-1}$
4621.72	6	8	$1.37 \cdot 10^{-1}$
5041.03	2	4	$7.00 \cdot 10^{-1}$
5055.98	4	6	1.45
5466.43	4	6	$2.16 \cdot 10^{-1}$
5466.87	6	6	$1.54 \cdot 10^{-2}$
5632.97	8	8	$4.00 \cdot 10^{-2}$
5660.66	6	6	$7.00 \cdot 10^{-2}$
5669.56	10	8	$5.00 \cdot 10^{-1}$
5681.44	4	4	$1.00 \cdot 10^{-1}$
5688.81	8	6	$4.60 \cdot 10^{-1}$
5701.37	6	4	$4.50 \cdot 10^{-1}$
5706.37	4	2	$6.10 \cdot 10^{-1}$
5957.56	2	2	$5.60 \cdot 10^{-1}$
5978.93	4	2	1.13
6347.10	2	4	$5.84 \cdot 10^{-1}$
6371.36	2	2	$6.80 \cdot 10^{-1}$
6660.52	4	6	$3.64 \cdot 10^{-1}$
6665.00	2	4	$2.16 \cdot 10^{-1}$
6671.88	6	8	$4.80 \cdot 10^{-1}$
6699.38	2	2	$4.20 \cdot 10^{-1}$
6750.28	6	6	$1.49 \cdot 10^{-1}$
6818.45	2	4	$1.08 \cdot 10^{-1}$
6829.82	4	4	$2.16 \cdot 10^{-2}$
7848.80	4	6	$3.73 \cdot 10^{-1}$
7849.72	6	8	$3.99 \cdot 10^{-1}$
9412.72	8	8	$4.65 \cdot 10^{-2}$

Si III

λ Å	Weights g_i	g_k	A 10^8 s^{-1}
566.61	1	3	1.21
652.22	3	3	2.67
653.33	5	3	4.40
673.48	5	5	$8.44 \cdot 10^{-2}$
673.48	5	3	$9.33 \cdot 10^{-3}$
800.07	3	1	4.13
823.41	3	5	5.85
883.40	5	7	$5.79 \cdot 10^{2}$
939.09	5	3	$1.11 \cdot 10^{1}$
967.95	5	7	4.08
993.52	1	3	2.68
994.79	3	3	8.05
997.39	5	3	$1.35 \cdot 10^{1}$
1005.37	5	5	$7.06 \cdot 10^{-1}$
1031.16	3	5	2.72
1033.92	5	5	8.11
1037.05	5	3	4.46
1083.22	5	3	3.65
1108.37	1	3	$1.54 \cdot 10^{1}$
1109.97	3	5	$2.07 \cdot 10^{1}$
1113.23	5	7	$2.74 \cdot 10^{1}$
1140.55	1	3	$1.96 \cdot 10^{1}$
1141.58	3	5	$2.67 \cdot 10^{1}$
1142.28	3	3	$1.75 \cdot 10^{1}$
1144.31	5	7	$3.86 \cdot 10^{1}$
1144.96	5	5	$1.19 \cdot 10^{1}$
1145.18	5	5	1.38
1155.00	1	3	9.34
1155.96	3	1	$2.22 \cdot 10^{1}$
1158.10	3	5	7.88
1160.26	5	3	8.65
1161.58	5	5	$1.42 \cdot 10^{1}$
1174.37	3	3	1.00
1174.43	3	5	1.00
1206.51	1	3	$2.55 \cdot 10^{1}$
1206.53	3	5	$4.17 \cdot 10^{1}$
1207.52	5	5	$2.43 \cdot 10^{1}$
1210.46	5	7	$1.60 \cdot 10^{1}$
1235.43	1	3	$2.77 \cdot 10^{1}$
1280.35	5	7	$1.04 \cdot 10^{1}$
1294.54	3	5	5.35
1296.73	1	3	7.10
1298.89	3	3	5.29
1298.96	5	5	$1.59 \cdot 10^{1}$

λ	Weights		A
Å	g_i	g_k	10^8 s^{-1}
1301.15	3	1	$2.11\cdot10^{1}$
1303.32	5	3	8.71
1312.59	3	1	6.66
1341.47	7	7	8.96
1342.39	5	5	7.54
1343.39	3	3	8.16
1361.60	3	5	$1.06\cdot10^{1}$
1362.37	3	1	$1.06\cdot10^{1}$
1363.47	5	3	8.53
1365.26	7	5	9.40
1367.05	3	3	$1.05\cdot10^{1}$
1369.44	3	1	$1.04\cdot10^{1}$
1373.03	5	5	8.00
1387.99	3	3	$2.09\cdot10^{-1}$
1417.24	3	1	$2.17\cdot10^{1}$
1433.69	5	7	8.67
1435.77	5	7	4.22
1436.17	1	3	7.79
1457.25	5	3	$1.07\cdot10^{1}$
1500.24	7	9	$2.10\cdot10^{1}$
1501.19	5	7	$1.86\cdot10^{1}$
1501.87	3	5	$1.76\cdot10^{1}$
1506.06	3	3	$1.24\cdot10^{1}$
1673.32	5	7	6.70
1842.55	5	3	2.99
2176.89	9	7	1.80
2295.48	5	7	$6.69\cdot10^{-1}$
2300.93	7	7	$5.80\cdot10^{-2}$
2308.19	9	11	$7.16\cdot10^{-1}$
2449.48	5	5	$1.81\cdot10^{-1}$
2483.20	7	9	$1.33\cdot10^{-1}$
2541.82	3	5	$3.22\cdot10^{-1}$
2546.09	5	5	$6.1\cdot10^{-1}$
2559.21	5	7	1.63
2640.79	5	3	1.83
2655.51	7	9	1.35
2817.11	9	7	$2.22\cdot10^{-1}$
2831.49	7	5	$2.12\cdot10^{-1}$
2839.62	5	3	$2.37\cdot10^{-1}$
2959.15	5	3	$1.48\cdot10^{-1}$
2980.52	5	3	$2.37\cdot10^{-3}$
3013.09	5	5	$5.67\cdot10^{-2}$
3034.73	5	7	$4.50\cdot10^{-1}$
3037.29	7	7	$3.93\cdot10^{-2}$
3040.93	9	11	$4.87\cdot10^{-1}$
3043.93	7	5	$2.51\cdot10^{-1}$
3045.08	5	3	$2.24\cdot10^{-1}$
3068.24	5	3	$5.15\cdot10^{-1}$
3086.24	7	5	1.46
3086.46	5	5	$2.61\cdot10^{-1}$
3093.42	5	3	1.30
3093.65	3	3	$4.33\cdot10^{-1}$
3096.83	3	1	1.73
3126.27	5	5	$4.67\cdot10^{-1}$
3147.37	5	3	$2.53\cdot10^{-1}$
3161.61	5	3	$4.91\cdot10^{-1}$
3185.13	3	1	4.04
3186.02	5	7	1.13
3196.50	7	9	1.14
3210.55	9	11	1.20

λ	Weights		A
Å	g_i	g_k	10^8 s^{-1}
3216.25	5	7	$1.90\cdot10^{-1}$
3230.50	1	3	$4.75\cdot10^{-1}$
3233.95	3	3	1.42
3241.62	5	3	2.35
3253.40	5	5	$1.82\cdot10^{-1}$
3253.74	7	5	$1.75\cdot10^{-1}$
3254.80	9	7	$1.81\cdot10^{-1}$
3258.66	7	5	1.01
3270.46	3	3	$2.99\cdot10^{-1}$
3276.26	5	3	$8.91\cdot10^{-1}$
3279.26	3	1	1.18
3486.91	5	7	1.54
3525.94	3	3	$2.48\cdot10^{-1}$
3569.67	3	1	$5.78\cdot10^{-1}$
3590.47	3	5	2.53
3622.54	7	7	$1.25\cdot10^{-1}$
3639.45	7	9	$2.24\cdot10^{-1}$
3645.12	5	5	$9.58\cdot10^{-2}$
3681.40	5	3	$2.51\cdot10^{-1}$
3682.15	3	3	$1.50\cdot10^{-1}$
3791.41	1	3	1.76
3796.11	3	5	2.36
3806.54	5	7	3.14
3842.46	3	5	$1.73\cdot10^{-1}$
3924.47	7	9	3.47
3947.49	5	7	$5.59\cdot10^{-2}$
3963.84	7	5	$3.63\cdot10^{-2}$
3981.24	3	3	$2.09\cdot10^{-3}$
4102.42	5	3	$2.46\cdot10^{-1}$
4115.50	7	9	$4.07\cdot10^{-1}$
4338.50	1	3	$1.47\cdot10^{-1}$
4341.40	3	1	$5.25\cdot10^{-3}$
4377.63	7	5	$4.14\cdot10^{-2}$
4405.90	5	5	$4.84\cdot10^{-2}$
4406.72	7	9	$3.10\cdot10^{-1}$
4494.05	3	3	$4.19\cdot10^{-1}$
4552.62	3	5	1.26
4554.00	5	3	$6.75\cdot10^{-1}$
4567.82	3	3	1.25
4574.76	3	1	1.24
4619.66	3	5	$2.89\cdot10^{-1}$
4638.28	1	3	$3.79\cdot10^{-1}$
4665.87	3	3	$2.79\cdot10^{-1}$
4683.02	5	5	$8.30\cdot10^{-1}$
4683.80	3	1	1.10
4716.65	5	7	1.32
4730.52	5	3	$4.47\cdot10^{-1}$
4800.43	1	3	$2.76\cdot10^{-1}$
4813.33	5	7	2.10
4819.72	7	9	2.12
4828.97	9	11	2.25
5113.76	5	7	$3.82\cdot10^{-1}$
5114.12	9	11	$4.17\cdot10^{-1}$
5197.26	3	5	$2.47\cdot10^{-1}$
5451.46	3	5	$4.60\cdot10^{-1}$
5473.05	5	7	$6.07\cdot10^{-1}$
5704.60	7	5	$1.86\cdot10^{-1}$
5716.29	9	7	$1.91\cdot10^{-1}$
5739.73	1	3	$5.41\cdot10^{-1}$
5898.79	7	9	$4.86\cdot10^{-1}$

λ	Weights		A
Å	g_i	g_k	10^8 s^{-1}
6314.46	3	1	1.18
6524.36	7	9	$3.82\cdot10^{-1}$
6831.56	5	3	$5.83\cdot10^{-1}$
7461.89	3	1	$4.99\cdot10^{-1}$
7462.62	5	3	$3.75\cdot10^{-1}$
7466.32	7	5	$4.19\cdot10^{-1}$
7612.36	3	5	$9.94\cdot10^{-1}$
8190.43	5	7	$8.09\cdot10^{-1}$
8191.16	7	7	$7.05\cdot10^{-2}$
8191.68	9	11	$8.78\cdot10^{-1}$
8262.57	5	7	$7.18\cdot10^{-1}$
8265.64	5	5	$1.80\cdot10^{-1}$
8269.32	3	5	$5.36\cdot10^{-1}$
8271.38	3	3	$2.98\cdot10^{-1}$
8271.94	1	3	$3.96\cdot10^{-1}$

Si IV

λ	Weights		A
Å	g_i	g_k	10^8 s^{-1}
457.82	2	4	3.90
458.16	2	2	4.05
515.12	2	2	4.66
516.35	4	2	9.32
645.76	6	6	$4.61\cdot10^{-1}$
645.76	6	8	6.92
645.76	4	6	6.46
749.94	6	6	$9.33\cdot10^{-1}$
749.94	6	8	$1.40\cdot10^{1}$
815.05	2	2	$1.18\cdot10^{1}$
818.13	4	2	$2.37\cdot10^{1}$
1066.63	6	6	2.54
1122.49	2	4	$2.14\cdot10^{1}$
1128.34	4	6	$2.53\cdot10^{1}$
1393.76	2	4	8.80
1402.77	2	2	8.63
1722.53	6	4	4.92
1727.38	4	2	5.47
2120.18	2	2	3.06
2127.47	4	2	6.14
2287.04	6	6	$4.27\cdot10^{-1}$
2287.04	6	8	6.41
2287.04	4	6	5.98
2366.76	2	2	$5.18\cdot10^{-1}$
2370.99	4	2	1.04
2482.82	2	4	$6.62\cdot10^{-2}$
2485.38	2	2	$7.06\cdot10^{-2}$
2672.19	2	4	$2.49\cdot10^{-2}$
2675.12	6	4	$2.74\cdot10^{-1}$
2675.25	8	6	$2.61\cdot10^{-1}$
2677.57	4	4	$4.51\cdot10^{-3}$
2723.81	6	8	1.10
2723.81	6	6	$7.34\cdot10^{-2}$
2971.52	6	6	$4.62\cdot10^{-3}$
2971.52	6	4	$9.70\cdot10^{-2}$
2971.52	8	6	$9.24\cdot10^{-2}$
3149.56	2	4	4.02
3165.71	4	6	4.77
3244.19	4	2	$4.21\cdot10^{-1}$
3762.44	4	4	$2.33\cdot10^{-1}$
3773.15	4	2	2.33
4031.39	2	2	$2.29\cdot10^{-1}$
4038.06	4	2	$4.59\cdot10^{-1}$

λ Å	Weights g_i	g_k	A 10^8 s^{-1}
4088.85	2	4	1.56
4116.10	2	2	1.53
4212.41	6	8	1.63
4212.41	6	6	$1.09{\cdot}10^{-1}$
4314.10	2	2	1.06
4328.18	4	2	2.12
4403.73	6	8	$4.09{\cdot}10^{-1}$
4403.73	6	6	$2.73{\cdot}10^{-2}$
4403.73	4	6	$3.82{\cdot}10^{-1}$
4411.65	2	4	$1.74{\cdot}10^{-2}$
4611.27	4	6	$2.23{\cdot}10^{-2}$
4611.27	4	4	$3.71{\cdot}10^{-3}$
4950.11	6	6	$9.46{\cdot}10^{-3}$
4950.11	8	6	$1.89{\cdot}10^{-1}$
4950.11	6	4	$1.99{\cdot}10^{-1}$
5304.97	6	4	$1.90{\cdot}10^{-1}$
5304.97	4	4	$2.11{\cdot}10^{-2}$
5309.49	4	2	$2.10{\cdot}10^{-1}$
6667.56	2	4	1.15
6701.21	4	6	1.36
6998.36	6	6	$3.65{\cdot}10^{-2}$
6998.36	6	8	$5.48{\cdot}10^{-1}$
6998.36	4	6	$5.11{\cdot}10^{-1}$
7047.94	6	4	$9.05{\cdot}10^{-1}$
7068.41	4	2	1.00
7620.50	2	2	$1.10{\cdot}10^{-1}$
7654.56	4	2	$8.82{\cdot}10^{-1}$
8240.61	6	6	$5.85{\cdot}10^{-3}$
8240.61	6	4	$1.23{\cdot}10^{-1}$
8240.61	8	6	$1.17{\cdot}10^{-1}$
8957.25	2	4	$4.26{\cdot}10^{-1}$
9018.16	2	2	$4.17{\cdot}10^{-1}$

Si V

λ Å	g_i	g_k	A 10^8 s^{-1}
96.44	1	3	$2.36{\cdot}10^{3}$
97.14	1	3	$2.42{\cdot}10^{2}$
98.21	1	3	6.88
117.86	1	3	$3.57{\cdot}10^{2}$
118.97	1	3	$3.84{\cdot}10^{1}$

Silver

Ag I

λ Å	g_i	g_k	A 10^8 s^{-1}
2061.2	2	4	$3.1{\cdot}10^{-2}$
2069.9	2	2	$1.5{\cdot}10^{-2}$
3280.7	2	4	1.4
3382.9	2	2	1.3
5209.1	2	4	$7.5{\cdot}10^{-1}$
5465.5	4	6	$8.6{\cdot}10^{-1}$
5471.6	4	4	$1.4{\cdot}10^{-1}$

Sodium

Na I

λ Å	g_i	g_k	A 10^8 s^{-1}
2543.84	2	4	$4.46{\cdot}10^{-4}$
2543.87	2	2	$4.35{\cdot}10^{-4}$
2593.87	2	4	$8.13{\cdot}10^{-4}$
2593.92	2	2	$7.96{\cdot}10^{-4}$
2680.34	2	4	$1.84{\cdot}10^{-3}$
2680.43	2	2	$1.81{\cdot}10^{-3}$
2852.81	2	4	$5.38{\cdot}10^{-3}$
2853.01	2	2	$5.31{\cdot}10^{-3}$
3302.37	2	4	$2.75{\cdot}10^{-2}$
3302.98	2	2	$2.73{\cdot}10^{-2}$
4238.99	2	4	$2.90{\cdot}10^{-3}$
4242.08	4	4	$5.8{\cdot}10^{-4}$
4242.08	4	6	$3.46{\cdot}10^{-3}$
4249.41	2	2	$8.7{\cdot}10^{-4}$
4252.52	4	2	$1.73{\cdot}10^{-3}$
4273.64	2	4	$3.91{\cdot}10^{-3}$
4276.79	4	4	$7.8{\cdot}10^{-4}$
4276.79	4	6	$4.69{\cdot}10^{-3}$
4287.84	2	2	$1.19{\cdot}10^{-3}$
4291.01	4	2	$2.38{\cdot}10^{-3}$
4321.40	2	4	$5.5{\cdot}10^{-3}$
4324.62	4	4	$1.09{\cdot}10^{-3}$
4324.62	4	6	$6.6{\cdot}10^{-3}$
4341.49	2	2	$3.26{\cdot}10^{-3}$
4344.74	4	2	$6.50{\cdot}10^{-3}$
4390.03	2	4	$9.83{\cdot}10^{-3}$
4393.34	4	4	$1.95{\cdot}10^{-3}$
4393.34	4	6	$1.17{\cdot}10^{-2}$
4419.88	2	2	$2.82{\cdot}10^{-3}$
4423.25	4	2	$5.61{\cdot}10^{-3}$
4494.18	2	4	$1.23{\cdot}10^{-2}$
4497.66	4	4	$2.44{\cdot}10^{-3}$
4497.66	4	6	$1.46{\cdot}10^{-2}$
4541.63	2	2	$3.76{\cdot}10^{-3}$
4545.19	4	2	$7.50{\cdot}10^{-3}$
4664.811	2	4	$2.08{\cdot}10^{-2}$
4668.560	4	4	$4.14{\cdot}10^{-3}$
4668.560	4	6	$2.49{\cdot}10^{-2}$
4747.941	2	2	$6.19{\cdot}10^{-3}$
4751.822	4	2	$1.23{\cdot}10^{-2}$
4978.541	2	4	$4.09{\cdot}10^{-2}$
4982.813	4	6	$4.88{\cdot}10^{-2}$
5148.838	2	2	$1.14{\cdot}10^{-2}$
5153.402	4	2	$2.27{\cdot}10^{-2}$
5682.633	2	4	$1.01{\cdot}10^{-1}$
5688.193	4	4	$2.02{\cdot}10^{-2}$
5688.205	4	6	$1.21{\cdot}10^{-1}$
5889.950	2	4	$6.16{\cdot}10^{-1}$
5895.924	2	2	$6.14{\cdot}10^{-1}$
6154.225	2	2	$2.50{\cdot}10^{-2}$
6160.747	4	2	$4.98{\cdot}10^{-2}$
7373.23	2	4	$5.42{\cdot}10^{-4}$
7373.49	2	2	$5.30{\cdot}10^{-4}$
7809.78	2	4	$9.91{\cdot}10^{-4}$
7810.24	2	2	$9.72{\cdot}10^{-4}$
8183.256	2	4	$4.29{\cdot}10^{-1}$
8194.790	4	4	$8.57{\cdot}10^{-2}$
8194.824	4	6	$5.14{\cdot}10^{-1}$
8649.92	2	4	$2.25{\cdot}10^{-3}$
8650.89	2	2	$2.21{\cdot}10^{-3}$
8942.96	6	6	$2.47{\cdot}10^{-4}$
8942.96	6	8	$3.71{\cdot}10^{-3}$
9153.88	6	6	$3.5{\cdot}10^{-4}$
9153.88	6	8	$5.3{\cdot}10^{-3}$
9465.94	6	6	$6.38{\cdot}10^{-4}$
9465.94	6	8	$9.57{\cdot}10^{-3}$
9961.28	6	6	$8.45{\cdot}10^{-4}$
9961.28	6	8	$1.27{\cdot}10^{-2}$

Na II

λ Å	g_i	g_k	A 10^8 s^{-1}
300.15	1	3	$1.18{\cdot}10^{1}$
300.20	1	3	$1.17{\cdot}10^{1}$
301.44	1	3	$3.33{\cdot}10^{1}$
302.45	1	3	1.4
372.08	1	3	$3.13{\cdot}10^{1}$
376.38	1	3	1.70
2315.65	3	1	$1.05{\cdot}10^{-1}$
2493.15	3	1	4.15
2506.30	3	3	$2.73{\cdot}10^{-1}$
2515.46	3	5	$2.25{\cdot}10^{-1}$
2531.54	3	1	$8.44{\cdot}10^{-1}$
2594.96	3	3	$3.71{\cdot}10^{-1}$
2661.00	3	5	1.65
2671.83	3	3	2.64
2678.09	3	1	3.01
2808.71	7	7	$1.93{\cdot}10^{-2}$
2829.87	5	5	$3.36{\cdot}10^{-1}$
2839.56	5	7	1.11
2872.95	3	5	$2.63{\cdot}10^{-1}$
2881.15	3	1	2.50
2886.26	3	5	1.07
2893.95	3	1	1.48
2901.14	7	7	$2.89{\cdot}10^{-1}$
2904.72	7	5	$1.04{\cdot}10^{-1}$
2919.05	7	5	$7.94{\cdot}10^{-2}$
2920.95	1	3	$6.66{\cdot}10^{-1}$
2930.88	5	3	$9.83{\cdot}10^{-1}$
2934.08	5	7	3.17
2937.74	5	5	1.28
2945.70	7	7	$2.50{\cdot}10^{-2}$
2951.24	7	9	4.33
2952.40	5	5	1.29
2960.12	5	5	$5.60{\cdot}10^{-1}$
2970.73	5	7	$1.03{\cdot}10^{-1}$
2974.24	5	5	$2.02{\cdot}10^{-1}$
2974.99	1	3	$6.47{\cdot}10^{-1}$
2977.13	3	3	$1.05{\cdot}10^{-1}$
2979.66	5	7	1.96
2980.63	7	5	$3.30{\cdot}10^{-3}$
3004.15	3	3	$9.44{\cdot}10^{-2}$
3009.14	3	3	$1.73{\cdot}10^{-1}$
3015.40	3	5	$1.51{\cdot}10^{-1}$
3053.67	5	7	2.99
3055.35	1	3	$8.80{\cdot}10^{-2}$
3057.38	5	5	$4.51{\cdot}10^{-1}$
3058.72	5	3	$2.18{\cdot}10^{-1}$
3064.38	3	5	$7.62{\cdot}10^{-2}$
3066.54	3	3	$3.74{\cdot}10^{-1}$
3080.25	3	5	2.81
3087.06	3	1	$1.66{\cdot}10^{-1}$
3094.45	5	5	$1.91{\cdot}10^{-3}$
3095.55	3	5	$3.97{\cdot}10^{-3}$
3104.40	3	1	$5.63{\cdot}10^{-1}$
3124.42	5	7	2.56
3125.21	3	3	$9.72{\cdot}10^{-2}$
3135.48	1	3	$7.42{\cdot}10^{-1}$
3137.86	3	3	1.41

λ Å	Weights g_i	g_k	A 10^8 s^{-1}
3145.71	3	5	$7.23 \cdot 10^{-3}$
3163.74	5	5	1.05
3179.06	5	3	$3.94 \cdot 10^{-1}$
3234.93	3	5	$1.83 \cdot 10^{-1}$
3257.96	5	5	1.03
3260.21	3	1	$4.65 \cdot 10^{-1}$
3274.22	5	3	$4.29 \cdot 10^{-1}$
3301.35	3	5	$4.54 \cdot 10^{-2}$
3304.96	1	3	$3.60 \cdot 10^{-1}$
3318.04	3	3	$4.14 \cdot 10^{-1}$
3327.69	3	1	$9.45 \cdot 10^{-1}$
3711.07	1	3	$1.02 \cdot 10^{-1}$
4123.08	1	3	$3.71 \cdot 10^{-1}$
4344.11	1	3	$4.32 \cdot 10^{-1}$
4368.60	1	3	$3.59 \cdot 10^{-1}$

Na III

λ Å	Weights g_i	g_k	A 10^8 s^{-1}
202.15	4	4	$1.40 \cdot 10^{1}$
202.49	4	2	$1.59 \cdot 10^{2}$
202.71	2	4	$6.40 \cdot 10^{1}$
202.72	4	2	$6.72 \cdot 10^{1}$
202.76	4	4	$1.26 \cdot 10^{2}$
203.28	2	2	$9.24 \cdot 10^{1}$
215.86	2	4	$1.72 \cdot 10^{1}$
216.12	2	4	$1.57 \cdot 10^{1}$
229.87	4	2	$2.91 \cdot 10^{1}$
267.87	2	2	$7.56 \cdot 10^{1}$
378.14	4	2	$8.42 \cdot 10^{1}$
380.10	2	2	$4.13 \cdot 10^{1}$
1336.76	4	2	4.27
1337.36	4	4	1.61
1340.67	6	4	2.29
1342.39	2	2	$8.44 \cdot 10^{-1}$
1342.73	4	2	1.06
1355.28	6	6	3.44
1361.90	2	2	2.24
1372.34	2	4	$5.81 \cdot 10^{-1}$
1420.89	8	6	5.73
1444.19	6	4	4.07
1449.31	8	6	5.12
1562.87	4	4	2.98
1565.29	4	4	$3.95 \cdot 10^{-1}$
1598.18	4	6	$7.96 \cdot 10^{-1}$
1711.12	6	4	1.52
1728.27	4	2	4.10
1731.11	2	4	1.87
1835.22	6	6	3.31
1838.94	8	8	3.31
1844.36	6	6	3.00
1849.56	6	8	6.87
1850.38	4	4	4.37
1855.92	2	2	5.77
1856.71	4	6	4.22
1861.21	2	4	2.34
1887.47	10	12	$1.25 \cdot 10^{1}$
1918.45	4	4	2.29
1926.26	8	10	8.94
1927.24	2	4	5.66
1932.74	4	6	5.37
1933.89	6	8	6.33
1946.43	6	8	8.28
1950.91	8	10	8.55
1951.24	6	4	2.38
1977.16	4	6	8.46
1985.57	4	4	1.78
1995.68	6	6	$8.12 \cdot 10^{-1}$
2004.21	2	4	2.07
2005.22	2	4	$9.67 \cdot 10^{-1}$
2008.47	4	6	1.45
2011.87	6	8	6.37
2014.17	4	6	4.88
2017.03	6	8	5.44
2028.56	8	8	1.93
2031.13	4	6	3.08
2035.90	2	2	6.23
2041.66	6	8	1.64
2043.29	4	2	4.02
2044.82	4	4	4.00
2045.44	6	6	1.05
2051.48	6	6	4.36
2060.36	4	6	$8.12 \cdot 10^{-1}$
2066.60	4	4	1.53
2140.72	6	6	3.80
2144.54	4	4	3.88
2202.83	6	6	$8.53 \cdot 10^{-1}$
2225.93	4	4	1.82
2230.33	6	8	3.64
2232.19	4	4	2.34
2246.70	4	6	2.72
2251.47	2	4	1.66
2278.42	2	2	2.98
2285.66	2	4	$7.73 \cdot 10^{-1}$
2309.99	4	2	3.08
2386.99	6	8	3.05
2394.03	4	6	2.80
2406.59	4	4	$5.1 \cdot 10^{-1}$
2459.31	4	6	3.0
2468.85	2	4	2.4
2474.73	6	4	1.38
2497.03	6	6	1.99
2510.26	4	2	2.19
2530.25	4	4	$3.65 \cdot 10^{-1}$
2542.80	2	2	$3.39 \cdot 10^{-1}$

Na IV

λ Å	Weights g_i	g_k	A 10^8 s^{-1}
136.551	5	5	$2.08 \cdot 10^{2}$
136.854	5	7	$1.56 \cdot 10^{2}$
142.232	5	7	$3.06 \cdot 10^{2}$
142.359	5	5	$2.77 \cdot 10^{2}$
146.064	5	5	$6.69 \cdot 10$
146.302	3	5	$2.00 \cdot 10^{2}$
150.298	5	7	$2.38 \cdot 10^{2}$
150.543	3	5	$1.71 \cdot 10^{2}$
150.687	5	5	$9.87 \cdot 10$
151.299	5	5	$6.70 \cdot 10$
155.083	5	3	$1.33 \cdot 10^{2}$
155.240	5	3	$1.88 \cdot 10^{2}$
156.537	5	7	$2.92 \cdot 10^{2}$
162.448	5	7	$6.04 \cdot 10^{2}$
163.190	5	5	$4.32 \cdot 10^{2}$
168.411	3	3	$1.37 \cdot 10^{2}$
168.546	1	3	$1.81 \cdot 10^{2}$
319.644	5	3	$2.52 \cdot 10^{2}$
360.76	1	3	$1.38 \cdot 10^{1}$
408.684	5	3	$2.66 \cdot 10^{1}$
409.614	3	1	$6.34 \cdot 10^{1}$
410.372	5	5	$4.73 \cdot 10^{1}$
411.334	1	3	$2.09 \cdot 10^{1}$
412.242	3	5	$1.55 \cdot 10^{1}$
1580.50	7	9	6.27
1582.18	3	3	7.16
1582.33	3	5	3.34
1583.98	5	5	5.55
1584.14	5	7	6.34
1587.05	7	9	9.46
1613.95	5	7	8.74
1615.92	7	9	8.89
1618.57	9	11	9.23
1655.47	7	9	8.72
1701.97	5	7	7.93
1702.41	3	5	5.95
1960.76	5	7	3.71
1965.08	5	5	3.68
1967.60	5	3	3.67
2018.39	3	5	3.44

Strontium

Sr I

λ Å	Weights g_i	g_k	A 10^8 s^{-1}
2206.2	1	3	$6.6 \cdot 10^{-3}$
2211.3	1	3	$8.5 \cdot 10^{-3}$
2217.8	1	3	$1.2 \cdot 10^{-2}$
2226.3	1	3	$1.6 \cdot 10^{-2}$
2237.7	1	3	$2.3 \cdot 10^{-2}$
2253.3	1	3	$3.7 \cdot 10^{-2}$
2275.3	1	3	$6.7 \cdot 10^{-2}$
2307.3	1	3	$1.2 \cdot 10^{-1}$
2354.3	1	3	$1.8 \cdot 10^{-1}$
2428.1	1	3	$1.7 \cdot 10^{-1}$
2569.5	1	3	$5.3 \cdot 10^{-2}$
2931.8	1	3	$1.9 \cdot 10^{-2}$
4607.3	1	3	2.01

Sr II

λ Å	Weights g_i	g_k	A 10^8 s^{-1}
2018.7	2	2	$1.2 \cdot 10^{-1}$
2051.9	4	2	$2.4 \cdot 10^{-1}$
2282.0	2	4	$8.3 \cdot 10^{-1}$
2322.4	4	6	$9.1 \cdot 10^{-1}$
2324.5	4	4	$1.5 \cdot 10^{-1}$
2423.5	2	2	$2.4 \cdot 10^{-1}$
2471.6	4	2	$4.8 \cdot 10^{-1}$
3464.5	4	6	3.1
3474.9	4	4	$5.1 \cdot 10^{-1}$
4077.7	2	4	1.42
4161.8	2	2	$6.5 \cdot 10^{-1}$
4215.5	2	2	1.27
4305.5	4	2	1.4
4414.8	4	6	$1.1 \cdot 10^{-1}$
4417.5	4	4	$1.8 \cdot 10^{-2}$
4585.9	4	2	$7.0 \cdot 10^{-2}$

λ Å	Weights g_i	g_k	A 10^8 s^{-1}
5303.1	2	4	$1.9\cdot10^{-1}$
5379.1	4	6	$2.2\cdot10^{-1}$
5385.5	4	4	$3.7\cdot10^{-2}$
5723.7	2	2	$7.1\cdot10^{-2}$
5819.0	4	2	$1.4\cdot10^{-1}$
8688.9	4	6	$5.5\cdot10^{-1}$
8719.6	4	4	$9.7\cdot10^{-2}$

Sulfur

S I

λ Å	Weights g_i	g_k	A 10^8 s^{-1}
1295.7	5	5	4.9
1296.2	5	3	2.7
1302.3	3	5	1.8
1302.9	3	3	1.6
1303.1	3	1	6.6
1303.4	5	3	1.9
1305.9	1	3	2.4
1401.5	5	3	$9.1\cdot10^{-1}$
1409.3	3	3	$5.0\cdot10^{-1}$
1412.9	1	3	$1.6\cdot10^{-1}$
1425.0	5	7	4.5
1425.2	5	5	1.2
1433.3	3	5	3.3
1433.3	3	3	1.9
1437.0	1	3	2.4
1448.2	5	3	7.3
1473.0	5	7	$4.2\cdot10^{-1}$
1474.0	5	7	1.6
1474.4	5	5	$5.0\cdot10^{-1}$
1474.6	5	3	$6.2\cdot10^{-2}$
1481.7	3	5	$1.7\cdot10^{-1}$
1483.0	3	5	1.2
1483.2	3	3	$7.5\cdot10^{-1}$
1487.2	1	3	$8.7\cdot10^{-1}$
1666.7	5	5	6.3
1687.5	1	3	$9.4\cdot10^{-1}$
1782.3	1	3	1.9
1807.3	5	3	3.8
1820.3	3	3	2.2
1826.2	1	3	$7.2\cdot10^{-1}$
4694.1	5	7	$6.7\cdot10^{-3}$
4695.4	5	5	$6.7\cdot10^{-3}$
4696.2	5	3	$6.5\cdot10^{-3}$
6403.6	3	5	$5.7\cdot10^{-3}$
6408.1	5	5	$9.5\cdot10^{-3}$
6415.5	7	5	$1.3\cdot10^{-2}$
*6751.2	15	25	$7.9\cdot10^{-2}$
7679.6	3	5	$1.2\cdot10^{-2}$
7686.1	5	5	$2.0\cdot10^{-2}$
7696.7	7	5	$2.8\cdot10^{-2}$

S II

λ Å	Weights g_i	g_k	A 10^8 s^{-1}
1124.4	2	4	1.0
1125.0	4	4	4.6
1131.0	2	2	3.5
1131.6	4	2	1.4
1250.5	4	2	$4.6\cdot10^{-1}$
1253.8	4	4	$4.2\cdot10^{-1}$
1259.5	4	6	$3.4\cdot10^{-1}$
4463.6	8	6	$5.3\cdot10^{-1}$
4483.4	6	4	$3.1\cdot10^{-1}$
4486.7	4	2	$6.6\cdot10^{-1}$
4524.7	4	4	$9.3\cdot10^{-2}$
4525.0	6	4	1.2
4552.4	4	2	1.2
4656.7	2	4	$9.0\cdot10^{-2}$
4716.2	4	4	$2.9\cdot10^{-1}$
4815.5	6	4	$8.8\cdot10^{-1}$
4885.6	2	4	$1.7\cdot10^{-1}$
4917.2	2	2	$6.6\cdot10^{-1}$
4924.1	4	6	$2.2\cdot10^{-1}$
4925.3	2	4	$2.4\cdot10^{-1}$
4942.5	2	2	$1.5\cdot10^{-1}$
4991.9	4	4	$1.5\cdot10^{-1}$
5009.5	4	2	$7.0\cdot10^{-1}$
5014.0	4	4	$8.4\cdot10^{-1}$
5027.2	4	2	$2.6\cdot10^{-1}$
5032.4	6	6	$8.1\cdot10^{-1}$
5047.3	4	2	$3.6\cdot10^{-1}$
5103.3	6	4	$5.0\cdot10^{-1}$
5142.3	2	2	$1.9\cdot10^{-1}$
5201.0	4	4	$7.5\cdot10^{-1}$
5201.3	6	4	$6.5\cdot10^{-2}$
5212.6	4	6	$9.8\cdot10^{-2}$
5212.6	6	6	$8.5\cdot10^{-1}$
5320.7	6	8	$9.2\cdot10^{-1}$
5345.7	4	6	$8.8\cdot10^{-1}$
5345.7	6	6	$1.1\cdot10^{-1}$
5428.6	2	4	$4.2\cdot10^{-1}$
5432.8	4	6	$6.8\cdot10^{-1}$
5453.8	6	8	$8.5\cdot10^{-1}$
5473.6	2	2	$7.3\cdot10^{-1}$
5509.7	4	4	$4.0\cdot10^{-1}$
5526.2	8	8	$8.1\cdot10^{-2}$
5536.8	4	6	$6.6\cdot10^{-2}$
5556.0	4	2	$1.1\cdot10^{-1}$
5564.9	6	6	$1.7\cdot10^{-1}$
5578.8	6	6	$1.1\cdot10^{-1}$
5606.1	10	8	$5.4\cdot10^{-1}$
5616.6	4	4	$1.2\cdot10^{-1}$
5640.0	4	6	$6.6\cdot10^{-1}$
5645.6	6	4	$1.8\cdot10^{-2}$
5647.0	2	4	$5.7\cdot10^{-1}$
5659.9	6	4	$4.6\cdot10^{-1}$
5664.7	4	2	$5.8\cdot10^{-1}$
5819.2	4	4	$8.5\cdot10^{-2}$
6305.5	8	6	$1.8\cdot10^{-1}$
6312.7	6	4	$3.0\cdot10^{-1}$

S III

λ Å	Weights g_i	g_k	A 10^8 s^{-1}
2496.2	7	5	2.5
2508.2	5	3	2.3
2636.9	3	5	$4.5\cdot10^{-1}$
2665.4	5	5	1.4
2680.5	1	3	$6.2\cdot10^{-1}$
2691.8	3	3	$4.6\cdot10^{-1}$
2702.8	3	1	1.9
2718.9	3	3	1.2
2721.4	5	3	$7.7\cdot10^{-1}$
2726.8	3	5	$6.0\cdot10^{-1}$
2731.1	5	5	1.1
2756.9	7	7	1.4
2785.5	3	3	$6.1\cdot10^{-1}$
2856.0	5	7	5.1
2863.5	7	9	5.7
2872.0	3	5	4.7
2950.2	3	5	3.0
2964.8	5	7	4.0
3662.0	3	3	$6.4\cdot10^{-1}$
3717.8	5	3	1.0
3778.9	3	5	$4.4\cdot10^{-1}$
3831.8	1	3	$5.6\cdot10^{-1}$
3837.8	3	3	$4.2\cdot10^{-1}$
3838.3	5	5	1.3
3860.6	3	1	1.6
3899.1	5	3	$6.7\cdot10^{-1}$
4253.6	5	7	1.2
4285.0	3	5	$9.0\cdot10^{-1}$

Tantalum

Ta I

λ Å	Weights g_i	g_k	A 10^8 s^{-1}
3170.3	8	10	$8.5\cdot10^{-2}$
3406.9	4	6	$6.8\cdot10^{-2}$
3419.7	8	8	$1.01\cdot10^{-2}$
3463.8	4	6	$2.62\cdot10^{-2}$
3497.9	6	8	$4.9\cdot10^{-2}$
3505.0	8	6	$2.72\cdot10^{-2}$
3607.4	6	8	$4.6\cdot10^{-2}$
3626.6	8	10	$7.1\cdot10^{-2}$
3642.1	10	12	$5.5\cdot10^{-2}$
3784.3	4	6	$4.3\cdot10^{-2}$
3848.1	10	8	$1.30\cdot10^{-2}$
3922.8	4	4	$3.98\cdot10^{-2}$
3996.2	2	4	$3.35\cdot10^{-2}$
4026.9	4	4	$3.60\cdot10^{-2}$
4029.9	10	10	$2.8\cdot10^{-2}$
4040.9	10	12	$7.3\cdot10^{-3}$
4061.4	2	4	$6.5\cdot10^{-2}$
4064.6	4	4	$3.83\cdot10^{-2}$
4136.2	8	6	$1.82\cdot10^{-2}$
4147.9	10	8	$1.79\cdot10^{-2}$
4175.2	6	8	$2.8\cdot10^{-2}$
4205.9	8	10	$8.9\cdot10^{-3}$
4303.0	6	6	$2.08\cdot10^{-2}$
4386.1	4	6	$1.0\cdot10^{-2}$
4402.5	6	6	$2.28\cdot10^{-2}$
4415.7	2	4	$2.53\cdot10^{-2}$
4441.7	10	8	$9.0\cdot10^{-3}$
4473.5	6	8	$1.36\cdot10^{-2}$
4511.0	10	12	$1.56\cdot10^{-2}$
4530.9	4	6	$2.42\cdot10^{-2}$
4553.7	6	8	$9.5\cdot10^{-3}$
4565.9	8	8	$2.5\cdot10^{-2}$
4619.5	6	4	$5.3\cdot10^{-2}$
4669.1	6	4	$2.85\cdot10^{-2}$
4681.9	6	6	$1.5\cdot10^{-2}$
4691.9	2	4	$4.08\cdot10^{-2}$
4706.1	6	6	$1.4\cdot10^{-2}$

λ Å	Weights g_i	g_k	A 10^8 s^{-1}
4740.2	4	4	$5.0 \cdot 10^{-2}$
4758.0	4	6	$7.5 \cdot 10^{-3}$
4769.0	8	8	$2.8 \cdot 10^{-2}$
4780.9	10	8	$2.16 \cdot 10^{-2}$
4812.8	4	4	$1.2 \cdot 10^{-2}$
4825.4	6	6	$2.63 \cdot 10^{-2}$
4832.2	4	4	$1.7 \cdot 10^{-2}$
4852.2	4	4	$1.7 \cdot 10^{-2}$
4884.0	6	8	$1.1 \cdot 10^{-2}$
4904.6	12	10	$1.95 \cdot 10^{-2}$
4921.3	2	4	$1.2 \cdot 10^{-2}$
4926.0	4	4	$1.5 \cdot 10^{-2}$
4936.4	8	6	$4.5 \cdot 10^{-2}$
4969.7	4	4	$1.0 \cdot 10^{-2}$
5012.5	4	4	$1.9 \cdot 10^{-2}$
5037.4	10	8	$4.4 \cdot 10^{-2}$
5043.3	6	4	$2.73 \cdot 10^{-2}$
5067.9	8	6	$2.92 \cdot 10^{-2}$
5087.4	6	4	$1.5 \cdot 10^{-2}$
5090.7	8	6	$9.5 \cdot 10^{-3}$
5136.5	2	2	$4.5 \cdot 10^{-2}$
5143.7	6	4	$1.7 \cdot 10^{-2}$
5147.6	6	4	$9.0 \cdot 10^{-3}$
5161.8	4	6	$6.3 \cdot 10^{-3}$
5218.7	8	6	$8.2 \cdot 10^{-3}$
5295.0	6	6	$7.5 \cdot 10^{-3}$
5336.1	6	8	$5.5 \cdot 10^{-3}$
5349.6	6	4	$2.2 \cdot 10^{-2}$
5435.3	4	6	$1.1 \cdot 10^{-2}$
5499.4	10	10	$6.1 \cdot 10^{-3}$
5518.9	8	10	$3.8 \cdot 10^{-2}$
5620.7	8	10	$6.0 \cdot 10^{-3}$
5640.2	6	8	$4.9 \cdot 10^{-3}$
5645.9	6	8	$1.43 \cdot 10^{-2}$
5811.1	8	6	$5.7 \cdot 10^{-3}$
5877.4	10	12	$2.3 \cdot 10^{-2}$
5939.8	2	4	$1.6 \cdot 10^{-2}$
5944.0	4	6	$2.13 \cdot 10^{-2}$
5997.2	10	10	$2.4 \cdot 10^{-2}$
6020.7	2	4	$1.0 \cdot 10^{-2}$
6045.4	6	8	$2.6 \cdot 10^{-2}$
6047.3	8	10	$9.0 \cdot 10^{-3}$
6249.8	6	6	$3.5 \cdot 10^{-3}$
6258.7	6	8	$3.3 \cdot 10^{-3}$
6309.6	4	6	$1.83 \cdot 10^{-2}$
6360.8	6	8	$4.6 \cdot 10^{-3}$
6428.6	6	6	$6.0 \cdot 10^{-3}$
6430.8	8	8	$2.9 \cdot 10^{-2}$
6450.4	8	10	$2.2 \cdot 10^{-2}$
6485.4	10	10	$5.8 \cdot 10^{-2}$
6514.4	6	4	$2.2 \cdot 10^{-2}$
6516.1	6	8	$1.25 \cdot 10^{-2}$
6612.0	6	4	$1.9 \cdot 10^{-2}$
6673.7	2	4	$9.0 \cdot 10^{-3}$
6771.7	4	4	$5.8 \cdot 10^{-3}$
6866.2	8	6	$2.58 \cdot 10^{-2}$
6927.4	10	12	$1.01 \cdot 10^{-2}$
6928.5	10	8	$1.69 \cdot 10^{-2}$
6951.3	10	10	$3.7 \cdot 10^{-3}$
6953.9	6	8	$8.3 \cdot 10^{-3}$
6966.1	8	8	$1.2 \cdot 10^{-2}$
6969.5	10	10	$2.9 \cdot 10^{-3}$
7407.9	6	4	$2.0 \cdot 10^{-2}$

Thallium

Tl I

λ Å	g_i	g_k	A 10^8 s^{-1}
2104.6	2	4	$4.0 \cdot 10^{-2}$
2118.9	2	2	$2.0 \cdot 10^{-2}$
2129.3	2	4	$5.8 \cdot 10^{-2}$
2151.9	2	2	$3.1 \cdot 10^{-2}$
2168.6	2	4	$9.8 \cdot 10^{-2}$
2237.8	2	4	$1.9 \cdot 10^{-1}$
2316.0	2	2	$7.8 \cdot 10^{-2}$
2379.7	2	4	$4.4 \cdot 10^{-1}$
2507.9	4	2	$1.1 \cdot 10^{-2}$
2538.2	4	2	$1.6 \cdot 10^{-2}$
2580.1	2	2	$1.8 \cdot 10^{-1}$
2609.0	4	6	$1.0 \cdot 10^{-1}$
2609.8	4	4	$1.9 \cdot 10^{-2}$
2665.6	4	2	$5.7 \cdot 10^{-2}$
2709.2	4	6	$1.7 \cdot 10^{-1}$
2710.7	4	4	$3.7 \cdot 10^{-2}$
2767.9	2	4	1.26
2826.2	4	2	$8.0 \cdot 10^{-2}$
2918.3	4	6	$4.2 \cdot 10^{-1}$
2921.5	4	4	$7.6 \cdot 10^{-2}$
3229.8	4	2	$1.73 \cdot 10^{-1}$
3519.2	4	6	1.24
3529.4	4	4	$2.20 \cdot 10^{-1}$
3775.7	2	2	$6.25 \cdot 10^{-1}$
5350.5	4	2	$7.05 \cdot 10^{-1}$

Thulium

Tm I

λ Å	g_i	g_k	A 10^8 s^{-1}
2513.8	8	10	$6.9 \cdot 10^{-2}$
2527.0	8	8	$1.7 \cdot 10^{-1}$
2596.5	8	10	$1.6 \cdot 10^{-1}$
2601.1	8	6	$1.7 \cdot 10^{-1}$
2622.5	8	10	$6.1 \cdot 10^{-2}$
2841.1	6	6	$2.0 \cdot 10^{-1}$
2854.2	8	6	$2.7 \cdot 10^{-1}$
2914.8	8	8	$7.7 \cdot 10^{-2}$
2933.0	8	6	$1.0 \cdot 10^{-1}$
2973.2	8	8	$2.3 \cdot 10^{-1}$
3046.9	8	8	$1.8 \cdot 10^{-1}$
3081.1	8	8	$1.9 \cdot 10^{-1}$
3122.5	6	6	$5.2 \cdot 10^{-1}$
3142.4	6	6	$8.8 \cdot 10^{-2}$
3172.7	8	8	$1.8 \cdot 10^{-1}$
3233.7	8	10	$5.1 \cdot 10^{-2}$
3247.0	6	8	$3.0 \cdot 10^{-1}$
3251.8	6	4	$5.2 \cdot 10^{-1}$
3380.7	6	8	$2.0 \cdot 10^{-1}$
3406.0	6	8	$1.5 \cdot 10^{-1}$
3410.1	8	10	$1.0 \cdot 10^{-1}$
3416.6	8	8	$5.7 \cdot 10^{-2}$
3418.6	6	6	$1.1 \cdot 10^{-1}$
3563.9	8	6	$9.8 \cdot 10^{-2}$
3567.4	8	10	$4.2 \cdot 10^{-2}$
3744.1	8	8	$9.5 \cdot 10^{-1}$
3751.8	8	10	$1.9 \cdot 10^{-1}$
3798.5	6	4	1.2
3807.7	6	6	$3.9 \cdot 10^{-1}$
3883.1	8	6	1.0
3887.4	8	8	$3.8 \cdot 10^{-1}$
3916.5	6	8	1.5
3949.3	6	6	1.0
4022.6	6	8	$4.0 \cdot 10^{-2}$
4044.5	6	4	$2.9 \cdot 10^{-1}$
4094.2	8	6	$9.0 \cdot 10^{-1}$
4105.8	8	10	$6.0 \cdot 10^{-1}$
4138.3	6	4	$7.0 \cdot 10^{-1}$
4158.6	6	8	$5.5 \cdot 10^{-2}$
4187.6	8	8	$6.1 \cdot 10^{-1}$
4203.7	8	10	$2.5 \cdot 10^{-1}$
4222.7	6	8	$1.5 \cdot 10^{-1}$
4271.7	6	6	$1.1 \cdot 10^{-1}$
4359.9	8	6	$1.3 \cdot 10^{-1}$
4386.4	8	8	$4.2 \cdot 10^{-2}$
4394.4	6	4	$1.1 \cdot 10^{-1}$
4643.1	6	6	$3.4 \cdot 10^{-2}$
4681.9	6	8	$3.9 \cdot 10^{-2}$
4691.1	6	6	$3.9 \cdot 10^{-2}$
5307.1	8	10	$2.3 \cdot 10^{-2}$
5658.3	6	8	$1.0 \cdot 10^{-2}$
5675.8	8	10	$1.3 \cdot 10^{-2}$
5760.2	6	6	$1.3 \cdot 10^{-2}$

Tin

Sn I

λ Å	g_i	g_k	A 10^8 s^{-1}
2073.1	1	3	$3.6 \cdot 10^{-2}$
2199.3	3	5	$2.9 \cdot 10^{-1}$
2209.7	5	5	$5.6 \cdot 10^{-1}$
2246.1	1	3	1.6
2268.9	5	7	1.2
2286.7	5	5	$3.1 \cdot 10^{-1}$
2317.2	5	7	2.0
2334.8	3	3	$6.6 \cdot 10^{-1}$
2354.8	3	5	1.7
2380.7	3	5	$3.1 \cdot 10^{-2}$
2408.2	5	3	$1.8 \cdot 10^{-1}$
2421.7	5	7	2.5
2429.5	5	7	1.5
2433.5	5	3	$8.0 \cdot 10^{-3}$
2455.2	5	5	$1.1 \cdot 10^{-2}$
2476.4	5	3	$1.1 \cdot 10^{-2}$
2483.4	5	5	$2.1 \cdot 10^{-1}$
2491.8	1	3	$1.7 \cdot 10^{-1}$
2495.7	5	5	$6.2 \cdot 10^{-1}$
2523.9	5	3	$7.4 \cdot 10^{-2}$
2546.6	1	3	$2.1 \cdot 10^{-1}$
2558.0	1	3	$3.4 \cdot 10^{-1}$
2571.6	5	7	$4.5 \cdot 10^{-1}$
2594.4	5	5	$3.0 \cdot 10^{-1}$
2636.9	1	3	$1.1 \cdot 10^{-1}$
2661.2	3	3	$1.1 \cdot 10^{-1}$
2706.5	3	5	$6.6 \cdot 10^{-1}$
2761.8	5	5	$3.7 \cdot 10^{-3}$

λ Å	Weights g_i	g_k	A $10^8\ s^{-1}$
2779.8	5	7	$1.8\cdot10^{-1}$
2785.0	5	3	$1.4\cdot10^{-1}$
2788.0	1	3	$1.4\cdot10^{-1}$
2812.6	1	3	$2.3\cdot10^{-1}$
2813.6	5	5	$1.2\cdot10^{-1}$
2840.0	5	5	1.7
2850.6	5	5	$3.3\cdot10^{-1}$
2863.3	1	3	$5.4\cdot10^{-1}$
2913.5	1	3	$8.3\cdot10^{-1}$
3009.1	3	3	$3.8\cdot10^{-1}$
3032.8	1	3	$6.2\cdot10^{-1}$
3034.1	3	1	2.0
3141.8	1	3	$1.9\cdot10^{-1}$
3175.1	5	3	1.0
3218.7	1	3	$4.7\cdot10^{-2}$
3223.6	5	5	$1.2\cdot10^{-3}$
3262.3	5	3	2.7
3330.6	5	5	$2.0\cdot10^{-1}$
3655.8	1	3	$4.1\cdot10^{-2}$
3801.0	5	3	$2.8\cdot10^{-1}$
4524.7	1	3	$2.6\cdot10^{-1}$
5631.7	1	3	$2.4\cdot10^{-2}$
5970.3	5	3	$9.6\cdot10^{-2}$
6037.7	5	5	$5.0\cdot10^{-2}$
6069.0	1	3	$4.6\cdot10^{-2}$
6073.5	3	1	$6.3\cdot10^{[illegible]}$
6171.5	3	3	$4.9\cdot10^{-2}$

Sn II

λ Å	g_i	g_k	A $10^8\ s^{-1}$
2368.3	4	2	$4.4\cdot10^{-3}$
2449.0	4	6	$3.7\cdot10^{-1}$
2487.0	6	8	$5.5\cdot10^{-1}$
3283.2	4	6	1.0
3352.0	6	8	1.0
3472.5	2	4	$1.6\cdot10^{-1}$
3575.5	4	6	$1.3\cdot10^{-1}$
5332.4	2	4	$8.6\cdot10^{-1}$
5562.0	4	6	1.2
5588.9	4	6	$8.5\cdot10^{-1}$
5596.2	4	4	$1.5\cdot10^{-1}$
5797.2	6	6	$2.8\cdot10^{-1}$
5799.2	6	8	$8.1\cdot10^{-1}$
6453.5	2	4	1.2
6761.5	2	2	$3.2\cdot10^{-1}$
6844.1	2	2	$6.6\cdot10^{-1}$

Titanium

Ti I

λ Å	g_i	g_k	A $10^8\ s^{-1}$
2644.28	7	5	1.4
2646.65	9	7	1.5
2733.27	5	5	1.9
2912.07	5	7	1.3
2942.00	5	5	1.0
2948.26	7	7	$9.3\cdot10^{-1}$
2956.13	9	9	$9.7\cdot10^{-1}$
3186.45	5	7	$8.0\cdot10^{-1}$
3191.99	7	9	$8.5\cdot10^{-1}$
3199.92	9	11	$9.4\cdot10^{-1}$
3341.88	5	7	$6.5\cdot10^{-1}$

λ Å	Weights g_i	g_k	A $10^8\ s^{-1}$
3354.63	7	9	$6.9\cdot10^{-1}$
3371.45	9	11	$7.2\cdot10^{-1}$
3377.58	7	5	$6.9\cdot10^{-1}$
3385.94	9	7	$5.0\cdot10^{-1}$
3635.46	5	7	$8.04\cdot10^{-1}$
3642.68	7	9	$7.74\cdot10^{-1}$
3653.50	9	11	$7.54\cdot10^{-1}$
3724.57	9	9	$9.1\cdot10^{-1}$
3741.06	7	7	$4.17\cdot10^{-1}$
3752.86	9	9	$5.04\cdot10^{-1}$
3786.04	5	3	1.4
3958.21	9	7	$4.05\cdot10^{-1}$
3981.76	5	5	$3.76\cdot10^{-1}$
3989.76	7	7	$3.79\cdot10^{-1}$
3998.64	9	9	$4.08\cdot10^{-1}$
4186.12	9	9	$2.10\cdot10^{-1}$
4266.23	5	5	$3.1\cdot10^{-1}$
4284.99	5	5	$3.2\cdot10^{-1}$
4289.07	5	5	$3.0\cdot10^{-1}$
4393.93	9	11	$3.3\cdot10^{-1}$
4417.27	11	9	$3.6\cdot10^{-1}$
4449.14	11	11	$9.7\cdot10^{-1}$
4450.90	9	9	$9.6\cdot10^{-1}$
4453.31	5	5	$5.98\cdot10^{-1}$
4453.71	7	7	$4.7\cdot10^{-1}$
4455.32	7	7	$4.8\cdot10^{-1}$
4457.43	9	9	$5.6\cdot10^{-1}$
4465.81	5	7	$3.28\cdot10^{-1}$
4481.26	7	7	$5.7\cdot10^{-1}$
4496.15	7	5	$4.4\cdot10^{-1}$
4518.02	7	9	$1.72\cdot10^{-1}$
4522.80	5	7	$1.9\cdot10^{-1}$
4533.24	11	11	$8.83\cdot10^{-1}$
4534.78	9	9	$6.87\cdot10^{-1}$
4548.76	7	5	$2.85\cdot10^{-1}$
4552.45	9	7	$2.1\cdot10^{-1}$
4563.43	9	11	$2.1\cdot10^{-1}$
4617.27	7	9	$8.51\cdot10^{-1}$
4623.10	5	7	$5.74\cdot10^{-1}$
4639.94	3	3	$6.64\cdot10^{-1}$
4742.79	9	9	$5.3\cdot10^{-1}$
4758.12	11	11	$7.13\cdot10^{-1}$
4759.27	13	13	$7.40\cdot10^{-1}$
4778.26	9	9	$2.0\cdot10^{-1}$
4805.42	5	7	$5.8\cdot10^{-1}$
4856.01	13	15	$5.2\cdot10^{-1}$
4885.08	11	13	$4.90\cdot10^{-1}$
4913.62	7	9	$4.44\cdot10^{-1}$
4928.34	3	5	$6.2\cdot10^{-1}$
4981.73	11	13	$6.60\cdot10^{-1}$
4989.14	7	5	$3.25\cdot10^{-1}$
4991.07	9	11	$5.84\cdot10^{-1}$
4999.50	7	9	$5.27\cdot10^{-1}$
5000.99	9	7	$3.52\cdot10^{-1}$
5007.21	5	7	$4.92\cdot10^{-1}$
5014.28	3	5	$6.8\cdot10^{-1}$
5036.47	7	9	$3.94\cdot10^{-1}$
5038.40	5	7	$3.87\cdot10^{-1}$
5224.30	11	11	$3.6\cdot10^{-1}$
5259.98	5	7	$2.3\cdot10^{-1}$

λ Å	Weights g_i	g_k	A $10^8\ s^{-1}$
5351.07	7	7	$3.4\cdot10^{-1}$
5503.90	11	9	$2.6\cdot10^{-1}$
5774.04	9	11	$5.5\cdot10^{-1}$
5785.98	11	13	$6.1\cdot10^{-1}$
5804.27	13	15	$6.8\cdot10^{-1}$
6098.66	9	7	$2.5\cdot10^{-1}$
6220.46	9	7	$1.8\cdot10^{-1}$

Ti II

λ Å	g_i	g_k	A $10^8\ s^{-1}$
2635.44	4	4	1.9
2638.56	6	6	1.7
2642.02	8	8	1.9
2645.86	10	10	2.7
2746.54	6	8	2.6
2751.59	8	10	3.7
2752.68	8	10	1.1
2757.62	6	8	$7.2\cdot10^{-1}$
2758.35	4	6	$9.9\cdot10^{-1}$
2804.82	6	8	4.6
2810.30	8	10	5.1
2817.83	10	12	3.8
2819.87	8	8	$6.5\cdot10^{-1}$
2821.26	6	8	$7.9\cdot10^{-1}$
2827.12	8	10	1.0
2828.06	12	14	4.4
2828.64	6	6	1.2
2828.83	10	10	$9.1\cdot10^{-1}$
2834.02	10	12	$7.9\cdot10^{-1}$
2836.47	8	8	1.2
2839.64	12	12	$8.3\cdot10^{-1}$
2845.93	10	10	1.2
2856.10	12	12	1.5
2877.47	8	8	$5.7\cdot10^{-1}$
2884.13	10	10	$5.2\cdot10^{-1}$
2926.64	10	8	$8.9\cdot10^{-1}$
2931.10	6	6	3.2
2936.02	4	6	2.7
2938.57	6	8	2.4
2941.90	8	10	1.8
2942.97	8	8	1.1
2945.30	10	12	2.7
2954.59	10	12	4.0
2958.80	8	10	4.0
2979.06	4	6	1.2
2990.06	6	8	$5.6\cdot10^{-1}$
3017.17	12	12	$3.6\cdot10^{-1}$
3022.64	10	10	1.2
3023.67	8	8	1.0
3075.23	6	4	1.13
3078.65	8	6	1.09
3081.52	10	8	1.1
3088.04	10	8	1.25
3089.44	8	6	1.3
3103.81	10	8	1.1
3106.26	6	6	$7.8\cdot10^{-1}$
3127.86	6	6	1.6
3128.50	8	8	1.1
3168.55	10	8	$4.1\cdot10^{-1}$
3181.73	6	8	$4.6\cdot10^{-1}$
3189.49	4	4	$9.2\cdot10^{-1}$

λ Å	Weights g_i	g_k	A 10^8 s^{-1}
3190.91	6	8	1.3
3202.56	4	6	1.1
3224.25	12	10	$7.0 \cdot 10^{-1}$
3228.62	4	2	2.0
3232.29	8	6	$6.0 \cdot 10^{-1}$
3234.51	10	10	1.38
3236.58	8	8	1.11
3239.04	6	6	$9.87 \cdot 10^{-1}$
3239.66	6	4	$9.4 \cdot 10^{-1}$
3241.99	4	4	1.16
3278.28	4	4	$9.6 \cdot 10^{-1}$
3278.91	6	4	1.0
3282.32	2	2	1.6
3287.66	8	10	1.4
3321.70	4	4	$7.2 \cdot 10^{-1}$
3322.94	10	10	$3.96 \cdot 10^{-1}$
3332.11	6	4	1.1
3361.23	8	10	1.1
3372.80	6	8	1.11
3383.77	4	6	1.09
3456.40	4	4	$8.2 \cdot 10^{-1}$
3483.63	10	8	$9.7 \cdot 10^{-1}$
3492.37	8	6	$9.8 \cdot 10^{-1}$
3504.90	10	10	$8.2 \cdot 10^{-1}$
3510.86	8	8	$9.3 \cdot 10^{-1}$
3535.41	4	6	$5.5 \cdot 10^{-1}$
3741.64	6	6	$6.2 \cdot 10^{-1}$
3759.30	8	8	$9.4 \cdot 10^{-1}$
3761.33	6	6	$9.9 \cdot 10^{-1}$
4911.18	6	4	$3.2 \cdot 10^{-1}$

Ti III

λ Å	Weights g_i	g_k	A 10^8 s^{-1}
865.79	5	3	$6.6 \cdot 10^{1}$
1002.37	5	5	7.6
1004.67	7	5	$4.3 \cdot 10^{1}$
1005.80	3	3	$1.3 \cdot 10^{1}$
1007.16	5	3	$3.8 \cdot 10^{1}$
1008.12	3	1	$5.1 \cdot 10^{1}$
1286.37	9	9	2.0
1289.30	7	7	2.2
1291.62	5	5	2.4
1293.23	9	7	1.0
1298.97	7	5	4.9
1327.59	5	3	3.2
1420.44	1	3	1.2
1421.63	3	1	4.0
1422.41	5	5	3.0
1424.14	5	3	1.6
1455.19	9	7	6.4
1498.70	5	5	2.8
2007.36	3	3	3.4
2007.60	1	3	1.2
2010.80	5	3	5.4
2097.30	5	7	3.3
2099.86	3	5	2.5
2104.86	3	3	1.1
2105.09	1	3	1.7
2199.22	3	3	5.7
2237.77	7	7	2.4
2331.35	3	1	4.3
2331.66	3	3	1.2
2339.00	5	3	3.0
2346.79	7	5	3.3
2374.99	5	3	4.0
2413.99	5	7	3.8
2516.05	7	9	3.4
2567.56	3	3	2.3
2984.75	5	5	1.9
3066.51	3	3	2.5
3228.89	3	3	1.5
3278.31	7	9	3.4
3320.94	3	5	2.8
3340.20	7	9	3.7
3346.18	9	11	3.7
3354.71	11	13	4.4
3397.24	3	1	1.8
3404.46	3	3	1.8
3417.62	3	5	1.9
3915.47	9	11	2.1
4119.14	5	5	$9.9 \cdot 10^{-1}$
4213.26	9	11	2.2
4215.53	9	11	2.2
4247.62	11	13	1.1
4248.54	5	7	2.3
4250.09	3	5	$9.5 \cdot 10^{-1}$
4259.01	11	13	$9.4 \cdot 10^{-1}$
4269.84	9	11	1.7
4285.61	13	15	3.0
4288.66	11	13	1.1
4296.70	11	13	1.6
4319.56	9	11	1.1
4343.25	3	1	1.0
4378.94	3	5	1.6
4433.91	11	13	1.8
4440.66	1	3	1.2
4533.26	3	5	1.5
4576.53	9	7	1.3
4628.07	3	1	1.5
4652.86	7	9	2.6
4874.00	5	7	1.5
4914.32	3	3	1.1
4971.19	9	11	2.1
5083.80	5	3	$9.7 \cdot 10^{-1}$
5278.33	3	3	$9.4 \cdot 10^{-1}$
7506.87	11	13	1.1

Ti IV

λ Å	Weights g_i	g_k	A 10^8 s^{-1}
423.49	4	6	$4.9 \cdot 10^{1}$
424.16	6	8	$5.3 \cdot 10^{1}$
433.63	4	2	5.5
433.76	6	4	5.0
729.36	4	2	5.7
1183.64	2	2	6.9
1195.21	4	2	$1.4 \cdot 10^{1}$
1451.74	2	4	$1.8 \cdot 10^{1}$
1467.34	4	6	$2.1 \cdot 10^{1}$
2067.56	2	4	5.1
2103.16	2	2	5.0
2541.79	4	6	6.9
2546.88	6	8	7.4
2862.60	4	2	4.1
3576.44	4	6	4.6

Tungsten

W I

λ Å	Weights g_i	g_k	A 10^8 s^{-1}
2879.4	1	3	$2.4 \cdot 10^{-1}$
2911.0	1	3	$7.7 \cdot 10^{-2}$
2923.5	7	9	$1.54 \cdot 10^{-2}$
2935.0	3	5	$1.5 \cdot 10^{-1}$
3013.8	7	9	$6.4 \cdot 10^{-2}$
3016.5	9	11	$9.27 \cdot 10^{-2}$
3017.4	7	9	$1.21 \cdot 10^{-1}$
3024.9	3	3	$1.4 \cdot 10^{-1}$
3046.4	3	5	$5.8 \cdot 10^{-2}$
3049.7	7	5	$1.7 \cdot 10^{-1}$
3064.9	5	7	$1.1 \cdot 10^{-2}$
3084.9	5	5	$1.3 \cdot 10^{-2}$
3093.5	7	9	$4.4 \cdot 10^{-2}$
3107.2	5	7	$2.33 \cdot 10^{-2}$
3108.0	7	9	$1.58 \cdot 10^{-2}$
3145.5	9	9	$4.8 \cdot 10^{-3}$
3170.2	7	5	$6.0 \cdot 10^{-3}$
3176.6	3	5	$2.12 \cdot 10^{-2}$
3183.5	7	7	$2.64 \cdot 10^{-3}$
3184.4	5	3	$2.3 \cdot 10^{-2}$
3191.6	1	3	$3.2 \cdot 10^{-2}$
3198.8	7	9	$4.6 \cdot 10^{-2}$
3207.3	7	9	$3.0 \cdot 10^{-2}$
3208.3	5	5	$4.4 \cdot 10^{-2}$
3215.6	9	11	$2.1 \cdot 10^{-1}$
3221.9	5	7	$1.61 \cdot 10^{-2}$
3223.1	5	3	$3.53 \cdot 10^{-3}$
3232.5	9	9	$2.4 \cdot 10^{-2}$
3235.1	7	5	$2.68 \cdot 10^{-3}$
3259.7	7	7	$1.3 \cdot 10^{-2}$
3300.8	7	9	$8.1 \cdot 10^{-2}$
3311.4	7	5	$5.6 \cdot 10^{-2}$
3363.3	9	7	$6.6 \cdot 10^{-3}$
3371.0	7	5	$1.0 \cdot 10^{-2}$
3371.4	3	3	$6.7 \cdot 10^{-3}$
3386.1	7	7	$2.64 \cdot 10^{-3}$
3413.0	7	9	$9.7 \cdot 10^{-3}$
3459.5	9	9	$2.04 \cdot 10^{-3}$
3510.0	7	9	$5.2 \cdot 10^{-3}$
3545.2	1	3	$3.2 \cdot 10^{-2}$
3570.6	5	3	$6.7 \cdot 10^{-3}$
3606.1	3	5	$9.6 \cdot 10^{-3}$
3617.5	7	7	$1.1 \cdot 10^{-1}$
3631.9	3	5	$1.3 \cdot 10^{-2}$
3675.6	9	11	$1.20 \cdot 10^{-2}$
3682.1	9	11	$2.0 \cdot 10^{-2}$
3707.9	7	7	$2.9 \cdot 10^{-2}$
3757.9	7	9	$1.38 \cdot 10^{-2}$
3760.1	5	7	$1.99 \cdot 10^{-2}$
3768.5	3	3	$3.47 \cdot 10^{-2}$
3780.8	7	5	$4.2 \cdot 10^{-2}$
3809.2	7	5	$9.0 \cdot 10^{-3}$
3817.5	7	7	$3.1 \cdot 10^{-2}$
3829.1	3	3	$3.83 \cdot 10^{-3}$

λ Å	Weights g_i	g_k	A $10^8\ s^{-1}$
3835.1	5	5	$5.2\cdot10^{-2}$
3846.3	3	5	$2.14\cdot10^{-2}$
3847.5	1	3	$8.3\cdot10^{-3}$
3864.3	5	5	$5.6\cdot10^{-3}$
3868.0	7	9	$4.6\cdot10^{-2}$
3881.4	7	7	$3.6\cdot10^{-2}$
3968.5	1	3	$5.07\cdot10^{-3}$
3975.5	9	11	$4.1\cdot10^{-3}$
4001.4	9	9	$5.6\cdot10^{-3}$
4008.8	7	9	$1.63\cdot10^{-1}$
4019.3	5	3	$6.7\cdot10^{-3}$
4028.8	1	3	$2.0\cdot10^{-2}$
4045.6	7	5	$2.88\cdot10^{-2}$
4055.2	7	9	$1.79\cdot10^{-3}$
4070.0	7	5	$3.60\cdot10^{-2}$
4070.6	3	5	$5.6\cdot10^{-3}$
4074.4	7	7	$1.0\cdot10^{-1}$
4088.3	5	3	$4.13\cdot10^{-3}$
4102.7	9	7	$4.9\cdot10^{-2}$
4115.6	11	11	$4.8\cdot10^{-3}$
4137.5	5	7	$8.4\cdot10^{-3}$
4171.2	7	9	$8.6\cdot10^{-3}$
4203.8	9	7	$4.9\cdot10^{-3}$
4219.4	9	7	$6.1\cdot10^{-3}$
4244.4	9	11	$1.38\cdot10^{-2}$
4269.4	7	5	$3.04\cdot10^{-2}$
4283.8	9	7	$1.69\cdot10^{-3}$
4294.6	7	5	$1.2\cdot10^{-1}$
4302.1	7	7	$3.6\cdot10^{-2}$
4355.2	9	9	$5.1\cdot10^{-3}$
4361.8	9	7	$1.64\cdot10^{-3}$
4378.5	7	5	$3.48\cdot10^{-3}$
4458.1	3	5	$4.2\cdot10^{-3}$
4466.3	7	5	$1.5\cdot10^{-2}$
4472.5	13	11	$1.55\cdot10^{-3}$
4484.2	3	5	$5.6\cdot10^{-3}$
4492.3	9	11	$3.6\cdot10^{-3}$
4495.3	11	11	$3.3\cdot10^{-3}$
4504.8	9	7	$7.0\cdot10^{-3}$
4552.5	9	9	$1.42\cdot10^{-2}$
4586.8	1	3	$4.20\cdot10^{-3}$
4592.6	7	9	$3.4\cdot10^{-3}$
4609.9	7	9	$1.42\cdot10^{-2}$
4613.3	9	9	$2.9\cdot10^{-3}$
4634.8	9	9	$8.8\cdot10^{-3}$
4659.9	1	3	$1.0\cdot10^{-2}$
4680.5	7	7	$1.4\cdot10^{-2}$
4720.4	3	5	$3.22\cdot10^{-3}$
4729.6	7	5	$7.8\cdot10^{-3}$
4752.6	3	3	$5.20\cdot10^{-3}$
4757.5	7	5	$2.72\cdot10^{-3}$
4757.8	11	9	$4.1\cdot10^{-3}$
4788.4	9	11	$2.6\cdot10^{-3}$
4843.8	5	5	$1.9\cdot10^{-2}$
4886.9	9	11	$8.1\cdot10^{-3}$
4924.6	13	11	$1.75\cdot10^{-3}$
4931.6	7	5	$1.0\cdot10^{-2}$
4948.6	9	11	$1.36\cdot10^{-3}$
4972.6	9	11	$3.9\cdot10^{-3}$
4982.6	1	3	$4.17\cdot10^{-3}$

λ Å	Weights g_i	g_k	A $10^8\ s^{-1}$
4986.9	11	9	$6.3\cdot10^{-3}$
5006.2	9	7	$1.2\cdot10^{-2}$
5015.3	7	9	$5.4\cdot10^{-3}$
5040.4	3	5	$5.2\cdot10^{-3}$
5053.3	3	3	$1.9\cdot10^{-2}$
5071.5	13	11	$3.4\cdot10^{-3}$
5117.6	11	11	$1.61\cdot10^{-3}$
5124.2	5	5	$4.0\cdot10^{-3}$
5141.2	7	9	$1.12\cdot10^{-3}$
5224.7	7	5	$1.2\cdot10^{-2}$
5243.0	9	7	$1.1\cdot10^{-2}$
5254.5	7	5	$3.86\cdot10^{-3}$
5268.6	9	9	$1.4\cdot10^{-3}$
5500.5	11	9	$6.9\cdot10^{-3}$
5514.7	5	3	$7.3\cdot10^{-3}$
5537.7	9	11	$2.2\cdot10^{-3}$
5617.1	7	7	$1.47\cdot10^{-3}$
5631.9	9	7	$1.43\cdot10^{-3}$
5660.7	13	11	$6.8\cdot10^{-3}$
5675.4	5	5	$2.20\cdot10^{-3}$
5796.5	9	7	$2.21\cdot10^{-3}$
5891.6	7	7	$1.47\cdot10^{-3}$
5947.6	5	7	$2.40\cdot10^{-3}$
5965.9	7	5	$1.0\cdot10^{-2}$
6021.5	5	3	$8.7\cdot10^{-3}$
6081.4	5	3	$4.7\cdot10^{-3}$
6203.5	7	7	$3.0\cdot10^{-3}$
6285.9	7	5	$6.6\cdot10^{-3}$
6292.0	3	5	$2.26\cdot10^{-3}$
6303.2	9	9	$1.84\cdot10^{-3}$
6404.2	5	7	$1.50\cdot10^{-3}$
6439.7	9	9	$1.29\cdot10^{-3}$
6445.1	7	5	$6.4\cdot10^{-3}$
6532.4	3	5	$4.6\cdot10^{-3}$
6538.1	11	9	$2.7\cdot10^{-3}$
6563.2	5	5	$2.04\cdot10^{-3}$
6814.9	9	9	$1.46\cdot10^{-3}$
7285.8	13	11	$1.47\cdot10^{-3}$
7569.9	5	3	$3.73\cdot10^{-3}$
7664.9	5	3	$3.80\cdot10^{-3}$
8017.2	5	7	$1.6\cdot10^{-3}$
8358.7	5	7	$1.89\cdot10^{-3}$
9381.4	9	7	$1.53\cdot10^{-3}$

Uranium

U I

λ Å	Weights g_i	g_k	A $10^8\ s^{-1}$
3553.0	13	13	$2.0\cdot10^{-2}$
3553.0	9	7	$1.4\cdot10^{-2}$
3553.4	15	13	$2.2\cdot10^{-2}$
3554.5	11	9	$8.4\cdot10^{-3}$
3554.9	15	17	$7.9\cdot10^{-3}$
3555.3	13	15	$2.7\cdot10^{-2}$
3555.8	13	11	$4.1\cdot10^{-3}$
3556.9	13	11	$7.5\cdot10^{-3}$
3557.8	13	13	$2.9\cdot10^{-2}$
3558.0	11	13	$1.6\cdot10^{-2}$
3558.6	9	7	$3.9\cdot10^{-2}$
3559.4	7	9	$1.5\cdot10^{-2}$
3560.3	9	7	$6.4\cdot10^{-2}$
3561.4	15	13	$5.5\cdot10^{-2}$
3561.5	9	9	$2.5\cdot10^{-2}$
3561.8	13	11	$5.7\cdot10^{-2}$
3563.7	13	13	$2.9\cdot10^{-2}$
3563.8	7	7	$1.1\cdot10^{-2}$
3565.0	13	11	$2.9\cdot10^{-2}$
3566.0	13	15	$1.7\cdot10^{-2}$
3566.6	11	11	$2.4\cdot10^{-1}$
3568.8	13	13	$3.8\cdot10^{-2}$
3569.1	17	15	$1.1\cdot10^{-1}$
3569.4	9	9	$1.5\cdot10^{-2}$
3570.1	13	11	$1.3\cdot10^{-2}$
3570.2	11	9	$5.3\cdot10^{-3}$
3570.6	13	15	$2.7\cdot10^{-2}$
3570.7	15	15	$1.2\cdot10^{-2}$
3571.2	11	11	$6.3\cdot10^{-3}$
3571.6	17	15	$1.3\cdot10^{-1}$
3572.9	13	15	$1.5\cdot10^{-2}$
3573.9	13	11	$4.0\cdot10^{-2}$
3574.1	13	15	$3.5\cdot10^{-2}$
3574.8	13	15	$1.9\cdot10^{-2}$
3577.1	17	15	$4.3\cdot10^{-2}$
3577.5	15	13	$7.8\cdot10^{-3}$
3577.8	11	11	$8.3\cdot10^{-3}$
3577.9	13	13	$2.3\cdot10^{-2}$
3578.3	13	11	$2.0\cdot10^{-2}$
3580.0	9	9	$1.2\cdot10^{-2}$
3580.2	11	9	$2.9\cdot10^{-2}$
3580.4	11	13	$7.5\cdot10^{-3}$
3580.9	13	13	$2.1\cdot10^{-2}$
3582.6	13	13	$2.9\cdot10^{-2}$
3584.6	7	5	$2.4\cdot10^{-2}$
3584.9	13	15	$1.8\cdot10^{-1}$
3585.4	11	11	$1.9\cdot10^{-2}$
3585.8	11	9	$2.8\cdot10^{-2}$
3587.8	9	11	$1.3\cdot10^{-2}$
3588.3	7	9	$1.8\cdot10^{-2}$
3589.7	11	13	$2.1\cdot10^{-2}$
3589.8	15	13	$5.9\cdot10^{-2}$
3590.7	9	7	$2.2\cdot10^{-1}$
3591.7	11	9	$5.3\cdot10^{-2}$
3593.0	11	11	$1.4\cdot10^{-2}$
3593.2	13	15	$4.2\cdot10^{-2}$
3593.7	11	11	$7.2\cdot10^{-2}$

Vanadium

V I

λ Å	Weights g_i	g_k	A $10^8\ s^{-1}$
3053.65	4	4	1.3
3056.33	6	6	1.3
3060.46	8	8	1.4
3066.37	10	10	2.1
3183.41	6	8	2.4
3183.96	8	10	2.5
3183.98	4	6	2.4
3185.38	10	12	2.7
3205.58	8	10	1.3
3212.43	10	12	1.4
3377.62	6	6	$6.0\cdot10^{-1}$
3533.68	6	8	$5.2\cdot10^{-1}$

λ Å	Weights g_i	g_k	A 10^8 s^{-1}
3663.60	4	6	3.1
3667.74	6	8	2.7
3672.41	12	12	$9.2 \cdot 10^{-1}$
3673.41	8	10	2.7
3676.70	14	14	1.3
3680.12	10	12	2.2
3686.26	10	12	$2.3 \cdot 10^{-1}$
3687.50	12	14	2.9
3688.07	8	8	$3.5 \cdot 10^{-1}$
3692.22	6	6	$5.4 \cdot 10^{-1}$
3695.34	14	16	2.8
3695.86	4	4	$6.6 \cdot 10^{-1}$
3703.57	10	8	$9.2 \cdot 10^{-1}$
3704.70	8	6	$6.6 \cdot 10^{-1}$
3706.03	10	10	$5.2 \cdot 10^{-1}$
3708.71	12	12	$4.4 \cdot 10^{-1}$
3794.96	10	10	$2.3 \cdot 10^{-1}$
3806.79	10	10	$2.5 \cdot 10^{-1}$
3840.75	8	6	$5.48 \cdot 10^{-1}$
3855.85	10	8	$5.78 \cdot 10^{-1}$
3871.07	10	8	$2.8 \cdot 10^{-1}$
3902.26	10	10	$2.68 \cdot 10^{-1}$
3930.02	10	10	$3.3 \cdot 10^{-1}$
3934.01	8	8	$6.2 \cdot 10^{-1}$
3992.80	12	10	1.2
3998.73	14	12	1.0
4050.96	10	10	1.4
4051.35	12	12	1.3
4090.57	8	10	$8.5 \cdot 10^{-1}$
4092.68	8	10	$2.30 \cdot 10^{-1}$
4095.48	6	8	$7.2 \cdot 10^{-1}$
4099.78	6	8	$4.10 \cdot 10^{-1}$
4102.15	4	6	$7.1 \cdot 10^{-1}$
4104.77	10	8	2.1
4105.16	4	6	$4.9 \cdot 10^{-1}$
4109.78	2	4	$5.00 \cdot 10^{-1}$
4111.78	10	10	1.01
4115.18	8	8	$5.80 \cdot 10^{-1}$
4116.47	6	6	$3.2 \cdot 10^{-1}$
4123.50	4	2	1.00
4128.06	6	4	$7.70 \cdot 10^{-1}$
4131.99	8	6	$5.5 \cdot 10^{-1}$
4134.49	10	8	$2.90 \cdot 10^{-1}$
4232.46	10	10	$9.8 \cdot 10^{-1}$
4232.95	8	8	$7.7 \cdot 10^{-1}$
4268.64	14	14	1.2
4271.55	12	12	$9.6 \cdot 10^{-1}$
4276.95	10	10	$9.4 \cdot 10^{-1}$
4284.05	8	8	1.2
4291.82	12	14	$8.8 \cdot 10^{-1}$
4296.10	10	12	$7.7 \cdot 10^{-1}$
4297.67	8	10	$7.0 \cdot 10^{-1}$
4298.03	6	8	$7.8 \cdot 10^{-1}$
4379.23	10	12	1.1
4384.71	8	10	1.1
4389.98	6	8	$6.9 \cdot 10^{-1}$
4395.22	4	6	$5.5 \cdot 10^{-1}$
4406.64	10	10	$2.2 \cdot 10^{-1}$
4407.63	8	8	$4.4 \cdot 10^{-1}$
4408.20	6	6	$6.0 \cdot 10^{-1}$

λ Å	Weights g_i	g_k	A 10^8 s^{-1}
4452.01	14	16	$9.2 \cdot 10^{-1}$
4457.75	10	12	$2.7 \cdot 10^{-1}$
4460.33	10	8	$3.0 \cdot 10^{-1}$
4462.36	12	14	$7.6 \cdot 10^{-1}$
4468.00	8	10	$2.3 \cdot 10^{-1}$
4469.71	10	12	$6.2 \cdot 10^{-1}$
4474.04	10	8	$4.7 \cdot 10^{-1}$
4496.06	8	6	$4.0 \cdot 10^{-1}$
4524.21	12	10	$3.0 \cdot 10^{-1}$
4529.58	10	8	$2.4 \cdot 10^{-1}$
4545.40	10	12	$7.6 \cdot 10^{-1}$
4560.72	8	10	$7.0 \cdot 10^{-1}$
4571.79	6	8	$6.0 \cdot 10^{-1}$
4578.73	4	6	$6.8 \cdot 10^{-1}$
4757.47	4	2	$7.6 \cdot 10^{-1}$
4766.62	6	4	$5.6 \cdot 10^{-1}$
4776.36	8	6	$5.1 \cdot 10^{-1}$
4786.50	10	8	$4.7 \cdot 10^{-1}$
4796.92	12	10	$4.8 \cdot 10^{-1}$
4807.52	14	12	$5.8 \cdot 10^{-1}$
5193.00	12	12	$4.0 \cdot 10^{-1}$
5195.39	8	8	$2.3 \cdot 10^{-1}$
5234.08	10	10	$4.9 \cdot 10^{-1}$
5240.87	12	12	$4.3 \cdot 10^{-1}$
5415.25	12	14	$3.1 \cdot 10^{-1}$
5487.91	12	10	$2.9 \cdot 10^{-1}$
5507.75	10	8	$3.5 \cdot 10^{-1}$
6090.21	8	6	$2.60 \cdot 10^{-1}$

VII

λ Å	Weights g_i	g_k	A 10^8 s^{-1}
2527.90	13	13	$6.1 \cdot 10^{-1}$
2528.47	9	9	$5.2 \cdot 10^{-1}$
2528.83	11	11	$5.3 \cdot 10^{-1}$
2554.04	9	9	$5.4 \cdot 10^{-1}$
2589.10	9	9	$7.7 \cdot 10^{-1}$
2640.86	5	7	1.2
2677.80	3	5	$3.4 \cdot 10^{-1}$
2679.33	7	7	$3.4 \cdot 10^{-1}$
2683.09	1	3	$3.4 \cdot 10^{-1}$
2687.96	9	9	$7.6 \cdot 10^{-1}$
2689.88	3	1	$9.2 \cdot 10^{-1}$
2690.25	7	5	$3.4 \cdot 10^{-1}$
2690.79	5	3	$5.2 \cdot 10^{-1}$
2700.94	9	11	$3.5 \cdot 10^{-1}$
2706.17	7	9	$3.4 \cdot 10^{-1}$
2734.22	9	7	$6.2 \cdot 10^{-1}$
2753.41	13	11	$4.2 \cdot 10^{-1}$
2784.20	9	9	1.3
2787.91	7	9	$5.0 \cdot 10^{-1}$
2825.86	9	7	1.2
2843.82	7	5	$9.9 \cdot 10^{-1}$
2847.57	9	7	$4.6 \cdot 10^{-1}$
2854.34	11	9	$5.0 \cdot 10^{-1}$
2862.31	11	11	$3.6 \cdot 10^{-1}$
2868.11	5	3	2.1
2869.13	13	11	$4.8 \cdot 10^{-1}$
2882.49	5	5	$4.2 \cdot 10^{-1}$
2884.78	3	3	$5.6 \cdot 10^{-1}$
2889.61	3	1	1.9
2891.64	5	3	1.4
2892.43	9	9	$3.6 \cdot 10^{-1}$
2892.65	7	5	1.3
2893.31	9	7	1.2
2903.07	3	5	$3.4 \cdot 10^{-1}$
2906.45	7	7	$7.8 \cdot 10^{-1}$
2908.81	11	9	1.6
2910.01	5	5	1.1
2910.38	3	3	1.2
2911.05	7	9	$3.7 \cdot 10^{-1}$
2912.46	11	9	$5.0 \cdot 10^{-1}$
2915.88	9	7	$4.9 \cdot 10^{-1}$
2924.02	11	11	1.7
2924.63	9	9	1.2
2930.80	7	7	$5.8 \cdot 10^{-1}$
2941.37	11	9	$3.5 \cdot 10^{-1}$
2944.57	9	7	$7.6 \cdot 10^{-1}$
2948.08	9	11	$4.0 \cdot 10^{-1}$
2952.07	7	5	$7.2 \cdot 10^{-1}$
2955.58	7	9	$3.3 \cdot 10^{-1}$
2968.37	7	9	$7.0 \cdot 10^{-1}$
2972.26	5	7	$5.2 \cdot 10^{-1}$
2973.98	9	11	$3.5 \cdot 10^{-1}$
2985.18	7	9	$4.4 \cdot 10^{-1}$
3001.20	7	7	$7.5 \cdot 10^{-1}$
3014.82	5	3	$8.9 \cdot 10^{-1}$
3016.78	7	5	$5.0 \cdot 10^{-1}$
3020.21	9	7	$5.0 \cdot 10^{-1}$
3048.21	11	13	$7.0 \cdot 10^{-1}$
3063.25	9	11	1.0
3100.94	7	7	$5.8 \cdot 10^{-1}$
3113.56	11	11	$5.0 \cdot 10^{-1}$
3122.89	11	13	$7.6 \cdot 10^{-1}$
3134.93	13	13	$5.9 \cdot 10^{-1}$
3136.50	11	11	$5.3 \cdot 10^{-1}$
3139.73	9	9	$5.2 \cdot 10^{-1}$
3151.32	3	5	$4.4 \cdot 10^{-1}$
3190.69	9	9	$3.3 \cdot 10^{-1}$
3250.78	11	9	$5.2 \cdot 10^{-1}$
3251.87	5	7	$3.5 \cdot 10^{-1}$
3271.12	7	9	$6.9 \cdot 10^{-1}$
3276.12	9	11	$5.2 \cdot 10^{-1}$
3279.84	9	11	$5.8 \cdot 10^{-1}$
3287.71	5	7	$7.5 \cdot 10^{-1}$
3337.85	5	7	$5.3 \cdot 10^{-1}$
3517.30	9	7	$3.8 \cdot 10^{-1}$
3530.77	5	3	$4.5 \cdot 10^{-1}$
3545.19	7	5	$4.3 \cdot 10^{-1}$
3556.80	9	7	$5.1 \cdot 10^{-1}$
3592.01	7	5	$4.4 \cdot 10^{-1}$
3618.92	3	5	$3.3 \cdot 10^{-1}$

VIII

λ Å	Weights g_i	g_k	A 10^8 s^{-1}
2318.06	8	10	4.6
2323.82	6	8	3.8
2330.42	10	10	3.2
2331.75	8	8	2.5
2334.21	6	6	2.2
2337.13	4	4	2.7
2343.10	6	8	3.6
2358.73	6	8	4.2

λ Å	Weights g_i	g_k	A 10^8 s^{-1}
2366.31	8	10	4.2
2371.06	10	12	5.2
2373.06	4	6	2.9
2382.46	8	10	5.0
2393.58	6	8	4.3
2404.18	4	6	2.5
2516.14	10	10	3.7
2521.55	8	8	3.5
2548.21	6	4	2.0
2554.22	8	6	1.2
2593.05	6	6	2.8
2595.10	8	8	2.8

V IV

λ Å	Weights g_i	g_k	A 10^8 s^{-1}
677.345	9	9	6.7
680.632	9	7	$1.2\cdot10^{1}$
681.145	7	5	$1.1\cdot10^{1}$
682.455	7	7	6.5
682.923	5	5	6.9
684.450	7	5	7.7
691.530	5	3	$1.1\cdot10^{1}$
723.537	3	1	$1.5\cdot10^{1}$
724.068	5	5	$1.1\cdot10^{1}$
724.809	5	3	5.6
737.854	9	7	$2.4\cdot10^{1}$
750.110	5	5	$1.0\cdot10^{1}$
884.146	1	3	4.7
1071.05	5	5	6.1
1110.72	3	3	5.0
1112.20	7	7	6.3
1112.44	5	5	5.0
1127.84	7	5	8.9
1131.26	9	7	9.4
1194.46	7	5	$1.0\cdot10^{1}$
1226.52	5	5	$1.5\cdot10^{1}$
1243.72	3	1	9.4
1247.07	5	3	4.7
1272.97	3	1	$2.7\cdot10^{1}$
1304.17	3	5	$1.5\cdot10^{1}$
1305.42	5	7	7.0
1308.06	7	9	7.9
1309.50	5	5	8.7
1312.72	7	7	8.6
1317.57	5	7	8.7
1321.92	7	9	9.9
1326.81	3	5	4.0
1329.29	5	5	$1.5\cdot10^{1}$
1329.97	3	3	4.8
1330.36	1	3	6.0
1331.67	3	1	$1.7\cdot10^{1}$
1332.46	5	3	7.5
1334.49	9	9	8.3
1355.13	7	9	$2.5\cdot10^{1}$
1356.53	5	3	4.9
1395.00	5	7	$1.4\cdot10^{1}$
1400.42	5	7	7.5
1403.62	7	9	8.4
1412.69	3	3	$1.1\cdot10^{1}$
1414.41	5	7	$1.2\cdot10^{1}$
1414.84	5	5	4.6
1418.53	7	7	5.2
1419.58	7	9	$1.3\cdot10^{1}$
1423.72	3	5	7.1
1426.65	9	11	$2.2\cdot10^{1}$
1429.11	5	5	5.0
1434.84	7	7	5.4
1451.04	3	3	7.0
1454.00	5	3	$1.1\cdot10^{1}$
1520.14	5	7	7.2
1522.49	3	5	5.5
1601.92	3	3	$1.2\cdot10^{1}$
1611.88	7	7	5.2
1806.18	5	3	7.3
1809.85	3	1	7.2
1817.68	5	3	4.8
1825.84	7	5	5.3
1861.56	5	7	6.6
1939.07	7	9	5.8
1951.43	5	7	5.0
1963.10	3	5	4.8
1997.72	7	7	4.7
2084.43	5	5	4.0
2120.05	7	9	8.1
2141.20	3	5	7.0
2146.83	7	9	6.6
2149.85	5	7	5.1
2151.09	7	9	4.3
2155.34	11	13	$1.2\cdot10^{1}$
2446.80	9	11	5.3
2570.72	9	11	7.6
3284.56	7	9	5.3
3496.42	7	9	4.4
3514.25	9	11	4.7

Xenon

Xe I

λ Å	Weights g_i	g_k	A 10^8 s^{-1}
1043.8	1	3	$5.9\cdot10^{-1}$
1047.1	1	3	1.3
1050.1	1	3	$8.5\cdot10^{-2}$
1056.1	1	3	2.45
1061.2	1	3	$1.9\cdot10^{-1}$
1068.2	1	3	3.99
1085.4	1	3	$4.10\cdot10^{-1}$
1099.7	1	3	$4.34\cdot10^{-1}$
1110.7	1	3	1.5
1129.3	1	3	$4.4\cdot10^{-2}$
1170.4	1	3	1.6
1192.0	1	3	6.2
1250.2	1	3	$1.4\cdot10^{-1}$
1295.6	1	3	2.46
1469.6	1	3	2.81
4501.0	5	3	$6.2\cdot10^{-3}$
4524.7	5	5	$2.1\cdot10^{-3}$
4624.3	5	5	$7.2\cdot10^{-3}$
4671.2	5	7	$1.0\cdot10^{-2}$
4807.0	3	1	$2.4\cdot10^{-2}$
7119.6	7	9	$6.6\cdot10^{-2}$
7967.3	1	3	$3.0\cdot10^{-3}$
8409.2	5	3	$1.0\cdot10^{-2}$

Xe II

λ Å	Weights g_i	g_k	A 10^8 s^{-1}
4180.1	4	4	2.2
4330.5	6	8	1.4
4414.8	6	6	1.0
4603.0	4	4	$8.2\cdot10^{-1}$
4844.3	6	8	1.1
4876.5	6	8	$6.3\cdot10^{-1}$
5260.4	2	4	$2.2\cdot10^{-1}$
5262.0	4	4	$8.5\cdot10^{-1}$
5292.2	6	6	$8.9\cdot10^{-1}$
5372.4	4	2	$7.1\cdot10^{-1}$
5419.2	4	6	$6.2\cdot10^{-1}$
5439.0	4	2	$7.4\cdot10^{-1}$
5472.6	8	8	$9.9\cdot10^{-2}$
5531.1	8	6	$8.8\cdot10^{-2}$
5719.6	4	6	$6.1\cdot10^{-2}$
5976.5	4	4	$2.8\cdot10^{-1}$
6036.2	6	6	$7.5\cdot10^{-2}$
6051.2	8	6	$1.7\cdot10^{-1}$
6097.6	6	4	$2.6\cdot10^{-1}$
6270.8	4	6	$1.8\cdot10^{-1}$
6277.5	4	6	$3.6\cdot10^{-2}$
6805.7	8	6	$6.1\cdot10^{-2}$
6990.9	10	8	$2.7\cdot10^{-1}$

Ytterbium

Yb I

λ Å	Weights g_i	g_k	A 10^8 s^{-1}
2464.5	1	3	$9.1\cdot10^{-1}$
2672.0	1	3	$1.18\cdot10^{-1}$
3464.4	1	3	$6.2\cdot10^{-1}$
3988.0	1	3	1.76
5556.5	1	3	$1.14\cdot10^{-2}$

Yb II

λ Å	Weights g_i	g_k	A 10^8 s^{-1}
3289.4	2	4	1.8
3694.2	2	2	1.4

Yttrium

Y I

λ Å	Weights g_i	g_k	A 10^8 s^{-1}
2984.25	6	8	$4.8\cdot10^{-1}$
4077.36	4	6	1.1
4102.36	6	8	1.3
4128.30	6	6	1.6
4142.84	4	4	1.6
4167.51	6	6	$2.38\cdot10^{-1}$
4235.93	6	4	$3.0\cdot10^{-1}$
4379.33	6	4	$7.83\cdot10^{-1}$
4476.95	8	6	$2.8\cdot10^{-1}$
4514.01	4	6	$3.34\cdot10^{-1}$
4527.78	8	6	$8.33\cdot10^{-1}$
4544.31	6	6	$4.10\cdot10^{-1}$
4559.36	2	4	$4.0\cdot10^{-1}$
4643.70	4	6	$1.8\cdot10^{-1}$
4653.78	4	6	$1.6\cdot10^{-1}$
4674.85	6	8	$1.3\cdot10^{-1}$
4781.03	8	10	$1.0\cdot10^{-1}$
4799.30	6	8	$1.6\cdot10^{-1}$
4804.31	6	4	$2.6\cdot10^{-1}$

λ Å	Weights g_i	g_k	A $10^8\ s^{-1}$
4804.80	4	4	$3.84 \cdot 10^{-1}$
4845.67	8	8	$6.8 \cdot 10^{-1}$
4852.68	6	6	$6.2 \cdot 10^{-1}$
4856.71	6	6	$2.0 \cdot 10^{-1}$
4859.84	4	4	$7.26 \cdot 10^{-1}$
4893.44	6	4	$2.2 \cdot 10^{-1}$
4900.08	8	6	$2.0 \cdot 10^{-1}$
4906.11	10	8	$1.2 \cdot 10^{-1}$
5380.63	6	4	$3.2 \cdot 10^{-1}$
5424.36	6	4	$3.47 \cdot 10^{-1}$
5466.47	10	12	$6.3 \cdot 10^{-1}$
5513.65	6	6	$2.39 \cdot 10^{-1}$
5527.56	8	10	$5.4 \cdot 10^{-1}$
5606.34	10	10	$5.84 \cdot 10^{-2}$
5630.14	4	6	$4.9 \cdot 10^{-1}$
5675.27	6	6	$9.3 \cdot 10^{-2}$
5732.09	6	6	$7.5 \cdot 10^{-2}$
6087.94	6	4	$1.1 \cdot 10^{-1}$
6437.17	10	8	$4.8 \cdot 10^{-2}$
6538.57	10	10	$1.5 \cdot 10^{-1}$
6815.15	2	4	$7.18 \cdot 10^{-2}$
7035.15	4	4	$6.3 \cdot 10^{-2}$

Y II

λ Å	Weights g_i	g_k	A $10^8\ s^{-1}$
3112.03	1	3	$1.3 \cdot 10^{-2}$
3179.42	3	5	$3.8 \cdot 10^{-2}$
3195.62	3	3	$8.23 \cdot 10^{-1}$
3200.27	5	5	$4.8 \cdot 10^{-1}$
3203.32	3	1	2.77
3216.69	5	3	2.0
3242.28	7	5	2.0
3448.81	5	5	$4.1 \cdot 10^{-2}$
3467.88	5	3	$2.7 \cdot 10^{-2}$
3496.08	1	3	$3.49 \cdot 10^{-1}$
3549.01	5	7	$3.97 \cdot 10^{-1}$
3584.51	3	5	$4.02 \cdot 10^{-1}$
3600.74	7	7	1.4
3601.91	3	3	1.13
3611.04	5	5	1.04
3628.70	5	3	$3.3 \cdot 10^{-1}$
3664.62	7	5	$3.7 \cdot 10^{-1}$
3710.29	7	9	1.5
3747.55	3	3	$1.9 \cdot 10^{-1}$
3774.34	5	7	1.1
3776.56	5	3	$2.42 \cdot 10^{-1}$
3788.70	3	5	$8.1 \cdot 10^{-1}$
3818.34	5	5	$9.70 \cdot 10^{-2}$
3832.90	7	7	$3.0 \cdot 10^{-1}$
3878.29	7	5	$2.9 \cdot 10^{-2}$
3930.66	5	5	$2.1 \cdot 10^{-2}$
3950.36	3	5	$2.80 \cdot 10^{-1}$
3951.59	5	3	$1.5 \cdot 10^{-2}$
3982.60	5	5	$2.7 \cdot 10^{-1}$
4124.91	5	7	$1.8 \cdot 10^{-2}$
4177.54	5	5	$5.27 \cdot 10^{-1}$
4199.27	3	5	$5.36 \cdot 10^{-3}$
4204.69	1	3	$2.20 \cdot 10^{-2}$
4235.73	5	5	$2.3 \cdot 10^{-2}$
4309.62	7	5	$1.29 \cdot 10^{-1}$
4358.73	3	3	$5.55 \cdot 10^{-2}$
4374.95	5	5	$9.97 \cdot 10^{-1}$
4398.01	5	3	$1.16 \cdot 10^{-1}$
4422.59	3	1	$1.83 \cdot 10^{-1}$
4682.33	5	5	$1.9 \cdot 10^{-2}$
4786.58	7	7	$2.1 \cdot 10^{-2}$
4823.31	5	5	$4.3 \cdot 10^{-2}$
4854.87	5	3	$3.9 \cdot 10^{-1}$
4881.44	5	3	$1.5 \cdot 10^{-3}$
4883.69	9	7	$4.7 \cdot 10^{-1}$
4900.11	7	5	$4.51 \cdot 10^{-1}$
4982.13	7	9	$1.5 \cdot 10^{-2}$
5087.42	9	9	$2.0 \cdot 10^{-1}$
5119.11	5	7	$1.6 \cdot 10^{-2}$
5200.41	5	5	$1.3 \cdot 10^{-1}$
5205.73	7	7	$1.6 \cdot 10^{-1}$
5289.82	7	5	$6.7 \cdot 10^{-3}$
5320.78	9	7	$3.9 \cdot 10^{-3}$
5473.39	3	5	$4.3 \cdot 10^{-2}$
5480.73	1	3	$7.62 \cdot 10^{-2}$
5497.41	5	5	$1.2 \cdot 10^{-1}$
5509.90	5	5	$4.24 \cdot 10^{-2}$
5544.61	3	1	$1.8 \cdot 10^{-1}$
5546.01	5	3	$5.8 \cdot 10^{-2}$
5728.89	5	5	$3.0 \cdot 10^{-2}$
6613.74	5	7	$1.7 \cdot 10^{-2}$
6832.48	5	5	$3.3 \cdot 10^{-3}$
7264.16	5	3	$1.3 \cdot 10^{-2}$

Zinc

Zn I

λ Å	Weights g_i	g_k	A $10^8\ s^{-1}$
748.29	1	3	$6.0 \cdot 10^{-2}$
765.60	1	3	$7.6 \cdot 10^{-2}$
792.05	1	3	$5.7 \cdot 10^{-2}$
793.85	1	3	$1.8 \cdot 10^{-1}$
809.92	1	3	$2.6 \cdot 10^{-1}$
1109.1	1	3	$3.05 \cdot 10^{-1}$
2138.6	1	3	7.09
3075.9	1	3	$3.29 \cdot 10^{-4}$
3282.3	1	3	$9.0 \cdot 10^{-1}$
3302.6	3	5	1.2
3302.9	3	3	$6.7 \cdot 10^{-1}$
3345.0	5	7	1.7
3345.6	5	5	$4.0 \cdot 10^{-1}$
3345.9	5	3	$4.5 \cdot 10^{-2}$
6362.3	3	5	$4.74 \cdot 10^{-1}$
11054	3	1	$2.43 \cdot 10^{-1}$

Zn II

λ Å	Weights g_i	g_k	A $10^8\ s^{-1}$
2025.5	2	4	3.3
2064.2	2	4	4.6
2099.9	4	6	5.6
2102.2	4	4	$9.3 \cdot 10^{-1}$
4911.6	4	6	1.6

ELECTRON AFFINITIES

Thomas M. Miller

Electron affinity is defined as the energy difference between the lowest (ground) state of the neutral and the lowest state of the corresponding negative ion. The accuracy of electron affinity measurements has been greatly improved since the advent of laser photodetachment experiments with negative ions. Electron affinities can be determined with optical precision, though a detailed understanding of atomic and molecular states and splittings is required to specify the photodetachment threshold corresponding to the electron affinity.

Atomic and molecular electron affinities are discussed in two excellent articles reviewing photodetachment studies which appear in *Gas Phase Ion Chemistry*, Vol. 3, Bowers, M. T., Ed., Academic Press, Orlando, 1984: Chapter 21 by Drzaic, P. S., Marks, J., and Brauman, J. I., "Electron Photodetachment from Gas Phase Negative Ions," p. 167, and Chapter 22 by Mead, R. D., Stevens, A. E., and Lineberger, W. C., "Photodetachment in Negative Ion Beams," p. 213. Persons interested in photodetachment details should consult these articles and the critical reviews of Andersen, T., Haugen, H. K., and Hotop, H., *J. Phys. Chem. Ref. Data* 28, 1511, 1999, Hotop, H., and Lineberger, W. C., J. *Phys. Chem. Ref. Data* 14, 731, 1985, and Andersen, T., Haugen, H. K., and Hotop, H. *J. Phys. Chem. Ref. Data* 28, 1511, 1999. For simplicity in the tables below, any electron affinity which was discussed in the articles by Drzaic et al. or Hotop and Lineberger is referenced to these sources, where original references are given. The development of cluster-ion photodetachment apparatuses has brought an explosion of electron affinity estimates for atomic and molecular clusters. The policy in this tabulation is to list the electron affinities for the atoms, diatoms, and triatoms, if adiabatic electron affinities have been determined, but to refer the reader to original sources for higher-order clusters. Additional data on molecular electron affinities may be found in Lias, S. G., Bartmess, J. E., Liebman, J. F., Holmes, J. L., Levin, R. D., and Mallard, W. G., Gas Phase Ion and Neutral Thermochemistry, *J. Phys. Chem. Ref. Data* 17, (Supplement No. 1), 1988 and on the NIST WebBook at the Internet address http://webbook.nist.gov/.

For the present tabulation, the 2006 CODATA value $e/(hc)$ = 8065.54465 ± 0.00020 cm^{-1} eV^{-1} (http://physics.nist.gov/constants/) has been used to convert electron affinities from the units used in spectroscopic work, cm^{-1}, into eV for these tables. Experimental measurements have improved to the level that the 25 ppb uncertainty in $e/(hc)$ will make a difference in a few cases. For this reason, very accurate electron affinities will be given in cm^{-1} with the relevant references.

Abbreviations used in the tables: calc = calculated value; PT = photodetachment threshold using a lamp as a light source; LPT = laser photodetachment threshold; LPES = laser photoelectron spectroscopy; DA = dissociative attachment; attach = electron attachment/detachment equilibrium; e-scat = electron scattering; kinetic = dissociation kinetics; Knud = Knudsen cell; CT = charge transfer; CD = collisional detachment; and ZEKE = zero electron kinetic energy spectroscopy.

TABLE 1. Atomic Electron Affinities

Atomic number	Atom	Electron affinity in eV	Uncertainty in eV	Method	Ref.	
1	H	0.754195	0.000019	LPT	89	
		0.75420812	—	calc	205	
	D	0.754593	0.000074	LPT	89	deuterium
	D	0.75465624	—	calc	205	deuterium
	T	0.75480540	—	calc	205	tritium
2	He	not stable	—	calc	1	
3	Li	0.618049	0.000020	LPT	185	
4	Be	not stable	—	calc	1	
5	B	0.279723	0.000025	LPES	191	
6	C	1.262119	0.000020	LPT	28	
7	N	not stable	—	DA	1	
8	O	1.4611135	0.0000012	LPT	334	
9	F	3.4011895	0.0000025	LPT	227	
10	Ne	not stable	—	calc	1	
11	Na	0.547926	0.000025	LPT	1	
12	Mg	not stable	—	e-scat	1	
13	Al	0.43283	0.00005	LPES	208	
14	Si	1.3895213	0.0000013	LPES	335	
15	P	0.7465	0.0003	LPT	1	

Atomic number	Atom	Electron affinity in eV	Uncertainty in eV	Method	Ref.
16	S	2.07710418	0.00000071	LPT	334
17	Cl	3.612724	0.000027	LPT	52
18	Ar	not stable	—	calc	1
19	K	0.50147	0.00010	LPT	1
20	Ca	0.02455	0.00010	LPT	44
21	Sc	0.188	0.020	LPES	1
22	Ti	0.079	0.014	LPES	1
23	V	0.525	0.012	LPES	1
24	Cr	0.666	0.012	LPES	1
25	Mn	not stable	—	calc	1
26	Fe	0.151	0.003	LPES	27
27	Co	0.662	0.003	LPES	27
28	Ni	1.156	0.010	LPES	1
29	Cu	1.235	0.005	LPES	37
30	Zn	not stable	—	e-scat	1
31	Ga	0.43	0.03	LPES	183
32	Ge	1.232712	0.000015	LPES	28
33	As	0.804	0.002	LPES	352
34	Se	2.020670	0.000025	LPT	1
35	Br	3.363588	0.000002	LPT	74
36	Kr	not stable	—	calc	1
37	Rb	0.48592	0.00002	LPT	1
38	Sr	0.048	0.006	LPT	122
39	Y	0.307	0.012	LPES	1
40	Zr	0.426	0.014	LPES	1
41	Nb	0.916	0.005	LPES	311
42	Mo	0.748	0.002	LPES	127
43	Tc	0.55	0.20	calc	1
44	Ru	1.05	0.15	calc	1
45	Rh	1.137	0.008	LPES	1
46	Pd	0.562	0.005	LPES	116
47	Ag	1.302	0.007	LPES	1
48	Cd	not stable	—	e-scat	1
49	In	0.3	0.2	PT	1
50	Sn	1.112067	0.000015	LPES	28
51	Sb	1.046	0.005	LPES	108
52	Te	1.970876	0.000007	LPT	261
53	I	3.059037	0.000010	LPT	92
54	Xe	not stable	—	calc	1
55	Cs	0.471626	0.000025	LPT	1
56	Ba	0.14462	0.00006	LPT	195
57	La	0.47	0.02	LPT	184
58	Ce	0.65	0.03	LPT	269
59	Pr	0.962	0.024	LPES	225
60	Nd	>1.916	—	LPES	342
63	Eu	0.864	0.024	LPES	268
65	Tb	>1.165	—	LPES	342
66	Dy	>0	—	LPES	342
69	Tm	1.029	0.022	LPES	264
70	Yb	−0.020	—	calc	196
71	Lu	0.34	0.01	LPT	223
72	Hf	0.017	—	calc	343
73	Ta	0.322	0.012	LPES	1
74	W	0.815	0.002	LPES	37
75	Re	0.15	0.15	calc	1
76	Os	1.1	0.2	calc	1
77	Ir	1.5638	0.0005	LPT	141
78	Pt	2.128	0.002	LPT	1
79	Au	2.30863	0.00003	LPT	1

Atomic number	Atom	Electron affinity in eV	Uncertainty in eV	Method	Ref.
80	Hg	not stable	—	e-scat	1
81	Tl	0.377	0.013	LPES	341
82	Pb	0.364	0.008	LPES	1
83	Bi	0.942362	0.000013	LPT	262
84	Po	1.9	0.3	calc	1
85	At	2.8	0.2	calc	1
86	Rn	not stable	—	calc	1
87	Fr	0.486	0.002	calc	82
88	Ra	0.10	—	calc	273
89	Ac	0.35	—	calc	207
118	ekaradon	0.056	0.01	calc	140
114	—	<0	—	calc	354
121	ekaactinium	0.57	—	calc	207
57–71	lanthanides	—	—	calc	355
89–103	actinides	—	—	calc	355

TABLE 2. Electron Affinities for Diatomic Molecules

Molecule	Electron affinity in eV	Uncertainty in eV	Method	Ref.
Ag_2	1.023	0.007	LPES	37
AgO	1.654	0.002	LPES	233
Al_2	1.10	0.15	LPES	68
AlO	2.60	0.02	LPES	143
AlP	2.043	0.020	LPES	218
AlS	2.60	0.03	LPES	129
As_2	0.739	0.008	LPES	200
AsH	1.0	0.1	PT	2
AsO	1.286	0.008	LPES	198
Au_2	1.938	0.007	LPES	37
AuO	2.374	0.007	LPES	282
AuPd	1.88	—	LPES	220
AuS	2.469	0.006	LPES	282
BN	3.160	0.005	LPES	189
BO	2.508	0.008	LPES	6
BeH	0.7	0.1	PT	2
Bi_2	1.271	0.008	LPES	119
Br_2	2.55	0.10	CT	2
BrO	2.353	0.006	LPES	88
C_2	3.269	0.006	LPES	87
CH	1.238	0.008	LPES	2
CN	3.862	0.004	LPES	111
CRh	1.46	0.02	LPES	206
CS	0.205	0.021	LPES	2
CaH	0.93	0.05	PT	2
Cl_2	2.38	0.10	CT	2
ClO	2.2775	0.0013	LPES	339
Co_2	1.110	0.008	LPES	27
CoD	0.680	0.010	LPES	29
CoH	0.671	0.010	LPES	29
Cr_2	0.505	0.005	LPES	114
CrD	0.568	0.010	LPES	29
CrH	0.563	0.010	LPES	29
CrO	1.221	0.006	LPES	5
Cs_2	0.469	0.015	LPES	104
CsCl	0.455	0.010	LPES	30
CsO	0.273	0.012	LPES	133
Cu_2	0.836	0.006	LPES	37
CuD	0.439	0.006	LPES	308
CuD_2	2.60	0.05	LPES	308
CuH	0.444	0.006	LPES	308
CuH_2	2.60	0.05	LPES	308
CuO	1.777	0.006	LPES	118
F_2	3.01	0.07	kinetic	331
FO	2.272	0.006	LPES	88
Fe_2	0.902	0.008	LPES	27
FeD	0.932	0.015	LPES	9
FeH	0.934	0.011	LPES	9
FeO	1.493	0.005	LPES	45
GaAs	1.949	0.020	LPES	218
GaO	2.612	0.008	LPES	279
GaP	1.988	0.020	LPES	218
Ge_2	2.035	0.001	LPES	123
I_2	2.524	0.015	LPES	305
IBr	2.512	0.003	LPES	350
IO	2.370	0.006	LPES	88
InP	1.845	0.020	LPES	218
K_2	0.497	0.012	LPES	104
KBr	0.642	0.010	LPES	30
KCl	0.582	0.010	LPES	30
KCs	0.471	0.020	LPES	104
KI	0.728	0.010	LPES	30
KRb	0.486	0.020	LPES	104
LiCl	0.593	0.010	LPES	30
LiD	0.337	0.012	LPES	102
LiH	0.342	0.012	LPES	102
MgCl	1.589	0.011	LPES	31
MgH	1.05	0.06	PT	2
MgI	1.899	0.018	LPES	31
MgO	1.630	0.025	LPES	178
MnD	0.866	0.010	LPES	9
MnH	0.869	0.010	LPES	9
MnO	1.375	0.010	LPES	158
MoO	1.290	0.006	LPES	127
NH	0.370	0.004	LPT	32
NO	0.026	0.005	LPES	73

Molecule	Electron affinity in eV	Uncertainty in eV	Method	Ref.
NRh	1.51	0.02	LPES	206
NS	1.194	0.011	LPES	2
Na_2	0.430	0.015	LPES	104
NaBr	0.788	0.010	LPES	30
NaCl	0.727	0.010	LPES	30
NaF	0.520	0.010	LPES	30
NaI	0.865	0.010	LPES	30
NaK	0.465	0.030	LPES	104
NbO	1.29	0.02	LPES	174
Ni_2	0.926	0.010	LPES	112
NiCu	0.889	0.010	LPES	128
NiAg	0.979	0.010	LPES	128
NiD	0.477	0.007	LPES	29
NiH	0.481	0.007	LPES	29
NiO	1.470	0.003	LPES	146
O_2	0.450	0.002	LPES	222
OD	1.825533	0.000037	LPT	142
OH	1.8276487	0.0000011	LPT	226
ORh	1.58	0.02	LPES	206
P_2	0.589	0.025	LPES	42
PH	1.027	0.006	LPES	281
PO	1.092	0.010	LPES	2
Pb_2	1.366	0.010	LPES	117
PbO	0.722	0.006	LPES	105
PbS	1.049	0.010	LPES	228
Pd_2	1.685	0.008	LPES	112
PdCO	0.604	0.010	LPES	160
PdO	1.570	0.006	LPES	290
Pt_2	1.898	0.008	LPES	112
PtN	1.240	0.010	LPES	46
Rb_2	0.498	0.015	LPES	104
RbCl	0.544	0.010	LPES	30
RbCs	0.478	0.020	LPES	104
Re_2	1.571	0.008	LPES	33
S_2	1.670	0.015	LPES	53
SD	2.315	0.002	LPES	10
SF	2.285	0.006	LPES	93
SH	2.3147282	0.0000017	LPT	47
SO	1.125	0.005	LPES	84
Sb_2	1.282	0.008	LPES	108
ScO	1.35	0.02	LPES	171
Se_2	1.94	0.07	LPES	38
SeH	2.212519	0.000025	LPT	48
SeO	1.456	0.020	LPES	41
Si_2	2.201	0.010	LPES	100
SiF	0.81	0.02	LPES	278
SiH	1.277	0.009	LPES	2
SiN	2.949	0.008	LPES	274
Sn_2	1.962	0.010	LPES	117
SnO	0.598	0.006	LPES	168
SnPb	1.569	0.008	LPES	117
TaO	1.07	0.06	LPES	360
Te_2	1.92	0.07	LPES	38
TeH	2.102	0.015	LPES	39
TeO	1.697	0.022	LPES	40
TiO	1.30	0.03	LPES	172
VO	1.229	0.008	LPES	170
YO	1.35	0.02	LPES	171
ZnF	1.974	0.008	LPES	179
ZnH	<0.95	—	PT	2
ZnO	2.087	0.008	LPES	179
ZrO	1.3	0.3	LPES	173

TABLE 3. Electron Affinities for Triatomic Molecules

Molecule	Electron affinity in eV	Uncertainty in eV	Method	Ref.	
Ag_3	2.32	0.05	LPES	37	
AgCN	1.588	0.010	LPES	163	
Al_3	1.916	0.004	LPES	332	
AlO_2	4.23	0.02	LPES	143	
AlP_2	1.933	0.007	LPES	217	
Al_2N	2.571	0.008	LPES	297	
Al_2P	2.513	0.020	LPES	217	
Al_2S	0.80	0.12	LPES	129	
As_3	1.45	0.03	LPES	200	
AsH_2	1.27	0.03	PT	2	
Au_3	3.7	0.3	LPES	37	
AuBO	1.46	0.02	LPES	336	
$AuBr_2$	4.46	0.07	LPES	294	
$AuCl_2$	4.60	0.07	LPES	294	
AuI_2	4.18	0.07	LPES	294	
AuOH	1.771	0.015	LPES	312	
Au_2H	3.55	0.03	LPES	276	
Au_2Pd	3.80	—	LPES	220	
BO_2	4.46	0.03	LPES	338	
B_2N	3.098	0.005	LPES	193	
B_3	2.82	0.02	LPES	221	
Bi_3	1.60	0.03	LPES	119	
C_3	1.981	0.020	LPES	11	
CBr_2	1.78	0.10	LPES	249	
CCl_2	1.593	0.006	LPES	249	
CD_2	0.645	0.006	LPES	12	
CDF	0.535	0.005	LPES	95	
CF_2	0.180	0.020	LPES	235	
CH_2	0.652	0.006	LPES	12	
CHBr	1.454	0.005	LPES	95	
CHCl	1.210	0.005	LPES	95	
CHF	0.542	0.005	LPES	95	
CHI	1.42	0.17	LPES	95	
CI_2	2.09	0.07	LPES	235	
C_2Cr	2.30	1.617	0.015	271	
C_2H	2.969	0.006	LPES	87	
C_2N	2.7489	0.0010	LPES	346	
C_2Nb	1.380	0.025	LPES	243	
C_2O	2.3107	0.0006	LPES	323	CCO
C_2S	2.7475	0.0006	LPES	323	CCS
COS	−0.04	—	LPES	272	
CS_2	0.58	0.05	LPES	278	
C_2Ti	1.542	0.020	LPES	147	
ClO_2	1.6600	0.0002	LPES	339	ClOO

Molecule	Electron affinity in eV	Uncertainty in eV	Method	Ref.	
CoD_2	1.465	0.013	LPES	34	
CoH_2	1.450	0.014	LPES	34	
CrH_2	>2.5	—	LPES	34	
Cr_2D	1.464	0.005	LPES	107	
Cr_2H	1.474	0.005	LPES	107	
Cr_2O	0.9	0.1	LPES	306	
CrO_2	2.413	0.008	LPES	144	OCrO
CrO_2	1.5	0.06	LPES	241	$Cr(O_2)$
Cs_3	0.864	0.030	LPES	18	
Cu_3	2.11	0.05	LPES	37	
CuCN	1.466	0.010	LPES	163	
$CuCl_2$	4.35	0.05	LPES	237	
$CuBr_2$	4.35	0.05	LPES	237	
DCO	0.301	0.005	LPES	35	
DNO	0.330	0.015	LPES	14	
DO_2	1.077	0.005	LPES	15	
DS_2	1.918	0.015	LPES	347	
Fe_3	1.43	0.06	LPES	149	
FeC_2	1.9782	0.0006	LPES	254	
FeCO	1.157	0.005	LPES	103	
FeD_2	1.038	0.013	LPES	34	
FeH_2	1.049	0.014	LPES	34	
FeO_2	2.358	0.030	LPES	130	
Fe_2H	0.564	0.019	LPES	254	
Fe_2O	1.60	0.02	LPES	152	
$GaAs_2$	1.891	0.033	LPES	192	
GaP_2	1.666	0.041	LPES	192	
Ga_2As	2.428	0.020	LPES	192	
Ga_2N	2.506	0.008	LPES	302	
Ga_2P	2.481	0.015	LPES	192	
Ge_3	2.23	0.01	LPES	123	
GeH_2	1.097	0.015	LPES	28	
HCO	0.313	0.005	LPES	35	
HCl_2	4.896	0.005	LPES	69	
HNO	0.338	0.015	LPES	14	
HO_2	1.078	0.006	LPES	15	
HS_2	1.916	0.015	LPES	347	
HfO_2	2.14	0.03	LPES	319	
I_3	4.226	0.013	LPES	162	
InP_2	1.61	0.05	LPES	137	
In_2P	2.36	0.05	LPES	137	
K_3	0.956	0.050	LPES	18	
MnD_2	0.465	0.014	LPES	34	
MnH_2	0.444	0.016	LPES	34	
MnO_2	2.06	0.03	LPES	158	
N_3	2.70	0.12	PT	2	
N_3	2.68	0.01	LPT	255	
NCN	2.484	0.006	LPES	154	
NCO	3.609	0.005	LPES	111	
NCS	3.537	0.005	LPES	111	

Molecule	Electron affinity in eV	Uncertainty in eV	Method	Ref.	
NH_2	0.771	0.005	LPES	58	
N_2O	−0.03	0.10	calc	59	
NO_2	2.273	0.005	LPES	63	
(NO)R	R=Ar,Kr,Xe	—	LPES	90	
Na_3	1.019	0.060	LPES	18	
Nb_3	1.032	0.010	LPES	175	
Ni_3	1.41	0.05	LPES	55	
NiCN	1.771	0.010	LPES	287	
NiCO	0.804	0.012	LPES	2	
NiD_2	1.926	0.007	LPES	34	
NiH_2	1.934	0.008	LPES	34	
NiO_2	3.05	0.01	LPES	214	ONiO
NiO_2	0.82	0.03	LPES	214	$Ni(O_2)$
O_3	2.1028	0.0025	LPT	2	
O_4Ar	0.52	0.02	LPES	75	
OClO	2.1451	0.0025	LPES	339	
OIO	2.577	0.008	LPES	88	
PH_2	1.263	0.006	LPES	281	
P_2H	1.514	0.010	LPES	281	
PO_2	3.42	0.01	LPES	124	
Pb_3	1.70	0.09	LPES	345	
Pd_3	<1.5	0.1	LPES	55	
PdCN	2.543	0.007	LPES	287	
PdCO	0.606	0.010	LPES	293	
Pt_3	1.87	0.02	LPES	55	
PtCN	3.191	0.003	LPES	287	
PtCO	1.212	0.010	LPES	293	
Rb_3	0.920	0.030	LPES	18	
ReO_2	2.5	0.1	LPES	216	
S_3	2.093	0.025	LPES	16	
SO_2	1.107	0.008	LPES	16	
S_2O	1.877	0.008	LPES	16	
Sb_3	1.85	0.03	LPES	108	
ScO_2	2.32	0.02	LPES	171	
SeO_2	1.823	0.050	LPES	38	
SiF_2	0.10	0.10	LPES	278	
Si_2F	1.99	0.28	LPES	17	
SiH_2	1.124	0.020	LPES	2	
Si_2H	2.31	0.01	LPES	182	
Si_3	2.29	0.02	LPES	110	
Sn_3	2.24	0.01	LPES	289	
SnCN	1.922	0.006	LPES	292	
Ta_3	1.36	0.03	LPES	169	
TaO_2	2.40	0.06	LPES	360	
TiO_2	1.59	0.03	LPES	172	
V_3	1.107	0.010	LPES	176	
VO_2	2.3	0.2	CT	101	
WO_2	1.998	0.010	LPES	299	
YO_2	2.00	0.03	LPES	171	
ZrO_2	1.64	0.03	LPES	319	

TABLE 4. Electron Affinities for Larger Polyatomic Molecules

Molecule	Electron Affinity in eV	Uncertainty in eV	Method	Ref.	Name
Ag_n	n=1–60	—	LPES	37	
Al_n	n=3–32	—	LPES	68	
Al_5	2.23	0.05	LPES	238	
Al_2C_2	0.64	0.05	LPES	239	acetylide
Al_3C	2.56	0.06	LPES	161	
Al_3C_2	2.19	0.03	LPES	244	
Al_3Ge_2	2.43	0.03	LPES	244	
Al_3Si_2	2.36	0.03	LPES	244	
Al_3O	1.00	0.15	LPES	68	
$Al_5H_2O_4$	3.10	0.10	LPES	283	
Al_5O_4	3.50	0.05	LPES	283	
Al_6N	2.58	0.04	LPES	337	
Al_8N	2.75	0.05	LPES	348	
Al_nO_m	n=1,2	m=1–5	LPES	143	
Al_nO_m	n=3–7	m=2–5	LPES	267	
Al_nP_m	n=1–4	m=1–4	LPES	217	
Al_nS_m	n=1–5	m=1–3	LPES	129	
$Ar(H_2O)_n$	n=2,6,7	—	LPES	77	
Ar_nBr	n=2–9	—	ZEKE	212	
Ar_nI	n=2–19	—	ZEKE	212	
As_4	<0.8	—	LPES	200	
As_5	≈1.7	—	LPES	200	
As_5	≈3.5	—	LPES	253	
Au_2BO	4.32	0.02	LPES	336	
Au_3BO	3.08	0.02	LPES	336	
Au_n	n=1–233	—	LPES	37	
Au_6	2.06	0.02	LPES	288	
$Au_6(CO)$	2.04	0.05	LPES	288	
$Au_6(CO)_2$	2.03	0.05	LPES	288	
$Au_6(CO)_3$	1.95	0.05	LPES	288	
AuF_6	7.5	estimate	CT	98	
Au_3Pd	2.51	—	LPES	220	
Au_4Pd	2.69	—	LPES	220	
$Au_{12}Nb$	3.70	0.03	LPES	275	
$Au_{12}Ta$	3.77	0.03	LPES	275	
$Au_{12}V$	3.76	0.03	LPES	275	
B_5	2.33	0.02	LPES	245	
BD_3	0.027	0.014	LPES	62	
BH_3	0.038	0.015	LPES	62	
B_6Li	2.3	0.1	LPES	298	
B_3N	2.098	0.035	LPES	193	
Bi_n	n=2–9	—	LPES	213	
Bi_4	1.05	0.010	LPES	119	
Bi_5	2.87	0.02	LPES	253	
$Br(CO_2)$	3.582	0.017	LPES	131	
$Br(H_2O)_n$	n=1–4	—	LPES	250	
Br_7Au_2	3.52	0.02	LPES	301	
C_n	n=2–84	—	LPES	70	
C_nCr	n=2–8	—	LPES	271	
C_nNb	n=2–7	—	LPES	243	
$(CO_2)_n$	n=1,2	—	LPES	75	
$(CS)_n$	n=2	—	LPES	75	
$(CS_2)_n$	n=1,2	—	LPES	75	
CAl_3Ge	2.70	0.06	LPES	224	
CAl_3Si	2.77	0.06	LPES	224	
CCl_4	≤1.14	—	CT	266	
$CCoNO_3$	1.73	0.03	LPES	199	$Co(CO_2)NO$
CDO_2	3.510	0.015	LPES	109	

Molecule	Electron Affinity in eV	Uncertainty in eV	Method	Ref.	Name
CF_3	1.82	0.05	LPES	187	
CF_3Br	0.91	0.2	CD	2	
CF_3I	1.57	0.2	CD	2	
CFO_2	4.277	0.030	LPES	131	
$CHCl_3$	≤0.78	—	CT	266	
CHO_2	3.498	0.015	LPES	109	
CH_2O_4	2.1	0.2	PT	2	$CO_3(H_2O)$
CH_2S	0.465	0.023	LPES	53	
CD_3O	1.5546	0.0019	LPES	194	
CD_3O_2	1.154	0.004	LPES	188	d_3-methyl peroxyl radical
CD_3S	1.856	0.006	LPT	2	
CD_3S_2	1.748	0.022	LPES	53	
CH_3	0.08	0.03	LPES	2	
CH_3I	0.11	0.02	LPES	277	
CH_3NO_2	0.172	0.006	LPES	211	
CH_3O	1.5690	0.0019	LPES	194	
CH_3O_2	1.161	0.005	LPES	188	methyl peroxyl radical
CH_3S	1.867	0.004	LPES	166	
CH_3S_2	1.757	0.022	LPES	53	
CH_3Si	0.852	0.010	LPES	97	CH_3-Si
CH_3Si	2.010	0.010	LPES	97	CH_2=SiH
CH_4N	0.432	0.015	LPES	215	
CH_5Si	1.19	0.04	LPT	65	CH_3SiH_2
CO_3	2.69	0.14	LPES	2	
C_2F_2	2.255	0.006	LPES	106	difluorovinylidene
C_2DN	2.009	0.020	LPES	219	DCCN
C_2DN	1.877	0.010	LPES	219	DCNC
C_2DO	2.350	0.020	LPES	13	
C_2HF	1.718	0.006	LPES	106	monofluorovinylidene
C_2HN	2.003	0.014	LPES	219	HCCN
C_2HN	1.883	0.013	LPES	219	HCNC
C_2HO	2.338	0.008	LPES	190	
C_2HNPd	2.17	0.03	LPES	291	
C_2HPd	1.98	0.03	LPES	287	
C_2HPt	2.650	0.010	LPES	287	
C_2D_2	0.492	0.006	LPES	83	vinylidene-d_2
C_2HD	0.489	0.006	LPES	83	vinylidene-d_1
C_2HFe	1.4512	0.0025	LPES	254	
C_2HNi	1.063	0.019	LPES	254	
C_2D_2N	1.538	0.012	LPES	21	cyanomethyl-d_2 radical
C_2D_2N	1.070	0.024	LPES	21	isocyanomethyl-d_2 radical
C_2H_2	0.490	0.006	LPES	83	vinylidene
$C_2H_2BrO_2$	3.97	0.03	LPES	310	
$C_2H_2ClO_2$	3.93	0.03	LPES	310	
C_2H_2FO	2.22	0.09	PT	2	acetyl fluoride enolate
$C_2H_2FO_2$	3.80	0.03	LPES	310	
C_2H_2Fe	1.328	0.019	LPES	254	
C_2H_2N	1.548	0.005	LPES	316	cyanomethyl radical
C_2H_2N	1.059	0.024	LPES	21	isocyanomethyl radical
$C_2H_2N_3$	3.447	0.004	LPES	316	1,2,3-triazolyl
C_2H_2Ni	2.531	0.005	LPES	287	$HNiC_2H$
C_2H_3	0.667	0.024	LPES	90	vinyl
C_2H_3Fe	1.587	0.019	LPES	254	
C_2H_3Ni	1.103	0.019	LPES	254	
C_2D_3O	1.8191	0.0012	LPT	22	vinoxy-d_3
C_2H_3O	1.8249	0.0012	LPT	22	vinoxy
C_2D_5O	1.699	0.004	LPES	194	ethoxide-d_3
C_2H_5N	0.56	0.01	PT	2	ethyl nitrine
$C_2H_5NO_2$	0.3	0.2	LPES	322	nitroethane

Molecule	Electron Affinity in eV	Uncertainty in eV	Method	Ref.	Name
$(C_2H_5NO_2)_n$	n=0–4	—	LPES	322	$(nitroethane)_n$
C_2H_5O	1.712	0.004	LPES	194	ethoxide
$C_2H_5O_2$	1.186	0.004	LPES	188	ethyl peroxyl radical
C_2H_5S	1.953	0.006	LPT	2	ethyl sulfide
C_2H_5S	0.868	0.051	LPES	53	CH_3SCH_2
$C_2H_7O_2$	2.26	0.08	PT	50	MeOHOMe
C_3D	1.997	0.005	LPES	315	c-C_3D
C_3Fe	1.69	0.08	LPES	132	
C_3H	1.999	0.003	LPES	315	c-C_3H
C_3HFe	1.58	0.06	LPES	132	
C_3H_2	1.794	0.008	LPES	153	
$C_3H_2F_3O$	2.625	0.010	LPT	113	1,1,1-trifluoroacetone enolate
C_3D_3	2.7300	0.0010	LPES	311	1-propynyl-d_3
C_3H_3	2.7355	0.0010	LPES	311	1-propynyl
C_3H_3	0.918	0.008	LPES	153	2-propynyl
C_3H_2D	0.88	0.15	LPES	24	2-propynyl-d_1
C_3D_2H	0.907	0.023	LPES	24	2-propynyl-d_2
C_3H_3N	1.247	0.012	LPES	21	CH_3CH-CN
C_3D_5	0.464	0.006	LPES	138	allyl-d5
C_3H_5	0.481	0.008	LPES	138	allyl
C_3H_5	0.397	0.069	kinetic	155	cyclopropyl
C_3H_4D	0.373	0.019	LPES	25	allyl-d_1
C_3H_5O	1.758	0.019	LPT	113	acetone enolate
C_3H_5O	1.621	0.006	LPT	113	propionaldehyde enolate
$C_3H_5O_2$	1.80	0.06	PT	2	methyl acetate enolate
C_3H_7O	1.789	0.033	LPES	23	n-propyl oxide
C_3H_7O	1.847	0.004	LPES	194	isopropyl oxide
C_3H_7S	2.00	0.02	PT	2	n-propyl sulfide
C_3H_7S	2.02	0.02	PT	2	isopropyl sulfide
C_3O	1.237	0.003	LPES	351	
C_3O_2	0.85	0.15	LPES	11	
C_3S	1.5957	0.0010	LPES	351	
C_3Ti	1.561	0.015	LPES	147	
C_4D	3.5308	0.0012	LPES	314	
$C_4F_4Cl_2$	0.87	0.08	attach	258	1,2-dichlorotetrafluorocyclobutene
$C_4F_4O_3$	0.5	0.2	CD	2	tetrafluorosuccinic anhydride
C_4F_8	0.63	0.05	attach	256	octafluorocyclobutane
C_4Fe	<2.2	0.2	LPES	132	
C_4H	3.5332	0.0010	LPES	314	
C_4HFe	1.67	0.06	LPES	132	
C_4H_2Fe	1.633	0.019	LPES	254	
$C_4H_2O_3$	1.44	0.10	CT	61	maleic anhydride
C_4H_3Fe	1.182	0.019	LPES	254	
C_4H_3Ni	0.824	0.019	LPES	254	
C_4D_4	0.909	0.015	LPES	125	vinylvinylidene-d_4
C_4H_4	0.914	0.015	LPES	125	vinylvinylidene
C_4H_4N	2.145	0.010	LPES	265	pyrrolyl
$C_4H_4N_3O$	0.75	—	LPES	285	NO pyrimidine
$C_4H_4N_3O$	3.037	0.015	LPES	309	dehydrogenated cytosine
C_4H_5O	1.801	0.008	LPT	113	cyclobutanone enolate
C_4H_6	0.431	0.006	LPES	135	trimethylenemethane
$C_4H_6O_2$	0.69	0.10	CT	61	2,3-butanedione
C_4H_6D	0.493	0.008	LPES	138	2-methylallyl-d_7
C_4H_7	0.505	0.006	LPES	138	2-methylallyl
C_4H_7O	1.67	0.05	PT	2	butyraldehyde enolate
C_4H_6DO	1.67	0.05	PT	2	2-butanone-3-d_1 enolate
$C_4H_5D_2O$	1.75	0.06	PT	2	2-butanone-3,3-d_2 enolate
C_4H_9O	1.909	0.004	LPES	194	t-butoxyl
C_4H_9S	2.03	0.02	PT	2	n-butyl sulfide

Molecule	Electron Affinity in eV	Uncertainty in eV	Method	Ref.	Name
C_4H_9S	2.07	0.02	PT	2	t-butyl sulfide
C_4N	3.1113	0.0010	LPES	346	
C_4O	2.05	0.15	LPES	11	
C_4O_2	2.0	0.2	LPES	11	
C_4Ti	1.494	0.020	LPES	147	
C_5	2.853	0.001	LPT	99	
C_5F_5N	0.70	0.05	attach	259	pentafluoropyridine
$C_5F_6O_3$	1.5	0.2	CD	2	hexafluoroglutaric anhydride
C_5H	2.421	0.019	LPES	317	linear
C_5H	2.857	0.028	LPES	317	cyclic
C_5HF_4N	0.40	0.08	attach	259	tetrafluoropyridine
C_5D_5	1.790	0.008	LPES	11	cyclopentadienyl-d5
C_5H_5	1.804	0.007	LPES	11	cyclopentadienyl
$(C_5H_5N)_nCo_m$	n=1–4	m=1–3	LPES	327	
$C_5H_5NO_2$	1.39	—	LPES	285	O_2·pyridine
$C_5H_5N_2O$	0.62	—	LPES	285	NO·pyridine
$C_5H_5N_2O_2$	3.250	0.015	LPES	309	dehydrogenated thymine
C_5H_7	0.91	0.03	PT	2	pentadienyl
$C_5H_7NO_3$	1.87	—	LPES	285	O_2·pyridine·H_2O
C_5H_7O	1.598	0.007	LPT	113	cyclopentanone enolate
C_5H_9O	1.69	0.05	PT	2	3-penanone enolate
$C_5H_{11}O$	1.93	0.05	LPT	2	neopentoxyl
$C_5H_{11}S$	2.09	0.02	PT	2	n-pentyl sulfide
C_5O_2	1.2	0.2	LPES	11	
C_5Ti	1.748	0.050	LPES	147	
C_6	1.180	0.001	LPT	8	
$C_6Br_4O_2$	2.44	0.20	CT	2	tetrabromobenzoquinone
$C_6Cl_4O_2$	2.78	0.10	CT	61	tetrachlorobenzoquinone
$C_6F_4O_2$	2.70	0.10	CT	61	tetrafluorobenzoquinone
C_6F_5Br	1.15	0.11	CT	67	pentafluorobromobenzene
C_6F_5Cl	0.75	0.05	attach	260	pentafluorochlorobenzene
C_6F_5I	1.41	0.11	CT	67	pentafluoroiodobenzene
$C_6F_5NO_2$	1.52	0.11	CT	67	pentafluoronitrobenzene
C_6F_6	0.53	0.05	attach	257	hexafluorobenzene
C_6F_{10}	>1.4	0.3	CT	2	perfluorocyclohexane
$C_6H_2Cl_2O_2$	2.48	0.10	CT	61	2,6-dichlorobenzoquinone
$C_6H_2O_2$	1.859	0.005	LPES	232	dehydrobenzoquinone
$C_6H_3F_2NO_2$	1.17	0.10	CT	61	2,4-difluoronitrobenzene
$C_6H_3O_2$	<2.18	—	LPES	232	benzoquinonide
C_6D_4	0.551	0.010	LPES	36	o-benzyne-d_4
C_6H_4	0.560	0.010	LPES	36	o-benzyne
$C_6H_4BrNO_2$	1.16	0.10	CT	61	o-bromonitrobenzene
$C_6H_4BrNO_2$	1.32	0.10	CT	61	m-bromonitrobenzene
$C_6H_4BrNO_2$	1.29	0.10	CT	61	p-bromonitrobenzene
$C_6H_4ClNO_2$	1.14	0.10	CT	61	o-chloronitrobenzene
$C_6H_4ClNO_2$	1.28	0.10	CT	61	m-chloronitrobenzene
$C_6H_4ClNO_2$	1.26	0.10	CT	61	p-chloronitrobenzene
C_6H_4ClO	≤2.58	0.08	PT	2	o-chlorophenoxide
$C_6H_4FNO_2$	1.07	0.10	CT	61	o-fluoronitrobenzene
$C_6H_4FNO_2$	1.23	0.10	CT	61	m-fluoronitrobenzene
$C_6H_4FN_2$	1.12	0.10	CT	61	p-fluoronitrobenzene
$C_6H_4N_2O_4$	1.65	0.10	CT	61	o-dinitrobenzene
$C_6H_4N_2O_4$	1.65	0.10	CT	61	m-dinitrobenzene
$C_6H_4N_2O_4$	2.00	0.10	CT	61	p-dinitrobenzene
$C_6H_4O_2$	1.860	0.005	LPES	284	1,4-benzoquinone
C_6D_5	1.092	0.020	LPES	26	phenyl-d_5
C_6D_5N	1.44	0.02	LPES	96	phenylnitrene-d_5
C_6H_5	1.096	0.006	LPES	26	phenyl
C_6H_5N	1.429	0.011	LPT	115	phenylnitrene

Molecule	Electron Affinity in eV	Uncertainty in eV	Method	Ref.	Name
$C_6H_5NO_2$	1.00	0.01	LPES	164	nitrobenzene
C_6H_5O	2.253	0.006	LPES	26	phenoxyl
C_6H_5S	<2.47	0.06	PT	2	thiophenoxide
C_6H_5NH	1.70	0.03	PT	2	anilide
$(C_6H_6)_nCo_m$	n=1–4	m=1–5	LPES	326	
$(C_6H_6)_nFe_m$	n=1–4	m=1–7	LPES	329	
C_6H_6NO	0.44	—	LPES	285	NO·(benzene)
C_6H_6Nb	0.893	0.006	LPES	311	
$C_6H_6O_2$	1.06	—	LPES	285	O_2·(benzene)
C_6H_7	<1.67	0.04	PT	2	methylchylopentadienyl
C_6H_8	0.855	0.010	LPES	203	$(CH_2)_2C\text{-}C(CH_2)_2$
C_6H_8Si	1.435	0.004	LPT	65	$C_6H_5SiH_3$
C_6H_9	0.654	0.010	LPES	203	$CH_2{=}C(CH_3)\text{-}C(CH_2)_2$
C_6H_9O	1.526	0.010	LPT	113	cyclohexanone enolate
C_6H_{10}	0.645	0.015	LPES	126	t-butyl vinylidene
$C_6H_{11}O$	1.755	+0.05/−0.005	LPT	113	pinacolone enolate
$C_6H_{11}O$	1.82	0.06	PT	2	3,3-dimethylbutananl enolate
C_6N	3.3715	0.0010	LPES	346	
C_6N_4	2.3	0.3	PT	2	TCNE
C_6O_6	2.54	0.05	LPES	333	
C_7F_5N	1.11	0.11	CT	67	pentafluorobenzonitrile
C_7F_8	0.86	0.11	CT	67	octafluorotoluene
C_7F_{14}	1.06	0.10	CT	56	perfluoromethylcyclohexane
C_7HF_5O	1.10	0.11	CT	67	pentafluorobenzaldehyde
$C_7H_3N_3O_4$	2.16	0.10	CT	61	3,5-dinitrobenzonitrile
$C_7H_4F_3NO_2$	1.41	0.10	CT	61	m-trifluoromethylnitrobenzene
$C_7H_4N_2O_2$	1.61	0.10	CT	61	o-cyanonitrobenzene
$C_7H_4N_2O_2$	1.56	0.10	CT	61	m-cyanonitrobenzene
$C_7H_4N_2O_2$	1.72	0.10	CT	61	p-cyanonitrobenzene
C_7H_6Br	1.308	0.008	LPES	167	o-bromobenzyl
C_7H_6Br	1.307	0.008	LPES	167	m-bromobenzyl
C_7H_6Br	1.229	0.008	LPES	167	p-bromobenzyl
C_7H_6Cl	1.257	0.008	LPES	167	o-chlorobenzyl
C_7H_6Cl	1.272	0.008	LPES	167	m-chlorobenzyl
C_7H_6Cl	1.174	0.008	LPES	167	p-chlorobenzyl
C_7H_6F	1.091	0.008	LPES	167	o-fluorobenzyl
C_7H_6F	1.173	0.008	LPES	167	m-fluorobenzyl
C_7H_6F	0.937	0.008	LPES	167	p-fluorobenzyl
C_7H_6FO	2.218	0.010	LPT	2	m-fluoroacetophenone enolate
C_7H_6FO	2.176	0.010	LPT	2	p-fluoroacetophenone enolate
$C_7H_6FeO_3$	0.990	0.10	CT	120	h_4-1,3-butadiene-$Fe(CO)_3$
$C_7H_6N_2O_4$	1.77	0.05	PT	60	3,4-dintrotoluene
$C_7H_6N_2O_4$	1.77	0.05	PT	60	2,3-dinitrotoluene
$C_7H_6N_2O_4$	1.60	0.05	PT	60	2,4-dinitrotoluene
$C_7H_6N_2O_4$	1.55	0.05	PT	60	2,6-dinitrotoluene
$C_7H_6O_2$	1.85	0.10	CT	61	o-CH_3-BQ
C_7H_7	0.912	0.006	LPES	26	benzyl
C_7H_7	0.868	0.006	LPES	136	1-quadricyclanide
C_7H_7	0.962	0.006	LPES	136	2-quadricyclanide
C_7H_7	1.286	0.006	LPES	136	norbornadienide
C_7H_7	0.39	0.04	LPES	136	cycloheptatrienide
C_7H_7	3.046	0.006	LPES	136	1-(1,6-heptadiynide)
C_7H_7	>1.140	0.006	LPES	136	3-(1,6-heptadiynide)
$C_7H_7NO_2$	0.92	0.10	CT	61	2-nitrotoluene
$C_7H_7NO_2$	0.99	0.10	CT	61	3-nitrotoluene
$C_7H_7NO_2$	0.95	0.10	CT	61	4-nitrotoluene
$C_7H_7NO_3$	1.04	0.10	CT	61	3-nitroanisole
$C_7H_7NO_3$	0.91	0.10	CT	61	4-nitroanisole
C_7H_7O	<2.36	0.06	PT	2	o-methyl phenoxide

Molecule	Electron Affinity in eV	Uncertainty in eV	Method	Ref.	Name
C_7H_7O	2.14	0.02	PT	50	benzyloxide
C_7H_8FO	<3.05	0.06	PT	50	$PhCH_2OHF$
C_7H_9	1.27	0.03	PT	2	heptatrienyl
C_7H_9O	1.61	0.05	PT	2	2-norbornanone enolate
C_7H_9Si	1.33	0.04	LPT	65	$C_6H_5(CH_3)SiH$
$C_7H_{11}O$	1.598	0.007	LPT	113	cycloheptanone enolate
$C_7H_{11}O$	1.49	0.04	PT	2	2,5-dimethyl-cyclopentanone enolate
$C_7H_{13}O$	1.72	0.06	PT	2	4-heptanone enolate
$C_7H_{13}O$	1.46	0.04	PT	2	di-isopropyl ketone enolate
$C_8F_{14}N_2$	1.89	0.10	CT	51	1,4-$(CN)_2C_6F_4$
$C_8H_3F_5O$	0.88	0.11	CT	67	pentafluoroacetophenone
$C_8H_3F_6NO_2$	1.79	0.10	CT	61	3,5-$(CF_3)_2$-nitrobenzene
$C_8H_4F_3N$	0.70	0.05	attach	263	o-trifluoromethylbenzonitrile
$C_8H_4F_3N$	0.67	0.05	attach	263	m-trifluoromethylbenzonitrile
$C_8H_4F_3N$	0.83	0.05	attach	263	p-trifluoromethylbenzonitrile
$C_8H_4O_3$	1.21	0.10	CT	61	phthalic anhydride
C_8H_6	1.044	0.008	LPES	148	
C_8H_7	1.091	0.008	LPES	134	
C_8H_7O	2.057	0.010	PT	2	acetophenone enolate
C_8H_7O	2.10	0.08	LPT	2	phenylacetaldehyde enolate
C_8H_8	0.55	0.02	CT	134	cyclooctatetraene
C_8H_8	0.919	0.008	LPES	139	m-xylylene
$C_8H_9NO_2$	1.21	0.05	PT	60	1,3-dimethyl-5-nitrobenzene
$C_8H_9NO_2$	2.61	0.05	PT	60	1,3-dimethyl-2-nitrobenzene
$C_8H_9NO_2$	0.86	0.10	CT	61	1,2-dimethyl-3-nitrobenzene
$C_8H_{13}O$	1.63	0.06	PT	2	cyclooctanone enolate
$C_8N_4NiS_4$	4.56	0.04	LPES	307	Ni-bis(dithiolene)
$C_8N_4PdS_4$	4.55	0.04	LPES	307	Ni-bis(dithiolene)
$C_8N_4PtS_4$	4.45	0.04	LPES	307	Ni-bis(dithiolene)
C_8S_2	0.049	0.005	LPES	230	bithiophene
$C_9H_8FeO_3$	0.76	0.10	CT	120	h_4-1,3-cyclohexadiene-$Fe(CO)_3$
C_9H_9O	2.030	0.010	LPT	2	m-methylacetophenone enolate
C_9H_9SiN	1.43	0.10	PT	2	trimethylsilylnitrene
$C_9H_{11}NO_2$	0.70	0.10	CT	61	2,4,6-trimethyl-NB
$C_9H_{15}O$	1.69	0.06	PT	2	cyclononanone enolate
$C_{10}H_4C_{12}O_2$	2.19	0.10	CT	61	2,3-dichloro-1,4-naphthoquinone
$C_{10}H_6N_2O_4$	1.78	0.10	CT	61	1,3-dinitronaphthalene
$C_{10}H_6N_2O_4$	1.77	0.10	CT	61	1,5-dinitronaphthalene
$C_{10}H_6O_2$	1.81	0.10	CT	61	1,4-naphthoquinone
$C_{10}H_7$	1.403	0.015	LPES	197	1-naphthyl radical
$C_{10}H_7NO_2$	1.23	0.10	CT	61	1-nitronaphthalene
$C_{10}H_7NO_2$	1.18	0.10	CT	61	2-nitronaphthalene
$C_{10}H_8$	0.790	0.008	LPES	230	azulene
$C_{10}H_8CrO_3$	0.93	0.10	CT	120	h_4-1,3,5-cycloheptatriene $Cr(CO)_3$
$C_{10}H_8FeO_3$	0.98	0.10	CT	120	h_4-1,3,5-cycloheptatriene-$Fe(CO)_3$
$C_{10}H_8NO$	0.66	—	LPES	285	NO·naphthlene
$C_{10}H_8O_2$	1.41	—	LPES	285	O_2·naphthlene
$C_{10}H_{10}O_3$	2.09	—	LPES	285	O_2·naphthalene·H_2O
$C_{10}H_{12}O_4$	2.72	—	LPES	285	O_2·naphthalene·$(H_2O)_2$
$C_{10}H_{17}O$	1.83	0.06	PT	2	cyclodecanone enolate
$C_{11}H_8FeO_3$	1.29	0.10	CT	120	h_4-1,3-butadiene-$Fe(CO)_3$
$C_{12}F_{10}$	0.82	0.11	CT	67	decafluorobiphenyl
$C_{12}H_4N_4$	2.8	0.3	CD	2	TCNQ
$C_{12}H_9$	1.07	0.10	PT	2	perinaphthenyl
$C_{12}H_{12}NO$	0.79	—	LPES	285	NO·$(benzene)_2$
$C_{12}H_{15}O$	2.032	0.010	LPT	2	t-butylacetophenone enolate
$C_{12}H_{21}O$	1.90	0.07	PT	2	cyclododecanone enolate
$C_{13}F_{10}O$	1.52	0.11	CT	67	decafluorobenzophenone
$C_{13}H_9FO$	0.64	0.10	CT	61	4-fluorobenzophenone

Molecule	Electron Affinity in eV	Uncertainty in eV	Method	Ref.	Name
$C_{13}H_{10}O$	0.62	0.10	CT	61	benzophenone
$C_{14}H_9NO_2$	1.43	0.10	CT	61	9-nitroanthracene
$C_{14}H_{10}$	0.530	0.005	LPES	286	anthracene
$C_{14}H_{12}O$	0.770	0.005	LPES	286	anthracene·H_2O
$(C_{14}H_{10})_n$	n=1–16	—	LPES	231	anthracene clusters
$C_{16}H_{10}$	0.406	0.010	LPES	270	pyrene
$(C_{16}H_{10})_nCo_m$	n=1,2	m=1,2	LPES	330	
$C_{18}H_{12}$	1.058	0.005	LPES	313	tetracene
$C_{18}H_{12}$	0.32	0.01	LPES	303	chrysene
$C_{20}H_{12}$	0.79	0.10	CT	66	benz[a]pyrene
$C_{20}H_{12}$	0.973	0.005	LPES	236	perylene
$C_{20}H_{16}NO$	1.06	—	LPES	285	NO·(naphthalene)$_2$
$C_{22}H_{14}$	1.35	0.10	CT	66	pentacene
$C_{24}H_{12}$	0.47	0.09	LPES	328	coronene
$C_{24}H_{12}Co$	1.15	0.15	LPES	324	Co·(coronene)
$C_{24}H_{12}Co_2$	1.15	0.10	LPES	324	Co_2·(coronene)
$C_{24}H_{12}Fe$	1.06	—	LPES	324	Fe·(coronene)
$C_{24}H_{12}Fe_2$	1.59	—	LPES	324	Fe_2·(coronene)
$C_{44}Cl_{28}FeN_4$	2.59	0.11	CT	186	$FeTPPCl_{28}$
$C_{44}Cl_8F_{20}FeN_4$	3.21	0.03	CT	186	$FeTPP\beta Cl_8$
$C_{44}Cl_9F_{20}FeN_4$	3.35	0.03	CT	186	$FeTPPF_{20}\beta Cl_8Cl$
$C_{44}H_8F_{20}FeN_4$	2.15	0.15	CT	186	$FeTPPF_{20}$
$C_{44}H_8ClF_{20}FeN_4$	3.14	0.03	CT	186	$FeTPPF_{20}Cl$
$C_{44}H_8Cl_{21}FeN_4$	2.93	0.23	CT	186	$FeTPPoCl_{20}Cl$
$C_{44}H_{12}Cl_{17}FeN_4$	3.14	0.03	CT	186	$FeTPPoCl_8\beta Cl_8Cl$
$C_{44}H_{20}Cl_8FeN_4$	1.86	0.03	CT	186	$FeTPPoCl_8$
$C_{44}H_{20}Cl_9FeN_4$	2.10	0.19	CT	186	$FeTPPoCl_8Cl$
$C_{44}H_{28}FeN_4$	1.87	0.03	CT	186	iron tetraphenylporphyrin (FeTPP)
$C_{44}H_{28}NiN_4$	1.51	0.01	CT	186	nickel tetraphenylporphyrin (NiTPP)
$C_{44}H_{28}ClFeN_4$	2.15	0.15	CT	186	FeTPPCl
$C_{44}H_{30}N_4$	1.69	0.01	CT	186	H_2 tetraphenylporphyrin
$C_{45}H_{29}NiN_4O$	1.74	0.01	CT	186	NiTPPCHO
$C_{48}H_{24}$	0.67	0.09	LPES	328	coronene dimer
$C_{48}H_{24}Fe$	1.50	—	LPES	324	Fe·(coronene)$_2$
$C_{48}H_{24}Fe_2$	1.48	—	LPES	324	Fe_2·(coronene)$_2$
$C_{52}H_{39}FeN_7O$	1.97	0.03	CT	186	FeTPP-val
C_{60}	2.683	0.008	LPES	201	
$C_{60}F_2$	2.74	0.07	Knud	202	
$C_{64}H_{64}FeN_8O_4$	2.07	0.03	CT	186	FeTPP-piv
$C_{70}F_2$	2.80	0.07	Knud	202	
$(benzene)_n$	n=53–124	—	LPES	248	
$(toluene)_n$	n=33–139	—	LPES	248	
CeF_4	3.8	0.4	CT	98	
$Cl(CO_2)$	3.907	0.010	LPES	131	
$Cl(H_2O)_n$	n=1–4	—	LPES	250	
ClO_3	4.25	0.10	LPES	340	
ClO_4	5.25	0.10	LPES	340	
Co_n	n=1–108	—	LPES	251	
$CoBr_3$	4.6	0.1	LPES	249	
$CoCl_3$	4.7	0.1	LPES	249	
CoF_4	6.4	0.3	CT	98	
$Cr(CO)_3$	1.349	0.006	LPES	94	
CrO_3	3.66	0.02	LPES	241	
CrO_4	4.98	0.09	LPES	241	
CrO_5	4.4	0.1	LPES	241	
Cr_2O_n	n=1–7		LPES	306	
CsO_4	2.5	0.2	LPES	252	
Cu_n	n=1–411	—	LPES	37	
$Cu_n(CN)_m$	n=1–6	m=1–6	LPES	159	

Molecule	Electron Affinity in eV	Uncertainty in eV	Method	Ref.	Name
$EuSi_n$	n=3–17	—	LPES	321	
$F(H_2O)_n$	n=1–4	—	LPES	242	
$F(H_2O)_n$	n=1–4	—	LPES	250	
Fe_n	n=3–34	—	LPES	149	
$Fe(CO)_2$	1.22	0.02	LPES	2	
$Fe(CO)_3$	1.8	0.2	LPES	2	
$Fe(CO)_4$	2.4	0.3	LPES	2	
$FeBr_3$	4.26	0.06	LPES	249	
$FeBr_4$	5.50	0.08	LPES	249	
$FeCl_3$	4.22	0.06	LPES	249	
$FeCl_4$	6.00	0.08	LPES	249	
FeF_3	3.6	0.1	CT	98	
FeF_4	6.0	estimate	CT	98	
Fe_2H_2	0.942	0.019	LPES	254	
Fe_nO_m	n=1–4	m=1–6	LPES	152	
Ga_2As_3	2.783	0.024	LPES	192	
Ga_xAs_y	n=2–50	n=x+y	LPES	229	
Ga_2P_3	2.991	0.026	LPES	192	
Ge_n	n=3–15	—	LPES	71	
Ge_xAs_y	n=5–30	n=x+y	LPES	72	
GeH_3	<1.74	0.04	PT	2	
$H(NH_3)_n$	n=1,2	—	LPES	76	
HNO_3	0.57	0.15	CD	2	
$(H_2O)_n$	n=2–19	—	LPES	77	
$I(CO_2)$	3.225	0.001	LPES	131	
$I(H_2O)_n$	n=1–4		LPES	250	
$I(CO_2)_n$	n=1–3	—	LPES	350	
In_xP_y	n=2–8	n=x+y	LPES	137	
IrF_4	4.7	0.3	CT	98	
IrF_6	6.5	0.4	CT	98	
K_n	n=2–7	—	LPES	18	
KO_4	2.8	0.2	LPES	252	
$LaCl_4$	7.03	0.01	LPES	145	
LiO_4	3.3	0.2	LPES	252	
$MnBr_3$	5.03	0.06	LPES	249	
$MnCl_3$	5.07	0.06	LPES	249	
MnF_4	5.5	0.2	CT	98	
MnO_3	3.335	0.010	LPES	158	
$Mo(CO)_3$	1.337	0.006	LPES	94	
MoF_5	3.5	0.2	CT	98	
MoF_6	3.8	0.2	CT	98	
MoO_3	3.17	0.02	LPES	280	
MoO_4	5.20	0.07	LPES	86	
MoO_5	5.10	0.07	LPES	86	
MoW_2O_6	2.76	0.05	LPES	353	
Mo_2O_2	2.24	0.02	LPES	280	
Mo_2O_3	2.33	0.07	LPES	280	
Mo_2O_4	2.13	0.04	LPES	280	
Mo_2WO_6	2.85	0.05	LPES	353	
Mo_3O_6	2.75	0.02	LPES	353	
N_2CD	2.622	0.005	LPES	154	NCND
N_2CH	2.622	0.005	LPES	154	NCNH
$(NH_3)_n$	n=41–1100	—	LPES	77	
$NH_2(NH_3)_n$	n=1,2	—	LPES	78	
$NO(H_2O)_n$	n=1,2	—	LPES	75	
NO_3	3.937	0.014	LPES	85	
$NO_3(H_2O)_n$	n=0–6	—	LPES	240	
$NO(N_2O)_n$	n=1–5	—	LPES	79	
$(NO)_2$	>2.1	—	LPES	75	

Molecule	Electron Affinity in eV	Uncertainty in eV	Method	Ref.	Name
$(N_2O)_n$	n=1,2	—	LPES	81	
Na_n	n=2–5	—	LPES	18	
$NaCS_2$	0.80	0.05	LPES	278	
Na_2CS_2	0.25	0.05	LPES	278	
$(NaF)_n$	n=1–7,12	—	LPES	64	
$Na(NaF)_n$	n=5,7–12	—	LPES	64	
NaO_4	3.1	0.2	LPES	252	
NaO_5	3.2	0.2	LPES	252	
$NaSO_3$	2.3	0.2	LPES	252	
Nb_n	n=6–17	—	LPES	181	
Nb_8	1.513	0.008	LPES	157	
Nb_3O	1.393	0.006	LPES	169	
Ni_n	n=1–100	—	LPES	247	
$Ni_n(benzene)_m$	n=1–3	m=1,2	LPES	295	
$NiBr_3$	4.94	0.08	LPES	249	
$NiCl_3$	5.20	0.08	LPES	249	
$Ni(CO)_2$	0.643	0.014	LPES	2	
$Ni(CO)_3$	1.077	0.013	LPES	2	
Ni(CO)H	1.126	0.010	LPES	293	HNiCO
$OH(H_2O)$	<2.95	0.15	PT	2	
$OH(NH_3)$	2.35	0.07	LPES	234	
$OH(N_2O)$	2.14	0.02	LPES	209	
$OH(N_2O)_n$	n=1–5	—	LPES	209	
OsF_4	3.9	0.3	CT	98	
OsF_6	6.0	0.3	CT	98	
P_5	3.88	0.03	LPES	253	
PBr_3	1.59	0.15	CD	2	
PBr_2Cl	1.63	0.20	CD	2	
PCl_2Br	1.52	0.20	CD	2	
PCl_3	0.82	0.10	CD	2	
PF_5	0.75	0.15	CT	121	
PO_3	4.95	0.06	LPES	156	
$POCl_2$	3.83	0.25	CD	2	
$POCl_3$	1.41	0.20	CD	2	
P_2H_2	1.00	0.01	LPES	281	trans-P_2H_2
P_2H_2	1.03	0.01	LPES	281	cis-P_2H_2
Pb_4	1.55	0.09	LPES	345	
PtF_4	5.5	0.3	CT	98	
PtF_6	7.0	0.4	CT	98	
ReF_6	4.7	estimate	CT	98	
ReO_3	3.6	0.1	LPES	216	
RhF_4	5.4	0.3	CT	98	
RuF_4	4.8	0.3	CT	98	
RuF_5	5.2	0.4	CT	98	
RuF_6	7.5	0.3	CT	98	
SF_4	1.5	0.2	CT	91	
SF_5	4.23	0.12	e-scat	204	
SF_6	1.20	0.05	attach	318	
SO_3	1.97	0.10	LPES	165	
$(SO_2)_2$	0.6	0.2	LPES	80	
Sb_n	n=2–9	—	LPES	213	
Sb_5	3.46	0.03	LPES	253	
$ScBr_4$	6.13	0.08	LPES	249	
$ScCl_4$	6.84	0.01	LPES	145	
SeF_6	2.9	0.2	CD	2	
Si_4	2.13	0.01	LPES	110	
Si_5	2.59	0.02	LPES	110	
Si_7	1.85	0.02	LPES	110	
Si_n	n=3–20	—	LPES	71	

Molecule	Electron Affinity in eV	Uncertainty in eV	Method	Ref.	Name
Si_2C_3	1.766	0.012	LPES	296	linear Si-C_3-Si
SiD_3	1.386	0.022	LPES	43	
SiF_3	2.41	0.22	LPES	17	
SiF_4	≤0	—	LPES	17	
SiF_5	≥4.66	—	LPES	17	
Si_nF	n=2–11	—	LPES	17	
SiH_3	1.406	0.014	LPES	43	
Si_3H	2.53	0.01	LPES	182	
Si_4H	2.68	0.01	LPES	182	
Si_nNa_m	n=4–11	m=1–3	LPES	210	
Sn_n	n=1–12	—	LPES	289	
$SnCH_2CN$	1.57	0.02	LPES	292	
$Sn(CN)_2$	2.622	0.004	LPES	292	
Sn(CN) (CH_2CN)	2.29	0.05	LPES	292	
Ta_3O	1.583	0.010	LPES	169	
TeF_6	3.34	0.17	CD	2	
Ti_n	n=1–130	—	LPES	151	
TiO_3	4.2	—	LPES	172	
UF_5	3.7	0.2	CT	98	
UF_6	5.1	0.2	CT	98	
UO_3	<2.1	—	CT	98	
V_n	n=3–65	—	LPES	150	
VF_4	3.5	0.2	CT	98	
V_2O_n	n=3–7	—	LPES	246	
V_3O	1.218	0.008	LPES	169	
V_4O_{10}	4.2	0.6	CT	101	
$W(CO)_3$	1.859	0.006	LPES	94	
WF_5	1.25	0.3	CD	18	
WF_6	3.5	0.1	CT	19	
WO_3	3.62	0.05	LPES	86	
$(WO_3)_n$	n=7–10	—	LPES	300	
WO_4	5.30	0.05	LPES	86	
WO_5	5.1	0.1	LPES	86	
W_3O_6	2.85	0.02	LPES	353	
YCl_4	7.02	0.01	LPES	145	

REFERENCES

1. Hotop, H., and Lineberger, W. C., *J. Phys. Chem. Ref. Data* 14, 731, 1985.
2. Drzaic, P. S., Marks, J., and Brauman, J. I., in *Gas Phase Ion Chemistry*, Vol. 3, Bowers, M. T., Ed., Academic Press, Orlando, 1984, p. 167. The reference for C_6H_4ClO should read "Richardson et al., 1975c."
3. Schulz, P. A., Mead, R. D., Jones, P. L., and Lineberger, W. C., *J. Chem. Phys.* 77, 1153, 1982.
4. Neumark, D. M., Lykke, K. R., Anderson, T., and Lineberger, W. C., *Phys. Rev. A* 32, 1890, 1985. EA(O) = 11,784.645 ± 0.008 cm^{-1}.
5. Wenthold, P. G., Gunion, R. F., and Lineberger, W. C., *Chem. Phys. Lett.* 258, 101, 1996.
6. Wenthold, P. G., Kim, J. B., Jonas, K. L., and Lineberger, W. C., *J. Phys. Chem.* A 101, 4472, 1997.
7. Klein, R., McGinnis, R. P., and Leone, S. R., *Chem. Phys. Lett.* 100, 475, 1983.
8. Arnold, D. W., Bradforth, S. E., Kitsopoulos, T. N., and Neumark, D. M., *J. Chem. Phys.* 95, 8753, 1991; linear C_n.
9. Stevens, A. E., Fiegerle, C. S., and Lineberger, W. C., *J. Chem. Phys.* 78, 5420, 1983.
10. Breyer, F., Frey, P., and Hotop, H., *Z. Phys.* A 300, 7, 1981.
11. Oakes, J. M., and Ellison, G. B., *Tetrahedron* 42, 6263, 1986.
12. Leopold, D. G., Murray, K. K., Miller, A. E. S., and Lineberger, W. C., *J. Chem. Phys.* 83, 4849, 1985.
13. Oakes, J. M., Jones, M.E., Bierbaum, V. M., and Ellison, G. B., *J. Phys. Chem.* 87, 4810, 1983.
14. Ellis, H. B., Jr., and Ellison, G. B., *J. Chem. Phys.* 78, 6541, 1983.
15. Ramond, T. M., Blanksby, S. J., Kato, S., Bierbaum, V. M., Davico, G. E., Schwartz, R. L., Lineberger, W. C., and Ellison, G. B., *J. Phys. Chem. A* 106, 9641, 2002.
16. Nimlos, M. E., and Ellison, G. B., *J. Chem. Phys.* 90, 2574, 1986.
17. Kawamata, H., Negishi, Y., Kishi, R., Iwata, S., Nakajima, A., and Kaya, K., *J. Chem. Phys.* 105, 5369, 1996.
18. McHugh, K. M., Eaton, J. G., Lee, G. H., Sarkas, H. W., Kidder, L. H., Snodgrass, J. T., Manaa, M. R., and Bowen, K. H., *J. Chem. Phys.* 91, 3792, 1989. See also Ref. 104.
19. George, P. M., and Beauchamp, J. L., *Chem. Phys.* 36, 345, 1979. The lower limit given in this paper (3.4 eV) may be increased to 3.5 eV, as rapid charge transfer from HCO_2^- to WF_6 has since been observed (Miller, T. M., and Viggiano, A. A., unpublished).
20. Burnett, S. M., Stevens, A. E., Fiegerle, C. S., and Lineberger, W. C., *Chem. Phys. Lett.* 100, 124, 1983.
21. Moran, S., Ellis, H. B., DeFrees, D. J., McLean, A. D., and Ellison, G. B., *J. Am. Chem. Soc.* 109, 5996, 1987; Moran, S., Ellis, H. B., DeFrees, D. J., McLean, A. D., Paulson, S. E., and Ellison, G. B., *J. Am. Chem. Soc.* 109, 6004, 1987; see also Lykke, K. R., Neumark, D. M., Andersen, T., Trapa, V. J., and Lineberger, W. C., *J. Chem. Phys.* 87, 6842, 1987.
22. Yacovitch, T. I., Garand, E., and Neumark, D. M., *J. Chem. Phys.* 130, 244309, 2009 for C_2H_3O (14,719 ± 9 cm^{-1}). EA(C_2D_3O) was deduced

from this and the difference measured by Mead, R. D., Lykke, K. R., Lineberger, W. C., Marks, J., and Brauman, J. I., *J. Chem. Phys.* 81, 4883, 1984.

23. Ellison, G. B., Engelking, P. C., and Lineberger, W. C., *J. Chem. Phys.* 86, 4873, 1982.
24. Oakes, J. M., and Ellison, G. B., *J. Am. Chem. Soc.* 105, 2969, 1983.
25. Ellison, G. B., and Oakes, J. M., *J. Am. Chem. Soc.* 106, 7734, 1984. EA(allyl) and EA(allyl-d_5) are 0.119 and 0.083 eV too low, respectively, in this work, according to Ref. 138. Therefore, EA(allyl-d_1) is likely too low by a similar amount.
26. Gunion, R. F., Gilles, M. K., Polak, M. L., and Lineberger, W. C., *Int. J. Mass Spectrom. Ion Process.* 117, 601, 1992.
27. Leopold, D. G., and Lineberger, W. C., *J. Chem. Phys.* 85, 51, 1986.
28. Scheer, M., Bilodeau, R. C., Brodie, C. A., and Haugen, H. K., *Phys. Rev. A* 58, 2844, 1998.
29. Miller, A. E. S., Fiegerle, C. S., and Lineberger, W. C., *J. Chem. Phys.* 87, 1549, 1987.
30. Miller, T. M., Leopold, D. G., Murray, K. K., and Lineberger, W. C., *J. Chem. Phys.* 85, 2368, 1986.
31. Miller, T. M., and Lineberger, W. C., *Chem. Phys. Lett.* 146, 364, 1988.
32. Neumark, D. M., Lykke, K. R., Andersen, T., and Lineberger, W. C., *J. Chem. Phys.* 83, 4364, 1985.
33. Leopold, D. G., Miller, T. M., and Lineberger, W. C., *J. Am. Chem. Soc.* 108, 178, 1986.
34. Miller, A. E. S., Fiegerle, C. S., and Lineberger, W. C., *J. Chem. Phys.* 84, 4127, 1986.
35. Murray, K. K., Miller, T. M., Leopold, D. G., and Lineberger, W. C., *J. Chem. Phys.* 84, 2520, 1986.
36. Leopold, D. G., Miller, A. E. S., and Lineberger, W. C., *J. Am. Chem. Soc.* 108, 1379, 1986.
37. Li, J., Li, X., Zhai, H. J., and Wang, L.-S., *Science* 299, 864, 2003; Hakkinen, H., Yoon, B., Landman, U., Li, X., Zhai, H. J., and Wang, L.-S., *J. Phys. Chem.* 107, 6168, 2003; Taylor, K. J., Pettiette-Hall, C. L., Cheshnovsky, O., and Smalley, R. E., *J. Chem. Phys.* 96, 3319, 1992; Handschuh, H., Cha, C.-Y., Bechthold, P. S., Ganteför, G., and Eberhardt, W., J., *Chem. Phys.* 102, 6406, 1995; Cha, C.-Y., Ganteför, G., and Eberhardt, W., J., *Chem. Phys.* 99, 6308, 1993; Ho, J., Ervin, K. M., and Lineberger, W. C., J., *Chem. Phys.* 93, 6987, 1990; Leopold, D. G., Ho, J., and Lineberger, W. C., *J. Chem. Phys.* 86, 1715, 1987; Pettiette, C. L., Yang, S. H., Craycraft, M. J., Conceicao, J., Laaksonen, R. T., Cheshnovsky, O., and Smalley, R. E., *J. Chem. Phys.* 88, 5377, 1988.
38. Snodgrass, J. T., Coe, J. V., McHugh, K. M., Friedhoff, C. B., and Bowen, K. H., *J. Phys. Chem.* 93, 1249, 1989.
39. Friedhoff, C. B., Snodgrass, J. T., Coe, J. V., McHugh, K. M., and Bowen, K. H., *J. Chem. Phys.* 84, 1051, 1986.
40. Friedhoff, C. B., Coe, J. V., Snodgrass, J. T., McHugh, K. M., and Bowen, K. H., *Chem. Phys. Lett.* 124, 268, 1986.
41. Coe, J. V., Snodgrass, J. T., Friedhoff, C. B., McHugh, K. M., and Bowen, K. H., *J. Chem. Phys.* 84, 619, 1986.
42. Snodgrass, J. T., Coe, J. V., Friedhoff, C. B., McHugh, K. M., and Bowen, K. H., *Chem. Phys. Lett.* 122, 352, 1985.
43. Nimlos, M. R., and Ellison, G. B., *J. Am. Chem. Soc.* 108, 6522, 1986.
44. Petrunin, V., Andersen, H., Balling, P., and Andersen, T., *Phys. Rev. Lett.* 76, 744, 1996.
45. Andersen, T., Lykke, K. R., Neumark, D. M., and Lineberger, W. C., *J. Chem. Phys.* 86, 1858, 1987.
46. Murray, K. K., Lykke, K. R., and Lineberger, W. C., *Phys. Rev. A* 36, 699, 1987.
47. Chaibi, W., Delsart, C., Drag, C., and Blondel, C., *J. Mol. Spectrosc.* 239, 11, 2006. EA(^{32}SH) = 18669.543(12) cm^{-1}.
48. Stonemann, R. C., and Larson, D. J., *Phys. Rev. A* 35, 2928, 1987. EA(SeH) = 17,845.17 ± 0.20 cm^{-1}.
49. Nimlos, M. R., Harding, L. B., and Ellison, G. B., *J. Chem. Phys.* 87, 5116, 1987.
50. Moylan, C. R., Dodd, J. A., Han, C.-C., and Braumann, J. I., *J. Chem. Phys.* 86, 5350, 1987.
51. Chowdhury, S., Grimsrud, E. P., Heinis, T., and Kebarle, P., *J. Am. Chem. Soc.* 108, 3630, 1986.
52. Berzinsh, U., Gustafsson, M., Hanstorp, D., Klinkmueller, A. E., Ljungblad, and U., Maartensson-Pendrill, A.-M., *Phys. Rev. A* 51, 231, 1995. EA(Cl) = 29138.59 ± 0.22 cm^{-1}.
53. Moran, S., and Ellison, G. B., *J. Phys. Chem.* 92, 1794, 1988.
54. Murray, K. K., Leopold, D. G., Miller, T. M., and Lineberger, W. C., *J. Chem. Phys.* 89, 5442, 1988.
55. Ervin, K. M., Ho, J., and Lineberger, W. C., *J. Chem. Phys.* 89, 4514, 1988.
56. Grimsrud, E. P., Chowdhury, S., and Kebarle, P., *J. Chem. Phys.* 83, 1059, 1985.
57. Fischer, C. F., *Phys. Rev. A* 39, 963, 1989.
58. Wickham-Jones, C. T., Ervin, K. M., Ellision, G. B., and Lineberger, W. C., *J. Chem. Phys.* 91, 2762, 1989.
59. Kryachko, E. S., Vinckier, C., and Nguyen, M. T., *J. Chem. Phys.* 114, 7911, 2001.
60. Mock, R. S., and Grimsrud, E. P., *J. Am. Chem. Soc.* 111, 2861, 1989.
61. Chowdhury, S., Heinis, T., Grimsrud, E. P., and Kebarle, P., *J. Phys. Chem.* 90, 2747, 1986. The uncertainty and other results are quoted in Ref. 60.
62. Wickham-Jones, C. T., Moran, S., and Ellison, G. B., *J. Chem. Phys.* 90, 795, 1989.
63. Ervin, K. M., Ho, J., and Lineberger, W. C., *J. Phys. Chem.* 92, 5405, 1988.
64. Miller, T. M., and Lineberger, W. C., *Int. J. Mass Spectrom. Ion Process.* 102, 239, 1990.
65. Wetzel, D. M., Salomon, K. E., Berger, S., and Brauman, J. I., *J. Am. Chem. Soc.* 111, 3835, 1989.
66. Crocker, L., Wang, T., and Kebarle, P., *J. Am. Chem. Soc.* 115, 7818, 1993.
67. Dillow, G. W., and Kebarle, P., *J. Am. Chem. Soc.* 111, 5592, 1989.
68. Gantefor, G., Gausa, M., Meiwes-Broer, K. H., and Lutz, H. O., *Z. Phys. D* 9, 253, 1988; Taylor, K. J., Petteitte, C. L., Craycraft, M. J., Chesnovsky, O., and Smalley, R. E., *Chem. Phys. Lett.* 152, 347, 1988.
69. Metz, R. B., Kitsopoulos, T., Weaver, A., and Neumark, D. M., *J. Chem. Phys.* 88, 1463, 1988.
70. Yang, S., Pettiette, C. L., Conceicao, J., Cheshnovsky, O., and Smalley, R. E., *Chem. Phys. Lett.* 139, 233, 1987; Yang, S., Taylor, K. J., Craycraft, M. J., Conceicao, J., Pettiette, C. L., Cheshnovsky, O., and Smalley, R. E., *Chem. Phys. Lett.* 144, 431, 1988; Arnold, D. W., Bradforth, S. E., Kitsopoulos, T. N., and Neumark, D. M., *J. Chem. Phys.* 95, 5479, 1991.
71. Cheshnovsky, O., Yang, S., Pettiette, C. L., Craycraft, M. J., Liu, Y., and Smalley, R. E., *Chem. Phys. Lett.* 138, 119, 1987.
72. Liu, Y., Zhang, Q.-L., Tittel, F. K., Curl, R. F., and Smalley, R. E., *J. Chem. Phys.* 85, 7434, 1986.
73. Travers, M. J., Cowles, D. C., and Ellison, G. B., *Chem. Phys. Lett.* 164, 449, 1989.
74. Blondel, C., Cacciani, P., Delsart, C., and Trainham, R., *Phys. Rev. A* 40, 3698, 1989. EA(Br) = 27,129.170 ± 0.015 cm^{-1} and EA(F) = 27,432.440 ± 0.025 cm^{-1}.
75. Bowen, K. H., and Eaton, J. G., in *The Structure of Small Molecules and Ions*, Naaman, R., and Vager, Z., Eds., Plenum, New York, 1988, pp. 147-169; Arnold, S. T., Eaton, J. G., Patel-Mistra, D., Sarkas, H. W., and Bowen, K. H., in *Ion and Cluster Ion Spectroscopy and Structure*, Maier, J. P., Ed., Elsevier Science, New York, 1989, p. 417.
76. Snodgrass, J. T., Coe, J. V., Friedhoff, C. B., McHugh, K. M., and Bowen, K. H., *Faraday Disc. Chem. Soc.* 88, 1988.
77. Lee, G. H., Arnold, S. T., Eaton, J. G., Sarkas, H. W., Bowen, K. H., Ludewigt, C., and Haberland, H., *Z. Phys. D – At., Mol. Clusters* 20, 9, 1991; Coe, J. V., Lee, G. H., Eaton, J. G., Arnold, S. T., Sarkas, H. W., Bowen, K. H., Ludewigt, C., Haberland, H., and Worsnop, D. R., *J. Chem. Phys.* 92, 3980, 1990.
78. Snodgrass, J. T., Coe, J. V., Freidhoff, C. B., McHugh, K. M., Arnold, S. T., and Bowen, K. H., *J. Phys. Chem.* **99**, 9675, 1995.
79. Hendricks, J. H., de Clercq, H. L., Freidhoff, C. B., Arnold, S. T., Eaton, J. G., Fancher, C., Lyapustina, S. A., Snodgrass, J. T., and Bowen, K. H., *J. Chem. Phys.* **116**, 7926, 2002.
80. Snodgrass, J. T., Coe, J. V., Friedhoff, C. B., McHugh, K. M., and Bowen, K. H., *J. Chem. Phys.* 88, 8014, 1988.
81. Coe, J. V., Snodgrass, J. T., Friedhoff, C. B., McHugh, K. M., and Bowen, K. H., *Chem. Phys. Lett.* 124, 274, 1986.

82. Eliav, E., Vilkas, M. J., Ishikawa, Y., and Kaldor, U., *J. Chem. Phys.* 123, 224113(5), 2005.
83. Ervin, K. M., Ho, J., and Lineberger, W. C., *J. Chem. Phys.* 91, 5974, 1989.
84. Polak, M. L., Fiala, B. L., Ervin, K. M., and Lineberger, W. C., *J. Chem. Phys.* 94, 6924, 1991.
85. Weaver, A., Arnold, D. W., Bradforth, S. E., Neumark, D. M., *J. Chem. Phys.* 94, 1740, 1991.
86. Zhai, H.-J., Kirian, B., Cui, L.-F., Li, X., Dixon, D. A., and Wang, L.-S, *J. Am. Chem. Soc.* 126, 16134, 2004.
87. Ervin, K. M., and Lineberger, W. C., *J. Phys. Chem.* 95, 1167, 1991.
88. Gilles, M. K., Polak, M. L., and Lineberger, W. C., *J. Chem. Phys.* 96, 8012, 1992.
89. Lykke, K. R., Murray, K. K., and Lineberger, W. C., *Phys. Rev. A* 43, 6104, 1991. EA(H) = 6082.99 ± 0.15 cm^{-1} and EA(D) = 6086.2 ± 0.6 cm^{-1}.
90. Ervin, K. M., Gronert, S., Barlow, S. E., Gilles, M. K., Harrison, A. G., Bierbaum, V. M., DePuy, C. H., Lineberger, W. C., and Ellison, G. B., *J. Am. Chem. Soc.* 112, 5750, 1990.
91. Viggiano, A. A., Miller, T. M., Miller, A. E. S., Morris, R. A., Van Doren, J. M., and Paulson, J. F., *Int. J. Mass Spectrom. Ion Process.* 109, 327, 1991.
92. Hanstorp, D., and Gustafsson, M., *J. Phys. B: At. Mol. Opt. Phys.* 25, 1773, 1992. EA(I) = 24,672.7956 ± 0.079 cm^{-1}.
93. Polak, M. L., Gilles, M. K., and Lineberger, W. C., *J. Chem. Phys.* 96, 7191, 1992.
94. Bengali, A. A., Casey, S. M., Cheng, C.-L., Dick, J. P., Fenn, P. T., Villalta, P. W., and Leopold, D. G., *J. Am. Chem. Soc.* 114, 5257, 1992.
95. Gilles, M. K., Ervin, K. M., Ho, J., and Lineberger, W. C., *J. Phys. Chem.* 96, 1130, 1992.
96. Travers, M. J., Cowles, D. C., Clifford, E. P., and Ellison, G. B., *J. Am. Chem. Soc.* 114, 8699, 1992.
97. Bengali, A. A., and Leopold, D. G., *J. Am. Chem. Soc.* 114, 9192, 1992.
98. Rudnyi, E. B., Kaibicheva, E. A., and Sidorov, L. N., *Rapid Commun. Mass Spectrom.* 6, 356, 1992; Sidorov, L. N., *High Temp. Sci.* 29, 153, 1990. See also Srivastava, R. D., Uy, O. M., and Farber, M., *Trans. Faraday Soc.* 67, 2941, 1971.
99. Kitsopoulos, T. N., Chick, C. J., Zhao, Y., and Neumark, D. M., *J. Chem. Phys.* 95, 5479, 1991.
100. Arnold, C. C., Kitsopoulos, T. N., and Neumark, D. M., *J. Chem. Phys.* 99, 766, 1993.
101. Rudnyi, E. B., Kaibicheva, E. A., and Sidorov, L. N., *J. Chem. Thermodyn.* 25, 929, 1993.
102. Sarkas, H. W., Hendricks, J. H., Arnold, S. T., and Bowen, K. H., *J. Chem. Phys.* 100, 1884, 1994.
103. Villalta, P. W., and Leopold, D. G., *J. Chem. Phys.* 98, 7730, 1993.
104. Eaton, J. G., Sarkas, H. W., Arnold, S. T., McHugh, K. M., and Bowen, K. H., *Chem. Phys. Lett.* 193, 141, 1992. See also Ref. 18.
105. Polak, M. L., Gilles, M. K., Gunion, R. F., and Lineberger, W. C., *Chem. Phys. Lett.* 210, 55, 1993.
106. Gilles, M. K., Lineberger, W. C., and Ervin, K. M., *J. Am. Chem. Soc.* 115, 1031, 1993.
107. Casey, S. M., and Leopold, D. G., *Chem. Phys. Lett.* 201, 205, 1993.
108. Polak, M. L., Gerber, G., Ho, J., and Lineberger, W. C., *J. Chem. Phys.* 97, 8990, 1992.
109. Kim, E. H., Bradforth, S. E., Arnold, D.W., Metz, R. B., and Neumark, D. M., *J. Chem. Phys.* 103, 7801, 1995.
110. Xu, C., Taylor, T. R., Burton, G. R., and Neumark, D. M., *J. Chem. Phys.* 108, 1395, 1998.
111. Bradforth, S. E., Kim, E. H., Arnold, D. W., and Neumark, D. M., *J. Chem. Phys.* 98, 800, 1993.
112. Ho, J., Polak, M. L., Ervin, K. M., and Lineberger, W. C., *J. Chem. Phys.* 99, 8542, 1993.
113. Brinkman, E. A., Berger, S., Marks, J., and Brauman, J. I., *J. Chem. Phys.* 99, 7586, 1993.
114. Casey, S. M., and Leopold, D. G., *J. Phys. Chem.* 97, 816, 1993.
115. McDonald, R. N., and Davidson, S. J., *J. Am. Chem. Soc.* 115, 10857, 1993.
116. Ho, J., Ervin, K. M., Polak, M. L., Gilles, M. K., and Lineberger, W. C., *J. Chem. Phys.* 95, 4845, 1991.
117. Ho, J., Polak, M. L., and Lineberger, W. C., *J. Chem. Phys.* 96, 144, 1992. See also Ref. 289.
118. Polak, M. L., Gilles, M. K., Ho, J., and Lineberger, W. C., *J. Phys. Chem.* 95, 3460, 1991.
119. Polak, M. L., Ho, J., Gerber, G., and Lineberger, W. C., *J. Chem. Phys.* 95, 3053, 1991.
120. Sharpe, P., and Kebarle, P., *J. Am. Chem. Soc.* 115, 782, 1993.
121. Miller, T. M., Miller, A. E. S., Viggiano, A. A., Morris, R. A., and Paulson, J. F., *J. Chem. Phys.* 100, 7200, 1994. Accurate calculations have yielded a higher result (0.90 eV); see Lau, J. K.-C. and Li, W.-K., *J. Mol. Struct.* (*Theochem*) 578, 221, 2002.
122. Berkovits, D., Boaretto, E., Gehlberg, S., Heber, O., and Paul, M., *Phys. Rev. Lett.* 75, 414, 1995.
123. Arnold, C. C., Xu, C., Burton, G. R., and Neumark, D. M., *J. Chem. Phys.* 102, 6982, 1995; Burton, G. R., Xu, C., Arnold, C. C., and Neumark, D. M., *J. Chem. Phys.* 104, 2757 1996.
124. Xu, C., de Beer, E., and Neumark, D. M., *J. Chem. Phys.* 104, 2749, 1996.
125. Gunion, R. F., Koppel, H., Leach, G. W., and Lineberger, W. C., *J. Chem. Phys.* 103, 1250, 1995.
126. Gunion, R. F., and Lineberger, W. C., *J. Phys. Chem.* 100, 4395, 1996.
127. Gunion, R. F., Dixon Warren, St. J., and Lineberger, W. C., *J. Chem. Phys.* 104, 1765, 1996.
128. Dixon-Warren, St. J., Gunion, R. F., and Lineberger, W. C., *J. Chem. Phys.* 104, 4902, 1996.
129. Nakajima, A., Zhang, N., Kawamata, H., Hayase, T., Nakao, K., and Kaya, K., *Chem. Phys. Lett.* 241, 295, 1995; Nakajima, A., Taguwa, T., Nakao, K., Hoshino, K., Iwata, S., and Kaya, K., *J. Chem. Phys.* 102, 660, 1995.
130. Fan, J., and Wang, L.-S., *J. Chem. Phys.* 102, 8714, 1995.
131. Arnold, D. W., Bradforth, S. E., Kim, E. H., and Neumark, D. M., *J. Chem. Phys.* 102, 3493, 1995; Zhao, Y., Arnold, C. C., and Neumark, D. M., *J. Chem. Soc. Faraday Trans.* 2 89, 1449, 1992.
132. Fan, J., Lou, L., and Wang, L.-S., *J. Chem. Phys.* 102, 2701, 1995.
133. Sarkas, H. W., Hendricks, J. H., Arnold, S. T., Slager, V. L., and Bowen, K. H., *J. Chem. Phys.* 100, 3358, 1994.
134. Kato, S., Lee, H. S., Gareyev, R., Wenthold, P. G., Lineberger, W. C., DePuy, C. H., and Bierbaum, V. M., *J. Am. Chem. Soc.* 119, 7863, 1997. See also Miller, T. M., Viggiano, A. A., and Miller, A. E. S., *J. Phys. Chem. A* 106, 10200, 2002.
135. Wenthold, P. G., Hu, J., Squires, R. R., and Lineberger, W. C., *J. Am. Chem. Soc.* 118, 475, 1996.
136. Gunion, R. F., Karney, W., Wenthold, P. G., Borden, W. T., and Lineberger, W. C., *J. Am. Chem. Soc.* 118, 5074, 1996. The numbers in the abstract for 1,6-heptadiyne were misprinted. EA(cycloheptatrienide) quoted here derives from the LPES data combined with other thermochemical data in Ref. 136.
137. Xu, C., de Beer, E., Arnold, D. W., Arnold, C. C., and Neumark, D. M., *J. Chem. Phys.* 101, 5406, 1996.
138. Wenthold, P. G., Polak, M. L., and Lineberger, W. C., *J. Phys. Chem.* 100, 6920, 1996.
139. Wenthold, P. G., Kim, J. B., and Lineberger, W. C., *J. Am. Chem. Soc.* 119, 1354, 1997.
140. Eliav, E., Kaldor, U., Ishikawa, Y., and Pyykko, P., *Phys. Rev. Lett.* 77, 5350, 1996.
141. Davies, B. J., Ingram, C. W., Larson, D. J., and Ljungblad, U., *J. Chem. Phys.* 106, 5783, 1997. EA(Ir) = 12,613 ± 4 cm^{-1}.
142. Smith, J. R., Kim, J. B., and Lineberger, W. C., *Phys. Rev. A* 55, 2036, 1997. EA(OH) = 14,741.02 ± 0.03 cm^{-1}. Schulz, P. A., Mead, R. D., Jones, P. L., and Lineberger, W. C., *J. Chem. Phys.* 77, 1153, 1982. EA(OD) = 14,723.92 ± 0.30 cm^{-1}. See also Rudmin, J. D., Ratliff, L. P., Yukich, J. N., and Larson, D. J., *J. Phys. B: At. Mol. Opt. Phys.* 29, L881, 1996.
143. Desai, S. R., Wu, H., Rohlfing, C. M., and Wang, L.-S., *Int. J. Chem. Phys.* 106, 1309, 1997.
144. Wenthold, P. G., Jonas, K.-L., and Lineberger, W. C., *J. Chem. Phys.* 106, 9961, 1997.
145. Yang, J., Wang, X.-B., Xing, X.-P., and Wang, L.-S., *J. Chem. Phys.* 128, 201102, 2008.
146. Moravec, V. D., and Jarrold, C. C., *J. Chem. Phys.* 108, 1804, 1998.
147. Wang, X.-B., Ding, C.-F., and Wang, L.-S., *J. Phys. Chem. A* 101, 7699, 1997.

148. Wenthold, P. G., and Lineberger, W. C., *J. Am. Chem. Soc.* 19, 7772, 1997.
149. Wang, L.-S., Li, X., and Zhang, H.-F., *Chem. Phys.* 262, 53, 2000; Wang, L.-S., Cheng, H.-S., and Fan, J., *J. Chem. Phys.* 102, 9480, 1995.
150. Wu, H., Desai, S. R., and Wang, L.-S., *Phys. Rev. Lett.* 77, 2436, 1996.
151. Liu, S.-R., Zhai, H.-J., Castro, M., and Wang, L.-S., *J. Chem. Phys.* 118, 2108, 2003.
152. Gutsev, G. L., Bauschlicher, C. W., Zhai, H.-J., and Wang, L.-S., *J. Chem. Phys.* 119, 11135, 2003; Wang, L.-S., Wu, H., and Desai, S. R., *Phys. Rev. Lett.* 76, 4853, 1996.
153. Robinson, M. S., Polak, M. L., Bierbaum, V. M., DePuy, C. H., and Lineberger, W. C., *J. Am. Chem. Soc.* 117, 6766, 1995.
154. Clifford, E. P., Wenthold, P. G., Lineberger, W. C., Petersson, G. A., and Ellison, G. B., *J. Phys. Chem.* 101, 4338, 1997.
155. Seburg, R. A., and Squires, R. R., *Int. J. Mass Spectrom. Ion Process.* 167/168, 541, 1997.
156. Wang, X.-B., and Wang, L.-S., *Chem. Phys. Lett.* 313, 179, 1999.
157. Marcy, T. P., and Leopold, D. G., *Int. J. Mass Spectrom.* 195/196, 653, 2000.
158. Gutsev, G. L., Rao, B. K., Jena, P., Li, X., and Wang, L.-S., *J. Chem. Phys.* 113, 1473, 2000.
159. Negishi, Y., Yasuike, T., Hayakawa, F., Kizawa, M., Yabushita, S., and Nakajima, A., *J. Chem. Phys.* 113, 1725, 2000.
160. Klopcic, S. A., Moravec, V. D., and Jarrold, C. C., *J. Chem. Phys.* 110, 8986, 1999.
161. Boldyrev, A. I., Simons, J., Li, X., Chen, W., and Wang, L.-S., *J. Chem. Phys.* 110, 8980, 1999.
162. Taylor, T. R., Asmis, K. R., Zanni, M. T., and Neumark, D. M., *J. Chem. Phys.* 110, 7607, 1999.
163. Boldyrev, A., Li, X., and Wang, L.-S., *J. Chem. Phys.* 112, 3627, 2000.
164. Defrançois, C., Périquet, V., Lyapustina, S. A., Lippa, T. P., Robinson, D. W., Bowen, K. H., Nonaka, H., and Compton, R. N., *J. Chem. Phys.* 111, 4569, 1999.
165. Dobrin, S., Boo, B. H., Alconcel, L. S., and Continetti, R. E., *J. Phys. Chem. A* 104, 10695, 2000.
166. Schwartz, R. L., Davico, G. E., and Lineberger, W. C., *J. Electron Spectrosc. Relat. Phenom.*108, 163, 2000.
167. Kim, J. B., Wenthold, P. G., and Lineberger, W. C., *J. Phys. Chem.* 103, 10833, 1999.
168. Davico, G. E., Ramond, T. M., and Lineberger, W. C., *J. Chem. Phys.* 113, 8852, 2000.
169. Green, S. M. E., Alex, S., Fleischer, N. L., Millam, E. L., Marcy, T. P., and Leopold, D. G., *J. Chem. Phys.* 114, 2653, 2001.
170. Wu, H., and Wang, L.-S., *J. Chem. Phys.* 108, 5310, 1998.
171. Wu, H., and Wang, L.-S., *J. Phys. Chem. A* 102, 9129, 1998.
172. Wu, H., and Wang, L.-S., *J. Chem. Phys.* 107, 8221, 1997.
173. Thomas, O. C., Xu, S. J., Lippa, T. P., and Bowen, K. H., *J. Clust. Sci.*10, 525, 1999.
174. Wang, L.-S., private communication quoted in Ref. 169.
175. Marcy, T. P., Ph.D. dissertation, quoted in Ref. 169.
176. Alex, S., Green, M. E., and Leopold, D. G., unpublished, quoted in Ref. 169.
177. Wang, X.-B., Wang, L.-S., Brown, R., Schwerdtfeger, P., Schröder, D., and Schwarz, H., *J. Chem. Phys.* 114, 7388, 2001.
178. Kim, J. H., Li, X., Wang, L.-S., de Clercq, H. L., Fancher, C. A., Thomas, O. C., and Bowen, K. H., J. *Phys. Chem. A* 105, 5709, 2001.
179. Moravec, V. D., Klopcic, S. A., Chatterjee, B., and Jarrold, C. C., *Chem. Phys. Lett.* 341, 313, 2001.
180. Zengin, V., Persson, B. J., Strong, K. M., and Continetti, R. E., *J. Chem. Phys.* 105, 9740, 1996.
181. Kietzmann, H., Morenzin, J., Bechthold, P. S., Ganteför, G., and Eberhardt, W., *J. Chem. Phys.* 109, 2275, 1998.
182. Xu, C., Taylor, T. R., Burton, G. R., and Neumark, D. M., *J. Chem. Phys.* 108, 7645, 1998.
183. Williams, W. W., Carpenter, D. L., Covington, A. M., Koepnick, M. C., Calabrese, D., and Thompson, J. S., *J. Phys. B: At. Mol. Opt. Phys.* 31, L341, 1998.
184. Covington, A. M., Calabrese, D., Thompson, J. S., and Kvale, T. J., *J. Phys. B: At. Mol. Opt. Phys.* 31, L855, 1998.
185. Haeffler, G., Hanstrorp, D., Kiyan, I., Klinkmueller, A. E., Ljungblad, U., and Pegg, D. J., *Phys. Rev. A* 53, 4127, 1996. EA(Li) = 4984.90 ± 0.17 cm^{-1}.
186. Chen, H. L., Ellis, Jr., P. E., Wijesekera, T., Hagan, T. E., Groh, S. E., Lyons, J. E., and Ridge, D. P., *J. Am. Chem. Soc.* 116, 1086, 1994.
187. Deyerl, H.-J., Alconcel, L. S., and Continetti, R. E., *J. Phys. Chem. A* 105, 552, 2001.
188. Blanksby, S. J., Ramond, T. M., Davico, G. E., Nimlos, M. R., Kato, S., Bierbaum, V. M., Lineberger, W. C., Ellison, G. B., and Okumura, M., *J. Am. Chem. Soc.* 123, 9585, 2001.
189. Asmis, K. R., Taylor, T. R., Xu, C., and Neumark, D. M., *Chem. Phys. Lett.* 295, 75, 1998.
190. Schäfer-Bung, B., Engels, B., Taylor, T. R., Neumark, D. M., Botschwina, P., and Peric, M., *J. Chem. Phys.* 115, 1777, 2001.
191. Scheer, M., Bilodeau, R. C., and Haugen, H. K., *Phys. Rev. Lett.* 80, 2562, 1998.
192. Taylor, T. R., Gómez, H., Asmis, K. R., and Neumark, D. M., *J. Chem. Phys.* 115, 4620, 2001.
193. Asmis, K. R., Taylor, T. R., and Neumark, D. M., *J. Chem. Phys.* 111, 8838, 1999 and 111, 10491, 1999.
194. Ramond, T. M., Davico, G. E., Schwartz, R. L., and Lineberger, W. C., *J. Chem. Phys.* 112, 1158, 2000.
195. Petrunin, V. V., Voldstad, J. D., Balling, P., Kristensen, P., Andersen, T., and Haugen, H. K., *Phys. Rev. Lett.* 75, 1911, 1995.
196. Dzuba, V. A., and Gribakin, G. F., *J. Phys. B: At. Mol. Opt. Phys.* 31, L483, 1998.
197. Ervin, K. M., Ramond, T. M., Davico, G. E., G. E., Schwartz, R. L., Casey, S. M., and Lineberger, W. C., *J. Phys. Chem.* 105, 10822, 2001.
198. Lippa, T. P., Xu, S.-J., Lyapustina, S. A., and Bowen, K. H., *J. Chem. Phys.* 109, 9263, 1998.
199. Turner, N. J., Martel, A. A., and Waller, I. M., *J. Phys. Chem.* 98, 474, 1994.
200. Lippa, T. P., Xu, S.-J., Lyapustina, S. A., Nilles, J. M., and Bowen, K. H., *J. Chem. Phys.* 109, 10727, 1998.
201. Wang, X.-B., Woo, H.-K., Wang, L.-S., *J. Chem. Phys.* 123, 051106, 2005.
202. Boltalina, O. V., Sidorov, L. N., Sukhanova, E. V., and Sorokin, I. D., *Chem. Phys. Lett.* 230, 567, 1994.
203. Clifford, E. P., Wenthold, P. G., Lineberger, W. C., Ellison, G. B., Wang, C. X., Grabowski, J. J., Vila, F., and Jordan, K. D., *J. Chem. Soc. Perkin Trans.* 2, 1015, 1998.
204. Spanel, P., Matejcik, S., and Smith, D., *J. Phys. B: At. Mol. Phys.* 28, 2941, 1995. See Miller, A. E. S., Miller, T. M., Viggiano, A. A., Morris, R. A., Van Doren, J. M., Arnold, S. T., and Paulson, J. F., *J. Chem. Phys.* 102, 8865, 1995 for interpretation in terms of EA(SF_5).
205. Kinghom, D. B., and Adamowicz, L., J., *Chem. Phys.* 106, 4589, 1997. EA(H) = 6083.0994 cm^{-1}, EA(D) = 6086.7137 cm^{-1}, and EA(T) = 6087.9168 cm^{-1}.
206. Xi, L., and Wang, L.-S., *J. Chem. Phys.* 109, 5264, 1998.
207. Ephraim, E., Shmulyian, S., Kaldor, U., and Isikawa, Y., *J. Chem. Phys.* 109, 3954, 1998. Also EA(La) = 0.35 eV.
208. Scheer, M., Bilodeau, R. C., Thogersen, J., and Haugen, H. K., *Phys. Rev. A* 57, R1493, 1998.
209. Kim, J. B., Wenthold, P. G., and Lineberger, W. C., *J. Chem. Phys.* 108, 830, 1998.
210. Kishi, R., Kawamata, H., Negishi, Y., Iwata, S., Nakajima, A., and Kaya, K., *J. Chem. Phys.* 107, 10029, 1997.
211. Adams, C. L., Schneider, H., Ervin, K. M. and Weber, J. M., *J. Chem. Phys.* 130, 074307, 2009. Earlier work by Compton, R. N., Carman, Jr., H. S., Desfrançois, C., Abdoul-Carmine, J., Schermann, J. P., Hendricks, J. H., Lyapustina, S. A., and Bowen, K. H., *J. Chem. Phys.* 105, 3472, 1996, found EA(CD_3NO_2) to be 0.02 eV lower than EA(CH_3NO_2).
212. Yourshaw, I., Zhao, Y., and Neumark, D. M., *J. Chem. Phys.* 105, 351, 1996.
213. Gausa, M., Kaschner, R., Seifert, G., Faehrmann, J. H., Lutz, H. O., and Meiwes-Broer, K., *J. Chem. Phys.* 104, 9719, 1996.
214. Wu, H., and Wang, L.-S., *J. Chem. Phys.* 107, 16, 1997.
215. Radisic, D., Xu, S., and Bowen, K. H., *Chem. Phys. Lett.* 354, 9, 2002.
216. Pramann, A., and Rademann, K., *Chem. Phys. Lett.* 343, 99, 2001.
217. Gómez, H., Taylor, T. R., and Neumark, D. M., *J. Phys. Chem. A* 105, 6886, 2001.

218. Gómez, H., Taylor, T. R., Zhao, Y., and Neumark, D. M., *J. Chem. Phys.* 117, 8644, 2002.
219. Nimlos, M. R., Davico, G., Geise, C. M., Wenthold, P. G., Lineberger, W. C., Blansksby, S. J., Hadad, C. M., Petersson, G. A., and Ellison, G. B., *J. Chem. Phys.* 117, 4323, 2002.
220. Koyasu, K., Mitsui, M., Nakajima, A., and Kaya, K., *Chem. Phys. Lett.* 358, 224, 2002.
221. Zhai, H.-J., Wang, L.-S., Alexandrova, A. N., Boldyrev, A. I., and Zakrzewski, V. G., *J. Phys. Chem. A* 107, 9319, 2003.
222. Schiedt, J., and Weinkauf, *R., Z. Naturforsch. A* 50, 1041, 1995. See also Ervin, K. M., Anusiewicz, I., Skurski, P., Simons, J., and Lineberger, W. C., *J. Phys. Chem. A* 107, 8521, 2003 [EA(O_2) = 0.448 ± 0.006 eV].
223. Davis, V. T., and Thompson, J. S., *J. Phys. B* 34, L433, 2001. Recent high-level calculations give EA(Lu) = 0.029 eV; see Ref. 343.
224. Li, X., Zhai, H.-J., and Wang, L.-S., *Chem. Phys. Lett.* 357, 415, 2002.
225. Davis, V. T., and Thompson, J. S., *J. Phys. B: At. Mol. Opt. Phys.* 35, L11, 2002.
226. Goldfarb, F., Drag, C., Chaibi, W., Kröger, S., Blondel, C., and Delsart, C., *J. Chem. Phys.* 122, 014308, 2005. EA(OH) = 14740.982(7) cm^{-1}.
227. Blondel, C., Delsart, C., and Goldfarb, F., *J. Phys. B* 34, L281, 2001. EA(F) = 27432.446(19) cm^{-1}.
228. Fancher, C. A., de Clercq, H. L., and Bowen, K. H., *Chem. Phys. Lett.* 366, 197, 2002.
229. Jin, C., Taylor, K. J., Conceicao, J., and Smalley, R. E., *Chem. Phys. Lett.* 175, 17, 1990.
230. Schiedt, J., Knott, W. J., Le Barbu, K., Schlag, E. W., and Weinkauf, R., *J. Chem. Phys.* 113, 9470, 2000.
231. Song, J. K., Lee, N. K., Kim, J. H., Han, S. Y., and Kim, S. K., *J. Chem. Phys.* 119, 3071, 2003.
232. Davico, G. E., Schwartz, R. L., Ramond, T. M., and Lineberger, W. C., *J. Amer. Chem. Soc.* 121, 6047, 1999.
233. Andrews, D. H., Gianola, A. J., and Lineberger, W. C., *J. Chem. Phys.* 117, 4074, 2002.
234. Schwartz, R. L., Davico, G. E., Kim, J. B., and Lineberger, W. C., *J. Chem. Phys.* 112, 4966, 2000.
235. Schwartz, R. L., Davico, G. E., Ramond, T. M., and Lineberger, W. C., *J. Phys. Chem. A* 103, 8213, 1999.
236. Schiedt, J., and Weinkauf, R., *Chem. Phys. Lett.* 274, 18, 1997.
237. Wang, X.-B., Wang, L.-S., Brown, R., Schwerdtfeger, P., Schröder, D., and Schwarz, H., *J. Chem. Phys.* 114, 7388, 2001.
238. Geske, G. D., Boldyrev, A. I., Li, X., and Wang, L.-S., *J. Chem. Phys.* 113, 5130, 2000.
239. Cannon, N. A., Boldyrev, A. I., Li, X., and Wang, L.–S., *J. Chem. Phys.* 113, 2671, 2000.
240. Wang, X.-B., Yang, X., and Wang, L.-S., *J. Chem. Phys.* 116, 561, 2002.
241. Gutsev, G. L., Jena, P., Zhai, H.-J., and Wang, L.-S., *J. Chem. Phys.* 115, 7935, 2001.
242. Yang, X., Wang, X.-B., and Wang, L.-S., *J. Chem. Phys.* 115, 2889, 2001.
243. Zhai, H.-J., Liu, S.-R., Li, X., and Wang, L.-S., *J. Chem. Phys.* 115, 5170, 2001.
244. Li, X., Wang, L.-S., Cannon, N. A., and Boldyrev, A. I., *J. Chem. Phys.* 116, 1330, 2002.
245. Zhai, H.-J., Wang, L.-S., Alexandrova, A. N., and Boldyrev, A. I., *J. Chem. Phys.* 117, 7917, 2002.
246. Zhai, H.-J., and Wang, L.-S., *J. Chem. Phys.* 117, 7882, 2002.
247. Liu, S.-R., Zhai, H.-J., and Wang, L.-S., *J. Chem. Phys.* 117, 9758, 2002.
248. Mitsui, M., Nakajima, A., and Kaya, K., *J. Chem. Phys.* 117, 9740, 2002.
249. Yang, X., Wang, X.-B., Wang, L.-S., Niu, S., and Ichiye, T., *J. Chem. Phys.* 119, 8311, 2003.
250. Kim, J., Lee, H. M., Suh, S. B., Majumdar, D., and Kim, K. S., *J. Chem. Phys.* 113, 5259, 2000.
251. Liu, S. R., Zhai, H. J., and Wang, L.-S., *Phys. Rev. B* 64, 153402, 2001.
252. Zhai, H. J., Yang, X., Wang, X. B., Wang, L.-S., Elliott, B., and Boldyrev, A. I., *J. Am. Chem. Soc.* 124, 6742, 2002.
253. Zhai, H. J., Wang, L.-S., Kuznetsov, A. E., and Boldyrev, A. I., *J. Phys. Chem. A* 106, 5600, 2002.
254. Drechsler, G., and Boesl, U., *Int. J. Mass Spectrom.* 228, 1067, 2003.
255. Illenberger, E, Comita, P. B., Brauman, J. I., Fenzlaff, H. P., Heni, M., Heinrich, N., Koch, W., and Fenking, G., *Ber. Bunsen-Ges. Phys. Chem.* 89, 1026, 1985; Jackson, R. L., Pellerite, M. J., and Brauman, J. I., *J. Am. Chem. Soc.* 103, 1802, 1981.
256. Miller, T. M., Friedman, J. F., and Viggiano, A. A., *J. Chem. Phys.* 120, 7024, 2004.
257. Miller, T. M., Van Doren, J. M., and Viggiano, A. A., *Int. J. Mass Spectrom.* 233, 67, 2004.
258. Van Doren, J. M., McSweeney, S. A., Hargus, M. D., Kerr, D. M., Miller, T. M., Arnold, S. T., and Viggiano, A. A., *Int. J. Mass Spectrom.* 228, 541, 2003.
259. Van Doren, J. M., Miller, T. M., and Viggiano, A. A., *J. Chem. Phys.* 123, 114303, 2005.
260. Miller, T. M., and Viggiano, A. A., *Phys. Rev. A* 71, 012702, 2005.
261. Haeffler, G., Klinkmueller, A. E., Rangell, J., Berzinsh, U., and Hanstorp, D., *Z. Phys. D* 38, 211, 1996.
262. Bilodeau, R. C., and Haugen, H. K., *Phys. Rev. A* 64, 024501, 2001.
263. Miller, T. M., Viggiano, A. A., Friedman, J. F., and Van Doren, J. M., *J. Chem. Phys.* 121, 9993, 2004.
264. Davis, V. T., and Thompson, J. S., *Phys. Rev A* 65, 010501, 2001. Theoretical work implies that the measured EA(Tm) was actually for a long-lived excited anion state, and that Tm does not form a stable anion. See O'Malley, S. M., and Beck, D. R., *Phys. Rev. A* 70, 022502, 2004.
265. Gianola, A. J., Ichino, T., Hoenigman, R. L., Kato, S., Bierbaum, V. M., and Lineberger, W. C., *J. Phys. Chem. A* 108, 10326, 2004.
266. Staneke, P. O., Groothuis, G., Ingemann, S., and Nibbering, N. M. M., *Int. J. Mass Spectrom. Ion Process.* 142, 83, 1995. A Gaussian-3 calculation yields EA(CCl_4) = 0.994 eV [Ed.].
267. Meloni, G., Ferguson, M. J., and Neumark, D. M., *Phys. Chem. Chem. Phys.* 5, 4073, 2003.
268. Davis, V. T., and Thompson, J. S., *J. Phys. B: At. Mol. Opt. Phys.* 37, 1961, 2004.
269. Walter, C. W., Gibson, N. D., Janczak, C. M., Starr, K. A., Snedden, A. P., Field III, R. L., and Andersson, P., *Phys. Rev. A*76, 052702, 2007.
270. Ando, N., Kokubo, S., Mitsui, M., and Nakajima, A., *Chem. Phys. Lett.* 389, 279, 2004.
271. Zhai, H.-J., and Wang, L.-S., *J. Chem. Phys.* 120, 8996, 2004.
272. Surber, E., and Sanov, A., *J. Chem. Phys.* 116, 5921, 2002.
273. Dzuba, V. A., and Gribakin, G. F., *Phys. Rev. A* 55, 2443, 1997.
274. Meloni, G., Sheehan, S. M., Ferguson, M. J., and Neumark, D. M., *J. Phys. Chem. A* 108, 9750, 2004.
275. Zhai, H.-J., Li, J., and Wang, L.-S., *J. Chem. Phys.* 121, 8369, 2004.
276. Zhai, H.-J., Kiran, B., and Wang, L.-S., *J. Chem. Phys.* 121, 8231, 2004.
277. Kim, J., Kelley, J. A., Ayotte, P., Nielsen, S. B., Weddle, G. H., and Johnson, M. A., *J. Am. Soc. Mass Spectrom.* 10, 810, 1999.
278. Misaizu, F., Tsunoyama, H., Yasumura, Y., Ohshimo, K., and Ohno, K., *Chem. Phys. Lett.* 389, 241, 2004. See also Ref. 236 for comments on EA(CS_2).
279. Meloni, G., Sheehan, S. M., and Neumark, D. M., *J. Chem. Phys.* 122, 074317, 2005.
280. Yoder, B. L., Maze, J. T., Raghavachari, K., and Jarrold, C. C., *J. Chem. Phys.* 122, 094313, 2005.
281. Ervin, K. M., and Lineberger, W. C., *J. Chem. Phys.* 122, 194303, 2005.
282. Ichino, T., Gianola, A. J., Andrews, D. H., and Lineberger, W. C., *J. Phys. Chem. A*108, 11307, 2004.
283. Das, U., Raghavachari, K., and Jarrold, C. C., *J. Chem. Phys.* 122, 014313, 2005.
284. Schiedt, J., and Weinkauf, R., *J. Chem. Phys.* 110, 304, 1999.
285. Le Barbu, K., Schiedt, J., Weinkauf, R., Schlag, E. W., Nilles, J. M., Xu, S.-J., Thomas, O. C., and Bowen, J. H., *J. Chem. Phys.* 116, 9663, 2002. Uncertainties not stated.
286. Schiedt, J., and Weinkauf, R., *Chem. Phys. Lett.* 266, 201, 1997. The uncertainty for EA(anthracene) quoted as ±0.008 eV in a later paper (Ref. 230).
287. Chatterjee, B., Akin, F. A., Jarrold, C. C., and Raghavachari, K., *J. Phys. Chem. A* 109, 6880, 2005.
288. Zhai, H.-J., Kiran, B., Dai, B., Li, J., and Wang, L.-S., *J. Am. Chem. Soc.* 127, 12098, 2005.
289. Moravec, V. D., Klopcic, S. A., and Jarrold, C. C., *J. Chem. Phys.* 110, 5079, 1999.

290. Klopcic, S. A., Moravec, V. D., and Jarrold, C. C., *J. Chem. Phys.* 110, 10216, 1999.
291. Moravec, V. D., and Jarrold, C. C., *J. Chem. Phys.* 112, 792, 2000.
292. Moravec, V. D., and Jarrold, C. C., *J. Chem. Phys.* 113, 1035, 2000.
293. Chatterjee, B., Akin, F. A., Jarrold, C. C., and Raghavachari, K., *J. Chem. Phys.* 119, 10591, 2003.
294. Schröder, D., Brown, R., Schwerdtfeger, P., Wang, X. B., Yang, X., Wang, L. S., and Schwarz, H., *Angew. Chem. Int. Ed.* 42, 311, 2003.
295. Zheng, W., Nilles, J. M., Thomas, O. C., and Bowen, K. H., *J. Chem. Phys.* 122, 044306, 2005.
296. Duan, X., Burggraf, L. W., Weeks, D. E., Davico, G. E., Schwartz, R. L., and Lineberger, W. C., *J. Chem. Phys.* 116, 3601, 2002.
297. Meloni, G., Sheehan, S. M., Parsons, B. F., and Neumark, D. M., *J. Phys. Chem. A* 110, 3527, 2006.
298. Alexandrova, A. N., Boldyrev, A. I., Zhai, H.-J., and Wang, L.-S., *J. Chem. Phys.* 122, 054313, 2005.
299. Davico, G. E., Schwartz, R. L., Raymond, T. M., and Lineberger, W. C., *J. Phys. Chem. A* 103, 6167, 1999. EA(WO_2) may be lower than 1.998 eV by either 0.040 eV or 0.080 eV depending on the precise assignment of structures in the photoelectron spectrum.
300. Huang, X., Zhai, H.-J., Li, J., and Wang, L.-S., *J. Phys. Chem. A* 110, 85, 2006.
301. Zhai, H.-J., Wang, L.-S., Zubarev, D. Yu., and Boldyrev, A. I., *J. Phys. Chem. A* 110, 1689, 2006.
302. Sheehan, S. M., Meloni, G., Parsons, B. F., Wehres, N., and Neumark, D. M., *J. Chem. Phys.* 124, 064303, 2006.
303. Tschurl, M., and Boesl, U., *Int. J. Mass Spectrom.* 249/250, 364, 2006.
304. Nee, M. J., Osterwalder, A., Zhou, J., and Neumark, D. M., *J. Chem. Phys.* 125, 014306, 2006.
305. Zanni, M. T., Taylor, T. R., Greenblatt, B. J., Soep, B., and Neumark, D. M., *J. Chem. Phys.* 110, 3748, 1997.
306. Zhai, H.-J., and Wang, L.-S., *J. Chem. Phys.* 125, 164315, 2006.
307. Waters, T., Woo, H.-K., Wang, X.-B., and Wang, L.-S., *J. Am. Chem. Soc.* 128, 4282, 2006.
308. Calvi, R. M. D., Andrews, D. H., and Lineberger, W. C., *Chem. Phys. Lett.* 442, 12, 2007.
309. Parsons, B. F., Sheehan, S. M., Yen, T. A., Neumark, D. M., Wehres, N., and Weinkauf, R., *Phys. Chem. Chem. Phys.* 9, 3291, 2007.
310. Yu, W., Lin, Z., and Ding, C., *J. Chem. Phys.* 126, 114301, 2007.
311. Zhou, J., Garand, E., Eisfeld, W., and Neumark, D. M., *J. Chem. Phys.* 127, 034304, 2007.
312. Zheng, W., Li, X., Eustis, S., Grubisic, A., Thomas, O., de Clercq, H., and Bowen, K., *Chem. Phys. Lett.* 444, 232, 2007.
313. Mitsui, M., Ando, N., and Nakajima, A., *J. Phys. Chem. A* 111, 9644, 2007.
314. Zhou, J. Garand, E., and Neumark, D. M., *J. Chem. Phys.* 127, 154320, 2007. The measured values are EA(C_4D) = 28478(10) cm^{-1} and EA(C_4H) = 28497(8) cm^{-1}.
315. Sheehan, S. M., Parsons, B. F., Zhou, J., Garand, E., Yen, T. A., Moore, D. T., and Neumark, D. M., J., *Chem. Phys.* 128, 034301, 2008.
316. Ichino, T., Andrews, D. H., Rathbone, G. J., Misaizu, F., Calvi, R. M. D., Wren, S. W., Kato, S., Bierbaum, V. M., and Lineberger, W. C., *J. Phys. Chem. B* 112, 545, 2008.
317. Sheehan, S. M., Parsons, B. F., Yen, T. A., Furlanetto, M. R., and Neumark, D. M., *J. Chem. Phys.* 128, 174301, 2008.
318. Viggiano, A. A., Miller, T. M., Friedman, J. F., and Troe, J., *J. Chem. Phys.* 127, 244305, 2007.
319. Zheng, W., Bowen, K. H., Li, J., Dąbkowska, I., Gutowski, M., *J. Phys. Chem. A* 109, 482, 2005. See also Mok, D. K. W., Chau, F.-t., Dyke, J. M., and Lee, E. P. F., *Chem. Phys. Lett.* 458, 11, 2008 and Mok, D. K. W., Lee, E. P. F., Chau, F.-t., and Dyke, J. M., *Phys. Chem. Chem. Phys.* 10, 7270, 2008.
320. Zheng, W., Li, X., Eustis, S., and Bowen, K., *Chem. Phys. Lett.* 460, 68, 2008.
321. Grubisic, A., Wang, H., Ko, Y. J., and Bowen, K. H., *J. Chem. Phys.* 129, 054302, 2008.
322. Stokes, S. T., Bowen, K. H., Sommerfeld, T., Ard, S., Mirsaleh-Kohan, N., Steill, J. D., and Compton, R. N., *J. Chem. Phys.* 129, 064308, 2008.
323. Garand, E., Yacovitch, T. I., and Neumark, D. M., *J. Chem. Phys.* 129, 074312, 2008.
324. Li, X., Eustis, S., Bowen, K. H., Kandalam, A. K., and Jena, P., *J. Chem. Phys.* 129, 074313, 2008. Uncertainties not stated.
325. Kandalam, A. K., Kiran, B., Jena, P., Li, X., Grubisic, A., and Bowen, K. H., *J. Chem. Phys.* 126, 084306, 2007.
326. Gerhards, M., Thomas, O. C., Nilles, J. M., Zhang, W.-J., and Bowen, K. H., *J. Chem. Phys.* 116, 10247, 2002.
327. Edmonds, B. D., Kandalam, A. K., Khanna, S. N., Li, X., Grubisic, A., Khanna, I., and Bowen, K. H., J. *Chem. Phys.* 124, 074316, 2006.
328. Duncan, M. A., Knight, A. M., Negishi, Y., Nagao, S., Nakamura, Y., Kato, A., Nakajima, A., and Kaya, K., *Chem. Phys. Lett.* 309, 49, 1999.
329. Zheng, W., Eustis, S. N., Li, X., Nilles, J. M., Thomas, O. C., Bowen, K. H., and Kandalam, A. K., *Chem. Phys. Lett.* 462, 35, 2008.
330. Kandalam, A. K., Jena, P., Li, X., Eustis, S. N., and Bowen, K. H., *J. Chem Phys.* 129, 134308, 2008.
331. Wenthold, P. G., and Squires, R. R., *J. Phys. Chem.* 99, 2002, 1995. See also Artau, A., Nizzi, K. E., Hill, B. T., Sunderlin, L. S., and Wenthold, P. G., *J. Am. Chem. Soc.* 122, 10667, 2000.
332. Villalta, P. W., and Leopold, D. G., *J. Chem. Phys.* 130, 024303, 2009.
333. Wyrwas, R. B., and Jarrold, C. C., *J. Am. Chem. Soc.* 128, 13688, 2008.
334. Blondel, C., Chaibi, W., Delsart, C., and Drag, C., *J. Phys. B*, 39, 1409, 2006. EA(^{32}S) = 16752.9760(42) cm^{-1}.
335. Blondel, C., Chaibi, W., Delsart, C., Drag, C., Goldfarb, F., and Kröger, J., *Eur. Phys. D* 33, 335, 2005. EA(^{28}Si) = 11207.246(8) cm^{-1}.
336. Zubarev, D. Yu., Boldyrev, A. I., Li, J., Zhai, H.-J., and Wang, L.-S., *J. Phys. Chem. A* 111, 1648, 2007.
337. Averkiev, B. B., Boldyrev, A. I., Li, Y., and Wang, L.-S., *J. Phys. Chem. A* 111, 34, 2007.
338. Zhai, H.-J., Wang, L.-M., Li, S.-D., and Wang, L.-S., *J. Phys. Chem. A* 111, 1030, 2007.
339. Distelrath, V., and Boesl, U., *Faraday Discuss.* 115, 161, 2000.
340. Wang, X.-B., and Wang, L.-S., *J. Chem. Phys.* 113, 10928, 2000.
341. Carpenter, D. L., Covington, A. M., and Thompson, J. S., *Phys. Rev. A* 61, 042501, 2000.
342. Davis, V. T., Thompson, J., and Covington, A., *Nucl. Instrum. Methods Phys. Res. B* 241, 118, 2005. High-level calculations give a lower value, EA(Nd) = 0.162 eV (Ref. 343) and 0.169 eV (O'Malley, S. M., and Beck, D. R., *Phys. Rev. A* 77, 012505, 2008).
343. Felfli, Z., Msezane, A. Z., and Sokolovski, D., *Phys. Rev. A* 78, 030703, 2008.
344. Sobhy, M. A., Reveless, J. U., Gupta, U., Khanna, S. N., and Castleman, A. W., *J. Chem. Phys.* 130, 054304, 2009.
345. Ganteför, G., Gausa, M., Meiwes-Broer, K. H., and Lutz, H. O., *Z. Phys. D* 12, 405, 1989.
346. Garand, E., Yacovitch, T. I., and Neumark, D. M., *J. Chem. Phys.* 130, 064304, 2009.
347. Entfellner, M., and Boesl, U., *Phys. Chem. Chem. Phys.* 11, 2657, 2009.
348. Wang, L.-M., Huang, W., Wang, L.-S., Averkiev, B. B., and Boldyrev, A. I., *J. Chem. Phys.* 130, 134303, 2009.
349. Wren, S. W., Vogelhuber, K. M., Ervin, K. M., and Lineberger, W. C., *Phys. Chem. Chem. Phys.* 11, 4745, 2009.
350. Sheps, L., Miller, E. M., and Lineberger, W. C., *J. Chem. Phys.* 131, 064304, 2009.
351. Garand, E., Yacovitch, T. I., and Neumark, D. M., *J. Chem. Phys.* 131, 054312, 2009.
352. Walter, C. W., Gibson, N. D., Field III, R. L., Snedden, A. P., Shapiro, J. Z., Janczak, C. M., and Hanstrorp, D., *Phys. Rev. A* 80, 014501, 2009.
353. Rothgeb, D. W., Hossain, E., Kuo, A. T., Troyer, J. L., and Jarrold, C. C., *J. Chem. Phys.* 131, 044310, 2009.
354. Borschevsky, A., Pershina, V., Bliav, E., and Kaldor, U., *Chem. Phys. Lett.* 480, 49, 2009.
355. O'Malley, S. M., and Beck, D. M., *Phys. Rev. A* 79, 012511, 2009 and 80, 032514, 2009.

PROTON AFFINITIES

Proton affinity is a useful parameter for describing gas phase ion-molecule reactions in fields such as atmospheric chemistry, plasma chemistry, mass spectrometry, and astrophysics. The proton affinity E_{pa} (often designated in the literature as PA) of a molecular species M is defined as the negative of the enthalpy change for the gas phase reaction

$$M + H^+ \rightarrow MH^+.$$

A closely related quantity is the gas phase basicity $\Delta_{base}G°$ (often designated as GB), which is the negative of the Gibbs energy change for the same reaction. Thus the two are related by

$$\Delta_{base}G° = E_{pa} + T\Delta S,$$

where T is the temperature and ΔS is the entropy change in the reaction (which can be calculated if the molecular structure of M and M^+ is known).

Direct measurement of the proton affinity is possible for only a few molecules, mainly olefins and carbonyl compounds. However, these measurements have been used to establish a scale of E_{pa} values that permits proton affinities to be determined for many other molecules, including unstable species and reaction intermediates. The basis for this scale is described by Hunter and Lias in Reference 1.

The E_{pa} and $\Delta_{base}G°$ values at a temperature of 298 K are tabulated below for selected molecules. Many values are given to one decimal place, but the majority are not accurate to better than one or two kilojoules per mole. The methods of measurement are described in Reference 1, which contains a much more extensive and detailed tabulation.

Compounds are listed by molecular formula in the Hill order, but with all compounds that do not contain carbon appearing before those that do contain carbon.

References

1. Hunter, E. P. L., and Lias, S. G., *J. Phys. Chem. Ref. Data* 27, 413, 1998.
2. Hunter, E. P., and Lias, S. G., "Proton Affinity Evaluation", in *NIST Chemistry WebBook*, NIST Standard Reference Database No. 69, Linstrom, P. J., and Mallard, W. G., Eds., March 2003, National Institute of Standards and Technology, Gaithersburg, MD 20899, <http://webbook.nist.gov>.
3. Do, K., Klein, T. P., Pommerening, C. A., Bachrach, S. M., and Sunderlin, L. S., *J. Am. Chem. Soc.* 120, 6093, 1998.
4. Kim, H.-T., Green, R. J., Qian, J., and Anderson, S. L., *J. Chem. Phys.* 112, 5717, 2000.
5. Park, S. T., Kim, S. K., and Kim, M. S., *J. Chem. Phys.* 114, 5568, 2001.
6. Hiraoka, K., Mizuno, T., Eguchi, D., Takao, T., and Ino, S., *J. Chem. Phys.* 116, 7574, 2002.
7. Oresmaa, L. O., Haukka, M., Vainiotalo, P., and Pakkanen, T. A., *J. Org. Chem.* 67, 8216, 2002.
8. Wang, F., Ma, S., Zhang, D., and Cooks, R. G., *J. Phys. Chem. A* 102, 2988, 1998.
9. Bouchoux, G., Gal, J.-F., Szulejko, J. E., McMahon, T. B., Tortajada, J., Luna, A., Yanez, M., and Mo, O., *J. Phys. Chem. A* 102, 9183, 1998.
10. van Beelen, E., Koblenz, T. A., Ingemann, S. and Hammerum, S., *J. Phys. Chem. A* 108, 2728, 2004.

Molecular formula	Name	E_{pa} kJ/mol	$\Delta_{base}G°$ kJ/mol	Notes
Ar	Argon	369.2	346.3	
AsF_3	Arsenic(III) fluoride	636.7	604.2	
AsH_3	Arsine	747.9	712.0	
BHO_2	Metaboric acid	763.0	730.5	
BH_3O_3	Boric acid	728.1	698.4	
B_2H_6	Diborane	615	586.0	
$B_3H_6N_3$	Borazine	802.5	772.8	
B_4H_{10}	Tetraborane(10)	605	572.5	
B_5H_9	Pentaborane(9)	699.4	666.9	
BaO	Barium oxide	1215.4	1187.6	
Br	Bromine (atomic)	554.4	531.2	
BrH	Hydrogen bromide	584.2	557.7	
BrLi	Lithium bromide	819	792.5	
CaO	Calcium oxide	1190.6	1162.3	
Cl	Chlorine (atomic)	513.6	490.1	
ClH	Hydrogen chloride	556.9	530.1	
ClLi	Lithium chloride	827	800.5	
Co	Cobalt	742.7	719.8	
Cr	Chromium	791.3	768.4	
CsHO	Cesium hydroxide	1117.9	1092.2	
Cs_2O	Cesium oxide	1442.9	1412.2	
Cu	Copper	655.3	632.4	
F	Fluorine (atomic)	340.1	315.1	
FH	Hydrogen fluoride	484	456.7	
FO	Fluorine oxide	508.7	482.2	
F_2	Fluorine	332	305.5	
F_2O_2S	Sulfuryl fluoride	605.5	580.5	

Molecular formula	Name	E_{pa} kJ/mol	$\Delta_{base}G°$ kJ/mol	Notes
F_3N	Nitrogen trifluoride	568.4	538.6	
F_3OP	Phosphoryl fluoride	694.0	664.2	
F_3P	Phosphorus(III) fluoride	695.3	662.8	
F_4Si	Tetrafluorosilane	502.9	476.6	
F_6S	Sulfur hexafluoride	575.3	550.7	
Fe	Iron	754	731.1	
FeO	Iron(II) oxide	907	880.5	
GeH_4	Germane	713.4	687.1	
HI	Hydrogen iodide	627.5	601.3	
HKO	Potassium hydroxide	1101.8	1075.4	
HLi	Lithium hydride	1021.7	996.4	
HLiO	Lithium hydroxide	1000.1	972.1	
HNO_3	Nitric acid	751.4	731.5	
HN_3	Hydrazoic acid	756.0	723.5	
HNa	Sodium hydride	1095	1070.6	
HNaO	Sodium hydroxide	1071.8	1044.8	
HO	Hydroxyl	593.2	564.0	
HO_2	Hydroperoxy	660	627.5	
HP	Phosphorus monohydride	670.3	639.6	
H_2	Hydrogen	422.3	394.7	
$H_2N_2O_2$	Nitramide	757.4	725.0	
H_2O	Water	691	660.0	
H_2O_2	Hydrogen peroxide	674.5	643.8	
H_2O_4S	Sulfuric acid	717	681	Ref. 3
H_2P	Phosphino	709.2	675.7	
H_2S	Hydrogen sulfide	705	673.8	
H_2Se	Hydrogen selenide	707.8	676.4	
H_2Si	Silylene	839.2	804.1	
H_2Te	Hydrogen telluride	735.9	704.5	
H_3N	Ammonia	853.6	819.0	
H_3P	Phosphine	785	750.9	
H_4N_2	Hydrazine	853.2	822.4	
H_4Si	Silane	639.7	613.4	
H_6OSi_2	Disiloxane	749	718.3	
He	Helium	177.8	148.5	
I	Iodine (atomic)	608.2	583.5	
K_2O	Potassium oxide	1342.5	1311.8	
Kr	Krypton	424.6	402.4	
La	Lanthanum	1013	991.9	
Li_2	Dilithium	1162	1133.1	
Li_2O	Lithium oxide	1206	1175.3	
Lu	Lutetium	992	970.6	
Mg	Magnesium	819.6	797.3	
MgO	Magnesium oxide	988	959.4	
Mg_2	Dimagnesium	919	886.5	
Mn	Manganese	797.3	774.4	
N	Nitrogen (atomic)	342.2	318.7	
NO	Nitric oxide	531.8	505.3	
NO_2	Nitrogen dioxide	591.0	560.3	
NP	Phosphorus nitride	789.4	757.0	
N_2	Nitrogen	493.8	464.5	
N_2O	Nitrous oxide	549.8	523.3	Protonation at N
N_2O	Nitrous oxide	575.2	548.7	Protonation at O
Na_2	Disodium	1146.8	1118.2	
Na_2O	Sodium oxide	1375.9	1345.2	
Ne	Neon	198.8	174.4	
Ni	Nickel	737	714.1	
O	Oxygen (atomic)	485.2	459.6	
OP	Phosphorus monoxide	682	649.5	
OSi	Silicon monoxide	777.8	750.4	Protonation at O

Molecular formula	Name	E_{pa} kJ/mol	$\Delta_{base}G°$ kJ/mol	Notes
OSi	Silicon monoxide	533	500.5	Protonation at Si
OSr	Strontium oxide	1209	1180.7	
O_2	Oxygen	421	396.3	
O_2S	Sulfur dioxide	672.3	643.3	
O_3	Ozone	625.5	595.9	
O_3S	Sulfur trioxide	588.3	560.3	
O_4Os	Osmium(VIII) oxide	676.9	650.6	
P	Phosphorus	626.8	604.8	
Pd	Palladium	696	673.4	
Rh	Rhodium	768	745.4	
Ru	Ruthenium	774	751.4	
S	Sulfur	664.3	640.2	
SSi	Silicon monosulfide	627	596.6	Protonation at Si
SSi	Silicon monosulfide	683	660.2	Protonation at S
Sc	Scandium	914	892.0	
Si	Silicon	837	814.1	
Ti	Titanium	876	853.7	
U	Uranium	995.2	973.2	
V	Vanadium	859.4	836.8	
Xe	Xenon	499.6	478.1	
Y	Yttrium	967	945.9	
Zn	Zinc	608.6	586.0	
$CBrF_3$	Bromotrifluoromethane	580.0	550.3	
CBrN	Cyanogen bromide	749.8	719.2	
$CClF_3$	Chlorotrifluoromethane	571.3	541.5	
CClN	Cyanogen chloride	722.1	691.5	
CCl_2	Dichloromethylene	861	828.5	
CCl_2S	Carbonothioic dichloride	752.5	721.8	
CFN	Cyanogen fluoride	632	601.3	
CF_2	Difluoromethylene	765	732.5	
CF_2O	Carbonyl fluoride	666.7	637.0	
CF_3I	Trifluoroiodomethane	628.0	598.2	
CF_3NO	Trifluoronitrosomethane	703.3	670.8	
CF_4	Tetrafluoromethane	529.3	503.7	
CHCl	Chloromethylene	874.1	839.9	
CHF	Fluoromethylene	797.9	763.8	
CHF_3	Trifluoromethane	619.5	589.7	
CHF_3O_3S	Trifluoromethanesulfonic acid	699.4	666.9	
CHN	Hydrogen cyanide	712.9	681.6	
CHN	Hydrogen isocyanide	772.3	739.8	
CHNO	Isocyanic acid (HNCO)	753	718.8	
CHNO	Fulminic acid	758	725.5	
CHO	Oxomethyl (HCO)	636	601.8	
CHO_2	Formyloxyl	623.4	590.9	
CH_2F_2	Difluoromethane	620.5	589.7	
CH_2N_2	Diazomethane	858.9	826.7	
CH_2N_2	Cyanamide	805.6	774.9	
CH_2O	Formaldehyde	712.9	683.3	
CH_2O_2	Formic acid	742.0	710.3	
CH_2S	Thioformaldehyde	759.7	730.5	
CH_2Se	Selenoformaldehyde	764.0	734.9	
CH_3Br	Bromomethane	664.2	638.0	
CH_3Cl	Chloromethane	647.3	621.1	
CH_3F	Fluoromethane	598.9	571.5	
CH_3I	Iodomethane	691.7	665.5	
CH_3NO	Formamide	822.2	791.2	
CH_3NO_2	Nitromethane	754.6	721.6	
CH_3NO_2	Methyl nitrite	798.9	766.4	
CH_3NO_3	Methyl nitrate	733.6	714.8	
CH_3N_3	Methyl azide	833	800.5	

Molecular formula	Name	E_{pa} kJ/mol	$\Delta_{base}G°$ kJ/mol	Notes
CH_4	Methane	543.5	520.6	
CH_4N	Methylamidogen	832.8	801.6	
CH_4N_2O	Urea	873.5	841.6	Protonation at O; Ref. 8
CH_4N_2S	Thiourea	893.7	863.9	
CH_4O	Methanol	754.3	724.5	
CH_4O_3S	Methanesulfonic acid	761.3	728.9	
CH_4S	Methanethiol	773.4	742	
CH_5N	Methylamine	899.0	864.5	
CH_5NO	*O*-Methylhydroxylamine	844.8	812.3	
CH_5N_3	Guanidine	986.3	949.4	
CH_5P	Methylphosphine	851.5	817.6	
CH_6N_2	Methylhydrazine	898.8	866.4	
CN	Cyanide	>595	>564	Protonation at N
CNS	Thiocyanate	751	718.5	
CO	Carbon monoxide	594	562.8	Protonation at C
CO	Carbon monoxide	426.3	402.2	Protonation at O
COS	Carbon oxysulfide	628.5	602.6	Protonation at S
COSe	Carbon oxyselenide	670	644.1	Protonation at Se
CO_2	Carbon dioxide	540.5	515.8	
CS	Carbon monosulfide	791.5	760	
CS_2	Carbon disulfide	681.9	657.7	
CSe	Carbon monoselenide	831.8	800.2	Protonation at C
CSe_2	Carbon diselenide	725	700.9	
C_2ClF_3O	Trifluoroacetyl chloride	681.6	649.8	
C_2Cl_3N	Trichloroacetonitrile	723.2	692.6	
C_2F_3N	Trifluoroacetonitrile	688.4	657.7	
C_2H	Ethynyl	753	720.8	
C_2HCl_3O	Trichloroacetaldehyde	722.3	690.5	
$C_2HCl_3O_2$	Trichloroacetic acid	770.0	739.1	
C_2HF	Fluoroacetylene	686	661.3	
C_2HF_3	Trifluoroethene	699.4	666.9	
$C_2HF_3O_2$	Trifluoroacetic acid	711.7	680.7	
C_2H_2	Acetylene	641.4	616.7	
C_2H_2ClN	Chloroacetonitrile	745.7	715.1	
$C_2H_2F_2$	1,1-Difluoroethene	734	705.1	
$C_2H_2F_2$	*trans*-1,2-Difluoroethene	688.6	657.9	
C_2H_2O	Ketene	825.3	793.6	
$C_2H_3ClO_2$	Chloroacetic acid	765.4	734.5	
$C_2H_3Cl_3O$	2,2,2-Trichloroethanol	729.3	698.9	
C_2H_3F	Fluoroethene	729	700.1	
$C_2H_3FO_2$	Fluoroacetic acid	765.4	734.5	
$C_2H_3F_3O$	2,2,2-Trifluoroethanol	700.2	669.9	
$C_2H_3F_3O$	Methyl trifluoromethyl ether	719.2	690.0	
C_2H_3N	Acetonitrile	779.2	748	
C_2H_3N	Isocyanomethane	839.1	806.6	
C_2H_3NO	Methyl isocyanate	764.4	732.0	
C_2H_3NS	Methyl thiocyanate	796.7	766.1	
C_2H_3NS	Methyl isothiocyanate	799.2	766.7	
$C_2H_3N_3$	1*H*-1,2,3-Triazole	879.3	847.4	
$C_2H_3N_3$	1*H*-1,2,4-Triazole	886.0	855.9	
C_2H_4	Ethylene	680.5	651.5	
$C_2H_4F_2O$	2,2-Difluoroethanol	727.4	697.0	
$C_2H_4F_3N$	2,2,2-Trifluoroethylamine	846.8	812.9	
$C_2H_4N_2$	Aminoacetonitrile	824.9	791.0	
C_2H_4O	Acetaldehyde	768.5	736.5	
C_2H_4O	Oxirane	774.2	745.3	
$C_2H_4O_2$	Acetic acid	783.7	752.8	
$C_2H_4O_2$	Methyl formate	782.5	751.5	
C_2H_4S	Thiirane	807.4	777.6	
C_2H_5Br	Bromoethane	696.2	669.7	

Molecular formula	Name	E_{pa} kJ/mol	$\Delta_{base}G°$ kJ/mol	Notes
C_2H_5BrO	2-Bromoethanol	766.1	735.7	
C_2H_5Cl	Chloroethane	693.4	666.9	
C_2H_5ClO	2-Chloroethanol	766.1	735.7	
C_2H_5F	Fluoroethane	683.4	655.8	
C_2H_5FO	2-Fluoroethanol	715.6	685.2	
C_2H_5I	Iodoethane	724.8	698.3	
C_2H_5N	Ethenamine	898.9	866.5	
C_2H_5N	Ethyleneimine	905.5	872.5	
C_2H_5NO	Acetamide	863.6	832.6	
C_2H_5NO	*N*-Methylformamide	851.3	820.3	
$C_2H_5NO_2$	Nitroethane	765.7	733.2	
$C_2H_5NO_2$	Ethyl nitrite	818.9	786.4	
$C_2H_5NO_2$	Glycine	886.5	852.2	
$C_2H_5NO_2$	Acetohydroxamic acid	854.0	823.0	
C_2H_5NS	Thioacetamide	884.6	852.8	
C_2H_6	Ethane	596.3	569.9	
C_2H_6Hg	Dimethyl mercury	771.6	740.8	
$C_2H_6N_2$	Ethanimidamide	970.7	938.2	
$C_2H_6N_2$	trans-Dimethyldiazene	865.1	834.4	
$C_2H_6N_2O$	2-Aminoacetamide		882.3	
$C_2H_6N_2O_2$	*N*-Methyl-*N*-nitromethanamine	828.3	795.8	
C_2H_6O	Ethanol	776.4	746	
C_2H_6O	Dimethyl ether	792	764.5	
C_2H_6OS	Dimethyl sulfoxide	884.4	853.7	
$C_2H_6O_2$	1,2-Ethanediol	815.9	773.6	
C_2H_6S	Ethanethiol	789.6	758.4	
C_2H_6S	Dimethyl sulfide	830.9	801.2	
$C_2H_6S_2$	Dimethyl disulfide	815.3	782.8	
C_2H_7N	Ethylamine	912.0	878	
C_2H_7N	Dimethylamine	929.5	896.5	
C_2H_7NO	Ethanolamine	930.3	896.8	
$C_2H_7O_3P$	Dimethyl hydrogen phosphite	894.8	862.4	
C_2H_7P	Dimethylphosphine	912.0	877.9	
$C_2H_8N_2$	1,2-Ethanediamine	951.6	912.5	
$C_2H_8N_2$	1,1-Dimethylhydrazine	927.1	894.7	
C_2N_2	Cyanogen	674.7	645.8	
C_2O	Dicarbon monoxide	774.7	747.0	
C_3	Carbon trimer	767.0	736.3	
C_3F_6O	Perfluoroacetone	670.4	639.7	
C_3HN	Cyanoacetylene	751.2	720.5	
$C_3H_2F_6O$	1,1,1,3,3,3-Hexafluoro-2-propanol	686.6	656.2	
$C_3H_2N_2$	Malononitrile	723.0	694.1	
C_3H_3	2-Propynyl	741	708.5	
$C_3H_3Cl_3O$	1,1,1-Trichloro-2-propanone	768.3	736.3	
$C_3H_3F_3O$	1,1,1-Trifluoroacetone	723.9	692.0	
$C_3H_3F_3O_2$	Methyl trifluoroacetate	740.5	709.6	
C_3H_3N	Acrylonitrile	784.7	753.7	
C_3H_3NO	Oxazole	876.4	844.5	
C_3H_3NO	Isoxazole	848.6	816.8	
C_3H_3NO	2-Oxopropanenitrile	746.9	716.2	
C_3H_3NS	Thiazole	904	872.1	
$C_3H_3N_3$	1,3,5-Triazine	848.8	819.6	
C_3H_4	Allene	775.3	745.8	
C_3H_4	Propyne	748.2	723.0	
C_3H_4	Cyclopropene	818.5	787.8	
C_3H_4ClN	3-Chloropropanenitrile	773.1	742.4	
$C_3H_4N_2$	1*H*-Pyrazole	894.1	860.5	
$C_3H_4N_2$	Imidazole	942.8	909.2	
$C_3H_4N_2S$	2-Thiazolamine	930.6	898.7	
C_3H_4O	Acrolein	797.0	765.1	

Molecular formula	Name	E_{pa} kJ/mol	$\Delta_{base}G°$ kJ/mol	Notes
C_3H_4O	1-Propen-1-one	834.1	803.4	
$C_3H_4O_3$	Ethylene carbonate	814.2	784.4	
C_3H_5	Allyl	736	707.4	
C_3H_5	Cyclopropyl	738.9	702.0	
$C_3H_5ClO_2$	Ethyl chloroformate	764.8	733.8	
C_3H_5FO	1-Fluoro-2-propanone	795.4	763.5	
$C_3H_5F_3O$	2,2,2-Trifluoroethyl methyl ether	747.6	718.4	
C_3H_5N	Propanenitrile	794.1	763.0	
C_3H_5N	2-Propyn-1-amine	887.4	853.5	
C_3H_5N	Ethyl isocyanide	851.3	818.9	
C_3H_5NO	Acrylamide	870.7	839.8	
C_3H_5NO	Methoxyacetonitrile	758.1	727.4	
C_3H_5NO	2-Azetidinone	852.6	821.7	
C_3H_5NS	(Methylthio)acetonitrile	784.8	754.1	
$C_3H_5N_3$	1*H*-Pyrazol-3-amine	921.5	889.6	
$C_3H_5N_3$	1*H*-Pyrazol-4-amine	907.6	874.0	
C_3H_6	Propene	741.6		Ref. 5
C_3H_6	Cyclopropane	750.3	722.2	
$C_3H_6N_2$	3-Aminopropanenitrile	866.4	832.5	
$C_3H_6N_2$	Dimethylcyanamide	852.1	821.4	
$C_3H_6N_2S$	2-Imidazolidinethione	921.9	891.2	
C_3H_6O	Methyl vinyl ether	859.2	830.3	
C_3H_6O	Propanal	786.0	754.0	
C_3H_6O	Acetone	812	782.1	
C_3H_6O	Oxetane	801.3	773.9	
$C_3H_6O_2$	Propanoic acid	797.2	766.2	
$C_3H_6O_2$	Ethyl formate	799.4	768.4	
$C_3H_6O_2$	Methyl acetate	821.6	790.7	
$C_3H_6O_3$	Dimethyl carbonate	830.2	799.2	
C_3H_6S	(Methylthio)ethene	858.2	829.3	
C_3H_6S	Thietane	834.8	805.0	
C_3H_6S	Methylthiirane	833.3	801.5	
C_3H_7N	Allylamine	909.5	875.5	
C_3H_7N	Cyclopropylamine	904.7	869.9	
C_3H_7N	Azetidine	943.4	908.6	
C_3H_7N	1-Methylaziridine	934.8	904.1	
C_3H_7N	Propyleneimine	925.1	892.1	
C_3H_7NO	*N,N*-Dimethylformamide	887.5	856.6	
C_3H_7NO	*N*-Methylacetamide	888.5	857.6	
C_3H_7NO	Propanamide	876.2	845.3	
$C_3H_7NO_2$	Isopropyl nitrite	845.5	813.0	
$C_3H_7NO_2$	*L*-Alanine	901.6	867.7	
$C_3H_7NO_2$	Sarcosine	921.2	888.7	
$C_3H_7NO_2S$	*L*-Cysteine	903.2	869.3	
$C_3H_7NO_3$	*L*-Serine	914.6	880.7	
C_3H_8	Propane	625.7	607.8	
$C_3H_8N_2O$	*N,N'*-Dimethylurea	903.3	873.5	
$C_3H_8N_2S$	*N,N'*-Dimethylthiourea	926.0	895.1	
C_3H_8O	1-Propanol	786.5	756.1	
C_3H_8O	2-Propanol	793.0	762.6	
C_3H_8O	Ethyl methyl ether	808.6	781.2	
$C_3H_8O_2$	1,3-Propanediol	876.2	825.9	
$C_3H_8O_2$	2-Methoxyethanol	768.8	729.8	
$C_3H_8O_3$	Glycerol	874.8	820	
C_3H_8S	1-Propanethiol	794.9	763.6	
C_3H_8S	2-Propanethiol	803.6	772.3	
C_3H_8S	Ethyl methyl sulfide	846.5	815.3	
C_3H_9As	Trimethylarsine	897.3	864.9	
$C_3H_9BO_3$	Trimethyl borate	815.8	783.4	
C_3H_9N	Propylamine	917.8	883.9	

Molecular formula	Name	E_{pa} kJ/mol	$\Delta_{base}G^{\circ}$ kJ/mol	Notes
C_3H_9N	Isopropylamine	923.8	889.0	
C_3H_9N	Ethylmethylamine	942.2	909.2	
C_3H_9N	Trimethylamine	948.9	918.1	
C_3H_9NO	2-Methoxyethylamine	928.6	894.6	
C_3H_9NO	Trimethylamine oxide	983.2	953.5	
C_3H_9NO	3-Amino-1-propanol	962.5	917.3	
$C_3H_9O_3P$	Trimethyl phosphite	929.7	899.9	
$C_3H_9O_4P$	Trimethyl phosphate	890.6	860.8	
C_3H_9P	Trimethylphosphine	958.8	926.3	
$C_3H_{10}N_2$	1,3-Propanediamine	987.0	940.0	
$C_3H_{10}OSi$	Trimethylsilanol	814.0	781.5	
C_4F_8	Perfluorocyclobutane	>544		Ref. 6
C_4H_2	1,3-Butadiyne	737.2	712.8	
$C_4H_4F_6O$	Bis(2,2,2-trifluoroethyl) ether	702.3	674.9	
$C_4H_4N_2$	Pyrazine	877.1	847.0	
$C_4H_4N_2$	Pyrimidine	885.8	855.7	
$C_4H_4N_2$	Pyridazine	907.2	877.1	
$C_4H_4N_2O_2$	Uracil	872.7	841.7	
$C_4H_4N_2S_2$	2,4(1*H*,3*H*)-Pyrimidinedithione	911.4	880.5	
C_4H_4O	Furan	812	781	Ref. 10
$C_4H_4O_3$	Succinic anhydride	797		Ref. 9
C_4H_4S	Thiophene	815.0	784.3	
$C_4H_5Cl_3O_2$	Ethyl trichloroacetate	790.4	759.4	
$C_4H_5F_3O_2$	Ethyl trifluoroacetate	758.8	727.9	
C_4H_5N	Pyrrole	875.4	843.8	
C_4H_5N	Cyclopropanecarbonitrile	808.2	777.5	
$C_4H_5NO_2$	Ethyl cyanoformate	745.7	714.7	
C_4H_5NS	2-Methylthiazole	930.6	898.7	
$C_4H_5N_3O$	Cytosine	949.9	918	
C_4H_6	1,2-Butadiene	778.9	749.8	
C_4H_6	1,3-Butadiene	783.4	757.6	
C_4H_6	2-Butyne	775.8	745.1	
C_4H_6	Cyclobutene	784.4	753.6	
$C_4H_6F_3NO$	2,2,2-Trifluoro-*N*,*N*-dimethylacetamide	849.0	818.0	
$C_4H_6N_2$	1-Methylimidazol	959.6	927.7	
$C_4H_6N_2$	2-Methyl-1*H*-imidazole	963.4	929.6	
$C_4H_6N_2$	4-Methyl-1*H*-imidazole	952.8	920.9	
$C_4H_6N_2$	1-Methyl-1*H*-pyrazole	912.0	880.1	
$C_4H_6N_2$	3-Methyl-1*H*-pyrazole	906.0	874.2	
$C_4H_6N_2$	4-Methyl-1*H*-pyrazole	906.8	873.4	
C_4H_6O	2-Methylpropenal	808.7	776.8	
C_4H_6O	3-Buten-2-one	834.7	802.8	
C_4H_6O	Cyclobutanone	802.5	772.7	
C_4H_6O	2,3-Dihydrofuran	866.9	834.4	
C_4H_6O	2,5-Dihydrofuran	823.4	796	
$C_4H_6O_2$	*trans*-2-Butenoic acid	824.0	793	
$C_4H_6O_2$	Methacrylic acid	816.7	785.7	
$C_4H_6O_2$	Cyclopropanecarboxylic acid	821.4	790.4	
$C_4H_6O_2$	Vinyl acetate	813.9	782.9	
$C_4H_6O_2$	Methyl acrylate	825.8	794.8	
$C_4H_6O_2$	2,3-Butanedione	801.9	770.1	
$C_4H_6O_2$	γ-Butyrolactone	840.0	808.1	
$C_4H_6O_2$	2,3-Dihydro-1,4-dioxin	823.5	792.8	
$C_4H_6O_3$	Acetic anhydride	844		Ref. 9
C_4H_7	2-Methylallyl	778	747.3	
C_4H_7N	Butanenitrile	798.4	767.7	
C_4H_7N	2-Methylpropanenitrile	803.6	772.8	
C_4H_7N	1-Isocyanopropane	856.8	824.3	
C_4H_7NO	2-Butenamide	887.1	856.1	

Molecular formula	Name	E_{pa} kJ/mol	$\Delta_{base}G°$ kJ/mol	Notes
C_4H_7NO	2-Methyl-2-propenamide	880.4	849.4	
$C_4H_7NO_4$	*L*-Aspartic acid	908.9	875	
C_4H_8	*trans*-2-Butene	747	719.9	
C_4H_8	Isobutene	802.1	775.6	
$C_4H_8N_2$	(Dimethylamino)acetonitrile	884.5	853.7	
$C_4H_8N_2O_3$	*L*-Asparagine	929	891.5	
$C_4H_8N_2O_3$	*N*-Glycylglycine		882	
C_4H_8O	Ethyl vinyl ether	870.1	840.4	
C_4H_8O	2-Methoxy-1-propene	894.9	866.1	
C_4H_8O	Butanal	792.7	760.8	
C_4H_8O	Isobutanal	797.3	765.5	
C_4H_8O	2-Butanone	827.3	795.5	
C_4H_8O	Tetrahydrofuran	822.1	794.7	
$C_4H_8O_2$	Propyl formate	804.9	773.9	
$C_4H_8O_2$	Isopropyl formate	811.3	780.3	
$C_4H_8O_2$	Ethyl acetate	835.7	804.7	
$C_4H_8O_2$	Methyl propanoate	830.2	799.2	
$C_4H_8O_2$	1,3-Dioxane	825.4	796.2	
$C_4H_8O_2$	1,4-Dioxane	797.4	770.0	
$C_4H_8O_3$	Ethyl methyl carbonate	842.7	810.8	
C_4H_8S	Tetrahydrothiophene	849.1	819.3	
C_4H_9N	Pyrrolidine	948.3	915.3	
C_4H_9NO	*N*-Methylpropanamide	920.4	889.4	
C_4H_9NO	2-Methylpropanamide	878.6	846.7	
C_4H_9NO	*N*-Ethylacetamide	898.0	867.0	
C_4H_9NO	*N,N*-Dimethylacetamide	908.0	877.0	
C_4H_9NO	Morpholine	924.3	891.2	
$C_4H_9NO_2$	*tert*-Butyl nitrite	863.9	831.4	
$C_4H_9NO_2$	Ethyl *N*-methylcarbamate	888.8	857.8	
$C_4H_9NO_3$	*L*-Threonine	922.5	888.5	
C_4H_9NS	*N,N*-Dimethylthioacetamide	925.3	894.4	
C_4H_{10}	Isobutane	677.8	671.3	
$C_4H_{10}N_2$	Piperazine	943.7	914.7	
$C_4H_{10}N_2$	3-Ethyl-3-methyldiaziridine	903.8	871.3	
$C_4H_{10}O$	1-Butanol	789.2	758.9	
$C_4H_{10}O$	2-Butanol	815.7	784.6	
$C_4H_{10}O$	2-Methyl-1-propanol	793.7	762.2	
$C_4H_{10}O$	2-Methyl-2-propanol	802.6	772.2	
$C_4H_{10}O$	Diethyl ether	828.4	801	
$C_4H_{10}O$	Methyl propyl ether	814.9	785.7	
$C_4H_{10}O$	Isopropyl methyl ether	826.3	797.1	
$C_4H_{10}O_2$	1,4-Butanediol	915.6	854.9	
$C_4H_{10}O_2$	1,2-Dimethoxyethane	858.0	820.2	
$C_4H_{10}O_3$	1,2,4-Butanetriol	905.9	841	
$C_4H_{10}S$	1-Butanethiol	801.7	770.5	
$C_4H_{10}S$	2-Methyl-1-propanethiol	802.6	771.4	
$C_4H_{10}S$	2-Methyl-2-propanethiol	816.4	785.1	
$C_4H_{10}S$	Diethyl sulfide	856.7	827.0	
$C_4H_{11}N$	Butylamine	921.5	886.6	
$C_4H_{11}N$	*tert*-Butylamine	934.1	899.9	
$C_4H_{11}N$	Isobutylamine	924.8	890.8	
$C_4H_{11}N$	Diethylamine	952.4	919.4	
$C_4H_{11}N$	Isopropylmethylamine	952.4	919.4	
$C_4H_{11}N$	Ethyldimethylamine	960.1	929.1	
$C_4H_{11}NO$	*N*-Ethyl-*N*-hydroxyethanamine	914.7	882.2	
$C_4H_{11}NO$	4-Amino-1-butanol	984.5	932.1	
$C_4H_{11}NO_2$	Diethanolamine	953	920	
$C_4H_{12}N_2$	1,4-Butanediamine	1005.6	954.3	
$C_4H_{12}N_2$	*N,N'*-Dimethyl-1,2-ethanediamine	989.2	946.9	
$C_4H_{12}Sn$	Tetramethylstannane	823.7	797.4	

Molecular formula	Name	E_{pa} kJ/mol	$\Delta_{base}G°$ kJ/mol	Notes
$C_4H_{14}OSi_2$	1,1,3,3-Tetramethyldisiloxane	845.3	814.6	
C_4NiO_4	Nickel carbonyl	742.3	716.0	
C_5F_5N	Perfluoropyridine	764.9	733.0	
C_5FeO_5	Iron pentacarbonyl	833.0	798.5	
$C_5H_3ClN_4$	6-Chloro-1*H*-purine	873.6	841.7	
C_5H_4BrN	2-Bromopyridine	904.8	873.0	
C_5H_4BrN	3-Bromopyridine	910.0	878.2	
C_5H_4BrN	4-Bromopyridine	917.8	886.0	
C_5H_4ClN	2-Chloropyridine	900.9	869	
C_5H_4ClN	3-Chloropyridine	903.4	871.5	
C_5H_4ClN	4-Chloropyridine	916.1	884.2	
C_5H_4FN	3-Fluoropyridine	902.0	870.1	
C_5H_4FN	2-Fluoropyridine	884.6	852.7	
$C_5H_4N_2O_2$	4-Nitropyridine	874.3	842.5	
$C_5H_4N_2O_3$	4-Nitropyridine 1-oxide	868.0	837.3	
$C_5H_4N_4$	1*H*-Purine	920.1	888.2	
$C_5H_4N_4O$	Hypoxanthine	912.3	880.5	
C_5H_5	Cyclopentadienyl	831.5	799.1	
C_5H_5N	Pyridine	930	898.1	
C_5H_5NO	3-Pyridinol	929.5	897.7	
C_5H_5NO	Pyridine-1-oxide	923.6	892.9	
$C_5H_5N_5$	Adenine	942.8	912.5	
$C_5H_5N_5O$	Guanine	959.5	927.6	
C_5H_6	1,3-Cyclopentadiene	821.6	798.4	
$C_5H_6N_2$	2-Pyridinamine	947.2	915.3	
$C_5H_6N_2$	3-Pyridinamine	954.4	922.6	
$C_5H_6N_2$	4-Pyridinamine	979.7	947.8	
$C_5H_6N_2O_2$	Thymine	880.9	850.0	
C_5H_6O	2-Methylfuran	865.9	833.5	
C_5H_6O	3-Methylfuran	854.0	821.5	
$C_5H_6O_3$	Glutaric anhydride	816		Ref. 9
$C_5H_6O_3$	3-Methylsuccinic anhydride	807		Ref. 9
C_5H_6S	2-Methylthiophene	859.0	826.5	
$C_5H_7F_3O_2$	Propyl trifluoroacetate	763.9	732.9	
C_5H_8	*trans*-1,3-Pentadiene	834.1	804.4	
C_5H_8	2-Methyl-1,3-butadiene	826.4	797.6	
C_5H_8	2-Pentyne	810.2	778.0	
C_5H_8	3-Methyl-1-butyne	814.9	787.8	
C_5H_8	Cyclopentene	766.3	733.8	
C_5H_8	1-Methylcyclobutene	841.5	807.3	
C_5H_8	Vinylcyclopropane	816.3	787.5	
C_5H_8	3,3-Dimethylcyclopropene	847.8	817.1	
$C_5H_8N_2$	1,3-Dimethyl-1*H*-pyrazole	933.9	902.3	
$C_5H_8N_2$	1,4-Dimethyl-1*H*-imidazole	976.7	944.9	
$C_5H_8N_2$	1,5-Dimethyl-1*H*-pyrazole	934.3	902.8	
$C_5H_8N_2$	3,4-Dimethyl-1*H*-pyrazole	927.3	895.4	
$C_5H_8N_2$	3,5-Dimethyl-1*H*-pyrazole	933.5	900.1	
$C_5H_8N_2$	1,2-Dimethyl-1*H*-imidazole	984.7	952.6	
$C_5H_8N_2$	1,5-Dimethyl-1*H*-imidazole	977.6	945.8	
C_5H_8O	*trans*-2-Pentenal	839.0	807.2	
C_5H_8O	3-Methyl-2-butenal	856.9	825.0	
C_5H_8O	3-Methyl-3-buten-2-one	843.1	811.3	
C_5H_8O	Cyclopropyl methyl ketone	854.9	823	
C_5H_8O	Cyclopentanone	823.7	794.0	
C_5H_8O	3,4-Dihydro-2H-pyran	865.8	833.4	
$C_5H_8O_2$	3-Methyl-2-butenoic acid	822.9	791.9	
$C_5H_8O_2$	*cis*-2-Methyl-2-butenoic acid	822.5	791.5	
$C_5H_8O_2$	Cyclobutanecarboxylic acid	817.4	786.4	
$C_5H_8O_2$	Methyl trans-2-butenoate	851.3	820.4	
$C_5H_8O_2$	Methyl methacrylate	831.4	800.5	

Molecular formula	Name	E_{pa} kJ/mol	$\Delta_{base}G°$ kJ/mol	Notes
$C_5H_8O_2$	Methyl cyclopropanecarboxylate	842.1	811.2	
$C_5H_8O_2$	2,4-Pentanedione	873.5	836.8	
C_5H_9N	2-Isocyano-2-methylpropane	870.7	838.3	
C_5H_9N	3-(Dimethylamino)-1-propyne	940.3	909.5	
C_5H_9N	Pentanenitrile	802.4	771.7	
C_5H_9N	2,2-Dimethylpropanenitrile	810.9	780.2	
C_5H_9NO	3-Ethoxypropanenitrile	807.2	776.5	
C_5H_9NO	*N,N*-Dimethyl-2-propenamide	904.3	873.4	
C_5H_9NO	*N*-Methyl-2-pyrrolidone	923.5	891.6	
$C_5H_9NO_2$	*L*-Proline	920.5	886.0	
$C_5H_9NO_4$	*L*-Glutamic acid	913.0	879.1	
$C_5H_9N_3$	Histamine	999.8	961.9	
C_5H_{10}	2-Methyl-2-butene	808.8	779.9	
$C_5H_{10}N_2O$	1,3-Dimethyl-2-imidazolidinone	918.4	886.0	
$C_5H_{10}N_2O_3$	*L*-Glutamine	937.8	900	
$C_5H_{10}O$	Allyl ethyl ether	833.7	804.5	
$C_5H_{10}O$	Pentanal	796.6	764.8	
$C_5H_{10}O$	2-Pentanone	832.7	800.9	
$C_5H_{10}O$	3-Pentanone	836.8	807	
$C_5H_{10}O$	3-Methyl-2-butanone	836.3	804.4	
$C_5H_{10}O$	Tetrahydropyran	822.8	795.4	
$C_5H_{10}O$	2-Methyltetrahydrofuran	840.8	811.6	
$C_5H_{10}O_2$	Butyl formate	806.0	775	
$C_5H_{10}O_2$	Propyl acetate	836.6	805.6	
$C_5H_{10}O_2$	Isopropyl acetate	836.6	805.6	
$C_5H_{10}O_2$	Methyl butanoate	836.4	805.4	
$C_5H_{10}O_2$	Methyl isobutanoate	836.6	805.7	
$C_5H_{10}O_2$	*cis*-1,2-Cyclopentanediol	885.6	853.1	
$C_5H_{10}S$	Thiacyclohexane	855.8	826.0	
$C_5H_{11}N$	Allyldimethylamine	957.8	926.8	
$C_5H_{11}N$	Piperidine	954.0	921	
$C_5H_{11}N$	*N*-Methylpyrrolidine	965.6	934.8	
$C_5H_{11}NO$	2,2-Dimethylpropanamide	889.0	857.2	
$C_5H_{11}NO_2$	*L*-Valine	910.6	876.7	
$C_5H_{11}NO_2S$	*L*-Methionine	935.4	901.5	
$C_5H_{12}N_2O$	Tetramethylurea	930.6	899.6	
$C_5H_{12}N_2S$	Tetramethylthiourea	947.6	916.6	
$C_5H_{12}O$	2,2-Dimethyl-1-propanol	795.5	765.2	
$C_5H_{12}O$	Butyl methyl ether	820.3	791.2	
$C_5H_{12}O$	Methyl tert-butyl ether	841.6	812.4	
$C_5H_{12}O$	Ethyl isopropyl ether	842.7	813.5	
$C_5H_{12}S$	2,2-Dimethyl-1-propanethiol	809.5	778.2	
$C_5H_{12}Si$	Vinyltrimethylsilane	833	804.1	
$C_5H_{13}N$	Pentylamine	923.5	889.5	
$C_5H_{13}N$	2-Methyl-2-butanamine	937.8	903.6	
$C_5H_{13}N$	2,2-Dimethylpropylamine	928.3	894.0	
$C_5H_{13}N$	Ethylisopropylamine	960.0	926.7	
$C_5H_{13}N$	*N,N*-Dimethyl-1-propanamine	962.8	931.9	
$C_5H_{13}N$	Diethylmethylamine	971.0	940.0	
$C_5H_{13}N_3$	1,1,3,3-Tetramethylguanidine	1031.6	997.4	
$C_5H_{14}N_2$	*N,N,N',N'*-Tetramethylmethanediamine	952.2	919.8	
$C_5H_{14}N_2$	*N,N*-Dimethyl-1,3-propanediamine	1025.0	975.3	
$C_5H_{14}N_2$	1,5-Pentanediamine	999.6	946.2	
C_6CrO_6	Chromium carbonyl	739.2	714.6	
C_6F_6	Hexafluorobenzene	648.0	624.4	
C_6HF_5	Pentafluorobenzene	690.4	662.7	
$C_6H_2F_4$	1,2,3,4-Tetrafluorobenzene	700.4	672.7	
$C_6H_2F_4$	1,2,3,5-Tetrafluorobenzene	747.3	719.6	
$C_6H_2F_4$	1,2,4,5-Tetrafluorobenzene	746.5	718.8	
$C_6H_3F_3$	1,2,3-Trifluorobenzene	724.3	696.6	

Molecular formula	Name	E_{pa} kJ/mol	$\Delta_{base}G°$ kJ/mol	Notes
$C_6H_3F_3$	1,2,4-Trifluorobenzene	729.5	699.4	
$C_6H_3F_3$	1,3,5-Trifluorobenzene	741.9	715.4	
C_6H_4	Benzyne	841	808.5	
$C_6H_4F_2$	*o*-Difluorobenzene	731.2	703.5	
$C_6H_4F_2$	*m*-Difluorobenzene	749.7	722	
$C_6H_4F_2$	*p*-Difluorobenzene	718.7	692.8	
$C_6H_4N_2$	2-Pyridinecarbonitrile	872.9	841	
$C_6H_4N_2$	3-Pyridinecarbonitrile	877.0	845.1	
$C_6H_4N_2$	4-Pyridinecarbonitrile	880.6	848.8	
$C_6H_4O_2$	*p*-Benzoquinone	799.1	769.3	
C_6H_5	Phenyl	884	851.5	
C_6H_5Br	Bromobenzene	754.1	725.8	
C_6H_5Cl	Chlorobenzene	753.1	724.6	
C_6H_5F	Fluorobenzene	755.9	726.6	
C_6H_5NO	Nitrosobenzene	854.3	823.6	
C_6H_5NO	4-Pyridinecarboxaldehyde	904.6	872.8	
$C_6H_5NO_2$	Nitrobenzene	800.3	769.5	
$C_6H_5N_3$	Azidobenzene	820	787.5	
C_6H_5O	Phenoxy	873.2		Ref. 4
C_6H_6	Benzene	750.4	725.4	
C_6H_6BrN	3-Bromoaniline	873.2	841.4	
C_6H_6ClN	3-Chloroaniline	868.1	836.3	
C_6H_6ClN	4-Chloroaniline	873.8	842.0	
C_6H_6ClN	2-Chloro-4-methylpyridine	921.2	889.4	
C_6H_6ClN	2-Chloro-6-methylpyridine	908.0	876.2	
C_6H_6ClNO	2-Chloro-6-methoxypyridine	909.9	878.0	
C_6H_6FN	3-Fluoroaniline	867.3	835.5	
C_6H_6FN	4-Fluoroaniline	871.5	839.7	
C_6H_6IN	3-Iodoaniline	878.7	846.8	
C_6H_6N	Anilino	949.8	917.4	
$C_6H_6N_2O$	3-Pyridinecarboxamide	918.3	886.4	
$C_6H_6N_2O_2$	4-Nitroaniline	866.0	834.2	
$C_6H_6N_4$	6-Methyl-1*H*-purine	939.2	907.3	
C_6H_6O	Bis(2-propynyl) ether	783.9	756.5	
C_6H_6O	Phenol	817.3	786.3	
C_6H_7N	Bis(2-propynyl)amine	910.0	876.9	
C_6H_7N	Aniline	882.5	850.6	
C_6H_7N	2-Methylpyridine	949.1	917.3	
C_6H_7N	3-Methylpyridine	943.4	911.6	
C_6H_7N	4-Methylpyridine	947.2	915.3	
C_6H_7NO	1-Methyl-2(1*H*)-pyridinone	925.8	894.8	
C_6H_7NO	2-Aminophenol	898.8	866.9	
C_6H_7NO	3-Aminophenol	898.8	866.9	
C_6H_7NO	2-Methoxypyridine	934.7	902.8	
C_6H_7NO	3-Methoxypyridine	942.7	910.9	
C_6H_7NO	4-Methoxypyridine	961.7	929.8	
C_6H_7NO	3-Methylpyridine-1-oxide	935.2	902.8	
C_6H_8	1,3-Cyclohexadiene	837	804.5	
C_6H_8	1,4-Cyclohexadiene	837	808.0	
$C_6H_8N_2$	1,2-Benzenediamine	896.5	865.8	
$C_6H_8N_2$	1,3-Benzenediamine	929.9	899.2	
$C_6H_8N_2$	1,4-Benzenediamine	905.9	874.0	
$C_6H_8N_2O$	Bis(2-cyanoethyl) ether	813.8	786.4	
C_6H_8O	2,4-Dimethylfuran	894.7	862.3	
C_6H_8O	2,5-Dimethylfuran	865.9	835.2	
C_6H_8O	3,4-Dimethylfuran	869.0	838.3	
$C_6H_8O_2$	1,3-Cyclohexanedione	881.2	849.4	
$C_6H_8O_2$	1,4-Cyclohexanedione	812.5	782.7	
$C_6H_8O_2$	1,2-Cyclohexanedione	849.6	818.9	
$C_6H_8O_3$	4-Methylglutaric anhydride	820		Ref. 9

Molecular formula	Name	E_{pa} kJ/mol	$\Delta_{base}G°$ kJ/mol	Notes
$C_6H_9F_3O_2$	Butyl trifluoroacetate	764.8	733.8	
C_6H_9N	2,5-Dimethylpyrrole	918.7	887.1	
$C_6H_9N_3O_2$	*L*-Histidine	988	950.2	
C_6H_{10}	Methylenecyclopentane	832.4	803.5	
C_6H_{10}	(1-Methylvinyl)cyclopropane	871.6	842.7	
C_6H_{10}	2-Methyl-1,3-pentadiene	864.9	836	
C_6H_{10}	3-Methyl-1,3-pentadiene	852.3	823.4	
C_6H_{10}	2,3-Dimethyl-1,3-butadiene	835.0	807.8	
C_6H_{10}	1-Hexyne	799.8	774.8	
C_6H_{10}	2-Hexyne	806.1	781.1	
C_6H_{10}	Cyclohexene	784.5	752.0	
C_6H_{10}	1-Methylcyclopentene	816.5	787.1	
$C_6H_{10}N_2$	1,3,5-Trimethyl-1*H*-pyrazole	949.3	917.4	
$C_6H_{10}N_2$	3,4,5-Trimethyl-1*H*-pyrazole	949.3	916.0	
$C_6H_{10}O$	7-Oxabicyclo[2.2.1]heptane	844.2	816.8	
$C_6H_{10}O$	7-Oxabicyclo[4.1.0]heptane	848.1	815.6	
$C_6H_{10}O$	*trans*-3-Hexen-2-one	865.6	833.8	
$C_6H_{10}O$	Diallyl ether	827.4	800.0	
$C_6H_{10}O$	Cyclohexanone	841.0	811.2	
$C_6H_{10}O$	Mesityl oxide	878.7	846.9	
$C_6H_{10}O_2$	Cyclopentanecarboxylic acid	817.4	786.4	
$C_6H_{10}O_2$	2,5-Hexanedione	892.0	851.8	
$C_6H_{11}N$	*N*-Allyl-2-propen-1-amine	949.3	916.3	
$C_6H_{11}NO$	1-Methyl-2-piperidinone	924.4	892.6	
$C_6H_{11}N_3O_4$	*N*-(*N*-Glycylglycyl)glycine	966.8	916.8	
C_6H_{12}	1-Hexene	805.2	776.3	
C_6H_{12}	2-Methyl-2-pentene	812	783.1	
C_6H_{12}	2,3-Dimethyl-2-butene	813.9	785.9	
C_6H_{12}	Cyclohexane	686.9	666.9	
$C_6H_{12}N_2$	Triethylenediamine	963.4	934.6	
$C_6H_{12}N_2O_3$	*N*-*L*-Alanyl-*L*-alanine		905.6	
$C_6H_{12}O$	Oxepane	834.2	806.8	
$C_6H_{12}O$	3-Hexanone	843.2	811.3	
$C_6H_{12}O$	3,3-Dimethyl-2-butanone	840.1	808.2	
$C_6H_{12}O_2$	*cis*-1,3-Cyclohexanediol	882.2	849.7	
$C_6H_{12}O_2$	*trans*-1,3-Cyclohexanediol	828.6	797.9	
$C_6H_{12}O_2$	Methyl 2,2-dimethylpropanoate	845.2	814.2	
$C_6H_{12}O_2$	Diacetone alcohol	822.9	791.1	
$C_6H_{12}O_6$	α-*D*-Glucose		778.9	
$C_6H_{12}O_6$	β-*D*-Glucose		778.9	
$C_6H_{13}N$	*N*,*N*,2-Trimethylpropenylamine	967.0	934.5	
$C_6H_{13}N$	Cyclohexylamine	934.4	899.6	
$C_6H_{13}N$	1-Methylpiperidine	971.1	940.1	
$C_6H_{13}N$	Hexahydro-1*H*-azepine	956.7	923.5	
$C_6H_{13}NO$	*N*,*N*-Dimethylbutanamide	921.7	890.8	
$C_6H_{13}NO$	*N*,*N*-Diethylacetamide	925.4	894.4	
$C_6H_{13}NO_2$	*L*-Leucine	914.6	880.6	
$C_6H_{13}NO_2$	*L*-Isoleucine	917.4	883.5	
$C_6H_{14}N_2O_2$	*L*-Lysine	996	951.0	
$C_6H_{14}N_4O_2$	*L*-Arginine	1051.0	1006.6	
$C_6H_{14}O$	Dipropyl ether	837.9	810.5	
$C_6H_{14}O$	Diisopropyl ether	855.5	828.1	
$C_6H_{14}O$	*tert*-Butyl ethyl ether	856.0	826.9	
$C_6H_{14}O_3$	Diethylene glycol dimethyl ether	918.8	870.9	
$C_6H_{14}S$	Dipropyl sulfide	864.7	834.9	
$C_6H_{14}S$	Diisopropyl sulfide	876.4	846.6	
$C_6H_{15}N$	Butyldimethylamine	969.2	938.2	
$C_6H_{15}N$	Isobutyldimethylamine	968.7	937.8	
$C_6H_{15}N$	Hexylamine	927.5	893.5	
$C_6H_{15}N$	Dipropylamine	962.3	929.3	

Molecular formula	Name	E_{pa} kJ/mol	$\Delta_{base}G°$ kJ/mol	Notes
$C_6H_{15}N$	Diisopropylamine	971.9	938.6	
$C_6H_{15}N$	Triethylamine	981.8	951	
$C_6H_{15}NO$	6-Amino-1-hexanol	969.0	915.7	
$C_6H_{15}OP$	Triethylphosphine oxide	936.6	906.8	
$C_6H_{15}O_4P$	Triethyl phosphate	909.3	879.6	
$C_6H_{15}P$	Triethylphosphine	984.5	952.0	
$C_6H_{16}N_2$	1,6-Hexanediamine	999.5	946.2	
$C_6H_{16}N_2$	*N,N,N',N'*-Tetramethyl-1,2-ethanediamine	1012.8	970.6	
$C_6H_{16}OSi$	Triethylsilanol	822.1	794.8	
$C_6H_{18}N_3OP$	Hexamethylphosphoric triamide	958.6	928.7	
$C_6H_{18}N_3P$	Hexamethylphosphorous triamide	930.1	897.7	
$C_6H_{18}OSi_2$	Hexamethyldisiloxane	846.4	816.2	
C_6MoO_6	Molybdenum hexacarbonyl	762.6	738.1	
C_6O_6W	Tungsten carbonyl	758.0	733.4	
$C_7H_4N_2O_2$	3-Nitrobenzonitrile	781.4	750.7	
$C_7H_4N_2O_2$	4-Nitrobenzonitrile	775.7	745.1	
C_7H_5ClO	3-Chlorobenzaldehyde	813.0	781.1	
C_7H_5ClO	4-Chlorobenzaldehyde	831.3	799.4	
C_7H_5FO	3-Fluorobenzaldehyde	814.3	782.5	
C_7H_5FO	4-Fluorobenzaldehyde	827.1	795.3	
C_7H_5N	Benzonitrile	811.5	780.9	
C_7H_5N	Isocyanobenzene	868.4	836.0	
C_7H_5NO	Benzoxazole	891.6	859.8	
$C_7H_5NO_3$	4-Nitrobenzaldehyde	795.1	763.2	
C_7H_6ClNO	3-Chlorobenzamide	877.2	846.3	
C_7H_6ClNO	4-Chlorobenzamide	877.2	846.3	
C_7H_6F	*m*-Fluorobenzyl	836.5	804	
C_7H_6FNO	3-Fluorobenzamide	877.2	846.3	
C_7H_6FNO	4-Fluorobenzamide	877.2	846.3	
$C_7H_6F_3N$	3-(Trifluoromethyl)aniline	856.9	825.1	
$C_7H_6N_2$	1*H*-Benzimidazole	953.8	920.5	
$C_7H_6N_2$	1*H*-Indazole	900.8	868.9	
$C_7H_6N_2$	3-Aminobenzonitrile	842.3	810.4	
$C_7H_6N_2$	1*H*-Pyrrolo[2,3-b]pyridine	940.2	908.3	
$C_7H_6N_2O_3$	4-Nitrobenzamide	845.3	814.4	
$C_7H_6N_2O_3$	3-Nitrobenzamide	854.2	823.2	
C_7H_6O	Benzaldehyde	834.0	802.1	
C_7H_6O	2,4,6-Cycloheptatrien-1-one	920.8	891.0	
$C_7H_6O_2$	Benzoic acid	821.1	790.1	
C_7H_7	Benzyl	831.4	800.7	
C_7H_7Br	2-Bromotoluene	775.3	745.8	
C_7H_7Br	3-Bromotoluene	782.0	752.5	
C_7H_7Br	4-Bromotoluene	775.3	745.8	
C_7H_7Cl	2-Chlorotoluene	790.5	761.1	
C_7H_7Cl	3-Chlorotoluene	783.9	754.5	
C_7H_7Cl	4-Chlorotoluene	762.9	735.2	
C_7H_7F	2-Fluorotoluene	773.3	743.8	
C_7H_7F	3-Fluorotoluene	785.4	756.0	
C_7H_7F	4-Fluorotoluene	763.8	736.1	
C_7H_7I	1-Iodo-2-methylbenzene	780.3	750.8	
C_7H_7N	4-Vinylpyridine	944.1	912.3	
C_7H_7NO	1-(3-Pyridinyl)ethanone	916.2	884.3	
C_7H_7NO	1-(4-Pyridinyl)ethanone	914.7	882.9	
C_7H_7NO	4-Aminobenzaldehyde	910.4	878.6	
C_7H_7NO	Benzamide	892.1	861.2	
$C_7H_7NO_2$	Methyl 3-pyridinecarboxylate	925.6	893.8	
$C_7H_7NO_2$	Methyl 4-pyridinecarboxylate	926.6	894.7	
$C_7H_7NO_2$	Aniline-2-carboxylic acid	901.5	869.0	
$C_7H_7NO_2$	Aniline-3-carboxylic acid	864.7	832.3	

Molecular formula	Name	E_{pa} kJ/mol	$\Delta_{base}G°$ kJ/mol	Notes
$C_7H_7NO_2$	Aniline-4-carboxylic acid	864.7	832.3	
$C_7H_7NO_2$	4-Nitrotoluene	815.2	782.7	
$C_7H_7NO_3$	4-Nitrobenzenemethanol	810.5	778.0	
$C_7H_7N_3$	1-Methyl-1*H*-benzotriazole	931.2	898.7	
C_7H_7O	2-Methylphenoxy	874.5	842	
C_7H_8	Toluene	784.0	756.3	
C_7H_8	2,5-Norbornadiene	849.3	820.3	
$C_7H_8N_2O$	4-Aminobenzamide	927.9	896.9	
$C_7H_8N_2O$	3-Aminobenzamide	900.9	869.9	
$C_7H_8N_2O_2$	*N*-Methyl-4-nitroaniline	891.6	865.1	
C_7H_8O	*o*-Cresol	832	800	Ref. 10
C_7H_8O	*m*-Cresol	841	809	Ref. 10
C_7H_8O	*p*-Cresol	814	782	Ref. 10
C_7H_8O	Benzyl alcohol	778.3	748.0	
C_7H_8O	Anisole	839.6	807.2	
$C_7H_8O_2$	2,6-Dimethyl-4*H*-pyran-4-one	941.5	907.3	
$C_7H_8O_2S$	Methyl phenyl sulfone	812.7	780.3	
C_7H_8S	(Methylthio)benzene	872.6	843.7	
C_7H_9N	Benzylamine	913.3	879.4	
C_7H_9N	2-Methylaniline	890.9	859.1	
C_7H_9N	3-Methylaniline	895.8	864.0	
C_7H_9N	4-Methylaniline	896.7	864.8	
C_7H_9N	*N*-Methylaniline	916.6	890.1	
C_7H_9N	2-Ethylpyridine	952.4	920.6	
C_7H_9N	3-Ethylpyridine	947.4	915.5	
C_7H_9N	4-Ethylpyridine	951.1	919.2	
C_7H_9N	2,3-Dimethylpyridine	958.9	927.0	
C_7H_9N	2,4-Dimethylpyridine	962.9	930.8	
C_7H_9N	2,5-Dimethylpyridine	958.8	926.9	
C_7H_9N	2,6-Dimethylpyridine	963.0	931.1	
C_7H_9N	3,4-Dimethylpyridine	957.3	925.5	
C_7H_9N	3,5-Dimethylpyridine	955.4	923.5	
C_7H_9NO	2-Methoxyaniline	905.2	873.3	
C_7H_9NO	3-Methoxyaniline	913.0	881.1	
C_7H_9NO	4-Methoxyaniline	900.3	868.5	
C_7H_{10}	Bicyclo[2.2.1]hept-2-ene	836.5	804.0	
$C_7H_{10}N_2$	*N,N*-Dimethyl-2-pyridinamine	968.2	941.6	
$C_7H_{10}N_2$	*N,N*-Dimethyl-4-pyridinamine	997.6	971.1	
$C_7H_{10}O$	Dicyclopropyl ketone	880.4	850.6	
$C_7H_{10}O$	Bicyclo[2.2.1]heptan-2-one	847.4	815.5	
$C_7H_{11}N$	Cyclohexanecarbonitrile	815.0	784.4	
C_7H_{12}	2,4-Dimethyl-1,3-pentadiene	886.5	857.6	
C_7H_{12}	1-Methylcyclohexene	825.1	792.6	
C_7H_{12}	1,2-Dimethylcyclopentene	822.6	791.9	
$C_7H_{12}N_2$	2,3,4,6,7,8-Hexahydropyrrolo[1,2-a]pyrimidine	1038.3	1005.9	
$C_7H_{12}O$	Cycloheptanone	845.6	815.9	
$C_7H_{12}O$	4-Methylcyclohexanone	844.9	813.0	
$C_7H_{12}O_2$	Cyclohexanecarboxylic acid	823.8	792.8	
$C_7H_{13}N$	1-Azabicyclo[2.2.2]octane	983.3	952.5	
C_7H_{14}	2,4-Dimethyl-2-pentene	812	783.1	
$C_7H_{14}O$	Methoxycyclohexane	840.5	811.3	
$C_7H_{14}O$	4-Heptanone	845.0	815.3	
$C_7H_{14}O$	2,4-Dimethyl-3-pentanone	850.3	820.5	
$C_7H_{14}O$	Cyclohexanemethanol	802.1	771.7	
$C_7H_{15}N$	Cyclohexanemethanamine	926.6	895.8	
$C_7H_{16}O$	*tert*-Butyl isopropyl ether	870.7	841.5	
$C_7H_{17}N$	Heptylamine	923.2	889.3	
$C_7H_{17}N$	Methyldipropylamine	983.5	950.9	
$C_7H_{17}N$	Diethylpropylamine	978.8	947.9	

Molecular formula	Name	E_{pa} kJ/mol	$\Delta_{base}G°$ kJ/mol	Notes
$C_7H_{18}N_2$	1,7-Heptanediamine	998.5	944.9	
$C_7H_{18}N_2$	*N,N,N',N'*-Tetramethyl-1,3-propanediamine	1035.2	985.4	
$C_8H_4F_3N$	3-(Trifluoromethyl)benzonitrile	791.4	760.8	
$C_8H_4F_3N$	4-(Trifluoromethyl)benzonitrile	787.2	758.3	
$C_8H_4N_2$	*m*-Dicyanobenzene	779.3	750.4	
$C_8H_4N_2$	*p*-Dicyanobenzene	779.0	751.8	
C_8H_5Cl	1-Chloro-4-ethynylbenzene	832.4	801.7	
$C_8H_5Cl_3O$	2,2,2-Trichloro-1-phenylethanone	818.9	787.0	
$C_8H_5F_3O$	2,2,2-Trifluoro-1-phenylethanone	799.2	767.4	
$C_8H_5F_3O$	4-(Trifluoromethyl)benzaldehyde	805.6	773.8	
C_8H_5NO	4-Formylbenzonitrile	796.9	766.3	
C_8H_6	Phenylacetylene	832.0	801.3	
C_8H_6ClN	4-(Chloromethyl)benzonitrile	812.8	782.1	
C_8H_6ClN	3-(Chloromethyl)benzonitrile	811.2	780.6	
$C_8H_6N_2$	Quinoxaline	903.8	873.7	
$C_8H_6N_2$	Cinnoline	936.3	904.4	
C_8H_7Br	1-Bromo-4-vinylbenzene	838.7	809.8	
C_8H_7Br	1-Bromo-3-vinylbenzene	822.4	793.5	
C_8H_7ClO	1-(3-Chlorophenyl)ethanone	846.9	815.1	
C_8H_7ClO	1-(4-Chlorophenyl)ethanone	856.6	824.8	
$C_8H_7ClO_2$	Methyl 4-chlorobenzoate	842.1	811.1	
$C_8H_7ClO_2$	Methyl 3-chlorobenzoate	835.4	804.4	
C_8H_7FO	1-(4-Fluorophenyl)ethanone	858.6	826.8	
C_8H_7N	Benzeneacetonitrile	805.5	774.8	
C_8H_7N	1*H*-Indole	933.4	901.9	
$C_8H_7NO_3$	1-(4-Nitrophenyl)ethanone	824.3	792.5	
$C_8H_7NO_3$	1-(3-Nitrophenyl)ethanone	826.0	794.1	
$C_8H_7NO_4$	Methyl 3-nitrobenzoate	815.7	784.7	
$C_8H_7NO_4$	Methyl 4-nitrobenzoate	813.2	782.3	
C_8H_8	Styrene	839.5	809.2	
$C_8H_8N_2$	1-Methyl-1*H*-benzimidazole	967.0	935.2	
$C_8H_8N_2$	2-Methyl-2*H*-indazole	941.4	909.6	
$C_8H_8N_2$	1-Methyl-1*H*-indazole	922.4	890.5	
C_8H_8O	3-Methylbenzaldehyde	840.0	808.1	
C_8H_8O	4-Methylbenzaldehyde	851.8	820.0	
C_8H_8O	Acetophenone	861.1	829.3	
$C_8H_8O_2$	*o*-Toluic acid	838.8	807.8	
$C_8H_8O_2$	*m*-Toluic acid	829.8	798.8	
$C_8H_8O_2$	*p*-Toluic acid	836.7	805.7	
$C_8H_8O_2$	Methyl benzoate	850.5	819.5	
$C_8H_8O_2$	3-Methoxybenzaldehyde	844.1	812.2	
$C_8H_8O_2$	4-Methoxybenzaldehyde	881.1	849.3	
$C_8H_8O_2$	1-(3-Hydroxyphenyl)ethanone	863.6	831.8	
$C_8H_8O_2$	1-(4-Hydroxyphenyl)ethanone	883.7	851.9	
$C_8H_8O_3$	Methyl 4-hydroxybenzoate	863.4	832.5	
$C_8H_8O_3$	Methyl 3-hydroxybenzoate	850.0	819.1	
C_8H_9N	2,3-Dihydro-1*H*-indole	957.1	926.3	
C_8H_9NO	3-Methylbenzamide	900.9	869.9	
C_8H_9NO	4-Methylbenzamide	900.9	869.9	
C_8H_9NO	1-(4-Aminophenyl)ethanone	908.8	877.0	
$C_8H_9NO_2$	3-Methoxybenzamide	900.9	869.9	
$C_8H_9NO_2$	2,4-Dimethyl-1-nitrobenzene	831.0	798.5	
$C_8H_9NO_2$	Methyl 4-aminobenzoate	883.9	853.0	
$C_8H_9NO_2$	4-Methoxybenzamide	900.3	869.4	
C_8H_{10}	Ethylbenzene	788.0	760.3	
C_8H_{10}	*o*-Xylene	796.0	768.3	
C_8H_{10}	*m*-Xylene	812.1	786.2	
C_8H_{10}	*p*-Xylene	794.4	766.8	
$C_8H_{10}ClN$	4-Chloro-*N,N*-dimethylaniline	922.9	896.4	

Molecular formula	Name	E_{pa} kJ/mol	$\Delta_{base}G°$ kJ/mol	Notes
$C_8H_{10}N_2O_2$	*N,N*-Dimethyl-4-nitroaniline	896.7	870.2	
$C_8H_{10}N_2O_2$	*N,N*-Dimethyl-3-nitroaniline	894.1	867.6	
$C_8H_{10}O$	Benzyl methyl ether	816.7	787.5	
$C_8H_{10}O$	2-Methylanisole	850	818	Ref. 10
$C_8H_{10}O$	3-Methylanisole	860	828	Ref. 10
$C_8H_{10}O$	4-Methylanisole	841	809	Ref. 10
$C_8H_{11}N$	4-Isopropylpyridine	955.7	923.8	
$C_8H_{11}N$	3-Ethylaniline	897.9	866.1	
$C_8H_{11}N$	*N*-Ethylaniline	924.8	892.9	
$C_8H_{11}N$	*N,N*-Dimethylaniline	941.1	909.2	
$C_8H_{11}N$	2,6-Dimethylaniline	901.7	869.8	
$C_8H_{11}N$	Benzeneethanamine	936.2	902.3	
$C_8H_{11}N$	2-Propylpyridine	955.7	923.8	
C_8H_{12}	2-Methyl-2-norbornene	845	812.5	
$C_8H_{12}N_2$	*N,N*-Dimethyl-1,4-benzenediamine	955.0	928.4	
$C_8H_{12}N_2O_2$	Ethyl 1,5-dimethyl-1*H*-pyrazole-3-carboxylate	933.4	901.5	
$C_8H_{14}O$	Cyclooctanone	849.4	819.6	
$C_8H_{14}O$	1-Cyclohexylethanone	841.4	809.5	
$C_8H_{14}O_2$	Methyl cyclohexanecarboxylate	846.2	815.3	
$C_8H_{16}O$	2,2,4-Trimethyl-3-pentanone	856.9	825.0	
$C_8H_{17}N$	Cyclohexyldimethylamine	983.6	952.6	
$C_8H_{18}O$	Dibutyl ether	845.7	818.3	
$C_8H_{18}O$	Di-*sec*-butyl ether	865.9	838.5	
$C_8H_{18}O$	Di-*tert*-butyl ether	887.4	860.0	
$C_8H_{18}O_4$	Triethylene glycol dimethyl ether	946.6	892.4	
$C_8H_{18}O_5$	Tetraethylene glycol		>910	
$C_8H_{18}S$	Dibutyl sulfide	871.8	842.1	
$C_8H_{18}S$	Di-*tert*-butyl sulfide	893.8	864.0	
$C_8H_{19}N$	*N*-Ethyl-*N*-isopropyl-2-propanamine	994.3	963.5	
$C_8H_{19}N$	Octylamine	928.9	895.0	
$C_8H_{19}N$	Dibutylamine	968.5	935.3	
$C_8H_{19}N$	Di-*sec*-butylamine	980.7	947.5	
$C_8H_{19}N$	Diisobutylamine	958.1	925.1	
$C_8H_{20}N_2$	*N,N,N',N'*-Tetramethyl-1,4-butanediamine	1046.3	992.7	
$C_8H_{20}N_2$	Tetraethylhydrazine	964.3	935.3	
$C_9H_7MnO_3$	Manganese 2-methylcyclopentadienyl tricarbonyl	833.8	801.3	
C_9H_7N	Quinoline	953.2	921.4	
C_9H_7N	Isoquinoline	951.7	919.9	
C_9H_7NO	4-Acetylbenzonitrile	826.8	795.0	
C_9H_8	Indene	848.8	819.6	
C_9H_8O	2-Methylbenzofuran	859.6	827.2	
$C_9H_8O_3$	Methyl 4-formylbenzoate	832.9	801.9	
C_9H_9Cl	1-Chloro-4-isopropenylbenzene	854.3	825.4	
$C_9H_9ClO_2$	3-Chloro-4-methoxyacetophenone	883.7	851.9	
C_9H_{10}	2-Methylstyrene	855.2	826.3	
C_9H_{10}	3-Methylstyrene	849.4	820.5	
C_9H_{10}	4-Methylstyrene	861.7	832.8	
C_9H_{10}	*cis*-1-Propenylbenzene	836.4	807.5	
C_9H_{10}	*trans*-1-Propenylbenzene	834.2	805.3	
C_9H_{10}	Isopropenylbenzene	864.2	835.3	
C_9H_{10}	Cyclopropylbenzene	834.9	802.4	
$C_9H_{10}N_2$	4-(Dimethylamino)benzonitrile	889.1	862.6	
$C_9H_{10}O$	1-(3-Methylphenyl)ethanone	868.2	836.4	
$C_9H_{10}O$	1-Phenyl-1-propanone	867.4	835.6	
$C_9H_{10}O$	1-Phenyl-2-propanone	842.6	810.8	
$C_9H_{10}O$	4-Methylacetophenone	875.5	843.6	
$C_9H_{10}OS$	4-Acetylthioanisole	888.2	856.3	
$C_9H_{10}O_2$	Methyl 2-methylbenzoate	858.3	827.3	

Molecular formula	Name	E_{pa} kJ/mol	$\Delta_{base}G°$ kJ/mol	Notes
$C_9H_{10}O_2$	Methyl 3-methylbenzoate	857.7	826.8	
$C_9H_{10}O_2$	1-(3-Methoxyphenyl)ethanone	871.2	839.3	
$C_9H_{10}O_2$	Methyl 4-methylbenzoate	861.5	830.6	
$C_9H_{10}O_2$	4-Acetylanisole	895.6	863.7	
$C_9H_{10}O_3$	Methyl 4-methoxybenzoate	870.6	839.6	
$C_9H_{10}O_3$	Methyl 3-methoxybenzoate	856.7	825.8	
$C_9H_{11}N$	5,6,7,8-Tetrahydroquinoline	966.0	934.1	
$C_9H_{11}N$	5,6,7,8-Tetrahydroisoquinoline	966.6	934.7	
$C_9H_{11}NO$	4-(Dimethylamino)benzaldehyde	924.8	898.3	
$C_9H_{11}NO$	*N,N*-Dimethylbenzamide	932.7	901.8	
$C_9H_{11}NO_2$	1,3,5-Trimethyl-2-nitrobenzene	823.8	793.1	
$C_9H_{11}NO_2$	*L*-Phenylalanine	922.9	888.9	
$C_9H_{11}NO_3$	*L*-Tyrosine	926	892.1	
C_9H_{12}	Propylbenzene	790.1	762.4	
C_9H_{12}	Isopropylbenzene	791.6	763.9	
C_9H_{12}	1,3,5-Trimethylbenzene	836.2	808.6	
$C_9H_{12}N_2$	3-(2-Pyrrolidinyl)pyridine, (*S*)-	964.0	931.0	
$C_9H_{12}N_2O_6$	Uridine	947.6	916.6	
$C_9H_{12}O_3$	1,3,5-Trimethoxybenzene	926.7	898.2	
$C_9H_{13}N$	*N*-Ethyl-*N*-methylaniline	939.0	912.4	
$C_9H_{13}N$	2,6-Diethylpyridine	972.3	940.4	
$C_9H_{13}N$	4-*tert*-Butylpyridine	957.7	925.8	
$C_9H_{13}N$	2-*tert*-Butylpyridine	961.7	929.8	
$C_9H_{13}N$	2-Methyl-*N,N*-dimethylaniline	951.8	925.3	
$C_9H_{13}N$	3-Methyl-*N,N*-dimethylaniline	942.1	915.7	
$C_9H_{13}N$	4-Methyl-*N,N*-dimethylaniline	950.0	918.1	
$C_9H_{13}N$	*N,N*-Dimethylbenzylamine	968.4	937.4	
$C_9H_{13}NO$	4-Methoxy-*N,N*-dimethylaniline	949.1	922.4	
$C_9H_{13}N_3O_5$	Cytidine	982.5	950.0	
$C_9H_{14}O$	Isophorone	893.5	861.6	
$C_9H_{15}N$	*N,N*-Diallyl-2-propen-1-amine	972.3	941.3	
$C_9H_{15}N$	*N*-(1-Cyclopenten-1-yl)pyrrolidine	1019.2	988.4	
$C_9H_{16}O$	Cyclononanone	852.6	822.8	
$C_9H_{17}N_3O_4$	*N*-(*N*-*L*-Alanyl-*L*-alanyl)-*L*-alanine		924.1	
$C_9H_{18}O$	5-Nonanone	853.7	821.9	
$C_9H_{18}O$	Di-*tert*-butyl ketone	861.3	831.5	
$C_9H_{19}N$	1-Isobutylpiperidine	974.5	943.5	
$C_9H_{19}N$	2,2,6,6-Tetramethylpiperidine	987.0	953.9	
$C_9H_{21}N$	Tripropylamine	991.0	960.1	
$C_{10}H_8$	Naphthalene	802.9	779.4	
$C_{10}H_8$	Azulene	925.2	896	
$C_{10}H_8N_2$	2,2'-Bipyridine	965		Ref. 7
$C_{10}H_9N$	1-Naphthylamine	907.0	875.1	
$C_{10}H_{10}Fe$	Ferrocene	863.6	841.3	
$C_{10}H_{10}N_2$	1,8-Naphthalenediamine	944.5	912.1	
$C_{10}H_{10}N_2$	1-Methyl-3-phenyl-1*H*-pyrazole	932.6	900.8	
$C_{10}H_{10}N_2$	1-Methyl-5-phenyl-1*H*-pyrazole	932.4	900.5	
$C_{10}H_{10}Ni$	Nickelocene	935.7	907.3	
$C_{10}H_{10}O_2$	1,4-Diacetylbenzene	850.8	821.0	
$C_{10}H_{10}O_2$	1,3-Diacetylbenzene	852.0	822.3	
$C_{10}H_{10}O_3$	4-Acetylphenyl acetate	853.2	821.3	
$C_{10}H_{10}O_4$	Dimethyl isophthalate	843.5	814.3	
$C_{10}H_{10}O_4$	Dimethyl terephthalate	843.2	812.3	
$C_{10}H_{10}Ru$	Ruthenocene	899.1	876.8	
$C_{10}H_{12}$	1-Methyl-3-(1-methylvinyl)benzene	867.6	838.7	
$C_{10}H_{12}$	1-Methyl-4-(1-methylvinyl)benzene	881.8	852.9	
$C_{10}H_{12}$	1-Methyl-2-(1-methylvinyl)benzene	857.8	828.9	
$C_{10}H_{12}$	1,2,3,4-Tetrahydronaphthalene	809.7	782.1	
$C_{10}H_{12}O$	1-(2,4-Dimethylphenyl)ethanone	882.6	850.8	
$C_{10}H_{12}O$	1-(2,5-Dimethylphenyl)ethanone	873.5	841.6	

Molecular formula	Name	E_{pa} kJ/mol	$\Delta_{base}G°$ kJ/mol	Notes
$C_{10}H_{12}O$	1-(3,4-Dimethylphenyl)ethanone	882.8	851.0	
$C_{10}H_{12}O_2$	Methyl 2,5-dimethylbenzoate	864.7	833.7	
$C_{10}H_{12}O_2$	Methyl 2,4-dimethylbenzoate	868.2	837.2	
$C_{10}H_{12}O_2$	Methyl 3,5-dimethylbenzoate	864.3	833.4	
$C_{10}H_{13}N$	1-Phenylpyrrolidine	941.6	915.1	
$C_{10}H_{13}NO$	4'-(Dimethylamino)acetophenone	932.8	906.3	
$C_{10}H_{13}NO$	*N,N*,3-Trimethylbenzamide	927.0	896.0	
$C_{10}H_{13}NO$	*N,N*,4-Trimethylbenzamide	927.0	896.0	
$C_{10}H_{13}NO$	1-[3-(Dimethylamino)phenyl]ethanone	928.0	901.5	
$C_{10}H_{13}N_5O_3$	2'-Deoxyadenosine	991.5	959.1	
$C_{10}H_{13}N_5O_4$	Adenosine	989.3	956.8	
$C_{10}H_{13}N_5O_5$	Guanosine	993.4	960.9	
$C_{10}H_{14}$	Butylbenzene	791.9	764.2	
$C_{10}H_{14}$	1,2,3,5-Tetramethylbenzene	845.6	816.5	
$C_{10}H_{14}ClN$	4-Chloro-*N,N*-diethylaniline	931.0	899.2	
$C_{10}H_{14}N_2$	*L*-Nicotine	963.4	932.6	
$C_{10}H_{14}N_2O$	*N,N*-Diethyl-3-pyridinecarboxamide	940.9	909.0	
$C_{10}H_{14}N_2O_5$	Thymidine	948.3	915.9	
$C_{10}H_{15}N$	*N,N*,2,6-Tetramethylaniline	954.1	923.2	
$C_{10}H_{15}N$	*N,N*,3,5-Tetramethylaniline	956.1	924.3	
$C_{10}H_{15}N$	*N,N*-Diethylaniline	959.8	927.9	
$C_{10}H_{16}N_2$	*N,N,N',N'*-Tetramethyl-1,2-benzenediamine	982.6	950.2	
$C_{10}H_{17}N$	Tricyclo[3.3.1.1^{3,7}]decan-1-amine	948.8	916.3	
$C_{10}H_{22}O$	Dipentyl ether	852.7	825.3	
$C_{10}H_{22}O_5$	Tetraethylene glycol dimethyl ether	953.8	897.8	
$C_{10}H_{23}N$	Decylamine	930.4	896.5	
$C_{10}H_{24}N_2$	*N,N,N',N'*-Tetramethyl-1,6-hexanediamine	1035.8	982.2	
$C_{11}H_9N$	4-Phenylpyridine	939.7	907.8	
$C_{11}H_{10}$	1-Methylnaphthalene	834.8	805.3	
$C_{11}H_{10}$	2-Methylnaphthalene	831.9	802.4	
$C_{11}H_{12}N_2O_2$	*L*-Tryptophan	948.9	915	
$C_{11}H_{14}O_2$	Methyl 2,4,6-trimethylbenzoate	866.3	835.3	
$C_{11}H_{15}N$	1-Phenylpiperidine	952.9	926.4	
$C_{11}H_{15}N$	Tricyclo[3.3.1.1^{3,7}]decane-1-carbonitrile	834.4	803.8	
$C_{11}H_{16}$	Pentamethylbenzene	850.7	823.5	
$C_{11}H_{17}N$	*N,N*-Diethyl-4-methylaniline	962.8	931.0	
$C_{11}H_{17}N$	2-Hexylpyridine	963.6	931.7	
$C_{11}H_{18}O$	1,4,7,7-Tetramethylbicyclo[2.2.1]heptan-2-one	863.3	831.4	
$C_{11}H_{24}O_4$	2,6,10,14-Tetraoxapentadecane		895.1	
$C_{12}H_8N_2$	Phenazine	938.4	908.3	
$C_{12}H_9NO$	Phenyl-3-pyridinylmethanone	934.1	902.3	
$C_{12}H_{10}$	Acenaphthene	851.7	821.0	
$C_{12}H_{10}$	Biphenyl	813.6	782.9	
$C_{12}H_{16}O$	1-(4-tert-Butylphenyl)ethanone	882.5	850.6	
$C_{12}H_{18}$	Hexamethylbenzene	860.6	836.0	
$C_{12}H_{18}O$	1-Tricyclo[3.3.1.1^{3,7}]dec-1-ylethanone	864.9	833.1	
$C_{12}H_{19}N$	*N,N*-Dipropylaniline	963.0	931.1	
$C_{12}H_{20}O$	2,5-Di-*tert*-butylfuran	894.7	863.9	
$C_{12}H_{27}N$	Tributylamine	998.5	967.6	
$C_{13}H_9N$	Acridine	972.6	940.7	
$C_{13}H_{10}$	9*H*-Fluorene	831.5	803.8	
$C_{13}H_{10}O$	Benzophenone	882.3	852.5	
$C_{13}H_{12}$	2-Methylbiphenyl	815.9	783.4	
$C_{13}H_{12}$	3-Methylbiphenyl	828.0	795.5	
$C_{13}H_{12}$	4-Methylbiphenyl	817.9	785.4	
$C_{13}H_{12}$	Diphenylmethane	802.0	769.5	
$C_{13}H_{13}P$	Methyldiphenylphosphine	972.1	939.7	

Molecular formula	Name	E_{pa} kJ/mol	$\Delta_{base}G°$ kJ/mol	Notes
$C_{13}H_{21}N$	2,4-Di-*tert*-butylpyridine	983.8	952.0	
$C_{13}H_{21}N$	2,6-Di-*tert*-butylpyridine	982.9	951	
$C_{13}H_{21}NO$	*N,N*-Dimethyltricyclo[3.3.1.1^{3,7}]decane-1-carboxamide	949.4	917.6	
$C_{14}H_{10}$	Anthracene	877.3	846.6	
$C_{14}H_{10}$	Phenanthrene	825.7	795.0	
$C_{14}H_{12}$	1,1-Diphenylethene	885.7	856.9	
$C_{14}H_{14}$	1,2-Diphenylethane	801.8	774.1	
$C_{14}H_{18}$	1,2,3,4,5,6,7,8-Octahydroanthracene	845.4	814.7	
$C_{14}H_{18}$	1,2,3,4,5,6,7,8-Octahydrophenanthrene	846.2	815.5	
$C_{14}H_{23}N$	4-Octylaniline	894.5	862	
$C_{15}H_{12}$	2-Methylanthracene	887.5	855.1	
$C_{15}H_{12}$	9-Methylanthracene	896.5	865.8	
$C_{15}H_{12}N_2$	3,5-Diphenyl-1*H*-pyrazole	946.3	912.7	
$C_{15}H_{16}$	1,3-Diphenylpropane	820.1	787.6	
$C_{15}H_{18}$	1,4-Dimethyl-7-isopropylazulene	983.1	950.6	
$C_{15}H_{24}$	1,3-Di-*tert*-butyl-5-methylbenzene	853.7	826.0	
$C_{16}H_{10}$	Fluoranthene	828.6	800.9	
$C_{16}H_{10}$	Pyrene	869.2	840.1	
$C_{16}H_{18}$	1,4-Diphenylbutane	822.0	779.8	
$C_{17}H_{20}$	1,5-Diphenylpentane	824.7	782.4	
$C_{18}H_{12}$	Chrysene	840.9	810.1	
$C_{18}H_{12}$	Naphthacene	905.5	876.5	
$C_{18}H_{12}$	Triphenylene	819.2	791.2	
$C_{18}H_{15}As$	Triphenylarsine	908.9	876.4	
$C_{18}H_{15}AsO$	Triphenylarsine oxide	906.2	876.4	
$C_{18}H_{15}N$	Triphenylamine	908.9	876.4	
$C_{18}H_{15}OP$	Triphenylphosphine oxide	906.2	876.4	
$C_{18}H_{15}P$	Triphenylphosphine	972.8	940.4	
$C_{18}H_{15}PS$	Triphenylphosphine sulfide	906.2	876.4	
$C_{18}H_{15}Sb$	Triphenylstibine	845.5	813.1	
$C_{18}H_{22}$	1,6-Diphenylhexane	826.1	783.8	
$C_{18}H_{30}$	1,3,5-Tri-*tert*-butylbenzene	848.8	822.3	
$C_{20}H_{12}$	Perylene	888.6	859.6	
$C_{22}H_{12}$	Benzo[ghi]perylene	876.0	845.2	
$C_{22}H_{14}$	Picene	851.3	820.6	
$C_{24}H_{12}$	Coronene	861.3	835.0	
C_{60}	Carbon (fullerene-C_{60})		827.5	
C_{70}	Carbon (fullerene-C_{70})		827.5	

ATOMIC AND MOLECULAR POLARIZABILITIES

Thomas M. Miller

The *polarizability* of an atom or molecule describes the response of the electron cloud to an external field. The atomic or molecular energy shift ΔW due to an external electric field E is proportional to E^2 for external fields that are weak compared to the internal electric fields between the nucleus and electron cloud. The *electric dipole polarizability* α is the constant of proportionality defined by $\Delta W = -\alpha E^2/2$. The induced electric dipole moment is αE. *Hyperpolarizabilities*, coefficients of higher powers of E, are less often required. Technically, the polarizability is a tensor quantity but for spherically symmetric charge distributions reduces to a single number. In any case, an *average polarizability* is usually adequate in calculations. Frequency-dependent or *dynamic polarizabilities* are needed for electric fields that vary in time, except for frequencies that are much lower than electron orbital frequencies, where *static polarizabilities* suffice.

Polarizabilities for atoms and molecules in excited states are found to be larger than for ground states and may be positive or negative. Molecular polarizabilities are very slightly temperature dependent since the size of the molecule depends on its rovibrational state. Only in the case of dihydrogen has this effect been studied enough to warrant consideration in Table 3.

Polarizabilities are normally expressed in c.g.s. units of cm³. Ground state polarizabilities are in the range of 10^{-24} cm³ = 1 Å³ and hence are often given in Å³ units. Theorists tend to use atomic units of a_o^3 where a_o is the Bohr radius. The conversion is $\alpha(\text{cm}^3) = 0.148184 \times 10^{-24} \times \alpha(a_o^3)$. Polarizabilities are only recently encountered in SI units, C m²/V = J/(V/m)². The conversion from c.g.s. units to SI units is $\alpha(\text{C m}^2/\text{V}) = 4\pi\varepsilon_o \times 10^{-6}\alpha(\text{cm}^3)$, where ε_o is the permittivity of free space in SI units and the factor 10^{-6} simply converts cm³ into m³. Thus, $\alpha(\text{C m}^2/\text{V}) = 1.11265 \times 10^{-16} \times \alpha(\text{cm}^3)$. Persons measuring excited state polarizabilities by optical methods tend to use units of MHz/(V/cm)², where the energy shift, ΔW, is expressed in frequency units with a factor of h understood. The polarizability is $-2\,\Delta W/E^2$. The conversion into c.g.s. units is $\alpha(\text{cm}^3) = 5.955214 \times 10^{-16} \times \alpha[\text{MHz}/(\text{V/cm})^2]$.

The polarizability appears in many formulas for low-energy processes involving the valence electrons of atoms or molecules. These formulas are given below in c.g.s. units: the polarizability α is in cm³; masses m or μ are in grams; energies are in ergs; and electric charges are in esu, where $e = 4.8032 \times 10^{-10}$ esu. The symbol $\alpha(\nu)$ denotes a frequency (ν) dependent polarizability, where $\alpha(\nu)$ reduces to the static polarizability a for $\nu = 0$. For further information, see Bonin, K. D., and Kresin, V. V., *Electric Dipole Polarizabilities of Atoms, Molecules, and Clusters*, World Scientific, Singapore, 1997; Bonin, K. D., and Kadar-Kallen, *Int. J. Mod. Phys. B*, 24, 3313, 1994; and Miller, T. M., and Bederson, B., *Advances in Atomic and Molecular Physics*, 13, 1, 1977, and Gould, H., and Miller, T. M., *Advances in Atomic, Molecular, and Optical Physics*, 51, 243, 2005. Details on polarizability-related interactions, especially in regard to hyperpolarizabilities and nonlinear optical phenomena, are given by Bogaard, M. P., and Orr, B. J., in *Physical Chemistry, Series Two, Vol. 2, Molecular Structure and Properties*, Buckingham, A. D., Ed., Butterworths, London, 1975, pp. 149–194. A tabulation of tensor and hyperpolarizabilities is included. The gas number density, n, in Table 1 is usually taken to be that of 1 atm at 0 °C in reporting experimental data.

TABLE 1. Formulas Involving Polarizability

Description	Formula	Remarks
Lorentz-Lorenz relation	$\alpha(\nu) = \frac{3}{4\pi n}\left[\frac{\eta^2(\nu)-1}{\eta^2(\nu)+2}\right]$	For a gas of atoms or nonpolar molecules; the index of refraction is $\eta(\nu)$
Refraction by polar molecules	$\alpha(\nu) + \frac{d^2}{3kT} = \frac{3}{4\pi n}\left[\frac{\eta^2(\nu)-1}{\eta^2(\nu)+2}\right]$	The dipole moment is d, in esu·cm (= 10^{-18} D)
Dielectric constant (dimensionless)	$\kappa(\nu) = 1 + 4\pi n \;\; \alpha(\nu)$	From the Lorentz-Lorenz relation for the usual case of $\kappa(\nu) \approx 1$
Index of refraction (dimensionless)	$\eta(\nu) = 1 + 2\pi n \;\; \alpha(\nu)$	From $\eta^3(\nu) = \kappa(\nu)$
Diamagnetic susceptibility	$\chi_m = e^2(a_o N\alpha)^{1/2} / 4m_e c^2$	From the approximation that the static polarizability is given by the variational formula $\alpha = (4/9a_o)\Sigma(N_i r_i^2)^2$; N is the number of electrons, m_e is the electron mass; a crude approximation is $\chi_m = (E_i/4m_e c^2)\alpha$, where E_i is the ionization energy
Long-range electron- or ion-molecule interaction energy	$V(r) = -e^2\alpha / 2r^4$	The target molecule polarizability is α
Ion mobility in a gas	$\kappa = -13.87 / (\alpha\mu)^{1/2}\, \text{cm}^2 / \text{V}\cdot\text{s}$	This one formula is not in c.g.s. units. Enter α in Å³ or 10^{-24} cm³ units and the reduced mass μ of the ion-molecule pair in amu. Classical limit; pure polarization potential
Langevin capture cross section	$\sigma(\nu_o) = (2\pi e / \nu_o)(\alpha / \mu)^{1/2}$	The relative velocity of approach for an ion-molecule pair is ν_o; the target molecular polarizability is α and the reduced mass of the ion-molecule pair is μ
Langevin reaction rate coefficient	$k = 2\pi e(\alpha / \mu)^{1/2}$	Collisional rate coefficient for an ion-molecule reaction
Rate coefficient for polar molecules	$k_d = 2\pi e\left[(\alpha / \mu)^{1/2} + cd(2 / \mu\pi kT)^{1/2}\right]$	The dipole moment of the neutral is d in esu cm; the number c is a "locking factor" that depends on α and d, and is between 0 and 1

Description	Formula	Remarks
Modified effective range cross section for electron-neutral scattering	$\sigma(k) = 4\pi A^2 + 32\pi^4 \mu e^2 \alpha A k / 3h^2 + \ldots$	Here, k is the electron momentum divided by $h/2\pi$, where h is Planck's constant; A is called the "scattering length"; the reduced mass is μ
van der Waals constant between two systems A, B	$C_6 = \frac{3}{2}\left[\frac{\alpha^A \alpha^B E^A E^B}{E^A + E^B}\right]$	For the interaction potential term $V_6(r) = -C_6 r^6$; $E^{A,B}$ represents average dipole transition energies and $\alpha^{A,B}$ the respective polarizabilities of A, B
Dipole-quadrupole constant between two systems A, B	$C_8 = \frac{15}{4}\left[\frac{\alpha^A \alpha_q{}^B E^A E_q{}^B}{E^A + E_q{}^B}\right] + \frac{15}{4}\left[\frac{\alpha_q{}^A \alpha^B E_q{}^A E^B}{E_q{}^A + E^B}\right]$	For the interaction potential term $V_8(r) = -C_8 r^8$; $E_q{}^{A,B}$ represents average quadrupole transition energies and $\alpha_q{}^{A,B}$ are the respective quadrupole polarizabilities of A, B
van der Waals constant between an atom and a surface	$C_3 = \frac{\alpha g E^A E^S}{8(E^A + E^S)}$	For an interaction potential $V_3(r) = -C_3 r^3$; $E^{A,S}$ are characteristic energies of the atom and surface; $g = 1$ for a free-electron metal and $g = (\varepsilon_\infty - 1)/(\varepsilon_\infty + 1)$ for an ionic crystal
Relationship between $\alpha(\nu)$ and oscillator strengths	$\alpha(\nu) = \frac{e^2 h^2}{4\pi^2 m_e} \Sigma \frac{f_k}{E_k^2 - (h\nu)^2}$	Here, f_k is the oscillator strength from the ground state to an excited state k, with excitation energy E_k. This formula is often used to estimate static polarizabilities ($\nu = 0$)
Dynamic polarizability	$\alpha(\nu) = \frac{\alpha E_r{}^2}{E_r{}^2 - (h\nu)^2}$	Approximate variation of the frequency-dependent polarizability $\alpha(\nu)$ from $\nu = 0$ up to the first dipole-allowed electronic transition, of energy E_r; the static dipole polarizability is $\alpha(0)$; infrared contributions ignored
Rayleigh scattering cross section	$\alpha(\nu) = \frac{8\pi}{9c^4}(2\pi\nu)^4 \times [3\alpha^2(\nu) + 2\gamma^2(\nu)/3]$	The photon frequency is ν; the polarizability anisotropy (the difference between polarizabilities parallel and perpendicular to the molecular axis) is $\gamma(\nu)$
Verdet constant	$V(\nu) = \frac{\nu n}{2m_e c^2}\left[\frac{d\alpha(\nu)}{d\nu}\right]$	Defined from $\theta = V(\nu)B$, where θ is the angle of rotation of linearly polarized light through a medium of number density n, per unit length, for a longitudinal magnetic field strength B (Faraday effect)

TABLE 2. Static Average Electric Dipole Polarizabilities for Ground State Atoms (in Units of 10^{-24} cm³)

Atomic number	Atom	Polarizability	Estimated accuracy (%)	Method	Ref.
1	H	0.666793	"exact"	calc	MB77
2	He	0.2050522	"exact"	calc	LJS04
		0.2050	0.1	index/diel	NB65/OC67
3	Li	24.33	0.16	beam	MJBTV06
4	Be	5.60	2	calc	MB77
5	B	3.03	2	calc	MB77
6	C	1.67	2	calc	TASSS08
7	N	1.10	2	calc/index	MB77
8	O	0.802	2	calc/index	MB77
9	F	0.557	2	calc	MB77
10	Ne	0.3956	0.1	diel	OC67
11	Na	24.11	0.12	inter-ferom	ESCHP94
12	Mg	10.6	2	calc	MB77
		11.1	5	calc	S71
		10.6	5	calc	BM02
13	Al	6.8	4.4	beam	MMD90
14	Si	5.53	2	calc	TASSS08
15	P	3.63	2	calc	MB77
16	S	2.90	2	calc	MB77
17	Cl	2.18	2	calc	MB77
18	Ar	1.6411	0.05	index/diel	NB65/OC67
19	K	43.4	2	beam	MB77
20	Ca	22.8	2	calc	MB77
		29.4	6	calc	BM02
		25.0	0	beam	MB77
21	Sc	17.8	25	calc	D84
22	Ti	14.6	25	calc	D84
23	V	12.4	25	calc	D84
24	Cr	11.6	25	calc	D84
25	Mn	9.4	25	calc	D84
26	Fe	8.4	25	calc	D84
27	Co	7.5	25	calc	D84
28	Ni	6.8	25	calc	D84
29	Cu	6.2	6	calc	BM02
		6.1	25	calc	D84
30	Zn	5.75	2	index	GHM96
		6.1	6	calc	BM02
		5.6	25	calc	D84
31	Ga	8.12	2	calc	MB77
32	Ge	5.84	2	calc	TASSS08
33	As	4.31	2	calc	MB77
34	Se	3.77	2	calc	MB77
35	Br	3.05	2	calc	MB77
36	Kr	2.4844	0.05	diel	OC67
37	Rb	47.3	2	beam	MB77
38	Sr	27.6	8	beam	MB77

Atomic number	Atom	Polarizability	Estimated accuracy (%)	Method	Ref.
		23.5	6	calc	BM02
39	Y	22.7	25	calc	D84
40	Zr	17.9	25	calc	D84
41	Nb	15.7	25	calc	D84
42	Mo	12.8	25	calc	D84
43	Tc	11.4	25	calc	D84
44	Ru	9.6	25	calc	D84
45	Rh	8.6	25	calc	D84
46	Pd	4.8	25	calc	D84
47	Ag	7.2	25	calc	D84
48	Cd	7.36	3	index	GH95
		7.4	6	calc	BM02
		7.2	25	calc	D84
49	In	10.2	12	beam	GMBSJ84
		9.1	25	calc	D84
50	Sn	6.28	26	beam	TASSS08
		7.84	2	calc	TASSS08
51	Sb	6.6	25	calc	D84
52	Te	5.5	25	calc	D84
53	I	5.35	25	index	A56
		4.7	25	calc	D84
54	Xe	4.044	0.5	diel	MB77
55	Cs	59.42	0.13	beam	AG03
56	Ba	39.7	8	beam	MB77
57	La	31.1	25	calc	D84
58	Ce	29.6	25	calc	D84
59	Pr	28.2	25	calc	D84
60	Nd	31.4	25	calc	D84
61	Pm	30.1	25	calc	D84
62	Sm	28.8	25	calc	D84
63	Eu	27.7	25	calc	D84
64	Gd	23.5	25	calc	D84
65	Tb	25.5	25	calc	D84
66	Dy	24.5	25	calc	D84
67	Ho	23.6	25	calc	D84
68	Er	22.7	25	calc	D84
69	Tm	21.8	25	calc	D84
70	Yb	20.9	2	calc	TS09
71	Lu	21.9	25	calc	D84
72	Hf	16.2	25	calc	D84
73	Ta	13.1	25	calc	D84
74	W	11.1	25	calc	D84
75	Re	9.7	25	calc	D84
76	Os	8.5	25	calc	D84
77	Ir	7.6	25	calc	D84
78	Pt	6.5	25	calc	D84
79	Au	5.8	25	calc	D84
80	Hg	5.02	1	index	GH96
		5.7	25	calc	D84
81	Tl	7.6	15	beam	NYU84
		7.5	25	calc	D84
82	Pb	6.98	15	beam	TASSS08
		7.01	2	calc	TASSS08
83	Bi	7.4	25	calc	D84
84	Po	6.8	25	calc	D84
85	At	6.0	25	calc	D84
86	Rn	5.3	25	calc	D84
87	Fr	48.60	2	calc	LSMS05
		47.1	5	calc	DJSB99
88	Ra	38.3	25	calc	D84
89	Ac	32.1	25	calc	D84
90	Th	32.1	25	calc	D84
91	Pa	25.4	25	calc	D84
92	U	24.9	6	beam	KB94
93	Np	24.8	25	calc	D84
94	Pu	24.5	25	calc	D84
95	Am	23.3	25	calc	D84
96	Cm	23.0	25	calc	D84
97	Bk	22.7	25	calc	D84
98	Cf	20.5	25	calc	D84
99	Es	19.7	25	calc	D84
100	Fm	23.8	25	calc	D84
101	Md	18.2	25	calc	D84
102	No	16.4	1	calc	TS09
112	E112	4.06	2	calc	PBEK08
114	E114	4.59	2	calc	TASSS08
		4.37	2	calc	PBEK08
119	ekafrancium	24.26	2	cal	LSMS05

[a] Methods: calc = calculated value; beam = atomic beam deflection technique; interferom = atomic beam interference; index = determination based on the measured index of refraction; diel = determination based on the measured dielectric constant.

References

A56. Atoji, M., *J. Chem. Phys.* 25, 174, 1956. Semiempirical method based on molecular polarizabilities and atomic radii.

AG03. Amini, J. M., and Gould, H., *Phys. Rev. Lett.* 91, 153001, 2003.

BM02. Bromley, M. W. J., and Mitroy, J., *Phys. Rev. A* 65, 062505, 2002; 062506, 2002.

D84. Doolen, G. D., Los Alamos National Laboratory, unpublished. A relativistic linear response method was used. The method is that described by Zangwill, A., and Soven, P., *Phys. Rev. A* 21, 1561, 1980. Adjustments of less than 10% across the periodic table have been made to these results to bring them into agreement with accurate experimental values where available, for the purpose of presenting "recommended" polarizabilities in Table 2.

DJSB99. Derevianko, A., Johnson, W. R., Safronova, M. S., and Babb, J. F., *Phys. Rev. Lett.* 82, 3589, 1999.

ESCHP94. Ekstrom, C. R., Schmiedmayer, J., Chapman, M. S., Hammond, T. D., and Pritchard, D. E., *Phys. Rev. A* 51, 3883, 1995. See theoretical work by Thakkar, A. J., and Lupinetti, C., *Chem. Phys. Lett.* 402, 270, 2005.

GH95. Goebel, D., and Holm, U., *Phys. Rev. A* 52, 3691, 1995.

GH96. Goebel, D., and Holm, U., *J. Chem. Phys.* 100, 7710, 1996.

GHM96. Goebel, D., Holm, U., and Maroulis, G., *Phys. Rev. A* 54, 1973, 1996.

GMBSJ84. Guella, T. P., Miller, T. M., Bederson, B., Stockdale, J. A. D., and Jaduszliwer, B., *Phys. Rev. A* 29, 2977, 1984.

KB94. Kadar-Kallen, M. A., and Bonin, K. D., *Phys. Rev. Lett.* 72, 828, 1994.

LJS04. Łach, G., Jezionski, B., and Szalewicz, K., *Phys. Rev. Lett.* 92, 233001, 2004.

LS04. Lim, I. S., and Schwerdfeger, P., *Phys. Rev. A* 70, 062501, 2004.

LSMS05. Lim, L. S., Schwerdfeger, P., Metz, B., Stoll, H., *J. Chem. Phys.* 122, 104103, 2005.

MB77. Miller, T. M., and Bederson, B., *Adv. At. Mol. Phys.* 13, 1, 1977. For simplicity, any value in Table 2 which has not changed since this 1977 review is referenced as MB77. Persons interested in original references and further details should consult MB77.

MJBTV06. Miffre, A., Jacquey, M., Büchner, M., Trénec, G., and Vigué, J., *Phys. Rev. A* 73, 011603(R), 2006.

MMD90. Milani, P., Moullet, I., and de Heer, W. A., *Phys. Rev. A* 42, 5150, 1990. See theoretical comments on this result, in Fuentealba, P., *Chem. Phys. Lett.* 397, 459, 2004, and in Lupinetti, C., and Thakkar, A. J., *J. Chem. Phys.* 122, 044301, 2005.

NB65. Newell, A. C., and Baird, R. D., *J. Appl. Phys.* 36, 3751, 1965.

NYU84. Preliminary value from the New York University group. See GMBSJ84.

OC67. Orcutt, R. H., and Cole, R. H., *J. Chem. Phys.* 46, 697, 1967; see also the later references from this group, given following the tables.

PBEK08. Pershina, V., Borschevsky, A., Eliav, E., and Kaldor, U., *J. Chem. Phys.* 128, 024707, 2008. Relativistic coupled-cluster calculations.

S71. Stwalley, W. C., *J. Chem. Phys.* 54, 4517, 1971.

TASSS08. Thierfelder, C., Assadollahzadeh, B., Schwerdtfeger, P., Schäfer, S., and Schäfer, R., *Phys. Rev. A* 78, 052506, 2008. Relativistic calculations, Dirac-Coulomb.

TS09. Thierfelder, C., and Schwerdtfeger, P., *Phys. Rev. A* 79, 032512, 2009. Relativistic coupled-cluster calculations.

TABLE 3. Average Electric Dipole Polarizabilities for Ground State Diatomic Molecules (in Units of 10^{-24} cm^3)

Molecule	Polarizability	Ref.	Molecule	Polarizability	Ref.
Al_2	19	23		5.35	2
BH	3.32*	1	HgCl	7.4*	9
Br_2	7.02	2	ICl	12.3	2
CO	1.95	3	K_2	77	22
Cl_2	4.61	3		72	21
Cs_2	104	22	Li_2	32.8	29
CsK	89	22		34	22
D_2 (v=0,J=0)	0.7921*	5	LiCl	3.46*	10
D_2 (293 K)	0.7954	6	LiF	10.8*	11
DCl	2.84	2	LiH	3.84*	12
F_2	1.38*	7		3.68*	13
H_2 (v=0,J=0)	0.8023*	5		3.88*	14
H_2 (293 K)	0.8045*	5	N_2	1.7403	6,8
H_2 (293 K)	0.8042	6	NO	1.70	2
H_2 (322 K)	0.8059	8	Na_2	40	22
HBr	3.61	3		38	21
HCl	2.63	3	NaK	51	22
	2.77	2	NaLi	40	4
HD (v=0,J=0)	0.7976*	5	O_2	1.5689	34
HF	0.80	37	Rb_2	79	22
HI	5.44	3			

TABLE 4. Average Electric Dipole Polarizabilities for Ground State Triatomic Molecules (in Units of 10^{-24} cm^3)

Molecule	Polarizability	Ref.	Molecule	Polarizability	Ref.
BeH_2	4.34*	14	HgI_2	19.1	2
CO_2	2.911	8	Li_3	34.5	29
CS_2	8.74	3	$LiNa_2$	61.2	30
	8.86	2	Li_2Na	35.4	30
D_2O	1.26	2	N_2O	3.03	8
H_2O	1.45	2	NO_2	3.02	2†
H_2S	3.782	3	Na_3	70	21
	3.95	2	O_3	3.21	2
HCN	2.59	3	OCS	5.71	2
	2.46	2		5.2	15
$HgBr_2$	14.5	2	SO_2	3.72	3
$HgCl_2$	11.6	2		4.28	2

TABLE 5. Average Electric Dipole Polarizabilities for Ground State Inorganic Polyatomic Molecules (Larger than Triatomic) (in Units of 10^{-24} cm^3)

Molecule	Polarizability	Ref.	Molecule	Polarizability	Ref.
$AsCl_3$	14.9	2	GeH_3Cl	6.7	2†
AsN_3	5.75	2	$(HgCl)_2$	14.7	9
BCl_3	9.38	20	K_n	n=2,5,7-9,11,20	21
BF_3	3.31	2	$(KBr)_2$	42.0	16
$(BN_3)_2$	5.73	2	$(KCl)_2$	32.1	16
$(BH_2N)_3$	8.0	2†	$(KF)_2$	21.0	16
ClF_3	6.32	2	$(KI)_2$	36.3	16
$(CsBr)_2$	54.5	16	Li_n	n=2-22	29
$(CsCl)_2$	42.4	16	$(LiBr)_2$	18.9	16
$(CsF)_2$	28.4	16	$(LiCl)_2$	13.1	16
$(CsI)_2$	51.8	16	$(LiF)_2$	6.9	16
Ga_nAs_m	$n+m$=4-30	28	$(LiI)_2$	23.4	16
$GeCl_4$	15.1	2	$LiNa_3$	75.6	30

Molecule	Polarizability	Ref.
(Li_2Na_2)	60.0	30
Li_3Na	54.8	30
ND_3	1.70	2
NF_3	3.62	2
NH_3	2.81	20
	2.10	2
	2.26	3
	2.22*	33
$(NO_2)_2$	6.69	2
Na_n	n=1-40	21
$(NaBr)_2$	26.8	16
$(NaCl)_2$	23.4	16
$(NaF)_2$	20.7	16
$(NaI)_2$	26.9	16
OsO_4	8.17	2
PCl_3	12.8	2
PF_5	6.10	2
PH_3	4.84	2
$(RbBr)_2$	48.2	16
$(RbCl)_2$	43.2	16
$(RbF)_2$	40.7	16
$(RbI)_2$	46.3	16
SF_6	6.54	8
$(SF_5)_2$	13.2	2
SO_3	4.84	2
SO_2Cl_2	10.5	2
SeF_6	7.33	2
SiF_4	5.45	2
SiH_4	5.44	2
$(SiH_3)_2$	11.1	2
$SiHCl_3$	10.7	2
SiH_2Cl_2	8.92	2
SiH_3Cl	7.02	2
$SnBr_4$	22.0	2
$SnCl_4$	18.0	2
	13.8	15
SnI_4	32.3	2
TeF_6	9.00	2
$TiCl_4$	16.4	2
UF_6	12.5	2

TABLE 6. Average Electric Dipole Polarizabilities for Ground State Hydrocarbon Molecules (in Units of 10^{-24} cm^3)

Molecule	Name	Polarizability	Ref.
CH_4	methane	2.593	8
C_2H_2	acetylene	3.33	3
		3.93	2
C_2H_4	ethylene	4.252	8
C_2H_6	ethane	4.47	3
		4.43	2
C_3H_4	propyne	6.18	2
C_3H_6	propene	6.26	2
	cyclopropane	5.66	2
C_3H_8	propane	6.29	3
		6.37	2
C_4H_6	1-butyne	7.41	2†
	1,3-butadiene	8.64	2
C_4H_8	1-butene	7.97	2
		8.52	2
	trans-2-butene	8.49	2
	2-methylpropene	8.29	2
C_4H_{10}	butane	8.20	2
	isobutane	8.14	27
C_5H_6	1,3-cyclopentadiene	8.64	2
C_5H_8	1-pentyne	9.12	2
	trans-1,3-pentadiene	10.0	2
	isoprene	9.99	2
C_5H_{10}	cyclopentane	9.15	18
	1-pentene	9.65	27
	2-pentene	9.84	27
C_5H_{12}	pentane	9.99	2
	neopentane	10.20	18
C_6H_6	benzene	10.0	25
		10.32	3
		10.74	2
C_6H_{10}	1-hexyne	10.9	2†
	2-ethyl-1,3-butadiene	11.8	2†
	3-methyl-1,3-pentadiene	11.8	2†
	2-methyl-1,3-pentadiene	12.1	2†
	2,3-dimethyl-1,3-butadiene	11.8	2†
	cyclohexene	10.7	2†
C_6H_{12}	cyclohexane	11.0	18
		10.87	15
	1-hexene	11.65	27
C_6H_{14}	hexane	11.9	2
C_7H_8	toluene	11.8	25
		12.26	15
		12.3	2
C_7H_{12}	1-heptyne	12.8	2†
C_7H_{14}	methylcyclohexane	13.1	2
	1-heptene	13.51	27
C_7H_{16}	heptane	13.61	2
C_8H_8	styrene	15.0	2
		14.41	27
C_8H_{10}	ethylbenzene	14.2	2
	o-xylene	14.9	2
		14.1	15
	p-xylene	13.7	25
		14.2	15
		14.9	2
	m-xylene	14.2	15
C_8H_{16}	ethylcyclohexane	15.9	2
C_8H_{18}	*n*-octane	15.9	2
	3-methylheptane	15.44	27
	2,2,4-trimethylpentane	15.44	27
C_9H_{10}	α- methylstyrene	16.05	27
C_9H_{12}	isopropylbenzene	16.0	2†
	1,3,5-trimethylbenzene	15.5	25
		16.14	27
C_9H_{18}	isopropylcyclohexane	17.2	2
C_9H_{20}	nonane	17.36	27
$C_{10}H_8$	naphthalene	16.5	17
		17.48	27
$C_{10}H_{14}$	durene	17.3	25
	tert-butylbenzene	17.2	25
		17.8	2†
$C_{10}H_{20}$	*tert*-butylcyclohexane	19.8	2
$C_{10}H_{22}$	decane	19.10	27
$C_{11}H_{10}$	α-methylnaphthalene	19.35	27
	β-methylnaphthalene	19.52	27

Molecule	Name	Polarizability	Ref.
$C_{11}H_{14}$	α,β,β-trimethylstyrene	19.64	27
$C_{11}H_{16}$	pentamethylbenzene	19.1	25
$C_{11}H_{24}$	undecane	21.03	27
$C_{12}H_{10}$	acenaphthene	20.61	27
$C_{12}H_{12}$	α-ethylnaphthalene	21.19	27
	β-ethylnaphthalene	21.36	27
$C_{12}H_{18}$	hexamethylbenzene	20.9	25
$C_{12}H_{26}$	dodecane	22.75	27
$C_{13}H_{10}$	fluorene	21.68	27
$C_{14}H_{10}$	anthracene	25.4	17
		25.93	27
	phenanthrene	36.8*	17
		24.70	27
$C_{14}H_{22}$	*p*-di-*tert*-butylbenzene	24.5	25
$C_{16}H_{10}$	pyrene	28.22	27
$C_{17}H_{12}$	2,3-benzfluorene	30.21	27
$C_{18}H_{12}$	naphthacene	32.27	27
	1,2-benzanthracene	32.86	27
	chrysene	33.06	27
	triphenylene	31.07	27
$C_{18}H_{30}$	1,3,5-tri-*tert*-butylbenzene	31.8	25
$C_{24}H_{12}$	coronene	42.50	27

TABLE 7. Average Electric Dipole Polarizabilities for Ground State Organic Halides (in Units of 10^{-24} cm^3)

Molecule	Name	Polarizability	Ref.
CBr_2F_2	dibromodifluoromethane	9.0	2†
$CClF_3$	chlorotrifluoromethane	5.72	20
		5.59	8
CCl_2F_2	dichlorodifluoromethane	7.93	20
		7.81	2
CCl_2O	phosgene	7.29	2
CCl_2S	thiophosgene	10.2	2
CCl_3F	trichlorofluoromethane	9.47	2
CCl_3NO_2	trichloronitromethane	10.8	2†
CCl_4	carbon tetrachloride	11.2	2
		10.5	3
CF_4	carbon tetrafluoride	3.838	8
CF_2O	carbonylfluoride	1.88*	17
$CHBr_3$	bromoform	11.8	27
$CHBrF_2$	bromodifluoromethane	5.7	2†
$CHClF_2$	chlorodifluoromethane	6.38	20
		5.91	2
$CHCl_2F$	dichlorofluoromethane	6.82	2
$CHCl_3$	chloroform	9.5	8
		8.23	27
CHF_3	fluoroform	3.52	20
		3.57	8
CHFO	fluoroformaldehyde	1.76*	17
CHI_3	iodoform	18.0	27
CH_2Br_2	dibromomethane	9.32	2
		8.68	27
CH_2ClNO_2	chloronitromethane	6.9	2†
CH_2Cl_2	dichloromethane	6.48	3
		7.93	2
CH_2I_2	diiodomethane	12.90	27
CH_3Br	bromomethane	5.87	20
		6.03	2
		5.55	15
CH_3Cl	chloromethane	5.35	20
		4.72	8
CH_3F	fluoromethane	2.97	8
CH_3I	iodomethane	7.97	2
C_2ClF_5	chloropentafluoroethane	6.3	2†
$C_2Cl_2F_4$	1,2-dichlorotetrafluoroethane	8.5	2†
C_2Cl_3N	trichloroacetonitrile	10.42	18
C_2F_6	hexafluoroethane	6.82	2
C_2HBr	bromoacetylene	7.39	2
C_2HCl	chloroacetylene	6.07	2
C_2HCl_3	trichloroethlyene	10.03	27
C_2HCl_5	pentachloroethane	14.0	2
$C_2H_2Cl_2$	1,1-dichloroethylene	7.83	27
	trans-dichloroethylene	8.15	27
	cis-dichloroethylene	8.03	27
$C_2H_2Cl_2F_2$	1,1-dichloro-2,2-difluoroethane	8.4	2†
$C_2H_2Cl_2O$	chloroacetyl chloride	8.92	2
$C_2H_2Cl_3F$	1,2,2-trichloro-1-fluoroethane	10.2	2†
$C_2H_2Cl_4$	1,1,2,2-tetrachloroethane	12.1	2†
C_2H_2ClN	chloroacetonitrile	6.10	18
$C_2H_2F_2$	1,1-difluoroethylene	5.01	20
C_2H_3Br	bromoethylene	7.59	2
C_2H_3Cl	chloroethylene	6.41	2
$C_2H_3ClF_2$	1-chloro-1,1-difluoroethane	8.05	2
C_2H_3ClO	acetyl chloride	6.62	2
$C_2H_3ClO_2$	methyl chloroformate	7.1	2†
$C_2H_3Cl_3$	1,1,1-trichloroethane	10.7	2
$C_2H_3F_3$	1,1,1-trifluoroethane	4.4	2†
C_2H_3I	iodoethylene	9.3	2†
C_2H_4BrCl	1-bromo-2-chloroethane	9.5	2†
$C_2H_4Br_2$	1,2-dibromoethane	10.7	2†
C_2H_4ClF	1-chloro-2-fluoroethane	6.5	2†
$C_2H_4ClNO_2$	1-chloro-1-nitroethane	10.9	2
$C_2H_4Cl_2$	1,1-dichloroethane	8.64	2
	1,2-dichloroethane	8.0	2†
C_2H_5Br	bromoethane	8.05	2
		7.28	27
C_2H_5Cl	chloroethane	7.27	20
		8.29	2
		6.4	15
C_2H_5ClO	2-chloroethanol	7.1	2†
		6.88	27
	chloromethyl methyl ether	7.1	2†
C_2H_5F	fluoroethane	4.96	2
C_2H_5I	iodoethane	10.0	2
$C_3H_4Cl_2$	dichloropropene	10.1	2†
C_3H_5Cl	chloropropene	8.3	2
C_3H_5ClO	chloroacetone	8.4	2†
$C_3H_5ClO_2$	ethyl chloroformate	9.0	2†
$C_3H_6ClNO_2$	1-chloro-1-nitropropane	10.4	2†
$C_3H_6Cl_2$	dichloropropane	10.9	2†
C_3H_7Br	1-bromopropane	9.4	2†
		9.07	27
	2-bromopropane	9.6	2†
C_3H_7Cl	chloropropane	10.0	2
C_3H_7ClO	β-chloroethyl methyl ether	8.71	27
	2-chloro-1-propanol	8.89	27
	3-chloro-1-propanol	8.84	27

Molecule	Name	Polarizability	Ref.
C_3H_7I	1-iodopropane	11.5	2†
C_4H_5Cl	4-chloro-1,2-butadiene	10.0	2†
C_4H_7Cl	1-chloro-2-methylpropene	10.8	2
$C_4H_7ClO_2$	2-chlorobutyric acid	10.87	27
	3-chlorobutyric acid	10.80	27
	4-chlorobutyric acid	10.69	27
$C_4H_8Cl_2$	1,4-dichlorobutane	12.0	2†
C_4H_9Br	bromobutane	13.9	2
		10.86	27
C_4H_9Cl	1-chlorobutane	11.3	2
	1-chloro-2-methylpropane	11.1	2
	2-chloro-2-methylpropane	12.5	2†
	2-chlorobutane	12.4	2
C_4H_9ClO	β-chloroethyl ethyl ether	10.56	27
	2-chloro-1-butanol	10.70	27
	3-chloro-1-butanol	10.38	27
C_4H_9I	1-iodobutane	13.3	2†
		12.65	27
$C_5H_9ClO_2$	methyl 2-chlorobutanoate	12.33	27
	methyl 3-chlorobutanoate	12.31	27
	methyl 4-chlorobutanoate	12.27	27
	2-chloropentanoic acid	12.69	27
	3-chloropentanoic acid	12.57	27
	4-chloropentanoic acid	12.53	27
$C_5H_{11}Br$	1-bromopentane	13.1	2†
$C_5H_{11}Cl$	1-chloropentane	12.0	2†
$C_5H_{11}F$	fluoropentane	9.95	27
C_6F_6	hexafluorobenzene	9.58	27
C_6HF_5	pentafluorobenzene	9.63	27
$C_6H_2Cl_2O_2$	2,5-dichloro-1,4-benzoquinone	18.4	2
$C_6H_2F_4$	1,2,3,4-tetrafluorobenzene	9.69	27
	1,2,4,5-tetrafluorobenzene	9.69	27
$C_6H_3F_3$	1,3,5-trifluorobenzene	9.74	27
C_6H_4BrF	*p*-bromofluorobenzene	13.4	2†
$C_6H_4ClNO_2$	chloronitrobenzene	14.6	2†
$C_6H_4Cl_2$	*o*-dichlorobenzene	14.17	27
	m-dichlorobenzene	14.23	27
	p-dichlorobenzene	14.20	27
C_6H_4FI	*p*-fluoroiodobenzene	15.5	2†
$C_6H_4FNO_2$	*p*-fluoronitrobenzene	12.8	2†
$C_6H_4F_2$	*o*-difluorobenzene	9.80	27
	m-difluorobenzene	10.3	2†
	p-difluorobenzene	9.80	27
C_6H_5Br	bromobenzene	14.7	2
		13.62	27
C_6H_5Cl	chlorobenzene	14.1	2
		12.3	15
C_6H_5ClO	chlorophenol	13.0	2†
C_6H_5F	fluorobenzene	10.3	2
C_6H_5I	iodobenzene	15.5	2†
$C_6H_{11}ClO_2$	ethyl 2-chlorobutanoate	14.16	27
	ethyl 3-chlorobutanoate	14.13	27
	ethyl 4-chlorobutanoate	14.11	27
$C_6H_{13}Br$	bromohexane	14.44	27
$C_6H_{13}F$	fluorohexane	11.80	27
C_7H_7Br	*p*-bromotoluene	14.80	27
C_7H_7Cl	*p*-chlorotoluene	13.70	27
C_7H_7F	*p*-fluorotoluene	11.70	27
C_7H_7I	*p*-iodotoluene	17.10	27
$C_7H_{15}Br$	1-bromoheptane	16.8	2†
		16.23	27
$C_7H_{15}F$	fluoroheptane	13.66	27
$C_8H_{17}Br$	bromooctane	18.02	27
$C_8H_{17}F$	fluorooctane	15.46	27
$C_9H_{19}Br$	bromononane	19.81	27
$C_9H_{19}F$	fluorononane	17.34	27
$C_{10}F_8$	octafluoronaphthalene	17.64	27
$C_{10}H_7Br$	α-bromonaphthalene	20.34	27
$C_{10}H_7Cl$	α-chloronaphthalene	19.30	27
	β-chloronaphthalene	19.58	27
$C_{10}H_7I$	α-iodonaphthalene	22.41	27
	β-iodonaphthalene	22.95	27
$C_{10}H_{21}Br$	bromodecane	21.60	27
$C_{10}H_{21}F$	fluorodecane	19.18	27
$C_{11}H_{23}F$	fluoroundecane	21.00	27
$C_{12}H_{25}Br$	bromododecane	25.18	27
$C_{12}H_{25}F$	fluorododecane	22.83	27
$C_{12}H_8Br_2O$	4,4'-dibromodiphenyl ether	27.8	2†
$C_{12}H_9BrO$	4-bromodiphenyl ether	24.2	2†
$C_{13}H_{11}BrO$	*p*-bromophenyl-*p*-tolyl ether	26.6	2†
$C_{14}H_9Br$	9-bromoanthracene	28.32	27
$C_{14}H_9Cl$	9-chloroanthracene	27.35	27
$C_{14}H_9F$	fluoranthracene	28.34	27
$C_{14}H_{29}F$	fluorotetradecane	26.57	27
$C_{16}H_{33}Br$	bromohexadecane	32.34	27
$C_{18}H_{37}Br$	bromooctadecane	35.92	27

TABLE 8. Static Average Electric Dipole Polarizabilities for Other Ground State Organic Molecules (in Units of 10^{-24} cm³)

Molecule	Name	Polarizability	Ref.
CN_4O_8	tetranitromethane	15.3	2
CH_2O	formaldehyde	2.8	2†
		2.45	18
CH_2O_2	formic acid	3.4	2†
CH_3NO	formamide	4.2	2†
		4.08	18
CH_3NO_2	nitromethane	7.37	2
CH_4O	methanol	3.29	2
		3.23	15
		3.32	18
CH_5N	methyl amine	4.7	2
		4.01	19
		4.01*	33
C_2N_2	cyanogen	7.99	2
C_2H_2O	ketene	4.4	2†
C_2H_3N	acetonitrile	4.40	2†
		4.48	18
C_2H_4O	acetaldehyde	4.6	2†
		4.59	18
	ethylene oxide	4.43	18
$C_2H_4O_2$	acetic acid	5.1	2†
	methyl formate	5.05	27
$C_2H_4O_4$	formic acid dimer	12.7	2
C_2H_5NO	acetamide	5.67	18
	N-methyl formamide	5.91	18
$C_2H_5NO_2$	nitroethane	9.63	2
	ethyl nitrite	7.0	15
C_2H_6O	ethanol	5.41	2
		5.11	18
	methyl ether	5.29	20

Molecule	Name	Polarizability	Ref.
		5.84	2
		5.16	15
$C_2H_6O_2$	ethylene glycol	5.7	2†
		5.61	27
$C_2H_6O_2S$	dimethyl sulfone	7.3	2†
C_2H_6S	ethanethiol	7.41	2
C_2H_7N	ethyl amine	7.10	2
	dimethyl amine	6.37	2
		5.90*	33
$C_2H_8N_2$	ethylene diamine	7.2	2†
$C_3H_2N_2$	malononitrile	5.79	18
C_3H_3N	acrylonitrile	8.05	2
$C_3H_4N_2$	pyrazole	7.23	27
C_3H_4O	propenal	6.38	2†
C_3H_5N	propionitrile	6.70	2
		6.24	18
		6.27*	32
C_3H_6O	acetone	6.33	15
		6.4	2†
		6.39	18
	allyl alcohol	7.65	2
	propionaldehyde	6.50	2
$C_3H_6O_2$	propionic acid	6.9	2†
	ethyl formate	8.01	2
		6.88	27
	methyl acetate	6.94	2
		6.81	27
$C_3H_6O_3$	dimethyl carbonate	7.7	2†
C_3H_7NO	*N*-methyl acetamide	7.82	18
	N,N-dimethyl formamide	7.01	18
$C_3H_7NO_2$	nitropropane	8.5	2†
C_3H_8O	2-propanol	7.61	2
		6.97	18
	1-propanol	6.74	2
	ethyl methyl ether	7.93	2
$C_3H_8O_2$	dimethoxymethane	7.7	2†
	ethylene glycol monomethyl ether	7.44	27
C_3H_9N	propylamine	7.70	27
		9.20	2
	isopropylamine	7.77	27
	trimethylamine	8.15	2
		7.78*	33
$C_4H_2N_2$	fumaronitrile	11.8	2
$C_4H_4N_2$	succinonitrile	8.1	2†
	pyrimidine	8.53*	17
	pyridazine	9.27*	17
$C_4H_4O_2$	diketene	8.0	2†
C_4H_4S	thiophene	9.67	2
C_4H_5N	methacrylonitrile	8.0	2†
	trans-crotononitrile	8.2	2†
$C_4H_6N_2$	*N*-methylpyrazole	8.99	27
C_4H_6O	crotonaldehyde	8.5	2†
	methacrylaldehyde	8.3	2†
$C_4H_6O_2$	biacetyl	8.2	2†
$C_4H_6O_3$	acetic anhydride	8.9	2†
C_4H_6S	divinyl sulfide	10.9	2†
C_4H_7N	butyronitrile	8.4	2†
	isobutyronitrile	8.05	18
		8.05*	32
C_4H_8O	butanal	8.2	2†
	methyl ethyl ketone	8.13	15
	trans-2,3-epoxy butane	8.22*	17
$C_4H_8O_2$	ethyl acetate	9.7	2
		8.62	27
	1,4-dioxane	10.0	2
	p-dioxane	8.60	18
	2-methyl-1,3-dioxolane	9.44	15
	butyric acid	8.58	27
	methyl propionate	8.97	27
$C_4H_9NO_2$	1-nitrobutane	10.4	2†
	2-methyl-2-nitropropane	10.3	2†
$C_4H_{10}O$	ethyl ether	10.2	2
		8.73	15
	1-butanol	8.88	2
	2-methylpropanol	8.92	2
	methyl propyl ether	8.86	27
	ethylene glycol monoethyl ether	9.28	27
$C_4H_{10}S$	ethyl sulfide	10.8	2
$C_4H_{11}N$	butylamine	13.5	2
	diethylamine	10.2	2
		9.61	27
C_5H_5N	pyridine	9.5	15
		9.18	27
	4-cyano-1,3-butadiene	10.5	2†
$C_5H_8N_2$	1,5-dimethylpyrazole	10.72	27
$C_5H_8O_2$	acetyl acetone	10.5	2†
C_5H_9N	valeronitrile	10.4	2
	22-DMPN	9.59	18
$C_5H_{10}O$	diethyl ketone	9.93	15
	methyl propyl ketone	9.93	15
$C_5H_{10}O_2$	ethyl propionate	10.41	27
	methyl butanoate	10.41	27
$C_5H_{10}O_3$	diethyl carbonate	11.3	2
$C_5H_{12}O$	ethyl propyl ether	10.68	27
$C_5H_{12}O_4$	tetramethyl orthocarbonate	13.0	2†
$C_6H_4N_2O_4$	*p*-dinitrobenzene	18.4	2
$C_6H_4O_2$	*p*-benzoquinone	14.5	2
$C_6H_5NO_2$	nitrobenzene	14.7	2
		12.92	15
C_6H_6O	phenol	11.1	2†
		9.94*	17
C_6H_7N	aniline	12.1	2†
$C_6H_8N_2$	phenylenediamine	13.8	2†
	phenylhydrazine	12.91	27
$C_6H_{10}N_2$	1-ethyl-5-methylpyrazole	12.50	27
$C_6H_{10}O_3$	ethyl acetoacetate	12.9	2†
$C_6H_{12}N_2$	dimethylketazine	15.6	2
$C_6H_{12}O$	cyclohexanol	11.56	18
$C_6H_{12}O_2$	amyl formate	14.2	2
$C_6H_{12}O_3$	paraldehyde	17.9	2
$C_6H_{14}O$	propyl ether	12.8	2
		12.5	15
$C_6H_{14}O_2$	1,1-diethoxyethane	13.2	2†
	1,2-diethoxyethane	11.3	2†
$C_6H_{15}N$	triethylamine	13.1	2
		13.38	27
	dipropylamine	13.29	27
$C_7H_4N_2O_2$	*p*-cyanonitrobenzene	19.0	2
C_7H_5N	benzonitrile	12.5	2†
$C_7H_7NO_3$	nitroanisole	15.7	2†
C_7H_8O	anisole	13.1	2†
C_7H_9NO	*o*-anisidine	14.2	2†
$C_7H_{10}N_2$	1,1-methylphenylhydrazine	14.81	27
$C_7H_{14}O$	cyclohexyl methyl ether	13.4	2†
	2,4-dimethyl-3-pentanone	13.5	15
$C_7H_{14}O_2$	pentyl acetate	14.9	2
$C_8H_4N_2$	*p*-dicyanobenzene	19.2	2

Molecule	Name	Polarizability	Ref.
$C_8H_6N_2$	quinoxaline	15.13	27
C_8H_8O	acetophenone	15.0	2
$C_8H_8O_2$	2,5-dimethyl-1,4-benzoquinone	18.8	2
$C_8H_{10}O$	phenetole	14.9	2
$C_8H_{11}N$	*N*-dimethylaniline	16.2	2†
$C_8H_{12}N_2$	1,1-ethylphenylhydrazine	16.62	27
$C_8H_{12}O_2$	ethyl sorbate	17.2	2†
	tetramethylcyclobutane-1,3-dione	18.6	2
$C_8H_{14}O_4$	diethyl succinate	16.8	2†
$C_8H_{18}O$	butyl ether	17.2	2
C_9H_7N	quinoline	15.70	27
	isoquinoline	16.43	27
$C_9H_{10}O_2$	ethyl benzoate	16.9	2†
$C_9H_{21}N$	tripropylamine	18.87	27
$C_{10}H_9N$	α-naphthylamine	19.50	27
	β-naphthylamine	19.73	27
	2-methylquinoline	18.65	27
	1-methylisoquinoline	18.28	27
$C_{10}H_{10}Fe$	ferrocene	17.1	26
$C_{10}H_{10}N_2$	2,3-dimethylquinoxaline	18.70	27
$C_{10}H_{14}BeO_4$	beryllium acetylacetonate	34.1	2
$C_{11}H_8O$	1-naphthaldehyde	19.75	27
	2-naphthaldehyde	20.06	27
$C_{12}H_8N_2$	phenazine	23.43	27
$C_{12}H_9NO_3$	4-nitrodiphenyl ether	24.7	2†
$C_{14}H_8O_2$	anthraquinone	24.46	27
$C_{14}H_{14}O$	di-*p*-tolyl ether	24.9	2†
$C_{15}H_{21}AlO_6$	aluminum acetylacetonate	51.9	2
$C_{15}H_{21}CrO_6$	chromium acetylacetonate	53.7	2
$C_{15}H_{21}FeO_6$	ferric acetylacetonate	58.1	2
$C_{20}H_{28}O_8Th$	thorium acetylacetonate	79.0	2
C_{60}	buckminsterfullerene	76.5	24
		79	31

Note: All polarizabilities in the tables are experimental values except those values marked by an asterisk (*), which indicates a calculated result. The experimental polarizabilities are mostly determined by measurements of a dielectric constant or refractive index that are quite accurate (0.5% or better). However, one should treat many of the results with several percent of caution because of the age of the data and because some of the results refer to optical frequencies rather than static. Comments given with the references are intended to allow one to judge the degree of caution required. Interested persons should consult these references. In many cases, the reference given is to a theoretical paper in which the experimental results are quoted. These papers, noted in the References, contain valuable information on polarizability calculations and experimental data which often includes the tensor components of the polarizability.
An empirical additive formula for molecular polarizabilities at 589 nm frequency has been given in Bosque, R., and Sales, J., *J. Chem. Inf. Comput. Sci.* 42, 1154, 2002: α=0.32 +1.51#C +0.17#H +0.57#O +1.05#N +2.99#S +2.48#P +2.16#Cl +3.29#Br +5.45#I, where "#C" denotes the number of carbon atoms in the molecule, etc.

References

1. McCullough, E. A., Jr., *J. Chem. Phys.*, 63, 5050, 1975. This calculation is for the parallel component, not the average polarizability.
2. Maryott, A. A., and Buckley, F., *U. S. National Bureau of Standards Circular No. 537*, 1953. A tabulation of dipole moments, dielectric constants, and molar refractions measured between 1910 and 1952, and used here to determine polarizabilities if no more recent result exists. The polarizability is $3/(4\pi N_A)$ times the molar polarization or molar refraction, where N_A is Avogadro's number. The value $3/(4\pi N_A)$ = 0.3964308 x 10^{-24} cm^3 was used for this conversion. A dagger (†) following the reference number in the tables indicates that the polarizability was derived from the molar refraction and hence may not include some low-frequency contributions to the static polarizability; these "static" polarizabilities are therefore low by 1 to 30%.
3. Hirschfelder, J. O., Curtis, C. F., and Bird, R. B., *Molecular Theory of Gases and Liquids,* Wiley, New York, 1954, p. 950. Fundamental information on molecular polarizabilities.
4. Miller, T. M., and Bederson, B., *Adv. At. Mol. Phys.*, 13, 1, 1977. Review emphasizing atomic polarizabilities and measurement techniques. The data quoted in Table 3 are accurate to 8 to 12%.
5. Kolos, W., and Wolniewicz, L., *J. Chem. Phys.*, 46, 1426, 1967. Highly accurate molecular hydrogen calculations. See also recent work by Machado, A.M., and Masilli, M., *J. Chem. Phys.* 120, 7505, 2004.
6. Newell, A. C., and Baird, R. C., *J. Appl. Phys.*, 36, 3751, 1965. Highly accurate refractive index measurements at 47.7 GHz (essentially static).
7. Jao, T. C., Beebe, N. H. F., Person, W. B., and Sabin, J. R., *Chem. Phys. Lett.*, 26, 474, 1974. Tensor polarizabilities, derivatives, and other results are reported.
8. Orcutt, R. H., and Cole, R. H., *J. Chem. Phys.*, 46, 697, 1967 (He, Ne, Ar, Kr, H_2, N_2); Sutter, H., and Cole, R. H., *J. Chem. Phys.*, 52, 132, 1970 (CF_3H, CFH_3, $CClF_3$, $CClH_3$); Bose, T. K., and Cole, R. H., *J. Chem. Phys.*, 52, 140, 1970 (CO_2), and 54, 3829, 1971 (C_2H_4); Nelson, R. D., and Cole, R. H., *J. Chem. Phys.*, 54, 4033, 1971 (SF_6, $CClF_3$); Bose, T. K., Sochanski, J. S., and Cole, R. H., *J. Chem. Phys.*, 57, 3592, 1972 (CH_4, CF_4); Kirouac, S., and Bose, T. K., *J. Chem. Phys.*, 59, 3043, 1973 (N_2O), and 64, 1580, 1976 (He). Highly accurate dielectric constant measurements. These modern data give the most accurate polarizabilities available. A criticism of the interpretation of these data in the case of polar molecules is given in Ref. 20, p. 2905.
9. Huestis, D. L., Technical Report #MP 78-25, SRI International (project PYU 6158), Menlo Park, CA 94025. Molar refractions for mercury-chlorine compounds are analyzed.
10. Bounds, D. G., Clarke, J. H. R., and Hinchliffe, A., *Chem. Phys. Lett.*, 45, 367, 1977. Theoretical tensor polarizability for LiCl.
11. Kolker, H. J., and Karplus, M., *J. Chem. Phys.*, 39, 2011, 1963. Theoretical.
12. Cutschick, V. P., and McKoy, V., *J. Chem. Phys.*, 58, 2397, 1973. Theoretical tensor polarizabilities.
13. Gready, J. E., Bacskay, G. B., and Hush, N. S., *Chem. Phys.*, 22, 141, 1977, and 23, 9, 1977. Theoretical.
14. Amos, A. T., and Yoffe, J. A., *J. Chem. Phys.*, 63, 4723, 1975. Theoretical.
15. Stuart, H. A., *Landolt-Börnstein Zahlenwerte and Funktionen,* Vol. 1, Part 3, Eucken, A., and Hellwege, K. H., Eds., Springer-Verlag, Berlin, 1951, p. 511. Tabulation of molecular polarizabilities. Two misprints in the chemical symbols have been corrected.
16. Guella, T., Miller, T. M., Stockdale, J. A. D., Bederson, B., and Vuskovic, L., *J. Chem. Phys.*, 94, 6857, 1991. Beam measurements with accuracies between 12 and 24%.
17. Marchese, F. T., and Jaff, H. H., *Theoret. Chim. Acta* (Berlin), 45, 241, 1977. Theoretical and experimental tensor polarizabilities are tabulated in this paper.
18. Applequist, J., Carl, J. R., and Fung, K.-K., *J. Am. Chem. Soc.*, 94, 2952, 1972. Excellent reference on the calculation of molecular polarizabilities, including extensive tables of tensor polarizabilities, both theoretical and experimental, at 589.3 nm wavelength.
19. Bridge, N. J., and Buckingham, A. D., *Proc. Roy. Soc.* (London), A295, 334, 1966. Measured tensor polarizabilities at 633 nm wavelength.
20. Barnes, A. N. M., Turner, D. J., and Sutton, L. E., *Trans. Faraday Soc.*, 67, 2902, 1971. Dielectric constants yielding polarizabilities accurate from 0.3–8%.
21. Rayane, D., Allouche, A. R., Benichou, E., Antoine, R., Aubert-Frecon, M., Dugourd, Ph., Broyer, M., Ristori, C., Chandezon, F., Huber, B. A., and Guet, C., *Eur. Phys. J.* D 9, 243, 1999. See also Knight, W. D., Clemenger, K., de Heer, W. A., and Saunders, W. A., *Phys. Rev. B* 31, 2539, 1985. These data probably correspond to a very low internal temperature.

22. Tarnovsky, V., Bunimovicz, M., Vuskovic, L., Stumpf, B., and Bederson, B., *J. Chem. Phys.*, 98, 3894, 1993. These data correspond to internal temperatures 480-948 K.
23. Milani, P., Moullet, I., and de Heer, W. A., *Phys. Rev. A*, 42, 5150, 1990. Beam measurements accurate to 11%.
24. Antoine, R., Dugourd, P., Rayane, D., Benichou, E., Broyer, M., Chandezon, F., and Guet, C., *J. Chem. Phys.*, 110, 9771, 1999.
25. Aroney, M. J., and Pratten, S. J., *J. Chem. Soc., Faraday Trans.*, 1, 80, 1201, 1984. Uncertainties in the range 1–3%.
26. Le Fevre, R. J. W., Murthy, D. S. N., and Saxby, J. D., *Aust. J. Chem.*, 24, 1057, 1971. Kerr effect.
27. No, K. T., Cho, K. H., Jhon, M. S., and Scheraga, H. A., J. Am. Chem. Soc., 115, 2005, 1993. Theoretical; these results are quoted in numerous valuable papers on calculated polarizabilities, e.g., Miller, K. J., and Savchik, J. A., *J. Am. Chem. Soc.*, 101, 7206, 1979.
28. Schlecht, S., Schäfer, R., Woenckhaus, J., Becker, J. A., *Chem. Phys. Lett.*, 246, 315, 1995.
29. Benichou, E., Antoine, R., Rayane, D., Vezin, B., Dalby, F. W., Dugourd, P., Ristori, C., Chandezon, F., Huber, B. A., Rocco, J. C., Blundell, S. A., and Guet, C., *Phys. Rev. A* 59, R1, 1999. See also Rayane, D., Allouche, A. R., Benichou, E., Antoine, R., Aubert-Frecon, M., Dugourd, Ph., Broyer, M., Ristori, C., Chandezon, F., Huber, B. A., and Guet, C., *Eur. Phys. J.* D 9, 243, 1999.
30. Antoine, R., Rayane, D., Allouche, A. R., Aubert-Frécon, M., Benichou, E., Dalby, F. W., Dugourd, P., Broyer, M., and Guet, C., *J. Chem. Phys.*, 110, 5568, 1999.
31. Ballard, A., Bonin, K., and Louderback, J., *J. Chem. Phys.* 113, 5732, 2000.
32. Ritchie, G.L.D., and Watson, J.N., *J. Phys. Chem. A* 108, 4515, 2004. These measurements are at 632.8 nm frequency, and are stated accurate to 0.4%.
33. Ritchie, G.L.D., and Blanch, E.W., *J. Phys. Chem. A* 107, 2093, 2003. These measurements are at 632.8 nm frequncy, and are stated accurate to better than 1%.
34. May, E. F., Moldover, M. R., and Schmidt, J. W., *Phys. Rev. A* 78, 032522 (2008). The uncertainty is 0.0003 Å^3.

IONIZATION ENERGIES OF ATOMS AND ATOMIC IONS

The ionization energies (often called ionization potentials) of neutral and partially ionized atoms are listed in this table. Data were obtained from the compilations cited below, supplemented by results from the recent research literature. Values for the first and second ionization energies come from Reference 6. All values are given in electron volts (eV).

Following the traditional spectroscopic notation, columns are headed I, II, III, etc. up to XXX, where I indicates the neutral atom, II the singly ionized atom, III the doubly ionized atom, etc. The first section of the table includes spectra I to VIII of all the elements through rutherfordium; subsequent sections cover higher spectra (ionization stages) for those elements for which data are available.

References

1. Moore, C. E., *Ionization Potentials and Ionization Limits Derived from the Analysis of Optical Spectra,* Natl. Stand. Ref. Data Ser. — Natl. Bur. Stand. (U.S.) No. 34, 1970.
2. Martin, W. C., Zalubas, R., and Hagan, L., *Atomic Energy Levels — The Rare Earth Elements,* Natl. Stand. Ref. Data Ser. — Natl. Bur. Stand. (U.S.), No. 60, 1978.
3. Sugar, J. and Corliss, C., *Atomic Energy Levels of the Iron Period Elements: Potassium through Nickel, J. Phys. Chem. Ref. Data,* Vol.14, Suppl. 2, 1985.
4. References to papers in *J. Phys. Chem. Ref. Data,* in the period 1973–91 covering other elements may be found in the cumulative index to that journal.
5. Martin, W.C., and Wiese, W.L., in *Atomic, Molecular, and Optical Physics Handbook,* Drake, G.W.F., Ed., AIP Press, New York, 1996.
6. Sansonetti, J. E., Martin, W. C., and Young, S. L., *Handbook of Basic Atomic Spectroscopic Data* (version 1.1), NIST Physical Data web site <http://physics.nist.gov/PhysRefData/Handbook> (October 2004); *J. Phys. Chem. Ref. Data,* 34, 1559, 2005.

Neutral Atoms to +7 Ions

Z	Element	I	II	III	IV	V	VI	VII	VIII
1	H	13.598443							
2	He	24.587387	54.417760						
3	Li	5.391719	75.6400	122.45429					
4	Be	9.32270	18.21114	153.89661	217.71865				
5	B	8.29802	25.1548	37.93064	259.37521	340.22580			
6	C	11.26030	24.3833	47.8878	64.4939	392.087	489.99334		
7	N	14.5341	29.6013	47.44924	77.4735	97.8902	552.0718	667.046	
8	O	13.61805	35.1211	54.9355	77.41353	113.8990	138.1197	739.29	871.4101
9	F	17.4228	34.9708	62.7084	87.1398	114.2428	157.1651	185.186	953.9112
10	Ne	21.56454	40.96296	63.45	97.12	126.21	157.93	207.2759	239.0989
11	Na	5.139076	47.2864	71.6200	98.91	138.40	172.18	208.50	264.25
12	Mg	7.646235	15.03527	80.1437	109.2655	141.27	186.76	225.02	265.96
13	Al	5.985768	18.82855	28.44765	119.992	153.825	190.49	241.76	284.66
14	Si	8.15168	16.34584	33.49302	45.14181	166.767	205.27	246.5	303.54
15	P	10.48669	19.7695	30.2027	51.4439	65.0251	220.421	263.57	309.60
16	S	10.36001	23.33788	34.79	47.222	72.5945	88.0530	280.948	328.75
17	Cl	12.96763	23.8136	39.61	53.4652	67.8	97.03	114.1958	348.28
18	Ar	15.759610	27.62966	40.74	59.81	75.02	91.009	124.323	143.460
19	K	4.3406633	31.63	45.806	60.91	82.66	99.4	117.56	154.88
20	Ca	6.11316	11.87172	50.9131	67.27	84.50	108.78	127.2	147.24
21	Sc	6.56149	12.79977	24.75666	73.4894	91.65	110.68	138.0	158.1
22	Ti	6.82812	13.5755	27.4917	43.2672	99.30	119.53	140.8	170.4
23	V	6.74619	14.618	29.311	46.709	65.2817	128.13	150.6	173.4
24	Cr	6.76651	16.4857	30.96	49.16	69.46	90.6349	160.18	184.7
25	Mn	7.43402	15.6400	33.668	51.2	72.4	95.6	119.203	194.5
26	Fe	7.9024	16.1877	30.652	54.8	75.0	99.1	124.98	151.06
27	Co	7.88101	17.084	33.50	51.3	79.5	102.0	128.9	157.8
28	Ni	7.6398	18.16884	35.19	54.9	76.06	108	133	162
29	Cu	7.72638	20.2924	36.841	57.38	79.8	103	139	166
30	Zn	9.394199	17.96439	39.723	59.4	82.6	108	134	174
31	Ga	5.999301	20.51515	30.7258	63.241	86.01	112.7	140.9	169.9
32	Ge	7.89943	15.93461	34.2241	45.7131	93.5			
33	As	9.7886	18.5892	28.351	50.13	62.63	127.6		
34	Se	9.75239	21.19	30.8204	42.9450	68.3	81.7	155.4	
35	Br	11.8138	21.591	36	47.3	59.7	88.6	103.0	192.8
36	Kr	13.99961	24.35984	36.950	52.5	64.7	78.5	111.0	125.802
37	Rb	4.177128	27.2895	40	52.6	71.0	84.4	99.2	136
38	Sr	5.69485	11.0301	42.89	57	71.6	90.8	106	122.3
39	Y	6.2173	12.224	20.52	60.597	77.0	93.0	116	129
40	Zr	6.63390	13.1	22.99	34.34	80.348			
41	Nb	6.75885	14.0	25.04	38.3	50.55	102.057	125	

Neutral Atoms to +7 Ions

Z	Element	I	II	III	IV	V	VI	VII	VIII
42	Mo	7.09243	16.16	27.13	46.4	54.49	68.8276	125.664	143.6
43	Tc	7.28	15.26	29.54					
44	Ru	7.36050	16.76	28.47					
45	Rh	7.45890	18.08	31.06					
46	Pd	8.3369	19.43	32.93					
47	Ag	7.57623	21.47746	34.83					
48	Cd	8.99382	16.90831	37.48					
49	In	5.78636	18.8703	28.03	54				
50	Sn	7.34392	14.6322	30.50260	40.73502	72.28			
51	Sb	8.60839	16.63	25.3	44.2	56	108		
52	Te	9.0096	18.6	27.96	37.41	58.75	70.7	137	
53	I	10.45126	19.1313	33					
54	Xe	12.12984	20.9750	32.1230					
55	Cs	3.893905	23.15744						
56	Ba	5.211664	10.00383						
57	La	5.5769	11.059	19.1773	49.95	61.6			
58	Ce	5.5387	10.85	20.198	36.758	65.55	77.6		
59	Pr	5.473	10.55	21.624	38.98	57.53			
60	Nd	5.5250	10.72	22.1	40.4				
61	Pm	5.582	10.90	22.3	41.1				
62	Sm	5.6437	11.07	23.4	41.4				
63	Eu	5.67038	11.25	24.92	42.7				
64	Gd	6.14980	12.09	20.63	44.0				
65	Tb	5.8638	11.52	21.91	39.79				
66	Dy	5.9389	11.67	22.8	41.47				
67	Ho	6.0215	11.80	22.84	42.5				
68	Er	6.1077	11.93	22.74	42.7				
69	Tm	6.18431	12.05	23.68	42.7				
70	Yb	6.25416	12.176	25.05	43.56				
71	Lu	5.42586	13.9	20.9594	45.25	66.8			
72	Hf	6.82507	15	23.3	33.33				
73	Ta	7.54957							
74	W	7.86403	16.1						
75	Re	7.83352							
76	Os	8.43823							
77	Ir	8.96702							
78	Pt	8.9588	18.563						
79	Au	9.22553	20.20						
80	Hg	10.4375	18.7568	34.2					
81	Tl	6.108194	20.4283	29.83					
82	Pb	7.41663	15.03248	31.9373	42.32	68.8			
83	Bi	7.2855	16.703	25.56	45.3	56.0	88.3		
84	Po	8.414							
85	At								
86	Rn	10.7485							
87	Fr	4.072741							
88	Ra	5.278423	10.14715						
89	Ac	5.17	11.75						
90	Th	6.3067	11.9	20.0	28.8				
91	Pa	5.89							
92	U	6.1941	10.6						
93	Np	6.2657							
94	Pu	6.0260	11.2						
95	Am	5.9738							
96	Cm	5.9914							
97	Bk	6.1979							
98	Cf	6.2817	11.8						
99	Es	6.42	12.0						
100	Fm	6.50							
101	Md	6.58							
102	No	6.65							
103	Lr	4.9							
104	Rf	6.0							

+8 Ions to +15 Ions

Z	Element	IX	X	XI	XII	XIII	XIV	XV	XVI
9	F	1103.1176							
10	Ne	1195.8286	1362.1995						
11	Na	299.864	1465.121	1648.702					
12	Mg	328.06	367.50	1761.805	1962.6650				
13	Al	330.13	398.75	442.00	2085.98	2304.1410			
14	Si	351.12	401.37	476.36	523.42	2437.63	2673.182		
15	P	372.13	424.4	479.46	560.8	611.74	2816.91	3069.842	
16	S	379.55	447.5	504.8	564.44	652.2	707.01	3223.78	3494.1892
17	Cl	400.06	455.63	529.28	591.99	656.71	749.76	809.40	3658.521
18	Ar	422.45	478.69	538.96	618.26	686.10	755.74	854.77	918.03
19	K	175.8174	503.8	564.7	629.4	714.6	786.6	861.1	968
20	Ca	188.54	211.275	591.9	657.2	726.6	817.6	894.5	974
21	Sc	180.03	225.18	249.798	687.36	756.7	830.8	927.5	1009
22	Ti	192.1	215.92	265.07	291.500	787.84	863.1	941.9	1044
23	V	205.8	230.5	255.7	308.1	336.277	896.0	976	1060
24	Cr	209.3	244.4	270.8	298.0	354.8	384.168	1010.6	1097
25	Mn	221.8	248.3	286.0	314.4	343.6	403.0	435.163	1134.7
26	Fe	233.6	262.1	290.2	330.8	361.0	392.2	457	489.256
27	Co	186.13	275.4	305	336	379	411	444	511.96
28	Ni	193	224.6	321.0	352	384	430	464	499
29	Cu	199	232	265.3	369	401	435	484	520
30	Zn	203	238	274	310.8	419.7	454	490	542
31	Ga	210.8	244.0	280.7	319.2	357.2	471.2	508.8	548.3
36	Kr	230.85	268.2	308	350	391	447	492	541
37	Rb	150	277.1						
38	Sr	162	177	324.1					
39	Y	146.2	191	206	374.0				
42	Mo	164.12	186.4	209.3	230.28	279.1	302.60	544.0	570

+16 Ions to +23 Ions

Z	Element	XVII	XVIII	XIX	XX	XXI	XXII	XXIII	XXIV
17	Cl	3946.2960							
18	Ar	4120.8857	4426.2296						
19	K	1033.4	4610.8	4934.046					
20	Ca	1087	1157.8	5128.8	5469.864				
21	Sc	1094	1213	1287.97	5674.8	6033.712			
22	Ti	1131	1221	1346	1425.4	6249.0	6625.82		
23	V	1168	1260	1355	1486	1569.6	6851.3	7246.12	
24	Cr	1185	1299	1396	1496	1634	1721.4	7481.7	7894.81
25	Mn	1224	1317	1437	1539	1644	1788	1879.9	8140.6
26	Fe	1266	1358	1456	1582	1689	1799	1950	2023
27	Co	546.58	1397.2	1504.6	1603	1735	1846	1962	2119
28	Ni	571.08	607.06	1541	1648	1756	1894	2011	2131
29	Cu	557	633	670.588	1697	1804	1916	2060	2182
30	Zn	579	619	698	738	1856			
36	Kr	592	641	786	833	884	937	998	1051
42	Mo	636	702	767	833	902	968	1020	1082

+24 Ions to +29 Ions

Z	Element	XXV	XXVI	XXVII	XXVIII	XXIX	XXX
25	Mn	8571.94					
26	Fe	8828	9277.69				
27	Co	2219.0	9544.1	10012.12			
28	Ni	2295	2399.2	10288.8	10775.40		
29	Cu	2308	2478	2587.5	11062.38	11567.617	
36	Kr	1151	1205.3	2928	3070	3227	3381
42	Mo	1263	1323	1387	1449	1535	1601

IONIZATION ENERGIES OF GAS-PHASE MOLECULES

Sharon G. Lias

This table presents values for the first ionization energies (IP) of approximately 1000 molecules and atoms. Substances are listed by molecular formula in the modified Hill order (see Preface). Values enclosed in parentheses are considered not to be well established. Data appearing in the 1988 reference were updated in 1996 for inclusion in the database of ionization energies available at the Internet site of the Standard Reference Data program of the National Institute of Standards and Technology (http://webbook.nist.gov). The list appearing here includes these updates.

The list also includes values for enthalpies of formation of the ions at 298 K, $\Delta_f H_{ion}$, given according to the ion convention used by mass spectrometrists; to convert these values to the electron convention used by thermodynamicists, add 6 kJ/mol. Details on the calculation of $\Delta_f H_{ion}$, as well as data for a much larger number of molecules, may be found in the reference and on the Internet site.

Reference

Lias, S. G., Bartmess, J. E., Liebman, J. F., Holmes, J. L., Levin, R. D., and Mallard, W.G., *Gas-Phase Ion and Neutral Thermochemistry, J. Phys. Chem. Ref. Data*, Vol. 17, Suppl. No. 1, 1988.

Mol. form.	Name	IP/eV	$\Delta_f H_{ion}$ kJ/mol
Substances not containing carbon			
Ac	Actinium	5.17	905
Ag	Silver	7.57624	1016
AgCl	Silver(I) chloride	(≤ 10.08)	≤ 1065
AgF	Silver(I) fluoride	(11.0 ± 0.3)	1071
Al	Aluminum	5.98577	905
AlBr	Aluminum monobromide	(9.3)	913
$AlBr_3$	Aluminum tribromide	(10.4)	593
AlCl	Aluminum monochloride	9.4	855
$AlCl_3$	Aluminum trichloride	(12.01)	670
AlF	Aluminum monofluoride	9.73 ± 0.01	673
AlF_3	Aluminum trifluoride	≤ 15.45	≤ 282
AlI	Aluminum monoiodide	9.3 ± 0.3	965
AlI_3	Aluminum triiodide	(9.1)	673
Am	Americium	5.9738 ± 0.0002	860
Ar	Argon	15.75962	1521
As	Arsenic	9.8152	1250
$AsCl_3$	Arsenic(III) chloride	(10.55 ± 0.025)	754
AsF_3	Arsenic(III) fluoride	(12.84 ± 0.05)	452
AsH_3	Arsine	(9.89)	1021
Au	Gold	9.22567	1254
B	Boron	8.29803	1363
BBr_3	Boron tribromide	(10.51)	809
BCl_3	Boron trichloride	11.60 ± 0.02	718
BF	Fluoroborane	11.12 ± 0.01	957
	Difluoroborane	(9.4)	317
BF_3	Boron trifluoride	15.7 ± 0.3	365
BH	Boron monohydride	(9.77)	1385
BH_3	Borane	12.026 ± 0.024	1261
BI_3	Boron triiodide	(9.25 ± 0.03)	964
BO_2	Boron dioxide	(13.5 ± 0.3)	1001
B_2H_6	Diborane	11.38 ± 0.05	1134
B_2O_3	Boron oxide	13.5 ± 0.15	460
B_4H_{10}	Tetraborane	10.76 ± 0.04	1105
B_5H_9	Pentaborane(9)	9.90 ± 0.04	1028
B_6H_{10}	Hexaborane	(9.0)	965
Ba	Barium	5.21170	683
BaO	Barium oxide	6.91 ± 0.06	543
Be	Beryllium	9.32263	1224
BeO	Beryllium oxide	(10.1 ± 0.4)	1111
Bi	Bismuth	7.2855	908
$BiCl_3$	Bismuth trichloride	(10.4)	736

Mol. form.	Name	IP/eV	$\Delta_f H_{ion}$ kJ/mol
Bk	Berkelium	6.23	911
Br	Bromine (atomic)	11.81381	1252
BrCl	Bromine chloride	11.01	1079
BrF	Bromine fluoride	11.86	1086
BrF_5	Bromine pentafluoride	13.172 ± 0.002	840
BrH	Hydrogen bromide	11.66 ± 0.03	1087
BrH_3Si	Bromosilane	10.6	943
BrI	Iodine bromide	9.790 ± 0.004	986
BrK	Potassium bromide	7.85 ± 0.1	578
BrLi	Lithium bromide	(8.7)	685
BrNO	Nitrosyl bromide	10.17 ± 0.03	1065
BrNa	Sodium bromide	8.31 ± 0.1	660
BrO	Bromine monoxide	10.46 ± 0.02	1135
BrRb	Rubidium bromide	7.94 ± 0.03	583
BrTl	Thallium(I) bromide	9.14 ± 0.02	844
Br_2	Bromine	10.516 ± 0.005	1046
Br_2Hg	Mercury(II) bromide	10.560 ± 0.003	935
Br_2Sn	Tin(II) bromide	9.0	839
Br_3Ga	Gallium(III) bromide	10.40	711
Br_3P	Phosphorus(III) bromide	9.7	798
Br_4Hf	Hafnium(IV) bromide	(10.9)	366
Br_4Sn	Tin(IV) bromide	10.6	709
Br_4Ti	Titanium(IV) bromide	10.3	375
Br_4Zr	Zirconium(IV) bromide	(10.7)	388
Ca	Calcium	6.11316	768
CaCl	Calcium monochloride	5.86 ± 0.07	462
CaO	Calcium oxide	6.66 ± 0.18	668
Cd	Cadmium	8.99367	980
Ce	Cerium	5.5387	957
Cf	Californium	6.30	805
Cl	Chlorine (atomic)	12.96764	1373
ClCs	Cesium chloride	(7.84 ± 0.05)	510
ClF	Chlorine fluoride	12.66 ± 0.01	1171
$ClFO_3$	Perchloryl fluoride	(12.945 ± 0.005)	1224
ClF_2	Chlorine difluoride	(12.77 ± 0.05)	1128
ClF_3	Chlorine trifluoride	(12.65 ± 0.05)	1057
ClF_5S	Sulfur chloride pentafluoride	(12.335 ± 0.005)	144
ClH	Hydrogen chloride	12.749 ± 0.009	1137
ClHO	Hypochlorous acid	(11.12 ± 0.01)	993
ClH_3Si	Chlorosilane	11.4	899
ClI	Iodine chloride	10.088 ± 0.01	991
ClIn	Indium(I) chloride	(9.51)	842
ClK	Potassium chloride	(8.0 ± 0.4)	557
ClLi	Lithium chloride	9.57	727
ClNO	Nitrosyl chloride	10.87 ± 0.01	1099
$ClNO_2$	Nitryl chloride	(11.84)	1155
ClNa	Sodium chloride	8.92 ± 0.06	681
ClO	Chlorine monoxide	10.95	1159
ClO_2	Chlorine dioxide	10.33 ± 0.02	1093
ClRb	Rubidium chloride	(8.50 ± 0.03)	590
ClTl	Thallium(I) chloride	9.70 ± 0.03	869
Cl_2	Chlorine	11.480 ± 0.005	1108
Cl_2CrO_2	Chromyl chloride	11.6	580
Cl_2Ge	Germanium(II) chloride	(10.20 ± 0.05)	813
Cl_2H_2Si	Dichlorosilane	11.4	765
Cl_2Hg	Mercury(II) chloride	11.380 ± 0.003	952
Cl_2O	Chlorine oxide	10.94	1135
Cl_2OS	Thionyl chloride	10.96	844
Cl_2O_2S	Sulfuryl chloride	12.05	807
Cl_2Pb	Lead(II) chloride	(10.2)	791

Mol. form.	Name	IP/eV	$\Delta_f H_{ion}$ kJ/mol
Cl_2S	Sulfur dichloride	9.45 ± 0.03	895
Cl_2Si	Dichlorosilylene	(10.93 ± 0.10)	887
Cl_2Sn	Tin(II) chloride	(10.0)	760
Cl_3Ga	Gallium(III) chloride	11.52	664
Cl_3HSi	Trichlorosilane	(11.7)	648
Cl_3N	Nitrogen trichloride	(10.12 ± 0.1)	1244
Cl_3OP	Phosphorus(V) oxychloride	11.36 ± 0.02	540
Cl_3OV	Vanadyl trichloride	(11.6)	425
Cl_3P	Phosphorus(III) chloride	9.91	668
Cl_3PS	Phosphorus(V) sulfide trichloride	9.71 ± 0.03	573
Cl_3Sb	Antimony(III) chloride	(≤ 10.7)	s719
Cl_4Ge	Germanium(IV) chloride	11.68 ± 0.05	629
Cl_4Hf	Hafnium(IV) chloride	(11.7)	246
Cl_4Si	Tetrachlorosilane	11.79 ± 0.01	527
Cl_4Sn	Tin(IV) chloride	11.7 ± 0.2	656
Cl_4Ti	Titanium(IV) chloride	(11.5)	349
Cl_4V	Vanadium(IV) chloride	(9.2)	361
Cl_4Zr	Zirconium(IV) chloride	(11.2)	210
Cl_5Mo	Molybdenum(V) chloride	(8.7)	392
Cl_5Nb	Niobium(V) chloride	(10.97)	356
Cl_5P	Phosphorus(V) chloride	(10.2)	608
Cl_5Ta	Tantalum(V) chloride	(11.08)	303
Cl_6W	Tungsten(VI) chloride	(9.5)	348
Cm	Curium	6.02	966
Co	Cobalt	7.8810	1187
Cr	Chromium	6.76664	1050
Cs	Cesium	3.89390	452
CsF	Cesium fluoride	(8.80 ± 0.10)	489
CsNa	Cesium sodium	(4.05 ± 0.04)	535
Cu	Copper	7.72638	1084
CuF	Copper(I) fluoride	10.15 ± 0.02	984
Dy	Dysprosium	5.9389	862
Er	Erbium	6.1078	907
Es	Einsteinium	6.42	753
Eu	Europium	5.6704	723
F	Fluorine (atomic)	17.42282	1761
FGa	Gallium monofluoride	(9.6 ± 0.5)	700
FH	Hydrogen fluoride	16.044 ± 0.003	1276
FHO	Hypofluorous acid	12.71 ± 0.01	1130
FH_3Si	Fluorosilane	11.7	752
FI	Iodine fluoride	10.54 ± 0.01	922
FIn	Indium monofluoride	(9.6 ± 0.5)	740
FNO	Nitrosyl fluoride	12.63 ± 0.03	1152
FNO_2	Nitryl fluoride	(13.09)	1154
FNS	Thionitrosyl fluoride (NSF)	11.51 ± 0.04	1090
FO	Fluorine monoxide	12.78 ± 0.03	1342
FO_2	Fluorine superoxide (FOO)	(12.6 ± 0.2)	1228
FS	Sulfur fluoride	10.09	986
FTl	Thallium(I) fluoride	10.52	835
F_2	Fluorine	15.697 ± 0.003	1515
F_2Ge	Germanium(II) fluoride	(≤ 11.65)	551
F_2HN	Difluoramine	(11.53 ± 0.08)	1046
F_2H_2Si	Difluorosilane	(12.2)	386
F_2Mg	Magnesium fluoride	(13.6 ± 0.3)	588
F_2N	Difluoroamidogen	11.628 ± 0.01	1155
F_2N_2	*trans*-Difluorodiazine	(12.8)	1315
F_2O	Fluorine monoxide	13.11 ± 0.01	1290
F_2OS	Thionyl fluoride	12.25	688
F_2O_2S	Sulfuryl fluoride	13.04 ± 0.01	501
F_2Pb	Lead(II) fluoride	(11.5)	679

Mol. form.	Name	IP/eV	$\Delta_f H_{ion}$ kJ/mol
F_2S	Sulfur difluoride	(10.08)	676
F_2Si	Difluorosilylene	10.78 ± 0.05	450
F_2Sn	Tin(II) fluoride	(11.1)	586
F_2Xe	Xenon difluoride	12.35 ± 0.01	1083
F_3HSi	Trifluorosilane	(14.0)	150
F_3N	Nitrogen trifluoride	13.00 ± 0.02	1125
F_3NO	Trifluoramine oxide	13.31 ± 0.06	1121
F_3OP	Phosphorus(V) oxyfluoride	12.76 ± 0.01	−24
F_3P	Phosphorus(III) fluoride	11.60 ± 0.05	161
F_3PS	Phosphorus(V) sulfide trifluoride	≤ 11.05 ± 0.035	≤ 58
F_3Si	Trifluorosilyl	(9.99)	− 32
F_4Ge	Germanium(IV) fluoride	(15.5)	307
F_4N_2	Tetrafluorohydrazine	11.94 ± 0.03	1119
F_4S	Sulfur tetrafluoride	12.0 ± 0.3	399
F_4Si	Tetrafluorosilane	15.24 ± 0.14	−144
F_4Xe	Xenon tetrafluoride	12.65 ± 0.1	1016
F_5I	Iodine pentafluoride	12.943 ± 0.005	408
F_5P	Phosphorus(V) fluoride	(15.1)	−137
F_5S	Sulfur pentafluoride	9.60 ± 0.05	10
F_6Mo	Molybdenum(VI) fluoride	(14.5 ± 0.1)	−159
F_6S	Sulfur hexafluoride	15.32 ± 0.02	258
F_6U	Uranium(VI) fluoride	14.00 ± 0.10	−796
Fe	Iron	7.9024	1177
Fm	Fermium	6.50	627
Ga	Gallium	5.99930	851
GaI_3	Gallium(III) iodide	9.40	765
Gd	Gadolinium	6.1500	991
Ge	Germanium	7.900	1139
GeH_4	Germane	≤ 10.53	≤ 1108
GeI_4	Germanium(IV) iodide	(9.42)	850
GeO	Germanium(II) oxide	11.25 ± 0.01	1044
GeS	Germanium(II) sulfide	(9.98)	1055
H	Hydrogen (atomic)	13.59844	1530
HI	Hydrogen iodide	10.386 ± 0.001	1028
HLi	Lithium hydride	7.7	882
HN	Imidogen	≤ 13.49 ± 0.01	1678
HNO	Nitrosyl hydride	(10.1)	1075
HNO_2	Nitrous acid	≤ 11.3	≤ 1011
HNO_3	Nitric acid	11.95 ± 0.01	1019
HN_3	Hydrazoic acid	10.72 ± 0.025	1328
HO	Hydroxyl	13.0170 ± 0.0002	1294
HO_2	Hydroperoxy	11.35 ± 0.01	1106
HS	Mercapto	10.4219 ± 0.0004	1145
H_2	Hydrogen	15.42593 ± 0.00005	1488
H_2N	Amidogen	11.14 ± 0.01	1264
H_2O	Water	12.6206 ± 0.0020	976
H_2O_2	Hydrogen peroxide	10.58 ± 0.04	885
H_2S	Hydrogen sulfide	10.457 ± 0.012	989
H_2Se	Hydrogen selenide	9.892 ± 0.005	984
H_2Si	Silylene	8.244 ± 0.025	1084
H_3N	Ammonia	10.070 ± 0.020	925
H_3NO	Hydroxylamine	(10.00)	923
H_3P	Phosphine	9.869 ± 0.002	958
H_3Sb	Stibine	9.54 ± 0.03	1067
H_4N_2	Hydrazine	8.1 ± 0.15	877
H_4Si	Silane	11.00 ± 0.02	1095
H_4Sn	Stannane	(10.75)	1200
H_6Si_2	Disilane	9.74 ± 0.02	1019
H_8Si_3	Trisilane	(9.2)	1009
He	Helium	24.58741	2372

Mol. form.	Name	IP/eV	$\Delta_f H_{ion}$ kJ/mol
Hf	Hafnium	6.82507 ± 0.00004	1278
Hg	Mercury	10.43750	1069
HgI_2	Mercury(II) iodide	9.5088 ± 0.0022	900
Ho	Holmium	6.0216	882
I	Iodine (atomic)	10.45126	1115
IK	Potassium iodide	(7.21 ± 0.3)	570
ILi	Lithium iodide	(7.5)	633
INa	Sodium iodide	7.64 ± 0.02	659
ITl	Thallium(I) iodide	8.47 ± 0.02	826
I_2	Iodine	9.3074 ± 0.0002	960
I_4Ti	Titanium(IV) iodide	(9.1)	602
I_4Zr	Zirconium(IV) iodide	(9.3)	500
In	Indium	5.78636	802
Ir	Iridium	9.1	1543
K	Potassium	4.34066	508
KLi	Lithium potassium	4.57 ± 0.04	512
KNa	Potassium sodium	4.41636 ± 0.00017	561
K_2	Dipotassium	4.0637 ± 0.0002	519
Kr	Krypton	13.99961	1351
La	Lanthanum	5.5770	969
Li	Lithium	5.39172	680
LiNa	Lithium sodium	5.05 ± 0.04	571
LiO	Lithium monoxide	(8.44)	894
LiRb	Lithium rubidium	4.3 ± 0.1	486
Li_2	Dilithium	5.1127 ± 0.0003	709
Lu	Lutetium	5.42585	950
Md	Mendelevium	6.58	635
Mg	Magnesium	7.64624	885
MgO	Magnesium oxide	(8.76 ± 0.22)	901
Mn	Manganese	7.43402	998
Mo	Molybdenum	7.09243	1343
N	Nitrogen (atomic)	14.53414	1875
NO	Nitric oxide	9.26438 ± 0.00005	985
NO_2	Nitrogen dioxide	9.586 ± 0.002	958
NP	Phosphorus nitride	11.84 ± 0.04	1247
NS	Nitrogen sulfide	8.87 ± 0.01	1119
N_2	Nitrogen	15.5808	1503
N_2O	Nitrous oxide	12.886	1325
N_2O_4	Nitrogen tetroxide	(10.8)	1050
N_2O_5	Nitrogen pentoxide	(11.9)	1161
Na	Sodium	5.13908	603
NaRb	Rubidium sodium	4.32 ± 0.04	480
Na_2	Disodium	4.894 ± 0.003	614
Nb	Niobium	6.75885	1384
Nd	Neodymium	5.5250	859
Ne	Neon	21.56454	2081
Ni	Nickel	7.6398	1167
No	Nobelium	6.65	642
Np	Neptunium	6.2657 ± 0.0003	1069
O	Oxygen (atomic)	13.61806	1563
OPb	Lead(II) oxide	9.08 ± 0.10	939
OS	Sulfur monoxide	10.294 ± 0.004	998
OS_2	Sulfur oxide (SSO)	10.584 ± 0.005	971
OSi	Silicon monoxide	11.49 ± 0.20	1008
OSn	Tin(II) oxide	9.60 ± 0.02	944
OSr	Strontium oxide	6.6 ± 0.2	623
O_2	Oxygen	12.0697 ± 0.0002	1165
O_2S	Sulfur dioxide	12.349 ± 0.001	894
O_2Th	Thorium(IV) oxide	(8.7 ± 0.15)	342
O_2Ti	Titanium(IV) oxide	(9.54 ± 0.1)	623

Mol. form.	Name	IP/eV	$\Delta_f H_{ion}$ kJ/mol
O_2U	Uranium(IV) oxide	(5.4 ± 0.1)	57
O_3	Ozone	12.43	1342
O_3S	Sulfur trioxide	12.82 ± 0.03	841
O_3U	Uranium(VI) oxide	(10.5 ± 0.5)	214
O_4Os	Osmium(VIII) oxide	(12.32)	850
O_4Ru	Ruthenium(VIII) oxide	12.15 ± 0.03	988
O_7Re_2	Rhenium(VII) oxide	(12.7 ± 0.2)	125
Os	Osmium	8.7	1630
P	Phosphorus	10.48669	1328
P_2	Diphosphorus	10.53	1160
Pa	Protactinium	5.89	1133
Pb	Lead	7.41666	911
PbS	Lead(II) sulfide	(8.5 ± 0.5)	954
Pd	Palladium	8.3367	1181
Pm	Promethium	5.55	536
Pr	Praseodymium	5.464	883
Pt	Platinum	9.0	1433
Pu	Plutonium	6.025	926
Ra	Radium	5.27892	668
Rb	Rubidium	4.17713	484
Re	Rhenium	7.88	1530
Rh	Rhodium	7.45890	1276
Rn	Radon	10.74850	1037
Ru	Ruthenium	7.36050	1355
S	Sulfur	10.36001	1277
SSn	Tin(II) sulfide	(8.8)	966
S_2	Disulfur	9.356 ± 0.002	1031
Sb	Antimony	8.64	1096
Sc	Scandium	6.56144	1010
Se	Selenium	9.75238	1168
Si	Silicon	8.15169	1238
Sm	Samarium	5.6437	751
Sn	Tin	7.34381	1011
Sr	Strontium	5.69484	713
Ta	Tantalum	7.89	1544
Tb	Terbium	5.8639	955
Tc	Technetium	7.28	1380
Te	Tellurium	9.0096	1066
Th	Thorium	6.308 ± 0.003	1207
Ti	Titanium	6.8282	1127
Tl	Thallium	6.10829	771
Tm	Thulium	6.18431	827
U	Uranium	6.19405	1129
V	Vanadium	6.746 ± 0.002	1166
W	Tungsten	7.98	1621
Xe	Xenon	12.12987	1170
Y	Yttrium	6.217	1022
Yb	Ytterbium	6.25416	754
Zn	Zinc	9.39405	1037
Zr	Zirconium	6.63390	1251
Substances containing carbon			
C	Carbon	11.26030	1803
$CBrClF_2$	Bromochlorodifluoromethane	(11.21)	642
$CBrCl_3$	Bromotrichloromethane	(10.6)	980
$CBrF_3$	Bromotrifluoromethane	(11.40)	451
CBr_2F_2	Dibromodifluoromethane	11.03 ± 0.04	683
CBr_4	Tetrabromomethane	(10.31 ± 0.02)	1079
CCl	Chloromethylidyne	(8.9 ± 0.2)	1244
$CClF_3$	Chlorotrifluoromethane	12.6 ± 0.2	505

Mol. form.	Name	IP/eV	$\Delta_f H_{ion}$ kJ/mol
$CClN$	Cyanogen chloride	12.34 ± 0.01	1329
CCl_2	Dichloromethylene	(9.27)	1058
CCl_2F_2	Dichlorodifluoromethane	12.05 ± 0.24	685
CCl_2O	Carbonyl chloride	(11.5)	888
CCl_3F	Trichlorofluoromethane	11.77 ± 0.02	868
CCl_4	Tetrachloromethane	11.47 ± 0.01	1010
CF	Fluoromethylidyne	9.11 ± 0.01	1134
CFN	Cyanogen fluoride	13.34 ± 0.02	1325
CF_2	Difluoromethylene	11.44 ± 0.03	899
CF_2O	Carbonyl fluoride	13.035 ± 0.030	617
CF_3	Trifluoromethyl	8.7 ± 0.2	379
CF_3I	Trifluoroiodomethane	10.23	397
CH	Methylidyne	10.64 ± 0.01	1622
$CHBrCl_2$	Bromodichloromethane	10.6	973
$CHBr_2Cl$	Chlorodibromomethane	10.59 ± 0.01	1030
$CHBr_3$	Tribromomethane	10.48 ± 0.02	1035
$CHCl$	Chloromethylene	9.84	1247
$CHClF_2$	Chlorodifluoromethane	(12.2)	693
$CHCl_2F$	Dichlorofluoromethane	(11.5)	829
$CHCl_3$	Trichloromethane	11.37 ± 0.02	992
CHF	Fluoromethylene	10.06 ± 0.05	1121
CHF_3	Trifluoromethane	(13.86)	643
CHI_3	Triiodomethane	9.25 ± 0.02	1010
CHN	Hydrogen cyanide	13.60 ± 0.01	1447
CHN	Hydrogen isocyanide	(12.5 ± 0.1)	1407
$CHNO$	Isocyanic acid	11.595 ± 0.005	1016
$CHNO$	Fulminic acid	(10.83)	1263
CHO	Oxomethyl (HCO)	(8.55)	826
CH_2	Methylene	10.396 ± 0.003	1392
CH_2BrCl	Bromochloromethane	10.77 ± 0.01	1085
CH_2Br_2	Dibromomethane	(10.50 ± 0.02)	1013
CH_2ClF	Chlorofluoromethane	11.71 ± 0.01	870
CH_2Cl_2	Dichloromethane	11.32 ± .01	996
CH_2F_2	Difluoromethane	12.71	774
CH_2I_2	Diiodomethane	9.46 ± 0.02	1030
CH_2N_2	Diazomethane	8.999 ± 0.001	1098
CH_2N_2	Cyanamide	(10.4)	1137
CH_2O	Formaldehyde	10.88 ± 0.01	941
CH_2O_2	Formic acid	11.33 ± 0.01	715
CH_3	Methyl	9.843 ± 0.002	1095
CH_3BO	Borane carbonyl	11.14 ± 0.02	962
CH_3Br	Bromomethane	10.541 ± 0.003	979
CH_3Cl	Chloromethane	11.22 ± 0.01	1001
CH_3Cl_3Si	Methyltrichlorosilane	(11.36 ± 0.03)	548
CH_3F	Fluoromethane	12.47 ± 0.02	956
CH_3I	Iodomethane	9.538	936
CH_3NO	Formamide	10.16 ± 0.06	796
CH_3NO_2	Nitromethane	11.08 ± 0.07	994
CH_3N_3	Methyl azide	9.81 ± 0.02	1227
CH_3O	Methoxy	(10.72)	1050
CH_4	Methane	12.61 ± 0.01	1143
CH_4N_2O	Urea	9.7	690
CH_4O	Methanol	10.85 ± 0.01	845
CH_4S	Methanethiol	9.44 ± 0.005	888
CH_5N	Methylamine	(8.80)	826
CH_6N_2	Methylhydrazine	7.7 ± 0.15	835
CH_6Si	Methylsilane	(10.7)	1003
CN	Cyanide	13.5984	1748
CNO	Cyanate	11.76 ± 0.01	1290
CO	Carbon monoxide	14.014 ± 0.0003	1242

Mol. form.	Name	IP/eV	$\Delta_f H_{ion}$ kJ/mol
COS	Carbon oxysulfide	11.18 ± 0.01	936
$COSe$	Carbon oxyselenide	10.36 ± 0.01	929
CO_2	Carbon dioxide	13.773 ± 0.002	935
CS	Carbon sulfide	11.33 ± 0.01	1361
CS_2	Carbon disulfide	10.0685 ± 0.0020	1089
C_2	Dicarbon	(11.4 ± 0.3)	2000
$C_2Br_2F_4$	1,2-Dibromotetrafluoroethane	(11.1)	280
C_2ClF_3	Chlorotrifluoroethylene	9.81 ± 0.03	373
C_2ClF_5	Chloropentafluoroethane	(12.6)	99
C_2Cl_2	Dichloroacetylene	9.9	1165
$C_2Cl_2F_4$	1,2-Dichlorotetrafluoroethane	12.2	252
$C_2Cl_3F_3$	1,1,1-Trichlorotrifluoroethane	11.5	386
$C_2Cl_3F_3$	1,1,2-Trichlorotrifluoroethane	11.99 ± 0.02	429
C_2Cl_4	Tetrachloroethylene	9.326 ± 0.001	887
$C_2Cl_4F_2$	1,1,2,2-Tetrachloro-1,2-difluoroethane	(11.3)	563
C_2Cl_4O	Trichloroacetyl chloride	(11.0)	827
C_2Cl_6	Hexachloroethane	(11.1)	920
C_2F_3N	Trifluoroacetonitrile	13.93 ± 0.07	845
C_2F_4	Tetrafluoroethylene	10.12 ± 0.02	315
C_2F_6	Hexafluoroethane	(13.6)	-30
C_2H	Ethynyl	(11.61 ± 0.07)	1685
C_2HBr	Bromoacetylene	10.31 ± 0.02	1242
$C_2HBrClF_3$	2-Bromo-2-chloro-1,1,1-trifluoroethane	(11.0)	363
C_2HCl	Chloroacetylene	10.58 ± 0.02	1276
C_2HClF_2	1-Chloro-2,2-difluoroethylene	9.80 ± 0.04	628
C_2HCl_3	Trichloroethylene	9.46 ± 0.02	894
C_2HCl_3O	Dichloroacetyl chloride	(10.9)	809
C_2HCl_5	Pentachloroethane	(11.0)	919
C_2HF	Fluoroacetylene	11.26	1195
C_2HF_3	Trifluoroethylene	10.14	489
$C_2HF_3O_2$	Trifluoroacetic acid	11.46	75
C_2H_2	Acetylene	11.400 ± 0.002	1328
$C_2H_2Cl_2$	1,1-Dichloroethylene	9.81 ± 0.04	949
$C_2H_2Cl_2$	*cis*-1,2-Dichloroethylene	9.66 ± 0.01	936
$C_2H_2Cl_2$	*trans*-1,2-Dichloroethylene	9.64 ± 0.02	934
$C_2H_2Cl_2O$	Chloroacetyl chloride	(≤ 10.3)	815
$C_2H_2Cl_4$	1,1,1,2-Tetrachloroethane	(11.1)	920
$C_2H_2Cl_4$	1,1,2,2-Tetrachloroethane	(≤ 11.62)	≤ 971
$C_2H_2F_2$	1,1-Difluoroethylene	10.29 ± 0.01	650
$C_2H_2F_2$	*cis*-1,2-Difluoroethylene	10.23 ± 0.02	690
C_2H_2O	Ketene	9.617 ± 0.003	880
$C_2H_2O_2$	Glyoxal	10.2	773
$C_2H_2S_2$	Thiirene	8.61	892
C_2H_3Br	Bromoethylene	9.83 ± 0.02	1028
C_2H_3Cl	Chloroethylene	9.99 ± 0.02	985
$C_2H_3ClF_2$	1-Chloro-1,1-difluoroethane	(11.98)	626
C_2H_3ClO	Acetyl chloride	10.82 ± 0.04	801
C_2H_3ClO	Chloroacetaldehyde	(10.48)	815
$C_2H_3ClO_2$	Chloroacetic acid	(10.7)	597
$C_2H_3Cl_3$	1,1,1-Trichloroethane	(11.0)	917
$C_2H_3Cl_3$	1,1,2-Trichloroethane	(11.0)	911
C_2H_3F	Fluoroethylene	10.36 ± 0.01	861
C_2H_3FO	Acetyl fluoride	(11.5)	667
$C_2H_3F_3$	1,1,1-Trifluoroethane	13.3 ± 0.5	536
C_2H_3N	Acetonitrile	12.20 ± 0.01	1253
C_2H_3NO	Methylisocyanate	(10.67)	900
C_2H_4	Ethylene	10.5138 ± 0.0006	1067
$C_2H_4Br_2$	1,2-Dibromoethane	10.35 ± 0.04	961
$C_2H_4Cl_2$	1,1-Dichloroethane	11.04 ± 0.02	935
$C_2H_4Cl_2$	1,2-Dichloroethane	11.04 ± 0.02	931

Mol. form.	Name	IP/eV	$\Delta_f H_{ion}$ kJ/mol
$C_2H_4F_2$	1,1-Difluoroethane	(11.87)	643
C_2H_4O	Acetaldehyde	10.229 ± 0.0007	821
C_2H_4O	Ethylene oxide	10.56 ± 0.01	966
$C_2H_4O_2$	Acetic acid	10.65 ± 0.02	595
$C_2H_4O_2$	Methyl formate	10.835 ± 0.005	690
C_2H_5Br	Bromoethane	10.29 ± 0.01	931
C_2H_5Cl	Chloroethane	10.98 ± 0.02	947
C_2H_5ClO	2-Chloroethanol	(10.5)	756
C_2H_5F	Fluoroethane	(11.78)	873
C_2H_5I	Iodoethane	9.3492 ± 0.0006	893
C_2H_5N	Ethyleneimine	(9.5 ± 0.3)	1044
C_2H_5NO	Acetamide	9.65 ± 0.03	693
C_2H_5NO	*N*-Methylformamide	9.83 ± 0.04	760
$C_2H_5NO_2$	Nitroethane	10.88 ± 0.05	948
C_2H_6	Ethane	11.56 ± 0.02	1031
$C_2H_6Cl_2Si$	Dichlorodimethylsilane	(10.7)	576
C_2H_6O	Ethanol	10.43 ± 0.05	772
C_2H_6O	Dimethyl ether	10.025 ± 0.025	783
C_2H_6OS	Dimethyl sulfoxide	9.10 ± 0.03	727
$C_2H_6O_2$	Ethylene glycol	10.16	593
C_2H_6S	Ethanethiol	9.31 ± 0.03	851
C_2H_6S	Dimethyl sulfide	8.69 ± 0.02	801
$C_2H_6S_2$	Dimethyl disulfide	(7.4 ± 0.3)	690
C_2H_7N	Ethylamine	8.86 ± 0.02	808
C_2H_7N	Dimethylamine	8.24 ± 0.08	777
C_2H_7NO	Ethanolamine	8.96	664
$C_2H_8N_2$	1,2-Ethanediamine	(8.6)	812
$C_2H_8N_2$	1,1-Dimethylhydrazine	7.29 ± 0.05	787
C_2N_2	Cyanogen	13.37 ± 0.01	1597
C_3F_6	Perfluoropropene	10.60 ± 0.03	−103
C_3F_6O	Perfluoroacetone	(11.57 ± 0.13)	−282
C_3F_8	Perfluoropropane	(13.38)	−491
C_3HN	Cyanoacetylene	11.64 ± 0.01	1475
C_3H_2O	2-Propynal	(10.7 ± 0.1)	1145
$C_3H_3F_3$	3,3,3-Trifluoropropene	(10.9)	437
C_3H_3N	2-Propenenitrile	10.91 ± 0.01	1237
C_3H_3NO	Oxazole	(9.9)	940
C_3H_3NO	Isoxazole	(9.93)	1038
C_3H_4	Allene	9.692 ± 0.004	1126
C_3H_4	Propyne	10.37 ± 0.01	1187
C_3H_4	Cyclopropene	9.67 ± 0.01	1209
$C_3H_4N_2$	Imidazole	(8.81)	997
C_3H_4O	Propargyl alcohol	10.49 ± 0.02	1060
C_3H_4O	Acrolein	10.103 ± 0.006	900
C_3H_4O	Cyclopropanone	(9.1 ± 0.1)	895
$C_3H_4O_2$	Propenoic acid	10.60	701
$C_3H_4O_2$	2-Oxetanone	(9.70 ± 0.01)	653
C_3H_5Br	3-Bromopropene	(9.96)	1008
C_3H_5Cl	3-Chloropropene	10.04 ± 0.01	965
C_3H_5ClO	Epichlorohydrin	(10.64)	919
$C_3H_5ClO_2$	Methyl chloroacetate	(10.3)	575
C_3H_5F	3-Fluoropropene	(10.11)	821
C_3H_5N	Propanenitrile	11.84 ± 0.02	1194
C_3H_5NO	Acrylamide	(9.5)	720
C_3H_6	Propene	9.73 ± 0.02	959
C_3H_6	Cyclopropane	9.86	1005
$C_3H_6Br_2$	1,2-Dibromopropane	(10.1)	903
$C_3H_6Br_2$	1,3-Dibromopropane	(≤ 10.2)	≤ 919
$C_3H_6Cl_2$	1,2-Dichloropropane	10.8 ± 0.1	886
$C_3H_6Cl_2$	1,3-Dichloropropane	10.89 ± 0.04	892

Mol. form.	Name	IP/eV	$\Delta_f H_{ion}$ kJ/mol
C_3H_6O	Allyl alcohol	9.67 ± 0.05	808
C_3H_6O	Methyl vinyl ether	8.95 ± 0.01	763
C_3H_6O	Propanal	9.96 ± 0.01	772
C_3H_6O	Acetone	9.703 ± 0.006	719
C_3H_6O	Methyloxirane	(10.22)	892
C_3H_6O	Oxetane	9.65 ± 0.01	851
$C_3H_6O_2$	Propanoic acid	10.525 ± 0.003	568
$C_3H_6O_2$	Ethyl formate	10.61 ± 0.01	639
$C_3H_6O_2$	Methyl acetate	10.25 ± 0.02	579
$C_3H_6O_2$	1,3-Dioxolane	(9.9)	658
$C_3H_6O_3$	1,3,5-Trioxane	(10.3)	528
C_3H_7Br	1-Bromopropane	10.18 ± 0.01	898
C_3H_7Br	2-Bromopropane	10.10 ± 0.03	877
C_3H_7Cl	1-Chloropropane	10.81 ± 0.01	911
C_3H_7Cl	2-Chloropropane	10.79 ± 0.02	896
C_3H_7F	1-Fluoropropane	(11.3)	806
C_3H_7F	2-Fluoropropane	(11.08)	776
C_3H_7I	1-Iodopropane	9.25 ± 0.01	860
C_3H_7I	2-Iodopropane	9.19 ± 0.02	845
C_3H_7N	Allylamine	(8.76)	891
C_3H_7N	Cyclopropylamine	(8.8)	926
C_3H_7N	Propyleneimine	(9.0)	960
C_3H_7NO	*N,N*-Dimethylformamide	(9.12)	688
$C_3H_7NO_2$	1-Nitropropane	(10.81)	919
$C_3H_7NO_2$	2-Nitropropane	(10.71)	894
C_3H_8	Propane	10.95 ± 0.05	952
C_3H_8O	1-Propanol	10.18 ± 0.06	727
C_3H_8O	2-Propanol	10.17 ± 0.02	709
C_3H_8O	Ethyl methyl ether	9.72 ± 0.07	722
$C_3H_8O_2$	Dimethoxymethane	9.7	588
C_3H_8S	1-Propanethiol	9.20 ± 0.01	819
C_3H_8S	2-Propanethiol	9.145 ± 0.005	806
C_3H_8S	Ethyl methyl sulfide	(8.55)	765
$C_3H_9BO_3$	Trimethyl borate	(10.0)	65
C_3H_9ClSi	Trimethylchlorosilane	(10.15)	624
C_3H_9N	Propylamine	(8.78)	777
C_3H_9N	Isopropylamine	(8.72)	758
C_3H_9N	Trimethylamine	7.82 ± 0.06	731
C_3H_9NO	3-Amino-1-propanol	(9.0)	651
$C_4H_2O_3$	Maleic anhydride	(10.8)	645
C_4H_4	1-Buten-3-yne	9.58 ± 0.02	1230
$C_4H_4N_2$	Succinonitrile	(12.1 ± 0.25)	1377
$C_4H_4N_2$	Pyrimidine	9.23	1087
$C_4H_4N_2$	Pyridazine	8.67 ± 0.03	1112
C_4H_4O	Furan	8.883 ± 0.003	822
$C_4H_4O_2$	Diketene	(9.6 ± 0.02)	736
$C_4H_4O_3$	Succinic anhydride	(10.6)	500
$C_4H_4O_4$	Fumaric acid	(10.7)	355
C_4H_4S	Thiophene	8.86 ± 0.02	970
C_4H_5N	Methylacrylonitrile	10.34	1127
C_4H_5N	Pyrrole	8.207 ± 0.005	900
C_4H_5N	Cyclopropanecarbonitrile	(10.25)	1173
C_4H_6	1,2-Butadiene	(9.03)	1034
C_4H_6	1,3-Butadiene	9.082 ± 0.004	986
C_4H_6	1-Butyne	10.19 ± 0.02	1148
C_4H_6	2-Butyne	9.59 ± 0.03	1071
C_4H_6	Cyclobutene	9.43 ± 0.02	1067
C_4H_6O	Divinyl ether	(8.7)	827
C_4H_6O	*trans*-2-Butenal	9.73 ± 0.01	835
C_4H_6O	2-Methylpropenal	(9.92)	834

Mol. form.	Name	IP/eV	$\Delta_f H_{ion}$ kJ/mol
C_4H_6O	Cyclobutanone	(9.35)	815
$C_4H_6O_2$	*cis*-Crotonic acid	(10.08)	625
$C_4H_6O_2$	*trans*-Crotonic acid	(9.9)	604
$C_4H_6O_2$	Methacrylic acid	(10.15)	611
$C_4H_6O_2$	Vinyl acetate	9.19 ± 0.05	572
$C_4H_6O_2$	Methyl acrylate	(9.9)	641
$C_4H_6O_3$	Acetic anhydride	(10.0)	398
$C_4H_6O_4$	Dimethyl oxalate	(10.0)	287
C_4H_6S	2,5-Dihydrothiophene	(8.4)	898
C_4H_7N	Butanenitrile	(11.2)	1110
C_4H_7N	2-Methylpropanenitrile	(11.3)	1115
C_4H_7NO	2-Pyrrolidone	(9.2)	674
C_4H_8	1-Butene	9.55 ± 0.06	921
C_4H_8	*cis*-2-Butene	9.11 ± 0.01	871
C_4H_8	*trans*-2-Butene	9.10 ± 0.01	866
C_4H_8	Isobutene	9.239 ± 0.003	875
C_4H_8	Cyclobutane	(9.82 ± 0.05)	976
C_4H_8	Methylcyclopropane	(9.46)	936
$C_4H_8Br_2$	1,4-Dibromobutane	(10.15)	879
C_4H_8O	Ethyl vinyl ether	(8.98)	709
C_4H_8O	1,2-Epoxybutane	(≤ 10.15)	862
C_4H_8O	Butanal	9.84 ± 0.02	742
C_4H_8O	Isobutanal	9.71 ± 0.01	721
C_4H_8O	2-Butanone	9.52 ± 0.04	678
C_4H_8O	Tetrahydrofuran	9.38 ± 0.05	721
$C_4H_8O_2$	Butanoic acid	10.17 ± 0.05	509
$C_4H_8O_2$	2-Methylpropanoic acid	10.33 ± 0.03	516
$C_4H_8O_2$	Propyl formate	10.52 ± 0.02	555
$C_4H_8O_2$	Ethyl acetate	10.01 ± 0.05	522
$C_4H_8O_2$	Methyl propanoate	10.15 ± 0.03	548
$C_4H_8O_2$	1,3-Dioxane	9.8	607
$C_4H_8O_2$	1,4-Dioxane	9.19 ± 0.01	571
$C_4H_8O_2S$	Sulfolane	(9.8)	577
C_4H_8S	Tetrahydrothiophene	8.38	774
C_4H_9Br	1-Bromobutane	(10.12)	869
C_4H_9Br	2-Bromobutane	10.01 ± 0.02	845
C_4H_9Br	1-Bromo-2-methylpropane	10.09 ± 0.02	861
C_4H_9Br	2-Bromo-2-methylpropane	9.92 ± 0.03	823
C_4H_9Cl	1-Chlorobutane	10.67 ± 0.03	875
C_4H_9Cl	2-Chlorobutane	10.53	857
C_4H_9Cl	1-Chloro-2-methylpropane	10.73 ± 0.07	877
C_4H_9Cl	2-Chloro-2-methylpropane	(10.61)	842
C_4H_9I	1-Iodobutane	9.23 ± 0.01	840
C_4H_9I	2-Iodobutane	9.10 ± 0.02	815
C_4H_9I	1-Iodo-2-methylpropane	9.19 ± 0.01	824
C_4H_9I	2-Iodo-2-methylpropane	(9.02)	798
C_4H_9N	Pyrrolidine	(8.0)	769
C_4H_9NO	*N,N*-Dimethylacetamide	8.81 ± 0.03	616
C_4H_9NO	Morpholine	(8.2)	841
C_4H_{10}	Butane	10.53 ± 0.10	890
C_4H_{10}	Isobutane	(10.57)	886
$C_4H_{10}O$	1-Butanol	9.99 ± 0.05	689
$C_4H_{10}O$	2-Butanol	9.88 ± 0.03	658
$C_4H_{10}O$	2-Methyl-1-propanol	10.02 ± 0.04	683
$C_4H_{10}O$	2-Methyl-2-propanol	9.90 ± 0.02	642
$C_4H_{10}O$	Diethyl ether	9.51 ± 0.03	666
$C_4H_{10}O$	Methyl propyl ether	9.41 ± 0.07	670
$C_4H_{10}O$	Isopropyl methyl ether	9.45 ± 0.04	661
$C_4H_{10}O_2$	Ethylene glycol monoethyl ether	(9.6)	529
$C_4H_{10}O_2$	Ethylene glycol dimethyl ether	(9.3)	558

Mol. form.	Name	IP/eV	$\Delta_f H_{ion}$ kJ/mol
$C_4H_{10}S$	1-Butanethiol	9.14 ± 0.01	794
$C_4H_{10}S$	2-Butanethiol	(9.10)	781
$C_4H_{10}S$	2-Methyl-1-propanethiol	(9.12)	783
$C_4H_{10}S$	2-Methyl-2-propanethiol	(9.03)	762
$C_4H_{10}S$	Diethyl sulfide	(8.43)	730
$C_4H_{10}S$	Methyl propyl sulfide	(8.8)	767
$C_4H_{10}S$	Isopropyl methyl sulfide	(8.7)	749
$C_4H_{10}S_2$	Diethyl disulfide	(8.27)	724
$C_4H_{11}N$	Butylamine	8.7 ± 0.1	748
$C_4H_{11}N$	*sec*-Butylamine	8.46 ± 0.1	711
$C_4H_{11}N$	*tert*-Butylamine	8.46 ± 0.1	695
$C_4H_{11}N$	Isobutylamine	8.50 ± 0.1	721
$C_4H_{11}N$	Diethylamine	7.85 ± 0.1	684
$C_4H_{12}Si$	Tetramethylsilane	9.80 ± 0.04	713
$C_4H_{12}Sn$	Tetramethylstannane	8.89 ± 0.05	837
C_4NiO_4	Nickel carbonyl	8.27 ± 0.04	200
$C_5H_4O_2$	Furfural	9.22 ± 0.01	739
C_5H_5N	Pyridine	9.25	1031
C_5H_6	1-Penten-3-yne	9.00 ± 0.01	1119
C_5H_6	*cis*-3-Penten-1-yne	9.14 ± 0.04	1137
C_5H_6	*trans*-3-Penten-1-yne	9.05 ± 0.01	1128
C_5H_6	2-Methyl-1-buten-3-yne	9.25 ± 0.02	1152
C_5H_6	1,3-Cyclopentadiene	8.55 ± 0.02	955
C_5H_6O	2-Methylfuran	8.38 ± 0.02	729
C_5H_6O	3-Methylfuran	(8.64)	763
C_5H_6S	2-Methylthiophene	(8.14)	867
C_5H_6S	3-Methylthiophene	(8.40)	893
C_5H_8	*cis*-1,3-Pentadiene	8.63 ± 0.03	914
C_5H_8	*trans*-1,3-Pentadiene	8.59 ± 0.02	905
C_5H_8	1,4-Pentadiene	9.60 ± 0.02	1032
C_5H_8	2-Methyl-1,3-butadiene	8.84 ± 0.01	928
C_5H_8	1-Pentyne	10.10 ± 0.01	1119
C_5H_8	Cyclopentene	9.01 ± 0.01	905
C_5H_8	Spiropentane	(9.26)	1078
C_5H_8O	Cyclopropyl methyl ketone	(≤ 9.46)	796
C_5H_8O	Cyclopentanone	9.26 ± 0.01	701
C_5H_8O	3,4-Dihydro-2H-pyran	8.35 ± 0.01	681
$C_5H_8O_2$	Ethyl acrylate	(≤ 10.3)	617
$C_5H_8O_2$	Methyl methacrylate	(9.7)	589
$C_5H_8O_2$	2,4-Pentanedione	8.85 ± 0.01	469
C_5H_9NO	*N*-Methyl-2-pyrrolidone	(≤ 9.17)	≤ 676
C_5H_{10}	1-Pentene	9.51 ± 0.01	896
C_5H_{10}	*cis*-2-Pentene	9.01 ± 0.03	843
C_5H_{10}	*trans*-2-Pentene	9.04 ± 0.01	841
C_5H_{10}	2-Methyl-1-butene	9.12 ± 0.01	844
C_5H_{10}	3-Methyl-1-butene	9.52 ± 0.01	891
C_5H_{10}	2-Methyl-2-butene	8.69 ± 0.01	796
C_5H_{10}	Cyclopentane	(10.33 ± 0.15)	918
$C_5H_{10}O$	2,2-Dimethylpropanal	9.51 ± 0.01	675
$C_5H_{10}O$	Cyclopentanol	(9.72)	695
$C_5H_{10}O$	Pentanal	9.74 ± 0.04	709
$C_5H_{10}O$	2-Pentanone	9.38 ± 0.01	646
$C_5H_{10}O$	3-Pentanone	9.31 ± 0.01	640
$C_5H_{10}O$	3-Methyl-2-butanone	9.30 ± 0.01	635
$C_5H_{10}O$	Tetrahydropyran	9.25 ± 0.01	670
$C_5H_{10}O_2$	Pentanoic acid	(≤ 10.53)	≤ 527
$C_5H_{10}O_2$	3-Methylbutanoic acid	(≤ 10.51)	≤ 499
$C_5H_{10}O_2$	Butyl formate	10.52 ± 0.02	584
$C_5H_{10}O_2$	Propyl acetate	(≤ 9.92)	501
$C_5H_{10}O_2$	Isopropyl acetate	9.99 ± 0.03	482

Mol. form.	Name	IP/eV	$\Delta_f H_{ion}$ kJ/mol
$C_5H_{10}O_2$	Ethyl propanoate	(10.00)	500
$C_5H_{10}O_2$	Methyl butanoate	(10.07)	520
$C_5H_{10}S$	Thiacyclohexane	(8.2)	728
$C_5H_{11}Br$	1-Bromopentane	10.10 ± 0.01	846
$C_5H_{11}I$	1-Iodopentane	9.20 ± 0.01	817
$C_5H_{11}N$	Piperidine	8.03 ± 0.11	726
$C_5H_{11}N$	*N*-Methylpyrrolidine	≤ 8.41 ± 0.02	≤ 809
C_5H_{12}	Pentane	10.28 ± 0.10	845
C_5H_{12}	Isopentane	10.32 ± 0.05	843
C_5H_{12}	Neopentane	(≤ 10.2)	≤ 818
$C_5H_{12}O$	1-Pentanol	(10.00)	668
$C_5H_{12}O$	2-Pentanol	(9.78)	630
$C_5H_{12}O$	3-Pentanol	9.78	628
$C_5H_{12}O$	2-Methyl-1-butanol	(9.86)	649
$C_5H_{12}O$	2-Methyl-2-butanol	(9.8)	615
$C_5H_{12}O$	3-Methyl-2-butanol	(9.88 ± 0.13)	637
$C_5H_{12}O$	Butyl methyl ether	(9.4 ± 0.1)	648
$C_5H_{12}O$	Methyl *tert*-butyl ether	(9.24)	608
$C_5H_{12}O$	Ethyl propyl ether	(9.45)	640
$C_5H_{12}S$	*tert*-Butyl methyl sulfide	(8.38)	687
$C_5H_{12}S$	Ethyl propyl sulfide	(8.50)	716
$C_5H_{12}S$	Ethyl isopropyl sulfide	(8.35)	689
C_6BrF_5	Bromopentafluorobenzene	(9.67)	222
C_6ClF_5	Chloropentafluorobenzene	(9.72)	126
C_6Cl_6	Hexachlorobenzene	(8.98)	822
C_6F_6	Hexafluorobenzene	9.89 ± 0.04	8
C_6F_{12}	Perfluorocyclohexane	(13.2)	−1095
C_6HF_5	Pentafluorobenzene	(9.63)	122
C_6HF_5O	Pentafluorophenol	(9.20)	−71
$C_6H_2F_4$	1,2,3,4-Tetrafluorobenzene	(9.53)	284
$C_6H_2F_4$	1,2,3,5-Tetrafluorobenzene	(9.53)	263
$C_6H_2F_4$	1,2,4,5-Tetrafluorobenzene	(9.35)	254
$C_6H_3Cl_3$	1,2,4-Trichlorobenzene	(9.04)	880
$C_6H_3Cl_3$	1,3,5-Trichlorobenzene	9.32 ± 0.02	899
$C_6H_4ClNO_2$	1-Chloro-3-nitrobenzene	(9.92 ± 0.1)	995
$C_6H_4ClNO_2$	1-Chloro-4-nitrobenzene	(9.96 ± 0.1)	999
$C_6H_4Cl_2$	*o*-Dichlorobenzene	9.06 ± 0.02	907
$C_6H_4Cl_2$	*m*-Dichlorobenzene	9.10 ± 0.02	906
$C_6H_4Cl_2$	*p*-Dichlorobenzene	8.92 ± 0.02	885
$C_6H_4FNO_2$	1-Fluoro-4-nitrobenzene	(9.90)	826
$C_6H_4F_2$	*o*-Difluorobenzene	9.29 ± 0.01	602
$C_6H_4F_2$	*m*-Difluorobenzene	9.33 ± 0.01	591
$C_6H_4F_2$	*p*-Difluorobenzene	9.1589 ± 0.0003	577
$C_6H_4O_2$	*p*-Benzoquinone	10.01 ± 0.06	844
C_6H_5Br	Bromobenzene	9.00 ± 0.02	971
C_6H_5Cl	Chlorobenzene	9.07 ± 0.02	930
C_6H_5ClO	*m*-Chlorophenol	8.655 ± 0.001	680
C_6H_5ClO	*p*-Chlorophenol	(≤ 8.69)	≤ 692
C_6H_5F	Fluorobenzene	9.20 ± 0.01	772
C_6H_5I	Iodobenzene	8.685	1003
$C_6H_5NO_2$	Nitrobenzene	9.86 ± 0.02	1019
$C_6H_5NO_3$	*o*-Nitrophenol	(9.1)	782
$C_6H_5NO_3$	*m*-Nitrophenol	(9.0)	755
$C_6H_5NO_3$	*p*-Nitrophenol	(9.1)	761
C_6H_6	Benzene	9.24378 ± 0.00007	975
C_6H_6	Fulvene	(8.36)	1031
C_6H_6ClN	*o*-Chloroaniline	(8.50)	883
C_6H_6ClN	*m*-Chloroaniline	(8.09)	835
C_6H_6ClN	*p*-Chloroaniline	(≤ 8.18)	≤ 844
$C_6H_6N_2O_2$	*o*-Nitroaniline	(8.27)	861

Mol. form.	Name	IP/eV	$\Delta_f H_{ion}$ kJ/mol
$C_6H_6N_2O_2$	*m*-Nitroaniline	(8.31)	865
$C_6H_6N_2O_2$	*p*-Nitroaniline	(8.34)	859
C_6H_6O	Phenol	8.49 ± 0.02	723
$C_6H_6O_2$	*p*-Hydroquinone	7.94 ± 0.01	503
C_6H_6S	Benzenethiol	(8.32)	915
C_6H_7N	Aniline	7.720 ± 0.002	832
C_6H_7N	2-Methylpyridine	(9.02)	970
C_6H_7N	3-Methylpyridine	(9.04)	979
C_6H_7N	4-Methylpyridine	(9.04)	976
$C_6H_8N_2$	*o*-Phenylenediamine	(7.2)	787
$C_6H_8N_2$	*m*-Phenylenediamine	(7.14)	777
$C_6H_8N_2$	*p*-Phenylenediamine	(6.87 ± 0.05)	759
C_6H_{10}	1,5-Hexadiene	9.27 ± 0.05	978
C_6H_{10}	1-Hexyne	10.03 ± 0.05	1089
C_6H_{10}	3,3-Dimethyl-1-butyne	9.90 ± 0.04	1060
C_6H_{10}	Cyclohexene	8.945 ± 0.01	859
$C_6H_{10}O$	Cyclohexanone	9.14 ± 0.01	656
$C_6H_{10}O$	Mesityl oxide	9.10 ± 0.01	694
$C_6H_{10}O_4$	Diethyl oxalate	(9.8)	205
$C_6H_{11}NO$	Caprolactam	(9.07 ± 0.02)	629
C_6H_{12}	1-Hexene	9.44 ± 0.04	869
C_6H_{12}	*cis*-2-Hexene	(8.97 ± 0.01)	818
C_6H_{12}	*trans*-2-Hexene	(8.97 ± 0.01)	814
C_6H_{12}	2-Methyl-1-pentene	(9.08 ± 0.01)	817
C_6H_{12}	4-Methyl-1-pentene	9.45 ± 0.01	862
C_6H_{12}	2-Methyl-2-pentene	(8.58)	761
C_6H_{12}	4-Methyl-*cis*-2-pentene	8.98 ± 0.01	809
C_6H_{12}	4-Methyl-*trans*-2-pentene	(8.97 ± 0.01)	804
C_6H_{12}	2-Ethyl-1-butene	(9.06 ± 0.02)	818
C_6H_{12}	2,3-Dimethyl-1-butene	(9.07 ± 0.01)	812
C_6H_{12}	2,3-Dimethyl-2-butene	8.27 ± 0.01	729
C_6H_{12}	Cyclohexane	9.86 ± 0.03	828
C_6H_{12}	Methylcyclopentane	(9.85)	845
$C_6H_{12}O$	Hexanal	9.72 ± 0.05	691
$C_6H_{12}O$	2-Hexanone	9.3 ± 0.1	626
$C_6H_{12}O$	3-Hexanone	9.12 ± 0.02	600
$C_6H_{12}O$	3-Methyl-2-pentanone	9.21 ± 0.01	600
$C_6H_{12}O$	4-Methyl-2-pentanone	9.30 ± 0.01	609
$C_6H_{12}O$	2-Methyl-3-pentanone	9.10 ± 0.01	592
$C_6H_{12}O$	3,3-Dimethyl-2-butanone	9.12 ± 0.02	589
$C_6H_{12}O$	Cyclohexanol	(9.75)	651
$C_6H_{12}O_2$	Hexanoic acid	≤ 10.12	≤ 463
$C_6H_{12}O_2$	Butyl acetate	(9.92 ± .05)	471
$C_6H_{12}O_2$	*sec*-Butyl acetate	9.90	453
$C_6H_{12}O_2$	Methyl 2,2-dimethylpropanoate	(9.90 ± 0.04)	466
$C_6H_{13}I$	1-Iodohexane	9.179	794
$C_6H_{13}N$	Cyclohexylamine	(8.86)	750
C_6H_{14}	Hexane	10.13	810
C_6H_{14}	2-Methylpentane	(10.12)	802
C_6H_{14}	3-Methylpentane	(10.08)	801
C_6H_{14}	2,2-Dimethylbutane	(10.06)	787
C_6H_{14}	2,3-Dimethylbutane	(10.02)	791
$C_6H_{14}O$	1-Hexanol	(9.89)	639
$C_6H_{14}O$	2-Hexanol	(9.80 ± 0.03)	611
$C_6H_{14}O$	3-Hexanol	(9.63 ± 0.03)	599
$C_6H_{14}O$	Dipropyl ether	(9.27)	602
$C_6H_{14}O$	Diisopropyl ether	9.20 ± 0.05	569
$C_6H_{14}O$	Butyl ethyl ether	(9.36)	610
$C_6H_{14}O$	Methyl pentyl ether	(≤ 9.67)	≤ 657
$C_6H_{14}O_2$	1,1-Diethoxyethane	(9.2)	434

Mol. form.	Name	IP/eV	$\Delta_f H_{ion}$ kJ/mol
$C_6H_{14}O_3$	Diethylene glycol dimethyl ether	≤ 9.8	≤ 448
$C_6H_{14}S$	Dipropyl sulfide	8.30 ± 0.02	676
$C_6H_{14}S$	Diisopropyl sulfide	(8.2 ± 0.2)	649
$C_6H_{15}N$	Hexylamine	(8.63 ± 0.05)	699
$C_6H_{15}N$	Dipropylamine	(7.84 ± 0.02)	641
$C_6H_{15}N$	Diisopropylamine	(7.73 ± 0.03)	602
$C_6H_{15}N$	Triethylamine	(7.50 ± 0.02)	631
$C_6H_{15}NO_3$	Triethanolamine	(7.9)	206
$C_7H_3F_5$	2,3,4,5,6-Pentafluorotoluene	(9.4)	64
C_7H_5ClO	Benzoyl chloride	(9.53)	815
$C_7H_5Cl_3$	(Trichloromethyl)benzene	(≤ 9.60)	≤ 914
$C_7H_5F_3$	(Trifluoromethyl)benzene	9.685 ± 0.005	335
C_7H_5N	Benzonitrile	9.70 ± 0.01	1154
C_7H_6O	Benzaldehyde	9.49 ± 0.02	878
$C_7H_6O_2$	Benzoic acid	(9.3)	604
C_7H_7Br	*p*-Bromotoluene	8.67 ± 0.02	908
C_7H_7Cl	*o*-Chlorotoluene	(8.7 ± 0.1)	856
C_7H_7Cl	*m*-Chlorotoluene	(8.83)	869
C_7H_7Cl	*p*-Chlorotoluene	(8.69)	855
C_7H_7Cl	(Chloromethyl)benzene	9.10 ± 0.02	897
C_7H_7F	*o*-Fluorotoluene	8.91 ± 0.01	709
C_7H_7F	*m*-Fluorotoluene	8.91 ± 0.01	709
C_7H_7F	*p*-Fluorotoluene	8.79 ± 0.01	701
C_7H_7NO	Benzamide	(9.25)	792
$C_7H_7NO_2$	*o*-Nitrotoluene	9.24	946
$C_7H_7NO_2$	*m*-Nitrotoluene	9.45 ± 0.1	941
$C_7H_7NO_2$	*p*-Nitrotoluene	9.46 ± 0.05	942
C_7H_8	Toluene	8.8276 ± 0.0006	901
C_7H_8O	*o*-Cresol	(8.24)	670
C_7H_8O	*m*-Cresol	8.29 ± 0.07	668
C_7H_8O	*p*-Cresol	(8.3)	675
C_7H_8O	Benzyl alcohol	(8.3)	701
C_7H_8O	Anisole	8.22 ± 0.03	725
C_7H_9N	Benzylamine	(8.64)	917
C_7H_9N	*o*-Methylaniline	(7.44 ± 0.02)	772
C_7H_9N	*m*-Methylaniline	(7.50 ± 0.02)	778
C_7H_9N	*p*-Methylaniline	(7.24 ± 0.02)	753
C_7H_9N	*N*-Methylaniline	7.34 ± 0.04	792
C_7H_9N	2,3-Dimethylpyridine	(8.85 ± 0.02)	922
C_7H_9N	2,4-Dimethylpyridine	(8.85 ± 0.03)	918
C_7H_9N	2,5-Dimethylpyridine	(≤ 8.80 ± 0.05)	≤ 916
C_7H_9N	2,6-Dimethylpyridine	8.86 ± 0.03	913
C_7H_9N	3,4-Dimethylpyridine	(≤ 9.15)	≤ 953
C_7H_9N	3,5-Dimethylpyridine	(≤ 9.25)	≤ 965
$C_7H_{10}O$	Dicyclopropyl ketone	(9.1)	1041
C_7H_{14}	1-Heptene	9.34 ± 0.10	839
C_7H_{14}	*trans*-3-Heptene	(8.92)	790
C_7H_{14}	Cycloheptane	9.97	844
C_7H_{14}	Methylcyclohexane	9.64	775
C_7H_{14}	*cis*-1,2-Dimethylcyclopentane	(9.92 ± 0.05)	828
C_7H_{14}	*trans*-1,2-Dimethylcyclopentane	9.7 ± 0.2	799
$C_7H_{14}O$	1-Heptanal	(9.65)	668
$C_7H_{14}O$	2-Heptanone	9.28 ± 0.10	594
$C_7H_{14}O$	3-Heptanone	9.18 ± 0.08	589
$C_7H_{14}O$	4-Heptanone	9.10 ± 0.06	577
$C_7H_{14}O$	5-Methyl-2-hexanone	(9.28)	586
$C_7H_{14}O$	2,4-Dimethyl-3-pentanone	8.95 ± 0.01	552
$C_7H_{14}O$	1-Methylcyclohexanol	(9.8 ± 0.2)	586
C_7H_{16}	Heptane	9.93 ± 0.10	771
$C_7H_{16}O$	1-Heptanol	(9.84)	614

Mol. form.	Name	IP/eV	$\Delta_f H_{ion}$ kJ/mol
$C_7H_{16}O$	2-Heptanol	(9.70)	580
$C_7H_{16}O$	3-Heptanol	(9.68)	578
$C_7H_{16}O$	4-Heptanol	(9.61)	572
$C_7H_{16}O$	Ethyl pentyl ether	(≤ 9.49)	≤ 602
$C_8H_4O_3$	Phthalic anhydride	(10.1)	603
$C_8H_6O_4$	Isophthalic acid	(9.98)	268
$C_8H_6O_4$	Terephthalic acid	(9.86)	232
C_8H_7N	2-Methylbenzonitrile	(≤ 9.38)	1085
C_8H_7N	3-Methylbenzonitrile	(≤ 9.34)	1085
C_8H_7N	4-Methylbenzonitrile	9.32 ± 0.02	1083
C_8H_7N	Indole	7.7602 ± 0.0006	908
C_8H_8	Styrene	8.464 ± 0.001	964
C_8H_8O	*p*-Tolualdehyde	(9.33)	825
C_8H_8O	Acetophenone	9.29 ± 0.03	810
$C_8H_8O_2$	*o*-Toluic acid	(9.1)	558
$C_8H_8O_2$	*m*-Toluic acid	(9.43)	579
$C_8H_8O_2$	*p*-Toluic acid	(9.23)	560
$C_8H_8O_2$	Benzeneacetic acid	(8.26)	479
$C_8H_8O_2$	Methyl benzoate	9.32 ± 0.03	611
C_8H_{10}	Ethylbenzene	8.77 ± 0.01	876
C_8H_{10}	*o*-Xylene	8.56 ± 0.01	844
C_8H_{10}	*m*-Xylene	8.56 ± 0.01	843
C_8H_{10}	*p*-Xylene	8.44 ± 0.01	832
$C_8H_{10}O$	*p*-Ethylphenol	(7.84)	613
$C_8H_{10}O$	2,3-Xylenol	(8.26)	640
$C_8H_{10}O$	2,4-Xylenol	(8.0)	609
$C_8H_{10}O$	2,6-Xylenol	(8.05)	615
$C_8H_{10}O$	3,4-Xylenol	(8.09)	624
$C_8H_{10}O$	Phenetole	(8.13)	683
$C_8H_{11}N$	2,4,6-Trimethylpyridine	(≤ 8.9)	≤ 880
$C_8H_{11}N$	*N*-Ethylaniline	(≤ 7.67)	≤ 794
$C_8H_{11}N$	*N,N*-Dimethylaniline	7.12 ± 0.02	787
C_8H_{14}	1-Octyne	(9.95 ± 0.02)	1040
C_8H_{14}	2-Octyne	9.31 ± 0.01	961
C_8H_{14}	3-Octyne	9.22 ± 0.01	952
C_8H_{14}	4-Octyne	9.20 ± 0.01	946
C_8H_{16}	1-Octene	9.43 ± 0.01	829
C_8H_{16}	Cyclooctane	9.75 ± 0.05	816
C_8H_{16}	Ethylcyclohexane	(9.54)	748
C_8H_{16}	1,1-Dimethylcyclohexane	(9.42)	728
C_8H_{16}	*cis*-1,2-Dimethylcyclohexane	(<9.78)	772
C_8H_{16}	*trans*-1,2-Dimethylcyclohexane	9.41	728
C_8H_{16}	*cis*-1,3-Dimethylcyclohexane	(<9.98)	778
C_8H_{16}	*trans*-1,3-Dimethylcyclohexane	9.53	743
C_8H_{16}	*cis*-1,4-Dimethylcyclohexane	(<9.93)	782
C_8H_{16}	*trans*-1,4-Dimethylcyclohexane	(9.56)	738
C_8H_{16}	Propylcyclopentane	(9.34)	753
$C_8H_{16}O$	2,2,4-Trimethyl-3-pentanone	(8.80)	511
C_8H_{18}	Octane	9.80 ± 0.10	737
C_8H_{18}	2-Methylheptane	(9.84)	734
C_8H_{18}	2,2,4-Trimethylpentane	(9.86)	713
C_8H_{18}	2,2,3,3-Tetramethylbutane	9.8	720
$C_8H_{18}O$	Dibutyl ether	(9.28)	s 560
$C_8H_{18}O$	Di-*sec*-butyl ether	(9.11)	511
$C_8H_{18}O$	Di-*tert*-butyl ether	8.88 ± 0.07	493
$C_8H_{18}S$	Dibutyl sulfide	(8.2)	624
$C_8H_{18}S$	Di-*tert*-butyl sulfide	(8.0)	583
$C_8H_{18}S$	Diisobutyl sulfide	(8.34)	625
$C_8H_{19}N$	Dibutylamine	(7.69)	586
$C_8H_{19}N$	Diisobutylamine	(7.8)	574

Mol. form.	Name	IP/eV	$\Delta_f H_{ion}$ kJ/mol
$C_8H_{20}Si$	Tetraethylsilane	(8.9)	595
C_9H_7N	Quinoline	8.62 ± 0.01	1041
C_9H_7N	Isoquinoline	8.53 ± 0.03	1032
C_9H_8	Indene	8.14 ± 0.01	949
C_9H_{10}	*o*-Methylstyrene	(8.20)	908
C_9H_{10}	*m*-Methylstyrene	(8.15)	899
C_9H_{10}	*p*-Methylstyrene	(8.1)	895
C_9H_{10}	Cyclopropylbenzene	(8.35)	956
C_9H_{10}	Indan	(8.3)	864
$C_9H_{10}O_2$	Ethyl benzoate	(8.9)	537
C_9H_{12}	Propylbenzene	8.713 ± 0.010	848
C_9H_{12}	Isopropylbenzene	8.73 ± 0.01	847
C_9H_{12}	1,2,3-Trimethylbenzene	8.42 ± 0.02	803
C_9H_{12}	1,2,4-Trimethylbenzene	8.27 ± 0.01	784
C_9H_{12}	1,3,5-Trimethylbenzene	8.41 ± 0.01	796
$C_9H_{13}N$	*N,N*-Dimethyl-*o*-toluidine	7.40 ± 0.02	814
$C_9H_{14}O$	Isophorone	(≤ 9.07)	≤ 670
C_9H_{18}	Butylcyclopentane	(9.95)	793
C_9H_{18}	Propylcyclohexane	(9.46)	720
C_9H_{18}	Isopropylcyclohexane	(9.33)	704
$C_9H_{18}O$	2-Nonanone	(9.16)	545
$C_9H_{18}O$	5-Nonanone	(9.07)	530
$C_9H_{18}O$	2,6-Dimethyl-4-heptanone	9.01 ± 0.06	512
C_9H_{20}	Nonane	9.71 ± 0.10	709
$C_{10}F_8$	Perfluoronaphthalene	8.85	−368
$C_{10}H_7Br$	1-Bromonaphthalene	8.08 ± 0.03	955
$C_{10}H_7Cl$	1-Chloronaphthalene	(8.13)	906
$C_{10}H_8$	Naphthalene	8.1442 ± 0.0009	936
$C_{10}H_8$	Azulene	7.38 ± 0.05	1001
$C_{10}H_8O$	1-Naphthol	7.76 ± 0.03	719
$C_{10}H_8O$	2-Naphthol	7.87 ± 0.06	729
$C_{10}H_{10}O_4$	Dimethyl phthalate	(9.64 ± 0.07)	277
$C_{10}H_{12}$	1,2,3,4-Tetrahydronaphthalene	8.46 ± 0.02	841
$C_{10}H_{14}$	Butylbenzene	8.69 ± 0.02	826
$C_{10}H_{14}$	*sec*-Butylbenzene	8.68 ± 0.02	820
$C_{10}H_{14}$	*tert*-Butylbenzene	8.68 ± 0.05	816
$C_{10}H_{14}$	Isobutylbenzene	8.69 ± 0.02	817
$C_{10}H_{14}$	*p*-Cymene	(8.29)	771
$C_{10}H_{14}$	*o*-Diethylbenzene	(≤ 8.51)	≤ 804
$C_{10}H_{14}$	*m*-Diethylbenzene	(8.49)	798
$C_{10}H_{14}$	*p*-Diethylbenzene	(8.40)	790
$C_{10}H_{14}$	1,2,4,5-Tetramethylbenzene	8.04 ± 0.02	730
$C_{10}H_{14}O$	*p*-tert-Butylphenol	(7.8)	552
$C_{10}H_{16}$	α-Pinene	(8.07)	808
$C_{10}H_{16}O$	Camphor	(8.76)	577
$C_{10}H_{18}$	*cis*-Decahydronaphthalene	9.36 ± 0.04	734
$C_{10}H_{18}$	*trans*-Decahydronaphthalene	9.34 ± 0.04	720
$C_{10}H_{20}$	1-Decene	9.42 ± 0.05	786
$C_{10}H_{20}$	Butylcyclohexane	(9.41)	695
$C_{10}H_{22}$	Decane	(9.65)	682
$C_{11}H_{10}$	1-Methylnaphthalene	7.97 ± 0.03	882
$C_{11}H_{10}$	2-Methylnaphthalene	7.91 ± 0.08	877
$C_{11}H_{16}$	*p-tert*-Butyltoluene	(8.12)	730
$C_{11}H_{24}$	Undecane	(9.56)	650
$C_{11}H_{24}$	2-Methyldecane	(9.7)	658
$C_{12}H_8$	Acenaphthylene	(8.22)	1053
$C_{12}H_9N$	Carbazole	(7.57)	961
$C_{12}H_{10}$	Acenaphthene	7.75 ± 0.07	903
$C_{12}H_{10}$	Biphenyl	8.23 ± 0.10	977
$C_{12}H_{10}N_2O$	*trans*-Azoxybenzene	(8.1)	1123

Mol. form.	Name	IP/eV	$\Delta_f H_{ion}$ kJ/mol
$C_{12}H_{10}O$	Diphenyl ether	(8.09)	766
$C_{12}H_{11}N$	Diphenylamine	7.16 ± 0.04	908
$C_{12}H_{18}$	5,7-Dodecadiyne	(8.67)	1079
$C_{12}H_{18}$	Hexamethylbenzene	7.85 ± 0.01	670
$C_{12}H_{22}$	Cyclohexylcyclohexane	(9.41)	690
$C_{12}H_{27}N$	Tributylamine	(7.4)	492
$C_{13}H_{10}$	9*H*-Fluorene	7.91 ± 0.02	952
$C_{13}H_{10}O$	Benzophenone	9.08 ± 0.05	926
$C_{13}H_{12}$	Diphenylmethane	(8.55)	963
$C_{14}H_{10}$	Anthracene	7.439 ± 0.006	948
$C_{14}H_{10}$	Phenanthrene	7.8914 ± 0.0006	966
$C_{14}H_{10}$	Diphenylacetylene	7.94 ± 0.03	1168
$C_{14}H_{12}$	*cis*-Stilbene	(7.80)	1005
$C_{14}H_{12}$	*trans*-Stilbene	7.656 ± 0.001	973
$C_{14}H_{14}$	1,2-Diphenylethane	8.9 ± 0.1	1002
$C_{16}H_{10}$	Fluoranthene	7.9 ± 0.1	1052
$C_{16}H_{10}$	Pyrene	7.4256 ± 0.0006	935
$C_{18}H_{12}$	Chrysene	7.60 ± 0.01	1017
$C_{18}H_{14}$	*o*-Terphenyl	(7.99)	1056
$C_{18}H_{14}$	*m*-Terphenyl	(8.01)	1057
$C_{18}H_{14}$	*p*-Terphenyl	7.80 ± 0.03	1037
$C_{20}H_{12}$	Perylene	6.960 ± 0.001	981
$C_{24}H_{12}$	Coronene	7.29 ± 0.01	1026

X-RAY ATOMIC ENERGY LEVELS

The energy levels in this table are the values recommended by Bearden and Burr on the basis of a thorough review of the literature on x-ray wavelengths and related data. All values are in electron volts (eV). Values in parentheses are interpolated, and an asterisk * indicates a level that is not resolved from the level above it. See Reference 1 for uncertainties in the levels and a complete description of how the recommended values were obtained.

References

1. Bearden, J. A., and Burr, A. F., *Rev. Mod. Phys.*, 39, 125, 1967; also published as *X-Ray Wavelengths and X-Ray Atomic Energy* Levels, Natl. Stand. Ref. Data Sys. — Natl. Bur. Standards (U.S.), No. 14, 1967.
2. Gray, D. E., Ed., *American Institute of Physics Handbook, Third Edition*, pp. 7-158 to 7-167, McGraw-Hill, New York, 1972.

Level	^{1}H	^{2}He	^{3}Li	^{4}Be	^{5}B	^{6}C	^{7}N	^{8}O
K	13.59811	24.58678	54.75	111.0	188.0	283.8	401.6	532.0
L_{I}								23.7
$L_{II,III}$					4.7	6.4	9.2	7.1

Level	^{9}F	^{10}Ne	^{11}Na	^{12}Mg	^{13}Al	^{14}Si	^{15}P	^{16}S
K	685.4	866.9	1072.1	1305.0	1559.6	1838.9	2145.5	2472.0
L_{I}	(31)	(45)	63.3	89.4	117.7	148.7	189.3	229.2
$L_{II,III}$	8.6	18.3	31.1	51.4	73.1	99.2	132.2	164.8

Level	^{17}Cl	^{18}Ar	^{19}K	^{20}Ca	^{21}Sc	^{22}Ti	^{23}V	^{24}Cr
K	2822.4	3202.9	3607.4	4038.1	4492.8	4966.4	5465.1	5989.2
L_{I}	270.2	320	377.1	437.8	500.4	563.7	628.2	694.6
L_{II}	201.6	247.3	296.3	350.0	406.7	461.5	520.5	583.7
L_{III}	200.0	245.2	293.6	346.4	402.2	455.5	512.9	574.5
M_{I}	17.5	25.3	33.9	43.7	53.8	60.3	66.5	74.1
$M_{II,III}$	6.8	12.4	17.8	25.4	32.3	34.6	37.8	42.5
$M_{IV,V}$					6.6	3.7	2.2	2.3

Level	^{25}Mn	^{26}Fe	^{27}Co	^{28}Ni	^{29}Cu	^{30}Zn	^{31}Ga	^{32}Ge
K	6539.0	7112.0	7708.9	8332.8	8978.9	9658.6	10367.1	11103.1
L_{I}	769.0	846.1	925.6	1008.1	1096.1	1193.6	1297.7	1414.3
L_{II}	651.4	721.1	793.6	871.9	951.0	1042.8	1142.3	1247.8
L_{III}	640.3	708.1	778.6	854.7	931.1	1019.7	1115.4	1216.7
M_{I}	83.9	92.9	100.7	111.8	119.8	135.9	158.1	180.0
M_{II}	48.6	54.0	59.5	68.1	73.6	86.6	106.8	127.9
M_{III}	48.6*	54.0*	59.5*	68.1*	73.6*	86.6*	102.9	120.8
$M_{IV,V}$	3.3	3.6	2.9	3.6	1.6	8.1	17.4	28.7

Level	^{33}As	^{34}Se	^{35}Br	^{36}Kr	^{37}Rb	^{38}Sr	^{39}Y	^{40}Zr
K	11866.7	12657.8	13473.7	14325.6	15199.7	16104.6	17038.4	17997.6
L_{I}	1526.5	1653.9	1782.0	1921.0	2065.1	2216.3	2372.5	2531.6
L_{II}	1358.6	1476.2	1596.0	1727.2	1863.9	2006.8	2155.5	2306.7
L_{III}	1323.1	1435.8	1549.9	1674.9	1804.4	1939.6	2080.0	2222.3
M_{I}	203.5	231.5	256.5		322.1	357.5	393.6	430.3
M_{II}	146.4	168.2	189.3	222.7	247.4	279.8	312.4	344.2
M_{III}	140.5	161.9	181.5	213.8	238.5	269.1	300.3	330.5
M_{IV}	41.2	56.7	70.1	88.9	111.8	135.0	159.6	182.4
M_{V}	41.2*	56.7*	69.0	88.9*	110.3	133.1	157.4	180.0
N_{I}			27.3	24.0	29.3	37.7	45.4	51.3
N_{II}	2.5	5.6	5.2	10.6	14.8	19.9	25.6	28.7
N_{III}	2.5*	5.6*	4.6	10.6*	14.0	19.9*	25.6*	28.7*

Level	^{41}Nb	^{42}Mo	^{43}Tc	^{44}Ru	^{45}Rh	^{46}Pd	^{47}Ag	^{48}Cd
K	18985.6	19999.5	21044.0	22117.2	23219.9	24350.3	25514.0	26711.2
L_{I}	2697.7	2865.5	3042.5	3224.0	3411.9	3604.3	3805.8	4018.0
L_{II}	2464.7	2625.1	2793.2	2966.9	3146.1	3330.3	3523.7	3727.0
L_{III}	2370.5	2520.2	2676.9	2837.9	3003.8	3173.3	3351.1	3537.5
M_{I}	468.4	504.6		585.0	627.1	669.9	717.5	770.2
M_{II}	378.4	409.7	444.9	482.8	521.0	559.1	602.4	650.7
M_{III}	363.0	392.3	425.0	460.6	496.2	531.5	571.4	616.5
M_{IV}	207.4	230.3	256.4	283.6	311.7	340.0	372.8	410.5

Level	^{41}Nb	^{42}Mo	^{43}Tc	^{44}Ru	^{45}Rh	^{46}Pd	^{47}Ag	^{48}Cd
M_V	204.6	227.0	252.9	279.4	307.0	334.7	366.7	403.7
N_I	58.1	61.8		74.9	81.0	86.4	95.2	107.6
N_{II}	33.9	34.8	38.9	43.1	47.9	51.1	62.6	66.9
N_{III}	33.9*	34.8*	38.9*	43.1*	47.9*	51.1*	55.9	66.9*
$N_{IV,V}$	3.2	1.8		2.0	2.5	1.5	3.3	9.3

Level	^{49}In	^{50}Sn	^{51}Sb	^{52}Te	^{53}I	^{54}Xe	^{55}Cs	^{56}Ba
K	27939.9	29200.1	30491.2	31813.8	33169.4	34561.4	35984.6	37440.6
L_I	4237.5	4464.7	4698.3	4939.2	5188.1	5452.8	5714.3	5988.8
L_{II}	3938.0	4156.1	4380.4	4612.0	4852.1	5103.7	5359.4	5623.6
L_{III}	3730.1	3928.8	4132.2	4341.4	4557.1	4782.2	5011.9	5247.0
M_I	825.6	883.8	943.7	1006.0	1072.1		1217.1	1292.8
M_{II}	702.2	756.4	811.9	869.7	930.5	999.0	1065.0	1136.7
M_{III}	664.3	714.4	765.6	818.7	874.6	937.0	997.6	1062.2
M_{IV}	450.8	493.3	536.9	582.5	631.3		739.5	796.1
M_V	443.1	484.8	527.5	572.1	619.4	672.3	725.5	780.7
N_I	121.9	136.5	152.0	168.3	186.4		230.8	253.0
N_{II}	77.4	88.6	98.4	110.2	122.7	146.7	172.3	191.8
N_{III}	77.4*	88.6*	98.4*	110.2*	122.7*	146.7*	161.6	179.7
N_{IV}	16.2	23.9	31.4	39.8	49.6		78.8	92.5
N_V	16.2*	23.9*	31.4*	39.8*	49.6*		76.5	89.9
O_I	0.1	0.9	6.7	11.6	13.6		22.7	39.1
O_{II}	0.8	1.1	2.1	2.3	3.3		13.1	16.6
O_{III}	0.8*	1.1*	2.1*	2.3*	3.3*		11.4	14.6

Level	^{57}La	^{58}Ce	^{59}Pr	^{60}Nd	^{61}Pm	^{62}Sm	^{63}Eu	^{64}Gd
K	38924.6	40443.0	41990.6	43568.9	45184.0	46834.2	48519.0	50239.1
L_I	6266.3	6548.8	6834.8	7126.0	7427.9	7736.8	8052.0	8375.6
L_{II}	5890.6	6164.2	6440.4	6721.5	7012.8	7311.8	7617.1	7930.3
L_{III}	5482.7	5723.4	5964.3	6207.9	6459.3	6716.2	6976.9	7242.8
M_I	1361.3	1434.6	1511.0	1575.3		1722.8	1800.0	1880.8
M_{II}	1204.4	1272.8	1337.4	1402.8	1471.4	1540.7	1613.9	1688.3
M_{III}	1123.4	1185.4	1242.2	1297.4	1356.9	1419.8	1480.6	1544.0
M_{IV}	848.5	901.3	951.1	999.9	1051.5	1106.0	1160.6	1217.2
M_V	831.7	883.3	931.0	977.7	1026.9	1080.2	1130.9	1185.2
N_I	270.4	289.6	304.5	315.2		345.7	360.2	375.8
N_{II}	205.8	223.3	236.3	243.3	242	265.6	283.9	288.5
N_{III}	191.4	207.2	217.6	224.6	242*	247.4	256.6	270.9
$N_{IV,V}$	98.9	110.0	113.2	117.5	120.4	129.0	133.2	140.5
$N_{VI,VII}$		0.1	2.0	1.5		5.5	0.0	0.1
O_I	32.3	37.8	37.4	37.5		37.4	31.8	36.1
$O_{II,III}$	14.4	19.8	22.3	21.1		21.3	22.0	20.3

Level	^{65}Tb	^{66}Dy	^{67}Ho	^{68}Er	^{69}Tm	^{70}Yb	^{71}Lu	^{72}Hf
K	51995.7	53788.5	55617.7	57485.5	59389.6	61332.3	63313.8	65350.8
L_I	8708.0	9045.8	9394.2	9751.3	10115.7	10486.4	10870.4	11270.7
L_{II}	8251.6	8580.6	8917.8	9264.3	9616.9	9978.2	10348.6	10739.4
L_{III}	7514.0	7790.1	8071.1	8357.9	8648.0	8943.6	9244.1	9560.7
M_I	1967.5	2046.8	2128.3	2206.5	2306.8	2398.1	2491.2	2600.9
M_{II}	1767.7	1841.8	1922.8	2005.8	2089.8	2173.0	2263.5	2365.4
M_{III}	1611.3	1675.6	1741.2	1811.8	1884.5	1949.8	2023.6	2107.6
M_{IV}	1275.0	1332.5	1391.5	1453.3	1514.6	1576.3	1639.4	1716.4
M_V	1241.2	1294.9	1351.4	1409.3	1467.7	1527.8	1588.5	1661.7
N_I	397.9	416.3	435.7	449.1	471.7	487.2	506.2	538.1
N_{II}	310.2	331.8	343.5	366.2	385.9	396.7	410.1	437.0
N_{III}	385.0	292.9	306.6	320.0	336.6	343.5	359.3	380.4
N_{IV}	147.0	154.2	161.0	176.7	179.6	198.1	204.8	223.8
N_V	147.0*	154.2*	161.0*	167.6	179.6*	184.9	195.0	213.7
$N_{VI,VII}$	2.6	4.2	3.7	4.3	5.3	6.3	6.9	17.1
O_I	39.0	62.9	51.2	59.8	53.2	54.1	56.8	64.9
O_{II}	25.4	26.3	20.3	29.4	32.3	23.4	28.0	38.1
O_{III}	25.4*	26.3*	20.3*	29.4*	32.3*	23.4*	28.0*	30.6

Level	^{73}Ta	^{74}W	^{75}Re	^{76}Os	^{77}Ir	^{78}Pt	^{79}Au	^{80}Hg
K	67416.4	69525.0	71676.4	73870.8	76111.0	78394.8	80724.9	83102.3
L_{I}	11681.5	12099.8	12526.7	12968.0	13418.5	13879.9	14352.8	14839.3
L_{II}	11136.1	11544.0	11958.7	12385.0	12824.1	13272.6	13733.6	14208.7
L_{III}	9881.1	10206.8	10535.3	10870.9	11215.2	11563.7	11918.7	12283.9
M_{I}	2708.0	2819.6	2931.7	3048.5	3173.7	3296.0	3424.9	3561.6
M_{II}	2468.7	2574.9	2681.6	2792.2	2908.7	3026.5	3147.8	3278.5
M_{III}	2194.0	2281.0	2367.3	2457.2	2550.7	2645.4	2743.0	2847.1
M_{IV}	1793.2	1871.6	1948.9	2030.8	2116.1	2201.9	2291.1	2384.9
M_{V}	1735.1	1809.2	1882.9	1960.1	2040.4	2121.6	2205.7	2294.9
N_{I}	565.5	595.0	625.0	654.3	690.1	722.0	758.8	800.3
N_{II}	464.8	491.6	517.9	546.5	577.1	609.2	643.7	676.9
N_{III}	404.5	425.3	444.4	468.2	494.3	519.0	545.4	571.0
N_{IV}	241.3	258.8	273.7	289.4	311.4	330.8	352.0	378.3
N_{V}	229.3	245.4	260.2	272.8	294.9	313.3	333.9	359.8
N_{VI}	25.0	36.5	40.6	46.3	63.4	74.3	86.4	102.2
N_{VII}	25.0*	33.6	40.6*	46.3*	60.5	71.1	82.8	98.5
O_{I}	71.1	77.1	82.8	83.7	95.2	101.7	107.8	120.3
O_{II}	44.9	46.8	45.6	58.0	63.0	65.3	71.7	80.5
O_{III}	36.4	35.6	34.6	45.4	50.5	51.7	53.7	57.6
$O_{IV,V}$	5.7	6.1	3.5		3.8	2.2	2.5	6.4

Level	^{81}Tl	^{82}Pb	^{83}Bi	^{84}Po	^{85}At	^{86}Rn	^{87}Fr	^{88}Ra
K	85530.4	88004.5	90525.9	93105.0	95729.9	98404	101137	103921.9
L_{I}	15346.7	15860.8	16387.5	16939.3	17493	18049	18639	19236.7
L_{II}	14697.9	15200.0	15711.1	16244.3	16784.7	17337.1	17906.5	18484.3
L_{III}	12657.5	13035.2	13418.6	13813.8	14213.5	14619.4	15031.2	15444.4
M_{I}	3704.1	3850.7	3999.1	4149.4	(4317)	(4482)	(4652)	4822.0
M_{II}	3415.7	3554.2	3696.3	3854.1	4008	4159	4327	4489.5
M_{III}	2956.6	3066.4	3176.9	3301.9	3426	3538	3663	3791.8
M_{IV}	2485.1	2585.6	2687.6	2798.0	2908.7	3021.5	3136.2	3248.4
M_{V}	2389.3	2484.0	2579.6	2683.0	2786.7	2892.4	2999.9	3104.9
N_{I}	845.5	893.6	938.2	995.3	(1042)	(1097)	(1153)	1208.4
N_{II}	721.3	763.9	805.3	851	886	929	980	1057.6
N_{III}	609.0	644.5	678.9	705	740	768	810	879.1
N_{IV}	406.6	435.2	463.6	500.2	533.2	566.6	603.3	635.9
N_{V}	386.2	412.9	440.0	473.4			577	602.7
N_{VI}	122.8	142.9	161.9					298.9
N_{VII}	118.5	138.1	157.4					298.9*
O_{I}	136.3	147.3	159.3					254.4
O_{II}	99.6	104.8	116.8					200.4
O_{III}	75.4	86.0	92.8					152.8
O_{IV}	15.3	21.8	26.5	31.4				67.2
O_{V}	13.1	19.2	24.4	31.4*				67.2*
P_{I}		3.1						43.5
$P_{II,III}$		0.7	2.7					18.8

Level	^{89}Ac	^{90}Th	^{91}Pa	^{92}U	^{93}Np	^{94}Pu	^{95}Am	^{96}Cm
K	106755.3	109650.9	112601.4	115606.1	118678	121818	125027	128220
L_{I}	19840	20472.1	21104.6	21757.4	22426.8	23097.2	23772.9	24460
L_{II}	19083.2	19693.2	20313.7	20947.6	21600.5	22266.2	22944.0	23779
L_{III}	15871.0	16300.3	16733.1	17166.3	17610.0	18056.8	18504.1	18930
M_{I}	(5002)	5182.3	5366.9	5548.0	5723.2	5932.9	6120.5	6288
M_{II}	4656	4830.4	5000.9	5182.2	5366.2	5541.2	5710.2	5895
M_{III}	3909	4046.1	4173.8	4303.4	4434.7	4556.6	4667.0	4797
M_{IV}	3370.2	3490.8	3611.2	3727.6	3850.3	3972.6	4092.1	4227
M_{V}	3219.0	3332.0	3441.8	3551.7	3665.8	3778.1	3886.9	3971
N_{I}	(1269)	1329.5	1387.1	1440.8	1500.7	1558.6	1617.1	1643
N_{II}	1080	1168.2	1224.3	1272.6	1327.7	1372.1	1411.8	1440
N_{III}	890	967.3	1006.7	1044.9	1086.8	1114.8	(1135.7)	1154
N_{IV}	674.9	714.1	743.4	780.4	815.9	848.9	878.7	

Level	^{89}Ac	^{90}Th	^{91}Pa	^{92}U	^{93}Np	^{94}Pu	^{95}Am	^{96}Cm
N_V		676.4	708.2	737.7	770.3	801.4	827.6	
N_{VI}		344.4	371.2	391.3	415.0	445.8		
N_{VII}		335.2	359.5	380.9	404.4	432.4		
O_I		290.2	309.6	323.7	351.9			385
O_{II}		229.4	222.9	259.3	283.4	274.1		
O_{III}		181.8	222.9*	195.1	206.1	206.5		
O_{IV}		94.3	94.1	105.0	109.3	116.0	115.8	
O_V		87.9	94.1*	96.3	101.3	105.4	103.3	
P_I		59.5		70.7				
P_{II}		49.0		42.3				
P_{III}		43.0		32.3				

Level	^{97}Bk	^{98}Cf	^{99}Es	^{100}Fm	^{101}Md	^{102}No	^{103}Lr
K	131590	135960	139490	143090	146780	150540	154380
L_I	25275	26110	26900	27700	28530	29380	30240
L_{II}	24385	25250	26020	26810	27610	28440	29280
L_{III}	19452	19930	20410	20900	21390	21880	22360
M_I	6556	6754	6977	7205	7441	7675	7900
M_{II}	6147	6359	6574	6793	7019	7245	7460
M_{III}	4977	5109	5252	5397	5546	5688	5710
M_{IV}	4366	4497	4630	4766	4903	5037	5150
M_V	4132	4253	4374	4498	4622	4741	4860
N_I	1755	1799	1868	1937	2010	2078	2140
N_{II}	1554	1616	1680	1747	1814	1876	1930
N_{III}	1235	1279	1321	1366	1410	1448	1480
O_I	398	419	435	454	472	484	490

ELECTRON BINDING ENERGIES OF THE ELEMENTS

Gwyn P. Williams

This table gives the binding energies in electron volts (eV) for selected electronic levels of the elements. For metallic elements the binding energy is referred to the Fermi level; for semiconductors, to the valence band maximum; and for gases and insulators, to the vacuum level. The atomic number is listed after the element name.

References

1. Fluggle, J. C., and Martensson, N., *J. Elect. Spect.*, 21, 275, 1980.
2. Cardona, M. and Ley, L., *Photoemission from Solids*, Springer-Verlag, Heidelberg, 1978.
3. Bearden, J. A. and Burr, A. F., *Rev. Mod. Phys.*, 39, 125, 1967.

Actinium (89)

Level	Orbital	Energy (eV)
K	1s	106755
L I	2s	19840
L II	$2p_{1/2}$	19083
L III	$2p_{3/2}$	15871
M I	3s	5002
M II	$3p_{1/2}$	4656
M III	$3p_{3/2}$	3909
M IV	$3d_{3/2}$	3370
M V	$3d_{5/2}$	3219
N I	4s	1269[a]
N II	$4p_{1/2}$	1080[a]
N III	$4p_{3/2}$	890[a]
N IV	$4d_{3/2}$	675[a]
N V	$4d_{5/2}$	639[a]
N VI	$4f_{5/2}$	319[a]
N VII	$4f_{7/2}$	319[a]
O I	5s	272[a]
O II	$5p_{1/2}$	215[a]
O III	$5p_{3/2}$	167[a]
O IV	$5d_{3/2}$	80[a]
O V	$5d_{5/2}$	80[a]
P I	6s	—
P II	$6p_{1/2}$	—
P III	$6p_{3/2}$	—

Aluminum (13)

Level	Orbital	Energy (eV)
K	1s	1559.0
L I	2s	117.8[a]
L II	$2p_{1/2}$	72.9[a]
L III	$2p_{3/2}$	72.5[a]

Antimony (51)

Level	Orbital	Energy (eV)
K	1s	30419
L I	2s	4698
L II	$2p_{1/2}$	4380
L III	$2p_{3/2}$	4132
M I	3s	946[b]
M II	$3p_{1/2}$	812.7[b]
M III	$3p_{3/2}$	766.4[b]
M IV	$3d_{3/2}$	537.5[b]
M V	$3d_{5/2}$	528.2[b]
N I	4s	153.2[b]
N II	$4p_{1/2}$	95.6[b,c]
N III	$4p_{3/2}$	95.6[b]
N IV	$4d_{3/2}$	33.3[b]
N V	$4d_{5/2}$	32.1[b]

Argon (18)

Level	Orbital	Energy (eV)
K	1s	3205.9[a]
L I	2s	326.3[a]
L II	$2p_{1/2}$	250.6[a]
L III	$2p_{3/2}$	248.4[a]
M I	3s	29.3[a]
M II	$3p_{1/2}$	15.9[a]
M III	$3p_{3/2}$	15.7[a]

Arsenic (33)

Level	Orbital	Energy (eV)
K	1s	11867
L I	2s	1527.0[a,d]
L II	$2p_{1/2}$	1359.1[a,d]
L III	$2p_{3/2}$	1323.6[a,d]
M I	3s	204.7[a]
M II	$3p_{1/2}$	146.2[a]
M III	$3p_{3/2}$	141.2[a]
M IV	$3d_{3/2}$	41.7[a]
M V	$3d_{5/2}$	41.7[a]

Astatine (85)

Level	Orbital	Energy (eV)
K	1s	95730
L I	2s	17493
L II	$2p_{1/2}$	16785
L III	$2p_{3/2}$	14214
M I	3s	4317
M II	$3p_{1/2}$	4008
M III	$3p_{3/2}$	3426
M IV	$3d_{3/2}$	2909
M V	$3d_{5/2}$	2787
N I	4s	1042[a]
N II	$4p_{1/2}$	886[a]
N III	$4p_{3/2}$	740[a]
N IV	$4d_{3/2}$	533[a]
N V	$4d_{5/2}$	507[a]
N VI	$4f_{5/2}$	210[a]
N VII	$4f_{7/2}$	210[a]
O I	5s	195[a]
O II	$5p_{1/2}$	148[a]
O III	$5p_{3/2}$	115[a]
O IV	$5d_{3/2}$	40[a]
O V	$5d_{5/2}$	40[a]

Barium (56)

Level	Orbital	Energy (eV)
K	1s	37441
L I	2s	5989
L II	$2p_{1/2}$	5624
L III	$2p_{3/2}$	5247
M I	3s	1293[a,d]
M II	$3p_{1/2}$	1137[a,d]
M III	$3p_{3/2}$	1063[a,d]
M IV	$3d_{3/2}$	795.7[a]
M V	$3d_{5/2}$	780.5[a]
N I	4s	253.5[b]
N II	$4p_{1/2}$	192
N III	$4p_{3/2}$	178.6[b]
N IV	$4d_{3/2}$	92.6[b]
N V	$4d_{5/2}$	89.9[b]
N VI	$4f_{5/2}$	—
N VII	$4f_{7/2}$	—
O I	5s	30.3[b]
O II	$5p_{1/2}$	17.0[b]
O III	$5p_{3/2}$	14.8[b]

Beryllium (4)

Level	Orbital	Energy (eV)
K	1s	111.5[a]

Bismuth (83)

Level	Orbital	Energy (eV)
K	1s	90526
L I	2s	16388
L II	$2p_{1/2}$	15711
L III	$2p_{3/2}$	13419
M I	3s	3999
M II	$3p_{1/2}$	3696
M III	$3p_{3/2}$	3177
M IV	$3d_{3/2}$	2688
M V	$3d_{5/2}$	2580
N I	4s	939[b]
N II	$4p_{1/2}$	805.2[b]
N III	$4p_{3/2}$	678.8[b]
N IV	$4d_{3/2}$	464.0[b]
N V	$4d_{5/2}$	440.1[b]
N VI	$4f_{5/2}$	162.3[b]
N VII	$4f_{7/2}$	157.0[b]
O I	5s	159.3[a,d]
O II	$5p_{1/2}$	119.0[b]
O III	$5p_{3/2}$	92.6[b]
O IV	$5d_{3/2}$	26.9[b]
O V	$5d_{5/2}$	23.8[b]

Boron (5)

Level	Orbital	Energy (eV)
K	1s	188[a]

Bromine (35)

Level	Orbital	Energy (eV)
K	1s	13474
L I	2s	1782[a]
L II	$2p_{1/2}$	1596[a]
L III	$2p_{3/2}$	1550[a]
M I	3s	257[a]
M II	$3p_{1/2}$	189[a]
M III	$3p_{3/2}$	182[a]
M IV	$3d_{3/2}$	70[a]
M V	$3d_{5/2}$	69[a]

Cadmium (48)

Level	Orbital	Energy (eV)
K	1s	26711
L I	2s	4018
L II	$2p_{1/2}$	3727
L III	$2p_{3/2}$	3538
M I	3s	772.0[b]
M II	$3p_{1/2}$	652.6[b]
M III	$3p_{3/2}$	618.4[b]
M IV	$3d_{3/2}$	411.9[b]
M V	$3d_{5/2}$	405.2[b]
N I	4s	109.8[b]
N II	$4p_{1/2}$	63.9[b,c]
N III	$4p_{3/2}$	63.9[b,c]
N IV	$4d_{3/2}$	11.7[b]
N V	$4d_{5/2}$	10.7[b]

Calcium (20)

Level	Orbital	Energy (eV)
K	1s	4038.5[a]
L I	2s	438.4[b]
L II	$2p_{1/2}$	349.7[b]
L III	$2p_{3/2}$	346.2[b]
M I	3s	44.3[b]
M II	$3p_{1/2}$	25.4[b]
M III	$3p_{3/2}$	25.4[b]

Carbon (6)

Level	Orbital	Energy (eV)
K	1s	284.2[a]

Cerium (58)

Level	Orbital	Energy (eV)
K	1s	40443
L I	2s	6548
L II	$2p_{1/2}$	6164
L III	$2p_{3/2}$	5723
M I	3s	1436[a,d]
M II	$3p_{1/2}$	1274[a,d]
M III	$3p_{3/2}$	1187[a,d]
M IV	$3d_{3/2}$	902.4[a]
M V	$3d_{5/2}$	883.8[a]
N I	4s	291.0[a]
N II	$4p_{1/2}$	223.3
N III	$4p_{3/2}$	206.5[a]
N IV	$4d_{3/2}$	109[a]
N V	$4d_{5/2}$	—
N VI	$4f_{5/2}$	0.1
N VII	$4f_{7/2}$	0.1
O I	5s	37.8
O II	$5p_{1/2}$	19.8[a]
O III	$5p_{3/2}$	17.0[a]

Cesium (55)

Level	Orbital	Energy (eV)
K	1s	35985
L I	2s	5714
L II	$2p_{1/2}$	5359
L III	$2p_{3/2}$	5012
M I	3s	1211[a,d]
M II	$3p_{1/2}$	1071[a]

M III	$3p_{3/2}$	1003[a]
M IV	$3d_{3/2}$	740.5[a]
M V	$3d_{5/2}$	726.6[a]
N I	4s	232.3[a]
N II	$4p_{1/2}$	172.4[a]
N III	$4p_{3/2}$	161.3[a]
N IV	$4d_{3/2}$	79.8[a]
N V	$4d_{5/2}$	77.5[a]
N VI	$4f_{5/2}$	—
N VII	$4f_{7/2}$	—
O I	5s	22.7
O II	$5p_{1/2}$	14.2[a]
O III	$5p_{3/2}$	12.1[a]

Chlorine (17)

K	1s	2822.0
L I	2s	270[a]
L II	$2p_{1/2}$	202[a]
L III	$2p_{3/2}$	200[a]

Chromium(24)

K	1s	5989
L I	2s	696.0[b]
L II	$2p_{1/2}$	583.8[b]
L III	$2p_{3/2}$	574.1[b]
M I	3s	74.1[b]
M II	$3p_{1/2}$	42.2[b]
M III	$3p_{3/2}$	42.2[b]

Cobalt (27)

K	1s	7709
L I	2s	925.1[b]
L II	$2p_{1/2}$	793.2[b]
L III	$2p_{3/2}$	778.1[b]
M I	3s	101.0[b]
M II	$3p_{1/2}$	58.9[b]
M III	$3p_{3/2}$	58.9[b]

Copper (29)

K	1s	8979
L I	2s	1096.7[b]
L II	$2p_{1/2}$	952.3[b]
L III	$2p_{3/2}$	932.5[b]
M I	3s	122.5[b]
M II	$3p_{1/2}$	77.3[b]
M III	$3p_{3/2}$	75.1[b]

Dysprosium (66)

K	1s	53789
L I	2s	9046
L II	$2p_{1/2}$	8581
L III	$2p_{3/2}$	7790
M I	3s	2047
M II	$3p_{1/2}$	1842
M III	$3p_{3/2}$	1676
M IV	$3d_{3/2}$	1333
M V	$3d_{5/2}$	1292[a]
N I	4s	414.2[a]
N II	$4p_{1/2}$	333.5[a]
N III	$4p_{3/2}$	293.2[a]
N IV	$4d_{3/2}$	153.6[a]
N V	$4d_{5/2}$	153.6[a]
N VI	$4f_{5/2}$	8.0[a]
N VII	$4f_{7/2}$	4.3[a]
O I	5s	49.9[a]
O II	$5p_{1/2}$	26.3
O III	$5p_{3/2}$	26.3

Erbium (68)

K	1s	57486
L I	2s	9751
L II	$2p_{1/2}$	9264
L III	$2p_{3/2}$	8358
M I	3s	2206
M II	$3p_{1/2}$	2006
M III	$3p_{3/2}$	1812
M IV	$3d_{3/2}$	1453
M V	$3d_{5/2}$	1409
N I	4s	449.8[a]
N II	$4p_{1/2}$	366.2
N III	$4p_{3/2}$	320.2[a]
N IV	$4d_{3/2}$	167.6[a]
N V	$4d_{5/2}$	167.6[a]
N VI	$4f_{5/2}$	—
N VII	$4f_{7/2}$	4.7[a]
O I	5s	50.6[a]
O II	$5p_{1/2}$	31.4[a]
O III	$5p_{3/2}$	24.7[a]

Europium (63)

K	1s	48519
L I	2s	8052
L II	$2p_{1/2}$	7617
L III	$2p_{3/2}$	6977
M I	3s	1800
M II	$3p_{1/2}$	1614
M III	$3p_{3/2}$	1481
M IV	$3d_{3/2}$	1158.6[a]
M V	$3d_{5/2}$	1127.5[a]
N I	4s	360
N II	$4p_{1/2}$	284
N III	$4p_{3/2}$	257
N IV	$4d_{3/2}$	133
N V	$4d_{5/2}$	1227[a]
N VI	$4f_{5/2}$	0
N VII	$4f_{7/2}$	0
O I	5s	32
O II	$5p_{1/2}$	22
O III	$5p_{3/2}$	22

Fluorine (9)

K	1s	696.7[a]

Francium (87)

K	1s	101137
L I	2s	18639
L II	$2p_{1/2}$	17907
L III	$2p_{3/2}$	15031
M I	3s	4652
M II	$3p_{1/2}$	4327
M III	$3p_{3/2}$	3663
M IV	$3d_{3/2}$	3136
M V	$3d_{5/2}$	3000
N I	4s	1153[a]
N II	$4p_{1/2}$	980[a]
N III	$4p_{3/2}$	810[a]
N IV	$4d_{3/2}$	603[a]
N V	$4d_{5/2}$	577[a]
N VI	$4f_{5/2}$	268[a]
N VII	$4f_{7/2}$	268[a]
O I	5s	234[a]
O II	$5p_{1/2}$	182[a]
O III	$5p_{3/2}$	140[a]
O IV	$5d_{3/2}$	58[a]
O V	$5d_{5/2}$	58[a]
P I	6s	34
P II	$6p_{1/2}$	15
P III	$6p_{3/2}$	15

Gadolinium (64)

K	1s	50239
L I	2s	8376
L II	$2p_{1/2}$	7930
L III	$2p_{3/2}$	7243
M I	3s	1881
M II	$3p_{1/2}$	1688
M III	$3p_{3/2}$	1544
M IV	$3d_{3/2}$	1221.9[a]
M V	$3d_{5/2}$	1189.6[a]
N I	4s	378.6[a]
N II	$4p_{1/2}$	286
N III	$4p_{3/2}$	271
N IV	$4d_{3/2}$	—
N V	$4d_{5/2}$	142.6[a]
N VI	$4f_{5/2}$	8.6[a]
N VII	$4f_{7/2}$	8.6[a]
O I	5s	36
O II	$5p_{1/2}$	20
O III	$5p_{3/2}$	20

Gallium (31)

K	1s	10367
L I	2s	1299.0[a,d]
L II	$2p_{1/2}$	1143.2[b]
L III	$2p_{3/2}$	1116.4[b]
M I	3s	159.5[b]
M II	$3p_{1/2}$	103.5[b]
M III	$3p_{3/2}$	100.0[b]
M IV	$3d_{3/2}$	18.7[b]
M V	$3d_{5/2}$	18.7[b]

Germanium (32)

K	1s	11103
L I	2s	1414.6[a,d]
L II	$2p_{1/2}$	1248.1[a,d]
L III	$2p_{3/2}$	1217.0[a,d]
M I	3s	180.1[a]
M II	$3p_{1/2}$	124.9[a]
M III	$3p_{3/2}$	120.8[a]
M IV	$3d_{3/2}$	29.8[a]
M V	$3d_{5/2}$	29.2[a]

Gold (79)

K	1s	80725
L I	2s	14353
L II	$2p_{1/2}$	13734
L III	$2p_{3/2}$	11919
M I	3s	3425
M II	$3p_{1/2}$	3148
M III	$3p_{3/2}$	2743
M IV	$3d_{3/2}$	2291
M V	$3d_{5/2}$	2206
N I	4s	762.1[b]
N II	$4p_{1/2}$	642.7[b]
N III	$4p_{3/2}$	546.3[b]
N IV	$4d_{3/2}$	353.2[b]
N V	$4d_{5/2}$	335.1[b]
N VI	$4f_{5/2}$	87.6[b]
N VII	$4f_{7/2}$	83.9[b]
O I	5s	107.2[a,d]
O II	$5p_{1/2}$	74.2[b]
O III	$5p_{3/2}$	57.2[b]

Hafnium (72)

K	1s	65351
L I	2s	11271
L II	$2p_{1/2}$	10739
L III	$2p_{3/2}$	9561
M I	3s	2601
M II	$3p_{1/2}$	2365
M III	$3p_{3/2}$	2107
M IV	$3d_{3/2}$	1176
M V	$3d_{5/2}$	1662
N I	4s	538[a]
N II	$4p_{1/2}$	438.2[b]
N III	$4p_{3/2}$	380.7[b]
N IV	$4d_{3/2}$	220.0[b]
N V	$4d_{5/2}$	211.5[b]
N VI	$4f_{5/2}$	15.9[b]
N VII	$4f_{7/2}$	14.2[b]
O I	5s	64.2[b]
O II	$5p_{1/2}$	38[a]
O III	$5p_{3/2}$	29.9[b]

Helium (2)

K	1s	24.6[a]

Holmium (67)

K	1s	55618
L I	2s	9394
L II	$2p_{1/2}$	8918
L III	$2p_{3/2}$	8071
M I	3s	2128
M II	$3p_{1/2}$	1923
M III	$3p_{3/2}$	1741
M IV	$3d_{3/2}$	1392
M V	$3d_{5/2}$	1351
N I	4s	432.4[a]
N II	$4p_{1/2}$	343.5
N III	$4p_{3/2}$	308.2[a]
N IV	$4d_{3/2}$	160[a]
N V	$4d_{5/2}$	160[a]
N VI	$4f_{5/2}$	8.6[a]
N VII	$4f_{7/2}$	5.2[a]
O I	5s	49.3[a]
O II	$5p_{1/2}$	30.8[a]
O III	$5p_{3/2}$	24.1[a]

Hydrogen (1)

K	1s	13.6

Indium (49)

K	1s	27940
L I	2s	4238
L II	$2p_{1/2}$	3938
L III	$2p_{3/2}$	3730
M I	3s	827.2[b]
M II	$3p_{1/2}$	703.2[b]
M III	$3p_{3/2}$	665.3[b]
M IV	$3d_{3/2}$	451.4[b]

M V	$3d_{5/2}$	443.9[b]
N I	4s	122.9[b]
N II	$4p_{1/2}$	73.5[b,c]
N III	$4p_{3/2}$	73.5[b,c]
N IV	$4d_{3/2}$	17.7[b]
N V	$4d_{5/2}$	16.9[b]
Iodine (53)		
K	1s	33169
L I	2s	5188
L II	$2p_{1/2}$	4852
L III	$2p_{3/2}$	4557
M I	3s	1072[a]
M II	$3p_{1/2}$	931[a]
M III	$3p_{3/2}$	875[a]
M IV	$3d_{3/2}$	631[a]
M V	$3d_{5/2}$	620[a]
N I	4s	186[a]
N II	$4p_{1/2}$	123[a]
N III	$4p_{3/2}$	123[a]
N IV	$4d_{3/2}$	50[a]
N V	$4d_{5/2}$	50[a]
Iridium (77)		
K	1s	76111
L I	2s	13419
L II	$2p_{1/2}$	12824
L III	$2p_{3/2}$	11215
M I	3s	3174
M II	$3p_{1/2}$	2909
M III	$3p_{3/2}$	2551
M IV	$3d_{3/2}$	2116
M V	$3d_{5/2}$	2040
N I	4s	691.1[b]
N II	$4p_{1/2}$	577.8[b]
N III	$4p_{3/2}$	495.8[b]
N IV	$4d_{3/2}$	311.9[b]
N V	$4d_{5/2}$	296.3[b]
N VI	$4f_{5/2}$	63.8[b]
N VII	$4f_{7/2}$	60.8[b]
O I	52	95.2[a,d]
O II	$5p_{1/2}$	63.0[a,d]
O III	$5p_{3/2}$	48.0[b]
Iron (26)		
K	1s	7112
L I	2s	844.6[b]
L II	$2p_{1/2}$	719.9[b]
L III	$2p_{3/2}$	706.8[b]
M I	3s	91.3[b]
M II	$3p_{1/2}$	52.7[b]
M III	$3p_{3/2}$	52.7[b]
Krypton (36)		
K	1s	14326
L I	2s	1921
L II	$2p_{1/2}$	1730.9[a]
L III	$2p_{3/2}$	1678.4[a]
M I	3s	292.8[a]
M II	$3p_{1/2}$	222.2[a]
M III	$3p_{3/2}$	214.4[a]
M IV	$3d_{3/2}$	95.0[a]
M V	$3d_{5/2}$	93.8[a]
N I	4s	27.5[a]
N II	$4p_{1/2}$	14.1[a]
N III	$4p_{3/2}$	14.1[a]
Lanthanum (57)		
K	1s	38925
L I	2s	6266
L II	$2p_{1/2}$	5891
L III	$2p_{3/2}$	5483
M I	3s	1362[a,d]
M II	$3p_{1/2}$	1209[a,d]
M III	$3p_{3/2}$	1128[a,d]
M IV	$3d_{3/2}$	853[a]
M V	$3d_{5/2}$	836[a]
N I	4s	247.7[a]
N II	$4p_{1/2}$	205.8
N III	$4p_{3/2}$	196.0[a]
N IV	$4d_{3/2}$	105.3[a]
N V	$4d_{5/2}$	102.5[a]
N VI	$4f_{5/2}$	—
N VII	$4f_{7/2}$	—
O I	5s	34.3[a]
O II	$5p_{1/2}$	19.3[a]
O III	$5p_{3/2}$	16.8[a]
Lead (82)		
K	1s	88005
L I	2s	15861
L II	$2p_{1/2}$	15200
L III	$2p_{3/2}$	13055
M I	3s	3851
M II	$3p_{1/2}$	3554
M III	$3p_{3/2}$	3066
M IV	$3d_{3/2}$	2586
M V	$3d_{5/2}$	2484
N I	4s	891.8[b]
N II	$4p_{1/2}$	761.9[b]
N III	$4p_{3/2}$	643.5[b]
N IV	$4d_{3/2}$	434.3[b]
N V	$4d_{5/2}$	412.2[b]
N VI	$4f_{5/2}$	141.7[b]
N VII	$4f_{7/2}$	136.9[b]
O I	5s	147[a,d]
O II	$5p_{1/2}$	106.4[b]
O III	$5p_{3/2}$	83.3[b]
O IV	$5d_{3/2}$	20.7[b]
O V	$5d_{5/2}$	18.1[b]
Lithium (3)		
K	1s	54.7[a]
Lutetium		
K	1s	63314
L I	2s	10870
L II	$2p_{1/2}$	10349
L III	$2p_{3/2}$	9244
M I	3s	2491
M II	$3p_{1/2}$	2264
M III	$3p_{3/2}$	2024
M IV	$3d_{3/2}$	1639
M V	$3d_{5/2}$	1589
N I	4s	506.8[a]
N II	$4p_{1/2}$	412.4[a]
N III	$4p_{3/2}$	359.2[a]
N IV	$4d_{3/2}$	206.1[a]
N V	$4d_{5/2}$	196.3[a]
N VI	$4f_{5/2}$	8.9[a]
N VII	$4f_{7/2}$	7.5[a]
O I	5s	57.3[a]
O II	$5p_{1/2}$	33.6[a]
O III	$5p_{3/2}$	26.7[a]
Magnesium (12)		
K	1s	1303.0[b]
L I	2s	88.6[a]
L II	$2p_{1/2}$	49.6[b]
L III	$2p_{3/2}$	49.2[a]
Manganese (25)		
K	1s	6539
L I	2s	769.1[b]
L II	$2p_{1/2}$	649.9[b]
L III	$2p_{3/2}$	638.7[b]
M I	3s	82.3[b]
M II	$3p_{1/2}$	47.2[b]
M III	$3p_{3/2}$	47.2[b]
Mercury (80)		
K	1s	83102
L I	2s	14839
L II	$2p_{1/2}$	14209
L III	$2p_{3/2}$	12284
M I	3s	3562
M II	$3p_{1/2}$	3279
M III	$3p_{3/2}$	2847
M IV	$3d_{3/2}$	2385
M V	$3d_{5/2}$	2295
N I	4s	802.2[b]
N II	$4p_{1/2}$	680.2[b]
N III	$4p_{3/2}$	576.6[b]
N IV	$4d_{3/2}$	378.2[b]
N V	$4d_{5/2}$	358.8[b]
N VI	$4f_{5/2}$	104.0[b]
N VII	$4f_{7/2}$	99.9[b]
O I	5s	127[b]
O II	$5p_{3/2}$	83.1[b]
O III	$5p_{3/2}$	64.5[b]
O IV	$5d_{3/2}$	9.6[b]
O V	$5d_{5/2}$	7.8[b]
Molybdenum (42)		
K	1s	20000
L I	2s	2866
L II	$2p_{1/2}$	2625
L III	$2p_{3/2}$	2520
M I	3s	506.3[b]
M II	$3p_{1/2}$	411.6[b]
M III	$3p_{3/2}$	394.0[b]
M IV	$3d_{3/2}$	231.1[b]
M V	$3d_{5/2}$	227.9[b]
N I	4s	63.2[b]
N II	$4p_{1/2}$	37.6[b]
N III	$4p_{3/2}$	35.5[b]
Neodymium (60)		
K	1s	43569
L I	2s	7126
L II	$2p_{1/2}$	6722
L III	$2p_{3/2}$	6208
M I	3s	1575
M II	$3p_{1/2}$	1403
M III	$3p_{3/2}$	1297
M IV	$3d_{3/2}$	1003.3[a]
M V	$3d_{5/2}$	980.4[a]
N I	4s	319.2[a]
N II	$4p_{1/2}$	243.3
N III	$4p_{3/2}$	224.6
N IV	$4d_{3/2}$	120.5[a]
N V	$4d_{5/2}$	120.5[a]
N VI	$4f_{5/2}$	1.5
N VII	$4f_{7/2}$	1.5
O I	5s	37.5
O II	$5p_{1/2}$	21.1
O III	$5p_{3/2}$	21.1
Neon (10)		
K	1s	870.2[a]
L I	2s	48.5[a]
L II	$2p_{1/2}$	21.7[a]
L III	$2p_{3/2}$	21.6[a]
Nickel (28)		
K	1s	8333
L I	2s	1008.6[b]
L II	$2p_{1/2}$	870.0[b]
L III	$2p_{3/2}$	852.7[b]
M I	3s	110.8[b]
M II	$3p_{1/2}$	68.0[b]
M III	$3p_{3/2}$	66.2[b]
Niobium (41)		
K	1s	18986
L I	2s	2698
L II	$2p_{1/2}$	2465
L III	$2p_{3/2}$	2371
M I	3s	466.6[b]
M II	$3p_{1/2}$	376.1[b]
M III	$3p_{3/2}$	360.6[b]
M IV	$3d_{3/2}$	205.0[b]
M V	$3d_{5/2}$	202.3[b]
N I	4s	56.4[b]
N II	$4p_{1/2}$	32.6[b]
N III	$4p_{3/2}$	30.8[b]
Nitrogen (7)		
K	1s	409.9[a]
L I	2s	37.3[a]
Osmium (76)		
K	1s	73871
L I	2s	12968
L II	$2p_{1/2}$	12385
L III	$2p_{3/2}$	10871
M I	3s	3049
M II	$3p_{1/2}$	2792
M III	$3p_{3/2}$	2457
M IV	$3d_{3/2}$	2031
M V	$3d_{5/2}$	1960
N I	4s	658.2[b]
N II	$4p_{1/2}$	549.1[b]
N III	$4p_{3/2}$	470.7[b]
N IV	$4d_{3/2}$	293.1[b]
N V	$4d_{5/2}$	278.5[b]
N VI	$4f_{5/2}$	53.4[b]
N VII	$4f_{7/2}$	50.7[b]
O I	5s	84[a]
O II	$5p_{1/2}$	58[a]

O III	$5p_{3/2}$	44.5[b]
Oxygen (8)		
K	1s	543.1[a]
L I	2s	41.6[a]
Palladium (46)		
K	1s	24350
L I	2s	3604
L II	$2p_{1/2}$	3330
L III	$2p_{3/2}$	3173
M I	3s	671.6[b]
M II	$3p_{1/2}$	559.9[b]
M III	$3p_{3/2}$	532.3[b]
M IV	$3d_{3/2}$	340.5[b]
M V	$3d_{5/2}$	335.2[b]
N I	4s	87.1[a,d]
N II	$4p_{1/2}$	55.7[b,c]
N III	$4p_{3/2}$	50.9[b,c]
Phosphorus (15)		
K	1s	2145.5
L I	2s	189[a]
L II	$2p_{1/2}$	136[a]
L III	$2p_{3/2}$	135[a]
Platinum (78)		
K	1s	78395
L I	2s	13880
L II	$2p_{1/2}$	13273
L III	$2p_{3/2}$	11564
M I	3s	3296
M II	$3p_{1/2}$	3027
M III	$3p_{3/2}$	2645
M IV	$3d_{3/2}$	2202
M V	$3d_{5/2}$	2122
N I	4s	725.4[b]
N II	$4p_{1/2}$	609.1[b]
N III	$4p_{3/2}$	519.4[b]
N IV	$4d_{3/2}$	331.6[b]
N V	$4d_{5/2}$	314.6[b]
N VI	$4f_{5/2}$	74.5[b]
N VII	$4f_{7/2}$	71.2[b]
O I	5s	101.7[a,d]
O II	$5p_{1/2}$	65.3[a,b]
O III	$5p_{3/2}$	51.7[b]
Polonium (84)		
K	1s	93105
L I	2s	16939
L II	$2p_{1/2}$	16244
L III	$2p_{3/2}$	13814
M I	3s	4149
M II	$3p_{1/2}$	3854
M III	$3p_{3/2}$	3302
M IV	$3d_{3/2}$	2798
M V	$3d_{5/2}$	2683
N I	4s	995[a]
N II	$4p_{1/2}$	851[a]
N III	$4p_{3/2}$	705[a]
N IV	$4d_{3/2}$	500[a]
N V	$4d_{5/2}$	473[a]
N VI	$4f_{5/2}$	184[a]
N VII	$4f_{7/2}$	184[a]
O I	5s	177[a]
O II	$5p_{1/2}$	132[a]
O III	$5p_{3/2}$	104[a]
O IV	$5d_{3/2}$	31[a]
O V	$5d_{5/2}$	31[a]
Potassium (19)		
K	1s	3608.4[a]
L I	2s	378.6[a]
L II	$2p_{1/2}$	297.3[a]
L III	$2p_{3/2}$	294.6[a]
M I	3s	34.8[a]
M II	$3p_{1/2}$	18.3[a]
M III	$3p_{3/2}$	18.3[a]
Praseodymium (59)		
K	1s	41991
L I	2s	6835
L II	$2p_{1/2}$	6440
L III	$2p_{3/2}$	5964
M I	3s	1511
M II	$3p_{1/2}$	1337
M III	$3p_{3/2}$	1242
M IV	$3d_{3/2}$	948.3[a]
M V	$3d_{5/2}$	928.8[a]
N I	4s	304.5
N II	$4p_{1/2}$	236.3
N III	$4p_{3/2}$	217.6
N IV	$4d_{3/2}$	115.1[a]
N V	$4d_{5/2}$	115.1[a]
N VI	$4f_{5/2}$	2.0
N VII	$4f_{7/2}$	2.0
O I	5s	37.4
O II	$5p_{1/2}$	22.3
O III	$5p_{3/2}$	22.3
Promethium (61)		
K	1s	45184
L I	2s	7428
L II	$2p_{1/2}$	7013
L III	$2p_{3/2}$	6459
M I	3s	—
M II	$3p_{1/2}$	1471.4
M III	$3p_{3/2}$	1357
M IV	$3d_{3/2}$	1052
M V	$3d_{5/2}$	1027
N I	4s	—
N II	$4p_{1/2}$	242
N III	$4p_{3/2}$	242
N IV	$4d_{3/2}$	120
N V	$4d_{5/2}$	120
Protactinium (91)		
K	1s	112601
L I	2s	21105
L II	$2p_{1/2}$	20314
L III	$2p_{3/2}$	16733
M I	3s	5367
M II	$3p_{1/2}$	5001
M III	$3p_{3/2}$	4174
M IV	$3d_{3/2}$	3611
M V	$3d_{5/2}$	3442
N I	4s	1387[a]
N II	$4p_{1/2}$	1224[a]
N III	$4p_{3/2}$	1007[a]
N IV	$4d_{3/2}$	743[a]
N V	$4d_{5/2}$	708[a]
N VI	$4f_{5/2}$	371[a]
N VII	$4f_{7/2}$	360[a]
O I	5s	310[a]
O II	$5p_{1/2}$	232[a]
O III	$5p_{3/2}$	232[a]
O IV	$5d_{3/2}$	94[a]
O V	$5d_{5/2}$	94[a]
P I	6s	—
P II	$6p_{1/2}$	—
P III	$6p_{3/2}$	—
Radium (88)		
K	1s	103922
L I	2s	19237
L II	$2p_{1/2}$	18484
L III	$2p_{3/2}$	15444
M I	3s	4822
M II	$3p_{1/2}$	4490
M III	$3p_{3/2}$	3792
M IV	$3d_{3/2}$	3248
M V	$3d_{5/2}$	3105
N I	4s	1208[a]
N II	$4p_{1/2}$	1058
N III	$4p_{3/2}$	879[a]
N IV	$4d_{3/2}$	636[a]
N V	$4d_{5/2}$	603[a]
N VI	$4f_{5/2}$	299[a]
N VII	$4f_{7/2}$	299[a]
O I	5s	254[a]
O II	$5p_{1/2}$	200[a]
O III	$5p_{3/2}$	153[a]
O IV	$5d_{3/2}$	68[a]
O V	$5d_{5/2}$	68[a]
P I	6s	44
P II	$6p_{1/2}$	19
P III	$6p_{3/2}$	19
Radon (86)		
K	1s	98404
L I	2s	18049
L II	$2p_{1/2}$	17337
L III	$2p_{3/2}$	14619
M I	3s	4482
M II	$3p_{1/2}$	4159
M III	$3p_{3/2}$	3538
M IV	$3d_{3/2}$	3022
M V	$3d_{5/2}$	2892
N I	4s	1097[a]
N II	$4p_{1/2}$	929[a]
N III	$4p_{3/2}$	768[a]
N IV	$4d_{3/2}$	567[a]
N V	$4d_{5/2}$	541[a]
N VI	$4f_{5/2}$	238[a]
N VII	$4f_{7/2}$	238[a]
O I	5s	214[a]
O II	$5p_{1/2}$	164[a]
O III	$5p_{3/2}$	127[a]
O IV	$5d_{3/2}$	48[a]
O V	$5d_{5/2}$	48[a]
P I	6s	26
Rhenium (75)		
K	1s	71676
L I	2s	12527
L II	$2p_{1/2}$	11959
L III	$2p_{3/2}$	10535
M I	3s	2932
M II	$3p_{1/2}$	2682
M III	$3p_{3/2}$	2367
M IV	$3d_{3/2}$	1949
M V	$3d_{5/2}$	1883
N I	4s	625.4[b]
N II	$4p_{1/2}$	518.7[b]
N III	$4p_{3/2}$	446.8[b]
N IV	$4d_{3/2}$	273.9[b]
N V	$4d_{5/2}$	260.5[b]
N VI	$4f_{5/2}$	42.9[a]
N VII	$4f_{7/2}$	40.5[a]
O I	5s	83[b]
O II	$5p_{1/2}$	45.6[b]
O III	$5p_{3/2}$	34.6[a,d]
Rhodium (45)		
K	1s	23220
L I	2s	3412
L II	$2p_{1/2}$	3146
L III	$2p_{3/2}$	3004
M I	3s	628.1[b]
M II	$3p_{1/2}$	521.3[b]
M III	$3p_{3/2}$	496.5[b]
M IV	$3d_{3/2}$	311.9[b]
M V	$3d_{5/2}$	307.2[b]
N I	4s	81.4[a,d]
N II	$4p_{1/2}$	50.5[b]
N III	$4p_{3/2}$	47.3[b]
Rubidium (37)		
K	1s	15200
L I	2s	2065
L II	$2p_{1/2}$	1864
L III	$2p_{3/2}$	1804
M I	3s	326.7[a]
M II	$3p_{1/2}$	248.7[a]
M III	$3p_{3/2}$	239.1[a]
M IV	$3d_{3/2}$	113.0[a]
M V	$3d_{5/2}$	112[a]
N I	4s	30.5[a]
N II	$4p_{1/2}$	16.3[a]
N III	$4p_{3/2}$	15.3[a]
Ruthenium (44)		
K	1s	22117
L I	2s	3224
L II	$2p_{1/2}$	2967
L III	$2p_{3/2}$	2838
M I	3s	586.2[b]
M II	$3p_{1/2}$	483.3[b]
M III	$3p_{3/2}$	461.5[b]
M IV	$3d_{3/2}$	284.2[b]
M V	$3d_{5/2}$	280.0[b]
N I	4s	75.0[b]
N II	$4p_{1/2}$	46.5[b]
N III	$4p_{3/2}$	43.2[b]
Samarium (62)		
K	1s	46834
L I	2s	7737
L II	$2p_{1/2}$	7312

L III	$2p_{3/2}$	6716
M I	$3s$	1723
M II	$3p_{1/2}$	1541
M III	$3p_{3/2}$	1419.8
M IV	$3d_{3/2}$	1110.9[a]
M V	$3d_{5/2}$	1083.4[a]
N I	$4s$	347.2[a]
N II	$4p_{1/2}$	265.6
N III	$4p_{3/2}$	247.4
N IV	$4d_{3/2}$	129.0
N V	$4d_{5/2}$	129.0
N VI	$4f_{5/2}$	5.2
N VII	$4f_{7/2}$	5.2
O I	$5s$	37.4
O II	$5p_{1/2}$	21.3
O III	$5p_{3/2}$	21.3

Scandium (21)

K	$1s$	4492
L I	$2s$	498.0[a]
L II	$2p_{1/2}$	403.6[a]
L III	$2p_{3/2}$	389.7[a]
M I	$3s$	51.1[a]
M II	$3p_{1/2}$	28.3[a]
M III	$3p_{3/2}$	28.3[a]

Selenium (34)

K	$1s$	12658
L I	$2s$	1652.0[a,d]
L II	$2p_{1/2}$	1474.3[a,d]
L III	$2p_{3/2}$	1433.9[a,d]
M I	$3s$	229.6[a]
M II	$3p_{1/2}$	166.5[a]
M III	$3p_{3/2}$	160.7[a]
M IV	$3d_{3/2}$	55.5[a]
M V	$3d_{5/2}$	54.6[a]

Silicon (14)

K	$1s$	1839
L I	$2s$	149.7[a,d]
L II	$2p_{1/2}$	99.8[a]
L III	$2p_{3/2}$	99.2[a]

Silver (47)

K	$1s$	25514
L I	$2s$	3806
L II	$2p_{1/2}$	3524
L III	$2p_{3/2}$	3351
M I	$3s$	719.0[b]
M II	$3p_{1/2}$	603.8[b]
M III	$3p_{3/2}$	573.0[b]
M IV	$3d_{3/2}$	374.0[b]
M V	$3d_{5/2}$	368.0[b]
N I	$4s$	97.0[b]
N II	$4p_{1/2}$	63.7[b]
N III	$4p_{3/2}$	58.3[b]

Sodium (11)

K	$1s$	1070.8[b]
L I	$2s$	63.5[b]
L II	$2p_{1/2}$	30.4[b]
L III	$2p_{3/2}$	30.5[a]

Strontium (38)

K	$1s$	16105
L I	$2s$	2216
L II	$2p_{1/2}$	2007
L III	$2p_{3/2}$	1940
M I	$3s$	358.7[b]
M II	$3p_{1/2}$	280.3[b]
M III	$3p_{3/2}$	270.0[b]
M IV	$3d_{3/2}$	136.0[b]
M V	$3d_{5/2}$	134.2[b]
N I	$4s$	38.9[b]
N II	$4p_{1/2}$	21.6[b]
N III	$4p_{3/2}$	20.1[b]

Sulfur (16)

K	$1s$	2472
L I	$2s$	230.9[a,d]
L II	$2p_{1/2}$	163.6[a]
L III	$2p_{3/2}$	162.5[a]

Tantalum (73)

K	$1s$	67416
L I	$2s$	11682
L II	$2p_{1/2}$	11136
L III	$2p_{3/2}$	9881
M I	$3s$	2708
M II	$3p_{1/2}$	2469
M III	$3p_{3/2}$	2194
M IV	$3d_{3/2}$	1793
M V	$3d_{5/2}$	1735
N I	$4s$	563.4[b]
N II	$4p_{1/2}$	463.4[b]
N III	$4p_{3/2}$	400.9[b]
N IV	$4d_{3/2}$	237.9[b]
N V	$4d_{5/2}$	226.4[b]
N VI	$4f_{5/2}$	23.5[b]
N VII	$4f_{7/2}$	21.6[b]
O I	$5s$	69.7[b]
O II	$5p_{1/2}$	42.2[a]
O III	$5p_{3/2}$	32.7[b]

Technetium (43)

K	$1s$	21044
L I	$2s$	3043
L II	$2p_{1/2}$	2793
L III	$2p_{3/2}$	2677
M I	$3s$	586.1[a]
M II	$3p_{1/2}$	447.6[a]
M III	$3p_{3/2}$	417.7[a]
M IV	$3d_{3/2}$	257.6[a]
M V	$3d_{5/2}$	253.9[a]
N I	$4s$	69.5[a]
N II	$4p_{1/2}$	42.3[a]
N III	$4p_{3/2}$	39.9[a]

Tellurium (52)

K	$1s$	31814
L I	$2s$	4939
L II	$2p_{1/2}$	4612
L III	$2p_{3/2}$	4341
M I	$3s$	1006[b]
M II	$3p_{1/2}$	870.8[b]
M III	$3p_{3/2}$	820.0[b]
M IV	$3d_{3/2}$	583.4[b]
M V	$3d_{5/2}$	573.0[b]
N I	$4s$	169.4[b]
N II	$4p_{1/2}$	103.3[b,c]
N III	$4p_{3/2}$	103.3[b,c]
N IV	$4d_{3/2}$	41.9[b]
N V	$4d_{5/2}$	40.4[b]

Terbium (65)

K	$1s$	51996
L I	$2s$	8708
L II	$2p_{1/2}$	8252
L III	$2p_{3/2}$	7514
M I	$3s$	1968
M II	$3p_{1/2}$	1768
M III	$3p_{3/2}$	1611
M IV	$3d_{3/2}$	1267.9[a]
M V	$3d_{5/2}$	1241.1[a]
N I	$4s$	396.0[a]
N II	$4p_{1/2}$	322.4[a]
N III	$4p_{3/2}$	284.1[a]
N IV	$4d_{3/2}$	150.5[a]
N V	$4d_{5/2}$	150.5[a]
N VI	$4f_{5/2}$	7.7[a]
N VII	$4f_{7/2}$	2.4[a]
O I	$5s$	45.6[a]
O II	$5p_{1/2}$	28.7[a]
O III	$5p_{3/2}$	22.6[a]

Thallium (81)

K	$1s$	85530
L I	$2s$	15347
L II	$2p_{1/2}$	14698
L III	$2p_{3/2}$	12658
M I	$3s$	3704
M II	$3p_{1/2}$	3416
M III	$3p_{3/2}$	2957
M IV	$3d_{3/2}$	2485
M V	$3d_{5/2}$	2389
N I	$4s$	846.2[b]
N II	$4p_{1/2}$	720.5[b]
N III	$4p_{3/2}$	609.5[b]
N IV	$4d_{3/2}$	405.7[b]
N V	$4d_{5/2}$	385.0[b]
N VI	$4f_{5/2}$	122.2[b]
N VII	$4f_{7/2}$	117.8[b]
O I	$5s$	136[a,d]
O II	$5p_{1/2}$	94.6[b]
O III	$5p_{3/2}$	73.5[b]
O IV	$5d_{3/2}$	14.7[b]
O V	$5d_{5/2}$	12.5[b]

Thorium (90)

K	$1s$	109651
L I	$2s$	20472
L II	$2p_{1/2}$	19693
L III	$2p_{3/2}$	16300
M I	$3s$	5182
M II	$3p_{1/2}$	4830
M III	$3p_{3/2}$	4046
M IV	$3d_{3/2}$	3491
M V	$3d_{5/2}$	3332
N I	$4s$	1330[a]
N II	$4p_{1/2}$	1168[a]
N III	$4p_{3/2}$	966.4[b]
N IV	$4d_{3/2}$	712.1[b]
N V	$4d_{5/2}$	675.2[b]
N VI	$4f_{5/2}$	342.4[b]
N VII	$4f_{7/2}$	333.1[b]
O I	$5s$	290[a,c]
O II	$5p_{1/2}$	229[a,c]
O III	$5p_{3/2}$	182[a,c]
O IV	$5d_{3/2}$	92.5[b]
O V	$5d_{5/2}$	85.4[b]
P I	$6s$	41.4[b]
P II	$6p_{1/2}$	24.5[b]
P III	$6p_{3/2}$	16.6[b]

Thulium (69)

K	$1s$	59390
L I	$2s$	10116
L II	$2p_{1/2}$	9617
L III	$2p_{3/2}$	8648
M I	$3s$	2307
M II	$3p_{1/2}$	2090
M III	$3p_{3/2}$	1885
M IV	$3d_{3/2}$	1515
M V	$3d_{5/2}$	1468
N I	$4s$	470.9[a]
N II	$4p_{1/2}$	385.9[a]
N III	$4p_{3/2}$	332.6[a]
N IV	$4d_{3/2}$	175.5[a]
N V	$4d_{5/2}$	175.5[a]
N VI	$4f_{5/2}$	—
N VII	$4f_{7/2}$	4.6
O I	$5s$	54.7[a]
O II	$5p_{1/2}$	31.8[a]
O III	$5p_{3/2}$	25.0[a]

Tin (50)

K	$1s$	29200
L I	$2s$	4465
L II	$2p_{1/2}$	4156
L III	$2p_{3/2}$	3929
M I	$3s$	884.7[b]
M II	$3p_{1/2}$	756.5[b]
M III	$3p_{3/2}$	714.6[b]
M IV	$3d_{3/2}$	493.2[b]
M V	$3d_{5/2}$	484.9[b]
N I	$4s$	137.1[b]
N II	$4p_{1/2}$	83.6[b,c]
N III	$4p_{3/2}$	83.6[b,c]
N IV	$4d_{3/2}$	24.9[b]
N V	$4d_{5/2}$	23.9[b]

Titanium (22)

K	$1s$	4966
L I	$2s$	560.9[b]
L II	$2p_{1/2}$	460.2[b]
L III	$2p_{3/2}$	453.8[b]
M I	$3s$	58.7[b]
M II	$3p_{1/2}$	32.6[b]
M III	$3p_{3/2}$	32.6[b]

Tungsten (74)

K	$1s$	69525
L I	$2s$	12100
L II	$2p_{1/2}$	11544
L III	$2p_{3/2}$	10207
M I	$3s$	2820
M II	$3p_{1/2}$	2575
M III	$3p_{3/2}$	2281
M IV	$3d_{3/2}$	1949

M V	$3d_{5/2}$	1809
N I	4s	594.1[b]
N II	$4p_{1/2}$	490.4[b]
N III	$4p_{3/2}$	423.6[b]
N IV	$4d_{3/2}$	255.9[b]
N V	$4d_{5/2}$	243.5[b]
N VI	$4f_{5/2}$	33.6[a]
N VII	$4f_{7/2}$	31.4[b]
O I	5s	75.6[b]
O II	$5p_{1/2}$	453[a,d]
O III	$5p_{3/2}$	36.8[b]
Uranium (92)		
K	1s	115606
L I	2s	21757
L II	$2p_{1/2}$	20948
L III	$2p_{3/2}$	17166
M I	3s	5548
M II	$3p_{1/2}$	5182
M III	$3p_{3/2}$	4303
M IV	$3d_{3/2}$	3728
M V	$3d_{5/2}$	3552
N I	4s	1439[a,d]
N II	$4p_{1/2}$	1271[a,d]
N III	$4p_{3/2}$	1043[b]
N IV	$4d_{3/2}$	778.3[b]
N V	$4d_{5/2}$	736.2[b]
N VI	$4f_{5/2}$	388.2[a]
N VII	$4f_{7/2}$	377.4[b]
O I	5s	321[a,c,d]
O II	$5p_{1/2}$	257[a,c,d]
O III	$5p_{3/2}$	192[a,c,d]
O IV	$5d_{3/2}$	102.8[b]
O V	$5d_{5/2}$	94.2[b]
P I	6s	43.9[b]
P II	$6p_{1/2}$	26.8[b]
P III	$6p_{3/2}$	16.8[b]
Vanadium (23)		
K	1s	5465
L I	2s	626.7[b]
L II	$2p_{1/2}$	519.8[b]
L III	$2p_{3/2}$	521.1[b]
M I	3s	66.3[b]
M II	$3p_{1/2}$	37.2[b]
M III	$3p_{3/2}$	37.2[b]
Xenon (54)		
K	1s	34561
L I	2s	5453
L II	$2p_{1/2}$	5107
L III	$2p_{3/2}$	4786
M I	3s	1148.7[a]
M II	$3p_{1/2}$	1002.1[a]
M III	$3p_{3/2}$	940.6[a]
M IV	$3d_{3/2}$	689.0[a]
M V	$3d_{5/2}$	676.4[a]
N I	4s	213.2[a]
N II	$4p_{1/2}$	146.7
N III	$4p_{3/2}$	145.5[a]
N IV	$4d_{3/2}$	69.5[a]
N V	$4d_{5/2}$	67.5[a]
N VI	$4f_{5/2}$	—
N VII	$4f_{7/2}$	—
O I	5s	23.3[a]
O II	$5p_{1/2}$	13.4[a]
O III	$5p_{3/2}$	12.1[a]
Ytterbium (70)		
K	1s	61332
L I	2s	10486
L II	$2p_{1/2}$	9978
L III	$2p_{3/2}$	8944
M I	3s	2398
M II	$3p_{1/2}$	2173
M III	$3p_{3/2}$	1950
M IV	$3d_{3/2}$	1576
M V	$3d_{5/2}$	1528
N I	4s	480.5[a]
N II	$4p_{1/2}$	388.7[a]
N III	$4p_{3/2}$	339.7[a]
N IV	$4d_{3/2}$	191.2[a]
N V	$4d_{5/2}$	182.4[a]
N VI	$4f_{5/2}$	2.5[a]
N VII	$4f_{7/2}$	1.3[a]
O I	5s	52.0[a]
O II	$5p_{1/2}$	30.3[a]
O III	$5p_{3/2}$	24.1[a]
Yttrium (39)		
K	1s	17038
L I	2s	2373
L II	$2p_{1/2}$	2156
L III	$2p_{3/2}$	2080
M I	3s	392.0[a,d]
M II	$3p_{1/2}$	310.6[a]
M III	$3p_{3/2}$	298.8[a]
M IV	$3d_{3/2}$	157.7[b]
M V	$3d_{5/2}$	155.8[b]
N I	4s	43.8[a]
N II	$4p_{1/2}$	24.4[a]
N III	$4p_{3/2}$	23.1[a]
Zinc (30)		
K	1s	9659
L I	2s	1196.2[a]
L II	$2p_{1/2}$	1044.9[a]
L III	$2p_{3/2}$	1021.8[a]
M I	3s	139.8[a]
M II	$3p_{1/2}$	91.4[a]
M III	$3p_{3/2}$	88.6[a]
M IV	$3d_{3/2}$	10.2[a]
M V	$3d_{5/2}$	10.1[a]
Zirconium (40)		
K	1s	17998
L I	2s	2532
L II	$2p_{1/2}$	2307
L III	$2p_{3/2}$	2223
M I	3s	430.3[b]
M II	$3p_{1/2}$	343.5[b]
M III	$3p_{3/2}$	329.8[b]
M IV	$3d_{3/2}$	181.1[b]
M V	$3d_{5/2}$	178.8[b]
N I	4s	50.6[b]
N II	$4p_{1/2}$	28.5[b]
N III	$4p_{3/2}$	27.1[b]

[a] Reference 1.
[b] Reference 2 (remaining values from Reference 3).
[c] One-particle approximation not valid.
[d] Derived using energy differences from Reference 3.

NATURAL WIDTH OF X-RAY LINES

Natural widths of K X-ray lines in eV:

Element	$K\alpha_1$	$K\alpha_2$	$K\beta_1$	$K\beta_3$
Ca	1.00	0.98		
Ti	1.45	2.13		
Cr	2.05	2.64		
Fe	2.45	3.20		
Ni	3.00	3.70		
Zn	3.40	3.96		
Ge	3.75	4.18		
Se	4.10	4.43		
Kr	4.23	4.62		
Sr	5.17	4.97		
Zr	5.70	5.25		
Mo	6.82	6.80		
Ru	7.41	7.96		
Pd	8.80	9.20		
Cd	9.80	10.40		
Sn	11.20	12.40	11.80	11.00
Te	12.80	14.20	13.30	13.10
Xc	14.20	15.10	15.30	14.50
Ba	16.10	16.80	18.15	16.70
Ce	18.60	19.50	20.60	18.60
Nd	21.50	21.50	23.25	21.33
Sm	26.00	24.70	25.65	24.65
Gd	29.50	28.00	29.37	28.00
Dy	33.90	32.20	32.73	32.00
Er	35.00	35.50	36.20	35.70
Yb	38.80	40.60	41.43	41.15
Hf	42.70	44.30	46.00	46.10
W	46.80	48.00	51.83	51.50
Os	49.00	49.40	55.90	55.95
Pt	54.10	54.30	59.98	62.13
Hg	64.75	68.20	65.75	68.95
Pb	67.10	72.30	72.20	73.80
Po	73.20	75.10	78.60	80.10
Rn	80.00	81.50	85.50	86.50
Ra	87.00	88.20	94.20	95.50
Th	94.70	95.00	99.70	101.00
U	103.00	104.30	105.00	107.30

From Salem, S. I. and Lee, P. L., *At. Data Nucl. Data Tables*, 18, 233, 1976.

Natural widths of L X-ray lines in eV:

Element	$L\alpha_1$	$L\alpha_2$	$L\beta_1$	$L\beta_2$	$L\beta_3$	$L\beta_4$	$L\gamma_1$
Zr	1.68	1.52	1.87	5.13	5.50	5.60	3.34
Mo	1.86	1.80	2.03	5.30	5.90	5.78	3.76
Ru	2.03	1.98	2.18	5.45	6.35	5.96	4.15
Pd	2.21	2.16	2.36	5.63	6.80	6.18	4.50
Gd	2.43	2.40	2.54	5.82	7.23	6.28	4.83
Sn	2.62	2.62	2.75	6.10	7.70	6.60	5.23
Tc	2.88	2.88	2.96	6.25	8.22	6.82	5.60
Xe	3.15	3.15	3.20	6.43	8.70	7.15	5.95
Ba	3.39	3.45	3.45	6.70	9.20	7.42	6.35
Ce	3.70	3.78	3.73	6.86	9.70	7.82	6.75
Nd	3.93	4.08	4.00	7.18	10.30	8.15	7.16
Sm	4.13	4.50	4.33	7.42	10.80	8.60	7.50
Cd	1.16	1.90	4.60	7.70	11.20	9.08	7.83
Dy	4.81	5.35	5.03	7.90	11.50	9.60	8.30
Er	5.17	5.73	5.45	8.28	11.85	10.03	8.75
Yb	5.40	6.22	5.90	8.58	12.20	11.00	9.20
Hf	5.83	6.70	6.36	8.92	12.40	12.80	9.63
W	6.50	7.20	6.90	9.06	13.10	14.60	10.20
Os	7.04	7.70	7.42	9.60	14.60	16.50	10.65
Pt	7.60	8.28	8.00	9.95	16.10	18.00	11.20
Hg	8.10	8.80	8.70	10.40	17.40	19.70	11.80
Pb	8.82	9.35	9.35	10.75	18.65	21.30	12.30
Po	9.50	9.95	10.10	11.25	19.90	22.70	13.05
Rn	10.03	10.50	10.65	11.65	21.00	24.00	13.55
Ra	11.00	11.20	11.60	12.20	22.00	25.20	14.30
Th	11.90	11.80	12.40	12.80	22.85	26.35	15.00
U	12.40	12.40	13.50	13.30	23.70	27.50	15.70
Pu	13.20	13.00	14.10	13.90	24.10	28.30	16.40
Cm	14.80	13.60	15.70	14.60	25.00	29.40	17.10

PHOTON ATTENUATION COEFFICIENTS

Martin J. Berger and John H. Hubbell

This table gives mass attenuation coefficients for photons for all elements at energies between 1 keV (soft x-rays) and 1 GeV (hard gamma rays). The mass attenuation coefficient μ describes the attenuation of radiation as it passes through matter by the relation

$$I(x)/I_o = e^{-\mu\rho x}$$

where I_o is the initial intensity, $I(x)$ the intensity after path length x, and ρ is the mass density of the element in question. To a high approximation the mass attenuation coefficient is additive for the elements present, independent of the way in which they are bound in chemical compounds.

The power of ten is indicated beside each number in the table; i.e., 7.41 + 03 means 7.41×10^3. A vertical line between two columns indicates that an absorption edge lies between those energy values. The various edges are labeled at the bottom of the table.

The attenuation coefficients were calculated with the computer program XCOM (Reference 1), which uses a cross-section database compiled at the Photon and Charged Particle Data Center at the National Institute of Standards and Technology. Their accuracy has been confirmed at all energies by extensive comparisons with experimental attenuation coefficients. Such comparisons for X-ray energies up to 100 keV can be found in Reference 2.

References

1. Berger, M. J. and Hubbell, J. H., *National Bureau of Standards Report* NBSIR-87-3597, 1987.
2. Saloman, E. B., Hubbell, J. H., and Scofield, J. H., *Atomic Data and Nuclear Data Tables*, 38, 1, 1988.

	Atomic no.	Mass attenuation coefficient, cm²/g Photon energy, MeV 0.001	0.002	0.005	0.01	0.02	0.05	0.1	0.2	0.5
H	1	7.21 + 00	1.06 + 00	4.19-01	3.85-01	3.69-01	3.36-01	2.94-01	2.43-01	1.73-01
He	2	6.08 + 01	6.86 + 00	5.77-01	2.48-01	1.96-01	1.70-01	1.49-01	1.22-01	8.71-02
Li	3	2.34 + 02	2.71 + 01	1.62 + 00	3.40-01	1.86-01	1.49-01	1.29-01	1.06-01	7.53-02
Be	4	6.04 + 02	7.47 + 01	4.37 + 00	6.47-01	2.25-01	1.55-01	1.33-01	1.09-01	7.74-02
B	5	1.23 + 03	1.60 + 02	9.68 + 00	1.25 + 00	3.01-01	1.66-01	1.39-01	1.14-01	8.07-02
C	6	2.21 + 03	3.03 + 02	1.91 + 01	2.37 + 00	4.42-01	1.87-01	1.51-01	1.23-01	8.72-02
N	7	3.31 + 03	4.77 + 02	3.14 + 01	3.88 + 00	6.18-01	1.98-01	1.53-01	1.23-01	8.72-02
O	8	4.59 + 03	6.95 + 02	4.79 + 01	5.95 + 00	8.65-01	2.13-01	1.55-01	1.24-01	8.73-02
F	9	5.65 + 03	9.05 + 02	6.51 + 01	8.21 + 00	1.13 + 00	2.21-01	1.50-01	1.18-01	8.27-02
Ne	10	7.41 + 03	1.24 + 03	9.34 + 01	1.20 + 01	1.61 + 00	2.58-01	1.60-01	1.24-01	8.66-02
Na	11	6.54 + 02	1.52 + 03	1.19 + 02	1.56 + 01	2.06 + 00	2.80-01	1.59-01	1.20-01	8.37-02
Mg	12	9.22 + 02	1.93 + 03	1.58 + 02	2.11 + 01	2.76 + 00	3.29-01	1.69-01	1.24-01	8.65-02
Al	13	1.19 + 03	2.26 + 03	1.93 + 02	2.62 + 01	3.44 + 00	3.68-01	1.70-01	1.22-01	8.44-02
Si	14	1.57 + 03	2.78 + 03	2.45 + 02	3.39 + 01	4.46 + 00	4.38-01	1.84-01	1.28-01	8.75-02
P	15	1.91 + 03	3.02 + 02	2.86 + 02	4.04 + 01	5.35 + 00	4.92-01	1.87-01	1.25-01	8.51-02
S	16	2.43 + 03	3.85 + 02	3.49 + 02	5.01 + 01	6.71 + 00	5.85-01	2.02-01	1.30-01	8.78-02
Cl	17	2.83 + 03	4.52 + 02	3.90 + 02	5.73 + 01	7.74 + 00	6.48-01	2.05-01	1.27-01	8.45-02
Ar	18	3.18 + 03	5.12 + 02	4.23 + 02	6.32 + 01	8.63 + 00	7.01-01	2.04-01	1.20-01	7.96-02
K	19	4.06 + 03	6.59 + 02	5.19 + 02	7.91 + 01	1.09 + 01	8.68-01	2.34-01	1.32-01	8.60-02
Ca	20	4.87 + 03	8.00 + 02	6.03 + 02	9.34 + 01	1.31 + 01	1.02 + 00	2.57-01	1.38-01	8.85-02
Sc	21	5.24 + 03	8.70 + 02	6.31 + 02	9.95 + 01	1.41 + 01	1.09 + 00	2.58-01	1.31-01	8.31-02
Ti	22	5.87 + 03	9.86 + 02	6.84 + 02	1.11 + 02	1.59 + 01	1.21 + 00	2.72-01	1.31-01	8.19-02
V	23	6.50 + 03	1.11 + 03	9.29 + 01	1.22 + 02	1.77 + 01	1.35 + 00	2.88-01	1.32-01	8.07-02
Cr	24	7.40 + 03	1.28 + 03	1.08 + 02	1.39 + 02	2.04 + 01	1.55 + 00	3.17-01	1.38-01	8.28-02
Mn	25	8.09 + 03	1.42 + 03	1.21 + 02	1.51 + 02	2.25 + 01	1.71 + 00	3.37-01	1.39-01	8.19-02
Fe	26	9.09 + 03	1.63 + 03	1.40 + 02	1.71 + 02	2.57 + 01	1.96 + 00	3.72-01	1.46-01	8.41-02
Co	27	9.80 + 03	1.78 + 03	1.54 + 02	1.84 + 02	2.80 + 01	2.14 + 00	3.95-01	1.48-01	8.32-02
Ni	28	9.86 + 03	2.05 + 03	1.79 + 02	2.09 + 02	3.22 + 01	2.47 + 00	4.44-01	1.58-01	8.70-02
Cu	29	1.06 + 04	2.15 + 03	1.90 + 02	2.16 + 02	3.38 + 01	2.61 + 00	4.58-01	1.56-01	8.36-02
Zn	30	1.55 + 03	2.37 + 03	2.12 + 02	2.33 + 02	3.72 + 01	2.89 + 00	4.97-01	1.62-01	8.45-02
Ga	31	1.70 + 03	2.52 + 03	2.27 + 02	3.42 + 01	3.93 + 01	3.08 + 00	5.20-01	1.62-01	8.24-02
Ge	32	1.89 + 03	2.71 + 03	2.47 + 02	3.74 + 01	4.22 + 01	3.34 + 00	5.55-01	1.66-01	8.21-02
As	33	2.12 + 03	2.93 + 03	2.71 + 02	4.12 + 01	4.56 + 01	3.63 + 00	5.97-01	1.72-01	8.26-02
Se	34	2.32 + 03	3.10 + 03	2.90 + 02	4.41 + 01	4.82 + 01	3.86 + 00	6.28-01	1.74-01	8.13-02
Br	35	2.62 + 03	3.41 + 03	3.21 + 02	4.91 + 01	5.27 + 01	4.26 + 00	6.86-01	1.84-01	8.33-02
Kr	36	2.85 + 03	3.60 + 03	3.43 + 02	5.26 + 01	5.55 + 01	4.52 + 00	7.22-01	1.87-01	8.23-02
Rb	37	3.17 + 03	3.41 + 03	3.74 + 02	5.77 + 01	5.98 + 01	4.92 + 00	7.80-01	1.96-01	8.36-02
Sr	38	3.49 + 03	2.59 + 03	4.06 + 02	6.27 + 01	6.39 + 01	5.31 + 00	8.37-01	2.01-01	0.44-02
Y	39	3.86 + 03	7.42 + 02	4.42 + 02	6.87 + 01	6.86 + 01	5.76 + 00	9.05-01	2.15-01	8.61-02
Zr	40	4.21 + 03	8.12 + 02	4.76 + 02	7.42 + 01	7.24 + 01	6.17 + 00	9.66-01	2.24-01	8.69-02
Nb	41	4.60 + 03	8.89 + 02	5.13 + 02	8.04 + 01	7.71 + 01	6.64 + 00	1.04 + 00	2.34-01	8.83-02
Mo	42	4.94 + 03	9.60 + 02	5.45 + 02	8.58 + 01	1.31 + 01	7.04 + 00	1.10 + 00	2.42-01	8.85-02
Tc	43	5.36 + 03	1.04 + 03	5.84 + 02	9.23 + 01	1.41 + 01	7.52 + 00	1.17 + 00	2.53-01	8.97-02
Ru	44	5.72 + 03	1.12 + 03	6.17 + 02	9.80 + 01	1.50 + 01	7.92 + 00	1.23 + 00	2.62-01	8.99-02
Rh	45	6.17 + 03	1.21 + 03	6.59 + 02	1.05 + 02	1.61 + 01	8.45 + 00	1.31 + 00	2.74-01	9.13-02
Pd	46	6.54 + 03	1.29 + 03	6.91 + 02	1.11 + 02	1.70 + 01	8.85 + 00	1.38 + 00	2.83-01	9.13-02
Ag	47	7.04 + 03	1.40 + 03	7.39 + 02	1.19 + 02	1.84 + 01	9.45 + 00	1.47 + 00	2.97-01	9.32-02
Cd	48	7.35 + 03	1.47 + 03	7.69 + 02	1.24 + 02	1.92 + 01	9.78 + 00	1.52 + 00	3.04-01	9.25-02
In	49	7.81 + 03	1.58 + 03	8.13 + 02	1.32 + 02	2.04 + 01	1.03 + 01	1.61 + 00	3.17-01	9.37-02
Sn	50	8.16 + 03	1.66 + 03	8.47 + 02	1.38 + 02	2.15 + 01	1.07 + 01	1.68 + 00	3.26-01	9.37-02

L_3 L_1 L_2 K EDGE

	Atomic no.	Mass attenuation coefficient, cm²/g								
		Photon energy, MeV								
		1.0	2.0	5.0	10.0	20.0	50.0	100.0	500.0	1000.0
H	1	1.26-01	8.77-02	5.05-02	3.25-02	2.15-02	1.42-02	1.19-02	1.14-02	1.16-02
He	2	6.36-02	4.42-02	2.58-02	1.70-02	1.18-02	8.61-03	7.78-03	7.79-03	7.95-03
Li	3	5.50-02	3.83-02	2.26-02	1.53-02	1.11-02	8.68-03	8.21-03	8.61-03	8.87-03
Be	4	5.65-02	3.94-02	2.35-02	1.63-02	1.23-02	1.02-02	9.94-03	1.08-02	1.12-02
B	5	5.89-02	4.11-02	2.48-02	1.76-02	1.37-02	1.19-02	1.19-02	1.32-02	1.37-02
C	6	6.36-02	4.44-02	2.71-02	1.96-02	1.58-02	1.43-02	1.46-02	1.64-02	1.70-02
N	7	6.36-02	4.45-02	2.74-02	2.02-02	1.67-02	1.57-02	1.63-02	1.85-02	1.92-02
O	8	6.37-02	4.46-02	2.78-02	2.09-02	1.77-02	1.71-02	1.79-02	2.06-02	2.13-02
F	9	6.04-02	4.23-02	2.66-02	2.04-02	1.77-02	1.75-02	1.86-02	2.14-02	2.21-02
Ne	10	6.32-02	4.43-02	2.82-02	2.20-02	1.95-02	1.96-02	2.11-02	2.43-02	2.51-02
Na	11	6.10-02	4.28-02	2.75-02	2.18-02	1.97-02	2.03-02	2.19-02	2.53-02	2.62-02
Mg	12	6.30-02	4.43-02	2.87-02	2.31-02	2.13-02	2.23-02	2.42-02	2.81-02	2.90-02
Al	13	6.15-02	4.32-02	2.84-02	2.32-02	2.17-02	2.31-02	2.52-02	2.93-02	3.03-02
Si	14	6.36-02	4.48-02	2.97-02	2.46-02	2.34-02	2.52-02	2.76-02	3.23-02	3.34-02
P	15	6.18-02	4.36-02	2.91-02	2.45-02	2.36-02	2.58-02	2.84-02	3.33-02	3.45-02
S	16	6.37-02	4.50-02	3.04-02	2.59-02	2.53-02	2.79-02	3.08-02	3.62-02	3.75-02
Cl	17	6.13-02	4.33-02	2.95-02	2.55-02	2.52-02	2.81-02	3.11-02	3.67-02	3.80-02
Ar	18	5.76-02	4.07-02	2.80-02	2.45-02	2.45-02	2.76-02	3.07-02	3.62-02	3.75-02
K	19	6.22-02	4.40-02	3.05-02	2.70-02	2.74-02	3.11-02	3.46-02	4.09-02	4.24-02
Ca	20	6.39-02	4.52-02	3.17-02	2.84-02	2.90-02	3.32-02	3.71-02	4.40-02	4.56-02
Sc	21	5.98-02	4.24-02	3.00-02	2.72-02	2.80-02	3.23-02	3.62-02	4.30-02	4.45-02
Ti	22	5.89-02	4.18-02	2.98-02	2.73-02	2.84-02	3.30-02	3.71-02	4.40-02	4.56-02
V	23	5.79-02	4.11-02	2.96-02	2.74-02	2.88-02	3.36-02	3.78-02	4.49-02	4.65-02
Cr	24	5.93-02	4.21-02	3.06-02	2.86-02	3.03-02	3.56-02	4.01-02	4.76-02	4.93-02
Mn	25	5.85-02	4.16-02	3.04-02	2.87-02	3.07-02	3.63-02	4.09-02	4.86-02	5.04-02
Fe	26	5.99-02	4.26-02	3.15-02	2.99-02	3.22-02	3.83-02	4.33-02	5.15-02	5.33-02
Co	27	5.91-02	4.20-02	3.13-02	3.00-02	3.26-02	3.88-02	4.40-02	5.23-02	5.41-02
Ni	28	6.16-02	4.39-02	3.29-02	3.18-02	3.48-02	4.17-02	4.73-02	5.61-02	5.81-02
Cu	29	5.90-02	4.20-02	3.18-02	3.10-02	3.41-02	4.10-02	4.66-02	5.53-02	5.72-02
Zn	30	5.94-02	4.24-02	3.22-02	3.18-02	3.51-02	4.24-02	4.82-02	5.72-02	5.91-02
Ga	31	5.77-02	4.11-02	3.16-02	3.13-02	3.48-02	4.22-02	4.80-02	5.70-02	5.89-02
Ge	32	5.73-02	4.09-02	3.16-02	3.16-02	3.53-02	4.30-02	4.89-02	5.80-02	6.00-02
As	33	5.73-02	4.09-02	3.19-02	3.21-02	3.60-02	4.40-02	5.01-02	5.95-02	6.15-02
Se	34	5.62-02	4.01-02	3.14-02	3.19-02	3.60-02	4.41-02	5.03-02	5.97-02	6.17-02
Br	35	5.73-02	4.09-02	3.23-02	3.29-02	3.74-02	4.60-02	5.24-02	6.22-02	6.43-02
Kr	36	5.63-02	4.02-02	3.20-02	3.28-02	3.74-02	4.61-02	5.26-02	6.25-02	6.46-02
Rb	37	5.69-02	4.06-02	3.25-02	3.36-02	3.85-02	4.75-02	5.43-02	6.45-02	6.67-02
Sr	38	5.71-02	4.08-02	3.29-02	3.41-02	3.93-02	4.87-02	5.56-02	6.61-02	6.83-02
Y	39	5.80-02	4.14-02	3.35-02	3.50-02	4.05-02	5.03-02	5.75-02	6.83-02	7.06-02
Zr	40	5.81-02	4.15-02	3.38-02	3.55-02	4.12-02	5.13-02	5.87-02	6.98-02	7.22-02
Nb	41	5.87-02	4.18-02	3.44-02	3.63-02	4.22-02	5.27-02	6.03-02	7.17-02	7.42-02
Mo	42	5.84-02	4.16-02	3.44-02	3.65-02	4.26-02	5.33-02	6.10-02	7.26-02	7.51-02
Tc	43	5.88-02	4.19-02	3.48-02	3.71-02	4.35-02	5.45-02	6.24-02	7.43-02	7.68-02
Ru	44	5.85-02	4.16-02	3.48-02	3.73-02	4.39-02	5.50-02	6.30-02	7.51-02	7.77-02
Rh	45	5.89-02	4.20-02	3.53-02	3.80-02	4.48-02	5.63-02	6.45-02	7.69-02	7.94-02
Pd	46	5.85-02	4.16-02	3.52-02	3.80-02	4.50-02	5.66-02	6.49-02	7.73-02	8.00-02
Ag	47	5.92-02	4.21-02	3.58-02	3.88-02	4.61-02	5.81-02	6.67-02	7.93-02	8.20-02
Cd	48	5.83-02	4.14-02	3.54-02	3.85-02	4.59-02	5.79-02	6.64-02	7.91-02	8.18-02
In	49	5.85-02	4.15-02	3.56-02	3.90-02	4.65-02	5.88-02	6.75-02	8.04-02	8.32-02
Sn	50	5.80-02	4.11-02	3.55-02	3.90-02	4.66-02	5.90-02	6.78-02	8.07-02	8.35-02

	Atomic no.	Mass attenuation coefficient, cm²/g — Photon energy, MeV								
		0.001	0.002	0.005	0.01	0.02	0.05	0.1	0.2	0.5
Sb	51	8.58 + 03	1.77 + 03	8.85 + 02	1.46 + 02	2.27 + 01	1.12 + 01	1.76 + 00	3.38-01	9.45-02
Te	52	8.43 + 03	1.83 + 03	9.01 + 02	1.50 + 02	2.34 + 01	1.14 + 01	1.80 + 00	3.43-01	9.33-02
I	53	9.10 + 03	2.00 + 03	8.43 + 02	1.63 + 02	2.54 + 01	1.23 + 01	1.94 + 00	3.66-01	9.70-02
Xe	54	9.41 + 03	2.09 + 03	6.39 + 02	1.69 + 02	2.65 + 01	1.27 + 01	2.01 + 00	3.76-01	9.70-02
Cs	55	9.37 + 03	2.23 + 03	2.30 + 02	1.79 + 02	2.82 + 01	1.34 + 01	2.12 + 00	3.94-01	9.91-02
ZB	56	8.54 + 03	2.32 + 03	2.41 + 02	1.86 + 02	2.94 + 01	1.38 + 01	2.20 + 00	4.05-01	9.92-02
La	57	9.09 + 03	2.46 + 03	2.58 + 02	1.97 + 02	3.12 + 01	1.45 + 01	2.32 + 00	4.24-01	1.01-01
Ce	58	9.71 + 03	2.61 + 03	2.74 + 02	2.08 + 02	3.31 + 01	1.52 + 01	2.45 + 00	4.45-01	1.04-01
Pr	59	1.06 + 04	2.77 + 03	2.92 + 02	2.21 + 02	3.53 + 01	1.60 + 01	2.59 + 00	4.69-01	1.07-01
Nd	60	6.63 + 03	2.88 + 03	3.06 + 02	2.30 + 02	3.68 + 01	1.65 + 01	2.69 + 00	4.84-01	1.08-01
Pm	61	2.06 + 03	3.05 + 03	3.26 + 02	2.44 + 02	3.92 + 01	1.73 + 01	2.84 + 00	5.10-01	1.12-01
Sm	62	2.11 + 03	3.12 + 03	3.36 + 02	2.50 + 02	4.03 + 01	1.77 + 01	2.90 + 00	5.19-01	1.11-01
Eu	63	2.22 + 03	3.28 + 03	3.54 + 02	2.63 + 02	4.24 + 01	1.85 + 01	3.04 + 00	5.43-01	1.14-01
Gd	64	2.29 + 03	3.36 + 03	3.65 + 02	2.69 + 02	4.36 + 01	3.86 + 00	3.11 + 00	5.54-01	1.14-01
Tb	65	2.40 + 03	3.51 + 03	3.84 + 02	2.82 + 02	4.59 + 01	4.06 + 00	3.25 + 00	5.77-01	1.17-01
Dy	66	2.49 + 03	3.47 + 03	3.99 + 02	2.90 + 02	4.76 + 01	4.23 + 00	3.36 + 00	5.95-01	1.18-01
Ho	67	2.62 + 03	3.59 + 03	4.17 + 02	3.01 + 02	4.98 + 01	4.43 + 00	3.49 + 00	6.18-01	1.20-01
Er	68	2.75 + 03	3.52 + 03	4.36 + 02	3.13 + 02	5.20 + 01	4.63 + 00	3.63 + 00	6.41-01	1.23-01
Tm	69	2.90 + 03	3.69 + 03	4.57 + 02	2.83 + 02	5.45 + 01	4.87 + 00	3.78 + 00	6.68-01	1.26-01
Yb	70	3.02 + 03	3.80 + 03	4.72 + 02	2.94 + 02	5.63 + 01	5.04 + 00	3.88 + 00	6.86-01	1.27-01
Lu	71	3.19 + 03	3.45 + 03	4.94 + 02	2.21 + 02	5.88 + 01	5.28 + 00	4.03 + 00	7.13-01	1.30-01
Hf	72	3.34 + 03	3.60 + 03	5.11 + 02	2.30 + 02	6.09 + 01	5.48 + 00	4.15 + 00	7.34-01	1.32-01
Ta	73	3.51 + 03	3.77 + 03	5.33 + 02	2.38 + 02	6.33 + 01	5.72 + 00	4.30 + 00	7.60-01	1.35-01
W	74	3.68 + 03	3.92 + 03	5.53 + 02	9.69 + 01	6.57 + 01	5.95 + 00	4.44 + 00	7.84-01	1.38-01
Re	75	3.87 + 03	3.77 + 03	5.76 + 02	1.01 + 02	6.84 + 01	6.21 + 00	4.59 + 00	8.12-01	1.41-01
Os	76	4.03 + 03	2.22 + 03	5.93 + 02	1.04 + 02	7.04 + 01	6.41 + 00	4.70 + 00	8.33-01	1.43-01
Ir	77	4.24 + 03	1.03 + 03	6.18 + 02	1.09 + 02	7.32 + 01	6.69 + 00	4.86 + 00	8.63-01	1.46-01
Pt	78	4.43 + 03	1.08 + 03	6.40 + 02	1.13 + 02	7.57 + 01	6.95 + 00	4.99 + 00	8.90-01	1.49-01
Au	79	4.65 + 03	1.14 + 03	6.66 + 02	1.18 + 02	7.88 + 01	7.26 + 00	5.16 + 00	9.22-01	1.53-01
Hg	80	4.83 + 03	1.18 + 03	6.87 + 02	1.22 + 02	8.12 + 01	7.50 + 00	5.28 + 00	9.46-01	1.56-01
Tl	81	5.01 + 03	1.23 + 03	7.07 + 02	1.26 + 02	8.36 + 01	7.75 + 00	5.40 + 00	9.69-01	1.58-01
Pb	82	5.21 + 03	1.29 + 03	7.30 + 02	1.31 + 02	8.64 + 01	8.04 + 00	5.55 + 00	9.99-01	1.61-01
Bi	83	5.44 + 03	1.35 + 03	7.58 + 02	1.36 + 02	8.95 + 01	8.38 + 00	5.74 + 00	1.03 + 00	1.66-01
Po	84	5.72 + 03	1.42 + 03	7.93 + 02	1.43 + 02	9.35 + 01	8.80 + 00	5.99 + 00	1.08 + 00	1.71-01
At	85	5.87 + 03	1.49 + 03	8.25 + 02	1.49 + 02	9.70 + 01	9.19 + 00	6.17 + 00	1.12 + 00	1.77-01
Rn	86	5.83 + 03	1.49 + 03	8.16 + 02	1.48 + 02	9.56 + 01	9.12 + 00	6.09 + 00	1.10 + 00	1.73-01
Fr	87	6.08 + 03	1.56 + 03	8.49 + 02	1.54 + 02	9.93 + 01	9.52 + 00	1.66 + 00	1.14 + 00	1.78-01
Ra	88	6.20 + 03	1.62 + 03	8.74 + 02	1.59 + 02	1.02 + 02	9.85 + 00	1.71 + 00	1.17 + 00	1.82-01
Ac	89	6.47 + 03	1.70 + 03	8.69 + 02	1.65 + 02	1.06 + 02	1.03 + 01	1.79 + 00	1.21 + 00	1.87-01
Th	90	6.61 + 03	1.74 + 03	8.88 + 02	1.69 + 02	9.37 + 01	1.05 + 01	1.83 + 00	1.23 + 00	1.90-01
Pa	91	6.53 + 03	1.83 + 03	8.76 + 02	1.77 + 02	7.03 + 01	1.10 + 01	1.92 + 00	1.29 + 00	1.97-01
U	92	6.63 + 03	1.86 + 03	8.89 + 02	1.79 + 02	7.11 + 01	1.12 + 01	1.95 + 00	1.30 + 00	1.98-01
Np	93	6.95 + 03	1.96 + 03	9.32 + 02	1.87 + 02	7.45 + 01	1.18 + 01	2.05 + 00	1.35 + 00	2.05-01
Pu	94	7.19 + 03	2.04 + 03	9.65 + 02	1.94 + 02	7.71 + 01	1.22 + 01	2.13 + 00	1.39 + 00	2.10-01
Am	95	7.37 + 03	2.10 + 03	9.90 + 02	1.98 + 02	7.93 + 01	1.25 + 01	2.19 + 00	1.42 + 00	2.14-01
Cm	96	7.54 + 03	2.15 + 03	1.02 + 03	2.03 + 02	8.14 + 01	1.28 + 01	2.25 + 00	1.44 + 00	2.18-01
Bk	97	7.84 + 03	2.25 + 03	1.06 + 03	2.10 + 02	8.39 + 01	1.34 + 01	2.35 + 00	1.50 + 00	2.25-01
Cf	98	7.89 + 03	2.31 + 03	9.27 + 02	2.15 + 02	8.58 + 01	1.37 + 01	2.41 + 00	1.52 + 00	2.29-01
Es	99	7.79 + 03	2.40 + 03	9.59 + 02	2.22 + 02	4.01 + 01	1.42 + 01	2.51 + 00	1.57 + 00	2.36-01
Fm	100	7.13 + 03	2.46 + 03	9.77 + 02	2.26 + 02	4.09 + 01	1.45 + 01	2.57 + 00	1.59 + 00	2.39-01

Edge lines: N_5, N_4, N_3, N_2, N_1; M_5, M_4; M_3, M_2, M_1; L_3, L_2, L_1; K EDGE

	Atomic no.	Mass attenuation coefficient, cm²/g — Photon energy, MeV 1.0	2.0	5.0	10.0	20.0	50.0	100.0	500.0	1000.0
Sb	51	5.80-02	4.10-02	3.56-02	3.92-02	4.70-02	5.96-02	6.85-02	8.16-02	8.44-02
Te	52	5.67-02	4.01-02	3.49-02	3.86-02	4.64-02	5.89-02	6.77-02	8.07-02	8.35-02
I	53	5.84-02	4.12-02	3.61-02	4.00-02	4.82-02	6.13-02	7.04-02	8.40-02	8.69-02
Xe	54	5.78-02	4.08-02	3.58-02	3.99-02	4.82-02	6.12-02	7.04-02	8.40-02	8.69-02
Cs	55	5.85-02	4.12-02	3.64-02	4.06-02	4.91-02	6.25-02	7.19-02	8.58-02	8.88-02
ZB	56	5.80-02	4.08-02	3.61-02	4.04-02	4.90-02	6.25-02	7.19-02	8.58-02	8.88-02
La	57	5.88-02	4.12-02	3.66-02	4.11-02	5.00-02	6.37-02	7.34-02	8.76-02	9.06-02
Ce	58	5.96-02	4.18-02	3.73-02	4.19-02	5.10-02	6.52-02	7.50-02	8.96-02	9.27-02
Pr	59	6.07-02	4.24-02	3.80-02	4.29-02	5.23-02	6.68-02	7.69-02	9.19-02	9.50-02
Nd	60	6.07-02	4.24-02	3.81-02	4.30-02	5.26-02	6.72-02	7.74-02	9.25-02	9.56-02
Pm	61	6.19-02	4.31-02	3.88-02	4.40-02	5.38-02	6.89-02	7.94-02	9.48-02	9.81-02
Sm	62	6.11-02	4.24-02	3.83-02	4.35-02	5.34-02	6.84-02	7.88-02	9.41-02	9.73-02
Eu	63	6.19-02	4.28-02	3.88-02	4.42-02	5.42-02	6.96-02	8.02-02	9.57-02	9.90-02
Gd	64	6.12-02	4.23-02	3.84-02	4.38-02	5.38-02	6.91-02	7.97-02	9.51-02	9.83-02
Tb	65	6.20-02	4.27-02	3.89-02	4.45-02	5.47-02	7.03-02	8.11-02	9.67-02	1.00-01
Dy	66	6.20-02	4.26-02	3.90-02	4.46-02	5.49-02	7.06-02	8.15-02	9.72-02	1.00-01
Ho	67	6.26-02	4.29-02	3.93-02	4.50-02	5.55-02	7.14-02	8.24-02	9.83-02	1.02-01
Er	68	6.32-02	4.32-02	3.96-02	4.55-02	5.61-02	7.23-02	8.34-02	9.95-02	1.03-01
Tm	69	6.40-02	4.36-02	4.01-02	4.61-02	5.70-02	7.35-02	8.48-02	1.01-01	1.04-01
Yb	70	6.40-02	4.35-02	4.00-02	4.61-02	5.70-02	7.35-02	8.49-02	1.01-01	1.04-01
Lu	71	6.48-02	4.39-02	4.05-02	4.66-02	5.77-02	7.45-02	8.60-02	1.02-01	1.06-01
Hf	72	6.50-02	4.39-02	4.05-02	4.68-02	5.80-02	7.48-02	8.64-02	1.03-01	1.06-01
Ta	73	6.57-02	4.41-02	4.08-02	4.72-02	5.85-02	7.56-02	8.73-02	1.04-01	1.07-01
W	74	6.62-02	4.43-02	4.10-02	4.75-02	5.89-02	7.62-02	8.80-02	1.05-01	1.08-01
Re	75	6.69-02	4.46-02	4.14-02	4.79-02	5.95-02	7.70-02	8.89-02	1.06-01	1.09-01
Os	76	6.71-02	4.46-02	4.13-02	4.79-02	5.96-02	7.71-02	8.90-02	1.06-01	1.10-01
Ir	77	6.79-02	4.50-02	4.17-02	4.84-02	6.02-02	7.80-02	9.01-02	1.07-01	1.11-01
Pt	78	6.86-02	4.52-02	4.20-02	4.87-02	6.06-02	7.86-02	9.08-02	1.08-01	1.12-01
Au	79	6.95-02	4.57-02	4.24-02	4.93-02	6.14-02	7.95-02	9.19-02	1.09-01	1.13-01
Hg	80	6.99-02	4.57-02	4.25-02	4.94-02	6.15-02	7.98-02	9.22-02	1.10-01	1.13-01
Tl	81	7.03-02	4.58-02	4.25-02	4.94-02	6.16-02	8.00-02	9.24-02	1.10-01	1.14-01
Pb	82	7.10-02	4.61-02	4.27-02	4.97-02	6.21-02	8.06-02	9.31-02	1.11-01	1.15-01
Bi	83	7.21-02	4.66-02	4.32-02	5.03-02	6.28-02	8.15-02	9.42-02	1.12-01	1.16-01
Po	84	7.39-02	4.75-02	4.40-02	5.12-02	6.40-02	8.32-02	9.61-02	1.15-01	1.18-01
At	85	7.54-02	4.82-02	4.46-02	5.20-02	6.49-02	8.44-02	9.76-02	1.16-01	1.20-01
Rn	86	7.30-02	4.65-02	4.30-02	5.01-02	6.26-02	8.14-02	9.42-02	1.12-01	1.16-01
Fr	87	7.45-02	4.72-02	4.36-02	5.08-02	6.35-02	8.26-02	9.56-02	1.14-01	1.18-01
Ra	88	7.53-02	4.75-02	4.38-02	5.10-02	6.38-02	8.31-02	9.61-02	1.15-01	1.19-01
Ac	89	7.69-02	4.82-02	4.44-02	5.17-02	6.47-02	8.43-02	9.75-02	1.16-01	1.20-01
Th	90	7.71-02	4.81-02	4.42-02	5.15-02	6.45-02	8.40-02	9.72-02	1.16-01	1.20-01
Pa	91	7.94-02	4.93-02	4.52-02	5.26-02	6.59-02	8.60-02	9.95-02	1.19-01	1.23-01
U	92	7.90-02	4.88-02	4.46-02	5.19-02	6.51-02	8.49-02	9.83-02	1.17-01	1.21-01
Np	93	8.13-02	4.99-02	4.56-02	5.30-02	6.65-02	8.68-02	1.01-01	1.20-01	1.24-01
Pu	94	8.26-02	5.05-02	4.60-02	5.34-02	6.71-02	8.76-02	1.01-01	1.21-01	1.25-01
Am	95	8.33-02	5.06-02	4.60-02	5.34-02	6.70-02	8.77-02	1.02-01	1.21-01	1.25-01
Cm	96	8.41-02	5.08-02	4.60-02	5.34-02	6.70-02	8.77-02	1.02-01	1.21-01	1.26-01
Bk	97	8.62-02	5.18-02	4.68-02	5.42-02	6.81-02	8.92-02	1.03-01	1.24-01	1.28-01
Cf	98	8.70-02	5.20-02	4.68-02	5.42-02	6.81-02	8.92-02	1.04-01	1.24-01	1.28-01
Es	99	8.89-02	5.28-02	4.74-02	5.48-02	6.89-02	9.04-02	1.05-01	1.25-01	1.29-01
Fm	100	8.94-02	5.28-02	4.72-02	5.45-02	6.86-02	9.00-02	1.05-01	1.25-01	1.29-01

CLASSIFICATION OF ELECTROMAGNETIC RADIATION

Hans Dolezalek

Basic Conversions:

$c = \lambda\nu = \nu/k$ $\qquad$ $\lambda = c/\nu = 1/k$

$\nu = c/\lambda = ck$ $\qquad$ $k = \nu/c = 1/\lambda$

c = speed of light = 2.99792458×10^8 m/s

Frequency (ν)	Wavelength (λ)	Wave number (k)	Names of bands	Approximate photon energies
$3 \times 10^0 - 3 \times 10^1$ Hz 3 – 30 Hz	$10^8 - 10^7$ m 100 – 10 Mm	$10^{-8} - 10^{-7}$ m^{-1} 10 – 100 Gm^{-1}	ELF-(ELF 1), ITU band no. 1	
$3 \times 10^1 - 3 \times 10^2$ Hz 30 – 300 Hz	$10^7 - 10^6$ m 10 – 1 Mm	$10^{-7} - 10^{-6}$ m^{-1} 100 Gm^{-1} – 1Mm^{-1}	SLF-(ELF 2), ITU band no. 2, megameter waves	
$3 \times 10^2 - 3 \times 10^3$ Hz 300 Hz – 3 kHz	$10^6 - 10^5$ m 1 Mm – 100 km	$10^{-6} - 10^{-5}$ m^{-1} 1 – 10 Mm^{-1}	ULF-(ELF 3), ITU band no. 3	
$3 \times 10^3 - 3 \times 10^4$ Hz 3 – 30 kHz	$10^5 - 10^4$ m 100 – 10 km	$10^{-5} - 10^{-4}$ m^{-1} 10 – 100 Mm^{-1}	VLF, ITU band no. 4, myriameter waves	
$3 \times 10^4 - 3 \times 10^5$ Hz 30 – 300 kHz	$10^4 - 10^3$ m 10 – 1 km	$10^{-4} - 10^{-3}$ m^{-1} 100 Mm^{-1} – 1 km^{-1}	LF, ITU band no. 5, kilometer waves	
$3 \times 10^5 - 3 \times 10^6$ Hz 300 kHz – 3 MHz	$10^3 - 10^2$ m 1 km – 100 m	$10^{-3} - 10^{-2}$ m^{-1} 1 – 10 km^{-1}	MF, ITU band no. 6, hectometer waves	
$3 \times 10^6 - 3 \times 10^7$ Hz 3 – 30 MHz	$10^2 - 10^1$ m 100 – 10 m	$10^{-2} - 10^{-1}$ m^{-1} 10 – 100 km^{-1}	HF, ITU band no. 7, decameter waves	
$3 \times 10^7 - 3 \times 10^8$ Hz 30 – 300 MHz	$10^1 - 10^0$ m 10 – 1 m	$10^{-1} - 10^0$ m^{-1} 100 km^{-1} – 1 mm^{-1}	VHF, ITU band no. 8, meter waves	
$3 \times 10^8 - 3 \times 10^9$ Hz 300 MHz – 3 GHz	$10^0 - 10^{-1}$ m 1 m – 100 mm	$10^0 - 10^1$ m^{-1} 1 – 10 m^{-1}	UHF, ITU band no. 9, decimeter waves[a]	
$3 \times 10^9 - 3 \times 10^{10}$ Hz 3 – 30 GHz	$10^{-1} - 10^{-2}$ m 100 –10 mm	$10^1 - 10^2$ m^{-1} 10 – 100 m^{-1}	SHF, ITU band no. 10, centimeter waves[a]	
$3 \times 10^{10} - 3 \times 10^{11}$ Hz 30 – 300 GHz	$10^{-2} - 10^{-3}$ m 10 –1 mm	$10^2 - 10^3$ m^{-1} 100 m^{-1} – 1 mm^{-1} (1 – 10 cm^{-1})	EHF, ITU band no. 11, millimeter waves	
$3 \times 10^{11} - 3 \times 10^{12}$ Hz 300 GHz – 3 THz	$10^{-3} - 10^{-4}$ m 1 mm – 100 μm	$10^3 - 10^4$ m^{-1} (1 – 10 mm^{-1}) (10 – 100 cm^{-1})	Part of micrometer waves, includes part of far or thermal infrared; ITU band no. 12	
$3 \times 10^{12} - 3 \times 10^{13}$ Hz 3 – 30 THz	$10^{-4} - 10^{-5}$ m 100 – 10 μm	$10^4 - 10^5$ m^{-1} 10 – 100 mm^{-1} (100 – 1000 cm^{-1})	Part of micrometer waves includes part of far (thermal) infrared	
$3 \times 10^{13} - 3 \times 10^{14}$ Hz 30 – 300 THz	$10^{-5} - 10^{-6}$ m 10 – 1 μm (100,000 – 10,000 Å)	$10^5 - 10^6$ m^{-1} 100 mm^{-1} – 1 μm^{-1}	Part of μm waves, part of infrared	$(1.6 - 16) \times 10^{-20}$ joule {0.1 – 1 eV}
$3 \times 10^{14} - 3 \times 10^{15}$ Hz 300 THz – 3 PHz	$10^{-6} - 10^{-7}$ m 1 μm – 100 m (10,000 – 1000 Å)	$10^6 - 10^7$ m^{-1} 1 – 10 μm^{-1}	Near infrared, visible, near ultraviolet	$(1.6 - 16) \times 10^{-19}$ joule {1 – 10 eV}
$3 \times 10^{15} - 3 \times 10^{16}$ Hz 3 – 30 PHz	$10^{-7} - 10^{-8}$ m 100 – 10 nm (1000 – 100 Å)	$10^7 - 10^8$ m^{-1} 10 – 100 μm^{-1}	Part of vacuum ultraviolet	$(1.6 - 16) \times 10^{-18}$ joule {10 – 100 eV}
$3 \times 10^{16} - 3 \times 10^{17}$ Hz 30 – 300 PHz	$10^{-8} - 10^{-9}$ m 10 –1 nm (100 – 10 Å)	$10^8 - 10^9$ m^{-1} 100 μm^{-1} – 1 nm^{-1}	Part of soft X-rays	$(1.6 - 16) \times 10^{-17}$ joule {100 – 1000 eV}
$3 \times 10^{17} - 3 \times 10^{18}$ Hz 300 PHz – 3 EHz	$10^{-9} - 10^{-10}$ m 1 nm – 100 pm (10 – 1 Å)	$10^9 - 10^{10}$ m^{-1} 1 – 10 nm^{-1}	Part of soft X-rays	$(1.6 - 16) \times 10^{-16}$ joule {1 – 10 keV}
$3 \times 10^{18} - 3 \times 10^{19}$ Hz 3 – 30 EHz	$10^{-10} - 10^{-11}$ m 100 – 10 pm (1 – 0.1 Å)	$10^{10} - 10^{11}$ m^{-1} 10 – 100 nm^{-1}	Hard X-rays and part of soft γ-rays	$(1.6 - 16) \times 10^{-15}$ joule {10 – 100 keV}
$3 \times 10^{19} - 3 \times 10^{20}$ Hz 30 – 300 EHz	$10^{-11} - 10^{-12}$ m 10 – 1 pm (0.1 – 0.01 Å)	$10^{11} - 10^{12}$ m^{-1} 100 nm^{-1} – 1 pm^{-1}	Part of soft and part of hard γ-rays (limit at 510 keV)	$(1.6 - 16) \times 10^{-14}$ joule {100 keV – 1 MeV}
$3 \times 10^{20} - 3 \times 10^{21}$ Hz 300 – 3000 EHz	$10^{-12} - 10^{-13}$ m 1 pm – 100 fm (0.01 – 0.001 Å)	$10^{12} - 10^{13}$ m^{-1} 1 – 10 pm^{-1}	Part of hard γ-rays and part of "cosmic" γ-rays	$(1.6 - 16) \times 10^{-13}$ joule {1 – 10 MeV}
$3 \times 10^{21} - 3 \times 10^{22}$ Hz 3000 – 30,000 EHz	$10^{-13} - 10^{-14}$ m 100 – 10 fm (0.001 – 0.0001 Å)	$10^{13} - 10^{14}$ m^{-1} 10 – 100 pm^{-1}	γ-rays produced by cosmic rays	$(1.6 - 16) \times 10^{-12}$ joule {10 – 100 MeV}

Note: Abbreviations used in this table: Å— ångstrom (1 Å=10^{-10} m); EHz—exahertz (10^{18} hertz); EHF—extremely high frequency; ELF— extremely low frequency; eV—electron volt (1 eV = 1.60218 × 10^{-19} joule); fm—femtometer (10^{-15} m); GHz—gigahertz (10^{9} hertz); Gm—gigameter (10^{9} m); HF—high frequency; Hz—hertz (s^{-1}); ITU—International Telecommunications Union; keV— kiloelectron volt (10^{3} eV); km—kilometer (10^{3} m); LF—low frequency; m—meter; MeV— megaelectron volt (10^{6} eV); MF—medium frequency; MHz—megahertz (10^{6} hertz); Mm—megameter (10^{6} meter); mm—millimeter (10^{-3} meter); µm—micrometer (10^{-6} meter); nm—nanometer (10^{-9}meter); PHz—petahertz (10^{15} hertz); pm—picometer (10^{-12} meter); SHF—super high frequency; SLF—super low frequency; THz—terahertz; UHF— ultra high frequency; ULF—ultra low frequency; VHF—very high frequency; VLF—very low frequency.

[a] Also called "microwaves"; not to be confused with "micrometer waves".

Letter Designations of Microwave Bands

Frequency (GHz)	Wavelength (cm)	Wavenumber (cm^{-1})	Band
1—2	30—15	0.033—0.067	L-Band
2—4	15—7.5	0.067—0.133	S-Band
4—8	7.5—3.7	0.133—0.267	C-Band
8—12	3.7—2.5	0.267—0.4	X-Band
12—18	2.5—1.7	0.4—0.6	Ku-Band
18—27	1.7—1.1	0.6—0.9	K-Band
27—40	1.1—0.75	0.9—1.33	Ka-Band

SENSITIVITY OF THE HUMAN EYE TO LIGHT OF DIFFERENT WAVELENGTHS

The human eye responds to electromagnetic radiation in the wavelength range from about 360 nm (violet) to 820 nm (red), with a peak sensitivity near 555 nm (green). While the detailed shape of this response curve depends on the individual person, studies on representative samples of human subjects have led to adoption of a standard function relating the perceived brightness (luminous flux) to the actual power of the spectral radiation. This function is referred to as $V(\lambda)$, the photopic spectral luminous efficiency function, and it plays an important role in photometry.

The function $V(\lambda)$, as adopted by the International Commission on Illumination (CIE), is tabulated and plotted below.

References

1. *The Basis for Physical Photometry*, CIE Publication #18.2, 1983.
2. *CIE Standard Colorimetric Observers*, ISO/CIE #10527, 1991.
3. *Kaye and Laby Tables of Physical and Chemical Constants, Sixteenth Edition*, Longman Group Ltd., Harlow, Essex, 1995.

λ/nm	V(λ)	λ/nm	V(λ)	λ/nm	V(λ)
360	0.000004	520	0.710000	670	0.032000
370	0.000012	530	0.862000	680	0.017000
380	0.000039	540	0.954000	690	0.008210
390	0.000120	550	0.994950	700	0.004102
400	0.000396	555	1.000000	710	0.002091
410	0.001210	560	0.995000	720	0.001047
420	0.004000	570	0.952000	730	0.000520
430	0.011600	580	0.870000	740	0.000249
440	0.023000	590	0.757000	750	0.000120
450	0.038000	600	0.631000	760	0.000060
460	0.060000	610	0.503000	770	0.000030
470	0.090980	620	0.381000	780	0.000015
480	0.139020	630	0.265000	790	0.000007
490	0.208020	640	0.175000	800	0.000004
500	0.323000	650	0.107000	810	0.000002
510	0.503000	660	0.061000	820	0.000001

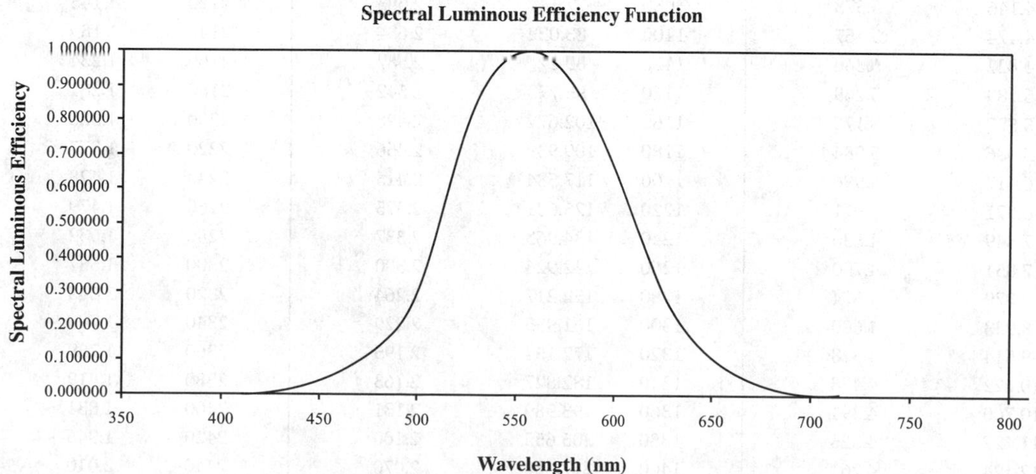

BLACK BODY RADIATION

The total power radiated from an ideal black body and the wavelength corresponding to maximum power are given here as a function of absolute temperature. Constants used in the calculation are taken from the table "Fundamental Physical Constants" in Section 1. The radiated power in a band $\Delta\lambda$ at λ_{max} may be calculated from:

$$P_{max} = 0.657548\ (\Delta\lambda/\lambda_{max})\ P_{tot}$$

T/K	P_{tot}	λ_{max}/μm	*T*/K	P_{tot}	λ_{max}/μm	*T*/K	P_{tot}	λ_{max}/μm
50	0.354 W/m^2	57.955	740	17.004	3.916	1520	302.689	1.906
100	5.671	28.978	750	17.942	3.864	1540	318.937	1.882
150	28.707	19.318	760	18.918	3.813	1560	335.831	1.858
200	90.728	14.489	770	19.934	3.763	1580	353.387	1.834
250	221.504	11.591	780	20.989	3.715	1600	371.623	1.811
273	314.973	10.614	790	22.087	3.668	1620	390.555	1.789
280	348.541	10.349	800	23.226	3.622	1640	410.202	1.767
290	401.064	9.992	810	24.410	3.577	1660	430.581	1.746
300	459.311	9.659	820	25.638	3.534	1680	451.710	1.725
310	523.684	9.348	830	26.911	3.491	1700	473.607	1.705
320	594.596	9.055	840	28.232	3.450	1720	496.290	1.685
330	672.478	8.781	850	29.600	3.409	1740	519.779	1.665
340	757.771	8.523	860	31.018	3.369	1760	544.093	1.646
350	850.931	8.279	870	32.486	3.331	1780	569.249	1.628
360	952.428	8.049	880	34.006	3.293	1800	595.267	1.610
370	1.063 kW/m^2	7.832	890	35.578	3.256	1820	622.168	1.592
380	1.182	7.626	900	37.204	3.220	1840	649.970	1.575
390	1.312	7.430	910	38.886	3.184	1860	678.694	1.558
400	1.452	7.244	920	40.623	3.150	1880	708.359	1.541
410	1.602	7.068	930	42.418	3.116	1900	738.987	1.525
420	1.764	6.899	940	44.272	3.083	1920	770.597	1.509
430	1.939	6.739	950	46.187	3.050	1940	803.210	1.494
440	2.125	6.586	960	48.162	3.018	1960	836.848	1.478
450	2.325	6.439	970	50.201	2.987	1980	871.531	1.464
460	2.539	6.299	980	52.303	2.957	2000	907.282	1.449
470	2.767	6.165	990	54.471	2.927	2020	944.121	1.435
480	3.010	6.037	1000	56.705	2.898	2040	982.071	1.420
490	3.269	5.914	1020	61.379	2.841	2060	1.021 MW/m^2	1.407
500	3.544	5.796	1040	66.337	2.786	2080	1.061	1.393
510	3.836	5.682	1060	71.589	2.734	2100	1.103	1.380
520	4.146	5.573	1080	77.147	2.683	2120	1.145	1.367
530	4.474	5.467	1100	83.022	2.634	2140	1.189	1.354
540	4.822	5.366	1120	89.227	2.587	2160	1.234	1.342
550	5.189	5.269	1140	95.773	2.542	2180	1.281	1.329
560	5.577	5.175	1160	102.672	2.498	2200	1.328	1.317
570	5.986	5.084	1180	109.939	2.456	2220	1.377	1.305
580	6.417	4.996	1200	117.584	2.415	2240	1.428	1.294
590	6.871	4.911	1220	125.621	2.375	2260	1.479	1.282
600	7.349	4.830	1240	134.063	2.337	2280	1.532	1.271
610	7.851	4.750	1260	142.924	2.300	2300	1.587	1.260
620	8.379	4.674	1280	152.217	2.264	2320	1.643	1.249
630	8.933	4.600	1300	161.955	2.229	2340	1.700	1.238
640	9.514	4.528	1320	172.154	2.195	2360	1.759	1.228
650	10.122	4.458	1340	182.827	2.163	2380	1.819	1.218
660	10.760	4.391	1360	193.989	2.131	2400	1.881	1.207
670	11.427	4.325	1380	205.655	2.100	2420	1.945	1.197
680	12.124	4.261	1400	217.838	2.070	2440	2.010	1.188
690	12.853	4.200	1420	230.556	2.041	2460	2.077	1.178
700	13.615	4.140	1440	243.822	2.012	2480	2.145	1.168
710	14.410	4.081	1460	257.652	1.985	2500	2.215	1.159
720	15.239	4.025	1480	272.063	1.958	2550	2.398	1.136
730	16.103	3.970	1500	287.070	1.932	2600	2.591	1.115

T/K	P_{tot}	λ_{max}/μm	T/K	P_{tot}	λ_{max}/μm	T/K	P_{tot}	λ_{max}/μm
2650	2.796	1.093	3600	9.524	0.805	5100	38.362	0.568
2700	3.014	1.073	3650	10.065	0.794	5200	41.461	0.557
2750	3.243	1.054	3700	10.627	0.783	5300	44.743	0.547
2800	3.485	1.035	3750	11.214	0.773	5400	48.217	0.537
2850	3.741	1.017	3800	11.824	0.763	5500	51.889	0.527
2900	4.011	0.999	3850	12.458	0.753	5600	55.767	0.517
2950	4.294	0.982	3900	13.118	0.743	5700	59.858	0.508
3000	4.593	0.966	3950	13.804	0.734	5800	64.170	0.500
3050	4.907	0.950	4000	14.517	0.724	5900	68.712	0.491
3100	5.237	0.935	4100	16.024	0.707	6000	73.490	0.483
3150	5.583	0.920	4200	17.645	0.690	6500	101.222	0.446
3200	5.946	0.906	4300	19.386	0.674	7000	136.149	0.414
3250	6.326	0.892	4400	21.254	0.659	7500	179.418	0.386
3300	6.725	0.878	4500	23.253	0.644	8000	232.264	0.362
3350	7.142	0.865	4600	25.389	0.630	8500	296.004	0.341
3400	7.578	0.852	4700	27.670	0.617	9000	372.042	0.322
3450	8.033	0.840	4800	30.101	0.604	9500	461.867	0.305
3500	8.509	0.828	4900	32.689	0.591	10000	567.051	0.290
3550	9.006	0.816	5000	35.441	0.580			

The curves below show, for various temperatures, the fraction of radiant power as a function of wavelength. The function plotted is $P_\lambda/\Delta\lambda\ P_{tot}$, where P_λ is the power at wavelength λ in a small interval $\Delta\lambda$ (in μm), and P_{tot} is the total power.

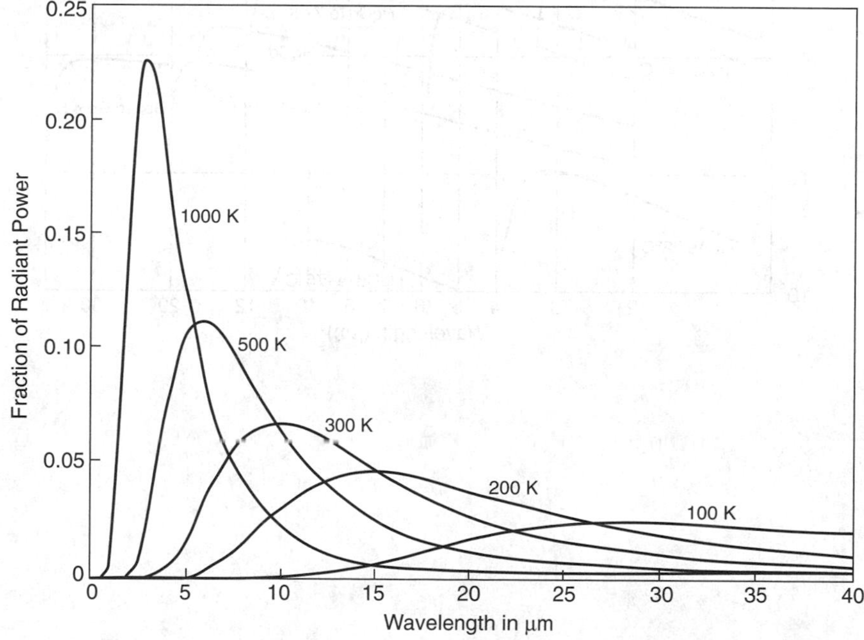

CHARACTERISTICS OF INFRARED DETECTORS

This graph summarizes the wavelength response of some semiconductors used as detectors for infrared radiation. The quantity D*(λ) is the signal-to-noise ratio for an incident radiant power density of 1 W/cm² and a bandwidth of 1 Hz (60° field of view). The Ge, InAs, and InSb detectors are photovoltaics, while the HgCdTe series are photoconductive devices. The cutoff wavelength of the latter can be varied by adjusting the relative amounts of Hg, Cd, and Te (three examples are shown at 77 K). The graph also shows the theoretical background limited sensitivity for ideal detectors which introduce no intrinsic noise.

Reference

Infrared Detectors 1995, EG&G Judson, Montgomeryville, PA.

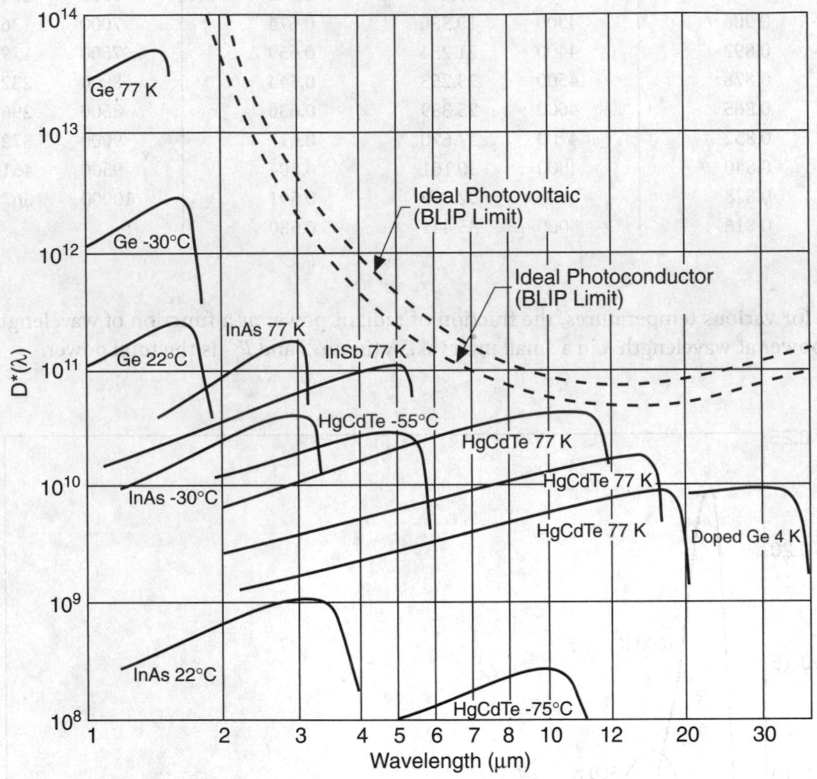

INDEX OF REFRACTION OF INORGANIC CRYSTALS

This table lists the index of refraction of selected crystalline inorganic compounds. When available, values are given as a function of wavelength in the range from the ultraviolet to the far infrared region. For each compound a value at 589 nm, the wavelength of the principal sodium line, is given. The data have been taken from the references indicated; in many cases, data from a reference have been refitted to generate the index of refraction at the wavelengths used in this table. All values refer to ambient temperature. Entries marked by * are based on extrapolation beyond the range of available experimental data.

Compounds belonging to the cubic crystal system have only a single refractive index value, but other systems are anisotropic, so that the crystal is characterized by two or three unique indexes. Hexagonal, rhombohedral, and tetragonal crystals have two unique indexes that are traditionally labeled n_o and n_e for "ordinary ray" and "extraordinary ray." Orthorhombic, monoclinic, and triclinic crystals are characterized by three indexes that here are called n_x, n_y, and n_z. The table indicates the crystal system for each entry in order to identify the material uniquely.

The refractive index and other optical properties for metals, semiconductors, and certain other compounds can be found in the tables "Optical Properties of Selected Elements" and "Optical Properties of Selected Inorganic and Organic Solids" in Section 12 of this *Handbook*.

References

1. Li, H. H., "Refractive Index of Alkali Halides and its Wavelength and Temperature Derivatives," *J. Phys. Chem. Ref. Data* 5, 329, 1976.
2. Li, H. H., "Refractive Index of Alkaline Earth Halides and its Wavelength and Temperature Derivatives," *J. Phys. Chem. Ref. Data* 9, 161, 1980.
3. Li, H. H., "Refractive Index of ZnS, ZnSe, and ZnTe and its Wavelength and Temperature Derivatives," *J. Phys. Chem. Ref. Data* 13, 103, 1984.
4. Shannon, R. D., Shannon, R. C., Medenbach, O., and Fischer, R. X., "Refractive Index and Dispersion of Fluorides and Oxides," *J. Phys. Chem. Ref. Data* 31, 931, 2002.
5. Gray, D. E., ed., *American Institute of Physics Handbook*, Sec. 6b, pp. 6-12, McGraw-Hill, New York, 1972.
6. *Landolt-Börnstein Numerical Data and Functional Relationships in Science and Technology, III/11, Elastic, Piezoelectric, Pyroelectric, Piezooptic, Electrooptic Constants, and Nonlinear Dielectric Susceptibilities of Crystals,* Springer-Verlag, Berlin, 1979.
7. *Landolt-Börnstein Numerical Data and Functional Relationships in Science and Technology, III/30A, High Frequency Properties of Dielectric Crystals. Piezooptic and Electrooptic Constants,* Springer-Verlag, Berlin, 1996.
8. Weber, M. J., *CRC Handbook of Laser Science and Technology*, Vol. IV. Optical Materials. Part 2: Properties, CRC Press, Boca Raton, FL, 1986.

Compound	Crystal system	Ray	Index of Refraction at the Indicated Wavelength								Ref.
			300 nm	589 nm	750 nm	1 µm	2 µm	5 µm	10 µm	20 µm	
AgCl	cub	n		2.0668	2.0401	2.0224	2.0062	1.9975	1.9803	1.9069	5
$AlPO_4$	rhomb	n_o		1.5247	1.5203	1.5161	1.5034				6
	rhomb	n_e		1.5338	1.5290	1.5245	1.5116				6
Al_2O_3	hex	n_o		1.7673							4
	hex	n_e		1.7598							4
As_2O_3[a]	cub	n		1.7537							4
BaF_2	cub	n	1.5010	1.4744	1.4712	1.4686	1.4647	1.4511	1.4014		2
BaO	cub	n		1.9841							4
$BaSO_4$	orth	n_x		1.6362							4
	orth	n_y		1.6374							4
	orth	n_z		1.6480							4
$BaTiO_3$	tetr	n_o		2.4405							4
	tetr	n_e		2.3831							4
$BaWO_4$	tetr	n_o		1.8426							4
	tetr	n_e		1.8405							4
BeO	hex	n_o		1.7184							4
	hex	n_e		1.7342							4
$BeSO_4 \cdot 4H_2O$	tetr	n_o		1.4713							4
	tetr	n_e		1.4328							4
$CaCO_3$[b]	hex	n_o	1.7216	1.6584	1.6503	1.6436	1.6249				5
	hex	n_e	1.5145	1.4864	1.4828	1.4801	1.4753				5
CaF_2	cub	n	1.4540	1.4338	1.4311	1.4289	1.4239	1.3990	1.299		2
CaO	cub	n		1.8396							4
$CaSO_4$	orth	n_x		1.5698							4
	orth	n_y		1.5755							4
	orth	n_z		1.6137							4
$CaSO_4 \cdot 2H_2O$	monocl	n_x		1.5207							4
	monocl	n_y		1.5227							4
	monocl	n_z		1.5304							4
$CaWO_4$	tetr	n_o		1.9195							4
	tetr	n_e		1.9355							4
CdS	hex	n_o		2.507	2.390	2.334					5

Compound	Crystal system	Ray	Index of Refraction at the Indicated Wavelength 300 nm	589 nm	750 nm	1 μm	2 μm	5 μm	10 μm	20 μm	Ref.
	hex	n_e		2.525	2.409	2.352					5
CdSe	hex	n_o			2.68*	2.5502	2.4682	2.4483	2.4331		7
	hex	n_e			2.69*	2.5696	2.4873	2.4676	2.4514		7
CdTe	cub	n							2.6724	2.6302	7
CeF_3	hex	n_o		1.6183							4
	hex	n_e		1.6113							4
CsBr	cub	n	1.8047	1.6974	1.6861	1.6784	1.6711	1.6678	1.6630	1.6439	1
CsCl	cub	n	1.712	1.640	1.631	1.626	1.620	1.616	1.606	1.563	1
$CsClO_4$	orth	n_x		1.4752							4
	orth	n_y		1.4788							4
	orth	n_z		1.4804							4
CsF	cub	n	1.506	1.477	1.474	1.472	1.469*	1.461*	1.436*	1.32*	1
CsI	cub	n	1.9790	1.7873	1.7694	1.7576	1.7465	1.7428	1.7396	1.7280	1
Cs_2SO_4	orth	n_x		1.5598							4
	orth	n_y		1.5644							4
	orth	n_z		1.5662							4
CuBr	cub	n		2.117							7
CuCl	cub	n		1.9727	1.9391				1.9245		7
$CuSO_4 \cdot 5H_2O$	tricl	n_x		1.5140							4
	tricl	n_y		1.5367							4
	tricl	n_z		1.5436							4
Dy_2O_3	cub	n		1.9757							4
FeF_2	tetr	n_o		1.514							4
	tetr	n_e		1.524							4
Gd_2O_3	cub	n		1.96							4
HgS	rhomb	n_o		2.9413	2.7770	2.7120	2.6305		2.6018		6
	rhomb	n_e		3.3072	3.0896	3.0050	2.8776		2.8522		6
KBr	cub	n	1.6482	1.5598	1.5498	1.5444	1.5383	1.5345	1.5264	1.4924	1
KCl	cub	n	1.5455	1.4902	1.4840	1.4798	1.4753	1.4704	1.4564	1.3946	1
$KClO_4$	orth	n_x		1.4730							4
	orth	n_y		1.4736							4
	orth	n_z		1.4768							4
KF	cub	n	1.380	1.362	1.360	1.358	1.355	1.344	1.304*	1.09*	1
KH_2AsO_4	tetr	n_o		1.5674							7
	tetr	n_e		1.5179							7
KH_2PO_4	tetr	n_o	1.5450	1.5093	1.5030	1.4957					5
	tetr	n_e	1.4977	1.4682	1.4641	1.4606					5
KI	cub	n	1.834*	1.665	1.650	1.640	1.631	1.627	1.620	1.593	1
KIO_3	tricl	n_x		1.6959							7
	tricl	n_y		1.8317							7
	tricl	n_z		1.8343							7
KIO_4	tetr	n_o		1.6205							4
	tetr	n_e		1.6476							4
$KNbO_3$	orth	n_x		2.2480	2.3395	2.2612					7
	orth	n_y		2.3464	2.2959	2.2622					7
	orth	n_x		2.1803	2.1457	2.1288					7
K_2SO_4	orth	n_x		1.4934							4
	orth	n_y		1.4947							4
	orth	n_z		1.4973							4
LaF_3	hex	n_o		1.605							4
	hex	n_e		1.599							4
LiBr	cub	n	1.810	1.783	1.781	1.778	1.774*	1.756*	1.68*	1.33*	1
LiCl	cub	n	1.677	1.662	1.660	1.658	1.654*	1.62*	1.53*		1
$LiClO_4 \cdot 3H_2O$	hex	n_o		1.4832							4
	hex	n_e		1.4384							4
LiF	cub	n	1.4087	1.3921	1.3895	1.3871	1.3786	1.3266	1.1005		1
LiI	cub	n	1.979	1.955	1.952	1.950	1.948*	1.940*	1.91*	1.77*	1
$LiIO_3$	hex	n_o		1.8875	1.8713	1.8589	1.8410				6
	hex	n_e		1.7400	1.7268	1.7179	1.7062				6
$LiNbO_3$	rhomb	n_o		2.3007	2.2632	2.2370					7

Compound	Crystal system	Ray	Index of Refraction at the Indicated Wavelength								Ref.
			300 nm	589 nm	750 nm	1 μm	2 μm	5 μm	10 μm	20 μm	
	rhomb	n_e		2.2116	2.1804	2.1567					7
$LiTaO_3$	rhomb	n_o		2.1864	2.1590	2.1391	2.1066				7
	rhomb	n_e		2.1908	2.1634	2.1432	2.1115				7
$Li_2SO_4{\cdot}H_2O$	monocl	n_x		1.4615							4
	monocl	n_y		1.4765							4
	monocl	n_z		1.4863							4
Lu_2O_3	cub	n		1.9349							4
MgF_2	tetr	n_o	1.3930	1.3776	1.375	1.373	1.368	1.34	1.21		2
	tetr	n_e	1.4055	1.3894	1.387	1.385	1.379	1.34	1.21		2
MgO	cub	n		1.7355	1.7283	1.7228	1.7084	1.6361			5
$MgSO_4{\cdot}7H_2O$	orth	n_x		1.4326							4
	orth	n_y		1.4555							4
	orth	n_z		1.4607							4
MnF_2	tetr	n_o		1.472							4
	tetr	n_e		1.501							4
$NH_4H_2AsO_4$	tetr	n_o	1.6401	1.5777	1.5704	1.5583					7
	tetr	n_e	1.5754	1.5232	1.5179	1.5101					7
$NH_4H_2PO_4$	tetr	n_o	1.5668	1.5247	1.5187	1.5084					7
	tetr	n_e	1.5137	1.4797	1.4754	1.4694					7
NaBr	cub	n	1.748	1.642	1.631	1.623	1.616	1.609	1.593*	1.520*	1
$NaBrO_3$	cub	n		1.6168							4
NaCl	cub	n	1.6066	1.5441	1.5369	1.5320	1.5265	1.5188	1.4947	1.382*	1
$NaClO_3$	cub	n		1.5151							7
NaF	cub	n	1.3424	1.3252	1.3231	1.3214	1.3179	1.3017	1.2400		1
$NaH_2PO_4{\cdot}2H_2O$	orth	n_x		1.4400							7
	orth	n_y		1.4020							7
	orth	n_z		1.4814							7
NaI	cub	n	1.93*	1.774	1.758	1.74	1.73*	1.73*	1.71*	1.66*	1
$NaNO_2$	orth	n_x		1.6547							7
	orth	n_y		1.3455							7
	orth	n_z		1.4125							7
$NaNO_3$	rhomb	n_o		1.5840							5
	rhomb	n_e		1.3340							5
$Na_2HPO_4{\cdot}7H_2O$	monocl	n_x		1.4411							4
	monocl	n_y		1.4423							4
	monocl	n_z		1.4525							4
Na_2SO_4	orth	n_x		1.4669							4
	orth	n_y		1.4730							4
	orth	n_z		1.4809							4
NdF_3	hex	n_o		1.6191							4
	hex	n_e		1.6132							4
Nd_2O_3	cub	n		1.92							4
NiF_2	tetr	n_o		1.526							4
	tetr	n_e		1.561							4
$NiSO_4{\cdot}6H_2O$	tetr	n_o		1.5107							4
	tetr	n_e		1.4870							4
PbF_2	cub	n	1.94*	1.767	1.754	1.745	1.73	1.70	1.66	1.32	5
$PbSO_4$	orth	n_x		1.8780							4
	orth	n_y		1.8834							4
	orth	n_z		1.8945							4
PrF_3	hex	n_o		1.6207							4
	hex	n_e		1.6146							4
RbBr	cub	n	1.639	1.553	1.544	1.538	1.532	1.530	1.525	1.505*	1
RbCl	cub	n	1.549	1.493	1.487	1.483	1.479	1.475	1.465	1.424*	1
$RbClO_4$	orth	n_x		1.4691							4
	orth	n_y		1.4701							4
	orth	n_z		1.4732							4
RbF	cub	n	1.428*	1.397	1.394	1.391	1.388	1.379	1.346	1.19*	1
RbH_2AsO_4	tetr	n_o	1.6183	1.5603	1.5538	1.5432					7
	tetr	n_e	1.5718	1.5232	1.5184	1.5121					7

Compound	Crystal system	Ray	Index of Refraction at the Indicated Wavelength 300 nm	589 nm	750 nm	1 μm	2 μm	5 μm	10 μm	20 μm	Ref.
RbH_2PO_4	tetr	n_o	1.5434	1.5078	1.5021	1.4941					7
	tetr	n_e	1.5106	1.4791	1.4754	1.4704					7
RbI	cub	n	1.808	1.647	1.633	1.623	1.615	1.612	1.608	1.595	1
Rb_2SO_4	orth	n_x		1.5131							4
	orth	n_y		1.5133							4
	orth	n_z		1.5144							4
Sb_2O_3[c]	cub	n		2.8017							4
Sc_2O_3	cub	n		1.9943							4
SiO_2[d]	hex	n_o	1.5733	1.5442	1.5394	1.5350	1.5209				5
	hex	n_e	1.5882	1.5534	1.5484	1.5438	1.5291				5
SnO_2	tetr	n_o		1.993							4
	tetr	n_e		2.088							4
SrF_2	cub	n	1.459	1.4380	1.435	1.433	1.429	1.412	1.35		2
SrO	cub	n		1.8710							4
$SrSO_4$	orth	n_x		1.6214							4
	orth	n_y		1.6231							4
	orth	n_z		1.6303							4
$SrTiO_3$	cub	n		2.4082	2.3525	2.3160	2.2676	2.1205			5
$SrWO_4$	tetr	n_o		1.8618							4
	tetr	n_e		1.8719							4
TbF_3	hex	n_o		1.6034							4
	hex	n_e		1.5603							4
TeO_2	tetr	n_o		2.2738		2.2080					7
	tetr	n_e		2.4295		2.3520					7
ThO_2	cub	n		2.1113							4
TiO_2[e]	tetr	n_o		2.612	2.533	2.485	2.399	2.220			5
	tetr	n_e		2.910	2.805	2.748					5
	tetr	n_o		2.562							4
	tetr	n_e		2.489							4
TlBr	cub	n		2.418	2.350	2.289	2.103	1.984	2.339	2.322	5
TlCl	cub	n		2.247	2.198	2.145	1.986	1.891	2.193		5
$TlClO_4$	orth	n_x		1.6427							4
	orth	n_y		1.6446							4
	orth	n_z		1.6542							4
Tl_2SO_4	orth	n_x		1.8604							4
	orth	n_y		1.8676							4
	orth	n_z		1.8857							4
Y_2O_3	cub	n		1.930							4
Yb_2O_3	cub	n		1.9468							4
ZnF_2	tetr	n_o		1.495							4
	tetr	n_e		1.525							4
ZnO	hex	n_o		2.0036	1.9662	1.9435	1.9197				7
	hex	n_e		2.0199	1.9821	1.9589	1.9330				7
ZnS[f]	cub	n		2.3691	2.3232	2.2932	2.2633				7
ZnS[g]	hex	n_o		2.372	2.331	2.303	2.26	2.25	2.20		3,5
	hex	n_e		2.368	2.327	2.301					5
ZnSe	cub	n		2.6222	2.5384	2.4888	2.4462	2.4296	2.4065		3
ZnTe	cub	n		3.060	2.880	2.789	2.719	2.698	2.684		3
$ZrSiO_4$[h]	tetr	n_o		1.9255							4
	tetr	n_e		1.9843							4

* Provisional value based on extrapolation beyond the range of experimental data.
[a] Arsenolite
[b] Calcite
[c] Senarmontite
[d] α-Quartz
[e] Rutile
[f] Sphalerite
[g] Wurtzite
[h] Zircon

REFRACTIVE INDEX AND TRANSMITTANCE OF REPRESENTATIVE GLASSES

Typical values of the index of refraction and internal transmittance (fraction of light transmitted through a one centimeter thickness) are tabulated here for selected types of glasses, as well as for synthetic fused (vitreous) silica. Nominal compositions are given in the first part of the table. The second part gives the index of refraction, relative to air, and the internal transmittance for representative samples of each glass at wavelengths in the infrared, visible, and near-ultraviolet regions. It should be emphasized that a wide variation of these parameters may be found among subtypes of each glass. More detailed data may be found in Reference 3.

Assuming that the Lambert-Beer Law is followed, the transmittance of a glass plate of thickness d (in centimeters) can be obtained by raising the transmittance value in the table to the power d.

References

1. Weber, M. J., *CRC Handbook of Laser Science and Technology*, Vol. IV, Part 2, CRC Press, Boca Raton, FL, 1988.
2. Gray, D. E., Ed., *American Institute of Physics Handbook, Third Edition*, McGraw Hill, New York, 1972.
3. *Schott Optical Glass*, Schott Glass Technologies, Inc., 400 York Ave., Duryea, PA.
4. Kaye, G. W. C., and Laby, T. H., *Tables of Physical and Chemical Constants, 15th Edition*, Longman, London, 1986.

		Composition in percent by mass									
Type	Name	SiO_2	B_2O_3	Al_2O_3	Na_2O	K_2O	CaO	BaO	ZnO	PbO	P_2O_5
PK	Phosphate crown		3	10		12	5				70
PSK	Dense phosphate crown		3	5			4	28			60
BK	Borosilicate crown	70	10		8	8	1	3			
K	Crown	74			9	11	6				
ZK	Zinc crown	71			17				12		
BaK	Barium crown	60	3		3	10		19	5		
SK	Dense crown	39	15	5				41			
KF	Crown flint	67		2	16				3	12	
BaLF	Barium light flint	51			6	5		20	14	4	
SSK	Extra dense crown	35	10	5				42	8		
LLF	Extra light flint	63			5	8				24	
BaF	Barium flint	46				8		16	8	22	
LF	Light flint	53			5	8				34	
F	Flint	47			2	7				44	
BaSF	Dense barium flint	43			1	7		11	5	33	
SF	Dense flint	33				5				62	
KzFS	Short flint										
SiO_2	Fused silica	100									

	Index of refraction				Transmittance of 1 cm plate			
Type	1.060 μm	546.1 nm	365.0 nm	312.6 nm	1.060 μm	546.1 nm	365.0 nm	310 nm
PK	1.51519	1.52736	1.54503	1.5574	0.997	0.998	0.987	0.46
PSK	1.54154	1.55440	1.57342	1.5868	0.996	0.998	0.984	0.46
BK	1.50669	1.51872	1.53627	1.5406	0.999	0.998	0.987	0.35
K	1.50091	1.51314	1.53189	1.5454	0.998	0.998	0.988	0.40
ZK	1.52220	1.53534	1.55588	1.5708	0.996	0.998	0.976	0.27
BaK	1.55695	1.57124	1.59407	1.6108	0.998	0.997	0.986	0.28
SK	1.59490	1.60994	1.63398		0.998	0.998	0.959	0.28
KF	1.50586	1.51978	1.54251	1.5600	0.998	0.996	0.989	0.49
BaLF	1.57579	1.59166	1.61804		0.996	0.998	0.933	0.010
SSK	1.60402	1.61993	1.64595		0.999	0.998	0.915	0.010
LaK	1.69710	1.71616	1.74573		0.999	0.998	0.882	0.17
LLF	1.52775	1.54344	1.57038		0.998	0.997	0.990	0.32
BaF	1.56873	1.58565	1.61524		0.999	0.997	0.992	0.004
LF	1.56594	1.58482	1.61926		0.999	0.998	0.981	0.008
F	1.58636	1.60718	1.64606		0.997	0.998	0.959	
BaSF	1.60889	1.62987	1.66926		0.999	0.998	0.857	
SF	1.71350	1.74620	1.8145		0.998	0.997	0.650	
KzFS	1.59680	1.61639	1.64849	1.6739		0.998	0.672	0.012
SiO_2	1.44968	1.46008	1.47435[a]	1.53430[b]				

[a] At 366.3 nm.

[b] At 213.9 nm.

INDEX OF REFRACTION OF WATER

This table gives the index of refraction of liquid water at atmospheric pressure, relative to a vacuum, at several temperatures and wavelengths. It is generated from the formulation in Reference 1, which covers a wide range of temperature, pressure, and wavelength. The wavelengths listed here correspond to prominent lines of cadmium (226.50 and 361.05 nm), potassium (404.41 nm), sodium (589.00 nm), Ne (632.80 nm, from a helium - neon laser), and mercury (1.01398 μm).

References

1. Schiebener, P., Straub, J., Levelt Sengers, J. M. H., and Gallagher, J. S., *J. Phys. Chem. Ref. Data*, 19, 677, 1990; 19, 1617, 1990.
2. Marsh, K. N., Editor, *Recommended Reference Materials for the Realization of Physicochemical Properties*, Blackwell Scientific Publications, Oxford, 1987.

t/°C	226.50 nm	361.05 nm	404.41 nm	589.00 nm	632.80 nm	1.01398 μm
0	1.39450	1.34896	1.34415	1.33432	1.33306	1.32612
10	1.39422	1.34870	1.34389	1.33408	1.33282	1.32591
20	1.39336	1.34795	1.34315	1.33336	1.33211	1.32524
30	1.39208	1.34682	1.34205	1.33230	1.33105	1.32424
40	1.39046	1.34540	1.34065	1.33095	1.32972	1.32296
50	1.38854	1.34373	1.33901	1.32937	1.32814	1.32145
60	1.38636	1.34184	1.33714	1.32757	1.32636	1.31974
70	1.38395	1.33974	1.33508	1.32559	1.32438	1.31784
80	1.38132	1.33746	1.33284	1.32342	1.32223	1.31576
90	1.37849	1.33501	1.33042	1.32109	1.31991	1.31353
100	1.37547	1.33239	1.32784	1.31861	1.31744	1.31114

INDEX OF REFRACTION OF LIQUIDS FOR CALIBRATION PURPOSES

This table gives the index of refraction of six liquids that are available in highly pure form and whose index of refraction has been accurately measured as a function of wavelength and temperature. They are therefore useful for calibration of refractometers. The estimated uncertainty in the values is:

2,2,4-Trimethylpentane	±0.00003
Hexadecane	±0.00008
trans-Bicyclo[4.0.0]decane	±0.00008
1-Methylnaphthalene	±0.00008
Toluene	±0.00003
Methylcyclohexane	±0.00003

Full details are given in the references. This table is reprinted from Reference 1 by permission of the International Union of Pure and Applied Chemistry.

References

1. Marsh, K. N., Editor, *Recommended Reference Materials for the Realization of Physicochemical Properties*, Blackwell Scientific Publications, Oxford, 1987.
2. Tilton, L. W., *J. Opt. Soc. Am.*, 32, 71, 1941.

λ	2,2,4-Trimethylpentane			Hexadecane		
nm	20 °C	25 °C	30 °C	20 °C	25 °C	30 °C
667.81	1.38916	1.38670	1.38424	1.43204	1.43001	1.42798
656.28	1.38945	1.38698	1.38452	1.43235	1.43032	1.42829
589.26	1.39145	1.38898	1.38650	1.43453	1.43250	1.43047
546.07	1.39316	1.39068	1.38820	1.43640	1.43436	1.43232
501.57	1.39544	1.39294	1.39044	1.43888	1.43684	1.43480
486.13	1.39639	1.39389	1.39138	1.43993	1.43788	1.43583
435.83	1.40029	1.39776	1.39523	1.44419	1.44213	1.44007

λ	*trans*-Bicyclo[4.4.0]decane			1-Methylnaphthalene		
nm	20 °C	25 °C	30 °C	20 °C	25 °C	30 °C
667.81	1.46654	1.46438	1.46222	1.60828	1.60592	1.60360
656.28	1.46688	1.46472	1.46256	1.60940	1.60703	1.60471
589.26	1.46932	1.46715	1.46498	1.61755	1.61512	1.61278
546.07	1.47141	1.46923	1.46705	1.62488	1.62240	1.62005
501.57	1.47420	1.47200	1.46980	1.63513	1.63259	1.63022
486.13	1.47535	1.47315	1.47095	1.63958	1.63701	1.63463
435.83	1.48011	1.47789	1.47567		1.65627	1.65386

λ	Toluene			Methylcyclohexane		
nm	20 °C	25 °C	30 °C	20 °C	25 °C	30 °C
667.81	1.49180	1.48903	1.48619	1.42064	1.41812	1.41560
656.28	1.49243	1.48966	1.48682	1.42094	1.41842	1.41591
589.26	1.49693	1.49413	1.49126	1.42312	1.42058	1.41806
546.07	1.50086	1.49803	1.49514	1.42497	1.42243	1.41989
501.57	1.50620	1.50334	1.50011	1.42744	1.42488	1.42233
486.13	1.50847	1.50559	1.50265	1.42847	1.42590	1.42334
435.83	1.51800	1.51506	1.51206	1.43269	1.43010	1.42752

INDEX OF REFRACTION OF AIR

This is a table of the index of refraction n of dry air at 15 °C and a pressure of 101.325 kPa and containing 0.045% by volume of carbon dioxide ("standard air"). The index of refraction is defined by $n = \lambda_{vac}/\lambda_{air}$ where λ is the wavelength of the radiation. The index is calculated from the expression

$$(n-1) \times 10^8 = 8342.54 + 2406147(130 - \sigma^2)^{-1} + 15998(38.9 - \sigma^2)^{-1}$$

where $\sigma = 1/\lambda_{vac}$ and λ_{vac} has units of μm. The equation is valid for λ_{vac} from 200 nm to 2 μm. The table also gives the correction $(n-1)\lambda_{air}$ that must be added to the wavelength in air to obtain λ_{vac}.

If the air is at a temperature t in °C (ITS-90) and a pressure p in pascals, a value of $(n - 1)$ from this table should be multiplied by

$$p[1 + p(60.1 - 0.972t) \times 10^{-10}]/96095.43(1 + 0.003661t)$$

References

1. Birch, K. P., and Downs, M. J., *Metrologia*, 31, 315, 1994.
2. Edlen, B., *Metrologia* 2, 71, 1966.

λ_{vac}	$(n-1) \times 10^8$	$\lambda_{vac} - \lambda_{air}$	λ_{vac}	$(n-1) \times 10^8$	$\lambda_{vac} - \lambda_{air}$	λ_{vac}	$(n-1) \times 10^8$	$\lambda_{vac} - \lambda_{air}$
200 nm	32409	0.06480 nm	540	27804	0.15010	880	27462	0.24160
210	31748	0.06665	550	27784	0.15277	890	27458	0.24431
220	31226	0.06868	560	27765	0.15544	900	27454	0.24701
230	30801	0.07082	570	27747	0.15811	910	27449	0.24972
240	30447	0.07305	580	27730	0.16079	920	27445	0.25243
250	30148	0.07535	590	27714	0.16347	930	27441	0.25513
260	29892	0.07769	600	27698	0.16614	940	27437	0.25784
270	29670	0.08009	610	27684	0.16882	950	27434	0.26055
280	29477	0.08251	620	27670	0.17151	960	27430	0.26326
290	29307	0.08497	630	27657	0.17419	970	27427	0.26597
300	29157	0.08745	640	27644	0.17688	980	27423	0.26868
310	29023	0.08995	650	27632	0.17956	990	27420	0.27138
320	28904	0.09247	660	27621	0.18225			
330	28796	0.09500	670	27610	0.18494	1.00 μm	27417	0.0002741
340	28700	0.09755	680	27600	0.18763	1.05	27402	0.0002876
350	28612	0.10011	690	27590	0.19032	1.10	27390	0.0003012
360	28532	0.10269	700	27581	0.19301	1.15	27379	0.0003148
370	28460	0.10527	710	27572	0.19570	1.20	27370	0.0003283
380	28393	0.10786	720	27563	0.19840	1.25	27361	0.0003419
390	28332	0.11046	730	27555	0.20109	1.30	27354	0.0003555
400	28276	0.11307	740	27547	0.20379	1.35	27347	0.0003691
410	28224	0.11569	750	27539	0.20649	1.40	27341	0.0003827
420	28177	0.11831	760	27532	0.20918	1.45	27336	0.0003963
430	28132	0.12094	770	27525	0.21188	1.50	27331	0.0004099
440	28091	0.12357	780	27518	0.21458	1.55	27327	0.0004234
450	28053	0.12620	790	27511	0.21728	1.60	27323	0.0004370
460	28018	0.12885	800	27505	0.21998	1.65	27319	0.0004506
470	27985	0.13149	810	27499	0.22268	1.70	27316	0.0004642
480	27954	0.13414	820	27493	0.22538	1.75	27313	0.0004778
490	27925	0.13679	830	27488	0.22808	1.80	27310	0.0004914
500	27897	0.13945	840	27482	0.23079	1.85	27307	0.0005050
510	27872	0.14211	850	27477	0.23349	1.90	27305	0.0005187
520	27848	0.14477	860	27472	0.23619	1.95	27303	0.0005323
530	27825	0.14743	870	27467	0.23890	2.00	27301	0.0005459

INDEX OF REFRACTION OF GASES

This table gives the index of refraction of several gases at selected wavelengths ranging from the blue to the red region of the spectrum. The entries at 0.5893 μm correspond to the prominent sodium D line in the yellow region. All values refer to gas at a pressure of one atmosphere (101.325 kPa) and at a temperature of 0 °C.

References

1. Gray, D. E., ed., *American Institute of Physics Handbook, 3rd Edition,* p. 6-110, McGraw-Hill, New York, 1972.
2. Forsythe, W. E., *Smithsonian Physical Tables, Ninth Edition,* p. 533, Smithsonian Institution, Washington, 1954.
3. *Kaye and Laby Tables of Physical and Chemical Constants, Sixteenth Edition,* p. 131, Longman Group Ltd., Harlow, Essex, 1995.

	Wavelength						
Gas	**0.4360 μm**	**0.4861 μm**	**0.5461 μm**	**0.5790 μm**	**0.5893 μm**	**0.6563 μm**	**0.6709 μm**
Air	1.0002966	1.0002947	1.0002932	1.0002926	1.0002924	1.0002915	1.0002913
Ar					1.000281		
CH_4					1.000444		
Cl_2					1.000773		
CO_2	1.0004563		1.0004506		1.0004493		1.0004471
H_2	1.0001418	1.0001406	1.0001397	1.0001393	1.0001392	1.0001387	1.0001385
He					1.000036		
N_2		1.0003012	1.0002998		1.0002990	1.0002982	
N_2O					1.000516		
NO					1.000297		
O_2	1.0002743	1.0002734	1.0002717	1.0002710	1.0002709	1.0002698	1.0002683
SO_2					1.000686		

CHARACTERISTICS OF LASER SOURCES

William F. Krupke

Light amplification by stimulated emission of radiation was first demonstrated by Maiman in 1960, the result of a population inversion produced between energy levels of chromium ions in a ruby crystal when irradiated with a xenon flashlamp. Since then population inversions and coherent emission have been generated in literally thousands of substances (neutral and ionized gases, liquids, and solids) using a variety of incoherent excitation techniques (optical pumping, electrical discharges, gas-dynamic flow, electron-beams, chemical reactions, nuclear decay).

The extrema of laser output parameters that have been demonstrated to date and the laser media used are summarized in Table 1. Note that the extreme power and energy parameters listed in this table were attained with laser systems rather than with simple laser oscillators.

Laser sources are commonly classified in terms of the state-of-matter of the active medium: gas, liquid, and solid. Each of these classes is further subdivided into one or more types as shown in Table 2. A well-known representative example of each type of laser is also given in Table 2 together with its nominal operation wavelength and the methods by which it is pumped.

The various lasers together cover a wide spectral range from the far ultraviolet to the far infrared. The particular wavelength of emission (usually a narrow line) is presented for some six dozen lasers in Figures 1A and 1B.

By suitably designing the excitation source and/or by controlling the laser resonator structure, laser systems can provide continuous or pulsed radiation as shown in Table 3.

Besides the method of excitation and the temporal behavior of a laser, there are many other parameters that characterize its operation and efficiency, as shown in Tables 4 and 5.

Although many lasers only emit in one or more narrow spectral "lines," an increasing number of lasers can be tuned by changing the composition or the pressure of the medium, or by varying the wavelength of the pump bands. The spectral regions in which these tunable lasers operate are presented in Figure 2.

Reference

Krupke, W. F., in *Handbook of Laser Science and Technology*, Vol. I, Weber, M. J., Ed., CRC Press, Boca Raton, FL, 1986.

TABLE 1. Extrema of Output Parameters of Laser Devices or Systems

Parameter	Value	Laser medium
Peak power	1×10^{14} W(collimated)	Nd:glass
Peak power density	10^{18} W/cm^2 (focused)	Nd:glass
Pulse energy	$>10^5$ J	CO_2, Nd:glass
Average power	10^5 W	CO_2
Pulse duration	3×10^{-15} s continuous wave (cw)	Rh6G dye; various gases, liquids, solids
Wavelength	60 nm ↔ 385 μm	Many required
Efficiency (nonlaser pumped)	70%	CO
Beam quality	Diffraction limited	Various gases, liquids, solids
Spectral linewidth	20 Hz (for 10^{-1} s)	Neon-helium
Spatial coherence	10 m	Ruby

TABLE 2. Classes, Types, and Representative Examples of Laser Sources

Class	Type (characteristic)	Representative example	Nominal operating wavelength (nm)	Method(s) of excitation
Gas	Atom, neutral (electronic transition)	Neon-Helium (Ne-He)	633	Glow discharge
	Atom, ionic (electronic transition)	Argon (Ar^+)	488	Arc discharge
	Molecule, neutral (electronic transition)	Krypton fluoride (KrF)	248	Glow discharge; e-beam
	Molecule, neutral (vibrational transition)	Carbon dioxide (CO_2)	10600	Glow discharge; gasdynamic flow
	Molecule, neutral (rotational transition)	Methyl fluoride (CH_3F)	496000	Laser pumping
	Molecule, ionic (electronic transition)	Nitrogen ion (N_2^+)	420	E-beam
Liquid	Organic solvent (dye-chromophore)	Rhodamine dye (Rh6G)	580–610	Flashlamp; laser pumping
	Organic solvent (rare earth chelate)	Europium:TTF	612	Flashlamp
	Inorganic solvent (trivalent rare earth ion)	Neodymium:$POCl_4$	1060	Flashlamp
Solid	Insulator, crystal (impurity)	Neodymium:YAG	1064	Flashlamp, arc lamp
	Insulator, crystal (stoichiometric)	Neodymium:UP(NdP_5O_{14})	1052	Flashlamp
	Insulator, crystal (color center)	F_2^-:LiF	1120	Laser pumping
	Insulator, amorphous (impurity)	Neodymium:glass	1061	Flashlamp
	Semiconductor (p-n junction)	GaAs	820	Injection current
	Semiconductor (electron-hole plasma)	GaAs	890	E- beam, laser pumping

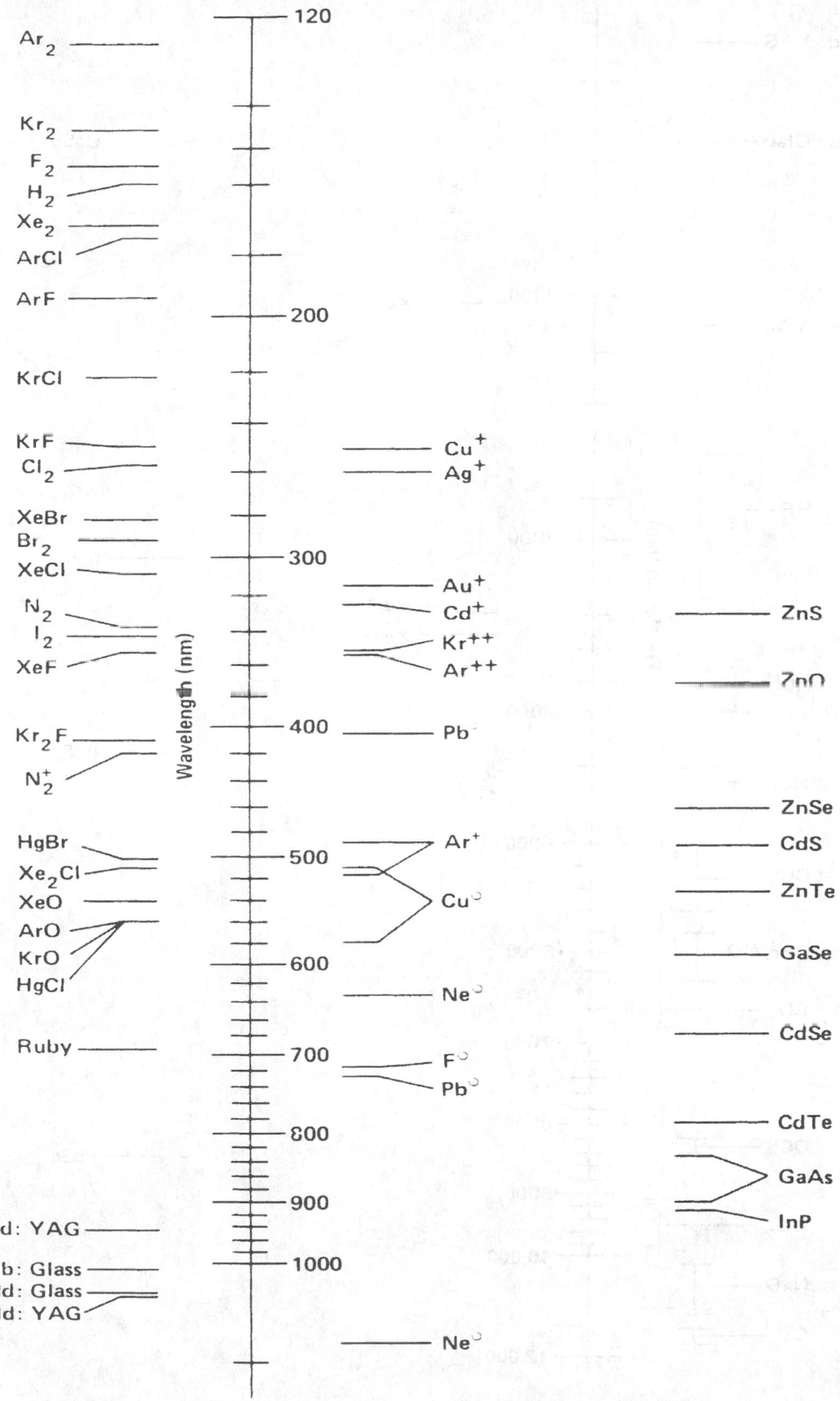

FIGURE 1A. Wavelengths of lasers operating in the 120 to 1200 nm spectral region.

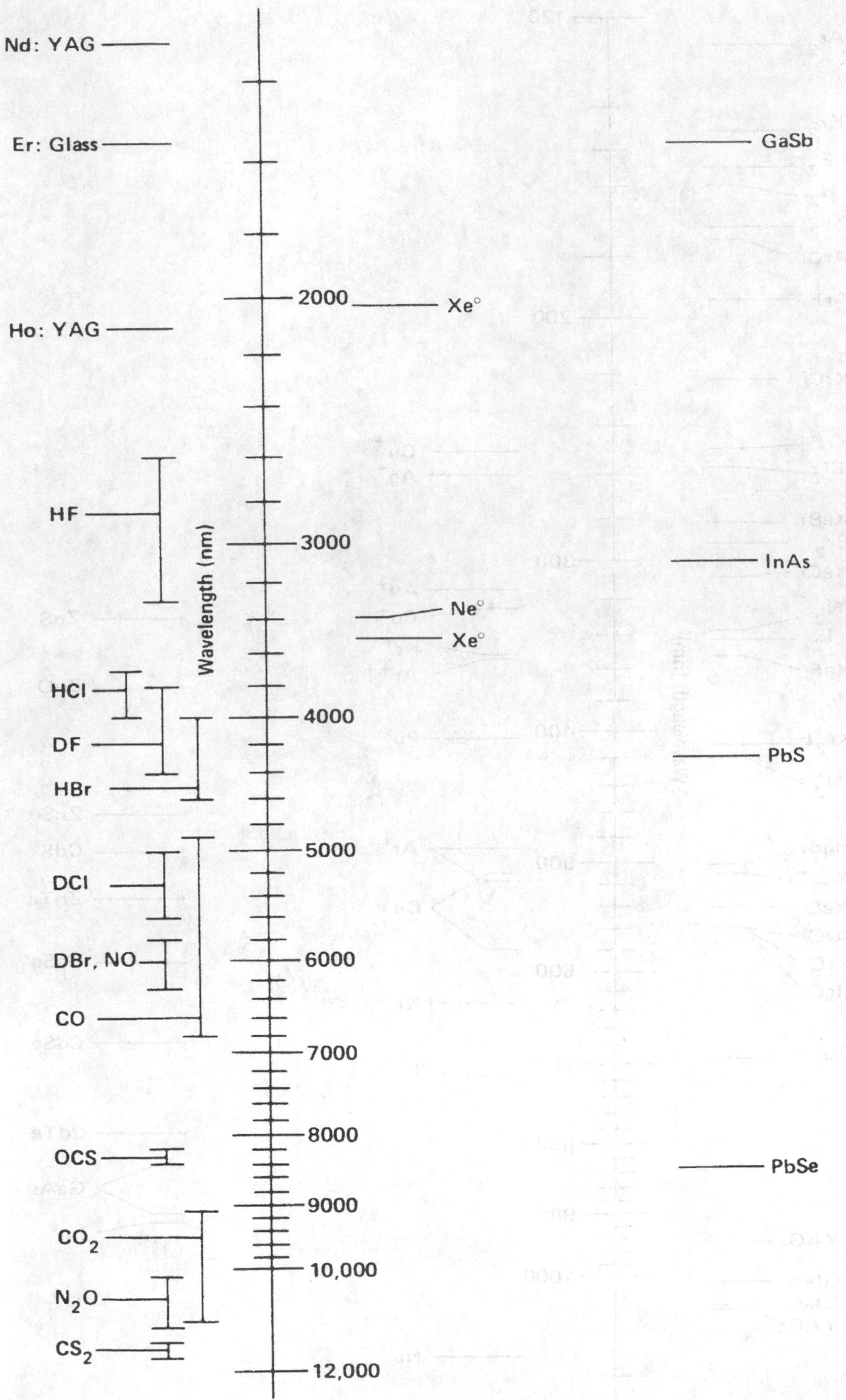

FIGURE 1B. Wavelength of lasers operating in the 1300 to 12,000 nm spectral region.

TABLE 3. Temporal Characteristics of Lasers and Laser Systems

Form	Technique	Pulse width range (s)
Continuous wave	Excitation is continuous; resonator Q is held constant at some moderate value	∞
Pulsed	Excitation is pulsed; resonator Q is held constant at some moderate value	$10^{-8} - 10^{-3}$
Q-Switched	Excitation is continuous or pulsed; resonator Q is switched from a very low value to a moderate value	$10^{-8} - 10^{-6}$
Cavity dumped	Excitation is continuous or pulsed; resonator Q is switched from a very high value to a low value	$10^{-7} - 10^{-5}$
Mode locked	Excitation is continuous or pulsed; phase or loss of the resonator modes is modulated at a rate related to the resonator transit time	$10^{-12} - 10^{-9}$

TABLE 4. Properties and Performance of Some Continuous Wave (CW) Lasers

		Gas			Liquid	Solid	
Parameter	**Unit**	**Neon helium**	**Argon ion**	**Carbon dioxide**	**Rhodamine 6G dye**	**Nd:YAG**	**GaAs**
Excitation method		DC discharge	DC discharge	DC discharge	Ar^+ laser pump	Krypton arc lamp	DC injection
Gain medium composition		Neon:helium	Argon	CO_2:N_2:He	Rh 6G:H_2O	Nd:YAG	p:n:GaAs
Gain medium density	Torr	0.1:1.0	0.4	0.4:0.8:5.0			
	ions/cm^3				2(18):2(22)	1.5(20):2(22)	2(19):3(18):3(22)
Wavelength	nm	633	488	10600	590	1064	810
Laser cross-section	cm^{-2}	3(-13)	1.6(-12)	1.5(-16)	1.8(-16)	7(-19)	~6(-15)
Radiative lifetime (upper level)	s	~1(-7)	7.5(-9)	4(-3)	6.5(-9)	2.6(-4)	~1(-9)
Decay lifetime (upper level)	s	~1(-7)	~5.0(-9)	~4(-3)	6.0(-9)	2.3(-4)	~1(-9)
Gain bandwidth	nm	2(-3)	5(-3)	1.6(-2)	80	0.5	10
Type, gain saturation		Inhomogeneous	Inhomogeneous	Homogeneous	Homogeneous	Homogeneous	Homogeneous
Homogeneous saturation flux	W cm^{-2}			~20	3(5)	2.3(3)	~2(4)
Decay lifetime (lower level)	s	~ 1(-8)	~4(-10)	~5(-6)	<1(-12)	< 1(-7)	<1(-12)
Inversion density	cm^{-3}	~ 1(9)	2(10)	2(15)	2(16)	6(16)	1(16)
Small signal gain coefficient	cm^{-1}	~ 1(-3)	~3(-2)	1(-2)	4	5(-2)	40
Pump power density	W cm^{-3}	3	900	0.15	1(6)	150	7(7)
Output power density	W cm^{-3}	2.6(-3)	~1	2(-2)	3(5)	95	5(6)
Laser size (diameter: length)	cm:cm	0.5:100	0.3:100	5.0:600	1(-3):0.3	0.6:10	5(-4):7(-3);2(-2)[a]
Excitation current/ voltage	A/V	3(-2):2(3)	30:300	0.1:1.5(4)		90:125	1.0/1.7
Excitation current density	A cm^{-2}	0.15	600	6(-3)		140	4.5(3)
Excitation power	W	60	9(3)	1.5(3)	4	1.1(4)	1.7
Output power	W	0.06	10	240	0.3	300	0.12
Efficiency[b]	%	0.1	0.1	13	7	2.6	7

[a] Junction thickness:width:length.
[b] Pressure dependent.

TABLE 5. Properties and Performance of Some Pulsed Lasers

		Gas				Liquid	Solid	
Parameter	**Unit**	**Carbon dixoxide**		**Krypton fluoride**		**Rhodamine 6G**	**Nd:YAG**	**Nd:glass**
Excitation method		TEA-discharge	E- beam/sust.	Glow discharge	E-beam	Xenon flashlamp	Xenon flashlamp	Xenon flashlamp
Gain medium composition		CO_2:N_2:He	CO_2:N_2:He	He:Kr:F_2	Ar:Kr:F_2	Rh6G:alcohol	Nd:YAG	Nd:Glass
Gain medium density	torr	100:50:600	240:240:320	1070:70:3	1235:52:3			–
	ions/cm³					1(18):1.5(22)	1.5(20):1(22)	3(20):2(22)
Wavelength	nm	10600	10600	249	249	590	1064	1061
Laser cross-section	cm^{-2}	2(-18)	2(-18)	2(-16)	2(-16)	1.8(-16)	7(-19)	2.8(-20)
Radiative lifetime (upper level)	s	4(-3)	4(-3)	7(-9)	7(-9)	6.5(-9)	2.6(-4)	4.1(-4)
Decay lifetime (upper level)	s	~1(-4)	5(-5)	2(-9)	3(-9)	6.0(-9)	2.3(-4)	3.7(-4)
Gain bandwidth	nm	1	1	2	2	80	0.5	26
Homogeneous saturation fluence	J/cm²	0.2	0.2	4(-3)	4(-3)	2(-3)	0.6	~5
Decay lifetime (lower level)	s	5(-8)[a]	1(-8)[a]	<1(-12)	<1(-12)	<1(-12)	<1(-7)	<1(-8)
Inversion density	cm^{-3}	3(17)	6(17)	4(14)	2(14)	2(16)	4(17)	3(18)
Small signal gain coefficient	cm^{-1}	2(-2)	4(-2)	8–92)	4(-2)	4	0.3	8(-2)
Medium excitation energy density	J/cm³	0.1	0.36	0.15	0.13	2.8	0.15	0.6
Output energy density	J/cm³	2(-2)	1.8(-2)	1.5(-3)	1.2(-2)	0.85	5(-2)	2(-2)
Laser dimensions	cm: cm: cm	4.5:4.5:87	10:10:100	1.5:4.5:100	8.5:10:100	1.2:25	0.6:7.5	0.6:8.3
Excitation current/voltage	A/V	6(4)/3.3(3)	2.4(4)/4(4)	2.5(4)/1.5(5)	1.2(4)/2.5(5)	2(5)/2.5(4)		
Excitation current density	A cm^{-2}	8.5	22	170	11.5	2.6(3)		
Excitation peak power	W	2(8)	9(8)	4(9)	3(9)	5.4(9)	4(4)	9(4)
Output pulse energy	J	35	180	1	102	32	0.1	1.0
Output pulse length	s	1(-6)	4(-6)	2.5(-8)	6(-7)	3.2(-6)	2(-8)	1(-4)
Output pulse power	W	3.5(7)	4(7)	4(7)	2(8)	1(7)	5(6)	1(4)
Efficiency	%	17	5	1	10[b]	0.2	1.5	3.7

[a] Pressure dependent.

[b] Intrinsic efficiency ≡ energy output/energy deposited in gas.

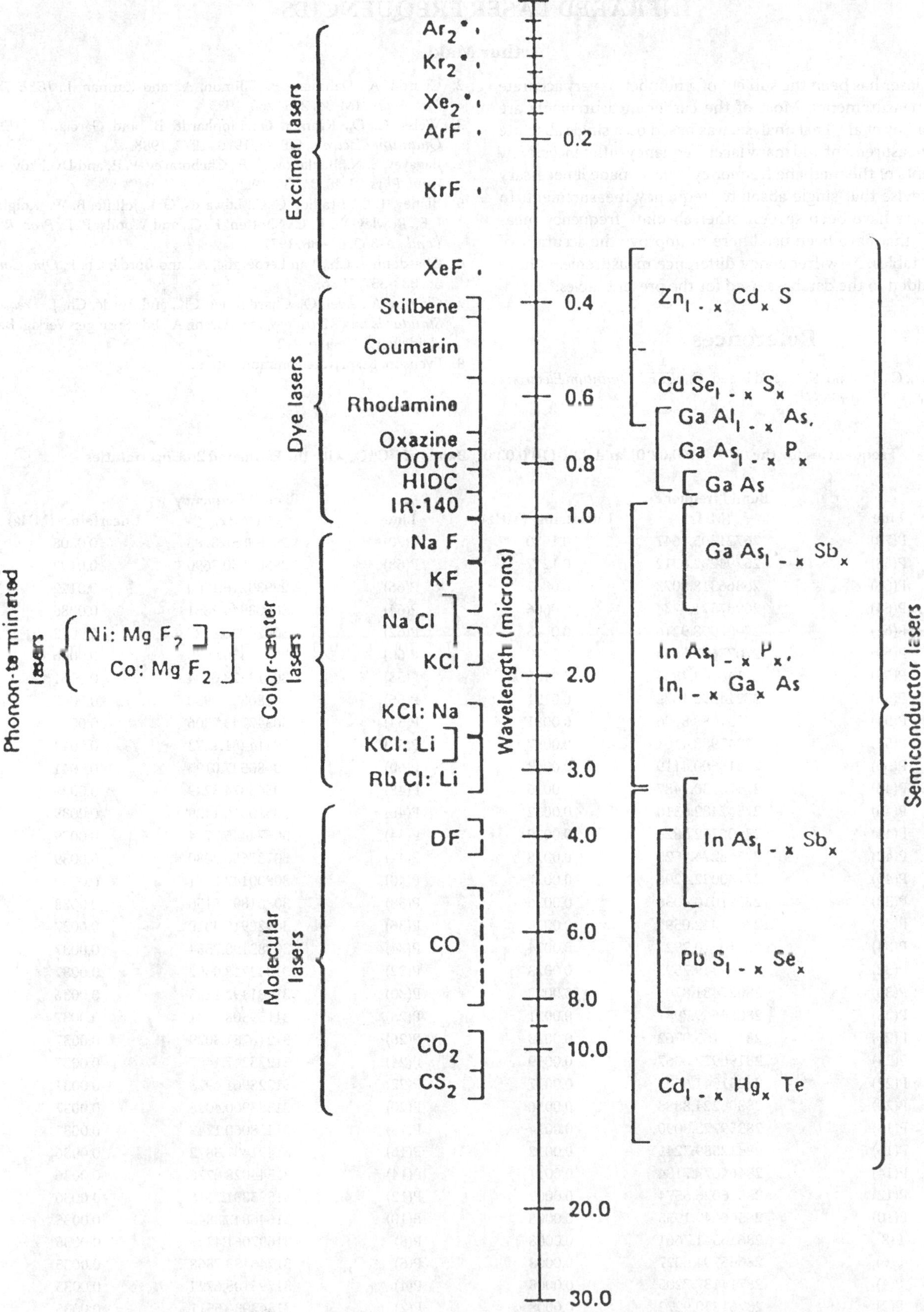

FIGURE 2. Spectral tuning ranges of various types of tunable lasers.

INFRARED LASER FREQUENCIES

Arthur Maki

The CO_2 laser has been the subject of a number of very accurate frequency measurements. Most of the earlier measurements are given by Bradley et al.[1] That analysis was based on a single absolute frequency measurement and many laser frequency differences. New measurements of the methane frequency[2-4] have made it necessary to slightly revise that single absolute frequency measurement. In addition, there have been several other absolute frequency measurements[5-7] that have been used here to improve the accuracy of the present tables. New frequency difference measurements have also been added to the database used for the present tables.[8]

References

1. Bradley, L. C., Soohoo, K. L., and Freed, C., *IEEE J. Quantum Electron.*, QE-22, 234–267, 1986.
2. Clairon, A., Dahmani, B., Filimon, A., and Rutman, J., *IEEE Trans. Inst. Meas.*, IM-34, 265–268, 1985.
3. Weiss, C. O., Kramer, G., Lipphardt, B., and Garcia, E., *IEEE J. Quantum Electron.*, QE-24, 1970–1972, 1988.
4. Bagayev, S. N., Baklanov, A. E., Chebotayev, V. P., and Dychkov, A. S., *Appl. Phys.*, B-48, 31–35, 1989.
5. Blaney, T. G., Bradley, C. C., Edwards, G. J., Jolliffe, B. W., Knight, D. J. E., Rowley, W. R. C., Shotten, K. C., and Woods, P. T., *Proc. R. Soc. Lond.*, A-355, 61–88, 1977.
6. Chardonnet, Ch., Van Lerberghe, A., and Bordé, Ch. J., *Opt. Comm.*, 58, 333-337, 1986.
7. Clairon, A., Acef, O., Chardonnet, Ch., and Bordé, Ch. J., *Frequency Standards and Metrology*, De Marchi, A., Ed., Springer-Verlag, Berlin, Heidelberg, 1989, p. 212.
8. Evenson, K., private communication.

Frequencies for the 00°1-(10°0,02°0)$_{I}$ and 00°1-(10°0,02°0)$_{II}$ Bands of $^{12}C^{16}O_2$ with the Estimated 2σ Uncertainties

Line	Band I frequency (MHz)	Uncertainty (MHz)	Line	Band II frequency (MHz)	Uncertainty (MHz)
P(70)	26721305.4647	0.1680	P(70)	29789856.3783	0.0308
P(68)	26794232.6712	0.1217	P(68)	29861850.7690	0.0192
P(66)	26866318.8073	0.0867	P(66)	29933216.1760	0.0122
P(64)	26937571.7234	0.0606	P(64)	30003944.2861	0.0086
P(62)	27007998.9216	0.0415	P(62)	30074026.9127	0.0072
P(60)	27077607.5643	0.0279	P(60)	30143456.0039	0.0066
P(58)	27146404.4834	0.0185	P(58)	30212223.6504	0.0061
P(56)	27214396.1873	0.0121	P(56)	30280322.0930	0.0055
P(54)	27281588.8696	0.0081	P(54)	30347743.7306	0.0049
P(52)	27347988.4161	0.0057	P(52)	30414481.1273	0.0044
P(50)	27413600.4119	0.0043	P(50)	30480527.0196	0.0041
P(48)	27478430.1487	0.0036	P(48)	30545874.3239	0.0039
P(46)	27542482.6310	0.0032	P(46)	30610516.1429	0.0039
P(44)	27605762.5826	0.0030	P(44)	30674445.7724	0.0039
P(42)	27668274.4525	0.0028	P(42)	30737656.7080	0.0039
P(40)	27730022.4206	0.0027	P(40)	30800142.6511	0.0039
P(38)	27791010.4036	0.0026	P(38)	30861897.5150	0.0038
P(36)	27851242.0594	0.0025	P(36)	30922915.4310	0.0037
P(34)	27910720.7927	0.0024	P(34)	30983190.7534	0.0037
P(32)	27969449.7593	0.0023	P(32)	31042718.0652	0.0037
P(30)	28027431.8708	0.0022	P(30)	31101492.1833	0.0036
P(28)	28084669.7981	0.0021	P(28)	31159508.1631	0.0037
P(26)	28141165.9762	0.0020	P(26)	31216761.3029	0.0037
P(24)	28196922.6067	0.0019	P(24)	31273247.1487	0.0037
P(22)	28251941.6622	0.0017	P(22)	31328961.4978	0.0037
P(20)	28306224.8888	0.0016	P(20)	31383900.4028	0.0037
P(18)	28359773.8090	0.0014	P(18)	31438060.1749	0.0037
P(16)	28412589.7245	0.0012	P(16)	31491437.3872	0.0036
P(14)	28464673.7184	0.0011	P(14)	31544028.8776	0.0036
P(12)	28516026.6574	0.0009	P(12)	31595831.7516	0.0036
P(10)	28566649.1935	0.0008	P(10)	31646843.3843	0.0035
P(8)	28616541.7661	0.0008	P(8)	31697061.4225	0.0035
P(6)	28665704.6027	0.0008	P(6)	31746483.7868	0.0035
P(4)	28714137.7205	0.0008	P(4)	31795108.6724	0.0035
P(2)	28761840.9272	0.0008	P(2)	31842934.5511	0.0035
R(0)	28832026.2198	0.0008	R(0)	31913172.5691	0.0035
R(2)	28877902.4382	0.0007	R(2)	31958996.0621	0.0034
R(4)	28923046.4303	0.0006	R(4)	32004017.3822	0.0034

Line	Band I frequency (MHz)	Uncertainty (MHz)	Line	Band II frequency (MHz)	Uncertainty (MHz)
R(6)	28967457.0657	0.0005	R(6)	32048236.2498	0.0034
R(8)	29011133.0054	0.0003	R(8)	32091652.6619	0.0034
R(10)	29054072.7010	0.0001	R(10)	32134266.8917	0.0034
R(12)	29096274.3935	0.0003	R(12)	32176079.4878	0.0034
R(14)	29137736.1129	0.0005	R(14)	32217091.2721	0.0035
R(16)	29178455.6759	0.0007	R(16)	32257303.3386	0.0036
R(18)	29218430.6852	0.0009	R(18)	32296717.0510	0.0037
R(20)	29257658.5269	0.0010	R(20)	32335334.0408	0.0038
R(22)	29296136.3689	0.0011	R(22)	32373156.2044	0.0039
R(24)	29333861.1583	0.0012	R(24)	32410185.7003	0.0041
R(26)	29370829.6191	0.0011	R(26)	32446424.9459	0.0042
R(28)	29407038.2491	0.0011	R(28)	32481876.6140	0.0042
R(30)	29442483.3168	0.0011	R(30)	32516543.6293	0.0042
R(32)	29477160.8582	0.0012	R(32)	32550429.1641	0.0042
R(34)	29511066.6733	0.0013	R(34)	32583536.6340	0.0042
R(36)	29544196.3221	0.0015	R(36)	32615869.6937	0.0041
R(38)	29576545.1205	0.0017	R(38)	32647432.2320	0.0040
R(40)	29608108.1360	0.0019	R(40)	32678228.3665	0.0039
R(42)	29638880.1831	0.0022	R(42)	32708262.4386	0.0038
R(44)	29668855.8183	0.0024	R(44)	32737539.0081	0.0039
R(46)	29698029.3350	0.0027	R(46)	32766062.8469	0.0041
R(48)	29726394.7582	0.0032	R(48)	32793838.9334	0.0045
R(50)	29753945.8385	0.0037	R(50)	32820872.4463	0.0055
R(52)	29780676.0464	0.0042	R(52)	32847168.7576	0.0071
R(54)	29806578.5659	0.0047	R(54)	32872733.4269	0.0099
R(56)	29831646.2878	0.0052	R(56)	32897572.1935	0.0141
R(58)	29855071.0032	0.0050	R(58)	32921090.9701	0.0202
R(60)	29879247.3960	0.0074	R(60)	32945095.8355	0.0288
R(62)	29901765.0357	0.0113	R(62)	32967793.0268	0.0407
R(64)	29923416.3695	0.0186	R(64)	32989788.9322	0.0567
R(66)	29944192.7145	0.0302	R(66)	33011090.0831	0.0780
R(68)	29964085.0488	0.0475	R(68)	33031703.1467	0.1060
R(70)	29983084.0036	0.0720	R(70)	33051634.9172	0.1423

Frequencies for the 00°1-(10°0,02°0)$_{I}$ and 00°1-(10°0,02°0)$_{II}$ Bands of $^{13}C^{16}O_2$ with the Estimated 2σ Uncertainties

Line	Band I frequency (MHz)	Uncertainty (MHz)	Line	Band II frequency (MHz)	Uncertainty (MHz)
P(66)	25523832.1808	0.7836	P(66)	28512082.5283	1.2894
P(64)	25590013.4703	0.5415	P(64)	28585121.9396	0.9194
P(62)	25655543.6502	0.3629	P(62)	28657449.4180	0.6420
P(60)	25720428.2487	0.2339	P(60)	28729056.6374	0.4375
P(58)	25784672.4840	0.1430	P(58)	28799935.4147	0.2897
P(56)	25848281.2771	0.0810	P(56)	28870077.7187	0.1853
P(54)	25911259.2627	0.0405	P(54)	28939475.6771	0.1135
P(52)	25973610.8005	0.0157	P(52)	29008121.5846	0.0659
P(50)	26035339.9857	0.0045	P(50)	29076007.9109	0.0357
P(48)	26096450.6582	0.0079	P(48)	29143127.3077	0.0180
P(46)	26156946.4123	0.0101	P(46)	29209472.6164	0.0090
P(44)	26216830.6053	0.0101	P(44)	29275036.8754	0.0058
P(42)	26276106.3655	0.0090	P(42)	29339813.3270	0.0050
P(40)	26334776.6003	0.0077	P(40)	29403795.4243	0.0044
P(38)	26392844.0030	0.0068	P(38)	29466976.8383	0.0037
P(36)	26450311.0599	0.0063	P(36)	29529351.4635	0.0032
P(34)	26507180.0565	0.0061	P(34)	29590913.4252	0.0029
P(32)	26563453.0836	0.0060	P(32)	29651657.0844	0.0028
P(30)	26619132.0428	0.0058	P(30)	29711577.0447	0.0028
P(28)	26674218.6515	0.0055	P(28)	29770668.1566	0.0031
P(26)	26728714.4479	0.0054	P(26)	29828925.5239	0.0035
P(24)	26782620.7952	0.0054	P(24)	29886344.5074	0.0041
P(22)	26835938.8858	0.0054	P(22)	29942920.7308	0.0046
P(20)	26888669.7451	0.0055	P(20)	29998650.0838	0.0051
P(18)	26940814.2347	0.0055	P(18)	30053528.7271	0.0054

Line	Band I frequency (MHz)	Uncertainty (MHz)	Line	Band II frequency (MHz)	Uncertainty (MHz)
P(16)	26992373.0555	0.0055	P(16)	30107553.0955	0.0055
P(14)	27043346.7508	0.0054	P(14)	30160719.9016	0.0055
P(12)	27093735.7083	0.0052	P(12)	30213026.1388	0.0054
P(10)	27143540.1624	0.0051	P(10)	30264469.0839	0.0054
P(8)	27192760.1962	0.0049	P(8)	30315046.2994	0.0054
P(6)	27241395.7431	0.0048	P(6)	30364755.6359	0.0055
P(4)	27289446.5880	0.0047	P(4)	30413595.2335	0.0056
P(2)	27336912.3682	0.0046	P(2)	30461563.5231	0.0057
R(0)	27407012.8882	0.0045	P(0)	30531879.5415	0.0057
R(2)	27453013.4589	0.0043	P(2)	30577664.6138	0.0056
R(4)	27498426.5430	0.0040	P(4)	30622575.1885	0.0054
R(6)	27543251.1200	0.0037	P(6)	30666611.0128	0.0051
R(8)	27587486.0225	0.0034	P(8)	30709772.1257	0.0047
R(10)	27631129.9356	0.0031	P(10)	30752058.8571	0.0045
R(12)	27674181.3963	0.0029	P(12)	30793471.8269	0.0044
R(14)	27716638.7917	0.0029	P(14)	30834011.9425	0.0043
R(16)	27758500.3577	0.0029	P(16)	30873680.3976	0.0044
R(18)	27799764.1770	0.0029	P(18)	30912478.6694	0.0044
R(20)	27840428.1773	0.0030	P(20)	30950408.5159	0.0044
R(22)	27880490.1283	0.0029	P(22)	30987471.9732	0.0043
R(24)	27919947.6395	0.0029	P(24)	31023671.3517	0.0042
R(26)	27958798.1567	0.0028	P(26)	31059009.2327	0.0042
R(28)	27997038.9591	0.0028	P(28)	31093488.4642	0.0042
R(30)	28034667.1551	0.0027	P(30)	31127112.1569	0.0043
R(32)	28071679.6785	0.0027	P(32)	31159883.6793	0.0045
R(34)	28108073.2842	0.0026	P(34)	31191806.6529	0.0046
R(36)	28143844.5432	0.0026	P(36)	31222884.9469	0.0048
R(38)	28178989.8377	0.0026	P(38)	31253122.6730	0.0053
R(40)	28213505.3554	0.0028	P(40)	31282524.1795	0.0061
R(42)	28247387.0838	0.0033	P(42)	31311094.0452	0.0077
R(44)	28280630.8035	0.0046	P(44)	31338837.0736	0.0108
R(46)	28313232.0818	0.0083	P(46)	31365758.2858	0.0173
R(48)	28345186.2652	0.0161	P(48)	31391862.9147	0.0295
R(50)	28376488.4720	0.0301	P(50)	31417156.3972	0.0505
R(52)	28407133.5839	0.0531	P(52)	31441644.3679	0.0845
R(54)	28437116.2372	0.0887	P(54)	31465332.6516	0.1366
R(56)	28466430.8141	0.1419	P(56)	31488227.2557	0.2138
R(58)	28495071.4324	0.2188	P(58)	31510334.3631	0.3247
R(60)	28523031.9357	0.3271	P(60)	31531660.3243	0.4800
R(62)	28550305.8819	0.4763	P(62)	31552211.6497	0.6932
R(64)	28576886.5323	0.6781	P(64)	31571995.0017	0.9805
R(66)	28602766.8393	0.9467	P(66)	31591017.1868	1.3619

Frequencies for the 00°1-(10°0,02°0)$_I$ and 00°1-(10°0,02°0)$_{II}$ Bands of $^{12}C^{18}O_2$ with the Estimated 2σ Uncertainties

P(70)	27045326.3119	0.4540	P(70)	30695237.5856	0.0858
P(68)	27114914.0922	0.3324	P(68)	30755520.2231	0.0570
P(66)	27183635.7945	0.2392	P(66)	30815311.4928	0.0364
P(64)	27251496.4118	0.1688	P(64)	30874607.2084	0.0223
P(62)	27318500.7361	0.1165	P(62)	30933403.2309	0.0131
P(60)	27384653.3618	0.0783	P(60)	30991695.4724	0.0075
P(58)	27449958.6881	0.0510	P(58)	31049479.9009	0.0049
P(56)	27514420.9224	0.0319	P(56)	31106752.5446	0.0041
P(54)	27578044.0828	0.0191	P(54)	31163509.4964	0.0040
P(52)	27640832.0010	0.0108	P(52)	31219746.9183	0.0040
P(50)	27702788.3248	0.0059	P(50)	31275461.0455	0.0039
P(48)	27763916.5206	0.0035	P(48)	31330648.1908	0.0039
P(46)	27824219.8762	0.0028	P(46)	31385304.7490	0.0039
P(44)	27883701.5029	0.0026	P(44)	31439427.2006	0.0039
P(42)	27942364.3379	0.0025	P(42)	31493012.1163	0.0038
P(40)	28000211.1464	0.0024	P(40)	31546056.1605	0.0038

Line	Band I frequency (MHz)	Uncertainty (MHz)	Line	Band II frequency (MHz)	Uncertainty (MHz)
P(38)	28057244.5242	0.0022	P(38)	31598556.0954	0.0037
P(36)	28113466.8992	0.0021	P(36)	31650508.7847	0.0037
P(34)	28168880.5335	0.0020	P(34)	31701911.1970	0.0037
P(32)	28223487.5256	0.0019	P(32)	31752760.4093	0.0037
P(30)	28277289.8118	0.0017	P(30)	31803053.6105	0.0037
P(28)	28330289.1679	0.0016	P(28)	31852788.1043	0.0038
P(26)	28382487.2111	0.0015	P(26)	31901961.3125	0.0038
P(24)	28433885.4012	0.0013	P(24)	31950570.7773	0.0038
P(22)	28484485.0420	0.0012	P(22)	31998614.1649	0.0038
P(20)	28534287.2828	0.0011	P(20)	32046089.2669	0.0037
P(18)	28583293.1193	0.0010	P(18)	32092994.0036	0.0037
P(16)	28631503.3952	0.0010	P(16)	32139326.4254	0.0036
P(14)	28678918.8025	0.0009	P(14)	32185084.7154	0.0036
P(12)	28725539.8830	0.0010	P(12)	32230267.1907	0.0036
P(10)	28771367.0288	0.0010	P(10)	32274872.3041	0.0037
P(8)	28816400.4829	0.0010	P(8)	32318898.6155	0.0038
P(6)	28860640.3403	0.0011	P(6)	32362344.9434	0.0039
P(4)	28904086.5477	0.0011	P(4)	32405210.0652	0.0041
P(2)	28946738.9048	0.0011	P(2)	32447493.0185	0.0041
R(0)	29009228.1702	0.0010	P(0)	32509824.0580	0.0042
R(2)	29049894.0586	0.0010	P(2)	32550648.1723	0.0042
R(4)	29089764.2368	0.0009	P(4)	32590887.7542	0.0042
R(6)	29128837.8426	0.0008	P(6)	32630542.4457	0.0041
R(8)	29167113.8668	0.0008	P(8)	32669612.0295	0.0041
R(10)	29204591.1529	0.0009	P(10)	32708096.4282	0.0040
R(12)	29241268.3964	0.0010	P(12)	32745995.7040	0.0040
R(14)	29277144.1444	0.0011	P(14)	32783310.0573	0.0040
R(16)	29312216.7955	0.0012	P(16)	32820039.8258	0.0040
R(18)	29346484.5984	0.0012	P(18)	32856185.4827	0.0040
R(20)	29379945.6517	0.0013	P(20)	32891747.6358	0.0040
R(22)	29412597.9024	0.0013	P(22)	32926727.0254	0.0040
R(24)	29444439.1458	0.0013	P(24)	32961124.5220	0.0040
R(26)	29475467.0236	0.0014	P(26)	32994941.1249	0.0040
R(28)	29505679.0230	0.0015	P(28)	33028177.9594	0.0040
R(30)	29535072.4755	0.0016	P(30)	33060836.2743	0.0040
R(32)	29563644.5557	0.0018	P(32)	33092917.4394	0.0041
R(34)	29591392.2794	0.0020	P(34)	33124422.9429	0.0043
R(36)	29618312.5023	0.0023	P(36)	33155354.3878	0.0046
R(38)	29644401.9182	0.0028	P(38)	33185713.4894	0.0049
R(40)	29669657.0575	0.0036	P(40)	33215502.0716	0.0056
R(42)	29694074.2853	0.0053	P(42)	33244722.0637	0.0068
R(44)	29717649.7992	0.0082	P(44)	33273375.4969	0.0092
R(46)	29740379.6276	0.0128	P(46)	33301464.5003	0.0134
R(48)	29762259.6274	0.0200	P(48)	33328991.2976	0.0199
R(50)	29783285.4820	0.0307	P(50)	33355958.2027	0.0294
R(52)	29803452.6988	0.0461	P(52)	33382367.6161	0.0427
R(54)	29822756.6072	0.0681	P(54)	33408222.0209	0.0607
R(56)	29841192.3558	0.0985	P(56)	33433523.9780	0.0848
R(58)	29858754.9100	0.1401	P(58)	33458276.1228	0.1165
R(60)	29875439.0495	0.1960	P(60)	33482481.1601	0.1576
R(62)	29891239.3658	0.2702	P(62)	33506141.8605	0.2104
R(64)	29906150.2589	0.3673	P(64)	33529261.0556	0.2775
R(66)	29920165.9352	0.4930	P(66)	33551841.6335	0.3621
R(68)	29933280.4042	0.6540	P(68)	33573886.5352	0.4679
R(70)	29945487.4756	0.8581	P(70)	33595398.7493	0.5992

Frequencies for the 00°1-(10°0,02°0)$_I$ and 00°1-(10°0,02°0)$_{II}$ Bands of $^{13}C^{18}O_2$ with the Estimated 2σ Uncertainties

Line	Band I frequency (MHz)	Uncertainty (MHz)	Line	Band II frequency (MHz)	Uncertainty (MHz)
P(70)	25967863.7652	1.1146	P(70)	28960476.2278	0.4069
P(68)	26033448.2798	0.8152	P(68)	29022326.9578	0.2861
P(66)	26098273.9159	0.5860	P(66)	29083661.3546	0.1961

Line	Band I frequency (MHz)	Uncertainty (MHz)	Line	Band II frequency (MHz)	Uncertainty (MHz)
P(64)	26162346.4813	0.4129	P(64)	29144473.5795	0.1303
P(62)	26225671.5466	0.2844	P(62)	29204757.8761	0.0833
P(60)	26288254.4494	0.1906	P(60)	29264508.5768	0.0507
P(58)	26350100.2984	0.1237	P(58)	29323720.1086	0.0290
P(56)	26411213.9778	0.0772	P(56)	29382386.9988	0.0152
P(54)	26471600.1504	0.0459	P(54)	29440503.8809	0.0073
P(52)	26531263.2618	0.0258	P(52)	29498065.4997	0.0038
P(50)	26590207.5442	0.0138	P(50)	29555066.7172	0.0032
P(48)	26648437.0195	0.0077	P(48)	29611502.5178	0.0031
P(46)	26705955.5026	0.0057	P(46)	29667368.0132	0.0031
P(44)	26762766.6051	0.0055	P(44)	29722658.4475	0.0034
P(42)	26818873.7378	0.0055	P(42)	29777369.2022	0.0039
P(40)	26874280.1143	0.0056	P(40)	29831495.8006	0.0044
P(38)	26928988.7531	0.0056	P(38)	29885033.9125	0.0049
P(36)	26983002.4809	0.0056	P(36)	29937979.3584	0.0053
P(34)	27036323.9351	0.0055	P(34)	29990328.1139	0.0054
P(32)	27088955.5657	0.0054	P(32)	30042076.3132	0.0055
P(30)	27140899.6384	0.0051	P(30)	30093220.2534	0.0055
P(28)	27192158.2363	0.0049	P(28)	30143756.3978	0.0054
P(26)	27242733.2620	0.0047	P(26)	30193681.3793	0.0053
P(24)	27292626.4396	0.0044	P(24)	30242992.0038	0.0052
P(22)	27341839.3165	0.0042	P(22)	30291685.2529	0.0051
P(20)	27390373.2651	0.0040	P(20)	30339758.2870	0.0049
P(18)	27438229.4843	0.0037	P(18)	30387208.4477	0.0048
P(16)	27485409.0008	0.0035	P(16)	30434033.2603	0.0046
P(14)	27531912.6704	0.0033	P(14)	30480230.4356	0.0045
P(12)	27577741.1795	0.0031	P(12)	30525797.8725	0.0044
P(10)	27622895.0455	0.0031	P(10)	30570733.6593	0.0043
P(8)	27667374.6182	0.0031	P(8)	30615036.0750	0.0043
P(6)	27711180.0803	0.0033	P(6)	30658703.5912	0.0044
P(4)	27754311.4480	0.0034	P(4)	30701734.8727	0.0045
P(2)	27796768.5718	0.0036	P(2)	30744128.7785	0.0045
R(0)	27859189.3155	0.0036	P(0)	30806522.5414	0.0045
R(2)	27899959.0889	0.0035	P(2)	30847319.2956	0.0044
R(4)	27940052.7921	0.0033	P(4)	30887476.2168	0.0043
R(6)	27979469.5315	0.0031	P(6)	30926993.0424	0.0042
R(8)	28018208.2478	0.0028	P(8)	30965869.7046	0.0041
R(10)	28056267.7161	0.0026	P(10)	31004106.3298	0.0040
R(12)	28093646.5448	0.0025	P(12)	31041703.2379	0.0040
R(14)	28130343.1757	0.0025	P(14)	31078660.9408	0.0040
R(16)	28166355.8825	0.0025	P(16)	31114980.1420	0.0040
R(18)	28201682.7706	0.0025	P(18)	31150661.7340	0.0041
R(20)	28236321.7757	0.0025	P(20)	31185706.7976	0.0042
R(22)	28270270.6628	0.0024	P(22)	31220116.5992	0.0043
R(24)	28303527.0249	0.0024	P(24)	31253892.5891	0.0043
R(26)	28336088.2817	0.0023	P(26)	31287036.3991	0.0044
R(28)	28367951.6781	0.0024	P(28)	31319549.8396	0.0043
R(30)	28399114.2823	0.0025	P(30)	31351434.8973	0.0043
R(32)	28429572.9843	0.0026	P(32)	31382693.7318	0.0042
R(34)	28459324.4940	0.0028	P(34)	31413328.6728	0.0042
R(36)	28488365.3390	0.0029	P(36)	31443342.2165	0.0041
R(38)	28516691.8625	0.0029	P(38)	31472737.0219	0.0040
R(40)	28544300.2211	0.0031	P(40)	31501515.9074	0.0039
R(42)	28571186.3823	0.0032	P(42)	31529681.8467	0.0040
R(44)	28597346.1222	0.0032	P(44)	31557237.9646	0.0042
R(46)	28622775.0223	0.0038	P(46)	31584187.5329	0.0046
R(48)	28647468.4672	0.0071	P(48)	31610533.9656	0.0057
R(50)	28671421.6417	0.0148	P(50)	31636280.8146	0.0088
R(52)	28694629.5272	0.0286	P(52)	31661431.7650	0.0151
R(54)	28717086.8993	0.0510	P(54)	31685990.6298	0.0261
R(56)	28738788.3239	0.0852	P(56)	31709961.3449	0.0434

Line	Band I frequency (MHz)	Uncertainty (MHz)	Line	Band II frequency (MHz)	Uncertainty (MHz)
R(58)	28759728.1540	0.1355	P(58)	31733347.9642	0.0693
R(60)	28779900.5263	0.2075	P(60)	31756154.6537	0.1068
R(62)	28799299.3572	0.3078	P(62)	31778385.6867	0.1594
R(64)	28817918.3393	0.4447	P(64)	31800045.4375	0.2317
R(66)	28835750.9374	0.6283	P(66)	31821138.3761	0.3291
R(68)	28852790.3843	0.8707	P(68)	31841669.0622	0.4581
R(70)	28869029.6768	1.1863	P(70)	31861642.1394	0.6268

Frequencies for the $01^{1e}1$-$(11^{1e}0,03^{1e}0)_I$ and $01^{1e}1$-$(11^{1e}0,03^{1e}0)_{II}$ Bands of $^{12}C^{16}O_2$ with the Estimated 2σ Uncertainties

Line	Band I frequency (MHz)	Uncertainty (MHz)	Line	Band II frequency (MHz)	Uncertainty (MHz)
P(59)	26125213.2723	1.6633	P(59)	30427055.2899	0.1962
P(57)	26191576.6703	1.0880	P(57)	30494640.3229	0.1332
P(55)	26257240.7898	0.6844	P(55)	30561557.5929	0.0865
P(53)	26322208.2302	0.4094	P(53)	30627802.0344	0.0530
P(51)	26386481.4313	0.2286	P(51)	30693368.7014	0.0306
P(49)	26450062.6783	0.1155	P(49)	30758252.7710	0.0175
P(47)	26512954.1076	0.0498	P(47)	30822449.5469	0.0123
P(45)	26575157.7109	0.0191	P(45)	30885954.4624	0.0114
P(43)	26636675.3402	0.0160	P(43)	30948763.0834	0.0109
P(41)	26697508.7115	0.0182	P(41)	31010871.1119	0.0100
P(39)	26757659.4084	0.0177	P(39)	31072274.3882	0.0091
P(37)	26817128.8857	0.0160	P(37)	31132968.8940	0.0091
P(35)	26875918.4726	0.0144	P(35)	31192950.7549	0.0102
P(33)	26934029.3751	0.0131	P(33)	31252216.2430	0.0118
P(31)	26991462.6787	0.0119	P(31)	31310761.7788	0.0134
P(29)	27048219.3509	0.0106	P(29)	31368583.9339	0.0147
P(27)	27104300.2431	0.0096	P(27)	31425679.4328	0.0155
P(25)	27159706.0925	0.0093	P(25)	31482045.1550	0.0157
P(23)	27214437.5237	0.0097	P(23)	31537678.1367	0.0154
P(21)	27268495.0505	0.0104	P(21)	31592575.5725	0.0147
P(19)	27321879.0769	0.0108	P(19)	31646734.8172	0.0137
P(17)	27374589.8987	0.0108	P(17)	31700153.3868	0.0127
P(15)	27426627.7040	0.0104	P(15)	31752828.9602	0.0119
P(13)	27477992.5747	0.0098	P(13)	31804759.3803	0.0113
P(11)	27528684.4867	0.0096	P(11)	31855942.6551	0.0113
P(9)	27578703.3113	0.0101	P(9)	31906376.9582	0.0116
P(7)	27628048.8151	0.0113	P(7)	31956060.6304	0.0122
P(5)	27676720.6609	0.0127	P(5)	32004992.1796	0.0129
P(3)	27724718.4080	0.0141	P(3)	32053170.2819	0.0136
R(1)	27841759.7696	0.0152	P(1)	32170312.0391	0.0149
R(3)	27887393.2105	0.0146	P(3)	32215845.0845	0.0151
R(5)	27932349.2934	0.0135	P(5)	32260620.8121	0.0152
R(7)	27976627.0108	0.0124	P(7)	32304638.8261	0.0152
R(9)	28020225.2521	0.0115	P(9)	32347898.8990	0.0150
R(11)	28063142.8031	0.0110	P(11)	32390400.9714	0.0148
R(13)	28105378.3457	0.0109	P(13)	32432145.1513	0.0145
R(15)	28146930.4576	0.0109	P(15)	32473131.7137	0.0142
R(17)	28187797.6116	0.0107	P(17)	32513361.0997	0.0140
R(19)	28227978.1750	0.0103	P(19)	32552833.9153	0.0140
R(21)	28267470.4088	0.0099	P(21)	32591550.9309	0.0141
R(23)	28306272.4666	0.0099	P(23)	32629513.0796	0.0143
R(25)	28344382.3939	0.0107	P(25)	32666721.4564	0.0144
R(27)	28381798.1267	0.0122	P(27)	32703177.3164	0.0142
R(29)	28418517.4902	0.0141	P(29)	32738882.0732	0.0136
R(31)	28454538.1976	0.0165	P(31)	32773837.2976	0.0136
R(33)	28489857.8477	0.0213	P(33)	32808044.7156	0.0174
R(35)	28524473.9240	0.0312	P(35)	32841506.2063	0.0279
R(37)	28558383.7917	0.0486	P(37)	32874223.8000	0.0462
R(39)	28591584.6963	0.0754	P(39)	32906199.6761	0.0735
R(41)	28624073.7602	0.1131	P(41)	32937436.1606	0.1114
R(43)	28655847.9806	0.1644	P(43)	32967935.7238	0.1624
R(45)	28686904.2261	0.2328	P(45)	32997700.9775	0.2292

Line	Band I frequency (MHz)	Uncertainty (MHz)	Line	Band II frequency (MHz)	Uncertainty (MHz)
R(47)	28717239.2334	0.3239	P(47)	33026734.6728	0.3151
R(49)	28746849.6038	0.4465	P(49)	33055039.6965	0.4238
R(51)	28775731.7988	0.6142	P(51)	33082619.0689	0.5595
R(53)	28803882.1361	0.8465	P(53)	33109475.9403	0.7272

Frequencies for the $01^{1f}1\text{-}(11^{1f}0,03^{1f}0)_I$ and $01^{1f}1\text{-}(11^{1f}0,03^{1f}0)_{II}$ Bands of $^{12}C^{16}O_2$ with the Estimated 2σ Uncertainties

Line	Band I frequency (MHz)	Uncertainty (MHz)	Line	Band II frequency (MHz)	Uncertainty (MHz)
P(60)	26051570.0104	4.4521	P(60)	30355115.0204	0.2752
P(58)	26120964.4932	3.0629	P(58)	30425283.5969	0.1926
P(56)	26189552.8496	2.0516	P(56)	30494732.8293	0.1301
P(54)	26257339.6006	1.3305	P(54)	30563455.6325	0.0840
P(52)	26324329.0344	0.8289	P(52)	30631445.1076	0.0512
P(50)	26390525.2136	0.4901	P(50)	30698694.5456	0.0292
P(48)	26455931.9824	0.2698	P(48)	30765197.4310	0.0163
P(46)	26520552.9722	0.1334	P(46)	30830947.4444	0.0111
P(44)	26584391.6075	0.0551	P(44)	30895938.4662	0.0104
P(42)	26647451.1105	0.0181	P(42)	30960164.5794	0.0105
P(40)	26709734.5057	0.0151	P(40)	31023620.0723	0.0105
P(38)	26771244.6242	0.0174	P(38)	31086299.4415	0.0107
P(36)	26831984.1067	0.0157	P(36)	31148197.3941	0.0114
P(34)	26891955.4069	0.0126	P(34)	31209308.8510	0.0126
P(32)	26951160.7945	0.0105	P(32)	31269628.9481	0.0138
P(30)	27009602.3576	0.0096	P(30)	31329153.0395	0.0147
P(28)	27067282.0045	0.0092	P(28)	31387876.6994	0.0151
P(26)	27124201.4662	0.0090	P(26)	31445795.7236	0.0149
P(24)	27180362.2977	0.0089	P(24)	31502906.1318	0.0141
P(22)	27235765.8792	0.0090	P(22)	31559204.1695	0.0128
P(20)	27290413.4182	0.0093	P(20)	31614686.3091	0.0113
P(18)	27344305.9494	0.0096	P(18)	31669349.2515	0.0098
P(16)	27397444.3368	0.0097	P(16)	31723189.9280	0.0086
P(14)	27449829.2733	0.0096	P(14)	31776205.5007	0.0081
P(12)	27501461.2824	0.0096	P(12)	31828393.3642	0.0085
P(10)	27552340.7179	0.0101	P(10)	31879751.1463	0.0095
P(8)	27602467.7649	0.0111	P(8)	31930276.7092	0.0107
P(6)	27651842.4399	0.0125	P(6)	31979968.1497	0.0120
P(4)	27700464.5912	0.0139	P(4)	32028823.8002	0.0131
P(2)	27748333.8988	0.0148	P(2)	32076842.2290	0.0139
R(2)	27864709.8633	0.0146	P(2)	32193218.1935	0.0150
R(4)	27909939.2762	0.0135	P(4)	32238298.4853	0.0151
R(6)	27954412.3294	0.0122	P(6)	32282538.0393	0.0153
R(8)	27998127.7801	0.0112	P(8)	32325936.7244	0.0153
R(10)	28041084.2173	0.0107	P(10)	32368494.6458	0.0154
R(12)	28083280.0620	0.0108	P(12)	32410212.1438	0.0155
R(14)	28124713.5668	0.0110	P(14)	32451089.7941	0.0155
R(16)	28165382.8151	0.0112	P(16)	32491128.4063	0.0154
R(18)	28205285.7213	0.0111	P(18)	32530329.0234	0.0152
R(20)	28244420.0302	0.0110	P(20)	32568692.9211	0.0149
R(22)	28282783.3158	0.0114	P(22)	32606221.6061	0.0147
R(24)	28320372.9812	0.0129	P(24)	32642916.8154	0.0144
R(26)	28357186.2574	0.0149	P(26)	32678780.5147	0.0140
R(28)	28393220.2023	0.0168	P(28)	32713814.8971	0.0133
R(30)	28428471.6994	0.0175	P(30)	32748022.3813	0.0120
R(32)	28462937.4565	0.0165	P(32)	32781405.6101	0.0106
R(34)	28496614.0042	0.0142	P(34)	32813967.4482	0.0122
R(36)	28529497.6934	0.0163	P(36)	32845710.9809	0.0212
R(38)	28561584.6939	0.0309	P(38)	32876639.5111	0.0385
R(40)	28592870.9914	0.0593	P(40)	32906756.5580	0.0646
R(42)	28623352.3850	0.1054	P(42)	32936065.8540	0.1015
R(44)	28653024.4839	0.1779	P(44)	32964571.3426	0.1518
R(46)	28681882.7038	0.2907	P(46)	32992277.1760	0.2184
R(48)	28709922.2632	0.4635	P(48)	33019187.7118	0.3048
R(50)	28737138.1785	0.7231	P(50)	33045307.5105	0.4152

INFRARED AND FAR-INFRARED ABSORPTION FREQUENCY STANDARDS

Arthur Maki

Aside from the CO_2 laser transitions, the absorption spectrum of CO has been more accurately and thoroughly measured than any other spectrum. A bibliography of earlier measurements on CO is given by Maki and Wells,[1] and the present tables were calculated from the measurements referred to in that work. In addition, some new and very accurate frequency measurements[2,3] have been made and were incorporated in the present tables. The frequencies of the rotational transitions of HF and HCl were calculated from constants obtained from fitting the measurements of Evenson et al.[4,5] and Jennings and Wells.[6]

A new report on infrared wavenumber standards from the International Union of Pure and Applied Chemistry, Commission on Molecular Structure and Spectroscopy, may be found in Reference 7.

References

1. Maki, A. G. and Wells, J. S., *Wavenumber Calibration Tables from Heterodyne Frequency Measurements*, NIST Special Publication 821, U.S. Dept. of Commerce, Washington, D.C., 1991.
2. Evenson, K. and Stroh, F., private communication.
3. George, T., Urban, W., and co-workers, private communication.
4. Jennings, D. A., Evenson, K. M., Zink, L. R., Demuynck, C., Destombes, J. L., Lemoine, B., and Johns, J. *W. C., J. Mol. Spectrosc.*, 122, 477–480, 1987.
5. Nolt, I. G., Radostitz, J. V., DiLonardo, G., Evenson, K. M., Jennings, D. A., Leopold, K. R., Vanek, M. D., Zink, L. R., Hinz, A., and Chance, K. V., *J. Mol. Spectrosc.*, 125, 274–287, 1987.
6. Jennings, D. A. and Wells, J. S., *J. Mol. Spectrosc.*, 130, 267–268, 1988.
7. High Resolution Wavenumber Standards for the Infrared, *Pure Appl. Chemistry*, 68, 193, 1996.

Wavenumbers for the v = 1 – 0 Band of CO

Wavenumber (unc)*/cm^{-1}	Transition	Wavenumber (unc)/cm^{-1}	Transition
		2147.081132(01)	R(0)
2139.426071(01)	P(1)	2150.856006(01)	R(1)
2135.546178(01)	P(2)	2154.595581(01)	R(2)
2131.631574(01)	P(3)	2158.299710(01)	R(3)
2127.682404(01)	P(4)	2161.968245(01)	R(4)
2123.698816(01)	P(5)	2165.601011(01)	R(5)
2119.680957(01)	P(6)	2169.197949(01)	R(6)
2115.628973(01)	P(7)	2172.758824(01)	R(7)
2111.543012(01)	P(8)	2176.283519(01)	R(8)
2107.423221(01)	P(9)	2179.771887(01)	R(9)
2103.269746(01)	P(10)	2183.223782(01)	R(10)
2099.082734(01)	P(11)	2186.639057(01)	R(11)
2094.862333(01)	P(12)	2190.017565(01)	R(12)
2090.608688(01)	P(13)	2193.359161(01)	R(13)
2086.321947(01)	P(14)	2196.663698(01)	R(14)
2082.002256(01)	P(15)	2199.931030(01)	R(15)
2077.649762(01)	P(16)	2203.161010(01)	R(16)
2073.264612(01)	P(17)	2206.353492(01)	R(17)
2068.846952(01)	P(18)	2209.508331(02)	R(18)
2064.396929(01)	P(19)	2212.625379(02)	R(19)
2059.914688(02)	P(20)	2215.704492(02)	R(20)
2055.400377(02)	P(21)	2218.745522(02)	R(21)
2050.854140(02)	P(22)	2221.748326(03)	R(22)
2046.276126(03)	P(23)	2224.712755(03)	R(23)
2041.666479(03)	P(24)	2227.638666(03)	R(24)
2037.025345(03)	P(25)	2230.525912(04)	R(25)
2032.352870(04)	P(26)	2233.374349(04)	R(26)
2027.649200(04)	P(27)	2236.183829(04)	R(27)
2022.914480(04)	P(28)	2238.954210(05)	R(28)
2018.148857(05)	P(29)	2241.685344(05)	R(29)
2013.352474(05)	P(30)	2244.377088(06)	R(30)
2008.525477(06)	P(31)	2247.029296(07)	R(31)
2003.668012(06)	P(32)	2249.641824(08)	R(32)
1998.780224(07)	P(33)	2252.214527(10)	R(33)
1993.862257(09)	P(34)	2254.747262(14)	R(34)
1988.914257(11)	P(35)	2257.239883(18)	R(35)
1983.936367(14)	P(36)	2259.692248(24)	R(36)
1978.928733(18)	P(37)	2262.104213(33)	R(37)
1973.891500(25)	P(38)	2264.475634(45)	R(38)

Wavenumbers for the $v = 1 - 0$ Band of CO

Wavenumber (unc)*/cm^{-1}	Transition	Wavenumber (unc)/cm^{-1}	Transition
1968.824811(34)	P(39)	2266.806368(61)	R(39)
1963.728813(46)	P(40)	2269.096273(81)	R(40)
1958.603648(61)	P(41)	2271.345206(106)	R(41)
1953.449462(82)	P(42)	2273.553027(139)	R(42)

* The uncertainty in the last digits (twice the standard error) is given in parentheses.

Wavenumbers for the $v = 2 - 0$ Band of CO

Wavenumber (unc)*/cm^{-1}	Transition	Wavenumber (unc)/cm^{-1}	Transition
		4263.837198(02)	R(0)
4256.217140(02)	P(1)	4267.542066(02)	R(1)
4252.302244(02)	P(2)	4271.176630(02)	R(2)
4248.317633(02)	P(3)	4274.740746(02)	R(3)
4244.263453(02)	P(4)	4278.234264(02)	R(4)
4240.139852(02)	P(5)	4281.657039(02)	R(5)
4235.946975(02)	P(6)	4285.008924(02)	R(6)
4231.684972(02)	P(7)	4288.289772(02)	R(7)
4227.353987(02)	P(8)	4291.499437(02)	R(8)
4222.954169(02)	P(9)	4294.637773(02)	R(9)
4218.485665(02)	P(10)	4297.704631(02)	R(10)
4213.948620(02)	P(11)	4300.699868(02)	R(11)
4209.343182(02)	P(12)	4303.623334(02)	R(12)
4204.669499(02)	P(13)	4306.474886(02)	R(13)
4199.927716(02)	P(14)	4309.254375(02)	R(14)
4195.117980(02)	P(15)	4311.961657(02)	R(15)
4190.240439(02)	P(16)	4314.596584(02)	R(16)
4185.295239(02)	P(17)	4317.159011(02)	R(17)
4180.282526(02)	P(18)	4319.648791(02)	R(18)
4175.202447(02)	P(19)	4322.065779(03)	R(19)
4170.055149(03)	P(20)	4324.409829(03)	R(20)
4164.840777(03)	P(21)	4326.680794(03)	R(21)
4159.559478(03)	P(22)	4328.878530(03)	R(22)
4154.211398(03)	P(23)	4331.002889(04)	R(23)
4148.796683(04)	P(24)	4333.053728(04)	R(24)
4143.315479(04)	P(25)	4335.030899(05)	R(25)
4137.767932(04)	P(26)	4336.934259(06)	R(26)
4132.154187(05)	P(27)	4338.763661(07)	R(27)
4126.474391(06)	P(28)	4340.518961(09)	R(28)
4120.728689(07)	P(29)	4342.200014(11)	R(29)
4114.917226(09)	P(30)	4343.806675(16)	R(30)
4109.040148(12)	P(31)	4345.338799(21)	R(31)
4103.097600(16)	P(32)	4346.796243(29)	R(32)
4097.089728(21)	P(33)	4348.178862(40)	R(33)
4091.016676(29)	P(34)	4349.486513(54)	R(34)
4084.878591(40)	P(35)	4350.719052(73)	R(35)
4078.675618(54)	P(36)	4351.876336(96)	R(36)
4072.407901(73)	P(37)	4352.958224(127)	R(37)
4066.075588(97)	P(38)	4353.964572(166)	R(38)
4059.678822(127)	P(39)	4354.895240(214)	R(39)

* The uncertainty in the last digits (twice the standard error) is given in parentheses.

Wavenumbers for the $v = 3 - 0$ Band of CO

Wavenumber (unc)*/cm^{-1}	Transition	Wavenumber (unc)/cm^{-1}	Transition
		6354.179057(13)	R(0)
6346.594000(13)	P(1)	6357.813923(13)	R(1)
6342.644103(13)	P(2)	6361.343487(13)	R(2)
6338.589491(13)	P(3)	6364.767599(13)	R(3)
6334.430309(13)	P(4)	6368.086115(13)	R(4)

Wavenumbers for the $v = 3 - 0$ Band of CO

Wavenumber (unc)*/cm⁻¹	Transition	Wavenumber (unc)/cm⁻¹	Transition
6330.166705(13)	P(5)	6371.298887(13)	R(5)
6325.798826(13)	P(6)	6374.405768(12)	R(6)
6321.326819(13)	P(7)	6377.406611(12)	R(7)
6316.750831(12)	P(8)	6380.301271(12)	R(8)
6312.071008(12)	P(9)	6383.089600(12)	R(9)
6307.287498(12)	P(10)	6385.771452(12)	R(10)
6302.400447(12)	P(11)	6388.346680(13)	R(11)
6297.410003(12)	P(12)	6390.815139(13)	R(12)
6292.316311(13)	P(13)	6393.176681(13)	R(13)
6287.119520(13)	P(14)	6395.431160(13)	R(14)
6281.819775(13)	P(15)	6397.578430(13)	R(15)
6276.417224(13)	P(16)	6399.618344(13)	R(16)
6270.912012(13)	P(17)	6401.550757(13)	R(17)
6265.304287(13)	P(18)	6403.375523(13)	R(18)
6259.594194(13)	P(19)	6405.092495(14)	R(19)
6253.781880(13)	P(20)	6406.701527(14)	R(20)
6247.867492(14)	P(21)	6408.202474(14)	R(21)
6241.851176(14)	P(22)	6409.595189(15)	R(22)
6235.733077(14)	P(23)	6410.879527(15)	R(23)
6229.513342(15)	P(24)	6412.055343(16)	R(24)
6223.192117(15)	P(25)	6413.122491(17)	R(25)
6216.769547(16)	P(26)	6414.080825(19)	R(26)
6210.245778(17)	P(27)	6414.930201(23)	R(27)
6203.620957(19)	P(28)	6415.670474(28)	R(28)
6196.895229(23)	P(29)	6416.301500(37)	R(29)
6190.068739(28)	P(30)	6416.823133(50)	R(30)
6183.141633(37)	P(31)	6417.235231(67)	R(31)
6176.114058(50)	P(32)	6417.537649(90)	R(32)
6168.986159(67)	P(33)		
6161.758082(90)	P(34)		

* The uncertainty in the last digits (twice the standard error) is given in parentheses.

Frequencies and Wavenumbers for the Rotational Lines of CO

Frequency/MHz	Uncertainty*/MHz	J′	J″	Wavenumber/cm⁻¹	Uncertainty*/cm⁻¹
115271.2029	0.0004	1	0	3.84503345	0.00000001
230538.0016	0.0008	2	1	7.68991999	0.00000003
345795.9923	0.0012	3	2	11.53451273	0.00000004
461040.7712	0.0016	4	3	15.37866477	0.00000005
576267.9350	0.0019	5	4	19.22222923	0.00000006
691473.0809	0.0021	6	5	23.06505926	0.00000007
806651.8065	0.0023	7	6	26.90700800	0.00000008
921799.7104	0.0025	8	7	30.74792863	0.00000008
1036912.3919	0.0027	9	8	34.58767438	0.00000009
1151985.4515	0.0029	10	9	38.42609848	0.00000010
1267014.4906	0.0031	11	10	42.26305422	0.00000010
1381995.1119	0.0034	12	11	46.09839491	0.00000011
1496922.9195	0.0038	13	12	49.93197392	0.00000013
1611793.5189	0.0042	14	13	53.76364468	0.00000014
1726602.5173	0.0047	15	14	57.59326065	0.00000016
1841345.5237	0.0052	16	15	61.42067535	0.00000017
1956018.1486	0.0057	17	16	65.24574239	0.00000019
2070616.0050	0.0061	18	17	69.06831542	0.00000020
2185134.7075	0.0065	19	18	72.88824816	0.00000022
2299569.8733	0.0069	20	19	76.70539441	0.00000023
2413917.1217	0.0071	21	20	80.51960806	0.00000024
2528172.0747	0.0073	22	21	84.33074306	0.00000024
2642330.3567	0.0074	23	22	88.13865346	0.00000025
2756387.5949	0.0075	24	23	91.94319341	0.00000025

Frequencies and Wavenumbers for the Rotational Lines of CO

Frequency/MHz	Uncertainty*/MHz	J′	J″	Wavenumber/cm⁻¹	Uncertainty*/cm⁻¹
2870339.4194	0.0077	25	24	95.74421713	0.00000026
2984181.4631	0.0080	26	25	99.54157896	0.00000027
3097909.3621	0.0085	27	26	103.33513334	0.00000028
3211518.7558	0.0090	28	27	107.12473480	0.00000030
3325005.2869	0.0096	29	28	110.91023800	0.00000032
3438364.6013	0.0102	30	29	114.69149772	0.00000034
3551592.3489	0.0107	31	30	118.46836884	0.00000036
3664684.1829	0.0111	32	31	122.24070637	0.00000037
3777635.7608	0.0118	33	32	126.00836545	0.00000039
3890442.7435	0.0137	34	33	129.77120137	0.00000046
4003100.7965	0.0179	35	34	133.52906952	0.00000060
4115605.5892	0.0254	36	35	137.28182546	0.00000085
4227952.7954	0.0370	37	36	141.02932487	0.00000123
4340138.0932	0.0531	38	37	144.77142361	0.00000177
4452157.1657	0.0746	39	38	148.50797766	0.00000249
4564005.7001	0.1025	40	39	152.23884318	0.00000342

* The uncertainty given is twice the standard error.

Frequencies and Wavenumbers for the Rotational Lines of HF

Frequency/MHz	Uncertainty*/MHz	J′	J″	Wavenumber/cm⁻¹	Uncertainty*/cm⁻¹
1232476.21	0.12	1	0	41.110981	0.000004
2463428.09	0.19	2	1	82.171116	0.000006
3691334.81	0.25	3	2	123.129676	0.000008
4914682.58	0.51	4	3	163.936165	0.000017
6131968.11	1.10	5	4	204.540439	0.000037
7341702.00	2.00	6	5	244.892818	0.000067
8542412.1	3.21	7	6	284.944197	0.000107
9732646.8	4.72	8	7	324.646153	0.000157
10910978.2	6.51	9	8	363.951056	0.000217
12076004.8	8.55	10	9	402.81216	0.000285
13226355.2	10.81	11	10	441.18372	0.000361
14360689.8	13.25	12	11	479.02105	0.00044
15477704.4	15.86	13	12	516.28065	0.00053
16576131.8	18.61	14	13	552.92024	0.00062
17654744.4	21.48	15	14	588.89888	0.00072
18712356.5	24.44	16	15	624.17703	0.00082
19747825.6	27.43	17	16	658.71656	0.00092
20760054.3	30.32	18	17	692.4809	0.00101
21747991.7	32.91	19	18	725.4349	0.00110
22710634.7	34.94	20	19	757.5452	0.00117
23647028.7	36.08	21	20	788.7800	0.00120
24556268.8	35.93	22	21	819.1090	0.00120
25437499.9	34.12	23	22	848.5037	0.00114
26289917.4	30.32	24	23	876.9373	0.00101
27112767.2	24.41	25	24	904.38457	0.00081
27905345.6	16.88	26	25	930.82214	0.00056
28666999.3	10.80	27	26	956.22817	0.00036
29397124.8	14.65	28	27	980.58253	0.00049
30095168.2	24.62	29	28	1003.86676	0.00082
30760624.2	33.36	30	29	1026.0640	0.00111
31393035.7	36.17	31	30	1047.1590	0.00121

* The uncertainty given is twice the standard error.

Frequencies and Wavenumbers for the Rotational Lines of $H^{35}Cl$

Frequency/MHz	Uncertainty*/MHz	J′	J″	Wavenumber/cm⁻¹	Uncertainty*/cm⁻¹
1876226.517	0.065	3	2	62.584180	0.000002
2499864.439	0.066	4	3	83.386502	0.000002
3121986.563	0.064	5	4	104.138262	0.000002
3742216.601	0.076	6	5	124.826909	0.000003
4360180.042	0.098	7	6	145.439951	0.000003
4975504.51	0.11	8	7	165.964966	0.000004
5587820.10	0.12	9	8	186.389615	0.000004
6196759.76	0.22	10	9	206.701656	0.000007
6801959.63	0.50	11	10	226.888951	0.000017
7403059.41	1.02	12	11	246.939481	0.000034
7999702.7	1.8	13	12	266.841359	0.000062
8591537.3	3.1	14	13	286.582837	0.000103
9178215.8	4.8	15	14	306.152324	0.000161

* The uncertainty given is twice the standard error.

Frequencies and Wavenumbers for the Rotational Lines of $H^{37}Cl$

Frequency/MHz	Uncertainty*/MHz	J′	J″	Wavenumber/cm⁻¹	Uncertainty*/cm⁻¹
1873410.72	0.05	3	2	62.490255	0.000002
2496115.33	0.05	4	3	83.261445	0.000002
3117308.69	0.05	5	4	103.982225	0.000002
3736615.64	0.06	6	5	124.640082	0.000002
4353662.84	0.08	7	6	145.222561	0.000003
4968079.04	0.09	8	7	165.717279	0.000003
5579495.53	0.10	9	8	186.111938	0.000003
6187546.42	0.19	10	9	206.394332	0.000006
6791869.04	0.45	11	10	226.552365	0.000015
7392104.3	0.9	12	11	246.574057	0.000030
7987896.9	1.6	13	12	266.447561	0.000054
8578896.1	2.7	14	13	286.161170	0.000089

* The uncertainty given is twice the standard error.

Section 11
Nuclear and Particle Physics

Summary Tables of Particle Properties 11-1
Table of the Isotopes 11-2
Neutron Scattering and Absorption Properties 11-170
Cosmic Radiation 11-183

SUMMARY TABLES OF PARTICLE PROPERTIES

Details of particle properties may be found in the following publication:

Amsler, C., et al. (Particle Data Group), Review of Particle Physics, *Physics Letters* B667, 1, 2008.

Abstract. This biennial *Review* summarizes much of particle physics. Using data from previous editions, plus 2778 new measurements from 645 papers, we list, evaluate, and average measured properties of gauge bosons, leptons, quarks, mesons, and baryons. We also summarize searches for hypothetical particles such as Higgs bosons, heavy neutrinos, and supersymmetric particles. All the particle properties and search limits are listed in Summary Tables. We also give numerous tables, figures, formulae, and reviews of topics such as the Standard Model, particle detectors, probability, and statistics. Among the 108 reviews are many that are new or heavily revised, including those on CKM quark-mixing matrix, V_{ud} & V_{us}, V_{cb} & V_{ub}, top quark, muon anomalous magnetic moment, extra dimensions, particle detectors, cosmic background radiation, dark matter, cosmological parameters, and big bang cosmology. A booklet is available containing the Summary Tables and abbreviated versions of some of the other sections of this full *Review*. All tables, listings, and reviews (and errata) are also available on the Particle Data Group website: http://pdg.lbl.gov.

An electronic version of this publication, which includes updates made in 2009, is available on the web site of the Particle Data Group: <pdg.lbl.gov>. A full update of the review will be published in 2010. The Summary Tables are included in the CD-ROM and Internet versions of the 91st Edition of the *CRC Handbook of Chemistry and Physics*.

TABLE OF THE ISOTOPES

Norman E. Holden

This table presents an evaluated set of values for the experimental quantities that characterize the decay of radioactive nuclides. A list of the major references used in this evaluation is given below. When uncertainties are not listed, they are assumed to be five or less in the last digit quoted. If the uncertainty in the value exceeds five in the last digit, the value is preceded by an approximate sign. For quasi-stable nuclides, the measured width, Γ, of the resonance is given. To estimate the approximate half-life, the Heisenberg relationship may be used, the half-life = 4.56×10^{-22} seconds / Γ (MeV). The effective literature cutoff date for data in this edition of the table is December 2007.

Table Layout

Column number	Column title	Description
1	Isotope or Element	For elements, the atomic number and chemical symbol are listed. For nuclides, the mass number and chemical symbol are listed. Isomers are indicated by the addition of m, m1, or m2.
2	Isotopic Abundance	The abundance of an isotope in normal terrestrial samples of an element, listed in atom percent.
3	Atomic Mass or Atomic Weight	Atomic mass relative to ^{12}C = 12. Atomic weight of elements is given on the same scale.
4	Half-life/Resonance Width	Half-life in decimal notation. μs = microseconds; ms = milliseconds; s = seconds; m = minutes; h = hours; d = days; and a = years. For quasi-stable nuclides, the measured width at half maximum of the energy resonance is given.
5	Decay Mode/Energy	Decay modes are α = alpha particle emission; β = negative beta emission; β+ = positron emission; EC = orbital electron capture; IT = isomeric transition from upper to lower isomeric state; n = neutron emission; p = proton emission; sf = spontaneous fission; ββ, ECEC, 2p, 2n, 3n = double beta, double EC, two proton and multiple neutron decay. Total disintegration energy in MeV units.
6	Particle Energy/Intensity	End point energies of beta transitions and discrete energies of alpha particles, neutrons and protons are given in MeV units and their intensities in percent.
7	Spin and Parity	Nuclear spin or angular momentum of the nuclides in units of $h/2\pi$; parity is positive or negative.
8	Magnetic Dipole Moment	Magnetic dipole moments in nuclear magneton units. An absolute value is indicated in the absence of a positive or a negative sign.
9	Electric Quadrupole Moment	Electric quadrupole moments in barn units (10^{-24} cm^2). An absolute value is indicated in the absence of a positive or a negative sign.
10	Gamma Ray Energy/ Intensity	Gamma ray energies are given in MeV units and intensities in percent. Annihilation radiation (Ann.Rad.) refers to the 511.006 keV photons emitted in the annihilation of positrons in matter.

General Nuclear Data References

The following references represent the major sources of the nuclear data presented, along with subsequent published journal articles and reports:

1. G. Audi, O. Bersillon, J. Blachot, A.H. Wapstra, *The Nubase Evaluation of Nuclear and Decay Properties*, Nuclear Physics A729, 3 (2003).
2. G. Audi, A.H. Wapstra, C. Thibault, *The AME2003 Atomic Mass Evaluation (II)*, Nuclear Physics A729, 337 (2003).
3. M.E. Wieser, *Atomic Weights of the Elements - 2005*, Pure & Applied Chemistry 78, 2051 (2006).
4. E.M. Baum, H.D. Knox, T.R. Miller, *Chart of the Nuclides, 16th Edition*, Knolls Atomic Power Lab (2002).
5. N.E. Holden, *Total and Spontaneous Fission Half-lives for Uranium, Plutonium, Americium and Curium Nuclides*, Pure & Applied Chemistry 61, 1483 (1989).
6. N.E. Holden, *Half-lives of Selected Nuclides*, Pure & Applied Chemistry 62, 941 (1990).
7. N.E. Holden, *Review of Thermal Neutron Cross Sections and Isotopic Composition of the Elements*, BNL-NCS-42224 (March 1989).
8. P. Raghavan, *Table of Nuclear Moments*, Atomic Data Nuclear Data Tables 42, 189 (1989).
9. E. Brown, R. Firestone, *Radioactivity Handbook*, Wiley Interscience Press (1986).
10. J.K. Tuli, *Nuclear Wallet Cards*, Brookhaven National Laboratory (April 2005).
11. N.E. Holden, D.C. Hoffman, *Spontaneous Fission Half-lives for Ground State Nuclides*, Pure & Applied Chemistry 72, 1525 (2000).
12. N.J. Stone, *Table of Nuclear Magnetic Dipole and Electrical Quadrupole Moments*, Atomic Data Nuclear Data Tables 90, 75 (2005).

This research was carried out under the auspices of the US Department of Energy Contract No. DE-AC02-98CH10886. The author is at Brookhaven National Laboratory, Upton, NY, and can be contacted at holden@bnl.gov.

Elem. or Isot.	Natural Abundance (Atom %)	Atomic Mass or Weight	Half-life/ Resonance Width (MeV)	Decay Mode/ Energy (/MeV)	Particle Energy/ Intensity (MeV/%)	Spin ($h/2\pi$)	Nuclear Magnetic Mom. (nm)	Elect. Quadr. Mom. (b)	γ-Energy / Intensity (MeV/%)
$_0$n		1.008664916	614. s	β- /0.78235	0.782/100.	1/2+	-1.913043		
				β-, γ	/ 0.0031				
$_1$H		**1.00794(7)**							
^{1}H	99.9885(70)	1.007825032	>2.8 × 10^{23} a			1/2+	+2.792847		
^{2}H	0.0115(70)	2.014101778				1+	+0.857438	+2.86 mb	
^{3}H		3.016049278	12.31 a	β-/0.01859	0.01860/100.	1/2+	+2.978962		
^{4}H		4.0278	Γ = 1.2	n/	/100	2-			
^{5}H		5.0353	Γ < 0.5	n/	/100	(1/2+)			
^{6}H		6.0449	Γ = 6.	n/	/100	(2-)			
^{7}H		7.053	Γ ~ 0.1						
$_2$He		**4.002602(2)**							
^{3}He	0.000134(3)	3.016029319				1/2+	-2.127750		
^{4}He	99.999866(3)	4.002603254				0+			
^{5}He		5.01222	Γ = 0.60(2)	n, α		3/2-			
^{6}He		6.018889	0.807 s	β-/3.508	3.510/100.	0+			
				β, d	/0.00076				
^{7}He		7.02802	Γ = 0.15(2)	n		(3/2)-			
^{8}He		8.03392	0.119 s	β-/10.65	/84.	0+			0.9807/84.
				n/	/16.				0.4776/5.
				β, t	/0.82				
^{9}He		9.04395	Γ = 0.10(6)	n	/100	(1/2-)			
^{10}He		10.0524	Γ = 0.3(2)	2n	/100	0+			
$_3$Li		**6.941(2)**							
^{4}Li		4.0272	Γ = 6.0	p/	/100	2-			
^{5}Li		5.01254	Γ = 1.2	p/α		3/2-			
^{6}Li	7.59(4)	6.01512280				1+	+0.822047	-0.8 mb	
^{7}Li	92.41(4)	7.0160046				3/2-	+3.25644	-0.0406	
^{8}Li		8.0224874	0.840 s	β-/16.004	12.5/100.	2+	+1.65334	+0.0314	
				α/	α(1.6)				
^{9}Li		9.026790	0.178 s	β-/13.606	13.5/75.	3/2-	3.439	-0.0306	
				β-/	11/25.				
^{10}Li		10.03548	Γ = 0.11(5)	n	/7.	1+			
^{11}Li		11.04380	8.8 ms	β-/20.6	/8.3	3/2(-)	3.668	-0.031	3.368/33.
				β-, n	/85.7				0.320/7.
				β-, 2n	/4.1				2.590/8.
				β-, 3n	/1.9				5.958/3.
				β-, d	/>0.01				2.895/1.5
				β-, t	/0.02				2.811/1.1
^{12}Li		12.054	< 0.01 μs						
$_4$Be		**9.012182(3)**							
^{5}Be		5.041		p, ^{3}He		(1/2+)			
^{6}Be		6.01973	Γ = 0.092(6)	2p,α		0+			
^{7}Be		7.0169298	53.28 d	EC/0.8618		3/2-	-1.40		0.4776/10.4
^{8}Be		8.00530510	Γ = 6.8(17) eV	2α/0.046		0+			
^{9}Be	100.	9.0121822				3/2-	-1.1776	+0.0529	
^{10}Be		10.0135338	1.52 × 10^{6} a	β-/0.5559	0.555/100.	0+			
^{11}Be		11.02166	13.8 s	β-, β-α/11.51	11.48/61.	1/2+	-1.681		2.125/35.5
^{12}Be		12.02692	22.0 ms	β-, (n)/11.71	n//0.5	0+			(0.95 - 4.4)
^{13}Be		13.0357	Γ ~ 1.						
^{14}Be		14.0429	4.6 ms	β-/16.2		0+			3.5346/0.9
				β-, n	0.288/94.				3.6845/7.
				β-, 2n	/6.				
				β-, α	/<0.012				

Elem. or Isot.	Natural Abundance (Atom %)	Atomic Mass or Weight	Half-life/ Resonance Width (MeV)	Decay Mode/ Energy (/MeV)	Particle Energy/ Intensity (MeV/%)	Spin ($h/2\pi$)	Nuclear Magnetic Mom. (nm)	Elect. Quadr. Mom. (b)	γ-Energy / Intensity (MeV/%)
				β-, t	/<0.04				
^{15}Be		15.053	< 0.2 μs	β-					
^{16}Be		16.062	< 0.2 μs	β-		0+			
$_5$B		**10.811(7)**							
^{7}B		7.0299	Γ = 1.4(2)	p, α		(3/2-)			
^{8}B		8.024607	0.770 s	β+, 2α/17.979	13.7(β+)/93.	2+	1.0358	0.065	ann.rad.
^{9}B		9.013329	Γ = 0.5(2) keV	p, 2α/		3/2-			
^{10}B	19.9(7)	10.0129370				3+	+1.800645	+0.085	
^{11}B	80.1(7)	11.0093054				3/2-	+2.688649	+0.0407	
^{12}B		12.014352	0.0202 s	β-/13.369		1+	+1.0027	0.0132	4.438/1.3
				β- α/1.6/					3.215/0.00065
^{13}B		13.017780	0.0174 s	β- /13.437	13.4	3/2-	+3.1778	0.037	3.68/7.6
				β- n/0.25/	2.43(n)/0.09				
					3.55(n)/0.16				
^{14}B		14.02540	14. ms	β-/20.64		2-	1.185	0.0298	6.094/90.
^{15}B		15.03110	9.9 ms	β-, (n)/19.09	n//99.7	(3/2-)	2.66	0.038	
^{16}B		16.0398	Γ < 0.1	n					
^{17}B		17.0470	5.1 ms	β-, (n)/22.7			2.55	0.039	
^{18}B		18.056	< 0.026 μs			0-			
^{19}B		19.0637	2.9 ms	β-, (n)/26.5	1n//72.	(3/2-)			
					2n//16.				
					3n// < 9.				
$_6$C		**12.0107(8)**							
^{8}C		8.03768	Γ = 0.25(4)	p		0+			
^{9}C		9.031037	127. ms	β+, p, 2α/16.498		(3/2-)	1.391		ann.rad.
^{10}C		10.0168532	19.3 s	β+/3.648	1.865	0+			ann.rad.
									0.71829/100.
^{11}C		11.011434	20.3 m	β+, EC/1.982	0.9608/99.	3/2-	-0.964	0.032	ann.rad.
^{12}C	98.93(8)	12.000000000				0+			
^{13}C	1.07(8)	13.003354838				½-	+0.702412		
^{14}C		14.003241989	5715. a	β-/0.15648	0.1565/100.	0+			
^{15}C		15.010599	2.45 s	β-/9.772	4.51/68.	½+	1.72		5.298/68.
					9.82/32.				(7.30–9.05)
^{16}C		16.014701	~ 0.750 s	β-/8.012	β/3.3, 4.3/84, 16	0+			
				β, n	n/0.8, 1.7/84, 16				
^{17}C		17.02259	0.19 s	β-/13.17		3/2+	0.758		1.375
				β-, n	n/1.6–3.7/11.				1.849
									1.906
^{18}C		18.02676	0.092 s	β-/11.81		0+			
				β-, n	n/0.88–4.59/21.				
^{19}C		19.0348	0.05 s	n		½+			
^{20}C		20.0403	0.02 s	β,n	1n// ~ 65.	0+			
					2n// < 19.				
^{21}C		21.049	< 0.03 μs						
^{22}C		22.057	6 ms	β-, n	1n// ~ 61.	0+			
					2n// < 37.				
$_7$N		**14.0067(2)**							
^{10}N		10.0417	Γ = 2.3(16)						
^{11}N		11.02609	Γ ~ 1.			½+			
^{12}N		12.018613	11.00 ms	β+, β+α/17.338	16.38/95.	1+	+0.457	+10. mb	ann.rad.
									4.438/2.
^{13}N		13.0057386	9.97 m	β+ /2.2204	1.190/100.	½-	0.3222		
^{14}N	99.636(20)	14.003074005				1+	+0.403761	+0.0200	

Elem. or Isot.	Natural Abundance (Atom %)	Atomic Mass or Weight	Half-life/ Resonance Width (MeV)	Decay Mode/ Energy (/MeV)	Particle Energy/ Intensity (MeV/%)	Spin ($h/2\pi$)	Nuclear Magnetic Mom. (nm)	Elect. Quadr. Mom. (b)	γ-Energy / Intensity (MeV/%)
^{15}N	0.364(20)	15.00010898				½-	-0.283189		
^{16}N		16.006102	7.13 s	β- /10.419	4.27/68.	2-			6.129/68.8
					10.44/26.		1.986	18 mb	7.115/4.7
				β-, α	1.85/.0012				(0.99–8.87)
					(0.6-2.3)				
^{17}N		17.00845	4.17 s	β-, β- n/8.68	3.7/100.	½-	0.352		0.871/3.
				0.4–1.7n/95.					2.1842/0.3
				β- α/	8.0, 8.2				
^{18}N		18.01408	0.62 s	β- /13.90	9.4/100.	1-	0.328	0.012	0.822/48.
				β-, α	(1.08-5.23)/12.				1.65/47.
				β-, n	/14				1.982/77.
									(0.535–7.13)
^{19}N		19.01703	0.336 s	β-/12.53			0.305		(0.096–3.14)
				β-, n	0.45-4.51/42.				
^{20}N		20.0234	0.136 s	β- /17.97					
				β-, n	1.10-3.68/43.				
^{21}N		21.0271	0.08 s						
^{22}N		22.0344	0.02 s	β-,n	1n// ~ 41.				
					2n// < 13.				
^{23}N		23.0412	15. ms	β-, n	n// ~ 42.				
					2n// ~ 8.				
					3n// < 3.4				
^{24}N		24.0510	< 0.052 μs						
^{25}N		25.061	< 0.26 μs						
$_8$O		**15.9994(3)**							
^{12}O		12.03441	Γ = 0.51(16)	2p		0+			
^{13}O		13.02481	8.9 ms	β+, p/17.77	1.560 (p)	(3/2-)	1.389	0.011	ann.rad.
					p/(1.00 - 13.5)				4.438/0.56
^{14}O		14.0085963	70.63 s	β+ /5.1430	1.81/99.	0+			ann.rad.
									2.312/99.4
^{15}O		15.0030656	122.2 s	β+ /2.754	1.723/100.	1/2-	0.7195		ann.rad.
^{16}O	99.757(16)	15.9949146196				0+			
^{17}O	0.038(1)	16.9991317				5/2+	-1.8938	-0.026	
^{18}O	0.205(14)	17.999161				0+			
^{19}O		19.003580	26.9 s	β- /4.820	3.25/60.	5/2+	1.5320	3.7 mb	0.197/95.9
					4.60/40.				1.3569/50.4
									(0.11–4.18)
^{20}O		20.004077	13.5 s	β- /3.814		0+			1.057/100.
^{21}O		21.00866	3.4 s	β- /8.11					(0.28–4.6)
^{22}O		22.0100	2.2 s	β- /6.5		0+			0.072/100
									0.638/98
									1.862/63
									(0.918-2.499)
^{23}O		23.0157	0.097 s	β-, n	/7.2%				2.234/51.5
									4.066/17.1
									(0.911-3.868)
^{24}O		24.0205	~ 65. ms	β-, n	n//18.	0+			1.83/28.
									0.52/14.
									1.31/12.
^{25}O		25.0295	< 0.05 μs						
^{26}O		26.0383	< 0.04 μs			0+			
^{27}O		27.048	< 0.026 μs						
^{28}O		28.058	< 0.10 μs			0+			

Elem. or Isot.	Natural Abundance (Atom %)	Atomic Mass or Weight	Half-life/ Resonance Width (MeV)	Decay Mode/ Energy (/MeV)	Particle Energy/ Intensity (MeV/%)	Spin ($h/2\pi$)	Nuclear Magnetic Mom. (nm)	Elect. Quadr. Mom. (b)	γ-Energy / Intensity (MeV/%)
$_9$F		**18.9984032(5)**							
^{14}F		14.0351							
^{15}F		15.0180	Γ = 0.8(3)	p		(1/2+)			
^{16}F		16.01147	Γ = 0.037(14)	p		0-			
^{17}F		17.0020952	64.5 s	β+ /2.761	1.75/	5/2+	+4.721	- 0.08	ann.rad.
^{18}F		18.000938	1.829 h	β+, EC/1.656	0.635/97.	1+			ann.rad.
^{19}F	100.	18.9984032				½+	+2.62887		
^{20}F		19.9999813	11.00 s	β- /7.0245	5.398/100.	2+	+2.0934	0.042	1.634/100.
									3.33/0.009
^{21}F		20.999949	4.16 s	β- /5.684	3.7/8.	5/2+	3.9		0.3507/90.
					5.0/63.				1.395/15.
					5.4/29.				(1.746–4.684)
^{22}F		22.00300	4.23 s	β- /10.82	3.48/15.	4+			1.2746/100.
					4.67/7.				2.0826/82.
					5.50/62.				(0.82–4.37)
^{23}F		23.0036	2.2 s	β- /8.5		5/2+			1.701/48.
									2.129/34.
									(0.493–3.83)
^{24}F		24.0081	0.3 s	β- /13.5					1.9816/
^{25}F		25.0121	~ 50. ms	β-, (n)	n//14.				1.70/39.
									(0.57–2.19)
^{26}F		26.0196	10. ms	β-, (n)	n//11.				2.02/67.
									1.67/19.
^{27}F		27.0268	5.0 ms	β-, (n)	n//90.				2.02/18.
^{28}F		28.036	< 0.04 μs						
^{29}F		29.043	2.5 ms	β-, (n)	n//100.				
^{30}F		30.053	< 0.26 μs						
^{31}F		31.060	> 0.26 μs						
$_{10}$Ne		**20.1797(6)**							
^{16}Ne		16.02576	Γ = 0.12(4)	2p		0+			
^{17}Ne		17.01767	109. ms	β+, p/14.53	1.4–10.6/6.9	1/2-	0.787		ann.rad./
				β+, α	/0.014				0.495
^{18}Ne		18.0057082	1.668 s	β+ /4.446	3.416/92.	0+			ann.rad./
									1.0413/7.8
									(0.658–1.70)
^{19}Ne		19.0018802	17.22 s	β+ /3.238	2.24/99.	1/2+	-1.885		ann.rad./
									(0.11–1.55)
^{20}Ne	90.48(3)	19.992440175				0+			
^{21}Ne	0.27(1)	20.99384668				3/2+	-0.66180	+0.103	
^{22}Ne	9.25(3)	21.99138511				0+			
^{23}Ne		22.9944669	37.2 s	β- /4.376	3.95/32.	5/2+	-1.08	+0.15	0.440/33.
					4.39/67.				(1.64–2.98)
^{24}Ne		23.9936108	3.38 m	β- /2.47	1.10/8.	0+			0.4723/100.
					1.98/92.				0.874/7.9
^{25}Ne		24.99774	0.61 s	β- /7.30	6.3/	1/2+	-1.006		0.0895/96.
					7.3/				(0.98–3.69)
^{26}Ne		26.00046	197 ms	β-, n/7.3	n //0.13	0+			0.082/100
									1.278/6
									0.233/5
									0.151/3
									1.211/1
									2.489/1
^{27}Ne		27.0076	31. ms	β-, n/12.7	n //3.	(3/2+)			
^{28}Ne		28.0121	19. ms	β-, n/12.3	n//12.	0+			2.06/19.

Elem. or Isot.	Natural Abundance (Atom %)	Atomic Mass or Weight	Half-life/ Resonance Width (MeV)	Decay Mode/ Energy (/MeV)	Particle Energy/ Intensity (MeV/%)	Spin ($h/2\pi$)	Nuclear Magnetic Mom. (nm)	Elect. Quadr. Mom. (b)	γ-Energy / Intensity (MeV/%)
					2n//3.				0.86/3.
^{29}Ne		29.0194	15. ms	β-, (n)/15.4	n//29.	(3/2+)			2.92/55.
				β-, 2n	2n//4.				(0.22–1.18)
^{30}Ne		30.025	7. ms	β-, (n)	n//9.	0+			0.151/9.
^{31}Ne		31.033	3. ms						
^{32}Ne		32.040	~ 3.5 ms			0+			
^{33}Ne		33.049	< 0.26 μs						
^{34}Ne		34.057	> 1.5 μs			0+			
$_{11}$**Na**		**22.98976928(2)**							
^{18}Na		18.02597	Γ = 0.34(9)						
^{19}Na		19.01388	0.03 s	β+, p/11.18					
^{20}Na		20.00735	0.446 s	β+ /13.89		2+	+0.3694	~ + 0.04	ann.rad./
				α	2.15/				1.634/79.
^{21}Na		20.997655	22.48 s	β+ /3.547	2.50/95.	3/2+	+2.3863	~ +0.05	ann.rad./
									0.351/5.
^{22}Na		21.9944364	2.605 a	β+ /90/2.842	0.545/90.	3+	+1.746	+0.19	ann.rad./
				EC/10/					1.2745/99.9
^{23}Na	100.	22.989769281				3/2+	+2.21752	+0.106	
^{24m}Na			20.2 ms	I.T., β$^-$		1+	-1.93		0.4723/100.
^{24}Na		23.9909628	14.96 h	β- /5.5158	1.389/>99.	4+	+1.690		1.3686/100.
									2.754/100.
									(0.997 4.238)
^{25}Na		24.989954	59.3 s	β- /3.835	2.6/7.	5/2+	+3.683	-0.10	0.3897/12.7
					3.15/25.				0.5850/13.
					4.0/65.				0.9747/14.9
									(0.836–2.80)
^{26}Na		25.99263	1.071 s	β- /9.31		3+	+2.851	-5.3 mb	1.809/98.9
									(0.24-7.37)
^{27}Na		26.994077	0.290 s	β- /9.01	7.95/	5/2+	+3.90	-7.2 mb	0.9847/87.4
				β-, n/					1.698/11.9
^{28}Na		27.99894	31. ms	β- /14.0	12.3/	1+	+2.43	+0.040	1.473/37.
				β-, n/					2.389/18.6
^{29}Na		29.00286	44. ms	β-, n/13.3	11.5/	3/2+	+2.45	+86. mb	2.560/36.
									(1.04–3.99)
^{30}Na		30.00898	50. ms	β-, n/17.5	n//30.	2+	+2.08		1.483/46.
^{31}Na		31.0136	17.2 ms	β-, n/15.9	n//37.	3/2-	+2.30		1.483/14.
									(0.05–3.54)
^{32}Na		32.0205	13.5 ms	β- /19.1					0.240–3.935
				β$^-$, n					0.171-0.894
				β$^-$, 2n					1.483/4.2
^{33}Na		33.027	8.0 ms	β- /20.	/ ~ 38				0.886/16
				β-, n	0.8,1.02/47(6)				0.546/6.4
				β-, 2n	/13(3)				0.050–2.55
^{34}Na		34.035	5. ms	β- /24.					
^{35}Na		35.042	1.5 ms	β- /24					
^{36}Na		36.051	< 0.26 μs						
^{37}Na		37.059	> 1.5 μs						
$_{12}$**Mg**		**24.3050(6)**							
^{19}Mg		19.0355	4 x 10^{-12} s	2p					
^{20}Mg		20.01886	96. ms	β+ /10.73	/70	0+			
				β+, p	/30				
^{21}Mg		21.01171	122. ms	β+, p/13.10		5/2+			0.332/51.
^{22}Mg		21.999574	3.876 s	β+ /4.786	3.05/	0+			0.0729/60.
									0.5820/100.

Elem. or Isot.	Natural Abundance (Atom %)	Atomic Mass or Weight	Half-life/ Resonance Width (MeV)	Decay Mode/ Energy (/MeV)	Particle Energy/ Intensity (MeV/%)	Spin $(h/2\pi)$	Nuclear Magnetic Mom. (nm)	Elect. Quadr. Mom. (b)	γ-Energy / Intensity (MeV/%)
									(1.28–1.93)
^{23}Mg		22.994124	11.32 s	β+ /4.057	3.09/92.	3/2+	0.536	1.25	0.440/8.2
^{24}Mg	78.99(4)	23.98504170				0+			
^{25}Mg	10.00(1)	24.98583692				5/2+	-0.85545	+0.200	
^{26}Mg	11.01(3)	25.98259293				0+			
^{27}Mg		26.98434059	9.45 m	β- /2.6103	1.59/41.	1/2+			0.17068/0.9
					1.75/58.				0.84376/72.
					2.65/0.3				1.01443/28.
^{28}Mg		27.983877	20.9 h	β- /1.832	0.459/95.	0+			0.0306/95.
									0.4006/36.
									0.9418/36.
									1.342/54.
^{29}Mg		28.98860	1.3 s	β-/7.55	5.4/	3/2+			0.960/15.
									1.398/16.
									2.224/36.
^{30}Mg		29.99043	0.32 s	β- /7.0		0+			0.224/85.
^{31}Mg		30.99655	0.24 s	β- /11.7	8.4/29.9	1/2+	-0.8836		1.613/47.
									0.947/37.
									(0.666-4.640)
				β-, n	/1.7				
^{32}Mg		31.99898	0.12 s	β- /10.3		0+			2.765/25.
^{33}Mg		33.00525	91. ms	β- /13.7	/83.				1.848/
				β-, n	/17.				
^{34}Mg		34.0095	0.02 s	β- /11.3		0+			
^{35}Mg		35.0173	0.07 s			(7/2-)			
^{36}Mg		36.023	4. ms			0+			
^{37}Mg		37.031	> 0.26 μs			(7/2-)			
^{38}Mg		38.038	> 0.26 μs			0+			
^{39}Mg		39.047	< 0.26 μs						
^{40}Mg		40.054				0+			
$_{13}$Al		**26.9815386(8)**							
^{21}Al		21.0280	< 0.035 μs						
^{22}Al		22.0195	59. ms	β+ /18.6	p/1.3/18.	4+			ann.rad./
					p/0.48-7.52				0.584/47.
									(1.25-2.15)
				β+, p, 2p, α/	α/3.3/0.3				
^{23m}Al			~ 0.35 s	β+, p/0.17					0.554
									0.839
^{23}Al		23.00727	0.47 s	β+ /12.24		5/2+	3.9		ann.rad./
				β+, p/					
^{24m}Al			0.129 s	I.T./0.4259					
				β+	13.3	1+			1.3686/5.3
^{24}Al		23.999939	2.07 s	β+ /13.878,p	3.40/48.	4+			1.078(2)/16.
					4.42/41.				1.368(2)/96.
					6.80/3.				2.753(2)/43.
					8.74/8.				4.315(3)/15.
									5.392(3)/20.
									7.0662(2)/41.
^{25}Al		24.9904281	7.17 s	β+ /4.277	3.27/	5/2+	3.646		ann.rad./
									1.6115(2)/100.
									0.975(2)/5.
^{26m}Al			6.345 s	β+ /	3.2/	0+			ann.rad./
^{26}Al		25.9868917	7.1 × 10^5 a	β+ /82/4.0042	1.16/	5+	+2.804	+0.27	ann.rad./
				EC/18					1.8087/99.8

Elem. or Isot.	Natural Abundance (Atom %)	Atomic Mass or Weight	Half-life/ Resonance Width (MeV)	Decay Mode/ Energy (/MeV)	Particle Energy/ Intensity (MeV/%)	Spin ($h/2\pi$)	Nuclear Magnetic Mom. (nm)	Elect. Quadr. Mom. (b)	γ-Energy / Intensity (MeV/%)
^{27}Al	100.	26.9815386				5/2+	+3.641507	+0.147	
^{28}Al		27.9819103	2.25 m	β- /4.6422	2.865/100.	3+	3.24	0.18	1.7778(6)/100.
^{29}Al		28.980445	6.5 m	β- /3.680	1.4/30.	5/2+			1.2732(8)/89.
					2.5/70.				2.0282(8)/4.
									2.4262(8)/7.
^{30}Al		29.98296	3.68 s	β- /8.56	5.05/	3+	3.01		1.26313(3)/35.
									2.23525(5)/65.
^{31}Al		30.98395	0.64 s	β- /8.00	6.25/	5/2+	3.8	0.08	0.75223(3)/18.
									1.69473(3)/59.
									2.31664(4)/73.
^{32}Al		31.9881	33. ms	β- /13.0		1+	1.96	0.024	
^{33}Al		32.9908	41.7 ms	β- /12.0	/91.5				1.940/2.5
				β-, n	/8.5				(1.01–4.34)
^{34}Al		33.9969	56. ms	β- /17.1	4.255/44	4			0.929/57
				β-, n	/26.				(0.12–4.26)
^{35}Al		34.9999	38. ms	β-/14.3	0.974/48	5/2+			0.064/45.
				β-, n	/ 38.				(0.12–5.63)
^{36}Al		36.0062	0.09 s	β-/18.3					
				β-, n	/<31.				
^{37}Al		37.0107	11. ms	β-/16.					
^{38}Al		38.017	> 7.6 ms						
^{39}Al		39.023	> 8. ms						
^{40}Al		40.031	> 0.26 μs						
^{41}Al		41.038	> 0.26 μs						
^{42}Al									
^{43}Al									
$_{14}$Si		**28.0855(3)**							
^{22}Si		22.0345	29. ms	β+, p	1.99/20	0+			
^{23}Si		23.0255	40.7 ms	β+, p/5.9	1.32,(0.6–11.6)				
^{24}Si		24.01155	0.14 s	β+, p/10.81	1.44,3.92,1.09	0+			ann.rad./
					(1.66–4.47)				
^{25}Si		25.00411	221 ms	β+, p/12.74	p/4.25/9.5	5/2+			ann.rad./
					p/0.40/4.75				
					p/0.56-6.80				
^{26}Si		25.992330	2.23 s	β+ /5.066	3.282/	0+			ann.rad./
									0.8294(8)/22.
^{27}Si		26.9867049	4.14 s	β+ /4.8118	3.85/100.	5/2+	-0.865		ann.rad./
									2.211(5)/0.2
^{28}Si	92.223(19)	27.976926533				0+			
^{29}Si	4.685(8)	28.97649470				1/2+	-0.5553		
^{30}Si	3.092(11)	29.97377017				0+			
^{31}Si		30.97536323	2.62 h	β- /1.4920	1.471/99.9	3/2+			1.2662(5)/0.05
^{32}Si		31.97414808	1.6×10^2 a	β- /0.224	0.213/100.	0+			
^{33}Si		32.97800	6.1 s	β- /5.85	3.92	(3/2+)	1.21		1.4313(5)/13.
									1.8477/100.
									2.538(2)/10.
^{34}Si		33.97858	2.8 s	β- /4.60	3.09/	0+			0.42907(5)/60.
									1.17852(2)/64.
									1.60756(5)/36.
^{35}Si		34.98458	0.9 s	β- /10.50					
^{36}Si		35.9866	0.5 s	β- /7.9		0^+			
				β-, n	/~ 12.				
^{37}Si		36.9929	~ 0.09 s	β- /12.5					
				β-, n	/~ 17.				

Elem. or Isot.	Natural Abundance (Atom %)	Atomic Mass or Weight	Half-life/ Resonance Width (MeV)	Decay Mode/ Energy (/MeV)	Particle Energy/ Intensity (MeV/%)	Spin ($h/2\pi$)	Nuclear Magnetic Mom. (nm)	Elect. Quadr. Mom. (b)	γ-Energy / Intensity (MeV/%)
^{38}Si		37.9956	> 1 μs	β- /10.7		0+			
				β-, n					
^{39}Si		39.0021	48. ms	β- /14.8					
^{40}Si		40.006	33. ms			0+			
^{41}Si		41.015	20. ms						
^{42}Si		42.020	13. ms			0+			
^{43}Si		43.029	> 0.26 μs						
^{44}Si									
$_{15}$P		**30.973762(2)**							
^{24}P		24.034							
^{25}P		25.0203	< 0.03 μs						
^{26}P		26.0118	44. ms	β+, p/18.1	p/0.41/18.0	3+			
					p/1.98/2.4				
					p/0.78-7.49				
^{27}P		26.99923	0.3 s	β+, p/11.63	p/0.73, 0.61/0.07	1/2+			
^{28}P		27.992315	270. ms	β+ /14.332	3.94/13.	3+			ann.rad./
					5.25/13.				1.779(2)/98.
					6.96/16.				2.839(2)/2.8
					8.8/7.				3.040(2)/3.2
					11.49/52.				4.498(2)/12.
									7.537(2)/9.
^{29}P		28.981801	4.14 s	β+ /4.9431	3.945/98.	1/2+	1.2349		ann.rad./
									1.273/1.32
									2.426/0.39
^{30}P		29.9783138	2.50 m	β+ /4.2323	3.245/99.9	1+			ann.rad./
									2.230(3)/0.07
^{31}P	100.	30.9737616				1/2+	+1.13160		
^{32}P		31.9739073	14.28 d	β- /1.7106	1.710/100.	1+	-0.2524		
^{33}P		32.971726	25.3 d	β- /0.249	0.249/100.	1/2+			
^{34}P		33.973636	12.4 s	β- /5.374	3.2/15.	1+			1.78–4.1/
					5.1/85.				2.127(5)/15.
^{35}P		34.973314	47. s	β- /3.989	2.34/100.	1/2+			1.572(1)/100.
^{36}P		35.97826	5.7 s	β- /10.41					0.902/77.
									3.291/100.
^{37}P		36.97961	2.3 s	β- /7.90					0.6462/
									1.5829/
^{38}P		37.9842	0.6 s	β- /12.4					1.2923/
				β-, n	/~ 12.				2.224/
^{39}P		38.9862	0.3 s	β- /10.5					
				β-, n	/26				
^{40}P		39.9913	0.15 s	β- /14.5					
				β-,n	/~ 30.				
^{41}P		40.9943	0.10 s	β-/~ 13.8					
				β-, n	/~ 30.				
^{42}P		42.0010	49. ms	β-/17.					
				β-, n	/ ~ 50.				
^{43}P		43.006	36. ms	β-/16.					
				β-, n	/100.				
^{44}P		44.013	19. ms						
^{45}P		45.019	> 0.2 μs						
^{46}P		46.027	> 0.2 μs						
$_{16}$S		**32.065(5)**							
^{26}S		26.0279	~ 10 ms			0+			
^{27}S		27.0188	16. ms	β+, 2p/18.3	p/2.26, 7.80				

Elem. or Isot.	Natural Abundance (Atom %)	Atomic Mass or Weight	Half-life/ Resonance Width (MeV)	Decay Mode/ Energy (/MeV)	Particle Energy/ Intensity (MeV/%)	Spin ($h/2\pi$)	Nuclear Magnetic Mom. (nm)	Elect. Quadr. Mom. (b)	γ-Energy / Intensity (MeV/%)
^{28}S		28.0044	0.13 s			0+			
^{29}S		28.99661	0.188 s	β+ /13.79		5/2+			ann.rad./
				β+, p/					
^{30}S		29.984903	1.18 s	β+ /6.138	4.42/78.	0+			ann.rad./
					5.08/20.				0.678/79.
^{31}S		30.979555	2.56 s	β+ /5.396	4.39/99.	1/2+	0.48793		ann.rad./
									1.2662(5)/1.2
^{32}S	94.99(26)	31.9720710				0+			
^{33}S	0.75(2)	32.9714588				3/2+	+0.64382	-0.068	
^{34}S	4.25(24)	33.9678669				0+			
^{35}S		34.9690322	87.2 d	β- /0.1672	0.1674/100.	3/2+	+1.00	+0.047	
^{36}S	0.01(1)	35.9670808				0+			
^{37}S		36.9711256	5.05 m	β- /4.8653	1.64/94.	7/2-			0.9083(4)/0.06
					4.75/5.6				3.1033(2)/94.2
^{38}S		37.97116	2.84 h	β- /2.94	1.00/	0+			0.1962(4)/0.2
									1.9421(3)/84.
^{39}S		38.97513	11.5 s	β- /6.64					1.301/52.
									1.697/44.
^{40}S		39.9755	9. s	β- /4.7		0+			(0.2116-1.874)
^{41}S		40.9796	~ 2.6 s	β- /8.7					
				β-, n					
^{42}S		41.9810	1.03 s	β- /7.8		0+			(0.118-2.912)
				β-, n	/ < 4.				
^{43}S		42.9872	0.26 s	β-/12.					
				β-, n	/ ~ 40				
^{44}S		43.9902	0.10 s	β-/9.		0+			
				β-, n	/18.				
^{45}S		44.997	68. ms	β-/14.					
				β-, n	/54.				
^{46}S		46.001	0.05 s			0+			
^{47}S		47.009	> 0.2 μs						
^{48}S		48.014	> 0.2 μs			0+			
^{49}S		49.024	< 0.2 μs						
$_{17}Cl$		**35.453(2)**							
^{28}Cl		28.029							
^{29}Cl		29.0141	< 0.02 μs						
^{30}Cl		30.0048	< 0.03 μs						
^{31}Cl		30.99241	0.15 s	β+, p/11.98	0.986, 1.52/0.7	3/2+			ann.rad./
									2.234
					0.762-2.75				1.249-4.045
^{32}Cl		31.98569	297. ms	β+ /12.69	9.47/50.	1+	1.11		ann.rad./
				β+, α	/0.05				2.2305/92
				β+, p	/0.026				(1.55–4.77)
^{33}Cl		32.9774519	2.511 s	β+ /5.583	4.51/98.	3/2+	+0.752		ann.rad./
									0.8409/0.52
									1.966/0.45
									2.866/0.44
^{34m}Cl			32.2 m	β+/	1.35/24.	3+			ann.rad./
					2.47/28.				
				I.T./					0.1457(8)/42.
									2.1276(5)/42.
^{34}Cl		33.9737628	1.528 s	β+ /5.4922	4.50/100.	0+			ann.rad./
^{35}Cl	75.76(10)	34.96885268				3/2+	+0.82187	-0.082	

Elem. or Isot.	Natural Abundance (Atom %)	Atomic Mass or Weight	Half-life/ Resonance Width (MeV)	Decay Mode/ Energy (/MeV)	Particle Energy/ Intensity (MeV/%)	Spin ($h/2\pi$)	Nuclear Magnetic Mom. (nm)	Elect. Quadr. Mom. (b)	γ-Energy / Intensity (MeV/%)
^{36}Cl		35.9683070	3.01 × 10^5 a	β- /0.7086	0.7093/98.	0+	+1.28547	-0.018	
				β+, EC/1.1421	0.115/0.002				ann.rad./
^{37}Cl	24.24(10)	36.96590259				3/2+	+0.68412	-0.0646	
^{38m}Cl			0.715 s	I.T./		5-			0.6714/100
^{38}Cl		37.9680104	37.2 m	β- /4.9168	1.11/31.	2-	2.05		1.64216/33.3
					2.77/11.				2.16760/44.8
					4.91/58.				
^{39}Cl		38.968008	55.6 m	β- /3.442	1.91/85.	3/2+			0.25026(1)/47.
					2.18/8.				1.26720(5)/54.
					3.45/7.				0.986–1.517
^{40}Cl		39.97042	1.38 m	β- /7.48		2-			0.6431(3)/6.
									1.4608(1)/77.
									2.8402(2)/17.
^{41}Cl		40.9707	34. s	β- /5.7	3.8/				(0.167–1.359)
^{42}Cl		41.9733	6.8 s	β- /9.4					
^{43}Cl		42.9741	3.3 s	β- /8.0					(0.352-4.247)
^{44}Cl		43.9783	~ 0.43 s	β-/12.3					
				β-, n	/ < 8.				
^{45}Cl		44.9803	0.40 s	β-/11.					
				β-, n	/24.				
^{46}Cl		45.984	0.23 s	β-/14.9					
				β-, n	/ ~ 60				
^{47}Cl		46.989	0.10 s	β-/15.					
				β-, n	/ < 3.				
^{48}Cl		47.995	> 0.2 μs						
^{49}Cl		48.000	> 0.17 s						
^{50}Cl		50.008							
^{51}Cl		51.014	> 0.2 μs						
$_{18}$Ar		**39.948(1)**							
^{30}Ar		30.0216	< 0.02 μs			0+			
^{31}Ar		31.0121	~ 14.1 ms	β+ /18.4	p/2.08/100.	5/2+			
				β+, p	/55.				
				β+, 2p	/2.5				
				β+, 3p	/0.11				
^{32}Ar		31.997638	98. ms	β+ /11.2	p/3.35, 0.6–5.55	0+			ann.rad./
				β+, p	/43.				
^{33}Ar		32.9899257	174. ms	β+ /11.62	3.17,2.10	1/2+	-0.72		ann.rad./
				β+, p/	(1.32–5.72)				0.810(2)/48.
^{34}Ar		33.9802712	0.844 s	β+/6.061	5.0/95.	0+			ann.rad./
									0.6658(1)/2.5
									3.1290(1)/1.3
^{35}Ar		34.975258	1.77 s	β+/5.965	4.94/93.	3/2+	+0.6322	-0.08	ann.rad./
									1.2185(5)/1.22
									1.763(1)/0.25
									2.964(1)/0.2
^{36}Ar	0.3336(21)	35.96754511				0+			
^{37}Ar		36.9667763	35.0 d	EC/.813		3/2+	+1.15	+0.076	
^{38}Ar	0.0629(7)	37.9627324				0+			
^{39}Ar		38.964313	268. a	β-/0.565	0.565/100.	7/2-	-1.59	-0.12	
^{40}Ar	99.6035(25)	39.962383123				0+			
^{41}Ar		40.9645006	1.82 h	β-/2.492	1.198/	7/2-			1.29364(5)/99.
									1.6770(3)/0.05
^{42}Ar		41.96305	33. a	β-/0.60	0.60/100.	0+			
^{43}Ar		42.96564	5.4 m	β-/4.6					0.4791(2)/10.

Elem. or Isot.	Natural Abundance (Atom %)	Atomic Mass or Weight	Half-life/ Resonance Width (MeV)	Decay Mode/ Energy (/MeV)	Particle Energy/ Intensity (MeV/%)	Spin ($h/2\pi$)	Nuclear Magnetic Mom. (nm)	Elect. Quadr. Mom. (b)	γ-Energy / Intensity (MeV/%)
									0.7380(1)/43.
									0.9752(1)/100.
									1.4400(3)/39.
^{44}Ar		43.964924	11.87 m	β-/3.55		0+			0.182–1.866
^{45}Ar		44.968040	21.5 s	β-/6.9		7/2-			0.0610/25.
									1.020/35.
									3.707/34.
^{46}Ar		45.96809	8.4 s	β-/5.70		0+			1.944/
^{47}Ar		46.9722	1.23 s	β-					0.36/100
									1.66/53
									1.74/41
									(2.02 - 4.01)
^{48}Ar		47.9745	0.48 s			0+			
^{49}Ar		48.981	0.17 s	β- ,n	n// ~ 65.				
^{50}Ar		49.984	~ 0.085 s	β- ,n	n// ~ 35.	0+			
^{51}Ar		50.992	> 0.2 μs	β-					
^{52}Ar		51.997	10 ms	β-		0+			
^{53}Ar		53.005		β-					
$_{19}$K		**39.0983(1)**							
^{32}K		32.022							
^{33}K		33.0073	< 0.025 μs						
^{34}K		33.9984	< 0.04 μs						
^{35}K		34.98801	0.19 s	β+ /11.88		3/2+	0.36		ann.rad./
				β+, p/	/0.37				1.751/14.
									2.5698/26.
									2.9827/51.
^{36}K		35.98129	0.342 s	β+ /12.81	5.3/42.	2+	+0.548		ann.rad./
					9.9/44.				1.97044(5)/82.
				β+, p	/0.048				2.20783(5)/30.
									2.43343(2)/32.
^{37}K		36.9733759	1.23 s	β+ /6.149	5.13/	3/2+	+0.2032		ann.rad./
									2.7944(8)/2.
									3.602(2)/0.05
^{38m}K			0.924 s	β+ /6.742	5.02/100.	0+			ann.rad./
^{38}K		37.9690812	7.63 m	β+ /5.913	2.60/99.8	3+	+1.37		ann.rad./
									2.1675(3)/99.8
									3.9356(5)/0.2
^{39}K	93.2581(44)	38.9637067				3/2+	+0.39146	+0.049	
^{40}K	0.0117(1)	39.9639985	1.248×10^9 a	β- /1.3111	1.312/89.	4-	-1.29810	-0.074	ann.rad./
				β+, EC/1.505	1.50/10.7				1.4608/10.5
^{41}K	6.7302(44)	40.9618258				3/2+	+0.21487	+0.071	
^{42}K		41.9624028	12.36 h	β- /3.525	1.97/19.	2-	-1.1425		0.31260(2)/0.3
					3.523/81.				1.5246(3)/18.1
^{43}K		42.96072	22.3 h	β- /1.82	0.465/8.	3/2+	+0.163		0.2211(2)/4.
					0.825/87.				0.3729(2)/88.
					1.24/3.5				0.3971(2)/11.
					1.814/1.3				0.6178(2)/81.
^{44}K		43.96156	22.1 m	β- /5.66	5.66/34.	2-	-0.856		0.36821/2.2
									1.15700(1)/58.
									2.15079(2)/22.
^{45}K		44.96070	17.8 m	β- /4.20	1.1/23.	3/2+	+0.173		0.1743(5)/80.
					2.1/69.				1.2607(8)/7.
					4.0/8.				1.7056(6)/69.
									2.3542(5)/14.

Elem. or Isot.	Natural Abundance (Atom %)	Atomic Mass or Weight	Half-life/ Resonance Width (MeV)	Decay Mode/ Energy (/MeV)	Particle Energy/ Intensity (MeV/%)	Spin ($h/2\pi$)	Nuclear Magnetic Mom. (nm)	Elect. Quadr. Mom. (b)	γ-Energy / Intensity (MeV/%)
^{46}K		45.96198	1.8 m	β- /7.72	6.3/	2-	-1.05		1.347(1)/91.
									3.700(5)/28.
^{47}K		46.96168	17.5 s	β- /6.64	4.1/99.	½+	+1.93		0.56474(3)/15.
					6.0/1.				0.58575(3)/85.
									2.0131/100
^{48}K		47.96551	6.8 s	β- /12.09	5.0/	(2-)			0.67122(1)/4.
									0.6723(5)/20.
									0.78016(1)/32.
									3.83153(7)/80.
^{49}K		48.9675	1.26 s	β- /11.0					2.025/
									2.252/
^{50}K		49.9728	0.472 s	β- /14.2					
^{51}K		50.976	0.365 s	β- /					1.027/21.7
				β-, n	//68(10)				3.46/3.9
					/2.23/0.218				(1.976-4.035)
^{52}K		51.983	0.105 s	β-	//79(12)				2.563/25.2
				β-, n	/1.040/0.216				2.377/6.9
				β-, 2n					1.027/0.55
^{53}K		52.987	30. ms	β-		3/2+			2.22/15.3
				β-, n	//85(19)				2.56/51.5
^{54}K		53.994	10. ms	β-					
$_{20}$Ca		**40.078(4)**							
^{34}Ca		34.0141	< 0.035 μs			0+			
^{35}Ca		35.0049	25.7 ms	β+, p/15.6	p/1.43/49				
					1.9–8.8				
^{36}Ca		35.99309	0.100 s	β+, (p)/10.99	p/2.61/32	0+			ann.rad./
				β+, n/	p/1.71/7				
^{37}Ca		36.98587	0.182 s	β+ /11.64	p/3.19/40.7	3/2+			ann.rad./
				β+, n/	p/(0.899-2.00)				1.369
^{38}Ca		37.976318	0.44 s	β+ /6.74		0+			ann.rad./
									1.5677(5)/25.
									3.210(2)/1.
^{39}Ca		38.970720	0.861 s	β+ /6.531	5.49/100.	3/2+	1.02168	0.04	ann.rad./
^{40}Ca	96.941(156)	39.9625910	5.92 × 10^{21} a	EC-EC		0+			
^{41}Ca		40.9622781	1.02 × 10^{5} a	EC/0.4214		7/2-	-1.5948	-0.090	
^{42}Ca	0.647(23)	41.9586180				0+			
^{43}Ca	0.135(10)	42.9587666				7/2-	-1.31764	-0.055	
^{44}Ca	2.086(110)	43.9554818				0+			
^{45}Ca		44.9561866	162.7 d	β- /0.257	0.257/100.	7/2-	-1.327	+0.05	
^{46}Ca	0.004(3)	45.953693	>0.4 × 10^{16} a	β-β-		0+			
^{47}Ca		46.954546	4.536 d	β-/1.992	0.684/84.	7/2-	-1.38	+0.02	1.297/75
					1.98/16.				(0.041–1.88)
^{48}Ca	0.187(21)	47.952534	4.3 × 10^{19} a	β-β-		0+			
			>7.1 × 10^{19} a	β-					
^{49}Ca		48.955674	8.72 m	β- /5.262	0.89/7.	3/2-			3.0844(1)/90.7
					1.95/92.				4.0719(1)/8.12
									(0.143 - 4.738)
^{50}Ca		49.95752	14. s	β- /4.97	3.12/	0+			0.2569/98.
									(0.0715–1.59)
^{51}Ca		50.9615	10. s	β- /7.3		(3/2-)			
^{52}Ca		51.965	4.6 s	β- /8.0		0+			
^{53}Ca		52.9701	0.09 s	β- /10.9					
^{54}Ca		53.974	> 0.3 μs			0+			
^{55}Ca		54.981	> 0.3 μs						

Elem. or Isot.	Natural Abundance (Atom %)	Atomic Mass or Weight	Half-life/ Resonance Width (MeV)	Decay Mode/ Energy (/MeV)	Particle Energy/ Intensity (MeV/%)	Spin ($h/2\pi$)	Nuclear Magnetic Mom. (nm)	Elect. Quadr. Mom. (b)	γ-Energy / Intensity (MeV/%)
^{56}Ca		55.986	> 0.3 μs			0+			
$_{21}$Sc		**44.955912(6)**							
^{36}Sc		36.0149	0.102 s						
^{37}Sc		37.0031	0.181 s						
^{38}Sc		37.9947	< 0.3 μs						
^{39}Sc		38.98479	< 0.3 μs	p					
^{40}Sc		39.977967	0.182 s	β+ /14.320	5.73/50.	4-			ann.rad./
					7.53/15.				0.752/41.
					8.76/15.				3.732/99.5
					9.58/20.				(1.12–3.92)
^{41}Sc		40.9692511	0.596 s	β+ /6.4953	5.61/100.	7/2-	+5.431	-0.156	ann.rad./
^{42m}Sc			61.6 s	β+ /	2.82/	7+			ann.rad./
									0.4375(5)/100.
									1.2270(5)/100.
									1.5245(5)/100.
^{42}Sc		41.9655164	0.682 s	β+ /6.4259	5.32/100.	0+			ann.rad./
^{43}Sc		42.961151	3.89 h	β+, EC/2.221	0.82/22.	7/2-	+4.62	-0.26	ann.rad./
					1.22/78.				0.3729(1)/22.
^{44m}Sc			58.2 h	I.T./0.27		6+	+3.88	-0.19	0.27124(1)/87.
				EC/3.926					(1.00–1.16)
^{44}Sc		43.959403	3.93 h	β+, EC/3.653	1.47/	2+	+2.56	+0.10	ann.rad./
									1.157/100
^{45}Sc	100.	44.955912				7/2-	+4.75649	-0.220	
^{46m}Sc			18.7 s	I.T./0.14253		1-			0.14253(2)/62.
^{46}Sc		45.955172	83.81 d	β- /2.367	0.357/100.	4+	+3.03	+0.12	0.8893/100
									1.121/100
^{47}Sc		46.952408	3.349 d	β- /0.600	0.439/69.	7/2-	+5.34	-0.22	0.15938(1)/68.
					0.601/31.				
^{48}Sc		47.95223	43.7 h	β- /3.99	0.655/	6+			0.9835/100
									1.03750(1)/97.
									1.3121/100
^{49}Sc		48.950024	57.3 m	β- /2.006	2.00/99.9.	7/2-			1.7619(3)/0.05
^{50}Sc		49.95219	1.71 m	β- /6.89	3.05/76.	(5+)			0.5235(1)/88.
					3.60/24.				1.1210(1)/100.
									1.5537(2)/100.
^{51}Sc		50.95360	12.4 s	β- /6.51	4.4/	7/2-			1.4373(4)/52.
					5.0/				0.718–2.144
^{52}Sc		51.9567	8.2 s	β- /9.0		(3+)			
^{53}Sc		52.9596	> 3. ms	β- /8.1					
^{54m}Sc			~ 7 μs	I.T.		(5+)			0.110/IT
^{54}Sc		53.9633	0.27 s	β- /11.6					0.100/50
									1.70/40
									0.50/40
^{55}Sc		54.968	0.103 s	β- /13					0.593(1)/40
^{56m}Sc			0.06 s						1.161/21
									0.690/19
^{56}Sc		55.973	35. ms	β-		(1+)			1.129/48
^{57}Sc		56.978	13. ms	β-					
^{58}Sc		57.984	12. ms	β-					
$_{22}$Ti		**47.867(1)**							
^{38}Ti		38.0098	< 0.12 μs			0+			
^{39}Ti		39.0016	29. ms	β+ /15.4	p//94				
^{40}Ti		39.9905	52. ms	β+ /11.7	p/2.16/29	0+			2.467/8.5
				β+, p	3.73/23				

Elem. or Isot.	Natural Abundance (Atom %)	Atomic Mass or Weight	Half-life/ Resonance Width (MeV)	Decay Mode/ Energy (/MeV)	Particle Energy/ Intensity (MeV/%)	Spin ($h/2\pi$)	Nuclear Magnetic Mom. (nm)	Elect. Quadr. Mom. (b)	γ-Energy / Intensity (MeV/%)
					1.70/22				
					0.242–5.74				
^{41}Ti		40.9832	83. ms	β+, p/12.93	p/4.73/107	3/2+			ann.rad./
					3.10/67				
					3.75/39				
					0.744–6.73				
^{42}Ti		41.97303	0.20 s	β+ /7.000	6.0/	0+			ann.rad./
									0.6107(5)/56.
^{43}Ti		42.96852	0.50 s	β+ /6.87	5.80/	7/2-	0.85		ann.rad./
^{44}Ti		43.959690	59. a	EC/0.268		0+			0.06787/91
									0.07832/97
^{45}Ti		44.958126	3.078 h	β+/86/2.062	1.04	7/2-	0.095	~ 0.015	ann.rad./
				EC/14/					(0.36–1.66)
^{46}Ti	8.25(3)	45.952632				0+			
^{47}Ti	7.44(2)	46.951763				5/2-	-0.78848	+0.30	
^{48}Ti	73.72(3)	47.947946				0+			
^{49}Ti	5.41(2)	48.947870				7/2-	-1.10417	+0.25	
^{50}Ti	5.18(2)	49.944791				0+			
^{51}Ti		50.946615	5.76 m	β- /2.471	1.50/92.	3/2-			0.3197(2)/93.
					2.13/				0.6094–0.9291
^{52}Ti		51.94690	1.7 m	β- /1.97	1.8/100.	0+			0.0170(5)/100.
									0.1245/100
^{53}Ti		52.9497	33. s	β- /5.0	(2.2–3)/	3/2-			0.1008(1)/20.
									0.1276(1)/45.
									0.2284(1)/39.
									1.6755(5)/45.
									(1.72–2.8)/
^{54}Ti		53.9511	1.5 s	β- /4.3		0+			
^{55}Ti		54.9553	1.3 s	β- /7.4					0.672/44
									(0.32 - 1.83)
^{56}Ti		55.9582	0.20 s	β- /7.0		0+			
^{57}Ti		56.9640	98. ms	β- /11.					
^{58}Ti		57.967	53. ms	β-		0+			0.114
^{59}Ti		58.973	30. ms	β-					
^{60}Ti		59.978	22. ms	β-		0+			
^{61}Ti		60.983	> 0.3 μs						
$_{23}$V		**50.9415(1)**							
^{40}V		40.0111							
^{41}V		40.9998							
^{42}V		41.9912	< 0.055 μs						
^{43}V		42.9807	79. ms	β+ /11.3	p// < 2.5				
^{44}V		43.9741	0.09 s	β+, α/13.7					ann.rad./
^{45}V		44.96578	0.54 s	β+ /7.13		7/2-			
^{46}V		45.960201	0.4223 s	β+ /7.051	6.03/100.	0+			ann.rad./
^{47}V		46.954909	32.6 m	β+, EC/2.928	1.90/99.+	3/2-			ann.rad./
									1.7949(8)/0.19
									(0.2–2.16)
^{48}V		47.952254	15.98 d	β+ /4.012	0.698/50.	4+	2.01		ann.rad./
									0.9835/100
									(1.3–2.4)
^{49}V		48.948516	337. d	EC/0.602		7/2-	4.47		
^{50}V	0.250(4)	49.947159	1.4×10^{17} a	EC	/82.7	6+	+3.34569	+0.21	
				β-	/17.3				
^{51}V	99.750(4)	50.943960				7/2-	+5.148706	-0.04	

Elem. or Isot.	Natural Abundance (Atom %)	Atomic Mass or Weight	Half-life/ Resonance Width (MeV)	Decay Mode/ Energy (/MeV)	Particle Energy/ Intensity (MeV/%)	Spin (h/2 π)	Nuclear Magnetic Mom. (nm)	Elect. Quadr. Mom. (b)	γ-Energy / Intensity (MeV/%)
^{52}V		51.944776	3.76 m	β- /3.976	2.47/	3+			1.4341(1)/100.
^{53}V		52.944338	1.56 m	β- /3.436	2.52/	7/2-			1.0060(5)/90.
									1.2891(3)/10.
^{54m}V			0.9 μs	I.T.		(5+)			0.108/IT
^{54}V		53.94644	49.8 s	β- /7.04	1.00/5.	3+			0.8348/97.
					2.00/12.				0.9887/80.
					2.95/45.				2.259/46.
					5.20/11.				(0.56–3.38)
^{55}V		54.9472	6.5 s	β- /6.0	6.0/	(7/2-)			0.5177/73.
									(0.224–1.21)
^{56}V		55.9505	0.22 s	β- /9.1					1.01/30.
									0.688/26.
									(0.82 – 1.32)
^{57}V		56.9526	0.35 s	β- /8.1					0.268/52.
									0.692/20.
									(0.25 – 1.31)
^{58}V		57.9567	0.19 s	β- /11.6					0.880/62
									1.056/28
									2.217/13
									(1.04 – 1.57)
^{59}V		58.9602	97. ms	β- /9.9					0.90/80.
^{60m}V			0.12 s						
^{60}V		59.9650	0.07 s	β- /14.					0.102–0.208
^{61}V		60.9685	47. ms						(0.071-1.144)
^{62}V		61.9738	34. ms						
^{63}V		62.978	17. ms						
^{64}V		63.983	> 0.3 μs						
$_{24}$Cr		**51.9961(6)**							
^{42}Cr		42.0064	13. ms	β+, p	p/1.90/29	0+			1.623/35
					p/1.50–3.7				
^{43}Cr		42.9977	21. ms	β+, p	p/3.83/18				1.555/35
					p/4.29/15				0.838/6.2
					p/1.01–4.59				1.937/1.8
^{44}Cr		43.98555	43. ms	β+, (p)/10.3	p/0.95–3.1	0+			0.677/59.
^{45}Cr		44.9796	61. ms	β+, p/12.5	p/2.088/19.6	7/2-			ann.rad./
					p/(0.945-1.61)				(1.08-1.37)
^{46}Cr		45.96836	0.3 s	β+ /7.60		0+			ann.rad./
^{47}Cr		46.96290	0.51 s	β+ /7.45		3/2-			ann.rad./
^{48}Cr		47.95403	21.6 h	EC/1.66		0+			ann.rad./
									0.116(2)/95.
									0.305(10)/100.
^{49}Cr		48.951336	42.3 m	β+, EC/2.631	1.39/	5/2-	0.476		ann.rad./
					1.45/				0.09064(1)/51.
					1.54/				0.15293(1)/27.
									(0.062-1.6)
^{50}Cr	4.345(13)	49.946044	>1.3 × 10^{18} a	β+EC		0+			
^{51}Cr		50.944767	27.70 d	EC/0.7527		7/2-	-0.934		0.3201/9.8
^{52}Cr	83.789(18)	51.940508				0+			
^{53}Cr	9.501(17)	52.940649				3/2-	-0.47454	-0.22	
^{54}Cr	2.365(7)	53.938880				0+			
^{55}Cr		54.940840	3.497 m	β- /2.603	2.5/	3/2-			1.5282(2)/0.04
									(0.13–2.37)
^{56}Cr		55.940653	5.9 m	β- /1.62	1.50/100.	0+			0.026(2)/100.
									0.083(3)/100.

Elem. or Isot.	Natural Abundance (Atom %)	Atomic Mass or Weight	Half-life/ Resonance Width (MeV)	Decay Mode/ Energy (/MeV)	Particle Energy/ Intensity (MeV/%)	Spin ($h/2\pi$)	Nuclear Magnetic Mom. (nm)	Elect. Quadr. Mom. (b)	γ-Energy / Intensity (MeV/%)
^{57}Cr		56.943613	21. s	β- /5.1	3.3/	3/2-	0.0834		0.850/8.
					3.5/				(0.083-2.62)
^{58}Cr		57.9444	7.0 s	β- /4.0		0+			(0.131–0.683)
^{59m}Cr			0.10 ms	I.T.		(9/2+)			0.208/IT
									0.193
									0.102
^{59}Cr		58.9486	1.0 s	β- /7.7					1.236
^{60}Cr		59.9500	0.6 s	β- /6.0		0+			
^{61}Cr		60.9547	0.26 s	β- /8.8					0.354–1.860
^{62}Cr		61.9566	0.19 s	β- /7.3		0+			(0.156-1.215)
^{63}Cr		62.9619	0.129 s	β-					(0.250-3.454)
^{64}Cr		63.9644	0.043 s	β-		0+			0.188
^{65}Cr		64.9702	0.027 s	β-					0.272, 1.368
^{66}Cr		65.973	0.01 s	β-		0+			
^{67}Cr		66.980	> 0.3 μs						
$_{25}$**Mn**		**54.938045(5)**							
^{44}Mn		44.0069	< 0.105 μs						
^{45}Mn		44.9945	< 0.07 μs						
^{46}Mn		45.9867	36. ms	β+ /17.1	p/3.00/6.5				0.329/11.
				β+, p	// ~ 58				(0.0544-1.322)
^{47}Mn		46.9761	88. ms	β+ /12.3	p// <1.7				
^{48}Mn		47.9685	0.15 s	β+ /13.5	5.79/58.	4+			
					4.43/10.				
^{49}Mn		48.95962	0.38 s	β+ /7.72	6.69/	5/2-			ann.rad./
^{50m}Mn			1.74 m	β+ /7.887	3.54/	5+			ann.rad./
									1.0980/94.
									0.783/91.
									(0.66–3.11)
^{50}Mn		49.954238	0.283 s	β+ /7.6330	6.61/	0+			ann.rad./
^{51}Mn		50.948211	46.2 m	β+, EC/3.208	2.2/	5/2-	3.568	0.4	ann.rad./
									0.7491(1)/0.26
									(1.148–1.164)
^{52m}Mn			21.1 m	β+ /98/5.09	2.631/	2+	0.0077		ann.rad./
				I.T./2/0.378					0.3778 (I.T.)
									1.43406(1)/98.
									(0.7–4.8)
^{52}Mn		51.945566	5.591 d	β+ /4.712	0.575/	6+	+3.063	+0.5	ann.rad./
				EC/					0.74421(1)/90.
									1.4341/100
^{53}Mn		52.941290	3.7 × 10^6 a	EC/0.5970		7/2-	5.024		
^{54}Mn		53.940359	312.1 d	EC/1.377		3+	+3.282	+0.33	0.8340/100
			6.7 × 10^8 a	β+	//1.3 × 10^{-7}				
^{55}Mn	100.	54.938045				5/2-	+3.4687	+0.32	
^{56}Mn		55.938905	2.579 h	β- /3.6954	0.718/18.	3+	+3.2266		0.84675/98.9
					1.028/34.				1.81072(4)/26.3
									2.113/13.8
									(1.04 – 3.37)
^{57}Mn		56.938285	1.45 m	β- /2.691		5/2-			
^{58}Mn		57.93998	65 s	β- /6.25	3.8/	3+			0.45916(2)/20.
					5.1/				0.81076(1)/82.
									1.32309(5)/53.
^{59}Mn		58.94044	4.6 s	β- /5.19	4.5/	5/2-			0.726/
									0.473/

Elem. or Isot.	Natural Abundance (Atom %)	Atomic Mass or Weight	Half-life/ Resonance Width (MeV)	Decay Mode/ Energy (/MeV)	Particle Energy/ Intensity (MeV/%)	Spin ($h/2\pi$)	Nuclear Magnetic Mom. (nm)	Elect. Quadr. Mom. (b)	γ-Energy / Intensity (MeV/%)
									0.287–2.35
^{60m}Mn			1.77 s	β- /IT	5.7/	3+			0.824/
^{60}Mn		59.9429	0.28 s	β- /8.6	8.2/88.	0+			0.8234/12.2
					6.2/5.0				1.150/5.0
					7.4/4.2				1.523/3.0
^{61}Mn		60.9447	0.67 s	β- /7.4		(5/2)-			
^{62}Mn		61.9484	0.67 s	β- /10.4		(3+)			0.877/
									0.942–1.299
^{63}Mn		62.9502	0.28 s	β- /8.8					0.356,0.450
^{64m}Mn			> 0.1 ms						0.135/IT
^{64}Mn		63.9543	87 ms	β- /11.8					0.746
^{65}Mn		64.9563	0.092 s	β- /10.					0.366
^{66}Mn		65.9611	64 ms						0.471
^{67}Mn		66.9641	45 ms						
^{68}Mn		67.969	~ 28 ms						
^{69}Mn		68.973	14 ms						
$_{26}$**Fe**		**55.845(2)**							
^{45}Fe		45.0146	2.6 ms	2p /1.14	p// ~ 59.				
^{46}Fe		46.0008	13. ms	β+ /13.1	p// 79.	0+			0.493/23.
^{47}Fe		46.9929	21.9 ms	β+ /15.6	p//87.				0.892/76.
^{48}Fe		47.9805	45. ms	β+ /11.2	p//16.	0+			0.313/63.
^{49}Fe		48.9736	65. ms	β+ /13.0	p/1.977/34.5	(7/2-)			0.797/23.7
					p/(1.161-1.550)				(0.261-1.279)
^{50}Fe		49.9630	0.15 s	β+ /8.2		0+			0.651
^{51}Fe		50.95682	0.31 s	β+ /8.02		(5/2-)			ann.rad./
^{52m}Fe			46. s	β+ /4.4		(12+)			ann.rad./
									(0.622–2.286)/
^{52}Fe		51.94811	8.28 h	β+ /57/2.37	0.804/	0+			ann.rad./
				EC/43/					0.16868(1)/99.
				I.T./					0.377 (I.T.)/
^{53m}Fe			2.6 m	I.T./3.0407		19/2-			0.7011(1)/99.
									1.0115(1)/87.
									1.3281(1)/87.
									2.3396(1)/13.
^{53}Fe		52.945308	8.51 m	β+ /3.743	2.40/42.	7/2-			ann.rad./
					2.80/57.				0.3779(1)/42.
									(1.2–3.2)
^{54}Fe	5.845(35)	53.939611	>3.1 × 10^{22} a	EC-EC		0+			
^{55}Fe		54.938293	2.73 a	EC/0.2314		3/2-			Mn x-ray
^{56}Fe	91.754(36)	55.934938				0+			
^{57}Fe	2.119(10)	56.935394				½-	+0.0906	0.16	
^{58}Fe	0.282(4)	57.933276				0+			
^{59}Fe		58.934876	44.51 d	β-/1.565	0.273/48.	3/2-	-0.336		1.099/57
					0.475/51.				1.292/43.
									(0.14–1.48)
^{60}Fe		59.934072	1.5 × 10^{6} a	β- /0.237	0.184/100.	0+			0.0586/100
^{61m}Fe			0.25 µs	I.T.		(9/2+)			0.654/IT
									0.207
^{61}Fe		60.93675	6.0 m	β- /3.98	2.5/13.				1.205/44.
					2.63/54.				1.028/43.
					2.80/31.				(0.12–3.37)
^{62}Fe		61.93677	68. s	β- /2.53	2.5/100.	0+			0.5061(1)/100.
^{63}Fe		62.9404	6. s	β- /6.3		5/2-			0.995/
									(1.365–1.427)

Elem. or Isot.	Natural Abundance (Atom %)	Atomic Mass or Weight	Half-life/ Resonance Width (MeV)	Decay Mode/ Energy (/MeV)	Particle Energy/ Intensity (MeV/%)	Spin ($h/2\pi$)	Nuclear Magnetic Mom. (nm)	Elect. Quadr. Mom. (b)	γ-Energy / Intensity (MeV/%)
^{64}Fe		63.9412	2.0 s	β- /4.9		0+			
^{65m}Fe			0.4 μs	I.T.		(5/2-)			0.364/IT
^{65}Fe		64.9454	1.3 s	β- /7.9					
^{66}Fe		65.9468	0.44 s	β- /5.7		0+			0.471–1.425
^{67m}Fe			~ 0.04 ms	I.T.		(5/2-)			0.367/IT
^{67}Fe		66.9510	0.48 s	β- /8.8					0.189
^{68}Fe		67.954	0.19 s	β- / ~ 7.6		0+			
^{69}Fe		68.959	0.11 s						
^{70}Fe		69.961	0.10 s			0+			
^{71}Fe		70.967	> 0.3 μs						
^{72}Fe		71.970	> 0.3 μs			0+			
$_{27}$**Co**		**58.933195(5)**							
^{47}Co		47.0115							
^{48}Co		48.0018							
^{49}Co		48.9897	< 0.035 μs						
^{50}Co		49.9815	39. ms	β+ /17.0	p/2.770/41.				0.2614/64.
					p/(1.874-2.296)				(0.482-1.308)
^{51}Co		50.9707	69. ms	β+ /12.8	p// < 3.8				
^{52}Co		51.9636	0.12 s	β+ /14.0					0.849–1.942
^{53m}Co			0.25 s	β+, p/		19/2-			ann.rad./
^{53}Co		52.95422	0.24 s	β+ /8.30		7/2-			ann.rad./
^{54m}Co			1.46 m	β+ /8.44	4.25/100.	7+			ann.rad./
									0.411(1)/99.
									1.130(1)/100.
									1.408(1)/100.
^{54}Co		53.948460	0.1932 s	β+ /8.2430	7.34/100.	0+			ann.rad./
^{55}Co		54.941999	17.53 h	β+ /3.4513	0.53/	7/2-	+4.822		ann.rad./
				EC/	1.03/				0.9312/75.
					1.50/				0.4772/20.
									(0.092–3.11)
^{56}Co		55.939839	77.3 d	β+/4.566	1.459/18.	4+	3.85	~+0.25	ann.rad./
				EC/					0.8468/99.9
									1.2383/68.
									(0.26–3.61)
^{57}Co		56.936291	271.8 d	EC/0.8361		7/2-	+4.72	+0.5	0.12206/86
									(0.014–0.706)
^{58m}Co			9.1 h	I.T./		5+			0.02489/0.035
^{58}Co		57.935753	70.88 d	β+ /2.307		2+	+4.04	+0.22	ann.rad./
				EC/					0.81076/99
^{59}Co	100.	58.933195				7/2-	+4.63	+0.41	
^{60m}Co			10.47 m	I.T./99.8/0.059		2+	+4.4	~+0.3	0.0586/2.0
				β- /0.2/1.56					
^{60}Co		59.933817	5.271 a	β- /2.824	0.315/99.7	5+	+3.799	+0.44	1.1732/100
									1.3325/100
^{61}Co		60.932476	1.650 h	β- /1.322	1.22/95.	7/2-			0.0674/86.
									0.842–0.909
^{62m}Co			13.9 m	β- /	0.88/25.	5+			1.1635(3)/70.
					2.88/75.				1.1730(3)/98.
									2.0039(3)/19.
^{62}Co		61.93405	1.50 m	β- /5.32	1.03/10.	2+			1.1292(3)/13.
					1.76/5.				1.1730(3)/83.
					2.9/20.				1.9851(1)/3.
					4.05/60.				2.3020(1)/19.
^{63}Co		62.93361	27.5 s	β- /3.67	3.6/	7/2-			0.08713(1)/49.

Elem. or Isot.	Natural Abundance (Atom %)	Atomic Mass or Weight	Half-life/ Resonance Width (MeV)	Decay Mode/ Energy (/MeV)	Particle Energy/ Intensity (MeV/%)	Spin (h/2 π)	Nuclear Magnetic Mom. (nm)	Elect. Quadr. Mom. (b)	γ-Energy / Intensity (MeV/%)
									0.9817(3)/2.6
									0.156–2.17
^{64}Co		63.93581	0.30 s	β- /7.31	7.0/	1+			
^{65}Co		64.93648	1.14 s	β- /5.96		(7/2)-			
^{66m2}Co			> 0.1 ms	I.T.		(8-)			0.252/IT
									0.214
									0.175
^{66m1}Co			1.2 μs	I.T.		(5+)			0.175/IT
^{66}Co		65.9398	0.25 s	β- /10.0					(1.245–1.425)
^{67}Co		66.9409	0.43 s	β- /8.4					0.694
^{68}Co		67.9449	0.19 s	β- /11.7					
^{69}Co		68.9463	0.20 s	β- /9.3					
^{70}Co		69.951	0.12 s	β- 13.					1.26/102
									0.97/100
									(0.45 – 0.92)
^{71}Co		70.953	97. ms	β-					0.566/100
				β-,n	// > 3				(0.25 – 0.77)
^{72}Co		71.958	60. ms	β-					1.096/100
				β-,n	// > 6				0.845
									(0.455 – 1.197)
^{73}Co		72.960	41. ms	β-					0.524/100
				β-,n	// > 9				(0.24 – 0.76)
^{74}Co		73.965	30. ms	β-					0.739
				β-,n	// > 26				1.024
^{75}Co		74.968	> 0.3 μs						
$_{28}$Ni		**58.6934(4)**							
^{48}Ni		48.020	~ 2.1 ms	2p	p // ~ 25	0+			
^{49}Ni		49.0097	12. ms		p //~83				0.965/82
^{50}Ni		49.9959	12. ms	β+, p	p //70.	0+			0.063
					p /1.97/14				0.090
^{51}Ni		50.9877	24. ms	β+ /16.0	p/4.66/8.7				0.765/73
					p/1.08-5.66				1.087/29
					p//87.				1.546-1.743
^{52}Ni		51.9757	38. ms	β+ /11.7	p//31.	0+			2.418/38
					p/1.35/9				0.142
^{53}Ni		52.9685	55. ms	β+, p/13.3	p//23.	7/2-			ann.rad./
					p/1.93/5.4				0.849/13
^{54}Ni		53.95791	0.10 s	β+ /8.80		0+			0.937
^{55}Ni		54.95133	0.20 s	β+ /8.70	7.66/	7/2-			ann.rad./
^{56}Ni		55.94213	6.08 d	EC/2.14		0+			0.15838/99
				β+ /<10^{-6}					0.81185(3)/87.
									0.2695–0.7500
^{57}Ni		56.939794	35.6 h	β+ /3.264	0.712/10.	3/2-	-0.798		ann.rad./
				EC/	0.849/76.				1.3776/78.
									(0.127–3.177)
^{58}Ni	68.077(16)	57.935343	>4 × 10^{19} a	EC-EC		0+			
^{59}Ni		58.934347	~ 7.6 × 10^{4} a	EC/		3/2-			
^{60}Ni	26.223(13)	59.930786				0+			
^{61}Ni	1.1399(12)	60.931056				3/2-	-0.75002	+0.16	
^{62}Ni	3.6345(35)	61.928345				0+			
^{63}Ni		62.929669	101. a	β- /0.066945	0.065/	½-			
^{64}Ni	0.9255(17)	63.927966				0+			
^{65}Ni		64.930084	2.517 h	β- /2.137	0.65/30.	5/2-	0.69		0.36627(3)/5.
					1.020/11.				1.11553(4)/16.

Elem. or Isot.	Natural Abundance (Atom %)	Atomic Mass or Weight	Half-life/ Resonance Width (MeV)	Decay Mode/ Energy (/MeV)	Particle Energy/ Intensity (MeV/%)	Spin ($h/2\pi$)	Nuclear Magnetic Mom. (nm)	Elect. Quadr. Mom. (b)	γ-Energy / Intensity (MeV/%)
					2.140/58.				1.48184(5)/23.
^{66}Ni		65.929139	54.6 h	β- /0.23		0+			
^{67m}Ni			13.3 μs	I.T.		9/2+	0.56		0.313/IT
									0.694
^{67}Ni		66.931569	21. s	β- /3.56	3.8/	½-	+0.601		1.0722/100.
									1.6539/100.
									(0.10–1.98)
^{68m2}Ni			0.34 μs						0.511
^{68m1}Ni			0.86 ms	I.T.		(5-)			0.814/IT
									2.033
^{68}Ni		67.931869	29. s	β- /2.06		0+			
^{69m2}Ni			0.44 μs	I.T.		(17/2)			0.148/IT
									0.593
									1.959
^{69m1}Ni			3.5 s						
^{69}Ni		68.935610	11. s	β- /5.4					0.6807(3)/100.
									(0.207–1.213)
^{70m}Ni			0.21 μs	I.T.		(8+)			0.183/IT
									0.448
									0.970
									1.259
^{70}Ni		69.9365	6.0 s	β- /3.5		0+			
^{71}Ni		70.9407	2.56 s	β- /6.9					
^{72}Ni		71.9421	1.6 s	β- /5.2		0+			
^{73}Ni		72.9465	0.84 s	β- /9.					
^{74}Ni		73.9481	0.9 s	β- /7.		0+			
^{75}Ni		74.9529	0.34 s						
^{76}Ni		75.955	0.24 s			0+			
^{77}Ni		76.961	0.13 s						
^{78}Ni		77.963	~ 0.11 s			0+			
$_{29}$Cu		**63.546(3)**							
^{52}Cu		51.9972							
^{53}Cu		52.9856	< 0.3 μs						
^{54}Cu		53.9767	< 0.075 μs						
^{55}Cu		54.9661	~ 27. ms	β+ /13.2	p//15.				
^{56}Cu		55.9586	93. ms	β+ /15.3					0.511/233
									2.700/100
									0.9507–3.287
^{57}Cu		56.94921	196. ms	β+ /8.77		3/2-	2.00		0.77–3.01
^{58}Cu		57.944539	3.21 s	β+ /8.563	4.5/15.	1+			ann.rad./
				EC/	7.439/83.				0.0403(4)/5.
									1.4483(2)/11.
									1.4546(2)/16.
^{59}Cu		58.939498	1.36 m	β+ /4.800	1.9/	3/2-	+ 1.89		ann.rad./
					3.75/				0.3393(1)/8.
									0.8780(1)/12.
									1.3015(1)/15.
									(0.4–2.6)
^{60}Cu		59.937365	23.7 m	β+ /6.127	2.00/69.	2+	+1.219		ann.rad./
				EC/	3.00/18.				1.3325/88.
					3.92/6.				1.7915/45.
									(0.12–5.048)
^{61}Cu		60.933458	3.35 h	β+ /2.237	0.56/3.	3/2-	+2.14		ann.rad./
					0.94/5.				0.2830/13.

Elem. or Isot.	Natural Abundance (Atom %)	Atomic Mass or Weight	Half-life/ Resonance Width (MeV)	Decay Mode/ Energy (/MeV)	Particle Energy/ Intensity (MeV/%)	Spin ($h/2\pi$)	Nuclear Magnetic Mom. (nm)	Elect. Quadr. Mom. (b)	γ-Energy / Intensity (MeV/%)
					1.15/2.				0.6560/11.
					1.220/51.				(0.067–2.123)
^{62}Cu		61.932584	9.67 m	β+ /98/3.948	2.93/98.	1+	-0.380		ann.rad./
				EC/					1.17302(1)/0.6
									(0.87–3.37)
^{63}Cu	69.15(15)	62.929598				3/2-	+2.2273	-0.211	
^{64}Cu		63.929764	12.701 h	β- /38/0.579	0.578/	1+	-0.217		ann.rad./
				β+ /19/1.6751	0.65/				1.3459(3)/0.6
				EC/41/					
^{65}Cu	30.85(15)	64.927790				3/2-	+2.3817	-0.195	
^{66}Cu		65.928869	5.09 m	β- /2.642	1.65/6.	1+	-0.282		0.8330(1)/0.22
					2.7/94.				1.0392(2)/9.2
^{67}Cu		66.927730	2.580 d	β- /0.58	0.395/56.	3/2-	+ 2.54		0.09125(1)/7.
					0.484/23.				0.09325(1)/17.
					0.577/20.				0.18453(1)/47.
^{68m}Cu			3.79 m	I.T./86/		6-	+1.24		0.0843(5)/70.
				β- /14/1.8					0.1112(5)/18.
									0.5259(5)/74.
									(0.64–1.34)
^{68}Cu		67.929611	31. s	β- /4.46	3.5/40.	1+	+2.48		1.0774(5)/58.
					4.6/31.				1.2613(5)/17.
									(0.15–2.34)
^{69m}Cu			0.36 μs	I.T.		(13/2+)	+1.5		0.075/IT
									0.190/IT
									0.680
									1.871
^{69}Cu		68.929429	2.8 m	β- /2.68	2.48/80.	3/2-	+2.84		0.5307(3)/3.
									0.8340(5)/6.
									1.0065(8)/10.
^{70m2}Cu			6.6 s	β/93		1+	+1.9		0.8849/100
									1.072/19
				IT/7					0.141/ IT
^{70m1}Cu			33. s	β- /52	2.52/10.	3-	-3.5		0.8848(2)/100.
				IT/48					0.9017(2)/90.
									1.2517(5)/60.
									(0.39–3.06)
^{70}Cu		69.932392	44.5 s	β- /6.60	5.42/54.	6-	+1.5		0.8848(2)/100.
					6.09/46.				0.9017/99.7
									(0.438 - 3.062)
^{71m}Cu			0.28 μs	I.T.		(19/2)			0.133/IT
									0.494
									0.939
									1.189
^{71}Cu		70.932677	20. s	β- /4.56		3/2-	+2.28		0.490/
^{72m}Cu			1.76 μs	I.T.		(4-)			0.051/IT
									0.082
									0.138
^{72}Cu		71.935820	6.6 s	β- /8.2		(1+)			0.652/
^{73}Cu		72.936675	4.2 s	β- /6.3	5.8/43				0.450/100
					6.25/42				0.307–1.559
^{74}Cu		73.93988	1.59 s	β- /9.9					
^{75}Cu		74.942	1.2 s	β- /7.9					
^{76m}Cu			1.2 s						
^{76}Cu		75.94528	0.64 s	β- /11.					

Elem. or Isot.	Natural Abundance (Atom %)	Atomic Mass or Weight	Half-life/ Resonance Width (MeV)	Decay Mode/ Energy (/MeV)	Particle Energy/ Intensity (MeV/%)	Spin $(h/2\pi)$	Nuclear Magnetic Mom. (nm)	Elect. Quadr. Mom. (b)	γ-Energy / Intensity (MeV/%)
^{77}Cu		76.9479	0.46 s	β- / ~ 10.					
^{78}Cu		77.9520	0.33 s	β- /12.					
^{79}Cu		78.9546	0.19 s	β- /11.					
^{80}Cu		79.961	> 0.3 μs						
$_{30}$**Zn**		**69.38(2)**							
^{54}Zn		53.9930	~ 3.2 ms	2p	p//87	0+			
^{55}Zn		54.9840	20. ms		p//91.				
^{56}Zn		55.9724	30. ms		p//86.	0+			
^{57}Zn		56.9648	47. ms	β+, p/14.6		(7/2-)			ann.rad./
^{58}Zn		57.95459	0.09 s	β+		0+			
^{59}Zn		58.94926	183. ms	β+, p/9.09	8.1/	3/2-			ann.rad./
									(0.491–0.914)
^{60}Zn		59.94183	2.40 m	β+ /97/4.16		0+			ann.rad./
				EC/3/					0.669/47.
									(0.062–0.947)
^{61}Zn		60.93951	1.485 m	β+ /5.64	4.38/68.	3/2-			ann.rad./
									0.4748/17.
									(0.15–3.52)
^{62}Zn		61.93433	9.22 h	β+ /3/1.63	0.66/7.	0+			ann.rad./
				EC/93/					0.0408/25
									0.5967/26.
									(0.20–1.526)/
^{63}Zn		62.933212	38.5 m	β+ /93/3.367	1.02/	3/2-	-0.28164	+0.29	ann.rad./
				EC/7/	1.40/				0.66962(5)/8.4
					1.71/				0.96206(5)/6.6
					2.36/84.				(0.24–3.1)
^{64}Zn	49.17(75)	63.929142	>4.3 × 10^{18} a	EC-β+		0+			
^{65}Zn		64.929241	244. d	β+ /98/1.3514	0.325/	5/2-	+0.7690	-0.023	ann.rad./
				EC/1.5/					1.1155/50.2
^{66}Zn	27.73(98)	65.926033				0+			
^{67}Zn	4.04(16)	66.927127				5/2-	+0.8753	+0.15	
^{68}Zn	18.45(63)	67.924844				0+			
^{69m}Zn			13.76 h	I.T./99+/0.439		9/2+	1.157		0.4390(2)/95.
^{69}Zn		68.926550	56. m	β- /0.906	0.905/99.9	½-			0.318/
^{70}Zn	0.61(10)	69.925319	>1.3 × 10^{16} a	β-β-		0+			
^{71m}Zn			3.97 h	β- /	1.45/	9/2+	1.05		0.3864/93.
									0.4874/62.
									0.6203/57.
									(0.099–2.489)
^{71}Zn		70.92772	2.4 m	β- /2.81		½-			0.5116(1)/30.
									0.9103(1)/7.5
									(0.12–2.29)
^{72}Zn		71.92686	46.5 h	β- /0.46	0.25/14.	0+			0.0164(3)/8.
					0.30/86.				0.1447(1)/83.
									0.1915(2)/9.4
^{73m}Zn			6. s		I.T./0.196	(7/2+)			0.042
^{73}Zn		72.92978	24. s	β- /4.29	4.7/	(1/2-)			0.216(1)/100.
									0.496–0.911
^{74}Zn		73.92946	1.60 m	β- /2.3	2.1/	0+			0.0565/
									0.1401/
									(0.05–0.35)
^{75}Zn		74.9329	10.2 s	β- /6.0					0.229/
^{76}Zn		75.9333	5.7 s	β- /4.2	3.6/	0+			0.119/
^{77m}Zn			1.0 s	β- /		(1/2-)			0.772

Elem. or Isot.	Natural Abundance (Atom %)	Atomic Mass or Weight	Half-life/ Resonance Width (MeV)	Decay Mode/ Energy (/MeV)	Particle Energy/ Intensity (MeV/%)	Spin (*h*/2 π)	Nuclear Magnetic Mom. (nm)	Elect. Quadr. Mom. (b)	γ-Energy / Intensity (MeV/%)
^{77}Zn		76.9370	2.1 s	β- /7.3	4.8/				0.189/
^{78m}Zn			> 0.03 ms						1.070
^{78}Zn		77.9384	1.5 s	β- /6.4		0+			0.225/
^{79}Zn		78.9427	1.0 s	β- /8.6					0.702/
^{80}Zn		79.9443	0.54 s	β- /7.3		0+			0.713/
									0.2248/
^{81}Zn		80.9505	0.29 s	β- /11.9					
^{82}Zn		81.9544	> 0.15 μs			0+			
^{83}Zn		82.9610	> 0.15 μs						
$_{31}$**Ga**		**69.723(1)**							
^{56}Ga		55.9949							
^{57}Ga		56.9829							
^{58}Ga		57.9743							
^{59}Ga		58.9634	< 0.043 μs						
^{60}Ga		59.9571	0.07 s	β+					1.004
				β+, p	// ~ 1.6				3.848
				β+, α	// ~ 0.02				1.555–2.559
^{61}Ga		60.9495	0.17 s	β+ /9.0		3/2			0.088–1.362
^{62}Ga		61.94418	116.0 ms	β+ /9.17	8.3/	0+			ann.rad./
				EC/					0.954/0.0012
^{63}Ga		62.939294	32. s	β+ /5.5	4.5/				ann.rad./
				EC/					0.6271(2)/10.
									0.6370(2)/11.
									1.0652(4)/45.
^{64m}Ga			0.022 ms						0.0429
^{64}Ga		63.936839	2.63 m	β+ /7.165	2.79/	0+			ann.rad./
					6.05/				0.80785(1)/14.
									0.99152(1)/43.
									1.38727(1)/12.
									3.3659(1)/13.
^{65}Ga		64.932735	15.2 m	β+ /86/3.255	0.82/10.	3/2-			ann.rad./
				EC/	1.39/19.				0.1151(2)/55.
					2.113/56.				0.1530(2)/96.
					2.237/15.				0.2069(2)/39.
									(0.06–2.4)
^{66}Ga		65.931589	9.5 h	β+ /56/5.175	0.74/1.	0+			ann.rad./
				EC/43/	1.84/54.				1.03935(8)/38.
					4.153/51.				2.7523(1)/23.
									(0.28–5.01)
^{67}Ga		66.928202	3.261 d	EC/1.001		3/2-	+1.8507	0.20	0.09332/37.
									0.18459/20.
									0.30024/17.
									(0.091–0.89)
^{68}Ga		67.927980	1.130 h	β+ /90/2.921	1.83/	1+	0.01175	0.028	ann.rad./
				EC/10/					1.0774(1)/3.
									(0.57–2.33)/
^{69}Ga	60.108(9)	68.925574				3/2-	+2.01659	+0.165	
^{70}Ga		69.926022	21.1 m	EC/0.2/0.655		1+			0.1755(5)/0.15
				β- /99.8/1.656	1.65/99.				1.042(5)/0.48
^{71}Ga	39.892(9)	70.924701	>2.4 × 10^{26} a	β-		3/2-	+2.56227	+0.104	
^{72}Ga		71.926366	14.10 h	β- /4.001	0.64/40.	3-	-0.13224	+0.52	0.8340/95.53
					1.51/9.				2.202/26.9
					2.52/8.				0.630/26.2
					3.15/11.				(0.113–3.678)

Elem. or Isot.	Natural Abundance (Atom %)	Atomic Mass or Weight	Half-life/ Resonance Width (MeV)	Decay Mode/ Energy (/MeV)	Particle Energy/ Intensity (MeV/%)	Spin (h/2 π)	Nuclear Magnetic Mom. (nm)	Elect. Quadr. Mom. (b)	γ-Energy / Intensity (MeV/%)
^{73}Ga		72.925175	74.87 h	β- /1.59		3/2-			0.05344(5)/10.
									0.29732(5)/47.
									(0.01–1.00)/
^{74m}Ga			10. s	I.T./		1+			0.0565(1)/75.
^{74}Ga		73.926946	8.1 m	β- /5.4	2.6/	3-			0.5959/92.
									2.354/45.
									(0.23–3.99)
^{75}Ga		74.926500	2.10 m	β- /3.39	3.3/	3/2-			0.2529/
									0.5746/
									(0.12–2.10)
^{76}Ga		75.928828	29. s	β- /7.0		3-			0.5629/66.
									0.5455/26.
									(0.34–4.25)
^{77}Ga		76.929154	13.0 s	β- /5.3	5.2/				0.469/
									0.459/
^{78}Ga		77.931608	5.09 s	β- /8.2		3+			0.619/77.
									1.187/20.
^{79}Ga		78.9329	2.85 s	β- /7.0	4.6/				0.465/
^{80}Ga		79.9365	1.68 s	β- /10.4	10./				0.659/
^{81}Ga		80.9378	1.22 s	β- /8.3	5.1/				0.217/
^{82}Ga		81.9430	0.599 s	β- /12.6					1.348/
^{83}Ga		82.9470	0.308 s	β- /~ 11.5					
^{84}Ga		83.9527	~ 0.085 s	β- /14					
^{85}Ga		84.9570	> 0.3 μs						
^{86}Ga		85.963	> 0.3 μs						
$_{32}$**Ge**		**72.64(1)**							
^{58}Ge		57.9910				0+			
^{59}Ge		58.9818							
^{60}Ge		59.9702	> 0.11 μs			0+			
^{61}Ge		60.9638	~ 44. ms	β+ /13.6					
^{62}Ge		61.9547	0.13 s			0+			
^{63}Ge		62.9496	0.15 s	β- /9.8					
^{64}Ge		63.94165	1.06 m	β+ /4.4	3.0/	0+			ann.rad./
				EC/					0.1282(2)/11.
				β+, p					0.4270(3)/37.
									0.6671(3)/17.
^{65}Ge		64.9394	31. s	β+ /6.2	0.82/10.				ann.rad./
				EC/	1.39/19.				0.0620/27.
				EC, p	2.113/56.				0.6497/33.
					2.237/15.				0.8091/21.
				β+, p	//0.011				(0.19–3.28)
^{66}Ge		65.93384	2.26 h	β+ /27/2.10		0+			ann.rad./
				EC/73/					0.0438/29.
									0.3819/28.
									(0.022–1.77)
^{67}Ge		66.932734	19.0 m	β+ /96/4.225	1.6/	½-			ann.rad./
				EC/4/	2.3/				0.1670/84.
					3.15/				(0.25–3.73)
^{68}Ge		67.92809	270.8 d	EC/0.11		0+			Ga k x-ray/39.
^{69}Ge		68.927965	1.63 d	β+ /36/2.2273	0.70/	5/2-	0.735	0.02	ann.rad./
				EC/64/	1.2/				0.574/13.
									1.1068/36.
									(0.2–2.04)
^{70}Ge	20.38(18)	69.924247				0+			

Elem. or Isot.	Natural Abundance (Atom %)	Atomic Mass or Weight	Half-life/ Resonance Width (MeV)	Decay Mode/ Energy (/MeV)	Particle Energy/ Intensity (MeV/%)	Spin ($h/2\pi$)	Nuclear Magnetic Mom. (nm)	Elect. Quadr. Mom. (b)	γ-Energy / Intensity (MeV/%)
^{71m}Ge			20.4 ms		I.T./0.0234	9/2+	-1.041	~ 0.34	0.1749
^{71}Ge		70.924951	11.2 d	EC/0.229		½-	+0.547		
^{72}Ge	27.31(26)	71.922076				0+			
^{73}Ge	7.76(8)	72.923459	>1.8 × 10^{23} a	β-		9/2+	-0.879468	-0.17	
^{74}Ge	36.72(15)	73.921178				0+			
^{75m}Ge			48. s	I.T./		7/2+			0.13968(3)/39.
^{75}Ge		74.922859	1.380 h	β- /1.177	1.19/	½-	+0.510		0.26461(5)/11.
									0.41931(5)/0.2
^{76}Ge	7.83(7)	75.921403	1.6 × 10^{21} a	β-β-		0+			
^{77m}Ge			53. s	I.T./20/		½-			1.605/0.22
				β- /80/2.861	2.9/				1.676/0.16
									0.195–1.482
^{77}Ge		76.923549	11.25 h	β- /2.702	0.71/23.	7/2+			0.2110/29.
					1.38/35.				0.2155/27.
					2.19/42.				0.2644/51.
									(0.15–2.35)
^{78}Ge		77.922853	1.45 h	β- /0.95	0.70/	0+			0.2773(5)/96.
									0.2939(5)/4.
^{79m}Ge			39. s	β- /IT		7/2+			
^{79}Ge		78.9254	19.1 s	β- /4.2	4.0/20.	½-			0.1096/21.
					4.3/80.				(0.10–2.59)
									0.5427(4)/15.
^{80}Ge		79.92537	29.5 s	β- /2.67	2.4/	0+			0.1104(4)/6.
									0.2656(4)/25.
^{81m}Ge			~ 7.6 s	β- /	3.75/	½+			0.3362(4)/
									0.7935(4)/
^{81}Ge		80.9288	~ 7.6 s	β- /6.2	3.44/	9/2+			0.1976(4)/21.
									0.3362(4)/100.
^{82}Ge		81.9296	4.6 s	β- /4.7	1.093/80	0+			1.093/
^{83}Ge		82.9346	1.9 s	β- /8.9					
^{84}Ge		83.9375	0.98 s	β- /7.7		0+			
^{85}Ge		84.9430	0.54 s	β- /10.					
^{86}Ge		85.9465	> 0.3 μs			0+			
^{87}Ge		86.9525	> 0.3 μs						
^{88}Ge		87.957	> 0.3 μs			0+			
^{89}Ge		88.964	> 0.3 μs						
$_{33}$**As**		**74.92160(2)**							
^{60}As		59.993							
^{61}As		60.981							
^{62}As		61.9732							
^{63}As		62.9637	< 0.043 μs						
^{64}As		63.9576	0.02 s						
^{65}As		64.9496	0.13 s	β+ /9.4					
^{66m2}As			8. μs						
^{66m1}As			1.1 μs						
^{66}As		65.945	95.8 ms	β+ /9.55					
^{67}As		66.9392	42. s	β+ /6.0	5.0/	5/2-			0.121/
				EC/					0.123/
									0.244/
^{68}As		67.93677	2.53 m	β+ /8.1		3+			ann.rad./
									0.652/32.
									0.762/33.
									1.016/77.
									(0.61–3.55)

Elem. or Isot.	Natural Abundance (Atom %)	Atomic Mass or Weight	Half-life/ Resonance Width (MeV)	Decay Mode/ Energy (/MeV)	Particle Energy/ Intensity (MeV/%)	Spin ($h/2\pi$)	Nuclear Magnetic Mom. (nm)	Elect. Quadr. Mom. (b)	γ-Energy / Intensity (MeV/%)
^{69}As		68.93227	15.2 m	β+ /98/4.01	2.95/	5/2-	+ 1.623		ann.rad./
				EC/2/					0.0868(5)/1.5
									0.1458(3)/2.4
^{70}As		69.93092	52.6 m	β+ /84/6.22	1.44/	4+	+2.1061	+0.09	ann.rad./
				EC/16/2.14					1.0395(7)/82.
				/2.89					(0.17–4.4)/
^{71}As		70.927112	2.72 d	β+ /32/2.013		5/2-	+1.6735	-0.02	ann.rad./
				EC/68/					0.1749(2)/84.
									1.0957(2)/4.2
^{72}As		71.926752	26.0 h	β+ /77/4.356	0.669/5.	2-	-2.1566	-0.08	ann.rad./
					1.884/12.				0.83395(5)/80.
					2.498/62.				1.0507(1)/9.6
					3.339/19.				(0.1–4.0)
^{73}As		72.923825	80.3 d	EC/0.341		3/2-			0.0133/0.1
									0.0534/10.5
									Se k x-ray/90.
^{74}As		73.923829	17.78 d	β+ /31/2.562	0.94/26.	2-	-1.597		ann.rad./
				EC/37/	1.53/3.				0.59588(1)/60.
				β- /1.353	0.71/16.				0.6084(1)/0.6
					1.35/16.				0.6348(1)/15.
^{75m}As			0.017 s						
^{75}As	100.	74.921597				3/2-	+1.43947	+0.31	
^{76}As		75.922394	26.3 h	β- /2.962	0.54/3.	2-	-0.903		0.5591(1)/45.
					1.785/8.				0.65703(5)/6.2
					2.410/36.				1.21602(1)/3.4
					2.97/51.				(0.3–2.67)
^{77}As		76.920647	38.8 h	β- /0.683	0.70/98.	3/2-	+1.295		0.2391(2)/1.6
									0.2500(3)/0.4
									0.5208/0.43
^{78}As		77.92183	1.512 h	β- /4.21	3.00/12.	2-			0.6136(3)/54.
					3.70/17.				0.6954(3)/18.
					4.42/37.				1.3088(3)/10.
^{79m}As			1.21 μs	I.T.		9/2+			0.542/IT
									0.231
^{79}As		78.92095	9.0 m	β- /2.28	1.80/95.	3/2-			0.0955(5)/16.
									0.3645(5)/1.9
^{80}As		79.92253	16. s	β- /5.64	3.38/	1+			0.6662(2)/42.
									(2.5–3.0)
^{81}As		80.92213	33. s	β- /3.856		3/2-			0.4676(2)/20.
									0.4911(2)/8.
^{82m}As			13.7 s	β- /	3.6/	5-			0.6544(1)/77.
									0.344/65.
									(0.561 – 1.894)
^{82}As		81.9245	19. s	β- /7.4	7.2/80.	(2-)			0.6544(1)/54.
									(0.755 – 3.667)
^{83}As		82.9250	13.4 s	β- /5.5					0.7345/100.
									1.1131/34.
									2.0767/28.
^{84m}As			0.6 s	β-					
^{84}As		83.9291	4. s	β-, n/7.2		1-			0.6671(2)/21.
									1.4439(5)/49.
									(0.325–5.150)
^{85}As		84.9320	2.03 s	β-, n/8.9		3/2-			0.667(1)/42.

Elem. or Isot.	Natural Abundance (Atom %)	Atomic Mass or Weight	Half-life/ Resonance Width (MeV)	Decay Mode/ Energy (/MeV)	Particle Energy/ Intensity (MeV/%)	Spin ($h/2\pi$)	Nuclear Magnetic Mom. (nm)	Elect. Quadr. Mom. (b)	γ-Energy / Intensity (MeV/%)
									1.4551(2)/100.
^{86}As		85.9365	0.95 s	β-, n/11.4					0.704/
^{87}As		86.9399	0.49 s	β-, n/10.					0.704/
^{88}As		87.9449	> 0.3 μs						
^{89}As		88.9494	> 0.3 μs						
^{90}As		89.956	> 0.3 μs						
^{91}As		90.960	> 0.3 μs						
^{92}As		91.967	> 0.3 μs						
$_{34}$Se		**78.96(3)**							
^{64}Se			> 0.18 μs			0+			
^{65}Se		64.965	0.011 s	β+ /60/14.					
				β+, p	3.55/				
^{66}Se		65.9552	0.03 s			0+			
^{67}Se		66.9501	0.13 s	β+ /10.2					ann.rad./
				β+, (p)/					0.352
^{68}Se		67.94180	36. s	β+ /4.7		0+			ann.rad./
									(0.050–0.426)
^{69}Se		68.93956	27.4 s	β+ /6.78	5.006/				ann.rad./
				EC/					0.0664(4)/27.
				β+, p	// ~ 0.045				0.0982(4)/63.
^{70}Se		69.9334	41.1 m	β+ /2.4		0+			ann.rad
									0.04951(5)/35.
									0.4262(2)/29.
^{71}Se		70.93224	4.7 m	β+ /4.4	3.4/36.	5/2-			ann.rad
				EC/					0.1472(3)/47.
									0.8309(3)/13.
									1.0960(3)/10.
^{72}Se		71.92711	8.5 d	EC/0.34		0+			0.0460(2)/57.
^{73m}Se			40. m	I.T./73/0.0257	0.85	3/2-			ann.rad.
				β+ /27/2.77	1.45/				0.0257(2)/27.
					1.70/				0.2538(1)/2.5
^{73}Se		72.92677	7.1 h	β+ /65/2.74	0.80/	9/2+	0.86		ann.rad
				EC/35/	1.32/95.				0.0670(1)/72.
					1.68/1.				0.3609(1)/97.
									(0.6–1.5)
^{74}Se	0.89(4)	73.922476	> 5.5x10^{18} a	EC-EC		0+			
^{75}Se		74.922523	119.78 d	EC/0.864		5/2+	0.68	1.0	0.13600/55
									0.26465/58
									(0.024–0.821)
^{76}Se	9.37(29)	75.919214				0+			
^{77m}Se			17.4 s	I.T./		7/2+			0.1619(2)/52.
^{77}Se	7.63(16)	76.919914				½-	+0.53506		
^{78}Se	23.77(28)	77.917309				0+			
^{79m}Se			3.92 m	I.T./					0.09573(3)/9.5
^{79}Se		78.918499	3.6 × 10^{5} a	β- /0.151		7/2+	-1.02	+0.8	
^{80}Se	49.61(41)	79.916521				0+			
^{81m}Se			57.3 m	I.T./99/0.1031		7/2+			0.1031(3)/9.7
									0.2602(2)/0.06
									0.2760/0.06
^{81}Se		80.917993	18.5 m	β- /1.585	1.6/98.	½-			0.2759/0.85
									0.2901/0.75
									0.8283/0.32
^{82}Se	8.73(22)	81.916699	>9.5 × 10^{19} a	β-β-		0+			
^{83m}Se			1.17 m	β- /3.96	2.88/	½-			0.35666(6)/17.

Elem. or Isot.	Natural Abundance (Atom %)	Atomic Mass or Weight	Half-life/ Resonance Width (MeV)	Decay Mode/ Energy (/MeV)	Particle Energy/ Intensity (MeV/%)	Spin ($h/2\pi$)	Nuclear Magnetic Mom. (nm)	Elect. Quadr. Mom. (b)	γ-Energy / Intensity (MeV/%)
					3.92/				0.9879(1)/15.
									1.0305(1)/21.
									2.0514(2)/11.
									(0.19–3.1)
^{83}Se		82.919118	22.3 m	β- /3.668	0.93/	9/2+			0.22516(6)/33.
					1.51/				0.35666(6)/69.
									0.51004(8)/45.
									(0.21–2.42)
^{84}Se		83.91846	3.3 m	β- /1.83	1.41/100.	0+			0.4088(5)/100.
^{85}Se		84.92225	32. s	β- /6.18	5.9/	5/2+			0.3450(1)/22.
									0.6094(1)/41.
^{86}Se		85.92427	15. s	β- /5.10		0+			2.0124(1)/24.
									2.4433(8)/100.
									2.6619(1)/49.
^{87}Se		86.92852	5.4 s	β- /7.28		5/2+			0.468(1)/100.
				n/					1.4979(1)/23.
^{88}Se		87.93142	1.5 s	β-, n/6.85		0+			0.5346/
^{89}Se		88.9365	0.41 s	β-, n/9.0					
^{90}Se		89.9400	> 0.3 μs			0+			
^{91}Se		90.9460	0.27 s	β-, n/8.					
^{92}Se		91.950	> 0.3 μs			0+			
^{93}Se		92.956	> 0.3 μs						
^{94}Se		93.960	> 0.3 μs			0+			
$_{35}$Br		**79.904(1)**							
^{67}Br		66.9648							
^{68}Br		67.9585	< 1.5 μs						
^{69}Br		68.9501	< 0.024 μs	β+ /9.6					
^{70m}Br			2.2 s			9+			
^{70}Br		69.9446	~ 0.08 s	β+ /10.0	/0.75				
^{71}Br		70.939	21. s	β+ /6.9					
^{72}Br		71.9366	1.31 m	β+ /8.7		3	0.6		0.4547–1.317
^{73}Br		72.93169	3.4 m	β+ /4.7	3.7/	3/2-			ann.rad
									0.065–0.700
^{74m}Br			46. m	β+ /	4.5/	4-	1.82		ann.rad
									0.6348
									0.7285
									(0.2–4.38)
^{74}Br		73.92989	25.4 m	β+ /6.91					ann.rad
									0.6341
									0.6348
									(0.2–4.7)
^{75}Br		74.92578	1.62 h	β+ /76/3.03		3/2-	+0.75		ann.rad
									0.28650
									(0.1–1.56)
^{76m}Br			1.4 s	I.T./5.05		4+			0.104548
									0.05711
^{76}Br		75.92454	16.0 h	β+ /57/4.96	1.9/	1-	0.54821	0.270	ann.rad
					3.68/				0.55911
									1.85368
									(0.4–4.6)
^{77m}Br			4.3 m	I.T./0.1059		9/2+			0.1059
^{77}Br		76.921379	2.376 d	EC/99/1.365		3/2-	0.973	+0.53	ann.rad.
									0.23898
									0.52069

Elem. or Isot.	Natural Abundance (Atom %)	Atomic Mass or Weight	Half-life/ Resonance Width (MeV)	Decay Mode/ Energy (/MeV)	Particle Energy/ Intensity (MeV/%)	Spin (*h*/2 π)	Nuclear Magnetic Mom. (nm)	Elect. Quadr. Mom. (b)	γ-Energy / Intensity (MeV/%)
									(0.08–1.2)
^{78}Br		77.921146	6.45 m	β+ /92/3.574	1.2/	1+	0.13		ann.rad.
				EC/8/	2.5/				0.61363
									(0.7–3.0)
^{79m}Br			4.86 s	I.T./0.207		9/2+			0.2072
^{79}Br	50.69(7)	78.918337				3/2-	+2.106400	+0.31	
^{80m}Br			4.42 h	I.T./0.04885		5-	+1.3177	+0.70	Br k x-ray
									0.03705/39.1
									0.04885/0.3
^{80}Br		79.918529	17.66 m	β- /92/2.004	1.38 β-/7.6	1+	0.5140	+0.18	ann.rad.
				EC/5.7/1.8706	1.99 β-/82				0.6169/6.7
				β+ /2.6/	0.85 β+ /2.8				(0.64–1.45)
^{81}Br	49.31(7)	80.916291				3/2-	+2.270562	+0.26	
^{82m}Br			6.1 m	I.T./98/0.046		2-			0.046/0.24
				β- /2 /3.139					(0.62–2.66)
^{82}Br		81.916804	1.471 d	β- /3.093	0.444/	5-	+1.6270	0.69	0.5544/71
									0.61905/43
									0.77649/84
									(0.013–1.96)
^{83}Br		82.915180	2.40 h	β- /0.972	0.395/1	3/2-			0.52964
					0.925/99				(0.12–0.68)
^{84m}Br			6.0 m	β- /4.97	2.2/100	(6-)			0.4240/100
									0.8817/98
									1.4637/101
^{84}Br		83.91648	31.8 m	β- /4.65	2.70/11	2-	2.		0.8816/41
					3.81/20				1.8976/13
					4.63/34				(0.23–4.12)
^{85}Br		84.91561	2.87 m	β- /2.87	2.57	3/2-			0.80241/2.56
									0.92463/1.6
									(0.09–2.4)
^{86}Br		85.91880	55.5 s	β- /7.63	3.3	(2-)			1.56460/64
					7.4				2.75106/21
									(0.5–6.8)
^{87}Br		86.92071	55.6 s	β- /6.85	6.1/	3/2-			1.41983/55.
				n/					1.578/45.
									(0.173–1.76)
^{88m}Br			5.1 μs						
^{88}Br		87.92407	16.3 s	β- /8.96		1-			0.7649
				n/					0.7753
									0.8021
									(0.1–6.99)
^{89}Br		88.92640	4.35 s	β- /8.16		3/2-			0.7753
				n/					1.0978
^{90}Br		89.9306	1.91 s	β- /10.4	8.3/	2-			0.6555
				n/	9.8/				0.7071
									1.3626
^{91}Br		90.9340	0.54 s	β- /90 /9.80					0.263
				β- n/10 /					0.803
^{92}Br		91.93926	0.31 s	β- /12.20					0.740
				β- n/					
^{93}Br		92.9431	0.10 s	β- /11					0.117
				β- n	//11				(0.237–3.606)
^{94}Br		93.9487	0.07 s	β- n/					
^{95}Br		94.9529	> 0.3 μs						

Elem. or Isot.	Natural Abundance (Atom %)	Atomic Mass or Weight	Half-life/ Resonance Width (MeV)	Decay Mode/ Energy (/MeV)	Particle Energy/ Intensity (MeV/%)	Spin ($h/2\pi$)	Nuclear Magnetic Mom. (nm)	Elect. Quadr. Mom. (b)	γ-Energy / Intensity (MeV/%)
^{96}Br		95.959	> 0.3 μs						
^{97}Br		96.963	> 0.3 μs						
$_{36}$Kr		**83.798(2)**							
^{69}Kr		68.9652	0.03 s	β+, (p)	4.07/				
^{70}Kr		69.9553	0.06 s			0+			
^{71}Kr		70.950	100. ms	β+, EC/10.1					(0.198–0.207)
^{72}Kr		71.94209	17.1 s	β+ /5.0		0+			ann.rad
				EC/					0.3099/15.3
									0.4150/12.8
									(0.305 – 3.305)
^{73}Kr		72.93929	28. s	β+ /6.7		5/2-			ann.rad.
				EC/					0.1781/66
				β+, p/	/0.25				(0.06–0.86)
^{74}Kr		73.933084	11.5 m	β+ /3.1		0+			ann.rad.
				EC/					0.08970/31
									0.2030/20
									(0.010–1.06)
^{75}Kr		74.93095	4.3 m	β+ /4.90	3.2/	5/2+	-0.531	+1.1	ann.rad.
				EC/					0.1325/68
									0.1547/21
									(0.02–1.7)
^{76}Kr		75.925910	14.8 h	EC/1.31		0+			Br k x-ray
									0.270/21
									0.3158/39
									(0.03–1.07)
^{77}Kr		76.924670	1.24 h	β+ /80/3.06		5/2+	-0.583	+0.9	ann.rad.
				EC/20/	1.55/				0.1297/80
					1.70/				0.1465/38
					1.87/				(0.02–2.3)
^{78}Kr	0.355(3)	77.920365	>1.5 × 10^{21} a	EC-EC		0+			
^{79m}Kr			53. s	I.T./0.1299		7/2+	-0.786	+0.40	Kr x-ray
^{79}Kr		78.920082	1.455 d	β+ /7 /1.626		½-	+0.536		ann.rad.
				EC/93 /					0.2613/13
									0.39756/19
									0.6061/8
									(0.04–1.3)
^{80}Kr	2.286(10)	79.916379				0+			
^{81m}Kr			13.1 s	I.T./0.1904		½-	+0.586		0.1904
^{81}Kr		80.916592	2.1 × 10^5 a	EC/0.2807		7/2+	-0.908	+0.644	Br k x-ray
									0.2760
^{82}Kr	11.593(31)	81.913484				0+			
^{83m}Kr			1.86 h	I.T./0.0416		½-	+0.591		Kr k x-ray
									0.00940
									0.03215/0.055
^{83}Kr	11.500(19)	82.914136				9/2+	-0.970669	+0.259	
^{84}Kr	56.987(15)	83.911507				0+			
^{85m}Kr			4.48 h	β- /79 /	0.83/79	½-	+ 0.633		0.30487
				I.T./21 /0.305					0.15118
^{85}Kr		84.912527	10.73 a	β- /0.687	0.15/0.4	9/2+	1.005	+0.443	0.51399
^{86}Kr	17.279(41)	85.9106107				0+			
^{87}Kr		86.9133549	1.27 h	β- /3.887	1.33/8	5/2+	-1.023	-0.30	0.40258/49.6
					3.49/43				2.5548/9.2
					3.89/30				(0.13–3.31)
^{88}Kr		87.91445	2.84 h	β- /2.91		0+			0.19632/26.

Elem. or Isot.	Natural Abundance (Atom %)	Atomic Mass or Weight	Half-life/ Resonance Width (MeV)	Decay Mode/ Energy (/MeV)	Particle Energy/ Intensity (MeV/%)	Spin ($h/2\pi$)	Nuclear Magnetic Mom. (nm)	Elect. Quadr. Mom. (b)	γ-Energy / Intensity (MeV/%)
									2.392/34.6
									(0.03–2.8)
^{89}Kr		88.9176	3.15 m	β- /4.99	3.8/	5/2+	-0.330	+0.16	0.19746
					4.6/				0.2209/19.9
					4.9/				0.5858/16.4
									1.4728/6.8
									(0.2–4.7)
^{90}Kr		89.91952	32.3 s	β- /4.39	2.6/77	0+			0.12182/32.9
					2.8/6				0.5395/28.6
									1.1187/36.2
									(0.1–4.2)
^{91}Kr		90.9235	8.6 s	β- /6.4	4.33/	5/2+	-0.583	+0.30	0.10878/43.5
					4.59/				0.50658/19.
									(0.2–4.4)
^{92}Kr		91.92616	1.84 s	β- /5.99		0+			0.1424/66.
				n/					(0.14–3.7)
^{93}Kr		92.9313	1.29 s	β- /8.6	7.1/	½+	-0.413		0.1820
				n/					0.2534/42.
									0.32309/24.6
									(0.057–4.03)
^{94}Kr		93.9344	0.21 s	β- /7.3		0+			0.2196/67
				n	n//1.0				0.6293/100.
									(0.098–0.985)
^{95}Kr		94.9398	0.10 s	β- /9.7	n//2.9		- 0.410		
^{96}Kr		95.9431	~ 80 ms	β- ,n	n//3.7	0+			
^{97}Kr		96.9486	0.06 s	β- ,n	n//7.				
^{98}Kr		97.952	0.05 s	β- ,n	n//7.	0+			
^{99}Kr		98.958	0.04 s	β- ,n	n// ~ 11.				
^{100}Kr		99.9611	> 0.34 μs			0+			
$_{37}$**Rb**		**85.4678(3)**							
^{71}Rb		70.9653							
^{72}Rb		71.9591	< 1.5 μs						
^{73}Rb		72.9506	< 0.03 μs						
^{74}Rb		73.944265	64.8 ms	β+ /10.4					0.456/0.0025
									(0.053 – 4.244)
^{75}Rb		74.93857	19. s	β+ /7.02	2.31/				ann.rad
									0.179
^{76}Rb		75.935072	39. s	β+ /8.50	4.7/	1-	-0.372623	+0.4	ann.rad
									0.4240/92.
									(0.064–1.68)
^{77}Rb		76.93041	3.8 m	β+ /5.34	3.86/	3/2-	+0.654468	+0.70	ann.rad
									0.0665/59
									(0.04–2.82)
^{78m}Rb			5.7 m	I.T./0.1034		4-	+2.549	+0.81	ann.rad
				β+ /	3.4				0.4553/81.
				EC/					(0.103–4.01)
^{78}Rb		77.92814	17.7 m	β+ /7.22		0+			ann.rad
				EC/					0.4553/63.
									(0.42–5.57)
^{79}Rb		78.92399	23. m	β+ /84/3.65		5/2+	+0.3358	-0.10	ann.rad.
				EC/16 /					0.68812/23.
									(0.017–3.02)
^{80}Rb		79.92252	34. s	β+ /5.72	4.1/22	1+	-0.0836	+0.35	ann.rad.

Elem. or Isot.	Natural Abundance (Atom %)	Atomic Mass or Weight	Half-life/ Resonance Width (MeV)	Decay Mode/ Energy (/MeV)	Particle Energy/ Intensity (MeV/%)	Spin (h/2 π)	Nuclear Magnetic Mom. (nm)	Elect. Quadr. Mom. (b)	γ-Energy / Intensity (MeV/%)
					4.7/74				0.6167/25.
^{81m}Rb			30.5 m	I.T./0.85	1.4	9/2+	+5.598	-0.74	ann.rad.
				β+, EC/					(0.085–1.9)
^{81}Rb		80.91900	4.57 h	β+ /27/2.24	1.05/	3/2-	+2.060	+0.40	ann.rad./
				EC/73					0.19030/64.
									(0.05–1.9)
^{82m}Rb			6.47 h	β+/26/	0.80/	5-	+1.51001	+1.0	ann.rad./
				EC/74/					0.5544/63.
									0.7765/85.
									(0.092–2.3)
^{82}Rb		81.918209	1.258 m	β+/96/4.40	3.3/	1+	+0.554508	+0.19	ann.rad./
				EC/4/					0.7665/13.
									(0.47–3.96)
^{83}Rb		82.91511	86.2 d	EC/0.91		5/2-	+1.425	+0.20	Kr x-ray
									0.5205/46.
									(0.03–0.80)
^{84m}Rb			20.3 m	I.T./0.216		6-	+0.21293	+0.6	0.2163/34.
									0.2482/63.
									0.4645/32.
^{84}Rb		83.914385	32.9 d	β+/22/2.681	0.780/11	2-	-1.32412	-0.015	ann.rad./
				EC/75 /	1.658/11				0.8817/68.
				β-/3/0.894	0.893/				(1.02–1.9)
^{85}Rb	72.17(2)	84.91178974				5/2-	+1.353	+0.28	
^{86m}Rb			1.018 m	I.T./0.5560		6-	+1.815	+0.37	0.556/98.
^{86}Rb		85.9111674	18.65 d	β-/1.775	1.774/8.8	2-	-1.6920	+0.19	1.0768/8.8
^{87}Rb	27.83(2)	86.90918053	4.88 × 10^{10} a	β-/0.283	0.273/100	3/2-	+2.7512	+0.13	
^{88}Rb		87.9113156	17.7 m	β-/5.316	5.31	2-	0.508		0.8980/14.4
									1.8360/22.8
									(0.34–4.85)
^{89}Rb		88.91228	15.4 m	β-/4.50	1.26/38	3/2-	+2.38	+0.14	1.032/58.
					1.9/5				1.248/42.
					2.2/34				2.1960/13
					4.49/18				(0.12–4.09)
^{90m}Rb			4.3 m	β-/4.50	1.7/	4-	+1.616	+0.20	0.1069(IT)
					6.5/				0.8317/94
									(0.20–5.00)
^{90}Rb		89.91480	2.6 m	β-/6.59	6.6	1-			0.8317/28.
									(0.31–5.60)
^{91}Rb		90.91654	58.0 s	β-/5.861	5.9	3/2-	+2.182	+0.15	0.0936/34.
									(0.35–4.70)
^{92}Rb		91.91073	4.48 s	β-/8.11	8.1/94	1-			0.8148/8.
									(0.1–6.1)
^{93}Rb		92.92204	5.85 s	β-/7.46	7.4/	5/2-	+1.410	+0.18	0.2134/4.8
				n/1					0.4326/12.5
									0.9861/4.9
									(0.16–5.41)
^{94}Rb		93.92641	2.71 s	β-/10.31	9.5/	3	+1.498	+0.16	0.8369/87.
				n/10					1.5775/32.
									(0.12–6.35)
^{95}Rb		94.92930	0.377 s	β-/9.30	8.6/	5/2-	+1.334	+0.21	0.352/65.
				n/8					0.680/22.
									(0.20–2.27)
^{96m}Rb			1.7 μs						0.2999
									0.4612

Elem. or Isot.	Natural Abundance (Atom %)	Atomic Mass or Weight	Half-life/ Resonance Width (MeV)	Decay Mode/ Energy (/MeV)	Particle Energy/ Intensity (MeV/%)	Spin ($h/2\pi$)	Nuclear Magnetic Mom. (nm)	Elect. Quadr. Mom. (b)	γ-Energy / Intensity (MeV/%)
									0.2400
									0.093–0.369
^{96}Rb		95.93427	0.199 s	β-/11.76	10.8/	2+	+1.466	+0.25	0.815/76.
				n/13/					(0.20–5.42)
^{97}Rb		96.93735	0.169 s	β-/10.42	10.0	3/2+	+1.841	+0.58	0.167/100.
				n/27/					0.585/79.
									0.599/56.
									1.258/52.
									(0.14–2.08)
^{98}Rb		97.94179	0.107 s	β-/12.34	0.144/				
				n/13					(0.07–3.68)
^{99}Rb		98.9454	59. ms	β-/11.3					
^{100}Rb		99.9499	53. ms	β- /13.5					0.129
									(0.058–4.483)
^{101}Rb		100.9532	0.03 s	β- /11.8					
^{102}Rb		101.9589	0.09 s	β-					
$_{38}$Sr		**87.62(1)**							
^{73}Sr		72.966	> 25 ms						
^{74}Sr		73.9563	> 1.5 μs			0+			
^{75}Sr		74.9499	88. ms	β+ ,p	p//5.				0.144/4.5
^{76}Sr		75.94177	7.9 s	β+ /6.1		0+			
^{77}Sr		76.93795	9.0 s	β+ /6.9	5.6		-0.348	+1.4	0.147
				β+, p	//0.08				
^{78}Sr		77.93218	2.7 m	β+ /3.76		0+			(0.047–0.793)
^{79}Sr		78.92971	2.1 m	β+ /5.32	4.1	3/2-	-0.474	+0.71	ann.rad./
									0.039/28.
									0.105/22.
									(0.135–0.612)
^{80}Sr		79.92452	1.77 h	β+ /1.87		0+			ann.rad./
									0.174/10.
									0.589/39.
									(0.24–0.55)
^{81}Sr		80.92321	22.3 m	β+ /87/3.93	2.43/	1/2-	+0.544		ann.rad./
				EC/13/	2.68/				0.148/31.
									0.1534/35
									(0.06–1.7)
^{82}Sr		81.91840	25.36 d	EC/0.18		0+			Rb x-ray
^{83m}Sr			5.0 s	I.T./0.2591		½-	+0.582		0.2591/87.5
^{83}Sr		82.91756	1.350 d	β+/24/2.28	0.465/	7/2+	-0.8298	+0.76	ann.rad./
				EC/76/	0.803/				0.3816/12.
					1.227/				0.3816
									0.7627/30.
									(0.094–2.15)
^{84}Sr	0.56(1)	83.913425				0+			
^{85m}Sr			1.127 h	I.T./87/0.2387		½-	+0.600		0.2318/84.
				EC/13					(0.15–0.24)
^{85}Sr		84.912933	64.85 d	EC/1.065		9/2+	-1.001	+0.28	0.51399/99.3
^{86}Sr	9.86(1)	85.909260				0+			
^{87m}Sr			2.81 h	I.T./0.3884		½-	+0.63		0.3884(IT)
^{87}Sr	7.00(1)	86.908877				9/2+	-1.0936	+0.305	
^{88}Sr	82.58(1)	87.905612				0+			
^{89}Sr		88.907451	50.6 d	β-/1.497	1.492/100	5/2+	-1.148	-0.3	0.9092
^{90}Sr		89.907738	29.1 a	β-/0.546	0.546/100	0+			
^{91}Sr		90.910203	9.5 h	β-/2.70	0.61/7	5/2+	-0.885	+0.05	0.5556/61.

Elem. or Isot.	Natural Abundance (Atom %)	Atomic Mass or Weight	Half-life/ Resonance Width (MeV)	Decay Mode/ Energy (/MeV)	Particle Energy/ Intensity (MeV/%)	Spin ($h/2\pi$)	Nuclear Magnetic Mom. (nm)	Elect. Quadr. Mom. (b)	γ-Energy / Intensity (MeV/%)
					1.09/33				0.7498/24.
					1.36/29				1.0243/33.
					2.66/26				(0.12–2.4)
^{92}Sr		91.911038	2.64 h	β-/1.91	0.55/96	0+			1.3831/90.
					1.5/3				(0.24–1.1)
^{93}Sr		92.91403	7.4 m	β-/4.08	2.2/10	5/2+	-0.793	+0.26	0.5903/
					2.6/25				0.7104
					3.2/65				0.87573
									0.8883/
									(0.17–3.97)
^{94}Sr		93.91536	1.25 m	β-/3.511	2.1/	0+			0.6219
					3.3/				0.7043
									0.7241
									0.8064
									1.4283
^{95}Sr		94.91936	25.1 s	β-/6.08		½+	-0.537		0.6859
					6.1/50				0.8269
									2.7173
									2.9332
^{96}Sr		95.92170	1.06 s	β-/5.37	4.2/	0+			0.1222
									0.5305
									0.8094
									0.9318
^{97}Sr		96.92615	0.42 s	β-/7.47	5.3	(1/2+)	-0.498		0.2164
									0.3071
									0.6522
									0.9538
									1.2580
									1.9050
^{98}Sr		97.92845	0.65 s	β-/5.83	5.1	0+			0.0365
									0.1190
									0.4286
									0.4447
									0.5636
^{99}Sr		98.9332	0.27 s	β-/8.0			-0.26	0.8	
^{100}Sr		99.9354	0.201 s	β-/7.1		0+			
^{101}Sr		100.9405	0.115 s	β-/9.5					
^{102}Sr		101.9430	68. ms	β-/8.8		0+			
^{103}Sr		102.9490	> 0.3 μs						
^{104}Sr		103.952	> 0.3 μs			0+			
^{105}Sr		104.959	> 0.3 μs						
$_{39}$Y		**88.90585(2)**							
^{76}Y		75.9585	> 0.2 μs						
^{77}Y		76.9497	~ 57. ms						
^{78m}Y			5.8 s			(5+)			
^{78}Y		77.9436	53 ms	β+/10.5					0.279/100
									0.504/90
									0.713/40
^{79}Y		78.9374	15. s	β+/7.1					(0.152–1.106)
^{80m}Y			4.8 s						0.2285
^{80}Y		79.9343	30. s	β+/7.0	5.5	(4-)			ann.rad./
					5.0/				0.3858/100
									0.5951/42
									0.756–1.396

Elem. or Isot.	Natural Abundance (Atom %)	Atomic Mass or Weight	Half-life/ Resonance Width (MeV)	Decay Mode/ Energy (/MeV)	Particle Energy/ Intensity (MeV/%)	Spin ($h/2\pi$)	Nuclear Magnetic Mom. (nm)	Elect. Quadr. Mom. (b)	γ-Energy / Intensity (MeV/%)
^{81}Y		80.9291	1.21 m	β+/5.5	3.7/				ann.rad./
					4.2/				0.428
									0.469
^{82}Y		81.9268	9.5 s	β+/7.8	6.3/	1+			ann.rad./
									0.5736
									0.6017
									0.7375
^{83m}Y			2.85 m	β+/95/4.6	2.9	1/2-			ann.rad./
				EC/5 /					0.2591
									0.4218
									0.4945
^{83}Y		82.92235	7.1 m	β+/4.47	3.3	9/2+			ann.rad./
				EC/					0.0355
									0.4899
									0.8821
									(0.03–3.4)
^{84m}Y			4.6 s	β+/		1+			ann.rad./
				EC/					0.7930
^{84}Y		83.9204	40. m	β+/6.4	1.64/47	5-			ann.rad./
				EC/	2.24/25				0.4628
					2.64/21				0.6606
					3.15/7				0.7931
									0.9744
									1.0398
									(0.2–3.3)
^{85m}Y			4.9 h	β+/70/		9/2+	6.2		ann.rad./
				EC/30/					0.2317
									0.5356
									2.1238
									(0.1–3.1)
									0.7673
^{85}Y		84.91643	2.6 h	β+/55/3.26	1.54/	1/2-			ann.rad./
				EC/45/					0.2317
									0.5045
									0.9140
									(0.07–1.4)
^{86m}Y			48. m	I.T./99/		8+	4.8		ann.rad./
				β+/					0.0102(IT)
				EC/					0.2080
									(0.09–1.1)
^{86}Y		85.91489	14.74 h	β+/5.24		4-	<0.6		ann.rad./
				EC/					0.3070
									0.6277
									1.0766
									1.1531
									1.9207
									(0.1–3.8)
^{87m}Y			13. h	I.T./98/		9/2+	+ 6.24	- 0.5	0.3807
				β+/0.7/	1.15/0.7				
				EC/					
^{87}Y		86.910876	3.35 d	EC/99+/1.862	0.78/	1/2-	- 0.19		0.3880
									0.4870
^{88}Y		87.909501	106.6 d	EC/99+/3.623	0.76/	4-	- 0.42	+ 0.16	ann.rad./
				β+/0.2/					0.89802

Elem. or Isot.	Natural Abundance (Atom %)	Atomic Mass or Weight	Half-life/ Resonance Width (MeV)	Decay Mode/ Energy (/MeV)	Particle Energy/ Intensity (MeV/%)	Spin ($h/2\pi$)	Nuclear Magnetic Mom. (nm)	Elect. Quadr. Mom. (b)	γ-Energy / Intensity (MeV/%)
									1.83601
									2.73404
									3.2190
^{89m}Y			15.7 s	I.T./0.909		9/2+	+ 6.37	~ - 0.43	0.9092(IT)
^{89}Y	100.	88.905848				1/2-	-0.13742		
^{90m}Y			3.24 h	I.T./99+/	0.68204	7+	+ 5.28	~ - 0.65	0.2025
				β-/0.002/					0.4794
									0.6820
^{90}Y		89.907152	2.669 d	β-/2.282	2.28/	2-	-1.63	-0.155	
^{91m}Y			49.7 m	I.T./0.555		9/2+	5.96		0.5556(IT)
^{91}Y		90.907305	58.5 d	β-/1.544	1.545/	1/2-	0.164		1.208
^{92}Y		91.90895	3.54 h	β-/3.63	3.64/	2-	- 0.67		0.4485
									0.5611
									0.9345
									1.4054
									(0.4–3.3)
^{93m}Y			0.82 s	I.T./0.759		9/2+	+ 6.04	~ - 0.64	0.1686(IT)
									0.5902
^{93}Y		92.90958	10.2 h	β-/2.87	2.88/90	1/2-	- 0.12		0.2669
									0.9471
									1.9178
^{94m}Y			1.4 μs						0.4322
									0.7699
									1.2024
^{94}Y		93.91160	18.7 m	β-/4.919	4.92/	2-	- 0.24	~ 0.03	0.3816
									0.9188
									1.1389
									(0.3–4.1)
^{95}Y		94.91282	10.3 m	β-/4.42		1/2-	- 0.16		0.4324
									0.9542
									2.1760
									3.5770
^{96m}Y			9.6 s	β-/		(3+)			0.1467
									0.6174
									0.9150
									1.1071
									1.7507
^{96}Y		95.91589	6.2 s	β- /7.09	7.12/	0-			1.594
^{97m}Y			1.21 s	β- /7.4	4.8/	9/2+	+ 5.88	~ -0.76	0.1614
					6.0/				0.9700
									1.1030
^{97}Y		96.91813	3.76 s	β- /6.69	6.7	1/2-	- 0.12		0.2969
									1.9960
									3.2876
									3.4013
^{98m}Y			2.1 s	β- /9.8	5.5/	(4-)	+ 2.98	+ 1.7	0.2415
									0.6205
									0.6473
									1.2228
									1.8016
^{98}Y		97.92220	0.59 s	β- /8.83	8.7/	1+			0.2131
									1.2228
									1.5907
									2.9413

Elem. or Isot.	Natural Abundance (Atom %)	Atomic Mass or Weight	Half-life/ Resonance Width (MeV)	Decay Mode/ Energy (/MeV)	Particle Energy/ Intensity (MeV/%)	Spin (h/2 π)	Nuclear Magnetic Mom. (nm)	Elect. Quadr. Mom. (b)	γ-Energy / Intensity (MeV/%)
									4.4501
^{99m}Y			0.011 ms						
^{99}Y		98.92464	1.47 s	β- /7.57		1/2-			0.1218/43.8
				n	/2.5/				0.5362
									0.7242
									1.0130
^{100m}Y			0.94 s	β-, n /		3+			
^{100}Y		99.9278	0.73 s	β-, n/9.3	n/1.8/	1+			
^{101}Y		100.9303	0.43 s	β-, n/8.6	n/1.5/	(5/2+)	+ 3.22	- 1.5	
^{102}Y		101.9336	0.36 s	β-, n/9.9	n/4.0/				
^{103}Y		102.9367	0.23 s	β-, n	n/8.3/				
^{104}Y		103.9411	0.18 s						
^{105}Y		104.9449	> 0.15 μs						
^{106}Y		105.950	> 0.15 μs						
^{107}Y		106.9414	> 0.15 μs						
^{108}Y		107.959	> 0.15 μs						
$_{40}$Zr		**91.224(2)**							
^{78}Zr		77.9552	> 0.2 μs			0+			
^{79}Zr		78.9492	0.06 s						
^{80}Zr		79.940	~ 4.5 s	β+ /8.0		0+			0.290
									0.538
^{81}Zr		80.9372	5.3 s	β+ /7.2	6.1	(3/2-)			
^{82}Zr		81.9311	32. s	β+ /4.0	3.	0+			ann.rad./
^{83m}Zr			7. s	β+ /7.0		(7/2+)			ann.rad./
^{83}Zr		82.9287	44. s	β+ /5.9	4.8	(1/2-)			ann.rad./
				EC					0.0556
									0.1050
									0.2560
									0.474
									1.525
^{84}Zr		83.9233	26. m	β+ /2.7		0+			ann.rad./
				EC/					0.0449
									0.1125
									0.3729
									0.667
^{85m}Zr			10.9 s	I.T./0.2922		½-			ann.rad./
				β+, EC/					0.2922(IT)
									0.4165
^{85}Zr		84.9215	7.9 m	β+ /4.7	3.1	7/2+			ann.rad./
				EC/					0.2663
									0.4163
									0.4543
^{86}Zr		85.91647	16.5 h	EC/1.47		0+			0.0280
									0.243
									0.612
^{87m}Zr			14.0 s	I.T./0.3362		½-	+ 0.64		0.1352(IT)
									0.2010
^{87}Zr		86.91482	1.73 h	β+ /3.67	2.26	9/2+	- 0.895	+ 0.42	ann.rad./
				EC/					0.3811
									1.228
^{88m}Zr			1.4 μs			(8+)	- 1.81	+ 0.51	0.077
^{88}Zr		87.91023	83.4 d	EC/0.67		0+			0.3929

Elem. or Isot.	Natural Abundance (Atom %)	Atomic Mass or Weight	Half-life/ Resonance Width (MeV)	Decay Mode/ Energy (/MeV)	Particle Energy/ Intensity (MeV/%)	Spin ($h/2\pi$)	Nuclear Magnetic Mom. (nm)	Elect. Quadr. Mom. (b)	γ-Energy / Intensity (MeV/%)
^{89m}Zr			4.18 m	I.T./94/0.5877		½-	+ 0.80		ann.rad./
				β+ /1.5/					0.5877(IT)
				EC/4.7/					1.507
^{89}Zr		88.908889	3.27 d	β+ /23/2.832	0.9/	9/2+	-1.05	+ 0.3	ann.rad./
				EC/77/					0.9092
^{90m}Zr			0.809 s	I.T./		5-	6.3		0.1326
									2.1862
									2.3189(IT)
^{90}Zr	51.45(40)	89.904704				0+			
^{91}Zr	11.22(5)	90.905646				5/2+	-1.30362	-0.176	
^{92}Zr	17.15(8)	91.905041				0+			
^{93}Zr		92.906476	1.5 × 10^{6} a	β- /0.091		5/2+			0.0304
^{94}Zr	17.38(28)	93.906315	>10^{17} a	β-β-		0+			
^{95}Zr		94.908043	64.02 d	β- /1.125	0.366/55	5/2+	1.13	+0.22	0.7242
					0.400/44				0.7567
^{96}Zr	2.80(9)	95.908273	3 × 10^{19} a	β-β-		0+			
			>1.7 × 10^{18} a	β-					
^{97}Zr		96.910953	16.75 h	β- /2.658	1.91/	½-	- 0.937		0.7434
^{98}Zr		97.91274	30.7 s	β- /2.26	2.2/100	0+			
^{99}Zr		98.91651	2.2 s	β- /4.56	3.9/	½+	- 0.930		0.4692/55.2
					3.5/				0.5459/48
									0.028–1.321
^{100}Zr		99.91776	7.1 s	β- /3.34		0+			0.4006/20
									0.5043/30
^{101}Zr		100.92114	2.1 s	β- /5.49	6.2/	3/2-	- 0.27	+ 0.81	0.1194
									0.2057
									0.2089
^{102}Zr		101.92298	2.9 s	β- /4.61		0+			0.064/15
									0.599/14
									0.535/11
^{103}Zr		102.9266	1.3 s	β- /7.0					
^{104}Zr		103.9288	1.2 s	β- /5.9		0+			
^{105}Zr		104.9331	~ 1. s	β- /8.5					
^{106}Zr		105.9359	> 0.24 μs			0+			
^{107}Zr		106.9408	> 0.24 μs						
^{108}Zr		107.944	> 0.15 μs			0+			
^{109}Zr		108.9492	> 0.15 μs						
^{110}Zr		109.953	> 0.15 μs			0+			
$_{41}$Nb		**92.90638(2)**							
^{81}Nb		80.949	<0.08 μs						
^{82}Nb		81.9431	50 ms	β+ /11.					
^{83}Nb		82.9367	4.1 s	β+ /7.5					
^{84}Nb		83.9336	10. s	β+, EC/9.6		(3+)			0.540
									(0.456-1.427)
^{85m}Nb			3. s						0.069
^{85}Nb		84.9279	21. s	β+ /6.0					
^{86m}Nb			56. s	β+					
^{86}Nb		85.9250	1.46 m	β+ /8.0					ann.rad./
									0.751
									1.003
^{87m}Nb			3.7 m	β+ /		1/2-			ann.rad./
				EC/					0.1352
									0.2010
^{87}Nb		86.92036	2.6 m	β+ /5.2/		(9/2+)			ann.rad./

Elem. or Isot.	Natural Abundance (Atom %)	Atomic Mass or Weight	Half-life/ Resonance Width (MeV)	Decay Mode/ Energy (/MeV)	Particle Energy/ Intensity (MeV/%)	Spin (h/2 π)	Nuclear Magnetic Mom. (nm)	Elect. Quadr. Mom. (b)	γ-Energy / Intensity (MeV/%)
				EC/					0.2010
									0.4706
									0.6165
									1.0665
									1.8842
^{88m}Nb			7.7 m	β+ /		4-			ann.rad./
				EC/					0.2625
									0.3996
									1.0569
									1.0825
^{88}Nb		87.9183	14.3 m	β+ /7.6	3.2/	8+			ann.rad./
				EC/					1.0570
									1.0828
									(0.07–2.5)
^{89m}Nb			2.0 h	β+ /	3.3/	9/2+			0.5880/10(D)
				EC/					(0.17–4.0)
^{89}Nb		88.91342	1.10 h	β+ /74/4.29	2.8/	1/2-	+6.216		ann.rad./
				EC/26 /					0.5074
									0.5880
									0.7696
									1.2775
^{90m}Nb			18.8 s	I.T./0.1246		4-			0.002
									0.1225
^{90}Nb		89.911265	14.6 h	β/53 /6.111	0.86/5	8+	4.961	+ 0.05	ann.rad./
				EC/47 /	1.5/92				0.1412
									1.1292
									2.1862
									2.3189
									(0.1–3.3)
^{91m}Nb			62. d	I.T./97 /		1/2-			0.1045(IT)
				EC/3 /					1.2050
^{91}Nb		90.906996	7 × 10^{2} a	EC/1.253		9/2+			Mo k x-ray
^{92m}Nb			10.13 d	EC/99+/		2+	6.14		0.9126
									0.9345
									1.8475
^{92}Nb		91.907194	3.7 × 10^{7} a	EC/2.006		7+			0.5611
									0.9345
^{93m}Nb			16.1 a	I.T./0.0304		1/2-			Nb x-ray
									0.0304
^{93}Nb	100.	92.906378				9/2+	+6.1705	-0.32	
^{94m}Nb			6.26 m	I.T./99+	/2.086	3+			Nb k x-ray
				β- /0.5/					0.0409
									0.87109
^{94}Nb		93.907284	2.4 × 10^{4} a	β- /2.045	0.47/	6+			0.70263
									0.87109
^{95m}Nb			3.61 d	I.T./97.5/	0.2357	1/2-			0.2040
				β- /2.5 /					0.2356
^{95}Nb		94.906836	34.97 d	β- /0.926	0.160/	9/2+	6.14		0.76578
^{96}Nb		95.908101	23.4 h	β- /3.187	0.5/10	6+	4.976		0.7782
					0.75/90				0.2191–1.498
^{97m}Nb			58.1 s	I.T./0.7434	0.734/98	1/2-			0.7434
^{97}Nb		96.908099	1.23 h	β- /1.934	1.27/98	9/2+	6.15		0.4809
									0.6579
^{98m}Nb			51. m	β- /4.67		5+			0.7874

Elem. or Isot.	Natural Abundance (Atom %)	Atomic Mass or Weight	Half-life/ Resonance Width (MeV)	Decay Mode/ Energy (/MeV)	Particle Energy/ Intensity (MeV/%)	Spin (*h*/2 π)	Nuclear Magnetic Mom. (nm)	Elect. Quadr. Mom. (b)	γ-Energy / Intensity (MeV/%)
									0.1726–1.89
^{98}Nb		97.91033	2.9 s	β- /4.59	4.6/	1+			0.6451
									0.7874
									1.0243
^{99m}Nb			2.6 m	β- /	3.2/	1/2-			0.0978/100
									(0.138–3.010)
^{99}Nb		98.91162	15.0 s	β- /3.64	3.5/100	9/2+			0.0977
									0.1378/3.1
^{100m2}Nb			0.013 ms						
^{100m1}Nb			3.0 s	β- /6.74	5.8				Nb k x-ray
									0.159
									0.6364
									1.0637
^{100}Nb		99.91418	1.5 s	β- /6.25	6.2/				0.5354
					5.3/				0.6001–1.566
^{101}Nb		100.91525	7.1 s	β- /4.57	4.3/				0.1105–0.810
^{102m}Nb			4.3 s	β- /					
^{102}Nb		101.91804	1.3 s	β- /7.21	7.2/				0.2960–2.184
^{103}Nb		102.9191	1.5 s	β- /5.53	5.3/	5/2+			
^{104m}Nb			0.9 s	β-, n/	n/0.06				
^{104}Nb		103.9225	4.8 s	β-, n/8.1	n/0.05				
^{105}Nb		104.9239	3.0 s	β-, n/6.5	n/1.7				
^{106}Nb		105.9280	1.0 s	β-, n/9.3	n/4.5				
^{107}Nb		106.9303	0.30 s	β-, n/7.9	n/6.0				
^{108}Nb		107.9348	0.19 s	β, n/	n/6.2				(0.193–0.590)
^{109}Nb		108.9376	0.19 s	β, n/	n/31				
^{110}Nb		109.9424	0.17 s	β, n/	n/40				
^{111}Nb		110.9457	> 0.15 μs						
^{112}Nb		111.951	> 0.15 μs						
^{113}Nb		112.955	> 0.15 μs						
$_{42}$Mo		**95.96(2)**							
^{83}Mo		82.9487	~ 6. ms						
^{84}Mo		83.9401	~ 3.7 s	β+ /6.		0+			
^{85}Mo		84.9366	3.2 s	β+/8.1		½+			
^{86}Mo		85.9307	19. s	β+ /4.8		0+			
^{87}Mo		86.9273	14. s	EC, β+/6.5					(0.752–1.004)
^{88}Mo		87.92195	8.0 m	β+ /3.4		0+	+0.5		ann.rad./
				EC					0.0800
									0.1399
									0.1707
^{89m}Mo			0.19 s	I.T./0.118		½-			0.118(IT)
									0.268
^{89}Mo		88.91948	2.2 m	β+ /5.58		9/2+			ann.rad./
				EC/					0.659
									0.803
									1.155
									1.272
^{90m}Mo			1.2 μs				- 1.39	0.58	0.063
^{90}Mo		89.91394	5.7 h	β+ /25/2.489	1.085/	0+			ann.rad./
				EC/75 /					0.04274
									0.12237
									0.25734
^{91m}Mo			1.08 m	I.T./50/0.653		½-			ann.rad./
				β+, EC/50 /	2.5/				0.6529

Elem. or Isot.	Natural Abundance (Atom %)	Atomic Mass or Weight	Half-life/ Resonance Width (MeV)	Decay Mode/ Energy (/MeV)	Particle Energy/ Intensity (MeV/%)	Spin ($h/2\pi$)	Nuclear Magnetic Mom. (nm)	Elect. Quadr. Mom. (b)	γ-Energy / Intensity (MeV/%)
					2.8/				1.2081
					4.0/				1.5080
									2.2407
^{91}Mo		90.91175	15.5 m	β+ /94/4.43	3.44/94	9/2-			ann.rad./
				EC/6/					1.6373
									2.6321
									3.0286
									(0.1–4.2)
^{92}Mo	14.53(31)	91.906811	> 3 × 10^{17} a	β+-EC		0+			
^{93m}Mo			6.9 h	I.T./99+ /2.425		21/2+	+9.9		0.26306(IT)
									0.68461
									1.47711
^{93}Mo		92.906813	3.5 × 10^3 a	EC/0.405		5/2+			0.0304
^{94}Mo	9.15(9)	93.905088				0+			
^{95}Mo	15.84(11)	94.905842				5/2+	-0.9142	-0.02	
^{96}Mo	16.67(15)	95.904680				0+			
^{97}Mo	9.60(14)	96.906022				5/2+	-0.9335	+0.26	
^{98}Mo	24.39(37)	97.905408				0+			
^{99}Mo		98.907712	2.7476 d	β- /1.357	0.45/14	½+	0.375		0.144048
					0.84/2				0.18109
					1.21/84				0.36644
									0.73947
^{100}Mo	9.82(31)	99.90748	6 × 10^{20} a	β β		0+			
^{101}Mo		100.91035	14.6 m	β- /2.82	2.23/	1/2+			0.0063
					0.7/				0.19193
									0.5909
									(0.0809–2.405)
^{102}Mo		101.91030	11.3 m	β- /1.01	1.2/	0+			0.1493/89.
									0.2116/100.
									0.2243/32.
^{103}Mo		102.9132	1.13 m	β- /3.8		3/2+			0.1028(2)/
									0.1440(2)
									0.2511(2)
^{104}Mo		103.9138	1.00 m	β- /2.16		0+			0.0686(1)/100.
									0.4239(4)/21.
^{105}Mo		104.9170	36. s	β- /4.95		3/2+			0.0642/
									0.0856/
									0.2495/
^{106}Mo		105.91814	8.4 s	β- /3.52		0+			0.1894(2)/22.
									0.3644(2)/6.
									0.3723(2)/12.
^{107}Mo		106.9217	3.5 s	β- /6.2					
^{108}Mo		107.9235	1.1 s	β- /5.1		0+			(0.028–0.636)
^{109}Mo		108.9278	0.5 s	β- /7.2					
^{110}Mo		109.9297	0.27 s	β- /5.7		0+			Tc k x-ray
									0.142
									(0.039–0.599)
^{111}Mo		110.9344	> 0.15 μs						
^{112}Mo		111.937	> 0.15 μs			0+			
^{113}Mo		112.942	> 0.15 μs						
^{114}Mo		113.945	> 0.15 μs			0+			
^{115}Mo		114.950	> 0.15 μs						
^{116}Mo			> 0.15 μs			0+			
^{117}Mo			> 0.15 μs						

Elem. or Isot.	Natural Abundance (Atom %)	Atomic Mass or Weight	Half-life/ Resonance Width (MeV)	Decay Mode/ Energy (/MeV)	Particle Energy/ Intensity (MeV/%)	Spin ($h/2\pi$)	Nuclear Magnetic Mom. (nm)	Elect. Quadr. Mom. (b)	γ-Energy / Intensity (MeV/%)
$_{43}$Tc									
^{85}Tc		84.9488	< 0.1 ms						
^{86}Tc		85.9429	0.05 s	β+ /11.9					
^{87}Tc		86.9365	2.4 s	β+ /8.6					
^{88}Tc		87.9327	5.8 s	β+ /10.1					
^{89m}Tc			13. s						
^{89}Tc		88.9272	13. s	β+ /7.5					
^{90m}Tc			49.2 s	β$^+$	5.3/	6+			ann.rad./
									0.9479/
									1.0542/
^{90}Tc		89.9236	8.3 s	β+ /8.9	7.0/15	1+			ann.rad./
					7.9/95.				0.9479/
^{91m}Tc			3.3 m	β+		½+			ann.rad./170.
				EC					0.8110(5)/5.
									1.6052(1)/7.8
									1.6339(1)/9.1
									1.9023(1)/6.
									2.4509(1)/13.5
^{91}Tc		90.9184	3.14 m	β+ /6.2	5.2	9/2+			ann.rad./200.
^{92}Tc		91.91526	4.4 m	β+ /7.87	4.1	8+			ann.rad./200.
				EC					0.0850/
									0.1475
									0.3293
									0.7731
									1.5096
^{93m}Tc			43. m	I.T./13		½-			0.3924(IT)
				EC/20					0.9437
									2.6445
^{93}Tc		92.910249	2.73 h	β+ /13/3.201	0.81	9/2+	6.3		ann.rad./
				EC/87/					1.3629
									1.4771
									1.5203
									(0.1–3.0)
^{94m}Tc			52. m	β+ /72/4.33		2+			ann.rad./
				EC/28/					0.8710
									1.8686
^{94}Tc		93.909657	4.88 h	β+ /11/4.256		7+	5.12		ann.rad./
				EC/89/					0.4491
									0.7026
									0.8496
									0.8710
^{95m}Tc			62.0 d	I.T./4/		1/2-			ann.rad./
				β+ /0.3	0.5/				0.0389(IT)
				EC/96	0.7/				0.2041
									0.5821
									0.5821
									0.8351
^{95}Tc		94.90766	20.0 h	EC/100/1.691		9/2+	5.9		0.7657
									1.0738
^{96m}Tc			52. m	I.T./90/		4+			0.0342(IT)
				β+, EC/2/					0.7782
									1.2002
^{96}Tc		95.90787	4.3 d	EC/2.973		7+	+5.1		Mo k x-ray
									0.7782

Elem. or Isot.	Natural Abundance (Atom %)	Atomic Mass or Weight	Half-life/ Resonance Width (MeV)	Decay Mode/ Energy (/MeV)	Particle Energy/ Intensity (MeV/%)	Spin (h/2 π)	Nuclear Magnetic Mom. (nm)	Elect. Quadr. Mom. (b)	γ-Energy / Intensity (MeV/%)
									0.8125
									0.8498
									1.12168
^{97m}Tc			91. d	I.T./0.0965		1/2-			Tc k x-ray
				EC	/3.9				0.0965
^{97}Tc		96.906365	4.2 × 10^{6} a	EC/100/0.320		9/2+			Mo k x-ray
^{98}Tc		97.907216	~ 6.6 × 10^{6} a	β- /1.80	0.40/100	6+			0.65241/100
				EC	//<0.036				0.74535/100
^{99m}Tc			6.01 h	I.T./100/0.142		1/2-			Tc k x-ray
									0.14049
									0.14261
^{99}Tc		98.906255	2.13 × 10^{5} a	β- /0.294	0.293/100	9/2+	+5.6847	-0.129	
^{100}Tc		99.907658	15.8 s	β- /3.202	2.2/	1+			0.5396
				EC /1.8(10)$^{-3}$/	2.9/				0.5908
					3.3				(0.3 79–2.30)
^{101}Tc		100.90732	14.2 m	β- /1.61	1.32/	9/2+			0.1272
									0.1841
									0.3068
									0.5451
									(0.073–0.969)
^{102m}Tc			4.4 m	I.T./2/4.8	1.8/				0.4184
				β- /98/					0.4752
									0.6281
									0.6302
									1.0464
									1.1033
									1.6163
									2.2447
^{102}Tc		101.90922	5.3 s	β- /4.53	3.4/	1+			0.4686
					4.2				0.4751
					2.2/				1.1055
^{103}Tc		102.90918	54. s	β- /2.66	2.0/	5/2+			0.1361
					2.2/				0.1743
									0.2104
									0.3464
									0.5629
									(0.13–1.0)
^{104m}Tc			0.005 ms						
^{104}Tc		103.91145	18.2 m	β- /5.60	5.3/	(3+)			0.3483
									0.3580
									0.5305
									0.5351
									0.8844
									0.8931
									1.6768
									(0.3–3.7)
^{105}Tc		104.9117	7.6 m	β- /3.6	3.4/	5/2+			0.1079
									0.1432
									0.3215
^{106}Tc		105.91436	36. s	β- /6.55		2+			0.2703
									0.5222
									1.9694
									2.2393
									2.7893

Elem. or Isot.	Natural Abundance (Atom %)	Atomic Mass or Weight	Half-life/ Resonance Width (MeV)	Decay Mode/ Energy (/MeV)	Particle Energy/ Intensity (MeV/%)	Spin ($h/2\pi$)	Nuclear Magnetic Mom. (nm)	Elect. Quadr. Mom. (b)	γ-Energy / Intensity (MeV/%)
^{107}Tc		106.9151	21.2 s	β- /4.8					0.1027
									0.1063
									0.1770
									0.4587
^{108}Tc		107.9185	5.1 s	β- /7.72		(3)			0.2422
									0.4656
									0.7078
									0.7326
									1.5835
^{109}Tc		108.9200	1.4 s	β- /6.3	p/0.08				
^{110}Tc		109.9238	0.83 s	β- /8.8	p/0.04				0.2407
^{111}Tc		110.9257	0.30 s	β- .n/7.0	n/0.85				0.150/92.7
									0.063–1.435
^{112}Tc		111.9292	0.26 s	β, n	n/2.6				
^{113}Tc		112.9316	0.15 s	β-, n/8.	/2.1				0.0985/100
									0.0658–1.520
^{114}Tc		113.936	0.15 s	β-, n	/1.3				
^{115}Tc		114.939	0.07 s						
^{116}Tc		115.943	> 0.15 μs						
^{117}Tc		116.946	> 0.15 μs						
^{118}Tc		117.951	> 0.15 μs						
$_{44}$Ru		**101.07(2)**							
^{87}Ru		86.949	> 1.5 μs						
^{88}Ru		87.9403	1.2 s			0^+			
^{89}Ru		88.9361	1.4 s	β+ .p/8.					
^{90}Ru		89.9299	12. s	β+ /5.9		0^+			ann.rad./
									0.155–1.551
^{91}Ru		90.9263	7.9 s	β+, EC/7.4		9/2+			ann.rad./
									(0.205-1.998)
^{92}Ru		91.9201	3.7 m	β+ /53/4.5		0+			ann.rad./
				EC/47/					0.1346
									0.2138
									0.2593
^{93m}Ru			10.8 s	I.T./21/		1/2-			ann.rad./
				β+, EC/79/	5.3/				0.7344
									1.1112
									1.3962
									2.0931
^{93}Ru		92.9171	1.0 m	β+ /6.3		9/2+			ann.rad./
				EC/					0.6807
									1.4349
									(0.5–4.2)weak
^{94}Ru		93.91136	52. m	EC/100/1.59		0+			0.3672
									0.5247
									0.8922
^{95}Ru		94.91041	1.64 h	EC/85/2.57	1.20/	5/2+	0.86		ann.rad./
				β+ /15/	0.91/				0.3364/71
									1.097/16.4
									0.6268/15.9
									0.036–2.424
^{96}Ru	5.54(14)	95.90760	>3.1 × 10^{16} a	β+β+		0+			
^{97}Ru		96.90756	2.89 d	EC/1.12		5/2+	-0.79		Tc k x-ray
									0.2157
									0.3245

Elem. or Isot.	Natural Abundance (Atom %)	Atomic Mass or Weight	Half-life/ Resonance Width (MeV)	Decay Mode/ Energy (/MeV)	Particle Energy/ Intensity (MeV/%)	Spin (h/2 π)	Nuclear Magnetic Mom. (nm)	Elect. Quadr. Mom. (b)	γ-Energy / Intensity (MeV/%)
									0.4606
^{98}Ru	1.87(3)	97.90529				0+			
^{99}Ru	12.76(14)	98.905939				5/2+	-0.64	+0.079	
^{100}Ru	12.60(7)	99.904220				0+			
^{101}Ru	17.06(2)	100.905582				5/2+	-0.72	+0.46	
^{102}Ru	31.55(14)	101.904349				0+			
^{103}Ru		102.906324	39.27 d	β- /0.763	0.223	3/2+	0.206	+0.62	0.05329
									0.29498
									0.4438
									0.49708
									0.55704
									0.61033
									(0.04–1.6)
^{104}Ru	18.62(27)	103.905433				0+			
^{105}Ru		104.907753	4.44 h	β- /1.917	1.11/22	3/2+	-0.3		0.12968
					1.134/13				0.1491
					1.187/49				0.2629
									0.31664
									0.46943
									0.67634
									0.72420
									(0.1–1.8)
^{106}Ru		105.90733	1.020 a	β- /0.0394	0.0394/100	0+			
^{107}Ru		106.9099	3.8 m	β- /2.9	2.1/				0.1939
					3.2/				0.3741
									0.4625
									0.8488
^{108}Ru		107.9102	4.5 m	β- /1.4	1.2/	0^+			0.0923
									0.1651
									0.4339
									0.4975
									0.6189
^{109}Ru		108.9132	34.5 s	β- /4.2					0.1164
									0.3584
^{110}Ru		109.9141	15. s	β- /2.81		0+			0.1121
									0.3737
									0.4397
									0.7967
^{111}Ru		110.9177	1.5 s	β- /5.5					
^{112}Ru		111.9190	4.5 s	β- /4.5		0+			
^{113m}Ru			0.6 s						
^{113}Ru		112.9225	0.80 s	β- /7.					0.2632
									0.048–2.418
^{114}Ru		113.9243	0.57 s	β- /6.1		0+			0.127/24
									(0.053–0.180)
^{115}Ru		114.9287	~ 0.74 s	β- /8.					
^{116}Ru		115.931	0.20 s			0+			
^{117}Ru		116.936	0.14 s						
^{118}Ru		117.938	0.12 s			0+			
^{119}Ru		118.943	> 0.15 μs						
^{120}Ru		119.945	> 0.15 μs			0+			
$_{45}$Rh		**102.90550(2)**							
^{89}Rh		88.9488	> 0.15 μs						
^{90m}Rh			~ 12. ms						

Elem. or Isot.	Natural Abundance (Atom %)	Atomic Mass or Weight	Half-life/ Resonance Width (MeV)	Decay Mode/ Energy (/MeV)	Particle Energy/ Intensity (MeV/%)	Spin (h/2 π)	Nuclear Magnetic Mom. (nm)	Elect. Quadr. Mom. (b)	γ-Energy / Intensity (MeV/%)
^{90}Rh		89.9429	1.0 s						
^{91m}Rh			1.5 s	IT					0.387
^{91}Rh		90.9366	1.5 s						(0.438-0.973)
^{92m}Rh			0.5 s						0.866
^{92}Rh		91.9320	4.7 s	β+ /11.1					(0.163-0.991)
^{93}Rh		92.9257	12. s	β+ /8.1					(0.138–1.493)
^{94m}Rh			25.8 s	β+ /		8+			ann.rad./
									0.1264
									0.3117
									0.7562
									1.0752
									1.4307
^{94}Rh		93.9217	1.18 m	β+ /9.6	6.4/	4+			ann.rad./
									0.1461
									0.3117
									0.7562
									1.4307
^{95m}Rh			1.96 m	I.T./88/		½+			ann.rad./
				β+, EC/12/					0.5433(IT)
									0.7837
^{95}Rh		94.9159	5.0 m	β+ /5.1	3.2	9/2+			ann.rad./
									0.2293
									0.4103
									0.6610
									0.9416
									1.3520
									(0.2–3.8)
^{96m}Rh			1.51 m	I.T./60/0.052		2+			ann.rad./
				β+, EC/40/	4.70/				Tc,Ru x-rays
									0.8326
									1.0985
									1.6921
									(0.4–3.3)
^{96}Rh		95.91446	9.6 m	β+/6.45	3.3/	5+			ann.rad./
				EC/					0.4299
									0.6315
									0.6853
									0.7418
									0.8326
									(0.2–3.4)
^{97m}Rh			46. m	I.T./5 /	2.6/	1/2-			ann.rad./
				β+, EC/95/					0.1886
									0.4215
									2.2452
^{97}Rh		96.91134	31.0 m	β+ /3.52	2.1/	9/2+			ann.rad./
									0.1886
									0.3892
									0.4515
									0.8398
									0.8788
									(0.2–3.5)
^{98m}Rh			3.5 m	β+ /		5+			ann.rad./
									0.6154
									0.6524

Elem. or Isot.	Natural Abundance (Atom %)	Atomic Mass or Weight	Half-life/ Resonance Width (MeV)	Decay Mode/ Energy (/MeV)	Particle Energy/ Intensity (MeV/%)	Spin ($h/2\pi$)	Nuclear Magnetic Mom. (nm)	Elect. Quadr. Mom. (b)	γ-Energy / Intensity (MeV/%)
									0.7452
^{98}Rh		97.91071	8.7 m	β+ /90/5.06	3.4/	2+			ann.rad./
									0.6524
									0.7623
^{99m}Rh			4.7 h	β+ /8/	.74/	9/2+	5.67		ann.rad./
				EC/92/					0.2766/
									0.3408
									0.6178
									1.2612
^{99}Rh		98.90813	16. d	β+/4/2.10	0.54/	1/2-			ann.rad./
				EC/97/	0.68/				0.0894/
									0.3530
									0.5277
									(0.1–2.0)
^{100m}Rh			4.7 m	I.T./99/		5+			ann.rad./
				β+ /0.4/					0.0748/
									0.2647(IT)
^{100}Rh		99.90812	20.8 h	β+ /3.63	2.62/	1-			0.4462
				EC/	2.07/				0.5396
									0.5882
									0.8225
									1.5534
									2.3761
^{101m}Rh			4.35 d	EC/92/		9/2+	+5.47		Rh k x-ray
				I.T./8/0.1573					0.1272/
									0.3069
									0.5451
^{101}Rh		100.90616	3.3 a	EC/0.54		1/2-			Ru k x-ray
									0.1272
									0.1980
									0.3252
^{102m}Rh			3.74 a	EC/2.323		6^+	4.04		0.4751
				IT/0.0419					0.6313
			$> 1.2 \times 10^6$ a	β^+	/<0.00025				0.6975
									0.7668
									1.0466
									1.1032
^{102}Rh		101.906843	207. d	EC/62			~ 0.5		ann.rad./
				β- /19/					0.4686
				β+ /14/					0.4751
									0.5566
									0.6280
									1.1032
									(0.4–1.6)
^{103m}Rh			56.12 m	IT		7/2+	4.54		
^{103}Rh	100.	102.905504				1/2-	-0.0884		
^{104m}Rh			4.36 m	I.T./99+ /		5+			Rh k x-ray
				β^-	1.3/				0.0514
									0.0971
									0.5558
^{104}Rh		103.906656	42.3 s	β-/99+/2.441	1.88/2	1+			0.3581
				EC/0.4/1.141	2.44/98				0.5558
									1.2370
									(0.35–1.8)

Elem. or Isot.	Natural Abundance (Atom %)	Atomic Mass or Weight	Half-life/ Resonance Width (MeV)	Decay Mode/ Energy (/MeV)	Particle Energy/ Intensity (MeV/%)	Spin (h/2 π)	Nuclear Magnetic Mom. (nm)	Elect. Quadr. Mom. (b)	γ-Energy / Intensity (MeV/%)
^{105m}Rh			43. s	I.T./1.296		1/2-			Rh k x-ray
									0.1296
^{105}Rh		104.905694	35.4 h	β- /0.567	0.247/30	7/2+	+4.45		0.2801
					0.567/70				0.3061/4.8
									0.3189/17.0
^{106m}Rh			2.18 h	β- /	0.92/	6+			0.2217
									0.4510
									0.5119
									0.6162
									0.7173
									0.7484
									1.0458
									1.5277
^{106}Rh		105.90729	29.9 s	β- /3.54	2.4/2	1+	+2.58		0.51186/
					3.0/12				0.61612
					3.54/79				0.62187
									(0.05–3.04)
^{107}Rh		106.90675	21.7 m	β- /1.51	1.20/65	7/2+			0.2776
					1.5/17				0.3028
									0.3925
^{108m}Rh			6.0 m	β- /	1.57/				0.4339
									0.4973
									0.6189
^{108}Rh		107.9087	17. s	β- /4.5		1+			0.4046
									0.4339
									0.4973
									0.5811
									0.6146
									0.9014
									0.9471
^{109}Rh		108.90874	1.34 m	β- /2.59	2.25/	7/2+			0.1134
									0.1780
									0.2914
									0.3254
									0.3268
									0.4261
									(0.1–1.6)
^{110m}Rh			29. s	β- /	6/				0.3737
									0.4397
									0.7967
^{110}Rh		109.91114	3.1 s	β- /5.4	5.5/	1+			0.3737
									0.4400
									0.5463
									0.6877
									0.8381
									0.9045
^{111}Rh		110.91159	11. s	β- /3.7					0.275
^{112m}Rh			6.8 s	β- /					
^{112}Rh		111.9144	3.5 s	β- /6.2		1+			0.3489
^{113}Rh		112.91553	0.9 s	β- /4.9					0.1285
^{114m}Rh			1.9 s	β- /					(0.103-1.923)
^{114}Rh		113.9188	1.8 s	β- /6.5		1+			(0.276-0.783)
^{115}Rh		114.9203	0.99 s	β- /6.0					
^{116m}Rh			0.9 s	β- /					0.3405

Elem. or Isot.	Natural Abundance (Atom %)	Atomic Mass or Weight	Half-life/ Resonance Width (MeV)	Decay Mode/ Energy (/MeV)	Particle Energy/ Intensity (MeV/%)	Spin ($h/2\pi$)	Nuclear Magnetic Mom. (nm)	Elect. Quadr. Mom. (b)	γ-Energy / Intensity (MeV/%)
^{116}Rh		115.9241	0.7 s	β- /8.0		1+			0.340
									0.398–1.665
^{117}Rh		116.9260	0.42 s	β- /7.					0.0346
									0.1317
^{118}Rh		117.9301	0.27 s						0.379
									0.575
									0.370–1.037
^{119}Rh		118.932	0.17 s						
^{120}Rh		119.936	0.13 s						
^{121}Rh		120.939	0.15 s						
^{122}Rh		121.943							
$_{46}$Pd		**106.42(1)**							
^{91}Pd		90.949	> 1.5 μs						
^{92}Pd		91.9404	1.0 s			0+			
^{93}Pd		92.9359	1.2 s	β+, p		9/2+			0.240/81
									0.382–0.864
^{94}Pd		93.9288	9.6 s	EC, β+ /~ 6.6		0+			0.5582
									(0.055–0.798)
^{95m}Pd		94.92684	13.4 s	EC, β+ /10.2		21/2+			
^{95}Pd		94.9247							
^{96}Pd		95.9182	2.03 m	EC, β+ /3.5	1.15/	0+			0.1248
									0.4995
^{97}Pd		96.9165	3.1 m	β+, EC/4.8	3.5/	5/2+			ann.rad./
									0.2653
									0.4752
									0.7927
									(0.2–3.4)
^{98}Pd		97.91272	17.7 m	β+ /1.87		0+			ann.rad./
				EC/					0.0677
									0.1125
									0.6630
									0.8379
^{99}Pd		98.91177	21.4 m	β+ /49/3.37	2.18/	5/2+			ann.rad./
				EC/51/					0.1360
									0.2636
									0.6734
									(0.2–2.85)
^{100}Pd		99.90851	3.7 d	EC/0.36		0+			0.03271
									0.0748
									0.0840
^{101}Pd		100.90829	8.4 h	β+ /5/1.980	0.776/	5/2+	-0.66		ann.rad./
				EC/95/					0.0244
									0.2963
									0.5904
^{102}Pd	1.02(1)	101.905609				0+			
^{103}Pd		102.906087	16.99 d	EC/0.543		5/2+			Rh k x-ray
									0.03975
									0.3575
									0.4971
^{104}Pd	11.14(8)	103.904036				0+			
^{105}Pd	22.33(8)	104.905085				5/2+	-0.642	+0.66	
^{106}Pd	27.33(3)	105.903486				0+			
^{107m}Pd			20.9 s	I.T./0.2149		11/2-			Pd k x-ray
									0.2149(IT)

Elem. or Isot.	Natural Abundance (Atom %)	Atomic Mass or Weight	Half-life/ Resonance Width (MeV)	Decay Mode/ Energy (/MeV)	Particle Energy/ Intensity (MeV/%)	Spin (*h*/2 π)	Nuclear Magnetic Mom. (nm)	Elect. Quadr. Mom. (b)	γ-Energy / Intensity (MeV/%)
^{107}Pd		106.905133	6.5 × 10^6 a	β- /0.033	0.03/	5/2+			
^{108}Pd	26.46(9)	107.903893				0+			
^{109m}Pd			4.75 m	I.T./0.1889		11/2-			Pd x-ray
									0.1889(IT)
^{109}Pd		108.905950	13.5 h	β- /1.116	1.028	5/2+			0.0880
									(0.08–1.0)
^{110}Pd	11.72(9)	109.905153				0+			
^{111m}Pd			5.5 h	I.T./73/0.172		11/2-			0.0704
				β- /27/	0.35				0.1722
					0.77				0.3912
									(0.1–1.97)
^{111}Pd		110.90767	23.4 m	β- /2.19	2.2/95	5/2+			0.0598
									0.2454
									0.5800
									0.6504
									1.3885
									1.4590
^{112}Pd		111.90731	21.04 h	β- /0.29	0.28/	0+			0.018
^{113m}Pd			1.48 m	β- /		5/2+			0.0959
^{113}Pd		112.91015	1.64 m	β- /3.34					0.0958
									0.4824
									0.6436
									0.7394
^{114}Pd		113.91036	2.48 m	β- /1.45		0+			0.1266
									0.2320
									0.5582
									0.5760
^{115m}Pd			50. s			(9/2-)			0.089
^{115}Pd		114.9137	25. s	β- /4.58		(3/2+)			0.1255
									0.2554
									0.3428
^{116}Pd		115.9142	12.7 s	β- /2.61		0+			0.1015
									0.1147
									0.1778
^{117m}Pd			19. ms			(9/2-)			0.203
^{117}Pd		116.9178	4.4 s	β- /5.7		(3/2+)			0.2473
									0.077–0.403
^{118}Pd		117.9190	2.4 s	β- /4.1		0+			0.1254
									0.028–0.596
^{119}Pd		118.9231	0.9 s	β- /6.5					0.2566
									0.070–0.326
^{120}Pd		119.9247	0.49 s	β- /5.0		0+			0.1581
									0.053–0.595
^{121}Pd		120.9289	0.29 s						
^{122}Pd		121.9306	0.18 s			0+			
^{123}Pd		122.935	0.17 s						
^{124}Pd		123.9369	~ 0.04 s			0+			
$_{47}$Ag		**107.8682(2)**							
^{93}Ag		92.950							
^{94m2}Ag			0.40 s	β+	p/1.01/2.2	21+			(0.153-1.132)
				β+,p	p/0.79/1.9				
^{94m1}Ag			0.60 s	β+		7+			(0.659-0.905)
				β+,p	p//20.				

Elem. or Isot.	Natural Abundance (Atom %)	Atomic Mass or Weight	Half-life/ Resonance Width (MeV)	Decay Mode/ Energy (/MeV)	Particle Energy/ Intensity (MeV/%)	Spin ($h/2\pi$)	Nuclear Magnetic Mom. (nm)	Elect. Quadr. Mom. (b)	γ-Energy / Intensity (MeV/%)
^{94}Ag		93.9428	0.03 s	β+		0+			
				β+,p					
^{95}Ag		94.9355	2.0 s	β+, p/					(0.089–2.940)
^{96m}Ag			4.4 s	β^+		8+			
				β+, p	/8.				
^{96}Ag		95.9307	7. s	β+ /11.6		2+			ann.rad./
				EC/					0.1248
				β+, p	/18.				0.4995
									(0.1066–1.416)
^{97}Ag		96.9240	19. s	β+ /7.0					ann.rad./
				EC/					0.6862
									1.2941
									(0.352–3.294)
^{98}Ag		97.9216	47.6 s	β+ /8.4		5+			ann.rad./
				EC/	/36.				0.5711
				β+, p	/0.11				0.6786
									0.8631
									(0.153–1.185)
^{99m}Ag			11. s	I.T./100/		½-			Ag k x-ray
									0.1636(IT)
									0.3426
^{99}Ag		98.9176	2.07 m	β+ /87 5.4		9/2+			ann.rad./
				EC/13/					0.2199
									0.2645
									0.8056
									0.8323
									(0.2–3.5)
^{100m}Ag			2.3 m	β+ /		2+			ann.rad./
				EC/					0.6657
									1.6941
^{100}Ag		99.9161	2.0 m	β+/7.1	4.7/	5+			ann.rad./
				EC/					0.2807
									0.4503
									0.6657
									0.7508
									0.7732
^{101m}Ag			3.1 s	I.T./0.23		½-			Ag k x-ray
									0.0981
									0.176(IT)
^{101}Ag		100.9128	11.1 m	β+ /69/4.2	2.7/	9/2+	5.7		ann.rad./
				EC/31/					0.2610
					2.18/				0.2747
					2.73/				0.3269
					3.38/				0.4392
									0.6673
									1.1739
									(0.2–3.1)
^{102m}Ag			7.8 m	β+ /38/	3.4	2+	+4.1		ann.rad./
				EC/13/					0.5567
				I.T./49/					0.9777
									1.8347
									2.0545
									2.1594
									3.2386

Elem. or Isot.	Natural Abundance (Atom %)	Atomic Mass or Weight	Half-life/ Resonance Width (MeV)	Decay Mode/ Energy (/MeV)	Particle Energy/ Intensity (MeV/%)	Spin (h/2 π)	Nuclear Magnetic Mom. (nm)	Elect. Quadr. Mom. (b)	γ-Energy / Intensity (MeV/%)
^{102}Ag		101.91169	13.0 m	β+ /78/5.92	2.26/	5+	~ 4.6		ann.rad./
				EC/22/					0.5564
									0.7193
									0.163–2.242
^{103m}Ag			5.7 s	I.T./0.134		1/2-			Ag k x-ray
									0.1344
^{103}Ag		102.90897	1.10 h	β+ /28/2.69	1.7	7/2+	+4.47		ann.rad./
				EC/72/	1.3				0.1187
									0.1482
^{104m}Ag			33. m	β+ /64/	2.71/	2+	+3.7		ann.rad./
				EC/36/					0.5558
				I.T./0.07/					0.7657
									(0.5–3.4)
^{104}Ag		103.90863	69. m	β+ /16/4.28	0.99/	5+	3.92		ann.rad./
				EC/84/					0.5558
									0.9259
									0.9416
									(0.18–2.27)
^{105m}Ag			7.2 m	I.T./98/0.0255		7/2+	+4.41		Ag x-ray
				EC/2 /					0.3063
									0.3192
									(0.1–1.0)
^{105}Ag		104.90653	41.3 d	EC/1.35		1/2-	0.101		0.0640
									0.2804
									0.3445
									0.4434
^{106m}Ag			8.4 d	EC/		6+	3.71	+1.1	Pd k x-ray
									0.4510
									0.5118
									0.7173
									1.0458
^{106}Ag		105.90667	24.0 m	β+ /59/2.965	/1.96	1+	+2.9		ann.rad./
				EC/41 /					0.5119
^{107m}Ag			44.2 s	I.T./0.093		7/2+	+4.40	1.0	Ag x-ray
									0.0931
^{107}Ag	51.839(8)	106.905097				1/2-	-0.113680		
^{108m}Ag			418. a	EC/92/		6+	3.58	+1.3	Ag k x-ray
				I.T./8 /0.079					Pd k x-ray
									0.43392
									0.61427
									0.72290
^{108}Ag		107.905956	2.39 m	β- /97/1.65	1.02/1.7	1+	+2.6884		ann.rad./
				EC/2/	1.65/96				0.43392
				β- /1/1.92	0.88/0.3				0.61885
									0.63298
^{109m}Ag			39.8 s	I.T./0.088		7/2+	+4.40	+1.0	Ag k x-ray
									0.0880
^{109}Ag	48.161(8)	108.904752				1/2-	-0.13069		
^{110m}Ag			249.8 d	β- /99/	0.087	6+	+3.61	+1.4	0.65774
				I.T./1 /0.1164	0.530				0.76393
									0.88467
									0.93748
									1.38427
									(0.447–1.56)

Elem. or Isot.	Natural Abundance (Atom %)	Atomic Mass or Weight	Half-life/ Resonance Width (MeV)	Decay Mode/ Energy (/MeV)	Particle Energy/ Intensity (MeV/%)	Spin ($h/2\pi$)	Nuclear Magnetic Mom. (nm)	Elect. Quadr. Mom. (b)	γ-Energy / Intensity (MeV/%)
^{110}Ag		109.906107	24.6 s	β- /2.892	2.22/5	1+	+2.727	0.2	0.65774
					2.89/95				0.8154
									1.1257
^{111m}Ag			1.08 m	IT/99/0.0598		7/2+			Ag k x-ray
				β- /1/					0.0598
									0.2454
^{111}Ag		110.905294	7.47 d	β- /1.037	1.035/	1/2-	-0.146		0.2454
									0.3421
^{112}Ag		111.90701	3.13 h	β- /3.96	3.94/	2-	0.0547		0.6067
					3.4				0.6174
									1.3877
									(0.4–2.9)
^{113m}Ag			1.14 m	I.T./80 /0.043		7/2+			0.1422
				β- /20 /	1.5				0.2983
									0.3161
									0.3923
^{113}Ag		112.90657	5.3 h	β- /2.02	2.01/	1/2-	0.159		0.2588
									0.2986
^{114}Ag		113.90880	4.6 s	β- /5.08	4.9/	1+			0.5582
									0.5760
									1.9946
^{115m}Ag			18.7 s	β- /		7/2+			0.1134
									0.1315
									0.2288
									0.3887
^{115}Ag		114.90876	20. m	β- /3.10		1/2-			0.1316
									0.2128
									0.2291
									0.4727
									(0.13–2.49)
^{116m2}Ag			20. s	β-,IT/7	IT/0.0479				
^{116m}Ag			9.8 s	β-/92 /	3.2/	5+			0.5134
					2.9				0.7055
				I.T./8	IT/.0809				(0.255–2.838)
^{116}Ag		115.91136	2.68 m	β- /6.16	5.3	2-			0.5134
									0.6993
									2.4779
^{117m}Ag			5.3 s	β- /	3.2/	7/2+			0.1354
									0.2981
									0.3868
									0.1571
^{117}Ag		116.91168	1.22 m	β- /4.18	2.3	1/2-			0.1354
									0.3377
^{118m}Ag			2.8 s	β- /59/					0.1277
				I.T./41 /0.1277					0.4878
									0.6771
									0.7709
									(0.190-2.778)
^{118}Ag		117.9146	4.0 s	β- /7.1					0.4878
									0.6771
									3.2259
^{119}Ag		118.9157	2.1 s	β- /5.35		7/2+			0.0674
									0.3662
									0.3991

Elem. or Isot.	Natural Abundance (Atom %)	Atomic Mass or Weight	Half-life/ Resonance Width (MeV)	Decay Mode/ Energy (/MeV)	Particle Energy/ Intensity (MeV/%)	Spin (h/2 π)	Nuclear Magnetic Mom. (nm)	Elect. Quadr. Mom. (b)	γ-Energy / Intensity (MeV/%)
									0.6264
^{120m}Ag			0.40 s	β- /63.					0.2030
				I.T./37.					0.5059
									0.6978
									0.8300
									(0.115-1.644)
^{120}Ag		119.9188	1.23 s	β- /8.2					0.5059
				β-,n	n//<0.0030%				0.6978
									0.8171
									(0.442-3.044)
^{121}Ag		120.9199	0.78 s	β- /6.4					0.1150
									0.3148
									0.3537
									0.3696
									0.5007
									1.5105
									(0.11–2.5)
^{122m}Ag			1. s	β- /					
^{122}Ag		121.9235	0.44 s	β- /9.2					
^{123}Ag		122.9249	0.28 s	β- /7.4					
^{124}Ag		123.9286	0.17 s	β- /10.1					
^{125}Ag		124.9304	0.17 s	β-					
^{126}Ag		125.9345	0.11 s	β-					
^{127}Ag		126.9368	0.11 s	β-					
^{128}Ag		127.9412	58 ms	β-					
^{129m}Ag			0.16 s						
^{129}Ag		128.9437	~ 46. ms	β-, n					
^{130}Ag		129.9505	~ 35 ms						
$_{48}$Cd		**112.411(8)**							
^{95}Cd		94.950							
^{96}Cd		95.9398				0+			
^{97}Cd		96.9349	3. s	β+, (p)					
^{98}Cd		97.9274	9.2 s	β+ /5.4		0+			
				(p)	/0.025				
^{99}Cd		98.9250	16. s	β+, EC/6.9					ann.rad./
^{100}Cd		99.9203	1.1 m	β+, EC/3.9		0+			ann.rad./
									(0.090–1.043)
^{101}Cd		100.9187	1.2 m	β+ /83/5.5	4.5	5/2+			In k x-ray
				EC/17/					0.0985
									1.7225
									0.31–2.84)
^{102}Cd		101.91446	5.8 m	β+ /27/2.59		0+			ann.rad./
				EC/73					0.0974
									0.4810
									1.0366
									1.3598
^{103}Cd		102.91342	7.5 m	β+ /33/4.14		5/2+	-0.81	~ -0.8	ann.rad./
				EC/67/					Ag k x-ray
									1.0799
									1.4487
									1.4618
									(0.1–2.8)
^{104}Cd		103.90985	58. m	EC/1.14		0+			Ag k x-ray
									0.0835

Elem. or Isot.	Natural Abundance (Atom %)	Atomic Mass or Weight	Half-life/ Resonance Width (MeV)	Decay Mode/ Energy (/MeV)	Particle Energy/ Intensity (MeV/%)	Spin ($h/2\pi$)	Nuclear Magnetic Mom. (nm)	Elect. Quadr. Mom. (b)	γ-Energy / Intensity (MeV/%)
									0.7093
^{105}Cd		104.90947	55.5 m	β+ /26/2.739	1.69/	5/2+	-0.7393	+0.43	Ag k x-ray
				EC/74/					0.3469
									0.6072
									0.9618
									1.3025
									(0.25–2.4)
^{106}Cd	1.25(6)	105.90646	>1.9 × 10^{19} a	EC, EC		0+			
^{107}Cd		106.90662	6.52 h	EC/99+/1.417		5/2+	-0.615055	~ +0.68	Ag k x-ray
				β+ /					0.0931
									0.8289
^{108}Cd	0.89(3)	107.90418	>4.1 × 10^{17} a	EC EC		0+			
^{109}Cd		108.904982	462.0 d	EC/0.214		5/2+	-0.827846	~ +0.69	Ag k x-ray
									0.08804/.0366
^{110}Cd	12.49(18)	109.903002				0+			
^{111m}Cd			48.5 m	I.T./		11/2-	- 1.105	~ -0.85	Cd k x-ray
									0.1508(IT)
									0.2454
^{111}Cd	12.80(12)	110.904178				1/2+	-0.594886		
^{112}Cd	24.13(21)	111.902758				0+			
^{113m}Cd			14.1 y	β- /99.9/0.59	0.59/99.9	11/2-	-1.087784	~ -0.71	0.2637
^{113}Cd	12.22(12)	112.904402	8.04 × 10^{15} a	β-		1/2+	-0.622301		
^{114}Cd	28.73(42)	113.903359	>6.0 × 10^{17} y	β β		0+			
^{115m}Cd			44.6 d	β- /1.629	0.68/1.6	11/2-	-1.041034	~ -0.54	0.48450
					1.62/97				0.93381
									1.29064
^{115}Cd		114.905431	2.228 d	β- /1.446	0.593/42	1/2+	-0.648426		0.23141
					1.11/58				0.26085
									0.33624
									0.49227
									0.52780
^{116}Cd	7.49(18)	115.904756	3.8 × 10^{19} a	β-β-		0+			
^{117m}Cd			3.4 h	β- /2.66	0.72/	11/2-			0.1586
									0.5529
									0.37–2.42
^{117}Cd		116.907219	2.49 h	β- /2.52	0.67/51	1/2+			0.2209
					2.2/10				0.2733
									0.3445
									1.3033
^{118}Cd		117.90692	50.3 m	β- /0.52		0+			
^{119m}Cd			2.20 m	β- /		11/2-			0.1056
									0.7208
									1.0250
									2.0213
^{119}Cd		118.9099	2.69 m	β- /3.8	~ 3.5/	1/2+			0.1340
									0.2929
									0.3429
^{120}Cd		119.90985	50.8 s	β- /1.76	1.5/	0+			
^{121m}Cd			8. s	β- /		11/2-			0.1008
									0.9878
									1.0209
									1.1815
									2.0594
^{121}Cd		120.9130	13.5 s	β- /4.9		(3/2+)			0.2102

Elem. or Isot.	Natural Abundance (Atom %)	Atomic Mass or Weight	Half-life/ Resonance Width (MeV)	Decay Mode/ Energy (/MeV)	Particle Energy/ Intensity (MeV/%)	Spin ($h/2\pi$)	Nuclear Magnetic Mom. (nm)	Elect. Quadr. Mom. (b)	γ-Energy / Intensity (MeV/%)
									0.3242
									0.3492
									1.0403
^{122}Cd		121.91333	5.3 s	β- /3.0		0+			
^{123m}Cd			1.9 s	β- /					
^{123}Cd		122.91700	2.09 s	β- /6.12		3+			
^{124}Cd		123.9177	1.24 s	β- /4.17		0+			0.0365
									0.0628
									0.1799
^{125m}Cd			0.66 s	β- /					
^{125}Cd		124.9213	0.68 s	β- /7.16		3/2+			
^{126}Cd		125.9224	0.52 s	β- /5.49		0+			0.2601
^{127}Cd		126.9264	0.4 s	β- /8.5		3/2+			
^{128}Cd		127.9278	0.28 s	β- /7.1		0+			0.247
^{129}Cd		128.9322	0.24 s	β- /5.9					0.281
^{130}Cd		129.9339	0.162 s	β- /		0+			
				β-, n	/~ 3.5				
^{131}Cd		130.9407	68 ms						
^{132}Cd		131.9456	0.10 s	β-, n/	/60	0+			
^{133}Cd			0.06 s						
$_{49}$**In**		**114.818(3)**							
^{97}In		96.950							
^{98m}In			~ 0.03 s						
^{98}In		97.9421	1. s						
^{99}In		98.9342	~ 3.8 s	β+ /8.9					
^{100}In		99.9311	5.9 s	β+, (p)/10.5					(0.297-1.365)
^{101}In		100.9263	15. s	β+ /7.3					
^{102}In		101.9241	23. s	EC/8.9		(5)			0.1566
									0.7767
									(0.397–0.923)
^{103m}In			34. s						
^{103}In		102.91991	1.1 m	β+, EC/6.05	4.2	9/2+			ann.rad./
				EC	/45				0.1879
									(0.157–3.98)
^{104m}In			16. s	IT/0.0935					
^{104}In		103.9183	1.84 m	β+, EC/7.9	4.8	5+	+4.44	+0.7	ann.rad./
									0.6580
									0.8341
									0.8781
^{105m}In			43. s	I.T.		½-			In k x-ray
									0.6740
^{105}In		104.91467	5.1 m	β+, EC/4.85	3.7	9/2+	+5.675	+0.83	0.1310
									0.2600
									0.6038
^{106m}In			5.3 m	β+ /85/	4.90	3+			ann.rad./
				EC/15/					0.6326
									0.8611
									1.7164
^{106}In		105.91347	6.2 m	β+ /65/6.52	2.6	7+	+4.92	~ +0.97	ann.rad./
				EC/35/					0.2259
									0.6327
									0.8611
									0.9978
									1.0091

Elem. or Isot.	Natural Abundance (Atom %)	Atomic Mass or Weight	Half-life/ Resonance Width (MeV)	Decay Mode/ Energy (/MeV)	Particle Energy/ Intensity (MeV/%)	Spin (h/2 π)	Nuclear Magnetic Mom. (nm)	Elect. Quadr. Mom. (b)	γ-Energy / Intensity (MeV/%)
^{107m}In			51. s	I.T./0.6786		½-			In k x-ray
									0.6785
^{107}In		106.91030	32.4 m	β+ /35/3.43	2.20/	9/2+	+5.59	+0.81	ann.rad./
				E.C/65/					Cd k x-ray
									0.2050
									0.3209
									0.5055
									(0.2–2.99)
^{108m}In			57. m	β+ /53/	1.3	6+	+4.94	+0.47	ann.rad./
				EC/47/					Cd k x-ray
									0.6329
									1.9863
									3.4522
^{108}In		107.90970	40. m	β+ /33/5.15	3.49/	3+	+4.561	+1.01	ann.rad./
				EC/67/					Cd k x-ray
									0.2429
									0.6331
									0.8756
^{109m}In			1.3 m	I.T./0.650		½-			In k x-ray
									0.6498
^{109}In		108.90715	4.17 h	β+ /8/2.02	0.79/	9/2+	+5.54	+0.84	ann.rad./
				EC/92/					Cd k x-ray
									0.2035
									0.6235
^{110m}In			4.9 h	EC/		7+	+4.71	+1.00	Cd k x-ray
									0.6577
									0.8847
									0.9375
									(0.1–1.98)
^{110}In		109.90717	1.15 h	β+ /62/3.88	2.22/	2+	+4.37	+0.35	ann.rad./
				EC/38/					Cd k x-ray
									0.6577
									(0.6–3.6)
^{111m}In			7.7 m	I.T./0.537		½-	+5.53		In k x-ray
									0.537
^{111}In		110.905103	2.8049 d	EC/0.866		9/2+	+5.50	+0.80	Cd k x-ray
									0.1712
									0.2453
^{112m}In			20.8 m	I.T./0.155		4+	+ 5.227	+0.71	In k x-ray
									0.1555
^{112}In		111.90553	14.4 m	β+ /22/2.586		1+	+2.82	~ +0.087	ann.rad./
				EC/34/					Cd k x-ray
				β- /0.663					0.6171
^{113m}In			1.658 h	I.T./0.3917		½-	-0.21074		In k x-ray
									0.3917
^{113}In	4.29(5)	112.904058				9/2+	+5.5289	+0.80	
^{114m}In			49.51 d	I.T./97/0.190		5+	+4.65	+0.74	In k x-ray
				EC/3 /					0.19027
^{114}In		113.904914	1.198 m	β- /97/1.989	1.984/	1+	+2.82		Cd k x-ray
				EC/3/1.453					0.5584
									0.5727
									1.2998
^{115m}In			4.486 h	I.T./95/0.336		½-	-0.2440		In k x-ray
				β- /5 /0.83					0.3362

Elem. or Isot.	Natural Abundance (Atom %)	Atomic Mass or Weight	Half-life/ Resonance Width (MeV)	Decay Mode/ Energy (/MeV)	Particle Energy/ Intensity (MeV/%)	Spin ($h/2\pi$)	Nuclear Magnetic Mom. (nm)	Elect. Quadr. Mom. (b)	γ-Energy / Intensity (MeV/%)
									0.4974
^{115}In	95.71(5)	114.903878	4.4 × 10^{14} a	β- /0.495		9/2+	+5.5408	+0.81	
$^{116m2}In$			2.16 s	I.T./0.162		8-	+3.22	+0.31	In k x-ray
				EC	/0.023				0.1624
$^{116m1}In$			54.1 m	β- /	1.0	5+	+4.44	+0.80	0.13792
									0.41688/27
									1.09723/58.5
									1.29349/85
^{116}In		115.905260	14.1 s	β- /3.274	3.3/99	1+	2.788	0.11	0.46313
									1.2526
									1.29349
^{117m}In			1.94 h	β- /53/1.769	1.77/	½-	-0.25174		In k x-ray
				I.T./47 /					0.15855
									0.31531
									0.55294
^{117}In		116.90451	44. m	β- /1.455	0.74/	9/2+	+5.519	+0.83	0.15855
									0.3966
									0.55294
$^{118m2}In$			8.5 s	I.T./98/		(8-)	+3.32	+0.44	In k x-ray
				β- /2/					0.1382
$^{118m1}In$			4.40 m	β- /	1.3	5+	+4.23	+0.80	0.2086
					2.0				0.6833
									1.2295
^{118}In		117.90635	5.0 s	β- /4.42	4.2/	1+			0.5282
									1.1734
									1.2295
									2.0432
^{119m}In			17.9 m	β- /97/	2.7/	½-	-0.32		0.3114
				I.T./3/0.311					0.7631
^{119}In		118.90585	2.3 m	β- /2.36	1.6/	9/2+	+5.52	+0.85	0.0239
									0.6495
									0.7631
									1.2149
$^{120m2}In$			47 s	β- /6.1		8-	+3.692	+0.53	1.171
									1.023
$^{120m1}In$			46. s	β- /5.8	2.2/	5+	+4.30	+0.81	1.171
									1.023
^{120}In		119.90796	3.1 s	β- /5.37	5.6/	(1+)			0.4146
					3.1/				0.5924
									0.8637
									1.0232
									1.1714
									(0.4–2.7)
^{121m}In			3.8 m	β- /99/	3.7/	1/2-	-0.355		0.0601
				I.T./1/0.313					0.3136
									0.9256
									1.0412
									1.1022
									1.1204
^{121}In		120.90785	23. s	β- /3.36	2.5	9/2+	+5.50	+0.81	0.2620
									0.6573
									0.9256
^{122m}In			10. s	β- /	4.4/	8-	+3.78	+0.59	1.0014
									1.1403

Elem. or Isot.	Natural Abundance (Atom %)	Atomic Mass or Weight	Half-life/ Resonance Width (MeV)	Decay Mode/ Energy (/MeV)	Particle Energy/ Intensity (MeV/%)	Spin ($h/2\pi$)	Nuclear Magnetic Mom. (nm)	Elect. Quadr. Mom. (b)	γ-Energy / Intensity (MeV/%)
^{122}In		121.91028	1.5 s	β- /6.37	5.3/	(1+)			0.2391
									1.0014
									1.1403
									1.164
									1.1903
^{123m}In			47. s	β- /	4.6/	(1/2-)	-0.400		0.1258
									1.170
									3.234
^{123}In		122.91044	6.0 s	β- /4.39	3.3/	(9/2+)	+5.49	+0.76	0.6188
									1.0197
									1.1305
^{124m}In			3.4 s	β-		8-	+3.89	+0.66	0.1029
									0.9699
									1.0729
									1.1316
^{124}In		123.91318	3.18 s	β- /7.36	5/	3+	+4.04	+0.61	0.7070
									0.9978
									1.1316
									3.2142
									(0.3–4.6)
^{125m}In			12.2 s	β- /	5.5/	1/2-	-0.433		0.1876
^{125}In		124.91360	2.33 s	β- /5.42	4.1/	9/2+	+5.50	+0.71	0.4260
									1.0318
									1.3350
^{126m}In			1.53 s		4.9/	3+	+4.03	+0.49	0.9086
									0.9696
									1.1411
^{126}In		125.91646	1.63 s	β- /8.21	4.2/	8-	+4.061		0.1118
									0.9086
									1.1411
^{127m}In			3.73 s	β- /	6.4/	(1/2-)			0.2523
									3.074
^{127}In		126.91735	1.14 s	β- /6.51	4.9/	(9/2+)	+5.52	+0.59	0.4680
									0.6461
									0.8051
									1.5977
^{128m}In			0.7 s	β- /	5.4/	(8-)			1.8670
									1.9739
									(0.1205–2.12)
^{128}In		127.92017	0.80 s	β- /8.98	5.0/	3+			0.9352
									1.1688
									3.5198
									4.2970
^{129m}In			1.23 s	β- /98/	~ 7.5/	1/2-			0.3153
				n/2/					0.9067
									1.2220
^{129}In		128.9217	0.63 s	β- /7.66	5.5/	9/2+			0.2853
									0.7693
									1.8650
									2.1180
^{130m2}In			0.53 s	β- /	8.8/	5+			0.0892
									0.7744
									1.2212
^{130m1}In			0.51 s	β- /	6.1/	10-			0.0892

Elem. or Isot.	Natural Abundance (Atom %)	Atomic Mass or Weight	Half-life/ Resonance Width (MeV)	Decay Mode/ Energy (/MeV)	Particle Energy/ Intensity (MeV/%)	Spin ($h/2\pi$)	Nuclear Magnetic Mom. (nm)	Elect. Quadr. Mom. (b)	γ-Energy / Intensity (MeV/%)
									0.1298
									0.7744
									1.2212
									1.9052
^{130}In		129.92497	0.29 s	β- /10.25	10.0/	1-			
^{131m2}In			0.3 s	β- /		(21/2+)			
^{131m1}In			0.35 s	β- /		(1/2-)			
^{131}In		130.92685	0.28 s	β- /9.18	6.4/	(9/2+)			0.3328
									2.433
^{132}In		131.9330	~ 0.206 s	β- /13.6	6.0/	(7-)			0.1320
					8.8/				0.2992
									0.3747
									4.0406
^{133}In		132.9378	0.165 s	β-, (n)					
^{134}In		133.9442	0.14 s						(0.354–2.005)
^{135}In		134.9493	0.09 s						
$_{50}$**Sn**		**118.710(7)**							
^{99}Sn		98.949							
^{100}Sn		99.939	1.0 s	β+ /7.3	3.4/	0+			
^{101}Sn		100.9361	1.9 s	β+ /9.					0.352
									1.065
^{102}Sn		101.9303	3.8 s	β+ /5.8		0+			(0.069-1.425)
^{103}Sn		102.9281	7. s	β+ /7.7					1.3558
				β+,p	p//1.2				(0.351-2.813)
				EC	/ 20.				
^{104}Sn		103.9231	21. s	β+, EC/4.5		0+			(0.913-1.846)
^{105}Sn		104.9214	32.7 s	β+ /6.3	EC//42.				In-x-ray
				β$^+$, p	p//0.11				(0.288–3.819)
^{106}Sn		105.91688	2.0 m	β+ /20/3.18		0+			ann.rad./
				EC/80/					In k x-ray
									0.3865
									0.4772
^{107}Sn		106.9156	2.92 m	EC/5.0	1.2/				0.4218
				β+ /					0.6105
									0.6785
									1.0013
									1.1290
									1.542
^{108}Sn		107.91193	10.3 m	β+ /1/2.09	0.36/	0+			In k x-ray
				EC/99/					0.2724
									0.3965
									(0.105–1.68)
^{109}Sn		108.91128	18.0 m	β+ /9/3.85	1.52/	7/2+	-1.08	+0.3	ann.rad./
				EC/91/					In k x-ray
									0.6498
									1.0992
^{110}Sn		109.90784	4.17 h	EC/0.64		0+			In k x-ray
									0.283
^{111}Sn		110.90773	35. m	β+ /31/2.45	1.5/	7/2+	+0.61	+0.2	In k x-ray
				EC/69/					0.7620
									1.1530
									1.9147
^{112}Sn	0.97(1)	111.904818				0+			
^{113m}Sn			21.4 m	I.T./92/0.077		7/2+			Sn k x-ray

Elem. or Isot.	Natural Abundance (Atom %)	Atomic Mass or Weight	Half-life/ Resonance Width (MeV)	Decay Mode/ Energy (/MeV)	Particle Energy/ Intensity (MeV/%)	Spin (*h*/2 π)	Nuclear Magnetic Mom. (nm)	Elect. Quadr. Mom. (b)	γ-Energy / Intensity (MeV/%)
				EC/8/					In x-ray
									0.0774
^{113}Sn		112.905171	115.1 d	EC/1.036		½+	-0.879		In k x-ray
									0.25511
									0.39169
^{114}Sn	0.66(1)	113.902779				0+			
^{115}Sn	0.34(1)	114.903342				½+	-0.9188		
^{116}Sn	14.54(9)	115.901741				0+			
^{117m}Sn			14.0 d	I.T./0.3146		11/2-	-1.396	-0.4	Sn k x-ray
									0.15856
^{117}Sn	7.68(7)	116.902952				½+	-1.0010		
^{118}Sn	24.22(9)	117.901603				0+			
^{119m}Sn			293. d	I.T./0.0896		11/2-	-1.4	0.21	Sn k x-ray
									0.02387
^{119}Sn	8.59(4)	118.903308				½+	-1.0473		
^{120}Sn	32.58(9)	119.902195				0+			
^{121m}Sn			44. a	I.T./78/0.006		11/2-	-1.388	-0.14	Sn k x-ray
				β- /22/	0.354/				0.03715
^{121}Sn		120.904236	1.128 d	β /0.388	0.383/100	3/2+	0.698	~ -0.02	
^{122}Sn	4.63(3)	121.903439				0+			
^{123m}Sn			40.1 m	β- /1.428	1.26/99	3/2+			0.1603
									0.3814
^{123}Sn		122.905721	129.2 d	β /1.404	1.42/99.4	11/2-	-1.370	~ +0.03	0.1603
									1.0302
									1.0886
^{124}Sn	5.79(5)	123.905274	>2.2 × 10^{18} a	$\beta^-\beta^-$		0+			
^{125m}Sn			9.51 m	β- /2.387	2.03/98	3/2+	+0.764	+0.8	0.3321
									1.4040
^{125}Sn		124.907784	9.63 d	β- /2.364	2.35/82	11/2-	-1.348	~ +0.1	1.0671
									(0.2–2.3)
^{126}Sn		125.90765	2.34 × 10^5 a	β- /0.38	0.25/100	0+			0.0643
									0.0876
									0.4148
									0.6663
									0.6950
^{127m}Sn			4.15 m	β- /3.21	2.72/	3/2+	+0.757	+0.60	0.4909
									1.3480
									1.5640
^{127}Sn		126.91036	2.12 h	β- /3.20	2.42/	11/2-	-1.33	+0.3	0.8231
					3.2/				1.0956
									(0.120–2.84)
^{128m}Sn			6.5 s	IT/0.091		(7-)			
^{128}Sn		127.91054	59.1 m	β- /1.27	0.48/	0+			0.4823
					0.63/				0.5573
									0.6805
^{129m}Sn			6.9 m	β- /		11/2-	-1.30	~ -0.2	1.1611
^{129}Sn		128.91348	2.4 m	β- /4.0		3/2+	+0.754	~ +0.05	0.6456
^{130m}Sn			1.7 m	β- /		(7-)	-0.381	-0.4	0.1449
									0.8992
^{130}Sn		129.91397	3.7 m	β- /2.15	1.10/	0+			0.0700
									0.1925
									0.7798
^{131m}Sn			1.02 m	β- /	3.4/	11/2-	-1.28		0.3043
									0.4500

Elem. or Isot.	Natural Abundance (Atom %)	Atomic Mass or Weight	Half-life/ Resonance Width (MeV)	Decay Mode/ Energy (/MeV)	Particle Energy/ Intensity (MeV/%)	Spin ($h/2\pi$)	Nuclear Magnetic Mom. (nm)	Elect. Quadr. Mom. (b)	γ-Energy / Intensity (MeV/%)
									0.7985
									1.2260
									(0.08–3.21)
^{131}Sn		130.91700	39. s	β- /4.69	3.8/	3/2+	+0.747	~ -0.04	see ^{131m}Sn
^{132}Sn		131.91782	40. s	β- /3.12	1.8/	0+			0.0855
									0.2467
									0.3402
									0.8985
^{133}Sn		132.92383	1.44 s	β- /7.8	7.5/	7/2-			
^{134}Sn		133.9283	1.04 s	β- /6.8		0+			(0.053-2.417)
^{135}Sn		134.9347	0.53 s	β-					(0.053-0.830)
				β-,n	/21.				0.733–1.855
^{136}Sn		135.9393	0.25 s	β-, n	/30.	0+			
^{137}Sn		136.946	0.19 s	β-, n	/~ 58				
^{138}Sn			0.15 s			0+			
$_{51}$Sb		**121.760(1)**							
^{103}Sb		102.9397	> 1.5 μs						
^{104}Sb		103.9365	0.5 s						
^{105}Sb		104.9315	1.1 s	β+,p	p//<0.1				
^{106}Sb		105.9288	0.6 s	β+ /10.5					
^{107}Sb		106.9242	4.0 s	β+ /7.9					1.280
									0.1515
									0.6666
									0.553–2.046
^{108}Sb		107.9222	7.0 s	β+ /9.5					(0.151–1.280)
^{109}Sb		108.91813	17.3 s	β+ /6.38	4.42/	5/2+			0.925
				EC/	4.67/				1.062
					4.33/				0.261–2.127
^{110}Sb		109.9168	24. s	β+ /9.0	6.8/	3+			ann.rad./
				EC/					0.6365
									0.9847
									1.2117
									1.2433
^{111}Sb		110.91316	1.25 m	β+ /87/4.47	3.3/	5/2+			ann.rad./
				EC/13 /					0.1002
									0.1545
									0.4891
									1.0326
^{112}Sb		111.91240	51.4 s	β+ /90/7.06	4.75/	3+			ann.rad./
				EC/10/					0.6700
									0.9909
									1.2571
^{113}Sb		112.90937	6.7 m	β+ /65/3.91	2.42/	5/2+			ann.rad./
									(0.3–3.6)
				EC/35/					Sn k x-ray
									0.3324
									0.4980
^{114}Sb		113.90927	3.49 m	β+ /78/5.9	3.4/	3+	1.7		ann.rad./
				EC/22/					Sn k x-ray
									0.8876
									1.2999
^{115}Sb		114.90660	32.1 m	β+ /67/3.03	1.51/	5/2+	+3.46	-0.4	ann.rad./
				EC/33/					Sn k x-ray
									0.4973

Elem. or Isot.	Natural Abundance (Atom %)	Atomic Mass or Weight	Half-life/ Resonance Width (MeV)	Decay Mode/ Energy (/MeV)	Particle Energy/ Intensity (MeV/%)	Spin (h/2 π)	Nuclear Magnetic Mom. (nm)	Elect. Quadr. Mom. (b)	γ-Energy / Intensity (MeV/%)
^{116m}Sb			1.00 h	β+ /78/	1.16/	8-	2.6		ann.rad./
				EC/22/					Sn k x-ray
									0.4073
									0.5429
									0.9725
									1.2935
									(0.100–1.501)
^{116}Sb		115.90679	16. m	β+ /50/4.707	1.3/	3+	2.72		ann.rad./
				EC/50/	2.3/				Sn k x-ray
									0.93180
									1.29354
									(0.138–3.903)
^{117}Sb		116.90484	2.80 h	β+ /2/1.76	0.57/	5/2+	+3.4		Sn k x-ray
				EC/98/					0.1586
^{118m}Sb			5.00 h	EC/99/		8-	2.32		Sn k x-ray
									0.25368
									1.05069
									1.22964
^{118}Sb		117.905529	3.6 m	β+ /74/3.657	2.65/	1+	2.5		ann.rad./
				EC/26/					Sn k x-ray
									1.22964
^{119}Sb		118.90394	38.1 h	EC/0.59		5/2+	+3.45	-0.4	Sn k x-ray
									0.0239
^{120m}Sb			5.76 d	EC/		8-	2.34		Sn k x-ray
									0.0898
									0.19730
									1.02301
									1.17121
^{120}Sb		119.90507	15.89 m	β+ /41/2.68	1.72/	1+	+2.3		ann.rad./
				EC/59/					Sn k x-ray
									0.7038
									1.17121
^{121}Sb	57.21(5)	120.903816				5/2+	+3.3634	-0.4	
^{122m}Sb			4.19 m	I.T./0.162		8-			Sb x-ray
									0.0614
									0.0761
^{122}Sb		121.905174	2.72 d	β- /98/1.979	1.414/65	2-	-1.90	+0.9	0.56409
				β+ /2/1.620	1.980/26				0.69277
									1.14050
									1.2569
^{123}Sb	42.79(5)	122.904214				7/2+	+2.5498	-0.5	
^{124m2}Sb			20.3 m	I.T./0.035		8-			
^{124m1}Sb			1.6 m	I.T./80/	1.2/	5+			0.4984
				β- /20/	1.7/				0.6027
									0.6458
									1.1010
^{124}Sb		123.905936	60.20 d	β- /2.905	0.61/52	3-	1.20	+1.9	0.60271/97.8
					2.301/23				0.64583/7.4
									0.72277/10.5
									1.69094/48.2
									(0.027–2.808)
^{125}Sb		124.905254	2.758 a	β- /0.767	0.13/30	7/2+	+2.63		0.0355
					0.302/45				0.17632
					0.62/13				0.38044

Elem. or Isot.	Natural Abundance (Atom %)	Atomic Mass or Weight	Half-life/ Resonance Width (MeV)	Decay Mode/ Energy (/MeV)	Particle Energy/ Intensity (MeV/%)	Spin ($h/2\pi$)	Nuclear Magnetic Mom. (nm)	Elect. Quadr. Mom. (b)	γ-Energy / Intensity (MeV/%)
									0.42786
									0.46336
									0.60060
									0.63595
^{126m2}Sb			11. s	I.T./		3-			L x-ray
									0.0227
^{126m1}Sb			19.0 m	β- /86 /	1.9	5+			0.4148
				I.T./14 /					0.6663
									0.6950
^{126}Sb		125.90725	12.4 d	β- /3.67	1.9	8-	1.3		0.2786
									0.4148/83.3
									0.6663/99.7
									0.6950/99
									0.7205
^{127}Sb		126.90692	3.84 d	β- /1.581	0.89/	7/2+	2.70		0.2524
					1.10/				0.2908
					1.50/				0.4121
									0.4370
									0.6857
									0.7837
^{128m}Sb			10.1 m	β- /96/	2.6/	5+			0.3140
				I.T./4/					0.5941
									0.7432
									0.7539
^{128}Sb		127.90917	9.1 h	β- /4.38	2.3/	8-	1.3		0.2148
									0.3141
									0.5265
									0.7433
									0.7540
^{129m}Sb			17.7 m	β- /					0.4338
									0.6578
									0.7598
^{129}Sb		128.90915	4.40 h	β- /2.38	0.65/	7/2-	2.79		0.0278
									0.1808
									0.3594
									0.4596
									0.5447
									0.8128
									0.9146
									1.0301
^{130m}Sb			6.5 m	β- /2.6	2.12/		3.09		0.1023
									0.7934
									0.8394
^{130}Sb		129.91166	38.4 m	β- /4.96	2.9/	8-			0.1823
									0.3309
									0.4680
									0.7394
									0.8394
^{131}Sb		130.91198	23.0 m	β- /3.20	1.31/	7/2+	2.89		0.6423
					3.0/				0.6579
									0.9331
									0.9434
^{132m}Sb			2.8 m	β- /	3.9/	4+	3.18		0.1034
									0.3538

Elem. or Isot.	Natural Abundance (Atom %)	Atomic Mass or Weight	Half-life/ Resonance Width (MeV)	Decay Mode/ Energy (/MeV)	Particle Energy/ Intensity (MeV/%)	Spin ($h/2\pi$)	Nuclear Magnetic Mom. (nm)	Elect. Quadr. Mom. (b)	γ-Energy / Intensity (MeV/%)
									0.6968
									0.9739
									0.9896
^{132}Sb		131.91447	4.2 m	β- /5.49		8-			0.1034
									0.1506
									0.6968
									0.9739
^{133}Sb		132.91525	2.5 m	β- /4.00	1.20/	7/2+	3.00		0.4235
									0.6318
									0.8165
									1.0764
^{134m}Sb			10.4 s	β- /	6.1	7-			
^{134}Sb		133.92038	0.8 s	β- /8.4	8.4	0-			0.1152
									0.2970
									0.7063
									1.2791
^{135}Sb		134.9252	1.71 s	β- /8.12		7/2+			1.127
									1.279
^{136}Sb		135.9304	0.82 s	β- /9.3					
^{137}Sb		136.9353	> 0.15 μs						
^{138}Sb		137.9408	> 0.15 μs						
^{139}Sb		138.9460	> 0.15 μs						
$_{52}$Te		**127.60(3)**							
^{105}Te		104.9436	~ 0.63 μs	α	A/4.70				
^{106}Te		105.9375	0.07 ms	α/4.3	/100	0+			
^{107}Te		106.9350	3.1 ms	α/ 70/	3.86(1)/				(0.090-0.721)
				β+, EC/10.1					
^{108}Te		107.9294	2.1 s	α /68 /	3.314(4)/	0+			
				β+, EC/32 /6.8					
^{109}Te		108.9274	4.6 s	β+ EC/96 /8.7					0.7523
				α/4 /	3.107(4)/				0.287–2.045
^{110}Te		109.9224	19. s	β+, EC/4.5		0+			ann.rad./
									0.2191
									0.6059
^{111}Te		110.9211	19.3 s	β+, EC/8.0		(7/2+)			ann.rad./
									0.267
									0.322
									0.341
^{112}Te		111.9170	2.0 m	β+, EC/4.3		0+			ann.rad./
									0.2962
									0.3727
									0.4187
^{113}Te		112.9159	1.7 s	β+ /85/5.7	4.5/	(7/2+)			ann.rad./
				EC/15/					Sb k x-ray
									0.8144
									1.0181
									1.1812
^{114}Te		113.91209	15. m	β+ /40/3.2		0+			ann.rad./
				EC/60/					Sb k x-ray
									0.0838
									0.0903
^{115m}Te			6.7 m	β+ /45/		(1/2+)			ann.rad./
				EC/55/					Sb k x-ray
									0.7236

Elem. or Isot.	Natural Abundance (Atom %)	Atomic Mass or Weight	Half-life/ Resonance Width (MeV)	Decay Mode/ Energy (/MeV)	Particle Energy/ Intensity (MeV/%)	Spin ($h/2\pi$)	Nuclear Magnetic Mom. (nm)	Elect. Quadr. Mom. (b)	γ-Energy / Intensity (MeV/%)
									0.7704
^{115}Te		114.91190	5.8 m	β+ /45/4.6	2.7/	7/2+			ann.rad./
				EC/55/					Sb k x-ray
									0.7236
									1.3268
									1.3806
									(0.22–2.7)
^{116}Te		115.90846	2.49 h	EC/1.5		0+			Sb k x-ray
									0.0937
^{117}Te		116.90865	1.03 h	EC/75/3.54	1.78/	½+			ann.rad./
				β+ /25/					Sb k x-ray
									0.9197
									1.7164
									2.3000
^{118}Te		117.90583	6.00 d	EC/0.28		0+			Sb k x-ray
^{119m}Te			4.69 d	EC/		11/2-	0.89		Sb k x-ray
									0.15360
									0.2705
									1.21271
^{119}Te		118.90640	16.0 h	β+ /2/2.293	0.627/	½+	0.25		ann.rad.
				EC/98/					Sb k x-ray
									0.6440
									0.6998
^{120}Te	0.09(1)	119.90402	1.9x10^{17} a	β$^+$ EC		0+			
^{121m}Te			~ 154. d	I.T. (89%)		11/2-	0.90		Te k x-ray
				EC (11%)					0.2122
^{121}Te		120.90494	16.8 d	EC/1.04		½+			Sb k x-ray
									0.5076
									0.5731
^{122}Te	2.55(12)	121.903044				0+			
^{123m}Te			119.7 d	I.T./0.247		11/2-	-0.93		Te k x-ray
									0.1590/84.1
^{123}Te	0.89(3)	122.904270	>9.2 × 10^{16} a	EC/0.051		½+	-0.736948		
^{124}Te	4.74(14)	123.902818				0+			
^{125m}Te			58. d	I.T./0.145		11/2-	-0.99	-0.06	Te k x-ray
									0.0355
^{125}Te	7.07(15)	124.904431				½+	-0.8885		
^{126}Te	18.84(25)	125.903312				0+			
^{127m}Te			109. d	I.T./98/0.088		11/2-	-1.04		Te k x-ray
				β- /2/0.77					0.0883
^{127}Te		126.905226	9.4 h	β- /0.698	0.696/	3/2+	0.635		0.3603
^{128}Te	31.74(8)	127.904463	2.2 × 10^{24} a	β-β-		0+			
^{129m}Te			33.6 d	I.T./63/0.105		11/2-	-1.09		Te k x-ray
				β- /37/	1.60/				0.45984
									0.6959
^{129}Te		128.906598	1.16 h	β- /1.498	0.99/9	3/2+	0.702	0.06	0.0278
					1.45/89				0.45984
									0.48728
^{130}Te	34.08(62)	129.906224	8 × 10^{20} a	β-β-		0+			
^{131m}Te			1.35 d	β- /78/2.4	0.42/	11/2-	-1.04		0.0811
				I.T./22/0.18					0.1021
									0.14973
									0.77369
									0.79375

Elem. or Isot.	Natural Abundance (Atom %)	Atomic Mass or Weight	Half-life/ Resonance Width (MeV)	Decay Mode/ Energy (/MeV)	Particle Energy/ Intensity (MeV/%)	Spin ($h/2\pi$)	Nuclear Magnetic Mom. (nm)	Elect. Quadr. Mom. (b)	γ-Energy / Intensity (MeV/%)
									0.85225
^{131}Te		130.908524	25.0 m	β- /2.233	1.35/12	3/2+	0.70		0.14973
					1.69/22				0.45327
					2.14/60				0.49269
^{132}Te		131.90855	3.26 d	β- /0.51	0.215	0+			0.049725
									0.11198
									0.22830
^{133m}Te			55.4 m	β- /82/	2.4/30	11/2-	-1.13		Te k x-ray
				I.T./18/0.334					0.0949
									0.1689
									0.3121
									0.3341
^{133}Te		132.91096	12.4 m	β- /2.94	2.25/25	3/2+			0.3121
					2.65				0.4079
									1.3334
^{134}Te		133.91137	42. m	β- /1.51	0.6/	0+			0.7672/29
					0.7/				(0.079–0.926)
^{135}Te		134.9165	19.0 s	β- /6.0	5.4/				0.267
					6.0				0.603
									0.870
^{136}Te		135.92010	17.5 s	β- /5.1	2.5/	0+			2.0779/25
									(0.087–3.235)
^{137}Te		136.9253	2.5 s	β- /98 /6.9	6.8	7/2-			0.2436
				n/2 /					
^{138}Te		137.9292	1.4 s	β- /6.4		0+			
^{139}Te		138.9347	> 0.15 μs						
^{140}Te		139.9389	> 0.15 μs			0+			
^{141}Te		140.9447	> 0.15 μs						
^{142}Te		141.949	> 0.15 μs			0+			
$_{53}$I		**126.90447(3)**							
^{108}I		107.9435	0.04 s	α/91/4.	3.95				
^{109}I		108.9382	93.5 μs	p					0.593/100
				α	α//0.014				0.717/63
									(0.496–1.057)
^{110}I		109.9352	0.65 s	β+, EC/83/11.4					ann.rad./
				α/17/~ 3.6	3.457(10)/				
				p/11/					
^{111}I		110.9303	2.5 s	β+, E.C./8.5					ann.rad./
									0.2665
									0.3215
									0.3412
^{112}I		111.9280	3.4 s	β+, EC/10.2					ann.rad./
									0.6889
									0.7869
^{113}I		112.9236	5.9 s	β+, EC/7.6					ann.rad./
									0.4625/100
									0.6224/74
									0.0550–1.422
^{114}I		113.9219	2.1 s	β+, EC/8.7					ann.rad./
									0.6826
									0.7088
^{115}I		114.9181	1.3 m	β+, EC/6.7		5/2+			ann.rad./
									0.275
									0.284

Elem. or Isot.	Natural Abundance (Atom %)	Atomic Mass or Weight	Half-life/ Resonance Width (MeV)	Decay Mode/ Energy (/MeV)	Particle Energy/ Intensity (MeV/%)	Spin ($h/2\pi$)	Nuclear Magnetic Mom. (nm)	Elect. Quadr. Mom. (b)	γ-Energy / Intensity (MeV/%)
									0.460
									0.709
^{116}I		115.9168	2.9 s	β+ /97/7.8	6.7/	1+			ann.rad./
				EC/3/					0.5402
									0.6789
^{117}I		116.91365	2.22 m	β+, EC/4.7	3.2/	(5/2+)	3.1		ann.rad./
									0.2744
									0.3259
^{118m}I			8.5 m	β+, EC/	4.9/	7-	4.2		ann.rad./
				I.T.					0.104
									0.5998
									0.6052
									0.6138
^{118}I		117.91307	14. m	β+, EC/7.0		2-	2.0		ann.rad./
									0.5448
									0.6052
									1.3384
^{119}I		118.91007	19. m	β+ /54/3.5	2.4/	(5/2+)	+2.9		ann.rad./
				EC/46/					Te k x-ray
									0.2575
^{120m}I			53. m	β+ /80/	3.8		4.2		ann.rad.
				EC/20/					Te k x-ray
									0.4257
									0.5604
									0.6147
									1.3459
^{120}I		119.91005	1.36 h	β+ /56/5.62	4.03	2-	1.23		ann.rad./
				EC/	4.60				Te k x-ray
									0.5604
									0.6411
									1.5230
									(0.111–3.1)
^{121}I		120.90737	2.12 h	β+ /13/2.27	1.2/	5/2+	2.3		ann.rad./
				EC/87/					Te k x-ray
									0.2122
									(0.14–1.1)
^{122}I		121.90759	3.6 m	β+ /4.234	3.1/	1+	+0.94		ann.rad./
				EC/					Te k x-ray
									0.5641
^{123}I		122.905589	13.2 h	EC/1.242		5/2+	2.82		Te k x-ray
									0.1590
^{124}I		123.906210	4.18 d	β+ /23/3.160	1.54/	2-	1.446		ann.rad./
				EC/77/	2.14/				Te k x-ray
					0.75/				0.6027/62.9
									0.7228/10.3
									1.6910/11.2
									(0.31–1.73)
^{125}I		124.904630	59.4 d	EC/0.1861		5/2+	2.82	-0.78	Te k x-ray
									0.0355
^{126}I		125.905624	13.0 d	EC/		2-	1.438		ann.rad./
				β+ /2.155	1.13/				Te k x-ray
				β- /1.258/47	0.87/				0.3887
					1.25/				0.6622
^{127}I	100.	126.904473				5/2+	+2.8133	-0.71	

Elem. or Isot.	Natural Abundance (Atom %)	Atomic Mass or Weight	Half-life/ Resonance Width (MeV)	Decay Mode/ Energy (/MeV)	Particle Energy/ Intensity (MeV/%)	Spin ($h/2\pi$)	Nuclear Magnetic Mom. (nm)	Elect. Quadr. Mom. (b)	γ-Energy / Intensity (MeV/%)
^{128}I		127.905809	25.00 m	β- /2.118	2.13/	1+			Te k x-ray
				EC/1.251					0.44287
									0.52658
^{129}I		128.904988	1.7 × 10^7 a	β- /0.194	0.15/	7/2+	+2.6210	-0.50	Xe k x-ray
									0.0396
^{130m}I			9.0 m	I.T./83/0.048		2+			I k x-ray
				β- /17/					0.5361
^{130}I		129.906674	12.36 h	β- /2.949	1.04/	5+	3.35		0.4180
					0.62				0.5361
									0.6685
									0.7395
^{131}I		130.906125	8.021 d	β- /0.971	0.606/	7/2+	+2.742	-0.35	0.08017
									0.28431
									0.36446
									0.63699
^{132m}I			1.39 h	IT		8-			
^{132}I		131.90800	2.283 h	β- /14/3.58	0.80/	4+	3.09	0.08	I k x-ray
				I.T./86/	1.03/				0.0980
					1.2/				0.5059
					1.6/				0.52264
					2.16/				0.63019
									0.6506
									0.66768
									0.77260
									0.95457
^{133m}I			9. s	I.T./1.63		19/2-			I kx-ray
									0.0730
									0.6474
									0.9126
^{133}I		132.907797	20.8 h	β- /1.77	1.24/85	7/2+	+2.86	-0.24	0.51056
									0.52989
									0.87537
^{134m}I			3.7 m	I.T./98/0.316		8-			I k x-ray
				β- /2/					0.0444
									0.2719
^{134}I		133.90974	52.6 m	β- /4.05	1.2/	4+			0.1354
									0.84702
									0.88409
^{135}I		134.91005	6.57 h	β- /2.63	0.9/	7/2+	2.940		0.2884
					1.3/				0.41768
									0.52658
									1.13156
									1.26046
^{136m}I			47. s	β- /	4.7/	6-			0.1973
					5.2/				0.3468
									0.3701
									0.3814
									1.3130
									(0.16–2.36)
^{136}I		135.91465	1.39 m	β- /6.93	4.3/	2-			0.3447
					5.6/				1.3130
									1.3211
									2.2896
									(0.3–6.1)

Elem. or Isot.	Natural Abundance (Atom %)	Atomic Mass or Weight	Half-life/ Resonance Width (MeV)	Decay Mode/ Energy (/MeV)	Particle Energy/ Intensity (MeV/%)	Spin ($h/2\pi$)	Nuclear Magnetic Mom. (nm)	Elect. Quadr. Mom. (b)	γ-Energy / Intensity (MeV/%)
^{137}I		136.91787	24.5 s	β- /5.88	5.0/	(7/2+)			0.6010
									1.2180
									1.2201
									1.3026
									1.5343
									(0.25–4.4)
^{138}I		137.9224	6.5 s	β- /7.8	6.9/	2-			0.4836
					7.4/				0.5888
									0.8752
									(0.4–5.3)
^{139}I		138.92610	2.30 s	β- /6.81					0.192
				n/					0.198
									0.273
									0.382
									0.386
									0.468
									0.683
									1.313
^{140}I		139.9310	0.86 s	β- /8.8		(3)			0.372
				n/					0.377
									0.457
^{141}I		140.9350	0.45 s	β- /7.8					
^{142}I		141.9402	~ 0.2 s	β-					
^{143}I		142.9446	> 0.15 μs						
^{144}I		143.9500	> 0.15 μs						
$_{54}$**Xe**		**131.293(6)**							
^{109}Xe			13. ms	α	α/3.92				
					α/4.06				
^{110}Xe		109.9443	0.11 s	β+ /9.2		0+			
				α	/~ 64				
^{111m}Xe			0.9 s	EC, β+					
^{111}Xe		110.9416	0.7 s	EC, β+ /10.6					
				α/	3.58(1)/				
^{112}Xe		111.9356	3. s	EC, β+ /7.2	α/0.8/	0+			
^{113}Xe		112.9333	2.8 s	EC, β+ /9.1					
^{114}Xe		113.92798	10.0 s	β+, EC/5.9		0+			ann.rad./
									0.1031
									0.1616
									0.3085
									0.6826
									0.7088
^{115}Xe		114.92629	18. s	β+, EC/7.6		(5/2+)			ann.rad./
^{116}Xe		115.92158	56. s	β+, EC/4.3	3.3/	0+			ann.rad./
									0.1042
									0.1916
									0.2477
									0.3107
									0.4127
^{117}Xe		116.92036	1.02 m	β+, EC/6.5		(5/2+)	-0.594	+1.16	ann.rad./
									0.2214
									0.5190
									0.6389
									0.6613
^{118}Xe		117.91618	~ 4. m	β+, EC/3.	2.7/	0+			ann.rad./

Elem. or Isot.	Natural Abundance (Atom %)	Atomic Mass or Weight	Half-life/ Resonance Width (MeV)	Decay Mode/ Energy (/MeV)	Particle Energy/ Intensity (MeV/%)	Spin ($h/2\pi$)	Nuclear Magnetic Mom. (nm)	Elect. Quadr. Mom. (b)	γ-Energy / Intensity (MeV/%)
									0.0535
									0.0600
									0.1199
^{119}Xe		118.91541	5.8 m	β+, EC/5.0	3.5/	7/2+	-0.654	+1.31	0.0873
									0.1000
									0.2318
									0.4615
^{120}Xe		119.91178	46. m	β+, EC/97/1.96		0+			I k x-ray
				β+ /3/					0.0251
									0.0726
									0.1781
									(0.1–1.03)
^{121}Xe		120.91146	39. m	β+ /44/3.73	2.8/	5/2+	-0.701	+1.33	ann.rad./
				EC/56/					I k x-ray
									0.1328
									0.2527
									0.4452
									(0.1–3.1)
^{122}Xe		121.90837	20.1 h	EC/0.9		0+			I k x-ray
									0.3501
^{123}Xe		122.90848	2.00 h	β+ /23/2.68	1.51/	½+	-0.150		ann.rad./
				EC/77/					I k x ray
									0.1489
									0.1781
									(0.1–2.1)
^{124}Xe	0.0952(3)	123.905893	> 10^{17} a	β β		0+			
^{125m}Xe			57. s	I.T./0.252		(9/2-)	-0.745	+0.42	Xe k x-ray
									0.1111
									0.141
^{125}Xe		124.906395	17.1 h	EC/1.653	0.47/	½+	-0.269		I k x-ray
									0.1884
									0.2434
^{126}Xe	0.0890(2)	125.90427				0+			
^{127m}Xe			1.15 m	I.T./0.297		(9/2-)	-0.884	+0.69	Xe k x-ray
									0.1246
									0.1725
^{127}Xe		126.905184	36.34 d	EC/0.662		½+	-0.504		I k x-ray
									0.1721
									0.2029
									0.3750
^{128}Xe	1.9102(8)	127.903531				0+			
^{129m}Xe			8.89 d	I.T./0.236		11/2-	-0.89122	+0.64	Xe k x-ray
									0.0396
									0.1966
^{129}Xe	26.4006(82)	128.904779				½+	-0.77798		
^{130}Xe	4.0710(13)	129.903508				0+			
^{131m}Xe			11.9 d	I.T./0.164		11/2-	-0.99405	+0.73	Xe k x-ray
									0.16398
^{131}Xe	21.2324(30)	130.905082				3/2+	+0.69186	-0.12	
^{132}Xe	26.9086(33)	131.904153				0+			
^{133m}Xe			2.19 d	I.T./0.233		11/2-	-1.082	+0.77	Xe k x-ray
									0.23325
^{133}Xe		132.905911	5.243 d	β- /0.427	0.346/99	3/2+	+0.813	+0.14	Cs k x-ray
									0.080998

Elem. or Isot.	Natural Abundance (Atom %)	Atomic Mass or Weight	Half-life/ Resonance Width (MeV)	Decay Mode/ Energy (/MeV)	Particle Energy/ Intensity (MeV/%)	Spin ($h/2\pi$)	Nuclear Magnetic Mom. (nm)	Elect. Quadr. Mom. (b)	γ-Energy / Intensity (MeV/%)
									0.1606
^{134}Xe	10.4357(21)	133.905394	$>1.1 \times 10^{16}$ a	β- β-		0+			
^{135m}Xe			15.3 m	I.T./		11/2-	1.1030	+0.62	Xe k x-ray
									0.52658
^{135}Xe		134.907227	9.10 h	β-/1.15	0.91/	3/2+	0.903	+0.21	0.24975
									0.60807
^{136}Xe	8.8573(44)	135.90722	$>8.5 \times 10^{21}$ a	β-β-		0+			
^{137}Xe		136.91156	3.82 m	β- /4.17	4.1/	7/2-	-0.97	-0.48	0.45549
					3.6/				0.8489
									0.9822
									1.2732
									1.7834
									2.8498
^{138}Xe		137.91395	14.1 m	β- /2.77	0.8/	0+			0.1538
					2.4/				0.2426
									0.2583
									0.4345
									1.76826
									2.0158
^{139}Xe		138.91879	39.7 s	β- /5.06	4.5/		-0.30	+0.40	0.1750
					5.0/				0.2186
									0.2965
									(0.1–3.37)
^{140}Xe		139.9216	13.6 s	β- /4.1	2.6	0+			0.0801
									0.6220
									0.8055
									1.4137
									(0.04–2.3)
^{141}Xe		140.9266	1.72 s	β- /6.2	6.2/	5/2+	+0.010	-0.58	0.1187
									0.9095
									(0.05–2.55)
^{142}Xe		141.9297	1.22 s	β- /5.0	3.7/	0+			0.0338
					4.2/				0.0729
									0.2038
									0.3091
									0.4145
									0.5382
									0.5718
									0.6181
									0.6448
^{143m}Xe			0.96 s	β^-					
^{143}Xe		142.9351	0.30 s	β- /7.3			-0.460	+0.93	
^{144}Xe		143.9385	1.2 s	β- /6.1		0+			
^{145}Xe		144.9441	0.9 s	β-, (n)					
^{146}Xe		145.9478	> 0.15 μs			0+			
^{147}Xe		146.9536	> 0.15 μs						
$_{55}$Cs		**132.9054519(2)**							
^{112}Cs		111.9503	0.5 ms	p	0.81				
^{113}Cs		112.9445	17. μs	p	0.96				
^{114}Cs		113.9414	0.58 s	β+, EC/11.8		1+			ann.rad./
									0.6826
									0.7088
^{115}Cs		114.9359	~ 1.4 s	β+, EC/8.4					ann.rad./
^{116m}Cs			0.7 s	β, EC/					ann.rad./

Elem. or Isot.	Natural Abundance (Atom %)	Atomic Mass or Weight	Half-life/ Resonance Width (MeV)	Decay Mode/ Energy (/MeV)	Particle Energy/ Intensity (MeV/%)	Spin ($h/2\pi$)	Nuclear Magnetic Mom. (nm)	Elect. Quadr. Mom. (b)	γ-Energy / Intensity (MeV/%)
									0.3935
^{116}Cs		115.9334	3.8 s	β+, EC/10.8					ann.rad./
									0.3935
									0.5243
									0.6151
									0.6223
^{117m}Cs			6.5 s	β+, EC/					
^{117}Cs		116.9287	~ 8.4 s	β+, EC/7.5					ann.rad./
^{118m}Cs			17. s	β+, EC/			5.		
^{118}Cs		117.92656	14. s	β+, EC/9.		2	+3.88	+1.4	ann.rad./
									0.3372
									0.4727
									0.5865
									0.5906
^{119m}Cs			29. s			3/2	+0.84	+0.9	
^{119}Cs		118.92238	43. s	β+, EC/6.3		9/2+	+5.46	+2.8	ann.rad./
									0.169
									0.176
									0.224
									0.257
^{120m}Cs			60. s	β+, EC/					
^{120}Cs		119.92068	64. s	β+, EC/7.92		2+	+3.87	+1.45	ann.rad./
									0.3224
									0.4735
									0.5534
									(0.3–3.28)
^{121m}Cs			2.0 m	I.T./60/		(9/2+)	+5.41	+2.7	ann.rad./
				β+ /40/	4.4				0.1794
									0.1961
^{121}Cs		120.91723	2.3 m	β+, EC/5.40	4.38/	3/2+	+0.77	+0.84	ann.rad./
									0.1537
									(0.08–0.56)
^{122m2}Cs			4.4 m	β+, EC		8-	+4.77	+3.3	ann.rad./
^{122m1}Cs			0.36 s	IT					0.3311
									0.4971
									0.6385
									(0.27–2.22)
^{122}Cs		121.91611	21. s	β+, EC/7.1	5.8/	(1+)	-0.133	-0.19	ann.rad./
									0.3311
									0.5120
									0.8179
^{123m}Cs			1.6 s	I.T./		11/2-			Cs k x-ray
									0.0946
^{123}Cs		122.91300	5.87 m	β+ /75/4.20	3.0/	1/2+	+1.38		ann.rad./
				EC/25/					Xe k x-ray
									0.0974
									0.5964
^{124m}Cs			6.3 s	IT		7+			
^{124}Cs		123.91226	30. s	β+ /9 /5.92	~ 5.	1+	+0.673	-0.74	ann.rad./
				EC/8 /					Xe k x-ray
									0.3539
									0.4925
									0.9418
^{125}Cs		124.90973	45. m	β+ /40/3.09	2.06/	1/2+	+1.41		ann.rad./

Elem. or Isot.	Natural Abundance (Atom %)	Atomic Mass or Weight	Half-life/ Resonance Width (MeV)	Decay Mode/ Energy (/MeV)	Particle Energy/ Intensity (MeV/%)	Spin (h/2 π)	Nuclear Magnetic Mom. (nm)	Elect. Quadr. Mom. (b)	γ-Energy / Intensity (MeV/%)
				EC/60/					Xe k x-ray
									0.112
									0.526
^{126}Cs		125.90945	1.64 m.	β+ /81/4.83	3.4	1+	+0.78	-0.68	ann.rad./
				EC/19/	3.7/				Xe k x-ray
									0.3886
									0.4912
									0.9252
^{127}Cs		126.90742	6.2 h	β+ /96/2.08	0.65/	1/2+	+1.46		Xe k x-ray
				EC/4/	1.06				0.1247
									0.4119
^{128}Cs		127.90775	3.62 m	β+ /68/3.930	2.44/	1+	+0.97	-0.57	ann.rad./
				EC/32 /	2.88/				Xe k x-ray
									0.4429
^{129}Cs		128.90606	1.336 d	EC/1.195		1/2+	+1.49		Xe k x-ray
									0.3719
									0.4115
^{130m}Cs			3.5 m	IT, β+, EC		5-	+0.629	+1.45	
^{130}Cs		129.90671	29.21 m	β+ /55/2.98	1.98/	1+	+1.46	-0.06	ann.rad./
				EC/43/					Xe k x-ray
				β- /1.6/0.37	0.44/1.6				0.5361
^{131}Cs		130.90546	9.69 d	EC/0.352		5/2+	+3.543	-0.58	Xe k x-ray
^{132}Cs		131.906434	6.48 d	EC/98/		2-	+2.22	+0.51	Xe k x-ray
				β+ /0.3/2.120					0.4646
				β- / /1.280					0.6302
									0.66769
^{133}Cs	100.	132.90545193				7/2+	+2.58291	-0.00355	
^{134m}Cs			2.91 h	I.T./0.139		8-	+1.0978	+1.0	Cs k x-ray
									0.12749
^{134}Cs		133.90671848	2.065 a	β- /2.059	0.089/27	4+	+2.994	+0.389	0.56327
					0.658/70				0.56935
				EC/1.22					0.60473
									0.79584
^{135m}Cs			53. m	I.T./1.627		19/2-	+2.18	+0.9	0.7869
									0.8402
^{135}Cs		134.905977	2.3×10^{6} a	β- /0.269	0.205/100	7/2+	+2.7324	+0.050	
^{136m}Cs			19. s	I.T./		8-	+1.32	+0.7	
^{136}Cs		135.907312	13.16 d	β- /2.548	0.341/	5+	+3.71	+0.23	0.06691
									0.34057
									0.81850
									1.04807
^{137}Cs		136.907089	30.2 a	β- /1.176	0.514/95	7/2+	+2.851	+0.051	Ba k x-ray
									0.66164
^{138m}Cs			2.9 m	I.T./75	/0.080	6-	+1.71	-0.40	Cs k x-ray
				β- /25 /	3.3				0.0799
									0.1917
									0.4628
									1.43579
^{138}Cs		137.91102	32.2 m	β- /5.37	2.9/	3-	+0.700	+0.12	0.1381
									0.46269
									1.00969
									1.43579
									2.21788
^{139}Cs		138.913364	9.3 m	β- /4.213	4.21	7/2+	+2.70	-0.07	0.6272

Elem. or Isot.	Natural Abundance (Atom %)	Atomic Mass or Weight	Half-life/ Resonance Width (MeV)	Decay Mode/ Energy (/MeV)	Particle Energy/ Intensity (MeV/%)	Spin ($h/2\pi$)	Nuclear Magnetic Mom. (nm)	Elect. Quadr. Mom. (b)	γ-Energy / Intensity (MeV/%)
									1.2832
									(0.4–3.66)
^{140}Cs		139.91728	1.06 m	β- /6.22	5.7/	1-	+0.133895	-0.11	0.5283
					6.21/				0.6023
									0.9084
									(0.41–3.94)
^{141}Cs		140.92005	24.9 s	β- /5.26	5.20/	7/2+	+2.44	-0.4	Ba k x-ray
									0.0485
									0.5616
									0.5887
									1.1940
									(0.05–3.33)
^{142}Cs		141.92430	1.8 s	β- /7.31	6.9/				0.3596
					7.28				0.9668
									1.1759
									1.3265
^{143}Cs		142.92735	1.78 s	β- /6.24	6.1	(3/2+)	+0.87	+0.47	0.1955
									0.2324
									0.3064
									(0.17–1.98)
^{144}Cs		143.93208	1.01 s	β- /8.47	8.46/	1	-0.546	+0.30	0.1993
					7.9/				0.5598
									0.6392
									0.7587
^{145}Cs		144.93553	0.59 s	β- /7.89	7.4/	3/2+	+0.784	+0.6	0.1126
					7.9/				0.1755
									0.1990
^{146}Cs		145.9403	0.322 s	β-, (n)/9.38	~ 9.0	2-	-0.515	+0.22	
^{147}Cs		146.9442	0.227 s	β-, (n)/9.3					(0.024-2.2798)
^{148}Cs		147.9492	0.15 s	β-, (n)/10.5					
^{149}Cs		148.9529	> 50 ms						
^{150}Cs		149.9582	> 50 ms						
^{151}Cs		[illegible]0.9622	> 50 ms						
$_{56}$**Ba**		**137.327(7)**							
^{114}Ba		113.9507	0.43 s	β+, (p)	p/20	0+			
				α	/0.9				
^{115}Ba		114.947	0.45 s	β+, (p)	p/<15				
^{116}Ba		115.9414	1.3 s	β+, (p)	p/3	0+			
^{117}Ba		116.9385	1.8 s	β+, (p), EC/8.4	p/16	(3/2-)			(0.046–0.364)
^{118}Ba		117.9330	5.2 s	β+,		0+			(0.040–0.156)
^{119}Ba		118.9307	5.4 s	β+, EC/8.					
^{120}Ba		119.9260	24. s	β+, EC/5.0		0+			ann.rad./
									0.140
									(0.075–0.146)
^{121}Ba		120.9241	30. s	β+, EC/6.8		5/2+	+0.660	+1.8	ann.rad./
^{122}Ba		121.91990	2.0 m	β+, EC/3.8		0+			ann.rad./
^{123}Ba		122.91878	2.7 m	β+, EC/5.5			-0.680	+1.5	ann.rad./
									0.0306
									0.0927
									0.1161
									0.1235
^{124}Ba		123.91509	12. m	β+, EC/2.65		0+			ann.rad./
									0.1695
									0.1888

Elem. or Isot.	Natural Abundance (Atom %)	Atomic Mass or Weight	Half-life/ Resonance Width (MeV)	Decay Mode/ Energy (/MeV)	Particle Energy/ Intensity (MeV/%)	Spin ($h/2\pi$)	Nuclear Magnetic Mom. (nm)	Elect. Quadr. Mom. (b)	γ-Energy / Intensity (MeV/%)
									1.2160
^{125m}Ba			8. m	β+, EC/	4.5		0.174		
^{125}Ba		124.9145	3.5 m	β+, EC/4.6	3.4	½+	+0.18		ann.rad./
									0.0550
									0.0776
									0.0854
									0.1409
^{126}Ba		125.91125	1.65 h	β+ /2/1.67		0+			Cs k x-ray
				EC/98 /					0.2179
									0.2336
									0.2576
^{127m}Ba			1.9 s	IT		7/2-	-0.723	+1.6	
^{127}Ba		126.91109	12.9 m	β+ /54/3.5		1/2+	+0.083		ann.rad./
				EC/46/					Cs k x-ray
									0.1148
									0.1808
									(0.07–2.5)
^{128}Ba		127.90832	2.43 d	EC/0.52		0+			Cs k x-ray
									0.27344
^{129m}Ba			2.17 h	EC/98/		7/2+	+0.93	+1.6	Cs k x-ray
				β+ /2/					0.1769
									0.1823
									0.2023
									1.4593
^{129}Ba		128.90868	2.2 h	β+ /20/2.43	1.42/	1/2+	-0.40		ann.rad./
				EC/80/					Cs k x-ray
									0.1291
									0.2143
									0.2208
^{130m}Ba			9.5 ms	I.T./2.475	/100.	8-	-0.04	+2.8	0.080–0.802
^{130}Ba	0.106(1)	129.906321	2.2 × 10^{21} a	β+β+		0+			
^{131m}Ba			14.6 m	I.T./0.187		9/2-	-0.87	+1.5	Ba k x-ray
									0.1085
^{131}Ba		130.906941	11.7 d	EC/1.37		1/2+	0.70811		Cs k x-ray
									0.12381/28.4
									0.21608/21.3
									0.49636/42.9
									(0.055–1.171)
^{132}Ba	0.101(1)	131.905061	1.3 × 10^{21} a	EC EC		0+			
^{133m}Ba			1.621 d	I.T./0.288		11/2-	-0.91	+0.9	Ba k x-ray
									0.2761
^{133}Ba		132.906008	10.53 a	EC/0.517		1/2+	0.77167		Cs k x-ray
									0.08099
									0.35600
^{134}Ba	2.417(18)	133.904508				0+			
^{135m}Ba			1.20 d	I.T./0.2682		11/2-	-1.00	+1.0	Ba k x-ray
									0.2682
^{135}Ba	6.592(12)	134.9056886				3/2+	+0.83863	+0.160	
^{136m}Ba			0.308 s	I.T./2.0305		7-			Ba k x-ray
									0.8185
									1.0481
^{136}Ba	7.854(24)	135.9045759				0+			
^{137m}Ba			2.552 m	I.T./0.6617		11/2-	-0.99	+0.8	Ba k x-ray
									0.66164

Elem. or Isot.	Natural Abundance (Atom %)	Atomic Mass or Weight	Half-life/ Resonance Width (MeV)	Decay Mode/ Energy (/MeV)	Particle Energy/ Intensity (MeV/%)	Spin (*h*/2 π)	Nuclear Magnetic Mom. (nm)	Elect. Quadr. Mom. (b)	γ-Energy / Intensity (MeV/%)
^{137}Ba	11.232(24)	136.9058274				3/2+	+0.93737	+0.245	
^{138}Ba	71.698(42)	137.9052472				0+			
^{139}Ba		138.9088412	1.396 h	β- /2.317	2.14/27	7/2-	-0.97	-0.57	0.16585
					2.27/72				1.2544
									1.42033
^{140}Ba		139.91060	12.75 d	β- /1.05	0.48	0+			0.16268
					1.0/				0.30485
					1.02/				0.53727
^{141}Ba		140.91441	18.3 m	β- /3.22	2.59/	3/2-	-0.34	+0.45	0.1903
					2.73/				0.2770
									0.3042
									(0.1–2.5)
^{142}Ba		141.91645	10.7 m	β- /2.212	1.0/	0+			0.23152
					1.10/				0.25512
									0.3090
									1.2040
^{143}Ba		142.92063	14.3 s	β- /4.24	4.2/	5/2+	+0.44	-0.88	0.1786
									0.21148
									0.7988
									(0.17–2.4)
^{144}Ba		143.92295	11.4 s	β- /3.1	2.4/	0+			La k x-ray
					2.9/				0.10386
									0.1566
									0.1728
									0.3882
									0.43048
^{145}Ba		144.9276	4.0 s	β- /4.9	4.9/	(5/2-)	-0.28	+1.22	La k x-ray
									0.0918
									0.09709
^{146}Ba		145.9302	2.20 s	β- /4.12	3.9/	0+			0.0644
									0.2513
									0.3270
									0.3329
									0.3622
^{147}Ba		146.9349	0.892 s	β- /5.75	5.5/				
^{148}Ba		147.9377	0.64 s	β-, n/5.11		0+			
^{149}Ba		148.9426	0.36 s	β-, (n)/7.3					
^{150}Ba		149.9457	0.3 s			0+			
^{151}Ba		150.9508	> 0.15 μs						
^{152}Ba		151.9543				0+			
^{153}Ba		151.960							
$_{57}$**La**		**138.90547(7)**							
^{117}La		116.9501	23 ms	p	0.806/	3/2+			
^{118}La		117.9467							
^{119}La		118.9410							
^{120}La		119.9381	2.8 s	EC, β+ /11.					
^{121}La		120.9330	5.3 s						
^{122}La		121.9307	9. s	EC, β+/~ 9.7					
^{123}La		122.9262	17. s	EC/7.					
^{124}La		123.9246	30. s	EC/ ~ 8.8		(7+)			
^{125m}La			0.39 s						
^{125}La		124.92082	1.2 m	β+, EC/5.6		11/2-			ann.rad./
									0.0436
									0.0676

Elem. or Isot.	Natural Abundance (Atom %)	Atomic Mass or Weight	Half-life/ Resonance Width (MeV)	Decay Mode/ Energy (/MeV)	Particle Energy/ Intensity (MeV/%)	Spin ($h/2\pi$)	Nuclear Magnetic Mom. (nm)	Elect. Quadr. Mom. (b)	γ-Energy / Intensity (MeV/%)
^{126m}La			< 50. s						
^{126}La		125.9195	54. s	β+, EC/7.6					ann.rad./
									0.256
									0.455
									0.117–3.853
^{127}La		126.91638	3.8 m	β+, EC/4.7		3/2+			ann.rad./
									0.025
									0.0562
^{128}La		127.9156	5.0 m	β+ /80/6.7		(5-)			ann.rad./
				EC/20/					Ba k x-ray
									0.2841/87
									0.4793/54
									(0.315–2.212)
^{129m}La			0.56 s	IT		(11/2-)			
^{129}La		128.91269	11.6 m	β+ /58/3.72	2.42/	3/2+			ann.rad./
				EC/42/					Ba k x-ray
									0.1105
									0.2786
									(0.1–1.8)
^{130}La		129.91237	8.7 m	β+ /78/5.6		3+			ann.rad./
				EC/22/					Ba k x-ray
									0.3573/81
									0.5506/27
									(0.1965–1.989)
^{131}La		130.91007	59. m	β+ /76/3.0	1.42/	3/2+			ann.rad./
				EC/24/	1.94/				Ba k x-ray
									0.1085
									0.3658
									0.5263
^{132m}La			24. m	I.T./76/		6-			La k x-ray
				β+, EC/24/					0.1352
									0.4645
^{132}La		131.91010	4.8 h	β+ /40/4.71	2.6/	2-			ann.rad./
				EC/60/	3.2				Ba k x-ray
					3.7/				0.4645
									0.5671
^{133}La		132.90822	3.91 h	β+ /4/2.2	1.2/	5/2+			Ba k x-ray
				EC/96/					0.2788
									0.2901
									0.3024
^{134}La		133.90851	6.5 m	β+ /63/3.71	2.67/	1+			ann.rad./
				EC/37/					Ba k x-ray
									0.6047
									(0.5–1.9)
^{135}La		134.90698	19.5 h	EC/1.20		5/2+			Ba k x-ray
									0.4805
^{136}La		135.9076	9.87 m	β+ /36/2.9	1.8/	1+	+3.7	~ -0.4	ann.rad./
				EC/64/					Ba k x-ray
									0.8185
^{137}La		136.90649	6 × 10^{4} a	EC/0.60		7/2+	+2.70	+0.21	0.2836
^{138}La	0.0888(6)	137.907112	1.06 × 10^{11} a			5+	+3.71365	+0.44	1.4358/65
									0.7887/35
^{139}La	99.9112(6)	138.906353				7/2+	+2.783046	+0.20	
^{140}La		139.909478	1.678 d	β- /3.762	1.35	3-	+0.73	+0.09	

Elem. or Isot.	Natural Abundance (Atom %)	Atomic Mass or Weight	Half-life/ Resonance Width (MeV)	Decay Mode/ Energy (/MeV)	Particle Energy/ Intensity (MeV/%)	Spin ($h/2\pi$)	Nuclear Magnetic Mom. (nm)	Elect. Quadr. Mom. (b)	γ-Energy / Intensity (MeV/%)
					1.24/				
					1.67/				
^{141}La		140.910962	3.90 h	β- /2.502	2.43/	7/2+			
^{142}La		141.91408	1.54 h	β- /4.505	2.11/	2-			
					2.98/				
					4.52/				
^{143}La		142.91606	14.1 m	β- /3.43	3.3/	7/2-			
^{144}La		143.91960	40.7 s	β- /5.5	4.1/				
^{145}La		144.9216	24. s	β- /4.1	4.1/	3/2+			
^{146m}La			10.0 s	β- /6.7	5.5/	(6)			
^{146}La		145.9258	6.3 s	β- /6.6	6.2/	(2-)			
^{147}La		146.9282	4.02 s	β- /5.0	4.6/				
^{148}La		147.9322	1.1 s	β- /7.26		2-			
^{149}La		148.9347	1.10 s	β- /5.5					0.1335
									0.009–1.709
^{150}La		149.9388	0.51 s						x-ray
									(0.097–0.209)
^{151}La		150.9417	> 0.15 μs						
^{152}La		151.9462	> 0.15 μs						
^{153}La		152.950	> 0.15 μs						
^{154}La		153.955							
^{155}La		154.958							
$_{58}$Ce		**140.116(1)**							
^{119}Ce		118.953							
^{120}Ce		119.947				0+			
^{121}Ce		120.943	1.1 s	β+, p					
^{122}Ce		121.9379				0+			
^{123}Ce		122.9354	3.8 s	β+, EC/~ 8.6					ann.rad./
^{124}Ce		123.9304	6. s	EC/~ 5.6		0+			
^{125m}Ce			2. m						
^{125}Ce		124.9284	9.6 s	β+, EC/7.		7/2-			ann.rad./
									0.1346
									0.1666
									0.056–1.329
^{126}Ce		125.92397	50. s	EC/4.		0+			
^{127}Ce		126.9227	29. s	β+, EC/6.1					ann.rad./
									(0.058–1.961)
^{128}Ce		127.91891	4.1 m	β+, EC/3.2		0+			ann.rad./
									(0.023–0.880)
^{129}Ce		128.91810	3.5 m	β+, EC/5.6					ann.rad./
									(0.0675–1.015)
^{130}Ce		129.91474	26. m	β+, EC/2.2		0+			ann.rad./
									La k x-ray
									0.047–1.431
^{131m}Ce			5. m	β+ EC					ann.rad./
									0.2304
									0.3955
									0.4213
^{131}Ce		130.91442	10. m	β+, EC/4.0	2.8/				ann.rad.
									0.119
									0.169
									0.414
^{132m}Ce			9.4 ms	IT/2.340					0.3255
									0.10–0.955

Elem. or Isot.	Natural Abundance (Atom %)	Atomic Mass or Weight	Half-life/ Resonance Width (MeV)	Decay Mode/ Energy (/MeV)	Particle Energy/ Intensity (MeV/%)	Spin (h/2 π)	Nuclear Magnetic Mom. (nm)	Elect. Quadr. Mom. (b)	γ-Energy / Intensity (MeV/%)
^{132}Ce		131.91146	3.5 h	EC/1.3		0+			La k x-ray
									0.1554
									0.1821
^{133m}Ce			1.6 h	β+, EC/		½+			ann.rad.
									0.0769
									0.0973
									0.5577
^{133}Ce		132.91152	5.4 h	β+/8/2.9	1.3/	9/2-			ann.rad.
				EC/92/					0.0584
									0.1308
									0.4722
									0.5104
^{134}Ce		133.90892	3.16 d	EC/0.5		0+			La k x-ray
									0.1304
									0.1623
									0.6047
^{135m}Ce			20. s	I.T./0.446		11/2-			Ce k x-ray
									0.0826
									0.1497
									0.2134
^{135}Ce		134.90915	17.7 h	β+/1 /2.026	0.8/	1/2+			La k x-ray
				EC/99 /					0.0345
									0.2656
									0.3001
									0.6068
^{136}Ce	0.185(2)	135.90717	>0.7 × 10^{14} a	EC EC		0+			
^{137m}Ce			1.43 d	I.T./99 /0.254		11/2-	1.0		Ce k x-ray
				EC/0.8 /					0.1693
									0.2543
^{137}Ce		136.90781	9.0 h	β+/1.222		3/2+	0.96		La k x-ray
									0.4472
^{138}Ce	0.251(2)	137.90599	>0.9 × 10^{14} a	EC EC		0+			
^{139m}Ce			56.4 s	I.T./0.7542		11/2-			Ce k x-ray
									0.7542
^{139}Ce		138.90665	137.6 d	EC/0.28		3/2+	1.06		La k x-ray
									1.320/72.1
									0.255/59.6
									0.825/45.8
									(0.231-2.364)
^{140}Ce	88.450(51)	139.905439				0+			
^{141}Ce		140.908276	32.50 d	β-/0.581	0.436/69	7/2-	1.1		Pr k x-ray
					0.581/31				0.14544/48.0
^{142}Ce	11.114(51)	141.909244	>1.6 × 10^{17} a	β- β-		0+			
^{143}Ce		142.912386	1.38 d	β-/1.462	1.404/	3/2-	0.43		Pr k x-ray
					1.110/47				0.0574
									0.2933
^{144}Ce		143.913647	284.6 d	β-/0.319	0.185/20	0+			Pr k x-ray
					0.318/				0.0801
									0.1335
^{145}Ce		144.91723	3.00 m	β-/2.54	1.7/24	3/2-			Pr k x-ray
					1.3				0.0627
									0.7245
^{146}Ce		145.9188	13.5 m	β-/1.04	0.7/90	0+			Pr k x-ray
									0.0986

Elem. or Isot.	Natural Abundance (Atom %)	Atomic Mass or Weight	Half-life/ Resonance Width (MeV)	Decay Mode/ Energy (/MeV)	Particle Energy/ Intensity (MeV/%)	Spin (*h*/2 π)	Nuclear Magnetic Mom. (nm)	Elect. Quadr. Mom. (b)	γ-Energy / Intensity (MeV/%)
									0.2182
									0.3167
									0.0930
^{147}Ce		146.92267	56. s	β-/3.29	3.3/				0.2687
^{148}Ce		147.92443	56. s	β-/2.1	1.66/	0+			0.0904
									0.0985
									0.1212
									0.2918
^{149}Ce		148.9284	5.2 s	β-/4.2					0.0577
									0.0864
									0.3800
									0.1099
^{150}Ce		149.93041	4.4 s	β-/3.0		0+			0.0526
^{151}Ce		150.9340	1.8 s	β-/5.3					Pr k x-ray
									(0.035-0.637)
^{152}Ce		151.9365	1.4 s	β-/4.4		0+			0.098
									0.115
^{153}Ce		152.9406	> 0.15 μs						
^{154}Ce		153.9434	> 0.15 μs			0+			
^{155}Ce		154.948	> 0.15 μs						
^{156}Ce		155.951				0+			
^{157}Ce		156.956							
$_{59}$Pr		**140.90765(2)**							
^{121}Pr		120.955	0.01 s	p	p/0.882				
^{122}Pr		121.9518							
^{123}Pr		122.946							
^{124}Pr		123.943	1.2 s	β+, EC/12.					ann.rad./
^{125}Pr		124.9378	~ 3.3 s	β+					ann.rad./
									0.1358
^{126}Pr		125.9353	3.1 s	β+, EC/~ 10.4					ann.rad./
									(0.170–0.985)
^{127}Pr		126.9308	4.2 s	β+ /~ 7.5					ann.rad./
									(0.028–0.8949)
^{128}Pr		127.92879	3.0 s	β+, EC/~ 9.3					ann.rad./
									0.207/100
									0.400–1.373
^{129}Pr		128.92510	32 s	β+, EC/5.8					ann.rad./
									(0.0395–1.865)
^{130}Pr		129.9236	40. s	β+, EC/8.1					ann.rad./
^{131m}Pr			5.7 s						(0.06–0.16)
^{131}Pr		130.9203	1.7 m	β+, EC/5.3				~ 5.5	ann.rad./
									(0.059–0.980)
^{132}Pr		131.9193	1.6 m	β+, EC/7.1					ann.rad./
									0.325
									0.496
									0.533
^{133m}Pr			1.1 s	IT/0.192					0.1305
									0.0617
^{133}Pr		132.91633	6.5 m	β+, EC/4.3		5/2+			ann.rad./
									0.074
									0.1343
									0.2419
									0.3156
									0.3308

Elem. or Isot.	Natural Abundance (Atom %)	Atomic Mass or Weight	Half-life/ Resonance Width (MeV)	Decay Mode/ Energy (/MeV)	Particle Energy/ Intensity (MeV/%)	Spin (h/2 π)	Nuclear Magnetic Mom. (nm)	Elect. Quadr. Mom. (b)	γ-Energy / Intensity (MeV/%)
									0.4650
^{134m}Pr			~ 11. m	β+, EC/					ann.rad./
									0.294
									0.460
									0.495
									0.632
^{134}Pr		133.91571	17. m	β+, EC/6.2		2+			ann.rad./
									0.294
									0.495
^{135}Pr		134.91311	24. m	β+, EC/3.7	2.5/	3/2+			ann.rad./
									0.0826
									0.2135
									0.2961
									0.5832
^{136}Pr		135.91269	13.1 m	β+ /57 /5.13	2.98/	2+			ann.rad./
				EC/43					Ce k x-ray
									0.5398
									0.5522
^{137}Pr		136.91071	1.28 h	β+ /26 /2.70	1.68/	5/2+			ann.rad./
				EC/74 /					Ce k x-ray
									0.4339
									0.5140
									0.8367
									(0.16–1.8)
^{138m}Pr			2.1 h	β+ /24 /	1.65/	7-			ann.rad./
				EC/76 /					Ce k x-ray
									0.3027
									0.7887
									1.0378
									(0.07–2.0)
^{138}Pr		137.91075	1.45 m	β+ /75 /4.44	3.42/	1+			ann.rad./
				EC/25 /					Ce k x-ray
									0.7887
^{139}Pr		138.90894	4.41 h	β+ /8 /2.129	1.09/	5/2+			ann.rad./
				EC/92 /					Ce k x-ray
									0.2551
									1.3473
									1.6307
^{140}Pr		139.90908	3.39 m	β+ /51 /3.39	2.37/	1+			ann.rad./
				EC/49 /					Ce k x-ray
									0.3069
									1.5965
^{141}Pr	100.	140.907653				5/2+	+4.275	-0.08	
^{142m}Pr			14.6 m	I.T./0.004	c.e.	5-	2.2		
^{142}Pr		141.910045	19.12 h	β- /2.162	0.58/4	2-	+0.234	+0.03	0.5088
				EC/0.744	2.16/96				1.57580
^{143}Pr		142.910817	13.57 d	β- /0.934	0.933/	7/2+	+2.70	+0.8	0.7420
^{144m}Pr			7.2 m	IT/99+/0.059		3-			Pr k x-ray
				β- /					0.0590
									0.6965
									0.8142
^{144}Pr		143.913305	17.28 m	β- /2.998	0.807/1	0-			0.69649
					2.30/				1.48912
					2.996/98				2.18562

Elem. or Isot.	Natural Abundance (Atom %)	Atomic Mass or Weight	Half-life/ Resonance Width (MeV)	Decay Mode/ Energy (/MeV)	Particle Energy/ Intensity (MeV/%)	Spin ($h/2\pi$)	Nuclear Magnetic Mom. (nm)	Elect. Quadr. Mom. (b)	γ-Energy / Intensity (MeV/%)
^{145}Pr		144.91451	5.98 h	β- /1.81	1.80/97	7/2+			0.0725
									0.6758
									0.7483
^{146}Pr		145.9176	24.2 m	β- /4.2	2.2/30	2-			0.4539/48
					3.7/10				1.5247
					4.2/40				
^{147}Pr		146.91900	13.4 m	β- /2.69	1.5/	3/2+			0.3146/24.
					2.1/				0.5779/16
									0.6413/19.
^{148m}Pr			2.0 m	β- /	4.0/	(4)			0.3016
					3.8/				0.4506
									0.6975
^{148}Pr		147.92213	2.27 m	β- /4.9	4.8/	1-			0.3017
					4.5/				
^{149}Pr		148.9237	2.3 m	β- /3.40	3.0	(5/2+)			0.1085
									0.1385
									0.1651
^{150}Pr		149.92667	6.2 s	β- /5.7		1-			0.1302
					~ 5.5				0.8044
									0.8527
^{151}Pr		150.92832	22.4 s	β- /4.2					
^{152}Pr		151.9315	3.2 s	β- /6.7		4+			0.0726
									0.164
									0.285
^{153}Pr		152.9338	4.3 s	β- /5.5					
^{154}Pr		153.9375	2.3 s	β- /7.9					
^{155}Pr		154.9401	> 0.3 μs						
^{156}Pr		155.9443	> 0.3 μs						
^{157}Pr		156.9474							
^{158}Pr		157.952							
^{159}Pr		158.956							
$_{60}$Nd		**144.242(3)**							
^{124}Nd		123.952				0+			
^{125}Nd		124.9489	0.6 s	β+, p					
^{126}Nd		125.9432				0+			
^{127}Nd		126.9405	1.8 s	β+, EC/9.		(5/2)			ann.rad./
^{128}Nd		127.9354	4. s	β+, EC/6.					ann.rad./
^{129}Nd		128.9332	4.9 s	β+, EC/8.		5/2(-)			ann.rad./
									(0.091–0.875)
^{130}Nd		129.92851	28. s	β+, EC/5.		0+			ann.rad./
^{131}Nd		130.92725	0.5 m	β+, EC/6.6					ann.rad./
									(0.09–0.36)
^{132}Nd		131.92332	1.5 m	β+, EC/3.7		0+			ann.rad./
									(0.099–0.567)
^{133}Nd		132.92235	1.2 m	β+, EC/5.6					ann.rad./
									(0.06–0.37)
^{134}Nd		133.91879	~ 8.5 m	β+ /17 /2.8		0+			ann.rad./
				EC/83 /					Pr k x-ray
									0.1631/58
									(0.09–1.00)
^{135m}Nd			5.5 m	β+ /					
^{135}Nd		134.91818	12. m	β+ /65 /4.8		9/2-	-0.78	+1.9	ann.rad./
				EC/35 /					Pr k x-ray
									0.0415/23.

Elem. or Isot.	Natural Abundance (Atom %)	Atomic Mass or Weight	Half-life/ Resonance Width (MeV)	Decay Mode/ Energy (/MeV)	Particle Energy/ Intensity (MeV/%)	Spin ($h/2\pi$)	Nuclear Magnetic Mom. (nm)	Elect. Quadr. Mom. (b)	γ-Energy / Intensity (MeV/%)
									0.204/51.
									(0.11–1.8)
^{136}Nd		135.91498	50.6 m	EC/94 /2.21	1.04/	0+			Pr kx-ray
				β+ /6 /					0.0401/21.
									0.1091/35.
									(0.10–0.97)
^{137m}Nd			1.6 s	I.T./0.5196		11/2-			Nd k x-ray
									0.1084
									0.1775
									0.2337
^{137}Nd		136.91457	38. m	β+ /40 /3.69	1.7/20	1/2+	-0.63		ann.rad./
				EC/60 /	2.40/20				Pr k x-ray
									0.0755
									0.5806
^{138}Nd		137.91195	5.1 h	EC/1.1		0+			Pr k x-ray
									0.1995
									0.3258
^{139m}Nd			5.5 h	I.T./12 /0.231	1.17/	11/2-			Nd k x-ray
				β+ /88 /					Pr k x-ray
									0.1139/34.
									0.7382/30.
^{139}Nd		138.91198	30. m	β+ /25 /2.79	1.77/	3/2+	+0.91	+0.3	ann.rad./
				EC/75 /					Pr k x-ray
									0.4050
^{140}Nd		139.90955	3.37 d	EC /0.22		0+			Pr k x-ray
^{141m}Nd			1.04 m	IT/99+/0.756		11/2-			Nd k x-ray
									0.7565
^{141}Nd		140.909610	2.49 h	EC/98 /1.823	0.802/	3/2+	+1.01	+0.3	Pr k x-ray
				β+ /2 /					(0.15–1.7)
^{142}Nd	27.153(40)	141.907723				0+			
^{143}Nd	12.173(26)	142.909814				7/2-	-1.07	-0.60	
^{144}Nd	23.798(19)	143.910087	2.1×10^{15} a	α	1.83	0+			
^{145}Nd	8.293(12)	144.912574				7/2-	-0.66	-0.31	
^{146}Nd	17.189(32)	145.913117				0+			
^{147}Nd		146.916100	10.98 d	β- /0.896	0.805/	5/2-	0.58	0.9	Pr k x-ray
									0.53102
									0.09111–0.686
^{148}Nd	5.756(21)	147.916893				0+			
^{149}Nd		148.920149	1.73 h	β- /1.691	1.03/25	5/2-	0.35	1.3	Pr k x-ray
					1.13/26				0.1143/19.
					1.42/				0.2113/27.
									(0.026–1.6)
^{150}Nd	5.638(28)	149.920891	1.4×10^{20} a	β-β-		0+			
^{151}Nd		150.923829	12.4 m	β- /2.442	1.2/	(3/2+)			Pm k x-ray
									0.1168
									0.2557
									1.1806
									(0.10–1.9)m
^{152}Nd		151.92468	11.4 m	β- /1.1		0+			0.2785/29.
									0.2501/18.
									(0.016–0.66)
^{153}Nd		152.92770	28.9 s	β- /3.6					0.418
^{154}Nd		153.9295	25.9 s	β- /2.8		0+			0.1519
									0.7998

Elem. or Isot.	Natural Abundance (Atom %)	Atomic Mass or Weight	Half-life/ Resonance Width (MeV)	Decay Mode/ Energy (/MeV)	Particle Energy/ Intensity (MeV/%)	Spin ($h/2\pi$)	Nuclear Magnetic Mom. (nm)	Elect. Quadr. Mom. (b)	γ-Energy / Intensity (MeV/%)
^{155}Nd		154.9329	8.9 s	β- /5.0					0.1807
^{156}Nd		155.9350	5.5 s	β- /4.1		0+			0.0848
^{157}Nd		156.9390	> 0.3 μs						
^{158}Nd		157.9416	> 0.3 μs			0+			
^{159}Nd		158.946							
^{160}Nd		159.949				0+			
^{161}Nd		160.954							
$_{61}$**Pm**									
^{128}Pm		127.9484	1.0 s	β+, p					ann.rad.
^{129}Pm		128.9432	~ 2.4 s						
^{130}Pm		129.9405	2.5 s	β+, EC/11.					0.1589
									0.326–1.062
^{131}Pm		130.9359	~ 6.3 s	β+					0.185
									0.220
									0.146
^{132}Pm		131.9338	6. s	β+, EC/10.					ann.rad./
^{133}Pm		132.92978	12. s	β+, EC/~ 7.0					ann.rad./
^{134}Pm		133.9284	24. s	β+, EC/~ 8.9		(5+)			ann.rad./
									0.294
									0.495
^{135}Pm		134.9249	0.8 m	β+, EC/6.0		11/2-			(0.13–0.47)
^{136}Pm		135.9236	1.8 m	β+ /89 /7.9		(3+)			ann.rad./
				EC/11 /					Nd k x-ray
									0.3735
									0.6027
^{137}Pm		136.92048	2.4 m	β+, EC/5.6		(11/2-)			ann.rad./
									0.1086
									0.1775
^{138m}Pm			3.2 m	β+ /50 /~ 7.0	3.9/	3+	3.		ann.rad./
				EC/50 /					Nd k x-ray
									0.5209
									0.7290
^{138}Pm		137.91955	10. s	β+ /6.9	6.1/	1+			ann.rad./
^{139m}Pm			0.18 s	IT/		(11/2-)			0.1887
^{139}Pm		138.91680	4.14 m	β+ /68 /4.52	3.52/	(5/2+)			ann.rad./
				EC/32 /					Nd k x-ray
									0.4028
									(0.27–2.4)
^{140m}Pm			5.87 m	β+ /70 /	3.2	7/2-			ann.rad./
				EC/30 /					Nd k x-ray
									0.4199
									0.7738
									1.0283
^{140}Pm		139.91604	9.2 s	β+ /89 /6.09	5.07/74	1+			ann.rad./
				EC/11 /					Nd k x-ray
									0.7738
									1.4898
^{141}Pm		140.91356	20.9 m	β+ /52 /3.72	2.71	5/2+			ann.rad./
				EC/48 /					Nd k x-ray
									0.8862
									1.2233
^{142m}Pm			67 μs						(0.208-0.882)
^{142}Pm		141.91287	40.5 s	β+ /86 /4.87	3.8/	1+			ann.rad./
				EC/20 /					Nd k x-ray

Elem. or Isot.	Natural Abundance (Atom %)	Atomic Mass or Weight	Half-life/ Resonance Width (MeV)	Decay Mode/ Energy (/MeV)	Particle Energy/ Intensity (MeV/%)	Spin ($h/2\pi$)	Nuclear Magnetic Mom. (nm)	Elect. Quadr. Mom. (b)	γ-Energy / Intensity (MeV/%)
									0.6414
									1.5758
^{143}Pm		142.910933	265. d	EC/1.041		5/2+	3.8		Nd k x-ray
				β+ /< 6 × 10^{-6}/					0.7420
^{144}Pm		143.912591	360. d	EC/2.332		5-	1.7		Nd k x-ray
				β+ /7 × 10^{-6}/					0.6180
									0.6965
^{145}Pm		144.912749	17.7 a	EC/0.163		5/2+	+3.8	+0.2	Nd k x-ray
									0.0723
^{146}Pm		145.914696	5.53 a	EC/63 /1.472		3-			Nd k x-ray
				β- /37 /1.542	0.795/				0.4538
									0.7362
									0.7474
^{147}Pm		146.915139	2.623 a	β- /0.224	0.224/	7/2+	+2.6	+0.6	0.1213
									0.1974
^{148m}Pm			41.3 d	β- /95 /2.6	0.4/60	6-	1.8		0.5503/94.
				I.T./5 /0.137	0.5/17				0.6300/89.
					0.7/21				0.7257/33
^{148}Pm		147.91748	5.37 d	β- /2.47	1.02/	1-	+2.0	~ +0.2	0.5503
					2.47/				0.9149
									1.4651
^{149}Pm		148.918334	2.212 d	β- /1.071	0.78/9	7/2+	3.3		0.2859
					1.072/90				0.5909
									0.8594
^{150}Pm		149.92098	2.68 h	β- /3.45	1.6/	(1-)			0.3339/69.
					2.3/				1.1658/16.
					1.8/				1.3245/17.
									(0.25–2.9)
^{151}Pm		150.92121	1.183 d	β- /1.187	0.84/	5/2+	+1.8	1.9	0.1677/8
									0.2751/7
									0.3401/22
^{152m2}Pm			15. m	β-, I.T./		(>6)			(0.14–1.4)
^{152m1}Pm			7.5 m	β- /		(4-)			0.1218
									0.2447
									0.3404
									1.0971
									1.4375
^{152}Pm		151.92350	4.1 m	β- /3.5	3.5/20	1+			0.1218
					3.50/60				(0.12–2.1)
^{153}Pm		152.92412	5.4 m	β- /1.90	1.7/	(5/2-)			0.0910
									0.1198
									0.1273
^{154m}Pm			2.7 m	β- /	2.0/				0.0820
									0.1848
									1.4403
^{154}Pm		153.92646	1.7 m	β- /4.1	1.9/				0.0820
									0.8396
									1.3940
									2.0589
									(0.08–2.8)
^{155}Pm		154.92810	48. s	β- /3.2		(5/2-)			(0.05–0.78)
^{156}Pm		155.93106	26.7 s	β- /5.16					
^{157}Pm		156.9330	10.9 s	β- /4.6					
^{158}Pm		157.9366	5. s	β- /6.3					

Elem. or Isot.	Natural Abundance (Atom %)	Atomic Mass or Weight	Half-life/ Resonance Width (MeV)	Decay Mode/ Energy (/MeV)	Particle Energy/ Intensity (MeV/%)	Spin ($h/2\pi$)	Nuclear Magnetic Mom. (nm)	Elect. Quadr. Mom. (b)	γ-Energy / Intensity (MeV/%)
^{159}Pm		158.9390	1.5 s						(0.072-0.261)
^{160}Pm		159.9430							
^{161}Pm		160.9459							
^{162}Pm		161.950							
^{163}Pm		162.954							
$_{62}$Sm		**150.36(2)**							
^{129}Sm		128.954	~ 0.55 s	β+, p					
^{130}Sm		129.9489				0+			
^{131}Sm		130.9461	1.2 s	β+, EC/					ann.rad./
^{132}Sm		131.9407	4.0 s	β+		0+			
^{133}Sm		132.9387	2.9 s	β+, EC/~ 8.4		5/2+			ann.rad./
									0.3696
									0.0845
^{134}Sm		133.9340	11. s	β+, EC/5.		0+			ann.rad./
^{135}Sm		134.9325	10. s	β+, EC/7.		7/2+			ann.rad./
^{136}Sm		135.92828	42. s	β+, EC/4.5		0+			ann.rad./
^{137}Sm		136.92697	45. s	β+, EC/6.1					ann.rad./
^{138}Sm		137.92324	3.0 m	β+, EC/3.9		0+			ann.rad./
									0.0536
									0.0747
^{139m}Sm			10. s	I.T./94 /0.457		(11/2-)	1.1		Sm k x-ray
				β+ /6 /	4.7				0.1118
									0.1553
									0.1901
									0.2673
^{139}Sm		138.92230	2.6 m	β+ /75 /5.5	4.1/	½+	-0.53		Pm k x-ray
				EC/25 /					0.3678
									0.4028
									(0.27–2.4)
^{140}Sm		139.91900	14.8 m	β+, EC/3.4	1.9/	0+			ann.rad./
									Pm k x-ray
									0.1396
									0.2255
									(0.07–1.7)
^{141m}Sm			22.6 m	β+ /32 /	1.6/	11/2-	-0.84	+1.6	ann.rad./
				EC/68 /	2.19/				Pm k x-ray
				I.T./0.3 /0.1758					0.1966
									0.4318
									0.7774
^{141}Sm		140.91848	10.2 m	β+ /52 /4.54	3.2/	½+	-0.74		ann.rad./
				EC/48 /					Pm k x-ray
									0.4382
^{142}Sm		141.91520	1.208 h	β+ /6 /2.10	1.0/	0+			ann.rad./
				EC/94 /					Pm k x-ray
^{143m}Sm			1.10 m	IT/99/0.7540		11/2-			Sm k x-ray
									0.7540
^{143}Sm		142.914628	8.83 m	β+ /46 /3.443	2.47/	3/2+	+1.01	+0.4	ann.rad./
				EC/54 /					Pm k x-ray
									1.0565
^{144}Sm	3.083(20)	143.911999				0+			
^{145}Sm		144.913410	340. d	EC/0.617		7/2-	-1.12	-0.6	Pm k x-ray
									0.0613
									0.4924
^{146}Sm		145.913041	1.03×10^8 a	α/	2.50/	0+			

Elem. or Isot.	Natural Abundance (Atom %)	Atomic Mass or Weight	Half-life/ Resonance Width (MeV)	Decay Mode/ Energy (/MeV)	Particle Energy/ Intensity (MeV/%)	Spin ($h/2\pi$)	Nuclear Magnetic Mom. (nm)	Elect. Quadr. Mom. (b)	γ-Energy / Intensity (MeV/%)
^{147}Sm	15.017(75)	146.914898	1.06 × 10^{11} a	α/	2.23/	7/2-	-0.815	-0.26	
^{148}Sm	11.254(51)	147.914823	7 × 10^{15} a	α/	1.96/	0+			
^{149}Sm	13.830(56)	148.917185	10^{16} a	α/		7/2-	-0.672	+0.075	
^{150}Sm	7.351(36)	149.917276				0+			
^{151}Sm		150.919932	90. a	β- /0.0768	0.076/	5/2-	-0.363	+0.7	0.02154
^{152}Sm	26.735(48)	151.919732				0+			
^{153}Sm		152.922097	1.929 d	β- /0.808	0.64/	3/2+	-0.0216	+1.3	Eu k x-ray
					0.69/				0.0697/4.7
									0.10318/29
									0.075–0.714
^{154}Sm	22.730(78)	153.922209				0+			
^{155}Sm		154.924640	22.2 m	β- /1.627	1.52	3/2-		1.1	Eu k x-ray
									0.1043/75.
^{156}Sm		155.92553	9.4 h	β- /0.72	0.43/	0+			0.0872
					0.71/				0.1657
									0.2038
^{157}Sm		156.92836	8.0 m	β- /2.7	2.4/	3/2-			Eu k x-ray
									0.1964
									0.1978
									0.3942
^{158}Sm		157.9300	5.5 m	β- /2.0		0+			0.1894/100.
									0.3636/82.
^{159}Sm		158.9332	11.3 s	β- /3.8					0.1898
^{160}Sm		159.9351	9.6 s	β- /3.6		0+			0.110
^{161}Sm		160.9388	~ 4.8 s						0.264
^{162}Sm		161.941	2.4 s			0+			(0.036-0.741)
^{163}Sm		162.945							
^{164}Sm		163.948				0+			
^{165}Sm		164.953							
$_{63}$Eu		**151.964(1)**							
^{130}Eu		129.964	0.9 ms	p	1.027/				
^{131}Eu		130.9578	~ 26. ms	β+, p	p/0.95				
^{132}Eu		131.9544							
^{133}Eu		132.9492							
^{134}Eu		133.9465	0.5 s	EC, β+					ann.rad./
^{135}Eu		134.9418	1.5 s	EC, β+ /~ 8.7					ann.rad./
^{136m}Eu			~ 3.2 s			7+			0.255
^{136}Eu		135.9396	~ 3.9 s	EC, β+ /10.		1+			ann.rad./
^{137}Eu		136.9356	11. s	EC/~ 7.5		11/2-			ann.rad./
^{138}Eu		137.93371	12. s	EC, β+ /~ 9.2		7+	5		ann.rad./
^{139}Eu		138.92979	18. s	EC, β+ /6.7			6		ann.rad./
^{140m}Eu			0.125 s	EC, β+					ann.rad./
^{140}Eu		139.9281	1.51 s	EC, β+ /8.4		1-	+1.37	+0.31	ann.rad./
^{141m}Eu			3.0 s	β+ /58 /		11/2-			ann.rad./
				EC/9 /					Eu k x-ray
				I.T./33 /0.0964					(0.09–1.6)
^{141}Eu		140.92493	40. s	β+ /81 /5.6		5/2+	+3.49	+0.85	ann.rad./
				EC/15 /					Sm k x-ray
									0.3845
									0.3940
^{142m}Eu			1.22 m	β+ /83 /	4.8/	8-	+2.98	+1.4	ann.rad./
				EC/17 /					Sm k x-ray
									0.5566
									0.7680

Elem. or Isot.	Natural Abundance (Atom %)	Atomic Mass or Weight	Half-life/ Resonance Width (MeV)	Decay Mode/ Energy (/MeV)	Particle Energy/ Intensity (MeV/%)	Spin (*h*/2 π)	Nuclear Magnetic Mom. (nm)	Elect. Quadr. Mom. (b)	γ-Energy / Intensity (MeV/%)
									1.0233
^{142}Eu		141.92343	2.4 s	β- /94/7.4	7.0/	1+	+1.54	+0.12	ann.rad./
				EC/6 /					0.7680
^{143}Eu		142.92030	2.62 m	β+ /72/5.17	4.1/	5/2+	+3.67	+0.51	ann.rad./
				EC/28/	5.1/				Sm k x-ray
									0.1107/7
									1.5368/3.
									1.9127/2.
^{144}Eu		143.91882	10.2 s	β+ /86 /6.33	5.31/	1+	+1.89	+0.10	ann.rad./
				EC/13 /					Sm k x-ray
									1.6601
^{145}Eu		144.916265	5.93 d	β+ /2 /2.660	0.79/	5/2+	+4.00	+0.29	ann.rad./
				EC/98 /1.71					Sm k x-ray
									0.6535
									0.8937
									1.6587
^{146}Eu		145.91721	4.57 d	β+ /5 /3.88	1.47/	4-	+1.42	-0.18	ann.rad./
				EC/95 /					Sm k x-ray
									0.6336
									0.6341
									0.7470
									(0.27–2.64)
^{147}Eu		146.916746	24.4 d	EC/99 /1.722		5/2+	+3.73	+0.53	Sm k x-ray
				β+ /0.4 /					0.12113/20.6
									0.19725/24.0
									(0.601-1.077)
^{148}Eu		147.91809	54.5 d	EC/3.11	0.92	5-	+2.34	+0.35	Sm k x-ray
									0.5503/99.
									0.6299/71.
									(0.067–2.17)
^{149}Eu		148.917931	93.1 d	EC/0.692		5/2+	+3.57	+0.75	Sm k x-ray
									0.2770/4.1
									0.3275/4.8
^{150}Eu		149.91970	36. a	EC/2.26		5-	+2.71	+1.13	Sm k x-ray
									0.3340
									0.4394
									0.5843
									(0.25–1.8)
^{150m}Eu			12.8 h	β- /92 /	1.013/	0-			Sm k x-ray
				β+ /0.4 /	1.24/				0.3339
				EC/8 /					0.4065
^{151}Eu	47.81(6)	150.919850	>1.7x10^{18} a			5/2+	+3.472	+0.90	
^{152m2}Eu			1.60 h	I.T./0.1478		8-			Eu k x-ray
									0.0898
^{152m1}Eu			9.30 h	β- /72 /	1.85/	0-			Sm k x-ray
				EC/28 /	0.89/				0.12178
									0.84153
									0.96334
^{152}Eu		151.921745	13.5 a	EC/72 /1.874	0.69/	3-	-1.941	+2.71	Sm k x-ray
				β- /28 /1.818	1.47/				Gd k x-ray
									0.12178
									0.34427
									1.40802
									(0.252–1.528)

Elem. or Isot.	Natural Abundance (Atom %)	Atomic Mass or Weight	Half-life/ Resonance Width (MeV)	Decay Mode/ Energy (/MeV)	Particle Energy/ Intensity (MeV/%)	Spin ($h/2\pi$)	Nuclear Magnetic Mom. (nm)	Elect. Quadr. Mom. (b)	γ-Energy / Intensity (MeV/%)
^{153}Eu	52.19(6)	152.921230				5/2+	+1.533	+2.41	
^{154m}Eu			46.1 m	I.T./~ 0.16		8-			Eu k x-ray
									0.0682
									0.1009
^{154}Eu		153.922979	8.59 a	β- /99.9/1.969	0.27/29	3-	-2.01	+2.8	Gd k x-ray
				EC/0.02/0.717	0.58/38				0.12299/40.
					0.84/17				0.72331/20.
					0.98/4				1.2745/36
					1.87/11				(0.059-1.90)
^{155}Eu		154.922893	4.76 a	β- /0.252	0.15/	5/2+	+1.520	+2.50	Gd k x-ray
									0.0865/30
									0.1053/20
^{156}Eu		155.92475	15.2 d	β- /2.451	0.30/11	1+	≈1.1		0.08899/9.
					0.49/30				0.64623/7.
					1.2/12				0.723441/6.
					2.45/31				0.8118/10.
^{157}Eu		156.92542	15.13 h	β- /1.36	0.98/	(5/2+)	+1.50	+2.6	Gd k x-ray
					1.30/41				0.0639/100.
									0.3705/48.
									0.4107/76.
^{158}Eu		157.9279	45.9 m	β- /3.5	2.5/	(1-)	+1.44	+0.7	0.0795
									0.8976
									0.9442
									0.9771
^{159}Eu		158.92909	18.1 m	β- /2.51	2.4/	(5/2+)	+1.38	+2.7	0.0678
					2.57/				0.0786
									0.0957
^{160}Eu		159.9320	38. s	β- /4.1	2.7/	(0-)			0.0753
					4.1/				0.1735
									0.4131
									0.5155
									0.8217
									0.9110
									0.9246
^{161}Eu		160.9337	27. s	β- /3.7					0.0719
^{162}Eu		161.9370	11. s	β- /5.6					
^{163}Eu		162.9392							
^{164}Eu		163.943							
^{165}Eu		164.946							
^{166}Eu		165.950							
^{167}Eu		166.953							
$_{64}$Gd		**157.25(3)**							
^{135}Gd		134.953	1.1 s	β+					(0.163–0.360)
^{136}Gd		135.9473				0+			
^{137}Gd		136.9450	7. s	EC, β+ /~ 8.8					ann.rad./
^{138}Gd		137.9401	~ 4.7 s	EC, β+		0+			0.0647
^{139m}Gd			~ 4.8 s						0.1216
^{139}Gd		138.9382	5. s	EC, β+ /~ 7.7					0.104–0.323
^{140}Gd		139.93367	16. s	EC/4.8		0+			0.1748
^{141m}Gd			25. s	EC, β+ /		11/2-			ann.rad./
^{141}Gd		140.93213	21. s	β+ /7.3		½+			ann.rad./
^{142}Gd		141.92812	1.17 m	EC, β+ /4.2		0+			ann.rad./
^{143m}Gd			1.84 m	β+ /67 /		11/2-			ann.rad./
				EC/33 /					Eu k x-ray

Elem. or Isot.	Natural Abundance (Atom %)	Atomic Mass or Weight	Half-life/ Resonance Width (MeV)	Decay Mode/ Energy (/MeV)	Particle Energy/ Intensity (MeV/%)	Spin ($h/2\pi$)	Nuclear Magnetic Mom. (nm)	Elect. Quadr. Mom. (b)	γ-Energy / Intensity (MeV/%)
				I.T./					0.1176
									0.2719
									0.5880
									0.6681
									0.7999
^{143}Gd		142.9268	39. s	β+ /82 /6.0		1/2+			ann.rad./
				EC/18 /					Eu k x-ray
									0.2048
									0.2588
^{144}Gd		143.92296	4.5 m	β+ /45 /4.3	3.3/	0+			ann.rad./
				EC/55 /					Eu k x-ray
									0.3332
^{145m}Gd			1.44 m	I.T./95 /0.749		11/2-	-1.0		0.0273
				β+ /4 /5.7					0.3295
									0.3866
									0.7214
^{145}Gd		144.92171	23.4 m	β+ /33 /5.05	2.5/	1/2+	-0.74		ann.rad./
				EC/67 /					Eu k x-ray
									1.7579
									1.8806
									(0.32–3.69)
^{146}Gd		145.918311	48.3 d	EC/99.9 /1.03	0.35/	0+			Eu k x-ray
				β+ /0.2					0.1147
									0.1155
									0.1546
^{147}Gd		146.919094	1.588 d	EC/99.8 /2.188	0.93/	7/2-	1.0		Eu k x-ray
				EC/0.2 /					0.2293
									0.3699
									0.3960
									0.9289
									(0.1–1.8)
^{148}Gd		147.918115	71. a	α/3.27	3.1828/	0+			
^{149}Gd		148.919341	9.3 d	EC/1.32		7/2-	0.9		Eu k x-ray
									0.1496
									0.2985
									0.3465
^{150}Gd		149.91866	1.8 × 10^{6} a	α/2.80	2.73/	0+			
^{151}Gd		150.920348	124. d	EC/0.464		7/2-	0.8		Eu k x-ray
									0.1536
									0.2432
^{152}Gd	0.20(1)	151.919791				0+			
^{153}Gd		152.921750	240. d	EC/0.485		3/2-	0.4		Eu k x-ray
									0.09743
									0.10318
^{154}Gd	2.18(3)	153.920867				0+			
^{155}Gd	14.80(12)	154.922622				3/2-	-0.258	+1.30	
^{156}Gd	20.47(9)	155.922123				0+			
^{157}Gd	15.65(2)	156.923960				3/2-	-0.340	+1.36	
^{158}Gd	24.84(7)	157.924104				0+			
^{159}Gd		158.926389	18.6 h	β- 0.971	0.971/58	3/2-	-0.44		Tb k x-ray
					0.913/29				0.36351
					0.607/12				0.058-0.855
^{160}Gd	21.86(19)	159.927054	>1.9 × 10^{19} a	β- β-		0+			
^{161}Gd		160.929669	3.66 m	β- /1.956	1.56/85	5/2-			Tb k x-ray

Elem. or Isot.	Natural Abundance (Atom %)	Atomic Mass or Weight	Half-life/ Resonance Width (MeV)	Decay Mode/ Energy (/MeV)	Particle Energy/ Intensity (MeV/%)	Spin (h/2 π)	Nuclear Magnetic Mom. (nm)	Elect. Quadr. Mom. (b)	γ-Energy / Intensity (MeV/%)
									0.1023
									0.3149
									0.3609
^{162}Gd		161.930985	8.4 m	β- /1.39	1.0/	0+			0.4030
									0.4421
^{163}Gd		162.9340	1.13 m	β- /3.1					0.2868
									0.214
									1.685
^{164}Gd		163.9359	45. s	β- /2.3		0+			
^{165}Gd		164.9394	10 s	β-					
^{166}Gd		165.942	~ 4.8 s			0+			(0.040-1.015)
^{167}Gd		166.946							
^{168}Gd		167.948				0+			
^{169}Gd		168.953							
$_{65}$**Tb**		**158.92535(2)**							
^{135}Tb			0.9 ms	p	p/1.179				
^{138}Tb		137.9532							
^{139}Tb		138.9483	1.6 s						0.109
									0.120
^{140}Tb		139.946	2.4 s	β+, EC/11					0.329
									0.355–0.740
^{141}Tb		140.9415	3.5 s	β+, EC/~ 8.3					
^{142m2}Tb			25 μs						
^{142m}Tb			0.30 s	β+, EC/		4-			
^{142}Tb		141.9387	0.60 s	β+, EC/10.		0+			
^{143}Tb		142.9351	12. s	β+, EC/7.4		11/2-			
^{144m}Tb			4.1 s	IT		5-			
^{144}Tb		143.93305	< 1.5 s	β+, EC/8.4		1+			
^{145m}Tb			30. s	β+, EC/~ 6.6		11/2-			ann.rad./
									0.2577
									0.5370
									0.9876
^{145}Tb		144.9293		β+, EC/6.5		½+			
^{146m}Tb			23. s	β+ /76 /		(5-)			ann.rad./
				EC/24 /					Gd k x-ray
									1.0789
									1.5795
^{146}Tb		145.92725	~ 8. s	β+ /8.1		1+			
^{147m}Tb			1.8 m	β+ /35 /		11/2-			ann.rad./
				EC/65 /					Gd k x-ray
									1.3977
									1.7978
^{147}Tb		146.92405	1.6 h	β+ /42 /4.61		5/2+	+1.70		ann.rad./
				EC/58 /					Gd k x-ray
									0.6944
									1.1522
									(0.120–3.318)
^{148m}Tb			2.3 m	β+ /25 /		9+			ann.rad./
				EC/75 /					Gd k x-ray
									0.3945
									0.6319
									0.7845
									0.8824
^{148}Tb		147.92427	1.00 h	β+, EC/5.69		2-	-1.75	-0.3	ann.rad./

Elem. or Isot.	Natural Abundance (Atom %)	Atomic Mass or Weight	Half-life/ Resonance Width (MeV)	Decay Mode/ Energy (/MeV)	Particle Energy/ Intensity (MeV/%)	Spin ($h/2\pi$)	Nuclear Magnetic Mom. (nm)	Elect. Quadr. Mom. (b)	γ-Energy / Intensity (MeV/%)
									Gd k x-ray
									0.4888
									0.7845
									(0.14–3.8)
^{149m}Tb			4.16 m	EC/88 /		11/2-			ann.rad./
				β+ /12 /					Gd k x-ray
									0.1650
									0.7960
^{149}Tb		148.923246	4.13 h	β+ /4 /3.636	1.8/	½+	+1.35		Gd k x-ray
				α/16/	3.97/				0.1650
									0.3522
									0.3886
									(0.1–3.2)
^{150m}Tb			6.0 m	β+ /17 /					ann.rad./
				EC/83 /					Gd k x-ray
									0.4384
									0.6380
									0.6504
									0.8275
^{150}Tb		149.92366	3.3 h	β+, EC/4.66		2-	-0.90		ann.rad./
									0.4963
									0.6380
									(0.3–4.29)
^{151m}Tb			25. s	I.T./95 /		11/2-			0.0229
				β+, EC/7 /					0.0495
									0.3797
									0.8305
^{151}Tb		150.923103	17.61 h	β+/1 /2.565	0.70/	1/2+	+0.92		Gd k x-ray
				EC/99 /					0.1083
									0.2517
									0.2870
									(0.1–1.8)
^{152m}Tb			4.3 m	I.T./79 /0.5018		(8+)			Tb k x-ray
				EC/21 /4.35					Gd k x-ray
									0.2833
									0.3443
									0.4111
^{152}Tb		151.92407	17.5 h	β+ /20 /3.99	2.5/	2-	-0.58	+0.3	ann.rad./
				EC/80 /	2.8/				Gd k x-ray
									0.3443
									(0.2–2.88)
^{153}Tb		152.923435	2.34 d	EC/1.570		5/2+	+3.44	+1.1	Gd k x-ray
									0.2119
									(0.05–1.1)
^{154m2}Tb			23.1 h	EC/98 /		(7-)	0.9		Gd k x-ray
				I.T./2 /					0.1231
									0.2479
									0.3467
									1.4199
^{154m1}Tb			9. h	β+ /78 /		(3-)	+1.7	+3.	Gd k x-ray
				I.T./22 /					0.1231
									0.2479
									0.5401
									(0.12–2.57)

Elem. or Isot.	Natural Abundance (Atom %)	Atomic Mass or Weight	Half-life/ Resonance Width (MeV)	Decay Mode/ Energy (/MeV)	Particle Energy/ Intensity (MeV/%)	Spin (*h*/2 π)	Nuclear Magnetic Mom. (nm)	Elect. Quadr. Mom. (b)	γ-Energy / Intensity (MeV/%)
^{154}Tb		153.92468	21.5 h	EC/99 /3.56	1.86/	0-			Gd k x-ray
				β+ /1 /	2.45				0.1231
									1.2744
									2.1872
									(0.12–3.14)
^{155}Tb		154.92351	5.3 d	EC/0.82		3/2+	+2.01	+1.41	Gd k x-ray
									0.08654
									0.10530
^{156m2}Tb			1.02 d	I.T./		(7-)			Tb k x-ray
									0.0496
^{156m1}Tb			5.3 h	I.T./0.0884		(0+)			Tb k x-ray
									0.0884
^{156}Tb		155.924747	5.3 d	EC/2.444		3-	~ 1.7	+2.	Gd k x-ray
									0.08896
									0.19921
									0.53435
									1.22245
^{157}Tb		156.924025	1.1 × 10^2 a	EC/0.0601		3/2+	+2.01	+1.4	Gd k x-ray
									0.0545
^{158m}Tb			10.5 s	I.T./0.11		0-			Gd k x-ray
									0.0110
^{158}Tb		157.925413	1.8 × 10^2 a	EC/80 /1.220		3-	+1.76	+2.7	Gd k x-ray
				β- /20 /0.937					0.0795
									0.9442
									0.9621
^{159}Tb	100.	158.925347				3/2+	+2.014	+1.43	
^{160}Tb		159.927168	72.3 d	β- /1.835	0.57/47	3-	+1.79	3.8	Dy k x-ray
					0.86/27				0.08678
									0.29857
									0.87936
									0.96615
^{161}Tb		160.927570	6.91 d	β- /0.593	0.46/23	3/2+	2.2	+1.2	Dy k x-ray
					0.52/66				0.02565
					0.6/10				0.04892
									0.07458
^{162}Tb		161.92949	7.6 m	β- /2.51	1.4	(1/2-)			Dy k x-ray
									0.2600
									0.8075
									0.8882
^{163}Tb		162.930648	19.5 m	β- /1.785	0.80/	3/2+			Dy k x-ray
									0.3511
									0.3897
									0.4945
^{164}Tb		163.9334	3.0 m	β- /3.9	1.7/	(5+)			Dy k x-ray
									0.1689
									0.2157
									0.6110
									0.6885
									0.7548
^{165}Tb		164.9349	2.1 m	β- /3.0		3/2+			0.5389
									1.1785
									1.2920
									1.6648
^{166}Tb		165.9380	26 s	β-/					

Elem. or Isot.	Natural Abundance (Atom %)	Atomic Mass or Weight	Half-life/ Resonance Width (MeV)	Decay Mode/ Energy (/MeV)	Particle Energy/ Intensity (MeV/%)	Spin ($h/2\pi$)	Nuclear Magnetic Mom. (nm)	Elect. Quadr. Mom. (b)	γ-Energy / Intensity (MeV/%)
^{167}Tb		166.9401	19 s						0.057
									0.070
^{168}Tb		167.944	8 s						(0.075–0.227)
^{169}Tb		168.946							
^{170}Tb		169.950							
^{171}Tb		170.953							
$_{66}$Dy		**162.500(1)**							
^{139}Dy		138.960	0.6 s	β+, p					
^{140}Dy		139.954				0+			
^{141}Dy		140.9514	0.9 s	EC, β+ /9.					
^{142}Dy		141.9464	2.3 s	EC, β+ /7.1		0+			
^{143}Dy		142.9438	3.9 s	EC, β+ /~ 8.8					
^{144}Dy		143.93925	9.1 s	EC, β+ /~ 6.2		0+			
^{145m}Dy		144.9365	14. s	EC, β+		11/2-			
^{146m}Dy			0.15 s	I.T.		10+			
^{146}Dy		145.93285	30. s	EC, β+ /5.2		0+			
^{147m}Dy			56. s	I.T./40 /		(11/2-)	-0.66	+0.7	Dy k x-ray
				β+, EC/60 /					0.072
									0.6787
^{147}Dy		146.93109	75. s	EC, β+ /6.37		½+	-0.92		ann.rad./
									0.1007
									0.2534
									0.3653
^{148}Dy		147.92715	3.1 m	β+ /4 /2.68	1.2/	0+			ann.rad./
				EC/96 /					Tb k x-ray
									0.6202
^{149}Dy		148.92731	4.2 m	β+, EC/3.81		(7/2-)	-0.12	-0.62	ann.rad./
									0.1008
									0.1063
									0.2534
									0.6536
									0.7894
									1.7765
									1.8062
^{150}Dy		149.925585	7.18 m	β+, EC/67 /1.79		0+			Tb k x-ray
				α/33 /	4.233/				0.3967
^{151}Dy		150.926185	17. m	β+ /5 /2.871		7/2-	-0.95	-0.30	Tb k x-ray
				EC/89 /					0.1764
				α /6 /	4.067/				0.3030
									0.3861
									0.5463
									(0.16–2.09)
^{152}Dy		151.92472	2.37 h	EC/0.60		0+			Tb k x-ray
				α /	3.63/				0.2569
^{153}Dy		152.925765	6.3 h	β+ /1 /2.171	0.89/	(7/2-)	-0.78	~-0.15	Tb k x-ray
				EC/99 /					0.0807
				α /0.01 /	3.46/				0.0997
									0.2137
									(0.08–1.66)
^{154}Dy		153.92442	3. × 10^6 a	α/2.95	2.87/	0+			
^{155}Dy		154.92575	9.9 h	β+ /2 /2.095	0.845/	3/2-	-0.385	+1.04	Tb k x-ray
				EC/98 /					0.0655
									0.2269

Elem. or Isot.	Natural Abundance (Atom %)	Atomic Mass or Weight	Half-life/ Resonance Width (MeV)	Decay Mode/ Energy (/MeV)	Particle Energy/ Intensity (MeV/%)	Spin (h/2 π)	Nuclear Magnetic Mom. (nm)	Elect. Quadr. Mom. (b)	γ-Energy / Intensity (MeV/%)
^{156}Dy	0.056(3)	155.92428				0+			
^{157}Dy		156.92547	8.1 h	EC/1.34		3/2-	-0.301	+1.30	Tb k x-ray
									(0.061–1.319)
^{158}Dy	0.095(3)	157.924409				0+			
^{159}Dy		158.925739	144. d	EC/0.366		3/2-	-0.354	+1.37	Tb k x-ray
									0.3262
^{160}Dy	2.329(18)	159.925198				0+			
^{161}Dy	18.889(42)	160.926933				5/2+	-0.480	+2.51	
^{162}Dy	25.475(36)	161.926798				0+			
^{163}Dy	24.896(42)	162.928731				5/2-	+0.673	+2.65	
^{164}Dy	28.260(54)	163.929175				0+			
^{165m}Dy			1.26 m	I.T./98 /0.108		1/2-			Dy k x-ray
				β- /2 /					0.1082
									0.5155
^{165}Dy		164.931703	2.33 h	β- /1.286	1.29/	7/2+	-0.52	-3.5	Ho k x-ray
									0.09468/3.8
^{166}Dy		165.932807	3.400 d	β- /0.486	0.40/	0+			Ho k x-ray
									0.0282
									0.0825
^{167}Dy		166.9357	6.2 m	β- /~ 2.35	1.78	(1/2-)			Ho k x-ray
									0.2593
									0.3103
									0.5697
									(0.06–1.4)
^{168}Dy		167.9371	8.5 m	β- /1.6		0+			Ho k x-ray
									0.1925
									0.4867
^{169}Dy		168.9403	~ 39. s	β- /3.2					
^{170}Dy		169.9424				0+			
^{171}Dy		170.9462							
^{172}Dy		171.9488				0+			
^{173}Dy		172.953							
$_{67}$**Ho**		**164.93032(2)**							
^{140}Ho		139.969	6 ms	p/	p/1.09				
^{141m}Ho			8 μs	p/	p/1.23				
^{141}Ho		140.963	4.1 ms	β+, p	p/1.17/99.3				
					p/0.97/0.7				
^{142}Ho		141.960	0.4 s	EC/β+, p					0.307
^{143}Ho		142.9546	> 0.2 μs						
^{144}Ho		143.9515	0.7 s	β+, EC/12					
^{145}Ho		144.9472	2.4 s	β+					
^{146}Ho		145.9446	3.3 s	β+, EC/10.7		(10+)			ann.rad./
^{147}Ho		146.94006	5.8 s	β+, EC/8.2		11/2-			ann.rad./
^{148m}Ho			9. s	β+, EC/		4-			ann.rad./
^{148}Ho		147.9377	2. s	β+, EC/9.4		1+			ann.rad./
									0.6615
									1.6883
^{149m}Ho			21. s	β+, EC/		11/2-			ann.rad./
									1.0733
									1.0911
^{149}Ho		148.93378	> 30. s	β+, EC/6.01		1/2+			
^{150m}Ho			25. s	β+, EC/		(9+)			ann.rad./
									0.3939
									0.5511

Elem. or Isot.	Natural Abundance (Atom %)	Atomic Mass or Weight	Half-life/ Resonance Width (MeV)	Decay Mode/ Energy (/MeV)	Particle Energy/ Intensity (MeV/%)	Spin (*h*/2 π)	Nuclear Magnetic Mom. (nm)	Elect. Quadr. Mom. (b)	γ-Energy / Intensity (MeV/%)
									0.6534
									0.8034
^{150}Ho		149.93350	1.3 m	β+, EC/6.6					ann.rad./
									0.5913
									0.6534
									0.8034
^{151m}Ho			47. s	β+, EC/87 /					ann.rad./
				α/13	4.605/				0.2102
									0.4889
									0.6948
									0.7762
^{151}Ho		150.93169	35.2 s	β+, EC/80/5.13					ann.rad./
				α/20 /	4.519/				0.3522
									0.5274
									0.9676
									1.0471
^{152m}Ho			50. s	β+, EC/90/		(9+)	+5.9	-1.3	ann.rad./
				α/10/	4.453/				0.4929
									0.6138
									0.6474
									0.6835
^{152}Ho		151.93171	2.4 m	β+, EC/88/6.47		(3+)	-1.02	~ +0.1	ann.rad./
				α/12/	4.387/				0.6140
									0.6476
^{153m}Ho			9.3 m	β+, EC/99+/4.12		5/2	+1.19		ann.rad./
				α/	4.01/				0.0905
									0.1089
									0.1618
									0.2302
									0.2707
									0.3659
									0.4565
^{153}Ho		152.93020	2.0 m	β+, EC/99+/4.13		11/2-	+6.8	-1.1	ann.rad./
				α/	3.91/				0.2958
									0.3346
									0.4381
									0.6383
^{154m}Ho			3.3 m	β+, EC/		(8+)	5.7	-1.0	ann.rad./
									0.3346
									0.4124
									0.4771
^{154}Ho		153.93060	12. m	β+, EC/5.75		1-	-0.64	+0.2	ann.rad./
									Dy k x-ray
									0.3346
									0.5700
									0.8734
^{155}Ho		154.92910	48. m	β+/6/3.10		(5/2+)	+3.51	+1.5	ann.rad./
				EC/94 /					Dy k x-ray
									0.0474
									0.1363
									0.3254
									(0.06–2.24)
^{156m}Ho			5.8 m	I.T./0.0352			+2.99	+2.3	ann.rad./

Elem. or Isot.	Natural Abundance (Atom %)	Atomic Mass or Weight	Half-life/ Resonance Width (MeV)	Decay Mode/ Energy (/MeV)	Particle Energy/ Intensity (MeV/%)	Spin ($h/2\pi$)	Nuclear Magnetic Mom. (nm)	Elect. Quadr. Mom. (b)	γ-Energy / Intensity (MeV/%)
				β+ /25 /	1.8/				Dy k x-ray
				EC/75 /	2.9/				0.1378
									0.2666
									(0.28–2.9)
^{156}Ho		155.92984	56. m	β+, EC/4.4		(5+)			ann.rad./
									0.1378
									0.2665
^{157}Ho		156.92826	12.6 m	β+/5/2.54	1.18/	7/2-	+4.35	+3.0	ann.rad./
				EC/95/					Dy k x-ray
									0.2800
									0.3411
^{158m2}Ho			28. m	I.T./44/		2-	+2.44	+1.6	ann.rad./
				EC/56/					Dy k x-ray
									0.0989
									0.2182
^{158m1}Ho			21. m	β+, EC/		(9+)			ann.rad./
									0.0981
									0.1664
									0.2182
									0.3205
									0.4062
									0.9774
									1.0532
									0.4846
^{158}Ho		157.92894	11.3 m	β+/8/4.24	1.30/	5+	+3.77	+4.1	ann.rad./
				EC/92/					Dy k x-ray
									0.0989
									0.2182
									0.9488
^{159m}Ho			8.3 s	IT/0.206		1/2+			Ho k x-ray
									0.1660
									0.2059
^{159}Ho		158.927712	33.0 m	EC/1.838		7/2-	+4.28	+3.2	Dy k x-ray
									0.1210
									0.1320
									0.2529
									0.3096
									(0.06–1.2)
^{160m2}Ho			3. s			1+			
^{160m}Ho			5.0 h	IT/67/0.060		2-	+2.52	+1.8	0.0868
				EC/33/3.35					0.1970
									0.6464
									0.7281
									0.8791
									0.9619
									0.9658
^{160}Ho		159.92873	25.6 m	β+, EC/3.29	0.57/	5+	+3.71	+4.0	See Ho[166m]
									0.7282
									0.8794
^{161m}Ho			6.8 s	IT/0.211					Ho k x-ray
									0.2112
^{161}Ho		160.927855	2.48 h	EC/0.859		7/2-	+4.25	+3.2	Dy k x-ray
									0.0256
									0.0592

Elem. or Isot.	Natural Abundance (Atom %)	Atomic Mass or Weight	Half-life/ Resonance Width (MeV)	Decay Mode/ Energy (/MeV)	Particle Energy/ Intensity (MeV/%)	Spin ($h/2\pi$)	Nuclear Magnetic Mom. (nm)	Elect. Quadr. Mom. (b)	γ-Energy / Intensity (MeV/%)
									0.0774
									0.1031
^{162m}Ho			1.12 h	IT/61/		6-	+3.60	+4.	Dy k x-ray
				EC/39/					Ho k x-ray
									0.0807
									0.1850
									0.2828
									0.9372
									1.2200
^{162}Ho		161.929096	15. m	EC/96 /0.295		1+			Dy k x-ray
				β+ /4 /					0.0807
									1.3196
									1.3728
^{163m}Ho			1.09 s	I.T./0.298		(1/2+)			Ho k x-ray
									0.2798
^{163}Ho		162.928734	4.57 × 10^3 a	EC/0.00258		7/2-	+4.23	+3.6	Dy M x-rays
^{164m}Ho			38. m	I.T./0.140		(6-)			Ho k x-ray
									0.0373
									0.0566
									0.0940
^{164}Ho		163.930234	29. m	EC/58 /0.987		1+			Dy k x-ray
				β- /42 /0.963					0.0734
									0.0914
^{165}Ho	100.	164.930322				7/2-	+4.17	+3.6	
^{166m}Ho			1.2 × 10^3 a	β- /		7-	3.6	~ -3.	Er k x-ray
									0.18407
									0.71169
									0.81031
^{166}Ho		165.932284	1.117 d	β- /1.855	1.776/48	0-			Er k x-ray
					1.855/51				0.08057
									1.37943
^{167}Ho		166.93313	3.1 h	β- /1.007	0.31/43	(7/2-)			Er k x-ray
					0.61/21				0.0793
					0.96/15				0.0835
					0.97/15				0.2379
									0.3213
									0.3465
^{168m}Ho			2.2 m	I.T./					
^{168}Ho		167.93552	3.0 m	β- /2.91	2.0/	3+			Er k x-ray
									0.7413
									0.8159
									0.8211
									(0.08–2.34)
^{169}Ho		168.93687	4.7 m	β- /2.12	1.2/	(7/2-)			
					2.0/				0.1496
									0.7610
									0.7784
									0.7884
									0.8529
^{170m}Ho			43. s	β- /		1+			0.0787
									0.8123
									1.8940
									1.9726
^{170}Ho		169.93962	2.8 m	β- /3.87		6+			Er k x-ray

Elem. or Isot.	Natural Abundance (Atom %)	Atomic Mass or Weight	Half-life/ Resonance Width (MeV)	Decay Mode/ Energy (/MeV)	Particle Energy/ Intensity (MeV/%)	Spin (h/2 π)	Nuclear Magnetic Mom. (nm)	Elect. Quadr. Mom. (b)	γ-Energy / Intensity (MeV/%)
									0.1816
									0.2582
									0.8902
									0.9321
									0.9414
									1.1387
^{171}Ho		170.941	53 s	β- /					
^{172}Ho		171.9448	25. s	β- /					Er k x-ray
									(0.077–1.186)
^{173}Ho		172.9473							
^{174}Ho		173.951							
^{175}Ho		174.954							
$_{68}$Er		**167.259(3)**							
^{144}Er		143.9604	> 0.2 μs			0+			
^{145}Er		144.9574	0.9 s	β+					
^{146}Er		145.9520	~ 1.7 s	β+		0+			
^{147}Er		146.9495	2.5 s	E.C, β+ /~ 9.1					
^{148}Er		147.9446	4.5 s	β+, EC/6.8		0+			
^{149m}Er			10. s	IT		11/2-			
^{149}Er		148.94231	10.7 s	ECβ+ /8.1		½+			
^{150}Er		149.93791	18. s	β+ /36 /4.11		0+			ann.rad./
				EC/64 /					Ho k x-ray
									0.4758
^{151}Er		150.93745	23. s	β+, EC/5.2		7/2-			ann.rad./
^{152}Er		151.93505	10.2 s	β+, EC/10/3.11		0+			ann.rad./
				α/90/	4.804/				
^{153}Er		152.935063	37.1 s	α/	4.674		-0.934	-0.42	0.351
				β+, EC/47/4.56	4.35/				(0.0945–1.70)
^{154}Er		153.93278	3.7 m	β+, EC/99+/2.03		0+			ann.rad./
				α/0.5/	4.166/				
^{155}Er		154.93321	5.3 m	β+, EC/47/3.84		(7/2-)	-0.669	-0.27	ann.rad./
				EC/53 /					Ho k x-ray
									0.1101
									0.2415
^{156}Er		155.93107	20. m	β+, EC/1.7		0+			ann.rad./
									0.0298
									0.0352
									0.0522
									0.1336
^{157}Er		156.93192	25. m	β+, EC/3.5		3/2-	-0.412	+0.92	ann.rad./
									0.117
									0.385
									1.320
									1.660
									1.820
									2.000
^{158}Er		157.92989	2.2 h	EC/99.5 /1.78	0.74/	0+			Ho k x-ray
				β+ /0.5 /					0.0719
									0.2486
									0.3868
^{159}Er		158.930684	36. m	β+ /7 /2.769		3/2-	-0.304	+1.17	ann.rad./
				EC/93 /					Ho k x-ray
									0.6245

Elem. or Isot.	Natural Abundance (Atom %)	Atomic Mass or Weight	Half-life/ Resonance Width (MeV)	Decay Mode/ Energy (/MeV)	Particle Energy/ Intensity (MeV/%)	Spin ($h/2\pi$)	Nuclear Magnetic Mom. (nm)	Elect. Quadr. Mom. (b)	γ-Energy / Intensity (MeV/%)
									0.6493
									(0.07–2.5)
^{160}Er		159.92908	1.191 d	EC/0.33		0+			Ho k x-ray
									(0.05–0.96)
^{161}Er		160.93000	3.21 h	EC/2.00		3/2-	-0.37	+1.36	Ho k x-ray
									0.8265
									(0.07–1.74)
^{162}Er	0.139(5)	161.928778				0+			
^{163}Er		162.93003	1.25 h	EC/1.210		5/2-	+0.557	+2.55	Ho k x-ray
									0.4361
									0.4399
									1.1135
^{164}Er	1.601(3)	163.929200				0+			
^{165}Er		164.930726	10.36 h	EC/0.376		5/2-	+0.643	+2.71	Ho k x-ray
^{166}Er	33.503(36)	165.930293				0+			
^{167m}Er			2.27 s	I.T./0.208		½-			Er k x-ray
									0.2078
^{167}Er	22.869(9)	166.932048				7/2+	-0.5639	+3.57	
^{168}Er	26.978(18)	167.932370				0+			
^{169}Er		168.934590	9.40 d	β- /0.351	0.35/~ 100	½-	+0.485		Tm k x-ray
									0.1098
									0.1182
^{170}Er	14.910(36)	169.935464				0+			
^{171}Er		170.938030	7.52 h	β- /1.491		5/2-	0.66	2.9	Tm k x-ray
									0.11160
									0.29591
									0.30832
									(0.08–1.4)
^{172}Er		171.939356	2.05 d	β-/0.891	0.28/48	0+			Tm k x-ray
					0.36/46				0.0597
									0.4073
									0.6101
^{173}Er		172.9424	1.4 m	β- /2.6		(7/2-)			Tm k x-ray
									0.1928
									0.1992
									0.8952
^{174}Er		173.9442	3.1 m	β- /1.8		0+			Tm k x-ray
									(0.100–0.152)
^{175}Er		174.9478	1.2 m	β-					(0.0765–1.17)
^{176}Er		175.9501				0+			
^{177}Er		176.954							
$_{69}$Tm		**168.93421(2)**							
^{144}Tm			~ 1.9 μs	p	1.70, 1.43				
^{145}Tm		144.9701	3.1 μs	p// ~ 10	1.73/91				
					1.40/~9.6				
^{146m}Tm			0.198 s	β+, p	p/1.12/99				
					p/0.89/~1				
^{146}Tm		145.9664	73 ms	β+/14.	p/1.19/68				
				p	p/1.01/~18				
					p/0.94/~14				
^{147m}Tm			0.4 ms	β+, p	p/1.115				
^{147}Tm		146.9610	0.56 s	EC, β+/85	~ 10.7				
				p/15/	1.052/				
^{148m}Tm		147.9578	0.7 s	β+, EC/12.					ann.rad./

Elem. or Isot.	Natural Abundance (Atom %)	Atomic Mass or Weight	Half-life/ Resonance Width (MeV)	Decay Mode/ Energy (/MeV)	Particle Energy/ Intensity (MeV/%)	Spin (h/2 π)	Nuclear Magnetic Mom. (nm)	Elect. Quadr. Mom. (b)	γ-Energy / Intensity (MeV/%)
^{148}Tm									
^{149}Tm		148.9527	0.9 s	β+, EC/~ 9.2		11/2-			
^{150}Tm		149.9500	2.3 s	β+, EC/~ 11.5		6-			(0.101–2.177)
^{151}Tm		150.94548	4. s	β+, EC/7.5					ann.rad./
^{152m}Tm			8. s	β+, EC/		9+			
^{152}Tm		151.9444	5. s	β+, EC/8.8					ann.rad./
^{153}Tm		152.94201	1.6 s	β+,EC/10 /6.46		(11/2-)	6.9	~ +0.5	ann.rad./
				α/90 /	5.109/				
^{154m}Tm			3.3 s	β+, EC/15 /	α/5.031/100	(9+)	+5.9	~ -0.2	ann.rad./
				α/	4.84/0.24				0.4605–0.796
^{154}Tm		153.94157	8.1 s	β+, EC/56 /7.4	α/4.956/100	(2-)	-1.14	~ +0.4	ann.rad./
				α/44 /	4.83/0.45				
^{155}Tm		154.93920	30. s	β+, EC/5.58					0.0315
				α/	4.46/				0.0638
									0.0881
									0.2268
									0.5320
									0.6067
^{156m}Tm			19. s	α/	4.46/				
^{156}Tm		155.93898	1.40 m	β+, EC/7.6		2-	+0.40	-0.5	ann.rad./
				α/	4.23/				0.3446
									0.4529
									0.5860
^{157}Tm		156.93697	3.6 m	β+, EC/4.5	2.6	½+	+0.48		ann.rad./
				α/	3.97/				0.1104
									0.3484
									0.3855
									0.4550
									(0.1–1.58)
^{158}Tm		157.93698	4.0 m	β+, EC/74 /6.5		(2-)	+0.04	+0.7	ann.rad./
				EC/26 /					Er k x-ray
									0.1921
									0.3351
									0.6280
									1.1498
									(0.18–2.81)
^{159}Tm		158.93498	9.1 m	β+/23 /3.9		5/2+	+3.42	+1.9	ann.rad./
				EC/77 /					Er k x-ray
									0.0591
									0.0848
									0.2713
									(0.05–1.27)
^{160m}Tm			1.24 m	IT		(5)			
^{160}Tm		159.93526	9.4 m	β+/15 /5.9		1-	+0.16	+0.58	ann.rad./
				EC/85 /					Er k x-ray
									0.1264
									0.2642
									0.7285
									0.8544
									0.8614
									1.3685
^{161}Tm		160.93355	31. m	β+, EC/3.2		7/2+	+2.40	+2.9	ann.rad./
									Er k x-ray
									0.0595

Elem. or Isot.	Natural Abundance (Atom %)	Atomic Mass or Weight	Half-life/ Resonance Width (MeV)	Decay Mode/ Energy (/MeV)	Particle Energy/ Intensity (MeV/%)	Spin ($h/2\pi$)	Nuclear Magnetic Mom. (nm)	Elect. Quadr. Mom. (b)	γ-Energy / Intensity (MeV/%)
									0.0844
									1.6481
									(0.04–2.15)
^{162m}Tm			24. s	I.T./90 /		5+			Tm k x-ray
				β+, EC/10 /					Er k x-ray
									0.0669
									0.8115
									0.9003
^{162}Tm		161.93400	21.7 m	β+ /8 /4.81		1-	+0.07	+0.69	ann.rad./
				EC/92 /					Er k x-ray
									0.1020
									0.7987
									(0.1–3.75)
^{163}Tm		162.93265	1.81 h	EC/98 /2.439		½+	-0.082		Er k x-ray
				β+ /1 /					0.0692
									0.1043
									0.2414
^{164m}Tm			5.1 m	I.T./80 /		6-			0.0914
				β+, EC/20 /					0.1394
									0.2081
									0.2405
									0.3149
^{164}Tm		163.93356	2.0 m	β+ /36 /3.96	2.94/	1+	+2.38	+0.71	ann.rad./
				EC/64 /					Er k x-ray
									0.0914
^{165}Tm		164.932435	1.253 d	EC/1.593		½+	-0.139		Er k x-ray
									0.0472
									0.0544
									0.29728
									0.80636
^{166}Tm		165.93355	7.70 h	EC/98 /3.04		2+	+0.092	+2.14	Er k x-ray
				β+ /2 /					0.0806
									0.1844
									0.7789
									1.2734
									2.0524
^{167}Tm		166.932852	9.24 d	EC/0.748		½+	-0.197		Er k x-ray
									0.0571
									0.20778
^{168}Tm		167.934173	93.1 d	EC/1.679		3+	+0.23	+3.2	Er k x-ray
									0.19825
									0.4475
									0.81595
^{169}Tm	100	168.934213				½+	-0.231	-1.2	
^{170}Tm		169.935801	128.6 d	β- /99.8/0.968	0.883/24	1-	+0.246	+0.74	Yb k x-ray
				EC/0.2 /0.314	0.968/76				0.08425
^{171}Tm		170.936429	1.92 a	β- /0.096	0.03/2	½+	-0.228		0.06674
					0.096/98				
^{172}Tm		171.93840	2.65 d	β- /1.88	1.79/36	2-			Yb k x-ray
					1.88/29				0.07879
									1.38722
									1.46601
									1.52982
									1.60861

Elem. or Isot.	Natural Abundance (Atom %)	Atomic Mass or Weight	Half-life/ Resonance Width (MeV)	Decay Mode/ Energy (/MeV)	Particle Energy/ Intensity (MeV/%)	Spin ($h/2\pi$)	Nuclear Magnetic Mom. (nm)	Elect. Quadr. Mom. (b)	γ-Energy / Intensity (MeV/%)
^{173}Tm		172.939604	8.2 h	β- /1.298	0.80/21	½+			Yb k x-ray
					0.86/71				0.3988
									0.4613
^{174m}Tm			2.29 s						
^{174}Tm		173.94217	5.4 m	β- /3.08	0.70/14	(4-)			Yb k x-ray
					1.20/83				0.07664
									0.17669
									0.27332
									0.3666
									0.99205
									(0.08–1.6)
^{175}Tm		174.94384	15.2 m	β- /2.39	0.9/36	(1/2+)			Yb k x-ray
					1.9/23				0.36396
									0.51487
									0.94125
									0.98247
^{176}Tm		175.9470	1.9 m	β-/4.2	2.0/	(4+)			Yb k x-ray
					1.2/				0.1898
									0.3819
									1.0691
^{177}Tm		176.9490	1.4 m	β-		(7/2-)			
^{178}Tm		177.9526							
^{179}Tm		178.955							
$_{70}$**Yb**		**173.054(5)**							
^{148}Yb		147.967				0+			
^{149}Yb		148.964	0.7 s	β+, p	p/2.5–6.4/				0.647
^{150}Yb		149.9584	> 0.2 μs			0+			
^{151}Yb		150.9554	1.6 s	β+ /8.5		(1/2+)			
^{152}Yb		151.9503	3.2 s	β+ EC/5.5		0+			
^{153}Yb		152.9495	4. s	β+ EC/6.7					
^{154}Yb		153.94639	0.40 s	β+ EC/7 /4.49		0+			ann.rad./
				α/93 /	5.32/				
^{155}Yb		154.9458	1.7 s	β+, EC/16 /6.0		(7/2-)	-0.91	-0.5	ann.rad./
				α/84 /	5.19/				
^{156}Yb		155.94282	26. s	β+, EC/21/3.57		0+			ann.rad./
				α/79 /	4.69/				
^{157}Yb		156.94263	39. s	β+, EC/99+/5.5		7/2-	-0.64		ann.rad./
				α/0.5/	4.69/				0.231
									(0.035–0.670)
^{158}Yb		157.93987	1.5 m	β+, EC/1.9		0+			ann.rad./
									0.0741
									0.2526
^{159}Yb		158.94005	1.4 m	EC, β+/5.1		5/2-	-0.37	-0.22	Tm k x-ray
									0.1661
									0.1772
									0.3297
									0.3903
^{160}Yb		159.93755	4.8 m	β+, EC/2.0		0+			ann.rad./
									0.1404
									0.1737
									0.2158
^{161}Yb		160.93790	4.2 m	β+, EC/3.9		3/2-	-0.33	+1.03	ann.rad./
									Tm k x-ray
									0.0782

Elem. or Isot.	Natural Abundance (Atom %)	Atomic Mass or Weight	Half-life/ Resonance Width (MeV)	Decay Mode/ Energy (/MeV)	Particle Energy/ Intensity (MeV/%)	Spin ($h/2\pi$)	Nuclear Magnetic Mom. (nm)	Elect. Quadr. Mom. (b)	γ-Energy / Intensity (MeV/%)
									0.5999
									0.6315
^{162}Yb		161.93577	18.9 m	β+, EC/1.7		0+			ann.rad./
									Tm k x-ray
									0.1188
									0.1635
^{163}Yb		162.93633	11.1 m	β+ /26 /3.4	1.4/	3/2-	-0.37	+1.24	ann.rad./
									Tm k x-ray
									0.0636
									0.8603
									(0.06 –1.9)
^{164}Yb		163.93449	1.26 h	EC/1.0		0+			Tm k x-ray
									0.0914
									0.6752
^{165}Yb		164.93528	9.9 m	β+ /10 /2.76	1.58/	(5/2-)	+0.48	+2.48	ann.rad./
				EC/90 /					Tm k x-ray
									0.0801
									1.0903
^{166}Yb		165.93388	2.363 d	EC/0.30		0+			Tm k x-ray
									0.0828
									0.1844
									0.7789
									1.2734
									2.0524
^{167}Yb		166.934950	17.5 m	β+ /0.5 /1.954	0.639/	5/2-	+0.62	+2.70	Tm k x-ray
				EC/99.5 /					0.06296
									0.10616
									0.11337
									0.17633
^{168}Yb	0.1233(29)	167.933897				0+			
^{169m}Yb			46. s	I.T./0.0242		1/2-	+0.51		Yb L x-ray
									0.0242
^{169}Yb		168.935190	32.02 d	EC/0.909		7/2+	-0.63	+3.5	0.1979/35.9
									0.3078/10.05
									0.0207–0.261
^{170}Yb	2.982(38)	169.934762				0+			
^{171}Yb	14.09(13)	170.936326				1/2-	+0.49367		
^{172}Yb	21.68(13)	171.936382				0+			
^{173}Yb	16.103(62)	172.938211				5/2-	-0.67989	+2.80	
^{174}Yb	32.026(78)	173.938862				0+			
^{175}Yb		174.941277	4.19 d	β- /0.470	0.466/73	7/2-	0.77		Lu k x-ray
					0.071/21				0.3963/13
					0.353/6.2				(0.114–0.28)
^{176m}Yb			11.4 s	I.T./1.051		(8-)			Yb k x-ray
									0.0961
									0.1901
									0.2929
									0.3897
^{176}Yb	12.996(81)	175.942572	10^{26} a	β-β-		0+			
^{177m}Yb			6.41 s	I.T./0.3315		1/2-			Yb k x-ray
									0.1131
									0.2084
^{177}Yb		176.945261	1.9 h	β- /1.399	1.40	9/2+			Lu k x-ray
									0.1504

Elem. or Isot.	Natural Abundance (Atom %)	Atomic Mass or Weight	Half-life/ Resonance Width (MeV)	Decay Mode/ Energy (/MeV)	Particle Energy/ Intensity (MeV/%)	Spin ($h/2\pi$)	Nuclear Magnetic Mom. (nm)	Elect. Quadr. Mom. (b)	γ-Energy / Intensity (MeV/%)
^{178}Yb		177.94665	1.23 h	β- /0.65	0.25/	0+			0.1415
									0.3246
									0.3516
									0.3815
									0.6125
^{179}Yb		178.9502	8. m	β- /2.4		(1/2-)			
^{180}Yb		179.9523	2. m	β-		0+			0.1028–0.442
^{181}Yb		180.9562							
$_{71}$Lu		**174.9668(1)**							
^{150m}Lu			0.045 ms	p/1.29					
^{150}Lu		149.973	43. ms	p					
^{151m}Lu			16 μs	p/1.31		(3/2+)			
^{151}Lu		150.9676	0.08 s	p/1.231		(11/2-)			
^{152}Lu		151.9641	0.7 s						
^{153}Lu		152.9588	0.9 s			11/2-			
^{154}Lu		153.9575	1.0 s	β+, EC/10.8					
^{155m}Lu			2.6 ms	α/7.41		(25/2-)			
^{155}Lu		154.95432	0.07 s	EC/8.0		(11/2-)			
				α/	5.66/90				
^{156m}Lu			0.20 s	α/	5.57/	9+			
^{156}Lu		155.9530	~ 0.5 s	β+, EC/9.5		2-			ann.rad./
				α/	5.45/				
^{157m}Lu			~ 9.6 s	α	4.925/				
^{157}Lu		156.95010	4.8 s	β+, EC/94 /6.93					ann.rad./
				α/	5.00/				
^{158}Lu		157.94931	10.4 s	β+, EC/99 /8.0		2-			ann.rad./
				α/	4.67/				0.3682
									0.4770
^{159}Lu		158.94663	12.3 s	β+, EC/6.0					ann.rad./
									0.1505
									0.1875
									0.3693
^{160}Lu		159.9460	36.1 s	β+, EC/7.3					ann.rad./
									0.2434
									0.3957
									0.5773
^{161}Lu		160.94357	1.2 m	β+, EC/5.3		1/2+	+0.223		ann.rad./
									0.0437
									0.0671
									0.1003
									0.1108
									0.1562
									0.2562
^{162m}Lu			~ 1.5 m	EC/		4-			
^{162}Lu		161.9433	1.37 m	β+, EC/6.9		1-	+0.055	+0.52	ann.rad./
									0.1666
									0.6314
^{163}Lu		162.94118	4.1 m	β+, EC/4.6		1/2+	+0.077		ann.rad./
									0.0539
									0.0581
									0.1504
									0.1631
									0.3717

Elem. or Isot.	Natural Abundance (Atom %)	Atomic Mass or Weight	Half-life/ Resonance Width (MeV)	Decay Mode/ Energy (/MeV)	Particle Energy/ Intensity (MeV/%)	Spin ($h/2\pi$)	Nuclear Magnetic Mom. (nm)	Elect. Quadr. Mom. (b)	γ-Energy / Intensity (MeV/%)
^{164}Lu		163.94134	3.14 m	β+, EC/6.3	1.6/	1-	+0.059	+0.61	0.1238
					3.8/				0.2621
									0.7404
									0.8639
									0.8804
^{165}Lu		164.93941	10.7 m	β+, EC/3.9	2.06/	1/2+	-0.0245		ann.rad./
									0.1206
									0.1324
									0.1742
									0.2036
									(0.04–2.0)
^{166m2}Lu			2.1 m	β+ /35 /		(0-)			ann.rad./
				EC/65 /					Yb k x-ray
									1.0673
									1.2566
									2.0986
^{166m1}Lu			1.4 m	β+, EC/58 /		(3-)	+0.189	+2.72	ann.rad./
				I.T./42 /0.0344					0.1024
									0.2281
									0.2861
									0.8119
									0.8301
^{166}Lu		165.93986	2.8 m	β+ /25 /5.5		(6-)	+2.91	+4.33	ann.rad./
				EC/75 /					Yb k x-ray
									0.1024
									0.2281
									0.3375
									0.3679
^{167}Lu		166.93827	52. m	β+ /2 /3.1	2.1/	7/2+	+2.325	+3.28	Yb k x-ray
				EC/98 /					0.0297
									0.2392
									(0.03–2.0)
^{168m}Lu			6.7 m	β+ /12 /		3+	+1.221	+2.43	ann.rad./
				EC/88 /					Yb k x-ray
				IT/<0.8					0.1988/190
									0.8960/100
									0.9792/128
									0.018–2.65
^{168}Lu		167.93874	5.5 m	β+ /6 /4.5	1.2/	(6-)	+3.02	+4.8	ann.rad./
				EC/94 /					Yb k x-ray
									0.1114
									0.1124
									0.2286
									0.3483
									1.4836
^{169m}Lu			2.7 m	I.T./0.0290		1/2-			Lu L x-ray
									0.0290
^{169}Lu		168.93765	1.419 d	EC/2.293	1.271/	7/2+	2.30	3.5	Yb k x-ray
									0.19121
									0.9606
									(0.08–2.1)
^{170m}Lu			0.7 s	I.T./0.0929		4-			Lu L x-ray
									0.04449
									0.0484

Elem. or Isot.	Natural Abundance (Atom %)	Atomic Mass or Weight	Half-life/ Resonance Width (MeV)	Decay Mode/ Energy (/MeV)	Particle Energy/ Intensity (MeV/%)	Spin ($h/2\pi$)	Nuclear Magnetic Mom. (nm)	Elect. Quadr. Mom. (b)	γ-Energy / Intensity (MeV/%)
^{170}Lu		169.93848	2.01 d	EC/3.46	2.44/	0+			Yb k x-ray
									0.58711
									0.5908
									1.28029
									(0.1–3.38)
^{171m}Lu			1.31 m	I.T./0.0711		1/2-	+0.59		Lu k x-ray
									0.07119
^{171}Lu		170.937913	8.24 d	EC/1.479	0.362/	7/2+	+2.293	+3.48	Yb k x-ray
									0.01939
									0.66744
									(0.02–1.3)
^{172m}Lu			3.7 m	I.T./0.0419		1-	+1.98	+0.76	Lu L x-rays
									0.04186
^{172}Lu		171.939086	6.64 d	EC/2.519		4-	2.90	+3.80	Yb k x-ray
									0.18156
									1.09367
									(0.07–2.2)
^{173}Lu		172.938931	1.37 a	EC/0.671		7/2+	+2.281	+3.53	Yb k x-ray
									0.07860
									0.27198
^{174m}Lu			142. d	IT/99.3/	0.17086	6-	+1.49	+4.80	Lu k x-ray
				EC/0.7 /					0.067055
^{174}Lu		173.940338	3.3 a	EC/1.374		1-	+1.988	+0.773	Yb k x-ray
									0.07664
									1.2419
^{175}Lu	97.4013(12)	174.940772				7/2+	+2.232	+3.49	
^{176m}Lu			3.66 h	β- /1.315	1.229/	1-	+0.318	-1.47	Hf k x-ray
					1.317/				0.088372
^{176}Lu	2.5987(12)	175.942686	3.73 × 10^{10} a	β- /1.192		7-	+3.169	+4.92	Hf k x-ray
				β+/ < 0.9					0.20187
				EC/<0.36					0.30691
^{177m2}Lu			6. m	β-		39/2-			0.089
^{177m}Lu			160.7 d	IT/22/0.9702		23/2-	+2.32	+5.7	Lu k x-ray
				β- /78					Hf k x-ray
									0.11295
									0.20836
									0.37850
									0.41853
^{177}Lu		176.943758	6.65 d	β- /0.498	0.497/	7/2+	+2.238	+3.39	0.11295
									0.20836
^{178m}Lu			23.1 m	β- /		(9-)	+4.83	+5.39	0.2166
									0.3317
^{178}Lu		177.945955	28.5 m	β-/2.099	2.03/	1+	-1.38	+0.71	Hf k x-ray
									0.0932
									1.3099
									1.3408
									(0.09–1.7)
^{179}Lu		178.94733	4.6 h	β- /1.405	1.35/	7/2+	+2.38	+3.32	0.2143
									0.3377
^{180}Lu		179.9499	5.7 m	β- /3.1	1.49/				0.40795/50.
									(0.07–1.9)
^{181}Lu		180.9520	3.5 m	β- /2.5		(7/2+)			0.0458
									0.2059
									0.5749

Elem. or Isot.	Natural Abundance (Atom %)	Atomic Mass or Weight	Half-life/ Resonance Width (MeV)	Decay Mode/ Energy (/MeV)	Particle Energy/ Intensity (MeV/%)	Spin ($h/2\pi$)	Nuclear Magnetic Mom. (nm)	Elect. Quadr. Mom. (b)	γ-Energy / Intensity (MeV/%)
^{182}Lu		181.9550	2.0 m	β- /~ 4.1					0.0978
									0.7208
									0.8182
^{183}Lu		182.9576	58. s	β- /		7/2+			
^{184}Lu		183.9609	20 s	β-					
$_{72}$Hf		**178.49(2)**							
^{153}Hf		152.971	> 0.2 μs						
^{154}Hf		153.965	2. s	EC, β+/~ 6.7		0+			
^{155}Hf		154.9634	0.9 s	EC, β+/8.					
^{156}Hf		155.9594	25. ms	α/		0+			
^{157}Hf		156.9584	0.11 s	α/					
^{158}Hf		157.95480	2.9 s	EC/54 /5.1		0+			
				α/46 /	5.27/				
^{159}Hf		158.95400	5.6 s	β+, EC/88 /6.9					ann.rad./
				α/12 /	5.09/				
^{160}Hf		159.95068	~ 12. s	β+, EC/97 /4.9		0+			ann.rad./
				α/4.78					
^{161}Hf		160.95028	17. s	α/	4.60/				
^{162}Hf		161.94721	38. s	β+, EC/3.7		0+			ann.rad./
									0.1739
									0.1963
									0.4101
^{163}Hf		162.94709	40. s	β+, EC/5.5					ann.rad./
									0.0454
									0.0621
									0.0710
									0.6882
^{164}Hf		163.94438	2.8 m	EC, β+/3.0		0+			
^{165}Hf		164.94457	1.32 m	EC/4.6		11/2-			
^{166}Hf		165.94218	6.8 m	EC/93 /2.3		0+			ann.rad./
				β+ /7 /					Lu k x-ray
									0.0788
^{167}Hf		166.94260	2.0 m	β+ /40 /4.0		(5/2-)			ann.rad./
				EC/60 /					Lu k x-ray
									0.1754
									0.3152
^{168}Hf		167.94057	25.9 m	β+, EC/1.8		0+			ann.rad./
									(0.014–1.311)
^{169}Hf		168.94126	3.25 m	EC/85 /3.3		(5/2-)			ann.rad./
				β+ /15 /					Lu k x-ray
									0.3695
									0.4929
^{170}Hf		169.93961	16.0 h	EC/1.1		0+			Lu k x-ray
									0.0985
									0.1202
									0.1647
									0.5729
									0.6207
^{171m}Hf			30. s			(1/2-)	+0.53		
^{171}Hf		170.94049	12.2 h	EC, β+ /2.4		7/2+	-0.67	+3.46	ann.rad./
									Lu k x-ray
									0.1221
									0.6620
									1.0714

Elem. or Isot.	Natural Abundance (Atom %)	Atomic Mass or Weight	Half-life/ Resonance Width (MeV)	Decay Mode/ Energy (/MeV)	Particle Energy/ Intensity (MeV/%)	Spin ($h/2\pi$)	Nuclear Magnetic Mom. (nm)	Elect. Quadr. Mom. (b)	γ-Energy / Intensity (MeV/%)
^{172}Hf		171.93945	1.87 a	EC/0.35		0+			Lu k x-ray
									0.02399
									0.12582
									(0.082–0.123)
^{173}Hf		172.94051	23.6 h	EC/1.6		½-	+0.50		Lu k x-ray
									0.12367
									0.13963
									0.29697
									0.31124
									(0.1–2.1)
^{174}Hf	0.16(1)	173.940046	2.0 × 10^{15} a			0+			
^{175}Hf		174.941509	71. d	EC/0.686		5/2-	-0.68	+2.72	Lu k x-ray
									0.08936
									0.34340
^{176}Hf	5.26(7)	175.941409				0+			
^{177m2}Hf			51.4 m	I.T./2.740		37/2-			Hf k x-ray
									0.2140
									0.2951
									0.3115
									0.3267
^{177m1}Hf			1.1 s	I.T./		23/2+			Hf k x-ray
									0.20836
									0.22847
									0.37851
^{177}Hf	18.60(9)	176.943221				7/2-	+0.794	+0.337	
^{178m2}Hf			31. a	I.T./		16+	+8.16	+6.0	Hf k x-ray
									0.32555
									0.42635
									0.089–0.574
^{178m1}Hf			4.0 s	I.T./		8-	3		Hf k x-ray
									0.21342
									0.32555
									0.42635
^{178}Hf	27.28(7)	177.943699				0+			
^{179m2}Hf			25.1 d	I.T./1.1057		25/2-	7.4		Hf k x-ray
									0.1227
									0.1461
									0.3626
									0.4537
^{179m1}Hf			18.7 s	I.T./0.375		½-			Hf k x-ray
									0.1607
									0.2141
^{179}Hf	13.62(2)	178.945816				9/2+	-0.641	+3.86	
^{180m}Hf			5.54 h	I.T./1.1416		8-	+9.	+4.6	Hf k x-ray
									0.2152
									0.3323
									0.4432
^{180}Hf	35.08(16)	179.946550				0+			
^{181m}Hf			1.5 ms	/1.738		25/2-			
^{181}Hf		180.949101	42.4 d	β- /1.027	0.408/	1/2-			Ta k x-ray
									0.13294/54
									0.48200/100
									0.3459/20
^{182m}Hf			62. m	β- /54 /1.60	0.49/43	8-			Hf k x-ray

Elem. or Isot.	Natural Abundance (Atom %)	Atomic Mass or Weight	Half-life/ Resonance Width (MeV)	Decay Mode/ Energy (/MeV)	Particle Energy/ Intensity (MeV/%)	Spin ($h/2\pi$)	Nuclear Magnetic Mom. (nm)	Elect. Quadr. Mom. (b)	γ-Energy / Intensity (MeV/%)
				IT/46 /1.173	0.95/10				0.0509
									0.2244
									0.3441
									0.4558
									0.5066
									0.9428
^{182}Hf		181.95055	8.9 × 10^6 a	β- /0.37		0+			Ta k x-ray
									0.2704/79
									(0.098-0.270)
^{183}Hf		182.95353	1.07 h	β- /2.01	1.18/68	3/2-			Ta k x-ray
					1.54/25				0.0732
									0.4591
									0.7837
^{184}Hf		183.95545	4.1 h	β- /1.34	0.74/38	0+			Ta k x-ray
					0.85/16				0.0414
					1.10/46				0.1391
									0.3449
^{185}Hf		184.9588	~ 3.5 m	β- /					0.165
^{186}Hf		185.9609	~ 2.6 m			0+			0.738
^{187}Hf		186.9646	> 0.3 μs						
^{188}Hf		187.967	> 0.3 μs			0+			
$_{73}$Ta		**180.94788(2)**							
^{155}Ta		154.975	3. ms	p/1.44					
^{156}Ta		155.9723	0.11 s	β+ /~ 11.6					
				p/	1.02/~ 100				
^{157}Ta		156.9682	10 ms	α/	6.117				
				p/	0.927/3.4				
^{158}Ta		157.9667	37. ms	α/	6.05/100				
					5.97/100				
^{159}Ta		158.96302	0.6 s	β+, EC/20 /8.5	α/5.52/34				ann.rad./
				α/80 /	5.60/55				
^{160}Ta		159.9615	1.4 s	β+, EC/10.1					ann.rad./
				α	5.41/				
^{161}Ta		160.9584	3.16 s	β+, EC/7.5					ann.rad./
				α/	5.15				
^{162}Ta		161.9573	4. s	EC/8.6					
^{163}Ta		162.95433	10.6 s	EC/6.8					
^{164}Ta		163.95353	14.2 s	β+ /8.5		3+			ann.rad./
				α/	4.62/				0.2110
									0.3768
^{165}Ta		164.95077	31. s	ECβ+/5.9					
^{166}Ta		165.95051	34. s	β+ /82 /7.7					ann.rad./
				EC/18 /					Hf k x-ray
									0.1587
									0.3117
									0.8101
^{167}Ta		166.94809	1.4 m	β+, EC/5.6					ann.rad./
^{168}Ta		167.94805	2.4 m	β+ /77 /6.7		3+			ann.rad./
				EC/23 /					Hf k x-ray
									0.1241
									0.2619
									0.7518
									(0.307-1.985)
^{169}Ta		168.94601	4.9 m	β+, EC/4.4					ann.rad./

Elem. or Isot.	Natural Abundance (Atom %)	Atomic Mass or Weight	Half-life/ Resonance Width (MeV)	Decay Mode/ Energy (/MeV)	Particle Energy/ Intensity (MeV/%)	Spin (h/2 π)	Nuclear Magnetic Mom. (nm)	Elect. Quadr. Mom. (b)	γ-Energy / Intensity (MeV/%)
									0.0288
									0.1535
									0.1924
^{170}Ta		169.94618	6.8 m	β +/70 /6.0		(3+)			ann.rad./
				EC/35 /					Hf k x-ray
									0.1008
									0.2212
^{171}Ta		170.94448	23.3 m	β+, EC/3.7		(5/2-)			0.0496
									0.5018
									0.5064
									(0.05–1.02)
^{172}Ta		171.94490	36.8 m	β+ /25 /4.9		(3-)			ann.rad./
				EC/75 /					Hf k x-ray
									0.21396
									1.10923
									(0.09 –3.8)
^{173}Ta		172.94375	3.6 h	β+ /24 /3.7		(5/2-)	1.70	-1.9	ann.rad./
				EC/76 /					Hf k x-ray
									0.06972
									0.17219
									(0.06 –2.7)
^{174}Ta		173.94445	1.12 h	β+ /27 /3.8		(3+)			ann.rad./
				EC/73 /					Hf k x-ray
									0.09089
									0.20638
									(0.09–3.64)
^{175}Ta		174.94374	10.5 h	EC/2.0		7/2+	2.27	+3.6	Hf k x-ray
									0.2077
									0.2671
									0.3487
^{176}Ta		175.94486	8.1 h	EC/3.1		1-			Hf k x-ray
									0.08837
									1.15735
^{177}Ta		176.944472	2.356 d	EC/1.166		7/2+	2.25		Hf k x-ray
									0.11295
									(0.07–1.06)
^{178m}Ta			2.4 h	EC/		(7-)			Hf k x-ray
									0.08886
									0.21342
									0.32555
									0.42635
^{178}Ta		177.94578	9.29 m	EC/99 /1.9		1+	+2.74	+0.65	ann.rad./
				β+ /1 /					Hf k x-ray
									0.09316
^{179}Ta		178.945930	1.8 a	EC/0.110		7/2+	+2.29	3.37	Hf k x-ray
^{180m}Ta	0.0120(2)		>7.1 × 10^{15} a			(9-)	+4.82	+4.95	
			> 1.2 × 10^{16} α	β-					
			> 1.7 × 10^{16} α	EC					
^{180}Ta		179.947465	8.15 h	EC/87 /0.854		1+			Hf k x-ray
				β- /13 /0.708	0.61/3				W k x-ray
					0.71/10				0.09333
									0.10340
^{181}Ta	99.9880(2)	180.947996				7/2+	+2.370	+3.3	
^{182m}Ta			15.8 m	I.T./0.5198		10-			Ta k x-ray

Elem. or Isot.	Natural Abundance (Atom %)	Atomic Mass or Weight	Half-life/ Resonance Width (MeV)	Decay Mode/ Energy (/MeV)	Particle Energy/ Intensity (MeV/%)	Spin ($h/2\pi$)	Nuclear Magnetic Mom. (nm)	Elect. Quadr. Mom. (b)	γ-Energy / Intensity (MeV/%)
									0.14678
									0.17157
^{182}Ta		181.950152	114.43 d	β- /1.814	0.25/30	3-	+3.02	+2.6	W k x-ray
					0.44/20				1.12127/100
					0.52/40				1.22138/79
									0.085–1.289
^{183}Ta		182.951373	5.1 d	β- /1.070	0.45/5	7/2+	+2.36		W k x-ray
					0.62/91				0.0847
									0.0991
									0.1079
									0.2461
									0.3540
^{184}Ta		183.95401	8.7 h	β- /2.87	1.11/15	(5-)			W k x-ray
					1.17/81				0.2528/44.
									0.4140/74.
									(0.09–1.4)
^{185}Ta		184.95556	49. m	β- /1.99	1.21/5	(7/2+)			W k x-ray
					1.77/81				0.0697
									0.1739
									0.1776
^{186}Ta		185.9586	10.5 m	β- /3.9	2.2/	(3-)			W k x-ray
									0.1979
									0.2149
									0.5106
									(0.09–1.5)
^{187}Ta		186.9605	> 0.3 μs						
^{188}Ta		187.9637	5 μs						0.292
^{189}Ta		188.9658	> 0.3 μs						
$_{74}$**W**		**183.84(1)**							
^{158m}W			0.14 ms	α	8.28(3)/				
^{158}W		157.975	1.3 ms	α/	6.433/96	0+			
^{159}W		158.9729	7. ms	α/					
^{160}W		159.9685	0.08 s	α/	5.92/	0+			
^{161}W		160.9674	0.41 s	β+, EC/18 /8.1					
				α/82 /	5.78/				
^{162}W		161.9635	1.39 s	β+, EC/54 /5.8		0+			
				α/46 /	5.54/				
^{163}W		162.9625	2.8 s	β+, EC/59 /7.5					
				α/41 /	5.38/				
^{164}W		163.95895	6. s	β+, EC/97 /5.0		0+			ann.rad./
				α/3 /	5.15/				
^{165}W		164.95828	5.1 s	β+, EC/99 /7.0					ann.rad./
				α/1 /	4.91/				
^{166}W		165.95503	16. s	β+, EC/99 /4.2		0+			ann.rad./
				α/1 /	4.74/				
^{167}W		166.95482	20. s	EC/5.6					
^{168}W		167.95181	53. s	EC/3.8		0+			ann.rad./
				α/10^{-5}/	4.40(1)				Ta k x-ray
									0.1755
									(0.037–0.573)
^{169}W		168.95178	1.3 m	EC/5.4					ann.rad./
									Ta k x-ray
									0.123
									(0.097–0.699)

Elem. or Isot.	Natural Abundance (Atom %)	Atomic Mass or Weight	Half-life/ Resonance Width (MeV)	Decay Mode/ Energy (/MeV)	Particle Energy/ Intensity (MeV/%)	Spin ($h/2\pi$)	Nuclear Magnetic Mom. (nm)	Elect. Quadr. Mom. (b)	γ-Energy / Intensity (MeV/%)
^{170}W		169.94923	2.4 m	EC/2.2		0+			ann.rad./
									Ta k x-ray
									0.3162
									(0.060–0.144)
^{171}W		170.94945	2.4 m	EC/4.6					ann.rad./
									Ta k x-ray
									0.1842
									(0.052–0.479)
^{172}W		171.94729	6.6 m	β+, EC/2.5		0+			ann.rad./
									Ta k x-ray
									0.0389
									(0.034–0.674)
^{173}W		172.94769	6.3 m	EC/4.0					ann.rad./
									Ta k x-ray
									0.4576
									(0.035–0.623)
^{174}W		173.94608	35. m	EC/1.9		0+			ann.rad./
									Ta k x-ray
									0.3287
									0.4288
									(0.056–0.429)
^{175}W		174.94672	35. m	EC/2.9		½-			(0.015–0.27)
^{176}W		175.94563	2.5 h	β+, EC/0.8		0+			0.03358
									0.06129
									0.09487
									0.10020
^{177}W		176.94664	2.21 h	EC/2.0		(1/2-)			Ta k x-ray
									0.15505
									0.18569
									0.42694
^{178}W		177.94588	21.6 d	EC/0.091		0+			Ta k x-ray
^{179m}W			6.4 m	IT/99.7/0.222		(1/2-)			W k x-ray
				EC/0.3/					0.2220
^{179}W		178.94707	38. m	EC/1.06		(7/2-)			Ta k x-ray
									0.0307
^{180}W	0.12(1)	179.946704	1.8×10^{18} a	α/		0+			
^{181}W		180.948197	121.1 d	EC/0.188		9/2+			Ta k x-ray
									0.13617
									0.15221
^{182}W	26.50(16)	181.948204	$>7.7 \times 10^{21}$ a	α/		0+			
^{183m}W			5.15 s	I.T./		(11/2+)			W k x-ray
									0.0465
									0.0526
									0.0991
									0.1605
^{183}W	14.31(4)	182.950223	$>4.1 \times 10^{21}$ a	α/		½-	+0.1177848		
^{184}W	30.64(2)	183.950931	$>8.9 \times 10^{21}$ a	α/		0+			
^{185m}W			1.6 m	I.T./0.1974		11/2+			W k x-ray
									0.0659
									0.1315
									0.1737
^{185}W		184.953419	74.8 d	β- /0.433	0.433/99.9	3/2-			0.12536
^{186}W	28.43(19)	185.954364	$>8.2 \times 10^{21}$ a	α/		0+			
^{187m}W			1.6 μs	IT	0.411	11/2+			(0.014-0.287)

Elem. or Isot.	Natural Abundance (Atom %)	Atomic Mass or Weight	Half-life/ Resonance Width (MeV)	Decay Mode/ Energy (/MeV)	Particle Energy/ Intensity (MeV/%)	Spin ($h/2\pi$)	Nuclear Magnetic Mom. (nm)	Elect. Quadr. Mom. (b)	γ-Energy / Intensity (MeV/%)
^{187}W		186.957161	23.9 h	β- /1.311	0.624/66	3/2-	0.62		Re k x-ray
					1.315/16				0.68572/33
					0.081–1.18				0.134–0.773
^{188}W		187.958489	69.78 d	β- /0.349	0.349/99	0+			0.0636
									0.2271
									0.2907
^{189}W		188.9619	9.7 m	β- /2.5	1.4/	(3/2-)			0.2604
					2.5/				(0.1262-1.466)
^{190m}W			~ 0.06 ms						(0.0585-0.694)
^{190}W		189.9632	30. m	β- /1.3	0.95/	0+			Re k x-ray
									0.1576
									0.1621
^{191}W		190.9666	> 0.3 μs						
^{192}W		191.968	> 0.3 μs			0+			
$_{75}$**Re**		**186.207(1)**							
^{159}Re			~ 0.02 ms	α	α/6.78/7.5				
				p	p/1.80				
^{160}Re		159.9821	0.7 ms	p/	1.261(6)/91				
				α/	6.54/				
^{161}Re		160.9776	14 ms	α/	6.24				
				p	1.35				
^{162}Re		161.9760	0.10 s	α/	6.12/94				
					6.09/94				
^{163}Re		162.97208	0.26 s	β+, EC/9.0	α/5.87/32				
				α/	5.92/66				
^{164}Re		163.9703	0.9 s	β+, EC/10.7					
				α/	5.78/				
^{165m}Re			~ 2.37 s	α/	5.502/				
^{165}Re		164.96709	2.6 s	β+, EC/87 /8.1					
				α/	5.49/ < 5.				
^{166}Re		165.9658	2.5 s	β+, EC/9.4					
				α/	5.50/				
^{167m}Re			6.2 s	α, EC/					
^{167}Re		166.9626	3.4 s	β+, EC/7.4					
				α/	5.015/				
^{168}Re		167.96157	4.4 s	β+, EC/9.1					
				α/	4.833/				0.1117
^{169m}Re			8.1 s	α	4.70/				
					4.87/				
^{169}Re		168.95879	16. s						
^{170}Re		169.95822	9.2 s	β+, EC/9.0					0.1560
									0.3055
									0.4125
^{171}Re		170.95572	15.2 s	EC/~ 5.7					
^{172m}Re			55. s	β+, EC/		(2)			ann.rad./
									0.1234
									0.2537
									0.3504
^{172}Re		171.9554	15. s	β+, EC/7.3					ann.rad./
									0.1234
									0.2537
^{173}Re		172.95324	2.0 m	EC/~ 3.9					ann.rad./
^{174}Re		173.95312	2.4 m	β+, EC/5.6					ann.rad./
									0.1119

Elem. or Isot.	Natural Abundance (Atom %)	Atomic Mass or Weight	Half-life/ Resonance Width (MeV)	Decay Mode/ Energy (/MeV)	Particle Energy/ Intensity (MeV/%)	Spin (h/2 π)	Nuclear Magnetic Mom. (nm)	Elect. Quadr. Mom. (b)	γ-Energy / Intensity (MeV/%)
									0.2430
^{175}Re		174.95138	5.8 m	β+, EC/4.3					ann.rad./
^{176}Re		175.95162	5.3 m	β+, EC/5.6		(3+)			ann.rad./
									0.1089
									0.2406
^{177}Re		176.95033	14. m	EC/78 /3.4		(5/2-)			ann.rad./
				β+ /22 /					W k x-ray
									0.0797
									0.0843
									0.1968
^{178}Re		177.95099	13.2 m	β+ /11 /4.7	3.3/	(3+)			ann.rad./
				EC/89 /					W k x-ray
									0.1059
									0.2373
									0.9391
^{179m}Re			0.47 ms						
^{179}Re		178.94999	19.7 m	EC/99 /2.71	0.95/	(5/2+)	2.8		W k x-ray
				β+ /1 /					0.1199
									0.2900
									0.4154
									0.4302
									1.6803
^{180}Re		179.95079	2.45 m	EC/92 /3.80	1.76/	1-	1.6		ann.rad./
				β+ /8 /					W k x-ray
									0.1036
									0.9028
									(0.07–2.2)
^{181}Re		180.95007	20. h	EC /1.74		5/2+	3.19		W k x-ray
									0.3607
									0.3655
									0.6390
^{182m}Re			12.7 h	EC/	0.55/	2+	3.3	+1.8	W k x-ray
					1.74/				0.0677
									1.1214
									1.2215
									(0.06–2.2)
^{182}Re		181.9512	2.67 d	EC/2.8		(7+)	2.8	+4.1	W k x-ray
									0.0678
									0.2293
									1.1213
									1.2214
^{183}Re		182.95082	70. d	EC/0.56		(5/2+)	+3.16	+2.2	W k x-ray
									0.16232
^{184m}Re			165. d	I.T./75 /0.188		8+	+2.9		Re k x-ray
				EC/25 /					0.1047
									0.2165
									0.92093
									(0.10–1.1)
^{184}Re		183.952521	35. d	EC/1.48		3-	+2.53	+3.0	W k x-ray
									0.79207
									08948
									0.90328
^{185}Re	37.40(2)	184.952955				5/2+	+3.1871	+2.19	
^{186m}Re			2.0 × 10^5 a	I.T./0.150		8+			Re k x-ray

Elem. or Isot.	Natural Abundance (Atom %)	Atomic Mass or Weight	Half-life/ Resonance Width (MeV)	Decay Mode/ Energy (/MeV)	Particle Energy/ Intensity (MeV/%)	Spin ($h/2\pi$)	Nuclear Magnetic Mom. (nm)	Elect. Quadr. Mom. (b)	γ-Energy / Intensity (MeV/%)
									0.0590
^{186}Re		185.954986	3.718 d	β- /92 /1.070	0.973/21	1-	+1.739	+0.62	W k x-ray
				EC/8 /0.582	1.07/71				0.1227/0.6
									0.1372/9.5
									(0.63–0.77)
^{187}Re	62.60(2)	186.955753	4.2 × 10^{10} a	β- /0.00266	0.0025/	5/2+	+3.2197	+2.07	
^{188m}Re			18.6 m	I.T./0.172		(6-)			Re k x-ray
									0.0925
									0.1059
^{188}Re		187.958114	17.00 h	β- /2.120	1.962/20	1-	+1.788	+0.57	Os k x-ray
					2.118/79				0.15502
									0.309–2.022
^{189}Re		188.95923	24. h	β- /1.01	1.01/	(5/2+)			0.1471
									0.2167
									0.2194
									0.2451
^{190m}Re			3.0 h	β- /51 /		(6-)			Re k x-ray
				I.T./49 /					0.1191
									0.2238
									0.6731
									(0.1–1.79)
^{190}Re		189.9618	3.0 m	β- /3.2	1.8/	(2-)			Os k x-ray
									0.1867
									0.5580
									0.6051
^{191}Re		190.96313	9.7 m	β- /2.05	1.8/				
^{192m}Re			~ 0.12 ms						(0.0606-0.146)
^{192}Re		191.9660	16. s	β- /4.2	~ 2.5/				(0.2–0.75)
^{193m}Re			~ 0.08 ms						(0.061-0.146)
^{193}Re		192.9675	> 0.3 μs						
^{194}Re		193.9704	> 0.3 μs						
$_{76}$Os		**190.23(3)**							
^{162}Os		161.984	1.8 ms	α/	6.60	0+			
^{163}Os		162.9827	5.5 ms	α/	6.51				
^{164}Os		163.9780	0.04 s	α		0+			
^{165}Os		164.9768	0.07 s	α					
^{166}Os		165.97269	0.18 s	β+, EC/28 /6.3	6.27/	0+			ann. rad./
				α/72 /	5.98/				
^{167}Os		166.9716	0.7 s	β+, EC/76 /8.2					ann.rad./
				α/24 /	5.84/				
^{168}Os		167.96780	2.2 s	β+, EC/51 /5.7		0+			ann. rad./
				α/49 /					
^{169}Os		168.96702	3.3 s	β+, EC/89 /7.7	5.57/80				ann.rad./
				α/13 /	5.51/12				
					5.54/8				
^{170}Os		169.96358	7.1 s	β+, EC/5.0		0+			ann.rad./
				α/	5.40/				(0.162–0.216)
^{171}Os		170.96319	8.4 s	β+, EC/98 /7.1	α/5.24/93.5				ann.rad./
				α/19 /	5.17/6.5				0.190–0.705
^{172}Os		171.96002	19. s	β+, EC/99 /4.5		0+			ann.rad./
				α/1.1/	5.10/				(0.063–1.120)
^{173}Os		172.95981	16. s	β+, EC/6.3					ann.rad./
				α/0.4 /	4.94/				0.142–0.299
^{174}Os		173.95706	44. s	β+, EC/3.9		0+			0.118

Elem. or Isot.	Natural Abundance (Atom %)	Atomic Mass or Weight	Half-life/ Resonance Width (MeV)	Decay Mode/ Energy (/MeV)	Particle Energy/ Intensity (MeV/%)	Spin ($h/2\pi$)	Nuclear Magnetic Mom. (nm)	Elect. Quadr. Mom. (b)	γ-Energy / Intensity (MeV/%)
				α/0.02 /	4.76/				0.138 / 0.001
									0.158
									0.325
^{175}Os		174.95695	1.4 m	β+, EC/5.3					0.125
									0.181
									0.248
^{176}Os		175.95481	3.6 m	β+, EC/3.2		0+			0.8155
									0.7758
									0.8573
									1.2093
									1.2909
^{177}Os		176.95497	2.8 m	β+, EC/4.5		(1/2-)			0.0848
									0.1958
									0.3002
									1.2686
^{178}Os		177.95325	5.0 m	β+, EC/2.3		0+			ann.rad./
									0.5946
									0.6850
									0.9687
									1.3311
^{179}Os		178.95382	7. m	β+, EC/3.7					ann.rad./
									0.0654
									0.2186
									0.5938
^{180}Os		179.95238	21.5 m	β+, EC/1.5		0+			Re k x-ray
									0.0202–0.7174
^{181m}Os			1.75 h	EC/		(1/2-)			ann.rad./
									0.0489
^{181}Os		180.95324	2.7 m	EC/2.9		(7/2-)			ann.rad./
									0.11794
									0.23868
									0.8267
									(0.07–2.64)
^{182}Os		181.95211	21.5 h	EC/0.9		0+			Re k x-ray
									0.1802
									0.5100
^{183m}Os			9.9 h	EC/84 /		½-			Os k x-ray
				I.T./16 /					Re k x-ray
									1.1020
									1.1080
^{183}Os		182.95313	13. h	EC/2.1		9/2+	-0.79	+3.1	Re k x-ray
									0.1144
									0.3818
^{184}Os	0.02(1)	183.952489				0+			
^{185}Os		184.954042	93.6 d	EC/1.013		½-			Re k x-ray
									0.6461
									0.8748
									0.8805
^{186}Os	1.59(3)	185.953838	2. × 10^{15} a	α/	~ 2.75/	0+			
^{187}Os	1.96(2)	186.955750				½-	+0.0646519		
^{188}Os	13.24(8)	187.955838				0+			
^{189m}Os			5.8 h	I.T./0.0308		9/2-			Os L x-ray
									0.0308
^{189}Os	16.15(5)	188.958148				3/2+	+0.659933	+0.9	

Elem. or Isot.	Natural Abundance (Atom %)	Atomic Mass or Weight	Half-life/ Resonance Width (MeV)	Decay Mode/ Energy (/MeV)	Particle Energy/ Intensity (MeV/%)	Spin ($h/2\pi$)	Nuclear Magnetic Mom. (nm)	Elect. Quadr. Mom. (b)	γ-Energy / Intensity (MeV/%)
^{190m}Os			9.9 m	I.T./1.705		10-	-0.6		Os k x-ray
									0.1867
									0.3611
									0.5026
									0.6161
^{190}Os	26.26(2)	189.958447				0+			
^{191m}Os			13.1 h	I.T./0.0744		3/2-			Os k x-ray
									0.0744
^{191}Os		190.960930	15.4 d	β- /0.314	0.140/100	9/2-	+0.96	+2.5	Ir k x-ray
									0.1294
^{192m}Os			6.0 s	I.T./2.0154		(10-)			Os k x-ray
									0.2058/65.9
									0.5692/70
									(0.201–1.000)
^{192}Os	40.78(19)	191.961481				0+			
^{193}Os		192.964152	30.5 h	β- /1.141	1.04/20	3/2-	+0.730	+0.47	Ir k x-ray
									0.1389
									0.4605
^{194}Os		193.965182	6.0 a	β- /0.097	0.054/33	0+			Ir L x-ray
					0.096/67				0.0429
^{195}Os		194.968	6.5 m	β- /2.0	2.0/				
^{196}Os		195.96964	34.9 m	β- /1.16	0.84/	0+			0.1262/5
									0.4079/5.9
^{197}Os			2.8 m	β-					0.2239
									(0.0412-0.406)
$_{77}$**Ir**		**192.217(3)**							
^{164}Ir		163.9922	0.06 ms	p	1.78				
^{165}Ir		164.9875	0.3 ms	p/87	1.71				
				α/13	6.72				
^{166m}Ir			14.3 ms	α/98.2	6.545				
				p/1.8	1.32				
^{166}Ir		165.9858	0.010 s	α/93	6.56				
				p/6.9	1.15				
^{167m}Ir			26. ms	α/48, β+	6.39/90				
				p/32	1.25/0.42				
^{167}Ir		166.98167	32. ms	α/80, β+	6.35/48				
				p/0.4	1.06/39.3				
^{168}Ir		167.9799	0.16 s	α/82					
^{169m}Ir			280. ms	α/	6.12/59				
^{169}Ir		168.97630	353. ms	α/	5.99/42				
^{170}Ir		169.9750	0.81 s	α/	6.01/				0.175
					6.05/				
					5.95/				
					6.12/				
^{171}Ir		170.97163	1.3 s	α/	5.91/				
^{172}Ir		171.9705	2.1 s	α/	5.811/				0.228
									(0.379–0.475)
^{173}Ir		172.96750	3.0 s	α/	5.665/				0.0493
									(0.092–0.296)
^{174}Ir		173.96686	4. s	α/	5.478/				0.1587
									(0.276–1.33)
^{175}Ir		174.96411	~ 4.5 s	α/	5.393/				0.1056
^{176}Ir		175.96365	8. s	EC, β+/80					0.260
				α/3.2/	5.118/				(0.135–0.415)

Elem. or Isot.	Natural Abundance (Atom %)	Atomic Mass or Weight	Half-life/ Resonance Width (MeV)	Decay Mode/ Energy (/MeV)	Particle Energy/ Intensity (MeV/%)	Spin ($h/2\pi$)	Nuclear Magnetic Mom. (nm)	Elect. Quadr. Mom. (b)	γ-Energy / Intensity (MeV/%)
^{177}Ir		176.96130	30. s	EC, β+/5.7					0.184
				α/0.06/	5.011/				(0.062–0.194)
^{178}Ir		177.96108	12. s	β+, EC/6.3					
									0.1320
									0.2667
									0.3633
^{179}Ir		178.95912	4. m	EC/4.9					0.0975
									(0.045–0.220)
^{180}Ir		179.95923	1.5 m	EC/6.4					0.2765
									((0.132–1.106)
^{181}Ir		180.95763	4.9 m	β+, EC/4.1		(7/2+)			ann.rad./
									0.1076
									(0.020–1.715)
^{182}Ir		181.95808	15. m	β+ /44 /5.6			+2.6	~ -1.7	ann.rad./
				EC/56 /					Os k x-ray
									0.1273
									0.2370
^{183}Ir		182.95685	57. m	β+, EC/3.5			+2.4	~ -1.8	ann.rad./
									0.0877
									0.2285
									0.2824
^{184}Ir		183.95748	3.0 h	β+ /12 /4.6	2.3/	5-	0.70	+2.41	ann.rad./
				EC/88 /	2.9/				Os k x-ray
									0.11968
									0.2640
									0.3904
^{185}Ir		184.95670	14. h	β+ /3 /2.4		(5/2-)	2.60	-2.1	ann.rad./
				EC/97 /					Os k x-ray
									0.2543
									1.8288
^{186m}Ir			1.7 h	EC /		(2-)	0.64	+1.46	Os k x-ray
									0.1371
									0.7675
^{186}Ir		185.95795	15.7 h	EC/98 /3.83		(5+)	3.9	-2.55	Os k x-ray
				β+ /2 /					0.1372
									0.2968
									0.4348
									(0.13–3.0)
^{187}Ir		186.95736	10.5 h	EC/1.50		3/2+	+0.17	+0.94	Os k x-ray
									0.0743
									0.4009
									0.4271
									0.6109
									0.9128
^{188}Ir		187.95885	1.72 d	β+ /2.81	1.13/	(2-)	0.30	+0.48	Os k x-ray
				EC/99+ /	1.64/				0.1550
									0.4780
									0.6330
									2.2146
^{189}Ir		188.95872	13.2 d	EC/0.53		3/2+	+0.14	+0.88	Os k x-ray
									0.2449
^{190m2}Ir			3.09 h	β+, EC/95 /		(11-)			0.376
				I.T./5 /					
^{190m1}Ir			1.12 h	I.T. /0.0263		7+			Ir L x-ray

Elem. or Isot.	Natural Abundance (Atom %)	Atomic Mass or Weight	Half-life/ Resonance Width (MeV)	Decay Mode/ Energy (/MeV)	Particle Energy/ Intensity (MeV/%)	Spin ($h/2\pi$)	Nuclear Magnetic Mom. (nm)	Elect. Quadr. Mom. (b)	γ-Energy / Intensity (MeV/%)
^{190}Ir		189.960546	11.8 d	EC/2.0		(4+)	0.04	+2.8	Os k x-ray
									0.1867
									0.4072
									0.5186
									0.5580
									0.6051
									(0.2–1.4)
^{191m}Ir			4.93 s	I.T./0.1714		11/2-	+0.603		Ir k x-ray
									0.1294
^{191}Ir	37.3(2)	190.960594				3/2+	+0.151	+0.82	
^{192m2}Ir			241. a	I.T./0.161		(9+)			Ir k x-ray
^{192m1}Ir			1.44 m	I.T./0.0580		(1+)			Ir L x-ray
									0.0580
									0.3165
^{192}Ir		191.962605	73.83 d	β- /1.460		(4-)	+1.92	+2.15	Pt k x-ray
									0.31649/83.
									0.46806/48.
^{193m}Ir			10.53 d	I.T./0.0802		11/2-			Ir L x-ray
									0.0803
^{193}Ir	62.7(2)	192.962926				3/2+	+0.164	+0.75	
^{194m}Ir			170. d	β- /		11			Pt k x-ray
									0.3284
									0.4829
									0.5624
^{194}Ir		193.965078	19.3 h	β-/2.247	1.92/9	1-	+0.39	+0.34	0.2935
					2.25/86				0.3284
									0.6451
									(0.1–2.2)
^{195m}Ir			3.9 h	β- /	0.41/	(11/2-)			Pt k x-ray
					0.97/				0.3199/9.6
									0.3649/9.5
									0.4329/9.6
									0.6849/9.6
^{195}Ir		194.965980	2.8 h	β- /1.120	1.0/80	(3/2+)			Pt k x-ray
					1.11/13				0.0989/9.7
^{196m}Ir			1.40 h	β-/	1.16/				Pt k x-ray
									0.3557
									0.3935
									0.4471
									0.5214
									0.6473
^{196}Ir		195.96840	52. s	β- /3.21	2.1/15	0-			0.3329
					3.2/80				0.3557
									0.7796
^{197m}Ir			8.9 m	β- /		(11/2-)			0.3465
				I.T./					see Ir[197]
^{197}Ir		196.96965	5.8 m	β- /2.16	1.5/	(3/2+)			0.0531
					2.0/				0.1351
									0.4306
									0.4697
^{198}Ir		197.9723	8. s	β- /4.1					0.4074
									0.5070
^{199}Ir		198.97380							

Elem. or Isot.	Natural Abundance (Atom %)	Atomic Mass or Weight	Half-life/ Resonance Width (MeV)	Decay Mode/ Energy (/MeV)	Particle Energy/ Intensity (MeV/%)	Spin $(h/2\pi)$	Nuclear Magnetic Mom. (nm)	Elect. Quadr. Mom. (b)	γ-Energy / Intensity (MeV/%)
$_{78}$Pt		**195.084(9)**							
^{166}Pt		165.995	0.3 ms	α/	7.11/	0+			
^{167}Pt		166.930	0.9 ms	α/	6.98/				
^{168}Pt		167.9882	2.1 ms	α	6.82	0+			0.582/69
									0.594/69
									0.725/62
^{169}Pt		168.9867	7.0 ms	α	6.69				
^{170}Pt		169.98250	14.0 ms	α	6.55	0+			0.509/100
									0.662/86
									0.214–0.726
^{171}Pt		170.9812	0.05 s	α	6.45				0.4450
									(0.1564-1.208)
^{172}Pt		171.97735	0.10 s	α/	6.31/94	0+			
^{173}Pt		172.9764	0.36 s	β+, EC/8.2	6.23				
				α/	6.20/				
^{174}Pt		173.97282	0.89 s	β+, EC/17 /5.6		0+			
				α/83 /	6.040/				
^{175}Pt		174.97242	2.5 s	β+, EC/65 /7.6					0.0774
				α/35 /	5.831/5				0.1354
					5.96/54				0.2128
					6.038/				
^{176}Pt		175.96895	6.3 s	β+, EC/60 /5.1		0+			ann.rad./
				α/40 /	5.528/0.6				0.2277
					5.750/41				
^{177}Pt		176.96847	11. s	EC/91 /6.8	5.53/				0.0908
				α/9 /	5.485/3				
					5.525/6				
^{178}Pt		177.96565	21. s	EC/93 /4.5		0+			
				α/7 /	5.286/0.2				
					5.442/7				
^{179}Pt		178.96536	33. s	β+, EC/5.7			+0.43		
				α/	5.16/				
^{180}Pt		179.96303	52. s	β+, EC/99.7 /3.7	0+				
				α/0.3 /	5.140/				
^{181}Pt		180.96310	51. s	β+, EC/5.2			+0.48		
^{182}Pt		181.96117	2.7 m	β+, EC/2.9		0+			ann.rad./
									0.1360
									0.1460
									0.2100
^{183m}Pt			43. s	β+, EC/		(7/2-)	+0.78	+3.4	ann.rad./
				I.T./					0.3132/26
									0.3164/59
									0.6296/100
									0.058–1.75
^{183}Pt		182.96160	7. m	β+, EC/4.6			+0.50		ann.rad./
									0.119/100
									0.307/93
									0.260/90
									0.058–1.377
^{184}Pt		183.95992	17.3 m	β+, EC/2.3		0+			ann.rad./
									0.1549
									0.1919
									0.5484

Elem. or Isot.	Natural Abundance (Atom %)	Atomic Mass or Weight	Half-life/ Resonance Width (MeV)	Decay Mode/ Energy (/MeV)	Particle Energy/ Intensity (MeV/%)	Spin (h/2 π)	Nuclear Magnetic Mom. (nm)	Elect. Quadr. Mom. (b)	γ-Energy / Intensity (MeV/%)
^{185m}Pt			33. m	β+, EC/		½-	+0.50		
^{185}Pt		184.96062	1.18 h	β+, EC/3.8		(9/2+)	-0.75	+4.	ann.rad./
									0.1353
									0.1974
									0.2296
									0.2551
^{186}Pt		185.95935	2.0 h	β+, EC/1.38		0+			ann.rad./
									0.6115
									0.6892
^{187}Pt		186.96059	2.35 h	β+, EC/3.1		3/2-	-0.40	-1.0	ann.rad./
									Ir k x-ray
									0.1064
									0.1100
									0.2015
									0.2849
									0.7092
^{188}Pt		187.95940	10.2 d	EC/0.51		0+			Ir k x-ray
									0.1876
									0.1951
^{189}Pt		188.96083	10.9 h	β+, EC/1.97		3/2-	-0.43	-1.2	Ir k x-ray
									0.0943
									0.6076
									0.7214
									(0.09–1.47)
^{190}Pt	0.012(2)	189.95993	4.5 × 10^{11} a	α		0+			
^{191}Pt		190.961677	2.86 d	EC/1.02		(3/2-)	-0.50	-0.9	Ir k x-ray
									0.3599
									0.4094
									0.5389
^{192}Pt	0.782(24)	191.961038				0+			
^{193m}Pt			4.33 d	I.T./0.1498		13/2+	-0.75		Pt k x-ray
									0.1355
^{193}Pt		192.962988	60. a	EC/0.0566		(1/2-)	+0.60		Ir k x-rays
^{194}Pt	32.86(40)	193.962680				0+			
^{195m}Pt			4.01 d	I.T./0.2952		13/2+	-0.61	~ +1.4	Pt k x-ray
									0.0989
^{195}Pt	33.78(24)	194.964791				1/2-	+0.6095		
^{196}Pt	25.21(34)	195.964952				0+			
^{197m}Pt			1.590 h	I.T./97 /		13/2+			Pt k x-ray
				β- /3 /					0.0530
									0.3465
^{197}Pt		196.967340	19.9 h	β- /0.719		1/2-	0.51		Au k x-ray
									0.1914
									0.2688
^{198}Pt	7.356(130)	197.967893				0+			
^{199m}Pt			13.6 s	I.T./0.424		13/2+			Pt k x-ray
									0.3919
^{199}Pt		198.970593	30.8 m	β- /1.70	0.90/18	(5/2-)			0.3170/3.88
					1.14/14				0.49375/4.47
									0.5430/11.7
									(0.055–1.293)
^{200}Pt		199.971441	12.5 h	β- /~ 0.66		0+			Au k x-ray
									0.13590
									0.22747

Elem. or Isot.	Natural Abundance (Atom %)	Atomic Mass or Weight	Half-life/ Resonance Width (MeV)	Decay Mode/ Energy (/MeV)	Particle Energy/ Intensity (MeV/%)	Spin ($h/2\pi$)	Nuclear Magnetic Mom. (nm)	Elect. Quadr. Mom. (b)	γ-Energy / Intensity (MeV/%)
									0.24371
^{201}Pt		200.97451	2.5 m	β- /2.66		(5/2-)			0.070
									0.152
									0.222
									1.760
^{202m}Pt			0.3 ms						(0.535-0.719)
^{202}Pt		201.9757	1.8 d			0+			0.440
$_{79}$**Au**		**196.966569(4)**							
^{170m}Au			0.62 ms	p/58	1.74/				
				α/42	7.11/				
^{170}Au		169.9961	0.30 ms	p/89	1.46/				
				α/11	7.00/				
^{171m}Au			1.09 ms	α/66	6.995				
				p/34	1.694				
^{171}Au		170.99188	0.022 ms	p/100	1.437				
^{172}Au		171.9900	4 ms	α/7.02	6.86				
^{173m}Au			15 ms	α/92	6.732				
^{173}Au		172.98624	0.02 s	α/94	6.672				
^{174}Au		173.9848	0.14 s	α	6.54				
^{175}Au		174.98127	0.15 s	α					
^{176}Au		175.9801	0.9 s	β+, EC/10.5					
				α/	6.260/80				
					6.290/20				
^{177}Au		176.97687	1.2 s	α/	6.115/				
					6.150/				
^{178}Au		177.9760	2.6 s	α/	5.920/				
^{179}Au		178.97321	7.5 s	α/	5.85/				
^{180}Au		179.97252	8.1 s	EC/8.6	5.65				0.1522
				α/	5.61				0.2564
					5.50				0.5242
									0.6765
									0.8084
									0.8597
^{181}Au		180.97008	11.4 s	EC/97.5/6.3	5.482/				
				α/2.7/					
^{182}Au		181.96962	21. s	β+, EC/6.9		(2+)	1.3		ann.rad./
				α/0.13/					0.1549
									0.2649
									(0.13–1.4)
^{183}Au		182.96759	42. s	EC/5.5			+1.97		0.1630
				α/0.8/					0.2730
									0.3625
^{184m}Au			48 s	I.T.		(2+)	+1.44	+1.9	0.069(IT)
^{184}Au		183.96745	21. s	EC, β+/7.1		(5+)	+2.07	+4.7	
				α/0.013/					
^{185m}Au			6.8 m	β+, EC/					
				I.T./0.145					
^{185}Au		184.96579	4.3 m	β+, EC/4.71		(5/2-)	+2.17	-1.1	ann.rad./
				α/0.26/					
^{186m}Au			< 2. m	β+, EC/					0.1915
^{186}Au		185.96595	10.7 m	β+, EC/6.0		3-	-1.28	+3.1	ann.rad./
				α/8(10)$^{-4}$/					0.1915
									0.2988

Elem. or Isot.	Natural Abundance (Atom %)	Atomic Mass or Weight	Half-life/ Resonance Width (MeV)	Decay Mode/ Energy (/MeV)	Particle Energy/ Intensity (MeV/%)	Spin (h/2 π)	Nuclear Magnetic Mom. (nm)	Elect. Quadr. Mom. (b)	γ-Energy / Intensity (MeV/%)
^{187m}Au			2.3 s	IT		9/2-			
^{187}Au		186.96457	8.3 m	β+, EC/3.60		1/2+	+0.53		ann.rad./
									0.9152
									1.2668
									1.3321
									1.4081
^{188}Au		187.96532	8.8 m	β+, EC/5.3		(1-)	-0.07		ann.rad./
									0.2660
									0.3404
									0.6061
^{189m}Au			4.6 m	β+, EC/		11/2-	+6.19		0.1667
^{189}Au		188.96395	28.7 m	EC/96 /3.2		1/2+	+0.49		ann.rad./
				β+ /4 /					Pt k x-ray
									0.4478
									0.7133
									0.8128
^{190}Au		189.96470	43. m	β+ /2 /4.44		1-	-0.07		ann.rad./
				EC/98 /					Pt k x-ray
									0.2958
									0.3018
									0.5977
^{191m}Au			0.9 s	I.T./0.2663		(11/2-)	6.6		Au k x-ray
									0.2411
									0.2526
^{191}Au		190.96370	3.2 h	EC/1.83		3/2+	+0.137	+0.72	Pt k x-ray
									0.5864/16
									(0.088–1.30)
^{192}Au		191.96481	4.9 h	β+ /5 /3.52	2.19/	1-	-0.011	-0.23	ann.rad./
				EC/95 /	2.49/				Pt k x-ray
									0.2959
									0.3165
^{193m}Au			3.9 s	I.T./0.2901		11/2-	6.2	+1.98	Au k x-ray
									0.2580
^{193}Au		192.96415	17.6 h	EC/1.07		3/2+	+0.140	+0.66	Pt k x-ray
									0.1862
									0.2556
^{194}Au		193.96537	1.64 d	β+ /3 /2.49	1.49/	1-	+0.076	-0.24	ann.rad./
				EC/97 /					Pt k x-ray
									0.2935
									0.3284/61
^{195m}Au			30.5 s	I.T./0.3186		11/2-	6.2	+1.9	Au k x-ray
									0.2617
^{195}Au		194.965035	186.10 d	EC/0.227		3/2+	+0.149	+0.61	Pt k x-ray
^{196m2}Au			9.7 h	I.T./0.5954		12-	5.7		Au k x-ray
									0.1478
									0.1883
^{196m1}Au			8.1 s	I.T./0.0846		8+			0.0847
^{196}Au		195.966570	6.17 d	EC/92 /1.506		2-	+0.591	~ 0.81	Pt k x-ray
^{197m}Au			7.8 s	I.T./0.4094		11/2-	+6.0	+1.7	Au k x-ray
				β- /8 /0.686					0.1302
									0.2790
^{197}Au	100.	196.966569				3/2+	+0.14575	+0.55	
^{198m}Au			2.30 d	I.T./0.812		(12-)	+5.9		Au k x-ray
									0.0972

Elem. or Isot.	Natural Abundance (Atom %)	Atomic Mass or Weight	Half-life/ Resonance Width (MeV)	Decay Mode/ Energy (/MeV)	Particle Energy/ Intensity (MeV/%)	Spin ($h/2\pi$)	Nuclear Magnetic Mom. (nm)	Elect. Quadr. Mom. (b)	γ-Energy / Intensity (MeV/%)
									0.1803
									0.2419
^{198}Au		197.968242	2.695 d	β- /1.372	0.290/1	2-	+0.5934	+0.64	Hg k x-ray
					0.961/99				0.411794
^{199}Au		198.968765	3.14 d	β- /0.453	0.25/22	3/2+	~ +0.2715	+0.51	Hg k x-ray
					0.292/72				0.15837
					0.462/6				0.20820
^{200m}Au			18.7 h	β- /84 /1.0	0.56/	12-	5.9		Au k x-ray
				I.T./16 /					0.2559/71
									0.3680/77
									0.4978/73
									0.5793/72
									0.084–0.904)
^{200}Au		199.97073	48.4 m	β- /2.24	0.7/15	1-			0.3679/19
					2.2/77				1.2254/10.6
									(0.077–1.570)
^{201}Au		200.971657	26. m	β- /1.28	1.27/82	3/2+			(0.027–0.732)
^{202}Au		201.9738	29. s	β- /3.0		(1-)			0.4396
^{203}Au		202.975155	1.0 m	β- /2.14	~ 1.9/	3/2+			(0.04–0.37)
^{204}Au		203.9777	40. s	β- /4.5		(2-)			0.4366
									1.5113
^{205}Au		204.9799	31. s	β- /					(0.38–1.33)
$_{80}$**Hg**		**200.59(2)**							
^{171}Hg		171.0038	0.06 ms	α	7.49				
^{172}Hg		171.9988	0.3 ms	α	7.36	0+			
^{173}Hg		172.9972	0.8 ms	α	7.20				
^{174}Hg		173.99286	1.9 ms	α	7.07	0+			
^{175}Hg		174.9914	0.02 s	α					
^{176}Hg		175.98736	21 ms	α	6.74/94	0+			
^{177m}Hg			1.5 μs	IT					0.246
^{177}Hg		176.9863	0.13 s	α	6.58				
^{178}Hg		177.98248	0.26 s	EC/50 /6.1		0+			
				α/50 /	6.43/				
^{179}Hg		178.98183	1.05 s	EC/8.0					
				α/	6.29/				
^{180}Hg		179.97827	2.6 s	EC/5.5		0+			0.1250
				α/	6.12/33				0.3005
					5.69/.03				0.3812
^{181}Hg		180.97782	3.6 s	β+ EC/76 /~ 7.3		(1/2-)	+0.507		0.0663
				α/24 /					0.0811
									0.0924
									0.1474
									0.1587
									0.2142
									0.2398
^{182}Hg		181.97469	10.8 s	β+, EC/85/5.0		0+			0.129/122
				α/15/	5.87/8.6				0.2176/66
					5.45/0.03				0.0256–0.543
^{183}Hg		182.97445	9. s	β+, EC/77/6.3		½-	+0.524		0.0714
				α/	5.83/				0.0874
					5.91/				0.1538
^{184}Hg		183.97171	30.9 s	β+, EC/99/4.1		0+			0.1565/102
				α/1/	5.54/1.3				0.2367/100

Elem. or Isot.	Natural Abundance (Atom %)	Atomic Mass or Weight	Half-life/ Resonance Width (MeV)	Decay Mode/ Energy (/MeV)	Particle Energy/ Intensity (MeV/%)	Spin ($h/2\pi$)	Nuclear Magnetic Mom. (nm)	Elect. Quadr. Mom. (b)	γ-Energy / Intensity (MeV/%)
					5.07/0.002				0.2384/18
									(0.018-0.4227)
^{185m}Hg			21. s	β+, EC, IT, α/	5.37/	13/2+	-1.02	~ +0.2	0.211
									0.292
^{185}Hg		184.97190	51. s	β+, EC/95/5.8		½-	+0.509		0.02–0.55
^{186}Hg		185.96936	1.4 m	β+, EC/3.3		0+			0.1119
				α	5.09/0.02				0.2518
^{187m}Hg			1.9 m	β+, EC/		13/2+	-1.04	+0.5	see Hg187
^{187}Hg		186.96981	2.4 m	β+, EC/4.9		3/2-	-0.594	-0.8	0.1034/32
									0.2334/100
									0.2403/33
									0.27151/31
									0.3763/38
									0.5254/30
									0.10–2.18
^{188}Hg		187.96758	3.2 m	β+, EC/2.3		0+			0.0988
				α	4.61				0.1148
									0.1424
									0.1900
^{189m}Hg			8.6 m	EC/		13/2+	-1.06	+0.7	0.0780
									0.3210
									0.4345
									0.5655
									(0.08–2.170)
^{189}Hg		188.96819	7.6 m	EC/4.2		3/2-	-0.609	-0.8	0.2005
									0.2038
									0.2386
									0.2485
^{190}Hg		189.96632	20.0 m	EC/1.5		0+			0.1296
									0.1426
^{191m}Hg			51. m	β+ /6 /		13/2+	-1.07	+0.6	ann.rad./
				EC/94 /					Au k x-ray
									0.2741
									0.4203
									0.5787
									(0.07–1.9)
^{191}Hg		190.96716	50. m	β+, EC/3.2		(3/2-)	-0.62	-0.8	0.1963
									0.2247
									0.2524
^{192}Hg		191.96563	5.0 h	EC/~ 0.5		0+			Au k x-ray
									0.1572
									0.2748
									0.3065
^{193m}Hg			11.8 h	β+, EC/91 /		13/2+	-1.058430	~ +0.92	Hg k x-ray
				I.T./9 /0.2901					0.1866
									0.2580
									0.4076
									0.5733
									0.9324
									(0.1–1.96)
^{193}Hg		192.96667	3.8 h	EC, B+/2.34		3/2-	-0.6276	-0.7	0.1866
									0.2580
									0.8611
^{194}Hg		193.96544	520. a	EC/0.04		0+			Au L x-rays

Elem. or Isot.	Natural Abundance (Atom %)	Atomic Mass or Weight	Half-life/ Resonance Width (MeV)	Decay Mode/ Energy (/MeV)	Particle Energy/ Intensity (MeV/%)	Spin (h/2 π)	Nuclear Magnetic Mom. (nm)	Elect. Quadr. Mom. (b)	γ-Energy / Intensity (MeV/%)
^{195m}Hg			1.67 d	I.T./(54)/0.3186		13/2+	-1.04465	+1.1	Hg k x-ray
				EC/(46)/					Au k x-ray
									0.2617
									0.5603
									0.7798
^{195}Hg		194.96672	10.5 h	EC/1.51		1/2-	+0.541475		Au k x-ray
									0.0614
									0.7798
^{196}Hg	0.15(1)	195.965833	>2.5 × 10^{18} a	α		0+			
^{197m}Hg			23.8 h	I.T./(93)/0.2989		13/2+	-1.027684	+1.2	Hg k x-ray
									Au k x-ray
									0.13398
^{197}Hg		196.967213	2.69 d	EC/0.600		1/2-	+0.527374		Au k x-ray
									0.07735
^{198}Hg	9.97(20)	197.9667690				0+			
^{199m}Hg			42.7 m	I.T./0.532		13/2+	-1.014703	+1.2	Hg k x-ray
									0.15841
^{199}Hg	16.87(22)	198.9682799				1/2-	+0.505885		
^{200}Hg	23.10(19)	199.9683260				0+			
^{201}Hg	13.18(9)	200.970302				3/2-	-0.560226	+0.37	
^{202}Hg	29.86(26)	201.970643				0+			
^{203}Hg		202.972873	46.61 d	β- /0.492	0.213/100	5/2-	+0.8489	+0.34	Tl k x-ray
									0.279188
^{204}Hg	6.87(15)	203.9734939				0+			
^{205}Hg		204.976073	5.2 m	β- /1.531	1.33/4	1/2-	+0.6009		0.20378
									(0.2–1.4)
^{206}Hg		205.97751	8.2 m	β- /1.31	0.935/34	0+			Tl k x-ray
					1.3/63				0.3052
									0.6502
^{207}Hg		206.9826	2.9 m	β- /4.8		(9/2+)			
^{208}Hg		207.9859	41. m	β-		0+			0.474
^{209}Hg		208.9910	36 s	β-					0.324
^{210}Hg		209.9945	> 0.3 μs	β-		0+			
$_{81}$Tl		**204.3833(2)**							
^{176}Tl		176.0006	5 ms	p	1.26/~ 100				
^{177m}Tl			0.23 ms	p/51	1.95				
				α/49	7.48				
^{177}Tl		176.99643	0.017 s	α/73					
				p/27					
^{178}Tl		177.9949	0.25 s	α/	6.704				
					6.785				
					6.62				
					6.859				
^{179m}Tl			1.7 ms	α	/7.21/80				
				α	/7.10/20				
^{179}Tl		178.99109	0.3 s	α	6.57/				
^{180}Tl		179.9899	1.5 s	α//8	6.28/30				
					6.36/30				
					6.21/18				
					6.56/15				
					6.47/7				
^{181m}Tl			1.4 ms	α	6.58/100				
^{181}Tl		180.98626	3.2 ms	α/ < 10	6.19/100				
^{182}Tl		181.9857	3. s	β+, EC/10.9					0.351

Elem. or Isot.	Natural Abundance (Atom %)	Atomic Mass or Weight	Half-life/ Resonance Width (MeV)	Decay Mode/ Energy (/MeV)	Particle Energy/ Intensity (MeV/%)	Spin (h/2 π)	Nuclear Magnetic Mom. (nm)	Elect. Quadr. Mom. (b)	γ-Energy / Intensity (MeV/%)
									(0.26–0.41)
^{183m}Tl			53. ms	α//1.5	6.33/80	9/2-			0.0618
					6.38/16				(0.046-0.0894)
					6.46/4				
^{183}Tl		182.98219	5. s	β+, EC/7.7		½+			0.208
^{184}Tl		183.98187	11. s	β+, EC/(98)/9.2					0.2868
				α/(2)/	6.16/				0.3399
									0.3667
^{185m}Tl			1.8 s	I.T./0.453		(9/2-)			0.1688
				α/5.97	6.01				0.2840
^{185}Tl		184.9788	20. s	EC/β+/6.6					
^{186m}Tl			4. s	I.T./0.374					0.3738
^{186}Tl		185.9783	28. s	β+, EC/7.5					0.3567
									0.4026
									0.4053
^{187m}Tl			15.6 s	I.T./~ 0.33		(9/2+)	+3.79	-2.4	0.2995
^{187}Tl		186.97591	50. s	β+, EC/6.0		½+	1.6		
^{188m}Tl			1.18 m	β+, EC/		(7+)			Hg k x-ray
									0.4129
									0.5043
									0.5921
^{188}Tl		187.97601	1.2 m	β+, EC/7.8		(2-)	+0.48	+0.129	see Tl[188m]
									0.4129
^{189m}Tl			1.4 m	β+, EC/		(9/2-)	+3.88	-2.29	0.2156
									0.2284
									0.3175
									0.4452
^{189}Tl		188.97359	2.3 m	β+, EC/5.2		(1/2+)			0.3337
									0.4510
									0.5223
									0.9422
^{190m}Tl			3.7 m	β+, EC/	4.2/	(7+)	+0.495	+0.29	0.1968
									0.4164
									0.7311
^{190}Tl		189.97388	2.6 m	β+, EC/7.0	5.7/	(2-)	+0.254	-0.33	0.4164
									0.6254
									0.6838
									1.0999
^{191m}Tl			5.2 m	β+, EC/(98)/		(9/2+)	+3.903	-2.2	0.2157
									0.2647
									0.3256
									0.3359
^{191}Tl		190.97179	2.2 m			(1/2)	+1.588		
^{192m}Tl			10.8 m	β+, EC/		(7+)	+0.518	+0.46	0.1740
									0.4228
									0.6348
									0.7863
									0.7455
^{192}Tl		191.97223	9.6 m	β+, EC/6.4		(2-)	+0.200	-0.33	0.3975
									0.4228
									0.6908
^{193m}Tl			2.1 m	I.T./(75)/		(9/2-)	+3.948	-2.20	0.3650
^{193}Tl		192.9707	22. m	β+, EC/3.6		(1/2+)	+1.591		0.2077
									0.3244

Elem. or Isot.	Natural Abundance (Atom %)	Atomic Mass or Weight	Half-life/ Resonance Width (MeV)	Decay Mode/ Energy (/MeV)	Particle Energy/ Intensity (MeV/%)	Spin ($h/2\pi$)	Nuclear Magnetic Mom. (nm)	Elect. Quadr. Mom. (b)	γ-Energy / Intensity (MeV/%)
									0.3440
									0.6761
									1.0447
									1.5793
^{194m}Tl			32.8 m	β+ /(20)/~ 0.30		(7+)	+0.540	+0.61	ann.rad./
				EC/(80)/					Hg k x-ray
									0.4282
									0.6363
									0.7490
^{194}Tl		193.9712	33.0 m	β+, EC/5.3		2-	+0.140	-0.28	0.4279/75.2
									0.6452/10.8
									(0.395-1.623)
^{195m}Tl			3.6 s	I.T./0.483		9/2-			Tl k x-ray
									0.0990
									0.3836
^{195}Tl		194.96977	1.16 h	EC/97/2.8		1/2+	+1.58		ann.rad./
				β+ /(3)/					Hg k x-ray
									0.2422
									0.5635
									0.8845
									1.3639
									(0.13–2.5)
^{196m}Tl			1.41 h	β+, EC/95/4.9		(7+)	+0.55	+0.76	0.0840
									0.4261
									0.6353
									0.6954
									(0.08–1.0)
^{196}Tl		195.97048	1.84 h	β+ /(15)/4.4		2-	+0.072	-0.18	ann.rad./
				EC/(85)/					Hg k x-ray
									0.4257
									0.6105
									(0.03–2.4)
^{197m}Tl			0.54 s	IT/53/0.608		9/2-			Tl k x-ray
				β+, EC/47/					0.2262
									0.4118
									0.5872
									0.6367
^{197}Tl		196.96958	2.83 h	β+ /(1)/2.18		1/2+	+1.58		Hg k x-ray
				EC/(99)/					0.1522/8.2
									0.4258
^{198m}Tl			1.87 h	β+, EC/(53)/		7+	+0.64		Hg k x-ray
				IT/47/0.5347					Tl k x-ray
									0.4118
									0.5872
									0.6367
^{198}Tl		197.9405	5.3 h	EC, β+ /(1)/3.5	1.4/	2-			Hg k x-ray
					2.1/				0.4118
					2.4/				0.6367
									0.6759
									(0.23–2.8)
^{199}Tl		198.96988	7.4 h	EC/1.4		1/2+	+1.60		Hg k x-ray
									0.2082
									0.2473
									0.4555

Elem. or Isot.	Natural Abundance (Atom %)	Atomic Mass or Weight	Half-life/ Resonance Width (MeV)	Decay Mode/ Energy (/MeV)	Particle Energy/ Intensity (MeV/%)	Spin (h/2 π)	Nuclear Magnetic Mom. (nm)	Elect. Quadr. Mom. (b)	γ-Energy / Intensity (MeV/%)
^{200}Tl		199.97096	1.087 d	EC/2.46	1.07/	2-	0.04		Hg k x-ray
					1.44/				0.36799
									1.2057
									(0.11–2.3)
^{201}Tl		200.97082	3.038 d	EC/0.48		1/2+	+1.605		Hg k x-ray
									0.13528
									0.16740/10.0
^{202}Tl		201.97211	12.47 d	EC/1.36		2-	0.06		Hg k x-ray
									0.43957
^{203}Tl	29.524(14)	202.972344				1/2+	+1.622258		
^{204}Tl		203.973864	3.78 a	β- /97/0.7637	0.763/97	2-	0.09		Hg k x-ray
				EC/(3)/0.347					
^{205}Tl	70.476(14)	204.974428				1/2+	+1.638215		
^{206m}Tl			3.76 m	I.T./2.644		12-			Tl k x-ray
									0.2166
									0.2661
									0.4534
									0.6866
									1.0219
^{206}Tl		205.976110	4.20 m	β- /1.533	1.53/99.9	0-			Pb k x-ray
									0.80313
^{207m}Tl			1.3 s	I.T./1.350		11/2-			Tl k x-ray
									0.3501
									1.0000
^{207}Tl		206.97742	4.77 m	β- /1.423	1.43/99.8	1/2+	+1.88		0.89723
^{208}Tl		207.982019	3.053 m	β- /5.001	1.28/23	(5+)	+0.29		Pb k x-ray
					1.52/22				0.27728
					1.796/51				0.51061
									0.58302
									2.61448
^{209}Tl		208.98536	2.16 m	β-/3.98	1.8 /100	(1/2+)			Pb k x-ray
									1.5670/100
									0.4651/95
									(0.12–1.33)
^{210}Tl		209.99007	1.30 m	β- /5.48	1.3/25	(5+)			Pb k x-ray
					1.9/56				0.081
									0.2981
									0.79788
^{211}Tl		210.9935	> 0.3 μs	β-					
^{212}Tl		211.9982	> 0.3 μs	β-					
$_{82}$Pb		**207.2(1)**							
^{178}Pb		178.00383	~ 0.2 ms			0+			
^{179}Pb		179.0022							
^{180}Pb		179.99792	5 ms	α/	7.25	0+			
^{181}Pb		180.9966	0.05 s	α/	7.07				
^{182}Pb		181.99267	55 ms	α	6.90	0+			
^{183m}Pb			0.42 s	α	6.70/82.7	13/2+			
					6.86/1.9				
^{183}Pb		182.99187	0.54 s	α/	6.57/4.3	(3/2-)			
					6.78/11.0				
^{184}Pb		183.98814	0.48 s	α/~ 80	6.63/	0+			
^{185m}Pb			4.3 s	α	6.41/100	13/2+	-1.19		
^{185}Pb		184.98761	6.3 s	α/	6.29/56	3/2-	-1.10		0.205
					6.49/44				0.269

Elem. or Isot.	Natural Abundance (Atom %)	Atomic Mass or Weight	Half-life/ Resonance Width (MeV)	Decay Mode/ Energy (/MeV)	Particle Energy/ Intensity (MeV/%)	Spin (h/2 π)	Nuclear Magnetic Mom. (nm)	Elect. Quadr. Mom. (b)	γ-Energy / Intensity (MeV/%)
					6.55/<1.4				
^{186}Pb		185.98424	5. s	β+, EC/95/5.5		0+			
				α/(5)/	6.32/				
					6.34/<100				
					6.01/<0.2				
^{187m}Pb			15.2 s	β+, EC/	5.99/	(1/2-)			0.0674
				α/12	6.19/				0.2080
									0.2755
									0.2995
									0.4487
									0.7477
^{187}Pb		186.98392	18.3 s	EC/7.2		13/2+			0.1930
				α/7	6.08/				0.3314
									0.3435
									0.3934
^{188}Pb		187.98087	23. s	EC/(78)/4.8		0+			0.1850
				α/(22)/	5.98/<10				0.7582
					5.61/<0.1				
^{189}Pb		188.98081	51. s	EC/6.1					
				α/	5.58/				
^{190}Pb		189.97808	1.2 m	β+ (13)/4.1		0+			ann.rad./
				EC/(86)/					Tl k x-ray
				α/(0.9)/	5.58/				0.1415
									0.1512
									0.9422
^{191m}Pb			2.2 m	β+, EC/		13/2+	-1.17	+0.085	ann.rad./
									0.3871
									0.6135
									0.7122
^{191}Pb		190.97827	1.3 m	β+, EC/5.5					ann.rad./
									0.9368
^{192}Pb		191.97579	3.5 m	β+, EC/~ 3.4		0+			ann.rad./
				α/.006/	5.11				0.1675
									0.6082
									1.1954
^{193m}Pb			5.8 m	β+, EC/		13/2+	-1.15	+0.19	ann.rad./
									0.3650
									0.3922
^{193}Pb		192.97617	~ 2. m	EC/5.2		3/2 (-)			
^{194}Pb		193.97401	10. m	β+, EC/2.7		0+			ann.rad./
				α	4.64				0.2036
^{195m}Pb			15. m	β+ /(8)/		13/2+	-1.132	+0.31	ann.rad./
				EC/(92)/					Tl k x-ray
									0.3836
									0.3942
									0.8784
^{195}Pb		194.97454	~ 15. m	β+, EC/5.8					ann.rad./
									0.3836
									0.3937
									0.7776
^{196}Pb		195.97277	37. m	β+, EC/2.1		0+			Tl k x-ray
									0.2531
									0.5021
^{197m}Pb			43. m	EC/79/		13/2+	-1.105	+0.38	Tl k x-ray

Elem. or Isot.	Natural Abundance (Atom %)	Atomic Mass or Weight	Half-life/ Resonance Width (MeV)	Decay Mode/ Energy (/MeV)	Particle Energy/ Intensity (MeV/%)	Spin ($h/2\pi$)	Nuclear Magnetic Mom. (nm)	Elect. Quadr. Mom. (b)	γ-Energy / Intensity (MeV/%)
				β+ /2/					0.3079
				IT/19/0.3193					0.3877
									0.7743
									(0.2–2.2)
^{197}Pb		196.97343	~ 8. m	EC/97/3.6		(3/2-)	-1.075	~ -0.08	Tl k x-ray
				β+ /3/					0.3755
									0.3858
									0.7611
^{198}Pb		197.97203	2.4 h	EC/1.4		0+			Tl k x-ray
									0.1734
									0.2903
									0.3654
^{199m}Pb			12.2 m	IT/93/0.4248		13/2+			Pb k x-ray
				β+, EC/(7)/					0.4255
^{199}Pb		198.97292	1.5 h	EC/(99)/2.9		5/2-	-1.074	~ +0.08	Tl k x-ray
				β+ /(1)/					0.3534
									0.7202
									1.1350
									(0.22–2.4)
^{200}Pb		199.97183	21.5 h	EC/0.81		0+			Tl k x-ray
									0.14763
^{201m}Pb			1.02 m	I.T./0.6291		13/2+			Pb k x-ray
									0.6290
^{201}Pb		200.97289	9.33 h	EC/1.90		5/2-	+0.675	~ -0.009	Tl k x-ray
									0.33120
									0.36131
									(0.11–1.8)
^{202m}Pb			3.53 h	IT/90/2.170		9-	-0.228	~ +0.58	Pb k x-ray
				β+ /10/					Tl k x-ray
									0.42219
									0.78700
									0.96271
^{202}Pb		201.97216	5.3 × 10^{4} a	EC/0.05		0+			Tl L x-ray
^{203m}Pb			6.2 s	I.T./0.8252		13/2+			Pb k x-ray
									0.8203
									0.8252
^{203}Pb		202.97339	2.163 d	EC/0.98		5/2-	+0.686	+0.10	Tl k x-ray
									0.279188
^{204m}Pb			1.13 h	I.T./2.185		9-			Pb k x-ray
									0.37481
									0.89922
									0.91175
^{204}Pb	1.4(1)	203.973044				0+			
^{205}Pb		204.974482	1.51 × 10^{7} a	EC/0.0512		5/2-	+0.712	+0.23	Tl L x-ray
^{206}Pb	24.1(1)	205.974465				0+			
^{207m}Pb			0.80 s	I.T./1.632		13/2+			Pb k x-ray
									0.56915
									1.06310
^{207}Pb	22.1(1)	206.975897				1/2-	+0.59258		
^{208}Pb	52.4(1)	207.976652	> 2 × 10^{19} a	sf		0+			
^{209}Pb		208.981090	3.25 h	β- /0.644	0.645/100	9/2+	-1.474	-0.3	
^{210}Pb		209.984189	22.6 a	β- /0.0635	0.017/81	0+			
					0.061/19				
				α	3.72				

Elem. or Isot.	Natural Abundance (Atom %)	Atomic Mass or Weight	Half-life/ Resonance Width (MeV)	Decay Mode/ Energy (/MeV)	Particle Energy/ Intensity (MeV/%)	Spin ($h/2\pi$)	Nuclear Magnetic Mom. (nm)	Elect. Quadr. Mom. (b)	γ-Energy / Intensity (MeV/%)
^{211}Pb		210.988737	36.1 m	β- /1.37	0.57/5	(9/2+)	-1.404	+0.09	0.40486
					1.36/92				0.42700
									0.83186
									(0.09–1.27)
^{212}Pb		211.991898	10.64 h	β- /0.574	0.28/83	0+			Bi k x-ray
					0.57/12				0.23858
^{213}Pb		212.99658	10.2 m	β- /2.1					
^{214}Pb		213.999805	26.9 m	β- /1.0	0.67/48	0+			Bi k x-ray
					0.73/42				0.24192
									0.29509
									0.35187
^{215}Pb		215.0048	36 s						
$_{83}$Bi		**208.98040(1)**							
^{184m}Bi			0.007 s	α	(7.22-7.85)				0.449
^{184}Bi		184.0011	13. ms	α					0.124
^{185}Bi		184.9976	60. μs	p/90	1.55				
				α/10	8.03				
^{186m}Bi			15. ms	α	7.07-7.23				(0.087-0.520)
				p/<0.5					
^{186}Bi		185.9966	9.8 ms	α	7.26				0.1085
					7.37				
^{187m}Bi			0.37 ms	α/12	7.72/100	1/2+			
^{187}Bi		186.99316	40. ms	α/7	7.00/88.	9/2-			
					7.61/9.0				
					7.34/3.0				
^{188}Bi		187.99227	0.271 s	α	6.81				(0.071-0.320)
^{189m}Bi			7.0 ms	α	7.30				
^{189}Bi		188.9892	0.68 s	α					
^{190m}Bi			5.7 s	α/90	6.43				(0.105-0.314)
					(6.23-6.72)				
^{190}Bi		189.9883	~ 5.9 s	β+, EC/(30)/8.7	α/6.45				(0.089-0.374)
				α/70	(6.39-6.82)				
^{191m}Bi			0.12 ms	α/	6.87/100				
^{191}Bi		190.98579	12.4 s	β+, EC/(60)/7.3					
				α/(40)/	6.31				
^{192}Bi		191.98546	40. s	β+, EC/(80)/9.0					
				α/(20)/	6.06/				
^{193m}Bi			3.2 s	β+, EC/		1/2+			
				α/	6.48/				
^{193}Bi		192.98296	1.11 m	β+, EC/40/7.1		9/2+			
				α/(60)/	5.91/				
^{194}Bi		193.98283	1.8 m	β+, EC/99.9/8.2		(10-)			0.1661
				α/0.1/					0.1740
									0.2802
									0.421
									0.5754
									0.9650
^{195m}Bi			1.45 m	β+, EC/(94)/					
				α/(6)/	6.11/				
^{195}Bi		194.98065	2.9 m	β+, EC/99.8/5.8		3/2-			
				α/(0.2)	5.45/				
^{196}Bi		195.98067	5. m	EC/~ 7.4					0.1376
									0.3720
									0.6880

Elem. or Isot.	Natural Abundance (Atom %)	Atomic Mass or Weight	Half-life/ Resonance Width (MeV)	Decay Mode/ Energy (/MeV)	Particle Energy/ Intensity (MeV/%)	Spin (h/2 π)	Nuclear Magnetic Mom. (nm)	Elect. Quadr. Mom. (b)	γ-Energy / Intensity (MeV/%)
									1.0486
^{197}Bi		196.97886	5. m	β+, EC/5.2		1/2+			
^{198m}Bi			7.7 s	I.T./0.2485		(10-)			0.2485
^{198}Bi		197.97921	11.8 m	β+, EC/6.6		(7+)			0.0900
									0.1976
									0.5624
									1.0635
^{199m}Bi			24.7 m	β+, EC/					ann.rad./
^{199}Bi		198.97767	27. m	β+, EC/4.3		9/2-	4.6		0.7203
									0.8374
									0.8417
									0.9460
									1.0528
									1.3056
									(0.12–3.2)
^{200m}Bi			31. m	β+, EC/		(2+)			0.2453
									0.4198
									0.4624
									1.0265
^{200}Bi		199.97813	36. m	EC/(90)/5.9		7+			ann.rad./
				β+ /(10)/					Pb k x-ray
									0.4198
									0.4623
									1.0265
^{201m}Bi			59.1 m	I.T./0.846		(1/2+)			Bi k x-ray
				β+, EC/					0.8464
^{201}Bi		200.97701	1.8 h	EC/3.84		9/2-	4.8		Pb k x-ray
									0.6288
									0.9357
									1.0138
									(0.13–2.4)
^{202}Bi		201.97774	1.72 h	β+ /(3)/5.16		5+	+4.26	-0.9	ann.rad./
				EC/(97)/					Pb k x-ray
									0.57860
									0.92734
									(0.08–3.5)
^{203}Bi		202.97688	11.8 h	EC/99.8/3.25		9/2-	+4.02	-0.8	Pb k x-ray
				β+ /(0.2)/	1.35/				0.1865
									0.8203
									0.8969
									1.8475
									(0.1–2.9)
^{204}Bi		203.97781	11.2 h	EC/4.44		6+	+4.32	-0.43	Pb k x-ray
									0.37481
									0.89922
									0.98409
^{205}Bi		204.97739	14.9 d	EC/2.71		9/2-	+4.07	-0.7	Pb k x-ray
									0.70347
									1.76435
									(0.550-1.862)
^{206}Bi		205.97850	6.243 d	EC/3.76		6+	+4.36	-0.4	Pb k x-ray
									0.51619
									0.80313
									0.88100

Elem. or Isot.	Natural Abundance (Atom %)	Atomic Mass or Weight	Half-life/ Resonance Width (MeV)	Decay Mode/ Energy (/MeV)	Particle Energy/ Intensity (MeV/%)	Spin (h/2 π)	Nuclear Magnetic Mom. (nm)	Elect. Quadr. Mom. (b)	γ-Energy / Intensity (MeV/%)
^{207}Bi		206.978471	31.55 a	EC/2.399		9/2-	4.092	-0.6	Pb k x-ray
									0.56915
									1.06310
^{208m}Bi			2.58 ms	IT		10-	2.67		0.921
^{208}Bi		207.979742	3.68 × 10^5 a	EC/2.880		5+	4.60	-0.64	Pb k x-ray
									2.61435
^{209}Bi	100.	208.980399	1.9 × 10^{19} a	α	3.13	9/2-	+4.111	-0.6	
^{210m}Bi			3.0 × 10^6 a	α/	4.420(3)/0.29	9-	+2.73	-0.6	Tl k x-ray
					4.569(3)/3.9				0.2661
					4.584(3)/1.4				0.3052
					4.908(4)/39				0.6502
					4.946(3)/55				
^{210}Bi		209.984120	5.01 d	β- /1.163	1.16/99	1-	-0.0445	+0.136	0.2661
									0.3.52
^{211}Bi		210.98727	2.14 m	α/(99.7)/	6.279/16	9/2-			Tl k x-ray
				β- /(0.3)/0.58	6.623/84				0.3501
^{212m2}Bi			7. m	β- /		(15-)			
^{212m1}Bi			25.0 m	α/(93)/	6.300/40	(9-)			0.120
				β- /(7)/	6.340/53				0.233
									0.275
									0.404
									0.727
^{212}Bi		211.991286	1.009 h	β- /(64)/2.254		(1-)	+0.32	~ +0.1	Tl k x-ray
				α/(36)/	6.051/25				Po k x-ray
					6.090/9.6				0.2881
									0.72725
									0.78551
									1.62066
^{213}Bi		212.994385	45.6 m	β- /(98)/1.43	1.02/31	9/2-	+3.72	-0.71	Po k x-ray
				α/(2)/	1.42/66				0.4404
					5.549/0.16				(0.15–1.328)
					5.869/2.0				
									1.10006
^{214}Bi		213.99871	19.7 m	β- /3.27					0.60931
									1.12027
									1.76449
									(0.19–3.2)
^{215m}Bi			37. s	β					(0.158-0.498)
^{215}Bi		215.00177	7.7 m	β- /2.3					0.2937/35.2
									(0.271–1.399)
^{216}Bi		216.00631	2.3 m	β-/4.0					0.5498
									0.4192
^{217}Bi		217.0095	98 s	β/					0.2646/100
									(0.254-1.017)
^{218}Bi		218.0143	33. s	β-					0.5097/134
									0.3857/100
									(0.174-0.703)
$_{84}$Po									
^{187}Po			1.4 ms	α	7.53/100				
^{188}Po		187.99942	0.27 ms	α	7.91/80	0+			
					7.320				
^{189}Po		188.99848	5 ms	α	7.532/8				
					7.259/80				
					7.309/12				

Elem. or Isot.	Natural Abundance (Atom %)	Atomic Mass or Weight	Half-life/ Resonance Width (MeV)	Decay Mode/ Energy (/MeV)	Particle Energy/ Intensity (MeV/%)	Spin ($h/2\pi$)	Nuclear Magnetic Mom. (nm)	Elect. Quadr. Mom. (b)	γ-Energy / Intensity (MeV/%)
^{190}Po		189.99510	2.4 ms	α/	7.53/96.4	0+			
					7.01/3.3				
^{191m}Po			93. ms	α	7.376/50				
					6.888/46				
^{191}Po		190.99457	22 ms	α/	7.334/77				
					6.97/8				
^{192}Po		191.99134	32. ms	α/8.5	7.17/98.6	0+			
					6.59/1.4				
^{193m}Po			~ 0.07 s	α/	7.00				
^{193}Po		192.99103	0.45 s	α/	6.95				
^{194}Po		193.98819	0.2 s	α/	6.84/93	0+			
					6.19/0.22				
^{195m}Po			~ 2.8 s	α/	6.70/				
^{195}Po		194.98811	~ 3.9 s	α/	6.62/				
^{196}Po		195.98554	5. s	α/(95)/	6.53/94	0+			
				β+, EC/(5)/~4.6	5.77/0.02				
^{197m}Po			25.8 s	α/(84)/	6.385(3)/55	13/2+			
				β+, EC/(16)/					
^{197}Po		196.98566	53. s	α/(44)/	6.282(4)/76	(3/2-)			
				β+, EC/(56)/6.2					
^{198}Po		197.98339	1.76 m	α/(70)/	6.18/57	0+			
				β+, EC/(30)/4.0	5.27/7.6 × 10^{-4}				
^{199m}Po			4.2 m	β+, EC/(51)/		13/2+	~ 0.99		ann.rad./
				α/(39)/	6.059/24				0.2745
									0.4998
									1.0020
^{199}Po		198.98367	5.2 m	β+, EC/(88)/7.		(3/2-)			Bi k x-ray
				α/(12)/	5.952/7.5				0.1877
									0.3616
									1.0214
									1.0344
^{200}Po		199.981780	11.5 m	β+, EC/85/3.4		0+			0.14748
				α/(15)/	5.863/11.1				0.32792
									0.6176
									0.6709
^{201m}Po			8.9 m	β+, EC/(57)/		13/2+	~ 1.00		Bi k x-ray
				IT/40/0.418					Po k x-ray
				α/(3)/	5.786/~ 3.				0.2726
									0.4123
									0.4179
									0.9670
^{201}Po		200.98226	15.3 m	β+, EC/98/4.9		3/2-	~ 0.94		Bi k x-ray
				α/(2)/	5.683(3)/1.1				0.2056
									0.2250
									0.8483
									0.9048
^{202}Po		201.98076	45. m	β+, EC/98/2.8		0+			0.0410
				α/(2)/	5.588/1.9				0.1656
									0.3158
									0.6884
^{203m}Po			1.2 m	IT/96/0.6414		13/2+			Bi k x-ray
				β-EC/(4)/					Po k x-ray
									0.6414
^{203}Po		202.98142	35. m	β+, EC/4.2		5/2-	+0.74		0.17516

Elem. or Isot.	Natural Abundance (Atom %)	Atomic Mass or Weight	Half-life/ Resonance Width (MeV)	Decay Mode/ Energy (/MeV)	Particle Energy/ Intensity (MeV/%)	Spin ($h/2\pi$)	Nuclear Magnetic Mom. (nm)	Elect. Quadr. Mom. (b)	γ-Energy / Intensity (MeV/%)
									0.21477
									0.89350
									0.90863
									1.09095
^{204}Po		203.98032	3.53 h	EC/2.34		0+			Bi k x-ray
				α	5.377/0.66				0.2702
									0.8844
									1.0162
									(0.11–1.9)
^{205}Po		204.98120	1.7 h	β+, EC/3.53		5/2-	~ +0.76	+0.17	Bi k x-ray
									0.83681
									0.84983
									0.87241
									1.00124
									(0.12–2.7)
^{206}Po		205.98048	8.8 d	EC/(95)/1.85		0+			Bi k x-ray
				α/(5)/	5.223/5.5				0.28644
									0.31156
									0.51134
									0.80737
									1.03228
									(0.11–1.5)
^{207m}Po			2.8 s	I.T./1.383		19/2-			Po k x-ray
									0.2682
									0.30074
									0.81448
^{207}Po		206.98159	5.80 h	EC, β+/2.91		5/2-	~ +0.79	+0.28	Bi k x-ray
									0.74263
									0.91176
									0.99225
^{208}Po		207.981246	2.898 a	α/5.213	4.233/0.0002	0+			
					5.1158/100				
^{209}Po		208.982430	128. a	α/4.976	4.624/0.56	1/2-	+0.7		0.26049
					4.879/99.2				0.8964
^{210}Po		209.982874	138.4 d	α/5.407	4.516/0.001	0+			0.80313
					5.304/100				
^{211m}Po			25.2 s	α/	7.273/91	25/2+			Pb k x-ray
					7.994/1.7				0.32808
					8.316/0.25				0.56915
					8.875/7.0				0.89723
									1.06310
^{211}Po		210.986653	0.516 s	α/7.594	6.570/0.54	9/2+			0.56915
					6.892/0.55				0.89723
					7.450/98.9				
^{212m}Po			45. s	α/	8.514/2.0	16+			
					9.086/1.0				
					11.650/97				
^{212}Po		211.988868	0.298 μs	α/8.953	8.784/100	0+			
^{213}Po		212.992857	3.7 μs	α/8.537	7.614/0.003	9/2+			
					8.375/100				
^{214}Po		213.995201	163.7 μs	α/7.833	6.904/0.01	0+			0.7995
					7.686/99.99				0.298
^{215}Po		214.999420	1.780 ms	α/7.526	6.950/0.02	(9/2+)			
					6.957/0.03				

Elem. or Isot.	Natural Abundance (Atom %)	Atomic Mass or Weight	Half-life/ Resonance Width (MeV)	Decay Mode/ Energy (/MeV)	Particle Energy/ Intensity (MeV/%)	Spin ($h/2\pi$)	Nuclear Magnetic Mom. (nm)	Elect. Quadr. Mom. (b)	γ-Energy / Intensity (MeV/%)
					7.386/100				
^{216}Po		216.001915	0.145 s	α/6.906	5.895/0.002	0+			
					6.778/99.99				
^{217}Po		217.00634	1.53 s	α/6.662	6.539/				
^{218}Po		218.008973	3.04 m	α/6.114	6.003/99.999	0+			
					5.181/0.11				
^{219}Po		219.0137	~ 2 m						
^{220}Po		220.0166	> 0.3 μs			0+			
$_{85}$At									
^{191m}At			2.1 ms	α	7.65/98				
					7.72/2				
^{191}At			~ 1.7 ms	α	7.55/100				
^{193m}At			21 ms	α	7.33/98				
					7.42/2				
^{193}At		192.9998	28 ms	α/	7.24/100				
^{194}At		193.9987	40 ms	α/					
^{195m}At			147 ms	α	7.07-7.22				
^{195}At		194.99627	0.33 s	α/	6.95				
^{196m}At			8 μs						0.158
^{196}At		195.9958	0.39 s	α/	7.05/				
^{197m}At			2.0 s	α	6.707	(1/2+)			
^{197}At		196.99319	0.39 s	β+, EC/7.8		(9/2-)			
				α/	6.960/				
^{198m}At			1.0 s	β+, EC/(75)/					
				α/(25)/	6.856/86				
^{198}At		197.99284	4.1 s	α/	6.755/94				
^{199}At		198.99053	6.9 s	β+, EC/8/5.6		9/2-			
				α/(92)/	6.643/				
^{200m2}At			47. s	α	6.411/				
^{200m}At			3.5 s	β+, EC/(80)		10-			
				α/(20)/	6.538/12				
^{200}At		199.99035	43. s	β+, EC/65/~ 8.0		5+			
				α/(35)/	6.412/44				
					6.465/57				
^{201}At		200.98842	1.48 m	β+, EC/29/5.9		9/2-			
				α/(71)/6.474	6.344/				
^{202m}At			0.46 s	I.T./0.391					
^{202}At		201.98863	3.02 m	β+, EC/88/7.2		5+			ann.rad./
				α/(12)/	6.135/7.7				0.4413
					6.225/4.3				0.5697
									0.6753
^{203}At		202.98694	7.4 m	β+, EC/69/5.1		9/2-			0.1458
				α/(31)/6.210	6.088/				0.2459
									0.6414
									1.0020
									1.0340
^{204}At		203.98725	9.1 m	β+, EC/95/6.5		(5+)			Po k x-ray
				α/(5)/	5.951/				0.3271
									0.4254
									0.5156
									0.6837
^{205}At		204.98607	26. m	β+, EC/90/4.54		(9/2-)			Po k x-ray
				α/(10)/6.020	5.902/				0.1543

Elem. or Isot.	Natural Abundance (Atom %)	Atomic Mass or Weight	Half-life/ Resonance Width (MeV)	Decay Mode/ Energy (/MeV)	Particle Energy/ Intensity (MeV/%)	Spin ($h/2\pi$)	Nuclear Magnetic Mom. (nm)	Elect. Quadr. Mom. (b)	γ-Energy / Intensity (MeV/%)
									0.6696
									0.7194
^{206}At		205.98667	29.4 m	β+, EC/99/5.72		5+			Po k x-ray
				α/(1)/5.881	5.703/				0.20186
									0.39561
									0.47716
									0.70071
^{207}At		206.98578	1.81 h	β+, EC/90/3.91		9/2-			Po k x-ray
				α/(10)/5.873	5.758/				0.16801
									0.58842
									0.81448
^{208}At		207.98650	1.63 h	β+, EC/99/4.97		(6+)			Po k x-ray
				α/(1)/5.752	5.626/0.01				0.1770
					5.641/0.53				0.2060
									0.6601
									0.6852
									0.8450
									1.0281
^{209}At		208.98617	5.4 h	β+, EC/96/3.49		(6+)			Po k x-ray
				α/(4)/5.757	5.647/4.1				0.10422
									0.54503
									0.78189
									0.79020
									(0.1–2.6)
^{210}At		209.98715	8.1 h	EC/99.8/3.98		5+			Po k x-ray
				α/(0.2)/5.632	5.361/0.05				0.24535
					5.442/0.05				0.52758
									1.18143
									1.43678
									1.48335
									(0.04–2.4)
^{211}At		210.987496	7.21 h	EC/(58)/0.787		9/2-			Po k x-ray
				α/(42)/5.980	5.211/0.004				0.66956
					5.868/42				0.6870
									0.74263
^{212m}At			0.119 s	α/	7.837/65	(9-)			
					7.897/33				
^{212}At		211.99075	0.314 s	α/7.828	7.058/0.4	(1-)			
					7.088/0.6				
					7.618/15				
					7.681/84				
^{213}At		212.992937	0.11 μs	α/9.254	9.080/	9/2-			
^{214m}At			0.76 μs	α/8.762		(9-)			
^{214}At		213.996372	0.56 μs	α/8.987	8.819/100	(1-)			
^{215}At		214.99865	0.10 ms	α/8.178	7.626/0.045	(9/2-)			0.40486
					8.023/99.9				
^{216}At		216.002423	0.30 ms	α/7.947	7.595/0.2	(1-)			
					7.697/2.1				
					7.800/97				
^{217}At		217.004719	32. ms	α/7.202	6.812/0.06	(9/2-)	3.8		0.2595
					7.067/99.9				0.3345
									0.5940
^{218}At		218.00869	1.6 s	α/6.883	6.654/6				
					6.695/90				

Elem. or Isot.	Natural Abundance (Atom %)	Atomic Mass or Weight	Half-life/ Resonance Width (MeV)	Decay Mode/ Energy (/MeV)	Particle Energy/ Intensity (MeV/%)	Spin ($h/2\pi$)	Nuclear Magnetic Mom. (nm)	Elect. Quadr. Mom. (b)	γ-Energy / Intensity (MeV/%)
					6.748/4				
^{219}At		219.011162	50. s	α/6.390	6.275/				
^{220}At		220.0154	3.71 m	β- /3.7					(0.24–0.70)
^{221}At		221.0181	2.3 m	β					
^{222}At		222.0223	0.9 m	β					
^{223}At		223.0252	50. s	β					
$_{86}$**Rn**									
^{193}Rn			1.2 ms	α	7.69/74				
					7.88/26				
^{194}Rn			0.8 ms	α	7.70				
^{195m}Rn			5 ms	α	7.56				
^{195}Rn		195.00544	6 ms	α	7.54				
^{196}Rn		196.00212	4. ms	α/	7.46	0+			
^{197m}Rn			0.02 s	α	7.36				
^{197}Rn		197.0016	0.07 s	α/	7.26				
^{198}Rn		197.99868	64. ms	α	7.205	0+			
^{199m}Rn			0.32 s	α	7.060	(13/2+)			
^{199}Rn		198.9984	0.62 s	α/	6.989	3/2-			
^{200}Rn		199.99570	1.06 s	α/(98)/	6.901/	0+			0.4329
				EC/(2)/5.					0.5043
^{201m}Rn			3.8 s	EC/(10)/		13/2+			
				α/(90)/	6.773/				
^{201}Rn		200.9956	7.0 s	α/(80)/	6.725/	(3/2-)			
				EC/(20)/	α/6.778				
^{202}Rn		201.99326	9.9 s	α/(12)/	6.641/	0+			0.5695
				EC/(88)/					0.288–0.6255
^{203m}Rn			28. s	α/	6.551	13/2+	-0.96	+1.3	
^{203}Rn		202.99339	45. s	α/(66)/6.629	6.499/	3/2-			
				EC/(34)/~ 7.4					
^{204}Rn		203.99143	1.24 m	α/(68)/	6.420/	0+			
				EC/(32)/3.8					
^{205}Rn		204.99172	2.8 m	α/(23)/6.390	6.123(3)/0.02	(5/2-)	+0.80	+0.06	0.2652
				EC/(77)/5.2	6.262(3)/23				0.3553
									0.4648
									0.6205
									0.6753
									0.7300
^{206}Rn		205.99021	5.7 m	α/(68)/6.384	6.258(3)/	0+			0.06170
				EC/(32)/3.3					0.0968
									0.3245
									0.3862
									0.4822
									0.4973
									0.7728
^{207}Rn		206.99073	9.3 m	β+, EC/77/4.6		5/2-	+0.82	+0.22	At k x-ray
				α/(23)/6.252	5.995(4)/0.02				0.32947
					6.068(3)/0.15				0.34455
					6.126(3)/22.8				0.36767
									0.40267
									0.74723
									(0.18–1.4)
^{208}Rn		207.98964	24.3 m	α/(60)/6.260	5.469(2)/0.003	0+			
				EC/(40)/2.85	6.140(2)/60				

Elem. or Isot.	Natural Abundance (Atom %)	Atomic Mass or Weight	Half-life/ Resonance Width (MeV)	Decay Mode/ Energy (/MeV)	Particle Energy/ Intensity (MeV/%)	Spin (*h*/2 π)	Nuclear Magnetic Mom. (nm)	Elect. Quadr. Mom. (b)	γ-Energy / Intensity (MeV/%)
^{209}Rn		208.99042	29. m	β+ /(83)/3.93	2.16/2.3	5/2-	+0.8388	+0.31	At k x-ray
				α/(17)/	5.887(3)/0.04				0.27933
					5.898(3)/0.02				0.33753
					6.039(2)/16.9				0.40841
									0.68942
									0.74594
									(0.18–3.2)
^{210}Rn		209.98970	2.4 h	α/(96)/6.157	5.351(2)/0.005	0+			At k x-ray
				EC/(4)/2.37	6.039(2)/96				0.19625
									0.45824
									0.57104
									0.64868
									(0.14–1.7)
^{211}Rn		210.99060	14.6 h	β+, EC/74/2.89		1/2-	+0.60		At k x-ray
				α/(26)/5.964	5.619(1)/0.7				0.16877
					5.784(1)/16.4				0.25022
					5.851(1)/8.8				0.37049
									0.67412
									0.67839
									1.36298
									(0.11–2.7)
^{212}Rn		211.990704	24. m	α/6.385	5.587(4)/0.05	0+			
					6.260(4)/99.95				
^{213}Rn		212.99388	19. ms	α/8.243	7.552(8)/1.0	9/2+			0.540
					8.087(8)/98.2				
					7.254/0.8				
^{214}Rn		213.99536	0.27 μs	α/9.209	9.037(9)/	0+			
^{215}Rn		214.99875	2.3 μs	α/8.840	8.674(8)/	(9/2+)			
^{216}Rn		216.00027	45. μs	α		0+			
^{217}Rn		217.003928	0.6 ms	α/7.885	7.500/0.1	9/2+			
					7.742(4)/100				
^{218}Rn		218.005601	35. ms	α/7.267	6.534(1)/0.16	0+			0.6093
					7.133(1)/99.8				0.6653
^{219}Rn		219.009480	3.96 s	α/6.946(1)	6.3130(5)/0.05	(5/2+)	-0.44	+1.0	Po k x-ray
					6.425(3)/7.5				0.13057
					6.5309(4)/0.12				0.27113
					6.5531(3)/12.2				0.40170
					6.8193(3)/81				(0.1–1.05)
^{220}Rn		220.011394	55.6 s	α/6.404	5.7486(5)/0.07	0+			
					6.2883(1)/99.9				
^{221}Rn		221.01554	25. m	α/(22)/6.148	5.778(3)/1.8	7/2+	-0.020	-0.4	Fr L x-ray
				β- /(78)/1.2	5.788(3)/2.2				0.07384
					6.037(3)/18				0.08323
									0.0610
									0.18639
^{222}Rn		222.017578	3.823 d	α/5.590	4.987(1)/0.08	0+			0.510
					5.4897(3)/99.9				
^{223}Rn		223.0218	23. m	β- /			-0.78	~ +0.80	
^{224}Rn		224.0241	1.8 h	β- /		0+			0.1085
									0.2601
									0.2655
^{225}Rn		225.0284	4.5 m	β- /		7/2-	-0.70	~ +0.84	
^{226}Rn		226.0309	7.4 m	β- /		0+			
^{227}Rn		227.0354	2. s	β- /					

Elem. or Isot.	Natural Abundance (Atom %)	Atomic Mass or Weight	Half-life/ Resonance Width (MeV)	Decay Mode/ Energy (/MeV)	Particle Energy/ Intensity (MeV/%)	Spin ($h/2\pi$)	Nuclear Magnetic Mom. (nm)	Elect. Quadr. Mom. (b)	γ-Energy / Intensity (MeV/%)
^{228}Rn		228.0380	65. s	β-/		0+			
$_{87}$Fr									
^{199}Fr		199.00726	12 s	α	7.66				
^{200}Fr		200.0066	49 ms	α	7.47				
^{201m}Fr			~ 0.02 s	α/	7.454				
^{201}Fr		201.0039	~ 60 ms	α/	7.36/	(9/2-)			
^{202m}Fr			0.29 s	α	7.236/				
^{202}Fr		202.00337	0.30 s	α/7.590	7.24/100				
^{203}Fr		203.00093	0.54 s	α/7.280	7.132(5)/	(9/2-)			
^{204m2}Fr			0.8 s	α	7.01				
^{204m1}Fr			2. s	α	6.97				
^{204}Fr		204.00065	1.8 s	α/	7.03/96				
					6.97/90				
					7.01/74				
^{205}Fr		204.99859	3.9 s	α/7.050	6.914(5)/	(9/2-)			
^{206m}Fr			0.7 s	α/	6.93				0.531(IT)
^{206}Fr		205.99867	16.0 s	α/7.416	6.792(5)/84				
^{207}Fr		206.99695	14.8 s	α/6.900	6.766(5)/	9/2-	+3.9	-0.16	
^{208}Fr		207.99714	59.1 s	α/(77)/6.770	6.636(5)/	7+	+4.8		
				EC/(23)/6.99					
^{209}Fr		208.99595	50.0 s	α/(89)/5.1	6.646(3)/	9/2-	+3.9	-0.24	0.7978
				EC/(11)/5.16					(0.110–1.384)
^{210}Fr		209.99641	3.2 m	α/6.670/71	6.543(5)/99.07	6+	+4.4	+0.19	0.2030
				EC/6.26	(5.90-6.42)				0.6438
									0.8175
									0.9008
^{211}Fr		210.99554	3.10 m	α/6.660/87	6.534(5)/99.94	9/2-	+4.0	-0.19	0.220
				EC/4.61	(5.87-6.20)				0.2799
									0.5389
									0.9169
^{212}Fr		211.99620	20. m	EC/(57)/5.12	6.261(1)/16	(5+)	+4.6	-0.10	Rn x-ray
				α/(43)/6.529	6.335(1)/4				0.08107
					6.335(1)/4				0.08378
					6.343(1)/1.3				0.2277
					6.383(1)/10				1.1856
					6.406(1)/9.5				1.2748
					6.08–6.18				0.014–1.178
^{213}Fr		212.99619	34.6 s	α/6.905	8.476(4)/51	9/2-	+4.0	-0.14	(0.408-0.577)
^{214m}Fr			3.4 ms	α/	8.547(4)/46	9-			
					6.775–8.046				
^{214}Fr		213.99897	5.0 ms	α/8.587	7.409(3)/0.3	(1-)			(0.073-0.966)
					7.605(8)/1.0				
					7.940(3)/1.0				
					8.355(3)/4.7				
					8.427(3)/93				
^{215}Fr		215.00034	0.12 μs	α/9.537	9.360(8)/	(9/2-)			
^{216}Fr		216.00320	0.70 μs	α/9.175	9.005(10)/95				(0.045–0.160)
^{217}Fr		217.00463	0.016 ms	α/8.471	8.315(8)/	(9/2-)			
^{218m}Fr			22. ms	α					
^{218}Fr		218.007578	1. ms	α/8.014	7.384(10)/0.5	(1-)			
					7.542(15)/1.0				
					7.572(10)/5				
					7.732(10)/0.5				
					7.867(2)/93				

Elem. or Isot.	Natural Abundance (Atom %)	Atomic Mass or Weight	Half-life/ Resonance Width (MeV)	Decay Mode/ Energy (/MeV)	Particle Energy/ Intensity (MeV/%)	Spin ($h/2\pi$)	Nuclear Magnetic Mom. (nm)	Elect. Quadr. Mom. (b)	γ-Energy / Intensity (MeV/%)
^{219}Fr		219.00925	21. ms	α/8.132	6.802(2)/0.25	(9/2-)			
					6.967(2)/0.6				
					7.146(2)/0.25				
					7.313(2)/99				
^{220}Fr		220.012327	27.4 s	α/6.800	6.582(1)/10	1+	-0.67	+0.47	0.0450
					6.630(2)/6				0.061
					6.641(1)/12				0.1060
					6.686(1)/61				0.1539
					6.39–6.58				0.1617
^{221}Fr		221.014255	4.79 m	α/6.457	5.9393(7)/0.17	(5/2-)	+1.58	-1.0	At k x-ray
					5.9797(7)/0.49				0.0995
					6.0751(7)/0.15				0.21798
					6.1270(7)/				0.4091
					6.2433(3)/1.3				
					6.3410(7)/83.4				
^{222}Fr		222.01755	14.3 m	β- /2.03	1.78/	2-	+0.63	+0.51	
				α/5.850					
^{223}Fr		223.019736	22.0 m	β- /1.149	α/5.291	(3/2+)	+1.17	+1.17	0.1509
				α//0.006	5.314				0.0589
					5.403				0.1453
^{224}Fr		224.02325	3.0 m	β- /2.82		1-	+0.40	+0.517	0.13150
									0.21575
									0.8367
									(0.1–2.21)
^{225}Fr		225.02557	3.9 m	β- /1.87		3/2	+1.07	+1.3	
^{226}Fr		226.0294	49. s	β- /3.6		1	+0.071	-1.35	0.18606
									0.25373
^{227}Fr		227.0318	2.48 m	β- /2.5		1/2	+1.50		
^{228}Fr		228.0357	39. s	β- /~ 3.5		2-	-0.76	+2.4	
^{229}Fr		229.03845	50. s	β- /					
^{230}Fr		230.0425	19. s	β- /		(3)			
^{231}Fr		231.0454	17. s	β- /					
^{232}Fr		232.050	5. s	β- /					(0.0545-0.721)
$_{88}$**Ra**									
^{201}Ra			~ 1.6 ms	α	7.91/				
^{202}Ra		202.0099	~ 0.02 ms	α	7.74	0+			
^{203m}Ra			24 ms	α	7.61				
^{203}Ra		203.0093	~ 31 ms	α	7.59				
^{204}Ra		204.0065	0.06 s	α	7.48	0+			
^{205m}Ra			~ 0.17 s						
^{205}Ra		205.0063	0.22 s	α	7.34				
^{206}Ra		206.00383	0.4 s	α/7.416	7.272(5)/	0+			
^{207}Ra		207.0038	1.3 s	α/7.270	7.133(5)/				
^{208}Ra		208.00184	1.4 s	α/7.273	7.133(5)/	0+			
^{209}Ra		209.00199	4.6 s	α/7.150	(6.50-7.14)	5/2-	+0.87	+0.40	(0.387-0.634)
^{210m}Ra			2.34 μs						(0.0967-0.775)
^{210}Ra		210.00050	3.7 s	α/7.610	7.020(5)/	0+			574.9
^{211m}Ra			5.1 μs						(0.396-0.802)
^{211}Ra		211.00090	13. s	α/7.046	6.907/99.	(5/2-)	+0.878	+0.48	
				EC/5.0	(6.26-6.79)				(0.120-0.665)
^{212m}Ra			8.4 μs						(0.440-0.824)
^{212}Ra		211.99979	13.0 s	α/7.033	6.901(2)/	0+			
^{213m}Ra			2.20 ms	IT//99	α/8.47/63	17/2-	7.4		0.161
				α//0.6	8.36/33				(0.160-1.061)

Elem. or Isot.	Natural Abundance (Atom %)	Atomic Mass or Weight	Half-life/ Resonance Width (MeV)	Decay Mode/ Energy (/MeV)	Particle Energy/ Intensity (MeV/%)	Spin (h/2 π)	Nuclear Magnetic Mom. (nm)	Elect. Quadr. Mom. (b)	γ-Energy / Intensity (MeV/%)
					8.27/4				
^{213}Ra		213.00038	2.7 m	EC/(20)/3.88	α/6.225/49	(1/2-)	+0.613		0.1024
				α/(80)/6.860	6.733/45				0.110
					6.522/5.4				0.215
					6.413/0.22				(0.105-0.511)
^{214m}Ra			0.69 ms						(0.181-1.382)
^{214}Ra		214.00011	2.46 s	α/7.272	7.14/99.8/	0+			0.642
					6.51/0.2				
^{215m}Ra			7.6 μs						(0.196-1.048)
^{215}Ra		215.00272	1.64 ms	α/8.864	7.883(6)/2.8	(9/2+)			0.773/100
					8.171(3)/1.4				0.852/74
					8.700(3)/95.9				0.055–1.048
^{216}Ra		216.00353	0.18 μs	α/9.526	9.349(8)/	0+			
^{217}Ra		217.00632	1.6 μs	α/9.161	8.992(8)/	9/2-			
^{218}Ra		218.00714	26. μs	α/8.547	8.390(8)/	0+			
^{219}Ra		219.01009	0.010 s	α/8.132	7.680(10)/65				
					7.982(9)/35				
^{220}Ra		220.01103	18. ms	α/7.593	7.39/5	0+			0.465
					7.45/95				
^{221}Ra		221.013917	29. s	α/6.879	6.254(10)/0.7	5/2+	-0.180	+2.0	
					6.578(5)/3				
					6.585(3)/8				
					6.608(3)/35				
					6.669(3)/21				
					6.758(3)/31				
^{222}Ra		222.015375	36.2 s	α/5.590	6.237(2)/3.0	0+			0.324
					6.556(2)/97				0.145–0.8402
^{223}Ra		223.018502	11.43 d	α/5.979	5.287(1)/0.15	(3/2+)	+0.271	+1.25	Rn k x-ray
					5.338(1)/0.13				0.12231
					5.365(1)/0.13				0.14418
					5.433(5)/2.3				0.15418
					5.502(1)/1.0				0.15859
					5.540(1)/9.2				0.26939
					5.607(3)/24				0.32388
					5.716(3)/52				0.33328
					5.747(1)/9				0.44494
					5.857(1)/0.32				(0.10–0.7)
					5.872(1)/0.85				
^{224}Ra		224.020212	3.66 d	α/5.789	5.034(10)/0.003	0+			Rn k x-ray
					5.047(1)/0.007				0.2407
					5.164(5)/0.007				0.4093
					5.449(2)/4.9				0.6501
					5.685(2)/95				
^{225}Ra		225.023612	14.9 d	β- /0.36	0.32/100	(3/2+)	-0.734		Ac k x-ray
				α	5.01 × 10^{-5}				0.0434
					4.98 × 10^{-6}				
^{226}Ra		226.025410	1599. a	α/4.870	4.194(1)/0.001	0+			Rn k x-ray
			> 4 × 10^{18} a	sf/4 × 10^{-14}	4.343(1)/0.006				0.1861/3.64
					4.601(1)/6.16				0.2624
					4.784(1)/93.8				0.053–2.448
^{227}Ra		227.029178	42. m	β- /1.325	1.03/	(3/2+)	-0.404	+1.5	Ac L x-ray
					1.30/				Ac k x-ray
									0.02739
^{228}Ra		228.031070	5.76 a	β- /0.046	0.039/50	0+			0.0135

Elem. or Isot.	Natural Abundance (Atom %)	Atomic Mass or Weight	Half-life/ Resonance Width (MeV)	Decay Mode/ Energy (/MeV)	Particle Energy/ Intensity (MeV/%)	Spin ($h/2\pi$)	Nuclear Magnetic Mom. (nm)	Elect. Quadr. Mom. (b)	γ-Energy / Intensity (MeV/%)
				βf//5x10^{-12}	0.014/30				(0.006–0.031)
					0.026/20				
^{229}Ra		229.03496	4.0 m	β- /1.76	1.76/	(3/2+)	+0.503	+3.1	0.0145–0.172
^{230}Ra		230.03706	1.5 h	β- /1.0	0.7/	0+			0.0631
									0.0720
									0.2028
									0.4698
									0.4787
^{231}Ra		231.0412	1.7 m	β-					
^{232}Ra		232.0436	4. m	β-		0+			
^{233}Ra		233.0481	30. s	β-					
^{234}Ra		234.051	~ 30. s	β-/		0+			
$_{89}$Ac									
^{206m}Ac			0.04 s	α	7.79				
^{206}Ac		206.0145	~ 26 ms	α	7.75				
^{207}Ac		207.0120	27 ms	α/	7.69				
^{208m}Ac			~ 25. ms	α/	7.72				
^{208}Ac		208.0116	~ 0.1 s	α/	7.62				
^{209}Ac		209.00949	~ 0.10 s	α/	7.58				
^{210}Ac		210.0094	0.34 s	α/7.610	7.462(8)/				
^{211}Ac		211.0077	0.20 s	α/7.620	7.480(8)/				
^{212}Ac		212.0078	0.9 s	α/7.520	7.379(8)/				
^{213}Ac		213.0066	0.73 s	α/7.500	7.364(8)/	(9/2-)			
^{214}Ac		214.00690	8.2 s	α/(86)/7.350	7.215/54	(5+)			(0.0626-0.754)
				EC/(14)/6.34	7.081/42				
					(6.48-7.15)				
^{215}Ac		215.00645	0.17 s	α/7.750	7.60/99.57	(9/2-)			0.399
					7.21/0.46				0.582
					7.03/0.20				0.654
					6.96/0.14				
^{216m}Ac			0.44 ms	α/	8.198(8)/1.7	(9-)			(0.0826-1.375)
					8.283(8)/2.5				
					9.028(5)/49				
					9.106(5)/46				
^{216}Ac		216.00872	44. ms	α/9.241	8.990(2)/10	(1-)			
					9.070(8)/90				
^{217m}Ac			0.7 μs	α/	10.540/100				
^{217}Ac		217.00935	0.07 μs	α/9.832	9.650(10)/100	9/2-			
^{218}Ac		218.01164	1.1 μs	α/9.380	9.205(15)/				
^{219}Ac		219.01242	0.012 ms	α/8.830	8.664(10)/	(9/2-)			
^{220}Ac		220.01476	26. ms	α/8.350	7.610(20)/23				
					4.680(20)/21				
					7.790(10)/13				
					7.850(10)/24				
					7.985(10)/4				
					8.005(10)/5				
					8.060(10)/6				
					8.195(10)/3				
^{221}Ac		221.01559	52. ms	α/7.790	7.170(10)/2				
					7.375(10)/10				
					7.440(15)/20				
					7.645(10)/70				
^{222m}Ac			63. s	α/(>89)/	6.710(20)/7				
				EC/(1)/	6.750(20)/13				

Elem. or Isot.	Natural Abundance (Atom %)	Atomic Mass or Weight	Half-life/ Resonance Width (MeV)	Decay Mode/ Energy (/MeV)	Particle Energy/ Intensity (MeV/%)	Spin ($h/2\pi$)	Nuclear Magnetic Mom. (nm)	Elect. Quadr. Mom. (b)	γ-Energy / Intensity (MeV/%)
				I.T./(<10)/	6.810(20)/24				
					6.840(20)/9				
					6.890(20)/13				
					6.970(20)/7				
					7.000(20)/13				
^{222}Ac		222.01784	5. s	α/7.141	6.967(10)/6	1-			
					7.013(2)/94				
^{223}Ac		223.01914	2.1 m	α/(99)/6.783	6.131(2)/0.12	(5/2-)			0.0725
				EC/(1)/0.59	6.177(2)/0.94				0.0839
					6.293(1)/0.47				0.0927
					6.326(1)/0.3				0.0990
					6.332(2)/0.14				0.1917
					6.360(1)/0.22				0.2158
					6.397(1)/0.13				0.3588
					6.448(1)/0.2				0.4768
					6.473(1)/3.1				
					6.523(2)/0.6				
					6.528(1)/3.1				
					6.563(1)/13.6				
					6.582(3)/0.3				
					6.646(1)/44				
					6.661(1)/31				
^{224}Ac		224.021723	2.7 h	EC/(90)/1.403	5.811(1)/0.5	0			Ra L x-ray
				α/(10)/6.323	5.860(1)/0.75				Ra k x-ray
					5.875(1)/1.7				0.08426
					5.941(1)/4.4				0.13150
					6.000(1)/6.7				0.1571
					6.013(1)/1.4				0.21575
					6.056(1)/22				0.2619
					6.138(1)/26				(0.03–0.3)
					6.154(1)/1.0				
					6.204(1)/12				
					6.210(1)/20				
^{225}Ac		225.023230	10.0 d	α/5.935	5.286(1)/0.2	3/2			Fr k x-ray
					5.444(3)/0.1				0.06296/0.48
					5.554(1)/0.1				0.09982/1.36
					5.608(1)/1.1				0.1084
					5.636(1)/4.5				0.1116
					5.681(1)/1.4				0.1451
					5.722(1)/2.9				0.150/0.691
					5.731(1)/10				0.15724
					5.791(1)/9				0.18795/0.54
					5.793(1)/18				0.0075–0.809
^{226}Ac		226.026098	1.224 d	EC/(17)/0.640		(1-)			Ra k x-ray
				β- /(83)/1.116					Th k x-ray
				α/(0.006)/5.51	5.399(5)/0.006				0.07218
									0.15816
									0.23034
^{227}Ac		227.027752	21.77 a	β- /98.6/0.045	0.045/54	(3/2-)	+1.1	+1.7	0.0838/23.
				α/(1.4)/5.043	4.869(1)/0.09				0.0811/14.
					4.938(1)/0.52				0.2696/13.
					4.951(1)/0.65				(0.044–1.27)
^{228}Ac		228.031021	6.15 h	β- /2.127	1.11/32	(3+)			Th L x-ray
					1.85/12				Th k x-ray

Elem. or Isot.	Natural Abundance (Atom %)	Atomic Mass or Weight	Half-life/ Resonance Width (MeV)	Decay Mode/ Energy (/MeV)	Particle Energy/ Intensity (MeV/%)	Spin ($h/2\pi$)	Nuclear Magnetic Mom. (nm)	Elect. Quadr. Mom. (b)	γ-Energy / Intensity (MeV/%)
					2.18/11				0.12903
									0.33842
									0.91116
									0.96897
									(0.2–1.96)
^{229}Ac		229.03302	1.04 h	β- /1.10	1.1/	(3/2+)			0.09335/2.43
									0.16451/2.61
									0.56916/2.24
									0.0111–0.898
^{230}Ac		230.0363	2.03 m	β- /2.7	1.4/	1+			Th k x-ray
				β-, sf	/0.000119				0.45497
									0.50820
									(0.12–2.5)
^{231}Ac		231.0386	7.5 m	β- /2.1	2.1/100	(1/2+)			0.14379
									0.18574
									0.22140
									0.28250
									0.3070
^{232}Ac		232.0420	2.0 m	β- /3.7		(2-)			
^{233}Ac		233.0446	2.4 m	β- /		(1/2+)			
^{234}Ac		234.0484	40. s	β- /		(1+)			
$_{90}$Th		**232.03806(2)**							
^{209}Th		209.0177	~ 0.01 s	α	8.08				
^{210}Th		210.0158	~ 9 ms	α	7.90	0+			
^{211}Th		211.0149	0.04 s	α	7.79				
^{212}Th		212.01298	~ 30. ms	α/	7.80/	0+			
^{213}Th		213.0130	0.14 s	α/7.840	7.692(10)/				
^{214}Th		214.01150	0.10 s	α/7.825	7.677(10)/	0+			
^{215}Th		215.01173	1.2 s	α/7.660	7.33(10)/8	(1/2-)			0.134
					7.395(8)/52				0.192
					7.524(8)/40				(0.069-0.295)
^{216m}Th			0.14 ms	α	9.93/74				(0.0905-1.478)
					8.00, 9.31				
^{216}Th		216.01106	27. ms	α/8.071	7.92/99.46	0+			0.628
					7.30/0.54				
^{217}Th		217.01311	0.25 ms	α/9.424	9.27/94.6				(0.546-0.822)
					8.46/3.8				
					8.73/1.6				
^{218}Th		218.01328	0.11 μs	α/9.847	9.665(10)/	0+			
^{219}Th		219.01554	1.05 μs	α/9.510	9.340(20)/				
^{220}Th		220.01575	10. μs	α/8.953	8.790(20)/	0+			
^{221}Th		221.01818	2. ms	α/8.628	7.732/7				
					8.142/72				
					8.469/21				
^{222}Th		222.01847	2.24 ms	α/8.129	7.980/97.7	0+			
					7.599/2.3				
^{223}Th		223.02081	0.60 s	α/7.454	7.29(1)/41(5)				
					7.32(1)/29(5)				
					7.350(15)/20(5)				
					7.390(15)/10(4)				
^{224}Th		224.02147	1.05 s	α/7.305	6.768(5)/1.2	0+			
					6.997(5)/19				
					7.170(5)/7				
^{225}Th		225.023951	8.72 m	EC/(10)/0.68		(3/2+)			

Elem. or Isot.	Natural Abundance (Atom %)	Atomic Mass or Weight	Half-life/ Resonance Width (MeV)	Decay Mode/ Energy (/MeV)	Particle Energy/ Intensity (MeV/%)	Spin ($h/2\pi$)	Nuclear Magnetic Mom. (nm)	Elect. Quadr. Mom. (b)	γ-Energy / Intensity (MeV/%)
				α/(90)/6.920	6.441(2)/15				
					6.479(2)/43				
					6.501(3)/14				
					6.627(3)/3				
					6.650(5)/3				
					6.700(5)/2				
					6.743(3)/7				
					6.796(2)/9				
^{226}Th		226.024903	30.83 m	α/6.454	6.026(1)/0.2	0+			Ra k x-ray
					6.041(1)/0.19				0.1112
					6.098(1)/1.3				0.2421
					6.2283(4)/23				0.1310
					6.3375(4)/75				0.1733–0.9295
^{227}Th		227.027704	18.72 d	α/6.146		(3/2+)			Ra L x-ray
									Ra k x-ray
									0.05014
									0.23597
									0.25624
									(0.02–1.0)
^{228}Th		228.028741	1.913 a	α/5.520	5.1770(2)/0.18	0+			
					5.2114(1)/0.4				
					5.3405(1)/26.7				
					5.4233(1)/73				
^{229m}Th			13.9 h	α	4.03-5.08				
^{229}Th		229.031762	7.9×10^{3} a	α/5.168	4.814/9.3	5/2+	+0.46	+4.	0.1935/4.3
					4.845(5)/56				0.21089/277
					4.9008(5)/10.2				0.13697/1.21
					4.689–5.077				0.011–0.6036
^{230}Th		230.033134	7.54×10^{4} a	α/4.771	4.4383(6)/0.03	0+			0.0677/0.46
					4.4798(6)/0.12				0.1439/0.078
			> 2×10^{18} a	sf/< 4×10^{-12}	4.6211(6)/23.4				
					4.6876(6)/76.3				
^{231}Th		231.036304	1.063 d	β- /0.390	0.138/22	5/2+			Pa L x-ray
					0.218/20				Pa k x-ray
					0.305/52				0.02564
									0.084203/
									(0.02–0.3)
^{232}Th	100.	232.038055	1.40×10^{10} a	α/4.081	3.830(10)/0.2	0+			0.0590
			1.2×10^{21} a	sf/1.1×10^{-9}	3.952(5)/23				0.124
					4.010(5)/77				
^{233}Th		233.041582	22.3 m	β-/1.245	1.245/	½+			Pa L x-ray
									Pa k x-ray
									0.02938
									0.08653
									0.45930
									(0.02–1.2)
^{234}Th		234.043601	24.10 d	β- /0.273	0.102/20	0+			Pa L x-ray
					0.198/72				0.06329/4.1
									0.09235/2.4
									0.09278/2.4
^{235}Th		235.04751	7.2 m	β- /1.9					0.4162
									0.6594
									0.7272
									0.747

Elem. or Isot.	Natural Abundance (Atom %)	Atomic Mass or Weight	Half-life/ Resonance Width (MeV)	Decay Mode/ Energy (/MeV)	Particle Energy/ Intensity (MeV/%)	Spin ($h/2\pi$)	Nuclear Magnetic Mom. (nm)	Elect. Quadr. Mom. (b)	γ-Energy / Intensity (MeV/%)
									0.9318
^{236}Th		236.0499	37.5 m	β- /~ 1.0		0+			Pa k x-ray
									0.1107
^{237}Th		237.0539	5.0 m	β-					
^{238}Th		238.0565	9.4 m			0+			0.0890
$_{91}$Pa		**231.03588(2)**							
^{212}Pa		212.0232	~ 5 ms	α	8.27				
^{213}Pa		213.0211	7 ms	α	8.24				
^{214}Pa		214.0209	17 ms	α	8.12				
^{215}Pa		215.0192	15. ms	α	8.08/100				
^{216}Pa		216.0191	0.19 s	α/	7.95/51				0.134
					7.82/45				
					7.79/4				
^{217m}Pa			1.08 ms	α/	10.16/72				0.4504–0.821
					8.306/11				
					9.55/6				
					9.69/2				
^{217}Pa		217.0183	3.8 ms	α/8.490	8.337/99				0.0466–0.634
					7.873/0.4				
					7.728/0.3				
					7.710/0.3				
^{218}Pa		218.02004	0.12 ms	α/	9.54/31				0.092
					9.61/69				
^{219}Pa		219.0199	0.05 μs	α					
^{220}Pa		220.0219	0.8 μs	α					
^{221}Pa		221.0219	6. μs	α	9.08(3)				
^{222}Pa		222.0237	~ 4.3 ms	α/8.700	8.180/50				
					8.330/20				
					8.540/30				
^{223}Pa		223.0240	~ 6.5 ms	α/8.340	8.006(10)/55				
					8.196(10)/45				
^{224}Pa		224.02563	0.84 s	α/7.630	7.555(10)/75(3)				0.1945
					7.46(1)/25(3)				(0.028–0.412)
^{225}Pa		225.0261	1.8 s	α/7.380	7.195(10)/30				
					7.245(10)/70				
^{226}Pa		226.02795	1.8 m	α/(74)/6.987	6.728(10)/0.7				
				EC/(26)/2.83	6.823(10)/35				
					6.863(10)/39				
^{227}Pa		227.02881	38.3 m	α/(85)/6.582	6.357(4)/7	(5/2-)			0.0649
				EC/(15)/1.02	6.376(10)/2.2				0.0669
					6.401(4)/8				0.1100
					6.416(4)/13				
					6.423(10)/10				
					6.465(4)/43				
^{228}Pa		228.031051	22. h	EC/(98)/2.111		(3+)	+3.5		Th k x-ray
				α/(2)	5.779/0.23				0.409/100
					5.805/0.15				0.4631/222
					6.078/0.4				0.91116/242
					6.105/0.25				0.96464/120
					6.118/0.22				0.96897/149
									0.058–1.96
^{229}Pa		229.032097	1.5 d	EC/(99.8)/0.32		(5/2+)			0.04244
				α/(0.2)/5.836	5.536(2)/0.02				(0.024–0.18)
					5.579(2)/0.09				

Elem. or Isot.	Natural Abundance (Atom %)	Atomic Mass or Weight	Half-life/ Resonance Width (MeV)	Decay Mode/ Energy (/MeV)	Particle Energy/ Intensity (MeV/%)	Spin ($h/2\pi$)	Nuclear Magnetic Mom. (nm)	Elect. Quadr. Mom. (b)	γ-Energy / Intensity (MeV/%)
					5.668(2)/0.05				
^{230}Pa		230.034541	17.4 d	EC/(90)/1.310	0.51/	(2-)	2.0		Th L x-ray
				β-/(10)/0.563					Th k x-ray
									0.4437
									0.45477
									0.89876
									0.91856
									0.95199
									(0.053–1.07)
^{231}Pa	100	231.035884	3.25 × 10^{4} a	α/5.148	4.6781(5)/1.5	3/2-	2.01	-1.7	Ac L x-ray
					4.7102(5)/1.0				Ac k x-ray
			> 2 × 10^{17} a	sf/< 1.6 × 10^{-15}	4.7343(5)/8.4				0.01899
					4.8513(5)/1.4				0.027396
					4.9339(5)/3				0.03823
					4.9505(5)/22.8				0.04639
					4.9858(5)/1.4				0.25586
					5.0131(5)/25.4				0.26029
					5.0292(5)/20				0.28367
					5.0318(5)/2.5				0.30007
					5.0587(5)/11				0.30264
									0.33007
									(0.02–0.61)
^{232}Pa		232.03859	1.31 d	β- /1.34		(2-)			U k x-ray
									0.10900
									0.15009
									0.89439
									0.96934
									(0.10–1.17)
^{233}Pa		233.040247	26.97 d	β-/0.571	0.15/40	3/2-	+4.0	-3.0	U L x-ray
					0.256/60				U k x-ray
									0.30017
									0.31201/38.4
									(0.0286-0.456)
^{234m}Pa			1.17 m	β- /99.9/2.29		(0-)			U k x-ray
				IT/0.13/					0.25818/0.07
									0.76641/0.32
									1.0009/0.86
									(0.06–1.96)
^{234}Pa		234.043308	6.69 h	β- /2.197	0.51/	(4+)			U L x-ray
									U k x-ray
									0.1312/0.03
									0.5695/0.02
									0.9256/0.02
									(0.02–1.99)
^{235}Pa		235.04544	24.4 m	β- /1.41	1.4/97	(3/2-)			0.0308–0.659
^{236}Pa		236.0487	9.1 m	β- /2.9	1.1/40	(1-)			U k x-ray
					2.0/50				0.64235
					3.1/10				0.68759
									1.7630
									(0.04–2.18)
^{237}Pa		237.0512	8.7 m	β- /2.3	1.1/60	(1/2+)			0.4986
					1.6/30				0.5293
					2.3/10				0.5407
									0.8536

Elem. or Isot.	Natural Abundance (Atom %)	Atomic Mass or Weight	Half-life/ Resonance Width (MeV)	Decay Mode/ Energy (/MeV)	Particle Energy/ Intensity (MeV/%)	Spin ($h/2\pi$)	Nuclear Magnetic Mom. (nm)	Elect. Quadr. Mom. (b)	γ-Energy / Intensity (MeV/%)
									0.8650
									(0.04–1.4)
^{238}Pa		238.0545	2.3 m	β- /3.5	1.2/	(3-)			0.10350
					1.7/				0.1785
									0.4484
									0.6350
									0.6800
									1.01446
									(0.04–2.5)
^{239}Pa		239.0573	1.8 h						
$_{92}$U		**238.02891(3)**							
^{217}U		217.0244	~ 0.2 ms	α	8.02				
^{218m}U			~ 0.56 ms	α	10.68				
^{218}U		218.02354	0.5 ms	α	8.61	0+			
^{219}U		219.0249	~ 0.08 ms	α	9.77				
^{222}U		222.0261	~ 1.μs	α		0+			
^{223}U		223.0277	0.02 s	α/	8.78(4)/				
^{224}U		224.02761	~ 1. ms	α/	8.46/100	0+			
^{225}U		225.02939	84. ms	α/	7.87/83				
					7.82/15				
					7.63/2				
^{226}U		226.02934	0.26 s	α/7.560	7.56/86	0+			
					7.38/14				
^{227}U		227.03116	1.1 m	α/7.200	6.870/				
^{228}U		228.03137	9.1 m	α/6.803	6.404(6)/0.6	0+			0.095
					6.440(5)/0.7				0.152
					6.589(5)/29				0.187
					6.681(6)/70				0.246
^{229}U		229.03351	58. m	EC/(80)/1.31	6.223/3	(3/2+)			
				α/(20)/6.473	6.297(3)/11				
					6.332(3)/20				
					6.360(3)/64				
^{230}U		230.033940	20.8 d	α/5.992	5.5866(3)/0.01	0+			Th L x-ray
			> 4 × 10^{10} a	sf/< 10^{-10}	5.6624(3)/0.26				0.07218
					5.6663(3)/0.38				0.15421
					5.8178(3)/32				0.23034
					5.8887(3)/67				(0.081–0.8565)
^{231}U		231.036294	4.2 d	EC/0.36		(5/2-)			Pa L x-ray
				α/(10^{-3})	5.46/1.6 × 10^{-3}				Pa k x-ray
					5.47/1.4 × 10^{-3}				0.02564
					5.40/1. × 10^{-3}				0.08420
^{232}U		232.037156	70. a	α/5.414	4.9979(1)/0.003	0+			
			2.6 × 10^{15} a	sf/2.7 × 10^{-12}	5.1367(1)/0.3				
					5.2635(1)/31				
					5.3203(1)/69				
^{233}U		233.039635	1.592 ×10^{5} a	α/4.909	4.7830(8)/13.2	5/2+	+0.59	3.66	Th L x-ray
			>2.7 × 10^{17} a	sf/6 × 10^{-11}	4.8247(8)/84.4				0.04244
					4.510–4.804				0.09714
									(0.0252–1.119)
^{234}U	0.0054(5)	234.040952	2.455 ×10^{5} a	α/4.856	4.604(1)/0.24	0+			0.05323/0.156
			1.5 × 10^{16} a	sf/1.6 × 10^{-9}	4.7231(1)/27.5				0.12091
					4.776(1)/72.5				
^{235m}U			26. m	IT/0.0007		1/2+			
^{235}U	0.7204(6)	235.043930	7.04 × 10^{8} a	α/4.6793	4.1525(9)/0.9	7/2-	-0.38	4.94	Th L x-ray

Elem. or Isot.	Natural Abundance (Atom %)	Atomic Mass or Weight	Half-life/ Resonance Width (MeV)	Decay Mode/ Energy (/MeV)	Particle Energy/ Intensity (MeV/%)	Spin ($h/2\pi$)	Nuclear Magnetic Mom. (nm)	Elect. Quadr. Mom. (b)	γ-Energy / Intensity (MeV/%)
			1.0×10^{19} a	sf/7×10^{-9}	4.2157(9)/6.				Th k x-ray
					4.3237(9)/4.6				0.10917
					4.3641(9)/19.				0.14378/0.134
					4.370(4)/6				0.16338/0.067
					4.3952(9)/57.				0.18574/0.806
					4.4144(9)/2.1				0.1949/0.009
					4.5025(9)/1.7				0.20533/0.774
					4.5558(9)/4.2				0.2214/0.0014
					4.5970(9)/4.8				(0.03–0.79)
^{236}U		236.045568	2.342×10^{7} a	α/4.569	4.332(8)/0.26	0+			Th L x-ray
			2.5×10^{16} a	sf/9×10^{-8}	4.445(5)/26				0.04946/100
					4.494(3)/74				0.11279/24.1
									0.17115/0.080
^{237}U		237.048730	6.75 d	β- /0.519	0.21/	1/2+			Np L x-ray
					0.25/				Np k x-ray
									0.05953
									0.20801
^{238}U	99.2742(10)	238.050788	4.47×10^{9} a	α	4.0395/0.23	0+			Th L x-ray
			8.2×10^{15} a	sf/5×10^{-5}	4.147(5)/23				0.04955/.06
					4.196(5)/77				0.1135/.01
^{239}U		239.054293	23.5 m	β- /1.265	1.2/	5/2+			(0135-1.102)
					1.3/				
^{240}U		240.05659	14.1 h	β- /0.39	0.36/	0+			Np L x-ray
									0.04410
									0.05558
									0.06760
^{242}U		242.0629	16.8 m	β- /~ 1.2		0+			
$_{93}$**Np**									
^{225}Np		225.0339	> 2 μs						
^{226}Np		226.0352	0.03 s	α/	8.04(2)/				
^{227}Np		227.0350	0.51 s	α/	7.65(2)/				
					7.68(1)/				
^{228}Np		228.0362	61. s	EC/60(7)/					
				α/40(7)/, sf					
^{229}Np		229.0363	4.0 m	α/7.010	6.890(20)				
^{230}Np		230.0378	4.6 m	EC/97 /3.6					
				α/3	6.660(20)				
^{231}Np		231.03825	48.8 m	EC/98 /1.8		5/2			0.2629
				α/2 /6.368	6.280/2				0.3475
									0.3703
^{232}Np		232.0401	14.7 m	EC/99 /2.7		(4-)			U L x-ray
									U k x-ray
									0.3268
									0.81925
									0.86683
^{233}Np		233.04074	36.2 m	EC/1.2		(5/2+)			U L x-ray
									U k x-ray
									0.29887
									0.31201
^{234}Np		234.04290	4.4 d	β+, EC/1.81	0.79/	(0+)			U L x-ray
									U k x-ray
									1.5272
									1.5587
									1.6022

Elem. or Isot.	Natural Abundance (Atom %)	Atomic Mass or Weight	Half-life/ Resonance Width (MeV)	Decay Mode/ Energy (/MeV)	Particle Energy/ Intensity (MeV/%)	Spin ($h/2\pi$)	Nuclear Magnetic Mom. (nm)	Elect. Quadr. Mom. (b)	γ-Energy / Intensity (MeV/%)
^{235}Np		235.044063	1.085 a	EC/99.9 /0.124		5/2+			U k x-ray
				α/0.001/5.191					
^{236m}Np			22.5 h	EC/52 /		(1-)			U L x-ray
				β- /48 /					Pu L x-ray
									U k x-ray
									0.64235
									0.68759
^{236}Np		236.04657	1.55 × 10^{5} a	EC/91 /0.94		(6-)			U L x-ray
				β- /9 /0.49					U k x-ray
									0.10423
									0.16031
^{237}Np		237.048173	2.14 × 10^{6} a	α/4.957	4.6395(5)/6.5	5/2+	+3.14	+3.87	Pa L x-ray
			1 × 10^{18} a	sf/2.1 × 10^{-10}	4.766(5)/9.7				Pa k x-ray
					4.7715(5)/22.7				0.029378/15
					4.7884(5)/47.8				0.08653/12
					4.558–4.873				(0.03–0.28)
^{238}Np		238.050946	2.117 d	β- /1.292	1.2/	2+			Pu L x-ray
									Pu k x-ray
									0.98447/25.2
									1.02855/18.3
									(.044–1.026)
^{239}Np		239.052939	2.355 d	β- /0.722	0.341/30	5/2+			Pu L x-ray
					0.438/48				Pu k x-ray
									0.10613
									0.228186/11
									0.27760/15
									(0.04–0.50)
^{240m}Np			7.22 m	β- /99.9 /	2.18/	(1+)			0.25143
				IT/0.1 /					0.26333
									0.55454
									0.59735
^{240}Np		240.05616	1.032 h	β- /2.20	0.89/	5+			0.1471/
									0.5664
									0.6008
^{241}Np		241.0583	13.9 m	β- /1.3	1.3/	5/2+			0.1330/
									0.1740
									0.280
^{242m}Np			2.2 m	β- /		(1+)			0.15910
									0.2651/
									0.78570
									0.9448/
^{242}Np		242.0616	5.5 m	β- /2.7	2.7/	6+			0.6209
									0.73620
									0.78074
									1.47340
									(0.04–2.37)
^{243}Np		243.06428	1.9 m						
^{244}Np		244.0679	2.3 m						
$_{94}$Pu									
^{228}Pu		228.03874	~ 1.1 s	α/	7.81(2)/	0+			
^{229}Pu		229.0402	~ 1.5 m	α/	7.46/				
^{230}Pu		230.03965	1.7 m	α/	7.06/81	0+			
					7.00/19				
^{231}Pu		231.04110	8.6 m	EC/90					

Elem. or Isot.	Natural Abundance (Atom %)	Atomic Mass or Weight	Half-life/ Resonance Width (MeV)	Decay Mode/ Energy (/MeV)	Particle Energy/ Intensity (MeV/%)	Spin ($h/2\pi$)	Nuclear Magnetic Mom. (nm)	Elect. Quadr. Mom. (b)	γ-Energy / Intensity (MeV/%)
				α/10	6.72				
^{232}Pu		232.04119	34. m	EC/>80/1.1		0+			
				α/<20/6.716	6.542(10)/38				
					6.600(10)/62				
^{233}Pu		233.04300	20.9 m	EC(99.9)/1.9					0.1503
				α/0.1 /6.416	6.300(20)/0.1				0.1804
									0.2353
									0.5002
									0.5346/
									1.0352/
^{234}Pu		234.04332	8.8 h	EC/94 /0.39		0+			
				α/6 /6.310	6.035(3)/0.024				
					6.149(3)/1.9				
					6.200(3)/4.				
^{235}Pu		235.04529	25.3 m	EC/99+ /1.2		(5/2+)			
				α/0.003/5.957	5.850(20)/0.003				
^{236m}Pu			1.2 μs						
^{236}Pu		236.046058	2.87 a	α/5.867	5.611/0.21	0+			0.0476/0.07
			1.5 × 10^{9} a	sf/1.9 × 10^{-7}	5.7210/30.5				0.109/0.02
					5.7677(1)/69.3				(0.17–0.97)
^{237}Pu		237.048410	45.7 d	EC/99.9 /0.220		7/2-			Np L x-ray
				α/0.003 /5.747	5.334(4)/0.0015				Np k x-ray
					5.356(4)/0.0006				0.026344
					5.650(4)/0.0007				0.03319
									0.05954
									(0.03–0.5)
^{238}Pu		238.049560	87.7 a	α/5.593	5.3583(1)/0.10	0+			U k x-ray
			4.75 × 10^{10} a	sf/1.8 × 10^{-7}	5.465(1)/28.3				0.04347
					5.4992(1)/71.6				(0.04–1.1)
^{239}Pu		239.052163	2.410 ×10^{4} a	α/5.244	5.055/0.047	1/2+	+0.203		U k x-ray
			8. × 10^{15} a	sf/3 × 10^{-10}	5.076/0.078				0.05162
					5.106/11.9				0.05682
					5.144/17.1				0.12928
					5.157/70.8				0.37502
					(4.74 –5.03)				0.41369
^{240}Pu		240.053814	6.56 × 10^{3} a	α/5.255	5.0212(1)/0.07	0+			U L x-ray
			1.14 × 10^{11} a	sf/5.7 × 10^{-6}	5.1237(1)/26.4				0.04524
					5.1681(1)/73.5				0.10423
									(0.04–0.97)
^{241}Pu		241.056852	14.3 a	β-/99+/0.0208	4.853/3 × 10^{-4}	5/2+	-0.68	+6.	0.14854
				α/0.002 /5.139	4.897/0.002				0.1600
			< 6 × 10^{16} a	sf/> 2.4 × 10^{-14}					
^{242}Pu		242.058743	3.75 × 10^{5} a	α/4.983	4.7546(7)/0.098	0+			U L x-ray
			6.77 × 10^{10} a	sf/5.5 × 10^{-4}	4.8564(7)/22.4				0.04491
					4.9006(7)/78				0.10350
^{243}Pu		243.062003	4.956 h	β- /0.582	0.49/21	7/2+			Am L x-ray
					0.58/60				0.0417
									0.0839
^{244}Pu		244.064204	8.12 × 10^{7} a	α/99.9/4.665	4.546(1)/19.4	0+			U L x-ray
			6.6 × 10^{10} a	sf/0.12	4.589(1)/80.5				0.0439
^{245}Pu		245.06775	10.5 h	β- /1.21	0.93/57	(9/2-)			Am L x-ray
					1.21/11				Am k x-ray
									0.2804 /
									0.30832

Elem. or Isot.	Natural Abundance (Atom %)	Atomic Mass or Weight	Half-life/ Resonance Width (MeV)	Decay Mode/ Energy (/MeV)	Particle Energy/ Intensity (MeV/%)	Spin ($h/2\pi$)	Nuclear Magnetic Mom. (nm)	Elect. Quadr. Mom. (b)	γ-Energy / Intensity (MeV/%)
									0.32752
									0.56014
									(0.03–1.2)
^{246}Pu		246.07021	10.85 d	β- /0.40	0.150/85	0+			Am L x-ray
					0.35/10				Am k x-ray
									0.04379
									0.22371
^{247}Pu		247.0741	2.3 d						
$_{95}$**Am**									
^{232}Am		232.0466	0.9 m	EC/~ 5.0					
^{233}Am		233.0464	~ 3.2 m	α	6.78				
^{234}Am		234.0478	2.3 m	EC/4.2					
^{235}Am		235.0480	10.3 m	EC					Pu K x-ray
				α	6.46/0.4				0.291/100
									(0.170-0.828)
^{236m}Am			2.9 m			(1-)			(0.583-0.713)
^{236}Am		236.0496	3.6 m	EC		(5-)			(0.158-1.038)
^{237}Am		237.0500	1.22 h	EC/99.98 /1.7		(5/2-)			Pu k x-ray
				α/0.02 /6.20	6.042(5)/0.02				0.14559
									0.28026
									0.43845
^{238}Am		238.05198	1.63 h	EC/2.26		1+			Pu L x-ray
				α/0.0001 /6.04	5.940/0.0001				Pu k x-ray
									0.91870
									0.96278
^{239}Am		239.053025	11.9 h	EC/99.99/0.803		5/2-			Pu L x-ray
				α/0.01/5.924	5.734(2)/0.001				Pu k x-ray
					5.776(2)/0.008				0.18172
									0.22818
									0.27760
^{240}Am		240.05530	2.12 d	EC/1.38		(3-)			Pu L x-ray
				α/5.592	5.378/16 × 10^{-4}				Pu k x-ray
									0.88878
									0.98764
									(0.1–1.3)
^{241}Am		241.056829	432.7 a	α/5.637	5.2443(1)/0.002	5/2-	+1.58	+3.1	Np L x-ray
			1.2 × 10^{14} a	sf/3.6 × 10^{-10}	5.3221(1)/0.015				0.02634 /.024
					5.3884(1)/1.4				0.0332/.00126
					5.4431(1)/12.8				0.05954/0.359
					5.4857(1)/85.2				(0.03–1.128)
					5.5116(1)/0.20				
					5.5442(1)/0.34				
^{242m}Am			141. a	IT/99.5/0.048		5-	+1.0	+7.	Am L x-ray
				α/0.5/5.62	5.141(4)/0.026				0.04863
			> 3 × 10^{12} a	sf/< 4.7 × 10^{-9}	5.2070(2)/0.4				0.08648
									0.10944
									0.16304
^{242}Am		242.059549	16.02 h	β- /83 /0.665	0.63/46	1-	+0.388	~ -2.4	Pu L x-ray
				EC/17 /0.750	0.67/37				Cm L x-ray
									Pu k x-ray
									0.0422
									0.04453
^{243}Am		243.061381	7.37 × 10^3 a	α/5.438	5.1798(5)/1.1	5/2-	+1.50	+2.86	0.04354
			2. × 10^{14} a	sf/3.7 × 10^{-9}	5.2343(5)/11				0.07467

Elem. or Isot.	Natural Abundance (Atom %)	Atomic Mass or Weight	Half-life/ Resonance Width (MeV)	Decay Mode/ Energy (/MeV)	Particle Energy/ Intensity (MeV/%)	Spin ($h/2\pi$)	Nuclear Magnetic Mom. (nm)	Elect. Quadr. Mom. (b)	γ-Energy / Intensity (MeV/%)
					5.2766(5)/88				0.08657
					5.394(5)/0.12				0.11770
					5.3500(5)/0.16				0.14197
^{244m}Am			~ 26. m	β- /1.498		(1-)			0.0429
^{244}Am		244.064285	10.1 h	β- /1.428					Am L x-ray
									Cm k x-ray
									0.7460
									0.9000
^{245}Am		245.066452	2.05 h	β- /0.894	0.65/19	(5/2+)			Cm L x-ray
					0.90/77				Cm k x-ray
									0.25299
^{246m}Am			25.0 m	β- /	1.3/79.	2-			Cm L x-ray
					1.60/14				Cm k x-ray
					2.1/7				0.27002
									0.79881
									1.06201
									1.07885
									(0.04–2.29)
^{246}Am		246.06978	39. m	β- /2.38	1.2/	(7-)			Cm L x-ray
									Cm k x-ray
									0.1529
									0.2046
									0.6786
^{247}Am		247.0721	22. m	β- /1.7					Cm L x-ray
									Cm k x-ray
									0.2267 /
									0.2853 /
$_{96}$Cm									
^{233}Cm		233.0508		α/	7.34/				
^{234}Cm		234.05016	~ 51. s	α	7.24/	0+			
^{235}Cm		235.0514							
^{236}Cm		236.0514		EC/1.7		0+			
^{237}Cm		237.0529		EC/2.5					
				α/<1	6.66				
^{238}Cm		238.05303	2.4 h	EC/>90 /0.97	α/6.558/86	0+			
				α/<10 /6.632	6.520(50)/14				
^{239}Cm		239.0550	~ 3. h	EC/1.7					
									0.0407
									0.1466
									0.1874
^{240}Cm		240.055530	27. d	α/6.397	5.989/0.014	0+			
					6.147/0.05				
			1.9 × 10^6 a	sf/3.9 × 10^{-6}	6.2478(6) /28.8				
					6.2906(6) /70.6				
^{241}Cm		241.057653	32.8 d	EC/99 /0.768		1/2+			Am k x-ray
				α/1 /6.184	5.8842(4)/0.12				0.13241
					5.9291(4)/0.18				0.16505
					5.9389(4)/0.69				0.18028
									0.43063
									0.47181
^{242}Cm		242.058836	162.8 d	α/6.216	5.9694(1)/0.035	0+			Pu L x-ray
					6.069(1)/25				0.04408
			7.0 × 10^6 a	sf/6.4 × 10^{-6}	6.1129(1)/74				0.10189
									(0.04–1.2)

Elem. or Isot.	Natural Abundance (Atom %)	Atomic Mass or Weight	Half-life/ Resonance Width (MeV)	Decay Mode/ Energy (/MeV)	Particle Energy/ Intensity (MeV/%)	Spin ($h/2\pi$)	Nuclear Magnetic Mom. (nm)	Elect. Quadr. Mom. (b)	γ-Energy / Intensity (MeV/%)
^{243}Cm		243.061389	29.1 a	α/6.167	5.6815(5) /0.2	5/2+	0.40		Pu L x-ray
					5.6856(5)/1.6				Pu k x-ray
			5.5×10^{11} a	sf/5.3×10^{-9}	5.7420(5)/10.6				0.10612
					5.7859(5)/73.3				0.20975
					5.9922(5)/6.5				0.22819
					6.0103(5)/1.0				0.27760
					6.0589(5)/5				0.28546
					6.0666(5)/1.5				0.33431
									(0.04–0.7)
^{244}Cm		244.062753	18.1 a	α/5.902	5.6656/0.02	0+			Pu L x-ray
					5.7528/23				0.04282
			1.32×10^{7} a	sf/1.4×10^{-4}	5.8050/77				0.09885
					5.515/0.004				0.15262
^{245}Cm		245.065491	8.48×10^{3} a	α/5.623	5.235(10)/0.3	7/2+	0.5		Pu L x-ray
					5.3038(10)/5.0				Pu k x-ray
			1.4×10^{12} a	sf/6.1×10^{-7}	5.3620(7)/93				0.04195
					5.4927(11)/0.8				0.13299
					5.5331(11)/0.6				0.13606
									0.17494
^{246}Cm		246.067224	4.76×10^{3} a	α/5.476	5.343(3)/21	0+			Pu L x-ray
			1.8×10^{7} a	sf/0.026	5.386(3)/79				0.04453
^{247}Cm		247.070354	1.56×10^{7} a	α/5.352	4.818(4)/4.7	9/2-	0.36		Pu k x-ray
					4.8690(20)/71				0.2792
					4.941(4)/1.6				0.2886
					4.9820(20)/2.0				0.3471
					5.1436(20)/1.2				0.4035
					5.2104(20)/5.7				
					5.2659(20)/13.8				
^{248}Cm		248.072349	3.48×10^{5} a	α/99.92 /5.162	4.931(5)/0.07	0+			
					5.0349(2)/16.5				
			4.15×10^{6} a	sf/8.38	5.0784(2)/(75)/1				
^{249}Cm		249.075953	64.15 m	β- /0.900	0.9/	1/2+			Bk k x-ray
									0.56039/0.84
									0.63431/1.5
									(0.085-0.653)
^{250}Cm		250.07836	~ 9.7×10^{3} a	sf/85.8		0+			
				α/5.27					
^{251}Cm		251.08229	16.8 m	β- /1.42	0.90/16	(1/2+)			0.3896 /
									0.5299
									0.5425
^{252}Cm		252.0849	< 2 d			0+			
$_{97}$**Bk**									
^{238}Bk		238.0583	2.4 m	EC/5.0					
^{239}Bk		239.0583							
^{240}Bk		240.0598	~ 4.8 m						
^{241}Bk		241.0602	4.6 m	EC					(0.152-0.262)
^{242}Bk		242.0620	7.0 m	EC/3.0					
^{243}Bk		243.063008	4.5 h	EC/99.8 /1.508	6.542(4)/0.03	(3/2-)			0.1466
				α/0.15 /6.871	6.5738(2)/0.04				0.1874
					6.7180(22)/0.02				0.755
					6.7581(20)/0.02				0.840
									0.946
^{244}Bk		244.06518	4.4 h	EC/99.99 /2.26		(4-)			0.1445
				α/0.01 /6.778	6.625(4)/0.003				0.1876

Elem. or Isot.	Natural Abundance (Atom %)	Atomic Mass or Weight	Half-life/ Resonance Width (MeV)	Decay Mode/ Energy (/MeV)	Particle Energy/ Intensity (MeV/%)	Spin ($h/2\pi$)	Nuclear Magnetic Mom. (nm)	Elect. Quadr. Mom. (b)	γ-Energy / Intensity (MeV/%)
					6.667(4)/0.003				0.2176
									0.9815
									0.9215/
^{245}Bk		245.066362	4.94 d	EC/99.9 /0.810		3/2-			Cm L x-ray
				α/0.1 /6.453	5.8851(5)/0.03				Cm k x-ray
					6.1176(9)/0.01				0.25299
					6.1467(5)/0.02				0.3809
					6.3087(5)/0.014				0.3851
					6.3492(5)/0.018				
^{246}Bk		246.0687	1.80 d	EC/1.35		(2-)			Cm L x-ray
									Cm k x-ray
									0.79881
									1.08142
^{247}Bk		247.07031	1.4 × 10^3 a	α/5.889	5.465(5)/1.5	(3/2-)			0.04175
					5.501(5)/7				0.0839
					5.532(5)/45				0.268
					5.6535(20)/5.5				
					5.678(2)/13				
					5.712(2)/17				
					5.753(2)/4.3				
					5.794(2)/5.5				
^{248}Bk		248.07310	23.7 h	β- /70 /0.87	0.86/	(1-)			Cm L x-ray
				EC/30 /0.72					Cf L x-ray
									Cm k x-ray
									Cf k x-ray
									0.5507
^{249}Bk		249.074987	320. d	β- /0.125	0.125/100	7/2+	2.0		0.327/10^{-5}
				α/0.001 /5.525	5.390(1)/0.0002				0.308/10^{-6}
			1.8 × 10^9 a	sf/4.9 × 10^{-8}	5.4174(6)/0.001				
^{250}Bk		250.078317	3.217 h	β- /1.780	0.74/	2-			Cf L x-ray
									Cf k x-ray
									0.98912
									1.03184
									(0.04–1.6)
^{251}Bk		251.08076	56. m	β- /1.09		(3/2-)			0.02481
									0.1528
									0.1776
^{252}Bk		252.0843	1.8 m						
$_{98}$Cf									
^{237}Cf		237.062	2.1 s	α, sf/10					
^{238}Cf		238.0614	21 ms	sf/~ 100		0+			
				α/~ 0.2					
^{239}Cf		239.0624	~ 0.7 m	α					
^{240}Cf		240.0623	1.1 m	α/7.719	7.590(10)/	0+			
				sf/ ~ 2.1					
^{241}Cf		241.0637	4. m	EC/3.3					
				α/7.60	7.335(5)/				
^{242}Cf		242.06370	3.5 m	α/7.509	7.351(6)/20	0+			
				sf/<0.014	7.385(4)/80				
^{243}Cf		243.0654	11. m	EC/86 /2.2	7.060(6)/20	(1/2+)			
				α/14 /7.40	7.170/4				
^{244}Cf		244.066001	20. m	α/7.328	7.168(5)/25	0+			
					7.210(5)/75				
^{245}Cf		245.068049	44. m	α/36 /7.255	7.14/91.7				Cm K x-ray

Elem. or Isot.	Natural Abundance (Atom %)	Atomic Mass or Weight	Half-life/ Resonance Width (MeV)	Decay Mode/ Energy (/MeV)	Particle Energy/ Intensity (MeV/%)	Spin ($h/2\pi$)	Nuclear Magnetic Mom. (nm)	Elect. Quadr. Mom. (b)	γ-Energy / Intensity (MeV/%)
				EC/64 /1.569	6.983/0.31				0.5709
					7.09/7				0.6014
					7.065/0.68				0.6163
^{246}Cf		246.068805	1.49 d	α/6.869	6.6156(10)/0.18	0+			Cm L x-ray
					6.7086(7)/21.8				0.04221
			1.8×10^3 a	sf/2.3×10^{-4}	6.7501(7)/78.0				0.0945
									0.147
^{247}Cf		247.07100	3.11 h	EC/99.96 /0.65		7/2+			Bk k x-ray
				α/0.04 /6.55	6.301(5)/				0.2941
									0.4778
^{248}Cf		248.07219	334. d	α/6.369	6.220(5)/17	0+			
			3.2×10^4 a	sf/0.0029	6.262(5)/83				
^{249}Cf		249.074854	351. a	α/6.295	5.758/3.7	9/2-			Cm L x-ray
					5.812/85.7				Cm k x-ray
			$8. \times 10^{10}$ a	sf/4.4×10^{-7}	5.8488(2)/1.0				0.25299/2.5
					5.9029(2)/2.8				0.33351/13.6
					5.9451(2)/4.0				0.38832/63.6
					6.1401(2)/1.1				(0.0376–1.10)
					6.1940(2)/2.2				
^{250}Cf		250.076406	13.1 a	α/6.129	5.8913(4)/0.3	0+			Cm L x-ray
			1.7×10^4 a	sf/0.077	5.9889(4)/15				0.04285
					6.0310(4)/84.5				
^{251m}Cf			26.3 μs						
^{251}Cf		251.079587	9.0×10^2 a	α/6.172	5.56448(7)/1.5	1/2+			0.109/19.8
					5.632(1)/4.5				0.1775/17.3
					5.648(1)/3.5				(0.0385-0.354)
					5.6773(6)/35				
					5.762(3)/3.8				
					5.7937(7)/2.0				
					5.8124(8)/4.2				
					5.8514(6)/27				
					6.0140(7)/11.6				
					6.0744(7)/2.7				
^{252}Cf		252.081626	2.65 a	α/96.9 /6.217	5.7977(1)/0.23	0+			Cm L x-ray
			86. a	sf/3.1/	6.0756(4)/15.2				0.04339
					6.1184(4)/81.6				0.1002
^{253}Cf		253.08513	17.8 d	β- /99.7 /0.29	0.27/100	(7/2+)			
				α/0.3 /6.126	5.921(5)/0.02				
^{254}Cf		254.08732	60.5 d	sf/99.7/		0+			
				α/0.3/5.930	5.792(5)/0.05				
					5.834(5)/0.26				
^{255}Cf		255.0911	1.4 h	β- /0.7					
^{256}Cf		256.0934	12. m	sf		0+			
$_{99}$**Es**									
^{241}Es		241.0685	~ 8 s	α	8.11				
^{242}Es		242.0698	16 s	α	8.03				(0.087-0.122)
^{243}Es		243.0696	22. s	α/60	7.89				
				EC/40 /4.0	7.86				
^{244}Es		244.0709	37. s	EC/76 /4.6					
				α/4 /	7.57/4				
^{245}Es		245.0713	1.3 m	α/40 /7.858	7.74				
				EC/60 /3.1					
^{246}Es		246.0729	7.7 m	EC/90 /3.9					
				α/10 /	7.35				

Elem. or Isot.	Natural Abundance (Atom %)	Atomic Mass or Weight	Half-life/ Resonance Width (MeV)	Decay Mode/ Energy (/MeV)	Particle Energy/ Intensity (MeV/%)	Spin ($h/2\pi$)	Nuclear Magnetic Mom. (nm)	Elect. Quadr. Mom. (b)	γ-Energy / Intensity (MeV/%)
^{247}Es		247.07366	4.8 m	EC/93 /2.48					
				α/7 /	7.32				
^{248}Es		248.0755	26. m	EC/99.7 /3.1					
				α/0.3 /	6.87				
^{249}Es		249.07641	1.70 h	EC/99.4 /1.45		(7/2+)			0.3795
				α/0.6 /	6.77				0.8132
^{250m}Es			2.2 h	EC/		(1-)			Cf L x-ray
				β+					Cf k x-ray
									0.9891
									1.0319
^{250}Es		250.0786	8.6 h	EC/2.1		(6+)			Cf L x-ray
									Cf k x-ray
									0.30339
									0.34948
									0.82883
^{251}Es		251.07999	1.38 d	EC/99.5 /0.38		(3/2-)			
				α/0.5 /	6.462/0.05				
					6.492/0.4				
^{252}Es		252.08298	1.29 a	α/76 /	6.632/61.0	(5-)			
				EC/24 /1.26	6.562/10.3				
^{253}Es		253.084825	20.47 d	α/	6.633/89.8	7/2+	~ +4.10	7.	0.04180/5.6
			6.3 × 10^5 a	sf/8.9 × 10^{-6}	6.5916/6.6				0.3892/2.7
									(0.0389-1.106)
^{254m}Es			1.64 d	β- /99.6 /	0.475	2+	2.9	3.7	Fm L x-ray
				α/0.3 /6.67	6.382	2+			Fm k x-ray
			> 10. a	sf/0.045					0.6488
									0.6938
^{254}Es		254.088022	276. d	α/	6.429	(7+)			0.064
			> 2.5 × 10^7 a	sf/< 3 × 10^{-6}					
^{255}Es		255.09027	40. d	β- /92 /0.29		(7/2+)			
				α/8 /	6.26				
			2.6 × 10^3 a	sf/0.0042	6.300				
^{256m}Es			7.6 h	β- /		(8+)			0.218
									0.232
									0.862
^{256}Es		256.0936	25. m	β-/1.7		(1+)			
^{257}Es		257.0960	7.7 d	β-					
$_{100}$**Fm**									
^{242}Fm		242.0734	0.8 ms	sf/> 96		0+			
^{243}Fm		243.0744	0.2 s	α/	8.55				
				sf/< 0.4					
^{244}Fm		244.0741	3.3 ms	sf/> 97		0+			
^{245}Fm		245.0754	4. s	α/	8.15/				
				sf/<0.1					
^{246}Fm		246.07530	1.2 s	α/85/	8.24/	0+			
				sf/15/					
^{247m}Fm			5.1 s	α/	8.17/				
^{247}Fm		247.0769	30. s	α/8.20	7.87/70				(0.082-0.167)
				EC/2.9	7.93/30				
^{248}Fm		248.07720	33. s	α/99.9 /8.001	7.83/20	0+			
				sf/0.1/	7.87/80				
^{249}Fm		249.0790	1.6 m	EC/2.4		(7/2+)			
				α/	7.57				
^{250m}Fm			1.8 s	IT/					

Elem. or Isot.	Natural Abundance (Atom %)	Atomic Mass or Weight	Half-life/ Resonance Width (MeV)	Decay Mode/ Energy (/MeV)	Particle Energy/ Intensity (MeV/%)	Spin ($h/2\pi$)	Nuclear Magnetic Mom. (nm)	Elect. Quadr. Mom. (b)	γ-Energy / Intensity (MeV/%)
				sf/<8 × 10^{-5}					
^{250}Fm		250.07952	30. m	α/	7.43/	0+			
				EC/0.8					
				sf/0.007					
^{251}Fm		251.08158	5.3 h	EC/98 /1.47		(9/2-)			
				α/2 /	6.833				
^{252}Fm		252.08247	1.058 d	α/7.154	6.998/15	0+			
				sf/0.0023	7.039/85				
^{253}Fm		253.085185	3.0 d	EC/88/0.333	6.676/	½+			Es k x-ray
				α/12 /	6.943/				0.2719
^{254}Fm		254.086854	3.240 h	α/	7.150	0+			
				sf/0.059	7.192				
^{255}Fm		255.089962	20.1 h	α/	6.9635(5)/5.0	7/2+			0.08148/1.
			1.0 × 10^{4} a	sf/2.3 × 10^{-5}	7.0225(5)/93.4				(0.041-0.900)
^{256}Fm		256.09177	2.63 h	sf/91		0+			
				α/19	6.92/				
^{257}Fm		257.09511	100.5 d	α/99.79	6.519	(9/2+)			0.1794
				sf/0.21					0.2410
^{258}Fm		258.0971	0.37 ms	sf/		0+			
^{259}Fm		259.1006	1.5 s	sf/					
^{260}Fm		260.103	~ 4 ms	sf/		0+			
$_{101}$**Md**									
^{245m}Md			~ 0.4 s	α	8.64, 8.68				
^{245}Md		245.0808	0.9 ms	sf					
^{246m}Md			~ 4.4 s	Sf					
				α	8.18				
^{246}Md		246.0819	1.3 s	α	8.74				(0.169-0.396)
					8.50–8.56				
^{247m}Md			0.26 s	sf/23					
				α	8.78				
^{247}Md		247.0816	1.3 s	α	8.42				0.2096/100
									0.1575/11
^{248}Md		248.0828	7. s	EC/80 /5.3	8.32/15				
				α/20 /	8.36/5				
				sf/<0.05					
^{249}Md		249.0830	24. s	EC>/<80 /3.7					0.2532/100
				α/>20 /8.46	8.026(20)/				0.2004/20
									0.2232/10
^{250}Md		250.0844	50. s	EC/94 /4.6	7.75/4				
				α/6 /8.25	7.83/2				
^{251}Md		251.0848	4.3 m	EC/>94 /3.1	α/7.55/87	7/2-			0.293/100
				α/10	7.59/13				0.243/16
^{252}Md		252.0866	2. m	EC/>50 /3.9					
				α/<50 /	7.73/				
^{253}Md		253.0873	~ 6 m	EC/2.0	α/7.100				0.3532/100
^{254m}Md			30. m	EC/					
^{254}Md		254.0897	10. m	EC/2.7					
^{255}Md		255.09108	27. m	EC/92 /1.04	α/7.33/93	(7/2-)			0.121/100
				α/8 /	7.27/5				0.115/65
				sf/< 0.15	7.75/1				0.136/35
					7.71/1				0.141–0.453
^{256}Md		256.0941	1.30 h	EC/89 /2.13	7.21/71				Fm k x-ray
				α/11 /	7.14/22				0.121/409
				sf/< 2.6	7.68/2.5				0.115/266

Elem. or Isot.	Natural Abundance (Atom %)	Atomic Mass or Weight	Half-life/ Resonance Width (MeV)	Decay Mode/ Energy (/MeV)	Particle Energy/ Intensity (MeV/%)	Spin ($h/2\pi$)	Nuclear Magnetic Mom. (nm)	Elect. Quadr. Mom. (b)	γ-Energy / Intensity (MeV/%)
					7.25/2.5				0.136/143
					7.64/2.1				0.634/119
									0.141–1.37
^{257}Md		257.095541	5.5 h	EC/85 /0.41	7.074	(7/2-)			Fm k x-ray
				α/15, sf/< 1	7.014				(0.181–0.389)
^{258m}Md			57. m	EC/		(1-)			Fm k x-ray
				sf/< 30					
^{258}Md		258.098431	51.5 d	α/7.40	6.718(2)/	(8-)			0.3678
				sf/< 0.003	6.763(4)/				0.057–0.448
^{259}Md		259.1005	1.64 h	sf/>98.7		7/2+			
				α/<1.3					
^{260}Md		260.1037	~ 27.8 d	sf/ 73–100					
$_{102}$No									
^{248}No		248.0866	< 1.0 μs	sf		0+			
^{249}No		249.0878	0.05 ms	sf					
				α/ < 20					
^{250m}No			0.04 ms	Sf					
^{250g}No		250.0875	~ 0.005 ms	sf/		0+			
				α/ < 10					
^{251m}No			1.0 s	α	8.668/~98				
					8.625/~2				
^{251}No		251.0890	0.80 s	α/91	8.612/~98				
				sf/0.26	8.552/~1				
^{252m}No			0.11 s						0.710
^{252g}No		252.08898	2.44 s	α/75/8.551	8.42	0+			
				sf/24/	8.37				
				EC, β+/<1.6					
^{253m2}No			~ 0.97 ms						
^{253m1}No			31. μs	IT		5/2+			0.167
^{253}No		253.0907	1.7 m	α/	8.00	(9/2-)			0.222/100
				EC/3.2	α/8.28				(0.151-0.280)
^{254m2}No			0.17 ms						
^{254m1}No			0.27 s	I.T./					
				sf/< .2					
^{254}No		254.09096	49. s	α/	8.09	0+			0.102
				EC/1.1					0.152
				sf/0.17					
^{255}No		255.09324	3.1 m	α/62 /	8.095/58	½+			0.187
				EC/38/2.01	7.742/19				(0.163-0.358)
					7.903/11				
^{256}No		256.09428	2.9 s	α/	8.43	0+			
				sf/0.5					
^{257}No		257.09688	24.5 s	α/	8.222/83	(7/2+)			0.0770/100
				sf/<1.5	8.323/17				0.1018/57
					8.19/<4				0.1241/78
^{258}No		258.0982	~ 1.2 ms	sf/		0+			
^{259}No		259.1010	58. m	α/78 /7.794	7.52	(9/2+)			
				EC/22/0.5	7.55				
				sf/<9.7					
^{260}No		260.1026	0.11 s	sf/		0+			
^{262}No		262.1073	~ 8. ms	sf/		0+			
$_{103}$Lr									
^{251}Lr		251.0944	39 m	sf					
^{252}Lr		252.0954	~ 0.36 s	α	9.02/73				

Elem. or Isot.	Natural Abundance (Atom %)	Atomic Mass or Weight	Half-life/ Resonance Width (MeV)	Decay Mode/ Energy (/MeV)	Particle Energy/ Intensity (MeV/%)	Spin (h/2 π)	Nuclear Magnetic Mom. (nm)	Elect. Quadr. Mom. (b)	γ-Energy / Intensity (MeV/%)
				sf/<1	8.97/27				
^{253m}Lr			~ 0.57 s	α	8.79				
				sf/1.3					
^{253}Lr		253.0952	1.5 s	α/	8.72				
				sf/8					
^{254}Lr		254.0965	13. s	α/	8.45				
				EC/5.2					
				sf/<0.1					
^{255m}Lr			2.5 s	α	8.46				
^{255g}Lr		255.09669	29. s	α/	8.37/67	1/2-			
				EC/3.2	8.43/<3.6				
				sf/< 0.1	8.29/1.2				
^{256}Lr		256.0986	27. s	α/99.7 /8.554	8.43/				
				EC/4.2	8.39				
				sf/< 0.03					
^{257}Lr		257.0996	0.65 s	α/	8.80	7/2+			
				EC/2.5					
				sf/< 0.03					
^{258}Lr		258.1018	3.9 s	α/	8.60/46				
				EC/3.4	8.62/25				
				sf/< 5	8.56/20				
					8.65/9				
^{259}Lr		259.1029	6.1 s	α/80	8.44(1)				
				sf/20					
^{260}Lr		260.1055	3. m	α/	8.03				
^{261}Lr		261.1069	40. m	sf					
^{262}Lr		262.1096	3.6 h	EC/2.					
				sf/<10					
$_{104}$**Rf**									
^{253}Rf		253.1007	~ 48. μs	sf					
				α/<10					
^{254}Rf		254.1002	23. μs	sf/>98.5		0+			
				α/<1.5					
^{255}Rf		255.1013	1.6 s	α	8.716/92				0.2036/49
				sf/52	8.678/3				0.1433/51
					8.906/2.5				
					8.646/1.5				
					8.575/1				
^{256}Rf		256.10117	6.2 ms	sf/99.68		0+			
				α/0.32	8.81				
^{257}Rf		257.1030	4.7 s	α/9.22	8.77				0.117
				EC/11	9.01				
				sf/<1.4	8.95				
					8.62				
^{258}Rf		258.1035	12. ms	sf/87		0+			
				α/13					
^{259}Rf		259.1056	2. s	α/9.09/93	8.77(2)/				
				sf/7	8.86/				
^{260}Rf		260.1064	20. ms	sf/		0+			
^{261m}Rf			1.3 m	α	8.28				
^{261}Rf		261.10877	5. s	α/60, sf/40	8.52/				
^{262}Rf		262.1099	2.1 s	sf/>99.2		0+			
^{263}Rf		263.1126	~ 24. m	sf, α					
^{265}Rf		265.1167	~ 13 h	α					

Elem. or Isot.	Natural Abundance (Atom %)	Atomic Mass or Weight	Half-life/ Resonance Width (MeV)	Decay Mode/ Energy (/MeV)	Particle Energy/ Intensity (MeV/%)	Spin (h/2 π)	Nuclear Magnetic Mom. (nm)	Elect. Quadr. Mom. (b)	γ-Energy / Intensity (MeV/%)
^{267}Rf		267.122	~ 1. h	sf					
$_{105}$Db									
^{255}Db		255.1074	~ 1.5 s	α,					
				sf/~ 20					
^{256}Db		256.1081	1.6 s	α/64	9.02/67				
				EC/35	8.89/11				
				sf/0.05	9.08/11				
					9.12/11				
^{257m}Db			0.8 s	α	9.16				
				sf/<13					
^{257}Db		257.1077	1.5 s	α/	8.97/33				
				sf/<6	9.07/38				
					9.12/5.5				
					8.94/9				
					9.02/9				
					8.89/5.5				
^{258}Db		258.1092	4.2 s	α/	9.30/				
				EC/5.3/39	9.17/				
				sf/<33	9.08/				
^{259}Db		259.1096	~ 0.51 s	sf/					
				α/	9.47/				
^{260m}Db			0.3 m						
^{260}Db		260.1113	1.5 s	α/	9.05/				
				sf/<9.6	9.08/				
					9.13/				
^{261}Db		261.1121	1.8 s	α/	8.93/				
				sf/<18					
^{262}Db		262.1141	0.5 m	sf/<33					
				α/	8.45/				
					8.53/				
					8.67/				
^{263}Db		263.1150	~ 0.45 m	sf/57	8.36/				
				α/41	8.41/				
				EC/3					
^{266}Db			~ 0.4 h	EC,sf					
^{267}Db		267.1224	1.2 h	sf					
^{268}Db		268.125	1.2 d	sf, EC					
$_{106}$Sg									
^{258}Sg		258.1132	~ 2.9 ms	sf		0+			
				α/<20					
^{259}Sg		259.1145	0.5 s	α/	9.62				
				sf/<20	9.35				
					9.03				
^{260}Sg		260.11442	4. ms	α/50	9.76	0+			
				sf/50	9.72				
					9.81				
^{261}Sg		261.1161	0.3 s	α, sf/<10	9.56				0.107
^{262}Sg		262.1164	15. ms	sf		0+			
				α/<16					
^{263m}Sg			0.3 s	α	9.2				
^{263}Sg		263.1183	0.8 s	α	9.06				
				sf/~13	9.25				
^{264}Sg			~ 0.04 s	α/<36					
^{265}Sg		265.1211	~ 15. s	α/>65	8.84/46				

Elem. or Isot.	Natural Abundance (Atom %)	Atomic Mass or Weight	Half-life/ Resonance Width (MeV)	Decay Mode/ Energy (/MeV)	Particle Energy/ Intensity (MeV/%)	Spin (h/2π)	Nuclear Magnetic Mom. (nm)	Elect. Quadr. Mom. (b)	γ-Energy / Intensity (MeV/%)
				sf/<35	8.76/23				
					8.94/23				
					8.69/8				
^{266}Sg		266.1221	~ 0.4. s	Sf		0+			
^{271}Sg		271.133	2. m	α/70	~ 8.54				
				sf/30					
$_{107}$**Bh**									
^{260}Bh		260.122		α					
^{261}Bh		261.1217	12. ms	α/, sf <10	10.40				
					10.10				
					10.03				
^{262m}Bh			8. ms	α/	10.37				
				sf/<24	10.24				
^{262}Bh		262.1229	~ 0.09 s	α/	10.06				
				sf/<11	9.91				
					9.74				
^{264}Bh		264.1246	1.0 s	α/	9.3 – 9.8				
				sf/					
^{265}Bh		265.1252	0.9 s	α	9.24				
^{266}Bh		266.1269	~ 2 s	α	9.08				
^{267}Bh		267.1277	~ 17 s	α	8.83				
^{270}Bh			~ 10. s	α	9.0				
^{272}Bh			~ 1. m	α	8.9				
$_{108}$**Hs**									
^{263}Hs		263.1286		α/					
^{264}Hs		264.12839	~ 0.08 ms	α/, sf/~ 50	11.0	0+			
^{265m}Hs			~ 0.75 ms	α	10.57/63				
					10.73				
					10.52				
					10.34				
^{265}Hs		265.1301	2.0 ms	α/	10.30/90				
				sf/<1	10.43				
					10.37				
					10.25				
^{266}Hs		266.1301	~ 2.3 ms	α	10.2	0+			
^{267m}Hs			~ 0.8 s						
^{267}Hs		267.1318	0.05 s	α/>88	9.88				
					9.83				
					9.75				
^{269}Hs		269.1341	~ 10. s	α	9.07				
					8.92				
^{270}Hs		270.1347	~ 3.6 s	α	9.16	0+			
^{275}Hs		275.146	~ 0.19 s	α	9.3				
^{277}Hs		277.150	~ 11 m	sf					
$_{109}$**Mt**									
^{266m}Mt			~ 1.2 ms	α	10.46–10.81				
^{266}Mt		266.1373	~ 0.7 ms	α	10.48–11.31				
^{267}Mt		267.137	19 ms	α					
^{268}Mt		268.1387	~ 0.03 s	α/>68	10.3 - 10.8				
^{270}Mt		270.141	5 ms	α	10.0				
^{274}Mt			~ 0.4 s	α	9.8				
^{275}Mt		275.149	~ 9.7 ms	α	10.3				

Elem. or Isot.	Natural Abundance (Atom %)	Atomic Mass or Weight	Half-life/ Resonance Width (MeV)	Decay Mode/ Energy (/MeV)	Particle Energy/ Intensity (MeV/%)	Spin ($h/2\pi$)	Nuclear Magnetic Mom. (nm)	Elect. Quadr. Mom. (b)	γ-Energy / Intensity (MeV/%)
^{276}Mt		276.151	~ 0.7 s	α	~ 9.71				
$_{110}$**Ds**									
^{267}Ds		267.1443	~ 3 μs	α/>32	11.6				
^{269}Ds		269.1451	0.17 ms	α/>75	11.11				
^{270m}Ds			~ 6 ms	α	10.95				
					11.15				
					12.15				
^{270}Ds		270.1447	0.1 ms	α	11.03	0+			
^{271m}Ds			0.07 s	α	9.9				
^{271}Ds		271.1461	1.6 ms	α	10.8				
^{273m}Ds			0.076 ms	α	11.8				
^{273}Ds		273.1489	118 ms	α/	9.73				
^{279}Ds		279.159	0.20 s	sf/90					
				α/10	~ 9.70				
^{280}Ds		280.160	~ 7.6 s	sf/		0+			
^{281}Ds		281.162	11. s	sf					
^{282}Ds			0.5 ms	sf					
$_{111}$**Rg**									
^{272}Rg		272.1536	~ 2 ms	α/>68	10.82				
^{274}Rg		274.156	~ 65 ms	α	11.2				
^{278}Rg			~ 4. ms	α	10.7				
^{279}Rg		279.162	~ 0.17 s	α	10.4				
^{280}Rg		280.164	3.6 s	α	~ 9.75				
$_{112}$**Cn**									
^{277}Cn		277.1639	~ 0.7 ms	α	11.45				
					11.65				
^{282}Cn			0.8 ms	Sf					
^{283}Cn		283.172	4. s	sf/< 10					
				α/~ 100	9.52				
^{284}Cn		284.172	~ 97. ms	sf					
^{285}Cn		285.174	~ 29. s	α	9.15				
$_{113}$**113**									
^{278}113			~ 3. ms	α	11.7				
^{282}113			~ 0.07 s	α	10.6				
^{283}113		283.176	~ 0.1 s	α	10.1				
^{284}113		284.178	~ 0.48 s	α	10.0				
$_{114}$**114**									
^{286}114		286.184	0.13 s	α/50	10.2				
				sf/50					
^{287}114		287.186	0.5 s	α	10.0				
^{288}114		288.186	0.8 s	α	9.94				
^{289}114		289.187	~ 2.6 s	α	9.82				
$_{115}$**115**									
^{287}115		287.191	~ 0.03 s	α	10.6				
^{288}115		288.192	~ 0.09 s	α	10.5				
$_{116}$**116**									
^{290}116		290.199	7. ms	α	10.8				
^{291}116		291.200	~ 18. ms	α	~ 10.74				
^{292}116		292.200	~ 18. ms	α	~ 10.66				
^{293}116			0.06 s	α	10.5				
$_{118}$**118**									
^{294}118			~ 0.89 ms	α	11.7				

NEUTRON SCATTERING AND ABSORPTION PROPERTIES

Norman E. Holden

This table presents an evaluated set of values for experimental quantities that characterize the properties for scattering and absorption of neutrons. The neutron cross section is given for room temperature neutrons, 20.43 °C, corresponding to a thermal neutron energy of 0.0253 electron volts (eV) or a neutron velocity of 2200 meters/second. The neutron resonance integral is defined over the energy range from 0.5 eV to 0.1×10^6 eV, or 0.1 MeV.

Bound neutron scattering lengths and neutron cross sections averaged over a Maxwellian spectrum at 30 keV for astrophysical applications are also presented. A list of the major references used is given below. The literature cutoff date is January 2003. Uncertainties are given in parentheses. Parentheses with two or more numbers indicate values to the excited state(s) and to the ground state of the product nucleus.

Table Layout

Column Number	Column Title	Description
1	Isotope/Element	For elements, atomic number and chemical symbol are listed. For nuclides, mass number and chemical symbol are listed. Isomers are indicated by the addition of m, m1, or m2.
2	Isotopic Abundance	In atom percent.
3	Half-life	Half-life in decimal notation. μs = microsecond; ms = millisecond; s = second; m = minute; h = hour; d = day; y = year.
4	Thermal Neutron Cross Sections	Cross sections for neutron capture reactions in units of barns (10^{-24} cm^2) or millibarns (mb). Proton, alpha production and fission reactions are designated by σ_p, σ_α, σ_f, respectively. Separate values are listed for isomeric production.
5	Neutron Resonance Integrals	Resonance integrals for neutron capture reactions in barns (10^{-24} cm^2) or millibarns (mb). Proton, alpha production and fission reactions are designated by $R.I._p$, $R.I._\alpha$, $R.I._f$, respectively. Separate values are listed for isomeric production.
6	Neutron Scattering Lengths	Bound coherent scattering lengths for neutron scattering reactions in units of femtometers (fm), which is equal to fermis (10^{-13} cm).
7	Maxwellian Averaged Cross Section	Astrophysical cross sections, averaged over a stellar neutron maxwellian spectrum characterized by a thermal energy of 30 keV, expressed in barns (10^{-24} cm^2), millibarns (mb) or microbarns (μb).

General Nuclear Data References

The following references represent the major sources of the nuclear data presented:

Mughabghab, S.F., Divadeenam, M., Holden, N.E.; Neutron Cross Sections, Vol. 1 *Neutron Resonance Parameters and Thermal Cross Sections*, Part A, *Z* = 1-60. Academic Press Inc., New York, New York (1981); Mughabghab, S.F.; Part B, *Z* = 61-100. Academic Press Inc., Orlando, FL (1984).

Holden, N.E.; *Fifty Years with Nuclear Fission* Conference, Wash., D.C., Gaithersburg, Md. April 26-29, 1989, p. 946. American Nuclear Society, LaGrange Park, IL (1989).

Tuli, J.K.; *Nuclear Wallet Cards*, Brookhaven National Laboratory (Jan. 2000).

Holden, N.E.; Half-lives of Selected Nuclides, *Pure & Applied Chemistry* 62, 941 (1990).

Holden, N.E., Hoffman, D.C.; Spontaneous Fission Half-lives for Ground State Nuclides, *Pure & Applied Chemistry* 72, 1525 (2000).

Koester, L., Rauch, H., Seymann, E.; Neutron Scattering Lengths: A Survey of Experimental Data and Methods, *Atomic Data Nuclear Data Tables* 49, 65 (1991).

Sears, V.F.; Neutron Scattering Lengths and Cross Sections, *Neutron News* 3, (3), 26 (1992).

Bao, Z.Y., Beer, H., Käppeler, F., Voss, F., Wisshak, K., Raucher, T.; Neutron Cross Sections for Nucleo-synthesis Studies, *Atomic Data Nuclear Data Tables* 76, 70 (2000).

Elem. or Isot.	Natural Abundance (%)	Half-Life	Thermal Neut. Cross Section (barns)	Resonance Integral (barns)	Coh. Scat. Length (fm)	σ (30 keV) Maxw. Avg. (barns)
$_1$H			0.332(2)	0.149(1)	-3.739(1)	
^{1}H	99.9885(70)	>2.8×10^{23} y	0.332(2)	0.149(1)	- 3.741(1)	0.25(2) mb*
^{2}H	0.0115(70)		0.51(1)mb	0.23(2) mb	6.671(4)	2.1(4) μb
^{3}H		12.33 y	< 6. μb		4.79(3)	
$_2$He			< 0.05		3.26(3)	
^{3}He	0.000134(3)		σ_p = 5.33(1)×10^3	RI_p = 2.39(1)×10^3	5.74(7)	
			0.05(1) mb			8.(1) μb*
^{4}He	99.999867(3)				3.26(3)	
$_3$Li			71.(2)	32.(1)	- 1.90(2)	
^{6}Li	7.59(4)		σ_t = 9.4(1)×10^2	RI_t = 422.(4)	2.0(1)	σ_t ≈ 1.

*Extrapolated value.

Elem. or Isot.	Natural Abundance (%)	Half-Life	Thermal Neut. Cross Section (barns)	Resonance Integral (barns)	Coh. Scat. Length (fm)	σ (30 keV) Maxw. Avg. (barns)
			39.(5) mb	17.(2) mb		0.06(1) mb*
^{7}Li	92.41(4)		45.(5) mb	20.(2) mb	- 2.22(2)	42.(3) μb
^{8}Li		0.84 s				< ≈ 5.5 μb
$_{4}$Be			8.8(4) mb	3.9(2) mb	7.79(1)	
^{7}Be		53.28 d	$\sigma_p = 3.9(1)\times10^4$	$RI_p = 1.75(5)\times10^4$		$\sigma_p = 16(4)^*$
			$\sigma_\alpha \approx 0.1$			
^{9}Be	100.		8.8(4) mb	3.9(2) mb	7.79(1)	
^{10}Be		1.52×10^6 y	<1. mb			
$_{5}$B			$7.6(1)\times10^2$	$3.4(1)\times10^2$	5.30(4)	
^{10}B	19.9(7)		$\sigma_\alpha = 38.4(1)\times10^2$	$RI_\alpha = 17.3(1)\times10^2$	- 0.1(3)	
			0.3(1)	0.13(4)		
			σ_p = 7.(1) mb			
			σ_t = 8.(2) mb			
^{11}B	80.1(7)		5.(3) mb	2.(1) mb	6.65(4)	
$_{6}$C			3.5(1) mb	1.6(1) mb	6.646(1)	
^{12}C	98.93(8)		3.5(1) mb	1.6(1) mb	6.651(2)	16.(1) μb*
^{13}C	1.07(8)		1.4(1) mb	1.7(2) mb	6.19(9)	0.021(4) mb
^{14}C		5715. y	<1.4 μb			3.(1) μb*
$_{7}$N			2.00(6)	0.90(3)	9.36(2)	
^{14}N	99.636(20)		$\sigma_p = 1.93(5)$	$RI_p = 0.87(3)$	9.37(2)	σ_p – 1.8(2) mb*
			0.080(1)	0.034(1)		0.04(1) mb
^{15}N	0.364(20)		0.04(1) mb	0.11(3) mb	6.44(3)	6.(1) μb*
$_{8}$O			0.29(1) mb	0.40(4) mb	5.805(4)	
^{16}O	99.757(16)		0.19(1) mb	0.36(4) mb	5.805(5)	34.(4) μb
^{17}O	0.038(1)		$\sigma_\alpha = 0.257(10)$	0.11(1)	5.8(2)	σ_α = 3.9(5) mb*
			0.54(7) mb	0.39(5) mb		
^{18}O	0.205(14)		0.16(1) mb	0.81(4) mb	5.84(7)	9.(1) μb*
$_{9}$F			9.5(1) mb	21.(3) mb	5.65(1)	6.(1) mb
^{19}F	100.		9.5(1) mb	21.(3) mb	5.65(1)	6.(1) mb
$_{10}$Ne			42.(5) mb	19.(3) mb	4.566(6)	
^{20}Ne	90.48(3)		39.(5) mb	18.(3) mb	4.631(6)	0.12(1) mb
^{21}Ne	0.27(1)		0.7(1)	0.31(5)	6.7(2)	≈ 1.5 mb
			σ_α = 0.18(9) mb			
^{22}Ne	9.25(3)		51.(5) mb	23.(3) mb	3.87(1)	58.(4) μb*
$_{11}$Na			0.53(2)	0.32(2)	3.63(2)	2.1(2) mb
^{22}Na		2.605 y	$\sigma_p = 2.8(3)\times10^4$	$RI_p < 2.\times10^5$		
			$\sigma_\alpha = 2.6(4)\times10^2$	$RI_\alpha = 1.2(2)\times10^2$		
^{23}Na	100.		$\sigma_m = 0.43(3)$	$RI_m = 0.30(6)$	3.63(2)	2.1(2) mb
$_{12}$Mg			66.(6) mb	38.(5) mb	5.375(4)	
^{24}Mg	78.99(4)		0.053(6)	32.(4) mb	5.7(2)	3.3(4) mb
^{25}Mg	10.00(1)		0.20(1)	98.(15) mb	3.6(2)	6.4(4) mb
^{26}Mg	11.01(3)		0.038(1)	25.(2) mb	4.9(2)	0.13(1) mb*
^{27}Mg		9.45 m	0.07(2)	0.03(1)		
$_{13}$Al			0.230(2)	0.17(1)	3.45(1)	
^{26}Al		7.1×10^5 y	$\sigma_p = 1.97(10)$			0.14(2)
			$\sigma_\alpha = 0.34(1)$			
^{27}Al	100.		0.230(2)	0.17(1)	3.45(1)	2.9(3) mb
$_{14}$Si			0.166(9)	0.12(2)	4.15(1)	
^{28}Si	92.223(19)		0.17(1)	0.11(2)	4.11(1)	2.9(3) mb
^{29}Si	4.685(8)		0.12(1)	0.08(2)	4.7(1)	7.9(9) mb
^{30}Si	3.092(11)		0.107(3)	0.62(6)	4.61(1)	3.2(3) mb*
^{31}Si		2.62 h	73.(6) mb	33.(3) mb		
^{32}Si		1.6×10^2 y	< 0.5			
$_{15}$P			0.17(1)	0.08(1)	5.13(1)	
^{31}P	100.		0.17(1)	0.08(1)	5.13(1)	1.7(1) mb
$_{16}$S			0.54(2)	0.24(2)	2.847(1)	
^{32}S	94.93(31)		0.55(5)	0.25(2)	2.804(2)	4.1(2) mb
			σ_α < 0.5 mb			

*Extrapolated value.

Elem. or Isot.	Natural Abundance (%)	Half-Life	Thermal Neut. Cross Section (barns)	Resonance Integral (barns)	Coh. Scat. Length (fm)	σ (30 keV) Maxw. Avg. (barns)
^{33}S	0.76(2)		0.46(3)	0.21(2)	4.7(2)	7.4(15) mb
			σ_α = 0.12(1)	RI_α = 0.05(1)		σ_α = 0.18(1)
			σ_p = 2. mb			
^{34}S	4.29(28)		0.25(1)	0.13	3.48(3)	0.23(1) mb
^{36}S	0.02(1)		0.24(2)	0.26(3)		0.17(1) mb*
$_{17}Cl$			33.6(3)	15.(2)	9.58(1)	
^{35}Cl	75.78(4)		43.7(4)	20.(2)	11.7(1)	9.4(3) mb
			σ_p = 0.44(1)	RI_p = 0.2		σ_p = 1.7(2) mb*
			σ_α≈ 0.08 mb			
^{36}Cl		3.01×10^5 y	σ_p = 46.(2) mb	RI_p = 0.02		σ_p = 91.(8) mb
			<10.			
			σ_α = 0.59(7) mb			σ_α = 0.9(2) mb
^{37}Cl	24.22(4)		(0.05 + 0.38)	(0.04+0.26)	3.1(1)	2.0(2) mb
$_{18}Ar$			0.66(3)	0.42(5)	1.91(1)	
^{36}Ar	0.3365(30)		5.(1)	2.(1)	24.9(1)	
			σ_α = 5.4(3) mb			
			σ_p < 1.5 mb			
^{37}Ar		35.0 d	σ_α = 1.08(8)×10^3	RI_α = 900.		σ_α ≈ 1.3
			σ_p = 37.(4)	RI_p = 31.		σ_p ≈ 0.04
^{38}Ar	0.0632(5)		0.8(2)	0.4(1)	3.5(35)	
^{39}Ar		268. y	6.(2)×10^2			
			σ_α <0.29			
^{40}Ar	99.6003(30)		0.64(3)	0.41(5)	1.83(1)	2.5(3) mb
^{41}Ar		1.82 h	0.5(1)	0.2(1)		
$_{19}K$			2.1(1)	1.0(1)	3.67(2)	
^{39}K	93.2581(44)		2.1(2)	0.9(1)	3.74(2)	11.8(4) mb
			σ_α = 4.3(5) mb			
			σ_p < 0.05 mb			
^{40}K	0.0117(1)	1.26×10^9 y	30.(8)	13.(4)		σ_p = 7.(1) mb
			σ_p = 4.4(4)	2.0(2)		σ_α = 40.(6) mb
			σ_α = 0.42(8)			
^{41}K	6.7302(44)		1.46(3)	1.4(2)	2.69(8)	22.(1) mb
$_{20}Ca$			0.43(2)	0.23(2)	4.70(2)	
^{40}Ca	96.941(156)		0.41(3)	0.22(4)	4.80(2)	6.7(7) mb
			σ_α = 0.13(4) mb			
^{41}Ca		1.02×10^5 y	≈ 4.			
			σ_α = 0.18(3)			
			σ_p = 7.(2) mb			
^{42}Ca	0.647(23)		0.65(10)	0.39(4)	3.4(1)	16.(2) mb
^{43}Ca	0.135(10)		6.(1)	3.9(2)	- 1.56(9)	51.(6) mb
^{44}Ca	2.086(110)		0.8(2)	0.56(1)	1.42(6)	9.(1) mb
^{45}Ca		162.7 d	≈ 15.			
^{46}Ca	0.004(3)	>4×10^{15} y	0.70(3)	0.9(1)	3.6(2)	5.3(5) mb*
^{48}Ca	0.187(21)	4.3×10^{19} y	1.0(1)	0.5(1)	0.39(9)	0.8(1) mb*
$_{21}Sc$			27.2(2)	12.(1)	12.3(1)	
^{45}Sc	100.		(10.+17.)	(5.6+6.4)	12.3(1)	69.(5) mb
^{46}Sc		83.81 d	8.(1)	3.6(5)		
$_{22}Ti$			6.1(1)	2.8(2)	- 3.438(2)	
^{44}Ti		60 y	1.1(2)			
			σ_p<0.2			
^{46}Ti	8.25(3)		0.6(2)	0.4(1)	4.93(6)	27.(3) mb
^{47}Ti	7.44(2)		1.6(2)	1.6(2)	3.63(1)	64.(8) mb
^{48}Ti	73.72(3)		7.9(9)	3.6(2)	- 6.09(2)	32.(5) mb
^{49}Ti	5.41(2)		1.9(5)	1.2(2)	1.04(5)	22.(2) mb
^{50}Ti	5.18(2)		0.179(3)	0.12(2)	6.18(8)	3.6(4) mb
$_{23}V$			5.0(2)	2.8(1)	- 0.382(1)	
^{50}V	0.250(4)	1.4×10^{17} y	21.(4)	50.(20)	7.6(6)	
			σ_p = 0.7(4) mb			

*Extrapolated value.

Elem. or Isot.	Natural Abundance (%)	Half-Life	Thermal Neut. Cross Section (barns)	Resonance Integral (barns)	Coh. Scat. Length (fm)	σ (30 keV) Maxw. Avg. (barns)
^{51}V	99.750(4)		4.9(1)	2.7(2)	- 0.402(2)	38.(4) mb
$_{24}$Cr			3.0(2)	1.7(1)	3.635(7)	
^{50}Cr	4.345(13)	>1.8×10^{17} y	15.(1)	8.(1)	- 4.5(1)	0.05(1)
^{51}Cr		27.70 d	< 10.			
^{52}Cr	83.789(18)		0.8(1)	0.6(2)	4.91(2)	8.8(4) mb
^{53}Cr	9.501(17)		18.(2)	9.(1)	- 4.2(1)	0.06(1)
^{54}Cr	2.365(7)		0.36(4)	0.25(5)	4.6(1)	7.(2) mb
$_{25}$Mn			13.3(1)	14.0(3)	- 3.75(2)	
^{53}Mn		3.7×10^{6} y	70.(10)	32.(5)		
^{54}Mn		312.1 d	< 10.			
^{55}Mn	100.		13.3(1)	14.0(3)	- 3.75(2)	40.(3) mb
$_{26}$Fe			2.7(1)	1.4(2)	9.45(2)	
^{54}Fe	5.845(35)		2.3(2)	1.3(2)	4.2(1)	29.(2) mb
			σ_α = 10. µb	RI$_\alpha$ = 1.1(1) mb		
^{55}Fe		2.73 y	13.(2)	6.(1)		
			σ_α = 0.01			
^{56}Fe	91.754(36)		2.8(3)	1.4(2)	9.93(3)	11.7(5) mb
^{57}Fe	2.119(10)		1.4(2)	0.8(4)	2.3(1)	40.(4) mb
^{58}Fe	0.282(4)		1.3(1)	1.3(2)	15.(7)	12.(1) mb
^{59}Fe		44.51 d	13.(3)	6.(1)		
$_{27}$Co			37.19(8)	74.(2)	2.49(2)	
^{58m}Co		9.1 h	1.4(1)×10^{5}	2.5(10)×10^{5}		
^{58}Co		70.88 d	1.9(2)×10^{3}	7.(1)×10^{3}		
^{59}Co	100.		(20.7+16.5)	(39.+35.)	2.49(2)	38.(4) mb
^{60m}Co		10.47 m	58.(3)	230.(50)		
^{60}Co		5.271 y	2.0(2)	4.3(10)		
$_{28}$Ni			4.5(2)	2.3(2)	10.3(1)	
^{58}Ni	68.0769(89)	>4×10^{19} y	4.6(4)	2.3(2)	14.4(1)	41.(2) mb
			σ_α < 0.03 mb			
^{59}Ni		≈ 7.6×10^{4} y	σ_{abs} = 92.(4)	RI$_{abs}$ = 1.4(1)×10^{2}		
			σ_α = 14.(2)			
			σ_p = 2.(1)			
^{60}Ni	26.2231(77)		2.9(3)	1.5(2)	2.8(1)	25.(1) mb
^{61}Ni	1.1399(6)		2.5(5)	1.5(4)	7.60(6)	82.(8) mb
			σ_α = 0.03 mb			
^{62}Ni	3.6345(17)		15.(1)	6.8(3)	- 8.7(2)	13.(4) mb
^{63}Ni		100. y	20.(5)	9.(2)		
^{64}Ni	0.9256(9)		1.6(1)	1.2(2)	- 0.37(7)	9.(1) mb
^{65}Ni		2.517 h	22.(2)	10.(1)		
$_{29}$Cu			3.8(1)	4.1(4)	7.718(4)	
^{63}Cu	69.15(15)		4.5(2)	5.(1)	6.43(15)	0.09(1)
^{64}Cu		12.701 h	≈ 270.			
^{65}Cu	30.85(15)		2.17(3)	2.2(1)	10.61(19)	41.(5) mb
^{66}Cu		5.09 m	1.4(1)×10^{2}	60.(20)		
$_{30}$Zn			1.1(2)	2.8(4)	5.680(5)	
^{64}Zn	48.27(32)	>2.3×10^{18} y	0.74(5)	1.4(3)	5.23(4)	59.(5) mb
			σ_p < 12. µb			
			σ_α = 11.(3) µb			
^{65}Zn		243.8 d	66.(8)	30.(4)		
			σ_α = 2.0(2)			
^{66}Zn	27.977(77)		0.9(3)	1.8(2)	5.98(5)	35.(3) mb
			σ_α < 0.02 mb			
^{67}Zn	4.102(21)		6.9(1.4)	25.(5)	7.58(8)	0.15(2)
			σ_α = 0.4 mb			
^{68}Zn	19.02(12)		(0.072 + 0.8)	(0.2 + 2.9)	6.04(3)	19.(2) mb
			σ_α <0.02 mb			σ_m = 3.(1) mb
^{70}Zn	0.631(9)		(8.1+83.) mb	0.9(2)		0.02(1)
$_{31}$Ga			2.9(1)	22.(3)	7.288(2)	

*Extrapolated value.

Elem. or Isot.	Natural Abundance (%)	Half-Life	Thermal Neut. Cross Section (barns)	Resonance Integral (barns)	Coh. Scat. Length (fm)	σ (30 keV) Maxw. Avg. (barns)
^{69}Ga	60.108(9)		1.68(7)	16.(2)	7.88(4)	0.14(1)
^{71}Ga	39.892(9)	>2.4×10^{26} y	4.7(2)	31.(3)	6.40(3)	0.12(1)
			σ_m = 0.15(5)			
$_{32}$Ge			2.2(1)	6.(2)	8.19(2)	
^{68}Ge		270.8 d	1.0(5)			
^{70}Ge	20.370(89)		(0.3 + 2.7)	2.3(1)	10.0(1)	88.(5) mb
^{72}Ge	27.380(60)		0.9(2)	0.8(3)	8.5(1)	0.07(2)
^{73}Ge	7.759(78)	>1.8×10^{23} y	15.(1)	66.(20)	5.02(4)	0.3(1)
^{74}Ge	36.656(80)		(0.14 + 0.28)	(0.4+0.5)	7.6(1)	53.(7) mb
^{76}Ge	7.835(81)	1.6×10^{21} y	(0.09 + 0.06)	(1.3+0.6)	8.2(15)	0.03(2)
$_{33}$As			4.0(4)	61.(5)	6.58(1)	
^{75}As	100.		4.0(4)	61.(5)	6.58(1)	0.57(4)
$_{34}$Se			12.(1)	14.(3)	7.970(9)	
^{74}Se	0.89(4)		50.(2)	520(50)	0.8(3)	0.2(1)
^{75}Se		119.78 d	3.3(10)×10^{2}			
^{76}Se	9.37(29)		(22. + 63.)	(9.+31.)	12.2(1)	0.16(1)
^{77}Se	7.63(16)		42.(4)	30.(5)	8.25(8)	0.3(1)
			σ_α = 0.97(3) μb			
^{78}Se	23.77(28)		σ_m = 0.38(2)	RI_m = 4.3(4)	8.24(9)	0.1
^{80}Se	49.61(41)		(0.05+0.54)	(0.15+0.85)	7.48(3)	42.(3) mb
^{82}Se	8.73(22)	≈ 1×10^{20} y	(39.+ 5.2) mb	39.(4) mb	6.34(8)	0.04(2)
$_{35}$Br			6.8(2)	92.(8)	6.79(2)	
^{76}Br		16.0 h	224.(42)			
^{79}Br	50.69(7)		(2.5+8.3)	(36.+96.)	6.79(7)	0.63(4)
						σ_m = 0.08(1)
^{81}Br	49.31(7)		(2.4+0.24)	51.(5)	6.78(7)	0.31(2)
$_{36}$Kr			24.(1)	39.(6)	7.81(2)	
^{78}Kr	0.353(3)	>2.3×10^{20} y	(0.17+6.)	20.(1)		(0.11+0.19)
^{80}Kr	2.286(10)		(4.6+7.)	57.(6)		(0.09+0.18)
^{82}Kr	11.593(3)		(14.+7.)	130.(13)		90.(6) mb
^{83}Kr	11.500(19)		183.(30)	183.(20)		0.24(2)
^{84}Kr	56.987(15)		(σ_m+ σ_g) = 0.11	2.4(3)		(16.+33.) mb
			σ_m = 0.09			
^{85}Kr		10.73 y	1.7(2)	1.8(10)		0.07(2)
^{86}Kr	17.279(41)		3.(2) mb	≈ 1. mb	8.1(3)	3.2(4) mb
$_{37}$Rb			0.39(4)	6.(3)	7.08(2)	
^{84}Rb		32.9 d	σ_p = 12.(2)			
^{85}Rb	72.17(2)		(0.06+0.38)	(0.7+7.)	7.0(1)	0.24(1)
^{86}Rb		18.65 d	<20.			
^{87}Rb	27.83(2)	4.88×10^{10} y	0.10(1)	2.3(4)	7.3(1)	16.(1) mb
^{88}Rb		17.7 m	1.2(3)	0.5(1)		
$_{38}$Sr			1.2(1)	10.(1)	7.02(2)	
^{84}Sr	0.56(1)		(0.6+0.2)	(9.+1.)		0.4(1)
^{86}Sr	9.86(1)		σ_m = 0.81(4)	RI_m = 4.(1)	5.68(5)	(48.+22.) mb
^{87}Sr	7.00(1)		16.(3)	118.(30)	7.41(7)	97.(5) mb
^{88}Sr	82.58(1)		5.8(4) mb	0.07(3)	7.16(6)	6.0(2) mb
^{89}Sr		50.52 d	0.42(4)	0.2		
^{90}Sr		29.1 y	10.(1) mb	0.10(2)		
$_{39}$Y			1.25(5)	1.0(1)	7.75(2)	
^{89}Y	100.		(0.001+1.25)	(0.006+1.0)	7.75(2)	19.(1) mb
^{90}Y		2.67 d	<6.5			
^{91}Y		58.5 d	1.4(3)	0.6(1)		
$_{40}$Zr			0.19(1)	0.95(9)	7.16(3)	
			σ_α <0.1 mb			
^{90}Zr	51.45(40)		≈ 0.014	0.2(1)	6.4(1)	21.(2) mb
^{91}Zr	11.22(5)		1.2(3)	5.(2)	8.8(1)	60.(8) mb
^{92}Zr	17.15(8)		0.2(1)	0.6(2)	7.5(2)	33.(4) mb
^{93}Zr		1.5×10^{6} y	<4.	16.(5)		0.10(1)

*Extrapolated value.

Elem. or Isot.	Natural Abundance (%)	Half-Life	Thermal Neut. Cross Section (barns)	Resonance Integral (barns)	Coh. Scat. Length (fm)	σ (30 keV) Maxw. Avg. (barns)
^{94}Zr	17.38(28)	>10^{17} y	0.049(6)	0.25(3)	8.3(2)	26.(1) mb
^{96}Zr	2.80(9)	>1.7×10^{18} y	0.020(3)	5.0(5)	5.5(1)	11.(1) mb
$_{41}$Nb			1.11(1)	8.5(6)	7.14(3)	
			σ_α <0.1 mb			
^{93}Nb	100.		1.1	(6.3+2.2)	7.14(3)	266.(5) mb
			σ_m = 0.86			
^{94}Nb		2.4×10^4 y	(σ_m+ σ_g) = 15.(1)	126.(13)		
			σ_m = 0.6(1)			
^{95}Nb		34.97 d	<7.	<200.		
$_{42}$Mo			2.5(1)	26.(5)	6.72(2)	
			σ_α <0.1 mb			
^{92}Mo	14.77(31)	>3×10^{17} y	0.06	≈ 0.8	6.93(8)	0.07(1)
			σ_m = 0.2 μb			
^{94}Mo	9.226(99)		0.02	≈ 0.8	6.82(7)	0.10(2)
^{95}Mo	15.900(85)		13.4(3)	109.(5)	6.93(6)	0.29(1)
			σ_α = 30.(4) μb			
^{96}Mo	16.674(12)		0.5	17.(3)	6.22(6)	0.11(1)
^{97}Mo	9.560(50)		2.5(2)	14.(3)	7.26(8)	0.34(1)
			σ_α = 0.4(2) μb			
^{98}Mo	24.20(25)		0.14(1)	7.2(7)	6.60(7)	0.10(1)
^{100}Mo	9.67(20)	≈ 1×10^{19} y	0.19(1)	3.6(3)	6.75(7)	0.11(1)
$_{43}$Tc						
^{98}Tc		≈ 6.6×10^6 y	σ_m = 0.9(2)			
^{99}Tc		2.13×10^5 y	23.(2)	4.0(4)×10^2	6.8(3)	0.93(5)
$_{44}$Ru			2.6 (1)	48.(5)	7.03(3)	
^{96}Ru	5.54(14)	>3.1×10^{16} y	0.23(4)	7.(2)		0.21(1)
^{98}Ru	1.87(3)		< 8.			0.3(1)
^{99}Ru	12.76(14)		4.(1)	195.(20)		1.2(3)
^{100}Ru	12.60(7)		5.8(6)	11.(2)		0.21(1)
^{101}Ru	17.06(2)		5.(1)	1.1(3)×10^2		1.00(4)
			σ_α <0.15 μb			
^{102}Ru	31.55(14)		1.2(1)	4.3(5)		0.15(1)
^{103}Ru		39.27 d	<20.	≈ 30.		
^{104}Ru	18.62(27)		0.49(2)	6.(2)		0.15(1)
^{105}Ru		4.44 h	0.29(3)	0.13(1)		
^{106}Ru		1.020 y	0.15(4)	2.0(6)		
$_{45}$Rh			145.(2)	1.2(1)×10^3	5.88(4)	
^{103}Rh	100.		(11.+ 134.)	(0.08+1.1)×10^3	5.88(4)	0.81(1)
^{104m}Rh		4.36 m	800.(100)			
^{104}Rh		42.3 s	40.(30)			
^{105}Rh		35.4 h	1.1(3)×10^4	1.7(4)×10^4		
$_{46}$Pd			7.(1)	82.(8)	5.91(6)	
^{102}Pd	1.02(1)		3.2(10)	10.(2)		0.3(1)
^{104}Pd	11.14(8)			16.(2)		0.29(3)
^{105}Pd	22.33(8)		22.(2)	60.(20)	5.5(3)	1.20(6)
			σ_α = 0.5(2) μb			
^{106}Pd	27.33(3)		(0.013+0.28)	(0.2+5.5)	6.4(4)	0.25(3)
^{107}Pd		6.5×10^6 y	1.8(2)	108.(4)		1.34(6)
^{108}Pd	26.46(9)		(0.19+8.5)	(2.+240.)	4.1(3)	0.20(2)
^{110}Pd	11.72(9)		(0.033+0.7)	(0.7+8.)		0.15(2)
$_{47}$Ag			62.(1)	767.(60)	5.922(7)	
^{107}Ag	51.839(8)		(1.+35.)	(3.+105.)	7.56(1)	0.80(3)
^{109}Ag	48.161(8)		(4.1 + 87.)	(0.7+14.1)×10^2	4.17(1)	0.79(3)
^{110m}Ag		249.8 d	82.(11)	20.(4)		
^{111}Ag		7.47 d	3.(2)	105.(20)		
$_{48}$Cd			2.52(5)×10^3	73.(8)	4.87(5)	
^{106}Cd	1.25(6)	>2.6×10^{17} y	0.20(3)	4.(1)		0.30(2)
^{108}Cd	0.89(3)	>4.1×10^{17} y	1.	14.(3)	5.4(1)	0.20(1)

*Extrapolated value.

Elem. or Isot.	Natural Abundance (%)	Half-Life	Thermal Neut. Cross Section (barns)	Resonance Integral (barns)	Coh. Scat. Length (fm)	σ (30 keV) Maxw. Avg. (barns)
^{109}Cd		462.0 d	≈ 180.	6.7(12)×10^{3}		
			σ_α <0.05			
^{110}Cd	12.49(18)		(0.06+11.)	(6.+34.)	5.9(1)	(0.01+0.22)
^{111}Cd	12.80(12)		3.5(20)	51.(6)	6.5(1)	0.75(1)
^{112}Cd	24.13(21)		(0.012+2.2)	15.	6.4(1)	0.19(1)
^{113}Cd	12.22(12)	7.7×10^{15} y	2.06(4)×10^{4}	390.(40)	- 8.0(2)	0.67(1)
			σ_α <1. μb			
^{114}Cd	28.73(42)		(0.04+0.29)	16.(7)	7.5(1)	(0.01+0.12)
^{116}Cd	7.49(18)	3.8×10^{19} y	(26.+52.) mb	1.2	6.3(1)	(12.+47.) mb
$_{49}$In			197.(4)	3.3(2)×10^{3}	4.07(2)	
^{113}In	4.29(5)		(3.1+5.0+3.9)	(220.+90.)	5.39(6)	(0.48+0.31)
^{115}In	95.71(5)	4.4×10^{14} y	(88.+73.+44.)	(1.5+1.2+0.7)×10^{3}	4.01(2)	(0.69+0.02)
$_{50}$Sn			0.61(3)	8.(2)	6.225(2)	
^{112}Sn	0.97(1)		(0.15+0.40)	(8.+19.)		0.21(1)
^{113}Sn		115.1 d	≈ 9.	210.(50)		
^{114}Sn	0.66(1)		≈ 0.12	5.(1)	6.2(3)	134.(3) mb
^{115}Sn	0.34(1)		σ_α = 0.06 mb	29.(6)		0.34(1)
^{116}Sn	14.54(9)		(0.006+0.14)	(0.5+11.)	5.93(5)	91.(2) mb
^{117}Sn	7.68(7)		1.1(1)	16.(5)	6.48(5)	319.(7) mb
^{118}Sn	24.22(9)		σ_m = 4. mb	4.7(5)	6.07(5)	62.(1) mb
^{119}Sn	8.59(4)		2.(1)	2.9(5)	6.12(5)	0.18(1)
^{120}Sn	32.58(9)		(0.001+0.13)	1.2(3)	6.49(5)	(0.5+36.) mb
^{122}Sn	4.63(3)		(0.15+0.001)	0.81(4)	5.74(5)	(18.+4.) mb
^{124}Sn	5.79(5)	>2.2×10^{18} y	(0.13+0.004)	(8.0+0.08)	5.97(5)	12.(2) mb
$_{51}$Sb			5.2(2)	169.(20)	5.57(3)	
^{121}Sb	57.21(5)		(0.4+5.8)	(13.+192.)	5.71(6)	0.53(2)
^{123}Sb	42.79(5)		(0.02+0.04+4.0)	(1.+119.)	5.38(7)	0.30(1)
^{124}Sb		60.20 d	17.(3)	≈ 8.		
$_{52}$Te			4.2(1)	47.(3)	5.80(3)	
^{120}Te	0.09(1)		(1.+5.)	≈ 1.	5.3(5)	0.4(1)
^{122}Te	2.55(12)		(0.4+3.)	(5.+75.)	3.8(2)	295.(3) mb
^{123}Te	0.89(3)	>5.3×10^{16} y	370.(40)	4.5(3)×10^{3}	- 0.05	0.83(1)
			σ_α = 0.05 mb			
^{124}Te	4.74(14)		(1.+6.)	(1.4+4.)	8.0(1)	155.(2) mb
^{125}Te	7.07(15)		1.1(2)	21.(4)	5.02(8)	431.(4) mb
^{126}Te	18.84(25)		(0.12+0.8)	(0.6+7.4)	5.56(7)	(28.+53.) mb
^{128}Te	31.74(8)	2.2×10^{24} y	(0.03+0.2)	(0.2+1.6)	5.89(7)	(3.+41.) mb
^{130}Te	34.08(62)	8.×10^{20} y	(0.01+0.19)	(0.03+0.3)	6.02(7)	(4.+11.) mb
$_{53}$I			6.2(1)	1.5(1)×10^{2}	5.28(2)	
^{125}I		59.4 d	900.(100)	1.4(2)×10^{4}		
^{127}I	100.		6.2(1)	1.5(1)×10^{2}	5.28(2)	0.64(3)
^{128}I		25.00 m	22.(4)	≈ 10.		
^{129}I		1.7×10^{7} y	(20.7+10.3)	36.(4)		0.44(2)
^{130}I		12.36 h	18.(3)	≈ 8.		
^{131}I		8.021 d	≈ 0.7	8.(4)		
$_{54}$Xe			25.(1)	263.(50)	4.92(3)	
^{124}Xe	0.0953(27)	>10^{17} y	(28.+137.)	(0.6+3.0)×10^{3}		(0.13+0.51)
^{125}Xe		17.1 h	σ_α < 0.03			
^{126}Xe	0.0890(14)		(0.45+3.)	(8.+52.)		(0.04+0.32)
^{127}Xe		36.34 d	σ_α ≤ 0.01			
^{128}Xe	1.910(22)		σ_m = 0.48	RI_m = 38.(10)		0.26(1)
^{129}Xe	26.40(18)		22.(5)	250.(50)		0.62(2)
^{130}Xe	4.071(53)		σ_m = 0.45	RI_m = 16.(4)		0.132(3)
^{131}Xe	21.233(62)		90.(10)	9.(1)×10^{2}		0.45(8)
^{132}Xe	26.9087(680)		(0.05+0.4)	(0.9+3.7)		(5.+60.) mb
^{133}Xe		5.243 d	190.(90)			
^{134}Xe	10.436(29)	>1.1×10^{16} y	(0.003 + 0.26)	0.40(4)		20.(2) mb
^{135}Xe		9.10 h	2.65(11)×10^{6}	7.6(5)×10^{3}		

*Extrapolated value.

Elem. or Isot.	Natural Abundance (%)	Half-Life	Thermal Neut. Cross Section (barns)	Resonance Integral (barns)	Coh. Scat. Length (fm)	σ (30 keV) Maxw. Avg. (barns)
^{136}Xe	8.858(33)	>8×10^{20} y	0.26(2)	0.7(2)		0.9(1) mb
$_{55}$Cs			30.4(8)	422.(50)	5.42(2)	
^{132}Cs		6.48 d	σ_α < 0.15			
^{133}Cs	100.		(2.7+27.3)	(32.+360.)	5.42(2)	(0.04+0.47)
^{134}Cs		2.065 y	140.(10)	54.(9)		
^{135}Cs		2.3×10^{6} y	8.3(3)	38.(3)		
^{137}Cs		30.2 y	(0.20+0.07)	0.36(7)		
$_{56}$Ba			1.3(2)	10.(2)	5.07(3)	
^{130}Ba	0.106(1)	2.2×10^{21} y	(1.+8.)	(25.+200.)	-3.6(6)	0.76(11)
^{132}Ba	0.101(1)	1.3×10^{21} y	(0.84+9.7)	(4.7+24.)	7.8(3)	0.6(1)
^{133}Ba		10.53 y	4.(1)	85.(30)		
^{134}Ba	2.417(18)		(0.1+1.3)	(5.6+18.)	5.7(1)	0.18(1)
^{135}Ba	6.592(12)		(0.014+5.8)	(0.47+131.)	4.7(1)	0.46(2)
^{136}Ba	7.854(24)		(0.010+0.44)	(0.1+1.5)	4.91(8)	61.(2) mb
^{137}Ba	11.232(24)		5.(1)	4.(1)	6.8(1)	76.(3) mb
^{138}Ba	71.698(42)		0.41(2)	0.4(1)	4.84(8)	4.0(2) mb
^{139}Ba		1.396 h	5.(1)	2.2(5)		
^{140}Ba		12.75 d	1.6(3)	14.(1)		
$_{57}$La			9.2(2)	12.(1)	8.24(4)	
^{138}La	0.090(1)	1.06×10^{11} y	57.(6)	4.1(9)×10^{2}		
^{139}La	99.910(1)		9.2(2)	12.(1)	8.24(4)	38.(3) mb
^{140}La		1.678 d	2.7(3)	69.(4)		
$_{58}$Ce			0.64(4)	0.71(6)	4.84(2)	
^{136}Ce	0.185(2)		(1.0+6.5)	58.(12)	5.80(9)	(0.028+0.3)
^{138}Ce	0.251(2)		(0.025+1.0)	(1.5+5.2)	6.70(9)	179.(5) mb
^{140}Ce	88.450(51)		0.58(4)	0.50(5)	4.84(9)	11.0(4) mb
^{141}Ce		32.50 d	29.(3)	13.(2)		
^{142}Ce	11.114(51)	>1.6×10^{17} y	0.97(3)	1.3(3)	4.75(9)	28.(1) mb
^{143}Ce		1.38 d	6.1(7)	2.7(3)		
^{144}Ce		284.6 d	1.0(1)	2.6(3)		
$_{59}$Pr			11.5(4)	14.(3)	4.58(5)	
^{141}Pr	100.		(4.+7.5)	14.(3)	4.58(5)	111.(2) mb
^{142}Pr		19.12 h	20.(3)	9.(1)		
^{143}Pr		13.57 d	90.(10)	190.(25)		
$_{60}$Nd			51.(2)	49.(5)	7.69(5)	
^{142}Nd	27.2(5)		19.(1)	34.(11)	7.7(3)	35.(1) mb
^{143}Nd	12.2(2)		330.(10)	128.(30)		0.24(1)
			σ_α = 17. mb			
^{144}Nd	23.8(3)	2.1×10^{15} y	3.6(3)	3.9(5)	2.8(3)	81.(2) mb
^{145}Nd	8.3(1)		47.(6)	260.(40)		0.42(1)
			σ_α = 12. μb			
^{146}Nd	17.2(3)		1.5(2)	3.0(4)	8.7(2)	91.(1) mb
^{147}Nd		10.98 d	440.(150)	200.		
^{148}Nd	5.7(1)		2.4(1)	13.(2)	5.7(3)	147.(2) mb
^{150}Nd	5.6(2)	≈ 1×10^{19} y	1.0(1)	14.(2)	5.3(2)	0.16(1)
$_{61}$Pm						
^{146}Pm		5.53 y	8.4(1.7)×10^{3}			
^{147}Pm		2.623 y	(84.+96.)	(1000.+1280.)	12.6(4)	2.(1)
^{148m}Pm		41.3 d	10600.(800)			
^{148}Pm		5.37 d	≈ 10^{3}	2.6(2.4)×10^{3}		
^{149}Pm		2.212 d	1400.(200)			
^{151}Pm		1.183 d	≈ 150.			
$_{62}$Sm			5.6(1)×10^{3}	1.4(2)×10^{3}		
^{144}Sm	3.07(7)		1.6(1)	2.4(3)		92.(6) mb
^{145}Sm		340. d	280.(20)	600.(90)		
^{147}Sm	14.99(18)	1.06×10^{11} y	56.(4), σ_α = 0.6 mb	710.(50)	14.(3)	0.97(1)
^{148}Sm	11.24(10)	7×10^{15} y	2.4(6)	27.(14)		241.(2) mb
^{149}Sm	13.82(7)	10^{16} y	4.01(6)×10^{4}, σ_α = 31. mb	3.1(5)×10^{3}		1.82(2)

*Extrapolated value.

Elem. or Isot.	Natural Abundance (%)	Half-Life	Thermal Neut. Cross Section (barns)	Resonance Integral (barns)	Coh. Scat. Length (fm)	σ (30 keV) Maxw. Avg. (barns)
^{150}Sm	7.38(1)		102.(5)	290.(30)	14.(3)	422.(4) mb
^{151}Sm		90. y	1.52(3)$\times10^4$	3520.(60)		2.(1)
^{152}Sm	26.75(16)		206.(15)	3.0(3)$\times10^3$	- 5.0(6)	473.(4) mb
^{153}Sm		1.929 d	420.(180)			
^{154}Sm	22.75(29)		7.5(3)	32.(6)	9.(1)	0.21(1)
$_{63}$Eu			4570.(100)	3.8(5)$\times10^3$	5.3(3)	
^{151}Eu	47.81(6)		(4.+3150.+6000.)	(2.+4.)$\times10^3$		(1.6+2.2)
			σ_α = 8.7(3) μb			
^{152m1}Eu		9.30 h	6.8(15)$\times10^4$	< 10^5		
^{152}Eu		13.5 y	1.1(2)$\times10^4$	1.6(2)$\times10^3$		5.(2)
^{153}Eu	52.19(6)		300.(20), σ_α <1. μb	1.8(4)$\times10^3$	8.2(1)	2.8(1)
^{154}Eu		8.59 y	1.5(3)$\times10^3$	1.6(2)$\times10^3$		4.4(7)
^{155}Eu		4.76 y	3.9(2)$\times10^3$	1.6(2)$\times10^4$		1.3(1)
$_{64}$Gd			48.8(6)$\times10^3$	400.(10)	9.5(2)	
^{148}Gd		75. y	1.40(14)$\times10^4$			
^{152}Gd	0.20(1)	1.1$\times10^{14}$ y	700.(200), σ_α <7. mb	700.(200)		1.05(2)
^{153}Gd		240. d	2.(1)$\times10^4$, σ_α = 0.03			
^{154}Gd	2.18(3)		(0.035+60.)	230.(50)		1.03(1)
^{155}Gd	14.80(12)		61.(1)$\times10^3$, σ_α = .08 mb	1540.(100)		2.65(3)
^{156}Gd	20.47(9)		≈ 2.0	104.(15)	6.3(4)	615.(5) mb
^{157}Gd	15.65(2)		2.54(3)$\times10^5$, σ_α <0.05	800.(100)		1.37(2)
^{158}Gd	24.84(7)		2.3(3)	73.(7)	9.(2)	324.(3) mb
^{160}Gd	21.86(19)	>1.9$\times10^{19}$ y	1.5(7)	6.(1)	9.15(5)	0.15(2)
^{161}Gd		3.66 m	2.0(6)$\times10^4$			
$_{65}$Tb			23.2(5)	420.(50)	7.34(2)	
^{159}Tb	100.		23.2(5)	420.(50)	7.34(2)	1.6(2)
^{160}Tb		72.3 d	570.(110)			
$_{66}$Dy			9.5(2)$\times10^2$	1.5(2)$\times10^3$	16.9(3)	
^{156}Dy	0.056(3)		33.(3), σ_α < 9. mb	1000.(100)		1.6(2)
^{158}Dy	0.095(3)		43.(6), σ_α < 6. mb	120.(10)	6.1(5)	0.8(2)
^{159}Dy		144. d	8.(2)$\times10^3$			
^{160}Dy	2.39(18)		60.(10), σ_α < 0.3 mb	1100.(200)	6.7(4)	0.89(1)
^{161}Dy	18.889(42)		600.(50), σ_α < 1. μb	1100.(100)	10.3(4)	1.96(2)
^{162}Dy	25.475(36)		170.(20)	2755.(300)	- 1.4(5)	446.(4) mb
^{163}Dy	24.896(42)		120.(10), σ_α < 20. μb	1600.(400)	5.0(4)	1.11(1)
^{164}Dy	28.260(54)		(1.7+1.0)$\times10^3$	(4.+2.)$\times10^2$	49.4(2)	212.(3) mb
^{165m}Dy		1.26 m	2.0(6)$\times10^3$			
^{165}Dy		2.33 h	3.5(3)$\times10^3$	2.2(3)$\times10^4$		
$_{67}$Ho			61.(2)	670.(40)	8.01(8)	
^{163}Ho		4.57$\times10^3$ y				(0.4+1.7)
^{165}Ho	100.		(3.1+58.), σ_α < 20. μb	(?+670.)	8.01(8)	(0.8+0.5)
^{166m}Ho		1.2$\times10^3$ y	3.1(8)$\times10^3$	10.(3)$\times10^3$		
$_{68}$Er			1.5(2)$\times10^2$	730.(10)	7.79(2)	
^{162}Er	0.139(5)		19.(3), σ_α < 11. mb	480.(50)	8.8(2)	1.6(1)
^{164}Er	1.601(3)		13.(3), σ_α < 1.2 mb	105.(10)	8.2(2)	1.08(5)
^{166}Er	33.503(36)		(3.+14.), σ_α < 70. μb	96.(12)	10.6(2)	0.56(6)
^{167}Er	22.869(9)		6.5(8)$\times10^2$, σ_α = 3. μb	2970.(70)	3.0(3)	1.4(2)
^{168}Er	26.978(18)		2.3(3), σ_α = 0.09 mb	37.(5)	7.4(4)	0.34(4)
^{170}Er	14.910(36)		8.(2)	26.(4)	9.6(5)	0.17(1)
^{171}Er		7.52 h	370.(40)	170.(20)		
$_{69}$Tm			108.(4)	1.5(2)$\times10^3$	7.07(3)	
^{169}Tm	100		(8.+100.)	1.5(2)$\times10^3$	7.07(3)	1.13(6)
^{170}Tm		128.6 d	100.(20)	460.(50)		
^{171}Tm		1.92 y	≈ 160.	118.(6)		
$_{70}$Yb			52.(10)	1.7(2)$\times10^2$	12.43(3)	
^{168}Yb	0.13(1)		2.4(2)$\times10^3$, σ_α < 0.1 mb	2.0(5)$\times10^4$	-4.07(2)	0.7(4)
^{169}Yb		32.02 d	3.6(3)$\times10^3$	5200.(500)		
^{170}Yb	3.04(15)		12.(2), σ_α < 10. μb	320.(30)	6.8(1)	0.77(1)

*Extrapolated value.

Elem. or Isot.	Natural Abundance (%)	Half-Life	Thermal Neut. Cross Section (barns)	Resonance Integral (barns)	Coh. Scat. Length (fm)	σ (30 keV) Maxw. Avg. (barns)
^{171}Yb	14.28(57)		53.(5), σ_α < 1.5 μb	315.(30)	9.7(1)	1.21(1)
^{172}Yb	21.83(67)		≈ 1.3, σ_α < 1. μb	25.(3)	9.4(1)	0.34(1)
^{173}Yb	16.13(27)		16.(2), σ_α < 1. μb	380.(30)	9.56(7)	0.75(1)
^{174}Yb	31.83(92)		(46.+17.), σ_α < 0.02 mb	(13.+16.)	19.3(1)	151.(2) mb
^{176}Yb	12.76(41)		3.1(2), σ_α < 1. μb	8.(2)	8.7(1)	116.(2) mb
$_{71}$Lu			78.(7)	8.3(7)×10^2	7.21(3)	
^{175}Lu	97.41(2)		(16.+8.)	(550.+270.)	7.24(3)	(1.04+0.11)
^{176}Lu	2.59(2)	3.73×10^{10} y	(2.+2100.)	(3.+930.)	6.1(2)	1.53(7)
^{177m}Lu		160.7 d	3.2(3)	1.4(2)		
^{177}Lu		6.65 d	1000.(300)			
$_{72}$Hf			106.(3)	19.7(5)×10^2	7.8(1)	
^{174}Hf	0.16(1)	2.0×10^{15} y	600.(50)	400.(50)	11.(1)	0.8(2)
^{176}Hf	5.26(7)		23.(4)	700.(100)	6.6(2)	0.46(2)
^{177}Hf	18.60(9)		(1.+375.), σ_α < 20. μb	7170.(200)		1.5(1)
^{178m2}Hf		31. y	σ_{m2} = 45.(5)	RI_{m2} = 8(1)×10^2		
^{178}Hf	27.28(7)		(54.+32.)	(0.9+1.0)×10^3	5.9(2)	0.31(1)
^{179}Hf	13.62(2)		(0.43+46.)	(6.8+620.)	7.5(2)	(0.01+0.95)
^{180}Hf	35.08(16)		13.0(5), σ_α < 13. μb	32.(1)	13.2(3)	179.(5) mb
^{181}Hf		42.4 d	30.(25)			
$_{73}$Ta			20.(1)	650(20.)	6.91(7)	
^{179}Ta		1.8 y	9.3(6)×10^2	1.22(7)×10^3		
^{180m}Ta	0.012(2)	> 1.2×10^{15} y	≈ 560.	1350.(100)		
^{181}Ta	99.988(2)		(0.012 + 20.), σ_α <1. μb	(0.4+650.)	6.91(7)	0.77(2)
^{182}Ta		114.43 d	8200.(600)	900.(90)		
$_{74}$W			18.(1)	3.6(3)×10^2	4.86(2)	
^{180}W	0.12(1)	7.4×10^{16} y	≈ 4.	210.(30)		0.54(6)
^{182}W	26.50(16)	8.3×10^{18} y	20.(1)	600.(90)	6.97(4)	274.(8) mb
^{183}W	14.31(4)	1.9×10^{18} y	10.5(3)	340.(50)	6.53(4)	0.52(2)
^{184}W	30.64(2)	4.0×10^{18} y	(0.002 + 2.0)	15.(2)	7.48(6)	0.22(1)
^{185}W		74.8 d	≈ 3.3	300.(50)		
^{186}W	28.43(19)	6.5×10^{18} y	37.(2)	510.(50)	- 0.72(4)	176.(5) mb
^{187}W		23.9 h	70.(10)	2760.(550)		
^{188}W		69.78 h	12.(1)			
$_{75}$Re			90.(4)	8.4(2)×10^2	9.2(3)	
^{185}Re	37.40(2)		(0.33+110.)	1700.(50)	9.0(3)	1.54(6)
^{187}Re	62.60(2)	4.2×10^{10} y	(2.+72.)	(9.+310.)	9.3(3)	1.16(6)
$_{76}$Os			17.(1)	1.5(1)×10^2	10.7(2)	
^{184}Os	0.02(1)	>5.6×10^{13} y	3.3(3)×10^3, σ_α <10. mb	1.4(1)×10^3		0.4(2)
^{186}Os	1.59(3)	2.×10^{15} y	≈ 80., σ_α < 0.1 mb	3.8(9)×10^2	12(2)	0.42(2)
^{187}Os	1.96(2)		2.(1)×10^2, σ_α < 0.1 mb	5.0(7)×10^2		0.90(3)
^{188}Os	13.24(8)		≈ 5., σ_α < 30. μb	1.5(2)×10^2	7.6(3)	0.40(2)
^{189}Os	16.15(5)		(0.00026+40.),σ_α<10. μb	(0.013+670.)	10.7(3)	1.17(5)
^{190}Os	26.26(2)		(9.+4.), σ_α < 20. μb	(22.+8.)	11.0(3)	0.30(5)
^{191}Os		15.4 d	3.8(6)×10^2	1.7(3)×10^2		
^{192}Os	40.78(19)		3.(1), σ_α < 10. μb	7.(1)	11.5(4)	0.31(5)
^{193}Os		30.5 h	2.5(5)×10^2	1.1(2)×10^2		
$_{77}$Ir			4.2(1)×10^2	2.8(4)×10^3	10.6(3)	
^{191}Ir	37.3(2)		(0.14+660.+260.)	(1.0+4.2)×10^3		1.35(4)
^{192}Ir		73.83 d	1.4(3)×10^3	4.8(7)×10^3		
^{193}Ir	62.7(2)		(0.04+6.+109.)	1.4(2)×10^3		0.99(7)
^{194}Ir		19.3 h	1.6(3)×10^3	7.(2)×10^2		
$_{78}$Pt			10.(1)	1.3(1)×10^2	9.60(1)	
^{190}Pt	0.014(1)	4.5×10^{11} y	1.5(1)×10^2, σ_α < 8. mb	70.(10)	9.(1)	0.7(2)
^{192}Pt	0.782(7)		(2.0+6.), σ_α < 0.2 mb	115.(20)	9.9(5)	0.6(1)
^{194}Pt	32.967(99)		(0.1+1.1), σ_α < 5. μb	(4.+?)	10.55(8)	(0.03+0.34)
^{195}Pt	33.832(10)		28.(1), σ_α < 5. μb	365.(50)	8.8(1)	0.9(2)
^{196}Pt	25.242(41)		(0.045+0.55)	7.(2)	9.89(8)	(0.01+0.19)
^{198}Pt	7.163(55)		(0.3+3.1)	(5.+53.)	7.8(1)	(3.+79.) mb

*Extrapolated value.

Elem. or Isot.	Natural Abundance (%)	Half-Life	Thermal Neut. Cross Section (barns)	Resonance Integral (barns)	Coh. Scat. Length (fm)	σ (30 keV) Maxw. Avg. (barns)
^{199}Pt		30.8 m	≈ 15.	≈ 7.		
$_{79}$Au			98.7(1)	1.55(3)×10^{3}	7.63(6)	
^{197}Au	100.		σ_{m+g} = 98.7(1)	RI_{m+g} = 1.55(3)×10^{3}	7.63(6)	582.(9) mb
			σ_m = 8.(2) mb	RI_m = 0.06(2)		
^{198}Au		2.695 d	26.5(15)×10^{3}	≈ 4.×10^{4}		
^{199}Au		3.14 d	≈ 30.			
$_{80}$Hg			3.7(1)×10^{2}	87.(5)	12.69(2)	
^{196}Hg	0.15(1)	>2.5×10^{18} y	(105.+3000.)	(53.+410.)	30.(1)	0.4(2)
^{198}Hg	9.97(8)		(0.017+2.)	(1.7+70.)		0.17(2)
^{199}Hg	16.87(10)		2.1(2)×10^{3}	435(20)	16.9(4)	0.37(2)
^{200}Hg	23.10(16)		≈ 1.	2.1(5)		0.12(1)
^{201}Hg	13.18(8)		≈ 8.	30.(3)		0.26(1)
^{202}Hg	29.86(20)		4.9(5)	4.5(2)	11.(1)	74.(6) mb
^{204}Hg	6.87(4)		0.4(1)	0.8(2)		42.(4) mb
$_{81}$Tl			3.3(1)	12.5(8)	8.776(5)	
^{203}Tl	29.524(14)		11.(1), σ_α < 0.3 mb	41.(2)	7.0(2)	124.(8) mb
^{204}Tl		3.78 y	22.(2)	90.(20)		0.14(5)
^{205}Tl	70.476(14)		0.11(2)	0.6(2)	9.52(7)	54.(4) mb
$_{82}$Pb			0.172(2)	0.14(4)	9.402(2)	
^{204}Pb	1.4(1)		0.68(7)	2.0(2)	10.9(1)	90.(6) mb
^{205}Pb		1.51×10^{7} y	≈ 5.	≈ 2.		0.06(1)
^{206}Pb	24.1(1)		0.027(1)	0.10(1)	9.23(5)	16.(1) mb
^{207}Pb	22.1(1)		0.61(3)	0.38(1)	9.28(2)	10.(1) mb
^{208}Pb	52.4(1)	>2×10^{19} y	0.23(1) mb, σ_α < 8. μb	2.0(2) mb	9.50(3)	0.36(4) mb
^{210}Pb		22.6 y	< 0.5			
$_{83}$Bi			0.034(1)	0.19(2)	8.532(2)	
^{209}Bi	100.		(11.+23.) mb,σ_α<0.3 μb	0.19(2)	8.532(2)	2.7(5) mb
^{210m}Bi		3.0×10^{6} y	54.(4) mb	0.20(3)		
$_{84}$Po						
^{210}Po		138.4 d	σ_m<0.5 mb, σ_α < 2. mb			
			σ_g<30. mb, σ_f< 0.1			
$_{85}$At						
$_{86}$Rn						
^{220}Rn		55.6 s	<0.2			
^{222}Rn		3.823 d	0.74(5)			
$_{88}$Ra						
^{223}Ra		11.43 d	1.3(2)×10^{2}, σ_f< 0.7			
^{224}Ra		3.66 d	12.0(5)			
^{226}Ra		1599. y	≈ 13., σ_f< 7. μb	280.(50)	10.(1)	
^{228}Ra		5.76 y	36.(5), σ_f< 2.			
$_{89}$Ac						
^{227}Ac		21.77 y	8.8(7)×10^{2}, σ_f< 0.35 mb	1.5(4)×10^{3}		
$_{90}$Th			7.4	85.(3)	10.31(3)	
^{227}Th		18.72 d	σ_f = 2.0(2)×10^{2}			
^{228}Th		1.913 y	1.2(2)×10^{2}, σ_f<0.3	1014.(400)		
^{229}Th		7.9×10^{3} y	≈ 60.	1.0(2)×10^{3}		
			σ_f = 30.(3)	RI_f = 466.(75)		
^{230}Th		7.54×10^{4} y	23.4(5)	1.0(1)×10^{3}		
			σ_f < 0.5 mb			
^{232}Th	100.	1.40×10^{10} y	7.37(4)	85.(3)	10.31(3)	
			σ_f = 3.(1) μb			
			σ_α < 1. μb			
^{233}Th		22.3 m	1.5(1)×10^{3}	4.(1)×10^{2}		
			σ_f = 15.(2)			
^{234}Th		24.10 d	1.8(5)			
			σ_f < 0.01			

*Extrapolated value.

Elem. or Isot.	Natural Abundance (%)	Half-Life	Thermal Neut. Cross Section (barns)	Resonance Integral (barns)	Coh. Scat. Length (fm)	σ (30 keV) Maxw. Avg. (barns)
$_{91}$Pa						
^{230}Pa		17.4 d	$1.5(3)\times10^{3}$			
^{231}Pa		3.25×10^{4} y	$2.0(1)\times10^{2}$	750.(80)	9.1(3)	
			σ_f = 20.(1) mb	RI_f = 0.05(1)		
^{232}Pa		1.31 d	$4.6(10)\times10^{2}$	300.(70)		
			$\sigma_f = 1.5(5)\times10^{3}$	$RI_f = 1.0(1)\times10^{3}$		
^{233}Pa		27.0 d	39.(2)	(460.+440.)		
			σ_m = 20.(4)			
			σ_g = 19.(3)			
			σ_f < 0.1			
$_{92}$U			3.4(3); σ_f = 4.2(1)	280.(20),RI_f = 2.0	8.417(5)	
^{230}U		20.8 d	σ_f ≈ 25.			
^{231}U		4.2 d	σ_f ≈ 250.			
^{232}U		70. y	73.(2)	280.(15)		
			σ_f = 74.(8)	RI_f = 350.(30)		
^{233}U		1.592×10^{5} y	47.(2)	137.(6)	10.1(2)	
			$\sigma_f = 5.3(1)\times10^{2}$	RI_f = 760.(17)		
			σ_α < 0.2 mb			
^{234}U	0.0054(5)	2.455×10^{5} y	96.(2)	660.(70)	12.(4)	
			σ_f = 0.07(2)	RI_f = 6.5		
^{235}U	0.7204(6)	7.04×10^{8} y	95.(5)	144.(6)	10.47(4)	
			σ_f = 586.(2)	RI_f = 275(5)		
			σ_α < 0.1 mb			
^{236}U		2.342×10^{7} y	5.1(3)	360.(15)		
			σ_f < 1.3 mb	RI_f = 4.38(50)		
^{237}U		6.75 d	≈ 10^{2}	1200.(200)		
			σ_f < 0.35			
^{238}U	99.2742(10)	4.47×10^{9} y	2.7(1)	277.(3)	8.402(5)	
			σ_f ≈ 3. μb	1.54(15) mb		
			σ_α = 1.4(5) μb			
^{239}U		23.5 m	22.(2)			
			σ_f = 15.(3)			
$_{93}$Np						
^{234}Np		4.4 d	$\sigma_f = 9.(3)\times10^{2}$			
^{235}Np		1.085 y	$1.6(1)\times10^{2}$			
^{236m}Np		22.5 h	$\sigma_f = 2.7(2)\times10^{3}$	$7.(4)\times10^{2}$		
^{236}Np		1.55×10^{5} y	$\sigma_f = 3.0(2)\times10^{3}$	$1.35(30)\times10^{3}$		
^{237}Np		2.14×10^{6} y	$1.7(1)\times10^{2}$	$6.5(3)\times10^{2}$	10.6(1)	
			σ_f = 20.(1) mb	RI_f = 4.7		
^{238}Np		2.117 d	$\sigma_f = 2.6(3)\times10^{3}$	$1.4(3)\times10^{3}$		
^{239}Np		2.355 d	(32.+19.)			
			σ_f < 1.			
$_{94}$Pu						
^{236}Pu		2.87 y	$\sigma_f = 1.6(3)\times10^{2}$	1000.(60)		
^{237}Pu		45.7 d	$\sigma_f = 2.3(3)\times10^{3}$			
^{238}Pu		87.7 y	$5.1(2)\times10^{2}$	$1.6(2)\times10^{2}$	14.1(5)	
			σ_f = 17.(1)	RI_f = 26.(2)		
^{239}Pu		2.410 x 10^{4} y	$2.7(1)\times10^{2}$	$2.0(2)\times10^{2}$	7.7(1)	
			σ_f = 752.(3)	$3.0(1)\times10^{2}$		
			σ_α ≤ 0.3 mb			
^{240}Pu		6.56×10^{3} y	$2.9(1)\times10^{2}$	$8.4(3)\times10^{3}$	3.5(1)	
			σ_f ≈ 59. mb	RI_f = 3.2		
^{241}Pu		14.4 y	$3.7(1)\times10^{2}$, σ_α <0.2 mb	$1.6(1)\times10^{2}$		
			$\sigma_f = 1.01(1)\times10^{3}$	$5.7(4)\times10^{2}$		
^{242}Pu		3.75 x 10^{5} y	19.(1)	$1.1(1)\times10^{3}$	8.1(1)	
			σ_f<0.2	RI_f = 0.23		
^{243}Pu		4.956 h	<100.			
			$\sigma_f = 2.0(2)\times10^{2}$			

*Extrapolated value.

Elem. or Isot.	Natural Abundance (%)	Half-Life	Thermal Neut. Cross Section (barns)	Resonance Integral (barns)	Coh. Scat. Length (fm)	σ (30 keV) Maxw. Avg. (barns)
^{244}Pu		8.00×10^{7} y	1.7(1)	41.(3)		
^{245}Pu		10.5 h	1.5(3)×10^{2}	220.(40)		
$_{95}$Am						
^{241}Am		432.7 y	(0.6+6.4)×10^{2}	(1.+14.)×10^{2}		
			σ_f = 3.15(10)	14.(1)		
^{242m}Am		141. y	1.7(4)×10^{3}	≈ 200.		
			σ_f = 5.9(3)×10^{3}	RI$_f$ = 1.8(1)×10^{3}		
^{242}Am		16.02 h	σ_f = 2.1(2)×10^{3}	RI$_f$ = < 300.		
			3.3(5)×10^{2}	≈ 1.5×10^{2}		
^{243}Am		7.37×10^{3} y	(75.+5.)	(17.1+1.0)×10^{2}	8.3(2)	
			σ_f = 79.(2) mb	RI$_f$ = 0.056		
^{244m}Am		≈ 26. m	σ_f = 1.6(3)×10^{3}			
^{244}Am		10.1 h	σ_f = 2.2(3)×10^{3}			
$_{96}$Cm						
^{242}Cm		162.8 d	≈ 20.	120.(50)		
			σ_f ≈ 5.			
^{243}Cm		29.1 y	1.3(1)×10^{2}	214.(20)		
			σ_f = 6.2(2)×10^{2}	RI$_f$ = 1.6(1)×10^{3}		
^{244}Cm		18.1 y	15.(1)	640.(50)	9.5(3)	
			σ_f = 1.1(2)	RI$_f$ = 10.8(8)		
^{245}Cm		8.48×10^{3} y	3.5(2)×10^{2}	110.(10)		
			σ_f = 2.1(1)×10^{3}	RI$_f$ = 8.(1)×10^{2}		
^{246}Cm		4.76×10^{3} y	1.2(2)	120.(10)	9.3(2)	
			σ_f = 0.16(7)	13.(2)		
^{247}Cm		1.56×10^{7} y	60.(30)	5.(1)×10^{2}		
			σ_f = 82.(5)	7.3(7)×10^{2}		
^{248}Cm		3.48×10^{5} y	2.6(3)	270.(30)	7.7(2)	
			σ_f = 0.36(7)	13.(2)		
^{249}Cm		64.15 m	≈ 1.6			
^{250}Cm		≈ 9.7×10^{3} y	≈ 80.			
$_{97}$Bk						
^{249}Bk		320. d	7.(1)×10^{2}	9.(1)×10^{2}		
			σ_f ≈ 0.1			
^{250}Bk		3.217 h	σ_f = 1.0(2)×10^{3}			
$_{98}$Cf						
^{249}Cf		351. y	5.0(3)×10^{2}	7.7(4)×10^{2}		
			σ_f = 1.7(1)×10^{3}	RI$_f$ = 2.1(3)×10^{3}		
^{250}Cf		13.1 y	2.0(2)×10^{3}	12.(2)×10^{3}		
			σ_f = 110.(90)	RI$_f$ = 160.(40)		
^{251}Cf		9.0×10^{2} y	2.9(2)×10^{3}	1.6(1)×10^{3}		
			σ_f = 4.5(5)×10^{3}	RI$_f$ = 5.5(3)×10^{3}		
^{252}Cf		2.65 y	20.(2)	43.(3)		
			σ_f = 32.(4)	RI$_f$ = 1.1(3)×10^{2}		
^{253}Cf		17.8 d	18.(2)	8.(1)		
			σ_f = 1.3(2)×10^{3}			
^{254}Cf		60.5 d	4.5(10)	2.		
$_{99}$Es						
^{253}Es		20.47 d	(180.+5.8)	(37.5+1.1)×10^{2}		
^{254m}Es		1.64 d	σ_f = 1.8(1)×10^{3}			
^{254}Es		276. d	28.(3)	18.(2)		
			σ_f = 1.8(2)×10^{3}	RI$_f$ = 1.2(3)×10^{3}		
^{255}Es		40. d	≈ 55.			
$_{100}$Fm						
^{255}Fm		20.1 h	26.(3)	14.(2)		
			σ_f = 3.3(2)×10^{3}			
^{257}Fm		100.5 d	σ_f = 3.0(2)×10^{3}			

*Extrapolated value.

COSMIC RADIATION

A.G. Gregory and R.W. Clay

The Nature of Cosmic Rays

Primary cosmic radiation, in the form of high energy nuclear particles, electrons and photons from outside the solar system and from the Sun, continually bombards our atmosphere. Secondary radiation, resulting from the interaction of the primary cosmic rays with atmospheric gas, is present at sea-level and throughout the atmosphere.

Secondary radiation is collimated by absorption and scattering in the atmosphere and consists of a number of components associated with different particle species. High energy primary particles can produce large numbers of secondary particles forming an extensive air shower. Thus, a number of particles may then be detected simultaneously at sea-level.

Primary particle energies accessible in the vicinity of the Earth range from ~10^8 eV to ~10^{20} eV. At the lower energies, the limit is determined by the inability of charged particles to traverse the heliosphere to us through the outward-moving solar wind. The upper energy limit is set by the practicality of building detectors to record particles with the extremely low fluxes found at those energies (O.C. Allkofer, 1975a; J.G. Wilson, 1976).

Primary Cosmic Rays

Primary Particle Energy Spectrum

Figure 1 shows the spectrum of primary particle energies. This includes all particle species. In differential form it is roughly a power law of intensity versus energy with an index of ~ –3. There appears to be a knee (a steepening) at a little above 10^{15} eV and an ankle (a flattening) above ~10^{18} eV. Figure 2 emphasizes the features in the spectrum at the highest energies through multiplying the flux with a strongly rising power law of energy. This figure should be used with caution as errors for the two axes are not now independent.

Data on the high energy cosmic ray spectrum are uncertain largely because of limited event statistics due to the very low flux which might best be measured in particles per square kilometer per century. The highest energy event recorded to 1995 had an energy of 3×10^{20} eV (D.J. Bird et al., 1993).

It is expected that the highest energy cosmic rays will interact with the 2.7 K cosmic microwave background through photoproduction or photodisintegration. These interactions will appreciably reduce the observed flux of cosmic rays with energies above 5×10^{19} eV if they travel further than ~150 million light years. This process is known as the Greisen-Zatsepin-Kuz'min (GZK) cutoff (P. Sokolsky, 1989).

At energies below ~10^{13} eV, solar system magnetic fields and plasma can modulate the primary component and Figure 3 shows the extent of this modulation between solar maximum and minimum (E. Juliusson, 1975; J. Linsley, 1981).

Primary Particle Energy Density

If the above spectrum is corrected for solar effects, the energy density above a particle energy of 10^9 eV outside the solar system is found to be ~5×10^5 eV m^{-3}. As the threshold energy is increased, the energy density decreases rapidly, being 2×10^4 eV m^{-3} above 10^{12} eV and 10^2 eV m^{-3} above 10^{15} eV. The energy density at lower energies outside the heliosphere is unknown but may be substantially greater if the particle rest mass energy is included together with the kinetic energy (A. W. Wolfendale, 1979).

Primary Particle Isotropy

This is measured as an anisotropy $(I_{max} - I_{min})/(I_{max} + I_{min}) \times$ 100%, where I, the intensity ($m^{-2}s^{-1}sr^{-1}$), is usually measured with an angular resolution of a few degrees.

The measured anisotropy is small and energy dependent. It is roughly constant in amplitude at between 0.05 and 0.1% (with a phase of 0 to 6 hours in right ascension) for energies between 10^{11} eV and 10^{14} eV and appears to increase at higher energies roughly as $0.4 \times (\text{Energy(eV)}/10^{16})^{0.5}$ % up to ~10^{18} eV. The latter rise may well be an artifact of the progressively more limited statistics as the flux drops rapidly with energy. It appears possible that a real anisotropy has been observed at the highest energies (above a few times 10^{19} eV) with a directional preference for the supergalactic plane (this plane reflects the directions of galaxies within about 100 million light years) (A.W. Wolfendale, 1979; R.W. Clay, 1987; T. Stanev et al., 1995).

Primary Particle Composition

The composition of low energy cosmic rays is close to universal nuclear abundances except where propagation effects are present. For example, Li, Be, and B which are spallation products, are overabundant by about six orders of magnitude.

Composition at 10^{11} eV per nucleus

Charge	1	2	(3–5)	(6–8)	(10–14)	(16–24)	(26–28)	≥30
% Composition (10% uncertainty)	50	25	1	12	7	4	4	0.1

Measurements at higher energies indicate that there is an increase in the relative abundances of nuclei with charge greater than 6 at energies above 50 TeV/nucleus (K. Asakimori et al., 1993) (1 TeV = 10^{12} eV).

Cosmic ray composition at low energies is often quoted at a fixed energy per nucleon. When presented in this way, protons constitute roughly 90% of the flux, helium nuclei about 10% and the remainder sum to a total of about 1%.

Certain radioactive isotopic ratios show lifetime effects. The ratio of Be^{10}/B^9 abundances is used to measure an "age" of cosmic rays since Be^{10} is unstable with a half life of about 1.6×10^6 years. A ratio of 0.6 is expected in the absence of Be^{10} decay and a ratio of about 0.2 is found experimentally (E. Juliusson, 1975; P. Meyer, 1981).

At higher energies, composition determinations are indirect and are rather contradictory and controversial. Experiments aim to differentiate between broad composition models. The measurement technique is based on studies of cosmic ray shower development. A rather direct technique for such studies is to use fluorescence observations of the shower development to determine the atmospheric depth of maximum development of the shower. Such observations suggest a heavy composition (large atomic number) at energies ~10^{17} eV which changes with increasing energy to a light composition (perhaps protonic) above ~10^{19} eV (T. K. Gaisser et al., 1993).

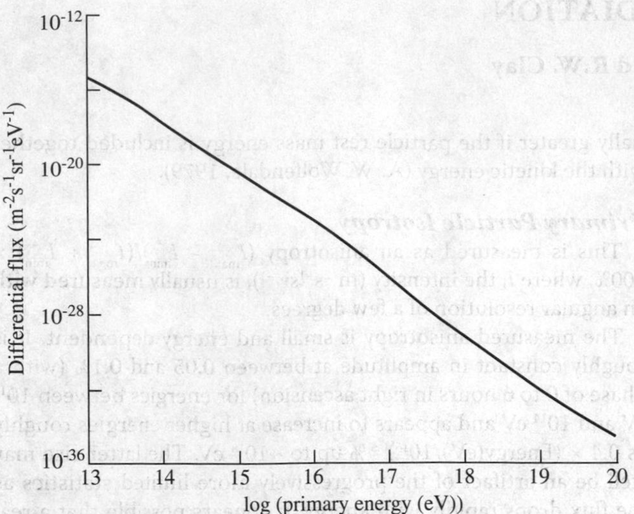

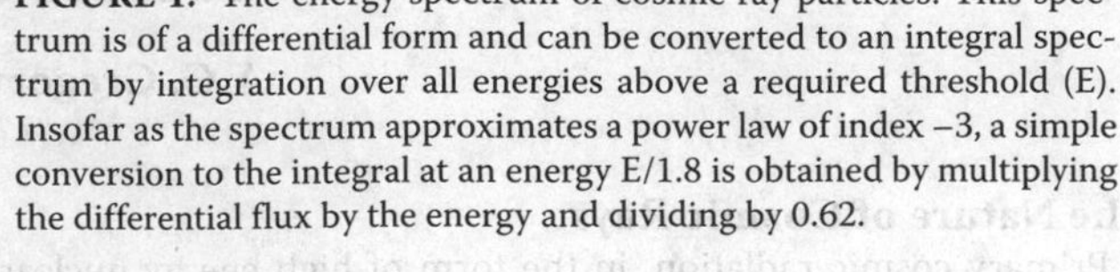

FIGURE 1. The energy spectrum of cosmic ray particles. This spectrum is of a differential form and can be converted to an integral spectrum by integration over all energies above a required threshold (E). Insofar as the spectrum approximates a power law of index −3, a simple conversion to the integral at an energy E/1.8 is obtained by multiplying the differential flux by the energy and dividing by 0.62.

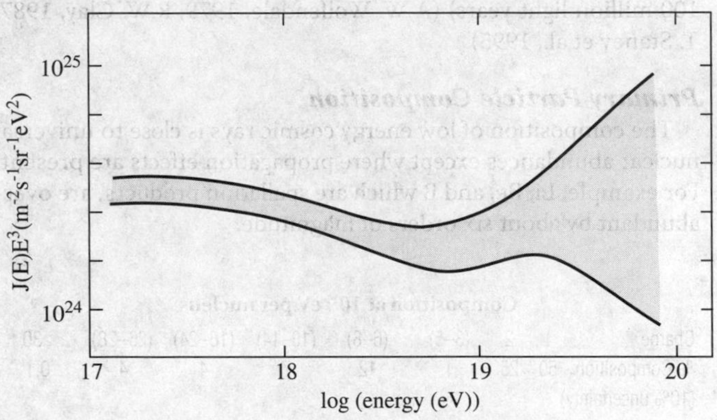

FIGURE 2. Energy spectrum at the highest energies. This spectrum (after Yoshida et al., 1995) has the differential spectrum multiplied by energy cubed. It is from a compilation of a number of measurements and indicates the good general agreement at the lower energies and a spread due to inadequate statistics at the highest energies.

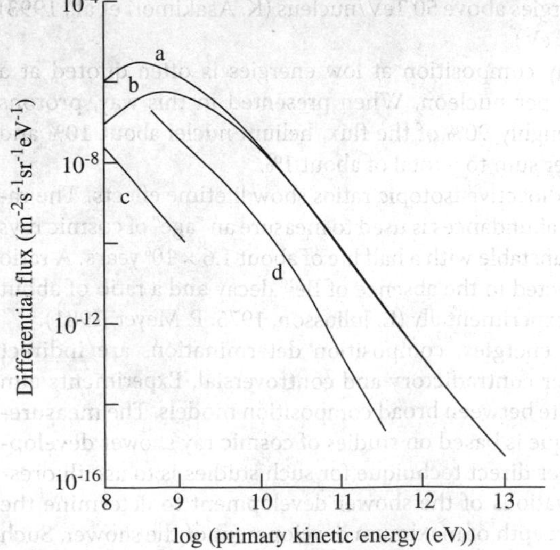

FIGURE 3. Energy spectrum of particles at lower energies. (a) Solar minimum proton energy spectrum. (b) Solar maximum proton energy spectrum. (c) Gamma-ray energy spectrum. (d) Local interstellar electron spectrum.

Primary Electrons

Primary electrons constitute about 1% of the cosmic ray beam. The positron to negative electron ratio is about 10% (J. M. Clem et al., 1995).

Antimatter in the Primary Beam

The ratio of antiprotons to protons in the primary cosmic ray beam (at about 400 MeV) is about 10^{-5}. At about 10 GeV the ratio is about 10^{-3}. At the highest measured energies (10 TeV), the upper limit to the ratio is about 20% (M. Amenomori et al., 1995; S. Orito et al., 1995).

Primary Gamma-Rays

The flux of primary gamma-rays is low at high energies. At 1 GeV the ratio of gamma-rays to protons is about 10^{-6}. The arrival directions of these gamma-rays are strongly concentrated in the plane of the Milky Way although there is a diffuse, near isotropic background flux and some point sources have been detected.

Since the absorption cross section for gamma-rays above 100 MeV is approximately 20 mbarn/electron, less than 10% of gamma-rays reach mountain altitudes (A. W. Wolfendale, 1979; P. F. Michelson, 1994).

Sea-Level Cosmic Radiation

The sea-level cosmic ray dose is 300 millirad·yr^{-1} and the sea-level ionization is 2.2×10^6 ion pairs m^{-3}s^{-1}. The sea-level flux has a soft component, which can be absorbed in about 100 mm of lead (about 100 g·cm^{-2} of absorber) and a more penetrating (largely muon) hard component. The sea-level radiation is largely produced in the atmosphere and is a secondary component from interactions of the primary particles. The steep primary energy spectrum means that most secondaries at sea level are from rather low energy primaries. Thus the secondary flux is dependent on the solar cycle and the geomagnetic latitude of the observer.

Absolute Flux of the Hard Component

Vertical Integral Intensity I(0) ~100 m^{-2}s^{-1}sr^{-1}
Angular dependence I(θ) ~ I(0) cos^2(θ)
Integrated Intensity ~200 m^{-2}s^{-1}
(O.C. Allkofer, 1975b).

Flux of the Soft Component

In free air, the soft component comprises about one third of the total cosmic ray flux.

Latitude Effect

The geomagnetic field influences the trajectories of lower energy cosmic rays approaching the Earth. As a result, the background flux is reduced by about 7% at the geomagnetic equator. The effect decreases toward the poles and is negligible at latitudes above about 40°.

Flux of Protons

The proton component is strongly attenuated by the atmosphere with an attenuation length (reduction by a factor of e) of about 120 g·cm^{-2}. It constitutes about 1% of the total vertical sea level flux.

Absorption

The soft component is absorbed in about 100 g·cm^{-2} of matter. The hard component is absorbed much more slowly:

Absorption in lead, 6% per 100 g·cm^{-2}
Absorption in rock, 8.5% per 100 g·cm^{-2}
Absorption in water, 10% per 100 g·cm^{-2}
(Absorption for depths less than 100 g·pd cm^{-2} is given by K. Greisen, 1943.)

Altitude Dependence

The cosmic ray background in the atmosphere has a maximum intensity of about 15 times that at sea level at a depth of about 150 g·cm^{-2} (15 km altitude). At maximum intensity, the soft and hard components contribute roughly equally but the hard component is then attenuated more slowly (S. Hayakawa, 1969).

Cosmic Ray Showers

High energy cosmic rays produce particle cascades in the atmosphere which can be detected at sea level provided that their energy exceeds about 100 GeV (such low energy cascades may be detected by using the most sensitive atmospheric Cerenkov detectors). The primary particle progressively loses energy which is transferred through the production of successive generations of secondary particles to a cascade of hadrons, an electromagnetic shower component (both positively and negatively charged electrons and gamma-rays) and muons. The secondary particles are relativistic and all travel effectively at the speed of light. As a result, they reach sea level at approximately the same time but, due to Coulomb scattering (for the electrons) and production angles (for the pions producing the muons), are spread laterally into a disk-like shower front with a characteristic lateral width of several tens of meters and thickness (near the central shower core) of 2 to 3 m. The number of particles at sea level is roughly proportional to the primary particle energy:

Number of particles at sea level ~10^{-10} × energy (eV).

At altitudes below a few kilometers, the number of particles in a shower attenuates with an *attenuation length* of about 200 g·cm^{-2}, i.e.,

particle number = original number × exp(−(depth increase)/200)

The above applies to an individual shower. The rate of observation of showers of a given size (particle number at the detector) at different depths of absorber attenuates with an *absorption length* of about 100 g·cm^{-2} (J.G. Wilson, 1976).

Atmospheric Background Light from Cosmic Rays

Cosmic ray particles produce Cerenkov light in the atmosphere and produce fluorescent light through the excitation of atmospheric molecules.

Cerenkov Light

High energy charged particles will cause the emission of Cerenkov light in air if their energies are above about 30 MeV (electrons). This threshold is pressure (and hence altitude) dependent. A typical Cerenkov light pulse (at sea level, 100 m from the central shower core) has a time spread of a few nanoseconds. Over this time, the photon flux between 430 and 530 nm would be ~10^{14} m^{-2}s^{-1} for a primary particle energy of 10^{16} eV. For comparison, the night sky background flux is ~6×10^{11} photons m^{-2}s^{-1}sr^{-1} in the same wavelength band (J.V. Jelley, 1967).

Fluorescence Light

Cosmic ray particles in the atmosphere excite atmospheric molecules which then emit fluorescence light. This is weak compared to the highly collimated Cerenkov component when viewed in the direction of the incident cosmic ray particle but is emitted isotropically. Typical pulse widths are longer than 50 ns and may be up to several microseconds for the total pulse from distant large showers (R.M. Baltrusaitis et al., 1985).

Effects of Cosmic Rays

Cerenkov Effects in Transparent Media

Background cosmic ray particles will produce Cerenkov light in transparent material with a photon yield between wavelengths λ_1 and λ_2

$$\sim (2\pi/137)\sin^2(\theta_c)\int_{\lambda_1}^{\lambda_2} d\lambda/\lambda^2 \text{ photons(unit length)}^{-1}$$

where θ_c (the Cerenkov angle) = $\cos^{-1}$ (1/refractive index).

This background light is known to affect light detectors, e.g., photomultipliers, and can be a major source of background noise (R.W. Clay and A.G. Gregory, 1977).

Effects on Electronic Components

If background cosmic ray particles pass through electronic components, they may deposit sufficient energy to affect the state of, e.g., a transistor flip-flop. This effect may be significant where reliability is of great importance or the background flux is high. For instance, it has been estimated that, in communication satellite operation, an error rate of about 2×10^{-3} per transistor per year may be found. Permanent damage may also result. A significant error rate may be found even at sea level in large electronic memories. This error rate is dependent on the sensitivity of the component devices to the deposition of electrons in their sensitive volumes (J.F. Ziegler, 1981).

Biophysical Significance

When cosmic rays interact with living tissue, they produce radiation damage. The amount of the damage depends on the total dose of radiation. At sea level, this dose is small compared with doses from other sources but both the quantity and quality of the radiation change rapidly with altitude. Approximate dose rates under various conditions are:

Dose rates (mrem·yr^{-1})
Sea level cosmic rays, 30
Cosmic rays at 10 km (subsonic jets), 2000
Cosmic rays at 18 km (supersonic transports), 10,000
(c.f., mean total sea level dose, 300)

Astronauts would be subject to radiation from galactic (0.05 rads per day) and solar (a few hundred rads per solar flare) cosmic rays as well as large fluxes of low energy radiation when passing through the Van Allen belts (about 0.3 rads per traverse).

Both astronauts and SST travellers would be subject to a small flux of low energy heavy nuclei stopping in the body. Such particles are capable of destroying cell nuclei and could be particularly harmful in the early stages of the development of an embryo. The rates of heavy nuclei stopping in tissue in supersonic transports and spacecraft are approximately as follows:

Stopping nuclei ((cm^3 tissue)$^{-1}$ hr^{-1})
Supersonic transport (16 km), 0.0005
Supersonic transport (20 km), 0.005
Spacecraft, 0.15
(O. C. Allkofer, 1975a; O. C. Allkofer et al., 1974).

Carbon Dating

Radiocarbon is produced in the atmosphere due to the action of cosmic ray slow neutrons. Solar cycle modulation of the very low energy cosmic rays causes an anticorrelation of the atmospheric ^{14}C activity with sunspot number with a mean amplitude of about 0.5%. In the long term, modulation of cosmic rays by a varying magnetic field may be important (A.A. Burchuladze et al., 1979).

Practical Uses of Cosmic Rays

There are few direct practical uses of cosmic rays. Their attenuation in water and snow have, however, enabled automatic monitors of water and snow depth to be constructed. A search for hidden cavities in pyramids has been carried out using a muon "telescope."

Other Effects

Stellar X-rays have been observed to affect the transmission times of radio signals between distant stations by altering the depth of the ionospheric reflecting layer. It has also been suggested that variations in ionization of the atmosphere due to solar modulation may have observable effects on climatic conditions.

References

O.C. Allkofer (1975a) *Introduction to Cosmic Radiation*, Verlag Karl Thiemig, Munchen, Germany.
O.O. Allkofer (1975b) *J. Phys. G: Nucl. Phys.*, 1, L51.
O.C. Allkofer and W. Heinrich (1974) *Health Phys.*, 27, 543.
M. Amenomori et al. (1995) *Proc. 24th Int. Cosmic Ray Conf. Rome*, 3, 85. Universita La Sapienza, Roma.
K. Asakimori et al. (1993) *Proc. 23rd Int. Cosmic Ray Conf. Calgary*, 2, 25, University of Calgary, Canada.
R.M. Baltrusaitis et al. (1985) *Nucl. Inst. Meth.*, A420, 410.
D.J. Bird et al. (1993) *Phys. Rev. Lett.*, 71, 3401.
A.A. Burchuladze, S.V. Pagava, P. Povinec, G. I. Togonidze, S. Usacev (1979) *Proc. 16th Int. Cosmic Ray Conf. Kyoto*, 3, 201, Univ. of Tokyo, Japan.
R.W. Clay (1987) *Aust. J. Phys.*, 40, 423.
R.W. Clay and A.G. Gregory (1977) *J. Phys. A: Math. Gen.*, 10, 135.
J.M. Clem et al. (1995) *Proc. 24th Int. Cosmic Ray Conf. Rome*, 3, 5, Universita La Sapienza, Roma.
T.K. Gaisser et al. (1993) *Phys. Rev. D*, 47, 1919.
K. Greisen (1943) *Phys. Rev.*, 63, 323.
S. Hayakawa (1969) *Cosmic Ray Physics*, Wiley-Interscience, New York.
J.V. Jelley (1967) *Prog. in Elementary Particle and Cosmic Ray Physics*, 9, 41.
E. Juliusson (1975) Proc. 14th Int. Cosmic Ray Conf. Munich, 8, 2689, Max Planck Institute fur Extraterrestriche Physik, Munchen, Germany.
J. Linsley (1981) *Origin of Cosmic Rays*, I.A.U. Symposium 94, 53, D. Reidel Publishing Co. Dordrecht, Holland.
P. Meyer (1981) *Origin of Cosmic Rays*, I.A.U. Symposium 94, 7, D. Reidel Publishing Co. Dordrecht, Holland.
P.F. Michelson (1994) in *Towards a Major Atmospheric Cerenkov Detector III*, 257, Ed. T. Kifune, Universal Academy Press Inc., Tokyo, Japan.
P. Sokolsky (1989) *Introduction to Ultrahigh Energy Cosmic Ray Physics*, Addison Wesley Publishing Company.
T. Stanev et al. (1995) *Phys. Rev. Lett.*, 75, 3056.
S. Orito et al. (1995) *Proc. 24th Int. Cosmic Ray Conf. Rome*, 3, 76. Universita La Sapienza, Roma.
J.G. Wilson (1976) *Cosmic Rays*, Wykeham Pub. (London) Lt., U.K.
A.W. Wolfendale (1979) *Pramana*, 12, 631.
S. Yoshida et al. (1995) *Astroparticle Phys.*, 3, 105.
J.F. Ziegler, (1981) *IEEE Trans. Electron Devices*, ED-28, 560.

Section 12
Properties of Solids

Solids

Techniques for Materials Characterization: Experimental Techniques Used to Determine the Composition, Structure, and Energy States of Solids and Liquids 12-1
Symmetry of Crystals 12-5
Ionic Radii in Crystals 12-11
Polarizabilities of Atoms and Ions in Solids 12-13
Crystal Structures and Lattice Parameters of Allotropes of the Elements 12-15
Phase Transitions in the Solid Elements at Atmospheric Pressure 12-19
Lattice Energies 12-21
The Madelung Constant and Crystal Lattice Energy 12-34
Elastic Constants of Single Crystals 12-35
Electrical Resistivity of Pure Metals 12-41
Electrical Resistivity of Selected Alloys 12-43
Electrical Resistivity of Graphite Materials 12-46
Permittivity (Dielectric Constant) of Inorganic Solids 12-47
Curie Temperature of Selected Ferroelectric Crystals 12-56
Properties of Antiferroelectric Crystals 12-57
Dielectric Constants of Glasses 12-58
Properties of Superconductors 12-59
High-Temperature Superconductors 12-75
Organic Superconductors 12-77
Properties of Semiconductors 12-80
Selected Properties of Semiconductor Solid Solutions 12-93
Properties of Organic Semiconductors 12-95
Diffusion Data for Semiconductors 12-99
Properties of Magnetic Materials 12-107
Organic Magnets 12-116
Electron Inelastic Mean Free Paths 12-119
Electron Stopping Powers 12-121
Electron Work Function of the Elements 12-123
Secondary Electron Emission 12-124
Optical Properties of Selected Elements 12-125
Optical Properties of Selected Inorganic and Organic Solids 12-150
Elasto-Optic, Electro-Optic, and Magneto-Optic Constants 12-169
Nonlinear Optical Constants 12-183
Phase Diagrams 12-186
Heat Capacity of Selected Solids 12-204
Thermal and Physical Properties of Pure Metals 12-205
Thermophysical Properties of Stainless Steel 310 12-207
Thermal Conductivity of Metals and Semiconductors as a Function of Temperature 12-208
Thermal Conductivity of Alloys as a Function of Temperature 12-210
Thermal Conductivity of Crystalline Dielectrics 12-211
Thermal Conductivity of Ceramics and Other Insulating Materials 12-213
Thermal Conductivity of Glasses 12-215
Thermoelectric Properties of Metals and Semiconductors 12-219
Fermi Energy and Related Properties of Metals 12-221
Properties of Commercial Metals and Alloys 12-223
Hardness of Minerals and Ceramics 12-224

Section 12
Properties of Solids

Techniques for Materials Characterization; Experimental Techniques Used to Determine the Composition, Structure, and Energy States of Solids and Liquids
Symmetry of Crystals
Ionic Radii in Crystals
Polarizabilities of Atoms and Ions in Solids
Crystal Structures and Lattice Parameters of Allotropes of the Elements
Phase Transitions in the Solid Elements at Atmospheric Pressure
Lattice Energies
The Madelung Constant and Crystal Lattice Energy
Elastic Constants of Single Crystals
Electrical Resistivity of Pure Metals
Electrical Resistivity of Selected Alloys
Electrical Resistivity of Graphite Materials
Permittivity (Dielectric Constant) of Inorganic Solids
Curie Temperature of Selected Ferroelectric Crystals
Properties of Antiferroelectric Crystals
Dielectric Constants of Glasses
Properties of Superconductors
High-Temperature Superconductors
Organic Superconductors
Properties of Semiconductors
Selected Properties of Semiconductor Solid Solutions
Diffusion Data for Semiconductors
Properties of Magnetic Materials
Organic Magnets
Electron Inelastic Mean Free Paths
Electron Stopping Powers
Electron Work Function of the Elements
Secondary Electron Emission
Optical Properties of Selected Elements
Optical Properties of Selected Inorganic and Organic Solids
Elasto-Optic, Electro-Optic, and Magneto-Optic Constants
Nonlinear Optical Constants
Phase Diagrams
Heat Capacity of Selected Solids
Mechanical and Physical Properties of Pure Metals
Thermophysical Properties of Stainless Steel 316
Thermal Conductivity of Metals and Semiconductors as a Function of Temperature
Thermal Conductivity of Alloys as a Function of Temperature
Thermal Conductivity of Crystalline Dielectrics
Thermal Conductivity of Ceramics and Other Insulating Materials
Thermal Conductivity of Glasses
Thermoelectric Properties of Metals and Semiconductors
Fermi Energy and Related Properties of Metals
Properties of Commercial Metals and Alloys
Hardness of Minerals and Ceramics

TECHNIQUES FOR MATERIALS CHARACTERIZATION

Experimental Techniques Used to Determine the Composition, Structure, and Energy States of Solids and Liquids

H. P. R. Frederikse

The many experimental methods, originally designed to study the chemical and physical behavior of solids and liquids, have grown into a new field known as Materials Characterization (or Materials Analysis). During the past 30 years a host of techniques aimed at the study of surfaces and thin films has been added to the many tools for the analysis of bulk samples. The field has benefitted particularly from the development of computers and microprocessors, which have vastly increased the speed and accuracy of the measuring devices and the recording of their output. Materials characterization was and is a very important tool in the search for new physical and chemical phenomena. It plays an essential role in new applications of solids and liquids in industry, communications, and medicine. Many of its techniques are used in quality control, in safety regulations, and in the fight against pollution.

In most Materials Characterization experiments the sample is subjected to some kind of radiation: electromagnetic, acoustic, thermal, or particles (electrons, ions, neutrons, etc.). The surface analysis techniques usually require a high vacuum. As a result of interactions between the solid (or liquid) and the incoming radiation a beam of a similar (or a different) nature will emerge from the sample. Measurement of the physical and/or chemical attributes of this emerging radiation will yield qualitative, and often quantitative, information about the composition and the properties of the material being probed.

The modern tendency of describing practically everything in this world by a combination of a few letters (acronyms) has also penetrated the field of Materials Characterization. The table below gives the meaning of the acronym for every technique listed, the form and size of the required sample (bulk, surface, film, liquid, powder, etc.), the nature of the incoming and of the emerging radiation, the depth and the lateral spatial resolution that can be probed, and the information obtained from the experiment. The last column lists one or two major references to the technique described.

	Technique	Sample	In	Out	Depth	Lateral resolution	Information obtained	Ref.
	Optical and Mass Spectroscopies for Chemical Analysis							
1	AAS Atomic Absorption Spectroscopy	Atomize (flame, electro thermal, etc.)	Light, r.f., glow discharge	Absorption spectrum		–	Concentration of atomic species (quantitative, using standards)	1,2
2.	ICP-AES Induct. Coupled Plasma – Atomic Emission Spectroscopy	Atomize (flame, electro, thermal, ICP, etc.)	–	Emission spectrum	–	–	Concentration of atomic species (quantitative, using standards)	3
3.	Dynamic SIMS Dynamic Secondary Ion Mass Spectroscopy	Surface	Ion beam (1–20 keV)	Secondary ions; analysis with mass spectrometer	2 nm–1 μm (or deeper: ion milling)	0.50 nm	Elemental and isotopic analysis; depth profile (all elements); detection limits: ppb-ppm	4
4.	Static SIMS Static Secondary Ion Mass Spectroscopy	Surface	Ion beam (0.5–20 keV)	Secondary ions, analysis with mass spectrometer	0.1–0.5 nm	10 μm	Elemental analysis of surface layers; molecular analysis; detection limits: ppb-ppm	4
5.	SNMS Sputtered Neutral Mass Spectroscopy	Surface, bulk	Plasma discharge; noble gases: 0.5–20 keV	Sputtered atoms ionized by atoms or electrons; then mass analyzed	0.1–0.5 nm (or deeper: ion milling)	1 cm	Elemental analysis Z ≥ 3; depth profile; detection limit: ppm	4,6
6.	SALI Surface Analysis by Laser Ionization	Surface	e-beam, ion-beam, or laser for sputtering	Sputtered atoms ionized by laser; then mass analyzed	0.1–0.5 nm up to 3 μm in milling mode	60 nm	Surface analysis; depth profiling	7
7.	LIMS Laser Ionization Mass Spectroscopy	Surface, bulk	u.v. laser (ns pulses)	Ionized species; analyzed with mass spectrometer	50–150 nm	5 μm–1 mm	Elemental (micro)analysis; detection limits: 1–100 ppm	8
8.	SSMS Spark Source Mass Spectroscopy	Sample in the form of two electrodes	High voltage R.F. spark produces ions	Ions – analyzed in mass spectrometer	1–5 μm	–	Survey of trace elements; detection limit: 0.01–0.05 ppm	9
9.	GDMS Glow Discharge Mass Spectroscopy	Sample forms the cathode for a D.C. glow discharge	Sputtered atoms ionized in plasma	Ions – analyzed in mass spectrometer	0.1–100 μm	3–4 mm	(Bulk) trace element analysis; detection limit: sub-ppb	9,10
10.	ICPMS Induct. Coupled Plasma Mass Spectroscopy	Liquid-dissolved sample carried by gas stream into R.F. induction coil	Ions produced in argon plasma	Ions – analyzed in quadrupole mass spectrometer	–	–	High sensitivity analysis of trace elements	11
	Photons — Absorption, Reflection and Electron Emission							
11.	IRS Infrared Spectroscopy	Thin crystal, glass, liquid	I.R. light (W-filament, globar, Hg-arc)	I.R. spectrum	–	–	Electronic transitions (mainly in semiconductors and superconductors); vibrational modes (in crystals and molecules)	12,13,14
12.	FTIR Fourier Transform I.R. Spectroscopy	Solid, liquid; transmission or reflection	White light (all frequencies)	Fourier Transform of spectrum (interferometer)	–	–	Spectra obtained at higher speed and resolution	15
13.	ATR Attenuated Total Reflection	Surface or thin crystal	–	–	μm's	–	Atomic or molecular spectra of surfaces and films	16
14.	(μ)-RS (Micro-) Raman Spectroscopy	Solid, liquid (1 μm–1 cm)	Laser beam, e.g., Ar-line, YAG-line	Raman spectra	0.5 μm	0.5 μm	Molecular and crystal vibrations	12,14,17

	Technique	Sample	In	Out	Depth	Lateral resolution	Information obtained	Ref.
15.	CARS Coherent Anti-Stokes Raman Spectroscopy	Solid, liquid (50 µm–3 cm)	Pump beam (ω_0)+ probe beam (ω_s)	Anti-Stokes spectrum	–	–	High resolution Raman spectra	14
16.	Ellipsometry	Transparent films, crystals, adsorbed layers	Polarized light	Change in polarization	0.05 nm–5 µm	25 µm (or sample thickness)	Refractive index and absorption	18,19
17.	UPS Ultraviolet Photo-electron Spectroscopy	Surfaces, adsorbed layers	u.v. light, 10–100 eV; 200 eV (synchrotron)	Electrons	0.2–10 nm	0.1–10 nm	Energies of electronic states of surfaces and free molecules	20,21
18.	PSD Photon Stimulated Desorption	Surfaces with adsorbed species	Far u.v. light E > 10 eV	Ions – analyzed with mass spectrometer	0.1–2 nm	–	Structure and desorption kinetics of adsorbed atoms and molecules	22
X-Rays								
19.	XRD X-Ray Diffraction	Single crystals, powders films	X-rays: λ = 0.05–0.2 nm (6–17 keV)	Diffracted X-ray beam	1–1000 µm	0.1–10 mm	Identification of crystallographic structures; all elements (low Z difficult)	23,24
20.	XRF/EDS X-Ray Fluorescence/Energy Dispersive Spectroscopy	Thin films, single layer	Prim. X-ray beam λ = 0.02–0.1 nm 12–80 keV	Fluorescent X-rays	1–100 µm	10 mm	Elemental analysis; all elements except H, He, Li – (EDS also used in XRD, SEM, TEM and EPMA)	25,26
21.	EXAFS Extended X-Ray Absorption Fine Structure	Films, foils	High intensity X-rays (synchrotron)	Spectrum near absorption edge	nm–µm	–	Local atomic structure: order/ disorder in vicinity of absorbing atom	27
22.	XPS/ESCA X-Ray Photo-electron Spectroscopy/ Electron Spect. for Chemical Analysis	Surfaces, thin films (≈20 atomic layers)	Soft X-rays (1–20 keV)	Core electrons; valence electrons	0.5–10 nm	5 nm–50 µm	(Quantitative) identification of all elements in surface layer or film	28,29
Electrons								
23.	CL Cathode Luminescence	Insulators, semiconductors	Electrons 5–50 keV	Photons 0.1–5 eV	1 nm–2 µn	1 or 2 µm	Energy levels of impurities and point defects	30
24.	APS Appearance Potential Spectroscopy	Surface (≈20 atomic layers)	Electrons (energy scan) 50–2000 eV	X-rays to pinpoint electron energy threshold	–	–	Identification of surface species	21, see also C
25.	AES Auger Electron Spectroscopy	Thin films, surfaces	Electrons 3–10 keV	Auger electrons 20–2000 eV	0.3–3 nm	≈30 nm	Elemental composition of surface (except H, He); detection limit 0.1–1%	28,29
26.	EELS Electron Energy Loss Spectroscopy	Very thin samples (<200 nm)	Electrons (100–400 keV)	(Retarded) electrons; minus 1–1000 eV	<200 nm	1–100 nm	Local elemental concentration; electronic structure, chem. bonding; interatomic distances	31
27.	EXELFS Extended Electron Energy Loss Fine Structure	Thin films	Electrons (100–400 keV)	Electrons energies 0–30 eV above edge	<200 nm	1–100 nm	Density of states of valence electrons (above Fermi level)	27,32
28.	ESD Electron Stimulated Desorption	Adsorbed species	Electrons E > 10 eV	Ions – analyzed with mass spectrometer	–	–	Structure and desorption properties of adsorbed atoms and molecules	22
29.	ESDIAD ESD-Ion Angular Distribution	(See ESD)	(See ESD)	Directional dependence of emitted ions	–	–	Geometries of adsorbed species (atoms or molecules)	22
30.	EPMA Electron Probe (X-Ray) Micro Analysis	Solid conductors and insulators <1 cm thick	Electrons 5–30 keV	Characteristic X-ray 0.1–15 keV	100 nm–5 µm	1 µm	Elemental analysis, $Z \leq 4$, major, minor and trace amounts	33,34
31.	LEED Low Energy Electron Diffraction	Surface	Mono-energetic electron beam 10–1000 eV	Diffracted electrons	0.4–2 nm	<5 µm	Crystallographic structure of surface; resolution: 0.01 nm	35
32.	RHEED Reflection High Energy Electron Diffraction	Surface	Electron beam at grazing angle 5–50 keV	Reflected electrons	0.2–10 nm	<5 µm	Surface symmetry	36,37
33.	SEM Scanning Electron Microscopy	Bulk, films (conducting)	High energy electrons usually ~30 keV	Secondary and backscattered electrons	1 nm–5 µm	1–20 nm	Surface image, defect structure; resolution 5–15 nm; magnification 300,000×	33,34
34.	(S)TEM (Scanning) Transmission Electron Microscopy	Thin specimen – <200 nm	High energy electrons typically 300 keV	Transmitted and diffracted electrons	(Sample thickness)	2–20 nm	(Defect) structure of cryst. solids; microchemistry; high resol.: 0.2 nm	33
35.	FEM Field Emission Microscopy	Metals, alloys (sharp point)	–	Electron emission (with appl. electric field – 50 kV)	≈0.5 nm	10–100 nm	Surface image, crystallographic structure	34
36.	STM Scanning Tunneling Microscopy	Polished or cleaved surface (conducting)	Tunneling current controls distance between sample and very sharp tip		1–5 nm	2–10 nm	Atomic-scale relief map of surface; resolution: vert. 0.002 nm, hor. 0.2 nm	39
37.	SPM Scanned Probe Microscopy	Very flat surface	Any field: e.g. mechan. vibration recorded with laser probe; same with magnetic, electric or thermal field		1–100 nm	1–100 nm	Surface-magnetic field, surface-thermal conductivity, etc.	39a
38.	AFM Atomic Force Microscopy	Very flat surface	Similar to STM; force measured with cantilever spring		0.5–5 nm	0.2–130 nm	Surface topography with atomic resolution; interatomic force	40

	Technique	Sample	In	Out	Depth	Lateral resolution	Information obtained	Ref.
Ions and Neutrons								
39.	ISS (or LEIS) Ion Scattering Spectroscopy (Low Energy Ion Scattering)	Surface	Ion beam He^+ or Ne^+ <3 keV	Sputtered ions (energy analysis)	0.1–0.5 nm	1–100 μm	Elemental analysis (better for low Z) detection limits: 0.01–1%	41
40.	FIM Field Ion Microscopy	Surface: metals, alloys; very sharp tip	(He gas above sample)	He ions + high electric field produce image	≈0.1 nm	0.1–2 nm	Atomic structure of surface	34,42
41.	RBS Rutherford Back Scattering	Solids, thin films	Mono-energetic ions (H^+ or He^{++}) 0.5–3 MeV	Backscattered ions	10 nm–1 μm	1 mm	Element identification (Li to U) detection limit: 0.01–1%	46
42.	NRA Nuclear Reaction Analysis	Solids, thin films	Mono-energetic ions (Li, Be, B, etc.) 200 keV–6 MeV	Protons, deuterons ^3He, α-particles, γ-rays	0.1–5 μm	10 μm–10 mm	Element identification (all) detection limit: 10^{-12}–10^{-2}	47
43.	PIXE Particle Induced X-ray Emission	Thin films, surface layers	High energy ions (H^+ or He^{++})	Characteristic X-rays	<10 μm	1 μm–2 mm	Trace impurities: $Z>3$ detection limit: 0.1–100 ppm (depending on sample thickness)	48
44.	INS Ion Neutralization Spectroscopy	Surface	He-ions (≈5 eV)	Electrons	–	–	Energies of valence electrons	49
45.	NAA Neutron Activation Analysis	Bulk, >0.5 g	Thermal neutrons	Characteristic γ-rays, (≈1 MeV)	Bulk	–	Trace concentrations (of isotopes) of elements: trans. metals, Pt-group; detection limit: 10^8–10^{14} atoms/cm^3	43
46.	N(P)D Neutron (Powder) Diffraction	Crystalline solids	Thermal neutrons E ≈0.0025 eV	Diffracted neutrons	Bulk	–	Crystallographic structure; porosity, particle size	44
47.	SANS Small Angle Neutron Scattering	Inhomogeneous solids; powders; porous samples	Thermal neutrons 2 θ = 10^{-2}–10^{-4}	Scattered neutrons	1–25 mm	–	Average size of inhomogeneities; range: 1 nm–1 mm	45
Acoustic								
48.	SLAM Scanning Laser Acoustic Microscopy	Bulk, film	Acoustic wave produced by laser 1 MHz–1 GHz	Reflected acoustic wave	μm–cm	0.1–20 mm	Defect structure; thickness measurement	50
Thermal								
49.	DTA Differential Thermal Analysis	Specimen and reference sample	Uniform heating	Temperature difference	Bulk	–	Phase transitions, crystallization	51
50.	DSC Differential Scanning Calorimetry	Specimen and ref. sample	Controlled heating	Measure heat required for equal temperature	Bulk	–	Phase transitions, crystallization; activation energies	51
51.	TGA Thermo Gravimetric Analysis	Bulk, 1–100 g	Controlled heating	Weight as function of temperature (and time)	Bulk	–	Decomposition, non-stoichiometry, kinetics of reaction	52
Resonance								
52.	EPR (ESR) Electron Paramagnetic (Spin) Resonance	Paramagnetic solids or liquids	Microwave radiation in magnetic field 3–300 GHz; 1–100 kG	Microwave absorption (at resonance)	Bulk	–	Local environment of paramagnetic ion; concentration of paramagnetic, species; detection limit: 10^{11} spins/cm^3	53,54
53.	ECR Electron Cyclotron Resonance	Semiconductors, metals; free electrons (low temperature)	Microwave radiation in magnetic field 10–30 GHz; 5–10 kG	Microwave absorption (at resonance)	Bulk	–	Electronic energy bands, effective masses	55
54.	Mössbauer Effect	Source and absorber	Mono-energetic γ-rays: 5–100 keV	Mössbauer spectrum (Doppler shifted (lines)	50 m	1 cm	Interaction between nucleus and its environment (local electric, magnetic fields; bonds; valency; diffusion, etc.)	56
55.	NMR (MRI) Nuclear Magnetic Resonance (Magnetic Resonance Imaging)	Solids, liquids	R.F. radiation + magnetic field; e.g. for protons: 60 MHz, 14 kG	R.F. absorption	<1 cm	1 cm	Quant. analysis; local magnetic environment; diffusion; imaging	58
56.	ENDOR Electron Nuclear Double Resonance	Solids, liquids	R.F. + microwave radiation in magn. field.	Microwave absorption	–	–	Hyperfine interaction → local atomic structure	54
57.	NQR Nuclear Quadrupole Resonance	Solids	R.F. radiation 0.5–1000 MHz	R.F. absorption	–	–	Asymmetry of the charge distribution at the nucleus	55,59
Other								
58.	BET Brunauer-Emmett-Teller	(Large) surface area 1–20 m^2/g	Adsorbed gas (e.g., N_2 at low temp.) as function of pressure (monolayer coverage)		–	–	Surface area measurement	60

References

General References

A. Wachtman, J. B., *Characterization of Materials*, Butterworth-Heinemann, Boston, 1993.
B. Brundle, C. R., Evans, C. A., and Wilson, S., Eds., *Encyclopedia of Materials*, Butterworth-Heinemann, Boston, 1992.
C. Woodruff, D. P. and Delchar, T. A., *Modern Techniques of Surface Science*, Cambridge University Press, Cambridge, 1986.
D. *Metals Handbook*, 9th Edition, Vol. 10, Materials Characterization, Whan, R. E., Coordinator, American Society for Metals, Metals Park, OH, 1986.

Specific References

1. Slavin, M., *Atomic Absorption Spectroscopy*, 2nd Edition, John Wiley & Sons, New York, 1978.
2. Schrenk, W. G., *Analytical Atomic Spectroscopy*, Plenum Press, New York, 1975.
3. Dean, J. A. and Rains, T. E., *Flame Emission and Atomic Absorption Spectroscopy*, Vols. 1–3, Marcel Dekker, New York, 1969.
4. Benninghoven, A., Rudenauer, F. G., and Werner, H. W., *Secondary Ion Mass Spectroscopy*, John Wiley & Sons, New York, 1987.
5. Bird, J. R. and Williams, J. S., Eds., in *Ion Beams for Materials Analysis*, Academic Press, New York, 1989, pp. 515–537.
6. Smith, G. C., *Quantitative Surface Analysis for Materials Science*, The Institute of Metals, London, 1991.
7. Becker, E. H., in *Ion Spectroscopies for Surface Analysis*, Czanderna, A. W. and Hercules, D. M., Eds., Plenum Press, New York, 1991, p. 273.
8. Simons, D. S., *Int. J. Mass Spectrometry and Ion Processes*, 55, 15, 1983.
9. White, F. A. and Wood, G. M., *Mass Spectrometry: Applications in Science and Engineering*, John Wiley & Sons, New York, 1986.
10. Harrison, W. W. and Bentz, B. L., *Prog. Anal. Spectrometry*, 11, 53, 1988.
11. Bowmans, P. W. J. M., *Inductively Coupled Plasma Emission Spectroscopy*, Parts I and II, John Wiley & Sons, New York, 1987.
12. Brame, Jr., E. G. and Grasselli, J., *Infrared and Raman Spectroscopy*, Practical Spectroscopy Series, Vol. I, Marcel Dekker, New York, 1976.
13. Hollas, J. M., *Modern Spectroscopy*, John Wiley & Sons, New York, 1987.
14. Turrell, G., *Infrared and Raman Spectroscopy of Crystals*, Academic Press, New York and London, 1972.
15. Griffith, P. R. and Haseth, J. A., *Fourier Transform Infrared Spectroscopy*, John Wiley & Sons, New York, 1986.
16. Barnowski, M. K., *Fundamentals of Optical Fiber Communications*, Academic Press, New York, 1976.
17. Long, D. A., *Raman Spectroscopy*, McGraw-Hill, New York, 1977.
18. Azzam, R. M. A., *Ellipsometry and Polarized Light*, Elsevier-North Holland, Amsterdam, 1977.
19. Hecht, E., *Optics*, 2nd Edition, Addison-Wesley, Reading MA, 1987.
20. Brundle, C. R., in *Molecular Spectroscopy*, West, A. R., Ed., Heyden, London, 1976.
21. Park, R. L., in *Experimental Methods in Catalytic Research*, Vol. III, Anderson, R. B. and Dawson, P. T., Academic Press, New York, 1976, pp. 1–39.
22. Madey, T. E. and Stockbauer, R., in *Solid State Physics: Surfaces*, Vol. 22 of Methods of Experimental Physics, Park, R.L. and Lagally, M. G., Eds., Academic Press, New York, 1985.
23. Cullity, B. D., *Elements of X-Ray Diffraction*, 2nd Edition, Addison-Wesley, Reading, MA, 1978.
24. Schwartz, L. H. and Cohen, J. B., *Diffraction from Materials*, Springer Verlag, Berlin, 1987.
25. deBoer, D. K. G., in *Advances in X-Ray Analysis*, Vol. 34, Barrett, C. S. et. al., Eds., Plenum Press, New York, 1991.
26. Birks, L. S., *X-Ray Spectrochemical Analysis*, 2nd Edition, John Wiley & Sons, New York, 1969.
27. Bonnelle, C. and Mande, C., *Advances in X-Ray Spectroscopy*, Pergamon Press, Oxford, 1982.
28. *Practical Surface Analysis by Auger and X-Ray Photo-Electric Spectroscopy*, Briggs, D. and Seah, M. P., Eds., John Wiley & Sons, New York, 1983.
29. Powell, C. J. and Seah, M. P., *J. Vac. Sci. Technol. A*, Vol. 8, 735, 1990.
30. Yacobi, G. G. and Holt, D. B., *Cathodeluminescence Microscopy of Inorganic Solids*, Plenum Press, New York, 1990.
31. Egerton, R. F., *Electron Energy Loss Spectroscopy in the Electron Microscope*, Plenum Press, New York, 1986.
32. Disko, M. M., Krivanek, O. L., and Rez, P., *Phys. Rev.*, B25, 4252, 1982.
33. Goldstein, J. I., et. al., *Scanning Electron Microscopy and X-Ray Microanalysis*, 2nd Edition, Plenum Press, New York, 1986.
34. Murr, L. E., *Electron and Ion Microscopy and Microanalysis*, Marcel Dekker, New York, 1982.
35. Armstrong, R. A., in *Experimental Methods in Catalytic Research*, Vol. III, Anderson, R. B., and Dawson, P. T., Eds., Academic Press, New York, 1976.
36. Dobson, P. J. et. al., *Vacuum*, 33, 593, 1983.
37. Rymer, T. B., *Electron Diffraction*, Methuen, London, 1970.
38. Reimer, L., *Transmission Electron Microscopy*, Springer-Verlag, Berlin, 1984.
39. *Scanning Tunneling Microscopy and Related Methods*, Behm, R. J., Garcia, N., and Rohrer, H., Eds., Kluwer Academic Publishers, Norwell, MA, 1990.

39a. Wikramasinghe, H.K., *Scientific American*, Vol. 261, No. 4, pp. 98–105, Oct. 1989.

40. Rugar, D. and Hansma, P., *Physics Today*, 43(10), pp. 23–30, 1990.
41. Feldman, C. C. and Mayer, J. W., *Fundamentals of Surface and Thin Film Analysis*, North-Holland, Amsterdam, 1986.
42. Muller, E. W. and Tsong, T. T., *Field Ion Microscopy*, Elsevier, Amsterdam, 1969.
43. Amiel, S., *Nondestructive Activation Analysis*, Elsevier, Amsterdam, 1981.
44. Bacon, G. E., *Neutron Diffraction*, 3rd Edition, Clarendon Press, Oxford, 1975.
45. Neutron Scattering, Part A., in *Methods of Experimental Physics*, Vol. 23, Skold, K. and Price, D. L., Eds., Academic Press, New York, 1986.
46. Chu, W. K., Mayer, J. W., and Nicolet, M. A., *Backscattering Spectroscopy*, Academic Press, New York, 1987.
47. Rickey, F. A., in *High Energy and Heavy Ion Beams in Materials Analysis*, Tesmer, J. R., et. al., Eds., MRS, 1990, pp. 3–26.
48. Johansson, S. A. E. and Campbell, J. L., *PIXE: A Novel Technique for Elemental Analysis*, John Wiley & Sons, New York, 1988.
49. Hagstrum, H. D., in *Inelastic Ion-Surface Collisions*, Tolk, N. H. et. al., Eds., Academic Press, New York, 1977, pp. 1–46.
50. Nikoonahad, M., in *Research Techniques in Nondestructive Testing*, Vol. VI, Sharpe, R.S., Ed., Academic Press, New York, 1984, pp. 217–257.
51. Gallagher, P. K., *Characterization of Materials by Thermoanalytical Techniques*, MRS - Bulletin, Vol. 13, No. 7, pp. 23–27, 1988.
52. Earnest, C. M., *Compositional Analysis by Thermogravimetry*, ASTM Special Technical Publication 997, 1988.
53. Poole, C. P., *Electron Spin Resonance – A Comprehensive Treatise on Experimental Techniques*, 2nd Edition, John Wiley & Sons, New York, 1983.
54. Atherton, N. M., *Principles of Electron Spin Resonance*, Ellis Horwood Ltd., Chichester, U.K., 1993.
55. Kittel, C., *Introduction to Solid State Physics, 6th Edition*, John Wiley & Sons, New York, 1986, p. 196.
56. Gibb, T. C., *Principles of Mössbauer Spectroscopy*, Chapman & Hall, London, 1976.
57. Slichter, C. P., *Principles of Magnetic Resonance*, 3rd Edition, Springer-Verlag, Berlin, 1990.
58. *NMR Spectroscopy Techniques*, Dybrowski, C. and Lichter, R. L., Eds., Marcel Dekker, New York, 1987.
59. Das, T. P. and Hahn, E. L., *Nuclear Quadrupole Resonance Spectroscopy*, Academic Press, New York, 1958.
60. Somorjai, G. A., *Principles of Surface Chemistry*, Prentice-Hall, Englewood Cliffs, NJ, 1972, p. 216

SYMMETRY OF CRYSTALS

L. I. Berger

The ability of a body to coincide with itself in its different positions regarding a coordinate system is called its symmetry. This property reveals itself in iteration of the parts of the body in space. The iteration may be done by reflection in mirror planes, rotation about certain axes, inversions and translations. These actions are called the symmetry operations. The planes, axes, points, etc., are known as symmetry elements. Essentially, mirror reflection is the only truly primitive symmetry operation. All other operations may be done by a sequence of reflections in certain mirror planes. Hence, the mirror plane is the only true basic symmetry element. But for clarity, it is convenient to use the other symmetry operations, and accordingly, the other aforementioned symmetry elements. The symmetry elements and operations are presented in Table 1.

The entire set of symmetry elements of a body is called its symmetry class. There are thirty-two symmetry classes that describe all crystals that have ever been noted in mineralogy or been synthesized (more than 150,000). The denominations and symbols of the symmetry classes are presented in Table 2.

There are several known approaches to classification of individual crystals in accordance with their symmetry and crystallochemistry. The particles that form a crystal are distributed in certain points in space. These points are separated by certain distances (translations) equal to each other in any chosen direction in the crystal. Crystal lattice is a diagram that describes the location of particles (individual or groups) in a crystal. The lattice parameters are three non-coplanar translations that form the crystal lattice. Three basic translations form the unit cell of a crystal. August Bravais (1848) has shown that all possible crystal lattice structures belong to one or another of fourteen lattice types (Bravais lattices). The Bravais lattices, both primitive and non-primitive, are the contents of Table 3.

Among the three-dimensional figures, there is a group of polyhedrons that are called regular, which have all faces of the same shape and all edges of the same size (regular polygons). It has been shown that there are only five regular polyhedrons. Because of their importance in crystallography and solid state physics, a brief description of these polyhedrons is included in Table 4.

The systematic description of crystal structures is presented primarily in the well-known *Structurbericht*. The classification of crystals by the Structurbericht does not reflect their crystal class, the Bravais lattice, but is based on the crystallochemical type. This makes it inconvenient to use the Structurbericht categories for comparison of some individual crystals. Thus, there have been several attempts to provide a more convenient classification of crystals. Table 5 presents a compilation of different classifications which allows the reader to correlate the Structurbericht type with the international and Schoenflies point and space groups and with Pearson's symbols, based on the Bravais lattice and chemical composition of the class prototype. The information included in Table 5 has been chosen as an introduction to a more detailed crystallophysical and crystallochemical description of solids.

TABLE 1. Symmetry Operations and Elements

Symmetry operation	Symmetry element: Name	Symbol: International (Hermann-Mauguin)	Symbol: Schoenflies	Presentation on the stereographic projection: Parallel	Presentation on the stereographic projection: Perpendicular
Reflection in a plane	Plane	m	C_s		
Rotation by angle α = 360°/n about an axis	Axis	n = 1, 2, 3, 4 or 6	C_n		
		n = 2	C_2		
		n = 3	C_3		
		n = 4	C_4		
		n = 6	C_6		
Rotation about an axis and inversion in a symmetry center lying on the axis	Inversion (improper) axis	$\bar{n} = \bar{3}, \bar{4}, \bar{6}$	C_{ni}		
		$\bar{n} = \bar{3}$	C_{3i}		
		$\bar{n} = \bar{4}$	C_{4i}		

TABLE 1. Symmetry Operations and Elements

Symmetry operation	Symmetry element: Name	Symbol: International (Hermann-Mauguin)	Symbol: Schoenflies	Presentation on the stereographic projection: Parallel	Presentation on the stereographic projection: Perpendicular
		$\bar{n} = \bar{6}$	C_{6i}		
Inversion in a point	Center	$\bar{1}$	C_i		
Parallel translation	Translation vector $\vec{a}, \vec{b}, \vec{c}$				
Reflection in a plane and translation parallel to the plane	Glide–plane	a, b, c, n, d			
Rotation about an axis and translation parallel to the axis	Screw axis	n_m (m = 1, 2, .., n – 1)			
Rotation about an axis and reflection in a plane perpendicular to the axis	Rotatory-reflection axis	$\bar{n}$ $\bar{n} = \bar{1}, \bar{2}, \bar{3}, \bar{4}, \bar{6}$	S_n		

TABLE 2. The Thirty-Two Symmetry Classes

Class name[a] and its symbol – International (Int) and Schoenflies (Sch)

Crystal symbol	Primitive Int	Primitive Sch	Central Int	Central Sch	Planal Int	Planal Sch	Axial Int	Axial Sch	Plane-axial Int	Plane-axial Sch	Inversion primitive Int	Inversion primitive Sch	Inversion-planal Int	Inversion-planal Sch
Triclinic	1	C_1	$\bar{1}$	C_i										
Monoclinic			—		m	C_s	2	C_2	2/m	C_{2h}				
Ortho-rhombic					mm2	C_{2v}	222	D_2	mmm	D_{2h}				
Trigonal	3	C_3	$\bar{3}$	C_{3i}	3m	C_{3v}	32	D_3	$\bar{3}$m	C_{3d}				
Tetragonal	4	C_4	4/m	C_{4h}	4mm	C_{4v}	422	D_4	4/mmm	D_{4h}	$\bar{4}$	S_4	$\bar{4}$2m	D_{2d}
Hexagonal	6	C_6	6/m	C_{6h}	6mm	C_{6v}	622	D_6	6/mmm	D_{6h}	$\bar{6}$	C_{3h}	$\bar{6}$m2	D_{3h}
Cubic	23	T	m3	T_h	$\bar{4}$3m	T_d	432	O	m3m	O_h				

[a] Per Fedorov Institute of Crystallography, Russian Academy of Sciences, nomenclature.

TABLE 3. The Fourteen Possible Space Lattices (Bravais Lattices)

Crystal system	Metric category of the system	No. of different lattices in the system	Lattice type[a] (marked by +): P	C	I	F	R	No. of identi-points per unit cell	Characteristic parameters (marked by +): a	b	c	α	β	γ	Description of characteristic parameters a⊂X, b⊂Y, c⊂Z α≡(b,c), β≡(a,c), γ≡(a,b)	Symmetry of the lattice: Int	Sch
Triclinic	Trimetric	1	+					1	+	+	+	+	+	+	$a \neq b \neq c, \alpha \neq \beta \neq \gamma$	1	C
Monoclinic	Trimetric	2	+	+				1 or 2	+	+	+		+		$a \neq b \neq c, \alpha = \gamma = 90° \neq \beta$	2/m	C_{2h}
Orthorhombic	Trimetric	4	+	+	+	+		1, 2 or 4	+	+	+				$a \neq b \neq c, \alpha = \beta = \gamma = 90°$	mmm	D_{2h}
Trigonal (rhombohedral)	Dimetric	1					+	1	+			+			$a = b = c, 120° > \alpha = \beta = \gamma \neq 90°$	3m	D_{3d}
Tetragonal	Dimetric	2	+		+			1 or 2	+		+				$a = b \neq c, \alpha = \beta = \gamma = 90°$	4/mmm	D_{4h}
Hexagonal	Dimetric	1	+					1	+		+				$a = b \neq c, \alpha = \beta = 90°, \gamma = 120°$	6/mmm	D_{6h}
Isometric (cubic)	Monometric	3	+		+	+		1, 2 or 4	+						$a = b = c, \alpha = \beta = \gamma = 90°$	m3m	O_h

[a] Designations of the space-lattice types: P – primitive, C – side-centered (base-centered), I – body-centered, F – face-centered, R – rhombohedral.

TABLE 4. The Five Possible Regular Polyhedrons

Polyhedron	Symmetry (Schoenflies): Class	Symmetry (Schoenflies): Elements	Form of faces	Number of[a]: Faces (F)	Number of[a]: Edges (E)	Number of[a]: Vertices (V)
Tetrahedron	T	$4C_3 3C_2$	Equilateral triangle	4	6	4

Cube (hexahedron)	O	$3C_4 4C_3 6C_2$	Square	6	12	8
Octahedron	O	$3C_4 4C_3 6C_2$	Equilateral triangle	8	12	6
Pentagonal dodecahedron	J	$6C_5 10C_3 15C_2$	Regular pentagon	12	30	20
Icosahedron	J	$6C_5 10C_3 15C_2$	Equilateral triangle	20	30	12

[a] Per formula by Leonhard Euler: F + V − E = 2

TABLE 5. Classification of Crystals

Strukturbericht symbol	Structure name	Symmetry group: International	Symmetry group: Schoenflies	Pearson symbol[a]	Standard ASTM E157-82a symbol[b]
1	2	3	4	5	6
A1	Cu	Fm3m	O_h^4	cF4	F
A2	W	Im3m	O_h^9	cI2	B
A3	Mg	$P6_3/mmc$	D_{6h}^4	hP2	H
A4	C	Fd3m	O_h^7	cF8	F
A5	Sn	If_1/amd	D_{4h}^{19}	tI4	U
A6	In	I4/mmm	D_{4h}^{17}	tI2	U
A7	As	$R\bar{3}m$	D_{3d}^5	hR2	R
A8	Se	$P3_121$ or $P3_221$	D_3^4 (D_3^6)	hP3	H
A10	Hg	$R\bar{3}m$	D_{3d}^5	hR1	R
A11	Ga	Cmca	D_{2h}^{18}	oC8	Q
A12	α Mn	$I\bar{4}3m$	T_d^3	cI58	B
A13	β-Mn	$P4_132$	O^7	cP20	C
A15	OW_3	Pm3n	O_h^3	cP8	C
A20	α-U	Cmcm	D_{2h}^{17}	oC4	Q
B1	ClNa	Fm3m	O_h^5	cF8	F
B2	ClCs	Pm3m	O_h^1	cP2	C
B3	SZn	$F\bar{4}3m$	T_d^2	cF8	F
B4	SZn	$P6_3mc$	C_{6v}^4	hP4	H
$B8_1$	AsNi	$P6_3/mmc$	D_{6h}^4	hP4	H
$B8_2$	$InNi_2$	$P6_3/mmc$	D_{6h}^4	hP6	H
B9	HgS	$P3_121$ or $P3_221$	D_3^4 or D_3^6	hP6	H
B10	OPb	P4/nmm	D_{4h}^7	tP4	T
B11	γ-CuTi	P4/nmm	D_{4h}^7	tP4	T
B13	NiS	$R\bar{3}m$	D_{3d}^5	hR6	R
B16	GeS	Pnma	D_{2h}^{16}	oP8	O
B17	PtS	$P4_2/mmc$	D_{4h}^9	tP4	T
B18	CuS	$P6_3/mmc$	D_{6h}^4	hP12	H
B19	AuCd	Pmma	D_{2h}^5	oP4	O
B20	FeSi	$P2_13$	T^4	cP8	C
B27	BFe	Pnma	D_{2h}^{16}	oP8	O
B31	MnP	Pnma	D_{2h}^{16}	oP8	O
B32	NaTl	Fd3m	O_h^7	cF16	F
B34	Pds	$P4_2/m$	C_{4h}^2	tP16	T
B35	CoSn	P6/mmm	D_{6h}^1	hP6	H
B37	SeTl	I4/mcm	D_{4h}^{18}	tI16	U
B_e	CdSb	Pbca	D_{2h}^{15}	oP16	O
B_f (B33)	ξ-BCr	Cmcm	D_{2h}^{17}	oC8	Q
B_g	BMo	$I4_1/amd$	D_{4h}^{19}	tI4	U
B_h	CW	P6m2	D_{3h}^1	hP2	H
B_i	γ′CMo (AsTi)	$P6_3/mmc$	D_{6h}^4	hP8	H
C1	CaF_2	$Fm\bar{3}m$	O_h^5	cF12	F
$C1_b$	AgAsMg	$F\bar{4}3m$	T_d^2	cF12	F
C2	FeS_2	Pa3	T_h^6	cP12	C
C3	Cu_2O	Pn3m	O_h^4	cP6	C
C4	O_2Ti	$P4_2/mnm$	D_{4h}^{14}	tP6	T
C6	CdI_2	P3m1	D_{3d}^3	hP3	H
C7	MoS_2	$P6_3/mmc$	D_{6h}^4	hP6	H
$C11_a$	C_2Ca	I4/mmm	D_{4h}^{17}	tI6	U
$C11_b$	$MoSi_2$	I4/mmm	D_{4h}^{17}	tI6	U

TABLE 5. Classification of Crystals

Strukturbericht symbol	Structure name	Symmetry group: International	Symmetry group: Schoenflies	Pearson symbol[a]	Standard ASTM E157-82a symbol[b]
1	2	3	4	5	6
C12	$CaSi_2$	$R\bar{3}m$	D^5_{3d}	hR6	R
C14	$MgZn_2$	$P6_3/mmc$	D^4_{6h}	hP12	H
C15	Cu_2Mg	Fd3m	O^7_h	cF24	F
$C15_b$	$AuBe_5$	$F\bar{4}3m$ or F23	T^2_d or T^2	cF24	F
C16	Al_2Cu	I4/mcm	D^{18}_{4h}	tI12	U
C18	FeS_2	Pnnm	D^{12}_{2h}	oP6	O
C19	$CdCl_2$	$R\bar{3}m$	D^5_{3d}	hR3	R
C22	Fe_2P	$P2\bar{6}m$	D^1_{3h}	hP9	H
C23	Cl_2Pb	Pnma	D^{16}_{2h}	oP12	O
C32	AlB_2	P6/mmm	D^1_{6h}	hP3	H
C33	Bi_2STe_2	$R\bar{3}m$	D^5_{3d}	hR5	R
C34	$AuTe_2$	C2/m (P2/m)	C^3_{2h} (C^1_{2h})	mC6	N
C36	$MgNi_2$	$P6_3/mmc$	D^4_{6h}	hP24	H
C38	Cu_2Sb	P4/nmm	D^7_{4h}	tP6	T
C40	$CrSi_2$	$P6_222$	D^4_6	hP9	H
C42	SiS_2	Ibam	D^{26}_{2h}	oI12	P
C44	GeS_2	Fdd2	C^{19}_{2v}	oF72	S
C46	$AuTe_2$	Pma2	C^4_{2v}	oP24	O
C49	Si_2Zr	Cmcm	D^{17}_{2h}	oC12	Q
C54	Si_2Ti	Fddd	D^{24}_{2h}	oF24	S
C_c	Si_2Th	$I4_1/amd$	D^{19}_{4h}	tI12	U
C_e	$CoGe_2$	Aba2	C^{17}_{2v}	oC23	Q
DO_2	As_3Co	Im3	T^5_h	cI32	B
DO_3	BiF_3	Fm3m	O^5_h	cF16	F
DO_9	O_3Re	Pm3m	O^1_h	cP4	C
DO_{11}	CFe_3	Pnma	D^{16}_{2h}	oP16	O
DO_{18}	$AsNa_3$	$P6_3/mmc$	D^4_{6h}	hP8	H
DO_{19}	Ni_3Sn	$P6_3/mmc$	D^4_{6h}	hP8	H
DO_{20}	Al_3Ni	Pnma	D^{16}_{2h}	oP16	O
DO_{21}	Cu_3P	$P\bar{3}c1$	D^4_{3d}	hP24	H
DO_{22}	Cu_3P	I4/mmm	D^{17}_{4h}	tI8	U
DO_{23}	Al_3Zr	I4/mmm	D^{17}_{4h}	tI16	U
DO_{24}	Ni_3Ti	$P6_3/mmc$	D^4_{6h}	hP16	H
DO_c	SiU_3	I4/mcm	D^{18}_{4h}	tI16	U
DO_e	Ni_3P	$I\bar{4}$	S^2_4	tI32	U
$D1_3$	Al_4Ba	I4/mmm	D^{17}_{4h}	tI10	U
$D1_a$	$MoNi_4$	I4/m	C^5_{4h}	tI10	U
$D1_b$	Al_4U	Imma	D^{28}_{2h}	oI20	P
$D1_c$	$PtSn_4$	Aba2	C^{17}_{2v}	oC20	Q
$D1_e$	B_4Th	P4/mbm	D^5_{4h}	tP20	T
$D1_f$	BMn_4	Fddd	D^{24}_{2h}	oF40	S
$D2_1$	B_6Ca	Pm3m	O^1_h	cP7	C
$D2_3$	$NaZn_{13}$	Fm3m	O^5_h	cF112	F
$D2_b$	$Mn_{12}Th$	I4/mmm	D^{17}_{4h}	tI26	U
$D2_c$	MnU_6	I4/mcm	D^{18}_{4h}	tI28	U
$D2_d$	$CaCu_5$	P6/mmm	D^1_{6h}	hP6	H
$D2_f$	$B_{12}U$	Fm3m	O^5_h	cF52	F
$D2_h$	Al_6Mn	Cmcm	D^{17}_{2h}	oC28	Q
$D5_1$	α-Al_2O_3	R3c	D^6_{3d}	hR10	R
$D5_2$	La_2O_3	$P\bar{3}m1$	D^3_{3d}	hP5	H
$D5_3$	Mn_2O_3	Ia3	T^7_h	cI80	B
$D5_8$	S_3Sb_2	Pnma	D^{16}_{2h}	oP20	O
$D5_9$	P_2Zn_3	$P4_2/mmc$	D^9_{4h}	tP40	T
$D5_{10}$	C_2C_3	Pnma	D^{16}_{2h}	oP20	O
$D5_{13}$	Al_3Ni_2	$P\bar{3}m1$	D^3_{3d}	hP5	H
$D5_a$	Si_2U_3	P4/mbm	D^5_{4h}	tP10	T
$D5_c$	C_3Pu_2	$I\bar{4}3d$	T^6_d	cI40	B

TABLE 5. Classification of Crystals

Strukturbericht symbol	Structure name	Symmetry group		Pearson symbol[a]	Standard ASTM E157-82a symbol[b]
		International	Schoenflies		
1	2	3	4	5	6
$D7_1$	Al_4C_3	$R\bar{3}m$	D^{5}_{3d}	hR7	R
$D7_3$	P_4Th_3	$I\bar{4}3d$	T^{6}_{d}	cI28	B
$D7_b$	B_4Ta_3	Immm	D^{25}_{2h}	oI14	P
$D8_1$	Fe_3Zn_{10}	Im3m	O^{9}_{h}	cI52	B
$D8_2$	Cu_5Zn_8	$I\bar{4}3m$	T^{3}_{d}	cI52	B
$D8_3$	Al_4Cu_9	P43m	T^{1}_{d}	cP52	C
$D8_4$	C_6Cr23	Fm3m	O^{5}_{h}	cF116	F
$D8_5$	Fe_7W_6	$R\bar{3}m$	D^{5}_{3d}	hR13	R
$D8_6$	$Cu_{15}Si_4$	$I\bar{4}3m$	T^{3}_{d}	cI76	B
$D8_8$	Mn_5Si_3	$P6_3/mcm$	D^{3}_{6h}	hP16	H
$D8_9$	Co_9S_8	Fm3m	O^{5}_{h}	cF68	F
$D8_{10}$	Al_8Cr_5	R3m	C^{5}_{3v}	hR26	R
$D8_{11}$	Al_5Co_2	$P6_3/mcm$	D^{3}_{6h}	hP28	H
$D8_a$	$Mn_{23}Th_6$	Fm3m	O^{5}_{h}	cF116	F
$D8_b$	σ-phase of Cr-Fe	$p\bar{4}_2/mnm$	D^{14}_{4h}	tP30	T
$D8_e$	$(Al,Zn)_{49}Mg_{32}$	Im3	T^{5}_{h}	cI162	B
$D8_f$	Ge_7Ir_3	Im3m	O^{9}_{h}	cI40	B
$D8_h$	B_5W_2	$P6_3/mmc$	D^{4}_{6h}	hP14	H
$D8_i$	B_5Mo_2	$R\bar{3}m$	D^{5}_{3d}	hR7	R
$D8_l$	B_3Cr_5	I4/mcm	D^{18}_{4h}	tI32	U
$D8_m$	Si_3W_5	I4/mcm	D^{18}_{4h}	tI32	U
$D10_1$	C_3Cr_7	P31c	C^{4}_{3v}	hP80	H
$D10_2$	Fe_3Th_7	$P6_3mc$	C^{4}_{6v}	hP20	H
$E0_1$	ClFPb	P4/nmm	D^{7}_{4h}	tP6	T
$E1_1$	$CuFeS_2$	$I\bar{4}2d$	D^{12}_{2d}	tI16	U
$E2_1$	CaO_3Ti	Pm3m	O^{1}_{h}	cP5	C
$E2_4$	S_3Sn_2	Pnma	D^{16}_{2h}	oP20	O
E3	Al_2CdS_4	$I\bar{4}$	S^{2}_{4}	tI14	U
$E9_3$	$SiFe_3W_3$	Fd3m	O^{7}_{h}	cF112	F
$E9_a$	Al_7Cu_2Fe	P4/mnc	D^{6}_{4h}	tP40	T
$E9_b$	$AlLi_3N_2$	Ia3	T^{7}_{h}	cI96	B
$F0_1$	NiSSb	$P2_13$	T^{4}	cP12	C
$F5_1$	$CrNaS_2$	R3m or R32	D^{5}_{3d} or D^{7}_{3}	hR4	R
$F5_6$	CuS_2Sb	Pnma	D^{16}_{2h}	oP16	O
$H1_1$	Al_2MgO_4	Fd3m	O^{7}_{h}	cF56	F
$H2_4$	Cu_3S_4V	P43m	T^{1}_{d}	cP8	C
$H2_5$	$AsCu_3S_4$	$Pmn2_1$	C^{7}_{2v}	oP16	O
$L1_0$	AuCu	P4/mmm	D^{1}_{4h}	tP4	T
$L1_2$	$AlCu_3$	Pm3m	O^{1}_{h}	cP4	C
$L2_1$	$AlCu_2Mn$	Fm3m	O^{5}_{h}	cF16	F
$L2_2$	Sb_2Tl_7	Im3m	O^{9}_{h}	cI54	B
$L'2_b$	H_2Th	I4/mmm	D^{17}_{4h}	tI6	U
L′3	Fe_2N	$P6_3/mmc$	D^{4}_{6h}	hP3	H
$L6_0$	$CuTi_3$	P4/mmm	D^{1}_{4h}	tP4	T

[a] The first letter denotes the crystal system: triclinic (a), monoclinic (m), orthorhombic (o), tetragonal (t), hexagonal (h) and cubic (c). Trigonal (rhombohedral) system is denoted by combination hR. The second letter of Pearson's symbol denotes lattice type: primitive (P), edge-(base-) centered (C), body-centered (I) or face-centered (F). The following number denotes number of atoms in the crystal unit cell.

[b] Standard ASTM E157-82a has the Bravais lattices designations as following: C – primitive cubic; B – body-centered cubic; F – face-centered cubic; T – primitive tetragonal; U – body-centered tetragonal; R – rhombohedral; H – hexagonal; O – primitive orthorhombic; P – body-centered orthorhombic; Q – base-centered orthorhombic; S – face-centered orthorhombic; M – primitive monoclinic; N – centered monoclinic; A – triclinic.

References

1. A. Schoenflies, *Kristallsysteme und Kristallstructur*, Leipzig, 1891.
2. E. S. Fedorow, Zusammenstellung der kristallographischen Resultate, *Zs. Krist.*, 20, 1892.

3. P. Groth, *Elemente der physikalischen und chemischen Krystallographie*, R. Oldenbourg, München/Berlin, 1921.
4. N. V. Belov, Class Method of Deriving Space Groups of Symmetry, *Trudy Instituta Kristallodraffi imeni Fedorova (Transactions of the Fedorov Inst. of Crystallography)*, 5, 25, 1951, in Russian.
5. W. B. Pearson, *Handbook of Lattice Spacings and Structures of Metals and Alloys*, Vol. 1, Pergamon Press, 1958; Vol. 2, 1967.
6. Ch. Kittel, *Introduction to Solid State Physics*, John Wiley & Sons, 1956.
7. G. S. Zhdanov, *Fizika Tverdogo Tela (Solid State Physics)*, Moscow University Press, 1962, in Russian.
8. M. J. Buerger, *Elementary Crystallography*, John Wiley & Sons, 1963.
9. F. D. Bloss, *Crystallography & Crystal Chemistry*, Holt, Rinehart & Winston, 1971.
10. T. Janssen, *Crystallographic Groups*, North-Holland/American Elsevier, 1973.
11. M. P. Shaskolskaya, *Kristallografiya (Crystallography)*, Vysshaya Shkola, Moscow, 1976, in Russian.
12. T. Hahn, Ed., Internat. *Tables for Crystallography*, Vol. A, D. Reidel Publishing, Boston, 1983.
13. Crystal Data. Determinative Tables, Volumes 1–6, 1966–1983, JCPDS-Intern Centre for Diffraction Data and U.S. Dept. of Commerce.
14. R. W. G. Wyckoff, *Crystal Structures*, 2nd ed., Volumes 1–6, Interscience, New York, 1963.
15. C. J. Bradley and A. P. Cracknell, *The Mathematical Theory of Symmetry in Solids*, Clarendon Press, Oxford, 1972.
16. International Tables for Crystallography. Volume A, *Space–Group Symmetry*, T. Hahn, Ed., 1989; Volume B, *Reciprocal Space*, U. Schmueli, Ed.; Volume C, *Mathematical, Physical and Chemical Tables*, A. J. C. Wilson, Ed., Kluwer Academic Publishers, Dordrecht, 1989.
17. G. R. Desiraju, *Crystal Engineering: The Design of Organic Solids*, Elsevier, Amsterdam, 1989.
18. M. Senechal, *Crystalline Symmetries: An Informal Mathematical Introduction*, Adam Hilger Publ., Bristol, 1990.
19. C. Hammond, *Introduction to Crystallography*, Oxford University Press, 1990.
20. N.W. Alcock, *Bonding and Structure: Structural Principles in Inorganic and Organic Chemistry*, Ellis Norwood Publ., 1990.
21. T. C. W. Mak and G. D. Zhou. *Crystallography in Modern Chemistry: A Resource Book of Crystal Structures*, Wiley–Interscience, New York, 1992.
22. S. C. Abrahams, K. Mirsky, and R. M. Nielson, *Acta Cryst*, B52, 806 (1996); B52, 1057 (1996).
23. C. Marcos, A. Panalague, D. B. Morciras, S. Garcia-Granda and M. R. Dias. *Acta Cryst*, B52, 899 (1996).
24. A. C. Larson, *Crystallographic Computing*, Manksgaard, Copenhagen, 1970.
25. G. M. Sheldrick, SHELXS86. Crystallographic Computing 3, Clarendon Press, Oxford, 1986; SHELXL93. Program for the Refinement of Crystal Structures, University of Göttingen Press, 1993.
26. Inorganic Crystal Structure Database, CD–ROM. Sci. Inf. Service. E-mail: SISI@Delphi.com.

IONIC RADII IN CRYSTALS

Ionic radii are a useful tool for predicting and visualizing crystal structures. This table lists a set of ionic radii R_i in Å units for the most common coordination numbers *CN* of positive and negative ions. The values are based on experimental crystal structure determinations, supplemented by empirical relationships, and theoretical calculations. The notation sq after the coordination number indicates a square configuration, while py indicates pyramidal.

The advice of Howard T. Evans and Marvin J. Weber in preparing this table is appreciated.

References

1. Shannon, R. D., *Acta Crystallogr.* A32, 751, 1976.
2. Jia, Y. Q., *J. Solid State Chem.* 95, 184, 1991.

Ion	*CN*	R_i/Å
Anions		
F^{-1}	6	1.33
Cl^{-1}	6	1.81
Br^{-1}	6	1.96
I^{-1}	6	2.20
OH^{-1}	4	1.35
	6	1.37
O^{-2}	2	1.21
	6	1.40
	8	1.42
S^{-2}	6	1.84
Se^{-2}	6	1.98
Te^{-2}	6	2.21
Cations		
Ac^{+3}	6	1.12
Ag^{+1}	4	1.00
	6	1.15
	8	1.28
Ag^{+2}	4sq	0.79
	6	0.94
Al^{+3}	4	0.39
	5	0.48
	6	0.54
Am^{+3}	6	0.98
	8	1.09
Am^{+4}	6	0.85
	8	0.95
As^{+3}	6	0.58
As^{+5}	4	0.34
	6	0.46
Au^{+1}	6	1.37
Au^{+3}	4sq	0.64
	6	0.85
Ba^{+2}	6	1.35
	8	1.42
	12	1.61
Be^{+2}	4	0.27
	6	0.45
Bi^{+3}	5	0.96
	6	1.03
	8	1.17
Bi^{+5}	6	0.76
Bk^{+3}	6	0.96
Bk^{+4}	6	0.83
	8	0.93
Br^{+5}	3py	0.31
Br^{+7}	4	0.25
	6	0.39
C^{+4}	4	0.15
	6	0.16
Ca^{+2}	6	1.00
	8	1.12
	10	1.23
	12	1.34
Cd^{+2}	4	0.78
	6	0.95
	8	1.10
	12	1.31
Ce^{+3}	6	1.01
	8	1.14
	10	1.25
	12	1.34
Ce^{+4}	6	0.87
	8	0.97
	10	1.07
	12	1.14
Cf^{+3}	6	0.95
Cf^{+4}	6	0.82
	8	0.92
Cl^{+5}	3py	0.12
Cl^{+7}	4	0.08
Cm^{+3}	6	0.97
Cm^{+4}	6	0.85
	8	0.95
Co^{+2}	4	0.56
	6	0.65
	8	0.90
Co^{+3}	6	0.55
Cr^{+2}	6	0.73
Cr^{+3}	6	0.62
Cr^{+4}	4	0.41
	6	0.55
Cr^{+6}	4	0.26
	6	0.44
Cs^{+1}	6	1.67
	8	1.74
	10	1.81
	12	1.88
Cu^{+1}	2	0.46
	4	0.60
	6	0.77
Cu^{+2}	4sq	0.57
	6	0.73
Dy^{+2}	6	1.07
	8	1.19
Dy^{+3}	6	0.91
	8	1.03
Er^{+3}	6	0.89
	8	1.00
Eu^{+2}	6	1.17
	8	1.25
	10	1.35
Eu^{+3}	6	0.95
	8	1.07
F^{+7}	6	0.08
Fe^{+2}	4	0.63
	6	0.61
	8	0.92
Fe^{+3}	4	0.49
	6	0.55
	8	0.78
Fr^{+1}	6	1.80
Ga^{+3}	4	0.47
	6	0.62
Gd^{+3}	6	0.94
	8	1.05
Ge^{+2}	6	0.73
Ge^{+4}	4	0.39
	6	0.53
Hf^{+4}	4	0.58
	6	0.71
	8	0.83
Hg^{+1}	6	1.19
Hg^{+2}	2	0.69
	4	0.96
	6	1.02
	8	1.14
I^{+5}	3py	0.44
	6	0.95
I^{+7}	4	0.42
	6	0.53
In^{+3}	4	0.62
	6	0.80
Ir^{+3}	6	0.68
Ir^{+4}	6	0.63
Ir^{+5}	6	0.57
K^{+1}	4	1.37
	6	1.38
	8	1.51
	12	1.64
La^{+3}	6	1.03
	8	1.16
	10	1.27
	12	1.36
Li^{+1}	4	0.59
	6	0.76
	8	0.92
Lu^{+3}	6	0.86
	8	0.97
Mg^{+2}	4	0.57
	6	0.72
	8	0.89
Mn^{+2}	4	0.66

Ion	*CN*	R_i/Å
	6	0.83
	8	0.96
Mn^{+3}	6	0.58
Mn^{+4}	4	0.39
	6	0.53
Mn^{+5}	4	0.33
Mn^{+6}	4	0.26
Mn^{+7}	4	0.25
Mo^{+3}	6	0.69
Mo^{+4}	6	0.65
Mo^{+5}	4	0.46
	6	0.61
Mo^{+6}	4	0.41
	6	0.59
	7	0.73
N^{+3}	6	0.16
N^{+5}	6	0.13
Na^{+1}	4	0.99
	6	1.02
	8	1.18
	9	1.24
	12	1.39
Nb^{+3}	6	0.72
	8	0.79
Nb^{+4}	6	0.68
Nb^{+5}	4	0.48
	6	0.64
	8	0.74
Nd^{+3}	6	0.98
	8	1.12
	9	1.16
	12	1.27
Ni^{+2}	4sq	0.49
	6	0.69
Ni^{+3}	6	0.56
Np^{+3}	6	1.01
Np^{+4}	6	0.87
Np^{+5}	6	0.75
Np^{+6}	6	0.72
Os^{+4}	6	0.63
Os^{+5}	6	0.58
Os^{+6}	6	0.55
Os^{+8}	4	0.39
P^{+5}	4	0.17
	6	0.38
Pa^{+3}	6	1.04
Pa^{+4}	6	0.90
Pa^{+5}	6	0.78
Pb^{+2}	6	1.19
	8	1.29
	10	1.40
	12	1.49
Pb^{+4}	4	0.65
	6	0.78
	8	0.94
Pd^{+2}	4sq	0.64
	6	0.86
Pd^{+3}	6	0.76
Pd^{+4}	6	0.62
Pm^{+3}	6	0.97
	8	1.09
Po^{+4}	6	0.97
Pr^{+3}	6	0.99
	8	1.13
Pr^{+4}	6	0.85
	8	0.96
Pt^{+2}	4sq	0.60
	6	0.80
Pt^{+4}	6	0.63
Pu^{+3}	6	1.00
Pu^{+4}	6	0.86
Pu^{+5}	6	0.74
Pu^{+6}	6	0.71
Ra^{+2}	8	1.48
	12	1.70
Rb^{+1}	6	1.52
	8	1.61
	10	1.66
	12	1.72
Re^{+4}	6	0.63
Re^{+5}	6	0.58
Re^{+6}	6	0.55
Re^{+7}	4	0.38
	6	0.53
Rh^{+3}	6	0.67
Rh^{+4}	6	0.60
Rh^{+5}	6	0.55
Ru^{+3}	6	0.68
Ru^{+4}	6	0.62
Ru^{+5}	6	0.57
Ru^{+7}	4	0.38
Ru^{+8}	4	0.36
S^{+4}	6	0.37
S^{+6}	4	0.12
	6	0.29
Sb^{+3}	4py	0.76
	6	0.76
Sb^{+5}	6	0.60
Sc^{+3}	6	0.75
	8	0.87
Se^{+4}	6	0.50
Se^{+6}	4	0.28
	6	0.42
Si^{+4}	4	0.26
	6	0.40
Sm^{+2}	6	1.19
	8	1.27
Sm^{+3}	6	0.96
	8	1.08
	12	1.24
Sn^{+4}	4	0.55
	6	0.69
	8	0.81
Sr^{+2}	6	1.18
	8	1.26
	10	1.36
	12	1.44
Ta^{+3}	6	0.72
Ta^{+4}	6	0.68
Ta^{+5}	6	0.64
Tb^{+3}	6	0.92
	8	1.04
Tb^{+4}	6	0.76
	8	0.88
Tc^{+4}	6	0.65
Te^{+4}	4	0.66
	6	0.97
Te^{+6}	4	0.43
	6	0.56
Th^{+4}	6	0.94
	8	1.05
	10	1.13
	12	1.21
Ti^{+2}	6	0.86
Ti^{+3}	6	0.67
Ti^{+4}	4	0.42
	6	0.61
	8	0.74
Tl^{+1}	6	1.50
	8	1.59
	12	1.70
Tl^{+3}	4	0.75
	6	0.89
	8	0.98
Tm^{+2}	6	1.01
	7	1.09
Tm^{+3}	6	0.88
	8	0.99
U^{+3}	6	1.03
U^{+4}	6	0.89
	8	1.00
	12	1.17
U^{+5}	6	0.76
U^{+6}	2	0.45
	4	0.52
	6	0.73
	8	0.86
V^{+2}	6	0.79
V^{+3}	6	0.64
V^{+4}	5	0.53
	6	0.58
	8	0.72
V^{+5}	4	0.36
	5	0.46
	6	0.54
W^{+4}	6	0.66
W^{+5}	6	0.62
W^{+6}	4	0.42
	5	0.51
	6	0.60
Y^{+3}	6	0.90
	8	1.02
	9	1.08
Yb^{+2}	6	1.02
	8	1.14
Yb^{+3}	8	0.99
	9	1.04
Zn^{+2}	4	0.60
	6	0.74
	8	0.90
Zr^{+4}	4	0.59
	6	0.72
	8	0.84
	9	0.89

POLARIZABILITIES OF ATOMS AND IONS IN SOLIDS

H. P. R. Frederikse

The polarization of a solid dielectric medium, ***P***, is defined as the dipole moment per unit volume averaged over the volume of a crystal cell. A component of ***P*** can be expanded as a function of the electric field ***E***:

$$P_i = \sum_j a_j E_j + \sum_{jk} b_{jk} E_j E_k$$

For relatively small electric fields in isotropic substances $\boldsymbol{P} = \chi_e \boldsymbol{E}$, where χ_e is the electric susceptibility. If the medium is made up of N atoms (or ions) per unit volume, the polarization is $\boldsymbol{P} = N\boldsymbol{p}_m$ where $\boldsymbol{p}_m$ is the average dipole moment per atom. The polarizability α can be defined as $\boldsymbol{p}_m = \alpha \boldsymbol{E}_0$, where $\boldsymbol{E}_0$ is the local field at the position of the atom. Using the Lorentz method to calculate the local field one finds:

$$\boldsymbol{P} = N\alpha(\boldsymbol{E} + 4\pi \boldsymbol{P}) = \chi_e \boldsymbol{E}$$

Together with the definition of the dielectric constant (relative permittivity), $\varepsilon = 1 + 4\pi\chi_e$, this leads to:

$$\alpha = \frac{3}{4\pi N}\left(\frac{\varepsilon - 1}{\varepsilon + 2}\right)$$

This expression is known as the Clausius-Mossotti equation.

The total polarization associated with atoms, ions, or molecules is due to three different sources:

1. Electronic polarization arises because the center of the local electronic charge cloud around the nucleus is displaced under the action of the field: $\boldsymbol{P}_e = N\alpha_e \boldsymbol{E}_0$ where α_e is the *electronic polarizability*.
2. Ionic polarization occurs in ionic materials because the electric field displaces cations and anions in opposite directions: $\boldsymbol{P}_i = N\alpha_i \boldsymbol{E}_0$, where α_i is the *ionic polarizability*.
3. Orientational polarization can occur in substances composed of molecules that have permanent electric dipoles. The alignment of these dipoles depends on temperature and leads to an *orientational polarizability* per molecule: $\alpha_{or} = p^2/3kT$, where p is the permanent dipole moment per molecule, k is the Boltzmann constant, and T is the temperature.

Because of the different nature of these three polarization processes the response of a dielectric solid to an applied electric field will strongly depend on the frequency of the field. The resonance of the electronic excitation in insulators (dielectrics) takes place in the ultraviolet part of the spectrum; the characteristic frequency of the lattice vibrations is located in the infrared, while the orientation of dipoles requires fields of much lower frequencies (below 10^{10} Hz). This response to electric fields of different frequencies is shown in Figure 1. Values of the electronic polarizabilities for selected atoms and ions are given in Table 1.

References

1. Kittel, C., *Introduction to Solid State Physics*, Fourth Edition, John Wiley & Sons, New York, 1971.
2. Lerner, R.G., and Trigg, G.L., Eds., *Encyclopedia of Physics, Second Edition*, VCH Publishers, New York, 1990.
3. Ralls, K.M., Courtney, T.H., and Wulff, J., *An Introduction to Materials Science and Engineering*, John Wiley & Sons, New York, 1976.

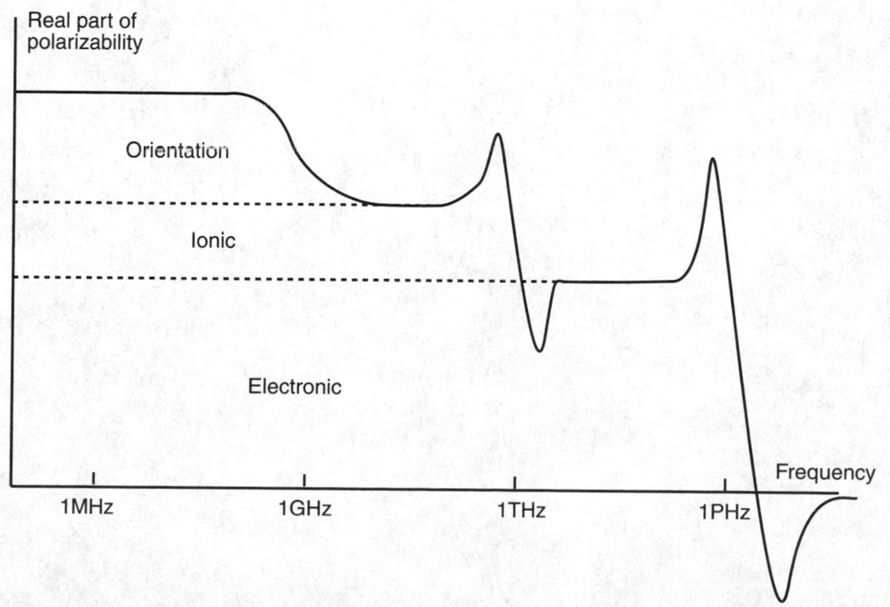

FIGURE 1. Schematic graph of the frequency dependence of the different contributions to polarizability.

TABLE 1. Electronic Polarizabilities in Units of 10^{-24} cm^3

						He 0.201
Li^+ 0.029	**Be^{2+}** 0.008	**B^{3+}** 0.003	**C^{4+}** 0.0013	**O^{2-}** 3.88	**F^-** 1.04	**Ne** 0.39
Na^+ 0.179	**Mg^{2+}** 0.094	**Al^{3+}** 0.052	**Si^{4+}** 0.0165	**S^{2-}** 10.2	**Cl^-** 3.66	**Ar** 1.62
K^+ 0.83	**Ca^{2+}** 0.47	**Sc^{3+}** 0.286	**Ti^{4+}** 0.185	**Se^{2-}** 10.5	**Br^-** 4.77	**Kr** 2.46
Rb^+ 1.40	**Sr^{2+}** 0.86	**Y^{3+}** 0.55	**Zr^{4+}** 0.37	**Te^{2-}** 14.0	**I^-** 7.1	**Xe** 3.99
Cs^+ 2.42	**Ba^{2+}** 1.55	**La^{3+}** 1.04	**Ce^{4+}** 0.73			

Data from Pauling, L., *Proc. R. Soc.* London, A114, 181, 1927. See also Jaswal, S.S. and Sharma, T.P., *J. Phys. Chem. Solids*, 34, 509, 1973.
Values are appropriate for cgs units. To convert to SI, use the relation $\alpha(SI)/C\ m^2V^{-1} = 1.11265.10^{-16}\ \alpha(cgs)/cm^3$.

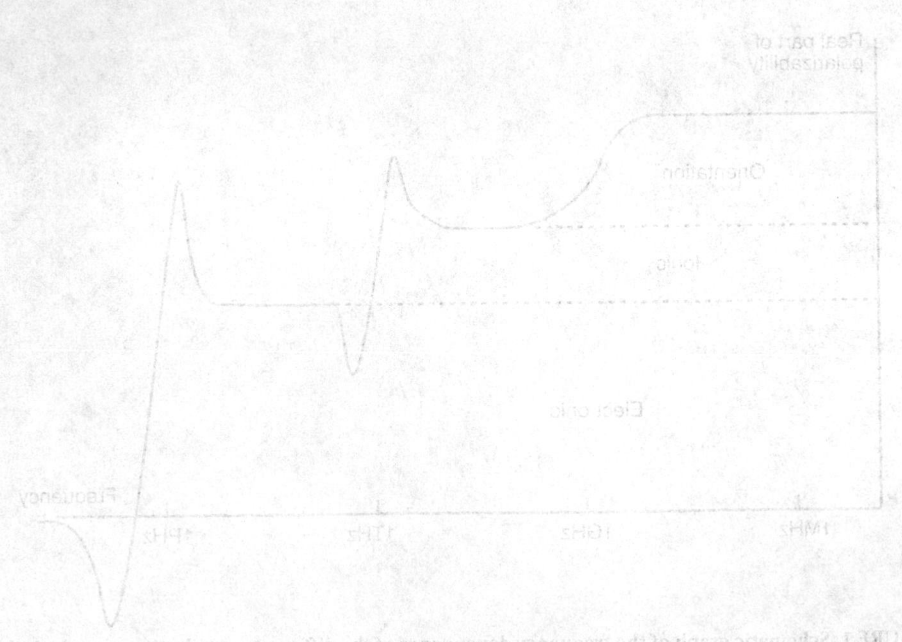

CRYSTAL STRUCTURES AND LATTICE PARAMETERS OF ALLOTROPES OF THE ELEMENTS

H. W. King

The crystal structures of the allotropic forms of the elements are presented in terms of the Pearson symbol, the Strukturbericht designation, and the prototype of the structure. The temperatures of the phase transformations are listed in degrees Celsius and the pressures are in GPa. A consistent nomenclature is used, whereby all allotropes are labeled by Greek letters. The lattice parameters of the units cells are given in nanometers (nm) and are considered to be accurate to ±2 in the last reported digit.

This compilation is restricted to changes in crystal structures that occur as a result of a change in temperature or pressure. Low-temperature structures are included for the diatomic and rare gases, which show many similarities with respect to the metallic elements. The elements identified with an asterisk (*) have polymorphic structures based on different molecular configurations. The crystal data given for these elements refer to the most stable structure at room temperature.

Reprinted with the permission of ASM International from T.B. Massalski, Ed., *Binary Alloy Phase Diagrams*, ASM International, Metals Park, Ohio, 1986; certain data on rare earth elements were provided by K.A. Gschneidner.

Element	Temper-ature, °C	Pressure, GPa	Pearson symbol	Space group	Struktur-bericht designation	Prototype	Lattice parameters, nm			Comment, c/a or α or β
							a	*b*	*c*	
Ac	25	atm	*cF4*	*Fm3m*	*A1*	Cu	0.5311	...	...	...
Ag	25	atm	*cF4*	*Fm3m*	*A1*	Cu	0.40857	...	...	...
αAl	25	atm	*cF4*	*Fm3m*	*A1*	Cu	0.40496	...	...	...
βAl	25	>20.5	*hP2*	$P6_3/mmc$	*A3*	Mg	0.2693	...	0.4398	1.6331
α′Am	25	atm	*hP4*	$P6_3/mmc$	*A3′*	αLa	0.34681	...	1.1241	2*1.621
αAm	>769	atm	*cF4*	*Fm3m*	*A1*	Cu	0.4894	...	...	...
βAm	>1074	atm	*cI2*	*Im3m*	*A2*	W	?	...	...	...
γAm	25	>15	*oC4*	*Cmcm*	*A20*	αU	0.3063	0.5968	0.5169	...
αAr	<-189.35	atm	*cF4*	*Fm3m*	*A1*	Cu	0.5316	...	...	...
(βAr)	<-189.40	atm	*hP2*	$P6_3/mmc$	*A3*	Mg	0.3760	...	0.6141	1.633
αAs	25	atm	*hR2*	$R\bar{3}m$	*A7*	αAs	0.41319	...	......	α = 54.12°
ϵAs	>448	atm	*oC8*	*Cmca*	...	P(black)	0.362	1.085	0.448	...
Au	25	atm	*cF4*	*Fm3m*	*A1*	Cu	0.40782	...	...	...
βB	25	atm	*hR105*	*R3m*	...	βB	1.017	...	...	α = 65.12°
αBa	25	atm	*cI2*	$Im\bar{3}m$	*A2*	W	0.50227	...	...	...
βBa	25	>5.33	*hP2*	$P6_3/mmc$	*A3*	Mg	0.3901	...	0.6154	1.5775
γBa	25	>23	?	?	...	...	...	...	...	...
αBe	25	atm	*hP2*	$P6_3/mmc$	*A3*	Mg	0.22859	...	0.35845	1.5681
βBe	>1270	atm	*cI2*	$Im\bar{3}m$	*A2*	W	0.25515	...	...	...
γBe	25	>9.3	?	...	...	...	...	...	...	...
αBi	25	atm	*hR2*	$r\bar{3}m$	*A7*	αAs	0.47460	...	...	α = 57.23°
βBi	25	>2.6	*mC4*	*C2/m*	...	βBi	0.6674	0.6117	0.3304	β = 110.33°
γBi	25	>3.0	*mP3*	?	...	...	0.605	0.42	0.465	β = 85.33°
σBi	25	>4.3	?	?	...	...	...	...	...	...
ϵBi	25	>6.5	?	?	...	...	...	...	...	...
ζBi	25	>9.0	*cI2*	*Im3m*	*A2*	W	0.3800	...	...	...
αBk	25	atm	*hP4*	$P6_3/mmc$	*A3′*	αLa	0.3416	...	1.1069	2*1.620
βBk	>977	atm	*cF4*	*Fm3m*	*A1*	Cu	0.4997	...	...	...
Br	<7.25	atm	*oC8*	*Cmca*	...	Cl	0.668	0.449	0.874	...
C (graphite)	25	atm	*hP4*	$P6_3/mmc$	*A9*	C (graphite)	0.24612	...	0.6709	2.7258
C (diamond)	25	>60	*cF8*	*Fd3m*	*A4*	C (diamond)	0.35669	...	...	...
C (hd)	25	HP	*hP4*	$P6_3/mmc$	...	C (hd)	0.2522	...	0.4119	1.633
αCa	25	atm	*cF4*	*Fm3m*	*A1*	Cu	0.55884	...	...	...
βCa	>443	atm	*cI2*	*Im3m*	*A2*	W	0.4480	...	...	...
γCa	25	>1.5	?	...	...	...	...	...	...	...
Cd	25	atm	*hP2*	$P6_3/mmc$	*A3*	Mg	0.29793	...	0.56196	1.8862
αCe	<-177	atm	*cF4*	*Fm3m*	*A1*	Cu	0.485	...	...	...
βCe	25	atm	*hP4*	$P6_3/mmc$	*A3′*	αLa	0.36810	...	1.1857	2*1.611
γCe	25	atm	*cF4*	*Fm3m*	*A1*	Cu	0.51610	...	...	...
δ-Ce	>726	atm	*cI2*	*Im3m*	*A2*	W	0.412	...	...	...
α′Ce	25	>5.4	*oC4*	*Cmcm*	*A20*	αU	0.3049	0.5998	0.5215	...

Element	Temperature, °C	Pressure, GPa	Pearson symbol	Space group	Strukturbericht designation	Prototype	Lattice parameters, nm			Comment, c/a or α or β
							a	*b*	*c*	
αCf	25	atm	*hP4*	$P6_3/mmc$	*A3*′	αLa	0.339	...	1.1015	2*1.625
βCf	>590	atm	*cF4*	*Fm3m*	*A1*	Cu	?	...	...	...
Cl	<-102	atm	*oC8*	*Cmca*	...	Cl	0.624	0.448	0.826	...
αCm	25	atm	*hP4*	$P6_3/mmc$	*A3*′	αLa	0.3496	...	1.1331	2*1.621
βCm	>1277	atm	*cF4*	*Fm3m*	*A1*	Cu	0.4382	...	...	...
εCo	25	atm	*hP2*	$P6_3/mmc$	*A3*	Mg	0.25071	...	0.40686	1.6228
αCo	>422	atm	*cF4*	*Fm3m*	*A1*	Cu	0.35447	...	...	...
αCr	25	atm	*cI2*	*Im3m*	*A2*	W	0.28848	...	...	...
α′Cr	25	HP	*tI2*	*I4/mmm*	...	α′Cr	0.2882	...	0.2887	1.002
αCs	25	atm	*cI2*	*Im3m*	*A2*	W	0.6141	...	...	...
βCs	25	>2.37	*cF4*	*Fm3m*	*A1*	Cu	0.6465	...	...	...
β′Cs	25	>4.22	*cF4*	*Fm3m*	*A1*	Cu	0.5800	...	...	...
γCs	25	>4.27	?	...	...	...	...	...	...	...
Cu	25	atm	*cF4*	*Fm3m*	*A1*	Cu	0.36146	...	...	...
α′Dy	<-187	atm	*oC4*	*Cmcm*	...	α′Dy	0.3595	0.6184	0.5678	...
αDy	25	atm	*hP2*	$P6_3/mmc$	*A3*	Mg	0.35915	...	0.56501	1.5732
βDy	>1381	atm	*cI2*	*Im3m*	*A2*	W	0.403	...	...	...
γDy	25	>7.5	*hR3*	$R\bar{3}m$	...	αSm	0.3436	...	2.483	4.5*1.606
Er	25	atm	*hP2*	$P6_3/mmc$	*A3*	Mg	0.35592	...	0.55850	1.5692
αEs	25	atm	*hP4*	$P6_3/mmc$	*A3*′	αLa	?	...	...	...
βEs	?	atm	*cF4*	*Fm3m*	*A1*	Cu	?	...	...	...
Eu	25	atm	*cI2*	*Im3m*	*A2*	W	0.45827	...	...	...
αF	<-227.6	atm	*mC8*	*C2/c*	...	αF	0.550	0.338	0.728	β = 102.17°
βF	<-219.67	atm	*cP16*	*Pm3n*	...	γO	0.667	...	...	...
αFe	25	atm	*cI2*	*Im3m*	*A2*	W	0.28665	...	...	...
γFe	>912	atm	*cF4*	*Fm3m*	*A1*	Cu	0.36467	...	...	...
σFe	>1394	atm	*cI2*	*Im3m*	*A2*	W	0.29315	...	...	...
εFe	25	>13	*hP2*	$P6_3/mmc$	*A3*	Mg	0.2468	...	0.396	1.603
αGa	25	atm	*oC8*	*Cmca*	*A11*	αGa	0.45186	0.76570	0.45258	...
βGa	25	>1.2	*tI2*	*I4/mmm*	*A6*	In	0.2808	...	0.4458	1.588
γGa	-53	>3.0	*oC40*	*Cmcm*	...	γGa	1.0593	1.3523	0.5203	...
αGd	25	atm	*hP2*	$P6_3/mmc$	*A3*	Mg	0.36336	...	0.57810	0.5910
βGd	>1235	atm	*cI2*	*Im3m*	*A2*	W	0.406	...	...	...
γGd	25	>3.0	*hR3*	*R3m*	...	αSm	0.361	...	2.603	4*1.60
αGe	25	atm	*cF8*	*Fd3m*	*A4*	C (diamond)	0.56574	...	...	...
βGe	25	>12	*tI4*	$I4_1/amd$	*A5*	βSn	0.4884	...	0.2692	0.551
γGe	25	>12→atm	*tP12*	$P4_12_12$	...	σGe	0.593	...	0.698	1.18
σGe	LT	>12	*cI16*	*Im3m*	...	γSi	0.692	...	...	...
αH	<-271.9	atm	*cF4*	*Fm3m*	*A1*	Cu	0.5338	...	...	...
βH	<-259.34	atm	*hP2*	$P6_3/mmc$	*A3*	Mg	0.3776	...	0.6162	1.632
αHe	<-268.94	atm	*hP2*	$P6_3/mmc$	*A3*	Mg	0.3555	...	0.5798	1.631
βHe	>-258	0.125	*cF4*	*Fm3m*	*A1*	Cu	0.4240	...	...	...
γHe	<-271.47	0.03	*cI2*	*Im3m*	*A2*	W	0.4110	...	...	...
αHf	25	atm	*hP2*	$P6_3/mmc$	*A3*	Mg	0.31946	...	0.50510	1.5811
βHf	>1995	atm	*cI2*	*Im3m*	*A2*	W	0.3610	...	...	...
αHg	<-38.84	atm	*hR1*	$R\bar{3}m$	*A10*	αHg	0.3005	...	...	α = 70.53°
βHg	<-194	HP	*tI2*	*I4/mmm*	...	βHg	0.3995	...	0.2825	0.707
γHg	<-194	c.w.	*hR1*	?	...	...	...	...	...	...
αHo	25	atm	*hP2*	$P6_3/mmc$	*A3*	Mg	0.35778	...	0.56178	1.5702
βHo	25	>7.5	*hR3*	$R\bar{3}m$	...	αSm	0.334	...	2.45	4.5*1.63
I	25	atm	*oC8*	*Cmca*	...	Cl	0.72697	0.47903	0.97942	...
In	25	atm	*tI2*	*I4/mmm*	*A6*	In	0.3253	...	0.49470	1.5210
Ir	25	atm	*cF4*	*Fm3m*	*A1*	Cu	0.38392	...	...	...
K	25	atm	*cI2*	*Im3m*	*A2*	W	0.5321	...	...	...
Kr	<-157.39	atm	*cF4*	*Fm3m*	*A1*	Cu	0.5810	...	...	...
αLa	25	atm	*hP4*	$P6_3/mmc$	*A3*′	αLa	0.37740	...	1.2171	2*1.6125
βLa	>310	atm	*cF4*	*Fm3m*	*A1*	Cu	0.5303	...	...	...
γLa	>865	atm	*cI2*	*Im3m*	*A2*	W	0.426	...	...	...
β′La	25	>2.0	*cF4*	*Fm3m*	*A1*	Cu	0.517	...	...	...
αLi	<-193	atm	*hP2*	$P6_3/mmc$	*A3*	Mg	0.3111	...	0.5093	1.637

Crystal Structures and Lattice Parameters of Allotropes of the Elements

Element	Temperature, °C	Pressure, GPa	Pearson symbol	Space group	Strukturbericht designation	Prototype	Lattice parameters, nm a	b	c	Comment, c/a or α or β
βLi	25	atm	cI2	Im3m	A2	W	0.35093	...	...	...
γLi	<-201	c.w.	cF4	Fm3m	A1	Cu	0.4388	...	...	...
Lu	25	atm	hP2	$P6_3/mmc$	A3	Mg	0.35052	...	0.55494	1.5832
Mg	25	atm	hP2	$P6_3/mmc$	A3	Mg	0.32094	...	0.52107	1.6236
αMn	25	atm	cI58	$I\bar{4}3m$	A12	αMn	0.89126	...	...	...
βMn	>710	atm	cP20	$P4_132$	A13	βMn	0.63152	...	...	...
γMn	>1079	atm	cF4	Fm3m	A1	Cu	0.3860	...	...	...
σMn	>1143	atm	cI2	Im3m	A2	W	0.3080	...	...	...
Mo	25	atm	cI2	Im3m	A2	W	0.31470	...	...	...
αN	<-237.6	atm	cP8	Pa3	...	αN	0.5661	...	...	...
βN	<-210.00	atm	hP4	$P6_3/mmc$	...	βN	0.4050	...	0.6604	1.631
γN	<-253	>3.3	tP4	$P4_2/mnm$	...	γN	0.3957	...	0.5109	1.291
αNa	<-233	atm	hP2	$P6_3/mmc$	A3	Mg	0.3767	...	0.6154	1.634
βNa	25	atm	cI2	Im3m	A2	W	0.42906	...	...	...
Nb	25	atm	cI2	Im3m	A2	W	0.33004	...	...	...
αNd	25	atm	hP4	$P6_3/mmc$	A3′	αLa	0.36582	...	1.17966	2*1.6124
βNd	>863	atm	cI2	Im3m	A2	W	0.413	...	...	...
γNd	25	>5.0	cF4	Fm3m	A1	Cu	0.480	...	...	...
Ne	<-243.59	atm	cF4	Fm3m	A1	Cu	0.4462	...	...	...
Ni	25	atm	cF4	Fm3m	A1	Cu	0.35240	...	...	...
αNp	25	atm	oP8	Pnma	A_c	αNp	0.6663	0.4723	0.4887	...
βNp	>280	atm	tP4	$P42_12$	A_d	βNp	0.4883	...	0.3389	0.694
γNp	>576	atm	cI2	Im3m	A2	W	0.352	...	...	...
αO	<-243.3	atm	mC4	C2m	...	αO	0.5403	0.3429	0.5086	β = 132.53°
βO	<-229.6	atm	hR2	$R\bar{3}m$	...	βO	0.4210	...	...	α = 46.27°
γO	<-218.79	atm	cP16	Pm3n	...	γO	0.683	...	...	...
Os	25	atm	hP2	$P6_3/mmc$	A3	Mg	0.27341	...	0.43918	1.6063
P (black)	25	atm	oC8	Cmca	...	P (black)	0.33136	1.0478	0.43763	...
αPa	25	atm	tI2	I4/mmm	A_a	αPa	0.3921	...	0.3235	0.825
βPa	>1170	atm	cI2	Im3m	A2	W	0.381	...	...	...
αPb	25	atm	cF4	Fm3m	A1	Cu	0.49502	...	...	...
βPb	25	>10.3	hP2	$P6_3/mmc$	A3	Mg	0.3265	...	0.5387	1.650
Pd	25	atm	cF4	Fm3m	A1	Cu	0.38903	...	...	...
αPm	25	atm	hP4	$P6_3/mmc$	A3′	αLa	0.365	...	1.165	2*1.60
βPm	>890	atm	cI2	Im3m	A2	W	(0.410)	...	...	...
αPo	25	atm	cP1	Pm3m	A_h	αPo	0.3366	...	...	...
βPo	>54	atm	hR1	$R\bar{3}m$	...	βPo	0.3373	...	...	α = 98.08°
αPr	25	atm	hP4	$P6_3/mmc$	A3′	αLa	0.36721	...	1.18326	2*1.6111
βPr	>795	atm	cI2	Im3m	A2	W	0.413	...	...	...
γPr	25	>4.0	cF4	Fm3m	A1	Cu	0.488	...	...	...
Pt	25	atm	cF4	Fm3m	A1	Cu	0.39236	...	...	...
αPu	25	atm	mP16	$P2_1/m$	...	αPu	0.6183	0.4822	1.0963	β = 101.97°
βPu	>125	atm	mI34	I2/m	...	βPu	0.9284	1.0463	0.7859	β = 92.13°
γPu	>215	atm	oF8	Fddd	...	γPu	0.31587	0.57682	1.0162	
σPu	>320	atm	cF4	Fm3m	A1	Cu	0.46371	...	...	...
σ′Pu	>463	atm	tI2	I4/mmm	A6	In	0.33261	...	0.44630	1.3418
εPu	>483	atm	cI2	Im3m	A2	W	0.36343	...	...	...
Ra	25	atm	cI2	Im3m	A2	W	0.5148	...	...	...
αRb	25	atm	cI2	Im3m	A2	W	0.5705	...	...	...
βRb	25	>1.08	?	...	...	...	...	...	...	...
γRb	25	>2.05	?	...	...	...	...	...	...	...
Re	25	atm	hP2	$P6_3/mmc$	A3	Mg	0.27609	...	0.4458	1.6145
Rh	25	atm	cF4	Fm3m	A1	Cu	0.38032	...	...	...
Ru	25	atm	hP2	$P6_3/mmc$	A3	Mg	0.27058	...	0.42816	1.5824
αS	25	atm	oF128	Fddd	A16	αS	1.0464	1.28660	2.44860	...
αSb	25	atm	hR2	$R\bar{3}m$	A7	αAs	0.45067	...	...	α = 57.11°
βSb	25	>5.0	cP1	Pm3m	A_h	αPo	0.2992	...	...	...
γSb	25	>7.5	hP2	$P6_3/mmc$	A3	Mg	0.3376	...	0.5341	1.582
σSb	25	>14.0	mP3	?	...	...	0.556	0.404	0.422	β = 86.0°
αSc	25	atm	hP2	$P6_3/mm$	A3	Mg	0.33088	...	0.52680	1.5921

Element	Temperature, °C	Pressure, GPa	Pearson symbol	Space group	Strukturbericht designation	Prototype	Lattice parameters, nm *a*	*b*	*c*	Comment, *c/a* or α or β
βSc	>1337	atm	*cI2*	*Im3m*	*A2*	W	(0.373)	...	...	...
γSe	25	atm	*hP3*	$P3_121$	*A8*	γSe	0.43659	...	0.49537	1.1346
αSi	25	atm	*cF8*	*Fd3m*	*A4*	C (diamond)	0.54306	...	...	...
βSi	25	>9.5	*tI4*	$I4_1/amd$	*A5*	βSn	0.4686	...	0.2585	0.552
γSi	25	>16.0	*cI16*	*Im3m*	...	γSi	0.6636	...	...	...
σSi	25	>16→atm	*hP4*	$P6_3/mmc$	*A3´*	αLa	0.380	...	0.628	1.653
αSm	25	atm	*hR3*	$R\bar{3}m$	...	αSm	0.36290	...	2.6207	4*1.6048
βSm	>734	atm	*hP2*	$P6_3/mmc$	*A3*	Mg	0.36630	...	0.58448	1.5956
γ´Sm	>922	atm	*cI2*	*Im3m*	*A2*	W	(0.410)	...	...	...
σSm	25	>4.0	*hP4*	$P6_3/mmc$	*A3´*	αLa	0.3618	...	1.166	2*1.611
αSn	<13	atm	*cF8*	*Fd3m*	*A4*	C (diamond)	0.64892	...	...	...
βSn	25	atm	*tI4*	$I4_1/amd$	*A5*	βSn	0.58318	...	0.31818	0.5456
γSn	25	>9.0	*tI2*	?	...	γSn	0.370	...	0.337	0.91
αSr	25	atm	*cF4*	*Fm3m*	*A1*	Cu	0.6084	...	...	...
βSr	>547	atm	*cI2*	*Im3m*	*A2*	W	0.487	...	...	...
β´Sr	25	>3.5	*cI2*	*Im3m*	*A2*	W	0.4437	...	...	...
Ta	25	atm	*cI2*	*Im3m*	*A2*	W	0.33030	...	...	...
α´Tb	<-53	atm	*oC4*	*Cmcm*	...	α´Dy	0.3605	0.6244	0.5706	...
aTb	25	atm	*hP2*	$P6_3/mmc$	*A3*	Mg	0.36055	...	0.56966	1.5800
βTb	>1289	atm	*cI2*	*Im3m*	*A2*	W	(0.407)	...	...	...
γTb	25	>6.0	*hR3*	$R\bar{3}m$	...	αSm	0.341	...	2.45	4*1.60
Tc	25	atm	*hP2*	$P6_3/mmc$	*A3*	Mg	0.2738	...	0.4393	1.604
αTe	25	atm	*hP3*	$P3_121$	*A8*	γSe	0.44566	...	0.59264	1.3298
βTe	25	>2.0	*hR2*	$R\bar{3}m$	*A7*	αAs	0.469	...	...	α = 53.30°
γTe	25	>7.0	*hR1*	$R\bar{3}m$	...	βPo	0.3002	...	...	α = 103.3°
αTh	25	atm	*cF4*	*Fm3m*	*A1*	Cu	0.50842	...	...	...
βTh	>1360	atm	*cI2*	*Im3m*	*A2*	W	0.411	...	...	...
αTi	25	atm	*hP2*	$P6_3/mmc$	*A3*	Mg	0.29506	...	0.46835	1.59873
βTi	>882	atm	*cI2*	*Im3m*	*A2*	W	0.33065	...	...	...
ωTi	25	HP→atm	*hP3*	*P6/mmm*	...	ωTi	0.4625	...	0.2813	0.6082
αTl	25	atm	*hP2*	$P6_3/mmc$	*A3*	Mg	0.34566	...	0.55248	1.5983
βTl	>230	atm	*cI2*	*Im3m*	*A2*	W	0.3879	...	...	...
γTl	25	HP	*cF4*	*Fm3m*	*A1*	Cu	?	...	...	...
Tm	25	atm	*hP2*	$P6_3/mmc$	*A3*	Mg	0.35375	...	0.55540	1.5700
αU	25	atm	*oC4*	*Cmcm*	*A20*	αU	0.28537	0.58695	0.49548	...
βU	>668	atm	*tP30*	$P4_2/mnm$	A_b	βU	1.0759	...	0.5656	0.526
γU	>776	atm	*cI2*	*Im3m*	*A2*	W	0.3524	...	...	...
V	25	atm	*cI2*	*Im3m*	*A2*	W	0.30240	...	...	...
W	25	atm	*cI2*	*Im3m*	*A2*	W	0.31652	...	...	...
Xe	<-111.76	atm	*cF4*	*Fm3m*	*A1*	Cu	0.6350	...	...	...
αY	25	atm	*hP2*	$P6_3/mmc$	*A3*	Mg	0.36482	...	0.57318	1.5711
βY	>1478	atm	*cI2*	*Im3m*	*A2*	W	(0.410)	...	...	...
αYb	<-3	atm	*hP2*	$P6_3/mmc$	*A3*	Mg	0.38799	...	0.63859	1.6459
βYb	25	atm	*cF4*	*Fm3m*	*A1*	Cu	0.54848	...	...	...
γYb	>795	atm	*cI2*	*Im3m*	*A2*	W	0.444	...	...	...
Zn	25	atm	*hP2*	$P6_3/mmc$	*A3*	Mg	0.26650	...	0.49470	1.8563
αZr	25	atm	*hP2*	$P6_3/mmc$	*A3*	Mg	0.32316	...	0.51475	1.5929
βZr	>863	atm	*cI2*	*Im3m*	*A2*	W	0.36090	...	...	...
ωZr	25	HP→atm	*hP2*	*P6/mmm*	...	ωTi	0.5036	...	0.3109	0.617

PHASE TRANSITIONS IN THE SOLID ELEMENTS AT ATMOSPHERIC PRESSURE

This table gives the phase transition temperatures for the elements that can exist in two or more crystalline forms (allotropes). The crystal phases are labeled by Greek letters in the most common conventions, although some variation is found. All data refer to normal atmospheric pressure.

References

1. Massalski, T. B., Ed., *Binary Alloy Phase Diagrams, Second Edition*, ASM International, Metals Park, OH, 1990.
2. Cordfunke, E. H. P., and Konings, R. J. M., Eds., *Thermochemical Data for Reactor Materials and Fission Products*, North-Holland, Amsterdam, 1990.
3. Greenwood, N. N., and Earnshaw, A., *Chemistry of the Elements, Second Edition*, Butterworth-Heinemann, Oxford, 1997.
4. Rhyne, J. J., "Magnetic Phase Transitions of the Elements", *Bull. Alloy Phase Diag.* 3, 402, 1982.

Element	Symbol	Transition	*t*/°C	Comments
Americium	Am	α→β	769	
	Am	β→γ	1077	
	Am	γ→liq	1176	
Beryllium	Be	α→β	1270	
	Be	β→liq	1287	
Boron	B	α→β	1100	
	B	β→γ	1500	
	B	γ→liq	2075	
Calcium	Ca	α→β	443	
	Ca	β→liq	842	
Californium	Cf	α→β	590	
	Cf	β→liq	900	
Cerium	Ce	α→β	−177	
	Ce	β→γ	61	β-Ce and γ-Ce are magnetic
	Ce	γ→δ	726	
	Ce	δ→liq	799	
Cobalt	Co	ε→α	422	magnetic transition at 1115 °C
	Co	α→liq	1495	
Curium	Cm	α→β	1277	magnetic transition at −221 °C
	Cm	β→liq	1345	
Dysprosium	Dy	α′→α	−187	
	Dy	α→β	1381	magnetic transitions in α-Dy at −184 °C and −94 °C
	Dy	β→liq	1412	
Fluorine	F_2	α→β	−227.60	
	F_2	β→liq	−219.67	
Gadolinium	Gd	α→β	1235	
	Gd	β→liq	1313	
Hafnium	Hf	α→β	1743	
	Hf	β→liq	2233	
Iron	Fe	α→γ	912	magnetic transition in α-Fe at 771 °C
	Fe	γ→δ	1394	
	Fe	δ→liq	1538	
Lanthanum	La	α→β	277	
	La	β→γ	860	
	La	γ→liq	920	
Lithium	Li	α→β	−193	
	Li	β→liq	180.50	
Manganese	Mn	α→β	727	magnetic transition in α-Mn at −100 °C
	Mn	β→γ	1100	
	Mn	γ→δ	1138	
	Mn	δ→liq	1246	
Neodymium	Nd	α→β	855	magnetic transition in α-Nd at −253 °C
	Nd	β→liq	1016	
Neptunium	Np	α→β	280	
	Np	β→γ	576	
	Np	γ→liq	644	

Element	Symbol	Transition	t/°C	Comments
Nitrogen	N_2	α→β	−237.54	
	N_2	β→liq	−210.0	
Oxygen	O_2	α→β	−249.29	
	O_2	β→γ	−229.35	
	O_2	γ→liq	−218.79	
Phosphorus	P	brown→β-white	−190	several amorphous phases (red, black, gray) exist (Ref. 3)
	P	β-white→α-white	−76.9	
	P	α-white→liq	44.15	
Plutonium	Pu	α→β	124.5	
	Pu	β→γ	214.8	
	Pu	γ→δ	320.0	
	Pu	δ→δ'	462.9	
	Pu	δ'→ε	482.6	
	Pu	ε→liq	640	
Polonium	Po	α→β	54	
	Po	β→liq	254	
Praseodymium	Pr	α→β	795	
	Pr	β→liq	931	
Promethium	Pm	α→β	890	magnetic transition in α-Pm at −175 °C
	Pm	β→liq	1042	
Protactinium	Pa	α→β	1170	
	Pa	β→liq	1572	
Samarium	Sm	α→β	734	magnetic transition in α-Sm at −167 °C
	Sm	β→γ	922	
	Sm	γ→liq	1072	
Scandium	Sc	α→β	1337	
	Sc	β→liq	1541	
Selenium	Se	α-red→gray	180	many allotropes exist (Ref. 3)
	Se	gray→liq	220.8	
Sodium	Na	α→β	−233	
	Na	β→liq	97.794	
Strontium	Sr	α→β	547	
	Sr	β→liq	777	
Sulfur	S	α→β	95.3	many allotropes exist (Ref. 3)
	S	β→liq	115.21	
Terbium	Tb	α'→α	−53	
	Tb	α→β	1289	magnetic transition in α-Tb at −230 °C
	Tb	β→liq	1359	
Thallium	Tl	α→β	230	
	Tl	β→liq	304	
Thorium	Th	α→β	1360	
	Th	β→liq	1750	
Tin	Sn	α (gray)→β (white)	13.2	
	Sn	β (white)→liq	231.928	defining fixed point on ITS-90
Titanium	Ti	α→β	882	
	Ti	β→liq	1668	
Uranium	U	α→β	669	
	U	β→γ	776	
	U	γ→liq	1135	
Ytterbium	Yb	α→β	3	
	Yb	β→γ	795	
	Yb	γ→liq	824	
Yttrium	Y	α→β	1478	
	Y	β→liq	1522	
Zirconium	Zr	α→β	866	
	Zr	β→liq	1854.7	

LATTICE ENERGIES

H. D. B. Jenkins and H. K. Roobottom

THERMOCHEMICAL CYCLE AND CALCULATED VALUES

Table 1 contains calculated values of the lattice energies (total lattice potential energies), U_{POT}, of crystalline salts, M_aX_b. U_{POT} is expressed in units of kilojoules per mole, kJ mol^{-1}. M and X can be either simple or complex ions. Substances are arranged by chemical class.

Also listed in the table is the lattice energy, U_{POT}^{BFHC}, obtained from the application of the Born–Fajans–Haber cycle (BHFC) described below, using the "Standard Thermochemical Properties of Chemical Substances" table in Section 5 of this *Handbook*, References 1 through 4, and certain other data that are given in Table 3 below.

The lattice enthalpy, ΔH_L, is given by the cycle:

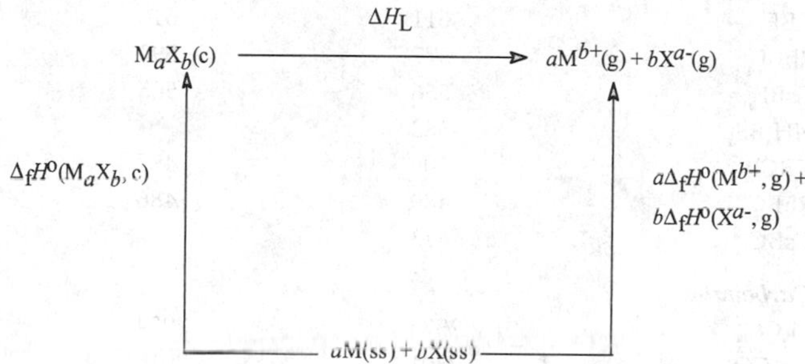

where (ss) is the standard state of the element concerned.

The lattice enthalpy, ΔH_L, is obtained using the equation:

$$\Delta H_L = a\Delta_f H°(M^{b+}, g) + b\Delta_f H°(X^{a-}, g) - \Delta_f H°(M_aX_b, c)$$

and is further related to the total lattice potential energy, U_{POT}, by the relationship:

$$\Delta H_L = U_{POT} + \left[a\left(\frac{n_M}{2} - 2\right) + b\left(\frac{n_X}{2} - 2\right)\right]RT$$

where n_M and n_X equal 3 for monatomic ions, 5 for linear polyatomic ions and 6 for polyatomic non-linear ions.

METHOD OF ESTIMATION OF VALUES NOT TABULATED

In cases where the lattice energy is not tabulated and we want to furnish an estimate, then the Kapustinskii equation (Ref. 5) can be used to obtain a value (in kJ mol^{-1}):

$$U_{POT} = \frac{121.4 z_a z_b \nu}{(r_a + r_b)}\left(1 - \frac{0.0345}{(r_a + r_b)}\right)$$

where z_a and z_b are the moduli of the charges on the ν ions in the lattice and r_a and r_b (in nm) are the thermochemical radii given in Table 2. The r_a for metal ions is taken to be the Goldschmidt radius (Ref. 6).

To cite an example, if we wish to estimate the lattice energy of the salt $[NH^+_4][HF^-_2]$ using the above procedure, we see that Table 2 gives the thermochemical radius (r_a) for NH^+_4 to be 0.136 nm and that for HF^-_2 (r_b) to be 0.172 nm. The lattice potential energy is then estimated to be 700 kJ mol^{-1} compared with the calculated value of 705 kJ mol^{-1} and the Born–Fajans–Haber cycle value of 658 kJ mol^{-1}.

References

1. Wagman, D. D., Evans, W. H., Parker, V. B., Schumm, R. H., Halow, I., Bailey, S. M., Churney, K. L., and Nuttall, R. L., The NBS Tables of Chemical Thermodynamic Properties, *J. Phys. Chem. Ref. Data*, Vol. 11, Suppl. 2, 1982.
2. Chase, M. W., Davies, C. A., Downey, J. R., Frurip, D. J., McDonald, R. A., and Syverud, A. N., JANAF Thermochemical Tables, Third Edition, *J. Phys. Chem. Ref. Data*, Vol. 14, Suppl. 1, 1985.
3. Lias, S. G., Bartmess, J. E., Liebman, J. F., Holmes, J. L., Levin, R. D., and Mallard, W. G., Gas-Phase Ion and Neutral Thermochemistry, *J. Phys. Chem. Ref. Data*, Vol. 17, Suppl. 1, 1988.
4. Jenkins, H. D. B., and Pratt, K. F., *Adv. Inorg. Chem. Radiochem.* 22, 1, 1978.
5. Kapustinskii, A. F., *Quart. Rev.* 10, 283, 1956.
6. Goldschmidt, V. M., *Skrifter Norske Videnskaps-Akad.* Oslo, I, Mat.-Naturn. Kl, 1926. See also Dasent, W. E., *Inorganic Energetics*, 2nd ed., Cambridge University Press, 1982.

Table 1. Lattice Energies (kJ mol^{-1})

Substance	Calc. U_{POT}	U_{POT}^{BHFC}
Acetates		
$Li(CH_3COO)$	–	843
$Na(CH_3COO)$	828	807
$K(CH_3COO)$	749	726
$Rb(CH_3COO)$	715	–
$Cs(CH_3COO)$	682	
Acetylides		
CaC_2	2911	2902
SrC_2	2788	2782
BaC_2	2647	2652
Azides		
LiN_3	861	875
NaN_3	770	784
KN_3	697	–
RbN_3	674	691
CsN_3	665	674
AgN_3	854	910
TlN_3	689	742
$Ca(N_3)_2$	2186	2316
$Sr(N_3)_2$	2056	2187
$Ba(N_3)_2$	2021	–
$Mn(N_3)_2$	2408	2348
$Cu(N_3)_2$	2730	2738
$Zn(N_3)_2$	2840	2970
$Cd(N_3)_2$	2446	2576
$Pb(N_3)_2$		2300
Bihalide Salts		
$LiHF_2$	821	847
$NaHF_2$	755	748
KHF_2	657	660
$RbHF_2$	627	631
$CsHF_2$	607	–
NH_4HF_2	705	658
$CsHCl_2$	601	–
Me_4NHCl_2	427	–
Et_4NHCl_2	346	–
Bu_4NHCl_2	290	
Bicarbonates		
$NaHCO_3$	820	656
$KHCO_3$	741	573
$RbHCO_3$	707	522
$CsHCO_3$	678	520
NH_4HCO_3	–	577
Borides		
CaB_6	5146	–
SrB_6	5104	–
BaB_6	5021	–
YB_6	7447	–
LaB_6	7406	–
CeB_6	10083	–
PrB_6	7447	–
NdB_6	7447	–
PmB_6	7406	–
SmB_6	7447	–
EuB_6	5104	–
GdB_6	7489	–
TbB_6	7489	–
DyB_6	7489	–
HoB_6	7489	–
ErB_6	7489	–
TmB_6	7489	–
YbB_6	5146	–
LuB_6	7489	–
ThB_6	10167	–
Borohydrides		
$LiBH_4$	778	–
$NaBH_4$	703	694
KBH_4	655	638
$RbBH_4$	648	–
$CsBH_4$	628	–
Borohalides		
$LiBF_4$	699	749
$NaBF_4$	657	674
KBF_4	611	616
$RbBF_4$	577	590
$CsBF_4$	556	565
NH_4BF_4	582	–
$KBCl_4$	506	497
$RbBCl_4$	489	486
$CsBCl_4$	473	–
Carbonates		
Li_2CO_3	2523	2254
Na_2CO_3	2301	2016
K_2CO_3	2084	1846
Rb_2CO_3	2000	1783
Cs_2CO_3	1920	1722
$MgCO_3$	3138	3122
$CaCO_3$	2804	2811
$SrCO_3$	2720	2688
$BaCO_3$	2615	2554
$MnCO_3$	3046	3092
$FeCO_3$	3121	3169
$CoCO_3$	3443	3235
$CuCO_3$	3494	–
$ZnCO_3$	3121	3273
$CdCO_3$	2929	3052
$SnCO_3$	2904	–
$PbCO_3$	2728	2750
Cyanates		
$LiNCO$	849	–
$NaNCO$	807	816
$KNCO$	726	734
$RbNCO$	692	–
$CsNCO$	661	–
NH_4NCO	724	–
Cyanides		
$LiCN$	874	
$NaCN$	766	759
KCN	692	686
$RbCN$	638	–
$CsCN$	601	–
$Ca(CN)_2$	2268	2240
$Sr(CN)_2$	2138	–
$Ba(CN)_2$	2001	2009
NH_4CN	617	691
$AgCN$	(741)	935

Substance	Calc. U_{POT}	U_{POT}^{BHFC}
$Zn(CN)_2$	2809	2817
$Cd(CN)_2$	2583	2591
Formates		
$Li(HCO_2)$	865	–
$Na(HCO_2)$	791	804
$K(HCO_2)$	713	722
$Rb(HCO_2)$	685	–
$Cs(HCO_2)$	651	–
$NH_4(HCO_2)$	715	–
Germanates		
Mg_2GeO_4	7991	–
Ca_2GeO_4	7301	7306
Sr_2GeO_4	6987	–
Ba_2GeO_4	6653	6643
Halates		
$LiBrO_3$	883	880
$NaBrO_3$	803	791
$KBrO_3$	740	722
$RbBrO_3$	720	705
$CsBrO_3$	694	681
$NaClO_3$	770	785
$KClO_3$	711	721
$RbClO_3$	690	703
$CsClO_3$	–	679
$LiIO_3$	975	974
$NaIO_3$	883	876
KIO_3	820	780
$RbIO_3$	791	–
$CsIO_3$	761	–
Halides		
LiF	1030	1049
LiCl	834	864
LiBr	788	820
LiI	730	764
NaF	910	930
NaCl	769	790
NaBr	732	754
NaI	682	705
KF	808	829
KCl	701	720
KBr	671	691
KI	632	650
RbF	774	795
RbCl	680	695
RbBr	651	668
RbI	617	632
CsF	744	759
CsCl	657	670
CsBr	632	647
CsI	600	613
FrF	715	–
FrCl	632	–
FrBr	611	–
FrI	582	–
CuCl	992	996
CuBr	969	978
CuI	948	966
AgF	953	974
AgCl	910	918
AgBr	897	905

Substance	Calc. U_{POT}	U_{POT}^{BHFC}
AgI	881	892
AuCl	1013	1066
AuBr	1029	1059
AuI	1027	1070
InCl	–	764
InBr	–	767
InI	–	733
TlF	–	850
TlCl	738	751
TlBr	720	734
TlI	692	710
Me_4NCl	566	–
Me_4NBr	553	–
Me_4NI	544	–
PH_4Br	616	–
PH_4I	590	–
BeF_2	3464	3526
$BeCl_2$	3004	3033
$BeBr_2$	2950	2914
BeI_2	2780	2813
MgF_2	2926	2978
$MgCl_2$	2477	2540
$MgBr_2$	2406	2451
MgI_2	2293	2340
CaF_2	2640	2651
$CaCl_2$	2268	2271
$CaBr_2$	2133	
CaI_2	1971	2087
SrF_2	2476	2513
$SrCl_2$	2142	2170
SrI_2	1984	1976
BaF_2	2347	2373
$BaCl_2$	2046	2069
$BaBr_2$	1971	1995
BaI_2	1862	1890
RaF_2	2284	–
$RaCl_2$	2004	–
$RaBr_2$	1929	–
RaI_2	1803	–
$ScCl_2$	2380	–
$ScBr_2$	2291	–
ScI_2	2201	–
TiF_2	2724	–
$TiCl_2$	2439	2514
$TiBr_2$	2360	2430
TiI_2	2259	2342
VCl_2	2607	2593
VBr_2	–	2534
VI_2	–	2470
CrF_2	2778	2939
$CrCl_2$	2540	2601
$CrBr_2$	2377	2536
CrI_2	2269	2440
$MoCl_2$	2737	2746
$MoBr_2$	2742	2753
MoI_2	2630	–
MnF_2	2644	–
$MnCl_2$	2510	2551
$MnBr_2$	2448	2482
MnI_2	2212	–
FeF_2	2849	2967

Substance	Calc. U_{POT}	U_{POT}^{BHFC}
$FeCl_2$	2569	2641
$FeBr_2$	2515	2577
FeI_2	2439	2491
CoF_2	3004	3042
$CoCl_2$	2707	2706
$CoBr_2$	2640	2643
CoI_2	2569	2561
NiF_2	3098	3089
$NiCl_2$	2753	2786
$NiBr_2$	2729	2721
NiI_2	2607	2637
$PdCl_2$	2778	2818
$PdBr_2$	2741	2751
PdI_2	2748	2760
CuF_2	3046	3102
$CuCl_2$	2774	2824
$CuBr_2$	2715	2774
CuI_2	2640	–
AgF_2	2942	2967
ZnF_2	3021	3053
$ZnCl_2$	2703	2748
$ZnBr_2$	2648	2689
ZnI_2	2581	2619
CdF_2	2809	2830
$CdCl_2$	2552	2565
$CdBr_2$	2507	2517
CdI_2	2441	2455
HgF_2	2757	–
$HgCl_2$	2657	2664
$HgBr_2$	2628	2639
HgI_2	2628	2624
SnF_2	2551	–
$SnCl_2$	2297	2310
$SnBr_2$	2251	2256
SnI_2	2193	2206
PbF_2	2535	2543
$PbCl_2$	2270	2282
$PbBr_2$	2219	2230
PbI_2	2163	2177
ScF_3	5492	5540
$ScCl_3$	4874	4901
$ScBr_3$	4729	4761
ScI_3	4640	–
YF_3	4983	–
YCl_3	4506	4524
YI_3	4240	4258
TiF_3	5644	–
$TiCl_3$	5134	5153
$TiBr_3$	5012	5023
TiI_3	4845	–
$ZrCl_3$	–	4791
$ZrBr_3$	–	4758
ZrI_3	–	4591
VF_3	5895	–
VCl_3	5322	5329
VBr_3	5214	5224
VI_3	5121	5136
$NbCl_3$	5062	–
$NbBr_3$	4980	–
NbI_3	4860	–
CrF_3	6033	6065
$CrCl_3$	5518	5529
$CrBr_3$	5355	–
CrI_3	5275	5294
MoF_3	6459	–
$MoCl_3$	5246	5253
$MoBr_3$	5156	–
MoI_3	5073	–
MnF_3	6017	–
$MnCl_3$	5544	–
$MnBr_3$	5448	–
MnI_3	5330	–
$TcCl_3$	5270	–
$TcBr_3$	5215	–
TcI_3	5188	–
FeF_3	5870	–
$FeCl_3$	5364	5436
$FeBr_3$	5333	5347
FeI_3	5117	–
$RuCl_3$	5245	5257
$RuBr_3$	5223	5232
RuI_3	5222	5235
CoF_3	5991	–
$RhCl_3$	5641	5665
IrF_3	(6112)	–
$IrBr_3$	(4794)	–
NiF_3	(6111)	–
AuF_3	(5777)	–
$AuCl_3$	(4605)	–
$ZnCl_3$	5832	–
$ZnBr_3$	5732	–
ZnI_3	5636	–
AlF_3	5924	6252
$AlCl_3$	5376	5513
$AlBr_3$	5247	5360
AlI_3	5070	5227
GaF_3	5829	6238
$GaCl_3$	5217	5665
$GaBr_3$	4966	5569
GaI_3	4611	5496
$InCl_3$	4736	5183
$InBr_3$	4535	5117
InI_3	4234	5001
TlF_3	5493	–
$TlCl_3$	5258	5278
$TlBr_3$	5171	–
TlI_3	5088	–
$AsBr_3$	5497	5365
AsI_3	4824	5295
SbF_3	5295	5324
$SbCl_3$	5032	4857
$SbBr_3$	4954	4776
SbI_3	4867	4692
$BiCl_3$	4689	4707
BiI_3	3774	–
LaF_3	4682	–
$LaCl_3$	4263	4242
$LaBr_3$	4209	–
LaI_3	3916	3986
$CeCl_3$	4394	4348
CeI_3	–	4061
$PrCl_3$	4322	4387

Substance	Calc. U_{POT}	U_{POT}^{BHFC}
PrI_3	–	4101
$NdCl_3$	4343	4415
$SmCl_3$	4376	4450
$EuCl_3$	4393	4490
$GdCl_3$	4406	4495
$DyCl_3$	4481	4529
$HoCl_3$	4501	4572
$ErCl_3$	4527	4591
$TmCl_3$	4548	4608
TmI_3	–	4340
$YbCl_3$	–	4651
$AcCl_3$	4096	–
UCl_3	4243	–
$NpCl_3$	4268	–
$PuCl_3$	4289	–
$PuBr_3$	(3959)	–
$AmCl_3$	4293	–
TiF_4	10012	9908
$TiCl_4$	9431	–
$TiBr_4$	9288	9059
TiI_4	9108	8918
ZrF_4	8853	8971
$ZrCl_4$	8021	8144
$ZrBr_4$	7661	7984
ZrI_4	7155	7801
MoF_4	8795	–
$MoCl_4$	8556	9603
$MoBr_4$	8510	9500
MoI_4	8427	–
$SnCl_4$	8355	8930
$SnBr_4$	7970	8852
PbF_4	9519	–
CrF_2Cl	5795	–
CrF_2Br	5753	–
CrF_2I	5669	–
$CrCl_2Br$	5448	–
$CrCl_2I$	5381	5429
$CrBr_2I$	5330	5370
CuFCl	2891	–
CuFBr	2853	–
CuFI	2803	–
CuClBr	2753	–
CuClI	2694	–
CuBrI	2669	–
FeF_2Cl	5711	–
FeF_2Br	5653	–
FeF_2I	5569	–
$FeCl_2Br$	5339	–
$FeCl_2I$	5272	–
$FeBr_2I$	5209	–
$LiIO_2F_2$	845	–
$NaIO_2F_2$	766	756
KIO_2F_2	699	689
$RbIO_2F_2$	674	–
$CsIO_2F_2$	636	–
$NH_4IO_2F_2$	678	–
$AgIO_2F_2$	736	685
Hydrides		
LiH	916	918
NaH	807	807
KH	711	713

Substance	Calc. U_{POT}	U_{POT}^{BHFC}
RbH	686	684
CsH	648	653
VH	1184	(1344)
NbH	1163	(1633)
PdH	979	1368
CuH	828	1254
TiH	996	1407
ZrH	916	1590
HfH	904	–
LaH	828	–
TaH	1021	–
CrH	1050	–
NiH	929	
PtH	937	–
AgH	941	–
AuH	1033	1108
TlH	745	–
GeH	950	–
PbH	778	–
BeH_2	3205	3306
MgH_2	2791	2718
CaH_2	2410	2406
SrH_2	2250	2265
BaH_2	2121	2133
ScH_2	2711	2744
YH_2	(2598)	2733
LaH_2	2380	2522
CeH_2	2414	2509
PrH_2	2448	2405
NdH_2	2464	2394
PmH_2	2519	–
SmH_2	2510	2389
GdH_2	2494	2651
AcH_2	2372	–
ThH_2	2711	2738
PuH_2	2519	–
AmH_2	2544	–
TiH_2	2866	2864
ZrH_2	2711	2999
CuH_2	2941	–
ZnH_2	2870	–
HgH_2	2707	–
AlH_3	5924	5969
FeH_3	5724	–
ScH_3	5439	–
YH_3	5063	4910
LaH_3	4895	4493
FeH_3	5724	–
GaH_3	5690	–
InH_3	5092	–
TlH_3	5092	–
Hydroselenides		
NaHSe	703	732
KHSe	644	712
RbHSe	623	689
CsHse	598	669
Hydrosulphides		
LiHS	768	862
NaHS	723	771
RbHS	655	682

Substance	Calc. U_{POT}	U_{POT}^{BHFC}
CsHS	628	657
NH_4HS	661	718
$Ca(HS)_2$	2184	(2171)
$Sr(HS)_2$	2063	–
$Ba(HS)_2$	1979	(1956)
Hydroxides		
LiOH	1021	1028
NaOH	887	892
KOH	789	796
RbOH	766	765
CsOH	721	732
$Be(OH)_2$	3477	3620
$Mg(OH)_2$	2870	2998
$Ca(OH)_2$	2506	2637
$Sr(OH)_2$	2330	2474
$Ba(OH)_2$	2142	2330
$Ti(OH)_2$	–	2953
$Mn(OH)_2$	2909	3008
$Fe(OH)_2$	2653	3044
$Co(OH)_2$	2786	3109
$Ni(OH)_2$	2832	3186
$Pd(OH)_2$	–	3189
$Cu(OH)_2$	2870	3229
CuOH	1006	–
AgOH	918	845
AuOH	1033	–
TlOH	705	874
$Zn(OH)_2$	2795	3151
$Cd(OH)_2$	2607	2909
$Hg(OH)_2$	2669	–
$Sn(OH)_2$	2489	2721
$Pb(OH)_2$	2376	–
$Sc(OH)_3$	5063	5602
$Y(OH)_3$	4707	–
$La(OH)_3$	4443	–
$Cr(OH)_3$	5556	6299
$Mn(OH)_3$	6213	–
$Al(OH)_3$	5627	–
$Ga(OH)_3$	5732	6368
$In(OH)_3$	5280	–
$Tl(OH)_3$	5314	–
$Ti(OH)_4$	9456	–
$Zr(OH)_4$	8619	–
$Mn(OH)_4$	10933	–
$Sn(OH)_4$	9188	9879
Imides		
CaNH	3293	–
SrNH	3146	–
BaNH	2975	–
Metavanadates		
Li_3VO_4	3945	–
Na_3VO_4	3766	–
K_3VO_4	3376	–
Rb_3VO_4	3243	–
Cs_3VO_4	3137	–
Nitrates		
$LiNO_3$	848	854
$NaNO_3$	755	763
KNO_3	685	694
$RbNO_3$	662	671
$CsNO_3$	648	650
$AgNO_3$	820	832
$TlNO_3$	690	707
$Mg(NO_3)_2$	2481	2521
$Ca(NO_3)_2$	2268	2247
$Sr(NO_3)_2$	2176	2151
$Ba(NO_3)_2$	2062	2035
$Mn(NO_3)_2$	2318	2478
$Fe(NO_3)_2$	–	(2580)
$Co(NO_3)_2$	2560	2647
$Ni(NO_3)_2$	–	2729
$Cu(NO_3)_2$	–	2739
$Zn(NO_3)_2$	2376	2649
$Cd(NO_3)_2$	2238	2462
$Sn(NO_3)_2$	2155	2254
$Pb(NO_3)_2$	2067	2208
Nitrides		
ScN	7547	7506
LaN	6876	6793
TiN	8130	8033
ZrN	7633	7723
VN	8283	8233
NbN	7939	8022
CrN	8269	8358
Nitrites		
$NaNO_2$	774	772
KNO_2	699	687
$RbNO_2$	724	765
$CsNO_2$	690	–
Oxides		
Li_2O	2799	2814
Na_2O	2481	2478
K_2O	2238	2232
Rb_2O	2163	2161
Cs_2O	2131	2063
Cu_2O	3273	3189
Ag_2O	3002	2910
Tl_2O	2659	2575
LiO_2	(878)	(872)
NaO_2	799	821
KO_2	741	751
RbO_2	706	721
CsO_2	679	696
Li_2O_2	2592	2557
Na_2O_2	2309	22717
K_2O_2	2114	2064
Rb_2O_2	2025	1994
Cs_2O_2	1948	1512
MgO_2	3356	3526
CaO_2	3144	3132
SrO_2	3037	2977
KO_3	697	707
BeO	4514	4443
MgO	3795	3791
CaO	3414	3401
SrO	3217	3223
BaO	3029	3054
TiO	3832	3811
VO	3932	3863
MnO	3724	3745

Substance	Calc. U_{POT}	U_{POT}^{BHFC}
FeO	3795	3865
CoO	3837	3910
NiO	3908	4010
PdO	3736	–
CuO	4135	4050
ZnO	4142	3971
CdO	3806	–
HgO	3907	–
GeO	3919	–
SnO	3652	–
PbO	3520	–
Sc_2O_3	13557	13708
Y_2O_3	12705	–
La_2O_3	12452	–
Ce_2O_3	12661	–
Pr_2O_3	12703	–
Nd_2O_3	12736	–
Pm_2O_3	12811	–
Sm_2O_3	12878	–
Eu_2O_3	12945	–
Gd_2O_3	12996	–
Tb_2O_3	13071	–
Dy_2O_3	13138	–
Ho_2O_3	13180	–
Er_2O_3	13263	–
Tm_2O_3	13322	–
Yb_2O_3	13380	–
Lu_2O_3	13665	–
Ac_2O_3	12573	–
Ti_2O_3	–	14149
V_2O_3	15096	14520
Cr_2O_3	15276	14957
Mn_2O_3	15146	15035
Fe_2O_3	14309	14774
Al_2O_3	15916	–
Ga_2O_3	15590	15220
In_2O_3	13928	–
Pb_2O_3	(14841)	–
CeO_2	9627	–
ThO_2	10397	–
PaO_2	10573	–
VO_2(g)	10644	–
NpO_2	10707	–
PuO_2	10786	–
AmO_2	10799	–
CmO_2	10832	–
TiO_2	12150	–
ZrO_2	11188	–
MoO_2	11648	–
MnO_2	12970	–
SiO_2	13125	–
GeO_2	12828	–
SnO_2	11807	–
PbO_2	11217	–
Perchlorates		
$LiClO_4$	709	715
$NaClO_4$	643	641
$KClO_4$	599	595
$RbClO_4$	564	576
$CsClO_4$	636	550
NH_4ClO_4	583	580

Substance	Calc. U_{POT}	U_{POT}^{BHFC}
$Ca(ClO_4)_2$	1958	1971
$Sr(ClO_4)_2$	1862	1862
$Ba(ClO_4)_2$	1795	1769
Permanganates		
$NaMnO_4$	661	–
$KMnO_4$	607	–
$RbMnO_4$	586	–
$CsMnO_4$	565	–
$Ca(MnO_4)_2$	1937	–
$Sr(MnO_4)_2$	1845	–
$Ba(MnO_4)_2$	1778	–
Phosphates		
$Mg_3(PO_4)_2$	11632	11407
$Ca_3(PO_4)_2$	10602	10479
$Sr_3(PO_4)_2$	10125	10075
$Ba_3(PO_4)_2$	9652	9654
$MnPO_4$	7397	–
$FePO_4$	7251	7300
BPO_4	8201	–
$AlPO_4$	7427	7507
$GaPO_4$	7381	–
Selenides		
Li_2Se	2364	–
Na_2Se	2130	–
K_2Se	1933	–
Rb_2Se	1837	–
Cs_2Se	1745	–
Ag_2Se	2686	–
Tl_2Se	2209	–
BeSe	3431	–
MgSe	3071	–
CaSe	2858	2862
SrSe	2736	–
BaSe	2611	–
MnSe	3176	–
Selenites		
Li_2SeO_3	2171	–
Na_2SeO_3	1950	1916
K_2SeO_3	1774	1749
Rb_2SeO_3	1715	1675
Cs_2SeO_3	1640	–
Tl_2SeO_3	1879	–
Ag_2SeO_3	2113	2148
$BeSeO_3$	3322	–
$MgSeO_3$	3012	2998
$CaSeO_3$	2732	–
$SrSeO_3$	2586	2588
$BaSeO_3$	2460	2451
Selenates		
Li_2SeO_4	2054	–
Na_2SeO_4	1879	–
K_2SeO_4	1732	–
Rb_2SeO_4	1686	–
Cs_2SeO_4	1615	–
Cu_2SeO_4	2201	–
Ag_2SeO_4	2033	–
Tl_2SeO_4	1766	–
Hg_2SeO_4	2163	–
$BeSeO_4$	3448	–

Substance	Calc. U_{POT}	U_{POT}^{BHFC}
$MgSeO_4$	2895	–
$CaSeO_4$	2632	–
$SrSeO_4$	2489	–
Sulphates		
Li_2SO_4	2229	2142
Na_2SO_4	1827	1938
K_2SO_4	1700	1796
Rb_2SO_4	1636	1748
Cs_2SO_4	1596	1658
$(NH_4)_2SO_4$	1766	1777
Cu_2SO_4	2276	2166
Ag_2SO_4	2104	1989
Tl_2SO_4	1828	1722
Hg_2SO_4	–	2127
$CaSO_4$	2489	2480
$SrSO_4$	2577	2484
$BaSO_4$	2469	2374
$MnSO_4$	2920	2825
Sulphides		
Li_2S	2464	2472
Na_2S	2192	2203
K_2S	1979	(2052)
Rb_2S	1929	1949
Cs_2S	1892	1850
$(NH_4)_2S$	2008	(2026)
Cu_2S	2786	2865
Ag_2S	2606	2677
Au_2S	2908	–
Tl_2S	2298	2258
Ternary Salts		
Cs_2CuCl_4	1393	–
Rb_2ZnCl_4	1529	–
Cs_2ZnCl_4	1492	–
Rb_2ZnBr_4	1498	–
Cs_2ZnBr_4	1454	–
Cs_2ZnI_4	1386	–
$CsGaCl_4$	494	–
$NaAlCl_4$	556	–
$CsAlCl_4$	486	–
$NaFeCl_4$	492	–
Rb_2CoCl_4	1447	–
Cs_2CoCl_4	1391	–
K_2PtCl_4	1574	1550
Cs_2GeF_6	1573	–
$(NH_4)_2GeF_6$	1657	–
Cs_2GeCl_6	1404	1419
K_2HfCl_6	1345	1461
K_2IrCl_6	1442	1440
Na_2MoCl_6	1526	1504
K_2MoCl_6	1418	1412
Rb_2MoCl_6	1399	1399
Cs_2MoCl_6	1347	1347
K_2NbCl_6	1375	1398
Rb_2NbCl_6	1371	1385
Cs_2NbCl_6	1381	1344
K_2OsCl_6	1447	1447
Cs_2OsCl_6	1409	–
K_2OsBr_6	1396	–
K_2PdCl_6	1481	1493
Rb_2PdCl_6	1449	–
Cs_2PdCl_6	1426	–
Rb_2PbCl_6	1343	1343
Cs_2PbCl_6	1344	–
$(NH_4)_2PbCl_6$	1355	–
K_2PtCl_6	1468	1471
Rb_2PtCl_6	1464	–
Cs_2PtCl_6	1444	–
$(NH_4)_2PtCl_6$	1468	–
Tl_2PtCl_6	1546	–
Ag_2PtCl_6	1773	1881
$BaPtCl_6$	2047	2070
K_2PtBr_6	1423	1392
Ag_2PtBr_6	1791	2276
K_2PtI_6	1421	–
K_2ReCl_6	1416	1442
Rb_2ReCl_6	1414	–
Cs_2ReCl_6	1398	–
K_2ReBr_6	1375	1375
K_2SiF_6	1670	1765
Rb_2SiF_6	1639	1673
Cs_2SiF_6	1604	1498
Tl_2SiF_6	1675	–
K_2SnCl_6	1363	1390
Rb_2SnCl_6	1361	1363
Cs_2SnCl_6	1358	–
Tl_2SnCl_6	1437	–
$(NH_4)_2SnCl_6$	1370	1344
Rb_2SnBr_6	1309	–
Cs_2SnBr_6	1306	–
Rb_2SnI_6	1226	–
Cs_2SnBr_6	1243	–
K_2TeCl_6	1318	1320
Rb_2TeCl_6	1321	–
Cs_2TeCl_6	1323	–
Tl_2TeCl_6	1392	–
$(NH_4)_2TeCl_6$	1318	–
K_2RuCl_6	1451	–
Rb_2CoF_6	1688	–
Cs_2CoF_6	1632	–
K_2NiF_6	1721	–
Rb_2NiF_6	1688	–
Rb_2SbCl_6	1357	–
Rb_2SeCl_6	1409	–
Cs_2SeCl_6	1397	–
$(NH_4)_2SeCl_6$	1420	–
$(NH_4)_2PoCl_6$	1338	–
Cs_2PoBr_6	1286	–
Cs_2CrF_6	1603	–
Rb_2MnF_6	1688	–
Cs_2MnF_6	1620	–
K_2MnCl_6	1462	–
Rb_2MnCl_6	1451	–
$(NH_4)_2MnCl_6$	1464	–
Cs_2TeBr_6	1306	–
Cs_2TeI_6	1246	–
K_2TiCl_6	1412	1447
Rb_2TiCl_6	1415	1416
Cs_2TiCl_6	1402	1384
Tl_2TiCl_6	1560	1553
K_2TiBr_6	1379	1379
Rb_2TiBr_6	1341	1331

Substance	Calc. U_{POT}	U_{POT}^{BHFC}
Cs_2TiBr_6	1339	1306
Na_2UBr_6	1504	–
K_2UBr_6	1484	–
Rb_2UBr_6	1473	–
Cs_2UBr_6	1459	–
K_2WCl_6	1398	1423
Rb_2WCl_6	1397	1434
Cs_2WCl_6	1392	1366
K_2WBr_6	1408	1408
Rb_2WBr_6	1361	1391
Cs_2WBr_6	1362	1332
K_2ZrCl_6	1339	1371
Rb_2ZrCl_6	1341	–
Cs_2ZrCl_6	1339	1307
Tellurides		
Li_2Te	2212	–
Na_2Te	1997	2095
K_2Te	1830	–
Rb_2Te	1837	–
Cs_2Te	1745	–
Cu_2Te	2706	2683
Ag_2Te	2607	2600
Tl_2Te	2084	2172
BeTe	3319	–
MgTe	2878	3081
CaTe	2721	–

Substance	Calc. U_{POT}	U_{POT}^{BHFC}
Thiocyanates		
LiCNS	764	(765)
NaCNS	682	682
KCNS	623	616
RbCNS	623	619
CsCNS	623	568
NH4CNS	605	611
Ca(CNS)2	2184	2118
Sr(CNS)2	2063	1957
Ba(CNS)2	1979	1852
Mn(CNS)2	2280	2351
Zn(CNS)2	2335	2560
Cd(CNS)2	2201	2374
Hg(CNS)2	2146	2492
Sn(CNS)2	2117	2142
Pb(CNS)2	2058	–
Vanadates		
LiVO3	810	–
NaVO3	761	–
KVO3	686	–
RbVO3	657	–
CsVO3	628	–

TABLE 2. Thermochemical Radii (nm)

Ion	Radius	
Singly Charged Anions		
AgF_4^-	0.231	± 0.019
$AlBr_4^-$	0.321	± 0.023
$AlCl_4^-$	0.317	± 0.019
AlF_4^-	0.214	± 0.023
AlH_4^-	0.226	± 0.019
AlI_4^-	0.374	± 0.019
AsF_6^-	0.243	± 0.019
AsO_2^-	0.211	± 0.019
$Au(CN)_2^-$	0.266	± 0.019
$AuCl_4^-$	0.288	± 0.019
AuF_4^-	0.240	± 0.019
AuF_6^-	0.235	± 0.038
$B(OH)_4^-$	0.229	± 0.019
BF_4^-	0.205	± 0.019
BH_4^-	0.205	± 0.019
Br^-	0.190	± 0.019
BrF_4^-	0.231	± 0.019
BrO_3^-	0.214	± 0.019
$CF_3SO_3^-$	0.230	± 0.049
$CH_3CO_2^-$	0.194	± 0.019
Cl^-	0.168	± 0.019
ClO_2^-	0.195	± 0.019
ClO_3^-	0.208	± 0.019
ClO_4^-	0.225	± 0.019
$ClS_2O_6^-$	0.260	± 0.049
CN^-	0.187	± 0.023
$Cr_3O_8^-$	0.276	± 0.019
$CuBr_4^-$	0.315	± 0.019
F^-	0.126	± 0.019
$FeCl_4^-$	0.317	± 0.019

Ion	Radius	
$GaCl_4^-$	0.328	± 0.019
H^-	0.148	± 0.019
$H_2AsO_4^-$	0.227	± 0.019
$H_2PO_4^-$	0.213	± 0.019
HCO_2^-	0.200	± 0.019
HCO_3^-	0.207	± 0.019
HF_2^-	0.172	± 0.019
HSO_4^-	0.221	± 0.019
I^-	0.211	± 0.019
I_2Br	0.261	± 0.019
I_3^-	0.272	± 0.019
I_4^-	0.300	± 0.019
IBr_2^-	0.251	± 0.019
ICl_2^-	0.235	± 0.019
ICl_4^-	0.307	± 0.019
$IO_2F_2^-$	0.233	± 0.019
IO_3^-	0.218	± 0.019
IO_4^-	0.231	± 0.019
IrF_6^-	0.242	± 0.019
MnO_4^-	0.220	± 0.019
MoF_6^-	0.241	± 0.019
$MoOF_5^-$	0.241	± 0.019
N_3^-	0.180	± 0.019
NCO^-	0.193	± 0.019
$NbCl_6^-$	0.338	± 0.049
NbF_6^-	0.254	± 0.019
$Nb_2F_{11}^-$	0.311	± 0.038
NbO_3^-	0.194	± 0.019
NH_2^-	0.168	± 0.019
$NH_2CH_2COO^-$	0.252	± 0.019
NO_2^-	0.187	± 0.019

Ion	Radius	
NO_3^-	0.200	± 0.019
O_2^-	0.165	± 0.019
O_3^-	0.199	± 0.034
OH^-	0.152	± 0.019
OsF_6^-	0.252	± 0.020
PaF_6^-	0.249	± 0.019
PdF_6^-	0.252	± 0.019
PF_6^-	0.242	± 0.019
PO_3^-	0.204	± 0.019
PtF_6^-	0.247	± 0.019
PuF_5^-	0.239	± 0.019
ReF_6^-	0.240	± 0.019
ReO_4^-	0.227	± 0.019
RuF_6^-	0.242	± 0.019
S_6^-	0.305	± 0.019
SCN^-	0.209	± 0.019
$SbCl_6^-$	0.320	± 0.019
SbF_6^-	0.252	± 0.019
$Sb_2F_{11}^-$	0.312	± 0.038
$Sb_3F_{14}^-$	0.374	± 0.038
$SeCl_5^-$	0.258	± 0.038
$SeCN^-$	0.230	± 0.019
SeH^-	0.195	± 0.019
SH^-	0.191	± 0.019
SO_3F^-	0.214	± 0.019
$S_3N_3^-$	0.231	± 0.038
$S_3N_3O_4^-$	0.252	± 0.038
$TaCl_6^-$	0.352	± 0.019
TaF_6^-	0.250	± 0.019
TaO_3^-	0.192	± 0.019
UF_6^-	0.301	± 0.019
VF_6^-	0.235	± 0.019
VO_3^-	0.201	± 0.019
WCl_6^-	0.337	± 0.019
WF_6^-	0.246	± 0.019
WOF_5^-	0.241	± 0.019
Doubly Charged Anions		
AmF_6^{2-}	0.255	± 0.019
$Bi_2Br_8^{2-}$	0.392	± 0.055
$Bi_6Cl_{20}^{2-}$	0.501	± 0.073
$CdCl_4^{2-}$	0.307	± 0.019
$CeCl_6^{2-}$	0.352	± 0.019
CeF_6^{2-}	0.249	± 0.019
CO_3^{2-}	0.189	± 0.019
$CoCl_4^{2-}$	0.306	± 0.019
CoF_4^{2-}	0.209	± 0.019
CoF_6^{2-}	0.256	± 0.019
$Cr_2O_7^{2-}$	0.292	± 0.019
CrF_6^{2-}	0.253	± 0.019
CrO_4^{2-}	0.229	± 0.019
$CuCl_4^{2-}$	0.304	± 0.019
CuF_4^{2-}	0.213	± 0.019
$GeCl_6^{2-}$	0.335	± 0.019
GeF_6^{2-}	0.244	± 0.019
HfF_6^{2-}	0.248	± 0.019
HgI_4^{2-}	0.377	± 0.019
$IrCl_6^{2-}$	0.332	± 0.019
$MnCl_6^{2-}$	0.314	± 0.031
MnF_4^{2-}	0.219	± 0.019
MnF_6^{2-}	0.241	± 0.019
$MoBr_6^{2-}$	0.364	± 0.019
$MoCl_6^{2-}$	0.338	± 0.019
MoF_6^{2-}	0.274	± 0.019
MoO_4^{2-}	0.231	± 0.019
$NbCl_6^{2-}$	0.343	± 0.019
NH^{2-}	0.128	± 0.019
$Ni(CN)_4^{2-}$	0.322	± 0.019
NiF_4^{2-}	0.211	± 0.019
NiF_6^{2-}	0.249	± 0.019
O^{2-}	0.141	± 0.019
O_2^{2-}	0.167	± 0.019
$OsBr_6^{2-}$	0.365	± 0.019
$OsCl_6^{2-}$	0.336	± 0.019
OsF_6^{2-}	0.276	± 0.019
$PbCl_4^{2-}$	0.279	± 0.019
$PbCl_6^{2-}$	0.347	± 0.019
PbF_6^{2-}	0.268	± 0.019
$PdBr_6^{2-}$	0.354	± 0.019
$PdCl_4^{2-}$	0.313	± 0.019
$PdCl_6^{2-}$	0.333	± 0.019
PdF_6^{2-}	0.252	± 0.019
$PoBr_6^{2-}$	0.380	± 0.019
PoI_6^{2-}	0.428	± 0.019
$Pt(NO_2)_3Cl_3^{2-}$	0.364	± 0.019
$Pt(NO_2)_4Cl_2^{2-}$	0.383	± 0.019
$Pt(OH)_2^{2-}$	0.333	± 0.019
$Pt(SCN)_6^{2-}$	0.451	± 0.019
$PtBr_4^{2-}$	0.324	± 0.019
$PtBr_6^{2-}$	0.363	± 0.019
$PtCl_4^{2-}$	0.307	± 0.019
$PtCl_6^{2-}$	0.333	± 0.019
PtF_6^{2-}	0.245	± 0.019
$PuCl_6^{2-}$	0.349	± 0.019
$ReBr_6^{2-}$	0.371	± 0.019
$ReCl_6^{2-}$	0.337	± 0.019
ReF_6^{2-}	0.256	± 0.019
ReF_8^{2-}	0.276	± 0.019
ReH_9^{2-}	0.257	± 0.019
ReI_6^{2-}	0.421	± 0.026
RhF_6^{2-}	0.240	± 0.019
$RuCl_6^{2-}$	0.336	± 0.019
RuF_6^{2-}	0.248	± 0.019
S^{2-}	0.189	± 0.019
$S_2O_3^{2-}$	0.251	± 0.019
$S_2O_4^{2-}$	0.262	± 0.019
$S_2O_5^{2-}$	0.270	± 0.019
$S_2O_6^{2-}$	0.283	± 0.019
$S_2O_7^{2-}$	0.275	± 0.019
$S_2O_8^{2-}$	0.291	± 0.019
$S_3O_6^{2-}$	0.302	± 0.019
$S_4O_6^{2-}$	0.325	± 0.019
$S_6O_6^{2-}$	0.382	± 0.019
ScF_6^{2-}	0.276	± 0.019
Se^{2-}	0.181	± 0.019
$SeBr_6^{2-}$	0.363	± 0.019
$SeCl_6^{2-}$	0.336	± 0.019
SeO_4^{2-}	0.229	± 0.019
SiF_6^{2-}	0.248	± 0.019
SiO_3^{2-}	0.195	± 0.019
SmF_4^{2-}	0.218	± 0.019
$Sn(OH)_6^{2-}$	0.279	± 0.020
$SnBr_6^{2-}$	0.374	± 0.019

Ion	Radius	
$SnCl_6^{2-}$	0.345	± 0.019
SnF_6^{2-}	0.265	± 0.019
SnI_6^{2-}	0.427	± 0.019
SO_3^{2-}	0.204	± 0.019
SO_4^{2-}	0.218	± 0.019
$TcBr_6^{2-}$	0.363	± 0.019
$TcCl_6^{2-}$	0.337	± 0.019
TcF_6^{2-}	0.244	± 0.019
TcH_9^{2-}	0.260	± 0.019
TcI_6^{2-}	0.419	± 0.019
Te^{2-}	0.220	± 0.019
$TeBr_6^{2-}$	0.383	± 0.019
$TeCl_6^{2-}$	0.353	± 0.019
TeI_6^{2-}	0.430	± 0.019
TeO_4^{2-}	0.238	± 0.019
$Th(NO_3)_6^{2-}$	0.424	± 0.019
$ThCl_6^{2-}$	0.360	± 0.019
ThF_6^{2-}	0.263	± 0.019
$TiBr_6^{2-}$	0.356	± 0.019
$TiCl_6^{2-}$	0.335	± 0.019
TiF_6^{2-}	0.252	± 0.019
UCl_6^{2-}	0.354	± 0.019
UF_6^{2-}	0.256	± 0.019
VO_3^{2-}	0.204	± 0.019
WBr_6^{2-}	0.363	± 0.019
WCl_6^{2}	0.339	± 0.019
WO_4^{2-}	0.237	± 0.019
$WOCl_5^{-}$	0.334	± 0.019
$ZnBr_4^{2-}$	0.335	± 0.019
$ZnCl_4^{2-}$	0.306	± 0.019
ZnF_4^{2-}	0.219	± 0.019
ZnI_4^{2-}	0.384	± 0.019
$ZrBr_4^{2-}$	0.334	± 0.019
$ZrCl_4^{2-}$	0.306	± 0.019
$ZrCl_6^{2-}$	0.348	± 0.019
ZrF_6^{2-}	0.258	± 0.019
Multi-Charged Anions		
AlH_6^{3-}	0.256	± 0.042
AsO_4^{3-}	0.237	± 0.042
$CdBr_6^{4-}$	0.374	± 0.038
$CdCl_6^{4-}$	0.352	± 0.038
CeF_6^{3-}	0.278	± 0.038
CeF_7^{3-}	0.282	± 0.038
$Co(CN)_6^{3-}$	0.349	± 0.038
$Co(NO_2)_6^{3-}$	0.343	± 0.038
$CoCl_5^{3-}$	0.320	± 0.038
CoF_6^{3-}	0.258	± 0.042
$Cr(CN)_6^{3-}$	0.351	± 0.038
CrF_6^{3-}	0.254	± 0.042
$Cu(CN)_4^{3-}$	0.312	± 0.038
$Fe(CN)_6^{3-}$	0.347	± 0.038
FeF_6^{3-}	0.298	± 0.042
HfF_7^{3-}	0.277	± 0.042
InF_6^{3-}	0.268	± 0.038
$Ir(CN)_6^{3-}$	0.347	± 0.038
$Ir(NO_2)_6^{3-}$	0.338	± 0.038
$Mn(CN)_6^{3-}$	0.350	± 0.038
$Mn(CN)_6^{5-}$	0.401	± 0.042
$MnCl_6^{4-}$	0.349	± 0.038
N^{3-}	0.180	± 0.042
$Ni(NO_2)_6^{3-}$	0.342	± 0.038

Ion	Radius	
$Ni(NO_2)_6^{4-}$	0.383	± 0.038
NiF_6^{3-}	0.250	± 0.042
O^{3-}	0.288	± 0.038
P^{3-}	0.224	± 0.042
PaF_8^{3-}	0.299	± 0.042
PO_4^{3-}	0.230	± 0.042
PrF_6^{3-}	0.281	± 0.038
$Rh(NO_2)_6^{3-}$	0.345	± 0.038
$Rh(SCN)_6^{3-}$	0.428	± 0.042
TaF_8^{3-}	0.284	± 0.042
TbF_7^{3-}	0.290	± 0.038
$Tc(CN)_6^{5-}$	0.410	± 0.042
ThF_7^{3-}	0.282	± 0.042
$TiBr_6^{3-}$	0.315	± 0.038
TlF_6^{3-}	0.271	± 0.038
UF_7^{3-}	0.285	± 0.042
YF_6^{3-}	0.275	± 0.038
ZrF_7^{3-}	0.273	± 0.038
Singly Charged Cations		
$N(CH_3)_4^+$	0.234	± 0.019
$N_2H_5^+$	0.158	± 0.019
$N_2H_6^{2+}$	0.158	± 0.029
$NH(C_2H_5)_3^+$	0.274	± 0.019
$NH_3C_2H_5^+$	0.193	± 0.019
$NH_3C_3H_7^+$	0.225	± 0.019
$NH_3CH_3^+$	0.177	± 0.019
NH_3OH^+	0.147	± 0.019
NH_4^+	0.136	± 0.019
$NH_3C_2H_4OH^+$	0.203	± 0.019
$As_3S_4^+$	0.244	± 0.027
$As_3Se_4^+$	0.253	± 0.027
$AsCl_4^+$	0.221	± 0.027
Br_2^+	0.155	± 0.027
Br_3^+	0.204	± 0.027
Br_3^-	0.238	± 0.027
Br_5^+	0.229	± 0.027
$BrClCNH_2^+$	0.175	± 0.027
BrF_2^+	0.183	± 0.027
BrF_4^+	0.172	± 0.027
$C_{10}F_8^+$	0.265	± 0.027
$C_6F_6^+$	0.228	± 0.027
$Cl(SNSCN)_2^+$	0.347	± 0.027
$Cl_2C{=}NH_2^+$	0.173	± 0.027
Cl_2F^+	0.165	± 0.027
Cl_3^+	0.182	± 0.027
ClF_2^+	0.147	± 0.027
ClO_2^+	0.118	± 0.027
$GaBr_4^-$	0.317	± 0.038
I_2^+	0.185	± 0.027
I_3^+	0.225	± 0.027
I_5^+	0.263	± 0.027
IBr_2^+	0.196	± 0.027
ICl_2^+	0.175	± 0.036
IF_6^+	0.209	± 0.027
$N(S_3N_2)_2^+$	0.258	± 0.027
$N(SCl)_2^+$	0.186	± 0.027
$N(SeCl)_2^+$	0.246	± 0.027
$N(SF_2)_2^+$	0.214	± 0.027
N_2F^+	0.156	± 0.027
NO^+	0.145	± 0.027
NO_2^+	0.153	± 0.027

Ion	Radius	
O_2^+	0.140	± 0.027
$O_2(SCCF_3Cl)_2^+$	0.275	± 0.027
$ONCH_3CF_3^+$	0.200	± 0.027
$OsOF_5^-$	0.246	± 0.038
$P(CH_3)_3Cl^+$	0.197	± 0.027
$P(CH_3)_3D^+$	0.196	± 0.027
PCl_4^+	0.235	± 0.027
$ReOF_5^-$	0.245	± 0.038
$S(CH_3)_2Cl^+$	0.207	± 0.027
$S(N(C_2H_5)_2)_3^+$	0.439	± 0.027
$S_2(CH_3)_2Cl^+$	0.265	± 0.027
$S_2(CH_3)_2CN^+$	0.223	± 0.027
$S_2(CH_3)_3^+$	0.233	± 0.027
$S_2Br_5^+$	0.267	± 0.027
S_2N^+	0.159	± 0.034
$S_2N_2C_2H_3^+$	0.211	± 0.027
$S_2NC_2(PhCH_3)_2^+$	0.310	± 0.027
$S_2NC_3H_4^+$	0.218	± 0.027
$S_2NC_4H_8^+$	0.225	± 0.027
$S_3(CH_3)_3^+$	0.239	± 0.027
$S_3Br_3^+$	0.245	± 0.027
$S_3C_3H_7^+$	0.199	± 0.027
$S_3C_4F_6^+$	0.261	± 0.027
$S_3CF_3CN^+$	0.263	± 0.027
$S_3Cl_3^+$	0.233	± 0.027
$S_3N_2^+$	0.201	± 0.027
$S_3N_2Cl^+$	0.232	± 0.027
$S_4N_3^+$	0.231	± 0.027
$S_4N_3(Ph)_2^+$	0.316	± 0.027
$S_4N_4H^+$	0.178	± 0.027
$S_5N_5^+$	0.257	± 0.027
S_7I^+	0.262	± 0.027
$Sb(NPPh_3)_4^+$	0.518	± 0.027
SBr_3^+	0.220	± 0.027
$SCH_3O_2^+$	0.183	± 0.027
$SCH_3P(CH_3)_3^+$	0.248	± 0.027
$SCH_3PCH_3Cl_2^+$	0.205	± 0.027
$SCl(C_2H_5)_2^+$	0.207	± 0.027
$SCl_2CF_3^+$	0.207	± 0.027
$SCl_2CH_3^+$	0.204	± 0.027
SCl_3^+	0.185	± 0.027
$Se_3Br_3^+$	0.253	± 0.027
$Se_3Cl_3^+$	0.245	± 0.027
$Se_3N_2^+$	0.288	± 0.042
$Se_3NC_{l2}^+$	0.163	± 0.027
Se_6I^+	0.260	± 0.027
$SeBr_3^+$	0.182	± 0.027
$SeCl_3^+$	0.192	± 0.027
SeF_3^+	0.179	± 0.027
SeI_3^+	0.238	± 0.027
SeN_2Cl^+	0.196	± 0.027
$SeNCl_2^+$	0.157	± 0.027
$(SeNMe_3)_3^+$	0.406	± 0.027
$SeS_2N_2^+$	0.282	± 0.042
$SF(C_6F_5)_2^+$	0.294	± 0.027
$SF_2CF_3^+$	0.198	± 0.027
$SF_2N(CH_3)_2^+$	0.210	± 0.027
SF_3^+	0.172	± 0.027
$SFS(C(CF_3)_2)_2^+$	0.275	± 0.027
$SH_2C_3H_7^+$	0.210	± 0.027
SN^+	0.158	± 0.027
$SNCl_5(CH_3CN)^-$	0.290	± 0.038
$(SNPMe_3)_3^+$	0.308	± 0.027
$SNSC(CH_3)N^+$	0.225	± 0.027
$SNSC(CN)CH^+$	0.209	± 0.027
$SNSC(Ph)N^+$	0.251	± 0.027
$SNSC(Ph)NS_3N_2^+$	0.327	± 0.027
$SNSC(PhCH_3)N^+$	0.264	± 0.027
$(Te(N(SiMe_3)_2)_2^+$	0.371	± 0.027
$Te(N_3)_3^+$	0.226	± 0.027
$Te_4Nb_3OTe_2I_6^+$	0.407	± 0.027
$TeBr_3^+$	0.235	± 0.027
$TeCl_3^+$	0.216	± 0.027
$TeCl_3$(15-crown-5)$^+$	0.282	± 0.027
TeI_3^+	0.243	± 0.027
$Xe_2F_{11}^+$	0.266	± 0.027
$Xe_2F_3^+$	0.221	± 0.027
XeF^+	0.174	± 0.027
XeF_3^+	0.183	± 0.027
XeF_5^+	0.186	± 0.027
$XeOF_3^+$	0.186	± 0.027

Doubly Charged Cations

Ion	Radius	
$Co_2S_2(CO)_6^{2+}$	0.263	± 0.035
$FeW(Se)_2(CO)^{2+}$	0.260	± 0.035
I_4^{2+}	0.207	± 0.035
$Mo(Te_3)(CO)_4^{2+}$	0.234	± 0.035
S_{19}^{2+}	0.292	± 0.035
$S_2(S(CH_3)_2)_2^{2+}$	0.230	± 0.035
$S_2I_4^{2+}$	0.231	± 0.035
$S_3N_2^{2+}$	0.184	± 0.035
$S_3NCCNS_3^{2+}$	0.220	± 0.035
S_3Se^{2+}	0.326	± 0.035
$S_4N_4^{2+}$	0.186	± 0.035
$S_6N_4^{2+}$	0.232	± 0.035
S_8^{2+}	0.182	± 0.035
Se_{10}^{2+}	0.253	± 0.035
Se_{17}^{2+}	0.236	± 0.035
Se_{19}^{2+}	0.296	± 0.035
$Se_2I_4^{2+}$	0.218	± 0.035
$Se_3N_2^{2+}$	0.182	± 0.035
Se_4^{2+}	0.152	± 0.035
$Se_4S_2N_4^{2+}$	0.224	± 0.035
Se_8^{2+}	0.186	± 0.035
$SeN_2S_2^{2+}$	0.182	± 0.035
$(SNP(C_2H_5)_3)_2^{2+}$	0.312	± 0.035
$TaBr_6^-$	0.351	± 0.049
$Te(trtu)_4^{2+}$	0.328	± 0.035
$Te(tu)_4^{2+}$	0.296	± 0.035
$Te_2(esu)_4Br_2^{2+}$	0.356	± 0.035
$Te_2(esu)_4Cl_2^{2+}$	0.361	± 0.035
$Te_2(esu)_4I_2^{2+}$	0.342	± 0.035
$Te_2Se_2^{2+}$	0.192	± 0.035
$Te_2Se_4^{2+}$	0.222	± 0.035
$Te_2Se_8^{2+}$	0.252	± 0.035
$Te_3S_3^{2+}$	0.217	± 0.035
Te_3Se^{2+}	0.193	± 0.035
Te_4^{2+}	0.169	± 0.035
Te_8^{2+}	0.187	± 0.035
$W(CO)_4$(h3-Te)$^{2+}$	0.234	± 0.035
$W_2(CO)_{10}Se_4^{2+}$	0.290	± 0.035

Multi-Charged Cations

Ion	Radius	
I_{15}^{3+}	0.442	± 0.051
$Te_2(su)_6^{4+}$	0.453	± 0.034

Ligand abbreviations: su = selenourea; esu = ethyleneselenourea; tu = thiourea; ph = phenyl.

TABLE 3. Ancillary Thermochemical Data (kJ mol⁻¹)

Species	State	$\Delta_f H^o$
AsO_4^{3-}	g	(289)
BrO_3^-	g	−145
ClO_4^-	g	−344
CN^-	g	66
CO_3^{2-}	g	−321
$Fe(NO_3)_2$	c	(−448)
HF_2^-	g	−774
$HfCl_6^{2-}$	g	−1640
$IO_2F_2^-$	g	−693
IO_3^-	g	−208
$IrCl_6^{2-}$	g	−785
$LiCH_3O_2$	c	(−745)
$NbCl_6^{2-}$	g	−1224
$NH_2CH_2CO_2^-$	g	−564
O_2^{2-}	g	553
$PdCl_6^{2-}$	g	−749
PO_4^{3-}	g	291
$PtCl_6^{2-}$	g	−774
$ReBr_6^{2-}$	g	−689
$ReCl_6^{2-}$	g	−919
$Ti(OH)_2$	c	−778

THE MADELUNG CONSTANT AND CRYSTAL LATTICE ENERGY

If U is the crystal lattice energy and M is the Madelung constant, then[a]

$$U = \frac{NMz_i z_j e^2}{r}(1 - 1/n)$$

Substance	Ion type	Crystal form[b]	M
Sodium chloride, NaCl	M^+, X^-	FCC	1.74756
Cesium chloride, CsCl	M^+, X^-	BCC	1.76267
Calcium chloride, $CaCl_2$	M^{++}, $2X^-$	Cubic	2.365
Calcium fluoride (fluorite), CaF_2	M^{++}, $2X^-$	Cubic	2.51939
Cadmium chloride, $CdCl_2$	M^{++}, $2X^-$	Hexagonal	2.244[c]
Cadmium iodide (α), CdI_2	M^{++}, $2X^-$	Hexagonal	2.355[c]
Magnesium fluoride, MgF_2	M^{++}, $2X^-$	Tetragonal	2.381[c]
Cuprous oxide (cuprite), Cu_2O	$2M^+$, X^{--}	Cubic	2.22124
Zinc oxide, ZnO	M^{++}, X^{--}	Hexagonal	1.4985[c]
Sphalerite (zinc blende), ZnS	M^{++}, X^{--}	FCC	1.63806
Wurtzite, ZnS	M^{++}, X^{--}	Hexagonal	1.64132[c]
Titanium dioxide (anatase), TiO_2	M^{4+}, $2X^{--}$	Tetragonal	2.400[c]
Titanium dioxide (rutile), TiO_2	M^{4+}, $2X^{--}$	Tetragonal	2.408[c]
β-Quartz, SiO_2	M^{4+}, $2X^{--}$	Hexagonal	2.2197[c]
Corundum, Al_2O_3	$2M^{3+}$, $3X^{--}$	Rhombohedral	4.1719

[a] N is Avogadro's number, z_i and z_j are the integral charges on the ions (in units of e), and e is the charge on the electron in electrostatic units ($e = 4.803 \times 10^{-10}$ esu). r is the shortest distance between cation–anion pairs in centimeters. Then U is in ergs (1 erg = 10^{-7} J).

[b] FCC = face centered cubic; BCC = body centered cubic.

[c] For tetragonal and hexagonal crystals the value of M depends on the details of the lattice parameters.

The Born Exponent, n is:

Ion type	n
He, Li^+	5
Ne, Na^+, F^-	7
Ar, K^+, Cu^+, Cl^-	9
Kr, Rb^+, Ag^+, Br^-	10
Xe, Cs^+, Au^+, I^-	12

For a crystal with a mixed-ion type, an average of the values of n in this table is to be used (6 for LiF, for example).

ELASTIC CONSTANTS OF SINGLE CRYSTALS

H. P. R. Frederikse

This table gives selected values of elastic constants for single crystals. The values believed most reliable were selected from the original literature. The substances are arranged by crystal system and, within each system, alphabetically by name. A reference to the original literature is given for each value; a useful compilation of published values from many sources may be found in Reference 1.

Data are given for the single-crystal density and for the elastic constants c_{ij}, in units of 10^{11} N/m^2, which is equivalent to 10^{12} dyn/cm^2.

General References

1. Simmons, G., and Wang, H., *Single Crystal Elastic Constants and Calculated Aggregate Properties: A Handbook, Second Edition*, The MIT Press, Cambridge, MA, 1971.
2. Gray, D. E., Ed., *American Institute of Physics Handbook, Third Edition*, McGraw-Hill, New York, 1972.

CUBIC CRYSTALS

Name	Formula	ρ/g cm^{-3}	*T*/K	Ref.	C_{11}	C_{12}	C_{44}
Aluminum	Al	2.6970	298	1	1.0675	0.6041	0.2834
Aluminum antimonide	AlSb	4.3600	300	2	0.8939	0.4427	0.4155
Ammonium bromide	NH_4Br	2.4314	300	3	0.3414	0.0782	0.0722
Ammonium chloride	NH_4Cl	1.5279	290	4	0.3814	0.0866	0.0903
Argon	Ar	1.7710	4.2	5	0.0529	0.0135	0.0159
Barium fluoride	BaF_2	4.8860	298	6	0.9199	0.4157	0.2568
Barium nitrate	$Ba(NO_3)_2$	3.2560	293	7	0.2925	0.2065	0.1277
Calcium fluoride	CaF_2	3.810	298	8	1.6420	0.4398	0.8406
Calcium telluride	CaTe	5.8544	298	9	0.5351	0.3681	0.1994
Cesium	Cs	1.9800	78	10	0.0247	0.0206	0.0148
Cesium bromide	CsBr	4.4560	298	11	0.3063	0.0807	0.0750
Cesium chloride	CsCl	3.9880	298	11	0.3644	0.0882	0.0804
Cesium iodide	CsI	4.5250	298	11	0.2446	0.0661	0.0629
Chromite	$FeCr_2O_4$	4.4500	RT	12	3.2250	1.4370	1.1670
Chromium	Cr	7.20	298	13	3.398	0.586	0.990
Cobalt oxide	CoO	6.44	298	14	2.6123	1.4699	0.8300
Cobalt zinc ferrite	$CoZnFeO_2$	5.43	303	12	2.660	1.530	0.780
Copper	Cu	8.932	298	15	1.683	1.221	0.757
Gallium antimonide	GaSb	5.6137	298	16	0.8839	0.4033	0.4316
Gallium arsenide	GaAs	5.3169	298	17	1.1877	0.5372	0.5944
Gallium phosphide	GaP	4.1297	300	18	1.4120	0.6253	0.7047
Garnet (yttrium-iron)	$Y_3Fe_2(FeO_4)_3$	5.17	298	19	2.680	1.106	0.766
Germanium	Ge	5.313	298	20	1.2835	0.4823	0.6666
Gold	Au	19.283	296.5	21	1.9244	1.6298	0.4200
Indium antimonide	InSb	5.7890	298	22	0.6720	0.3670	0.3020
Indium arsenide	InAs	5.6720	293	23	0.8329	0.4526	0.3959
Indium phosphide	InP	4.78	RT	24	1.0220	0.5760	0.4600
Iridium	Ir	22.52	300	25	5.80	2.42	2.56
Iron	Fe	7.8672	298	26	2.26	1.40	1.16
Lead	Pb	11.34	296	27	0.4966	0.4231	0.1498
Lead fluoride	PbF_2	7.79	300	28	0.8880	0.4720	0.2454
Lead nitrate	$Pb(NO_3)_2$	4.547	293	29	0.3729	0.2765	0.1347
Lead telluride	PbTe	8.2379	303.2	30	1.0795	0.0764	0.1343
Lithium	Li	0.5326	298	31	0.1350	0.1144	0.0878
Lithium bromide	LiBr	3.47	RT	32	0.3940	0.1880	0.1910
Lithium chloride	LiCl	2.068	295	33	0.4927	0.2310	0.2495
Lithium fluoride	LiF	2.638	RT	34	1.1397	0.4767	0.6364
Lithium iodide	LiI	4.061	RT	32	0.2850	0.1400	0.1350
Magnesium oxide	MgO	3.579	298	20	2.9708	0.9536	1.5613
Magnetite	Fe_3O_4	5.18	RT	32	2.730	1.060	0.971
Manganese oxide	MnO	5.39	298	35	2.23	1.20	0.79
Mercury telluride	HgTe	8.079	290	36	0.548	0.381	0.204
Molybdenum	Mo	10.2284	273	37	4.637	1.578	1.092
Nickel	Ni	8.91	298	15	2.481	1.549	1.242
Niobium	Nb	8.578	300	38	2.4650	1.3450	0.2873

CUBIC CRYSTALS

Name	Formula	ρ/g cm⁻³	T/K	Ref.	C_{11}	C_{12}	C_{44}
Palladium	Pd	12.038	300	39	2.2710	1.7604	0.7173
Platinum	Pt	21.50	300	40	3.4670	2.5070	0.7650
Potassium	K	0.851	295	41	0.0370	0.0314	0.0188
Potassium bromide	KBr	2.740	298	11	0.3468	0.0580	0.0507
Potassium chloride	KCl	1.984	298	11	0.4069	0.0711	0.0631
Potassium cyanide	KCN	1.553	RT	32	0.1940	0.1180	0.0150
Potassium fluoride	KF	2.480	295	33	0.6490	0.1520	0.1232
Potassium iodide	KI	3.128	300	42	0.2710	0.0450	0.0364
Pyrite	FeS_2	5.016	RT	43	3.818	0.310	1.094
Rubidium	Rb	1.58	170	44	0.0296	0.0250	0.0171
Rubidium bromide	RbBr	3.350	300	45	0.3152	0.0500	0.0380
Rubidium chloride	RbCl	2.797	300	45	0.3624	0.0612	0.0468
Rubidium iodide	RbI	3.551	300	45	0.2556	0.0382	0.0278
Silicon	Si	2.331	298	46	1.6578	0.6394	0.7962
Silver	Ag	10.50	300	47	1.2399	0.9367	0.4612
Silver bromide	AgBr	5.585	300	48	0.5920	0.3640	0.0616
Sodium	Na	0.971	299	49	0.0739	0.0622	0.0419
Sodium bromate	$NaBrO_3$	3.339	RT	32	0.5450	0.1910	0.1500
Sodium bromide	NaBr	3.202	300	33	0.3970	0.1001	0.0998
Sodium chlorate	$NaClO_3$	2.485	RT	50	0.4920	0.1420	0.1160
Sodium chloride	NaCl	2.163	298	11	0.4947	0.1288	0.1287
Sodium fluoride	NaF	2.804	300	51	0.9700	0.2380	0.2822
Sodium iodide	NaI	3.6689	300	52	0.3007	0.0912	0.0733
Spinel	$MgAl_2O_4$	3.6193	298	53	2.9857	1.5372	1.5758
Strontium fluoride	SrF_2	4.277	300	54	1.2350	0.4305	0.3128
Strontium nitrate	$Sr(NO_3)_2$	2.989	293	29	0.4255	0.2921	0.1590
Strontium oxide	SrO	4.99	300	55	1.601	0.435	0.590
Strontium titanate	$SrTiO_3$	5.123	RT	56	3.4817	1.0064	4.5455
Tantalum	Ta	16.626	298	57	2.6023	1.5446	0.8255
Tantalum carbide	TaC	14.65	RT	58	5.05	0.73	0.79
Thallium bromide	TlBr	7.4529	298	59	0.3760	0.1458	0.0757
Thorium	Th	11.694	300	60	0.7530	0.4890	0.4780
Thorium oxide	ThO_2	9.991	298	61	3.670	1.060	0.797
Tin telluride	SnTe	6.445	300	62	1.1250	0.0750	0.1172
Titanium carbide	TiC	4.940	RT	107	5.00	1.13	1.75
Tungsten	W	19.257	297	64	5.2239	2.0437	1.6083
Uranium carbide	UC	13.63	300	65	3.200	0.850	0.647
Uranium dioxide	UO_2	10.97	298	66	3.960	1.210	0.641
Vanadium	V	6.022	300	67	2.287	1.190	0.432
Zinc selenide	ZnSe	5.262	298	68	0.8096	0.4881	0.4405
Zinc sulfide	ZnS	4.088	298	68	1.0462	0.6534	0.4613
Zinc telluride	ZnTe	5.636	298	68	0.7134	0.4078	0.3115
Zirconium carbide	ZrC	6.606	298	63	4.720	0.987	1.593

TETRAGONAL CRYSTALS

Name	Formula	ρ/g cm^{-3}	*T*/K	Ref.	C_{11}	C_{12}	C_{13}	C_{16}	C_{33}	C_{44}	C_{66}
Ammonium dihydrogen arsenate (ADA)	$NH_4H_2AsO_4$	2.3110	298	69	0.6747	−0.106	0.1652		0.3022	0.0685	0.0639
Ammonium dihydrogen phosphate (ADP)	$NH_4H_2PO_4$	1.8030	293	69	0.6200	−0.050	0.1400		0.3000	0.0910	0.0610
Barium titanate	$BaTiO_3$	5.9988	298	70	2.7512	1.7897	1.5156		1.6486	0.5435	1.1312
Calcium molybdate	$CaMoO_4$	4.255	298	79	1.447	0.664	0.466	0.134	1.265	0.369	0.451
Indium	In	7.300	RT	71	0.4450	0.3950	0.4050		0.4440	0.0655	0.1220
Magnesium fluoride	MgF_2	3.177	RT	72	1.237	0.732	0.536		1.770	0.552	0.978
Nickel sulfate hexahydrate	$NiSO_4·6H_2O$	2.070	RT	73	0.3209	0.2315	0.0209		0.2931	0.1156	0.1779
Potassium dihydrogen arsenate (KDA)	KH_2AsO_4	2.867	RT	12	0.530	−0.060	−0.020		0.370	0.120	0.070
Potassium dihydrogen phosphate (KDP)	KH_2PO_4	2.388	RT	71	0.7140	−0.049	0.1290		0.5620	0.1270	0.0628
Rubidium dihydrogen phosphate (RDP)	RbH_2PO_4	2.800	298	74	0.5562	−0.064	0.0279		0.4398	0.1142	0.0350
Rutile	TiO_2	4.260	298	75	2.7143	1.7796	1.4957		4.8395	1.2443	1.9477
Tellurium oxide	TeO_2	5.99	RT	76	0.5320	0.4860	0.2120		1.0850	0.2440	0.5520
Tin (white)	Sn	7.29	288	77	0.7529	0.6156	0.4400		0.9552	0.2193	0.2336
Zircon	$ZrSiO_4$	4.70	RT	78	2.585	1.791	1.542		3.805	0.733	1.113

ORTHORHOMBIC CRYSTALS

Name	Formula	ρ/g cm^{-3}	*T*/K	Ref.	C_{11}	C_{12}	C_{13}	C_{22}	C_{23}	C_{33}	C_{44}	C_{55}	C_{66}
Acenaphthene	$C_{12}H_{10}$	1.220	293	80	0.1380	0.0210	0.0410	0.1262	0.0460	0.1117	0.0265	0.0290	0.0185
Ammonium sulfate	$(NH_4)_2SO_4$	1.774	293	81	0.3607	0.1651	0.1580	0.2981	0.1456	0.3534	0.1025	0.0717	0.0974
Aragonite	$CaCO_3$	2.93	RT	82	1.5958	0.3663	0.0197	0.8697	0.1597	0.8503	0.4132	0.2564	0.4274
Barite	$BaSO_4$	4.40	RT	82	0.8941	0.4614	0.2691	0.7842	0.2676	1.0548	0.1190	0.2874	0.2778
Benzene	C_6H_6	1.061	250	83	0.0614	0.0352	0.0401	0.0656	0.0390	0.0583	0.0197	0.0378	0.0153
Benzophenone	$(C_6H_5)_2CO$	1.219	RT	32	0.1070	0.0550	0.0169	0.1000	0.0321	0.0710	0.0203	0.0155	0.0353
Bronzite	$(MgFe)SiO_3$	3.38	RT	78	1.876	0.686	0.605	1.578	0.561	2.085	0.700	0.592	0.544
Calcium sulfate	$CaSO_4$	2.962	RT	84	0.9382	0.1650	0.1520	1.845	0.3173	1.1180	0.3247	0.2653	0.0926
Celestite	$SrSO_3$	3.96	RT	12	1.044	0.773	0.605	1.061	0.619	1.286	0.135	0.279	0.266
Cesium sulfate	Cs_2SO_4	4.243	293	81	0.4490	0.1958	0.1815	0.4283	0.1800	0.3785	0.1326	0.1319	0.1323
Fosterite	Mg_2SiO_4	3.224	298	85	3.2848	0.6390	0.6880	1.9980	0.7380	2.3530	0.6515	0.8120	0.8088
Iodic acid	HIO_3	4.630	RT	73	0.3030	0.1194	0.1169	0.5448	0.0548	0.4359	0.1835	0.2193	0.1736
Lithium ammonium tartrate	$LiNH_4C_4H_4O_6·4H_2O$	1.71	RT	12	0.3864	0.1655	0.0875	0.5393	0.2007	0.3624	0.1190	0.0667	0.2326
Magnesium sulfate heptahydrate	$MgSO_4·7H_2O$	1.68	RT	86	0.325	0.174	0.182	0.288	0.182	0.315	0.078	0.156	0.090
Natrolite	$(Na,Al)SiO_3$	2.25	RT	78	0.716	0.261	0.297	0.632	0.297	1.378	0.196	0.248	0.423
Nickel sulfate heptahydrate	$NiSO_4·7H_2O$	1.948	RT	86	0.353	0.198	0.201	0.311	0.201	0.335	0.091	0.172	0.099
Olivine	$(MgFe)SiO_4$	3.324	RT	87	3.240	0.590	0.790	1.980	0.780	2.490	0.667	0.810	0.793
Potassium pentaborate	$KB_5O_8·4H_2O$	1.74	RT	71	0.582	0.229	0.174	0.359	0.231	0.255	0.164	0.046	0.057
Potassium sulfate	K_2SO_4	2.665	293	81	0.5357	0.1999	0.2095	0.5653	0.1990	0.5523	0.195	0.1879	0.1424
Rochelle salt	$NaK(C_4H_4O_6)·4H_2O$	1.79	RT	71	0.255	0.141	0.116	0.381	0.146	0.371	0.134	0.032	0.098
Rubidium sulfate	Rb_2SO_4	3.621	293	81	0.5029	0.1965	0.1999	0.5098	0.1925	0.4761	0.1626	0.1589	0.1407
Sodium ammonium tartrate	$NaNH_4C_4H_4O_6·4H_2O$	1.587	RT	12	0.3685	0.2725	0.3083	0.5092	0.3472	0.5541	0.1058	0.0303	0.0870
Sodium tartrate	$Na_2C_4H_4O_6·2H_2O$	1.794	RT	12	0.461	0.286	0.320	0.547	0.352	0.665	0.124	0.031	0.098
Strontium formate dihydrate	$Sr(CHO_2)_2·2H_2O$	2.25	RT	12	0.4391	0.1037	−0.149	0.3484	−0.014	0.3746	0.1538	0.1075	0.1724
Sulfur	S	2.07	RT	12	0.240	0.133	0.171	0.205	0.159	0.483	0.043	0.087	0.076
Thallium sulfate	$TlSO_4$	6.776	293	81	0.4106	0.2573	0.2288	0.3885	0.2174	0.4268	0.1125	0.1068	0.0751
Topaz	$Al_2SiO_3(OH,F)_2$	3.52	RT	82	2.8136	1.2582	0.8464	3.8495	0.8815	2.9452	1.0811	1.3298	1.3089
Uranium (alpha)	U	19.0453	293	88	2.1486	0.4622	0.2176	1.9983	1.0764	2.6763	1.2479	0.7379	0.7454
Zinc sulfate heptahydrate	$ZnSO_4·7H_2O$	1.970	RT	86	0.3320	0.1720	0.2000	0.2930	0.1980	0.3200	0.0780	0.1530	0.0830

MONOCLINIC CRYSTALS

Name	Formula	ρ/g cm^{-3}	T/K	Ref.	C_{11}	C_{12}	C_{13}	C_{15}	C_{22}
Aegirine	$(NaFe)Si_2O_6$	3.50	RT	89	1.858	0.685	0.707	0.098	1.813
Anthracene	$C_{14}H_{10}$	1.258	RT	90	0.0852	0.0672	0.0590	−0.0192	0.1170
Cobalt sulfate heptahydrate	$CoSO_4{\cdot}7H_2O$	1.948	RT	86	0.335	0.205	0.158	0.016	0.378
Diopside	$(CaMg)Si_2O_6$	3.31	RT	91	2.040	0.884	0.0883	−0.193	1.750
Dipotassium tartrate	$KHC_4H_4O_6$	1.97	RT	12	0.4294	0.1399	0.3129	−0.0105	0.3460
Feldspar (microceine)	$KAlSi_3O_8$	2.56	RT	92	0.664	0.438	0.259	−0.033	1.710
Ferrous sulfate heptahydrate	$FeSO_4{\cdot}7H_2O$	1.898	RT	86	0.349	0.208	0.174	−0.020	0.376
Lithium sulfate monohydrate	$Li_2SO_4{\cdot}H_2O$	2.221	RT	32	0.5250	0.1715	0.1730	−0.0196	0.5060
Naphthalene	$C_{10}H_8$	1.127	RT	93	0.0780	0.0445	0.0340	−0.006	0.0990
Potassium tartrate	$K_2C_4H_4O_6$	1.987	RT	32	0.3110	0.1720	0.1690	0.0287	0.3900
Sodium thiosulfate	$Na_2S_2O_3$	1.7499	RT	12	0.3323	0.1814	0.1875	0.0225	0.2953
Stilbene	$(C_6H_5CH)_2$	1.60	RT	94	0.0930	0.0570	0.0670	−0.003	0.0920
Triglycine sulfate (TGS)	$(NH_2CH_2COOH)_3{\cdot}\ H_2SO_4$	1.68	RT	32	0.4550	0.1720	0.1980	−0.030	0.3210

Name	C_{23}	C_{25}	C_{33}	C_{35}	C_{44}	C_{46}	C_{55}	C_{66}
Aegirine	0.626	0.094	2.344	0.214	0.692	0.077	0.510	0.474
Anthracene	0.0375	−0.0170	0.1522	−0.0187	0.0272	0.0138	0.0242	0.0399
Cobalt sulfate heptahydrate	0.158	−0.018	0.371	−0.047	0.060	0.016	0.058	0.101
Diopside	0.482	−0.196	2.380	−0.336	0.675	−0.113	0.588	0.705
Dipotassium tartrate	0.1173	0.0176	0.6816	0.0294	0.0961	−0.0044	0.1270	0.0841
Feldspar (microceine)	0.192	−0.148	1.215	−0.131	0.143	−0.015	0.238	0.361
Ferrous sulfate heptahydrate	0.172	−0.019	0.360	−0.014	0.064	0.001	0.056	0.096
Lithium sulfate monohydrate	0.0368	0.0571	0.5400	−0.0254	0.1400	−0.0054	0.1565	0.2770
Naphthalene	0.0230	−0.0270	0.1190	0.0290	0.0330	−0.0050	0.0210	0.0415
Potassium tartrate	0.1330	0.0182	0.5540	0.0710	0.0870	0.0072	0.1040	0.0826
Sodium thiosulfate	0.1713	0.0983	0.4590	−0.0678	0.0569	−0.0268	0.1070	0.0598
Stilbene	0.0485	−0.005	0.0790	−0.005	0.0325	0.0050	0.0640	0.0245
Triglycine sulfate (TGS)	0.2080	−0.0036	0.2630	−0.0500	0.0950	−0.0026	0.1110	0.0620

HEXAGONAL CRYSTALS

Name	Formula	ρ/g cm⁻³	T/K	Ref.	C_{11}	C_{12}	C_{13}	C_{33}	C_{55}
Apatite	$Ca_5(PO_4)_3(OH,F,Cl)$	3.218	RT	12	1.667	0.131	0.655	1.396	0.663
Beryl	$Be_3Al_2Si_6O_{18}$	2.68	RT	12	2.800	0.990	0.670	2.480	0.658
Beryllium	Be	1.8477	300	95	2.923	0.267	0.140	3.364	1.625
Beryllium oxide	BeO	3.01	RT	96	4.70	1.68	1.19	4.94	1.53
Cadmium	Cd	8.652	300	97	1.1450	0.3950	0.3990	0.5085	0.1985
Cadmium selenide	CdSe	5.655	298	68	0.7046	0.4516	0.3930	0.8355	0.1317
Cadmium sulfide	CdS	4.824	298	98	0.8431	0.5208	0.4567	0.9183	0.1458
Cobalt	Co	8.836	298	99	3.071	1.650	1.027	3.581	0.755
Dysprosium	Dy	8.560	298	100	0.7466	0.2616	0.2233	0.7871	0.2427
Erbium	Er	9.064	298	100	0.8634	0.3050	0.2270	0.8554	0.2809
Gadolinium	Gd	7.888	298	101	0.6667	0.2499	0.2132	0.7191	0.2089
Hafnium	Hf	12.727	298	102	1.881	0.772	0.661	1.969	0.557
Ice	H_2O(solid)	0.920	250	103	0.1410	0.0660	0.0624	0.1515	0.0288
Indium	In	7.2788	300	104	0.4535	0.4006	0.4151	0.4515	0.0651
Magnesium	Mg	1.7364	298	105	0.5950	0.2612	0.2180	0.6155	0.1635
Rhenium	Re	21.024	298	100	6.1820	2.7530	2.0780	6.8350	1.6060
Ruthenium	Ru	12.3615	298	100	5.6260	1.8780	1.6820	6.2420	1.8060
Thallium	Tl	11.560	300	106	0.4080	0.3540	0.2900	0.5280	0.0726
Titanium	Ti	4.5063	298	102	1.6240	0.9200	0.6900	1.8070	0.4670
Titanium diboride	TiB_2	4.95	RT	107	6.90	4.10	3.20	4.40	2.50
Yttrium	Y	4.472	300	108	0.7790	0.2850	0.2100	0.7690	0.2431
Zinc	Zn	7.134	295	109	1.6368	0.3640	0.5300	0.6347	0.3879
Zinc oxide	ZnO	5.6760	298	110	2.0970	1.2110	1.0510	2.1090	0.4247
Zinc sulfide	ZnS	4.089	298	96	1.2420	0.6015	0.4554	1.4000	0.2864
Zirconium	Zr	6.505	298	102	1.434	0.728	0.653	1.648	0.320

TRIGONAL CRYSTALS

Name	Formula	ρ/g cm⁻³	T/K	Ref.	C_{11}	C_{12}	C_{13}	C_{14}	C_{33}	C_{44}
Aluminum oxide	Al_2O_3	3.986	300	111	4.9735	1.6397	1.1220	−0.2358	4.9911	1.4739
Aluminum phosphate	$AlPO_4$	2.556	RT	73	1.0503	0.2934	0.6927	−0.1271	1.3353	0.2314
Antimony	Sb	6.70	295	112	1.0130	0.3450	0.2920	0.2090	0.4500	0.3930
Bismuth	Bi	9.80	295	112	0.6370	0.2490	0.2470	0.0717	0.3820	0.1123
Calcite	$CaCO_3$	2.712	300	113	1.4806	0.5578	0.5464	−0.2058	0.8557	0.3269
Hematite	Fe_2O_3	5.240	RT	82	2.4243	0.5464	0.1542	−0.1247	2.2734	0.8569
Lithium niobate	$LiNbO_3$	4.70	RT	114	2.030	0.530	0.750	0.090	2.450	0.600
Lithium tantalate	$LiTaO_3$	7.45	RT	114	2.330	0.470	0.800	−0.110	2.750	0.940
Quartz	SiO_2	2.6485	298	115	0.8680	0.0704	0.1191	−0.1804	1.0575	0.5820
Selenium	Se	4.838	300	116	0.1870	0.0710	0.2620	0.0620	0.7410	0.1490
Sodium nitrate	$NaNO_3$	2.27	RT	12	0.8670	0.1630	0.1600	0.0820	0.3740	0.2130
Tourmaline		3.05	RT	82	2.7066	0.6927	0.0872	−0.0774	1.6070	0.6682

References

1. Thomas, J. F., *Phys. Rev.*, 175, 955–962, 1968.
2. Bolef, D. I. and M. Menes, *J. Appl. Phys.*, 31, 1426–1427, 1960.
3. Garland, C. W. and C. F. Yarnell, *J. Chem. Phys.*, 44, 1112–1120, 1966.
4. Garland, C. W. and R. Renard, *J. Chem. Phys.*, 44, 1130–1139, 1966.
5. Gsänger, M., H. Egger and E. Lüscher, *Phys. Letters*, 27A, 695–696, 1968.
6. Wong, C. and D. E. Schuele, *J. Phys. Chem. Solids*, 29, 1309–1330, 1968.
7. Haussühl, S., *Phys. Stat. Sol.*, 3, 1072–1076, 1963.
8. Wong, C. and D. E. Schuele, *J. Phys. Chem. Solids*, 28, 1225–1231, 1967.
9. McSkimin, H. J. and D. G. Thomas, *J. Appl. Phys.*, 33, 56–59, 1962.
10. Kollarits, F. J. and J. Trivisonno, *J. Phys. Chem. Solids*, 29, 2133–2139, 1968.
11. Slagle, D. D. and H. A. McKinstry, *J. Appl. Phys.*, 38, 446–458, 1967.
12. Hearmon, R. F. S., *Adv. Phys.*, 5, 323–382, 1956.
13. Sumer, A. and J. F. Smith, *J. Appl. Phys.*, 34, 2691–2694, 1963.
14. Alexandrov, K. S. et al., *Sov. Phys. Sol. State*, 10, 1316–1321, 1968.
15. Epstein, S. G. and O. N. Carlson, *Acta Metal.*, 13, 487–491, 1965.
16. McSkimin, H. J., et al., *J. Appl. Phys.*, 39, 4127–4128, 1968.
17. McSkimin, H. J., et al., *J. Appl. Phys.*, 38, 2362–2364, 1967.
18. Weil, R. and W. O. Groves, *J. Appl. Phys.*, 39, 4049–4051, 1968.
19. Bateman, T. B., *J. Appl. Phys.*, 37, 2194–2195, 1966.
20. Bogardus, E. H., *J. Appl. Phys.*, 36, 2504–2513, 1965.
21. Golding, B., S. C. Moss and B. L. Averbach, *Phys. Rev.*, 158, 637–645, 1967.
22. Bateman, T. B., H. J. McSkimin and J. M. Whelan, *J. Appl. Phys.*, 30, 544–545, 1959.
23. Gerlich, D., *J. Appl. Phys.*, 35, 3062, 1964.
24. Hickernell, F. S. and W. R. Gayton, *J. Appl. Phys.*, 37, 462, 1966.
25. MacFarlane, R. E., et al., *Phys. Letters*, 20, 234–235, 1966.
26. Leese, J. and A. E. Lord Jr., *J. Appl. Phys.*, 39, 3986–3988, 1968.
27. Miller, R. A. and D. E. Schuele, *J. Phys. Chem. Solids*, 30, 589–600, 1969.
28. Wasilik, J. H. and M. L. Wheat, *J. Appl. Phys.*, 36, 791–793, 1965.
29. Haussühl, S., *Phys. Stat. Sol.*, 3, 1072–1076, 1963.
30. Houston, B., et al., *J. Appl. Phys.*, 39, 3913–3916, 1968.
31. Trivisonno, J. and C. S. Smith, *Acta Metal.*, 9, 1064–1071, 1961.
32. Alexandrov, K. S. and T. V. Ryzhova, *Sov. Phys. Cryst.*, 6, 228–252, 1961.
33. Lewis, J. T., A. Lehoczky and C. V. Briscoe, *Phys. Rev.*, 161, 877–887, 1967.
34. Drabble, J. R. and R. E. B. Strathen, *Proc. Phys. Soc.*, 92, 1090–1995, 1967.
35. Oliver, D. W., *J. Appl. Phys.*, 40, 893, 1969.
36. Alper, T., and G. A. Saunders, *J. Phys. Chem. Solids*, 28, 1637–1642, 1967.
37. Dickinson, J. M. and P. E. Armstrong, *J. Appl. Phys.*, 38, 602–606, 1967.
38. Bolef, D. I., *J. Appl. Phys.*, 32, 100–105, 1961.
39. Rayne, J. A., *Phys. Rev.*, 112, 1125–1130, 1958.
40. MacFarlane, R. E., et al., *Phys. Letters*, 18, 91–92, 1965.
41. Smith, P. A. and C. S. Smith, *J. Phys. Chem. Solids*, 26, 279–289, 1965.
42. Norwood, M. H. and C. V. Briscoe, *Phys. Rev.*, 112, 45–48, 1958.
43. Simmons, G. and F. Birch, *J. Appl. Phys.*, 34, 2736–2738, 1963.
44. Gutman, E. J. and J. Trivisonno, *J. Phys. Chem. Sol.*, 28, 805–809, 1967.
45. Ghafelehbashi, M., et al., *J. Appl. Phys.*, 41, 652–666, 1970.
46. McSkimin, H. J. and P. Andreatch, Jr., *J. Appl. Phys.*, 35, 2161–2165, 1964.
47. Neighbours, J. R. and G. A. Alers, *Phys. Rev.*, 111, 707–712, 1958.
48. Hidshaw, W., J. T. Lewis, and C. V. Briscoe, *Phys. Rev.*, 163, 876–881, 1967.
49. Daniels, W. B., *Phys. Rev.*, 119, 1246–1252, 1960.
50. Viswanathan, R., *J. Appl. Phys.*, 37, 884–886, 1966.
51. Miller, R. A. and C. S. Smith, *J. Phys. Chem. Sol.*, 25, 1279–1292, 1964.
52. Claytor, R. N. and B. J. Marshall, *Phys. Rev.*, 120, 332–334, 1960.
53. Schreiber, E., *J. Appl. Phys.*, 38, 2508–2511, 1967.
54. Gerlich, D., *Phys. Rev.*, 136, A1366–A1368, 1964.
55. Johnston, D. L., P. H. Thrasher and R. J. Kearney, *J. Appl. Phys.*, 41, 427–428, 1970.
56. Poindexter, E. and A. A. Giardini, *Phys. Rev.*, 110, 1069, 1958.
57. Soga, N., *J. Appl. Phys.*, 37, 3416–3420, 1966.
58. Bartlett, R. W. and C. W. Smith, *J. Appl. Phys.*, 38, 5428–5429, 1967.
59. Morse, G. E. and A. W. Lawson, *J. Phys. Chem. Sol.*, 28, 939–950, 1967.
60. Armstrong, P. E., O. N. Carlson and J. F. Smith, *J. Appl. Phys.*, 30, 36–41, 1959.
61. Macedo, P. M., W. Capps and J. B. Wachtman, *J. Am. Cer. Soc.*, 47, 651, 1964.
62. Beattie, A. G., *J. Appl. Phys.*, 40, 4818–4821, 1969.
63. Chang, R. and L. J. Graham, *J. Appl Phys.*, 37, 3778–3783, 1966.
64. Lowrie, R. and A. M. Gonas, *J. Appl. Phys.*, 38, 4505–4509, 1967.
65. Graham, L. J., H. Nadler and R. Chang, *J. Appl. Phys.*, 34, 1572–1573, 1963.
66. Wachtman, J. B., Jr., et al., *J. Nucl. Mat.*, 16, 39–41, 1965.
67. Bolef, D. I., *J. Appl. Phys.*, 32, 100–105, 1961.
68. Berlincourt, D., H. Jaffe and L. R. Shiozawa, *Phys. Rev.*, 129, 1009–1017, 1963.
69. Adhav. R. S. J. *Acoust. Soc. Am.*, 43, 835–838, 1968.
70. Berlincourt, D. and H. Jaffe, *Phys. Rev.*, 111, 143–148, 1958.
71. Huntington, H. B., in *Solid State Pysics, Vol. 7*, Seitz, F., and Turnbull, D., Ed., pp. 213–285, Academic Press, New York 1958.
72. Cutler, H. R., J. J. Gibson and K. A. McCarthy, *Sol. State Comm.*, 6, 431–433, 1968.
73. Mason, W. P., *Piezoelectric Crystals and Their Application to Ultrasonics*, D. Van Nostrand Co., Inc., New York, 1950.
74. Adhav, R. S., *J. Appl. Phys.*, 40, 2725–2727, 1969.
75. Manghnani, M. H., *J. Geophys. Res.*, 74, 4317–4328, 1969.
76. Uchida, N. and Y. Ohmachi, *J. Appl Phys.*, 40, 4692–4695, 1969.
77. House, D. G. and E. Y. Vernon, *Br. J. Appl. Phys.*, 11, 254–259, 1960.
78. Ryzhova, T. V., et al., *Bull. Acad. Sci. USSR, Earth Phys. Ser.*, English Transl., no. 2, 111–113, 1966.
79. Alton, W. J. and A. J. Barlow, *J. Appl. Phys.*, 38, 3817–3820, 1967.
80. Michard, F., et al., *C. R. Acad. Sci., Paris*, 265, 565–567, 1967.
81. Haussühl, S., *Acta Cryst.*, 18, 839–842, 1965.
82. Hearmon, R. F. S., *Rev. Mod. Phys.*, 18, 409–440, 1946.
83. Heseltine, J. C. W., D. W. Elliott and O. B. Wilson, *J. Chem. Phys.*, 40, 2584–2587, 1964.
84. Schwerdtner, W. M., et al., *Canad. J. Earth Sci.*, 2, 673–683, 1965.
85. Kumazawa, M. and O. L. Anderson, *J. Geophys. Res.*, 74, 5961–5972, 1969.
86. Alexandrov, K. S., et al., *Sov. Phys. Cryst.*, 7, 753–755, 1963.
87. Verma, R. K., *J. Geophys. Soc.*, 65, 757–766, 1960.
88. McSkimin, H. J. and E. S. Fisher, *J. Appl. Phys.*, 31, 1627–1639, 1960.
89. Alexandrov, K. S. and T.V. Ryzhova, *Bull. Acad. Sci. USSR, Geophys. Ser.*, English Transl., no. 8, 871–875, 1961.
90. Afanaseva, G. K., et al, *Phys. Stat. Sol.*, 24, K61–K63, 1967.
91. Alexandrov, K. S., et al., *Sov. Phys. Cryst.*, 8, 589–591, 1964.
92. Alexandrov, K. S. and T. V Ryzhova, *Bull Acad. Sci. USSR, Geophys. Ser.*, English Transl., no. 2, 129–131, 1962.
93. Alexandrov, K. S., et al., *Sov. Phys. Cryst.*, 8, 164–166, 1963.
94. Teslenko, V. F., et al., *Sov. Phys. Cryst.*, 10, 744–747, 1966.
95. Smith, J. F. and C. L. Arbogast, *J. Appl. Phys.*, 31, 99–102, 1960.
96. Cline, C. F., H. L. Dunegan and G. M. Henderson, *J. Appl. Phys.*, 38, 1944–1948, 1967.
97. Chang, Y. A. and L. Himmel, *J. Appl. Phys.*, 37, 3787–3790, 1966.
98. Gerlich, D., *J. Phys. Chem. Solids*, 28, 2575–2579, 1967.
99. McSkimin, H. J., *J. Appl. Phys.*, 26, 406–409, 1955.
100. Fisher, E. S. and D. Dever, *Trans. Met. Soc. AIME*, 239, 48–57, 1967.
101. Fisher, E. S. and D. Dever, *Proc. Conf. Rare Earth Res.*, 6th, Gatlinburg, TN, 522–533, 1967.
102. Fisher, E. S. and C. J. Renken, *Phys. Rev.*, 135, A482–A494, 1964.
103. Proctor, T. M., Jr., *J. Acoust. Soc. Am.*, 39, 972–977, 1966.
104. Chandrasekhar, B. S. and J. A. Rayne, *Phys. Rev.*, 124, 1011–1041, 1961.
105. Wazzan, A. R. and L. B. Robinson, *Phys. Rev.*, 155, 586–594, 1967.
106. Ferris, R. W., et al., *J. Appl. Phys.*, 34, 768–770, 1963.
107. Gilman, J. J. and B. W. Roberts, *J. Appl. Phys.*, 32, 1405, 1961.
108. Smith, J. F. and J. A. Gjevre, *J. Appl. Phys.*, 31, 645–647, 1960.
109. Alers, G. A. and J. R. Neighbours, *J. Phys. Chem. Solids*, 7, 58–64, 1908.
110. Bateman, T. B., *J. Appl. Phys.*, 33, 3309–3312, 1962.
111. Tefft, W. E., *J. Res. Natl. Bur. Stand.*, 70A, 277–280, 1966.
112. DeBretteville, Jr., A. et al., *Phys. Rev.*, 148, 575–579, 1966.
113. Dandekar, D. P. and A. L. Ruoff, *J. Appl. Phys.*, 39, 6004–6009, 1968.
114. Warner, A. W., M. Onoe and G. A. Coquin, *J. Acoust. Soc. Am.*, 42, 1223–1231, 1967.
115. McSkimin, H. J., P. Andreatch and R. N. Thurston, *J. Appl. Phys.*, 36, 1624–1632, 1965.
116. Mort, J., *J. Appl. Phys.*, 38, 3414–3415, 1967.

ELECTRICAL RESISTIVITY OF PURE METALS

The first part of this table gives the electrical resistivity, in units of $10^{-8}\ \Omega$ m, for 28 common metallic elements as a function of temperature. The data refer to polycrystalline samples. The number of significant figures indicates the accuracy of the values. However, at low temperatures (especially below 50 K) the electrical resistivity is extremely sensitive to sample purity. Thus the low-temperature values refer to samples of specified purity and treatment. The references should be consulted for further information on this point, as well as for values at additional temperatures.

The second part of the table gives resistivity values in the neighborhood of room temperature for other metallic elements that have not been studied over an extended temperature range.

References

1. C. Y. Ho, et al., *J. Phys. Chem. Ref. Data*, 12, 183–322, 1983; 13, 1069–1096, 1984; 13, 1097–1130, 1984, 13, 1131–1172, 1984.
2. R. A. Matula, *J. Phys Chem. Ref. Data*, 8, 1147–1298, 1979.
3. T. C. Chi, *J. Phys. Chem. Ref. Data*, 8, 339–438, 1979; 8, 439–498, 1979.
4. K. H. Hellwege, Ed., *Landolt-Börnstein Numerical Data and Functional Relationships in Science and Technology*, Group III, Vol. 15, Subvolume a, Springer-Verlag, Heidelberg, 1982.
5. L. A. Hall, *Survey of Electrical Resistivity Measurements on 16 Pure Metals in the Temperature Range 0 to 273 K*, NBS Technical Note 365, U.S. Superintendent of Documents, 1968.

Electrical Resistivity in $10^{-8}\ \Omega$ m

T/K	Aluminum	Barium	Beryllium	Calcium	Cesium	Chromium	Copper
1	0.000100	0.081	0.0332	0.045	0.0026		0.00200
10	0.000193	0.189	0.0332	0.047	0.243		0.00202
20	0.000755	0.94	0.0336	0.060	0.86		0.00280
40	0.0181	2.91	0.0367	0.175	1.99		0.0239
60	0.0959	4.86	0.067	0.40	3.07		0.0971
80	0.245	6.83	0.075	0.65	4.16		0.215
100	0.442	8.85	0.133	0.91	5.28	1.6	0.348
150	1.006	14.3	0.510	1.56	8.43	4.5	0.699
200	1.587	20.2	1.29	2.19	12.2	7.7	1.046
273	2.417	30.2	3.02	3.11	18.7	11.8	1.543
293	2.650	33.2	3.56	3.36	20.5	12.5	1.678
298	2.709	34.0	3.70	3.42	20.8	12.6	1.712
300	2.733	34.3	3.76	3.45	21.0	12.7	1.725
400	3.87	51.4	6.76	4.7		15.8	2.402
500	4.99	72.4	9.9	6.0		20.1	3.090
600	6.13	98.2	13.2	7.3		24.7	3.792
700	7.35	130	16.5	8.7		29.5	4.514
800	8.70	168	20.0	10.0		34.6	5.262
900	10.18	216	23.7	11.4		39.9	6.041

T/K	Gold	Hafnium	Iron	Lead	Lithium	Magnesium	Manganese
1	0.0220	1.00	0.0225		0.007	0.0062	7.02
10	0.0226	1.00	0.0238		0.008	0.0069	18.9
20	0.035	1.11	0.0287		0.012	0.0123	54
40	0.141	2.52	0.0758		0.074	0.074	116
60	0.308	4.53	0.271		0.345	0.261	131
80	0.481	6.75	0.693	4.9	1.00	0.557	132
100	0.650	9.12	1.28	6.4	1.73	0.91	132
150	1.061	15.0	3.15	9.9	3.72	1.84	136
200	1.462	21.0	5.20	13.6	5.71	2.75	139
273	2.051	30.4	8.57	19.2	8.53	4.05	143
293	2.214	33.1	9.61	20.8	9.28	4.39	144
298	2.255	33.7	9.87	21.1	9.47	4.48	144
300	2.271	34.0	9.98	21.3	9.55	4.51	144
400	3.107	48.1	16.1	29.6	13.4	6.19	147
500	3.97	63.1	23.7	38.3		7.86	149
600	4.87	78.5	32.9			9.52	151
700	5.82		44.0			11.2	152
800	6.81		57.1			12.8	
900	7.86		71.3			14.4	

T/K	Molybdenum	Nickel	Palladium	Platinum	Potassium	Rubidium	Silver
1	0.00070	0.0032	0.0200	0.002	0.0008	0.0131	0.00100
10	0.00089	0.0057	0.0242	0.0154	0.0160	0.109	0.00115
20	0.00261	0.0140	0.0563	0.0484	0.117	0.444	0.0042
40	0.0457	0.068	0.334	0.409	0.480	1.21	0.0539
60	0.206	0.242	0.938	1.107	0.90	1.94	0.162
80	0.482	0.545	1.75	1.922	1.34	2.65	0.289
100	0.858	0.96	2.62	2.755	1.79	3.36	0.418
150	1.99	2.21	4.80	4.76	2.99	5.27	0.726
200	3.13	3.67	6.88	6.77	4.26	7.49	1.029
273	4.85	6.16	9.78	9.6	6.49	11.5	1.467
293	5.34	6.93	10.54	10.5	7.20	12.8	1.587
298	5.47	7.12	10.73	10.7	7.39	13.1	1.617
300	5.52	7.20	10.80	10.8	7.47	13.3	1.629
400	8.02	11.8	14.48	14.6			2.241
500	10.6	17.7	17.94	18.3			2.87
600	13.1	25.5	21.2	21.9			3.53
700	15.8	32.1	24.2	25.4			4.21
800	18.4	35.5	27.1	28.7			4.91
900	21.2	38.6	29.4	32.0			5.64

T/K	Sodium	Strontium	Tantalum	Tungsten	Vanadium	Zinc	Zirconium
1	0.0009	0.80	0.10	0.000016		0.0100	0.250
10	0.0015	0.80	0.102	0.000137	0.0145	0.0112	0.253
20	0.016	0.92	0.146	0.00196	0.039	0.0387	0.357
40	0.172	1.70	0.751	0.0544	0.304	0.306	1.44
60	0.447	2.68	1.65	0.266	1.11	0.715	3.75
80	0.80	3.64	2.62	0.606	2.41	1.15	6.64
100	1.16	4.58	3.64	1.02	4.01	1.60	9.79
150	2.03	6.84	6.19	2.09	8.2	2.71	17.8
200	2.89	9.04	8.66	3.18	12.4	3.83	26.3
273	4.33	12.3	12.2	4.82	18.1	5.46	38.8
293	4.77	13.2	13.1	5.28	19.7	5.90	42.1
298	4.88	13.4	13.4	5.39	20.1	6.01	42.9
300	4.93	13.5	13.5	5.44	20.2	6.06	43.3
400		17.8	18.2	7.83	28.0	8.37	60.3
500		22.2	22.9	10.3	34.8	10.82	76.5
600		26.7	27.4	13.0	41.1	13.49	91.5
700		31.2	31.8	15.7	47.2		104.2
800		35.6	35.9	18.6	53.1		114.9
900			40.1	21.5	58.7		123.1

Element	T/K	Electrical resistivity $10^{-8}\ \Omega$ m	Element	T/K	Electrical resistivity $10^{-8}\ \Omega$ m
Antimony	273	39	Osmium	273	8.1
Bismuth	273	107	Polonium	273	40
Cadmium	273	6.8	Praseodymium	290–300	70.0
Cerium (β, hex)	290–300	82.8	Promethium	290–300	75 est.
Cerium (γ, cub)	298	74.4	Protactinium	273	17.7
Cobalt	273	5.6	Rhenium	273	17.2
Dysprosium	290–300	92.6	Rhodium	273	4.3
Erbium	290–300	86.0	Ruthenium	273	7.1
Europium	290–300	90.0	Samarium	290–300	94.0
Gadolinium	290–300	131	Scandium	290–300	56.2
Gallium	273	13.6	Terbium	290–300	115
Holmium	290–300	81.4	Thallium	273	15
Indium	273	8.0	Thorium	273	14.7
Iridium	273	4.7	Thulium	290–300	67.6
Lanthanum	290–300	61.5	Tin	273	11.5
Lutetium	290–300	58.2	Titanium	273	39
Mercury	298	96.1	Uranium	273	28
Neodymium	290–300	64.3	Ytterbium	290–300	25.0
Niobium	273	15.2	Yttrium	290–300	59.6

ELECTRICAL RESISTIVITY OF SELECTED ALLOYS

These values were obtained by fitting all available measurements to a theoretical formulation describing the temperature and composition dependence of the electrical resistivity of metals. Some of the values listed here fall in regions of temperature and composition where no actual measurements exist. Details of the procedure may be found in the reference.

Values of the resistivity are given in units of 10^{-8} Ω m. General comments in the preceding table for pure metals also apply here.

Reference

C. Y. Ho, et al., *J. Phys. Chem. Ref. Data*, 12, 183–322, 1983.

Aluminum-Copper

Wt % Al	100 K	273 K	293 K	300 K	350 K	400 K
99[a]	0.531	2.51	2.74	2.82	3.38	3.95
95[a]	0.895	2.88	3.10	3.18	3.75	4.33
90[b]	1.38	3.36	3.59	3.67	4.25	4.86
85[b]	1.88	3.87	4.10	4.19	4.79	5.42
80[b]	2.34	4.33	4.58	4.67	5.31	5.99
70[b]	3.02	5.03	5.31	5.41	6.16	6.94
60[b]	3.49	5.56	5.88	5.99	6.77	7.63
50[b]	4.00	6.22	6.55	6.67	7.55	8.52
40[c]		7.57	7.96	8.10	9.12	10.2
30[c]		11.2	11.8	12.0	13.5	15.2
25[f]		16.3	17.2	17.6	19.8	22.2
15[h]			12.3			
10[g]	8.71	10.8	11.0	11.1	11.7	12.3
5[c]	7.92	9.43	9.61	9.68	10.2	10.7
1[b]	3.22	4.46	4.60	4.65	5.00	5.37

Aluminum-Magnesium

Wt % Al	100 K	273 K	293 K	300 K	350 K	400 K
99[c]	0.958	2.96	3.18	3.26	3.82	4.39
95[c]	3.01	5.05	5.28	5.36	5.93	6.51
90[c]	5.42	7.52	7.76	7.85	8.43	9.02
10[b]	14.0	17.1	17.4	17.6	18.4	19.2
5[b]	9.93	13.1	13.4	13.5	14.3	15.2
1[a]	2.78	5.92	6.25	6.37	7.20	8.03

Copper-Gold

Wt % Cu	100 K	273 K	293 K	300 K	350 K	400 K
99[c]	0.520	1.73	1.86	1.91	2.24	2.58
95[c]	1.21	2.41	2.54	2.59	2.92	3.26
90[c]	2.11	3.29	4.42	3.46	3.79	4.12
85[c]	3.01	4.20	4.33	4.38	4.71	5.05
80[c]	3.95	5.15	5.28	5.32	5.65	5.99
70[c]	5.91	7.12	7.25	7.30	7.64	7.99
60[c]	8.04	9.18	9.13	9.36	9.70	10.05
50[c]	9.88	11.07	11.20	11.25	11.60	11.94
40[c]	11.44	12.70	12.85	12.90	13.27	13.65
30[c]	12.43	13.77	13.93	13.99	14.38	14.78
25[c]	12.59	13.93	14.09	14.14	14.54	14.94
15[c]	11.38	12.75	12.91	12.96	13.36	13.77
10[c]	9.33	10.70	10.86	10.91	11.31	11.72
5[c]	5.91	7.25	7.41	7.46	7.87	8.28
1[c]	2.00	3.40	3.57	3.62	4.03	4.45

Copper-Nickel

Wt % Cu	100 K	273 K	293 K	300 K	350 K	400 K
99[c]	1.45	2.71	2.85	2.91	3.27	3.62
95[c]	6.19	7.60	7.71	7.82	8.22	8.62
90[c]	12.08	13.69	13.89	13.96	14.40	14.81
85[c]	18.01	19.63	19.83	19.90	20.32	20.70
80[c]	23.89	25.46	25.66	25.72	26.12	26.44
70[i]	35.73	36.67	36.72	36.76	36.85	36.89
60[i]	45.76	45.43	45.38	43.35	45.20	45.01
50[i]	50.22	50.19	50.05	50.01	49.73	49.50
40[c]	36.77	47.42	47.73	47.82	48.28	48.49
30[i]	26.73	40.19	41.79	42.34	44.51	45.40
25[c]	22.22	33.46	35.11	35.69	39.67	42.81
15[c]	13.49	22.00	23.35	23.85	27.60	31.38
10[c]	9.28	16.65	17.82	18.26	21.51	25.19
5[c]	5.20	11.49	12.50	12.90	15.69	18.78
1[c]	1.81	7.23	8.08	8.37	10.63	13.18

Copper-Palladium

Wt % Cu	100 K	273 K	293 K	300 K	350 K	400 K
99[c]	0.91	2.10	2.23	2.27	2.59	2.92
95[c]	2.99	4.21	4.35	4.40	4.74	5.08
90[c]	5.69	6.89	7.03	7.08	7.41	7.74
85[c]	8.30	9.48	9.61	9.66	10.01	10.36
80[c]	10.74	11.99	12.12	12.16	12.51	12.87
70[c]	15.67	16.87	17.01	17.06	17.41	17.78
60[c]	20.45	21.73	21.87	21.92	22.30	22.69
50[c]	26.07	27.62	27.79	27.86	28.25	28.64
40[c]	33.53	35.31	35.51	35.57	36.03	36.47
30[c]	45.03	46.50	46.66	46.71	47.11	47.47
25[c]	44.12	46.25	46.45	46.52	46.99	47.43
15[c]	31.79	36.52	36.99	37.16	38.28	39.35
10[c]	23.00	28.90	29.51	29.73	31.19	32.56
5[c]	13.09	20.00	20.75	21.02	22.84	24.54
1[c]	8.97	11.90	12.67	12.93	14.82	16.68

Copper-Zinc

Wt % Cu	100 K	273 K	293 K	300 K	350 K	400 K
99[b]	0.671	1.84	1.97	2.02	2.36	2.71
95[b]	1.54	2.78	2.92	2.97	3.33	3.69
90[b]	2.33	3.66	3.81	3.86	4.25	4.63
85[b]	2.93	4.37	4.54	4.60	5.02	5.44
80[b]	3.44	5.01	5.19	5.26	5.71	6.17
70[b]	4.08	5.87	6.08	6.15	6.67	7.19

Gold-Palladium

Wt % Au	100 K	273 K	293 K	300 K	350 K	400 K
99[c]	1.31	2.69	2.86	2.91	3.32	3.73
95[c]	3.88	5.21	5.35	5.41	5.79	6.17
90[i]	6.70	8.01	8.17	8.22	8.56	8.93
85[b]	9.14	10.50	10.66	10.72	11.10	11.48
80[b]	11.23	12.75	12.93	12.99	13.45	13.93
70[c]	16.44	18.23	18.46	18.54	19.10	19.67
60[b]	24.64	26.70	26.94	27.02	27.63	28.23
50[a]	23.09	27.23	27.63	27.76	28.64	29.42
40[a]	19.40	24.65	25.23	25.42	26.74	27.95
30[b]	14.94	20.82	21.49	21.72	23.35	24.92
25[b]	12.72	18.86	19.53	19.77	21.51	23.19
15[a]	8.54	15.08	15.77	16.01	17.80	19.61
10[a]	6.54	13.25	13.95	14.20	16.00	17.81
5[a]	4.58	11.49	12.21	12.46	14.26	16.07
1[a]	3.01	10.07	10.85	11.12	12.99	14.80

Gold-Silver

Wt % Au	100 K	273 K	293 K	300 K	350 K	400 K
99[b]	1.20	2.58	2.75	2.80	3.22	3.63
95[a]	3.16	4.58	4.74	4.79	5.19	5.59
90[j]	5.16	6.57	6.73	6.78	7.19	7.58
85[j]	6.75	8.14	8.30	8.36	8.75	9.15
80[j]	7.96	9.34	9.50	9.55	9.94	10.33
70[j]	9.36	10.70	10.86	10.91	11.29	11.68
60[j]	9.61	10.92	11.07	11.12	11.50	11.87
50[j]	8.96	10.23	10.37	10.42	10.78	11.14
40[j]	7.69	8.92	9.06	9.11	9.46	9.81
30[a]	6.15	7.34	7.47	7.52	7.85	8.19
25[a]	5.29	6.46	6.59	6.63	6.96	7.30
15[a]	3.42	4.55	4.67	4.72	5.03	5.34
10[a]	2.44	3.54	3.66	3.71	4.00	4.31
5[i]	1.44	2.52	2.64	2.68	2.96	3.25
1[b]	0.627	1.69	1.80	1.84	2.12	2.42

Iron-Nickel

Wt % Fe	100 K	273 K	293 K	300 K	400 K
99[a]	3.32	10.9	12.0	12.4	18.7
95[c]	10.0	18.7	19.9	20.2	26.8
90[c]	14.5	24.2	25.5	25.9	33.2
85[c]	17.5	27.8	29.2	29.7	37.3
80[c]	19.3	30.1	31.6	32.2	40.0
70[b]	20.9	32.3	33.9	34.4	42.4
60[c]	28.6	53.8	57.1	58.2	73.9
50[d]	12.3	28.4	30.6	31.4	43.7
40[d]	7.73	19.6	21.6	22.5	34.0
30[c]	5.97	15.3	17.1	17.7	27.4
25[b]	5.62	14.3	15.9	16.4	25.1
15[c]	4.97	12.6	13.8	14.2	21.1
10[c]	4.20	11.4	12.5	12.9	18.9
5[c]	3.34	9.66	10.6	10.9	16.1
1[b]	1.66	7.17	7.94	8.12	12.8

Silver-Palladium

Wt % Ag	100 K	273 K	293 K	300 K	350 K	400 K
99[b]	0.839	1.891	2.007	2.049	2.35	2.66
95[b]	2.528	3.58	3.70	3.74	4.04	4.34
90[b]	4.72	5.82	5.94	5.98	6.28	6.59
85[k]	6.82	7.92	8.04	8.08	8.38	8.68
80[k]	8.91	10.01	10.13	10.17	10.47	10.78
70[k]	13.43	14.53	14.65	14.69	14.99	15.30
60[i]	19.4	20.9	21.1	21.2	21.6	22.0
50[k]	29.3	31.2	31.4	31.5	32.0	32.4
40[m]	40.8	42.2	42.2	42.2	42.3	42.3
30[b]	37.1	40.4	40.6	40.7	41.3	41.7
25[k]	32.4	36.67	37.06	37.19	38.1	38.8
15[i]	21.0	27.08	26.68	27.89	29.3	30.6
10[i]	14.95	21.69	22.39	22.63	24.3	25.9
5[b]	8.91	15.98	16.72	16.98	18.8	20.5
1[a]	3.97	11.06	11.82	12.08	13.92	15.70

[a] Uncertainty in resistivity is ± 2%.
[b] Uncertainty in resistivity is ± 3%.
[c] Uncertainty in resistivity is ± 5%.
[d] Uncertainty in resistivity is ± 7% below 300 K and ± 5% at 300 and 400 K.
[e] Uncertainty in resistivity is ± 7%.
[f] Uncertainty in resistivity is ± 8%.
[g] Uncertainty in resistivity is ± 10%.
[h] Uncertainty in resistivity is ± 12%.
[i] Uncertainty in resistivity is ± 4%.
[j] Uncertainty in resistivity is ± 1%.
[k] Uncertainty in resistivity is ± 3% up to 300 K and ± 4% above 300 K.
[m] Uncertainty in resistivity is ± 2% up to 300 K and ± 4% above 300 K.

ELECTRICAL RESISTIVITY OF GRAPHITE MATERIALS

L. I. Berger

At normal conditions, the only stable crystallographic modification of carbon is graphite. The quasi-stable diamond turns into graphite starting from about 1000 °C in air. In industry, a graphitic material is commonly called either *carbon*, if it consists of small and low-oriented crystallites, or *graphite*, the material with highly ordered structure. In the 1970s, the first carbon filaments of about 7 nm in diameter were grown by Morinobu Endo at the University of Orleans, France, by the vapor-growth technique. In 1985, Sir Harold Walter Kroto of Sussex University, UK, and Richard E. Smalley and co-workers at Rice University discovered spherical carbon molecules, C_{60} (or C60), consisting of combinations of carbon atoms organized into hexagons and pentagons, named *buckminsterfullerenes* or *fullerenes* and possessing very promising mechanical and electrical properties. In 1991, Sumio Iijima, NEC Labs, Japan, and David S. Bethune, IBM Almaden Labs, observed the carbon atomic groups in the form of tubes capped by halves of the fullerene molecules and formed on the cathodes of carbon arc devices. The length of the tubes could be up to tens of micrometers and the diameter, naturally, is equal to that of the fullerene molecule. These tubes, called *nanotubes*, may be single wall (SWNT) or consist of several concentric tubes with a common axis (multi-walled nanotubes, MWNT). Two-dimensional *graphene* is another crystallographic modification of graphite (Saroj Nayak, Rensselaer U., 2004) that is a flat hexagonal network of carbon atoms with a thickness equal to the carbon atom size. The nanotube may be considered as formed by strips of graphenes turned into a cylinder. The character of the electrical conductivity (metallic or semiconductive) of a SWNT depends on orientation of the carbon hexagons of the nanotube surface regarding its axis (the chiral angle [Ref. 1]). The following table contains some typical data on electrical and electronic properties of graphite materials.

References

1. M. S. Dresselhaus, G. Dresselhaus, and Ph. Avouris (Eds.), *Carbon Nanotubes. Synthesis, Structure, Properties, and Applications*, Springer-Verlag, 2001.
2. ESPI Metals Catalog, 2007.
3. SPI Supplies Catalog, 2007.
4. F. L. Vogel, *J. Mater. Sci.* 12, 982–986, 1977.
5. K. S. Novoselov et al., *Nature* 438, 197–200, 2005.
6. Y. Zhang et al., *Nature* 438, 201–204, 2005.
7. N. Tombros et al., *Nature* 448, 571–574, 2007.
8. H. Dai, in Ref. 1, pp. 29–53.
9. CTI Carbon Nanotube Cat., 2007.
10. L. Matija et al., *Sci. Forum* 413, 49–52, 2003.

Material	Electrical resistivity ρ at R. T. mΩ cm [μΩ inch]	Energy gap at R. T. eV	Electron mobility cm²/V s	(1/ρ)dρ/d*t* near R. T. 10^{-4}°C⁻¹	Ref.
Bulk graphite					
Electromet graphite	1.90 [750]			–5	2
Electro graphite	1.60 [630]			–5	2
Aeromet graphite	1.47 [580]			–5	2
ESPI Superconductive	1.75 [690]			–5	2
Radioelectronics data	30 [11,800]			–5.6	3
Highly ordered pyrolytic graphite	Parallel 0.04 [15.7]				3
	Across 150 [59000]				
Single crystal graphite, normal to *c*-axis	$1 \cdot 10^{-6}$				4
Graphenes					
n-Graphene		≈5 (M); ≈10 (Γ)[c]	10^6		5,6
p-Graphene			10^4		7
Carbon nanotubes					
Metallic SWNT	12 kΩ[a]				1
Semiconducting SWNT		0.7 – 0.9[b]		128[d]	1
MWNT	10^2				9
Carbon fullerenes					
Fullerene (C_{60})	10^{12}	1.95			10

[a] Minimum resistance of individual nanotubes [Ref. 8]
[b] Est. from Ref. 1, p. 47
[c] Est. from Ref. 1, p. 116
[d] Est. from Ref. 1, p. 179.

PERMITTIVITY (DIELECTRIC CONSTANT) OF INORGANIC SOLIDS

H. P. R. Frederikse

This table lists the permittivity ε, frequently called the dielectric constant, of a number of inorganic solids. When the material is not isotropic, the individual components of the permittivity are given. A superscript S indicates a measurement made under constant strain ("clamped" dielectric constant). If the constraint is removed, the measurement yields ε^T, the "unclamped" or free dielectric constant.

The temperature of the measurement is given when available; the symbol r.t. indicates a value at nominal room temperature. The frequency of the measurement is given in the last column (i.r. indicates a measurement in the infrared).

Substances are listed in alphabetical order by chemical formula.

Reference

Young, K. F., and Frederikse, H. P. R., *J. Phys. Chem. Ref. Data* 2, 313, 1973.

Formula	Name	ε_{ijk}	*T*/K	ν/Hz
Ag_3AsS_3	Silver thioarsenate (Proustite)	$\varepsilon_{11}^T = 16.5, \varepsilon_{11}^S = 14.5$	r.t.	2×10^7
		$\varepsilon_{33}^T = 20.0, \varepsilon_{11}^S = 18.0$	r.t.	2×10^7
AgBr	Silver bromide	12.50	r.t.	
AgCN	Silver cyanide	5.6	r.t.	10^6
AgCl	Silver chloride	11.15	r.t.	
$AgNO_3$	Silver nitrate	9.0	293	5×10^5
$AgNa(NO_2)_2$	Silver sodium nitrite	4.5 ± 0.5	r.t.	9.4×10^9
Ag_2O	Silver oxide	8.8	r.t.	
$(AlF)_2SiO_4$	Aluminum fluosilicate (topaz)	$\varepsilon_{11} = 6.62$	297	7×10^3
		$\varepsilon_{22} = 6.58$	297	7×10^3
		$\varepsilon_{33} = 6.95$	297	7×10^3
Al_2O_3	Aluminum oxide (alumina)	$\varepsilon_{11} = \varepsilon_{22} = 9.34$	298	$10^2 - 8 \times 10^9$
		$\varepsilon_{33} = 11.54$	298	$10^2 - 8 \times 10^9$
$AlPO_4$	Aluminum phosphate	$\varepsilon_{11}^T = 6.05$	r.t.	10^3
AlSb	Aluminum antimonide	11.21	300	i.r.
AsF_3	Arsenic trifluoride	5.7	r.t.	
BN	Boron nitride	7.1	r.t.	i.r.
$BaCO_3$	Barium carbonate	8.53	291	2×10^5
$Ba(COOH)_2$	Barium formate	$\varepsilon_{11} = 7.9$	r.t.	10^3
		$\varepsilon_{22} = 5.9$	r.t.	10^3
		$\varepsilon_{33} = 7.5$	r.t.	10^3
$BaCl_2$	Barium chloride	9.81	r.t.	
$BaCl_2 \cdot 2H_2O$	Barium chloride dihydrate	9.00	r.t.	10^3
BaF_2	Barium fluoride	7.32	292	$5 \times 10^2 - 10^{11}$
$Ba(NO_3)_2$	Barium nitrate	4.95	292	2×10^5
$Ba_2NaNb_5O_{15}$	Barium sodium niobate ("Bananas")	$\varepsilon_{11}^S = 222, \varepsilon_{11}^T = 235$	296	10^4
		$\varepsilon_{22}^S = 227, \varepsilon_{22}^T = 247$	296	
		$\varepsilon_{33}^S = 32, \varepsilon_{33}^T = 51$	296	
BaO	Barium oxide (baria)	34 ± 1	248, 333	60×10^7
BaO_2	Barium peroxide	10.7	r.t.	2×10^6
BaS	Barium sulfide	19.23	r.t.	7.25×10^6
$BaSO_4$	Barium sulfate	11.4	288	10^8
$BaSnO_3$	Barium stannate	18	298	25×10^5
$BaTiO_3$	Barium titanate	$\varepsilon_{11}^T = 3600$	298	10^5
		$\varepsilon_{11}^S = 2300$	298	2.5×10^8
		$\varepsilon_{33}^T = 150$	298	10^5
		$\varepsilon_{33}^S = 80$	298	2.5×10^8
$Ba_6Ti_2Nb_8O_{30}$	Barium titanium niobate	$\varepsilon_{11} = \varepsilon_{22} \approx 190$	298	
		$\varepsilon_{33} \approx 220$	298	
$BaWO_4$	Barium tungstate	$\varepsilon_{11} = \varepsilon_{22} = 35.5 \pm 0.2$	297.5	1.6×10^3
		$\varepsilon_{33} = 37.2 \pm 0.2$	297.5	1.6×10^3
$BaZrO_3$	Barium zirconate	43	r.t.	

Formula	Name	ε_{ijk}	*T*/K	ν/Hz
$Be_3Al_2Si_6O_{18}$	Beryllium aluminum silicate (Beryl)	$\varepsilon_{33} = 5.95$	297	7×10^3
		$\varepsilon_{11} = \varepsilon_{22} = 6.86$	297	7×10^3
$BeCO_3$	Beryllium carbonate	9.7	291	2×10^5
BeO	Beryllium oxide (beryllia)	7.35 ± 0.2	293	2×10^6
$BiFeO_3$	Bismuth iron oxide	40 ± 3	300	9.4×10^9
		$\varepsilon^S_{11} = 38$		
$Bi_{12}GeO_{20}$	Bismuth germanite		r.t.	
$Bi(GeO_4)_3$	Bismuth germanate	16	293	
Bi_2O_3	Bismuth sesquioxide	18.2	r.t.	2×10^6
$Bi_4Ti_3O_{12}$	Bismuth titanate	112	r.t.	10^3
C	Diamond			
	Type I	5.87 ± 0.19	300	10^3
	Type IIa	5.66 ± 0.04	300	10^3
$C_4H_4O_6$	Tartaric acid	$\varepsilon_{11} = \varepsilon_{22} = 4.3$	298	
		$\varepsilon_{33} = 4.5$	298	
		$\varepsilon_{13} = 0.55$	298	
$C_6H_{14}N_2O_6$	Ethylene diamine tartrate (EDT)	$\varepsilon^T_{11} = 5.0$	293	
		$\varepsilon^T_{22} = 8.3$	293	
		$\varepsilon^T_{33} = 6.0$	293	
		$\varepsilon^T_{13} = 0.7$	293	
$C_6H_{12}O_6NaBr$	Dextrose sodium bromide	$\varepsilon^T_{11} = 4.0$	r.t.	10^3
$(CH_3NH_3)Al(SO_4)_2 \cdot 2H_2O$	Methyl ammonium alum (MASD)	19	197	
$Ca_2B_6O_{11} \cdot 5H_2O$	Colemanite	$\varepsilon_{11} = 20$	293	10^3
		$\varepsilon_{33} = 25$	293	10^3
$CaCO_3$	Calcium carbonate	$\varepsilon_{11} = 8.67$	r.t.	9.4×10^{10}
		$\varepsilon_{22} = 8.69$	r.t.	9.4×10^{10}
		$\varepsilon_{33} = 8.31$	r.t.	9.4×10^{10}
$CaCeO_3$	Calcium cerate	21	r.t.	
CaF_2	Calcium fluoride	6.81	300	5×10^2–10^{11}
$CaMoO_4$	Calcium molybdate	$\varepsilon_{11} = \varepsilon_{22} = 24.0 \pm 0.2$	297.5	<10
		$\varepsilon_{33} = 20.0 \pm 0.2$	297.5	<10
$Ca(NO_3)_2$	Calcium nitrate	6.54	292	2×10^5
$CaNb_2O_6$	Calcium niobate	$\varepsilon_{11} = 22.8 \pm 1.9$	r.t.	$(5–500) \times 10^3$
$Ca_2Nb_2O_7$	Calcium pyroniobate	~45	r.t.	5×10^7
CaO	Calcium oxide	11.8 ± 0.3	283	2×10^6
CaS	Calcium sulfide	6.699	r.t.	7.25×10^6
$CaSO_4 \cdot 2H_2O$	Calcium sulfate dihydrate	$\varepsilon_{11} = 5.10$	r.t.	
		$\varepsilon_{22} = 5.24$	r.t.	
		$\varepsilon_{33} = 10.30$	r.t.	
$CaTiO_3$	Calcium titanate	165	r.t.	
$CaWO_4$	Calcium tungstate	$\varepsilon_{11} = \varepsilon_{22} = 11.7 \pm 0.1$	297.5	1.59×10^3
		$\varepsilon_{33} = 9.5 \pm 0.2$	297.5	1.59×10^3
Cd_3As_2	Cadmium arsenide	$\varepsilon_{33} = 18.5$	4	
$CdBr_2$	Cadmium bromide	8.6	293	5×10^5
CdF_2	Cadmium fluoride	8.33 ± 0.08	300	10^5–10^7
CdS	Cadmium sulfide	$\varepsilon_{11} = \varepsilon_{22} = 8.7$	300	i.r.
		$\varepsilon_{33} = 9.25$	300	i.r.
		$\varepsilon_{11} = \varepsilon_{22} = 8.37$	8	i.r.
		$\varepsilon_{33} = 9.00$	8	i.r.
		$\varepsilon^T_{11} = 8.48$	77	10^4
		$\varepsilon^T_{33} = 9.48$	77	10^4
		$\varepsilon^S_{11} = 9.02, \varepsilon^T_{11} = 9.35$	298	10^4
		$\varepsilon^S_{33} = 9.53, \varepsilon^T_{33} = 10.33$	298	10^4
CdSe	Cadmium selenide	$\varepsilon^S_{11} = 9.53, \varepsilon^T_{11} = 9.70$	298	10^4
		$\varepsilon^S_{33} = 10.2, \varepsilon^T_{33} = 10.65$	298	10^4
CdTe	Cadmium telluride	$\varepsilon_{11} = \varepsilon_{22} = 10.60 \pm 0.15$	297	i.r.

Formula	Name	ε_{ijk}	T/K	ν/Hz
		$\varepsilon_{33} = 7.05 \pm 0.05$	297	i.r.
$Cd_2Nb_2O_7$	Cadmium pyroniobate	500–580	293	10^3
CeO_2	Cerium oxide	7.0	r.t.	2×10^6
$CoNb_2O_6$	Cobalt niobate	$\varepsilon_{11} = 18.4 \pm 1.1$	r.t.	$(5–500) \times 10^3$
		$\varepsilon_{22} = 21.4 \pm 1.1$	r.t.	$(5–500) \times 10^3$
		$\varepsilon_{33} = 33.0 \pm 0.7$	r.t.	$(5–500) \times 10^3$
CoO	Cobalt oxide	12.9	298	$10^2–10^{10}$
Cr_2O_3	Chromic sesquioxide	$\varepsilon_{11} = \varepsilon_{22} = 13.3$	298.5	10^3
		$\varepsilon_{33} = 11.9$	298.5	10^3
		8	315 (T_N)	6×10^{10}
$CsAl(SO_4)_2 \cdot 12H_2O$	Cesium alum	5.0	r.t.	$20–20 \times 10^3$
CsBr	Cesium bromide	6.38	298	1.6×10^3
Cs_2CO_3	Cesium carbonate	6.53	291	2×10^5
CsCl	Cesium chloride	7.2	298	
CsH_2AsO_4	Cesium dihydrogen arsenate (CDA)	4.8	273	9.5×10^9
CsH_2PO_4	Cesium dihydrogen phosphate (CDP)	6.15	285	9.5×10^9
$CsH_3(SeO_3)_2$	Cesium trihydrogen selenite	$\varepsilon_{11} = 80$	273	10^5
		$\varepsilon_{22} = 63$	273	10^5
		$\varepsilon_{33} = 12$	273	10^5
CsI	Cesium iodide	6.31	298	1.6×10^3
$CsNO_3$	Cesium nitrate	$\varepsilon_{11} = \varepsilon_{22} = 9.4$	r.t.	5×10^5
		$\varepsilon_{33} = 8.3$	r.t.	5×10^5
$CsPbCl_3$	Cesium lead chloride	14.37	300	$10^5–10^6$
CuBr	Cuprous bromide	8.0	293	5×10^5
CuCl	Cuprous chloride	9.8 ± 0.5	r.t.	10^3
CuO	Cupric oxide	18.1	r.t.	2×10^6
Cu_2O	Cuprous oxide (Cuprite)	7.60 ± 0.06	r.t.	10^5
$CuSO_4 \cdot 5H_2O$	Cupric sulfate pentahydrate	6.60	r.t.	
EuF_2	Europium fluoride	7.7 ± 0.2	298	$(1–300) \times 10^3$
$Eu_2(MoO_4)_3$	Europium molybdate	9.5	298	
EuS	Europium sulfide	13.10 ± 0.04	80	$5 \times 10^2–10^5$
FeO	Ferrous oxide	14.2	r.t.	2×10^6
Fe_2O_3	Ferric sesquioxide	4.5	r.t.	$10^5–10^7$
Fe_2O_3-α	Ferric sesquioxide (hematite)	12		6×10^{10}
Fe_3O_4	Ferrosoferric oxide (magnetite)	20	r.t.	$10^5–10^7$
GaAs	Gallium arsenide	13.13	300	
		12.90	4	i.r.
GaP	Gallium phosphide	11.1	r.t.	
		10.75 ± 0.1	1.6	i.r.
GaSb	Gallium antimonide	15.69	r.t.	
		15.7	4	i.r.
$Gd_2(MoO_4)_3$	Gadolinium molybdate	$\varepsilon^T = 10$	298	
		$\varepsilon^S = 9.5$	298	10^3
Ge	Germanium	16.0 ± 0.3	4	9.2×10^9
		15.8 ± 0.2	r.t.	$500–3 \times 10^{10}$
GeO_2	Germanium dioxide	$\varepsilon_{11} = \varepsilon_{22} = 7.44$	r.t.	i.r.
HIO_3	Iodic acid	$\varepsilon_{11} = 7.5$	r.t.	10^3
		$\varepsilon_{22} = 12.4$	r.t.	10^3
		$\varepsilon_{33} = 8.1$	r.t.	10^3
$HNH_4(ClCH_2COO)_2$	Hydrogen ammonium dichloroacetate	$\varepsilon_{[102]} = 5.9$	r.t.	10^5
H_2O	Ice I (P = 0 kbar)	99	243	
	Ice III (P = 3 kbar)	117	243	
	Ice V (P = 5 kbar)	114	243	
	Ice VI (P = 8 kbar)	193	243	
HgCl	Mercurous chloride (Calumel)	$\varepsilon_{11} = \varepsilon_{22} = 14.0$	r.t.	10^{12}
$HgCl_2$	Mercuric chloride	6.5	r.t.	10^{12}
HgS	Mercurous sulfide (Cinnabar)	$\varepsilon_{11} = \varepsilon_{22} = 18.0$	r.t.	i.r.
		$\varepsilon_{33} = 32.5$	r.t.	i.r.
HgSe	Mercurous selenide	25.6	r.t.	$10^4–10^6$
I_2	Iodine	$\varepsilon_{11} = 6$	r.t.	$5 \times 10^4–10^7$
		$\varepsilon_{22} = 3$	r.t.	$5 \times 10^4–10^7$

Formula	Name	ε_{ijk}	*T*/K	ν/Hz
		$\varepsilon_{33} = 40$	r.t.	5×10^4–10^7
InAs	Indium arsenide	14.55 ± 0.3	r.t.	i.r.
		15.15	4	i.r.
InP	Indium phosphide	12.61	r.t.	i.r.
InSb	Indium antimonide	17.88	4	i.r.
$KAl(SO_4)_2 \cdot 12H_2O$	Potassium alum	6.5	r.t.	20–20×10^3
KBr	Potassium bromide	4.88	300	
		4.53	4.2	
$KBrO_3$	Potassium bromate	7.3	r.t.	2×10^6
KCN	Potassium cyanide	6.15	r.t.	2×10^6
K_2CO_3	Potassium carbonate	4.96	291	2×10^5
$K_2C_4H_4O_6 \cdot {}^1/_2\,H_2O$	Dipotassium tartrate (DKT)	$\varepsilon_{11} = 6.44$	r.t.	
		$\varepsilon_{22} = 5.80$	r.t.	
		$\varepsilon_{33} = 6.49$	r.t.	
		$\varepsilon_{13} = 0.005$	r.t.	
KCl	Potassium chloride	4.86 ± 0.02	r.t.	5×10^3
		4.50	4.2	
$KClO_3$	Potassium chlorate	5.1	r.t.	2×10^6
$KClO_4$	Potassium perchlorate	5.9	r.t.	2×10^6
K_2CrO_4	Potassium chromate	7.3	r.t.	6×10^7
$KCr(SO_4)_2 \cdot 12H_2O$	Potassium chrome alum	6.5	100–240	175×10^3
KD_2AsO_4	Potassium dideuterium arsenate (KDDA)	$\varepsilon_{11} = 70$	298	
		$\varepsilon_{33} = 31$	298	
KD_2PO_4	Potassium dideuterium phosphate (KDDP)	50 ± 2	297	10^3
KF	Potassium fluoride	6.05	r.t.	2×10^6
KH_2AsO_4	Potassium dihydrogen arsenate (KDA)	$\varepsilon_{11} = 60$	298	
		$\varepsilon_{33} = 24$	298	
KH_2PO_4	Potassium dihydrogen phosphate (KDP)	46	298	10^3
		$\varepsilon_{11} = 42$	r.t.	
		$\varepsilon_{33} = 21$	r.t.	
K_2HPO_4	Dipotassium monohydrogen orthophosphate	9.05	r.t.	2×10^6
KI	Potassium iodide	5.00	r.t.	9.4×10^{10}
KIO_3	Potassium iodate	170	255	10^5
		10	293	10^5
		$\varepsilon_{[101]} \approx 40{,}70$	r.t.	10^5
		16.85	r.t.	2×10^6
$(K,H)Al_3(SiO_4)_3$	Mica (muscovite)	5.4	299	10^2–3×10^9
$(K,H)Mg_3Al(SiO_4)_3$	Mica (Canadian)	$\varepsilon_{11} = \varepsilon_{22} = 6.9$	298	10^2–10^4
		$\varepsilon_{33} = 7.3$	298	10^4
KNO_2	Potassium nitrite	25	305	
KNO_3	Potassium nitrate	4.37	293	2×10^5
$KNbO_3$	Potassium niobate	700	r.t.	
K_3PO_4	Potassium orthophosphate	7.75	r.t.	2×10^6
KSCN	Potassium thiocyanate	7.9	r.t.	2×10^6
K_2SO_4	Potassium sulfate	6.4	r.t.	2×10^6
$K_2S_3O_6$	Potassium trithionate	5.7	293	1.8×10^6
$K_2S_4O_6$	Potassium tetrathionate	5.5	293	1.8×10^6
$K_2S_5O_6 \cdot H_2O$	Potassium pentathionate	7.8	293	1.8×10^6
$K_2S_6O_6$	Potassium hexathionate	7.8	293	1.8×10^6
K_2SeO_4	Potassium selenate	$\varepsilon_{11} = 5.9$	r.t.	10^3
		$\varepsilon_{22} = 7.7$	r.t.	10^3
$KSr_2Nb_5O_{15}$	Potassium strontium niobate	$\varepsilon_{11} = \varepsilon_{11} \approx 1200$	298	
		$\varepsilon_{33} \approx 800$	298	
$KTaNbO_3$	Potassium tantalate niobate (KTN)	34,000	273	10^4
		6,000	293	10^4
$KTaO_3$	Potassium tantalate	242	298	2×10^5
$LaScO_3$	Lanthanum scandate	30	r.t.	
LiBr	Lithium bromide	12.1	r.t.	2×10^6
Li_2CO_3	Lithium carbonate	4.9	291	2×10^5
LiCl	Lithium chloride	11.05	r.t.	2×10^6

Formula	Name	ε_{ijk}	T/K	ν/Hz
LiD	Lithium deuteride	14.0 ± 0.5	r.t.	i.r.
LiF	Lithium fluoride	9.00	298	10^2–10^7
		9.11	353	10^2–10^7
$LiGaO_2$	Lithium metagallate	$\varepsilon_{11}^T = 7.0, \varepsilon_{22}^T = 6.0$	r.t.	
		$\varepsilon_{33}^T = 9.5$	r.t.	
		$\varepsilon_{11}^S = 6.8, \varepsilon_{22}^S = 5.8$	r.t.	
Li^6H	Lithium-6 hydride	13.2 ± 0.5	r.t.	
Li^7H	Lithium-7 hydride	12.9 ± 0.5	r.t.	
$LiH_3(SeO_3)_2$	Lithium trihydrogen selenite	29	298	10^4
		ε_{11} = 13.0	r.t.	
		ε_{22} = 12.9	r.t.	
		ε_{33} = 46	r.t.	
LiI	Lithium iodide	11.03	r.t.	2×10^6
$LiIO_3$	Lithium iodate	ε_{11} = ε_{22} = 65	294.5	10^3
		ε_{33} = 554	298	
$LiNH_4C_4O_6 \cdot H_2O$	Lithium ammonium tartrate (LAT)	$\varepsilon_{11}^T = 7.2$	298	
		$\varepsilon_{22}^T = 8.0$	298	
		$\varepsilon_{33}^T = 6.9$	298	
$LiNa_3CrO_4 \cdot 6H_2O$	Lithium trisodium chromate	8.0	r.t.	10^3
$LiNa_3MoO_4 \cdot 6H_2O$	Lithium trisodium molybdate	ε_{11} = 6.7	r.t.	10^3
		ε_{33} = 5.3	r.t.	10^3
$LiNbO_3$	Lithium niobate	ε_{11} = ε_{22} = 82	298	10^5
		ε_{33}= 30	298	10^5
$Li_2SO_4 \cdot H_2O$	Lithium sulfate monohydrate	ε_{11} = 5.6	298	
		ε_{22} = 10.3	298	
		ε_{33} = 6.5	298	
		ε_{13} = 0.07	298	
$LiTaO_3$	Lithium tantalate	ε_{11} = ε_{22} = 53	r.t.	10^5
		ε_{33} = 46	r.t.	10^5
		$\varepsilon_{11}^S = \varepsilon_{22}^S = 41$	r.t.	
		$\varepsilon_{33}^S = 43$	r.t.	
		$\varepsilon_{11}^T = \varepsilon_{22}^T = 51$	r.t.	
		$\varepsilon_{33}^T = 45$	r.t.	
$LiTlC_4O_6 \cdot H_2O$	Lithium thallium tartrate (LTT)	ε_{11} ≈ 20	80	
$Mg_3B_7O_{13}Cl$	Magnesium borate monochloride (boracite)	ε_{11} = 14.1	r.t.	5×10^5
$MgCO_3$	Magnesium carbonate	8.1	291	2×10^5
$MgNb_2O_6$	Magnesium niobate	ε_{11} = 16.4 ± 0.5	r.t.	(5–500) × 10^3
		ε_{22} = 20.9 ± 0.5	r.t.	(5–500) × 10^3
		ε_{33} = 32.4 ± 0.5	r.t.	(5–500) × 10^3
MgO	Magnesium oxide (Periclase)	9.65	298	10^2–10^8
$(MgO)_xAl_2O_3$	Spinel	8.6	r.t.	
$MgSO_4$	Magnesium sulfate	8.2	r.t.	
$MgSO_4 \cdot 7H_2O$	Magnesium sulfate septahydrate	5.46	r.t.	
$MgTiO_3$	Magnesium titanate	13.5	r.t.	
$MgWO_4$	Magnesium tungstate	ε_{11} = 18.0 ± 1	r.t.	(5–500) × 10^3
		ε_{22} = 18.0 ± 1	r.t.	(5–500) × 10^3
$MnNb_2O_6$	Manganese niobate	ε_{11} = 17.4 ± 2	r.t.	(5–500) × 10^3
		ε_{22} = 16.1 ± 0.5	r.t.	(5–500) × 10^3
		ε_{33} = 30.7 ± 1	r.t.	(5–500) × 10^3
MnO	Manganese oxide (Pyrolusite)	12.8	r.t.	6×10^{10}
MnO_2	Manganese dioxide	~10^4	298	10^4
Mn_2O_3	Manganese sesquioxide	8	r.t.	6×10^{10}
$MnWO_4$	Manganese tungstate	ε_{11} = 19.3 ± 1.3	r.t.	(5–500) × 10^3
		ε_{22} = 14.3 ± 0.5	r.t.	(5–500) × 10^3
		ε_{33} = 16.5 ± 1.1	r.t.	(5–500) × 10^3
$N(CH_3)_4HgBr_3$	Tetramethylammonium tribromomercurate (TTM)	~10	233–373	

Formula	Name	ε_{ijk}	T/K	ν/Hz
$N(CH_3)_4HgI_3$	Tetramethylammonium triiodo mercurate (TTM)	~10	233–373	
$N_4(CH_2)_6$	Hexamethylene tetramine (HMTA)	2.6 ± 0.2	r.t.	10^9–10^{10}
$(ND_4)_2BeF_4$	Deuteroammonium fluoberyllate	ε_{11} = 10	r.t.	
		ε_{22} = 9	r.t.	
		ε_{33} = 9	r.t.	
$(ND_4)_2SO_4$	Deuteroammonium sulfate	ε_{11} = 9	r.t.	
		ε_{22} = 10	r.t.	
		ε_{33} = 9	r.t.	
$(NH_2 \cdot CH_2COOH)_3 \cdot H_2SO_4$	Triglycine sulfate (TGS)	ε_{11} = 9	273	10^4
		ε_{22} = 30	273	10^4
		ε_{33} = 6.5	273	10^4
$(NH_2 \cdot CH_2COOH)_3 \cdot H_2SeO_4$	Triglycine selenate (TGSe)	200	293	1.6×10^3
$(NH_2 \cdot CH_2COOH)_3 \cdot H_2BeF_4$	Triglycine fluorberyllate (TGFB)	ε_{22} = 12	273	10^4
$NH_4Al(SO_4)_2 \cdot 12H_2O$	Ammonium alum	6	r.t.	10^{12}
$(NH_4)_2BeF_4$	Ammonium fluorberyllate	$\varepsilon_{11} = \varepsilon_{22}$ = 7.8	123	10^5
		ε_{33} = 7.1	123	10^5
		$\varepsilon_{11} = \varepsilon_{22}$ = 8.8	293	10^5
		ε_{33} = 9.2	293	10^5
NH_4Br	Ammonium bromide	7.1	r.t.	7×10^5
NH_4I	Ammonium iodide	9.8	r.t.	
$(NH_4)_2C_2H_6O_6$	Ammonium tartrate	ε_{11} = 6.45	r.t.	10^3
		ε_{22} = 6.8	r.t.	10^3
		ε_{33} = 6.0	r.t.	10^3
$(NH_4)_2Cd_2(SO_4)_3$	Ammonium cadmium sulfate	10.0	r.t.	10^4
NH_4Cl	Ammonium chloride	6.9	r.t.	7×10^5
$NH_4(ClCH_2COO)$	Ammonium monochloroacetate	5	r.t.	2×10^6
$NH_4Cr(SO_4)_2 \cdot 12H_2O$	Ammonium chrome alum	6.5	r.t.	175×10^3
NH_4HSO_4	Ammonium bisulfate	165	273	5×10^4
$NH_4H_2AsO_4$	Ammonium dihydrogen arsenate (ADA)	5.1	265	9.5×10^9
		$\varepsilon_{11} = \varepsilon_{22}$ = 85	298	10^3
		ε_{33} = 22	298	
$NH_4H_2PO_4$	Ammonium dihydrogen phosphate (ADP)	$\varepsilon_{11} = \varepsilon_{22}$ = 57.1 ± 0.6	294.5	10^5–35×10^9
		ε_{33} = 14.0 ± 0.3	294	10^5–36×10^9
$ND_4D_2PO_4$	Ammonium dideuterium phosphate (ADDP)	$\varepsilon_{11} = \varepsilon_{22}$ = 74, ε_{33} = 24	300	
NH_4NO_3	Ammonium nitrate	10.7	322	$(5–50) \times 10^3$
$(NH_4)_2SO_4$	Ammonium sulfate	$\varepsilon_{11} = \varepsilon_{22}$ = 8.0	123	10^5
		ε_{33} = 6.3	123	10^5
		$\varepsilon_{11} = \varepsilon_{22}$ = 10.0	293	10^5
		ε_{33} = 9.3	293	10^5
$(NH_4)_2UO_2(C_2O_4)_2$	Ammonium uranyl oxalate	8.03	r.t.	10^4–3.3×10^9
$(NH_4)_2UO_2(C_2O_4)_2 \cdot 3H_2O$	Ammonium uranyl oxalate trihydrate	6.06	r.t.	10^4–3.3×10^9
NaBr	Sodium bromide	6.44	298	1.6×10^3
$NaBrO_3$	Sodium bromate	ε^T_{11} = 5.70	298	10^3
NaCN	Sodium cyanide	7.55	293	10^5
$NaCO_3$	Sodium carbonate	8.75	291	2×10^5
$NaCO_3 \cdot 10H_2O$	Sodium carbonate decahydrate	5.3	r.t.	6×10^7
NaCl	Sodium chloride	5.9	298	10^2–10^7
		5.45	4.2	
$NaClO_3$	Sodium chlorate	ε^T_{11} = 5.76	301	10^3
		5.28	r.t.	10^3
$NaClO_4$	Sodium perchlorate	5.76	r.t.	10^3
NaF	Sodium fluoride	5.08 ± 0.02	r.t.	5×10^3
$NaH_3(SeO_3)_2$	Sodium trihydrogen selenite	ε_{11} ≈ 75	273	2×10^5
$NaD_3(SeO_3)_2$	Sodium trideuterium selenite	ε_{11} ≈ 220	273	2×10^5
NaI	Sodium iodide	7.28 ± 0.03	r.t.	

Formula	Name	ε_{ijk}	*T*/K	ν/Hz
$NaK(C_4H_2D_2O_6)\cdot 4D_2O$	Sodium potassium tartrate tetradeutrate	$\varepsilon_{11} = 70$	273	10^3
	(double deuterated Rochelle salt)	$\varepsilon_{22} = 8.9$	273	10^3
$NaK(C_4H_4O_6)\cdot 4H_2O$	Sodium potassium tartrate tetrahydrate	$\varepsilon_{11} = 170$	273	10^3
	(Rochelle salt)	$\varepsilon_{22} = 9.1$	273	10^3
$NaNH_4(C_4H_4O_6)\cdot 4H_2O$	Sodium ammonium tartrate	$\varepsilon_{11} = 8.4$	298	
	(Ammonium Rochelle salt)	$\varepsilon_{22} = 9.2$	298	
		$\varepsilon_{33} = 9.5$	298	
$NaNbO_3$	Sodium niobate	$\varepsilon_{33} = 670 \pm 13$	r.t.	
		$\varepsilon_{11} = \varepsilon_{22} = 76 \pm 2$	r.t.	
$NaNO_2$	Sodium nitrite	$\varepsilon_{11} = 7.4$	r.t.	5×10^5
		$\varepsilon_{22} = 5.5$	r.t.	5×10^5
		$\varepsilon_{33} = 5.0$	r.t.	5×10^5
$NaNO_3$	Sodium nitrate	6.85	292	2×10^5
$NaSO_4$	Sodium sulfate	7.90	r.t.	
$NaSO_4\cdot 10H_2O$	Sodium sulfate decahydrate	5.0	r.t.	
$Na_2SO_4\cdot 5H_2O$	Sodium sulfate pentahydrate	7	250–290	$300–10^4$
$Na_4UO_2(C_2O_4)_3$	Sodium uranyl oxalate	5.18	r.t.	
$NdAlO_3$	Neodymium aluminate	17.5	r.t.	
$NdScO_3$	Neodymium scandate	27	r.t.	
$Ni_3B_7O_{13}I$	Nickel iodine boracite	$\varepsilon_{11} = 14$	260	
$NiNb_2O_6$	Nickel niobate	$\varepsilon_{11} = 16.0 \pm 0.5$	r.t.	$(5–500)\times 10^3$
		$\varepsilon_{22} = 23.8 \pm 1.8$	r.t.	$(5–500)\times 10^3$
		$\varepsilon_{33} = 31.3 \pm 2.5$	r.t.	$(5–500)\times 10^3$
NiO	Nickel oxide	11.9	298	10^5
$NiSO_4\cdot 6H_2O$	Nickel sulfate hexahydrate	$\varepsilon_{11} = 6.2$	r.t.	
		$\varepsilon_{33} = 6.8$	r.t.	
$NiWO_4$	Nickel tungstate	$\varepsilon_{11} = 17.4 \pm 2.4$	r.t.	$(5–500)\times 10^3$
		$\varepsilon_{22} = 13.6 \pm 1.0$	r.t.	$(5–500)\times 10^3$
		$\varepsilon_{33} = 19.7 \pm 0.6$	r.t.	$(5–500)\times 10^3$
P	Phosphorous (red)	4.1	r.t.	10^8
	Phosphorous (yellow)	3.6	r.t.	10^8
$[P(CH_3)_4]HgBr_3$	Tetramethylphosphonium tribromomercurate (TTM)	~10	233–373	
$PbBr_2$	Lead bromide	>30	293	$(0.5–3)\times 10^6$
$PbCO_3$	Lead carbonate	18.6	288	10.8
$Pb(C_2H_3O_2)_2$	Lead acetate	2.6	290–295	10^6
$PbCl_2$	Lead chloride	33.5	273	$(0.5–3)\times 10^6$
Pb_2CoWO_6	Lead cobalt tungstate	~250	r.t.	
PbF_2	Lead fluoride	26.3	r.t.	
$PbHfO_3$	Lead hafnate	390	300	10^5
		185	400	
PbI_2	Lead iodide	20.8	293	$(0.5–3)\times 10^6$
$Pb_3MgNb_2O_9$	Lead magnesium niobate	10,000	297	
$PbMoO_4$	Lead molybdate	$\varepsilon_{11} = 34.0 \pm 0.4$	297.5	1.6 ± 10^3
		$\varepsilon_{33} = 40.6 \pm 0.2$	297.5	1.6 ± 10^3
$Pb(NO_3)_2$	Lead nitrate	16.8	r.t.	$(0.5–3)\times 10^6$
$PbNb_2O_6$	Lead niobate	$\varepsilon^T_{33} = 180$	298	
PbO	Lead oxide	25.9	r.t.	2×10^6
PbS	Lead sulfide (Galena)	190	77	i.r.
		200 ± 35	r.t.	i.r.
$PbSO_4$	Lead sulfate	14.3	290— 295	10^6
PbSe	Lead selenide	280	r.t.	i.r.
$PbTa_2O_6$	Lead metatantalate	$\varepsilon_{11} = \varepsilon_{22} \approx 300$	r.t.	10^4
		$\varepsilon_{33} = 150$	r.t.	10^4
PbTe	Lead telluride	450	r.t.	i.r.
		40	77	$10^4–15 \times 10^4$
		430	4.2	$10^4–15 \times 10^4$
$PbTiO_3$	Lead titanate	~200	r.t.	10^3
$PbWO_4$	Lead tungstate	$\varepsilon_{11} = \varepsilon_{22} = 23.6 \pm 0.3$	297.5	1.59×10^3
		$\varepsilon_{33} = 31.0 \pm 0.4$	297.5	1.59×10^3
$Pb(Zn_{1/3}Nb_{2/3})O_3$	Lead zinc niobate	7	300	10^3, 300×10^3

Formula	Name	ε_{ijk}	T/K	ν/Hz
$PbZrO_3$	Lead zirconate	200	400	
$RbAl(SO_4)_2 \cdot 12H_2O$	Rubidium alum	5.1	r.t.	10^{12}
RbBr	Rubidium bromide	4.83	300	
Rb_2CO_3	Rubidium carbonate	4.87 ± 0.02	r.t.	5×10^3
RbCl	Rubidium chloride	4.91 ± 0.02	r.t.	5×10^3
$RbCr(SO_4)_2 \cdot 12H_2O$	Rubidium chrome alum	5.0	r.t.	10^{12}
RbF	Rubidium fluoride	5.91	r.t.	2×10^6
$RbHSO_4$	Rubidium bisulfate	$\varepsilon_{11} = 7$	r.t.	10^5
		$\varepsilon_{22} = 8$	r.t.	10^5
		$\varepsilon_{33} = 10$	r.t.	10^5
RbH_2AsO_4	Rubidium dihydrogen arsenate (RDA)	3.90	273	9.5×10^9
RbH_2PO_4	Rubidium dihydrogen phosphate (RDP)	6.15	285	9.5×10^9
RbI	Rubidium iodide	4.94 ± 0.02	r.t.	5×10^3
$RbInSO_4$	Rubidium indium sulfate	6.85	r.t.	
$RbNO_3$	Rubidium nitrate	20—380	433–488	10^6
		30	488–538	10^6
S	Sulfur	$\varepsilon_{11} = 3.75$	298	10^2–10^3
		$\varepsilon_{22} = 3.95$	298	10^2–10^3
		$\varepsilon_{33} = 4.44$	298	10^2–10^3
	(sublimed)	3.69	298	10^2–10^3
$SC(NH_2)_2$	Thiourea	$\varepsilon_{11} = \varepsilon_{22} \approx 3$	77–300	10^3
		$\varepsilon_{22} = 35$	300	10^3
Sb_2O_3	Antimonous sesquioxide	12.8	r.t.	$(1.5–2) \times 10^3$
Sb_2S_3	Antimonous sulfide (stibnite)	$\varepsilon_{11} = \varepsilon_{33} = 15$	r.t.	10^3
		$\varepsilon_{33} = 180$	r.t.	10^3
Sb_2Se_3	Antimonous selenide	~110	r.t.	$(10–16.5) \times 10^9$
SbSI	Antimonous sulfide iodide	2000	273	10^5
		$\varepsilon_{11} = \varepsilon_{22} \approx 25$	r.t.	10^3–10^5
		$\varepsilon_{33} \approx 5 \times 10^4$	295	10^3–10^5
Se	Selenium (monocrystal)	$\varepsilon_{11} = \varepsilon_{22} = 11$	300	24×10^9
		$\varepsilon_{33} = 21$	300	24×10^9
	(amorphous)	6.0	298	10^2—10^{10}
Si	Silicon	12.1	4.2	10^7—10^9
SiC	Silicon carbide			
	cubic	9.72	r.t.	i.r.
	6H	$\varepsilon_{11} = \varepsilon_{22} = 9.66$	r.t.	i.r.
		$\varepsilon_{33} = 10.03$	r.t.	i.r.
		9.7 ± 0.1	1.8	i.r.
Si_3N_4	Silicon nitride	4.2 (film)	r.t.	10^3
SiO	Silicon monoxide	5.8	r.t.	10^3
SiO_2	Silicon dioxide	$\varepsilon_{11} = 4.42$	r.t.	9.4×10^{10}
		$\varepsilon_{22} = 4.41$	r.t.	9.4×10^{10}
		$\varepsilon_{33} = 4.60$	r.t.	9.4×10^{10}
$Sm_2(MoO_4)_3$	Samarium molybdate	12	298	
SnO_2	Stannic dioxide	$\varepsilon_{11} = \varepsilon_{22} = 14 \pm 2$	r.t.	10^4–10^{10}
		$\varepsilon_{33} = 9.0 \pm 0.5$	r.t.	10^4–10^{10}
SnSb	Tin antimonide	147	r.t.	10^4–10^6
SnTe	Tin telluride	1770 ± 300	r.t.	i.r.
$Sr(COOH)_2 \cdot 2H_2O$	Strontium formate dihydrate	6.1	r.t.	10^3
$SrCO_3$	Strontium carbonate	8.85	298	2×10^5
$SrCl_2$	Strontium chloride	9.19	r.t.	
$SrCl_2 \cdot 6H_2O$	Strontium chloride hexahydrate	8.52	r.t.	
SrF_2	Strontium fluoride	6.50	300	5×10^2–10^{11}
$SrMoO_4$	Strontium molybdate	$\varepsilon_{11} = \varepsilon_{22} = 31.7 \pm 0.2$	297.5	1.59×10^3
		$\varepsilon_{33} = 41.7 \pm 0.2$	297.5	1.59×10^3
$Sr(NO_3)_2$	Strontium nitrate	5.33	292	2×10^5
$Sr_2Nb_2O_7$	Strontium niobate	$\varepsilon_{11} = 75$	r.t.	10^3
		$\varepsilon_{22} = 46$	r.t.	10^3
		$\varepsilon_{33} = 43$	r.t.	10^3
SrO	Strontium oxide	13.3 ± 0.3	273	2×10^6
SrS	Strontium sulfide	11.3	r.t.	7.25×10^6

Formula	Name	ε_{ijk}	*T*/K	ν/Hz
$SrSO_4$	Strontium sulfate	11.5	r.t.	
$SrTiO_3$	Strontium titanate	332	298	10^3
		2080	78	10^3
$SrWO_4$	Strontium tungstate	$\varepsilon_{11} = \varepsilon_{22} = 25.7 \pm 0.2$	297.5	1.6×10^3
		$\varepsilon_{33} = 34.1 \pm 0.2$	297.5	1.6×10^3
Ta_2O_5	Tantalum pentoxide (tantala)			
	α phase	$\varepsilon_{11} = \varepsilon_{22} = 30$	77	10^3
		$\varepsilon_{33} = 65$	77	10^3
	β phase	24	292	10^3
$Tb(MoO_4)_3$	Terbium molybdate	11	298	
		$\varepsilon_{11} = \varepsilon_{22} = 33$	100–200	9.4×10^9
		$\varepsilon_{33} = 53$	100–200	9.4×10^9
Te	Tellurium	$\varepsilon_{11} = \varepsilon_{22} = 33$	r.t.	
		$\varepsilon_{33} = 54$	r.t.	
	polycrystalline	27.5	r.t.	i.r.
	monocrystalline	28.0	r.t.	i.r.
ThO_2	Thorium dioxide	18.9 ± 0.4	r.t.	3×10^5
TiO_2	Titanium dioxide (rutile)	$\varepsilon_{11} = \varepsilon_{22} = 86$	300	10^4–10^6
		$\varepsilon_{33} = 170$	300	10^4–10^6
Ti_2O_3	Titanium sesquioxide	30	77	6×10^{10}
TlBr	Thallium bromide	30	293	10^3–10^7
TlCl	Thallous chloride	32.2 ± 0.2	293	10^3–10^5
TlI	Thallous iodide (orthorhombic)	20.7 ± 0.2	293	10^4
		37.3	193	10^7
$TlNO_3$	Thallous nitrate	16.5	293	5×10^5
$TlSO_4$	Thallous sulfate	25.5	293	5×10^5
UO_2	Uranium dioxide	24	r.t.	3×10^5
WO_3	Tungsten trioxide	300		
$YMnO_3$	Yttrium manganate	20	r.t.	2×10^7
Y_2O_3	Yttrium sesquioxide	10	r.t.	10^6
$YbMnO_3$	Ytterbium manganate	20	r.t.	2×10^7
Yb_2O_3	Ytterbium sesquioxide	5.0 (film)	r.t.	10^3
ZnO	Zinc monoxide	$\varepsilon^S_{11} = 8.33$	r.t.	
		$\varepsilon^S_{33} = 8.84$	r.t.	
		$\varepsilon^T_{11} = 9.26$	r.t.	
		$\varepsilon^T_{33} = 11.0$	r.t.	
		$\varepsilon_{11} = 9.26$	r.t.	
		$\varepsilon_{33} = 8.2$	r.t.	
		8.15	r.t.	i.r.
ZnS	Zinc sulfide	$\varepsilon^S_{11} = 8.08 \pm 2\%$	77	10^4
		$\varepsilon^S_{11} = 8.32 \pm 2\%$	298	10^4
		$\varepsilon^T_{11} = 8.14 \pm 2\%$	77	10^4
		$\varepsilon^T_{11} = 8.37 \pm 2\%$	298	10^4
ZnSe	Zinc selenide	$\varepsilon^T_{11} = \varepsilon^S_{11} = 9.12 \pm 2\%$	298	10^4
ZnTe	Zinc telluride	$\varepsilon^T_{11} = \varepsilon^S_{11} = 10.10 \pm 2\%$	r.t.	
$ZnWO_4$	Zinc tungstate	$\varepsilon_{22} = 16.1 \pm 0.5$	r.t.	$(5–500) \times 10^3$
ZrO_2	Zirconium dioxide (zirconia)	12.5	r.t.	2×10^6

CURIE TEMPERATURE OF SELECTED FERROELECTRIC CRYSTALS

H. P. R. Frederikse

The following table lists the major ferroelectric crystals and their Curie temperatures, T_C.

Reference

Young, K. F. and Frederikse, H. P. R., *J. Phys. Chem. Ref. Data*, 2, 313, 1973.

Name or acronym	Formula	T_C/K
Potassium dihydrogen phosphate group		
KDP	KH_2PO_4	123
KDA	KH_2AsO_4	97
KDDP	KD_2PO_4	213
KDDA	KD_2AsO_4	162
RDP	RbH_2PO_4	146
RDA	RbH_2AsO_4	111
RDDP	RbD_2PO_4	218
RDDA	RbD_2AsO_4	178
CDP	CsH_2PO_4	159
CDA	CsH_2AsO_4	143
CDDA	CsD_2AsO_4	212
Rochelle salt group		
Rochelle salt	$NaKC_4H_4O_6{\cdot}4H_2O$	255–297
Deuterated Rochelle salt	$NaKC_4H_2D_2O_6{\cdot}4H_2O$	251–308
Ammonium Rochelle salt	$NaNH_4C_4H_4O_6{\cdot}4H_2O$	109
LAT	$LiNH_4C_4H_4O_6{\cdot}H_2O$	106
Triglycine sulfate group		
TGS	$(NH_2CH_2COOH)_3{\cdot}H_2SO_4$	322
TGSe	$(NH_2CH_2COOH)_3{\cdot}H_2SeO_4$	295
TGFB	$(NH_2CH_2COOH)_3{\cdot}H_2BeF_4$	346
AFB	$(NH_4)_2BeF_4$	176
HADA	$HNH_4(ClCH_2COO)_2$	128
Perovskites and related compounds		
Barium titanate	$BaTiO_3$	406, 278, 193
Lead titanate	$PbTiO_3$	765
Potassium niobate	$KNbO_3$	712
Potassium tantalate niobate	$KTa_{2/3}Nb_{1/3}O_3$	241, 220, 170
Lithium niobate	$LiNBO_3$	1483
Lithium tantalate	$LiTaO_3$	891
Barium titanium niobate	$Ba_6Ti_2Nb_8O_{30}$	521
Ba-Na niobate ("Bananas")	$Ba_2NaNb_5O_{15}$	833
Potassium iodate	KIO_3	485, 343, 257–263, 83
Lithium iodate	$LiIO_3$	529
Potassium nitrate	KNO_3	397
Sodium nitrate	$NaNO_3$	548
Rubidium nitrate	$RbNO_3$	437–487
Miscellaneous compounds		
Cesium trihydrogen selenite	$CsH_3(SeO_3)_2$	143
Lithium trihydrogen selenite	$LiH_3(SeO_3)_2$	$T_C > T_{mp}$
Potassium selenate	K_2SeO_4	93
Methyl ammonium alum (MASD)	$CH_3NH_3Al(SO_4)_2{\cdot}12H_2O$	177
Ammonium cadmium sulfate	$(NH_4)_2Cd_2(SO_4)_3$	95
Ammonium bisulfate	$(NH_4)HSO_4$	271
Ammonium sulfate	$(NH_4)_2SO_4$	224
Ammonium nitrate	NH_4NO_3	398, 357, 305, 255
Colemanite	$CaB_3O_4(OH)_3{\cdot}H_2O$	266
Cadmium pyroniobite	$Cd_2Nb_2O_7$	185
Gadolinium molybdate	$Gd_2(MoO_4)_3$	432

PROPERTIES OF ANTIFERROELECTRIC CRYSTALS

H. P. R. Frederikse

Some important antiferroelectric crystals are listed here with their Curie temperatures T_C. The last column gives the constant T_0 which appears in the Curie–Weiss law describing the dielectric constant of these materials above the Curie temperature:

$$\varepsilon = \text{const.}/(T - T_0)$$

Name or acronym	**Formula**	T_C/K	T_0/K
ADP	$NH_4H_2PO_4$	148	
ADA	$NH_4H_2AsO_4$	216	
ADDP	$NH_4D_2PO_4$	242, 245	
ADDA	$NH_4D_2AsO_4$	299	
A_dDDP	$ND_4D_2PO_4$	243	
A_dDDA	$ND_4D_2AsO_4$	304	
Sodium niobate	$NaNbO_3$	911, 793	
Lead hafnate	$PbHfO_3$	476	378
Lead zirconate	$PbZrO_3$	503	475
Lead metaniobate	$PbNb_2O_6$	843	530
Lead metatantalate	$PbTa_2O_6$	543	533
Tungsten trioxide	WO_3	1010	
Potassium strontium niobate	$KSr_2Nb_5O_{15}$	427	413
Sodium nitrite	$NaNO_2$	437	437
Sodium trihydrogen selenite	$NaH_3(SeO_3)_2$	193	192
Sodium trideuterium selenite	$NaD_3(SeO_3)_2$	271	245
Ammonium trihydrogen periodate	$(NH_4)_2H_3IO_6$	245	

DIELECTRIC CONSTANTS OF GLASSES

Type	Dielectric constant at 100 MHz (20 °C)	Volume resistivity (in MΩ cm at 350 °C)	Loss factor[a]
Corning 0010	6.32	10	0.015
Corning 0080	6.75	0.13	0.058
Corning 0120	6.65	100	0.012
Pyrex 1710	6.00	2,500	0.025
Pyrex 3320	4.71	–	0.019
Pyrex 7040	4.65	80	0.013
Pyrex 7050	4.77	16	0.017
Pyrex 7052	5.07	25	0.019
Pyrex 7060	4.70	13	0.018
Pyrex 7070	4.00	1,300	0.0048
Vycor 7230	3.83	–	0.0061
Pyrex 7720	4.50	16	0.014
Pyrex 7740	5.00	4	0.040
Pyrex 7750	4.28	50	0.011
Pyrex 7760	4.50	50	0.0081
Vycor 7900	3.9	130	0.0023
Vycor 7910	3.8	1,600	0.00091
Vycor 7911	3.8	4,000	0.00072
Corning 8870	9.5	5,000	0.0085
G. E. Clear (silica glass)	3.81	4,000–30,000	0.00038
Quartz (fused)	3.75 (4.1 at 1 MHz)	–	0.0002 (1 MHz)

[a] Power factor × dielectric constant equals loss factor.

PROPERTIES OF SUPERCONDUCTORS

L. I. Berger and B. W. Roberts

The following tables include superconductive properties of selected elements, compounds, and alloys. Individual tables are given for thin films, elements at high pressures, superconductors with high critical magnetic fields, and high critical temperature superconductors.

The historically first observed and most distinctive property of a superconductive body is the near total loss of resistance at a critical temperature (T_c) that is characteristic of each material. Figure 1(a) below illustrates schematically two types of possible transitions. The sharp vertical discontinuity in resistance is indicative of that found for a single crystal of a very pure element or one of a few well-annealed alloy compositions. The broad transition, illustrated by broken lines, suggests the transition shape seen for materials that are not homogeneous and contain unusual strain distributions. Careful testing of the resistivity limit for superconductors shows that it is less than 4×10^{-23} ohm cm, while the lowest resistivity observed in metals is of the order of 10^{-13} ohm cm. If one compares the resistivity of a superconductive body to that of copper at room temperature, the superconductive body is at least 10^{17} times less resistive.

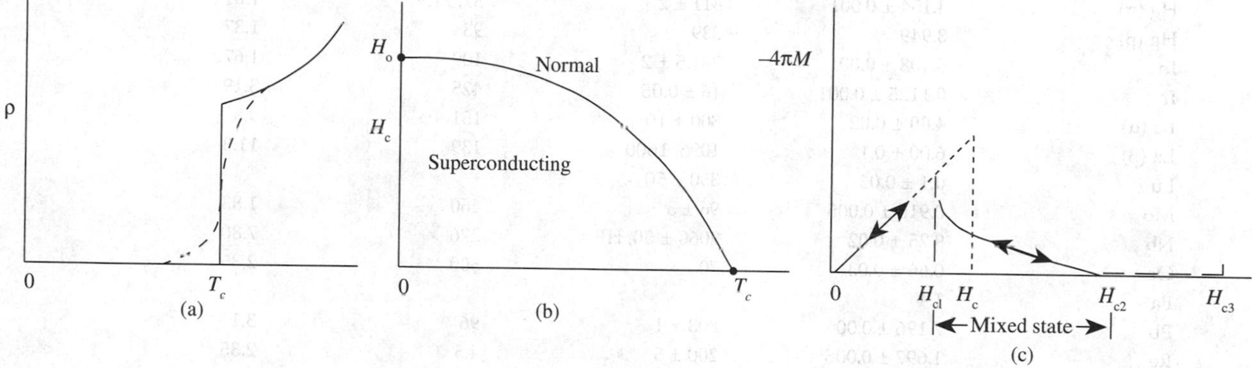

FIGURE 1. Physical properties of superconductors. (a) Resistivity vs. temperature for a pure and perfect lattice (solid line); impure and/or imperfect lattice (broken line). (b) Magnetic-field temperature dependence for Type-I or "soft" superconductors. (c) Schematic magnetization curve for "hard" or Type-II superconductors.

The temperature interval ΔT_c, over which the transition between the normal and superconductive states takes place, may be of the order of as little as 2×10^{-5} K or several K in width, depending on the material state. The narrow transition width was attained in 99.9999% pure gallium single crystals.

A Type-I superconductor below T_c, as exemplified by a pure metal, exhibits perfect diamagnetism and excludes a magnetic field up to some critical field H_c, whereupon it reverts to the normal state as shown in the *H-T* diagram of Figure 1(b).

The magnetization of a typical high-field superconductor is shown in Figure 1(c). The discovery of the large current-carrying capability of Nb_3Sn and other similar alloys has led to an extensive study of the physical properties of these alloys. In brief, a high-field superconductor, or Type-II superconductor, passes from the perfect diamagnetic state at low magnetic fields to a mixed state and finally to a sheathed state before attaining the normal resistive state of the metal. The magnetic field values separating the four stages are given as H_{c1}, H_{c2}, and H_{c3}. The superconductive state below H_{c1} is perfectly diamagnetic, identical to the state of most pure metals of the "soft" or Type-I superconductor. Between H_{c1} and H_{c2} a "mixed superconductive state" is found in which fluxons (a minimal unit of magnetic flux) create lines of normal flux in a superconductive matrix. The volume of the normal state is proportional to $-4\pi M$ in the "mixed state" region. Thus at H_{c2} the fluxon density has become so great as to drive the interior volume of the superconductive body completely normal. Between H_{c2} and H_{c3} the superconductor has a sheath of current-carrying superconductive material at the body surface, and above H_{c3} the normal state exists. With several types of careful measurement, it is possible to determine H_{c1}, H_{c2}, and H_{c3}. Table 6 contains some of the available data on high-field superconductive materials.

High-field superconductive phenomena are also related to specimen dimension and configuration. For example, the Type-I superconductor, Hg, has entirely different magnetization behavior in high magnetic fields when contained in the very fine sets of filamentary tunnels found in an unprocessed Vycor glass. The great majority of superconductive materials are Type-II. The elements in very pure form and a very few precisely stoichiometric and well annealed compounds are Type I with the possible exceptions of vanadium and niobium.

Metallurgical Aspects. The sensitivity of superconductive properties to the material state is most pronounced and has been used in a reverse sense to study and specify the detailed state of alloys. The mechanical state, the homogeneity, and the presence of impurity atoms and other electron-scattering centers are all capable of controlling the critical temperature and the current-carrying capabilities in high-magnetic fields. Well-annealed specimens tend to show sharper transitions than those that are strained or inhomogeneous. This sensitivity to mechanical state underlines a general problem in the tabulation of properties for superconductive materials. The occasional divergent values of the critical temperature and of the critical fields quoted for a Type-II superconductor may lie in the variation in sample preparation. Critical temperatures of materials studied early in the history of superconductivity must be evaluated in light of the probable metallurgical state of the material, as well as the availability of less pure starting elements. It has been noted that recent work has given extended consideration to the metallurgical aspects of sample preparation.

Symbols in tables: T_c: Critical temperature; H_o: Critical magnetic field in the $T = 0$ limit; θ_D: Debye temperature; and γ: Electronic specific heat.

TABLE 1. Selective Properties of Superconductive Elements

Element	T_c(K)	H_o(oersted)	θ_D(K)	γ(mJ mol^{-1}K^{-1})
Al	1.175 ± 0.002	104.9 ± 0.3	420	1.35
Am* (α,?)	0.6			
Am* (β,?)	1.0			
Be	0.026			0.21
Cd	0.517 ± 0.002	28 ± 1	209	0.69
Ga	1.083 ± 0.001	58.3 ± 0.2	325	0.60
Ga (β)	5.9, 6.2	560		
Ga (γ)	7	950, HF[a]		
Ga (Δ)	7.85	815, HF		
Hf	0.128	12.7		2.21
Hg (α)	4.154 ± 0.001	411 ± 2	87, 71.9	1.81
Hg (β)	3.949	339	93	1.37
In	3.408 ± 0.001	281.5 ± 2	109	1.672
Ir	0.1125 ± 0.001	16 ± 0.05	425	3.19
La (α)	4.88 ± 0.02	800 ± 10	151	9.8
La (β)	6.00 ± 0.1	1096, 1600	139	11.3
Lu	0.1 ± 0.03	350 ± 50		
Mo	0.915 ± 0.005	96 ± 3	460	1.83
Nb	9.25 ± 0.02	2060 ± 50, HF	276	7.80
Os	0.66 ± 0.03	70	500	2.35
Pa	1.4			
Pb	7.196 ± 0.006	803 ± 1	96	3.1
Re	1.697 ± 0.006	200 ± 5	4.5	2.35
Ru	0.49 ± 0.015	69 ± 2	580	2.8
Sn	3.722 ± 0.001	305 ± 2	195	1.78
Ta	4.47 ± 0.04	829 ± 6	258	6.15
Tc	7.8 ± 0.1	1410, HF	411	6.28
Th	1.38 ± 0.02	1.60 ± 3	165	4.32
Ti	0.40 ± 0.04	56	415	3.3
Tl	2.38 ± 0.02	178 ± 2	78.5	1.47
U	0.2			
V	5.40 ± 0.05	1408	383	9.82
W	0.0154 ± 0.0005	1.15 ± 0.03	383	0.90
Zn	0.85 ± 0.01	54 ± 0.3	310	0.66
Zr	0.61 ± 0.15	47	290	2.77
Zr (ω)	0.65, 0.95			

TABLE 2. Range of Critical Temperatures Observed for Superconductive Elements in Thin Films Condensed Usually at Low Temperatures

Element	T_c Range (K)	Comments
Al	1.15–5.7	HF[a]
Be	5–9.75	HF
Bi	6.17–6.6	
Cd		
(Disordered)	0.79–0.91	
(Ordered)	0.53–0.59	
Ga	2.5–8.5	HF
Hg	3.87–4.5	
In	2.2–5.6	HF
La	3.55–6.74	
Mo	3.3–8.0	
Nb	2.0–10.1	
Pb	1.8–7.5	
Re	1.7–7	
Sn	3.5–6	
Ta	<1.7–4.51	HF[a]
Tc	4.6–7.7	
Ti	1.3 Max	
Tl	2.33–2.96	
V	1.8–6.02	
W	<1.0–4.1	
Zn	0.77–1.9	

[a] HF denotes high magnetic field superconductive properties.

TABLE 3. Elements Exhibiting Superconductivity Under or After Application of High Pressure

Element	T_c Range (K)	Pressure (kbar)
Al	1.98–0.075	0–62
As	0.31–0.5	220–140
	0.2–0.25	140–100
Ba II	1–1.8	55–85
III	1.8–5	85–144
IV	4.5–5.4	144–190
Bi II	3.9	25–27
III	6.55–7.25	28–38
IV	7.0, 8.7–6.0	43, 43–62
V	6.7, 8.3	48–80
VI	8.55	90, 92–101
VII(?)	8.2	30
Ce (α)	0.020–0.045	20–35
Ce (α′)	1.9–1.3	45–125
Cs V	1.5	>125
Ga II	6.38	≥35
II′	7.5	≥35 then P removed
Ge	5.35	115
Lu	0.022–1.0	45–190
P	5.8	170
Pb II	3.55	160
Re II	2.3 Max.	"Plastic" compression
Sb (prepared 120 kbar, held below 77K)	2.6–2.7	
Sb II	3.55–3.40	85–150
Se II	6.75, 6.95	130
Si	6.7–7.1	120–130
Sn II	5.2–4.85	125–160
III	5.30	113
Te II	2.4–5.1	38–55
	4.1–4.2	53–62
IV	4.72–4	63–80
( )	3.3–2.8	100–260
Tl (cubic form)	1.45	35
(hexagonal form)	1.95	35
U	2.4–0.4	
Y	1.7–2.5	
Zr (omega form, metastable)	1–1.7	

TABLE 4. Superconductive Compounds and Alloys

All compositions are denoted on an atomic basis, i.e., AB, AB_2, or AB_3 for compounds, unless noted. Solid solutions or odd compositions may be denoted as A_zB_{1-z} or A_zB. A series of three or more alloys is indicated as A_xB_{1-x} or by actual indication of the atomic fraction range, such as $A_{0-0.6}B_{1-0.4}$. The critical temperature of such a series of alloys is denoted by a range of values or possibly the maximum value.

The selection of the critical temperature from a transition in the effective permeability, or the change in resistance, or possibly the incremental changes in frequency observed by certain techniques is not often obvious from the literature. Most authors choose the mid-point of such curves as the probable critical temperature of the idealized material, while others will choose the highest temperature at which a deviation from the normal state property is observed. In view of the previous discussion concerning the variability of the superconductive properties as a function of purity and other metallurgical aspects, it is recommended that appropriate literature be checked to determine the most probable critical temperature or critical field of a given alloy.

A very limited amount of data on critical fields, H_0, is available for these compounds and alloys; these values are given at the end of the table.

A. Superconductors with $T_c < 10$ K

Substance	T_c, K	Crystal structure type
$Ag_{3.3}Al$	0.34	A12-cI58 (Mn)
$Ag_xAl_yZn_{1-x-y}$	0.15	Cubic
$AgBi_2$	2.87–3.0	
$Ag_7F_{0.25}N_{0.75}O_{10.25}$	0.85–0.90	
Ag_2F	0.0.066	
Ag_7FO_8	0.3	Cubic
$Ag_{0.8-0.3}Ga_{0.2-0.7}$	6.5–8	
Ag_4Ge	0.85	Hex., c.p.
$Ag_{0.438}Hg_{0.562}$	0.64	$D8_2$
$AgIn_2$	~2.4	C16
$Ag_{0.1}In_{0.9}Te$ ($n = 1.4 \times 10^{22}$)*	1.2–1.89	B1
$Ag_{0.2}In_{0.8}Te$ ($n = 1.07 \times 10^{22}$)	0.77–1.00	B1
AgLa	0.94	B2-cP2 (CsCl)
AgLa (9.5 kbar)	1.2	B2
AgLu	0.33	B2-cP2
$AgMo_4S_5$	9.1	hR15 (Mo_6PbS_8)
$Ag_{1.2}Mo_6Se_8$	5.9	Same
Ag_7NO_{11}	1.04	Cubic
Ag_xPb_{1-x}	7.2 max.	
Ag_4Sn	0.1	h**
Ag_xSn_{1-x}	1.5–3.7	
Ag_xSn_{1-x} (film)	2.0–3.8	
$AgTe_3$	2.6	Cubic
AgTh	2.2	C16-tI12 (Al_2Cu)
$AgTh_2$	2.26	C16
$Ag_{0.03}Tl_{0.97}$	2.67	
$Ag_{0.94}Tl_{0.06}$	2.32	
AgY	0.33	B2-cP2 (CsCl)
Ag_xZn_{1-x}	0.5–0.845	
$AlAu_4$	0.4–0.7	Like A13
Al_2Au	0.1	C1-cF12 (CaF_2)
Al_2CMo_3	9.8–10.2	A13+trace 2nd. phase
Al_2CaSi	5.8	
$Al_{0.131}Cr_{0.088}V_{0.781}$	1.46	Cubic
$AlGe_2$	1.75	

Substance	T_c, K	Crystal structure type
Al_2Ge_2U	1.6	LI_2-cP4 (Cu_3Au)
$AlLa_3$	5.57	DO_{19}
Al_2La	3.23	C15
Al_2Lu	1.02	C15-cF24 (Cu_2Mg)
Al_3Mg_2	0.84	F.C.C.
$AlMo_3$	0.58	A15
$AlMo_6Pd$	2.1	
AlN	1.55	B4
Al_2NNb_3	1.3	A13
Al_3Nb	0.64	tI8 (Al_3Ti)
AlOs	0.39	B2
Al_3Os	5.90	C16
AlPb (film)	1.2–7	
Al_2Pt	0.48–0.55	C1
Al_5Re_{24}	3.35	A12
AlSb	2.8	B4-tI4 (Sn)
Al_2Sc	1.02	C15-cF24 (Cu_2Mg)
Al_2Si_2U	1.34	LI_2-cP4 (Cu_3[illegible]
[illegible]	[illegible]	[illegible]
[illegible]	10–85	[illegible]
[illegible]	110–160	[illegible]
[illegible]	60–130	[illegible]
[illegible]	[illegible]	[illegible]
[illegible]	[illegible]	[illegible]
[illegible]	[illegible]	[illegible]
Al[illegible]$_3$	[illegible]	[illegible]LI_2
AsBiPb	9.0	
AsBiPbSb	9.0	
AsHfOs	3.2	C22-hP9 (Fe_2P)
AsHfRu	4.9	same
$As_{0.33}InTe_{0.67}$ ($n = 1.24 \times 10^{22}$)	0.85–1.15	B1
$As_{0.5}InTe_{0.5}$ ($n = 0.97 \times 10^{22}$)	0.44–0.62	B1
As_4La_3	0.6	cI28 (Th_3P_4)
$AsNb_3$	0.3	$L1_2$-tP32
$As_{0.50}Ni_{0.06}Pd_{0.44}$	1.39	C2
$AsNi_{0.25}Pd_{0.75}$	1.6	$B8_1$-hP4 (NiAs)
AsOsZr	8.0	C22-hP9 (Fe_2P)
AsPb	8.4	
$AsPd_2$ (low-temp. phase)	0.60	Hexagonal
$AsPd_2$ (high-temp. phase)	1.70	C22
$AsPd_5$	0.46	Complex
As_3Pd_5	1.9	
AsRh	0.58	B31
$AsRh_{1.4–1.6}$	< 0.03–0.56	Hexagonal
AsSn	4.10	
AsSn ($n = 2.14 \times 10^{22}$)	3.41–3.65	B1
$As_{\sim 2}Sn_{\sim 3}$	3.5–3.6; 1.21–1.17	
As_3Sn_4 ($n = 0.56 \times 10^{22}$)	1.16–1.19	Rhombohedral
AsV_3	0.20	A15-cP8 (Cr_3Si)
Au_5Ba	0.4–0.7	$D2_d$
AuBe	2.64	B20
Au_2Bi	1.80	C15
Au_5Ca	0.34–0.38	$C15_b$
$AuGa_2$	1.6	C1-cF12 (CaF_2)
AuGa	1.2	B31
$Au_{0.40–0.92}Ge_{0.60–0.08}$	<0.32–1.63	Complex
$AuIn_2$	0.2	C1-cF12
AuIn	0.4–0.6	Complex
AuLu	<0.35	B2
$AuNb_3$	1.2	A2

Substance	T_c, K	Crystal structure type
$AuPb_2$	3.15	
$AuPb_2$ (film)	4.3	
$AuPb_3$	4.40	
$AuPb_3$ (film)	4.25	
Au_2Pb	1.18; 6–7	C15
$AuSb_2$	0.58	C2
AuSn	1.25	$B8_1$
Au_xSn_{1-x} (film)	2.0–3.8	
Au_5Sn	0.7–1.1	A3
$AuTa_{4.3}$	0.55	A15-cP8 (Cr_3Si)
Au_3Te_5	1.62	Cubic
$AuTh_2$	3.08	C16
AuTl	1.92	
AuV_3	0.74	A15
Au_xZn_{1-x}	0.50–0.845	
$AuZn_3$	1.21	Cubic
Au_xZr_y	1.7–2.8	A3
$AuZr_3$	0.92	A15
$B_2Ba_{0.67}Pt_3$	5.60	hP12 (B_2BaPt_3)
$BCMo_2$	5.4	Orthorhombic
$BCMo_2$	5.3–7.0	Same
$B_2Ca_{0.67}Pt_3$	1.57	hP12
B_4ErIr_4	2.1	tP18 (B_4CeCo_4)
B_4ErRh_4	4.3	oC108 (B_4LuRh_4)
B_4ErRh_4	8.7	tP18 (B_4CeCo_4)
BHf	3.1	Cubic
B_4HoIr_4	2.0	tP18
B_4HoRh_4	1.4	oC108
B_3Ir_3La	1.65	hP6 ($CaCu_5$)
B_2Ir_3Th	2.09	Same
B_4Ir_4Tm	1.6	tP18
B_6La	5.7	
B_2LaRh_3	2.82	hP6
$B_{12}Lu$	0.48	
B_2LuOs	2.66	oP16 (B_2LuRu)
B_2LuOs_3	4.62	hP6
B_4LuRh_4	6.2	oC108
B_2LuRu	9.86	oP16
B_4LuRu_4	2.0	tI72 (B_4LuRu_4)
BMo	0.5 (extrapol.)	
BMo_2	4.74	C16
BNb	8.25	B_f
B_4NdRh_4	5.3	tP18
B_2OsSc	1.34	oP16
B_2OsY	2.22	oP16
$B_2Pt_3Sr_{0.67}$	2.78	hP12 (B_2BaPt_3)
BRe_2	2.80; 4.6	
$B_4Rh_{3.4}Ru_{0.6}$	8.38	tI72
B_4Rh_4Sm	2.7	tP18
B_4Rh_4Th	4.3	Same
B_4Rh_4Tm	9.8	Same
B_4Rh_4Tm	5.4	oC108
$B_{0.3}Ru_{0.7}$	2.58	$D10_2$
B_4Ru_4Sc	7.2	tI72
B_2Ru_3Th	1.79	hP6
B_2Ru_3Y	2.85	Same
$B_2Ru\ Y$	7.80	oP16
B_4Ru_4Y	1.4	tI72
$B_{12}Sc$	0.39	
BTa	4.0	B_f

Substance	T_c, K	Crystal structure type
BTa_2	3.12	C16-tI12 (Al_2Cu)
B_6Th	0.74	
BW_2	3.1	C16
B_6Y	6.5–7.1	
$B_{12}Y$	4.7	
BZr	3.4	Cubic
$B_{12}Zr$	5.82	
$BaBi_3$	5.69	Tetragonal
$Ba_2Mo_{15}Se_{19}$	2.75	hP15 (Mo_6PbS_8)
$Ba_xO_3Sr_{1-x}Ti$ (n = 4.2 ×10[19])	<0.1–0.55	
$Ba_{0.13}O_3W$	1.9	Tetragonal
$Ba_{0.14}O_3W$	<1.25–2.2	Hexagonal
$BaRh_2$	6.0	C15
$Be_{22}Mo$	2.51	Cubic ($Be_{22}Re$)
$Be_8Nb_5Zr_2$	5.2	
$Be_{0.98-0.92}Re_{0.02-0.08}$ (quenched)	9.5–9.75	Cubic
$Be_{0.987}Re_{0.013}$	9.62	Cubic ($Be_{22}Re$)
BeTc	5.21	Cubic
$Be_{22}W$	4.12	Cubic ($Be_{22}Re$)
$Be_{13}W$	4.1	Tetragonal
Bi_3Ca	2.0	
$Bi_{0.5}Cd_{0.13}Pb_{0.25}Sn_{0.12}$ (weight fractions)	8.2	
BiCo	0.42–0.49	
Bi_2Cs	4.75	C15
Bi_xCu_{1-x} (electrodeposited)	2.2	
BiCu	1.33–1.40	
Bi_3Fe	1.0	m**
$Bi_{0.019}In_{0.981}$	3.86	
$Bi_{0.05}In_{0.95}$	4.65	α-phase
$Bi_{0.10}In_{0.90}$	5.05	Same
$Bi_{0.15-0.30}In_{0.85-0.70}$	5.3–5.4	α- and β-phases
$Bi_{0.34-0.48}In_{0.66-0.52}$	4.0–4.1	
Bi_3In_5	4.1	
$BiIn_2$	5.65	β-phase
Bi_2Ir	1.7–2.3	
Bi_2Ir (quenched)	3.0–3.96	
BiK	3.6	
Bi_2K	3.58	C15
BiLi	2.47	$L1_o$, α-phase
$Bi_{4-9}Mg$	0.7–~1.0	
Bi_3Mo	3–3.7	
BiNa	2.25	$L1_o$
$BiNb_3$	4.5	A15-cP8 (Cr_3Si)
$BiNb_3$ (high pressure and temperature)	3.05	A15
BiNi	4.25	$B8_1$
Bi_3Ni	4.06	Orthorhombic
$BiNi_{0.5}Rh_{0.5}$	3.0	$B8_1$-hP4 (AsNi)
$Bi_{0.5}NiSb_{0.5}$	2.0	Same
$Bi_{1-0}Pb_{0-1}$	7.26–9.14	
$Bi_{1-0}Pb_{0-1}$ (film)	7.25–8.67	
$Bi_{0.05-0.40}Pb_{0.95-0.60}$	7.35–8.4	H.C.P. to ε-phase
Bi_2Pb	4.25	t**
BiPbSb	8.9	
$Bi_{0.5}Pb_{0.31}Sn_{0.19}$ (weight fractions)	8.5	
$Bi_{0.5}Pb_{0.25}Sn_{0.25}$	8.5	
$BiPd_2$	4.0	
$Bi_{0.4}Pd_{0.6}$	3.7–4	Hexagonal, ordered
BiPd	3.7	Orthorhombic
Bi_2Pd	1.70	Monoclinic, α-phase
Bi_2Pd	4.25	Tetragonal, β-phase
$BiPd_{0.45}Pt_{0.55}$	3.7	$B8_1$-hP4 (NiAs)
BiPdSe	1.0	C2
BiPdTe	1.2	C2
BiPt	1.21	$B8_1$
$Bi_{0.1}PtSb_{0.9}$	2.05; 1.5	$B8_1$-hP4 (NiAs)
BiPtSe	1.45	C2
BiPtTe	1.15	C2
Bi_2Pt	0.155	Hexagonal
Bi_2Rb	4.25	C15
$BiRe_2$	1.9–2.2	
BiRh	2.06	$B8_1$
Bi_3Rh	3.2	Orthorhombic (NiB_3)
Bi_4Rh	2.7	Hexagonal
BiRu	5.7	m**
Bi_3Sn	3.6–3.8	
BiSn	3.8	
Bi_xSn_y	3.85–4.18	
Bi_3Sr	5.62	$L1_2$
Bi_3Te	0.75–1.0	
Bi_5Tl_3	6.4	
$Bi_{0.26}Tl_{0.74}$	4.4	Cubic, disordered
$Bi_{0.26}Tl_{0.74}$	4.15	$L1_2$, ordered (?)
Bi_2Y_3	2.25	
Bi_3Zn	0.8–0.9	
$Bi_{0.3}Zr_{0.7}$	1.51	
$BiZr_3$	2.4–2.8	
$BrMo_6Se_7$	7.1	hP15 (Mo_6PbS_8)
$Br_3Mo_6Se_5$	7.1	Same
CCs_x	0.020–0.135	Hexagonal
CFe_3	1.30	DO_{11}-oP16 (Fe_3C)
$CGaMo_2$	3.7–4.1	Hexagonal
$CHf_{0.5}Mo_{0.5}$	3.4	B1
$CHf_{0.3}Mo_{0.7}$	5.5	B1
$CHf_{0.25}Mo_{0.75}$	6.6	B1
$CHf_{0.7}Nb_{0.3}$	6.1	B1
$CHf_{0.6}Nb_{0.4}$	4.5	B1
$CHf_{0.5}Nb_{0.5}$	4.8	B1
$CHf_{0.4}Nb_{0.6}$	5.6	B1
$CHf_{0.25}Nb_{0.75}$	7.0	B1
$CHf_{0.2}Nb_{0.8}$	7.8	B1
$CHf_{0.9-0.1}Ta_{0.1-0.9}$	5.0–9.0	B1
CK (excess K)	0.55	Hexagonal
C_8K	0.39	Hexagonal
C_2La	1.66	tI6 (CaC_2)
C_2Lu	3.33	Same
$C_{0.40-0.44}Mo_{0.60-0.56}$	9–13	
C_3MoRe	3.8	B1-cF8
$C_{0.6}Mo_{4.8}Si_3$	7.6	$D8_8$
$CMo_{0.2}Ta_{0.8}$	7.5	B1
$CMo_{0.5}Ta_{0.5}$	7.7	B1
$CMo_{0.75}Ta_{0.25}$	8.5	B1
$CMo_{0.8}Ta_{0.2}$	8.7	B1
$CMo_{0.85}Ta_{0.15}$	8.9	B1
CMo_xV_{1-x}	2.9–9.3	B1
CMo_xZr_{1-x}	9.8	B1
$C_{0.984}Nb$	9.8	B1
CNb_2	9.1	
CNb_xTi_{1-x}	<4.2–8.8	B1
$CNb_{0.1-0.9}Zr_{0.9-0.1}$	4.2–8.4	B1

Substance	T_c, K	Crystal structure type
CRb_x (Au)	0.023–0.151	Hexagonal
$CRe_{0.06}W$	5.0	
CRu	2.00	hP2 (CW)
$C_{0.98}7Ta$	9.7	
$C_{0.848–0.987}$	2.04–9.7	
CTa (film)	5.09	B1
CTa_2	3.26	L'_3
$CTa_{0.4}Ti_{0.6}$	4.8	B1
$Cta_{1–0.4}W_{0–0.6}$	8.5–10.5	B1
$CTa_{0.2–0.9}Zr_{0.8–0.1}$	4.6–8.3	B1
CTc (excess C)	3.85	Cubic
$CTi_{0.5–0.7}W_{0.5–0.3}$	6.7–2.1	B1
CW	1.0	
CW_2	2.74	L'_3
CW_2	5.2	F.C.C.
C_2Y	3.88	tI6 (CaC_2)
$Ca_3Co_4Sn_{13}$	5.9	cP40 ($Pr_3Rh_2Sn_{13}$)
$Ca_3Ge_{13}Rh_4$	2.1	Same
CaHg	1.6	B2-cP2 (CsCl)
$CaHg_3$	1.6	hP8 (Ni_3Sn)
$CaIr_2$	6.15	C15
$Ca_3Ir_4Sn_{13}$	7.1	cP40
$Ca_xO_3Sr_{1-x}Ti$ (n = 3.7–11 × 10^{19})	< 0.1–0.55	
$Ca_{0.1}O_3W$	1.4–3.4	Hexagonal
CaPb	7.0	
$CaRh_2$	6.40	C15
$CaRh_{1.2}Sn_{4.5}$	8.7	cP40
$CaTl_3$	2.0	B2-cP2
$Cd_{0.3–0.5}Hg_{0.7–0.5}$	1.70–1.92	
CdHg	1.77; 2.15	Tetragonal
$Cd_{0.0075–0.05}In_{0.9925–0.95}$	3.24–3.36	Tetragonal
$Cd_{0.97}Pb_{0.03}$	4.2	
CdSn	3.65	
$Cd_{0.17}Tl_{0.83}$	2.3	
$Cd_{0.18}Tl_{0.82}$	2.54	
$CeCo_2$	0.84	C15
$CeCo_{1.67}Ni_{0.33}$	0.46	C15
$CeCo_{1.67}Rh_{0.33}$	0.47	C15
$Ce_xGd_{1-x}Ru_2$	3.2–5.2	C15
$CeIr_3$	3.34	
$CeIr_5$	1.82	
$Ce_{0.005}La_{0.995}$	4.6	
Ce_xLa_{1-x}	1.3–6.3	
$Ce_xPr_{1-x}Ru_2$	1.4–5.3	C15
Ce_xPt_{1-x}	0.7–1.55	
$CeRu_2$	6.0	C15
$Ce_3Mo_6Se_5$	5.7	hR15 (Mo_6PbS_8)
$Ce_2Mo_6Te_6$	1.7	Same
$Co_xFe_{1-x}Si_2$	1.4 (max.)	C1
$CoHf_2$	0.56	$E9_3$
$CoLa_3$	4.28	
$Co_4La_3Sn_{13}$	2.8	cP40
$CoLu_3$	~0.35	
Co_xLuSn_y	1.5	cP40
$Co_{0–0.01}Mo_{0.8}Re_{0.2}$	2–10	
$Co_{0.02–0.10}Nb_3Rh_{0.98–0.90}$	2.28–1.90	A15
$Co_xNi_{1-x}Si_2$	1.4 (max.)	C1
$Co_{0.5}Rh_{0.5}Si_2$	2.5	
$Co_xRh_{1-x}Si_2$	3.65 (max.)	
$Co_{\sim0.3}So_{\sim0.7}$	~0.35	
$Co_4Sc_5Si_{10}$	5.0	tP38 ($Co_4Sc_5Si_{10}$)
$CoSi_2$	1.40; 1.22	C1
Co_xSn_yYb	2.5	cP40
Co_3Th_7	1.83	$D10_2$
Co_xTi_{1-x}	2.8 (max.)	Co in α-Ti
Co_xTi_{1-x}	3.8 (max.)	Co in β-Ti
$CoTi_2$	3.44	$E9_3$
CoTi	0.71	A2
CoU	1.7	B2, distorted
CoU_6	2.29	$D2_c$
$Co_{0.28}Y_{0.72}$	0.34	
CoY_3	<0.34	
$CoZr_2$	6.3	C16
$Co_{0.1}Zr_{0.9}$	3.9	A3
$Cr_{0.6}Ir_{0.4}$	0.4	H.C.P.
$Cr_{0.65}Ir_{0.35}$	0.59	H.C.P.
$Cr_{0.7}Ir_{0.3}$	0.76	H.C.P.
$Cr_{0.72}Ir_{0.28}$	0.83	
Cr_3Ir	0.45	A15
$Cr_{0–0.1}Nb_{1–0.9}$	4.6–9.2	A2
$Cr_{0.80}Os_{0.20}$	2.5	Cubic
Cr_3Os	4.68	A15-cP8 (Cr_3Si)
Cr_xRe_{1-x}	1.2–5.2	
$Cr_{0.4}Re_{0.6}$	2.15	$D8_b$
$Cr_{0.8–0.6}Rh_{0.2–0.4}$	0.5–1.10	A3
Cr_3Rh	0.3	A15-cP8
Cr_3Ru (annealed)	3.3	A15
Cr_2Ru	2.02	$D8_b$
Cr_3Ru_2	2.10	$D8_b$-tP30 (CrFe)
$Cr_{0.1–0.5}Ru_{0.9–0.5}$	0.34–1.65	A3
Cr_xTi_{1-x}	3.6 (max.)	Cr in α-Ti
Cr_xTi_{1-x}	4.2 (max.)	Cr in β-Ti
$Cr_{0.1}Ti_{0.3}V_{0.6}$	5.6	
$Cr_{0.0175}U_{0.9825}$	0.75	β-phase
$Cs_{0.32}O_3W$	1.12	Hexagonal
$Cu_{0.15}In_{0.85}$ (film)	3.75	
$Cu_{0.04–0.08}In_{0.94–0.92}$	4.4	
CuLa	5.85	
$Cu_2Mo_6O_2S_6$	9	hR15 (Mo_6PbS_8)
$Cu_2Mo_6Se_8$	5.9	Same
Cu_xPb_{1-x}	5.7–7.7	
CuS	1.62	B18
CuS_2	1.48–1.53	C18
CuSSe	1.5–2.0	C18
$CuSe_2$	2.3–2.43	C18
CuSeTe	1.6–2.0	C18
Cu_xSn_{1-x}	3.2–3.7	
Cu_xSn_{1-x} (film, made at 10K)	3.6–7	
Cu_xSn_{1-x} (film, made at 300K)	2.8–3.7	
$CuTe_2$	<1.25–1.3	C18
$CuTh_2$	3.49	C16
$Cu_{0–0.027}V$	3.9–5.3	A2
CuY	0.33	B2-cP2 (CsCl)
Cu_xZn_{1-x}	0.5–0.845	
$DyMo_6S_8$	2.1	hR15
Er_xLa_{1-x}	1.4–6.3	
$ErMo_6S_8$	2.2	hR15
$ErMo_6Se_8$	6.2	hR15
$Fe_3Lu_2Si_5$	6.1	tP40 ($Fe_3Sc_2Si_5$)
$Fe_{0–0.04}Mo_{0.8}Re_{0.2}$	1–10	
$Fe_{0.05}Ni_{0.05}Zr_{0.90}$	~3.9	

Substance	T_c, K	Crystal structure type
Fe_3Re_2	6.55	$D8_b$-tP30 (FeCr)
$Fe_3Sc_2Si_5$	4.52	tP40
Fe_3Si_5Tm	1.3	Same
$Fe_3Si_5Y_2$	2.4	Same
Fe_3Th_7	1.86	D10
Fe_xTi_{1-x}	3.2 (max.)	Fe in α-Ti
Fe_xTi_{1-x}	3.7 (max.)	Fe in β-Ti
$Fe_xTi_{0.6}V_{1-x}$	6.8 (max.)	
FeU_6	3.86	$D2_c$
$Fe_{0.1}Zr_{0.9}$	1.0	A3
$Ga_{0.5}Ge_{0.5}Nb_3$	7.3	A15
Ga_2Ge_2U	0.87	B2-cP2
$GaHf_2$	0.21	C16-tI12 (Al_2Cu)
$GaLa_3$	5.84	
Ga_3Lu	2.3	B2-cP2
Ga_2Mo	9.5	
$GaMo_3$	0.76	A15
GaN (black)	5.85	B4
$Ga_{0.7}Pt_{0.3}$	2.9	C1
GaPt	1.74	B20
GaSb (120kbar, 77K, annealed)	4.24	A5
GaSb (unannealed)	~5.9	
$Ga_{0-1}Sn_{1-0}$ (quenched)	3.47–4.18	
$Ga_{0-1}Sn_{1-0}$ (annealed)	2.6–3.85	
GaTe	0.17	mC24 (GaTe)
Ga_5V_2	3.55	Tetragonal (Mn_2Hg_5)
$GaV_{4.5}$	9.15	
Ga_3Zr	1.38	
Ga_3Zr_5	3.8	$D8_8$-hP16 (Mn_5Si_3)
Gd_xLa_{1-x}	< 1.0–5.5	
$GdMo_6S_8$	3.5	hR15
$GdMo_6Se_8$	5.6	hR15
$Gd_xOs_2Y_{1-x}$	1.4–4.7	
$Gd_xRu_2Th_{1-x}$	3.6 (max.)	C15
$Ge_{10}As_4Y_5$	9.06	tP38 ($C0_4Sc_5Si_{10}$)
GeIr	4.7	B31
GeIrLa	1.64	tI12 (LaPtSi)
$Ge_{10}Ir_4Lu_5$	2.60	tP38
$Ge_{10}Ir_4Y_5$	2.62	tP38
Ge_2La	1.49; 2.2	Orthorhombic, distorted (Mn_2Hg_5)
GeLaPt	3.53	tI12
$Ge_{13}Lu_3Os_4$	3.6	cP40 ($Pr_3Rh_2Sn_{13}$)
$Ge_{10}Lu_5Rh_4$	2.79	tP38
$Ge_{13}Lu_3Ru_4$	2.3	cP40
$GeMo_3$	1.43	A15
$GeNb_2$	1.9	
$Ge_{0.29}Nb_{0.71}$	6	A15
GePt	0.40	B31
Ge_3Rh_5	2.12	Orthorhombic, related to $InNi_2$
GeRh	0.96	B31-oP8 (MnP)
$Ge_{13}Rh_4Sc_3$	1.9	c P40
$Ge_{10}Rh_4Y_5$	1.35	tP38
$Ge_{13}Ru_4Y_3$	1.7	cP40
Ge_2So	1.3	
$GeTa_3$	8.0	A15-cP8 (Cr_3Si)
Ge_3Te_4 (n = 1.06 × 10^{22})	1.55–1.80	Rhombohedral
Ge_xTe_{1-x} (n = 8.5–64 × 10^{20})	0.07–0.41	R1

Substance	T_c, K	Crystal structure type
GeV_3	6.01	A15
Ge_2Y	3.80	C_c
$Ge_{1.62}Y$	2.4	
Ge_2Zr	0.30	oC12 ($ZrSi_2$)
$GeZr_3$	0.4	$L1_2$-tP32 (Ti_3P)
$H_{0.33}Nb_{0.67}$	7.28	B.C.C.
$H_{0.1}Nb_{0.9}$	7.38	Same
$H_{0.05}Nb_{0.95}$	7.83	Same
$H_{0.12}Ta_{0.88}$	2.81	B.C.C.
$H_{0.08}Ta_{0.92}$	3.26	Same
$H_{0.04}Ta_{0.96}$	3.62	Same
HfIrSi	3.50	C37-cP12 (Co_2Si)
$HfMo_2$	0.05	hP24 (Ni_2Mn)
$HfN_{0.989}$	6.6	B1
$Hf_{0-0.5}Nb_{1-0.5}$	8.3–9.5	A2
$Hf_{0.75}Nb_{0.25}$	> 4.2	
$HfOs_2$	2.69	C14
HfOsP	6.1	C22-hP9 (Fe_2P)
HfPRu	9.9	Same
$HfRe_2$	4.80	C14
$Hf_{0.14}Rc_{0.86}$	5.86	A12
$Hf_{0.99-0.96}Rh_{0.01-0.04}$	0.85–1.51	
$Hf_{0-0.55}Ta_{1-0.45}$	4.4–6.5	A2
HfV_2	8.9–9.6	C15
Hg_xIn_{1-x}	3.14–4.55	
HgIn	3.81	
Hg_2K	1.20	Orthorhombic
Hg_3K	3.18	
Hg_4K	3.27	
Hg_8K	3.42	
Hg_3Li	1.7	Hexagonal
$HgMg_3$	0.17	hP8 (Na_3As)
Hg_2Mg	4.0	tI6 ($MoSi_2$)
Hg_3Mg_5	0.48	$D8_8$-hP16 (Mn_5Si_3)
Hg_2Na	1.62	Hexagonal
Hg_4Na	3.05	
Hg_xPb_{1-x}	4.14–7.26	
HgSn	4.2	
Hg_xTl_{1-x}	2.30–4.19	
Hg_5Tl_2	3.86	
Ho_xLa_{1-x}	1.3–6.3	
$Ho_{1.2}Mo_6Se_8$	6.1	$D10_2$-hR12 (Be_3Nb)
$In_{1-0.86}Mg_{0-0.14}$	3.395–3.363	
$In_2Mo_6Te_6$	2.6	hR15 (Mo_6PbS_8)
$InNb_3$ (high pressure and temp.)	4–8; 9.2	A15
$In_{0.5}Nb_3Zr_{0.5}$	6.4	
$In_{0.11}O_3W$	< 1.25–2.8	Hexagonal
$In_{0.95-0.85}Pb_{0.05-0.15}$	3.6–5.05	
$In_{0.98-0.91}Pb_{0.02-0.09}$	3.45–4.2	
InPb	6.65	
InPd	0.7	B2
InSb (quenched from 170 kbar into liquid N_2)	4.8	Like A5
InSb	2.1	B4
$(InSb)_{0.95-0.10}Sn_{0.05-0.90}$ (various heat treatments)	3.8–5.1	
$(InSb)_{0-0.07}Sn_{1-0.93}$	3.67–3.74	
In_3Sn	~5.5	
In_xSn_{1-x}	3.4–7.3	

Substance	T_c, K	Crystal structure type
$In_{0.82-1}Te$ (n = 0.83–1.71 × 10^{22})	1.02–3.45	B1
$In_{1.000}Te_{1.002}$	3.5–3.7	B1
In_3Te_4 (n = 4.7 × 10^{21})	1.15–1.25	Rhombohedral
In_xTl_{1-x}	2.7–3.374	
$In_{0.8}Tl_{0.2}$	3.223	
$In_{0.62}Tl_{0.38}$	2.760	
$In_{0.78-0.69}Tl_{0.22-0.31}$	3.18–3.32	Tetragonal
$In_{0.69-0.62}Tl_{0.31-0.38}$	2.98–3.3	F.C.C.
Ir_2La	0.48	C15
Ir_3La	2.32	$D10_2$
Ir_3La_7	2.24	$D10_2$
Ir_5La	2.13	
$IrLaSi_2$	2.03	oC16 ($CeNiSi_2$)
$IrLaSi_3$	2.7	tI10 ($BaNiSn_3$)
Ir_2Lu	2.47	C15
Ir_3Lu	2.89	C15
$Ir_4Lu_5Si_{10}$	3.9	tP38 ($Co_4Sc_5Si_{10}$)
IrMo	< 1.0	A3
$IrMo_3$	9.6	A15
$IrMo_3$	6.8	$D8_b$
$IrNb_3$	1.9	A15
$Ir_{0.4}Nb_{0.6}$	9.8	$D8_b$
$Ir_{0.37}Nb_{0.63}$	2.32	$D8_b$
IrNb	7.9	$D8_b$
$Ir_{1.15}Nb_{0.85}$	4.6	oP12 (IrTa)
$Ir_{0.02}Nb_3Rh_{0.98}$	2.43	A15
$Ir_{0.05}Nb_3Rh_{0.95}$	2.38	A15
$Ir_{0.287}O_{0.14}Ti_{0.573}$	5.5	$E9_3$
$Ir_{0.265}O_{0.035}Ti_{0.65}$	2.30	$E9_3$
Ir_xOs_{1-x}	0.3–0.98	
$Ir_{1.5}Os_{0.5}$	2.4	C14
IrOsY	2.6	C15
IrSiY	2.70	C37-oP12 (Co_2Si)
IrSiZr	2.04	Same
Ir_2Sc	2.07	C15
$Ir_{2.5}Sc$	2.46	C15
$Ir_4Sc_5Si_{10}$	8.46	tP38
Ir_2Si_2Th	2.14	tI10
$IrSi_3Th$	1.75	tI10
IrSiTh	6.50	tI12 (LaPtSi)
Ir_2Si_2Y	2.60	tI10 (Al4Ba)
$Ir_4Si_{10}Y_5$	3.10	tP38
$Ir_3Si_5Y_2$	2.83	oI40
$IrSn_2$	0.65–0.78	C1
Ir_2Sr	5.70	C15
Ir_7Ta_{13}	1.2	$D8_b$-tP30 (FeCr)
$Ir_{0.5}Te_{0.5}$	~3	
$IrTe_3$	1.18	C2
IrTh	< 0.37	B_f
Ir_2Th	6.50	C15
Ir_3Th	4.71	
Ir_3Th_7	1.52	$D10_2$
Ir_5Th	3.93	$D2_d$
$IrTi_3$	5.40	A15
IrV_2	1.39	A15
IrW_3	3.82	
$Ir_{0.28}W_{0.72}$	4.49	
Ir_2Y	2.18; 1.38	C15
$Ir_{0.69}Y_{0.31}$	1.98; 1.44	C15
$Ir_{0.70}Y_{0.30}$	2.16	C15
Ir_2Y_3	1.61	
Ir_3Y	3.50	$D10_2$-hR13 (Be_3Nb)
Ir_xY_{1-x}	0.3–3.7	
Ir_2Zr	4.10	C15
$Ir_{0.1}Zr_{0.9}$	5.5	A3
$K_2Mo_{15}S_{19}$	3.32	hR15
$K_{0.27-0.31}O_3W$	0.50	Hexagonal
$K_{0.40-0.57}O_3W$	1.5	Tetragonal
$La_{0.55}Lu_{0.45}$	2.2	Hexagonal, La type
$La_{0.8}Lu_{0.2}$	3.4	Same
$LaMg_2$	1.05	C15
$LaMo_6S_8$	7.1	hR15
LaN	1.35	
$LaOs_2$	6.5	C15
$LaPt_2$	0.46	C15
$La_{0.28}Pt_{0.72}$	0.54	C15
LaPtSi	3.48	tI12
$LaRh_3$	2.60	
$LaRh_5$	1.62	
La_7Rh_3	2.58	$D10_2$
$LaRhSi_2$	3.42	oC16 ($CeNiSi_2$)
$La_2Rh_3Si_5$	4.45	oI40 ($Co_3Si_5U_2$)
$LaRhSi_3$	2.7	tI10 ($BaNiSn_3$)
$LaRh_2Si_2$	3.90	tI10 (Al_4Ba)
$LaRu_2$	1.63	C15
La_3S_4	6.5	$D7_3$
La_3Se_4	8.6	$D7_3$
$LaSi_2$	2.3	C_c
La_xY_{1-x}	1.7–5.4	
LaZn	1.04	B2
$Li_2Mo_6S_8$	4.2	hR15
LiPb	7.2	
$LuOs_2$	3.49	C14
$Lu_{0.275}Rh_{0.725}$	1.27	C15
$LuRh_5$	0.49	
$Lu_5Rh_4Si_{10}$	3.95	tP38 ($Co_4So_5Si_{10}$)
$LuRu_2$	0.86	C14
$Mg_{1.14}Mo_{6.6}S_8$	3.5	hR15
Mg2Nb	5.6	
$Mg_{\sim 0.47}Tl_{\sim 0.53}$	2.75	B2
MgZn	0.9	A3-oP4 (AuCd)
Mn_xTi_{1-x}	2.3 (max.)	Mn in -Ti
Mn_xTi_{1-x}	1.1–3.0	Mn in -Ti
MnU_6	2.32	$D2_c$
Mo_2N	5.0	F.C.C.
$Mo_6Na_2S_8$	8.6	hR15
Mo_xNb_{1-x}	0.016–9.2	
$Mo_{5.25}Nb_{0.75}Se_8$	6.2	hR15
Mo_6NdSa_8	8.2	hR15
Mo_3Os	7.2	A15
$Mo_{0.62}Cs_{0.38}$	5.65	$D8_b$
Mo_3P	5.31	DO_e
$Mo_6Pb_{1.2}Se_8$	6.75	hR15
$Mo_{0.5}Pd_{0.5}$	3.52	A3
Mo_6PrSe_8	9.2	hR15
MoRe	7.8	$D8_b$-tP30
$MoRe_3$	9.25; 9.89	A12
Mo_xRe_{1-x}	1.2–12.2	
$Mo_{0.42}Re_{0.58}$	6.35	$D8_b$
MoRh	1.97	A3
Mo_xRh_{1-x}	1.5–8.2	B.C.C.

Substance	T_c, K	Crystal structure type
MoRu	9.5–10.5	A3
$Mo_{0.61}Ru_{0.39}$	7.18	$D8_b$
$Mo_{0.2}Ru_{0.8}$	1.66	A3
Mo_3Ru_2	7.0	$D8_b$-tP30
$Mo_4Ru_2Te_8$	1.7	hR15
Mo_6S_8	1.85	hR15
Mo_6S_8Sc	3.6	hR15
$Mo_6S_8Sm_{1.2}$	2.9	hR15
Mo_6S_8Tb	2.0	hR15
Mo_6S_8Tl	8.7	hR15
$Mo_6S_8Tm_{1.2}$	2.1	hR15
$Mo_6S_8Y_{1.2}$	3.0	hR15
Mo_6S_8Yb	9.2	hR15
$Mo_{6.6}S_8Zn_{11}$	3.6	hR15
Mo_3Sb_4	2.1	
Mo_6Se_8	6.3	hR15
$Mo_6Se_8Sm_{1.2}$	6.8	hR15
$Mo_6Se_8Sn_{1.2}$	6.8	hR15
Mo_6Se_8Tb	5.7	hR15
Mo_3Se_3Tl	4.0	hP14
$Mo_6Se_8Tm_{1.2}$	6.3	hR15
Mo_6Se_8Yb	6.2	hR15
Mo_3Si	1.30	A15
$MoSi_{0.7}$	1.34	
Mo_xSiV_{3-x}	4.54–16.0	A15
$Mo_{5.25}Ta_{0.75}Te_8$	1.7	hR15
Mo_6Te_8	1.7	hR15
$Mo_{0.16}Ti_{0.84}$	4.18; 4.25	
$Mo_{0.913}Ti_{0.087}$	2.95	
$Mo_{0.04}Ti_{0.96}$	2.0	Cubic
$Mo_{0.025}Ti_{0.975}$	1.8	
Mo_xU_{1-x}	0.7–2.1	
Mo_xV_{1-x}	0–~5.3	
Mo_2Zr	4.25–4.75	C15
NNb (film)	6–9	B1
$N_xO_yTi_z$	2.9–5.6	Cubic
$N_xO_yV_z$	5.8–8.2	Cubic
$N_{0.34}Re$	4–5	F.C.C.
NTa (film)	4.84	B1
$N_{0.6-0.987}Ti$	< 1.17–5.8	B1
$N_{0.82-0.99}V$	2.9–7.9	B1
NZr	9.8	B1
$N_{0.906-0.984}Zr$	3.0–9.5	B1
$Na_{0.28-0.35}O_3W$	0.56	Tetragonal
$Na_{0.28}Pb_{0.72}$	7.2	
NbO	1.25	
$NbOs_2$	2.52	A12
Nb_3Os	1.05	A15
$Nb_{0.6}Os_{0.4}$	1.89; 1.78	$D8_b$
$Nb_3Os_{0.02-0.10}Rh_{0.98-0.90}$	2.42–2.30	A15
Nb_3P	1.8	$L1_2$tP32 (Ti_3P)
NbPRh	4.08	C37-oP12 (Co_2Si)
$Nb_{0.6}Pd_{0.4}$	1.60	$D8_f$ plus cubic
$Nb_3Pd_{0.02-0.10}Rh_{0.92-0.90}$	2.49–2.55	A15
$Nb_{0.62}Pt_{0.38}$	4.21	$D8_b$
Nb_5Pt_3	3.73	$D8_b$
$Nb_3Pt_{0.02-0.98}Rh_{0.98-0.02}$	2.52–9.6	A15
$NbRe_3$	5.27	$D8_b$-tP30 (FeCr)
$Nb_{0.38-0.18}Re_{0.62-0.82}$	2.43–9.70	A15
NbRe	3.8	$D8_b$-tP30
NbReSi	5.1	oI36 (FeTiSi)
Nb_3Rh	2.64	A15
$Nb_{0.6}Rh_{0.40}$	4.21	$D8_b$ plus other
$Nb_{0.9}Rh_{1.1}$	3.07	A3-oP4 (AuCd)
$Nb_3Rh_{0.98-0.90}Ru_{0.02-0.10}$	2.42–2.44	A15
Nb_xRu_{1-x}	1.2–4.8	
NbRuSi	2.65	oI36
NbS_2	6.1–6.3	Hexagonal, $NbSe_2$ type
NbS_2	5.0–5.5	Hexagonal, three-layer type
Nb_3Sb	0.2	$L1_2$-tP32 (Ti_3P)
$Nb_3Sb_{0-0.7}Sn_{1-0.3}$	6.8–18	A15
$NbSe_2$	5.15–5.62	Hexagonal
$Nb_{1-1.05}Se_2$	2.2–7.0	Same
Nb_3Se_4	2.0	hP14
Nb_3Si	1.5	$L1_2$
Nb_3SiSnV_3	4.0	
$NbSn_2$	2.60	Orthorhombic
Nb_6Sn_5	2.8	oI44 (Sn_5Ti_6)
NbSnTaV	6.2	A15
$NbSnV_2$	5.5	A15
Nb_2SnV	9.8	A15
Nb_xTa_{1-x}	4.4–9.2	A2
Nb_3Te_4	1.8	hP14
Nb_xTi_{1-x}	0.6–9.8	
$Nb_{0.6}Ti_{0.4}$	9.8	
Nb_xU_{1-x}	1.95 (max.)	
$Nb_{0.88}V_{0.12}$	5.7	A2
$Nb_{0.5}V_{1.5}Zr$	4.3	C15-hP12 ($MgZn_2$)
$Ni_{0.3}Th_{0.7}$	1.98	$D10_2$
$NiZr_2$	1.52	
$Ni_{0.1}Zr_{0.9}$	1.5	A3
$O_3Rb_{0.27-0.29}W$	1.98	Hexagonal
OSn	3.81	tP4 (PbO)
O_3SrTi (n = 1.7–12.0 × 10^{19})	0.12–0.37	
O_3SrTi (n = 10^{18}–10^{21})	0.05–0.47	
O_3SrTi (n = 10^{20})	0.47	
$O_3Sr_{0.08}W$	2–4	Hexagonal
OTi	0.58	
$O_3Tl_{0.30}W$	2.0–2.14	Hexagonal
OV_3Zr_3	7.5	$E9_3$
OW_3 (film)	3.35; 1.1	A15
OsPTi	1.2	C22-hP9 (Fe_2P)
OsPZr	7.4	Same
OsReY	2.0	C14
Os_2Sc	4.6	C14
OsTa	1.95	A12
Os_3Th_7	1.51	$D10_2$
Os_xW_{1-x}	0.9–4.1	
OsW_3	~3	
Os_2Y	4.7	C14
Os_2Zr	3.0	C14
Os_xZr_{1-x}	1.5–5.6	
PPb	7.8	
OsW_2	3.81	$D8_b$-tP30 (FeCr)
$PPd_{3.0-3.2}$	<0.35–0.7	DO_{11}
P_3Pd_7 (high temperature)	1.0	Rhombohedral
P_3Pd_7 (low temperature)	0.70	Complex
PRh	1.22	
PRh_2	1.3	C1
P_4Rh_5	1.22	oP28 ($CaFe_2O_4$)
PRhTa	4.41	C37-oP12 (Co_2Si)

Substance	T_c, K	Crystal structure type
PRhZr	1.55	Same
PRuTi	1.3	C22-hP9 (Fe_2P)
PRuZr	3.46	C37-oP12
PW_3	2.26	DO_e
Pb_2Pd	2.95	C16
Pb_4Pt	2.80	Related to C16
Pb_2Rh	2.66	C16
PbSb	6.6	
PbTe (plus 0.1 w/o Pb)	5.19	
PbTe (plus 0.1 w/o Te)	5.24–5.27	
$PbTl_{0.27}$	6.43	
$PbTl_{0.17}$	6.73	
$PbTl_{0.12}$	6.88	
$PbTl_{0.075}$	6.98	
$PbTl_{0.04}$	7.06	
$Pb_{1-0.26}Tl_{0-0.74}$	7.20–3.68	
$PbTl_2$	3.75–4.1	
Pb_3Zr_5	4.60	$D8_8$
$PbZr_3$	0.76	A15
$Pd_{0.9}Pt_{0.1}Te_2$	1.65	C6
$Pd_{0.05}Ru_{0.05}Zr_{0.9}$	~9	
$Pd_{2.2}S$ (quenched)	1.63	Cubic
$PdSb_2$	1.25	C2
PdSb	1.5	$B8_1$
PdSbSe	1.0	C2
PdSbTe	1.2	C2
Pd_4Se	0.42	Tetragonal
$Pd_{6-7}Se$	0.66	Like Pd_4Te
$Pd_{2.8}Se$	2.3	
Pd_xSe_{1-x}	2.5 (max.)	
PdSi	0.93	B31
PdSn	0.41	B31
$PdSn_2$	3.34	
Pd_2Sn	0.41	C37
Pd_3Sn	0.47–0.64	$B8_2$
Pd_2SnTm	1.77	DO_3-cF16 (BiF_3)
Pd_2SnY	4.92	Same
Pd_2SnYb	1.79	Same
PdTe	2.3; 3.85	$B8_1$
$PdTe_{1.02-1.08}$	2.56–1.88	$B8_1$
$PdTe_2$	1.69	C6
$PdTe_{2.1}$	1.89	C6
$PdTe_{2.3}$	1.85	C6
$Pd_{1.1}Te$	4.07	$B8_1$
Pd_3Te	0.76	cI2 (W)
$PdTh_2$	0.85	C16
$Pd_{0.1}Zr_{0.9}$	7.5	A3
PtSb	2.1	$B8_1$
PtSi	0.88	B31
PtSn	0.37	$B8_1$
$PtSn_4$	2.38	C16-oC20 ($PdSn_4$)
Pt_3Ta_7	1.5	$D8_b$-tP30
$PtTa_3$	0.4	A15-cP8 (Cr_3Si)
PtTe	0.59	Orthorhombic
PtTh	0.44	B_f
Pt_3Th_7	0.98	$D10_2$
Pt_5Th	3.13	
$PtTi_3$	0.58	A15
$Pt_{0.02}U_{0.98}$	0.87	β-phase
$PtV_{2.5}$	1.36	A15
PtV_3	2.87–3.20	A15

Substance	T_c, K	Crystal structure type
$PtV_{3.5}$	1.26	A15
$Pt_{0.5}W_{0.5}$	1.45	A1
Pt_xW_{1-x}	0.4–2.7	
Pt_2Y_3	0.90	
Pt_2Y	1.57; 1.70	C15
Pt_3Y_7	0.82	$D10_2$
PtZr	3.0	A3
Re_2Sc	4.2	C15-hP12 ($MgZn_2$)
$Re_{24}Sc_5$	2.2	A12-cI58 (Mg)
ReSiTa	4.4	oI36 (FeTiSi)
$Re_3Si_5Y_2$	1.76	tP40 ($Fe_3Sc_2Si_5$)
Re_3Ta_2	1.4	$D8_b$-tP30 (FeCr)
$Re_{0.64}Ta_{0.36}$	1.46	A12
Re_3Ta	6.78	A12-cI58 (Mn)
$Re_{24}Ti_5$	6.60	A12
Re_xTi_{1-x}	6.6 (max.)	
$Re_{0.76}V_{0.24}$	4.52	$D8_b$
Re_3V	6.26	$D8_b$-tP30
$Re_{0.92}V_{0.08}$	6.8	A3
$Re_{0.6}W_{0.4}$	6.0	
$Re_{0.5}W_{0.5}$	5.12	$D8_b$
$Re_{13}W_{12}$	5.2	$D8_b$-tP30
Re_3W	9.0	A12-cI58
Re_2Y	1.83	C14
Re_2Zr	5.9	C14
Re_3Zr	7.40	A12-cI58
Re_6Zr	7.40	Same
$Rh_{17}S_{15}$	5.8	Cubic
$Rh_{\sim 0.24}Sc_{\sim 0.76}$	0.88; 0.92	
$Rh_4Sc_5Si_{10}$	8.54	tP38
$Rh_4Sc_3Sn_{13}$	4.5	cP40
Rh_xSe_{1-x}	6.0 (max.)	
$RhSi_3Th$	1.76	tI10
$Rh_{0.86}Sc_{1.04}Th$	6.45	tI12
Rh_2Si_2Y	3.11	tI10
$Rh_3Si_5Y_2$	2.70	oI40
$Rh_4Sn_{13}Sr_3$	4.3	cP40
Rh_xSn_yTh	1.9	cI2 (W)
Rh_xSn_yTm	2.3	cP40
$Rh_4Sn_{13}Y_3$	3.2	cP40
Rh_2Sr	6.2	C15
$Rh_{0.4}Ta_{0.6}$	2.35	$D8_b$
$RhTe_2$	1.51	C2
$Rh_{0.67}Te_{0.33}$	0.49	
Rh_xTe_{1-x}	1.51 (max.)	
RhTh	0.36	B_f
Rh_3Th_7	2.15	$D10_2$
Rh_5Th	1.07	
Rh_xTi_{1-x}	2.25–3.95	
$Rh_{0.02}U_{0.98}$	0.96	
RhV_3	0.38	A15
RhW	~3.4	A3
RhY_3	0.65	
Rh_2Y_3	1.48	
Rh_3Y	1.07	C15
Rh_5Y	0.56	
Rh_3Y_7	0.32	hP20 (Fe_3Th_7)
$Rh_{0.005}Zr$ (annealed)	5.8	
$Rh_{0-0.45}Zr_{1-0.55}$	2.1–10.8	
$Rh_{0.1}Zr_{0.9}$	9.0	H.C.P.
Ru_2Sc	1.67	C14
RuSiTa	3.15	oI36

Substance	T_c, K	Crystal structure type
Ru_3Si_2Th	3.98	hP12
Ru_3Si_2Y	3.51	hP12
$Ru_{1.1}Sn_{3.1}Y$	1.3	cP40
Ru_2Th	3.56	C15
RuTi	1.07	B2
$Ru_{0.05}Ti_{0.95}$	2.5	
$Ru_{0.1}Ti_{0.9}$	3.5	
$Ru_xTi_{0.6}V_y$	6.6 (max.)	
Ru_3U	0.15	$L1_2$-cP4
$Ru_{0.45}V_{0.55}$	4.0	B2
RuW	7.5	A3
Ru_2Y	1.52	C14
Ru_2Zr	1.84	C14
$Ru_{0.1}Zr_{0.9}$	5.7	A3
STh	0.5	B1-cF8 (NaCl)
SbSn	1.30–1.42	B1 or distorted
$SbTa_3$	0.72	A15-cP8 (Cr_3Si)
$SbTi_3$	5.8	Same
Sb_2Ti_7	5.2	
$Sb_{0.01-0.03}V_{0.99-0.97}$	3.76–2.63	A2
SbV_3	0.80	A15
SeTh	1.7	B1-cF8
$SiMo_3$	1.4	A15-cP8
Si_2Th	3.2	C_c, α-phase
Si_2Th	2.4	C32, β-phase
$SiV_{2.7}Ru_{0.3}$	2.9	A15
Si_2W_3	2.8; 2.84	
$SiZr_3$	0.5	$L1_2$-tP32 (Ti_3P)
$Sn_{0.174-0.104}Ta_{0.826-0.896}$	6.5–< 4.2	A15
$SnTa_3$	8.35	A15, highly ordered
$SnTa_3$	6.2	A15, partially ordered
$SnTaV_2$	2.8	A15
$SnTa_2V$	3.7	A15
Sn_xTe_{1-x} (n = 10.5–20 × 10^{20})	0.07–0.22	B1
Sn_3Th	3.33	$L1_2$-cP4
$SnTi_3$	5.80	A15-cP8
Sn_xTl_{1-x}	2.37–5.2	
SnV_3	3.8	A15
$Sn_{0.02-0.057}V_{0.98-0.943}$	2.87–~1.6	A2
$SnZr_3$	0.92	A15-cP8
$Ta_{0.025}Ti_{0.975}$	1.3	Hexagonal
$Ta_{0.05}Ti_{0.95}$	2.9	Hexagonal
$Ta_{0.05-0.75}V_{0.95-0.25}$	4.30–2.65	A2
$Ta_{0.8-1}W_{0.2-0}$	1.2–4.4	A2
$Tc_{0.1-0.4}W_{0.9-0.6}$	1.25–7.18	Cubic
$Tc_{0.50}W_{0.50}$	7.52	α plus
$Tc_{0.60}W_{0.40}$	7.88	plus α
Tc_6Zr	9.7	A12
TeY	1.02	B1-cF8
$ThTl_3$	0.87	$L1_2$-cP4
$Th_{0-0.55}Y_{1-0.45}$	1.2–1.8	
$Ti_{0.70}V_{0.30}$	6.14	Cubic
Ti_xV_{1-x}	0.2–7.5	
$Ti_{0.5}Zr_{0.5}$ (annealed)	1.23	
$Ti_{0.5}Zr_{0.5}$ (quenched)	2.0	
Tl_3Y	1.52	$L1_2$-cP4
V_2Zr	8.80	C15
$V_{0.26}Zr_{0.74}$	5.9	
W_2Zr	2.16	C15
YZn	0.33	B2-cP2 (CsCl)

* n denotes current carriers concentration in cm^{-3}.

B. Superconductors with T_c > 10K

Substance	T_c,K	Crystal structure type	
Al_2CMo_3	10.0	A13	
$Al_{0.5}Ge_{0.5}Nb$	12.6	A15	
$Al_{\sim0.8}Ge_{\sim0.2}Nb_3$	20.7	A15	
$AlNb_3$	18.0	A15	(Cr_3Si)
$AlNb_3$	12.0		(FeCr)
$A1_xNb_{1-x}$	<4.2–13.5	$D8_b$	
$A1_xNb_{1-x}$	12–17.5	A15	
$A1_{0.27}Nb_{0.73-0.48}V_{0-0.25}$	14.5–17.5	A15	
A1 Nb_xV_{1-x}	4.4–13.5		
$A1_{0.1}Si_{0.9}V_3$	14.05		
$A1V_3$	11.8	A15	(Cr_3Si)
$AuNb_3$	11.5	A15	
$Au_{0-0.3}Nb_{1-0.7}$	1.1–11.0		
$Au_{0.02-0.98}Nb_3Rh_{0.98-0.02}$	2.53–10.9	A15	
$AuNb_{3(1-x)}V_{3x}$	1.5–11.0	A15	
$B_{0.03}C_{0.51}Mo_{0.47}$	12.5		
B_4LuRh_4	11.7		(B_4CeCo_4)
B_2LuRu	10		
B_4Rh_4Y	11.3		(B_4CeCo_4)
$B_{0.1}Si_{0.9}V_3$	15.8	A15	
$BaBi_{0.2}O_3Pb_{0.8}$	13.2		
$Ba_2CaCu_2O_8T1_2$	120		
$Ba_2Cu_3LaO_6$	80		
$Ba_2Cu_3O_7Tm$	101		
$Ba_2Cu_3O_7Y$	90		
$(Ba,La)_2CuO_4$	36	A15	(K_2NiF_4)
$Bi_2CaCu_2O_8Sr_2$	110		
$Br_2Mo_6S_6$	13.8		(Mo_6PbS_8)
C_3La	11.0		(C_3Pu_2)
CMo	14.3	B1	(NaC1)
CMo_2	12.2		o**
$C_{0.5}Mo_xNb_{1-x}$	10.8–12.5	B1	
CMo_xTi_{1-x}	10.2(max)	B1	
$CMo_{0.83}Ti_{0.17}$	10.2	B1	
$C_{0-0.38}N_{1-0.62}Ta$	10.0–11.3		
CNb (whiskers)	7.5–10.5		
CNb	11.5	B1	
$C_{0.7-1.0}Nb_{0.3-0}$	6–11	B1	
CNb_xTa_{1-x}	8.2–13.9		
$CNb_{0.6-0.9}W_{0.4-0.1}$	12.5–11.6	B1	
$C_{0.1}Si_{0.9}V_3$	16.4	A15	
CTa	10.3	B1	
$CTa_{1-0.4}W_{0-0.6}$	8.5–10.5	B1	
$C_{0.66}Th_{0.13}Y_{0.21}$	17		(C_3Pu_2)
C_3Y_2	11.5		(C_3Pu_2)
CW	10	B1	
$(Ca,La)_2CuO_4$	18		(K_2NiF_4)
$Cu(La,Sr)_2O_4$	39		
$Cu_{1.8}Mo_6S_8$	10.8		(Mo_6PbS_8)
$Cr_{0.3}SiV_{2.7}$	11.3	A15	
$GaNb_3$	14.5	A15	(Cr_3Si)
$Ga_xNb_3Sn_{1-x}$	14–18.37	A15	
GaV_3	16.8	A15	
$GaV_{2.1-3.5}$	6.3–14.45	A15	
$GeNb_3$	23.2	A15	
$GeNb_3$ (quenched)	6–17	A15	
$Ge_xNb_3Sn_{1-x}$	17.6–18.0	A15	
$Ge_{0.5}Nb_3Sn_{0.5}$	11.3		
$Ge_{0.1}Si_{0.9}V_3$	14.0	A15	
GeV_3	11	A15	
$InLa_3$	9.83; 10.4	LI_2	($AuCu_3$)

Substance	T_c,K	Crystal structure type	
$In_{0-0.3}Nb_3Sn_{1-0.7}$	18.0–18.19	A15	
InV_3	13.9	A15	
$Ir_{0.4}Nb_{0.6}$	10		(FeCr)
$LaMo_6Se_8$	11.4		(Mo_6PbS_8)
LiO_4Ti_2	13.7		$(A1_2MgO_4)$
MgB_2	39.0±0.5	C32	
MoN	12; 14.8		h*
Mo_3Os	12.7	A15	
$Mo_6Pb_{0.9}S_{7.5}$	15.2		(Mo_6PbS_8)
Mo_3Re	10.0; 15	A15	
Mo_xRe_{1-x}	1.2–12.2		
$Mo_{0.52}Re_{0.48}$	11.1		
$Mo_{0.57}Re_{0.43}$	14.0		
$Mo_{\sim 0.60}Re_{0.395}$	10.6		
MoRu	9.5–10.5	A3	
Mo_3Ru	10.6	A15	
Mo_6Se_8T1	12.2		(Mo_6PbS_8)
$Mo_{0.3}SiV_{2.7}$	11.7	A15	
Mn_3Si	12.5	A15	
Mo_3Tc	15	A15	
$Mo_{0.3}Tc_{0.7}$	12.0	A15	
Mo_xTc_{1-x}	10.8–15.8		
$MoTc_3$	15.8		
NNb (whiskers)	10–14.5		
NNb (diffusion wires)	16.10		
$N_{0.988}Nb$	14.9; 17.3	B1	
$N_{0.824-0.988}Nb$	14.4–15.3	B1	
$N_{0.7-0.795}Nb$	11.3–12.9		
NNb_xO_y	13.5–17.0	B1	
NNb_xO_y	6.0–11		
$N_{100-42w/o}Nb_{0-58w/o}Ti$	15–16.8		
$N_{100-75w/o}Nb_{0-25w/o}Zr$	12.5–16.35		
NNb_xZr_{1-x}	9.8–13.8	B1	
$N_{0.93}Nb_{0.85}Zr_{0.15}$	13.8	B1	
NTa	12–14	B1	
NZr	10.7	B1	
Nb_3Pt	10.9	A15	
$Nb_{0.18}Re_{0.82}$	10		(Mn)
Nb_3Si	19	A15	
$Nb_{0.3}SiV_{2.7}$	12.8	A15	
Nb_3Sn	18.05	A15	
$Nb_{0.8}Sn_{0.2}$	18.18; 18.5	A15	
Nb_xSn_{1-x} (film)	2.6–18.5		o*
Nb_3Sn_2	16.6		t*
$NbSnTa_2$	10.8	A15	
Nb_2SnTa	16.4	A15	
$Nb_{2.5}SnTa_{0.5}$	17.6	A15	
$Nb_{2.75}SnTa_{0.25}$	17.8	A15	
$Nb_{3x}SnTa_{3(1-x)}$	6.0–18.0		
$Nb_2SnTa_{0.5}V_{0.5}$	12.2	A15	
$NbTc_3$	10.5	A12	
$Nb_{0.75}Zr_{0.25}$	10.8		
$Nb_{0.66}Zr_{0.33}$	10.8		
$PbTa_3$	17	A15	
$RhTa_3$	10	A15	
$RhZr_2$	10.8; 11.3	C16	$(A1_2Cu)$
$Rh_{0-0.45}Zr_{1-0.55}$	2.1–10.8		
$SiTi_{0.3}V_{2.7}$	10.9	A15	
SiV_3	17.1	A15	
$SiV_{2.7}Zr_{0.3}$	13.2	A15	

TABLE 5. Critical Field Data

Substance	H_o (oersteds)
Ag_2F	2.5
Ag_7NO_{11}	57
Al_2CMo_3	1700
$BaBi_3$	740
Bi_2Pt	10
Bi_3Sr	530
Bi_5Tl_3	>400
CdSn	>266
$CoSi_2$	105
$Cr_{0.1}Ti_{0.3}V_{0.6}$	1360
$In_{1-0.86}Mg_{0-0.14}$	272.4–259.2
InSb	1100
In_xTl_{1-x}	252–284
$In_{0.8}Tl_{0.2}$	252
$Mg_{0.47}Tl_{0.53}$	220
$Mo_{0.16}Ti_{0.84}$	<985
$NbSn_2$	620
$PbTl_{0.27}$	756
$PbTl_{0.17}$	796
$PbTl_{0.12}$	849
$PbTl_{0.075}$	880
$PbTl_{0.04}$	864

TABLE 6. High Critical Magnetic-Field Superconductive Compounds and Alloys

Substance	T_c, K	H_{c1}, kOe	H_{c2}, kOe	H_{c3}, kOe	T_{obs}, K[a]
Al_2CMo_3	9.8–10.2	0.091	156		1.2
$AlNb_3$		0.375			
$Ba_xO_3Sr_{1-x}Ti$	<0.1–0.55	0.0039 max.			
$Bi_{0.5}Cd_{0.1}Pb_{0.27}Sn_{0.13}$			>24		3.06
Bi_xPb_{1-x}	7.35–8.4	0.122 max.	30 max.		4.2
$Bi_{0.56}Pb_{0.44}$	8.8		15		4.2
$Bi_{7.5w/o}Pb_{92.5w/o}$ [b]			2.32		
$Bi_{0.099}Pb_{0.901}$		0.29	2.8		
$Bi_{0.02}Pb_{0.98}$		0.46	0.73		
$Bi_{0.53}Pb_{0.32}Sn_{0.16}$			>25		3.06
$Bi_{1-0.93}Sn_{0-0.07}$			0–0.032		3.7
Bi_5Tl_3	6.4		>5.6		3.35
C_8K (excess K)	0.55		0.160 (H⊥c)		0.32
			0.730 (H∥c)		0.32
C_8K	0.39		0.025 (H⊥c)		0.32
			0.250 (H∥c)		0.32
$C_{0.44}Mo_{0.56}$	12.5–13.5	0.087	98.5		1.2
CNb	8–10	0.12	16.9		4.2
$CNb_{0.4}Ta_{0.6}$	10–13.6	0.19	14.1		1.2
CTa	9–11.4	0.22	4.6		1.2
$Ca_xO_3Sr_{1-x}Ti$	<0.1–0.55	0.002–0.004			
$Cd_{0.1}Hg_{0.9}$ (by weight)		0.23	0.34		2.04
$Cd_{0.05}Hg_{0.95}$		0.28	0.31		2.16
$Cr_{0.10}Ti_{0.30}V_{0.60}$	5.6	0.071	84.4		0
GaN	5.85	0.725			4.2
Ga_xNb_{1-x}			>28		4.2
GaSb (annealed)	4.24		2.64		3.5
$GaV_{1.95}$	5.3		73[e]		
$GaV_{2.1-3.5}$	6.3–14.45		230–300[d]		0
GaV_3		0.4	350[e]		0
			500[d]		
$GaV_{4.5}$	9.15		121[c]		0
Hf_xNb_y			>52–>102		1.2
Hf_xTa_y			>28–>86		1.2
$Hg_{0.05}Pb_{0.95}$		0.235	2.3		
$Hg_{0.101}Pb_{0.899}$		0.23	4.3		4.2
$Hg_{0.15}Pb_{0.85}$	6.75		>13		2.93
$In_{0.98}Pb_{0.02}$	3.45	0.1		0.12	2.76
$In_{0.96}Pb_{0.04}$	3.68	0.1	0.12	0.25	2.94
$In_{0.94}Pb_{0.06}$	3.90	0.095	0.18	0.35	3.12
$In_{0.913}Pb_{0.087}$	4.2	~10.17	0.55	2.65	
$In_{0.316}Pb_{0.684}$		0.155	3.7		4.2
$In_{0.17}Pb_{0.83}$			2.8	5.5	4.2
$In_{1.000}Te_{1.002}$	3.5–3.7		1.2[c]		0
$In_{0.95}Tl_{0.05}$		0.263	0.263		3.3
$In_{0.90}Tl_{0.10}$		0.257	0.257		3.25
$In_{0.83}Tl_{0.17}$		0.242	0.39		3.21
$In_{0.75}Tl_{0.25}$		0.216	0.50		3.16
LaN	1.35	0.45			0.76
La_3S_4	6.5	≈0.15	>25		1.3
La_3Se_4	8.6	≈0.2	>25		1.25
$Mo_{0.52}Re_{0.48}$	11.1		14–21	22–33	4.2
			18–28	37–43	1.3
$Mo_{0.6}Re_{0.395}$	10.6		14–20	20–37	4.2
			19–26	26–37	1.3
$Mo_{0.5}Ti_{0.5}$			75[c]		0
$Mo_{0.16}Ti_{0.84}$	4.18	0.028	98.7[c]		0
			36–38		3.0
$Mo_{0.913}Ti_{0.087}$	2.95	0.060	15		4.2
$Mo_{0.1-0.3}U_{0.9-0.7}$	1.85–2.06		>25		
$Mo_{0.17}Zr_{0.83}$			30		

Substance	T_c, K	H_{c1}, kOe	H_{c2}, kOe	H_{c3}, kOe	T_{obs}, K[a]
$N_{(12.8\ w/o)}Nb$	15.2		>9.5		13.2
NNb (wires)	16.1		153[c]		0
			132		4.2
			95		8
			53		12
NNb_xO_{1-x}	13.5–17.0		38		
NNb_xZr_{1-x}	9.8–13.8		4- >130		4.2
$N_{0.93}Nb_{0.85}Zr_{0.15}$	13.8		>130		4.2
$Na_{0.086}Pb_{0.914}$		0.19	6.0		
$Na_{0.016}Pb_{0.984}$		0.28	2.05		
Nb	9.15		2.020		1.4
			1.710		4.2
Nb		0.4–1.1	3–5.5		4.2
Nb (unstrained)		1.1–1.8	3.40	6–9.1	4.2
Nb (strained)		1.25–1.92	3.44	6.0–8.7	4.2
Nb (cold-drawn wire)		2.48	4.10	≈10	4.2
Nb (film)			>25		4.2
NbSc			>30		
Nb_3Sn		0.170	221		4.2
			70		14.15
			54		15
			34		16
			17		17
$Nb_{0.1}Ta_{0.9}$		0.084	0.154		4.195
$Nb_{0.2}Ta_{0.8}$			10		4.2
$Nb_{0.65\text{-}0.73}Ta_{0.02\text{-}0.10}Zr_{0.25}$			>70–>90		4.2
Nb_xTi_{1-x}			148 max.		1.2
			120 max.		4.2
$Nb_{0.222}U_{0.778}$		1.98	23		1.2
Nb_xZr_{1-x}			127 max.		1.2
			94 max.		4.2
O_3SrTi	0.43	0.0049[c]	0.504[c]		0
O_3SrTi	0.33	0.00195[c]	0.420[c]		0
$PbSb_{1\ w/o}$(quenched)			>1.5		4.2
$PbSb_{1\ w/o}$(annealed)			>0.7		4.2
$PbSb_{2.8\ w/o}$(quenched)			>2.3		4.2
$PbSb_{2.8\ w/o}$(annealed)			>0.7		4.2
$Pb_{0.871}Sn_{0.129}$		0.45	1.1		
$Pb_{0.965}Sn_{0.035}$		0.53	0.56		
$Pb_{1\text{-}0.26}Tl_{0\text{-}0.74}$	7.20–3.68		2–6.9[c]		0
$PbTl_{0.17}$	6.73		4.5[c]		0
$Re_{0.26}W_{0.74}$			>30		
$Sb_{0.93}Sn_{0.07}$			0.12		3.7
SiV_3	17.0	0.55	156[e]		
Sn_xTe_{1-x}		0.00043–0.00236	0.005–0.0775		0.012–0.079
Ta (99.95%)		0.425	1.850		1.3
		0.325	1.425		2.27
		0.275	1.175		2.66
		0.090	0.375		3.72
$Ta_{0.5}Nb_{0.5}$			3.55		4.2
$Ta_{0.65\text{-}0}Ti_{0.35\text{-}1}$	4.4–7.8		>14–138		1.2
$Ta_{0.5}Ti_{0.5}$			138		1.2
Te	3.3	0.25[c]			0
Tc_xW_{1-x}	5.75–7.88		8–44		4.2
Ti				2.7	4.2
$Ti_{0.75}V_{0.25}$	5.3	0.029[c]	199[c]		0
$Ti_{0.775}V_{0.225}$	4.7	0.024[c]	172[c]		0
$Ti_{0.615}V_{0.385}$	7.07	0.050	34		4.2
$Ti_{0.516}V_{0.484}$	7.20	0.062	28		4.2
$Ti_{0.415}V_{0.585}$	7.49	0.078	25		4.2
$Ti_{0.12}V_{0.88}$			17.3	28.1	4.2
$Ti_{0.09}V_{0.91}$			14.3	16.4	4.2
$Ti_{0.06}V_{0.94}$			8.2	12.7	4.2

Substance	T_c, K	H_{c1}, kOe	H_{c2}, kOe	H_{c3}, kOe	T_{obs}, K[a]
$Ti_{0.03}V_{0.97}$			3.8	6.8	4.2
Ti_xV_{1-x}			108 max.		1.2
V	5.31	0.8	3.4		1.79
		0.75	3.15		2
		0.45	2.2		3
		0.30	1.2		4
$V_{0.26}Zr_{0.74}$	≈5.9	0.238			1.05
		0.227			1.78
		0.185			3.04
		0.165			3.5
W (film)	1.7–4.1		>34		1

[a] Temperature of critical field measurement.
[b] w/o denotes weight percent.
[c] Extrapolated.
[d] Linear extrapolation.
[e] Parabolic extrapolation.

References

1. B. W. Roberts, in *Superconductive Materials and Some of Their Properties. Progress in Cryogenics*, Vol. IV, 1964, pp. 160–231.
2. B. W. Roberts, Superconductive Materials and Some of Their Properties, NBS Technical Notes 408 and 482, U.S. Government Printing Office, 1966 and 1969; B. W. Roberts, *J. Phys. Chem. Ref. Data*, 5, 581, 1976.
3. B. W. Roberts, Properties of Selected Superconductive Materials, 1978 Supplement, NBS Technical Note 983, 1978.
4. T. Claeson, *Phys. Rev.*, 147, 340, 1966.
5. C. J. Raub, W. H. Zachariasen, T. H. Geballe, and B. T. Matthias, *J. Phys. Chem. Solids*, 24, 1093, 1963.
6. T. H. Geballe, B. T. Matthias, V. B. Compton, E. Corenzwit, G. W. Hull, Jr., and L. D. Longinotti, *Phys. Rev.*, 1A, 119, 1965.
7. C. J. Raub, V. B. Compton, T. H. Geballe, B. T. Matthias, J. P. Maita, and G. W. Hull, Jr., *J. Phys. Chem. Solids*, 26, 2051, 1965.
8. R. D. Blaugher, J. K. Hulm, and P. N. Yocom, *J. Phys. Chem. Solids*, 26, 2037, 1965.
9. T. Claeson and H. L. Luo, *J. Phys. Chem. Solids*, 27, 1081, 1966.
10. S. C. Ng and B. N. Brockhouse, *Solid State Comm.*, 5, 79, 1967.
11. O. I. Shulishova and I. A. Shcherbak, *Izv. AN SSSR, Neorg. Materials*, 3, 1495, 1967.
12. T. F. Smith and H. L. Luo, *J. Phys. Chem. Solids*, 28, 569, 1967.
13. A. C. Lawson, *J. Less-Common Metals*, 23, 103, 1971.
14. R. Chevrel, M. Sergent, and J. Prigent, *J. Solid State Chem.*, 3, 515, 1971.
15. M. Marezio, P. D. Dernier, J. P. Remeika, and B. T. Matthias, *Mat. Res. Bull.*, 8, 657, 1973.
16. J. K. Hulm and R. D. Blaugher in *Superconductivity in d- and f-Band Metals*, D. H. Douglass, Ed., American Institute of Physics, 4, 1, 1972.
17. R. N. Shelton, A. C. Lawson, and D. C. Johnston, *Mat. Res. Bull.*, 10, 297, 1975.
18. H. D. Wiesinger, *Phys. Status Sol.*, 41A, 465, 1977.
19. O. Fisher, *Applied Phys.*, 16, 1, 1978.
20. D. C. Johnston, *Solid State Comm.*, 24, 699, 1977.
21. H. C. Ku and R. H. Shelton, *Mat. Res. Bull.*, 15, 1441, 1980.
22. H. Barz, *Mat. Res. Bull.*, 15, 1489, 1980.
23. G. P. Espinosa, A. S. Cooper, H. Barz, and J. P. Remeika, *Mat. Res. Bull.*, 15, 1635, 1980.
24. E. M. Savitskii, V. V. Baron, Yu. V. Efimov, M. I. Bychkova, and L. F. Myzenkova, in *Superconducting Materials*, Plenum Press, 1981, p. 107.
25. R. Fluckiger and R. Baillif, in Topics in *Current Physics*, O. Fischer and M. B. Maple, Eds., Springer Verlag, 34, 113, 1982.
26. R. N. Shelton, in *Superconductivity in d- and f-Band Metals*, W. Buckel and W. Weber, Eds., Kernforschungszentrum, Karlsruhe, 1982, p. 123.
27. D. C. Johnston and H. F. Braun, *Topics in Current Phys.*, 32, 11, 1982.
28. R. Chevrel and M. Sergent, *Topics in Current Phys.*, 32, 25, 1982.
29. G. P. Espinosa, A. S. Cooper, and H. Barz, *Mat. Res. Bull.*, 17, 963, 1982.
30. R. Muller, R. N. Shelton, J. W. Richardson, Jr., and R. A. Jacobson, *J. Less-Comm. Met.*, 92, 177, 1983.
31. You-Xian Zhao and Shou-An He, in *High Pressure in Science and Technology*, North Holland, 22, 51, 1983.
32. You-Xian Zhao and Shou-An He, *Solid State Comm.*, 24, 699, 1983.
33. G. P. Meisner and H. C. Ku, *Appl. Phys.*, A31, 201, 1983.
34. R. J. Cava, D. W. Murphy, and S. M. Zahurak, *J. Electrochem. Soc.*, 130, 2345, 1983.
35. R. N. Shelton, *J. Less-Comm. Met.*, 94, 69, 1983.
36. B. Chevalier, P. Lejay, B. Lloret, Wang Xian–Zhong, J. Etourneau, and P. Hagenmuller, *Annales de Chemie*, 9, 191, 1984.
37. G. Venturini, M. Meot-Meyer, E. McRae, J. F. Mareche, and B. Rogues, *Mat. Res. Bull.*, 19, 1647, 1984.
38. J. M. Tarascon, F. G. DiSalvo, D. W. Murphy, G. Hull, and J. V. Waszczak, *Phys. Rev.*, 29B, 172, 1984.
39. G. V. Subba and G. Balakrishnan, *Bull. Mat. Sci.*, 6, 283, 1984.
40. B. Batlog, *Physica*, 126B, 275, 1984.
41. M. J. Johnson, Ames Lab (USA) Report IS-T-1140, 1984.
42. I. M. Chapnik, *J. Mat. Sci. Lett.*, 4, 370, 1985.
43. W. Rong-Yao, L. Q-Guang, and Z. Xiao, *Phys. Status Sol.*, 90A, 763, 1985.
44. W. Xian-Zhong, B. Chevalier, J. Etourneau, and P. Hagenmuller, *Mat. Res. Bull.*, 20, 517, 1985.
45. H. R. Ott, F. Hulliger, H. Rudigier, and Z. Fisk, *Phys. Rev.*, 31B, 1329, 1985.
46. P. Villars and L. D. Calver, *Pearson's Handbook of Crystallographic Data for Intermetallic Phases*, Vol. 1–3, ASM, 1985.
47. G. V. Subba Rao, K. Wagner, G. Balakhrishnan, J. Jakani, W. Paulus, and R. Scollhorn, *Bull. Mat. Sci.*, 7, 215, 1985.
48. J. G. Bednorz and K. A. Muller, *Zs. Physik*, B64, 189, 1986.
49. W. Rong-Yao, *Phys. Status Sol.*, 94A, 445, 1986.
50. H. D. Yang, R. N. Shelton, and H. F. Braun, *Phys. Rev.*, 33B, 5062, 1986.
51. G. Venturini, M. Kanta, E. McRae, J. F. Mareche, B. Malaman, and B. Roques, *Mat. Res. Bull.*, 21, 1203, 1986.
52. W. Rong-Yao, *J. Mat. Sci. Lett.*, 5, 87, 1986.
53. M. K. Wu, J. R. Ashburn, C. J. Torng, P. H. Hor, R. L. Meng, L. Gao, Z. J. Huang, Y. Q. Wang, and C. W. Chu, *Phys. Rev. Lett.*, 58, 908, 1987.
54. R. J. Cava, R. B. Van Dover, B. Batlog, and E. A. Rietman, *Phys. Rev. Lett.*, 58, 408, 1987.
55. L. C. Porter, T. J. Thorn, U. Geiser, A. Umezawa, H. H. Wang, W. K. Kwok, H-C. I. Kao, M. R. Monaghan, G. W. Crabtree, K. D. Carlson, and J. M. Williams, *Inorg. Chem.*, 26, 1645, 1987.
56. A. M. Kini, U. Geiser, H-C. I. Kao, K. D. Carlson, H. H. Wang, M. R. Monaghan, and K. M. Williams, *Inorg. Chem.*, 26, 1834, 1987.

57. T. Penney, S. von Molnar, D. Kaiser, F. Holtzberg, and A. W. Kleinsasser, *Phys. Rev.*, B38, 2918, 1988.
58. Y. K. Tao, J. S. Swinnea, A. Manthiram, J. S. Kim, J. B. Goodenoug, and H. Steinfink, *J. Mat. Res.*, 3, 248, 1988.
59. G. G. Peterson, B. R. Weinberger, L. Lynds, and H. A. Krasinski, *J. Mat. Res.*, 3, 605, 1988.
60. J. B. Torrance, Y. Tokura, A. Nazzai, and S. S. P. Parkin, *Phys. Rev. Lett.*, 60, 542, 1988.
61. K. Kourtakis, M. Robbins, P. K. Gallagher, and T. Teifel, *J. Mat. Res.*, 4, 1289, 1989.
62. J. C. Phillips, *Physics of High-T_c Superconductors*, Academic Press, 1989, p. 336.
63. Shui Wai Lin and L. I. Berger, *Rev. Sci. Instrum.*, 60, 507, 1989.
64. M. Tinkham, *Introduction to Superconductivity*, McGraw–Hill, New York, 1975.
65. O. Fischer and M.B. Maple, Eds., *Topics in Current Physics*, Volume 32: Superconductivity in Ternary Compounds I; Volume 34: Superconductivity in Ternary Compounds II, Springer–Verlag, Berlin, 1982.
66. K. J. Dunn and F. P. Bundy, *Phys. Rev.*, B25, 194, 1982.
67. A. Barone and G. Paterno, *Physics and Applications of the Josephson Effect*, Wiley, New York, 1982.
68. D. H. Douglass, Ed., *Superconductivity in d- and f-Band Metals*, Plenum Press, New York, 1976.
69. D. M. Ginsberg, Ed., *Physical Properties of High Temperature Superconductors*, (Volume II, 1990; Volume III, 1992; Volume V, 1996), World Scientific, Singapore.
70. T. Ishiguro and K. Yamji, *Organic Superconductors*, Springer-Verlag, Berlin, 1990.
71. Sh. Okada, K. Shimizu, T. C. Kobayashi, K. Amaya, and Sh. Endo., *J. Phys. Soc. Jpn.*, 65, 1924, 1996.
72. A. Bourdillon and N. X. Tan Bourdillon, *High Temperature Superconductors: Processing and Science*, Academic Press, 1994.
73. J. M. Williams, J. R. Ferraro, R. J. Thorn, K. Carlson, U. Geiser, H. H. Wang, A. M. Kini, and M.-H. Whangbo, *Organic Superconductors (Including Fullerenes): Synthesis, Structure, Properties, and Theory*, Prentice–Hall, 1992.
74. J. Nagamatsu, N. Nakagawa, T. Muranaka, Y. Zenitani, and J. Akimitsu, *Nature (London)*, 410, 63, 2001.
75. Y. Boguslavsky, G. K. Perkins, X. Qi, L. F. Cohen, and A. D. Caplin, *Nature (London)*, 410, 563, 2001.
76. B. Q. Fu, Y. Feng, G. Yan, Y. Zhao, A. K. Pradhan, C. H. Cheng, P. Ji, X. H. Liu, C. F. Liu, L. Zhou, and K. F. Yau, *J. Appl. Phys.*, 92, 7341, 2002.

HIGH-TEMPERATURE SUPERCONDUCTORS

C. N. R. Rao and A. K. Raychaudhuri

The following tables give properties of a number of high-temperature superconductors. Table 1 lists the crystal structure (space group and lattice constants) and the critical transition temperature T_c for the more important high-temperature superconductors studied so far. Table 2 gives the energy gap, critical current density, and penetration depth in the superconducting state. Table 3 gives electrical and thermal properties of some of these materials in the normal state. The tables were prepared in November 1992 and updated in November 1994.

References

1. Ginsburg, D. M., Ed., *Physical Properties of High-Temperature Superconductors*, Vols. I–III, World Scientific, Singapore, 1989–1992.
2. Rao, C. N .R., Ed., *Chemistry of High-Temperature Superconductors*, World Scientific, Singapore, 1991.
3. Shackelford, J. F., *The CRC Materials Science and Engineering Handbook*, CRC Press, Boca Raton, 1992, 98–99 and 122–123.
4. Kaldis, E., Ed., *Materials and Crystallographic Aspects of HT_c-Superconductivity*, Kluwer Academic Publ., Dordrecht, The Netherlands, 1992.
5. Malik, S. K. and Shah, S. S., Ed., *Physical and Material Properties of High Temperature Superconductors*, Nova Science Publ., Commack, N.Y., 1994.
6. Chmaissem, O. et al., *Physica*, C230, 231–238, 1994.
7. Antipov, E. V. et al., *Physica*, C215, 1–10, 1993.

TABLE 1. Structural Parameters and Approximate T_c Values of High-Temperature Superconductors

Material	Structure	T_c/K (maximum value)
$La_2CuO_{4+\delta}$	Bmab; a = 5.355, b = 5.401, c = 13.15 Å	39
$La_{2-x}Sr_x(Ba_x)CuO_4$	I4/mmm; a = 3.779, c = 13.23 Å	35
$La_2Ca_{1-x}Sr_xCu_2O_6$	I4/mmm; a = 3.825, c = 19.42 Å	60
$YBa_2Cu_3O_7$	Pmmm; a = 3.821, b = 3.885, c = 11.676 Å	93
$YBa_2Cu_4O_8$	Ammm; a = 3.84, b = 3.87, c = 27.24 Å	80
$Y_2Ba_4Cu_7O_{15}$	Ammm; a = 3.851, b = 3.869, c = 50.29 Å	93
$Bi_2Sr_2CuO_6$	Amaa; a = 5.362, b = 5.374, c = 24.622 Å	10
$Bi_2CaSr_2Cu_2O_8$	A_2aa; a = 5.409, b = 5.420, c = 30.93 Å	92
$Bi_2Ca_2Sr_2Cu_3O_{10}$	A_2aa; a = 5.39, b = 5.40, c = 37 Å	110
$Bi_2Sr_2(Ln_{1-x}Ce_x)_2Cu_2O_{10}$	P4/mmm; a = 3.888, c = 17.28 Å	25
$Tl_2Ba_2CuO_6$	A_2aa; a = 5.468, b = 5.472, c = 23.238 Å; I4/mmm; a = 3.866, c = 23.239 Å	92
$Tl_2CaBa_2Cu_2O_8$	I4/mmm; a = 3.855, c = 29.318 Å	119
$Tl_2Ca_2Ba_2Cu_3O_{10}$	I4/mmm; a = 3.85, c = 35.9 Å	128
$Tl(BaLa)CuO_5$	P4/mmm; a = 3.83, c = 9.55 Å	40
$Tl(SrLa)CuO_5$	P4/mmm; a = 3.7, c = 9 Å	40
$(Tl_{0.5}Pb_{0.5})Sr_2CuO_5$	P4/mmm; a = 3.738, c = 9.01 Å	40
$TlCaBa_2Cu_2O_7$	P4/mmm; a = 3.856, c = 12.754 Å	103
$(Tl_{0.5}Pb_{0.5})CaSr_2Cu_2O_7$	P4/mmm; a = 3.80, c = 12.05 Å	90
$TlSr_2Y_{0.5}Ca_{0.5}Cu_2O_7$	P4/mmm; a = 3.80, c = 12.10 Å	90
$TlCa_2Ba_2Cu_3O_8$	P4/mmm; a = 3.853, c = 15.913 Å	110
$(Tl_{0.5}Pb_{0.5})Sr_2Ca_2Cu_3O_9$	P4/mmm; a = 3.81, c = 15.23 Å	120
$TlBa_2(La_{1-x}Ce_x)_2Cu_2O_9$	I4/mmm; a = 3.8, c = 29.5 Å	40
$Pb_2Sr_2La_{0.5}Ca_{0.5}Cu_3O_8$	Cmmm; a = 5.435, b = 5.463, c = 15.817 Å	70
$Pb_2(Sr,La)_2Cu_2O_6$	$P22_12$; a = 5.333, b = 5.421, c = 12.609 Å	32
$(Pb,Cu)Sr_2(La,Ca)Cu_2O_7$	P4/mmm; a = 3.820, c = 11.826 Å	50
$(Pb,Cu)(Sr,Eu)(Eu,Ce)Cu_2O_x$	I4/mmm; a = 3.837, c = 29.01 Å	25
$Nd_{2-x}Ce_xCuO_4$	I4/mmm; a = 3.95, c = 12.07 Å	30
$Ca_{1-x}Sr_xCuO_2$	P4/mmm; a = 3.902, c = 3.35 Å	110
$Sr_{1-x}Nd_xCuO_2$	P4/mmm; a = 3.942, c = 3.393 Å	40
$Ba_{0.6}K_{0.4}BiO_3$	Pm3m; a = 4.287 Å	31
Rb_2CsC_{60}	a = 14.493 Å	31
$NdBa_2Cu_3O_7$	Pmmm; a = 3.878, b = 3.913, c = 11.753	58
$SmBaSrCu_3O_7$	I4/mmm; a = 3.854, c = 11.62	84
$EuBaSrCu_3O_7$	I4/mmm; a = 3.845, c = 11.59	88
$GdBaSrCu_3O_7$	I4/mmm; a = 3.849, c = 11.53	86
$DyBaSrCu_3O_7$	Pmmm; a = 3.802, b = 3.850, c = 11.56	90
$HoBaSrCu_3O_7$	Pmmm; a = 3.794, b = 3.849, c = 11.55	87
$ErBaSrCu_3O_7$ (multiphase)	Pmmm; a = 3.787, b = 3.846, c = 11.54	82
$TmBaSrCu_3O_7$ (multiphase)	Pmmm; a = 3.784, b = 3.849, c = 11.55	88
$YBaSrCu_3O_7$	Pmmm; a = 3.803, b = 3.842, c = 11.54	84
$HgBa_2CuO_4$	I4/mmm; a = 3.878, c = 9.507	94
$HgBa_2CaCu_2O_6$ (annealed in O_2)	I4/mmm; a = 3.862, c = 12.705	127
$HgBa_2Ca_2Cu_3O_8$	Pmmm; a = 3.85, c = 15.85	133
HgBa2Ca3Cu4O10	Pmmm; a = 3.854, c = 19.008	126

TABLE 2. Superconducting Properties

J_c (0): Critical current density extrapolated to 0 K
λ_{ab}: Penetration depth in *a-b* plane
k_B: Boltzmann constant

Material	Form	Energy gap (Δ) $2\Delta_{pp}/k_BT_c$*	Energy gap (Δ) $2\Delta_{fit}/k_BT_c$†	$10^{-6} \times J_c$ (0)/A cm^{-2}	λ_{ab}/Å
$YBa_2Cu_3O_7$	Single Crystal	5–6	4–5	30 (film)	1400
$Bi_2Sr_2CaCu_2O_8$	Single Crystal	8–9	5.5–6.5	2	2700
$Tl_2Ba_2CaCu_2O_8$	Ceramic	6–7	4–6	10 (film, 80 K)	2000
$La_{2-x}Sr_xCuO_4$, x = 0.15	Ceramic	7–9	4–6		
$Nd_{2-x}Ce_xCuO_4$	Ceramic	8	4–5	0.2 (film)	

* Obtained from peak to peak value.
† Obtained from fit to BCS-type relation.

TABLE 3. Normal State Properties

ρ_{ab}: Resistivity in the *a-b* plane
ρ_c: Resistivity along the *c* axis
+ve: ρ_c has positive temperature coefficient of resistivity
−ve: ρ_c has negative temperature coefficient of resistivity
n_H: Hall density
k: Thermal conductivity
in plane: Along *a-b* plane
out of plane: Perpendicular to *a-b* plane

Material	Form	$\rho_{ab}/\mu\Omega$ cm 300 K	$\rho_{ab}/\mu\Omega$ cm 100 K	ρ_c/mΩ cm 300 K	$d\rho_c/dT$	$10^{-21} \times n_H$/cm^{-3} 300 K	$10^{-21} \times n_H$/cm^{-3} 100 K	*k*/(mW/cm K) at 300 K in plane	*k*/(mW/cm K) at 300 K out of plane
$YBa_2Cu_3O_7$	Single crystal	110	35	5	+ve	11–16	4–6	120	3
	Film	200–300	60–100			5–9	2–3		
$YBa_2Cu_4O_8$	Single crystal	75	20	10	−ve	14			
	Film	100–200	20–50			22	17		
$Bi_2Sr_2CuO_6$	Single crystal	300	150	5000	−ve	6	5		
$Bi_2Sr_2CaCu_2O_8$	Single crystal	150	50	>1000	−ve	4	3	60	8
$Tl_2Ba_2CuO_6$	Single crystal	300–400	50–75	200–300	+ve	3.1	2.5		
$Tl_2Ba_2Ca_2Cu_3O_{10}$	Ceramic	***	**				≈ 2*		
$La_{2-x}Sr_xCuO_4$, *x* = 0.12	Single crystal	900	350	200	+ve for *T* >225 K	2.5			
$La_{2-x}Sr_xCuO_4$, *x* = 0.20	Single crystal	400	200	80	+ve for *T* >150 K	10		50 (for *x* = 0.04)	20
	Film	400	160			8.4	6.3		
$Nd_{2-x}Ce_xCuO_4$, *x* = 0.17	Single crystal	500	275			53	17		
x = 0.15	Film	140–180	35			32	11	250 (for *x* = 0.15)	

* At 200 K
** ρ ~0.4 mΩ cm at 120 K
*** ρ ~1.5 mΩ cm at 300 K

ORGANIC SUPERCONDUCTORS

H. P. R. Frederikse

Although the vast majority of organic compounds are insulators, a small number of organic solids show considerable electrical conductivity. Some of these materials appear to be superconductors. The superconducting organics fall primarily into two groups: those containing fulvalenes (pentagonal rings containing sulfur or selenium) and those based on fullerenes, involving the nearly spherical cluster C_{60}.

The transition temperatures T_c of the fulvalene derivatives are shown in Table 1. The abbreviations of the various molecular groups are listed in Table 2 and their chemical structures are depicted in Figure 1. Most of the T_c's are between 1 and 12 K. Several of the compounds only show superconductivity under pressure.

The fullerenes are A_3C_{60} compounds, where A represents a single or a combination of alkali atoms. The C_{60} cluster is shown in Figure 2a, while Figure 2b illustrates how the alkali atoms fit into the A_3C_{60} molecule to form the A15 crystallographic structure. Their superconducting transition temperatures range from 8 to 31.3 K (see Table 3).

References

1. Ishigura, T. and Yamaji, K., *Organic Superconductors*, Springer-Verlag, Berlin, 1990.
2. Williams, J. M. et al., *Organic Superconductors (Including Fullerenes)*, Prentice Hall, Englewood Cliffs, N.J., 1992.
3. *The Fullerenes*, Ed.: Krato, H. W., Fisher, J. E., and Cox, D. E., Pergammon Press, Oxford, 1993.
4. Schluter, M. et al., in *The Fullerenes* (Ref. 3), p. 303.

TABLE 1. Critical Pressure and Maximum Critical Temperature of Organic Superconductors

Material	P_c/kbar	T_c/K	Material	P_c/kbar	T_c/K
$(TMTSF)_2PF_6$	6.5	1.2	β-$(ET)_2IBr_2$	0	2.8
$(TMTSF)_2AsF_6$	9	1.3	β-$(ET)_2AuI_2$	0	4.8
$(TMTSF)_2SbF_6$	11	0.4	$(ET)_4Hg_{2.89}Cl_8$	0	4.2
$(TMTSF)_2TaF_6$	12	1.4	$(ET)_4Hg_{2.89}Br_8$	12	1.8
$(TMTSF)_2ClO_4$	0	1.4	$(ET)_3Cl_2(H_2O)_2$	16	2
$(TMTSF)_2ReO_4$	9.5	1.3	κ-$(ET)_2Cu(NCS)_2$	0	10.4
$(TMTSF)_2FSO_3$	5	3	κ-(d-ET)$_2$Cu(NCS)$_2$	0	11.4
$(ET)_4(ReO_4)_2$	4.5	2	$(DMET)_2Au(CN)_2$	1.5	0.9
$β_L$-$(ET)_2I_3$	0	1.4	$(DMET)_2AuI_2$	5	0.6
$β_H$-$(ET)_2I_3$	0	8.1	$(DMET)_2AuBr_2$	0	1.9
γ-$(ET)_3I_{2.5}$	0	2.5	$(DMET)_2AuCl_2$	0	0.9
ε-$(ET)_2I_3(I_8)_{0.5}$	0	2.5	$(DMET)_2I_3$	0	0.6
α-$(ET)_2I_3I_2$-doped	0	3.3	$(DMET)_2IBr_2$	0	0.7
$α_t$-$(ET)_2I_3$	0	8	(MDT-TTF)$_2$AuI$_2$	0	3.5
ε→β-$(ET)_2I_3$[a]	0	6	TTF[Ni(dmit)$_2$]$_2$	2	1.6[b]
θ-$(ET)_2I_3$	0	3.6	TTF[Pd(dmit)$_2$]$_2$	20	6.5
κ-$(ET)_2I_3$	0	3.6	$(CH_3)_4$N[Ni(dmit)$_2$]$_2$	7	5

[a] Converted form ε-type to β-type by thermal treatment.
[b] For 7 kbar.
From Ishigura, T. and Yamaji, K., *Organic Superconductors*, Springer-Verlag, Berlin, 1990. With permission.

TABLE 2. List of Symbols and Abbreviations

TTF	tetrathiafulvalene
TMTSF	tetramethyltetraselenafulvalene
BEDT-TTF or "ET"	bis(ethylenedithio)tetrathiafulvalene
MDT-TTF	methylenedithiotetrathiafulvalene
DMET	[dimethyl(ethylenedithio)diselenadithiafulvalene]
dmit	4,5-dimercapto-1,3-dithiole-2-thione
T_c	transition temperature to superconducting state
P_c	minimum pressure required for superconducting transition

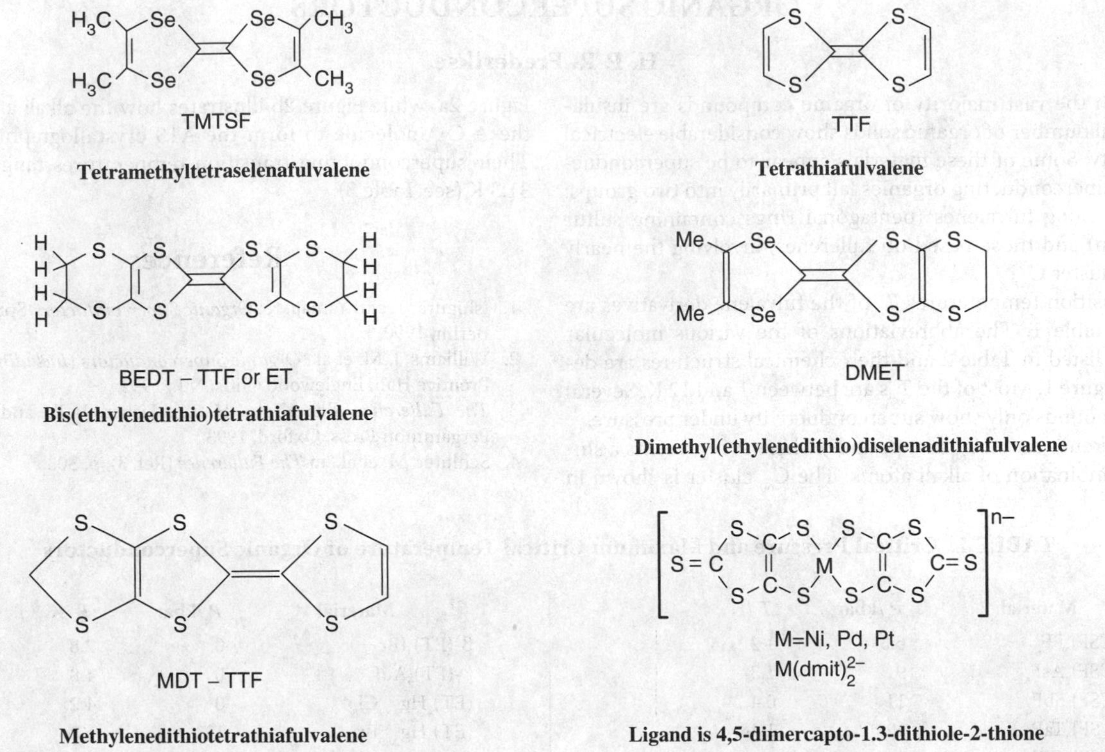

TMTSF

Tetramethyltetraselenafulvalene

TTF

Tetrathiafulvalene

BEDT – TTF or ET

Bis(ethylenedithio)tetrathiafulvalene

DMET

Dimethyl(ethylenedithio)diselenadithiafulvalene

MDT – TTF

Methylenedithiotetrathiafulvalene

M=Ni, Pd, Pt

$M(dmit)_2^{2-}$

Ligand is 4,5-dimercapto-1.3-dithiole-2-thione

FIGURE 1. Structures of various donor molecules and acceptor species.

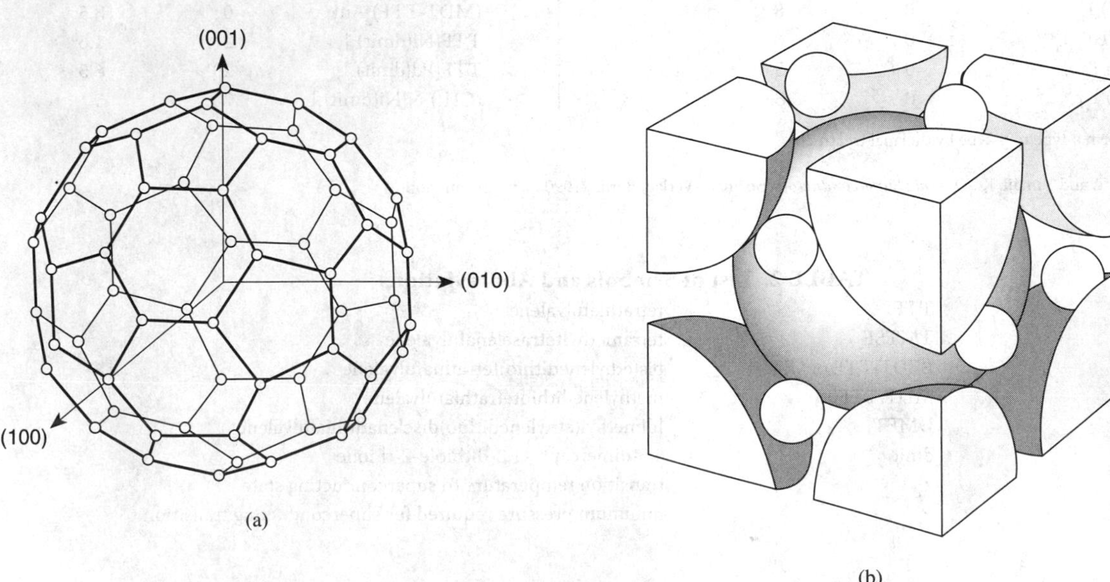

FIGURE 2. (a) C_{60} cluster placed in a fcc lattice. Each crystal axis crosses a double bond shared by two hexagons. (b) A hypothetical A_3C_{60} with the A15 structure. The structure can be seen to be an ordered defect structure of A_6C_{60}.

TABLE 3. Unit Cell and T_c for FCC-A_3 C_{60}

	Lattice parameter(s) (Å)	T_c/K
$Na_2Rb_{0.5}Cs_{0.5}C_{60}$	14.148(3)	8.0
Na_2CsC_{60} No. 1[a]	14.132(2)	10.5
Na_2CsC_{60} No. 2[a]	14.176(9)	14.0
K_3C_{60}	14.253(3)	19.3
K_2RbC_{60}	14.299(2)	21.8
Rb_2KC_{60} No. 1[a]	14.336(1)	24.4
Rb_2KC_{60} No. 2[a]	14.364(5)	26.4
Rb_3C_{60}	14.436(2)	29.4
Rb_2CsC_{60}	14.493(2)	31.3

[a] Samples labeled No. 1 and No. 2 have the same nominal composition.

From Schluter, M et al., *The Fullerenes*, Ed.: Krato, H.W., Fisher, J.E., and Cox, D.E., Pergamon Press, Oxford, 1993. With permission.

PROPERTIES OF SEMICONDUCTORS

L. I. Berger

The term *semiconductor* is applied to a material in which electric current is carried by electrons or holes and whose electrical conductivity, when extremely pure, rises exponentially with temperature and may be increased from its low "intrinsic" value by many orders of magnitude by "doping" with electrically active impurities.

Semiconductors are characterized by an energy gap in the allowed energies of electrons in the material that separates the normally filled energy levels of the *valence band* (where "missing" electrons behave like positively charged current carriers "holes") and the *conduction band* (where electrons behave rather like a gas of free negatively charged carriers with an effective mass dependent on the material and the direction of the electrons' motion). This energy gap depends on the nature of the material and varies with direction in anisotropic crystals. It is slightly dependent on temperature and pressure, and this dependence is usually almost linear at normal temperatures and pressures.

Data are presented in five tables. Table 1 lists the main crystallographic and semiconducting properties of a large number of semiconducting materials in three main categories: "Tetrahedral Semiconductors" in which every atom is tetrahedrally coordinated to four nearest neighbor atoms (or atomic sites) as for example in the diamond structure; "Octahedral Semiconductors" in which every atom is octahedrally coordinated to six nearest neighbor atoms—as for example the halite structure; and "Other Semiconductors."

Table 2 gives electrical, magnetic, and optical properties, while Tables 3 and 4 give more details on the semiconducting properties and band structures of the most common semiconductors. Table 5 lists semiconducting minerals with typical resistivity ranges.

TABLE 1. Physico-Chemical Properties of Semiconductors (Listed by Crystal Structure)

Substance	Molecular weight	Average atomic weight	Lattice parameters (Å, room temp.)	Density (g/cm³)	Melting point (K)	Microhardness, N/mm² (M-Mohs Scale)	Specific heat, J/kg·K (300 K)	Debye temp. (K)	Coefficient of thermal linear expansion [10^{-6} K^{-1} (300K)]	Thermal conductivity [mW/cm·K (300K)]
1.1. Tetrahedral (Adamantine) Semiconductors										
1.1.1. Diamond Structure Elements (Strukturbericht symbol A4, Space Group Fd3m-O_h^7)										
C (Diamond)		12.01	3.56683	3.513	≈4713 (12.4 GPa) Transition to graphite > 980	10 (M)	471.5	2340	1.18	9900(I) 23200(IIA) 13600(IIB)
Si		28.09	5.43072	2.329	1687	11270	702	645	2.6	1240
Ge		72.64	5.65754	5.323	1211.35	7644	321.9	374	5.8	640
α-Sn		118.71	6.4912	5.769	505.1 (Tr. 286.4)		213	230	5.4 (220 K)	
1.1.2. Sphalerite (Zinc Blende) Structure Compounds (Strukturbericht symbol B3 Space Group F $\bar{4}$ 3m-T_d^2)										
I-VII Compounds										
CuF	82.54	41.27	4.255		1181					
CuCl	98.99	49.49	5.4057	3.53	695	2.3 (M)	490	240	12.1	8.4
CuBr	143.45	71.73	5.6905	4.98	770	2.5 (M)	381	207	15.4	12.5
CuI	190.45	95.23	6.60427	5.63	878	192	276	181	19.2	16.8
AgBr	187.77	93.89		6.473	>1570 (Tr. 410)	2.5 (M)	270			
AgI	234.77	117.39	6.502	5.67	831	2.5 (M)	232	134	–2.5	4.2
II-VI Compounds										
BeS	41.08	20.54	4.865	2.36	dec.					
BeSe	87.97	43.99	5.139	4.315						
BeTe	136.61	68.31	5.626	5.090						
BePo	(2318)	(109)	5.838	7.3						
ZnO	81.39	40.69	4.63	5.675	2248	5.0 (M)	494	416	2.9	234
ZnS	97.46	48.72	5.4093	4.079	2100 (Tr. 1295)	1780	472	530	6.36	251

Substance	Molecular weight	Average atomic weight	Lattice parameters (Å, room temp.)	Density (g/cm³)	Melting point (K)	Microhardness, N/mm² (M-Mohs Scale)	Specific heat, J/kg·K (300 K)	Debye temp. (K)	Coefficient of thermal linear expansion [10^{-6} K^{-1} (300K)]	Thermal conductivity [mW/cm·K (300K)]
ZnSe	144.34	72.17	5.6676	5.42	1790	1350	339	400	7.2	140
ZnTe	192.99	96.5	6.101	6.34	1568	900	264	223	8.19	108
ZnPo	(274)	(137)	6.309							
CdS	144.48	72.24	5.832	4.826	1750	1250	330	219	4.7	200
CdSe	191.37	95.68	6.05	5.674	1512	1300	255	181	3.8	90
CdTe	240.01	120.00	6.477	5.86	1365	600	205	200	4.9	58.5
CdPo	(321)	(161)	6.665							
HgS	232.66	116.33	5.8517	7.73	1820	3 (M)	210			
HgSe	279.55	139.78	6.084	8.25	1070	2.5 (M)	178	151	5.46	10
HgTe	328.19	164.10	6.4623	8.17	943	300	164	242	4.6	20
III-V Compounds										
BN	24.82	12.41	3.615	3.49	3239	10 (M)	793	≈1900		200
BP(L.T.)	41.78	20.87	4.538	2.9	1398 (dec)	37000		≈980		
BAs	85.73	42.87	4.777		≈2300	19000		≈625		
AlP	57.95	28.98	5.451	2.42	≈2100	5.5 (M)		588		920
AlAs	101.90	50.95	5.6622	3.81	2013	5000		417	3.5	840
AlSb	148.74	74.37	6.1355	4.218	1330	4000		292	4.2	600
GaP	100.70	50.35	5.4905	4.13	1750	9450		446	5.3	752
GaAs	144.64	72.32	5.65315	5.316	1510	7500		344	5.4	560
GaSb	191.48	95.74	6.0954	5.619	980	4480	320	265	6.1	270
InP	145.79	72.90	5.86875	4.787	1330	4100		321	4.6	800
InAs	189.74	94.87	6.05838	5.66	1215	3300	268	249	4.7	290
InSb	236.58	118.29	6.47877	5.775	798	2200	144	202	4.7	160
Other sphalerite structure compounds										
MnS	87.00	43.5	5.011							
MnSe	133.90	66.95	5.82							
β-SiC (3-C SiC)	40.10	20.1	4.348	3.21	3070				2.9	4.9
Ga_2Se_3	376.32	75.26	5.429	4.92	1020	3160			8.9	50
Ga_2Te_3	522.24	104.45	5.899	5.75	1063	2370				47
In_2Te_3(H.T.)	608.44	121.7	6.173	5.8	940	1660				69
$MgGeP_2$	158.84	39.71	5.652							
$ZnSnP_2$	246.00	61.5	5.65		1200					
$ZnSnAs_2$(H.T.)	333.90	82.38	5.851	5.53	1050					76
$ZnSnSb_2$	427.56	106.89	6.281	5.67	870	2500				76
1.1.3. Wurtzite (Zincite) Structure Compounds (Strukturbericht symbol B4, Space Group P 6_3mc-C^4_{6v})										
I-VII Compounds										
CuCl	99.0	49.5	3.91 6.42		703					
CuBr	143.45	71.73	4.06 6.66		770					
Cul	190.45	95.23	4.31 7.09							
Agl	234.77	117.40	4.580 7.494							
II-VI Compounds										
BeO	25.01	12.51	2.698 4.380		2800					
MgTe	151.9	76.0	4.54 7.39	3.85	≈2800					
ZnO	81.37	40.69	3.24950 5.2069	5.66	2250					600
ZnS	97.43	48.72	3.8140 6.2576	4.1	2100					460

Substance	Molecular weight	Average atomic weight	Lattice parameters (Å, room temp.)		Density (g/cm³)	Melting point (K)	Microhardness, N/mm² (M-Mohs Scale)	Specific heat, J/kg·K (300 K)	Debye temp. (K)	Coefficient of thermal linear expansion [10^{-6} K^{-1} (300K)]	Thermal conductivity [mW/cm·K (300K)]
ZnTe	192.99	46.50	4.27	6.99		1568					
CdS	144.48	72.23	4.1348	6.7490	4.82	1748					401
CdSe	191.37	95.68	4.299	7.010	5.66	1512					316
CdTe	240.01	120.00	4.57	7.47							
III-V Compounds											
BP(H.T.)	41.79	20.90	3.562	5.900							
AlN	40.99	20.50	3.111	4.978	3.26	≈2500					823
GaN	83.73	41.87	3.190	5.189	6.10	1500					656
InN	128.83	64.42	3.533	5.693	6.88	1200					556
Other wurtzite structure compounds											
MnS	87.00	43.5	3.985	6.45	3.248						
MnSe	133.90	66.95	4.12	6.72							
SiC	40.10	20.1	3.076	5.048							
MnTe	182.54	91.27	4.078	6.701							
Al_2S_3	150.14	30.03	3.579	5.829	2.55	1400					
Al_2Se_3	290.84	58.17	3.890	6.30	3.91	1250					

1.1.4. Chalcopyrite Structure Compounds (Strukturbericht symbol E1$_1$, Space Group I $\bar{4}$ 2d-D^{12}_{24})

Substance	Molecular weight	Average atomic weight	Lattice parameters (Å, room temp.)		Density (g/cm³)	Melting point (K)	Microhardness, N/mm² (M-Mohs Scale)	Specific heat, J/kg·K (300 K)	Debye temp. (K)	Coefficient of thermal linear expansion [10^{-6} K^{-1} (300K)]	Thermal conductivity [mW/cm·K (300K)]
I-III-VI$_2$ Compounds											
$CuAlS_2$	154.65	38.66	5.323	10.44	3.47	2500					
$CuAlSe_2$	248.45	62.11	5.617	10.92	4.70	2260					
$CuAlTe_2$	345.73	86.43	5.976	11.80	5.50	2550					
$CuGaS_2$	197.39	49.53	5.360	10.49	4.35	2300					
$CuGaSe_2$	291.19	72.80	5.618	11.01	5.56	1970	4200		275	5.4	42
$CuGaTe_2$	388.47	97.12	6.013	11.93	5.99	2400	3500			6.9	27
$CuInS_2$	242.49	60.62	5.528	11.08	4.75	1400	2550				
$CuInSe_2$	336.29	84.07	5.785	11.56	5.77	1600	2050			6.6	37
$CuInTe_2$	433.57	108.39	6.179	12.365	6.10	1660	400		195	7.1	49
$CuTlS_2$	322.05	83.01	5.580	11.17	6.32						
$CuTlSe_2$(L.T.)	425.85	106.46	5.844	11.65	7.11	900					
$CuFeS_2$	183.51	45.88	5.29	10.32	4.088	1135					
$CuFeSe_2$	277.31	69.33				850					
$CuLaS_2$	266.58	66.65	5.65	10.86							
$AgAlS_2$	198.97	49.74	5.707	10.28	3.94						
$AgAlSe_2$	292.77	73.19	5.968	10.77	5.07	1220					
$AgAlTe_2$	390.05	97.51	6.309	11.85	6.18	1000					
$AgGaS_2$	241.71	60.43	5.755	10.28	4.72						
$AgGaSe_2$	335.51	83.88	5.985	10.90	5.84	1120	4400				
$AgGaTe_2$	432.79	108.2	6.301	11.96	6.05	990	1800		212		10
$AgInS_2$(L.T.)	286.87	71.70	5.828	11.19	5.00		2250				
$AgInSe_2$	380.61	95.15	6.102	11.69	5.81	1053	1850				30
$AgInTe_2$	477.89	119.47	6.42	12.59	6.12	965				9.49, 0.69	
$AgFeS_2$	227.83	56.96	5.66	10.30	4.53						
II-IV-V$_2$ Compounds											
$ZnSiP_2$	155.40	38.85	5.400	10.441	3.39	1640	1100				
$ZnGeP_2$	199.90	49.98	5.465	10.771	4.17	1295	8100				180
$ZnSnP_2$	246.00	61.5					6500				
$CdSiP_2$	202.43	50.61	5.678	10.431	4.00	≈1470	10500		282		

Substance	Molecular weight	Average atomic weight	Lattice parameters (Å, room temp.)		Density (g/cm³)	Melting point (K)	Microhardness, N/mm² (M-Mohs Scale)	Specific heat, J/kg·K (300 K)	Debye temp. (K)	Coefficient of thermal linear expansion [10^{-6} K^{-1} (300K)]	Thermal conductivity [mW/cm·K (300K)]
$CdGeP_2$	246.94	61.74	5.741	10.775	4.48	1049	5650				110
$CdSnP_2$	243.03	73.26	5.900	11.518		840	5000		195		140
$ZnSiAs_2$	242.20	60.55	5.61	10.88	4.70	1311	9200				
$ZnGeAs_2$	287.80	71.95	5.672	11.153	5.32	1150	6800		263		110
$ZnSnAs_2$	333.90	83.48	5.8515	11.704	5.53	1048	4550		271		150
$CdSiAs_2$	290.34	72.58	5.884	10.882		>1120	6850				
$CdGeAs_2$	334.83	83.71	5.9427	11.2172	5.60	938	4700				48
$CdSnAs_2$	380.93	95.23	6.0944	11.9182	5.72	880	3450				40

1.1.5. Other Ternary Semiconductors with Tetrahedral Coordination

I_2-IV-VI_3 Compounds

Substance	Molecular weight	Average atomic weight	Lattice parameters (Å, room temp.)		Density (g/cm³)	Melting point (K)	Microhardness, N/mm² (M-Mohs Scale)	Specific heat, J/kg·K (300 K)	Debye temp. (K)	Coefficient of thermal linear expansion [10^{-6} K^{-1} (300K)]	Thermal conductivity [mW/cm·K (300K)]
Cu_2SiS_3(H.T.)	251.36	41.89	3.684	6.004	3.81	1200					23
Cu_2SiS_3(L.T.)			5.290	10.156	3.63						
Cu_2SiTe_3	537.98	89.66	5.93		5.47						
Cu_2GeS_3(H.T.)	295.88	49.31	5.317		4.45	1210	4550	510	254	7.2	12
Cu_2GeS_3(L.T.)			5.327	5.215	4.46						
Cu_2GeSe_3	436.56	72.76	5.589	5.485	5.57	1030	3840	340	168	8.4	24
Cu_2GeTe_3	582.51	97.09	5.958	5.935	5.92		2890				130
Cu_2SnS_3	341.98	57.00	5.436		5.02	1110	2770	440	214	7.8	28
$CuSnSe_3$	482.66	80.44	5.687		5.94	960	2510	310	148	8.9	35
Cu_2SnTe_3	628.61	104.77	6.048		6.51	680	1970				144
Ag_2GeSe_3	525.21	87.54				810					
Ag_2SnSe_3	571.31	95.22									
Ag_2GeTe_3	671.13	111.86				600					
Ag_2SnTe_3	717.23	119.54									

I_3-V -VI_4-Compounds

Substance	Molecular weight	Average atomic weight	Lattice parameters (Å, room temp.)		Density (g/cm³)	Melting point (K)	Microhardness, N/mm² (M-Mohs Scale)	Specific heat, J/kg·K (300 K)	Debye temp. (K)	Coefficient of thermal linear expansion [10^{-6} K^{-1} (300K)]	Thermal conductivity [mW/cm·K (300K)]
Cu_3PS_4	349.85	40.73	7.44	6.19							
Cu_3AsS_4	393.79	49.22	6.43	6.14	4.37	931				3.2	30.2
Cu_3AsSe_4	581.37	72.67	5.570	10.957	5.61	733			169	9.5	19
Cu_3SbS_4	440.64	55.08	5.38	16.76	4.90	830					
Cu_3SbSe_4	628.22	78.53	5.654	11.256	6.0	700			131	12.4	14.6

I-IV_2-V_3 Compounds

Substance	Molecular weight	Average atomic weight	Lattice parameters (Å, room temp.)		Density (g/cm³)	Melting point (K)	Microhardness, N/mm² (M-Mohs Scale)	Specific heat, J/kg·K (300 K)	Debye temp. (K)	Coefficient of thermal linear expansion [10^{-6} K^{-1} (300K)]	Thermal conductivity [mW/cm·K (300K)]
$CuSi_2P_3$	212.64	35.44	5.25								
$CuGe_2P_3$	301.65	50.28	5.375		4.318	1113	8500	429	8.21	37.6	
$AgGe_2P_3$	345.97	57.66				1015	6150				

1.1.6. "Defect Chalcopyrite" Structure Compounds (Strukturbericht symbol E3, Space Group I $\bar{4}$-S_4^2)

Substance	Molecular weight	Average atomic weight	Lattice parameters (Å, room temp.)		Density (g/cm³)	Melting point (K)
$ZnAl_2Se_4$	435.18	62.17	5.503	10.90	4.37	
$ZnAl_2Te_4$(?)	629.74	84.96	5.904	12.05	4.95	
$ZnGa_2S_4$(?)	333.06	47.58	5.274	10.44	3.80	
$ZnGa_2Se_4$(?)	520.66	74.38	5.496	10.99	5.21	
$ZnGa_2Te_4$(?)	715.22	102.17	5.937	11.87	5.67	
$ZnIn_2Se_4$	610.86	87.27	5.711	11.42	5.44	1250
$ZnIn_2Te_4$	805.42	115.06	6.122	12.24	5.83	1075
$CdAl_2S_4$	294.61	42.09	5.564	10.32	3.06	
$CdAl_2Se_4$	482.21	68.89	5.747	10.68	4.54	
$CdAl_2Te_4$(?)	676.77	97.68	6.011	12.21	5.10	

Substance	Molecular weight	Average atomic weight	Lattice parameters (Å, room temp.)		Density (g/cm³)	Melting point (K)	Microhardness, N/mm² (M-Mohs Scale)	Specific heat, J/kg·K (300 K)	Debye temp. (K)	Coefficient of thermal linear expansion [10^{-6} K^{-1} (300K)]	Thermal conductivity [mW/cm·K (300K)]
$CdGa_2S_4$	380.09	54.30	5.577	10.08	4.03						
$CdGa_2Se_4$	567.69	81.10	5.743	10.73	5.32						
$CdGa_2Te_4$	762.25	108.89	6.093	11.81	5.77						
$CdIn_2Te_4$	852.45	121.78	6.205	12.41	5.9	1060					
$HgAl_2S_4$	382.79	54.68	5.488	10.26	4.11						
$HgAl_2Se_4$	570.39	82.48	5.708	10.74	5.05						
$HgAl_2Te_4$(?)	764.48	109.28	6.004	12.11	5.81						
$HgGa_2S_4$	468.27	66.90	5.507	10.23	5.00						
$HgGa_2Se_4$	655.87	93.70	5.715	10.78	6.18						
$HgIn_2Se_4$	746.07	106.58	5.764	11.80	6.3	1100					
$HgIn_2Te_4$(?)	940.63	134.38	6.186	12.37	6.3	980					
1.1.7. Other Adamantine Compounds											
α–SiC	40.10	20.10	3.0817	15.12	3.21	3070					
$Hg_5Ga_2Te_8$	2163.19	144.21	6.235								
$Hg_5In_2Te_8$	2253.39	150.23	6.328								
$CdIn_2Se_4$	657.89	93.98	a = c = 5.823								
1.2. Octahedral Semiconductors											
1.2.1. Halite Structure Semiconductors (Strukturbericht symbol B1, Space Group Fm3m-O_h^5)											
GeTe	200.21	100.10	5.98		6.14						
SnSe	197.67	98.83	6.020			1133					
SnTe	246.31	123.15	6.313		6.45	1080 (max)					91
PbS	239.3	119.63	5.9362		7.61	1390					23
PbSe	286.2	143.08	6.1243		8.15	1340					17
PbTe	334.8	167.4	6.454		8.16	1180					23
1.2.2. Selected Other Binary Halites											
BiSe	287.94	143.97	5.99		7.98	880					
BiTe	336.58	168.29	6.47								
EuSe	230.92	115.46	6.191			2300					2.4
GdSe	236.21	118.11	5.771			2400					
NiO	74.69	37.35	4.1684		6.6	2260					
CdO	128.41	64.21	4.6953			1700					7
SrS	119.69	59.84	6.0199		3.643	3000					
1.3. Other Semiconductors											
1.3.1. Antifluorite Structure Compounds (Fm3m–O_h^5)											
Mg_2Si	76.70	25.57	6.338		1.88	1375				11.5	
Mg_2Ge	121.22	40.4	6.380		3.08	1388				15.0	
Mg_2Sn	167.32	55.77	6.765		3.53	1051				9.9	92
Mg_2Pb	225.81	85.27	6.836		5.1	823				10.0	
1.3.2. Tetradymite Structure Compounds ($\overline{R}3m$–D_{3d}^5)											
Sb_2Te_3	626.3	125.26	4.25	30.3	6.44	895					
Bi_2Se_3	654.84	130.97	4.14	28.7	7.51	979	167				24
Bi_2Te_3	800.76	160.15	4.38	30.45	7.73	858	155	16			30

Substance	Molecular weight	Average atomic weight	Lattice parameters (Å, room temp.)	Density (g/cm³)	Melting point (K)	Microhardness, N/mm² (M-Mohs Scale)	Specific heat, J/kg·K (300 K)	Debye temp. (K)	Coefficient of thermal linear expansion [10^{-6} K^{-1} (300K)]	Thermal conductivity [mW/cm·K (300K)]
1.3.3. Skutterudite Structure Compounds ($Im3–T_h^5$)										
CoP_3	151.85	37.96	7.7073		>1270					
$CoAs_3$	286.70	71.65	8.2060	6.73	1230					
$CoSb_3$	424.18	106.05	9.0385		1123			307		50
$NiAs_3$	283.45	70.86	8.330	6.43						
RhP_3	195.83	48.96	7.9951		>1470					
$RhAs_3$	327.67	81.92	8.4427		>1270					100
$RhSb_3$	468.16	117.04	9.2322		1170					
IrP_3	285.14	71.29	8.0151	7.36	>1470					
$IrAs_3$	416.98	104.25	8.4673	9.12	>1470					90
$IrSb_3$	557.47	139.37	9.2533	9.35	1170			303		
1.3.4. Selected Multinary Compounds										
$AgSbSe_2$	387.54	96.88	5.786	6.60	910					10.5
$AgSbTe_2$ (or $Ag_{19}Sb_{29}Te_{52}$)	484.82	121.2	6.078	7.12	830					86
$AgBiS_2$(H.T.)	380.97	95.24	5.648							
$AgBiSe_2$(H.T.)	474.77	118.69	5.82							
$AgBiTe_2$(H.T.)	572.05	143.01	6.155							
Cu_2CdSnS_4	486.43	60.80	5.586	10.83						
1.3.5. Some Elemental Semiconductors										
B		10.81	4.91 12.6	2.34	2348	9.5 (M)	1277	1370	8.3	600
Se(gray)		78.96	4.36 4.95	4.81	493	350	292.6		(‖C) 17.89 (⊥C) 74.09	(‖C) 45.2 (⊥C) 13.1
Te		127.60	4.45 5.91	6.23	723		196.5		16.8	(‖C) 33.8 (⊥C) 19.7

TABLE 2. Basic Thermodynamic, Electrical, and Magnetic Properties of Semiconductors (Listed by Crystal Structure)

Substance	Heat of formation [kJ/mol (300K)]	Volume compressibility (10^{-10}m²/N)	Static dielectric constant	Atomic magnetic susceptibility (10^{-6} cgs)	Index of refraction	Minimum room temperature energy gap (eV)	Mobility (room temp.) (cm²/V·s) Electrons	Holes	Optical transition	Breakdown voltage kV/mm	Remarks
2.1. Adamantine Semiconductors											
2.1.1. Diamond Structure Elements (Strukturbericht symbol A4, Space Group Fd 3m–O_h^7)											
C	714.4	18	5.7	−5.88	2.419 (589 nm)	5.4	1800	1400	i*	500	
Si	324	0.306	11.9	−3.9	3.49 (589 nm)	1.12	1900	500	i	30	
Ge	291	0.768	16	−0.12	3.99 (589 nm)	0.67	3800	1820	i		
α-Sn	267.5		24		2.75 (589 nm)	0.0; 0.8	2500	2400			
2.1.2. Sphalerite (Zinc Blende) Structure Compounds (Strukturbericht symbol B3 Space Group $F\bar{4}3m–T_d^2$)											
I-VII Compounds											
CuF											
CuCl	481	0.26	7.9		1.93	3.17			d		Nantokite

Substance	Heat of formation [kJ/mol (300K)]	Volume compressibility ($10^{-10}m^2/N$)	Static dielectric constant	Atomic magnetic susceptibility (10^{-6} cgs)	Index of refraction	Minimum room temperature energy gap (eV)	Mobility (room temp.) ($cm^2/V \cdot s$) Electrons	Holes	Optical transition	Breakdown voltage kV/mm	Remarks
CuBr	481	0.26	7.9		2.12	2.91			d		
CuI	439	0.27	6.5		2.346	2.95			d		Marshite
AgBr	486		12.4		2.253	2.50	4000		i		Bromirite
AgI	389	0.41	10		2.22	2.22	30		d		Miersite
II-VI Compounds											
BeS					4.17				i		
BeSe					3.61				i		
BeTe					1.45		20		d		
BePo											
ZnO											See 2.1.3.
ZnS	477		8.9	−9.9	2.356	3.54	180	5(400°C)	d		See also 2.1.3.
ZnSe	422		9.2		2.89	2.58	540	28	d		
ZnTe	376		10.4		3.56	2.26	340	100	d		
ZnP											
CdS											See 2.1.3.
CdSe											See 2.1.3.
CdTe	339		7.2		2.50	1.44	1200	50	d		
CdPo											
HgS					2.85		250		d		Metacinna-barite
HgSe	247					2.10 (α)	20000	≈1.5	s		Tiemannite
HgTe	242					−0.06	25000	350	s		Coloradoite
III-V Compounds											
BN	815					4.6					Borazone
BP(L.T.)						≈2.1	500	70			Ignites 470K
BAs						≈1.5					
AlP						2.45	80		i		
AlAs	627		10.9			2.16	1200	420	i		
AlSb	585	0.571	11		3.2	1.60	200–400	550	i		
GaP	635	0.110	11.1	−13.8	3.2	2.24	300	150	i		
GaAs	535	0.771	13.2	−16.2	3.30	1.35	8800	400	d		
GaSb	493	0.457	15.7	−14.2	3.8	0.67	4000	1400	d		
InP	560	0.735	12.4	−22.8	3.1	1.27	4600	150	d		
InAs	477	0.549	14.6	−27.7	3.5	0.36	33000	460	d		
InSb	447	0.442	17.7	−32.9	3.96	0.163	78000	750	d		

* i = indirect, d = direct, s = semimetal.

Substance	Heat of formation	Volume compressibility	Static dielectric constant	Atomic magnetic susceptibility	Index of refraction	Energy gap	Electrons	Holes	Optical transition	Breakdown voltage	Remarks
Other sphalerite structure compounds											
MnS											See also 2.1.3.
MnSe											See also 2.1.3.
β-SiC					2.697	2.3	4000				
Ga_2Te_3	271			−13.5		1.35	50				
In_2Te_3(H.T.)	198			−13.6		1.04	50				
$MgGeP_2$											El–T^{d12}
$ZnSnP_2$						2.1					Same
$ZnSnAs_2$(H. T.)						≈0.7					Same
$ZnSnSb_2$						0.4					Same

2.1.3. Wurtzite (Zincite) Structure Compounds (Strukturbericht symbol B4, Space Group P 6_3 mc-C^4_{6v})

Substance	Heat of formation	Volume compressibility	Static dielectric constant	Atomic magnetic susceptibility	Index of refraction	Energy gap	Electrons	Holes	Optical transition	Breakdown voltage	Remarks
I-VII Compounds											
CuCl											
CuBr											
CuI											
AgI						2.63					Iodargirite
II-VI Compounds											
BeO											
MgTe											
ZnO	−350					3.2	180				
ZnS	−206					3.67					
ZnTe	−163										

Substance	Heat of formation [kJ/mol (300K)]	Volume compressibility ($10^{-10}m^2/N$)	Static dielectric constant	Atomic magnetic susceptibility (10^{-6}cgs)	Index of refraction	Minimum room temperature energy gap (eV)	Mobility (room temp.) ($cm^2/V{\cdot}s$) Electrons	Holes	Optical transition	Breakdown voltage kV/mm	Remarks
CdS				8.45; 9.12	2.32	2.42	350	40	d		Greenockide
CdSe						1.74	900	50	d		Cadmoselite
CdTe						1.50	650				
III-V Compounds											
BP(H.T.)											
AlN						6.02					
GaN						3.34					
InN						2.0					
Other wurtzite structure compounds											
MnS											
MnSe											
SiC					2.654						
MnTe						≈1.0					
Al_2S_3	426					4.1					
Al_2Se_3	367					3.1					

2.1.4 Chalcopyrite Structure Compounds (Strukturbericht symbol E1$_1$, Space Group I $\overline{4}$ 2d-D$^{12}_{2d}$)

Substance	Heat of formation [kJ/mol (300K)]	Volume compressibility ($10^{-10}m^2/N$)	Static dielectric constant	Atomic magnetic susceptibility (10^{-6}cgs)	Index of refraction	Minimum room temperature energy gap (eV)	Mobility (room temp.) ($cm^2/V{\cdot}s$) Electrons	Holes	Optical transition	Breakdown voltage kV/mm	Remarks
I-III-VI$_2$ Compounds											
$CuAlS_2$		0.106				2.5					
$CuAlSe_2$						2.67					
$CuAlTe_2$						0.88					
$CuCaS_2$		0.106				2.38					
$CuGaSe_2$		0.141				0.96, 1.63					
$CuGaTe_2$		0.227				0.82, 1.0					
$CuInS_2$		0.141				1.2					
$CuInSe_2$		0.187				0.86, 0.92					
$CuInTe_2$		0.278				0.95					
$CuTlS_2$											
$CuTlSe_2$ (L.T.)						1.07					
$CuFeS_2$						0.53					Chalcopyrite
$CuFeSe_2$						0.16					
$CuLaS_2$											
$AgAlS_2$											
$AgAlSe_2$						0.7					
$AgAlTe_2$						0.56					
$AgGaS_2$		0.150				1.66					
$AgGaSe_2$		0.182				1.1					
$AgGaTe_2$		0.280				1.9					
$AgInS_2$ (L.T.)		0.185				1.18					
$AgInSe_2$		0.238				0.96, 0.52					
$AgInTe_2$		0.338									
$AgFeS_2$											
II-IV-V$_2$ Compounds											
$ZnSiP_2$	312					2.3	1000				
$ZnGeP_2$	293					2.2					
$ZnSnP_2$	275					1.45					
$CdSiP_2$		0.103				2.2	1000				
$CdGeP_2$	289					1.8					
$CdSnP_2$	270					1.5					
$ZnSiAs_2$	290					1.7		50			
$ZnGeAs_2$	271			−14.4		0.85					
$ZnSnAs_2$	252			−18.4		0.65		300			Disorders at 910 K
$CdSiAs_2$		0.143				1.6					
$CdGeAs_2$	266			−23.4		0.53	70	25			Disorders at 903 K
$CdSnAs_2$	247		13.7	−21.5		0.26	22000	250			

Substance	Heat of formation [kJ/mol (300K)]	Volume compressibility (10^{-10} m^2/N)	Static dielectric constant	Atomic magnetic susceptibility (10^{-6} cgs)	Index of refraction	Minimum room temperature energy gap (eV)	Mobility (room temp.) (cm^2/V·s) Electrons	Holes	Optical transition	Breakdown voltage kV/mm	Remarks
2.1.5. Other Ternary Semiconductors with Tetrahedral Coordination											
II_2-IV-VI_3 Compounds											
Cu_2SiS_3 (H.T.)											Wurtzite
Cu_2SiS_3 (L.T.)											Tetragonal
Cu_2SiTe_3											Cubic
Cu_2GeS_3 (H.T.)				−18.7							Cubic
Cu_2GeS_3 (L.T.)							360				Tetragonal
Cu_2GeSe_3	211.5			−21.3		0.94	238				Same
Cu_2GeTe_3	190.2			−23.4							Same
Cu_2SnS_3				−18.2		0.91	405				Cubic
$CuSnSe_3$				−21.0		0.66	870				Cubic
Cu_2SnTe_3				−28.4							Cubic
Ag_2GeSe_3				−29.6		0.91 (77K)					
Ag_2SnSe_3				−29.5		0.81					
Ag_2GeTe_3				−31.4		0.25					
Ag_2SnTe_3				−31.0		0.08					
II_3-V-VI_4 Compounds											
Cu_3PS_4											Enargite
Cu_3AsS_4	269.6			−15.8		1.24					
Cu_3AsSe_4	161.3			−13.1		0.88					Famatinite
Cu_3SbS_4				−8.3		0.74					Famatinite
Cu_3SbSe_4	127.1			−20.5		0.31					
II-IV_2-V_3 Compounds											
$CuSi_2P_3$											El
$CuGe_2P_3$		0.12				0.90					El
$AgGe_2P_3$											
2.1.6. "Defect Chalcopyrite" Structure Compounds (Strukturbericht symbol E3, Space Group $I\,\bar{4}$-S_4^2)											
$ZnAl_2Se_4$											
$ZnAl_2Te_4$(?)											
$ZnGa_2S_4$ (?)						≈3.4					
$ZnGa_2Se_4$(?)						≈2.2					
$ZnGa_2Te_4$(?)						1.35					
$ZnIn_2Se_4$	206					1.82	35				
$ZnIn_2Te_4$	198					1.2					
$CdAl_2S_4$											
$CdAl_2Se_4$											
$CdAl_2Te_4$(?)											
$CdGa_2S_4$	256					3.44	60				
$CdGa_2Se_4$	216					2.43	33				
$CdGa_2Te_4$											
$CdIn_2Te_4$	195					(1.26 or 0.9)	4000				
$HgAl_2S_4$											
$HgAl_2Se_4$											
$HgAl_2Te_4$(?)											
$HgGa_2S_4$	249					2.84					
$HgGa_2Se_4$	204					1.95	400				
$HgIn_2Se_4$	196					0.6	290				
$HgIn_2Te_4$(?)	188					0.86	200				
2.1.7. Other Adamantine Compounds											
α–SiC			10.2	−6.4	2.67	2.86	400				6H structure
$Hg_5Ga_2Te_8$											B3 with superlattice
$Hg_5In_2Te_8$						0.7	2000				B3 with superlattice
$CdIn_2Se_4$						1.55					

Substance	Heat of formation [kJ/mol (300K)]	Volume compressibility (10^{-10}m²/N)	Static dielectric constant	Atomic magnetic susceptibility (10^{-6} cgs)	Index of refraction	Minimum room temperature energy gap (eV)	Mobility (room temp.) (cm²/V·s) Electrons	Holes	Optical transition	Breakdown voltage kV/mm	Remarks
2.2. Octahedral Semiconductors											
2.2.1. Halite Structure Semiconductors (Strukturbericht symbol B1, Space Group Fm3m-O_h^5)											
GeTe											
SnSe											
SnTe											
PbS	435					0.5	600	600			
PbSe	393		161			0.37	1000	900			
PbTe	393		280			0.26	1600	600			Altaite
			360			0.25					
2.2.2. Selected Other Binary Halites											
BiSe											
BiTe						0.4					
EuSe											
GdSe						1.8	4				
NiO						2.0 or 3.7	100				
CdO	531					2.5					
SrS						4.1					
2.3. Other Semiconductors											
2.3.1. Antifluorite Structure Compounds (Fm3m-O_h^5)											
Mg_2Si	79.08					0.77	405	70			
Mg_2Ge						0.74	520	110			
Mg_2Sn	76.57					0.36	320	260			
Mg_2Pb	52.72					0.1					
2.3.2. Tetradymite Structure Compounds ($R\bar{3}m$-D_{3d}^5)											
Sb_2Te_3						0.3		360			
Bi_2Se_3						0.35	600				
Bi_2Te_3						0.21	1140	680			R3m (166)
2.3.3. Skutterudite Structure Compounds (Im3-T_h^5)											
CoP_3						0.43					
$CoAs_3$						0.69		~4000			
$CoSb_3$						0.63	70	~3000			
RhP_3								700			
$RhAs_3$						0.85		~3000			
$RhSb_3$						0.80		~7000			
$IrSb_3$						1.18		1500			
2.3.4. Selected Multinary Compounds											
$AgSbSe_2$						0.58					
$AgSbTe_2$ (or$Ag_{19}Sb_{29}Te_{52}$)						0.7, 0.27					
$AgBiS_2$ (H.T.)											
$AgBiSe_2$ (H.T.)											
$AgBiTe_2$ (H.T.)											
Cu_2CdSnS_4						1.16	<2				
2.3.5. Some Elemental Semiconductors											
B	397.1			−6.7	3.4	1.55	10				
Se(gray)			6.6 (0.1 GHz)	−22.1	2.5	1.5		5			$P3_121(152)$
Te				−39.5	3.3	0.33	1700	1200			Same

TABLE 3. Semiconducting Properties of Selected Materials

Substance	Minimum energy gap (eV) R.T.	Minimum energy gap (eV) 0 K	$10^4 \cdot dE_g/dT$ eV/°C	$10^6 \cdot dE_g/dP$ eV·cm²/kg	Density of states electron effective mass m_{da} (m_o)	Electron mobility and temperature dependence μ_n (cm²/V·s)	Electron mobility and temperature dependence $-x$	Density of states hole effective Mass m_{dp} (m_n)	Hole mobility and temperature dependence μ_p (cm²/V·s)	Hole mobility and temperature dependence $-x$
Elements										
Si	1.107	1.153	−2.3	−2.0	1.1	1900	2.6	0.56	500	2.3
Ge	0.67	0.744	−3.7	+7.3	0.55	3800	1.66	0.3	1820	2.33
α–Sn	0.08	0.094	−0.5		0.02	2500	1.65	0.3	2400	2.0
Te	0.33				0.08	1100		0.19	560	
III-V Compounds										
AlAs	2.2	2.3				1200			420	
AlSb	1.6	1.7	−3.5	−1.6	0.09	200	1.5	0.4	500	1.8
GaP	2.24	2.40	−5.4	−1.7	0.35	300	1.5	0.5	150	1.5
GaAs	1.35	1.53	−5.0	+9.4	0.068	9000	1.0	0.5	500	2.1
GaSb	0.67	0.78	−3.5	+12	0.050	5000	2.0	0.23	1400	0.9
InP	1.27	1.41	−4.6	+4.6	0.067	5000	2.0		200	2.4
InAs	0.36	0.43	−2.8	+8	0.022	33,000	1.2	0.41	460	2.3
InSb	0.165	0.23	−2.8	+15	0.014	78,000	1.6	0.4	750	2.1
II-VI Compounds										
ZnO	3.2		−9.5	+0.6	0.38	180	1.5			
ZnS	3.54		−5.3	+5.7		180			5(400°C)	
ZnSe	2.58	2.80	−7.2	+6		540			28	
ZnTe	2.26			+6		340			100	
CdO	2.5 ± .1		−6		0.1	120				
CdS	2.4		−5	+3.3	0.165	400		0.8		
CdSe	1.74	1.85	−4.6		0.13	650	1.0	0.6		
GdTe	1.44	1.56	−4.1	+8	0.14	1200		0.35	50	
HgSe	0.30				0.030	20,000	2.0			
HgTe	0.15		−1		0.017	25,000		0.5	350	
Halite Structure Compounds										
PbS	0.37	0.28	+4		0.16	800		0.1	1000	2.2
PbSe	0.26	0.16	+4		0.3	1500		0.34	1500	2.2
PbTe	0.25	0.19	+4	−7	0.21	1600		0.14	750	2.2
Others										
ZnSb	0.50	0.56			0.15	10				1.5
CdSb	0.45	0.57	−5.4		0.15	300			2000	1.5
Bi_2S_3	1.3					200			1100	
Bi_2Se_3	0.27					600			675	
Bi_2Te_3	0.13		−0.95		0.58	1200	1.68	1.07	510	1.95
Mg_2Si		0.77	−6.4		0.46	400	2.5		70	
Mg_2Ge		0.74	−9			280	2		110	
Mg_2Sn	0.21	0.33	−3.5		0.37	320			260	
Mg_3Sb_2		0.32				20			82	
Zn_3As_2	0.93					10	1.1		10	
Cd_3As_2	0.55				0.046	100,000	0.88			
GaSe	2.05		−3.8						20	
GaTe	1.66	1.80	−3.6			14				
InSe	1.8					900				
TlSe	0.57		−3.9		0.3	30		0.6	20	1.5
$CdSnAs_2$	0.23				0.05	25,000	1.7			
Ga_2Te_2	1.1	1.55	−4.8							
α-In_2Te_2	1.1	1.2			0.7				50	1.1
β-In_2Te_2	1.0								5	
$Hg_5In_2Te_8$	0.5								11,000	
SnO_2									78	

TABLE 4. Band Properties of Semiconductors

4.1. Data on Valence Bands of Semiconductors (Room Temperature)

Substance	Band curvature effective mass (Expressed as fraction of free electron mass): Heavy holes	Light holes	"Split-off" band holes	Energy separation of "split-off" band (eV)	Measured (light) hole mobility (cm²/V·s)
4.1.1. Semiconductors with Valence Band Maximum at the Center of the Brillouin Zone ("Γ")					
Si	0.52	0.16	0.25	0.044	500
Ge	0.34	0.043	0.08	0.3	1820
Sn	0.3				2400
AlAs					
AlSb	0.4			0.7	550
GaP				0.13	100
GaAs	0.8	0.12	0.20	0.34	400
GaSb	0.23	0.06		0.7	1400
InP				0.21	150
InAs	0.41	0.025	0.083	0.43	460
InSb	0.4	0.015		0.85	750
CdTe	0.35				50
HgTe	0.5				350

4.1.2. Semiconductors with Multiple Band Maxima

Substance	Number of equivalent valleys and direction	Band curvature effective masses: Longitudinal m_L	Transverse m_T	Anistrophy $K = m_L/m_T$	Measured (light) hole mobility (cm²/V·s)
PbSe	4 "L" [111]	0.095	0.047	2.0	1500
PbTe	4 "L" [111]	0.27	0.02	10	750
Bi_2Te_3	6	0.207	~0.045	4.5	515

4.2. Data on Conduction Bands of Semiconductors (Room Temperature Data)

4.2.1. Single Valley Semiconductors

Substance	Energy gap (eV)	Effective mass (m_o)	Mobility (cm²/V·s)	Comments
GaAs	1.35	0.067	8500	3(or 6?) equivalent [100] valleys 0.36 eV above this maximum with a mobility of ~50.
InP	1.27	0.067	5000	3(or 6?) equivalent [100] valleys 0.4 eV above this minimum.
InAs	0.36	0.022	33,000	Equivalent valleys ~1.0 eV above this minimum.
InSb	0.165	0.014	78,000	
CdTe	1.44	0.11	1000	4(or 8?) equivalent [111] valleys 0.51 eV above this minimum.

4.2.2. Multivalley Semiconductors

Substance	Energy gap	Number of equivalent valleys and direction	Band curvature effective mass: Longitudinal m_L	Transverse m_T	Anisotropy $K = m_L/m_T$
Si	1.107	6 in [100] "Δ"	0.00	0.192	4.7
Ge	0.67	4 in [111] at "L"	1.588	0.0815	19.5
GaSb	0.67	as Ge (?)	~1.0	~0.2	~5
PbSe	0.26	4 in [111] at "L"	0.085	0.05	1.7
PbTe	0.25	4 in [111] at "L"	0.21	0.029	5.5
Bi_2Te_3	0.13	6			~0.05

TABLE 5. Resistivity of Semiconducting Minerals

Mineral	ρ (ohm · m)
Diamond (C)	2.7
Sulfides	
Argentite, Ag_2S	1.5 to 2.0×10^{-3}
Bismuthinite, Bi_2S_3	3 to 570
Bornite, $Fe_2S_3 \cdot nCu_2S$	1.6 to 6000×10^{-6}
Chalcocite, Cu_2S	80 to 100×10^{-6}
Chalcopyrite, $Fe_2S_3 \cdot Cu_2S$	150 to 9000×10^{-6}
Covellite, CuS	0.30 to 83×10^{-6}
Galena, PbS	6.8×10^{-6} to 9.0×10^{-2}
Haverite, MnS_2	10 to 20
Marcasite, FeS_2	1 to 150×10^{-3}
Metacinnabarite, HgS	2×10^{-6} to 1×10^{-3}
Millerite, NiS	2 to 4×10^{-7}
Pentlandite, $(Fe, Ni)_4S_4$	1 to 11×10^{-6}
Pyrrhotite, Fe_7S_4	2 to 160×10^{-6}
Pyrite, FeS_2	1.2 to 600×10^{-3}
Sphalerite, ZnS	2.7×10^{-3} to 1.2×10^{4}
Antimony-sulfur compounds	
Berthierite, $FeSb_2S_4$	0.0083 to 2.0
Boulangerite, $Pb_5Sb_3S_{11}$	2×10^{3} to 4×10^{4}
Cylindrite, $Pb_3Sn_4Sb_2S_{14}$	2.5 to 60
Franckeite, $Pb_5Sn_3Sb_2S_{14}$	1.2 to 4
Hauchecornite, $Ni_4(Bi, Sb)_2S_{14}$	1 to 83×10^{-6}
Jamesonite, $Pb_4FeSb_6S_{14}$	0.020 to 0.15
Tetrahedrite, Cu_3SbS_3	0.30 to 30,000
Arsenic-sulfur compounds	

Mineral	ρ (ohm · m)
Molybdenite, MoS_2	0.12 to 7.5
Cobaltite, CoAsS	6.5 to 130 × 10^{-3}
Enargite, Cu_3AsS_4	0.2 to 40 × 10^{-3}
Gersdorfite, NiAsS	1 to 160 × 10^{-6}
Glaucodote, (Co, Fe)AsS	5 to 100 × 10^{-6}
Antimonide	
Dyscrasite, Ag_3Sb	0.12 to 1.2 × 10^{-6}
Arsenides	
Allemonite, $SbAs_3$	70 to 60,000
Lollingite, $FeAs_2$	2 to 270 × 10^{-6}
Nicollite, NiAs	0.1 to 2 × 10^{-6}
Skutterudite, $CoAs_3$	1 to 400 × 10^{-6}
Smaltite, $CoAs_2$	1 to 12 × 10^{-6}
Tellurides	
Altaite, PbTe	20 to 200 × 10^{-6}
Calavarite, $AuTe_2$	6 to 12 × 10^{-6}
Coloradoite, HgTe	4 to 100 × 10^{-6}
Arsenopyrite, FeAsS	20 to 300 × 10^{-6}
Hessite, Ag_2Te	4 to 100 × 10^{-6}
Nagyagite, $Pb_6Au(S,Te)_{14}$	20 to 80 × 10^{-6}
Sylvanite, $AgAuTe_4$	4 to 20 × 10^{-6}
Oxides	
Braunite, Mn_2O_3	0.16 to 1.0
Cassiterite, SnO_2	4.5 × 10^{-4} to 10,000
Cuprite, Cu_2O	10 to 50
Hollandite, (Ba, Na, K) Mn_8O_{16}	2 to 100 × 10^{-3}
Ilmenite, $FeTiO_3$	0.001 to 4
Magnetite, Fe_3O_4	52 × 10^{-6}
Manganite, MnO · OH	0.018 to 0.5
Melaconite, CuO	6000
Psilomelane, $BaMn_9O_{18} \cdot 2H_2O$	0.04 to 6000
Pyrolusite, MnO_2	0.007 to 30
Rutile, TiO_2	29 to 910
Uraninite, UO_2	1.5 to 200

References

1. Beer, A. C., *Galvanomagnetic Effects in Semiconductors*, Academic Press, New York, 1963.
2. Goryunova, N. A., *The Chemistry of Diamond-Like Semiconductors*, The MIT Press, Cambridge, MA, 1965.
3. Abrikosov, N. Kh., Bankina, V. F., Poretskaya, L. E., Shelimova, L. E., and Skudnova, E.V., *Semiconducting II-VI, IV-VI, and V-VI Compounds*, Plenum Press, New York, 1969.
4. Berger, L. I. and Prochukhan, V. D., *Ternary Diamond-Like Semiconductors*, Cons. Bureau/Plenum Press, New York, 1969.
5. Shay, J. L. and Wernick, J. H., *Ternary Chalcopyrite Semiconductors: Growth, Electronic Properties, and Applications*, Pergammon Press, 1975.
6. Bergman, R., *Thermal Conductivity in Solids*, Clarendon, Oxford, 1976.
7. Handbook of Semiconductors, Vol. 1, Moss, T.S. and Paul, W., Eds., *Band Theory and Transport Properties*; Vol. 2, Moss, T.S. and Balkanski, M., Eds., *Optical Properties of Solids*; Vol. 3, Moss, T.S. and Keller, S.P., Eds., *Materials Properties and Preparation*, North Holland Publ. Co., Amsterdam, 1980.
8. Böer, K. W., *Survey of Semiconductor Physics*, Van Nostrand Reinhold, 1990.
9. Rowe, D. M., Ed., *CRC Handbook of Thermoelectrics*, CRC Press, Boca Raton, FL, 1995.
10. Berger, L. I., *Semiconductor Materials*, CRC Press, Boca Raton, FL, 1997.
11. Glazov, V. M., Chizhevskaya, S.N., and Glagoleva, N.N., *Liquid Semiconductors*, Plenum Press, New York, 1969.
12. Phillips, J. C., *Bonds and Bands in Semiconductors*, Academic Press, New York, 1973.
13. Harrison, W. A., *Electronic Structure and the Properties of Solids*, Freeman Publ. House, San Francisco, 1980.
14. Balkanski, M., Ed., *Optical Properties of Solids*, North-Holland, Amsterdam, 1980.
15. *Landolt-Börnstein. Numerical Data and Functional Relationships in Science and Technology, New Series, Group III: Crystal and Solid State Physics*, Hellwege, K.-H. and Madelung, O., Eds., Volumes 17 and 22, Springer Verlag, Berlin, 1984 (and further).
16. Shklovskii, B. L. and Efros, A.L., *Electronic Processes in Doped Semiconductors*, Springer Verlag, Berlin, 1984.
17. Cohen, M. L. and Chelikowsky, J. R., *Electronic Structure and Optical Properties of Semiconductors*, Springer Verlag, New York, 1988.
18. Glass, J.T., Messier, R.F., and Fujimori, N., Eds., *Diamond, Silicon Carbide, and Related Wide Bandgap Semiconductors*, MRS Symposia Proc. 1652, Mater. Res. Soc., Pittsburgh, 1990.
19. Palik, E., Ed., *Handbook of Optical Constants of Solids II*, Academic Press, New York, 1991.
20. Reed, M., Ed., *Semiconductors and Semimetals*, Volume 35, Academic Press, Boston, 1992.
21. Haug, H. and Koch, S. W., *Quantum Theory of the Optical and Electronic Properties of Semiconductors*, 2nd Edition, World Scientific, Singapore, 1993.
22. Lockwood, D. J., Ed., *Proc. 22nd Intl. Conf. on the Physics of Semiconductors, Vancouver, 1994*, World Scientific, Singapore, 1994.
23. Morelli, D. T., Caillat, T., Fleurial, J.-P., Borschchevsky, A., Vandersande, J., Chen, B., and Uher, C., *Phys. Rev.*, B51, 9622, 1995.
24. Caillat, T., Borshchevsky, A., and Fleurial, J.-P., *J. Appl. Phys.*, 80, 4442, 1996.
25. Fleurial, J.-P.,Caillat, T., and Borshchevsky, A., *Proc. XVI Intl. Conf. Thermoelectrics*, Dresden, Germany, August 26–29, 1997 (in print).
26. Borshchevsky, A. et al., U.S. Patents 5,610,366 (March 1977) and 5,831,286 (March 1998).
27. Jarrendahl, K. and Davis, R. F., *Semiconductors and Semimetals*, Vol. 52, Y.S. Park, Ed., 1998, pp. 1–20.
28. Bettini, M., *Solid State Comm.*,13, 599, 1973.
29. Chen, A. and Sher, A., *Semiconductor Alloys, Physics and Material Engineering*, Plenum Press, New York, 1995.
30. Holloway, P. H. and McGuire, G. E., Eds., *Handbook of Compound Semiconductors*, Noyes Publ., Park Ridge, NJ, 1995.
31. Madelung, O., Ed., *Semiconductors: Group IV Elements and III-V Compounds (Data in Science and Technology)*, Springer-Verlag, Berlin, Heidelberg, 1991.
32. Madelung, O., Ed., *Semiconductors: Other than Group IV Elements and III-V Compounds (Data in Science and Technology)*, Springer-Verlag, Berlin, Heidelberg, 1992.
33. Levinshtein, M., Rumyantsev, S., and Shur, M., Eds., Handbook Series on Semiconductor Parameters. Vol. 1, *Si, Ge, C (Diamond), GaAs, GaP, GaSb, InAs, InP, InSb*, World Scientific, Singapore, 1996.
34. Levinshtein, M., Rumyantsev, S., and Shur, M., Eds., Handbook Series on Semiconductor Parameters. Vol. 2, *Ternary and Quaternary III-V Compounds*, World Scientific, Singapore, 1996.
35. *Physical Properties of Semiconductors*, NSM Archive:http://www.ioffe.ru/SVA/NSM/Semicond/

SELECTED PROPERTIES OF SEMICONDUCTOR SOLID SOLUTIONS

L. I. Berger

Alloy system	Limits of solubility	Energy gap in eV (300 K)	Remarks, references
		Adamantine Semiconductors IV-IV	
Si_xGe_{1-x}	$0 \le x \le 1$	$0.8941+0.0421x+0.1691x^2$ $0.7596+1.0860x+0.3306x^2$	Transition Γ - X [Ref.1] Trans. Γ - L [Ref. 1]
		Adamantine Semiconductors III-V/III-V	
Common Anion			
$Al_xGa_{1-x}N$	$0 \le x \le 1$		Wurtzite Structure [Ref. 2 & 3]
$Al_xGa_{1-x}P$	$0 \le x \le 0.5$	$2.28+0.16x$	[Ref. 2]
$Al_xIn_{1-x}P$	$0 \le x \le 0.44$	at Γ: $134+2.23x$; at X: $2.24+0.18x$	[Ref. 2]
$Al_xGa_{1-x}As$	$0 \le x \le 0.5$	$1.42=0.75x$ [Ref.3]; $1.424+1.429x-0.14x^2$ [Ref.4]	
$Al_xIn_{1-x}As$	$0 \le x \le 1$	at Γ: $0.37+1.91x+0.74x^2$; at X: $1.8+0.4x$	[Ref. 2 and 6]
$Al_xGa_{1-x}Sb$	$0 \le x < \le 1$	$0.73+1.10x+0.47x^2$	Trans. Γ_{0v} - Γ_{0c} [Ref. 2]
$Al_xIn_{1-x}Sb$	$0 \le x \le 1$		[Ref. 6]
$Ga_xIn_{1-x}N$	$0 \le x \le 1$	$1.950+1.487x-1.000x(1-x)$	Wurtzite [Ref. 8 and 10]
$Ga_xIn_{1-x}P$	$0 \le x \le 1$		[Ref. 2]
$Ga_xIn_{1-x}As$	$0 \le x \le 1$	$0.360+0.629x+0.436x^2$	[Ref. 5]
$Ga_xIn_{1-x}Sb$	$0 < x < 1$	$0.235+0.1653x+0.413x^2$	[Ref. 2, see also Ref. 9]
Common Cation			
GaN_xAs_{1-x}	$0 \le x \le 0.05$	$1.42-9.9x$	[Ref. 2]
GaP_xAs_{1-x}	$0 < x < 1$	$2.270-0.846x$	[Ref. 2]
GaP_xAs_{1-x}	$0 \le x \le 0.05$	$1.515+1.172x+0.186x^2$ $1.9715+0.144x+0.211x^2$	(at 2K, Γ–Γ) [Ref. 7] [Ref. 2]
$GaAs_xSb_{1-x}$	$0 \le x \le 0.45$, $0.6 \le x \le 1$	$1.43-1.9x+1.2x^2$	[Ref. 5]
InP_xAs_{1-x}	$0 < x < 1$	$0.356+0.675x+0.32x^2$	[Ref. 2]
		Adamantine Binary Semiconductors II-VI/II-VI [Ref. 3 and 6]	
Common Anion			
$Zn_xCd_{1-x}S$	$0 \le x \le 1$		Wurtzite Structure
$Zn_xHg_{1-x}S$	$0 \le x \le 1$		
$Cd_xHg_{1-x}S$	$0 \le x \le 1$		Wurtzite Structure at $x<0.6$
$Zn_xCd_{1-x}Se$	$0.7 \le x \le 1$		
$Zn_xHg_{1-x}Se$	$0 \le x \le 1$		
$Cd_xHg_{1-x}Se$	$0 \le x \le 0.7$ and $0.75 \le x^* \le 1$		x^*- Wurtzite Structure
$Zn_xCd_{1-x}Te$	$0 \le x \le 1$		
$Zn_xHg_{1-x}Te$	$0 \le x \le 1$		
$Cd_xHg_{1-x}Te$	$0 \le x \le 1$		
Common Cation			
ZnS_xSe_{1-x}	$0 \le x \le 1$		
ZnS_xTe_{1-x}	$0 \le x \le 0.1$ and $0.9 \le x^* \le 1$		x^*- Wurtzite Structure
$ZnSe_xTe_{1-x}$	$0 \le x \le 1$		
CdS_xSe_{1-x}	$0 \le x \le 1$		Wurtzite Structure
CdS_xTe_{1-x}	$0 \le x \le 0.25$ and $0.8 \le x^* \le 1$		x^*- Wurtzite Structure
$CdSe_xTe_{1-x}$	$0 \le x \le 0.4$ and $0.6 \le x^* \le 1$		x^*- Wurtzite Structure
HgS_xSe_{1-x}	$0 \le x \le 1$		
HgS_xTe_{1-x}	$0 \le x \le 1$		
$HgSe_xTe_{1-x}$	$0 \le x \le 1$		
		Quaternary Adamantine Semiconductors II-VI/III-V [Ref. 6]	
$(ZnS)_x(AlP)_{1-x}$	$0.99 \le x \le 1$		
$(ZnSe)_x(GaAs)_{1-x}$	$0 \le x < 1$		
$(CdTe)_x(InAs)_{1-x}$	$0 < x \le 0.2$ and $0.7 \le x \le 1$		
$(CdTe)_x(AlSb)_{1-x}$	$0 \le x \le 1$		
$(HgTe)_x(InAs)_{1-x}$	$0 \le x \le 1$		

Alloy system	Limits of solubility	Energy gap in eV (300 K)	Remarks, references
	Quaternary Adamantine Semiconductors III_x-III_{1-x}-V_y - V_{1-y}		
$Ga_xIn_{1-x}As_yP_{1-y}$	$0 \le x \le 1$, $0 \le x \le 1$	$1.35+0.668x-1.068y+0.758x^2+0.078y^2-0.069xy-0.322x^2y+0.03xy^2$	[Ref. 2 and 6]
	Quaternary Adamantine Semiconductors III_{1-x-y}-III_x-III_y-V		
$Al_xGa_yIn_{1-x-y}Sb$	$0 \le x \le 1$, $0 \le y \le 1$	$0.095+1.76x+0.28y+0.345(x^2+y^2)+0.085(1-x-y)^2+xy(1-x-y)(23-28y)$	[Ref. 2 and 6]

References

1. Krishnamurti, S., Sher, A., and Chen, A. *Appl. Phys. Lett.* 47, 160, 1985.
2. Madelung, O., Ed., *Semiconductors Group IV Elements and III-V Compounds*, Springer, 1991; *Semiconductors Other than Group IV Elements and III-V Compounds*, Springer, 1992.
3. Goryunova, N. A., Kesamanly, F.P., and Nasledov, D.N., *Semiconductors and Semimetals*, Vol. 4, 1968, p. 413.
4. El Allali, M., Sorensen, C. B., Veje, E., and Tideman-Peterson, P., *Phys. Rev. B*, 48, 4398, 1993.
5. Nahorny, R. E., Pollack, M. A., Johnson, W. D., and Barns, R. L. *Appl. Phys. Lett.* 33, 695, 1978.
6. Goryunova, N. A. *Multicomponent Diamond-Like Semiconductors*, Sov. Radio, Moscow, 1968 (in Russian).
7. Capizzi, M., Modesti, S., Martelli, F., and Frova, A., *Solid State Comm.* 39, 333, 1981.
8. Nakamura, S., Pearton, S., and Fasol, G., *The Blue Laser Diode*, 2nd ed., Springer, 2000.
9. Roth, A. P., Keeler, W. J., and Fortin, E. *Canad. J. Phys.* 58, 560, 1980.
10. Wu, J., Walukiewicz, Yu, K. M., Ager, J. W., Haller, E. E., Lu, H., and Schaff, W. J., *Appl. Phys. Lett.* 80, 4741, 2002.

PROPERTIES OF ORGANIC SEMICONDUCTORS

L. I. Berger

Substance	Energy Gap, E, (in E/2 kT) eV	Room Temperature Electrical Resistivity, ohm · cm	Mobility, μ, cm²/V · s	Sign of Majority Carriers	Temperature Range, °C	Ref.
Metal-Free Molecular Crystals						
3-Acetylamino-*N*-methylphthalimide	3.46				67 to 100	1
3-Acetylamino-*N*-phenylphthalimide	3.50				54 to 124	1
4-Amino-*N*-cyclomethylphthalimide	2.90				73 to 100	1
4-Aminophthalimide	2.78				123 to 151	1
Acridine	3.90					1
Anthanthrene	1.67	$1.5 \cdot 10^{19}$			40 to 105	1
Anthanthrene	0.84	$1.5 \cdot 10^{19}$ (15°C)				2
Anthanthrone	1.70	$7.7 \cdot 10^{18}$			20 to 150	1
Anthracene	0.83	$1.3 \cdot 10^{14}$ (15°C)				2
Anthracene	2.50	$1.5 \cdot 10^{11}$	2.3	+	20 to 130	1
Anthracene	3.88 to 4.1	$>10^{15}$	1.74(n), 2.07(p)	+ & −		4
1,2-Benzanthracene	1.04	10^{16} (30°C)				2
Benzanthrone	3.12	$1.6 \cdot 10^{16}$				1
Benzene (liquid)	0.41					2
Benzene (amorphous)	0.84	10^{15}			−14 to 5	1
Benzene (cryst.)	7		2	−	−23	4
Benzimidazole	3.0 to 4.0	$5 \cdot 10^{13}$			84 to 144	1
Benzophenone	3.34				−23 to 14	1
Benzo[*f*]quinoline	2.77				30 to 50	1
Benzo[*h*]quinoline	2.72					1
3-Benzoylamino-*N*-methylphthalimide	3.28				84 to 112	1
Benzpentacene	1.72				0 to 150	1
Biphenyl	1.46	$1.7 \cdot 10^{15}$ (50°C)				2
Biphenyl	1.45				20 to 70	1
o-Chloranil	3.0	10^{15}				1
p-Chloranil	0.61					1
Chlorpromazine	2.1	10^{12} (32°C)		+ & −	32 to 80	1
Chrysene	1.1	$4 \cdot 10^{19}$ (15°C)				2
Chrysene	2.20	$4 \cdot 10^{19}$			25 to 90	1
Circumanthracene	1.8	$6 \cdot 10^{12}$				1
Coronene	1.7	$1.7 \cdot 10^{17}$			60 to 80	1
Coronene	0.85	$1.7 \cdot 10^{17}$ (15°C)				2
Cyananthrone	0.20	$1.2 \cdot 10^{7}$			30 to 125	1
1,6-Diaminopyrene	0.6	10^{8}				
Dibenzpentacene	1.50				0 to 150	1
Dinaphthopyrene	1.60				25 to 90	1
1,8-Diphenyl-1,3,5,7-octatetraene	1.7				72 to 191	1
Diphenylpentacene	1.60				0 to 150	1
4,4′-Diphenylstilbene	1.56				160 to 280	1
4,4′-Diphenylstilbene	0.80					2
Ferrocene	1.22	10^{14}		+		1
Flavanthrone	0.70	$1.4 \cdot 10^{11}$				1
Fluorene	2.7					2
Fluorene	1.4					2
Fluoridine	1.6	$6 \cdot 10^{13}$		+	20 to 140	1
Hexacene	0.57	$3.8 \cdot 10^{10}$ (50°C)				2
Hexacene	1.3					1
Hexamethylbenzene	1.86			+	20 to 140	1
3-Hydroxy-*N*-methylphthalimide	3.80				60 to 91	1
Imidazole	2.6	10^{11}			28 to 68	1

Substance	Energy Gap, *E*, (in *E*/2 *kT*) eV	Room Temperature Electrical Resistivity, ohm · cm	Mobility, μ, $cm^2/V \cdot s$	Sign of Majority Carriers	Temperature Range, °C	Ref.
Indanthrazine	0.66	$1.4 \cdot 10^{15}$			30 to 125	1
Indanthrone	0.64	$7.5 \cdot 10^{14}$			30 to 125	1
Indanthrone (black)	0.56	$2.5 \cdot 10^{8}$			30 to 125	1
Mesitylene (liquid)	0.19					2
Mesonaphthodianthracene	0.6	$4.0 \cdot 10^{18}$ (15°C)				2
Mesonaphthodianthrene	1.48				45 to 250	1
Mesonaphthodianthrone	0.86				5 to 110	1
3-Methoxy-*N*-methyl-phthalimide	3.18				54 to 78	1
Naphthacene	1.7	$1 \cdot 10^{15}$				1
Naphthalene	3.5	10^{14}			27 to 47	1
Naphthalene	1.15	$2.8 \cdot 10^{14}$ (50°C)				2
Naphthalene	4.9 to 5.1		0.64(n), 1.50(p)	+ & –		4
m-Naphthodianthrene	1.20	$4 \cdot 10^{18}$			40 to 150	1
m-Naphthodianthrone	1.30	$1.5 \cdot 10^{18}$			40 to 150	1
β-Naphthol	2.36	$2 \cdot 10^{5}$			60 to 110	1
β-Naphthoquinoline	2.77					1
1-Naphthylamine	2.2				25 to 42	1
1-Naphthylamine picrate	2.7				28 to 98	1
2-Naphthylphenyl sulphone	3.5				67 to 102	1
1-Nitronaphthalene	2.5				25 to 44	1
Ovalene	1.13	$2.3 \cdot 10^{15}$				1
Pentacene	0.58	$2.4 \cdot 10^{9}$ (50°C)				2
Pentacene	1.5	$6 \cdot 10^{13}$			20 to 140	1
Perylene	2.1	$4.1 \cdot 10^{13}$			40 to 100	1
Perylene	3.10		5.53(n), 87.4(p)	+ & –	–213	4
Phenanthrene	1.15	$1.3 \cdot 10^{14}$			12 to 72	1
Phenanthrene	0.65					2
1,10-Phenanthroline	2.73				50 to 90	1
Phenazine	2.1	$7 \cdot 10^{14}$ (100°C)			98 to 143	1
Phenazine	1.1			–		4
Phenothiazine	1.6	10^{11}			50 to 150	1
Phenothiazine			2.45(n), 0.02(p)	+ & –		4
Phenylanthranilic acid	3.30				87 to 119	1
4-Phenylstilbene	1.74				140 to 220	1
4-Phenylstilbene	0.86					2
Phosphonitrilic chloride trimer	1.68	10^{15}				1
Phthalocyanine, PcH_2	1.66	10^{13}	0.1 to 0.4	+	26 to 350	1
Phthalocyanine, PcH_2	2	10^{7}	1.2(n), 1.1(p)	+ & –	100	4
Pyranthrene	1.11	$1 \cdot 10^{15}$				1
Pyranthrene	0.51	$4.5 \cdot 10^{16}$ (15°C)				2
Pyranthrone	1.06	$3.9 \cdot 10^{15}$			40 to 150	1
Pyrene	2.02	$5 \cdot 10^{17}$		–		1
Pyrene			0.50	+		4
5,6-*N*-Pyridine-1,9-benzanthrone	1.60					2
p-Quaterphenyl	0.89	$1.0 \cdot 10^{15}$ (50°C)				2
Quaterrylene	0.6	10^{5}		–		1
p-Quinquiphenyl	0.91	$2.0 \cdot 10^{15}$ (50°C)				2
α-Resorcin	2.10	$2 \cdot 10^{16}$			30 to 94	1
β-Resorcin	3.27	$2 \cdot 10^{18}$			30 to 94	1
Salanil	4.1	10^{4}			20 to 40	1
p-Sexiphenyl	0.91	$7.0 \cdot 10^{14}$ (50°C)				2
cis-Stilbene	2.4			+	at 20	1
trans-Stilbene	1.80		2.4	+	70 to 120	1
trans-Stilbene	0.91					2
trans-Stilbene	1.4			+		4
o-Terphenyl		$3 \cdot 10^{-5}$		+		1
m-Terphenyl			10^{-5}	+		1
p-Terphenyl	0.6	10^{14} (25°C)				2
p-Terphenyl	1.2		0.025	+		1

Substance	Energy Gap, E, (in $E/2kT$) eV	Room Temperature Electrical Resistivity, ohm · cm	Mobility, μ, cm²/V · s	Sign of Majority Carriers	Temperature Range, °C	Ref.
p-Terphenyl			1.2(n), 0.80(p)	+ & –		4
Tetracene	0.66	$3.2 \cdot 10^{12}$ (50°C)				2
Tetracene	1.7					1
Tetracene	3.4		0.85	+		4
1,1,10,10-Tetracyanodecapentaene	2.24	10^{13}			>68	1
1,1,6,6-Tetracyanohexatriene	1.54	10^{14}		–		1
Tetracyanoethylene			0.26(max)	+		4
7,7,8,8-Tetracyanoquinodimethane			0.65	–		4
1,1,8,8-Tetracyanooctatetraene	1.42	10^{12}		–		1
Tetraphenylpentacene	1.62				0 to 150	1
Tetrathiotetracene	0.46	10^4				1
Triphenodioxazine	1.65	$5 \cdot 10^{14}$		–	20 to 140	1
Triphenyldiamine			$2 \cdot 10^{-2}$		at 20	1
Violanthrene	0.85	$2.1 \cdot 10^{14}$			40 to 105	1
Isoviolanthrene	0.82	$8.4 \cdot 10^{13}$			40 to 150	1
Violanthrone	0.78	$2.3 \cdot 10^{10}$			40 to 150	1
Isoviolanthrone	0.76	$5.7 \cdot 10^9$			40 to 150	1
o-Xylene (liquid)	0.45					2
m-Xylene (liquid)	0.41					2
p-Xylene (liquid)	0.41					2
Long-Chain Compounds and Polymers						
Acrylic acid-amylproparylaniline copolymers		10^9–10^{10}				3
Acrylic acid-methylproparylaniline copolymers		10^9–10^{10}				3
Acrylic acid-octylproparylaniline copolymers		10^9–10^{10}				3
Anthrone polymers		0.28 ≥100 at 1.8 kbar ≥2 at 33 kbar				3
$[CH(AsF_5)_{0.1}]_x$		0.0005		+		3
$[CH \cdot I_{0.22}]_x$	1.9	*trans* 10^5, *cis* 10^9				3
1,6-Diacetylenes (cyclopolymerized)		10^{10}–10^{14}				3
Ionene elastomers		$2.7 \cdot 10^7$ to $2.2 \cdot 10^8$			–80 to 60	3
1,3,4-Oxydiazole polymers	0.81	$3 \cdot 10^{12}$			20 to 140	3
Oxypyrrole polymer films		0.125				3
Phenylformaldehyde polymeric pyrolysates						3
a) Pyrolysis Temperature 600°C		27	0.0014	–		
b) 1200°C		0.0044	7.84	+		
Phenylthiocyanate polymers	0.5 to 0.8	10^5–10^8				3
Polyacetylene (undoped)		10^{10}				3
Polyacetylene (I_2 doped)		0.04				3
Polyacetylene (*cis*-rich, undoped)		10^7				3
trans-Polyacetylene (I_2 doped, 0.22 mole %)				+		3
Polyacrylonitrile (heat treated 700°C)			0.01	–	–100 to 100	3
Poly-5,5′-biisatyl	air 0.84	air $2.6 \cdot 10^9$		+	20 to 140	3
thiophene-indophene	vacuum 1.0	vacuum $3.1 \cdot 10^9$				
Poly bis(amino)-phosphazenes	1.75	$1.8 \cdot 10^{11}$			20 to 180	3
Poly-5,5′-diisatylmetane-thiophene-indophenine	0.45	$7.3 \cdot 10^4$		+	20 to 140	3
Polyethylene	2.74				20 to 70	3
Polyethylene (low density)	0.17	$4 \cdot 10^9$			above T_g	3
Polyimide	2.84					3
Polymalonitrile	1.72			–		3
Poly(metalphthalocyanines) :Cu	0.12	$7 \cdot 10^6$				3
:Fe	0.15	$1.1 \cdot 10^6$				3
:Ni	0.46	100				3
:Sb		$3.1 \cdot 10^6$				3
:Zn	0.12	$5.3 \cdot 10^3$				3
Poly-*N*-methylpyrrole		$2 \cdot 10^6$				3
Polyoxypyrrole (black)	0.044	1790			–173 to 27	3
Polyphthalocyanines	0.01	7 to 58				3
Polypyrrole	0.01				–193 to 250	3

Substance	Energy Gap, *E*, (in *E*/2 *kT*) eV	Room Temperature Electrical Resistivity, ohm · cm	Mobility, μ, $cm^2/V \cdot s$	Sign of Majority Carriers	Temperature Range, °C	Ref.
Polypyrroline II	1.74					3
Polyselenomethylene	0.7 to 2.62	$>10^{13}$			20 to 120	3
Poly(2-vinylpyridine):I_2 (1:2)	0.12	1000			−73 to 27	3
PVC (commercial)	2.84–3.04				$T<T_g$	3
PVC (commercial)	1.24–1.96				$T>T_g$	3
PVC (pure)	1.0±0.1				0 to 30	3
Salicylal-*N*-alkyliminate-Cu	1.62	$1.7 \cdot 10^{14}$				4
TTF-acetylacetonate polymers		$1.6 \cdot 10^4$				3
TTF-metal polymers		$1.6 \cdot 10^4$				3

References

1. F. Gutman and L. E. Lyons, *Organic Semiconductors*, John Wiley & Sons, New York, 1967.
2. Y. Okamoto and W. Brenner, *Organic Semiconductors*, Reinhold Publ. Corp., New York, 1964.
3. F. Gutman, H. Keyzer, L. E. Lyons, and R. B. Somoano, *Organic Semiconductors, Part B*, R. E. Krieger Publ. Co., Melbourne, FL, 1983.
4. L. I. Berger, *Semiconductor Materials*, CRC Press, Boca Raton, FL, 1997.

DIFFUSION DATA FOR SEMICONDUCTORS

B. L. Sharma

The diffusion coefficient D in many semiconductors may be expressed by an Arrhenius-type relation

$$D = D_o \exp(-Q/kT)$$

where D_o is a frequency factor, Q is the activation energy for diffusion, k is the Boltzmann constant, and T is the absolute temperature. This table lists D_o and Q for various diffusants in common semiconductors.

Abbreviations used in the table are

AES – Auger Electron Spectroscopy
DLTS – Deep Level Transient Spectroscopy
SEM – Scanning Electron Microscopy
SIMS – Secondary Ion Mass Spectrometry
$D(c)$ – Concentration Dependent Diffusion Coefficient
D_{max} – Maximum Diffusion Coefficient
(f) – Fast Diffusion Component
(i) – Interstitial Diffusion Component
(s) – Slow Diffusion Component
(‖) – Parallel to c Direction
(⊥) – Perpendicular to c Direction

Semiconductor	Diffusant	Frequency factor, D_o (cm²/s)	Activation energy, Q(eV)	Temperature range (°C)	Method of measurement	Ref.
Si	H	6×10^{-1}	1.03	120–1207	Electrical and SIMS	1
	Li	2.5×10^{-3}	0.65	25–1350	Electrical	2
	Na	1.65×10^{-3}	0.72	530–800	Electrical and flame photometry	3
	K	1.1×10^{-3}	0.76	740–800	Electrical and flame photometry	3
	Cu	4×10^{-2}	1.0	800–1100	Radioactive	4
		4.7×10^{-3}	0.43 (i)	300–700	Radioactive	5
	Ag	2×10^{-3}	1.6	1100–1350	Radioactive	6
	Au	2.4×10^{-4}	0.39 (i)	700–1300	Radioactive	7
		2.75×10^{-3}	2.05 (s)			
	Be	$(D \sim 10^{-7})$	–	1050	Electrical	8
	Ca	$(D \sim 6 \times 10^{-14})$	–	1100	Electrical and SIMS	1
	Zn	1×10^{-1}	1.4	980–1270	Electrical	9
	B	2.46	3.59	1100–1250	Electrical	10
		2.4×10^{1}	3.87	840–1250	Electrical	11
	Al	1.38	3.41	1119–1390	Electrical	12
		1.8	3.2	1025–1175	Electrical	13
	Ga	3.74×10^{-1}	3.39	1143–1393	Electrical	12
		6×10^{1}	3.89	900–1050	Radioactive	14
	In	7.85×10^{-1}	3.63	1180–1389	Electrical	12
		1.94×10^{1}	3.86	1150–1242	Radioactive	15
	Tl	1.37	3.7	1244–1338	Electrical	12
		1.65×10^{1}	3.9	1105–1360	Electrical	16
	Sc	8×10^{-2}	3.2	1100–1250	Radioactive	1
	Ce	$(D \sim 3.9 \times 10^{-13})$	–	1050	SIMS	1
	Pr	2.5×10^{-7}	1.74	1100–1280	Electrical	1
	Pm	7.5×10^{-9}	1.2 (s)	730–1270	Radioactive	1
		4.2×10^{-12}	0.13 (f)			
	Er	2×10^{-3}	2.9	1100–1250	Radioactive	1
	Tm	8×10^{-3}	3.0	1100–1280	Radioactive	1
	Yb	2.8×10^{-5}	0.95	947–1097	Neutron activation	1
	Ti	1.45×10^{-2}	1.79	950–1200	DLTS	17
	C	3.3×10^{-1}	2.92	1070–1400	Radioactive	18
	Si (self)	1.54×10^{2}	4.65	855–1175	SIMS	19
		1.6×10^{3}	4.77	1200–1400	Radioactive	20
	Ge	3.5×10^{-1}	3.92	855–1000	Radioactive	21
		2.5×10^{3}	4.97	1030–1302	Radioactive	21
		7.55×10^{3}	5.08	1100–1300	SIMS	22
	Sn	3.2×10^{1}	4.25	1050–1294	Neutron activation	23
	N	2.7×10^{-3}	2.8	800–1200	Out Diffusion; SIMS	1

Semiconductor	Diffusant	Frequency factor, D_o (cm²/s)	Activation energy, Q(eV)	Temperature range (°C)	Method of measurement	Ref.
	P	2.02×10^1	3.87	1100–1250	Electrical	10
		1.1	3.4	900–1200	Radioactive	24
		7.4×10^{-2}	3.3	1130–1405	Electrical	25
	As	6.0×10^1	4.2	950–1350	Radioactive	26
		6.55×10^{-2}	3.44	1167–1394	Electrical	27
		2.29×10^1	4.1	900–1250	Electrical	28
	Sb	1.29×10^1	3.98	1190–1398	Radioactive	29
		2.14×10^{-1}	3.65	1190–1405	Electrical	27
	Bi	1.03×10^3	4.64	1220–1380	Electrical	16
		1.08	3.85	1190–1394	Electrical	27
	Cr	1×10^{-2}	1	1100–1250	Radioactive	30
	Mo	$(D \sim 2 \times 10^{-10})$	–	1000	DLTS	1
	W	$(D \sim 10^{-12})$	–	1100	DLTS	1
	O	7×10^{-2}	2.44	700–1250	SIMS	31
		1.4×10^{-1}	2.53	700–1160	SIMS	32
	S	5.95×10^{-3}	1.83	975–1200	Radioactive	33
	Se	9.5×10^{-1}	2.6	1050–1250	Electrical	34
	Te	5×10^{-1}	3.34	900–1250	SIMS	1
	Mn	6.9×10^{-4}	0.63	900–1200	Radioactive	35
	Fe	1.3×10^{-3}	0.68	30–1250	Radioactive	36
	Co	2×10^{-3}	0.69	700–1300	Radioactive	37
	Ni	2×10^{-3}	0.47	800–1300	Radioactive	38
	Ru	$(D \sim 5 \times 10^{-7}$ $-5 \times 10^{-6})$	–	1000–1280	Electrical	1
	Rh	$(D \sim 10^{-6}–10^{-4})$	–	1000–1200	Electrical	39
	Pd	2.95×10^{-4}	0.22 (i)	702–1320	Nuclear Activation	1
	Pt	1.5×10^2	2.22	800–1000	Electrical	1
	Os	$(D \sim 2 \times 10^{-6})$	–	1280	Electrical	40
	Ir	4.2×10^{-2}	1.3	950–1250	Electrical	41
Ge	Li	1.3×10^{-3}	0.46	350–800	Electrical	42
		9.1×10^{-3}	0.57	800–500	Electrical	43
	Na	3.95×10^{-1}	2.03	700–850	Radioactive	44
	Cu	1.9×10^{-4}	0.18 (i)	750–900	Radioactive	45
		4×10^{-2}	0.99 (s)	600–700		
		4×10^{-3}	0.33 (i)	350–750	Radioactive	5
	Ag	4.4×10^{-2}	1.0 (i)	700–900	Radioactive	46, 47
		4×10^{-2}	2.23 (s)	800–900	Radioactive	48
	Au	2.25×10^2	2.5	600–900	Radioactive	49
	Be	5×10^{-1}	2.5	720–900	Electrical	50
	Mg	$(D \sim 8 \times 10^{-9})$	–	900	Electrical	1
	Zn	5	2.7	600–900	Radioactive and electrical	51
	Cd	1.75×10^9	4.4	760–915	Radioactive	52
	B	1.8×10^9	4.55	600–900	Electrical	51
	Al	1.0×10^3	3.45	554–905	SIMS	53
		$\sim 1.6 \times 10^2$	~3.24	750–850	Electrical	54
	Ga	1.4×10^2	3.35	554–916	SIMS	55
		3.4×10^1	3.1	600–900	Electrical	51
	In	1.8×10^4	3.67	554–919	SIMS	56
		3.3×10^1	3.02	700–855	Radioactive	57
	Tl	1.7×10^3	3.4	800–930	Radioactive	58
	Si	2.4×10^{-1}	2.9	650–900	(γ) resonance	59
	Ge (self)	2.48×10^1	3.14	549–891	Radioactive	60
		7.8	2.95	766–928	Radioactive	61
	Sn	1.7×10^{-2}	1.9	–	Radioactive	45
	P	3.3	2.5	600–900	Electrical	51
	As	2.1	2.39	700–900	Electrical	62
	Sb	3.2	2.41	700–855	Radioactive	57
		1.0×10^1	2.5	600–900	Radioactive and electrical	51

Semiconductor	Diffusant	Frequency factor, D_o (cm²/s)	Activation energy, Q(eV)	Temperature range (°C)	Method of measurement	Ref.
	Bi	3.3	2.57	650–850	–	63
	O	4×10^{-1}	2.08	–	Optical	64
	S	$(D \sim 10^{-9})$	–	920	–	65
	Se	$(D \sim 10^{-10})$	–	920	–	65
	Te	5.6	2.43	750–900	Radioactive	66
	Fe	1.3×10^{-1}	1.08	750–900	Radioactive	67
	Co	1.6×10^{-1}	1.12	750–850	Radioactive	47
	Ni	8×10^{-1}	0.9	670–900	Electrical	68
GaAs	Li	5.3×10^{-1}	1.0	250–500	Electrical and chemical	69
	Cu	3×10^{-2}	0.53	100–500	Radioactive	69
		6×10^{-2}	0.98	450–750	Ultrasonic	69
		1.5×10^{-3}	0.6	800–1000	Radioactive	69
	Ag	4×10^{-4}	0.8	500–1150	Radioactive	69
	Au	1×10^{-3}	1.0	740–1025	Radioactive	69
	Be	7.3×10^{-6}	1.2	800–990	Electrical	69
	Mg	4×10^{-5}	1.22	800–1200	Electrical	69
	Zn	1.5×10^{1}	2.49	600–980	Radioactive	69
		2.5×10^{-1}	3.0	750–1000	Radioactive	69
	Cd	1.3×10^{-3}	2.2	800–1100	Radioactive	69
		5×10^{-2}	2.43	868–1149	Radioactive	69
	Hg	$(D \sim 5 \times 10^{-14})$	–	1100	Radioactive	69
	Al	$(D \sim 4 \times 10^{-18}–10^{-14})$	4.3	850–1100	AES	70
	Ga (self)	4×10^{-5}	2.6	1025–1100	Radioactive	69
		1×10^{7}	5.6	1125–1230	Radioactive	69
	In	$(D \sim 7 \times 10^{-11})$	–	1000	Radioactive	69
	C	$(D \sim 1.04 \times 10^{-16})$	–	825	SIMS	69
	Si	1.1×10^{-1}	2.5	850–1050	SIMS	69
	Ge	1.6×10^{-5}	2.06	650–850	SIMS	69
	Sn	6×10^{-4}	2.5	1060–1200	Radioactive	69
		1×10^{-5}	2	800–1000	Radioactive	69
	P	$(D \sim 10^{-12}–10^{-10})$	2.9	800–1150	Reflectance measurements	69
	As (self)	7×10^{-1}	3.2	–	Radioactive	69
	Cr	2.04×10^{-6}	0.83 (f)	750–1000	SIMS	69
			1.7 (s)	700–900		
		7.9×10^{-3}	2.2	800–1100	Chemical analysis	69
	O	2×10^{-3}	1.1	700–900	Mass spectroscopy	69
	S	1.85×10^{-2}	2.6	1000–1300	Radioactive	69
		1.1×10^{1}	2.95	750–900	Electrical	69
	Se	3×10^{3}	4.16	1025–1200	Radioactive	69
	Te	1.5×10^{-1}	3.5	1000–1150	Radioactive	69
	Mn	6.5×10^{-1}	2.49	850–1100	Radioactive	69
	Fe	4.2×10^{-2}	1.8	850–1150	Radioactive	69
		2.2×10^{-3}	2.32	750–1050	Radioactive	69
	Co	5×10^{2}	2.5	800–1000	Radioactive	69
		1.2×10^{-1}	2.64	750–1050	Radioactive	69
	Tm	2.3×10^{-16}	1.0	800–1000	Radioactive	69
GaSb	Li	2.3×10^{-4}	1.9 (s)	527–657	Electrical and flame photometry	69
		1.2×10^{-1}	0.7 (f)	277–657		
	Cu	4.7×10^{-3}	0.9	470–650	Radioactive	69
	Zn	$(D \sim 2 \times 10^{-13} – 1 \times 10^{-11})$	2	510–600	Radioactive	69
	Cd	1.5×10^{-6}	0.72	640–800	Electrical	69
	Ga (self)	3.2×10^{3}	3.15	658–700	Radioactive	69
	In	1.2×10^{-7}	0.53	320–650	Radioactive	69
	Sn	2.4×10^{-5}	0.8	320–650	Radioactive	69
		1.3×10^{-5}	1.1	500–650	Radioactive	69
	Sb (self)	3.4×10^{4}	3.45	658–700	Radioactive	69
	Se	$(D \sim 2.4 \times 10^{-13} – 1.37 \times 10^{-11})$	–	400–500	Radioactive	69

Semiconductor	Diffusant	Frequency factor, D_o (cm²/s)	Activation energy, Q(eV)	Temperature range (°C)	Method of measurement	Ref.
	Te	3.8×10^{-4}	1.20	320–650	Radioactive	69
	Fe	5×10^{-2}	1.9 (I)	500–650	Radioactive	69
		5×10^{2}	2.3 (II)	500–650		
GaP	Ag	–	–	1000–1300	Radioactive	69
	Au	8	2.5 (I)	1050–1250	Radioactive	69
		20	2.4 (II)	1100–1250	Diffusion (I) A face and (II) B face	
	Be	($D_{max} \sim 2.4 \times 10^{-9}$– 8.5×10^{-8})	–	900–1000	Atomic absorption analysis	69
	Mg	5×10^{-5}	1.4	700–1050	Electrical	69
	Zn	1.0	2.1	700–1300	Radioactive	69
	Ge	–	–	900–1000	Radioactive	69
	Cr	6.2×10^{-4}	1.2	900–1130	Radioactive; ESR	69
	S	3.2×10^{3}	4.7	1120–1305	Radioactive	69
	Mn	2.1×10^{9}	4.7	T < 950	Radioactive; ESR	69
		1.1×10^{-6}	0.9	950–1130		
	Fe	1.6×10^{-1}	2.3	980–1180	Radioactive	69
	Co	2.8×10^{-3}	2.9	850–1100	Radioactive	69
InP	Cu	3.8×10^{-3}	0.69	600–900	Radioactive	69
	Ag	3.6×10^{-4}	0.59	500–900	Radioactive	69
	Au	1.32×10^{-5}	0.48	600–820	Radioactive	69
		1.37×10^{-4}	0.73	600–900	Radioactive	69
	Zn	1.6×10^{-8}	0.3	750–900	Electrical	69
		($D \sim 2 \times 10^{-9}$– 4×10^{-8})	–	700–900	Radioactive	69
	Cd	1.8	1.9	700–900	Radioactive	69
		1.1×10^{-7}	0.72	700–900	Electrical	69
		($D \sim 7 \times 10^{-13}$– 2×10^{-10})	–	450-650	Electrical	69
	In (self)	1×10^{5}	3.85	830–990	Radioactive	69
	Sn	($D \sim 3 \times 10^{-8}$)	–	550	Etching and cathodo-luminescence	69
	P (self)	7×10^{10}	5.65	900–1000	Radioactive	69
	Cr	–	–	600–900	Radioactive	69
	S	3.6×10^{-4}	1.94	585–708	Electrical	69
	Se	($D \sim 2 \times 10^{-8}$)	–	550	Cathodoluminescence	69
	Mn	–	2.9	650–750	SIMS	69
	Fe	3	2	600–950	Radioactive	69
		6.8×10^{5}	3.4	600–700	SIMS	69
	Co	9×10^{-1}	1.8	600–950	Radioactive	69
InAs	Cu	3.6×10^{-3}	0.52	342–875	Radioactive	69
		2.2×10^{-2}	0.54	525–890	Radioactive	69
	Ag	7.3×10^{-4}	0.26	450–900	Radioactive	69
	Au	5.8×10^{-3}	0.65	600–900	Radioactive	69
	Mg	1.98×10^{-6}	1.17	600–900	Electrical	69
	Zn	4.2×10^{-3}	0.96	600–900	Radioactive	69
		3.11×10^{-3}	1.17	600–900	Electrical	69
	Cd	7.4×10^{-4}	1.15	650–900	Radioactive	69
	Hg	1.45×10^{-5}	1.32	650–850	Radioactive	69
	In (self)	6×10^{5}	4.0	740–900	Radioactive	69
	Ge	3.74×10^{-6}	1.17	600–900	Electrical	69
	Sn	1.49×10^{-6}	1.17	600–900	Electrical	69
	As (self)	3×10^{7}	4.45	740–900	Radioactive	69
	S	6.78	2.2	600–900	Electrical	69
	Se	12.6	2.2	600–900	Electrical	69
	Te	3.43×10^{-5}	1.28	600–900	Electrical	69
InSb	Li	7×10^{-4}	0.28	0–210	Electrical	69
	Cu	9×10^{-4}	1.08	200–500	Radioactive	69
		3×10^{-5}	0.37	230–490	Radioactive	69
	Ag	1×10^{-7}	0.25	440–510	Radioactive	69
	Au	7×10^{-4}	0.32	140–510	Radioactive	69
	Zn	5×10^{-1}	1.35	362–508	Radioactive	69

Semiconductor	Diffusant	Frequency factor, D_o (cm²/s)	Activation energy, Q(eV)	Temperature range (°C)	Method of measurement	Ref.
		–	1.5	355–455	SIMS	69
	Cd	1×10^{-5}	1.1	250–500	Radioactive	69
		1.3×10^{-4}	1.2	360–500	Electrical	69
	Hg	4×10^{-6}	1.17	425–500	Radioactive	69
	In (self)	6×10^{-7}	1.45	400–500	Radioactive	69
		1.8×10^{13}	4.3	475–517	Radioactive	69
	Sn	5.5×10^{-8}	0.75	390–512	Radioactive	69
	Pb	$(D \sim 2.7 \times 10^{-15})$	–	500	Radioactive	71
	Sb (self)	5.35×10^{-4}	1.91	400–500	Radioactive	69
		3.1×10^{13}	4.3	475–517	Radioactive	69
	S	9×10^{-2}	1.4	360–500	Electrical	69
	Se	1.6	1.87	380–500	Electrical	69
	Te	1.7×10^{-7}	0.57	300–500	Radioactive	69
	Fe	1×10^{-7}	0.25	440–510	Radioactive	69
	Co	2.7×10^{-11}	0.39	420–500	Radioactive	69
AlAs	Ga	$(D \sim 2 \times 10^{-18}– 10^{-15})$	3.6	850–1100	AES	70
	Zn	$(D \sim 9 \times 10^{-11})$	–	557	SEM	69
AlSb	Cu	3.5×10^{-3}	0.36	150–500	Radioactive	69
	Zn	3.3×10^{-1}	1.93	660–860	Radioactive	69
	Cd	$D(c) \sim 4 \times 10^{-12}– 3 \times 10^{-10}$	–	900	Radioactive	69
	Al (self)	2	1.88	570–620	X-ray	69
	Sb (self)	1	1.7	570–620	X-ray	69
ZnS	Cu	2.6×10^{-3}	0.79	470–750	Radioactive	69
		4.3×10^{-4}	0.64	250–1200	Electroluminescence	69
		9.75×10^{-3}	1.04	400–800	Luminescence	69
	Au	1.75×10^{-4}	1.16	500–800	Radioactive	69
	Zn (self)	3×10^{-4}	1.5	925<T<940	Radioactive	69
		1.5×10^{4}	3.26	940<T<1030		
		1×10^{16}	6.5	1030<T<1075		
	Cd	$(D \sim 10^{-10})$	–	1100	Luminescence	72
	Al	5.69×10^{-4}	1.28	800–1000	Luminescence	69
	In	3×10^{1}	2.2	750–1000	Radioactive	69
	S (self)	2.16×10^{4}	3.15	600–800	Radioactive	69
		8×10^{-5}	2.2	740–1100	Radioactive	69
	Se	$(D \sim 5 \times 10^{-13})$	–	1070	X-ray microprobe	69
	Mn	2.3×10^{3}	2.46	500–800	Radioactive	69
ZnSe	Li	2.66×10^{-6}	0.49	950–980	Electrical	69
	Cu	1×10^{-4}	0.66	400–800	Luminescence	69
		1.7×10^{-5}	0.56	200–570	Radioactive	69
	Ag	2.2×10^{-2}	1.18	400–800	Luminescence	69
	Zn (self)	9.8	3.0	760–1150	Radioactive	69
	Cd	6.39×10^{-4}	1.87	700–950	Photoluminescence	69
	Al	2.3×10^{-2}	1.8	800–1100	Luminescence	69
	Ga	1.81×10^{2}	3.0	900–1100	Luminescence	69
		–	1.3	700–850	Electron probe	69
	In	$(D \sim 2 \times 10^{-12})$	–	940	–	69
	S	$(D \sim 8 \times 10^{-12})$	–	1060	X-ray microprobe	69
	Se (self)	1.3×10^{1}	2.5	860–1020	Radioactive	69
		2.3×10^{-1}	2.7	1000–1050	Radioactive	69
	Ni	$(D \sim 1.5 \times 10^{-8}– 1.7 \times 10^{-7})$	–	740–910	Luminescence	69
ZnTe	Li	2.9×10^{-2}	1.22 (s)	400–700	Nuclear and chemical analysis	69
		1.7×10^{-4}	0.78 (f)			
	Zn (self)	2.34	2.56	760–860	Radioactive	69
		1.4×10^{1}	2.69	667–1077	Radioactive	69
	Al	–	2.0	700–1000	Electrical and optical	69
	In	4	1.96	1100–1300	Radioactive	69
	Te (self)	2×10^{4}	3.8	727–977	Radioactive	69
CdS	Li	3×10^{-6}	0.68	610–960	Microhardness	69
	Na	$(D \sim 3 \times 10^{-7})$	–	800	Radioactive	69

Semiconductor	Diffusant	Frequency factor, D_o (cm²/s)	Activation energy, Q(eV)	Temperature range (°C)	Method of measurement	Ref.
	Cu	1.5×10^{-3}	0.76	400–700	Radioactive	69
		1.2×10^{-2}	1.05	300–700	Ultrasonic	69
		8×10^{-5}	0.72	20–200	Electrical	69
	Ag	2.5×10^{1}	1.2 (s)	300–500	Radioactive	69
		2.4×10^{-1}	0.8 (f)			
	Au	2×10^{2}	1.8	500–800	Radioactive	69
	Zn	1.27×10^{-9}	0.86 (s)	720–1000	Radioactive	69
		1.22×10^{-8}	0.66 (f)			
	Cd (self)	3.4	2.0	700–1100	Radioactive	69
	Ga	–	–	667–967	Optical and microprobe	69
	In	6×10^{1}	2.3 (∥)	650–930	Radioactive, optical and microprobe	69
		1×10^{1}	2.03 (⊥)			
	P	6.5×10^{-4}	1.6	800–1100	Radioactive	69
	S (self)	1.6×10^{-2}	2.05	800–900	Radioactive	69
		–	2.4	750–1050	Radioactive	69
	Se	($D \sim 1.2 \times 10^{-9}$)	–	900	Radioactive	69
	Te	1.3×10^{-7}	10.4	700–1000	Radioactive	69
	Cl	($D \sim 3 \times 10^{-10}$)	–	800	Electrical	69
	I	($D \sim 5 \times 10^{-12}$)	–	1000	Radioactive	69
	Ni	6.75×10^{-3}	10.9	570–900	Luminescence	69
	Yb	($D \sim 1.3 \times 10^{-9}$)	–	960	Photoluminescence	69
CdSe	Ag	2×10^{-4}	0.53	22–400	Ultrasonic	69
	Cd (self)	1.6×10^{-3}	1.5	700–1000	Radioactive	69
		6.3×10^{-2}	1.25 (I)	600–900	Radioactive;	69
		4.12×10^{-2}	2.18 (II)	600–900	(I) saturated Cd and (II) saturated Se pressure	
	P	($D \sim 5.3 \times 10^{-12}$– 6×10^{-11})	–	900–1000	Radioactive	69
	Se (self)	2.6×10^{3}	1.55	700–1000	Radioactive; saturated Se pressure	69
CdTe	Li	($D \sim 1.5 \times 10^{-10}$)	–	300	Ion microprobe	69
	Cu	3.7×10^{-4}	0.67	97–300	Radioactive	69
		8.2×10^{-8}	0.64	290–350	Ion backscattering	69
	Ag	–	–	700–800	Electrical and photo-luminescence	69
	Au	6.7×10^{1}	2.0	600–1000	Radioactive	69
	Cd (self)	1.26	2.07	700–1000	Radioactive	69
		3.26×10^{2}	2.67 (I)	650–900	Radioactive;	69
		1.58×10^{1}	2.44 (II)		(I) saturated Cd and (II) saturated Te pressure	
	In	8×10^{-2}	1.61	650–1000	Radioactive	69
		1.17×10^{2}	2.21 (I)	500–850	Radioactive; (I) saturated Cd and (II) saturated Te pressure	69
		6.48×10^{-4}	1.15 (II)			
	Sn	8.3×10^{-2}	2.2	700–925	Radioactive	69
	P	($D \sim 1.2 \times 10^{-10}$)	–	900	Radioactive	69
	As	–	–	850	–	69
	O	5.6×10^{-9}	1.22	200–650	Mass spectrometry	69
		6.0×10^{-10}	0.29	650–900		
	Se	1.7×10^{-4}	1.35	700–1000	Radioactive	69
	Te (self)	8.54×10^{-7}	1.42 (I)	600–900	Radioactive; (I) saturated Cd and (II) saturated Te pressure	69
		1.66×10^{-4}	1.38 (II)	500–800		
	Cl	7.1×10^{-2}	1.6	520–800	Radioactive	69
	Fe	($D \sim 4 \times 10^{-8}$)	0.77	900	Radioactive	69

Semiconductor	Diffusant	Frequency factor, D_o (cm²/s)	Activation energy, Q(eV)	Temperature range (°C)	Method of measurement	Ref.
HgSe	Sb	6.3×10^{-5}	0.85	540–630	Radioactive	69
	Se (self)	–	–	200–400	Radioactive	69
HgTe	Ag	6×10^{-4}	0.8	250–350	Radioactive	69
	Zn	5×10^{-8}	0.6	250–350	Radioactive	69
	Cd	3.1×10^{-4}	0.66	250–350	Radioactive	69
	Hg (self)	2×10^{-8}	0.6	200–350	Radioactive	69
	In	6×10^{-6}	0.9	200–300	Radioactive	69
	Sn	1.72×10^{-6}	0.66 (s)	200–300	Radioactive	69
		1.8×10^{-3}	0.80 (f)			
	Te (self)	10^{-6}	1.4	200–400	Radioactive	69
	Mn	1.5×10^{-4}	1.3	250–350	Radioactive	69
PbS	Cu	4.6×10^{-4}	0.36	150–450	Electrical	69
		5×10^{-3}	0.31	100–400	Electrical	69
	Pb (self)	8.6×10^{-5}	1.52	500–800	Radioactive	69
	S (self)	6.8×10^{-5}	1.38	500–750	Radioactive	69
	Ni	1.78×10^{1}	0.95	200–500	Electrical	69
PbSe	Na	1.5×10^{1}	1.74 (s)	400–850	Radioactive	69
		5.6×10^{-6}	0.4 (f)			
	Cu	2×10^{-5}	0.31	93–520	Radioactive	69
	Ag	7.4×10^{-4}	0.35	400–850	Radioactive	69
	Pb (self)	4.98×10^{-6}	0.83	400–800	Radioactive	69
	Sb	3.4×10^{-1}	2.0	650–850	Radioactive	69
	Se (self)	2.1×10^{-5}	1.2	650–850	Radioactive	69
	Cl	1.6×10^{-8}	0.45	400–850	Radioactive	69
	Ni	$(D \sim 1 \times 10^{-10})$	–	700	Radioactive	69
PbTe	Na	1.7×10^{-1}	1.91	600–850	Radioactive	69
	Sn	3.1×10^{-7}	[illegible]	600–800	Radioactive	69
	Pb (self)	2.9×10^{-5}	0.6	250–500	Radioactive	69
	Sb	4.9×10^{-2}	1.54	500–800	Radioactive	69
	Te	2.7×10^{-6}	0.75	500–800	Radioactive	69
	Cl	$(D > 2.3 \times 10^{-10})$	–	700	Radioactive	69
	Ni	$(D > 1 \times 10^{-6})$	–	700	Radioactive	69

References

1. N. A. Stolwijk and H. Bracht, in *Diffusion in Semiconductors and Non-Metallic Solids*, D. L. Beke, Ed., Springer-Verlag, Berlin, 1998, 2-1.
2. E. M. Pell, *Phys. Rev.*, 119, 1960; 119, 1014, 1960.
3. L. Svob, *Solid State Electron*, 10, 991, 1967.
4. B. I. Boltaks and I. I. Sosinov, *Zh. Tekh. Fiz.*, 28, 3, 1958.
5. R. N. Hall and J. N. Racette, *J. Appl. Phys.*, 35, 379, 1964.
6. B. I. Boltaks and Hsueh Shih-Yin, *Sov. Phys. Solid State*, 2, 2383, 1961.
7. W. R. Wilcox and T. J. LaChapelle, *J. Appl. Phys.*, 35, 240, 1964.
8. E. A. Taft and R. O. Carlson, *J. Electrochem. Soc.*, 117, 711, 1970.
9. R. Sh. Malkovich and N. A. Alimbarashvili, *Sov. Phys. Solid State*, 4, 1725, 1963.
10. R. N. Ghoshtagore, *Solid State Electron*, 15, 1113, 1972.
11. C. Hill, *Semiconductor Silicon 1981*, H. R. Huff, R. J. Kreiger, and Y. Takeishi, Eds., p. 988, *Electrochem. Soc.*, 1981.
12. R. N. Ghoshtagore, *Phys. Rev. B*, 3, 2507, 1971.
13. W. Rosnowski, *J. Electrochem. Soc.*, 125, 957, 1978.
14. J. S. Makris and B. J. Masters, *J. Appl. Phys.*, 42, 3750, 1971.
15. M. F. Millea, *J. Phys. Chem. Solids*, 27, 315, 1965 (refer Reference 2).
16. C. S. Fuller and J. A. Ditzenberger, *J. Appl. Phys.*, 27, 544, 1956.
17. S. Hocine and D. Mathiot, *Appl. Phys. Lett.*, 53, 1269, 1988.
18. R. C. Newman and J. Wakefield, *J. Phys. Chem. Solids*, 19, 230, 1961.
19. L. Kalinowski and R. Seguin, *Appl. Phys. Lett.*, 35, 211, 1979; *Appl. Phys. Lett.*, 36, 171, 1980.
20. R. F. Peart, *Phys. Stat. Sol.*, 15, K 119, 1966.
21. G. Hettich, H. Mehrer and K. Maler, *Inst. Phys. Conf. Ser.*, 46, 500, 1979.
22. M. Ogina, Y. Oana and M. Watanabe, *Phys. Stat. Sol. (a)*, 72, 535, 1982.
23. T. H. Yeh, S. M. Hu, and R. H. Kastl, *Appl. Phys.*, 39, 4266, 1968.
24. I. Franz and W. Langheinrich, *Solid State Electron*, 14, 835, 1971.
25. R. N. Ghoshtagore, *Hys. Rev. B*, 3, 389, 1971.
26. B. J. Masters and J. M. Fairfield, *J. Appl. Phys.*, 40, 2390, 1969.
27. R. N. Goshtagore, *Phys. Rev. B*, 3, 397, 1971.
28. R. S. Fair and J. C. C. Tsai, *J. Electrochem. Soc.*, 122, 1689, 1975.
29. J. J. Rohan, N. E. Pickering, and J. Kennedy, *J. Electrochem. Soc.*, 106, 705, 1969.
30. W. Wuerker, K. Roy, and J. Hesse, *Matsr. Res. Bull.*, 9, 971, 1974.
31. J. C. Mikkelsen, Jr., *Appl. Phys. Lett.*, 40, 336, 1982.
32. S. Tang Lee and D. Nicols, *Appl. Phys. Lett.*, 47, 1001, 1985.
33. P. L. Gruzin, S. V. Zemskii, A. D. Bullkin, and N. M. Makarov, *Sov. Phys. Sem.*, 7, 1241, 1974.
34. N. S. Zhdanovich and Yu. I. Kozlov, *Svoistva Legir, Poluprovodn.*, V. S. Zemskov, Ed., Nauka, Moscow, 1977, 115–120; *Fiz Tekh. Poluprovod.*, 9, 1594, 1975.
35. D. Gilles, W. Bergholze, and W. Schroeter, *J. Appl. Phys.*, 59, 3590, 1986.
36. E. R. Weber, *Appl. Phys.* A, 30, 1, 1983.
37. E. R. Weber, Properties of Silicon, EMIS Datareviews Ser. No. 4, INSPEC Publications, 1988, 409–451.
38. M. K. Bakhadyrkhanov, S. Zainabidinov, and A. Khamidov, *Sov. Phys. Sem.*, 14, 243, 1980.
39. S. A. Azimov, M. S. Yunosov, F. K. Khatamkulov, and G. Nasyrov, *Poluprovod.*, N. Kh. Abrikosov and V. S. Zemskov, Eds., Nauka, Moscow, 1975, 21–23.

40. S. A. Azimov, M. S. Yunosov, G. Nurkuziev, and F. R. Karimov, *Sov. Phys. Sem.*, 12, 981, 1978.
41. S. A. Azimov, B. V. Umarov, and M. S. Yunusov, *Sov. Phys. Sem.*, 10, 842, 1976.
42. C. S. Fuller and J. A. Ditzenberger, *Phys. Rev.*, 91, 193, 1953.
43. B. Pratt and F. Friedman, *J. Appl. Phys.*, 37, 1893, 1966.
44. M. Stojic, V. Spiric, and D. Kostoski, *Inst. Phys. Conf. Ser.*, 31, 304, 1976.
45. B. I. Boltaks, *Diffusion in Semiconductors*, Inforsearch, London, 1963, 162.
46. A. A. Bugai, V. E. Kosenko, and E. G. Miselyuk, *Zh. Tekh. Fiz.*, 27, 67, 1957.
47. L. Y. Wei, *J. Phys. Chem. Solids*, 18, 162, 1961.
48. V. E. Kosenko, *Sov. Phys. Solid State*, 4, 42, 1962.
49. W. C. Dunlap, Jr., *Phys. Rev.*, 97, 614, 1955
50. Yu. I. Belyaev and V. A. Zhidkov, *Sov. Phys. Solid State*, 3, 133, 1961.
51. W. C. Dunlap, Jr. *Phys. Rev.*, 94, 1531, 1954.
52. V. E. Kosenko, *Sov. Phys. Solid State*, 1, 1481, 1960.
53. P. Dorner, W. Gust, A. Lodding, H. Odelius, B. Predel, and U. Roll, *Acta Metall.*, 30, 941, 1982.
54. W. Meer and D. Pommerrening, *Z. Agnew. Phys.*, 23, 369, 1967.
55. U. Sodervall, H. Odelius, A. Lodding, U. Roll, B. Predel, W. Gust, and P. Dorner, *Phil. Mag. A*, 54, 539, 1986.
56. P. Dorner, W. Gust, A. Lodding, H. Odelius, B. Predel, and U. Roll, *Z. Metalkd.*, 73, 325, 1982.
57. P. V. Pavlov, *Sov. Phys. Solid State*, 8, 2377, 1967.
58. V. I. Tagirov and A. A. Kuliev, *Sov. Phys. Solid State*, 4, 196, 1962.
59. J. Raisanen, J. Hirvonen, and A. Anttila, *Solid State Electron.*, 24, 333, 1981.
60. C. Vogel, G. Hettich, and H. Mehrer, *J. Phys. C.*, 16, 6197, 1983.
61. H. Letaw, Jr., W. M. Portnoy, and L. Slifkin, *Phys. Rev.*, 102, 363, 1956.
62. W. Bosenberg, *Z. Naturforsch.*, 10a, 285, 1955.
63. V. M. Glazov and V. S. Zemskov, Physicochemical Principles of Semiconductor Doping, Israel Program for Scientific Translation, Jerusalem, 1968.
64. J. W. Corbett, R. S. McDonald, and G. D. Watkins, *J. Phys. Chem. Solids*, 25, 873, 1964.
65. W. W. Tyler, *J. Phys. Chem. Solids*, 8, 59, 1959.
66. V. D. Ignatkov and V. E. Kosenko, *Sov. Phys. Solid State*, 4, 1193, 1962.
67. A. A. Bugal, V. E. Kosenko, and E. G. Miseluk, *Zh. Tekh. Fiz.*, 27, 210, 1957.
68. F. van der Maesen and J. A. Brenkman, *Phillips Res. Rep.*, 9, 255, 1954.
69. M. B. Dutt and B. L. Sharma, in *Diffusion in Semiconductors and Non-Metallic Solids*, D. L. Beke, Ed., Springer-Verlag, Berlin, 1998, 3-1.
70. L. L. Chang and A. Koma, *Appl. Physics Lett.*, 29, 138, 1976.
71. D. L. Kendall, *Semiconductors and Semimetals*, Vol. 4, R. K. Willardson and A. C. Beer, Eds., Academic, 1968, 255.
72. H. J. Biter and F. Williams, *J. Luminescence*, 3, 395, 1971.

PROPERTIES OF MAGNETIC MATERIALS

H. P. R. Frederikse

Glossary of Symbols

Quantity	Symbol	Units SI	Units emu
Magnetic field	H	A m^{-1}	Oe (oersted)
Magnetic induction	B	T (tesla)	G (gauss)
Magnetization	M	A m^{-1}	emu cm^{-3}
Spontaneous magnetization	M_s	A m^{-1}	emu cm^{-3}
Saturation magnetization	M_0	A m^{-1}	emu cm^{-3}
Magnetic flux	Φ	Wb (weber)	maxwell
Magnetic moment	m, μ	A m^2	erg/G
Coercive field	H_c	A m^{-1}	Oe
Remanence	B_r	T	G
Saturation magnetic polarization	J_s	T	G
Magnetic susceptibility	χ		
Magnetic permeability	μ	H m^{-1} (henry/meter)	
Magnetic permeability of free space	μ_0	H m^{-1}	
Saturation magnetostriction	λ ($\Delta l/l$)		
Curie temperature	T_C	K	K
Néel temperature	T_N	K	K

Magnetic moment $\mu = \gamma\hbar J = g\,\mu_B\,J$
where

γ = gyromagnetic ratio; J = angular momentum; g = spectroscopic splitting factor (≈ 2)
μ_B = Bohr magneton = $9.2741 \cdot 10^{-24}$ J/T = $9.2741 \cdot 10^{-21}$ erg/G

Earth's magnetic field H = 56 A m^{-1} = 0.7 Oe
For iron: $M_0 = 1.7 \cdot 10^6$ A m^{-1}; $B_r = 0.8 \cdot 10^6$ A m^{-1}
1 Oe = $(1000/4\pi)$ A m^{-1}; 1 G = 10^{-4} T; 1 emu cm^{-3} = 10^3 A m^{-1}
1 Maxwell = 10^{-8} Wb
$\mu_0 = 4\pi\ 10^{-7}$ H m^{-1}

Relation between Magnetic Induction and Magnetic Field

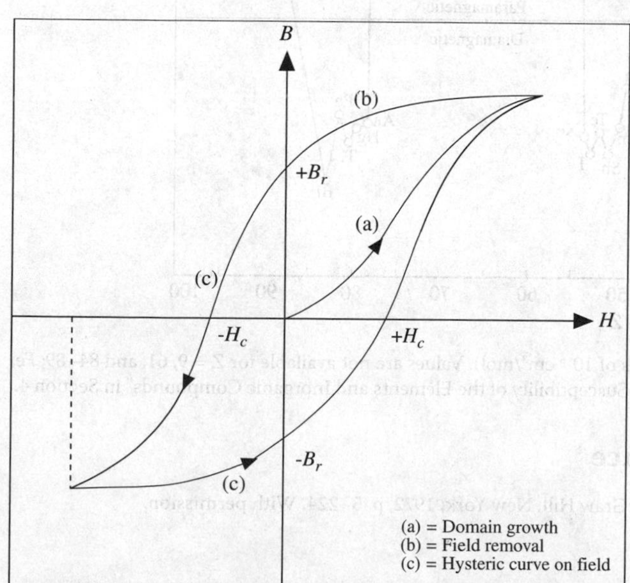

FIGURE 1. Typical curve representing the dependence of magnetic induction B on magnetic field H for a ferromagnetic material. When H is first applied, B follows curve (a) as the favorably oriented magnetic domains grow. This curve flattens as saturation is approached. When H is then reduced, B follows curve (b), but retains a finite value (the remanence B_r) at $H = 0$. In order to demagnetize the material, a negative field $-H_c$ (where H_c is called the coercive field or coercivity) must be applied. As H is further decreased and then increased to complete the cycle (curve c), a hysteresis loop is obtained. The area within this loop is a measure of the energy loss per cycle for a unit volume of the material.

12-107

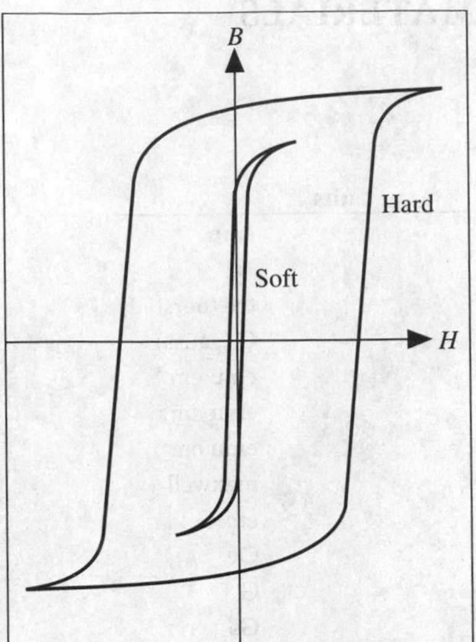

FIGURE 2. Schematic curve illustrating the *B* vs. *H* dependence for hard and soft magnetic materials. Hard materials have a larger remanence and coercive field, and a correspondingly large hysteresis loss.

Reference

Ralls, K. M., Courtney, T. H., and Wulff, J., *Introduction to Materials Science and Engineering*, J. Wiley & Sons, New York, 1976, p. 577, 582. With permission.

Magnetic Susceptibility of the Elements

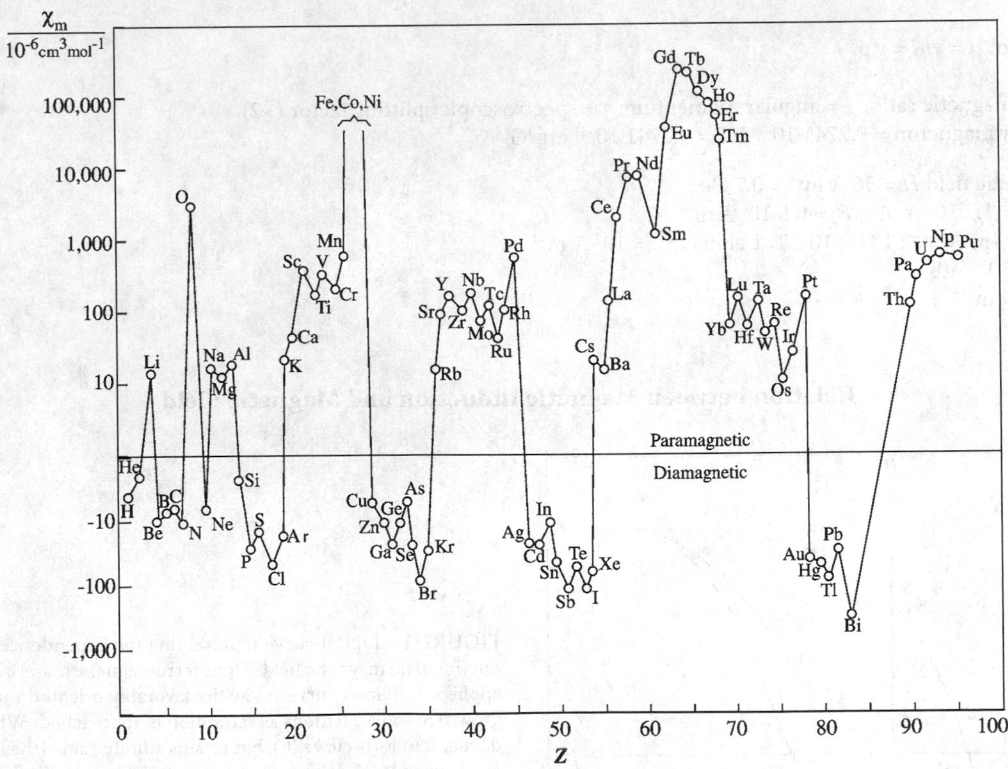

FIGURE 3. Molar susceptibility of the elements at room temperature (cgs units of 10^{-6} cm^3/mol). Values are not available for Z = 9, 61, and 84–89; Fe, Co, and Ni (Z = 26–28) are ferromagnetic. Data taken from the table “Magnetic Susceptibility of the Elements and Inorganic Compounds” in Section 4.

Reference

Gray, D. E., Ed., *American Institute of Physics Handbook, Third Edition*, McGraw Hill, New York, 1972, p. 5–224. With permission.

Ground State of Ions with Partly Filled *d* or *f* Shells

Z	Element	n	S	L	J	Gr. state	p_{calc} [a]	p_{calc} [b]	p_{meas}
22	Ti^{3+}	1	1/2	2	3/2	$^2D_{3/2}$	1.73	1.55	1.8
23	V^{4+}	1	1/2	2	3/2	$^2D_{3/2}$	1.73	1.55	1.8
23	V^{3+}	2	1	3	2	$^3F_2$	2.83	1.63	2.8
23	V^{2+}	3	3/2	3	3/2	$^4F_{3/2}$	3.87	0.77	3.8
24	Cr^{3+}	3	3/2	3	3/2	$^4F_{3/2}$	3.87	0.77	3.7
25	Mn^{4+}	3	3/2	3	3/2	$^4F_{3/2}$	3.87	0.77	4.0
24	Cr^{2+}	4	2	2	0	$^5D_0$	4.90	0	4.9
25	Mn^{3+}	4	2	2	0	$^5D_0$	4.90	0	5.0
25	Mn^{2+}	5	5/2	0	5/2	$^6S_{5/2}$	5.92	5.92	5.9
26	Fe^{3+}	5	5/2	0	5/2	$^6S_{5/2}$	5.92	5.92	5.9
26	Fe^{2+}	6	2	2	4	$^5D_4$	4.90	6.70	5.4
27	Co^{2+}	7	3/2	3	9/2	$^4F_{9/2}$	3.87	6.54	4.8
28	Ni^{2+}	8	1	3	4	$^3F_4$	2.83	5.59	3.2
29	Cu^{2+}	9	1/2	2	5/2	$^2D_{5/2}$	1.73	3.55	1.9
							p_{calc} [c]		
58	Ce^{3+}	1	1/2	3	5/2	$^2F_{5/2}$	2.54		2.4
59	Pr^{3+}	2	1	5	4	$^3H_4$	3.58		3.5
60	Nd^{3+}	3	3/2	6	9/2	$^4I_{9/2}$	3.62		3.5
61	Pm^{3+}	4	2	6	4	$^5I_4$	2.68		
62	Sm^{3+}	5	5/2	5	5/2	$^6H_{5/2}$	0.84		1.5
63	Eu^{3+}	6	3	3	0	$^7F_0$	0.0		3.4
64	Gd^{3+}	7	7/2	0	7/2	$^8S_{7/2}$	7.94		8.0
65	Tb^{3+}	8	3	3	6	$^7F_6$	9.72		9.5
66	Dy^{3+}	9	5/2	5	15/2	$^6H_{15/2}$	10.63		10.6
67	Ho^{3+}	10	2	6	8	$^5I_8$	10.60		10.4
68	Er^{3+}	11	3/2	6	15/2	$^4I_{15/2}$	9.59		9.5
69	Tm^{3+}	12	1	5	6	$^3H_6$	7.57		7.3
70	Yb^{3+}	13	1/2	3	7/2	$^2F_{7/2}$	4.54		4.5

[a] $p_{calc} = 2[S(S + 1)]^{1/2}$
[b] $p_{calc} = 2[J(J + 1)]^{1/2}$
[c] $p_{calc} = g[J(J + 1)]^{1/2}$

References

1. Jiles, D., *Magnetism and Magnetic Materials*, Chapman & Hall, London, 1991, p. 243.
2. Kittel, C., *Introduction to Solid State Physics, 6th Edition*, J. Wiley & Sons, New York, 1986, pp. 405–406.
3. Ashcroft, N. W. and Mermin, N. D., *Solid State Physics*, Holt, Rinehart, and Winston, New York, 1976, p. 652.

Ferro- and Antiferromagnetic Elements

M_0 is the saturation magnetization at $T = 0$ K
n_B is the number of Bohr magnetons per atom
T_C is the Curie temperature
T_N is the Néel temperature

	M_0/gauss	n_B	T_C/K	T_N/K	Comments
Fe	22020	2.22	1043		
Co	18170	1.72	1388		
Ni	6410	0.62	627		
Cr				311	
Mn				100	
Ce				12.5	*c*-Axis antiferromagnetic
Nd				19.2	Basal plane modulation on hexagonal sites
				7.8	Cubic sites order (periodicity different from high-T phase)
Sm				106	Ordering on hexagonal sites
				13.8	Cubic site order
Eu				90.5	Spiral along cube axis
Gd	24880	7	293		
Tb		9	220		Basal plane ferromagnet
				230.2	Basal plane spiral
Dy		10	87		Basal plane ferromagnet
				176	Basal plane spiral
Ho		10	20		Bunched cone structure
				133	Basal plane spiral
Er		9	32		*c*-Axis ferrimagnetic cone structure
				80	*c*-Axis modulated structure
Tm		7	32		*c*-Axis ferrimagnetic cone structure
				56	*c*-Axis modulated structure

References

1. Ashcroft, N. W., and Mermin, N. D., *Solid State Physics*, Holt, Rinehart, and Winston, New York, 1976, p.652.
2. Gschneidner, K. A., and Eyring, L., *Handbook on the Physics and Chemistry of Rare Earths*, North Holland Publishing Co., Amsterdam, 1978.

Selected Ferromagnetic Compounds

M_0 is the saturation magnetization at $T = 293$ K
T_C is the Curie temperature

Compound	M_0/gauss	T_C/K	Crystal system
MnB	152	578	orthorh(FeB)
MnAs	670	318	hex(FeB)
MnBi	620	630	hex(FeB)
MnSb	710	587	hex(FeB)
Mn_4N	183	743	
MnSi		34	cub(FeSi)
CrTe	247	339	hex(NiAs)
$CrBr_3$	270	37	hex(BiI_3)
CrI_3		68	hex(BiI_3)
CrO_2	515	386	tetr(TiO_2)
EuO	1910*	77	cub
EuS	1184*	16.5	cub
$GdCl_3$	550*	2.2	orthorh
FeB		598	orthorh
Fe_2B		1043	tetr ($CuAl_2$)
$FeBe_5$		75	cub($MgCu_2$)
Fe_3C		483	orthorh
FeP		215	orthorh (MnP)

* At $T = 0$ K

References

1. Kittel, C., *Introduction to Solid State Physics, 6th Edition*, J. Wiley & Sons, New York, 1986.
2. Ashcroft, N. W., and Mermin, N. D., *Solid State Physics*, Holt, Rinehart, and Winston, New York, 1976.

Magnetic Properties of High-Permeability Metals and Alloys (Soft)

μ_i is the initial permeability
μ_m is the maximum permeability
H_c is the coercive force
J_s is the saturation polarization
W_H is the hysteresis loss per cycle
T_C is the Curie temperature

Material	Composition (mass %)	μ_i/μ_0	μ_m/μ_0	H_c/A m^{-1}	J_s/T	W_H/J m^{-3}	T_C/K
Iron	Commercial 99Fe	200	6000	70	2.16	500	1043
Iron	Pure 99.9Fe	25000	350000	0.8	2.16	60	1043
Silicon-iron	96Fe-4Si	500	7000	40	1.95	50–150	1008
Silicon-iron (110) [001]	97Fe-3Si	9000	40000	12	2.01	35–140	1015
Silicon-iron {100} <100>	97Fe-3Si		100000	6	2.01		1015
Mild steel	Fe-0.1C-0.1Si-0.4Mn	800	1100	200			
Hypernik	50Fe-50Ni	4000	70000	4	1.60	22	753
Deltamax {100} <100>	50Fe-50Ni	500	200000	16	1.55		773
Isoperm {100} <100>	50Fe-50Ni	90	100	480	1.60		
78 Permalloy	78Ni-22Fe	4000	100000	4	1.05	50	651
Supermalloy	79Ni-16Fe-5Mo	100000	1000000	0.15	0.79	2	673
Mumetal	77Ni-16Fe-5Cu-2Cr	20000	100000	4	0.75	20	673
Hyperco	64Fe-35Co-0.5Cr	650	10000	80	2.42	300	1243
Permendur	50Fe-50Co	500	6000	160	2.46	1200	1253
2V-Permendur	49Fe-49Co-2V	800	4000	160	2.45	600	1253
Supermendur	49Fe-49Co-2V		60000	16	2.40	1150	1253
25Perminvar	45Ni-30Fe-25Co	400	2000	100	1.55		
7Perminvar	70Ni-23Fe-7Co	850	4000	50	1.25		
Perminvar (magnet. annealed)	43Ni-34Fe-23Co		400000	2.4	1.50		
Alfenol (or Alperm)	84Fe-16Al	3000	55000	3.2	0.8		723
Alfor	87Fe-13Al	700	3700	53	1.20		673
Aluminum-Iron	96.5Fe-3.5Al	500	19000	24	1.90		
Sendust	85Fe-10Si-5Al	36000	120000	1.6	0.89		753

References

1. McCurrie, R. A., *Structure and Properties of Ferromagnetic Materials*, Academic Press, London, 1994, p. 42.
2. Gray, D. E., Ed., *American Institute of Physics Handbook, Third Edition*, McGraw Hill, New York, 1972, p. 5–224.

Applications of High-Permeability Materials

Applications	Requirements
Power applications	
Distribution and power transformers	Low core losses, high permeability, high saturation magnetic polarization
High-quality motors and generators, stators and armatures, switched-mode power supplies	
Instrument transformers	
Audiofrequency transformers	Low core losses, high permeability, high magnetic polarization
Pulse transformers	High permeability
Cores for inductor coils	
Audiofrequency	Low hysteresis, high permeability
Carrier frequency	Very low hysteresis and eddy current loss
Radiofrequency	High permeability at low fields
Miscellaneous	
Relays, switches; Earth leakage circuit	High permeability, low remanence, low coercivity
Magnetic shielding	Low core loss for AC applications

Applications of High-Permeability Materials

Applications	Requirements
Magnetic recording heads	High initial permeability, low or zero remanence
Magnetic amplifiers Saturable reactors Saturable transformers Transformer cores	Rectangular hysteresis loops, low hysteresis loss
Magnetic shunts for temperature compensation in magnetic circuits	Low Curie temperature, appropriate decrease in permeability with increase in temperature
Electromagnets in indicating instruments, fire detection, quartz watches, electromechanical devices	High permeability, high saturation magnetic polarization
Magnetic yokes in permanent magnet devices, such as lifting and holding magnets, loudspeakers	High permeability, high saturation magnetic polarization

Reference

McCurrie, R. A., *Structure and Properties of Ferromagnetic Materials*, Academic Press, London, 1994. With permission.

Saturation Magnetostriction of Selected Materials

The tabulated parameter λ_s is related to the fractional change in length $\Delta l/l$ by $\Delta l/l = (3/2)\lambda_s(\cos^2\theta - 1/3)$, where θ is the angle of rotation.

Material	$\lambda_s \times 10^6$
Iron	−7
Fe - 3.2% Si	+9
Nickel	−33
Cobalt	−62
45 Permalloy, 45% Ni - 55% Fe	+27
Permalloy, 82% Ni - 18% Fe	0
Permendur, 49% Co - 49% Fe - 2% V	+70
Alfer, 87% Fe - 13% Al	+30
Magnetite, Fe_3O_4	+40
Cobalt ferrite, $CoFe_2O_4$	−110
$SmFe_2$	−1560
$TbFe_2$	+1753
$Tb_{0.3}Dy_{0.7}Fe_{1.93}$ (Terfenol D)	+2000
$Fe_{66}Co_{18}B_{15}Si$ (amorphous)	+35
$Co_{72}Fe_3B_6A_{13}$ (amorphous)	0

Reference

McCurrie, R.A., *Structure and Properties of Ferromagnetic Materials*, Academic Press, London, 1994, p. 91; additional data provided by A.E. Clark, Adelphi, MD.

Properties of Various Permanent Magnetic Materials (Hard)

B_r is the remanence
${}_BH_c$ is the flux coercivity
${}_iH_c$ is the intrinsic coercivity
$(BH)_{max}$ is the maximum energy product
T_C is the Curie temperature
T_{max} is the maximum operating temperature

Composition	B_r/T	${}_BH_c/10^3$ A m^{-1}	${}_iH_c/10^3$ A m^{-1}	$(BH)_{max}$/kJ m^{-3}	T_C/°C	T_{max}/°C
Alnico1 20Ni;12Al;5Co	0.72		35	25		
Alnico2 17Ni;10Al;12.5Co;6Cu	0.72		40–50	13–14		
Alnico3 24-30Ni;12-14Al;0-3Cu	0.5–0.6		40–54	10		
Alnico4 21-28Ni;11-13Al;3-5Co;2-4Cu	0.55–0.75		36–56	11–12		
Alnico5 14Ni;8Al;24Co;3Cu	1.25	53	54	40	850	520
Alnico6 16Ni;8Al;24Co;3Cu;2Ti	1.05		75	52		
Alnico8 15Ni;7Al;35Co;4Cu;5Ti	0.83	1.6	160	45		
Alnico9 15Ni;7Al;35Co;4Cu;5Ti	1.10	1.45	1.45	75	850	520
Alnico12 13.5Ni;8Al;24.5Co;2Nb	1.20		64	76.8		

Composition	B_r/T	$_BH_c/10^3$ A m^{-1}	$_iH_c/10^3$ A m^{-1}	$(BH)_{max}$/kJ m^{-3}	T_C/°C	T_{max}/°C
$BaFe_{12}O_{19}$ (Ferroxdur)	0.4	1.6	192	29	450	400
$SrFe_{12}O_{19}$	0.4	2.95	3.3	30	450	400
$LaCo_5$	0.91			164	567	
$CeCo_5$	0.77			117	380	
$PrCo_5$	1.20			286	620	
$NdCo_5$	1.22			295	637	
$SmCo_5$	1.00	7.9	696	196	700	250
$Sm(Co_{0.76}Fe_{0.10}Cu_{0.14})_{6.8}$	1.04	4.8	5	212	800	300
$Sm(Co_{0.65}Fe_{0.28}Cu_{0.05}Zr_{0.02})_{7.7}$	1.2	10	16	264	800	300
$Nd_2Fe_{14}B$ sintered	1.22	8.4	1120	280	300	100
Fe;52Co;14V (Vicalloy II)	1.0	42		28	700	500
Fe;24Cr;15Co;3Mo (anisotropic)	1.54	67		76	630	500
Fe;28Cr;10.5Co (Chromindur II)	0.98	32		16	630	500
Fe;23Cr;15Co;3V;2Ti	1.35	4		44	630	500
Cu;20Ni;20Fe (Cunife)	0.55	4		12	410	350
Cu;21Ni;29Fe (Cunico)	0.34	0.5		8		
Pt;23Co	0.64	4		76	480	350
Mn;29.5Al;0.5C (anisotropic)	0.61	2.16	2.4	56	300	120

References

1. McCurrie, R. A., *Structure and Properties of Ferromagnetic Materials*, Academic Press, London, 1994, p. 204.
2. Gray, D. E., Ed., *American Institute of Physics Handbook, Third Edition*, McGraw Hill, New York, 1972, p. 5–165.
3. Jiles, D., *Magnetism and Magnetic Materials*, Chapman & Hall, London, 1991.

Selected Ferrites

J_s is the saturation magnetic polarization
T_C is the Curie temperature
ΔH is the line width

Material	J_s/T	T_C/°C	ΔH/kA m^{-1}	Applications
Spinels				
γ-Fe_2O_3	0.52	575		
Fe_3O_4	0.60	585		
$NiFe_2O_4$	0.34	575	350	Microwave devices
$MgFe_2O_4$	0.14	440	70	
$NiZnFe_2O_4$	0.50	375	120	Transformer cores
$MnFe_2O_4$	0.50	300	50	Microwave devices
$NiCoFe_2O_4$	0.31	590	140	Microwave devices
$NiCoAlFe_2O_4$	0.15	450	330	Microwave devices
$NiAl_{0.35}Fe_{1.65}O_4$	0.12	430	67	Microwave devices
$NiAlFe_2O_4$	0.05	1860	32	Microwave devices
$Mg_{0.9}Mn_{0.1}Fe_2O_4$	0.25	290	56	Microwave devices
$Ni_{0.5}Zn_{0.5}Al_{0.8}Fe_{1.2}O_4$	0.14		17	Microwave devices
$CuFe_2O_4$	0.17	455		Electromechanical transducers
$CoFe_2O_4$	0.53	520		
$LiFe_5O_8$	0.39	670		Microwave devices
Garnets				
$Y_3Fe_5O_{12}$	0.178	280	55	Microwave devices
$Y_3Fe_5O_{12}$ (single crys.)	0.178	292	0.5	Microwave devices
$(Y,Al)_3Fe_5O_{12}$	0.12	250	80	Microwave devices
$(Y,Gd)_3Fe_5O_{12}$	0.06	250	150	Microwave devices
$Sm_3Fe_5O_{12}$	0.170	305		Microwave devices
$Eu_3Fe_5O_{12}$	0.116	293		Microwave devices
$GdFe_5O_{12}$	0.017	291		Microwave devices
Hexagonal crystals				
$BaFe_{12}O_{19}$	0.45	430	1.5	Permanent magnets
$Ba_3Co_2Fe_{24}O_{41}$	0.34	470	12	Microwave devices
$Ba_2Zn_2Fe_{12}O_{22}$	0.28	130	25	Microwave devices
$Ba_3Co_{1.35}Zn_{0.65}Fe_{24}O_{41}$		390	16	Microwave devices
$Ba_2Ni_2Fe_{12}O_{22}$	0.16	500	8	Microwave devices
$SrFe_{12}O_{19}$	0.4	450		Permanent magnets

Reference

McCurrie, R. A., *Structure and Properties of Ferromagnetic Materials*, Academic Press, London, 1994.

Spinel Structure (AB_2O_4)

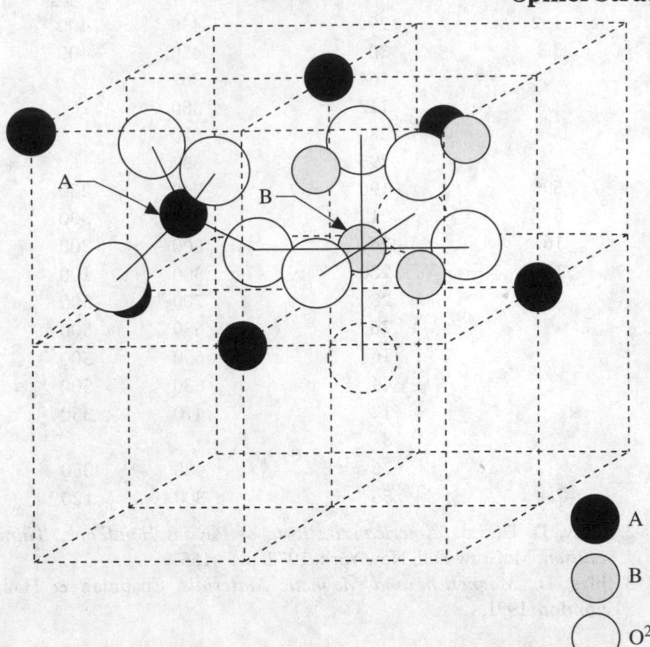

FIGURE 4. Arrangement of metal ions in the two octants A and B, showing tetrahedrally (A) and octahedrally (B) coordinated sites. (Reprinted from McCurrie, R.A., *Ferromagnetic Materials*, Academic Press, London, 1994. With permission.)

Selected Antiferromagnetic Solids

T_N is the Néel temperature

Material	Structure	T_N/K
Binary oxides		
MnO	cub(fcc)	122
FeO	cub(fcc)	198
CoO	cub(fcc)	291
NiO	cub(fcc)	525
α-Mn_2O_3	cub	90
CuO	monocl	230
UO_2	cub	30.8
Er_2O_3	cub	3.4
Gd_2O_3	cub	1.6
Perovskites		
$LaCrO_3$	orth	282
$LaMnO_3$	orth	100
$LaFeO_3$	orth	750
$NdCrO_3$	orth	224
$NdFeO_3$	orth	760
$YbCrO_3$	orth	118
$CaMnO_3$	cub	110
$EuTiO_3$	cub	5.3
$YCrO_3$	orth	141
$BiFeO_3$	cub*	673
$KCoF_3$	cub	125
$KMnF_3$	cub*	88.3
$KFeF_3$	cub	115
$KNiF_3$	cub	275
$NaMnF_3$	cub*	60
$NaNiF_3$	orth	149
$RbMnF_3$	cub	82
Spinels		
Co_3O_4	cub	40
$NiCr_2O_4$	tetr	65
$ZnCr_2O_4$	cub	15
$ZnFe_2O_4$	cub	9
$GeFe_2O_4$	cub	10
MgV_2O_4	cub	45
$MnGa_2O_4$	cub	33
NiAs and related structures		
CrAs	orth	300
CrSb	hex	705–723
CrSe	hex	300
MnTe	hex	320–323
NiS	hex	263
CrS	monocl	460
Rutile and related structures		
CoF_2	tetr	38
CrF_2	monocl	53
FeF_2	tetr	79
MnF_2	tetr	67
NiF_2	tetr	83
$CrCl_2$	orth	20
MnO_2	tetr	84
FeOF	tetr	315
Corundum and related structures		
Cr_2O_3	rhomb	318
α-Fe_2O_3	rhomb	948
$FeTiO_3$	rhomb	68
$MnTiO_3$	rhomb	41
$CoTiO_3$	rhomb	38
VF_3 and related structures		
CoF_3	rhomb	460
CrF_3	rhomb	80

Material	Structure	T_N/K
FeF_3	rhomb	394
MnF_3	monocl	43
MoF_3	rhomb	185
Miscellaneous		
K_2NiF_4	tetr	97
MnI_2	hex	3.4
$CoUO_4$	orth	12
$CaMn_2O_4$	orth	225
CrN	cub*	273
CeC_2	tetr	33
FeSn	hex	373
Mn_2P	hex	103

* Distorted.

References

1. Gray, D. E., Ed., *American Institute of Physics Handbook, Third Edition*, McGraw Hill, New York, 1972, p. 5–168 to 5–183.
2. Kittel, C., *Introduction to Solid State Physics, 6th Edition*, J. Wiley & Sons, New York, 1986.
3. Ashcroft, N. W., and Mermin, N. D., *Solid State Physics*, Holt, Rinehart, and Winston, New York, 1976, p. 697.

ORGANIC MAGNETS

J.S. Miller

Magnetic ordering, e.g., ferromagnetism, like superconductivity, is a property of a solid, not of an individual molecule or ion, and very rarely occurs for organic compounds. In contrast to superconductivity, where all electron spins pair to form a perfect diamagnetic material, magnetic ordering requires unpaired electron spins; hence, superconductivity and ferromagnetism are mutually exclusive.

The vast majority of organic compounds are diamagnetic (i.e., all electron spins are paired), and a relative few possess unpaired electrons (designated by an arrow, ↑) and are paramagnetic (PM), i.e., they are oriented in random directions. A few organic solids, however, exhibit strong magnetic behavior and magnetically order as ferromagnets (FO) with all spins aligned in the same direction. In some cases the spins align in the opposite direction and compensate to form an antiferromagnet (AF). In some cases these spins are not opposed to each other and do not compensate and lead to a canted antiferromagnet or weak ferromagnet (WF). If the number of spins that align in one direction differs from the number of spins that align in the opposite direction, the spins cannot compensate and a ferrimagnet (FI) results. Metamagnets (MM) are antiferromagnets in which all the spins become aligned like a ferromagnet in an applied magnetic field. Above the ordering or critical temperature, T_c, all magnets are paramagnets (PM). Organic magnets all possess electron spins in p-orbitals, but these may be in conjunction with metal ion-based spins.

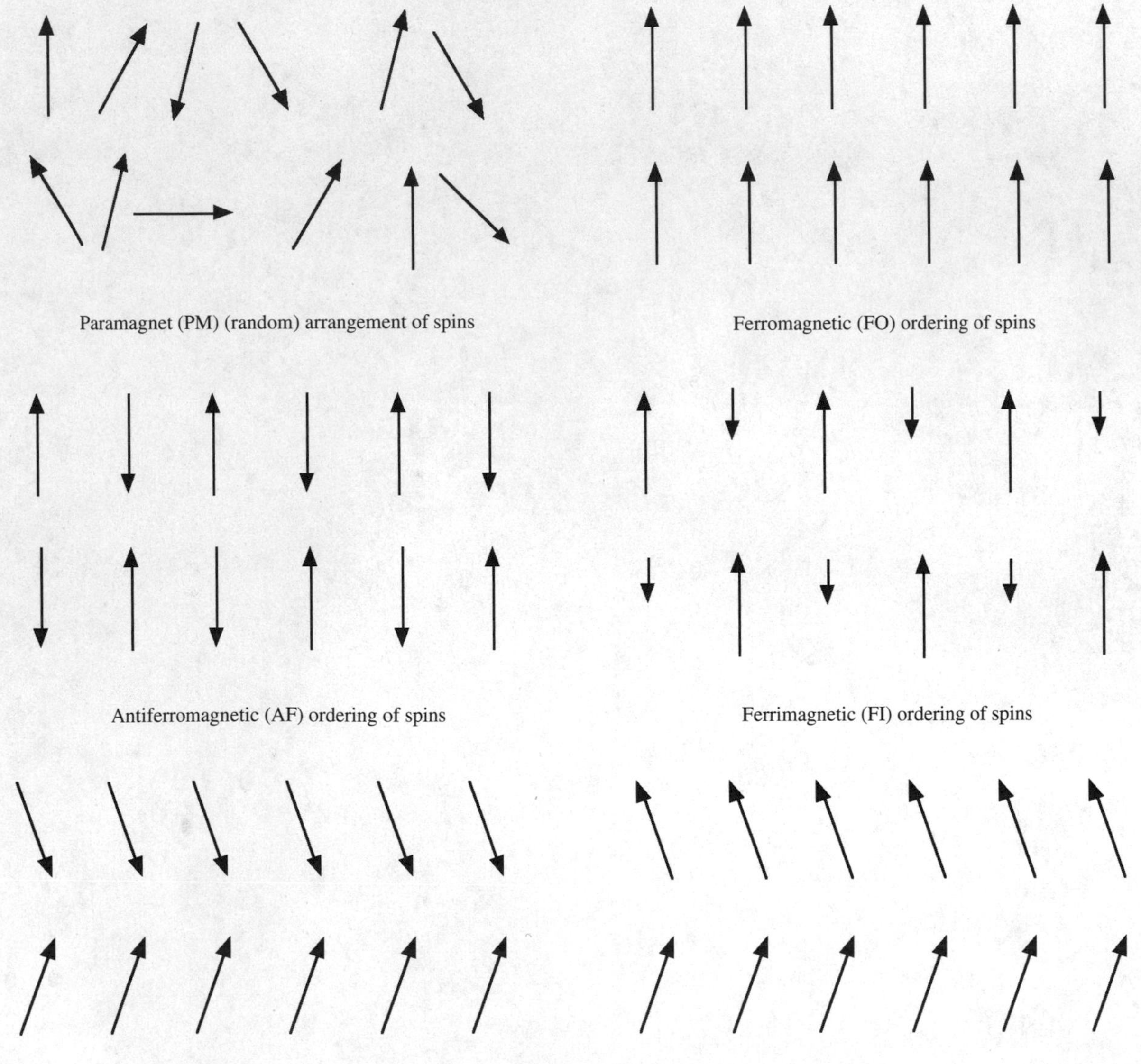

FIGURE 1. Schematic illustration of the different types of magnetic behavior.

Summary of the Critical Temperature, T_c, Saturation Magnetization, M_s, Coercive Field, H_{cr}, and Remanent Magnetization, M_r, for Selected Organic-Based Magnets

Magnet	Type	T_c/K	M_s/A m^{-1}	H_{cr}/T	M_r/A m^{-1}
α-1,3,5,7-Tetramethyl-2,6-diazaadamantane-*N,N'*-doxyl	FO	1.48	48,300	<0.00001	—
β-2-(4'-Nitrophenyl)-4,4,5,5-tetramethyl-4,5-dihydro-1H-imidazol-1-oxyl-3-*N*-oxide	FO	0.6	22,300	0.00008	<200
$\{Fe^{III}[C_5(CH_3)_5]_2\}$[TCNE]	FO	4.8	37,600	0.10	2,300
$\{Mn^{III}[C_5(CH_3)_5]_2\}$[TCNE]	FO	8.8	58,200	0.12	3,700
$\{Cr^{III}[C_5(CH_3)_5]_2\}$[TCNE]	FO	3.65	46,300	—	—
α-$\{Fe^{III}[C_5(CH_3)_5]_2\}$[TCNQ]	MM	2.55	34,200	—	—
β-$\{Fe^{III}[C_5(CH_3)_5]_2\}$[TCNQ]	FO	3.0	21,600	—	—
Tanol subarate	MM	0.38	20,700	—	—
$NCC_6F_4CN_2S_2$	WF	35.5	45	0.00009	—
$Mn^{II}(hfac)_2NITC_2H_5$	FI	7.8	39,400	0.03	27,600
$Mn^{II}(hfac)_2NIT(i\text{-}C_3H_8)$	FI	7.6	42,400	<0.0005	<420
$[Mn(hfac)_2]_3[\{ON[C_6H_3(t\text{-}C(CH_3)_3]_2NO]_2\}$	FI	46	24,400	—	—
$[MnTPP][TCNE]{\cdot}2C_6H_5CH_3$	FI	13	18,400	2.4	10,300
V[TCNE]$_x$·$y$$CH_2Cl_2$ (x ~ 2; y ~ 0.5)	FI	~400	28,200	0.0015 - 0.006	1,650
Mn[TCNE]$_x$·$y$$CH_2Cl_2$ (x ~ 2; y ~ 0.5)	FI	75	52,000	0.002	270
Fe[TCNE]$_x$·$y$$CH_2Cl_2$ (x ~ 2; y ~ 0.5)	FI	97	46,300	0.23	3
Co[TCNE]$_x$·$y$$CH_2Cl_2$ (x ~ 2; y ~ 0.5)	FI	44	22,000	0.65	—

List of Symbols and Abbreviations

M_s	Saturation magnetization at 2 K	hfac	Hexafluoroacetonate
H_{cr}	Coercive Field	NIT	Nitronyl nitroxide
T_c	Critical Temperature	FO	Ferromagnet
M_r	Remanent magnetization at 2 K	FI	Ferrimagnet
TCNE	Tetracyanoethylene	MM	Metamagnet
TCNQ	7,7,8,8-Tetracyano-*p*-quinodimethane	WF	Weak ferromagnet

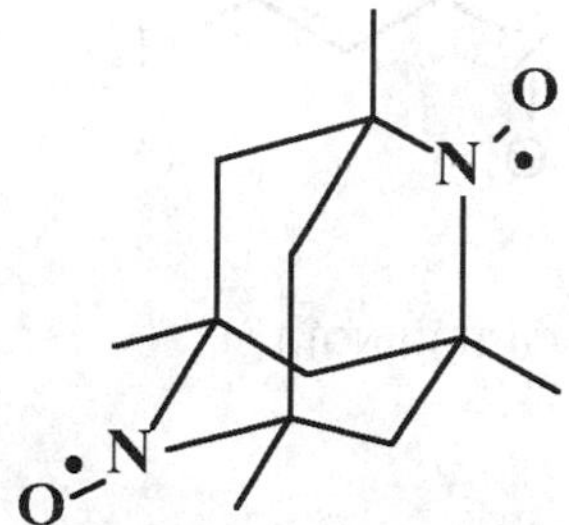

1,3,5,7-Tetramethyl-2,6-diazaadamantane-N,N'-doxyl

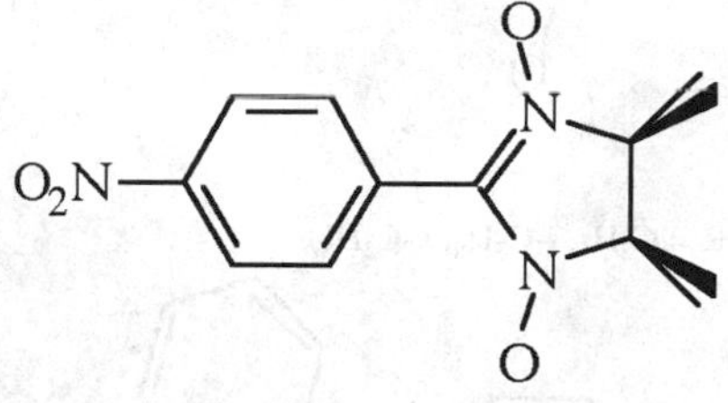

2-(4'-Nitrophyenyl)-4,4,5,5-tetramethyl-4,5-dihydro-1H-imidazol-1-oxyl-3-*N*-oxide

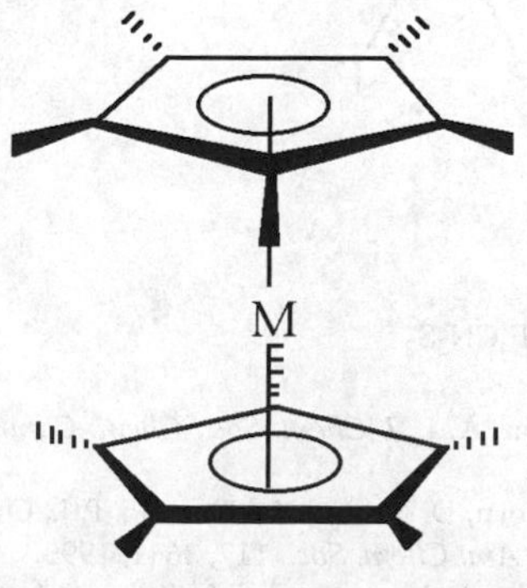

$M[C_5(CH_3)_5]_2$(M = Cr, Mn, Fe)

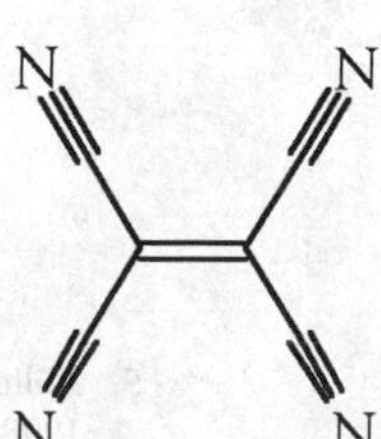

TCNE

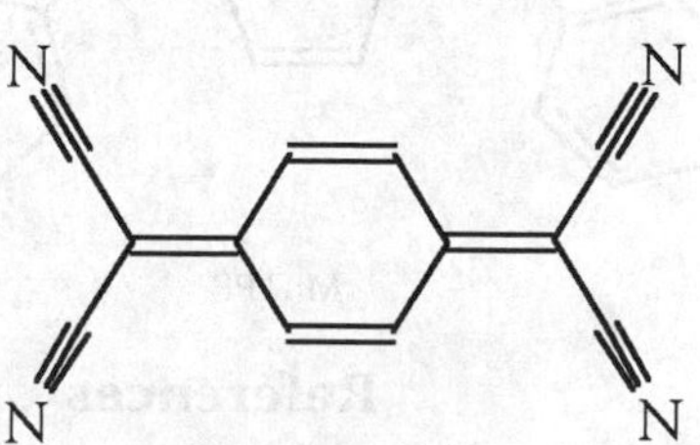

TCNQ

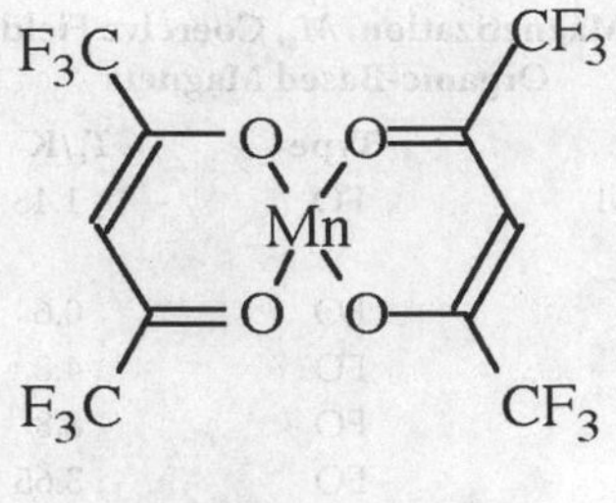

$Mn(hfac)_2$

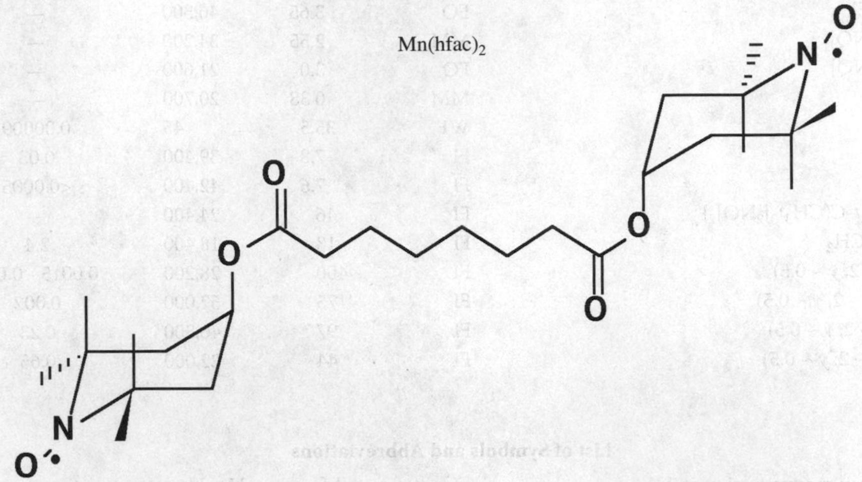

Tanol subarate

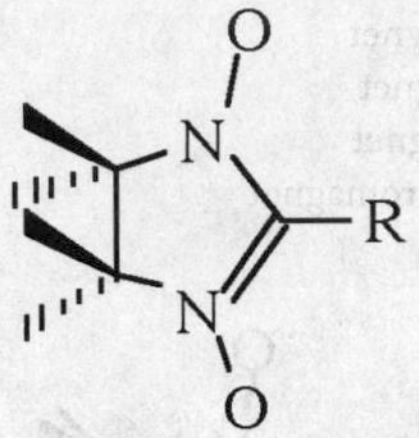

NITR ($R = C_2H_5$, $i\text{-}C_3H_8$, $n\text{-}C_3H_8$)

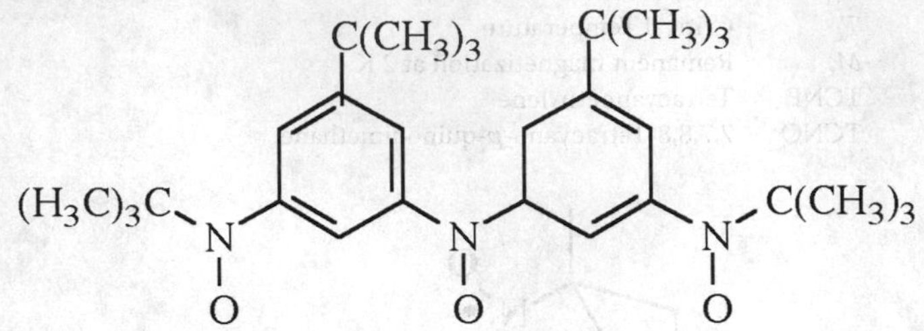

$\{ON[C_6H_3(t\text{-}C(CH_3)_3]_2NO]_2\}$

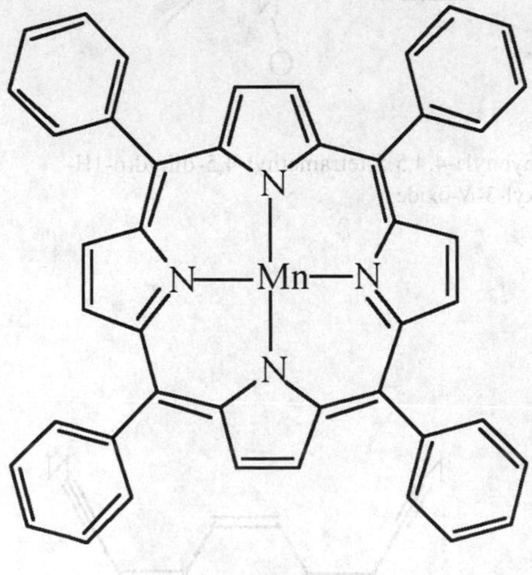

MnTPP

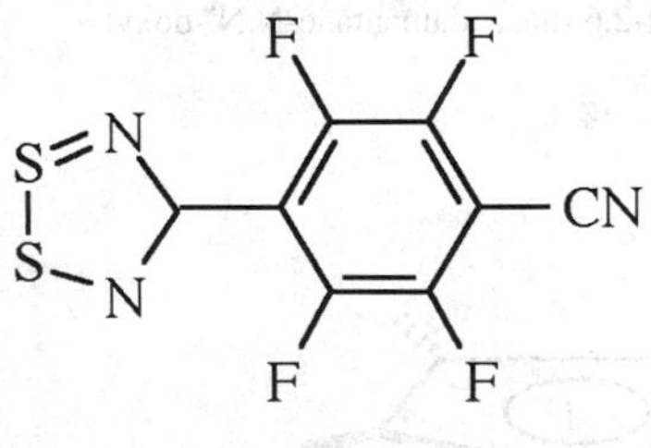

$NCC_6F_4CN_2S_2$

References

1. Miller, J. S. and Epstein, A. J., *Angew. Chem. Internat. Ed.*, 33, 385, 1994.
2. Chiarelli, R., Rassat, A., Dromzee, Y., Jeannin, Y., Novak, M. A., and Tholence, J. L., *Phys. Scrip.*, T49, 706, 1993.
3. Kinoshita, M., *Jap. J. Appl. Phys.*, 33, 5718, 1994.
4. Gatteschi, D., *Adv. Mat.*, 6, 635, 1994.
5. Miller, J. S. and Epstein, A. J., *J. Chem. Soc., Chem. Commun.*, 1319, 1998.
6. Broderick, W. E., Eichorn, D. M., Lu, X., Toscano, P. J., Owens, S. M. and Hoffman, B. M., *J. Am. Chem. Soc.*, 117, 3641, 1995.
7. Banister, A. J., Bricklebank, N., Lavander, I., Rawson, J., Gregory, C. I., Tanner, B. K., Clegg, W. J., Elsegood, M. R., and Palacio, F., *Angew. Chem. Internat. Ed.*, 35, 2533, 1996.

ELECTRON INELASTIC MEAN FREE PATHS

Cedric J. Powell

The inelastic mean free path (IMFP) of electrons is defined as the average of distances, measured along the trajectories, that electrons with a given energy travel between inelastic collisions in the substance. It is an important parameter in analyzing results from surface-characterization techniques such as Auger electron spectroscopy, x-ray photoelectron spectroscopy, low-energy electron diffraction, and others. IMFPs can be measured by the elastic-peak electron spectroscopy technique and other methods, and they can be calculated from optical data. A detailed analysis of the experimental and theoretical considerations in obtaining reliable IMPF values can be found in Reference 4.

The table below gives recommended IMFP values for several elements, simple inorganic compounds, and organic materials. Values are given in Ångström units (1 Å = 10^{-10} m) for a range of electron energies. Substances are given in alphabetical order by name, with elements and inorganic compounds listed before the organic materials.

References

1. Tanuma, S., Powell, C. J., and Penn, D. R., *Surf. Interface Anal.* 17, 911, 1991.
2. Tanuma, S., Powell, C. J., and Penn, D. R., *Surf. Interface Anal.* 17, 927, 1991.
3. Tanuma, S., Powell, C. J., and Penn, D. R., *Surf. Interface Anal.* 21, 165, 1994.
4. Powell, C. J., and Jablonski, A., *J. Phys. Chem. Ref. Data* 28, 19, 1999.
5. Tanuma, S., Powell, C. J., and Penn, D. R., *Surf. Interface Anal.*, 37, 1, 2005.

Electron Inelastic Mean Free Path in Å

Electron Energy in eV

Substance	Formula	50	100	150	200	300	400	600	800	1000	1200	1400	1600	1800	2000	Ref.
Aluminum	Al	3.5	4.6	5.7	6.8	8.9	10.9	14.5	18.0	21.3	24.6	27.8	31.0	34.1	37.2	5
Bismuth	Bi	4.9	5.5	6.3	7.2	8.8	10.6	14.0	17.2	20.2	23.2	26.1	28.9	31.6	34.4	1
Carbon	C	5.9	6.4	7.5	8.8	11.2	13.7	18.4	22.8	27.0	31.1	35.2	39.1	43.0	46.8	1
Chromium	Cr	4.4	4.3	5.0	5.7	7.2	8.6	11.4	14.0	16.6	19.1	21.6	24.0	26.3	28.6	1
Copper	Cu	5.2	5.0	5.4	6.0	7.3	8.6	11.3	13.9	16.5	18.9	21.3	23.7	26.0	28.2	1
Gallium phosphide	GaP	5.6	6.5	7.8	9.0	11.4	13.7	18.1	22.3	26.3	30.2	34.1	37.8	41.5	45.2	2
Gold	Au	6.3	4.7	4.7	5.1	6.1	7.2	9.5	11.7	13.8	15.8	17.8	19.7	21.6	23.4	4
Hafnium	Hf	5.8	6.2	7.1	8.0	10.2	12.0	15.6	19.0	22.2	25.3	28.4	31.4	34.4	37.3	1
Indium phosphide	InP	4.8	4.9	5.6	6.4	8.1	9.7	12.8	15.7	18.7	21.4	24.2	26.8	29.4	32.0	2
Iridium	Ir	5.3	4.3	4.7	5.2	6.4	7.5	9.7	11.8	13.8	15.7	17.6	19.4	21.2	22.9	1
Iron	Fe	4.3	4.4	5.1	5.8	7.2	8.5	11.2	13.7	16.2	18.6	20.9	23.2	25.5	27.7	1
Lead(II) sulfide	PbS	4.8	5.6	6.7	7.8	10.0	12.1	16.1	19.8	23.6	27.1	30.6	33.9	37.2	40.5	2
Lead(II) telluride	PbTe	4.3	5.5	6.6	7.7	9.8	11.9	15.8	19.6	23.4	26.9	30.3	33.7	37.0	40.2	2
Magnesium	Mg	4.0	5.4	6.8	8.2	10.7	13.0	17.5	21.7	25.9	29.9	33.9	37.7	41.6	45.3	1
Molybdenum	Mo	5.1	4.5	5.0	5.6	7.1	8.5	11.3	14.0	16.5	18.9	21.2	23.5	25.8	28.0	1
Nickel	Ni	4.9	4.5	4.9	5.4	6.5	7.7	10.1	12.4	14.6	16.7	18.8	20.8	22.8	24.7	4
Niobium	Nb	6.0	6.0	6.7	7.7	9.7	11.7	15.6	19.2	22.6	25.9	29.1	32.3	35.3	38.4	1
Osmium	Os	5.5	4.3	4.5	5.0	6.0	7.1	9.1	11.1	12.9	14.7	16.4	18.1	19.8	21.5	1
Palladium	Pd	4.8	4.8	5.4	6.2	7.8	9.4	12.5	15.4	18.2	20.9	23.5	26.0	28.4	30.9	1
Platinum	Pt	5.0	4.2	4.5	4.9	6.0	7.1	9.2	11.2	13.1	14.9	16.7	18.5	20.2	21.9	1
Potassium chloride	KCl	7.5	7.8	9.3	10.9	14.2	17.3	23.2	28.7	34.0	39.2	44.2	49.1	54.0	58.8	2
Rhenium	Re	5.2	3.8	3.9	4.3	5.1	6.0	7.7	9.4	10.9	12.5	14.0	15.4	16.9	18.3	1
Rhodium	Rh	4.8	4.1	4.5	5.0	6.1	7.3	9.7	12.0	14.1	16.2	18.2	20.1	22.0	23.9	1
Ruthenium	Ru	4.9	4.2	4.6	5.2	6.5	7.8	10.4	12.8	15.1	17.4	19.5	21.6	23.7	25.7	1
Silicon	Si	4.1	5.3	6.5	7.8	10.3	12.5	16.6	20.6	24.4	28.2	31.8	35.4	39.0	42.5	1
Silicon carbide	SiC	4.7	4.9	5.8	6.8	8.7	10.5	13.9	17.1	20.3	23.3	26.3	29.2	32.1	35.0	2
Silicon dioxide (vitreous)	SiO_2	8.0	7.7	8.8	10.0	12.6	15.2	20.0	24.7	29.3	33.7	38.0	42.2	46.4	50.5	2
Silver	Ag	6.4	4.7	4.8	5.2	6.4	7.7	10.2	12.6	14.9	17.2	19.3	21.4	23.5	25.5	4
Tantalum	Ta	4.8	4.5	5.0	5.5	6.8	8.0	10.4	12.7	14.8	16.9	19.0	21.0	22.9	24.9	1
Titanium	Ti	4.5	5.1	6.2	7.3	9.5	11.6	15.6	19.5	23.2	26.8	30.2	33.6	36.9	40.2	1
Tungsten	W	5.0	4.1	4.5	5.0	6.1	7.3	9.4	11.4	13.4	15.2	17.1	18.9	20.6	22.4	1
Vanadium	V	4.2	4.9	5.9	6.8	8.8	10.7	14.3	17.7	21.0	24.3	27.4	30.5	33.5	36.4	1
Yttrium	Y	5.0	5.5	6.4	7.5	9.8	11.9	16.0	19.8	23.4	27.0	30.4	33.8	37.1	40.4	1
Zinc sulfide	ZnS	5.8	6.5	7.7	8.9	11.3	13.6	17.9	22.0	26.0	29.8	33.6	37.3	40.9	44.5	2
Zirconium	Zr	4.4	4.8	5.7	6.6	8.6	10.5	14.1	17.5	20.7	23.8	26.9	29.9	32.8	35.7	1
Adenine	$C_5H_5N_5$	6.4	6.6	7.8	9.2	11.8	14.4	19.2	24.1	28.6	33.1	37.4	41.6	45.8	49.9	3
Bovine plasma albumin		7.3	7.2	8.5	9.9	12.7	15.4	20.7	25.8	30.8	35.6	40.2	44.8	49.4	53.8	3
β-Carotene	$C_{40}H_{56}$	6.4	7.0	8.5	10.0	13.0	15.9	21.4	26.9	32.0	37.0	41.9	46.6	51.3	56.0	3

Substance	Formula	50	100	150	200	300	400	600	800	1000	1200	1400	1600	1800	2000	Ref.
Deoxyribonucleic acid (DNA)		7.3	7.3	8.5	9.8	12.6	15.4	20.7	25.9	30.8	35.6	40.3	44.9	49.4	53.8	3
1,6-Diphenyl-1,3,5-hexatriene	$C_{18}H_{16}$	6.4	7.0	8.4	9.9	12.9	15.8	21.3	26.7	31.8	36.8	41.7	46.4	51.1	55.7	3
Guanine	$C_5H_5N_5O$	6.2	6.2	7.2	8.4	10.8	13.1	17.5	21.8	25.9	29.9	33.8	37.6	41.4	45.1	3
Hexacosane	$C_{26}H_{54}$	7.0	7.6	9.2	10.9	14.1	17.2	23.2	29.2	34.7	40.1	45.4	50.6	55.7	60.7	3
Kapton		7.0	6.8	7.9	9.2	11.7	14.2	19.0	23.7	28.2	32.5	36.7	40.9	44.9	49.0	3
Polyacetylene		5.3	5.7	6.8	7.9	10.2	12.5	16.9	21.1	25.1	29.0	32.8	36.5	40.2	43.8	3
Poly(butene-1-sulfone)		7.1	7.2	8.5	9.9	12.7	15.4	20.7	25.8	30.6	35.3	39.9	44.4	48.8	53.2	3
Polyethylene		6.9	7.2	8.6	10.1	13.0	15.9	21.4	26.8	31.8	36.8	41.6	46.3	51.0	55.6	3
Poly(methyl methacrylate)		7.8	7.9	9.3	10.8	13.9	16.9	22.7	28.3	33.7	38.8	43.9	48.9	53.8	58.6	3
Polystyrene		6.9	7.3	8.7	10.2	13.2	16.1	21.6	27.1	32.2	37.2	42.1	46.9	51.6	56.2	3
Poly(2-vinylpyridine)		6.9	7.3	8.7	10.3	13.3	16.2	21.8	27.3	32.5	37.5	42.4	47.3	52.0	56.7	3

ELECTRON STOPPING POWERS

Cedric J. Powell

Numerical data are given for collision electron stopping powers in 22 elemental solids for energies between 100 eV and 30 keV. These stopping powers were determined with an algorithm that utilizes experimental optical data for each solid.

The stopping power for electrons and other charged particles in matter is often needed in calculations of electron transport in a medium, particularly in radiation physics and in descriptions of signal generation in analytical techniques such as electron-probe microanalysis and Auger electron spectroscopy. The stopping power is defined as the average rate at which the charged particles lose energy at any point along their trajectories. For electrons, it is customary to separate the total stopping power into two components, the collision stopping power due to inelastic-scattering events of the electrons in a medium and the radiative stopping power due to the emission of bremsstrahlung in the electric field of the atomic nucleus and atomic electrons (Ref. 1). For electron energies less than 30 keV, the radiative stopping power is less than 1% of the collision stopping power (Ref. 1), and is neglected in the numerical data given here.

Numerical data for collision and radiative stopping powers at electron energies between 10 keV and 1 GeV have been published for materials of interest in radiation physics and dosimetry (Ref. 1). Similar data can also be obtained from a web site of the National Institute of Standards and Technology (Ref. 2). The collision stopping powers were calculated from the theory of Bethe (Refs. 3 and 4) and recommended values of the one material-dependent parameter, the mean excitation energy (Ref. 1). While the Bethe theory is expected to be valid for electron energies much larger than the largest K-shell binding energy of atoms in the particular material, the Bethe stopping-power equation is frequently utilized to calculate stopping powers for energies of 10 keV and above (Refs. 1 and 2). Detailed analyses of the Bethe stopping-power theory have been published (Refs. 1 and 4).

The table gives collision stopping powers for 22 elemental solids at energies between 100 eV and 30 keV (Refs. 5 and 6). These stopping powers were determined by interpolation with a clamped cubic spline from the published data (Refs. 5 and 6) which had been calculated with an algorithm that utilizes experimental optical data for each solid. Comparisons with stopping powers from the Bethe stopping-power equation at 30 keV showed a root-mean-square difference of about 10%. This level of agreement was considered satisfactory on account of uncertainties of the algorithm and optical data used for the calculations as well as uncertainties of the mean excitation energies used with the Bethe equation.

The stopping powers in the table are given in units of eV/Å (1 Å = 10^{-10} m) for a range of electron energies. The elemental solids are listed in order of atomic number. For some applications, the mass collision stopping power is desired and these values can be obtained by dividing the tabulated values by the density of the solid.

References

1. *Stopping Powers for Electrons and Positrons*, ICRU Report 37 (International Commission on Radiation Units and Measurements, Bethesda, 1984).
2. Berger, M. J., Coursey, J. S., Zucker, M. A., and Chang, J., *Stopping-Power and Range Tables for Electrons, Positrons, and Helium Ions*, Version 1.2.3, (http://physics.nist.gov/Star.html), (http://physics.nist.gov/PhysRefData/Star/Text/contents.html), 2005.
3. Bethe, H., *Ann. Physik* 5, 325, 1930.
4. Inokuti, M., *Rev. Mod. Phys.* 43, 297, 1971.
5. Tanuma, S., Powell, C. J., and Penn, D. R., *Surf. Interface Anal.* 37, 978, 2005.
6. Tanuma, S., Powell, C. J., and Penn, D. R., *J. Appl. Phys.* 103, 063707, 2008.

TABLE 1. Values of collision electron stopping powers in eV/Å for the indicated elemental solids and electron energies (Refs. 5 and 6)

	Electron energy in eV														
	100	200	300	400	500	1000	1500	2000	3000	4000	5000	10000	15000	20000	30000
Li	1.71	1.49	1.31	1.15	1.02	0.657	0.493	0.399	0.292	0.233	0.195	0.110	0.0787	0.0617	0.0437
Be	6.43	4.69	3.96	3.52	3.17	2.13	1.63	1.33	0.982	0.787	0.660	0.377	0.270	0.212	0.150
C[a]	4.84	4.09	3.39	2.90	2.54	1.73	1.39	1.16	0.881	0.715	0.606	0.353	0.255	0.201	0.143
C[b]	8.14	6.80	5.60	4.76	4.15	2.72	2.11	1.73	1.30	1.05	0.883	0.509	0.366	0.288	0.205
C[c]	10.02	9.02	7.57	6.49	5.69	3.80	2.97	2.46	1.85	1.50	1.27	0.734	0.528	0.416	0.296
Na	1.57	1.67	1.59	1.47	1.35	0.936	0.721	0.591	0.441	0.359	0.306	0.181	0.131	0.104	0.0747
Mg	3.40	3.23	3.07	2.82	2.61	1.84	1.44	1.18	0.887	0.718	0.610	0.362	0.263	0.209	0.150
Al	4.99	3.86	3.59	3.36	3.16	2.35	1.86	1.55	1.17	0.947	0.804	0.481	0.351	0.280	0.201
Si	4.66	3.34	3.02	2.87	2.72	2.06	1.64	1.37	1.04	0.845	0.716	0.426	0.312	0.248	0.178
K	1.79	1.37	1.11	0.944	0.825	0.572	0.472	0.404	0.316	0.261	0.223	0.133	0.0973	0.0778	0.0564
Sc	7.19	6.28	5.23	4.49	3.94	2.54	2.02	1.70	1.31	1.08	0.922	0.547	0.397	0.316	0.228
Ni	7.04	8.73	8.71	8.26	7.74	5.70	4.51	3.75	2.92	2.45	2.13	1.31	0.969	0.776	0.562
Cu	6.27	7.23	7.27	7.05	6.73	5.15	4.14	3.48	2.71	2.28	1.99	1.24	0.920	0.738	0.536
Ge	4.78	4.29	4.07	3.90	3.75	2.99	2.46	2.09	1.61	1.34	1.17	0.732	0.544	0.437	0.318
Zr	6.51	5.61	4.68	4.02	3.58	2.65	2.23	1.94	1.55	1.30	1.13	0.701	0.528	0.428	0.315
Ag	7.43	11.45	10.53	9.32	8.34	5.84	4.85	4.19	3.37	2.80	2.41	1.47	1.08	0.872	0.638
In	4.60	6.19	5.91	5.26	4.71	3.19	2.56	2.23	1.82	1.55	1.35	0.849	0.632	0.513	0.378

	Electron energy in eV														
	100	200	300	400	500	1000	1500	2000	3000	4000	5000	10000	15000	20000	30000
Sn	3.94	5.71	5.70	5.14	4.63	3.18	2.52	2.19	1.79	1.53	1.34	0.840	0.626	0.507	0.374
Cs	1.50	1.19	1.27	1.16	1.05	0.718	0.562	0.472	0.383	0.329	0.290	0.187	0.140	0.113	0.0839
Gd	6.28	5.58	4.85	4.62	4.32	3.20	2.58	2.19	1.70	1.43	1.26	0.829	0.629	0.512	0.378
Tb	7.87	7.25	6.24	5.73	5.28	3.82	3.06	2.58	1.99	1.65	1.44	0.926	0.697	0.563	0.413
Dy	7.43	7.12	6.25	5.79	5.38	3.97	3.20	2.71	2.10	1.74	1.52	0.978	0.736	0.596	0.437
Pt	9.35	10.81	9.88	8.98	8.33	6.54	5.60	4.95	4.06	3.45	3.02	1.95	1.50	1.23	0.920
Bi	4.79	6.16	5.81	5.21	4.70	3.41	2.85	2.50	2.05	1.75	1.53	0.968	0.744	0.605	0.450

[a] Glassy carbon
[b] Graphite
[c] Diamond

ELECTRON WORK FUNCTION OF THE ELEMENTS

The electron work function Φ is a measure of the minimum energy required to extract an electron from the surface of a solid. It is defined more precisely as the energy difference between the state in which an electron has been removed to a distance from the surface of a single crystal face that is large enough that the image force is negligible but small compared to the distance to any other face (typically about 10^{-4} cm) and the state in which the electron is in the bulk solid. In general, Φ differs for each face of a monocrystalline sample.

Since Φ is dependent on the cleanliness of the surface, measurements reported in the literature often cover a considerable range. This table contains selected values for the electron work function of the elements which may be regarded as typical values for a reasonably clean surface. The method of measurement is indicated for each value. The following abbreviations appear:

TE – Thermionic emission
PE – Photoelectric effect
FE – Field emission
CPD – Contact potential difference
polycr – Polycrystalline sample
amorp – Amorphous sample

Values in parentheses are only approximate.

References

1. Hölzl, J., and Schulte, F. K., Work Functions of Metals, in *Solid Surface Physics*, Höhler, G., Ed., Springer-Verlag, Berlin, 1979.
2. Riviere, J. C., Work Function: Measurements and Results, in *Solid State Surface Science, Vol.1*, Green, M., Ed., Decker, New York, 1969.
3. Michaelson, H. B., *J. Appl. Phys.*, 48, 4729, 1977.

Element	Plane	Φ/eV	Method	Element	Plane	Φ/eV	Method	Element	Plane	Φ/eV	Method
Ag	100	4.64	PE		210	5.00	PE	Ru	polycr	4.71	PE
	110	4.52	PE	K	polycr	2.29	PE	Sb	amorp	4.55	
	111	4.74	PE	La	polycr	3.5	PE		100	4.7	
Al	100	4.20	PE	Li	polycr	2.93	FE	Sc	polycr	3.5	PE
	110	4.06	PE	Lu	polycr	(3.3)	CPD	Se	polycr	5.9	PE
	111	4.26	PE	Mg	polycr	3.66	PE	Si	n	4.85	CPD
As	polycr	(3.75)	PE	Mn	polycr	4.1	PE		p 100	(4.91)	CPD
Au	100	5.47	PE	Mo	100	4.53	PE		p 111	4.60	PE
	110	5.37	PE		110	4.95	PE	Sm	polycr	2.7	PE
	111	5.31	PE		111	4.55	PE	Sn	polycr	4.42	CPD
B	polycr	(4.45)	TH		112	4.36	PE	Sr	polycr	(2.59)	TH
Ba	polycr	2.52	TH		114	4.50	PE	Ta	polycr	4.25	TH
Be	polycr	4.98	PE		332	4.55	PE		100	4.15	TH
Bi	polycr	4.34	PE	Na	polycr	2.36	PE		110	4.80	TH
C	polycr	(5.0)	CPD	Nb	001	4.02	TH		111	4.00	TH
Ca	polycr	2.87	PE		110	4.87	TH	Tb	polycr	3.0	PE
Cd	polycr	4.08	CPD		111	4.36	TH	Te	polycr	4.95	PE
Ce	polycr	2.9	PE		112	4.63	TH	Th	polycr	3.4	TH
Co	polycr	5.0	PE		113	4.29	TH	Ti	polycr	4.33	PE
Cr	polycr	4.5	PE		116	3.95	TH	Tl	polycr	(3.84)	CPD
Cs	polycr	1.95	PE		310	4.18	TH	U	polycr	3.63	PE
Cu	100	5.10	FE	Nd	polycr	3.2	PE		100	3.73	PE
	110	4.48	PE	Ni	100	5.22	PE		110	3.90	PE
	111	4.94	PE		110	5.04	PE		113	3.67	PE
	112	4.53	PE		111	5.35	PE	V	polycr	4.3	PE
Eu	polycr	2.5	PE	Os	polycr	5.93	PE	W	polycr	4.55	CPD
Fe	100	4.67	PE	Pb	polycr	4.25	PE		100	4.63	FE
	111	4.81	PE	Pd	polycr	5.22	PE		110	5.22	FE
Ga	polycr	4.32	PE		111	5.6	PE		111	4.45	FE
Gd	polycr	2.90	CPD	Pt	polycr	5.64	PE		113	4.46	FE
Ge	polycr	5.0	CPD		110	5.84	FE		116	4.32	TH
Hf	polycr	3.9	PE		111	5.93	FE	Y	polycr	3.1	PE
Hg	liquid	4.475	PE		320	5.22	FE	Zn	polycr	3.63	PE
In	polycr	4.09	PE		331	5.12	FE		polycr	(4.9)	CPD
Ir	100	5.67	PE	Rb	polycr	2.261	PE	Zr	polycr	4.05	PE
	110	5.42	PE	Re	polycr	4.72	TE				
	111	5.76	PE	Rh	polycr	4.98	PE				

SECONDARY ELECTRON EMISSION

The secondary emission yield, or secondary emission ratio, δ, is the average number of secondary electrons emitted from a bombarded material for every incident primary electron. It is a function of the primary electron energy E_p. The maximum yield δ_{max} corresponds to a primary electron energy E_{pmax} (see figure). The two primary electron energies corresponding to a yield of unity are denoted by the first and second crossovers (E_I and E_{II}). An insulating target, or a conducting target that is electrically floating, will charge positively or negatively depending on the primary electron energy. For $E_I < E_p < E_{II}$, $\delta > 1$ and the surface charges positively provided there is a collector present that is positive with respect to the target. For $E_p < E_I$ or $E_p > E_{II}$, $\delta < 1$, and the surface charges negatively with respect to the potential of the source of primary electrons.

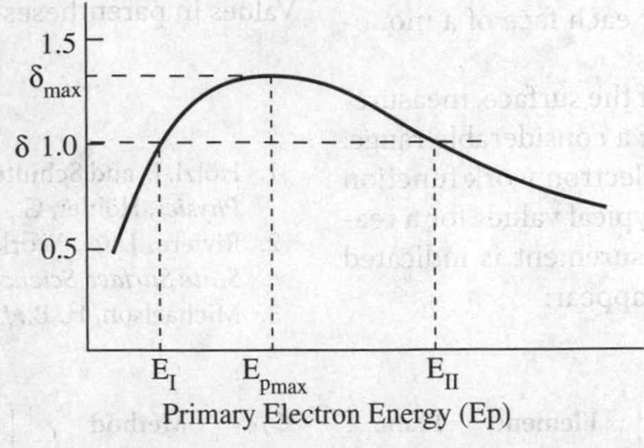

Element	δ_{max}	E_{pmax} (eV)	E_I (eV)	E_{II} (eV)
Ag	1.5	800	200	>2000
Al	1.0	300	300	300
Au	1.4	800	150	>2000
B	1.2	150	50	600
Ba	0.8	400	None	None
Bi	1.2	550	None	None
Be	0.5	200	None	None
C (diamond)	2.8	750	None	>5000
C (graphite)	1.0	300	300	300
C (soot)	0.45	500	None	None
Cd	1.1	450	300	700
Co	1.2	600	200	None
Cs	0.7	400	None	None
Cu	1.3	600	200	1500
Fe	1.3	400	120	1400
Ga	1.55	500	75	None
Ge	1.15	500	150	900
Hg	1.3	600	350	>1200
K	0.7	200	None	None
Li	0.5	85	None	None
Mg	0.95	300	None	None
Mo	1.25	375	150	1200
Na	0.82	300	None	None
Nb	1.2	375	150	1050
Ni	1.3	550	150	>1500
Pb	1.1	500	250	1000
Pd	>1.3	>250	120	None
Pt	1.8	700	350	3000
Rb	0.9	350	None	None
Sb	1.3	600	250	2000
Si	1.1	250	125	500
Sn	1.35	500	None	None
Ta	1.3	600	250	>2000
Th	1.1	800	None	None
Ti	0.9	280	None	None
Tl	1.7	650	70	>1500
W	1.4	650	250	>1500
Zr	1.1	350	None	None

Compound	δ_{max}	E_{pmax} (eV)
Alkali halides		
CsCl	6.5	
KBr (crystal)	14	1800
KCl (crystal)	12	1600
KCl (layer)	7.5	1200
KI (crystal)	10	1600
KI (layer)	5.6	
LiF (crystal)	8.5	
LiF (layer)	5.6	700
NaBr (crystal)	24	1800
NaBr (layer)	6.3	
NaCl (crystal)	14	1200
NaCl (layer)	6.8	600
NaF (crystal)	14	1200
NaF (layer)	5.7	
NaI (crystal)	19	1300
NaI (layer)	5.5	
RbCl (layer)	5.8	
Oxides		
Ag_2O	1.0	
Al_2O_3 (layer)	2–9	
BaO (layer)	2.3–4.8	400
BeO	3.4	2000
CaO	2.2	500
Cu_2O	1.2	400
MgO (crystal)	20–25	1500
MgO (layer)	3–15	400–1500
MoO_2	1.2	
SiO_2 (quartz)	2.1–4	400
SnO_2	3.2	640
Sulfides		
MoS_2	1.1	
PbS	1.2	500
WS_2	1.0	
ZnS	1.8	350
Others		
BaF_2 (layer)	4.5	
CaF_2 (layer)	3.2	
$BiCs_3$	6	1000
BiCs	1.9	1000
GeCs	7	700
Rb_3Sb	7.1	450
$SbCs_3$	6	700
Mica	2.4	350
Glasses	2–3	300–450

OPTICAL PROPERTIES OF SELECTED ELEMENTS

J. H. Weaver and H. P. R. Frederikse

These tables list the index of refraction n, the extinction coefficient k, and the normal incidence reflection R ($\phi = 0$) as a function of photon energy E, which is expressed in electron volts (eV). To convert the energy in eV to the wavelength in μm, use $\lambda = 1.2398/E$. To compute the dielectric function $\tilde{\varepsilon} = \varepsilon_1 + i\varepsilon_2$ from the complex index of refraction $\tilde{N} = n + ik$, use $\varepsilon_1 = n^2 - k^2$ and $\varepsilon_2 = 2nk$.

The optical constants in these tables are abridged from three more extensive tabulations:

- *Optical Properties of Metals* (OPM), Volumes I and II, *Physics Data, Nr.* 18-1 and 18-2, J. H. Weaver, C. Krafka, D. W. Lynch, and E. E. Koch, Fachinformationzentrum, Karlsruhe, Germany.
- *Handbook of Optical Constants* (HOC), Vol. I, 1985, and Vol. II, 1991. E. D. Palik, Ed., Academic Press, Inc., London.
- *American Institute of Physics Handbook* (AIPH), 3rd Edition, D. E. Gray, Ed., McGraw-Hill, New York, 1972.

The first two of these major sources provide detailed comparisons of all optical data available in the literature at the time of the compilation. For critical applications the reader should refer to the original work. References for individual metals and semiconductors are listed at the end of the tables. Generally, tabulated values for the optical properties are accurate to better than 10%. Data in parentheses are extrapolated or interpolated values. For most elements the spectral range covered is from the far infrared (0.010 or 0.10 eV) to the far ultraviolet (10, 30 or 300 eV). The intervals between successive energies in the tables are chosen in such a way that the major spectral features are preserved.

Very small values of k are expressed in exponential notation, e.g., 1.23E-5 means 1.23×10^{-5}.

The following table is convenient for associating the energy entries in these tables with the corresponding wavelengths:

λ	E/eV	λ	E/eV
1 mm	0.00124	6000 Å	2.066
500 μm	0.00248	5000 Å	2.480
100 μm	0.01240	4000 Å	3.100
50 μm	0.02480	3000 Å	4.133
10 μm	0.12398	2000 Å	6.199
5 μm	0.24797	1000 Å	12.398
1 μm	1.240	400 Å	30.996

Energy (eV)	n	k	R(φ = 0)
Aluminum[1]			
0.040	98.595	203.701	0.9923
0.050	74.997	172.199	0.9915
0.060	62.852	150.799	0.9906
0.070	53.790	135.500	0.9899
0.080	45.784	123.734	0.9895
0.090	39.651	114.102	0.9892
0.100	34.464	105.600	0.9889
0.125	24.965	89.250	0.9884
0.150	18.572	76.960	0.9882
0.175	14.274	66.930	0.9879
0.200	11.733	59.370	0.9873
0.250	8.586	48.235	0.9858
0.300	6.759	40.960	0.9844
0.350	5.438	35.599	0.9834
0.400	4.454	31.485	0.9826
0.500	3.072	25.581	0.9817
0.600	2.273	21.403	0.9806
0.700	1.770	18.328	0.9794
0.800	1.444	15.955	0.9778
0.900	1.264	14.021	0.9749
1.000	1.212	12.464	0.9697
1.100	1.201	11.181	0.9630
1.200	1.260	10.010	0.9521
1.300	1.468	8.949	0.9318
1.400	2.237	8.212	0.8852
1.500	2.745	8.309	0.8678
1.600	2.625	8.597	0.8794
1.700	2.143	8.573	0.8972
1.800	1.741	8.205	0.9069
1.900	1.488	7.821	0.9116
2.000	1.304	7.479	0.9148
2.200	1.018	6.846	0.9200
2.400	0.826	6.283	0.9228
2.600	0.695	5.800	0.9238
2.800	0.598	5.385	0.9242
3.000	0.523	5.024	0.9241
3.200	0.460	4.708	0.9243
3.400	0.407	4.426	0.9245
3.600	0.363	4.174	0.9246
3.800	0.326	3.946	0.9247
4.000	0.294	3.740	0.9248
4.200	0.267	3.552	0.9248
4.400	0.244	3.380	0.9249
4.600	0.223	3.222	0.9249
4.800	0.205	3.076	0.9249
5.000	0.190	2.942	0.9244
6.000	0.130	2.391	0.9257
6.500	0.110	2.173	0.9260
7.000	0.095	1.983	0.9262
7.500	0.082	1.814	0.9265
8.000	0.072	1.663	0.9269
8.500	0.063	1.527	0.9272
9.000	0.056	1.402	0.9277
9.500	0.049	1.286	0.9282
10.000	0.044	1.178	0.9286
10.500	0.040	1.076	0.9293
11.000	0.036	0.979	0.9298
11.500	0.033	0.883	0.9283
12.000	0.033	0.791	0.9224
12.500	0.034	0.700	0.9118
13.000	0.038	0.609	0.8960
13.500	0.041	0.517	0.8789
14.000	0.048	0.417	0.8486
14.200	0.053	0.373	0.8312
14.400	0.058	0.327	0.8102
14.600	0.067	0.273	0.7802
14.800	0.086	0.211	0.7202
15.000	0.125	0.153	0.6119
15.200	0.178	0.108	0.4903
15.400	0.234	0.184	0.3881
15.600	0.280	0.073	0.3182
15.800	0.318	0.065	0.2694
16.000	0.351	0.060	0.2326
16.200	0.380	0.055	0.2031
16.400	0.407	0.050	0.1789
16.750	0.448	0.045	0.1460
17.000	0.474	0.042	0.1278
17.250	0.498	0.040	0.1129
17.500	0.520	0.038	0.1005
17.750	0.540	0.036	0.0899
18.000	0.558	0.035	0.0809
18.500	0.591	0.032	0.0664
19.000	0.620	0.030	0.0554
19.500	0.646	0.028	0.0467
20.000	0.668	0.027	0.0398
20.500	0.689	0.025	0.0342
21.000	0.707	0.024	0.0296
21.500	0.724	0.023	0.0258
22.000	0.739	0.022	0.0226
22.500	0.753	0.021	0.0199
23.000	0.766	0.021	0.0177
23.500	0.778	0.020	0.0157
24.000	0.789	0.019	0.0140
24.500	0.799	0.018	0.0126
25.000	0.809	0.018	0.0113
25.500	0.817	0.017	0.0102
26.000	0.826	0.016	0.0092

Energy (eV)	n	k	R(φ = 0)
27.000	0.840	0.015	0.0076
28.000	0.854	0.014	0.0063
29.000	0.865	0.014	0.0053
30.000	0.876	0.013	0.0044
35.000	0.915	0.010	0.0020
40.000	0.940	0.008	0.0010
45.000	0.957	0.007	0.0005
50.000	0.969	0.006	0.0003
55.000	0.979	0.005	0.0001
60.000	0.987	0.004	0.0000
65.000	0.995	0.004	0.0000
70.000	1.006	0.004	0.0000
72.500	1.025	0.004	0.0002
75.000	1.011	0.024	0.0002
77.500	1.008	0.025	0.0002
80.000	1.007	0.024	0.0002
85.000	1.007	0.028	0.0002
90.000	1.005	0.031	0.0002
95.000	0.999	0.036	0.0003
100.000	0.991	0.030	0.0002
110.000	0.994	0.025	0.0002
120.000	0.991	0.024	0.0002
130.000	0.987	0.021	0.0001
140.000	0.989	0.016	0.0001
150.000	0.990	0.015	0.0001
160.000	0.989	0.014	0.0001
170.000	0.989	0.011	0.0001
180.000	0.990	0.010	0.0000
190.000	0.990	0.009	0.0000
200.000	0.991	0.007	0.0000
220.000	0.992	0.006	0.0000
240.000	0.993	0.005	0.0000
260.000	0.993	0.004	0.0000
280.000	0.994	0.003	0.0000
300.000	0.995	0.002	0.0000
Carbon (diamond) [2]			
0.06199	2.3741		0.166
0.06888	2.3741		0.166
0.07749	2.3745		0.166
0.08856	2.3750		0.166
0.1033	2.3757		0.166
0.1240	2.3765		0.166
0.1550	2.3772		0.166
0.1907		3.1 E-05	
0.2066	2.3779	5.7 E-05	0.166
0.22		1.21E-04	
0.23		2.36E-04	
0.24		3.82E-04	
0.25		5.21E-04	
0.26		2.96E-04	
0.27		4.39E-04	
0.28		2.75E-04	
0.29		7.82E-05	
0.30		1.32E-04	
0.31	2.3787	1.30E-04	0.167
0.32		1.11E-04	
0.33		2.99E-05	
0.34		1.89E-05	
0.35		2.11E-05	

Energy (eV)	n	k	R(φ = 0)
0.36		2.47E-05	
0.37		2.80E-05	
0.38		3.11E-05	
0.39		3.67E-05	
0.40		3.58E-05	
0.41		3.25E-05	
0.4133	2.3795		0.167
0.42		2.94E-05	
0.43		2.87E-05	
0.44		3.14E-05	
0.45		3.62E-05	
0.46		3.22E-05	
0.47		1.57E-05	
0.48		6.17E-06	
0.4959	2.3801		0.167
0.6199	2.3813		0.167
0.8266	2.3837		0.167
1.240	2.3905		0.168
1.378	2.3934		0.169
1.459	2.3953		0.169
1.550	2.3975		0.169
1.653	2.4003		0.170
1.771	2.4036		0.170
1.889	2.4073		0.171
1.926	2.4084		0.171
2.066	2.4133		0.171
2.105	2.4147		0.172
2.271	2.4210		0.173
2.480	2.4299		0.174
2.650	2.4380		0.175
2.845		3.82E-07	
3.100	2.4627		0.178
3.434	2.4849		0.182
3.576	2.4955		0.183
3.961		8.97E-07	
4.160	2.5465		0.190
4.511		1.29E-06	
4.8187	2.6205	1.47E-06	0.200
5.00	2.6383		0.203
5.30		2.98E-06	
5.35		6.45E-06	
5.40		1.04E-05	
5.50		3.41E-05	
5.55		5.48E-04	
5.60	2.740	1.48E-03	0.216
5.80	2.780	5.02E-03	0.222
6.00	2.826	7.99E-03	0.228
6.10	2.852	8.62E-03	0.231
6.20	2.879	9.30E-03	0.235
6.30	2.910	9.74E-03	0.239
6.40	2.944	9.87E-03	0.243
6.50	2.985	1.10E-02	0.248
6.60	3.031	1.47E-02	0.254
6.70	3.085	2.20E-02	0.261
6.80	3.146	3.44E-02	0.268
6.90	3.220	5.24E-02	0.277
7.00	3.322	9.35E-02	0.289
7.10	3.444	0.210	0.304
7.15	3.464	0.307	0.308
7.20	3.437	0.388	0.307

Energy (eV)	n	k	R(φ = 0)
7.30	3.376	0.473	0.303
7.40	3.335	0.515	0.300
7.50	3.321	0.533	0.299
7.60	3.306	0.592	0.300
7.80	3.276	0.659	0.300
8.00	3.251	0.712	0.300
8.25	3.232	0.765	0.301
8.50	3.228	0.806	0.303
8.75	3.247	0.855	0.308
9.00	3.272	0.910	0.314
9.25	3.308	0.978	0.322
9.50	3.348	1.055	0.331
9.75	3.398	1.147	0.342
10.00	3.453	1.258	0.355
10.25	3.514	1.403	0.371
10.50	3.565	1.581	0.389
10.75	3.600	1.813	0.411
11.00	3.582	2.078	0.434
11.25	3.507	2.380	0.460
11.50	3.346	2.693	0.488
11.75	3.090	2.986	0.518
12.00	2.736	3.228	0.551
12.20	2.383	3.354	0.580
12.40	1.983	3.382	0.610
12.60	1.532	3.265	0.641
12.80	1.312	2.953	0.627
13.00	1.223	2.722	0.604
13.50	1.129	2.379	0.557
14.00	1.070	2.178	0.526
14.50	1.018	2.034	0.504
15.00	0.972	1.929	0.489
15.50	0.917	1.845	0.482
16.00	0.861	1.767	0.477
16.50	0.805	1.692	0.474
17.00	0.753	1.619	0.471
17.50	0.707	1.546	0.467
18.00	0.665	1.476	0.463
18.50	0.626	1.408	0.459
19.00	0.589	1.341	0.455
19.50	0.557	1.273	0.449
20.00	0.527	1.203	0.442
21.00	0.487	1.052	0.413
22.00	0.518	0.888	0.330
23.00	0.597	0.850	0.270
24.00	0.586	0.829	0.268
25.00	0.562	0.787	0.265
26.00	0.538	0.736	0.260
27.00	0.516	0.679	0.252
28.00	0.501	0.616	0.239
29.00	0.494	0.552	0.221
30.00	0.493	0.490	0.201
Cesium (evaporated) [3]			
2.145	0.264	1.123	0.631
2.271	0.278	0.950	0.561
2.845	0.425	0.438	0.235
3.064	0.540	0.320	0.127
3.397	0.671	0.233	0.057
3.966	0.827	0.174	0.018
4.889	0.916	0.143	0.007

Energy (eV)	n	k	$R(\phi = 0)$
Chromium[4]			
0.06	21.19	42.00	0.962
0.10	11.81	29.76	0.955
0.14	15.31	26.36	0.936
0.18	8.73	25.37	0.53
0.22	5.30	20.62	0.954
0.26	3.91	17.12	0.951
0.30	3.15	14.28	0.943
0.42	3.47	8.97	0.862
0.54	3.92	7.06	0.788
0.66	3.96	5.95	0.736
0.78	4.13	5.03	0.680
0.90	4.43	4.60	0.650
1.00	4.47	4.43	0.639
1.12	4.53	4.31	0.631
1.24	4.50	4.28	0.629
1.36	4.42	4.30	0.631
1.46	4.31	4.32	0.632
1.77	3.84	4.37	0.639
2.00	3.48	4.36	0.644
2.20	3.18	4.41	0.656
2.40	2.75	4.46	0.677
2.60	2.22	4.36	0.698
2.80	1.80	4.06	0.703
3.00	1.54	3.71	0.695
3.20	1.44	3.40	0.670
3.40	1.39	3.24	0.657
3.60	1.26	3.12	0.661
3.80	1.12	2.95	0.660
4.00	1.02	2.76	0.651
4.20	0.94	2.58	0.639
4.40	0.90	2.42	0.620
4.50	0.89	2.35	0.607
4.60	0.88	2.28	0.598
4.70	0.86	2.21	0.586
4.80	0.86	2.13	0.572
4.90	0.86	2.07	0.557
5.00	0.85	2.01	0.542
5.10	0.86	1.94	0.523
5.20	0.87	1.87	0.503
5.40	0.93	1.80	0.466
5.60	0.95	1.74	0.443
5.80	0.97	1.74	0.437
6.00	0.94	1.73	0.444
6.20	0.89	1.69	0.446
6.40	0.85	1.66	0.447
6.60	0.80	1.59	0.444
6.80	0.75	1.51	0.439
7.00	0.74	1.45	0.425
7.20	0.71	1.39	0.414
7.40	0.69	1.33	0.404
7.60	0.66	1.23	0.378
7.80	0.67	1.15	0.347
8.00	0.68	1.07	0.315
8.20	0.71	1.00	0.278
8.50	0.74	0.92	0.235
9.0	0.83	0.81	0.170
9.50	0.92	0.74	0.132
10.00	0.98	0.73	0.120
10.50	1.01	0.72	0.112

Energy (eV)	n	k	$R(\phi = 0)$
11.00	1.05	0.69	0.103
11.50	1.09	0.69	0.100
12.00	1.13	0.70	0.101
12.50	1.15	0.73	0.108
13.00	1.15	0.77	0.119
13.50	1.12	0.80	0.128
14.00	1.09	0.82	0.135
14.50	1.03	0.82	0.142
15.00	1.00	0.82	0.143
15.50	0.96	0.80	0.141
16.00	0.92	0.77	0.139
16.50	0.31	0.75	0.134
17.00	0.90	0.73	0.132
17.50	0.88	0.72	0.130
18.00	0.87	0.70	0.129
18.50	0.84	0.69	0.130
19.00	0.82	0.68	0.131
20.00	0.77	0.64	0.130
20.5	0.76	0.63	0.129
21.0	0.74	0.58	0.121
21.5	0.72	0.55	0.116
22.0	0.71	0.52	0.112
22.5	0.70	0.50	0.109
23.0	0.69	0.48	0.105
23.5	0.68	0.45	0.101
24.0	0.68	0.43	0.096
24.5	0.67	0.39	0.089
25.0	0.68	0.36	0.080
25.5	0.68	0.33	0.072
26.0	0.70	0.31	0.063
26.5	0.71	0.28	0.055
27.0	0.72	0.26	0.048
27.5	0.73	0.25	0.043
28.0	0.75	0.23	0.037
29.0	0.77	0.22	0.032
30.0	0.78	0.21	0.030
Cobalt, single crystal, $\vec{E} \parallel \hat{c}$[5]			
0.10	6.71	37.87	0.982
0.15	4.66	25.47	0.973
0.20	3.55	18.78	0.962
0.25	3.98	14.59	0.933
0.30	4.04	12.16	0.907
0.40	4.24	9.13	0.847
0.50	4.41	7.19	0.782
0.60	4.91	6.13	0.729
0.70	5.24	5.85	0.713
0.80	5.17	5.89	0.716
0.90	4.94	5.95	0.720
1.00	4.46	5.86	0.722
1.10	4.07	5.61	0.715
1.20	3.81	5.36	0.706
1.30	3.60	5.20	0.701
1.40	3.37	5.09	0.701
1.50	3.10	4.96	0.701
1.60	2.84	4.77	0.697
1.70	2.66	4.57	0.690
1.80	2.45	4.41	0.687
1.90	2.31	4.18	0.675
2.00	2.21	4.00	0.664
2.10	2.13	3.85	0.654

Energy (eV)	n	k	$R(\phi = 0)$
2.20	2.07	3.70	0.642
2.30	2.01	3.59	0.634
2.40	1.95	3.49	0.627
2.50	1.88	3.40	0.622
2.60	1.81	3.32	0.618
2.70	1.73	3.24	0.615
2.80	1.66	3.13	0.607
2.90	1.61	3.05	0.600
3.00	1.55	2.96	0.594
3.20	1.46	2.80	0.579
3.40	1.38	2.64	0.563
3.60	1.31	2.48	0.544
3.80	1.28	2.33	0.519
4.00	1.26	2.20	0.495
4.20	1.25	2.10	0.471
4.40	1.24	2.01	0.152
4.60	1.24	1.94	0.435
4.80	1.23	1.88	0.423
5.00	1.22	1.83	0.411
5.20	1.21	1.79	0.403
5.40	1.19	1.77	0.399
5.60	1.16	1.75	0.400
5.80	1.10	1.73	0.406
6.00	1.03	1.68	0.407
6.20	0.97	1.62	0.401
6.40	0.94	1.53	0.386
6.60	0.91	1.46	0.368
6.80	0.91	1.38	0.345
7.00	0.91	1.32	0.326
7.00	0.91	1.26	0.305
7.40	0.92	1.21	0.286
7.60	0.93	1.17	0.269
7.80	0.94	1.13	0.253
8.00	0.95	1.09	0.239
Cobalt, single crystal, $\vec{E} \perp \hat{c}$[5]			
0.10	5.83	32.36	0.979
0.15	4.24	21.37	0.965
0.20	3.87	15.53	0.042
0.30	4.34	10.01	0.865
0.40	4.66	7.39	0.785
0.50	5.17	5.75	0.709
0.60	5.77	5.17	0.682
0.70	6.15	5.20	0.685
0.80	6.08	5.61	0.702
0.90	5.57	5.93	0.715
1.00	4.83	5.94	0.721
1.10	4.31	5.60	0.711
1.20	4.02	5.34	0.701
1.30	3.78	5.16	0.694
1.40	3.55	5.05	0.692
1.50	3.26	4.93	0.692
1.60	3.03	4.74	0.687
1.70	2.83	4.60	0.684
1.80	2.61	4.45	0.683
1.90	2.41	4.27	0.677
2.00	2.25	4.09	0.670
2.10	2.13	3.89	0.659
2.20	2.04	3.72	0.646
2.30	1.99	3.56	0.632
2.40	1.95	3.44	0.620

Energy (eV)	n	k	R(ϕ = 0)
2.50	1.90	3.34	0.611
2.60	1.86	3.26	0.605
2.70	1.79	3.19	0.602
2.80	1.72	3.11	0.596
2.90	1.66	3.03	0.591
3.00	1.60	2.94	0.586
3.20	1.50	2.78	0.571
3.40	1.42	2.62	0.553
3.60	1.36	2.47	0.533
3.80	1.33	2.33	0.511
4.00	1.31	2.21	0.488
4.20	1.28	2.12	0.471
4.40	1.27	2.03	0.452
4.60	1.26	1.95	0.435
4.80	1.25	1.90	0.423
5.00	1.24	1.84	0.411
5.20	1.22	1.80	0.403
5.40	1.21	1.78	0.399
5.60	1.17	1.76	0.400
5.80	1.11	1.74	0.406
6.00	1.04	1.69	0.407
6.20	0.98	1.62	0.401
6.40	0.94	1.54	0.386
6.60	0.92	1.46	0.368
6.80	0.91	1.38	0.345
7.00	0.91	1.32	0.326
7.20	0.91	1.26	0.305
7.40	0.92	1.21	0.285
7.60	0.93	1.17	0.269
7.80	0.94	1.13	0.253
Copper[6]			
0.10	29.69	71.57	0.980
0.50	1.71	17.63	0.979
1.00	0.44	8.48	0.976
1.50	0.26	5.26	0.965
1.70	0.22	4.43	0.958
1.75	0.21	4.25	0.956
1.80	0.21	4.04	0.952
1.85	0.22	3.85	0.947
1.90	0.21	3.67	0.943
2.00	0.27	3.24	0.910
2.10	0.47	2.81	0.814
2.20	0.83	2.60	0.673
2.30	1.04	2.59	0.618
2.40	1.12	2.60	0.602
2.60	1.15	2.50	0.577
2.80	1.17	2.36	0.545
3.00	1.18	2.21	0.509
3.20	1.23	2.07	0.468
3.40	1.27	1.95	0.434
3.60	1.31	1.87	0.407
3.80	1.34	1.81	0.387
4.00	1.34	1.72	0.364
4.20	1.42	1.64	0.336
4.40	1.49	1.64	0.329
4.60	1.52	1.67	0.334
4.80	1.53	1.71	0.345
5.00	1.47	1.78	0.366
5.20	1.38	1.80	0.380
5.40	1.28	1.78	0.389

Energy (eV)	n	k	R(ϕ = 0)
5.60	1.18	1.74	0.391
5.80	1.10	1.67	0.389
6.00	1.04	1.59	0.380
6.50	0.96	1.37	0.329
7.00	0.97	1.20	0.271
7.50	1.00	1.09	0.230
8.00	1.03	1.03	0.206
8.50	1.03	0.98	0.189
9.00	1.03	0.92	0.171
9.50	1.03	0.87	0.154
10.00	1.04	0.82	0.139
11.00	1.07	0.75	0.118
12.00	1.09	0.73	0.111
13.00	1.08	0.72	0.109
14.00	1.06	0.72	0.111
14.50	1.03	0.72	0.111
15.00	1.01	0.71	0.111
15.50	0.98	0.69	0.109
16.00	0.95	0.67	0.106
17.00	0.91	0.62	0.097
18.00	0.89	0.56	0.084
19.00	0.88	0.51	0.071
20.00	0.88	0.45	0.059
21.00	0.90	0.41	0.048
22.00	0.92	0.38	0.040
23.00	0.94	0.37	0.035
24.00	0.96	0.37	0.035
25.00	0.96	0.40	0.040
26.00	0.92	0.40	0.044
27.00	0.88	0.38	0.043
28.00	0.86	0.35	0.039
29.00	0.85	0.30	0.032
30.00	0.86	0.26	0.025
31.00	0.88	0.24	0.020
32.00	0.89	0.22	0.017
33.00	0.90	0.21	0.015
34.00	0.91	0.20	0.014
35.00	0.92	0.20	0.013
36.00	0.92	0.19	0.012
37.00	0.92	0.19	0.011
38.00	0.93	0.18	0.010
39.00	0.93	0.17	0.009
40.00	0.93	0.17	0.009
41.00	0.94	0.16	0.008
42.00	0.94	0.16	0.007
43.00	0.94	0.15	0.007
44.00	0.95	0.15	0.007
45.00	0.95	0.15	0.006
46.00	0.95	0.15	0.006
47.00	0.95	0.14	0.006
48.00	0.95	0.14	0.006
49.00	0.95	0.14	0.005
50.00	0.95	0.13	0.005
51.00	0.95	0.13	0.005
52.00	0.95	0.13	0.005
53.00	0.96	0.12	0.004
54.00	0.96	0.12	0.004
55.00	0.96	0.12	0.004
56.00	0.96	0.11	0.004
57.00	0.96	0.11	0.004

Energy (eV)	n	k	R(ϕ = 0)
58.00	0.96	0.11	0.004
59.00	0.97	0.11	0.003
60.00	0.97	0.11	0.003
61.00	0.97	0.11	0.003
62.00	0.97	0.11	0.003
63.00	0.96	0.10	0.003
64.00	0.96	0.10	0.003
65.00	0.97	0.10	0.003
66.00	0.97	0.10	0.003
67.00	0.97	0.09	0.003
68.00	0.97	0.09	0.002
69.00	0.97	0.09	0.002
70.00	0.97	0.09	0.002
75.00	0.98	0.09	0.002
80.00	0.98	0.09	0.002
85.00	0.97	0.09	0.002
90.00	0.96	0.08	0.002
Gallium (liquid)[7]			
1.425	2.40	9.20	0.900
1.550	2.09	8.50	0.898
1.771	1.65	7.60	0.898
2.066	1.25	6.60	0.897
2.480	0.89	5.60	0.898
3.100	0.59	4.50	0.896
Germanium, single crystal[8]			
0.01240	(4.0065)	3.00E-03	0.361
0.01364	4.0063	2.40E-03	0.361
0.01488	(4.0060)	1.70E-03	0.361
0.01612	(4.0060)	1.55E-03	0.361
0.01736	(4.0060)	1.50E-03	0.361
0.01860		1.50E-03	
0.01984		1.60E-03	
0.02108		1.60E-03	
0.02232		1.55E-03	
0.02356		1.53E-03	
0.02480		1.50E-03	
0.02604		1.25E-03	
0.02728		8.50E-04	
0.02852		6.50E-04	
0.02976		7.00E-04	
0.03100	3.9827	8.50E-04	0.358
0.03224		1.55E-03	
0.03348		2.75E-03	
0.03472		3.55E-03	
0.03596	(3.9900)	3.05E-03	0.359
0.03720		2.75E-03	
0.03844		2.70E-03	
0.03968	(3.9930)	2.90E-03	0.359
0.04092		2.95E-03	
0.04215		3.20E-03	
0.04339		6.30E-03	
0.04463		3.40E-03	
0.04587	(3.9955)	2.50E-03	0.360
0.04711		2.10E-03	
0.04835		2.00E-03	
0.04959		8.00E-04	
0.05083		1.40E-03	
0.05207		1.35E-03	
0.05331		1.10E-03	
0.05455		8.00E-04	

Energy (eV)	n	k	R(ϕ = 0)
0.05579		6.00E-04	
0.05703		9.0 E-04	
0.05827		6.5 E-04	
0.05951		4.6 E-04	
0.06075		4.0 E-04	
0.06199	3.9992	3.98E-04	0.360
0.06323		4.0 E-04	
0.06447		4.3 E-04	
0.06571		4.4 E-04	
0.06695	(4.0000)	4.3 E-04	0.360
0.06819		3.1 E-04	
0.06943		3.3 E-04	
0.07067		3.8 E-04	
0.07191		3.3 E-04	
0.07315		2.5 E-04	
0.07439		1.9 E-04	
0.07514		1.58E-04	
0.07749	4.0009	9.55E-05	0.360
0.07999	4.0011	1.71E-04	0.360
0.08266	4.0013	9.78E-05	0.360
0.08551	4.0015	5.77E-05	0.360
0.08920		3.98E-05	
0.09460		4.59E-05	
0.09840		3.51E-05	
0.1	4.0063	3.70E-05	0.361
0.2	4.0108		0.361
0.3	4.0246		0.362
0.4	4.0429		0.364
0.5	(4.074)		0.367
0.6	(4.104)	6.58E-07	0.370
0.7	4.180	1.27E-04	0.377
0.8	4.275	5.67E-03	0.385
0.9	4.285	7.45E-02	0.386
1.0	4.325	8.09E-02	0.390
1.1	4.385	0.103	0.395
1.2	4.420	0.123	0.398
1.3	4.495	0.167	0.405
1.4	4.560	0.190	0.411
1.5	4.635	0.298	0.418
1.6	4.763	0.345	0.428
1.7	4.897	0.401	0.439
1.8	5.067	0.500	0.453
1.9	5.380	0.540	0.475
2.0	5.588	0.933	0.495
2.1	5.748	1.634	0.523
2.2	5.283	2.049	0.516
2.3	5.062	2.318	0.519
2.4	4.610	2.455	0.508
2.5	4.340	2.384	0.492
2.6	4.180	2.309	0.480
2.7	4.082	2.240	0.471
2.8	4.035	2.181	0.464
2.9	4.037	2.140	0.461
3.0	4.082	2.145	0.463
3.1	4.141	2.215	0.471
3.2	4.157	2.340	0.482
3.3	4.128	2.469	0.490
3.4	4.070	2.579	0.497
3.5	4.020	2.667	0.502
3.6	3.985	2.759	0.509
3.7	3.958	2.863	0.517
3.8	3.936	2.986	0.527
3.9	3.920	3.137	0.539
4.0	3.905	3.336	0.556
4.1	3.869	3.614	0.579
4.2	3.745	4.009	0.612
4.3	3.338	4.507	0.659
4.4	2.516	4.669	0.705
4.5	1.953	4.297	0.713
4.6	1.720	3.960	0.702
4.7	1.586	3.709	0.690
4.8	1.498	3.509	0.677
4.9	1.435	3.342	0.664
5.0	1.394	3.197	0.650
5.1	1.370	3.073	0.636
5.2	1.364	2.973	0.622
5.3	1.371	2.897	0.609
5.4	1.383	2.854	0.600
5.5	1.380	2.842	0.598
5.6	1.360	2.846	0.602
5.7	1.293	2.163	0.479
5.8	1.209	2.873	0.632
5.9	1.108	2.813	0.641
6.0	1.30	2.34	0.517
6.5	1.10	2.05	0.489
7.0	1.00	1.80	0.448
7.5		1.60	
8.0	0.92	1.40	0.348
8.5	0.92	1.20	0.282
9.0	0.92	1.14	0.262
9.5		1.00	
10.0	0.93	0.86	0.167
20.0		0.237	
22.0		0.179	
24.0		0.144	
26.0		0.110	
28.0		0.0747	
30.0		0.1020	
32.0		0.0999	
34.0		0.0856	
36.0		0.0740	
38.0		0.0651	
40.0		0.0604	

Gold, electropolished, Au (110)[9]

Energy (eV)	n	k	R(ϕ = 0)
0.10	8.17	82.83	0.995
0.20	2.13	41.73	0.995
0.30	0.99	27.82	0.995
0.40	0.59	20.83	0.995
0.50	0.39	16.61	0.994
0.60	0.28	13.78	0.994
0.70	0.22	11.75	0.994
0.80	0.18	10.21	0.993
0.90	0.15	9.01	0.993
1.00	0.13	8.03	0.992
1.20	0.10	6.54	0.991
1.40	0.08	5.44	0.989
1.60	0.08	4.56	0.986
1.80	0.09	3.82	0.979
2.00	0.13	3.16	0.953
2.10	0.18	2.84	0.925
2.20	0.24	2.54	0.880
2.40	0.50	1.86	0.647
2.50	0.82	1.59	0.438
2.60	1.24	1.54	0.331
2.70	1.43	1.72	0.356
2.80	1.46	1.77	0.368
2.90	1.50	1.79	0.368
3.00	1.54	1.80	0.369
3.10	1.54	1.81	0.371
3.20	1.54	1.80	0.368
3.30	1.55	1.78	0.362
3.40	1.56	1.76	0.356
3.50	1.58	1.73	0.349
3.60	1.62	1.73	0.346
3.70	1.64	1.75	0.351
3.80	1.63	1.79	0.360
3.90	1.59	1.81	0.366
4.00	1.55	1.81	0.369
4.10	1.51	1.79	0.368
4.20	1.48	1.78	0.367
4.30	1.45	1.77	0.368
4.40	1.41	1.76	0.370
4.50	1.35	1.74	0.370
4.60	1.30	1.69	0.364
4.70	1.27	1.64	0.354
4.80	1.25	1.59	0.344
4.90	1.23	1.54	0.332
5.00	1.22	1.49	0.319
5.20	1.21	1.40	0.295
5.40	1.21	1.33	0.275
5.60	1.21	1.27	0.256
5.80	1.21	1.20	0.236
6.00	1.22	1.14	0.218
6.20	1.24	1.09	0.203
6.40	1.25	1.05	0.190
6.60	1.27	1.01	0.177
6.80	1.30	0.97	0.167
7.00	1.34	0.95	0.162
7.20	1.36	0.95	0.161
7.40	1.38	0.96	0.164
7.60	1.38	0.98	0.169
7.80	1.35	0.99	0.171
8.00	1.31	0.96	0.165
8.20	1.30	0.92	0.155
8.40	1.30	0.89	0.147
8.60	1.31	0.88	0.144
8.80	1.31	0.86	0.140
9.00	1.30	0.83	0.133
9.20	1.31	0.81	0.126
9.40	1.33	0.78	0.122
9.60	1.36	0.78	0.121
9.80	1.37	0.79	0.124
10.00	1.37	0.80	0.126
10.20	1.36	0.80	0.127
10.40	1.35	0.80	0.125
10.60	1.34	0.79	0.123
10.80	1.34	0.77	0.120
11.00	1.34	0.76	0.116
11.20	1.34	0.74	0.113
11.40	1.35	0.73	0.111

Energy (eV)	n	k	$R(\phi = 0)$
11.60	1.36	0.72	0.109
11.80	1.38	0.71	0.108
12.00	1.39	0.71	0.109
12.40	1.44	0.73	0.115
12.80	1.45	0.79	0.127
13.20	1.42	0.84	0.137
13.60	1.37	0.86	0.140
14.00	1.33	0.86	0.140
14.40	1.29	0.86	0.139
14.80	1.26	0.84	0.135
15.20	1.24	0.83	0.132
15.60	1.22	0.81	0.127
16.00	1.21	0.79	0.123
16.40	1.20	0.78	0.119
16.80	1.19	0.76	0.116
17.20	1.19	0.75	0.114
17.60	1.19	0.74	0.111
18.00	1.19	0.74	0.109
18.40	1.19	0.73	0.109
18.80	1.20	0.74	0.110
19.20	1.21	0.76	0.116
19.60	1.21	0.80	0.125
20.00	1.18	0.83	0.133
20.40	1.14	0.85	0.141
20.80	1.10	0.87	0.149
21.20	1.05	0.88	0.156
21.60	1.00	0.88	0.162
22.00	0.94	0.86	0.164
22.40	0.89	0.83	0.163
22.80	0.85	0.79	0.157
23.20	0.82	0.75	0.149
23.60	0.80	0.70	0.138
24.00	0.80	0.66	0.125
24.40	0.80	0.62	0.113
24.80	0.80	0.58	0.101
25.20	0.82	0.56	0.090
25.60	0.83	0.54	0.084
26.00	0.84	0.52	0.079
26.40	0.85	0.51	0.074
26.80	0.85	0.50	0.071
27.20	0.86	0.49	0.068
27.60	0.86	0.49	0.065
28.00	0.87	0.48	0.063
28.40	0.88	0.48	0.062
28.80	0.88	0.48	0.062
29.20	0.88	0.48	0.062
29.60	0.87	0.48	0.064
30.00	0.86	0.48	0.064

Hafnium, single crystal, $\vec{E} \parallel \hat{c}$[10]

Energy (eV)	n	k	$R(\phi = 0)$
0.52	1.48	4.11	0.747
0.56	1.84	3.29	0.615
0.60	2.34	2.62	0.486
0.66	3.21	2.13	0.428
0.70	3.70	2.03	0.441
0.76	4.31	2.10	0.476
0.80	4.61	2.31	0.504
0.86	4.71	2.70	0.533
0.90	4.64	2.85	0.541
0.95	4.54	2.96	0.545
1.00	4.45	3.00	0.545
1.10	4.28	3.08	0.547
1.20	4.08	3.10	0.544
1.30	3.87	3.04	0.536
1.40	3.72	2.95	0.525
1.50	3.60	2.85	0.514
1.60	3.52	2.73	0.500
1.70	3.52	2.61	0.488
1.80	3.57	2.56	0.485
1.90	3.63	2.59	0.489
2.00	3.65	2.67	0.498
2.10	3.64	2.81	0.511
2.20	3.53	2.99	0.526
2.30	3.34	3.09	0.534
2.40	3.15	3.11	0.537
2.50	2.99	3.13	0.540
2.60	2.83	3.12	0.542
2.70	2.68	3.10	0.542
2.80	2.54	3.08	0.543
2.90	2.40	3.04	0.544
3.00	2.27	3.00	0.544
3.10	2.14	2.95	0.544
3.20	2.00	2.89	0.544
3.30	1.87	2.79	0.538
3.40	1.78	2.68	0.528
3.50	1.71	2.58	0.517
3.60	1.66	2.48	0.503
3.70	1.63	2.40	0.491
3.80	1.60	2.33	0.481
3.90	1.56	2.27	0.473
4.00	1.52	2.21	0.466
4.10	1.48	2.14	0.455
4.20	1.45	2.07	0.442
4.30	1.43	2.01	0.431
4.40	1.41	1.95	0.420
4.50	1.39	1.89	0.407
4.60	1.39	1.83	0.394
4.70	1.39	1.79	0.382
4.80	1.38	1.75	0.373
4.90	1.38	1.71	0.364
5.00	1.37	1.68	0.356
5.20	1.36	1.61	0.341
5.40	1.35	1.55	0.324
5.60	1.35	1.51	0.314
5.80	1.32	1.48	0.308
6.00	1.28	1.41	0.295
6.20	1.26	1.35	0.278
6.40	1.26	1.28	0.258
6.60	1.27	1.22	0.240
6.80	1.28	1.16	0.224
7.00	1.31	1.13	0.212
7.20	1.33	1.10	0.204
7.40	1.34	1.07	0.197
7.60	1.36	1.05	0.191
7.80	1.37	1.02	0.183
8.00	1.40	1.01	0.179
8.20	1.43	1.01	0.178
8.40	1.45	1.01	0.180
8.60	1.47	1.02	0.183
8.80	1.48	1.04	0.186
9.00	1.49	1.07	0.193
9.20	1.50	1.10	0.201
9.40	1.48	1.14	0.211
9.60	1.46	1.18	0.222
9.80	1.41	1.21	0.230
10.00	1.36	1.22	0.235
10.20	1.32	1.22	0.238
10.40	1.28	1.22	0.240
10.60	1.24	1.21	0.241
10.80	1.20	1.20	0.242
11.00	1.16	1.19	0.242
11.20	1.13	1.17	0.241
11.40	1.10	1.16	0.241
11.60	1.07	1.14	0.239
11.80	1.04	1.12	0.238
12.00	1.02	1.10	0.236
12.40	0.96	1.06	0.232
12.80	0.92	1.01	0.225
13.20	0.88	0.96	0.218
13.60	0.84	0.90	0.205
14.00	0.83	0.83	0.186
14.40	0.83	0.80	0.172
14.80	0.81	0.76	0.167
15.20	0.79	0.70	0.153
15.60	0.79	0.64	0.132
16.00	0.83	0.60	0.111
16.40	0.81	0.60	0.114
16.80	0.79	0.55	0.105
17.20	0.79	0.50	0.089
17.60	0.80	0.46	0.077
18.00	0.81	0.42	0.064
18.40	0.84	0.38	0.051
18.80	0.87	0.34	0.040
19.00	0.89	0.33	0.036
19.60	0.93	0.32	0.030
20.00	0.94	0.31	0.027
20.60	0.97	0.30	0.023
21.00	0.99	0.29	0.022
21.60	1.01	0.28	0.020
22.00	1.03	0.28	0.020
22.60	1.06	0.28	0.020
23.00	1.07	0.28	0.021
23.60	1.09	0.29	0.022
24.00	1.09	0.30	0.023
24.60	1.10	0.31	0.024

Hafnium, single crystal, $\vec{E} \perp \hat{c}$[10]

Energy (eV)	n	k	$R(\phi = 0)$
0.52	2.25	4.65	0.723
0.56	2.34	3.66	0.623
0.60	2.84	2.89	0.512
0.66	3.71	2.35	0.469
0.70	4.26	2.21	0.482
0.76	4.97	2.33	0.521
0.80	5.41	2.62	0.554
0.86	5.46	3.36	0.593
0.90	5.22	3.62	0.601
0.95	4.95	3.72	0.602
1.00	4.76	3.76	0.602
1.10	4.43	3.80	0.601
1.20	4.07	3.74	0.594
1.30	3.79	3.55	0.578
1.40	3.61	3.36	0.561

Energy (eV)	n	k	R(φ = 0)
1.50	3.55	3.13	0.540
1.60	3.58	3.01	0.529
1.70	3.63	2.98	0.526
1.80	3.66	3.02	0.530
1.90	3.63	3.14	0.541
2.00	3.51	3.26	0.551
2.10	3.35	3.33	0.558
2.20	3.18	3.36	0.563
2.30	2.99	3.39	0.568
2.40	2.78	3.35	0.569
2.50	2.65	3.26	0.562
2.60	2.54	3.22	0.560
2.70	2.42	3.17	0.559
2.80	2.31	3.13	0.558
2.90	2.20	3.08	0.558
3.00	2.08	3.05	0.561
3.10	1.94	2.98	0.560
3.20	1.83	2.88	0.555
3.30	1.74	2.78	0.547
3.40	1.68	2.69	0.538
3.50	1.62	2.61	0.529
3.60	1.57	2.52	0.519
3.70	1.53	2.45	0.510
3.80	1.49	2.38	0.501
3.90	1.45	2.32	0.493
4.00	1.41	2.25	0.484
4.10	1.38	2.18	0.474
4.20	1.35	2.11	0.462
4.30	1.33	2.05	0.451
4.40	1.31	1.99	0.438
4.50	1.30	1.93	0.427
4.60	1.29	1.88	0.415
4.70	1.28	1.82	0.402
4.80	1.28	1.77	0.389
4.90	1.27	1.73	0.379
5.00	1.27	1.69	0.367
5.20	1.27	1.62	0.349
5.40	1.27	1.57	0.335
5.60	1.26	1.52	0.322
5.80	1.24	1.48	0.313
6.00	1.21	1.42	0.302
6.20	1.19	1.36	0.285
6.40	1.18	1.29	0.265
6.60	1.19	1.22	0.244
6.80	1.21	1.18	0.230
7.00	1.22	1.14	0.217
7.20	1.23	1.10	0.206
7.40	1.26	1.06	0.194
7.60	1.28	1.04	0.187
7.80	1.30	1.02	0.180
8.00	1.33	1.00	0.174
8.20	1.35	0.99	0.173
8.40	1.38	0.99	0.173
8.60	1.40	1.00	0.174
8.80	1.42	1.02	0.178
9.00	1.43	1.04	0.184
9.20	1.45	1.08	0.193
9.40	1.43	1.12	0.204
9.60	1.40	1.16	0.214
9.80	1.37	1.19	0.223
10.00	1.32	1.21	0.230
10.20	1.27	1.21	0.234
10.40	1.23	1.20	0.235
10.60	1.19	1.20	0.237
10.80	1.15	1.19	0.237
11.00	1.12	1.17	0.237
11.20	1.08	1.16	0.237
11.40	1.05	1.14	0.236
11.60	1.03	1.12	0.235
11.80	1.00	1.10	0.233
12.00	0.97	1.08	0.231
12.40	0.92	1.04	0.226
12.80	0.88	0.99	0.219
13.20	0.83	0.94	0.211
13.60	0.80	0.88	0.196
14.00	0.79	0.81	0.177
14.40	0.80	0.77	0.160
14.80	0.77	0.73	0.154
15.20	0.76	0.68	0.140
15.60	0.76	0.61	0.119
16.00	0.81	0.58	0.099
16.40	0.78	0.57	0.102
16.80	0.77	0.53	0.092
17.20	0.77	0.48	0.077
17.60	0.79	0.44	0.065
18.00	0.80	0.39	0.053
18.40	0.87	0.36	0.011
18.80	0.86	0.33	0.032
19.00	0.88	0.32	0.030
19.60	0.91	0.31	0.025
20.00	0.93	0.30	0.023
20.60	0.96	0.29	0.021
21.00	0.97	0.29	0.020
21.60	1.00	0.28	0.019
22.00	1.01	0.28	0.019
22.60	1.03	0.27	0.018
23.00	1.05	0.28	0.019
23.60	1.06	0.28	0.020
24.00	1.07	0.29	0.021
24.60	1.09	0.30	0.022

Iridium[11]

Energy (eV)	n	k	R(φ = 0)
0.10	28.49	60.62	0.975
0.15	15.32	45.15	0.973
0.20	9.69	35.34	0.972
0.25	6.86	28.84	0.969
0.30	5.16	24.25	0.967
0.35	4.11	20.79	0.964
0.40	3.42	18.06	0.960
0.45	3.05	15.82	0.954
0.50	2.98	14.06	0.944
0.60	2.79	11.58	0.925
0.70	2.93	9.78	0.895
0.80	3.14	8.61	0.862
0.90	3.19	7.88	0.840
1.00	3.15	7.31	0.822
1.10	3.04	6.84	0.808
1.20	2.96	6.41	0.791
1.30	2.85	6.07	0.779
1.40	2.72	5.74	0.767
1.50	2.65	5.39	0.750
1.60	2.68	5.08	0.728
1.70	2.69	4.92	0.716
1.80	2.64	4.81	0.710
1.90	2.57	4.68	0.704
2.00	2.50	4.57	0.699
2.10	2.40	4.48	0.697
2.20	2.29	4.38	0.695
2.30	2.18	4.26	0.692
2.40	2.07	4.14	0.689
2.50	1.98	4.00	0.682
2.60	1.91	3.86	0.673
2.70	1.85	3.73	0.665
2.80	1.81	3.61	0.655
2.90	1.77	3.51	0.646
3.00	1.73	3.43	0.640
3.20	1.62	3.26	0.629
3.40	1.53	3.05	0.610
3.60	1.52	2.81	0.573
3.80	1.61	2.69	0.541
4.00	1.64	2.68	0.535
4.20	1.58	2.71	0.549
4.40	1.45	2.68	0.561
4.60	1.31	2.60	0.567
4.80	1.18	2.49	0.570
5.00	1.10	2.35	0.559
5.20	1.04	2.22	0.543
5.40	1.00	2.09	0.522
5.60	0.98	1.98	0.499
5.80	0.96	1.86	0.474
6.00	0.95	1.78	0.454
6.20	0.94	1.68	0.427
6.40	0.94	1.59	0.401
6.60	0.94	1.50	0.375
6.80	0.95	1.42	0.345
7.00	0.97	1.34	0.318
7.20	0.99	1.27	0.290
7.40	1.02	1.20	0.262
7.60	1.03	1.14	0.241
7.80	1.08	1.06	0.208
8.00	1.13	1.03	0.191
8.20	1.18	1.00	0.179
8.40	1.22	0.98	0.171
8.60	1.26	0.96	0.164
8.80	1.29	0.95	0.160
9.00	1.33	0.94	0.157
9.20	1.36	0.95	0.159
9.40	1.39	0.95	0.161
9.60	1.42	0.97	0.163
9.80	1.44	0.99	0.169
10.00	1.45	1.01	0.175
10.20	1.45	1.04	0.182
10.40	1.44	1.07	0.187
10.60	1.43	1.09	0.193
10.80	1.41	1.12	0.200
11.00	1.38	1.13	0.206
11.20	1.34	1.14	0.208
11.40	1.31	1.13	0.208
11.60	1.28	1.12	0.206
11.80	1.25	1.10	0.203
12.00	1.24	1.08	0.199

Energy (eV)	n	k	R(ϕ = 0)
12.40	1.21	1.05	0.191
12.80	1.19	1.01	0.181
13.20	1.18	0.98	0.173
13.60	1.17	0.95	0.165
14.00	1.16	0.91	0.155
14.40	1.17	0.88	0.147
14.80	1.18	0.87	0.142
15.20	1.19	0.84	0.136
15.60	1.20	0.83	0.133
16.00	1.21	0.83	0.131
16.40	1.23	0.82	0.129
16.80	1.25	0.82	0.127
17.20	1.28	0.83	0.131
17.60	1.30	0.87	0.140
18.00	1.30	0.93	0.154
18.40	1.27	0.97	0.166
18.80	1.24	1.00	0.176
19.20	1.20	1.03	0.187
19.60	1.15	1.05	0.197
20.00	1.10	1.06	0.205
20.50	1.04	1.05	0.210
21.00	0.99	1.04	0.215
21.50	0.94	1.02	0.220
22.00	0.89	1.00	0.222
22.50	0.84	0.99	0.228
23.00	0.79	0.96	0.232
23.50	0.76	0.92	0.228
24.00	0.73	0.87	0.223
24.50	0.70	0.83	0.218
25.00	0.69	0.79	0.209
25.50	0.68	0.76	0.200
26.00	0.67	0.72	0.192
26.50	0.67	0.69	0.181
27.00	0.66	0.66	0.174
27.50	0.66	0.63	0.166
28.00	0.66	0.61	0.158
28.50	0.66	0.59	0.151
29.00	0.65	0.57	0.148
29.50	0.64	0.55	0.145
30.00	0.64	0.53	0.140
32.00	0.62	0.44	0.119
34.00	0.64	0.35	0.091
36.00	0.69	0.27	0.059
38.00	0.73	0.24	0.044
40.00	0.76	0.22	0.034

***Iron*[5]**

Energy (eV)	n	k	R(ϕ = 0)
0.10	6.41	33.07	0.978
0.15	6.26	22.82	0.956
0.20	3.68	18.23	0.958
0.26	4.98	13.68	0.911
0.30	4.87	12.05	0.892
0.36	4.68	10.44	0.867
0.40	4.42	9.75	0.858
0.50	4.14	8.02	0.817
0.60	3.93	6.95	0.783
0.70	3.78	6.17	0.752
0.80	3.65	5.60	0.725
0.90	3.52	5.16	0.700
1.00	3.43	4.79	0.678
1.10	3.33	4.52	0.660
1.20	3.24	4.26	0.641
1.30	3.16	4.07	0.626
1.40	3.12	3.87	0.609
1.50	3.05	3.77	0.601
1.60	3.00	3.60	0.585
1.70	2.98	3.52	0.577
1.80	2.92	3.46	0.573
1.90	2.89	3.37	0.563
2.00	2.85	3.36	0.563
2.10	2.80	3.34	0.562
2.20	2.74	3.33	0.563
2.30	2.65	3.34	0.567
2.40	2.56	3.31	0.567
2.50	2.46	3.31	0.570
2.60	2.34	3.30	0.576
2.70	2.23	3.25	0.575
2.80	2.12	3.23	0.580
2.90	2.01	3.17	0.580
3.00	1.88	3.12	0.583
3.10	1.78	3.04	0.580
3.20	1.70	2.96	0.576
3.30	1.62	2.87	0.572
3.40	1.55	2.79	0.565
3.50	1.50	2.70	0.556
3.60	1.47	2.63	0.548
3.70	1.43	2.56	0.542
3.83	1.38	2.49	0.534
4.00	1.30	2.39	0.527
4.17	1.26	2.27	0.510
4.33	1.23	2.18	0.494
4.50	1.20	2.10	0.482
4.67	1.16	2.02	0.470
4.83	1.14	1.93	0.451
5.00	1.14	1.87	0.435
5.17	1.12	1.81	0.425
5.33	1.11	1.75	0.408
5.50	1.09	1.17	0.401
5.67	1.09	1.65	0.383
5.83	1.10	1.61	0.373
6.00	1.09	1.59	0.366
6.17	1.08	1.57	0.365
6.33	1.04	1.55	0.365
6.50	1.02	1.51	0.358
6.67	1.00	1.47	0.351
6.83	0.97	1.43	0.346
7.00	0.96	1.39	0.333
7.17	0.94	1.35	0.327
7.33	0.94	1.30	0.311
7.50	0.94	1.26	0.298
7.67	0.94	1.23	0.288
7.83	0.94	1.21	0.279
8.00	0.94	1.18	0.272
8.17	0.94	1.16	0.265
8.33	0.94	1.14	0.258
8.50	0.94	1.12	0.251
8.67	0.94	1.10	0.246
8.83	0.92	1.08	0.240
9.00	0.93	1.07	0.236
9.17	0.92	1.06	0.233
9.33	0.91	1.04	0.231
9.50	0.90	1.02	0.226
9.67	0.90	1.00	0.221
9.83	0.89	0.99	0.218
10.00	0.88	0.97	0.213
10.17	0.87	0.94	0.203
10.33	0.87	0.91	0.196
10.50	0.87	0.89	0.189
10.67	0.88	0.87	0.179
10.83	0.89	0.85	0.170
11.00	0.91	0.83	0.162
11.17	0.92	0.83	0.159
11.33	0.93	0.84	0.159
11.50	0.93	0.84	0.160
11.67	0.93	0.84	0.162
11.83	0.92	0.84	0.163
12.00	0.91	0.84	0.163
12.17	0.90	0.84	0.165
12.33	0.89	0.83	0.164
12.50	0.98	0.83	0.165
12.67	0.87	0.82	0.166
12.83	0.86	0.81	0.166
13.00	0.85	0.80	0.162
13.17	0.84	0.79	0.161
13.33	0.84	0.78	0.160
13.50	0.83	0.77	0.159
13.67	0.82	0.76	0.157
13.83	0.81	0.75	0.154
14.00	0.81	0.73	0.151
14.17	0.80	0.72	0.149
14.33	0.80	0.71	0.146
14.50	0.79	0.79	0.144
14.67	0.79	0.69	0.141
14.83	0.78	0.67	0.138
15.00	0.78	0.66	0.135
15.17	0.78	0.65	0.131
15.33	0.78	0.64	0.238
15.50	0.77	0.63	0.126
15.67	0.77	0.62	0.123
15.83	0.77	0.61	0.119
16.00	0.77	0.60	0.116
16.17	0.78	0.58	0.112
16.33	0.78	0.58	0.110
16.50	0.78	0.57	0.107
16.67	0.77	0.56	0.106
16.83	0.78	0.55	0.103
17.00	0.78	0.55	0.102
17.17	0.78	0.54	0.100
17.33	0.78	0.54	0.098
17.50	0.77	0.53	0.097
17.67	0.77	0.52	0.095
17.83	0.78	0.51	0.092
18.00	0.78	0.51	0.091
18.17	0.78	0.51	0.090
18.33	0.78	0.50	0.089
18.50	0.77	0.50	0.089
18.67	0.77	0.50	0.088
18.83	0.77	0.49	0.087
19.00	0.77	0.49	0.087
19.17	0.76	0.49	0.088
19.33	0.76	0.48	0.087

Energy (eV)	n	k	$R(\phi = 0)$
19.50	0.75	0.47	0.086
19.67	0.75	0.47	0.085
19.83	0.75	0.46	0.084
20.00	0.74	0.45	0.083
20.17	0.74	0.44	0.081
20.33	0.74	0.44	0.081
20.50	0.74	0.42	0.080
20.67	0.73	0.43	0.079
20.83	0.73	0.42	0.078
21.00	0.73	0.41	0.077
21.17	0.72	0.40	0.076
21.33	0.72	0.39	0.074
21.50	0.72	0.38	0.073
21.67	0.72	0.38	0.071
21.83	0.72	0.37	0.070
22.00	0.72	0.36	0.068
22.17	0.71	0.35	0.067
22.33	0.72	0.34	0.064
22.50	0.72	0.34	0.063
22.67	0.72	0.33	0.062
22.83	0.72	0.32	0.059
23.00	0.72	0.31	0.058
23.17	0.72	0.30	0.056
23.33	0.72	0.29	0.054
23.50	0.73	0.28	0.050
23.67	0.73	0.28	0.049
23.83	0.74	0.27	0.047
24.00	0.74	0.27	0.045
24.17	0.74	0.26	0.044
24.33	0.74	0.26	0.043
24.50	0.74	0.25	0.042
24.67	0.75	0.25	0.040
24.83	0.75	0.24	0.039
25.00	0.75	0.24	0.038
26.00	0.76	0.21	0.031
27.00	0.78	0.18	0.026
28.00	0.79	0.16	0.021
29.00	0.81	0.14	0.017
30.00	0.82	0.13	0.014

Lithium[12]

Energy (eV)	n	k	$R(\phi = 0)$
0.14	0.659	38.0	0.998
0.54	0.661	12.6	0.984
0.75	0.561	7.68	0.963
1.05	0.448	5.58	0.946
1.35	0.338	4.36	0.935
1.65	0.265	3.55	0.925
1.95	0.221	2.94	0.913
2.25	0.206	2.48	0.892
2.55	0.217	2.11	0.854
2.85	0.247	1.82	0.797
3.15	0.304	1.60	0.715
3.45	0.334	1.45	0.656
3.75	0.345	1.32	0.611
4.05	0.346	1.21	0.578
4.35	0.333	1.11	0.557
4.65	0.317	1.01	0.540
4.95	0.302	0.906	0.520
5.25	0.299	0.795	0.484
5.55	0.310	0.688	0.434
5.85	0.342	0.594	0.365
6.15	0.376	0.522	0.306
6.45	0.408	0.460	0.256
6.75	0.440	0.407	0.214
7.05	0.466	0.364	0.183
7.35	0.492	0.320	0.155
7.65	0.517	0.282	0.131
7.95	0.545	0.246	0.109
8.25	0.572	0.214	0.091
8.55	0.601	0.189	0.075
8.85	0.624	0.163	0.063
9.15	0.657	0.144	0.050
9.45	0.680	0.130	0.042
9.75	0.708	0.119	0.034
10.1	0.726	0.108	0.029
10.4	0.743	0.102	0.025
10.6	0.753	0.080	0.022

Magnesium (evaporated)[13]

Energy (eV)	n	k	$R(\phi = 0)$
2.145	0.48	3.71	0.880
2.270	0.57	3.47	0.843
2.522	0.53	2.92	0.805
2.845	0.52	2.65	0.777
3.064	0.52	2.05	0.681
5.167	0.10	1.60	0.894
5.636	0.15	1.50	0.832
6.200	0.20	1.40	0.765
6.889	0.25	1.30	0.693
7.750	0.20	1.20	0.722
8.857	0.15	0.95	0.730
10.335	0.25	0.40	0.419

Manganese[14]

Energy (eV)	n	k	$R(\phi = 0)$
0.64	3.89	5.95	0.738
0.77	3.78	5.41	0.710
0.89	3.65	5.02	0.688
1.02	3.48	4.74	0.673
1.14	3.30	4.53	0.662
1.26	3.10	4.35	0.653
1.39	2.97	4.18	0.643
1.51	2.83	4.03	0.634
1.64	2.70	3.91	0.627
1.76	2.62	3.78	0.617
1.88	2.56	3.65	0.606
2.01	2.51	3.54	0.596
2.13	2.47	3.43	0.585
2.26	2.39	3.33	0.577
2.38	2.32	3.23	0.567
2.50	2.25	3.14	0.559
2.63	2.19	3.06	0.552
2.75	2.11	2.98	0.545
2.88	2.06	2.90	0.536
3.00	2.00	2.82	0.528
3.12	1.96	2.74	0.518
3.25	1.92	2.67	0.509
3.37	1.89	2.59	0.498
3.50	1.89	2.51	0.484
3.62	1.87	2.45	0.475
3.74	1.86	2.38	0.463
3.87	1.86	2.32	0.451
3.99	1.86	2.25	0.438
4.12	1.86	2.19	0.427
4.24	1.85	2.14	0.417
4.36	1.85	2.08	0.406
4.49	1.86	2.03	0.395
4.61	1.85	1.99	0.388
4.74	1.84	1.94	0.378
4.86	1.83	1.91	0.372
4.98	1.82	1.86	0.362
5.11	1.82	1.82	0.354
5.23	1.81	1.79	0.348
5.36	1.78	1.76	0.342
5.48	1.74	1.73	0.337
5.60	1.73	1.70	0.331
5.73	1.72	1.67	0.325
5.85	1.70	1.64	0.319
5.98	1.67	1.61	0.313
6.10	1.63	1.58	0.307
6.22	1.62	1.55	0.301
6.35	1.59	1.52	0.295
6.47	1.55	1.50	0.292
6.60	1.48	1.47	0.288

Mercury (liquid)[15]

Energy (eV)	n	k	$R(\phi = 0)$
0.2	13.99	14.27	0.869
0.3	11.37	11.95	0.846
0.4	9.741	10.65	0.830
0.5	8.528	9.805	0.818
0.6	7.574	9.195	0.808
0.8	6.086	8.312	0.796
1.0	4.962	7.643	0.789
1.2	4.050	7.082	0.786
1.4	3.324	6.558	0.785
1.6	2.746	6.054	0.783
1.8	2.284	5.582	0.782
2.0	1.910	5.150	0.782
2.2	1.620	4.751	0.780
2.4	1.384	4.407	0.779
2.6	1.186	4.090	0.779
2.8	1.027	3.802	0.779
3.0	0.898	3.538	0.777
3.2	0.798	3.294	0.773
3.4	0.713	3.074	0.770
3.6	0.644	2.860	0.763
3.8	0.589	2.665	0.755
4.0	0.542	2.502	0.749
4.2	0.507	2.341	0.738
4.4	0.477	2.195	0.727
4.6	0.452	2.058	0.715
4.8	0.431	1.929	0.701
5.0	0.414	1.806	0.685
5.2	0.401	1.687	0.666
5.4	0.394	1.569	0.642
5.6	0.386	1.454	0.617
5.7	0.386	1.396	0.601
5.8	0.386	1.341	0.585
5.9	0.385	1.287	0.569
6.0	0.386	1.232	0.551
6.1	0.388	1.176	0.531
6.2	0.390	1.118	0.510
6.3	0.399	1.058	0.481
6.4	0.412	1.002	0.450
6.5	0.428	0.949	0.418

Energy (eV)	n	k	$R(\phi = 0)$
6.6	0.436	0.898	0.392
6.7	0.438	0.836	0.367
6.8	0.459	0.756	0.320
6.9	0.510	0.676	0.255
7.0	0.585	0.617	0.191
7.1	0.663	0.589	0.148
7.2	0.717	0.584	0.128
7.3	0.769	0.575	0.111
7.4	0.817	0.574	0.100
7.5	0.860	0.580	0.094
7.6	0.893	0.597	0.093
7.8	0.929	0.623	0.096
8.0	0.946	0.639	0.098
8.2	0.952	0.645	0.099
8.4	0.953	0.638	0.097
8.6	0.956	0.624	0.093
8.8	0.965	0.607	0.087
9.0	0.975	0.588	0.082
9.2	0.988	0.568	0.076
9.4	1.009	0.548	0.069
9.6	1.044	0.541	0.066
9.8	1.061	0.557	0.069
10.0	1.062	0.567	0.071
10.2	1.054	0.569	0.072
10.4	1.045	0.561	0.070
10.6	1.041	0.550	0.068
10.8	1.039	0.537	0.065
11.0	1.039	0.523	0.062
11.5	1.050	0.491	0.055
12.0	1.064	0.467	0.050
12.5	1.078	0.445	0.045
13.0	1.092	0.430	0.042
13.5	1.104	0.416	0.040
14.0	1.115	0.404	0.038
14.5	1.125	0.394	0.037
15.0	1.135	0.383	0.035
15.5	1.146	0.374	0.034
16.0	1.159	0.368	0.034
16.5	1.170	0.367	0.034
17.0	1.177	0.367	0.034
17.5	1.184	0.366	0.034
18.0	1.191	0.367	0.035
18.5	1.195	0.367	0.035
19.0	1.200	0.366	0.035
19.5	1.208	0.364	0.035

Molybdenum[16]

Energy (eV)	n	k	$R(\phi = 0)$
0.10	18.53	68.51	0.985
0.15	8.78	47.54	0.985
0.20	5.10	35.99	0.985
0.25	3.36	28.75	0.984
0.30	2.44	23.80	0.983
0.34	2.00	20.84	0.982
0.38	1.70	18.44	0.980
0.42	1.57	16.50	0.978
0.46	1.46	14.91	0.975
0.50	1.37	13.55	0.971
0.54	1.35	12.36	0.966
0.58	1.34	11.34	0.960
0.62	1.38	10.44	0.952
0.66	1.43	9.67	0.942
0.70	1.48	8.99	0.932
0.74	1.51	8.38	0.921
0.78	1.60	7.83	0.906
0.82	1.64	7.35	0.892
0.86	1.70	6.89	0.876
9.90	1.74	6.48	0.859
1.00	1.94	5.58	0.805
1.10	2.15	4.85	0.743
1.20	2.44	4.22	0.671
1.30	2.77	3.74	0.608
1.40	3.15	3.40	0.562
1.50	3.53	3.30	0.550
1.60	3.77	3.41	0.562
1.70	3.84	3.51	0.570
1.80	3.81	3.58	0.576
1.90	3.74	3.58	0.576
2.00	3.68	3.52	0.571
2.10	3.68	3.45	0.565
2.20	3.76	3.41	0.562
2.30	3.79	3.61	0.578
2.40	3.59	3.78	0.594
2.50	3.36	3.73	0.591
2.60	3.22	3.61	0.582
2.70	3.13	3.51	0.573
2.80	3.08	3.42	0.565
2.90	3.05	3.33	0.566
3.00	3.04	3.27	0.550
3.10	3.03	3.21	0.544
3.20	3.05	3.18	0.540
3.30	3.06	3.18	0.540
3.40	3.06	3.19	0.541
3.50	3.06	3.21	0.543
3.60	3.05	3.23	0.546
3.70	3.04	3.27	0.550
3.80	3.04	3.31	0.554
3.90	3.04	3.40	0.564
4.00	3.01	3.51	0.576
4.20	2.77	3.77	0.610
4.40	2.39	3.88	0.640
4.60	2.06	3.84	0.658
4.80	1.75	3.76	0.678
5.00	1.46	3.62	0.695
5.20	1.22	3.42	0.706
5.40	1.07	3.20	0.706
5.60	0.96	2.99	0.700
5.80	0.89	2.80	0.688
6.00	0.85	2.64	0.674
6.20	0.81	2.50	0.660
6.40	0.79	2.36	0.641
6.60	0.78	2.24	0.619
6.80	0.78	2.13	0.592
7.00	0.80	2.04	0.568
7.20	0.81	1.98	0.548
7.40	0.81	1.95	0.542
7.60	0.75	1.90	0.552
7.80	0.71	1.81	0.542
8.00	0.69	1.73	0.530
8.20	0.67	1.65	0.512
8.40	0.66	1.57	0.495
8.60	0.65	1.49	0.475
8.80	0.65	1.41	0.450
9.00	0.65	1.33	0.420
9.20	0.67	1.25	0.385
9.40	0.69	1.19	0.355
9.60	0.71	1.12	0.320
9.80	0.74	1.05	0.285
10.00	0.77	0.99	0.250
10.20	0.81	0.93	0.217
10.40	0.86	0.88	0.188
10.60	0.91	0.83	0.162
10.80	0.98	0.79	0.138
11.00	1.05	0.77	0.125
11.20	1.12	0.78	0.123
11.40	1.18	0.80	0.125
11.60	1.23	0.85	0.135
11.80	1.25	0.89	0.145
12.00	1.26	0.92	0.154
12.40	1.25	0.98	0.168
12.80	1.23	1.00	0.178
13.20	1.20	1.02	0.185
13.60	1.17	1.02	0.187
14.00	1.15	1.01	0.185
14.40	1.13	1.00	0.182
14.80	1.13	0.99	0.179
15.00	1.14	0.99	0.179
15.60	1.15	1.01	0.184
16.00	1.14	1.04	0.194
16.60	1.10	1.10	0.216
17.00	1.04	1.12	0.233
17.60	0.94	1.14	0.257
18.00	0.87	1.12	0.270
18.60	0.77	1.08	0.283
19.00	0.71	1.02	0.284
19.60	0.66	0.94	0.275
20.00	0.64	0.89	0.264
20.60	0.62	0.81	0.245
21.00	0.61	0.77	0.234
21.60	0.61	0.71	0.215
22.00	0.60	0.69	0.207
22.60	0.59	0.63	0.195
23.00	0.58	0.60	0.185
23.60	0.58	0.53	0.166
24.00	0.58	0.49	0.151
24.60	0.60	0.43	0.124
25.00	0.62	0.39	0.106
25.60	0.66	0.35	0.085
26.00	0.68	0.33	0.072
26.50	0.71	0.31	0.060
27.00	0.73	0.29	0.050
27.50	0.76	0.28	0.041
28.00	0.79	0.27	0.036
28.50	0.81	0.26	0.031
29.00	0.83	0.26	0.028
29.50	0.86	0.26	0.025
30.00	0.88	0.26	0.023
31.00	0.92	0.29	0.024
32.00	0.92	0.32	0.030
33.00	0.90	0.33	0.032
34.00	0.91	0.34	0.034
35.00	0.87	0.37	0.043

Energy (eV)	n	k	R(ϕ = 0)
36.00	0.82	0.34	0.043
37.00	0.81	0.30	0.038
38.00	0.81	0.27	0.033
39.00	0.82	0.25	0.029
40.00	0.83	0.23	0.025

Nickel[17]

Energy (eV)	n	k	R(ϕ = 0)
0.10	9.54	45.82	0.983
0.15	5.45	30.56	0.978
0.20	4.12	22.48	0.969
0.25	4.25	17.68	0.950
0.30	4.19	15.05	0.934
0.35	4.03	13.05	0.918
0.40	3.84	11.43	0.900
0.50	4.03	9.64	0.864
0.60	3.84	8.35	0.835
0.70	3.59	7.48	0.813
0.80	3.38	6.82	0.794
0.90	3.18	6.23	0.774
1.00	3.06	5.74	0.753
1.10	2.97	5.38	0.734
1.20	2.85	5.10	0.721
1.30	2.74	4.85	0.708
1.40	2.65	4.63	0.695
1.50	2.53	4.47	0.688
1.60	2.43	4.31	0.679
1.70	2.28	4.18	0.677
1.80	2.14	4.01	0.670
1.90	2.02	3.82	0.659
2.00	1.92	3.65	0.649
2.10	1.85	3.48	0.634
2.20	1.80	3.33	0.620
2.30	1.75	3.19	0.605
2.40	1.71	3.06	0.590
2.50	1.67	2.93	0.575
2.60	1.65	2.81	0.557
2.70	1.64	2.71	0.542
2.80	1.63	2.61	0.525
2.90	1.62	2.52	0.509
3.00	1.61	2.44	0.495
3.10	1.61	2.36	0.480
3.20	1.61	2.30	0.467
3.30	1.61	2.23	0.454
3.40	1.62	2.17	0.441
3.50	1.63	2.11	0.428
3.60	1.64	2.07	0.416
3.70	1.66	2.02	0.405
3.80	1.69	1.99	0.397
3.90	1.72	1.98	0.393
4.00	1.73	1.98	0.392
4.20	1.74	2.01	0.396
4.40	1.71	2.06	0.409
4.60	1.63	2.09	0.421
4.80	1.53	2.11	0.435
5.00	1.40	2.10	0.449
5.20	1.27	2.04	0.454
5.40	1.16	1.94	0.449
5.60	1.09	1.83	0.435
5.80	1.04	1.73	0.417
6.20	1.00	1.54	0.371
6.40	1.01	1.46	0.345
6.60	1.01	1.40	0.325
6.80	1.02	1.35	0.308
7.00	1.03	1.30	0.291
7.20	1.03	1.27	0.282
7.40	1.03	1.24	0.273
7.60	1.02	1.22	0.265
7.80	1.01	1.18	0.256
8.00	1.01	1.15	0.248
8.20	1.00	1.13	0.242
8.40	0.99	1.11	0.235
8.60	0.98	1.08	0.228
8.80	0.97	1.05	0.220
9.00	0.97	1.01	0.211
9.20	0.96	0.99	0.203
9.40	0.95	0.96	0.194
9.60	0.95	0.93	0.185
9.80	0.95	0.89	0.175
10.00	0.95	0.87	0.166
10.20	0.95	0.83	0.155
10.40	0.95	0.80	0.145
10.60	0.97	0.76	0.129
10.80	0.99	0.75	0.123
11.00	1.01	0.73	0.115
11.25	1.04	0.72	0.111
11.50	1.05	0.71	0.109
11.75	1.07	0.71	0.108
12.00	1.07	0.71	0.108
12.25	1.07	0.71	0.107
12.50	1.08	0.71	0.106
12.75	1.08	0.71	0.106
13.00	1.08	0.71	0.105
13.25	1.08	0.71	0.105
13.50	1.07	0.70	0.105
13.75	1.07	0.70	0.105
14.00	1.07	0.71	0.106
14.25	1.06	0.70	0.106
14.50	1.05	0.70	0.106
14.75	1.04	0.70	0.107
15.00	1.03	0.70	0.107
15.25	1.02	0.69	0.106
15.50	1.01	0.69	0.105
15.75	1.00	0.68	0.104
16.00	0.99	0.67	0.103
16.50	0.98	0.66	0.101
17.00	0.96	0.64	0.098
17.50	0.94	0.63	0.096
18.00	0.92	0.61	0.092
18.50	0.91	0.58	0.087
19.00	0.90	0.56	0.082
19.50	0.90	0.54	0.077
20.00	0.89	0.51	0.071
20.50	0.89	0.49	0.066
21.00	0.90	0.47	0.061
21.50	0.91	0.46	0.057
22.00	0.91	0.45	0.055
22.50	0.91	0.44	0.053
23.00	0.92	0.44	0.051
23.50	0.91	0.44	0.052
24.00	0.90	0.43	0.051
24.50	0.90	0.43	0.051
25.00	0.89	0.42	0.050
26.00	0.88	0.39	0.046
27.00	0.87	0.37	0.042
28.00	0.87	0.35	0.040
29.00	0.86	0.34	0.037
30.00	0.86	0.32	0.034
35.00	0.86	0.24	0.022
40.00	0.87	0.18	0.014
45.00	0.88	0.13	0.008
50.00	0.92	0.10	0.004
60.00	0.96	0.08	0.002
65.00	0.98	0.09	0.002
68.00	0.96	0.12	0.004
70.00	0.94	0.11	0.004
75.00	0.94	0.09	0.003
80.00	0.94	0.07	0.002
90.00	0.94	0.06	0.002

Niobium[18]

Energy (eV)	n	k	R(ϕ = 0)
0.12	15.99	53.20	0.979
0.20	7.25	34.14	0.976
0.24	5.47	28.88	0.975
0.28	4.26	24.95	0.974
0.35	3.11	20.03	0.970
0.45	2.28	15.58	0.964
0.55	1.83	12.67	0.956
0.65	1.57	10.59	0.947
0.75	1.41	9.00	0.935
0.85	1.35	7.74	0.918
0.95	1.35	6.70	0.893
1.05	1.44	5.86	0.857
1.15	1.55	5.18	0.814
1.25	1.65	4.63	0.768
1.35	1.76	4.13	0.715
1.45	1.95	3.68	0.650
1.55	2.15	3.37	0.595
1.65	2.36	3.13	0.552
1.75	2.54	2.99	0.527
1.85	2.69	2.89	0.510
1.95	2.82	2.86	0.505
2.05	2.89	2.87	0.505
2.15	2.92	2.87	0.505
2.25	2.93	2.87	0.505
2.35	2.92	2.88	0.506
2.45	2.89	2.90	0.509
2.55	2.83	2.92	0.512
2.65	2.74	2.90	0.511
2.75	2.66	2.86	0.507
2.85	2.58	2.80	0.500
3.00	2.51	2.68	0.485
3.10	2.48	2.60	0.475
3.20	2.45	2.53	0.465
3.30	2.44	2.45	0.453
3.40	2.46	2.38	0.442
3.50	2.48	2.33	0.435
3.60	2.52	2.29	0.428
3.70	2.56	2.27	0.426
3.80	2.59	2.28	0.427
3.90	2.62	2.29	0.429
4.00	2.64	2.33	0.434
4.20	2.64	2.42	0.447

Energy (eV)	n	k	R(ϕ = 0)
4.40	2.53	2.56	0.467
4.60	2.39	2.56	0.470
4.80	2.32	2.52	0.465
5.00	2.26	2.57	0.475
5.20	2.16	2.62	0.487
5.40	2.00	2.68	0.505
5.60	1.81	2.67	0.518
5.80	1.63	2.60	0.522
6.00	1.49	2.49	0.520
6.20	1.38	2.38	0.512
6.40	1.31	2.25	0.496
6.60	1.26	2.14	0.480
6.80	1.24	2.04	0.460
7.00	1.23	1.96	0.441
7.20	1.22	1.91	0.430
7.40	1.20	1.88	0.427
7.60	1.14	1.85	0.430
7.80	1.07	1.78	0.428
8.00	1.02	1.69	0.412
8.20	1.00	1.60	0.390
8.40	0.99	1.51	0.365
8.60	0.99	1.43	0.340
8.70	0.99	1.39	0.328
8.80	1.00	1.36	0.315
9.00	1.01	1.29	0.290
9.20	1.04	1.22	0.265
9.40	1.07	1.18	0.245
9.60	1.10	1.13	0.227
9.80	1.13	1.09	0.209
10.00	1.18	1.05	0.194
10.20	1.23	1.04	0.187
10.40	1.27	1.04	0.185
10.60	1.30	1.06	0.190
10.80	1.32	1.08	0.195
11.00	1.32	1.10	0.200
11.20	1.31	1.12	0.204
11.40	1.30	1.13	0.207
11.60	1.28	1.13	0.209
11.80	1.27	1.13	0.210
12.00	1.25	1.12	0.209
12.40	1.24	1.10	0.204
12.80	1.24	1.09	0.200
13.20	1.24	1.09	0.201
13.60	1.23	1.12	0.208
14.00	1.20	1.13	0.216
14.40	1.16	1.15	0.225
14.80	1.11	1.16	0.234
15.00	1.08	1.16	0.238
15.60	0.99	1.14	0.247
16.00	0.92	1.11	0.250
16.60	0.85	1.04	0.245
17.00	0.80	0.99	0.240
17.20	0.79	0.96	0.236
17.40	0.77	0.93	0.230
17.80	0.75	0.87	0.217
18.00	0.74	0.85	0.209
18.60	0.73	0.77	0.185
19.00	0.72	0.72	0.170
19.60	0.72	0.66	0.150
20.00	0.72	0.62	0.137

Energy (eV)	n	k	R(ϕ = 0)
20.60	0.71	0.55	0.119
21.00	0.72	0.50	0.100
21.60	0.75	0.43	0.075
22.00	0.78	0.40	0.063
22.60	0.82	0.35	0.045
23.00	0.85	0.33	0.038
23.60	0.88	0.30	0.029
24.00	0.91	0.29	0.025
24.60	0.94	0.28	0.022
25.00	0.96	0.27	0.020
25.60	0.99	0.26	0.018
26.00	1.00	0.26	0.017
26.60	1.03	0.25	0.016
27.00	1.04	0.25	0.015
27.60	1.06	0.25	0.015
28.00	1.08	0.24	0.015
28.60	1.11	0.24	0.016
29.00	1.13	0.25	0.017
29.60	1.16	0.26	0.020
30.00	1.18	0.28	0.023
31.00	1.18	0.31	0.026
32.00	1.20	0.34	0.031
33.00	1.21	0.38	0.038
34.00	1.20	0.42	0.044
35.20	1.17	0.47	0.051
36.00	1.15	0.50	0.056
37.50	1.07	0.53	0.064
39.50	0.95	0.50	0.063
40.50	0.92	0.47	0.059

Osmium (Polycrystalline)[9]

Energy (eV)	n	k	R(ϕ = 0)
0.10	4.08	50.23	0.994
0.15	2.90	33.60	0.990
0.20	2.44	25.11	0.985
0.25	2.35	19.99	0.977
0.30	2.23	16.54	0.969
0.35	2.33	14.06	0.955
0.40	2.45	12.32	0.940
0.45	2.43	11.02	0.927
0.50	2.41	9.97	0.913
0.55	2.33	9.12	0.901
0.60	2.21	8.37	0.890
0.65	2.11	7.68	0.877
0.70	2.02	7.04	0.862
0.75	2.00	6.46	0.842
0.80	2.00	5.95	0.820
0.85	2.01	5.51	0.796
0.90	2.03	5.10	0.769
0.95	2.05	4.74	0.742
1.00	2.09	4.41	0.712
1.10	2.15	3.84	0.651
1.20	2.16	3.35	0.592
1.30	2.25	2.77	0.506
1.40	2.49	2.23	0.419
1.50	2.84	1.80	0.369
1.60	3.36	1.62	0.379
1.70	3.70	1.75	0.411
1.80	3.78	1.83	0.423
1.90	3.81	1.75	0.418
2.00	3.98	1.60	0.418
2.10	4.26	1.54	0.432

Energy (eV)	n	k	R(ϕ = 0)
2.20	4.58	1.62	0.457
2.30	4.84	1.76	0.479
2.40	5.10	2.01	0.506
2.50	5.28	2.38	0.532
2.60	5.36	2.82	0.557
2.70	5.30	3.29	0.580
2.80	5.07	3.78	0.603
2.90	4.65	4.18	0.624
3.00	4.05	4.40	0.639
3.20	3.29	3.96	0.614
3.40	2.93	3.79	0.607
3.60	2.75	3.45	0.577
3.80	2.73	3.32	0.562
4.00	2.71	3.34	0.565
4.20	2.53	3.44	0.584
4.40	2.24	3.44	0.599
4.60	2.01	3.31	0.598
4.80	1.88	3.19	0.592
5.00	1.74	3.12	0.596
5.20	1.58	3.00	0.597
5.40	1.46	2.88	0.593
5.60	1.36	2.77	0.589
5.80	1.27	2.65	0.582
6.00	1.20	2.54	0.575
6.20	1.13	2.44	0.571
6.40	1.06	2.33	0.562
6.60	1.01	2.21	0.548
6.80	0.97	2.11	0.532
7.00	0.95	2.00	0.514
7.20	0.92	1.91	0.497
7.40	0.91	1.81	0.476
7.60	0.90	1.72	0.451
7.80	0.90	1.63	0.426
8.00	0.91	1.55	0.400
8.20	0.91	1.48	0.375
8.40	0.94	1.40	0.344
8.60	0.96	1.34	0.319
8.80	0.98	1.29	0.296
9.00	1.01	1.24	0.274
9.20	1.04	1.19	0.255
9.40	1.08	1.16	0.238
9.60	1.10	1.14	0.229
9.80	1.13	1.11	0.217
10.00	1.16	1.10	0.209
10.20	1.19	1.08	0.203
10.30	1.20	1.08	0.201
10.40	1.22	1.08	0.200
10.50	1.23	1.09	0.201
10.60	1.24	1.10	0.203
10.80	1.25	1.11	0.206
11.00	1.24	1.13	0.213
11.20	1.23	1.14	0.217
11.40	1.19	1.15	0.223
11.60	1.17	1.12	0.216
11.80	1.16	1.10	0.211
12.00	1.15	1.08	0.205
12.40	1.14	1.03	0.191
12.80	1.15	1.01	0.183
13.20	1.16	0.98	0.174
13.60	1.17	0.97	0.170

Energy (eV)	n	k	R(ϕ = 0)
14.00	1.17	0.96	0.169
14.40	1.16	0.94	0.165
14.80	1.16	0.91	0.156
15.20	1.17	0.89	0.148
15.60	1.20	0.86	0.140
16.00	1.25	0.87	0.140
16.40	1.28	0.90	0.147
16.80	1.28	0.94	0.157
17.20	1.27	0.97	0.167
17.60	1.26	1.01	0.178
18.00	1.23	1.04	0.189
18.40	1.19	1.08	0.200
18.80	1.14	1.10	0.210
19.20	1.10	1.10	0.219
19.60	1.05	1.11	0.227
20.00	0.96	1.10	0.239
20.40	0.93	1.09	0.240
20.80	0.89	1.05	0.240
21.20	0.86	1.02	0.237
21.60	0.83	0.99	0.235
22.00	0.80	0.96	0.230
22.40	0.78	0.93	0.226
22.80	0.77	0.90	0.220
23.20	0.75	0.88	0.217
23.60	0.75	0.86	0.211
24.00	0.73	0.84	0.209
24.40	0.72	0.82	0.207
24.80	0.70	0.80	0.205
25.20	0.69	0.77	0.202
25.60	0.67	0.75	0.199
26.00	0.66	0.72	0.195
26.40	0.65	0.69	0.189
26.80	0.63	0.66	0.183
27.20	0.65	0.62	0.165
28.00	0.64	0.59	0.156
28.40	0.64	0.57	0.148
28.80	0.65	0.55	0.140
29.20	0.65	0.53	0.134
29.60	0.65	0.51	0.128
30.00	0.65	0.49	0.121
31.00	0.65	0.45	0.111
32.00	0.66	0.41	0.095
33.00	0.68	0.37	0.079
34.00	0.70	0.34	0.068
35.00	0.72	0.31	0.057
36.00	0.74	0.29	0.048
37.00	0.77	0.27	0.040
38.00	0.79	0.26	0.035
39.00	0.81	0.26	0.031
40.00	0.84	0.26	0.026

Palladium[19]

Energy (eV)	n	k	R(ϕ = 0)
0.10	4.13	54.15	0.994
0.15	3.13	35.82	0.990
0.20	3.07	26.59	0.983
0.26	3.11	20.15	0.971
0.30	3.56	17.27	0.955
0.36	3.98	14.41	0.932
0.40	4.27	13.27	0.916
0.46	4.27	12.11	0.902
0.50	4.10	11.44	0.896
0.56	3.92	10.49	0.883
0.60	3.80	9.96	0.876
0.72	3.51	8.70	0.854
0.80	3.35	8.06	0.840
1.00	2.99	6.89	0.811
1.10	2.81	6.46	0.800
1.20	2.65	6.10	0.790
1.30	2.50	5.78	0.781
1.40	2.34	5.50	0.774
1.50	2.17	5.22	0.767
1.60	2.08	4.95	0.755
1.70	2.00	4.72	0.745
1.80	1.92	4.54	0.737
1.90	1.82	4.35	0.729
2.00	1.75	4.18	0.721
2.10	1.67	4.03	0.714
2.20	1.60	3.88	0.707
2.30	1.53	3.75	0.700
2.40	1.47	3.61	0.693
2.50	1.41	3.48	0.685
2.60	1.37	3.36	0.676
2.70	1.32	3.25	0.668
2.80	1.29	3.13	0.658
2.90	1.26	3.03	0.648
3.00	1.23	2.94	0.639
3.10	1.20	2.85	0.630
3.20	1.17	2.77	0.622
3.30	1.14	2.68	0.613
3.40	1.12	2.60	0.602
3.50	1.10	2.52	0.591
3.60	1.08	2.45	0.581
3.70	1.07	2.38	0.570
3.80	1.06	2.31	0.558
3.90	1.05	2.25	0.547
4.00	1.03	2.19	0.537
4.20	1.04	2.09	0.510
4.40	1.03	2.01	0.493
4.60	1.03	1.94	0.476
4.80	1.01	1.90	0.470
5.00	0.96	1.86	0.472
5.20	0.90	1.79	0.474
5.40	0.85	1.70	0.463
5.60	0.81	1.62	0.449
5.80	0.78	1.54	0.437
6.00	0.76	1.45	0.418
6.20	0.74	1.37	0.397
6.40	0.73	1.29	0.375
6.60	0.72	1.21	0.350
6.80	0.73	1.13	0.316
7.00	0.73	1.05	0.287
7.20	0.75	0.98	0.255
7.40	0.77	0.91	0.223
7.60	0.79	0.85	0.195
7.80	0.83	0.78	0.163
8.00	0.88	0.73	0.133
8.20	0.94	0.70	0.117
8.40	0.96	0.70	0.114
8.60	1.00	0.65	0.097
8.80	1.04	0.65	0.094
9.00	1.07	0.64	0.090
9.50	1.12	0.65	0.089
10.00	1.14	0.65	0.088
10.50	1.16	0.65	0.087
11.00	1.18	0.64	0.086
11.50	1.19	0.65	0.087
12.00	1.20	0.66	0.089
12.50	1.19	0.67	0.091
13.00	1.18	0.67	0.091
13.50	1.18	0.67	0.092
14.00	1.17	0.67	0.093
14.50	1.15	0.68	0.095
15.00	1.13	0.69	0.098
15.50	1.10	0.68	0.096
16.00	1.08	0.66	0.092
16.50	1.06	0.63	0.086
17.00	1.07	0.61	0.081
17.50	1.06	0.61	0.080
18.00	1.07	0.59	0.077
18.50	1.07	0.59	0.077
19.00	1.08	0.59	0.077
19.50	1.08	0.61	0.080
20.00	1.07	0.65	0.090
20.50	1.03	0.67	0.098
21.00	0.99	0.67	0.103
21.50	0.95	0.66	0.103
22.00	0.91	0.64	0.103
22.50	0.88	0.62	0.101
23.00	0.86	0.59	0.097
23.50	0.85	0.56	0.091
24.00	0.84	0.54	0.086
25.00	0.81	0.51	0.084
26.40	0.80	0.43	0.066
27.80	0.81	0.38	0.052
29.20	0.82	0.35	0.046

Platinum[20]

Energy (eV)	n	k	R(ϕ = 0)
0.10	13.21	44.72	0.976
0.15	8.18	31.16	0.969
0.20	5.90	23.95	0.962
0.25	4.70	19.40	0.954
0.30	3.92	16.16	0.945
0.35	3.28	13.66	0.936
0.40	2.81	11.38	0.922
0.45	3.03	9.31	0.882
0.50	3.91	7.71	0.813
0.55	4.58	7.14	0.777
0.60	5.13	6.75	0.753
0.65	5.52	6.66	0.746
0.70	5.71	6.83	0.751
0.75	5.57	7.02	0.759
0.80	5.31	7.04	0.762
0.85	5.05	6.98	0.763
0.90	4.77	6.91	0.765
0.95	4.50	6.77	0.763
1.00	4.25	6.62	0.762
1.10	3.86	6.24	0.753
1.20	3.55	5.92	0.746
1.30	3.29	5.61	0.736
1.40	3.10	5.32	0.725
1.50	2.92	5.07	0.716
1.60	2.76	4.84	0.706

Energy (eV)	n	k	R(φ = 0)
1.70	2.63	4.64	0.697
1.80	2.51	4.43	0.686
1.90	2.38	4.26	0.678
2.00	2.30	4.07	0.664
2.10	2.23	3.92	0.654
2.20	2.17	3.77	0.642
2.30	2.10	3.67	0.636
2.40	2.03	3.54	0.626
2.50	1.96	3.42	0.616
2.60	1.91	3.30	0.605
2.70	1.87	3.20	0.595
2.80	1.83	3.10	0.585
2.90	1.79	3.01	0.575
3.00	1.75	2.92	0.565
3.20	1.68	2.76	0.546
3.40	1.63	2.62	0.527
3.60	1.58	2.48	0.507
3.80	1.53	2.37	0.491
4.00	1.49	2.25	0.472
4.20	1.45	2.14	0.452
4.40	1.43	2.04	0.432
4.60	1.39	1.95	0.415
4.80	1.38	1.85	0.392
5.00	1.36	1.76	0.372
5.20	1.36	1.67	0.350
5.40	1.36	1.61	0.332
5.60	1.36	1.54	0.315
5.80	1.36	1.47	0.295
6.00	1.38	1.40	0.276
6.20	1.39	1.35	0.261
6.40	1.42	1.29	0.246
6.60	1.45	1.26	0.236
6.80	1.48	1.24	0.231
7.00	1.50	1.24	0.230
7.20	1.50	1.25	0.231
7.40	1.49	1.23	0.228
7.60	1.48	1.22	0.225
7.80	1.48	1.20	0.221
8.00	1.47	1.18	0.216
8.20	1.47	1.17	0.212
8.40	1.47	1.15	0.209
8.60	1.47	1.14	0.205
8.80	1.47	1.13	0.202
9.00	1.48	1.12	0.200
9.20	1.49	1.11	0.198
9.40	1.49	1.12	0.200
9.60	1.49	1.13	0.203
9.80	1.48	1.15	0.207
10.00	1.46	1.15	0.209
10.20	1.43	1.16	0.211
10.40	1.40	1.15	0.210
10.60	1.37	1.14	0.207
10.80	1.35	1.12	0.203
11.00	1.33	1.10	0.199
11.20	1.31	1.08	0.194
11.40	1.30	1.06	0.188
11.60	1.29	1.04	0.183
11.80	1.29	1.01	0.177
12.00	1.29	1.00	0.173
12.40	1.29	0.97	0.165
12.80	1.29	0.94	0.158
13.20	1.31	0.93	0.155
13.60	1.31	0.93	0.155
14.00	1.31	0.93	0.155
14.40	1.30	0.93	0.156
14.80	1.27	0.93	0.157
15.20	1.27	0.93	0.155
15.60	1.25	0.92	0.151
16.00	1.24	0.89	0.146
16.50	1.24	0.87	0.142
17.00	1.25	0.86	0.138
17.50	1.27	0.85	0.135
18.00	1.31	0.88	0.142
18.50	1.30	0.94	0.157
19.00	1.28	0.99	0.171
19.50	1.23	1.03	0.184
20.00	1.18	1.06	0.197
20.50	1.11	1.09	0.212
21.00	1.03	1.10	0.226
21.50	0.94	1.08	0.238
22.00	0.87	1.04	0.240
22.50	0.81	0.98	0.235
23.00	0.77	0.92	0.226
23.50	0.75	0.87	0.213
24.00	0.74	0.82	0.201
24.50	0.73	0.77	0.187
25.00	0.73	0.73	0.174
25.50	0.73	0.70	0.162
26.00	0.74	0.67	0.150
26.50	0.74	0.65	0.142
27.00	0.74	0.63	0.136
27.50	0.74	0.62	0.130
28.00	0.75	0.60	0.125
28.50	0.75	0.59	0.121
29.00	0.75	0.58	0.118
29.50	0.74	0.58	0.120
30.00	0.73	0.58	0.124

***Potassium*[21]**

Energy (eV)	n	k	R(φ = 0)
0.55	0.139	7.10	0.989
0.58	0.119	6.72	0.990
0.63	0.106	6.32	0.990
0.67	0.091	5.79	0.990
0.73	0.079	5.30	0.989
0.81	0.066	4.75	0.989
0.92	0.056	4.19	0.988
1.05	0.044	3.58	0.987
1.23	0.040	3.04	0.985
1.44	0.040	2.56	0.979
1.65	0.044	2.19	0.970
1.87	0.050	1.84	0.955
2.07	0.053	1.62	0.943
2.27	0.049	1.43	0.938
2.45	0.046	1.28	0.933
2.64	0.043	1.14	0.928
2.82	0.043	1.02	0.919
2.95	0.041	0.898	0.913
3.06	0.041	0.799	0.905
3.40	0.052	0.549	0.852
3.71	0.089	0.288	0.719
3.97	0.287	0.091	0.310
4.00	0.34	0.08	0.245
4.065	0.38	0.07	0.204
4.133	0.41	0.07	0.177
4.203	0.45	0.06	0.145
4.275	0.48	0.06	0.125
4.350	0.52	0.05	0.101
4.428	0.55	0.05	0.085
4.509	0.58	0.05	0.072
4.592	0.61	0.05	0.060
4.679	0.64	0.04	0.049
4.769	0.66	0.04	0.043
4.862	0.68	0.04	0.037
4.959	0.70	0.04	0.032
5.061	0.72	0.04	0.027
5.166	0.74	0.04	0.023
5.276	0.76	0.04	0.019
5.391	0.78	0.04	0.016
5.510	0.79	0.05	0.015
5.637	0.81	0.05	0.012
5.767	0.83	0.05	0.009
6.048	0.85	0.05	0.007
6.199	0.87	0.05	0.006
6.358	0.88	0.05	0.005
6.526	0.90	0.06	0.004
6.702	0.91	0.06	0.003
6.888	0.92	0.06	0.003
7.085	0.92	0.06	0.003
7.293	0.93	0.06	0.002
7.514	0.93	0.06	0.002
7.749	0.94	0.06	0.002
7.999	0.94	0.06	0.002
8.260	0.94	0.06	0.002
8.551	0.94	0.06	0.002
8.856	0.94	0.05	0.002
9.184	0.94	0.05	0.002
9.537	0.94	0.04	0.001
9.919	0.94	0.04	0.001
10.33	0.94	0.03	0.001
11.0		0.03	
12.0		0.028	

***Rhenium, single crystal,* $\vec{E} \parallel \hat{c}$[9]**

Energy (eV)	n	k	R(φ = 0)
0.10	6.06	51.03	0.991
0.15	4.66	33.96	0.984
0.20	4.16	25.36	0.975
0.25	4.03	20.10	0.962
0.30	4.37	16.69	0.943
0.35	4.50	14.53	0.925
0.40	4.53	12.96	0.909
0.45	4.53	11.78	0.893
0.50	4.53	10.88	0.878
0.55	4.50	10.26	0.867
0.60	4.29	9.75	0.861
0.65	4.07	9.35	0.856
0.70	3.80	8.94	0.853
0.75	3.48	8.55	0.850
0.80	3.21	8.10	0.846
0.85	2.96	7.68	0.841
0.90	2.73	7.24	0.835
0.95	2.56	6.79	0.826
1.00	2.45	6.36	0.813

Energy (eV)	n	k	$R(\phi = 0)$
1.10	2.38	5.61	0.778
1.20	2.35	5.02	0.742
1.30	2.39	4.54	0.702
1.40	2.44	4.13	0.662
1.50	2.50	3.79	0.624
1.60	2.59	3.49	0.587
1.70	2.70	3.27	0.557
1.80	2.82	3.10	0.535
1.90	2.90	3.00	0.520
2.00	2.97	2.91	0.510
2.10	3.03	2.86	0.504
2.20	3.06	2.84	0.501
2.30	3.07	2.82	0.499
2.40	3.06	2.81	0.498
2.50	3.02	2.80	0.497
2.60	2.96	2.77	0.493
2.70	2.89	2.68	0.482
2.80	2.89	2.57	0.468
2.90	2.99	2.47	0.457
3.00	3.11	2.57	0.470
3.20	2.90	2.68	0.482
3.40	2.83	2.50	0.459
3.60	2.93	2.48	0.457
3.80	2.86	2.56	0.467
4.00	2.81	2.51	0.460
4.20	2.86	2.55	0.466
4.40	2.81	2.74	0.489
4.60	2.56	2.83	0.504
4.80	2.41	2.71	0.493
5.00	2.39	2.68	0.488
5.20	2.34	2.75	0.500
5.40	2.20	2.81	0.515
5.60	2.02	2.84	0.530
5.80	1.83	2.80	0.538
6.00	1.65	2.71	0.541
6.20	1.54	2.59	0.532
6.40	1.45	2.50	0.526
6.80	1.32	2.31	0.508
7.00	1.26	2.23	0.500
7.20	1.20	2.15	0.493
7.40	1.16	2.06	0.480
7.60	1.12	1.99	0.470
7.80	1.08	1.89	0.454
8.00	1.05	1.80	0.435
8.20	1.05	1.71	0.411
8.40	1.05	1.62	0.386
8.60	1.06	1.55	0.360
8.80	1.09	1.48	0.336
9.00	1.11	1.43	0.317
9.20	1.13	1.39	0.301
9.40	1.16	1.34	0.281
9.60	1.18	1.32	0.274
9.80	1.20	1.29	0.264
10.00	1.23	1.26	0.252
10.20	1.25	1.25	0.246
10.40	1.28	1.25	0.242
10.60	1.29	1.25	0.242
10.80	1.30	1.26	0.244
11.00	1.30	1.27	0.247
11.20	1.29	1.28	0.249

Energy (eV)	n	k	$R(\phi = 0)$
11.40	1.28	1.28	0.252
11.60	1.26	1.28	0.252
11.80	1.24	1.26	0.249
12.00	1.23	1.24	0.244
12.40	1.22	1.21	0.237
12.80	1.21	1.18	0.230
13.20	1.22	1.16	0.222
13.60	1.22	1.13	0.215
14.00	1.24	1.12	0.209
14.40	1.27	1.11	0.204
14.80	1.29	1.15	0.213
15.20	1.29	1.19	0.225
15.60	1.26	1.22	0.236
16.00	1.23	1.25	0.248
16.40	1.19	1.27	0.259
16.80	1.14	1.29	0.269
17.00	1.12	1.30	0.275
17.40	1.07	1.30	0.286
18.00	0.99	1.30	0.300
18.40	0.93	1.29	0.311
18.80	0.87	1.28	0.321
19.20	0.81	1.25	0.330
19.60	0.77	1.21	0.332
20.00	0.73	1.18	0.333
20.40	0.70	1.14	0.332
20.80	0.67	1.11	0.332
21.20	0.64	1.08	0.334
21.60	0.61	1.04	0.335
22.00	0.58	1.01	0.340
22.40	0.55	0.97	0.341
22.80	0.53	0.93	0.338
23.20	0.51	0.89	0.334
23.60	0.50	0.85	0.329
24.00	0.48	0.80	0.319
24.40	0.48	0.76	0.207
24.80	0.47	0.72	0.296
25.20	0.47	0.68	0.282
25.60	0.47	0.65	0.270
26.00	0.47	0.61	0.255
26.40	0.48	0.57	0.240
26.80	0.48	0.54	0.225
27.20	0.49	0.51	0.208
27.60	0.50	0.48	0.193
28.00	0.51	0.45	0.176
29.00	0.54	0.39	0.145
30.00	0.57	0.33	0.114
31.00	0.62	0.29	0.086
32.00	0.66	0.26	0.065
33.00	0.68	0.24	0.054
34.00	0.72	0.21	0.041
35.00	0.76	0.20	0.031
36.00	0.79	0.20	0.025
37.00	0.82	0.19	0.021
38.00	0.85	0.20	0.018
39.00	0.89	0.21	0.016
40.00	0.88	0.26	0.022
42.00	0.88	0.26	0.022
44.00	0.89	0.29	0.026
46.00	0.85	0.32	0.035
48.00	0.82	0.30	0.036

Energy (eV)	n	k	$R(\phi = 0)$
50.00	0.80	0.30	0.038
52.00	0.78	0.30	0.044
54.00	0.72	0.30	0.055
56.00	0.66	0.24	0.061
58.00	0.65	0.16	0.055

Rhenium, single crystal, $\vec{E} \perp \vec{c}$ [9]

Energy (eV)	n	k	$R(\phi = 0)$
0.10	4.25	42.83	0.991
0.15	3.28	28.08	0.984
0.20	3.28	20.66	0.971
0.25	3.47	16.27	0.951
0.30	3.73	13.44	0.926
0.35	3.93	11.54	0.900
0.40	3.99	10.15	0.875
0.45	4.17	9.03	0.846
0.50	4.34	8.26	0.821
0.55	4.45	7.73	0.801
0.60	4.53	7.40	0.788
0.65	4.44	7.26	0.784
0.70	4.13	7.09	0.784
0.75	3.77	6.75	0.779
0.80	3.55	6.32	0.766
0.85	3.39	5.95	0.752
0.90	3.26	5.61	0.737
0.95	3.17	5.27	0.719
1.00	3.09	4.96	0.701
1.10	3.05	4.39	0.658
1.20	3.08	3.89	0.613
1.30	3.20	3.56	0.578
1.40	3.23	3.38	0.559
1.50	3.23	3.12	0.532
1.60	3.29	2.88	0.507
1.70	3.38	2.72	0.491
1.80	3.47	2.59	0.480
1.90	3.54	2.50	0.473
2.00	3.63	2.43	0.469
2.10	3.74	2.40	0.470
2.20	3.83	2.38	0.472
2.30	3.93	2.44	0.481
2.40	4.00	2.55	0.492
2.50	4.01	2.70	0.505
2.60	3.90	2.84	0.514
2.70	3.74	2.92	0.517
2.80	3.57	2.88	0.511
2.90	3.49	2.75	0.497
3.00	3.53	2.71	0.493
3.20	3.55	2.84	0.506
3.40	3.34	2.88	0.508
3.60	3.25	2.83	0.501
3.80	3.24	2.84	0.502
4.00	3.19	2.94	0.513
4.20	3.05	3.06	0.526
4.40	2.88	3.15	0.539
4.60	2.67	3.18	0.548
4.80	2.44	3.17	0.554
5.00	2.25	3.12	0.556
5.20	2.10	3.04	0.555
5.40	1.96	2.96	0.553
5.60	1.84	2.88	0.551
5.80	1.73	2.81	0.549
6.00	1.61	2.74	0.549

Energy (eV)	n	k	R(φ = 0)
6.20	1.51	2.64	0.545
6.40	1.42	2.56	0.541
6.80	1.28	2.37	0.526
7.00	1.22	2.28	0.517
7.20	1.16	2.19	0.508
7.40	1.12	2.08	0.493
7.60	1.12	1.98	0.468
7.80	1.08	1.93	0.463
8.00	1.05	1.83	0.443
8.20	1.05	1.74	0.418
8.40	1.05	1.66	0.397
8.60	1.06	1.58	0.372
8.80	1.07	1.52	0.351
9.00	1.09	1.46	0.327
9.20	1.11	1.41	0.309
9.40	1.14	1.36	0.290
9.60	1.17	1.31	0.273
9.80	1.20	1.27	0.258
10.00	1.24	1.24	0.244
10.20	1.29	1.22	0.234
10.40	1.33	1.23	0.233
10.60	1.36	1.25	0.238
10.80	1.38	1.28	0.245
11.00	1.37	1.31	0.253
11.20	1.36	1.33	0.259
11.40	1.33	1.34	0.264
11.60	1.31	1.34	0.266
11.80	1.28	1.33	0.266
12.00	1.26	1.32	0.264
12.40	1.23	1.29	0.257
12.80	1.22	1.26	0.251
13.20	1.20	1.23	0.245
13.60	1.19	1.20	0.236
14.00	1.20	1.16	0.225
14.40	1.22	1.13	0.214
14.80	1.27	1.12	0.207
15.20	1.31	1.17	0.218
15.60	1.31	1.23	0.234
16.00	1.28	1.28	0.251
16.40	1.24	1.33	0.270
16.80	1.17	1.37	0.288
17.00	1.14	1.38	0.297
17.40	1.06	1.39	0.314
18.00	0.95	1.38	0.334
18.40	0.88	1.36	0.346
18.80	0.82	1.33	0.355
19.20	0.76	1.29	0.360
19.60	0.72	1.25	0.363
20.00	0.67	1.21	0.369
20.40	0.64	1.15	0.364
20.80	0.61	1.10	0.357
21.20	0.60	1.06	0.349
21.60	0.58	1.02	0.342
22.00	0.57	0.98	0.336
22.40	0.56	0.95	0.328
22.80	0.55	0.92	0.325
23.20	0.53	0.89	0.322
23.60	0.52	0.85	0.317
24.00	0.50	0.82	0.314
24.40	0.49	0.79	0.309
24.80	0.48	0.75	0.303
25.20	0.47	0.72	0.295
25.60	0.47	0.68	0.286
26.00	0.46	0.64	0.276
26.40	0.46	0.61	0.263
26.80	0.46	0.57	0.249
27.20	0.47	0.53	0.231
27.60	0.48	0.50	0.216
28.00	0.49	0.47	0.198
29.00	0.51	0.41	0.164
30.00	0.55	0.34	0.129
31.00	0.59	0.29	0.097
32.00	0.64	0.26	0.072
33.00	0.67	0.24	0.060
34.00	0.70	0.22	0.047
35.00	0.74	0.20	0.036
36.00	0.77	0.19	0.029
37.00	0.80	0.19	0.023
38.00	0.84	0.19	0.018
39.00	0.88	0.21	0.016
40.00	0.87	0.25	0.023
42.00	0.87	0.25	0.023
44.00	0.88	0.28	0.026
46.00	0.84	0.31	0.035
48.00	0.82	0.30	0.036
50.00	0.80	0.30	0.039
52.00	0.77	0.30	0.044
54.00	0.71	0.29	0.055
56.00	0.66	0.23	0.061
58.00	0.64	0.16	0.055

***Rhodium*[11]**

Energy (eV)	n	k	R(φ = 0)
0.10	18.48	69.43	0.986
0.20	8.66	37.46	0.977
0.30	5.85	25.94	0.967
0.40	4.74	19.80	0.955
0.50	4.20	16.07	0.941
0.60	3.87	13.51	0.925
0.70	3.67	11.72	0.908
0.80	3.63	10.34	0.887
0.90	3.62	9.36	0.867
1.00	3.71	8.67	0.848
1.10	3.67	8.26	0.837
1.20	3.51	7.94	0.832
1.30	3.26	7.63	0.829
1.40	3.01	7.31	0.827
1.50	2.78	6.97	0.823
1.60	2.60	6.64	0.818
1.70	2.42	6.33	0.813
1.80	2.30	6.02	0.805
1.90	2.20	5.76	0.798
2.00	2.12	5.51	0.789
2.10	2.05	5.30	0.780
2.20	2.00	5.11	0.772
2.30	1.94	4.94	0.765
2.40	1.90	4.78	0.756
2.50	1.88	4.65	0.748
2.60	1.85	4.55	0.743
2.70	1.80	4.49	0.742
2.90	1.63	4.36	0.748
3.00	1.53	4.29	0.753
3.10	1.41	4.20	0.760
3.20	1.30	4.09	0.764
3.30	1.20	3.97	0.767
3.40	1.11	3.84	0.769
3.50	1.04	3.71	0.768
3.60	0.99	3.58	0.764
3.70	0.95	3.45	0.759
3.80	0.91	3.34	0.753
3.90	0.88	3.23	0.747
4.00	0.86	3.12	0.739
4.20	0.83	2.94	0.722
4.40	0.80	2.76	0.706
4.60	0.78	2.60	0.684
4.80	0.79	2.46	0.659
5.00	0.79	2.34	0.635
5.20	0.79	2.23	0.613
5.40	0.80	2.14	0.591
5.60	0.80	2.06	0.573
5.80	0.79	2.00	0.561
6.00	0.76	1.93	0.556
6.20	0.73	1.85	0.544
6.40	0.70	1.77	0.534
6.60	0.68	1.69	0.518
6.80	0.67	1.60	0.498
7.00	0.66	1.52	0.476
7.20	0.66	1.43	0.452
7.40	0.66	1.35	0.423
7.60	0.67	1.27	0.394
7.80	0.68	1.20	0.363
8.00	0.69	1.12	0.329
8.20	0.71	1.04	0.288
8.40	0.74	0.97	0.252
8.60	0.78	0.89	0.212
8.80	0.83	0.83	0.179
9.00	0.88	0.77	0.148
9.20	0.95	0.73	0.125
9.40	1.01	0.71	0.110
9.60	1.07	0.69	0.102
9.80	1.12	0.69	0.098
10.00	1.17	0.69	0.098
10.60	1.26	0.73	0.106
11.00	1.29	0.76	0.113
11.60	1.32	0.80	0.124
12.00	1.32	0.82	0.127
12.60	1.32	0.82	0.129
13.00	1.32	0.83	0.131
13.60	1.32	0.85	0.134
14.00	1.32	0.86	0.138
14.60	1.30	0.89	0.144
15.00	1.28	0.90	0.147
15.60	1.25	0.90	0.147
16.00	1.24	0.89	0.147
16.50	1.23	0.88	0.145
17.00	1.22	0.88	0.144
17.50	1.22	0.87	0.143
18.00	1.23	0.88	0.145
18.50	1.25	0.92	0.155
19.00	1.24	0.98	0.172
19.50	1.18	1.05	0.193
20.00	1.10	1.09	0.213

Energy (eV)	n	k	R(ϕ = 0)
20.50	1.00	1.09	0.230
21.00	0.91	1.05	0.234
21.50	0.86	1.00	0.228
22.00	0.83	0.95	0.219
22.50	0.81	0.92	0.214
23.00	0.79	0.90	0.213
23.50	0.75	0.87	0.214
24.00	0.73	0.84	0.210
24.50	0.70	0.81	0.208
25.00	0.69	0.77	0.202
25.50	0.67	0.74	0.195
26.00	0.66	0.70	0.188
26.50	0.65	0.66	0.176
27.00	0.65	0.64	0.168
27.50	0.65	0.61	0.159
28.00	0.65	0.59	0.152
29.00	0.65	0.54	0.137
30.00	0.66	0.51	0.127
31.00	0.64	0.49	0.127
32.00	0.61	0.44	0.126
33.00	0.60	0.37	0.110
34.00	0.65	0.30	0.074
35.00	0.69	0.28	0.058
36.00	0.73	0.27	0.049
37.00	0.74	0.28	0.047
38.00	0.74	0.27	0.045
39.00	0.75	0.25	0.041

Ruthenium, single crystal, $\vec{E} \parallel \vec{c}$[9]

Energy (eV)	n	k	R(ϕ = 0)
0.10	11.50	51.38	0.984
0.20	5.93	27.14	0.970
0.30	4.33	18.50	0.953
0.40	3.60	13.97	0.933
0.50	3.18	11.04	0.909
0.60	3.28	8.89	0.865
0.70	3.62	7.73	0.822
0.80	3.42	7.02	0.801
0.90	3.25	6.12	0.766
1.00	3.39	5.33	0.715
1.10	3.66	4.83	0.675
1.20	3.84	4.57	0.654
1.30	3.94	4.38	0.638
1.40	4.02	4.19	0.624
1.50	4.16	4.07	0.614
1.60	4.33	4.08	0.615
1.70	4.42	4.21	0.624
1.80	4.40	4.38	0.636
1.90	4.29	4.61	0.651
2.00	4.04	4.81	0.667
2.10	3.69	4.90	0.679
2.20	3.35	4.82	0.683
2.30	3.09	4.70	0.681
2.40	2.89	4.55	0.677
2.50	2.74	4.40	0.671
2.60	2.64	4.25	0.663
2.70	2.58	4.14	0.656
2.80	2.54	4.05	0.650
2.90	2.48	4.03	0.650
3.00	2.38	4.03	0.656
3.10	2.26	4.00	0.661
3.20	2.13	3.96	0.666
3.30	2.00	3.91	0.671
3.40	1.87	3.83	0.673
3.50	1.76	3.74	0.674
3.60	1.66	3.65	0.675
3.70	1.57	3.55	0.673
3.80	1.49	3.45	0.672
3.90	1.42	3.35	0.668
4.00	1.37	3.24	0.661
4.20	1.29	3.08	0.649
4.40	1.22	2.93	0.639
4.60	1.16	2.79	0.628
4.80	1.11	2.67	0.617
5.00	1.06	2.56	0.607
5.20	1.01	2.46	0.600
5.40	0.95	2.35	0.593
5.60	0.92	2.23	0.576
5.80	0.90	2.14	0.559
6.00	0.88	2.05	0.545
6.20	0.87	1.98	0.531
6.40	0.84	1.91	0.521
6.60	0.82	1.84	0.510
6.80	0.79	1.77	0.500
7.00	0.76	1.69	0.489
7.20	0.75	1.61	0.472
7.40	0.73	1.54	0.455
7.60	0.73	1.46	0.433
7.80	0.73	1.39	0.411
8.00	0.72	1.33	0.391
8.20	0.72	1.26	0.366
8.40	0.73	1.20	0.342
8.60	0.74	1.14	0.318
8.80	0.74	1.08	0.295
9.00	0.75	1.02	0.267
9.20	0.77	0.97	0.243
9.40	0.79	0.91	0.217
9.60	0.82	0.86	0.190
9.80	0.85	0.81	0.167
10.00	0.88	0.76	0.144
10.20	0.92	0.72	0.125
10.40	0.96	0.69	0.110
10.60	1.01	0.67	0.100
10.80	1.05	0.66	0.094
11.00	1.09	0.65	0.090
11.20	1.12	0.65	0.088
11.40	1.15	0.65	0.087
11.60	1.18	0.65	0.088
11.80	1.21	0.66	0.090
12.00	1.23	0.67	0.092
12.40	1.26	0.69	0.098
12.80	1.27	0.72	0.104
13.20	1.28	0.74	0.108
13.60	1.28	0.75	0.111
14.00	1.28	0.76	0.114
14.40	1.27	0.76	0.114
14.80	1.27	0.76	0.114
15.00	1.27	0.76	0.114
15.60	1.28	0.77	0.115
16.00	1.30	0.78	0.118
16.50	1.32	0.80	0.123
17.00	1.34	0.85	0.136
17.50	1.32	0.93	0.155
18.00	1.26	0.99	0.173
18.50	1.18	1.02	0.185
19.00	1.11	1.02	0.192
19.50	1.05	1.02	0.199
20.00	0.99	1.02	0.208
20.50	0.92	0.99	0.212
21.00	0.86	0.94	0.209
21.50	0.83	0.90	0.203
22.00	0.81	0.86	0.193
23.00	0.77	0.79	0.182
24.00	0.74	0.74	0.171
25.00	0.71	0.69	0.163
26.00	0.68	0.63	0.154
27.00	0.67	0.57	0.140
28.00	0.66	0.51	0.124
29.00	0.67	0.46	0.107
30.00	0.67	0.43	0.097
31.00	0.67	0.37	0.084
32.00	0.69	0.33	0.070
33.00	0.71	0.30	0.058
34.00	0.73	0.27	0.048
35.00	0.75	0.25	0.039
36.00	0.77	0.24	0.035
37.00	0.79	0.23	0.039
38.00	0.80	0.22	0.027
39.00	0.82	0.22	0.024
40.00	0.83	0.22	0.022

Ruthenium, single crystal, $\vec{E} \perp \vec{c}$[5]

Energy (eV)	n	k	R(ϕ = 0)
0.10	11.85	50.81	0.983
0.20	6.68	27.18	0.966
0.30	4.94	18.92	0.950
0.40	3.90	14.51	0.933
0.50	3.27	11.63	0.915
0.60	2.98	9.54	0.888
0.70	2.82	7.99	0.856
0.80	2.73	6.71	0.815
0.90	2.82	5.54	0.751
1.00	3.17	4.59	0.670
1.10	3.69	3.91	0.604
1.20	4.28	3.66	0.585
1.30	4.66	3.72	0.593
1.40	4.86	3.79	0.601
1.50	4.99	3.89	0.609
1.60	5.08	4.03	0.618
1.70	5.12	4.22	0.629
1.80	5.10	4.45	0.642
1.90	4.96	4.78	0.660
2.00	4.61	5.06	0.677
2.10	4.21	5.09	0.682
2.20	3.94	5.00	0.681
2.30	3.69	4.97	0.684
2.40	3.44	4.88	0.684
2.50	3.27	4.77	0.681
2.60	3.14	4.66	0.677
2.70	3.06	4.59	0.674
2.80	2.99	4.59	0.676
2.90	2.87	4.64	0.686
3.00	2.64	4.69	0.701
3.10	2.40	4.64	0.710

Energy (eV)	n	k	$R(\phi = 0)$
3.20	2.18	4.55	0.717
3.30	2.00	4.43	0.721
3.40	1.84	4.30	0.723
3.50	1.71	4.16	0.723
3.60	1.60	4.03	0.722
3.70	1.50	3.90	0.721
3.80	1.41	3.77	0.718
3.90	1.35	3.64	0.713
4.00	1.29	3.53	0.707
4.20	1.21	3.31	0.694
4.40	1.16	3.13	0.679
4.60	1.13	2.97	0.662
4.80	1.09	2.86	0.652
5.00	1.03	2.75	0.648
5.20	0.97	2.64	0.643
5.40	0.91	2.52	0.635
5.60	0.88	2.40	0.622
5.80	0.86	2.29	0.605
6.00	0.84	2.20	0.591
6.20	0.82	2.11	0.576
6.40	0.81	2.04	0.564
6.60	0.78	1.97	0.556
6.80	0.76	1.89	0.545
7.00	0.73	1.82	0.538
7.20	0.70	1.75	0.527
7.40	0.68	1.67	0.513
7.60	0.67	1.59	0.496
7.80	0.66	1.51	0.476
8.00	0.66	1.44	0.454
8.20	0.65	1.36	0.430
8.40	0.66	1.29	0.403
8.60	0.66	1.22	0.378
8.80	0.68	1.15	0.346
9.00	0.69	1.09	0.317
9.20	0.70	1.02	0.286
9.40	0.73	0.95	0.251
9.60	0.77	0.89	0.216
9.80	0.82	0.84	0.185
10.00	0.86	0.81	0.163
10.20	0.90	0.77	0.143
10.40	0.94	0.74	0.127
10.60	0.99	0.72	0.115
10.80	1.04	0.71	0.108
11.00	1.08	0.70	0.104
11.20	1.11	0.70	0.102
11.40	1.14	0.70	0.101
11.60	1.17	0.71	0.102
11.80	1.20	0.72	0.104
12.00	1.22	0.73	0.107
12.40	1.25	0.76	0.113
12.80	1.26	0.78	0.118
13.20	1.27	0.81	0.124
13.60	1.27	0.83	0.129
14.00	1.26	0.84	0.132
14.40	1.25	0.84	0.132
14.80	1.25	0.84	0.133
15.00	1.25	0.84	0.133
15.60	1.25	0.85	0.134
16.00	1.27	0.85	0.134
16.50	1.28	0.89	0.145
17.00	1.28	0.94	0.158
17.50	1.25	1.00	0.175
18.00	1.19	1.04	0.190
18.50	1.12	1.05	0.200
19.00	1.07	1.05	0.205
19.50	1.02	1.04	0.212
20.00	0.97	1.04	0.219
20.50	0.91	1.03	0.228
21.00	0.85	1.01	0.234
21.50	0.80	0.97	0.234
22.00	0.77	0.94	0.233
23.00	0.71	0.87	0.229
24.00	0.67	0.79	0.218
25.00	0.64	0.73	0.205
26.00	0.61	0.66	0.194
27.00	0.60	0.59	0.177
28.00	0.60	0.53	0.155
29.00	0.61	0.48	0.134
30.00	0.62	0.45	0.123
31.00	0.61	0.40	0.114
32.00	0.63	0.34	0.093
33.00	0.65	0.31	0.077
34.00	0.67	0.28	0.065
35.00	0.70	0.26	0.054
36.00	0.72	0.25	0.047
37.00	0.73	0.23	0.041
38.00	0.75	0.22	0.035
39.00	0.77	0.22	0.031
40.00	0.79	0.22	0.028

Selenium, single crystal, $\vec{E} \parallel \hat{c}$[22]

Energy (eV)	n	k	$R(\phi = 0)$
0.01364	2.914	0.248	0.242
0.01488	3.175	9.95E-02	0.272
0.01612	3.263	2.13E-03	0.282
0.01736	3.306	3.81E-02	0.287
0.01860	3.330	7.04E-03	0.290
0.01984	3.346	4.23E-02	0.291
0.02108	3.358	3.40E-03	0.293
0.02232	3.366	5.31E-02	0.294
0.02356	3.372	1.96E-03	0.294
0.02480	3.377	2.39E-02	0.295
0.02604	3.380		0.295
0.02728		1.16E-02	
0.02976		7.96E-03	
0.03224		8.57E-03	
0.03472		2.70E-02	
0.03720	3.397	1.72E-02	0.297
0.04463		1.13E-02	
0.04959	3.403	2.79E-03	0.298
0.05703		1.56E-03	
0.06199	3.405	1.35E-03	0.298
0.06819		5.79E-04	
0.07439	3.407	4.44E-04	0.298
0.08059		4.41E-04	
0.08679	3.408	4.32E-04	0.298
0.09299		2.44E-04	
0.09919	3.409	3.23E-04	0.299
0.1116	3.409	2.87E-04	0.299
0.1240	3.410	2.71E-04	0.299
0.2480	3.417	2.67E-04	0.299
0.3720	3.427	1.90E-04	0.301
0.4959	3.442	1.41E-04	0.302
0.6199	3.462	1.12E-04	0.304
0.7439	3.486	9.42E-05	0.307
0.8679	3.516	8.07E-05	0.310
0.9919	3.551	7.11E-05	0.314
1.116	3.592	6.37E-05	0.319
1.240	3.640	5.81E-05	0.324
1.50		1.33E-04	
1.60		1.59E-04	
1.70		6.27E-04	
1.80	4.46	2.20E-02	0.402
2.0	4.79	0.76	0.438
2.2	4.49	1.19	0.431
2.4	4.28	1.21	0.417
2.6	4.40	1.32	0.430
2.8	4.59	1.70	0.462
3.0	4.44	2.29	0.490
3.2	3.92	2.59	0.493
3.4	3.69	2.76	0.502
3.6	3.39	3.01	0.521
3.8	(3.00)		
4.0	(2.65)		
4.2	(2.30)		
4.5	1.92	2.78	0.528
5.0	1.50	2.31	0.482
6.0	1.57	1.49	0.288
7.0	1.84	1.45	0.276
8.0	1.35	1.68	0.353
9.0	1.35	1.64	0.342
10.0	0.92	1.07	0.238
12.0	1.00	1.10	0.232
14.0	0.81	0.91	0.211
16.0	0.65	0.61	0.160
18.0	0.65	0.48	0.120
20.0	0.69	0.36	0.076
22.0	0.81	0.25	0.030
24.0	0.91	0.18	0.011
26.0	0.86	0.15	0.012
28.0	0.85	0.13	0.011
30.0	0.87	0.11	0.008

Selenium, single crystal, $\vec{E} \perp \hat{c}$[22]

Energy (eV)	n	k	$R(\phi = 0)$
0.01364	2.854	0.0239	0.231
0.01488	2.932	0.0325	0.241
0.01612	3.140	0.1750	0.269
0.01736	2.959	1.3300	0.321
0.01860	2.111	0.2550	0.133
0.01984	2.356	0.0746	0.164
0.02108	2.462	0.0276	0.178
0.02232	2.502	0.0442	0.184
0.02356	2.543	0.0097	0.190
0.02480	2.550	0.0239	0.191
0.02604	2.582		0.195
0.02728	2.600	0.0101	0.198
0.02976	2.576	9.95E-03	0.194
0.03224	2.598	1.16E-02	0.197
0.03472	2.607	1.68E-02	0.199
0.03720	2.613	1.54E-02	0.199
0.04463		1.17E-02	
0.04959	2.627	3.58E-03	0.201
0.05703		8.65E-04	

Energy (eV)	n	k	$R(\phi = 0)$
0.06199	2.632	2.07E-03	0.202
0.06819		2.89E-04	
0.07439	2.635	1.59E-04	0.202
0.08059		1.35E-04	
0.08679	2.636	1.42E-04	0.202
0.09299		1.04E-04	
0.09919	2.637	8.95E-05	0.203
0.1116	2.638	8.84E-05	0.203
0.1240	2.639	8.51E-05	0.203
0.2480	2.645	5.97E-05	0.204
0.3720	2.652	5.44E-05	0.205
0.4959	2.654	4.58E-05	0.205
0.6199	2.675	3.82E-05	0.208
0.7439	2.692	3.32E-05	0.210
0.8679	2.713	2.96E-05	0.213
0.9919	2.739	2.69E-05	0.216
1.116	2.772	2.48E-05	0.221
1.240	2.816	2.31E-05	0.226
1.50		7.37E-05	
1.60		8.63E-05	
1.70		3.60E-04	
1.80	3.32	0.11	0.289
2.00	3.38	0.65	0.310
2.20	3.07	0.73	0.282
2.40	2.93	0.61	0.259
2.60	3.00	0.53	0.263
2.80	3.12	0.58	0.279
3.00	3.30	0.70	0.305
3.20	3.35	1.01	0.328
3.40	3.22	1.24	0.334
3.60	3.06	1.47	0.344
3.80	2.84	1.66	0.351
4.00	2.51	1.81	0.356
4.20	2.18	1.83	0.352
4.50	1.75	1.94	0.382
5.00	1.25	1.50	0.316
6.00	1.32	0.73	0.107
7.00	1.62	0.61	0.105
8.00	1.81	0.69	0.135
9.00	1.66	1.02	0.182
10.00	1.72	0.95	0.171
12.00	1.25	1.02	0.181
14.00	0.98	0.92	0.178
16.00	0.68	0.96	0.274
18.00	0.61	0.65	0.191
20.00	0.73	0.48	0.094
22.00	0.78	0.39	0.060
24.00	0.78	0.32	0.046
26.00	0.78	0.26	0.036
28.00	0.80	0.19	0.023
30.00	0.79	0.14	0.020

Silicon, single crystal[23]

Energy (eV)	n	k	$R(\phi = 0)$
0.01240	3.4185	2.90E-04	0.300
0.01488	3.4190	2.30E-04	0.300
0.01736	3.4192	1.90E-04	0.300
0.01984	3.4195	1.70E-04	0.300
0.02480	3.4197		0.300
0.03100	3.4199		0.300
0.04092	3.4200		0.300
0.04463		1.08E-04	
0.04959	3.4201	9.15E-05	0.300
0.05703		1.56E-04	
0.06199	3.4204	2.86E-04	0.300
0.06943		3.84E-04	
0.07439		7.16E-04	
0.08059	(3.4207)	1.52E-04	0.300
0.08679		1.02E-04	
0.09299		2.59E-04	
0.09919		1.77E-04	
0.1054		1.53E-04	
0.1116		2.02E-04	
0.1178		1.22E-04	
0.1240	3.4215	6.76E-05	0.300
0.1364		5.49E-05	
0.1488		2.41E-05	
0.1612		2.49E-05	
0.1736	(3.4230)	1.68E-05	0.300
0.1798		2.45E-05	
0.1860		2.66E-06	
0.1922		1.74E-06	
0.1984		8.46E-07	
0.2046		5.64E-07	
0.2108	(3.4244)	4.17E-07	0.300
0.2170		4.05E-07	
0.2232		3.94E-07	
0.2294		3.26E-07	
0.2356		2.97E-07	
0.2418		2.82E-07	
0.2480	3.4261	1.99E-07	0.300
0.3100	3.4294		0.301
0.3626	3.4327		0.301
0.4568	3.4393	2.50E-09	0.302
0.6199	3.4490		0.303
0.8093	3.4784		0.306
1.033	3.5193		0.311
1.1	(3.5341)	1.30E-05	0.312
1.2		1.80E-04	
1.3		2.26E-03	
1.4		7.75E-03	
1.5	3.673	5.00E-03	0.327
1.6	3.714	8.00E-03	0.331
1.7	3.752	1.00E-02	0.335
1.8	3.796	0.013	0.340
1.9	3.847	0.016	0.345
2.0	3.906	0.022	0.351
2.1	3.969	0.030	0.357
2.2	4.042	0.032	0.364
2.3	4.123	0.048	0.372
2.4	4.215	0.060	0.380
2.5	4.320	0.073	0.390
2.6	4.442	0.090	0.400
2.7	4.583	0.130	0.412
2.8	4.753	0.163	0.426
2.9	4.961	0.203	0.442
3.0	5.222	0.269	0.461
3.1	5.570	0.387	0.486
3.2	6.062	0.630	0.518
3.3	6.709	1.321	0.561
3.4	6.522	2.705	0.592
3.5	5.610	3.014	0.575
3.6	5.296	2.987	0.564
3.7	5.156	3.058	0.563
3.8	5.065	3.182	0.568
3.9	5.016	3.346	0.577
4.0	5.010	3.587	0.591
4.1	5.020	3.979	0.614
4.2	4.888	4.639	0.652
4.3	4.086	5.395	0.703
4.4	3.120	5.344	0.726
4.5	2.451	5.082	0.740
4.6	1.988	4.678	0.742
4.7	1.764	4.278	0.728
4.8	1.658	3.979	0.710
4.9	1.597	3.749	0.693
5.0	1.570	3.565	0.675
5.1	1.571	3.429	0.658
5.2	1.589	3.354	0.646
5.3	1.579	3.353	0.647
5.4	1.471	3.366	0.663
5.5	1.340	3.302	0.673
5.6	1.247	3.206	0.675
5.7	1.180	3.112	0.673
5.8	1.133	3.045	0.672
5.9	1.083	2.982	0.673
6.0	1.010	2.909	0.677
6.5	0.847	2.73	0.688
7.0	0.682	2.45	0.691
7.5	0.563	2.21	0.693
8.0	0.478	2.00	0.691
8.5	0.414	1.82	0.688
9.0	0.367	1.66	0.683
9.5	0.332	1.51	0.672
10.0	0.306	1.38	0.661
12.0	0.257	0.963	0.590
14.0	0.275	0.641	0.460
16.0	0.345	0.394	0.297
18.0	0.455	0.219	0.159
20.0	0.567	0.0835	0.079
22.14	0.675	0.0405	0.038
24.31	0.752	0.0243	0.020
26.38	0.803	0.0178	0.012
28.18	0.834	0.0152	0.008
30.24	0.860	0.0138	0.006
31.79	0.877	0.0132	0.004
34.44	0.899	0.0121	0.003
36.47	0.913	0.0113	0.002
38.75	0.925	0.0104	0.002
40.00	0.930	0.0100	0.001

Silver[6]

Energy (eV)	n	k	$R(\phi = 0)$
0.10	9.91	90.27	0.995
0.20	2.84	45.70	0.995
0.30	1.41	30.51	0.994
0.40	0.91	22.89	0.993
0.50	0.67	18.32	0.992
1.00	0.28	9.03	0.987
1.50	0.27	5.79	0.969
2.00	0.27	4.18	0.944
2.50	0.24	3.09	0.914
3.00	0.23	2.27	0.864
3.25	0.23	1.86	0.816

Energy (eV)	n	k	R(ϕ = 0)
3.50	0.21	1.42	0.756
3.60	0.23	1.13	0.671
3.70	0.30	0.77	0.475
3.77	0.53	0.40	0.154
3.80	0.73	0.30	0.053
3.90	1.30	0.36	0.040
4.00	1.61	0.60	0.103
4.10	1.73	0.85	0.153
4.20	1.75	1.06	0.194
4.30	1.73	1.13	0.208
4.50	1.69	1.28	0.238
4.75	1.61	1.34	0.252
5.00	1.55	1.36	0.257
5.50	1.45	1.34	0.257
6.00	1.34	1.28	0.246
6.50	1.25	1.18	0.225
7.00	1.18	1.06	0.196
7.50	1.14	0.91	0.157
8.00	1.16	0.75	0.114
9.00	1.33	0.56	0.074
10.00	1.46	0.56	0.082
11.00	1.52	0.56	0.088
12.00	1.61	0.59	0.100
13.00	1.66	0.64	0.112
14.00	1.72	0.78	0.141
14.50	1.64	0.88	0.152
15.00	1.56	0.92	0.156
16.00	1.42	0.91	0.151
17.00	1.33	0.86	0.139
18.00	1.28	0.80	0.124
19.00	1.27	0.75	0.111
20.00	1.29	0.71	0.103
21.00	1.35	0.75	0.112
21.50	1.37	0.80	0.124
22.00	1.34	0.87	0.141
22.50	1.26	0.93	0.157
23.00	1.17	0.94	0.163
23.50	1.10	0.93	0.165
24.00	1.04	0.90	0.165
24.50	0.99	0.87	0.160
25.00	0.95	0.83	0.154
25.50	0.91	0.78	0.144
26.00	0.90	0.74	0.133
26.50	0.89	0.69	0.121
27.00	0.89	0.65	0.109
27.50	0.89	0.62	0.099
28.00	0.90	0.59	0.090
28.50	0.91	0.57	0.084
29.00	0.92	0.56	0.079
30.00	0.93	0.54	0.074
31.00	0.93	0.53	0.072
32.00	0.92	0.53	0.072
33.00	0.90	0.51	0.071
34.00	0.88	0.49	0.067
35.00	0.86	0.45	0.061
36.00	0.89	0.44	0.055
38.00	0.89	0.39	0.043
40.00	0.90	0.37	0.039
42.00	0.90	0.35	0.036
44.00	0.90	0.33	0.033
46.00	0.90	0.32	0.031
48.00	0.89	0.31	0.030
50.00	0.88	0.29	0.027
52.00	0.89	0.28	0.024
54.00	0.88	0.17	0.024
56.00	0.87	0.26	0.024
58.00	0.87	0.24	0.021
60.00	0.87	0.22	0.018
62.00	0.88	0.21	0.016
64.00	0.88	0.21	0.016
66.00	0.88	0.21	0.016
68.00	0.87	0.21	0.017
70.00	0.83	0.20	0.021
72.00	0.85	0.18	0.016
74.00	0.85	0.17	0.014
76.00	0.85	0.16	0.013
78.00	0.85	0.15	0.013
80.00	0.85	0.14	0.012
85.00	0.85	0.11	0.011
90.00	0.85	0.08	0.009
95.00	0.86	0.06	0.007
100.00	0.87	0.04	0.005

Sodium[24]

Energy (eV)	n	k	R(ϕ = 0)
0.55	0.262	9.97	0.990
0.58	0.241	9.45	0.989
0.63	0.207	8.80	0.990
0.67	0.175	8.09	0.990
0.73	0.147	7.42	0.990
0.81	0.123	6.67	0.989
0.92	0.099	5.82	0.989
1.05	0.078	5.11	0.989
1.23	0.064	4.35	0.987
1.44	0.053	3.72	0.986
1.65	0.050	3.22	0.983
1.87	0.049	2.76	0.978
2.07	0.053	2.48	0.971
2.27	0.059	2.23	0.961
2.45	0.063	2.07	0.953
2.64	0.066	1.88	0.943
2.82	0.068	1.76	0.936
2.95	0.068	1.63	0.928
3.06	0.069	1.54	0.921
3.20	0.065	1.47	0.921
3.40	0.061	1.33	0.916
3.71	0.055	1.13	0.908
3.97	0.049	1.01	0.908
6.199	0.390		0.193
6.358	0.454		0.141
6.526	0.485		0.120
6.702	0.533		0.093
6.888	0.574		0.073
7.130	0.616		0.056
7.328	0.641		0.048
7.583	0.674		0.038
7.847	0.700		0.031
8.015	0.710		0.029
8.634	0.762		0.018
9.143	0.800		0.012
9.709	0.819		0.010
10.20	0.843		0.007
11.08	0.870		0.005
11.83	0.887		0.004
12.73	0.907		0.002
13.05	0.913		0.002
13.42	0.914		0.002
13.73	0.917		0.002
14.07	0.922		0.002
14.83	0.934		0.001
15.05	0.936		0.001
15.46	0.942		0.001
16.21	0.948		0.001
18.10	0.964		0.000
21.12	0.979		0.000
25.51	0.993		0.000
26.95	1.00		0.000
27.68	1.01		0.000
28.37	1.01		0.000
29.52	1.02		0.000

Tantalum[16]

Energy (eV)	n	k	R(ϕ = 0)
0.10	10.14	66.39	0.984
0.15	9.45	46.41	0.9834
0.20	5.77	35.46	0.982
0.26	3.67	27.53	0.981
0.30	2.87	23.90	0.980
0.38	2.03	18.87	0.978
0.50	1.37	14.26	0.974
0.58	1.15	12.19	0.970
0.70	0.96	9.92	0.962
0.78	0.89	8.77	0.956
0.90	0.84	7.38	0.942
1.00	0.89	6.47	0.992
1.10	0.93	5.75	0.899
1.20	0.98	5.14	0.872
1.30	1.00	4.62	0.842
1.40	1.04	4.15	0.805
1.50	1.09	3.73	0.762
1.60	1.15	3.33	0.707
1.70	1.24	2.95	0.640
1.80	1.35	2.60	0.560
1.90	1.57	2.24	0.460
2.00	1.83	1.99	0.388
2.10	2.10	1.84	0.354
2.20	2.36	1.81	0.351
2.30	2.56	1.86	0.365
2.40	2.68	1.92	0.378
2.50	2.75	1.98	0.388
2.60	2.80	2.02	0.395
2.70	2.84	2.08	0.405
2.80	2.85	2.14	0.412
2.90	2.84	2.20	0.420
3.00	2.81	2.24	0.425
3.20	2.73	2.31	0.432
3.40	2.61	2.33	0.435
3.60	2.49	2.30	0.430
3.80	2.40	2.22	0.418
4.00	2.36	2.14	0.406
4.20	2.35	2.06	0.392
4.40	2.39	2.01	0.384
4.60	2.45	2.00	0.384
4.80	2.53	2.06	0.394

Energy (eV)	n	k	$R(\phi = 0)$
5.00	2.58	2.20	0.416
5.20	2.52	2.44	0.450
5.40	2.31	2.61	0.480
5.60	2.06	2.67	0.501
5.80	1.83	2.63	0.510
6.00	1.63	2.56	0.515
6.20	1.48	2.45	0.512
6.40	1.37	2.33	0.504
6.60	1.29	2.22	0.492
6.80	1.23	2.11	0.478
7.00	1.18	2.01	0.462
7.20	1.15	1.91	0.445
7.40	1.13	1.82	0.425
7.60	1.12	1.75	0.406
7.80	1.11	1.68	0.390
8.00	1.11	1.61	0.370
8.20	1.12	1.55	0.350
8.40	1.13	1.50	0.332
8.60	1.14	1.45	0.317
8.80	1.17	1.41	0.301
9.00	1.19	1.40	0.294
9.20	1.21	1.38	0.289
9.40	1.21	1.38	0.287
9.60	1.21	1.38	0.285
9.80	1.21	1.37	0.285
10.00	1.20	1.37	0.286
10.20	1.19	1.37	0.286
10.40	1.18	1.37	0.287
10.60	1.16	1.36	0.288
10.80	1.15	1.36	0.289
11.00	1.13	1.35	0.290
11.20	1.11	1.35	0.292
11.40	1.09	1.34	0.293
11.60	1.07	1.33	0.294
11.80	1.05	1.32	0.295
12.00	1.02	1.31	0.296
12.20	1.00	1.29	0.295
12.40	0.98	1.28	0.294
12.60	0.96	1.26	0.292
12.80	0.94	1.24	0.289
13.00	0.93	1.22	0.286
13.60	0.91	1.16	0.272
14.00	0.90	1.15	0.272
14.60	0.85	1.15	0.285
15.00	0.80	1.13	0.293
15.60	0.72	1.08	0.301
16.00	0.68	1.04	0.304
16.60	0.63	0.97	0.301
17.00	0.60	0.92	0.296
17.60	0.60	0.92	0.296
18.00	0.55	0.79	0.274
18.60	0.53	0.71	0.254
19.00	0.53	0.65	0.236
19.60	0.53	0.57	0.207
20.00	0.54	0.52	0.185
20.60	0.55	0.44	0.153
21.00	0.57	0.39	0.127
21.60	0.64	0.34	0.089
22.00	0.64	0.32	0.081
22.60	0.69	0.27	0.058
23.00	0.73	0.24	0.043
23.60	0.80	0.26	0.033
24.00	0.80	0.26	0.034
24.60	0.82	0.25	0.029
25.00	0.83	0.25	0.026
25.60	0.86	0.24	0.022
26.00	0.88	0.25	0.022
26.60	0.87	0.26	0.023
27.00	0.87	0.25	0.022
27.60	0.89	0.23	0.019
28.00	0.90	0.23	0.017
28.60	0.91	0.22	0.015
29.00	0.92	0.22	0.014
29.60	0.94	0.22	0.014
30.00	0.95	0.22	0.014
31.00	0.97	0.23	0.014
32.00	0.98	0.24	0.015
33.00	0.98	0.25	0.015
34.00	0.99	0.25	0.016
35.00	0.99	0.26	0.017
36.00	0.99	0.27	0.018
37.00	0.99	0.28	0.019
38.00	0.98	0.28	0.021
39.00	0.97	0.29	0.022
40.00	0.95	0.29	0.023

Tellurium, $\vec{E} \parallel \vec{c}$[25]

Energy (eV)	n	k	$R(\phi = 0)$
0.01364	4.82	0.118	0.431
0.01488	5.26	0.0505	0.463
0.01612	5.47	0.0278	0.477
0.01736	5.59	0.0174	0.485
0.01860		0.0796	
0.01984		0.0696	
0.02108		0.0749	
0.02232		0.1900	
0.02356		0.2220	
0.02480		0.0716	
0.02604		0.0682	
0.02728		0.0832	
0.02976		0.0149	
0.03224		2.14E-03	
0.03472		1.71E-02	
0.03720	5.94	3.71E-03	0.507
0.03968		2.44E-03	
0.04339	5.96	1.59E-03	0.508
0.04711		7.85E-04	
0.05083		7.38E-04	
0.05579		3.89E-04	
0.06199	5.98	3.09E-04	0.509
0.07439		2.52E-04	
0.08679		2.96E-04	
0.09919		3.68E-04	
0.12400	6.246	3.34E-04	0.524
0.15500	6.253		0.525
0.20660	6.286		0.526
0.24800	6.316	7.48E-05	0.528
0.31	6.372	1.18E-05	0.531
0.35		4.93E-04	
0.41		6.74E-03	
0.5	6.53	2.30E-02	0.539
0.6	6.71	7.50E-02	0.549
0.7	7.00	0.24	0.563
0.8	7.23	0.48	0.574
0.9	7.48	0.94	0.589
1.0	7.70	1.56	0.606
1.2	6.99	2.22	0.593
1.4	7.11	2.46	0.604
1.6	6.75	2.91	0.606
1.8	6.89	3.70	0.637
2.0	4.67	4.67	0.654
2.2	4.94	5.16	0.681
2.4	3.94	5.08	0.686
2.6	3.25	4.77	0.681
2.8	2.73	4.42	0.674
3.0	2.30	4.16	0.674
3.5	1.69	3.44	0.646
4.0	1.33	2.64	0.571
4.5	1.32	1.96	0.428
5.0	1.63	1.60	0.312
5.5	1.72	1.57	0.302
6.0	1.73	1.45	0.276
6.5	1.78	1.36	0.257
7.0	1.83	1.36	0.257
7.5	1.72	1.51	0.289
8.0	1.54	1.37	0.260
8.5	1.55	1.23	0.226
9.0	0.99	0.93	0.179
9.5	1.17	1.25	0.233
10.0	0.86	0.86	0.181
11.0	0.80	0.77	0.165
12.0	0.79	0.76	0.164
14.0	0.67	0.59	0.146
16.0	0.59	0.49	0.147
18.0	0.48	0.31	0.160
20.0	0.74	0.20	0.035
22.0	0.83	0.18	0.018
24.0	0.85	0.15	0.013
26.0	0.87	0.12	0.009
28.0	0.89	0.090	0.006
30.0	0.90	0.045	0.003

Tellurium, $\vec{E} \perp \vec{c}$[25]

Energy (eV)	n	k	$R(\phi = 0)$
0.01364	2.61	0.2980	0.204
0.01488	3.65	0.0894	0.325
0.01612	4.10	0.0535	0.370
0.01736	4.63	0.4990	0.420
0.01860		0.1170	
0.01984		0.0343	
0.02108	(4.42)	0.0421	0.398
0.02232		0.1060	
0.02356		0.0880	
0.02480		0.0458	
0.02604		0.0928	
0.02728		0.0886	
0.02976		0.0232	
0.03224		3.06E-03	
0.03472		1.25E-02	
0.03720	4.71	2.65E-03	0.422
0.03968		1.89E-03	
0.04339	4.74	1.41E-03	0.425
0.04711		8.38E-04	
0.05083		6.79E-04	

Energy (eV)	n	k	R(ϕ = 0)
0.05579		1.59E-04	
0.06199	4.77	1.16E-04	0.427
0.07439		7.23E-05	
0.08679		5.34E-05	
0.09919		4.28E-05	
0.1240	4.796	3.18E-05	0.429
0.1550	4.809		0.430
0.2066	4.838		0.432
0.2480	4.864	2.19E-05	0.434
0.31	4.929	3.18E-05	0.439
0.35		7.89E-02	
0.41		0.149	
0.5	4.90		0.437
0.6	4.93		0.439
0.7	4.95	0.11	0.441
0.8	5.10	0.13	0.452
0.9	5.22	0.22	0.461
1.0	5.35	0.45	0.472
1.2	5.17	0.63	0.462
1.4	5.56	0.63	0.488
1.6	5.88	1.15	0.517
1.8	6.10	1.80	0.545
2.0	5.94	2.69	0.571
2.2	5.10	3.61	0.594
2.4	4.24	3.77	0.593
2.6	3.57	3.75	0.591
2.8	3.03	3.63	0.588
3.0	2.51	3.39	0.578
3.5	1.72	2.70	0.532
4.0	1.32	2.01	0.440
4.5	1.28	1.28	0.251
5.0	1.47	0.82	0.132
5.5	1.74	0.51	0.104
6.0	1.94	0.39	0.118
6.5	2.19	0.32	0.148
7.0	2.48	0.40	0.192
7.5	2.60	0.69	0.226
8.0	2.59	0.91	0.245
8.5	2.39	1.00	0.235
9.0	1.11	1.24	0.259
9.5	2.08	1.11	0.224
10.0	0.99	1.04	0.215
11.0	0.84	1.01	0.237
12.0	0.87	0.87	0.182
14.0	0.59	0.87	0.282
16.0	0.64	0.55	0.144
18.0	0.52	0.41	0.161
20.0	0.50	0.38	0.165
22.0	0.56	0.29	0.110
24.0	0.54	0.25	0.113
26.0	0.50	0.20	0.127
28.0	0.48	0.17	0.135
30.0	0.46	0.088	0.140

Titanium (Polycrystalline)[14]

Energy (eV)	n	k	R(ϕ = 0)
0.10	5.03	23.38	0.965
0.15	3.00	15.72	0.954
0.20	2.12	11.34	0.939
0.25	2.05	8.10	0.890
0.30	6.39	9.94	0.833
0.35	2.74	6.21	0.792
0.40	2.49	4.68	0.708
0.45	3.35	3.25	0.545
0.50	4.43	3.22	0.555
0.60	4.71	3.77	0.597
0.70	4.38	3.89	0.603
0.80	4.04	3.82	0.596
0.90	3.80	3.65	0.582
1.00	3.62	3.52	0.570
1.10	3.47	3.40	0.560
1.20	3.35	3.30	0.550
1.30	3.28	3.25	0.546
1.40	3.17	3.28	0.549
1.50	2.98	3.32	0.557
1.60	2.74	3.30	0.559
1.70	2.54	3.23	0.557
1.80	2.36	3.11	0.550
1.90	2.22	2.99	0.540
2.00	2.11	2.88	0.530
2.10	2.01	2.77	0.520
2.20	1.92	2.67	0.509
2.30	1.86	2.56	0.495
2.40	1.81	2.47	0.483
2.50	1.78	2.39	0.471
2.60	1.75	2.34	0.462
2.70	1.71	2.29	0.456
2.80	1.68	2.25	0.451
2.90	1.63	2.21	0.447
3.00	1.59	2.17	0.444
3.10	1.55	2.15	0.442
3.20	1.50	2.12	0.442
3.30	1.44	2.09	0.442
3.40	1.37	2.06	0.443
3.50	1.30	2.01	0.443
3.60	1.24	1.96	0.441
3.70	1.17	1.90	0.436
3.80	1.11	1.83	0.430
3.85	1.08	1.78	0.423
3.90	1.06	1.73	0.413
4.00	1.04	1.62	0.389
4.20	1.05	1.45	0.333
4.40	1.13	1.33	0.284
4.60	1.17	1.29	0.265
4.80	1.21	1.23	0.244
5.00	1.24	1.21	0.236
5.20	1.27	1.20	0.228
5.40	1.17	1.16	0.228
5.60	1.24	1.21	0.234
5.80	1.21	1.22	0.241
6.00	1.15	1.21	0.244
6.20	1.11	1.18	0.240
6.40	1.08	1.14	0.232
6.60	1.04	1.06	0.212
6.80	1.05	1.02	0.198
7.00	1.06	0.97	0.182
7.20	1.07	0.95	0.175
7.40	1.11	0.94	0.167
7.60	1.09	0.92	0.165
7.80	1.11	0.93	0.165
8.00	1.10	0.94	0.169
8.20	1.10	0.95	0.171
8.40	1.08	0.95	0.175
8.60	1.04	0.96	0.181
8.80	1.02	0.95	0.181
9.00	1.00	0.94	0.182
9.20	0.97	0.93	0.182
9.40	0.95	0.91	0.181
9.60	0.94	0.90	0.179
9.80	0.91	0.88	0.179
10.00	0.89	0.88	0.180
10.20	0.86	0.85	0.178
10.40	0.85	0.83	0.175
10.60	0.81	0.79	0.167
10.80	0.80	0.76	0.162
11.00	0.79	0.72	0.152
11.20	0.81	0.69	0.139
11.40	0.81	0.69	0.139
11.60	0.79	0.68	0.139
11.80	0.78	0.67	0.137
12.00	0.77	0.65	0.132
12.80	0.76	0.55	0.106
13.20	0.76	0.52	0.097
13.60	0.76	0.48	0.087
14.00	0.77	0.45	0.077
14.40	0.77	0.42	0.069
14.80	0.79	0.38	0.058
15.20	0.79	0.36	0.052
15.60	0.79	0.32	0.045
16.00	0.83	0.31	0.037
16.40	0.84	0.28	0.030
16.80	0.87	0.27	0.025
17.20	0.90	0.25	0.020
17.60	0.93	0.25	0.017
18.00	0.94	0.24	0.165
18.40	0.94	0.23	0.017
18.80	0.95	0.24	0.016
19.20	0.96	0.25	0.016
19.60	0.97	0.25	0.017
20.00	0.98	0.27	0.018
20.40	0.98	0.27	0.019
20.60	1.00	0.29	0.020
21.20	0.99	0.31	0.023
21.60	0.99	0.31	0.024
22.00	0.98	0.32	0.025
22.40	0.98	0.33	0.027
22.80	0.97	0.33	0.028
23.20	0.96	0.34	0.030
23.60	0.95	0.35	0.031
24.00	0.92	0.35	0.033
24.5	0.91	0.34	0.032
25.0	0.91	0.33	0.032
25.5	0.89	0.33	0.032
26.0	0.89	0.33	0.032
26.5	0.88	0.32	0.032
27.0	0.86	0.31	0.032
27.5	0.85	0.30	0.033
28.0	0.84	0.29	0.033
28.5	0.82	0.26	0.029
29.0	0.83	0.25	0.027
30.0	0.84	0.22	0.022

Energy (eV)	n	k	$R(\phi = 0)$
Tungsten[27]			
0.10	14.06	54.71	0.983
0.20	3.87	28.30	0.981
0.25	2.56	22.44	0.980
0.30	1.83	18.32	0.979
0.34	1.71	15.71	0.973
0.38	1.86	13.88	0.963
0.42	1.92	12.63	0.954
0.46	1.69	11.59	0.952
0.50	1.40	10.52	0.952
0.54	1.23	9.45	0.948
0.58	1.17	8.44	0.938
0.62	1.28	7.52	0.917
0.66	1.45	6.78	0.888
0.70	1.59	6.13	0.856
0.74	1.83	5.52	0.810
0.78	2.12	5.00	0.759
0.82	2.36	4.61	0.710
0.86	2.92	4.37	0.661
0.90	3.11	4.44	0.660
0.94	3.15	4.43	0.658
0.98	3.15	4.36	0.653
1.00	3.14	4.32	0.649
1.10	3.05	4.04	0.627
1.20	3.00	3.64	0.590
1.30	3.12	3.24	0.545
1.40	3.29	2.96	0.515
1.50	3.48	2.79	0.500
1.60	3.67	2.68	0.494
1.70	3.84	2.79	0.507
1.80	3.82	2.91	0.518
1.90	3.70	2.94	0.518
2.00	3.60	2.89	0.512
2.10	3.54	2.84	0.506
2.20	3.49	2.76	0.497
2.30	3.49	2.72	0.494
2.40	3.45	2.72	0.493
2.50	3.38	2.68	0.487
2.60	3.34	2.62	0.480
2.70	3.31	2.55	0.472
2.80	3.31	2.49	0.466
2.90	3.32	2.45	0.461
3.00	3.35	2.42	0.459
3.10	3.39	2.41	0.460
3.20	3.43	2.45	0.465
3.30	3.45	2.55	0.476
3.40	3.39	2.66	0.485
3.50	3.24	2.70	0.488
3.60	3.13	2.67	0.482
3.70	3.05	2.62	0.476
3.80	2.99	2.56	0.468
3.90	2.96	2.50	0.460
4.00	2.95	2.43	0.451
4.20	3.02	2.33	0.440
4.40	3.13	2.32	0.442
4.60	3.24	2.41	0.455
4.80	3.33	2.57	0.475
5.00	3.40	2.85	0.505
5.20	3.27	3.27	0.548
5.40	2.92	3.58	0.586
5.60	2.43	3.70	0.618
5.80	2.00	3.61	0.637
6.00	1.70	3.42	0.643
6.20	1.47	3.24	0.646
6.40	1.32	3.04	0.640
6.60	1.21	2.87	0.631
6.80	1.12	2.70	0.619
7.00	1.06	2.56	0.607
7.20	1.01	2.43	0.593
7.40	0.98	2.30	0.573
7.60	0.95	2.18	0.556
7.80	0.93	2.06	0.533
8.00	0.94	1.95	0.505
8.20	0.94	1.86	0.481
8.40	0.96	1.76	0.449
8.60	0.99	1.70	0.422
8.80	1.01	1.65	0.401
9.00	1.01	1.60	0.388
9.20	1.02	1.55	0.369
9.40	1.03	1.50	0.352
9.60	1.05	1.44	0.329
9.80	1.09	1.38	0.307
10.00	1.13	1.34	0.287
10.20	1.19	1.33	0.274
10.40	1.24	1.34	0.270
10.60	1.27	1.36	0.274
10.80	1.29	1.39	0.282
11.00	1.28	1.42	0.290
11.20	1.27	1.44	0.297
11.40	1.25	1.46	0.305
11.60	1.22	1.48	0.313
11.80	1.20	1.48	0.318
12.00	1.16	1.48	0.323
12.40	1.10	1.47	0.329
12.80	1.04	1.44	0.333
13.20	0.98	1.40	0.332
13.60	0.94	1.35	0.325
14.00	0.91	1.28	0.312
14.40	0.90	1.23	0.296
14.80	0.90	1.17	0.276
15.20	0.93	1.13	0.255
15.60	0.97	1.12	0.246
16.00	0.98	1.14	0.249
16.40	0.97	1.17	0.260
16.80	0.94	1.19	0.273
17.20	0.90	1.21	0.289
17.60	0.85	1.21	0.304
18.00	0.80	1.20	0.317
18.40	0.74	1.18	0.330
18.80	0.69	1.15	0.340
19.20	0.64	1.11	0.347
19.60	0.60	1.07	0.353
20.00	0.56	1.02	0.354
20.40	0.54	0.97	0.350
20.80	0.52	0.92	0.342
21.20	0.50	0.87	0.331
21.60	0.50	0.82	0.318
22.00	0.49	0.77	0.303
22.40	0.49	0.73	0.287
22.80	0.49	0.69	0.272
23.20	0.49	0.66	0.263
23.60	0.48	0.62	0.252
24.00	0.49	0.57	0.234
24.40	0.50	0.53	0.213
24.80	0.51	0.49	0.191
25.20	0.53	0.46	0.171
25.60	0.55	0.43	0.150
26.00	0.57	0.40	0.132
26.40	0.59	0.38	0.117
26.80	0.61	0.37	0.105
27.00	0.62	0.36	0.099
27.50	0.64	0.34	0.085
28.00	0.67	0.32	0.073
28.50	0.69	0.31	0.065
29.00	0.71	0.30	0.057
29.50	0.73	0.30	0.052
30.00	0.75	0.29	0.047
31.00	0.78	0.29	0.042
32.00	0.79	0.29	0.040
33.00	0.82	0.28	0.033
34.00	0.84	0.29	0.032
35.00	0.85	0.31	0.033
36.00	0.85	0.32	0.036
37.00	0.84	0.33	0.039
38.00	0.83	0.33	0.040
39.00	0.81	0.33	0.042
40.00	0.80	0.33	0.045
Vanadium[9]			
0.10	12.83	45.89	0.978
0.20	3.90	24.30	0.975
0.28	2.13	17.35	0.973
0.36	1.54	13.32	0.966
0.44	1.28	10.74	0.957
0.52	1.16	8.93	0.945
0.60	1.10	7.59	0.929
0.68	1.07	6.54	0.909
0.76	1.08	5.67	0.882
0.80	1.10	5.30	0.864
0.90	1.18	4.50	0.811
1.00	1.34	3.80	0.730
1.10	1.60	3.26	0.632
1.20	1.93	2.88	0.543
1.30	2.25	2.71	0.498
1.40	2.48	2.72	0.491
1.50	2.57	2.79	0.499
1.60	2.57	2.84	0.507
1.70	2.52	2.88	0.512
1.80	2.45	2.88	0.515
1.90	2.36	2.85	0.514
2.00	2.34	2.81	0.509
2.10	2.31	2.78	0.506
2.20	2.28	2.80	0.510
2.30	2.23	2.83	0.516
2.40	2.15	2.88	0.528
2.50	2.02	2.91	0.540
2.60	1.89	2.92	0.552
2.70	1.74	2.89	0.561
2.80	1.61	2.85	0.569
2.90	1.48	2.80	0.577

Energy (eV)	n	k	R(ϕ = 0)
3.00	1.36	2.73	0.582
3.20	1.16	2.55	0.585
3.40	0.99	2.37	0.586
3.60	0.87	2.17	0.575
3.80	0.80	1.96	0.547
4.00	0.78	1.76	0.503
4.20	0.80	1.60	0.449
4.40	0.83	1.47	0.400
4.60	0.87	1.38	0.355
4.80	0.90	1.31	0.326
5.00	0.91	1.26	0.304
5.25	0.93	1.18	0.271
5.50	0.94	1.14	0.258
5.75	0.96	1.09	0.235
6.00	0.98	1.06	0.223
6.25	0.97	1.02	0.212
6.50	0.97	0.98	0.199
6.75	0.97	0.94	0.185
7.00	0.98	0.91	0.175
7.33	0.97	0.89	0.170
7.66	0.98	0.87	0.162
8.00	0.98	0.85	0.155
8.33	0.98	0.81	0.146
8.66	0.98	0.81	0.145
9.00	0.96	0.79	0.142
9.50	0.94	0.77	0.136
10.00	0.91	0.74	0.133
10.50	0.89	0.71	0.126
11.00	0.87	0.65	0.112
11.50	0.88	0.58	0.091
12.00	0.90	0.58	0.089
12.50	0.89	0.57	0.086
13.00	0.88	0.55	0.082
13.50	0.87	0.53	0.079
14.00	0.86	0.51	0.075
14.50	0.86	0.49	0.070
15.00	0.86	0.47	0.065
15.50	0.86	0.46	0.062
16.00	0.85	0.45	0.061
16.50	0.84	0.43	0.059
17.00	0.84	0.41	0.056
17.50	0.83	0.40	0.054
18.00	0.82	0.38	0.051
18.50	0.82	0.37	0.048
19.00	0.82	0.35	0.045
19.50	0.82	0.34	0.043
20.00	0.81	0.32	0.041
20.50	0.81	0.31	0.038
21.00	0.81	0.29	0.036
21.50	0.81	0.28	0.033
22.00	0.81	0.27	0.032
22.50	0.81	0.25	0.029
23.00	0.82	0.24	0.027
23.50	0.82	0.23	0.025
24.00	0.82	0.22	0.024
24.50	0.83	0.21	0.022
25.00	0.83	0.20	0.020
25.50	0.83	0.19	0.019
26.00	0.83	0.18	0.018
26.50	0.84	0.17	0.016
27.00	0.84	0.16	0.015
27.50	0.85	0.16	0.014
28.00	0.85	0.15	0.013
28.50	0.86	0.14	0.012
29.00	0.86	0.14	0.011
29.50	0.86	0.13	0.010
30.00	0.87	0.13	0.009
31.00	0.88	0.12	0.008
32.00	0.90	0.11	0.007
33.00	0.90	0.10	0.005
34.00	0.91	0.10	0.005
35.00	0.92	0.09	0.004
36.00	0.94	0.10	0.004
37.00	0.94	0.10	0.004
38.00	0.95	0.11	0.004
39.00	0.95	0.12	0.004
40.00	0.95	0.13	0.005

Zinc, $\vec{E} \parallel \hat{c}$[28]

Energy (eV)	n	k	R(ϕ = 0)
0.7514	1.9241	7.5619	0.883
0.827	1.7921	6.9973	0.874
0.866	1.5571	6.7753	0.881
0.952	1.4824	6.2296	0.868
0.992	1.5762	5.8843	0.847
1.033	1.5407	5.3192	0.823
1.078	1.5853	4.9013	0.793
1.127	1.7768	4.5307	0.748
1.181	1.9808	4.2004	0.701
1.240	2.8821	3.4766	0.575
1.305	3.2039	3.0042	0.520
1.377	2.9459	3.5761	0.584
1.459	3.2523	4.2447	0.640
1.550	3.8086	4.6212	0.657
1.653	3.7577	4.6239	0.659
1.722	3.5908	4.4614	0.650
1.823	3.4234	4.3232	0.642
1.937	3.0132	3.9974	0.624
1.984	1.8562	3.9706	0.690
2.066	1.4856	4.0555	0.737
2.094	1.2525	3.9961	0.762
2.119	1.0017	3.8683	0.789
2.275	0.7737	3.9129	0.832
2.445	0.6395	3.4013	0.821
2.666	0.4430	3.1379	0.851
2.917	0.3589	2.8140	0.853
3.220	0.3069	2.5088	0.847
3.594	0.2737	2.1737	0.828
4.065	0.2510	1.8528	0.799
4.678	0.2354	1.6357	0.776

Zinc, $\vec{E} \perp \hat{c}$[28]

Energy (eV)	n	k	R(ϕ = 0)
0.751	1.4469	7.4158	0.905
0.827	1.4744	6.9688	0.892
0.866	1.3628	6.6886	0.892
0.952	1.3165	6.2212	0.881
0.992	1.3835	5.8910	0.863
1.033	1.2889	5.4001	0.850
1.078	1.3095	4.9025	0.822
1.127	1.6897	4.4062	0.746
1.181	1.9701	4.0176	0.684
1.240	2.8717	3.2873	0.555
1.305	3.3991	2.7684	0.497
1.377	3.1807	3.4709	0.569
1.459	3.5064	4.1994	0.630
1.550	4.1241	4.7768	0.664
1.653	4.0269	4.8027	0.667
1.722	3.9369	4.6356	0.657
1.823	3.7549	4.3042	0.635
1.937	3.4512	4.1942	0.631
1.984	3.2515	4.2980	0.644
2.066	2.0802	4.7231	0.738
2.094	1.7084	4.7923	0.774
2.119	1.3329	4.4751	0.791
2.275	0.9725	4.2879	0.825
2.455	0.7568	3.7627	0.824
2.666	0.5470	3.4277	0.845
2.917	0.4774	3.0476	0.834
3.220	0.3911	2.7463	0.835
3.594	0.3147	2.3041	0.821
4.065	0.3013	2.0077	0.789
4.678	0.2806	1.7997	0.770

Zirconium (Polycrystalline)[28]

Energy (eV)	n	k	R(ϕ = 0)
0.10	6.18	1.76	0.300
0.15	3.37	1.30	0.123
0.20	2.34	1.08	0.058
0.26	2.24	1.06	0.052
0.30	2.59	1.14	0.073
0.36	3.17	1.26	0.110
0.40	3.09	1.24	0.105
0.46	3.36	1.30	0.123
0.50	4.13	1.44	0.175
0.56	5.01	1.58	0.231
0.60	5.18	1.61	0.242
0.70	4.54	1.51	0.202
0.80	4.03	1.42	0.168
0.90	3.74	1.37	0.149
0.96	3.69	1.36	0.145
1.00	3.66	1.35	0.143
1.10	3.65	1.35	0.142
1.20	3.53	1.33	0.134
1.30	3.25	1.27	0.116
1.40	3.10	1.25	0.106
1.50	3.02	1.23	0.100
1.60	2.88	1.20	0.091
1.70	2.68	1.16	0.078
1.80	2.49	1.12	0.067
2.00	2.14	1.03	0.047
2.10	1.99	1.00	0.040
2.20	1.87	0.97	0.034
2.30	1.78	0.94	0.030
2.40	1.71	0.92	0.027
2.50	1.62	0.90	0.024
2.60	1.54	0.88	0.022
2.70	1.46	0.86	0.019
2.80	1.40	0.84	0.018
2.90	1.34	0.82	0.016
3.00	1.30	0.81	0.016
3.10	1.26	0.80	0.015
3.30	1.19	0.77	0.014
3.40	1.16	0.76	0.013
3.50	1.13	0.75	0.013

Energy (eV)	n	k	$R(\phi = 0)$	Energy (eV)	n	k	$R(\phi = 0)$	Energy (eV)	n	k	$R(\phi = 0)$
3.60	1.10	0.74	0.013	9.00	1.65	0.91	0.025	17.20	1.09	0.74	0.013
3.70	1.07	0.73	0.013	9.20	1.63	0.90	0.025	17.60	1.13	0.75	0.013
3.80	1.04	0.72	0.012	9.40	1.60	0.89	0.024	18.00	1.17	0.76	0.014
3.90	1.01	0.71	0.012	9.60	1.57	0.89	0.023	18.40	1.21	0.78	0.014
4.00	0.98	0.70	0.012	9.80	1.52	0.87	0.021	18.80	1.24	0.79	0.014
4.20	0.94	0.68	0.013	10.00	1.47	0.86	0.020	19.20	1.27	0.80	0.015
4.40	0.89	0.67	0.013	10.20	1.42	0.84	0.018	19.60	1.29	0.80	0.015
4.60	0.85	0.65	0.014	10.40	1.35	0.82	0.016	20.00	1.30	0.81	0.015
4.80	0.81	0.64	0.014	10.50	1.32	0.81	0.016	20.60	1.29	0.80	0.015
5.00	0.78	0.63	0.015	10.60	1.28	0.80	0.015	21.00	1.27	0.80	0.015
5.20	0.77	0.62	0.016	10.80	1.23	0.78	0.014	21.60	1.23	0.78	0.014
5.40	0.77	0.62	0.016	11.00	1.19	0.77	0.014	22.00	1.20	0.77	0.014
5.60	0.80	0.63	0.014	11.20	1.16	0.76	0.013	22.60	1.15	0.76	0.013
5.80	0.87	0.66	0.013	11.40	1.13	0.75	0.013	23.00	1.12	0.75	0.013
6.00	1.00	0.71	0.012	11.60	1.11	0.74	0.013	23.60	1.08	0.73	0.013
6.20	1.11	0.75	0.013	11.80	1.09	0.74	0.013	24.00	1.05	0.73	0.013
6.40	1.23	0.78	0.014	12.00	1.08	0.73	0.013	24.60	1.02	0.71	0.012
6.60	1.33	0.81	0.016	12.40	1.05	0.72	0.012	25.00	1.00	0.71	0.012
6.80	1.42	0.84	0.018	12.80	1.01	0.71	0.012	25.60	0.97	0.69	0.012
7.00	1.49	0.86	0.020	13.20	0.98	0.70	0.012	26.00	0.95	0.69	0.013
7.20	1.54	0.88	0.022	13.60	0.95	0.69	0.013	26.60	0.91	0.67	0.013
7.40	1.58	0.89	0.023	14.00	0.92	0.68	0.013	27.00	0.88	0.66	0.013
7.60	1.61	0.90	0.024	14.40	0.89	0.67	0.013	27.60	0.84	0.65	0.014
7.80	1.63	0.90	0.025	14.80	0.90	0.67	0.013	28.00	0.83	0.64	0.014
8.00	1.66	0.91	0.026	15.20	0.92	0.68	0.013	28.60	0.82	0.64	0.014
8.20	1.67	0.91	0.026	15.60	0.95	0.69	0.013	29.00	0.81	0.64	0.014
8.40	1.68	0.92	0.026	16.00	0.98	0.70	0.012	29.60	0.82	0.64	0.014
8.60	1.68	0.92	0.026	16.40	1.01	0.71	0.012	30.00	0.82	0.64	0.014
8.80	1.66	0.91	0.026	16.80	1.04	0.72	0.012				

References

1. Shiles, E., Sasaki, T., Inokuti, M., and Smith, D. Y., *Phys. Rev. Sect. B*, 22, 1612, 1980.
2. Edwards, D. F., and Philipp, H. R., in *HOC-I*, p. 665.
3. Ives, H. E., and Briggs, N. B., *J. Opt. Soc. Am.*, 27, 395, 1937.
4. Bos, L. W., and Lynch, D. W., *Phys. Rev. Sect. B*, 2, 4567, 1970.
5. Weaver, J. H., Colavita, E., Lynch, D. W., and Rosei, R., *Phys. Rev. Sect. B*, 19, 3850, 1979.
6. Hagemann, H. J., Gudat, W., and Kunz, C., *J. Opt. Soc. Am.*, 65, 742, 1975.
7. Schulz, L. G., *J. Opt. Soc. Am.*, 47, 64, 1957.
8. Potter, R. F., in *HOC-I*, p. 465.
9. Olson, C. G., Lynch, D. W., and Weaver, J. H., unpublished.
10. Lynch, D. W., Olson, C. G., and Weaver, J. H., unpublished.
11. Weaver, J. H., Olson, C. G., and Lynch, D. W., *Phys. Rev. Sect. B*, 15, 4115, 1977.
12. Lynch, D. W., and Hunter, W. R., in *HOC-II*, p. 345.
13. Priol, M. A., Daudé, A., and Robin, S., *Compt. Rend.*, 264, 935, 1967.
14. Johnson, P. B., and Christy, R. W., *Phys. Rev. Sect. B*, 9, 5056, 1974.
15. Arakawn, E. T., and Inagaki, T., in *HOC-II*, p. 461.
16. Weaver, J. H., Lynch, D. W., and Olson, D. G., *Phys. Rev. Sect. B*, 10, 501, 1973.
17. Lynch, D. W., Rosei, R., and Weaver, J. H., *Solid State Commun.*, 9, 2195, 1971.
18. Weaver, J. H., Lynch, D. W., and Olson, C. G., *Phys. Rev. Sect. B*, 7, 4311, 1973.
19. Weaver, J. H., and Benbow, R. L., *Phys. Rev. Sect. B*, 12, 3509, 1975.
20. Weaver, J. H., *Phys. Rev., Sect. B*, 11, 1416, 1975.
21. Lynch, D. W., and Hunter, W. R., in *HOC-II*, p. 364.
22. Palik, E. D., in *HOC-II*, p. 691.
23. Edwards, D. F., in *HOC-I*, p. 547.
24. Lynch, D. W., and Hunter, W. R., in *HOC-II*, p. 354.
25. Palik, E. D., in *HOC-II*, p. 709.
26. Lynch, D. W., Olson, C. G., and Weaver, J. H., *Phys. Rev. Sect. B*, 11, 3671, 1975.
27. Weaver, J. H., Lynch, D. W., and Olson, C. G., *Phys. Rev. Sect. B*, 12, 1293, 1975.
28. Lanham, A. P., and Terherne, D. M., *Proc. Phys. Soc.*, 83, 1059, 1964.

OPTICAL PROPERTIES OF SELECTED INORGANIC AND ORGANIC SOLIDS

L. I. Berger

Optical properties of materials are closely related to their dielectric properties. The complex dielectric function (relative permittivity) of a material is equal to

$$\varepsilon(\omega) = \varepsilon'(\omega) - j\varepsilon''(\omega),$$

where $\varepsilon'(\omega)$ and $\varepsilon''(\omega)$ are its real and imaginary parts, respectively, and ω is the angular frequency of the applied electric field. For a non-absorbing medium, the index of refraction is $n = (\varepsilon\mu)^{1/2}$, where μ is the relative magnetic permeability of the medium (material); in the majority of dielectrics, $\mu \cong 1$.

For many applications, the most important optical properties of materials are the index of refraction, the extinction coefficient, k, and the reflectivity, R. The common index of refraction of a material is equal to the ratio of the phase velocity of propagation of an electromagnetic wave of a given frequency in vacuum to that in the material. Hence, $n \geqq 1$. The optical properties of highly conductive materials like metals and semiconductors (at photon energy range above the energy gap) differ from those of optically transparent media. Free electrons absorb the incident electromagnetic wave in a thin surface layer (a few hundred nanometers thick) and then release the absorbed energy in the form of secondary waves reflected from the surface. Thus, the light reflection becomes very strong; for example, highly conductive sodium reflects 99.8% of the incident wave (at 589 nm). Introduction of the effective index of refraction, $n_{eff} = (\varepsilon')^{1/2} = n - jk$, where $\varepsilon' = \varepsilon - j\delta/\omega\,\varepsilon_o$, δ is the electrical conductivity of the material in S/m, and $\varepsilon_o = 8.8542 \cdot 10^{-12}$ F/m is the permittivity of vacuum, allows one to apply the expressions of the optics of transparent media to the conductive materials. It is clear that the effective index of refraction may be smaller than 1. For example, $n = 0.05$ for pure sodium and $n = 0.18$ for pure silver (at 589.3 nm). At very high photon energies, the quantum effects, such as the internal photoeffect, start playing a greater role, and the optical properties of these materials become similar to those of insulators (low reflectance, existence of Brewster's angle, etc.).

The extinction coefficient characterizes absorption of the electromagnetic wave energy in the process of propagation of a wave through a material. The wave intensity, I, after it passes a distance x in an isotropic medium is equal to

$$I = I_0 \exp(-\alpha x),$$

where I_0 is the intensity at $x = 0$ and α is called the absorption coefficient. For many applications, the extinction coefficient, k, which is equal to

$$k = \alpha\frac{\lambda}{4\pi},$$

where λ is the wavelength of the wave in the medium, is more commonly used for characterization of the electromagnetic losses in materials.

Reflection of an electromagnetic wave from the interface between two media depends on the media indices of refraction and on the angle of incidence. It is characterized by the reflectivity, which is equal to the ratio of the intensity of the wave reflected back into the first medium to the intensity of the wave approaching the interface. For polarized light and two nonabsorbing media,

$$R = \frac{(N_1 - N_2)^2}{(N_1 + N_2)^2},$$

where $N_1 = n_1/\cos\theta_1$ and $N_2 = n_2/\cos\theta_2$ for the wave polarized in the plane of incidence, and $N_1 = n_1\cos\theta_1$ and $N_2 = n_2\cos\theta_2$ for the wave polarized normal to the plane of incidence; θ_1 and θ_2 are the angles between the normal to the interface in the point of incidence and the directions of the beams in the first and second medium, respectively. The reflectivity at normal incidence in this case is

$$R = [(n_1 - n_2)/(n_1 + n_2)]^2$$

For any two opaque (absorbing) media, the normal incidence reflectivity is

$$R = \frac{(n_1 - n_2)^2 + k_2^2}{(n_1 + n_2)^2 + k_2^2}.$$

In the majority of experiments, the first medium is air ($n \approx 1$), and hence,

$$R = \frac{(1 - n)^2 + k^2}{(1 + n)^2 + k^2}.$$

The data on n and k in the following table are abridged from the sources listed in the references. The reflectivity at normal incidence, R, has been calculated from the last equation. For convenience, the energy E, wavenumber $\bar{\nu}$, and wavelength λ are given for the incidence radiation.

E/eV	$\bar{\nu}$ /cm^{-1}	λ/μm	*n*	n_a	n_c	*k*	k_a	k_c	*R*	R_a	R_c
			Crystalline Arsenic Selenide (As_2Se_3) [Ref. 1]*								
2.194	17700	0.565					0.30				
2.168	17480	0.572					0.25				
2.141	17270	0.579					0.20				
2.123	17120	0.584					0.17				
2.098	16920	0.591					0.13				
2.094	16890	0.592						0.26			
2.091	16860	0.593						0.26			
2.073	16720	0.598					0.10	0.23			
2.060	16610	0.602						0.20			
2.049	16530	0.605					0.079	0.17			
2.036	16420	0.609						0.15			
2.023	16310	0.613						0.12			
2.013	16230	0.616					0.050				
2.009	16210	0.617						0.097			
2.000	16130	0.620						0.082			
1.987	16030	0.624						0.063			
1.977	15940	0.627					0.031				
1.974	15920	0.628						0.051			
1.962	15820	0.632						0.038			
1.953	15750	0.635						0.030			
1.949	15720	0.636					0.020				
1.937	15630	0.640						0.022			
1.925	15530	0.644						0.017			
1.922	15500	0.645					0.012				
1.905	15360	0.651					$8.6\cdot10^{-3}$				
1.893	15270	0.655					6.4				
1.881	15170	0.659					5.2				
1.859	14990	0.667					3.1				
1.848	14900	0.671						$1.7\cdot10^{-3}$			
1.845	14880	0.672					2.0				
1.842	14860	0.673						$1.2\cdot10^{-3}$			
1.831	14770	0.677					$1.3\cdot10^{-3}$	$9.0\cdot10^{-4}$			
1.826	14730	0.679						6.4			
1.821	14680	0.681						4.7			
1.818	14660	0.682					$8.6\cdot10^{-4}$				
1.815	14640	0.683						3.4			
1.807	14580	0.686					5.5				
1.802	14530	0.688					4.1				
0.06199	500.0	20.0		3.2	2.9		$1.7\cdot10^{-3}$	$1.8\cdot10^{-3}$		0.27	0.24
0.05904	476.2	21.0		3.1	2.9		$2.1\cdot10^{-3}$	$2.2\cdot10^{-3}$		0.26	0.24
0.05636	454.5	22.0		3.1	2.9		$2.5\cdot10^{-3}$	$2.6\cdot10^{-3}$		0.26	0.24
0.05391	434.8	23.0		3.1	2.9		$3.0\cdot10^{-3}$	$3.1\cdot10^{-3}$			
0.04592	370.4	27.0		3.0	2.8		$6.3\cdot10^{-3}$	$6.4\cdot10^{-3}$		0.25	0.22
0.04428	357.1	28.0		3.0	2.8		$7.6\cdot10^{-3}$	$7.7\cdot10^{-3}$		0.25	0.22
0.04275	344.8	29.0		3.0	2.8		0.0092	0.0093		0.25	0.22
0.04133	333.3	30.0		3.0	2.7		0.011	0.011		0.25	0.21
0.03542	285.7	35.0		2.7	2.5			0.037	0.034	0.21	0.18
0.03100	250.0	40.0		1.9	1.7			0.38	1.0	0.19	0.18
0.03061	247.0	40.5		2.0	2.6			0.33	0.95	0.12	0.25
0.03024	244.0	41.0		1.7	2.4			0.41	0.46	0.088	0.18
0.02883	232.6	43.0		1.2	1.3			2.2	0.94	0.50	0.16
0.02850	229.9	43.5		1.6	1.2			2.8	1.4	0.56	0.29
0.02818	227.3	44.0		2.3	1.2		3.3	2.0		0.58	0.48
0.02755	222.2	45.0		4.2	2.0		2.5	3.3		0.50	0.60
0.02480	200.0	50.0		6.5	4.0		3.6	0.26		0.62	0.36
0.02254	181.8	55.0		4.5	3.5		0.17	0.10		0.40	0.31
0.02066	166.7	60.0		4.0	3.2		0.089	0.10		0.36	0.27
0.01907	153.8	65.0		3.8	3.1		0.097	0.16		0.34	0.26
0.01771	142.9	70.0		3.6	3.0		0.19	0.30		0.32	0.25

E/eV	$\bar{\nu}$ /cm^{-1}	λ/μm	*n*	n_a	n_c	*k*	k_a	k_c	*R*	R_a	R_c
0.01653	133.3	75.0		3.7	3.0		0.41	0.44		0.34	0.26
0.01550	125.0	80.0		3.8	3.1		0.29	0.40		0.34	0.27
0.01459	117.6	85.0		3.6	2.9		0.20	0.34		0.32	0.24
0.01378	111.1	90.0		3.2	2.6		0.43	0.49		0.28	0.21
0.01305	105.3	95.0		4.7	3.0		1.5	1.5		0.46	0.34
0.01240	100.0	100.0		4.4	2.7		0.22	0.81		0.40	0.25
0.01181	95.24	105.0		4.2	3.0		0.094	3.9		0.38	0.62
0.01127	90.91	110.0		4.1	5.3		0.059	0.70		0.37	0.47
0.01033	83.33	120.0		3.9	4.2		0.034	0.13		0.35	0.38
0.009537	76.92	130.0		3.9	4.0		0.024	0.069		0.35	0.36
0.008856	71.43	140.0		3.9	3.8		0.019	0.048		0.35	0.34
0.007749	63.50	160.0		3.8	3.7		0.014	0.032		0.34	0.33
0.006888	55.55	180.0		3.8	3.7		0.011	0.024		0.34	0.33
0.006199	50.0	200.0		3.8	3.6		0.0091	0.019		0.34	0.32

*Indices a and c relate to the radiation electric field parallel to the a and c axes of the crystal, respectively.

Vitreous Arsenic Selenide (As_2Se_3) [Ref. 1]

E/eV	$\bar{\nu}$ /cm^{-1}	λ/μm	*n*	n_a	n_c	*k*	k_a	k_c	*R*	R_a	R_c
2.056	16580	0.603				0.12					
2.026	16340	0.612				0.11					
2.006	16180	0.618				0.099					
1.990	16050	0.623				9.0					
1.925	15530	0.644				5.6					
1.826	14730	0.679				1.4					
1.810	14600	0.685				0.012					
1.794	14470	0.691				0.0089					
1.771	14290	0.700				6.2					
1.715	13830	0.723				2.6					
1.701	13720	0.729				0.0022					
1.647	13280	0.753				0.00046					
1.629	13140	0.761	3.07			4.0			0.62		
1.596	12870	0.777	3.06			2.7			0.49		
1.579	12740	0.785	3.05			1.9			0.39		
1.562	12590	0.794	3.05			0.00013			0.26		
1.544	12450	0.803	3.04			0.000094			0.25		
1.529	12330	0.811	3.03			6.3			0.78		
1.512	12200	0.820	3.03			4.2			0.64		
1.494	12050	0.830	3.02			2.8			0.50		
1.476	11910	0.840	3.01			1.8			0.38		
1.378	11110	0.90	2.98								
1.240	10000	1.00	2.93								
1.127	9091	1.10	2.90								
1.051	8475	1.18	2.89								
1.033	8333	1.20	2.88								
0.2555	1980	5.05				$1.6 \cdot 10^{-7}$					
0.2380	1919	5.21				$9.9 \cdot 10^{-8}$					
0.2344	1890	5.29				$1.1 \cdot 10^{-7}$					
0.1345	1085	9.22				4.4					
0.1339	1080	9.26				3.7					
0.1333	1075	9.30				4.4					
0.1308	1055	9.48				4.5					
0.1215	980	10.20				8.9					
0.1203	970	10.31				$9.9 \cdot 10^{-7}$					
0.1196	965	10.36				$1.0 \cdot 10^{-6}$					
0.1178	950	10.53				1.1					
0.1116	900	11.11				1.8					
0.1004	810	12.35				4.9					
0.09919	800	12.50				$7.0 \cdot 10^{-6}$					
0.09795	790	12.66				$1.0 \cdot 10^{-5}$					
0.09671	780	12.82				1.5					
0.09299	750	13.33				3.7					
0.08555	690	14.49				6.9					

E/eV	$\bar{\nu}$ /cm^{-1}	λ/μm	n	n_a	n_c	k	k_a	k_c	R	R_a	R_c
0.08431	680	14.71				5.9					
0.08059	650	15.38				6.1					
0.07811	630	15.87				6.3					
0.07687	620	16.13				7.7					
0.07563	610	16.39				7.8					
0.07439	600	16.67				$9.3 \cdot 10^{-5}$					
0.07315	590	16.95	2.8			$1.2 \cdot 10^{-4}$			0.22		
0.07191	580	17.24	2.8			1.4			0.32		
0.07067	570	17.54	2.8			1.8			0.37		
0.06943	560	17.86	2.8			2.8			0.50		
0.06633	535	18.69	2.8			5.2			0.73		
0.06571	530	18.87	2.8			$7.2 \cdot 10^{-4}$			0.22		
0.06509	525	19.05	2.8			$1.2 \cdot 10^{-3}$			0.22		
0.06447	520	19.23	2.8			1.7			0.35		
0.06075	490	20.41	2.7			4.9			0.71		
0.06024	485.9	20.58	2.7			5.2			0.73		
0.05331	430	23.26	2.7			1.4			0.31		
0.05269	425	23.53	2.7			$1.1 \cdot 10^{-3}$			0.21		
0.05207	420	23.81	2.7			$8.5 \cdot 10^{-4}$			0.21		
0.05145	415	24.10	2.7			7.3			0.84		
0.05083	410	24.39	2.7			8.3			0.87		
0.05021	405	24.69	2.7			$9.4 \cdot 10^{-4}$			0.21		
0.04959	400	25.0	2.7			$1.2 \cdot 10^{-3}$			0.21		
0.04862	392.2	25.5	2.6			1.6			0.33		
0.04679	377.4	26.5	2.6			5.0			0.73		
0.04592	370.4	27.0	2.6			$8.0 \cdot 10^{-3}$			0.20		
0.04509	363.6	27.5	2.6			$1.2 \cdot 10^{-2}$			0.20		
0.04428	357.1	28.0	2.6			1.7			0.34		
0.03875	312.5	32.0	2.5			8.2			0.87		
0.03815	307.7	32.5	2.5			$9.3 \cdot 10^{-3}$			0.18		
0.03757	303.0	33.0	2.4			0.11			0.17		
0.02988	241.0	41.5	2.2			0.89			0.20		
0.02952	238.1	42.0	2.2			1.0			0.22		
0.02725	219.8	45.5	3.2			1.8			0.39		
0.02362	190.5	52.5	3.6			0.30			0.32		
0.01937	156.2	64.0	3.2			0.10			0.27		
0.01922	155.0	64.5	3.2			$9.6 \cdot 10^{-2}$			0.27		
0.01907	153.8	65.0	3.2			9.4			0.88		
0.01734	139.9	71.5	3.1			8.7			0.87		
0.01653	133.3	75.0	3.1			9.4			0.88		
0.01642	132.5	75.5	3.1			0.096			0.26		
0.01494	120.5	83.0	3.0			0.15			0.25		
0.01246	100.5	99.5	3.2			0.60			0.26		
0.007606	61.35	163.0	3.3			0.12			0.29		
0.006199	50.00	200.0	3.2								
0.004592	37.04	270.0	3.1			0.072			0.26		
0.002799	22.57	443.0	3.0			4.5			0.67		
0.001826	14.73	679.0	3.0			2.8			0.50		
0.001273	10.27	974.0	3.0			2.1			0.41		
0.0006491	5.236	1910.0	3.0			$1.1 \cdot 10^{-2}$			0.25		
0.0004376	3.530	2833.0	3.0			$7.5 \cdot 10^{-3}$			0.25		
0.0002903	2.341	4271.0	3.0			5.0			0.71		
0.0001716	1.384	7224.0	3.0			3.1			0.53		
0.00009047	0.7297	13704	3.0			$1.6 \cdot 10^{-3}$			0.25		
0.00005621	0.4534	22056	3.0			$9.9 \cdot 10^{-4}$			0.25		
0.00002774	0.2237	44699	3.0			5.2			0.72		
0.00001439	0.1161	86153	3.0			2.6			0.47		
Vitreous Arsenic Sulfide (As_2S_3) - [Ref. 2]											
4.959	40000	0.2500	2.48			1.21			0.27		
3.100	25000	0.40	3.09			0.34			0.27		

E/eV	$\bar{\nu}$ /cm^{-1}	λ/μm	n	n_a	n_c	k	k_a	k_c	R	R_a	R_c
2.48	20000	0.4999	2.83			0.013			0.23		
1.879	15150	0.66	2.59			$1.7 \cdot 10^{-6}$			0.20		
1.240	10000	1.0	2.48			$2.4 \cdot 10^{-7}$			0.18		
0.6199	5000	2.0	2.43						0.17		
0.3100	2500	4.0	2.41						0.17		
0.2480	2000	5.0	2.41						0.17		
0.1736	1400	7.143	2.40			$7.4 \cdot 10^{-7}$			0.17		
0.1240	1000	10.00	2.38			$1.3 \cdot 10^{-4}$			0.17		
0.09299	750	13.33	2.35			$3.0 \cdot 10^{-3}$			0.16		
0.07439	600	16.67	2.31			$4.6 \cdot 10^{-4}$			0.16		
0.04959	400.0	25.0	1.79			0.2			0.085		
0.03757	303.0	33.0	3.59			1.4			0.38		
0.03100	250.0	40.0	2.98			0.15			0.25		
0.02480	200.0	50	2.66			0.11			0.21		
0.02066	166.7	60	2.64			0.57			0.22		
0.01771	142.9	70	2.99			0.17			0.25		
0.01550	125.0	80	2.89			0.14			0.24		
0.01378	111.1	90	2.84			0.12			0.23		
0.01240	100	100	2.81			0.10			0.23		
0.008183	66	152	2.76			0.072			0.22		
0.004029	32.5	308	2.74			0.044			0.22		
0.002418	19.5	513	2.74			0.031			0.22		
0.001984	16	625	2.74			0.025			0.22		
0.001048	8.45	1180	2.73			$8.8 \cdot 10^{-3}$			0.22		
0.0001033	0.833	12000	2.73			$1.3 \cdot 10^{-3}$			0.22		
$4.129 \cdot 10^{-12}$	$3.33 \cdot 10^{-8}$	$3 \cdot 10^{11}$	2.73						0.22		
					Cadmium Telluride (CdTe) - [Ref. 3]						
4.9	39520	0.2530	2.48			2.04			0.39		
4.1	33070	0.3024	2.33			1.59			0.32		
3.9	31460	0.3179	2.57			1.90			0.37		
3.5	28230	0.3542	2.89			1.52			0.34		
3.1	25000	0.4000	3.43			1.02			0.34		
3.0	24200	0.4133	3.37			0.861			0.32		
2.755	22220	0.45	3.080			0.485			0.27		
2.75	22180	0.4509	3.23			0.636			0.29		
2.610	21050	0.475	3.045								
2.5	20160	0.4959	3.14			0.525			0.28		
2.25	18150	0.5510	3.05			0.411			0.26		
1.771	14290	0.70	2.861			0.210			0.23		
1.512	12200	0.82	2.880			0.040			0.23		
1.50	12100	0.8266	2.98			0.319			0.25		
1.475	11900	0.840	2.905			0.00134			0.24		
1.47	11860	0.8434				0.000671					
1.465	11820	0.8463				3.37					
1.46	11780	0.8492				1.89					
1.459	11760	0.850	2.948						0.24		
1.455	11740	0.8521				$1.08 \cdot 10^{-4}$					
1.45	11690	0.8551	2.9565			$5.10 \cdot 10^{-5}$			0.24		
1.445	11650	0.8580				2.73					
1.442	11630	0.860	2.952						0.24		
1.44	11610	0.8610	2.9479			1.37			0.32		
1.43	11530	0.8670	2.9402						0.24		
1.30	10490	0.9537	2.8720						0.23		
1.24	10000	1.0	2.840						0.23		
1.20	9679	1.033	2.8353						0.23		
1.10	8872	1.127	2.8050						0.23		
1.00	8065	1.240	2.7793						0.22		
0.90	7259	1.378	2.7537						0.22		
0.80	6452	1.550	2.7384						0.22		
0.70	5646	1.771	2.7223						0.21		

E/eV	$\bar{\nu}$ /cm^{-1}	λ/μm	n	n_a	n_c	k	k_a	k_c	R	R_a	R_c
0.60	4839	2.066	2.7086						0.21		
0.50	4033	2.480	2.6972						0.21		
0.40	3226	3.100	2.6878						0.21		
0.30	2420	4.133	2.6800						0.21		
0.20	1613	6.199	2.6722						0.21		
0.10	806.5	12.40	2.6535						0.21		
0.09	725.9	13.78	2.6482						0.20		
0.06819	550	18.18	2.623						0.20		
0.0573	462	21.6				$3.8 \cdot 10^{-6}$			0.20		
0.05	403.3	24.80	2.5801						0.19		
0.0469	378	26.5				$8.0 \cdot 10^{-5}$					
0.04592	370.3	27				$9.88 \cdot 10^{-5}$					
0.04133	333.3	30	2.55916			$2.86 \cdot 10^{-4}$			0.19		
0.04092	330	30.30	2.531			3.34			0.57		
0.03720	300	33.33	2.494			4.97			0.73		
0.03647	294.1	34.00				8.93					
0.03596	290	34.48	2.478			$5.77 \cdot 10^{-3}$			0.18		
0.03493	281.7	35.5				7.91					
0.03472	280	35.71	2.459			6.76			0.83		
0.03100	250	40	2.378			$1.18 \cdot 10^{-2}$			0.17		
0.02917	235.3	42.5				6.93					
0.02852	230	43.48	2.289			1.87			0.36		
0.02728	220	45.45	2.224			$2.47 \cdot 10^{-2}$			0.14		
0.02604	210	47.62	2.137			$3.4 \cdot 10^{-2}$			0.13		
0.02480	200	50.00	2.013			$4.97 \cdot 10^{-2}$			0.11		
0.02384	192.3	52.0				6.21					
0.01798	145	68.97	1.8			5.2			0.79		
0.01736	140	71.43	6.778			4.50			0.66		
0.01550	125	80.0	4.598			0.294			0.41		
0.01364	110	90.91	3.868			$9.47 \cdot 10^{-2}$			0.35		
0.01240	100	100	3.649			$5.68 \cdot 10^{-2}$			0.32		
0.009919	80	125	3.415			0.0262			0.30		
0.008679	70	142.9	3.348			0.0189			0.29		
0.007439	60	166.7	3.299			1.39			0.35		
0.006199	50	200	3.263			1.03			0.32		
0.004959	40	250	3.236			$7.52 \cdot 10^{-3}$			0.28		
0.003720	30	333.3	3.217						0.28		
0.023015	18.563		538.71			3.2096			0.28		
0.001550	12.50	800				6.18					
Gallium Arsenide (GaAs) - [Ref. 4]											
155		0.007999				0.0181					
145		0.008551				0.0203					
130		0.009537				0.0224					
110		0.01127				0.0278					
90		0.01378				0.0323					
70		0.01771				0.0376					
40		0.03100				0.0426					
23		0.05391	1.037			0.228					
7.0		0.1771	1.063			1.838					
6.0	48390	0.2066	1.264			2.472			0.61		
5.00	40330	0.2480	2.273			4.084			0.67		
4.00	32260	0.3100	3.601			1.920			0.42		
3.00	24200	0.4133	4.509			1.948			0.47		

E/eV	$\bar{\nu}$ /cm^{-1}	λ/μm	n	n_a	n_c	k	k_a	k_c	R	R_a	R_c
2.50	20160	0.4959	4.333			0.441			0.39		
2.00	16130	0.6199	3.878			0.211			0.35		
1.80	14520	0.8888	3.785			0.151			0.34		
1.60	12900	0.7749	3.700			0.091			0.33		
1.50	12100	0.8266	3.666			0.080			0.33		
1.40	11290	0.8856	3.6140			$1.69 \cdot 10^{-3}$			0.32		
1.20	9679	1.033	3.4920						0.31		
1.00	8065	1.240	3.4232						0.30		
0.80	6452	1.550	3.3737						0.29		
0.50	4033	2.480	3.3240						0.29		
0.25	2016	4.959	3.2978						0.29		
0.15	1210	8.266	3.2831						0.28		
0.100	806.5	12.40	3.2597			$4.93 \cdot 10^{-6}$			0.28		
0.090	725.9	13.78	3.2493			$1.64 \cdot 10^{-5}$			0.28		
0.070	564.6	17.71	3.2081			$2.32 \cdot 10^{-4}$			0.28		
0.060	483.9	20.66	3.1609			$3.45 \cdot 10^{-3}$			0.27		
0.0495	399.2	25.05	3.058			$2.07 \cdot 10^{-3}$			0.26		
0.03968	320	31.25	2.495			$2.43 \cdot 10^{-2}$			0.18		
0.03496	282	35.46	0.307			$294 \cdot 10^{-2}$					
0.02976	240	41.67	4.57			$4.26 \cdot 10^{-2}$			0.41		
0.02066	166.7	60	3.77			$3.89 \cdot 10^{-3}$			0.34		
0.01550	125	80	3.681			$1.84 \cdot 10^{-3}$			0.33		
0.008266	66.67	150	3.62			$2.14 \cdot 10^{-3}$			0.32		
0.002480	20	500	3.607			$1.3 \cdot 10^{-3}$			0.32		
0.001240	10	1000	3.606						0.32		

Gallium Phosphide (GaP) - [Ref. 5]

E/eV	$\bar{\nu}$ /cm^{-1}	λ/μm	n	n_a	n_c	k	k_a	k_c	R	R_a	R_c
154.0		0.00805				$1.7 \cdot 10^{-2}$					
110.0		0.0113				$2.15 \cdot 10^{-2}$					
100.0		0.0124				$215 \cdot 10^{-2}$					
80.0		0.0155				$3.0 \cdot 10^{-2}$					
50.0		0.0248				$4.7 \cdot 10^{-2}$					
27.0		0.0459				$9.3 \cdot 10^{-2}$					
25.0		0.0496				0.122					
20.0		0.0620				0.180					
15.0		0.0826	0.748			0.628					
5.5	44360	0.2254	1.543			3.556			0.68		
4.68	37750	0.2649	4.181			2.634			0.50		
3.50	28230	0.3542	5.050			0.819			0.46		
3.00	24200	0.4133	4.081			0.224			0.37		
2.78	22420	0.4460	3.904			0.103			0.35		
2.621	21140	0.473	3.73			$6.37 \cdot 10^{-3}$			0.33		
2.480	20000	0.500	3.590			$2.47 \cdot 10^{-3}$			0.32		
2.18	17580	0.5687	3.411			$2.8 \cdot 10^{-7}$			0.30		
2.000	16130	0.62	3.3254						0.29		
1.6	12900	0.7749	3.209						0.28		
1.240	10000	1.0	3.1192						0.26		
0.6888	5556	1.8	3.0439						0.26		
0.4769	3846	2.6	3.0271						0.25		
0.1907	1538	6.5	2.995			$4.29 \cdot 10^{-4}$			0.25		
0.1550	1250	8.0	2.984						0.25		
0.1240	1000	10	2.964						0.25		
0.06199	500	20	2.615			$7.16 \cdot 10^{-3}$			0.20		
0.03100	250	40	3.594			$1.81 \cdot 10^{-2}$			0.32		

E/eV	$\bar{\nu}$ /cm^{-1}	λ/μm	n	n_a	n_c	k	k_a	k_c	R	R_a	R_c
0.02480	200	50	3.461			$5.77 \cdot 10^{-3}$			0.30		
0.01727	139.27	71.80	3.3922			$4.34 \cdot 10^{-3}$			0.30		
0.01168	94.21	106.1	3.3621			$4.26 \cdot 10^{-3}$			0.29		
0.006199	50.00	200	3.3447			$1.3 \cdot 10^{-4}$			0.29		
0.004133	33.33	300	3.3413						0.29		
0.001240	10.00	1000	3.3319						0.29		
Indium Antimonide (InSb) - [Ref. 6]											
155		0.007999				$4.77 \cdot 10^{-3}$					
60		0.02066				$7.30 \cdot 10^{-2}$					
25		0.04959	1.15			.015					
24		0.05166	1.15			0.18					
15		0.08266	0.97			0.230					
10		0.1240	0.74			0.88					
5.00	40330	0.2480	1.307			2.441			0.53		
4.50	36290	0.2755	1.443			2.894			0.60		
4.00	32260	0.3100	2.632			3.694			0.61		
3.34	26940	0.3712	3.528			2.280			0.45		
2.84	22910	0.4366	3.340			2.021			0.45		
1.80	14520	0.6888	4.909			1.396			0.47		
1.50	12100	0.8266	4.418			0.643			0.41		
0.6	4839	2.066	4.03						0.36		
0.2480	2000	5.0	4.14			$9.1 \cdot 10^{-2}$			0.37		
0.1907	1538	6.5	4.30			$6.3 \cdot 10^{-2}$			0.39		
0.1653	1333	7.5	4.18			$2.7 \cdot 10^{-2}$			0.38		
0.06199	500	20.00	3.869			$2.0 \cdot 10^{-3}$			0.35		
0.03100	250	40.00	2.98			$2.6 \cdot 10^{-3}$			0.25		
0.02480	200	50.00	2.22			0.165			0.14		
0.02244	181	55.25	3.05			7.59			0.84		
0.02207	178	56.18	9.61			4.20			0.70		
0.02033	164	60.98	4.94			0.140			0.44		
0.01054	85	117.6	2.12			0.423			0.14		
0.005579	45	222.2	1.02			5.59			0.88		
0.001860	15	666.7	6.03			17.9			0.93		
0.001240	10	1000	10.7			24.0			0.94		
Indium Arsenide (InAs) - [Ref. 7]											
25		0.04959				1.139			0.168		
20		0.06199				1.125			0.225		
15		0.08266				0.894			0.336		
10		0.1240				0.835			1.071		
6	48390	0.2066	1.434			2.112			0.45		
5.0	40330	0.2480	1.524			2.871			0.58		
4.0	32260	0.3100	3.313			1.799			0.39		
3.5	28230	0.3542	3.008			1.754			0.37		
3.0	24200	0.4133	3.197			2.034			0.41		
2.5	20160	0.4959	4.364			1.786			0.45		
2.44	19680	0.5081	4.489			1.446			0.44		
1.86	15000	0.6666	3.889			0.554			0.36		
1.8	14520	0.6888	3.851			0.530			0.35		
1.7	13710	0.7293	3.798			0.493			0.35		
1.6	12900	0.7749	3.755			0.463			0.34		
1.5	12100	0.8266	3.714			0.432			0.34		
1.2	9679	1.033	3.613						0.32		
1.0	8065	1.240	3.548						0.31		
0.6	4839	2.066				0.161					
0.35	2823	3.542	3.608			$9.58 \cdot 10^{-3}$			0.32		
0.32	2581	3.875	3.512			$1.23 \cdot 10^{-4}$			0.31		
0.20	1613	6.199	3.427						0.30		
0.1240	1000	10.00	3.402						0.30		
0.06199	500	20.00	3.334						0.29		
0.04959	400	25.00	3.264						0.28		

E/eV	$\bar{\nu}$ /cm^{-1}	λ/μm	n	n_a	n_c	k	k_a	k_c	R	R_a	R_c
0.04339	350	28.57	3.182			$5.46\cdot10^{-3}$			0.27		
0.03720	300	33.33	2.988						0.25		
0.03100	250	40.00	1.970			$6.37\cdot10^{-2}$			0.11		
0.02765	222	44.84	5.90			6.53			0.74		
0.02480	200	50.00	6.91			0.30			0.56		
0.01984	160	62.50	5.27			0.41			0.47		
0.01860	150	66.67	5.27			0.51			0.47		
0.01736	140	71.43	3.99			$1.1\cdot10^{-2}$			0.36		
0.01488	120	83.33	3.91			$6.6\cdot10^{-3}$			0.35		
0.01240	100	100.0	3.85			$4.3\cdot10^{-3}$			0.35		
0.009919	80	125.0	3.817						0.34		
0.007439	60	166.7	3.793						0.34		
0.004959	40	250.0	3.778						0.34		
0.002480	20	500	3.769						0.37		
0.001240	10	1000	3.766						0.34		
Indium Phosphide (InP) - [Ref. 8]											
20		0.06199	0.793			0.494					
15		0.08266	0.695			0.574					
10		0.1240	0.806			1.154					
5.5	44360	0.2254	1.426			2.562			0.79		
5.0	40330	0.2480	2.131			3.495			0.61		
4.0	32260	0.3100	3.141			1.730			0.38		
3.0	24200	0.4133	4.395			1.247			0.43		
2.0	16130	0.6199	3.549			0.317			0.32		
1.5	12100	0.8266	3.456			0.203			0.31		
1.25	10085	0.9915	3.324						0.29		
1.00	8068	1.239	3.220						0.28		
0.50	4034	2.479	3.114						0.26		
0.30	2420	4.131	3.089						0.26		
0.10	806.8	12.39	3.012						0.25		
0.075	605.1	16.53	2.932						0.24		
0.060	484.1	20.66	2.780			$1.46\cdot10^{-2}$			0.22		
0.050	403.4	24.79	2.429			$3.35\cdot10^{-2}$			0.17		
0.03992	322	31.06	0.307			3.57					
0.03496	282	35.46	3.89			0.282			0.35		
0.03100	250	40.00	4.27			$3.0\cdot10^{-2}$			0.39		
0.02728	220	45.45	3.93			$1.3\cdot10^{-2}$			0.35		
0.02480	200	50.0	3.81			$8.7\cdot10^{-3}$			0.34		
0.02418	195	51.28	3.19						0.27		
0.02232	180	55.56	3.19						0.27		
0.01860	150	66.67	3.65						0.32		
0.01240	100	100	3.57						0.32		
0.009919	80	125.0	3.551						0.31		
0.007439	60	166.7	3.538						0.31		
0.004959	40	250.0	3.529						0.31		
0.002480	20	500	3.523						0.31		
0.001240	10	1000.0	3.522						0.31		
Lead Selenide (PbSe) - [Ref. 9]											
14.5		0.08551	0.72			0.20					
10		0.1240	0.68			0.50					
5	40330	0.2480	0.54			1.2					
2.0	16130	0.6199	3.65			2.9			0.51		
1.65	13310	0.7514	4.51			1.73			0.46		
1.5	12100	0.8266	4.64			2.64			0.52		
1.0	8065	1.240	4.65			1.1			0.44		
0.75	6049	1.653				0.269					
0.62	5001	2.000	4.59			0.770			0.42		
0.48	3871	2.583	4.90						0.44		
0.40	3226	3.100	4.91						0.44		
0.32	2581	3.875	4.98			0.173			0.44		

E/eV	$\bar{\nu}$ /cm^{-1}	λ/μm	n	n_a	n_c	k	k_a	k_c	R	R_a	R_c
0.20	1613	6.199	4.82						0.43		
0.1190	960	10.42	4.74			$1.20 \cdot 10^{-3}$			0.42		
0.09919	800	12.50	4.72			$2.09 \cdot 10^{-3}$			0.42		
0.07935	640	15.63	4.68			$4.12 \cdot 10^{-3}$			0.42		
0.05951	480	20.83	4.59			$1.00 \cdot 10^{-2}$			0.41		
0.04959	400	25.00	4.49			$1.77 \cdot 10^{-2}$			0.40		
0.03968	320	31.25	4.31			$3.62 \cdot 10^{-2}$			0.39		
0.02976	240	41.67	3.89			$9.61 \cdot 10^{-2}$			0.24		
0.01984	160	62.50	2.34			0.56			0.18		
0.009919	80	125.0	1.73			7.38			0.88		
0.007935	64	156.3	2.91			10.1			0.90		
0.004959	40	250.0	11.2			14.6			0.88		
0.002480	20	500.0	12.6			12.2					
0.001736	14	714.3	14.1			16.6					
0.001240	10	1000	17.4			21.1					
Lead Sulfide (PbS) - [Ref. 10]											
150		0.008266				$3.86 \cdot 10^{-3}$					
125		0.009919				$5.59 \cdot 10^{-3}$					
100		0.01240				$1.54 \cdot 10^{-2}$					
80		0.01550				$2.88 \cdot 10^{-2}$					
60		0.02066				$6.17 \cdot 10^{-2}$					
25		0.04959	0.845			0.171					
18.0		0.06888	0.846			0.294					
14.0		0.08856	0.651			0.665					
10.0		0.1240	0.879			1.050					
4.95	39920	0.2505	1.52			2.10			0.43		
4.0	32260	0.3100	1.73			2.83			0.55		
3.00	24200	0.4133	3.88			3.00			0.53		
2.90	23390	0.4275	4.12			2.70			0.51		
2.75	22180	0.4509	4.25			2.33			0.48		
2.55	20570	0.4862	4.35			2.00			0.47		
2.00	16130	0.6199	4.29			1.48			0.43		
1.60	12910	0.7749	4.62			0.94			0.43		
1.24	10000	1.00	4.43			0.597			0.41		
1.03	8333	1.2	4.30			0.458			0.39		
0.650	5263	1.9	4.24			0.318			0.39		
0.496	4000	2.5	4.30			0.235			0.39		
0.400	3226	3.1	4.30			$2.27 \cdot 10^{-2}$			0.39		
0.3100	2500	4.0	4.16			$6.38 \cdot 10^{-4}$			0.38		
0.2480	2000	5	4.115			$9.25 \cdot 10^{-4}$			0.37		
0.1240	1000	10	4.01			$6.32 \cdot 10^{-3}$			0.36		
0.1033	833.3	12	3.90			$1.14 \cdot 10^{-2}$			0.35		
0.08059	650	15.38	3.90						0.35		
0.06819	550	18.18	3.81						0.34		
0.04959	400	25.00	3.53						0.31		
0.03720	300	33.33	2.99						0.25		
0.02480	200.0	50	0.514			1.59					
0.01378	111.1	90	1.175			8.48			0.94		
0.01240	100.0	100	1.79			10.51			0.94		
0.008856	71.43	140	17.41			17.94			0.89		
0.006199	50.0	200	16.27			2.20			0.79		
0.003100	25.00	400	12.96			0.495			0.73		
0.001653	13.33	750	12.44			0.228			0.72		
0.001240	10.00	1000	12.35			0.167			0.72		
0.0006199	5.000	2000	12.27			0.0815			0.72		
Lead Telluride (PbTe) - [Ref. 11]											
150		0.008266				$2.37 \cdot 10^{-3}$					
125		0.009919				$9.71 \cdot 10^{-3}$					
100		0.01240				$4.39 \cdot 10^{-2}$					
75		0.01653				$6.43 \cdot 10^{-2}$					

E/eV	$\bar{\nu}$ /cm^{-1}	λ/μm	n	n_a	n_c	k	k_a	k_c	R	R_a	R_c
50		0.02480				$6.87\cdot10^{-2}$					
30		0.04133				$7.77\cdot10^{-2}$					
15		0.08266	0.72			0.17					
10		0.1240	0.66			0.60					
7.5		0.1653	0.8			0.92					
5.0	40330	0.2480	0.72			1.0					
3.0	24200	0.4133	1.0			2.2					
2.5	20160	0.4959	1.35			2.86			0.61		
1.5	12100	0.8266	3.8			3.1			0.53		
1.0	8065	1.240	4.55			2.2			0.49		
0.80	6452	1.550	6.25			0.71			0.53		
0.60	4839	2.066	6.10			0.521			0.52		
0.40	3226	3.100	6.075			0.331			0.52		
0.30	2420	4.133	5.95			$3.55\cdot10^{-2}$			0.51		
0.20	1613	6.199	5.77						0.50		
0.15	1210	8.266	5.76						0.50		
0.1017	820	12.20	5.47			$9.16\cdot10^{-3}$			0.48		
0.08927	720	13.89	5.38			$1.37\cdot10^{-2}$			0.47		
0.06943	560	17.86	5.13			$3.06\cdot10^{-2}$			0.45		
0.04959	400	25.00	4.50			$9.6\cdot10^{-2}$			0.40		
0.03968	320	31.25	3.58			0.23			0.32		
0.02976	240	41.67	1.01			1.9					
0.009919	80	125.0	2.95			16.6			0.96		
0.007439	60	166.7	4.9			22.5			0.96		
0.006199	50	200.0	6.9			27.2			0.97		
0.004959	40	250.0	11.6			34.8			0.97		
0.003720	30	333.3	27.7			35.7			0.95		
0.002480	20	500.0	27.6			39.1			0.95		
0.001240	10	1000	45.1			57.8			0.97		

Lithium Fluoride (LiF) - [Ref. 12]

E/eV	$\bar{\nu}$ /cm^{-1}	λ/μm	n	n_a	n_c	k	k_a	k_c	R	R_a	R_c
2000		$6.199\cdot10^{-4}$	0.9999347			$4.33\cdot10^{-6}$					
1496		$8.287\cdot10^{-4}$	0.999883			$1.28\cdot10^{-5}$					
1016		$1.220\cdot10^{-3}$	0.999757			$5.18\cdot10^{-5}$					
725		$1.710\cdot10^{-3}$	0.999643			$1.62\cdot10^{-4}$					
504		$2.460\cdot10^{-3}$	0.999162			$4.96\cdot10^{-5}$					
303		$4.092\cdot10^{-3}$	0.99752			$3.12\cdot10^{-4}$					
250		$4.959\cdot10^{-3}$	0.99632			$6.17\cdot10^{-5}$					
200		$6.199\cdot10^{-3}$				$2.12\cdot10^{-3}$					
150		$8.265\cdot10^{-3}$	0.9899			$3.54\cdot10^{-3}$					
100		$1.240\cdot10^{-2}$	0.9801			$1.32\cdot10^{-2}$					
75		$1.653\cdot10^{-2}$				$2.63\cdot10^{-2}$					
50		$2.480\cdot10^{-2}$				$7.89\cdot10^{-2}$					
25		$4.959\cdot10^{-2}$	0.558			0.521					
20		$6.199\cdot10^{-2}$	1.20			0.58			0.10		
15.1		$8.211\cdot10^{-2}$	1.08			0.68			0.10		
13		$9.537\cdot10^{-2}$	1.04			1.64					
12.0		0.1033	2.28			0.11			0.15		
11.0		0.1127	1.77			$8.07\cdot10^{-7}$			0.08		
10.00		0.12398	1.606			$7.70\cdot10^{-7}$			0.05		
9		0.1375	1.53						0.04		
7		0.1771	1.46								
4.959	40000	0.250	1.4189						0.03		
4.000	32260	0.31	1.4073						0.03		
2.952	23810	0.42	1.3978						0.03		
2.000	16130	0.62	1.3915						0.03		
0.9919	8000	1.25	1.3851								
0.7999	6452	1.55	1.3858						0.03		
0.4959	4000	2.5	1.3731						0.02		
0.4000	3226	3.1	1.3650								
0.3100	2500	4.0	1.3493								

E/eV	$\bar{\nu}$ /cm^{-1}	λ/µm	*n*	n_a	n_c	*k*	k_a	k_c	*R*	R_a	R_c
0.2480	2000	5.0	1.3266			$1.8\cdot10^{-6}$			0.02		
0.2000	1613	6.2	1.2912								
0.1698	1370	7.3	1.2499								
0.1494	1205	8.3	1.2036								
0.1240	1000	10.0	1.1005			$2.6\cdot10^{-3}$					
0.1127	909.1	11.0	1.0208			$8.0\cdot10^{-3}$					
0.1033	833.3	12.0				$1.9\cdot10^{-2}$					
0.09537	769.2	13.0				$3.7\cdot10^{-2}$					
0.08679	700	14.29	0.508			$7.74\cdot10^{-2}$					
0.07439	600	16.67	0.124			0.804					
0.06199	500	20.00	0.306			1.47			0.68		
0.05579	450	22.22	0.191			1.88			0.85		
0.04959	400	25.00	0.208			2.71			0.91		
0.03720	300	33.33	8.76			3.91			0.68		
0.03100	250	40.00	4.64			0.287			0.42		
0.02480	200	50.00	3.69			0.102			0.33		
0.01240	100.0	100	3.067			0.106			0.26		
0.06199	50.0	200	3.067			$4.0\cdot10^{-2}$			0.26		
0.04959	40.00	250	3.067			$2.2\cdot10^{-2}$			0.26		
0.02480	20.00	500	3.067			$6.3\cdot10^{-3}$					
0.01378	11.11	900				$3.1\cdot10^{-3}$					
$4.798\cdot10^{-4}$	3.870	2584	3.023			$1.19\cdot10^{-3}$			0.25		
$1.464\cdot10^{-4}$	1.181	8469	3.023			$6.20\cdot10^{-4}$			0.25		
$4.053\cdot10^{-5}$	0.3269	30590	3.023			$2.63\cdot10^{-4}$			0.25		
$1.861\cdot10^{-7}$	$1.501\cdot10^{-3}$	$6.662\cdot10^{6}$	3.018			$1.6\cdot10^{-5}$					
$3.718\cdot10^{-8}$	$2.999\cdot10^{-4}$	$3.335\cdot10^{7}$	3.018			$1.6\cdot10^{-5}$					
Potassium Chloride (KCl) - [Ref. 13]											
2860.3		$4.3347\cdot10^{-4}$				$3.93\cdot10^{-6}$					
2855.3		$4.3423\cdot10^{-4}$				$3.39\cdot10^{-6}$					
2849.3		$4.3514\cdot10^{-4}$				$4.61\cdot10^{-6}$					
2835.8		$4.3721\cdot10^{-4}$				$5.85\cdot10^{-6}$					
2832.3		$4.3775\cdot10^{-4}$				$5.85\cdot10^{-6}$					
2829.8		$4.3814\cdot10^{-4}$				$1.57\cdot10^{-6}$					
2828.3		$4.3837\cdot10^{-4}$				$4.19\cdot10^{-7}$					
219		$5.661\cdot10^{-3}$				$1.82\cdot10^{-3}$					
215		$5.767\cdot10^{-3}$				$1.84\cdot10^{-3}$					
212.5		$5.834\cdot10^{-3}$				$2.19\cdot10^{-3}$					
211		$5.876\cdot10^{-3}$				$1.82\cdot10^{-3}$					
185.1		$6.7\cdot10^{-3}$	0.99874						$1.01\cdot10^{-3}$		
109.7		$1.13\cdot10^{-2}$	0.99578						$4.22\cdot10^{-3}$		
43		0.02883	0.96			$3.0\cdot10^{-2}$					
40		0.03179	0.925			$1.8\cdot10^{-2}$					
29.9		0.04147	0.756			0.145					
20.1		0.06168	0.910			0.495					
15.1		0.08211	0.965			0.344					
10.0		0.1240	1.16			0.38			0.035		
9.0		0.1378	1.99			0.50			0.13		
8.0		0.1550	1.15			0.46			0.048		
7.0		0.1771	2.0			$8.46\cdot10^{-7}$			0.11		
6.199	50000	0.20	1.71739						0.070		
4.959	40000	0.25	1.58972								
3.999	32260	0.31	1.54005								
2.952	23810	0.42	1.50701								
2.695	21740	0.46	1.50115						0.040		
2.616	21100	0.474				$7.6\cdot10^{-11}$					
2.384	19230	0.52	1.49501								
2.066	16670	0.60	1.48969						0.039		
1.550	12500	0.80	1.48291						0.038		
1.033	8333	1.2	1.47813						0.037		
0.5166	4167	2.4	1.47464						0.037		

E/eV	$\bar{\nu}$ /cm^{-1}	λ/μm	*n*	n_a	n_c	*k*	k_a	k_c	*R*	R_a	R_c
0.2480	2000	5.0	1.47048						0.036		
0.2000	1.613	6.2	1.46796						0.036		
0.1512	1220	8.2	1.46260						0.035		
0.09999	806.5	12.4	1.44611						0.033		
0.07560	609.8	16.4	1.42295						0.030		
0.04959	400.0	25.0	1.34059			$6.57 \cdot 10^{-4}$			0.021		
0.03999	322.6	31.0	1.2431						0.012		
0.02976	240	41.67	0.85			0.16					
0.02728	220	45.45	0.53			0.35					
0.02232	180	55.56	0.31			1.05					
0.01860	150	66.67	0.44			4.0					
0.01612	130	76.92	4.1			0.32			0.37		
0.01240	100	100.0	2.7			0.11			0.21		
0.008679	70	142.9	2.4			$9.2 \cdot 10^{-2}$			0.17		
0.006199	50	200.0	2.2						0.14		
0.001240	10.00	1000				$9.0 \cdot 10^{-3}$					
0.0006199	5.000	2000				$3.7 \cdot 10^{-3}$					
0.0004133	3.333	3000				$2.0 \cdot 10^{-3}$					

Silicon Dioxide (Glass) - [Ref. 14]

E/eV	$\bar{\nu}$ /cm^{-1}	λ/μm	*n*	n_a	n_c	*k*	k_a	k_c	*R*	R_a	R_c
2000		$6.199 \cdot 10^{-4}$	0.99993			$1.503 \cdot 10^{-5}$					
1860		$6.665 \cdot 10^{-4}$	0.99991			$1.936 \cdot 10^{-5}$					
1609		$7.705 \cdot 10^{-4}$	0.99989			$9.941 \cdot 10^{-6}$					
1496		$8.287 \cdot 10^{-4}$	0.99987			$1.308 \cdot 10^{-5}$					
1204		$1.030 \cdot 10^{-3}$	0.99980			$2.916 \cdot 10^{-5}$					
1093		$1.134 \cdot 10^{-3}$	0.99975			$4.155 \cdot 10^{-5}$					
1016		$1.220 \cdot 10^{-3}$	0.99971			$5.423 \cdot 10^{-5}$					
798		$1.554 \cdot 10^{-3}$	0.99954			$1.289 \cdot 10^{-4}$					
597		$2.077 \cdot 10^{-3}$	0.99917			$3.560 \cdot 10^{-4}$					
396		$3.131 \cdot 10^{-3}$	0.99812			$4.04 \cdot 10^{-4}$					
303		$4.092 \cdot 10^{-3}$	0.99678			$9.91 \cdot 10^{-4}$					
201		$6.168 \cdot 10^{-3}$	0.99269			$3.63 \cdot 10^{-3}$					
151.2		$8.2 \cdot 10^{-3}$	0.9871			$7.3 \cdot 10^{-3}$					
99.99		$1.24 \cdot 10^{-2}$	0.9813			$7.0 \cdot 10^{-3}$					
49.59		$2.50 \cdot 10^{-2}$	0.9164			$6.5 \cdot 10^{-2}$					
40.00		$3.10 \cdot 10^{-2}$	0.907			$9.2 \cdot 10^{-2}$					
31.00		$4.00 \cdot 10^{-2}$	0.851			0.156					
25.00		0.04959	0.733			0.325					
20.00		0.06199	0.859			0.585					
15.00		0.08266	1.168			0.711			0.10		
13.00		0.09537	1.368			0.747			0.11		
11.00		0.1127	1.739			0.569			0.11		
10.00		0.1240	2.330			0.323			0.17		
9.00		0.1378	1.904			$1.89 \cdot 10^{-2}$			0.097		
7.00		0.1771	1.600						0.053		
6.00	48390	0.2066	1.543						0.046		
4.9939	40278.4	0.248272	1.50841						0.041		
4.1034	33096.1	0.302150	1.48719						0.038		
3.0640	24712.3	0.404656	1.46961						0.036		
2.5504	20570.5	0.486133	1.46313						0.035		
2.4379	19662.5	0.508582	1.46187						0.035		
2.2705	18312.5	0.546074	1.46008						0.035		
2.1489	17332.3	0.576959	1.45885						0.035		
2.1411	17269.2	0.579065	1.45877						0.035		
2.1102	17019.5	0.587561	1.45847						0.035		
2.1041	16970.4	0.589262	1.45841						0.035		
1.9257	15531.6	0.643847	1.45671						0.035		
1.8892	15237.6	0.656272	1.45637						0.035		
1.8566	14974.2	0.667815	1.45608						0.034		
1.7549	14153.9	0.706519	1.45515						0.034		
1.4550	11735.6	0.852111	1.45248						0.034		

E/eV	$\bar{\nu}$ /cm^{-1}	λ/μm	*n*	n_a	n_c	*k*	k_a	k_c	*R*	R_a	R_c
1.0985	8860.06	1.12866	1.44888						0.034		
0.60243	4858.9	2.0581	1.43722						0.032		
0.35354	2851.4	3.5070	1.40568						0.028		
0.2976	2400	4.176	1.383			$1.07\cdot10^{-4}$			0.026		
0.2728	2200	4.545	1.365			$2.56\cdot10^{-4}$			0.024		
0.2480	2000	5.000	1.342			$3.98\cdot10^{-3}$			0.021		
0.2232	1800	5.556	1.306			$5.63\cdot10^{-3}$					
0.1984	1600	6.250	1.239			$6.52\cdot10^{-3}$					
0.1736	1400	7.143	1.053			$1.06\cdot10^{-2}$					
0.1674	1350	7.407	0.9488			$1.48\cdot10^{-2}$					
0.1612	1300	7.692	0.7719			$3.72\cdot10^{-2}$					
0.1500	1210	8.265	0.4530			0.704			0.30		
0.1401	1130	8.850	0.3563			1.53			0.66		
0.1302	1050	9.524	2.760			1.65			0.35		
0.1209	975	10.26	2.448			0.231			0.18		
0.1091	880	11.36	1.784			$7.75\cdot10^{-2}$			0.079		
0.09919	800	12.50	1.753			0.343			0.089		
0.08989	725	13.79	1.698			0.175			0.071		
0.06943	560	17.86	1.337			0.298			0.036		
0.06199	500	20.00	0.6616						0.882		
0.04959	400	25.0	2.739			0.397			0.23		
0.03720	300	33.33	2.210			$6.7\cdot10^{-2}$			0.14		
0.01240	100	100.0	1.967			$1.59\cdot10^{-2}$			0.11		
0.007439	60	166.7	1.959			$8.62\cdot10^{-3}$			0.11		
0.002480	20	500.0	1.955			$7.96\cdot10^{-3}$			0.10		
			Silicon Monoxide (Noncrystalline) - [Ref. 15]								
25		0.04959	0.8090			0.2717					
20		0.06199	0.8853			0.4919					
17.5		0.07085	0.9825			0.5961					
15		0.08266	1.132			0.6651			0.092		
12.5		0.09919	1.283			0.6523			0.090		
10		0.1240	1.378			0.6843			0.10		
7.5		0.1653	1.593			0.7473			0.12		
5	40330	0.2480	2.001			0.6052			0.15		
4	32260	0.3100	2.141			0.4006			0.15		
3	24200	0.4133	2.116			0.1211			0.13		
2.8	22580	0.4428	2.085			0.08374			0.12		
2.6	20970	0.4769	2.053			0.05544			0.12		
2.4	19360	0.5166	2.021			0.03533			0.11		
2.2	17740	0.5636	1.994			0.02153			0.11		
2	16130	0.6199	1.969			0.01175			0.11		
1.8	14520	0.6888	1.948			0.00523			0.10		
1.6	12900	0.7749	1.929			0.00151			0.10		
1.240	10000	1.000	1.87						0.092		
0.6199	5000	2.000	1.84						0.087		
0.3100	2500	4.000	1.80						0.082		
0.2480	2000	5.000	1.75						0.074		
0.2066	1667	6.000	1.70						0.067		
0.1771	1492	7.000	1.60						0.053		
0.1653	1333	7.500	1.42								
0.1459	1176	8.500	0.90			0.18					
0.1305	1053	9.500	1.20			1.20			0.024		
0.1240	1000	10.00	2.00			1.38			0.27		
0.1181	952.4	10.50	2.85			0.90			0.27		
0.1153	930.2	10.75	2.86			0.58			0.25		
0.1127	909.1	11.00	2.82			0.40			0.24		
0.1078	869.6	11.50	2.50			0.20			0.19		
0.1033	833.3	12.00	2.13			0.14			0.13		
0.09537	769.2	13.00	2.04			0.20			0.12		
0.08856	714.3	14.00	2.01			0.30			0.12		

E/eV	$\bar{\nu}$ /cm^{-1}	λ/μm	n	n_a	n_c	k	k_a	k_c	R	R_a	R_c
			Noncrystalline Silicon Nitride (Si_3N_4) - [Ref. 16]								
24		0.05166	0.655			0.420			0.28		
23		0.05391	0.625			0.481			0.22		
22		0.05636	0.611			0.560			0.16		
21		0.05904	0.617			0.647			0.19		
20		0.06199	0.635			0.743			0.21		
19		0.06526	0.676			0.841			0.23		
18		0.06888	0.735			0.936			0.26		
17		0.07293	0.810			1.03			0.25		
16		0.07749	0.902			1.11			0.26		
15		0.08266	1.001			1.18			0.26		
14		0.08856	1.111			1.26			0.26		
13		0.09537	1.247			1.35			0.27		
12	96790	0.1033	1.417			1.43			0.28		
11	88720	0.1127	1.657			1.52			0.29		
10.5	84690	0.1181	1.827			1.53			0.29		
10	80650	0.1240	2.000			1.49			0.29		
9.5	76620	0.1305	2.162			1.44			0.28		
9	72590	0.1378	2.326			1.32			0.27		
8	64520	0.1550	2.651			0.962			0.26		
7	56460	0.1771	2.752			0.493			0.23		
6	48390	0.2066	2.541			0.102			0.19		
5	40330	0.2480	2.278			$4.9 \cdot 10^{-3}$			0.15		
4.75	38310	0.2610	2.234			$1.2 \cdot 10^{-3}$			0.15		
4.5	36290	0.2755	2.198			$2.2 \cdot 10^{-4}$			0.14		
4	32260	0.3100	2.141						0.13		
3.5	28230	0.3542	2.099						0.13		
3	24200	0.4133	2.066						0.12		
2.5	20160	0.4959	2.041						0.12		
2	16130	0.6199	2.022						0.11		
1.5	12100	0.8266	2.008						0.11		
1	8065	1.240	1.998						0.11		
			Sodium Chloride (NaCl) - [Ref. 17]								
209.5		$5.918 \cdot 10^{-3}$				$2.54 \cdot 10^{-3}$					
206		$6.019 \cdot 10^{-3}$				$2.62 \cdot 10^{-3}$					
203		$6.107 \cdot 10^{-3}$				$2.08 \cdot 10^{-3}$					
200		$6.199 \cdot 10^{-3}$				$1.92 \cdot 10^{-3}$					
26.0		0.04769	0.83			0.15			0.015		
25.0		0.04959	0.83			0.18			0.018		
22.0		0.05636	0.83			0.31			0.057		
20.0		0.06199	0.88			0.34			0.036		
18.0		0.06888	0.89			0.33			0.033		
16.1		0.07700	0.74			0.45			0.084		
14.0		0.08856	0.98			0.89			0.17		
12.0		0.1033	1.22			0.79			0.12		
10.0		0.1240	1.55			0.71			0.12		
8.00		0.1550	1.38			1.10			0.20		
6.00	48390	0.2066	1.75						0.074		
5.00	40330	0.2480	1.65						0.060		
2.952	23810	0.42	1.56324						0.048		
2.480	20000	0.50	1.55157						0.047		
2.214	17860	0.56	1.54613						0.046		
2.000	16130	0.62	1.54228						0.045		
1.771	14290	0.70	1.53865						0.045		
1.675	13510	0.74	1.53728						0.045		
1.550	12500	0.80	1.53560						0.045		
1.240	10000	1.00	1.53200						0.044		
1.033	8333	1.2	1.53000						0.044		
0.6888	5556	1.8	1.52712						0.043		
0.4959	4000	2.5	1.52531						0.043		

E/eV	$\bar{\nu}$ /cm^{-1}	λ/μm	n	n_a	n_c	k	k_a	k_c	R	R_a	R_c
0.4000	3226	3.1	1.52395						0.043		
0.3263	2632	3.8	1.52226			$(1.8\pm0.2)\cdot10^{-9}$			0.043		
0.2952	2381	4.2	1.52121						0.043		
0.2755	2222	4.5	1.52036						0.043		
0.2480	2000	5.0	1.51883						0.042		
0.1240	1000	10.0	1.49473						0.039		
0.1033	833.3	12.0	1.48000						0.037		
0.08856	714.3	14.0	1.46188						0.035		
0.07749	625.0	16.0	1.4399						0.033		
0.06888	555.5	18.0	1.41364						0.029		
0.06199	500.0	20.0	1.3822						0.026		
0.04959	400	25.0	1.27			$3.5\cdot10^{-3}$			0.014		
0.04215	340	29.41	1.12			$1.7\cdot10^{-2}$			0.0032		
0.03720	300	33.33	0.85			0.85			0.18		
0.03410	275	36.36	0.59			0.22			0.084		
0.03286	265	37.74	0.42			0.50			0.26		
0.03224	260	38.46	0.45			0.45			0.22		
0.02480	200	50.00	0.14			1.99			0.89		
0.02108	170	58.82	1.35			6.03			0.87		
0.01984	160	62.50	6.92			2.14			0.59		
0.01922	155	64.52	5.50			0.87			0.49		
0.01860	150	66.67	4.52			0.380			0.41		
0.01736	140	71.43	3.72			0.219			0.33		
0.01612	130	76.92	3.31			0.135			0.29		
0.01488	120	83.33	3.02			0.110			0.25		
0.01240	100	100.0	2.74			0.087			0.22		
0.009919	80	125.0	2.57			0.077			0.19		
0.07439	60	166.7	2.48			0.055			0.18		
0.04959	40	250.00	2.44			0.041			0.18		
0.002480	20	500.0	2.43			0.024			0.17		
0.001240	10	1000	2.43			0.006			0.17		
0.001033	8.333	1200				$8.8\cdot10^{-3}$					
0.0006888	5.556	1800				$5.4\cdot10^{-3}$					
0.0006199	5.000	2000	2.43						0.17		
0.0004959	4.000	2500				$4.4\cdot10^{-3}$					
0.0004797	3.869	2584	2.43			$2.1\cdot10^{-3}$			0.17		
0.0003875	3.125	3200				$3.3\cdot10^{-3}$					
0.0001464	1.181	8469	2.43			$5.8\cdot10^{-4}$			0.17		
0.00004053	0.3269	30590	2.43			$2.5\cdot10^{-4}$					
			Cubic Zinc Sulfide (ZnS) - [Ref. 18]								
2000		$6.199\cdot10^{-4}$	0.999904			$1.76\cdot10^{-5}$					
1204		$1.030\cdot10^{-3}$	0.999777			$1.00\cdot10^{-4}$					
1016		$1.220\cdot10^{-3}$	0.999838			$3.61\cdot10^{-5}$					
901		$1.376\cdot10^{-3}$	0.999647			$5.42\cdot10^{-5}$					
798		$1.554\cdot10^{-3}$	0.999520			$8.28\cdot10^{-5}$					
707		$1.754\cdot10^{-3}$	0.999372			$1.25\cdot10^{-4}$					
597		$2.077\cdot10^{-3}$	0.999160			$2.19\cdot10^{-4}$					
377		$9.50\cdot10^{-3}$	0.99789			$9.50\cdot10^{-4}$					
201		$6.168\cdot10^{-3}$	0.99553			$4.82\cdot10^{-3}$					
100		$1.240\cdot10^{-2}$	0.99061			$1.17\cdot10^{-2}$					
61.99		$2.000\cdot10^{-2}$	0.964			$3.32\cdot10^{-2}$			$6.2\cdot10^{-4}$		
41.33		$3.000\cdot10^{-2}$	0.941			$5.10\cdot10^{-2}$					
31.00		$4.000\cdot10^{-2}$	0.847			$9.95\cdot10^{-2}$					
24.80		$5.000\cdot10^{-2}$	0.796			0.171			$2.2\cdot10^{-2}$		
17.71		$7.000\cdot10^{-2}$	0.747			0.431			$7.7\cdot10^{-2}$		
13.78		$9.000\cdot10^{-2}$	0.758			0.824			0.20		
12.40		0.1000	0.862			0.876			0.19		
9.919		0.125	1.02			1.36			0.31		
8.266		0.150	1.41			1.47			0.29		

E/eV	$\bar{\nu}$ /cm^{-1}	λ/μm	n	n_a	n_c	k	k_a	k_c	R	R_a	R_c
6.199		0.200	2.32			1.62			0.32		
6.00	48390	0.2066	2.24			1.65			0.59		
4.00	32260	0.3100	2.70			0.44			0.22		
3.00	24200	0.4133	2.54			$4 \cdot 10^{-2}$			0.19		
2.50	20160	0.4959	2.42			$3 \cdot 10^{-2}$			0.17		
2.30	18550	0.5391	2.3950						0.17		
2.00	16130	0.6199	2.3576						0.16		
1.75	14110	0.7085	2.3319						0.16		
1.55	12500	0.7999	2.3146			$3.50 \cdot 10^{-6}$			0.16		
1.40	11290	0.8856	2.3033						0.16		
1.240	10000	1.000	2.2907			$3.02 \cdot 10^{-6}$			0.15		
1.00	8065	1.240	2.2795						0.15		
0.80	6452	1.550	2.2706						0.15		
0.6199	5000	2.000	2.2631			$6.2 \cdot 10^{-6}$			0.15		
0.45	3629	2.755	2.2587						0.15		
0.30	2420	4.133	2.2529						0.15		
0.20	1613	6.199	2.2443						0.15		
0.1550	1250	8.0	2.2213			$4.5 \cdot 10^{-6}$			0.14		
0.1240	1000	10.00	2.1986			$8.8 \cdot 10^{-6}$			0.14		
0.100	806.5	12.4	2.1969						0.14		
0.09	725.9	13.78	2.1793						0.14		
0.07999	645.2	15.5	2.1518			$3.82 \cdot 10^{-3}$			0.14		
0.07	564.6	17.71	2.1040						0.13		
0.06075	490	20.41	2.03			$8.0 \cdot 10^{-3}$			0.12		
0.05	403.3	24.80	1.6866						0.065		
0.03546	286	34.97	3.29			$8.3 \cdot 10^{-2}$			0.28		
0.03472	280	35.71	9.54			$5.2 \cdot 10^{-2}$			0.66		
0.02480	200	50.00	3.48			$3.1 \cdot 10^{-2}$			0.31		
0.01240	100	100.0	3.06			$5.8 \cdot 10^{-3}$			0.26		
0.004955	40	250.0	2.903			$6.2 \cdot 10^{-3}$			0.24		
0.004339	35	285.7	2.899			$7.0 \cdot 10^{-3}$			0.24		
0.003720	30	333.3	2.896						0.24		
0.003100	25	400.0	2.894						0.24		
0.002480	20	500.0	2.892						0.24		
0.001860	15	666.7	2.890						0.24		

Polytetrafluoroethylene (Teflon) - [Ref. 19]

E/eV	$\bar{\nu}$ /cm^{-1}	λ/μm	n	n_a	n_c	k	k_a	k_c	R	R_a	R_c
4.960	40000	0.250							0.970		
4.769	38462	0.260							0.972		
4.593	37037	0.270							0.975		
4.426	35714	0.280							0.978		
4.276	34483	0.290							0.980		
4.133	33333	0.300							0.983		
4.000	32258	0.310							0.986		
3.875	31250	0.320							0.988		
3.758	30303	0.330							0.990		
3.647	29412	0.340							0.991		
3.543	28571	0.350							0.992		
3.444	27778	0.360							0.992		
3.351	27027	0.370							0.993		
2.255	18182	0.550							0.993		
2.067	16667	0.600							0.992		
1.378	11111	0.900							0.992		
1.305	10526	0.950							0.991		
1.078	8696	1.150							0.991		
1.033	8333	1.200							0.990		
0.9920	8000	1.250							0.990		
0.9538	7692	1.300							0.989		
0.9185	7407	1.350							0.988		
0.8857	7143	1.400							0.988		
0.8552	6897	1.450							0.989		

E/eV	$\tilde{\nu}$ /cm^{-1}	λ/μm	n	n_a	n_c	k	k_a	k_c	R	R_a	R_c
0.8267	6667	1.500							0.989		
0.8000	6452	1.550							0.988		
0.7750	6250	1.600							0.988		
0.7515	6061	1.650							0.987		
0.7294	5882	1.700							0.986		
0.7086	5714	1.750							0.986		
0.6889	5556	1.800							0.985		
0.6703	5405	1.850							0.980		
0.6526	5263	1.900							0.978		
0.6359	51282	1.950							0.978		
0.6200	5000	2.000							0.970		
0.6049	4878	2.050							0.959		
0.5905	4762	2.100							0.951		
0.5767	4651	2.150							0.946		
0.5636	4545	2.200							0.966		
0.5511	44444	2.250							0.965		
0.5487	44247	2.260							0.964		
0.5439	4386	2.280							0.963		
0.5415	4367	2.290							0.961		
0.5368	4329	2.310							0.959		
0.5345	4310	2.320							0.957		
0.5322	4292	2.330							0.956		
0.5299	4274	2.340							0.954		
0.5277	4255	2.350							0.951		
0.5232	4219	2.370							0.950		
0.5188	4184	2.390							0.949		
0.5167	4167	2.400							0.947		
0.5061	4082	2.450							0.946		
0.4960	4000	2.500							0.945		

References

1. Arsenic Selenide
   D. J. Treacy in *Handbook of Optical Constants of Solids*, E. D. Palik, Editor, Academic Press, 1985, p. 623. (Hereafter abbreviated as *HOCS*.)
   R. Zallen, R. E. Drews, R. L. Emerald, and M. L. Slade, *Phys. Rev. Lett.* 26, 1564 (1971).
   R. Zallen, M. L. Slade, and A. T. Ward, *Phys. Rev.* B 3, 4257 (1971).
   U. Strom and P. C. Taylor, *Phys. Rev.* B 16, 5512 (1977).
   G. Lucovsky, *Phys. Rev.* B 6, 1480 (1972).
   C. T. Moynihan, P. B. Macedo, M. S. Maklad, R. K. Mohr, and R. E. Howard, *J. Non-Cryst. Solids*, 17, 369 (1975).
   Y. Ohmachi, *J. Opt. Soc. Am.* 63, 630 (1973).
2. Arsenic Sulfide
   D. J. Treacy in *HOCS*, 1985, p. 641.
   P. A. Young, *J. Phys.* C 4, 93 (1971).
   W. S. Rodny, I. H. Malitson, and T. A. King, *J. Opt. Soc. Am.* 48, 633 (1958).
   R. Zallen, R.E. Drew, R. L. Emerald, and M.L. Slade, *Phys. Rev. Lett.* 26, 1564 (1971).
   M. S. Maklad, R. K. Mohr, R. E. Howard, P. B. Macedo, and C. T. Moynihan, *Solid State Commun.* 15, 855 (1974).
   P. B. Klein, P. C. Taylor, and D. J. Treacy, *Phys. Rev.* B16, 4511 (1977).
   G. Lucovsky, *Phys. Rev.* B 6, 1480 (1972).
3. Cadmium Telluride
   E. D. Palik in *HOCS*, 1985, p. 409.
   D. T. F. Marple and H. Ehrenreich, *Phys. Lett.* 8, 87 (1962).
   T. H. Myers, S. W. Edwards, and J. F. Schetzina, *J. Appl. Phys.* 52, 4231 (1981).
   D. T. F. Marple, *Phys. Rev.* 150, 728 (1966).
   A. N. Pikhtin and A. D. Yas'kov, *Sov. Phys. Semicond.* 12, 622 (1978).
   L. S. Ladd, *Infrared Phys.* 6, 145 (1966).
   J. E. Harvey and W. L. Wolfe, *J. Opt. Soc. Am.* 65, 1267 (1975).
   A. Manabe, A. Mitsuishi, and H. Yoshinaga, *Jpn. J. Appl. Phys.* 6, 593 (1967).
   A. Manabe, A. Mitsuishi, H. Oshinaga, Y. Ueda, and H. Sei, *Technol. Rep. Osaka Univ. Jpn.* 17, 263 (1967).
   J. R. Birch and D. K. Murrey, *Infrared Phys.* 18, 283 (1978).
4. Gallium Arsenide
   E. D. Palik in *HOCS*, 1985, p. 429.
   M. Cardona, W. Gudat, B. Sonntag, and P. Y. Yu, in *Proc. Intl. Conf. Phys. Semicond.*, 10th. Cambridge, 1970, p. 208. US Atom. Energy Commission, Oak Ridge, TN, 1970.
   H. R. Philipp and H. Ehrenreich, *Phys. Rev.* 129, 1550 (1963).
   J. B. Theeten, D. E. Aspnes, and R. P. H. Chang, *J. Appl. Phys.* 49, 6097 (1978).
   H. C. Casey, D. D. Sell, and K. W. Wecht, *J. Appl. Phys.* 46, 250 (1975).
   A. H. Kachare, W. G. Spitzer, F. K. Euler, and A. Kahan, *J. Appl. Phys.* 45, 2938 (1974).
   R. T. Holm, J. W. Gibson, and E. D. Palik, *J. Appl. Phys.* 48, 212 (1977).
   W. Cochran, S. J. Fray, F. A. Johnson, J. E. Quarrington, and N. Williams, *J. Appl. Phys. Suppl.* 32, 2102 (1961).
   C. P. Christensen, R. Joiner, S. K. T. Nieh, and W. H. Steier, *J. Appl. Phys.* 45, 4957 (1974).
   R. H. Stolen, Phys. Rev. B 11, 767 (1975); *Appl. Phys. Lett.* 15, 74 (1969).
5. Gallium Phosphide
   A. Borghesi and G. Guizzetti in *HOCS*, 1985, p. 445.
   M. Cardona, W. Gudat, B. Sonntag, and P. Y. Yu, *Proc. Intl. Conf. Phys. Semicond.* Cambridge, 1970, p. 208. US Atom. Energy Commission, Oak Ridge, TN, 1970.
   M. Cardona, W. Gudat, E. E. Koch, M. Skibowski, B. Sonntag, and P. Yu. *Phys. Rev. Lett.* 25, 659 (1970).
   S. E. Stokowski and D. D. Sell, *Phys. Rev.* B 5, 1636 (1972).
   S. A. Abagyan, G. A. Ivanov, Y. E. Shanurin, and V. I. Amosov, *Sov. Phys. Semicond.* 5, 889 (1971).
   P. G. Dean, G. Kaminsky, and R. B. Zetterstorm, *J. Appl. Phys.* 38, 3551 (1967).

D. E. Aspnes and A. A. Studna, *Phys. Rev.* B 27, 985 (1983).

6. Indium Antimonide
R. T. Holm in *HOCS,* 1985, p. 491.
M. Cardona, W. Gudat, B. Sonntag, and P. Y. Yu, Proc. *Int. Conf. Phys. Semicond.,* 10th. Cambridge, 1970, p. 208. US Atom. Comm., Oak Ridge, TN, 1970.
H. R. Philipp and H. Ehrenreich, *Phys. Rev.* 129, 1550 (1963).
D. E. Aspnes and A. A. Studna, *Phys. Rev.* B 27, 985 (1983).
T. S. Moss, S. D. Smith, and T. D. F. Hawkins, *Proc. Phys. Soc. London* 70B, 776 (1957).
H. Yoshinaga and R. A. Oetjen, *Phys. Rev.* 101, 526 (1956).
R. B. Sanderson, *J. Phys. Chem. Solids* 26, 803 (1965).

7. Indium Arsenide
E. D. Palick and R. T. Holm in *HOCS,* 1985, p. 479.
H. R. Philipp and H. Ehrenreich, *Phys. Rev.* 129, 1550 (1963).
B. O. Seraphin and H. E. Bennett in *Semiconductors and Semimetals* (R. K. Willardson and A. C. Beer, Eds.), vol. 3, Academic, 1967, p. 499.
D. E. Aspnes and A. A. Studna, *Phys. Rev.* B 27, 985 (1983).
J. R. Dixon and J. M. Ellis, *Phys. Rev.* 123, 1560 (1961).
A. Memon, T. J. Parker, and J. R. Birch, *Proc. SPIE,* 289, 20 (1981).

8. Indium Phosphide
O. J. Glembocki and H. Piller in *HOCS,* 1985, p. 503.
M. Cardona, *J. Appl. Phys.* 32, 958 (1961); 36, 2181 (1965).
D. E. Aspnes and A. A. Studna, *Phys. Rev.* B 27, 985 (1983).
G. D. Pettit and W. J. Turner, *J. Appl. Phys.* 36, 2081 (1965).
R. Newman, *Phys. Rev.* 111, 1518 (1958).
W. N. Reynolds, M. T. Lilburne, and R. M. Dell, *Proc. Phys. Soc. London* 71, 416 (1958).
H. Jamshidi and T. J. Parker, Int. Meet. Infrared Mm. Waves, 7th., Marseilles, 1983.

9. Lead Selenide
G. Bauer and H. Krenn in *HOCS,* 1985, p. 517.
M. Cardona and D. L. Greenaway, *Phys. Rev.* A 133, 1685 (1964).
T. S. Moss, *Optical Properties of Semiconductors,* Butterworth, 1959, p. 189.
J. N. Zemel, J. D. Jensen, and R. B. Schoolar, *Phys. Rev.* A 140, 330 (1965).
W. W. Scanlon, *J. Phys. Chem. Solids,* 8, 423 (1959).
K. V. Vyatkin and A. P. Shotov, *Sov. Phys. Semicond.* 14, 785 (1980); Fiz. Tekh. Poluprovodn. 14, 1331 (1980).

10. Lead Sulfide
G. Guizzetti and A. Borghesi in *HOCS,* 1985, p. 525.
M. Cardona and R. Haensel, *Phys. Rev.* B 1, 2605 (1970).
M. Cardona and D. L. Greenaway, *Phys. Rev.* A 133, 1685 (1964).
M. Cardona, C. M. Penchina, E. E. Koch, and P. Y. Yu, *Phys. Status Solidi* B 53, 327 (1972).
P. R. Wessel, *Phys. Rev.* 153, 836 (1967).
C. E. Rossi and W. Paul, *J. Appl. Phys.* 38, 1803 (1967).
J. N. Zemel, J. D. Jensen, and R. B. Schoolar, *Phys. Rev.* A 140, 330 (1965).

11. Lead Telluride
G. Bauer and H. Krenn in *HOCS,* 1985, p. 535.
M. Cardona and R. Haensel, *Phys. Rev.* B 1, 2605 (1970).
M. Cardona and D. L. Greenaway, *Phys. Rev.* 133, A1685 (1964).
D. M. Korn and R. Braunstein, *Phys. Rev.* B 5, 4837 (1972).
W. W. Scanlon, *J. Phys. Chem. Solids* 8, 423 (1959).
J. N. Zemel, J. D. Jensen, and R. B. Schoolar, *Phys. Rev.* 140, A330 (1965).

12. Lithium Fluoride
E. D. Palik and W. R. Hunter in *HOCS,* 1985, p. 675.
B. L. Henke, P. Lee, T. J. Tanaka, R. L. Shimabukuro, and B. K. Fujikawa, *Low Energy X-ray Diagnostics-1981* (D. T. Attwood and B. L. Henke, Eds.), AIP Conf. Proc. No. 75, 1981.
A. P. Lukirskii, E. P. Savinov, O. A. Ershov, and Y. F. Shepelev, *Opt. Spektrosk.* 16, 168 (1964); 16, 310 (1964).
F. C. Brown, C. Gahwiller, A. B. Kunz, and N. O. Lipari, *Phys. Rev. Lett.* 25, 927 (1970).
A. Milgram and M. P. Givens, *Phys. Rev.* 125, 1506 (1962).
T. Tomiki and T. Miyata, *J. Phys. Soc. Jpn.* 27, 658 (1969).
A. Kachare, G. Andermann, and L. R. Brantley, *J. Phys. Chem. Solids* 33, 467 (1972).

13. Potassium Chloride
E. D. Palik in *HOCS,* 1985, p. 703.
O. Aita, I. Nagakura, and T. Sagawa, *J. Phys. Soc. Jpn.* 30, 1414 (1971).
A. P. Lukirskii, E. P. Savinov, O. A. Ershov, and Y. F. Shepelev, *Opt. Spectrosc.* 16, 168 (1964); Opt. Spektrosk. 16, 310 (1964).
T. Tomika, *J. Phys. Soc. Jpn.* 22, 463 (1967).
M. Antinori, A. Balzarotti, and M. Piacentini, *Phys. Rev.* B 7, 1541 (1973).
H. H. Li, *J. Phys. Chem. Ref. Data* 5, 329 (1976).
S. D. Allen and J. A. Harrington, *Appl. Opt.* 17, 1679 (1978).
K. W. Johnson and E. E. Bell, *Phys. Rev.* 139A, 1295 (1965).

14. Silicon Dioxide
H. R. Philipp in *HOCS,* 1985, p. 749.
J. Rife and J. Osantowski, *J. Opt. Soc. Am.* 70, 1513 (1980).
B. L. Henke, P. Lee, T. J. Tanaka, R. L. Shimabukuro, and B. K. Fujikawa, *Low Energy X-ray Diagnostics-1981* (D. T. Attwood and B. L. Henke, Eds.), AIP Conf. Proc. No. 75, 1981.
H. R. Philipp, *Solid State Commun.* 4, 73 (1966); *J. Phys. Chem. Solids,* 32, 1935 (1971).
P. L. Lamy, *Appl. Opt.* 16, 2212 (1977).
H. R. Philipp, *J. Appl. Phys.* 50 1053 (1979).
D. G. Drummond, *Proc. Roy. Soc. London,* 153, 328 (1935).

15. Silicon Monoxide
H. R. Philipp in *HOCS,* 1985, p. 765.
H. R. Philipp, *J. Phys. Chem. Solids,* 32, 1935 (1971).
G. Hass and C. D. Salzberg, *J. Opt. Soc. Am.* 44, 181 (1954).
E. Cremer, T. Kraus, and E. Ritter, *Zs. Electrochem.* 62, 939 (1958).
A. P. Bradford, G. Hass, M. McFarland, and E. Ritter, *Appl. Opt.* 4, 971 (1965).

16. Silicon Nitride
H. R. Philipp in *HOCS,* 1985, p. 771.
H. R. Philipp, *J. Electrochem. Soc.* 120, 295 (1973).
J. B. Theeten, D. E. Aspnes, F. Simondet, M. Errman, and P. C. Mürau, *J. Appl. Phys.* 52, 6788 (1981).
J. Bauer, *Phys. Status Solidi,* A 39, 411 (1977).

17. Sodium Chloride
J. E. Eldridge and E. D. Palik in *HOCS,* p. 775.
J. A. Harrington, C. J. Duthler, F. W. Patten, and M. Hass, *Solid State Commun.* 18, 1043 (1976).
T. Miyata and T. Tomiki, *J. Phys. Soc. Jpn.* 24, 1286 (1968); ibid., 22, 209 (1967).
D. M. Roessler and W. C. Walker, *J. Opt. Soc. Am.* 58, 279 (1968).
D. M. Roessler and W. C. Walker, *Phys. Rev.* 166, 599 (1968).
S. Allen and J. A. Harrington, *Appl. Opt.* 17, 1679 (1978).
O. Aita, I. Nagakura, and T. Sagawa, *J. Phys. Soc. Jpn.* 30, 1414 (1971).

18. Zinc Sulfide
E. D. Palik and A. Addamiano in *HOCS,* 1985, p. 597.
B. L. Henke, P. L. Lee, T. J. Tanaka, R. L. Shimabukuro, and B. F. Fujikawa, *Low Energy X-ray Diagnostics-1981* (D. T. Attwood and B. L. Henke, Eds.), *AIP Conf. Proc.* No. 75, 1981.
M. Cardona and G. Harbeke, *Phys. Rev.* 137, A1467 (1965).
Eastman Kodak, Publ. No. U-72, Rochester, New York (1981).
C. A. Klein and R. N. Donadio, *J. Appl. Phys.* 51, 797 (1980).
T. Deutsch, *Proc. Int. Conf. Phys. Semicond.,* 6th Exeter 1962, p. 505. The Inst. of Physics and the Physical Soc., London, 1962.
A. Manabe, A. Mitsuishi, and H. Yoshinaga, *Jpn. J. Appl. Phys.* 6, 593 (1967).
W. W. Piper, D. T. F. Marple, and P. D. Johnson, *Phys. Rev.* 110, 323 (1958).

19. Polytetrafluoroethylene
J. W. L. Thomas (NIST), Private communication.
NIST Certificate, STM 2044.
P. Y. Barnes, E. A. Early, and A. C. Parr, *NIST Special Publ. 250-48,* NIST Measurement Services: Spectral Reflectance.
Diffuse Reflectance Coatings and Materials Sections, Labsphere Catalog, 1996.
A. Arecchi and C. Ryder (Labsphere, North Sutten, NJ), private communication.

ELASTO-OPTIC, ELECTRO-OPTIC, AND MAGNETO-OPTIC CONSTANTS

When a crystal is subjected to a stress field, an electric field, or a magnetic field, the resulting optical effects are in general dependent on the orientation of these fields with respect to the crystal axes. It is useful, therefore, to express the optical properties in terms of the refractive index ellipsoid (or indicatrix):

$$\frac{x^2}{n_x^2}+\frac{y^2}{n_y^2}+\frac{z^2}{n_z^2}=1$$

or

$$\sum_{ij} B_{ij}x_i y_j = 1 \, (i,j=1,2,3)$$

where

$$B_{ij}=\left[\frac{1}{\varepsilon}\right]_{ij}=\left[\frac{1}{n^2}\right]_{ij}$$

ε is the dielectric constant or permeability; the quantity B_{ij} is called impermeability.

A crystal exposed to a *stress* **S** will show a change of its impermeability. The photo-elastic (or elasto-optic) constants, P_{ijkl}, are defined by

$$\Delta\left[\frac{1}{\varepsilon}\right]_{ij}=\Delta\left[\frac{1}{n^2}\right]_{ij}=\sum_{kl} P_{ijkl}S_{kl}$$

where n is the refractive index and S_{kl} are the strain tensor elements; the P_{ijkl} are the elements of a 4th rank tensor.

When a crystal is subjected to an *electric field* **E**, two possible changes of the refractive index may occur depending on the symmetry of the crystal.

1. All materials, including isotropic solids and polar liquids, show an electro-optic birefringence (Kerr effect) which is proportional to the square of the electric field, E:

$$\Delta\left[\frac{1}{n^2}\right]_{ij}=\sum_k K_{ijkl}E_kE_l=\sum_{k,l=1,2,3} g_{ijkl}P_kP_l$$

   where E_k and E_l are the components of the electric field and P_k and P_l the electric polarizations. The coefficients, K_{ijkl}, are the quadratic electro-optic coefficients, while the constants g_{ijkl} are known as the Kerr constants.
2. The other electro-optic effect only occurs in the 20 piezoelectric crystal classes (no center of symmetry). This effect is known as the Pockels effect. The optical impermeability changes linearly with the static field

$$\Delta\left[\frac{1}{n^2}\right]_{ij}=\sum_k r_{ij,k}E_k$$

The coefficients $r_{ij,k}$ have the name (linear) electro-optic coefficients.

The values of the electro-optic coefficients depend on the boundary conditions. If the superscripts T and S denote, respectively, the conditions of zero stress (free) and zero strain (clamped) one finds:

$$r_{ij}^{\mathrm{T}}=r_{ij}^{\mathrm{S}}+q_{ik}^{\mathrm{E}}e_{jk}=r_{ij}^{\mathrm{S}}+P_{ik}^{\mathrm{E}}d_{jk}$$

where $e_{jk}=(\partial T_k/\partial E_j)_S$ and $d_{jk}=(\partial S_k/\partial E_j)_T$ are the appropriate piezoelectric coefficients.

The interaction between a *magnetic field* and a light wave propagating in a solid or in a liquid gives rise to a rotation of the plane of polarization. This effect is known as *Faraday rotation*. It results from a difference in propagation velocity for left and right circular polarized light.

The Faraday rotation, θ_F, is linearly proportional to the magnetic field H:

$$\theta_F = VlH$$

where l is the light path length and V is the *Verdet* constant (minutes/oersted·cm).

For ferromagnetic, ferrimagnetic, and antiferromagnetic materials the magnetic field in the above expression is replaced by the magnetization M and the magneto-optic coefficient in this case is known as the Kund constant K:

Specific Faraday rotation $F=KM$

In the tables below the *Faraday rotation* is listed at the saturation magnetization per unit length, together with the absorption coefficient α, the temperature T, the critical temperature T_C (or T_N), and the wavelength of the measurement.

In the tables that follow, the properties are presented in groups:

- Elasto-optic coefficients (photoelastic constants)
- Linear electro-optic coefficients (Pockels constants)
- Quadratic electro-optic coefficients (Kerr constants)
- Magneto-optic coefficients:
  - Verdet constants
  - Faraday rotation parameters

Within each group, materials are classified by crystal system or physical state. References are given at the end of each group of tables.

ELASTO-OPTIC COEFFICIENTS (PHOTOELASTIC CONSTANTS)

Name Cubic (43m, 432, m3m)	Formula	λ/μm	p_{11}	p_{12}	p_{44}	p_{11}-p_{12}	Ref.
Sodium fluoride	NaF	0.633	0.08	0.20	−0.03	−0.12	1
Sodium chloride	NaCl	0.589	0.115	0.159	−0.011	−0.042	2
Sodium bromide	NaBr	0.589	0.148	0.184	−0.0036	−0.035	1
Sodium iodide	NaI	0.589	–	–	0.0048	−0.0141	3
Potassium fluoride	KF	0.546	0.26	0.20	−0.029	0.06	1
Potassium chloride	KCl	0.633	0.22	0.16	−0.025	0.06	4
Potassium bromide	KBr	0.589	0.212	0.165	−0.022	0.047	5
Potassium iodide	KI	0.590	0.212	0.171	–	0.041	6
Rubidium chloride	RbCl	0.589	0.288	0.172	−0.041	0.116	7,8
Rubidium bromide	RbBr	0.589	0.293	0.185	−0.034	0.108	7,8
Rubidium iodide	RbI	0.589	0.262	0.167	−0.023	0.095	7,8
Lithium fluoride	LiF	0.589	0.02	0.13	−0.045	−0.11	5
Lithium chloride	LiCl	0.589	–	–	−0.0177	−0.0407	3
Ammonium chloride	NH_4Cl	0.589	0.142	0.245	0.042	−0.103	9
Cadmium telluride	CdTe	1.06	−0.152	−0.017	−0.057	−0.135	10
Calcium fluoride	CaF_2	0.55–0.65	0.038	0.226	0.0254	−0.183	11
Copper chloride	CuCl	0.633	0.120	0.250	−0.082	−0.130	12
Copper bromide	CuBr	0.633	0.072	0.195	−0.083	−0.123	12
Copper iodide	CuI	0.633	0.032	0.151	−0.068	−0.119	12
Diamond	C	0.540–0.589	−0.278	0.123	−0.161	−0.385	13
Germanium	Ge	3.39	−0.151	−0.128	−0.072	−0.023	14
Gallium arsenide	GaAs	1.15	−0.165	−0.140	−0.072	−0.025	15
Gallium phosphide	GaP	0.633	−0.151	−0.082	−0.074	−0.069	15
Strontium fluoride	SrF_2	0.633	0.080	0.269	0.0185	−0.189	16
Strontium titanate	$SrTiO_3$	0.633	0.15	0.095	0.072	–	17
KRS-5	Tl(Br,I)	0.633	−0.140	0.149	−0.0725	−0.289	18,20
KRS-6	Tl(Br,Cl)	0.633	−0.451	−0.337	−0.164	−0.114	19,20
Zinc sulfide	ZnS	0.633	0.091	−0.01	0.075	0.101	15

Rare Gases	Formula	λ/μm	p_{11}	p_{12}	p_{44}	p_{11}-p_{12}	Ref.
Neon (T = 24.3 K)	Ne	0.488	0.157	0.168	0.004	−0.011	21
Argon (T = 82.3 K)	Ar	0.488	0.256	0.302	0.015	−0.046	22
Krypton (T = 115.6 K)	Kr	0.488	0.34	0.34	0.037	0	21
Xenon (T = 160.5 K)	Xe	0.488	0.284	0.370	0.029	−0.086	22

Garnets	Formula	λ/μm	p_{11}	p_{12}	p_{44}	p_{11}-p_{12}	Ref.
GGG	$Gd_3Ga_5O_{12}$	0.514	−0.086	−0.027	−0.078	−0.059	23
YIG	$Y_3Fe_5O_{12}$	1.15	0.025	0.073	0.041	–	15
YGG	$Y_3Ga_5O_{12}$	0.633	0.091	0.019	0.079	–	17
YAG	$Y_3Al_5O_{12}$	0.633	−0.029	0.0091	−0.0615	−0.038	15

Cubic (23, m3)	Formula	λ/μm	p_{11}	p_{12}	p_{44}	p_{13}	Ref.
Barium nitrate	$Ba(NO_3)_2$	0.589	–	$p_{11}-p_{22}$ = 0.992	−0.0205	$p_{11}-p_{13}$ = 0.713	13
Lead nitrate	$Pb(NO_3)_2$	0.589	0.162	0.24	−0.0198	0.20	24,25
Sodium bromate	$NaBrO_3$	0.589	0.185	0.218	−0.0139	0.213	26
Sodium chlorate	$NaClO_3$	0.589	0.162	0.24	−0.0198	0.20	26
Strontium nitrate	$Sr(NO_3)_2$	0.41	0.178	0.362	−0.014	0.316	27

Hexagonal (mmc, 6mm)	Formula	λ/μm	p_{11}	p_{12}	p_{13}	p_{31}	p_{33}	p_{44}	Ref.
Beryl	$Be_3Al_2Si_6O_{18}$	0.589	0.0099	0.175	0.191	0.313	0.023	−0.152	28
Cadmium sulfide	CdS	0.633	−0.142	−0.066	−0.057	−0.041	−0.20	−0.099	15,2
Zinc oxide	ZnO	0.633	±0.222	±0.099	−0.111	±0.088	−0.235	0.0585	30
Zinc sulfide	ZnS	0.633	−0.115	0.017	0.025	0.0271	−0.13	−0.0627	31

Trigonal (3m, 32, $\bar{3}$m)	Formula	λ/µm	p_{11}	p_{12}	p_{13}	p_{14}	p_{31}	p_{33}	p_{41}	p_{44}	Ref.
Sapphire	Al_2O_3	0.644	−0.23	−0.03	0.02	0.00	−0.04	−0.20	0.01	−0.10	15,32
Calcite	$CaCO_3$	0.514	0.062	0.147	0.186	−0.011	0.241	0.139	−0.036	−0.058	33
Lithium niobate	$LiNbO_3$	0.633	±0.034	±0.072	±0.139	±0.066	±0.178	±0.060	±0.154	±0.300	15,34
Lithium tantalate	$LiTaO_3$	0.633	−0.081	0.081	0.093	−0.026	0.089	−0.044	−0.085	0.028	15,35
Cinnabar	HgS	0.633			±0.445			±0.115	–	–	36
Quartz	SiO_2	0.589	0.16	0.27	0.27	−0.030	0.29	0.10	−0.047	−0.079	37
Proustite	Ag_3AsS_3	0.633	±0.10	±0.19	±0.22		±0.24	±0.20	–	–	38
Sodium nitrite	$NaNO_2$	0.633		±0.21	±0.215	±0.027	±0.25		0.055	−0.06	39
Tellurium	Te	10.6	0.155	0.130	–	–	–		–	–	15

Tetragonal (4/mmm, $\bar{4}$2m, 422)	Formula	λ/µm	p_{11}	p_{12}	p_{13}	p_{31}	p_{33}	p_{44}	p_{66}	Ref.
Ammonium dihydrogen phosphate	ADP	0.589	0.319	0.277	0.169	0.197	0.167	−0.058	−0.091	40
Barium titanate	$BaTiO_3$	0.633	0.425	–	–	–	–	–	–	41
Cesium dihydrogen arsenate	CDA	0.633	0.267	0.225	0.200	0.195	0.227	–	–	42
Magnesium fluoride	MgF_2	0.546	–	–	–	–	–	±0.0776	±0.0488	43
Calomel	Hg_2Cl_2	0.633	±0.551	±0.440	±0.256	±0.137	±0.010	–	±0.047	44
Potassium dihydrogen phosphate	KDP	0.589	0.287	0.282	0.174	0.241	0.122	−0.019	−0.064	45
Rubidium dihydrogen arsenate	RDA	0.633	0.227	0.239	0.200	0.205	0.182	–	–	41
Rubidium dihydrogen phosphate	RDP	0.633	0.273	0.240	0.218	0.210	0.208	–	–	41
Strontium barium niobate	$Sr_{0.75}Ba_{0.25}Nb_2O_6$	0.633	0.16	0.10	0.08	0.11	0.47	–	–	46
Strontium barium niobate	$Sr_{0.5}Ba_{0.5}Nb_2O_6$	0.633	0.06	0.08	0.17	0.09	0.23	–	–	46
Tellurium oxide	TeO_2	0.633	0.0074	0.187	0.340	0.090	0.240	−0.17	−0.046	47
Rutile	TiO_2	0.633	0.017	0.143	−0.139	−0.080	−0.057	−0.009	−0.060	48

Tetragonal (4, $\bar{4}$, 4/m)	Formula	λ/µm	p_{11}	p_{12}	p_{13}	p_{16}	p_{31}	p_{33}	p_{44}	p_{45}	p_{61}	p_{66}	Ref.
Cadmium molybdate	$CdMoO_4$	0.633	0.12	0.10	0.13	–	0.11	0.18	–	–	–	–	49
Lead molybdate	$PbMoO_4$	0.633	0.24	0.24	0.255	0.017	0.175	0.300	0.067	−0.01	0.013	0.05	52
Sodium bismuth molybdate	$NaBi(MoO_4)_2$	0.633	0.243	0.205	0.25	–	0.21	0.29	–	–	–	–	–

Orthorhombic (222, m22, mmm)	Formula	λ/µm	p_{11}	p_{12}	p_{13}	p_{21}	p_{22}	p_{23}	p_{31}	p_{32}	p_{33}	p_{44}	p_{55}	p_{66}	Ref.
Ammonium chlorate	NH_4ClO_3	0.633	–	0.24	0.18	0.23	–	0.20	0.19	0.18	±0.02	<±0.02	–	±0.04	51
Ammonium sulfate	$(NH_4)_2SO_4$	0.633	0.26	0.19	±0.260	±0.230	±0.27	±0.254	0.20	±0.26	0.26	0.015	±0.0015	0.012	52
Rochelle salt	$NaKC_4H_4O_6$	0.589	0.35	0.41	0.42	0.37	0.28	0.34	0.36	0.35	0.36	−0.030	0.0046	−0.025	53
Iodic acid (α)	HIO_3	0.633	0.302	0.496	0.339	0.263	0.412	0.304	0.251	0.345	0.336	0.084	−0.030	0.098	54
Sulfur (α)	S	0.633	0.324	0.307	0.268	0.272	0.301	0.310	0.203	0.232	0.270	0.143	0.019	0.118	54
Barite	$BaSO_4$	0.589	0.21	0.25	0.16	0.34	0.24	0.19	0.28	0.22	0.31	0.002	−0.012	0.037	55
Topaz	$Al_2SiO_4(OH,F)_2$	–	−0.085	0.069	0.052	0.095	−0.120	0.065	0.095	0.085	−0.083	−0.095	−0.031	0.098	28

Monoclinic (2, m, 2/m)	Formula	λ/µm			
Taurine	$C_2H_7NO_3S$	0.589	p_{11} = 0.313	p_{25} = −0.0025	p_{51} = −0.014
			p_{12} = 0.251	p_{31} = 0.362	p_{52} = 0.006
			p_{13} = 0.270	p_{32} = 0.275	p_{53} = 0.0048
			p_{15} = −0.10	p_{33} = 0.308	p_{55} = 0.047
			p_{21} = 0.281	p_{35} = −0.003	p_{64} = 0.0024
			p_{22} = 0.252	p_{44} = 0.0025	p_{66} = 0.0028
			p_{23} = 0.272	p_{46} = −0.0056	

Isotropic	Formula	λ/μm	p_{11}	p_{12}	p_{44}	Ref.
Fused silica	SiO_2	0.633	0.121	0.270	−0.075	15
Water	H_2O	0.633	±0.31	±0.31		15
Polystyrene		0.633	±0.30	±0.31		25
Lucite		0.633	±0.30	0.28		25
Orpiment	As_2S_3-glass	1.15	0.308	0.299	0.0045	15
Tellurium oxide	TeO_2-glass	0.633	0.257	0.241	0.0079	56
Laser glasses	LGS-247-2	0.488	±0.168	±0.230		57
	LGS-250-3		±0.135	±0.198		
	LGS-1		±0.214	±0.250		
	KGSS-1621		±0.205	±0.239		
Dense flint glasses	LaSF	0.633	0.088	0.147	−0.030	58
(examples)	SF_4		0.215	0.243	−0.014	
	U10502		0.172	0.179	−0.004	
	$TaFd_7$		0.099	0.138	−0.020	

References

A. Narasimhamurty, T. S., *Photoelastic and Electro-Optic Properties of Crystals*, Plenum Press, New York, 1981, pp. 290–293.

B. Weber, M. J., Ed., *CRC Handbook of Laser Science and Technology*, Volume IV, Part 2, CRC Press, Boca Raton, FL, 1986, pp. 324–331.

1. Petterson, H. E., *J. Opt. Soc. Am.*, 63, 1243, 1973.
2. Burstein, E. and Smith, P. L., *Phys. Rev.*, 74, 229, 1948.
3. Pakhnev, A. V., et al., *Sov. Phys. J. (transl.)*, 18, 1662, 1975.
4. Feldman, A., Horovitz, D., and Waxler, R. M., *Appl. Opt.*, 16, 2925, 1977.
5. Iyengar, K. S., *Nature (London)*, 176, 1119, 1955.
6. Bansigir, K. G. and Iyengar, K. S., *Acta Crystallogr.*, 14, 727, 1961.
7. Pakhev, A. V., et al., *Sov. Phys. J. (transl.)*, 20, 648, 1975.
8. Bansigir, K. G., *Acta Crystallogr.*, 23, 505, 1967.
9. Krishna Rao, K. V. and Krishna Murty, V. G., *Ind. J. Phys.*, 41, 150, 1967.
10. Weil, R. and Sun, M. J., *Proc. Int. Symp. CdTe (Detectors)*, Strasbourg Centre de Rech. Nucl., 1971, XIX-1 to 6, 1972.
11. Schmidt, E. D. D. and Vedam, K., *J. Phys. Chem. Solids*, 27, 1563, 1966.
12. Biegelsen, D. K., et al., *Phys. Rev. B*, 14, 3578, 1976.
13. Helwege, K. H., *Landolt-Börnstein, New Series Group III*, Vol. II, Springer-Verlag, Berlin, 1979.
14. Feldman, A., Waxler, R. M., and Horovitz, D., *J. Appl. Phys.*, 49, 2589, 1978.
15. Dixon, R. W., *J. Appl. Phys.*, 38, 5149, 1967.
16. Shabin, O. V., et al., *Sov. Phys. Sol. State (transl.)*, 13, 3141, 1972.
17. Reintjes, J. and Schultz, M. B., *J. Appl. Phys.*, 39, 5254, 1968.
18. Rivoallan, L. and Favre, F., *Opt. Commun.*, 8, 404, 1973.
19. Rivoallan, L. and Favre, F., *Opt. Commun.*, 11, 296, 1974.
20. Afanasev, I. I., et al., *Sov. J. Opt. Technol.*, 46, 663, 1979.
21. Rand, S. C., et al., *Phys. Rev. B*, 19, 4205, 1979.
22. Sipe, J. E., *Can J. Phys.*, 56, 199, 1978.
23. Christyi, I. L., et al., *Sov. Phys. Sol. State (transl.)*, 17, 922, 1975.
24. Narasimhamurty, T. S., *Curr. Sci. (India)*, 23, 149, 1954.
25. Smith, T. M. and Korpel, A., *IEEE J. Quant. Electron.*, QE-1, 283, 1965.
26. Narasimhamurty, T. S., *Proc. Ind. Acad. Sci.*, A40, 164, 1954.
27. Rabman, A., *Bhagarantam Commem. Vol.*, Bangalore Print. and Publ., 173, 1969.
28. Eppendahl, R., *Ann. Phys. (IV)*, 61, 591, 1920.
29. Laurenti, J. P. and Rouzeyre, M., *J. Appl. Phys.*, 52, 6484, 1981.
30. Sasaki, H., et al., *J. Appl. Phys.*, 47, 2046, 1976.
31. Uchida, N. and Saito, S., *J. Appl. Phys.*, 43, 971, 1972.
32. Waxler, R. M. and Farabaugh, E. M., *J. Res. Natl. Bur. Stand.*, A74, 215, 1970.
33. Nelson, D. F., Lazay, P. D., and Lax, M., *Phys. Rev.*, B6, 3109, 1972.
34. O'Brien, R. J., Rosasco, G. J., and Weber, A., *J. Opt. Soc. Am.*, 60, 716, 1970.
35. Avakyants, L. P., et al., *Sov. Phys.*, 18, 1242, 1976.
36. Sapriel, J., *Appl. Phys. Lett.*, 19, 533, 1971.
37. Narasimhamurty, T. S., *J. Opt. Soc. Am.*, 59, 682, 1969.
38. Zubrinov, I. I., et al., *Sov. Phys. Sol. State (transl.)*, 15, 1921, 1974.
39. Kachalov, O. V. and Shpilko, I. O., *Sov. Phys. JETP (transl.)*, 35, 957, 1972.
40. Narasimhamurty, T. S., et al., *J. Mater. Sci.*, 8, 577, 1973.
41. Tada, K. and Kikuchi, K., *Jpn. J. Appl. Phys.*, 19, 1311, 1980.
42. Aleksandrov, K. S., et al., *Sov. Phys. Sol. State (transl.)*, 19, 1090, 1977.
43. Afanasev, I. I., et al., *Sov. Phys. Sol. State (transl.)*, 17, 2006, 1975.
44. Silvestrova, I. M., et al., *Sov. Phys. Cryst. (transl.)*, 20, 649, 1975.
45. Veerabhadra Rao, K. and Narasimhamurty, T. S., *J. Mater. Sci.*, 10, 1019, 1975.
46. Venturini, E. L., et al., *J. Appl. Phys.*, 40, 1622, 1969.
47. Vehida, N. and Ohmachi, Y., *J. Appl. Phys.*, 40, 4692, 1969.
48. Grimsditch, M. H. and Ramdus, A. K., *Phys. Rev. B*, 22, 4094, 1980.
49. Schinke, D. P. and Viehman,W., unpublished data.
50. Coquin, G. A., et al., *J. Appl. Phys.*, 42, 2162, 1971.
51. Vasquez, F., et al., *J. Phys. Chem. Solids*, 37, 451, 1976.
52. Luspin, Y. and Hauret, G., *C.R.Ac. Sci. Paris*, B274, 995, 1972.
53. Narasimhamurty, T. S., *Phys. Rev.*, 186, 945, 1969.
54. Haussühl, S. and Weber, H. J., *Z. Kristall.*, 132, 266, 1970.
55. Vedam, K., *Proc. Ind. Ac. Sci.*, A34, 161, 1951.
56. Yano, T., Fukumoto, A., and Watanabe, A., *J. Appl. Phys.*, 42, 3674, 1971.
57. Manenkov, A. A. and Ritus, A. I., *Sov. J. Quant. Electr.*, 8, 78, 1978.
58. Eschler, H. and Weidinger, F., *J. Appl. Phys.*, 46, 65, 1975.

LINEAR ELECTRO-OPTIC COEFFICIENTS

Name Cubic ($\bar{4}3m$)	Formula	λ/μm	r_{41} pm/V
Cuprous bromide	CuBr	0.525	0.85
Cuprous chloride	CuCl	0.633	3.6
Cuprous iodide	CuI	0.55	–5.0
Eulytite (BSO)	$Bi_4Si_3O_{12}$	0.63	0.54
Germanium eulytite (BGO)	$Bi_4Ge_3O_{12}$	0.63	1.0
Gallium arsenide	GaAs	10.6	1.6
Gallium phosphide	GaP	0.56	–1.07
Hexamethylenetetramine	$C_6H_{12}N_4$	0.633	0.78
Sphalerite	ZnS	0.65	2.1
Zinc selenide	ZnSe	0.546	2.0
Zinc telluride	ZnTe	3.41	4.2
Cadmium telluride	CdTe	3.39	6.8

Cubic (23)	Formula	λ/μm	r_{41} pm/V
Ammonium chloride (77 K)	NH_4Cl	–	1.5
Ammonium cadmium langbeinite	$(NH_4)_2Cd_2(SO_4)_3$	0.546	0.70
Ammonium manganese langbeinite	$(NH_4)_2Mn_2(SO_4)_3$	0.546	0.53
Thallium cadmium langbeinite	$Tl_2Cd_2(SO_4)_3$	0.546	0.37
Potassium magnesium langbeinite	$K_2Mg_2(SO_4)_3$	0.546	0.40
Bismuth monogermanate	$Bi_{12}GeO_{20}$	–	3.3
Bismuth monosilicate	$Bi_{12}SiO_{20}$	–	3.3
Sodium chlorate	$NaClO_3$	0.589	0.4
Sodium uranyl acetate	$NaUO_2(CH_3COO)_3$	0.546	0.87
Trenhydrobromide	$N(CH_2CH_2NH_2)_3$ 3HBr	–	1.5
Trenhydrochloride	$N(CH_2CH_2NH_2)_3$ 3HCl	–	1.7

Tetragonal ($\bar{4}2m$)	Formula	T_{tran} K	r_{41} pm/V	r_{63} pm/V
Ammonium dihydrogen phosphate (ADP)	$NH_4H_2PO_4$	148	24.5	–8.5
Ammonium dideuterium phosphate (AD*P)	$NH_4D_2PO_4$	242	–	11.9
Ammonium dihydrogen arsenate (ADA)	$NH_4H_2AsO_4$	–	–	9.2
Cesium dihydrogen arsenate (CsDA)	CsH_2AsO_4	143	–	18.6
Cesium dideuterium arsenate (CsD*A)	CsD_2AsO_4	212	–	36.6
Potassium dihydrogen phosphate (KDP)	KH_2PO_4	123	8.6	–10.5
Potassium dideuterium phosphate (KD*P)	KD_2PO_4	222	8.8	23.8
Potassium dihydrogen arsenate (KDA)	KH_2AsO_4	97	12.5	10.9
Potassium dideuterium arsenate (KD*A)	KD_2AsO_4	162	–	18.2
Rubidium dihydrogen phosphate (RDP)	RbH_2PO_4	147	–	15.5
Rubidium dihydrogen arsenate (RDA)	RbH_2AsO_4	110	–	13.0
Rubidium dideuterium arsenate (RD*A)	RbD_2AsO_4	178	–	21.4

Tetragonal (4mm)	Formula	T_{tran} K	r_{13} pm/V	r_{33} pm/V	r_{51} pm/V
Barium titanate	$BaTiO_3$	406	8	28	–
Potassium lithium niobate	$K_3Li_2Nb_5O_{15}$	693	8.9	5.9	–
Lead titanate	$PbTiO_3$	765	13.8	5.9	–
Strontium barium niobate (SBN75)	$Sr_{0.75}Ba_{0.25}Nb_2O_6$	330	6.7	1340	42
Strontium barium niobate (SBN46)	$Sr_{0.46}Ba_{0.54}Nb_2O_6$	602	~180	35	–

Hexagonal (6mm)	Formula	r_{13} pm/V	r_{33} pm/V	r_{42} pm/V	r_{51} pm/V
Greenockite	CdS	3.1	2.9	2.0	3.7
Greenockite (const. strain)	CdS	1.1	2.4	–	–
Wurzite	ZnS	0.9	1.8	–	–
Zincite	ZnO	−1.4	+2.6	–	–

Hexagonal (6)	Formula	r_{13} pm/V	r_{33} pm/V	r_{42} pm/V	r_{51} pm/V
Lithium iodate	$LiIO_3$	4.1	6.4	1.4	3.3
Lithium potassium sulfate	$LiKSO_4$	$r_{13}-r_{33}$ = 1.6	–	–	–

Trigonal (3m)	Formula	T_{tran} K	r_{13} pm/V	r_{22} pm/V	r_{33} pm/V	r_{42} pm/V
Cesium nitrate	$CsNO_3$	425	–	0.43	–	–
Lithium niobate	$LiNbO_3$	1483	8.6	7.0	30.8	28
Lithium tantalate	$LiTaO_3$	890	8.4	–	30.5	–
Lithium sodium sulfate	$LiNaSO_4$	–	–	<0.02	–	–
Tourmaline	–	–	–	0.3	–	–

Trigonal (32)	Formula	T_{tran} K	r_{11} pm/V	r_{41} pm/V
Cesium tartrate	$Cs_2C_4H_4O_6$	–	1.0	
Cinnabar	HgS	659	3.1	1.5
Potassium dithionate	$K_2S_2O_6$	–	0.26	–
Strontium dithionate	$SrS_2O_6{\cdot}4H_2O$	–	0.1	–
Quartz	SiO_2	1140	−0.47	0.2
Selenium	Se	398	2.5	

Orthorhombic (222)	Formula	T_{tran} K	r_{41} pm/V	r_{52} pm/V	r_{63} pm/V
Ammonium oxalate	$(NH_4)_2C_2O_4{\cdot}4H_2O$	–	230	330	250
Rochelle salt	$KNaC_4H_4O_6{\cdot}4H_2O$	T_u = 297 T_l = 255	−2.0	−1.7	+0.32

Orthorhombic (mm2)	Formula	*T*trans K	r_{13} pm/V	r_{23} pm/V	r_{33} pm/V	r_{42} pm/V	r_{51} pm/V
Barium sodium niobate (BSN)	Ba_2NaNbO_{15}	833	15	13	48	92	90
Potassium niobate	$KNbO_3$	476	28	1.3	64	380	105

Monoclinic (2)	Formula	T_{trans} K	r_{22} pm/V	r_{32} pm/V
Calcium pyroniobate	$Ca_2Nb_2O_7$	–	0.33	13.7
Triglycine sulfate (TGS)	$(NH_2CH_2COOH)_3{\cdot}H_2SO_4$	322	7.2	13.6

References

1. Narasimhamurty, T. S., *Photoelastic and Electro-Optic Properties of Crystals*, Plenum Press, New York, 1981, pp. 405–407.
2. Weber, M. J., Ed., *CRC Handbook of Laser Science and Technology*, Vol. IV, CRC Press, Boca Raton, FL, 1986, pp. 258–278.

QUADRATIC ELECTRO-OPTIC COEFFICIENTS

Kerr Constants of Ferroelectric Crystals[1,2]

Name	Formula	T_{tran} K	λ μm	g_{11} 10^{10} esu	g_{12} 10^{10} esu	$g_{11}-g_{12}$ 10^{10} esu	g_{44} 10^{10} esu
Barium titanate	$BaTiO_3$	406	0.633	1.33	−0.11	1.44	
Strontium titanate	$SrTiO_3$	–	0.633	–	–	1.56	–
Potassium tantalate niobate	$KTa_{0.65}Nb_{0.35}O_3$	330	0.633	1.50	−0.42	1.92	1.63
Potassium tantalate	$KTaO_3$	13	0.633	–	–	1.77	1.33
Lithium niobate	$LiNbO_3$	1483	–	0.94	0.25	0.7	0.6
Lithium tantalate	$LiTaO_3$	938	–	1.0	0.17	0.8	0.7
Barium sodium niobate (BSN)	$Ba_{0.8}Na_{0.4}Nb_2O_6$	833	–	1.55	0.44	1.11	

Kerr Constants of Selected Liquids[2]

K is the Kerr constant at a wavelength of 589 nm and at room temperature; ε is the static dielectric constant; t_m is the melting point; and t_b is the normal boiling point

Name	Molecular formula	*K* 10^{-7} esu	ε	t_m °C	t_b °C
Carbon disulfide	CS_2	+3.23	2.63	−111.5	+46.3
Acetone	C_3H_6O	+16.3	21.0	−94.8	+56.1
Methyl ethyl ketone	C_4H_8O	+13.6	18.56	−86.67	+79.6
Pyridine	C_5H_5N	+20.4	13.26	−42	+115.23
Ethyl cyanoacetate	$C_5H_7NO_2$	+38.8	31.6	−22.5	205
o-Dichlorobenzene	$C_6H_4Cl_2$	+42.6	10.12	−16.7	180
Benzenesulfonyl chloride	$C_6H_5ClO_2S$	+89.9	28.90	+14.5	247
Nitrobenzene	$C_6H_5NO_2$	+326	35.6	+5.7	210.8
Ethyl 3-aminocrotonate	$C_6H_{11}NO_2$	+31.0	–	+33.9	210
Paraldehyde	$C_6H_{12}O_3$	−23.0	14.7	+12.6	124
			12.0[a]		
Benzaldehyde	C_7H_6O	+80.8	17.85	−26	179.05
			14.1[a]		
p-Chlorotoluene	C_7H_7Cl	+23.0	6.25	+7.5	162.4
o-Nitrotoluene	$C_7H_7NO_2$	+174	26.26	−10	222.3
m-Nitrotoluene	$C_7H_7NO_2$	+177	24.95	+15.5	232
p-Nitrotoluene	$C_7H_7NO_2$	+222	22.2	+51.6	238.3
Benzyl alcohol	C_7H_8O	−15.4	11.92	−15.3	205.8
			10.8[a]		
m-Cresol	C_7H_8O	+21.2	12.44	+11.8	202.27
			5.0[a]		
m-Chloroacetophenone	C_8H_7ClO	+69.1			
Acetophenone	C_8H_8O	+66.6	17.44	+19.7	202.3
			15.8[a]		
Quinoline	C_9H_7N	+15.0	9.16	−14.78	237.16
Ethyl salicylate	$C_9H_{10}O_3$	+19.6	8.48	+1.3	231.5
Carvone	$C_{10}H_{14}O$	+23.6	11.2	<0	230
Ethyl benzoylacetate	$C_{11}H_{12}O_3$	+16.0	13.50	<0	270
Water	H_2O	+4.0	80.10	0.00	100.0

[a] Dielectric constant at radio frequencies (108–109 Hz).

References

1. Narasimhamurty, T. S., *Photoelastic and Electro-Optic Properties of Crystals*, Plenum Press, New York, 1981, p. 408.
2. Gray, D. E., Ed., *AIP Handbook of Physics*, McGraw Hill, New York, 1972, p. 6–241.

MAGNETO–OPTIC CONSTANTS

Verdet Constants of Non-Magnetic Crystals[1]

V is the Verdet constant; *n* is the refractive index; and λ is the wavelength. "min" is minutes of angle.

Material	*T* K	λ nm	*n*	*V* min/Oe cm
Al_2O_3	300	546.1	1.771	0.0240
	300	589.3	1.768	0.0210
$BaTaO_3$	403	427		0.95
	403	496		0.38
	403	620		0.18
	403	826		0.072
$Bi_4Ge_3O_{12}$	300	442	2.077	0.289
	300	632.8	2.048	0.099
	300	1064	2.031	0.026
C (diamond)	300	589.3	2.417	0.0233
$CaCO_3$	300	589.3	1.658	0.019
CaF_2	300	589.3	1.434	0.0088
$Cd_{0.55}Mn_{0.45}Te$	300	632.8		6.87
CuCl	300	546.1	1.93	0.20
GaSe	298	632.8		0.80
$KAl(SO_4)_2{\cdot}12H_2O$	300	589.3	1.456	0.0124
KBr	300	546.1	1.564	0.0500
	300	589.3	1.560	0.0425
KCl	300	589.3	1.490	0.0275
KI	300	546.1	1.673	0.083
	300	589.3	1.666	0.070
$KTaO_3$	296	352		0.44
	296	413		0.19
	296	496		0.096
	296	620		0.051
	296	826		0.022
LaF_3	300	325	1.639	0.054
(H‖c)	300	442	1.615	0.028
	300	632.8	1.601	0.012
	300	1064	1.592	0.006
$MgAl_2O_4$	300	589.3	1.718	0.021
$NH_4AlSO_4{\cdot}12H_2O$	300	589.3	1.459	0.0128
NH_4Br	300	589.3	1.711	0.0504
NH_4Cl	300	546.1		0.0410
	300	589.3	1.643	0.0362
NaBr	300	546.1		0.0621
NaCl	300	546.1		0.0410
	300	589.3	1.544	0.0345
$NaClO_3$	300	546.1		0.0105
	300	589.3	1.515	0.0081
$NiSO_4{\cdot}6H_2O$	297	546.1		0.0256
	297	589.3	1.511	0.0221
SiO_2	300	546.1	1.546	0.0195
	300	589.3	1.544	0.0166
$SrTiO_3$	298	413	2.627	0.78
	298	496		0.31
	298	620		0.14
	298	826		0.066
ZnS	300	546.1		0.287
	300	589.3	2.368	0.226
ZnSe	300	476	2.826	1.50
	300	496	2.759	1.04
	300	514	2.721	0.839
	300	587	2.627	0.529
	300	632.8	2.592	0.406

Verdet Constants of Rare-Earth Aluminum Garnets at Various Wavelengths[1]

The absorption coefficient α for these materials ranges from 0.2 to 0.6 cm^{-1} at 300 K.

		V in min/Oe cm							
Material	***T*/K**	**λ = 405 nm**	**450 nm**	**480 nm**	**520 nm**	**546 nm**	**578 nm**	**635 nm**	**670 nm**
$Tb_2Al_5O_{12}$	300	−2.266	−1.565	−1.290	−1.039	−0.912	−0.787	−0.620	−0.542
	77		−102.16	−83.45	−3.425	−3.051	−2.603	−2.008	−1.815
	4.2				−64.80	−58.35	−53.77	48.39	−45.15
	1.45		−200.95	−172.52	−139.28	−125.07	−111.27	97.47	−93.42
$Dy_3Al_5O_{12}$	300	−1.241	−0.942	−0.803	−0.667	−0.592	−0.518	−0.411	−0.359
$Ho_3Al_5O_{12}$	300	−0.709	−0.320	−0.260	−0.335	−0.304	−0.299		−0.206
$Er_3Al_5O_{12}$	300	−0.189	−0.240	−0.154	−0.162	−0.157	−0.145	−0.105	−0.089
$Tm_3Al_5O_{12}$	300	+0.151	+0.103	+0.093	0.076	0.069	+0.059	+0.048	
$Yb_3Al_5O_{12}$	298	0.287	0.215	0.186	0.140	0.133	0.116	0.094	
	77	0.718	0.540	0.481	0.393	0.342	0.302	0.239	

Verdet Constants for KDP-Type Crystals[1]

Measurements refer to T = 298 K and λ = 632.8 nm, with $\mathbf{k} \parallel [001]$.

Material	***V* min/Oe cm**
KH_2PO_4 (KDP)	0.0124
$KH_{0.3}D_{1.7}PO_4$ (KD*P)	0.145
$NH_4H_2PO_4$ (ADP)	0.138
KH_2AsO_4 (KDA)	0.238
$KH_{0.1}D_{1.9}AsO_4$ (KD*A)	0.245
$NH_4H_2AsO_4$ (ADH)	0.244

Verdet Constants of Gases[2]

Values refer to T = 0 °C and P = 101.325 kPa (760 mmHg); n_D is the refractive index at a wavelength of 589 nm.

Gas	**$(n_D - 1) \times 10^3$**	**$10^6 \times V$ min/Oe cm**
He	0.036	+0.40
Ar	2.81	+9.36
H_2		+6.29
N_2	0.297	+6.46
O_2	0.272	+5.69
Air	0.293	+6.27
Cl_2	0.773	+31.9
HCl	0.447	+21.5
H_2S	0.63	+41.5
NH_3	0.376	+19.0
CO	0.34	+11.0
CO_2	0.45	+9.39
NO	0.297	−58
CH4	0.444	+17.4
n-C_4H_{10}		+44.0

Verdet Constants of Liquids[2]

n_D is the refractive index at a wavelength of 589 nm and a temperature of 20 °C, unless otherwise indicated. V is the Verdet constant.

Liquid	**λ/nm**	***t*/°C**	**$10^2 \times V$ min/Oe cm**	**n_D**
P	589	33	+13.3	
S	589	114	+8.1	1.929 (110 °C)
H_2O	589	20	+1.309	1.3328
D_2O	589	19.7	+1.257	1.3384
H_3PO_4	578	97.4	+1.35	
CS_2	589	20	+4.255	1.6255
CCl_4	578–589	25.1	+1.60	1.463 (15 °C)
$SbCl_5$	578	18	+7.45	1.601 (14 °C)
$TiCl_4$	578	17	−1.65	1.61
$TiBr_4$	578	46	−5.3	
Methanol	589	18.7	+0.958	1.3289
Acetone	578–589	20.0	+1.116	1.3585
Toluene	578–589	15.0	+2.71	1.4950
Benzene	578–589	15.0	+3.00	1.5005
Chlorobenzene	589	15	+2.92	1.5246
Nitrobenzene	589	15	+2.17	1.5523
Bromoform	589	17.9	+3.13	1.5960

Verdet Constants of Rare-Earth Paramagnetic Crystals[1]

n is the refractive index, and *V* is the Verdet constant at the wavelength and temperature indicated.

Rare Earth	Host	*T*/K	λ/nm	*n*	*V* min/Oe cm
Ce^{3+}(30%)	CaF_2	300	325	1.516	−0.956
		300	442	1.502	−0.297
		300	633	1.494	−0.111
		300	1064	1.489	−0.035
Ce^{3+}	CeF_3	300	442	1.613	−1.05
		300	633	1.598	−0.406
		77	633		−1.418
		300	1064	1.589	−0.113
Pr^{3+}(5%)	CaF_2	300	266	1.471	−0.172
		300	325	1.461	−0.0818
		300	442	1.451	−0.0089
		300	633	1.445	−0.0168
		300	1064	1.441	−0.0045
Nd^{3+}(2.9%)	CaF_2	4.2	426		−0.19
Nd^{3+}	NdF_3	300	442	1.60	−0.553
		290	633	1.59	−0.209
		77	633		−0.755
		300	1064	1.58	−0.097
Eu^{3+}(3%)	CaF_2	4.2	430		29
		4.2	440		22
Eu^{2+}	EuF_2	300	450		−4.5
		300	500		−2.6
		300	550		−1.6
		300	600		−1.1
		300	650		−0.8
		300	1064		−0.19
Tb^{3+}	KTb_3F_{10}	300	325	1.531	−2.174
		300	442	1.518	−0.933
		300	633	1.510	−0.386
		77	633		−1.94
		300	1064	1.505	−0.114
Tb^{3+}	$LiTbF_4$	300	325	1.493	−1.9
		300	442	1.481	−0.98
		300	633	1.473	−0.44
		300	1064	1.469	−0.13
Tb^{3+}	$Tb_3Ga_5O_{12}$	300	500	1.989	−0.749
		300	570	1.981	−0.581
		300	633	1.976	−0.461
		300	830	1.967	−0.21
		300	1060	1.954	−0.12

Verdet Constants of Paramagnetic Glasses[1]

The Verdet constant *V* is given at room temperature for the wavelengths indicated.

Rare-earth phosphate glasses of composition $R_2O_3 \cdot xP_2O_5$, where x is given in the second column

		Verdet constant *V* in min/Oe cm									
		λ = 405	λ = 436	λ = 480	λ = 500	λ = 520	λ = 546	λ = 578	λ = 600	λ = 635	λ = 670
R	***x***	**nm**	**nm**	**nm**	**nm**	**nm**	**nm**	**nm**	**nm**	**nm**	**nm**
La		0.037	0.030	0.024	0.022	0.020	0.018	0.015	–0.014	0.013	–
Ce	2.67	–0.672	0.510	–0.366	–0.326	–0.287	–0.253	–0.217	–0.197	–0.173	–0.150
Pr	3.09	–0.447	–0.332	–0.283	–0.261	–0.236	–0.208	–0.182	–0.170	–0.150	–0.132
Nd	2.92	–0.250	–0.209	–0.167	–0.155	–0.136	–0.134	–0.094	–0.080	–0.080	–0.071
Sm	2.87	0.026	0.024	0.020	0.020	0.017	0.015	0.014	0.012	0.011	0.010
Eu	2.93	–0.025	–0.017	–0.010	–0.006	–0.006	–0.005	–0.004	–0.003	–0.002	–0.002
Gd	3.01	0.018	0.015	0.014	0.012	0.012	0.011	0.011	0.010	0.009	0.009
Tb	2.94	–0.560	–0.458	–0.357	–0.323	–0.295	–0.261	–0.226	–0.206	–0.190	–0.164
Dy	2.51	–0.540	–0.453	–0.359	–0.331	–0.301	0.268	–0.237	–0.217	–0.197	–0.173
Ho	2.94	–0.299	–0.313	–0.156	–0.153	–0.138	–0.138	–0.119	–0.110	–0.098	–0.084
Er	3.01	–0.139	–0.121	–0.100	–0.111	–0.095	–0.062	–0.060	–0.057	–0.051	–0.044
Tm	2.79	0.019	0.013	0.012	0.009	0.008	0.006	0.005	0.004	0.004	0.007
Yb	3.01	0.087	0.072	0.056	0.050	0.045	0.041	0.036	0.032	0.029	0.024

The following are rare-earth borate glasses with composition:

for La and Pr: $R_2O_3 \cdot xP_2O_5$; for Tb–Pr and Dy–Pr: $R_2O_3 \cdot xB_2O_3$; and for other elements: $R_2O_3 \cdot 0.85La_2O_3 \cdot xB_2O_3$.

R	*x*	λ = 405 nm	λ = 436 nm	λ = 480 nm	λ = 500 nm	λ = 520 nm	λ = 546 nm	λ = 578 nm	λ = 600 nm	λ = 635 nm	λ = 670 nm
La	3.04	0.043	0.036	0.029	0.026	0.023	0.022	0.019	0.018	0.016	0.014
Pr-La	5.44	–0.380	–0.307	–0.230	–0.220	–0.201	–0.178	–0.153	–0.146	–0.128	–0.110
Nd-La	5.41	–0.180	–0.147	–0.120	–0.111	–0.096	–0.094	–0.100	–0.059	–0.056	–0.046
Sm-La	4.97	0.032	0.030	0.025	0.024	0.022	0.019	0.017	0.016	0.014	0.012
Eu-La	4.69	–0.081	–0.060	–0.038	–0.033	–0.029	–0.024	0.019	–0.016	0.014	–0.012
Gd-La	4.71	0.032	0.026	0.024	0.022	0.021	0.020	0.018	0.017	0.015	0.013
Tb-La	4.73	–0.512	–0.419	–0.319	–0.288	–0.262	–0.234	–0.205	–0.186	–0.167	0.112
Dy-La	4.88	–0.436	–0.361	–0.299	–0.273	–0.246	–0.220	–0.193	–0.177	–0.159	–0.138
Ho-La	4.36	–0.269	–0.252	–0.123	–0.131	–0.112	–0.128	–0.104	–0.096	–	–0.074
Er-La	4.50	–0.093	–0.078	–0.068	–0.082	–	–0.045	–0.042	–0.040	–0.035	–0.034
Tm-La	4.75	0.060	0.046	0.039	0.034	0.031	0.026	0.023	0.021	0.018	0.016
Yb-La	8.58	0.115	0.094	0.073	0.066	0.060	0.054	0.046	0.043	0.037	0.033
Tb-Pr	4.99	–0.940	–0.786	–0.560	–0.536	–0.489	–0.436	–0.380	–0.348	–0.306	–0.265
Dy-Pr	4.63	–0.850	–	–	–0.497	–0.465	–0.413	–0.358	–0.332	–0.290	–0.252
Pr	2.56	–0.843	–0.646	–0.471	–0.480	–0.432	–0.390	–0.334	–0.317	–0.271	–0.243

Verdet Constants of Diamagnetic Glasses[1]

The Verdet constant *V* is given at room temperature for the wavelengths indicated.

Glass		Verdet constant *V* in min/Oe cm			
type	**Composition (wt. %)**	**λ = 325 nm**	**λ = 442 nm**	**λ = 633 nm**	**λ = 1064 nm**
SiO_2	100% SiO_2			0.013	
B_2O_3	100% B_2O_3			0.010	
CdO	47.5% CdO, 52.5% P_2O_5	0.079	0.033	0.022	
ZnO	36.4% ZnO, 63.6% P_2O_5	0.072	0.044	0.020	
TeO_2	88.9% TeO_2, 11.1% P_2O_5		0.196	0.076	0.022
ZrF_4	63.1% ZrF_4, 14.9% BaF_2, 7.2% LaF_3, 1.9% AlF_3, 9.1% PbF_2, 3.8% LiF			0.011	

		λ = 700 nm	λ = 853 nm	λ = 1060 nm
Bi_2O_3	95% Bi2O3, 5% B2O3	0.086	0.051	0.033
PbO	95% PbO, 5% B2O3	0.093	0.061	0.031
	82% PbO, 18% SiO2	0.077	0.045	0.027
	50% PbO, 15% K2O, 35% SiO2	0.032	0.020	0.011
Tl_2O	95% Tl2O, 5% B2O3	0.092	0.061	0.032
	82% Tl2O, 18% SiO2	0.100	0.067	0.043
	50% Tl2O, 15% K2O, 35% SiO2	0.036	0.022	0.012
SnO	76% SnO, 13% B2O3, 11% SiO2	0.071	0.046	0.026
TeO_3	75% TeO2, 25% Sb2O3	0.076	0.052	0.032
	80% TeO2, 20% ZnCl2	0.073	0.046	0.025
	84% TeO2, 16% BaO	0.056	0.041	0.029
	70% TeO2, 30% WO3	0.052	0.035	0.022
	20% TeO2, 80% PbO	0.128	0.075	0.048
Sb_2O_3	25% Sb2O3, 75% TeO2	0.076	0.050	0.032
	75% Sb2O3, 75% Cs2O, 5% Al2O3	0.074	0.044	0.025
	75% Sb2O3, 10% Cs2O, 10% Rb2O, 5% Al2O3	0.078	0.052	0.030

Verdet Constants of Commercial Glasses[1]

This table gives the density, ρ, refractive index at 589 nm, n_D, and Verdet constant, V, for the wavelengths indicated; the data refer to room temperature.

Glass	ρ		*V* in min/Oe cm				
type	g/cm³	n_D	λ = 365.0 nm	λ = 404.7 nm	λ = 435.8 nm	λ = 546.1 nm	λ = 578.0 nm
BSC	2.49	1.5096	0.0499	0.0392	0.0333	0.02034	0.01798
HC	2.53	1.5189	0.0561	0.0440	0.0372	0.0225	0.01995
LBC	2.87	1.5406	0.0609	0.0477	0.0403	0.0245	0.0216
LF	3.23	1.5785	0.1143	0.0850	0.0693	0.0394	0.0344
BLF	3.48	1.6047	0.1112	0.0832	0.0685	0.0393	0.0344
DBC	3.56	1.6122	0.0662	0.0517	0.0435	0.0261	0.0231
DF	3.63	1.6203	0.1473	0.1076	0.0872	0.0485	0.0423
EDF	3.9	1.6533	0.1725	0.1248	0.1007	0.0556	0.0483

The composition of the glasses in weight percent is:

Glass type	SiO_2	B_2O_3	K_2O	CaO	Al_2O_3	As_2O_3	Na_2O	BaO	ZnO	PbO
BSC	69.6	6.7	20.5	2.9	0.3	0.1	–	–	–	–
HC	72.0	–	10.1	11.4	0.3	0.2	6.1	–	–	–
LBC	57.1	1.8	13.7	0.3	0.2	0.1	–	26.9	–	–
LF	52.5	–	9.5	0.3	0.2	0.1	–	–	–	37.6
BLF	45.2	–	7.8	–	–	0.4	–	16.0	8.3	22.2
DBC	36.2	7.7	0.2	0.2	3.5	0.7	–	44.6	6.7	–
DF	46.3	–	1.1	0.3	0.2	0.1	5.0	–	–	47.0
EDF	40.6	–	7.5	0.2	0.2	0.2	0.1	–	–	51.5

References

1. Weber, M. J., *CRC Handbook of Laser Science and Technology*, Vol. IV, Part 2, CRC Press, Boca Raton, FL, 1988, pp. 299–310.
2. Gray, D. E., Ed., *American Institute of Physics Handbook*, Third edition, McGraw Hill, New York, 1972, p. 6–230.

FARADAY ROTATION

Ferro-, Ferri-, and Antiferromagnetic Solids

Material	T_c K	$4\pi M_s$ gauss	F deg/cm	α cm^{-1}	$2F/\alpha$ deg	T K	λ nm
Fe	1043	21,800	4.4×10^5	6.5×10^5	1.4	300	500
			6.5×10^5	5.0×10^5	2.6	300	1000
			7×10^5	4.2×10^5	3.3	300	1500
			7×10^5	3.5×10^5	4.0	300	2000
Co	1390	18,200	2.9×10^5	–	–	300	500
			5.5×10^5	6.1×10^5	1.8	300	1000
			5.5×10^5	4.5×10^5	2.4	300	1500
			5.5×10^5	3.6×10^5	2.7	300	2000
Ni	633	6,400	0.8×10^5	–	–	300	500
			2.6×10^5	5.8×10^5	0.9	300	1000
			1.5×10^5	4.8×10^5	0.6	300	1500
			1×10^5	4.1×10^5	0.25	300	2000
Permalloy (Ni/Fe = 82/18)	803	10,700	1.2×10^5	6×10^5	0.4	300	500
Ni/Fe = 100/0		6,000	1.2×10^5	7.05×10^5	0.34	300	632.8
Ni/Fe = 80/20		10,800	2.2×10^5	7.10×10^5	0.62	300	632.8
Ni/Fe = 60/40		14,900	2.9×10^5	7.54×10^5	0.77	300	632.8
Ni/Fe = 40/60		14,400	2.2×10^5	8.17×10^5	0.54	300	632.8
Ni/Fe = 20/80		19,400	3.3×10^5	8.10×10^5	0.81	300	632.8
Ni/Fe = 0/100	639	21,600	3.5×10^5	8.13×10^5	0.86	300	632.8
MnBi		7,700	4.2×10^5	6.1×10^5	1.4	300	450
			7.5×10^5	4.2×10^5	3.6	300	900
MnAs	313		0.44×10^5	5.0×10^5	0.174	300	500
			0.62×10^5	4.4×10^5	0.28	300	900
CrTe	334	1015	0.5×10^5	2.0×10^5	0.5	300	550
			0.4×10^5	1.2×10^5	0.7	300	900
FeRh	333	–	0.9×10^5	3.3×10^5	0.56	348	700
$Y_3Fe_5O_{12}$ (YIG)	560	2500	2400	1500	3.2	300	555
			1250	1400	1.8	300	625
			750	450	3.3	300	770
			175	<0.06	$>3 \times 10^3$	300	5000 to 1500
$Gd_3Fe_5O_{12}$ (GdIG)	$T_n = 564$	7300	–2000	6000	0.6	300	500
	$T = 286$		–1050	900	2.3	300	600
			–300	100	6.0	300	800
			–80	70	2.3	300	1000
$NiFe_2O_4$	858	3350	2.0×10^4	5.9×10^4	0.7	300	286
			-1.0×10^4	10×10^4	0.2	300	500
			–120	38	6	300	1500
			+75	15	10	300	3000
			+110	32	7	300	5000
$CoFe_2O_4$	793	4930	2.75×10^4	12×10^4	0.5	300	286
			3.6×10^4	17×10^4	0.4	300	400
			-2.5×10^4	6×10^4	0.8	300	660
$MgFe_2O_4$	593-713[e]	1450[e]	–60	100	1	300	2500
			0	12	0	300	4000
			+35	6	11	300	6000
$Li_{0.5}Fe_{2.5}O_4$	863–953[e]	3240[e] to 3900	–440	150	6	300	1500
			+10	85	0.2	300	3000
			+110	44	5	300	5000
			+135	80	3	300	7000
$BaFe_{12}O_{19}$	723	–	–50	–38	3	300	2000
			+75	20	7.5	300	3000
			+150	20	15	300	5000
			+165	22	15	300	7000
$Ba_2Zn_2Fe_{12}O_{19}$	–	–	90	120	1.5	300	5000

Material	T_c K	$4\pi M_s$ gauss	F deg/cm	α cm^{-1}	$2F/\alpha$ deg	T K	λ nm
			75	65	2.0	300	7000
$RbNiF_3$	220	1250	360	35	20	77	450[a]
			70	10	14	77	600[a]
			310	70	9	77	800[a]
			75	25	6	77	1000[a]
$RbNi_{0.75}Co_{0.25}F_3$	109	–	180	9	40	77	600[b]
$RbFeF_3$	102	–	3400	7	900	82	300[c]
			1600	3	1100	82	400[c]
			620	1.5	830	82	600[c]
			300	2.5	240	82	800[c]
FeF_3	365	40	670	14	95	300	349[d]
		at 300 K	180	4.4	82	300	522.5[d]
$CrCl_3$	16.8	3880	2000	200	20	1.5	410
			–500	300	3	1.5	450
			–1000	70	30	1.5	590
$CrBr_3$	32.5	3390	3×10^5	3×10^3	200	1.5	478
			1.6×10^5	1.4×10^4	23	1.5	500
CrI_3	68	2690	1.1×10^5	6.3×10^3	35	1.5	970
			0.8×10^5	3×10^3	53	1.5	1000
$FeBO_3$	348	115	3200	140	45	300	500
		at 300 K	450	38	24	300	700
EuO	69	23700	-1.0×10^5	0.5×10^4	40	5	1100
			5×10^5	9.7×10^4	10	5	700
			0.5×10^5	7.8×10^4	1.3	5	500
			3×10^4	>0.5	~105	20	2500
			660	>1.0	1300	20	10600
EuS	16.3	–	-1.6×10^5	0	–	6	825
			-9.6×10^5	3.3×10^4	58	6	690
			$+5.5 \times 10^5$	1.2×10^5	9.2	6	563
EuSe	7.0	13,200	1.45×10^5	80	3600	4.2	750
			0.95×10^5	60	3170	4.2	800

[a] Measured along the C-axis (magnetic hard axis).
[b] Measured along the C-axis (magnetic easy axis).
[c] Measured along the C-axis ([100]-direction at room temperature).
[d] Strong natural birefringence interferes with the Faraday effect.
[e] Depends on heat treatment.

Reference

1. Weber, M. J., Ed., *CRC Handbook of Laser Science and Technology*, Vol. IV, Part 2, CRC Press, Boca Raton, FL, 1988, pp. 288–296.

NONLINEAR OPTICAL CONSTANTS

H. P. R. Frederikse

The relation between the polarization density P of a dielectric medium and the electric field E is linear when E is small, but becomes nonlinear as E acquires values comparable with interatomic electric fields (10^5 to 10^8 V/cm). Under these conditions the relation between P and E can be expanded in a Taylor's series

$$P = \varepsilon_0 \chi^{(1)} E + 2\chi^{(2)} E^2 + 4\chi^{(3)} E^3 + \cdots \ldots \quad (1)$$

where ε_o is the permittivity of free space, while $\chi^{(1)}$ is the linear and $\chi^{(2)}$, $\chi^{(3)}$ etc. the nonlinear optical susceptibilities.

If we consider two optical fields, the first $E_j^{\omega_1}$ (along the j-direction at frequency ω_1) and the second $E_k^{\omega_2}$ (along the k-direction at frequency ω_2) one can write the second term of the Taylor's series as follows

$$P_i(\omega_1 \omega_2) = 2\chi_{ijk}^{\omega_3 = \omega_1 \pm \omega_2} E_j^{\omega_1} E_k^{\omega_2}$$

When $\omega_1 \neq \omega_2$ the (parametric) mixing of the two fields gives rise to two new polarizations at the frequencies $\omega_3 = \omega_1 + \omega_2$ and $\omega_3' = \omega_1 - \omega_2$. When the two frequencies are equal, $\omega_1 = \omega_2 = \omega$, the result is Second Harmonic Generation (SHG): $\chi_{ijk}(2\omega, \omega, \omega)$, while equal and opposite frequencies, $\omega_1 = \omega$ and $\omega_2 = -\omega$ lead to Optical Rectification (OR): $\chi_{ijk}(0, \omega, -\omega)$. In the SHG case the following convention is adopted: the second order nonlinear coefficient d is equal to one half of the second order nonlinear susceptibility

$$d_{ijk} = 1/2\chi^{(2)}$$

Because of the symmetry of the indices j and k one can replace these two by a single index (subscript) m. Consequently the notation for the SHG nonlinear coefficient in reduced form is d_{im} where m takes the values 1 to 6. Only noncentrosymmetric crystals can possess a nonvanishing d_{ijk} tensor (third rank). The unit of the SHG coefficients is m/V (in the MKSQ/SI system).

In centrosymmetric media the dominant nonlinearity is of the third order. This effect is represented by the third term in the Taylor's series (Equation 1); it is the result of the interaction of a number of optical fields (one to three) producing a new frequency $\omega_4 = \omega_1 + \omega_2 + \omega_3$. The third order polarization is given by

$$P_j(\omega_1 \omega_2 \omega_3) = g_4 \chi_{jklm} E_k^{\omega_1} E_l^{\omega_2} E_m^{\omega_3}$$

Third Harmonic Generation (THG) is achieved when $\omega_1 = \omega_2 = \omega_3 = \omega$. In this case the constant $g_4 = 1/4$. The third order nonlinear coefficient C is related to the third order susceptibility as follows:

$$C_{jklm} = 1/4\chi_{jklm}$$

This coefficient is a fourth rank tensor. In the THG case the matrices must be invariant under permutation of the indices k, l, and m; as a result the notation for the third order nonlinear coefficient can be simplified to C_{jn}. The unit of C_{jn} is $m^2 \cdot V^{-2}$ (in the MKSQ/SI system).

Applications of second order nonlinear optical materials include the generation of higher (up to sixth) optical harmonics, the mixing of monochromatic waves to generate sum or difference frequencies (frequency conversion), the use of two monochromatic waves to amplify a third wave (parametric amplification) and the addition of feedback to such an amplifier to create an oscillation (parametric oscillation).

Third order nonlinear optical materials are used for THG, self-focusing, four wave mixing, optical amplification, and optical conjugation. Many of these effects — as well as the variation and modulation of optical propagation caused by mechanical, electric, and magnetic fields (see the preceding table on "Elasto-Optic, Electro-Optic, and Magneto-Optic Constants") are used in the areas of optical communication, optical computing, and optical imaging.

References

1. *Handbook of Laser Science and Technology*, Vol. 111, Part 1; Weber, M. J. Ed., CRC Press, Boca Raton, FL, 1986.
2. Dmitriev, V.G., Gurzadyan, G.G., and Nikogosyan, D., *Handbook of Nonlinear Optical Crystals*, Springer-Verlag, Berlin, 1991.
3. Shen, Y.R., *The Principles of Nonlinear Optics*, John Wiley, New York, 1984.
4. Yariv, A., *Quantum Electronics*, 3rd edition, John Wiley, New York, 1988.
5. Bloembergen, N., *Nonlinear Optics*, W.A. Benjamin, New York, 1965.
6. Zernike F. and Midwinter, J.E., *Applied Nonlinear Optics*, John Wiley, New York, 1973.
7. Hopf, F.A. and Stegeman, G.I., *Applied Classical Electrodynamics*, Volume 2: Nonlinear Optics, John Wiley, New York, 1986.
8. *Nonlinear Optical Properties of Organic Molecules and Crystals*, Chemla, D. S., and Zyss, J., Eds., Academic Press, Orlando, FL, 1987.
9. *Optical Phase Conjugation*, Fisher, R. A., Ed., Academic Press, New York, 1983.
10. Zyss, J., *Molecular Nonlinear Optics: Materials, Devices and Physics*, Academic Press, Boston, 1994.
11. Nonlinear Optics, 5 articles in *Physics Today, (Am. Inst. of Phys.)*, Vol. 47, No. 5, May, 1994.

Selected SHG Coefficients of NLO Crystals*

Material	Symmetry class	$d_{im} \times 10^{12}$ m/V	λ μm
GaAs	$\bar{4}$3 m	$d_{14} = 134.1 \pm 42$	10.6
GaP	$\bar{4}$3 m	$d_{14} = 71.8 \pm 12.3$	1.058
InAs	$\bar{4}$3 m	$d_{14} = 364 \pm 47$	1.058
		$d_{14} = 210$	10.6
ZnSe	$\bar{4}$3 m	$d_{14} = 78.4 \pm 29.3$	10.6
		$d_{36} = 26.6 \pm 1.7$	1.058
β-ZnS	$\bar{4}$3 m	$d_{14} = 30.6 \pm 8.4$	10.6
		$d_{36} = 20.7 \pm 1.3$	1.058
ZnTe	$\bar{4}$3 m	$d_{14} = 92.2 \pm 33.5$	10.6
		$d_{14} = 83.2 \pm 8.4$	1.058
		$d_{36} = 89.6 \pm 5.7$	1.058
CdTe	$\bar{4}$3 m	$d_{14} = 167.6 \pm 63$	10.6
Bi_4GeO_{12}	$\bar{4}$3 m	$d_{14} = 1.28$	1.064
$N_4(CH_2)_6$ (hexamine)	$\bar{4}$3 m	$d_{14} = 4.1$	1.06
$LiIO_3$	6	$d_{33} = -7.02$	1.06
		$d_{31} = -5.53 \pm 0.3$	1.064
ZnO	6 mm	$d_{33} = -5.86 \pm 0.16$	1.058
		$d_{31} = 1.76 \pm 0.16$	1.058
		$d_{15} = 1.93 \pm 0.16$	1.058
α-ZnS	6 mm	$d_{33} = 11.37 \pm 0.07$	1.058
		$d_{33} = 37.3 \pm 12.6$	10.6
		$d_{31} = -18.9 \pm 6.3$	10.6
		$d_{15} = 21.37 \pm 8.4$	10.6
CdS	6 mm	$d_{33} = 25.8 \pm 1.6$	1.058
		$d_{31} = -13.1 \pm 0.8$	1.058
		$d_{15} = 14.4 \pm 0.8$	1.058
CdSe	6 mm	$d_{33} = 54.5 \pm 12.6$	10.6
		$d_{31} = -26.8 \pm 2.7$	10.6
$BaTiO_3$	4 mm	$d_{33} = 6.8 \pm 1.0$	1.064
		$d_{31} = 15.7 \pm 1.8$	1.064
		$d_{15} = 17.0 \pm 1.8$	1.064
$PbTiO_3$	4 mm	$d_{33} = 7.5 \pm 1.2$	1.064
		$d_{31} = 37.6 \pm 5.6$	1.064
		$d_{15} = 33.3 \pm 5$	1.064
$K_3Li_2Nb_5O_{15}$	4 mm	$d_{33} = 11.2 \pm 1.6$	1.064
		$d_{31} = 6.18 \pm 1.28$	1.064
		$d_{15} = 5.45 \pm 0.54$	1.064
$K_{0.8}Na_{0.2}Ba_2Nb_5O_{15}$	4 mm	$d_{31} = 13.6 \pm 1.6$	1.064
$SrBaNb_5O_{15}$	4 mm	$d_{33} = 11.3 \pm 3.3$	1.064
		$d_{31} = 4.31 \pm 1.32$	1.064
		$d_{15} = 5.98 \pm 2$	1.064
$NH_4H_2PO_4$ (ADP)	$\bar{4}$2 m	$d_{36} = 0.53$	1.064
		$d_{36} = 0.85$	0.694
KH_2PO_4 (KDP)	$\bar{4}$2 m	$d_{36} = 0.44$	1.064
		$d_{36} = 0.47 \pm 0.07$	0.694
KD_2PO_4 (KD*P)	$\bar{4}$2 m	$d_{36} = 0.38 \pm 0.016$	1.058
		$d_{36} = 0.34 \pm 0.06$	0.694
		$d_{14} = 0.37$	1.058
KH_2AsO_4 (KDA)	$\bar{4}$2 m	$d_{36} = 0.43 \pm 0.025$	1.06
		$d_{36} = 0.39 \pm 0.4$	0.694
$CdGeAs_2$	$\bar{4}$2 m	$d_{36} = 351 \pm 105$	10.6
$AgGaS_2$	$\bar{4}$2 m	$d_{36} = 18 \pm 2.7$	10.6
$AgGaSe_2$	$\bar{4}$2 m	$d_{36} = 37.4 \pm 6.0$	10.6
$(NH_2)_2CO$ (urea)	$\bar{4}$2 m	$d_{36} = 1.3$	1.06
$AlPO_4$	32	$d_{11} = 0.35 \pm 0.03$	1.058
Se	32	$d_{11} = 97 \pm 25$	10.6
Te	32	$d_{11} = 650 \pm 30$	10.6
SiO_2 (quartz)	32	$d_{11} = 0.335$	1.064
HgS	32	$d_{11} = 50.3 \pm 17$	10.6
$(C_6H_5CO)_2$ [benzil]	32	$d_{11} = 3.6 \pm 0.5$	1.064
β-BaB_2O_4 [BBO]	3 m	$d_{22} = 2.22 \pm 0.09$	1.06
		$d_{31} = 0.16 \pm 0.08$	1.06
$LiNbO_3$	3 m	$d_{33} = 34.4$	1.06
		$d_{31} = -5.95$	1.06
		$d_{22} = 2.76$	1.06
$LiTaO_3$	3 m	$d_{33} = -16.4 \pm 2$	1.058
		$d_{31} = -1.07 \pm 0.2$	1.058
		$d_{22} = +1.76 \pm 0.2$	1.058
Ag_3AsS_3 [proustite]	3 m	$d_{31} = 11.3 \pm 2.5$	10.6
		$d_{22} = 18.0 \pm 2.5$	10.6
Ag_3SbS_3 [pyrargerite]	3m	$d_{31} = 12.6 \pm 4$	10.6
		$d_{22} = 13.4 \pm 4$	10.6
α-HIO_3	222	$d_{36} = 5.15 \pm 0.16$	1.064
$NO_2 \cdot CH_3NOC_5H_4 \cdot$ (POM)	222	$d_{36} = 6.4 \pm 1.0$	1.064
$Ba_2NaNb_5O_{15}$ [Banana]	mm 2	$d_{33} = -17.6 \pm 1.28$	1.064
		$d_{31} = -12.8 \pm 1.28$	1.064
$C_6H_4(NO_2)_2$ [MDB]	mm 2	$d_{33} = 0.74$	1.064
		$d_{32} = 2.7$	1.064
		$d_{31} = 1.78$	1.064
$Gd_2(MoO_4)_3$	mm 2	$d_{33} = -0.044 \pm 0.008$	1.064
		$d_{32} = +2.42 \pm 0.36$	1.064
		$d_{31} = -2.49 \pm 0.37$	1.064
$KNbO_3$	mm 2	$d_{33} = -19.58 \pm 1.03$	1.064
		$d_{32} = +11.34 \pm 1.03$	1.064
		$d_{31} = -12.88 \pm 1.03$	1.064
$KTiOPO_4$ [KTP]	mm 2	$d_{33} = 13.7$	1.06
		$d_{32} = \pm 5.0$	1.06
		$d_{31} = \pm 6.5$	1.06
$NO_2C_6H_4 \cdot NH_2$ [mNA]	mm 2	$d_{33} = 13.12 \pm 1.28$	1.064
		$d_{32} = 1.02 \pm 0.22$	1.064
		$d_{31} = 12.48 \pm 1.28$	1.064
$C_{10}H_{12}N_3O_6$ [MAP]	2	$d_{23} = 10.67 \pm 1.3$	1.064
		$d_{22} = 11.7 \pm 1.3$	1.064
		$d_{21} = 2.35 \pm 0.5$	1.064
		$d_{25} = -0.35 \pm 0.3$	1.064
$(NH_2CH_2COOH)_3H_2SO_4$ [TGS]	2	$d_{23} = 0.32$	0.694

* These data are taken from References 1 and 2.

Selected THG Coefficients of Some NLO Materials*

Material	NLO process	$C_{jm} \times 10^{20}$ m^2/V^{-2}	λ μm
$NH_4H_2PO_4$ [ADP]	(−3ω,ω,ω,ω)	$C_{11} = 0.0104$	1.06
		$C_{18} = 0.0098$	1.06
C_6H_6 [benzene]	(−3ω,ω,ω,ω)	$C_{11} = 0.0184 \pm 0.0042$	1.89
$CdGeAs_2$	(−3ω,ω,ω,ω)	$C_{11} = 182 \pm 84$	10.6
p-type: 5×10^{16} cm^{-3}		$C_{16} = 175$	10.6
		$C_{18} = -35$	10.6
$C_{40}H_{56}$ [β-carotene]	(−3ω,ω,ω,ω)	C_{11} 0.263 ± 0.08	1.89
GaAs high-resistivity	(−3ω,ω,ω,−ω)	$C_{11} = 62 \pm 31$	1.06
Ge	(−3ω,ω,ω,−ω)	$C_{11} = 23.5 \pm 12$	1.06
$LiIO_3$	(−3ω,ω,ω,−ω)	$C_{12} = 0.2285$	1.06
		$C_{35} = 6.66 \pm 1$	1.06
KBr	(−3ω,ω,ω,−ω)	$C_{11} = 0.0392$	1.06
		$C_{18}/C_{11} = 0.3667$	1.06
KCl	(−3ω,ω,ω,−ω)	$C_{11} = 0.0168$	1.06
		$C_{18}/C_{11} = 0.28$	1.06
KH_2PO_4 [KDP]	(−3ω,ω,ω,−ω)	$C_{11}-3C_{18} = 0.04$	1.06
Si p-type: 10^{14} cm^{-3}	(−3ω,ω,ω,−ω)	$C_{11} = 82.8 \pm 25$	1.06
NaCl	(−3,ω,ω,ω,−ω)	$C_{11} = 0.0168$	1.06
		$C_{18}/C_{11} - 0.4133$	1.06
NaF	(−3ω,ω,ω,−ω)	$C_{11} = 0.0035$	1.06

* These data are taken from Reference 1.

PHASE DIAGRAMS

H. P. R. Frederikse

A phase is a structurally homogeneous portion of matter. Regardless of the number of chemical constituents of a gas, there is only one vapor phase. This is true also for the liquid form of a pure substance, although a mixture of several liquid substances may exist as one or several phases, depending on the interactions among the substances. On the other hand a pure solid may exist in several phases at different temperatures and pressures because of differences in crystal structure (Reference 1). At the phase transition temperature, T_{tr}, the chemical composition of the solid remains the same, but a change in the physical properties often will take place. Such changes are found in ferroelectric crystals (example $BaTiO_3$) that develop a spontaneous polarization below T_{tr}, in superconductors (example Pb) that lose all electrical resistance below the transition point, and in many other classes of solids.

In quite a few cases it is difficult to bring about the phase transition, and the high- (or low-) temperature phase persists in its metastable form. Many liquids remain in the liquid state for shorter or longer periods of time when cooled below the melting point (supercooling). However, often the slightest disturbance will cause solidification. Persistence of the high temperature phase in solid–solid transitions is usually of much longer duration. An example of this behavior is found in white tin; although gray tin is the thermodynamically stable form below T_{tr} (286.4 K), the metal remains in its undercooled, white tin state all the way to $T = 0$ K, and crystals of gray tin are very difficult to produce.

A *phase diagram* is a map that indicates the areas of stability of the various phases as a function of external conditions (temperature and pressure). Pure materials, such as mercury, helium, water, and methyl alcohol are considered one-component systems and they have *unary* phase diagrams. The equilibrium phases in two-component systems are presented in *binary* phase diagrams. Because many important materials consist of three, four, and more components, many attempts have been made to deduce their multicomponent phase diagrams. However, the vast majority of systems with three or more components are very complex, and no overall maps of the phase relationships have been worked out.

It has been shown during the last 20 to 25 years that very useful partial phase diagrams of complex systems can be obtained by means of thermodynamic modeling (References 2, 3). Especially for complicated, multicomponent alloy systems the CALPHAD method has proved to be a successful approach for producing valuable portions of very intricate phase diagrams (Reference 4). With this method thermodynamic descriptions of the free energy functions of various phases are obtained that are consistent with existing (binary) phase diagram information and other thermodynamic data. Extrapolation methods are then used to extend the thermodynamic functions into a ternary system. Comparison of the results of this procedure with available experimental data is then used to fine-tune the phase diagram and add ternary interaction functions if necessary. In principle this approximation strategy can be extended to four, five, and more component systems.

The nearly two dozen phase diagrams shown below present the reader with examples of some important types of single and multicomponent systems, especially for ceramics and metal alloys. This makes it possible to draw attention to certain features like the kinetic aspects of phase transitions (see Figure 22, which presents a time–temperature–transformation, or TTT, diagram for the precipitation of α-phase particles from the β-phase in a Ti-Mo alloy; Reference 1, pp. 358–360). The general references listed below and the references to individual figures contain phase diagrams for many additional systems.

General References

1. Ralls, K. M., Courtney, T. H., and Wulff, J., *Introduction to Materials Science and Engineering*, Chapters 16 and 17, John Wiley & Sons, New York, 1976.
2. Kaufman, L., and Bernstein, H., *Computer Calculation of Phase Diagrams*, Academic Press, New York, 1970.
3. Kattner, U. R., Boettinger, W. J. B., and Coriell, S. R., *Z. Metallkd.*, 87, 9, 1996.
4. Dinsdale, A. T., Ed., *CALPHAD*, Vol. 1–20, Pergamon Press, Oxford, 1977–1996 and continuing.
5. Baker, H., Ed., *ASM Handbook, Volume 3: Alloy Phase Diagrams*, ASM International, Materials Park, OH, 1992.
6. Massalski, T. B., Ed., *Binary Alloy Phase Diagrams, Second Edition*, ASM International, Materials Park, OH, 1990.
7. Roth. R. S., Ed., *Phase Diagrams for Ceramists*, Vol. I (1964) to Volume XI (1995), American Ceramic Society, Waterville, OH.

References to Individual Phase Diagrams

Figure 1. Carbon: Reference 7, Vol. X (1994), Figure 8930. Reprinted with permission.

Figure 2. Si-Ge : Ref. 5, p. 2.231. Reprinted with permission.

Figure 3. H_2O (ice): See figure.

Figure 4. SiO_2: Reference 7, Vol. XI (1995), Figure 9174. Reprinted with permission.

Figure 5. Fe-O: Darken, L.S., and Gurry, R.W., *J. Am. Chem. Soc.*, 68, 798, 1946. Reprinted with permission.

Figure 6. Ti-O: Reference 5, p. 2.324. Reprinted with permission.

Figure 7. BaO-TiO_2: Reference 7, Vol. III (1975), Figure 4302. Reprinted with permission.

Figure 8. MgO-Al_2O_3: Reference 7, Vol. XI (1995), Figure 9239. Reprinted with permission.

Figure 9. Y_2O_3-ZrO_2: Reference 7, Vol. XI (1995), Figure 9348. Reprinted with permission.

Figure 10. Si-N-Al-O (Sialon): Reference 7, Vol. X (1994), Figure 8759. Reprinted with permission.

Figure 11. PbO-ZrO_2-TiO_2 (PZT): Reference 7, Vol. III (1975), Figure 4587. Reprinted with permission.

Figure 12. Al-Si-Ca-O: Reference 7 (1964), Vol. I, Figure 630. Reprinted with permission.

Figure 13. Y-Ba-Cu-O: Whitler, J.D., and Roth, R.S., *Phase Diagrams for High T_c Superconductors*, Figure S-082, American Ceramic Society, Waterville, OH, 1990. Reprinted with permission.

Figure 14. Al-Cu: Reference 5, p. 2.44. Reprinted with permission.

Figure 15. Fe-C: Ralls, K.M., Courtney, T.H., and Wulff, J., *Introduction to Materials Science and Engineering*, Figure 16.13, John Wiley & Sons, New York, 1976. Reprinted with permission.

Figure 16. Fe-Cr: Reference 5, p. 2.152. Reprinted with permission.

Figure 17. Cu-Sn: Reference 5, p. 2.178. Reprinted with permission.

Figure 18. Cu-Ni: Reference 5, p. 2.173. Reprinted with permission.

Figure 19. Pb-Sn (solder): Reference 5, p. 2.335. Reprinted with permission.

Figure 20. Cu-Zn (brass): Subramanian, P.R., Chakrabarti, D.J., and Laughlin, D.E., Eds., *Phase Diagrams of Binary Copper Alloys*, p. 487, ASM International, Materials Park, OH, 1994. Reprinted with permission.

Figure 21. Co-Sm: Reference 5, p. 2.148. Reprinted with permission.

Figure 22. Ti-Mo: Reference 5, p. 2.296; Reference 1, p. 359. Reprinted with permission.

Figure 23. Fe-Cr-Ni: Reference 5, Figure 48. Reprinted with permission.

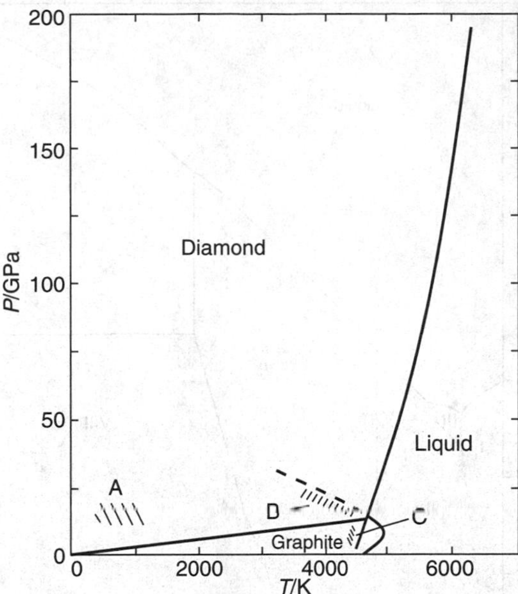

FIGURE 1. Phase diagram of carbon. (A) Martensitic transition: hex graphite → hex diamond. (B) Fast graphite-to-diamond transition. (C) Fast diamond-to-graphite transition.

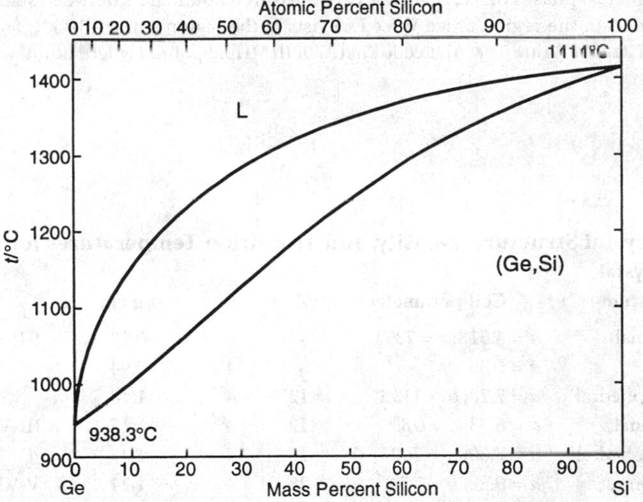

FIGURE 2. Si-Ge system.

Phase	Composition, mass % Si	Pearson symbol	Space group
(Ge,Si)	0 to 100	*cF*8	$Fd\bar{3}m$
High-pressure phases			
GeII	–	*tI*4	$I4_1/amd$
SiII	–	*tI*4	$I4_1/amd$

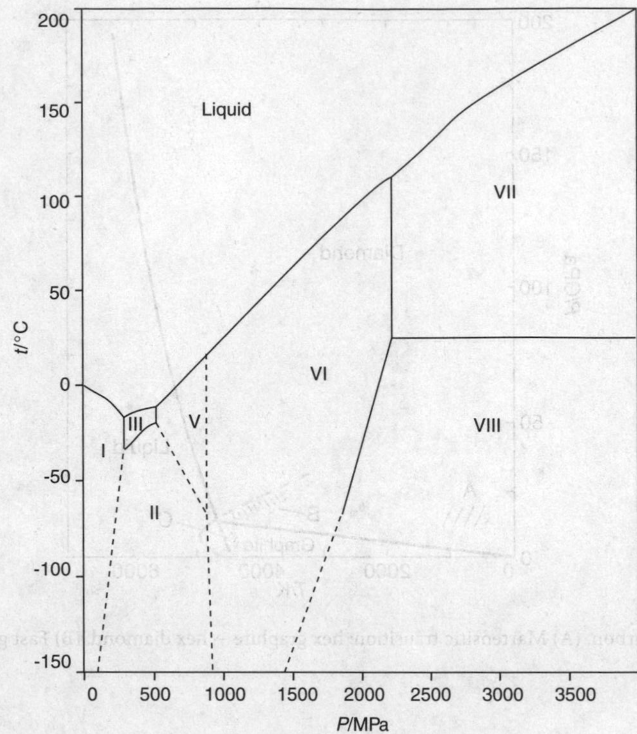

FIGURE 3. Diagram of the principal phases of ice. Solid lines are measured boundaries between stable phases; dotted lines are extrapolated. Ice IV is a metastable phase that exists in the region of ice V. Ice IX exists in the region below −100 °C and pressures in the range 200–400 MPa. Ice X exists at pressures above 44 GPa. See Table 1 for the coordinates of the triple points, where liquid water is in equilibrium with two adjacent solid phases.

TABLE 1. Crystal Structure, Density, and Transition Temperatures for the Phases of Ice

Phase	Crystal system	Cell parameters	Z	n	ρ/g cm^{-3}	Triple points
Ih	Hexagonal	a = 4.513; c = 7352	4	4	0.93	I-III: −21.99 °C, 209.9 MPa
Ic	Cubic	a = 6.35	8	4	0.94	
II	Rhombohedral	a = 7.78; α = 113.1°	12	4	1.18	
III	Tetragonal	a = 6.73; c = 6.83	12	4	1.15	III-V: −16.99 °C, 350.1 MPa
IV	Rhombohedral	a = 7.60; α = 70.1°	16	4	1.27	
V	Monoclinic	a = 9.22; b = 7.54, c = 10.35; β = 109.2°	28	4	1.24	V-VI: 0.16 °C, 632.4 MPa
VI	Tetragonal	a = 6.27; c = 5.79	10	4	1.31	VI-VII: 82 °C, 2216 MPa
VII	Cubic	a = 3.41	2	8	1.56	
VIII	Tetragonal	a = 4.80; c = 6.99	8	8	1.56	
IX	Tetragonal	a = 6.73; c = 6.83	12	4	1.16	
X	Cubic	a = 2.83	2	8	2.51	

References

1. Wagner, W., Saul, A., and Pruss, A., *J. Phys. Chem. Ref. Data*, 23, 515, 1994.
2. Lerner, R.G. and Trigg, G.L., Eds., *Encyclopedia of Physics*, VCH Publishers, New York, 1990.
3. Donnay, J.D.H. and Ondik, H.M, *Crystal Data Determinative Tables, Third Edition, Volume 2, Inorganic Compounds,* Joint Committee on Powder Diffraction Standards, Swarthmore, PA, 1973.
4. Hobbs, P.V., *Ice Physics*, Oxford University Press, Oxford, 1974.
5. Glasser, L., *J. Chem. Edu.*, 81, 414, 2004.

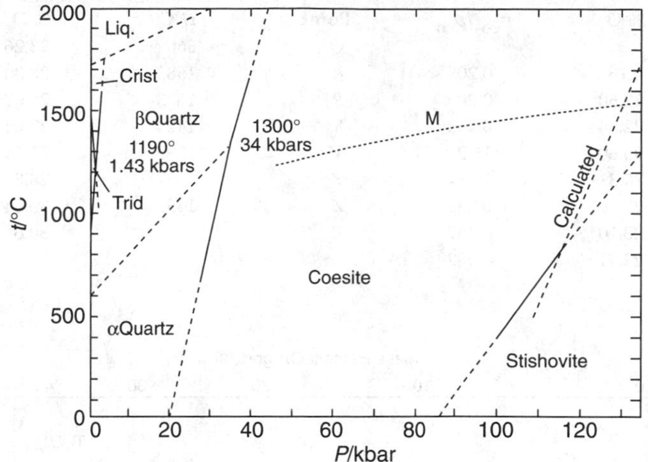

FIGURE 4. SiO_2 system. Crist = cristobalite; Trid = tridymite.

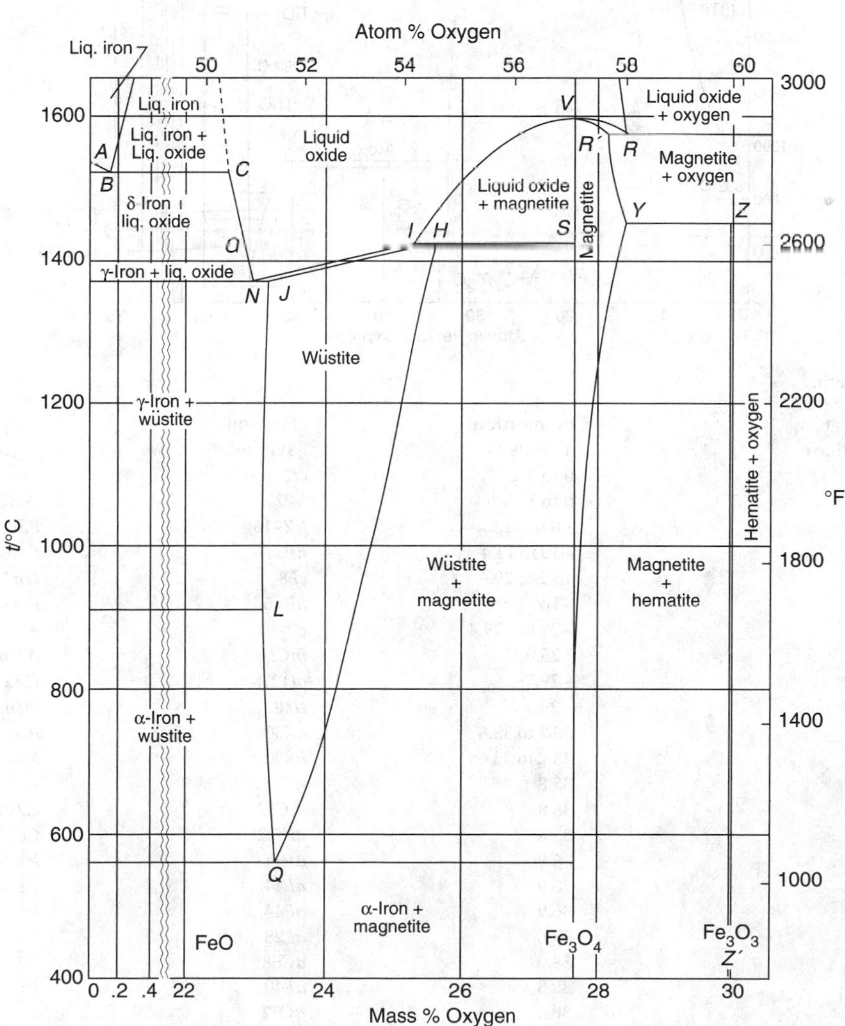

FIGURE 5. Fe-O system.

Point	t/°C	% O	p_{CO_2}/p_{CO}	Point	t/°C	% O	p_{CO_2}/p_{CO}	p_{O_2}/atm
A	1539			*Q*	560	23.26	1.05	
B	1528	0.16	0.209	*R*	1583	28.30		1
C	1528	22.60	0.209	*R′*	1583	28.07		1
G	1400[a]	22.84	0.263	*S*	1424	27.64	16.2	
H	1424	25.60	16.2	*V*	1597	27.64		0.0575
I	1424	25.31	16.2	*Y*	1457	28.36		1
J	1371	23.16	0.282	*Z*	1457	30.04		1
L	911[a]	23.10	0.447	*Z′*		30.6		
N	1371	22.91	0.282					

[a] Values for pure iron.

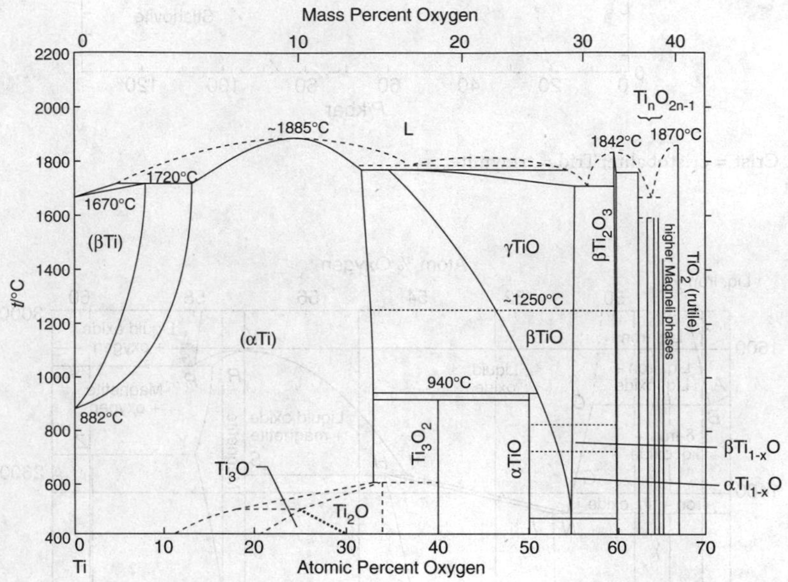

FIGURE 6. Ti-O system.

Phase	Composition, mass % O	Pearson symbol	Space group
(βTi)	0 to 3	*cI*2	*Im*$\bar{3}$*m*
(αTi)	0 to 13.5	*hP*2	*P*63/*mmc*
Ti_3O	~8 to ~13	*hP*~16	*P*$\bar{3}$*c*
Ti_2O	~10 to 14.4	*hP*3	*P*$\bar{3}$*m*1
γTiO	15.2 to 29.4	*cF*8	*Fm*$\bar{3}$*m*
Ti_3O_2	~18	*hP*~5	*P*6/*mmm*
βTiO	~24 to ~29.4	*c***	–
αTiO	~25.0	*mC*16	*A*2/*m* or *B**/*
$\beta Ti_{1-x}O$	~29.5	*oI*12	*I*222
$\alpha Ti_{1-x}O$	~29.5	*tI*18	*I*4/*m*
βTi_2O_3	33.2 to 33.6	*hR*30	*R*$\bar{3}$*c*
αTi_2O_3	33.2 to 33.6	*hR*30	*R*$\bar{3}$*c*
βTi_3O_5	35.8	*m***	–
αTi_3O_5	35.8	*mC*32	*C*2/*m*
$\alpha' Ti_3O_5$	35.8	*mC*32	*Cc*
γTi_4O_7	36.9	*aP*44	*P*$\bar{1}$
βTi_4O_7	36.9	*aP*44	*P*$\bar{1}$
$\alpha Ti4O_7$	36.9	*aP*44	*P*$\bar{1}$
γTi_5O_9	37.6	*aP*28	*P*$\bar{1}$
βTi_6O_{11}	38.0	*aC*68	*A*$\bar{1}$
Ti_7O_{13}	38.3	*aP*40	*P*$\bar{1}$
Ti_8O_{15}	38.5	*aC*92	*A*$\bar{1}$
Ti_9O_{17}	38.7	*aP*52	*P*$\bar{1}$
Rutile TiO_2	40.1	*tP*6	*P*4$_2$/*mnm*
Metastable phases			
Anatase	–	*tI*12	*I*4$_1$/*amd*
Brookite	–	*oP*24	*Pbca*
High-pressure phases			
TiO_2-II	–	*oP*12	*Pbcn*
TiO_2-III	–	*hP*~48	–

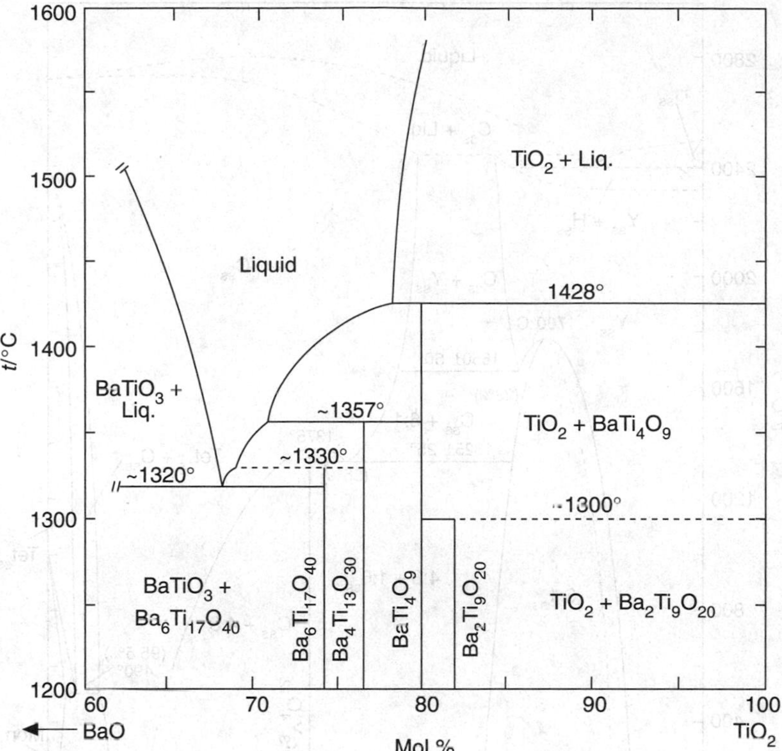

FIGURE 7. BaO-TiO_2 system.

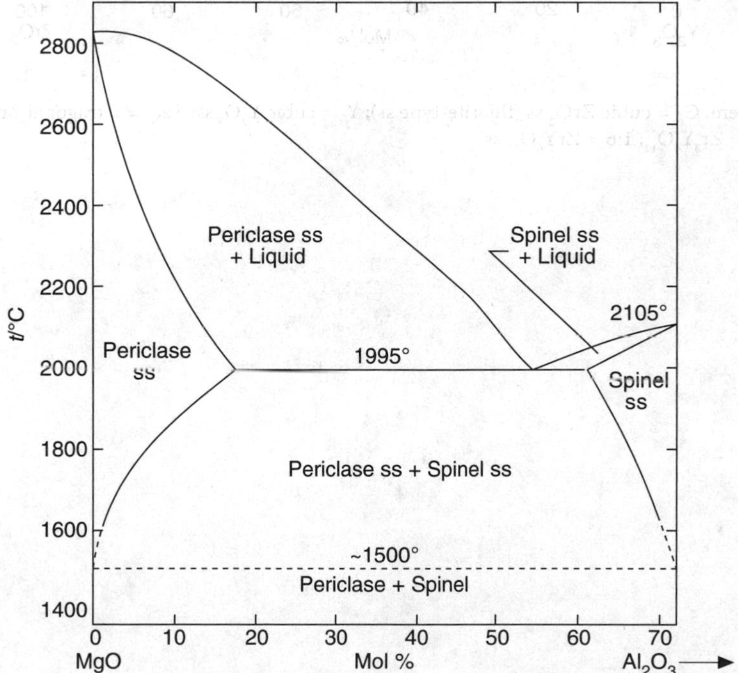

FIGURE 8. MgO-Al_2O_3 system.

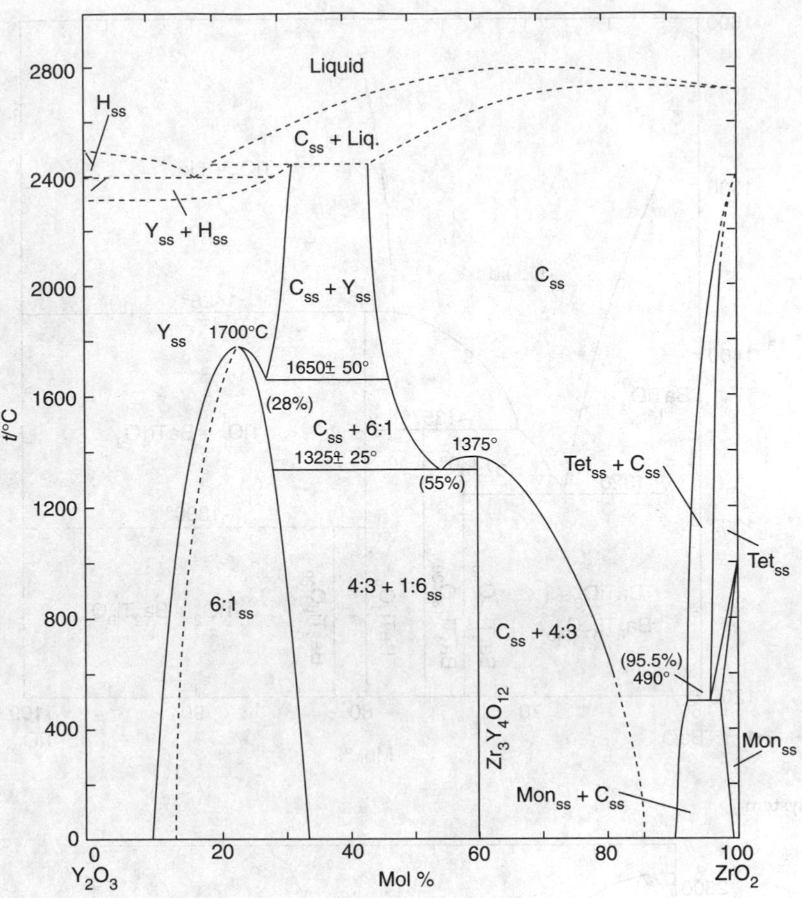

FIGURE 9. Y_2O_3-ZrO_2 system. C_{ss} = cubic ZrO_2 ss (fluorite-type ss); Y_{ss} = cubic Y_2O_3 ss; Tet_{ss} = tetragonal ZrO_2 ss; Mon_{ss} = monoclinic ZrO_2 ss; H_{ss} = hexagonal Y_2O_3 ss; 3:4 = $Zr_3Y_4O_{12}$; 1:6 = ZrY_6O_{11} ss.

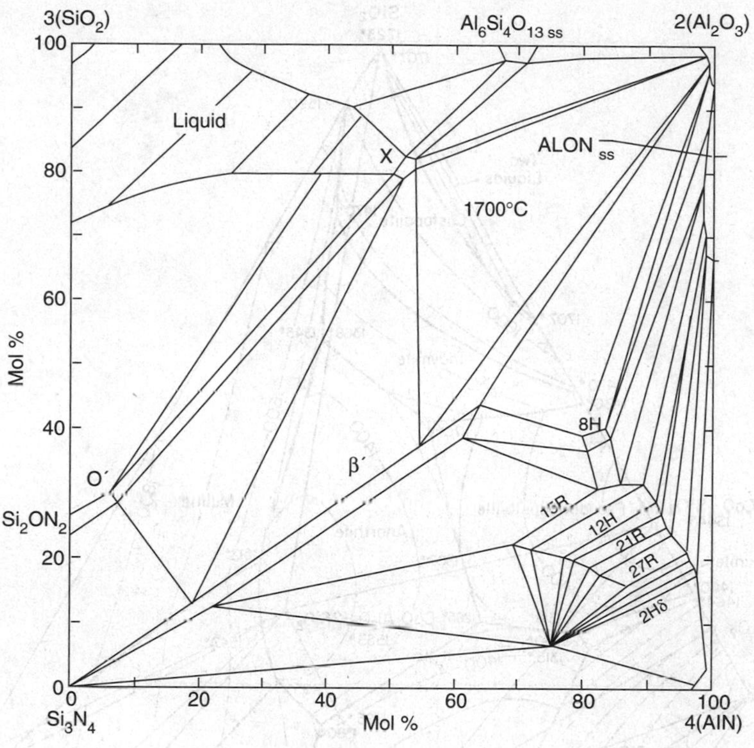

FIGURE 10. $3(SiO_2)$-Si_3N_4-$4(AlN)$-$2(Al_2O_3)$ system. "Behavior" diagram at 1700 °C. The labels 8H, 15R, 12H, 21R, 27R, $2H^{\delta}$ indicate defect AlN polytypes. β´ = 3-sialon ($Si_{6-x}Al_xO_xN_{8-x}$); O´ = sialon of Si_2ON_2 type; X = $SiAlO_2N$ ("nitrogen mullite"). ALON ss = aluminum oxynitride ss extending from approximately Al_7O_9N to Al_3O_3N.

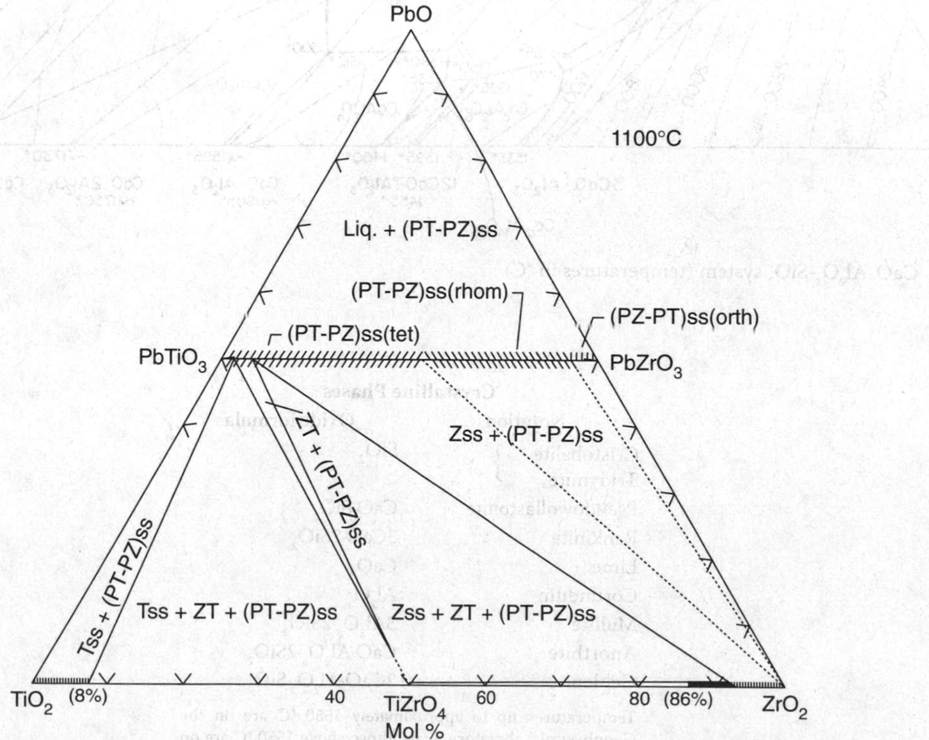

FIGURE 11. PbO-ZrO_2-TiO_2 (PZT) system, subsolidus at 1100 °C. *P* = PbO; *T* = TiO_2; *Z* = ZrO_2.

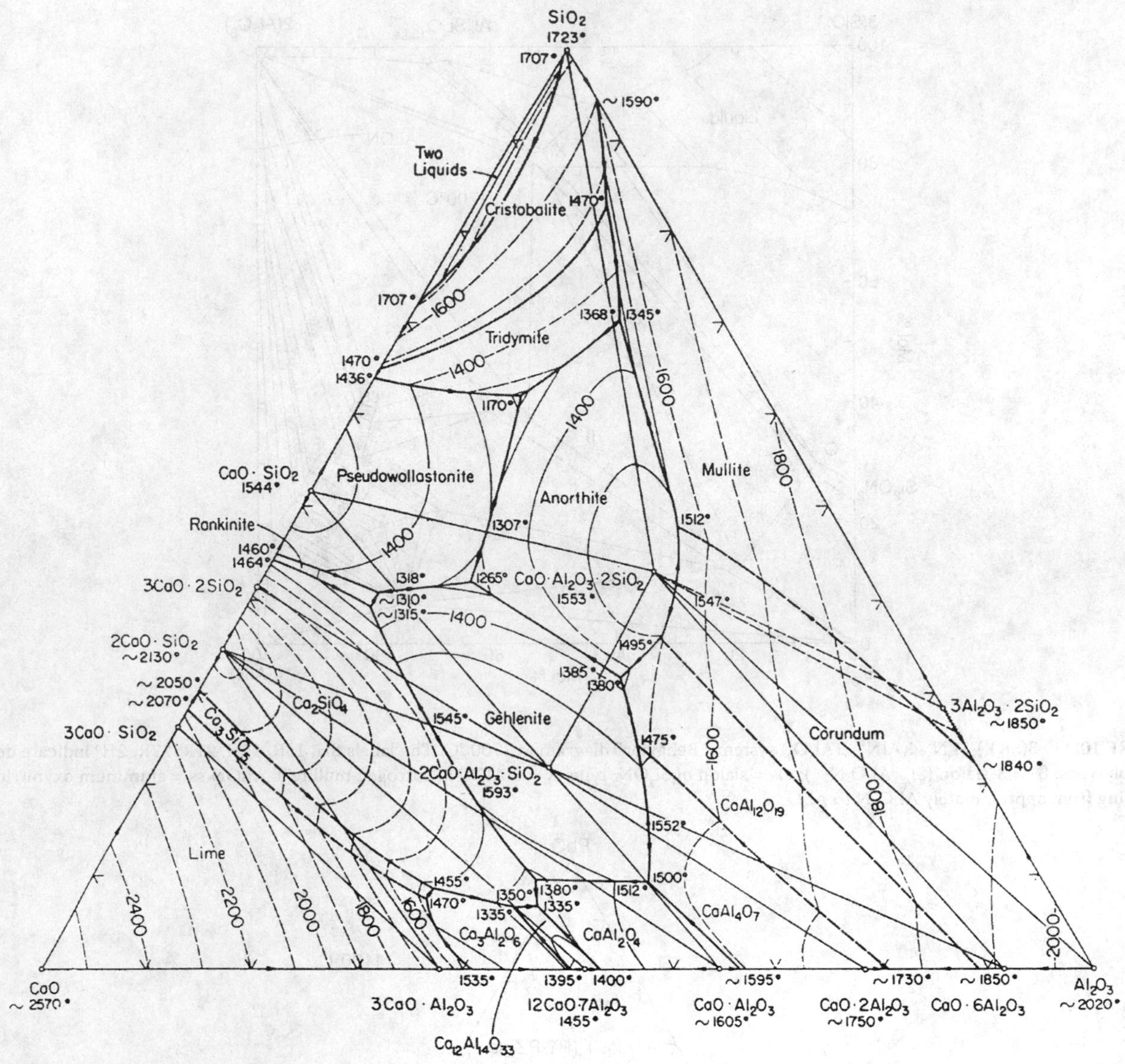

FIGURE 12. CaO-Al_2O_3-SiO_2 system (temperatures in °C).

Crystalline Phases

Notation	Oxide formula
Cristobalite }	SiO_2
Tridymite }	
Pseudowollastonite	$CaO{\cdot}SiO_2$
Rankinite	$3CaO{\cdot}2SiO_2$
Lime	CaO
Corundum	Al_2O_3
Mullite	$3Al_2O_3{\cdot}2SiO_2$
Anorthite	$CaO{\cdot}Al_2O_3{\cdot}2SiO_2$
Gehlenite	$2CaO{\cdot}Al_2O_3{\cdot}SiO_2$

Temperatures up to approximately 1550 °C are on the Geophysical Laboratory Scale; those above 1550 °C are on the 1948 International Scale.

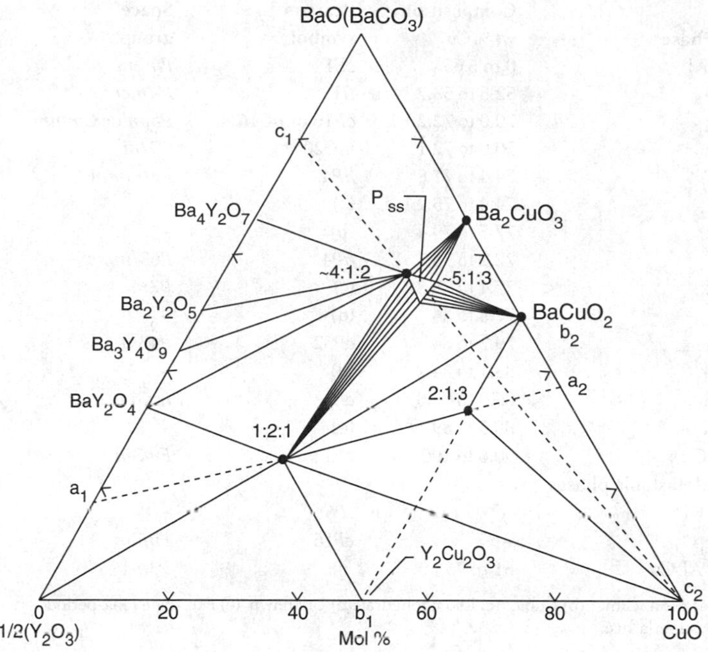

FIGURE 13. BaO-Y_2O_3-CuO system. 2:1:3 = $Ba_2YCu_3O_{7-x}$; 1:2:1 = BaY_2CuO_5; 4:1:2 = $Ba_4YCu_2O_{7.5+x}$; and 5:1:3 = $Ba_5YCu_3O_{9.5+x}$. The superconducting 2:1:3 phase was prepared using barium peroxide.

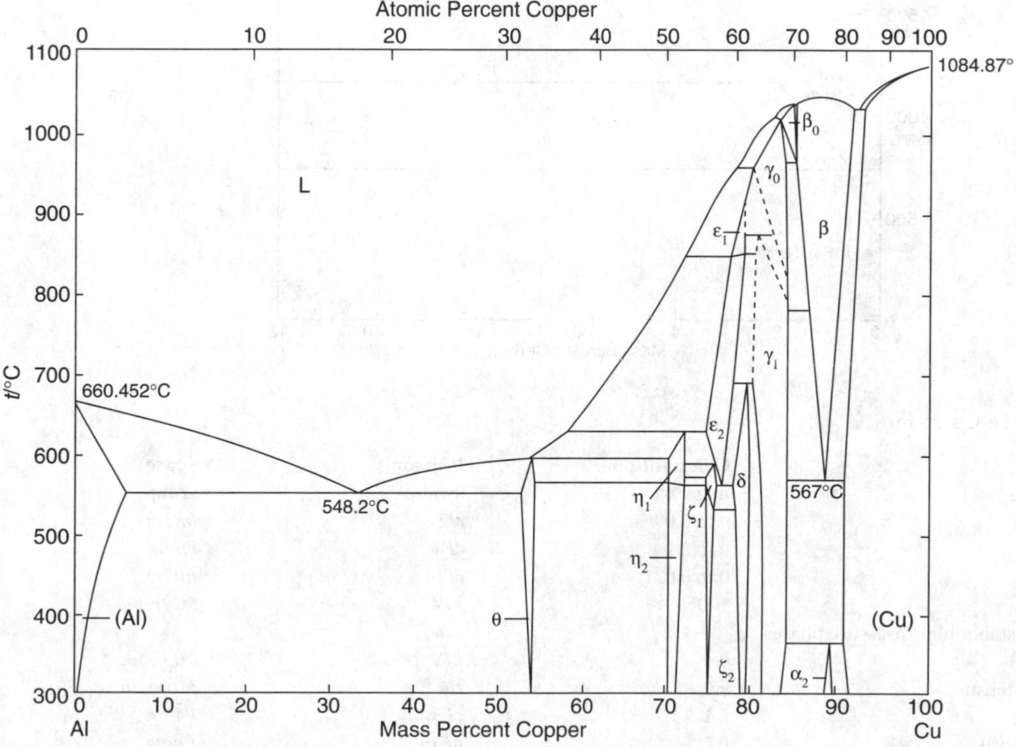

FIGURE 14. Al-Cu system.

Phase	Composition, wt % Cu	Pearson symbol	Space group
(Al)	0 to 5.65	*cF*4	$Fm\bar{3}m$
θ	52.5 to 53.7	*tI*12	*I*4/*mcm*
η_1	70.0 to 72.2	*oP*16 or *oC*16	*Pban* or *Cmmm*
η_2	70.0 to 72.1	*mC*20	*C*2/*m*
ζ_1	74.4 to 77.8	*hP*42	*P*6/*mmm*
ζ_2	74.4 to 75.2	(a)	–
ε_1	77.5 to 79.4	(b)	–
ε_2	72.2 to 78.7	*hP*4	*P*63/*mmc*
δ	77.4 to 78.3	(c)	$R\bar{3}m$
γ_0	77.8 to 84	(d)	–
γ_1	79.7 to 84	*cP*52	$P\bar{4}3m$
β_0	83.1 to 84.7	(d)	–
β	85.0 to 91.5	*cI*2	$Im\bar{3}m$
α_2	88.5 to 89	(e)	–
(Cu)	90.6 to 100	*cF*4	$Fm\bar{3}m$
Metastable phases			
θ′	–	*tP*6	–
β′	–	*cF*16	$Fm\bar{3}m$
Al_3Cu_2	61 to 70	*hp*5	$P\bar{3}m1$

(a) Monoclinic? (b) Cubic? (c) Rhombohedral. (d) Unknown. (e) $D0_{22}$-type long-period superlattice.

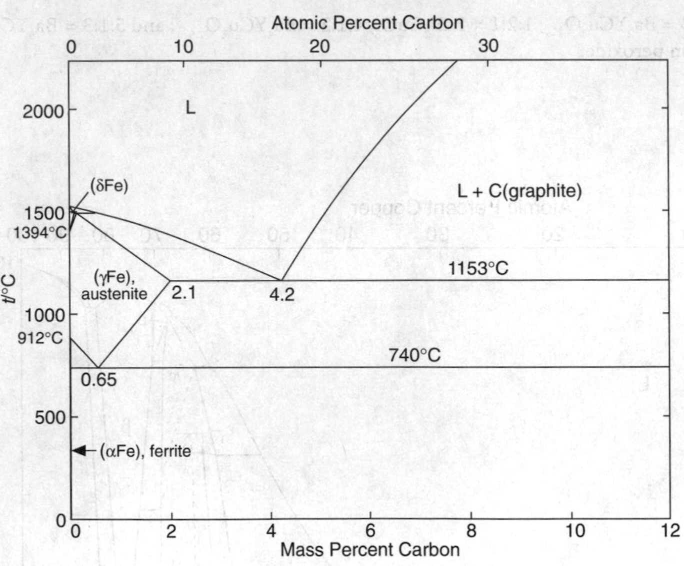

FIGURE 15. Fe-C system.

Phase	Composition, mass % C	Pearson symbol	Space group
(δFe)	0 to 0.09	*cI*2	$Im\bar{3}m$
(γFe)	0 to 2.1	*cF*4	$Fm\bar{3}m$
(αFe)	0 to 0.021	*cI*2	$Im\bar{3}m$
(C)	100	*hP*4	$P6_3/mmc$
Metastable/high-pressure phases			
(εFe)	0	*hP*2	$P6_3/mmc$
Martensite	< 2.1	*tI*4	*I*4/*mmm*
Fe_4C	5.1	*cP*5	$P\bar{4}3m$
Fe_3C (θ)	6.7	*oP*16	*Pnma*
Fe_5C_2 (χ)	7.9	*mC*28	*C*2/*c*
Fe_7C_3	8.4	*hP*20	$P6_3mc$
Fe_7C_3	8.4	*oP*40	*Pnma*
Fe_2C (η)	9.7	*oP*6	*Pnnm*
Fe_2C (ε)	9.7	*hP**	$P6_322$
Fe_2C	9.7	*hP**	$P\bar{3}m1$
(C)	100	*cF*8	$Fd\bar{3}m$

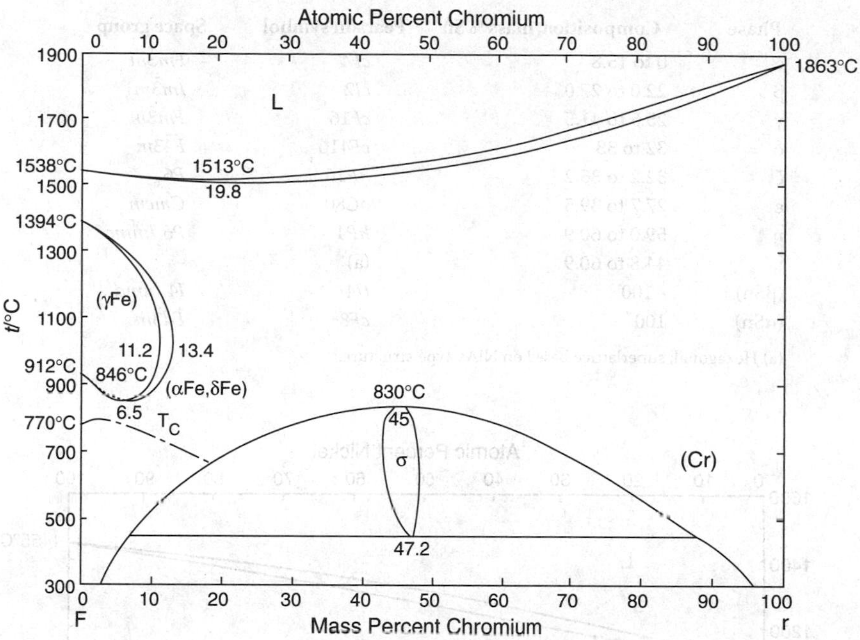

FIGURE 16. Fe-Cr system.

Phase	Composition, mass % Cr	Pearson symbol	Space group
(αFe, Cr)	0 to 100	*cI*2	$Im\bar{3}m$
(γFe)	0 to 11.2	*cF*4	$Fm\bar{3}m$
σ	42.7 to 48.2	*tP*30	$P4_2/mnm$

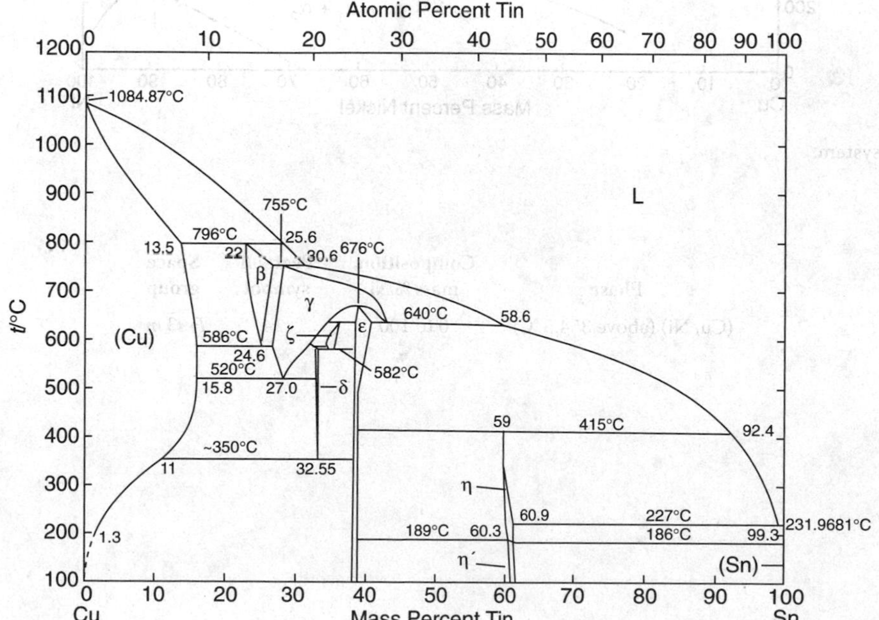

FIGURE 17. Cu-Sn system.

Phase	Composition, mass % Sn	Pearson symbol	Space group
α	0 to 15.8	*cF*4	$Fm\bar{3}m$
β	22.0 to 27.0	*cI*2	$Im\bar{3}m$
γ	25.5 to 41.5	*cF*16	$Fm\bar{3}m$
δ	32 to 33	*cF*416	$F\bar{4}3m$
ζ	32.2 to 35.2	*hP*26	$P6_3$
ε	27.7 to 39.5	*oC*80	*Cmcm*
η	59.0 to 60.9	*hP*4	$P6_3/mmc$
η′	44.8 to 60.9	(a)	–
(βSn)	~100	*tI*4	$I4_1/amd$
(αSn)	100	*cF*8	$Fd\bar{3}m$

(a) Hexagonal; superlattice based on NiAs-type structure.

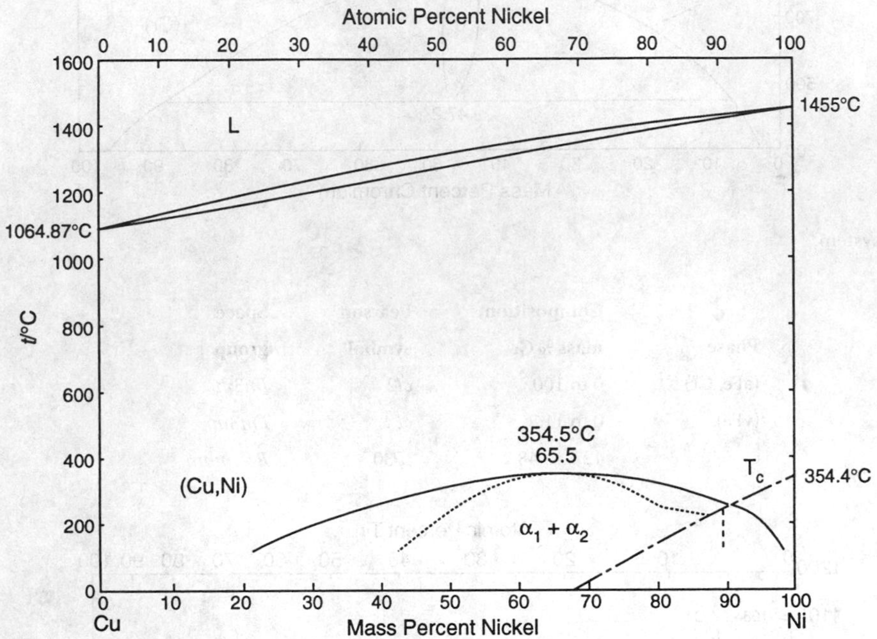

FIGURE 18. Cu-Ni system.

Phase	Composition, mass % Ni	Pearson symbol	Space group
(Cu, Ni) (above 354.5 °C)	0 to 100	*cF*4	$Fm\bar{3}m$

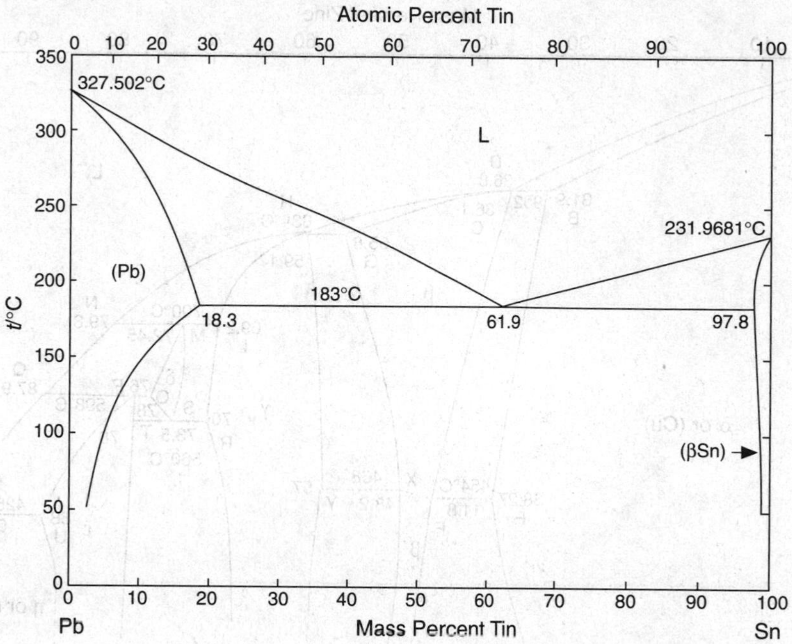

FIGURE 19. Pb-Sn system.

Phase	Composition, mass % Sn	Pearson symbol	Space group
(Pb)	0 to 18.3	*cF4*	*Fm*$\bar{3}$*m*
(βSn)	97.8 to 100	*tI4*	*I*4_1*/amd*
(αSn)	100	*cF8*	*Fd*$\bar{3}$*m*
High-pressure phases			
ε(a)	52 to 74	*hP1*	*P6/mmm*
ε′(b)	52	*hP2*	*P*6_3*/mmc*

(a) From phase diagram calculated at 2500 MPa. (b) This phase was claimed for alloys at 350 °C and 5500 MPa.

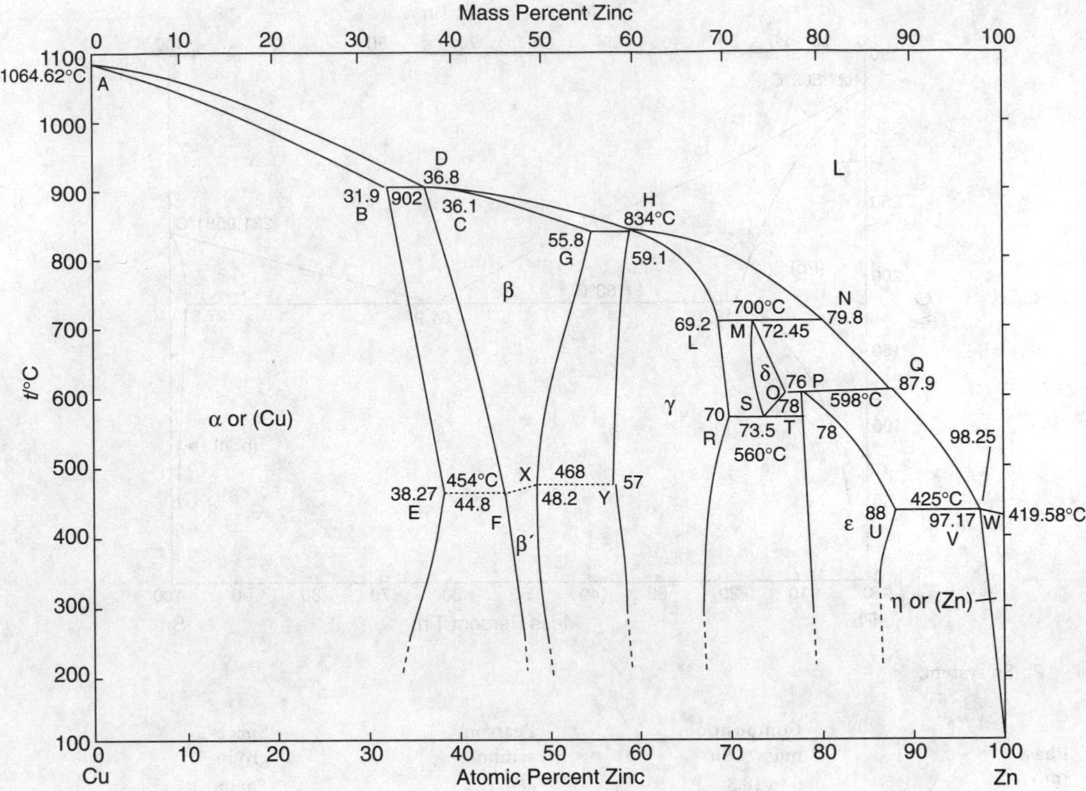

FIGURE 20. Cu-Zn system.

Phase	Composition, mass % Zn	Pearson symbol	Space group
α or (Cu)	0 to 38.95	*cF*4	$Fm\bar{3}m$
β	36.8 to 56.5	*cI*2	$Im\bar{3}m$
β′	45.5 to 50.7	*cP*2	$Pm\bar{3}m$
γ	57.7 to 70.6	*cI*52	$I\bar{4}3m$
δ	73.02 to 76.5	*hP*3	$P\bar{6}$
ε	78.5 to 88.3	*hP*2	$P6_3/mmc$
η or (Zn)	97.25 to 100	*hP*2	$P6_3/mmc$

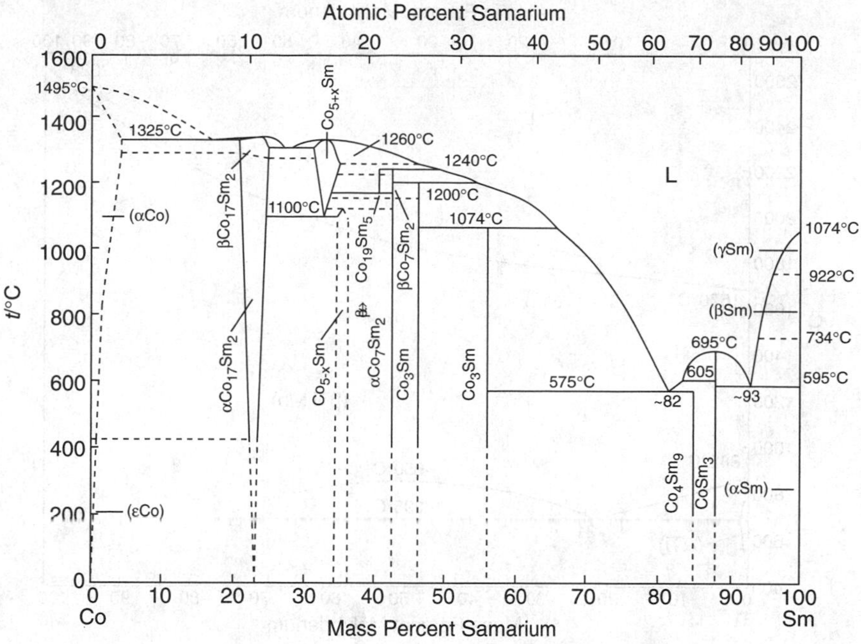

FIGURE 21. Co-Sm system.

Phase	Composition, mass % Sm	Pearson symbol	Space group
(αCo)	0 to ~3.7	*cF4*	*Fm*$\bar{3}$*m*
(εCo)	~0	*hP2*	*P*6_3*/mmc*
β$Co_{17}Sm_2$	~23.0	*hP38*	*P*6_3*/mmc*
α$Co_{17}Sm_2$	~23.0	*hR19*	*R*$\bar{3}$*m*
		hP8	*P6/mmm*
$Co_{5+x}Sm$	~33 to 34	–	–
$Co_{5-x}Sm$	~34 to 35	–	–
$Co_{19}Sm_5$	~40.1	*hR24*	*R*$\bar{3}$*m*
		hP48	*P*6_3*/mmc*
αCo_7Sm_2	~42.1	*hR18*	*R*$\bar{3}$*m*
βCo_7Sm_2	~42.1	*hP36*	*P*6_3*/mmc*
Co_3Sm	46	*hR12*	*R*$\bar{3}$*m*
Co_2Sm	56.0	*hR4*	*R*$\bar{3}$*m*
		cF24	*Fd*$\bar{3}$*m*
Co_4Sm_9	~85.1	*o***	–
$CoSm_3$	88	*oP16*	*Pnma*
(γSm)	~100	*cI2*	*Im*$\bar{3}$*m*
(βSm)	~100	*hP2*	*P*6_3*/mmc*
(αSm)	~100	*hR3*	*R*$\bar{3}$*m*
Other reported phases			
Co_5Sm	~33.8	*hP6*	*P6/mmm*
Co_2Sm_5	~86.4	*mC28*	*C2/c*

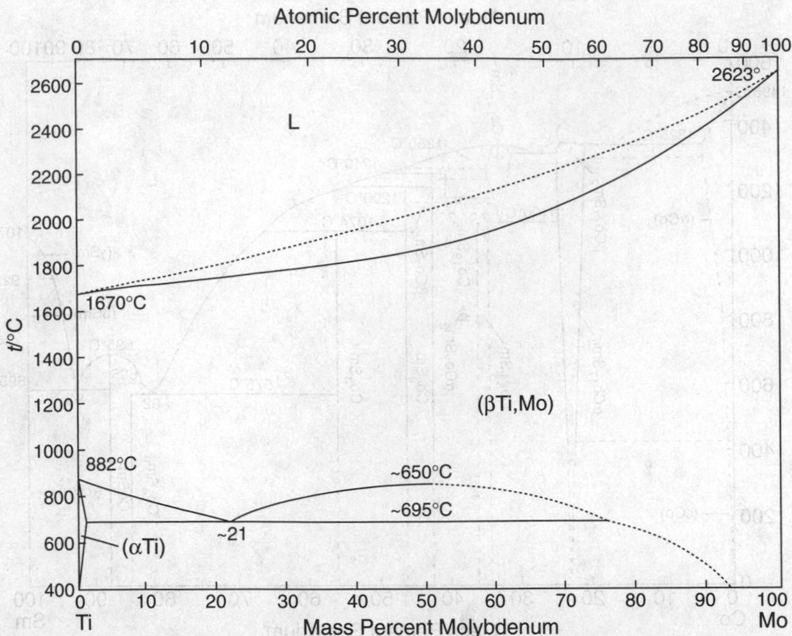

FIGURE 22. Ti-Mo system.

Phase	Composition, mass % Mo	Pearson symbol	Space group
(βTi, Mo)	0 to 100	*cI*2	*Im*$\bar{3}$*m*
(αTi)	0 to 0.8	*hP*2	$P6_3/mmc$
α′	(a)	*hP*2	$P6_3/mmc$
α″	(a)	*oC*4	*Cmcm*
ω	(a)	*hP*3	*P*6/*mmm*

(a) Metastable.

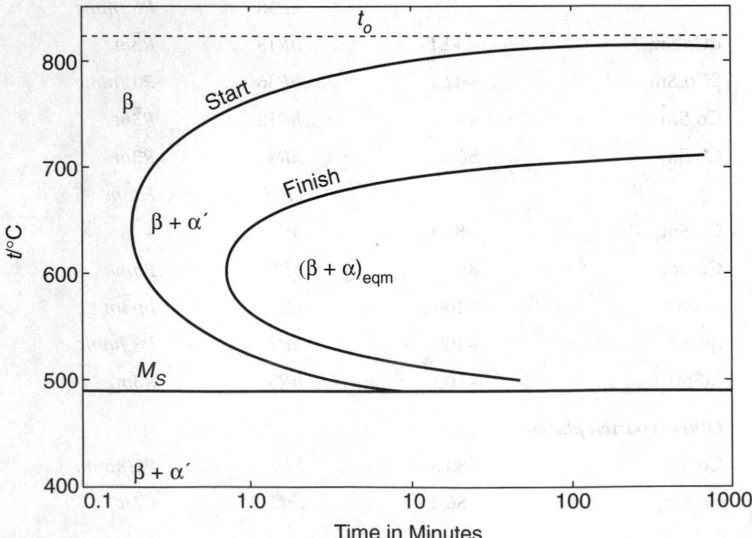

Experimental time–temperature–transformation (TTT) diagram for Ti-Mo. The start and finish times of the isothermal precipitation reaction vary with temperature as a result of the temperature dependence of the nucleation and growth processes. Precipitation is complete, at any temperature, when the equilibrium fraction of α is established in accordance with the lever rule. The solid horizontal line represents the athermal (or nonthermally activated) martensitic transformation that occurs when the β phase is quenched.

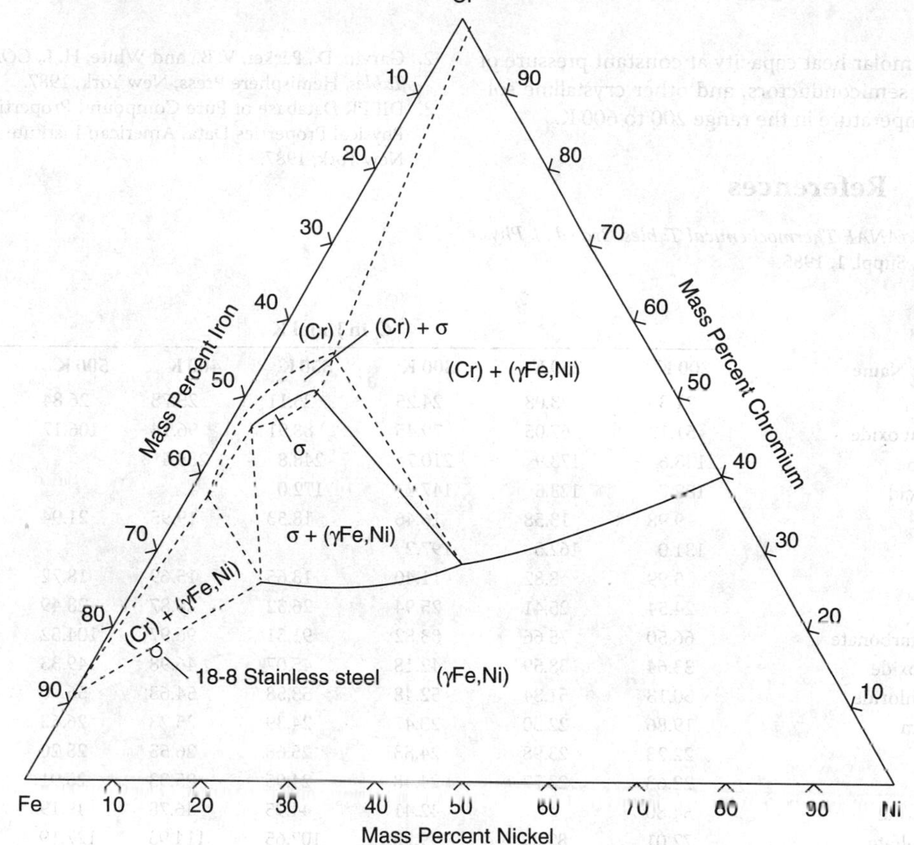

FIGURE 23. The isothermal section at 900 °C (1652 °F) of the iron-chromium-nickel ternary phase diagram, showing the nominal composition of 18-8 stainless steel.

HEAT CAPACITY OF SELECTED SOLIDS

This table gives the molar heat capacity at constant pressure of representative metals, semiconductors, and other crystalline solids as a function of temperature in the range 200 to 600 K.

References

1. Chase, M. W., et al., *JANAF Thermochemical Tables, 3rd ed., J. Phys. Chem. Ref. Data,* 14, Suppl. 1, 1985.
2. Garvin, D., Parker, V. B., and White, H. J., *CODATA Thermodynamic Tables,* Hemisphere Press, New York, 1987.
3. DIPPR Database of Pure Compound Properties, Design Institute for Physical Properties Data, American Institute of Chemical Engineers, New York, 1987.

	C_p in J/mol K						
Name	**200 K**	**250 K**	**300 K**	**350 K**	**400 K**	**500 K**	**600 K**
Aluminum	21.33	23.08	24.25	25.11	25.78	26.84	27.89
Aluminum oxide	51.12	67.05	79.45	88.91	96.14	106.17	112.55
Anthracene	138.6	173.9	210.7	248.8	288.4		
Benzoic acid	102.7	123.5	147.4	172.0			
Beryllium	9.98	13.58	16.46	18.53	19.95	21.94	23.34
Biphenyl	131.0	162.5	197.2				
Boron	5.99	8.82	11.40	13.65	15.69	18.72	20.78
Calcium	24.54	25.41	25.94	26.32	26.87	28.49	30.38
Calcium carbonate	66.50	75.66	83.82	91.51	96.97	104.52	109.86
Calcium oxide	33.64	38.59	42.18	45.07	46.98	49.33	50.72
Cesium chloride	50.13	51.34	52.48	53.58	54.68	56.90	59.10
Chromium	19.86	22.30	23.47	24.39	25.23	26.63	27.72
Cobalt	22.23	23.98	24.83	25.68	26.53	28.20	29.66
Copper	22.63	23.77	24.48	24.95	25.33	25.91	26.48
Copper oxide	34.80		42.41	44.95	46.78	49.19	50.83
Copper sulfate	77.01	89.25	99.25	107.65	114.93	127.19	136.31
Germanium			23.25	23.85	24.31	24.96	25.45
Gold			25.41	25.37	25.51	26.06	26.65
Graphite	5.01	6.82	8.58	10.24	11.81	14.62	16.84
Hexachlorobenzene	162.7	183.6	202.4				
Iodine	51.57	53.24	54.51	58.60			
Iron	21.59	23.74	25.15	26.28	27.39	29.70	32.05
Lead	25.87	26.36	26.85	27.30	27.72	28.55	29.40
Lithium	21.57	23.42	24.64	25.96	27.60	29.28	
Lithium chloride	43.35	46.08	48.10	49.66	50.97	53.34	55.59
Magnesium	22.72	24.02	24.90	25.57	26.14	27.17	28.18
Magnesium oxide			37.38	40.59	42.77	45.56	47.30
Manganese	23.05	24.95	26.35	27.52	28.53	30.29	31.90
Naphthalene	105.8	134.1	167.8	204.1			
Potassium	27.00	28.01	29.60				
Potassium chloride	48.44	50.10	51.37	52.31	53.08	54.71	56.35
Silicon	15.64	18.22	20.04	21.28	22.14	23.33	24.15
Silicon dioxide	32.64	39.21	44.77	49.47	53.43	59.64	64.42
Silver			25.36	25.55	25.79	26.36	26.99
Sodium	22.45	27.01	28.20	30.14			
Sodium chloride	46.89	48.85	50.21	51.25	52.14	53.96	55.81
Tantalum	24.08	24.86	25.31	25.60	25.84	26.35	26.84
Titanium	22.37	24.07	25.28	26.17	26.86	27.88	28.60
Tungsten	22.49	23.69	24.30	24.65	24.92	25.36	25.79
Vanadium	21.88	23.70	24.93	25.68	26.23	26.94	27.49
Zinc	24.05	25.02	25.45	25.88	26.35	27.39	28.59
Zirconium	23.87	24.69	25.22	25.61	25.93	26.56	27.28

THERMAL AND PHYSICAL PROPERTIES OF PURE METALS

This table gives the following properties for the metallic elements:

- t_m: Melting point in °C
- t_b: Normal boiling point in °C, at a pressure of 101.325 kPa (760 Torr)
- $\Delta_{fus}H$: Enthalpy of fusion at the melting point in J g^{-1}
- ρ: Density at 25 °C in g cm^{-3}
- α: Coefficient of linear expansion at 25 °C in K^{-1} (the quantity listed is $10^6 \times \alpha$)
- c_p: Specific heat capacity at constant pressure at 25 °C in J g^{-1} K^{-1}
- λ: Thermal conductivity at 27 °C in W m^{-1} K^{-1}

References

1. Dinsdale, A. T., *CALPHAD* 15, 317, 1991 [melting point, enthalpy of fusion].
2. Touloukian, Y. S., *Thermophysical Properties of Matter*, Vol. 12, Thermal Expansion, IFI/Plenum, New York, 1975 [coefficient of expansion, density].
3. Ho, C. Y., Powell, R. W., and Liley, P. E., *J. Phys. Chem. Ref. Data* 3, Suppl. 1, 1974 [thermal conductivity].
4. Cox, J. D., Wagman, D. D., and Medvedev, V. A., *CODATA Key Values for Thermodynamics*, Hemisphere Publishing Corp., New York, 1989 [heat capacity].
5. Glushko, V. P., Ed., *Thermal Constants of Substances*, VINITI, Moscow, 1965-1981 [enthalpy of fusion, heat capacity].
6. Wagman, D. D., Evans, W. H., Parker, V. B., Schumm, R. H., Halow, I., Bailey, S. M., Churney, K. L., and Nuttall, R. L., *The NBS Tables of Chemical Thermodynamic Properties, J. Phys. Chem. Ref. Data* 11, Suppl. 2, 1982 [heat capacity].
7. Gschneidner, K. A., *Bull. Alloy Phase Diagrams* 11, 216–224, 1990 [various properties of the rare earth metals].
8. Hellwege, K. H., Ed., *Landolt Börnstein, Numerical Values and Functions in Physics, Chemistry, Astronomy, Geophysics, and Technology*, Vol. 2, Part 1, Mechanical-Thermal Properties of State, 1971 [density].
9. *Physical Encyclopedic Dictionary*, Vol. 1–5, Encyclopedy Publishing House, Moscow, 1960–66.

Metal (symbol)	Atomic weight	t_m °C	t_b °C	$\Delta_{fus}H$ J g^{-1}	ρ g cm^{-3}	$\alpha \times 10^6$ K^{-1}	c_p J g^{-1} K^{-1}	λ W m^{-1} K^{-1}
Actinium (Ac)	227	1050	3198	52.9	10		0.12	
Aluminum (Al)	26.98	660.32	2519	397.0	2.70	23.1	0.897	237
Antimony (Sb)	121.76	630.628	1587	162.5	6.68	11.0	0.207	24.3
Barium (Ba)	137.33	727	1897	51.8	3.62	20.6	0.204	18.4
Beryllium (Be)	9.01	1287	2471	876.1	1.85	11.3	1.825	200
Bismuth (Bi)	208.98	271.406	1564	53.1	9.79	13.4	0.122	7.87
Cadmium (Cd)	112.41	321.069	767	55.2	8.69	30.8	0.231	96.8
Calcium (Ca)	40.08	842	1484	213.1	1.54	22.3	0.647	200
Cerium (Ce)	140.12	799	3443	39.0	6.770	6.3	0.192	11.3
Cesium (Cs)	132.91	28.5	671	15.7	1.93	97	0.242	35.9
Chromium (Cr)	52.00	1907	2671	403.9	7.15	4.9	0.449	93.7
Cobalt (Co)	58.93	1495	2927	274.9	8.86	13.0	0.421	100
Copper (Cu)	63.55	1084.62	2562	208.7	8.96	16.5	0.385	401
Dysprosium (Dy)	162.50	1412	2567	69.8	8.55	9.9	0.173	10.7
Erbium (Er)	167.26	1529	2868	119	9.07	12.2	0.168	14.5
Europium (Eu)	151.96	822	1529	60.6	5.24	35.0	0.182	13.9
Gadolinium (Gd)	157.25	1313	3273	61.5	7.90	9.4	0.235	10.5
Gallium (Ga)	69.72	29.7666	2204	80.1	5.91	18	0.373	40.6
Gold (Au)	196.97	1064.18	2856	63.7	19.3	14.2	0.129	317
Hafnium (Hf)	178.49	2233	4603	152.4	13.3	5.9	0.144	23.0
Holmium (Ho)	164.93	1472	2700	71.3	8.80	11.2	0.165	16.2
Indium (In)	114.82	156.60	2072	28.7	7.31	32.1	0.233	81.6
Iridium (Ir)	192.22	2446	4428	213.9	22.56	6.4	0.131	147
Iron (Fe)	55.85	1538	2861	247.3	7.87	11.8	0.449	80.2
Lanthanum (La)	138.91	920	3464	44.6	6.15	12.1	0.195	13.4
Lead (Pb)	207.2	327.462	1749	23.0	11.3	28.9	0.130	35.3
Lithium (Li)	6.94	180.50	1342	432	0.534	46	3.582	84.7
Lutetium (Lu)	174.97	1663	3402	106.6	9.84	9.9	0.154	16.4
Magnesium (Mg)	24.31	650	1090	348.9	1.74	24.8	1.023	156
Manganese (Mn)	54.94	1246	2061	235.0	7.3	21.7	0.479	7.82
Mercury (Hg)	200.59	−38.8290	356.62	11.4	13.5336	60.4	0.140	8.34
Molybdenum (Mo)	95.96	2623	4639	390.6	10.2	4.8	0.251	138

Metal (symbol)	Atomic weight	t_m °C	t_b °C	$\Delta_{fus}H$ J g^{-1}	ρ g cm^{-3}	$\alpha \times 10^6$ K^{-1}	c_p J g^{-1} K^{-1}	λ W m^{-1} K^{-1}
Neodymium (Nd)	144.24	1016	3074	49.5	7.01	9.6	0.190	16.5
Neptunium (Np)	237	644		13.5	20.2			6.3
Nickel (Ni)	58.69	1455	2913	297.8	8.90	13.4	0.444	90.7
Niobium (Nb)	92.91	2477	4744	323	8.57	7.3	0.265	53.7
Osmium (Os)	190.23	3033	5012	304.1	22.59	5.1	0.130	87.6
Palladium (Pd)	106.42	1554.8	2963	157.3	12.0	11.8	0.244	71.8
Platinum (Pt)	195.08	1768.2	3825	113.7	21.5	8.8	0.133	71.6
Plutonium (Pu)	244	640	3228	11.6	19.7	46.7		6.74
Polonium (Po)	209	254	962	48	9.20	23.5		20
Potassium (K)	39.10	63.5	759	59.7	0.89	83.3	0.757	102.4
Praseodymium (Pr)	140.91	931	3520	48.9	6.77	6.7	0.193	12.5
Promethium (Pm)	145	1042	3000		7.26	11	0.19	15
Protactinium (Pa)	231.04	1572		53.4	15.4			
Radium (Ra)	226	696		34	5			
Rhenium (Re)	186.21	3185	5596	183.0	20.8	6.2	0.137	47.9
Rhodium (Rh)	102.91	1964	3695	258.4	12.4	8.2	0.243	150
Rubidium (Rb)	85.47	39.30	688	25.6	1.53		0.363	58.2
Ruthenium (Ru)	101.07	2333	4150	381.8	12.1	6.4	0.238	117
Samarium (Sm)	150.36	1072	1794	57.3	7.52	12.7	0.196	13.3
Scandium (Sc)	44.96	1541	2836	313.6	2.99	10.2	0.568	15.8
Silver (Ag)	107.87	961.78	2162	104.8	10.5	18.9	0.235	429
Sodium (Na)	22.99	97.794	882.940	113.1	0.97	71	1.228	141
Strontium (Sr)	87.62	777	1382	84.8	2.64	22.5	0.306	35.3
Tantalum (Ta)	180.95	3017	5458	202.1	16.4	6.3	0.140	57.5
Technetium (Tc)	98	2157	4265	339.7	11			50.6
Terbium (Tb)	158.93	1359	3230	63.9	8.23	10.3	0.182	11.1
Thallium (Tl)	204.38	304	1473	20.3	11.8	29.9	0.129	46.1
Thorium (Th)	232.04	1750	4788	59.5	11.7	11.0	0.118	54.0
Thulium (Tm)	168.93	1545	1950	99.7	9.32	13.3	0.160	16.9
Tin (Sn)	118.71	231.93	2602	60.2	7.265	22.0	0.227	66.6
Titanium (Ti)	47.87	1668	3287	295.6	4.506	8.6	0.524	21.9
Tungsten (W)	183.84	3422	5555	284.5	19.3	4.5	0.132	174
Uranium (U)	238.03	1135	4131	38.4	19.1	13.9	0.116	27.6
Vanadium (V)	50.94	1910	3407	422	6.0	8.4	0.489	30.7
Ytterbium (Yb)	173.05	824	1196	44.3	6.90	26.3	0.155	38.5
Yttrium (Y)	88.91	1522	3345	128.1	4.47	10.6	0.298	17.2
Zinc (Zn)	65.38	419.53	907	108.1	7.14	30.2	0.388	116
Zirconium (Zr)	91.22	1854.7	4409	230.2	6.52	5.7	0.278	22.7

THERMOPHYSICAL PROPERTIES OF STAINLESS STEEL 310

Stainless steel is used in a wide variety of applications, especially at high temperatures. This table gives properties of a reference standard Stainless Steel 310 certified by the U. K. National Physical Laboratory. The properties are:

- a: thermal diffusivity in $mm^2 s^{-1}$
- c_p: specific heat capacity at constant pressure in $J g^{-1} K^{-1}$
- ρ: density in $g cm^{-3}$
- k: thermal conductivity in $W m^{-1} K^{-1}$

With kind permission from Springer Science+Business Media: International Journal of Thermophysics, 28, 674, 2007, Table II.

Reference

Blumm, J., Lindemann, A. Niedrig, B., and Campbell, R., *Int. J. Thermophys.* 28, 674, 2007.

t/°C	a/mm² s⁻¹	c_p/J g⁻¹ K⁻¹	ρ/g cm⁻³	k/W m⁻¹ K⁻¹
−125	3.170	0.376	7.878	9.39
−100	3.130	0.411	7.871	10.12
−75	3.145	0.435	7.863	10.76
−50	3.170	0.451	7.855	11.23
−25	3.210	0.464	7.846	11.69
0	3.256	0.475	7.838	12.12
25	3.352	0.483	7.829	12.67
50	3.439	0.490	7.820	13.18
101	3.611	0.501	7.801	14.11
150	3.763	0.512	7.782	14.99
200	3.917	0.518	7.762	15.75
250	4.075	0.525	7.742	16.56
300	4.205	0.533	7.722	17.31
350	4.331	0.541	7.701	18.04
400	4.455	0.548	7.681	18.75
450	4.571	0.555	7.660	19.43
500	4.686	0.562	7.639	20.12
550	4.806	0.570	7.618	20.86
600	4.920	0.595	7.596	22.24
651	5.058	0.601	7.574	23.02
701	5.179	0.607	7.551	23.74
750	5.207	0.611	7.529	23.95
800	5.288	0.617	7.506	24.49
850	5.404	0.624	7.483	25.23
901	5.506	0.633	7.460	26.00
950	5.618	0.645	7.436	26.94
1000	5.707	0.655	7.411	27.70

THERMAL CONDUCTIVITY OF METALS AND SEMICONDUCTORS AS A FUNCTION OF TEMPERATURE

This table gives the temperature dependence of the thermal conductivity of several metals and of carbon, germanium, and silicon. For graphite, separate entries are given for the thermal conductivity parallel (||) and perpendicular (⊥) to the layer planes. The thermal conductivity of all these materials is very sensitive to impurities at low temperatures, especially below 100 K. Therefore, the values given here should be regarded as typical values for a highly purified specimen; the thermal conductivity of different specimens can vary by more than an order of magnitude in the low-temperature range. See Reference 2 for details.

References

1. Ho, C. Y., Powell, R. W., and Liley, P. E., *J. Phys. Chem. Ref. Data*, 1, 279, 1972.
2. White, G. K., and Minges, M. L., *Thermophysical Properties of Some Key Solids*, CODATA Bulletin No. 59, 1985.

Thermal Conductivity in W/cm K

				Carbon (C)						
				Diamond (type)			Pyrolytic graphite			
T/K	Ag	Al	Au	I	IIa	IIb	∥	⊥	Cr	Cu
1	39.4	41.1	5.46						0.402*	42.2
2	78.3	81.8	10.9	0.0138*	0.033*	0.0200*			0.803	84.0
3	115	121	16.1	0.0461	0.111	0.0676			1.20	125
4	147	157	20.9	0.108	0.261	0.160			1.60	162
5	172	188	25.2	0.206	0.494	0.307			2.00	195
6	187	213	28.5	0.344	0.820	0.510			2.39	222
7	193	229	30.9	0.523	1.24	0.778			2.27	239
8	190	237	32.3	0.762	1.77	1.12			3.14	248
9	181	239	32.7	1.05	2.41	1.53			3.50	249
10	168	235	32.4	1.40	3.17	2.03	0.811	0.0116	3.85	243
15	96.0	176	24.6	3.96	8.65	5.66			5.24	171
20	51.0	117	15.8	7.87	16.8	11.2	4.20	0.0397	5.93	108
30	19.3	49.5	7.55	18.8	38.9	26.5	9.86	0.0786	5.49	44.5
40	10.5	24.0	5.15	29.4	65.9	44.0	16.4	0.120	4.25	21.7
50	7.0	13.5	4.21	35.3	92.1	59.1	23.1	0.152	3.17	12.5
60	5.5	8.5	3.74	37.4	112	67.5	29.8	0.173	2.48	8.29
70	4.97	5.85	3.48	36.9	119	69.1	36.6	0.181	2.07	6.47
80	4.71	4.32	3.32	35.1	117	65.7	42.8	0.181	1.84	5.57
90	4.60	3.42	3.28	32.7	109	60.0	47.5	0.176	1.69	5.08
100	4.50	3.02	3.27	30.0	100	54.2	49.7	0.168	1.59	4.82
150	4.32	2.48	3.25	19.5	60.2	32.5	45.1	0.125	1.29	4.29
200	4.30	2.37	3.23	14.1	40.3	22.6	32.3	0.0923	1.11	4.13
250	4.29	2.35	3.21	11.0	29.7	17.0	24.4	0.0711	1.00	4.06
300	4.29	2.37	3.17	8.95	23.0	13.5	19.5	0.0570	0.937	4.01
350	4.27	2.40	3.14	7.55*	18.5*	11.1*	16.2	0.0477	0.929	3.96
400	4.25	2.40	3.11	6.5*	15.4*	9.32*	13.9	0.0409	0.909	3.93
500	4.19	2.36	3.04				10.8	0.0322	0.860	3.86
600	4.12	2.31	2.98				8.92	0.0268	0.807	3.79
800	3.96	2.18	2.84				6.67	0.0201	0.713	3.66
1000	3.79		2.70				5.34	0.0160	0.654	3.52
1200	3.61*		2.55				4.48	0.0134	0.619	3.39
1400							3.84	0.0116	0.588	
1600							3.33	0.0100	0.556	
1800							2.93	0.00895	0.526*	
2000							2.62	0.00807	0.494*	

T/K	Fe	Ge[a]	Mg	Ni	Pb	Pt	Si[a]	Sn	Ti	W
1	1.71	0.274	9.86	2.17	27.9	2.31	0.0693*	183	0.0144*	14.4
2	3.42	2.06	19.6	4.34	44.6	4.60	0.454	323	0.0288*	28.7
3	5.11	5.35	29.0	6.49	35.8	6.79	1.38	297	0.0432	42.8
4	6.77	8.77	37.6	8.59	22.2	8.8	2.97	181	0.0575	56.3
5	8.39	11.6	45.0	10.6	13.8	10.5	5.27	117	0.0719	68.7
6	9.93	13.9	50.8	12.5	8.10	11.8	8.23	76	0.0863	79.5
7	11.4	15.5	54.7	14.2	4.86	12.6	11.7	52	0.101	88.0
8	12.7	16.6	56.7	15.8	3.20	12.9	15.5	36	0.115	93.8
9	13.9	17.3	57.0	17.1	2.30	12.8	19.5	26	0.129	96.8
10	14.8	17.7	55.8	18.1	1.78	12.3	23.3	19.3	0.143	97.1
15	17.0	17.3	41.1	19.5	0.845	8.41	41.6	6.3	0.212	72.0
20	15.4	14.9	27.2	16.5	0.591	4.95	49.8	3.2	0.275	40.5
30	10.0	10.8	12.9	9.56	0.477	2.15	48.1	1.79	0.365	14.4
40	6.23	7.98	7.19	5.82	0.451	1.39	35.3	1.33	0.390	6.92
50	4.05	6.15	4.65	4.00	0.436	1.09	26.8	1.15	0.374	4.27
60	2.85	4.87	3.27	3.08	0.425	0.947	21.1	1.04	0.355	3.14
70	2.16	3.93	2.49	2.50	0.416	0.862	16.8	0.96	0.340	2.58
80	1.75	3.25	2.02	2.10	0.409	0.815	13.4	0.915	0.326	2.29
90	1.50	2.70	1.78	1.83	0.403	0.789	10.8	0.880	0.315	2.17
100	1.34	2.32	1.69	1.64	0.397	0.775	8.84	0.853	0.305	2.08
150	1.04	1.32	1.61	1.22	0.379	0.740	4.09	0.779	0.270	1.92
200	0.94	0.968	1.59	1.07	0.367	0.726	2.64	0.733	0.245	1.85
250	0.865	0.749	1.57	0.975	0.360	0.718	1.91	0.696	0.229	1.80
300	0.802	0.599	1.56	0.907	0.353	0.716	1.48	0.666	0.219	1.74
350	0.744	0.495	1.55	0.850	0.347	0.717	1.19	0.642	0.210	1.67
400	0.695	0.432	1.53	0.802	0.340	0.718	0.989	0.622	0.204	1.59
500	0.613	0.338	1.51	0.722	0.328	0.723	0.762	0.596	0.197	1.46
600	0.547	0.273	1.49	0.656	0.314	0.732	0.619		0.194	1.37
800	0.433	0.198	1.46*	0.676		0.756	0.422		0.197	1.25
1000	0.323	0.174		0.718		0.787	0.312		0.207	1.18
1200	0.283	0.174		0.762		0.826	0.257		0.220	1.12
1400	0.312			0.804		0.871	0.235		0.236	1.08
1600	0.330					0.919	0.221		0.253	1.04
1800	0.345*					0.961			0.270*	1.01
2000						0.994*				0.98

[a] Values below 300 K are typical values.

* Extrapolated.

THERMAL CONDUCTIVITY OF ALLOYS AS A FUNCTION OF TEMPERATURE

This table lists the thermal conductivity of selected alloys at various temperatures. The indicated compositions refer to weight percent. Since the thermal conductivity is sensitive to exact composition and processing history, especially at low temperatures, these values should be considered approximate.

References

1. Powell, R. L., and Childs, G. E., in *American Institute of Physics Handbook, 3rd Edition*, Gray, D. E., Ed., McGraw-Hill, New York, 1972.
2. Ho, C. Y., et al., *J. Phys. Chem. Ref. Data*, 7, 959, 1978.

Thermal conductivity in W/m K

	Alloy	4 K	20 K	77 K	194 K	273 K	373 K	573 K	973 K
Aluminum:	1100	50	240	270	220	220			
	2024	3.2	17	56	95	130			
	3003	11	58	140	150	160			
	5052	4.8	25	77	120	140			
	5083, 5086	3	17	55	95	120			
	Duralumin	5.5	30	91	140	160	180		
Bismuth:	Rose metal		5.5	8.3	14	16			
	Wood's metal	4	17	23					
Copper:	electrolytic tough pitch	330	1300	550	400	390	380	370	350
	free cutting, leaded	200	800	460	380	380			
	phosphorus, deoxidized	7.5	42	120	190	220			
	brass, leaded	2.3	12	39	70	120			
	bronze, 68% Cu; 32% Zn	2.3	16	48	92	110			
	beryllium	2	17	36	70	90	113	172	
	german silver	0.75	7.5	17	20	23	25	30	40
	silicon bronze A		3.4	11	23	30			
	manganin	0.48	3.2	14	17	22			
	constantan	0.9	8.6	17	19	22			
Ferrous:	commercial pure iron	15	72	106	82	76	66	54	34
	plain carbon steel(AISI 1020)	13	20	58	65	65			
	plain carbon steel(AISI 1095)		8.5	31	41	45			
	3% Ni; 0.7% Cr; 0.6% Mo		6	22		33	35	36	30
	4% Si					20	24	28	26
	stainless steel	0.3	2	8	13	14	16	19	25
	27% Ni; 15% Cr		1.7	55		11	12	16	21
Gold:	colbalt thermocouple	1.2	8.6	20					
	65% Au; 35% Ag		12	24		61	89		
Indium:	85.5% In; 14.5% Pb	1.9	7.8	24	41				
Lead:	60% Pb; 40% Sn (soft solder)		28	44					
	64.35% Pb; 35.65% In	0.8	3.26	9.1		20.2			
Nickel:	80% Ni; 20% Cr					12	14	17	23
	contracid	0.2	2	7.3	9.5	13			
	inconel	0.5	4.2	12.5	13	15	16	19	26
	monel	0.9	7.1	15	20	21	24	30	43
Platinum:	90% Pt; 10% Ir					31	31.4		
	90% Pt; 10% Rh					30.1	30.5		
Silver:	silver solder		12	34	58				
	normal Ag thermocouple	48	230	310					
Tin:	60% Sn; 40% Pb	16	55	51					
Titanium:	5.5% Al; 2.5% Sn;0.2% Fe		1.8	4.3	6.4	7.8	8.4	10.8	
	4.7% Mn; 3.99% Al; 0.14% C		1.7	4.5	6.5	8.5			

THERMAL CONDUCTIVITY OF CRYSTALLINE DIELECTRICS

This table lists the thermal conductivity of a number of crystalline dielectrics, including some that find use as optical materials. Values are given at temperatures for which data are available.

Reference

Powell, R. L., and Childs, G. E., in *American Institute of Physics Handbook, 3rd Edition*, Gray, D. E., Ed., McGraw-Hill, New York, 1972.

Material	*T*/K	Ther. cond. W/m K
AgCl	223	1.3
	273	1.2
	323	1.1
	373	1.1
Al,B silicate (tourmaline)	398	2.9
‖ to *c* axis	540	3.2
	723	3.5
Al,Be silicate (beryl)	315	6.4
Al,F silicate (topaz)	315	17.7
‖ to *c* axis	358	15.6
	417	13.3
Al,Fe silicate (garnet)	315	35.8
	358	35.4
	377	35.6
Al_2O_3 (sapphire):		
36° to *c* axis	4.2	110
	20	3500
	35	6000
	77	1100
⊥ to *c* axis	373	2.6
	523	3.9
	773	5.8
Al_2O_3 (sintered)	4.2	0.5
	20	23
	77	150
	194	48
	273	35
	373	26
	973	8
Ar	8	6.0
	10	3.7
	20	1.4
	77	0.31
As_2S_3 (glass)	283	0.16
	323	0.21
	373	0.27
BN	1047	36.2
	1475	22.7
	1928	21.9
	2111	18.5
BaF_2	225	20
	260	13.4
	305	10.9
	370	10.5
$BaTiO_3$	5	4.2
	30	24.0
	40	25.0
	100	12.0
	250	4.8
	300	6.2
BeO	4.2	0.3
	20	16
	77	270
	373	210
	573	120
	1273	29
Bi_2Te_3	80	6.4
	204	2.8
	303	3.6
	370	4.6
C (diamond)	4.2	13
type I	20	800
	77	3550
	194	1450
	273	1000
$CaCO_3$		
‖ to *c* axis	83	25
	273	5.5
⊥ to *c* axis	83	17
	194	6.5
	273	4.6
	373	3.6
CaF_2	83	39
	223	18
	273	10
	323	9.2
	373	9
$CaWO_4$ (scheelite)	422	11.3
CdTe	160	7.0
	297	3.6
	422	2.9
CsBr	223	1.2
	273	0.94
	323	0.81
	373	0.77
CsI	223	1.4
	273	1.2
	323	1
	373	0.95
Cu_2O (cuprite)	102	3.74
	163	7.76
	299	5.58
	360	4.86
Fe_3O_4 (magnetite)	4.5	27.4
	20.5	293.0
	126.5	7.4
	304	7.0
Glass:		
phoenix	4.2	0.095
	20	0.13
	77	0.37
plastic perspex	4.2	0.058

Material	T/K	Ther. cond. W/m K
	20	0.074
pyrex	77	0.44
	194	0.88
	273	1
H_2 (para + 0.5% ortho)	2.5	100
	3	150
	4	200
	6	30
	10	3
H_2O (ice)	173	3.5
	223	2.8
	273	2.2
He^3 (high pressure)	0.6	25
	1	2
	1.5	0.57
	2	0.21
He^4 (high pressure)	0.5	42
	0.8	120
	1	24
	2	0.18
I_2	300	0.45
	325	0.42
	350	0.4
KBr	2	150
	4.2	360
	100	12
	273	5
	323	4.8
	373	4.8
KCl	4.2	500
	25	140
	80	35
	194	10
	273	7.0
	323	6.5
	373	6.3
KI	4.2	700
	80	13
	194	4.6
	273	3.1
Kr	4.2	0.48
	10	1.7
	20	1.2
	77	0.36
LaF_3	78	7.8
	197	5.0
	274	5.4
LiF	4.2	620
	20	1800
	77	150
$MgO{\cdot}Al_2O_3$ (spinel)	373	13
	773	8.5
MnO	4.2	0.25
	40	55
	120	8
	573	3.5
NaCl	4.2	440
	20	300
	77	30
	273	6.4
	323	5.6
	373	5.4
NaF	5	1100
	50	250
	100	90
Ne	2	3.0
	3	4.6
	4.2	4.2
	10	0.8
	20	0.3
NH_4Cl	77	17
	194	23
	230	38
	273	27
$NH_4H_2PO_4$		
‖ to optic axis	315	0.71
	339	0.71
⊥ to optic axis	313	1.26
	342	1.34
NiO	4.2	5.9
	40	400
	194	82
SiO_2 (quartz)		
‖ to *c* axis	20	720
	194	20
	273	12
⊥ to *c* axis	20	370
	194	10
	273	6.8
SiO_2 (fused silica)	4.2	0.25
	20	0.7
	77	0.8
	194	1.2
	273	1.4
	373	1.6
	673	1.8
$SrTiO_3$	5	2.4
	30	21.0
	40	19.2
	100	18.5
	250	12.5
	300	11.2
TlBr	316	0.59
TlCl	311	0.75
TiO_2 (rutile)		
‖ to optic axis	4.2	200
	20	1000
	273	13
⊥ to optic axis	4.2	160
	20	690
	273	9

THERMAL CONDUCTIVITY OF CERAMICS AND OTHER INSULATING MATERIALS

Thermal conductivity values for ceramics, refractory oxides, and miscellaneous insulating materials are given here. The thermal conductivity refers to samples with the density indicated in the second column. Since most of these materials are highly variable, the values should only be considered as a rough guide.

References

1. Powell, R. L., and Childs, G. E., in *American Institute of Physics Handbook, 3rd Edition*, Gray, D. E., Ed., McGraw-Hill, New York, 1972.
2. Perry, R. H., and Green, D., *Perry's Chemical Engineers' Handbook, Sixth Edition*, McGraw-Hill, New York, 1984.

Material	Dens. g/cm³	*t* °C	Ther. cond. W/m K
Alumina (Al_2O_3)	3.8	100	30
		400	13
		1300	6
		1800	7.4
	3.5	100	17
		800	7.6
Al_2O_3 + MgO		100	15
		400	10
		1000	5.6
Asbestos	0.4	−100	0.07
		0	0.09
		100	0.10
Asbestos + 85% MgO	0.3	30	0.08
Asphalt	2.1	20	0.06
Beryllia (BeO)	2.8	100	210
		400	90
		1000	20
		1800	15
	1.85	50	64
		200	40
		600	23
Brick, dry	1.54	0	0.04
Brick, refractory:			
alosite		1000	1.3
aluminous	1.99	400	1.2
		1000	1.3
diatomaceous	0.77	100	0.2
		500	0.24
	0.4	100	0.08
		500	0.1
fireclay	2	400	1
		1000	1.2
silicon carbide	2	200	2
		600	2.4
vermiculite	0.77	200	0.26
		600	0.31
Calcium oxide		100	16
		400	9
		1000	7.5
Cement mortar	2	90	0.55
Charcoal	0.2	20	0.055
Coal	1.35	20	0.26
Concrete	1.6	0	0.8
Cork	0.05	0	0.03
		100	0.04
	0.35	0	0.06
		100	0.08
Cotton wool	0.08	30	0.04
Diatomite	0.2	0	0.05
		400	0.09
	0.5	0	0.09
		400	0.16
Ebonite	1.2	0	0.16
Felt, flax	0.2	30	0.05
	0.3	30	0.04
Fuller's earth	0.53	30	0.1
Glass wool	0.2	−200 to 20	0.005
		50	0.04
		100	0.05
		300	0.08
Graphite			
100 mesh	0.48	40	0.18
20-40 mesh	0.7	40	1.29
Linoleum cork	0.54	20	0.08
Magnesia (MgO)		100	36
		400	18
		1200	5.8
		1700	9.2
MgO + SiO_2		100	5.3
		400	3.5
		1500	2.3
Mica:			
muscovite		100	0.72
		300	0.65
		600	0.69
phlogopite		100	0.66
Canadian		300	0.19
		600	0.2
Micanite		30	0.3
Mineral wool	0.15	30	0.04
Perlite, expanded	0.1	−200 to 20	0.002
Plastics:			
bakelite	1.3	20	1.4
celluloid	1.4	30	0.02
polystyrene foam	0.05	−200 to 20	0.033
mylar foil	0.05	−200 to 20	0.0001
nylon		−253	0.10
		−193	0.23
		25	0.30
polytetrafluoroethylene		−253	0.13
		−193	0.16
		25	0.26
		230	2.5
urethane foam	0.07	20	0.06
Porcelain		90	1
Rock:			
basalt		20	2
chalk		20	0.92

Material	Dens. g/cm³	t °C	Ther. cond. W/m K
granite	2.8	20	2.2
limestone	2	20	1
sandstone	2.2	20	1.3
slate, ⊥		95	1.4
slate, ‖		95	2.5
Rubber:			
sponge	0.2	20	0.05
92 percent		25	0.16
Sand, dry	1.5	20	0.33
Sawdust	0.2	30	0.06
Shellac		20	0.23
Silica aerogel	0.1	−200 to 20	0.003
Snow	0.25	0	0.16
Steel wool	0.1	55	0.09
Thoria (ThO_2)		100	10
		400	5.8
		1500	2.4
Titanium dioxide		100	6.5
		400	3.8
		1200	3.3
Uranium dioxide		100	9.8
		400	5.5
		1000	3.4
Wood:			
balsa, ⊥	0.11	30	0.04
fir, ⊥	0.54	20	0.14
fir, ‖	0.54	20	0.35
oak		20	0.16
plywood		20	0.11
pine, ⊥	0.45	60	0.11
pine, ‖	0.45	60	0.26
walnut, ⊥	0.65	20	0.14
Wool	0.09	30	0.04
Zinc oxide		200	17
		800	5.3
Zirconia (ZrO_2)		100	2
		400	2
		1500	2.5
Zirconia + silica		200	5.6
		600	4.6
		1500	3.7

THERMAL CONDUCTIVITY OF GLASSES

This table gives the composition of various types of glasses and the thermal conductivity k as a function of temperature. Because of the variability of glasses, the data should be regarded as only approximate.

Type of glass	Composition SiO$_2$ (wt%)	Other oxides (wt%)		t °C	k W/m K
Vitreous silica	100			−150	0.85
				−100	1.05
				−50	1.20
				0	1.30
				50	1.40
				100	1.50
Vycor glass	96	B_2O_3	3	−100	1.00
				0	1.25
				100	1.40
Pyrex type chemically resistant borosilicate glasses	80–81	B_2O_3	12–13	−100	0.90
		Na_2O	4	0	1.10
		Al	2	100	1.25
Borosilicate crown glasses	60–65	B_2O_3	15–20	−100	0.65–0.75
				0	0.90–0.95
				100	1.00–1.05
	65–70	B_2O_3	10–15	−100	0.75–0.80
				0	0.95–1.00
				100	1.05–1.15
	70–75	B_3O_3	5–10	−100	0.80–0.85
				0	1.05–1.10
				100	1.15–1.20
Zinc crown glasses (i)	55–65	ZnO	5–15	−100	0.88–0.92
		Remainder:		0	1.10–1.15
		B_2O_3, Al_2O_3		100	1.15–1.25
		ZnO	5–15	−100	0.60–0.70
		Remainder:		0	0.70–0.90
		Na_2O, K_2O		100	0.85–0.95
		ZnO	15–25	−100	0.88–0.92
		Remainder:		0	1.10–1.15
		B_2O_3, Al_2O_3		100	1.15–1.20
		ZnO	15–25	−100	0.65–0.80
		Remainder:		0	0.85–0.95
		Na_2O, K_2O		100	0.90–1.05
Zinc crown glasses (ii)	65–75	ZnO	5–15	−100	0.88–0.92
		Remainder:		0	1.15–1.15
		B_2O_3, Al_2O_3		100	1.20–1.30
		ZnO	5–15	−100	0.70–0.85
		Remainder:		0	0.90–1.05
		Na_2O, K_2O		100	1.00–1.15
		ZnO	15–25	−100	0.90–0.95

Type of glass	Composition: SiO_2 (wt%)	Composition: Other oxides (wt%)		t °C	k W/m K
		Remainder:		0	1.15–1.15
		B_2O_3, Al_2O_3		100	1.20–1.25
		ZnO	15–25	−100	0.65–0.85
		Remainder:		0	0.85–1.00
		Na_2O, K_2O		100	1.05–1.20
Barium crown glasses	31	B_2O_3	12	−100	0.55
		Al_2O_3	8	0	0.70
		BaO	48	100	0.80
	41	B_2O_3	6	−100	0.60
		Al_2O_3	2	0	0.75
		ZnO	8	100	0.85
		BaO	43		
	47	B_2O_3	4	−100	0.65
		Na_2O	1	0	0.75
		K_2O	7	100	0.90
		ZnO	8		
		BaO	32		
	65	B_2O_3	2	−100	0.70
		Na_2O	5	0	0.90
		K_2O	15	100	1.00
		ZnO	2		
		BaO	10		
Borate glasses					
Borate flint glass	9	B_2O_3	36	−100	0.55
		Na_2O	1	0	0.65
		K_2O	2	100	0.80
		PbO	36		
		Al_2O_3	10		
		ZnO	6		
Borate flint glass	0	B_2O_3	56	−100	0.50
		Al_2O_3	12	0	0.65
		PbO	32	100	0.85
Borate flint glass	0	B_2O_3	43	−100	0.40
		Al_2O_3	5	0	0.55
		PbO	52	100	0.70
Borate glass	4	B_2O_3	55	−100	0.65
		Al_2O_3	14	0	0.80
		PbO	11	100	0.90
		K_2O	4		
		ZnO	12		
Borate crown glass	0	B_2O_3	64	−100	0.50
		Na_2O	8	0	0.65
		K_2O	3	100	0.85
		BaO	4		
		PbO	3		
		Al_2O_3	18		
Light borate crown glass	0	B_2O_3	69	−100	0.55
		Na_2O	8	0	0.70

Type of glass	Composition SiO$_2$ (wt%)	Other oxides (wt%)		*t* °C	*k* W/m K
		BaO	5	100	0.90
		Al_2O_3	18		
Zinc borate glass	0	B_2O_3	40	−100	0.65
		ZnO	60	0	0.75
				100	0.85
Phosphate crown glasses					
Potash phosphate glass	0	P_2O_5	70	0	0.75
		B_2O_3	3	100	0.85
		K_2O	12		
		Al_2O_3	10		
		MgO	4		
Baryta phosphate glass	0	P_2O_5	60	45	0.75
		B_2O_3	3		
		Al_2O_3	8		
		BaO	28		
Soda-lime glasses	75	Na_2O	17	−100	0.75
		CaO	8	0	0.95
				100	1.10
	75	Na_2O	12	−100	0.90
		CaO	13	0	1.10
				100	1.15
	72	Na_2O	15	−100	0.80
		CaO	11	0	1.00
		Al_2O_3	2	100	1.15
	65	Na_2O	25	−100	0.65
		CaO	10	0	0.85
				100	0.95
	65	Na_2O	15	−100	0.85
		CaO	20	0	1.00
				100	1.10
	60	Na_2O	20	−100	0.75
		CaO	20	0	0.90
				100	1.00
Other crown glasses					
Crown glass	75	Na_2O	9	−100	0.80
		K_2O	11	0	1.00
		CaO	5	100	1.10
High dispersion crown glass	68	Na_2O	16	−100	0.65
		ZnO	3	0	0.85
		PbO	13	100	1.00
Miscellaneous flint glasses					
(i) Silicate flint glasses					
Light flint glasses	65	PbO	25	−100	0.65–0.70
		Others	10	0	0.88–0.92
				100	1.00–1.05
	55	PbO	35	−100	0.60–0.65

Type of glass	Composition SiO$_2$ (wt%)	Other oxides (wt%)		t °C	k W/m K
		Others	10	0	0.75–0.85
				100	0.88–0.92
Ordinary flint glass	45	PbO	45	−100	0.50–0.60
		Others	10	0	0.65–0.75
				100	0.80–0.85
Heavy flint glass	35	PbO	60	−100	0.45–0.50
		Others	5	0	0.60–0.65
				100	0.70–0.75
Very heavy flint glasses	25	PbO	73	−100	0.40–0.45
		Others	2	0	0.55–0.60
				100	0.63–0.67
	20	PbO	80	−100	0.40
				0	0.50
				100	0.60
(ii) Borosilicate flint glass	33	B_2O_3	31	−100	0.65
		PbO	25	0	0.85
		Al_2O_3	7	100	0.95
		K_2O	3		
		Na_2O	1		
(iii) Barium flint glass	50	BaO	24	−100	0.60
		PbO	6	0	0.70
		K_2O	8	100	0.85
		Na_2O	3		
		ZnO	8		
		Sb_2O_3	1		
Other glasses					
Potassium glass	59	K_2O	33	50	0.88–0.92
		CaO	8		
Iron glasses	63	Fe_2O_3	10	−100	0.80
		Na_2O	17	0	0.95
		MgO	4	100	1.05
		CaO	3		
		Al_2O_3	2		
	67	Fe_2O_3	15	0	0.88–0.92
		Na_2O_3	18	100	1.00–1.05
	62	Fe_2O_3	20	0	0.85–0.90
		Na_2O	18	100	0.95–1.00
Rock glasses					
Obsidian				0	1.35
Artificial diabase				100	1.25

THERMOELECTRIC PROPERTIES OF METALS AND SEMICONDUCTORS

Lev I. Berger

There are three thermoelectric phenomena that result from correlation between propagation of heat through a conductor and displacement of the current carriers in the conductor. The Seebeck effect (Ref. 1) consists of formation of an electric current in an electrical circuit formed by two dissimilar conductors if the contacts between the conductors are held at different temperatures. A reverse phenomenon, the Peltier effect (Ref. 2), consists of formation of a temperature difference between the contacts in a circuit of this type if an electric current is created in the circuit by an external current source to which the circuit is connected. W. Thomson (Lord Kelvin), who explained both effects (Refs. 3,4), predicted and experimentally confirmed the existence of another thermoelectric phenomenon, named the Thomson effect, which consists of absorption or release of heat in a uniform conductor with a current passing through it when a temperature gradient (positive or negative) is present along the current direction.

The electromotive force, ΔU, which creates the Seebeck current in the circuit, is the algebraic sum of the emf's created in each of the conductors, and is proportional to the temperature difference, ΔT, between the electrical contact points: $\Delta U = \Delta U_1 + \Delta U_2 = \alpha_1 \Delta T + \alpha_2 \Delta T$. The coefficient of proportionality, α, called the Seebeck coefficient or thermoelectric power or thermal electromotive force (thermal emf), of each of the two materials depends on the electrical properties and temperature of the material. The Peltier effect is measured by the amount of heat, ΔQ, released or absorbed in a unit of time (in addition to the Joule heat) at a contact of two dissimilar conductors with electric current ΔI passing through the contact: $\Delta Q = \Pi \cdot \Delta I$. Thomson showed that $\Pi = \alpha T$. The Thomson effect's heat, dQ, released or absorbed in a unit of time along a part of a conductor of length dx is proportional to the current magnitude I, the temperature gradient along the conductor $\partial T/\partial x$, and the increment dx: $dQ = \tau I(\partial T/\partial x)dx$. Thomson showed that the magnitude of the coefficient of proportionality, τ, later named the Thomson coefficient, depends on only the properties of the conductor and the ambient temperature and correlates with the other thermoelectric parameters of a material through the equation $\tau = T(\partial\alpha/\partial T)$.

Another thermoelectric phenomenon, called the Bridgman effect or the internal Peltier effect (Ref. 5), occurs when an electric current passes through an anisotropic crystal, resulting in absorption or liberation of heat because of non-uniformity in current distribution.

In view of the correlations between α, Π, and τ, we need only to present data for one of these parameters, namely, thermal emf α and its dependence on temperature. These values are presented below, first for metals and then for semiconductors. In accordance with modern theory of solids, thermal emf in semiconductors is up to three or even four orders of magnitude higher than that in metals (Ref. 9).

References

1. Seebeck, T. J., *Abhand. Deut. Akad. Wiss. Berlin*, 265–373, 1822.
2. Peltier, J. C. A, *Ann. Chem.*, LVI, 371–387, 1834.
3. Thomson, W., *Proc. Roy. Soc. Edinburgh*, 91–98, 1851.
4. Thomson, W., *Math. and Phys. Papers, Cambridge*, 1, 558, 1882; 2, 306, 1882.
5. Bridgman, P. W., *Proc. Natl. Acad. Sci. USA*, 13(2), 46–50, 1927; *Phys. Rev.*30, 911–921 (1927).
6. Blatt, F. J., *Thermoelectric Power of Metals*. Plenum Press, NY 1976.
7. Foiles, C. L., Thermopower of Pure Metals and Dilute Alloys, in *Landolt–Bornstein. Numerical Data and Functional Relationships in Science and Technology. New Series.* Group III, v. 15, Metals. Springer-Verlag, NY, 1985.
8. Burkov, A. T., and Vedernikov, M. V., in *CRC Handbook of Thermoelectrics*, D. M. Rowe, Ed., CRC Press, Boca Raton, FL, 1995, pp. 387–399.
9. Ioffe, A. F., *Semiconductor Thermoelements and Thermoelectric Cooling*, Infosearch Ltd., 1957.
10. Berger, L. I., and Prochuchan, V. D., *Ternary Diamond-Like Semiconductors*, Cons. Bureau, Plenum Press, New York, 1969.
11. Rowe, D. M., Ed., *Thermoelectrics Handbook Macro to Nano*, Taylor & Francis, Boca Raton, 2006.
12. Berger, L. I., *Semiconductor Materials*, CRC Press, Boca Raton, FL, 1996.
13. Glazov, V. M., Tshizhevskaya, S. N., and Glagoleva, N. N., *Liquid Semiconductors*, Nauka Publ. House, Moscow, 1967.
14. Shay, J. L., and Wernick, J. H., *Ternary Chalcopyrite Semiconductors: Growth, Electronic Properties and Applications*, Pergamon Press, New York, 1975.
15. Heikes, R. R., and Ure, R. W., *Thermoelectricity: Science and Engineering*, Interscience Publ., New York, 1961.
16. Goland, A. N., and Ewald, A. W., *Phys. Rev.* 104, 948 (1956).
17. Tauc J., *Photo and Thermoelectric Effects in Semiconductors*, Pergamon, New York, 1962.
18. Dugdale, J. S., *The Electrical Properties of Metals and Alloys*, Edward Arnold, London, 1977.
19. Rowe, D. M., Ed., *CRC Handbook of Thermoelectrics*, CRC Press, Boca Raton, FL, 1994.

Thermoelectric Properties of Elemental Metals

	Thermal emf α(T) in μV/K at Temperature T				
	100 K	300 K	500 K	1000 K	1500 K
Ag	0.73	1.51	2.82	7.95	
Al	−2.2	−1.66	1.96		
Au	0.82	1.94	2.86	3.85	
Ba	−4	12.1	28.5		
Be	−2.5	1.7	2.7	7.9	
Ca	1.05	10.3	17.1		
Cd	−0.05	2.55			
Ce	13.6	6.2	5.2	−4.8	
Co	−8.43	−30.8	−44.8	−35.9	−7.8
Cr	5	21.8	16.6	17.9	5.7
Cs		−0.9			
Cu	1.19	1.83	2.83	5.36	
Dy	−4.1	−1.8	0.9	2.3	
Er	−3.8	−0.1	1.9	4.2	
Eu	5.3	24.5	46		
Fe	11.6	15	3	0.4	
Ga	0.5				
Gd	−4.6	−1.6	−0.5	−0.8	
Hf	0	5.5	5.7	−0.5	
Ho	−6.7	−1.6	1.4	2.8	

	Thermal emf $\alpha(T)$ in μV/K at Temperature T				
	100 K	300 K	500 K	1000 K	1500 K
In	0.56	1.68			
Ir	1.42	0.86	−0.1	−2.7	−5.7
K	−5.2	−13.7			
La	0.1	1.7	2	−1.7	
Li	4.3				
Lu	−6.9	−4.3	−2.6	0	
Mg	−2.1	−1.46			
Mn	−2.5	−9.8	−8.4	−1.5	
Mo	0.1	5.6	11.4	17.4	13.7
Na	−2.6	−6.3			
Nb	1.05	−0.44	−1.1	0.45	3.2
Nd	−4	−2.3	0	−1.2	
Ni	−8.5	−19.5	−25.8	−29.9	
Np	8.9	−3.1			
Os	−3.2	−4.4	−4.7	−6.3	−8.5
Pb	−0.58	−1.05	−1.5		
Pd	1.1	−10.7	−16.3	−32.3	−46.4
Pt	4.1	−5.3	−7.9	−8.2	
Pu	12				
Rb	−3.6	−10			

	Thermal emf $\alpha(T)$ in μV/K at Temperature T				
	100 K	300 K	500 K	1000 K	1500 K
Re	−1.4	−5.9	−5.9	−1.9	1.8
Rh	0.8	0.6	0.5	−1.5	
Ru	0.3	−1.4	−1.8	−4.2	−7.5
Sc	−14.3	−19	−17.5	−5.4	10.2
Sm	0.7	1.2	0.6	−3	
Sn	−0.04	−1			
Sr	−3	1.1	4.2		
Ta	0.7	−1.9	−2.3	1.6	7.2
Tb	−1.6	−1	0.3	0.6	
Th	0.6	−3.2	−9.2	−14.3	−10.4
Ti	−2	9.1	5.3	−3.1	−0.5
Tl	0.6	0.3	−1.5		
Tm	−1.3	1.9	2.7	2.2	
U	3	7.1	11	16.7	
V	2.9	0.23	1.1	4.6	
W	−4.4	0.9	9	19.8	21.3
Y	−5.1	−0.7	0.3	2.9	6.6
Yb	5.1	30	20.3	12.3	
Zn	0.7	2.4			
Zr	4.4	8.9	4.6	−3	1.1

Thermoelectric Properties of Selected Semiconductors; Values Near Room Temperature unless Otherwise Indicated

Material	α/μV K⁻¹	Material	α/μV K⁻¹	Material	α/μV K⁻¹
		Elemental Semiconductors			
B	600 (500 K)	n-Si	300	p-Si	−500
n-Ge	600	p-Ge	−830	α-Sn	−40 (250 K)
		I-VI Compounds			
Cu_2S	327	Cu_2Se	135	Cu_2Te	40
Ag_2Te	120				
		II-VI Compounds			
ZnO	300	CdS	700	ZnSe	55
CdSe	200				
		III-V Compounds			
GaN	70	GaP	1200	InP	−400
AlAs	70	n-GaAs	380	p-GaAs	−310
InAs	200	AlSb	500	n-GaSb	250
p-GaSb	−55	n-InSb	240	p-InSb	200
		V-VI Compounds			
Sb_2Te_3	110	n-Bi_2Te_3	224	p-Bi_2Te_3	−227
		I-III-VI Compounds			
$CuAlS_2$	50	$AgInSe_2$	−370	$CuTlTe_2$	80
CuGaSe	40	$AgTlSe_2$	800	$AgAlTe_2$	321
$CuInSe_2$	340	$CuGaTe_2$	340	$AgGaTe_2$	950
$CuTlSe_2$	−5	$CuInTe_2$	260	$AgInTe_2$	298
$AgGaSe_2$	90	$CuTlTe_2$	80		
		I-IV-VI Compounds			
Cu_2GeS_3	300	Cu_2GeSe_3	100	Cu_2GeTe_3	10
Cu_2SnS_3	600	Cu_2SnSe_3	250	Cu_2SnTe_3	30
		I-V-VI Compounds			
Cu_3AsS_4	130	Cu_3AsSe_4	120	Cu_3SbSe_4	200
		II-IV-V Compounds			
$ZnGeP_2$	1200	$ZnSiAs_2$	1100	$CdGeAs_2$	190
$CdSnAs_2$	600				

FERMI ENERGY AND RELATED PROPERTIES OF METALS

Lev I. Berger

In the classical Drude theory of metals, the Maxwell-Boltzmann velocity distribution of electrons is used. It states that the number of electrons per unit volume with velocities in the range of $d\vec{v}$ about any magnitude $\vec{v}$ at temperature T is

$$f_B(\vec{v})d\vec{v} = n\left(\frac{m}{2\pi k_B T}\right)\exp\left(-\frac{mv^2}{2k_B T}\right)d\vec{v}$$

where n is the total number of conduction electrons in a unit volume of a metal, m is the free electron mass, and k_B is the Boltzmann constant. In an attempt to explain a substantial discrepancy between the experimental data on the specific heat of metals and the values calculated on the basis of the Drude model, Sommerfeld suggested a model of the metal in which the Pauli exclusion principle is applied to free electrons. In this case, the Maxwell-Boltzmann distribution is replaced by the Fermi-Dirac distribution:

$$f(\vec{v})d\vec{v} = 2\left(\frac{m}{h}\right)^3 d\vec{v}\left\{\exp\left[\left(\frac{mv^2}{2} - k_B T_0\right)\Big/ k_B T\right] + 1\right\}^{-1}$$

Here h is the Planck constant and T_0 is a characteristic temperature which is determined by the normalization condition

$$n = \int d\vec{v} \cdot f(\vec{v})$$

The magnitude of T_0 is quite high; usually, $T_0 > 10^4$ K. So, at common temperatures ($T < 10^3$ K), the free electron density of a metal is much smaller than in the case of the Maxwell-Boltzmann distribution. This allows us to explain why the experimental data on specific heat for metals are close to those for insulators.

The maximum kinetic energy the electrons of a metal may possess at $T = 0$ K is called the Fermi energy, e.g.,

$$E_F = \frac{\hbar^2 k_F^2}{2m} = \left(\frac{e^2}{2k_B}\right)(k_F r_B)^2$$

where k_F is the Fermi momentum or the Fermi wave vector

$$k_F = (3\pi^2 n)^{1/3}$$

e is the electron charge, and r_B is the Bohr radius

$$r_B = \hbar^2/me^2 = 0.529 \cdot 10^{-10}\ \mathrm{m}$$

Another, more common expression for the Fermi energy is

$$E_F = \tfrac{1}{2} m v_F^2$$

where $v_F = \hbar k_F/m$ is the Fermi velocity which can be expressed using the concept of the electron radius, r_s. It is equal to radius of a sphere occupied by one free electron. If the total volume of a metal sample is V and the number of conduction electrons in this volume is N, then the volume per electron is equal to

$$\frac{V}{N} = \frac{1}{n} = \frac{4}{3}\pi r_S^3$$

and

$$r_S = \left(\frac{3}{4\pi n}\right)^{1/3}$$

The following table contains information pertinent to the Sommerfeld model for some metals. The magnitudes of T_0 are calculated using the expression

$$T_0 = \frac{E_F}{k_B} = \frac{58.2 \cdot 10^4}{(r_S / r_B)^2}\ \mathrm{K}$$

Ground State Properties of the Electron Gas in Some Metals

Metal	Valency	$n/10^{28}$ m^{-3}	r_S/pm	r_S/r_B	E_F/eV	$T_0/10^4$ K	$k_F/10^{10}$ m^{-1}	$v_F/10^6$ m s^{-1}
Li[a]	1	4.70	172	3.25	4.74	5.51	1.12	1.29
Na[b]	1	2.65	208	3.93	3.24	3.77	0.92	1.07
K[b]	1	1.40	257	4.86	2.12	2.46	0.75	0.86
Rb[b]	1	1.15	275	5.20	1.85	2.15	0.70	0.81
Cs[b]	1	0.91	298	5.62	1.59	1.84	0.65	0.75
Cu	1	8.47	141	2.67	7.00	8.16	1.36	1.57
Ag	1	5.86	160	3.02	5.49	6.38	1.20	1.39
Au	1	5.90	159	3.01	5.53	6.42	1.21	1.40
Be	2	24.7	99	1.87	14.3	16.6	1.94	2.25
Mg	2	8.61	141	2.66	7.08	8.23	1.36	1.58
Ca	2	4.61	173	3.27	4.69	5.44	1.11	1.28
Sr	2	3.55	189	3.57	3.93	4.57	1.02	1.18
Ba	2	3.15	196	3.71	3.64	4.23	0.98	1.13
Nb	1	5.56	163	3.07	5.32	6.18	1.18	1.37
Fe	2	17.0	112	2.12	11.1	13.0	1.71	1.98
Mn[c]	2	16.5	113	2.14	10.9	12.7	1.70	1.96
Zn	2	13.2	122	2.30	9.47	11.0	1.58	1.83
Cd	2	9.27	137	2.59	7.47	8.68	1.40	1.62
Hg[a]	2	8.65	140	2.65	7.13	8.29	1.37	1.58

Metal	Valency	$n/10^{28}$ m^{-3}	r_S/pm	r_S/r_B	E_F/eV	$T_0/10^4$ K	$k_F/10^{10}$ m^{-1}	$v_F/10^6$ m s^{-1}
Al	3	18.1	110	2.07	11.7	13.6	1.75	2.03
Ga	3	15.4	116	2.19	10.4	12.1	1.66	1.92
In	3	11.5	127	2.41	8.63	10.0	1.51	1.74
Tl	3	10.5	131	2.48	8.15	9.46	1.46	1.69
Sn	4	14.8	117	2.22	10.2	11.8	1.64	1.90
Pb	4	13.2	122	2.30	9.47	11.0	1.58	1.83
Bi	5	14.1	119	2.25	9.90	11.5	1.61	1.87
Sb	5	16.5	113	2.14	10.9	12.7	1.70	1.96

[a] At 78 K.
[b] At 5 K.
[c] α-phase.
The data in the table are for atmospheric pressure and room temperature unless otherwise noted.

References

1. Drude, P., *Ann. Physik*, 1, 566, 1900; *ibid*., 3, 369, 1900.
2. Sommerfeld, A. and Bethe, H., *Handbuch der Physik*, Chapter 3, Springer, 1933.
3. Wyckoff, R. W. G., *Crystal Structures*, 2nd. ed., Interscience, 1963.
4. Ashcroft, N. W. and Mermin, N. D., *Solid State Physics*, Holt, Rinehart and Winston, 1976.

PROPERTIES OF COMMERCIAL METALS AND ALLOYS

This table gives typical values of mechanical, thermal, and electrical properties of several common commercial metals and alloys. Values refer to ambient temperature (0 to 25 °C). All values should be regarded as typical, since these properties are dependent on the particular type of alloy, heat treatment, and other factors. Values for individual specimens can vary widely.

References

1. *ASM Metals Reference Book, Second Edition*, American Society for Metals, Metals Park, OH, 1983.
2. Lynch, C. T., *CRC Practical Handbook of Materials Science*, CRC Press, Boca Raton, FL, 1989.
3. Shackelford, J. F., and Alexander, W., *CRC Materials Science and Engineering Handbook*, CRC Press, Boca Raton, FL, 1991.

Common name	Thermal conductivity W/cm K	Density g/cm³	Coeff. of linear expansion 10^{-6}/°C	Electrical resistivity μΩ cm	Modulus of elasticity GPa	Tensile strength MPa	Approx. melting point °C
Ingot iron	0.7	7.86	11.7	9.7	205	–	1540
Plain carbon steel AISI-SAE 1020	0.52	7.86	11.7	10	205	450	1515
Stainless steel type 304	0.15	7.9	17.3	72	195	550	1425
Cast gray iron	0.47	7.2	10.5	67	90	180	1175
Malleable iron		7.3	12	30	170	345	1230
Hastelloy C	0.12	8.94	11.3	125	200	780	1350
Inconel	0.15	8.25	11.5	103	200	800	1370
Aluminum alloy 3003, rolled	1.9	2.73	23.2	3.7	70	110	650
Aluminum alloy 2014, annealed	1.9	2.8	23.0	3.4	70	185	650
Aluminum alloy 360	1.5	2.64	21.0	7.5	70	325	565
Copper, electrolytic (ETP)	3.9	8.94	16.5	1.7	120	300	1080
Yellow brass (high brass)	1.2	8.47	20.3	6.4	100	300-800	930
Aluminum bronze	0.7	7.8	16.4	12	120	400-600	1050
Beryllium copper 25	0.8	8.23	17.8	7	130	500-1400	925
Cupronickel 30%	0.3	8.94	16.2		150	400-600	1200
Red brass, 85%	1.6	8.75	18.7	11	90	300-700	1000
Chemical lead	0.35	11.34	29.3	21	13	17	327
Antimonial lead (hard lead)	0.3	10.9	26.5	23	20	47	290
Solder 50-50	0.5	8.89	23.4	15	–	42	215
Magnesium alloy AZ31B	1.0	1.77	26	9	45	260	620
Monel	0.3	8.84	14.0	58	180	545	1330
Nickel (commercial)	0.9	8.89	13.3	10	200	460	1440
Cupronickel 55-45 (constantan)	0.2	8.9	18.8	49	160	–	1260
Titanium (commercial)	1.8	4.5	8.5	43	110	330-500	1670
Zinc (commercial)	1.1	7.14	32.5	6	–	130	419
Zirconium (commercial)	0.2	6.5	5.85	41	95	450	1855

HARDNESS OF MINERALS AND CERAMICS

There are several hardness scales for describing the resistance of a material to indentation or scratching. This table lists a number of common materials in order of increasing hardness. Values are given, when available, on three different hardness scales: the original Mohs Scale (range 1 to 10); the modified Mohs Scale (range 1 to 15), and the Knoop Hardness Scale. In the last case, a load of 100 g is assumed.

Reference

Shackelford, J. F. and Alexander, W., *CRC Materials Science and Engineering Handbook*, CRC Press, Boca Raton, FL, 1991.

Material	Formula	Mohs	Modified mohs	Knoop
Graphite	C	0.5		
Talc	$3MgO{\cdot}4SiO_2{\cdot}H_2O$	1	1	
Alabaster	$CaSO_4{\cdot}2H_2O$	1.7		
Gypsum	$CaSO_4{\cdot}2H_2O$	2	2	32
Halite (rock salt)	NaCl	2		
Stibnite (antimonite)	Sb_2S_3	2.0		
Galena	PbS	2.5		
Mica		2.8		
Calcite	$CaCO_3$	3	3	135
Barite	$BaSO_4$	3.3		
Marble		3.5		
Aragonite	$CaCO_3$	3.5		
Dolomite	$CaMg(CO_3)_2$	3.5		
Fluorite	CaF_2	4	4	163
Magnesia	MgO	5		370
Apatite	$CaF_2{\cdot}3Ca_3(PO_4)_2$	5	5	430
Opal		5		
Feldspar (orthoclase)	$K_2O{\cdot}Al_2O{\cdot}6SiO_2$	6	6	560
Augite		6		
Hematite	Fe_2O_3	6		750
Magnetite	Fe_3O_4	6		
Rutile	TiO_2	6.2		
Pyrite	FeS_2	6.3		
Agate	SiO_2	6.5		
Uranium dioxide	UO_2	6.7		600
Silica (fused)	SiO_2		7	
Quartz	SiO_2	7	8	820
Flint		7		
Silicon	Si	7		
Andalusite	Al_2OSiO_4	7.5		
Zircon	$ZrSiO_4$	7.5		
Zirconia	ZrO_2			1200
Aluminum nitride	AlN			1225
Beryl	$Be_3Al_2Si_6O_{18}$	7.8		
Beryllia	BeO			1300
Topaz	$Al_2SiO_4(OH,F)_2$	8	9	1340
Garnet	$Al_2O_3{\cdot}3FeO{\cdot}3SiO_2$		10	1360
Emery	Al_2O_3 (impure)	8		
Zirconium nitride	ZrN	8+		1510
Zirconium boride	ZrB_2			1560
Titanium nitride	TiN	9		1770
Zirconia (fused)	ZrO_2		11	
Tantalum carbide	TaC			1800
Tungsten carbide	WC			1880
Corundum (alumina)	Al_2O_3	9		2025
Zirconium carbide	ZrC			2150
Alumina (fused)	Al_2O_3		12	
Beryllium carbide	Be_2C			2400
Titanium carbide	TiC			2470
Carborundum (silicon carbide)	SiC	9.3	13	2500
Aluminum boride	AlB			2500
Tantalum boride	TaB_2			2600
Boron carbide	B_4C		14	2800
Boron	B	9.5		
Titanium boride	TiB_2			2850
Diamond	C	10	15	7000

Section 13
Polymer Properties

Abbreviations Used in Polymer Science and Technology ... 13-1
Physical Properties of Selected Polymers ... 13-3
Nomenclature for Organic Polymers ... 13-5
Solvents for Common Polymers ... 13-9
Glass Transition Temperature for Selected Polymers ... 13-10
Dielectric Constant of Selected Polymers ... 13-17
Pressure–Volume–Temperature Relationships for Polymer Melts ... 13-18
Upper Critical (UCST) and Lower Critical (LCST) Solution Temperatures of Binary Polymer Solutions ... 13-23
Vapor Pressures (Solvent Activities) for Binary Polymer Solutions ... 13-41
Specific Enthalpies of Solution of Polymers and Copolymers ... 13-46
Solubility Parameters of Selected Polymers ... 13-74

ABBREVIATIONS USED IN POLYMER SCIENCE AND TECHNOLOGY

ABA	triblock copolymers; acrylonitrile-butadiene acrylate
ABS	copolymer of acrylonitrile, butadiene, and styrene
ACS	acrylonitrile-chlorinated polyethylene styrene terpolymer
AIBN	2,2'-azobisisobutyronitrile
AMA	acrylate maleic anhydride terpolymer
AMMA	acrylate-methyl methacrylate copolymer
AN	acrylonitrile
AP	ethylene-propylene copolymers
APO	amorphous polyolefin
AS	acrylonitrile styrene copolymer
ASA	acrylonitrile-styrene-acrylonitrile block
ATR	attenuated total reflectance spectroscopy
AU	polyurethane
BMC	bulk molding compound
BMI	bis maleimide
BPO	benzoyl peroxide
CA	cellulose acetate
CAB	cellulose acetate butyrate
CAP	cellulose acetate proprionate
CAR	carbon fiber
CED	cohesive energy density
CFRP	carbon reinforced plastics
CMC	carboxymethylcellulose
CN	cellulose nitrate
COC	cycloolefin copolymer
COP	copolyester thermoplastic elastomer
CPE	chloronated polyethylene
CPVC	chlorinated poly(vinyl chloride)
CR	neoprene
CTA	cellulose triacetate
CTFE	chlorotrifluoroethylene
C_s	chain transfer constant
DAIP	diallyl isophthalate plasticizer
DAP	dially phthalate plasticizer
DNA	deoxyribonucleic acid
DP	degree of polymerization
DRS	dynamic reflectance spectroscopy
DS	degree of substitution
EAA	ethylene acrylic acid copolymer
EC	ethyl cellulose
ECTFE	ethylene-chlorotrifluoroethylene copolymer
EEA	ethylene-ethyl acetate copolymer
EGG	Einstein-Guth-Gold equation
EMAC	ethylene-methyl acrylate copolymer
EnBA	ethylene n-butyl acetate
EP	epoxy resin
EPDM	poly(ethylene-co-propylene) crosslinked
EPM	ethylene-propylene copolymer
EPR	ethylene propylene rubber
EPS	expanded polystyrene
ET	thiokol
ETFE	ethylene tetrafluoroethylene polymer
EU	polyether polyurethane
EVA	ethylene-vinyl acetate copolymer
EVOH	ethylene-vinyl alcohol copolymer
FEP	fluorinated ethylene propylene
FRP	fibrous glass reinforced polyester; fiber reinforced plastic

G	molar attraction constant
GF	glass reinforced
GRS	poly(butadiene-co-styrene)
HDPE	high-density polyethylene
HIPS	high-impact polystyrene
HMC	high strength molding compound
HMWHDPE	high-molecular-weight high-density polyethylene
I	ionomer
IIR	butyl rubber
IPN	interpenetrating polymer network
K	constant in Mark-Houwink equation
LC	liquid crystal
LCP	liquid crystal polymer
LDPE	low-density polyethylene
LLDPE	linear low-density polyethylene
LPE	linear polyethylene
MA	maleic anhydride
MABS	methyl methacrylate ABS copolymer
MBS	methyl methacrylate butadiene styrene terpolymer
MDPE	medium density polyethylene
MDI	methylene diphenylisocyanate
MF	melamine-formaldehyde resin
MP	melamine phenolic
MWD	molecular weight distribution
M_n	number-average molecular weight
M_v	viscosity-average molecular weight
M_w	weight-average molecular weight
M_z	Z-average molecular weight
NBR	poly(butadiene-co-acrylonitrile); nitrile butadiene rubber
NR	natural rubber
OSA	olefin modified styrene acrylonitrile
P	phenolic
PA	polyamide; nylon
PA6	polyamide 6, nylon6
PA11	polyamide 11, nylon11
PA12	polyamide 12, nylon12
PA46	polyamide 46, nylon46
PA66	polyamide 66, nylon6-6
PA66/6T	polyamide 66/6T
PA610	polyamide 610, nylon6-10
PA612	polyamide 612, nylon6-12
PA666	polyamide 666
PAA	poly(acrylic acid)
PAEK	polyaryletherketone
PAI	polyamide-imide
PAK	polyester alkyd
PAL	polyanaline
PAN	polyacrylonitrile
PARA	polyaryl amide
PAS	polyarylsulfone
PB	polybutylene
PBAN	polybutylene-acrylonitrile copolymer
PBD	polybutadine
PBI	polybenzimidazole
PBN	poly(butylene napthalate)
PBS	polybutadiene-styrene copolymer
PBT	poly(butylene terephthalate)
PC	polycarbonate

PC/ABS polycarbonate/acrylonitrile butadiene styrene blend
PCB polychlorinated biphenyl
PCCE poly(cyclohexylene dimethylene cyclohexanedicarboxylate), glycol and acid comonomer
PCL polycaprolactone
PCT poly(cyclohexylene terephthalate)
PCTA poly(cyclohexylene dimethylene terephthalate) copolyester
PCTFE polychlorotrifluoroethylene
PCTG poly(cyclohexylene dimethylene terephthalate) copolyester
PCT-G glycol modified polycyclohexyl terephthallate
PE polyethylene
PEBA polyether block amide or polyester block amide
PEEK poly(ether ether ketone)
PEG poly(ethylene glycol)
PEI polyetherimide
PEK polyetherketone
PEKEKK polyetherketone etherketone ketone
PEKK polyetherketoneketone
PEN poly(ethylene napthalene)
PEO poly(ethylene oxide)
PES polyethersulfone
PET poly(ethylene terephthalate)
PET-G glycol modified poly(ethylene terephthalate)
PEX crosslinked polyethylene
PF phenol-formaldehyde resin
PFA perfluoroalkoxy
PI polyimide, polyisoprene
PIB polyisobutylene
PIR polyisocyanurate
PK polyketone
PLGA poly(lactic-*co*-glycolic acid)
PMAN polymethactylonitrile
PMMA poly(methyl methacrylate)
PMP polymethylpentene
PMS polymethylstyrene
PNF poly(phosphonitrilic fluorides)
PO polyolefin
POM polyoxymethylene, polyformaldehyde, acetals
PP polypropylene
PPA polyphthalamide
PPC chlorinated polypropylene, polyphthalate carbonate
PPE poly(phenylene ether)
PPI polymeric polyisocyanate
PPO poly(phenylene oxide)
PPOX poly(propylene oxide)
PPS poly(phenylene sulfide)
PPSU poly(phenylene sulfone)
PPT poly(propylene terephthalate)
PS polystyrene
PS-b-PI polystyrene/polyisoprene block copolymer
PSO, PSU polysulfone
PTFE polytetrafluoroethylene, Teflon
PTME poly(tetramethylene terephthalate)
PTMT poly(tetramethylene terephthalate)
PU polyurethane
PUR polyurethane rubber
PVA poly(vinyl alcohol); sometimes poly(vinyl acetate)
PVAc poly(vinyl acetate)
PVB poly(vinyl butyral)
PVC poly(vinyl chloride)
PVCA copolymer of vinyl chloride and vinyl acetate
PVDA polyvinylidene acetate
PVDC poly(vinylidene chloride)
PVDF poly(vinylidene fluoride)
PVF poly(vinyl fluoride)
PVK poly(vinyl carbazole)
PVOH poly(vinyl alcohol)
PVP poly(vinyl pyrrolidone)
RIM reaction injection molding
RNA ribonucleic acid
ROMP ring opening metathesis polymerization
ROP ring opening polymerization
S radius of gyration
SAN poly(styrene-co-acrylonitrile)
SB styrene butadiene copolymer
SBR poly(butadiene-co-styrene) elastomer
SBS styrene butadiene styrene block copolymer
SEBS styrene ethylene butylene styrene block copolymer
SI silicon
SIS styrene isoprene styrene block copolymer
SMA poly(styrene-co-maleic anhydride)
SMC sheet molding compound
SMMA styrene methyl methacrylate copolymer
SMS styrene/a-methyl styrene
SN sulfur nitride
SR synthetic rubber
SRP styrene-rubber plastics
SVA styrene vinyl acrylonitrile
TDI toluenediisocyanate
TEO thermoplastic elastic olefin
TGA, TG thermal gravimetric analysis
TMC thick molding compound
TMMV threshold molecular weight value
TPA polyamide thermoplastic elastomer
TPC copolyester thermoplastic elastomer
TPE thermoplastic elastomer
TPE-O, TPO thermoplastic elastomer - olefinic
TPE-S, TPS thermoplastic elastomer - styrenic
TPU thermoplastic urethane
TPX poly-4-methylpentene
TVO thermoplastic vulcanites
T_c ceiling temperature; cloud-point temperature
T_g glass transition temperature
T_m melting point temperature
UF urea-formaldehyde resin
UHMWPE ultrahigh molecular weight polyethylene
ULDPE ultra low-density polyethylene
ULPE ultra linear polyethylene
UP, UPE unsaturated polyester (thermoset)
VA vinyl acetate
VAE vinyl acetate ethylene
VLDPE very low-density polyethylene
WLF Williams-Landel-Ferry equation
WS polyurethane
XLPE crosslinked polyethylene
XPS expandable polystyrene

PHYSICAL PROPERTIES OF SELECTED POLYMERS

The physical properties of polymers are important parameters in determining their behavior and performance in a wide range of applications. This table lists some examples of general representative physical properties (including mechanical properties) of representative polymeric compounds. For glass transition temperatures of selected polymers, see pages 13-10 through 13-16 in this section. Some of the properties in this table are defined as follows:

The **heat deflection temperature** (HDT), or heat distortion temperature, is the temperature at which a polymer or plastic sample deforms under a specified load (normally either 0.455 MPa or 1.82 MPa).

The **crystalline melting point** is the temperature (or temperature range) at which a crystalline solid changes its state from solid to liquid. Although the phrase would suggest a specific temperature, most crystalline compounds actually melt over a range of a few degrees or less.

The **coefficient of linear thermal expansion** is the fractional change in length per °C change in temperature at constant pressure.

The **compressive strength** of a material is the maximum uniaxial compressive stress (compressive force per unit area) reached when the material fails completely on being subjected to a load that pushes it together.

The **tensile strength** is a measure of the ability of a material to withstand pulling stresses. It is defined as the stress (stretching force per unit area) required to break a specimen. Polymers are approximately 20 % stronger in compression than in tension.

The **flexural strength**, or cross-breaking strength, of a material is a measure of the bending strength or stiffness of a specimen expressed as the stress required to break a specimen by exerting a torque on it.

The **impact strength** is a measure of the energy needed to break a sample. The term toughness is sometimes used to describe the impact strength of a material. The notched izod impact test is a single point test that measures the resistance of a material to impact from a swinging pendulum. Izod impact is defined as the kinetic energy needed to initiate fracture and continue the fracture until the specimen is broken. Izod specimens are notched to prevent deformation of the specimen upon impact. This test can be used as a quick and easy quality control check to determine if a material meets specific impact properties or to compare materials for general toughness.

The **ultimate elongation** is a measure of how far a material will stretch before breaking, expressed as a percentage of its original length.

The properties of the following polymers are presented in this table:

PET	poly(ethylene terephthalate)
PBT	poly(butylene terephthalate)
PC	polycarbonate
Nylon 6,6	poly(iminoadipoyliminohexamethylene)
Nylon 6	poly[imino(1-oxohexamethylene)]
PPO	poly(phenylene ether)
POM	polyoxymethylene
LDPE	low-density polyethylene
HDPE	high-density polyethylene
UHMWPE	ultrahigh molecular weight polyethylene
iPP	isotactic polypropylene
ABS	copolymer of acrylonitrile, butadiene, and styrene (extrusion grade)
PTFE	polytetrafluoroethylene, Teflon
PCTFE	polymonochlorotrifluoroethylene
PVDF	poly(vinylidene fluoride)
PVF	poly(vinyl fluoride)
PVC (rigid)	poly(vinyl chloride)
PVC (plasterized)	poly(vinyl chloride)
PMMA	poly(methyl methacrylate)

The assistance of Charles E. Carraher, Jr. in providing these data is gratefully acknowledged.

Reference

Carraher, Jr., C.E., *Seymour/Carraher's Polymer Chemistry,* 7th Edition, CRC Press, Taylor & Francis Group, Boca Raton, FL, 2008.

	PET	PBT	PC	Nylon 6,6	Nylon 6	PPO
Heat deflection temperature at 1820 kPa (°C)	100	65	130	75	80	100
Maximum resistance to continuous heat (°C)	100	60	115	120	125	80
Crystalline melting point (°C)	—	—	225	265	225	215
Coefficient of linear expansion (10^{-5}/°C)	6.5	7.0	6.8	8.0	8.0	5.0
Compressive strength (kPa)	8.6×10^4	7.5×10^4	8.6×10^4	1×10^5	9.7×10^4	9.6×10^4
Flexural strength (kPa)	1.1×10^5	9.6×10^4	9.3×10^4	1×10^5	9.7×10^4	8.9×10^4
Impact strength (Izod: cm N/cm of notch)	26	53	530	80	160	270
Tensile strength (kPa)	6.2×10^4	5.5×10^4	7.2×10^4	8.3×10^4	6.2×10^4	5.5×10^4
Ultimate elongation (%)	100	100	110	30	—	50
Density (g cm^{-3})	1.35	1.35	1.2	1.2	1.15	1.1

	POM	LDPE	HDPE	UHMWPE	iPP	ABS
Heat deflection temperature at 1820 kPa (°C)	125	40	50	85	55	90
Maximum resistance to continuous heat (°C)	100	40	80	80	100	90
Crystalline melting point (°C)	180	—				
Coefficient of linear expansion (10^{-5}/°C)	10.0	10	12	12	9	9.5
Compressive strength (kPa)	1.1×10^5	—	3×10^4	—	—	4.8×10^4
Flexural strength (kPa)	9.7×10^4	—	—	—	5×10^4	6.2×10^4
Impact strength (Izod: cm N/cm of notch)	80	No break	30	No break	27	320
Tensile strength (kPa)	6.9×10^4	5×10^3	2×10^4	6×10^4	3.5×10^4	3.4×10^4
Ultimate elongation (%)	30	—	—	—	100	60
Density (g cm^{-3})	1.4	0.91	0.96	0.93	0.90	1.0

	PTFE	PCTFE	PVDF	PVF	Rigid PVC	Plasticized PVC	PMMA
Heat deflection temperature at 1820 kPa (°C)	100	100	80	90	75	—	95
Maximum resistance to continuous heat (°C)	250	200	150	125	60	35	75
Crystalline melting point (°C)	—	—	—	—	170	—	—
Coefficient of linear expansion (10^{-5}/°C)	10	14	8.5	10	6	12	7.0
Compressive strength (kPa)	2.7×10^4	3.8×10^4	—	—	6.8×10^4	6×10^3	1×10^5
Flexural strength (kPa)	—	6×10^4	—	—	9×10^4	—	9.6×10^4
Impact strength (Izod: cm N/cm of notch)	160	130	—	—	27	—	21
Tensile strength (kPa)	2.4×10^4	3.4×10^4	5.5×10^4	—	4.4×10^4	1×10^4	6.5×10^4
Ultimate elongation (%)	200	100	200	—	50	200	4
Density (g cm^{-3})	2.16	2.1	1.76	1.4	1.4	1.3	1.2

NOMENCLATURE FOR ORGANIC POLYMERS

Robert B. Fox and Edward S. Wilks

Organic polymers have traditionally been named on the basis of the monomer used, a hypothetical monomer or a semi-systematic structure. Alternatively, they may be named in the same way as organic compounds, i.e., on the basis of a structure as drawn. The former method, often called "source-based nomenclature" or "monomer-based nomenclature," sometimes results in ambiguity and multiple names for a single material. The latter method, termed "structure-based nomenclature," generates a sometimes cumbersome unique name for a given polymer, independent of its source. Within their limitations, both types of names are acceptable and well-documented.[1] The use of stereochemical descriptors with both types of polymer nomenclature has been published.[2]

Traditional Polymer Names

Monomer-Based Names

"Polystyrene" is the name of a homopolymer made from the single monomer styrene. When the name of a monomer comprises two or more words, the name should be enclosed in parentheses, as in "poly(methyl methacrylate)" or "poly(4-bromostyrene)" to identify the monomer more clearly. This method can result in several names for a given polymer; thus, "poly(ethylene glycol)," "poly(ethylene oxide)," and "poly(oxirane)" describe the same polymer. Sometimes, the name of a hypothetical monomer is used, as in "poly(vinyl alcohol)." Even though a name like "polyethylene" covers a multitude of materials, the system does provide understandable names when a single monomer is involved in the synthesis of a single polymer. When one monomer can yield more than one polymer, e.g., 1,3-butadiene or acrolein, some sort of structural notation must be used to identify the product, and one is not far from a formal structure-based name.

Copolymers, Block Polymers, and Graft Polymers. When more than one monomer is involved, monomer-based names are more complex. Some common polymers have been given names based on an apparent structure, as with "poly(ethylene terephthalate)." A better system has been approved by the IUPAC.[1] With this method, the arrangement of the monomeric units is introduced through use of an italicized connective placed between the names of the monomers. For monomer names represented by A, B, and C, the various types of arrangements are shown in Table 1.

Table 2 contains examples of common or semi-systematic names of copolymers. The systematic names of comonomers may also be used; thus, the polyacrylonitrile-*block*-polybutadiene-*block*-polystyrene polymer in Table 2 may also be named poly(prop-2-enenitrile)-*block*-polybuta-1,3-diene-*block*-poly(ethenylbenzene). IUPAC does not require alphabetized names of comonomers within a polymer name; many names are thus possible for some copolymers.

These connectives may be used in combination and with small, non-repeating (i.e. non-polymeric) junction units; see, for example, Table 2, line 8. A long dash may be used in place of the connective *-block-*; thus, in Table 2, the polymers of lines 7 and 8 may also be written as shown on lines 9 and 10.

IUPAC also recommends an alternative scheme for naming copolymers that comprises use of "copoly" as a prefix followed by the names of the comonomers, a solidus (an oblique stroke) to separate comonomer names, and the addition before "copoly" of any applicable connectives listed in Table 2 except *-co-*.

Table 3 gives the same examples shown in Table 2 but with the alternative format. Comonomer names need not be parenthesized.

TABLE 1. IUPAC Source-Based Copolymer Classification

No.	Copolymer type	Connective	Example
1	Unspecified or unknown	*-co-*	poly(A-*co*-B)
2	Random (obeys Bernoullian distribution)	*-ran-*	poly(A-*ran*-B)
3	Statistical (obeys known statistical laws)	*-stat-*	poly(A-*stat*-B)
4	Alternating (for two monomeric units)	*-alt-*	poly(A-*alt*-B)
5	Periodic (ordered sequence for 2 or more monomeric units)	*-per-*	poly(A-*per*-B-*per*-C)
6	Block (linear block arrangement)	*-block-*	polyA-*block*-polyB
7	Graft (side chains connected to main chains)	*-graft-*	polyA-*graft*-polyB

TABLE 2. Examples of Source-Based Copolymer Nomenclature

No.	Copolymer name
1	poly(propene-*co*-methacrylonitrile)
2	poly[(acrylic acid)-*ran*-(ethyl acrylate)]
3	poly(butene-*stat*-ethylene-*stat*-styrene)
4	poly[(sebacic acid)-*alt*-butanediol]
5	poly[(ethylene oxide)-*per*-(ethylene oxide)-*per*-tetrahydrofuran]
6	polyisoprene-*graft*-poly(methacrylic acid)
7	polyacrylonitrile-*block*-polybutadiene-*block*-polystyrene
8	polystyrene-*block*-dimethylsilylene-*block*-polybutadiene
9	polyacrylonitrile—polybutadiene—polystyrene
10	polystyrene—dimethylsilylene—polybutadiene

TABLE 3. Examples of Source-Based Copolymer Nomenclature (Alternative Format)

No.	Polymer name
1	copoly(propene/methacrylonitrile)
2	*ran*-copoly(acrylic acid/ethyl acrylate)
3	*stat*-copoly(butene/ethylene/styrene)
4	*alt*-copoly(sebacic acid/butanediol)
5	*block*-copoly(acrylonitrile/butadiene/styrene)
6	*per*-copoly(ethylene oxide/ethylene oxide/tetrahydrofuran)
7	*graft*-copoly(isoprene/methacrylic acid)

Source-based nomenclature for non-linear macromolecules and macromolecular assemblies is covered by a 1997 IUPAC document.[11] The types of polymers in these classes, together with their connectives, are given in Table 4; the terms shown may be used as connectives, prefixes, or both to designate the features present.

TABLE 4. Connectives for Non-Linear Macromolecules and Macromolecular Assemblies

No.	Type	Connective
1	Branched (type unspecified)	branch
2	Branched with branch point of functionality f	f-branch
3	Comb	comb
4	Cross-link	ι (Greek iota)
5	Cyclic	cyclo
6	Interpenetrating polymer network	ipn
7	Long-chain branched	l-branch
8	Network	net
9	Polymer blend	blend
10	Polymer-polymer complex	compl
11	Semi-interpenetrating polymer network	sipn
12	Short-chain branched	sh-branch
13	Star	star
14	Star with f arms	f-star

Non-linear polymers are named by using the italicized connective as a *prefix* to the source-based name of the polymer component or components to which the prefix applies; some examples are listed in Table 5.

TABLE 5. Non-Linear Macromolecules

No.	Polymer name	Polymer structural features
1	poly(methacrylic acid)-*comb*-polyacrylonitrile	Comb polymer with a poly(methacrylic acid) backbone and polyacrylonitrile side chains
2	*comb*-poly[ethylene-*stat*-(vinyl chloride)]	Comb polymer with unspecified backbone composition and statistical ethylene/vinyl chloride copolymer side chains
3	polybutadiene-*comb*-(polyethylene; polypropene)	Comb polymer with butadiene backbone and side chains of polyethylene and polypropene
4	*star*-(polyA; polyB; polyC; polyD; polyE)	Star polymer with arms derived from monomers A, B, C, D, and E, respectively
5	*star*-(polyA-*block*-polyB-*block*-polyC)	Star polymer with every arm comprising a tri-block segment derived from comonomers A, B, and C
6	*star*-poly(propylene oxide)	A star polymer prepared from propylene oxide
7	5-*star*-poly(propylene oxide)	A 5-arm star polymer prepared from propylene oxide
8	*star*-(polyacrylonitrile; polypropylene) (M_r 10000: 25000)	A star polymer containing polyacrylonitrile arms of MW 10000 and polypropylene arms of MW 25000

Macromolecular assemblies held together by forces other than covalent bonds are named by inserting the appropriate italicized connective between names of individual components; Table 6 gives examples.

TABLE 6. Examples of Polymer Blends and Nets

No.	Polymer name
1	polyethylene-*blend*-polypropene
2	poly(methacrylic acid)-*blend*-poly(ethyl acrylate)
3	*net*-poly(4-methylstyrene-ι-divinylbenzene)
4	*net*-poly[styrene-*alt*-(maleic anhydride)]-ι-(polyethylene glycol; polypropylene glycol)
5	*net*-poly(ethyl methacrylate)-*sipn*-polyethylene
6	[*net*-poly(butadiene-*stat*-styrene)]-*ipn*-[*net*-poly(4-methylstyrene-ι-divinylbenzene)]

Structure-Based Polymer Nomenclature

Regular Single-Strand Polymers

Structure-based nomenclature has been approved by the IUPAC[4] and is currently being updated; it is used by *Chemical Abstracts*.[5] Monomer names are not used. To the extent that a polymer chain can be described by a repeating unit in the chain, it can be named "poly(repeating unit)." For regular single-strand polymers, "repeating unit" is a bivalent group; for regular double-strand (ladder and spiro) polymers, "repeating unit" is usually a tetravalent group.[9]

Since there are usually many possible repeating units in a given chain, it is necessary to select one, called the "constitutional repeating unit" (CRU) to provide a unique and unambiguous name, "poly(CRU)," where "CRU" is a recitation of the names of successive units as one proceeds through the CRU from left to right. For this purpose, a portion of the main chain structure that includes at least two repeating sequences is written out. These sequences will typically be composed of bivalent subunits such as $-CH_2-$, -O-, and groups from ring systems, each of which can be named by the usual nomenclature rules.[6,7]

Where a chain is simply one long sequence comprising repetition of a single subunit, that subunit is itself the CRU, as in "poly(methylene)" or "poly(1,4-phenylene)." In chains having more than one kind of subunit, a seniority system is used to determine the beginning of the CRU and the direction in which to move along the main chain atoms (following the shortest path in rings) to complete the CRU. Determination of the first, most senior, subunit is based on a descending order of seniority: (1) heterocyclic rings, (2) hetero atoms, (3) carbocyclic rings, and lowest, (4) acyclic carbon chains.

Within each of these classes, there is a further order of seniority that follows the usual rules of nomenclature.

Heterocycles: A nitrogen-containing ring system is senior to a ring system not containing nitrogen.[4,9] Further descending order of seniority is determined by:

(i) the highest number of rings in the ring system
(ii) the largest individual ring in the ring system
(iii) the largest number of hetero atoms
(iv) the greatest variety of hetero atoms

Hetero atoms: The senior bivalent subunit is the one nearest the top right-hand corner of the Periodic Table; the order of seniority is: O, S, Se, Te, N, P, As, Sb, Bi, Si, Ge, Sn, Pb, B, Hg.

Carbocycles: Seniority[4] is determined by:

(i) the highest number of rings in the ring system
(ii) the largest individual ring in the ring system
(iii) degree of ring saturation; an unsaturated ring is senior to a saturated ring of the same size

Carbon chains: Descending order of seniority is determined by:

(i) chain length (longer is senior to shorter)
(ii) highest degree of unsaturation
(iii) number of substituents (higher number is senior to lower number)
(iv) ascending order of locants
(v) alphabetical order of names of substituent groups

Among equivalent ring systems, preference is given to the one having lowest locants for the free valences in the subunit, and among otherwise identical ring systems, the one having least hydrogenation is senior. Lowest locants in unsaturated chains are also given preference. Lowest locants for substituents are the final determinant of seniority.

Direction within the repeating unit depends upon the shortest path, which is determined by counting main chain atoms, both cyclic and acyclic, from the most senior subunit to another subunit of the same kind or to a subunit next lower in seniority. When identification and orientation of the CRU have been accomplished, the CRU is named by writing, in sequence, the names of the largest possible subunits within the CRU from left to right. For example, the main chain of the polymer traditionally named "poly(ethylene terephthalate)" has the structure shown in Figure 1.

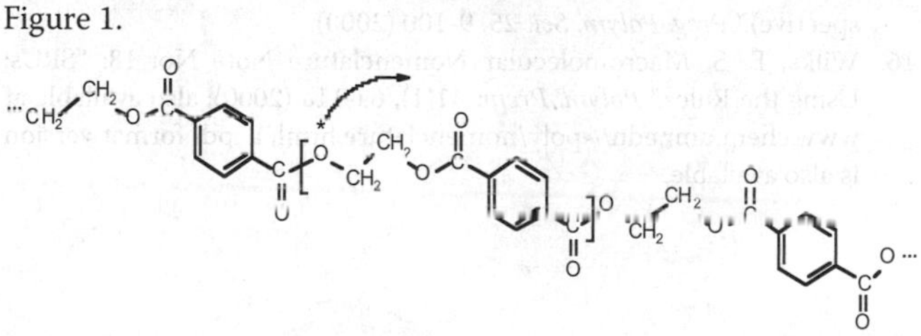

Figure 1. Structure-based name: poly(oxyethyleneoxyterephthaloyl); traditional name: poly(ethylene terephthalate).

The CRU in Figure 1 is enclosed in brackets and read from left to right. It is selected because (1) either backbone oxygen atom qualifies as the "most senior subunit," (2) the shortest path length from either -O- to the other -O- is via the ethylene subunit. Orientation of the CRU is thus defined by (1) beginning at the -O- marked with an asterisk, and (2) reading in the direction of the arrow. The structure-based name of this polymer is therefore "poly(oxyethyleneoxyterephthaloyl)," not much longer than the traditional name and much more adaptable to the complexities of substitution. As organic nomenclature evolves, more systematic names may be used for subunits, e.g., "ethane-1,2-diyl" instead of "ethylene." IUPAC still prefers "ethylene" for the $-CH_2-CH_2-$ unit, however, but also accepts "ethane-1,2-diyl."

Structure-based nomenclature can also be used when the CRU backbone has no carbon atoms. An example is the polymer traditionally named "poly(dimethylsiloxane)," which on the basis of structure would be named "poly(oxydimethylsilylene)" or "poly(oxydimethylsilanediyl)." This nomenclature method has also been applied to inorganic and coordination polymers[8] and to double-strand (ladder and spiro) organic polymers.[9]

Irregular Single-Strand Polymers

Polymers that cannot be described by the repetition of a single CRU or comprise units not all connected identically in a directional sense can also be named on a structure basis.[10] These include copolymers, block and graft polymers, and star polymers. They are given names of the type "poly(A/B/C...)," where A, B, C, etc. are the names of the component constitutional units, the number of which are minimized. The constitutional units may include regular or irregular blocks as well as atoms or atomic groupings, and each is named by the method described above or by the rules of organic nomenclature.

The solidus denotes an unspecified arrangement of the units within the main chain.[10] For example, a statistical copolymer derived from styrene and vinyl chloride with the monomeric units joined head-to-tail is named "poly(1-chloroethylene/1-phenylethylene)." A polymer obtained by 1,4-polymerization and both head-to-head and head-to-tail 1,2-polymerization of 1,3-butadiene would be named "poly(but-1-ene-1,4-diyl/1-vinylethylene/2-vinylethylene)."[12] In graphic representations of these polymers, shown in Figure 2, the hyphens or dashes at each end of each CRU depiction are shown *completely within* the enclosing parentheses; this indicates that they are not necessarily the terminal bonds of the macromolecule.

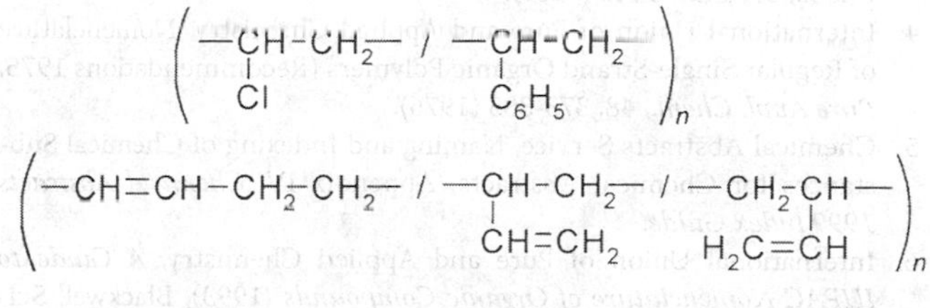

Figure 2. Graphic representations of copolymers.

A long hyphen is used to separate components in names of block polymers, as in "poly(A)—poly(B)—poly(C)," or "poly(A)—X—poly(B)" in which X is a non polymeric junction unit, e.g., dimethylsilylene.

In graphic representations of these polymers, the blocks are shown connected when the bonding is known (Figure 3, for example); when the bonding between the blocks is unknown, the blocks are separated by solidi and are shown *completely within* the outer set of enclosing parentheses (Figure 4, for example).[10,13]

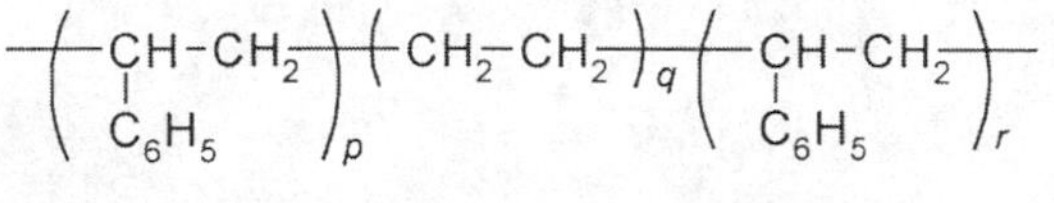

Figure 3. polystyrene—polyethylene—polystyrene.

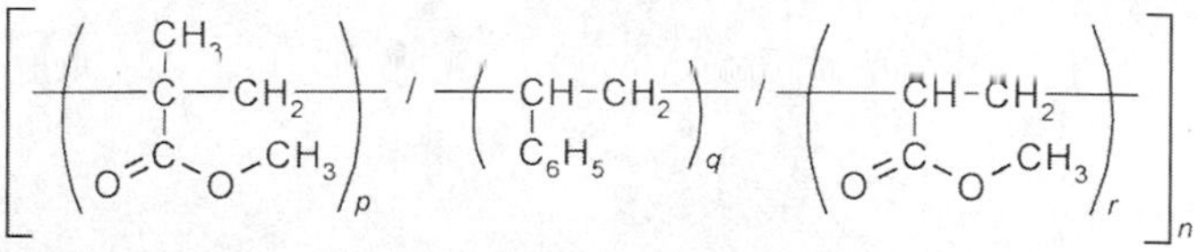

Figure 4. poly[poly(methyl methacrylate)—polystyrene—poly(methyl acrylate)].

Graft polymers are named in the same way as a substituted polymer but without the ending "yl" for the grafted chain; the name of a regular polymer, comprising Z units in which some have grafts of "poly(A)," is "poly[Z/poly(A)Z]." Star polymers are treated as a central unit with substituent blocks, as in "tetrakis(polymethylene)silane."[10,13]

Other Nomenclature Articles and Publications

In addition to the *Chemical Abstracts* and IUPAC documents cited above and listed below, other articles on polymer nomenclature are available. A 1999 article lists significant documents on polymer nomenclature published during the last 50 years in books, encyclopedias, and journals by *Chemical Abstracts*,

IUPAC, and individual authors.[14] A comprehensive review of source-based and structure-based nomenclature for all of the major classes of polymers,[15] and a short tutorial on the correct identification, orientation, and naming of most commonly encountered constitutional repeating units were both published in 2000.[16]

References and Notes

1. International Union of Pure and Applied Chemistry, *Compendium of Macromolecular Nomenclature*, Blackwell Scientific Publications, Oxford, 1991.
2. International Union of Pure and Applied Chemistry, Stereochemical Definitions and Notations Relating to Polymers (Recommendations 1980), *Pure Appl. Chem.*, 53, 733-752 (1981).
3. International Union of Pure and Applied Chemistry, Source-Based Nomenclature for Copolymers (Recommendations 1985), *Pure Appl. Chem.*, 57, 1427-1440 (1985).
4. International Union of Pure and Applied Chemistry, Nomenclature of Regular Single-Strand Organic Polymers (Recommendations 1975, *Pure Appl. Chem.*, 48, 373-385 (1976).
5. Chemical Abstracts Service, Naming and Indexing of Chemical Substances for Chemical Abstracts, Appendix IV, *Chemical Abstracts 1999 Index Guide.*
6. International Union of Pure and Applied Chemistry, *A Guide to IUPAC Nomenclature of Organic Compounds* (1993), Blackwell Scientific Publications, Oxford, 1993.
7. International Union of Pure and Applied Chemistry, *Nomenclature of Organic Chemistry, Sections A, B, C, D, E, F, and H*, Pergamon Press, Oxford, 1979.
8. International Union of Pure and Applied Chemistry, Nomenclature of Regular Double-Strand and Quasi-Single-Strand Inorganic and Coordination Polymers (Recommendations 1984), *Pure Appl. Chem.*, 57, 149-168 (1985).
9. International Union of Pure and Applied Chemistry, Nomenclature of Regular Double-Strand (Ladder and Spiro) Organic Polymers (Recommendations 1993), *Pure Appl. Chem.*, 65, 1561-1580 (1993).
10. International Union of Pure and Applied Chemistry, Structure-Based Nomenclature for Irregular Single-Strand Organic Polymers (Recommendations 1994), *Pure Appl. Chem.*, 66, 873-889 (1994).
11. International Union of Pure and Applied Chemistry, "Source-Based Nomenclature for Non-Linear Macromolecules and Macromolecular Assemblies (Recommendations 1997)." *Pure Appl. Chem.*, 69, 2511-2521 (1997).
12. Poly(1,3-butadiene) obtained by polymerization of 1,3-butadiene in the so-called 1,4- mode is frequently drawn incorrectly in publications as $\text{-(CH}_2\text{-CH=CH-CH}_2)_n\text{-}$; the double bond should be assigned the lowest locant possible, i.e., the structure should be drawn as $\text{–(CH=CH-CH}_2\text{-CH}_2)_n\text{-}$.
13. International Union of Pure and Applied Chemistry, "Graphic Representations (Chemical Formulae) of Macromolecules (Recommendations 1994)." *Pure Appl. Chem.*, 66, 2469-2482 (1994).
14. Wilks, E. S. Macromolecular Nomenclature Note No. 17: "Whither Nomenclature?" *Polym. Prepr.* 40(2), 6-11 (1999); also available at www.chem.umr.edu/~poly/nomenclature.html.
15. Wilks, E. S. "Polymer Nomenclature: The Controversy Between Source-Based and Structure-Based Representations (A Personal Perspective)." *Prog. Polym. Sci.* 25, 9-100 (2000).
16. Wilks, E. S. Macromolecular Nomenclature Note No. 18: "SRUs: Using the Rules." *Polym. Prepr.* 41(1), 6a-11a (2000); also available at www.chem.umr.edu/~poly/nomenclature.html; a .pdf format version is also available.

SOLVENTS FOR COMMON POLYMERS

Abbreviations:
- HC: hydrocarbons
- MEK: methyl ethyl ketone
- THF: tetrahydrofuran
- DMF: dimethylformamide
- DMSO: dimethylsulfoxide

Polymer	Solvents
Polyethylene (HDPE)	HC and halogenated HC
Polypropylene (atactic)	HC and halogenated HC
Polybutadiene	HC, THF, ketones
Polystyrene	ethylbenzene, $CHCl_3$, CCl_4, THF, MEK
Polyacrylates	aromatic HC, chlorinated HC, THF, esters, ketones
Polymethacrylates	aromatic HC, chlorinated HC, THF, esters, MEK
Polyacrylamide	water
Poly(vinyl ethers)	halogenated HC, MEK, butanol
Poly(vinyl alcohol)	glycols (hot), DMF
Poly(vinyl acetate)	aromatic HC, chlorinated HC, THF, esters, DMF
Poly(vinyl chloride)	THF, DMF, DMSO
Poly(vinylidene chloride)	THF (hot), dioxane, DMF
Poly(vinyl fluoride)	DMF, DMSO (hot)
Polyacrylonitrile	DMF, DMSO
Poly(oxyethylene)	aromatic HC, $CHCl_3$, alcohols, esters, DMF
Poly(2,6-dimethylphenylene oxide)	aromatic HC, halogenated HC
Poly(ethylene terephthalate)	phenol, DMSO (hot)
Polyurethanes (linear)	aromatic HC, THF, DMF
Polyureas	phenol, formic acid
Polysiloxanes	HC, THF, DMF
Poly[bis(2,2,2-trifluoroethoxy)-phosphazene]	THF, ketones, ethyl acetate

GLASS TRANSITION TEMPERATURE FOR SELECTED POLYMERS

Robert B. Fox

Polymer names are based on the IUPAC structure-based nomenclature system described in the table "Naming Organic Polymers." Within each category, names are listed in alphabetical order. Source-based and trivial names are also given (in italics) for the most common polymers. The table does not include polymers for which T_g is not clearly defined because of variability of structure or because of reactions taking place near the glass transition.

All values of T_g cited in this table have been determined by differential scanning calorimetry (DSC) except those values indicated by:

(D)	dynamic method
(Dil)	dilatometry
(M)	mechanical method

Polymer name	Glass transition temperature (T_g/K)
ACYCLIC CARBON CHAINS	
Polyalkadienes	
Poly(alkenylene) *Polyalkadiene* –[$CH{=}CHCH_2CH_2$]–	
Poly(*cis*-1-butenylene)	171
cis-1,3-polybutadiene [PBD]	
Poly(*trans*-1-butenylene)	215
trans-1,3-polybutadiene [PBD]	
Poly(1-chloro-*cis*-1-butenylene)	253
cis-1,3-polychloroprene	
Poly(1-chloro-*trans*-1-butenylene)	233
trans-1,3-polychloroprene	
Poly(1-methyl-*cis*-1-butenylene)	200
cis-1,3-polyisoprene	
Poly(1-methyl-*trans*-1-butenylene)	207
trans-1,3-polyisoprene	
Poly(1,4,4-trifluoro-1-butenylene)	238
Polyalkenes	
Poly(alkylethylene) *Poly(alkylethylene)* -[$RCHCH_2$]-	
Poly(1-benzylethylene)	333
Poly(1-butylethylene)	223
Poly(1-cyclohexylethylene) (atactic)	393
Poly(1-cyclohexylethylene) (isotactic)	406 (D)
Poly(1,1-dimethylethylene)	200
Polyisobutylene [PIB]	
Poly(ethylene)	148
Poly(methylene)	155
Poly(1-phenethylethylene)	283
Poly(propylene) (isotactic)	272
Poly(propylene) (syndiotactic)	ca. 265
Poly[1-(2-pyridyl)ethylene]	377
Poly[1-(4-pyridyl)ethylene]	415
Poly(1-vinylethylene)	273
Polyacrylics	
Poly[1-(alkoxycarbonyl)ethylene] *Poly(alkyl acrylate)* –[$(ROCO)CHCH_2$]–	
Poly[1-(benzyloxycarbonyl)ethylene]	279
Poly[1-(butoxycarbonyl)ethylene]	219 (M)
Poly(butyl acrylate) [PBA]	
Poly[1-(*sec*-butoxycarbonyl)ethylene]	251
Poly[1-(butoxycarbonyl)-1-cyanoethylene]	358
Poly[1-(butylcarbamoyl)ethylene]	319 (M)
Poly(1-carbamoylethylene)	438
Polyacrylamide [PAM]	
Poly(1-carboxyethylene)	379

Polymer name	Glass transition temperature (T_g/K)
Poly(acrylic acid) [PAA]	
Poly[1-(2-chlorophenoxycarbonyl)ethylene]	326
Poly[1-(4-chlorophenoxycarbonyl)ethylene]	331
Poly[1-(4-cyanobenzyloxycarbonyl)ethylene]	317
Poly[1-(2-cyanoethoxycarbonyl)ethylene]	277
Poly[1-(cyanomethoxycarbonyl)ethylene)]	433 Dil
Poly[1-(4-cyanophenoxycarbonyl)ethylene]	363
Poly[1-(cyclohexyloxycarbonyl)ethylene]	292
Poly[1-(2,4-dichlorophenoxycarbonyl)ethylene]	333
Poly[1-(dimethylcarbamoyl)ethylene]	362
Poly[1-(ethoxycarbonyl)ethylene]	249
Poly(ethyl acrylate) [PEA]	
Poly[1-(ethoxycarbonyl)-1-fluoroethylene]	316
Poly[1-(2-ethoxycarbonylphenoxycarbonyl)ethylene]	303
Poly[1-(3-ethoxycarbonylphenoxycarbonyl)ethylene]	297
Poly[1-(4-ethoxycarbonylphenoxycarbonyl)ethylene]	310
Poly[1-(2-ethoxyethoxycarbonyl)ethylene]	223
Poly[1-(3-ethoxypropoxycarbonyl)ethylene]	218
Poly[1-(isopropoxycarbonyl)ethylene]	267-270
Poly[1-(methoxycarbonyl)ethylene]	283
Poly(methyl acrylate) [PMA]	
Poly[1-(2-methoxycarbonylphenoxycarbonyl)ethylene]	319
Poly[1-(3-methoxycarbonylphenoxycarbonyl)ethylene]	311
Poly[1-(4-methoxycarbonylphenoxycarbonyl)ethylene]	340
Poly[1-(2-methoxyethoxycarbonyl)ethylene]	223
Poly[1-(4-methoxyphenoxycarbonyl)ethylene]	324
Poly[1-(3-methoxypropoxycarbonyl)ethylene]	198
Poly[1-(2-naphthyloxycarbonyl)ethylene]	358
Poly[1-(pentachlorophenoxycarbonyl)ethylene]	420
Poly[1-(phenethoxycarbonyl)ethylene]	270
Poly[1-(phenoxycarbonyl)ethylene]	330
Poly[1-(*m*-tolyloxycarbonyl)ethylene]	298
Poly[1-(*o*-tolyloxycarbonyl)ethylene]	325
Poly[1-(*p*-tolyloxycarbonyl)ethylene]	316
Poly[1-(2,2,2-trifluorethoxycarbonyl)ethylene]	263
Polymethacrylics	
Poly[1-(alkoxycarbonyl)-1-methylethylene] *Poly(alkyl methacrylate)* $-[(ROCO)(Me)CCH_2]-$	
Poly[1-(benzyloxycarbonyl)-1-methylethylene]	327
Poly[1-(2-bromoethoxycarbonyl)-1-methylethylene]	325
Poly[(1-(butoxycarbonyl)-1-methylethylene]	293
Poly(butyl methacrylate) [PBMA]	
Poly[1-(*sec*-butoxycarbonyl)-1-methylethylene]	333
Poly[1-(*tert*-butoxycarbonyl)-1-methylethylene)]	391
Poly[1-(2-chloroethoxycarbonyl)-1-methylethylene]	ca 315
Poly[1-(2-cyanoethoxycarbonyl)-1-methylethylene]	364
Poly[1-(4-cyanophenoxycarbonyl)-1-methylethylene]	428
Poly[1-(cyclohexyloxycarbonyl)-1-methylethylene] (atactic)	356
Poly[1-(cyclohexyloxycarbonyl)-1-methylethylene)] (isotactic)	324
Poly[1-(dimethylaminoethoxycarbonyl)-1-methylethylene]	292
Poly[1-(ethoxycarbonyl)-1-ethylethylene]	300
Poly[1-(ethoxycarbonyl)-1-methylethylene] (atactic) *Poly(ethyl methacrylate)* [PEMA]	338
Poly[1-(ethoxycarbonyl)-1-methylethylene] (isotactic)	285
Poly[1-(ethoxycarbonyl)-1-methylethylene)] (syndiotactic)	339
Poly[1-(hexyloxycarbonyl)-1-methylethylene]	268
Poly[1-(isobutoxycarbonyl)-1-methylethylene]	326
Poly[1-(isopropoxycarbonyl)-1-methylethylene]	354
Poly[1-(methoxycarbonyl)-1-methylethylene] (atactic) *Poly(methyl methacrylate)* [PMMA]	378
Poly[1-(methoxycarbonyl)-1-methylethylene)] (isotactic)	311
Poly[1-(methoxycarbonyl)-1-methylethylene)] (syndiotactic)	378
Poly[1-(4-methoxycarbonylphenoxy)-1-methylethylene]	379

Polymer name	Glass transition temperature (T_g/K)
Poly[1-(methoxycarbonyl)-1-phenylethylene)] (atactic)	391
Poly[1-(methoxycarbonyl)-1-phenylethylene)] (isotactic)	397
Poly[1-methyl-1-(phenethoxycarbonyl)ethylene]	299
Poly[1-methyl-1-(phenoxycarbonyl)ethylene]	383
Polyvinyl ethers, alcohols, and ketones	
Poly(1-alkoxyethylene) *Poly(alkyl vinyl ether)* –[ROCHCH$_2$]–	
Poly(1-hydroxyethylene) *Poly(vinyl alcohol)* –[HOCHCH$_2$]–	
Poly(1-alkanoylethylene) *Poly(alkyl vinyl ketone)* –[RCOCHCH$_2$]–	
Poly(1-butoxyethylene)	218
Poly(1-*sec*-butoxyethylene)	253
Poly(1-*tert*-butoxyethylene)	361
Poly[1-(butylthio)ethylene]	253
Poly(1-ethoxyethylene)	230
Poly[1-(4-ethylbenzoyl)ethylene]	325
Poly(1-hydroxyethylene)	358 (D)
Poly(vinyl alcohol) [PVA]	
Poly(hydroxymethylene)	407
Poly(1-isopropoxyethylene)	270
Poly[1-(4-methoxybenzoyl)ethylene]	319 (M)
Poly(1-methoxyethylene)	242
Poly(methyl vinyl ether) [PMVE]	
Poly[1-(methylthio)ethylene]	272
Poly(1-propoxyethylene)	224
Poly[1-(trifluoromethoxy)trifluoroethylene]	268
Polyvinyl halides and nitriles	
Poly(1-haloethylene) *Poly(vinyl halide)* –[XCHCH$_2$]–	
Poly(1-cyanoethylene) *Poly(acrylonitrile)* –[NCCHCH$_2$]–	
Poly(1-chloroethylene)	354
Poly(vinyl chloride) [PVC]	
Poly(chlorotrifluoroethylene)	373
Poly(1-cyanoethylene)	370
Polyacrylonitrile [PAN]	
Poly(1-cyano-1-methylethylene)	393
Polymethacrylonitrile	
Poly(1,1-dichloroethylene)	255
Poly(vinylidene chloride)	
Poly(1,1-difluoroethylene)	ca 233
Poly(vinylidene fluoride)	
Poly(1-fluoroethylene)	314 (M)
Poly(vinyl fluoride)	
Poly(1-hexafluoropropylene)	425
Poly[1-(2-iodoethyl)ethylene]	343
Poly(tetrafluoroethylene)	(160)
Poly[1-(trifluoromethyl)ethylene]	300
Polyvinyl esters	
Poly[1-(alkanoyloxy)ethylene] *Poly(vinyl alkanoate)* –[RCOOCHCH$_2$]–	
Poly(1-acetoxyethylene)	305
Poly(vinyl acetate) [PVAc]	
Poly[1-(benzoyloxy)ethylene]	344
Poly[1-(4-bromobenzoyloxy)ethylene]	365
Poly[1-(2-chlorobenzoyloxy)ethylene]	335
Poly[1-(3-chlorobenzoyloxy)ethylene]	338
Poly[1-(4-chlorobenzoyloxy)ethylene]	357
Poly[1-(cyclohexanoyloxy)ethylene]	349 (M)
Poly[1-(4-ethoxybenzoyloxy)ethylene]	343
Poly[1-(4-ethylbenzoyloxy)ethylene]	326

Polymer name	Glass transition temperature (T_g/K)
Poly[1-(4-isopropylbenzoyloxy)ethylene]	342
Poly[1-(2-methoxybenzoyloxy)ethylene]	338
Poly[1-(3-methoxybenzoyloxy)ethylene]	ca 317
Poly[1-(4-methoxybenzoyloxy)ethylene]	360
Poly[1-(4-methylbenzoyloxy)ethylene]	343
Poly[1-(4-nitrobenzoyloxy)ethylene]	395
Poly[1-(propionoyloxy)ethylene]	283 (M)
Polystyrenes	
Poly(1-phenylethylene) *Polystyrene* –[$C_6H_5CHCH_2$]–	
Poly[1-(4-acetylphenyl)ethylene]	389 (M)
Poly[1-(4-benzoylphenyl)ethylene]	371 (M)
Poly[1-(4-bromophenyl)ethylene]	391
Poly[1-(4-butoxyphenyl)ethylene]	ca 320 (M)
Poly[1-(4-butoxycarbonylphenyl)ethylene]	349 (M)
Pol[(1-(4-butylphenyl)ethylene]	279
Poly[1-(4-carboxyphenyl)ethylene]	386 (M)
Poly[1-(2-chlorophenyl)ethylene]	392
Poly[1-(3-chlorophenyl)ethylene]	363
Poly[1-(4-chlorophenyl)ethylene]	383
Poly[1-(2,4-dichlorophenyl)ethylene]	406
Poly[1-(2,5-dichlorophenyl)ethylene]	379
Poly[1-(2,6-dichlorophenyl)ethylene]	440
Poly[1-(3,4-dichlorophenyl)ethylene]	401
Poly[1-(2,4-dimethylphenyl)ethylene]	385
Poly[1-(4-(dimethylamino)phenyl)ethylene]	398 (M)
Poly[1-(4-ethoxyphenyl)ethylene]	ca 359 (M)
Poly[1-(4-ethoxycarbonylphenyl)ethylene]	367 (M)
Poly[1-(4-fluorophenyl)ethylene]	368
Poly[1-(4-iodophenyl)ethylene]	429
Poly[1-(4-methoxyphenyl)ethylene]	386
Poly[1-(4-methoxycarbonylphenyl)ethylene]	386 (M)
Poly(1-methyl-1-phenylethylene) *Poly(α-methylstyrene)*	373
Poly[1-(2-(methylamino)phenyl)ethylene]	462 (M)
Poly(1-phenylethylene) *Polystyrene* [PS]	373
Poly[1-(4-propoxyphenyl)ethylene]	343 (M)
Poly[1-(4-propoxycarbonylphenyl)ethylene]	365 (M)
Poly(1-*o*-tolylethylene)	409
CHAINS WITH CARBOCYCLIC UNITS	
Poly(arylenealkylene) –[–Ar–(CH_2)$_n$]–	
Poly[1-(2-bromo-1,4-phenylene)ethylene]	353 (M)
Poly[1-(2-chloro-1,4-phenylene)ethylene]	343 (M)
Poly[1-(2-cyano-1,4-phenylene)ethylene]	363 (M)
Poly[1-(2,5-dimethyl-1,4-phenylene)ethylene]	373 (M)
Poly[1-(2-ethyl-1,4-phenylene)ethylene]	298 (M)
Poly[1-(1,4-naphthylene)ethylene]	433 (M)
Poly[1-(1,4-phenylene)ethylene]	ca 353 (M)
CHAINS WITH HETEROATOM UNITS	
Main chain oxide units	
Poly(oxyalkylene) *Poly(alkylene oxide)* –[O(CH_2)$_n$]–	
Poly[oxy(1,1-bis(chloromethyl)trimethylene)]	265
Poly[oxy(1-(bromomethyl)ethylene)]	259
Poly[oxy(1-(butoxymethyl)ethylene)]	194
Poly[oxy(1-butylethylene)]	203
Poly[oxy(1-*tert*-butylethylene)]	308
Poly[oxy(1-(chloromethyl)ethylene)]	251

Polymer name	Glass transition temperature (T_g/K)
Poly(epichlorohydrin)	
Poly[oxy(2,6-dimethoxy-1,4-phenylene)]	440
Poly[oxy(1,1-dimethylethylene)]	264
Poly[oxy(2,6-dimethyl-1,4-phenylene)]	482
Poly[oxy(2,6-diphenyl-1,4-phenylene)]	493
Poly[oxy(1-ethylethylene)]	203
Poly(oxyethylidene)	243
Polyacetaldehyde	
Poly[oxy(1-(methoxymethyl)ethylene)]	211
Poly[oxy(2-methyl-6-phenyl-1,4-phenylene)]	428
Poly[oxy(1-methyltrimethylene)]	223 (D)
Poly[oxy(2-methyltrimethylene)]	218
Poly(oxy-1,4-phenylene)	358
Poly(phenylene oxide) [PPO]	
Poly[oxy(1-phenylethylene)]	313
Poly(oxytetramethylene)	189
Poly(tetrahydrofuran) [PTMO]	
Poly(oxytrimethylene)	195
Main-chain ester or anhydride units	
Poly(oxyalkyleneoxyalkanedioyl) *Poly(alkylene alkanedioate)*–[O(CH$_2$)$_m$OCO(CH$_2$)$_n$CO]–	
Poly(oxyadipoyloxydecamethylene)	217
Poly(oxyadipoyloxy-1,4-phenyleneisopropylidene-1,4-phenylene)	341
Poly(oxycarbonyloxy-1,4-phenylene-isopropylidene-1,4-phenylene)	422
Bisphenol A polycarbonate	
Poly(oxycarbonylpentamethylene)	213
Poly(oxycarbonyl-1,4-phenylenemethylene-1,4-phenylene)	395
Poly(oxycarbonyl-1,4-phenyleneisopropylidene-1,4-phenylene)	333
Poly[oxy(2,6-dimethyl-1,4-phenyleneisopropylidene-3,5-dimethyl-1,4-phenylene)oxysebacoyl]	318
Poly(oxyethylenecarbonyl-1,4-cyclohexylenecarbonyl) (trans)	291
Poly(oxyethyleneoxycarbonyl-1,4-naphthylenecarbonyl)	337
Poly(oxyethyleneoxycarbonyl-1,5-naphthylenecarbonyl)	344
Poly(oxyethyleneoxycarbonyl-2,6-naphthylenecarbonyl)	386
Poly(oxyethyleneoxycarbonyl-2,7-naphthylenecarbonyl)	392
Poly(oxyethyleneoxyterephthaloyl)	342
Poly(ethylene terephthalate) [PET]	
Poly(oxyisophthaloyl)	403 (D)
Poly(oxy(1-oxo-2,2-dimethyltrimethylene))	263
Poly(pivalolactone)	
Poly(oxy-1,4-phenyleneisopropylidene-1,4-phenyleneoxysebacoyl)	280
Poly(oxy-1,4-phenyleneoxy-1,4-phenyleneoxy-carbonyl-1,4-phenylene) [PEEK]	416
Poly(oxypropyleneoxyterephthaloyl)	341
Poly[oxyterephthaloyloxy(2,6-dimethyl-1,4-phenyleneisopropylidene-3,5-dimethyl-1,4-(D)phenylene)]	498
Poly(oxyterephthaloyloxyoctamethylene)	318 (D)
Poly(oxyterephthaloyloxy-1,4-phenyleneisopropylidene-1,4-phenylene)	478
Poly(bisphenol A terephthalate)	
Poly(oxytetramethyleneoxyterephthaloyl)	323
Poly(butylene terephthalate) [PBT]	
Main-chain amide units	
Poly(iminoalkyleneiminoalkanedioyl) *Poly(alkylene alkanediamide)*–[NH(CH$_2$)$_m$NHCO(CH$_2$)$_n$CO]–	
Poly(iminoadipoyliminodecamethylene)	313
Nylon 10,6	
Poly(iminoadipoyliminohexamethylene)	ca 323
Nylon 6,6	
Poly(iminoadipoyliminooctamethylene)	318
Nylon 8,6	
Poly[iminoadipoyliminotrimethylene(methylimino)trimethylene]	278
Poly(iminocarbonyl-1,4-cyclohexylenemethylene)	466
Poly[iminocarbonyl-1,4-phenylene(2-oxoethylene)iminohexamethylene]	377
Poly(iminoethylene-1,4-phenyleneethyleneiminosebacoyl)	378 (D)

Polymer name	Glass transition temperature (T_g/K)
Poly(iminohexamethyleneiminoazelaoyl)	331
Nylon 6,9	
Poly(iminohexamethyleneiminododecanedioyl)	319
Nylon 6, 12	
Poly(iminohexamethyleneiminopimeloyl)	331
Nylon 6,7	
Poly(iminohexamethyleneiminosebacoyl)	323
Nylon 6,10	
Poly(iminohexamethyleneiminosuberoyl)	330
Nylon 6,8	
Poly(iminoisophthaloylimino-4,4′-biphenylylene)	558
Poly(iminoisophthaloyliminohexamethylene)	390
Poly(iminoisophthaloyliminomethylene-1,4-cyclohexylenemethylene)	481
Poly(iminoisophthaloyliminomethylene-1,3-phenylenemethylene)	438 (M)
Poly[iminomethylene(2,5-dimethyl-1,4-phenylene)methyleneiminosuberoyl]	351
Poly(imino-1,5-naphthyleneiminoisophthaloyl)	598
Poly(imino-1,5-naphthyleneiminoterephthaloyl)	578
Poly(iminooctamethyleneiminodecanedioyl)	333
Nylon 8,10	
Poly(iminooxalyliminohexamethylene)	430
Nylon 6,2	
Poly[imino(1-oxohexamethylene)]	326
Nylon 6	
Poly[imino(1-oxodecamethylene)]	315
Nylon 10	
Poly[imino(1-oxoheptamethylene)]	325
Nylon 7	
Poly[imino(1-oxo-3-methyltrimethylene]	369
Poly[imino(1-oxononamethylene)]	319
Nylon 9	
Poly[imino(1-oxooctamethylene)]	323
Nylon 8	
Poly[imino(1-oxotrimethylene)]	384
Nylon 3	
Poly(iminopentamethyleneiminoadipoyl)	318
Nylon 5,6	
Poly[iminopentamethyleneiminocarbonyl-1,4-phenylene(2-oxoethylene)]	376
Poly(imino-1,3-phenyleneiminoisophthaloyl)	553 (M)
Poly(imino-1,4-phenyleneiminoterephthaloyl)	618
Poly(iminopimeloyliminoheptamethylene)	328
Nylon 7,7	
Poly(iminoterephthaloylimino-4,4′-biphenylylene)	613
Poly(iminotetramethyleneiminoadipoyl)	316
Nylon 4,6	
Poly[iminotetramethyleneiminocarbonyl-1,4-phenylene(2-oxoethylene)]	357
Poly(iminotrimethyleneiminoadipoyliminotrimethylene)	307
Poly[iminotrimethyleneiminocarbonyl-1,4-phenylene(2-oxoethylene)]	382
Poly(oxy-1,4-phenyleneiminoterephthaloyl-imino-1,4-phenylene)	613
Poly(sulfonylimino-1,4-phenyleneiminoadipoylimino-1,4-phenylene)	467
Main-chain urethane units	
Poly(oxyalkyleneoxycarbonyliminoalkyleneiminocarbonyl)–[O(CH$_2$)$_m$OCONH(CH$_2$)$_n$NHCO]–	
Poly(oxyethyleneoxycarbonyliminohexamethyleneiminocarbonyl)	329
Poly[oxyethyleneoxycarbonylimino(6-methyl-1,3-phenylene)iminocarbonyl]	325
Poly(oxyethyleneoxycarbonylimino-1,4-phenylenemethylene-1,4-phenyleneiminocarbonyl)	412
Poly(oxyhexamethyleneoxycarbonyliminohexamethyleneiminocarbonyl)	332
Poly[oxyhexamethyleneoxycarbonylimino(6-methyl-1,3-phenylene)iminocarbonyl]	305
Poly(oxyhexamethyleneoxycarbonylimino-1,4-phenylenemethylene-1,4-phenyleneiminocarbonyl)	364
Poly(oxyoctamethyleneoxycarbonyliminohexamethyleneiminocarbonyl)	331
Poly[oxyoctamethyleneoxycarbonylimino(6-methyl-1,3-phenylene)iminocarbonyl]	337
Poly(oxyoctamethyleneoxycarbonylimino-1,4-phenylenemethylene-1,4-phenyleneiminocarbonyl)	352

Polymer name	Glass transition temperature (T_g/K)
Poly(oxytetramethyleneoxycarbonyliminohexamethyleneiminocarbonyl)	332
Poly[oxytetramethyleneoxycarbonylimino(6-methyl-1,3-phenylene)iminocarbonyl]	315
Poly(oxytetramethyleneoxycarbonylimino-1,4-phenylenemethylene-1,4-phenyleneiminocarbonyl)	382
Main-chain siloxanes	
Poly[oxy(dialkylsilylene)] *Poly(dialkylsiloxane)* –[O(R_2Si)]–	
Poly[oxy(dimethylsilylene)]	148
Poly(dimethylsiloxane) [PDMS]	
Poly[oxy(dimethylsilylene)oxy-1,4-phenylene]	363 (M)
Poly[oxy(dimethylsilylene)oxy-1,4-phenyleneisopropylidene-1,4-phenylene]	318 (M)
Poly[oxy(diphenylsilylene)]	238
Poly(diphenylsiloxane)	
Poly[oxy(diphenylsilylene)-1,3-phenylene]	ca 331
Poly[oxy((methyl)phenylsilylene)]	187
Poly[oxy((methyl)-3,3,3-trifluoropropylsilylene]	<193
Main-chain sulfur-containing units	
Poly(dithioethylene)	223
Poly(dithiomethylene-1,4-phenylenemethylene)	296
Poly(oxy-4,4′-biphenylylene-1,4-phenylenesulfonyl-1,4-phenylene)	503 (M)
Poly(oxycarbonyloxy-1,4-phenylenethio-1,4-phenylene)	ca 383
Poly(oxyethylenedithioethylene)	220 (M)
Poly[oxy(2-hydroxytrimethylene)oxy-1,4-phenylenesulfonyl-1,4-phenylene]	428
Poly(oxymethyleneoxyethylenedithioethylene)	214
Poly(oxy-1,4-phenylenesulfinyl-1,4-phenyleneoxy-1,4-phenylenecarbonyl-1,4-phenylene)	478 (M)
Poly(oxy-1,4-phenylenesulfinyl-1,4-phenyleneoxy-1,4-phenyleneisopropylidene-1,4-phenylene)	438 (M)
Poly(oxy-1,4-phenylenesulfonyl-1,4-phenylene)	487
Poly(oxy-1,4-phenylenesulfonyl-4,4′-biphenylylenesulfonyl-1,4-phenylene)	533
Poly[oxy-1,4-phenylenesulfonyl-1,4-phenyleneoxy(2,6-dimethyl-1,4-phenylene)isopropylidene (3,5-dimethyl-1,4-phenylene)]	508 (M)
Poly(oxy-1,4-phenylenesulfonyl-1,4-phenyleneoxy-1,4-phenylenecarbonyl-1,4-phenylene)	478 (M)
Poly[oxy-1,4-phenylenesulfonyl-1,4-phenyleneoxy-1,4-phenylene(hexafluoroisopropylidene)1,4-phenylene]	478 (M)
Poly(oxy-1,4-phenylenesulfonyl-1,4-phenyleneoxy-1,4-phenyleneisopropylidene-1,4-phenylene)	449
Poly(oxy-1,4-phenylenesulfonyl-1,4-phenyleneoxy-1.4-phenylenemethylene-1,4-phenylene)	453 (M)
Poly(oxy-1,4-phenylenesulfonyl-1,4-phenyleneoxy-1.4-phenylenethio-1,4-phenylene)	448 (M)
Poly(oxy-1,4-phenylenesulfonyl-1,4-phenyleneoxyterephthaloyl)	522
Poly(oxytetramethylenedithiotetramethylene)	197
Poly(sulfonyl-1,2-cyclohexylene)	401
Poly(sulfonyl-1,3-cyclohexylene)	381
Poly(sulfonyl-1,4-phenylenemethylene-1,4-phenylene)	497
Poly(thio-1,3-cyclohexylene)	221
Poly[thio(difluoromethylene)]	155
Poly(thioethylene)	223
Poly[thio(1-ethylethylene]	218
Poly[thio(1-methyl-3-oxotrimethylene)]	285
Poly[thio(1-methyltrimethylene)]	214
Pol[(thio(1-oxohexamethylene)]	292
Poly(thio-1,4-phenylene)	370
Poly(thiopropylene)	226
Main-chain heterocyclic units	
Poly(1,3-dioxa-4,6-cyclohexylenemethylene)	378
Poly(vinyl formal)	
Poly[(2,6-dioxopiperidine-1,4-diyl)trimethylene]	363
Poly[(2-methyl-1,3-dioxa-4,6-cyclohexylene)methylene]	355
Poly(vinyl acetal)	
Poly(1,4-piperazinediylcarbonyloxyethyleneoxycarbonyl)	333
Poly(1,4-piperazinediylisophthaloyl)	465 (M)
Poly[(2-propyl-1,3-dioxa-4,6-cyclohexylene)methylene]	322
Poly(vinyl butyral)	
Poly(3,6-pyridazinediyloxy-1,4-phenyleneisopropylidene-1,4-phenyleneoxy)	453 (M)
Poly(2,5-pyridinediylcarbonyliminohexamethyleneiminocarbonyl)	322

DIELECTRIC CONSTANT OF SELECTED POLYMERS

This table lists typical values of the dielectric constant (more properly called relative permittivity) of some important polymers. Values are given for frequencies of 1 kHz, 1 MHz, and 1 GHz; in most cases the dielectric constant at frequencies below 1 kHz does not differ significantly from the value at 1 kHz. Since the dielectric constant of a polymeric material can vary with density, degree of crystallinity, and other details of a particular sample, the values given here should be regarded as only typical or average values.

References

1. Gray, D. E., Ed., *American Institute of Physics Handbook, Third Edition*, p. 5-132, McGraw Hill, New York, 1972.
2. Anderson, H. L., Ed., *A Physicist's Desk Reference*, American Institute of Physics, New York, 1989.
3. Brandrup, J., and Immergut, E. H., *Polymer Handbook, Third Edition*, John Wiley & Sons, New York, 1989.

Name	t/°C	1 kHz	1 MHz	1 GHz
Polyacrylonitrile	25	5.5	4.2	
Polyamides (nylons)	25	3.50	3.14	2.8
	84	11	4.4	2.8
Polybutadiene	25	2.5		
Polycarbonate	23	2.92	2.8	
Polychloroprene (neoprene)	25	6.6	6.3	4.2
Polychlorotrifluoroethylene	23	2.65	2.46	2.39
Polyethylene	23	2.3		
Poly(ethylene terephthalate) (Mylar)	23	3.25	3.0	2.8
Polyisoprene (natural rubber)	27	2.6	2.5	2.4
Poly(methyl methacrylate)	27	3.12	2.76	2.6
	80	3.80	2.7	2.6
Polyoxymethylene (polyformaldehyde)	25	3.8		
Poly(phenylene oxide)	23	2.59	2.59	
Polypropylene	25	2.3	2.3	2.3
Polystyrene	25	2.6	2.6	2.6
Polysulfones	25	3.13	2.10	
Polytetrafluoroethylene (teflon)	25	2.1	2.1	2.1
Poly(vinyl acetate)	50		3.5	
	150		8.3	
Poly(vinyl chloride)	25	3.39	2.9	2.8
	100	5.3	3.3	2.7
Poly(vinylidene chloride)	23	4.6	3.2	2.7
Poly(vinylidene fluoride)	23	12.2	8.9	4.7

PRESSURE–VOLUME–TEMPERATURE RELATIONSHIPS FOR POLYMER MELTS

Christian Wohlfarth

Numerous theoretical equations of state for polymer liquids have been developed. These, at the minimum, have to provide accurate fitting functions to experimental data. However, for the purpose of this table, the empirical Tait equation along with a polynomial expression for the zero pressure isobar is used. This equation is able to represent the experimental data for the melt state within the limits of experimental errors, i.e., the maximum deviations between measured and calculated specific volumes are about 0.001-0.002 cm³/g.

The general form of the Tait equation is:

$$V(P,T) = V(0,T)\{1 - C\ \ln[1 + P/B(T)]\} \tag{1}$$

where the coefficient C is usually taken to be a universal constant equal to 0.0894. T is the absolute temperature in K and P the pressure in MPa. The volume V is the specific volume in cm³/g. The Tait parameter $B(T)$ has the very simple meaning that it is inversely proportional to the compressibility κ at constant temperature and zero pressure:

$$\kappa(0,T) = -[1/V(0,T)](dV/dP) = C/B(T) \tag{2}$$

The $B(T)$ function is usually given by:

$$B(T) = B_0 \exp[-B_1(T\text{-}273.15)] \tag{3}$$

but, sometimes a polynomial expression is used:

$$B(T) = b_0 + b_1(T\text{-}273.15) + b_2(T\text{-}273.15)^2 \tag{4}$$

The zero-pressure isobar $V(0,T)$ is usually given by:

$$V(0,T) = A_0 + A_1(T\text{-}273.15) + A_2(T\text{-}273.15)^2 \tag{5}$$

where A_0, A_1, A_2 are specific constants for a given polymer (the expression T-273.15 is used because fitting to the zero-pressure isobar is usually done in terms of Celsius temperature). Other forms for $V(0,T)$ are also found in the literature, such as

$$V(0,T) = A_3 \exp[A_4(T\text{-}273.15)] \tag{6}$$

or

$$V(0,T) = A_5 \exp(A_6 T^{1.5}) \tag{7}$$

where A_3 and A_4 or A_5 and A_6 are again specific constants for a given polymer.

The Tait equation is particularly useful to calculate derivative quantities, such as the isothermal compressibility and the thermal expansivity coefficients. The isothermal compressibility $\kappa(P,T)$ is derived from equation (1) as:

$$\kappa(P,T) = -(1/V)(dV/dP) = 1/\{[P + B(T)][1/C - \ln(1 + P/B(T))]\} \tag{8}$$

and the thermal expansivity $\alpha(P,T)$ as:

$$\alpha(P,T) = (1/V)(dV/dT) = \alpha(0,T) - PB_1\kappa(P,T) \tag{9}$$

where $\alpha(0,T)$ represents the thermal expansivity at zero (atmospheric) pressure and is calculated from any suitable fit for the zero-pressure volume, such as equations (5) through (7) above.

Because polymer melt PVT-behavior depends only slightly on polymer molar mass above the oligomeric region, usually no information is given in the original literature for the average molar mass of the polymers.

Table 1 summarizes the polymers or copolymers considered here and the experimental ranges of pressure and temperature over which data are available. In Table 2 the Tait-equation functions, with parameters obtained from the fit, are given for 90 polymer or copolymer melts.

References

1. Zoller, P., *J. Appl. Polym. Sci.*, 23, 1051–1056, 1979.
2. Starkweather, H. W., Jones, G. A., and Zoller, P., *J. Polym. Sci., Pt. B Polym. Phys.*, 26, 257–266,1988.
3. Fakhreddine, Y. A., and Zoller, P., *J. Polym. Sci., Pt. B Polym. Phys.*, 29, 1141–1146, 1991.
4. Rodgers, P. A., *J. Appl. Polym. Sci.*, 48, 1061–1080, 1993.
5. Rodgers, P. A., *J. Appl. Polym. Sci.*, 48, 2075–2083, 1993.
6. Yi, Y. X., and Zoller, P., *J. Polym. Sci., Pt. B Polym. Phys.*, 31, 779–788, 1993.
7. Callaghan, T. A., and Paul, D. R., *Macromolecules*, 26, 2439–2450, 1993.
8. Wang, Y. Z., Hsieh, K. H., Chen, L. W.,and Tseng, H. C., *J. Appl. Polym. Sci.*, 53, 1191–1201, 1994.
9. Privalko, V. P., Arbuzova, A. P., Korskanov, V. V., and Zagdanskaya, N. E., *Polym. Intern.*, 35, 161–169, 1994.
10. Sachdev, V. K., Yashi, U., and Jain, R. K., *J. Polym. Sci., Pt. B Polym. Phys.*, 36, 841–850, 1998.

TABLE 1. Names of the Polymers, Abbreviation Used, and Range of Experimental Data Applied in the Determination of the Equation Constants

Polymer	Symbol	*T*/K	*P*/MPa	Ref.
Ethylene/propylene copolymer (50 wt%)	EP50	413-523	0.1-63	4
Ethylene/vinyl acetate copolymer				
18 wt% vinyl acetate	EVA18	385-491	0.1-177	4
25 wt% vinyl acetate	EVA25	367-506	0.1-177	4
28 wt% vinyl acetate	EVA28	367-508	0.1-177	4
40 wt% vinyl acetate	EVA40	348-508	0.1-177	4
Polyamide-6	PA6	509-569	0.1-196	4
Polyamide-11	PA11	478-542	0.1-200	5
Polyamide-66	PA66	519-571	0.1-196	4
cis-1,4-Polybutadiene	cPBD	277-328	0.1-284	4
Polybutadiene, 8% 1,2-content	PBD-8	298-473	0.1-200	6
Polybutadiene, 24% 1,2-content	PBD-24	298-473	0.1-200	6
Polybutadiene, 40% 1,2-content	PBD-40	298-473	0.1-200	6
Polybutadiene, 50% 1,2-content	PBD-50	298-473	0.1-200	6
Polybutadiene, 87% 1,2-content	PBD-87	298-473	0.1-200	6
Poly(1-butene), isotactic	iPB	406-519	0.1-196	4
Poly(butyl methacrylate)	PnBMA	307-473	0.1-200	4
Poly(butylene terephthalate)	PBT	508-576	0.1-200	3
Poly(ε-caprolactone)	PCL	373-421	0.1-200	4
Polycarbonate-bisphenol-A	PC	424-613	0.1-177	4
Polycarbonate-bisphenol-chloral	BCPC	428-557	0.1-200	4
Polycarbonate-hexafluorobisphenol-A	HFPC	432-553	0.1-200	4
Polycarbonate-tetramethylbisphenol-A	TMPC	491-563	0.1-160	4
Poly(cyclohexyl methacrylate)	PcHMA	396-471	0.1-200	4
Poly(2,5-dimethylphenylene oxide)	PPO	473-593	0.1-177	4
Poly(dimethyl siloxane)	PDMS	298-343	0.1-100	4
Poly(dimethyl siloxane) $M_n = 1000$	PDMS-10	304-420	0.1-250	10
Poly(dimethyl siloxane) $M_n = 4000$	PDMS-40	298-418	0.1-250	10
Poly(dimethyl siloxane) $M_n = 6000$	PDMS-60	291-423	0.1-250	10
Poly(epichlorohydrin)	PECH	333-413	0.1-200	4
Poly(ether ether ketone)	PEEK	619-671	0.1-200	4
Poly(ethyl acrylate)	PEA	310-490	0.1-196	4
Poly(ethyl methacrylate)	PEMA	386-434	0.1-196	4
Polyethylene, high density	HDPE	413-476	0.1-196	4
Polyethylene, linear	LPE	415-473	0.1-200	4
Polyethylene, linear, high MW	HMLPE	410-473	0.1-200	4
Polyethylene, branched	BPE	398-471	0.1-200	4
Polyethylene, low density	LDPE	394-448	0.1-196	4
Polyethylene, low density, type A	LDPE-A	385-498	0.1-196	1
Polyethylene, low density, type B	LDPE-B	385-498	0.1-196	1
Polyethylene, low density, type C	LDPE-C	385-498	0.1-196	1
Poly(ethylene oxide)	PEO	361-497	0.1- 68	4
Poly(ethylene terephthalate)	PET	547-615	0.1-196	4
Poly(4-hexylstyrene)	P4HS	303-403	30-100	4
Polyisobutylene	PIB	326-383	0.1-100	4
Polyisoprene, 8% 3,4-content	PI-8	298-473	0.1-200	6
Polyisoprene, 14% 3,4-content	PI-14	298-473	0.1-200	6
Polyisoprene, 41% 3,4-content	PI-41	298-473	0.1-200	6
Polyisoprene, 56% 3,4-content	PI-56	298-473	0.1-200	6
Poly(methyl acrylate)	PMA	310-493	0.1-196	4
Poly(methyl methacrylate)	PMMA	387-432	0.1-200	4
Poly(4-methyl-1-pentene)	P4MP	514-592	0.1-196	4
Poly(α-methylstyrene)	PαMS	473-533	0.1-170	7
Poly(*o*-methylstyrene)	PoMS	412-471	0.1-180	4
Polyoxymethylene	POM	463-493	0.1-196	2
Phenoxy[a]	PH	341-573	0.1-177	4
Polysulfone[b]	PSF	475-644	0.1-196	4
Polyarylate[c]	PAr	450-583	0.1-177	4
Polypropylene, atactic	aPP	353-393	0.1-100	4

Polymer	Symbol	*T*/K	*P*/MPa	Ref.
Polypropylene, isotactic	iPP	443-570	0.1-196	4
Polystyrene	PS	388-469	0.1-200	4
Poly(tetrafluoroethylene)	PTFE	603-645	0.1- 39	4
Poly(tetrahydrofuran)	PTHF	335-439	0.1- 78	4
Poly(vinyl acetate)	PVAc	308-373	0.1- 80	4
Poly(vinyl chloride)	PVC	373-423	0.1-200	4
Poly(vinyl methyl ether)	PVME	303-471	0.1-200	4
Poly(vinylidene fluoride)	PVdF	451-521	0.1-200	5
Styrene/acrylonitrile copolymer				
2.7 wt% acrylonitrile	SAN3	378-539	0.1-200	4
5.7 wt% acrylonitrile	SAN6	370-540	0.1-200	4
15.3 wt% acrylonitrile	SAN15	405-531	0.1-200	4
18.0 wt% acrylonitrile	SAN18	377-528	0.1-200	4
40 wt% acrylonitrile	SAN40	373-543	0.1-200	4
70 wt% acrylonitrile	SAN70	373-544	0.1-200	4
Styrene/butadiene copolymer				
10 wt% styrene	SBR10	393-533	0.1-196	8
23.5 wt% styrene	SBR23	393-533	0.1-196	8
60 wt% styrene	SBR60	393-533	0.1-196	8
85 wt% styrene	SBR85	393-533	0.1-196	8
Styrene/methyl methacrylate copolymer				
20 wt% methyl methacrylate	SMMA20	383-543	0.1-200	4
60 wt% methyl methacrylate	SMMA60	383-543	0.1-200	4
N-Vinylcarbazole/4-ethylstyrene copolymer				
50 mol% ethylstyrene	VCES50	393-443	30-100	9
N-Vinylcarbazole/4-hexylstyrene copolymer				
80 mol% hexylstyrene	VCHS80	313-423	30-100	9
67 mol% hexylstyrene	VCHS67	333-423	30-100	9
60 mol% hexylstyrene	VCHS60	383-453	30-100	9
50 mol% hexylstyrene	VCHS50	373-443	30-100	9
40 mol% hexylstyrene	VCHS40	423-493	30-100	9
33 mol% hexylstyrene	VCHS33	463-523	30-100	9
20 mol% hexylstyrene	VCHS20	473-523	30-100	9
N-Vinylcarbazole/4-octylstyrene copolymer				
50 mol% octylstyrene	VCOS50	403-453	30-100	9
N-Vinylcarbazole/4-pentylstyrene copolymer				
50 mol% pentylstyrene	VCPS50	383-443	30-100	9

[a] Phenoxy = Poly(oxy-2-hydroxytrimethyleneoxy-1,4-phenyleneisopropylidene-1,4-phenylene)

[b] Polysulfone = Poly(oxy-1,4-phenylenesulfonyl-1,4-phenyleneoxy-1,4-phenyleneisopropylidene-1,4-phenylene)

[c] Polyarylate = Poly(oxyterephthaloyl/isophthaloyl T/I=50/50)oxy-1,4-phenyleneisopropylidene-1,4-phenylene

TABLE 2. Tait Equation Parameter Functions for Polymer Melts

Polymer	$V(0,T)/\mathrm{cm^3g^{-1}}$	$B(T)$/MPa
EP50	$1.2291 + 5.799\cdot10^{-5}(T-273.15) + 1.964\cdot10^{-6}(T-273.15)^2$	$487.0\exp[-8.103\cdot10^{-3}(T-273.15)]$
EVA18	$1.02391\exp(2.173\cdot10^{-5}T^{1.5})$	$188.2\exp[-4.537\cdot10^{-3}(T-273.15)]$
EVA25	$1.00416\exp(2.244\cdot10^{-5}T^{1.5})$	$184.4\exp[-4.734\cdot10^{-3}(T-273.15)]$
EVA28	$1.00832\exp(2.241\cdot10^{-5}T^{1.5})$	$183.5\exp[-4.457\cdot10^{-3}(T-273.15)]$
EVA40	$1.06332\exp(2.288\cdot10^{-5}T^{1.5})$	$205.1\exp[-4.989\cdot10^{-3}(T-273.15)]$
PA6	$0.7597\exp[4.701\cdot10^{-4}(T-273.15)]$	$376.7\exp[-4.660\cdot10^{-3}(T-273.15)]$
PA11	$0.9581\exp[6.664\cdot10^{-4}(T-273.15)]$	$254.7\exp[-4.178\cdot10^{-3}(T-273.15)]$
PA66	$0.7657\exp[6.600\cdot10^{-4}(T-273.15)]$	$316.4\exp[-5.040\cdot10^{-3}(T-273.15)]$
cPBD	$1.0970\exp[6.600\cdot10^{-4}(T-273.15)]$	$177.7\exp[-3.593\cdot10^{-3}(T-273.15)]$
PBD-8	$1.1004 + 6.718\cdot10^{-4}(T-273.15) + 6.584\cdot10^{-7}(T-273.15)^2$	$200.0\exp[-4.606\cdot10^{-3}(T-273.15)]$
PBD-24	$1.1049 + 6.489\cdot10^{-4}(T-273.15) + 7.099\cdot10^{-7}(T-273.15)^2$	$193.0\exp[-4.519\cdot10^{-3}(T-273.15)]$
PBD-40	$1.1013 + 6.593\cdot10^{-4}(T-273.15) + 5.776\cdot10^{-7}(T-273.15)^2$	$188.0\exp[-4.437\cdot10^{-3}(T-273.15)]$
PBD-50	$1.1037 + 5.955\cdot10^{-4}(T-273.15) + 7.789\cdot10^{-7}(T-273.15)^2$	$183.0\exp[-4.425\cdot10^{-3}(T-273.15)]$
PBD-87	$1.1094 + 6.729\cdot10^{-4}(T-273.15) + 4.470\cdot10^{-7}(T-273.15)^2$	$175.0\exp[-4.538\cdot10^{-3}(T-273.15)]$
iPB	$1.1417\exp[6.751\cdot10^{-4}(T-273.15)]$	$167.5\exp[-4.533\cdot10^{-3}(T-273.15)]$
PnBMA	$0.9341 + 5.5254\cdot10^{-4}(T-273.15) + 6.5803\cdot10^{-6}(T-273.15)^2 + 1.5691\cdot10^{-10}(T-273.15)^3$	$226.7\exp[-5.344\cdot10^{-3}(T-273.15)]$
PBT	$0.9640 - 1.017\cdot10^{-3}(T-273.15) + 3.065\cdot10^{-6}(T-273.15)^2$	$263.0\exp[-3.444\cdot10^{-3}(T-273.15)]$
PCL	$0.9049\exp[6.392\cdot10^{-4}(T-273.15)]$	$189.0\exp[-3.931\cdot10^{-3}(T-273.15)]$
PC	$0.73565\exp(1.859\cdot10^{-5}T^{1.5})$	$310.0\exp[-4.078\cdot10^{-3}(T-273.15)]$
BCPC	$0.6737 + 3.634\cdot10^{-4}(T-273.15) + 2.370\cdot10^{-7}(T-273.15)^2$	$363.4\exp[-4.921\cdot10^{-3}(T-273.15)]$
HFPC	$0.6111 + 4.898\cdot10^{-4}(T-273.15) + 1.730\cdot10^{-7}(T-273.15)^2$	$236.6\exp[-5.156\cdot10^{-3}(T-273.15)]$
TMPC	$0.8497 + 5.073\cdot10^{-4}(T-273.15) + 3.832\cdot10^{-7}(T-273.15)^2$	$231.4\exp[-4.242\cdot10^{-3}(T-273.15)]$
PcHMA	$0.8793 + 4.0504\cdot10^{-4}(T-273.15) + 7.774\cdot10^{-7}(T-273.15)^2 - 7.7534\cdot10^{-10}(T-273.15)^3$	$295.2\exp[-5.220\cdot10^{-3}(T-273.15)]$
PPO	$0.78075\exp(2.151\cdot10^{-5}T^{1.5})$	$227.8\exp[-4.290\cdot10^{-3}(T-273.15)]$
PDMS	$1.0079\exp[9.121\cdot10^{-4}(T-273.15)]$	$89.4\exp[-5.701\cdot10^{-3}(T-273.15)]$
PDMS-10	$0.8343 + 5.991\cdot10^{-4}(T-273.15) + 5.734\cdot10^{-7}(T-273.15)^2$	$542.63\exp[-6.69\cdot10^{-3}(T-273.15)]$
PDMS-40	$0.8018 + 7.072\cdot10^{-4}(T-273.15) + 3.635\cdot10^{-7}(T-273.15)^2$	$482.73\exp[-6.09\cdot10^{-3}(T-273.15)]$
PDMS-60	$0.8146 + 5.578\cdot10^{-4}(T-273.15) + 5.774\cdot10^{-7}(T-273.15)^2$	$482.73\exp[-6.09\cdot10^{-3}(T-273.15)]$
PECH	$0.7216\exp[5.825\cdot10^{-4}(T-273.15)]$	$238.3\exp[-4.171\cdot10^{-3}(T-273.15)]$
PEEK	$0.7158\exp[6.690\cdot10^{-4}(T-273.15)]$	$388.0\exp[-4.124\cdot10^{-3}(T-273.15)]$
PEA	$0.8756\exp[7.241\cdot10^{-4}(T-273.15)]$	$193.2\exp[-4.839\cdot10^{-3}(T-273.15)]$
PEMA	$0.8614\exp[7.468\cdot10^{-4}(T-273.15)]$	$260.9\exp[-5.356\cdot10^{-3}(T-273.15)]$
HDPE	$1.1595 + 8.0394\cdot10^{-4}(T-273.15)$	$179.9\exp[-4.739\cdot10^{-3}(T-273.15)]$
LPE	$0.9172\exp[7.806\cdot10^{-4}(T-273.15)]$	$176.7\exp[-4.661\cdot10^{-3}(T-273.15)]$
HMLPE	$0.8992\exp[8.502\cdot10^{-4}(T-273.15)]$	$168.3\exp[-4.292\cdot10^{-3}(T-273.15)]$
BPE	$0.9399\exp[7.341\cdot10^{-4}(T-273.15)]$	$177.1\exp[-4.699\cdot10^{-3}(T-273.15)]$
LDPE	$1.1944 + 2.841\cdot10^{-4}(T-273.15) + 1.872\cdot10^{-6}(T-273.15)^2$	$202.2\exp[-5.243\cdot10^{-3}(T-273.15)]$
LDPE-A	$1.1484\exp[6.950\cdot10^{-4}(T-273.15)]$	$192.9\exp[-4.701\cdot10^{-3}(T-273.15)]$
LDPE-B	$1.1524\exp[6.700\cdot10^{-4}(T-273.15)]$	$196.6\exp[-4.601\cdot10^{-3}(T-273.15)]$
LDPE-C	$1.1516\exp[6.730\cdot10^{-4}(T-273.15)]$	$106.7\exp[-4.391\cdot10^{-3}(T-273.15)]$
PEO	$0.8766\exp[7.087\cdot10^{-4}(T-273.15)]$	$207.7\exp[-3.947\cdot10^{-3}(T-273.15)]$
PET	$0.6883 + 5.90\cdot10^{-4}(T-273.15)$	$369.7\exp[-4.150\cdot10^{-3}(T-273.15)]$
P4HS	$0.8251 + 6.77\cdot10^{-4}T$	$103.1\exp[-2.417\cdot10^{-3}(T-273.15)]$
PIB	$1.0750\exp[5.651\cdot10^{-4}(T-273.15)]$	$200.3\exp[-4.329\cdot10^{-3}(T-273.15)]$
PI-8	$1.1030 + 6.488\cdot10^{-4}(T-273.15) + 5.125\cdot10^{-7}(T-273.15)^2$	$188.0\exp[-4.541\cdot10^{-3}(T-273.15)]$
PI-14	$1.0943 + 6.293\cdot10^{-4}(T-273.15) + 6.231\cdot10^{-7}(T-273.15)^2$	$202.0\exp[-4.653\cdot10^{-3}(T-273.15)]$
PI-41	$1.0951 + 6.188\cdot10^{-4}(T-273.15) + 6.629\cdot10^{-7}(T-273.15)^2$	$199.0\exp[-4.622\cdot10^{-3}(T-273.15)]$
PI-56	$1.0957 + 6.655\cdot10^{-4}(T-273.15) + 5.661\cdot10^{-7}(T-273.15)^2$	$200.0\exp[-4.644\cdot10^{-3}(T-273.15)]$
PMA	$0.8365\exp[6.795\cdot10^{-4}(T-273.15)]$	$235.8\exp[-4.493\cdot10^{-3}(T-273.15)]$
PMMA	$0.8254 + 2.8383\cdot10^{-4}(T-273.15) + 7.792\cdot10^{-7}(T-273.15)^2$	$287.5\exp[-4.146\cdot10^{-3}(T-273.15)]$
P4MP	$1.4075 - 9.095\cdot10^{-4}(T-273.15) + 3.497\cdot10^{-6}(T-273.15)^2$	$37.67 + 0.2134(T-273.15)] - 7.0445\cdot10^{-4}(T-273.15)^2$
PαMS	$0.89365 + 3.4864\cdot10^{-4}(T-273.15) + 5.0184\cdot10^{-7}(T-273.15)^2$	$297.7\exp[-4.074\cdot10^{-3}(T-273.15)]$
PoMS	$0.9396\exp[5.306\cdot10^{-4}(T-273.15)]$	$261.9\exp[-4.114\cdot10^{-3}(T-273.15)]$
POM	$0.7484\exp[6.770\cdot10^{-4}(T-273.15)]$	$305.6\exp[-4.326\cdot10^{-3}(T-273.15)]$
PH	$0.76644\exp(1.921\cdot10^{-5}T^{1.5})$	$359.9\exp[-4.378\cdot10^{-3}(T-273.15)]$
PSF	$0.7644 + 3.419\cdot10^{-4}(T-273.15) + 3.126\cdot10^{-7}(T-273.15)^2$	$365.9\exp[-3.757\cdot10^{-3}(T-273.15)]$
PAr	$0.73381\exp(1.626\cdot10^{-5}T^{1.5})$	$296.9\exp[-3.375\cdot10^{-3}(T-273.15)]$
aPP	$1.1841 - 1.091\cdot10^{-4}(T-273.15) + 5.286\cdot10^{-6}(T-273.15)^2$	$162.1\exp[-6.604\cdot10^{-3}(T-273.15)]$
iPP	$1.1606\exp[6.700\cdot10^{-4}(T-273.15)]$	$149.1\exp[-4.177\cdot10^{-3}(T-273.15)]$

Polymer	$V(0,T)$/cm^{3}g^{-1}	$B(T)$/MPa
PS	$0.9287 \exp[5.131\cdot10^{-4}(T-273.15)]$	$216.9 \exp[-3.319\cdot10^{-3}(T-273.15)]$
PTFE	$0.3200 + 9.5862\cdot10^{-4}(T-273.15)$	$425.2 \exp[-9.380\cdot10^{-3}(T-273.15)]$
PTHF	$1.0043 \exp[6.691\cdot10^{-4}(T-273.15)]$	$178.6 \exp[-4.223\cdot10^{-3}(T-273.15)]$
PVAc	$0.82496 + 5.820\cdot10^{-4}(T-273.15) + 2.940\cdot10^{-7}(T-273.15)^2$	$204.9 \exp[-4.346\cdot10^{-3}(T-273.15)]$
PVC	$0.7196 + 5.581\cdot10^{-5}(T-273.15) + 1.468\cdot10^{-6}(T-273.15)^2$	$294.2 \exp[-5.321\cdot10^{-3}(T-273.15)]$
PVME	$0.9585 \exp[6.653\cdot10^{-4}(T-273.15)]$	$215.8 \exp[-4.588\cdot10^{-3}(T-273.15)]$
PVdF	$0.5790 \exp[8.051\cdot10^{-4}(T-273.15)]$	$244.0 \exp[-5.210\cdot10^{-3}(T-273.15)]$
SAN3	$0.9233 + 3.936\cdot10^{-4}(T-273.15) + 5.685\cdot10^{-7}(T-273.15)^2$	$239.8 \exp[-4.376\cdot10^{-3}(T-273.15)]$
SAN6	$0.9211 + 4.370\cdot10^{-4}(T-273.15) + 5.846\cdot10^{-7}(T-273.15)^2$	$226.9 \exp[-4.286\cdot10^{-3}(T-273.15)]$
SAN15	$0.9044 + 4.207\cdot10^{-4}(T-273.15) + 4.077\cdot10^{-7}(T-273.15)^2$	$238.4 \exp[-3.943\cdot10^{-3}(T-273.15)]$
SAN18	$0.9016 + 4.036\cdot10^{-4}(T-273.15) + 4.206\cdot10^{-7}(T-273.15)^2$	$240.4 \exp[-3.858\cdot10^{-3}(T-273.15)]$
SAN40	$0.8871 + 3.406\cdot10^{-4}(T-273.15) + 4.938\cdot10^{-7}(T-273.15)^2$	$289.3 \exp[-4.431\cdot10^{-3}(T-273.15)]$
SAN70	$0.8528 + 3.616\cdot10^{-4}(T-273.15) + 2.634\cdot10^{-7}(T-273.15)^2$	$335.4 \exp[-3.923\cdot10^{-3}(T-273.15)]$
SBR10	$0.9053 \exp(2.437\cdot10^{-5}T^{1.5})$	$530.3 \exp[-3.99\cdot10^{-3}(T-273.15)]$
SBR23	$0.8986 \exp(2.317\cdot10^{-5}T^{1.5})$	$551.6 \exp[-4.17\cdot10^{-3}(T-273.15)]$
SBR60	$0.8812 \exp(2.031\cdot10^{-5}T^{1.5})$	$486.0 \exp[-4.34\cdot10^{-3}(T-273.15)]$
SBR85	$0.8704 \exp(1.846\cdot10^{-5}T^{1.5})$	$356.7 \exp[-4.24\cdot10^{-3}(T-273.15)]$
SMMA20	$0.9063 + 3.570\cdot10^{-4}(T-273.15) + 6.532\cdot10^{-7}(T-273.15)^2$	$232.0 \exp[-4.143\cdot10^{-3}(T-273.15)]$
SMMA60	$0.8610 + 3.350\cdot10^{-4}(T-273.15) + 6.980\cdot10^{-7}(T-273.15)^2$	$261.0 \exp[-4.611\cdot10^{-3}(T-273.15)]$
VCES50	$0.6676 + 6.63\cdot10^{-4}T$	$5281.7 \exp[-9.264\cdot10^{-3}(T-273.15)]$
VCHS80	$0.7753 + 6.17\cdot10^{-4}T$	$247.6 \exp[-2.604\cdot10^{-3}(T-273.15)]$
VCHS67	$0.8028 + 6.50\cdot10^{-4}T$	$581.7 \exp[-4.553\cdot10^{-3}(T-273.15)]$
VCHS60	$0.8213 + 6.23\cdot10^{-4}T$	$229.1 \exp[-2.133\cdot10^{-3}(T-273.15)]$
VCHS50	$0.7827 + 5.05\cdot10^{-4}T$	$136.0 \exp[-1.083\cdot10^{-3}(T-273.15)]$
VCHS40	$0.7805 + 4.92\cdot10^{-4}T$	$155.0 \exp[-1.605\cdot10^{-3}(T-273.15)]$
VCHS33	$0.7710 + 4.86\cdot10^{-4}T$	$460.4 \exp[-3.453\cdot10^{-3}(T-273.15)]$
VCHS20	$0.6416 + 5.42\cdot10^{-4}T$	$489.8 \exp[-3.193\cdot10^{-3}(T-273.15)]$
VCOS50	$0.7081 + 7.40\cdot10^{-4}T$	$666.5 \exp[-4.503\cdot10^{-3}(T-273.15)]$
VCPS50	$0.7814 + 4.36\cdot10^{-4}T$	$880.1 \exp[-4.393\cdot10^{-3}(T-273.15)]$

UPPER CRITICAL (UCST) AND LOWER CRITICAL (LCST) SOLUTION TEMPERATURES OF BINARY POLYMER SOLUTIONS

Christian Wohlfarth

Liquid–liquid demixing in solutions of polymers in low molar mass solvents is not a rare phenomenon. Demixing depends on concentration, temperature, pressure, molar mass and molar mass distribution function of the polymer, chain branching and end groups of the polymer, the chemical nature of the solvent, isotope substitution in solvents or polymers, chemical composition of copolymers and its distributions, and other variables. Phase diagrams of polymer solutions can therefore show quite complicated behavior when they have to be considered in detail (see Ref. 1a).

Polymer solutions can undergo demixing when cooling a homogeneous solution as well as when heating such a solution. The corresponding cloud-point curves show a maximum (UCST behavior) or a minimum (LCST behavior). For common polymer solutions, the LCST region is at higher temperatures (in many cases near the critical temperature of the solvent) than the UCST region. The temperature range between both extrema provides the essential information where the one-phase region of a polymer solution can be found. In the case of monodisperse polymers the extrema are equal to the critical points. However, in the case of polydisperse polymers with distribution functions, these extrema are threshold temperatures whereas the critical point shifts to higher concentrations on the shoulder of the cloud-point curve. Usually, the critical concentration is much more strongly influenced than the critical temperature. Thus, the table below does not distinguish between threshold and critical temperatures.

UCST and LCST values depend somewhat on pressure. LCST values in the table are usually given at the vapor pressure of the solvent at this temperature. UCST values are measured in most cases at normal pressure; data at higher pressures are neglected here. The interested reader can find such information, for example, in Refs. 76, 84, 104, 157, 165, 177, 185–187, or 192.

However, UCST and LCST values of a given polymer/solvent pair depend strongly on the molar mass of the polymer. In the case of monodisperse polymers, this dependency can be described in good approximation by the so-called Shultz-Flory plot (see Refs. 6 and 8):

$$\frac{1}{T_{\text{crit}}} = \frac{1}{\theta}\left[1 + const.\left(\frac{1}{\sqrt{r}} - \frac{1}{2r}\right)\right] \tag{1}$$

where r denotes the number of segments of a polymer (being proportional to the degree of polymerization or to the molar mass or molar volume of the polymer). Extrapolation to $r \rightarrow \infty$, i.e., to infinite molar mass, leads to the value of the θ-temperature. This θ-temperature is the highest temperature for UCST behavior or the lowest temperature for LCST behavior and a given polymer/solvent pair. In the case of polydisperse polymers, the segment number in equation (1) is to be replaced by its weight average, r_w (related to M_w). The constant in equation (1) reflects further thermodynamic properties of the given polymer/solvent pair, but should not depend on molar mass. A detailed discussion can be found in Ref. 1b.

The printed table in the *Handbook* provides only one data line for a given polymer/solvent pair and does not show the molar mass dependence of UCST or LCST data. The entire table with all data at different molar masses for many of the systems is given in the electronic version, however. Nevertheless, the necessary molar mass information for a system is always provided in the table by the corresponding number average, M_n, mass average, M_w, or viscosity average, M_η, values of the polymer as given in the original sources.

Polymer	M_n/g mol^{-1}	M_w/g mol^{-1}	M_η/g mol^{-1}	Solvent	UCST/K	LCST/K	Ref.
Acrylonitrile/butadiene copolymer							
(18% Acrylonitrile)			840000	Ethyl acetate		427	220
(26% Acrylonitrile)			1000000	Ethyl acetate		412	220
Butadiene/α-methylstyrene copolymer							
(10% α-Methylstyrene)			100000	Ethyl acetate	387	393	220
Carbon monoxide/ethylene copolymer (1:1, alternating)							
		1000000		1,1,1,3,3,3-Hexafluoro-2-propanol		453	159
Cellulose diacetate			120000	Benzyl alcohol	372		86
	59900	75500		2-Butanone	279.7	471.5	111
	59300			2-Propanone	216.2	438.2	42
Cellulose diacetate/styrene graft copolymer (77.4 wt% grafted polystyrene)							
		750000		*N,N*-Dimethylformamide	262	399	106
		750000		Tetrahydrofuran		363	106
Cellulose nitrate (13.3 wt% N)							
	unknown			2-Propanone	328	182	148
Cellulose triacetate							

Polymer	M_n/g mol^{-1}	M_w/g mol^{-1}	M_η/g mol^{-1}	Solvent	UCST/K	LCST/K	Ref.
			20000	Benzyl alcohol	322		86
		100500		2-Propanone	290.0	472.0	42
Cellulose tricaprylate							
	infinite			*N,N*-Dimethylformamide	413		5
	infinite			3-Phenyl-1-propanol	321		5
Decamethyltetrasiloxane							
	310.69			Tetradecafluorohexane	332.59		195
N,N-Dimethylacrylamide/2-butoxyethyl acrylate copolymer (50 wt% 2-butoxyethyl acrylate)							
				Water		<273.2	164
N,N-Dimethylacrylamide/butyl acrylate copolymer							
(15 wt% Butyl acrylate)				Water		346.2	164
(20 wt% Butyl acrylate)				Water		323.2	164
(30 wt% Butyl acrylate)				Water		294.2	164
(35 wt% Butyl acrylate)				Water		281.2	164
N,N-Dimethylacrylamide/2-ethoxyethyl acrylate copolymer							
(50 wt% 2-Ethoxyethyl acrylate)				Water		319.2	164
(75 wt% 2-Ethoxyethyl acrylate)				Water		285.2	164
N,N-Dimethylacrylamide/ethyl acrylate copolymer							
(25 wt% Ethyl acrylate)				Water		347.2	164
(30 wt% Ethyl acrylate)				Water		334.2	164
(50 wt% Ethyl acrylate)				Water		287.2	164
(55 wt% Ethyl acrylate)				Water		<273.2	164
N,N-Dimethylacrylamide/2-methoxyethyl acrylate copolymer							
(38 mol% 2-Methoxyethyl acrylate)				Water		353	184
(45 mol% 2-Methoxyethyl acrylate)				Water		333	184
(55 mol% 2-Methoxyethyl acrylate)				Water		315	184
(68 mol% 2-Methoxyethyl acrylate)				Water		305	184
(82 mol% 2-Methoxyethyl acrylate)				Water		288	184
(92 mol% 2-Methoxyethyl acrylate)				Water		283	184
N,N-Dimethylacrylamide/methyl acrylate copolymer							
(30 wt% Methyl acrylate)				Water		371.2	164
(40 wt% Methyl acrylate)				Water		338.2	164
(50 wt% Methyl acrylate)				Water		314.2	164
(55 wt% Methyl acrylate)				Water		294.2	164
(60 wt% Methyl acrylate)				Water		279.2	164
(70 wt% Methyl acrylate)				Water		<273.2	164
N,N-Dimethylacrylamide/propyl acrylate copolymer							
(20 wt% Propyl acrylate)				Water		353.2	164
(30 wt% Propyl acrylate)				Water		337.2	164
(40 wt% Propyl acrylate)				Water		294.2	164
(50 wt% Propyl acrylate)				Water		281.2	164
Dimethylsiloxane/methylphenylsiloxane copolymer (15 wt% methylphenylsiloxane)							
	9100	41200		Anisole	291.45		198
	9100	41200		2-Propanone	282.45		198

Polymer	M_n/g mol⁻¹	M_w/g mol⁻¹	M_η/g mol⁻¹	Solvent	UCST/K	LCST/K	Ref.
Ethylene/propylene copolymer (33 mol% ethylene)							
			145000	Cyclohexane		534	101
			145000	Cyclopentane		490	101
			145000	2,2-Dimethylbutane		428	101
			145000	2,3-Dimethylbutane		452	101
			145000	3,4-Dimethylhexane		541	101
			145000	2,2-Dimethylpentane		472	101
			145000	2,3-Dimethylpentane		500	101
			145000	2,4-Dimethylpentane		464	101
			145000	3-Ethylpentane		511	101
			145000	Heptane		502	101
			145000	Hexane		455	101
			145000	2-Methylbutane		396	101
			145000	Methylcyclohexane		558	101
			145000	Methylcyclopentane		512	101
			145000	2-Methylhexane		486	101
			145000	Nonane		558	101
			145000	Octane		528	101
			145000	Pentane		409	101
			145000	2,2,4,4-Tetramethylpentane		539	101
			145000	2,2,3-Trimethylbutane		500	101
			145000	2,2,4-Trimethylpentane		503	101
Ethylene/propylene copolymer (43 mol% ethylene)							
	70000	140000		Hexane		436	127
	70000	140000		2-Methylpentane		474	127
	70000	140000		Pentane		441	127
Ethylene/propylene copolymer (53 mol% ethylene)							
			154000	2,2-Dimethylbutane		407	101
			154000	2,3-Dimethylbutane		437	101
			154000	2,2-Dimethylpentane		453	101
			154000	2,3-Dimethylpentane		488	101
			154000	2,4-Dimethylpentane		445	101
			154000	3-Ethylpentane		500	101
			154000	Heptane		493	101
			154000	Hexane		443	101
			154000	Pentane		395	101
			154000	2,2,3-Trimethylbutane		488	101
			154000	2,3,4-Trimethylhexane		565	101
			154000	2,2,4-Trimethylpentane		484	101
Ethylene/propylene copolymer (63 mol% ethylene)							
			236000	Cyclohexane		526	101
			236000	Cyclopentane		481	101
			236000	2,3-Dimethylbutane		429	101
			236000	3,4-Dimethylhexane		530	101
			236000	2,2-Dimethylpentane		444	101
			236000	2,3-Dimethylpentane		482	101
			236000	2,4-Dimethylpentane		434	101
			236000	3-Ethylpentane		492	101

Polymer	M_n/g mol^{-1}	M_w/g mol^{-1}	M_η/g mol^{-1}	Solvent	UCST/K	LCST/K	Ref.
			236000	Heptane		485	101
			236000	Hexane		436	101
			236000	2-Methylbutane		348	101
			236000	Methylcyclopentane		498	101
			236000	Nonane		547	101
			236000	Octane		512	101
			236000	Pentane		387	101
			236000	2,2,4,4-Tetramethylpentane		528	101
			236000	2,2,3-Trimethylbutane		479	101
			236000	2,2,4-Trimethylpentane		479	101
Ethylene/propylene copolymer (75 mol% ethylene)							
			109000	2,2-Dimethylpentane		431	101
			109000	2,4-Dimethylpentane		425	101
			109000	Heptane		475	101
			109000	Hexane		427	101
			109000	Nonane		542	101
			109000	Octane		509	101
			109000	Pentane		378	101
			109000	2,2,4,4-Tetramethylpentane		523	101
			109000	2,2,4-Trimethylpentane		469	101
Ethylene/propylene copolymer (81 mol% ethylene)							
			195000	Cyclohexane		522	101
			195000	Cyclopentane		474	101
			195000	2,2-Dimethylbutane		381	101
			195000	2,3-Dimethylbutane		413	101
			195000	2,4-Dimethylhexane		478	101
			195000	2,5-Dimethylhexane		466	101
			195000	3,4-Dimethylhexane		522	101
			195000	2,2-Dimethylpentane		425	101
			195000	2,3-Dimethylpentane		471	101
			195000	2,4-Dimethylpentane		420	101
			195000	3-Ethylpentane		478	101
			195000	Heptane		468	101
			195000	Hexane		425	101
			195000	2-Methylbutane		327	101
			195000	Methylcyclohexane		541	101
			195000	Methylcyclopentane		493	101
			195000	2-Methylhexane		453	101
			195000	3-Methylhexane		459	101
			195000	Nonane		540	101
			195000	Octane		506	101
			195000	Pentane		370	101
			195000	2,2,4,4-Tetramethylpentane		519	101
			195000	2,2,3-Trimethylbutane		461	101
			195000	2,2,4-Trimethylpentane		460	101
Ethylene/vinyl acetate copolymer							
(2.3 wt% vinyl acetate)	52000	465000		Diphenyl ether	404.2		143
(4.0 wt% vinyl acetate)	47000	280000		Diphenyl ether	392.5		143
(7.1 wt% Vinyl acetate)	34000	460000		Diphenyl ether	378.2		143

Polymer	M_n/g mol^{-1}	M_w/g mol^{-1}	M_η/g mol^{-1}	Solvent	UCST/K	LCST/K	Ref.
(9.5 wt% Vinyl acetate)	53000	350000		Diphenyl ether	367.3		143
(9.7 wt% Vinyl acetate)	55000	490000		Diphenyl ether	370.8		143
(12.1 wt% Vinyl acetate)	66000	300000		Diphenyl ether	360.4		143
(42.6 mol% Vinyl acetate)	14800	41500		Methyl acetate	307.0		130
Ethylene/vinyl alcohol copolymer							
(87.2 mol% Vinyl alcohol)			infinite	Water	463.55	285.65	44
(88.9 mol% Vinyl alcohol)			infinite	Water	449.15	290.75	44
(91.0 mol% Vinyl alcohol)			infinite	Water	428.45	302.95	44
(94.1 mol% Vinyl alcohol)			infinite	Water	389.25	324.45	44
Ethylene oxide/propylene oxide copolymer							
(20.0 mol% Ethylene oxide)	3400			Water		303	211
(27.0 mol% Ethylene oxide)	3000			Water		309	210
(30.0 mol% Ethylene oxide)	5400			Water		313	211
(38.5 mol% Ethylene oxide)	5000			Water		309	210
(50.0 mol% Ethylene oxide)	3900			Water		323	211
(58.8 mol% Ethylene oxide)	3000			Water		326.65	210
(72.4 mol% Ethylene oxide)		36000		Water		333	153
(79.5 mol% Ethylene oxide)		30800		Water		345	153
(86.6 mol% Ethylene oxide)		30100		Water		355.5	153
Gutta Percha			194000	Propyl acetate	318.95		7
Hydroxypropylcellulose							
		75000		Water		318.45	43
		300000		Water		331.25	43
N-Isopropylacrylamide/acrylamide copolymer							
(15 mol% Acrylamide)		3100000		Water		315.15	172
(30 mol% Acrylamide)		4500000		Water		326.15	172
(45 mol% Acrylamide)		3900000		Water		347.15	172
N-Isopropylacrylamide/1-deoxy-1-methacrylamido-*D*-glucitol							
(12.9 mol% Glucitol)	78000	170000		Water		311.3	218
(13.7 mol% Glucitol)	51600	110000		Water		314.9	218
(14.0 mol% Glucitol)	145000	432000		Water		307.5	218
N-Isopropylacrylamide/*N*-isopropylmethacrylamide copolymer							
(10.56 mol% *N*-Isopropylmethacrylamide)	55300	177000		Water		307.15	212
(30.00 mol% *N*-Isopropylmethacrylamide)	28800	92000		Water		309.75	212
(39.99 mol% *N*-Isopropylmethacrylamide)	23100	74000		Water		311.05	212
(59.89 mol% *N*-Isopropylmethacrylamide)	23100	74000		Water		314.65	212
(79.81 mol% *N*-Isopropylmethacrylamide)	16600	53000		Water		317.35	212
(89.99 mol% *N*-Isopropylmethacrylamide)	14700	47000		Water		318.75	212
Methylcellulose (about 30 mol% methyl substitution)			70000	Water		324.75	47
Methylcellulose/hydroxypropylcellulose copolymer (25 mol% methyl, 8 mol% hydroxypropyl substitution)			80000	Water		340.15	63
Natural rubber		300000		Pentane		403	10
			74500	2-Pentanone	274.45		7
Phenol-formaldehyde resin (acetylated)				2-Ethoxyethanol	378.2		200

Polymer	M_n/g mol^{-1}	M_w/g mol^{-1}	M_η/g mol^{-1}	Solvent	UCST/K	LCST/K	Ref.
Poly(acrylic acid)		120000		Tetrahydrofuran		268.3	189
Poly[bis(2,3-dimethoxypropanoxy) phosphazene]							
	1070000	1500000		Water		317.15	183
Poly[bis(2-(2′-methoxyethoxy)ethoxy) phosphazene]							
	667000	1000000		Water		338.15	183
Poly[bis(2,3-bis(2-methoxyethoxy) propanoxy)phosphazene]							
	714000	1000000		Water		311.15	183
Poly[bis(2,3-bis(2-(2′-methoxyethoxy) ethoxy)propanoxy)phosphazene]							
	1420000	1700000		Water		322.65	183
Poly[bis(2,3-bis(2-(2′-(2″-dimethoxyethoxy) ethoxy)ethoxy)propanoxy)phosphazene]							
	857000	1200000		Water		334.65	183
Poly(1-butene) (atactic)							
	infinite			Anisole	359.4		11
	infinite			Toluene	356.2		28
Poly(1-butene) (isotactic)							
	infinite			Anisole	362.3		11
			530000	Cyclopentane		498	102
			530000	2,2-Dimethylbutane		444	102
			530000	2,5-Dimethylhexane		519	102
			530000	3,4-Dimethylhexane		559	102
			530000	2,3-Dimethylpentane		517	102
			530000	2,4-Dimethylpentane		480	102
			530000	3-Ethylpentane		523	102
			530000	Heptane		509	102
	infinite			Hexane		464	102
			530000	2-Methylbutane		416	102
	infinite			Nonane		564	102
			530000	Octane		540	102
	infinite			Pentane		421	102
			530000	2,2,3-Trimethylbutane		507	102
Poly(butyl methacrylate)							
	278000	470000		1-Butanol	287.15		132
	278000	470000		Decane	357.25		132
	278000	470000		Ethanol	315.25		132
	278000	470000		Heptane	342.55		132
	278000	470000		Octane	345.80		132
	278000	470000		1-Pentanol	286.30		132
	278000	470000		2-Propanol	294.90		132
	278000	470000		2,2,4-Trimethylpentane	347.50		132
Poly(2-chlorostyrene)							
	infinite			Benzene		298	40
Poly(4-chlorostyrene)							
	infinite			Benzene	274.0		22
	infinite			2-(Butoxyethoxy)ethanol		323.25	46
	infinite			Butyl acetate		502.4	22
	infinite			*tert*-Butyl acetate		338.55	46
	infinite			Chlorobenzene	128.8		22

Polymer	M_n/g mol⁻¹	M_w/g mol⁻¹	M_η/g mol⁻¹	Solvent	UCST/K	LCST/K	Ref.
	infinite			2-(Ethoxyethoxy)ethanol		300.95	46
	infinite			Ethyl acetate		613.2	22
	infinite			Ethylbenzene	283.2		22
	infinite			Ethylbenzene	258.45		46
	infinite			Ethyl chloroacetate	271.35		46
	infinite			Isopropyl acetate		348.65	46
	infinite			Isopropylbenzene	332.15		46
	infinite			Isopropyl chloroacetate	264.95		46
	infinite			Methyl chloroacetate	337.75		46
	infinite			Propyl acetate		908.7	22
	infinite			Tetrachloroethene	317.55		46
	infinite			Tetrachloromethane	323.85		46
	infinite			Toluene	236.8		22
Poly(decyl methacrylate)							
	390000	468000		1-Butanol	304.85		113
	390000	468000		1-Pentanol	278.40		113
	220000	252000		2-Propanol	346.85		132
Polydimethylsiloxane (cyclic)							
	9810	10300		2,2-Dimethylpropane		433	133
	9810	10300		Tetramethylsilane		448	133, 171
Polydimethylsiloxane							
			626000	Butane		392.95	53
	infinite			Decane		603	30
	14750	16370		2,2-Dimethylpropane		428	133
	infinite			Dodecane		643	30
			626000	Ethane		259.65	53
			100000	Ethoxybenzene	341.99		108
	infinite			Heptane		528	30
	infinite			Hexadecane		708	30
	infinite			Hexane		493	30
	infinite			Octane		553	30
	infinite			Pentane		453	30
			203000	Propane		340.15	53
	14750	16370		Tetramethylsilane		443	133, 171
Poly(ethyl acrylate)							
			48000	1-Butanol	310.05		27
			48000	Ethanol	301.15		27
			380000	Methanol	287.25		27
			48000	1-Propanol	305.15		27
Polyethylene (branched)							
	8400	32000		Diphenyl ether	384.7		95, 98
	24000	123000		Diphenyl ether	396.7		95, 98
	65000	425000		Diphenyl ether	415.3		95, 98
Polyethylene (linear)							
			20000	Anisole	368.15		24
			20000	Benzyl acetate	459.65		24
			20000	Benzyl phenyl ether	437.15		24
			20000	Benzyl propionate	436.15		24
			50900	Biphenyl	383.55		25
			61100	Butyl acetate	448	497	70
			20000	4-*tert*-Butylphenol	466.15		24

Polymer	M_n/g mol^{-1}	M_w/g mol^{-1}	M_η/g mol^{-1}	Solvent	UCST/K	LCST/K	Ref.
			134000	Cyclohexane		518	101
			20000	Cyclohexanone	389.65		24
			134000	Cyclopentane		472	101
	36700	49300		Decane		563.75	91
			20000	1-Decanol	400.15		24
			20000	Dibenzyl ether	448.65		24
			134000	3,4-Dimethylhexane		515	101
			134000	2,2-Dimethylpentane		399	101
			134000	2,3-Dimethylpentane		463	101
			134000	2,4-Dimethylpentane		395	101
	12000	150000		Diphenyl ether	416.2		95, 98
			97200	Diphenylmethane	400.25		25
	60400	82600		Dodecane		610.85	91
		218000		1-Dodecanol	405.15		141
			134000	3-Ethylpentane		471	101
	36700	49300		Heptane		464.70	91
			20000	1-Heptanol	440.15		24
	36700	49300		Hexane		414.65	91
	7900	92000		1-Hexanol	458.15		154
			20000	2-Methoxynaphthalene	427.65		24
			20000	3-Methylbutyl acetate	407.15		24
			134000	Methylcyclohexane		537	101
			134000	Methylcyclopentane		488	101
	60400	82600		Nonane		531.90	91
			20000	1-Nonanol	431.15		24
			20000	4-Nonylphenol	410.15		24
	36700	49300		Octane		502.40	91
	7900	92000		1-Octanol	426.65		154
			20000	4-Octylphenol	424.65		24
			134000	Pentane		353	101
			20000	1-Pentanol	445.15		24
			175000	Pentyl acetate	421	528	70
			20000	4-*tert*-Pentylphenol	443.65		24
			20000	Phenetole	366.65		24
			134000	2,2,4,4-Tetramethylpentane		513	101
	60400	82600		Tridecane		639.30	91
			134000	2,2,3-Trimethylbutane		444	101
			134000	2,3,4-Trimethylhexane		545	101
			134000	2,2,4-Trimethylpentane		495	101
	97700	135900		Undecane		583.95	91
Poly(ethylene glycol)							
			8000	*tert*-Butyl acetate	321.2	464.2	83
			21200	*tert*-Butyl acetate	353.2	431.2	83
	6100	6200		Water		404.79	185
	10457	11615		Water		394.33	205
	40800	151000		Water		378.25	205
Poly(ethylene oxide)-b-poly[bis(methoxyethoxyethoxy)-phosphazene] block copolymer (about 67 mol% Ethylene oxide)							
	22000	31500		Water		338	222

Polymer	M_n/g mol⁻¹	M_w/g mol⁻¹	M_η/g mol⁻¹	Solvent	UCST/K	LCST/K	Ref.
Poly(ethylene oxide)-b-poly(propylene oxide)-b-poly(ethylene oxide) triblock copolymer (about 30 mol% Ethylene oxide)							
		4400		Water		286.65	209
Polyethylethylene							
	48000	52000		Diphenyl ether	411.2		95, 98
Poly(*p*-hexylstyrene)							
	infinite			2-Butanone	302.6		135
Poly(2-hydroxyethyl methacrylate)							
			77400	1-Butanol	337.25		35
			233600	2-Butanol	287		35
			233600	2-Metyl-1-propanol	342		35
			77400	1,2,3-Propanetriol	345		35
			77400	1-Propanol	311		35
Polyisobutylene							
	infinite			Anisole	377		3
			72000	Benzene		540.5	39
			703000	Butane		264.75	53
	infinite			Cycloheptane		572	34
			1500000	Cyclohexane		412	10
	infinite			Cyclooctane		637	34
			1500000	Cyclopentane		344	10
	infinite			Decane		535	34
			1500000	2,2-Dimethylbutane		376	10
			1500000	2,3-Dimethylbutane		404	10
	infinite			2,2-Dimethylhexane		454	34
	infinite			2,4-Dimethylhexane		458	34
	infinite			2,5-Dimethylhexane		446	34
	infinite			3,4-Dimethylhexane		497	34
	infinite			2,2-Dimethylpentane		404	34
	infinite			2,3-Dimethylpentane		451	34
	infinite			2,4-Dimethylpentane		403	34
	infinite			3,3-Dimethylpentane		451	34
	infinite			Diphenyl ether	306		3
	infinite			Decane		585	30
	infinite			Dodecane		582	34
	infinite			Ethylbenzene	249		3
	infinite			Ethylcyclopentane		524	34
	infinite			Ethyl heptanoate	306		3
	infinite			Ethyl hexanoate	330		3
	infinite			3-Ethylpentane		458	34
	infinite			Heptane		442	34
			72000	Hexane		428.5	39
			6030	2-Methylbutane		357.85	53
	infinite			Methylcyclohexane		526	34
	infinite			Methylcyclopentane		478	34
	infinite			2-Methylheptane		466	34
	infinite			3-Methylheptane		478	34
	infinite			2-Methylhexane		426	34
	infinite			3-Methylhexane		446	34
	infinite			2-Methylpentane		376	34

Polymer	M_n/g mol^{-1}	M_w/g mol^{-1}	M_η/g mol^{-1}	Solvent	UCST/K	LCST/K	Ref.
	infinite			3-Methylpentane		405	34
			470	2-Methylpropane		387	10
			72000	Octane		506.0	39
			6030	Pentane		403.55	53
			72000	Pentane		373.5	39
	infinite			Phenetole	357		3
			470	Propane		358	10
	infinite			Propylcyclopentane		547	34
	infinite			Toluene	260		3
	infinite			2,2,3-Trimethylbutane		445	34
	infinite			2,2,4-Trimethylpentane		435	34
1,4-*cis*-Polyisoprene							
			780000	2,5-Dimethylhexane		474.15	140
			780000	3,4-Dimethylhexane		520.15	140
			780000	2,2-Dimethylpentane		445.15	140
			780000	2,3-Dimethylpentane		484.15	140
			780000	2,4-Dimethylpentane		442.15	140
			780000	3-Methylpentane		483.15	140
			780000	Heptane		488.15	140
			780000	Hexane		434.15	140
			780000	Nonane		541.15	140
			780000	Octane		509.15	140
			780000	2,2,4,4-Tetramethylpentane		518.15	140
			780000	2,3,4-Trimethylhexane		548.15	140
			780000	2,2,4-Trimethylpentane		471.15	140
1,4-*trans*-Polyisoprene							
			180000	2,5-Dimethylhexane		451.15	140
			180000	3,4-Dimethylhexane		521.15	140
			180000	2,2-Dimethylpentane		405.15	140
			180000	2,3-Dimethylpentane		460.15	140
			180000	2,4-Dimethylpentane		404.15	140
			180000	3-Methylpentane		473.15	140
			180000	Heptane		467.15	140
			180000	Hexane		407.15	140
			180000	Nonane		540.15	140
			180000	Octane		503.15	140
			180000	2,2,4,4-Tetramethylpentane		519.15	140
			180000	2,3,4-Trimethylhexane		548.15	140
Poly(*N*-isopropylacrylamide)							
	5400	14000		Water		307.45	146
	146000	530000		Water		305.85	146
Poly(*N*-isopropylacrylamide)-poly[(*N*-acetylimino)ethylene] block copolymer (80 wt% *N*-Isopropylacrylamide)							
	5500			Water		306.2	223
Poly(*N*-isopropylacrylamide)-poly[(*N*-acetylimino)ethylene] graft copolymer (75 wt% *N*-Isopropylacrylamide)							
	6030			Water		306.2	223
Poly(*N*-isopropylmethacrylamide)							
	6250	20000		Water		319.95	212
Poly(methyl methacrylate)							

Polymer	M_n/g mol^{-1}	M_w/g mol^{-1}	M_η/g mol^{-1}	Solvent	UCST/K	LCST/K	Ref.
			127000	Acetonitrile	267.15		16
			970000	Acetonitrile	303.15		16
			50000	1-Butanol	353.25		2
	infinite			2-Butanone		482	80
	infinite			1-Chlorobutane	320	463	80
			970000	2,2-Dimethyl-3-pentanone	301.55		16
			127000	2,4-Dimethyl-3-pentanone	280.15		16
	200000	264000		2-Ethoxyethanol	312.15		196
		77000		Ethyl acetate	290	533	190
			127000	2-Ethylbutanal	264.65		16
	infinite			3-Heptanone	307.7		126
			970000	4-Heptanone	299.95		16
	infinite			3-Hexanone		522	80
	infinite			Methyl acetate		451	80
			50000	1-Methyl-4-isopropylbenzene	400.15		2
			1400000	2-Octanone	321.15		16
	572400	595300		3-Octanone	329.88		166
	infinite			3-Pentanone		506	80
			50000	1-Propanol	349.95		2
	infinite			2-Propanone		439	80
	200000	264000		Tetra(ethylene glycol)	390.15		196
			400000	Toluene	225.35		2
			50000	Trichloromethane	231.15		2
	200000	264000		Tri(ethylene glycol)	407.15		196
Poly(methyl methacrylate) (isotactic)							
	infinite			Acetonitrile	301	461	80
	infinite			2-Butanone		464	80
	infinite			1-Chlorobutane	309	454	80
	infinite			4-Heptanone	319	522	80
	infinite			3-Hexanone	279	511	80
	infinite			Methyl acetate		441	80
	infinite			3-Pentanone		497	80
	infinite			2-Propanone		428	80
Poly(4-methyl-1-pentene) (isotactic)							
			152000	Butane		388	102
			152000	Cyclopentane		505	102
			152000	2,2-Dimethylbutane		462	102
			152000	2,2-Dimethylpentane		499	102
			152000	2,4-Dimethylpentane		499	102
	infinite			Diphenyl	467.8		62
	infinite			Diphenyl ether	483.2		62
	infinite			Diphenylmethane	449.8		62
			152000	3-Ethylpentane		532	102
			152000	Heptane		522	102
			152000	Hexane		487	102
			152000	2-Methylbutane		431	102
			152000	Nonane		579	102
			152000	Octane		553	102
			152000	Pentane		441	102
			152000	2,2,3-Trimethylbutane		521	102

Polymer	M_n/g mol^{-1}	M_w/g mol^{-1}	M_η/g mol^{-1}	Solvent	UCST/K	LCST/K	Ref.
Poly(α-methylstyrene)							
	58500	61400		Butyl acetate	262.05	457.15	181
	99100	113000		Cyclohexane	293.55		152
	26000	31200		Cyclopentane	276.7	435.95	181
		289000		*trans*-Decahydronaphthalene	273		181
	69500	76500		Hexyl acetate	285.05	508.15	181
	72000	75600		Methylcyclohexane	328.9		203
	58500	61400		Pentyl acetate	287.1	484.6	181
Poly(2-methyl-5-vinylpyridine)							
			600000	Butyl acetate	287.95		20
			263000	Ethyl butyrate	319.05		20
			335000	Ethyl propionate	293.55		20
			275000	3-Methylbutyl acetate	314.75		20
			335000	4-Methyl-2-pentanone	299.95		20
			170000	2-Methylpropyl acetate	312.35		20
			165000	Pentyl acetate	316.95		20
			284000	Propionitrile	262.35		20
			152000	Propyl acetate	282.65		20
			181000	Propyl propionate	312.15		20
			233000	Tetrahydronaphthalene	316.95		20
Poly(1-pentene) (isotactic)							
			4500000	Cyclopentane		502	102
			4500000	2,2-Dimethylbutane		457	102
			4500000	3,4-Dimethylhexane		>569	102
			4500000	2,2-Dimethylpentane		502	102
			4500000	2,3-Dimethylpentane		529	102
			4500000	2,4-Dimethylpentane		493	102
			4500000	3-Ethylpentane		537	102
			4500000	Heptane		522	102
			4500000	Hexane		482	102
			4500000	2-Methylbutane		422	102
			4500000	Octane		556	102
			4500000	Pentane		433	102
			4500000	2,2,4-Trimethylpentane		527	102
Polypropylene (atactic)							
	infinite			Diphenyl ether	426.5		9
	infinite			Diethyl ether		383	68
			242000	Heptane		511	101
	infinite			Hexane		441	68
			242000	2-Methylbutane		413	101
			242000	Methylcyclohexane		564	101
	infinite			Pentane		397	68
Polypropylene (isotactic)							
			28000	Benzyl phenyl ether	429.2		31
			28000	Benzyl propionate	405.2		31
			28000	1-Butanol	395.2		31
			28000	4-*tert*-Butylphenol	413.2		31
			242000	Cyclohexane		540	101
			242000	Cyclopentane		495	101
			28000	Dibenzyl ether	433.2		31

Polymer	M_n/g mol⁻¹	M_w/g mol⁻¹	M_η/g mol⁻¹	Solvent	UCST/K	LCST/K	Ref.
			242000	2,2-Dimethylbutane		441	101
			242000	2,3-Dimethylbutane		465	101
			242000	3,4-Dimethylhexane		553	101
			242000	2,2-Dimethylpentane		489	101
			242000	2,3-Dimethylpentane		513	101
			242000	2,4-Dimethylpentane		481	101
			28000	Diphenyl	388.2		31
			28000	Diphenyl ether	395.2		31
			28000	Diphenylmethane	389.7		31
			242000	3-Ethylpentane		520	101
			28000	4-Ethylphenol	457.2		31
			242000	Heptane		511	101
			242000	Hexane		470	101
			242000	2-Methylbutane		413	101
			28000	3-Methylbutyl benzyl ether	384.2		31
			242000	Methylcyclohexane		564	101
			242000	Methylcyclopentane		518	101
			28000	4-Methylphenol	479.2		31
			28000	2-Methyl-1-propanol	395.2		31
			242000	Nonane		571	101
			242000	Octane		542	101
			28000	4-Octylphenol	379.2		31
			28000	4-Isooctylphenol	383.2		31
			242000	Pentane		422	101
			242000	2,2,4,4-Tetramethylpentane		548	101
			242000	2,2,3-Trimethylbutane		511	101
			242000	2,3,4-Trimethylhexane		585	101
			242000	2,2,4-Trimethylpentane		510	101
Poly(propylene glycol)							
	1000			Hexane	288.15		88
	575			Water		318.2	65
Polystyrene							
	34900	37000		Benzene		538.7	61
			62600	Butanedioic acid dimethyl ester	335.15		2
	3700	4000		1-Butanol	383.45		154
	91700	97200		2-Butanone		448.8	61
	545500	600000		Butyl acetate		489	181
	104000	110000		*tert*-Butyl acetate	250.0	417.9	74
			62600	Butyl stearate	387.15		2
	18400	19200		1-Chlorododecane	274.65		154
	18400	19200		1-Chlorohexadecane	337.05		154
	18400	19200		1-Chlorooctadecane	365.55		154
	18400	19200		1-Chlorotetradecane	309.35		154
	46400	51000		Cyclodecane	278.9		128
	46400	51000		Cycloheptane	276.2		128
	34900	37000		Cyclohexane	285.6	510.9	60
			236000	Cyclohexanol	353.5		8
	46400	51000		Cyclooctane	275.2		128
	91700	97200		Cyclopentane	275.2	445.5	61

Polymer	M_n/g mol⁻¹	M_w/g mol⁻¹	M_η/g mol⁻¹	Solvent	UCST/K	LCST/K	Ref.
	91500	97000		*trans*-Decahydronaphthalene	281.95		81
		4800		Decane	360.95		154
	3700	4000		1-Decanol	375.15		154
			570000	Decyl acetate		650	64
	18700	19800		Diethyl ether	235.6	314.5	51
	187000	200000		Diethyl malonate	285.8	589.6	74
	47200	50000		Diethyl oxalate	280.05		131
	151000	160000		Dimethoxymethane		401.2	51
		240000		1,4-Dimethylcyclohexane	387	482	116
			62600	Dimethyl malonate	409.15		2
			62600	Dimethyl oxalate	453.15		2
	116000	123000		Dodecadeuterocyclohexane	298.10		224
		25000		Dodecadeuteromethyl-cyclopentane	310.07		180
		4800		Dodecane	368.65		154
	3700	4000		1-Dodecanol	379.75		154
	infinite			Dodecyl acetate	285.2		206
	104000	110000		Ethyl acetate	213.9	435.4	72
	104000	110000		Ethyl butanoate		490.8	74
	221000	239000		Ethylcyclohexane	330.52		18
	9440	10000		Ethyl formate	272	451	74
		900000		Bis(2-ethylhexyl) phthalate	283.05		136
	4530	4800		Heptane	359	477	112
	3700	4000		1-Dexadecanol	386.25		154
	5500	5770		1,1,1,3,3,3-Hexadeutero-2-propanone	270	436	157
	1920	2030		Hexane	318	470	112
			62600	Hexanoic acid	448.15		2
	3700	4000		1-Hexanol	372.15		154
			62600	3-Hexanol	396.65		2
			90000	Hexyl acetate		578	64
	104000	110000		Methyl acetate	284.2	415.7	72
	104000	110000		3-Methyl-1-butyl acetate	210.1	510.1	72
	91700	97200		Methylcyclohexane	321.8	505.9	60
	10750	11500		Methylcyclopentane	295	480	157
	104000	110000		2-Methyl-1-propyl acetate	210.4	468.5	72
		48000		Nitroethane	303.1		151
		4800		Octadecane	403.55		154
	3700	4000		1-Octadecanol	390.55		154
	4530	4800		Octane	353	527	112
	3700	4000		1-Octanol	372.35		154
			62600	1-Octene	355.15		2
		4800		Pentadecane	385.25		154
		1100		Pentane	292		137
	3700	4000		1-Pentanol	375.05		154
	219800	233000		Pentyl acetate		519	181
		100000		1-Phenyldecane	283.60		105
	5500	5770		2-Propanone	251	452	157
	12750	13500		Propionitrile	312		187
	104000	110000		Propyl acetate	183.7	469.0	72
	104000	110000		2-Propyl acetate	220.9	414.2	72

Polymer	M_n/g mol^{-1}	M_w/g mol^{-1}	M_η/g mol^{-1}	Solvent	UCST/K	LCST/K	Ref.
	3700	4000		1-Tetradecanol	383.25		154
	34900	37000		Toluene		567.2	60
			62600	Vinyl acetate	384.15		2
Polystyrene (three-arm star)							
		230000		Cyclohexane	297.1	496.8	93
Polystyrene (four-arm star)							
		155000		Cyclohexane	294.13		199
Poly(trimethylene oxide)							
		infinite		Cyclohexane	300		79
Poly(vinyl alcohol)							
		40000		Water		514	45
Poly(*N*-vinyl caprolactam)							
		150000		Water		306.45	217
Poly(vinyl chloride)							
	55000			Dibutyl phthalate	353		114
	55000			Tricresyl phosphate	383		114
			85000	Dimethyl phthalate	355		219
Poly(*N*-vinylisobutyramide)							
	66000	105600		Water		313.25	208
Poly(vinyl methyl ether)							
	46500	98600		Deuterium oxide		307.2	173
	83000	155000		Water		306.95	146
Poly(*N*-vinyl-*N*-propylacetamide)							
			30000	Water		313.5	176
Styrene/acrylonitrile copolymer							
(21.1 wt% acrylonitrile)	infinite			Toluene	325.4		52
(23.2 wt% Acrylonitrile)	infinite			Toluene	355.1		52
(25.0 wt% Acrylonitrile)	90000	147000		Toluene	313.15		198
(51.0 wt% Acrylonitrile)		347000		Ethyl acetate		344.15	107
Styrene/methyl methacrylate copolymer (52.0 mol% Styrene)							
		infinite		Cyclohexanol	334.65		38
Styrene/α-methylstyrene copolymer (20.0 mol% Styrene)							
	100000	114000		Butyl acetate	288.85	453.05	181
	100000	114000		Cyclohexane	285.85	484.85	181
	100000	114000		Cyclopentane	290.95	421.05	181
	100000	114000		*trans*-Decahydronaphthalene	264.15		181
	100000	114000		Hexyl acetate	288.55	514.15	181
	100000	114000		Pentyl acetate	303.15	480.65	181
Trifluoronitrosomethane/tetrafluoroethylene copolymer (1:1) alternating							
	infinite			1,1,2-Trichloro-1,2,2-trifluoroethane	301.6		12
N-Vinylacetamide/vinyl acetate copolymer							
(58 mol% Vinyl acetate)	30000	57000		Water		340.15	225
(63 mol% Vinyl acetate)	27000	48600		Water		323.15	225
(78 mol% Vinyl acetate)	26000	46800		Water		282.15	225
Vinyl alcohol/vinyl butyrate copolymer							
(7.5 mol% Butyralized PVA)	infinite			Water	408.0	298.25	121

Polymer	M_n/g mol^{-1}	M_w/g mol^{-1}	M_η/g mol^{-1}	Solvent	UCST/K	LCST/K	Ref.
N-Vinylcaprolactam/*N*-vinylamine copolymer (3.8 mol% Vinyl amine)			160000	Water		308.8	176
N-Vinylformamide/vinyl acetate copolymer							
(60 mol% Vinyl acetate)	24000	45600		Water		310.15	225
(66 mol% Vinyl acetate)	25000	47500		Water		291.15	225
(73 mol% Vinyl acetate)	23000	50600		Water		277.15	225

References

1a. Koningsveld, R., Stockmayer, W.H., and Nies, E., *Polymer Phase Diagrams*, Oxford University Press, Oxford, 2001.
1b. Kamide, K., *Thermodynamics of Polymer Solutions*, Elsevier, Amsterdam, 1990.
2. Jenckel, E. and Gorke, K., *Z. Naturforsch.*, 5a, 317, 556, 1950.
3. Fox, T.G. and Flory, P.J., *J. Amer. Chem. Soc.*, 73, 1909, 1951.
4. Fox, T.G. and Flory, P.J., *J. Amer. Chem. Soc.*, 73, 1915, 1951.
5. Mandelkern, L. and Flory, P.J., *J. Amer. Chem. Soc.*, 74, 2517, 1952.
6. Shultz, A.R. and Flory, P.J., *J. Amer. Chem. Soc.*, 74, 4760, 1952.
7. Wagner, H.L. and Flory, P.J., *J. Amer. Chem. Soc.*, 74, 195, 1952.
8. Shultz, A.R.and Flory, P.J., *J. Amer. Chem. Soc.*, 75, 3888, 1953.
9. Kinsinger, J.B. and Wessling, R.A., *J. Amer. Chem. Soc.*, 81, 2908, 1959.
10. Freeman, P.I. and Rowlinson, J.S., *Polymer*, 1, 20, 1960.
11. Krigbaum, W.R., Kurz, J.E., and Smith, P., *J. Phys. Chem.*, 65, 1984, 1961.
12. Morneau, G.A., Roth, P.I., and Shultz, A.R., *J. Polym. Sci.*, 55, 609, 1961.
13. Debye, P., Coll, H., and Woermann, D., *J. Chem. Phys.*, 32, 939, 1960.
14. Debye, P., Coll, H., and Woermann, D., *J. Chem. Phys.*, 33, 1746, 1960.
15. Debye, P., Chu, B., and Woermann, D., *J. Chem. Phys.*, 36, 1803, 1962.
16. Fox, T.G., *Polymer*, 3, 111, 1962.
17. Ham, J.S., Bolen, M.C., and Hughes, J.K., *J. Polym. Sci.*, 57, 25, 1962.
18. Debye, P., Woermann, D., and Chu, B., *J. Polym. Sci.: Part A*, 1, 255, 1963.
19. Allen, G. and Baker, C.H., *Polymer*, 6, 181, 1965.
20. Gechele, G.B., Crescentini, L., *J. Polym. Sci.: Part A*, 3, 3599, 1965.
21. Myrat, C.D. and Rowlinson, J.S., *Polymer*, 6, 645, 1965.
22. Kubo, K. and Ogino, K., *Sci. Pap. Coll. Art. Sci. Univ. Tokyo*, 16, 193, 1966.
23. Rehage, G., Moeller, D., and Ernst, O., *Makromol.Chem.*, 88, 232, 1965.
24. Nakajima, A., Fujiwara, H., and Hamada, F., *J. Polym. Sci.: Part A-2*, 4, 507, 1966.
25. Nakajima, A., Hamada, F., and Hayashi, S., *J. Polym. Sci.: Part C*, 15, 285, 1966.
26. Koningsveld, R., *Proefschrift Univ. Leiden*, Heerlen, 1967.
27. Llopis, J., Albert, A., and Usobinaga P., *Eur. Polym. J.*, 3, 259, 1967.
28. Moraglio, G., Gianotti, G., and Danusso, F., *Eur. Polym. J.*, 3, 251, 1967.
29. Orwoll, R.A. and Flory, P.J., *J. Amer. Chem. Soc.*, 89, 6822, 1967.
30. Patterson, D., Delmas, G., and Somcynsky, T., *Polymer*, 8, 503, 1967.
31. Nakajima, A. and Fujiwara, H., *J. Polym. Sci.: Part A-2*, 6, 723, 1968.
32. Rehage, G. and Koningsveld, R., *J. Polym. Sci.: Polym. Lett.*, 6, 421, 1968.
33. Andreeva, V. M., et al., *Vysokomol. Soedin., Ser. B*, 11, 555, 1969.
34. Bardin, J.-M. and Patterson, D., *Polymer*, 10, 247, 1969.
35. Dusek, K., *Coll. Czech. Chem. Commun.*, 34, 3309, 1969.
36. Delmas, G. and Patterson, D., *J. Polym. Sci.: Part C*, 30, 1, 1970.
37. Koningsveld, R., Kleintjens, L.A., and Shultz, A.R., *J. Polym. Sci.: Part A-2*, 8, 1261, 1970.
38. Kotaka, T., et al., *Polym. J.*, 1, 245, 1970.
39. Liddell, A.H. and Swinton, F.L., *Discuss. Faraday Soc.*, 49, 115, 1970.
40. Matsumura, K., *Polym. J.*, 1, 322, 1970.
41. Nakayama, H., *Bull. Chem. Soc. Japan*, 43, 1683, 1970.
42. Cowie, J.M.G., Maconnachie, A., and Ranson, R.J., *Macromolecules*, 4, 57, 1971.
43. Kagemoto, A. and Baba, Y., *Kobunshi Kagaku*, 28, 784, 1971.
44. Shibatani, K. and Oyanagi, Y., *Kobunshi Kagaku*, 28 (1971) 361-367.
45. Tager, A.A., et al., *Vysokomol. Soedin., Ser. A*, 13, 659, 1971.
46. Izumi, Y. and Miyake, Y., *Polym. J.*, 3, 647, 1972.
47. Kagemoto, A., Baba, Y., and Fujishiro, R., *Makromol. Chem.*, 154, 105, 1972.
48. Kennedy, J.W., Gordon, M., and Koningsveld, R., *J. Polym. Sci.: Part C*, 39, 43, 1972.
49. Lirova, B.I., et al., *Vysokomol. Soedin., Ser. B*, 14,265, 1972.
50. Nakayama, H., *Bull. Chem. Soc. Japan*, 45, 1371, 1972.
51. Siow, K.S., Delmas, G., and Patterson, D., *Macromolecules*, 5, 29, 1972.
52. Teramachi, S. and Fujikawa, T., *J. Macromol. Sci.-Chem. A*, 6, 1393, 1972.
53. Zeman, L., Biros, J., Delmas, G., and Patterson, D., *J. Phys. Chem.*, 76, 1206, 1972.
54. Zeman, L. and Patterson, D., *J. Phys. Chem.*, 76, 1214, 1972.
55. Baba, Y., Fujita, Y., and Kagemoto, A., *Makromol. Chem.*, 164, 349, 1973.
56. Candau, F., Strazielle, C., and Benoit, H., *Makromol. Chem.*, 170, 165, 1973.
57. Hamada, F., Fujisawa, K., and Nakajima, A., *Polym. J.*, 4, 316, 1973.
58. Kuwahara, N., Nakata, M., and Kaneko, M., *Polymer*, 14, 415, 1973.
59. Kuwahara, N., Kojima, J., and Kaneko, M., *J. Polym. Sci.: Polym. Phys. Ed.*, 11, 2307, 1973.
60. Saeki, S., Kuwahara, N., Konno, S., and Kaneko, M., *Macromolecules*, 6, 246, 1973.
61. Saeki, S., Kuwahara, N., Konno, S., and Kaneko, M., *Macromolecules*, 6, 589, 1973.
62. Tani, S., Hamada, F., and Nakajima, A., *Polym.J.*, 5, 86, 1973.
63. Baba, Y. and Kagemoto, A., *Kobunshi Ronbunshu*, 31, 446, 1974.
64. Bataille, P., *J. Chem. Eng. Data*, 19, 224, 1974.
65. Bessonov, Yu.S. and Tager, A.A., *Trud. Khim. Khim. Tekhnol.*, 1, 150, 1974.
66. Cowie, J.M.G. and McEwen, L.J., *J. Chem. Soc., Faraday Trans. I*, 70, 171, 1974.
67. Cowie, J.M.G. and McEwen, I.J., *Macromolecules*, 7, 291, 1974.
68. Cowie, J.M.G. and McEwen, I.J., *J. Polym. Sci.: Polym. Phys. Ed.*, 12, 441, 1974.
69. Derham, K.W., Goldsbrough, J., and Gordon, M., *Pure Appl. Chem.*, 38, 97, 1974.
70. Kuwahara, N., Saeki, S., Chiba, T., and Kaneko, M., *Polymer*, 15, 777, 1974.
71. Nakajima, A., et al., *Makromol. Chem.*, 175, 197, 1974.
72. Saeki, S., Konno, S., Kuwahara, N., Nakata, M., and Kaneko, M., *Macromolecules*, 7, 521, 1974.
73. Ver Strate, G. and Philippoff, W., *J. Polym. Sci.: Polym. Lett. Ed.*, 12, 267, 1974.
74. Konno, S., et al., *Macromolecules*, 8, 799, 1975.
75. Nakata, M., Kuwahara, N., and Kaneko, M., *J. Chem. Phys.*, 62, 4278, 1975.
76. Saeki, S., Kuwahara, N., Nakata, M., and Kaneko, M., *Polymer*, 16, 445, 1975.

77. Strazielle, C. and Benoit, H., *Macromolecules*, 8, 203, 1975.
78. Tager, A. A., et al., *Vysokomol. Soedin., Ser. B*, 17, 61, 1975.
79. Chiu, D.S., Takahashi, Y., and Mark, J.E., *Polymer*, 17, 670, 1976.
80. Cowie, J.M.G. and McEwen, I.J., *J. Chem. Soc., Faraday Trans. I*, 72, 526, 1976.
81. Nakata, M., et al., *J. Chem. Phys.*, 64, 1022, 1976.
82. Nose, T. and Tan, T.V., *J. Polym. Sci.: Polym. Lett. Ed.*, 14, 705, 1976.
83. Saeki, S., Kuwahara, N., Nakata, M., and Kaneko, M., *Polymer*, 17, 685, 1976.
84. Saeki, S., Kuwahara, N., and Kaneko, M., *Macromolecules*, 101, 1976.
85. Slagowski, E., Tsai, B., and McIntyre, D., *Macromolecules*, 9, 687, 1976.
86. Panina, N.I., Lozgacheva, V.P., and Aver'yanova, V.M., *Vysokomol. Soedin., Ser. B*, 19, 786,.1977.
87. Rigler, J.K., Wolf, B.A., and Breitenbach, J.W., *Angew. Makromol. Chem.*, 57, 15, 1977.
88. Vshivkov, S.A., et al., *Prots. Studneobras. Polimern. Sistem.*, (2), 3, 1977.
89. Wolf, B.A. and Jend, R., *Makromol. Chem.*, 178, 1811, 1977.
90. Wolf, B.A. and Sezen, M.C., *Macromolecules*, 10, 1010, 1977.
91. Kodama, Y. and Swinton, F.L., *Brit. Polym. J.*, 10, 191, 1978.
92. Nakata, M., Dobashi, T., Kuwahara, N., Kaneko, M., and Chu, B., *Phys. Rev. A*, 18, 2683, 1978.
93. Cowie, J.M.G., Horta, A., McEwen, I.J., and Prochazka, K., *Polym. Bull.*, 1, 329, 1979.
94. Hamano, K., Kuwahara, N., and Kaneko, M., *Phys. Rev. A*, 20, 1135, 1979.
95. Kleintjens, L.A.L., *Ph.D. Thesis*, Univ. Essex, U.K., 1979.
96. Dobashi, T., Nakata, M., and Kaneko, M., *J. Chem. Phys.*, 72, 6685, 1980.
97. Irvine, P. and Gordon, M., *Macromolecules*, 13, 761, 1980.
98. Kleintjens, L.A., Koningsveld, R., and Gordon, M., *Macromolecules*, 13, 303, 1980.
99. Lang, J.C. and Morgan, R.D., *J. Chem. Phys.*, 73, 5849, 1980.
100. Richards, R.W., *Polymer*, 21, 715, 1980.
101. Charlet, G. and Delmas, G., *Polymer*, 22, 1181, 1981.
102. Charlet, G., Ducasse, R., and Delmas, G., *Polymer*, 22, 1190, 1981.
103. Hashizume, J., Teramoto, A., and Fujita, H., *J. Polym. Sci.: Polym. Phys. Ed.*, 19, 1405, 1981.
104. Wolf, B.A. and Geerissen, H., *Colloid Polym. Sci.*, 259, 1214, 1981.
105. Geerissen, H. and Wolf, B.A., *Makromol. Chem., Rapid Commun.*, 3, 17, 1982.
106. Goloborod'ko, V.I., Valatin, S.M., and Tashmukhamedov, I.P., *Uzb. Khim. Zh.*, (3), 33, 1982.
107. Mangalam, P. V. and Kalpagam, V., *J. Polym. Sci.: Polym. Phys. Ed.*, 20, 773, 1982.
108. Shinozaki, K., Abe, M., and Nose, T., *Polymer*, 23, 722, 1982.
109. Shinozaki, K., Van Tan, T., Saito, Y., and Nose, T., *Polymer*, 23, 728, 1982.
110. Suzuki, H., Kamide, K., and Saitoh, M., *Eur. Polym. J.*, 18, 123, 1982.
111. Suzuki, H., Muraoka, Y., Saitoh, M., and Kamide, K., *Brit. Polym. J.*, 14, 23, 1982.
112. Cowie, J.M.G. and McEwen, I.J., *Polymer*, 24, 1445, 1983.
113. Herold, F.K., Schulz, G.V., and Wolf, B.A., *Materials Chem. Phys.*, 8, 243, 1983.
114. Tager, A.A., et al., *Vysokomol. Soedin., Ser. A*, 25, 1444, 1983.
115. Corti, M., Minero, C., and Degiorgio, V., *J. Phys. Chem.*, 88, 309, 1984.
116. Cowie, J.M.G. and McEwen, I.J., *Polymer*, 25, 1107, 1984.
117. Dobashi, T., Nakata, M., and Kaneko, M., *J. Chem. Phys.*, 80, 948, 1984.
118. Florin, E., Kjellander, R., and Eriksson, J.C., *J. Chem. Soc., Faraday Trans. I*, 80, 2889, 1984.
119. Gilluck, M., *Dissertation*, TH Leuna-Merseburg, 1984.
120. Rangel-Nafaile, C., Metzner, A.B., and Wissbrun, K.F., *Macromolecules*, 17, 1187, 1984.
121. Shiomi, T., et al., *J. Polym. Sci.: Polym. Phys. Ed.*, 22, 1305, 1984.
122. Tsuyumoto, M., Einaga, Y., and Fujita, H., *Polym. J.*, 16, 229, 1984.
123. Varennes, S., Charlet, G., and Delmas, G., *Polym. Eng. Sci.*, 24, 98, 1984.
124. Hamano, K., Kuwahara, N., Koyama, T., and Harada, S., *Phys. Rev. A*, 32, 3168, 1985.
125. Kraemer, H. and Wolf, B.A., *Makromol. Chem., Rapid Commun.*, 6, 21, 1985.
126. Herold, F.K. and Wolf, B.A., *Mater. Chem. Phys.*, 14, 311, 1986.
127. Irani, C.A. and Cozewith, C., *J. Appl. Polym. Sci.*, 31, 1879, 1986.
128. Cowie, J.M.G. and McEwen, I.J., *Brit. Polym. J.*, 18, 387, 1986.
129. Krüger, B., *Dissertation*, TH Leuna-Merseburg, 1986.
130. Rätzsch, M.T., et al., *J. Macromol. Sci.-Chem. A*, 23, 1349, 1986.
131. Saeki, S., et al., *Macromolecules*, 19, 2353, 1986.
132. Sander, U. and Wolf, B.A., *Angew. Makromol. Chem.*, 139, 149, 1986.
133. Barbarin-Castillo, J.-M., et al., *Polym.Commun.*, 28, 212, 1987.
134. Gruner, K. and Greer, S.C., *Macromolecules*, 20, 2238, 1987.
135. Magarik, S.Ya., Filippov, A.P., and D'yakonova, N.V., *Vysokomol. Soedin., Ser. A*, 29, 698, 1987.
136. Rangel-Nafaile, C. and Munoz-Lara, J.J., *Chem. Eng. Commun.*, 53, 177, 1987.
137. Kiepen, F. and Borchard, W., *Macromolecules*, 21, 1784, 1988.
138. Schuster, R., *Diploma Paper*, TH Leuna-Merseburg, 1988.
139. Tveekrem, J.L., Greer, S.C., and Jacobs, D.T., *Macromolecules*, 21, 147, 1988.
140. Bohossian, T., Charlet, G., and Delmas, G., *Polymer*, 30, 1695, 1989.
141. Chiu, G. and Mandelkern, L., *Macromolecules*, 23, 5356, 1990.
142. Goedel, W.A., et al., *Ber. Bunsenges. Phys. Chem.*, 94, 17, 1990.
143. Van der Haegen, R. and Van Opstal, L., *Makromol. Chem.*, 191, 1871, 1990.
144. Iwai, Y., et al., *Sekiyu Gakkaishi*, 33, 117, 1990.
145. Raetzsch, M.T., Krueger, B., and Kehlen, H., *J. Macromol. Sci.-Chem. A*, 27, 683, 1990.
146. Schild, H.G. and Tirrell, D.A., *J. Phys. Chem.*, 94, 4352, 1990.
147. Stafford, S.G., Ploplis, A.C., and Jacobs, D.T., *Macromolecules*, 23, 470, 1990.
148. Akhmadeev, I.R., et al., *Vysokomol. Soedin., Ser. B*, 33, 543, 1991.
149. Bae, Y.C., Lambert, S.M., Soane, D.S., and Prausnitz, J.M., *Macromolecules*, 24, 4403, 1991.
150. Chu, B., Linliu, K., Xie, P., Ying, Q., Wang, Z., and Shook, J.W., *Rev. Sci. Instr.*, 62, 2252, 1991.
151. Kawate, K., Imagawa, I., and Nakata, M., *Polym. J.*, 23, 233, 1991.
152. Lee, K.D. and Lee, D.C., *Pollimo*, 15, 274, 1991.
153. Louai, A., Sarazin, D., Pollet, G., Francois, J., and Moreaux, F., *Polymer*, 32, 703, 1991.
154. Van Opstal, L., Koningsveld, R., and Kleintjens, L.A., *Macromolecules*, 24, 161, 1991.
155. Schubert, K.-V., Strey, R., and Kahlweit, M., *J. Colloid Interface Sci.*, 141, 21, 1991.
156. Shen, W., Smith, G.R., Knobler, C.M., and Scott, R.L., *J. Phys. Chem.*, 95, 3376, 1991.
157. Szydlowski, J. and Van Hook, W.A., Macromolecules, 24, 4883, 1991.
158. Tager, A.A., et al., *Vysokomol. Soedin., Ser. B*, 33, 572, 1991.
159. Wakker, A., *Polymer*, 32, 279, 1991.
160. Yokoyama, H., Takano, A., Okada, M., and Nose, T., *Polymer*, 32, 3218, 1991.
161. Heinrich, M. and Wolf, B.A., *Polymer*, 33, 1926, 1992.
162. Heinrich, M. and Wolf, B.A., *Macromolecules*, 25, 3817, 1992.
163. Lecointe, J.P., Pascault, J.P., Suspene, L., and Yang, Y.S., *Polymer*, 33, 3226, 1992.
164. Mueller, K.F., *Polymer*, 33, 3470, 1992.
165. Szydlowski, J., Rebelo, L., and Van HooK, W.A., *Rev. Sci. Instrum.*, 63, 1717, 1992.
166. Xia, K.-Q., Franck, C., and Widom, B., *J. Chem. Phys.*, 97, 1446, 1992.
167. Arnauts, J., Berghmans, H., and Koningsveld, R., *Makromol. Chem.*, 194, 77, 1993.
168. Iwai, Y., Shigematsu, Y., Furuya, T., Fukuda, H., Arai, Y., *Polym. Eng. Sci.*, 33, 480, 1993.
169. Wakker, A., Van Dijk, F., and Van Dijk, M.A., *Macromolecules*, 26, 5088, 1993.

170. Wells, P.A., de Loos, Th.W., and Kleintjens, L.A., *Fluid Phase Equil.*, 83, 383, 1993.
171. Barbarin-Castillo, J.-M. and McLure, I.A., *Polymer*, 35, 3075, 1994.
172. Mumick, P.S. and McCormick, C.L., *Polym. Eng. Sci.*, 34, 1419, 1994.
173. Okano, K., Takada, M., Kurita, K., and Furusaka, M., *Polymer*, 35, 2284, 1994.
174. Sato, H., Kuwahara, N., and Kubota, K., *Phys. Rev. E*, 50, 1752, 1994.
175. Song, S.-W. and Torkelson, J.M., *Macromolecules*, 27, 6389, 1994.
176. Tager, A.A., et al., *Colloid Polym. Sci.*, 272, 1234, 1994.
177. Vanhee, S., et al., *Makromol. Chem. Phys.*, 195, 759, 1994.
178. Haas, C.K. and Torkelson, J.M., *Phys. Rev. Lett.*, 75, 3134, 1995.
179. Ikier, C. and Klein, H., *Macromolecules*, 28, 1003, 1995.
180. Luszczyk, M., Rebelo, L.P.N., and Van Hook, W.A., *Macromolecules*, 28, 745, 1995.
181. Pfohl, O., Hino, T., and Prausnitz, J.M., Polymer, 36, 2065, 1995.
182. Vshivkov, S.A. and Safronov, A.P., *Vysokomol. Soedin., Ser. B*, 37, 1779, 1995.
183. Allcock, H.R. and Dudley, G.K., *Macromolecules*, 29, 1313, 1996.
184. El-Ejmi, A.A.S. and Huglin, M.B., *Polym. Int.* 39, 113, 1996.
185. Fischer, V., Borchard, W., and Karas, M., *J. Phys. Chem.*, 100, 15992, 1996.
186. Imre, A. and Van Hook, W.A., *J. Polym. Sci.: Part B: Polym. Sci.*, 34, 751, 1996.
187. Luszczyk, M. and Van Hook, W.A., *Macromolecules*, 29, 6612, 1996.
188. Rong, Z., Wang, H., Ying, X., and Hu, Y., *J. East China Univ. Sci. Technol.*, 22, 754, 1996.
189. Safronov, A.P., Tager, A.A., and Koroleva, E.V., *Vysokomol. Soedin., Ser. B*, 38, 900, 1996.
190. Vshivkov, S.A. and Rusinova, E.V., *Vysokomol. Soedin., Ser. A*, 38, 1746, 1996.
191. Xia, K.-Q., An, X.-Q., and Shen, W.-G., *J. Chem.Phys.*, 105, 6018, 1996.
192. Imre, A. and Van Hook, W.A., *J. Polym. Sci.: Part B: Polym. Phys.*, 35, 1251, 1997.
193. Kita, R., Dobashi, T., Yamamoto, T., Nakata, M., and Kamide, K., *Phys. Rev. E*, 55, 3159, 1997.
194. Li, M., Zhu, Z.-Q., and Mei, L.-H., *Biotechnol. Progr.*, 13, 105, 1997.
195. McLure, I.A., Mokhtari, A., and Bowers, J., *J. Chem. Soc., Faraday Trans.*, 93, 249, 1997.
196. Chalykh, A.E., Dement'eva, O.V., and Gerasimov, V.K., *Vysokomol. Soedin., Ser. A*, 40, 815,.1998.
197. Kubota, K., Kita, R., and Dobashi, T., *J. Chem. Phys.*, 109, 711, 1998.
198. Schneider, A., *Dissertation*, Johannes Gutenberg Universität Mainz. 1998.
199. Terao, K., et al., *Macromolecules*, 31, 6885, 1998.
200. Yamagishi, T.-A., et al., *Macromol. Chem. Phys.*, 199, 423, 1998.
201. Lau, A.C.W. and Wu, C., *Macromolecules*, 32, 581, 1999.
202. Nakata, M., Dobashi, T., Inakuma, Y.-I., and Yamamura, K., *J. Chem. Phys.*, 111, 6617, 1999.
203. Pruessner, M.D., Retzer, M.E., and Greer, S.C., *J. Chem. Eng. Data*, 44, 1419, 1999.
204. Shimofure, S., Kubota, K., Kita, R., and Dobashi, T., *J.Chem.Phys.*, 111, 4199, 1999.
205. Fischer, V. and Borchard, W., *J. Phys. Chem. B*, 104, 4463, 2000.
206. Imre, A., and Van Hook, W.A., *Macromolecules*, 33, 5308, 2000.
207. Koizumi, J., et al., *J. Phys. Soc. Japan*, 69, 2543, 2000.
208. Kunugi, S., Tada, T., Yamazaki, Y., Yamamoto, K., and Akashi, M., *Langmuir*, 16, 2042, 2000.
209. La Mesa, C., *J. Therm. Anal. Calorim.*, 61, 493, 2000.
210. Persson, J., et al., *Bioseparation*, 9, 105, 2000.
211. Persson, J., Kaul, A., and Tjerneld, F., *J. Chromatogr. B*, 743, 115, 2000.
212. Djokpe, E. and Vogt, W., *Macromol. Chem. Phys.*, 202, 750, 2001.
213. Kujawa, P. and Winnik, F.M., *Macromolecules*, 34, 4130, 2001.
214. Pendyala, K.S., Greer, S.C., and Jacobs, D.T., *J. Chem. Phys.*, 115, 9995, 2001.
215. Berlinova, I. V., Nedelcheva, A. N., Samchikov, V., and Ivanov, Ya., *Polymer*, 43, 7243, 2002.
216. Freitag, R. and Garret-Flaudy, F., *Langmuir*, 18, 3434, 2002.
217. Maeda, Y., Nakamura, T., and Ikeda, I., *Macromolecules*, 35, 217, 2002.
218. Rebelo, L.P.N., et al., J., *Macromolecules*, 35, 1887, 2002.
219. Safronov, A.P. and Somova, T.V., *Vysokomol. Soedin., Ser. A*, 44, 2014, 2002.
220. Vshvikov, S.A., Rusinova, E.V., and Gur'ev, A.A., *Vysokomol. Soedin., Ser. B*, 44, 504, 2002.
221. Zhou, C.-S., An, X.-Q., Xia, K.-Q., Yin, X.-L., and Shen, W.-G., *J. Chem. Phys.*, 117, 4557, 2002.
222. Chang, Y., Powell, E.S., Allcock, H.R., Park, S.M., and Kim, C., *Macromolecules*, 36, 2568, 2003.
223. David, G., et al., *Eur. Polym. J.*, 39, 1209, 2003.
224. Siporska, A., Szydlowski, J., and Rebelo, L.P.N., *Phys. Chem. Chem. Phys.*, 5, 2996, 2003.
225. Yamamoto, K., Serizawa, T., and Akashi, M., *Macromol. Chem. Phys.*, 204, 1027, 2003.

VAPOR PRESSURES (SOLVENT ACTIVITIES) FOR BINARY POLYMER SOLUTIONS

Christian Wohlfarth

The vapor pressure of a binary polymer solution is given by the activity of the solvent A, a_A. Solvent activities in polymer solutions are measured either by the isopiestic method applying a reference system whose solvent activity is precisely known or by determining the solvent partial pressure, P_A, and calculating the activity of the solvent by equation (1):

$$a_A = (P_A/P_A^s)\exp\left[\frac{(B_{AA} - V_A^L)(P - P_A^s)}{RT}\right] \quad (1)$$

where B_{AA} is the second virial coefficient, P_A^s is the saturation vapor pressure, and V_A^L is the molar volume of the pure solvent A at the measuring temperature T. The exponential term is neglected in quite a lot of original papers, however, and only the reduced vapor pressures are given (such data are indicated by an asterisk in the table below). Vapor pressures of polymer solutions have been measured since the 1940s, but the amount of experimental data for polymer solutions is still relatively small in comparison to low-molecular mixtures and solutions. The data scatter with respect to temperature, concentration, molar mass, and other polymer characterization variables. Furthermore, the concentration range for measuring vapor pressures in good thermodynamic quality is often limited to the polymer mass fraction range between 0.4 and 0.85. A recent review on methods for the measurement of vapor pressures/solvent activities of polymer solutions and on related problems is given in Ref. [1]. Experimental data have been collected in several books [2-6].

The table in this *Handbook* provides data for a number of polymer solutions as smoothed values over the complete range of solvent activities between 0 (polymer mass fraction = 1) and 1 (polymer mass fraction = 0). For this purpose, the data were selected from data books [4–6] as well as from a number of original sources [7–22] which are not included in these books. The appropriate data were smoothed. The final table provides then the polymer mass fractions at given fixed solvent activities between 0.1 and 0.9. Of course, the user must keep in mind that the activity vs. concentration range of the experimental data is sometimes smaller than the below given complete range, thus the smoothed data should be used with sufficient care.

Generally, vapor pressures or solvent activities of binary polymer solutions depend on molar mass. However, for high molecular weight polymers (well above the oligomer region), this molar-mass dependence can be neglected in many cases. Therefore, the table below presents only data for polymer solutions where the number average molar mass, M_n, is in the order of 10^5 g/mol or even higher, therefore, the molar mass is not specified. The temperature is stated, even though the temperature dependence of a_A is relatively small for the temperature ranges where most of the experimental data exist.

References

1. Wohlfarth, C., Methods for the measurement of solvent activity of polymer solutions, in *Handbook of Solvents*, Wypych, G., Ed., ChemTec Publishing, Toronto, 2000, 146.
2. Wen, H., Elbro, H. S., and Alessi, P., *Polymer Solution Data Collection.* I. Vapor-liquid equilibrium; II. Solvent activity coefficients at infinite dilution; III. Liquid-liquid equilibrium, Chemistry Data Series, Vol. 15, DECHEMA, Frankfurt am Main, 1992.
3. Danner, R. P. and High, M. S., *Handbook of Polymer Solution Thermodynamics*, American Institute of Chemical Engineers, New York, 1993.
4. Wohlfarth, C., *Vapour-Liquid Equilibrium Data of Binary Polymer Solutions: Physical Science Data*, 44, Elsevier, Amsterdam, 1994.
5. Wohlfarth, C., *CRC Handbook of Thermodynamic Data of Copolymer Solutions*, CRC Press, Boca Raton, FL, 2001.
6. Wohlfarth, C., *CRC Handbook of Thermodynamic Data of Aqueous Polymer Solutions*, CRC Press, Boca Raton, FL, 2003.
7. Wang, K., Hu, Y., and Wu, D. T., *J. Chem. Eng. Data*, 39, 916, 1994.
8. Choi, J. S., Tochigi, K., and Kojima, K., *Fluid Phase Equil.*, 111, 143, 1995.
9. Tochigi, K., Kurita, S., Ohashi, M., and Kojima, K., *Kagaku Kogaku Ronbunshu*, 23, 720, 1997.
10. Wong, H. C., Campbell, S. W., and Bhethanabotla, V. R., *Fluid Phase Equil.*, 139, 371, 1997.
11. Kim, J., Joung, K. C., Hwang, S., Huh, W., Lee, C. S., and Yoo, K.-P., *Korean J. Chem. Eng.*, 15, 199, 1998.
12. Kim, N. H., Kim, S.J., Won, Y. S., and Choi, J. S., *Korean J. Chem. Eng.*, 15, 141, 1998.
13. Feng, W., Wang, W., and Feng, Z., *J. Chem. Ind. Eng. (China)*, 49, 271, 1998.
14. French, R. N. and Koplos, G. J., *Fluid Phase Equil.*, 160, 879, 1999.
15. Striolo, A. and Praunsitz, J. M., *Polymer*, 41, 1109, 2000.
16. Fornasiero, F., Halim, M., and Prausnitz, J. M., *Macromolecules*, 33, 8435, 2000.
17. Wong, H. C., Campbell, S. W., and Bhethanabotla, V. R., *Fluid Phase Equil.*, 179, 181, 2001.
18. Wibawa, G., Takahashi, M., Sato, Y., Takishima, S., and Masuoka, H., *J. Chem. Eng. Data*, 47, 518, 2002.
19. Wibawa, G., Hatano, R., Sato, Y., Takishima, S., and Masuoka, H., *J. Chem. Eng. Data*, 47, 1022, 2002.
20. Pfohl, O., Riebesell, C., and Dohrn, R., *Fluid Phase Equil.*, 202, 289, 2002.
21. Jung, J. K., Joung, S. N., Shin, H.Y., Kim, S. Y., Yoo, K.-P., Huh, W., Lee, C. S., *Korean J. Chem. Eng.*, 19, 296, 2002.
22. Kang, S., Huang, Y., Fu, J., Liu, H., and Hu, Y., *J. Chem. Eng. Data*, 47, 788, 2002.

Solvent Activity a_A as Function of Temperature and Mass Fraction

Polymer/ solvents	a_A: T/K	0.1	0.2	0.3	0.4	0.5	0.6	0.7	0.8	0.9
					Mass Fraction of the Polymer					
Acrylonitrile/Styrene Copolymer (28 wt% Acrylonitrile)										
Benzene*)	343.15	0.982	0.962	0.940	0.915	0.886	0.851	0.809	0.753	0.670
1,2-Dimethylbenzene*)	398.15	0.983	0.964	0.942	0.918	0.890	0.857	0.817	0.764	0.685
1,3-Dimethylbenzene*)	398.15	0.983	0.965	0.944	0.921	0.893	0.861	0.821	0.769	0.690
1,4-Dimethylbenzene*)	398.15	0.983	0.964	0.942	0.918	0.890	0.857	0.817	0.763	0.684
Propylbenzene*)	398.15	0.987	0.972	0.955	0.935	0.913	0.885	0.851	0.804	0.732
Toluene*)	343.15	0.982	0.962	0.940	0.915	0.886	0.851	0.809	0.753	0.669

Polymer/ solvents	a_A: T/K	0.1	0.2	0.3	0.4	0.5	0.6	0.7	0.8	0.9
					Mass Fraction of the Polymer					
Butadiene/Styrene Copolymer (41 wt% Styrene)										
Benzene*)	343.15	0.968	0.934	0.896	0.853	0.805	0.748	0.680	0.591	0.461
Cyclohexane*)	343.15	0.978	0.953	0.925	0.893	0.856	0.811	0.754	0.678	0.556
Ethylbenzene*)	398.15	0.974	0.945	0.912	0.875	0.831	0.779	0.713	0.625	0.491
Mesitylene*)	398.15	0.977	0.950	0.921	0.887	0.847	0.799	0.738	0.656	0.526
Toluene*)	343.15	0.970	0.936	0.899	0.857	0.808	0.751	0.682	0.591	0.456
Cellulose Triacetate										
Dichloromethane	298.15	0.979	0.956	0.930	0.899	0.863	0.819	0.762	0.683	0.554
Trichloromethane	298.15	0.978	0.953	0.924	0.892	0.853	0.806	0.747	0.665	0.533
Dextran										
Water	313.15	0.988	0.975	0.960	0.942	0.921	0.894	0.860	0.810	0.725
Hydroxyethylcellulose										
Water	368.15	0.988	0.974	0.958	0.939	0.915	0.884	0.841	0.775	0.650
Hydroxypropylstarch										
Water	293.15	0.989	0.977	0.963	0.947	0.927	0.903	0.872	0.827	0.749
Nitrocellulose										
Ethyl acetate	293.15	0.938	0.885	0.835	0.786	0.737	0.685	0.627	0.560	0.471
Ethyl formate	293.15	0.958	0.916	0.873	0.828	0.780	0.728	0.668	0.595	0.494
Ethyl propionate	293.15	0.941	0.889	0.839	0.789	0.739	0.685	0.625	0.555	0.460
Methyl acetate	293.15	0.890	0.820	0.763	0.711	0.660	0.609	0.554	0.490	0.406
2-Propanone	293.15	0.922	0.861	0.807	0.756	0.706	0.653	0.596	0.530	0.443
Propyl acetate	293.15	0.937	0.881	0.827	0.775	0.722	0.665	0.602	0.528	0.426
Polybutadiene (random cis-trans-vinyl)										
Benzene	298.15	0.964	0.925	0.884	0.839	0.788	0.731	0.663	0.578	0.455
Cyclohexane	298.15	0.974	0.945	0.913	0.876	0.833	0.782	0.719	0.635	0.507
Dichloromethane	298.15	0.951	0.902	0.852	0.800	0.745	0.684	0.616	0.532	0.415
Hexane	298.15	0.984	0.965	0.943	0.916	0.881	0.837	0.775	0.683	0.534
Tetrachloromethane	298.15	0.932	0.865	0.799	0.731	0.660	0.585	0.503	0.409	0.288
Toluene	298.15	0.969	0.935	0.898	0.856	0.809	0.754	0.688	0.603	0.476
Trichloromethane	298.15	0.925	0.855	0.788	0.720	0.650	0.578	0.498	0.406	0.289
1,4-cis-Polybutadiene										
Benzene	298.15	0.966	0.930	0.890	0.846	0.796	0.738	0.668	0.580	0.450
Cyclohexane	298.15	0.977	0.951	0.922	0.888	0.849	0.803	0.747	0.677	0.581
Dichloromethane	298.15	0.948	0.898	0.848	0.796	0.742	0.683	0.616	0.536	0.424
Hexane	298.15	0.983	0.963	0.941	0.916	0.886	0.850	0.804	0.741	0.639
Tetrachloromethane	298.15	0.936	0.871	0.805	0.736	0.665	0.588	0.505	0.409	0.287
Toluene	298.15	0.969	0.936	0.900	0.860	0.815	0.763	0.701	0.622	0.506
Trichloromethane	298.15	0.915	0.840	0.770	0.702	0.634	0.562	0.485	0.396	0.283
Poly(butyl acrylate)										
Benzene	298.15	0.964	0.926	0.887	0.845	0.799	0.749	0.691	0.619	0.519
Dichloromethane	298.15	0.868	0.801	0.744	0.690	0.636	0.577	0.511	0.430	0.318
Tetrachloromethane	298.15	0.932	0.868	0.805	0.742	0.677	0.607	0.529	0.438	0.317
Toluene	298.15	0.967	0.932	0.893	0.849	0.801	0.744	0.676	0.590	0.463
Trichloromethane	298.15	0.901	0.811	0.733	0.662	0.595	0.529	0.459	0.381	0.282
Poly(butyl methacrylate)										
Benzene	313.15	0.971	0.939	0.902	0.861	0.813	0.756	0.685	0.592	0.453
1-Butanol	313.15	0.991	0.980	0.968	0.953	0.936	0.914	0.885	0.842	0.762
2-Butanol	313.15	0.992	0.982	0.969	0.953	0.933	0.906	0.869	0.815	0.719
2-Butanone	313.15	0.982	0.963	0.940	0.914	0.884	0.846	0.799	0.732	0.623
Butyl acetate*)	308.15	0.982	0.961	0.936	0.908	0.875	0.836	0.789	0.730	0.652
Cyclohexane	313.15	0.985	0.968	0.948	0.925	0.899	0.866	0.823	0.764	0.666
Cyclopentane	313.15	0.984	0.965	0.944	0.918	0.886	0.846	0.792	0.714	0.579
Diethyl ether*)	298.15	0.987	0.973	0.956	0.937	0.914	0.885	0.848	0.795	0.703
1,4-Dimethylbenzene	333.15	0.971	0.940	0.905	0.866	0.822	0.770	0.706	0.622	0.497
Ethylbenzene	333.15	0.969	0.935	0.899	0.859	0.815	0.764	0.704	0.627	0.517
Methyl acetate	313.15	0.984	0.965	0.944	0.920	0.891	0.856	0.811	0.748	0.645

Polymer/ solvents	a_A: T/K	0.1	0.2	0.3	0.4	0.5	0.6	0.7	0.8	0.9
					Mass Fraction of the Polymer					
2-Methyl-1-propanol	333.15	0.988	0.974	0.958	0.940	0.919	0.893	0.860	0.815	0.744
Octane	313.15	0.988	0.974	0.959	0.942	0.921	0.896	0.865	0.823	0.758
1-Propanol	333.15	0.990	0.980	0.967	0.952	0.934	0.911	0.881	0.834	0.746
2-Propanol	313.15	0.991	0.981	0.970	0.956	0.939	0.918	0.889	0.845	0.755
2-Propanone	313.15	0.989	0.976	0.961	0.944	0.921	0.892	0.850	0.783	0.647
Propyl acetate	313.15	0.980	0.957	0.932	0.903	0.870	0.830	0.780	0.714	0.612
Toluene	313.15	0.971	0.939	0.903	0.863	0.818	0.764	0.698	0.613	0.485
Poly(ε-caprolacton)										
Tetrachloromethane*)	338.15	0.956	0.910	0.864	0.815	0.762	0.704	0.637	0.554	0.438
Poly(dimethylsiloxane)										
Chlorodifluoromethane	298.15	0.976	0.950	0.921	0.888	0.850	0.805	0.750	0.677	0.565
Cyclohexane	303.15	0.979	0.955	0.928	0.898	0.863	0.822	0.770	0.702	0.596
Hexane	303.15	0.982	0.962	0.939	0.912	0.880	0.842	0.793	0.724	0.611
Pentane	308.15	0.982	0.962	0.940	0.913	0.881	0.842	0.791	0.720	0.600
Pentane	423.15	0.984	0.966	0.946	0.922	0.893	0.858	0.813	0.749	0.641
Poly(ethyl acrylate)										
Benzene	298.15	0.970	0.939	0.904	0.866	0.823	0.774	0.716	0.641	0.533
Dichloromethane	298.15	0.900	0.830	0.768	0.709	0.648	0.584	0.512	0.427	0.313
Tetrachloromethane	298.15	0.950	0.900	0.848	0.794	0.736	0.672	0.598	0.509	0.385
Toluene	298.15	0.972	0.942	0.910	0.874	0.833	0.786	0.730	0.659	0.555
Trichloromethane	298.15	0.866	0.776	0.701	0.632	0.566	0.499	0.428	0.349	0.248
Poly(ethylene oxide)										
Benzene	323.15	0.972	0.942	0.908	0.869	0.824	0.771	0.706	0.620	0.490
2-Butanone	353.15	0.981	0.959	0.934	0.902	0.863	0.813	0.746	0.651	0.503
Cyclohexane	353.15	0.989	0.976	0.960	0.943	0.921	0.893	0.855	0.798	0.688
Methanol	303.15	0.964	0.927	0.887	0.844	0.797	0.744	0.682	0.604	0.494
2-Propanone	353.15	0.979	0.947	0.896	0.815	0.719	0.625	0.532	0.434	0.315
Water	293.15	0.977	0.951	0.923	0.890	0.852	0.806	0.748	0.671	0.550
Poly(ethylenimine)										
Water	353.15	0.975	0.947	0.917	0.883	0.845	0.801	0.748	0.680	0.581
Poly(ethyl methacrylate)										
Benzene	298.15	0.970	0.938	0.903	0.864	0.821	0.771	0.712	0.637	0.529
Dichloromethane	298.15	0.912	0.838	0.769	0.703	0.636	0.567	0.491	0.404	0.292
Tetrachloromethane	298.15	0.935	0.873	0.812	0.750	0.686	0.616	0.540	0.449	0.328
Toluene	298.15	0.974	0.945	0.913	0.877	0.836	0.787	0.727	0.647	0.527
Trichloromethane	298.15	0.859	0.760	0.678	0.604	0.533	0.464	0.392	0.313	0.217
Polyisobutylene										
Benzene	313.15	0.984	0.965	0.945	0.921	0.892	0.858	0.813	0.751	0.645
Cyclohexane	313.15	0.976	0.950	0.921	0.888	0.850	0.805	0.749	0.676	0.563
Cyclopentane	313.15	0.977	0.952	0.924	0.892	0.855	0.812	0.758	0.687	0.579
1,4-Dimethylbenzene	313.15	0.979	0.955	0.929	0.899	0.863	0.821	0.767	0.694	0.579
2,2-Dimethylbutane	298.15	0.983	0.964	0.942	0.917	0.887	0.852	0.806	0.743	0.640
Ethylbenzene	313.15	0.979	0.955	0.927	0.895	0.857	0.810	0.750	0.668	0.535
Heptane	298.15	0.983	0.964	0.942	0.917	0.887	0.851	0.804	0.741	0.637
Hexane	298.15	0.980	0.959	0.934	0.906	0.873	0.834	0.784	0.715	0.606
Octane	298.15	0.983	0.963	0.940	0.914	0.883	0.845	0.797	0.729	0.617
Tetrachloromethane	298.15	0.962	0.921	0.877	0.829	0.776	0.715	0.643	0.552	0.423
Toluene	313.15	0.984	0.966	0.944	0.918	0.884	0.840	0.779	0.688	0.537
Trichloromethane	298.15	0.969	0.935	0.899	0.858	0.813	0.761	0.698	0.619	0.503
2,4,4-Trimethylpentane	298.15	0.981	0.961	0.937	0.911	0.879	0.842	0.794	0.730	0.628
1,4-cis-Polyisoprene										
Benzene	313.15	0.982	0.962	0.937	0.908	0.873	0.827	0.766	0.679	0.537
2-Butanone	353.15	0.986	0.970	0.953	0.933	0.910	0.883	0.850	0.808	0.746
Cyclohexane	313.15	0.978	0.954	0.928	0.899	0.865	0.825	0.778	0.716	0.625
Dichloromethane	298.15	0.969	0.935	0.898	0.857	0.811	0.757	0.693	0.610	0.488
1,4-Dimethylbenzene	313.15	0.977	0.951	0.923	0.892	0.857	0.816	0.767	0.704	0.613

Polymer/ solvents	a_A: T/K	0.1	0.2	0.3	0.4	0.5	0.6	0.7	0.8	0.9
		Mass Fraction of the Polymer								
Ethylbenzene	313.15	0.978	0.954	0.928	0.898	0.864	0.823	0.774	0.709	0.612
Methyl acetate	313.15	0.968	0.935	0.900	0.862	0.820	0.773	0.717	0.649	0.554
Octane	313.15	0.984	0.967	0.948	0.926	0.901	0.871	0.834	0.785	0.711
Propyl acetate	333.15	0.983	0.964	0.942	0.916	0.886	0.850	0.803	0.738	0.633
Tetrachloromethane	298.15	0.929	0.864	0.800	0.737	0.672	0.602	0.526	0.435	0.316
Toluene	313.15	0.978	0.954	0.927	0.898	0.865	0.827	0.782	0.725	0.645
Trichloromethane	298.15	0.930	0.867	0.807	0.747	0.685	0.620	0.547	0.462	0.346
Poly(methyl acrylate)										
Benzene	298.15	0.979	0.956	0.930	0.901	0.867	0.826	0.776	0.710	0.608
Dichloromethane	298.15	0.917	0.851	0.791	0.732	0.671	0.605	0.532	0.444	0.326
Tetrachloromethane	298.15	0.963	0.924	0.882	0.838	0.788	0.733	0.668	0.586	0.470
Toluene	298.15	0.981	0.960	0.936	0.909	0.878	0.840	0.792	0.727	0.626
Trichloromethane	298.15	0.912	0.830	0.753	0.678	0.603	0.527	0.446	0.357	0.248
Poly(methyl methacrylate)										
Benzene	298.15	0.982	0.961	0.938	0.912	0.881	0.843	0.795	0.729	0.622
2-Butanone*)	308.15	0.989	0.976	0.961	0.945	0.925	0.900	0.869	0.825	0.751
Cyclohexanone*)	323.15	0.978	0.954	0.928	0.899	0.866	0.827	0.781	0.723	0.640
Dichloromethane	298.15	0.939	0.882	0.825	0.766	0.704	0.637	0.560	0.468	0.343
Ethyl acetate*)	308.15	0.986	0.969	0.950	0.928	0.902	0.869	0.826	0.763	0.649
Toluene	298.15	0.981	0.959	0.935	0.908	0.877	0.841	0.795	0.736	0.646
Trichloromethane	298.15	0.924	0.848	0.771	0.694	0.616	0.536	0.451	0.358	0.246
Poly(α-methylstyrene)										
Cumene	338.15	0.984	0.965	0.944	0.918	0.887	0.848	0.796	0.721	0.593
α-Methylstyrene	338.15	0.978	0.954	0.927	0.896	0.859	0.816	0.761	0.687	0.570
Poly(propylene oxide)										
Benzene	333.15	0.967	0.932	0.893	0.850	0.801	0.744	0.675	0.588	0.460
Metvhanol	298.15	0.992	0.982	0.970	0.955	0.936	0.910	0.872	0.812	0.689
Polystyrene										
Benzene	333.15	0.978	0.953	0.924	0.891	0.852	0.804	0.742	0.657	0.521
2-Butanone*)	298.15	0.986	0.971	0.954	0.935	0.912	0.885	0.851	0.804	0.724
Cyclohexane	313.15	0.990	0.978	0.965	0.949	0.931	0.908	0.877	0.833	0.754
Cyclohexanone*)	313.15	0.970	0.937	0.900	0.858	0.810	0.753	0.684	0.593	0.459
Dichloromethane	298.15	0.949	0.899	0.849	0.797	0.743	0.684	0.617	0.536	0.423
1,3-Dimethylbenzene*)	323.15	0.980	0.956	0.926	0.891	0.846	0.791	0.723	0.638	0.524
1,4-Dimethylbenzene	423.15	0.974	0.944	0.911	0.872	0.826	0.770	0.698	0.601	0.452
Ethyl acetate*)	313.15	0.976	0.948	0.918	0.882	0.841	0.791	0.728	0.642	0.507
Hexane	423.15	0.980	0.958	0.933	0.904	0.869	0.827	0.772	0.697	0.574
2-Propanone	323.15	0.991	0.980	0.969	0.955	0.938	0.918	0.892	0.854	0.788
Propyl acetate	343.15	0.983	0.965	0.943	0.919	0.891	0.858	0.815	0.758	0.667
Tetrachloromethane	298.15	0.961	0.917	0.869	0.814	0.751	0.678	0.592	0.486	0.344
Toluene	313.15	0.981	0.959	0.933	0.901	0.861	0.809	0.738	0.638	0.481
Trichloromethane	298.15	0.949	0.898	0.847	0.793	0.736	0.675	0.604	0.519	0.400
Poly(tetramethylene glycol)										
Methanol	303.15	0.981	0.961	0.938	0.913	0.883	0.849	0.806	0.751	0.671
Poly(vinyl acetate)										
Benzene	313.15	0.985	0.967	0.945	0.919	0.886	0.844	0.784	0.696	0.548
1-Butanol	313.15	0.992	0.982	0.971	0.958	0.942	0.923	0.896	0.856	0.779
2-Butanol	313.15	0.987	0.972	0.956	0.937	0.915	0.889	0.856	0.813	0.747
2-Butanone	313.15	0.980	0.958	0.934	0.906	0.873	0.835	0.787	0.724	0.626
1,2-Dichloroethane*)	300.15	0.955	0.906	0.851	0.790	0.722	0.644	0.556	0.450	0.315
1,4-Dimethylbenzene	313.15	0.990	0.978	0.964	0.948	0.928	0.903	0.868	0.814	0.705
Ethylbenzene	313.15	0.990	0.979	0.966	0.950	0.932	0.910	0.880	0.836	0.759
Methanol	333.15	0.990	0.978	0.965	0.949	0.931	0.908	0.877	0.834	0.757
Methyl acetate	313.15	0.976	0.949	0.919	0.886	0.849	0.805	0.752	0.684	0.583
2-Methyl-1-propanol	353.15	0.984	0.966	0.946	0.924	0.899	0.868	0.832	0.784	0.715
1-Propanol	353.15	0.987	0.972	0.955	0.936	0.914	0.888	0.856	0.815	0.753
2-Propanol	353.15	0.988	0.974	0.958	0.940	0.919	0.894	0.863	0.820	0.754

Polymer/ solvents	a_A: T/K	0.1	0.2	0.3	0.4	0.5	0.6	0.7	0.8	0.9
		Mass Fraction of the Polymer								
2-Propanone	333.15	0.983	0.963	0.940	0.913	0.880	0.838	0.784	0.707	0.578
Propyl acetate	333.15	0.979	0.955	0.930	0.901	0.869	0.831	0.786	0.728	0.645
Tetrahydrofuran	323.15	0.973	0.943	0.911	0.874	0.831	0.781	0.720	0.640	0.519
Toluene	333.15	0.983	0.965	0.944	0.920	0.891	0.857	0.815	0.756	0.664
Poly(vinyl chloride)										
2-Butanone*)	313.15	0.976	0.949	0.920	0.887	0.849	0.804	0.749	0.676	0.566
Cyclohexanone*)	333.15	0.971	0.934	0.889	0.839	0.781	0.714	0.635	0.536	0.397
Poly(vinyl methyl ether)										
Benzene*)	298.15	0.969	0.935	0.897	0.855	0.807	0.751	0.683	0.596	0.466
Chlorobenzene*)	343.15	0.972	0.941	0.906	0.867	0.822	0.769	0.705	0.620	0.494
1,2-Dimethylbenzene*)	363.15	0.973	0.943	0.910	0.871	0.826	0.772	0.705	0.616	0.478
Ethylbenzene*)	343.15	0.978	0.954	0.927	0.895	0.857	0.811	0.753	0.672	0.542
Propylbenzene*)	373.15	0.977	0.951	0.923	0.890	0.852	0.808	0.752	0.678	0.563
Poly(4-vinylpyridine)										
Methanol	343.15	0.986	0.971	0.953	0.931	0.905	0.871	0.825	0.756	0.627
2-Propanol	343.15	0.989	0.977	0.964	0.948	0.928	0.904	0.872	0.826	0.743
Poly(1-vinyl-2-pyrrolidinone)										
Water	368.15	0.984	0.966	0.946	0.924	0.899	0.870	0.835	0.790	0.727
Starch (amorphous)										
Water	383.15	0.991	0.981	0.970	0.956	0.939	0.918	0.889	0.845	0.754
Styrene/Methyl methacrylate Copolymer (41.45 wt% Styrene)										
Benzene*)	308.15	0.982	0.963	0.940	0.913	0.881	0.841	0.789	0.716	0.590
Vinyl acetate/Vinyl chloride Copolymer (12 wt% Vinyl acetate)										
Benzene	398.15	0.976	0.949	0.918	0.883	0.841	0.791	0.728	0.643	0.509
Chlorobenzene	398.15	0.984	0.965	0.944	0.920	0.891	0.856	0.810	0.746	0.638
1,4-Dimethylbenzene	398.15	0.989	0.977	0.963	0.946	0.926	0.899	0.863	0.807	0.692
Ethylbenzene	398.15	0.989	0.976	0.961	0.944	0.924	0.899	0.866	0.818	0.735
Octane	398.15	0.992	0.982	0.971	0.958	0.942	0.922	0.893	0.847	0.739

*) $a_A = P_A / P_A^S$

SPECIFIC ENTHALPIES OF SOLUTION OF POLYMERS AND COPOLYMERS

Christian Wohlfarth

Enthalpies of solution or mixing, expressed as the enthalpy change per unit mass of polymer, are given in the table at infinite dilution, i.e., a very small amount of polymer and a large excess of solvent were mixed isothermally to form a homogeneous solution. By thermodynamics, $\Delta_{sol}H_B^{\infty}$ or $\Delta_M H_B^{\infty}$ are obtained from the following derivatives:

$$\Delta_{sol}H_B^{\infty} = \lim_{m_B \to 0}\left(\partial \Delta_{sol}h / \partial m_B\right)_{P,T,m_{j\neq B}} \quad (1)$$

$$\Delta_{M}H_B^{\infty} = \lim_{m_B \to 0}\left(\partial \Delta_{M}h / \partial m_B\right)_{P,T,m_{j\neq B}} \quad (2)$$

with a unit of J/g. Thus, they are the partial specific enthalpies of solution or mixing of the polymer B at infinite dilution where $\Delta_{sol}h$ or $\Delta_M h$ is the extensive enthalpy of the solution or mixing process.

The state of the polymer before dissolution can significantly affect the enthalpy of solution. The dissolving of a semicrystalline polymer requires an additional amount of heat associated with the disordering of crystalline regions. Consequently, its enthalpy of solution is usually positive and depends on the degree of crystallinity of the given polymer sample. An amorphous polymer below its glass transition temperature, T_g (see the T_g-table of this Section), often dissolves with the release of heat. The enthalpy of solution of a glassy polymer is additionally dependent to some extent on the thermal history of the glass-forming process. An amorphous polymer above T_g can show endothermic or exothermic dissolution behavior depending on the nature of the solvent and the interaction energies involved as is the case for any enthalpy of mixing. This enthalpy of mixing is then independent of any crystalline or glassy aspects of the polymer. It can be obtained without difficulties for liquid/molten polymers mixed with a solvent. Therefore, the enthalpies given in the table are either enthalpies of solution or enthalpies of mixing, depending on the state of the polymer.

The enthalpies depend on temperature and molar mass. The necessary molar mass information for a system is provided in the table (if available) by the corresponding number average, M_n, mass average, M_w, or viscosity average, M_η, values of the polymer as given in the original sources. Outside the oligomer range, specific enthalpies of solution or mixing do not remarkably depend on molar mass, however. More enthalpy data of polymer-solvent systems can be found in Ref. 106.

Polymer	M_n/ g/mol	M_w/ g/mol	M_η/ g/mol	Solvent	T/K	ΔH_B^{∞}/ J/g	Ref.
Acrylonitrile/butadiene copolymer							
(18 wt% Acrylonitrile)				Benzene	298.15	0.0	18
(26 wt% Acrylonitrile)				Benzene	298.15	−1.9	18
(40 wt% Acrylonitrile)				Benzene	298.15	−2.9	18
Acrylonitrile/isoprene copolymer							
(15 mol% Isoprene)				*N,N*-Dimethylformamide	323.15	−32	66
Acrylonitrile/vinyl chloride copolymer							
(13 wt% Acrylonitrile)				*N,N*-Dimethylformamide	293.15	−38	35
(13 wt% Acrylonitrile)				*N,N*-Dimethylformamide	308.15	−22	35
(13 wt% Acrylonitrile)				*N,N*-Dimethylformamide	323.15	−18	35
(13 wt% Acrylonitrile)				*N,N*-Dimethylformamide	338.15	−15	35
(13 wt% Acrylonitrile)				*N,N*-Dimethylformamide	353.15	−12	35
(29 wt% Acrylonitrile)				*N,N*-Dimethylformamide	295.15	−42	35
(29 wt% Acrylonitrile)				*N,N*-Dimethylformamide	308.15	−27	35
(29 wt% Acrylonitrile)				*N,N*-Dimethylformamide	323.15	−21	35
(29 wt% Acrylonitrile)				*N,N*-Dimethylformamide	338.15	−19	35
(29 wt% Acrylonitrile)				*N,N*-Dimethylformamide	353.15	−16	35
(40 wt% Acrylonitrile)				*N,N*-Dimethylformamide	295.15	−47	35
(40 wt% Acrylonitrile)				*N,N*-Dimethylformamide	308.15	−30	35
(40 wt% Acrylonitrile)				*N,N*-Dimethylformamide	323.15	−28	35
(40 wt% Acrylonitrile)				*N,N*-Dimethylformamide	338.15	−18	35
(40 wt% Acrylonitrile)				*N,N*-Dimethylformamide	353.15	−17	35
Benzylcellulose							
				Benzene	298.15	−11	8
				Cyclohexanone	298.15	−15	25
				Trichloromethane	298.15	−38	25

Polymer	M_n/ g/mol	M_w/ g/mol	M_η/ g/mol	Solvent	T/K	ΔH_B^∞/ J/g	Ref.
Bisphenol A-isophthaloyl chloride/terephthaloyl chloride							
(50/50 Iso/terephtaloyl chloride)				*N,N*-Dimethylacetamide	298.15	−56	69
(50/50 Iso/terephtaloyl chloride)				1,1,2,2-Tetrachloroethane	298.15	+72	69
Butadiene/styrene copolymer							
(10 wt% Styrene)				Benzene	293.65	4.9	17
(30 wt% Styrene)				Benzene	293.65	3.0	17
(30 wt% Styrene)				Benzene	298.15	3.0	25
(50 wt% Styrene)				Benzene	293.65	1.8	17
(60 wt% Styrene)				Benzene	293.65	0.0	17
(70 wt% Styrene)				Benzene	293.65	0.0	17
(75 wt% Styrene)				Benzene	298.15	1.5	7
(80 wt% Styrene)				Benzene	293.65	−0.6	17
(90 wt% Styrene)				Benzene	293.65	−4.9	17
Butyl methacrylate/isobutyl methacrylate copolymer (50 wt%/50 wt%)							
Glass		150000		Cyclohexanone	303.15	5.9	98
Liquid		150000		Cyclohexanone	303.15	14.0	98
Butyl methacrylate/methyl methacrylate copolymer (45 wt%/55 wt%)							
Glass	107000	250000		Cyclohexanone	304.15	−5.4	99
Liquid	107000	250000		Cyclohexanone	304.15	19.1	99
Cellulose acetate							
(52.2 wt% Acetate)				Formic acid	298.15	−30	10
(55.8 wt% Acetate)				Formic acid	298.15	−44	10
(52.5 wt% Acetate)				Methyl acetate	298.15	−80	1
(48 wt% Acetate)				2-Propanone	298.15	−35	25
(52.2 wt% Acetate)				2-Propanone	298.15	−30	10
(55.8 wt% Acetate)				2-Propanone	298.15	−26	10
(56 wt% Acetate)				2-Propanone	298.15	−45	25
(56 wt% Acetate)				2-Propanone	298.15	−30	4
Cellulose triacetate							
				2-Propanone	298.15	−29	4
				Trichloromethane	298.15	−47	4
Dextran							
	8200	10400		Dimethylsulfoxide	298.15	−185	75
	75900	101000		Dimethylsulfoxide	298.15	−187	70
	75900	101000		1,2-Ethanediol	298.15	−98	70
	75900	101000		Formamide	298.15	−228	70
	8200	10400		Water	298.15	−140	75
	75900	101000		Water	298.15	−150	75
(amorph)				Water	298.15	−123	65
Ethylene/propylene copolymer							
(33 mol% Ethylene)				Cyclohexane	298.15	1.4	74
(63 mol% Ethylene)				Cyclohexane	298.15	8.1	74
(75 mol% Ethylene)				Cyclohexane	298.15	11.8	74
(33 mol% Ethylene)				Cyclooctane	298.15	1.2	74

Polymer	M_n/ g/mol	M_w/ g/mol	M_η/ g/mol	Solvent	T/K	ΔH_B^∞/ J/g	Ref.
(63 mol% Ethylene)				Cyclooctane	298.15	6.9	74
(75 mol% Ethylene)				Cyclooctane	298.15	8.6	74
(33 mol% Ethylene)				Cyclopentane	298.15	−3.5	74
(63 mol% Ethylene)				Cyclopentane	298.15	1.1	74
(33 mol% Ethylene)				*cis*-Decahydronaphthalene	298.15	−2.4	74
(63 mol% Ethylene)				*cis*-Decahydronaphthalene	298.15	2.4	74
(75 mol% Ethylene)				*cis*-Decahydronaphthalene	298.15	3.9	74
(33 mol% Ethylene)				*trans*-Decahydronaphthalene	298.15	−4.8	74
(63 mol% Ethylene)				*trans*-Decahydronaphthalene	298.15	−1.3	74
(75 mol% Ethylene)				*trans*-Decahydronaphthalene	298.15	−0.3	74
(63 mol% Ethylene)				3,3-Diethylpentane	298.15	−1.4	74
(75 mol% Ethylene)				3,3-Diethylpentane	298.15	<0.1	74
(63 mol% Ethylene)				2,2-Dimethylpentane	298.15	5.3	74
(75 mol% Ethylene)				2,2-Dimethylpentane	298.15	2.3	74
(63 mol% Ethylene)				2,3-Dimethylpentane	298.15	0.7	74
(75 mol% Ethylene)				2,3-Dimethylpentane	298.15	0.4	74
(33 mol% Ethylene)				2,4-Dimethylpentane	298.15	−1.2	74
(63 mol% Ethylene)				2,4-Dimethylpentane	298.15	3.0	74
(75 mol% Ethylene)				2,4-Dimethylpentane	298.15	0.2	74
(33 mol% Ethylene)				3,3-Dimethylpentane	298.15	−2.7	74
(63 mol% Ethylene)				3,3-Dimethylpentane	298.15	0.3	74
(33 mol% Ethylene)				Dodecane	298.15	−0.1	73
(63 mol% Ethylene)				Dodecane	298.15	0.8	73
(75 mol% Ethylene)				Dodecane	298.15	−4.0	73
(63 mol% Ethylene)				3-Ethylpentane	298.15	2.6	74
(75 mol% Ethylene)				3-Ethylpentane	298.15	−0.6	74
(33 mol% Ethylene)				2,2,4,4,6,8,8-Heptamethylnonane	298.15	−0.5	73
(63 mol% Ethylene)				2,2,4,4,6,8,8-Heptamethylnonane	298.15	2.2	73
(75 mol% Ethylene)				2,2,4,4,6,8,8-Heptamethylnonane	298.15	−0.9	73
(33 mol% Ethylene)				Hexadecane	298.15	0.7	73
(63 mol% Ethylene)				Hexadecane	298.15	−1.1	73
(75 mol% Ethylene)				Hexadecane	298.15	−4.6	73
(63 mol% Ethylene)				3-Methylhexane	298.15	0.7	74
(75 mol% Ethylene)				3-Methylhexane	298.15	1.7	74
(33 mol% Ethylene)				Octane	298.15	−1.6	73
(63 mol% Ethylene)				Octane	298.15	3.6	73
(75 mol% Ethylene)				Octane	298.15	0.3	73
(33 mol% Ethylene)				2,2,4,6,6-Pentamethylheptane	298.15	−0.3	73
(63 mol% Ethylene)				2,2,4,6,6-Pentamethylheptane	298.15	3.6	73
(75 mol% Ethylene)				2,2,4,6,6-Pentamethylheptane	298.15	0.0	73
(63 mol% Ethylene)				2,2,4,4-Tetramethylpentane	298.15	2.7	74
(75 mol% Ethylene)				2,2,4,4-Tetramethylpentane	298.15	3.1	74
(33 mol% Ethylene)				2,2,4-Trimethylpentane	298.15	−0.2	73
(63 mol% Ethylene)				2,2,4-Trimethylpentane	298.15	1.9	73
(75 mol% Ethylene)				2,2,4-Trimethylpentane	298.15	3.5	73

Polymer	M_n/ g/mol	M_w/ g/mol	M_η/ g/mol	Solvent	T/K	ΔH_B^∞/ J/g	Ref.
Ethylene/vinylacetate copolymer							
(85 wt% Vinyl acetate)				Cyclopentanone	298.15	−0.5	104
(70 wt% Vinyl acetate)		220000		Tetrahydrofuran	304.65	−1.3	93
Gelatine							
				Water	293.15	−92	29
				Water	323.15	−63	29
Guttapercha							
				Trichloromethane	303.15	47	22
Isobutyl methacrylate/methyl methacrylate copolymer (51 wt%/49 wt%)							
Glass		150000		Cyclohexanone	303.15	−11	98
Liquid		150000		Cyclohexanone	303.15	15	98
Natural rubber							
				Benzene	298.15	10	25
				Benzene	298.15	12	20
Nitrocellulose							
			16600	2-Butanone	298.15	−80	4
			23000	2-Butanone	298.15	−81	4
			40000	2-Butanone	298.15	−81	4
				Butyl acetate	293.15	−75	23
				Butyl acetate	298.15	−75	23
				Butyl acetate	298.15	−73	26
				Butyl acetate	303.15	−75	23
				Butyl acetate	308.15	−71	23
				Butyl acetate	313.15	−65	23
				Butyl acetate	313.15	−67	26
				Butyl acetate	318.15	−59	23
				Butyl acetate	323.15	−54	23
				Butyl acetate	328.15	−50	23
				Butyl acetate	333.15	−59	26
				Butyl acetate	343.15	−55	26
				Butyl acetate	353.15	−47	26
				Diethyl ether	295.15	−62	3
				Dibutyl phthalate	273.15	−45	26
				Dibutyl phthalate	298.15	−46	26
				Dibutyl phthalate	313.15	−46	26
				Dibutyl phthalate	333.15	−42	26
				Ethanol	295.15	−46	3
				Ethyl acetate	293.15	−76	23
				Ethyl acetate	298.15	−75	23
				Ethyl acetate	303.15	−69	23
				Ethyl acetate	308.15	−61	23
				Ethyl acetate	313.15	−54	23
				Ethyl acetate	318.15	−50	23
				Ethyl acetate	323.15	−50	23
				Ethyl acetate	328.15	−50	23
				Methanol	293.15	−69	23

Polymer	M_n/ g/mol	M_w/ g/mol	M_η/ g/mol	Solvent	T/K	ΔH_B^∞/ J/g	Ref.
				Methanol	298.15	−56	23
				Methanol	303.15	−50	23
				Methanol	308.15	−50	23
				Methanol	313.15	−50	23
				Methanol	318.15	−50	23
				Methanol	323.15	−50	23
				Methanol	328.15	−50	23
				2,4-Pentanedione	298.15	−74	4
				2-Pentanone	298.15	−64	4
				2-Propanone	273.15	−75	26
				2-Propanone	293.15	−75	23
				2-Propanone	298.15	−83	2
				2-Propanone	298.15	−68	4
				2-Propanone	298.15	−71	8
				2-Propanone	298.15	−74	23
				2-Propanone	298.15	−79	25
				2-Propanone	298.15	−75	26
				2-Propanone	303.15	−60	23
				2-Propanone	308.15	−51	23
				2-Propanone	313.15	−50	23
				2-Propanone	313.15	−65	26
				2-Propanone	318.15	−50	23
				2-Propanone	323.15	−50	23
				2-Propanone	323.15	−50	26
				2-Propanone	328.15	−50	23
				Pyridine	298.15	−106	2
				Tri(4-methylphenyl) phosphate	298.15	−16	26
				Tri(4-methylphenyl) phosphate	313.15	−28	26
				Tri(4-methylphenyl) phosphate	333.15	−41	26
				Tri(4-methylphenyl) phosphate	343.15	−44	26
				Tri(4-methylphenyl) phosphate	353.15	−47	26
Nylon-6 (unoriented)							
				Formic acid	295.15	−53	24
				Tricresol	323.55	−66	22
				Tricresol	345.55	−66	22
Poly(acrylonitrile)							
				Benzene	298.15	0.0	18
				N,N-Dimethylformamide	295.15	−23	35
				N,N-Dimethylformamide	298.15	−21	18
				N,N-Dimethylformamide	298.15	−43	42
				N,N-Dimethylformamide	308.15	−17	35
				N,N-Dimethylformamide	323.15	−13	35

Polymer	M_n/ g/mol	M_w/ g/mol	M_η/ g/mol	Solvent	T/K	ΔH_B^∞/ J/g	Ref.
				N,N-Dimethylformamide	323.15	–15	66
				N,N-Dimethylformamide	338.15	–10	35
				Dimethylsulfoxide	298.15	–70	42
Poly(γ-benzyl-L-glutamate)							
		160000		Dichloroacetic acid	303.15	–35	46
		160000		1,2-Dichloroethane	303.15	–1.6	46
Polybutadiene							
				Benzene	298.15	6.1	7
				Benzene	298.15	7.1	25
				Benzene	298.15	10.5	32
				2,2,4-Trimethylpentane	298.15	1.1	32
1,4-cis-Polybutadiene							
	low			Cyclohexane	298.15	5.4	74
	low			Cyclooctane	298.15	5.8	74
	low			Cyclopentane	298.15	<0.1	74
	low			*cis*-Decahydronaphthalene	298.15	4.2	74
	low			*trans*-Decahydronaphthalene	298.15	2.6	74
	low			3,3-Diethylpentane	298.15	5.2	74
	low			2,2-Dimethylpentane	298.15	4.1	74
	low			2,3-Dimethylpentane	298.15	4.5	74
	low			2,4-Dimethylpentane	298.15	3.2	74
	low			3,3-Dimethylpentane	298.15	3.2	74
	low			Dodecane	298.15	4.2	74
	low			3-Ethylpentane	298.15	3.7	74
	low			2,2,4,4,6,8,8-Heptamethylnonane	298.15	4.8	74
	low			Hexadecane	298.15	4.9	74
	low			3-Methylhexane	298.15	3.6	74
	low			Octane	298.15	4.3	74
	low			2,2,4,6,6-Pentamethylheptane	298.15	5.0	74
	low			2,2,4,4-Tetramethylpentane	298.15	5.8	74
	low			2,3,3,4-Tetramethylpentane	298.15	5.1	74
Poly(1-butene)							
			20000	Cyclohexane	298.15	1.0	74
			20000	Cyclooctane	298.15	1.8	74
			20000	Cyclopentane	298.15	–2.9	74
			20000	*cis*-Decahydronaphthalene	298.15	<0.1	74
			20000	*trans*-Decahydronaphthalene	298.15	–2.0	74
			20000	Decane	298.15	1.2	62
			20000	3,3-Diethylpentane	298.15	–2.6	74
			20000	2,2-Dimethylpentane	298.15	–4.0	74
			20000	2,3-Dimethylpentane	298.15	–2.8	74
			20000	2,4-Dimethylpentane	298.15	–2.3	74
			20000	3,3-Dimethylpentane	298.15	–2.2	74
			20000	Dodecane	298.15	2.1	74

Polymer	M_n/ g/mol	M_w/ g/mol	M_η/ g/mol	Solvent	T/K	ΔH_B^∞/ J/g	Ref.
			20000	3-Ethylpentane	298.15	−2.8	74
			20000	Heptane	298.15	0.0	73
			20000	Hexadecane	298.15	3.0	62
			20000	Hexane	298.15	−1.2	62
			20000	3-Methylhexane	298.15	−2.1	74
			20000	Nonane	298.15	0.9	73
			20000	Octane	298.15	0.4	73
			20000	2,2,4,6,6-Pentamethylheptane	298.15	0.6	73
			20000	Pentane	298.15	−2.6	62
			20000	Tetradecane	298.15	2.7	62
			20000	2,2,4,4-Tetramethylpentane	298.15	−1.4	74
			20000	2,3,3,4-Tetramethylpentane	298.15	−2.2	74
			20000	2,2,4-Trimethylpentane	298.15	−0.5	73
Poly(butyl acrylate)							
				2-Propanone	298.15	0.8	25
Poly(butyl methacrylate)							
Glass	91300	210000		Cyclohexanone	304.15	7.7	99
Liquid	91300	210000		Cyclohexanone	304.15	8.2	99
				2-Propanone	298.15	19.5	25
Polychloroprene							
				Benzene	298.15	0.5	7
Poly(2,6-dimethyl phenylene oxide)							
	17000	46400		1,2-Dichlorobenzene	303.05	55	89
Poly(dimethylsiloxane)							
	13000			Benzene	298.15	11.2	50
			20000	Benzene	298.15	13.5	61
			100000	Benzene	298.15	14.2	40
			170000	Bromocyclohexane	303.15	10.2	51
			80000	2-Butanone	293.15	14.4	77
	30900			2-Butanone	303.15	14.2	44
			170000	2-Butanone	303.15	14.7	51
			80000	2-Butanone	308.15	14.3	77
			80000	2-Butanone	323.15	14.3	77
			80000	Butyl acetate	298.15	6.1	41
			80000	Butyl propanoate	298.15	4.9	41
			100000	Chlorobenzene	298.15	7.5	40
	13000			Cyclohexane	298.15	3.0	50
			20000	Cyclohexane	298.15	5.2	74
			100000	Cyclohexane	298.15	5.2	40
			20000	Cyclooctane	298.15	6.8	74
			20000	Cyclopentane	298.15	1.0	74
			20000	*cis*-Decahydronapthalene	298.15	7.1	74
			20000	*trans*-Decahydronapthalene	298.15	4.3	74
			20000	Decamethyltetrasiloxane	297.65	0.45	37
			20000	Decane	298.15	3.8	37
			20000	Decane	298.15	3.9	61

Polymer	M_n/ g/mol	M_w/ g/mol	M_η/ g/mol	Solvent	T/K	ΔH_B^∞/ J/g	Ref.
			80000	Decane	298.15	3.8	41
			80000	Decyl acetate	298.15	4.5	41
			80000	Dibutyl ether	298.15	0.6	41
			80000	Diethoxymethane	298.15	1.8	41
			80000	Diethyl ether	298.15	−1.3	41
			20000	3,3-Diethylpentane	298.15	1.9	74
			80000	Dihexyl ether	298.15	3.0	41
			80000	1,2-Dimethoxyethane	298.15	12.2	41
			80000	Dimethoxymethane	298.15	7.4	41
	13000			1,2-Dimethylbenzene	298.15	4.3	50
	13000			1,3-Dimethylbenzene	298.15	3.0	50
	13000			1,4-Dimethylbenzene	298.15	3.2	50
			20000	1,4-Dimethylbenzene	298.15	4.2	61
			80000	2,6-Dimethyl-4-heptanone	293.15	6.1	77
			20000	2,2-Dimethylpentane	298.15	0.8	74
			20000	2,3-Dimethylpentane	298.15	1.4	74
			20000	2,4-Dimethylpentane	298.15	1.6	74
			20000	3,3-Dimethylpentane	298.15	0.5	74
			80000	Dipentyl ether	298.15	2.1	41
			80000	Dipropyl ether	298.15	−1.2	41
			20000	Dodecamethylpentasiloxane	297.65	−0.3	37
			20000	Dodecane	297.65	4.5	37
			20000	Dodecane	298.15	4.4	73
			80000	Dodecane	298.15	4.5	41
			80000	Ethyl acetate	298.15	12.7	41
			170000	Ethyl acetate	303.15	13.7	51
	13000			Ethylbenzene	298.15	6.4	50
			20000	Ethylbenzene	298.15	6.2	61
			80000	Ethyl butanoate	298.15	6.0	41
			80000	Ethyl decanoate	298.15	3.8	41
			80000	Ethyl dodecanoate	298.15	3.8	41
			80000	Ethyl heptanoate	298.15	4.1	41
			80000	Ethyl hexanoate	298.15	4.3	41
			80000	Ethyl nonanoate	298.15	3.7	41
			80000	Ethyl octanoate	298.15	3.8	41
			20000	3-Ethylpentane	298.15	0.6	74
			80000	Ethyl propanoate	298.15	8.0	41
			20000	2,2,4,4,6,8,8-Heptamethylnonane	298.15	3.5	74
	13000			Heptane	298.15	1.8	50
			20000	Heptane	298.15	1.9	73
			20000	Heptane	297.65	2.0	37
			20000	Heptane	298.15	2.0	61
			80000	Heptane	298.15	2.0	41
			100000	Heptane	298.15	2.1	40
			170000	Heptane	303.15	1.9	51
			80000	3-Heptanone	308.15	8.8	77
			80000	3-Heptanone	323.15	8.8	77

Polymer	M_n/ g/mol	M_w/ g/mol	M_η/ g/mol	Solvent	T/K	ΔH_B^∞/ J/g	Ref.
			20000	Hexadecane	297.65	5.5	37
			20000	Hexadecane	298.15	5.5	73
			20000	Hexamethyldisiloxane	298.15	−1.2	37
			20000	Hexamethyldisiloxane	298.15	−1.6	61
			170000	Hexamethyldisiloxane	303.15	−1.5	51
			20000	Hexane	297.65	0.7	37
			20000	Hexane	298.15	0.7	61
			80000	Hexane	298.15	0.7	41
			170000	Hexane	303.15	0.3	51
			80000	Hexyl acetate	298.15	5.0	41
	13000			Isopropylbenzene	298.15	4.1	50
			80000	Methyl butanoate	298.15	8.6	41
	13000			Methylcyclohexane	298.15	2.9	50
			100000	Methylcyclohexane	298.15	1.9	40
			80000	Methyl decanoate	298.15	4.8	41
			20000	3-Methylhexane	298.15	1.3	74
			80000	Methyl hexanoate	298.15	5.3	41
			80000	Methyl octanoate	298.15	5.0	41
			80000	4-Methyl-2-pentanone	293.15	9.9	77
			80000	4-Methyl-2-pentanone	308.15	9.0	77
			80000	Methyl propanoate	298.15	12.1	41
			20000	Nonane	297.65	3.4	37
			20000	Nonane	298.15	3.3	73
			80000	Nonane	298.15	3.4	41
			80000	Octamethylcyclotetrasiloxane	293.15	−0.4	78
			20000	Octamethyltrisiloxane	297.65	−0.6	37
			20000	Octamethyltrisiloxane	298.15	−0.8	61
			170000	Octamethyltrisiloxane	303.15	−1.0	51
			20000	Octane	297.65	2.6	37
			20000	Octane	298.15	2.4	73
			20000	Octane	298.15	2.6	61
			80000	Octane	298.15	2.6	41
			20000	2,2,4,6,6-Pentamethylheptane	298.15	2.7	73
			20000	Pentane	298.15	−0.9	37
			20000	Pentane	298.15	−0.9	61
			80000	Pentane	298.15	−1.0	41
			80000	Pentyl acetate	298.15	5.8	41
			80000	Pentyl propanoate	298.15	3.8	41
			80000	Propyl acetate	298.15	8.6	41
			170000	Propyl acetate	303.15	9.9	51
			80000	Propyl propanoate	298.15	5.6	41
			100000	Tetrachloromethane	298.15	2.4	40
			20000	Tetradecane	297.65	5.1	37
			20000	Tetradecane	298.15	5.1	61
			80000	Tetradecane	298.15	5.1	41
			20000	2,2,4,4-Tetramethylpentane	298.15	2.1	73
			20000	2,2,4,4-Tetramethylpentane	298.15	2.3	74

Polymer	M_n/ g/mol	M_w/ g/mol	M_η/ g/mol	Solvent	T/K	ΔH_B^∞/ J/g	Ref.
			20000	2,3,3,4-Tetramethylpentane	298.15	1.9	74
	13000			Toluene	298.15	5.5	50
			20000	Toluene	298.15	6.7	61
			20000	Tridecane	297.65	4.8	37
			80000	Tridecane	298.15	4.8	41
	13000			1,3,5-Trimethylbenzene	298.15	3.7	50
			20000	2,2,4-Trimethylpentane	298.15	1.4	73
			20000	Undecane	297.65	4.2	37
			80000	Undecane	298.15	4.3	41
Polyethylene							
Semicrystalline		65000		1-Chloronaphthalene	373.15	780	47
Semicrystalline		65000		1-Chloronaphthalene	383.15	980	47
Semicrystalline		65000		1-Chloronaphthalene	393.15	800	47
Liquid		65000		1-Chloronaphthalene	403.15	49	47
Semicrystalline		144000		1-Chloronaphthalene	383.15	920	47
Semicrystalline		144000		1-Chloronaphthalene	393.15	990	47
Semicrystalline		144000		1-Chloronaphthalene	403.15	690	47
Liquid		144000		1-Chloronaphthalene	413.15	67	47
Liquid		144000		1-Chloronaphthalene	423.15	85	47
Semicrystalline		670000		1-Chloronaphthalene	363.15	380	47
Semicrystalline		670000		1-Chloronaphthalene	373.15	430	47
Semicrystalline		670000		1-Chloronaphthalene	383.15	165	47
Liquid		670000		1-Chloronaphthalene	393.15	39	47
Liquid		670000		1-Chloronaphthalene	403.15	36	47
Semicrystalline		900000		1-Chloronaphthalene	391.80	245	105
Semicrystalline		900000		Cyclohexane	379.50	205	105
Semicrystalline		900000		Cyclopentane	380.00	190	105
Alkathene				Decahydronaphthalene	349.85	142	56
Rigidex-3				Decahydronaphthalene	366.65	180	56
Rigidex-50				Decahydronaphthalene	374.05	233	56
Semicrystalline		900000		Decahydronaphthalene	384.00	260	105
Semicrystalline				1,2-Dichloroethane	333.15	30	27
Semicrystalline				1,2-Dichloroethane	338.15	38	27
Semicrystalline				1,2-Dichloroethane	343.15	54	27
Semicrystalline				1,2-Dichloroethane	348.15	65	27
Semicrystalline			10000	1,4-Dimethylbenzene	354.15	139	5
Semicrystalline			11800	1,4-Dimethylbenzene	352.15	139	5
Semicrystalline			15600	1,4-Dimethylbenzene	353.65	154	5
Semicrystalline			15600	1,4-Dimethylbenzene	363.65	113	5
Semicrystalline			15600	1,4-Dimethylbenzene	368.65	104	5
Semicrystalline		900000		2,4-Dimethylpentane	393.00	230	105
Semicrystalline		900000		2,2,4,4,6,8,8-Heptamethylnonane	399.50	170	105
Semicrystalline		900000		Hexadecane	399.50	262	105
Semicrystalline		900000		2-Methylbutane	394.20	165	105
Semicrystalline		65000		1,2,3,4-Tetrahydronaphthalene	373.15	940	47

Polymer	M_n/ g/mol	M_w/ g/mol	M_η/ g/mol	Solvent	T/K	ΔH_B^∞/ J/g	Ref.
Semicrystalline		65000		1,2,3,4-Tetrahydronaphthalene	383.15	990	47
Semicrystalline		65000		1,2,3,4-Tetrahydronaphthalene	393.15	790	47
Liquid		65000		1,2,3,4-Tetrahydronaphthalene	403.15	58	47
Semicrystalline		84000		1,2,3,4-Tetrahydronaphthalene	373.15	835	47
Semicrystalline		130000		1,2,3,4-Tetrahydronaphthalene	353.15	630	47
Semicrystalline		130000		1,2,3,4-Tetrahydronaphthalene	373.15	520	47
Liquid		130000		1,2,3,4-Tetrahydronaphthalene	393.15	69	47
Semicrystalline		144000		1,2,3,4-Tetrahydronaphthalene	373.15	610	47
Semicrystalline		144000		1,2,3,4-Tetrahydronaphthalene	383.15	1200	47
Semicrystalline		144000		1,2,3,4-Tetrahydronaphthalene	393.15	1130	47
Semicrystalline		144000		1,2,3,4-Tetrahydronaphthalene	403.15	800	47
Liquid		144000		1,2,3,4-Tetrahydronaphthalene	413.15	136	47
Liquid		144000		1,2,3,4-Tetrahydronaphthalene	423.15	88	47
Semicrystalline		310000		1,2,3,4-Tetrahydronaphthalene	343.15	485	47
Semicrystalline		670000		1,2,3,4-Tetrahydronaphthalene	353.15	560	47
Semicrystalline		670000		1,2,3,4-Tetrahydronaphthalene	363.15	560	47
Semicrystalline		670000		1,2,3,4-Tetrahydronaphthalene	373.15	460	47
Semicrystalline		670000		1,2,3,4-Tetrahydronaphthalene	383.15	155	47
Liquid		670000		1,2,3,4-Tetrahydronaphthalene	393.15	67	47
Liquid		670000		1,2,3,4-Tetrahydronaphthalene	403.15	39	47
Semicrystalline			16000	Toluene	353.15	110	22
Semicrystalline			22000	Toluene	358.35	118	22
Semicrystalline			22000	Toluene	367.35	106	22
Semicrystalline		900000		1,2,4-Trichlorobenzene	386.50	255	105
Poly(ethylene glycol)							
	180			Benzene	303.15	110	71
	385			Benzene	303.15	60	71
	560			Benzene	303.15	40	71
	1050			Benzene	303.15	90	71
	1610			Benzene	303.15	140	71
	1940			Benzene	303.15	215	71
	3200			Benzene	303.15	195	71
	4330			Benzene	303.15	195	71

Polymer	M_n/ g/mol	M_w/ g/mol	M_η/ g/mol	Solvent	T/K	ΔH_B^∞/ J/g	Ref.
	5850			Benzene	303.15	190	71
	9950			Benzene	303.15	195	71
			43400	Benzene	303.15	190	71
	400	420		Trichloromethane	303.15	–79	95
	590	615		Trichloromethane	303.15	–88	95
	180			Water	303.15	–136	71
	200			Water	321.35	–125	83
	355			Water	303.15	–159	71
	400			Water	321.35	–150	83
	560			Water	303.15	–150	71
	990			Water	321.35	–101	83
	1050			Water	303.15	–106	71
	1460			Water	321.35	–137	83
	1610			Water	303.15	–6	71
	1940			Water	303.15	57	71
	3200			Water	303.15	58	71
	4330			Water	303.15	28	71
	5850			Water	303.15	39	71
	9950			Water	303.15	30	71
			14000	Water	303.15	7	55
			14000	Water	313.15	27	55
			20300	Water	303.15	45	71
			34500	Water	303.15	34	71
			43300	Water	303.15	+40	71
Poly(ethylene glycol) dimethyl ether							
	250			Tetrachloromethane	303.15	–12	95
	250			Tetrachloromethane	318.15	–7.6	95
	400			Tetrachloromethane	303.15	–12	95
	520	550		Tetrachloromethane	303.15	–12	95
	520	550		Tetrachloromethane	303.15	–7.6	95
	250			Trichloromethane	303.15	–184	95
	520	550		Trichloromethane	303.15	–135	95
Poly(ethylene glycol) monododecyl ether							
	230			Dodecane	302.15	42	68
	274			Dodecane	302.15	23	68
	318			Dodecane	302.15	34	68
	362			Dodecane	302.15	37	68
	406			Dodecane	302.15	42	68
Poly(ethylene glycol) monomethyl ether							
	353	377		Trichloromethane	303.15	–125	95
	550	580		Trichloromethane	303.15	–117	95
Poly(ethylene oxide)							
Semicrystalline	6000			Dichloromethane	303.15	+84	58
Liquid	6000			Dichloromethane	303.15	–160	58

Polymer	M_n/ g/mol	M_w/ g/mol	M_η/ g/mol	Solvent	T/K	ΔH_B^∞/ J/g	Ref.
Semicrystalline	6000			Trichloromethane	303.15	+52	58
Liquid	6000			Trichloromethane	303.15	−186	58
Quenched	1520	1720		Water	293.15	−403	64
Annealed	1520	1720		Water	293.15	−392	64
Quenched	1520	1720		Water	298.15	−180	64
Annealed	1520	1720		Water	298.15	−150	64
Quenched	1520	1720		Water	303.15	+68	64
Annealed	1520	1720		Water	303.15	+109	64
Liquid	6000			Water	303.15	−50	58
Quenched	6840	7525		Water	293.15	−28	64
Annealed	6840	7525		Water	293.15	+209	64
Quenched	6840	7525		Water	298.15	+241	64
Annealed	6840	7525		Water	298.15	+540	64
Quenched	16600	19600		Water	293.15	−160	64
Annealed	16600	19600		Water	293.15	−143	64
Quenched	16600	19600		Water	298.15	+59	64
Annealed	16600	19600		Water	298.15	+155	64
Quenched	16600	19600		Water	303.15	+353	64
Annealed	16600	19600		Water	303.15	+490	64
Semicrystalline		20000		Water	298.15	+10	84
Polyindene							
	765	1023		Anisole	299.15	2.1	102
	765	1023		Benzene	299.15	−0.04	102
	765	1023		Benzonitrile	299.15	−4.4	102
	765	1023		Bromobenzene	299.15	−3.9	102
	765	1023		2-Butanone	299.15	1.9	102
	765	1023		Chlorobenzene	299.15	−3.9	102
	765	1023		1-Chlorobutane	299.15	−4.0	102
	765	1023		1-Chloroheptane	299.15	1.5	102
	765	1023		Cyclohexane	299.15	15	102
	765	1023		*N*,*N*-dimethylaniline	299.15	−8.2	102
	765	1023		Ethyl acetate	299.15	4.2	102
	765	1023		Ethylbenzene	299.15	−1.4	102
	765	1023		Ethyl benzoate	299.15	−0.7	102
	765	1023		Nitrobenzene	299.15	4.6	102
	765	1023		1-Nitropropane	299.15	8.4	102
	765	1023		Pyridine	299.15	−6.7	102
	765	1023		1,1,2,2-Tetrachloroethane	299.15	−19	102
	765	1023		Tetrachloromethane	299.15	−2.5	102
	765	1023		1,1,1-Trichloroethane	299.15	−1.8	102
	765	1023		Trichloromethane	299.15	−20	102
Polyisobutylene							
	360		700	Benzene	298.15	30	67
	1000		2000	Benzene	298.15	25	67
	1300		2500	Benzene	298.15	23	67
			30000	Benzene	297.65	19	38
			30000	Benzene	298.15	19	40

Polymer	M_n/ g/mol	M_w/ g/mol	M_η/ g/mol	Solvent	T/K	ΔH_B^∞/ J/g	Ref.
	44700			Benzene	303.15	16	44
			48000	Benzene	303.4	19	72
			50000	Benzene	303.15	16	45
			72000	Benzene	300.15	19	53
			72000	Benzene	323.15	18	53
			72000	Benzene	343.15	16	53
			72000	Benzene	375.15	13	53
			72000	Benzene	394.15	9.2	53
			72000	Benzene	423.15	3.5	53
			72000	Benzene	437.15	−0.5	53
			72000	Benzene	453.15	−4.7	53
			90000	Benzene	298.15	6.7	32
			160000	Benzene	303.15	16	51
			560000	Benzene	298.15	18	34
				Benzene	298.15	6.8	7
				Benzene	298.15	6.8	10
			30000	Chlorobenzene	297.65	12	38
			30000	Chlorobenzene	298.15	13	40
			50000	Chlorobenzene	303.15	12	45
			160000	Chlorobenzene	303.15	12	51
			560000	Chlorobenzene	298.15	12	34
	360		700	Cyclohexane	298.15	3.8	67
	1000		2000	Cyclohexane	298.15	1.2	67
	1300		2500	Cyclohexane	298.15	1.1	67
			4500	Cyclohexane	298.15	−0.6	74
			30000	Cyclohexane	297.65	−0.7	38
			30000	Cyclohexane	298.15	−0.6	40
			50000	Cyclohexane	303.15	−0.7	45
			160000	Cyclohexane	303.15	−0.6	51
			1990000	Cyclohexane	298.15	−0.7	39
			4500	Cyclooctane	298.15	+0.3	74
			4500	Cyclopentane	298.15	−5.9	74
			4500	*cis*-Decahydronaphthalene	298.15	0.2	74
			4500	*trans*-Decahydronaphthalene	298.15	−0.8	74
			30000	Decane	297.65	−0.5	38
			50000	Decane	303.15	−0.5	45
			30000	Dibutyl ether	297.65	1.2	37
			30000	Diethyl ether	297.65	2.8	37
			30000	Diethyl ether	297.65	2.8	38
			4500	3,3-Diethylpentane	298.15	−1.4	74
			30000	Dihexyl ether	297.65	0.9	37
			4500	2,2-Dimethylpentane	298.15	−1.1	74
			4500	2,3-Dimethylpentane	298.15	−1.9	74
			4500	2,4-Dimethylpentane	298.15	−1.1	74
			4500	3,3-Dimethylpentane	298.15	−1.7	74
			30000	Dipentyl ether	297.65	1.0	37
			30000	Dipropyl ether	297.65	1.8	37

Polymer	M_n/ g/mol	M_w/ g/mol	M_η/ g/mol	Solvent	T/K	ΔH_B^∞/ J/g	Ref.
	360		700	Dodecane	298.15	1.9	67
	1000		2000	Dodecane	298.15	0.7	67
	1300		2500	Dodecane	298.15	0.5	67
			4500	Dodecane	298.15	0.2	73
			30000	Dodecane	297.65	−0.1	38
			30000	Dodecane	298.15	−0.1	40
			48000	Ethylbenzene	291.15	9.5	72
			48000	Ethylbenzene	343.15	3.6	72
			30000	Ethyl decanoate	297.65	3.0	37
			30000	Ethyl heptanoate	297.65	5.6	37
			30000	Ethyl hexadecanoate	297.65	1.3	37
			30000	Ethyl hexanoate	297.65	6.7	37
			30000	Ethyl nonanoate	297.65	3.7	37
			30000	Ethyl octanoate	297.65	4.6	37
			4500	3-Ethylpentane	298.15	−2.0	74
			30000	Ethyl tetradecanoate	297.65	1.8	37
			4500	2,2,4,4,6,8,8-Heptamethylnonane	298.15	−0.5	74
	360		700	Heptane	298.15	−0.5	67
	1000		2000	Heptane	298.15	−1.0	67
	1300		2500	Heptane	298.15	−1.4	67
			4500	Heptane	298.15	−1.7	73
			30000	Heptane	297.65	−1.8	38
			30000	Heptane	298.15	−2.0	40
			50000	Heptane	303.15	−1.8	45
			160000	Heptane	303.15	−1.6	51
				Heptane	298.15	−1.4	7
				Heptane	298.15	−1.4	10
	360		700	Hexadecane	298.15	4.5	67
	1000		2000	Hexadecane	298.15	2.1	67
	1300		2500	Hexadecane	298.15	1.0	67
			4500	Hexadecane	298.15	0.9	73
			30000	Hexadecane	297.65	0.04	38
			30000	Hexane	297.65	−2.5	38
			30000	Hexane	298.15	−2.6	40
			50000	Hexane	303.15	−2.5	45
			72000	Hexane	303.15	−1.8	53
			72000	Hexane	324.15	−2.3	53
			72000	Hexane	348.15	−2.9	53
			72000	Hexane	373.15	−3.7	53
			72000	Hexane	393.15	−5.3	53
			72000	Hexane	408.15	−6.7	53
			72000	Hexane	423.15	−9.0	53
			72000	Hexane	433.15	−9.9	53
			160000	Hexane	303.15	−2.5	51
			30000	2-Methylbutane	297.65	−3.1	38
			30000	Methylcyclohexane	297.65	−1.2	38
			50000	Methylcyclohexane	303.15	−1.2	45

Polymer	M_n/ g/mol	M_w/ g/mol	M_η/ g/mol	Solvent	T/K	ΔH_B^∞/ J/g	Ref.
			160000	Methylcyclohexane	303.15	−1.2	51
			4500	3-Methylhexane	298.15	−1.0	74
			30000	3-Methylpentane	297.65	−2.8	38
			30000	Nonane	297.65	−0.8	38
			4500	Nonane	298.15	−0.8	73
			4500	Octane	298.15	−1.1	73
			30000	Octane	297.65	−1.2	38
			72000	Octane	303.15	−0.3	53
			72000	Octane	324.15	−0.8	53
			72000	Octane	348.15	−0.9	53
			72000	Octane	373.15	−1.1	53
			72000	Octane	393.15	−1.3	53
			72000	Octane	423.15	3.6	53
			4500	2,2,4,6,6-Pentamethylheptane	298.15	−0.1	73
	360		700	Pentane	298.15	−1.9	67
	1000		2000	Pentane	298.15	−2.9	67
	1300		2500	Pentane	298.15	−3.2	67
			30000	Pentane	297.65	−3.6	38
			72000	Pentane	303.15	−2.8	53
			72000	Pentane	333.15	−3.4	53
			72000	Pentane	352.15	−4.5	53
			72000	Pentane	365.15	−5.5	53
	360		700	Tetrachloromethane	298.15	5.9	67
	1000		2000	Tetrachloromethane	298.15	5.8	67
	1300		2500	Tetrachloromethane	298.15	5.0	67
			1990000	Tetrachloromethane	298.15	4.1	39
			30000	Tetradecane	297.65	0.0	38
			4500	2,2,4,4-Tetramethylpentane	298.15	−0.6	74
			4500	2,3,3,4-Tetramethylpentane	298.15	−2.3	74
			50000	Toluene	303.15	7.4	45
			160000	Toluene	303.15	7.4	51
			1990000	Toluene	298.15	8.8	39
				Toluene	298.15	1.8	7
				Toluene	298.15	1.8	10
			30000	Tridecane	297.65	−0.04	38
			4500	2,2,4-Trimethylpentane	298.15	−0.4	73
			30000	2,2,4-Trimethylpentane	297.65	−0.6	38
			1990000	2,2,4-Trimethylpentane	298.15	0.0	39
				2,2,4-Trimethylpentane	298.15	0.0	7
				2,2,4-Trimethylpentane	298.15	0.0	10
			30000	Undecane	297.65	−0.4	38
Poly(isobutyl methacrylate)							
Glass		260000		Cyclohexanone	303.15	−5.2	98
Liquid		260000		Cyclohexanone	303.15	13	98
Poly(methyl acrylate)							
				2-Propanone	298.15	0.0	25

Polymer	M_n/ g/mol	M_w/ g/mol	M_η/ g/mol	Solvent	T/K	ΔH_B^∞/ J/g	Ref.
Poly(methyl methacrylate)							
Glass	73900	170000		Cyclohexanone	304.15	−14	98
Liquid	73900	170000		Cyclohexanone	304.15	17	98
	1930			1,2-Dichloroethane	298.15	−20	32
	240000			1,2-Dichloroethane	298.15	−27	32
			53000	Ethylbenzene	298.15	−31	28
			180000	Ethylbenzene	298.15	−29	28
	28900	35900		4-Methyl-2-pentanone	303.15	−21	76
	93940	101000		4-Methyl-2-pentanone	303.15	−24	76
	137000	215000		4-Methyl-2-pentanone	303.15	−28	76
				2-Propanone	298.15	−30	25
	93940	101000		Toluene	303.15	−22	76
	689000	782000		Toluene	303.15	−24	76
			12000	Trichloromethane	298.15	−65	52
			54000	Trichloromethane	298.15	−80	52
			80000	Trichloromethane	298.15	−81	52
			100000	Trichloromethane	298.15	−84	52
			320000	Trichloromethane	298.15	−83	52
	93940	101000		Trichloromethane	303.15	−71	76
	689000	782000		Trichloromethane	303.15	−72	76
		2320000		Trichloromethane	303.15	−73	76
Poly(4-methyl-1-pentene)							
Semicrystalline			350000	Cyclohexane	303.15	30	79
Poly(α-methylstyrene)							
	1030	1180		Toluene	298.15	−7.1	43
		1430		Toluene	298.15	−30	43
	1820	2230		Toluene	298.15	−34	43
	1920			Toluene	298.15	−37	43
	2700	3300		Toluene	298.15	−39	43
	3280			Toluene	298.15	−46	43
	5260			Toluene	298.15	−46	43
	8600			Toluene	298.15	−45	43
	12200			Toluene	298.15	−46	43
		10500		Toluene	310.15	−8.4	92
		53000		Toluene	310.15	−13	92
		55000		Toluene	333.15	−16	96
		87000		Toluene	298.15	−17	90
		87000		Toluene	310.15	−16	92
		87000		Toluene	333.15	−11	90
Poly(2-methyl-5-vinyltetrazole)							
				Acetic acid	298.15	47	100
				Acetonitrile	298.15	14	100
				1,2-Dichloroethane	298.15	17	100
				N,N-Diethylacetamide	298.15	17	100
				N,N-Dimethylformamide	298.15	33	100
				Dimethylsulfoxide	298.15	10	100

Polymer	M_n/ g/mol	M_w/ g/mol	M_η/ g/mol	Solvent	T/K	ΔH_B^∞/ J/g	Ref.
				Formamide	298.15	12	100
				Formic acid	298.15	110	100
				Nitromethane	298.15	10	100
				Pyridine	298.15	16	100
Poly(octamethylene oxide)							
	7000			Benzene	298.15	20	64
	7000			Benzene	303.15	22	64
	7000			Benzene	308.15	25	64
Polypentenamer							
			50000	Cyclohexane	298.15	4.6	74
			50000	Cyclooctane	298.15	5.1	74
			50000	Cyclopentane	298.15	−2.3	74
			50000	*cis*-Decahydronaphthalene	298.15	2.6	74
			50000	*trans*-Decahydronaphthalene	298.15	<0.1	74
			50000	3,3-Diethylpentane	298.15	2.4	74
			50000	2,2-Dimethylpentane	298.15	3.3	74
			50000	2,3-Dimethylpentane	298.15	2.2	74
			50000	2,4-Dimethylpentane	298.15	3.3	74
			50000	3,3-Dimethylpentane	298.15	2.7	74
			50000	Dodecane	298.15	2.9	74
			50000	3-Ethylpentane	298.15	2.1	74
			50000	2,2,4,4,6,8,8-Heptamethylnonane	298.15	3.2	74
			50000	Hexadecane	298.15	2.6	74
			50000	3-Methylhexane	298.15	2.4	74
			50000	Octane	298.15	2.2	74
			50000	2,2,4,6,6-Pentamethylheptane	298.15	3.8	74
			50000	2,2,4,4-Tetramethylpentane	298.15	4.5	74
			50000	2,3,3,4-Tetramethylpentane	298.15	2.4	74
			50000	2,2,4-Trimethylpentane	298.15	4.3	74
Poly(m-phenyleneisophthalamide)							
Glass				*N,N*-Dimethylacetamide	298.15	−171	60
Semicrystalline				*N,N*-Dimethylacetamide	298.15	−128	60
Glass				*N,N*-Dimethylformamide	298.15	−149	60
Semicrystalline				*N,N*-Dimethylformamide	298.15	−125	60
Glass				1-Methyl-2-pyrrolidone	298.15	−177	60
Semicrystalline				1-Methyl-2-pyrrolidone	298.15	−118	60
Polypropylene (atactic)							
			18000	Benzene	298.15	31	80
			6000	Cyclohexane	298.15	2.3	74
			18000	Cyclohexane	298.15	3.9	80
			6000	Cyclooctane	298.15	3.0	74
			6000	Cyclopentane	298.15	−2.3	74
			6000	*cis*-Decahydronaphthalene	298.15	0.5	74

Polymer	M_n/ g/mol	M_w/ g/mol	M_η/ g/mol	Solvent	T/K	ΔH_B^∞/ J/g	Ref.
			6000	*trans*-Decahydronaphthalene	298.15	−2.4	74
			18000	Decane	298.15	3.1	80
			6000	3,3-Diethylpentane	298.15	−3.9	74
			18000	1,2-Dimethylbenzene	298.15	13	80
			18000	1,3-Dimethylbenzene	298.15	12	80
			18000	1,4-Dimethylbenzene	298.15	10	80
			6000	2,2-Dimethylpentane	298.15	−2.2	74
			6000	2,3-Dimethylpentane	298.15	−2.5	74
			6000	2,4-Dimethylpentane	298.15	−1.8	74
			6000	3,3-Dimethylpentane	298.15	−3.0	74
			6000	Dodecane	298.15	1.7	73
			18000	Ethylbenzene	298.15	14	80
			6000	3-Ethylpentane	298.15	−2.5	74
			6000	2,2,4,4,6,8,8-Heptamethylnonane	298.15	−0.7	73
			6000	Heptane	298.15	−1.6	73
			18000	Heptane	298.15	0.5	80
			6000	Hexadecane	298.15	2.3	73
			18000	Hexane	298.15	−1.4	80
			6000	3-Methylhexane	298.15	−1.8	74
			6000	Nonane	298.15	0.8	73
			18000	Nonane	298.15	2.4	80
			6000	Octane	298.15	−1.2	73
			18000	Octane	298.15	1.0	80
			6000	2,2,4,6,6-Pentamethylheptane	298.15	−0.2	73
			18000	Pentane	298.15	−4.7	80
			18000	Tetrachloromethane	298.15	6.6	80
			6000	2,2,4,4-Tetramethylpentane	298.15	−0.8	74
			6000	2,3,3,4-Tetramethylpentane	298.15	−3.1	74
			18000	Toluene	298.15	17	80
			18000	Trichloromethane	298.15	17	80
			6000	2,2,4-Trimethylpentane	298.15	−1.0	73
Polypropylene (isotactic)							
				1-Chloronaphthalene	383.15	26	47
				1-Chloronaphthalene	393.15	170	47
				1-Chloronaphthalene	403.15	245	47
				1-Chloronaphthalene	423.15	275	47
				1,2,3,4-Tetrahydronaphthalene	373.15	140	47
				1,2,3,4-Tetrahydronaphthalene	383.15	215	47
				1,2,3,4-Tetrahydronaphthalene	393.15	330	47
				1,2,3,4-Tetrahydronaphthalene	403.15	330	47
				1,2,3,4-Tetrahydronaphthalene	413.15	335	47
				1,2,3,4-Tetrahydronaphthalene	423.15	290	47

Polymer	M_n/ g/mol	M_w/ g/mol	M_η/ g/mol	Solvent	T/K	ΔH_B^∞/ J/g	Ref.
Poly(propylene glycol)							
	150			Benzene	321.35	200	103
	425			Benzene	321.35	80	103
	2025			Benzene	321.35	45	103
	150			Ethanol	321.35	40	103
	425			Ethanol	321.35	60	103
	2025			Ethanol	321.35	65	103
	396	412		Tetrachloromethane	303.15	4.7	95
	396	412		Tetrachloromethane	318.15	5.2	95
	1900			Tetrachloromethane	303.15	–8.2	95
	1900			Tetrachloromethane	318.15	11	95
	1900			Trichloromethane	303.15	–81	95
		400		Water	298.15	–165	97
	150			Water	321.35	–90	103
	425			Water	321.35	–95	103
Polystyrene							
		600		Benzene	298.15	–1.3	54
		600		Benzene	313.15	–2.5	54
		900		Benzene	291.15	–10	54
		900		Benzene	318.15	–5.8	54
		2000		Benzene	291.15	–16	54
		2000		Benzene	318.15	–6.8	54
		5000		Benzene	291.15	–23	54
		5000		Benzene	318.15	–12	54
		10300		Benzene	291.15	–26	54
		10300		Benzene	318.15	–18	54
			18000	Benzene	298.15	–4.1	19
	20000			Benzene	296.15	–15	9
			29000	Benzene	298.15	–5.0	19
			30000	Benzene	298.15	–7.5	19
			59000	Benzene	298.15	–13	19
			91000	Benzene	298.15	–15	19
		97200		Benzene	318.15	–21	54
			142000	Benzene	298.15	–17	19
			190000	Benzene	303.15	–18	51
	214000			Benzene	300.15	–16	85
			216000	Benzene	298.15	–18	19
			272000	Benzene	298.15	–21	19
			300000	Benzene	298.15	–21	12
				Benzene	298.15	–27	25
				Benzene	298.15	–10	7
	20000			Butyl acetate	296.15	–13	9
	20000			2-Butanone	296.15	–15	9
			142000	2-Butanone	296.15	–17	30
			190000	Butylbenzene	303.15	–14	51
	150000			Chlorobenzene	293.15	–32	49
			266000	Chlorobenzene	298.15	5.4	34

Polymer	M_n/ g/mol	M_w/ g/mol	M_η/ g/mol	Solvent	T/K	ΔH_B^∞/ J/g	Ref.
				Chlorobenzene	293.15	−39	21
	1260			Cyclohexane	298.15	10	16
	1910			Cyclohexane	298.15	5.4	16
	3160			Cyclohexane	298.15	−5.4	16
	3980			Cyclohexane	298.15	−6.9	16
	5630			Cyclohexane	298.15	−9.3	16
	9070			Cyclohexane	298.15	−11	16
	20000			Cyclohexane	296.15	2.5	9
			190000	Cyclohexane	303.15	−2.1	51
				Cyclohexane	293.15	−14	21
	22400			Cyclohexanone	298.15	−29	25
	20000			Cyclohexene	296.15	−9.4	9
			190000	Decahydronaphthalene	303.15	3.8	51
	110000	115000		1,2-Dichlorobenzene	303.05	26	89
	20000			1,2-Dimethylbenzene	296.15	−13	9
	20000			1,3-Dimethylbenzene	296.15	−12	9
			190000	1,3-Dimethylbenzene	303.15	−15	51
			190000	1,4-Dioxane	303.15	−12	51
	20000			Ethyl acetate	296.15	−11	9
			142000	Ethyl acetate	296.15	−13	30
			785	Ethylbenzene	298.15	0.0	14
			18000	Ethylbenzene	298.15	−3.8	14
			18000	Ethylbenzene	298.15	−3.9	19
			30000	Ethylbenzene	298.15	−5.7	19
			35000	Ethylbenzene	298.15	−6.5	19
			45000	Ethylbenzene	298.15	−8.4	19
			91000	Ethylbenzene	298.15	−11	19
			142000	Ethylbenzene	298.15	−13	19
			216000	Ethylbenzene	298.15	−17	19
	60000			Ethylbenzene	303.15	−22	57
	113000	122000		Ethylbenzene	306.65	−24	63
	113000	122000		Ethylbenzene	317.15	−19	63
	113000	122000		Ethylbenzene	337.15	−11	63
	113000	122000		Ethylbenzene	347.15	−6.4	63
	113000	122000		Ethylbenzene	350.65	−4.9	63
	113000	122000		Ethylbenzene	366.65	−2.3	63
	113000	122000		Ethylbenzene	367.15	−2.1	63
	113000	122000		Ethylbenzene	368.65	−2.6	63
	113000	122000		Ethylbenzene	372.15	−4.2	63
	113000	122000		Ethylbenzene	378.15	−4.6	63
	113000	122000		Ethylbenzene	385.15	−5.7	63
	150000			Ethylbenzene	293.15	−34	49
			190000	Ethylbenzene	303.15	−15	51
			272000	Ethylbenzene	298.15	−18	14
			272000	Ethylbenzene	298.15	−18	19
			413000	Ethylbenzene	298.15	−24	39
				Ethylbenzene	293.15	−34	21

Polymer	M_n/ g/mol	M_w/ g/mol	M_η/ g/mol	Solvent	T/K	ΔH_B^∞/ J/g	Ref.
				Ethylbenzene	298.15	−17	11
				Ethylbenzene	298.15	−30	28
	1260			2-Propanone	298.15	−0.6	16
	1910			2-Propanone	298.15	−7.7	16
	3160			2-Propanone	298.15	−16	16
	3980			2-Propanone	298.15	−17	16
	5630			2-Propanone	298.15	−19	16
	9070			2-Propanone	298.15	−21	16
	20000			2-Propanone	296.15	−11	9
			190000	Propylbenzene	303.15	−14	51
	20000			Styrene	296.15	−18	9
				Styrene	296.15	−35	6
			413000	Tetrachloromethane	298.15	22	39
	600			Toluene	296.15	−2.1	16
	600			Toluene	309.15	−1.8	16
	600			Toluene	318.15	−1.5	16
	1260			Toluene	296.15	−11	16
	1260			Toluene	303.15	−8.0	16
	1260			Toluene	309.15	−5.9	16
	1260			Toluene	318.15	−3.4	16
	1260			Toluene	328.15	−2.3	16
	1260			Toluene	338.15	−1.9	16
	1260			Toluene	346.65	−1.3	16
	1910			Toluene	298.15	−16	16
	1910			Toluene	318.15	−6.7	16
	1910			Toluene	338.15	−3.4	16
	1910			Toluene	348.15	−2.5	16
	3160			Toluene	298.15	−23	16
	3980			Toluene	298.15	−24	16
	3980			Toluene	318.15	−17	16
	3980			Toluene	338.15	−10	16
	5630			Toluene	298.15	−26	16
	9070			Toluene	298.15	−28	16
	270000			Toluene	298.15	−33	16
		600		Toluene	298.15	−1.4	54
		600		Toluene	313.15	−3.2	54
		900		Toluene	291.15	−7.3	54
		900		Toluene	318.15	−6.6	54
		2000		Toluene	291.15	−11	54
		2000		Toluene	318.15	−7.2	54
		5000		Toluene	291.15	−21	54
		5000		Toluene	318.15	−11	54
		10300		Toluene	291.15	−24	54
		10300		Toluene	318.15	−15	54
		97200		Toluene	318.15	−17	54
		9000		Toluene	310.15	−9.2	92
	20000			Toluene	296.15	−17	9

Polymer	M_n/ g/mol	M_w/ g/mol	M_η/ g/mol	Solvent	T/K	ΔH_B^∞/ J/g	Ref.
	20400			Toluene	298.15	−8.2	90
	20400			Toluene	310.15	−8.4	92
	20400			Toluene	333.15	−6.4	90
	47000			Toluene	310.15	−5.0	92
	50000			Toluene	333.15	−6.8	96
	60000			Toluene	303.15	−21	57
	113000	122000		Toluene	304.15	−29	63
	113000	122000		Toluene	306.15	−27	63
	113000	122000		Toluene	306.65	−26	63
	113000	122000		Toluene	316.15	−23	63
	113000	122000		Toluene	333.15	−15	63
	113000	122000		Toluene	337.15	−12	63
	113000	122000		Toluene	346.15	−8.3	63
	113000	122000		Toluene	347.15	−7.8	63
	113000	122000		Toluene	348.15	−8.2	63
	113000	122000		Toluene	350.65	−6.5	63
	113000	122000		Toluene	359.15	−4.3	63
	113000	122000		Toluene	362.15	−2.7	63
	113000	122000		Toluene	369.15	−3.3	63
	113000	122000		Toluene	372.15	−2.8	63
	115000			Toluene	310.15	−5.0	92
	150000			Toluene	293.15	−34	49
			190000	Toluene	303.15	−18	51
	214000			Toluene	300.15	−19	85
			250000	Toluene	303.15	−18	51
				Toluene	293.15	−34	21
				Toluene	298.65	−39	27
				Toluene	308.15	−34	27
				Toluene	318.15	−30	27
				Toluene	333.15	−23	27
				Toluene	343.15	−13	27
				Toluene	353.15	−13	27
		600		Trichloromethane	298.15	−13	54
		600		Trichloromethane	313.15	−9.9	54
		900		Trichloromethane	291.15	−22	54
		900		Trichloromethane	313.15	−15	54
		2000		Trichloromethane	291.15	−28	54
		2000		Trichloromethane	313.15	−16	54
		5000		Trichloromethane	291.15	−30	54
		5000		Trichloromethane	313.15	−18	54
		10300		Trichloromethane	291.15	−33	54
		10300		Trichloromethane	313.15	−23	54
	22400			Trichloromethane	298.15	−17	25
		97200		Trichloromethane	313.15	−25	54
	20000			1,3,5-Trimethylbenzene	296.15	−11	9
			190000	1,3,5-Trimethylbenzene	303.15	−13	51

Polymer	M_n/ g/mol	M_w/ g/mol	M_η/ g/mol	Solvent	T/K	ΔH_B^∞/ J/g	Ref.
Poly(tetramethylene oxide)							
	650			Benzene	313.15	4.0	86
	1000			Benzene	313.15	2.0	86
	2000			Benzene	313.15	1.1	86
	650			1,2-Dichloroethane	313.35	3.1	88
	2000			1,2-Dichloroethane	313.35	0.3	88
	650			1,2-Dimethylbenzene	313.15	5.9	87
	1000			1,2-Dimethylbenzene	313.15	1.8	87
	2000			1,2-Dimethylbenzene	313.15	0.9	87
	650			1,3-Dimethylbenzene	313.15	6.4	87
	1000			1,3-Dimethylbenzene	313.15	0.6	87
	2000			1,3-Dimethylbenzene	313.15	0.8	87
	650			1,4-Dimethylbenzene	313.15	4.3	87
	1000			1,4-Dimethylbenzene	313.15	1.8	87
	2000			1,4-Dimethylbenzene	313.15	0.7	87
	650			1,4-Dioxane	321.35	4.0	81
	1000			1,4-Dioxane	321.35	2.3	81
	2000			1,4-Dioxane	321.35	1.0	81
	650			Ethylbenzene	313.15	6.9	86
	1000			Ethylbenzene	313.15	3.4	86
	2000			Ethylbenzene	313.15	−0.05	86
	650			Propylbenzene	313.15	5.8	86
	1000			Propylbenzene	313.15	1.3	86
	2000			Propylbenzene	313.15	0.9	86
	650			Tetrachloromethane	313.15	3.3	88
	1000			Tetrachloromethane	313.15	1.4	88
	2000			Tetrachloromethane	321.35	0.7	82
	650			Toluene	313.15	4.3	86
	1000			Toluene	313.15	2.0	86
	2000			Toluene	313.15	0.9	86
	650			1,3,5-Trimethylbenzene	313.15	6.1	87
	1000			1,3,5-Trimethylbenzene	313.15	2.7	87
	2000			1,3,5-Trimethylbenzene	313.15	0.6	87
Poly(vinyl acetate)							
			140000	Benzene	298.15	2.3	13
			350000	2-Butanone	303.15	−1.7	51
			350000	Butyl acetate	303.15	1.0	51
			135000	Chlorobenzene	298.15	5.0	34
			350000	Ethyl acetate	303.15	−6.7	51
				Ethyl acetate	303.15	0.0	11
			26000	3-Heptanone	303.15	7.0	44
			350000	3-Heptanone	303.15	4.9	51
			140000	Methanol	298.15	−45	13
			350000	Methyl acetate	303.15	−9.7	51
			350000	2-Pentanone	303.15	0.0	51
			93000	2-Propanone	303.15	−0.4	32

Polymer	M_n/ g/mol	M_w/ g/mol	M_η/ g/mol	Solvent	T/K	ΔH_B^∞/ J/g	Ref.
			350000	2-Propanone	303.15	−3.9	51
				2-Propanone	303.15	−2.9	25
			350000	Propyl acetate	303.15	−2.7	51
		150000		Tetrahydrofuran	304.65	4.5	93
			140000	Trichloromethane	298.15	28	13
Poly(vinyl alcohol)							
				Ethanol	298.15	3.8	11
				Ethanol	298.15	9.6	31
	7260			Water	303.15	−34	15
	17000			Water	303.15	−18	32
	61600			Water	303.15	−41	15
				Water	303.15	−8.4	11
Poly(vinyl chloride)							
				Chlorobenzene	298.15	−17	36
Glass	23200			Cyclohexanone	303.15	−27	59
Liquid	23200			Cyclohexanone	303.15	−7.5	59
Glass	38700			Cyclohexanone	303.15	−29	59
Liquid	38700			Cyclohexanone	303.15	−6.6	59
Glass	53500			Cyclohexanone	303.15	−28	59
Liquid	53500			Cyclohexanone	303.15	−6.3	59
Glass	66700			Cyclohexanone	303.15	−29	59
Liquid	66700			Cyclohexanone	303.15	−6.1	59
Glass	136000			Cyclohexanone	303.15	−31	59
Liquid	136000			Cyclohexanone	303.15	−5.8	59
Glass	155400			Cyclohexanone	303.15	−32	59
Liquid	155400			Cyclohexanone	303.15	−5.8	59
	48000			Cyclopentanone	298.15	−28	104
				1,2-Dichloroethane	323.65	24	27
				1,2-Dichloroethane	328.15	34	27
				1,2-Dichloroethane	333.15	38	27
				1,2-Dichloroethane	368.15	44	27
				1,2-Dichloroethane	373.15	46	27
				1,2-Dichloroethane	378.15	46	27
				N,N-Dimethylformamide	293.15	−28	35
				N,N-Dimethylformamide	308.15	−19	35
				N,N-Dimethylformamide	323.15	−14	35
				N,N-Dimethylformamide	338.15	−7.5	35
				N,N-Dimethylformamide	353.15	2.4	35
Glass	23200			Tetrahydrofuran	303.15	−34	59
Liquid	23200			Tetrahydrofuran	303.15	−14	59
Glass	38700			Tetrahydrofuran	303.15	−35	59
Liquid	38700			Tetrahydrofuran	303.15	−14	59
Glass	53500			Tetrahydrofuran	303.15	−39	59
Liquid	53500			Tetrahydrofuran	303.15	−14	59
Glass	66700			Tetrahydrofuran	303.15	−36	59
Liquid	66700			Tetrahydrofuran	303.15	−14	59
Glass	136000			Tetrahydrofuran	303.15	−39	59

Polymer	M_n/ g/mol	M_w/ g/mol	M_η/ g/mol	Solvent	T/K	ΔH_B^∞/ J/g	Ref.
Liquid	136000			Tetrahydrofuran	303.15	−14	59
Glass	155400			Tetrahydrofuran	303.15	−39	59
Liquid	155400			Tetrahydrofuran	303.15	−14	59
Poly(1-vinyl-3,5-dimethyl-1,2,4-triazole)							
				N,N-Dimethylformamide	298.15	−28	94
				Water	298.15	−139	94
Poly(1-vinylimidazole)							
	20700			Acetic acid	298.15	−393	94
	20700			Butanoic acid	298.15	−322	94
	20700			*N,N*-Dimethylacetamide	298.15	−48	94
	20700			*N,N*-Dimethylformamide	298.15	−48	94
	20700			1-Methyl-2-pyrrolidinone	298.15	−54	91
	20700			Pentanoic acid	298.15	−325	94
	20700			Propanoic acid	298.15	−278	94
	20700			Water	298.15	−119	91
Poly(1-vinylpyrazole)							
	18900			Acetic acid	298.15	−88	94
	18900			Butanoic acid	298.15	−60	94
	18900			*N,N*-Dimethylacetamide	298.15	−26	94
	18900			*N,N*-Dimethylformamide	298.15	−28	94
	18900			Pentanoic acid	298.15	−52	94
	18900			Propanoic acid	298.15	−36	94
Poly(1-vinyl-2-pyrrolidone)							
			32000	Trichloromethane	298.15	−75	48
			32000	Water	298.15	−150	48
Poly(1-vinyl-1,2,4-triazole)							
	69500			Acetic acid	298.15	−85	94
	69500			Butanoic acid	298.15	−74	94
	69500			*N,N*-Dimethylacetamide	298.15	−49	94
	69500			*N,N*-Dimethylformamide	298.15	−47	94
	69500			1-Methyl-2-pyrrolidinone	298.15	−55	91
	69500			Pentanoic acid	298.15	−75	94
	69500			Propanoic acid	298.15	−72	94
	69500			Water	298.15	−68	91
Vinyl acetate/vinyl alcohol copolymer							
(9 wt% Vinyl acetate)				2-Propanone	298.15	6.3	33
(44 wt% Vinyl acetate)				2-Propanone	298.15	4.6	33
(57 wt% Vinyl acetate)				2-Propanone	298.15	0.0	33
(67 wt% Vinyl acetate)				2-Propanone	298.15	−1.3	33
(4.2 mol% Vinyl acetate) 7560				Water	303.15	−41	15
(4.3 mol% Vinyl acetate) 64300				Water	303.15	−49	15
(9.0 mol% Vinyl acetate) 66900				Water	303.15	−55	15
(10.3 mol% Vinyl acetate) 7970				Water	303.15	−41	15

Polymer	M_n/ g/mol	M_w/ g/mol	M_η/ g/mol	Solvent	T/K	ΔH_B^∞/ J/g	Ref.
(15.3 mol% Vinyl acetate)							
8300				Water	303.15	−60	15
(15.4 mol% Vinyl acetate)							
70700				Water	303.15	−65	15
(19.5 mol% Vinyl acetate)							
73100				Water	303.15	−66	15
(22.1 mol% Vinyl acetate)							
8800				Water	303.15	−60	15
(26.2 mol% Vinyl acetate)							
77000				Water	303.15	−64	15
(30.6 mol% Vinyl acetate)							
9370				Water	303.15	−53	15
(34.0 mol% Vinyl acetate)							
81600				Water	303.15	−60	15
(34.7 mol% Vinyl acetate)							
9670				Water	303.15	−44	15
Vinyl acetate/vinyl chloride copolymer (90 wt% Vinyl chloride)							
Glass	12400	26000		Cyclohexanone	304.15	−37	101
Liquid	12400	26000		Cyclohexanone	304.15	−16	101
Vinyl chloride/vinylidene chloride copolymer							
				Trichloromethane	297.15	−17	2

References

1. Liepatoff, S. and Preobagenskaja, S., *Kolloid Z. Z. Polym.*, 68, 324, 1934.
2. Kargin, V. and Papkov, S., *Acta Physicochim. URSS*, 3, 839, 1935.
3. Papkov, S. and Kargin, V., *Acta Physicochim. URSS*, 7, 667, 1937.
4. Tager, A. and Kargin, V., *Acta Physicochim. URSS*, 14, 713, 1941.
5. Raine, H.C., Richards, R.B., and Ryder, H., *Trans Faraday Soc.*, 41, 56, 1945.
6. Roberts, D.E., Walton, W.W., and Jessup, R.S., *J. Polym. Sci.*, 2, 420, 1947.
7. Tager, A. and Sanatina, V., *Kolloidn. Zhur.*, 12, 474, 1950.
8. Glikman, S.A. and Root, L.A., *Zh. Obshch. Khim.*, 21, 58, 1951.
9. Hellfritz, H., *Makromol. Chem.*, 7, 191, 1951.
10. Tager, A. and Vershkain, R., *Kolloidn. Zhur.*, 13, 123, 1951.
11. Tager, A.A., and Kargin, V.A., *Kolloidn. Zhur.*, 14, 367, 1952.
12. Tager, A.A. and Dombek, Zh.S., *Kolloidn. Zhur.*, 15, 69, 1953.
13. Daoust, H. and Rinfret, M., *Can. J. Chem.*, 32, 492, 1954.
14. Gatovskaya, T.V., Kargin, V.A., and Tager, A.A., *Zh. Fiz. Khim.*, 29, 883, 1955.
15. Oya, S., *Chem. High Polym. Japan*, 12, 122, 1955.
16. Schulz, G.V., Guenner, K. von, and Gerrens, H., *Z. Phys. Chem., N. F.*, 4, 192, 1955.
17. Tager, A.A., Kosova, L.K., Karlinskaya, D.Yu., and Yurina, I.A., *Kolloid. Zhur.*, 17, 315, 1955.
18. Tager, A.A. and Kosova, L.K., *Kolloid. Zhur.*, 17, 391, 1955.
19. Tager, A.A., Krivokorytova, R.V., and Khodorov, P.M., *Dokl. Akad. Nauk SSSR*, 100, 741, 1955.
20. Glikman, S.A. and Root, L.A., *Kolloidn. Zhur.*, 18, 523, 1956.
21. Jenckel, E. and Gorke, K., *Z. Elektrochem.*, 60, 579, 1956.
22. Lipatov, Yu.S., Kargin, V.A., and Slonimskii, G.L., *Zh. Fiz. Khim.*, 30, 1202, 1956.
23. Meerson, S.I. and Lipatov, S.M., *Kolloidn., Zh.*, 18, 447, 1956.
24. Mikhailov, N.V. and Fainberg, E.Z., *Kolloidn. Zhur.*, 18, 44, 1956.
25. Struminskii, G.V. and Slonimskii, G.L., *Zh. Fiz. Khim.*, 30, 1941, 1956.
26. Gal'perin, D.I. and Moseev, L.I., *Kolloidn. Zhur.*, 19, 167, 1957.
27. Akhmedov, K.S., *Uzb. Khim. Zh.* (1), 19, 1958.
28. Kargin, V.A. and Lipatov, Yu.S., *Zh. Fiz. Khim.*, 32, 326, 1958.
29. Meerson, S.I. and Lipatov, S.M., *Kolloidn. Zhur.*, 20, 353, 1958.
30. Tager, A.A. and Galkina, L.A., *Nauchn. Dokl. Vyssh. Shkol., Khim. Khim. Tekhnol.*, (2), 357, 1958.
31. Tager, A. A. and Kargin, V. A., *Zh. Fiz. Khim.*, 32, 1362, 1958.
32. Tager, A. A. and Kargin, V. A., *Zh. Fiz. Khim.*, 32, 2694, 1958.
33. Tager, A.A. and Iovleva, M., *Zh. Fiz. Khim.*, 32, 1774, 1958.
34. Horth, A., Patterson, D., and Rinfret, M., *J. Polym. Sci.*, 39. 189, 1959.
35. Zelikman, S.G. and Mikhailov, N.V., *Vysokomol. Soedin.*, 1, 1077, 1959.
36. Mueller, F.H. and Engelter, A., *Kolloid Z.*, 171, 152, 1960.
37. Delmas, G., Patterson, D., and Boehme, A., *Trans. Faraday Soc.*, 58, 2116, 1962.
38. Delmas, G., Patterson, D., and Somcynsky, T., *J. Polym. Sci.*, 57, 79, 1962.
39. Tager, A.A. and Podlesnyak, A.I., *Vysokomol. Soedin., Ser. A*, 5, 87, 1963.
40. Delmas, G., Patterson, D., and Bhattacharyya, S.N., *J. Phys. Chem.*, 68, 1468, 1964.
41. Patterson, D., *J. Polym. Sci.: Part A*, 2, 5177, 1964.
42. Zverev, M.P., Barash, A.N., and Zubov, P.I., *Vysokomol. Soedin., Ser. A*, 6, 1012, 1964.
43. Cottam, B.J., Cowie, J.M.G., and Bywater, S., *Makromol. Chem.*, 86, 116, 1965.
44. Bianchi, U., Pedemonte, E., and Rossi, C., *Makromol. Chem.*, 92, 114, 1966.
45. Cuniberti, C. and Bianchi, U., *Polymer*, 7, 151, 1966.
46. Giacommeti, G. and Turolla, A., *Z. Phys. Chem., N.F.*, 51, 108, 1966.
47. Schreiber, H.P. and Waldman, M.H., *J. Polym. Sci.: Part A-2*, 5, 555, 1967.
48. Goldfarb, J. and Rodriguez, S., *Makromol. Chem.*, 116, 96, 1968.
49. Maron, S.H. and Daniels, C.A., *J. Macromol. Sci.-Phys. B*, 2, 769, 1968.
50. Morimoto, S., *J. Polym. Sci.: Part A-1*, 6, 1547, 1968.
51. Bianchi, U., Cuniberti, C., Pedemonte, E., and Rossi, C., *J. Polym. Sci.: Part A-2*, 7, 855, 1969.
52. Gerth, Ch. and Mueller, F.H., *Kolloid-Z. Z. Polym.*, 241, 1071, 1970.
53. Liddell, A.H. and Swinton, F.L., *Discuss. Faraday Soc.*, 49, 115, 1970.
54. Morimoto, S., *Nippon Kagaku Zasshi*, 91, 31, 1970.
55. Nakayama, H., *Bull. Chem. Soc. Japan*, 43, 1683, 1970.

56. Blackadder, D.A. and Roberts, T.L., *Angew. Makromol. Chem.*, 27, 165, 1972.
57. Maron, S.H. and Filisko, F.E., *J. Macromol. Sci.-Phys. B*, 6, 57, 1972.
58. Maron, S.H. and Filisko, F.E., *J. Macromol. Sci.-Phys. B*, 6, 79, 1972.
59. Maron, S.H. and Filisko, F.E., *J. Macromol. Sci.-Phys. B*, 6, 413, 1972.
60. Sokolova, D.F., Sokolov, L.B., and Gerasimov, V.D., *Vysokomol. Soedin., Ser. B*, 14, 580, 1972.
61. Chahal, R.S., Kao, W.-P., and Patterson, D., *J. Chem. Soc., Faraday Trans. I*, 69, 1834, 1973.
62. Delmas, G. and Tancrede, P., *Eur. Polym. J.*, 9, 199, 1973.
63. Filisko, F.E., Raghava, R.S., and Yeh, G.S.Y., *J. Macromol. Sci.-Phys. B*, 10, 371, 1974.
64. Ikeda, M., Suga, H., and Seki, S., *Polymer*, 16, 634, 1975.
65. Kiselev, V.P., Shakhova, E.M., Fainberg, E.Z., Virnik, A.D, and Rogovin, Z.A., *Vysokomol. Soedin., Ser. B*, 18, 847, 1976.
66. Petrosyan, V.A., Gabrielyan, G.A., and Rogovin, Z.A., *Arm. Khim. Zhur.*, 29, 516, 1976.
67. Deshpande, D.D. and Prabhu, C.S., *Macromolecules*, 10, 433, 1977.
68. Miura, T. and Nakamura, M., *Bull. Chem. Soc. Japan*, 50, 2528, 1977.
69. Sokolova, D.F., Kudim, T.V., Sokolov, L.B., Zhegalova, N.I., and Zhuravlev, N.D., *Vysokomol. Soedin., Ser. B*, 20, 596, 1978.
70 Basedow, A.M. and Ebert, K.H., *J. Polym. Sci.: Polym. Symp.*, 66, 101, 1979.
71 Koller, J., *Dissertation*, TU München, 1979.
72. Lee, J.-O., Ono, M., Hamada, F., and Nakajima, A., *Polym. Bull.*, 1, 763, 1979.
73. Phuong-Nguyen, H. and Delmas, G., *Macromolecules*, 12, 740, 1979.
74. Phuong-Nguyen, H. and Delmas, G., *Macromolecules*, 12, 746, 1979.
75. Basedow, A.M., Ebert, K.H., and Feigenbutz, W., *Makromol. Chem.*, 181, 1071, 1980.
76. Graun, K., *Dissertation*, TU München, 1980.
77. Shiomi, T., Izumi, Z., Hamada, F., and Nakajima, A., *Macromolecules*, 13, 1149, 1980.
78. Shiomi, T., Kohra, Y., Hamada, F., and Nakajima, A., *Macromolecules*, 13, 1154, 1980.
79. Aharoni, S.M., Charlet, G., and Delmas, G., *Macromolecules*, 14, 1390, 1981.
80. Ochiai, H., Ohashi, T., Tadokoro, Y., and Murakami, I., *Polym. J.*, 14, 457, 1982.
81. Sharma, S.C., Mahajan, R., Sharma, V.K., and Lakhanpal, M.L., *Indian J. Chem.*, 21A, 682, 1982.
82. Sharma, S.C., Mahajan, R., Sharma, V.K., and Lakhanpal, M.L., *Indian J. Chem.*, 21A, 685, 1982.
83. Lakhanpal, M.L. and Parashar, R.N., *Indian J. Chem.*, 22A, 48, 1983.
84. Daoust, H. and St-Cyr, D., *Macromolecules*, 17, 596, 1984.
85. Aeleni, N., *Mater. Plast. (Bucharest)*, 22, 92, 1985.
86. Sharma, S.C. and Sharma, V.K., *Indian J. Chem.*, 24A, 292, 1985.
87. Sharma, S.C., Bhalla, S., and Sharma, V.K., *Indian J. Chem.*, 25A, 131, 1986.
88. Sharma, S.C., Syngal, M., and Sharma, V.K., *Indian J. Chem.*, 26A, 285, 1987.
89. Aukett, P.N. and Brown, C.S., *J. Therm. Anal.*, 33, 1079, 1988.
90. Lanzavecchia, L. and Pedemonte, E., *Thermochim. Acta*, 137, 123, 1988.
91. Tager, A.A., Safronov, A.P., Voit, V.V., Lopyrev, V.A., Ermakova, T.G., Tatarova, L.A., and Shagelaeva, N.S., *Vysokomol. Soedin., Ser. A*, 30, 2360, 1988.
92. Pedemonte, E. and Lanzavecchia, L., *Thermochim. Acta*, 162, 223, 1990.
93. Shiomi, T., Ishimatsu, H., Eguchi, T., and Imai, K., *Macromolecules*, 23, 4970, 1990.
94. Tager, A.A. and Safronov, A.P., *Vysokomol. Soedin., Ser. A*, 33, 67, 1991.
95. Zellner, H., *Dissertation*, TU München, 1993.
96. Brunacci, A., Pedemonte, E., Cowie, J.M.G., and McEwen, I. J., *Polymer*, 35, 2893, 1994.
97. Carlsson, M., Hallen, D., and Linse, P., *J. Chem. Soc. Faraday Trans.*, 91, 2081, 1995.
98. Sato, T., Tohyama, M., Suzuki, M., Shiomi, T., and Imai, K., *Macromolecules*, 29, 8231, 1996.
99. Shiomi, T., Tohyama, M., Endo, M., Sato, T., and Imai, K., *J. Polym. Sci.: Part B: Polym. Phys.*, 34, 2599, 1996.
100. Kizhnyaev, V.N., Gorkovenko, O.P., Bazhenov, D.N., and Smirnov, A.I., *Vysokomol. Soedin., Ser. A*, 39, 856, 1997.
101. Sato, T., Suzuki, M., Tohyama, M., Endo, M., Shiomi, T., and Imai, K., *Polym. J.*, 29, 417, 1997.
102. Vanderryn, J. and Zettlemoyer, A.C., *Ind. Eng. Chem., Chem. Eng. Data Ser.*, 2, 56, 1957.
103. Parashar, R. and Sharma, S.C., *Indian J. Chem.*, 27A, 1092, 1988.
104. Righetti, M.C., Cardelli, C., Scalari, M., Tombari, E., and Conti, G., *Polymer*, 43, 5035, 2002.
105. Phuong-Nguyen, H. and Delmas, G., *J. Solution Chem.*, 23, 249, 1994.
106. Wohlfarth, C., *CRC Handbook of Enthalpy Data of Polymer-Solvent Systems*, CRC Press, Boca Raton, 2006.

SOLUBILITY PARAMETERS OF SELECTED POLYMERS

Christian Wohlfarth

The concept of cohesive energy density and solubility parameter was introduced by Hildebrand:

$$\delta^2 = \frac{\Delta U_m^{LV}}{V_m} = \frac{\Delta H_m^{LV} - RT}{V_m} \qquad (1)$$

V_m is the molar volume, ΔU_m^{LV} is the molar energy of vaporization, and ΔH_m^{LV} is the molar enthalpy of vaporization. Units for the solubility parameter are $(MPa)^{1/2} = (J/cm^3)^{1/2} = 0.4887(cal/cm^3)^{1/2}$. The energy of vaporization is not accessible for polymers, but cohesive energy density of polymers can be determined from *PVT*-data. However, common ways for determining polymer solubility parameters use thermodynamic properties of polymer solutions and their relations to excess enthalpy or excess Gibbs energy per unit volume. These excess quantities are related to the (square) difference between the solubility parameters of solvents and polymers, i.e. $(\delta_1 - \delta_2)^2$.

$$\frac{H^E}{V} \sim (\delta_1 - \delta_2)^2 \quad \text{or} \quad \frac{G^E}{V} \sim (\delta_1 - \delta_2)^2 \qquad (2)$$

Often, the Flory-Huggins solvent-polymer interaction parameter is applied instead of H^E or G^E. There are some books (Refs. 1–3) giving details for such procedures as well as extensive tables of polymer solubility parameters from which the table below is extracted. Methods for calculating solubility parameters can be found in Refs. 4–7.

References

1. Barton, A.F.M., *CRC Handbook of Polymer-Liquid Interaction Parameters and Solubility Parameters*, CRC Press, Boca Raton, 1991.
2. Brandrup, J., Immergut, E.H., Grulke, E.A. (eds.), *Polymer Handbook*, 4th ed., J. Wiley & Sons, New York, 1999.
3. Wohlfarth, C., *Thermodynamic Properties of Polymer Solutions*, in Landolt-Börnstein, New Series, Group VIII, Vol. 6D, Lechner, M.D., (ed.), Springer Verlag, Berlin, Heidelberg, 2010.
4. Hildebrand, J.H., Prausnitz, J.M., Scott, R.L., *Regular and Related Solutions*, Van Nostrand Reinhold Co., New York, 1970.
5. [Van] Krevelen, D.W., *Properties of Polymers*, 3rd ed., Elsevier, Amsterdam, 1990.
6. Bicerano, J., *Prediction of Polymer Properties*, 3rd ed., CRC Press, Boca Raton, 2002.
7. Hansen, C.M., *Hansen Solubility Parameters: A User's Handbook*, 2nd ed., CRC Press, Boca Raton, 2007.

Polymer	*T*/K	$\delta/(J/cm^3)^{1/2}$
Benzyl cellulose	298	25.2
Butadiene/acrylonitrile copolymer		
(25 wt % acrylonitrile)	298	19.4
(30 wt % acrylonitrile)	298	19.2
(34 wt % acrylonitrile)	298	20.4
(39 wt % acrylonitrile)	298	21.35
Butadiene/styrene copolymer		
(6 wt % styrene)	298	16.5
(12.5 wt % styrene)	298	16.5
(15 wt% styrene)	298	17.5
(25 wt% styrene)	298	17.6
(40 wt% styrene)	298	17.8
Cellulose	298	26.0
Cellulose acetate	298	25.1
Cellulose diacetate	298	22.3
Cellulose nitrate	298	22.0
Cellulose triacetate	298	19.0
Ethyl cellulose	298	21.1
Ethylene/1-octene copolymer		
(2.0 wt% 1-octene)	473	17.2
(7.5 wt% 1-octene)	473	16.7
(12.0 wt% 1-octene)	473	16.5
(25.0 wt% 1-octene)	473	16.5
(39.4 wt% 1-octene)	473	16.4
(55.0 wt% 1-octene)	473	16.3
(64.0 wt% 1-octene)	473	16.3
Ethylene/vinyl acetate copolymer		
(11 wt% vinyl acetate)	323	16.6
(16 wt% vinyl acetate)	323	17.1
(24 wt% vinyl acetate)	323	16.8
(37 wt% vinyl acetate)	323	17.0
(42 wt% vinyl acetate)	323	17.2
Gelatine	298	24.6
Hydroxypropyl cellulose	298	26.8
Natural rubber	298	16.6
Poly(acrylonitrile)	298	26.0
Polyamide 4	298	24.0
Polyamide 6	298	21.7
Polyamide 66	298	22.9
Polyamide 7	298	24.1
Polyamide 8	298	20.3
Polyamide 9	298	22.6
Polyamide 10	298	19.4
Polyamide 11	298	22.9
Polyamide 12	298	20.8
Poly(*p*-benzamide)	298	23.0
Poly(benzyl methacrylate)	298	15.3
Polybutadiene	298	16.6
1,2-Polybutadiene	298	16.5
1,4-*cis*-Polybutadiene	298	16.5
Poly(1-butene), *isotactic*	298	16.0
Poly(butyl acrylate)	298	18.0
Poly(butyl methacrylate)	298	17.9
Poly(2-butyl methacrylate)	413	14.7

Polymer	T/K	δ/(J/cm^3)$^{1/2}$
Poly(ε-caprolactone)	298	19.9
Polycarbonate bisphenol-A	298	20.0
Polycarbonate hexafluorobisphenol-A	298	20.1
Poly(chloroprene)	298	17.6
Poly(4-chlorostyrene)	298	19.3
Poly(cyclohexyl methacrylate)	298	19.8
Poly(3,3-diethyloxetane)	298	16.2
Poly(3,3-dimethyloxetane)	298	16.2
Poly(2,6-dimethyl-1,4-phenylene ether)	298	18.1
Poly(dimethylsiloxane)	298	15.4
Poly(1,3-dioxepane)	298	18.8
Poly(1,3-dioxolane)	298	20.7
Poly(dodecyl methacrylate)	298	16.8
Poly(epichlorohydrin)	298	16.2
Poly(ethoxyethyl methacrylate)	298	18.4
Poly(ethyl acrylate)	298	19.2
Polyethylene, branched	298	16.2
Polyethylene, linear	298	16.2
Poly(ethylene adipate)	298	19.8
Poly(ethylene glycol)	298	23.7
Poly(ethylene oxide)	298	20.5
Poly(ethylene terephthalate)	298	21.9
Poly(ethyl methacrylate)	298	18.3
Poly(hexyl methacrylate)	298	17.7
Poly(4-hydroxystyrene)	298	24.0
Poly(isobornyl acrylate)	298	16.8
Poly(isobornyl methacrylate)	298	17.0
Poly(isobutylene)	298	16.2
Poly(isobutyl methacrylate)	413	14.6
1,4-*cis*-Poly(isoprene)	298	16.5
Poly(*N*-isopropylacrylamide)	298	23.5
Poly(DL-lactic acid)	298	20.5
Poly(L-lactide)	298	19.3
Poly(methacrylonitrile)	298	21.0
Poly(methyl acrylate)	298	20.5
Poly(methyl methacrylate)	298	19.3
Poly(4-methyl-1-pentene)	298	15.3
Poly(2-methylpropene)	298	17.7
Poly(2-methylstyrene)	298	18.4
Poly(4-methylstyrene)	298	19.3
Poly(methylvinylsiloxane)	298	15.65
Polynorbornene	298	14.0
Poly(1-octene)	298	16.6
Poly(octyl methacrylate)	298	18.0
Poly(propyl acrylate)	298	18.4
Poly(propyl methacrylate)	298	16.0
Polypropylene, *atactic*	298	15.5
Polypropylene, *isotactic*	298	17.5
Polypropylene, *syndiotactic*	298	17.6
Poly(propylene glycol)	298	19.5
Poly(propylene oxide)	298	18.5
Poly(propyl methacrylate)	298	20.0
Polystyrene	298	19.0
Polysulfone	298	19.9
Poly(tetrafluoroethylene)	298	19.6
Poly(tetramethylene oxide)	298	16.8
Poly(thioethylene)	298	18.8
Poly(trimethylene sulfide)	298	23.0
Poly(vinyl acetate)	298	20.6
Poly(vinyl alcohol)	298	22.0
Poly(vinyl bromide)	298	19.4
Poly(vinyl butyl ether)	298	19.2
Poly(*N*-vinylcarbazole)	298	19.2
Poly(vinyl chloride)	298	19.6
Poly(vinyl ethyl ether)	298	19.5
Poly(vinylidene fluoride)	298	23.2
Poly(vinyl methyl ether)	298	21.0
Poly(vinyl phenyl ether)	298	20.2
Poly(vinyl propionate)	298	18.1
Poly(vinyl propyl ether)	298	19.3
Poly(1-vinyl-2-pyrrolidinone)	298	25.6
Vinyl acetate/vinyl alcohol copolymer		
(43.4 mol% vinyl acetate)	298	21.8
(60.9 mol% vinyl acetate)	298	21.4
(74.4 mol% vinyl acetate)	298	20.9
(94.8 mol% vinyl acetate)	298	20.2
Vinyl acetate/vinyl chloride copolymer		
(3 wt% vinyl acetate)	298	18.8
(10 wt% vinyl acetate)	298	17.3
(17 wt% vinyl acetate)	298	19.1

Section 14
Geophysics, Astronomy, and Acoustics

Astronomical Constants ... 14-1
Properties of the Solar System ... 14-2
Satellites of the Planets ... 14-4
Interstellar Molecules ... 14-7
Mass, Dimensions, and Other Parameters of the Earth ... 14-10
Geological Time Scale ... 14-12
Acceleration Due to Gravity ... 14-13
Density, Pressure, and Gravity as a Function of Depth within the Earth ... 14-14
Ocean Pressure as a Function of Depth and Latitude ... 14-15
Properties of Seawater ... 14-16
Abundance of Elements in the Earth's Crust and in the Sea ... 14-18
Solar Irradiance at the Earth ... 14-19
U.S. Standard Atmosphere (1976) ... 14-20
Geographical and Seasonal Variations in Solar Radiation ... 14-26
Major World Earthquakes ... 14-27
Weather-Related Scales ... 14-31
Infrared Absorption by the Earth's Atmosphere ... 14-33
Atmospheric Concentration of Carbon Dioxide, 1958–2008 ... 14-34
Mean Temperatures in the United States, 1900–1992 ... 14-36
Global Temperature Trend, 1880–2009 ... 14-38
Global Warming Potential of Greenhouse Gases ... 14-39
Atmospheric Electricity ... 14-41
Speed of Sound in Various Media ... 14-48
Attenuation and Speed of Sound in Air as a Function of Humidity and Frequency ... 14-50
Speed of Sound in Dry Air ... 14-51
Musical Scales ... 14-52
Characteristics of Human Hearing ... 14-53

ASTRONOMICAL CONSTANTS

Victor Abalakin

The constants in this table are based originally on the set of constants adopted by the International Astronomical Union (IAU) in 1976. Updates have been made when new data were available. All values are given in SI Units; thus masses are expressed in kilograms and distances in meters.

The astronomical unit of time is a time interval of one day (1 d) equal to 86400 s. An interval of 36525 d is one Julian century (1 cy).

Defining constants	
Gaussian gravitational constant	$k = 0.01720209895$ m^3 kg^{-1} s^{-2}
Speed of light	$c = 299792458$ m s^{-1}
Primary constants	
Light-time for unit distance (1 ua)	$\tau_A = 499.004786$ s
Equatorial radius of Earth	$a_e = 6378140$ m
Equatorial radius of Earth (IUGG value)	$a_e = 6378136$ m
Dynamical form-factor for Earth	$J_2 = 0.001082636$
Geocentric gravitational constant	$GE = 3.986004 \times 10^{14}$ m^3 s^{-2}
Constant of gravitation	$G = 6.67428 \times 10^{-11}$ m^3 kg^{-1} s^{-2}
Ratio of mass of moon to that of Earth	$\mu = 0.01230002$
	$1/\mu = 81.300587$
General precession in longitude, per Julian century, at standard epoch J2000	$\rho = 5028''.796$
Obliquity of the ecliptic at standard epoch J2000	$\varepsilon = 23°26'21''.448$
Derived constants	
Constant of nutation at standard epoch J2000	$N = 9''.2025$
Unit distance (ua = $c\tau_A$)	ua = $1.49597871464 \times 10^{11}$ m
Solar parallax ($\pi_0 = \arcsin(a_e/\text{ua})$)	$\pi_0 = 8''.794143$
Constant of aberration for standard epoch J2000	$\kappa = 20''.49552$
Flattening factor for the Earth	$f = 1/298.256 = 0.00335282$
Heliocentric gravitational constant ($GS = A^3k^2/D^2$)	$GS = 1.32712438 \times 10^{20}$ m^3 s^{-2}
Ratio of mass of sun to that of the Earth (S/E) = $(GS)/(GE)$)	$S/E = 332946.0$
Ratio of mass of sun to that of Earth + moon	$(S/E)/(1 + \mu) = 328900.56$
Mass of the sun ($S = (GS)/G$)	$S = 1.98844 \times 10^{30}$ kg
Ratios of mass of sun to masses of the planets	
Mercury	6023600
Venus	408523.7
Earth + moon	328900.56
Mars	3098708
Jupiter	1047.349
Saturn	3497.898
Uranus	22902.98
Neptune	19412.24

References

1. Seidelmann, P. K., *Explanatory Supplement to the Astronomical Almanac*, University Science Books, Mill Valley, CA, 1990.
2. Lang, K. R., *Astrophysical Data: Planets and Stars*, Springer-Verlag, New York, 1992.
3. *The Astronomical Almanac for the Year 2007*, U.S. Government Printing Office, Washington, and Her Majesty's Stationary Office, London (2005).

PROPERTIES OF THE SOLAR SYSTEM

The following tables give various properties of the planets and characteristics of their orbits in the solar system. Certain properties of the sun and of the earth's moon are also included.

Explanations of the column headings:

- *Mass*: mass of the planet in units of 10^{24} kg
- *Radius*: radius at the equator in km
- *Density*: mean density in g/cm^3
- *Flattening*: degree of oblateness, defined as $(r_e - r_p)/r_e$, where r_e and r_p are the equatorial and polar radii, respectively
- *Potential coefficients*: coefficients in the spherical harmonic representation of the gravitational potential U by the equation

$$U(r, \phi) = (GM/r)\,[1 - \Sigma J_n (a/r)^n \mathrm{P}_n (\sin \phi)],$$

where G is the gravitational constant, r the distance from the center of the planet, a the radius of the planet, M the mass, ϕ the latitude, and P_n the Legendre polynomial of degree n.

- *Gravity*: acceleration due to gravity at the surface
- *Escape vel.*: velocity needed at the surface of the planet to escape the gravitational pull
- *Dist. to sun*: semi-major axis of the elliptical orbit in astronomical units (1 ua ≈ 1.496·10^8 km)
- ε: eccentricity of the orbit
- *Ecliptic angle*: angle between the planetary orbit and the plane of the earth's orbit around the sun
- *Inclin.*: angle between the equatorial plane of the planet and the plane of the planetary orbit
- *Orbit period*: period of revolution around the sun measured in years
- *Rotation period*: period of rotation of the planet measured in hours. A negative value indicates retrograde rotation.
- *Albedo*: ratio of the light reflected from the planet to the light incident on it
- *No. of satellites*: Number of confirmed satellites; this includes satellites that have not been named.
- T_{sur}: mean temperature at the surface
- P_{sur}: pressure of the atmosphere at the surface

The last four entries in the table are *dwarf planets* as defined by the International Astronomical Union. These are bodies in orbit around the sun that are massive enough to adopt a near-spherical shape as a result of their self-gravity, but are appreciably smaller than the major planets. *Plutoids* form a subset of the dwarf planets; their orbits are larger than that of Neptune (see Ref. 9). As of 2008, the IAU has recognized the names for three plutoids: Pluto, Eris, and Makemake.

The following general information on the solar system is of interest:

Mass of the earth = M_e = 5.9736·10^{24} kg
Total mass of planetary system = 2.669·10^{27} kg = 447 M_e
Total angular momentum of planetary system = 3.148·10^{43} kg m^2 s^{-1}
Total kinetic energy of the planets = 1.99·10^{35} J
Total rotational energy of planets = 0.7·10^{35} J

Properties of the sun:

Mass = 1.9884·10^{30} kg = 332943 M_e
Radius = 6.9551·10^8 m
Surface area = 6.079·10^{18} m^2
Volume = 1.409·10^{27} m^3
Mean density = 1.411 g/cm^3
Gravity at surface = 27398 cm/s^2
Escape velocity at surface = 6.177·10^5 m/s
Effective temperature = 5780 K
Total radiant power emitted (luminosity) = 3.8427·10^{26} W
Surface flux of radiant energy = 6.322·10^7 W/m^2
Flux of radiant energy at the earth (Solar Constant) = 1366.4 W/m^2 (Ref. 8)

References

1. *Planetary Fact Sheet*, NASA Goddard Space Flight Center, <nssdc.gsfc.nasa.gov/planetary/factsheet>, November 2007.
2. The Planetary Society, <www.planetary.org/explore/topics/groups/our_solar_system/>.
3. Arnet, B., *The Nine Planets*, <www.nineplanets.org>.
4. Onasch, B., *Our Solar System*, <www.onasch.de/astro/>.
5. *The Astronomical Almanac for the year 2007*, U.S. Government Printing Office, Washington, 2005; available online at <asa.usno.navy.mil/>.
6. Lang, K. R., *Astrophysical Data: Planets and Stars*, Springer-Verlag, New York, 1992.
7. Cox, A. N., *Allen's Astrophysical Quantities, Fourth Edition*, Springer-Verlag, New York, 2000; this is a revision of Allen, C. W., *Astrophysical Quantities, Third Edition*, 1983.
8. Amsler, C., et al, *Physics Letters* B667, 1, 2008; section on astrophysical constants available at <pdg.lbl.gov/2008/reviews/astrorpp.pdf>.
9. IAU Press Release, <www.iau.org/public_press/news/release/iau0804>, 11 June 2008.
10. Seidelmann, P. K., Editor, *Explanatory Supplement to the Astronomical Almanac*, University Science Books, Mill Valley, CA, 1992.

Planet	Mass 10^{24} kg	Radius km	Density g/cm³	Flattening	Potential coefficients $10^3 J_2$	$10^6 J_3$	$10^6 J_4$	Gravity cm/s²	Escape vel. km/s
Mercury	0.33022	2439.7	5.43	0.0000				370	4.25
Venus	4.8685	6051.8	5.24	0.000	0.027			887	10.36
Earth	5.9736	6378.14	5.52	0.00335364	1.08263	−2.54	−1.61	980	11.18
(Moon)	0.07349	1738.1	3.35	0.0012	0.2027			162	2.38
Mars	0.64185	3396.2	3.93	0.0064763	1.964	36		371	5.03
Jupiter	1898.6	71492	1.33	0.0648744	14.75	−580		2312	59.54
Saturn	568.46	60268	0.69	0.09796	16.45	−1000		896	35.5
Uranus	86.832	25559	1.27	0.02293	3.343			869	21.29
Neptune	102.43	24764	1.64	0.01708	3.41			1100	23.5
Dwarf planets:									
Pluto	0.0125	1195	1.75	0				58	1.2
Eris	0.0166	1225						≈80	
Makemake	≈0.004	750	≈2					≈47	≈0.84
Ceres	0.000943	471	2.08					27	0.51

Planet	Dist. to Sun ua	ε	Ecliptic angle	Inclin.	Orbit period (yr)	Rotation period (hr)	Albedo	No. of satellites
Mercury	0.38710	0.2056	7.00°	0.01°	0.2408467	1407.6	0.106	0
Venus	0.72333	0.0068	3.39°	177.36°	0.61519726	−5832.5	0.65	0
Earth	1.00000	0.0167	0	23.45°	1.0000174	23.9345	0.367	1
Mars	1.52366	0.0934	1.85°	25.19°	1.8808476	24.6229	0.150	2
Jupiter	5.20336	0.0484	1.305°	3.12°	11.862615	9.9250	0.52	63
Saturn	9.53707	0.0542	2.48°	26.73°	29.447498	10.656	0.47	60
Uranus	19.19126	0.0472	0.77°	97.86°	84.016846	−17.24	0.51	27
Neptune	30.06896	0.0086	1.77°	28.32°	164.79132	16.11	0.41	13
Dwarf planets:								
Pluto	39.48168	0.2488	17.14°	118°	247.92065	−153.29	0.6	3
Eris	67.7	0.44		44.19°	557		0.8	1
Makemake	45.79	0.159		28.96°	309.88			
Ceres	2.77	0.08		10.59°	4.599	9.074	0.090	

Planet	T_{sur} K	P_{sur} bar	Atmospheric composition CO_2	N_2	O_2	H_2O	H_2	He	A	Ne	CO	CH_4
Mercury	440	$1 \cdot 10^{-15}$										
Venus	737	90	96.5%	3.5%	69 ppm	20 ppm		12 ppm	70 ppm	7 ppm	17 ppm	
Earth	288	1.014	0.038%	78.084%	20.946%	0 to 3%	0.55 ppm	5.24 ppm	0.934%	18.18 ppm	1 ppm	1.7 ppm
Mars	210	0.007	95.32%	2.7%	0.13%	0.021%			1.6%	2.5 ppm	0.08%	
Jupiter	165	>>1000				4 ppm	89.8%	10.2%				0.30%
Saturn	134	>>1000					96.3%	3.25%				0.45%
Uranus	76	>>1000					82.5%	15.2%				2.3%
Neptune	56						80%	19%				1%
Pluto	50	$3 \cdot 10^{-6}$										

SATELLITES OF THE PLANETS

This table gives characteristics of the known satellites of the planets. The parameters covered are:

- Orbital period in units of earth days. An R following the value indicates a retrograde motion.
- Distance from the planet, as measured by the semi-major axis of the orbit
- Eccentricity of the orbit
- Inclination of the satellite orbit with respect to the equator of the planet
- Mass of the satellite in kilograms
- Radius of the satellite in kilometers
- Geometric albedo, which is a measure of the fraction of incident sunlight reflected by the satellite.

Since this is a very active field of research, the Internet sites listed below should be consulted for the most recent data.

References

1. *Solar System Dynamics*, Jet Propulsion Laboratory, California Institute of Technology, <ssd.jpl.nasa.gov/?phys_data>, June 2008.
2. The Planetary Society, <www.planetary.org/explore/topics/groups/our_solar_system/>.
3. Arnet, B., *The Nine Planets*, <www.nineplanets.org>.
4. Onasch, B., *Our Solar System*, <www.onasch.de/astro/>.
5. Sheppard, S. S., *The Giant Satellite and Moon Page*, <www.dtm.ciw.edu/sheppard/satellites/>.
6. *Gazetteer of Planetary Nomenclature*, U. S. Geological Survey, <planetarynames.wr.usgs.gov/append7.html>.
7. Seidelmann, P. K., Editor, *Explanatory Supplement to the Astronomical Almanac*, University Science Books, Mill Valley, CA, 1992.
8. Lang, K. R., *Astrophysical Data: Planets and Stars*, Springer-Verlag, New York, 1992.
9. Allen, C. W., *Astrophysical Quantities, Second Edition*, Athlone Press, London, 1955.

Planet		Satellite	Orb. period Earth days	Distance 10^3 km	Eccentricity	Inclination	Mass kg	Radius km	Albedo
Earth		Moon	27.321661	384.400	0.054900489	18.28–28.58°	$7.3483 \cdot 10^{22}$	1737.5	0.12
Mars	I	Phobos	0.31891023	9.378	0.0151	1.0°	$1.06 \cdot 10^{16}$	13.5×10.8×9.4	0.07
	II	Deimos	1.2624407	23.460	0.0005	0.9–2.7°	$2.4 \cdot 10^{15}$	7.5×6.1×5.5	0.07
Jupiter	I	Io	1.769137786	421.8	0.0041	0.04°	$8.932 \cdot 10^{22}$	1821.6	0.63
	II	Europa	3.551181041	671.1	0.0101	0.47°	$4.8 \cdot 10^{22}$	1560.8	0.67
	III	Ganymede	7.15455296	1070.4	0.0015	0.21°	$1.4819 \cdot 10^{23}$	2631.2	0.43
	IV	Callisto	16.6890184	1882.7	0.007	0.51°	$1.0759 \cdot 10^{23}$	2410.3	0.17
	V	Amalthea	0.49817905	181.4	0.003	0.40°	$7.17 \cdot 10^{18}$	131×73×67	0.09
	VI	Himalia	250.5662	11460	0.162	27.63°	$9.56 \cdot 10^{18}$	85	0.04
	VII	Elara	259.6528	11737	0.217	24.77°	$7.77 \cdot 10^{17}$	40	0.04
	VIII	Pasiphae	743.63 R	23620	0.409	145°	$1.91 \cdot 10^{17}$	18	0.04
	IX	Sinope	758.90 R	23940	0.250	153°	$7.77 \cdot 10^{16}$	14	0.04
	X	Lysithea	259.20	11720	0.112	29.02°	$7.77 \cdot 10^{16}$	12	0.04
	XI	Carme	734.17 R	23400	0.253	164°	$9.56 \cdot 10^{16}$	15	0.04
	XII	Ananke	629.77 R	21280	0.244	147°	$3.82 \cdot 10^{16}$	10	0.04
	XIII	Leda	240.92	11170	0.164	26.07°	$5.68 \cdot 10^{15}$	5	0.04
	XIV	Thebe	0.6745	221.9	0.018	0.8°	$7.77 \cdot 10^{17}$	55×45	0.05
	XV	Adrastea	0.29826	129	0.0015		$1.91 \cdot 10^{16}$	13×10×8	0.10
	XVI	Metis	0.29478	128	0.0002		$9.56 \cdot 10^{16}$	20	0.06
	XVII	Callirrhoe	758.77	24100	0.283		$8.7 \cdot 10^{14}$	4	0.04
	XVIII	Themisto	130.02	7507	0.242		$6.9 \cdot 10^{14}$	4	0.04
	XIX	Megaclite	752.86	23810	0.425		$2.1 \cdot 10^{14}$	2.7	0.04
	XX	Taygete	732.41	23360	0.251		$1.6 \cdot 10^{14}$	2.5	0.04
	XXI	Chaldene	723.72	23180	0.238		$7.5 \cdot 10^{13}$	1.9	0.04
	XXII	Harpalyke	623.32	21110	0.227		$1.2 \cdot 10^{14}$	2.2	0.04
	XXIII	Kalyke	742.06	23580	0.243		$1.9 \cdot 10^{14}$	2.6	0.04
	XXIV	Iocaste	631.60	21270	0.218		$1.9 \cdot 10^{14}$	2.6	0.04
	XXV	Erinome	728.46	23280	0.270		$4.5 \cdot 10^{13}$	1.6	0.04
	XXVI	Isonoe	726.63	23220	0.261		$7.5 \cdot 10^{13}$	1.9	0.04
	XVII	Praxidike	625.39	21150	0.220		$4.3 \cdot 10^{14}$	3.4	0.04
	XXVIII	Autonoe	760.95	23039	0.334		$9.0 \cdot 10^{13}$	2.0	0.04
	XXIX	Thyone	627.21	20940	0.229		$9.0 \cdot 10^{13}$	2.0	0.04
	XXX	Hermippe	633.90	21131	0.210		$9.0 \cdot 10^{13}$	2.0	0.04
	XXXI	Aitne	730.18	23231	0.264		$4.5 \cdot 10^{13}$	1.5	0.04
	XXXII	Eurydome	717.33	22685	0.276		$4.5 \cdot 10^{13}$	1.5	0.04
	XXXIII	Euanthe	620.49	20721	0.232		$4.5 \cdot 10^{13}$	1.5	0.04
	XXXIV	Euporie	550.74	19302	0.144		$1.5 \cdot 10^{13}$	1	0.04
	XXXV	Orthosie	622.56	20721	0.281		$1.5 \cdot 10^{13}$	1	0.04
	XXXVI	Sponde	748.34	23487	0.312		$1.5 \cdot 10^{13}$	1	0.04

Planet		Satellite	Orb. period Earth days	Distance 10^3 km	Eccentricity	Inclination	Mass kg	Radius km	Albedo
	XXXVII	Kale	729.47	23217	0.260		$1.5\cdot10^{13}$	1	0.04
	XXXVIII	Pasithee	719.44	23096	0.267		$1.5\cdot10^{13}$	1	0.04
	XXXIX	Hegemone	739.6	23947	0.328		$4.5\cdot10^{13}$	1.5	
	XL	Mneme	620.0	21069	0.227		$1.5\cdot10^{13}$	1	
	XLI	Aoede	761.5	23981	0.432		$9.0\cdot10^{13}$	2.0	
	XLII	Thelxinoe	628.1	21162	0.221		$1.5\cdot10^{13}$	1	
	XLIII	Arche	723.9	22931	0.259		$4.5\cdot10^{13}$	1.5	
	XLIV	Kallichore	764.7	24043	0.264		$1.5\cdot10^{13}$	1	
	XLV	Helike	634.8	21263	0.156		$9.0\cdot10^{13}$	2.0	
	XLVI	Carpo	456.1	16989	0.430		$4.5\cdot10^{13}$	1.5	
	XLVII	Eukelade	746.4	23661	0.272		$9.0\cdot10^{13}$	2.0	
	XLVIII	Cyllene	737.8	24349	0.319		$1.5\cdot10^{13}$	1	
	XLIX	Kore	779.2	24543	0.325			1	
Saturn	I	Mimas	0.942421813	185.52	0.0202	1.53°	$3.75\cdot10^{19}$	196	0.5
	II	Enceladus	1.370217855	238.02	0.00452	1.86°	$6.50\cdot10^{19}$	250	1.0
	III	Tethys	1.887802160	294.66	0.00000	1.86°	$6.27\cdot10^{20}$	530	0.9
	IV	Dione	2.736914742	377.40	0.002230	0.02°	$1.10\cdot10^{21}$	560	0.7
	V	Rhea	4.517500436	527.04	0.00100	0.35°	$2.31\cdot10^{21}$	765	0.7
	VI	Titan	15.94542068	1221.83	0.029192	0.33°	$1.3455\cdot10^{23}$	2575	0.21
	VII	Hyperion	21.2766088	1481.1	0.104	0.43°	$1.59\cdot10^{19}$	205×130×110	0.3
	VIII	Iapetus	79.3301825	3561.3	0.02828	14.72°	$1.59\cdot10^{21}$	730	0.6
	IX	Phoebe	550.31 R	12952	0.16326	177°	$7.2\cdot10^{18}$	110	0.08
	X	Janus	0.6945	151.472	0.007	0.14°	$1.92\cdot10^{18}$	110×100×80	0.6
	XI	Epimetheus	0.6942	151.422	0.009	0.34°	$5.4\cdot10^{17}$	70×60×50	0.5
	XII	Helene	2.7369	377.40	0.005	0.0°	$2.5\cdot10^{16}$	18×16×15	0.6
	XIII	Telesto	1.8878	294.66			$7.2\cdot10^{15}$	17×14×13	1.0
	XIV	Calypso	1.8878	294.66			$3.6\cdot10^{15}$	17×11×11	0.7
	XV	Atlas	0.6019	137.670		0.3°	$1.1\cdot10^{16}$	20×10	0.4
	XVI	Prometheus	0.6130	139.353	0.0024	0.0°	$3.3\cdot10^{17}$	70×50×40	0.6
	XVII	Pandora	0.6285	141.70	0.0042	0.0°	$1.9\cdot10^{17}$	55×45×35	0.5
	XVIII	Pan	0.5750	133.583			$2.7\cdot10^{15}$	10	0.5
	XIX	Ymir	1315.14	23096	0.470		$4.9\cdot10^{15}$	8	0.06
	XX	Paaliaq	686.95	15199	0.364		$8.2\cdot10^{15}$	9.5	0.06
	XXI	Tarvos	926.23	18247	0.536		$2.7\cdot10^{15}$	6.5	0.06
	XXII	Ijiraq	451.42	11440	0.322		$1.2\cdot10^{15}$	5	0.06
	XXIII	Suttungr	1016.67	19463	0.114		$2.1\cdot10^{14}$	2.8	0.06
	XXIV	Kiviuq	449.22	11365	0.334		$3.3\cdot10^{15}$	7	0.06
	XXV	Mundilfari	952.77	18709	0.208		$2.1\cdot10^{14}$	2.8	0.06
	XXVI	Albiorix	783.45	16404	0.478		$2.1\cdot10^{16}$	13	0.06
	XXVII	Skathi	728.20	15647	0.270		$3.1\cdot10^{14}$	3.2	0.06
	XXVIII	Erriapus	871.19	17616	0.474		$7.6\cdot10^{14}$	4.3	0.06
	XXIX	Siarnaq	895.53	18160	0.295		$3.9\cdot10^{16}$	16	0.06
	XXX	Thrymr	1094.11	20382	0.470		$2.1\cdot10^{14}$	2.8	0.06
	XXXI	Narvi	1003.86	19007	0.431		$4.9\cdot10^{15}$	3.3	0.04
	XXXII	Methone	1.010	194			$1.65\cdot10^{13}$	1.5	
	XXXIII	Pallene	1.154	211			$3.92\cdot10^{13}$	2	
	XXXIV	Polydeuces	2.737	377.4				4	
	XXXV	Daphnis	0.594	136.5				3.5	
	XXXVI	Aegir	1117.52	20735				3.5	
	XXXVII	Bebhionn	834.84	17119				3	
	XXXVIII	Bergelmir	1005.74	19338				3	
	XXXIX	Bestla	1088.72	20129				3.5	
	XL	Farbauti	1085.55	20390				2.5	
	XLI	Fenrir	1260.35	22453				2	
	XLII	Fornjot	1494.20	25108				3	
	XLIII	Hati	1038.61	19856				3	
	XLIV	Hyrrokkin	931.86	18437				4	
	XLV	Kari	1230.97	22118				3.5	
	XLVI	Loge	1311.36	23065				3	
	XLVII	Skoll	878.29	17665				3	

Planet		Satellite	Orb. period Earth days	Distance 10^3 km	Eccentricity	Inclination	Mass kg	Radius km	Albedo
	XLVIII	Surtur	1297.36	22707				3	
	XLIX	Anthe		197.7				1	
	L	Jarnsaxa		18600				3	
	LI	Greip		18105				3	
	LII	Tarqeq		19720				3.5	
Uranus	I	Ariel	2.52037935	191.02	0.0034	0.3°	$1.35 \cdot 10^{21}$	579	0.39
	II	Umbriel	4.1441772	266.30	0.0050	0.36°	$1.17 \cdot 10^{21}$	584.7	0.21
	III	Titania	8.7058717	435.91	0.0022	0.14°	$3.52 \cdot 10^{21}$	788.9	0.27
	IV	Oberon	13.4632389	583.52	0.0008	0.10°	$3.01 \cdot 10^{21}$	761.4	0.23
	V	Miranda	1.41347925	129.39	0.0027	4.2°	$6.59 \cdot 10^{19}$	236	0.32
	VI	Cordelia	0.335033	49.77	0.0003	0.1°	$5.4 \cdot 10^{16}$	20.1	0.07
	VII	Ophelia	0.376409	53.79	0.0099	0.1°	$5.4 \cdot 10^{16}$	21.4	0.07
	VIII	Bianca	0.434577	59.17	0.0009	0.2°	$9.3 \cdot 10^{16}$	25.7	0.07
	IX	Cressida	0.463570	61.78	0.0004	0.0°	$3.4 \cdot 10^{17}$	39.8	0.07
	X	Desdemona	0.473651	62.68	0.0001	0.2°	$1.8 \cdot 10^{17}$	32.0	0.07
	XI	Juliet	0.493066	64.35	0.0007	0.1°	$5.6 \cdot 10^{17}$	46.8	0.07
	XII	Portia	0.513196	66.09	0.0001	0.1°	$1.7 \cdot 10^{18}$	67.6	0.07
	XIII	Rosalind	0.558459	69.94	0.0001	0.3°	$2.6 \cdot 10^{17}$	36.0	0.07
	XIV	Belinda	0.623525	75.26	0.0001	0.0°	$3.6 \cdot 10^{17}$	40.3	0.07
	XV	Puck	0.761832	86.01	0.0001	0.31°	$2.9 \cdot 10^{18}$	81.0	0.07
	XVI	Caliban	579.73	7231	0.1587		$7.4 \cdot 10^{17}$	49	0.07
	XVII	Sycorax	1288.30	12179	0.5224		$5.4 \cdot 10^{18}$	95	0.07
	XVIII	Prospero	1978.29	16256	0.4448		$2.1 \cdot 10^{16}$	15	0.07
	XIX	Setebos	2225.21	17418	0.5914		$2.1 \cdot 10^{16}$	15	0.07
	XX	Stephano	677.36	8004	0.2292		$6.0 \cdot 10^{15}$	10	0.07
	XXI	Trinculo	749.24	8504	0.2200		$7.5 \cdot 10^{14}$	5	0.04
	XXII	Francisco	266.56	4276	0.146		$1.3 \cdot 10^{15}$	11	
	XXIII	Margaret	1687.01	14345	0.661		$1.0 \cdot 10^{15}$	5.5	
	XXIV	Ferdinand	2887.21	20901	0.368		$1.3 \cdot 10^{15}$	6	
	XXV	Perdita	0.638	76.42	0.0		$4.0 \cdot 10^{17}$	40	
	XXVI	Mab	0.923	97.73	0.0		$4.0 \cdot 10^{15}$	8	
	XXVII	Cupid	0.613	74.8	0.0		$1.2 \cdot 10^{15}$	6	
Neptune	I	Triton	5.8768541 R	354.76	0.000016	157.345°	$2.147 \cdot 10^{22}$	1353.4	0.76
	II	Nereid	360.13619	5513.4	0.7512	27.6°	$3.1 \cdot 10^{19}$	170	0.15
	III	Naiad	0.294396	48.227	0.0003	4.74°	$1.3 \cdot 10^{17}$	33	0.07
	IV	Thalassa	0.311485	50.075	0.0002	0.21°	$3.5 \cdot 10^{17}$	41	0.09
	V	Despina	0.334655	52.526	0.0001	0.07°	$2.3 \cdot 10^{18}$	75	0.09
	VI	Galatea	0.428745	61.953	0.0001	0.05°	$2.7 \cdot 10^{18}$	88	0.08
	VII	Larissa	0.554654	73.548	0.0014	0.20°	$4.8 \cdot 10^{18}$	104×89	0.09
	VIII	Proteus	1.122315	117.647	0.0004	0.55°	$4.9 \cdot 10^{19}$	218×208×201	0.10
	IX	Halimede	1879.08	16611	0.2646				
	X	Psamathe	9074.30	48096	0.3809		$1.5 \cdot 10^{16}$	14	
	XI	Sao	2912.72	22228	0.1365				
	XII	Laomedeia	3171.33	23567	0.3969				
	XIII	Neso	9740.73	49285	0.5714				
Plutoids*									
Pluto	I	Charon	6.387	17.536	0.0022	99°	$1.6 \cdot 10^{21}$	593	0.37
	II	Nix	24.86	48.708	0.0030		$5 \cdot 10^{16}$	22–65	
	III	Hydra	38.20	64.749	0.0051		$5 \cdot 10^{16}$	22–65	
Eris	I	Dysnomia		30				100–200	

* In June 2008 the International Astronomical Union decided on the name *plutoid* for the category of transneptunian dwarf planets. Plutoids are celestial bodies in orbit around the sun at a semimajor axis greater than that of Neptune and sufficiently massive to adopt a near-spherical shape. See <www.iau.org/public_press/news/release/iau0804/>.

INTERSTELLAR MOLECULES

Frank J. Lovas and Lewis E. Snyder

A number of molecules have been detected in the interstellar medium, in circumstellar envelopes around evolved stars, and comae and tails of comets through observation of their microwave, infrared, or optical spectra. The following list gives the molecules and the particular isotopic species that have been reported so far. Molecules are listed by molecular formula in the Hill order. All species not footnoted otherwise are observed in interstellar clouds, while some are also found in comets and circumstellar clouds. The list was last updated in October 2008 and lists 162 molecules (298 isotopic forms).

References

1. Lovas, F. J., Recommended Rest Frequencies for Observed Interstellar Molecule Microwave Transitions — 2002 Revision, *J. Phys. Chem. Ref. Data* 33, 177–355 (2004); and update appearing at http://physics.nist.gov/PhysRefData/micro/html/contents.html
2. Snyder, L. E., Cometary Molecules, Internat. Astron. Union Symposium No. 150, *Astrochemistry of Cosmic Phenomena*, ed. P. D. Singh, Kluwer Academic Publishers, Dordrecht, The Netherlands, pp. 427–434 (1992).

Molecular formula	Name	Isotopic species
AlCl	Aluminum monochloride	$Al^{35}Cl$[a]
		$Al^{37}Cl$[a]
AlF	Aluminum monofluoride	AlF[a]
CAlN	Aluminum isocyanide	AlNC[a]
CF^+	Fluoromethylidynium ion	CF^+
CH	Methylidyne	CH
CH^+	Methyliumylidene	CH^+
CHN	Hydrogen cyanide	HCN
		$H^{13}CN$
		$HC^{15}N$
		DCN
CHN	Hydrogen isocyanide	HNC
		$H^{15}NC$
		$HN^{13}C$
		DNC
		$D^{15}NC$
CHNO	Isocyanic acid	HNCO
		DNCO
CHNO	Hydroxyl cyanide	HOCN
CHNS	Isothiocyanic acid	HNCS
CHO	Oxomethyl	HCO
CHO^+	Oxomethylium	HCO^+
		$H^{13}CO^+$
		$HC^{17}O^+$
		$HC^{18}O^+$
		DCO^+
		$D^{13}CO^+$
CHO^+	Hydroxymethylidyne	HOC^+
CHO_2^+	Hydroxyoxomethylium	$HOCO^+$
CHP	Phosphaethyne	HCP[a]
CHS^+	Thiooxomethylium	HCS^+
CH_2	Methylene	CH_2
CH_2N^+	Iminomethylium	$HCNH^+$
CH_2N	Methylene amidogen	CH_2N
CH_2N_2	Cyanamide	NH_2CN
CH_2O	Formaldehyde	H_2CO
		$H_2{}^{13}CO$
		$H_2C^{18}O$
		HDCO
		D_2CO
CH_2O_2	Formic acid	HCOOH
		$H^{13}COOH$
		HCOOD
		DCOOH
CH_2S	Thioformaldehyde	H_2CS
		$H_2{}^{13}CS$
		$H_2C^{34}S$
		HDCS
		D_2CS
CH_3	Methyl	CH_3[a]
CH_3N	Methanimine	CH_2NH
		$^{13}CH_2NH$
CH_3NO	Formamide	NH_2CHO
		$NH_2{}^{13}CHO$
CH_3O^+	Hydroxymethylium ion	H_2COH^+
CH_4	Methane	CH_4
CH_4O	Methanol	CH_3OH
		$^{13}CH_3OH$
		$CH_3{}^{18}OH$
		CH_2DOH
		CH_3OD
		CHD_2OH
		CD_3OH
CH_4S	Methanethiol	CH_3SH
CH_5N	Methylamine	CH_3NH_2
CMgN	Magnesium cyanide	MgCN[a]
CMgN	Magnesium isocyanide	$^{24}MgNC$[a]
		$^{25}MgNC$[a]
		$^{26}MgNC$[a]
CN	Cyanide radical	CN
		^{13}CN
		$C^{15}N$
CN^+	Cyanide radical ion	CN^+[b]
CNNa	Sodium cyanide	NaCN[a]
CNSi	Silicon cyanide	SiCN[a]
CNSi	Silicon isocyanide	SiNC[a]
CN_2	Cyanoimidogen	NCN[b]
CO	Carbon monoxide	CO
		^{13}CO
		$C^{17}O$
		$C^{18}O$
		$^{13}C^{17}O$
		$^{13}C^{18}O$
		^{14}CO

Molecular formula	Name	Isotopic species
CO^+	Carbon monoxide ion	CO^+
COS	Carbon oxysulfide	OCS
		$OC^{34}S$
		$O^{13}CS$
		^{18}OCS
CO_2	Carbon dioxide	CO_2
CO_2^+	Carbon dioxide ion	CO_2^+[b]
CP	Carbon phosphide	CP[a]
CS	Carbon monosulfide	CS
		$C^{33}S$
		$C^{34}S$
		$C^{36}S$
		^{13}CS
		$^{13}C^{34}S$
CSi	Silicon carbide	SiC[a]
C_2	Dicarbon	C_2
C_2H	Ethynyl	C_2H
		^{13}CCH
		$C^{13}CH$
		C_2D
C_2HN	Cyanomethylene	$HCCN$
C_2HNO	Cyanoformaldehyde	$CNCHO$
C_2H_2	Acetylene	$HCCH$
C_2H_2N	Cyanomethyl	CH_2CN
C_2H_2O	Ketene	H_2CCO
C_2H_3N	Acetonitrile	CH_3CN
		$^{13}CH_3CN$
		$CH_3{}^{13}CN$
		$CH_3C^{15}N$
		CH_2DCN
C_2H_3N	Isocyanomethane	CH_3NC
C_2H_3N	Keteneimine	CH_2CNH
C_2H_4	Ethylene	H_2CCH_2
$C_2H_4N_2$	Aminoacetonitrile	NH_2CH_2CN
C_2H_4O	Acetaldehyde	CH_3CHO
C_2H_4O	Ethylene oxide	*c*-C_2H_4O
C_2H_4O	*anti*-Ethenol	*a*-CH_2CHOH
C_2H_4O	*syn*-Ethenol	*s*-CH_2CHOH
$C_2H_4O_2$	Methyl formate	CH_3OCHO
$C_2H_4O_2$	Acetic acid	CH_3COOH
$C_2H_4O_2$	Glycolaldehyde	CH_2OHCHO
C_2H_5NO	Acetamide	CH_3CONH_2
C_2H_6	Ethane	CH_3CH_3[b]
C_2H_6O	*trans*-Ethanol	*t*-CH_3CH_2OH
C_2H_6O	*gauche*-Ethanol	*g*-CH_3CH_2OH
C_2H_6O	Dimethyl ether	CH_3OCH_3
$C_2H_6O_2$	Ethylene glycol	$HOCH_2CH_2OH$
C_2O	Oxoethenylidene	CCO
C_2P	Phosphaethenylidene	CCP[a]
C_2S	Thioxoethenylidene	CCS
		$CC^{34}S$
		^{13}CCS
		$C^{13}CS$
C_2Si	Silicon dicarbide	*c*-SiC_2
		c-$^{29}SiC_2$
		c-$^{30}SiC_2$
		c-$Si^{13}CC$
C_3	Tricarbon	C_3
C_3H	Cyclopropenylidyne	*c*-C_3H
		c-$CC^{13}CH$
C_3H	Propenylidyne	*l*-C_3H
C_3HN	Cyanoacetylene	$HCCCN$
		$H^{13}CCCN$
		$HC^{13}CCN$
		$HCC^{13}CN$
		$HCCC^{15}N$
		$DCCCN$
C_3HN	Isocyanoacetylene	$HCCNC$
C_3HN	3-Imino-1,2-propadienylidene	$HNCCC$
C_3H_2	Cyclopropenylidene	*c*-C_3H_2
		c-$H^{13}CCCH$
		c-$HC^{13}CCH$
		c-C_3HD
C_3H_2	Propadienylidene	*l*-H_2CCC
$C_3H_2N^+$	Protonated cyanoacetylene	$HCCCNH^+$
C_3H_2O	2-Propynal	$HCCCHO$
C_3H_2O	Cyclopropenone	*c*-C_3H_2O
C_3H_3N	Acrylonitrile (vinyl cyanide)	CH_2CHCN
		$^{13}CH_2CHCN$
		$CH_2{}^{13}CHCN$
C_3H_4	Propyne	CH_3CCH
		$CH_3C^{13}CH$
		$^{13}CH_3CCH$
		CH_2DCCH
		CH_3CCD
C_3H_4O	Propenal	CH_2CHCHO
C_3H_5N	Propanenitrile (ethyl cyanide)	CH_3CH_2CN
		$^{13}CH_3CH_2CN$
		$CH_3{}^{13}CH_2CN$
		$CH_3CH_2{}^{13}CN$
C_3H_6	Propylene	CH_2CHCH_3
C_3H_6O	Acetone	$(CH_3)_2CO$
C_3H_6O	Propanal	CH_3CH_2CHO
C_3N	Cyanoethynyl	$CCCN$
		$^{13}CCCN$
		$C^{13}CCN$
		$CC^{13}CN$
C_3N^-	Cyanoethynyl anion	$CCCN^-$
C_3O	1,2-Propadienylidene, 3-oxo	$CCCO$
C_3S	1,2-Propadienylidene, 3-thioxo	$CCCS$
		$CCC^{34}S$
		$C^{13}CCS$
C_3Si	Silicon tricarbon	SiC_3
C_4H	1,3-Butadiynyl radical	$HCCCC$
		$H^{13}CCCC$
		$HC^{13}CCC$
		$HCC^{13}CC$
		$HCCC^{13}C$
		$DCCCC$
C_4H^-	1,3-Butadiynyl anion	$HCCCC^-$
C_4HN	3-Cyano-1,2-propadienylidene	$HCCCCN$
C_4H_2	Butatrienylidene	H_2CCCC
C_4H_2	1,3-Butadiyne	$HCCCCH$[a]
C_4H_3N	2-Butynenitrile	CH_3CCCN

Molecular formula	Name	Isotopic species
C_4H_3N	Cyanoallene	CH_2CCHCN
C_4Si	Silicon tetracarbide	SiC_4[a]
		$SiCCC^{13}C$
C_5	Pentacarbon	C_5[a]
C_5H	2,4-Pentadiynylidyne	$HCCCCC$
C_5HN	2,4-Pentadiynenitrile	$HCCCCCN$
		$H^{13}CCCCCN$
		$HC^{13}CCCCN$
		$HCC^{13}CCCN$
		$HCCC^{13}CCN$
		$HCCCC^{13}CN$
		$DCCCCCN$
C_5H_4	1,3-Pentadiyne	CH_3C_4H
C_5N	1,3-Butadiynylium, 4-cyano	C_5N
C_6H	1,3,5-Hexatriynyl	$HCCCCCC$
C_6H^-	1,3,5-Hexatriynyl anion	$HCCCCCC^-$
C_6H_2	1,3,5-Hexatriyne	$HCCCCCCH$[a]
C_6H_2	1,2,3,4,5-Hexapentaenylidene	$H_2CCCCCC$
C_6H_3N	Methylcyanodiacetylene	CH_3C_4CN
C_6H_6	Benzene	C_6H_6
C_7H	2,4,6-Heptatriynylidyne	$HCCCCCCC$
C_7HN	2,4,6-Heptatriynenitrile	HC_7N
C_7H_4	Methyltriacetylene	CH_3C_6H
C_8H	1,3,5,7-Octatetraynyl	HC_8
C_8H^-	1,3,5,7-Octatetraynyl anion	HC_8^-
C_9HN	2,4,6,8-Nonatetraynenitrile	HC_9N
$C_{11}HN$	2,4,6,8,10-Undecapentaynenitrile	$HC_{11}N$
ClH	Hydrogen chloride	$H^{35}Cl$
		$H^{37}Cl$
ClK	Potassium chloride	$K^{35}Cl$[a]
		$K^{37}Cl$[a]
$ClNa$	Sodium chloride	$Na^{35}Cl$[a]
		$Na^{37}Cl$[a]
FH	Hydrogen fluoride	HF
FeO	Iron monoxide	FeO
HLi	Lithium hydride	^7LiH
HN	Imidogen	HN
HNO	Nitrosyl hydride	HNO
HN_2^+	Hydrodinitrogen(1+)	N_2H^+
		$^{15}NNH^+$
		$N^{15}NH^+$
		N_2D^+
HO	Hydroxyl	OH
		^{17}OH
		^{18}OH
HO^+	Oxoniumylidene	OH^+[b]
HS	Mercapto	SH
H_2	Hydrogen	H_2
H_2N	Amidogen	NH_2

Molecular formula	Name	Isotopic species
H_2O	Water	H_2O
		$H_2^{18}O$
		HDO
		D_2O
H_2O^+	Oxoniumyl	H_2O^+[b]
H_2S	Hydrogen sulfide	H_2S
		$H_2^{34}S$
		HDS
		D_2S
H_3^+	Trihydrogen ion	H_3^+
		H_2D^+
		D_2H^+
H_3N	Ammonia	NH_3
		$^{15}NH_3$
		NH_2D
		NHD_2
		ND_3
H_3O^+	Oxonium hydride	H_3O^+
H_4Si	Silane	SiH_4[a]
NO	Nitric oxide	NO
NP	Phosphorous nitride	PN
NS	Nitrogen sulfide	NS
		$N^{34}S$
NSi	Silicon nitride	SiN
N_2	Nitrogen	N_2
N_2^+	Nitrogen ion	N_2^+[b]
N_2O	Nitrous oxide	N_2O
OP	Phosphorus monoxide	PO[a]
OS	Sulfur monoxide	SO
		^{34}SO
		^{33}SO
		$S^{18}O$
OS^+	Sulfur monoxide ion	SO^+
OSi	Silicon monoxide	SiO
		$Si^{18}O$
		^{29}SiO
		^{30}SiO
O_2	Oxygen	O_2
O_2S	Sulfur dioxide	SO_2
		$^{33}SO_2$
		$^{34}SO_2$
		$OS^{17}O$
		$OS^{18}O$
SSi	Silicon monosulfide	SiS
		$Si^{33}S$
		$Si^{34}S$
		^{29}SiS
		$^{29}Si^{34}S$
		^{30}SiS
		$^{30}Si^{34}S$
S_2	Sulfur	S_2[b]

l- before the isotopic species indicates a linear configuration, while *c*- indicates a cyclic molecule.

[a] Reported only in circumstellar clouds.

[b] Reported only in comets.

MASS, DIMENSIONS, AND OTHER PARAMETERS OF THE EARTH

This table is a collection of data on various properties of the Earth. Most of the values are given in SI units. Note that 1 ua (astronomical unit) = 149,597,870 km.

References

1. Seidelmann, P. K., Ed., *Explanatory Supplement to the Astronomical Almanac*, University Science Books, Mill Valley, CA, 1992.
2. Lang, K. R., *Astrophysical Data: Planets and Stars*, Springer-Verlag, New York, 1992.

Quantity	Symbol	Value	Unit
Mass	M	$5.9723 \cdot 10^{27}$	g
Major orbital semi-axis	a_{orb}	1.000000	ua
		$1.4959787 \cdot 10^{8}$	km
Distance from Sun at perihelion	r_{π}	0.9833	ua
Distance from Sun at aphelion	r_{α}	1.0167	ua
Moment of perihelion passage	T_{π}	Jan. 2, 4 h 52 min	
Moment of aphelion passage	T_{α}	July 4, 5 h 05 min	
Siderial rotation period around Sun	P_{orb}	$31.5581 \cdot 10^{6}$	s
		365.25636	d
Mean rotational velocity	U_{orb}	29.78	km/s
Mean equatorial radius	$\bar{a}$	6378.140	km
Mean polar compression (flattening factor)	α	1/298.257	
Difference in equatorial and polar semi-axes	$a - c$	21.385	km
Compression of meridian of major equatorial axis	α_a	1/295.2	
Compression of meridian of minor equatorial axis	α_b	1/298.0	
Equatorial compression	ε	1/30 000	
Difference in equatorial semi-axes	$a - b$	213	m
Difference in polar semi-axes	$c_N - c_S$	~70	m
Polar asymmetry	η	$\sim 1.10^{-5}$	
Mean acceleration of gravity at equator	g_e	9.78036	m/s^2
Mean acceleration of gravity at poles	g_p	9.83208	m/s^2
Difference in acceleration of gravity at pole and at equator	$g_p - g_e$	5.172	cm/s^2
Mean acceleration of gravity for entire surface of terrestrial ellipsoid	g	9.7978	m/s^2
Mean radius	R	6371.0	km
Area of surface	S	$5.10 \cdot 10^{8}$	km^2
Volume	V	$1.0832 \cdot 10^{12}$	km^3
Mean density	ρ	5.515	g/cm^3
Siderial rotational period	P	86,164.09	s
Rotational angular velocity	ω	$7.292116 \cdot 10^{-5}$	rad/s
Mean equatorial rotational velocity	v	0.46512	km/s
Rotational angular momentum	L	$5.861 \cdot 10^{33}$	J s
Rotational energy	E	$2.137 \cdot 10^{29}$	J
Ratio of centrifugal force to force of gravity at equator	q_c	0.0034677 = 1/288	
Moment of inertia	I	$8.070 \cdot 10^{37}$	kg m^2
Relative braking of Earth's rotation due to tidal friction	$\Delta\omega_e/\omega$	$-4.2 \cdot 10^{-8}$	$century^{-1}$
Relative secular acceleration of Earth's rotation	$\Delta\omega_i/\omega$	$+1.4 \cdot 10^{-8}$	$century^{-1}$
Not secular braking of Earth's rotation	$\Delta\omega/\omega$	$-2.8 \cdot 10^{-8}$	$century^{-1}$
Probable value of total energy of tectonic deformation of Earth	E_t	$\sim 1 \cdot 10^{23}$	J/century
Secular loss of heat of Earth through radiation into space	$\Delta' E_k$	$1 \cdot 10^{23}$	J/century
Portion of Earth's kinetic energy transformed into heat as a result of lunar and solar tides in the hydrosphere	$\Delta'' E_k$	$1.3 \cdot 10^{23}$	J/century
Differences in duration of days in March and August	ΔP	0.0025 (March-August)	s
Corresponding relative annual variation in Earth's rotational velocity	$\Delta^* \omega/\omega$	$2.9 \cdot 10^{-8}$ (Aug.-March)	
Presumed variation in Earth's radius between August and March	$\Delta^* R$	−9.2 (Aug.-March)	cm
Annual variation in level of world ocean	Δh_o	~10 (Sept.-March)	cm
Area of continents	S_C	$1.49 \cdot 10^{8}$	km^2
		29.2	% of surface

Quantity	Symbol	Value	Unit
Area of world ocean	S_o	$3.61 \cdot 10^{8}$	km^2
		70.8	% of surface
Mean height of continents above sea level	h_C	875	m
Mean depth of world ocean	h_o	3794	m
Mean thickness of lithosphere within the limits of the continents	$h_{c.l.}$	35	km
Mean thickness of lithosphere within the limits of the ocean	$h_{o.l.}$	4.7	km
Mean rate of thickening of continental lithosphere	$\Delta h/\Delta t$	10 – 40	$m/10^6$ y
Mean rate of horizontal extension of continental lithosphere	$\Delta l/\Delta t$	0.75 – 20	$km/10^6$ y
Mass of crust	m_l	$2.36 \cdot 10^{22}$	kg
Mass of mantle		$4.05 \cdot 10^{24}$	kg
Amount of water released from the mantle and core in the course of geological time		$3.40 \cdot 10^{21}$	kg
Total reserve of water in the mantle		$2 \cdot 10^{23}$	kg
Present content of free and bound water in the Earth's lithosphere		$2.4 \cdot 10^{21}$	kg
Mass of hydrosphere	m_h	$1.664 \cdot 10^{21}$	kg
Amount of oxygen bound in the Earth's crust		$1.300 \cdot 10^{21}$	kg
Amount of free oxygen		$1.5 \cdot 10^{18}$	kg
Mass of atmosphere	m_a	$5.136 \cdot 10^{18}$	kg
Mass of biosphere	m_b	$1.148 \cdot 10^{16}$	kg
Mass of living matter in the biosphere		$3.6 \cdot 10^{14}$	kg
Density of living matter on dry land		0.1	g/cm^2
Density of living matter in ocean		$15 \cdot 10^{-8}$	g/cm^3
Age of the Earth		$4.55 \cdot 10^{9}$	y
Age of oldest rocks		$4.0 \cdot 10^{9}$	y
Age of most ancient fossils		$3.4 \cdot 10^{9}$	y

GEOLOGICAL TIME SCALE

Period or epoch	Beginning and end, in 10^6 years	Key events
Cenozoic era		
Quaternary		
Holocene	0–0.0115	
Pleistocene*	0.0115–1.81	Homo Erectus breakout
Tertiary		
Pliocene*	1.81–5.3	Ape man fossils
Miocene	5.3–23	Origin of grass
Oligocene	23–34	Rise of cats, dogs, pigs
Eocene	34–56	Debut of hoofed mammals
Paleocene	56–65	Earliest primates
Mesozoic era		
Cretaceous	65–145	Demise of dinosaurs
Jurassic	145–200	First birds
Triassic	200–251	Appearance of dinosaurs
Paleozoic era		
Permian	251–299	Flowers, insect pollination
Carboniferous	299–359	First conifers
Devonian	359–416	First vertebrates ashore
Silurian	416–444	Spore-bearing plants
Ordovician	444–488	First animals ashore
Cambrian	488–542	Vertebrates appear
Pre-Cambrian		
Pre-Cambrian III (Proterozoic)	542–2500	First plants, jellyfish
Pre-Cambrian II (Archean)	2500–3850	Photosynthetic bacteria
Pre-Cambrian I (Hadean)	3850–4600	Earth formed 4600 million years ago

* Some authorities place the boundary between the Pleistocene and Pliocene at $2.6 \cdot 10^6$ years.

References

1. U.S. Geological Survey Geologic Names Committee, 2007, Divisions of geologic time—Major chronostratigraphic and geochronologic units: U.S. Geological Survey Fact Sheet 2007-3015. Available on the Internet at <pubs.usgs.gov/fs/2007/3015/index.html>.
2. Walker, J. D., and Geissman, J. W., compilers, 2009, Geologic Time Scale: Geological Society of America, <www.geosociety.org/science/timescale/timescl.pdf>.
3. Calder, N., *Timescale - An Atlas of the Fourth Dimension,* Viking Press, New York, 1983.

ACCELERATION DUE TO GRAVITY

The acceleration due to gravity is tabulated here as a function of latitude and height above the Earth's surface. Values were calculated from the expression

$$g/(\mathrm{m/s^2}) = 9.780356\ (1 + 0.0052885 \sin^2 \phi - 0.0000059 \sin^2 2\phi) - 0.003086\ H$$

where ϕ is the latitude and H is the height in kilometers.

Reference

Jursa, A. S., Ed., *Handbook of Geophysics and the Space Environment*, 4th ed., Air Force Geophysics Laboratory, 1985, p. 14–17.

ϕ	$H = 0$	$H = 1$ km	$H = 5$ km	$H = 10$ km
0	9.78036	9.77727	9.76493	9.74950
5	9.78075	9.77766	9.76532	9.74989
10	9.78191	9.77882	9.76648	9.75105
15	9.78381	9.78072	9.76838	9.75295
20	9.78638	9.78330	9.77095	9.75552
25	9.78956	9.78647	9.77413	9.75870
30	9.79324	9.79016	9.77781	9.76238
35	9.79732	9.79424	9.78189	9.76646
40	9.80167	9.79858	9.78624	9.77081
45	9.80616	9.80307	9.79073	9.77530
50	9.81065	9.80757	9.79522	9.77979
55	9.81501	9.81193	9.79958	9.78415
60	9.81911	9.81602	9.80368	9.78825
65	9.82281	9.81972	9.80738	9.79195
70	9.82601	9.82292	9.81058	9.79515
75	9.82860	9.82551	9.81317	9.79774
80	9.83051	9.82710	9.81500	9.79965
85	9.83168	9.82860	9.81625	9.80082
90	9.83208	9.82899	9.81665	9.80122

DENSITY, PRESSURE, AND GRAVITY AS A FUNCTION OF DEPTH WITHIN THE EARTH

This table gives the density ρ, pressure *p*, and acceleration due to gravity *g* as a function of depth below the Earth's surface, as calculated from the model of the structure of the Earth in Reference 1. The model assumes a radius of 6371 km for the Earth. The boundary between the crust and mantle (the Mohorovicic discontinuity) is taken as 21 km, while in reality it varies considerable with location.

References

1. Anderson, D. L., and Hart, R. S., *J. Geophys. Res.*, 81, 1461, 1976.
2. Carmichael, R. S., *CRC Practical Handbook of Physical Properties of Rocks and Minerals*, p. 467, CRC Press, Boca Raton, FL, 1989.

Depth km	ρ g/cm³	*p* kbar	*g* cm/s²
Crust			
0	1.02	0	981
3	1.02	3	982
3	2.80	3	982
21	2.80	5	983
Mantle (solid)			
21	3.49	5	983
41	3.51	12	983
61	3.52	19	984
81	3.48	26	984
101	3.44	33	984
121	3.40	39	985
171	3.37	56	987
221	3.34	73	989
271	3.37	89	991
321	3.47	106	993
371	3.59	124	994
571	3.95	199	999
871	4.54	328	997
1171	4.67	466	992
1471	4.81	607	991
1771	4.96	752	994
2071	5.12	903	1002
2371	5.31	1061	1017
2671	5.45	1227	1042
2886	5.53	1352	1069
Outer core (liquid)			
2886	9.96	1352	1069
2971	10.09	1442	1050
3371	10.63	1858	953
3671	11.00	2154	874
4071	11.36	2520	760
4471	11.69	2844	641
4871	11.99	3116	517
5156	12.12	3281	427
Inner core (solid)			
5156	12.30	3281	427
5371	12.48	3385	355
5771	12.52	3529	218
6071	12.53	3592	122
6371	12.58	3617	0

OCEAN PRESSURE AS A FUNCTION OF DEPTH AND LATITUDE

The following table is based upon an ocean model which takes into account the equation of state of standard seawater and the dependence on latitude of the acceleration of gravity. The tabulated pressure value is the excess pressure over the ambient atmospheric pressure at the surface.

References

1. *International Oceanographic Tables, Volume 4*, Unesco Technical Papers in Marine Science No. 40, Unesco, Paris, 1987.
2. Saunders, P. M., and Fofonoff, N. P., *Deep-Sea Res.* 23, 109–111, 1976.

Pressure in MPa at the Specified Latitude

Depth (meters)	0°	15°	30°	45°	60°	75°	90°
0	0.0000	0.0000	0.0000	0.0000	0.0000	0.0000	0.0000
500	5.0338	5.0355	5.0404	5.0471	5.0537	5.0586	5.0605
1000	10.0796	10.0832	10.0930	10.1064	10.1198	10.1296	10.1333
1500	15.1376	15.1431	15.1577	15.1778	15.1980	15.2127	15.2182
2000	20.2076	20.2148	20.2344	20.2613	20.2882	20.3080	20.3153
2500	25.2895	25.2985	25.3231	25.3560	25.3905	25.4153	25.4244
3000	30.3831	30.3940	30.4236	30.4641	30.5047	30.5345	30.5453
3500	35.4886	35.5012	35.5358	35.5832	35.6307	35.6654	35.6782
4000	40.6056	40.6201	40.6598	40.7140	40.7683	40.8082	40.8229
4500	45.7342	45.7505	45.7952	45.8564	45.9176	45.9626	45.9791
5000	50.8742	50.8924	50.9421	51.0102	51.0785	51.1285	51.1469
5500	56.0255	56.0456	56.1004	56.1755	56.2508	56.3059	56.3262
6000	61.1882	61.2100	61.2700	61.3521	61.4344	61.4947	61.5168
6500	66.3619	66.3857	66.4508	66.5399	66.6292	66.6947	66.7187
7000	71.5467	71.5724	71.6427	71.7388	71.8352	71.9059	71.9318
7500	76.7426	76.7701	76.8456	76.9488	77.0523	77.1282	77.1560
8000	81.9493	81.9788	82.0594	82.1697	82.2804	82.3614	82.3911
8500	87.1669	87.1983	87.2841	87.4016	87.5193	87.6057	87.6373
9000	92.3950	92.4284	92.5194	92.6440	92.7689	92.8606	92.8941
9500	97.6346	97.6698	97.7661	97.8978	98.0300	98.1269	98.1624
10000	102.8800	102.9170	103.0185	103.1572	103.2961	103.3981	103.4355

PROPERTIES OF SEAWATER

In addition to the dependence on temperature and pressure, the physical properties of seawater vary with the concentration of the dissolved constituents. A convenient parameter for describing the composition is the salinity, S, which is defined in terms of the electrical conductivity of the seawater sample. The defining equation for the practical salinity is:

$$S = a_0 + a_1K^{1/2} + a_2K + a_3K^{3/2} + a_4K^2 + a_5K^{5/2},$$

where K is the ratio of the conductivity of the seawater sample at 15 °C and atmospheric pressure to the conductivity of a potassium chloride solution in which the mass fraction of KCl is 0.0324356, at the same temperature and pressure. The values of the coefficients are:

$a_0 = 0.0080$ $a_3 = 14.0941$
$a_1 = -0.1692$ $a_4 = -7.0261$
$a_2 = 25.3851$ $a_5 = 2.7081$
$\Sigma a_i = 35.0000$

Thus when $K = 1$, $S = 35$ exactly (S is normally quoted in units of ‰, i.e., parts per thousand). The value of S can be roughly equated with the mass of dissolved material in grams per kilogram of seawater. Salinity values in the open oceans at midlatitudes typically fall between 34 and 36.

It is customary in oceanography to define the pressure at a given point as the pressure due to the column of water between that point and the surface. Thus by convention $P = 0$ at the sea surface. To a good approximation the pressure in decibars (dbar) can be equated to the depth in meters. Thus at 45° latitude the pressure is 5000 dbar at 4902 m, 10000 dbar at 9700 m.

The first table below gives several properties of seawater as a function of temperature for a salinity of 35. The second and third give density and electrical conductivity as a function of salinity at several temperatures, and the fourth lists typical concentrations of the main constituents of seawater as a function of salinity. The final table gives the freezing point as a function of salinity and pressure.

References

1. *The Practical Salinity Scale 1978 and the International Equation of State of Seawater 1980*, Unesco Technical Papers in Marine Science No. 36, Unesco, Paris, 1981; sections No. 37, 38, 39, and 40 in this series give background papers and detailed tables.
2. Kennish, M. J., *CRC Practical Handbook of Marine Science*, CRC Press, Boca Raton, FL, 1989.
3. Poisson, A. *IEEE J. Ocean. Eng.* OE-5, 50, 1981.
4. Webster, F., in *AIP Physics Desk Reference*, E. R. Cohen, D. R. Lide and G. L. Trigg, Eds., Springer-Verlag, New York, 2003.

Properties of Seawater as a Function of Temperature at Salinity $S = 35$ and Normal Atmospheric Pressure

ρ = density in g/cm³
β = $(1/\rho)(d\rho/dS)$ = fractional change in density per unit change in salinity
α = $-(1/\rho)(d\rho/dt)$ = fractional change in density per unit change in temperature (°C⁻¹)
κ = electrical conductivity in S/cm
η = viscosity in mPa s (equal to cP)
c_p = specific heat in J/kg °C
v = speed of sound in m/s

t/°C	ρ/g cm⁻³	$10^7\beta$	$10^7\,\alpha$/°C⁻¹	κ/S cm⁻¹	η/mPa s	c_p/J kg⁻¹°C⁻¹	v/m s⁻¹
0	1.028106	7854	526	0.029048	1.892	3986.5	1449.1
5	1.027675	7717	1136	0.033468	1.610		
10	1.026952	7606	1668	0.038103	1.388	3986.3	1489.8
15	1.025973	7516	2141	0.042933	1.221		
20	1.024763	7444	2572	0.047934	1.085	3993.9	1521.5
25	1.023343	7385	2970	0.053088	0.966		
30	1.021729	7338	3341	0.058373	0.871	4000.7	1545.6
35	1.019934	7300	3687				
40		7270	4004			4003.5	1563.2

Density of Surface Seawater in g/cm³ as a Function of Temperature and Salinity

t/°C	S = 0	S = 5	S = 10	S = 15	S = 20	S = 25	S = 30	S = 35	S = 40
0	0.999843	1.003913	1.007955	1.011986	1.016014	1.020041	1.024072	1.028106	1.032147
5	0.999967	1.003949	1.007907	1.011858	1.015807	1.019758	1.023714	1.027675	1.031645
10	0.999702	1.003612	1.007501	1.011385	1.015269	1.019157	1.023051	1.026952	1.030862
15	0.999102	1.002952	1.006784	1.010613	1.014443	1.018279	1.022122	1.025973	1.029834
20	0.998206	1.002008	1.005793	1.009576	1.013362	1.017154	1.020954	1.024763	1.028583
25	0.997048	1.000809	1.004556	1.008301	1.012050	1.015806	1.019569	1.023343	1.027128
30	0.995651	0.999380	1.003095	1.006809	1.010527	1.014252	1.017985	1.021729	1.025483
35	0.994036	0.997740	1.001429	1.005118	1.008810	1.012509	1.016217	1.019934	1.023662
40	0.992220	0.995906	0.999575	1.003244	1.006915	1.010593	1.014278	1.017973	1.021679

Electrical Conductivity of Seawater in S/cm as a Function of Temperature and Salinity

t/°C	S = 5	S = 10	S = 15	S = 20	S = 25	S = 30	S = 35	S = 40
0	0.004808	0.009171	0.013357	0.017421	0.021385	0.025257	0.029048	0.032775
5	0.005570	0.010616	0.015441	0.020118	0.024674	0.029120	0.033468	0.037734
10	0.006370	0.012131	0.017627	0.022947	0.028123	0.033171	0.038103	0.042935
15	0.007204	0.013709	0.019905	0.025894	0.031716	0.037391	0.042933	0.048355
20	0.008068	0.015346	0.022267	0.028948	0.035438	0.041762	0.047934	0.053968
25	0.008960	0.017035	0.024703	0.032097	0.039276	0.046267	0.053088	0.059751
30	0.009877	0.018771	0.027204	0.035330	0.043213	0.050888	0.058373	0.065683

Composition of Seawater and Ionic Strength at Various Salinities (Ref. 2)

	Expressed as molality			As grams per kilogram of seawater		
Constituent	S = 30	S = 35	S = 40	S = 30	S = 35	S = 40
Cl^-	0.482	0.562	0.650	16.58	19.33	22.36
Br^-	0.00074	0.00087	0.00100	0.057	0.067	0.078
F^-		0.00007			0.001	
SO_4^{2-}	0.0104	0.0114	0.0122	0.97	1.06	1.14
HCO_3^-	0.00131	0.00143	0.00100	0.078	0.085	0.059
$NaSO_4^-$	0.0085	0.0108	0.0139	0.98	1.25	1.60
KSO_4^-	0.00010	0.00012	0.00015	0.013	0.016	0.020
Na^+	0.405	0.472	0.544	9.03	10.53	12.13
K^+	0.00892	0.01039	0.01200	0.338	0.394	0.455
Mg^{2+}	0.0413	0.0483	0.0561	0.974	1.139	1.323
Ca^{2+}	0.00131	0.00143	0.00154	0.051	0.056	0.060
Sr^{2+}	0.00008	0.00009	0.00011	0.007	0.008	0.009
$MgHCO_3^+$	0.00028	0.00036	0.00045	0.023	0.030	0.037
$MgSO_4$	0.00498	0.00561	0.00614	0.582	0.655	0.717
$CaSO_4$	0.00102	0.00115	0.00126	0.135	0.152	0.166
$NaHCO_3$	0.00015	0.00020	0.00024	0.012	0.016	0.020
H_3BO_3	0.00032	0.00037	0.00042	0.019	0.022	0.025
Ionic strength	0.5736	0.6675	0.7701			

Freezing Point of Seawater in °C as a Function of Salinity and Pressure

P/dbar	S = 0	5	10	15	20	25	30	35	40
0	0.000	−0.274	−0.542	−0.812	−1.083	−1.358	−1.638	−1.922	−2.212
50	−0.038	−0.311	−0.580	−0.849	−1.121	−1.396	−1.676	−1.960	−2.250
100	−0.075	−0.349	−0.618	−0.887	−1.159	−1.434	−1.713	−1.998	−2.287
500	−0.377	−0.650	−0.919	−1.188	−1.460	−1.735	−2.014	−2.299	−2.589

ABUNDANCE OF ELEMENTS IN THE EARTH'S CRUST AND IN THE SEA

This table gives the estimated abundance of the elements in the continental crust (in mg/kg, equivalent to parts per million by mass) and in seawater near the surface (in mg/L). Values represent the median of reported measurements. The concentrations of the less abundant elements may vary with location by several orders of magnitude.

References

1. Carmichael, R. S., Ed., *CRC Practical Handbook of Physical Properties of Rocks and Minerals*, CRC Press, Boca Raton, FL, 1989.
2. Bodek, I., et al., *Environmental Inorganic Chemistry*, Pergamon Press, New York, 1988.
3. Ronov, A. B., and Yaroshevsky, A. A., "Earth's Crust Geochemistry," in *Encyclopedia of Geochemistry and Environmental Sciences*, Fairbridge, R. W., Ed., Van Nostrand, New York, 1969.

	Abundance			Abundance	
	Crust	Sea		Crust	Sea
Element	**mg/kg**	**mg/L**	**Element**	**mg/kg**	**mg/L**
Ac	5.5×10^{-10}		N	1.9×10^{1}	5×10^{-1}
Ag	7.5×10^{-2}	4×10^{-5}	Na	2.36×10^{4}	1.08×10^{4}
Al	8.23×10^{4}	2×10^{-3}	Nb	2.0×10^{1}	1×10^{-5}
Ar	3.5	4.5×10^{-1}	Nd	4.15×10^{1}	2.8×10^{-6}
As	1.8	3.7×10^{-3}	Ne	5×10^{-3}	1.2×10^{-4}
Au	4×10^{-3}	4×10^{-6}	Ni	8.4×10^{1}	5.6×10^{-4}
B	1.0×10^{1}	4.44	O	4.61×10^{5}	8.57×10^{5}
Ba	4.25×10^{2}	1.3×10^{-2}	Os	1.5×10^{-3}	
Be	2.8	5.6×10^{-6}	P	1.05×10^{3}	6×10^{-2}
Bi	8.5×10^{-3}	2×10^{-5}	Pa	1.4×10^{-6}	5×10^{-11}
Br	2.4	6.73×10^{1}	Pb	1.4×10^{1}	3×10^{-5}
C	2.00×10^{2}	2.8×10^{1}	Pd	1.5×10^{-2}	
Ca	4.15×10^{4}	4.12×10^{2}	Po	2×10^{-10}	1.5×10^{-14}
Cd	1.5×10^{-1}	1.1×10^{-4}	Pr	9.2	6.4×10^{-7}
Ce	6.65×10^{1}	1.2×10^{-6}	Pt	5×10^{-3}	
Cl	1.45×10^{2}	1.94×10^{4}	Ra	9×10^{-7}	8.9×10^{-11}
Co	2.5×10^{1}	2×10^{-5}	Rb	9.0×10^{1}	1.2×10^{-1}
Cr	1.02×10^{2}	3×10^{-4}	Re	7×10^{-4}	4×10^{-6}
Cs	3	3×10^{-4}	Rh	1×10^{-3}	
Cu	6.0×10^{1}	2.5×10^{-4}	Rn	4×10^{-13}	6×10^{-16}
Dy	5.2	9.1×10^{-7}	Ru	1×10^{-3}	7×10^{-7}
Er	3.5	8.7×10^{-7}	S	3.50×10^{2}	9.05×10^{2}
Eu	2.0	1.3×10^{-7}	Sb	2×10^{-1}	2.4×10^{-4}
F	5.85×10^{2}	1.3	Sc	2.2×10^{1}	6×10^{-7}
Fe	5.63×10^{4}	2×10^{-3}	Se	5×10^{-2}	2×10^{-4}
Ga	1.9×10^{1}	3×10^{-5}	Si	2.82×10^{5}	2.2
Gd	6.2	7×10^{-7}	Sm	7.05	4.5×10^{-7}
Ge	1.5	5×10^{-5}	Sn	2.3	4×10^{-6}
H	1.40×10^{3}	1.08×10^{5}	Sr	3.70×10^{2}	7.9
He	8×10^{-3}	7×10^{-6}	Ta	2.0	2×10^{-6}
Hf	3.0	7×10^{-6}	Tb	1.2	1.4×10^{-7}
Hg	8.5×10^{-2}	3×10^{-5}	Te	1×10^{-3}	
Ho	1.3	2.2×10^{-7}	Th	9.6	1×10^{-6}
I	4.5×10^{-1}	6×10^{-2}	Ti	5.65×10^{3}	1×10^{-3}
In	2.5×10^{-1}	2×10^{-2}	Tl	8.5×10^{-1}	1.9×10^{-5}
Ir	1×10^{-3}		Tm	5.2×10^{-1}	1.7×10^{-7}
K	2.09×10^{4}	3.99×10^{2}	U	2.7	3.2×10^{-3}
Kr	1×10^{-4}	2.1×10^{-4}	V	1.20×10^{2}	2.5×10^{-3}
La	3.9×10^{1}	3.4×10^{-6}	W	1.25	1×10^{-4}
Li	2.0×10^{1}	1.8×10^{-1}	Xe	3×10^{-5}	5×10^{-5}
Lu	8×10^{-1}	1.5×10^{-7}	Y	3.3×10^{1}	1.3×10^{-5}
Mg	2.33×10^{4}	1.29×10^{3}	Yb	3.2	8.2×10^{-7}
Mn	9.50×10^{2}	2×10^{-4}	Zn	7.0×10^{1}	4.9×10^{-3}
Mo	1.2	1×10^{-2}	Zr	1.65×10^{2}	3×10^{-5}

SOLAR IRRADIANCE AT THE EARTH

The solar luminosity (total radiant power emitted by the sun) is $3.86 \cdot 10^{26}$ W, of which about 1366 W m^{-2} (the solar irradiance or "solar constant") reaches the top of the earth's atmosphere. To a zeroth approximation the sun can be considered a black body with an effective temperature of 5780 K, which implies a peak in the radiation at around 0.520 μm (5200 Å). The actual solar spectral emission is more complex, especially at ultraviolet and shorter wavelengths. The graph in Fig. 1, which was taken from Ref. 1, summarizes the solar irradiance at the top of the atmosphere in the range 0.3 μm to 10 μm.

While the solar irradiance has been known for some time to undergo both long-term and short-term variations, accurate measurements have become possible only recently. Fig. 2, which is taken from Ref. 4, shows the variation over the last three decades.

References

1. Jursa, A. S., Ed., *Handbook of Geophysics and the Space Environment*, Air Force Geophysics Laboratory, 1985.
2. Pierce, A. K, and Allen, R. G., "The Solar Spectrum between 0.3 and 10 μm", in *The Solar Output and Its Variation*, White, O. R., Ed., Colorado Associated University Press, Boulder, CO, 1977.
3. Lang, K. R., *Astrophysical Data. Planets and Stars*, Springer-Verlag, New York, 1992.
4. GISS Surface Temperature Analysis, <data.giss.nasa.gov/gistemp/2008/>, Goddard Institute for Space Studies, Jan. 13, 2009.
5. Frohlich, C., and Lean, J., *Astron. Astrophys. Rev.* 12, 273, 2004.

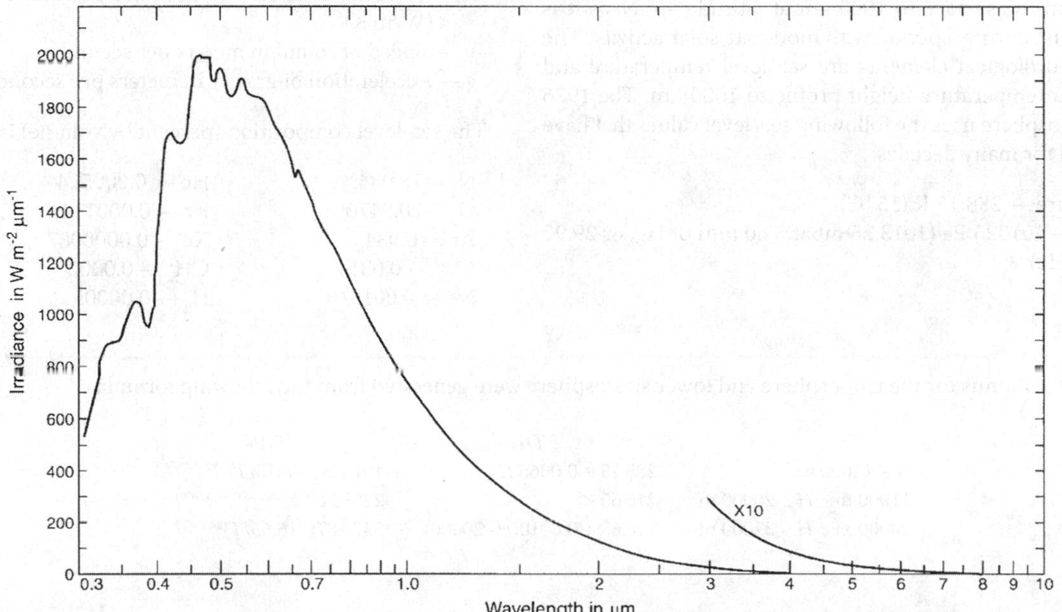

FIGURE 1. Wavelength dependence of solar irradiance.

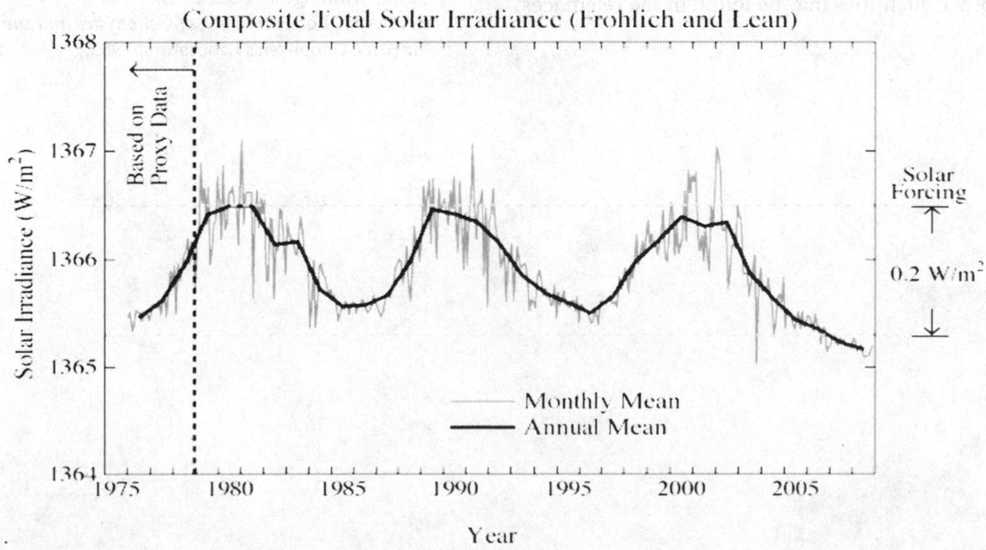

FIGURE 2. Variation of solar irradiance in the period 1976–2008.

U.S. STANDARD ATMOSPHERE (1976)

A Standard Atmosphere is a hypothetical vertical distribution of atmospheric temperature, pressure, and density that is roughly representative of year-round, midlatitude conditions. Typical uses are to serve as a basis for pressure altimeter calibrations, aircraft performance calculations, aircraft and rocket design, ballistic tables, meteorological diagrams, and various types of atmospheric modeling. The air is assumed to be dry and to obey the perfect gas law and the hydrostatic equation which, taken together, relate temperature, pressure, and density with vertical position. The atmosphere is considered to rotate with the Earth and to be an average over the diurnal cycle, the semiannual variation, and the range from active to quiet geomagnetic and sunspot conditions.

The U.S. Standard Atmosphere (1976) is an idealized, steady-state representation of mean annual conditions of the Earth's atmosphere from the surface to 1000 km at latitude 45° N, as it is assumed to exist during a period with moderate solar activity. The defining meteorological elements are sea-level temperature and pressure and a temperature-height profile to 1000 km. The 1976 Standard Atmosphere uses the following sea-level values that have been standard for many decades:

Temperature — 288.15 K (15 °C)
Pressure — 101325 Pa (1013.25 mbar, 760 mm of Hg, or 29.92 in. of Hg)
Density — 1225 g/m^3 (1.225 g/L)
Mean molar mass — 28.964 g/mol

The parameters included in this condensed version of the U.S. Standard Atmosphere are:

Z — Height (geometric) above mean sea level in meters
T — Temperature in kelvins
P — Pressure in pascals (1 Pa = 0.01 millibars)
ρ — Density in kilograms per cubic meter (1 kg/m^3 = 1 g/L)
n — Number density in molecules per cubic meter
ν — Mean collision frequency in collisions per second
l — Mean free path in meters
η — Absolute viscosity in pascal seconds (1 Pa s = 1000 cP)
k — Thermal conductivity in joules per meter second kelvin (W/m K)
v_s — Speed of sound in meters per second
g — Acceleration of gravity in meters per second square

The sea-level composition (percent by volume) is taken to be:

N_2 — 78.084%	He — 0.000524
O_2 — 20.9476	Kr — 0.000114
Ar — 0.934	Xe — 0.0000087
CO_2 — 0.0314	CH_4 — 0.0002
Ne — 0.001818	H_2 — 0.00005

The T and P columns for the troposphere and lower stratosphere were generated from the following formulas:

	T/K	P/Pa
$H \le 11000$ m	$288.15 - 0.0065\,H$	$101325(288.15/T)^{-5.25577}$
$11000 \text{ m} < H \le 20000$ m	216.65	$22632\ e^{-0.00015768832(H-11000)}$
$20000 \text{ m} < H \le 32000$ m	$216.65 + 0.0010(H-20000)$	$5474.87(216.65/T)^{34.16319}$

where $H = rZ/(r + Z)$ is the geopotential height in meters and r is the mean Earth radius at 45° N latitude, taken as 6356766 m. For altitudes up to 32 km, $\rho = 0.003483677(P/T)$ in the units used here. Formulas for the other quantities may be found in the references.

References

1. COESA, *U.S. Standard Atmosphere*, 1976, U.S. Government Printing Office, Washington, D.C., 1976.
2. Jursa, A. S., Ed., *Handbook of Geophysics and the Space Environment*, Air Force Geophysics Laboratory, 1985.

Z/m	T/K	P/Pa	ρ/kg m^{-3}	n/m^{-3}	ν/s^{-1}	l/m	η/Pa s	W m^{-1} K^{-1}	v_s/m s^{-1}	g/m s^{-2}
−5000	320.68	1.778E+05	1.931	4.015E+25	1.151E+10	4.208E−08	1.942E−05	0.02788	359.0	9.822
−4500	317.42	1.685E+05	1.849	3.845E+25	1.096E+10	4.395E−08	1.927E−05	0.02763	357.2	9.830
−4000	314.17	1.596E+05	1.770	3.680E+25	1.044E+10	4.592E−08	1.912E−05	0.02738	355.3	9.819
−3500	310.91	1.511E+05	1.693	3.520E+25	9.933E+09	4.800E−08	1.897E−05	0.02713	353.5	9.818
−3000	307.66	1.430E+05	1.619	3.366E+25	9.448E+09	5.019E−08	1.882E−05	0.02688	351.6	9.816
−2500	304.41	1.352E+05	1.547	3.217E+25	8.982E+09	5.252E−08	1.867E−05	0.02663	349.8	9.814
−2000	301.15	1.278E+05	1.478	3.102E+25	8.623E+09	5.447E−08	1.852E−05	0.02638	347.9	9.813
−1500	297.90	1.207E+05	1.411	2.935E+25	8.106E+09	5.757E−08	1.836E−05	0.02613	346.0	9.811
−1000	294.65	1.139E+05	1.347	2.801E+25	7.693E+09	6.032E−08	1.821E−05	0.02587	344.1	9.810
−500	291.40	1.075E+05	1.285	2.672E+25	7.298E+09	6.324E−08	1.805E−05	0.02562	342.2	9.808
0	288.15	1.013E+05	1.225	2.547E+25	6.919E+09	6.633E−08	1.789E−05	0.02533	340.3	9.807
500	284.90	9.546E+04	1.167	2.427E+25	6.556E+09	6.961E−08	1.774E−05	0.02511	338.4	9.805
1000	281.65	8.988E+04	1.112	2.311E+25	6.208E+09	7.310E−08	1.758E−05	0.02485	336.4	9.804
1500	278.40	8.456E+04	1.058	2.200E+25	5.874E+09	7.680E−08	1.742E−05	0.02459	334.5	9.802
2000	275.15	7.950E+04	1.007	2.093E+25	5.555E+09	8.073E−08	1.726E−05	0.02433	332.5	9.801
2500	271.91	7.469E+04	0.957	1.990E+25	5.250E+09	8.491E−08	1.710E−05	0.02407	330.6	9.799
3000	268.66	7.012E+04	0.909	1.891E+25	4.959E+09	8.937E−08	1.694E−05	0.02381	328.6	9.797
3500	265.41	6.579E+04	0.863	1.795E+25	4.680E+09	9.411E−08	1.678E−05	0.02355	326.6	9.796
4000	262.17	6.166E+04	0.819	1.704E+25	4.414E+09	9.917E−08	1.661E−05	0.02329	324.6	9.794
4500	258.92	5.775E+04	0.777	1.616E+25	4.160E+09	1.046E−07	1.645E−05	0.02303	322.6	9.793
5000	255.68	5.405E+04	0.736	1.531E+25	3.918E+09	1.103E−07	1.628E−05	0.02277	320.6	9.791
5500	252.43	5.054E+04	0.697	1.450E+25	3.687E+09	1.165E−07	1.612E−05	0.02250	318.5	9.790
6000	249.19	4.722E+04	0.660	1.373E+25	3.467E+09	1.231E−07	1.595E−05	0.02224	316.5	9.788
6500	245.94	4.408E+04	0.664	1.299E+25	3.258E+09	1.302E−07	1.578E−05	0.02197	314.4	9.787
7000	242.70	4.111E+04	0.590	1.227E+25	3.058E+09	1.377E−07	1.561E−05	0.02170	312.3	9.785
7500	239.46	3.830E+04	0.557	1.159E+25	2.869E+09	1.458E−07	1.544E−05	0.02144	310.2	9.784
8000	236.22	3.565E+04	0.526	1.093E+25	2.689E+09	1.545E−07	1.527E−05	0.02117	308.1	9.782
8500	232.97	3.315E+04	0.496	1.031E+25	2.518E+09	1.639E−07	1.510E−05	0.02090	306.0	9.781
9000	229.73	3.080E+04	0.467	9.711E+24	2.356E+09	1.740E−07	1.493E−05	0.02063	303.9	9.779
9500	226.49	2.858E+04	0.440	9.141E+24	2.202E+09	1.848E−07	1.475E−05	0.02036	301.7	9.777
10000	223.25	2.650E+04	0.414	8.598E+24	2.056E+09	1.965E−07	1.458E−05	0.02009	299.5	9.776
10500	220.01	2.454E+04	0.389	8.079E+24	1.918E+09	2.091E−07	1.440E−05	0.01982	297.4	9.774
11000	216.77	2.270E+04	0.365	7.585E+24	1.787E+09	2.227E−07	1.422E−05	0.01954	295.2	9.773
11500	216.65	2.098E+04	0.337	7.016E+24	1.653E+09	2.408E−07	1.422E−05	0.01953	295.1	9.771
12000	216.65	1.940E+04	0.312	6.486E+24	1.528E+09	2.605E−07	1.422E−05	0.01953	295.1	9.770
12500	216.65	1.793E+04	0.288	5.996E+24	1.412E+09	2.818E−07	1.422E−05	0.01953	295.1	9.768
13000	216.65	1.658E+04	0.267	5.543E+24	1.306E+09	3.048E−07	1.422E−05	0.01953	295.1	9.767
13500	216.65	1.533E+04	0.246	5.124E+24	1.207E+09	3.297E−07	1.422E−05	0.01953	295.1	9.765
14000	216.65	1.417E+04	0.228	4.738E+24	1.116E+09	3.566E−07	1.422E−05	0.01953	295.1	9.764
14500	216.65	1.310E+04	0.211	4.380E+24	1.032E+09	3.857E−07	1.422E−05	0.01953	295.1	9.762
15000	216.65	1.211E+04	0.195	4.049E+24	9.538E+08	4.172E−07	1.422E−05	0.01953	295.1	9.761
16000	216.65	1.035E+04	0.166	3.461E+24	8.153E+08	4.881E−07	1.422E−05	0.01953	295.1	9.758
17000	216.65	8.850E+03	0.142	2.959E+24	6.969E+08	5.710E−07	1.422E−05	0.01953	295.1	9.754
18000	216.65	7.565E+03	0.122	2.529E+24	5.958E+08	6.680E−07	1.422E−05	0.01953	295.1	9.751
19000	216.65	6.467E+03	0.104	2.162E+24	5.093E+08	7.814E−07	1.422E−05	0.01953	295.1	9.748
20000	216.65	5.529E+03	8.891E−02	1.849E+24	4.354E+08	9.139E−07	1.422E−05	0.01953	295.1	9.745
21000	217.58	4.729E+03	7.572E−02	1.574E+24	3.716E+08	1.073E−06	1.427E−05	0.01961	295.1	9.742
22000	218.57	4.048E+03	6.451E−02	1.341E+24	3.173E+08	1.260E−06	1.432E−05	0.01970	296.4	9.739
23000	219.57	3.467E+03	5.501E−02	1.144E+24	2.712E+08	1.477E−06	1.438E−05	0.01978	297.1	9.736
24000	220.56	2.972E+03	4.694E−02	9.759E+23	2.319E+08	1.731E−06	1.443E−05	0.01986	297.7	9.733
25000	221.55	2.549E+03	4.008E−02	8.334E+23	1.985E+08	2.027E−06	1.448E−05	0.01995	298.4	9.730
26000	222.54	2.188E+03	3.426E−02	7.123E+23	1.700E+08	2.372E−06	1.454E−05	0.02003	299.1	9.727
27000	223.54	1.880E+03	2.930E−02	6.092E+23	1.458E+08	2.773E−06	1.459E−05	0.02011	299.7	9.724
28000	224.53	1.610E+03	2.508E−02	5.214E+23	1.250E+08	3.240E−06	1.465E−05	0.02020	300.4	9.721
29000	225.52	1.390E+03	2.148E−02	4.466E+23	1.073E+08	3.783E−06	1.470E−05	0.02028	301.1	9.718
30000	226.51	1.197E+03	1.841E−02	3.828E+23	9.219E+07	4.414E−06	1.475E−05	0.02036	301.7	9.715
31000	227.50	1.031E+03	1.579E−02	3.283E+23	7.925E+07	5.146E−06	1.481E−05	0.02044	302.4	9.712
32000	228.49	8.891E+02	1.356E−02	2.813E+23	6.818E+07	5.995E−06	1.486E−05	0.02053	303.0	9.709
33000	230.97	7.673E+02	1.157E−02	2.406E+23	5.852E+07	7.021E−06	1.499E−05	0.02073	304.7	9.706
34000	233.74	6.634E+02	9.887E−03	2.056E+23	5.030E+07	8.218E−06	1.514E−05	0.02096	306.5	9.703
35000	236.51	5.746E+02	8.463E−03	1.760E+23	4.331E+07	9.601E−06	1.529E−05	0.02119	308.3	9.700

Z/m	T/K	P/Pa	ρ/kg m^{-3}	n/m^{-3}	ν/s^{-1}	l/m	η/Pa s	W m^{-1} K^{-1}	v_s/m s^{-1}	g/m s^{-2}
36000	239.28	4.985E+02	7.258E−03	1.509E+23	3.736E+07	1.120E−05	1.543E−05	0.02142	310.1	9.697
38000	244.82	3.771E+02	5.367E−03	1.116E+23	2.794E+07	1.514E−05	1.572E−05	0.02188	313.7	9.690
40000	250.35	2.871E+02	3.996E−03	8.308E+22	2.104E+07	2.034E−05	1.601E−05	0.02233	317.2	9.684
42000	255.88	2.200E+02	2.995E−03	6.227E+22	1.594E+07	2.713E−05	1.629E−05	0.02278	320.7	9.678
44000	261.40	1.695E+02	2.259E−03	4.697E+22	1.215E+07	3.597E−05	1.657E−05	0.02323	324.1	9.672
46000	266.93	1.313E+02	1.714E−03	3.564E+22	9.318E+06	4.740E−05	1.685E−05	0.02376	327.5	9.666
48000	270.65	1.023E+02	1.317E−03	2.738E+22	7.208E+06	6.171E−05	1.704E−05	0.02397	329.8	9.660
50000	270.65	7.978E+01	1.027E−03	2.135E+22	5.620E+06	7.913E−05	1.703E−05	0.02397	329.8	9.654
52000	269.03	6.221E+01	8.056E−04	1.675E+22	4.397E+06	1.009E−04	1.696E−05	0.02384	328.8	9.648
54000	263.52	4.834E+01	6.390E−04	1.329E+22	3.452E+06	1.272E−04	1.660E−05	0.02340	325.4	9.642
56000	258.02	3.736E+01	5.045E−04	1.049E+22	2.696E+06	1.611E−04	1.640E−05	0.02296	322.0	9.636
58000	252.52	2.872E+01	3.963E−04	8.239E+21	2.095E+06	2.051E−04	1.612E−05	0.02251	318.6	9.632
60000	247.02	2.196E+01	3.097E−04	6.439E+21	1.620E+06	2.624E−04	1.584E−05	0.02206	315.1	9.624
65000	233.29	1.093E+01	1.632E−04	3.393E+21	8.294E+05	4.979E−04	1.512E−05	0.02093	306.2	9.609
70000	219.59	5.221	8.283E−05	1.722E+21	4.084E+05	9.810E−04	1.438E−05	0.01978	297.1	9.594
75000	208.40	2.388	3.992E−05	8.300E+20	1.918E+05	2.035E−03	1.376E−05	0.01883	289.4	9.579
80000	198.64	1.052	1.846E−05	3.838E+20	8.656E+04	4.402E−03	1.321E−05	0.01800	282.5	9.564
85000	188.89	4.457E−01	8.220E−06	1.709E+20	3.766E+04	9.886E−03	1.265E−05	0.01716	275.5	9.550
90000	186.87	1.836E−01	3.416E−06	7.116E+19	1.560E+04	2.370E−02				9.535
95000	188.42	7.597E−02	1.393E−06	2.920E+19	6.440E+03	5.790E−02				9.520
100000	195.08	3.201E−02	5.604E−07	1.189E+19	2.680E+03	1.420E−01				9.505
110000	240.00	7.104E−03	9.708E−08	2.144E+18	5.480E+02	7.880E−01				9.476
120000	360.00	2.538E−03	2.222E−08	5.107E+17	1.630E+02	3.310				9.447
130000	469.27	1.251E−03	8.152E−09	1.930E+17	7.100E+01	8.800				9.418
140000	559.63	7.203E−04	3.831E−09	9.322E+16	3.800E+01	1.800E+01				9.389
150000	634.39	4.542E−04	2.076E−09	5.186E+16	2.300E+01	3.300E+01				9.360
160000	696.29	3.040E−04	1.233E−09	3.162E+16	1.500E+01	5.300E+01				9.331
170000	747.57	2.121E−04	7.815E−10	2.055E+16	1.000E+01	8.200E+01				9.302
180000	790.07	1.527E−04	5.194E−10	1.400E+16	7.200	1.200E+02				9.274
190000	825.16	1.127E−04	3.581E−10	9.887E+15	5.200	1.700E+02				9.246
200000	854.56	8.474E−05	2.541E−10	7.182E+15	3.900	2.400E+02				9.218
220000	899.01	5.015E−05	1.367E−10	4.040E+15	2.300	4.200E+02				9.162
240000	929.73	3.106E−05	7.858E−11	2.420E+15	1.400	7.000E+02				9.106
260000	950.99	1.989E−05	4.742E−11	1.515E+15	9.300E−01	1.100E+03				9.051
280000	965.75	1.308E−05	2.971E−11	9.807E+14	6.100E−01	1.700E+03				8.997
300000	976.01	8.770E−06	1.916E−11	6.509E+14	4.200E−01	2.600E+03				8.943
320000	983.16	5.980E−06	1.264E−11	4.405E+14	2.900E−01	3.800E+03				8.889
340000	988.15	4.132E−06	8.503E−12	3.029E+14	2.000E−01	5.600E+03				8.836
360000	991.65	2.888E−06	5.805E−12	2.109E+14	1.400E−01	8.000E+03				8.784
380000	994.10	2.038E−06	4.013E−12	1.485E+14	1.000E−01	1.100E+04				8.732
400000	995.83	1.452E−06	2.803E−12	1.056E+14	7.200E−02	1.600E+04				8.680
450000	998.22	6.447E−07	1.184E−12	4.678E+13	3.300E−02	3.600E+04				8.553
500000	999.24	3.024E−07	5.215E−13	2.192E+13	1.600E−02	7.700E+04				8.429
550000	999.67	1.514E−07	2.384E−13	1.097E+13	8.400E−03	1.500E+05				8.307
600000	999.85	8.213E−08	1.137E−13	5.950E+12	4.800E−03	2.800E+05				8.188
650000	999.93	4.887E−08	5.712E−14	3.540E+12	3.100E−03	4.800E+05				8.072
700000	999.97	3.191E−08	3.070E−14	2.311E+12	2.200E−03	7.300E+05				7.958
750000	999.98	2.260E−08	1.788E−14	1.637E+12	1.700E−03	1.000E+06				7.846
800000	999.99	1.704E−08	1.136E−14	1.234E+12	1.400E−03	1.400E+06				7.737
850000	1000.00	1.342E−08	7.824E−15	9.717E+11	1.200E−03	1.700E+06				7.630
900000	1000.00	1.087E−08	5.759E−15	7.876E+11	1.000E−03	2.100E+06				7.525
950000	1000.00	8.982E−09	4.453E−15	6.505E+11	8.700E−04	2.600E+06				7.422
1000000	1000.00	7.514E−09	3.561E−15	5.442E+11	7.500E−04	3.100E+06				7.322

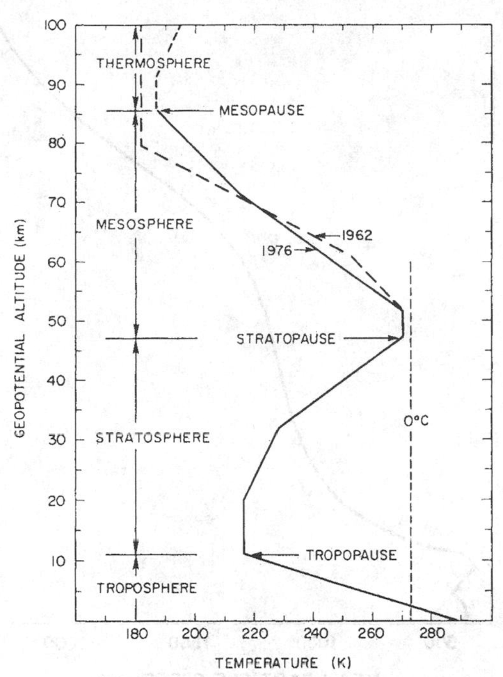

FIGURE 1. Temperature-height profile for U.S. Standard Atmosphere.

FIGURE 3. Mean molecular weight as a function of geometric altitude.

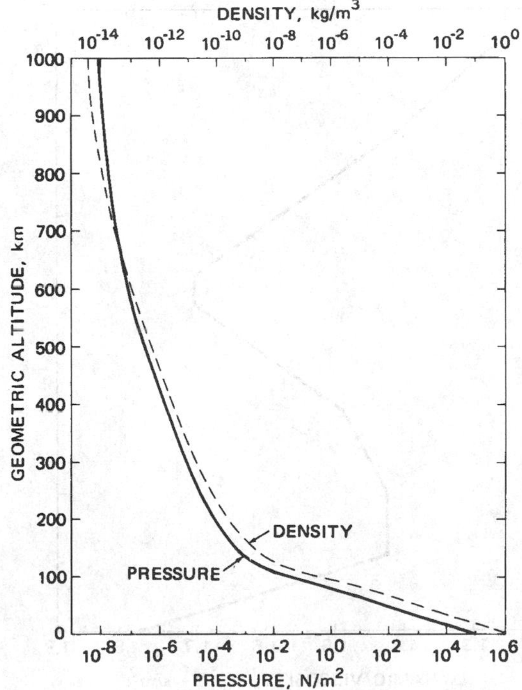

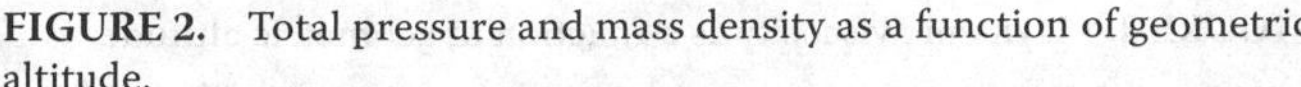

FIGURE 2. Total pressure and mass density as a function of geometric altitude.

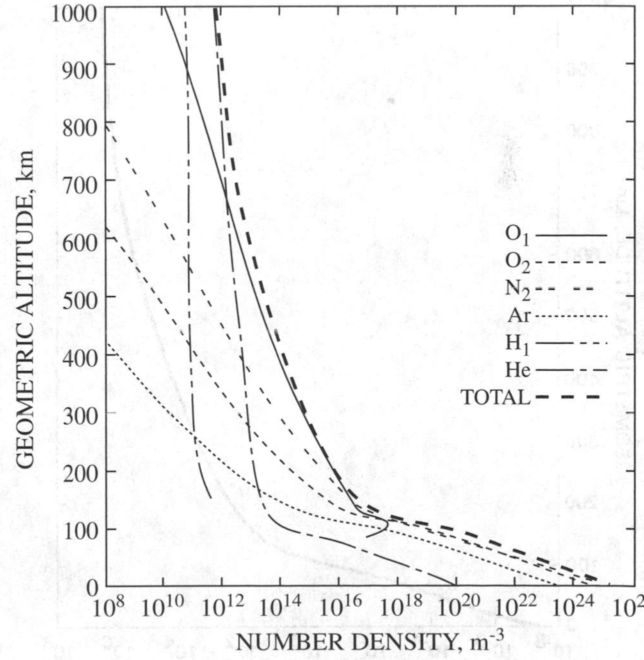

FIGURE 4. Number density of individual species and total number density as a function of geometric altitude.

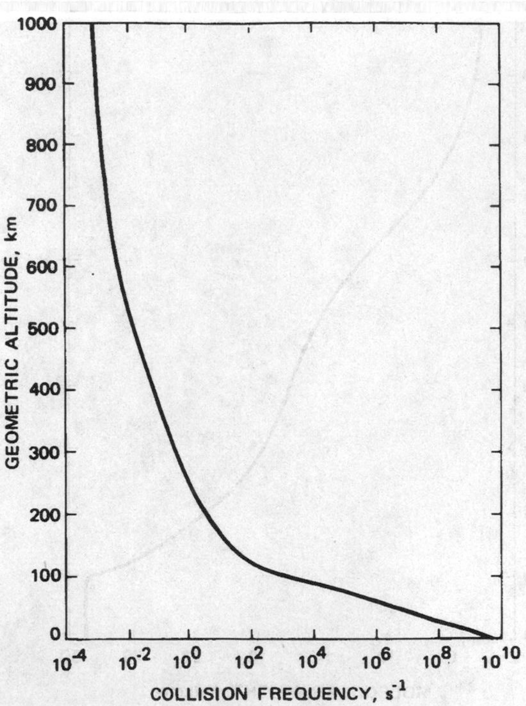

FIGURE 5. Collision frequency as a function of geometric altitude.

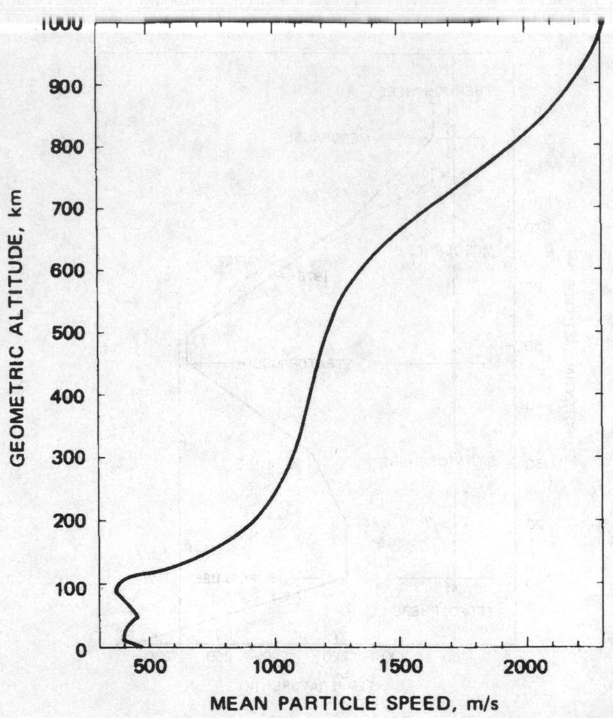

FIGURE 7. Mean air-particle speed as a function of geometric altitude.

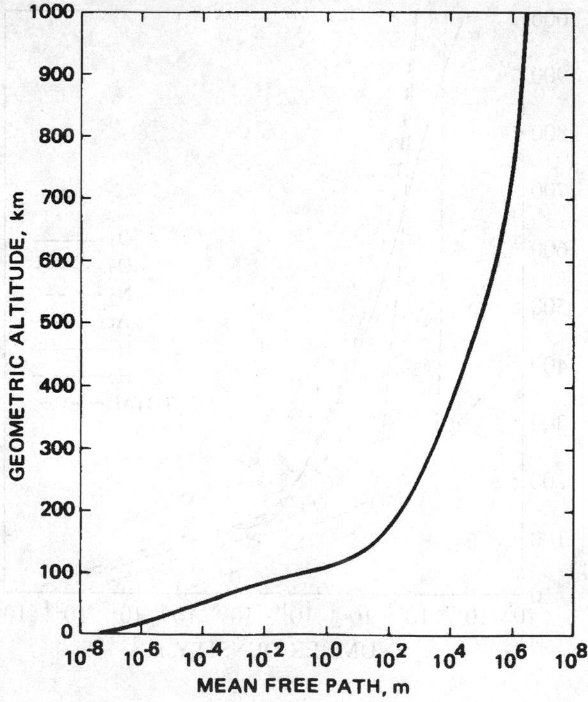

FIGURE 6. Mean free path as a function of geometric altitude.

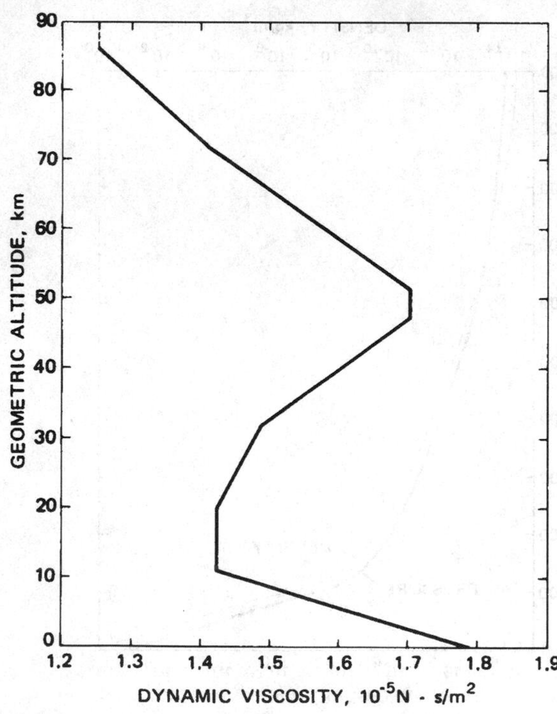

FIGURE 8. Dynamic viscosity as a function of geometric altitude.

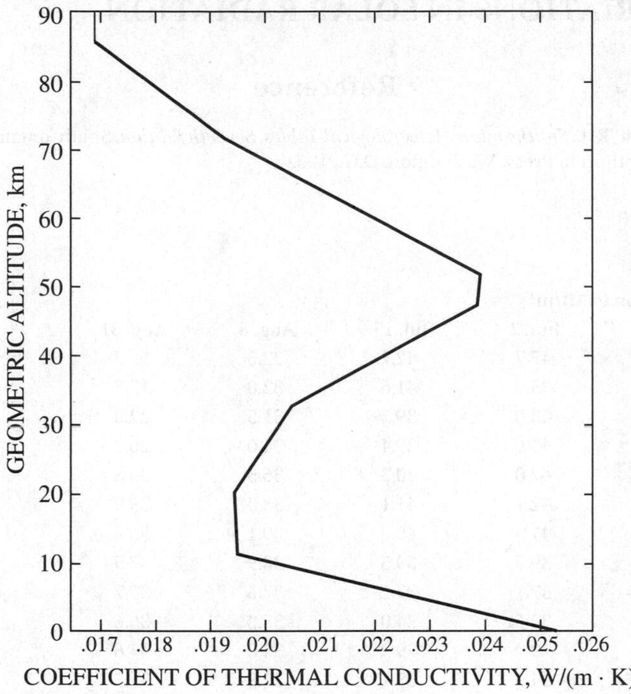

FIGURE 9. Coefficient of thermal conductivity as a function of geometric altitude.

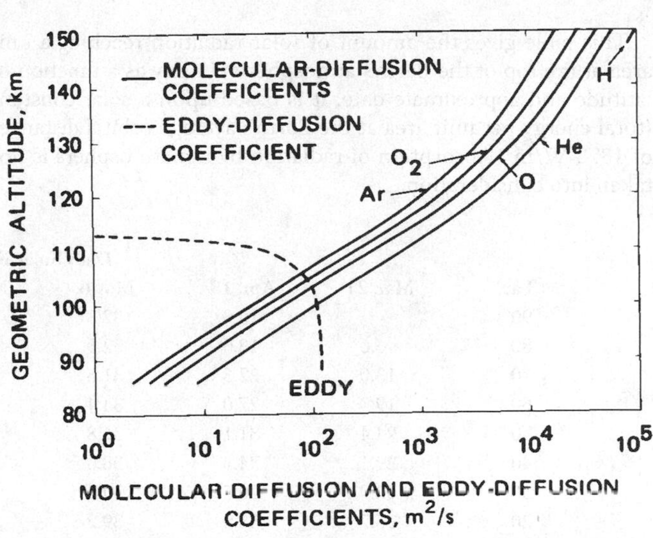

FIGURE 11. Molecular-diffusion and eddy-diffusion coefficients as a function of geometric altitude.

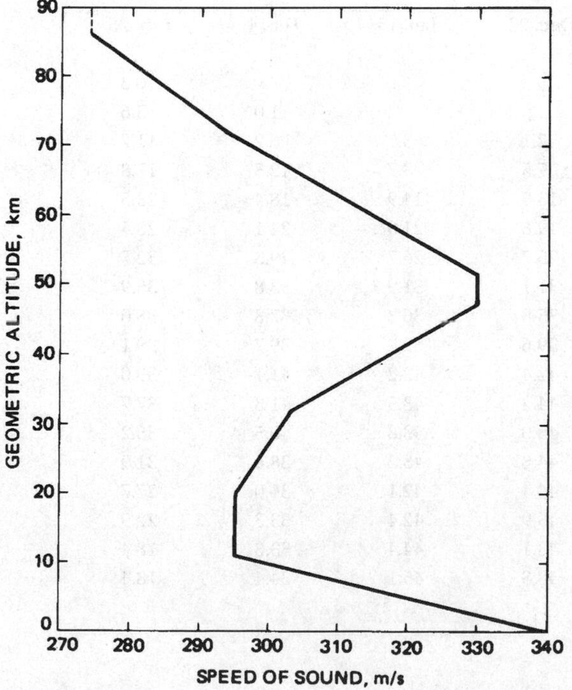

FIGURE 10. Speed of sound as a function of geometric altitude.

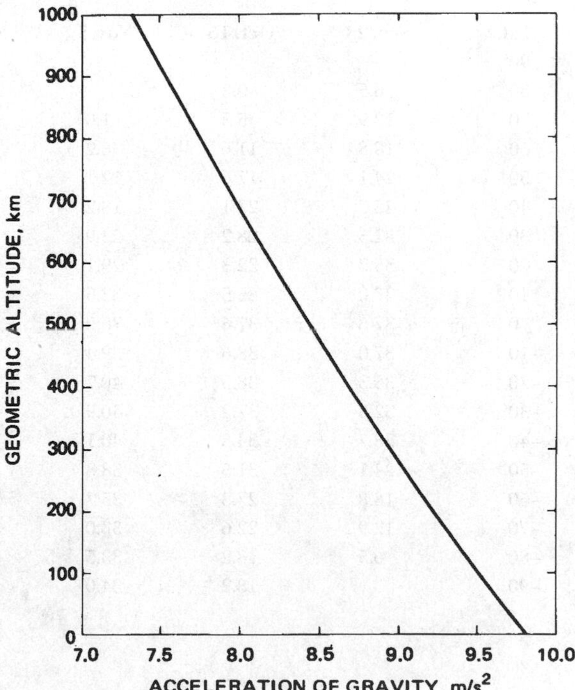

FIGURE 12. Acceleration of gravity as a function of geometric altitude.

GEOGRAPHICAL AND SEASONAL VARIATIONS IN SOLAR RADIATION

This table gives the amount of solar radiation reaching a unit area at the top of the Earth's atmosphere per day as a function of latitude and approximate date. It is based upon a solar constant (total energy per unit area at the Earth's average orbital distance) of 1373 W/m^2. Absorption of radiation by the atmosphere is not taken into consideration.

Reference

List, R. J., *Smithsonian Meteorological Tables, Seventh Edition*, Smithsonian Institution Press, Washington, D.C., 1962.

Daily Solar Radiation in MJ/m^2

Lat.	Mar. 21	Apr. 13	May 6	May 29	Jun. 2	Jul. 15	Aug. 8	Aug. 31
90°		18.0	32.8	42.4	45.7	42.2	32.5	17.7
80	6.6	18.0	32.3	41.8	45.0	41.6	32.0	17.7
70	13.0	22.3	31.8	39.9	43.0	39.7	31.5	22.0
60	19.0	27.0	34.4	39.7	41.6	39.4	34.0	26.7
50	24.4	31.1	36.8	40.7	42.0	40.5	36.5	30.8
40	29.1	34.3	38.6	41.3	42.1	41.1	38.3	33.9
30	32.9	36.7	39.4	41.1	41.4	40.8	39.1	36.3
20	35.7	38.0	39.2	39.7	39.7	39.5	38.9	37.5
10	37.4	38.1	37.9	37.4	37.1	37.2	37.6	37.7
0	38.0	37.1	35.5	34.1	33.5	34.0	35.2	36.6
−10	37.4	35.0	32.3	30.0	29.2	29.9	32.0	34.6
−20	35.7	31.8	28.0	25.2	24.1	25.1	27.8	31.5
−30	32.9	27.8	23.1	19.7	18.5	19.7	22.8	27.4
−40	29.1	22.8	17.5	14.0	12.6	13.9	17.4	22.6
−50	24.4	17.3	11.7	8.2	7.0	8.2	11.6	17.2
−60	19.0	11.4	5.9	2.9	2.0	2.9	5.9	11.3
−70	13.0	5.4	1.0				1.0	5.3
−80	6.6	0.3						0.3
−90								

Lat.	Sep. 23	Oct. 16	Nov. 8	Nov. 30	Dec. 22	Jan. 13	Feb. 4	Feb. 26
90°								
80	6.5	0.3						0.3
70	12.9	5.5	1.0				1.0	5.6
60	18.8	11.6	6.2	3.1	2.1	3.1	6.2	11.7
50	24.1	17.6	12.1	8.7	7.5	8.7	12.3	17.8
40	28.7	23.1	18.2	14.8	13.5	14.9	18.4	23.5
30	32.5	28.2	23.9	20.9	19.8	21.0	24.1	28.4
20	35.3	32.3	29.1	26.6	25.7	26.7	29.3	32.7
10	37.0	35.5	33.5	31.8	31.1	31.9	33.8	35.9
0	37.6	37.6	36.9	36.1	35.8	36.3	37.3	38.0
−10	37.0	38.6	39.4	39.5	39.6	39.7	39.7	39.1
−20	35.3	38.5	40.7	42.0	42.4	42.2	41.1	39.0
−30	32.5	37.2	40.9	43.3	44.2	43.5	41.3	37.7
−40	28.7	34.8	40.1	43.6	45.0	43.8	40.5	35.2
−50	24.1	31.5	38.3	43.1	44.8	43.2	38.6	31.9
−60	18.8	27.3	35.7	41.9	44.4	42.1	36.0	27.7
−70	12.9	22.6	33.0	42.2	45.9	42.4	33.3	22.9
−80	6.5	18.2	33.5	44.2	48.1	44.4	33.8	18.4
−90		18.2	34.0	44.8	48.8	45.1	34.4	18.4

MAJOR WORLD EARTHQUAKES

The United States Geological Survey maintains a database of historic earthquakes throughout the world (Reference 1). The table below is extracted from that database; it includes about 300 major earthquakes, based upon the magnitude and the degree of destruction. All recorded earthquakes of magnitude 7.5 or greater are listed, even if the fatalities are unknown or small. The death toll is often a rough estimate; in many cases the true toll could be much greater. More details on the exact location and degree of destruction can be found in References 2 and 3.

The magnitude is given on the Richter scale, which was developed in 1935 by Charles F. Richter of the California Institute of Technology as a mathematical device to compare the size of earthquakes. The magnitude of an earthquake is measured by the logarithm of the amplitude of waves recorded by seismographs. Adjustments are included for the variation in the distance between the various seismographs and the epicenter of the earthquake. On the Richter Scale, magnitude is expressed in whole numbers and decimal fractions, e.g. 6.3. Because of the logarithmic basis of the scale, each whole number increase in magnitude represents a tenfold increase in measured amplitude; as an estimate of energy, each whole number step in the magnitude scale corresponds to the release of about 31 times more energy than the amount associated with the preceding whole number value.

References

1. Historic Worldwide Earthquakes, http://earthquake.usgs.gov/regional/world/historical.php.
2. Most Destructive Known Earthquakes on Record in the World, http://earthquake.usgs.gov/regional/world/most_destructive.php.
3. Earthquakes with 1,000 or More Deaths since 1900, http://earthquake.usgs.gov/regional/world/world_deaths.php.

Date	Location	Magnitude	Fatalities
856/12/22	Damghan, Iran		200,000
893/03/23	Ardabil, Iran		150,000
1138/08/09	Aleppo, Syria		230,000
1268	Silicia, Asia Minor		60,000
1290/09/27	Chihli, China		100,000
1556/01/23	Shensi, China	8.0	830,000
1619/02/14	Trujillo, Peru	7.7	350
1667/11	Shemakha, Caucasia		80,000
1668/08/17	Anatolia, Turkey	8.0	
1687/10/20	Lima, Peru	8.5	
1693/01/11	Sicily, Italy	7.5	60,000
1700/01/26	Cascadia Subduction Zone (Oregon to British Columbia)	9.0	
1727/11/18	Tabriz, Iran		77,000
1730/07/08	Valparasio, Chile	8.7	
1755/11/01	Lisbon, Portugal	8.7	70,000
1783/02/04	Calabria, Italy		50,000
1787/05/02	Puerto Rico	8.0	
1811/12/16	New Madrid Region, Missouri	8.1	
1812/02/07	New Madrid Region, Missouri	8.0	
1812/03/26	Caracas, Venezuela	7.7	
1812/12/08	Southwest of San Bernardino County, California	6.9	40
1812/12/21	West of Ventura, California	7.1	1
1821/07/10	Camana, Peru	8.2	162
1835/02/20	Concepcion, Chile	8.2	500
1843/02/08	Leeward Islands	8.3	
1855/01/23	Wellington, New Zealand	8.0	4
1857/01/09	Fort Tejon, California	7.9	1
1868/04/03	Ka'u District, Island of Hawaii	7.9	77
1868/08/13	Arica, Peru (now Chile)	9.0	400
1872/03/26	Owens Valley, California	7.4	27
1877/05/10	Offshore Tarapaca, Chile	8.3	34
1886/09/01	Charleston, South Carolina	7.3	60
1887/05/03	Northern Sonora, Mexico	7.4	51
1891/10/27	Mino-Owari, Japan	8.0	
1892/02/24	Imperial Valley, California	7.8	
1896/06/15	Sanriku, Japan	8.5	28,000
1897/06/12	Assam, India	8.3	1,500
1899/09/04	Cape Yakataga, Alaska	7.9	
1899/09/10	Yakutat Bay, Alaska	8.0	
1900/10/09	Kodiak Island, Alaska	7.7	
1902/04/19	Quezaltenango and San Marcos, Guatemala	7.5	2,000
1902/12/16	Eastern Uzbekistan (Turkestan)	6.4	4,700
1903/04/28	Malazgirt, Turkey	7.0	3,500
1903/05/28	Gole, Turkey (Ottoman Empire)	5.8	1,000
1903/08/11	Southern Greece	8.3	
1905/04/04	Kangra, India	7.5	19,000
1905/07/09	Mongolia	8.4	
1905/09/08	Calabria, Italy	7.9	557
1906/01/31	Off the Coast of Esmeraldas, Ecuador	8.8	1,000
1906/03/16	Chia-i, Taiwan	6.8	1,250
1906/04/18	San Francisco, California	7.8	3,000
1906/08/17	Valparaiso, Chile	8.2	20,000
1907/01/14	Kingston, Jamaica	6.5	800-1,000
1907/04/15	Guerrero, Mexico	7.7	
1907/10/21	Qaratog, Tajikistan, Russia	8.0	12,000
1908/12/12	Off the Coast of Central Peru	8.2	
1908/12/28	Messina, Italy	7.2	72,000
1909/01/23	Silakhor, Iran (Persia)	7.3	5,000-6,000

Date	Location	Magnitude	Fatalities
1910/04/12	Taiwan region	7.6	
1911/01/03	Chong-Kemin, Kyrgyzstan	7.8	450
1911/02/18	Sarez, Tajikistan	7.4	90
1911/06/07	Off Guerrero, Mexico	7.7	45
1912/08/09	Murefte, Turkey (Ottoman Empire)	7.8	2,800
1914/10/03	Burdur, Turkey (Ottoman Empire)	7.0	4,000
1915/01/13	Avezzano, Italy	7.0	32,610
1917/01/20	Bali, Indonesia		1,500
1917/07/30	Daguan, Yunnan, China	7.5	1,800
1918/02/13	Nan'ao, Guangdong, (Kwangtung), China	7.3	1,000
1918/10/11	Mona Passage	7.5	116
1920/06/05	Taiwan region	8.0	
1920/12/16	Haiyuan, Ningxia, China	7.8	200,000
1922/11/11	Chile-Argentina Border	8.5	100
1923/02/03	Kamchatka Peninsula	8.5	
1923/03/24	Near Luhuo, Sichuan, China	7.3	3,500
1923/05/25	Torbat-e Heydariyeh, Iran	5.7	2,200
1923/09/01	Kanto (Kwanto), Japan	7.9	142,800
1925/03/16	Yunnan, China	7.1	5,800
1927/03/07	Tango, Japan	7.6	3,020
1927/05/22	Tsinghai (Kansu), China	7.6	40,900
1928/12/01	Talca, Chile	7.6	225
1929/03/07	Fox Islands, Aleutian Islands, Alaska	7.8	
1929/05/01	Koppeh Dagh, Iran (Persia)	7.4	3,800
1930/05/06	Salmas, Iran (Persia)	7.2	2,500
1930/07/23	Irpinia, Italy	6.5	1,400
1931/01/15	Oaxaca, Mexico	7.8	114
1931/02/02	Hawke's Bay, New Zealand	7.9	256
1931/03/31	Managua, Nicaragua	6.0	2,500
1931/04/27	Zangezur Mountains, Armenia — Azerbaijan border	5.7	2,800
1931/08/10	Xinjiang, China	8.0	10,000
1932/06/03	Jalisco, Mexico	8.1	45
1932/06/18	Colima, Mexico	7.8	
1932/12/25	Gansu, China	7.6	275
1933/03/02	Sanriku, Japan	8.4	3,000
1933/03/11	Long Beach, California	6.4	115
1933/08/25	Sichuan, China	7.4	9,300
1934/01/15	Bihar, India — Nepal	8.1	10,700
1935/04/20	Taiwan (Formosa)	7.1	3,270
1935/05/30	Quetta, Pakistan	7.5	30,000
1935/07/16	Taiwan (Formosa)	6.5	2,740
1938/02/01	Banda Sea, Indonesia	8.5	
1938/11/10	Shumagin Islands, Alaska	8.2	
1939/01/25	Chillan, Chile	7.8	28,000
1939/12/26	Erzincan, Turkey	7.8	32,700
1940/05/19	Imperial Valley, California	7.1	9
1940/05/24	Callao, Peru	8.2	249
1940/11/10	Vrancea, Romania	7.3	1,000
1942/08/06	Guatemala	7.9	38
1942/08/24	Off the coast of central Peru	8.2	30
1942/11/26	Turkey	7.6	
1942/12/20	Erbaa, Turkey	7.3	1,100
1943/04/06	Illapel — Salamanca, Chile	8.2	25
1943/09/10	Tottori, Japan	7.4	1,190
1943/11/26	Ladik, Turkey	7.6	4,000
1944/01/15	San Juan, Argentina	7.4	8,000
1944/02/01	Gerede, Turkey	7.4	2,790
1944/12/07	Tonankai, Japan	8.1	998
1945/01/12	Mikawa, Japan	7.1	1,961
1945/11/27	Makran Coast, Pakistan	8.0	4,000
1946/04/01	Unimak Island, Alaska	8.1	165
1946/05/31	Ustukran, Turkey	5.9	840-1,300
1946/08/04	Samana, Dominican Republic	8.0	100
1946/11/10	Ancash, Peru	7.3	1,400
1946/12/20	Nankaido, Japan	8.1	1,362
1947/11/01	Satipo, Peru	7.3	233
1948/05/11	Moquegua, Peru	7.4	70
1948/05/25	Sichuan, China	7.3	800
1948/06/28	Fukui, Japan	7.3	3,769
1948/10/05	Ashgabat, Turkmenistan	7.3	110,000
1949/04/13	Puget Sound, Washington	7.1	8
1949/07/10	Khait, Tajikistan	7.5	12,000
1949/08/05	Ambato, Ecuador	6.8	5,050
1949/08/22	Queen Charlotte Islands, British Columbia, Canada	8.1	
1950/08/15	Assam — Tibet	8.6	1,526
1951/08/02	Cosiguina, Nicaragua	5.8	1,000
1952/07/21	Kern County, California	7.3	12
1952/11/04	Kamchatka Peninsula	9.0	
1953/02/12	Torud, Iran	6.5	970
1953/03/18	Yenice-Gonen, Turkey	7.3	1,070
1953/08/12	Kefallinia, Greece	7.1	455
1953/12/12	Tumbes, Peru	7.4	7
1954/03/29	Spain	7.9	
1954/04/30	Greece	7.1	31
1954/09/09	Orleansville, Algeria	6.8	1,250
1957/03/09	Andreanof Islands, Alaska	8.6	
1957/04/25	Fethiye, Turkey	7.1	15
1957/05/26	Bolu Province, Turkey	7.1	66
1957/06/27	Stanovoy Mountains, Russia (USSR)	7.6	1,200
1957/07/02	Mazandaran, Iran	7.1	1,200
1957/07/28	Guerrero, Mexico	7.9	68
1957/12/04	Gobi-Altay, Mongolia	8.1	30
1957/12/13	Sahneh, Iran	7.1	1,130

Date	Location	Magnitude	Fatalities
1958/01/15	Arequipa, Peru	7.3	26
1958/07/10	Lituya Bay, Alaska	7.7	5
1958/11/06	Kuril Islands	8.3	
1959/04/26	Taiwan region	7.5	2
1959/08/18	Hebgen Lake, Montana	7.3	28
1960/01/13	Arequipa, Peru	7.5	57
1960/02/29	Agadir, Morocco	5.7	12,000-15,000
1960/05/21	Arauco Peninsula, Chile	7.9	
1960/05/22	Chile (off coast)	9.5	1,655
1962/05/11	Guerrero, Mexico	7.0	4
1962/05/19	Guerrero, Mexico	7.1	3
1962/09/01	Qazvin, Iran	7.1	12,225
1963/07/26	Skopje, Macedonia	6.0	1,100
1963/10/13	Kuril Islands	8.5	
1964/03/28	Prince William Sound, Alaska	9.2	128
1964/06/16	Niigata, Japan	7.5	26
1964/10/06	Western Turkey	7.0	36
1965/01/24	Sanana, Indonesia (Ceram Sea)	7.6	71
1965/02/04	Rat Islands, Alaska	8.7	
1965/03/14	Hindu Kush, Afghanistan	7.8	
1965/03/28	La Ligua, Chile	7.4	400
1965/03/31	Central Greece	7.1	6
1965/08/23	Oaxaca, Mexico	7.3	6
1966/03/07	Hebei, China	7.0	1,000
1966/03/22	Hebei, China	6.9	1,000
1966/08/19	Varto, Turkey	6.8	2,529
1966/10/17	Near the Coast of Peru	8.1	125
1967/07/22	Mudurnu Valley, Turkey	7.3	173
1968/05/23	Inangahua, New Zealand	7.1	2
1968/08/02	Oaxaca, Mexico	7.1	18
1968/08/31	Dasht-e Bayaz, Iran	7.3	12,000
1969/02/28	Portugal-Morocco area	7.8	13
1969/07/25	Guangdong, China	5.9	3,000
1970/01/04	Yunnan Province, China	7.5	10,000
1970/03/28	Gediz, Turkey	6.9	1,086
1970/05/31	Chimbote, Peru	7.9	70,000
1970/07/31	Colombia	8.0	1
1971/02/09	San Fernando, California	6.6	65
1971/05/22	Eastern Turkey	6.9	1,000
1971/07/09	Valparaiso region, Chile	7.5	90
1972/01/25	Taiwan region	7.5	
1972/04/10	Southern Iran	7.1	5,054
1972/04/24	Taiwan region	7.2	4
1972/07/30	Sitka, Alaska	7.6	
1972/12/23	Nicaragua	6.2	5,000
1974/05/10	Near Zhaotong, China	6.8	20,000
1974/07/13	Panama-Colombia border region	7.3	11
1974/10/03	Near the Coast of Central Peru	8.1	78
1974/10/08	Leeward Islands	7.5	
1974/12/28	Northern Pakistan	6.2	5,300
1975/02/02	Near Islands, Alaska	7.6	
1975/02/04	Haicheng, China	7.0	2,000
1975/09/06	Diyarbakir Province, Turkey	6.7	2,300
1975/11/29	Kalapana, Hawaii	7.2	2
1976/02/04	Guatemala	7.5	23,000
1976/05/06	Northeastern Italy	6.5	1,000
1976/06/25	Papua, Indonesia	7.1	422
1976/07/27	Tangshan, China	7.5	255,000
1976/08/16	Mindanao, Philippines	7.9	8,000
1976/11/24	Turkey-Iran border region	7.3	5,000
1977/03/04	Romania	7.2	1,500
1978/09/16	Iran	7.8	15,000
1979/02/28	Mt. St. Elias, Alaska	7.5	
1980/10/10	El Asnam (formerly Orleansville), Algeria	7.7	5,000
1981/02/24	Greece	6.8	3,000
1981/06/11	Southern Iran	6.9	3,000
1981/07/28	Southern Iran	7.3	1,500
1982/12/13	Yemen	6.0	2,800
1983/10/30	Erzurum Province, Turkey	6.9	1,342
1985/03/03	Offshore Valparaiso, Chile	7.8	177
1985/09/19	Michoacan, Mexico	8.0	9,500
1986/05/07	Andreanof Islands, Alaska	7.9	
1986/10/10	El Salvador	5.5	1,000
1987/03/06	Colombia-Ecuador	7.0	1,000
1987/11/30	Gulf of Alaska	7.8	
1988/03/06	Gulf of Alaska	7.7	
1988/08/20	Nepal-India border region	6.8	1,000
1988/12/07	Spitak, Armenia	6.8	25,000
1989/10/18	Loma Prieta, California	6.9	63
1990/06/20	Western Iran	7.4	50,000
1990/07/16	Luzon, Philippine Islands	7.7	1,621
1991/04/22	Costa Rica	7.6	47
1991/10/19	Northern India	6.8	2,000
1992/09/02	Nicaragua	7.6	116
1992/12/12	Flores Region, Indonesia	7.8	2,500
1993/08/08	South of the Mariana Islands	7.8	
1993/09/29	Latur-Killari, India	6.2	9,748
1994/01/17	Northridge, California	6.7	60
1994/06/09	Bolivia	8.2	10
1995/01/16	Kobe, Japan	6.9	5,502
1995/05/27	Sakhalin Island	7.1	1,989
1996/06/10	Andreanof Islands, Alaska	7.9	
1997/05/10	Northern Iran	7.3	1,567
1997/10/14	South of Fiji Islands	7.8	
1997/12/05	Near East Coast of Kamchatka	7.8	
1998/01/04	Loyalty Islands Region	7.5	

Date	Location	Magnitude	Fatalities
1998/02/04	Afghanistan-Tajikistan Border Region	5.9	2,323
1998/03/25	Balleny Islands Region (off Antarctica)	8.1	
1998/05/03	Southeast of Taiwan	7.5	
1998/05/30	Afghanistan-Tajikistan Border Region	6.6	4,000
1998/07/17	Near North Coast of New Guinea, Papua New Guinea	7.0	2,183
1999/01/25	Colombia	6.1	1,185
1999/08/17	Izmit, Turkey	7.6	17,118
1999/09/20	Taiwan	7.6	2,400
1999/09/30	Oaxaca, Mexico	7.5	
1999/11/12	Duzce, Turkey	7.2	894
2000/06/04	Southern Sumatera, Indonesia	7.9	103
2000/06/18	South Indian Ocean	7.9	
2000/11/16	New Ireland Region, Papua New Guinea	8.0	2
2001/01/01	Mindanao, Philippines	7.5	
2001/01/13	El Salvador	7.7	852
2001/01/26	Gujarat, India	7.6	20,085
2001/02/13	El Salvador	6.6	315
2001/06/23	Near the Coast of Peru	8.4	138
2002/03/03	Hindu Kush Region, Afghanistan	7.4	166
2002/03/05	Mindanao, Philippines	7.5	15
2002/03/25	Hindu Kush Region, Afghanistan	6.1	1,000
2002/03/31	Taiwan region	7.1	5
2002/08/19	Fiji Islands	7.7	
2002/09/08	New Guinea, Papua New Guinea	7.6	
2002/10/10	Irian Jaya, Indonesia	7.6	8
2002/11/02	Northern Sumatera, Indonesia	7.4	3
2002/11/03	Denali Fault, Alaska	7.9	
2003/01/22	Offshore Colima, Mexico	7.6	29
2003/05/21	Northern Algeria	6.8	2,226
2003/05/26	Halmahera, Indonesia	7.0	1
2003/07/15	Carlsberg Ridge	7.6	
2003/08/04	Scotia Sea	7.6	
2003/09/25	Hokkaido, Japan Region	8.3	
2003/09/27	Southwestern Siberia, Russia	7.3	3
2003/11/17	Rat Islands, Aleutian Islands, Alaska	7.8	
2003/12/26	Southeastern Iran	6.6	31,000
2004/02/05	Irian Jaya, Indonesia	7.0	37
2004/11/11	Kepulauan Alor, Indonesia	7.5	34
2004/11/26	Papua, Indonesia	7.1	32
2004/12/23	North of Macquarie Island, New Zealand	8.1	
2004/12/26	Sumatra-Andaman Islands	9.1	227,898
2005/03/28	Northern Sumatra, Indonesia	8.6	1,313
2005/06/13	Tarapaca, Chile	7.8	11
2005/09/09	New Ireland Region, Papua New Guinea	7.6	
2005/09/26	Northern Peru	7.5	5
2005/10/08	Pakistan	7.6	86,000
2006/01/27	Banda Sea	7.6	
2006/02/22	Mozambique	7.0	4
2006/04/20	Koryakia, Russia	7.6	
2006/05/03	Tonga	8.0	
2006/05/26	Java, Indonesia	6.3	5,749
2006/07/17	South of Java, Indonesia	7.7	730
2006/11/15	Kuril Islands	8.3	
2006/12/26	Taiwan Region	7.1	2
2007/01/13	East of the Kuril Islands	8.1	
2007/01/21	Molucca Sea	7.5	
2007/04/01	Solomon Islands	8.1	40
2007/08/08	Java, Indonesia	7.5	
2007/08/15	Near the Coast of Central Peru	8.0	514
2007/09/12	Southern Sumatra, Indonesia	8.5	25
2007/09/12	Kepulauan Mentawai region, Indonesia	7.9	
2007/09/28	Mariana Islands region	7.5	
2007/11/14	Antofagasta, Chile	7.7	2
2007/12/09	South of the Fiji Islands	7.8	
2008/05/12	Eastern Sichuan, China	7.9	87,652
2008/07/05	Sea of Okhotsk	7.7	
2009/01/03	Near the North Coast of Papua, Indonesia	7.7	5
2009/03/19	Tonga region	7.6	
2009/04/06	Central Italy	6.3	295
2009/07/15	Off West Coast of the South Island, New Zealand	7.8	
2009/08/10	Andaman Islands, India region	7.5	
2009/09/29	Samoa Islands region	8.1	192
2009/09/30	Southern Sumatra, Indonesia	7.5	1117
2009/10/07	Santa Cruz Islands	7.8	
2009/10/07	Vanuatu, Coral Sea	7.7	
2010/01/12	Haiti region	7.0	>220,000
2010/02/27	Offshore Maule, Chile	8.8	>800

WEATHER-RELATED SCALES

Saffir-Simpson Hurricane Scale

- **Tropical Storm**
  Winds 39–73 mph
- **Category 1 Hurricane** — winds 74–95 mph (64–82 knots); pressure greater than 980 mbar; storm surge 3–5 ft (1.0–1.7 m)
  No real damage to buildings. Damage to unanchored mobile homes. Some damage to poorly constructed signs. Also, some coastal flooding and minor pier damage.
  — Examples: Irene 1999 and Allison 1995
- **Category 2 Hurricane** — winds 96–110 mph (83–95 knots); pressure 979–965 mbar; storm surge 6–8 ft (1.8–2.6 m)
  Some damage to building roofs, doors and windows. Considerable damage to mobile homes. Flooding damages piers and small craft in unprotected moorings may break their moorings. Some trees blown down.
  — Examples: Bonnie 1998, Georges (FL & LA) 1998 and Gloria 1985
- **Category 3 Hurricane** — winds 111–130 mph (96–113 knots); pressure 964–945 mbar; storm surge 9–12 ft (2.7–3.8 m)
  Some structural damage to small residences and utility buildings. Large trees blown down. Mobile homes and poorly built signs destroyed. Flooding near the coast destroys smaller structures with larger structures damaged by floating debris. Terrain may be flooded well inland.
  — Examples: Keith 2000, Fran 1996, Opal 1995, Alicia 1983 and Betsy 1965
- **Category 4 Hurricane** — winds 131–155 mph (114–135 knots); pressure 944–920 mbar; storm surge 13–18 ft (3.9–5.6 m)
  More extensive curtainwall failures with some complete roof structure failure on small residences. Major erosion of beach areas. Terrain may be flooded well inland.
  — Examples: Hugo 1989 and Donna 1960
- **Category 5 Hurricane** — winds 156 mph and up (135+ knots); pressure less than 920 mbar; storm surge 19+ ft (5.7+ m)
  Complete roof failure on many residences and industrial buildings. Some complete building failures with small utility buildings blown over or away. Flooding causes major damage to lower floors of all structures near the shoreline. Massive evacuation of residential areas may be required.
  — Examples: Andrew (FL) 1992, Camille 1969 and Labor Day 1935

Fujita Tornado Damage Scale

The original Fujita Scale was modified by NOAA in February 2007 and is now called the Enhanced Fujita Scale (EF). It is an operational scale based on the estimated speed of three-second wind gusts, as indicated by typical damage levels. The table below describes the damage levels according to the original scale. In the enhanced scale, the damage is measured by a more elaborate set of criteria (see http://www.spc.noaa.gov/efscale/ef-scale.html).

EF Number	3 s Gusts (mph)	Typical damage (according to the original Fujita Scale)
0	65–85	**Light damage.** Some damage to chimneys; branches broken off trees; shallow-rooted trees pushed over; sign boards damaged.
1	86–110	**Moderate damage.** Peels surface off roofs; mobile homes pushed off foundations or overturned; moving autos blown off roads.
2	111–135	**Considerable damage.** Roofs torn off frame houses; mobile homes demolished; boxcars overturned; large trees snapped or uprooted; light-object missiles generated; cars lifted off ground.
3	136–165	**Severe damage.** Roofs and some walls torn off well-constructed houses; trains overturned; most trees in forest uprooted; heavy cars lifted off the ground and thrown.
4	166–200	**Devastating damage.** Well-constructed houses leveled; structures with weak foundations blown away some distance; cars thrown and large missiles generated.
5	Over 200	**Incredible damage.** Strong frame houses leveled off foundations and swept away; automobile-sized missiles fly through the air in excess of 100 meters (109 yd); trees debarked; incredible phenomena will occur.

Beaufort Wind Scale

The Beaufort Wind Scale was devised by British Rear-Admiral Sir Francis Beaufort in 1805 based on observations of the effects of the wind.

Force	Wind (knots)	WMO classification	Appearance of wind effects: On the water	Appearance of wind effects: On land
0	< 1	Calm	Sea surface smooth and mirror-like	Calm, smoke rises vertically
1	1–3	Light Air	Scaly ripples, no foam crests	Smoke drift indicates wind direction, still wind vanes
2	4–6	Light Breeze	Small wavelets, crests glassy, no breaking	Wind felt on face, leaves rustle, vanes begin to move
3	7–10	Gentle Breeze	Large wavelets, crests begin to break, scattered whitecaps	Leaves and small twigs constantly moving, light flags extended

Force	Wind (knots)	WMO classification	Appearance of wind effects: On the water	Appearance of wind effects: On land
4	11–16	Moderate Breeze	Small waves 1–4 ft. becoming longer, numerous whitecaps	Dust, leaves, and loose paper lifted, small tree branches move
5	17–21	Fresh Breeze	Moderate waves 4–8 ft taking longer form, many whitecaps, some spray	Small trees in leaf begin to sway
6	22–27	Strong Breeze	Larger waves 8–13 ft, whitecaps common, more spray	Larger tree branches moving, whistling in wires
7	28–33	Near Gale	Sea heaps up, waves 13–20 ft, white foam streaks off breakers	Whole trees moving, resistance felt walking against wind
8	34–40	Gale	Moderately high (13–20 ft) waves of greater length, edges of crests begin to break into spindrift, foam blown in streaks	Whole trees in motion, resistance felt walking against wind
9	41–47	Strong Gale	High waves (20 ft), sea begins to roll, dense streaks of foam, spray may reduce visibility	Slight structural damage occurs, slate blows off roofs
10	48–55	Storm	Very high waves (20–30 ft) with overhanging crests, sea white with densely blown foam, heavy rolling, lowered visibility	Seldom experienced on land, trees broken or uprooted, "considerable structural damage"
11	56–63	Violent Storm	Exceptionally high (30–45 ft) waves, foam patches cover sea, visibility more reduced	
12	64+	Hurricane	Air filled with foam, waves over 45 ft, sea completely white with driving spray, visibility greatly reduced	

Wind Chill

The following chart prepared by the U. S. National Weather Service gives the temperature perceived by an average person as a function of the real air temperature and the wind speed. The current scale was adopted in 2001.

Temperature (°F)

Wind (mph) / Calm	40	35	30	25	20	15	10	5	0	-5	-10	-15	-20	-25	-30	-35	-40	-45
5	36	31	25	19	13	7	1	-5	-11	-16	-22	-28	-34	-40	-46	-52	-57	-63
10	34	27	21	15	9	3	-4	-10	-16	-22	-28	-35	-41	-47	-53	-59	-66	-72
15	32	25	19	13	6	0	-7	-13	-19	-26	-32	-39	-45	-51	-58	-64	-71	-77
20	30	24	17	11	4	-2	-9	-15	-22	-29	-35	-42	-48	-55	-61	-68	-74	-81
25	29	23	16	9	3	-4	-11	-17	-24	-31	-37	-44	-51	-58	-64	-71	-78	-84
30	28	22	15	8	1	-5	-12	-19	-26	-33	-39	-46	-53	-60	-67	-73	-80	-87
35	28	21	14	7	0	-7	-14	-21	-27	-34	-41	-48	-55	-62	-69	-76	-82	-89
40	27	20	13	6	-1	-8	-15	-22	-29	-36	-43	-50	-57	-64	-71	-78	-84	-91
45	26	19	12	5	-2	-9	-16	-23	-30	-37	-44	-51	-58	-65	-72	-79	-86	-93
50	26	19	12	4	-3	-10	-17	-24	-31	-38	-45	-52	-60	-67	-74	-81	-88	-95
55	25	18	11	4	-3	-11	-18	-25	-32	-39	-46	-54	-61	-68	-75	-82	-89	-97
60	25	17	10	3	-4	-11	-19	-26	-33	-40	-48	-55	-62	-69	-76	-84	-91	-98

Frostbite Times 30 minutes 10 minutes 5 minutes

$$\text{Wind Chill } (°F) = 35.74 + 0.6215T - 35.75(V^{0.16}) + 0.4275T(V^{0.16})$$

Where, T= Air Temperature (°F) V= Wind Speed (mph)

Effective 11/01/01

Reference

National Oceanic and Atmospheric Administration, http://www.noaa.gov

INFRARED ABSORPTION BY THE EARTH'S ATMOSPHERE

Several constituents of the Earth's atmosphere absorb infrared radiation. At ground level the strongest absorbers are H_2O and CO_2, but 30 to 40 other compounds can make significant contributions. The centers of the most important absorption bands are listed below:

Molecule	Vibrational mode	Band center in cm^{-1}
H_2O	Bend	1595
H_2O	Symmetric O-H stretch	3657
H_2O	Antisymmetric O-H stretch	3756
CO_2	Bend	667
CO_2	Antisymmetric C-O stretch	2349
O_3	Bend	701
O_3	Antisymmetric O-O stretch	1042
O_3	Symmetric O-O stretch	1103
N_2O	Bend	589
N_2O	N-O stretch	1285
N_2O	N-N stretch	2224
CO	C-O stretch	2143
CH_4	Degenerate deformation	1306
CH_4	Degenerate stretch	3019

The HITRAN Molecular Spectroscopy Database (References 1 and 2) is a compilation of wavenumbers and intensities of more than 1.7 million spectral lines of atmospheric constituents. It is a valuable resource for calculating transmission of the atmosphere, radiative energy transfer, and other phenomena. The graph below, which was supplied by Walter J. Lafferty (Reference 3), gives the transmittance of the atmosphere for one set of conditions.

References

1. Rothman, L. S., et al., *J. Quant. Spectros. Radiat. Transfer* 82, 5, 2003; *ibid.*, to be published, 2005.
2. HITRAN Molecular Spectroscopy Database, <http://cfa-www.Harvard.edu/HITRAN/hitrandata04/>.
3. Lafferty, W. J., Some Aspects of High Resolution Molecular Spectroscopy, in *Lectures on Molecular Physics*, Institute for the Structure of Matter, Centro de Fisica Miguel A. Catalan, Madrid, 1997.

Transmittance of U.S. Standard Atmosphere at Ground Level for a Path of 1 km at 296 K

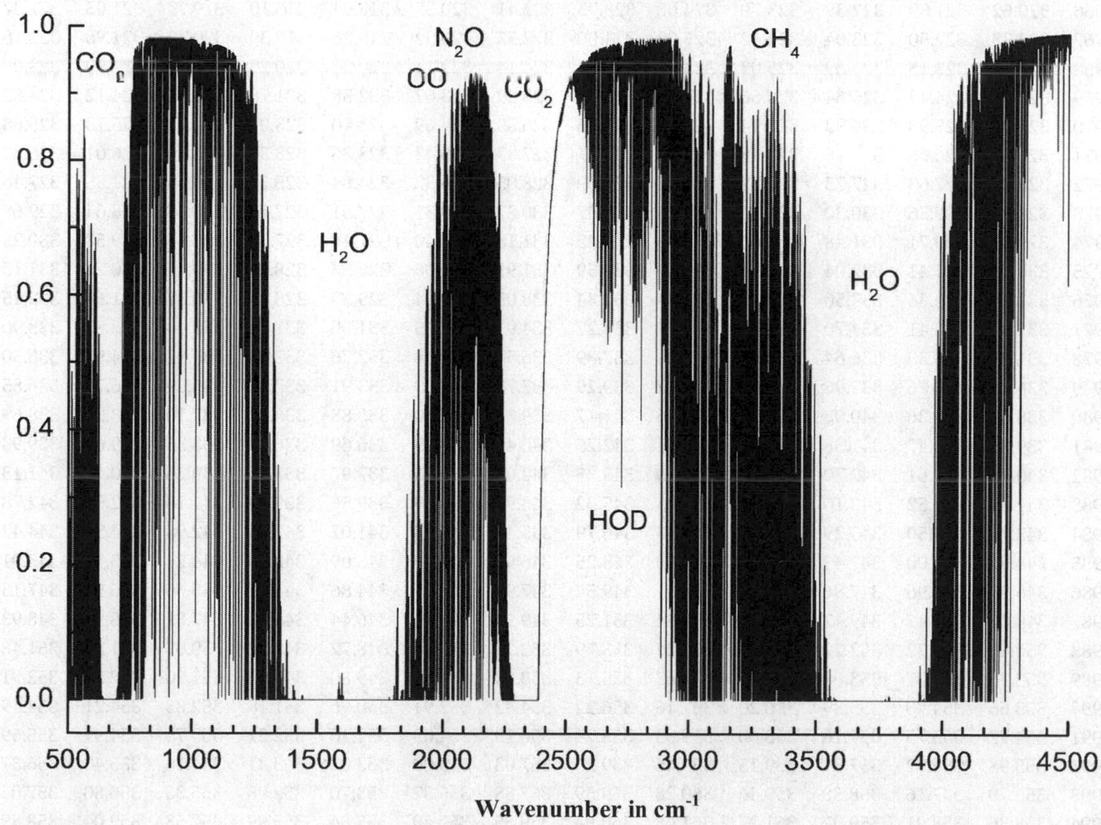

ATMOSPHERIC CONCENTRATION OF CARBON DIOXIDE, 1958–2008

The data in this table were taken at the Mauna Loa Observatory in Hawaii and represent averages adjusted to the 15th of each month. The last column gives the average over the year. The concentration of CO_2 is given in parts per million by volume. Data from other measurement sites may be found in Ref. 1.

The first graph illustrates the seasonal variation of CO_2 concentration and the steady increase over the last 50 years. The second graph summarizes the growth in global emissions of CO_2 into the atmosphere as a result of burning of fossil fuels (Ref. 2).

References

1. Keeling, C. D., Piper, S. C., Bollenbacher, A. F., and Walker, S. J., Atmospheric carbon dioxide record from Mauna Loa, Carbon Dioxide Information Analysis Center, Oak Ridge National Laboratory, U.S. Department of Energy, Oak Ridge, TN, <cdiac.ornl.gov/trends/co2/sio-mlo.html>, February 2009.
2. Boden, T. A., Marland, G., and Andres, R. J., Global, Regional, and National Fossil-Fuel CO_2 Emissions, Carbon Dioxide Information Analysis Center, Oak Ridge National Laboratory, U.S. Department of Energy, Oak Ridge, TN, <cdiac.ornl.gov/trends/emis/tre_glob.html>, 2009.

CO_2 Concentration in ppm at Mauna Loa

Year	Jan.	Feb.	March	April	May	June	July	Aug.	Sept.	Oct.	Nov.	Dec.	Annual
1958			315.71	317.45	317.50		315.85	314.93	313.19		313.34	314.67	
1959	315.58	316.47	316.65	317.72	318.29	318.16	316.55	314.80	313.84	313.34	314.82	315.59	315.98
1960	316.43	316.97	317.58	319.03	320.03	319.59	318.18	315.91	314.16	313.84	315.00	316.19	316.91
1961	316.89	317.70	318.54	319.48	320.58	319.77	318.58	316.79	314.99	315.31	316.10	317.01	317.65
1962	317.94	318.56	319.69	320.58	321.01	320.61	319.61	317.40	316.26	315.42	316.69	317.69	318.45
1963	318.74	319.08	319.86	321.39	322.24	321.47	319.74	317.77	316.21	315.99	317.06	318.36	318.99
1964	319.57				322.24	321.89	320.44	318.70	316.70	316.87	317.68	318.71	
1965	319.44	320.44	320.89	322.13	322.16	321.87	321.21	318.87	317.81	317.30	318.87	319.42	320.03
1966	320.62	321.59	322.39	323.70	324.07	323.75	322.41	320.37	318.64	318.10	319.79	321.03	321.37
1967	322.33	322.50	323.04	324.42	325.00	324.09	322.55	320.92	319.26	319.39	320.72	321.96	322.18
1968	322.57	323.15	323.89	325.03	325.57	325.36	324.14	322.11	320.33	320.25	321.33	322.90	323.05
1969	324.00	324.42	325.64	326.66	327.38	326.70	325.89	323.67	322.38	321.78	322.85	324.12	324.62
1970	325.06	325.98	326.93	328.14	328.07	327.66	326.35	324.69	323.10	323.07	324.01	325.13	325.68
1971	326.17	326.68	327.18	327.78	328.92	328.57	327.37	325.43	323.36	323.57	324.80	326.01	326.32
1972	326.77	327.63	327.75	329.72	330.07	329.09	328.05	326.32	324.84	325.20	326.50	327.55	327.46
1973	328.54	329.56	330.30	331.50	332.48	332.07	330.87	329.31	327.51	327.18	328.16	328.64	329.68
1974	329.35	330.71	331.48	332.65	333.08	332.25	331.18	329.40	327.44	327.37	328.46	329.58	330.25
1975	330.40	331.41	332.04	333.31	333.96	333.59	331.91	330.06	328.56	328.34	329.49	330.76	331.15
1976	331.74	332.56	333.50	334.58	334.87	334.34	333.05	330.94	329.30	328.94	330.31	331.68	332.15
1977	332.92	333.41	334.70	336.07	336.74	336.27	334.93	332.75	331.58	331.16	332.40	333.85	333.90
1978	334.97	335.39	336.64	337.76	338.01	337.89	336.54	334.68	332.76	332.54	333.92	334.95	335.50
1979	336.23	336.76	337.96	338.89	339.47	339.29	337.73	336.09	333.91	333.86	335.29	336.73	336.85
1980	338.01	338.36	340.08	340.77	341.46	341.17	339.56	337.60	335.88	336.02	337.10	338.21	338.69
1981	339.23	340.47	341.38	342.51	342.91	342.25	340.49	338.43	336.69	336.85	338.36	339.61	339.93
1982	340.75	341.61	342.70	343.57	344.13	343.35	342.06	339.82	337.97	337.86	339.26	340.49	341.13
1983	341.37	342.52	343.10	344.94	345.75	345.32	343.99	342.39	339.86	339.99	341.16	342.99	342.78
1984	343.70	344.50	345.29	347.08	347.43	346.79	345.40	343.28	341.07	341.35	342.98	344.22	344.42
1985	344.97	346.00	347.43	348.35	348.93	348.25	346.56	344.69	343.09	342.80	344.24	345.56	345.91
1986	346.29	346.96	347.86	349.55	350.21	349.54	347.94	345.91	344.86	344.17	345.66	346.90	347.15
1987	348.02	348.47	349.42	350.99	351.84	351.25	349.52	348.11	346.44	346.36	347.81	348.96	348.93
1988	350.43	351.72	352.22	353.59	354.22	353.79	352.39	350.44	348.72	348.88	350.07	351.34	351.48
1989	352.76	353.07	353.68	355.42	355.67	355.13	353.90	351.67	349.80	349.99	351.30	352.53	352.91
1990	353.66	354.70	355.39	356.20	357.16	356.22	354.82	352.91	350.96	351.18	352.83	354.21	354.19
1991	354.72	355.75	357.16	358.60	359.33	358.24	356.18	354.03	352.16	352.21	353.75	354.99	355.59
1992	355.98	356.72	357.81	359.15	359.66	359.25	357.03	355.00	353.01	353.31	354.16	355.40	356.37
1993	356.70	357.16	358.38	359.46	360.28	359.59	357.58	355.52	353.70	353.98	355.33	356.80	357.04
1994	358.36	358.91	359.97	361.27	361.68	360.94	359.55	357.49	355.84	355.99	357.58	359.04	358.89
1995	359.96	361.00	361.64	363.45	363.79	363.26	361.90	359.46	358.06	357.75	359.56	360.70	360.88
1996	362.05	363.25	364.03	364.72	365.41	364.97	363.65	361.49	359.46	359.60	360.76	362.33	362.64
1997	363.18	364.00	364.57	366.35	366.80	365.62	364.47	362.51	360.19	360.77	362.43	364.28	363.76
1998	365.32	366.15	367.31	368.61	369.29	368.87	367.64	365.77	363.90	364.23	365.46	366.97	366.63
1999	368.15	368.87	369.59	371.14	371.00	370.35	369.27	366.94	364.63	365.12	366.67	368.01	368.31
2000	369.14	369.46	370.52	371.66	371.82	371.70	370.12	368.12	366.62	366.73	368.29	369.53	369.48

Year	Jan.	Feb.	March	April	May	June	July	Aug.	Sept.	Oct.	Nov.	Dec.	Annual
2001	370.28	371.50	372.12	372.87	374.02	373.30	371.62	369.55	367.96	368.09	369.68	371.24	371.02
2002	372.43	373.09	373.52	374.86	375.55	375.40	374.02	371.49	370.71	370.24	372.08	373.78	373.10
2003	374.68	375.63	376.11	377.65	378.35	378.13	376.62	374.50	372.99	373.00	374.35	375.70	375.64
2004	376.79	377.37	378.41	380.52	380.63	379.57	377.79	375.86	374.06	374.24	375.86	377.48	377.38
2005	378.37	379.69	380.41	382.10	382.28	382.13	380.66	378.71	376.42	376.88	378.32	380.04	379.67
2006	381.38	382.03	382.64	384.62	384.95	384.06	382.29	380.47	378.67	379.06	380.14	381.74	381.84
2007	382.45	383.68	384.23	386.26	386.39	385.87	384.39	381.78	380.73	380.81	382.33	383.69	383.55
2008	385.07	385.72	385.85	386.71	388.45	387.64	386.10	383.95	382.91	382.73	383.96	385.02	385.34

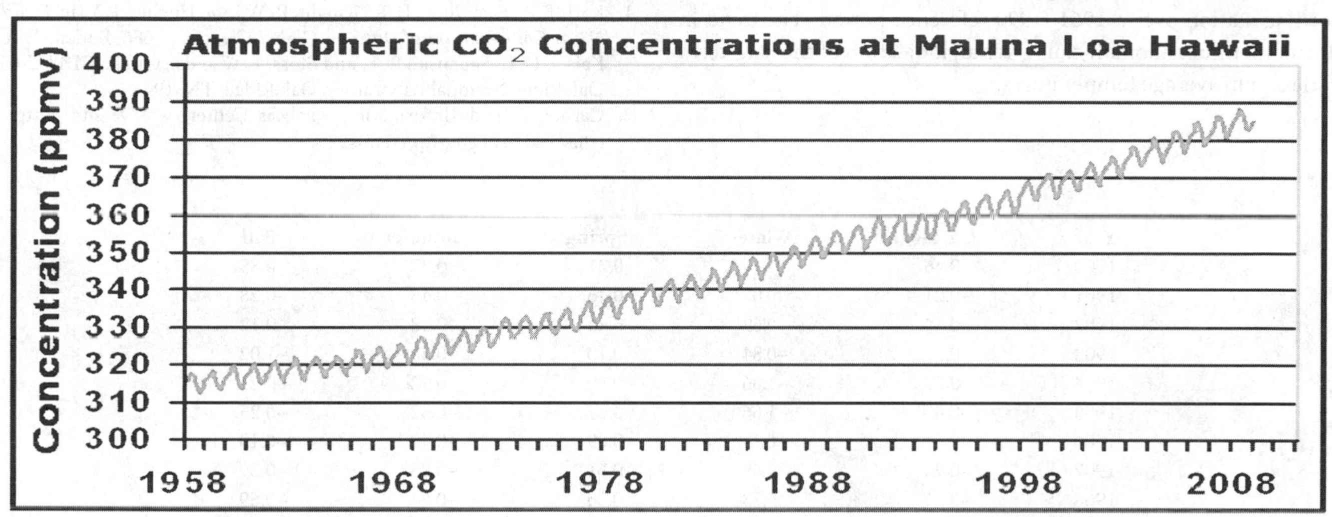

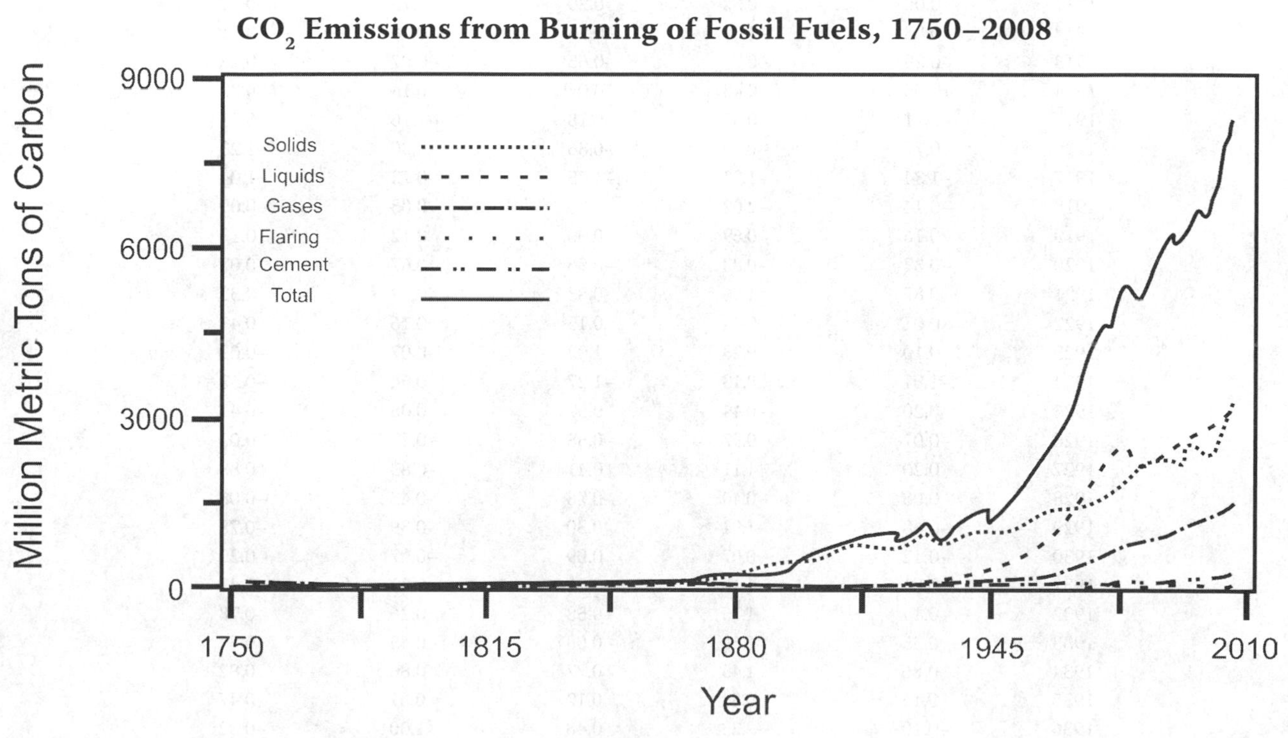

MEAN TEMPERATURES IN THE UNITED STATES, 1900–1992

Historical records of atmospheric temperatures have been analyzed to obtain mean temperatures in °C for 23 climatically distinct regions of the United States. The table below gives the average over these 23 regions, which cover completely the contiguous 48 states. Data for the individual regions and for other parts of the world may be found in the references.

The data are presented as temperature anomalies, i.e., as deviations (in °C) from the average temperature at each individual recording station over a 1961–1990 reference period. The trend in the temperature anomaly thus gives an indication of the long-term variation in average temperatures.

CY Mean: Calendar year mean (January–December)
Winter: December–February
Spring: March–May
Summer: June–August
Fall: September–November

References

1. Karl, T. R., Easterling, D. R., Knight, R. W., and Hughes, P. Y., in *Trends '93: A Compendium of Data on Global Change*, p. 686, Boden, T. A., Kaiser, D. P., Sepanski, R. J., and Stoss, F. W., Eds., ORNL/CDIAC-65, Oak Ridge National Laboratory, Oak Ridge, TN, 1994.
2. Carbon Dioxide Information Analysis Center, WWW site <http://cdiac.esd.ornl.gov/ftp/trends93>.

Year	CY Mean	Winter	Spring	Summer	Fall
1900	0.46		0.21	0.27	0.85
1901	−0.21	0.07	−0.46	0.48	−0.28
1902	−0.14	−0.69	0.48	−0.58	0.12
1903	−0.77	−0.84	0.10	−0.83	−1.02
1904	−0.72	−1.86	−0.39	−0.92	−0.21
1905	−0.45	−1.86	0.67	−0.32	−0.25
1906	−0.04	0.23	−0.69	−0.42	−0.10
1907	−0.23	1.09	−0.61	−0.85	−0.39
1908	−0.11	0.73	0.36	−0.64	−0.59
1909	−0.36	0.82	−1.06	0.06	0.01
1910	−0.14	−2.08	0.95	−0.55	0.13
1911	0.02	0.52	0.20	−0.19	−0.62
1912	−0.88	−1.50	−0.75	−0.76	−0.51
1913	−0.23	−0.74	−0.65	−0.07	0.22
1914	−0.05	0.48	0.04	0.14	0.32
1915	−0.11	−0.37	−0.18	−1.16	0.31
1916	−0.77	−0.29	−0.36	−0.30	−1.23
1917	−1.34	−1.93	−1.75	−0.73	−1.04
1918	−0.14	−2.02	0.30	0.03	−0.09
1919	−0.16	0.69	0.00	0.12	−0.13
1920	−0.37	−0.83	−0.96	−0.67	0.02
1921	0.87	1.56	0.83	0.52	0.32
1922	0.01	−0.44	0.15	0.16	0.41
1923	−0.10	0.23	−1.02	−0.07	−0.09
1924	−1.01	0.13	−1.27	−0.64	−0.53
1925	0.20	−0.44	0.57	0.05	−0.41
1926	−0.01	0.97	−0.58	−0.27	0.02
1927	0.20	1.11	0.41	−0.83	0.83
1928	−0.08	−0.40	−0.28	−0.43	−0.08
1929	−0.68	−1.94	0.30	−0.39	−0.78
1930	−0.12	0.07	0.09	−0.07	−0.25
1931	0.81	1.16	−0.71	0.51	1.41
1932	−0.14	1.75	−0.58	0.26	−0.78
1933	0.35	−0.60	−0.08	0.45	0.54
1934	0.86	1.45	0.77	0.86	0.82
1935	0.12	0.84	0.12	0.31	−0.47
1936	−0.10	−2.23	0.48	1.00	−0.32
1937	−0.13	−0.65	−0.24	0.66	−0.03
1938	0.71	1.31	0.98	0.39	0.08
1939	0.38	0.36	0.34	0.18	0.15
1940	0.06	0.03	−0.10	0.18	0.12
1941	0.79	1.56	0.44	0.28	0.85

Year	CY Mean	Winter	Spring	Summer	Fall
1942	0.01	0.35	0.22	0.08	0.06
1943	−0.17	0.20	−0.46	0.54	−0.79
1944	0.04	0.61	−0.38	−0.20	0.43
1945	−0.02	0.30	0.25	−0.47	0.08
1946	0.53	−0.26	1.21	−0.21	0.18
1947	0.10	0.47	−0.42	0.03	0.85
1948	−0.22	−0.67	−0.01	0.02	−0.15
1949	0.05	−0.77	0.39	0.33	0.23
1950	−0.37	0.39	−1.02	−0.86	0.13
1951	−0.45	0.05	−0.69	−0.30	−0.70
1952	0.03	0.68	−0.40	0.49	−1.22
1953	0.61	1.91	0.03	0.31	0.31
1954	0.57	1.47	−0.22	0.36	0.70
1955	−0.25	−0.33	0.12	0.19	−0.60
1956	−0.05	−0.16	−0.42	0.01	−0.37
1957	0.45	1.02	0.44	0.25	−0.13
1958	0.14	0.93	0.06	0.11	0.45
1959	0.12	−0.60	0.16	0.50	−0.52
1960	−0.27	0.43	−0.98	0.01	0.56
1961	−0.01	−0.02	−0.17	0.18	−0.29
1962	−0.04	−0.52	−0.09	−0.41	0.65
1963	−0.06	−1.35	0.61	0.12	1.38
1964	−0.27	−1.30	−0.24	−0.08	−0.40
1965	−0.03	−0.07	−0.56	−0.42	0.28
1966	−0.34	−0.31	−0.25	−0.07	−0.13
1967	−0.13	0.23	−0.05	−0.37	−0.32
1968	−0.28	−0.31	−0.16	−0.12	0.01
1969	0.06	0.36	0.51	0.12	0.16
1970	−0.12	−0.17	−0.44	0.32	−0.07
1971	−0.04	−0.08	−0.99	0.01	0.56
1972	−0.13	0.20	0.26	−0.19	0.05
1973	0.48	−0.23	0.47	0.22	0.90
1974	0.16	0.52	0.74	−0.32	−0.39
1975	−0.09	0.63	−0.79	0.03	−0.20
1976	−0.62	0.88	−0.09	−0.54	−1.72
1977	0.32	−1.95	1.07	0.60	0.68
1978	−0.37	−1.31	0.23	0.09	0.21
1979	−0.53	−2.92	0.09	−0.21	−0.07
1980	0.18	0.72	−0.25	0.43	−0.04
1981	0.64	0.90	0.80	0.57	0.36
1982	−0.08	−0.86	0.03	−0.11	0.06
1983	0.40	2.33	−0.36	0.58	0.94
1984	0.21	−0.78	−0.30	0.31	0.07
1985	−0.26	−0.78	1.24	−0.23	0.05
1986	0.93	0.22	1.22	0.45	0.49
1987	0.67	1.52	0.97	0.33	−0.17
1988	−0.07	−0.26	−0.06	0.57	0.04
1989	−0.30	−0.28	0.36	0.12	−0.27
1990	0.72	0.41	0.72	0.41	0.66
1991	0.77	0.32	1.36	0.56	−0.31
1992		2.48	0.82	−0.70	

GLOBAL TEMPERATURE TREND, 1880–2009

This table and graph summarize the trend in annual mean global surface temperature from 1880 to 2009. The values were calculated from mean temperature anomalies by assuming an absolute global mean of 14.00°C, which is the best estimate for the 1951–1980 period. The 95% confidence interval for comparing the annual mean temperature values for recent years is 0.05 °C.

Reference

GISS Surface Temperature Analysis, Goddard Institute for Space Studies, <data.giss.nasa.gov/gistemp/2008>, January 2009.

Year	*t*/°C	Year	*t*/°C	Year	*t*/°C	Year	*t*/°C	Year	*t*/°C
1880	13.92	1907	13.61	1934	14.02	1961	14.09	1988	14.42
1881	13.89	1908	13.70	1935	13.91	1962	14.05	1989	14.28
1882	14.03	1909	13.69	1936	14.01	1963	14.02	1990	14.48
1883	13.97	1910	13.80	1937	14.11	1964	13.75	1991	14.44
1884	13.61	1911	13.75	1938	14.14	1965	13.85	1992	14.15
1885	13.80	1912	13.67	1939	13.98	1966	13.92	1993	14.18
1886	13.76	1913	13.72	1940	14.14	1967	13.98	1994	14.32
1887	13.57	1914	13.98	1941	14.11	1968	13.91	1995	14.45
1888	13.78	1915	14.05	1942	14.11	1969	14.00	1996	14.36
1889	14.10	1916	13.79	1943	14.06	1970	14.04	1997	14.40
1890	13.81	1917	13.53	1944	14.11	1971	13.90	1998	14.70
1891	13.48	1918	13.66	1945	13.99	1972	13.95	1999	14.43
1892	13.64	1919	13.91	1946	14.00	1973	14.18	2000	14.40
1893	13.63	1920	13.82	1947	14.11	1974	13.94	2001	14.56
1894	13.69	1921	13.96	1948	13.97	1975	13.98	2002	14.67
1895	13.69	1922	13.90	1949	13.91	1976	13.79	2003	14.65
1896	13.75	1923	13.84	1950	13.83	1977	14.16	2004	14.59
1897	13.86	1924	13.88	1951	13.98	1978	14.07	2005	14.77
1898	13.83	1925	13.85	1952	14.03	1979	14.14	2006	14.64
1899	13.77	1926	14.04	1953	14.11	1980	14.28	2007	14.72
1900	13.96	1927	13.95	1954	13.91	1981	14.40	2008	14.54
1901	13.97	1928	13.99	1955	13.92	1982	14.08	2009	14.71
1902	13.72	1929	13.78	1956	13.82	1983	14.34		
1903	13.65	1930	13.96	1957	14.08	1984	14.15		
1904	13.60	1931	14.02	1958	14.09	1985	14.12		
1905	13.76	1932	14.03	1959	14.05	1986	14.19		
1906	13.86	1933	13.88	1960	13.98	1987	14.34		

Annual Mean Global Temperature in °C

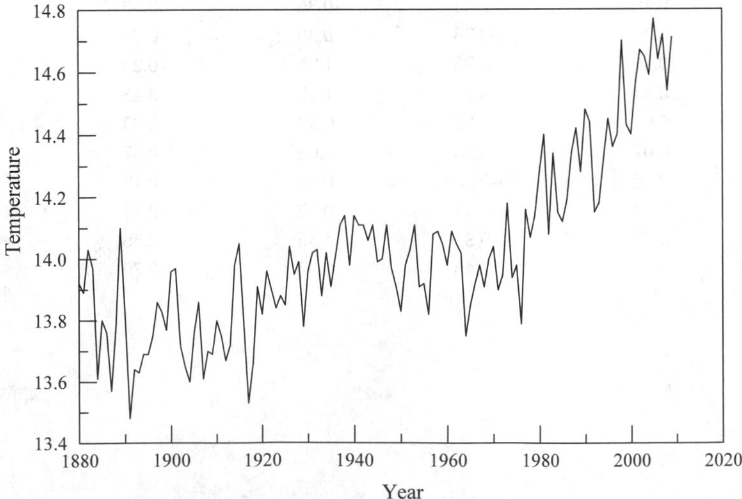

GLOBAL WARMING POTENTIAL OF GREENHOUSE GASES

The Global Warming Potential (GWP) of a gas is a measure of the degree, relative to carbon dioxide, to which the presence of that gas in the atmosphere will contribute to a long-term increase in global temperature. The calculation of the GWP for a given gas takes into account the efficiency of the gas in absorbing solar radiation (primarily determined by the infrared spectrum of the compound) and the time the compound will remain in the atmosphere before it is removed by natural processes. Thus if a pulse of 1 kg of the gas is emitted to the atmosphere at the same time as a pulse of 1 kg of CO_2, the GWP compares the warming effect of the gas relative to CO_2 over various time horizons.

This table, which is taken from the 2007 report of the Intergovernmental Panel of Climate Change (IPCC), gives the lifetime in years and the radiative efficiency in watts per square meter for a concentration of one part per billion for the major compounds identified in the Kyoto Protocol as contributing to global climate change. Radiative efficiency is a measure of the radiative forcing that influences the energy balance in the Earth–atmosphere system. The last four columns of the table give the Global Warming Potential, first as estimated in 1995 for a 100 year time horizon, and then as estimated with improved data in 2007 for 20-, 100-, and 500-year horizons. The calculation of a GWP involves a number of assumptions, and other measures have been proposed (see Reference).

The list of compounds includes those identified in the Montreal Protocol as contributing to ozone depletion, since these compounds also contribute to global warming. It also includes compounds used or proposed as replacements for the ozone-depleting compounds but which still have global warming potential.

Reference

Forster, P., V. Ramaswamy, P. Artaxo, T. Berntsen, R. Betts, D. W. Fahey, J. Haywood, J. Lean, D.C. Lowe, G. Myhre, J. Nganga, R. Prinn, G. Raga, M. Schulz, and R. Van Dorland, 2007: Changes in Atmospheric Constituents and in Radiative Forcing. In: *Climate Change 2007: The Physical Science Basis. Contribution of* Working *Group I to the Fourth Assessment Report of the Intergovernmental Panel on Climate Change* [Solomon, S., D. Sin, M. Manning, Z. Chen, M. Marquis, K. B. Averyt, M. Tignor and H.L. Miller (eds.)], Cambridge University Press, Cambridge, United Kingdom and New York, NY, USA. Available on the Internet at < http://ipcc-wg1.ucar.edu/wg1/wg1-report.html>.

Compound	Synonym/Code	Formula	Lifetime (years)	Rad. eff. $W\ m^{-2}ppb^{-1}$	GWP for given time horizon: 1995 100 yr	Current estimate 20 yr	Current estimate 100 yr	Current estimate 500 yr
Natural atmospheric constituents								
Carbon dioxide		CO_2		1.4×10^{-5}	1	1	1	1
Methane		CH_4	12	3.7×10^{-4}	21	72	25	7.6
Nitrous oxide		N_2O	114	3.03×10^{-3}	310	289	298	153
Substances controlled by the Montreal Protocol								
Trichlorofluoromethane	CFC-11	CCl_3F	45	0.25	3800	6730	4750	1620
Dichlorodifluoromethane	CFC-12	CCl_2F_2	100	0.32	8100	11000	10900	5200
Chlorotrifluoromethane	CFC-13	CCl_3F	640	0.25		10800	14400	16400
1,1,2-Trichloro-1,2,2-trifluoroethane	CFC-113	CCl_2FCClF_2	85	0.3	4800	6540	6130	2700
1,2-Dichloro-1,1,2,2-tetrafluoroethane	CFC-114	$CClF_2CClF_2$	300	0.31		8040	10000	8730
Chloropentafluoroethane	CFC-115	$CClF_2CF_3$	1700	0.18		5310	7370	9990
Bromotrifluoromethane	Halon-1301	$CBrF_3$	65	0.32	5400	8480	7140	2760
Bromochlorodifluoromethane	Halon-1211	$CBrClF_2$	16	0.3		4750	1890	575
1,2-Dibromotetrafluoroethane	Halon-2402	$CBrF_2CBrF_2$	20	0.33		3680	1640	503
Tetrachloromethane	Carbon tetrachloride	CCl_4	26	0.13	1400	2700	1400	435
Bromomethane	Methyl bromide	CH_3Br	0.7	0.01		17	5	1
1,1,1-Trichloroethane	Methyl chloroform	CH_3CCl_3	5	0.06		506	146	45
Chlorodifluoromethane	HCFC-22	$CHClF_2$	12	0.2	1500	5160	1810	549
2,2-Dichloro-1,1,1-trifluoroethane	HCFC-123	$CHCl_2CF_3$	1.3	0.14	90	273	77	24
1-Chloro-1,2,2,2-tetrafluoroethane	HCFC-124	$CHClFCF_3$	5.8	0.22	470	2070	609	185
1,1-Dichloro-1-fluoroethane	HCFC-141b	CH_3CCl_2F	9.3	0.14		2250	725	220
1-Chloro-1,1-difluoroethane	HCFC-142b	CH_3CClF_2	17.9	0.2	1800	5490	2310	705
3,3-Dichloro-1,1,1,2,2-pentafluoropropane	HCFC-225ca	$CHCl_2CF_2CF_3$	1.9	0.2		429	122	37
1,3-Dichloro-1,1,2,2,3-pentafluoropropane	HCFC-225cb	$CHClFCF_2CClF_2$	5.8	0.32		2030	595	181
Hydrofluorocarbons								
Trifluoromethane	HFC-23	CHF_3	270	0.19	11700	12000	14800	12200
Difluoromethane	HFC-32	CH_2F_2	4.9	0.11	650	2330	675	205
Pentafluoroethane	HFC-125	CHF_2CF_3	29	0.23	2800	6350	3500	1100

Compound	Synonym/Code	Formula	Lifetime (years)	Rad. eff. W m^{-2}ppb^{-1}	GWP for given time horizon: 1995 100 yr	Current estimate 20 yr	Current estimate 100 yr	Current estimate 500 yr
1,1,1,2-Tetrafluoroethane	HFC-134a	CH_2FCF_3	14	0.16	1300	3830	1430	435
1,1,1-Trifluoroethane	HFC-143a	CH_3CF_3	52	0.13	3800	5890	4470	1590
1,1-Difluoroethane	HFC-152a	CH_3CHF_2	1.4	0.09	140	437	124	38
1,1,1,2,3,3,3-Heptafluoropropane	HFC-227ea	CF_3CHFCF_3	34.2	0.26	2900	5310	3220	1040
1,1,1,3,3,3-Hexafluoropropane	HFC-236fa	$CF_3CH_2CF_3$	240	0.28	6300	8100	9810	7660
1,1,1,3,3-Pentafluoropropane	HFC-245fa	$CHF_2CH_2CF_3$	7.6	0.28		3380	1030	314
1,1,1,3,3-Pentafluorobutane	HFC-365mfc	$CH_3CF_2CH_2CF_3$	8.6	0.21		2520	794	241
1,1,1,2,3,4,4,5,5,5-Decafluoropentane	HFC-43-10mee	$CF_3CHFCHFCF_2CF_3$	15.9	0.4	1300	4140	1640	500
Perfluorinated compounds								
Sulfur hexafluoride		SF_6	3200	0.52	23900	16300	22800	32600
Nitrogen trifluoride		NF_3	740	0.21		12300	17200	20700
Tetrafluoromethane	PFC-14	CF_4	50000	0.10	6500	5210	7390	11200
Hexafluoroethane	PFC-116	C_2F_6	10000	0.26	9200	8630	12200	18200
Perfluoropropane	PFC-218	C_3F_8	2600	0.26	7000	6310	8830	12500
Perfluorocyclobutane	PFC-318	c-C_4F_8	3200	0.32	8700	7310	10300	14700
Perfluorobutane	PFC-3-1-10	C_4F_{10}	2600	0.33	7000	6330	8860	12500
Perfluoropentane	PFC-4-1-12	C_5F_{12}	4100	0.41		6510	9160	13300
Perfluorohexane	PFC-5-1-14	C_6F_{14}	3200	0.49	7400	6600	9300	13300
Perfluorodecalin	PFC-9-1-18	$C_{10}F_{18}$	>1,000	0.56		>5500	>7500	>9500
(Trifluoromethyl)sulfur pentafluoride		SF_5CF_3	800	0.57		13200	17700	21200
Fluorinated ethers								
Trifluoromethyl difluoromethyl ether	HFE-125	CHF_2OCF_3	136	0.44		13800	14900	8490
Bis(difluoromethyl) ether	HFE-134	CHF_2OCHF_2	26	0.45		12200	6320	1960
Methyl trifluoromethyl ether	HFE-143a	CH_3OCF_3	4.3	0.27		2630	756	230
2-Chloro-2-(difluoromethoxy)-1,1,1-trifluoroethane	HCFE-235da2	$CHF_2OCHClCF_3$	2.6	0.38		1230	350	106
Methyl 1,1,2,2-tetrafluoroethyl ether	HFE-245cb2	$CH_3OCF_2CHF_2$	5.1	0.32		2440	708	215
2-(Difluoromethoxy)-1,1,1-trifluoroethane	HFE-245fa2	$CHF_2OCH_2CF_3$	4.9	0.31		2280	659	200
Methyl pentafluoroethyl ether	HFE-254cb2	$CH_3OCF_2CHF_2$	2.6	0.28		1260	359	109
Perfluoropropyl methyl ether	HFE-347mcc3	$CH_3OCF_2CF_2CF_3$	5.2	0.34		1980	575	175
1,1,2,2-Tetrafluoroethyl 1,1,1-trifluoroethyl ether	HFE-347pcf2	$CHF_2CF_2OCH_2CF_3$	7.1	0.25		1900	580	175
1-Methoxy-1,1,2,2,3,3-hexafluoropropane	HFE-356pcc3	$CH_3OCF_2CF_2CHF_2$	0.33	0.93		386	110	33
Methyl nonafluorobutyl ether	HFE-449sl (HFE-7100)	$C_4F_9OCH_3$	3.8	0.31		1040	297	90
1-Ethoxy-1,1,2,2,3,3,4,4,4-nonafluorobutane	HFE-569sf2 (HFE-7200)	$C_4F_9OC_2H_5$	0.77	0.3		207	59	18
1-(Difluoromethoxy)-2-[(difluoromethoxy)difluoromethoxy]-1,1,2,2-tetrafluoroethane	HFE-43-10pccc124 (H-Galden 1040x)	$CHF_2OCF_2OC_2F_4OCHF_2$	6.3	1.37		6320	1870	569
Bis(difluoromethoxy)difluoromethane	HFE-236ca12 (HG-10)	$CHF_2OCF_2OCHF_2$	12.1	0.66		8000	2800	860
1,2-Bis(difluoromethoxy)-1,1,2,2-tetrafluoroethane	HFE-338pcc13 (HG-01)	$CHF_2OCF_2CF_2OCHF_2$	6.2	0.87		5100	1500	460
Perfluoropolymethylisopropyl ether	PFPMIE	$CF_3OCF(CF_3)CF_2OCF_2OCF_3$	800	0.65		7620	10300	12400
Other compounds - Direct effects								
Dimethyl ether	Methyl ether	CH_3OCH_3	0.015	0.02		1	1	<<1
Dichloromethane	Methylene chloride	CH_2Cl_2	0.38	0.03		31	8.7	2.7
Chloromethane	Methyl chloride	CH_3Cl	1	0.01		45	13	4

ATMOSPHERIC ELECTRICITY

Hans Dolezalek, Hannes Tammet, John Latham, and Martin A. Uman

I. SURVEY AND GLOBAL CIRCUIT

Hans Dolezalek

The science of atmospheric electricity originated in 1752 by an experimental proof of a related earlier hypothesis (that lightning is an electrical event). In spite of a large effort, in part by such eminent physicists as Coulomb, Lord Kelvin, and many others, an overall, proven theory able to generate models with sufficient resolution is not yet available. Generally accepted and encompassing textbooks are now more than 20 years old. The voluminous proceedings of the, so far, nine international atmospheric electricity conferences (1954 to 1992) give much valuable detail and demonstrate impressive progress, as do a number of less comprehensive textbooks published in the last 20 years, but a general theory as indicated above is not yet created. Only now, are certain related measuring techniques and mathematical possibilities emerging.

Applications to practical purposes do exist in the field of lightning research (including the electromagnetic radiation emanating from lightning) by the establishment of lightning-location networks and by the now developing possibility of detecting electrified clouds that pose hazards to aircraft. Application of atmospheric electricity to other parts of meteorology seems to be promising but so far has seldom been instituted. Because some atmospheric electric signals propagate around the Earth and because of the existence of a global circuit, applications for the monitoring of global change processes and conditions are now being proposed. Significant secular changes in the global circuit would indicate a change in the global climate; the availability of many old data (about a span of 100 years) could help detect a long-term trend.

The concept of the "global circuit" is based on the theory of the global spherical capacitor: both the solid (and liquid) Earth as one electrode, and the high atmospheric layers (about the ionosphere) as the other, are by orders of magnitude more electrically conductive than the atmosphere between them. According to the "classical picture of atmospheric electricity," this capacitor is continuously charged by the common action of all thunderstorms to a d.c. voltage difference of several hundred kilovolts, the Earth being negative. The much smaller but still existing conductivity of the atmosphere allows a current flowing from the ionosphere to the ground, integrated for all sink areas of the whole Earth, of the order of 1.5 kA. In this way, a global circuit is created with many generators and sink-areas both interspaced and distributed over the whole globe, all connected to two nodes: ionosphere and ground. Within the scope of the global circuit, for each location, the current density (order of several pA/m^2) is determined by the voltage difference between ionosphere and ground (which is the same for all locations but varying in time) and the columnar resistance reaching from the ground up to the ionosphere (in the order of 10^{17} Ωm^2).

Natural processes, especially meteorological processes and some human activity, which produce or move electric charges ("space charges") or affect the ion distribution, constitute local generators and thereby "local circuits," horziontally and/or in parallel or antiparallel to the local part of the global circuit. In many cases, the local currents are much stronger than the global ones, making the measurement of the global current at a given location and/or during a period of time very difficult or, often, impossible. The strongest local circuits usually occur with certain weather conditions (precipitation, fog, high wind, blown-up dust or snow, heavy cloudiness) that make measurement of the global circuit impossible everywhere; but even in their absence local generators exist in varying magnitudes and of different characters. The separation of the local and global shares in the measured values of current density is a central problem of the science of atmospheric electricity. Aerological measurements are of high value in this regard.

The above description is within the "classical picture" of atmospheric electricity, a group of hypotheses to explain the electrification of the atmosphere. It is probably fundamentally correct but certainly not complete; it has not yet been confirmed by systems of measurements resulting in no inner contradictions. In particular, extraterrestrial influences must be permitted; their general significance is still under debate.

Within this "classical picture" a kind of electric standard atmosphere may be constructed as shown in Table 1.

Values with a star, *, are rough average values from measurement. A star in parentheses, (*), points to a typical value from one or a few measurements. All other values have been calculated from starred values, under the assumption that at 2 km 50% and at 12 km 90% of the columnar resistance is reached. Voltage drop along one of the partial columns can be calculated by subtracting the value for the lower column from that of the upper one. Columnar resistances, conductances, and capacitances are valid for that particular part of the column which is indicated at the left. Capacitances are calculated with the formula for plate capacitors, and this fact must be considered also for the time constants for columns.

According to measurements, U, the potential difference between 0 m and 65 km may vary by a factor of approximately 2. The total columnar resistance, R_c, is estimated to vary up to a factor of 3, the variation being due to either reduction of conductivity in the exchange layer (about lowest 2 km of this table) or to the presence of high mountains; in both cases the variation is caused in the troposphere. Smaller variations in the stratosphere and mesosphere are being discussed because of aerosols there. The air–Earth current density in fair weather varies by a factor of 3 to 6 accordingly. Conductivity near the ground varies by a factor of about 3 but only decreasing; increase of conductivity due to extraordinary radioactivity is a singular event. The field strength near the ground varies as a consequence of variations of air–Earth current density and conductivity from about 1/3 to about 10 times of the value quoted in the table. Conductivity near the ground shows a diurnal and an annual variation which depends strongly on the locality: air–Earth current density shows a diurnal and annual variation because the Earth–ionosphere potential difference undergoes such variations, and also because the columnar resistance is supposed to have a diurnal and probably an annual variation.

Conductivities and air–earth current densities on high mountains are greater than at sea level by factors of up to 10. Conductivity decreases when atmospheric humidity increases. Values for space charges are not quoted because measurements are too few to allow calculation of average values. Values of parameters over the oceans are still rather uncertain.

Theoretically, in fair weather conditions, Ohm's law must be fulfilled for the electric field, the conduction current density, and the electrical conductivity of the atmosphere. Deviations point to shortcomings in the applied measuring techniques. Data which are representative for a large area (in the extreme, "globally representative data," i.e., data on the global circuit), can be obtained on the ground only by stations on an open plane and only if local generators are either small or constant or are independently measured. Certain measurements with instrumented aircraft provide globally representative information valid for the period of the actual measurement.

TABLE 1. Electrical Parameters of the Clear (Fair Weather) Atmosphere, Pertinent to the Classical Picture of Atm. Electricity (Electric Standard Atmosphere)

Part of atmosphere for which the values are calculated (elements are in free, cloudless atmosphere)	Currents, I, in A; and current densities, i, in A/m²	Potential differences, U, in V; field strength E in V/m; U = 0 at sea level	Resistances, R, in Ω; columnar resistances, R_c, in Ω m² and resistivities, ρ, in Ω m	Conductances, G, in Ω^{-1}; columnar conductances G_c, in $\Omega^{-1}m^2$; total conductivities, γ, in $\Omega^{-1}m^{-1}$	Capacitances, C, in F; columnar capacitances, C_c, in F m^{-2} and capacitivities, ε, in F m^{-1}	Time constants τ, in seconds
Volume element at about sea level, 1 m³	$i = 3 \times 10^{-12*}$	$E_0 = 1.2 \times 10^{2*}$	$\rho_0 = 4 \times 10^{13}$	$\gamma_0 = 2.5 \times 10^{-14}$	$\varepsilon_0 = 8.9 \times 10^{-12*}$	$\tau_0 = 3.6 \times 10^2$
Lower column of 1 m² cross section from sea level to 2 km height	$i = 3 \times 10^{-12}$	At upper end: $U_1 = 1.8 \times 10^5$	$R_{c1} = 6 \times 10^{16}$	$G_{c1} = 1.7 \times 10^{-17}$	$C_{cl} = 4.4 \times 10^{-15}$	$\tau_{c1} = 2.6 \times 10^2$
Volume element at about 2 km height, 1 m³	$i = 3 \times 10^{-12}$	$E_2 = 6.6 \times 10^1$	$\rho_2 = 2.2 \times 10^{13(*)}$	$\gamma_2 = 4.5 \times 10^{-14}$	$\varepsilon_2 = 8.9 \times 10^{-12*}$	$\tau_2 = 2 \times 10^2$
Center column of 1 m² cross section from 2 to 12 km	$i = 3 \times 10^{-12}$	At upper end: $U_m = 3.15 \times 10^5$	$R_{cm} = 4.5 \times 10^{16}$	$G_{cm} = 5 \times 10^{-17}$	$C_{cm} = 8.8 \times 10^{-16}$	$\tau_{cm} = 1.8 \times 10^1$
Volume element at about 12 km height, 1 m³	$i = 3 \times 10^{-12}$	$E_{12} = 3.9 \times 10^0$	$\rho_{12} = 1.3 \times 10^{12(*)}$	$\gamma_{12} = 7.7 \times 10^{-13}$	$\varepsilon_{12} = 8.9 \times 10^{-12}$	$\tau_{12} = 1.2 \times 10^1$
Upper column of 1 m² cross section from 12 to 65 km height	$i = 3 \times 10^{-12}$	At upper end: $U_u = 3.5 \times 10^5$	$R_{cu} = 1.5 \times 10^{16}$	$G_{cu} = 2.5 \times 10^{-17}$	$C_{cu} = 1.67 \times 10^{-16}$	$\tau_{cm} = 6.7 \times 10^0$
Whole column of 1 m² cross section from 0 to 65 km height	$i = 3 \times 10^{-12}$	At upper end: $U = 3.5 \times 10^5$	$R_c = 1.2 \times 10^{17}$	$G_c = 8.3 \times 10^{-18}$	$C_c = 1.36 \times 10^{-16}$	$\tau_c = 1.64 \times 10^1$
Total spherical capacitor area: 5×10^{14} m²	$i = 1.5 \times 10^3$	$U = 3.5 \times 10^{5*}$	$R = 2.4 \times 10^2$	$G = 4.2 \times 10^{-3}$	$C = 6.8 \times 10^{-2}$	$\tau = 1.64 \times 10^1$

Note: All currents and fields listed are part of the global circuit, i.e., circuits of local generators are not included. Values are subject to variations due to latitude and altitude of the point of observation above sea level, locality with respect to sources of disturbances, meteorological and climatological factors, and man-made changes. For more explanations, see text.

II. AIR IONS

Hannes Tammet

The term "air ions" signifies all airborne particles that are the carriers of the electrical current in the air and have drift velocities determined by the electric field.

The probability of electrical dissociation of molecules in the atmospheric air under thermodynamic equilibrium is near to zero. The average ionization at the ground level over the ocean is $2 \cdot 10^6$ ion pairs $m^{-3}s^{-1}$. This ionization is produced mainly by cosmic rays. Over the continents the ionizing radiation from soil and from radioactive substances in the air each add about $4 \cdot 10^6$ $m^{-3}s^{-1}$. The total average ionization rate of 10^7 m^3s^{-1} is equivalent to 17 μR/h, which is a customary expression of the background level of the ionizing radiations. The ionization rate over the ground varies in space due to the radioactivity of soil, and in time depending on the exchange of air between the atmosphere and radon-containing soil. Radioactive pollution increases the ionization rate. A temporary increase of about 10 times was registered in Sweden after the Chernobyl accident in 1986. The emission of Kr^{85} from nuclear power plants can noticeably increase the global ionization rate in the 21st century. The ionization rate decreases with altitude near the ground and increases at higher altitudes up to 15 km, where it has a maximum of about $5 \cdot 10^7$ $m^{-3}s^{-1}$. Solar X-ray and extreme UV radiation cause a new increase at altitudes over 60 km.

Local sources of air ions are point discharges in strong electric fields, fluidization of charged drops from waves, etc.

The enhanced chemical activity of an ion results in a chain of ion-molecule reactions with the colliding neutrals, and, in the first microsecond of the life of an air ion, a charged molecular cluster called the *cluster ion* is formed. According to theoretical calculations in the air free from exotic trace gases the following cluster ions should be dominant:

$NO_3^-{\cdot}(HNO_3){\cdot}H_2O$, $NO_2^-{\cdot}(H_2O)_2$, $NO_3^-{\cdot}H_2O$, $O_2^-{\cdot}(H_2O)_4$, $O_2^-{\cdot}(H_2O)_5$,

$H_3O^+{\cdot}(H_2O)_6$, $NH_4^+{\cdot}(H_2O)_2$, $NH_4^+{\cdot}(H_2O)$, $H_3O^+{\cdot}(H_2O)_5$, $NH_4^+{\cdot}NH_3$

A measurable parameter of air ions is the electrical mobility k, characterizing the drift velocity in the unit electric field. The mobility is inversely proportional to the density of air, and the results of measurements are as a rule reduced to normal conditions. According to mobility the air ions are called fast or small or light ions with mobility $k > 5 \cdot 10^{-5}$ $m^2V^{-1}s^1$, intermediate ions, and slow or large or heavy ions with mobility $k > 10^{-6}$ $m^2V^{-1}s^1$. The boundary between intermediate and slow ions is conventional.

Cluster ions are fast ions. The masses of cluster ions may be measured with mass spectrometers, but the possible ion-molecule reactions during the passage of the air through nozzles to the vacuum chamber complicate the measurement. Mass and mobility of cluster ions are highly correlated. The experimental results[5] can be expressed by the empirical formula

$$m \approx \frac{850\ \mathrm{u}}{[0.3 + k/(10^{-4}\,\mathrm{m^2V^{-1}s^{-1}})]^3}$$

where u is the unified atomic mass unit.

The value of the transport cross-section of a cluster ion is needed to calculate its mobility according to the kinetic theory of Chapman and Enskog. The theoretical estimation of transport cross-sections is rough and cannot be used to identify the chemi-

ous components, among which are typically three or four high-current pulses called *strokes*. Each stroke lasts about a millisecond, the separation time between strokes being typically several tens of milliseconds. Such lightning often appears to "flicker" because the human eye can just resolve the individual light pulse associated with each stroke. A drawing of the components of a negative cloud-to-ground flash is found in Figure 3. Some values for salient parameters are found in Table 1. The negatively charged *stepped leader* initiates the first stroke in a flash by propagating from cloud to ground through virgin air in a series of discrete steps. Photographically observed leader steps in clear air are typically 1 μs in duration and tens of meters in length, with a pause time between steps of about 50 μs. A fully developed stepped leader lowers up to 10 or more coulombs of negative cloud charge toward ground in tens of milliseconds with an average downward speed of about 2×10^5 m/s. The average leader current is in the 100 to 1000 A range. The steps have pulse currents of at least 1 kA. Associated with these currents are electric- and magnetic-field pulses with widths of about 1 μs or less and risetimes of about 0.1 μs or less. The stepped leader, during its trip toward ground, branches in a downward direction, resulting in the characteristic downward-branched geometrical structure commonly observed. The electric potential of the bottom of the negatively charged leader channel with respect to ground has a magnitude in excess of 10^7 V. As the leader tip nears ground, the electric field at sharp objects on the ground or at irregularities of the ground itself exceeds the breakdown value of air, and one or more upward-moving discharges (often called upward leaders) are initiated from those points, thus beginning the *attachment process*. An understanding of the physics of the attachment process is central to an understanding of the operation of lightning protection of ground-based objects and the effects of lightning on humans and animals, since it is the attachment process that determines where the lightning connects to objects on the ground and the value of the early currents which flow. When one of the upward-moving discharges from the ground (or from a lightning rod or an individual) contacts the tip of the downward-moving stepped leader, typically some tens of meters above the ground, the leader tip is effectively connected to ground potential. The negatively charged leader channel is then discharged to earth when a ground potential wave, referred to as the first *return stroke*, propagates continuously up the leader path. The upward speed of a return stroke near the ground is typically near one third the speed of light, and the speed decreases with height. The first return stroke produces a peak current near ground of typically 30 kA, with a time from zero to peak of a few microseconds. Currents measured at the ground fall to half of the peak value in about 50 μs, and currents of the order of hundreds of amperes may flow for times of a few milliseconds up to several hundred milliseconds. The longer-lasting currents are known as *continuing currents*. The rapid release of return stroke energy heats the leader channel to a temperature near 30,000 K and creates a high-pressure channel which expands and generates the shock waves that eventually become thunder, as further discussed later. The return stroke effectively lowers to ground the charge originally deposited onto the stepped-leader channel and additionally initiates the lowering of other charges that may be available to the top of its channel. First return-stroke electric fields exhibit a microsecond scale rise to peak with a typical peak value of 5 V/m, normalized to a distance of 100 km by an inverse distance relationship. Roughly half of the field rise to peak, the so-called "fast transition," takes place in tenths of a microsecond, an observation that can only be made if the field propagation is over a highly conducting surface such as salt water.

After the first return-stroke current has ceased to flow, the flash, including charge motion in the cloud, may end. The lightning is then called a single-stroke flash. On the other hand, if additional charge is made available to the top of the channel, a continuous or *dart leader* may propagate down the residual first-stroke channel at a typical speed of about 1×10^7 m/s. The dart leader lowers a charge of the order of 1 C by virtue of a current of about 1 kA. The dart leader then initiates the second (or any subsequent) return stroke. Subsequent return-stroke currents generally have faster zero-to-peak rise times than do first-stroke currents, but similar maximum rates of change, about 100 kA/μs. Some leaders begin as dart leaders, but toward the end of their trip toward ground become stepped leaders. These leaders are known as *dart-stepped leaders* and may have different ground termination points (and separate upward leaders) from the first stroke. Most often the dart-stepped leaders are associated with the second stroke of the flash. Nearly half of all flashes exhibit more than one termination point on ground with the distance between separate terminations being up to several kilometers. Subsequent return-stroke radiated electric and magnetic fields are similar to, but usually a factor of two or so smaller than first return-stroke fields. About one third of all multiple-stroke flashes has at least one subsequent stroke that is larger than the first stroke.

Cloud-to-ground flashes that lower positive charge, though not common, are of considerable practical interest because their peak currents and total charge transfer can be much larger than for the more common negative ground flash. The largest recorded peak currents, those in the 200- to 300-kA range, are due to the return strokes of positive lightning. Such positive flashes to ground are initiated by downward-moving leaders that do not exhibit the distinct steps of their negative counterparts. Rather, they show a luminosity that is more or less continuous but modulated in intensity. Positive flashes are generally composed of a single stroke followed by a period of continuing current. Positive flashes are probably initiated from the upper positive charge in the thundercloud charge dipole when that cloud charge is horizontally separated from the negative charge beneath it, the source of the usual negative cloud-to-ground lightning. Positive flashes are relatively common in winter thunderstorms (snow storms), which produce few flashes overall, and are relatively uncommon in summer thunderstorms. The fraction of positive lightning in summer thunderstorms apparently increases with increasing latitude and with increasing height of the ground above sea level.

Distant lightning return stroke fields are often referred to as sferics (called "atmospherics" in the older literature). The peak in the sferics frequency spectrum is near 5 kHz due to the bipolar or ringing nature of the distant return-stroke electromagnetic signal and to the effects of propagation.

Thunder, the acoustic radiation associated with lightning, is sometimes divided into the categories "audible," sounds that one can hear, and "infrasonic," below a few tens of hertz, a frequency range that is inaudible. This division is made because it is thought that the mechanisms that produce audible and infrasonic thunder are different. Audible thunder is thought to be due to the expansion of a rapidly heated return stroke channel, as noted earlier, whereas infrasonic thunder is thought to be associated with the conversion to sound of the energy stored in the electrostatic field of the thundercloud when lightning rapidly reduces that cloud field.

The technology of artificially initiating lightning by firing upward small rocket trailing grounded wire of a few hundred meters length has been well-developed during the past decade. Such "triggered" flashes are similar to natural upward-initiated discharges from tall structure. They often contain subsequent strokes which,

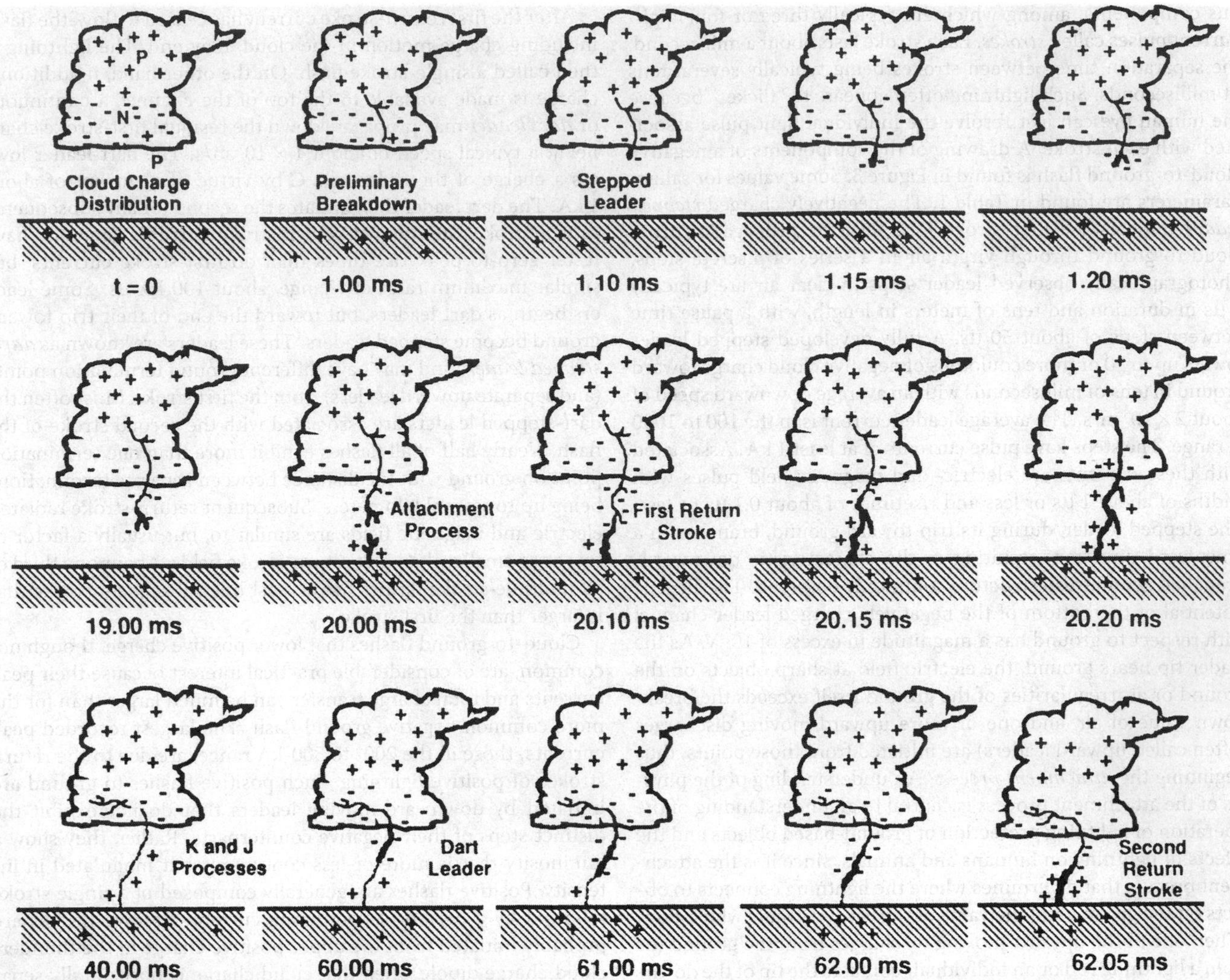

FIGURE 3. Sequence of steps in cloud-to-ground lightning.

when they occur, are similar to the subsequent strokes in natural lightning. These triggered subsequent strokes have been the subject of considerable recent research.

Also in the past 10 years or so sophisticated lightning locating equipment has been installed throughout the world. For example, all ground flashes in the U.S. are now centrally monitored for research, for better overall weather prediction, and for hazard warning for aviation, electric utilities and other lightning-sensitive facilities.

Information on lightning physics can be found in M. A. Uman, *The Lightning Discharge*, Academic Press, San Diego, 1987; on lightning death and injury in *Medical Aspects of Lightning Injury*, C. Andrews, M. A. Cooper, M. Darveniza, and D. Mackerras, Eds., CRC Press, Boca Raton, FL, 1992. Ground flash location information for the U.S., in real time or archived, is available from Geomet Data Service of Tucson, AZ, which is also a source of the names of providers of those data in other countries.

Table 2 has data for cloud-to-ground lightning discharges bringing negative charge to earth. The values listed are intended to convey a rough feeling for the various physical parameters of lightning. No great accuracy is claimed since the results of different investigators are often not in good agreement. These values may, in fact, depend on the particular environment in which the lightning discharge is generated. The choice of some of the entries in the table is arbitrary.

TABLE 2. Data for Cloud-to-Ground Lightning Discharges

	Minimum[a]	Representative values	Maximum[a]
Stepped leader			
Length of step, m	3	50	200
Time interval between steps, μs	30	50	125
Average speed of propagation of stepped leader, m/s[b]	1.0×10^5	2.0×10^5	3.0×10^6
Charged deposited on stepped-leader channel, coulombs	3	5	20
Dart leader			
Speed of propagation, m/s[b]	1.0×10^6	1.0×10^7	2.4×10^7
Charged deposited on dart-leader channel, coulombs	0.2	1	6
Return stroke[c]			
Speed of propagation, m/s[b]	2.0×10^7	1.0×10^8	2.0×10^8
Maximum current rate of increase, kA/μs	<1	100	400
Time to peak current, μs	<1	2	30
Peak current, kA	2	30	200
Time to half of peak current, μs	10	50	250
Charge transferred excluding continuing current, coulombs	0.02	3	20
Channel length, km	2	5	15
Lightning flash			
Number of strokes per flash	1	4	26
Time interval between strokes in absence of continuing current, ms	3	60	100
Time duration of flash, s	10^{-2}	0.5	2
Charge transferred including continuing current, coulombs	3	30	200

[a] The words maximum and minimum are used in the sense that most measured values fall between these limits.
[b] Speeds of propagation are generally determined from photographic data and are "two-dimensional." Since many lightning flashes are not vertical, values stated are probably slight underestimates of actual values.
[c] First return strokes have longer times to current peak and generally larger charge transfer than do subsequent return strokes.

Adapted from Uman, M.A., *Lightning*, Dover Paperbook, New York, 1986, and Uman, M.A., *The Lightning Discharge*, Academic Press, San Diego, 1987.

SPEED OF SOUND IN VARIOUS MEDIA

The speed of sound in various solids, liquids, and gases is given in these tables. While only a single parameter v is needed for liquids and gases, sound propagation in isotropic solids is characterized by three velocity parameters. For a solid of infinite extent (or of finite extent if all dimensions are much larger than a wavelength), there are two relevant quantities,

v_l: velocity of longitudinal waves
v_s: velocity of shear waves.

For a cylindrical rod with diameter much smaller than a wavelength,

v_{ext}: velocity of extensional waves along the rod. (Torsional waves in the rod are propagated at the same speed as sheer waves in an infinite solid.)

Table 1 lists values for a variety of solid materials. Table 2 covers liquids and gases; values for cryogenic liquids are given at the normal boiling point. Table 3 gives the speed of sound in pure water and in seawater of salinity $S = 3.5\%$ as a function of temperature. All values are in meters per second and are given for normal atmospheric pressure.

References

1. Gray, D. E., Ed., *American Institute of Physics Handbook, Third Edition*, McGraw Hill, New York, 1972.
2. Anderson, H.L., Ed., *A Physicist's Desk Reference*, American Institute of Physics, New York, 1989.
3. Younglove, B. A., Thermophysical Proeprties of Fluids. Part I, *J. Phys. Chem. Ref. Data*, 11, Suppl. 1, 1982.
4. Younglove, B. A., and Ely, J. F., Thermophysical Properties of Fluids. Part II, *J. Phys. Chem. Ref. Data*, 16, 577, 1987.
5. Haar, L., Gallagher, J. S., and Kell, G. S., *NBS/NRC Steam Tables*, Hemisphere Publishing Corp., New York, 1984.
6. Mason, W. P., *Physical Acoustics and the Properties of Solids*, D. Van Nostrand Co., Princeton, N.J., 1958.
7. *Landolt-Börnstein, Numerical Data and Functional Relationships in Science and Technology, New Series, II/5, Molecular Acoustics*, Springer-Verlag, Heidelberg, 1967.

TABLE 1. Speed of Sound in Solids at Room Temperature

Name	v_l/m s^{-1}	v_s/m s^{-1}	v_{ext}/m s^{-1}
Metals			
Aluminum, rolled	6420	3040	5000
Beryllium	12890	8880	12870
Brass (70 Cu, 30 Zn)	4700	2110	3480
Constantan	5177	2625	4270
Copper, annealed	4760	2325	3810
Copper, rolled	5010	2270	3750
Duralumin 17S	6320	3130	5150
Gold, hard-drawn	3240	1200	2030
Iron, cast	4994	2809	4480
Iron, electrolytic	5950	3240	5120
Iron, Armco	5960	3240	5200
Lead, annealed	2160	700	1190
Lead, rolled	1960	690	1210
Magnesium, annealed	5770	3050	4940
Molybdenum	6250	3350	5400
Monel metal	5350	2720	4400
Nickel	6040	3000	4900
Platinum	3260	1730	2800
Silver	3650	1610	2680
Steel (1% C)	5940	3220	5180
Steel, 347 Stainless	5790	3100	5000
Steel, K9	5940	3250	5250
Tin, rolled	3320	1670	2730
Titanium	6070	3125	5090
Tungsten, annealed	5220	2890	4620
Tungsten, drawn	5410	2640	4320
Zinc, rolled	4210	2440	3850
Other materials			
Fused silica	5968	3764	5760
Glass, heavy silicate flint	3980	2380	3720
Glass, light borate crown	5100	2840	4540
Glass, pyrex	5640	3280	5170
Lucite	2680	1100	1840
Nylon 6-6	2620	1070	1800
Polyethylene	1950	540	920
Polystyrene	2350	1120	1840
Rubber, butyl	1830		
Rubber, gum	1550		
Rubber, neoprene	1600		
Tungsten carbide	6655	3980	6220

TABLE 2. Speed of Sound in Liquids and Gases

Name	t/°C	v/m s⁻¹
Liquids		
Acetone	20	1203
Argon	−185.9	813
Benzene	25	1310
Bromobenzene	20	1169
Butane	−0.5	1034
1-Butanol	20	1258
Carbon disulphide	25	1140
Chlorobenzene	20	1311
Cyclohexane	19	1280
1-Decene	20	1250
Diethyl ether	25	976
Ethane	−88.6	1326
Ethanol	20	1162
Ethylene	−103.8	1309
Ethylene glycol	25	1658
Fluorobenzene	20	1183
Glycerol	25	1904
Helium	−268.9	180
Heptane	20	1162
1-Heptene	20	1128
Hexane	20	1083
Hydrogen	−252.9	1101
Iodobenzene	20	1114
Mercury	25	1450
Methane	−161.5	1337
Methanol	20	1121
Nitrobenzene	25	1463
Nitrogen	−195.8	939
1-Nonene	20	1218
Octane	20	1197
1-Octene	20	1184
Oxygen	−183.0	906
1-Pentadecene	20	1351
Pentane	20	1008
Propane	−42.1	1158
1-Propanol	20	1223
Tetrachloromethane	25	930
Trichloromethane	25	987
1-Undecene	20	1275
Water	25	1497
Water (sea, *S* = 3.5%)	25	1535
Gases at 1 atm		
Air, dry	25	346
Ammonia	0	415
Argon	27	323
Carbon monoxide	0	338
Carbon dioxide	0	259
Chlorine	0	206
Deuterium	0	890
Ethane	27	312
Ethylene	27	331
Helium	0	965
Hydrogen	27	1310
Hydrogen bromide	0	200
Hydrogen chloride	0	296
Hydrogen iodide	0	157
Hydrogen sulfide	0	289
Methane	27	450
Neon	0	435
Nitric oxide	10	325
Nitrogen	27	353
Nitrous oxide	0	263
Oxygen	27	330
Sulfur dioxide	0	213
Water (steam)	100	473

TABLE 3. Speed of Sound in Water and Seawater (S = 3.5%) at Different Temperatures

	v/m s⁻¹	
t/°C	Water	Seawater
0	1401.0	1449.4
10	1447.8	1490.4
20	1483.2	1522.2
25	1497.4	1535.1
30	1509.5	1546.2
40	1528.4	
50	1541.4	
60	1549.5	
70	1553.2	
80	1552.8	

ATTENUATION AND SPEED OF SOUND IN AIR AS A FUNCTION OF HUMIDITY AND FREQUENCY

This table gives the attenuation and speed of sound as a function of frequency at various values of relative humidity. All values refer to still air at 20 °C.

References

1. Tables of Absorption and Velocity of Sound in Still Air at 68 °F (20 °C), AD-738576, National Technical Information Service, Springfield, VA.
2. Evans, L. B., Bass, H. E., and Sutherland, L. C., *J. Acoust. Soc. Am.*, 51, 1565, 1972.

Frequency (Hz)	Attenuation (dB/km)	Speed (m/s)
Relative humidity 0%		
20	0.51	343.477
40	1.07	343.514
50	1.26	343.525
63	1.43	343.536
100	1.67	343.550
200	1.84	343.559
400	1.96	343.561
630	2.11	343.562
800	2.27	343.562
1250	2.82	343.562
2000	4.14	343.562
4000	8.84	343.564
6300	14.89	343.565
10000	26.28	343.566
12500	35.81	343.566
16000	52.15	343.567
20000	75.37	343.567
40000	267.01	343.567
63000	644.66	343.567
80000	1032.14	343.567
Relative humidity 30%		
20	0.03	343.807
40	0.11	343.808
50	0.17	343.810
63	0.25	343.810
100	0.50	343.814
200	1.01	343.821
400	1.59	343.826
630	2.24	343.827
800	2.85	343.828
1250	5.09	343.828
2000	10.93	343.829
4000	38.89	343.831
6300	90.61	343.836
10000	204.98	343.846
12500	294.08	343.854
16000	422.51	343.865
20000	563.66	343.877
40000	1110.97	343.911
63000	1639.47	343.924
80000	2083.08	343.929
Relative humidity 60%		
20	0.02	344.182
40	0.06	344.183
50	0.09	344.183
63	0.15	344.184
100	0.34	344.185
200	0.99	344.190
400	1.94	344.197
630	2.57	344.200
800	2.94	344.201
1250	4.01	344.202
2000	6.55	344.203
4000	18.73	344.204
6300	42.51	344.204
10000	101.84	344.206
12500	155.67	344.208
16000	247.78	344.211
20000	373.78	344.215
40000	1195.37	344.238
63000	2220.64	344.262
80000	2951.71	344.274
Relative humidity 100%		
20	0.01	344.685
40	0.04	344.685
50	0.06	344.685
63	0.09	344.685
100	0.22	344.686
200	0.77	344.689
400	2.02	344.695
630	3.05	344.699
800	3.57	344.701
1250	4.59	344.704
2000	6.29	344.705
4000	13.58	344.706
6300	27.72	344.706
10000	63.49	344.706
12500	96.63	344.707
16000	154.90	344.708
20000	237.93	344.709
40000	884.28	344.718
63000	1973.62	344.731
80000	2913.01	344.742

SPEED OF SOUND IN DRY AIR

Eric W. Lemmon

These values were calculated from the equation of state for dry air (average molecular weight 28.96) treated as a real gas. Values refer to standard atmospheric pressure. The speed of sound varies only slightly with pressure; at two atmospheres and −100 °C the value decreases by 0.16%, while at two atmospheres and 80 °C the speed increases by 0.05%. For additional values, see the table in Section 6 labeled "Thermophysical Properties of Air."

Reference

Lemmon, E.W., Jacobsen, R.T, Penoncello, S.G., and Friend, D.G., Thermodynamic Properties of Air and Mixtures of Nitrogen, Argon, and Oxygen from 60 to 2000 K at Pressures to 2000 MPa, *J. Phys. Chem. Ref. Data* 29, 331, 2000.

t/°C	v_s/m s^{-1}	t/°C	v_s/m s^{-1}	t/°C	v_s/m s^{-1}
−100	263.5	−30	312.7	40	354.9
−95	267.3	−25	315.9	45	357.7
−90	271.1	−20	319.1	50	360.4
−85	274.8	−15	322.2	55	363.2
−80	278.5	−10	325.4	60	365.9
−75	282.1	−5	328.4	65	368.7
−70	285.7	0	331.5	70	371.3
−65	289.2	5	334.5	75	374.0
−60	292.7	10	337.5	80	376.7
−55	296.1	15	340.5	85	379.3
−50	299.5	20	343.4	90	381.9
−45	302.9	25	346.3	95	384.5
−40	306.2	30	349.2	100	387.0
−35	309.5	35	352.0		

MUSICAL SCALES

Equal Tempered Chromatic Scale

A_4 = 440 Hz

International Concert Pitch

Note	Frequency	Note	Frequency	Note	Frequency	Note	Frequency
C_0	16.35	C_2	65.41	C_4	261.63	C_6	1046.50
$C\#_0$	17.32	$C\#_2$	69.30	$C\#_4$	277.18	$C\#_6$	1108.73
D_0	18.35	D_2	73.42	D_4	293.66	D_6	1174.66
$D\#_0$	19.45	$D\#_2$	77.78	$D\#_4$	311.13	$D\#_6$	1244.51
E_0	20.60	E_2	82.41	E_4	329.63	E_6	1318.51
F_0	21.83	F_2	87.31	F_4	349.23	F_6	1396.91
$F\#_0$	23.12	$F\#_2$	92.50	$F\#_4$	369.99	$F\#_6$	1479.98
G_0	24.50	G_2	98.00	G_4	392.00	G_6	1567.98
$G\#_0$	25.96	$G\#_2$	103.83	$G\#_4$	415.30	$G\#_6$	1661.22
A_0	27.50	A_2	110.00	A_4	440.00	A_6	1760.00
$A\#_0$	29.14	$A\#_2$	116.54	$A\#_4$	466.16	$A\#_6$	1864.66
B_0	30.87	B_2	123.47	B_4	493.88	B_6	1975.53
C_1	32.70	C_3	130.81	C_5	523.25	C_7	2093.00
$C\#_1$	34.65	$C\#_3$	138.59	$C\#_5$	554.37	$C\#_7$	2217.46
D_1	36.71	D_3	146.83	D_5	587.33	D_7	2349.32
$D\#_1$	38.89	$D\#_3$	155.56	$D\#_5$	622.25	$D\#_7$	2489.02
E_1	41.20	E_3	164.81	E_5	659.26	E_7	2637.02
F_1	43.65	F_3	174.61	F_5	698.46	F_7	2793.83
$F\#_1$	46.25	$F\#_3$	185.00	$F\#_5$	739.99	$F\#_7$	2959.96
G_1	49.00	G_3	196.00	G_5	783.99	G_7	3135.96
$G\#_1$	51.91	$G\#_3$	207.65	$G\#_5$	830.61	$G\#_7$	3322.44
A_1	55.00	A_3	220.00	A_5	880.00	A_7	3520.00
$A\#_1$	58.27	$A\#_3$	233.08	$A\#_5$	932.33	$A\#_7$	3729.31
B_1	61.74	B_3	246.94	B_5	987.77	B_7	3951.07
						C_8	4186.01

Scientific or Just Scale

C_4 = 256 Hz

Note	Frequency	Note	Frequency	Note	Frequency	Note	Frequency
C_0	16	C_2	64	C_4	256	C_6	1024
D_0	18	D_2	72	D_4	288	D_6	1152
E_0	20	E_2	80	E_4	320	E_6	1280
F_0	21.33	F_2	85.33	F_4	341.33	F_6	1365.33
G_0	24	G_2	96	G_4	384	G_6	1536
A_0	26.67	A_2	106.67	A_4	426.67	A_6	1706.67
B_0	30	B_2	120	B_4	480	B_6	1920
C_1	32	C_3	128	C_5	512	C_7	2048
D_1	36	D_3	144	D_5	576	D_7	2304
E_1	40	E_3	160	E_5	640	E_7	2560
F_1	42.67	F_3	170.67	F_5	682.67	F_7	2730.67
G_1	48	G_3	192	G_5	768	G_7	3072
A_1	53.33	A_3	213.33	A_5	853.33	A_7	3413.33
B_1	60	B_3	240	B_5	960	B_7	3840
						C_8	4096

CHARACTERISTICS OF HUMAN HEARING

The human ear is sensitive to sound waves with frequencies in the range from a few hertz to almost 20 kHz. Auditory response is usually expressed in terms of the *loudness level* of a sound, which is a measure of the sound pressure. The reference level, which is given in the unit *phon,* is a pure tone of frequency 1000 Hz with sound pressure of 20 μPa (in cgs units, $2 \cdot 10^{-4}$ dyn/cm^2); loudness level is usually expressed in decibels (dB) relative to this reference level. If a normal observer perceives an arbitrary sound to be equally loud as this reference sound, the sound is said to have the loudness level of the reference. The sensitivity of the typical human ear ranges from about 0 dB, the threshold loudness level, to about 140 dB, the level at which pain sets in. The minimum detectable level thus represents a sound wave of pressure 20 μPa and intensity (power density) 10^{-16} W/cm^2.

The following figure illustrates the frequency dependence of the threshold for an average young adult.

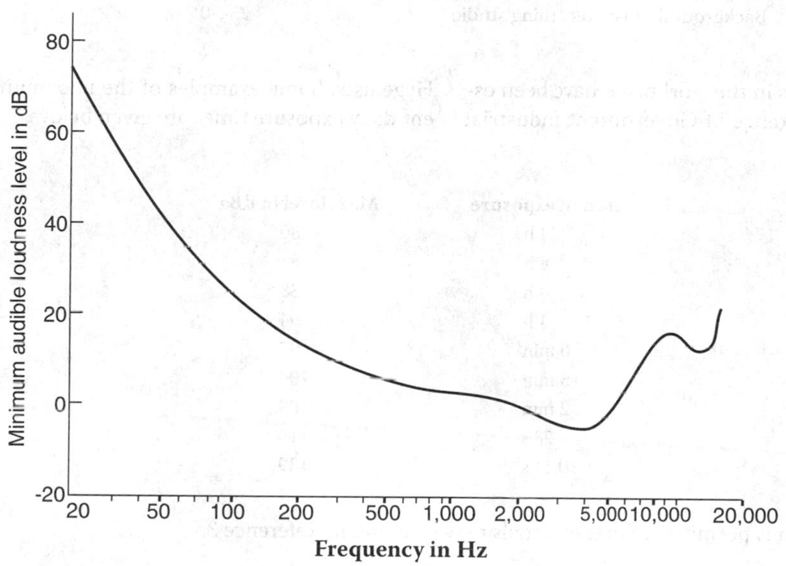

The relation between loudness level and frequency for a typical person is expressed by the following table:

Sound pressure level in dB relative to 20 μPa	Frequency in Hz 125	500	1000	4000	8000	10000
10			10	18		
20		16	20	28	11	
30	4	27	30	37	21	17
40	17	39	40	45	30	26
50	34	52	50	54	38	35
60	52	65	60	64	47	44
70	70	76	70	73	56	54
80	86	86	80	83	66	64
90	98	96	90	94	77	74
100	108	105	100	106	88	86

Thus, a 10,000 Hz tone at a pressure level of 50 dB seems equally loud as a 1000 Hz tone at a pressure of 35 dB.

The term *noise* refers to any unwanted sound, either a pure tone or a mixture of frequencies. Since the sensitivity of the ear is frequency dependent, as illustrated by the above table, noise level is expressed in a frequency-weighted scale, known as A-weighting. Decibel readings on this scale are designated as dBa. Typical noise levels from various sources are illustrated in this table:

Source	Noise level in dBa
Rocket engine	200
Jet aircraft engine	160
Light aircraft, cruising	140
Tractor, 150 hp	115
Electric motor, 100 hp at 2600 rpm	105
Pneumatic drill	100
Subway train	90
Vacuum cleaner	85
Heavy automobile traffic	75
Conversational speech	65
Whispered speech	40
Background noise, recording studio	25-30

Recommended noise thresholds in the workplace have been established by the American Conference of Government Industrial Hygenists. Some examples of the maximum safe levels for different daily exposure times are given below.

Duration of exposure	Max. level in dBa
24 h	80
8 h	85
4 h	88
1 h	94
30 min	97
15 min	100
2 min	109
28 s	115
0.11 s	139

No exposure greater than 140 dBa is permitted. Further details may be found in Reference 3.

References

1. Anderson, H. L., Ed., *A Physicist's Desk Reference*, American Institute of Physics, New York, 1989, chap. 2.
2. Gray, D. E., Ed., *American Institute of Physics Handbook, Third Edition*, McGraw Hill, New York, 1972, chap. 3.
3. *Threshold Limit Values for Chemical Substances and Physical Agents; Biological Exposure Indices*, 2008 Edition, American Conference of Governmental Industrial Hygienists, 1330 Kemper Meadow Drive, Cincinnati, OH 45240-1634; <www.acgih.org>.

Section 15
Practical Laboratory Data

Standard ITS-90 Thermocouple Tables ... 15-1
Secondary Reference Points on the ITS-90 Temperature Scale ... 15-10
Relative Sensitivity of Bayard-Alpert Ionization Gauges to Various Gases ... 15-12
Laboratory Solvents and Other Liquid Reagents ... 15-13
Miscibility of Organic Solvents ... 15-23
Density of Solvents as a Function of Temperature ... 15-25
Dependence of Boiling Point on Pressure ... 15-26
Ebullioscopic Constants for Calculation of Boiling Point Elevation ... 15-27
Cryoscopic Constants for Calculation of Freezing Point Depression ... 15-28
Freezing Point Lowering by Electrolytes in Aqueous Solution ... 15-29
Correction of Barometer Readings to 0 °C Temperature ... 15-30
Determination of Relative Humidity from Dew Point ... 15-31
Determination of Relative Humidity from Wet and Dry Bulb Temperatures ... 15-32
Constant Humidity Solutions ... 15-33
Standard Salt Solutions for Humidity Calibration ... 15-34
Low-Temperature Baths for Maintaining Constant Temperature ... 15-35
Metals and Alloys with Low Melting Temperature ... 15-36
Wire Tables ... 15-37
Characteristics of Particles and Particle Dispersoids ... 15-38
Density of Various Solids ... 15-39
Density of Sulfuric Acid ... 15-40
Density of Ethanol–Water Mixtures ... 15-41
Dielectric Strength of Insulating Materials ... 15-42
Coefficient of Friction ... 15-47
Flame Temperatures ... 15-49
Allocation of Frequencies in the Radio Spectrum ... 15-50

STANDARD ITS-90 THERMOCOUPLE TABLES

The Instrument Society of America (ISA) has assigned standard letter designations to a number of thermocouple types having specified emf-temperature relations. These designations and the approximate metal compositions that meet the required relations, as well as the useful temperature ranges, are given below:

Type B	(Pt + 30% Rh) vs. (Pt + 6% Rh)	0 to 1820 °C
Type E	(Ni + 10% Cr) vs. (Cu + 43% Ni)	−270 to 1000 °C
Type J	Fe vs. (Cu + 43% Ni)	−210 to 1200 °C
Type K	(Ni + 10% Cr) vs. (Ni + 2% Al + 2% Mn + 1% Si)	−270 to 1372 °C
Type N	(Ni + 14% Cr + 1.5% Si) vs. (Ni + 4.5% Si + 0.1% Mg)	−270 to 1300 °C
Type R	(Pt + 13% Rh) vs. Pt	−50 to 1768 °C
Type S	(Pt + 10% Rh) vs. Pt	−50 to 1768 °C
Type T	Cu vs. (Cu + 43% Ni)	−270 to 400 °C

The compositions are given in weight percent, and the positive leg is listed first. It should be emphasized that the standard letter designations do not imply a precise composition but rather that the specified emf-temperature relation is satisfied.

The first set of tables below lists, for each thermocouple type, the emf as a function of temperature on the International Temperature Scale of 1990 (ITS-90). The coefficients in the equation used to generate the table are also given. The second set of tables gives the inverse relationships, i.e., the coefficients in the polynomial equation that expresses the temperature as a function of thermocouple emf. The accuracy of these equations is also stated.

Further details and tables at closer intervals may be found in Reference 1.

References

1. Burns, G. W., Seroger, M. G., Strouse, G. F., Croarkin, M. C., and Guthrie, W. F., *Temperature-Electromotive Force Reference Functions and Tables for the Letter-Designated Thermocouple Types Based on the ITS-90*, Natl. Inst. Stand. Tech. (U.S.) Monogr. 175, 1993.
2. Schooley, J. F., *Thermometry*, CRC Press, Boca Raton, FL, 1986.

Type B Thermocouples: emf-Temperature (°C) Reference Table and Equations

Thermocouple emf as a Function of Temperature in Degrees Celsius (ITS-90)

emf in Millivolts — **Reference junctions at 0 °C**

°C	0	10	20	30	40	50	60	70	80	90	100
0	0.000	−0.002	−0.003	−0.002	−0.000	0.002	0.006	0.011	0.017	0.025	0.033
100	0.033	0.043	0.053	0.065	0.078	0.092	0.107	0.123	0.141	0.159	0.178
200	0.178	0.199	0.220	0.243	0.267	0.291	0.317	0.344	0.372	0.401	0.431
300	0.431	0.462	0.494	0.527	0.561	0.596	0.632	0.669	0.707	0.746	0.787
400	0.787	0.828	0.870	0.913	0.957	1.002	1.048	1.095	1.143	1.192	1.242
500	1.242	1.293	1.344	1.397	1.451	1.505	1.561	1.617	1.675	1.733	1.792
600	1.792	1.852	1.913	1.975	2.037	2.101	2.165	2.230	2.296	2.363	2.431
700	2.431	2.499	2.569	2.639	2.710	2.782	2.854	2.928	3.002	3.078	3.154
800	3.154	3.230	3.308	3.386	3.466	3.546	3.626	3.708	3.790	3.873	3.957
900	3.957	4.041	4.127	4.213	4.299	4.387	4.475	4.564	4.653	4.743	4.834
1000	4.834	4.926	5.018	5.111	5.205	5.299	5.394	5.489	5.585	5.682	5.780
1100	5.780	5.878	5.976	6.075	6.175	6.276	6.377	6.478	6.580	6.683	6.786
1200	6.786	6.890	6.995	7.100	7.205	7.311	7.417	7.524	7.632	7.740	7.848
1300	7.848	7.957	8.066	8.176	8.286	8.397	8.508	8.620	8.731	8.844	8.956
1400	8.956	9.069	9.182	9.296	9.410	9.524	9.639	9.753	9.868	9.984	10.099
1500	10.099	10.215	10.331	10.447	10.563	10.679	10.796	10.913	11.029	11.146	11.263
1600	11.263	11.380	11.497	11.614	11.731	11.848	11.965	12.082	12.199	12.316	12.433
1700	12.433	12.549	12.666	12.782	12.898	13.014	13.130	13.246	13.361	13.476	13.591
1800	13.591	13.706	13.820								

NOTE—Temperature ranges and coefficients of equations used to compute the above table: The equations are of the form: $E = c_0 + c_1t + c_2t^2 + c_3t^3 + \ldots c_nt^n$, where E is the emf in millivolts, t is the temperature in degrees Celsius (ITS- 90), and c_0, c_1, c_2, c_3, etc. are the coefficients. These coefficients are extracted from Reference 1.

		0 °C to 630.615 °C	630.615 °C to 1820 °C
c_0	=	0.000 000 000 0	−3.893 816 862 1 …
c_1	=	− 2.465 081 834 6 × 10^{-4}	2.857 174 747 0 × 10^{-2}
c_2	=	5.904 042 117 1 × 10^{-6}	−8.488 510 478 5 × 10^{-5}
c_3	=	− 1.325 793 163 6 × 10^{-9}	1.578 528 016 4 × 10^{-7}
c_4	=	1.566 829 190 1 × 10^{-12}	−1.683 534 486 4 × 10^{-10}
c_5	=	− 1.694 452 924 0 × 10^{-15}	1.110 979 401 3 × 10^{-13}
c_6	=	6.299 034 709 4 × 10^{-19}	−4.451 543 103 3 × 10^{-17}
c_7	=	…………	9.897 564 082 1 × 10^{-21}
c_8	=	…………	−9.379 133 028 9 × 10^{-25}

Type E Thermocouples: emf-Temperature (°C) Reference Table and Equations

Thermocouple emf as a Function of Temperature in Degrees Celsius (ITS-90)

emf in Millivolts — Reference junctions at 0 °C

°C	0	-10	-20	-30	-40	-50	-60	-70	-80	-90	-100
-200	-8.825	-9.063	-9.274	-9.455	-9.604	-9.718	-9.797	-9.835			
-100	-5.237	-5.681	-6.107	-6.516	-6.907	-7.279	-7.632	-7.963	-8.273	-8.561	-8.825
0	0.000	-0.582	-1.152	-1.709	-2.255	-2.787	-3.306	-3.811	-4.302	-4.777	-5.237
°C	0	10	20	30	40	50	60	70	80	90	100
0	0.000	0.591	1.192	1.801	2.420	3.048	3.685	4.330	4.985	5.648	6.319
100	6.319	6.998	7.685	8.379	9.081	9.789	10.503	11.224	11.951	12.684	13.421
200	13.421	14.164	14.912	15.664	16.420	17.181	17.945	18.713	19.484	20.259	21.036
300	21.036	21.817	22.600	23.386	24.174	24.964	25.757	26.552	27.348	28.146	28.946
400	28.946	29.747	30.550	31.354	32.159	32.965	33.772	34.579	35.387	36.196	37.005
500	37.005	37.815	38.624	39.434	40.243	41.053	41.862	42.671	43.479	44.286	45.093
600	45.093	45.900	46.705	47.509	48.313	49.116	49.917	50.718	51.517	52.315	53.112
700	53.112	53.908	54.703	55.497	56.289	57.080	57.870	58.659	59.446	60.232	61.017
800	61.017	61.801	62.583	63.364	64.144	64.922	65.698	66.473	67.246	68.017	68.787
900	68.787	69.554	70.319	71.082	71.844	72.603	73.360	74.115	74.869	75.621	76.373
1000	76.373										

NOTE—Temperature ranges and coefficients of equations used to compute the above table: The equations are of the form: $E = c_0 + c_1t + c_2t^2 + c_3t^3 + \ldots c_nt^n$, where E is the emf in millivolts, t is the temperature in degrees Celsius (ITS- 90), and c_0, c_1, c_2, c_3, etc. are the coefficients. These coefficients are extracted from Reference 1.

		−270 to 0 °C	0 °C to 1000 °C
c_0	=	0.000 000 000 0 ...	0.000 000 000 0 ...
c_1	=	5.866 550 870 8 × 10^{-2}	5.866 550 871 0 × 10^{-2}
c_2	=	4.541 097 712 4 × 10^{-5}	4.503 227 558 2 × 10^{-5}
c_3	=	−7.799 804 868 6 × 10^{-7}	2.890 840 721 2 × 10^{-8}
c_4	=	−2.580 016 084 3 × 10^{-8}	−3.305 689 665 2 × 10^{-10}
c_5	=	−5.945 258 305 7 × 10^{-10}	6.502 440 327 0 × 10^{-13}
c_6	=	−9.321 405 866 7 × 10^{-12}	−1.919 749 550 4 × 10^{-16}
c_7	=	−1.028 760 553 4 × 10^{-13}	−1.253 660 049 7 × 10^{-18}
c_8	=	−8.037 012 362 1 × 10^{-16}	2.148 921 756 9 × 10^{-21}
c_9	=	−4.397 949 739 1 × 10^{-18}	−1.438 804 178 2 × 10^{-24}
c_{10}	=	−1.641 477 635 5 × 10^{-20}	3.596 089 948 1 × 10^{-28}
c_{11}	=	−3.967 361 951 6 × 10^{-23}	…………
c_{12}	=	−5.582 732 872 1 × 10^{-26}	…………
c_{13}	=	−3.465 784 201 3 × 10^{-29}	…………

Type J Thermocouples: emf-Temperature (°C) Reference Table and Equations

Thermocouple emf as a Function of Temperature in Degrees Celsius (ITS-90)

emf in Millivolts — Reference junctions at 0 °C

°C	0	-10	-20	-30	-40	-50	-60	-70	-80	-90	-100
-200	-7.890	-8.095									
-100	-4.633	-5.037	-5.426	-5.801	-6.159	-6.500	-6.821	-7.123	-7.403	-7.659	-7.890
0	0.000	-0.501	-0.995	-1.482	-1.961	-2.431	-2.893	-3.344	-3.786	-4.215	-4.633

emf in Millivolts — Reference junctions at 0 °C

°C	0	10	20	30	40	50	60	70	80	90	100
0	0.000	0.507	1.019	1.537	2.059	2.585	3.116	3.650	4.187	4.726	5.269
100	5.269	5.814	6.360	6.909	7.459	8.010	8.562	9.115	9.669	10.224	10.779
200	10.779	11.334	11.889	12.445	13.000	13.555	14.110	14.665	15.219	15.773	16.327
300	16.327	16.881	17.434	17.986	18.538	19.090	19.642	20.194	20.745	21.297	21.848
400	21.848	22.400	22.952	23.504	24.057	24.610	25.164	25.720	26.276	26.834	27.393
500	27.393	27.953	28.516	29.080	29.647	30.216	30.788	31.362	31.939	32.519	33.102
600	33.102	33.689	34.279	34.873	35.470	36.071	36.675	37.284	37.896	38.512	39.132
700	39.132	39.755	40.382	41.012	41.645	42.281	42.919	43.559	44.203	44.848	45.494
800	45.494	46.141	46.786	47.431	48.074	48.715	49.353	49.989	50.622	51.251	51.877
900	51.877	52.500	53.119	53.735	54.347	54.956	55.561	56.164	56.763	57.360	57.953
1000	57.953	58.545	59.134	59.721	60.307	60.890	61.473	62.054	62.634	63.214	63.792
1100	63.792	64.370	64.948	65.525	66.102	66.679	67.255	67.831	68.406	68.980	69.553
1200	69.553										

NOTE—Temperature ranges and coefficients of equations used to compute the above table: The equations are of the form: $E = c_0 + c_1t + c_2t^2 + c_3t^3 + \ldots c_nt^n$, where E is the emf in millivolts, t is the temperature in degrees Celsius (ITS-90), and c_0, c_1, c_2, c_3, etc. are the coefficients. These coefficients are extracted from Reference 1.

		−260 °C to 760 °C	760 °C to 1200 °C
c_0	=	0.000 000 000 0 …	$2.964\ 562\ 568\ 1 \times 10^{2}$
c_1	=	$5.038\ 118\ 781\ 5 \times 10^{-2}$	−1.497 612 778 6 …
c_2	=	$3.047\ 583\ 693\ 0 \times 10^{-5}$	3.178 7[illegible] × 10^{-3}
c_3	=	$-8.568\ 106\ 572\ 0 \times 10^{-8}$	−3.184 [illegible]0 1 × 10^{-6}
c_4	=	$1.322\ 819\ 529\ 5 \times 10^{-10}$	1.572 [illegible] 900 4 × 10^{-9}
c_5	=	$-1.705\ 295\ 833\ 7 \times 10^{-13}$	$-3.069\ 136\ 905\ 6 \times 10^{-13}$
c_6	=	$2.094\ 809\ 069\ 7 \times 10^{-16}$	…………
c_7	=	$-1.253\ 839\ 533\ 6 \times 10^{-19}$	…………
c_8	=	$1.563\ 172\ 569\ 7 \times 10^{-23}$	…………

Type K Thermocouples: emf-Temperature (°C) Reference Table and Equations

Thermocouple emf as a Function of Temperature in Degrees Celsius (ITS-90)

emf in Millivolts — Reference junctions at 0 °C

°C	0	-10	-20	-30	-40	-50	-60	-70	-80	-90	-100
-200	-5.891	-6.035	-6.158	-6.262	-6.344	-6.404	-6.441	-6.458			
-100	3.554	-3.852	-4.138	-4.411	-4.669	-4.913	-5.141	-5.354	-5.550	-5.730	-5.891
0	0.000	-0.392	-0.778	-1.156	-1.527	-1.889	-2.243	-2.587	-2.920	-3.243	-3.554
°C	**0**	**10**	**20**	**30**	**40**	**50**	**60**	**70**	**80**	**90**	**100**
0	0.000	0.397	0.798	1.203	1.612	2.023	2.436	2.851	3.267	3.682	4.096
100	4.096	4.509	4.920	5.328	5.735	6.138	6.540	6.941	7.340	7.739	8.138
200	8.138	8.539	8.940	9.343	9.747	10.153	10.561	10.971	11.382	11.795	12.209
300	12.209	12.624	13.040	13.457	13.874	14.293	14.713	15.133	15.554	15.975	16.397
400	16.397	16.820	17.243	17.667	18.091	18.516	18.941	19.366	19.792	20.218	20.644
500	20.644	21.071	21.497	21.924	22.350	22.776	23.203	23.629	24.055	24.480	24.905
600	24.905	25.330	25.755	26.179	26.602	27.025	27.447	27.869	28.289	28.710	29.129
700	29.129	29.548	29.965	30.382	30.798	31.213	31.628	32.041	32.453	32.865	33.275
800	33.275	33.685	34.093	34.501	34.908	35.313	35.718	36.121	36.524	36.925	37.326
900	37.326	37.725	38.124	38.522	38.918	39.314	39.708	40.101	40.494	40.885	41.276
1000	41.276	41.665	42.053	42.440	42.826	43.211	43.595	43.978	44.359	44.740	45.119
1100	45.119	45.497	45.873	46.249	46.623	46.995	47.367	47.737	48.105	48.473	48.838
1200	48.838	49.202	49.565	49.926	50.286	50.644	51.000	51.355	51.708	52.060	52.410
1300	52.410	52.759	53.106	53.451	53.795	54.138	54.479	54.819			

NOTE—Temperature ranges and coefficients of equations used to compute the above table: The equations are of the form $E = c_0 + c_1t + c_2t^2 + c_3t^3 + \ldots c_nt^n$, where E is the emf in millivolts, t is the temperature in degrees Celsius (ITS-90), and c_0, c_1, c_2, c_3, etc. are the coefficients. In the 0 °C to 1372 °C range there is also an exponential term that must be evaluated and added to the equation. The exponential term is of the form $c_0 e^{c_1 (t-126.9686)^2}$, where t is the temperature in °C, e is the natural logarithm base, and c_0 and c_1 are the coefficients. These coefficients are extracted from Reference 1.

		-270 °C to 0 °C	0 °C to 1372 °C	0 °C to 1372 °C (Exponential term)
c_0	=	0.000 000 000 0	$-1.760\,041\,368\,6 \times 10^{-2}$	$1.185\,976 \times 10^{-1}$
c_1	=	$3.945\,012\,802\,5 \times 10^{-2}$	$3.892\,120\,497\,5 \times 10^{-2}$	$-1.183\,432 \times 10^{-4}$
c_2	=	$2.362\,237\,359\,8 \times 10^{-5}$	$1.855\,877\,003\,2 \times 10^{-5}$	....
c_3	=	$-3.285\,890\,678\,4 \times 10^{-7}$	$-9.945\,759\,287\,4 \times 10^{-8}$	....
c_4	=	$-4.990\,482\,877\,7 \times 10^{-9}$	$3.184\,094\,571\,9 \times 10^{-10}$	....
c_5	=	$-6.750\,905\,917\,3 \times 10^{-11}$	$-5.607\,284\,488\,9 \times 10^{-13}$	....
c_6	=	$-5.741\,032\,742\,8 \times 10^{-13}$	$5.607\,505\,905\,9 \times 10^{-16}$	....
c_7	=	$-3.108\,887\,289\,4 \times 10^{-15}$	$-3.202\,072\,000\,3 \times 10^{-19}$	....
c_8	=	$-1.045\,160\,936\,5 \times 10^{-17}$	$9.715\,114\,715\,2 \times 10^{-23}$	....
c_9	=	$-1.988\,926\,687\,8 \times 10^{-20}$	$-1.210\,472\,127\,5 \times 10^{-26}$	....
c_{10}	=	$-1.632\,269\,748\,6 \times 10^{-23}$	......	....

Type N Thermocouples: emf-Temperature (°C) Reference Table and Equations

Thermocouple emf as a Function of Temperature in Degrees Celsius (ITS-90)

emf in Millivolts — Reference junctions at 0 °C

°C	0	-10	-20	-30	-40	-50	-60	-70	-80	-90	-100
-200	3.990	-4.083	-4.162	-4.226	-4.277	-4.313	-4.336	-4.345			
-100	-2.407	-2.612	2.808	2.994	3.171	3.336	-3.491	3.634	-3.766	-3.884	-3.990
0	0.000	-0.260	-0.518	-0.772	-1.023	-1.269	-1.509	1.744	-1.972	-2.193	-2.407
°C	**0**	**10**	**20**	**30**	**40**	**50**	**60**	**70**	**80**	**90**	**100**
0	0.000	0.261	0.525	0.793	1.065	1.340	1.619	1.902	2.189	2.480	2.774
100	2.774	3.072	3.374	3.680	3.989	4.302	4.618	4.937	5.259	5.585	5.913
200	5.913	6.245	6.579	6.916	7.255	7.597	7.941	8.288	8.637	8.988	9.341
300	9.341	9.696	10.054	10.413	10.774	11.136	11.501	11.867	12.234	12.603	12.974
400	12.974	13.346	13.719	14.094	14.469	14.846	15.225	15.604	15.984	16.366	16.748
500	16.748	17.131	17.515	17.900	18.286	18.672	19.059	19.447	19.835	20.224	20.613
600	20.613	21.003	21.393	21.784	22.175	22.566	22.958	23.350	23.742	24.134	24.527
700	24.527	24.919	25.312	25.705	26.098	26.491	26.883	27.276	27.669	28.062	28.455
800	28.455	28.847	29.239	29.632	30.024	30.416	30.807	31.199	31.590	31.981	32.371
900	32.371	32.761	33.151	33.541	33.930	34.319	34.707	35.095	35.482	35.869	36.256
1000	36.256	36.641	37.027	37.411	37.795	38.179	38.562	38.944	39.326	39.706	40.087
1100	40.087	40.466	40.845	41.223	41.600	41.976	42.352	42.727	43.101	43.474	43.846
1200	43.846	44.218	44.588	44.958	45.326	45.694	46.060	46.425	46.789	47.152	47.513
1300	47.513										

NOTE—Temperature ranges and coefficients of equations used to compute the above table: The equations are of the form $E = c_0 + c_1t + c_2t^2 + c_3t^3 + \ldots c_nt^n$, where E is the emf in millivolts, t is the temperature in degrees Celsius (ITS-90), and c_0, c_1, c_2, c_3, etc. are the coefficients. These coefficients are extracted from Reference 1.

		–270 °C to 0 °C	0 °C to 1300 °C
c_0	=	0.000 000 000 0 ...	0.000 000 000 0...
c_1	=	$2.615\,910\,596\,2 \times 10^{-2}$	$2.592\,939\,460\,1 \times 10^{-2}$
c_2	=	$1.095\,748\,422\,8 \times 10^{-5}$	$1.571\,014\,188\,0 \times 10^{-5}$
c_3	=	$-9.384\,111\,155\,4 \times 10^{-8}$	$4.382\,562\,723\,7 \times 10^{-8}$
c_4	=	$-4.641\,203\,975\,9 \times 10^{-11}$	$-2.526\,116\,979\,4 \times 10^{-10}$
c_5	=	$-2.630\,335\,771\,6 \times 10^{-12}$	$6.431\,181\,933\,9 \times 10^{-13}$
c_6	=	$-2.265\,343\,800\,3 \times 10^{-14}$	$-1.006\,347\,151\,9 \times 10^{-15}$
c_7	=	$-7.608\,930\,079\,1 \times 10^{-17}$	$9.974\,533\,899\,2 \times 10^{-19}$
c_8	=	$-9.341\,966\,783\,5 \times 10^{-20}$	$-6.086\,324\,560\,7 \times 10^{-22}$
c_9	=	....	$2.084\,922\,933\,9 \times 10^{-25}$
c_{10}	=	....	$-3.068\,219\,615\,1 \times 10^{-29}$

Type R Thermocouples: emf-Temperature (°C) Reference Table and Equations

Thermocouple emf as a Function of Temperature in Degrees Celsius (ITS-90)

emf in Millivolts — Reference junctions at 0 °C

°C	0	-10	-20	-30	-40	-50	-60	-70	-80	-90	-100
0	0.000	-0.051	-0.100	-0.145	-0.188	-0.226					
°C	**0**	**10**	**20**	**30**	**40**	**50**	**60**	**70**	**80**	**90**	**100**
0	0.000	0.054	0.111	0.171	0.232	0.296	0.363	0.431	0.501	0.573	0.647
100	0.647	0.723	0.800	0.879	0.959	1.041	1.124	1.208	1.294	1.381	1.469
200	1.469	1.558	1.648	1.739	1.831	1.923	2.017	2.112	2.207	2.304	2.401
300	2.401	2.498	2.597	2.696	2.796	2.896	2.997	3.099	3.201	3.304	3.408
400	3.408	3.512	3.616	3.721	3.827	3.933	4.040	4.147	4.255	4.363	4.471
500	4.471	4.580	4.690	4.800	4.910	5.021	5.133	5.245	5.357	5.470	5.583
600	5.583	5.697	5.812	5.926	6.041	6.157	6.273	6.390	6.507	6.625	6.743
700	6.743	6.861	6.980	7.100	7.220	7.340	7.461	7.583	7.705	7.827	7.950
800	7.950	8.073	8.197	8.321	8.446	8.571	8.697	8.823	8.950	9.077	9.205
900	9.205	9.333	9.461	9.590	9.720	9.850	9.980	10.111	10.242	10.374	10.506
1000	10.506	10.638	10.771	10.905	11.039	11.173	11.307	11.442	11.578	11.714	11.850
1100	11.850	11.986	12.123	12.260	12.397	12.535	12.673	12.812	12.950	13.089	13.228
1200	13.228	13.367	13.507	13.646	13.786	13.926	14.066	14.207	14.347	14.488	14.629
1300	14.629	14.770	14.911	15.052	15.193	15.334	15.475	15.616	15.758	15.899	16.040
1400	16.040	16.181	16.323	16.464	16.605	16.746	16.887	17.028	17.169	17.310	17.451
1500	17.451	17.591	17.732	17.872	18.012	18.152	18.292	18.431	18.571	18.710	18.849
1600	18.849	18.988	19.126	19.264	19.402	19.540	19.677	19.814	19.951	20.087	20.222
1700	20.222	20.356	20.488	20.620	20.749	20.877	21.003				

NOTE—Temperature ranges and coefficients of equations used to compute the above table: The equations are of the form $E = c_0 + c_1t + c_2t^2 + c_3t^3 + \ldots c_nt^n$, where E is the emf in millivolts, t is the temperature in degrees Celsius (ITS-90), and c_0, c_1, c_2, c_3, etc. are the coefficients. These coefficients are extracted from Reference 1.

		−50 °C to 1064.18 °C	1064.18 °C to 1664.5 °C	1664.5 °C to 1768.1 °C
c_0	=	0.000 000 000 00 ...	2.951 579 253 16 ...	$1.522\ 321\ 182\ 09 \times 10^{2}$
c_1	=	$5.289\ 617\ 297\ 65 \times 10^{-3}$	$-2.520\ 612\ 513\ 32 \times 10^{-3}$	$-2.688\ 198\ 885\ 45 \times 10^{-1}$
c_2	=	$1.391\ 665\ 897\ 82 \times 10^{-5}$	$1.595\ 645\ 018\ 65 \times 10^{-5}$	$1.712\ 802\ 804\ 71 \times 10^{-4}$
c_3	=	$-2.388\ 556\ 930\ 17 \times 10^{-8}$	$-7.640\ 859\ 475\ 76 \times 10^{-9}$	$-3.458\ 957\ 064\ 53 \times 10^{-8}$
c_4	=	$3.569\ 160\ 010\ 63 \times 10^{-11}$	$2.053\ 052\ 910\ 24 \times 10^{-12}$	$-9.346\ 339\ 710\ 46 \times 10^{-15}$
c_5	=	$-4.623\ 476\ 662\ 98 \times 10^{-14}$	$-2.933\ 596\ 681\ 73 \times 10^{-16}$	....
c_6	=	$5.007\ 774\ 410\ 34 \times 10^{-17}$	....	....
c_7	=	$-3.731\ 058\ 861\ 91 \times 10^{-20}$	....	....
c_8	=	$1.577\ 164\ 823\ 67 \times 10^{-23}$	....	....
c_9	=	$-2.810\ 386\ 252\ 51 \times 10^{-27}$	....	....

Type S Thermocouples: emf-Temperature (°C) Reference Table and Equations.

Thermocouple emf as a Function of Temperature in Degrees Celsius (ITS-90)

emf in Millivolts — Reference junctions at 0 °C

°C	0	-10	-20	-30	-40-	-50	-60	-70	-80	-90	-100
0	0.000	-0.053	-0.103	-0.150	-0.194	-0.236					
°C	**0**	**10**	**20**	**30**	**40**	**50**	**60**	**70**	**80**	**90**	**100**
0	0.000	0.055	0.113	0.173	0.235	0.299	0.365	0.433	0.502	0.573	0.646
100	0.646	0.720	0.795	0.872	0.950	1.029	1.110	1.191	1.273	1.357	1.441
200	1.441	1.526	1.612	1.698	1.786	1.874	1.962	2.052	2.141	2.232	2.323
300	2.323	2.415	2.507	2.599	2.692	2.786	2.880	2.974	3.069	3.164	3.259
400	3.259	3.355	3.451	3.548	3.645	3.742	3.840	3.938	4.036	4.134	4.233
500	4.233	4.332	4.432	4.532	4.632	4.732	4.833	4.934	5.035	5.137	5.239
600	5.239	5.341	5.443	5.546	5.649	5.753	5.857	5.961	6.065	6.170	6.275
700	6.275	6.381	6.486	6.593	6.699	6.806	6.913	7.020	7.128	7.236	7.345
800	7.345	7.454	7.563	7.673	7.783	7.893	8.003	8.114	8.226	8.337	8.449
900	8.449	8.562	8.674	8.787	8.900	9.014	9.128	9.242	9.357	9.472	9.587
1000	9.587	9.703	9.819	9.935	10.051	10.168	10.285	10.403	10.520	10.638	10.757
1100	10.757	10.875	10.994	11.113	11.232	11.351	11.471	11.590	11.710	11.830	11.951
1200	11.951	12.071	12.191	12.312	12.433	12.554	12.675	12.796	12.917	13.038	13.159

Type S Thermocouples: emf-Temperature (°C) Reference Table and Equations.

Thermocouple emf as a Function of Temperature in Degrees Celsius (ITS-90)

emf in Millivolts **Reference junctions at 0 °C**

°C	0	-10	-20	-30	-40	-50	-60	-70	-80	-90	-100
1300	13.159	13.280	13.402	13.523	13.644	13.766	13.887	14.009	14.130	14.251	14.373
1400	14.373	14.494	14.615	14.736	14.857	14.978	15.099	15.220	15.341	15.461	15.582
1500	15.582	15.702	15.822	15.942	16.062	16.182	16.301	16.420	16.539	16.658	16.777
1600	16.777	16.895	17.013	17.131	17.249	17.366	17.483	17.600	17.717	17.832	17.947
1700	17.947	18.061	18.174	18.285	18.395	18.503	18.609				

NOTE—Temperature ranges and coefficients of equations used to compute the above table: The equations are of the form $E = c_0 + c_1t + c_2t^2 + c_3t^3 + \ldots c_nt^n$, where E is the emf in millivolts, t is the temperature in degrees Celsius (ITS-90), and c_0, c_1, c_2, c_3, etc. are the coefficients. These coefficients are extracted from Reference 1.

		−50 °C to 1064.18 °C	1064.18 °C to 1664.5 °C	1664.5 °C to 1768.1 °C
c_0	=	0.000 000 000 00 ...	1.329 004 440 85 ...	$1.466\,282\,326\,36 \times 10^{2}$
c_1	=	$5.403\,133\,086\,31 \times 10^{-3}$	$3.345\,093\,113\,44 \times 10^{-3}$	$-2.584\,305\,167\,52 \times 10^{-1}$
c_2	=	$1.259\,342\,897\,40 \times 10^{-5}$	$6.548\,051\,928\,18 \times 10^{-6}$	$1.636\,935\,746\,41 \times 10^{-4}$
c_3	=	$-2.324\,779\,686\,89 \times 10^{-8}$	$-1.648\,562\,592\,09 \times 10^{-9}$	$-3.304\,390\,469\,87 \times 10^{-8}$
c_4	=	$3.220\,288\,230\,36 \times 10^{-11}$	$1.299\,896\,051\,74 \times 10^{-14}$	$-9.432\,236\,906\,12 \times 10^{-15}$
c_5	=	$-3.314\,651\,963\,89 \times 10^{-14}$	....	....
c_6	=	$2.557\,442\,517\,86 \times 10^{-17}$	....	....
c_7	=	$-1.250\,688\,713\,93 \times 10^{-20}$	....	....
c_8	=	$2.714\,431\,761\,45 \times 10^{-24}$	....	....

Type T Thermocouples: emf-Temperature (°C) Reference Table and Equations.

Thermocouple emf as a Function of Temperature in Degrees Celsius (ITS-90)

emf in Millivolts **Reference junctions at 0 °C**

°C	0	-10	-20	-30	-40	-50	-60	-70	-80	-90	-100
-200	-5.603	-5.753	-5.888	-6.007	-6.105	-6.180	-6.232	-6.258			
-100	-3.379	-3.657	-3.923	-4.177	-4.419	-4.648	-4.865	-5.070	-5.261	5.439	-5.603
0	0.000	-0.383	-0.757	-1.121	-1.475	-1.819	-2.153	-2.476	-2.788	-3.089	-3.379
°C	**0**	**10**	**20**	**30**	**40**	**50**	**60**	**70**	**80**	**90**	**100**
0	0.000	0.391	0.790	1.196	1.612	2.036	2.468	2.909	3.358	3.814	4.279
100	4.279	4.750	5.228	5.714	6.206	6.704	7.209	7.720	8.237	8.759	9.288
200	9.288	9.822	10.362	10.907	11.458	12.013	12.574	13.139	13.709	14.283	14.862
300	14.862	15.445	16.032	16.624	17.219	17.819	18.422	19.030	19.641	20.255	20.872
400	20.872										

NOTE—Temperature ranges and coefficients of equations used to compute the above table: The equations are of the form $E = c_0 + c_1t + c_2t^2 + c_3t^3 + \ldots c_nt^n$, where E is the emf in millivolts, t is the temperature in degrees Celsius (ITS-90), and c_0, c_1, c_2, c_3, etc. are the coefficients. These coefficients are extracted from Reference 1.

		270 °C to 0 °C	0 °C to 400 °C
c_0	=	0.000 000 000 0 ...	0.000 000 000 0 ...
c_1	=	$3.874\,810\,636\,4 \times 10^{-2}$	$3.874\,810\,636\,4 \times 10^{-2}$
c_2	=	$4.419\,443\,434\,7 \times 10^{-5}$	$3.329\,222\,788\,0 \times 10^{-5}$
c_3	=	$1.184\,432\,310\,5 \times 10^{-7}$	$2.061\,824\,340\,4 \times 10^{-7}$
c_4	=	$2.003\,297\,355\,4 \times 10^{-8}$	$-2.188\,225\,684\,6 \times 10^{-9}$
c_5	=	$9.013\,801\,955\,9 \times 10^{-10}$	$1.099\,688\,092\,8 \times 10^{-11}$
c_6	=	$2.265\,115\,659\,3 \times 10^{-11}$	$-3.081\,575\,877\,2 \times 10^{-14}$
c_7	=	$3.607\,115\,420\,5 \times 10^{-13}$	$4.547\,913\,529\,0 \times 10^{-17}$
c_8	=	$3.849\,393\,988\,3 \times 10^{-15}$	$-2.751\,290\,167\,3 \times 10^{-20}$
c_9	=	$2.821\,352\,192\,5 \times 10^{-17}$	....
c_{10}	=	$1.425\,159\,477\,9 \times 10^{-19}$	....
c_{11}	=	$4.876\,866\,228\,6 \times 10^{-22}$	....
c_{12}	=	$1.079\,553\,927\,0 \times 10^{-24}$	....
c_{13}	=	$1.394\,502\,706\,2 \times 10^{-27}$	....
c_{14}	=	$7.979\,515\,392\,7 \times 10^{-31}$	....

Type B Thermocouples: Coefficients (c_i) of Polynomials for the Computation of Temperatures in °C as a Function of the Thermocouple emf in Various Temperature and emf Ranges

Temperature range:	250 °C to 700 °C	700 °C to 1820 °C
emf range:	0.291 mV to 2.431 mV	2.431 mV to 13.820 mV
c_0 =	$9.842\,332\,1 \times 10^{1}$	$2.131\,507\,1 \times 10^{2}$
c_1 =	$6.997\,150\,0 \times 10^{2}$	$2.851\,050\,4 \times 10^{2}$
c_2 =	$-8.476\,530\,4 \times 10^{2}$	$-5.274\,288\,7 \times 10^{1}$
c_3 =	$1.005\,264\,4 \times 10^{3}$	9.916 080 4 ...
c_4 =	$-8.334\,595\,2 \times 10^{2}$	−1.296 530 3 ...
c_5 =	$4.550\,854\,2 \times 10^{2}$	$1.119\,587\,0 \times 10^{-1}$
c_6 =	$-1.552\,303\,7 \times 10^{2}$	$-6.062\,519\,9 \times 10^{-3}$
c_7 =	$2.988\,675\,0 \times 10^{1}$	$1.866\,169\,6 \times 10^{-4}$
c_8 =	−2.474 286 0 ...	$-2.487\,858\,5 \times 10^{-6}$

NOTE— The above coefficients are extracted from Reference 1 and are for an expression of the form shown in Section 10.3.2. They yield approximate values of temperature that agree within ±0.03 °C with the values given in Table 10.2.

Type E Thermocouples: Coefficients (c_i) of Polynomials for the Computation of Temperatures in °C as a Function of the Thermocouple emf in Various Temperature and emf Ranges

Temperature range:	-200 °C TO 0 °C	0 °C to 1000 °C
emf range:	-8.825 mV to 0.0 mV	0.0 mV to 76.373 mV
c_0 =	0.000 000 0 ...	0.000 000 0 ...
c_1 =	$1.697\,728\,8 \times 10^{1}$	$1.705\,703\,5 \times 10^{1}$
c_2 =	$-4.351\,497\,0 \times 10^{-1}$	$-2.330\,175\,9 \times 10^{-1}$
c_3 =	$-1.585\,969\,7 \times 10^{-1}$	$6.543\,558\,5 \times 10^{-3}$
c_4 =	$-9.250\,287\,1 \times 10^{-2}$	$-7.356\,274\,9 \times 10^{-5}$
c_5 =	$-2.608\,431\,4 \times 10^{-2}$	$-1.789\,600 \times 10^{-6}$
c_6 =	$-4.136\,019\,9 \times 10^{-3}$	$8.403\,616\,5 \times 10^{-8}$
c_7 =	$-3.403\,403\,0 \times 10^{-4}$	$-1.373\,587\,9 \times 10^{-9}$
c_8 =	$-1.156\,489\,0 \times 10^{-5}$	$1.062\,982\,3 \times 10^{-11}$
c_9 =	....	$-3.244\,708\,7 \times 10^{-14}$

NOTE— The above coefficients are extracted from Reference 1 and are for an expression of the form shown in Section 10.3.2. They yield approximate values of temperature that agree within ±0.02 °C with the values given in Table 10.4

Type J Thermocouples: Coefficients (c_i) of Polynomials for the Computation of Temperatures in °C as a Function of the Thermocouple emf in Various Temperature and emf Ranges

Temperature range:	-210 °C to 0 °C	0 °C to 760 °C	760 °C to 1200 °C
emf Range:	-8.095 mVto 0.0 mV	0.0 mV to 42.919 mV	42.919 mV to 69.553 mV
c_0 =	0.000 000 0...	0.000 000 ...	$-3.113\,581\,87 \times 10^{3}$
c_1 =	$1.952\,826\,8 \times 10^{1}$	$1.978\,425 \times 10^{1}$	$3.005\,436\,84 \times 10^{2}$
c_2 =	−1.228 618 5 ...	$-2.001\,204 \times 10^{-1}$	−9.947 732 30 ...
c_3 =	−1.075 217 8 ...	$1.036\,969 \times 10^{-2}$	$1.702\,766\,30 \times 10^{-1}$
c_4 =	$-5.908\,693\,3 \times 10^{-1}$	$-2.549\,687 \times 10^{-4}$	$1.430\,334\,68 \times 10^{-3}$
c_5 =	$-1.725\,671\,3 \times 10^{-1}$	$3.585\,153 \times 10^{-6}$	$4.438\,860\,84 \times 10^{-6}$
c_6 =	$-2.813\,151\,3 \times 10^{-2}$	$-5.344\,285 \times 10^{-8}$	....
c_7 =	$-2.396\,337\,0 \times 10^{-3}$	$5.099\,890 \times 10^{-10}$	....
c_8 =	$-8.382\,332\,1 \times 10^{-5}$	....	....

NOTE— The above coefficients are extracted from Reference 1 and are for an expression of the form shown in Section 10.3.2. They yield approximate values of temperature that agree within ± 0.5 °C with the values given in Table 10.6.

Type K Thermocouples: Coefficients (c_i) of Polynomials for the Computation of Temperatures in °C as a Function of the Thermocouple emf in Various Temperature and emf Ranges

Temperature range:	– 200 °C to 0 °C	0 °C to 500 °C	500 °C to 1372 °C
emf Range:	– 5.891 mV to 0.0 mV	0.0 mV to 20.644 mV	20.644 mV to 54.886 mV
c_0 =	0.000 000 0 ...	0.000 000 ...	$-1.318\ 058 \times 10^{2}$
c_1 =	$2.517\ 346\ 2 \times 10^{1}$	$2.508\ 355 \times 10^{1}$	$4.830\ 222 \times 10^{1}$
c_2 =	−1.166 287 8 ...	$7.860\ 106 \times 10^{-2}$	−1.646 031 ...
c_3 =	– 1.083 363 8 ...	$-2.503\ 131 \times 10^{-1}$	$5.464\ 731 \times 10^{-2}$
c_4 =	$-8.977\ 354\ 0 \times 10^{-1}$	$8.315\ 270 \times 10^{-2}$	$-9.650\ 715 \times 10^{-4}$
c_5 =	$-3.734\ 237\ 7 \times 10^{-1}$	$-1.228\ 034 \times 10^{-2}$	$8.802\ 193 \times 10^{-6}$
c_6 =	$-8.663\ 264\ 3 \times 10^{-2}$	$9.804\ 036 \times 10^{-4}$	$3.110\ 810 \times 10^{-8}$
c_7 =	$-1.045\ 059\ 8 \times 10^{-2}$	$-4.413\ 030 \times 10^{-5}$	....
c_8 =	$-5.192\ 057\ 7 \times 10^{-4}$	$1.057\ 734 \times 10^{-6}$	....
c_9 =	....	$-1.052\ 755 \times 10^{-8}$	....

NOTE—The above coefficients are extracted from Reference 1 and are for an expression of the form shown in Section 10.3.2. They yield approximate values of temperature that agree within ±0.05 °C with the values given in Table 10.8.

Type N Thermocouples: Coefficients (c_i) of Polynomials for the Computation of Temperatures in °C as a Function of the Thermocouple emf in Various Temperature and emf Ranges

Temperature range:	−200 °C to 0 °C	0 °C to 600 °C	600 °C to 1300 °C
emf Range:	−3.990 mV to 0.0 mV	0.0 mV to 20.613 mV	20.613 mV to 47.513 mV
c_0 =	0.000 000 0 ...	0.000 00 ...	$1.972\ 485 \times 10^{1}$
c_1 =	$3.843\ 684\ 7 \times 10^{1}$	$3.868\ 96 \times 10^{1}$	$3.300\ 943 \times 10^{1}$
c_2 =	1.101 048 5 ...	−1.082 67 ...	$-3.915\ 159 \times 10^{-1}$
c_3 =	5.222 931 2 ...	$4.702\ 05 \times 10^{-2}$	$9.855\ 391 \times 10^{-3}$
c_4 =	7.206 052 5 ...	$-2.121\ 69 \times 10^{-6}$	$-1.274\ 371 \times 10^{-4}$
c_5 =	5.848 858 6 ...	$-1.172\ 72 \times 10^{-4}$	$7.767\ 022 \times 10^{-7}$
c_6 =	2.775 491 6 ...	$5.392\ 80 \times 10^{-6}$	....
c_7 =	$7.707\ 516\ 6 \times 10^{-1}$	$-7.981\ 56 \times 10^{-8}$	....
c_8 =	$1.158\ 266\ 5 \times 10^{-1}$	....	....
c_9 =	$7.313\ 886\ 8 \times 10^{-3}$	....	....

NOTE—The above coefficients are extracted from Reference 1 and are for an expression of the form shown in Section 10.3.2. They yield approximate values of temperature that agree within ± 0.04 °C with the values given in Table 10.10.

Type R Thermocouples: Coefficients (c_i) of Polynomials for the Computation of Temperatures in °C as a Function of the Thermocouple emf in Various Temperature and emf Ranges

Temperature range:	– 50 °C to 250 °C	250 °C to 1200 °C	1064 °C to 1664.5 °C	1664.5 °C to 1768.1 °C
emf Range:	– 0.226 mV to 1.923 mV	1.923 mV to 13.228 mV	11.361 mV to 19.739 mV	19.739 mV to 21.103 mV
c_0 =	0.000 000 0 ...	$1.334\ 584\ 505 \times 10^{1}$	$-8.199\ 599\ 416 \times 10^{1}$	$3.406\ 177\ 836 \times 10^{4}$
c_1 =	$1.889\ 138\ 0 \times 10^{2}$	$1.472\ 644\ 573 \times 10^{2}$	$1.553\ 962\ 042 \times 10^{2}$	$-7.023\ 729\ 171 \times 10^{3}$
c_2 =	$-9.383\ 529\ 0 \times 10^{1}$	$-1.844\ 024\ 844 \times 10^{1}$	−8.342 197 663	$5.582\ 903\ 813 \times 10^{2}$
c_3 =	$1.306\ 861\ 9 \times 10^{2}$	4.031 129 726 ...	$4.279\ 433\ 549 \times 10^{-1}$	$-1.952\ 394\ 635 \times 10^{1}$
c_4 =	$-2.270\ 358\ 0 \times 10^{2}$	$-6.249\ 428\ 360 \times 10^{-1}$	$-1.191\ 577\ 910 \times 10^{-2}$	$2.560\ 740\ 231 \times 10^{-1}$
c_5 =	$3.514\ 565\ 9 \times 10^{2}$	$6.468\ 412\ 046 \times 10^{-2}$	$1.492\ 290\ 091 \times 10^{-4}$	....
c_6 =	$-3.895\ 390\ 0 \times 10^{2}$	$-4.458\ 750\ 426 \times 10^{-3}$	....	....
c_7 =	$2.823\ 947\ 1 \times 10^{2}$	$1.994\ 710\ 149 \times 10^{-4}$	....	....
c_8 =	$-1.260\ 728\ 1 \times 10^{2}$	$-5.313\ 401\ 790 \times 10^{-6}$	....	....
c_9 =	$3.135\ 361\ 1 \times 10^{1}$	$6.481\ 976\ 217 \times 10^{-8}$	....	....
c_{10} =	−3.318 776 9 ...	....	....	....

NOTE—The above coefficients are extracted from Reference 1 and are for an expression of the form shown in Section 10.3.2. They yield approximate values of temperature that agree within ±0.02 °C with the values given in Table 10.12.

Type S Thermocouples: Coefficients (c_i) of Polynomials for the Computation of Temperatures in °C as a Function of the Thermocouple emf in Various Temperature and emf Ranges

Temperature range:	−50 °C to 250 °C	250 °C to 1200 °C	1064 °C to 1664.5 °C	1664.5 °C to 1768.1 °C
emf Range:	−0.235 mV to 1.874 mV	1.874 mV to 11.950 mV	10.332 mV to 17.536 mV	17.536 mV to 18.693 mV
c_0 =	0.000 000 00 . . .	1.291 507 177 × 10^1	-8.087 801 117 ×10^1	5.333 875 126 × 10^4
c_1 =	1.849 494 60 × 10^2	1.466 298 863 × 10^2	1.621 573 104 × 10^2	-1.235 892 298 × 10^4
c_2 =	-8.005 040 62 × 10^1	-1.534 713 402 ×10^1	-8.536 869 453 . . .	1.092 657 613 × 10^3
c_3 =	1.022 374 30 × 10^2	3.145 945 973 . . .	4.719 686 976 × 10^{-1}	-4.265 693 686 × 10^1
c_4 =	-1.522 485 92 × 10^2	-4.163 257 839 × 10^{-1}	-1.441 693 666 × 10^{-2}	6.247 205 420 × 10^{-1}
c_5 =	1.888 213 43 × 10^2	3.187 963 771 × 10^{-2}	2.081 618 890 × 10^{-4}	....
c_6 =	-1.590 859 41 × 10^2	-1.291 637 500 × 10^{-3}	....	....
c_7 =	8.230 278 80 × 10^1	2.183 475 087 × 10^{-5}	....	....
c_8 =	-2.341 819 44 × 10^1	-1.447 379 511 × 10^{-7}	....	....
c_9 =	2.797 862 60 . . .	8.211 272 125 × 10^{-9}	....	....

NOTE—The above coefficients are extracted from Reference 1 and are for an expression of the form shown in Section 10.3.2. They yield approximate values of temperature that agree within ± 0.02 °C with the values given in Table 10.14.

Type T Thermocouples: Coefficients (c_i) of Polynomials for the Computation of temperatures in °C as a Function of the Thermocouple emf in Various Temperature and emf Ranges

Temperature range:	−200 °C to 0 °C	0 °C to 400 °C
emf Range:	−5.603 mV to 0.0 mV	0.0 mV to 20.872 mV
c_0 =	0.000 000 0 . . .	0.000 000 . . .
c_1 =	2.594 919 2 × 10^1	2.592 800 × 10^1
c_2 =	-2.131 696 7 × 10^{-1}	-7.602 961 × 10^{-1}
c_3 =	7.901 869 2 × 10^{-1}	4.637 791 × 10^{-2}
c_4 =	4.252 777 7 × 10^{-1}	-2.165 394 × 10^{-3}
c_5 =	1.330 447 3 × 10^{-1}	6.048 144 × 10^{-5}
c_6 =	2.024 144 6 × 10^{-2}	-7.293 422 × 10^{-7}
c_7 =	1.266 817 1 × 10^{-3}	....

NOTE—The above coefficients are extracted from Reference 1 and are for an expression of the form shown in Section 10.3.2. They yield approximate values of temperature that agree within ± 0.04 °C with the values given in Table 10.16.

SECONDARY REFERENCE POINTS ON THE ITS-90 TEMPERATURE SCALE

The International Temperature Scale of 1990 is described in Section 1 of this *Handbook*, where the defining fixed points are listed. The Consultative Committee on Thermometry (CCT) of the International Committee on Weights and Measures (CIPM), which oversees the temperature scale, has recommended a number of secondary reference points whose values have been accurately determined with respect to the primary fixed points. The most accurate of these, referred to as "first quality points," satisfy several criteria involving purity of the material, reproducibility, and documentation of the measurements. The CCT also lists "second quality points" that do not yet satisfy all the criteria but are still useful. Taken together, these secondary reference points, help fill in the gaps between the primary fixed points.

The table below describes these secondary reference points. The best values resulting from the CCT evaluation are listed on both the Kelvin and Celsius scales, along with an estimate of uncertainty. Full details are given in the reference.

The entries within each quality group are listed in order of increasing temperature.

Reference

Bedford, R. E., Bonnier, G., Maas, H., and Pavese, F., *Metrologia* 33, 133, 1996.

Substance	Type of Transition	T_{90}/K	t_{90}/°C	Uncert.
First quality points				
Zinc	Superconductive transition	0.8500	-272.300	0.0030
Aluminum	Superconductive transition	1.1810	-271.9690	0.0025
Helium (^{4}He)	Superfluid transition	2.1768	-270.9732	0.0001
Indium	Superconductive transition	3.4145	-269.7355	0.0025
Lead	Superconductive transition	7.1997	-265.9503	0.0025
Niobium	Superconductive transition	9.2880	-263.8620	0.0025
Deuterium ($^2H_2$)	Triple point (equilibrium D_2)	18.689	-254.461	0.001
Deuterium ($^2H_2$)	Triple point (normal D_2)	18.724	-254.426	0.001
Neon (^{20}Ne)	Triple point	24.541	-248.609	0.001
Neon	Boiling point	27.097	-246.053	0.001
Nitrogen	Triple point	63.151	-209.999	0.001
Nitrogen	Boiling point	77.352	-195.798	0.002
Argon	Boiling point	87.303	-185.847	0.001
Oxygen	Condensation point	90.197	-182.953	0.001
Methane	Triple point	90.694	-182.456	0.001
Xenon	Triple point	161.405	-111.745	0.001
Carbon dioxide	Triple point	216.592	-56.558	0.001
Mercury	Freezing point	234.3210	-38.8290	0.0005
Water	Ice point	273.15	0	
Gallium	Triple point	302.9166	29.7666	0.0001
Water	Boiling point	373.124	99.974	0.001
Indium	Triple point	429.7436	156.5936	0.0002
Bismuth	Freezing point	544.552	271.402	0.001
Cadmium	Freezing point	594.219	321.069	0.001
Lead	Freezing point	600.612	327.462	0.001
Antimony	Freezing point	903.778	630.628	0.001
Copper/71.9% silver	Eutectic melting point	1052.78	779.63	0.05
Palladium	Freezing point	1828.0	1554.8	0.1
Platinum	Freezing point	2041.3	1768.2	0.4
Rhodium	Freezing point	2236	1963	3
Iridium	Freezing point	2719	2446	6
Molybdenum	Melting point	2895	2622	4
Tungsten	Melting point	3687	3414	7
Second quality points				
Hydrogen	Triple point (normal H_2)	13.952	-259.198	0.002
Hydrogen	Boiling point (normal H_2)	20.388	-252.762	0.002
Oxygen	α-β transition	23.868	-249.282	0.005
Nitrogen	α-β transition	35.614	-237.536	0.006
Oxygen	β-γ transition	43.796	-229.354	0.001
Krypton	Triple point	115.775	-157.375	0.001
Carbon dioxide	Sublimation point	194.686	-78.464	0.003

Substance	Type of Transition	T_{90}/K	t_{90}/°C	Uncert.
Sulfur hexafluoride	Triple point	223.554	-49.596	0.005
Gallium/20% indium	Eutectic melting point	288.800	15.650	0.001
Gallium/8% tin	Eutectic melting point	293.626	20.476	0.002
Diphenyl ether	Triple point	300.014	26.864	0.001
Ethylene carbonate	Triple point	309.465	36.315	0.001
Succinonitrile	Triple point	331.215	58.065	0.002
Sodium	Freezing point	370.944	97.794	0.005
Benzoic acid	Triple point	395.486	122.336	0.002
Benzoic acid	Freezing point	395.502	122.352	0.007
Mercury	Boiling point	629.769	356.619	0.004
Sulfur	Boiling point	717.764	444.614	0.002
Copper/66.9% aluminum	Eutectic melting point	840.957	567.807	0.010
Silver/30% aluminum	Eutectic melting point	840.957	567.807	0.002
Sodium chloride	Freezing point	1075.168	802.018	0.011
Sodium	Boiling point	1156.090	882.940	0.005
Nickel	Freezing point	1728	1455	1
Cobalt	Freezing point	1768	1495	3
Iron	Freezing point	1811	1538	3
Titanium	Melting point	1943	1670	2
Zirconium	Melting point	2127	1854	8
Aluminum oxide	Melting point	2326	2053	2
Ruthenium	Melting point	2606	2333	10

RELATIVE SENSITIVITY OF BAYARD-ALPERT IONIZATION GAUGES TO VARIOUS GASES

Paul Redhead

The ion current I_+ in a hot-cathode ionization gauge is given by $I_+ = KI_eP$. The gauge constant is $K = (I_+/I_e)(1/P)$, where I_e is the electron current, and P the pressure. The sensitivity is given by $S = KI_e = I_+/P$. The constant K is independent of pressure below about 10^{-3} Pa.

Relative sensitivities for different Bayard-Alpert ionization gauges may differ by as much as ± 15% as a result of differences in applied voltages, electron current, and electrode structure. The table below presents the average of the measurements of 12 experimenters on Bayard-Alpert ionization gauges in various gases. The sensitivity relative to nitrogen is tabulated.

Gas		Relative sensitivity $S/S(N_2)$
Helium	He	0.18
Neon	Ne	0.31
Argon	Ar	1.4
Krypton	Kr	1.9
Xenon	Xe	2.7
Nitrogen	N_2	1.00
Hydrogen	H_2	0.43
Oxygen	O_2	0.96
Carbon monoxide	CO	1.0
Carbon dioxide	CO_2	1.4
Water	H_2O	0.93
Sulfur hexafluoride	SF_6	2.3
Mercury	Hg	3.5
Methane	CH_4	1.6
Ethane	C_2H_6	2.6
Propane	C_3H_8	3.5
Butane	C_4H_{10}	4.3
Ethene	C_2H_4	1.3
Propene	C_3H_6	1.8
Acetylene	C_2H_2	0.61
Allene	C_3H_4	1.3
1-Propyne (Methyl acetylene)	C_3H_4	1.4
Benzene	C_6H_6	3.8

References

1. Hollanda, R., *J. Vac. Sci. Technol.*, **10**, 1133, 1973.
2. Nakayama, K., and Hojo, H., *Jap. J. Appl. Phys.*, Suppl. 2, part 1, p. 113, 1974.
3. Tilford, C. R., *J. Vac. Sci. Technol.A*, **1**, 152, 1983.
4. Tilford, C. R., in *Physical Methods of Chemistry, vol.6, Determination of Thermodynamic Properties*, B. W. Rossiter and R. C. Baetzoid, Eds., pp. 101-173, John Wiley, New York, 1992.

LABORATORY SOLVENTS AND OTHER LIQUID REAGENTS

This table summarizes the properties of 575 liquids that are commonly used in the laboratory as solvents or chemical reagents.

The properties tabulated are:

- M_r: Molecular weight
- t_m: Melting point in °C
- t_b: Normal boiling point in °C
- ρ: Density in g/mL at the temperature in °C indicated by the superscript
- η: Viscosity in mPa s (1 mPa s = 1 centipoise) at 25 °C
- ε: Dielectric constant at ambient temperature (15 to 30 °C)
- μ: Dipole moment in D
- c_p: Specific heat capacity of the liquid at constant pressure at 25 °C in J/g K
- vp: Vapor pressure at 25 °C in kPa (1 kPa = 7.50 mmHg)
- FP: Flash point in °C
- Fl. lim.: Flammable (explosive) limit in air in percent by volume
- IT: Autoignition temperature in °C
- TLV: Threshold limit for allowable airborne concentration in parts per million by volume at 25 °C and atmospheric pressure

Data on the temperature dependence of viscosity, dielectric constant, and vapor pressure can be found in the pertinent tables in this *Handbook*.

References

1. Lide, D. R., *Handbook of Organic Solvents,* CRC Press, Boca Raton, FL, 1994.
2. Lide, D. R., and Kehiaian, H. V., *Handbook of Thermophysical and Thermochemical Data,* CRC Press, Boca Raton, FL, 1994.
3. Riddick, J. A., Bunger, W. B., and Sakano, T. K., *Organic Solvents, Fourth Edition,* John Wiley & Sons, New York, 1986.
4. *Fire Protection Guide to Hazardous Materials, 11th Edition,* National Fire Protection Association, Quincy, MA, 1994.
5. Urben, P. G., Ed., *Bretherick's Handbook of Reactive Chemical Hazards, 5th Edition,* Butterworth-Heinemann, Oxford, 1995.
6. *2004 TLV's and BEI's,* American Conference of Governmental Industrial Hygienists, Cincinnati, OH, 2004.

Name	Mol. form.	M_r	t_m/°C	t_b/°C	ρ/g mL^{-1}	η/mPa s	ε	μ/D	c_p/J g^{-1}K^{-1}	vp/kPa	FP/°C	Fl. lim.	IT/°C	TLV/ppm
Acetaldehyde	C_2H_4O	44.052	-123.37	20.1	0.7834^{18}		21.0	2.750	2.020	120	-39	4-60%	175	25
Acetic acid	$C_2H_4O_2$	60.052	16.64	117.9	1.0446^{25}	1.056	6.20	1.70	2.053	2.07	39	4-20%	463	10
Acetic anhydride	$C_4H_6O_3$	102.089	-74.1	139.5	1.082^{20}	0.843	22.45	≈ 2.8	1.648	0.680	49	2.7-10.3%	316	5
Acetone	C_3H_6O	58.079	-94.7	56.05	0.7845^{25}	0.306	21.01	2.88	2.175	30.8	-20	3-13%	465	500
Acetone cyanohydrin	C_4H_7NO	85.105	-19	95	0.932^{19}						74	2.2-12%	688	4.6
Acetonitrile	C_2H_3N	41.052	-43.82	81.65	0.7857^{20}	0.369	36.64	3.92	2.229	11.9	6	3-16%	524	20
Acetophenone	C_8H_8O	120.149	20.5	202	1.0281^{20}	1.681	17.44	3.02	1.703	0.049	77		570	10
Acetyl bromide	C_2H_3BrO	122.948	-96	76	1.6625^{16}					16.2				
Acetyl chloride	C_2H_3ClO	78.497	-112.8	50.7	1.1051^{20}	0.368	15.8	2.72	1.491	38.4	4		390	
Acrolein	C_3H_4O	56.063	-87.7	52.6	0.840^{20}			3.1		36.2	-26	2.8-31%	220	0.1
Acrylic acid	$C_3H_4O_2$	72.063	12.5	141	1.0511^{20}				2.022	0.53	50	2.4-8%	438	2
Acrylonitrile	C_3H_3N	53.063	-83.48	77.3	0.8007^{25}		33.0	3.87	2.05	14.1	0	3-17%	481	2
Allyl alcohol	C_3H_6O	58.079	-129	97.0	0.8540^{20}	1.218	19.7	1.60	2.392	3.14	21	3-18%	378	0.5
Allylamine	C_3H_7N	57.095	-88.2	53.3	0.758^{20}			1.2		33.1	-29	2-22%	374	
2-Amino-2-methyl-1-propanol	$C_4H_{11}NO$	89.136	25.5	165.5	0.934^{20}						67			
3-Amino-1-propanol	C_3H_9NO	75.109	12.4	187.5	0.9824^{26}						80			
Aniline	C_6H_7N	93.127	-6.02	184.17	1.0217^{20}	3.85	7.06	1.13	2.061	0.090	70	1.3-11%	615	2
Anisole	C_7H_8O	108.138	-37.13	153.7	0.9940^{20}	1.056	4.30	1.38	1.840	0.472	52		475	
Antimony(V) chloride	Cl_5Sb	299.024	4	140 dec	2.34		3.222							
Antimony(V) fluoride	F_5Sb	216.752	8.3	141	3.10									
Arsenic(III) chloride	$AsCl_3$	181.280	-16	130	2.150			1.59		5.38				
Benzaldehyde	C_7H_6O	106.122	-57.1	178.8	1.0401^{25}		17.85	3.0	1.621	0.169	63		192	
Benzene	C_6H_6	78.112	5.49	80.09	0.8765^{20}	0.604	2.2825	0	1.741	12.7	-11	1-8%	498	0.5
Benzeneacetonitrile	C_8H_7N	117.149	-23.8	233.5	1.0205^{15}		17.87	3.5		0.012	113			
Benzeneethanamine	$C_8H_{11}N$	121.180	<0	195	0.9640^{25}									
Benzeneethanol	$C_8H_{10}O$	122.164	-27	218.2	1.0202^{20}		12.31		2.068	0.01	96			
Benzenemethanethiol	C_7H_8S	124.204	-30	194.5	1.058^{20}		4.705							
Benzenesulfonyl chloride	$C_6H_5ClO_2S$	176.621	14.5	251 dec	1.3470^{15}		28.90			0.008				
Benzenethiol	C_6H_6S	110.177	-14.93	169.1	1.0775^{20}		4.26	1.23	1.572	0.26				0.1
Benzonitrile	C_7H_5N	103.122	-13.99	191.1	1.0093^{15}	1.267	25.9	4.18	1.602	0.11				
Benzoyl chloride	C_7H_5ClO	140.567	-0.4	197.2	1.2120^{20}		23.0			0.084	72			0.5
Benzyl acetate	$C_9H_{10}O_2$	150.174	-51.3	213	1.0550^{20}		5.34	1.22	0.989	0.022	90		460	10
Benzyl alcohol	C_7H_8O	108.138	-15.4	205.31	1.0419^{24}	5.47	11.916	1.71	2.015	0.015	93		436	
Benzylamine	C_7H_9N	107.153		185	0.9813^{20}	1.624	5.18			0.096				
2,2'-Bioxirane	$C_4H_6O_2$	86.090	2.0	144	1.113^{20}									

Name	Mol. form.	M_r	t_m/°C	t_b/°C	ρ/g mL^{-1}	η/mPa s	ε	μ/D	c_p/J g^{-1}K^{-1}	vp/kPa	FP/ °C	Fl. lim.	IT/°C	TLV/ppm
Bis(2-aminoethyl)amine	$C_4H_{13}N_3$	103.166	-39	207	0.9569^{20}		12.62	1.9	2.462	0.03	98	2-7%	358	1
N,N'-Bis(2-aminoethyl)-1,2-ethanediamine	$C_6H_{18}N_4$	146.234	12	266.5			10.76							
Bis(2-chloroethyl) ether	$C_4H_8Cl_2O$	143.012	-51.9	178.5	1.22^{20}		21.20	2.6	1.545	0.143	55	3%-	369	5
Bis(chloromethyl) ether	$C_2H_4Cl_2O$	114.958	-41.5	106	1.323^{15}		3.51							0.001
Bis(2-ethylhexyl) phthalate	$C_{24}H_{38}O_4$	390.557	-55	384	0.981^{25}		5.3	2.84	1.804	0.00000005	218			0.3
Bis(2-hydroxyethyl) sulfide	$C_4H_{10}O_2S$	122.186	-10.2	282	1.1793^{25}		28.61			0.08	160		298	
Boron tribromide	BBr_3	250.523	-45	91	2.6			0						1
Boron trichloride	BCl_3	117.169	-107	12.65				0	0.911	156				
Bromine	Br_2	159.808	-7.2	58.8	3.1028	0.944	3.1484	0	0.474	28.2				0.1
Bromobenzene	C_6H_5Br	157.008	-30.72	156.06	1.4950^{20}	1.074	5.45	1.70	0.983	0.556	51		565	
1-Bromobutane	C_4H_9Br	137.018	-112.6	101.6	1.2758^{20}	0.606	7.315	2.08	0.798	5.26	18	2.6-6.6%	265	
2-Bromobutane, (±)-	C_4H_9Br	137.018	-112.65	91.3	1.2585^{20}		8.64	2.23		9.32	21			
Bromochloromethane	CH_2BrCl	129.384	-87.9	68.0	1.9344^{20}			1.7	0.41	19.5				200
Bromodichloromethane	$CHBrCl_2$	163.829	-57	90	1.980^{20}									
Bromoethane	C_2H_5Br	108.965	-118.6	38.5	1.4604^{20}	0.374	9.01	2.03	0.925	62.5		7-8%	511	5
Bromoethene	C_2H_3Br	106.949	-139.54	15.8	1.4933^{20}		5.63	1.42	1.007	141		9-15%	530	0.5
2-Bromo-2-methylpropane	C_4H_9Br	137.018	-16.2	73.3	1.4278^{20}		10.98	2.17	1.102	17.7				
1-Bromopentane	$C_5H_{11}Br$	151.045	-88.0	129.8	1.2182^{20}		6.31	2.20	0.875	1.68	32			
1-Bromopropane	C_3H_7Br	122.992	-110.3	71.1	1.3537^{20}	0.489	8.09	2.18	0.702	18.6			490	
2-Bromopropane	C_3H_7Br	122.992	-89.0	59.5	1.3140^{20}	0.458	9.46	2.21	1.075	28.9				
3-Bromopropene	C_3H_5Br	120.976	-119	70.1	1.398^{20}	0.471	7.0	≈ 1.9		18.6	-1	4.4-7.3%	295	
2-Bromotoluene	C_7H_7Br	171.035	-27.8	181.7	1.4232^{20}		4.641			0.17	79			
Bromotrichloromethane	$CBrCl_3$	198.274	-5.65	105	2.012^{25}		2.405			5.35				
Butanal	C_4H_8O	72.106	-96.86	74.8	0.8016^{20}		13.45	2.72	2.270	15.7	-22	2-12.5%	218	
1,3-Butanediol	$C_4H_{10}O_2$	90.121	-77	207.5	1.0053^{20}		28.8		2.521	0.008	121		395	
1,4-Butanediol	$C_4H_{10}O_2$	90.121	20.4	235	1.0171^{20}		31.9	2.58	2.220	0.002	121			
2,3-Butanediol	$C_4H_{10}O_2$	90.121	7.6	182.5	1.0033^{20}				2.363	0.02			402	
2,3-Butanedione	$C_4H_6O_2$	86.090	-1.2	88	0.9808^{18}		4.04			7.45	27			
Butanenitrile	C_4H_7N	69.106	-111.9	117.6	0.7936^{20}	0.553	24.83	3.9	2.301	2.55	24	>1.6%	501	
1-Butanethiol	$C_4H_{10}S$	90.187	-115.7	98.5	0.8416^{20}		5.204	1.53	1.898	6.07	2			0.5
2-Butanethiol	$C_4H_{10}S$	90.187	-165	85.0	0.8295^{20}		5.645			10.8	-23			
Butanoic acid	$C_4H_8O_2$	88.106	-5.1	163.75	0.9528^{25}	1.426	2.98	1.65	2.027	0.221	72	2-10%	443	
Butanoic anhydride	$C_8H_{14}O_3$	158.195	-75	200	0.9668^{20}		12.8		1.793	0.07	54	0.9-5.8%	279	
1-Butanol	$C_4H_{10}O$	74.121	-88.6	117.73	0.8095^{20}	2.54	17.84	1.66	2.391	0.86	37	1-11%	343	20
2-Butanol	$C_4H_{10}O$	74.121	-88.5	99.51	0.8063^{20}	3.10	17.26	1.8	2.656	2.32	24	2-10%	405	100
2-Butanone	C_4H_8O	72.106	-86.64	79.59	0.7999^{25}	0.405	18.56	2.78	2.201	12.6	-9	1-11%	404	200
trans-2-Butenal	C_4H_6O	70.090	-76	102.2	0.8516^{20}			3.67	1.361	4.92	13	2.1-15.5%	232	0.3
cis-2-Butenoic acid	$C_4H_6O_2$	86.090	15	169	1.0267^{20}					0.06				
2-Butoxyethanol	$C_6H_{14}O_2$	118.174	-74.8	168.4	0.9015^{20}		9.30	2.1	2.378	0.15	69	4-13%	238	20
Butyl acetate	$C_6H_{12}O_2$	116.158	-78	126.1	0.8825^{20}	0.685	5.07	1.9	1.961	1.66	22	2-8%	425	150
sec-Butyl acetate	$C_6H_{12}O_2$	116.158	-98.9	112	0.8748^{20}		5.135	1.87			31	1.7-9.8%		200
Butyl acrylate	$C_7H_{12}O_2$	128.169	-64.6	145	0.8898^{20}		5.25		1.958	0.731	29	1.7-9.9%	292	2
Butylamine	$C_4H_{11}N$	73.137	-49.1	77.00	0.7414^{20}	0.574	4.71	1.0	2.450	12.2	-12	2-10%	312	5
sec-Butylamine	$C_4H_{11}N$	73.137	<-72	62.73	0.7246^{20}			1.28			-9			
tert-Butylamine	$C_4H_{11}N$	73.137	-66.94	44.04	0.6958^{20}		58.5	1.3	2.627	48.4	-9	2-9%	380	
Butylbenzene	$C_{10}H_{14}$	134.218	-87.85	183.31	0.8601^{20}	0.950	2.359	≈ 0	1.813	0.150	71	0.8-5.8%	410	
tert-Butylbenzene	$C_{10}H_{14}$	134.218	-57.8	169.1	0.8665^{20}		2.359	≈ 0.83	1.773	0.280	60	0.7-5.7%	450	
Butyl benzoate	$C_{11}H_{14}O_2$	178.228	-22.4	250.3	1.000^{20}		5.52			0.005	107			
tert-Butyl ethyl ether	$C_6H_{14}O$	102.174	-94	72.6	0.736^{25}				2.13	16.5				5
tert-Butyl hydroperoxide	$C_4H_{10}O_2$	90.121	6	89 dec	0.8960^{20}						27			
1-*tert*-Butyl-4-methylbenzene	$C_{11}H_{16}$	148.245	-52	190	0.8612^{20}			≈ 0		0.09	68			1
Butyl vinyl ether	$C_6H_{12}O$	100.158	-92	94	0.7888^{20}			1.25	2.316	6.65	-9		255	
γ-Butyrolactone	$C_4H_6O_2$	86.090	-43.61	204	1.1296^{20}		39.0	4.27	1.642	0.43	98			
Carbon disulfide	CS_2	76.141	-112.1	46	1.2632^{20}	0.352	2.6320	0	1.003	48.2	-30	1-50%	90	10
Chloroacetaldehyde	C_2H_3ClO	78.497	-16.3	85.5	1.19									1
Chloroacetone	C_3H_5ClO	92.524	-44.5	119	1.15^{20}					2				1
Chloroacetyl chloride	$C_2H_2Cl_2O$	112.942	-22	106	1.4202^{20}			2.23		3.33				0.05
2-Chloroaniline	C_6H_6ClN	127.572	-1.9	208.8		3.32	13.40	1.77		0.034				
3-Chloroaniline	C_6H_6ClN	127.572	-10.28	230.5	1.2161^{20}		13.3		1.558	0.0156			705	
Chlorobenzene	C_6H_5Cl	112.557	-45.31	131.72	1.1058^{20}	0.753	5.6895	1.69	1.334	1.6	28	1-10%	593	10
2-Chloro-1,3-butadiene	C_4H_5Cl	88.536	-130	59.4	0.956^{20}		4.914			29.5	-20	4-20%		10
1-Chlorobutane	C_4H_9Cl	92.567	-123.1	78.4	0.8857^{20}	0.422	7.276	2.05	1.891	13.7	-12	2-10%	240	
2-Chlorobutane	C_4H_9Cl	92.567	-131.3	68.2	0.8732^{20}		8.564	2.04		21.0	-10			
Chlorocyclohexane	$C_6H_{11}Cl$	118.604	-43.81	142	1.000^{20}		7.9505	2.1		1.0	32			
Chlorodibromomethane	$CHBr_2Cl$	208.280	-20	120	2.451^{20}									
Chloroethane	C_2H_5Cl	64.514	-138.4	12.3	0.9239^{0}		9.45	2.05	1.617	160	-50	4-15%	519	100

Name	Mol. form.	M_r	t_m/°C	t_b/°C	ρ/g mL^{-1}	η/mPa s	ε	μ/D	c_p/J g^{-1}K^{-1}	vp/kPa	FP/ °C	Fl. lim.	IT/°C	TLV/ppm
2-Chloroethanol	C_2H_5ClO	80.513	-67.5	128.6	1.2019^{20}		25.80	1.78		1.2	60	5-16%	425	1
2-Chloroethyl vinyl ether	C_4H_7ClO	106.551	-70	108	1.0495^{20}						27			
(Chloromethyl)benzene	C_7H_7Cl	126.584	-45	179	1.1004^{20}		6.854	1.8	1.44	0.164	67	1%-	585	1
Chloromethyl methyl ether	C_2H_5ClO	80.513	-103.5	59.5	1.063^{10}					24.9				
1-Chloro-2-methylpropane	C_4H_9Cl	92.567	-130.3	68.5	0.8773^{20}		7.027	2.00	1.713	19.9	-6	2-8.7%		
2-Chloro-2-methylpropane	C_4H_9Cl	92.567	-25.60	50.9	0.8420^{20}		9.663	2.13	1.867	42.7	0			
1-Chloronaphthalene	$C_{10}H_7Cl$	162.616	-2.5	259	1.1880^{25}		5.04	1.57	1.307	0.003	121		>558	
1-Chlorooctane	$C_8H_{17}Cl$	148.674	-57.8	183.5	0.8734^{20}		5.05	2.00	1.335	0.11	70			
1-Chloropentane	$C_5H_{11}Cl$	106.594	-99.0	108.4	0.8820^{20}		6.654	2.16		4.36	13	1.6-8.6%	260	
2-Chlorophenol	C_6H_5ClO	128.556	9.4	174.9	1.2634^{20}	3.59	7.40		1.468	0.308	64			
1-Chloropropane	C_3H_7Cl	78.541	-122.9	46.5	0.8899^{20}	0.334	8.588	2.05	1.683	45.8	<-18	2.6-11%	520	
2-Chloropropane	C_3H_7Cl	78.541	-117.18	35.7	0.8617^{20}	0.303		2.17		68.9	-32	2.8-11%	593	
3-Chloro-1,2-propanediol	$C_3H_7ClO_2$	110.540		213 dec	1.325^{18}		31.0							
3-Chloropropanenitrile	C_3H_4ClN	89.524	-51	175.5	1.1573^{20}						76			
2-Chloropropene	C_3H_5Cl	76.525	-137.4	22.6	0.9017^{20}		8.92	1.647		110	-37	4.5-16%		
3-Chloropropene	C_3H_5Cl	76.525	-134.5	45.1	0.9376^{20}	0.314	8.2	1.94	1.635	48.9	-32	2.9-11%	485	1
Chlorosulfonic acid	$ClHO_3S$	116.525	-80	152	1.75					0.42				
2-Chlorotoluene	C_7H_7Cl	126.584	-35.8	159.0	1.0825^{20}	0.964	4.721	1.56	1.318	0.482				50
4-Chlorotoluene	C_7H_7Cl	126.584	7.5	162.4	1.0697^{20}	0.837	6.25	2.21		0.4				
Chromyl chloride	Cl_2CrO_2	154.900	-96.5	117	1.91									0.025
trans-Cinnamaldehyde	C_9H_8O	132.159	-7.5	246	1.0497^{20}		17.72			0.005				
o-Cresol	C_7H_8O	108.138	31.03	191.04	1.0327^{35}		6.76	1.45	2.160	0.041	81	>1.4%	599	5
m-Cresol	C_7H_8O	108.138	12.24	202.27	1.0339^{20}	12.91	12.44	1.40	2.080	0.019	86	>1.1%	558	5
p-Cresol	C_7H_8O	108.138	34.77	201.98	1.0185^{40}		13.05	1.48	2.044	0.017	86	>1.1%	558	5
Cyanogen chloride	CClN	61.471	-6.5	13	1.186^{20}			2.8331						0.3
Cyclobutane	C_4H_8	56.107	-90.7	12.6	0.7038^{0}			0		157	<10	>1.8%		
Cyclohexane	C_6H_{12}	84.159	6.59	80.73	0.7739^{25}	0.894	2.0243	≈ 0	1.841	13.0	-20	1-8%	245	100
Cyclohexanol	$C_6H_{12}O$	100.158	25.93	160.84	0.9624^{20}	57.5	16.40		2.079	0.10	68	1-9%	300	50
Cyclohexanone	$C_6H_{10}O$	98.142	-27.9	155.43	0.9478^{20}	2.02	16.1	2.87	1.856	0.53	44	1-9%	420	20
Cyclohexene	C_6H_{10}	82.143	-103.5	82.98	0.8110^{20}	0.625	2.2176	0.33	1.805	11.8	-12	>1.2%	310	300
Cyclohexylamine	$C_6H_{13}N$	99.174	-17.8	134	0.8191^{20}	1.944	4.547	1.3		1.20	31	1-9%	293	10
1,3-Cyclopentadiene	C_5H_6	66.102	-85	41	0.8021^{20}			0.419		58.5				75
Cyclopentane	C_5H_{10}	70.133	-93.4	49.3	0.7457^{20}	0.413	1.9687	≈ 0	1.837	42.3	-25	2%-	361	600
Cyclopentanol	$C_5H_{10}O$	86.132	-17.5	140.42	0.9488^{20}		18.5		2.119	0.294	51			
Cyclopentanone	C_5H_8O	84.117	51.90	130.57	0.9487^{20}		13.58	3.3	1.84	1.55	26			
cis-Decahydronaphthalene	$C_{10}H_{18}$	138.250	-42.9	195.8	0.8965^{20}	3.04	2.219	≈ 0	1.678	0.10				
trans-Decahydronaphthalene	$C_{10}H_{18}$	138.250	-30.4	187.3	0.8659^{25}	1.948	2.184	≈ 0	1.653	0.164	54	1-5%	255	
Decamethylcyclopenta-siloxane	$C_{10}H_{30}O_5Si_5$	370.770	-38	210	0.9593^{20}		2.50			0.02				
Decanal	$C_{10}H_{20}O$	156.265	-4.0	208.5	0.830^{15}					0.02				
Decane	$C_{10}H_{22}$	142.282	-29.6	174.15	0.7266^{25}	0.838	1.9853	≈ 0	2.210	0.170	51	0.8-5.4%	210	
Decanoic acid	$C_{10}H_{20}O_2$	172.265	31.4	268.7	0.8858^{40}				2.761					
1-Decanol	$C_{10}H_{22}O$	158.281	6.9	231.1	0.8297^{20}	10.91	7.93		2.341	0.009	82		288	
1-Decene	$C_{10}H_{20}$	140.266	-66.3	170.5	0.7408^{20}	0.756	2.136	≈ 0	2.144	0.210	<55		235	
Diacetone alcohol	$C_6H_{12}O_2$	116.158	-44	167.9	0.9387^{20}	2.80	18.2	3.2	1.905	0.224	58	2-7%	643	50
Dibenzyl ether	$C_{14}H_{14}O$	198.260	1.8	298	1.0428^{20}		3.821				135			
Dibromodifluoromethane	CBr_2F_2	209.816	-110.1	22.76				0.66		110				100
1,2-Dibromoethane	$C_2H_4Br_2$	187.861	9.84	131.6	2.1683^{25}	1.595	4.9612	1.2	0.724	1.55				
Dibromomethane	CH_2Br_2	173.835	-52.5	97	2.4969^{20}	0.980	7.77	1.43	0.61	6.12				
1,2-Dibromotetrafluoroethane	$C_2Br_2F_4$	259.823	-110.32	47.35	2.149^{25}		2.34		0.69	43.4				
Dibutylamine	$C_8H_{19}N$	129.244	-62	159.6	0.7670^{20}	0.918	2.765	1.0	2.266	0.34	47	1-6%		
Dibutyl ether	$C_8H_{18}O$	130.228	-95.2	140.28	0.7684^{20}	0.637	3.0830	1.17	2.136	0.898	25	1.5-7.6%	194	
Di-*tert*-butyl peroxide	$C_8H_{18}O_2$	146.228	-40	111	0.704^{20}					3.43	18			
Dibutyl phthalate	$C_{16}H_{22}O_4$	278.344	-35	340	1.0465^{20}	16.63	6.58	2.82	1.789		157	>0.5%	402	0.4
Dibutyl sebacate	$C_{18}H_{34}O_4$	314.461	-10	344.5	0.9405^{15}		4.54	2.48	1.968		178	>0.4%	365	
Dibutyl sulfide	$C_8H_{18}S$	146.294	-79.7	185	0.8386^{20}		4.29	1.61	1.943	0.09	76			
Dichloroacetic acid	$C_2H_2Cl_2O_2$	128.942	10	194	1.5634^{20}		8.33			0.03				
o-Dichlorobenzene	$C_6H_4Cl_2$	147.002	-17.0	180	1.3059^{20}	1.324	10.12	2.50	1.105	0.18	66	2-9%	648	25
m-Dichlorobenzene	$C_6H_4Cl_2$	147.002	-24.8	173	1.2884^{20}	1.044	5.02	1.72	1.163	0.252	72			
trans-1,4-Dichloro-2-butene	$C_4H_6Cl_2$	124.997	1.0	155.4	1.183^{25}									0.005
Dichlorodimethylsilane	$C_2H_6Cl_2Si$	129.061	-16	70.3	1.064^{25}					18.9	<21	3.4-9.5%		
1,1-Dichloroethane	$C_2H_4Cl_2$	98.959	-96.9	57.3	1.1757^{20}	0.464	10.10	2.06	1.276	30.5	-17	5-11%	458	100
1,2-Dichloroethane	$C_2H_4Cl_2$	98.959	-35.7	83.5	1.2454^{25}	0.779	10.42	1.8	1.298	10.6	13	6-16%	413	10
1,1-Dichloroethene	$C_2H_2Cl_2$	96.943	-122.56	31.6	1.213^{20}		4.60	1.34	1.148	80.0	-28	7-16%	570	5
cis-1,2-Dichloroethene	$C_2H_2Cl_2$	96.943	-80.0	60.1	1.2837^{20}	0.445	9.20	1.90	1.201	26.8	6	3-15%	460	200
trans-1,2-Dichloroethene	$C_2H_2Cl_2$	96.943	-49.8	48.7	1.2565^{20}	0.317	2.14	0	1.205	44.2	2	6-13%	460	200
Dichloromethane	CH_2Cl_2	84.933	-97.2	40	1.3266^{20}	0.413	8.93	1.60	1.192	58.2		13-23%	556	50

Name	Mol. form.	M_r	t_m/°C	t_b/°C	ρ/g mL^{-1}	η/mPa s	ε	μ/D	c_p/J g^{-1}K^{-1}	vp/kPa	FP/ °C	Fl. lim.	IT/°C	TLV/ppm
(Dichloromethyl)benzene	$C_7H_6Cl_2$	161.029	-17	205	1.26^{25}		6.9	2.1		0.06				
1,1-Dichloropropane	$C_3H_6Cl_2$	112.986		88.1	1.1321^{20}					9.09				
1,2-Dichloropropane, (±)-	$C_3H_6Cl_2$	112.986	-100.53	96.4	1.1560^{20}		8.37	1.8	1.320	6.62	21	3-15%	557	75
1,3-Dichloropropane	$C_3H_6Cl_2$	112.986	-99.5	120.9	1.1785^{25}		10.27	2.08		2.44				
2,3-Dichloropropene	$C_3H_4Cl_2$	110.970	10	94	1.211^{20}						15	2.6-7.8%		
2,4-Dichlorotoluene	$C_7H_6Cl_2$	161.029	-13.5	201	1.2476^{20}		5.68	1.70		0.055				
Dicyclohexylamine	$C_{12}H_{23}N$	181.318	-0.1	256 dec	0.9123^{20}					0.003	>99			
Diethanolamine	$C_4H_{11}NO_2$	105.136	28	268.8	1.0966^{20}		25.75	2.8	2.22	<0.01	172	2-13%	662	0.5
1,1-Diethoxyethane	$C_6H_{14}O_2$	118.174	-100	102.25	0.8254^{20}		3.80	1.4	2.01	3.68	-21	2-10%	230	
1,2-Diethoxyethane	$C_6H_{14}O_2$	118.174	-74.0	121.2	0.8351^{25}		3.90		2.195	4.33	27		205	
Diethylamine	$C_4H_{11}N$	73.137	-49.8	55.5	0.7056^{20}	0.319	3.680	0.92	2.313	30.1	-23	2-10%	312	5
N,N-Diethylaniline	$C_{10}H_{15}N$	149.233	-38.8	216.3	0.9307^{20}		5.15			0.025	85		630	
o-Diethylbenzene	$C_{10}H_{14}$	134.218	-31.2	184	0.8800^{20}		2.594			0.13	57		395	
m-Diethylbenzene	$C_{10}H_{14}$	134.218	-83.9	181.1	0.8602^{20}		2.369			0.14	56		450	
p-Diethylbenzene	$C_{10}H_{14}$	134.218	-42.83	183.7	0.8620^{20}		2.259			0.13	55	0.7-6%	430	
Diethyl carbonate	$C_5H_{10}O_3$	118.131	-43	126	0.9692^{25}		2.820	1.10	1.80	1.63	25			
Diethylene glycol	$C_4H_{10}O_3$	106.120	-10.4	245.8	1.1197^{15}	30.2	31.82	2.3	2.307	0.001	124	2-17%	224	
Diethylene glycol diethyl ether	$C_8H_{18}O_3$	162.227	-45	188	0.9063^{20}		5.70		2.104	0.10	82			
Diethylene glycol dimethyl ether	$C_6H_{14}O_3$	134.173	-68	162	0.9434^{20}	0.989	7.23	2.0	2.043	0.315	67			
Diethylene glycol monobutyl ether	$C_8H_{18}O_3$	162.227	-68	231	0.9553^{20}				2.188	0.0032				
Diethylene glycol monoethyl ether	$C_6H_{14}O_3$	134.173		196	0.9885^{20}			1.6	2.243	0.017	96			
Diethylene glycol monoethyl ether acetate	$C_8H_{16}O_4$	176.211	-25	218.5	1.0096^{20}			1.8		0.029	110		425	
Diethylene glycol monomethyl ether	$C_5H_{12}O_3$	120.147		193	1.035^{20}			1.6	2.256	0.024	96	1-23%	240	
Diethyl ether	$C_4H_{10}O$	74.121	-116.2	34.5	0.7138^{20}	0.224	4.2666	1.15	2.369	71.7	-45	2-36%	180	400
Diethyl maleate	$C_8H_{12}O_4$	172.179	-8.8	223	1.0662^{20}		7.560			0.015	121		350	
Diethyl malonate	$C_7H_{12}O_4$	160.168	-50	200	1.0551^{20}		7.550	2.54	1.779	0.048	93			
Diethyl oxalate	$C_6H_{10}O_4$	146.141	-40.6	185.7	1.0785^{20}		8.266	2.49	1.784	0.030	76			
Diethyl phthalate	$C_{12}H_{14}O_4$	222.237	-40.5	295	1.232^{14}		7.86		1.647	0.002	161	>0.7%	457	0.6
Diethyl succinate	$C_8H_{14}O_4$	174.195	-21	217.7	1.0402^{20}		6.098			0.15	90			
Diethyl sulfate	$C_4H_{10}O_4S$	154.185	-24	208	1.172^{25}		29.2			0.05	104		436	
Diethyl sulfide	$C_4H_{10}S$	90.187	-103.91	92.1	0.8362^{20}	0.422	5.723	1.54	1.900	7.78				
Diiodomethane	CH_2I_2	267.836	6.1	182	3.3211^{20}		5.32	1.08	0.500	0.172				
Diiodosilane	H_2I_2Si	283.911	-1	150										
Diisobutylamine	$C_8H_{19}N$	129.244	-73.5	139.6		0.723				0.972	29			
Diisopentyl ether	$C_{10}H_{22}O$	158.281		172.5	0.7777^{20}		2.817	1.23	2.394	0.210				
Diisopropylamine	$C_6H_{15}N$	101.190	-61	83.9	0.7153^{20}	0.393		1.15		10.7	-1	1.1-7.1%	316	5
Diisopropyl ether	$C_6H_{14}O$	102.174	-85.4	68.4	0.7192^{25}	0.379	3.805	1.13	2.122	19.9	-28	1-8%	443	250
1,2-Dimethoxyethane	$C_4H_{10}O_2$	90.121	-69.20	84.5	0.8637^{25}	0.455	7.30		2.145	9.93	-2		202	
Dimethoxymethane	$C_3H_8O_2$	76.095	-105.1	42	0.8593^{20}		2.644	0.7	2.129	53.1	-32	2-14%	237	1000
Dimethylacetal	$C_4H_{10}O_2$	90.121	-113.2	64.5	0.8501^{20}					22.9				
N,N-Dimethylacetamide	C_4H_9NO	87.120	-18.59	165	0.9372^{25}	1.927	38.85	3.7	2.016	0.075	70	2-12%	490	10
2,3-Dimethylaniline	$C_8H_{11}N$	121.180	<-15	221.5	0.9931^{20}						97	>1%		
2,6-Dimethylaniline	$C_8H_{11}N$	121.180	11.2	215	0.9842^{20}			1.63	1.971	0.45	96			
N,N-Dimethylaniline	$C_8H_{11}N$	121.180	2.42	194.15	0.9557^{20}	1.300	4.90	1.68	1.771	0.107	63		371	5
2,2-Dimethylbutane	C_6H_{14}	86.175	-98.8	49.73	0.6444^{25}	0.351	1.869	≈ 0	2.227	42.5	-48	1.2-7%	405	500
2,3-Dimethylbutane	C_6H_{14}	86.175	-128.10	57.93	0.6616^{20}	0.361	1.889	≈ 0	2.201	31.3	-29	1.2-7%	405	500
3,3-Dimethyl-2-butanone	$C_6H_{12}O$	100.158	-52.5	106.1	0.7229^{25}		12.73			4.27				
Dimethylcarbamic chloride	C_3H_6ClNO	107.539	-33	167	1.168^{25}									
Dimethyl disulfide	$C_2H_6S_2$	94.199	-84.67	109.74	1.0625^{20}		9.6	1.8	1.551	3.82	24			
N,N-Dimethylethanolamine	$C_4H_{11}NO$	89.136	-59	134	0.8866^{20}					0.9				
N,N-Dimethylformamide	C_3H_7NO	73.094	-60.48	153	0.9445^{25}	0.794	38.25	3.82	2.060	0.439	58	2-15%	445	10
2,6-Dimethyl-4-heptanone	$C_9H_{18}O$	142.238	-41.5	169.4	0.8062^{20}		9.91	2.7	2.090	0.23	49	1-7%	396	25
1,1-Dimethylhydrazine	$C_2H_8N_2$	60.098	-57.20	63.9	0.791^{22}				2.731	20.9	-15	2-95%	249	0.01
Dimethyl phthalate	$C_{10}H_{10}O_4$	194.184	5.5	283.7	1.1905^{20}	14.36	8.66		1.561	0.001	146	>0.9%	490	0.6
2,6-Dimethylpyridine	C_7H_9N	107.153	-6.1	144.01	0.9226^{20}		7.33	1.7	1.728	0.746				
Dimethyl sulfate	$C_2H_6O_4S$	126.132	-31.7	188 dec	1.3322^{20}		55.0			0.13	83		188	0.1
Dimethyl sulfide	C_2H_6S	62.134	-98.24	37.33	0.8483^{20}	0.284	6.70	1.554	1.901	64.4	-37	2.2-20%	206	10
Dimethyl sulfoxide	C_2H_6OS	78.133	17.89	189	1.1010^{25}	1.987	47.24	3.96	1.958	0.084	95	3-42%	215	
1,4-Dioxane	$C_4H_8O_2$	88.106	11.85	101.5	1.0337^{20}	1.177	2.2189	0	1.726	4.95	12	2-22%	180	20
1,3-Dioxolane	$C_3H_6O_2$	74.079	-97.22	78	1.060^{20}			1.19	1.593	14.6	2			20
Dipentyl ether	$C_{10}H_{22}O$	158.281	-69	190	0.7833^{20}		2.798	1.20	1.579	0.13	57		170	
Dipropylamine	$C_6H_{15}N$	101.190	-63	109.3	0.7400^{20}	0.517	2.923	1.03	2.500	3.21	17		299	

Name	Mol. form.	M_r	t_m/°C	t_b/°C	ρ/g mL⁻¹	η/mPa s	ε	μ/D	c_p/J g⁻¹K⁻¹	vp/kPa	FP/ °C	Fl. lim.	IT/°C	TLV/ppm
Dipropylene glycol monomethyl ether	$C_7H_{16}O_3$	148.200	-80	188.3	0.95									
Dipropyl ether	$C_6H_{14}O$	102.174	-114.8	90.08	0.7466^{20}	0.396	3.38	1.21	2.169	8.35	21	1.3-7%	188	
Dodecane	$C_{12}H_{26}$	170.334	-9.57	216.32	0.7495^{20}	1.383	2.0120	≈ 0	2.206	0.016	74	>0.6%	203	
1-Dodecanol	$C_{12}H_{26}O$	186.333	23.9	260	0.8309^{24}		5.82		2.351	0.000016	127		275	
1-Dodecene	$C_{12}H_{24}$	168.319	-35.2	213.8	0.7584^{20}	1.20	2.152	≈ 0	2.143	0.019	79			
Epichlorohydrin	C_3H_5ClO	92.524	-26	118	1.1812^{20}	1.073	22.6	1.8	1.422	2.2	31	4-21%	411	0.5
1,2-Epoxybutane	C_4H_8O	72.106	-150	63.4	0.8297^{20}			1.891	2.039	31.7	-22	1.7-19%	439	
1,2-Epoxy-4-(epoxyethyl)cyclohexane	$C_8H_{12}O_2$	140.180	<-55	227	1.0966^{20}									0.1
1,2-Ethanediamine	$C_2H_8N_2$	60.098	11.14	117	0.8979^{20}		13.82	1.99	2.872	1.62	40	3-12%	385	10
1,2-Ethanediol	$C_2H_6O_2$	62.068	-12.69	197.3	1.1135^{20}	16.06	41.4	2.28	2.394	0.01	111	3-22%	398	40
1,2-Ethanediol, diacetate	$C_6H_{10}O_4$	146.141	-31	190	1.1043^{20}		7.7	2.34	2.121	0.030	88	1.6-8.4%	482	
1,2-Ethanediol, dinitrate	$C_2H_4N_2O_6$	152.062	-22.3	198.5	1.4918^{20}		28.26			0.009				0.05
1,2-Ethanedithiol	$C_2H_6S_2$	94.199	-41.2	146.1	1.234^{20}		7.26	2.03						
Ethanethiol	C_2H_6S	62.134	-147.88	35.0	0.8315^{25}	0.287	6.667	1.60	1.898	70.3	-17	2.8-18%	300	0.5
Ethanol	C_2H_6O	46.068	-114.14	78.29	0.7893^{20}	1.074	25.3	1.69	2.438	7.87	13	3-19%	363	1000
Ethanolamine	C_2H_7NO	61.083	10.5	171	1.0180^{20}	21.1	31.94	2.3	3.201	0.05	86	3-24%	410	3
4-Ethoxyaniline	$C_8H_{11}NO$	137.179	4.6	254	1.0652^{16}		7.43			0.0007	116			
Ethoxybenzene	$C_8H_{10}O$	122.164	-29.43	169.81	0.9651^{20}	1.197	4.216	1.45	1.870	0.204	63			
2-Ethoxyethanol	$C_4H_{10}O_2$	90.121	-70	135	0.9253^{25}		13.38	2.1	2.339	0.71	43	3-18%	235	5
2-Ethoxyethyl acetate	$C_6H_{12}O_3$	132.157	-61.7	156.4	0.9740^{20}		7.567	2.2	2.845	0.24	56	2-8%	379	5
Ethyl acetate	$C_4H_8O_2$	88.106	-83.8	77.11	0.9003^{20}	0.423	6.0814	1.78	1.937	12.6	-4	2-12%	426	400
Ethyl acetoacetate	$C_6H_{10}O_3$	130.141	-45	180.8	1.0368^{10}		14.0		1.906	0.095	57	1-10%	295	
Ethyl acrylate	$C_5H_8O_2$	100.117	-71.2	99.4	0.9234^{20}		6.05	1.96		5.14	10	1.4-14%	372	5
Ethylamine	C_2H_7N	45.084	-80.5	16.5	0.689^{15}		8.7	1.22	2.884	141	-16	4-14%	385	5
N-Ethylaniline	$C_8H_{11}N$	121.180	-63.5	203.0	0.9625^{20}	2.05	5.87			0.039	85			
Ethylbenzene	C_8H_{10}	106.165	-94.96	136.19	0.8626^{25}	0.631	2.4463	0.59	1.726	1.28	21	1-7%	432	100
Ethyl benzoate	$C_9H_{10}O_2$	150.174	-34	212	1.0415^{25}		6.20	2.00	1.638	0.04	88		490	
Ethyl butanoate	$C_6H_{12}O_2$	116.158	-98	121.3	0.8735^{25}	0.639	5.18	1.74	1.963	2.01	24		463	
2-Ethyl-1-butanol	$C_6H_{14}O$	102.174	<-15	147	0.8326^{20}		6.19			0.206	57			
Ethyl chloroacetate	$C_4H_7ClO_2$	122.551	-21	144.3	1.1585^{20}					0.640	64			
Ethyl chloroformate	$C_3H_5ClO_2$	108.524	-80.6	95	1.1352^{20}		9.736				16		500	
Ethyl cyanoacetate	$C_5H_7NO_2$	113.116	-22.5	205	1.0654^{20}		31.62	2.17	1.947	0.003	110			
Ethyleneimine	C_2H_5N	43.068	-77.9	56	0.832^{25}		18.3	1.90		28.9	-11	3.3-55%	320	0.5
Ethyl formate	$C_3H_6O_2$	74.079	-79.6	54.4	0.9208^{20}	0.380	8.57	1.9	2.015	32.3	-20	3-16%	455	100
2-Ethylhexanal	$C_8H_{16}O$	128.212	<-100	163	0.8540^{20}						44	0.9-7.2%	190	
2-Ethyl-1,3-hexanediol	$C_8H_{18}O_2$	146.228	-40	244	0.9325^{22}		18.73				127		360	
2-Ethyl-1-hexanol	$C_8H_{18}O$	130.228	-70	184.6	0.8319^{25}	6.27	7.58	1.74	2.438	0.019	73	0.8-9.7%	231	
2-Ethylhexyl acetate	$C_{10}H_{20}O_2$	172.265	-80	199	0.8718^{20}			1.8		0.09	71	1-8%	268	
Ethyl lactate	$C_5H_{10}O_3$	118.131	-26	154.5	1.0328^{20}		15.4	2.4	2.150		46	>1.5%	400	
Ethyl 3-methylbutanoate	$C_7H_{14}O_2$	130.185	-99.3	135.0	0.8656^{20}		4.71			1.07				
Ethyl 2-methylpropanoate	$C_6H_{12}O_2$	116.158	-88.2	110.1	0.868^{20}					3.25	13			
Ethyl nitrite	$C_2H_5NO_2$	75.067		18	0.899^{15}					135	-35	4-50%	90	
Ethyl propanoate	$C_5H_{10}O_2$	102.132	-73.9	99.1	0.8843^{25}	0.501	5.76	1.74	1.920	4.97	12	1.9-11%	440	
Ethyl silicate	$C_8H_{20}O_4Si$	208.329	-82.5	168.8	0.9320^{20}		2.50		1.749	1.17	52			10
Eucalyptol	$C_{10}H_{18}O$	154.249	0.8	176.4	0.9267^{20}		4.57			0.260	48			
Fluorobenzene	C_6H_5F	96.102	-42.18	84.73	1.0225^{20}	0.550	5.465	1.60	1.523	10.4	-15			
Fluorosulfonic acid	FHO_3S	100.070	-89	163	1.726					0.08				
Formamide	CH_3NO	45.041	2.49	220	1.1334^{20}	3.34	111.0	3.73	2.389	0.01	154			10
Formic acid	CH_2O_2	46.026	8.3	101	1.220^{20}	1.607	51.1	1.425	2.151	5.75	50	18-57%	434	5
Furan	C_4H_4O	68.074	-85.61	31.5	0.9514^{20}	0.361	2.94	0.66	1.686	80.0	-36	2-14%		
Furfural	$C_5H_4O_2$	96.085	-38.1	161.7	1.1594^{20}	1.587	42.1	3.5	1.698	0.29	60	2-19%	316	2
Furfuryl alcohol	$C_5H_6O_2$	98.101	-14.6	171	1.1296^{20}		16.85	1.9	2.079	0.097	75	2-16%	491	10
Germanium(IV) chloride	Cl_4Ge	214.42	-51.50	86.55	1.88			0						
Glycerol	$C_3H_8O_3$	92.094	18.1	290	1.2613^{20}	934	46.53	2.6	2.377	<0.01	199	3-19%	370	2.7
Glycerol triacetate	$C_9H_{14}O_6$	218.203	-78	259	1.1583^{20}		7.11		1.763	<0.01	138	1%-	433	
Glycerol trioleate	$C_{57}H_{104}O_6$	885.432	-4		0.915^{15}		3.109							
Heptanal	$C_7H_{14}O$	114.185	-43.4	152.8	0.8132^{25}		9.07		2.015	0.46				
Heptane	C_7H_{16}	100.202	-90.55	98.4	0.6795^{25}	0.387	1.9209	≈ 0	2.242	6.09	-4	1-7%	204	400
Heptanoic acid	$C_7H_{14}O_2$	130.185	-7.17	222.2	0.9124^{25}	3.84	3.04		2.039	0.001			275	
1-Heptanol	$C_7H_{16}O$	116.201	-33.2	176.45	0.8219^{20}	5.81	11.75		2.342	0.0044				
2-Heptanone	$C_7H_{14}O$	114.185	-35	151.05	0.8111^{20}	0.714	11.95	2.6	2.037	0.49	39	1-8%	393	50
3-Heptanone	$C_7H_{14}O$	114.185	-39	147	0.8183^{20}		12.7	2.78		0.5	46			50
4-Heptanone	$C_7H_{14}O$	114.185	-33	144	0.8174^{20}		12.60			0.164	49			50
1-Heptene	C_7H_{14}	98.186	-118.9	93.64	0.6970^{20}	0.340	2.092	≈ 0	2.157	7.52	-1		260	
Hexachloro-1,3-butadiene	C_4Cl_6	260.761	-21	215	1.556^{25}		2.55			0.13			610	0.02

Name	Mol. form.	M_r	t_m/°C	t_b/°C	ρ/g mL⁻¹	η/mPa s	ε	μ/D	c_p/J g⁻¹K⁻¹	vp/kPa	FP/°C	Fl. lim.	IT/°C	TLV/ppm
Hexachloro-1,3-cyclopentadiene	C_5Cl_6	272.772	-9	239	1.7019^{25}									0.01
Hexafluorobenzene	C_6F_6	186.054	5.03	80.26	1.6184^{20}	2.79	2.029	0	1.191	11.3				
Hexamethyldisiloxane	$C_6H_{18}OSi_2$	162.377	-66	99	0.7638^{20}		2.179		1.918	5.57				
Hexamethylphosphoric triamide	$C_6H_{18}N_3OP$	179.200	7.2	232.5	1.03^{20}		31.3	5.5	1.791					
Hexanal	$C_6H_{12}O$	100.158	-56	131	0.8335^{20}				2.101	1.48	32			
Hexane	C_6H_{14}	86.175	-95.35	68.73	0.6606^{25}	0.300	1.8865	≈ 0	2.270	20.2	-22	1-8%	225	50
Hexanedinitrile	$C_6H_8N_2$	108.141	1	295	0.9676^{20}				1.190	<0.01	93	2-5%	550	2
Hexanoic acid	$C_6H_{12}O_2$	116.158	-3	205.2	0.9212^{25}		2.600	1.13	1.937	0.005	102		380	
1-Hexanol	$C_6H_{14}O$	102.174	-47.4	157.6	0.8136^{20}	4.58	13.03		2.353	0.11	63		290	
2-Hexanone	$C_6H_{12}O$	100.158	-55.5	127.6	0.8113^{20}	0.583	14.56	2.7	2.130	1.54	25	1-8%	423	5
1-Hexene	C_6H_{12}	84.159	-139.76	63.48	0.6685^{25}	0.252	2.077	≈ 0	2.178	24.8	-26	1.2-6.9%	253	50
Hexyl acetate	$C_8H_{16}O_2$	144.212	-80.9	171.5	0.8779^{15}		4.42		1.961	0.185	45			
Hydrazine	H_4N_2	32.045	1.4	113.55	1.0036	0.876	51.7	1.75	3.086	1.91	38	5-100%		0.01
Hydrazoic acid	HN_3	43.028	-80	35.7				1.70		68.2				0.11
Hydrogen cyanide	CHN	27.026	-13.29	26	0.6876^{20}	0.183	114.9	2.985	2.612	98.8	-18	6-40%	538	4.7
Hydrogen peroxide	H_2O_2	34.015	-0.43	150.2	1.44		74.6	1.573	2.619	0.26				1
3-Hydroxypropanenitrile	C_3H_5NO	71.078	-46	221	1.0404^{25}			3.2		0.010	129			
Indan	C_9H_{10}	118.175	-51.38	177.97	0.9639^{20}	1.357			1.609	0.2				
Indene	C_9H_8	116.160	-1.5	182	0.9960^{25}				1.609	0.220				10
Iodine bromide	BrI	206.808	40	116 dec	4.3			0.726						
Iodine chloride	ClI	162.357	27.39	100 dec	3.24			1.24		3.59				
Iodobenzene	C_6H_5I	204.008	-31.3	188.4	1.8308^{20}	1.554	4.59	1.70	0.778	0.133				
1-Iodobutane	C_4H_9I	184.018	-103	130.5	1.6154^{20}		6.27	1.93		1.85				
Iodoethane	C_2H_5I	155.965	-111.1	72.3	1.9357^{20}	0.556	7.82	1.976	0.738	18.2				
Iodomethane	CH_3I	141.939	-66.4	42.43	2.2789^{20}	0.469	6.97	1.62	0.888	53.9				2
1-Iodopropane	C_3H_7I	169.992	-101.3	102.5	1.7489^{20}	0.703	7.07	2.04	0.746	5.75				
2-Iodopropane	C_3H_7I	169.992	-90	89.5	1.7042^{20}	0.653	8.19	1.95	0.535	9.36				
Iron pentacarbonyl	C_5FeO_5	195.896	-20	103	1.5^{20}		2.602		1.228	4				0.1
Isobutanal	C_4H_8O	72.106	-65.9	64.5	0.7891^{20}			2.75		23.0	-18	1.6-10.6%	196	
Isobutyl acetate	$C_6H_{12}O_2$	116.158	-98.8	116.5	0.8712^{20}	0.676	5.068	1.9	2.013	2.39	18	1-11%	421	150
Isobutyl acrylate	$C_7H_{12}O_2$	128.169	-61	132	0.8896^{20}						30		427	
Isobutylamine	$C_4H_{11}N$	73.137	-86.7	67.75	0.724^{25}	0.571	4.43	1.3	2.505	19.0	-9	2-12%	378	
Isobutylbenzene	$C_{10}H_{14}$	134.218	-51.4	172.79	0.8532^{20}		2.318	≈ 0	1.793	0.257	55	0.8-6%	427	
Isobutyl formate	$C_5H_{10}O_2$	102.132	-95.8	98.2	0.8776^{20}		6.41	1.88		5.34	5	2-9%	320	
Isobutyl isobutanoate	$C_8H_{16}O_2$	144.212	-80.7	148.6	0.8542^{20}			1.9		0.552	38	1-8%	432	
Isopentane	C_5H_{12}	72.149	-159.77	27.88	0.6201^{20}	0.214	1.845	0.13	2.284	91.7	-51	1.4-7.6%	420	600
Isopentyl acetate	$C_7H_{14}O_2$	130.185	-78.5	142.5	0.876^{15}		4.72	1.9	1.909	0.728	25	1-8%	360	50
Isophorone	$C_9H_{14}O$	138.206	-8.1	215.2	0.9255^{20}	2.33			1.834	0.06	84	1-4%	460	5
Isopropenyl acetate	$C_5H_8O_2$	100.117	-92.9	94	0.9090^{20}					6.02	26		432	
Isopropenylbenzene	C_9H_{10}	118.175	-23.2	165.4	0.9106^{20}		2.28		1.711	0.40	54	1.9-6.1%	574	50
Isopropyl acetate	$C_5H_{10}O_2$	102.132	-73.4	88.6	0.8718^{20}				1.952	7.88	2	2-8%	460	100
Isopropylamine	C_3H_9N	59.110	-95.13	31.76	0.6891^{20}	0.325	5.6268	1.19	2.771	78.0	-37		402	5
Isopropylbenzene	C_9H_{12}	120.191	-96.02	152.41	0.8640^{25}	0.737	2.381	0.79	1.753	0.61	36	1-7%	424	50
Isopropylbenzene hydroperoxide	$C_9H_{12}O_2$	152.190		153	1.03^{20}					0.004				
1-Isopropyl-2-methylbenzene	$C_{10}H_{14}$	134.218	-71.5	178.1	0.8766^{20}					0.2				
1-Isopropyl-3-methylbenzene	$C_{10}H_{14}$	134.218	-63.7	175.1	0.8610^{20}					0.22				
1-Isopropyl-4-methylbenzene	$C_{10}H_{14}$	134.218	-67.94	177.1	0.8573^{20}		2.2322	≈0	1.761	0.19	47	1-6%	436	
Isoquinoline	C_9H_7N	129.159	26.47	243.22	1.0910^{30}		11.0	2.73	1.519	0.007				
d-Limonene	$C_{10}H_{16}$	136.234	-74.0	178	0.8411^{20}	1.47	2.3746		1.828	0.277	45	0.7-6.1%	237	
l-Limonene	$C_{10}H_{16}$	136.234		178	0.843^{20}		2.3738			0.254				
Mesityl oxide	$C_6H_{10}O$	98.142	-59	130	0.8653^{20}	0.602	15.6	2.8	2.165	1.47	31	1-7%	344	15
Methacrylic acid	$C_4H_6O_2$	86.090	16	162.5	1.0153^{20}			1.65	1.871	0.12	77	1.6-8.8%	68	20
Methanol	CH_4O	32.042	-97.53	64.6	0.7914^{20}	0.544	33.0	1.70	2.531	16.9	11	6-36%	464	200
2-Methoxyaniline	C_7H_9NO	123.152	6.2	224	1.0923^{20}		5.230			0.013	118			0.1
4-Methoxybenzaldehyde	$C_8H_8O_2$	136.149	0	248	1.119^{15}		22.0			0.004				
2-Methoxyethanol	$C_3H_8O_2$	76.095	-85.1	124.1	0.9647^{20}		17.2	2.36	2.249	1.31	39	2-14%	285	5
2-Methoxyethyl acetate	$C_5H_{10}O_3$	118.131	-70	143	1.0074^{19}		8.25	2.1	2.624	0.67	49	2-12%	392	5
Methyl acetate	$C_3H_6O_2$	74.079	-98.25	56.87	0.9342^{20}	0.364	7.07	1.72	1.916	28.8	-10	3-16%	454	200
Methyl acrylate	$C_4H_6O_2$	86.090	<-75	80.7	0.9535^{20}		7.03	1.77	1.845	11.0	-3	2.8-25%	468	2
2-Methylacrylonitrile	C_4H_5N	67.090	-35.8	90.3	0.8001^{20}			3.69	1.883	8.26	1	2-6.8%		1
2-Methylaniline	C_7H_9N	107.153	-14.41	200.3	0.9984^{20}	3.82	6.138	1.6	1.96	0.043	85		482	2
3-Methylaniline	C_7H_9N	107.153	-31.3	203.3	0.9889^{20}	3.31	5.816	1.45	2.118	0.036				2
N-Methylaniline	C_7H_9N	107.153	-57	196.2	0.9891^{20}	2.04	5.96		1.933	0.05				0.5
Methyl benzoate	$C_8H_8O_2$	136.149	-12.4	199	1.0837^{25}	1.857	6.642	1.9	1.625	0.052	83			

Name	Mol. form.	M_r	t_m/°C	t_b/°C	ρ/g mL^{-1}	η/mPa s	ε	μ/D	c_p/J g^{-1}K^{-1}	vp/kPa	FP/ °C	Fl. lim.	IT/°C	TLV/ppm
2-Methyl-1,3-butadiene	C_5H_8	68.118	-145.9	34.0	0.679^{20}		2.098	0.25	2.240	73.4	-54	1.5-8.9%	395	
Methyl butanoate	$C_5H_{10}O_2$	102.132	-85.8	102.8	0.8984^{20}	0.541	5.48		1.941	4.30	14			
3-Methylbutanoic acid	$C_5H_{10}O_2$	102.132	-29.3	176.5	0.931^{20}			0.63	1.930	0.067			416	
3-Methyl-1-butanol	$C_5H_{12}O$	88.148	-117.2	131.1	0.8104^{20}	3.69	15.63		2.382	0.315	43	1.2-9%	350	100
2-Methyl-2-butanol	$C_5H_{12}O$	88.148	-9.1	102.4	0.8096^{20}	3.55	5.78	1.82	2.803	2.19	19	1.2-9%	437	
3-Methyl-2-butanol, (±)-	$C_5H_{12}O$	88.148		112.9	0.8180^{20}		12.1			1.20	38			
3-Methyl-2-butanone	$C_5H_{10}O$	86.132	-93.1	94.33	0.8051^{20}		10.37		2.089	6.99				200
2-Methyl-1-butene	C_5H_{10}	70.133	-137.53	31.2	0.6504^{20}		2.180		2.241	81.4	-20			
2-Methyl-2-butene	C_5H_{10}	70.133	-133.72	38.56	0.6623^{20}	0.203	1.979		2.179	62.1	-20			
Methyl *tert*-butyl ether	$C_5H_{12}O$	88.148	-108.6	55.0	0.7353^{25}				2.127	33.6				50
Methyl chloroacetate	$C_3H_5ClO_2$	108.524	-32.1	129.5	1.236^{20}		12.0			1.0	57	7.5-18.5%		
Methylcyclohexane	C_7H_{14}	98.186	-126.6	100.93	0.7694^{20}	0.679	2.024	≈ 0	1.882	6.18	-4	1-7%	250	400
Methylcyclopentane	C_6H_{12}	84.159	-142.42	71.8	0.7486^{20}	0.479	1.9853	≈ 0	1.886	18.3	-29	1-8%	258	
N-Methylformamide	C_2H_5NO	59.067	-3.8	199.51	1.011^{19}	1.678	189.0	3.83	2.096	0.03				
Methyl formate	$C_2H_4O_2$	60.052	-99	31.7	0.9713^{20}	0.325	9.20	1.77	1.983	78.1	-19	5-23%	449	100
5-Methyl-2-hexanone	$C_7H_{14}O$	114.185		144	0.888^{20}		13.53			0.691	36	1-8%	191	50
Methylhydrazine	CH_6N_2	46.072	-52.36	87.5					2.928	6.61	-8	2.5-92%	194	0.01
Methyl isocyanate	C_2H_3NO	57.051	-45	39.5	0.9230^{27}		21.75	≈ 2.8		57.7	-7	5.3-26%	534	0.02
Methyl lactate, (±)-	$C_4H_8O_3$	104.105		144.8	1.0928^{20}					0.62	49	>2.2%	385	
Methyl methacrylate	$C_5H_8O_2$	100.117	-47.55	100.5	0.9377^{25}		6.32	1.67	1.910	5.10	10	1.7-8.2%		50
1-Methylnaphthalene	$C_{11}H_{10}$	142.197	-30.43	244.7	1.0202^{20}		2.915	≈ 0	1.578	0.009			529	
Methyloxirane	C_3H_6O	58.079	-111.9	35	0.859^{0}			2.01	2.073	71.7	-37	3.1-27.5%	449	2
2-Methylpentane	C_6H_{14}	86.175	-153.6	60.26	0.650^{25}	0.286	1.886	≈ 0	2.248	28.2	<-29	1-7%	264	500
3-Methylpentane	C_6H_{14}	86.175	-162.90	63.27	0.6598^{25}	0.306	1.886	≈ 0	2.213	25.3	-7	1.2-7%	278	500
2-Methyl-2,4-pentanediol	$C_6H_{14}O_2$	118.174	-50	197.1	0.923^{15}		23.4	2.9	2.843	<0.01	102	1-9%	306	25
2-Methyl-1-pentanol	$C_6H_{14}O$	102.174		149	0.8263^{20}				2.427	0.236	54	1.1-9.65%	310	
4-Methyl-2-pentanol	$C_6H_{14}O$	102.174	-90	131.6	0.8075^{20}	4.07			2.672	0.698	41	1-6%		25
4-Methyl-2-pentanone	$C_6H_{12}O$	100.158	-84	116.5	0.7965^{25}	0.545	13.11		2.130	2.64	18	1-8%	448	50
2-Methylpropanenitrile	C_4H_7N	69.106	-71.5	103.9	0.7704^{20}		24.42	4.29			8		482	
2-Methyl-2-propanethiol	$C_4H_{10}S$	90.187	-0.5	64.2	0.7943^{25}		5.475	1.66		24.2	<-29			
Methyl propanoate	$C_4H_8O_2$	88.106	-87.5	79.8	0.9150^{20}	0.431	6.200		1.943	11.5	-2	2.5-13%	469	
2-Methylpropanoic acid	$C_4H_8O_2$	88.106	-46	154.45	0.9681^{20}	1.226	2.58	1.08	1.964	0.17	56	2-9.2%	481	
2-Methyl-1-propanol	$C_4H_{10}O$	74.121	-101.9	107.89	0.8018^{20}	3.33	17.93	1.64	2.449	1.39	28	2-11%	415	50
2-Methyl-2-propanol	$C_4H_{10}O$	74.121	25.69	82.4	0.7887^{20}	4.31	12.47	1.7	2.949	5.52	11	2-8%	478	100
2-Methylpyridine	C_6H_7N	93.127	-66.68	129.38	0.9443^{20}		10.18	1.85	1.703	1.5	39		538	
3-Methylpyridine	C_6H_7N	93.127	-18.14	144.14	0.9566^{20}		11.10	2.40	1.704	0.795				
4-Methylpyridine	C_6H_7N	93.127	3.67	145.36	0.9548^{20}		12.2	2.70	1.707	0.759	57			
N-Methyl-2-pyrrolidone	C_5H_9NO	99.131	-23.09	202	1.0230^{25}		32.55	4.1	3.105	0.04	96	1-10%	346	
Methyl salicylate	$C_8H_8O_3$	152.148	-8	222.9	1.181^{25}		8.80	2.47	1.637	0.015	96		454	
4-Methylstyrene	C_9H_{10}	118.175	-34.1	172.8	0.9173^{25}					0.245	53	0.8-11%	538	50
Morpholine	C_4H_9NO	87.120	-4.8	128	1.0005^{20}	2.02	7.42	1.55	1.892	1.34	37	1-11%	290	20
β-Myrcene	$C_{10}H_{16}$	136.234		167	0.8013^{15}		2.3			0.280				
Nickel carbonyl	C_4NiO_4	170.734	-19.3	43 (exp 60)	1.31^{25}				1.198					0.05
L-Nicotine	$C_{10}H_{14}N_2$	162.231	-79	247	1.0097^{20}		8.937							0.1
Nitric acid	HNO_3	63.013	-41.6	83	1.55			2.17	1.744	8.31				2
2-Nitroanisole	$C_7H_7NO_3$	153.136	10.5	272	1.2540^{20}		45.75	5.0		0.002				
Nitrobenzene	$C_6H_5NO_2$	123.110	5.7	210.8	1.2037^{20}	1.863	35.6	4.22	1.509	0.03	88	2-9%	482	1
Nitroethane	$C_2H_5NO_2$	75.067	-89.5	114.0	1.0448^{25}	0.688	29.11	3.23	1.790	2.79	28	3-17%	414	100
Nitromethane	CH_3NO_2	61.041	-28.38	101.19	1.1371^{20}	0.630	37.27	3.46	1.746	4.79	35	7-22%	418	20
1-Nitropropane	$C_3H_7NO_2$	89.094	-108	131.1	0.9961^{25}	0.798	24.70	3.66	1.97	1.36	36	2%-	421	25
2-Nitropropane	$C_3H_7NO_2$	89.094	-91.3	120.2	0.9821^{25}		26.74	3.73	1.911	2.3	24	3-11%	428	10
N-Nitrosodiethylamine	$C_4H_{10}N_2O$	102.134		176.9	0.9422^{20}									
N-Nitrosodimethylamine	$C_2H_6N_2O$	74.081		152	1.0048^{20}					0.73				
2-Nitrotoluene	$C_7H_7NO_2$	137.137	-10.4	222	1.1611^{19}		26.26		1.474	0.0014	106			2
3-Nitrotoluene	$C_7H_7NO_2$	137.137	15.5	232	1.1581^{20}		24.95		1.474	0.03	106			2
Nonane	C_9H_{20}	128.255	-53.46	150.82	0.7192^{20}	0.665	1.9722	≈ 0	2.217	0.570	31	0.8-2.9%	205	200
Nonanoic acid	$C_9H_{18}O_2$	158.238	12.4	254.5	0.9052^{20}	7.01	2.475	0.79	2.290	0.00005				
1-Nonanol	$C_9H_{20}O$	144.254	-5	213.37	0.8280^{20}	9.12	8.83		2.470	0.00050			260	
1-Nonene	C_9H_{18}	126.239	-81.3	146.9	0.7253^{25}	0.586	2.180	≈ 0	2.142	0.714	26			
4-Nonylphenol	$C_{15}H_{24}O$	220.351	42	≈ 295	0.950^{20}									
cis,cis-9,12-Octadecadienoic acid	$C_{18}H_{32}O_2$	280.446	-7		0.9022^{20}		2.754							
cis-9-Octadecenoic acid	$C_{18}H_{34}O_2$	282.462	13.4	360	0.8935^{20}		2.336	1.18	2.043	0.000001	189		363	
Octane	C_8H_{18}	114.229	-56.82	125.67	0.6986^{25}	0.508	1.948	≈0	2.229	1.86	13	1-7%	206	300
Octanoic acid	$C_8H_{16}O_2$	144.212	16.5	239	0.9073^{25}	5.02	2.85	1.15	2.066	0.0002				
1-Octanol	$C_8H_{18}O$	130.228	-14.8	195.16	0.8262^{25}	7.29	10.30	1.8	2.344	0.01	81		270	
2-Octanol	$C_8H_{18}O$	130.228	-31.6	179.3	0.8193^{20}	6.49	8.13	1.71	2.535		88		265	

Name	Mol. form.	M_r	t_m/°C	t_b/°C	ρ/g mL^{-1}	η/mPa s	ε	μ/D	c_p/J g^{-1}K^{-1}	vp/kPa	FP/ °C	Fl. lim.	IT/°C	TLV/ppm
2-Octanone	$C_8H_{16}O$	128.212	-16	172.5	0.820^{20}		9.51	2.7	2.132	0.12	52			
1-Octene	C_8H_{16}	112.213	-101.7	121.29	0.7149^{20}	0.447	2.113	≈ 0	2.148	2.30	21		230	
Oxetane	C_3H_6O	58.079	-97	47.6	0.8930^{25}			1.94						
2-Oxetanone	$C_3H_4O_2$	72.063	-33.4	162	1.1460^{20}			4.18	1.694	0.3	74	>2.9%		0.5
Oxirane	C_2H_4O	44.052	-112.5	10.6	0.8821^{10}		12.42	1.89	1.998	175	-20	3-100%	429	1
Oxiranemethanol, (±)-	$C_3H_6O_2$	74.079	-45	167 dec	1.1143^{25}									2
Paraldehyde	$C_6H_{12}O_3$	132.157	12.6	124.3	0.9943^{20}	1.079		1.43		1.6	36	>1.3%	238	
Parathion	$C_{10}H_{14}NO_5PS$	291.261	6.1	375	1.2681^{20}									0.01
Pentachloroethane	C_2HCl_5	202.294	-28.78	162.0	1.6796^{20}	2.25	3.716	0.92	0.859	0.478				
cis-1,3-Pentadiene	C_5H_8	68.118	-140.8	44.1	0.6910^{20}		2.319	0.500		50.6				
trans-1,3-Pentadiene	C_5H_8	68.118	-87.4	42	0.6710^{25}			0.585		54.7				
Pentanal	$C_5H_{10}O$	86.132	-91.5	103	0.8095^{20}		10.00			4.58	12		222	50
Pentane	C_5H_{12}	72.149	-129.67	36.06	0.6262^{20}	0.224	1.8371	≈ 0	2.317	68.3	-40	2-8%	260	600
Pentanedial	$C_5H_8O_2$	100.117	-14	188 dec										0.05
1,5-Pentanediol	$C_5H_{12}O_2$	104.148	-18	239	0.9914^{20}		26.2	2.5	3.08	0.001	129		335	
2,4-Pentanedione	$C_5H_8O_2$	100.117	-23	138	0.9721^{25}		26.524	2.8	2.08	1.02	34		340	
1-Pentanethiol	$C_5H_{12}S$	104.214	-75.65	126.6	0.850^{20}		4.847			1.83	18			
Pentanoic acid	$C_5H_{10}O_2$	102.132	-33.6	186.1	0.9339^{25}		2.661	1.61	2.059	0.024	96		400	
1-Pentanol	$C_5H_{12}O$	88.148	-77.6	137.98	0.8144^{20}	3.62	15.13	1.7	2.361	0.259	33	1-10%	300	
2-Pentanol	$C_5H_{12}O$	88.148	-73	119.3	0.8094^{20}	3.47	13.71	1.66	2.716	0.804	34	1.2-9%	343	
3-Pentanol	$C_5H_{12}O$	88.148	-69	116.25	0.8203^{20}	4.15	13.35	1.64	2.719	1.10	41	1.2-9%	435	
2-Pentanone	$C_5H_{10}O$	86.132	-76.8	102.26	0.809^{20}	0.470	15.45	2.7	2.137	4.97	7	2-8%	452	200
3-Pentanone	$C_5H_{10}O$	86.132	-39	101.7	0.8098^{25}	0.444	17.00	2.82	2.216	4.72	13	>1.6%	450	200
1-Pentene	C_5H_{10}	70.133	-165.12	29.96	0.6405^{20}	0.195	2.011	≈ 0.5	2.196	85.0	-18	1.5-8.7%	275	
cis-2-Pentene	C_5H_{10}	70.133	-151.36	36.93	0.6556^{20}			≈ 0	2.163	66.0	<-20			
trans-2-Pentene	C_5H_{10}	70.133	-140.21	36.34	0.6431^{25}			≈ 0	2.239	67.4	<-20			
Pentyl acetate	$C_7H_{14}O_2$	130.185	-70.8	149.2	0.8756^{20}		4.79	1.75	2.005	0.60	16	1-8%	360	50
Pentylamine	$C_5H_{13}N$	87.164	-55	104.3	0.7544^{20}	0.702	4.27		2.501	4.00	-1	2.2-22%		
Perchloric acid	$ClHO_4$	100.459	-112	≈ 90 dec	1.77									
Peroxyacetic acid	$C_2H_4O_3$	76.051	-0.2	110	1.226^{15}					1.93	41			
Phenol	C_6H_6O	94.111	40.89	181.87	1.0545^{45}		12.40	1.224	2.123	0.055	79	1.8-8.6%	715	5
2-Phenoxyethanol	$C_8H_{10}O_2$	138.164	14	245	1.102^{22}					0.001	121			
Phenylhydrazine	$C_6H_8N_2$	108.141	20.6	243.5	1.0986^{20}	13.03	7.15		2.007	0.003	88			0.1
1-Phenyl-2-propylamine, (±)-	$C_9H_{13}N$	135.206		203	0.9306^{25}					0.06	<100			
Phosphinic acid	H_3O_2P	65.997	26.5	130	1.49									
Phosphoric acid	H_3O_4P	97.995	42.4	407					1.480					0.25
Phosphorothioc trichloride	Cl_3PS	169.398	-36.2	125	1.635		4.94							
Phosphorus(III) bromide	Br_3P	270.686	-41.5	173.2	2.8					0.38				
Phosphorus(III) chloride	Cl_3P	137.332	-93.6	76.1	1.574	0.529	3.498	0.56		16.1				0.2
Phosphoryl chloride	Cl_3OP	153.331	1.18	105.5	1.645		14.1	2.54	0.905	4.97				0.1
α-Pinene	$C_{10}H_{16}$	136.234	-64	156.2	0.8539^{25}		2.1787			0.64	33		255	
β-Pinene	$C_{10}H_{16}$	136.234	-61.5	166	0.860^{25}		2.4970			0.61	38		275	
Piperidine	$C_5H_{11}N$	85.148	-11.02	106.22	0.8606^{20}	1.573	4.33	1.2	2.113	4.28	16	1-10%		
Propanal	C_3H_6O	58.079	-80	48	0.8657^{25}	0.321	18.5	2.72	2.362	42.2	-30	2.6-17%	207	20
1,2-Propanediol	$C_3H_8O_2$	76.095	-60	187.6	1.0361^{20}	40.4	27.5	2.2	2.507	0.02	99	3-13%	371	
1,3-Propanediol	$C_3H_8O_2$	76.095	-27.7	214.4	1.0538^{20}		35.1	2.5		0.007			400	
Propanenitrile	C_3H_5N	55.079	-92.78	97.14	0.7818^{20}	0.294	29.7	4.05	2.166	6.14	2	3-14%	512	
Propanoic acid	$C_3H_6O_2$	74.079	-20.5	141.15	0.9882^{25}	1.030	3.44	1.75	2.063	0.553	52	2.9-12.1%	465	10
Propanoic anhydride	$C_6H_{10}O_3$	130.141	-45	170	1.0110^{20}		18.30		1.806	0.45	63	1.3-9.5%	285	
1-Propanol	C_3H_8O	60.095	-124.39	97.2	0.7997^{25}	1.945	20.8	1.55	2.395	2.76	23	2-14%	412	200
2-Propanol	C_3H_8O	60.095	-87.9	82.3	0.7809^{25}	2.04	20.18	1.56	2.604	6.02	12	2-13%	399	200
Propargyl alcohol	C_3H_4O	56.063	-51.8	113.6	0.9478^{20}		20.8	1.13			36			1
Propyl acetate	$C_5H_{10}O_2$	102.132	-93	101.54	0.8878^{20}	0.544	5.62	1.8	1.921	4.49	13	2-8%	450	200
Propylamine	C_3H_9N	59.110	-84.75	47.22	0.7173^{20}	0.376	5.08	1.17	2.776	42.1	-37	2-10%	318	
Propylbenzene	C_9H_{12}	120.191	-99.6	159.24	0.8593^{25}		2.370	≈ 0	1.786	0.45	30	1-6%	450	
Propyl butanoate	$C_7H_{14}O_2$	130.185	-95.2	143.0	0.8730^{20}		4.3			0.618	37			
Propylene carbonate	$C_4H_6O_3$	102.089	-48.8	242	1.2047^{20}		66.14	4.9	2.141	0.05	135			
Propyl formate	$C_4H_8O_2$	88.106	-92.9	80.9	0.9073^{20}	0.485	6.92	1.89	1.945	10.9	-3		455	
Propyl propanoate	$C_6H_{12}O_2$	116.158	-75.9	122.5	0.8809^{20}		5.249			1.88	79			
Pyridine	C_5H_5N	79.101	-41.70	115.23	0.9819^{20}	0.879	13.260	2.21	1.678	2.76	20	2-12%	482	1
Pyrrole	C_4H_5N	67.090	-23.39	129.79	0.9698^{20}	1.225	8.00	1.74	1.903	1.10	39			
Pyrrolidine	C_4H_9N	71.121	-57.79	86.56	0.8586^{20}	0.704	8.30	1.6	2.202	8.40	3			
2-Pyrrolidone	C_4H_7NO	85.105	25	251	1.120^{20}		28.18	3.5	1.99		129			
Quinoline	C_9H_7N	129.159	-14.78	237.16	1.0977^{15}	3.34	9.16	2.29	1.51	0.011			480	
Safrole	$C_{10}H_{10}O_2$	162.185	11.2	234.5	1.1000^{20}					0.01	100			
Salicylaldehyde	$C_7H_6O_2$	122.122	-7	197	1.1674^{20}		18.35	2.86	1.818	0.075	78			

Name	Mol. form.	M_r	t_m/°C	t_b/°C	ρ/g mL⁻¹	η/mPa s	ε	μ/D	c_p/J g⁻¹K⁻¹	vp/kPa	FP/ °C	Fl. lim.	IT/°C	TLV/ppm
Selenium chloride	Cl_2Se_2	228.83	-85	130 dec	2.774									
Selenium oxychloride	Cl_2OSe	165.86	8.5	177	2.44		46.2			0.02				
Selenium oxyfluoride	F_2OSe	132.96	15	125	2.8					0.56				
Styrene	C_8H_8	104.150	-30.65	145	0.9016^{25}	0.695	2.4737	0.123	1.747	0.81	31	1-7%	490	20
Sulfolane	$C_4H_8O_2S$	120.171	27.6	287.3	1.2723^{18}		43.26	4.8	1.498	<0.01	177			
Sulfur chloride	Cl_2S_2	135.037	-77	137	1.69		4.79			1.27				1
Sulfur dichloride	Cl_2S	102.971	-122	59.6	1.62		2.915	0.36		17.9				
Sulfuric acid	H_2O_4S	98.080	10.31	337	1.8				1.416					0.05
Sulfuryl chloride	Cl_2O_2S	134.970	-51	69.4	1.680		9.1	1.81	0.993	18.7				
α-Terpinene	$C_{10}H_{16}$	136.234		174	0.8375^{19}		2.4526							
1,1,2,2-Tetrabromoethane	$C_2H_2Br_4$	345.653	0	243.5	2.9655^{20}		6.72	1.38	0.479	0.003			335	1
Tetrabromosilane	Br_4Si	347.702	5.39	154	2.8			0						
1,1,2,2-Tetrachloro-1,2-difluoroethane	$C_2Cl_4F_2$	203.830	24.8	92.8	1.5951^{50}		2.52		0.852	7.51				500
1,1,1,2-Tetrachloroethane	$C_2H_2Cl_4$	167.849	-70.2	130.2	1.5406^{20}	1.437			0.92	1.6	47	5-12%		
1,1,2,2-Tetrachloroethane	$C_2H_2Cl_4$	167.849	-42.4	145.2	1.5953^{20}		8.50	1.32	0.967	0.622	62	20-54%		1
Tetrachloroethene	C_2Cl_4	165.833	-22.3	121.3	1.6230^{20}	0.844	2.268	0	0.865	2.42	45			25
Tetrachloromethane	CCl_4	153.823	-22.62	76.8	1.5940^{20}	0.908	2.2379	0	0.850	15.2				5
Tetrachlorosilane	Cl_4Si	169.897	-68.74	57.65	1.5	99.4		0	0.855	31.3				
Tetradecane	$C_{14}H_{30}$	198.388	5.82	253.58	0.7596^{20}	2.13	2.0343	≈ 0		0.002	112	>0.5%	200	
Tetraethylene glycol	$C_8H_{18}O_5$	194.226	-6.2	328	1.1285^{15}		20.44		2.208	0.000001	182			
Tetrafluoroboric acid	BF_4H	87.813		130 dec	~1.8									
Tetrahydrofuran	C_4H_8O	72.106	-108.44	65	0.8833^{25}	0.456	7.52	1.75	1.720	21.6	-14	2-12%	321	200
Tetrahydrofurfuryl alcohol	$C_5H_{10}O_2$	102.132	< 80	178	1.0524^{20}		13.48	2.1	1.774	0.100	75	1.5-9.7%	282	
1,2,3,4-Tetrahydronaphthalene	$C_{10}H_{12}$	132.202	-35.7	207.6	0.9645^{25}	2.14	2.771	≈ 0	1.645	0.05	71	1 5%	385	
Tetrahydropyran	$C_5H_{10}O$	86.132	-49.1	88	0.8814^{20}		5.66	1.74	1.82	9.54	-20			
Tetrahydrothiophene	C_4H_8S	88.172	-96.2	121.1	0.9987^{20}	0.973		1.90		2.45				
Tetramethylsilane	$C_4H_{12}Si$	88.224	-99.06	26.6	0.648^{19}		1.921	0	2.313	94.2				
Tetramethylurea	$C_5H_{12}N_2O$	116.161	-0.6	176.5	0.9687^{20}		23.10	3.5		0.138	77			
Tetranitromethane	CN_4O_8	196.033	13.8	126.1	1.6380^{20}		2.317	0		1.13				0.005
Thionyl bromide	Br_2OS	207.873	-50	140			9.06			0.84				
Thionyl chloride	Cl_2OS	118.970	-101	75.6	1.631		8.675	1.45	1.017	16.0				1
Thiophene	C_4H_4S	84.140	-38.21	84.0	1.0649^{20}		2.739	0.55	1.471	10.6	-1			
Tin(IV) chloride	Cl_4Sn	260.521	-34.07	114.15	2.234			0	0.634					
Titanium(IV) chloride	Cl_4Ti	189.678	-24.12	136.45	1.73				0.766					
Toluene	C_7H_8	92.139	-94.95	110.63	0.8668^{20}	0.560	2.379	0.37	1.707	3.79	4	1-7%	480	50
Toluene-2,4-diisocyanate	$C_9H_6N_2O_2$	174.156	20.5	251	1.2244^{20}		8.433		1.653	0.003	127	0.9-9.5%		0.005
Tribromomethane	$CHBr_3$	252.731	8.69	149.1	2.8788^{25}	1.857	4.404	0.99	0.517	0.726	83			0.5
Tributylamine	$C_{12}H_{27}N$	185.349	-70	216.5	0.7770^{20}		2.340	0.8		0.01	63	1-5%		
Tributyl borate	$C_{12}H_{27}BO_3$	230.151	<-70	234	0.8567^{20}		2.23	0.77			93			
Tributyrin	$C_{15}H_{26}O_6$	302.363	-75	307.5	1.0350^{20}		5.72		1.837		180	>0.5%	407	
Trichloroacetaldehyde	C_2HCl_3O	147.387	-57.5	97.8	1.512^{20}		6.8		1.025	6.66				
1,2,4-Trichlorobenzene	$C_6H_3Cl_3$	181.447	16.92	213.5	1.459^{25}					0.057	105	2.5-6.6%	571	5
1,1,1-Trichloroethane	$C_2H_3Cl_3$	133.404	-30.01	74.09	1.3390^{20}	0.793	7.243	1.76	1.082	16.5	-1	8-13%	500	350
1,1,2-Trichloroethane	$C_2H_3Cl_3$	133.404	-36.3	113.8	1.4397^{20}		7.1937	1.4	1.131	3.1	32	6-28%	400	10
Trichloroethene	C_2HCl_3	131.388	-84.7	87.21	1.4642^{20}	0.545	3.390	0.8	0.947	9.91	32	8-11%	420	50
Trichloroethylsilane	$C_2H_5Cl_3Si$	163.506	-105.6	100.5	1.2373^{20}			2.04		6.29	22			
Trichlorofluoromethane	CCl_3F	137.368	-110.44	23.7	1.4879^{20}	0.421	3.00	0.46	0.885	106				1000
Trichloromethane	$CHCl_3$	119.378	-63.41	61.17	1.4788^{25}	0.537	4.8069	1.04	0.957	26.2				10
(Trichloromethyl)benzene	$C_7H_5Cl_3$	195.474	-4.42	221	1.3723^{20}		6.9	2.03		0.35	127		211	0.1
Trichloromethylsilane	CH_3Cl_3Si	149.480	-90	65.6	1.273^{20}			1.91	1.091	22.5	-9	7.6->20%	>404	
Trichloronitromethane	CCl_3NO_2	164.376	-64	112	1.6558^{20}		7.319			3.18				0.1
1,2,3-Trichloropropane	$C_3H_5Cl_3$	147.431	-14.7	157	1.3889^{20}		7.5		1.245	0.492	71	3.2-12.6%		10
Trichlorosilane	Cl_3HSi	135.452	-128.2	33	1.331	0.326		0.86			-50		104	
1,1,2-Trichloro-1,2,2-trifluoroethane	$C_2Cl_3F_3$	187.375	-36.22	47.7	1.5635^{25}	0.656	2.41		0.908	44.8				1000
Tri-*o*-cresyl phosphate	$C_{21}H_{21}O_4P$	368.363	11	410	1.1955^{20}		6.7	2.87	1.57	0.0000002	225		385	0.01
Tridecane	$C_{13}H_{28}$	184.361	-5.4	235.47	0.7564^{20}	1.724	2.0213	≈ 0	2.206	0.005	79			
1-Tridecene	$C_{13}H_{26}$	182.345	-13	232.8	0.7658^{20}	1.50	2.139	≈ 0	2.149	0.0047	79			
Triethanolamine	$C_6H_{15}NO_3$	149.188	20.5	335.4	1.1242^{20}	609	29.36	3.6	2.61	<0.01	179	1-10%		0.8
Triethylamine	$C_6H_{15}N$	101.190	-114.7	89	0.7275^{20}	0.347	2.418	0.66	2.173	7.70	-7	1-8%	249	1
Triethylene glycol	$C_6H_{14}O_4$	150.173	-7	285	1.1274^{15}		23.69		2.18	0.0002	177	1-9%	371	
Triethylene glycol dimethyl ether	$C_8H_{18}O_4$	178.227	-45	216	0.986^{20}		7.62				111			
Triethyl phosphate	$C_6H_{15}O_4P$	182.154	-56.4	215.5	1.0695^{20}		13.20	3.1			115		454	
Trifluoroacetic acid	$C_2HF_3O_2$	114.023	-15.2	73	1.5351^{25}	0.808	8.42	2.28		15.1				
(Trifluoromethyl)benzene	$C_7H_5F_3$	146.110	-28.95	102.1	1.1884^{20}		9.22	2.86	1.289	5.14	12			

Name	Mol. form.	M_r	t_m/°C	t_b/°C	ρ/g mL^{-1}	η/mPa s	ε	μ/D	c_p/J g^{-1}K^{-1}	vp/kPa	FP/ °C	Fl. lim.	IT/°C	TLV/ppm
1,2,3-Trimethylbenzene	C_9H_{12}	120.191	-25.4	176.12	0.8944^{20}		2.656	≈ 0	1.800	0.20	44	0.8-6.6%	470	25
1,2,4-Trimethylbenzene	C_9H_{12}	120.191	-43.77	169.38	0.8758^{20}		2.377	≈ 0	1.789	0.30	44	1-6%	500	25
1,3,5-Trimethylbenzene	C_9H_{12}	120.191	-44.72	164.74	0.8615^{25}		2.279	0	1.741	0.33	50	1-5%	559	25
Trimethyl borate	$C_3H_9BO_3$	103.912	-29.3	67.5	0.915^{25}		2.2762		1.828	17.2	-8			
Trimethylchlorosilane	C_3H_9ClSi	108.642	-40	60	0.856^{25}					30.7	-28		395	
2,2,4-Trimethylpentane	C_8H_{18}	114.229	-107.3	99.22	0.6878^{25}		1.943	≈ 0	2.093	6.50	-12		418	300
2,3,3-Trimethylpentane	C_8H_{18}	114.229	-100.9	114.8	0.7262^{20}		1.9780	≈ 0	2.150	3.60	<21		425	300
Trimethyl phosphate	$C_3H_9O_4P$	140.074	-46	197.2	1.2144^{20}		20.6	3.2		0.11	107			
2,4,6-Trimethylpyridine	$C_8H_{11}N$	121.180	-46	170.6	0.9166^{22}		7.807	2.05		4.1				
Trinitroglycerol	$C_3H_5N_3O_9$	227.087	13.5	exp 218	1.5931^{20}		19.25			0.00005			270	0.05
Undecane	$C_{11}H_{24}$	156.309	-25.5	195.9	0.7402^{20}	1.098	1.9972	≈ 0	2.207	0.05	69			
Vanadium(IV) chloride	Cl_4V	192.753	-25.7	148	1.816		3.05							
Vanadyl trichloride	Cl_3OV	173.299	-79	127	1.829		3.4							
Vinyl acetate	$C_4H_6O_2$	86.090	-93.2	72.8	0.9256^{25}			1.79	1.969	15.4	-8	2.6-13.4%	402	10
4-Vinylcyclohexene	C_8H_{12}	108.181	-108.9	128	0.8299^{20}					1.87	16		269	0.1
Water	H_2O	18.015	0.00	100.0	0.9970	0.890	80.100	1.8546	4.180	3.17				
o-Xylene	C_8H_{10}	106.165	-25.2	144.5	0.8802^{10}	0.760	2.562	0.64	1.753	0.88	32	1-7%	463	100
m-Xylene	C_8H_{10}	106.165	-47.8	139.12	0.8596^{25}	0.581	2.359	≈ 0	1.724	1.13	27	1-7%	527	100
p-Xylene	C_8H_{10}	106.165	13.25	138.37	0.8566^{25}	0.603	2.2735	0	1.710	1.19	27	1-7%	528	100
2,4-Xylenol	$C_8H_{10}O$	122.164	24.5	210.98	0.9650^{20}		5.060	1.4		0.022				

MISCIBILITY OF ORGANIC SOLVENTS

The chart below gives qualitative information on the miscibility of pairs of organic liquids. Two liquids are considered miscible (indicated by **M** in the chart) if mixing equal volumes produces a single liquid phase. If two phases separate, they are considered immiscible (**I**). An entry of **P** indicates two phases whose volumes differ appreciably, suggesting a partial miscibility of the components. The symbol **R** indicates a reaction between the components. All data refer to room temperature.

The codes for the columns are:

A Acetone
B Benzaldehyde
C Benzene
D Butyl acetate
E Butyl alcohol
F Carbon tetrachloride
G 2-Chloroethanol
H Chloroform
I *o*-Cresol
J Diethyl ether
K *N,N*-Dimethylaniline
L Dipentylamine
M Ethyl alcohol
N Ethylene glycol
O Ethylene glycol monoethyl ether
P Formamide
Q Furfuryl alcohol
R Glycerol
S Methyl isopropyl ketone
T Nitromethane
U 1-Octanol
V 1,3-Propanediol
W Pyridine
X Triethylenetetramine
Y Triethyl phosphate

References

1. Drury, J. S., *Ind. Eng. Chem.* 44, 2744, 1959.
2. Jackson, W. M., and Drury, J. S., *Ind. Eng. Chem.* 51, 1491, 1959.

	A	B	C	D	E	F	G	H	I	J	K	L	M	N	O	P	Q	R	S	T	U	V	W	X	Y
Acetone	-	M	M	M	M	M	M	M		M	M	M	M	M	M	M	M	I	M	M	M	M	M	M	M
Adiponitrile	M		M		M	I				I	M		M	I		M	M	I				M	M		
2-Amino-2-methyl-1- propanol	M	M	M	M	M	M				M	M		M	M		M	M	M		M	M	M	M		
p-Anisaldehyde							M							I	M			I				I			M
Benzaldehyde	M	-	M	M	M	M				M	M		M	P		M	M	P		M	M	M	M		
Benzene	M	M	-	M	M	M	M	M		M	M	M	M	I	M	I	M	I	M	I	M	I	M	M	M
Benzonitrile	M		M		M	M				M	M		M	I		I	M	I				I	M		
Benzothiazole	M		M		M	M				M	M		M	M		I	M	I				M	M		
Benzyl alcohol	M	M	M	M	M	M				M	M		M	M		M	M	M		M	M	M	M		
Benzyl mercaptan	M		M		M	M				M	M		M	I		I	M	I				I	M		
2-Bromoethyl acetate	M		P					M				R	M						M					R	
1,3-Butanediol	M		I					M	M	P		M	M						M				M	M	
2,3-Butanediol	M		P					M	M	M		M	M						M				M	M	
Butyl acetate	M	M	M	-	M	M				M	M		M	P		I	M	I		M	M	P	M		
Butyl alcohol	M	M	M	M	-	M				M	M		M	M		M	M	M		M	M	M	M		
Carbon tetrachloride	M	M	M	M	M	-				M	M		M	I		I	M	I		M	M	I	M		
2-Chloroethanol	M		M				-		M	M		M							M				M	M	M
Chloroform	M		M				M	-		M		M	M	P	M			I	M			M		M	M
3-Chloro-1,2-propanediol	M		I					M	M	M		R	M						M				M	R	
Cinnamaldehyde	M		M				M	M				M	M	I	M			I	M			I		R	M
o-Cresol							M		-					M	M			M				M			M
Diacetone alcohol	M	M	P	M	M	P				M	M		M	M		M	M	I		M	M	M	M		
Dibenzyl ether	M		M				M	M				M		I	M			I	M					M	M
Dibutylamine							R							M	M			P							M
Dibutyl carbonate	M		M					M				M	M						M					I	
Dibutyl ether	M	M	M	M	M	M				M	M		M	I		I	M	I		I	M	I	M		
Diethanolamine	M	I	I	I	M	I				I	P		M	M		M	M	M		I	M	M	M		
Diethylacetic acid	M		M				M	M				R	M	M	M			I	M			M		R	M
Diethylene glycol dibutyl ether	M		M					M	M	M		R	M						M				M		
Diethylene glycol diethyl ether	M		M					M	M	M		M	M						M				M	M	
Diethylene glycol monobutyl ether	M		M					M	M	M		M	M						M				M	M	
Diethylene glycol monoethyl ether	M		M					M	M	M		M	M						M				M	M	
Diethylene glycol monomethyl ether	M		M					M		M		M	M						M				M	M	
Diethylenetriamine	M		M				R	M				I	M	M	M			M	R					M	M
Diethyl ether	M	M	M	M	M	M	M	M		-	M	M	M	I	M	I	M	I	M	M	M	I	M	M	M
Diethylformamide	M		M				M	M				R	M	M	M			M	M			M		R	M
Dihexyl ether	M		M				M	M				M	M	I	M			I	M			I		I	

	A	B	C	D	E	F	G	H	I	J	K	L	M	N	O	P	Q	R	S	T	U	V	W	X	Y
Diisobutyl ketone	M		M				M	M				M	M	I	M			I	M			I		M	M
Diisopropylamine	M		M				R	M				M	M	M	M			M	M			M		M	M
N,N-Dimethylaniline	M	M	M	M	M	M				M	-		M	I		I	M	I		M	M	I	M		
Dipentylamine	M		M					M		M		-	M	P	M			P	M			M		I	M
N,N-Dipropylaniline	M		M		M	M	M	M		M	M	M	M	I	M	I	M	I	M			I	M	M	M
Dipropylene glycol	M		M				M	M	M	M			M						M				M	M	
Ethyl alcohol	M	M	M	M	M	M		M		M	M	M	-	M	M	M	M	M	M	M	M	M	M	M	M
Ethyl benzoate	M	M	M	M	M	M		M		M	M	M	M	I		I	M	I	M	M	M	P	M	M	
Ethyl chloroacetate	M		M				M	M				M	M	I	M			I	M			I		R	M
Ethyl cinnamate	M		M					M				M	M	I	M			I	M			I		M	M
Ethylene glycol	M	P	I	P	M	I		P	M	I	I	P	M	-		M	M	M	I	I	M	M	M	M	
Ethylene glycol monobutyl ether	M		M					M	M	M		M	M						M				M	M	
Ethylene glycol monoethyl ether	M		M					M	M	M		M	M		-				M				M	M	
Ethylene glycol monomethyl ether	M		M						M	M		M	M										M	M	
2-Ethyl-1-hexanol	M	M	M	M	M	M				M	M		M	M		I	M	I		I	M	M	M		
Ethyl phenylacetate	M		M				M	M				M		I	M			I	M			I		M	M
Ethyl thiocyanate	M		M		M	M				M	M		M	I		I	M	I				I	M		
Formamide	M	M	I	I	M	I				I	I		M	M		-	M	M		M	I	M	M		
Furfuryl alcohol	M	M	M	M	M	M				M	M		M	M		M	-	M		M	M	M	M		
Glycerol	I	P	I	I	M	I		I	M	I	I	P	M	M		M	M	-	I	I	I	M	M	M	
1-Heptadecanol	M		M					M				M	M						M					M	
3-Heptanol	M		M				M	M				M	M	M	M			I				M		M	M
Heptyl acetate	M		M				M	M				M		I	M			I	M			I		R	M
Hexanenitrile	M		M				M	M				M	M	I	M			I	M			I		M	M
Isobutyl mercaptan	M		M		M	M				M	M		M	I		I	M	I				R	M		
Isopentyl acetate	M		M				M	M				M	M	I	M			I	M			I		M	M
Isopentyl alcohol	M	M	M	M	M	M				M	M		M	M		M	M	I		M	M	M	M		
Isopentyl sulfide	M		M		M	M				M	M		M	I		I	I	I				I	M		
Methyl disulfide	M		M		M	M				M	M		M	I		I	M	I				R	M		
Methyl isobutyl ketone	M	M	M	M	M	M				M	M		M	I		P	M	I		M	M	I	M		
Methyl isopropyl ketone	M		M				M	M		M		M	M	I	M			I	-			M		R	M
4-Methylpentanoic acid	M		M					M				M	M	M	M			I	M			M		R	M
Nitromethane	M	M	I	M	M	M				M	M		M	I		M	M	I		-	P	I	M		
1-Octanol	M	M	M	M	M	M				M	M		M	M		I	M	I		P	-	M	M		
o-Phenetidine	M		M				M	M				M	M	M	M				M			M		M	
1,2-Propanediol	M		I					M	M	P		M	M						M				M	M	
1,3-Propanediol	M	M	I	P	M	I		M	M	I	I	M	M	M		M	M	M	M	I	M	-	M	M	
Pyridine	M	M	M	M	M	M	M			M	M		M	M	M	M	M	M		M	M	M	-		M
Tetradecanol	M		M				M	M				M	M	I	M			I	M			P		M	M
Tributyl phosphate	M		M				M	M				M	M	P	M			I	M			M		M	M
Triethylene glycol	M		P					M	M	I		P	M						M				M	M	
Triethylenetetramine	M		M				M	M		M		I	M	M	M			M	R			M		-	M
Triethyl phosphate	M		M					M	M	M		M	M						M				M	M	-
2,6,8-Trimethyl-4-nonanone	M		M				M	M				M	M	I	M			I	M			I		I	M

DENSITY OF SOLVENTS AS A FUNCTION OF TEMPERATURE

The table below lists the density of several common solvents in the temperature range from 0 °C to 100 °C. The values have been calculated from the Rackett Equation using parameters in the reference. Density values refer to the liquid at its saturation vapor pressure; thus entries for temperatures above the normal boiling point are for pressures greater than atmospheric.

Reference

Lide, D. R., and Kehiaian, H. V., *Handbook of Thermophysical and Thermochemical Data*, CRC Press, Boca Raton, FL, 1994.

	Density in g/mL										
Solvent	**0 °C**	**10 °C**	**20 °C**	**30 °C**	**40 °C**	**50 °C**	**60 °C**	**70 °C**	**80 °C**	**90 °C**	**100 °C**
Acetic acid			1.051	1.038	1.025	1.012	0.9993	0.9861	0.9728	0.9592	0.9454
Acetone	0.8129	0.8016	0.7902	0.7785	0.7666	0.7545	0.7421	0.7293	0.7163	0.7029	0.6890
Acetonitrile			0.7825	0.7707	0.7591	0.7473	0.7353	0.7231	0.7106	0.6980	0.6851
Aniline	1.041	1.033	1.025	1.016	1.008	1.000	0.9909	0.9823	0.9735	0.9646	0.9557
Benzene		0.8884	0.8786	0.8686	0.8584	0.8481	0.8376	0.8269	0.8160	0.8049	0.7935
1-Butanol	0.8293	0.8200	0.8105	0.8009	0.7912	0.7812	0.7712	0.7609	0.7504	0.7398	0.7289
Butylamine	0.7606	0.7512	0.7417	0.7320	0.7221	0.7120	0.7017	0.6911	0.6803	0.6693	0.6579
Carbon disulfide	1.290	1.277	1.263	1.248	1.234						
Chlorobenzene	1.127	1.116	1.106	1.096	1.085	1.074	1.064	1.053	1.042	1.030	1.019
Cyclohexane		0.7872	0.7784	0.7694	0.7602	0.7509	0.7414	0.7317	0.7218	0.7117	0.7013
Decane	0.7447	0.7374	0.7301	0.7226	0.7151	0.7074	0.6997	0.6919	0.6839	0.6758	0.6676
1-Decanol			0.8294	0.8229	0.8162	0.8093	0.8024	0.7955	0.7884	0.7813	0.7740
Dichloromethane	1.362	1.344	1.326	1.307	1.289	1.269	1.250	1.229	1.208	1.187	1.165
Diethyl ether	0.7368	0.7254	0.7137	0.7018	0.6896	0.6770	0.6639	0.6505	0.6366	0.6220	0.6068
N,N-Dimethylaniline		0.9638	0.9562	0.9483	0.9401	0.9318	0.9234	0.9150	0.9064	0.8978	0.8890
Ethanol	0.8121	0.8014	0.7905	0.7793	0.7680	0.7564	0.7446	0.7324	0.7200	0.7073	0.6942
Ethyl acetate	0.9245	0.9126	0.9006	0.8884	0.8759	0.8632	0.8503	0.8370	0.8234	0.8095	0.7952
Ethylbenzene	0.8836	0.8753	0.8668	0.8582	0.8495	0.8407	0.8318	0.8228	0.8136	0.8043	0.7948
Ethyl formate	0.9472	0.9346	0.9218	0.9087	0.8954	0.8818	0.8678	0.8535	0.8389	0.8238	0.8082
Ethyl propanoate	0.9113	0.9005	0.8895	0.8784	0.8671	0.8556	0.8439	0.8319	0.8197	0.8072	0.7944
Heptane	0.7004	0.6921	0.6837	0.6751	0.6664	0.6575	0.6485	0.6393	0.6298	0.6202	0.6102
Hexane	0.6774	0.6685	0.6594	0.6502	0.6407	0.6311	0.6212	0.6111	0.6006	0.5899	0.5789
1-Hexanol	0.8359	0.8278	0.8195	0.8111	0.8027	0.7941	0.7854	0.7766	0.7676	0.7585	0.7492
Isopropylbenzene	0.8769	0.8696	0.8615	0.8533	0.8450	0.8366	0.8280	0.8194	0.8106	0.8017	0.7927
Methanol	0.8157	0.8042	0.7925	0.7807	0.7685	0.7562	0.7435	0.7306	0.7174	0.7038	0.6898
Methyl acetate	0.9606	0.9478	0.9346	0.9211	0.9074	0.8933	0.8790	0.8643	0.8491	0.8336	0.8176
N-Methylaniline	1.0010	0.9933	0.9859	0.9785	0.9709	0.9633	0.9556	0.9478	0.9399	0.9319	0.9239
Methylcyclohexane	0.7858	0.7776	0.7693	0.7608	0.7522	0.7435	0.7346	0.7255	0.7163	0.7069	0.6973
Methyl formate	1.003	0.9887	0.9739	0.9588	0.9433	0.9275	0.9112	0.8945	0.8772	0.8594	0.8409
Methyl propanoate	0.9383	0.9268	0.9150	0.9030	0.8907	0.8783	0.8656	0.8526	0.8393	0.8257	0.8117
Nitromethane			1.139	1.125	1.111	1.097	1.083	1.069	1.055	1.040	1.026
Nonane	0.7327	0.7252	0.7176	0.7099	0.7021	0.6941	0.6861	0.6779	0.6696	0.6611	0.6525
Octane	0.7185	0.7106	0.7027	0.6945	0.6863	0.6779	0.6694	0.6608	0.6520	0.6430	0.6338
Pentanoic acid	0.9563	0.9476	0.9389	0.9301	0.9211	0.9121	0.9029	0.8937	0.8843	0.8748	0.8652
1-Propanol	0.8252	0.8151	0.8048	0.7943	0.7837	0.7729	0.7619	0.7506	0.7391	0.7273	0.7152
2-Propanol	0.8092	0.7982	0.7869	0.7755	0.7638	0.7519	0.7397	0.7272	0.7143	0.7011	0.6876
Propyl acetate	0.9101	0.8994	0.8885	0.8775	0.8662	0.8548	0.8432	0.8313	0.8192	0.8069	0.7942
Propylbenzene	0.8779	0.8700	0.8619	0.8538	0.8456	0.8373	0.8289	0.8204	0.8117	0.8030	0.7943
Propyl formate	0.9275	0.9166	0.9053	0.8938	0.8821	0.8702	0.8581	0.8457	0.8330	0.8201	0.8068
Tetrachloromethane	1.629	1.611	1.593	1.575	1.557	1.538	1.518	1.499	1.479	1.458	1.437
Toluene	0.8846	0.8757	0.8667	0.8576	0.8483	0.8389	0.8294	0.8197	0.8098	0.7998	0.7896
Trichloromethane	1.524	1.507	1.489	1.471	1.452	1.433	1.414	1.394			
2,2,4-Trimethylpentane			0.6921	0.6836	0.6750	0.6663	0.6574	0.6484	0.6391	0.6296	0.6199
o-Xylene			0.8801	0.8717	0.8633	0.8547	0.8460	0.8372	0.8282	0.8191	0.8099
m-Xylene	0.8813	0.8729	0.8644	0.8558	0.8470	0.8382	0.8292	0.8201	0.8109	0.8015	0.7920
p-Xylene			0.8609	0.8523	0.8436	0.8347	0.8258	0.8167	0.8075	0.7981	0.7886

DEPENDENCE OF BOILING POINT ON PRESSURE

The normal boiling point of a liquid is defined as the temperature at which the vapor pressure reaches standard atmospheric pressure, 101.325 kPa. The change in boiling point with pressure may be calculated from the representation of the vapor pressure by the Antoine Equation,

$$\ln p = A_1 - A_2/(T + A_3)$$

where p is the vapor pressure, T the absolute temperature, and A_1, A_2, and A_3 are constants. This table, which has been calculated using the Antoine constants in Reference 1, gives values of $\Delta t/\Delta p$ for a number of liquids, in units of both °C/kPa and °C/mmHg. The correction to the boiling point is generally accurate to 0.1 to 0.2 °C as long as the pressure is within 10% of standard atmospheric pressure.

A slightly less accurate estimate of $\Delta t/\Delta p$ may be obtained from the Clausius-Clapeyron equation, with the assumption that the change in volume upon vaporization equals the ideal-gas volume of the vapor. This leads to the equation

$$\Delta t/\Delta p = RT_b^2/p_0\Delta_{vap}H(T_b)$$

where R is the molar gas constant, p_0 is 101.325 kPa, T_b is the normal boiling point temperature (absolute), and $\Delta_{vap}H(T_b)$ is the molar enthalpy of vaporization at the normal boiling point. Values of the last quantity may be obtained from the table "Enthalpy of Vaporization" in Section 6.

Reference

1. Lide, D. R., and Kehiaian, H. V., *CRC Handbook of Thermophysical and Thermochemical Data*, CRC Press, Boca Raton, FL, 1994, pp. 49-59.

Compound	t_b	$\Delta t/\Delta p$	
	°C	°C/kPa	°C/mmHg
Acetaldehyde	20.1	0.261	0.0348
Acetic acid	117.9	0.324	0.0432
Acetone	56.0	0.289	0.0385
Acetonitrile	81.6	0.316	0.0421
Ammonia	-33.33	0.198	0.0264
Aniline	184.1	0.378	0.0504
Anisole	153.7	0.367	0.0489
Benzaldehyde	179.0	0.392	0.0523
Benzene	80.0	0.321	0.0428
Bromine	58.8	0.300	0.0400
Butane	-0.5	0.267	0.0356
1-Butanol	117.7	0.278	0.0371
Carbon disulfide	46.2	0.304	0.0405
Chlorine	-34.04	0.224	0.0299
Chlorobenzene	131.7	0.365	0.0487
1-Chlorobutane	78.6	0.321	0.0428
Chloroethane	12.3	0.262	0.0349
Chloroethylene	-13.3	0.241	0.0321
Cyclohexane	80.7	0.328	0.0437
Cyclohexanol	160.8	0.344	0.0459
Cyclohexanone	155.4	0.382	0.0509
Decane	174.1	0.388	0.0517
Dibutyl ether	140.2	0.363	0.0484
Dichloromethane	39.6	0.276	0.0368
Diethyl ether	34.5	0.278	0.0371
Dimethyl sulfoxide	189.0	0.379	0.0505
1,4-Dioxane	101.5	0.321	0.0428
Dipropyl ether	90.0	0.326	0.0435
Ethanol	78.2	0.249	0.0332
Ethyl acetate	77.1	0.300	0.0400
Ethylene glycol	197.3	0.331	0.0441
Heptane	98.5	0.336	0.0448
Hexafluorobenzene	80.2	0.305	0.0407
Hexane	68.7	0.314	0.0419
1-Hexanol	157.6	0.318	0.0424
Hydrogen fluoride	20.1	0.276	0.0368
Iodomethane	42.5	0.291	0.0388
Isobutane	-11.7	0.254	0.0339
Methanol	64.6	0.251	0.0335
Methyl acetate	56.8	0.282	0.0376
Methyl formate	31.7	0.582	0.0776
N-Methylaniline	196.2	0.396	0.0528
N-Methylformamide	199.5	0.371	0.0495
Nitrobenzene	210.8	0.418	0.0557
Nitromethane	101.1	0.320	0.0427
1-Octanol	195.1	0.360	0.0480
Pentane	36.0	0.289	0.0385
1-Pentanol	137.9	0.296	0.0395
Phenol	181.8	0.349	0.0465
Propane	-42.1	0.224	0.0299
1-Propanol	97.2	0.261	0.0348
2-Propanol	82.3	0.247	0.0329
Pyridine	115.2	0.340	0.0453
Pyrrole	129.7	0.330	0.0440
Pyrrolidine	86.5	0.309	0.0412
Styrene	145.1	0.369	0.0492
Sulfur dioxide	-10.05	0.221	0.0295
Tetrachloroethylene	121.3	0.354	0.0472
Tetrachloromethane	76.8	0.325	0.0433
Toluene	110.6	0.353	0.0471
Trichloroethylene	87.2	0.330	0.0440
Trichloromethane	61.1	0.302	0.0403
Trimethylamine	2.8	0.248	0.0331
Water	100.0	0.276	0.0368
o-Xylene	144.5	0.373	0.0497
m-Xylene	139.1	0.368	0.0491
p-Xylene	138.3	0.369	0.0492

EBULLIOSCOPIC CONSTANTS FOR CALCULATION OF BOILING POINT ELEVATION

The boiling point T_b of a dilute solution of a non-volatile, non-dissociating solute is elevated relative to that of the pure solvent. If the solution is ideal (i.e., follows Raoult's Law), the amount of elevation depends only on the number of particles of solute present. Hence the change in boiling point ΔT_b can be expressed as

$$\Delta T_b = E_b\, m_2$$

where m_2 is the molality (moles of solute per kilogram of solvent) and E_b is the Ebullioscopic Constant, a characteristic property of the solvent. The Ebullioscopic Constant may be calculated from the relation

$$E_b = R\, T_b^{\,2}\, M/\Delta_{vap}H$$

where R is the molar gas constant, T_b is the normal boiling point temperature (absolute) of the solvent, M the molar mass of the solvent, and $\Delta_{vap}H$ the molar enthalpy (heat) of vaporization of the solvent at its normal boiling point.

This table lists E_b values for some common solvents, as calculated from data in the table "Enthalpy of Vaporization" in Section 6.

Compound	E_b/K kg mol^{-1}
Acetic acid	3.22
Acetone	1.80
Acetonitrile	1.44
Aniline	3.82
Anisole	4.20
Benzaldehyde	4.24
Benzene	2.64
1-Butanol	2.17
Carbon disulfide	2.42
Chlorobenzene	4.36
1-Chlorobutane	3.13
Cyclohexane	2.92
Cyclohexanol	3.5
Decane	6.10
Dichloromethane	2.42
Diethyl ether	2.20
Dimethyl sulfoxide	3.22
1,4 Dioxane	3.01
Ethanol	1.23
Ethyl acetate	2.82
Ethylene glycol	2.26
Heptane	3.62
Hexane	2.90
Iodomethane	4.31
Methanol	0.86
Methyl acetate	2.21
N-Methylaniline	4.3
N-Methylformamide	2.2
Nitrobenzene	5.2
Nitromethane	2.09
1-Octanol	5.06
Phenol	3.54
1-Propanol	1.66
2-Propanol	1.58
Pyridine	2.83
Pyrrole	2.33
Pyrrolidine	2.32
Tetrachloroethylene	6.18
Tetrachloromethane	5.26
Toluene	3.40
Trichloroethylene	4.52
Trichloromethane	3.80
Water	0.513
o-Xylene	4.25

CRYOSCOPIC CONSTANTS FOR CALCULATION OF FREEZING POINT DEPRESSION

The freezing point T_f of a dilute solution of a non-volatile, non-dissociating solute is depressed relative to that of the pure solvent. If the solution is ideal (i.e., follows Raoult's Law), this lowering is a function only of the number of particles of solute present. Thus the absolute value of the lowering of freezing point ΔT_f can be expressed as

$$\Delta T_f = E_f m_2$$

where m_2 is the molality (moles of solute per kilogram of solvent) and E_f is the Cryoscopic Constant, a characteristic property of the solvent. The Cryoscopic Constant may be calculated from the relation

$$E_f = R\,T_f^2\,M/\Delta_{fus}H$$

where R is the molar gas constant, T_f is the freezing point temperature (absolute) of the solvent, M the molar mass of the solvent, and $\Delta_{fus}H$ the molar enthalpy (heat) of fusion of the solvent.

This table lists cryscopic constants for selected substances, as calculated from data in the table "Enthalpy of Fusion" in Section 6.

Compound	E_f/K kg mol^{-1}
Acetamide	3.92
Acetic acid	3.63
Acetophenone	5.16
Aniline	5.23
Benzene	5.07
Benzonitrile	5.35
Benzophenone	8.58
(+)-Camphor	37.8
1-Chloronaphthalene	7.68
o-Cresol	5.92
m-Cresol	7.76
p-Cresol	7.20
Cyclohexane	20.8
Cyclohexanol	42.2
cis-Decahydronaphthalene	6.42
trans-Decahydronaphthalene	4.70
Dibenzyl ether	6.17
p-Dichlorobenzene	7.57
Diethanolamine	3.16
Dimethyl sulfoxide	3.85
1,4-Dioxane	4.63
Diphenylamine	8.38
Ethylene glycol	3.11
Formamide	4.25
Formic acid	2.38
Glycerol	3.56
Methylcyclohexane	2.60
Naphthalene	7.45
Nitrobenzene	6.87
Phenol	6.84
Pyridine	4.26
Quinoline	6.73
Succinonitrile	19.3
1,1,2,2-Tetrabromoethane	21.4
1,1,2,2-Tetrachloro-1,2-difluoroethane	41.0
Toluene	3.55
p-Toluidine	4.91
Tribromomethane	15.0
Water	1.86
p-Xylene	4.31

FREEZING POINT LOWERING BY ELECTROLYTES IN AQUEOUS SOLUTION

Reference

Forsythe, W. E., *Smithsonian Physical Tables, Ninth Edition*, Smithsonian Institution, Washington, D.C., 1956.

	Lowering of freezing point of water (in °C) as function of molality (mol/kg)									
Compound	**0.05**	**0.10**	**0.25**	**0.50**	**0.75**	**1.00**	**1.50**	**2.00**	**2.50**	**3.00**
$CaCl_2$	0.25	0.49	1.27	2.66	4.28	6.35	10.78	15.27	20.42	28.08
$CuSO_4$	0.13	0.23	0.47	0.96						
HCl	0.18	0.36	0.90	1.86	2.90	4.02	6.63	9.94		
HNO_3	0.18	0.35	0.88	1.80	2.78	3.80	5.98	8.34	10.95	13.92
H_2SO_4	0.20	0.39	0.96	1.95	3.04	4.28	7.35	11.35	16.32	
KBr	0.18	0.36	0.92	1.78						
KCl	0.17	0.35	0.86	1.68	2.49	3.29	4.88	6.50	8.14	9.77
KNO_3	0.17	0.33	0.78	1.47	2.11	2.66				
K_2SO_4	0.23	0.43	1.01	1.87						
LiCl	0.18	0.35	0.88	1.80	2.78					
$MgSO_4$	0.13	0.24	0.55	1.01	1.50	2.08	3.41			
NH_4Cl	0.17	0.34	0.85	1.70	2.55					
NaCl	0.18	0.35	0.85	1.68	2.60					
$NaNO_3$	0.18	0.36	0.80	1.62	2.63	3.10				

CORRECTION OF BAROMETER READINGS TO 0 °C TEMPERATURE

The following corrections are used to reduce the reading of a mercury barometer with a brass scale to 0 °C. The number in the table should be subtracted from the observed height of the mercury column to give the true pressure in mmHg (1mmHg = 133.322 Pa). The table is calculated from the formula

$$\Delta h = -0.0001634\, ht/(1+0.0001818\, t),$$

where h is the observed column height in mm and t the Celsius temperature. This relation is based on thermal expansion coefficients of $181.8 \cdot 10^{-6}$ °C^{-1} for mercury and $18.4 \cdot 10^{-6}$ °C^{-1} for brass.

	Observed Height in mm																		
t/°C	**620**	**630**	**640**	**650**	**660**	**670**	**680**	**690**	**700**	**710**	**720**	**730**	**740**	**750**	**760**	**770**	**780**	**790**	**800**
0	0.00	0.00	0.00	0.00	0.00	0.00	0.00	0.00	0.00	0.00	0.00	0.00	0.00	0.00	0.00	0.00	0.00	0.00	0.00
1	0.10	0.10	0.10	0.11	0.11	0.11	0.11	0.11	0.11	0.12	0.12	0.12	0.12	0.12	0.12	0.13	0.13	0.13	0.13
2	0.20	0.21	0.21	0.21	0.22	0.22	0.22	0.23	0.23	0.23	0.24	0.24	0.24	0.25	0.25	0.25	0.25	0.26	0.26
3	0.30	0.31	0.31	0.32	0.32	0.33	0.33	0.34	0.34	0.35	0.35	0.36	0.36	0.37	0.37	0.38	0.38	0.39	0.39
4	0.40	0.41	0.42	0.42	0.43	0.44	0.44	0.45	0.46	0.46	0.47	0.48	0.48	0.49	0.50	0.50	0.51	0.52	0.52
5	0.51	0.51	0.52	0.53	0.54	0.55	0.56	0.56	0.57	0.58	0.59	0.60	0.60	0.61	0.62	0.63	0.64	0.64	0.65
6	0.61	0.62	0.63	0.64	0.65	0.66	0.67	0.68	0.69	0.70	0.71	0.71	0.72	0.73	0.74	0.75	0.76	0.77	0.78
7	0.71	0.72	0.73	0.74	0.75	0.77	0.78	0.79	0.80	0.81	0.82	0.83	0.85	0.86	0.87	0.88	0.89	0.90	0.91
8	0.81	0.82	0.84	0.85	0.86	0.87	0.89	0.90	0.91	0.93	0.94	0.95	0.97	0.98	0.99	1.01	1.02	1.03	1.04
9	0.91	0.92	0.94	0.95	0.97	0.98	1.00	1.01	1.03	1.04	1.06	1.07	1.09	1.10	1.12	1.13	1.15	1.16	1.17
10	1.01	1.03	1.04	1.06	1.08	1.09	1.11	1.13	1.14	1.16	1.17	1.19	1.21	1.22	1.24	1.26	1.27	1.29	1.30
11	1.11	1.13	1.15	1.17	1.18	1.20	1.22	1.24	1.26	1.27	1.29	1.31	1.33	1.35	1.36	1.38	1.40	1.42	1.44
12	1.21	1.23	1.25	1.27	1.29	1.31	1.33	1.35	1.37	1.39	1.41	1.43	1.45	1.47	1.49	1.51	1.53	1.55	1.57
13	1.31	1.34	1.36	1.38	1.40	1.42	1.44	1.46	1.48	1.50	1.53	1.55	1.57	1.59	1.61	1.63	1.65	1.67	1.70
14	1.41	1.44	1.46	1.48	1.51	1.53	1.55	1.57	1.60	1.62	1.64	1.67	1.69	1.71	1.73	1.76	1.78	1.80	1.83
15	1.52	1.54	1.56	1.59	1.61	1.64	1.66	1.69	1.71	1.74	1.76	1.78	1.81	1.83	1.86	1.88	1.91	1.93	1.96
16	1.62	1.64	1.67	1.69	1.72	1.75	1.77	1.80	1.82	1.85	1.88	1.90	1.93	1.96	1.98	2.01	2.03	2.06	2.09
17	1.72	1.74	1.77	1.80	1.83	1.86	1.88	1.91	1.94	1.97	1.99	2.02	2.05	2.08	2.10	2.13	2.16	2.19	2.22
18	1.82	1.85	1.88	1.91	1.93	1.96	1.99	2.02	2.05	2.08	2.11	2.14	2.17	2.20	2.23	2.26	2.29	2.32	2.35
19	1.92	1.95	1.98	2.01	2.04	2.07	2.10	2.13	2.17	2.20	2.23	2.26	2.29	2.32	2.35	2.38	2.41	2.44	2.48
20	2.02	2.05	2.08	2.12	2.15	2.18	2.21	2.25	2.28	2.31	2.34	2.38	2.41	2.44	2.47	2.51	2.54	2.57	2.60
21	2.12	2.15	2.19	2.22	2.26	2.29	2.32	2.36	2.39	2.43	2.46	2.50	2.53	2.56	2.60	2.63	2.67	2.70	2.73
22	2.22	2.26	2.29	2.33	2.36	2.40	2.43	2.47	2.51	2.54	2.58	2.61	2.65	2.69	2.72	2.76	2.79	2.83	2.86
23	2.32	2.36	2.40	2.43	2.47	2.51	2.54	2.58	2.62	2.66	2.69	2.73	2.77	2.81	2.84	2.88	2.92	2.96	2.99
24	2.42	2.46	2.50	2.54	2.58	2.62	2.66	2.69	2.73	2.77	2.81	2.85	2.89	2.93	2.97	3.01	3.05	3.08	3.12
25	2.52	2.56	2.60	2.64	2.68	2.72	2.77	2.81	2.85	2.89	2.93	2.97	3.01	3.05	3.09	3.13	3.17	3.21	3.25
26	2.62	2.66	2.71	2.75	2.79	2.83	2.88	2.92	2.96	3.00	3.04	3.09	3.13	3.17	3.21	3.26	3.30	3.34	3.38
27	2.72	2.77	2.81	2.85	2.90	2.94	2.99	3.03	3.07	3.12	3.16	3.20	3.25	3.29	3.34	3.38	3.42	3.47	3.51
28	2.82	2.87	2.91	2.96	3.00	3.05	3.10	3.14	3.19	3.23	3.28	3.32	3.37	3.41	3.46	3.51	3.55	3.60	3.64
29	2.92	2.97	3.02	3.06	3.11	3.16	3.21	3.25	3.30	3.35	3.39	3.44	3.49	3.54	3.58	3.63	3.68	3.72	3.77
30	3.02	3.07	3.12	3.17	3.22	3.27	3.32	3.36	3.41	3.46	3.51	3.56	3.61	3.66	3.71	3.75	3.80	3.85	3.90
31	3.12	3.17	3.22	3.27	3.32	3.37	3.43	3.48	3.53	3.58	3.63	3.68	3.73	3.78	3.83	3.88	3.93	3.98	4.03
32	3.22	3.28	3.33	3.38	3.43	3.48	3.54	3.59	3.64	3.69	3.74	3.79	3.85	3.90	3.95	4.00	4.05	4.11	4.16
33	3.32	3.38	3.43	3.48	3.54	3.59	3.64	3.70	3.75	3.81	3.86	3.91	3.97	4.02	4.07	4.13	4.18	4.23	4.29
34	3.42	3.48	3.53	3.59	3.64	3.70	3.75	3.81	3.87	3.92	3.98	4.03	4.09	4.14	4.20	4.25	4.31	4.36	4.42
35	3.52	3.58	3.64	3.69	3.75	3.81	3.86	3.92	3.98	4.03	4.09	4.15	4.21	4.26	4.32	4.38	4.43	4.49	4.55
36	3.62	3.68	3.74	3.80	3.86	3.92	3.97	4.03	4.09	4.15	4.21	4.27	4.32	4.38	4.44	4.50	4.56	4.62	4.68
37	3.72	3.78	3.84	3.90	3.96	4.02	4.08	4.14	4.20	4.26	4.32	4.38	4.44	4.50	4.56	4.62	4.68	4.74	4.80
38	3.82	3.88	3.95	4.01	4.07	4.13	4.19	4.25	4.32	4.38	4.44	4.50	4.56	4.62	4.69	4.75	4.81	4.87	4.93
39	3.92	3.99	4.05	4.11	4.18	4.24	4.30	4.37	4.43	4.49	4.56	4.62	4.68	4.75	4.81	4.87	4.94	5.00	5.06
40	4.02	4.09	4.15	4.22	4.28	4.35	4.41	4.48	4.54	4.61	4.67	4.74	4.80	4.87	4.93	5.00	5.06	5.13	5.19

DETERMINATION OF RELATIVE HUMIDITY FROM DEW POINT

The relative humidity of a water vapor–air mixture is defined as 100 times the partial pressure of water divided by the saturation vapor pressure of water at the same temperature. The relative humidity may be determined from the dew point t_{dew}, which is the temperature at which liquid water first condenses when the mixture is cooled from an initial temperature t. This table gives relative humidity as a function of the dew point depression $t - t_{dew}$ for several values of the dew point. Values are calculated from the vapor pressure table in Section 6.

	t_{dew}/°C				
$t - t_{dew}$	−10	0	10	20	30
0.0	100	100	100	100	100
0.2	99	99	99	99	99
0.4	97	97	97	98	98
0.6	95	96	96	96	97
0.9	94	94	95	95	96
1.0	92	93	94	94	94
1.2	91	92	92	93	93
1.4	90	90	91	92	92
1.6	88	89	90	91	91
1.8	87	88	89	90	90
2.0	86	87	88	88	89
2.2	84	85	86	87	89
2.4	83	84	85	86	87
2.6	82	83	84	85	86
2.8	80	82	83	84	85
3.0	79	81	82	83	84
3.2	78	80	81	82	83
3.4	77	79	80	81	82
3.6	76	77	79	80	82
3.8	75	76	78	79	81
4.0	73	75	77	78	80
4.2	72	74	76	77	79
4.4	71	73	75	77	78
4.6	70	72	74	76	77
4.8	69	71	73	75	76
5.0	69	70	72	74	75
5.2	67	69	71	73	75
5.4	66	68	70	72	74
5.6	65	67	69	71	73
5.9	64	66	69	70	72
6.0	63	66	68	70	71
6.2	62	65	67	69	71
6.4	61	64	66	68	70
6.6	60	63	65	67	69
6.8	60	62	64	66	68
7.0	59	61	63	66	68
7.2	58	60	63	65	67
7.4	57	60	62	64	66
7.6	56	59	61	63	65
7.8	55	58	60	63	65
8.0	54	57	60	62	64
8.2	54	56	59	61	63
8.4	53	56	58	60	63
8.6	52	55	57	60	62
8.8	51	54	57	59	61
9.0	51	53	56	58	61
9.2	50	53	55	58	60
9.4	49	52	55	57	59
9.6	48	51	54	56	59
9.8	48	51	53	56	58
10.0	47	50	53	55	57
10.5	45	48	51	54	56
11.0	44	47	49	52	55
11.5	42	45	48	51	53
12.0	41	44	47	49	52
12.5	39	42	45	48	50
13.0	38	41	44	46	49
13.5	37	40	43	45	48
14.0	35	38	41	44	47
14.5	34	37	40	43	45
15.0	33	36	39	42	44
15.5	32	35	38	40	
16.0	31	34	37	39	
16.5	30	33	36	38	
17.0	29	32	35	37	
17.5	28	31	34	36	
18.0	27	30	33	35	
18.5	26	29	32	34	
19.0	25	28	31	33	
19.5	24	27	30	33	
20.0	24	26	29	32	
21.0	22	25	27	30	
22.0	21	23	26	29	
23.0	19	22	24	27	
24.0	18	21	23	26	
25.0	17	19	22	24	
26.0	16	18	21	23	
27.0	15	17	20	22	
28.0	14	16	19	21	
29.0	13	15	18	20	
30.0	12	14	17	19	

DETERMINATION OF RELATIVE HUMIDITY FROM WET AND DRY BULB TEMPERATURES

Relative humidity may be determined by comparing temperature readings of wet and dry bulb thermometers. The following table, extracted from more extensive U.S. National Weather Service tables, gives the relative humidity as a function of air temperature t_d (dry bulb) and the difference $t_d - t_w$ between dry and wet bulb temperatures. The data assume a pressure near normal atmospheric pressure and an instrumental configuration with forced ventilation.

	$(t_d - t_w)$/°C											
t_d/°C	**0.5**	**1.0**	**1.5**	**2.0**	**2.5**	**3.0**	**3.5**	**4.0**	**4.5**	**5.0**	**5.5**	**6.0**
−10	83	67	51	35	19							
−8	86	71	57	43	29	15						
−6	88	74	61	49	37	25	8					
−4	89	77	66	55	44	33	23	12				
−2	90	79	69	60	50	40	31	22	12			
0	91	81	72	64	55	46	38	29	21	13	5	
2	91	84	76	68	60	52	44	37	29	22	14	7
4	92	85	78	71	63	57	49	43	36	29	22	16
6	93	86	79	73	66	60	54	48	41	35	29	24
8	93	87	81	75	69	63	57	51	46	40	35	29
10	94	88	82	77	71	66	60	55	50	44	39	34
12	94	89	83	78	73	68	63	58	53	48	43	39
14	95	90	85	79	75	70	65	60	56	51	47	42
16	95	90	85	81	76	71	67	63	58	54	50	46
18	95	91	86	82	77	73	69	65	61	57	53	49
20	96	91	87	83	78	74	70	66	63	59	55	51
22	96	92	87	83	80	76	72	68	64	61	57	54
24	96	92	88	84	80	77	73	69	66	62	59	56
26	96	92	88	85	81	78	74	71	67	64	61	58
28	96	93	89	85	82	78	75	72	69	65	62	59
30	96	93	89	86	83	79	76	73	70	67	64	61
35	97	94	90	87	84	81	78	75	72	69	67	64
40	97	94	91	88	85	82	80	77	74	72	69	67

	$(t_d - t_w)$/°C											
t_d/°C	**6.5**	**7.0**	**7.5**	**8.0**	**8.5**	**9.0**	**10.0**	**11.0**	**12.0**	**13.0**	**14.0**	**15.0**
4	9											
6	17	11	5									
8	24	19	14	8								
10	29	24	20	15	10	6						
12	34	29	25	21	16	12	5					
14	38	34	30	26	22	18	10					
16	42	38	34	30	26	23	15	8				
18	45	41	38	34	30	27	20	14	7			
20	48	44	41	37	34	31	24	18	12	6		
22	50	47	44	40	37	34	28	22	17	11	6	
24	53	49	46	43	40	37	31	26	20	15	10	5
26	54	51	49	46	43	40	34	29	24	19	14	10
28	56	53	51	48	45	42	37	32	27	22	18	13
30	58	55	52	50	47	44	39	35	30	25	21	17
32	60	57	54	51	49	46	41	37	32	28	24	20
34	61	58	56	53	51	48	43	39	35	30	26	23
36	62	59	57	54	52	50	45	41	37	33	29	25
38	63	61	58	56	54	51	47	43	39	35	31	27
40	64	62	59	57	54	53	48	44	40	36	33	29

CONSTANT HUMIDITY SOLUTIONS

Anthony Wexler

An excess of a water soluble salt in contact with its saturated solution and contained within an enclosed space produces a constant relative humidity and water vapor pressure according to

$$RH = A \exp(B/T)$$

where RH is the percent relative humidity (generally accurate to ±2 %), T is the temperature in kelvin, and the constants A and B and the range of valid temperatures are given in the table below. The vapor pressure, p, can be calculated from

$$p = (RH/100) \times p_0$$

where p_0 is the vapor pressure of pure water at temperature T as given in the table in Section 6 titled "Vapor Pressure of Water from 0 to 370 °C".

References

1. Wexler, A. S. and Seinfeld, J. H., *Atmospheric Environment*, 25A, 2731, 1991.
2. Greenspan, L., *J. Res. National Bureau of Standards*, 81A, 89, 1977.
3. Broul, et al., *Solubility of Inorganic Two-Component Systems*, Elsevier, New York, 1981.
4. Wagman, D. D. et al., *J. Phys. Chem. Ref. Data, Vol. 11, Suppl.* 2, 1982.

Compound	Temperature range (°C)	RH (25 °C)	A	B
$NaOH \cdot H_2O$	15–60	6	5.48	27
$LiBr \cdot 2H_2O$	10–30	6	0.23	996
$ZnBr_2 \cdot 2H_2O$	5–30	8	1.69	455
$KOH \cdot 2H_2O$	5–30	9	0.014	1924
$LiCl \cdot H_2O$	20–65	11	14.53	–75
$CaBr_2 \cdot 6H_2O$	11–22	16	0.17	1360
$LiI \cdot 3H_2O$	15–65	18	0.15	1424
$CaCl_2 \cdot 6H_2O$	15–25	29	0.11	1653
$MgCl_2 \cdot 6H_2O$	5–45	33	29.26	34
$NaI \cdot 2H_2O$	5–45	38	3.62	702
$Ca(NO_3)_2 \cdot 4H_2O$	10–30	51	1.89	981
$Mg(NO_3)_2 \cdot 6H_2O$	5–35	53	25.28	220
$NaBr \cdot 2H_2O$	0–35	58	20.49	308
NH_4NO_3	10–40	62	3.54	853
KI	5–30	69	29.35	254
$SrCl_2 \cdot 6H_2O$	5–30	71	31.58	241
$NaNO_3$	10–40	74	26.94	302
NaCl	10–40	75	69.20	25
NH_4Cl	10–40	79	35.67	235
KBr	5–25	81	40.98	203
$(NH_4)_2SO_4$	10–40	81	62.06	79
KCl	5–25	84	49.38	159
$Sr(NO_3)_2 \cdot 4H_2O$	5–25	85	28.34	328
$BaCl_2 \cdot 2H_2O$	5–25	90	69.99	75
CsI	5–25	91	70.77	75
KNO_3	0–50	92	43.22	225
K_2SO_4	10–50	97	86.75	34

STANDARD SALT SOLUTIONS FOR HUMIDITY CALIBRATION

Saturated aqueous solutions of inorganic salts are convenient secondary standards for calibration of instruments for measurement of relative humidity. The International Union of Pure and Applied Chemistry has recommended salt solutions for calibrations in the range of 10% to 90% relative humidity, and the American Society for Testing and Materials has published similar standards. The data in this table are taken from the IUPAC recommendations, except for K_2CO_3 and K_2SO_4, which are ASTM recommendations.

Details on the preparation and use of these standards may be found in References 1 and 2. Data for other salts are given in Reference 3.

References

1. Marsh, K. N., Editor, *Recommended Reference Materials for the Realization of Physicochemical Properties*, Blackwell Scientific Publications, Oxford, 1987, pp. 157–162.
2. *Standard Practice for Maintaining Constant Relative Humidity by Means of Aqueous Solutions*, ASTM Standard E 104-85, Reapproved 1991.
3. Greenspan, L., *J. Res. Nat. Bur. Stand.*, 81A, 89, 1977.

	Relative Humidity in %						
t/°C	LiCl	$MgCl_2$	K_2CO_3	$Mg(NO_3)_2$	NaCl	KCl	K_2SO_4
0		33.66 ± 0.33	43.1 ± 0.7	60.35 ± 0.55	75.51 ± 0.34	88.61 ± 0.53	98.8 ± 2.1
5		33.60 ± 0.28	43.1 ± 0.5	58.86 ± 0.43	75.65 ± 0.27	87.67 ± 0.45	98.5 ± 0.9
10		33.47 ± 0.24	43.1 ± 0.4	57.36 ± 0.33	75.67 ± 0.22	86.77 ± 0.39	98.2 ± 0.8
15		33.30 ± 0.21	43.2 ± 0.3	55.87 ± 0.27	75.61 ± 0.18	85.92 ± 0.33	97.9 ± 0.6
20	11.31 ± 0.31	33.07 ± 0.18	43.2 ± 0.3	54.38 ± 0.23	75.47 ± 0.14	85.11 ± 0.29	97.6 ± 0.5
25	11.30 ± 0.27	32.78 ± 0.16	43.2 ± 0.4	52.89 ± 0.22	75.29 ± 0.12	84.34 ± 0.26	97.3 ± 0.5
30	11.28 ± 0.24	32.44 ± 0.14	43.2 ± 0.5	51.40 ± 0.24	75.09 ± 0.11	83.62 ± 0.25	97.0 ± 0.4
35	11.25 ± 0.22	32.05 ± 0.13		49.91 ± 0.29	74.87 ± 0.12	82.95 ± 0.25	96.7 ± 0.4
40	11.21 ± 0.21	31.60 ± 0.13		48.42 ± 0.37		82.32 ± 0.25	96.4 ± 0.4
45	11.16 ± 0.21	31.10 ± 0.13		46.93 ± 0.47		81.74 ± 0.28	96.1 ± 0.4
50	11.10 ± 0.22	30.54 ± 0.14		45.44 ± 0.60		81.20 ± 0.31	95.8 ± 0.5
55	11.03 ± 0.23	29.93 ± 0.16				80.70 ± 0.35	
60	10.95 ± 0.26	29.26 ± 0.18				80.25 ± 0.41	
65	10.86 ± 0.29	28.54 ± 0.21				79.85 ± 0.48	
70	10.75 ± 0.33	27.77 ± 0.25				79.49 ± 0.57	
75	10.64 ± 0.38	26.94 ± 0.29				79.17 ± 0.66	
80	10.51 ± 0.44	26.05 ± 0.34				78.90 ± 0.77	

LOW-TEMPERATURE BATHS FOR MAINTAINING CONSTANT TEMPERATURE

A liquid–solid slurry is a convenient means of maintaining a constant temperature environment below room temperature. The following is a list of readily available organic liquids suitable for this purpose, arranged in order of their melting (freezing) points t_m. The normal boiling points t_b are also given.

Compound	t_m/°C	t_b/°C
Isopentane (2-Methylbutane)	−159.9	27.8
Methylcyclopentane	−142.5	71.8
3-Chloropropene (Allyl chloride)	−134.5	45.1
Pentane	−129.7	36.0
Allyl alcohol	−129	97.0
Ethanol	−114.1	78.2
Carbon disulfide	−111.5	46
Isobutyl alcohol	−108	107.8
Toluene	−94.9	110.6
Acetone	−94.8	56.0
Ethyl acetate	−83.6	77.1
Dry ice + acetone	−78	
p-Cymene	−68.9	177.1
Trichloromethane (Chloroform)	−63.6	61.1
N-Methylaniline	−57	196.2
Chlorobenzene	−45.2	131.7
Anisole	−37.5	153.7
Bromobenzene	−30.6	156.0
Tetrachloromethane (Carbon tetrachloride)	−23	76.8
Benzonitrile	−12.7	191.1

METALS AND ALLOYS WITH LOW MELTING TEMPERATURE

L. I. Berger

Metal or alloy system	Composition, % * Weight	Atomic	Melting temperature (°C)	Comments	Ref.
Hg	100	100	−38.84		
Cs–K	77.0–23.0	50.0–50.0	−37.5	Eutectic (?)	1
Cs–Na	94.5–5.5	75.0–25.0	−30.0	Eutectic	2
K–Na	76.7–23.3	65.9–34.1	−12.65	Eutectic	3
Na–Rb	8.0–92.0	24.4–75.6	−5	Eutectic	4
Ga–In–Sn	62.5–21.5–16.0	73.6–15.3–11.1	11	Eutectic	5
Ga–Sn–Zn	82.0–12.0–6.0	86.0–7.3–6.7	17	Eutectic	5
Cs	100	100	28.44		
Ga	100	100	29.77		
K–Rb	32.0–68.0	50–50	33	Eutectic	4
Bi–Cd–In–Pb–Sn	44.7–5.3–19.1–22.6–8.3	35.1–8.2–27.3–17.9–11.5	46.7	Eutectic	6
Bi–In–Pb–Sn	49.5–21.3–17.6–11.6	39.2–30.7–14.0–16.2	58.2	Eutectic	6
Bi–In–Sn	32.5–51.0–16.5	21.1–60.1–18.8	60.5	Eutectic	7
K	100	100	63.38		
Bi–Cd–Pb–Sn	50.0–12.5–25.0–12.5	41.5–19.3–21.0–18.2	70	Wood's alloy	6
Bi–In	33.0–67.0	21.3–78.7	72	Eutectic	8
Bi–Cd–Pb	51.6–8.2–40.2	48.1–14.2–37.7	91.5	Eutectic	6
Bi–Pb–Sn	52.5–32.0–15.5	46.8–28.7–24.5	95	Eutectic	6
Na	100	100	97.8		
Bi–Cd–Sn	54.0–20.0–26.0	39.4–27.2–33.4	102.5	Eutectic	6
In–Sn	51.8–48.2	52.6–47.4	119	Eutectic	9
Cd–In	25.3–74.7	25.7–74.3	120	Eutectic	10
Bi–Pb	55.5–44.5	55.3–44.7	124	Eutectic	11
Bi–Sn–Zn	56.0–40.0–4.0	40.2–50.6–9.2	130	Eutectic	6, 7
Bi–Sn	70–30	57.0–43.0	138.5	Eutectic	6, 12
Bi–Cd	60.3–39.7	45.0–55.0	145.5	Eutectic	13, 14
In	100	100	156.6		
Li	100	100	180.5		
Pb–Sn	38.1–61.9	26.1–73.9	183	Eutectic	6,15
Bi–Tl	48.0–52.0	47.5–52.5	185	Eutectic	13
Sn–Zn	91.0–9.0	85.0–15.0	198	Eutectic	14
Sb–Sn	8.0–92.0	7.8–92.2	199	White Metal	16
Au–Pb	14.6–85.4	15.2–84.8	212	Eutectic	17
Ag–Sn	3.5–96.5	3.8–96.2	221	Eutectic	13,18
Bi–Pb–Sb–Sn	48.0–28.5–9.0–14.5	40.8–24.5–13.1–21.6	226	Matrix Alloy	6
Cu–Sn	0.75–99.25	1.3–98.7	227	Eutectic	13, 19
Sn	100	100	231.9		

* The useful expressions for correlations between the atomic and weight concentrations of an alloy components are:

$$f(\mathrm{a}, A_k) = \frac{f(\mathrm{w}, A_k)}{M_k \sum_{i=1}^{N} \frac{f(\mathrm{w}, A_i)}{M_i}} \quad \text{and} \quad f(\mathrm{w}, A_k) = \frac{M_k \cdot f(\mathrm{a}, A_k)}{\sum_{i=1}^{N} M_i \cdot f(\mathrm{a}, A_i)} \quad (i = 1, \ldots, k, \ldots, N)$$

where $f(\mathrm{a}, A_i)$ and $f(\mathrm{w}, A_i)$ are the atomic and weight concentrations of component A_i, respectively, and M_i is the atomic weight of this component.

References

1. Zintle, E. and Hauke, W., *Z. Electrochem.*, 44, 104, 1938.
2. Rinck, E., *Compt. Rend.*, 199, 1217, 1934.
3. Krier, C. A., Craign, R. S., and Wallace, W. E., *J. Phys. Chem.*, 61, 522, 1957.
4. Goria, C., *Gazz. Chim. Ital.*, 65, 865, 1935.
5. Baker, H., Ed., *ASM Handbook, Volume 3: Alloy Phase Diagrams*, ASM Intl., Materials Park, OH, 1992.
6. Sedlacek, V., *Non-Ferrous Metals and Alloys*, Elsevier, 1986.
7. Villars, P., Prince, A., Okamoto, H., Eds., *Handbook of Ternary Alloy Phase Diagrams*, ASM Intl., 1994.
8. Palatnik, L. S., Kosevich, V. M., and Tyrina, L. V., *Phys. Metals Metallog. (USSR)*, 11, 75, 1961.
9. Neumann, T. and Alpout, O., *J. Less-Common Metals*, 6, 108, 1964.
10. Neumann, T. and Predel, B., *Z. Metallk.*, 50, 309, 1959.
11. Roy, P., Orr, R. L., and Hultgren, R., *J. Phys. Chem.*, 64, 1034, 1960.
12. Dobovicek, B. and Smajic, N., *Rudarsko–Met. Zbornik*, 4, 353, 1962.
13. Massalski, T. B., Okamoto, H., Subramanian, P. R., and Kacprzak, L., Eds., *Binary Alloy Phase Diagrams*, 2nd ed., ASM Intl., 1990.
14. Dobovicek, B. and Straus, B., *Rudarsko–Met. Zbornik*, 3, 273, 1960.
15. Schurmann, E. and Gilhaus, F. J., *Arch. Eisenhuettenw.*, 32, 867, 1961.
16. Rosenblatt, G. M. and Birchenall, C. E., *Trans. AIME*, 224, 481, 1962.
17. Evans, D. S. and Prince, A., in *Alloy Phase Diagrams*, MRS Simposia Proc., Vol. 19, North Holland, 1983, p. 383.
18. Umanskiy, M. M., *Zh. Fiz. Khim.*, 14, 846, 1940.
19. Homer, C. E. and Plummer, H., *J. Inst. Met.*, 64, 169, 1939.

WIRE TABLES

The resistance per unit length of wires of various metals is tabulated here. Values were calculated from resistivity values in the tables "Electrical Resistivity of Pure Metals" and "Electrical Resistivity of Selected Alloys", which appear in Section 12. In practice, resistance may vary because of differing heat treatments and metal composition. The values in the table refer to 20 °C, but values at other temperatures may be calculated from the following resistivity data:

Metal	Resistivity in 10^{-8} Ω m at temperature			
	0 °C	20 °C	25 °C	100 °C
Aluminum	2.417	2.650	2.709	3.56
Brass (70% Cu, 30% Zn)	5.87	6.08	6.13	6.91
Constantan (60% Cu, 40% Ni)	45.43	45.38	45.35	45.11
Copper	1.543	1.678	1.712	2.22
Nichrome (79% Ni, 21% Cr)	107.3	107.5	107.6	108.3
Platinum	9.6	10.5	10.7	13.6
Silver	1.467	1.587	1.617	2.07
Tungsten	4.82	5.28	5.39	7.18

Resistance per unit length at 20 °C in Ω/m

AWG Gauge[a]	Diameter (mm)	Aluminum	Brass	Constantan	Copper	Nichrome	Platinum	Silver	Tungsten
0	8.252	0.000495	0.00114	0.00848	0.000314	0.0201	0.00196	0.000297	0.00099
2	6.543	0.000788	0.00181	0.0135	0.000499	0.0320	0.00312	0.000472	0.00157
4	5.189	0.00125	0.00287	0.0214	0.000793	0.0508	0.00496	0.000750	0.00250
6	4.115	0.00199	0.00457	0.0341	0.00126	0.0808	0.00789	0.00119	0.00397
8	3.264	0.00317	0.00727	0.0542	0.00200	0.128	0.0125	0.00190	0.00631
10	2.588	0.00504	0.0115	0.0863	0.00319	0.204	0.0200	0.00302	0.0100
12	2.053	0.00800	0.0184	0.137	0.00507	0.325	0.0317	0.00479	0.0159
14	1.628	0.0127	0.0292	0.218	0.00806	0.516	0.0504	0.00762	0.0254
16	1.291	0.0202	0.0464	0.347	0.0128	0.821	0.0802	0.0121	0.0403
18	1.024	0.0322	0.0738	0.551	0.0204	1.30	0.127	0.0193	0.0641
20	0.8118	0.0512	0.117	0.877	0.0324	2.08	0.203	0.0307	0.102
22	0.6439	0.0814	0.187	1.39	0.0515	3.30	0.322	0.0487	0.162
24	0.5105	0.129	0.297	2.22	0.0820	5.25	0.513	0.0775	0.258
26	0.4049	0.206	0.472	3.52	0.130	8.35	0.815	0.123	0.410
28	0.3211	0.327	0.751	5.60	0.207	13.3	1.30	0.196	0.652
30	0.2548	0.520	1.19	8.90	0.329	21.1	2.06	0.311	1.03
32	0.2019	0.828	1.90	14.2	0.524	33.6	3.28	0.496	1.65
34	0.1601	1.32	3.02	22.5	0.833	53.4	5.22	0.788	2.62
36	0.1270	2.09	4.80	35.8	1.32	84.9	8.29	1.25	4.17
38	0.1007	3.33	7.63	57.0	2.11	135	13.2	1.99	6.63
40	0.07988	5.29	12.1	90.5	3.35	214	20.9	3.17	10.5

[a] Often called Brown & Sharpe Gauge.

CHARACTERISTICS OF PARTICLES AND PARTICLE DISPERSOIDS

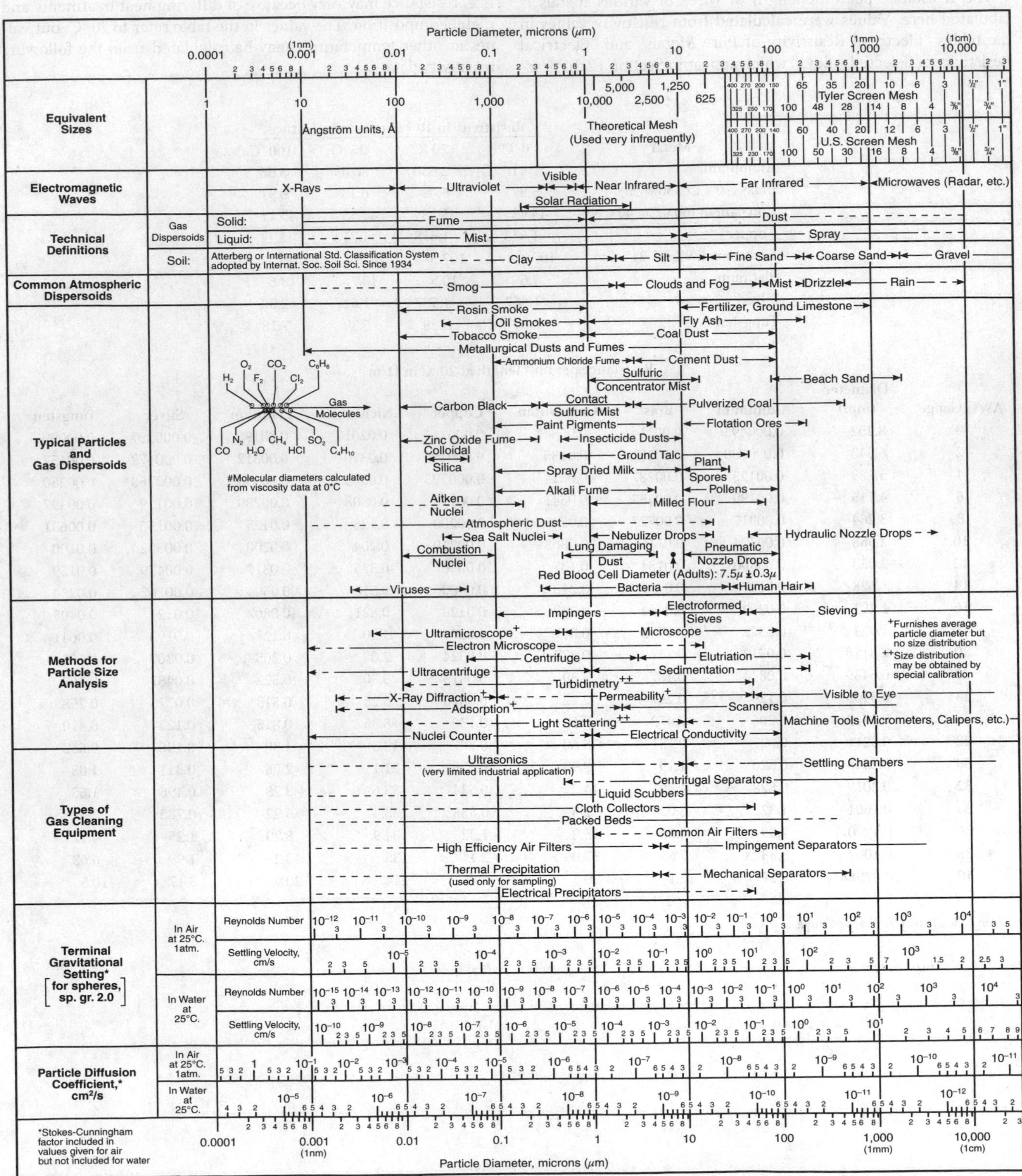

From C.E. Lapple, *Stanford Research Institute Journal*, Vol. 5, p. 95 (Third Quarter) 1961.

DENSITY OF VARIOUS SOLIDS

This table gives the range of density for miscellaneous solid materials whose characteristics depend on the source or method of preparation.

References

1. Forsythe, W. E., *Smithsonian Physical Tables, Ninth Edition*, Smithsonian Institution, Washington, D.C., 1956.
2. Kaye, G. W. C., and Laby, T. H., *Tables of Physical and Chemical Constants, 16th Edition*, Longman, London, 1995.
3. Brandrup, J., and Immergut, E. H., *Polymer Handbook, Third Edition*, John Wiley & Sons, New York, 1989.

Material	ρ/ g cm^{-3}
Agate	2.5-2.7
Alabaster,	
carbonate	2.69-2.78
sulfate	2.26-2.32
Albite	2.62-2.65
Amber	1.06-1.11
Amphiboles	2.9-3.2
Anorthite	2.74-2.76
Asbestos	2.0-2.8
Asbestos slate	1.8
Asphalt	1.1-1.5
Basalt	2.4-3.1
Beeswax	0.96-0.97
Beryl	2.69-2.70
Biotite	2.7-3.1
Bone	1.7-2.0
Brasses	8.44-8.75
Brick	1.4-2.2
Bronzes	0.74-0.89
Butter	0.86-0.87
Calamine	4.1-4.5
Calcspar	2.6-2.8
Camphor	0.99
Cardboard	0.69
Celluloid	1.4
Cement, set	2.7-3.0
Chalk	1.9-2.8
Charcoal,	
oak	0.57
pine	0.28-0.44
Cinnabar	8.12
Clay	1.8-2.6
Coal,	
anthracite	1.4-1.8
bituminous	1.2-1.5
Coke	1.0-1.7
Copal	1.04-1.14
Cork	0.22-0.26
Corundum	3.9-4.0
Diamond	3.51
Dolomite	2.84
Ebonite	1.15
Emery	4.0
Epidote	3.25-3.50
Feldspar	2.55-2.75
Flint	2.63
Fluorite	3.18
Galena	7.3-7.6
Garnet	3.15-4.3
Gelatin	1.27
Glass,	
common	2.4-2.8
lead	3-4
Pyrex	2.23
Granite	2.64-2.76
Graphite	2.30-2.72
Gum arabic	1.3-1.4
Gypsum	2.31-2.33
Hematite	4.9-5.3
Hornblende	3.0
Ice	0.917
Iron, cast	7.0-7.4
Ivory	1.83-1.92
Kaolin	2.6
Leather, dry	0.86
Lime, slaked	1.3-1.4
Limestone	2.68-2.76
Linoleum	1.18
Magnetite	4.9-5.2
Malachite	3.7-4.1
Marble	2.6-2.84
Meerschaum	0.99-1.28
Mica	2.6-3.2
Muscovite	2.76-3.00
Ochre	3.5
Opal	2.2
Paper	0.7-1.15
Paraffin	0.87-0.91
Peat blocks	0.84
Pitch	1.07
Polyamides	1.15-1.25
Polyethylene	0.92-0.97
Poly(methyl methacrylate)	1.19
Polypropylene	0.91-0.94
Polystyrene	1.06-1.12
Polytetrafluoroethylene	2.28-2.30
Poly(vinyl acetate)	1.19
Poly(vinyl chloride)	1.39-1.42
Porcelain	2.3-2.5
Porphyry	2.6-2.9
Pyrite	4.95-5.10
Quartz (α)	2.65
Resin	1.07
Rock salt	2.18
Rubber,	
hard	1.19
soft	1.1
pure gum	0.91-0.93
Neoprene	1.23-1.25
Sandstone	2.14-2.36
Serpentine	2.50-2.65
Silica, fused,	2.21
Silicon carbide	3.16
Slag	2.0-3.9
Slate	2.6-3.3
Soapstone	2.6-2.8
Solder	8.7-9.4
Starch	1.53
Steel, stainless	7.8
Sugar	1.59
Talc	2.7-2.8
Tallow, beef	0.94
Tar	1.02
Topaz	3.5-3.6
Tourmaline	3.0-3.2
Tungsten carbide	14.0-15.0
Wax, sealing	1.8
Wood (seasoned)	
alder	0.42-0.68
apple	0.66-0.84
ash	0.65-0.85
balsa	0.11-0.14
bamboo	0.31-0.40
basswood	0.32-0.59
beech	0.70-0.90
birch	0.51-0.77
blue gum	1.00
box	0.95-1.16
butternut	0.38
cedar	0.49-0.57
cherry	0.70-0.90
dogwood	0.76
ebony	1.11-1.33
elm	0.54-0.60
hickory	0.60-0.93
holly	0.76
juniper	0.56
larch	0.50-0.56
locust	0.67-0.71
logwood	0.91
mahogany	0.66-0.85
maple	0.62-0.75
oak	0.60-0.90
pear	0.61-0.73
pine, pitch	0.83-0.85
white	0.35-0.50
yellow	0.37-0.60
plum	0.66-0.78
poplar	0.35-0.50
satinwood	0.95
spruce	0.48-0.70
sycamore	0.40-0.60
teak, Indian	0.66-0.98
walnut	0.64-0.70
water gum	1.00
willow	0.40-0.60
Wood's metal	9.70

DENSITY OF SULFURIC ACID

This table gives the density of aqueous sulfuric acid solutions as a function of concentration (in mass percent of H_2SO_4) and temperature.

	Density in g/mL										
Mass %	0 °C	10 °C	15 °C	20 °C	25 °C	30 °C	40 °C	50 °C	60 °C	80 °C	100 °C
1	1.0074	1.0068	1.0060	1.0051	1.0038	1.0022	0.9986	0.9944	0.9895	0.9779	0.9645
2	1.0147	1.0138	1.0129	1.0118	1.0104	1.0087	1.0050	1.0006	0.9956	0.9839	0.9705
3	1.0219	1.0206	1.0197	1.0184	1.0169	1.0152	1.0113	1.0067	1.0017	0.9900	0.9766
4	1.0291	1.0275	1.0264	1.0250	1.0234	1.0216	1.0176	1.0129	1.0078	0.9961	0.9827
5	1.0364	1.0344	1.0332	1.0317	1.0300	1.0281	1.0240	1.0192	1.0140	1.0022	0.9888
6	1.0437	1.0414	1.0400	1.0385	1.0367	1.0347	1.0305	1.0256	1.0203	1.0084	0.9950
7	1.0511	1.0485	1.0469	1.0453	1.0434	1.0414	1.0371	1.0321	1.0266	1.0146	1.0013
8	1.0585	1.0556	1.0539	1.0522	1.0502	1.0481	1.0437	1.0386	1.0330	1.0209	1.0076
9	1.0660	1.0628	1.0610	1.0591	1.0571	1.0549	1.0503	1.0451	1.0395	1.0273	1.0140
10	1.0735	1.0700	1.0681	1.0661	1.0640	1.0617	1.0570	1.0517	1.0460	1.0338	1.0204
12	1.0886	1.0846	1.0825	1.0802	1.0780	1.0756	1.0705	1.0651	1.0593	1.0469	1.0335
14	1.1039	1.0994	1.0971	1.0947	1.0922	1.0897	1.0844	1.0788	1.0729	1.0603	1.0469
16	1.1194	1.1145	1.1120	1.1094	1.1067	1.1040	1.0985	1.0927	1.0868	1.0740	1.0605
18	1.1351	1.1298	1.1271	1.1243	1.1215	1.1187	1.1129	1.1070	1.1009	1.0879	1.0744
20	1.1510	1.1453	1.1424	1.1394	1.1365	1.1335	1.1275	1.1215	1.1153	1.1021	1.0885
22	1.1670	1.1609	1.1579	1.1548	1.1517	1.1486	1.1424	1.1362	1.1299	1.1166	1.1029
24	1.1832	1.1768	1.1736	1.1704	1.1672	1.1640	1.1576	1.1512	1.1448	1.1313	1.1176
26	1.1996	1.1929	1.1896	1.1862	1.1829	1.1796	1.1730	1.1665	1.1599	1.1463	1.1325
28	1.2160	1.2091	1.2057	1.2023	1.1989	1.1955	1.1887	1.1820	1.1753	1.1616	1.1476
30	1.2326	1.2255	1.2220	1.2185	1.2150	1.2115	1.2046	1.1977	1.1909	1.1771	1.1630
32	1.2493	1.2421	1.2385	1.2349	1.2314	1.2278	1.2207	1.2137	1.2068	1.1928	1.1787
34	1.2661	1.2588	1.2552	1.2515	1.2479	1.2443	1.2371	1.2300	1.2229	1.2088	1.1946
36	1.2831	1.2757	1.2720	1.2684	1.2647	1.2610	1.2538	1.2466	1.2394	1.2251	1.2109
38	1.3004	1.2929	1.2891	1.2855	1.2818	1.2780	1.2707	1.2635	1.2561	1.2418	1.2276
40	1.3179	1.3103	1.3065	1.3028	1.2991	1.2953	1.2880	1.2806	1.2732	1.2589	1.2446
42	1.3357	1.3280	1.3242	1.3205	1.3167	1.3129	1.3055	1.2981	1.2907	1.2762	1.2619
44	1.3538	1.3461	1.3423	1.3384	1.3346	1.3308	1.3234	1.3160	1.3086	1.2939	1.2796
46	1.3724	1.3646	1.3608	1.3569	1.3530	1.3492	1.3417	1.3343	1.3269	1.3120	1.2976
48	1.3915	1.3835	1.3797	1.3758	1.3719	1.3680	1.3604	1.3528	1.3455	1.3305	1.3159
50	1.4110	1.4029	1.3990	1.3951	1.3911	1.3872	1.3795	1.3719	1.3644	1.3494	1.3348
52	1.4310	1.4228	1.4188	1.4148	1.4109	1.4069	1.3991	1.3914	1.3837	1.3687	1.3540
54	1.4515	1.4431	1.4391	1.4350	1.4310	1.4270	1.4191	1.4113	1.4036	1.3884	1.3735
56	1.4724	1.4640	1.4598	1.4557	1.4516	1.4475	1.4396	1.4317	1.4239	1.4085	1.3934
58	1.4937	1.4852	1.4809	1.4768	1.4726	1.4685	1.4604	1.4524	1.4446	1.4290	1.4137
60	1.5154	1.5067	1.5024	1.4983	1.4940	1.4898	1.4816	1.4735	1.4656	1.4497	1.4344
62	1.5375	1.5287	1.5243	1.5200	1.5157	1.5115	1.5031	1.4950	1.4869	1.4708	1.4554
64	1.5600	1.5510	1.5465	1.5421	1.5378	1.5335	1.5250	1.5167	1.5086	1.4923	1.4766
66	1.5828	1.5736	1.5691	1.5646	1.5602	1.5558	1.5472	1.5388	1.5305	1.5140	1.4981
68	1.6059	1.5965	1.5920	1.5874	1.5829	1.5785	1.5697	1.5611	1.5528	1.5359	1.5198
70	1.6293	1.6198	1.6151	1.6105	1.6059	1.6014	1.5925	1.5838	1.5753	1.5582	1.5417
72	1.6529	1.6433	1.6385	1.6338	1.6292	1.6246	1.6155	1.6067	1.5981	1.5806	1.5637
74	1.6768	1.6670	1.6622	1.6574	1.6526	1.6480	1.6387	1.6297	1.6209	1.6031	1.5857
76	1.7008	1.6908	1.6858	1.6810	1.6761	1.6713	1.6619	1.6526	1.6435	1.6252	1.6074
78	1.7247	1.7144	1.7093	1.7043	1.6994	1.6944	1.6847	1.6751	1.6657	1.6469	1.6286
80	1.7482	1.7376	1.7323	1.7272	1.7221	1.7170	1.7069	1.6971	1.6873	1.6680	1.6493
82	1.7709	1.7599	1.7544	1.7491	1.7437	1.7385	1.7281	1.7180	1.7080	1.6882	1.6692
84	1.7916	1.7804	1.7748	1.7693	1.7639	1.7585	1.7479	1.7375	1.7274	1.7072	1.6878
86	1.8095	1.7983	1.7927	1.7872	1.7818	1.7763	1.7657	1.7552	1.7449	1.7245	1.7050
88	1.8243	1.8132	1.8077	1.8022	1.7968	1.7914	1.7809	1.7705	1.7602	1.7397	1.7202
90	1.8361	1.8252	1.8198	1.8144	1.8091	1.8038	1.7933	1.7829	1.7729	1.7525	1.7331
91	1.8410	1.8302	1.8248	1.8195	1.8142	1.8090	1.7986	1.7883	1.7783	1.7581	1.7388
92	1.8453	1.8346	1.8293	1.8240	1.8188	1.8136	1.8033	1.7932	1.7832	1.7633	1.7439
93	1.8490	1.8384	1.8331	1.8279	1.8227	1.8176	1.8074	1.7974	1.7876	1.7681	1.7485
94	1.8520	1.8415	1.8363	1.8312	1.8260	1.8210	1.8109	1.8011	1.7914		
95	1.8544	1.8439	1.8388	1.8337	1.8286	1.8236	1.8137	1.8040	1.7944		
96	1.8560	1.8457	1.8406	1.8355	1.8305	1.8255	1.8157	1.8060	1.7965		
97	1.8569	1.8466	1.8414	1.8364	1.8314	1.8264	1.8166	1.8071	1.7977		
98	1.8567	1.8463	1.8411	1.8361	1.8310	1.8261	1.8163	1.8068	1.7976		
99	1.8551	1.8445	1.8393	1.8342	1.8292	1.8242	1.8145	1.8050	1.7958		
100	1.8517	1.8409	1.8357	1.8305	1.8255	1.8205	1.8107	1.8013	1.7922		

DENSITY OF ETHANOL–WATER MIXTURES

This table gives the density of mixtures of ethanol and water as a function of composition and temperature. The composition is specified in weight percent of ethanol, i.e., mass of ethanol per 100 g of solution. Values from the reference have been converted to true densities.

Reference

Washburn, E. W., Ed., *International Critical Tables of Numerical Data of Physics, Chemistry, and Technology*, Vol. 3, McGraw-Hill, New York, 1926–1932.

	Density in g/cm³						
Weight % Ethanol	**10 °C**	**15 °C**	**20 °C**	**25 °C**	**30 °C**	**35 °C**	**40 °C**
0	0.99970	0.99910	0.99820	0.99705	0.99565	0.99403	0.99222
5	0.99095	0.99029	0.98935	0.98814	0.98667	0.98498	0.98308
10	0.98390	0.98301	0.98184	0.98040	0.97872	0.97682	0.97472
15	0.97797	0.97666	0.97511	0.97331	0.97130	0.96908	0.96667
20	0.97249	0.97065	0.96861	0.96636	0.96392	0.96131	0.95853
25	0.96662	0.96421	0.96165	0.95892	0.95604	0.95303	0.94988
30	0.95974	0.95683	0.95379	0.95064	0.94738	0.94400	0.94052
35	0.95159	0.94829	0.94491	0.94143	0.93787	0.93422	0.93048
40	0.94235	0.93879	0.93515	0.93145	0.92767	0.92382	0.91989
45	0.93223	0.92849	0.92469	0.92082	0.91689	0.91288	0.90881
50	0.92159	0.91773	0.91381	0.90982	0.90577	0.90165	0.89747
55	0.91052	0.90656	0.90255	0.89847	0.89434	0.89013	0.88586
60	0.89924	0.89520	0.89110	0.88696	0.88275	0.87848	0.87414
65	0.88771	0.88361	0.87945	0.87524	0.87097	0.86664	0.86224
70	0.87599	0.87184	0.86763	0.86337	0.85905	0.85467	0.85022
75	0.86405	0.85985	0.85561	0.85131	0.84695	0.84254	0.83006
80	0.85194	0.84769	0.84341	0.83908	0.83470	0.83027	0.82576
85	0.83948	0.83522	0.83093	0.82658	0.82218	0.81772	0.81320
90	0.82652	0.82225	0.81795	0.81360	0.80920	0.80476	0.80026
95	0.81276	0.80850	0.80422	0.79989	0.79553	0.79112	0.78668
100	0.79782	0.79358	0.78932	0.78504	0.78073	0.77639	0.77201

DIELECTRIC STRENGTH OF INSULATING MATERIALS

L. I. Berger

The loss of the dielectric properties by a sample of a gaseous, liquid, or solid insulator as a result of application to the sample of an electric field* greater than a certain critical magnitude is called *dielectric breakdown*. The critical magnitude of electric field at which the breakdown of a material takes place is called the *dielectric strength* of the material (or *breakdown voltage*). The dielectric strength of a material depends on the specimen thickness (as a rule, thin films have greater dielectric strength than that of thicker samples of a material), the electrode shape**, the rate of the applied voltage increase, the shape of the voltage vs. time curve, and the medium surrounding the sample, e.g., air or other gas (or a liquid — for solid materials only).

Breakdown in Gases

The current carriers in gases are free electrons and ions generated by external radiation. The equilibrium concentration of these particles at normal pressure is about 10^3 cm^{-3}, and hence the electrical conductivity is very small, of the order of 10^{-16} – 10^{-15} S/cm. But in a strong electric field, these particles acquire kinetic energy along their free path, large enough to ionize the gas molecules. The new charged particles ionize more molecules; this avalanche-like process leads to formation between the electrodes of channels of conducting plasma (streamers), and the electrical resistance of the space between the electrodes decreases virtually to zero.

Because the dielectric strength (breakdown voltage) of gases strongly depends on the electrode geometry and surface condition and the gas pressure, it is generally accepted to present the data for a particular gas as a fraction of the dielectric strength of either nitrogen or sulfur hexafluoride measured at the same conditions. In Table 1, the data are presented in comparison with the dielectric strength of nitrogen, which is considered equal to 1.00. For convenience to the reader, a few average magnitudes of the dielectric strength of some gases are expressed in kilovolts per millimeter. The data in the table relate to the standard conditions, unless indicated otherwise.

Breakdown in Liquids

If a liquid is pure, the breakdown mechanism in it is similar to that in gases. If a liquid contains liquid impurities in the form of small drops with greater dielectric constant than that of the main liquid, the breakdown is the result of formation of ellipsoids from these drops by the electric field. In a strong enough electric field, these ellipsoids merge and form a high-conductivity channel between the electrodes. The current increases the temperature in the channel, liquid boils, and the current along the steam canal leads to breakdown. Formation of a conductive channel (bridge) between the electrodes is observed also in liquids with solid impurities. If a liquid contains gas impurities in the form of small bubbles, breakdown is the result of heating of the liquid in strong electric fields. In the locations with the highest current density, the liquid boils, the size of the gas bubbles increases, they merge and form gaseous channels between the electrodes, and the breakdown medium is again the gas plasma.

Breakdown in Solids

It is known that the current in solid insulators does not obey Ohm's law in strong electric fields. The current density increases almost exponentially with the electric field, and at a certain field magnitude it jumps to very high magnitudes at which a specimen of a material is destroyed. The two known kinds of electric breakdown are thermal and electrical breakdowns. The former is the result of material heating by the electric current. Destruction of a sample of a material happens when, at [illegible] amount of heat produced by the current [illegible] through the sample surface; the breakdown voltage in this case is proportional to the square root of the ratio of the thermal conductivity and electrical conductivity of the material. A semi-empirical expression for dependence of the breakdown voltage, V_B, on the physical properties and geometry of a sample of a solid material for the one-dimensional case is

$$V_B = [A\rho\kappa / a\varphi(d)]^{1/2}$$

where A is a numerical constant related to the system of units used, ρ and κ are the volume resistivity and thermal conductivity of the sample material, a is a constant related to the chemical bond nature and crystal structure of the sample material, and $\phi(d)$ is a function of the sample geometry, first of all, thickness, d (see, e.g., Ref. R6). In the majority of materials, $\phi(d)$ increases with d, hence, the magnitude of V_B is greater in the thinner samples of a particular material.

The electrical breakdown results from the tunneling of the charge carriers from electrodes or from the valence band or from the impurity levels into the conduction band, or by the impact ionization. The tunnel effect breakdown happens mainly in thin layers, e.g., in thin p-n junctions. Otherwise, the impact ionization mechanism dominates. For this mechanism, the dielectric strength of an insulator can be estimated using Boltzmann's kinetic equation for electrons in a crystal.

In the following tables, the dielectric strength values are for room temperature and normal atmospheric pressure, unless indicated otherwise.

* The unit of electric field in the SI system is newton per coulomb or volt per meter.

** For example, the U.S. standard ASTM D149 is based on use of symmetrical electrodes, while per U.K. standard BS2918 one electrode is a plane and the other is a rod with the axis normal to the plane.

TABLE 1. Dielectric Strength of Gases

Material	Dielectric* strength	Ref.
Nitrogen, N_2	1.00	
Hydrogen, H_2	0.50	1,2
Helium, He	0.15	1
Oxygen, O_2	0.92	2
Air	0.97	6
Air (flat electrodes), kV/mm	3.0	3
Air, kV/mm	0.4-0.7	4
Air, kV/mm	1.40	5
Neon, Ne	0.25	1
	0.16	2
Argon, Ar	0.18	2
Chlorine, Cl_2	1.55	1
Carbon monoxide, CO	1.02	1
	1.05	2
Carbon dioxide, CO_2	0.88	1
	0.82	2
	0.84	6
Nitrous oxide, N_2O	1.24	2
Sulfur dioxide, SO_2	2.63	2
	2.68	6
Sulfur monochloride, S_2Cl_2 (at 12.5 Torr)	1.02	1
Thionyl fluoride, SOF_2	2.50	1
Sulfur hexafluoride, SF_6	2.50	1
	2.63	2
Sulfur hexafluoride, SF_6, kV/mm	8.50	7
	9.8	8
Perchloryl fluoride, ClO_3F	2.73	1
Tetrachloromethane, CCl_4	6.33	1
	6.21	2
Tetrafluoromethane, CF_4	1.01	1
Methane, CH_4	1.00	1
	1.13	2
Bromotrifluoromethane, CF_3Br	1.35	1
	1.97	2
Bromomethane, CH_3Br	0.71	2
Chloromethane, CH_3Cl	1.29	2
Iodomethane, CH_3I	3.02	2
Iodomethane, CH_3I, at 370 Torr	2.20	7
Dichloromethane, CH_2Cl_2	1.92	2
Dichlorodifluoromethane, CCl_2F_2	2.42	1
	2.63	2,6
Chlorotrifluoromethane, $CClF_3$	1.43	1
	1.53	2
Trichlorofluoromethane, CCl_3F	3.50	1
	4.53	2
Trichloromethane, $CHCl_3$	4.2	1
	4.39	2
Methylamine, CH_3NH_2	0.81	1
Difluoromethane, CH_2F_2	0.79	2
Trifluoromethane, CHF_3	0.71	2
Bromochlorodifluoromethane, CF_2ClBr	3.84	2
Chlorodifluoromethane, $CHClF_2$	1.40	1
	1.11	2
Dichlorofluoromethane, $CHCl_2F$	1.33	1
	2.61	2
Chlorofluoromethane, CH_2ClF	1.03	1
Hexafluoroethane, C_2F_6	1.82	1
	2.55	2
Ethyne (Acetylene), C_2H_2	1.10	1
	1.11	2
Chloropentafluoroethane, C_2ClF_5	2.3	1
	3.0	6
Dichlorotetrafluoroethane, $C_2Cl_2F_4$	2.52	1
Chlorotrifluoroethylene, C_2ClF_3	1.82	2
1,1,1-Trichloro-2,2,2-trifluoroethane	6.55	2
1,1,2-Trichloro-1,2,2-trifluoroethane	6.05	2
Chloroethane, C_2H_5Cl	1.00	1
1,1-Dichloroethane	2.66	2
Trifluoroacetonitrile, CF_3CN	3.5	1
Acetonitrile, CH_3CN	2.11	2
Dimethylamine, $(CH_3)_2NH$	1.04	1
Ethylamine, $C_2H_5NH_2$	1.01	1
Ethylene oxide (oxirane), CH_3CHO	1.01	1
Perfluoropropene, C_3F_6	2.55	2
Octafluoropropane, C_3F_8	2.19	1
	2.47	2
3,3,3-Trifluoro-1-propene, CH_2CHCF_3	2.11	2
Pentafluoroisocyanoethane, C_2F_5NC	4.5	1
1,1,1,4,4,4-Hexafluoro-2-butyne, CF_3CCCF_3	5.84	2
Octafluorocyclobutane, C_4F_8	3.34	2
1,1,1,2,3,4,4,4-Octafluoro-2-butene	2.8	1
Decafluorobutane, C_4F_{10}	3.08	1
Perfluorobutanenitrile, C_3F_7CN	5.5	1
Perfluoro-2-methyl-1,3-butadiene, C_5F_8	5.5	1
Hexafluorobenzene, C_6F_6	2.11	2
Perfluorocyclohexane, C_6F_{12}, (saturated vapor)	6.18	2

* Relative to nitrogen, unless units of kV/mm are indicated.

TABLE 2. Dielectric Strength of Liquids

Material	Dielectric strength kV/mm	Ref.
Helium, He, liquid, 4.2 K	10	9
Static	10	11
Dynamic	5	11
	23	12
Nitrogen, N_2, liquid, 77K		
Coaxial cylinder electrodes	20	10
Sphere to plane electrodes	60	10
Water, H_2O, distilled	65-70	13
Carbon tetrachloride, CCl_4	5.5	14
	16.0	15
Hexane, C_6H_{14}	42.0	16
Two 2.54 cm diameter spherical electrodes, 50.8 μm space	156	17,18
Cyclohexane, C_6H_{12}	42-48	16
2-Methylpentane, C_6H_{14}	149	17,18
2,2-Dimethylbutane, C H	[illegible]	[illegible]
[illegible], C_6H_6	163	17,18
Chlorobenzene, C_6H_5Cl	7.1	14
	18.8	15
2,2,4-Trimethylpentane, C_8H_{18}	140	17,18
Phenylxylylethane	23.6	19
Heptane, C_7H_{16}	166	17,18
2,4-Dimethylpentane, C_7H_{16}	133	17,18
Toluene, $C_6H_5CH_3$	199	17,18
	46	16
	12.0	14
	20.4	15
Octane, C_8H_{18}	16.6	14
	20.4	15
	179	17,18
Ethylbenzene, C_8H_{10}	226	17,18
Propylbenzene, C_9H_{12}	250	17,18
Isopropylbenzene, C_9H_{12}	238	17,18
Decane, $C_{10}H_{22}$	192	17,18
Synthetic Paraffin Mixture		
Synfluid 2cSt PAO	29.5	37
Butylbenzene, $C_{10}H_{14}$	275	17,18
Isobutylbenzene, $C_{10}H_{14}$	222	17,18
Silicone oils—polydimethylsiloxanes, $(CH_3)_3Si\text{-}O\text{-}[Si(CH_3)_2]_x\text{-}O\text{-}Si(CH_3)_3$		
Polydimethylsiloxane silicone fluid	15.4	20
Dimethyl silicone	24.0	[illegible]
[illegible]	10-15	23
Mineral insulating oils	11.8	6
Polybutene oil for capacitors	13.8	6
Transformer dielectric liquid	28-30	6
Isopropylbiphenyl capacitor oil	23.6	6
Transformer oil	110.7	24
Transformer oil Agip ITE 360	9-12.6	23
Perfluorinated hydrocarbons		
Fluorinert FC 6001	8.0	23
Fluorinert FC 77	10.7	23
Perfluorinated polyethers		
Galden XAD (Mol. wt. 800)	10.5	23
Galden D40 (Mol. wt. 2000)	10.2	23
Castor oil	65	25

TABLE 3. Dielectric Strength of Solids

Material	Dielectric strength kV/mm	Ref
Sodium chloride, NaCl, crystalline	150	26
Potassium bromide, KBr, crystalline	80	26
Ceramics		
Alumina (99.9% Al_2O_3)	13.4	6,27a
Aluminum silicate, Al_2SiO_5	5.9	6
Berillia (99% BeO)	13.8	6,27b
Boron nitride, BN	37.4	6
Cordierite, $Mg_2Al_4Si_5O_{18}$	7.9	6,27c
Forsterite, Mg_2SiO_4	9.8	28
Porcelain	35-160	26
Steatite, $Mg_3Si_4O_{11}\cdot H_2O$	9.1-15.4	6
Titanates of Mg, Ca, Sr, Ba, and Pb	20-120	3
Barium titanate, glass bonded	>30	36
Zirconia, ZrO_2	11.4	29
Glasses		
Fused silica, SiO_2	470-670	26
Alkali-silicate glass	200	26
Standard window glass	9.8-13.8	28
Micas		
Muscovite, ruby, natural	118	6
Phlogopite, amber, natural	118	6
Fluorophlogopite, synthetic	118	6
Glass-bonded mica	14.0-15.7	6
Thermoplastic Polymers		
Polypropylene	23.6	6
Amide polymer nylon 6/6, dry	23.6	6
Polyamide-imide copolymer	22.8	6
Modified polyphenylene oxide	21.7	6
Polystyrene	19.7	6
Polymethyl methacrylate	19.7	6
Polyetherimide	18.9	6
Amide polymer nylon 11(dry)	16.7	6
Polysulfone	16.7	6
Styrene-acrylonitrile copolymer	16.7	6
Acrylonitrile-butadiene-styrene	16.7	6
Polyethersulfone	15.7	6
Polybutylene terephthalate	15.7	6
Polystyrene-butadiene copolymer	15.7	6
Acetal homopolymer	15.0	6
Acetal copolymer	15.0	6
Polyphenylene sulfide	15.0	6

Material	Dielectric strength kV/mm	Ref
Polycarbonate	15.0	6
Acetal homopolymer resin (molding resin)	15.0	6
Acetal copolymer resin	15.0	6
Thermosetting Molding Compounds		
Glass-filled allyl (Type GDI-30 per MIL-M-14G)	15.7	6
Glass-filled epoxy, electrical grade	15.4	6
Glass-filled phenolic (Type GPI-100 per MIL-M-14G)	15.0	6
Glass-filled alkyd/polyester (Type MAI-60 per MIL-M-14G)	14.8	6
Glass-filled melamine (Type MMI-30 per MIL-M-14G)	13.4	6
Extrusion Compounds for High-Temperature Insulation		
Polytetrafluoroethylene	19.7	6
Perfluoroalkoxy polymer	21.7	6
Fluorinated ethylene-propylene copolymer	19.7	6
Ethylene-tetrafluoroethylene copolymer	15.7	6
Polyvinylidene fluoride	10.2	6
Ethylene-chlorotrifluoroethylene copolymer	19.3	6
Polychlorotrifluoroethylene	19.7	6
Extrusion Compounds for Low-Temperature Insulation		
Polyvinyl chloride		
Flexible	11.8-15.7	30
Rigid	13.8-19.7	30
Polyethylene	18.9	28
Polyethylene, low-density	21.7	6
	300	31
Polyethylene, high-density	19.7	6
Polypropylene/polyethylene copolymer	23.6	6
Embedding Compounds		
Basic epoxy resin: bisphenol-A/epichlorohydrin polycondensate	19.7	6
Cycloaliphatic epoxy: alicyclic diepoxy carboxylate	19.7	6
Polyetherketone	18.9	30
Polyurethanes		
Two-component, polyol-cured	25.4	6
Two-part solventless, polybutylene-based	24.0	6
Silicones		
Clear two-part heat curing eletrical grade silicone embedding resin	21.7	6
Red insulating enamel (MIL-E-22118)		
Dry	47.2	6
Wet	11.8	6
Enamels		
Red enamel, fast cure		
Standard conditions	78.7	6
Immersion conditions	47.2	6
Black enamel		
Standard conditions	70.9	6
Immersion conditions	47.2	6
Varnishes		
Vacuum-pressure impregnated baking type solventless polyester varnish		
Rigid, two-part	70.9	6
Semiflexible high-bond thixotropic	78.7	6
Rigid high-bond high-flash freon-resistant	68.9	6
Baking type epoxy varnish		
Solventless, rigid, low viscosity, one-part	90.6	6
Solventless, semiflexible, one-part	82.7	6
Solventless, semirigid, chemical resistant, low dielectric constant	106.3	6
Solvable, for hermetic electric motors	181.1	6
Polyurethane coating		
Clear conformal, fast cure		
Standard conditions	78.7	6
Immersion conditions	47.2	6
Insulating Films and Tapes		
Low-density polyethylene film (40 μm thick)	300	31
Poly-*p*-xylylene film	410-590	32
Aromatic polymer films		
Kapton H (Du Pont)	389-430	33
Ultem (GE Plastic and Roem AG)	437-565	33
Hostaphan (Hoechst AG)	338-447	33
Amorphous Stabar K2000 (ICI film)	404-422	33
Stabar S100 (ICI film)	353-452	33
Polyetherimide film (26 μm)	486	34
Parylene N/D (poly-*p*-xylylene/poly-dichloro-*p*-xylylene) 25 μm film	275	6
Cellulose acetate film	157	6
Cellulose triacetate film	157	6
Polytetrafluoroethylene film	87-173	6
Perfluoroalkoxy film	157-197	6
Fluorinated ethylene-propylene copolymer film	197	6
Ethylene-tetrafluoroethylene film	197	6
Ethylene-chlorotrifluoroethylene copolymer film	197	6
Polychlorotrifluoroethylene film	118-153.5	6
High-voltage rubber insulating tape	28	6
Composites		
Isophthalic polyester (vinyl toluene monomer) filled with		
Calcium carbonate, $CaCO_3$	15.0	38
Gypsum, $CaSO_4$	14.4	38
Alumina trihydrate	15.4	38
Clay	14.4	38
BPA fumarate polyester (vinyl toluene monomer) filled with		
Calcium carbonate	6.1	38
Gypsum	5.9	38
Alumina trihydrate	11.8	38
Clay	12.6	38
Polysulfone resin—30% glass fiber	16.5-18.7	38
Polyamid resin (Nylon 66)—30% carbon fiber	13.0	38
Polyimide thermoset resin, glass reinforced	12.0	39
Polyester resin (thermoplastic)—		

Material	Dielectric strength kV/mm	Ref
40% glass fiber	20.0	38
Epoxy resin (diglycidyl ether of bisphenol A), glass reinforced	16.0	40
Various Insulators		
Rubber, natural	100-215	26
Butyl rubber	23.6	6
Neoprene	15.7-27.6	6
Silicone rubber	26-36	6

Material	Dielectric strength kV/mm	Ref
Room-temperature vulcanized silicone rubber	9.2-10.9	35
Ureas (from carbamide to tetraphenylurea)	11.8-15.7	28
Dielectric papers		
Aramid paper, calendered	28.7	6
Aramid paper, uncalendered	12.2	6
Aramid with Mica	39.4	6

References

1. Vijh, A. K. *IEEE Trans.*, EI-12, 313, 1997.
2. Brand, K. P., *IEEE Trans.*, EI-17, 451, 1982.
3. *Encyclopedic Dictionary in Physics*, Vedensky, B. A. and Vul, B. M., Eds., Vol. 4, Soviet Encyclopedia Publishing House, Moscow, 1965.
4. Kubuki, M., Yoshimoto, R., [illegible]
5. [illegible] M. I., *IEEE Trans.*, DEI-1, 305, 1994.
6. Shugg, W. T., *Handbook of Electrical and Electronic Insulating Materials*, Van Nostrand Reinhold, New York, 1986.
7. Devins, J. C., *IEEE Trans.*, EI-15, 81, 1980.
8. Xu, X., Jayaram, S., and Boggs, S. A., *IEEE Trans.*, DEI-3, 836, 1996.
9. Okubo, H., Wakita, M., Chigusa, S., Nayakawa, N., and Hikita, M., *IEEE Trans.*, DEI-4, 120, 1997.
10. Hayakawa, H., Sakakibara, H., Goshima, H., Hikita, M., and Okubo, H., *IEEE Trans.*, DEI-4, 127, 1997.
11. Okubo, H., Wakita, M., Chigusa, S., Hayakawa, N., and Hikita, M., *IEEE Trans.*, DEI-4, 220, 1997.
12. Von Hippel, A. R., *Dielectric Materials and Applications*, MIT Press, Cambridge, MA, 1954.
13. Jones, H. M. and Kunhards, E. E., *IEEE Trans.*, DEI-1, 1016, 1994.
14. Nitta, Y. and Ayhara, Y., *IEEE Trans.*, EI-11, 91, 1976.
15. Gallagher, T. J., *IEEE Trans.*, EI-12, 249, 1977.
16. Wong, P. P. and Forster, E. O., in *Dielectric Materials. Measurements and Applications*, IEE Conf. Publ. 177, 1, 1979.
17. Kao, K. C. *IEEE Trans.*, EI-11, 121, 1976.
18. Sharbaugh, A. H., Crowe, R. W., and Cox, E. B., *J. Appl. Phys.*, 27, 806, 1956.
19. Miller, R. L., Mandelcorn, L., and Mercier, G. E., in *Proc. Intl. Conf. on Properties and Applications of Dielectric Materials*, Xian, China, June 24-28, 1985; cited in Ref. 6, p. 492.
20. Hakim, R. M., Oliver, R. G., and St-Onge, H., *IEEE Trans.*, EI-12, 360, 1977.
21. Hosticka, C., *IEEE Trans.*, 389, 1977.
22. Yasufuku, S., Umemura, T., and Ishioka, Y., *IEEE Trans.*, EI-12, 402, 1977.
23. Forster, E. O., Yamashita, H., Mazzetti, C., Pompini, M., Caroli, L., and Patrissi, S., *IEEE Trans.*, DEI-1, 440, 1994.
24. Bell, W. R., *IEEE Trans.*, 281, 1977.
25. Ramu, T. C. and Narayana Rao, Y., in *Dielectric Materials. Measurements and Applications*, IEE Conf. Publ. 177, 37.
26. Skanavi, G. I., *Fizika Dielektrikov; Oblast Silnykh Polei* (Physics of Dielectrics; Strong Fields). Gos. Izd. Fiz. Mat. Nauk (State Publ. House for Phys. and Math. Scis.), Moscow, 1958.
27. Kleiner, R. N., in *Practical Handbook of* [illegible]
28. [illegible] Methods, Reinhold Publ., New York, 1973.
29. Flinn, R. A. and Trojan, P. K., *Engineering Materials and Their Applications*, 2nd ed., Houghton Mifflin, 1981, p. 614.
30. Lynch, C. T., Ed., *Practical Handbook of Materials Science*, CRC Press, Boca Raton, FL, 1989.
31. Suzuki, H., Mukai, S., Ohki, Y., Nakamichi, Y., and Ajiki, K., *IEEE Trans.*, DEI-4, 238, 1997.
32. Mori, T., Matsuoka, T., and Muzitani, T., *IEEE Trans.*, DEI-1, 71, 1994.
33. Bjellheim, P. and Helgee, B., *IEEE Trans.*, DEI-1, 89, 1994.
34. Zheng, J. P., Cygan, P. J., and Jow, T. R., *IEEE Trans.*, DEI-3, 144, 1996.
35. Danukas, M. G., *IEEE Trans.*, DEI-1, 1196, 1994.
36. Burn, I. and Smithe, D. H., *J. Mater. Sci.*, 7, 339, 1972.
37. Hope, K.D., Chevron Chemical, Private Communication.
38. *Engineering Materials Handbook*, Vol. 1, Composites, C.A. Dostal, Ed., ASM Intl., 1987.
39. 1985 Materials Selector, *Mater. Eng.*, (12) 1984.
40. *Modern Plastics Encyclopedia*, McGraw-Hill, v. 62 (No. 10A) 1985–1986.

Review Literature on the Subject

R1. Kuffel, E. and Zaengl, W. S., *HV Engineering Fundamentals*, Pergamon, 1989.
R2. Kok, J. A., *Electrical Breakdown of Insulating Liquids*, Phillips Tech. Library, Cleaver-Hum, London, 1961.
R3. Gallagher, T. J., *Simple Dielectric Liquids*, Clarendon, Oxford, 1975.
R4. Meek, J. M. and Craggs, J. D., Eds., *Electric Breakdown in Gases*, John Wiley & Sons, 1976.
R5. Von Hippel, A. R., *Dielectric Materials and Applications*, MIT Press, Cambridge, MA, 1954.
R6. O'Dwyer, J. J. *The Theory of Dielectric Breakdown of Solids*, Clarendon Press, 1964.

COEFFICIENT OF FRICTION

The coefficient of friction between two surfaces is the ratio of the force required to move one over the other to the force pressing the two together. Thus if F is the minimum force needed to move one surface over the other, and W is the force pressing the surfaces together, the coefficient of friction μ is given by $\mu = F/W$. A greater force is generally needed to initiate movement from rest than to continue the motion once sliding has started. Thus the static coefficient of friction μ (static) is usually larger that the sliding or kinetic coefficient μ (sliding).

This table gives characteristic values of both the static and sliding coefficients of friction for a number of material combinations. In each case Material 1 is moving over the surface of Material 2. The type of lubrication or any other special condition is indicated in the third column. All values refer to room temperature unless otherwise indicated. It should be emphasized that the coefficient of friction is very sensitive to the condition of the surface, so that these values represent only a rough guide.

References

1. Minshall, H., in *CRC Handbook of Chemistry and Physics, 73rd Edition*, Lide, D. R., Ed., CRC Press, Boca Raton, FL, 1992.
2. Fuller, D. D., in *American Institute of Physics Handbook, 3rd Edition*, Gray, D. E., Ed., McGraw-Hill, New York, 1972.

Material 1	Material 2	Conditions	μ (static)	μ (sliding)
Metals				
Hard steel	Hard steel	Dry	0.78	0.42
		Castor oil	0.15	0.081
		Steric acid	0.005	0.029
		Lard	0.11	0.084
		Light mineral oil	0.23	
		Graphite		0.058
Hard steel	Graphite	Dry	0.21	
Mild steel	Mild steel	Dry	0.74	0.57
		Oleic acid		0.09
Mild steel	Phosphor bronze	Dry		0.34
Mild steel	Cast iron	Dry		0.23
Mild steel	Lead	Dry	0.95	0.95
		Mineral oil	0.5	0.3
Mild steel	Brass	Dry	0.35	
Cast iron	Cast iron	Dry	1.10	0.15
Aluminum	Aluminum	Dry	1.05	1.4
Aluminum	Mild steel	Dry	0.61	0.47
Brass	Mild steel	Dry	0.51	0.44
		Castor oil	0.11	
Brass	Cast iron	Dry		0.30
Bronze	Cast iron	Dry		0.22
Cadmium	Mild steel	Dry		0.46
Copper	Copper	Dry	1.6	
Copper	Mild steel	Dry	0.53	0.36
		Oleic acid		0.18
Copper	Cast iron	Dry	1.05	0.29
Copper	Glass	Dry	0.68	0.53
Lead	Cast iron	Dry		0.43
Magnesium	Magnesium	Dry	0.6	
Magnesium	Mild steel	Dry		0.42
Magnesium	Cast iron	Dry		0.25
Nickel	Nickel	Dry	1.10	0.53
Nickel	Mild steel	Dry		0.64
Tin	Cast iron	Dry		0.32
Zinc	Cast iron	Dry	0.85	0.21
Nonmetals				
Diamond	Diamond	Dry	0.1	
Diamond	Metals	Dry	0.12	
Garnet	Mild steel	Dry		0.39
Glass	Glass	Dry	0.94	0.4
Glass	Nickel	Dry	0.78	0.56

Material 1	Material 2	Conditions	μ (static)	μ (sliding)
Graphite	Graphite	Dry	0.1	
Mica	Mica	Freshly cleaved	1.0	
Nylon	Nylon	Dry	0.2	
Nylon	Steel	Dry	0.40	
Polyethylene	Polyethylene	Dry	0.2	
Polyethylene	Steel	Dry	0.2	
Polystyrene	Polystyrene	Dry	0.5	
Polystyrene	Steel	Dry	0.3	
Sapphire	Sapphire	Dry	0.2	
Teflon	Teflon	Dry	0.04	0.04
Teflon	Steel	Dry	0.04	0.04
Tungsten carbide	Tungsten carbide	Dry, room temp.	0.17	
		Dry, 1000 °C	0.45	
		Dry, 1600 °C	1.8	
		Oleic acid	0.12	
Tungsten carbide	Graphite	Dry	0.15	
Tungsten carbide	Steel	Dry	0.5	
		Oleic acid	0.08	
Miscellaneous materials				
Cotton	Cotton	Threads	0.3	
Leather	Cast iron	Dry	0.6	0.56
Leather	Oak	Parallel to grain	0.61	0.52
Oak	Oak	Parallel to grain	0.62	0.48
		Perpendicular to grain	0.54	0.32
Silk	Silk	Clean	0.25	
Wood	Wood	Dry	0.35	
		Wet	0.2	
Wood	Brick	Dry	0.6	
Wood	Leather	Dry	0.35	
Various materials on ice and snow				
Ice	Ice	Clean, 0 °C	0.1	0.02
		Clean, -12 °C	0.3	0.035
		Clean, -80 °C	0.5	0.09
Aluminum	Snow	Wet, 0 °C	0.4	
		Dry, 0 °C	0.35	
Brass	Ice	Clean, 0 °C		0.02
		Clean, -80 °C		0.15
Nylon	Snow	Wet, 0 °C	0.4	
		Dry, -10 °C	0.3	
Teflon	Snow	Wet, 0 °C	0.05	
		Dry, 0 °C	0.02	
Wax, ski	Snow	Wet, 0 °C	0.1	
		Dry, 0 °C	0.04	
		Dry, -10 °C	0.2	

FLAME TEMPERATURES

This table gives the adiabatic flame temperatures for stoichemetric mixtures of various fuels and oxidizers. The temperatures are calculated from thermodynamic and transport properties under ideal adiabatic conditions, using methods described in the reference.

Reference

Fristrom, R. M., *Flame Structures and Processes*, Oxford University Press, New York, 1995.

Adiabatic Flame Temperature in K for Various Fuel-Oxidizer Combinations

	Oxidizer					
Fuel	**Air**	**O_2**	**F_2**	**Cl_2**	**N_2O**	**NO**
Organic liquids and gases						
Acetaldehyde	2288					
Acetone	2253					
Acetylene	2607					
Benzene	2363					
Butane	2248					
Carbon disulfide	2257					
Cyanogen	2596	4855				
Cyclohexane	2250					
Cyclopropane	2370					
Decane	2286					
Ethane	2244					
Ethanol	2238					
Ethylene	2375					
Hexane	2238					
Methane	2236					
Methanol	2222					
Oxirane	2177					
Pentane	2250					
Propane	2250					
Toluene	2344					
Solids						
Aluminum		4005				
Lithium		2711				
Phosphorus (white)		3242				
Zirconium		4278				
Other						
Ammonia		2845				
Carbon monoxide	1388					
Diborane		3350				
Hydrazine		3037				
Hydrogen	2169	3000	4006	2493	2965	3127
Hydrogen sulfide	2091	3414				
Phosphine		3139				
Silane		3043				

ALLOCATION OF FREQUENCIES IN THE RADIO SPECTRUM

In the United States the National Telecommunications and Information Administration (NTIA) has responsibility for assigning each portion of the radio spectrum (9 kHz to 300 GHz) for different uses. These assignments must be compatible with the rules of the International Telecommunications Union (ITU), to which the United States is bound by treaty. The current assignments are given in a wall chart (Reference 1) and may also be found on the NTIA web site (Reference 2). The list below summarizes the broad features of the spectrum allocation, with particular attention to those sections of scientific interest. The references should be consulted for details of the allocations in the frequency bands listed here, which in some cases are quite complex.

References

1. *United States Frequency Allocations*, 1996 Spectrum Wall Chart, Stock No. 003-000-00652-2, U. S. Government Printing Office, P. O. Box 371954, Pittsburgh, PA 15250-7954.
2. http://www.ntia.doc.gov/osmhome/allochrt.html

Frequency range	Allocation
9–19.95 kHz	Maritime communication, navigation
19.95–20.05 kHz	Standard frequency and time signal (also at 60 kHz and 2.5, 5, 10, 15, 20, 25 MHz)
20.05–535 kHz	Maritime and aeronautical communication, navigation
535–1605 kHz	AM radio broadcasting
1605–3500 kHz	Mobile communication and navigation, amateur radio (1800–1900 kHz)
3.5–4.0 MHz	Amateur radio
4.0–5.95 MHz	Mobile communication
5.95–13.36 MHz	Mobile communication, amateur, short-wave broadcasting
13.36–13.41 MHz	Radioastronomy
13.41–25.55 MHz	Mobile communication, amateur, short-wave broadcasting
25.55–25.67 MHz	Radioastronomy
25.67–37.5 MHz	Mobile communication, amateur, short-wave broadcasting
37.5–38.25 MHz	Radioastronomy
38.25–50.0 MHz	Mobile communication
50.0–54.0 MHz	Amateur
54.0–72.0 MHz	TV channels 2–4
72.0–73.0 MHz	Mobile communication
73.0–74.6 MHz	Radioastronomy
74.6–76.0 MHz	Mobile communication
76.0–88.0 MHz	TV channels 5–6
88.0–108.0 MHz	FM radio broadcasting
108.0–118.0 MHz	Aeronautical navigation
118.0–174.0 MHz	Mobile communication, space research, meteorological satellites
174.0–216.0 MHz	TV channels 7–13
216.0–400.05 MHz	Mobile communication
400.05–400.15 MHz	Standard frequency and time satellite (also 20 and 25 GHz)
400.15–406.1 MHz	Meteorological aids (radiosonde)
406.1–410.0 MHz	Radioastronomy
410.0–470.0 MHz	Mobile communication, amateur
470.0–512.0 MHz	TV channels 14–20
512.0–608.0 MHz	TV channels 21–36
608.0–614.0 MHz	Radioastronomy
614.0–806.0 MHz	TV channels 38–69
806–1400 MHz	Mobile communication, navigation
1400–1427 MHz	Radioastronomy, space research
1427–1660 MHz	Various navigation and satellite applications
1660–1710 MHz	Radioastronomy, space research, meteorology
1710–2655 MHz	Various navigation and satellite applications
2655–2700 MHz	Radioastronomy, space research
2.7–4.99 GHz	Various navigation and satellite applications
4.99–5.0 GHz	Radioastronomy, space research
5.0–10.6 GHz	Various navigation and satellite applications
10.6–10.7 GHz	Radioastronomy, space research
10.7–15.35 GHz	Various navigation and satellite applications
15.35–15.4 GHz	Radioastronomy, space research
15.4–22.21 GHz	Various navigation and satellite applications

Frequency range	Allocation
22.21–22.5 GHz	Radioastronomy, space research
22.25–23.6 GHz	Various navigation and satellite applications
23.6–24.0 GHz	Radioastronomy, space research
24.0–31.3 GHz	Various navigation and satellite applications
31.3–31.8 GHz	Radioastronomy, space research
31.8–42.5 GHz	Various navigation and satellite applications
42.5–43.5 GHz	Radioastronomy
43.5–51.4 GHz	Various navigation and satellite applications
51.4–54.25 GHz	Radioastronomy, space research
54.25–58.2 GHz	Space research
58.2–59.0 GHz	Radioastronomy, space research
59.0–64.0 GHz	Satellite applications
64.0–65.0 GHz	Radioastronomy, space research
65.0–72.77 GHz	Various navigation and satellite applications
72.77–72.91 GHz	Radioastronomy, space research
72.91–86.0 GHz	Various navigation and satellite applications
86.0–92.0 GHz	Radioastronomy, space research
92.0–105.0 GHz	Various navigation and satellite applications
105.0–116.0 GHz	Radioastronomy, space research
116.0–164.0 GHz	Various navigation and satellite applications
164.0–168.0 GHz	Radioastronomy, space research
168.0–182.0 GHz	Various navigation and satellite applications
182.0–185.0 GHz	Radioastronomy, space research
185.0–217.0 GHz	Various navigation and satellite applications
217.0–231.0 GHz	Radioastronomy, space research
231.0–265.0 GHz	Various navigation and satellite applications
265.0–275.0 GHz	Radioastronomy
275.0–300.0 GHz	Mobile communications

Section 16
Health and Safety Information

Handling and Disposal of Chemicals in Laboratories.................................... 16-1
Flammability of Chemical Substances... 16-13
Threshold Limits for Airborne Contaminants..................................... 16-29
Octanol–Water Partition Coefficients.. 16-43
Protection against Ionizing Radiation 16-48
Annual Limits on Intakes of Radionuclides 16-49
Chemical Carcinogens 16-53

Safety

HANDLING AND DISPOSAL OF CHEMICALS IN LABORATORIES

Robert Joyce and Blaine C. McKusick

The following material has been extracted from two books prepared under the auspices of the Committee on Hazardous Substances in the Laboratory of the National Academy of Sciences – National Research Council. Readers are referred to these books for full details:

Prudent Practices for Handling Hazardous Chemicals in Laboratories, National Academy Press, Washington, D.C., 1981.

Prudent Practices for Disposal of Chemicals from Laboratories, National Academy Press, Washington, D.C., 1983.

The permission of the National Academy Press to use these extracts is gratefully acknowledged.

INCOMPATIBLE CHEMICALS

The term "incompatible chemicals" refers to chemicals that can react with each other

- Violently
- With evolution of substantial heat
- To produce flammable products
- To produce toxic products

Good laboratory safety practice requires that incompatible chemicals be stored, transported, and disposed of in ways that will prevent their coming together in the event of an accident. Tables 1 and 2 give some basic guidelines for the safe handling of acids, bases, reactive metals, and other chemicals. Neither of these tables is exhaustive, and additional information on incompatible chemicals can be found in the following references.

1. Urben, P. G., Ed., *Bretherick's Handbook of Reactive Chemical Hazards,* 5th ed., Butterworth-Heinemann, Oxford, 1995.
2. Luxon, S. G., Ed., *Hazards in the Chemical Laboratory,* 5th ed., Royal Society of Chemistry, Cambridge, 1992.
3. *Fire Protection Guide to Hazardous Materials,* 11th ed., National Fire Protection Association, Quincy, MA, 1994.

TABLE 1. General Classes of Incompatible Chemicals

A	B
Acids	Bases, reactive metals
Oxidizing agents[a]	Reducing agents[a]
Chlorates	Ammonia, anhydrous and aqueous
Chromates	Carbon
Chromium trioxide	Metals
Dichromates	Metal hydrides
Halogens	Nitrites
Halogenating agents	Organic compounds
Hydrogen peroxide	Phosphorus
Nitric acid	Silicon
Nitrates	Sulfur
Perchlorates	
Peroxides	
Permanganates	
Persulfates	

[a] The examples of oxidizing and reducing agents are illustrative of common laboratory chemicals; they are not intended to be exhaustive.

TABLE 2. Examples of Incompatible Chemicals

Chemical	Is incompatible with
Acetic acid	Chromic acid, nitric acid, hydroxyl compounds, ethylene glycol, perchloric acid, peroxides, permanaganates
Acetylene	Chlorine, bromine, copper, fluorine, silver, mercury
Acetone	Concentrated nitric and sulfuric acid mixtures
Alkali and alkaline earth metals (such as powdered aluminum or magnesium, calcium, lithium, sodium, potassium)	Water, carbon tetrachloride or other chlorinated hydrocarbons, carbon dioxide, halogens
Ammonia (anhydrous)	Mercury (in manometers, for example), chlorine, calcium hypochlorite, iodine, bromine, hydrofluoric acid (anhydrous)
Ammonium nitrate	Acids, powdered metals, flammable liquids, chlorates, nitrites, sulfur, finely divided organic or combustible materials
Aniline	Nitric acid, hydrogen peroxide
Arsenical materials	Any reducing agent
Azides	Acids

Chemical	Is incompatible with
Bromine	See Chlorine
Calcium oxide	Water
Carbon (activated)	Calcium hypochlorite, all oxidizing agents
Carbon tetrachloride	Sodium
Chlorates	Ammonium salts, acids, powdered metals, sulfur, finely divided organic or combustible materials
Chromic acid and chromium troixide	Acetic acid, naphthalene, camphor, glycerol, alcohol, flammable liquids in general
Chlorine	Ammonia, acetylene, butadiene, butane, methane, propane (or other petroleum gases), hydrogen, sodium carbide, benzene, finely divided metals, turpentine
Chlorine dioxide	Ammonia, methane, phosphine, hydrogen sulfide
Copper	Acetylene, hydrogen peroxide
Cumene hydroperoxide	Acids (organic or inorganic)
Cyanides	Acids
Flammable liquids	Ammonium nitrate, chromic acid, hydrogen peroxide, nitric acid, sodium peroxide, halogens
Fluorine	Everything
Hydrocarbons (such as butane, propane, benzene)	Fluorine, chlorine, bromine, chromic acid, sodium peroxide
Hydrocyanic acid	Nitric acid, alkali
Hydrofluoric acid (anhydrous)	Ammonia (aqueous or anhydrous)
Hydrogen peroxide	Copper, chromium, iron, most metals or their salts, alcohols, acetone, organic materials, aniline, nitro-methane, combustible materials
Hydrogen sulfide	Fuming nitric acid, oxidizing gases
Hypochlorites	Acids, activated carbon
Iodine	Acetylene, ammonia (aqueous or anhydrous), hydrogen
Mercury	Acetylene, fulminic acid, ammonia
Nitrates	Sulfuric acid
Nitric acid (concentrated)	Acetic acid, aniline, chromic acid, hydrocyanic acid, hydrogen sulfide, flammable liquids, flammable gases, copper, brass, any heavy metals
Nitrites	Acids
Nitroparaffins	Inorganic bases, amines
Oxalic acid	Silver, mercury
Oxygen	Oils, grease, hydrogen, flammable liquids, solids, or gases
Perchloric acid	Acetic anhydride, bismuth and its alloys, alcohol, paper, wood, grease, oils
Peroxides, organic	Acids (organic or mineral), avoid friction, store cold
Phosphorus (white)	Air, oxygen, alkalis, reducing agents
Potassium	Carbon tetrachloride, carbon dioxide, water
Potassium chlorate	Sulfuric and other acids
Potassium perchlorate (see also chlorates)	Sulfuric and other acids
Potassium permanganate	Glycerol, ethylene glycol, benzaldehyde, surfuric acid
Selenides	Reducing agents
Silver	Acetylene, oxalic acid, tartartic acid, ammonium compounds, fulminic acid
Sodium	Carbon tetrachloride, carbon dioxide, water
Sodium nitrite	Ammonium nitrate and other ammonium salts
Sodium peroxide	Ethyl or methyl alcohol, glacial acetic acid, acetic anhydride, benzaldehyde, carbon disulfide, glycerin, ethylene glycol, ethyl acetate, methyl acetate, furfural
Sulfides	Acids
Sulfuric acid	Potassium chlorate, potassium perchlorate, potassium permanganate (similar compounds of light metals, such as sodium, lithium)
Tellurides	Reducing agents

EXPLOSION HAZARDS

Table 3 lists some common classes of laboratory chemicals that have potential for producing a violent explosion when subjected to shock or friction. These chemicals should never be disposed of as such, but should be handled by procedures given in *Prudent Practices for Disposal of Chemicals from Laboratories*, National Academy Press, 1983, chapters 6 and 7. Additional information on these, as well as on some less common classes of explosives, can be found in L. Bretherick, *Handbook of Reactive Chemical Hazards*, 5th ed., Butterworth-Heinemann, Oxford, 1995.

Table 4 lists some illustrative combinations of common laboratory reagents that can produce explosions when they are brought together or that form reaction products that can explode without any apparent external initiating action. This list is not exhaustive, and additional information on potentially explosive reagent combinations can be found in *Manual of Hazardous Chemical Reactions, A Compilation of Chemical Reactions Reported to be Potentially Hazardous,* National Fire Protection Association, NFPA 491M, 1991, NFPA, Quincy, MA.

WATER-REACTIVE CHEMICALS

Table 5 lists some common laboratory chemicals that react violently with water and that should always be stored and handled so that they do not come into contact with liquid water or water vapor. Procedures for decomposing laboratory quantities are given in *Prudent Practices for Disposal of Chemicals from Laboratories*, chapter 6; the pertinent section of that chapter is given in parentheses.

PYROPHORIC CHEMICALS

Many members of the classes of readily oxidized, common laboratory chemicals listed in Table 6 ignite spontaneously in air. A more extensive list can be found in L. Bretherick, *Handbook of Reactive Chemical Hazards*, 3rd ed., Butterworths, London-Boston, 1985. Pyrophoric chemicals should be stored in tightly closed containers under an inert atmosphere (or, for some, an inert liquid), and all transfers and manipulations of them must be carried out under an inert atmosphere or liquid. Suggested procedures for decomposing them are given in *Prudent Practices for Disposal of Chemicals from Laboratories*, chapter 6; the pertinent section of that chapter is given in parentheses.

TABLE 3. Shock-Sensitive Compounds

Acetylenic compounds, especially polyacetylenes, haloacetylenes, and heavy metal salts of acetylenes (copper, silver, and mercury salts are particularly sensitive)

Acyl nitrates

Alkyl nitrates, particularly polyol nitrates such as nitrocellulose and nitroglycerine

Alkyl and acyl nitrites

Alkyl perchlorates

Amminemetal oxosalts: metal compounds with coordinated ammonia, hydrazine, or similar nitrogenous donors and ionic perchlorate, nitrate, permanganate, or other oxidizing group

Azides, including metal, nonmetal, and organic azides

Chlorite salts of metals, such as $AgClO_2$ and $Hg(ClO_2)_2$

Diazo compounds such as CH_2N_2

Diazonium slats, when dry

Fulminates (silver fulminate, AgCNO, can form in the reaction mixture from the Tollens' test for aldehydes if it is allowed to stand for some time; this can be prevented by adding dilute nitric acid to the test mixture as soon as the test has been completed)

Hydrogen peroxide becomes increasingly treacherous as the concentration rises above 30%, forming explosive mixtures with organic materials and decomposing violently in the presence of traces of transition metals

N–Halogen compounds such as difluoroamino compounds and halogen azides

N–Nitro compounds such as *N*–nitromethylamine, nitrourea, nitroguanidine, and nitric amide

Oxo salts of nitrogenous bases: perchlorates, dichromates, nitrates, iodates, chlorites, chlorates, and permanganates of ammonia, amines, hydroxylamine, guanidine, etc.

Perchlorate salts. Most metal, nonmetal, and amine perchlorates can be detonated and may undergo violent reaction in contact with combustible materials

Peroxides and hydroperoxides, organic (see Chapter 6, Section II.P)

Peroxides (solid) that crystallize from or are left from evaporation of peroxidizable solvents (see Chapter 6 and Appendix I)

Peroxides, transition–metal salts

Picrates, especially salts of transition and heavy metals, such as Ni, Pb, Hg, Cu, and Zn; picric acid is explosive but is less sensitive to shock or friction than its metal salts and is relatively safe as a water–wet paste (see Chapter 7)

Polynitroalkyl compounds such as tetranitromethane and dinitroacetonitrile

Polynitroaromatic compounds, especially polynitro hydrocarbons, phenols, and amines

TABLE 4. Potentially Explosive Combinations of Some Common Reagents

Acetone + chloroform in the presence of base
Acetylene + copper, silver, mercury, or their salts
Ammonia (including aqueous solutions) + Cl_2, Br_2, or I_2
Carbon disulfide + sodium azide
Chlorine + an alcohol
Chloroform or carbon tetrachloride + powdered Al or Mg
Decolorizing carbon + an oxidizing agent
Diethyl ether + chlorine (including a chlorine atmosphere)
Dimethyl sulfoxide + an acyl halide, $SOCl_2$ or $POCl_3$
Dimethyl sulfoxide + CrO_3
Ethanol + calcium hypochlorite
Ethanol + silver nitrate
Nitric acid + acetic anhydride or acetic acid
Picric acid + a heavy–metal salt, such as of Pb, Hg, or Ag
Silver oxide + ammonia + ethanol
Sodium + a chlorinated hydrocarbon
Sodium hypochlorite + an amine

TABLE 5. Water-Reactive Chemicals

Alkali metals (III.D)
Alkali metal hydrides (III.C.2)
Alkali metal amides (III.C.7)
Metal alkyls, such as lithium alkyls and aluminum alkyls (IV.A)
Grignard reagents (IV.A)
Halides of nonmetals, such as BCl_3, BF_3, PCl_3, PCl_5, $SiCl_4$, S_2Cl_2 (III.F)
Inorganic acid halides, such as $POCl_3$, $SOCl_2$, SO_2Cl_2 (III.F)
Anhydrous metal halides, such as $AlCl_3$, $TiCl_4$, $ZrCl_4$, $SnCl_4$ (III.E)
Phosphorus pentoxide (III.I)
Calcium carbide (IV.E)
Organic acid halides and anhydrides of low molecular weight (II.J)

TABLE 6. Classes of Pyrophoric Chemicals

Grignard reagents, RMgX (IV.A)
Metal alkyls and aryls, such as RLi, RNa, R_3Al, R_2Zn (IV.A)
Metal carbonyls, such as $Ni(CO)_4$, $Fe(CO)_5$, $Co_2(CO)_8$ (IV.B)
Alkali metals such as Na, K (III.D.1)
Metal powders, such as Al, Co, Fe, Mg, Mn, Pd, Pt, Ti, Sn, Zn, Zr (III.D.2)
Metal hydrides, such as NaH, $LiAlH_4$ (IV.C.2)
Nonmetal hydrides, such as B_2H_6 and other boranes, PH_3, AsH_3 (III.G)
Nonmetal alkyls, such as R_3B, R_3P, R_3As (IV.C)
Phosphorus (white) (III.H)

HAZARDS FROM PEROXIDE FORMATION

Many common laboratory chemicals can form peroxides when allowed access to air over a period of time. A single opening of a container to remove some of the contents can introduce enough air for peroxide formation to occur. Some types of compounds form peroxides that are treacherously and violently explosive in concentrated solution or as solids. Accordingly, peroxide-containing liquids should never be evaporated near to or to dryness. Peroxide formation can also occur in many polymerizable unsaturated compounds, and these peroxides can initiate a runaway, sometimes explosive, polymerization reaction. Procedures for testing for peroxides and for removing small amounts from laboratory chemicals are given in *Prudent Practices for Disposal of Chemicals from Laboratories*, chapter 6, Section II.P.

Table 7 provides a list of structural characteristics in organic compounds that can peroxidize. These structures are listed in approximate order of decreasing hazard. Reports of serious incidents involving the last five structural types are extremely rare, but these structures are listed because laboratory workers should be aware that they can form peroxides that can influence the course of experiments in which they are used.

Table 8 gives examples of common laboratory chemicals that are prone to form peroxides on exposure to air. The lists are not exhaustive, and analogous organic compounds that have any of the structural features given in Table 7 should be tested for peroxides before being used as solvents or reagents, or before being distilled. The recommended retention times begin with the date of synthesis or of opening the original container.

DISPOSAL OF TOXIC CHEMICALS

It is often desirable to precipitate toxic cations or hazardous anions from solution to facilitate recovery or disposal. Table 9 lists precipitants for many common cations, and Table 10 gives precipitants for some hazardous anions. Many cations can be precipitated as sulfides by adding sodium sulfide solution (preferable to the highly toxic hydrogen sulfide) to a neutral solution of the cation (Table 11). Control of pH is important because some sulfides will redissolve in excess sulfide ion. After precipitation, excess sulfide can be destroyed by addition of hypochlorite.

Most metal cations are precipitated as hydroxides or oxides at high pH. Since many of these precipitates will redissolve in excess base, it is often necessary to control pH. Table 12 shows the recommended pH range for precipitating many cations in their most common oxidation state. The notation "1 N" in the right-hand column indicates that the precipitate will not dissolve in 1 N sodium hydroxide (pH 14).

The distinctions between high and low toxicity or hazard are based on toxicological and other data, and are relative. There is no implication of a sharp distinction between high and low, or that any cations or anions are totally without hazard.

TABLE 7. Types of Chemicals That Are Prone to Form Peroxides

A. Organic structures (in approximate order of decreasing hazard)

	Structure	Type
1.	$\overset{H}{\overset{\vert}{C}}-O-$	Ethers and acetals with α hydrogen atoms
2.	$C=C-\overset{H}{\overset{\vert}{C}}$	Olefins with allylic hydrogen atoms
3.	$C=\overset{X}{\overset{\vert}{C}}-$	Chloroolefins and fluoroolefins
4.	$CH_2=C<$	Vinyl halides, esters, and ethers
5.	$C=\overset{\vert}{C}-\overset{\vert}{C}=C$	Dienes
6.	$C=\overset{H}{\overset{\vert}{C}}-C\equiv CH$	Vinylacetylenes with α hydrogen atoms
7.	$\overset{H}{\overset{\vert}{C}}-C\equiv CH$	Alkylacetylenes with α hydrogen atoms
8.	$\overset{H}{\overset{\vert}{C}}-Ar$	Alkylarenes that contain tertiary hydrogen atoms
9.	$-\overset{\vert}{\underset{\vert}{C}}-H$	Alkanes and cycloalkanes that contain tertiary hydrogen atoms

10. $\text{C=C—CO}_2\text{R}$ Acrylates and methacrylates

11. H—C—OH Secondary alcohols

12. O=C—C—H Ketones that contain α hydrogen atoms

13. H—C=O Aldehydes

14. O=C—N(H)—C—H Ureas, amides, and lactams that have a hydrogen atom on a carbon atom attached to nitrogen

B. Inorganic substances

1. Alkali metals, especially potassium, rubidium, and cesium (see Chapter 6, Section III.D)
2. Metal amides (see Chapter 6, Section III.C.7)
3. Organometallic compounds with a metal atom bonded to carbon (see Chapter 6, Section IV)
4. Metal alkoxides

TABLE 8. Common Peroxide-Forming Chemicals

LIST A

Severe Peroxide Hazard on Storage with Exposure to Air

Discard within 3 months

- Diisopropyl ether (isopropyl ether)
- Divinylacetylene (DVA)[a]
- Potassium metal
- Potassium amide
- Sodium amide (sodamide)
- Vinylidene chloride (1,1-dichloroethylene)[a]

LIST B

Peroxide Hazard on Concentration; Do Not Distill or Evaporate without First Testing for the Presence of Peroxides

Discard or Test for Peroxides after 6 Months

- Acetaldehyde diethyl acetal (acetal)
- Cumene (isopropylbenzene)
- Cyclohexene
- Cyclopentene
- Decalin (decahydronaphthalene)
- Diacetylene
- Dicyclopentadiene
- Diethyl ether (ether)
- Diethylene glycol dimethyl ether (diglyme)
- Dioxane
- Ethylene glycol dimethyl ether (glyme)
- Ethylene glycol ether acetates
- Ethylene glycol monoethers (cellosolves)
- Furan
- Methylacetylene
- Methylcyclopentane
- Methyl isobutyl ketone
- Tetrahydrofuran (THF)
- Tetralin (tetrahydronaphthalene)
- Vinyl ethers[a]

LIST C

Hazard of Rapid Polymerization Initiated by Internally Formed Peroxides[a]

a. Normal Liquids; discard or test for peroxides after 6 months[b]

- Chloroprene (2-chloro-1,3-butadiene)[c]
- Styrene
- Vinyl acetate
- Vinylpyridine

b. Normal Gases; discard after 12 months[d]

- Butadiene[c]
- Tetrafluoroethylene (TFE)[c]
- Vinylacetylene (MVA)[c]
- Vinyl chloride

[a] Polymerizable monomers should be stored with a polymerization inhibitor from which the monomer can be separated by distillation just before use.

[b] Although common acrylic monomers such as acrylonitrile, acrylic acid, ethyl acrylate, and methyl methacrylate can form peroxides, they have not been reported to develop hazardous levels in normal use and storage.

[c] The hazard from peroxides in these compounds is substantially greater when they are stored in the liquid phase, and if so stored without an inhibitor they should be considered as in LIST A.

[d] Although air will not enter a gas cylinder in which gases are stored under pressure, these gases are sometimes transferred from the original cylinder to another in the laboratory, and it is difficult to be sure that there is no residual air in the receiving cylinder. An inhibitor should be put into any such secondary cylinder before one of these gases is transferred into it; the supplier can suggest inhibitors to be used. The hazard posed by these gases is much greater if there is a liquid phase in such a secondary container, and even inhibited gases that have been put into a secondary container under conditions that create a liquid phase should be discarded within 12 months.

Note: Laboratory workers should label all containers of peroxidizable solvents or reagents with one of the following:

[LIST A]

Peroxidizable compound

	Received	Opened
Date	________	________

Discard 3 months after opening

[LISTS B AND C]

Peroxidizable compound

	Received	Opened
Date	________	________

Discard or test for peroxides
6 months after opening

TABLE 9. Relative Toxicity of Cations

High toxic hazard	Precipitant[a]	Low toxic hazard	Precipitant[a]
Antimony	OH^-, S^{2-}	Aluminum	OH^-
Arsenic	S^{2-}	Bismuth	OH^-, S^{2-}
Barium	SO_4^{2-}, CO_3^{2-}	Calcium	SO_4^{2-}, CO_3^{2-}
Beryllium	OH^-	Cerium	OH^-
Cadmium	OH^-, S^{2-}	Cesium	
Chromium (III)[b]	OH^-	Copper[c]	OH^-, S^{2-}
Cobalt (II)[b]	OH^-, S^{2-}	Gold	OH^-, S^{2-}
Gallium	OH^-	Iron[c]	OH^-, S^{2-}
Germanium	OH^-, S^2	Lanthanides	OH
Hafnium	OH^-	Lithium	
Indium	OH^-, S^{2-}	Magnesium	OH^-
Iridium	OH^-, S^{2-}	Molybdenum (VI)[b,d]	
Lead	OH^-, S^{2-}	Niobium (V)	OH^-
Manganese (II)[b]	OH^-, S^{2-}	Palladium	OH^-, S^{2-}
Mercury	OH^-, S^{2-}	Potassium	
Nickel	OH^-, S^{2-}	Rubidium	
Osmium (IV)[b,e]	OH^-, S^{2-}	Scandium	OH^-
Platinum (II)[b]	OH^-, S^{2-}	Sodium	
Rhenium (VII)[b]	S^{2-}	Strontium	SO_4^{2-} CO_3^{2-}
Rhodium (III)[b]	OH^-, S^{2-}	Tantalum	OH^-
Ruthenium (III)[b]	OH^-, S^{2-}	Tin	OH^-, S^{2-}
Selenium	S^{2-}	Titanium	OH^-
Silver	Cl^-, OH^-, S^{2-}	Yttrium	OH^-
Tellurium	S^{2-}	Zinc[c]	OH^-, S^{2-}
Thallium	OH^-, S^2	Zirconium	OH^-
Tungsten (VI)[b,d]			
Vanadium	OH^-, S^{2-}		

[a] Precipitants are listed in order of preference:
OH^- = base (sodium hydroxide or sodium carbonate)
S^{2-} = sulfide
Cl^- = chloride
SO_4^{2-} = sulfate
CO_3^{2-} = carbonate

[b] The precipitant is for the indicated valence state.

[c] Maximum tolerance levels have been set for these low-toxicity ions by the U.S. Public Health Service, and large amounts should not be put into public sewer systems. The small amounts typically used in laboratories will not normally affect water supplies.

[d] These ions are best precipitated as calcium molybdate or calcium tungstate.

[e] CAUTION: OsO_4, a volatile, extremely poisonous substance, is formed from almost any osmium compound under acid conditions in the presence of air.

TABLE 10. Relative Hazard of Anions

High-hazard anions			
Ion	**Hazard type**[a]	**Precipitant**	**Low-hazard anions**
Aluminum hydride, AlH_4^-	F	—	Bisulfite, HSO_3^-
Amide, NH_2^-	F,E[b]	—	Borate, BO_3^{3-}, $B_4O_7^{2-}$
Arsenate, AsO_3^-, AsO_4^{3-}	T	Cu^{2+}, Fe^{2+}	Bromide, Br^-
Arsenite, AsO_2^-, AsO_3^{3-}	T	Pb^{2+}	Carbonate, CO_3^{2-}
Azide, N_3^-	E, T	—	Chloride, Cl^-
Borohydride, BH_4^-	F	—	Cyanate, OCN^-
Bromate, BrO_3^-	O, E	—	Hydroxide, OH^-
Chlorate, ClO_3^-	O, E	—	Iodide, I^-
Chromate, CrO_4^{2-}, $Cr_2O_7^{2-}$	T, O	[c]	Oxide, O^{2-}
Cyanide, CN^-	T	—	Phosphate, PO_4^{3-}
Ferricyanide, $Fe(CN)_6^{3-}$	T	Fe^{2+}	Sulfate, SO_4^{2}
Ferrocyanide, $Fe(CN)_6^{4-}$	T	Fe^{3+}	Sulfite, SO_3^{2-}
Fluoride, F^-	T	Ca^{2+}	Thiocyanate, SCN^-
Hydride, H^-	F	—	
Hydroperoxide, O_2H^-	O, E	—	
Hydrosulfide, SH^-	T	—	
Hypochlorite, OCl^-	O	—	
Iodate, IO_3^-	O, E	—	
Nitrate, NO_3^-	O	—	
Nitrite, NO_2^-	T, O	—	
Perchlorate, ClO_4^-	O, E	—	
Permanganate, MnO_4^-	T, O	[d]	
Peroxide, O_2^{2-}	O, E	—	
Persulfate, $S_2O_8^{2-}$	O	—	
Selenate, SeO_4^{2-}	T	Pb^{2+}	
Selenide, Se^{2-}	T	Cu^{2+}	
Sulfide, S^{2-}	T	[e]	

[a] Toxic, T: oxidant, O; flammable, F; explosive, E.
[b] Metal amides readily form explosive peroxides on exposure to air.
[c] Reduce and precipitate as Cr(III); see Table 9.
[d] Reduce and precipitate as Mn(II); see Table 9.
[e] See Table 11.

TABLE 11. Precipitation of Sulfides

Precipitated at pH 7	Not precipitated at low pH	Forms a soluble complex at high pH
Ag^+		
As^{3+}[a]		X
Au^+[a]		X
Bi^{3+}		
Cd^{2+}		
Co^{2+}	X	
Cr^{3+}[a]		
Cu^{2+}		
Fe^{2+}[a]	X	
Ge^{2+}		X
Hg^{2+}		X
In^{3+}	X	
Ir^{4+}		X
Mn^{2+}[a]	X	
Mo^{3+}		X
Ni^{2+}	X	
Os^{4+}		
Pb^{2+}		
Pd^{2+}[a]		
Pt^{2+}[a]		X
Re^{4+}		
Rh^{2+}[a]		
Ru^{4+}		

TABLE 11. Precipitation of Sulfides

Precipitated at pH 7	Not precipitated at low pH	Forms a soluble complex at high pH
Sb^{3+a}		X
Se^{2+}		X
Sn^{2+}		X
Te^{4+}		X
Tl^{+a}	X	
V^{4+a}		
Zn^{2+}	X	

[a] Higher oxidation states of this ion are reduced by sulfide ion and precipitated as this sulfide.

TABLE 12. pH Range for Precipitation of Metal Hydroxides and Oxides

Ion	pH range (scale 1–10)	
Ag^{1+}	9 →	1N
Al^{3+}	7–8	
As^{3+}	Not precipitated (precipitate as sulfide)	
As^{5+}	Not precipitated (precipitate as sulfide)	
Au^{3+}	7–8	
Be^{2+}	7–8	
Bi^{3+}	7 →	1 N
Cd^{2+}	7 →	1 N
Co^{2+}	8 →	1 N
Cr^{3+}	7 →	1 N
Cu^{1+}	9 →	1 N
Cu^{2+}	7 →	1 N
Fe^{2+}	7 →	1 N
Fe^{3+}	7 →	1 N
Ga^{3+}	7–8	
Ge^{4+}	6–8	
Hf^{4+}	6–7	
Hg^{1+}	8 →	1 N
Hg^{2+}	8 →	1 N
In^{3+}	6 →	pH 13
Ir^{4+}	6–8	
Mg^{2+}	9 →	1 N
Mn^{2+}	8 →	1 N
Mn^{4+}	7 →	1 N
Mo^{6+}	Not precipitated (precipitate as Ca salt)	
Nb^{5+}	1–10	
Ni^{2+}	8 →	1 N
Os^{4+}	7–8	
Pb^{2+}	7–8	
Pd^{2+}	7–8	
Pd^{4+}	7–8	
Pt^{2+}	7–8	
Re^{3+}	6 →	1 N
Re^{7+}	Not precipitated (precipitate as sulfide)	
Rh^{3+}	7–8	
Ru^{3+}	7 →	1 N
Sb^{3+}	7–8	
Sb^{5+}	7–8	
Sc^{3+}	8 →	1 N
Se^{4+}	Not precipitated (precipitate as sulfide)	
Se^{6+}	Not precipitated (precipitate as sulfide)	
Sn^{2+}	7–8	
Sn^{4+}	7–8	
Ta^{5+}	1–10	
Te^{4+}	Not precipitated (precipitate as sulfide)	
Te^{6+}	Not precipitated (precipitate as sulfide)	
Th^{4+}	6 →	1 N
Ti^{3+}	8 →	1 N
Ti^{4+}	8 →	1 N

TABLE 12. pH Range for Precipitation of Metal Hydroxides and Oxides

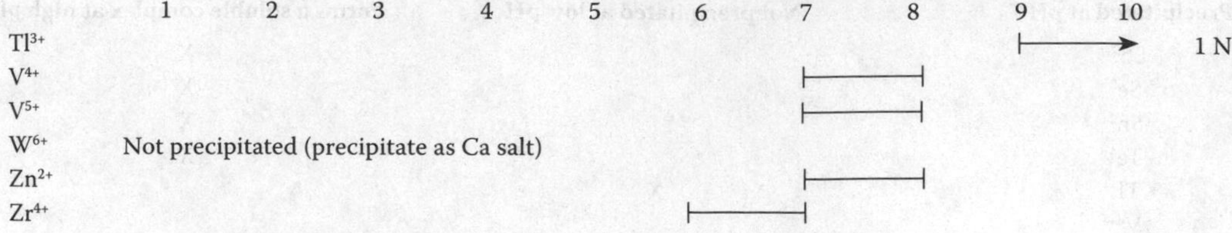

References

L. Erdey, *Gravimetric Analysis,* Part II, Pergamon Press, New York, 1965.
D. T. Burns, A. Towsend, and A. H. Carter, *Inorganic Reaction Chemistry,* Vol. 2, Ellis Horwood, New York, 1981.

FIRE HAZARDS

Flammable solvents are a common source of laboratory fires. The relative ease with which some common laboratory solvents can be ignited is indicated by the following properties.

Flash Point — The lowest temperature, as determined by standard tests, at which a liquid emits vapor in sufficient concentration to form an ignitable mixture with air near the surface of the liquid in a test vessel. Note that many of these common chemicals have flash points below room temperature.

Ignition Temperature — The minimum temperature required to initiate self-sustained combustion, regardless of the heat source.

Flammable Limits — The lower flammable limit is the minimum concentration (percent by volume) of a vapor in air below which a flame is not propagated when an ignition source is present. Below this concentration the mixture is too lean to burn. The upper flammable limit is the maximum concentration (percent by volume) of the vapor in air above which a flame is not propagated. Above this concentration the mixture is too rich to burn. The flammable range comprises all concentrations between these two limits. This range becomes wider with increasing temperature and in oxygen-rich atmospheres. Table 13 lists these properties for a few common laboratory chemicals.

GLOVE MATERIALS

It is good safety practice (and mandated in some laboratories) to wear rubber gloves while handling chemicals that can cause injury when in contact with, or absorbed through, the skin. The various common rubbers are not equally resistant to all chemicals. Table 14 provides guidelines for selecting the best, and avoiding the poorest, glove material for handling a given chemical.

RESPIRATORS

In the event of a laboratory accident or spill, it will be necessary for someone to enter the contaminated area for cleanup. If significant quantities of a chemical are spilled, or even minor quantities of a known toxic material, it is essential to wear the correct kind of respirator equipment when entering the area. If it is not known whether the contamination is of a chemical "immediately dangerous to life or health," the prudent course is to assume that it is, and to use the corresponding type of respirator. Guidelines are presented in Table 15.

TABLE 13. Flash Points, Boiling Points, Ignition Temperatures, and Flammable Limits of Some Common Laboratory Chemicals

				Flammable limit (percent by volume in air)	
Chemical	**Flash point (°C)**	**Boiling point (°C)**	**Ignition temp. (°C)**	**Lower**	**Upper**
Acetaldehyde	−37.8	21.1	175.0	4.0	60.0
Acetone	−19.0	56.0	538.0	2.6	12.8
Benzene	−11.1	80.1	560.0	1.4	8.0
Carbon disulfide	−30.0	45.8	90.0	1.0	44.0
Cyclohexane	−18.0	80.7	260.0	1.3	8.0
Diethyl ether	−45.0	34.4	160.0	1.8	48.0
Ethanol	12.0	78.3	363.0	3.3	19.0
n-Heptane	−3.9	98.4	204.0	1.0	6.7
n-Hexane	−21.7	68.7	223.0	1.2	7.5
Isopropyl alcohol	11.7	82.2	398.9	2.0	12.0
Methanol	11.1	64.5	385.0	6.0	36.5
Methyl ethyl ketone	−6.1	79.6	515.6	1.9	11.0

TABLE 13. Flash Points, Boiling Points, Ignition Temperatures, and Flammable Limits of Some Common Laboratory Chemicals

Chemical	Flash point (°C)	Boiling point (°C)	Ignition temp. (°C)	Flammable limit (percent by volume in air)	
				Lower	Upper
Pentane	−40.0	36.1	260.0	1.4	7.8
Styrene	31.0	145.0	490.0	1.1	6.1
Toluene	4.4	110.6	530.0	1.3	7.0
p-Xylene	25.0	132.4	529.0	1.1	7.0

Note: For a more extensive listing, see the table "Properties of Common Solvents" in Section 15.

TABLE 14. Resistance to Chemicals of Common Glove Materials (E = Excellent, G = Good, F = Fair, P = Poor)

Chemical	Natural rubber	Neoprene	Nitrile	Vinyl
Acetaldehyde	G	G	E	G
Acetic acid	E	E	E	E
Acetone	G	G	G	F
Acrylonitrile	P	G	—	F
Ammonium hydroxide (sat)	G	E	E	E
Aniline	F	G	E	G
Benzaldehyde	F	F	E	G
Benzene[a]	P	F	G	F
Benzyl chloride[a]	F	P	G	P
Bromine	G	G	—	G
Butane	P	E	—	P
Butyraldehyde	P	G	—	G
Calcium hypochlorite	P	G	G	G
Carbon disulfide	P	P	G	F
Carbon tetrachloride[a]	P	F	G	F
Chlorine	G	G	—	G
Chloroacetone	F	E	—	P
Chloroform[a]	P	F	G	P
Chromic acid	P	F	F	E
Cyclohexane	F	E	—	P
Dibenzyl ether	F	G	—	P
Dibutyl phthalate	F	G	—	P
Diethanolamine	F	E	—	E
Diethyl ether	F	G	E	P
Dimethyl sulfoxide[b]	—	—	—	—
Ethyl acetate	F	G	G	F
Ethylene dichloride[a]	P	F	G	P
Ethylene glycol	G	G	E	E
Ethylene trichloride[a]	P	P	—	P
Fluorine	G	G	—	G
Formaldehyde	G	E	E	E
Formic acid	G	E	E	E
Glycerol	G	G	E	E
Hexane	P	E	—	P
Hydrobromic acid (40%)	G	E	—	E
Hydrochloric acid (conc)	G	G	G	E
Hydrofluoric acid (30%)	G	G	G	E
Hydrogen peroxide	G	G	G	E
Iodine	G	G	—	G
Methylamine	G	G	E	E
Methyl cellosolve	F	E	—	P
Methyl chloride[a]	P	E	—	P
Methyl ethyl ketone	F	G	G	P
Methylene chloride[a]	F	F	G	F
Monoethanolamine	F	E	—	E
Morpholine	F	E	—	E
Naphthalene[a]	G	G	E	G
Nitric acid (conc)	P	P	P	G
Perchloric acid	F	G	F	E

TABLE 14. Resistance to Chemicals of Common Glove Materials (E = Excellent, G = Good, F = Fair, P = Poor)

Chemical	Natural rubber	Neoprene	Nitrile	Vinyl
Phenol	G	E	—	E
Phosphoric acid	G	E	—	E
Potassium hydroxide (sat)	G	G	G	E
Propylene dichloride[a]	P	F	—	P
Sodium hydroxide	G	G	G	E
Sodium hypochlorite	G	P	F	G
Sulfuric acid (conc)	G	G	F	G
Toluene[a]	P	F	G	F
Trichloroethylene[a]	P	F	G	F
Tricresyl phosphate	P	F	—	F
Triethanolamine	F	E	E	E
Trinitrotoluene	P	E	—	P

[a] Aromatic and halogenated hydrocarbons will attack all types of natural and synthetic glove materials. Should swelling occur, the user should change to fresh gloves and allow the swollen gloves to dry and return to normal.

[b] No data on the resistance to dimethyl sulfoxide of natural rubber, neoprene, nitrile rubber, or vinyl materials are available; the manufacturer of the substance recommends the use of butyl rubber gloves.

TABLE 15. Guide for Selection of Respirators

Type of hazard	Type of respirator
Oxygen deficiency	Self-contained breathing apparatus
	Hose mask with blower
	Combination of air-line respirator and auxiliary self-contained air supply or air-storage receiver with alarm
Gas and vapor contaminants	Self-contained breathing apparatus
Immediately dangerous to life or health	Hose mask with blower
	Air-purifying full-facepiece respirator with chemical canister (gas mask)
	Self-rescue mouthpiece respirator (for escape only)
	Combination of air-line respirator and auxiliary self-contained air supply or air-storage receiver with alarm
Not immediately dangerous to life or health	Air-line respirator
	Hose mask with blower
	Air-purifying half-mask or mouthpiece respirator with chemical cartridge
Particulate contaminants	Self-contained breathing apparatus
Immediately dangerous to life or health	Hose mask with blower
	Air-purifying full-facepiece respirator with appropriate filter
	Self-rescue mouthpiece respirator (for escape only)
	Combination of air-line respirator and auxiliary self-contained air supply or air-storage receiver with alarm
Not immediately dangerous to life or health	Air-purifying half-mask or mouthpiece respirator with filter pad or cartridge
	Air-line respirator
	Air-line abrasive-blasting respirator
	Hose mask with blower
Combination of gas, vapor, and particulate contaminants	Self-contained breathing apparatus
Immediately dangerous to life or health	Hose mask with blower
	Air-purifying full-facepiece respirator with chemical canister and appropriate filter (gas mask with filter)
	Self-rescue mouthpiece respirator (for escape only)
	Combination of air-line respirator and auxiliary self-contained air supply or air-storage receiver with alarm
Not immediately dangerous to life or health	Air-line respirator
	Hose mask without blower
	Air-purifying half-mask or mouthpiece respirator with chemical cartridge and appropriate filter

Source: ANSI Standard Z88.2 (1969).

FLAMMABILITY OF CHEMICAL SUBSTANCES

This table gives properties related to the flammability of about 900 chemical substances. The properties listed are:

- t_B: Normal boiling point in °C (at 101.325 kPa pressure).
- FP: Flash point, which is the minimum temperature at which the vapor pressure of a liquid is sufficient to form an ignitable mixture with air near the surface of the liquid. Flash point is not an intrinsic physical property but depends on the conditions of measurement (see Reference 1).
- Fl. limits: Flammable limits (often called explosive limits), which specify the range of concentration of the vapor in air (in percent by volume) for which a flame can propagate. Below the lower flammable limit, the gas mixture is too lean to burn; above the upper flammable limit, the mixture is too rich. Values refer to ambient temperature and pressure and are dependent on the precise test conditions. A ? indicates that one of the limits is not known.
- IT: Ignition temperature (sometimes called autoignition temperature), which is the minimum temperature required for self-sustained combustion in the absence of an external ignition source. As in the case of flash point, the value depends on specified test conditions.

Even in cases where very careful measurements of flash point have been replicated in several laboratories, observed values can differ by 3 to 6 °C (Reference 4). For more typical measurements, larger uncertainties should be assumed in both flash points and autoignition temperatures. The absence of a flash point entry in this table does not mean that the substance is nonflammable, but only that no reliable value is available.

Compounds are listed by molecular formula following the Hill convention. Substances not containing carbon are listed first, followed by those that contain carbon. To locate an organic compound by name or CAS Registry Number when the molecular formula is not known, use the table "Physical Constants of Organic Compounds" in Section 3 and its indexes to determine the molecular formula.

References

1. *Fire Protection Guide to Hazardous Materials, 11th Edition*, National Fire Protection Association, Quincy, MA, 1994.
2. Urben, P. G., Ed., *Bretherick's Handbook of Reactive Chemical Hazards, 5th Edition*, Butterworth-Heinemann, Oxford, 1995.
3. Daubert, T. E., Danner, R. P., Sibul, H. M., and Stebbins, C. C., *Physical and Thermodynamic Properties of Pure Compounds: Data Compilation*, extant 1994 (core with 4 supplements), Taylor & Francis, Bristol, PA.
4. *Report of Investigation: Flash Point Reference Materials*, National Institute of Standards and Technology, Standard Reference Materials Program, Gaithersburg, MD, 1995.

Mol. form.	Name	t_B/°C	FP/°C	Fl. limits	IT/°C
	Compounds not containing carbon				
B_2H_6	Diborane	−92.4	−90	1–98%	≈40
B_5H_9	Pentaborane(9)	60	30	0.4–?	35
BrH_3Si	Bromosilane	1.9	<0		≈20
Br_3HSi	Tribromosilane	109			≈20
Cl_2H_2Si	Dichlorosilane	8.3		4.1–99%	36
Cl_3HSi	Trichlorosilane	33	−50		104
GeH_4	Germane	−88.1			≈20
Ge_2H_6	Digermane	29			≈50
H_2	Hydrogen	−252.8		4–74%	
H_2S	Hydrogen sulfide	−59.55		4–44%	260
H_2S_2	Hydrogen disulfide	70.7	<22		
H_2Te	Hydrogen telluride	−2			−50
H_3N	Ammonia	−33.33		16–25%	
H_3P	Phosphine	−87.75		1.8–?	
H_4N_2	Hydrazine	113.55	38	5–100%	
H_4P_2	Diphosphine	63.5			≈20
H_4Si	Silane	−111.9	−112	1.4–?	≈20
H_6Si_2	Disilane	−14.3	−14		≈20
H_8Si_3	Trisilane	52.9	<0		≈20
P	Phosphorus (white)	280.5			38
	Compounds containing carbon				
CHN	Hydrogen cyanide	26	−18	6–40%	538
CH_2Cl_2	Dichloromethane	40		13–23%	556
CH_2N_2	Cyanamide		141		
CH_2O	Formaldehyde	−19.1	85	7.0–73%	424
$(CH_2O)_x$	Paraformaldehyde		70	7.0–73%	300
CH_2O_2	Formic acid	101	50	18–57%	434
CH_3Br	Bromomethane	3.5		10–16%	537
CH_3Cl	Chloromethane	−24.0		8.1–17.4%	632

Mol. form.	Name	t_B/°C	FP/°C	Fl. limits	IT/°C
CH_3Cl_3Si	Methyltrichlorosilane	65.6	–9	7.6–>20%	>404
CH_3NO	Formamide	220	154		
CH_3NO_2	Nitromethane	101.1	35	7.3–?	418
CH_4	Methane	–161.5		5.0–15.0%	537
CH_4Cl_2Si	Dichloromethylsilane	41	–9	6.0–55%	316
CH_4O	Methanol	64.6	11	6.0–36%	464
CH_4S	Methanethiol	5.9	–18	3.9–21.8%	
CH_5N	Methylamine	–6.3	0	4.9–20.7%	430
CH_6N_2	Methylhydrazine	87.5	–8	2.5–92%	194
CO	Carbon monoxide	–191.5		12.5–74%	609
COS	Carbon oxysulfide	–50		12–29%	
CS_2	Carbon disulfide	46	–30	1.3–50.0%	90
C_2ClF_3	Chlorotrifluoroethylene	–27.8		8.4–16.0%	
C_2F_4	Tetrafluoroethylene	–75.9		10.0–50.0%	200
C_2HCl_3	Trichloroethylene	87.2		8–10.5%	420
C_2HCl_3O	Dichloroacetyl chloride	108	66		
C_2H_2	Acetylene	–84.7		2.5–100%	305
$C_2H_2Cl_2$	1,1-Dichloroethylene	31.6	–28	6.5–15.5%	570
$C_2H_2Cl_2$	*cis*-1,2-Dichloroethylene	60.1	6	3–15%	460
$C_2H_2Cl_2$	*trans*-1,2-Dichloroethylene	48.7	2	6–13%	460
$C_2H_2F_2$	1,1-Difluoroethylene	–85.7		5.5–21.3%	
C_2H_3Br	Bromoethylene	15.8		9–15%	530
C_2H_3Cl	Chloroethylene	–13.3	–78	3.6–33.0%	472
$C_2H_3ClF_2$	1-Chloro-1,1-difluoroethane	–9.7		6–18%	632
C_2H_3ClO	Acetyl chloride	50.7	4		390
$C_2H_3Cl_2NO_2$	1,1-Dichloro-1-nitroethane	123.5	76		
$C_2H_3Cl_3$	1,1,1-Trichloroethane	74.0		8–10.5%	500
$C_2H_3Cl_3$	1,1,2-Trichloroethane	113.8	32	6–28%	460
$C_2H_3Cl_3Si$	Trichlorovinylsilane	91.5	21		
C_2H_3F	Fluoroethylene	–72		2.6–21.7%	
C_2H_3N	Acetonitrile	81.6	6	3.0–16.0%	524
C_2H_3NO	Methyl isocyanate	39.5	–7	5.3–26%	534
C_2H_4	Ethylene	–103.7		2.7–36%	450
$C_2H_4ClNO_2$	1-Chloro-1-nitroethane	124.5	56		
$C_2H_4Cl_2$	1,1-Dichloroethane	57.4	–17	5.4–11.4%	458
$C_2H_4Cl_2$	1,2-Dichloroethane	83.5	13	6.2–16%	413
C_2H_4O	Acetaldehyde	20.1	–39	4.0–60%	175
C_2H_4O	Ethylene oxide	10.6	–20	3.0–100%	429
$C_2H_4O_2$	Acetic acid	117.9	39	4.0–19.9%	463
$C_2H_4O_2$	Methyl formate	31.7	–19	4.5–23%	449
$C_2H_4O_3$	Ethaneperoxoic acid	110	41		
C_2H_5Br	Bromoethane	38.5		6.8–8.0%	511
C_2H_5Cl	Chloroethane	12.3	–50	3.8–15.4%	519
C_2H_5ClO	Ethylene chlorohydrin	128.6	60	4.9–15.9%	425
$C_2H_5Cl_3Si$	Trichloroethylsilane	100.5	22		
C_2H_5N	Ethyleneimine	56	–11	3.3–54.8%	320
$C_2H_5NO_2$	Nitroethane	114.0	28	3.4–17%	414
$C_2H_5NO_2$	Ethyl nitrite	18	–35	4.0–50%	90
$C_2H_5NO_3$	Ethyl nitrate	87.2	10	4–?	
C_2H_6	Ethane	–88.6		3.0–12.5%	472
$C_2H_6Cl_2Si$	Dichlorodimethylsilane	70.3	<21	3.4–9.5%	
C_2H_6O	Ethanol	78.2	13	3.3–19%	363
C_2H_6O	Dimethyl ether	–24.8	–41	3.4–27.0%	350
C_2H_6OS	2-Mercaptoethanol	158	74		
C_2H_6OS	Dimethyl sulfoxide	189	95	2.6–42%	215
$C_2H_6O_2$	Ethylene glycol	197.3	111	3.2–22%	398
$C_2H_6O_4S$	Dimethyl sulfate		83		188
C_2H_6S	Ethanethiol	35.1	–17	2.8–18.0%	300
C_2H_6S	Dimethyl sulfide	37.3	–37	2.2–19.7%	206
$C_2H_6S_2$	Dimethyl disulfide	109.8	24		
C_2H_7N	Ethylamine	16.5	–16	3.5–14%	385

Mol. form.	Name	t_B/°C	FP/°C	Fl. limits	IT/°C
C_2H_7N	Dimethylamine	6.8	20	2.8–14.4%	400
C_2H_7NO	Ethanolamine	171	86	3.0–23.5%	410
$C_2H_8N_2$	1,2-Ethanediamine	117	40	2.5–12.0%	385
$C_2H_8N_2$	1,1-Dimethylhydrazine	63.9	−15	2–95%	249
C_2N_2	Cyanogen	−21.1		6.6–32%	
C_3H_3Br	3-Bromo-1-propyne	89	10	3.0–?	324
C_3H_3N	2-Propenenitrile	77.3	0	3.0–17.0%	481
C_3H_4	Propyne	−23.2		2.1–12.5%	
C_3H_4ClN	3-Chloropropanenitrile	175.5	76		
$C_3H_4Cl_2$	2,3-Dichloropropene	94	15	2.6–7.8%	
C_3H_4O	Propargyl alcohol	113.6	36		
C_3H_4O	Acrolein	52.6	−26	2.8–31%	220
$C_3H_4O_2$	Propenoic acid	141	50	2.4–8.0%	438
$C_3H_4O_2$	2-Oxetanone	162	74	2.9–?	
$C_3H_4O_3$	Ethylene carbonate	248	143		
C_3H_5Br	3-Bromopropene	70.1	−1	4.4–7.3%	295
C_3H_5Cl	2-Chloropropene	22.6	−37	4.5–16%	
C_3H_5Cl	3-Chloropropene	45.1	−32	2.9–11.1%	485
C_3H_5ClO	Epichlorohydrin	118	31	3.8–21.0%	411
C_3H_5ClO	Propanoyl chloride	80	12		
$C_3H_5ClO_2$	2-Chloropropanoic acid	185	107		500
$C_3H_5ClO_2$	Ethyl chloroformate	95	16		500
$C_3H_5ClO_2$	Methyl chloroacetate	129.5	57	7.5–18.5%	
$C_3H_5Cl_2NO_2$	1,1-Dichloro-1-nitropropane	145	66		
$C_3H_5Cl_3$	1,2,3-Trichloropropane	157	71	3.2–12.6%	
$C_3H_5Cl_3Si$	Trichloro-2-propenylsilane	117.5	35		
C_3H_5N	Propanenitrile	97.1	2	3.1–14%	512
C_3H_5NO	3-Hydroxypropanenitrile	221	129		
$C_3H_5N_3O_9$	Trinitroglycerol				270
C_3H_6	Propene	−47.6		2.0–11.1%	455
C_3H_6	Cyclopropane	−32.8		2.4–10.4%	498
$C_3H_6ClNO_2$	1-Chloro-1-nitropropane	142	62		
$C_3H_6ClNO_2$	2-Chloro-2-nitropropane		57		
$C_3H_6Cl_2$	1,2-Dichloropropane	96.4	21	3.4–14.5%	557
$C_3H_6Cl_2O$	1,3-Dichloro-2-propanol	176	74		
$C_3H_6N_2$	Dimethylcyanamide	163.5	71		
C_3H_6O	Allyl alcohol	97.0	21	2.5–18.0%	378
C_3H_6O	Methyl vinyl ether	5.5			287
C_3H_6O	Propanal	48	−30	2.6–17%	207
C_3H_6O	Acetone	56.0	−20	2.5–12.8%	465
C_3H_6O	Methyloxirane	35	−37	3.1–27.5%	449
$C_3H_6O_2$	Propanoic acid	141.1	52	2.9–12.1%	465
$C_3H_6O_2$	Ethyl formate	54.4	−20	2.8–16.0%	455
$C_3H_6O_2$	Methyl acetate	56.8	−10	3.1–16%	454
$C_3H_6O_2$	1,3-Dioxolane	78	2		
$C_3H_6O_3$	Dimethyl carbonate	90.5	19		
$C_3H_6O_3$	1,3,5-Trioxane	114.5	45	3.6–29%	414
C_3H_7Br	1-Bromopropane	71.1			490
C_3H_7Cl	1-Chloropropane	46.5	<−18	2.6–11.1%	520
C_3H_7Cl	2-Chloropropane	35.7	−32	2.8–10.7%	593
C_3H_7ClO	2-Chloro-1-propanol	133.5	52		
C_3H_7ClO	1-Chloro-2-propanol	127	52		
$C_3H_7Cl_3Si$	Trichloropropylsilane	123.5	37		
C_3H_7N	Allylamine	53.3	−29	2.2–22%	374
C_3H_7NO	*N,N*-Dimethylformamide	153	58	2.2–15.2%	445
$C_3H_7NO_2$	1-Nitropropane	131.1	36	2.2–?	421
$C_3H_7NO_2$	2-Nitropropane	120.2	24	2.6–11.0%	428
$C_3H_7NO_3$	Propyl nitrate	110	20	2–100%	175
C_3H_8	Propane	−42.1	−104	2.1–9.5%	450
C_3H_8O	1-Propanol	97.2	23	2.2–13.7%	412
C_3H_8O	2-Propanol	82.3	12	2.0–12.7%	399

Mol. form.	Name	t_B/°C	FP/°C	Fl. limits	IT/°C
C_3H_8O	Ethyl methyl ether	7.4	−37	2.0–10.1%	190
$C_3H_8O_2$	1,2-Propylene glycol	187.6	99	2.6–12.5%	371
$C_3H_8O_2$	1,3-Propylene glycol	214.4			400
$C_3H_8O_2$	Ethylene glycol monomethyl ether	124.1	39	1.8–14%	285
$C_3H_8O_2$	Dimethoxymethane	42	−32	2.2–13.8%	237
$C_3H_8O_3$	Glycerol	290	199	3–19%	370
$C_3H_9BO_3$	Trimethyl borate	67.5	−8		
C_3H_9ClSi	Trimethylchlorosilane	60	−28		395
C_3H_9N	Propylamine	47.2	−37	2.0–10.4%	318
C_3H_9N	Isopropylamine	31.7	−37		402
C_3H_9N	Trimethylamine	2.8	−5	2.0–11.6%	190
C_3H_9NO	3-Amino-1-propanol	187.5	80		
C_3H_9NO	1-Amino-2-propanol	159.4	77		374
C_3H_9NO	*N*-Methyl-2-ethanolamine	158	74		
$C_3H_9O_3P$	Trimethyl phosphite	111.5	54		
$C_3H_9O_4P$	Trimethyl phosphate	197.2	107		
$C_3H_{10}N_2$	1,3-Propanediamine	139.8	24		
C_4Cl_6	Hexachloro-1,3-butadiene	215			610
$C_4H_2O_3$	Maleic anhydride	202	102	1.4–7.1%	477
C_4H_4	1-Buten-3-yne	5.1		21–100%	
$C_4H_4N_2$	Succinonitrile	266	132		
C_4H_4O	Furan	31.5	−36	2.3–14.3%	
$C_4H_4O_2$	Diketene	126.1	34		
C_4H_4S	Thiophene	84.0	−1		
C_4H_5Cl	2-Chloro-1,3-butadiene	59.4	−20	4.0–20.0%	
C_4H_5N	2-Butenenitrile	120.5	16		
C_4H_5N	Methylacrylonitrile	90.3	1	2–6.8%	
C_4H_5N	Pyrrole	129.7	39		
C_4H_6	1,3-Butadiene	−4.4		2.0–12.0%	420
C_4H_6	2-Butyne	26.9	−31	1.4–?	
C_4H_6O	Divinyl ether	28.3	<−30	1.7–27%	360
C_4H_6O	Ethoxyacetylene	50	<−7		
C_4H_6O	*trans*-2-Butenal	102.2	13	2.1–15.5%	232
C_4H_6O	3-Buten-2-one	81.4	−7	2.1–15.6%	491
C_4H_6O	Vinyloxirane	68	<−50		
$C_4H_6O_2$	Methacrylic acid	162.5	77	1.6–8.8%	68
$C_4H_6O_2$	Vinyl acetate	72.5	−8	2.6–13.4%	402
$C_4H_6O_2$	Methyl acrylate	80.7	−3	2.8–25%	468
$C_4H_6O_2$	2,3-Butanedione	88	27		
$C_4H_6O_2$	γ-Butyrolactone	204	98		
$C_4H_6O_3$	Acetic anhydride	139.5	49	2.7–10.3%	316
$C_4H_6O_3$	Propylene carbonate	242	135		
$C_4H_6O_6$	*L*-Tartaric acid		210		425
C_4H_7Br	1-Bromo-2-butene	104.5		4.6–12.0%	
$C_4H_7BrO_2$	Ethyl bromoacetate	168.5	48		
C_4H_7Cl	2-Chloro-1-butene	58.5	−19	2.3–9.3%	
C_4H_7Cl	3-Chloro-2-methylpropene	71.5	−12	3.2–8.1%	
C_4H_7ClO	2-Chloroethyl vinyl ether	108	27		
$C_4H_7ClO_2$	Ethyl chloroacetate	144.3	64		
C_4H_7N	Butanenitrile	117.6	24	1.6–?	501
C_4H_7N	2-Methylpropanenitrile	103.9	8		482
C_4H_7NO	Acetone cyanohydrin		74	2.2–12.0%	688
C_4H_7NO	2-Pyrrolidone	251	129		
C_4H_8	1-Butene	−6.2		1.6–10.0%	385
C_4H_8	*cis*-2-Butene	3.7		1.7–9.0%	325
C_4H_8	*trans*-2-Butene	0.8		1.8–9.7%	324
C_4H_8	Isobutene	−6.9		1.8–9.6%	465
C_4H_8	Cyclobutane	12.6	<10	1.8–?	
$C_4H_8Cl_2$	1,2-Dichlorobutane	124.1			275
$C_4H_8Cl_2$	1,4-Dichlorobutane	161	52		
$C_4H_8Cl_2O$	Bis(2-chloroethyl) ether	178.5	55	2.7–?	369

Mol. form.	Name	t_B/°C	FP/°C	Fl. limits	IT/°C
C_4H_8O	2-Buten-1-ol	121.5	27	4.2–35.3%	349
C_4H_8O	2-Methyl-2-propenol	114.5	33		
C_4H_8O	Ethyl vinyl ether	35.5	<-46	1.7–28%	202
C_4H_8O	1,2-Epoxybutane	63.4	−22	1.7–19%	439
C_4H_8O	Butanal	74.8	−22	1.9–12.5%	218
C_4H_8O	Isobutanal	64.5	−18	1.6–10.6%	196
C_4H_8O	2-Butanone	79.5	−9	1.4–11.4%	404
C_4H_8O	Tetrahydrofuran	65	−14	2–11.8%	321
C_4H_8OS	1,4-Oxathiane	147	42		
$C_4H_8O_2$	Butanoic acid	163.7	72	2.0–10.0%	443
$C_4H_8O_2$	2-Methylpropanoic acid	154.4	56	2.0–9.2%	481
$C_4H_8O_2$	Propyl formate	80.9	−3		455
$C_4H_8O_2$	Isopropyl formate	68.2	−6		485
$C_4H_8O_2$	Ethyl acetate	77.1	−4	2.0–11.5%	426
$C_4H_8O_2$	Methyl propanoate	79.8	−2	2.5–13%	469
$C_4H_8O_2$	3-Hydroxybutanal		66		250
$C_4H_8O_2$	1,4-Dioxane	101.5	12	2.0–22%	180
$C_4H_8O_2S$	Sulfolane	287.3	177		
$C_4H_8O_3$	Methyl lactate	144.8	49	2.2–?	385
$C_4H_8O_3$	Ethylene glycol monoacetate	188	102		
C_4H_9Br	1-Bromobutane	101.6	18	2.6–6.6%	265
C_4H_9Br	2-Bromobutane	91.2	21		
C_4H_9Cl	1-Chlorobutane	78.6	−12	1.9–10.1%	240
C_4H_9Cl	2-Chlorobutane	68.2	−10		
C_4H_9Cl	1-Chloro-2-methylpropane	68.5	−6	2.0–8.7%	
C_4H_9Cl	2-Chloro-2-methylpropane	50.9	0		
$C_4H_9Cl_3Si$	Butyltrichlorosilane	148.5	54		
C_4H_9N	Pyrrolidine	86.5	3		
C_4H_9NO	*N*-Ethylacetamide	205	110		
C_4H_9NO	*N,N*-Dimethylacetamide	165	70	1.8–11.5%	490
C_4H_9NO	Butanal oxime	154	58		
C_4H_9NO	2-Butanone oxime	152.5	≈70		
C_4H_9NO	Morpholine	128	37	1.4–11.2%	290
$C_4H_9NO_2$	*N*-Acetylethanolamine		179		460
$C_4H_9NO_3$	Butyl nitrate	133	36		
C_4H_{10}	Butane	−0.5	−60	1.9–8.5%	287
C_4H_{10}	Isobutane	−11.7	−87	1.8–8.4%	460
$C_4H_{10}N_2$	Piperazine	146	81		
$C_4H_{10}O$	1-Butanol	117.7	37	1.4–11.2%	343
$C_4H_{10}O$	2-Butanol	99.5	24	1.7–9.8%	405
$C_4H_{10}O$	2-Methyl-1-propanol	107.8	28	1.7–10.6%	415
$C_4H_{10}O$	2-Methyl-2-propanol	82.4	11	2.4–8.0%	478
$C_4H_{10}O$	Diethyl ether	34.5	−45	1.9–36.0%	180
$C_4H_{10}O$	Methyl propyl ether	39.1	−20	2.0–14.8%	
$C_4H_{10}O_2$	1,2-Butanediol	190.5	40		
$C_4H_{10}O_2$	1,3-Butanediol	207.5	121		395
$C_4H_{10}O_2$	1,4-Butanediol	235	121		
$C_4H_{10}O_2$	2,3-Butanediol	182.5			402
$C_4H_{10}O_2$	Ethylene glycol monoethyl ether	135	43	3–18%	235
$C_4H_{10}O_2$	Ethylene glycol dimethyl ether	85	−2		202
$C_4H_{10}O_2$	*tert*-Butyl hydroperoxide		27		
$C_4H_{10}O_2S$	2,2′-Thiodiethanol	282	160		298
$C_4H_{10}O_3$	Diethylene glycol	245.8	124	2–17%	224
$C_4H_{10}O_4S$	Diethyl sulfate	208	104		436
$C_4H_{10}S$	1-Butanethiol	98.5	2		
$C_4H_{10}S$	2-Butanethiol	85	−23		
$C_4H_{10}S$	2-Methyl-1-propanethiol	88.5	2		
$C_4H_{10}S$	2-Methyl-2-propanethiol	64.3	<-29		
$C_4H_{10}Se$	Diethyl selenide	108		2.5–?	
$C_4H_{11}N$	Butylamine	77.0	−12	1.7–9.8%	312
$C_4H_{11}N$	*sec*-Butylamine	63.5	−9		

Mol. form.	Name	t_B/°C	FP/°C	Fl. limits	IT/°C
$C_4H_{11}N$	*tert*-Butylamine	44.0	–9	1.7–8.9%	380
$C_4H_{11}N$	Isobutylamine	67.7	–9	2–12%	378
$C_4H_{11}N$	Diethylamine	55.5	–23	1.8–10.1%	312
$C_4H_{11}NO$	2-Amino-1-butanol	178	74		
$C_4H_{11}NO$	2-Amino-2-methyl-1-propanol	165.5	67		
$C_4H_{11}NO_2$	Diethanolamine	268.8	172	2–13%	662
$C_4H_{12}Sn$	Tetramethylstannane	78	–12	1.9–?	
$C_4H_{13}N_3$	Diethylenetriamine	207	98	2–6.7%	358
$C_5H_4O_2$	Furfural	161.7	60	2.1–19.3%	316
C_5H_5N	Pyridine	115.2	20	1.8–12.4%	482
C_5H_6	2-Methyl-1-buten-3-yne	32	<–7		
$C_5H_6N_2$	2-Methylpyrazine	137	50		
C_5H_6O	3-Methylfuran	66	–30		
$C_5H_6O_2$	Furfuryl alcohol	171	75	1.8–16.3%	491
C_5H_7N	1-Methylpyrrole	115	16		
C_5H_7NO	2-Furanmethanamine	145.5	37		
$C_5H_7NO_2$	Ethyl cyanoacetate	205	110		
C_5H_8	2-Methyl-1,3-butadiene	34.0	–54	1.5–8.9%	395
C_5H_8	1-Pentyne	40.1	<–20		
C_5H_8	Cyclopentene	44.2	–29		395
C_5H_8O	3-Methyl-3-buten-2-one	98		1.8–9.0%	
C_5H_8O	Cyclopentanone	130.5	26		
C_5H_8O	3,4-Dihydro-2H-pyran	86	–18		
$C_5H_8O_2$	Allyl acetate	103.5	22		374
$C_5H_8O_2$	Isopropenyl acetate	94	26		432
$C_5H_8O_2$	Vinyl propanoate	91.2	1		
$C_5H_8O_2$	Ethyl acrylate	99.4	10	1.4–14%	372
$C_5H_8O_2$	Methyl methacrylate	100.5	10	1.7–8.2%	
$C_5H_8O_2$	2,4-Pentanedione	138	34		340
$C_5H_8O_3$	Methyl acetoacetate	171.7	77		280
C_5H_9NO	*N*-Methyl-2-pyrrolidone	202	96	1–10%	346
C_5H_{10}	1-Pentene	29.9	–18	1.5–8.7%	275
C_5H_{10}	*cis*-2-Pentene	36.9	<–20		
C_5H_{10}	*trans*-2-Pentene	36.3	<–20		
C_5H_{10}	2-Methyl-1-butene	31.2	–20		
C_5H_{10}	3-Methyl-1-butene	20.1	–7	1.5–9.1%	365
C_5H_{10}	2-Methyl-2-butene	38.5	–20		
C_5H_{10}	Cyclopentane	49.3	–25	1.5–?	361
$C_5H_{10}Cl_2$	1,5-Dichloropentane	179	>27		
$C_5H_{10}N_2$	3-(Dimethylamino)propanenitrile	173	65		
$C_5H_{10}O$	Cyclopentanol	140.4	51		
$C_5H_{10}O$	Pentanal	103	12		222
$C_5H_{10}O$	2-Pentanone	102.2	7	1.5–8.2%	452
$C_5H_{10}O$	3-Pentanone	101.9	13	1.6–?	450
$C_5H_{10}O$	Tetrahydropyran	88	–20		
$C_5H_{10}O$	2-Methyltetrahydrofuran	78	–11		
$C_5H_{10}O_2$	Pentanoic acid	186.1	96		400
$C_5H_{10}O_2$	3-Methylbutanoic acid	176.5			416
$C_5H_{10}O_2$	Butyl formate	106.1	18	1.7–8.2%	322
$C_5H_{10}O_2$	Isobutyl formate	98.2	5	2–9%	320
$C_5H_{10}O_2$	Propyl acetate	101.5	13	1.7–8%	450
$C_5H_{10}O_2$	Isopropyl acetate	88.6	2	1.8–8%	460
$C_5H_{10}O_2$	Ethyl propanoate	99.1	12	1.9–11%	440
$C_5H_{10}O_2$	Methyl butanoate	102.8	14		
$C_5H_{10}O_2$	3-Ethoxypropanal	135.2	38		
$C_5H_{10}O_2$	Tetrahydrofurfuryl alcohol	178	75	1.5–9.7%	282
$C_5H_{10}O_3$	Diethyl carbonate	126	25		
$C_5H_{10}O_3$	Ethylene glycol monomethyl ether acetate	143	49	1.5–12.3%	392
$C_5H_{10}O_3$	Ethyl lactate	154.5	46	1.5–?	400
$C_5H_{11}Br$	1-Bromopentane	129.8	32		

Mol. form.	Name	t_B/°C	FP/°C	Fl. limits	IT/°C
$C_5H_{11}Cl$	1-Chloropentane	107.8	13	1.6–8.6%	260
$C_5H_{11}Cl$	2-Chloro-2-methylbutane	85.6		1.5–7.4%	345
$C_5H_{11}Cl$	1-Chloro-3-methylbutane	98.9	<21	1.5–7.4%	
$C_5H_{11}Cl_3Si$	Trichloropentylsilane	172	63		
$C_5H_{11}N$	Piperidine	106.2	16		
$C_5H_{11}N$	*N*-Methylpyrrolidine	81	−14		
$C_5H_{11}NO$	4-Methylmorpholine	116	24		
$C_5H_{11}NO_2$	Isopentyl nitrite	99.2			210
C_5H_{12}	Pentane	36.0	−40	1.4–8.0%	260
C_5H_{12}	Isopentane	27.8	−51	1.4–7.6%	420
C_5H_{12}	Neopentane	9.4	−65	1.4–7.5%	450
$C_5H_{12}N_2$	1-Methylpiperazine	138	42		
$C_5H_{12}N_2O$	Tetramethylurea	176.5	77		
$C_5H_{12}O$	1-Pentanol	137.9	33	1.2–10.0%	300
$C_5H_{12}O$	2-Pentanol	119.3	34	1.2–9.0%	343
$C_5H_{12}O$	3-Pentanol	116.2	41	1.2–9.0%	435
$C_5H_{12}O$	2-Methyl-1-butanol	128	50		385
$C_5H_{12}O$	3-Methyl-1-butanol	131.1	43	1.2–9.0%	350
$C_5H_{12}O$	2-Methyl-2-butanol	102.4	19	1.2–9.0%	437
$C_5H_{12}O$	3-Methyl-2-butanol	112.9	38		
$C_5H_{12}O$	2,2-Dimethyl-1-propanol	113.5	37		
$C_5H_{12}O$	Ethyl propyl ether	63.2	<−20	1.7–9.0%	
$C_5H_{12}O_2$	1,5-Pentanediol	239	129		335
$C_5H_{12}O_2$	2-Isopropoxyethanol	145	33		
$C_5H_{12}O_2$	2,2-Dimethyl-1,3-propanediol	208	129		399
$C_5H_{12}O_3$	Diethylene glycol monomethyl ether	193	96	1.38–22.7%	240
$C_5H_{12}S$	1-Pentanethiol	126.6	18		
$C_5H_{12}S$	3-Methyl-2-butanethiol		3		
$C_5H_{13}N$	Pentylamine	104.3	−1	2.2–22%	
$C_5H_{13}N$	Butylmethylamine	91	13		
$C_6H_2Cl_4$	1,2,4,5-Tetrachlorobenzene	244.5	155		
$C_6H_3ClN_2O_4$	1-Chloro-2,4-dinitrobenzene	315	194	2.0–22%	
$C_6H_3Cl_3$	1,2,4-Trichlorobenzene	213.5	105	2.5–6.6%	571
$C_6H_4ClNO_2$	1-Chloro-4-nitrobenzene	242	127		
$C_6H_4Cl_2$	*o*-Dichlorobenzene	180	66	2.2–9.2%	648
$C_6H_4Cl_2$	*m*-Dichlorobenzene	173	72		
$C_6H_4Cl_2$	*p*-Dichlorobenzene	174	66		
$C_6H_4Cl_2O$	2,4-Dichlorophenol	210	114		
C_6H_5Br	Bromobenzene	156.0	51		565
C_6H_5Cl	Chlorobenzene	131.7	28	1.3–9.6%	593
C_6H_5ClO	*o*-Chlorophenol	174.9	64		
C_6H_5ClO	*p*-Chlorophenol	220	121		
$C_6H_5Cl_2N$	3,4-Dichloroaniline	272	166		
$C_6H_5Cl_3Si$	Trichlorophenylsilane	201	91		
C_6H_5F	Fluorobenzene	84.7	−15		
$C_6H_5NO_2$	Nitrobenzene	210.8	88	1.8–?	482
$C_6H_5N_3O_4$	2,4-Dinitroaniline		224		
C_6H_6	1,5-Hexadien-3-yne	85	<−20	1.5–?	
C_6H_6	Benzene	80.0	−11	1.2–7.8%	498
$C_6H_6N_2O_2$	*p*-Nitroaniline	332	199		
C_6H_6O	Phenol	181.8	79	1.8–8.6%	715
$C_6H_6O_2$	1,2-Benzenediol	245	127		
$C_6H_6O_2$	Resorcinol		127	1.4–?	608
$C_6H_6O_2$	*p*-Hydroquinone	287	165		516
C_6H_7N	Aniline	184.1	70	1.3–11%	615
C_6H_7N	2-Methylpyridine	129.3	39		538
C_6H_7N	4-Methylpyridine	145.3	57		
C_6H_8ClN	Aniline, hydrochloride		193		
$C_6H_8Cl_2O_2$	Hexanedioyl dichloride		72		
$C_6H_8N_2$	Adiponitrile	295	93	1.0–?	550
$C_6H_8N_2$	*o*-Phenylenediamine	257	156	1.5–?	

Mol. form.	Name	t_B/°C	FP/°C	Fl. limits	IT/°C
$C_6H_8N_2$	Phenylhydrazine	243.5	88		
$C_6H_8N_2$	2,5-Dimethylpyrazine	155	64		
C_6H_8O	2,5-Dimethylfuran	93.5	7		
$C_6H_8O_4$	Dimethyl maleate	202	113		
C_6H_{10}	1,4-Hexadiene	65	−21	2.0–6.1%	
C_6H_{10}	2-Methyl-1,3-pentadiene	75.8	−12		
C_6H_{10}	4-Methyl-1,3-pentadiene	76.5	−34		
C_6H_{10}	2-Hexyne	84.5	−10		
C_6H_{10}	Cyclohexene	82.9	−12	1.2–?	310
$C_6H_{10}O$	Diallyl ether	94	−7		
$C_6H_{10}O$	Cyclohexanone	155.4	44	1.1–9.4%	420
$C_6H_{10}O$	Mesityl oxide	130	31	1.4–7.2%	344
$C_6H_{10}O_2$	Vinyl butanoate	116.7	20	1.4–8.8%	
$C_6H_{10}O_2$	Ethyl 2-butenoate	136.5	2		
$C_6H_{10}O_2$	Ethyl methacrylate	117	20		
$C_6H_{10}O_2$	2,5-Hexanedione	194	79		499
$C_6H_{10}O_3$	Ethyl acetoacetate	180.8	57	1.4–9.5%	295
$C_6H_{10}O_3$	Propanoic anhydride	170	63	1.3–9.5%	285
$C_6H_{10}O_4$	Adipic acid	337.5	196		420
$C_6H_{10}O_4$	Diethyl oxalate	185.7	76		
$C_6H_{10}O_4$	Ethylene glycol diacetate	190	88	1.6–8.4%	482
$C_6H_{11}Cl$	Chlorocyclohexane	142	32		
$C_6H_{11}NO$	Caprolactam	270	125		
$C_6H_{11}NO_2$	Nitrocyclohexane	205	88		
$C_6H_{11}NO_2$	4-Acetylmorpholine		113		
C_6H_{12}	1-Hexene	63.4	−26	1.2–6.9%	253
C_6H_{12}	*cis*-2-Hexene	68.8	−21		
C_6H_{12}	2-Methyl-1-pentene	62.1	−28		300
C_6H_{12}	4-Methyl-1-pentene	53.9	−7		300
C_6H_{12}	4-Methyl-*cis*-2-pentene	56.3	−32		
C_6H_{12}	4-Methyl-*trans*-2-pentene	58.6	−29		
C_6H_{12}	2-Ethyl-1-butene	64.7	<−20		315
C_6H_{12}	2,3-Dimethyl-1-butene	55.6	<−20		360
C_6H_{12}	2,3-Dimethyl-2-butene	73.3	<−20		401
C_6H_{12}	Cyclohexane	80.7	−20	1.3–8%	245
C_6H_{12}	Methylcyclopentane	71.8	−29	1.0–8.35%	258
C_6H_{12}	Ethylcyclobutane	70.8	−15	1.2–7.7%	210
C_6H_{12}	2-Methyl-2-pentene	67.3	<−7		
$C_6H_{12}Cl_2O_2$	1,2-Bis(2-chloroethoxy)ethane	232	121		
$C_6H_{12}O$	*cis*-3-Hexen-1-ol	156.5	54		
$C_6H_{12}O$	Butyl vinyl ether	94	−9		255
$C_6H_{12}O$	Isobutyl vinyl ether	83	−9		
$C_6H_{12}O$	Hexanal	131	32		
$C_6H_{12}O$	2-Ethylbutanal		21	1.2–7.7%	
$C_6H_{12}O$	2-Methylpentanal	117	17		199
$C_6H_{12}O$	2-Hexanone	127.6	25	1–8%	423
$C_6H_{12}O$	3-Hexanone	123.5	35	1–8%	
$C_6H_{12}O$	4-Methyl-2-pentanone	116.5	18	1.2–8.0%	448
$C_6H_{12}O$	Cyclohexanol	160.8	68	1–9%	300
$C_6H_{12}O_2$	Hexanoic acid	205.2	102		380
$C_6H_{12}O_2$	2-Methylpentanoic acid	195.6	107		378
$C_6H_{12}O_2$	Diethylacetic acid	194	99		400
$C_6H_{12}O_2$	Pentyl formate	130.4	26		
$C_6H_{12}O_2$	Butyl acetate	126.1	22	1.7–7.6%	425
$C_6H_{12}O_2$	*sec*-Butyl acetate	112	31	1.7–9.8%	
$C_6H_{12}O_2$	Isobutyl acetate	116.5	18	1.3–10.5%	421
$C_6H_{12}O_2$	Propyl propanoate	122.5	79		
$C_6H_{12}O_2$	Ethyl butanoate	121.5	24		463
$C_6H_{12}O_2$	Ethyl 2-methylpropanoate	110.1	13		
$C_6H_{12}O_2$	Diacetone alcohol	167.9	58	1.8–6.9%	643

Mol. form.	Name	t_B/°C	FP/°C	Fl. limits	IT/°C
$C_6H_{12}O_3$	Ethylene glycol monoethyl ether acetate	156.4	56	2–8%	379
$C_6H_{12}O_3$	Paraldehyde	124.3	36	1.3–?	238
$C_6H_{12}S$	Cyclohexanethiol	158.9	43		
$C_6H_{13}Cl$	1-Chlorohexane	135	35		
$C_6H_{13}N$	Cyclohexylamine	134	31	1.9–9.4%	293
$C_6H_{13}NO$	*N*-Butylacetamide	229	116		
$C_6H_{13}NO$	2,6-Dimethylmorpholine	146.6	44		
$C_6H_{13}NO$	*N*-Ethylmorpholine	138.5	32		
$C_6H_{13}NO_2$	4-Morpholineethanol	227	99		
C_6H_{14}	Hexane	68.7	–22	1.1–7.5%	225
C_6H_{14}	2-Methylpentane	60.2	<–29	1.0–7.0%	264
C_6H_{14}	3-Methylpentane	63.2	–7	1.2–7.0%	278
C_6H_{14}	2,2-Dimethylbutane	49.7	–48	1.2–7.0%	405
C_6H_{14}	2,3-Dimethylbutane	57.9	–29	1.2–7.0%	405
$C_6H_{14}N_2O$	1-Piperazineethanol	246	124		
$C_6H_{14}O$	1-Hexanol	157.6	63		
$C_6H_{14}O$	2-Methyl-1-pentanol	149	54	1.1–9.65%	310
$C_6H_{14}O$	4-Methyl-2-pentanol	131.6	41	1.0–5.5%	
$C_6H_{14}O$	2-Ethyl-1-butanol	147	57		
$C_6H_{14}O$	Dipropyl ether	90.0	21	1.3–7.0%	188
$C_6H_{14}O$	Diisopropyl ether	68.5	–28	1.4–7.9%	443
$C_6H_{14}O$	Butyl ethyl ether	92.3	4		
$C_6H_{14}O_2$	2,5-Hexanediol	218	110		
$C_6H_{14}O_2$	2-Methyl-2,4-pentanediol	197.1	102	1–9%	306
$C_6H_{14}O_2$	Ethylene glycol monobutyl ether	168.4	69	4–13%	238
$C_6H_{14}O_2$	1,1-Diethoxyethane	102.2	–21	1.6–10.4%	230
$C_6H_{14}O_2$	Ethylene glycol diethyl ether	119.4	27		205
$C_6H_{14}O_3$	1,2,6-Hexanetriol		191		
$C_6H_{14}O_3$	Diethylene glycol monoethyl ether	196	96		
$C_6H_{14}O_3$	Diethylene glycol dimethyl ether	162	67		
$C_6H_{14}O_3$	Trimethylolpropane		149		
$C_6H_{14}O_4$	Triethylene glycol	285	177	0.9–9.2%	371
$C_6H_{15}N$	Hexylamine	132.8	29		
$C_6H_{15}N$	Butylethylamine	107.5	18		
$C_6H_{15}N$	Dipropylamine	109.3	17		299
$C_6H_{15}N$	Diisopropylamine	83.9	–1	1.1–7.1%	316
$C_6H_{15}N$	Triethylamine	89	–7	1.2–8.0%	249
$C_6H_{15}NO_2$	Diisopropanolamine	250	127		374
$C_6H_{15}NO_3$	Triethanolamine	335.4	179	1–10%	
$C_6H_{15}N_3$	1-Piperazineethanamine	220	93		
$C_6H_{15}O_4P$	Triethyl phosphate	215.5	115		454
$C_6H_{16}N_2$	*N,N*-Diethylethylenediamine	144	46		
$C_7H_3ClF_3NO_2$	1-Chloro-4-nitro-2-(trifluoromethyl)benzene	232	135		
$C_7H_4ClF_3$	1-Chloro-2-(trifluoromethyl)benzene	152.2	59		
$C_7H_4F_3NO_2$	1-Nitro-3-(trifluoromethyl)benzene	202.8	103		
C_7H_5ClO	Benzoyl chloride	197.2	72		
C_7H_5ClO	4-Chlorobenzaldehyde	213.5	88		
$C_7H_5Cl_3$	(Trichloromethyl)benzene	221	127		211
$C_7H_5F_3$	(Trifluoromethyl)benzene	102.1	12		
$C_7H_6N_2O_4$	1-Methyl-2,4-dinitrobenzene		207		
C_7H_6O	Benzaldehyde	179.0	63		192
$C_7H_6O_2$	Benzoic acid	249.2	121		570
$C_7H_6O_2$	Salicylaldehyde	197	78		
$C_7H_6O_3$	Salicylic acid		157	1.1–?	540
C_7H_7Br	*o*-Bromotoluene	181.7	79		
C_7H_7Br	*p*-Bromotoluene	184.3	85		
C_7H_7Cl	(Chloromethyl)benzene	179	67	1.1–?	585
$C_7H_7NO_2$	*o*-Nitrotoluene	222	106		
$C_7H_7NO_2$	*m*-Nitrotoluene	232	106		
$C_7H_7NO_2$	*p*-Nitrotoluene	238.3	106		

Mol. form.	Name	t_B/°C	FP/°C	Fl. limits	IT/°C
C_7H_8	Toluene	110.6	4	1.1–7.1%	480
C_7H_8	Bicyclo[2.2.1]hepta-2,5-diene	89.5	–21		
C_7H_8O	*o*-Cresol	191.0	81	1.4–?	599
C_7H_8O	*m*-Cresol	202.2	86	1.1–?	558
C_7H_8O	*p*-Cresol	201.9	86	1.1–?	558
C_7H_8O	Benzyl alcohol	205.3	93		436
C_7H_8O	Anisole	153.7	52		475
$C_7H_8O_2$	4-Methoxyphenol	243	132		421
$C_7H_8O_3S$	*p*-Toluenesulfonic acid		184		
C_7H_9N	*o*-Methylaniline	200.3	85		482
C_7H_9N	*p*-Methylaniline	200.4	87		482
C_7H_9NO	*o*-Anisidine	224	118		
$C_7H_{10}O$	3-Cyclohexene-1-carboxaldehyde	105	57		
$C_7H_{10}O_4$	3,3-Diacetoxy-1-propene	180	82		
C_7H_{12}	4-Methylcyclohexene	102.7	–1		
$C_7H_{12}O_2$	Butyl acrylate	145	29	1.7–9.9%	292
$C_7H_{12}O_2$	Isobutyl acrylate	132	30		427
$C_7H_{12}O_2$	Cyclohexyl formate	162	51		
$C_7H_{12}O_4$	Diethyl malonate	200	93		
C_7H_{14}	1-Heptene	93.6	–1		260
C_7H_{14}	*trans*-2-Heptene	98	<0		
C_7H_{14}	Cycloheptane	118.4	<21	1.1–6.7%	
C_7H_{14}	Methylcyclohexane	100.9	–4	1.2–6.7%	250
C_7H_{14}	Ethylcyclopentane	103.5	<21	1.1–6.7%	260
$C_7H_{14}O$	2-Heptanone	151.0	39	1.1–7.9%	393
$C_7H_{14}O$	3-Heptanone	147	46		
$C_7H_{14}O$	4-Heptanone	144	49		
$C_7H_{14}O$	5-Methyl-2-hexanone	144	36	1.0–8.2%	191
$C_7H_{14}O$	*cis*-2-Methylcyclohexanol	165	65		296
$C_7H_{14}O$	*trans*-2-Methylcyclohexanol	167.5	65		296
$C_7H_{14}O$	*cis*-3-Methylcyclohexanol	174.5	70		295
$C_7H_{14}O$	*trans*-3-Methylcyclohexanol	174.5	70		295
$C_7H_{14}O$	*cis*-4-Methylcyclohexanol	173	70		295
$C_7H_{14}O$	*trans*-4-Methylcyclohexanol	174	70		295
$C_7H_{14}O_2$	Pentyl acetate	149.2	16	1.1–7.5%	360
$C_7H_{14}O_2$	Isopentyl acetate	142.5	25	1.0–7.5%	360
$C_7H_{14}O_2$	*sec*-Pentyl acetate	130.5	32		
$C_7H_{14}O_2$	Butyl propanoate	146.8	32		426
$C_7H_{14}O_2$	Propyl butanoate	143.0	37		
$C_7H_{15}NO_2$	Ethyl *N*-butylcarbamate	202	92		
C_7H_{16}	Heptane	98.5	–4	1.05–6.7%	204
C_7H_{16}	2-Methylhexane	90.0	–1	1.0–6.0%	280
C_7H_{16}	3-Methylhexane	92	–4		280
C_7H_{16}	2,3-Dimethylpentane	89.7	–56	1.1–6.7%	335
C_7H_{16}	2,4-Dimethylpentane	80.4	–12		
C_7H_{16}	2,2,3-Trimethylbutane	80.8	<0		412
$C_7H_{16}N_2O$	4-Morpholinepropanamine	220	104		
$C_7H_{16}O$	2-Heptanol	159	71		
$C_7H_{16}O$	3-Heptanol	157	60		
$C_7H_{16}O$	2,4-Dimethyl-3-pentanol	138.7	49		
$C_7H_{16}O$	2,3,3-Trimethyl-2-butanol	131	<0		375
$C_7H_{17}N$	Heptylamine	156	54		
$C_7H_{18}N_2$	*N,N*-Diethyl-1,3-propanediamine	168.5	59		
$C_8H_4O_3$	Phthalic anhydride	295	152	1.7–10.5%	570
$C_8H_6O_4$	Phthalic acid		168		
$C_8H_6O_4$	Terephthalic acid		260		496
C_8H_7ClO	α-Chloroacetophenone	247	118		
C_8H_7N	Benzeneacetonitrile	233.5	113		
C_8H_8	Styrene	145	31	0.9–6.8%	490
C_8H_8O	Phenyloxirane	194.1	74		498
C_8H_8O	Benzeneacetaldehyde	195	71		

Mol. form.	Name	t_B/°C	FP/°C	Fl. limits	IT/°C
C_8H_8O	Acetophenone	202	77		570
$C_8H_8O_2$	Benzeneacetic acid	265.5	>100		
$C_8H_8O_2$	Phenyl acetate	196	80		
$C_8H_8O_2$	Methyl benzoate	199	83		
$C_8H_8O_2$	2-Methoxybenzaldehyde	243.5	118		
$C_8H_8O_3$	Methyl salicylate	222.9	96		454
C_8H_9Cl	1-Chloro-4-ethylbenzene	184.4	64		
C_8H_9NO	Acetanilide	304	169		530
$C_8H_9NO_2$	Methyl 2-aminobenzoate	256	>100		
C_8H_{10}	Ethylbenzene	136.1	21	0.8–6.7%	432
C_8H_{10}	*o*-Xylene	144.5	32	0.9–6.7%	463
C_8H_{10}	*m*-Xylene	139.1	27	1.1–7.0%	527
C_8H_{10}	*p*-Xylene	138.3	27	1.1–7.0%	528
$C_8H_{10}O$	*p*-Ethylphenol	217.9	104		
$C_8H_{10}O$	Benzeneethanol	218.2	96		
$C_8H_{10}O$	α-Methylbenzyl alcohol	205	93		
$C_8H_{10}O$	Phenetole	169.8	63		
$C_8H_{10}O$	Benzyl methyl ether	170	135		
$C_8H_{10}O$	4-Methylanisole	175.5	60		
$C_8H_{10}O_2$	2-Phenoxyethanol	245	121		
$C_8H_{11}N$	*N*-Ethylaniline	203.0	85		
$C_8H_{11}N$	*N,N*-Dimethylaniline	194.1	63		371
$C_8H_{11}N$	2,3-Xylidine	221.5	97	1.0–?	
$C_8H_{11}N$	2,6-Xylidine	215	96		
$C_8H_{11}N$	α-Methylbenzylamine	187	79		
$C_8H_{11}N$	5-Ethyl-2-picoline	178.3	68	1.1–6.6%	
$C_8H_{11}NO$	*N*-Phenylethanolamine	279.5	152		
$C_8H_{11}NO$	*o*-Phenetidine	232.5	115		
$C_8H_{11}NO$	*p*-Phenetidine	254	116		
C_8H_{12}	1,5-Cyclooctadiene	150.8	35		
C_8H_{12}	4-Vinylcyclohexene	128	16		269
$C_8H_{12}O_4$	Diethyl maleate	223	121		350
$C_8H_{12}O_4$	Diethyl fumarate	214	104		
$C_8H_{14}O_2$	Cyclohexyl acetate	173	58		335
$C_8H_{14}O_2$	Butyl methacrylate	160	52		
$C_8H_{14}O_3$	Butanoic anhydride	200	54	0.9–5.8%	279
$C_8H_{14}O_3$	2-Methylpropanoic anhydride	183	59	1.0–6.2%	329
$C_8H_{14}O_3$	Butyl acetoacetate		85		
$C_8H_{14}O_4$	Ethyl succinate	217.7	90		
$C_8H_{14}O_5$	Diethylene glycol diacetate	200	135		
$C_8H_{14}O_6$	Diethyl tartrate	281	93		
$C_8H_{15}ClO$	Octanoyl chloride	195.6	82		
C_8H_{16}	1-Octene	121.2	21		230
C_8H_{16}	2,4,4-Trimethyl-1-pentene	101.4	–5	0.8–4.8%	391
C_8H_{16}	2,4,4-Trimethyl-2-pentene	104.9	2		305
C_8H_{16}	Ethylcyclohexane	131.9	35	0.9–6.6%	238
C_8H_{16}	*cis*-1,2-Dimethylcyclohexane	129.8	16		304
C_8H_{16}	*trans*-1,2-Dimethylcyclohexane	123.5	11		304
C_8H_{16}	*cis*-1,4-Dimethylcyclohexane	124.4	16		
C_8H_{16}	Propylcyclopentane	131			269
$C_8H_{16}O$	Octanal	171	52		
$C_8H_{16}O$	2-Ethylhexanal	163	44	0.85–7.2%	190
$C_8H_{16}O$	2-Octanone	172.5	52		
$C_8H_{16}O_2$	Hexyl acetate	171.5	45		
$C_8H_{16}O_2$	sec-Hexyl acetate	147.5	45		
$C_8H_{16}O_2$	2-Ethylbutyl acetate	162.5	54		
$C_8H_{16}O_2$	Pentyl propanoate	168.6	41		378
$C_8H_{16}O_2$	Butyl butanoate	166	53		
$C_8H_{16}O_2$	Isobutyl butanoate	156.9	50		
$C_8H_{16}O_2$	Isobutyl isobutanoate	148.6	38	0.96–7.59%	432
$C_8H_{16}O_2$	Ethyl hexanoate	167	49		

Mol. form.	Name	t_B/°C	FP/°C	Fl. limits	IT/°C
$C_8H_{16}O_2$	1,4-Cyclohexanedimethanol	283	167		316
$C_8H_{16}O_3$	Pentyl lactate		79		
$C_8H_{16}O_4$	Diethylene glycol monoethyl ether acetate	218.5	110		425
$C_8H_{17}Cl$	1-Chlorooctane	181.5	70		
$C_8H_{17}Cl$	3-(Chloromethyl)heptane	172	60		
C_8H_{18}	Octane	125.6	13	1.0–6.5%	206
C_8H_{18}	2,3-Dimethylhexane	115.6	7		438
C_8H_{18}	2,4-Dimethylhexane	109.5	10		
C_8H_{18}	3-Ethyl-2-methylpentane	115.6	<21		460
C_8H_{18}	2,2,3-Trimethylpentane	110	<21		346
C_8H_{18}	2,2,4-Trimethylpentane	99.2	–12		418
C_8H_{18}	2,3,3-Trimethylpentane	114.8	<21		425
$C_8H_{18}O$	1-Octanol	195.1	81		
$C_8H_{18}O$	2-Octanol	180	88		
$C_8H_{18}O$	2-Ethyl-1-hexanol	184.6	73	0.88–9.7%	231
$C_8H_{18}O$	Dibutyl ether	140.2	25	1.5–7.6%	194
$C_8H_{18}O_2$	2-Ethyl-1,3-hexanediol	244	127		360
$C_8H_{18}O_2$	2,2,4-Trimethyl-1,3-pentanediol	235	113		346
$C_8H_{18}O_2$	Di-*tert*-butyl peroxide	111	18		
$C_8H_{18}O_3$	Diethylene glycol diethyl ether	188	82		
$C_8H_{18}O_4$	2,5,8,11-Tetraoxadodecane	216	111		
$C_8H_{18}O_5$	Tetraethylene glycol	328	182		
$C_8H_{18}S$	1-Octanethiol	199.1	69		
$C_8H_{18}S$	Dibutyl sulfide	185	76		
$C_8H_{19}N$	Octylamine	179.6	60		
$C_8H_{19}N$	Dibutylamine	159.6	47	1.1–6%	
$C_8H_{19}N$	Diisobutylamine	139.6	29		
$C_8H_{19}N$	2-Ethylhexylamine	169.2	60		
$C_8H_{20}O_4Si$	Ethyl silicate	168.8	52		
$C_8H_{23}N_5$	Tetraethylenepentamine	341.5	163		321
$C_9H_6N_2O_2$	Toluene-2,4-diisocyanate	251	127	0.9–9.5%	
C_9H_7N	Quinoline	237.1			480
C_9H_{10}	*o*-Methylstyrene	169.8	53	0.8–11.0%	538
C_9H_{10}	*m*-Methylstyrene	164	53	0.8–11.0%	538
C_9H_{10}	*p*-Methylstyrene	172.8	53	0.8–11.0%	538
C_9H_{10}	Isopropenylbenzene	165.4	54	1.9–6.1%	574
$C_9H_{10}O$	1-Phenyl-1-propanone	217.5	99		
$C_9H_{10}O$	4-Methylacetophenone	226	96		
$C_9H_{10}O_2$	Ethyl benzoate	212	88		490
$C_9H_{10}O_2$	Benzyl acetate	213	90		460
$C_9H_{10}O_2$	Methyl 2-phenylacetate	216.5	91		
$C_9H_{11}NO$	4-Methylacetanilide	307	168		
C_9H_{12}	Propylbenzene	159.2	30	0.8–6.0%	450
C_9H_{12}	Isopropylbenzene	152.4	36	0.9–6.5%	424
C_9H_{12}	*o*-Ethyltoluene	165.2			440
C_9H_{12}	*m*-Ethyltoluene	161.3			480
C_9H_{12}	*p*-Ethyltoluene	162			475
C_9H_{12}	1,2,3-Trimethylbenzene	176.1	44	0.8–6.6%	470
C_9H_{12}	1,2,4-Trimethylbenzene	169.3	44	0.9–6.4%	500
C_9H_{12}	1,3,5-Trimethylbenzene	164.7	50	1–5%	559
$C_9H_{12}O$	α–Ethylbenzyl alcohol	219	100		
$C_9H_{12}O_2$	Ethylene glycol monobenzyl ether	256	129		352
$C_9H_{12}O_3S$	Ethyl *p*-toluenesulfonate		158		
$C_9H_{13}N$	Amphetamine	203	<100		
$C_9H_{14}O$	Phorone	197.5	85		
$C_9H_{14}O$	Isophorone	215.2	84	0.8–3.8%	460
$C_9H_{14}O_6$	Triacetin	259	138	1.0–?	433
C_9H_{16}	Octahydroindene	167			296
$C_9H_{16}O_2$	Allyl hexanoate	186	66		
C_9H_{18}	1-Nonene	146.9	26		

Mol. form.	Name	t_B/°C	FP/°C	Fl. limits	IT/°C
C_9H_{18}	Propylcyclohexane	156.7			248
C_9H_{18}	Isopropylcyclohexane	154.8			283
C_9H_{18}	Butylcyclopentane	156.6			250
$C_9H_{18}O$	2-Nonanone	195.3	60	0.9–5.9%	360
$C_9H_{18}O$	Diisobutyl ketone	169.4	49	0.8–7.1%	396
$C_9H_{18}O_2$	Pentyl butanoate	186.4	57		
$C_9H_{18}O_2$	Isopentyl butanoate	179	59		
$C_9H_{18}O_2$	Butyl 3-methylbutanoate		53		
C_9H_{20}	Nonane	150.8	31	0.8–2.9%	205
C_9H_{20}	3-Ethyl-4-methylhexane	140	24		
C_9H_{20}	4-Ethyl-2-methylhexane	133.8	<21	0.7–?	280
C_9H_{20}	2,2,5-Trimethylhexane	124.0	13		
C_9H_{20}	3,3-Diethylpentane	146.3		0.7–5.7%	290
C_9H_{20}	3-Ethyl-2,4-dimethylpentane	136.7	390		
C_9H_{20}	2,2,3,3-Tetramethylpentane	140.2	<21	0.8–4.9%	430
C_9H_{20}	2,2,3,4-Tetramethylpentane	133.0	<21		
$C_9H_{21}BO_3$	Triisopropyl borate	140	28		
$C_9H_{21}N$	Tripropylamine	156	41		
$C_9H_{21}NO_3$	Triisopropanolamine		160		320
$C_{10}H_7Cl$	1-Chloronaphthalene	259	121		>558
$C_{10}H_8$	Naphthalene	217.9	79	0.9–5.9%	526
$C_{10}H_8O$	2-Naphthol	285	153		
$C_{10}H_9N$	1-Naphthalenamine	300.8	157		
$C_{10}H_{10}O_2$	Safrole	234.5	100		
$C_{10}H_{10}O_4$	Dimethyl phthalate	283.7	146	0.9–?	490
$C_{10}H_{10}O_4$	Dimethyl isophthalate	282	138		
$C_{10}H_{10}O_4$	Dimethyl terephthalate	288	153		518
$C_{10}H_{11}NO_2$	Acetoacetanilide		185		
$C_{10}H_{12}$	1,2,3,4-Tetrahydronaphthalene	207.6	71	0.8–5.0%	385
$C_{10}H_{12}O_2$	Isopropyl benzoate	216	99		
$C_{10}H_{12}O_2$	Ethyl phenylacetate	227	99		
$C_{10}H_{14}$	Butylbenzene	183.3	71	0.8–5.8%	410
$C_{10}H_{14}$	*sec*-Butylbenzene	173.3	52	0.8–6.9%	418
$C_{10}H_{14}$	*tert*-Butylbenzene	169.1	60	0.7–5.7%	450
$C_{10}H_{14}$	Isobutylbenzene	172.7	55	0.8–6.0%	427
$C_{10}H_{14}$	*p*-Cymene	177.1	47	0.7–5.6%	436
$C_{10}H_{14}$	1,2,3,4-Tetramethylbenzene	205	74		427
$C_{10}H_{14}$	1,2,3,5-Tetramethylbenzene	198	71		427
$C_{10}H_{14}$	1,2,4,5-Tetramethylbenzene	196.8	54		
$C_{10}H_{14}$	*o*-Diethylbenzene	184	57		395
$C_{10}H_{14}$	*m*-Diethylbenzene	181.1	56		450
$C_{10}H_{14}$	*p*-Diethylbenzene	183.7	55	0.7–6.0%	430
$C_{10}H_{14}O$	Butyl phenyl ether	210	82		
$C_{10}H_{14}O_2$	4-*tert*-Butyl-1,2-benzenediol	285	130		
$C_{10}H_{15}N$	*N*-Butylaniline	243.5	107		
$C_{10}H_{15}N$	*N,N*-Diethylaniline	216.3	85		630
$C_{10}H_{15}NO_2$	*N*-Phenyl-*N,N*-diethanolamine		196	0.7–?	387
$C_{10}H_{16}$	Dipentene	178	45		237
$C_{10}H_{16}$	*d*-Limonene	178	45	0.7–6.1%	237
$C_{10}H_{16}$	α-Pinene	156.2	33		255
$C_{10}H_{16}$	β-Pinene	166	38		275
$C_{10}H_{16}$	β-Phellandrene	171.5	49		
$C_{10}H_{16}O$	Camphor	207.4	66	0.6–3.5%	466
$C_{10}H_{18}$	*trans*-Decahydronaphthalene	187.3	54	0.7–5.4%	255
$C_{10}H_{18}O$	Borneol		66		
$C_{10}H_{18}O$	Linalol	198	71		
$C_{10}H_{18}O$	α-Terpineol	220	90		
$C_{10}H_{18}O$	Cineole	176.4	48		
$C_{10}H_{18}O$	*trans*-Geraniol	230	>100		
$C_{10}H_{18}O_4$	Dibutyl oxalate	241	104		
$C_{10}H_{19}NO_2$	*N-tert*-Butylaminoethyl methacrylate		96		

Mol. form.	Name	t_B/°C	FP/°C	Fl. limits	IT/°C
$C_{10}H_{20}$	1-Decene	170.5	<55		235
$C_{10}H_{20}$	Butylcyclohexane	180.9			246
$C_{10}H_{20}$	Isobutylcyclohexane	171.3			274
$C_{10}H_{20}$	*tert*-Butylcyclohexane	171.5			342
$C_{10}H_{20}O$	Citronellol	224	96		
$C_{10}H_{20}O_2$	2-Ethylhexyl acetate	199	71	0.76–8.14%	268
$C_{10}H_{20}O_2$	Ethyl octanoate	208.5	79		
$C_{10}H_{21}N$	*N*-Butylcyclohexanamine		93		
$C_{10}H_{22}$	Decane	174.1	51	0.8–5.4%	210
$C_{10}H_{22}$	2-Methylnonane	167.1			210
$C_{10}H_{22}$	3-Ethyloctane	166.5			230
$C_{10}H_{22}$	4-Ethyloctane	163.7			229
$C_{10}H_{22}O$	1-Decanol	231.1	82		288
$C_{10}H_{22}O$	Dipentyl ether	190	57		170
$C_{10}H_{22}O_2$	Ethylene glycol dibutyl ether	203.3	85		
$C_{10}H_{22}O_5$	Tetraethylene glycol dimethyl ether	275.3	141		
$C_{10}H_{22}S$	Dipentyl sulfide		85		
$C_{10}H_{23}N$	Decylamine	220.5	99		
$C_{10}H_{23}N$	Dipentylamine	202.5	51		
$C_{11}H_{10}$	1-Methylnaphthalene	244.7			529
$C_{11}H_{12}O_3$	Ethyl benzoylacetate		141		
$C_{11}H_{14}O_2$	Butyl benzoate	250.3	107		
$C_{11}H_{16}$	*p-tert*-Butyltoluene	190	68		
$C_{11}H_{16}$	Pentylbenzene	205.4	66		
$C_{11}H_{16}$	1,3-Diethyl-5-methylbenzene	205			455
$C_{11}H_{16}$	Pentamethylbenzene	232	93		427
$C_{11}H_{16}O$	4-*tert*-Butyl-2-methylphenol	237	118		
$C_{11}H_{17}N$	*p-tert*-Pentylaniline	260.5	102		
$C_{11}H_{20}O_2$	2-Ethylhexyl acrylate		82		252
$C_{11}H_{22}$	Pentylcyclohexane	203.7			239
$C_{11}H_{22}O$	2-Undecanone	231.5	89		
$C_{11}H_{22}O_2$	Nonyl acetate	210	68		
$C_{11}H_{24}$	Undecane	195.9	69		
$C_{11}H_{24}$	2-Methyldecane	189.3			225
$C_{11}H_{24}O$	2-Undecanol	228	113		
$C_{12}H_9Br$	4-Bromo-1,1'-biphenyl	310	144		
$C_{12}H_{10}$	Biphenyl	256.1	113	0.6–5.8%	540
$C_{12}H_{10}Cl_2Si$	Dichlorodiphenylsilane	305	142		
$C_{12}H_{10}O$	*o*-Phenylphenol	286	124		530
$C_{12}H_{10}O$	Diphenyl ether	258.0	112	0.8–1.5%	618
$C_{12}H_{11}N$	2-Aminobiphenyl	299			450
$C_{12}H_{11}N$	Diphenylamine	302	153		634
$C_{12}H_{12}$	1-Ethylnaphthalene	258.6			480
$C_{12}H_{14}O_4$	Diethyl phthalate	295	161	0.7–?	457
$C_{12}H_{14}O_4$	Diethyl terephthalate	302	117		
$C_{12}H_{16}$	Cyclohexylbenzene	240.1	99		
$C_{12}H_{16}O_3$	Pentyl salicylate	270	132		
$C_{12}H_{17}NO$	*N*-Butyl-*N*-phenylacetamide	281	141		
$C_{12}H_{18}$	1,5,9-Cyclododecatriene	240	71		
$C_{12}H_{20}O_4$	Dibutyl maleate	280	141		
$C_{12}H_{22}O_4$	Dimethyl sebacate		145		
$C_{12}H_{22}O_6$	Dibutyl tartrate	320	91		284
$C_{12}H_{23}N$	Dicyclohexylamine		>99		
$C_{12}H_{24}$	1-Dodecene	213.8	79		
$C_{12}H_{24}O_2$	Ethyl decanoate	241.5	>100		
$C_{12}H_{25}Br$	1-Bromododecane	276	144		
$C_{12}H_{26}$	Dodecane	216.3	74	0.6–?	203
$C_{12}H_{26}O$	1-Dodecanol	259	127		275
$C_{12}H_{26}O$	2-Butyl-1-octanol	246.5	110		
$C_{12}H_{26}O_3$	Diethylene glycol dibutyl ether	256	118		310
$C_{12}H_{26}S$	1-Dodecanethiol	277	128		

Mol. form.	Name	t_B/°C	FP/°C	Fl. limits	IT/°C
$C_{12}H_{27}BO_3$	Tributyl borate	234	93		
$C_{12}H_{27}N$	Tributylamine	216.5	63		
$C_{12}H_{27}O_4P$	Tributyl phosphate	289	146		
$C_{13}H_{12}$	2-Methylbiphenyl	255.5	137		502
$C_{13}H_{12}$	Diphenylmethane	265.0	130		485
$C_{13}H_{14}N_2$	*p,p′*-Diaminodiphenylmethane	398	220		
$C_{13}H_{26}$	1-Tridecene	232.8	79		
$C_{13}H_{26}O$	2-Tridecanone	263	107		
$C_{13}H_{28}$	Tridecane	235.4	79		
$C_{13}H_{28}O$	1-Tridecanol		121		
$C_{14}H_8O_2$	9,10-Anthracenedione	377	185		
$C_{14}H_{10}$	Anthracene	339.9	121	0.6–?	540
$C_{14}H_{10}$	Phenanthrene	340	171		
$C_{14}H_{12}O_2$	Benzyl benzoate	323.5	148		480
$C_{14}H_{12}O_3$	Benzyl salicylate	320	>100		
$C_{14}H_{14}$	1,1-Diphenylethane	272.6	>100		440
$C_{14}H_{14}O$	Dibenzyl ether	298	135		
$C_{14}H_{16}$	1-Butylnaphthalene	289.3	360		
$C_{14}H_{16}N_2O_2$	*o*-Dianisidine		206		
$C_{14}H_{23}N$	*N,N*-Dibutylaniline	274.8	110		
$C_{14}H_{28}$	1-Tetradecene	233	110		235
$C_{14}H_{30}$	Tetradecane	253.5	112	0.5–?	200
$C_{14}H_{30}O$	1-Tetradecanol	289	141		
$C_{15}H_{18}$	1-Pentylnaphthalene	307	124		
$C_{15}H_{24}$	Nonylbenzene	280.5	99		
$C_{15}H_{24}O$	2,6 Di-*tert*-butyl-4-methylphenol	265	127		
$C_{15}H_{26}O_6$	Tributyrin	307.5	180	0.5–?	407
$C_{15}H_{33}N$	Tripentylamine	242.5	102		
$C_{16}H_{14}O$	1,3-Diphenyl-2-buten-1-one	342.5	177		
$C_{16}H_{18}$	2-Butyl-1,1′-biphenyl		>100		430
$C_{16}H_{22}O_4$	Dibutyl phthalate	340	157	0.5–?	402
$C_{16}H_{26}$	Decylbenzene	298	107		
$C_{16}H_{34}$	Hexadecane	286.8	136		202
$C_{16}H_{34}O$	Dioctyl ether	283	>100		205
$C_{16}H_{35}N$	Bis(2-ethylhexyl)amine		132		
$C_{17}H_{20}N_2O$	*N,N′*-Diethylcarbanilide		150		
$C_{17}H_{34}O$	2-Heptadecanone	320	120		
$C_{17}H_{36}O$	1-Heptadecanol	333	154		
$C_{18}H_{14}$	*o*-Terphenyl	332	163		
$C_{18}H_{14}$	*m*-Terphenyl	363	191		
$C_{18}H_{15}O_3P$	Triphenyl phosphite	360	218		
$C_{18}H_{15}O_4P$	Triphenyl phosphate		220		
$C_{18}H_{15}P$	Triphenylphosphine		180		
$C_{18}H_{30}$	Dodecylbenzene	328	140		
$C_{18}H_{32}O_7$	Butyl citrate		157		368
$C_{18}H_{34}O_2$	Oleic acid	360	189		363
$C_{18}H_{34}O_4$	Dibutyl sebacate	344.5	178	0.4–?	365
$C_{18}H_{36}O_2$	Stearic acid		196		395
$C_{18}H_{37}Cl_3Si$	Trichlorooctadecylsilane		89		
$C_{18}H_{38}$	Octadecane	316.3	>100		227
$C_{18}H_{38}O$	1-Octadecanol				450
$C_{19}H_{16}$	Triphenylmethane	359	>100		
$C_{19}H_{38}O$	2-Nonadecanone		124		
$C_{19}H_{38}O_2$	Methyl stearate	443	153		
$C_{19}H_{40}$	Nonadecane	329.9	>100		230
$C_{20}H_{14}O_4$	Diphenyl phthalate		224		
$C_{20}H_{28}$	1-Decylnaphthalene	379	177		
$C_{20}H_{42}$	Eicosane	343	>100		232
$C_{21}H_{21}O_4P$	Tri-*o*-cresyl phosphate	410	225		385
$C_{21}H_{26}O_3$	4-Octylphenyl salicylate		216		416
$C_{21}H_{32}O_2$	Methyl abietate		180		

Mol. form.	Name	t_B/°C	FP/°C	Fl. limits	IT/°C
$C_{22}H_{42}O_2$	Butyl oleate		180		
$C_{22}H_{42}O_4$	Bis(2-ethylhexyl) adipate		206	0.4–?	377
$C_{22}H_{44}O_2$	Butyl stearate	343	160		355
$C_{23}H_{46}O_2$	Pentyl stearate		185		
$C_{24}H_{20}Sn$	Tetraphenylstannane	420	232		
$C_{24}H_{38}O_4$	Bis(2-ethylhexyl) phthalate	384	218		
$C_{25}H_{48}O_4$	Bis(2-ethylhexyl) azelate		227	0.3–?	374

THRESHOLD LIMITS FOR AIRBORNE CONTAMINANTS

Several organizations recommend limits of exposure to airborne contaminants in the workplace. These include the Occupational Safety and Health Administration (OSHA), the National Institute for Occupational Safety and Health (NIOSH), and the non-governmental organization, American Conference of Governmental Industrial Hygienists (ACGIH). The threshold limit value (TLV) for a substance is defined as the concentration level under which the majority of workers may be repeatedly exposed, day after day, without adverse effects. The TLV recommendations are given in two forms:

- Time-weighted average (TWA) concentration for a normal 8-hr workday and 40-hr workweek.
- Short-term exposure limit (STEL), which should not be exceeded for more than 15 min.

Both kinds of limits are specified for some substances.

The following table gives threshold limit values for a number of substances that may be encountered in the atmosphere of a chemical laboratory or industrial facility. All values refer to the concentration in air at 25 °C and normal atmospheric pressure. Data for gases are given in parts per million by volume (ppm). Values for liquids refer to mists or aerosols, and those for solids to dusts or fumes; both are stated in mass concentration (mg/m^3). A "C" preceding a value indicates a ceiling limit, which should not be exceeded even for very brief periods because of acute toxic effects of the substance. The notation "levels as low as possible" in the Notes column indicates such a high level of hazard that no safe limit can be recommended.

Substances are listed alphabetically by systematic name. The Notes column gives further information on the form of the substance and the basis to which the TLV is referred. This column also includes common synonyms and acronyms in brackets (e.g., [MTBE]). The Formula column gives the molecular formula in the Hill convention for organic compounds and the customary line formula for inorganic compounds. The TWA and STEL limits appear in the last two columns.

Proposed changes in the limits are given in the Notes column with the notation "NIC", which stands for Notice of Intended Changes. These are considered trial values in 2008, which may be adopted permanently in subsequent years.

References

1. *2008 TLV's and BEI's*, American Conference of Governmental Industrial Hygienists, 1330 Kemper Meadow Drive, Cincinnati, OH 45240-1634, 2008 (www.acgih.org).
2. *NIOSH Pocket Guide to Chemical Hazards*, U.S. Department of Health and Human Services, National Institute for Occupational Health and Safety, U.S. Government Printing Office, Washington, DC, 1994.
3. *Chemical Information Manual*, U.S. Department of Labor, Occupational Safety and Health Administration, Washington, DC, 1991.

Substance	Notes	Formula	CAS Reg. No.	Time-weighted average	Short-term exposure limit
Abate	[Temephos]	$C_{16}H_{20}O_6P_2S_3$	3383-96-8	10 mg/m^3	
Acetaldehyde	[Ethanal]	C_2H_4O	75-07-0		C 25 ppm
Acetic acid	[Ethanoic acid]	$C_2H_4O_2$	64-19-7	10 ppm	15 ppm
Acetic anhydride	[Acetyl acetate]	$C_4H_6O_3$	108-24-7	5 ppm	
Acetone	[2-Propanone]	C_3H_6O	67-64-1	500 ppm	750 ppm
Acetone cyanohydrin	as CN	C_4H_7NO	75-86-5		C 5 mg/m^3
Acetonitrile	[Methyl cyanide]	C_2H_3N	75-05-8	20 ppm	
Acetophenone	[Methyl phenyl ketone]	C_8H_8O	98-86-2	10 ppm	
2-(Acetyloxy)benzoic acid	[Aspirin]	$C_9H_8O_4$	50-78-2	5 mg/m^3	
Acrolein	[2-Propenal]	C_3H_4O	107-02-8		C 0.1 ppm
Acrylamide	[2-Propenamide]	C_3H_5NO	79-06-1	0.03 mg/m^3	
Acrylic acid	[2-Propenoic acid]	$C_3H_4O_2$	79-10-7	2 ppm	
Acrylonitrile	[Propenenitrile]	C_3H_3N	107-13-1	2 ppm	
Alachlor	[Acetamide, 2-chloro-*N*-(2,6-diethylphenyl)-*N*-(methoxymethyl)-]	$C_{14}H_{20}ClNO_2$	15972-60-8	1 mg/m^3	
Aldrin		$C_{12}H_8Cl_6$	309-00-2	0.05 mg/m^3	
Allyl alcohol	[2-Propen-1-ol]	C_3H_6O	107-18-6	0.5 ppm	
Allyl glycidyl ether	[AGE]	$C_6H_{10}O_2$	106-92-3	1 ppm	
Allyl propyl disulfide		$C_6H_{12}S_2$	2179-59-1	0.5 ppm	
Aluminum	metal dust and insoluble compounds	Al	7429-90-5	1 mg/m^3	
4-Aminobiphenyl	[*p*-Biphenylamine] levels as low as possible	$C_{12}H_{11}N$	92-67-1		
4-Amino-3,5,6-trichloro-2-pyridinecarboxlic acid	[Picloram]	$C_6H_3Cl_3N_2O_2$	1918-02-1	10 mg/m^3	
Ammonia		NH_3	7664-41-7	25 ppm	35 ppm
Ammonium chloride	fume	NH_4Cl	12125-02-9	10 mg/m^3	20 mg/m^3
Ammonium perfluorooctanoate		$C_8H_4F_{15}NO_2$	3825-26-1	0.01 mg/m^3	
Ammonium sulfamate		$NH_4NH_2SO_3$	7773-06-0	10 mg/m^3	

Substance	Notes	Formula	CAS Reg. No.	Time-weighted average	Short-term exposure limit
Aniline	[Benzenamine]	C_6H_7N	62-53-3	2 ppm	
Antimony	and compounds, as Sb	Sb	7440-36-0	0.5 mg/m^3	
Antimony trioxide	levels as low as possible	Sb_2O_3	1309-64-4		
Arsenic	and inorganic compounds, as As	As	7440-38-2	0.01 mg/m^3	
Arsine	[Arsenic hydride]	AsH_3	7784-42-1	0.005 ppm	
Asbestos	all forms		1332-21-4	0.1 fiber/cm^3	
Asphalt	fume, as aerosol		8052-42-4	0.5 mg/m^3	
Atrazine	[6-Chloro-*N*-ethyl-*N*′-(1-methylethyl)-1,3,5-triazine-2,4-diamine]	$C_8H_{14}ClN_5$	1912-24-9	5 mg/m^3	
Azinphos-methyl		$C_{10}H_{12}N_3O_3PS_2$	86-50-0	0.2 mg/m^3	
Barium	and soluble compounds, as Ba	Ba	7440-39-3	0.5 mg/m^3	
Barium sulfate	[Barite]	$BaSO_4$	7727-43-7	10 mg/m^3	
Benomyl		$C_{14}H_{18}N_4O_3$	17804-35-2	1 mg/m^3	
Benz[a]anthracene	levels as low as possible	$C_{18}H_{12}$	56-55-3		
Benzene		C_6H_6	71-43-2	0.5 ppm	2.5 ppm
1,2-Benzenediamine	[*o*-Phenylenediamine]	$C_6H_8N_2$	95-54-5	0.1 mg/m^3	
1,3-Benzenediamine	[*m*-Phenylenediamine]	$C_6H_8N_2$	108-45-2	0.1 mg/m^3	
1,4-Benzenediamine	[*p*-Phenylenediamine]	$C_6H_8N_2$	106-50-3	0.1 mg/m^3	
1,3-Benzenedimethanamine	[*m*-Xylene diamine]	$C_8H_{12}N_2$	1477-55-0		C 0.1 mg/m^3
Benzenethiol	[Phenyl mercaptan]	C_6H_6S	108-98-5	0.1 ppm	
p-Benzidine	levels as low as possible	$C_{12}H_{12}N_2$	92-87-5		
Benzo[b]fluoranthene	levels as low as possible	$C_{20}H_{12}$	205-99-2		
Benzo[a]pyrene	levels as low as possible	$C_{20}H_{12}$	50-32-8		
p-Benzoquinone	[Quinone]	$C_6H_4O_2$	106-51-4	0.1 ppm	
Benzoyl chloride	[Benzoic acid, chloride]	C_7H_5ClO	98-88-4		C 0.5 ppm
Benzoyl peroxide		$C_{14}H_{10}O_4$	94-36-0	5 mg/m^3	
Benzyl acetate	[(Acetoxymethyl)benzene]	$C_9H_{10}O_2$	140-11-4	10 ppm	
Beryllium	and compounds, as Be. NIC in 2008 to TWA 0.00005 mg/m^3, STEL 0.0002 mg/m^3.	Be	7440-41-7	0.002 mg/m^3	0.01 mg/m^3
Biphenyl	[Diphenyl]	$C_{12}H_{10}$	92-52-4	0.2 ppm	
Bis(2-aminoethyl)amine	[Diethylenetriamine]	$C_4H_{13}N_3$	111-40-0	1 ppm	
Bis(2-chloroethyl) ether	[2,2′-Dichlorethyl ether]	$C_4H_8Cl_2O$	111-44-4	5 ppm	10 ppm
Bis(chloromethyl) ether	[Chloromethyl ether]	$C_2H_4Cl_2O$	542-88-1	0.001 ppm	
Bis(2-dimethylaminoethyl) ether	[DMAEE]	$C_8H_{20}N_2O$	3033-62-3	0.05 ppm	0.15 ppm
Bis(2-ethylhexyl) phthalate	[Di-*sec*-octyl phthalate; DEHP]	$C_{24}H_{38}O_4$	117-81-7	5 mg/m^3	
Bismuth telluride	[Tetradymite]	Bi_2Te_3	1304-82-1	10 mg/m^3	
Boric acid	and inorganic borate compounds	H_3BO_3	10043-35-3	2 mg/m^3	6 mg/m^3
Boron oxide	[Boric oxide]	B_2O_3	1303-86-2	10 mg/m^3	
Boron tribromide	[Tribromoborane]	BBr_3	10294-33-4		C 1 ppm
Boron trifluoride	[Trifluoroborane]	BF_3	7637-07-2		C 1 ppm
Bromacil	[5-Bromo-3-*sec*-butyl-6-methyluracil]	$C_9H_{13}BrN_2O_2$	314-40-9	10 mg/m^3	
Bromine		Br_2	7726-95-6	0.1 ppm	0.2 ppm
Bromine pentafluoride		BrF_5	7789-30-2	0.1 ppm	
Bromochloromethane	[Halon 1011]	CH_2BrCl	74-97-5	200 ppm	
2-Bromo-2-chloro-1,1,1-trifluoroethane	[Halothane]	$C_2HBrClF_3$	151-67-7	50 ppm	
Bromoethane	[Ethyl bromide]	C_2H_5Br	74-96-4	5 ppm	
Bromoethene	[Vinyl bromide]	C_2H_3Br	593-60-2	0.5 ppm	
Bromomethane	[Methyl bromide]	CH_3Br	74-83-9	1 ppm	
1-Bromopropane	[Propyl bromide]	C_3H_7Br	106-94-5	10 ppm	
Bromotrifluoromethane	[Trifluorobromomethane]	$CBrF_3$	75-63-8	1000 ppm	
1,3-Butadiene	[Divinyl]	C_4H_6	106-99-0	2 ppm	
Butane		C_4H_{10}	106-97-8	1000 ppm	
1-Butanethiol	[Butyl mercaptan]	$C_4H_{10}S$	109-79-5	0.5 ppm	
1-Butanol	[Butyl alcohol]	$C_4H_{10}O$	71-36-3	20 ppm	

Substance	Notes	Formula	CAS Reg. No.	Time-weighted average	Short-term exposure limit
2-Butanol	[*sec*-Butyl alcohol]	$C_4H_{10}O$	78-92-2	100 ppm	
2-Butanone	[Methyl ethyl ketone; MEK]	C_4H_8O	78-93-3	200 ppm	300 ppm
2-Butanone peroxide	[Methyl ethyl ketone peroxide]	$C_8H_{16}O_4$	1338-23-4		C 0.2 ppm
trans-2-Butenal	[Crotonaldehyde]	C_4H_6O	4170-30-3		C 0.3 ppm
1-Butene	[1-Butylene]	C_4H_8	106-98-9	250 ppm	
cis-2-Butene		C_4H_8	590-18-1	250 ppm	
trans-2-Butene		C_4H_8	624-64-6	250 ppm	
3-Buten-2-one	[Methyl vinyl ketone]	C_4H_6O	78-94-4		C 0.2 ppm
2-Butoxyethanol	[EGBE]	$C_6H_{14}O_2$	111-76-2	20 ppm	
2-Butoxyethyl acetate	[EGBEA]	$C_8H_{16}O_3$	112-07-2	20 ppm	
Butyl acetate		$C_6H_{12}O_2$	123-86-4	150 ppm	200 ppm
sec-Butyl acetate	[1-Methylpropyl acetate]	$C_6H_{12}O_2$	105-46-4	200 ppm	
tert-Butyl acetate		$C_6H_{12}O_2$	540-88-5	200 ppm	
Butyl acrylate	[Butyl 2-propenoate]	$C_7H_{12}O_2$	141-32-2	2 ppm	
Butylamine	[1-Butanamine]	$C_4H_{11}N$	109-73-9		C 5 ppm
tert-Butyl chromate	as CrO_3	$C_8H_{18}CrO_4$	1189-85-1		C 0.1 mg/m^3
tert-Butyl ethyl ether	[ETBE]	$C_6H_{14}O$	637-92-3	5 ppm	
Butyl glycidyl ether	[BGE]	$C_7H_{14}O_2$	2426-08-6	3 ppm	
Butyl lactate		$C_7H_{14}O_3$	138-22-7	5 ppm	
1-*tert*-Butyl-4-methylbenzene	[*p*-*tert*-Butyltoluene]	$C_{11}H_{16}$	98-51-1	1 ppm	
2-*sec*-Butylphenol		$C_{10}H_{14}O$	89-72-5	5 ppm	
Cadmium	metal	Cd	7440-43-9	0.01 mg/m^3	
Cadmium	compounds, as Cd	Cd	7440-43-9	0.002 mg/m^3	
Calcium chromate	as Cr	$CaCrO_4$	13765-19-0	0.001 mg/m^3	
Calcium cyanamide	[Calcium carbimide]	$CaCN_2$	156-62-7	0.5 mg/m^3	
Calcium hydroxide	[Portlandite]	$Ca(OH)_2$	1305-62-0	5 mg/m^3	
Calcium metasilicate	synthetic, nonfibrous	$CaSiO_3$	1344-95-2	10 mg/m^3	
Calcium oxide	[Lime]	CaO	1305-78-8	2 mg/m^3	
Calcium sulfate	[Anhydrite]	$CaSO_4$	7778-18-9	10 mg/m^3	
Camphor	[1,7,7-Trimethylbicyclo[2.2.1]heptan-2-one], synthetic	$C_{10}H_{16}O$	76-22-2	2 ppm	3 ppm
Caprolactam	[Hexahydro-2-azepinone]	$C_6H_{11}NO$	105-60-2	5 mg/m^3	
Captafol		$C_{10}H_9Cl_4NO_2S$	2425-06-1	0.1 mg/m^3	
Captan		$C_9H_8Cl_3NO_2S$	133-06-2	5 mg/m^3	
Carbaryl		$C_{12}H_{11}NO_2$	63-25-2	0.5 mg/m^3	
Carbofuran	[7-Benzofuranol, 2,3-dihydro-2,2-dimethyl-, methylcarbamate]	$C_{12}H_{15}NO_3$	1563-66-2	0.1 mg/m^3	
Carbon black	[Carbon (amorphous)]	C	1333-86-4	3.5 mg/m^3	
Carbon dioxide		CO_2	124-38-9	5000 ppm	30,000 ppm
Carbon disulfide	[Carbon bisulfide]	CS_2	75-15-0	1 ppm	
Carbon monoxide		CO	630-08-0	25 ppm	
Carbonyl chloride	[Phosgene]	CCl_2O	75-44-5	0.1 ppm	
Carbonyl fluoride		CF_2O	353-50-4	2 ppm	5 ppm
Cellulose			9004-34-6	10 mg/m^3	
Cesium hydroxide		CsOH	21351-79-1	2 mg/m^3	
Chlordane	[1,2,4,5,6,7,8,8-Octachloro-2,3,3a,4,7,7a-hexahydro-4,7-methano-1*H*-indene]	$C_{10}H_6Cl_8$	57-74-9	0.5 mg/m^3	
o-Chlorinated diphenyl oxide			31242-93-0	0.5 mg/m^3	
Chlorine		Cl_2	7782-50-5	0.5 ppm	1 ppm
Chlorine dioxide		ClO_2	10049-04-4	0.1 ppm	0.3 ppm
Chlorine trifluoride		ClF_3	7790-91-2		C 0.1 ppm
Chloroacetaldehyde	[2-Chloro-1-ethanal]	C_2H_3ClO	107-20-0		C 1 ppm
Chloroacetic acid		$C_2H_3ClO_2$	79-11-8	0.5 ppm	
Chloroacetone		C_3H_5ClO	78-95-5		C 1 ppm
α-Chloroacetophenone	[ω-Chloroacetophenone]	C_8H_7ClO	532-27-4	0.05 ppm	

Substance	Notes	Formula	CAS Reg. No.	Time-weighted average	Short-term exposure limit
Chloroacetyl chloride		$C_2H_2Cl_2O$	79-04-9	0.05 ppm	0.15 ppm
Chlorobenzene	[Phenyl chloride]	C_6H_5Cl	108-90-7	10 ppm	
o-Chlorobenzylidene malononitrile		$C_{10}H_5ClN_2$	2698-41-1		C 0.05 ppm
2-Chloro-1,3-butadiene	[Chloroprene]	C_4H_5Cl	126-99-8	10 ppm	
Chlorodifluoromethane	[HCFC-22]	$CHClF_2$	75-45-6	1000 ppm	
Chloroethane	[Ethyl chloride]	C_2H_5Cl	75-00-3	100 ppm	
2-Chloroethanol	[Ethylene chlorohydrin]	C_2H_5ClO	107-07-3		C 1 ppm
Chloroethene	[Vinyl chloride]	C_2H_3Cl	75-01-4	1 ppm	
Chloromethane	[Methyl chloride]	CH_3Cl	74-87-3	50 ppm	100 ppm
(Chloromethyl)benzene	[Benzyl chloride]	C_7H_7Cl	100-44-7	1 ppm	
Chloromethyl methyl ether	levels as low as possible	C_2H_5ClO	107-30-2		
1-Chloro-4-nitrobenzene	[*p*-Nitrochlorobenzene]	$C_6H_4ClNO_2$	100-00-5	0.1 ppm	
1-Chloro-1-nitropropane		$C_3H_6ClNO_2$	600-25-9	2 ppm	
Chloropentafluoroethane	[CFC-115]	C_2ClF_5	76-15-3	1000 ppm	
2-Chloropropanoic acid	[2-Chloropropionic acid]	$C_3H_5ClO_2$	598-78-7	0.1 ppm	
2-Chloro-1-propanol	[Propylene chlorohydrin]	C_3H_7ClO	78-89-7	1 ppm	
1-Chloro-2-propanol	[*sec*-Propylene chlorohydrin]	C_3H_7ClO	127-00-4	1 ppm	
3-Chloropropene	[Allyl chloride]	C_3H_5Cl	107-05-1	1 ppm	2 ppm
2-Chlorostyrene		C_8H_7Cl	2039-87-4	50 ppm	75 ppm
2-Chlorotoluene	[1-Chloro-2-methylbenzene]	C_7H_7Cl	95-49-8	50 ppm	
Chlorpyrifos	[Phosphorothioic acid, *O,O*-diethyl *O*-(3,5,6-trichloro-2-pyridinyl) ester]	$C_9H_{11}Cl_3NO_3PS$	2921-88-2	0.1 mg/m^3	
Chromium	metal	Cr	7440-47-3	0.5 mg/m^3	
Chromium	Cr(III) compounds, as Cr	Cr	7440-47-3	0.5 mg/m^3	
Chromium	soluble Cr(VI) compounds, as Cr	Cr	7440-47-3	0.05 mg/m^3	
Chromium	insoluble Cr(VI) compounds, as Cr	Cr	7440-47-3	0.01 mg/m^3	
Chromium(VI) dichloride dioxide	[Chromyl chloride]	CrO_2Cl_2	14977-61-8	0.025 ppm	
Chrysene	levels as low as possible	$C_{18}H_{12}$	218-01-9		
Clopidol		$C_7H_7Cl_2NO$	2971-90-6	10 mg/m^3	
Coal	dust, anthracite			0.4 mg/m^3	
Coal	dust, bituminous			0.9 mg/m^3	
Coal tar	volatiles, as aerosol		65966-93-2	0.2 mg/m^3	
Cobalt	metal and inorganic compounds, as Co	Co	7440-48-4	0.02 mg/m^3	
Cobalt carbonyl	as Co	$Co_2(CO)_8$	10210-68-1	0.1 mg/m^3	
Cobalt hydrocarbonyl	as Co	C_4HCoO_4	16842-03-8	0.1 mg/m^3	
Copper	fume	Cu	7440-50-8	0.2 mg/m^3	
Copper	dusts & mists, as Cu	Cu	7440-50-8	1 mg/m^3	
Coumaphos		$C_{14}H_{16}ClO_5PS$	56-72-4	0.05 mg/m^3	
Cresol	all isomers. NIC in 2008 to 20 mg/m^3	C_7H_8O	1319-77-3	5 ppm	
Crufomate		$C_{12}H_{19}ClNO_3P$	299-86-5	5 mg/m^3	
Cyanamide	[Cyanogenamide]	CH_2N_2	420-04-2	2 mg/m^3	
Cyanide ion	cyanide salts, as CN	CN^-	57-12-5		C 5 mg/m^3
Cyanogen		C_2N_2	460-19-5	10 ppm	
Cyanogen chloride	[Chlorine cyanide]	CClN	506-77-4		C 0.3 ppm
Cyclohexane	[Hexahydrobenzene]	C_6H_{12}	110-82-7	100 ppm	
Cyclohexanol	[Cyclohexyl alcohol]	$C_6H_{12}O$	108-93-0	50 ppm	
Cyclohexanone	[Pimelic ketone]	$C_6H_{10}O$	108-94-1	20 ppm	50 ppm
Cyclohexene	[Tetrahydrobenzene]	C_6H_{10}	110-83-8	300 ppm	
Cyclohexylamine	[Cyclohexanamine]	$C_6H_{13}N$	108-91-8	10 ppm	
Cyclonite	[Hexahydro-1,3,5-trinitro-1,3,5-triazine]	$C_3H_6N_6O_6$	121-82-4	0.5 mg/m^3	
1,3-Cyclopentadiene	[Pyropentylene]	C_5H_6	542-92-7	75 ppm	
Cyclopentane	[Pentamethylene]	C_5H_{10}	287-92-3	600 ppm	
Cyhexatin	[Tricyclohexylhydroxystannane]	$C_{18}H_{34}OSn$	13121-70-5	5 mg/m^3	

Substance	Notes	Formula	CAS Reg. No.	Time-weighted average	Short-term exposure limit
Decaborane(14)		$B_{10}H_{14}$	17702-41-9	0.05 ppm	0.15 ppm
Demeton	[Systox]	$C_8H_{19}O_3PS_2$	8065-48-3	0.05 mg/m^3	
Demeton-*S*-methyl	[Phosphorothioic acid, *S*-[2-(ethylthio) ethyl] *O,O*-dimethyl ester]	$C_6H_{15}O_3PS_2$	919-86-8	0.05 mg/m^3	
Diacetone alcohol	[4-Hydroxy-4-methyl-2-pentanone]	$C_6H_{12}O_2$	123-42-2	50 ppm	
4,4'-Diaminodiphenylmethane	[4,4-Methylene dianiline]	$C_{13}H_{14}N_2$	101-77-9	0.1 ppm	
Diazinon		$C_{12}H_{21}N_2O_3PS$	333-41-5	0.01 mg/m^3	
Diazomethane		CH_2N_2	334-88-3	0.2 ppm	
Diborane		B_2H_6	19287-45-7	0.1 ppm	
Dibromodifluoromethane		CBr_2F_2	75-61-6	100 ppm	
2-Dibutylaminoethanol		$C_{10}H_{23}NO$	102-81-8	0.5 ppm	
2,6-Di-*tert*-butyl-4-methylphenol	[Butylated hydroxytoluene; BHT]	$C_{15}H_{24}O$	128-37-0	2 mg/m^3	
Dibutylphenyl phosphate		$C_{14}H_{23}O_4P$	2528-36-1	0.3 ppm	
Dibutyl phosphate	NIC in 2008 to TWA 5 mg/m^3	$C_8H_{19}O_4P$	107-66-4	1 ppm	2 ppm
Dibutyl phthalate	[Butyl phthalate]	$C_{16}H_{22}O_4$	84-74-2	5 mg/m^3	
Dichloroacetic acid		$C_2H_2Cl_2O_2$	79-43-6	0.5 ppm	
Dichloroacetylene		C_2Cl_2	7572-29-4		C 0.1 ppm
o-Dichlorobenzene	[1,2-Dichlorobenzene]	$C_6H_4Cl_2$	95-50-1	25 ppm	50 ppm
p-Dichlorobenzene	[1,4-Dichlorobenzene]	$C_6H_4Cl_2$	106-46-7	10 ppm	
3,3'-Dichloro-*p*-benzidine	levels as low as possible	$C_{12}H_{10}Cl_2N_2$	91-94-1		
1,4-Dichloro-2-butene	both isomers	$C_4H_6Cl_2$	764-41-0	0.005 ppm	
Dichlorodifluoromethane	[CFC-12]	CCl_2F_2	75-71-8	1000 ppm	
1,3-Dichloro-5,5-dimethyl hydantoin		$C_5H_6Cl_2N_2O_2$	118-52-5	0.2 mg/m^3	0.4 mg/m^3
Dichlorodiphenyltrichloroethane	[DDT]	$C_{14}H_9Cl_5$	50-29-3	1 mg/m^3	
1,1-Dichloroethane	[Ethylidene dichloride]	$C_2H_4Cl_2$	75-34-3	100 ppm	
1,2-Dichloroethane	[Ethylene dichloride]	$C_2H_4Cl_2$	107-06-2	10 ppm	
1,1-Dichloroethene	[Vinylidene chloride]	$C_2H_2Cl_2$	75-35-4	5 ppm	
1,2-Dichloroethylene	both isomers	$C_2H_2Cl_2$	540-59-0	200 ppm	
Dichlorofluoromethane	[CFC 21]	$CHCl_2F$	75-43-4	10 ppm	
Dichloromethane	[Methylene chloride]	CH_2Cl_2	75-09-2	50 ppm	
1,1-Dichloro-1-nitroethane	[Ethide]	$C_2H_3Cl_2NO_2$	594-72-9	2 ppm	
(2,4-Dichlorophenoxy)acetic acid	[2,4-D]	$C_8H_6Cl_2O_3$	94-75-7	10 mg/m^3	
1,2-Dichloropropane	[Propylene dichloride]	$C_3H_6Cl_2$	78-87-5	10 ppm	
2,2-Dichloropropanoic acid	[2,2-Dichloropropionic acid]	$C_3H_4Cl_2O_2$	75-99-0	5 mg/m^3	
1,3-Dichloropropene	both isomers	$C_3H_4Cl_2$	542-75-6	1 ppm	
1,2-Dichloro-1,1,2,2-tetrafluoroethane	[CFC-114]	$C_2Cl_2F_4$	76-14-2	1000 ppm	
Dichlorvos	[Phosphoric acid, 2,2-dichloroethenyl dimethyl ester]	$C_4H_7Cl_2O_4P$	62-73-7	0.1 mg/m^3	
Dicrotophos		$C_8H_{16}NO_5P$	141-66-2	0.05 mg/m^3	
m-Dicyanobenzene	[*m*-Phthalodinitrile]	$C_8H_4N_2$	626-17-5	5 mg/m^3	
Dicyclopentadiene		$C_{10}H_{12}$	77-73-6	5 ppm	
Dieldrin		$C_{12}H_8Cl_6O$	60-57-1	0.25 mg/m^3	
Diesel fuel	as total hydrocarbons		68334-30-5	100 mg/m^3	
Diethanolamine	[Bis(2-hydroxyethyl)amine]. NIC in 2008 to 1 mg/m^3.	$C_4H_{11}NO_2$	111-42-2	2 mg/m^3	
Diethylamine	[*N*-Ethylethanamine]	$C_4H_{11}N$	109-89-7	5 ppm	15 ppm
2-Diethylaminoethanol		$C_6H_{15}NO$	100-37-8	2 ppm	
Diethyl ether	[Ethyl ether]	$C_4H_{10}O$	60-29-7	400 ppm	500 ppm
Diethyl phthalate		$C_{12}H_{14}O_4$	84-66-2	5 mg/m^3	
1,1-Difluoroethene	[Vinylidene fluoride]	$C_2H_2F_2$	75-38-7	500 ppm	
Diglycidyl ether	[DGE]	$C_6H_{10}O_3$	2238-07-5	0.1 ppm	
Diisopropylamine	[*N*-Isopropyl-2-propanamine]	$C_6H_{15}N$	108-18-9	5 ppm	
Diisopropyl ether	[Isopropyl ether]	$C_6H_{14}O$	108-20-3	250 ppm	310 ppm

Substance	Notes	Formula	CAS Reg. No.	Time-weighted average	Short-term exposure limit
Dimethoxymethane	[Methylal]	$C_3H_8O_2$	109-87-5	1000 ppm	
N,N-Dimethylacetamide	[*N,N*-Dimethylethanamide]	C_4H_9NO	127-19-5	10 ppm	
Dimethylamine	[*N*-Methylmethanamine]	C_2H_7N	124-40-3	5 ppm	15 ppm
N,N-Dimethylaniline		$C_8H_{11}N$	121-69-7	5 ppm	
Dimethylaniline	other isomers	$C_8H_{11}N$	1300-73-8	0.5 ppm	
2,2-Dimethylbutane	[Neohexane]	C_6H_{14}	75-83-2	500 ppm	1000 ppm
2,3-Dimethylbutane	[Diisopropyl]	C_6H_{14}	79-29-8	500 ppm	1000 ppm
Dimethylcarbamic chloride	[Dimethyl carbamoyl chloride]	C_3H_6ClNO	79-44-7	0.005 mg/m³	
Dimethyl disulfide	[Methyl disulfide]	$C_2H_6S_2$	624-92-0	0.5 mg/m³	
N,N-Dimethylformamide	[DMF]	C_3H_7NO	68-12-2	10 ppm	
2,6-Dimethyl-4-heptanone	[Diisobutyl ketone]	$C_9H_{18}O$	108-83-8	25 ppm	
1,1-Dimethylhydrazine	[UDMH]	$C_2H_8N_2$	57-14-7	0.01 ppm	
Dimethyl mercury	[Mercury dimethyl]	C_2H_6Hg	593-74-8	0.01 mg/m³	
Dimethyl phthalate	[Methyl phthalate]	$C_{10}H_{10}O_4$	131-11-3	5 mg/m³	
2,2-Dimethyl-1-propanol acetate		$C_7H_{14}O_2$	926-41-0	50 ppm	100 ppm
Dimethyl sulfate		$C_2H_6O_4S$	77-78-1	0.1 ppm	
Dimethyl sulfide	[2-Thiapropane]	C_2H_6S	75-18-3	10 ppm	
Dinitrobenzene	all isomers	$C_6H_4N_2O_4$	25154-54-5	0.15 ppm	
Dinitrotoluene	all isomers	$C_7H_6N_2O_4$	25321-14-6	0.2 mg/m³	
1,4-Dioxane	[1,4-Dioxacyclohexane]	$C_4H_8O_2$	123-91-1	20 ppm	
Dioxathion		$C_{12}H_{26}O_6P_2S_4$	78-34-2	0.1 mg/m³	
1,3-Dioxolane	[1,3-Dioxacyclopentane]	$C_3H_6O_2$	646-06-0	20 ppm	
Diphenylamine	[*N*-Phenylbenzenamine]	$C_{12}H_{11}N$	122-39-4	10 mg/m³	
Diphenyl ether	[Phenyl ether]	$C_{12}H_{10}O$	101-84-8	1 ppm	2 ppm
4,4′-Diphenylmethane diisocyanate	[Methylene diphenyl isocyanate; MDI]	$C_{15}H_{10}N_2O_2$	101-68-8	0.005 ppm	
Dipropylene glycol monomethyl ether	[Bis(2-Methoxypropyl) ether; DPGME]	$C_7H_{16}O_3$	34590-94-8	100 ppm	150 ppm
Diquat		$C_{12}H_{12}N_2$	231-36-7	0.5 mg/m³	
Disulfiram		$C_{10}H_{20}N_2S_4$	97-77-8	2 mg/m³	
Disulfoton	[Phosphorodithioic acid, *O,O*-diethyl *S*-[2-(ethylthio)ethyl] ester]	$C_8H_{19}O_2PS_3$	298-04-4	0.05 mg/m³	
Diuron		$C_9H_{10}Cl_2N_2O$	330-54-1	10 mg/m³	
Divinyl benzene	all isomers	$C_{10}H_{10}$	1321-74-0	10 ppm	
1-Dodecanethiol		$C_{12}H_{26}S$	112-55-0	0.1 ppm	
Endosulfan		$C_9H_6Cl_6O_3S$	115-29-7	0.1 mg/m³	
Endrin		$C_{12}H_8Cl_6O$	72-20-8	0.1 mg/m³	
Enflurane		$C_3H_2ClF_5O$	13838-16-9	75 ppm	
Epichlorohydrin	[(Chloromethyl)oxirane]	C_3H_5ClO	106-89-8	0.5 ppm	
1,2-Epoxy-4-(epoxyethyl) cyclohexane	[Vinylcyclohexene dioxide]	$C_8H_{12}O_2$	106-87-6	0.1 ppm	
Ethane		C_2H_6	74-84-0	1000 ppm	
1,2-Ethanediamine	[Ethylenediamine]	$C_2H_8N_2$	107-15-3	10 ppm	
1,2-Ethanediol	[Ethylene glycol]	$C_2H_6O_2$	107-21-1		C 100 mg/m³
1,2-Ethanediol, dinitrate	[Ethylene glycol dinitrate; EGDN]	$C_2H_4N_2O_6$	628-96-6	0.05 ppm	
Ethanethiol	[Ethyl mercaptan]	C_2H_6S	75-08-1	0.5 ppm	
Ethanol	[Ethyl alcohol]	C_2H_6O	64-17-5	1000 ppm	
Ethanolamine	[2-Aminoethanol]	C_2H_7NO	141-43-5	3 ppm	6 ppm
Ethion	[Phosphorodithioic acid, *S,S*′-methylene *O,O,O′,O′*-tetraethyl ester]	$C_9H_{22}O_4P_2S_4$	563-12-2	0.05 mg/m³	
Ethoxydimethylsilane	[Dimethylethoxysilane]	$C_4H_{12}OSi$	14857-34-2	0.5 ppm	1.5 ppm
2-Ethoxyethanol	[Ethylene glycol monoethyl ether; EGEE]	$C_4H_{10}O_2$	110-80-5	5 ppm	
2-Ethoxyethyl acetate	[Ethylene glycol monoethyl ether acetate; EGEEA]	$C_6H_{12}O_3$	111-15-9	5 ppm	
Ethyl acetate		$C_4H_8O_2$	141-78-6	400 ppm	
Ethyl acrylate	[Ethyl propenoate]	$C_5H_8O_2$	140-88-5	5 ppm	15 ppm

Substance	Notes	Formula	CAS Reg. No.	Time-weighted average	Short-term exposure limit
Ethylamine	[Ethanamine]	C_2H_7N	75-04-7	5 ppm	15 ppm
Ethylbenzene	[Phenylethane]	C_8H_{10}	100-41-4	100 ppm	125 ppm
Ethyl 2-cyanoacrylate	[Ethyl 2-cyano-2-propenoate]	$C_6H_7NO_2$	7085-85-0	0.2 ppm	
Ethylene	[Ethene]	C_2H_4	74-85-1	200 ppm	
Ethyleneimine	[Aziridine]	C_2H_5N	151-56-4	0.5 ppm	
Ethyl formate		$C_3H_6O_2$	109-94-4	100 ppm	
2-Ethylhexanoic acid		$C_8H_{16}O_2$	149-57-5	5 mg/m³	
5-Ethylidene-2-norbornene	[5-Ethylidenebicyclo[2.2.1]hept-2-ene]	C_9H_{12}	16219-75-3		C 5 ppm
N-Ethylmorpholine		$C_6H_{13}NO$	100-74-3	5 ppm	
O-Ethyl *O*-*p*-nitrophenyl benzenethiophosphonate	[EPN]	$C_{14}H_{14}NO_4PS$	2104-64-5	0.1 mg/m³	
Ethyl silicate		$C_8H_{20}O_4Si$	78-10-4	10 ppm	
Fenamiphos		$C_{13}H_{22}NO_3PS$	22224-92-6	0.05 mg/m³	
Fensulfothion	[Phosphorothioic acid, *O,O*-diethyl *O*-[4-(methylsulfinyl)phenyl] ester]	$C_{11}H_{17}O_4PS_2$	115-90-2	0.01 mg/m³	
Fenthion	[Phosphorothioic acid, *O,O*-dimethyl *O*-[3-methyl-4-(methylthio)phenyl] ester]	$C_{10}H_{15}O_3PS_2$	55-38-9	0.05 mg/m³	
Ferbam	[Iron, tris(dimethylcarbamodithioato-*S,S'*)-, (OC-6-11)-]	$C_9H_{18}FeN_3S_6$	14484-64-1	10 mg/m³	
Ferrocene	[Dicyclopentadienyl iron]	$C_{10}H_{10}Fe$	102-54-5	10 mg/m³	
Ferrovanadium	dust		12604-58-9	1 mg/m³	3 mg/m³
Flour	dust			0.5 mg/m³	
Fluoride ion	fluoride salts, as F	F^-	16984-48-8	2.5 mg/m³	
Fluorine		F_2	7782-41-4	1 ppm	2 ppm
Fluorine monoxide	[Oxygen difluoride]	F_2O	7783-41-7		C 0.05 ppm
Fluoroethene	[Vinyl fluoride]	C_2H_3F	75-02-5	1 ppm	
Fonofos	[Phosphonodithioic acid, ethyl-, *O*-ethyl *S*-phenyl ester]	$C_{10}H_{15}OPS_2$	944-22-9	0.01 mg/m³	
Formaldehyde	[Methanal]	CH_2O	50-00-0		C 0.3 ppm
Formamide	[Methanamide]	CH_3NO	75-12-7	10 ppm	
Formic acid	[Methanoic acid]	CH_2O_2	64-18-6	5 ppm	10 ppm
Furfural	[2-Furaldehyde]	$C_5H_4O_2$	98-01-1	2 ppm	
Furfuryl alcohol	[2-Furanmethanol]	$C_5H_6O_2$	98-00-0	10 ppm	15 ppm
Gallium arsenide		GaAs	1303-00-0	0.0003 mg/m³	
Gasoline			8006-61-9	300 ppm	500 ppm
Germane	[Germanium tetrahydride]	GeH_4	7782-65-2	0.2 ppm	
Glycerol	[1,2,3-Propanetriol], mist	$C_3H_8O_3$	56-81-5	10 mg/m³	
Glyoxal		$C_2H_2O_2$	107-22-2	0.1 mg/m³	
Grain	dust			4 mg/m³	
Graphite	except fibers	C	7440-44-0	2 mg/m³	
Hafnium	metal and compounds, as Hf	Hf	7440-58-6	0.5 mg/m³	
Heptachlor		$C_{10}H_5Cl_7$	76-44-8	0.05 mg/m³	
Heptachlor epoxide		$C_{10}H_5Cl_7O$	1024-57-3	0.05 mg/m³	
Heptane	all isomers	C_7H_{16}	142-82-5	400 ppm	500 ppm
2-Heptanone	[Methyl pentyl ketone]	$C_7H_{14}O$	110-43-0	50 ppm	
3-Heptanone	[Ethyl butyl ketone]	$C_7H_{14}O$	106-35-4	50 ppm	75 ppm
4-Heptanone	[Dipropyl ketone]	$C_7H_{14}O$	123-19-3	50 ppm	
Hexachlorobenzene	[Perchlorobenzene]	C_6Cl_6	118-74-1	0.002 mg/m³	
Hexachloro-1,3-butadiene	[Perchlorobutadiene]	C_4Cl_6	87-68-3	0.02 ppm	
1,2,3,4,5,6-Hexachlorocyclohexane	[Lindane]	$C_6H_6Cl_6$	58-89-9	0.5 mg/m³	
Hexachloro-1,3-cyclopentadiene	[Perchlorocyclopentadiene]	C_5Cl_6	77-47-4	0.01 ppm	
Hexachloroethane	[Perchloroethane]	C_2Cl_6	67-72-1	1 ppm	
Hexachloronaphthalene	all isomers	$C_{10}H_2Cl_6$	1335-87-1	0.2 mg/m³	

Substance	Notes	Formula	CAS Reg. No.	Time-weighted average	Short-term exposure limit
Hexahydro-1,3-isobenzofurandione	[Hexahydrophthalic anhydride]	$C_8H_{10}O_3$	85-42-7		C 0.005 mg/m³
Hexamethylene diisocyanate		$C_8H_{12}N_2O_2$	822-06-0	0.005 ppm	
Hexane		C_6H_{14}	110-54-3	50 ppm	
1,6-Hexanediamine	[Hexamethylenediamine]	$C_6H_{16}N_2$	124-09-4	0.5 ppm	
Hexanedinitrile	[Adiponitrile]	$C_6H_8N_2$	111-69-3	2 ppm	
1,6-Hexanedioic acid	[Adipic acid]	$C_6H_{10}O_4$	124-04-9	5 mg/m³	
2-Hexanone	[Butyl methyl ketone]	$C_6H_{12}O$	591-78-6	5 ppm	10 ppm
1-Hexene		C_6H_{12}	592-41-6	50 ppm	
sec-Hexyl acetate	[4-Methyl-2-pentyl acetate]	$C_8H_{16}O_2$	108-84-9	50 ppm	
Hydrazine		N_2H_4	302-01-2	0.01 ppm	
Hydrazoic acid	vapor	HN_3	7782-79-8		C 0.11 ppm
Hydrogen bromide	[Hydrobromic acid]	HBr	10035-10-6		C 2 ppm
Hydrogen chloride	[Hydrochloric acid]	HCl	7647-01-0		C 2 ppm
Hydrogen cyanide	[Hydrocyanic acid]	HCN	74-90-8		C 4.7 ppm
Hydrogen fluoride	[Hydrofluoric acid]	HF	7664-39-3	0.5 ppm	C 2 ppm
Hydrogen peroxide		H_2O_2	7722-84-1	1 ppm	
Hydrogen selenide		H_2Se	7783-07-5	0.05 ppm	
Hydrogen sulfide		H_2S	7783-06-4	10 ppm	15 ppm
p-Hydroquinone	[1,4-Benzenediol]	$C_6H_6O_2$	123-31-9	1 mg/m³	
2-Hydroxypropyl acrylate		$C_6H_{10}O_3$	999-61-1	0.5 ppm	
Indene	[Indonaphthene]	C_9H_8	95-13-6	5 ppm	
Indium	metal and compounds, as In	In	7440-74-6	0.1 mg/m³	
Iodine	and volatile iodides	I_2	7553-56-2	0.01 ppm	C 0.1 ppm
Iodomethane	[Methyl iodide]	CH_3I	74-88-4	2 ppm	
Iron ion [Fe^{+2}]	soluble ferrous salts, as Fe	Fe^{+2}	15438-31-0	1 mg/m³	
Iron ion [Fe^{+3}]	soluble ferric salts, as Fe	Fe^{+3}	20074-52-6	1 mg/m³	
Iron(III) oxide	dust and fume, as Fe	Fe_2O_3	1309-37-1	5 mg/m³	
Iron pentacarbonyl	as Fe	C_5FeO_5	13463-40-6	0.1 ppm	0.2 ppm
Isobutane	[2-Methylpropane]	C_4H_{10}	75-28-5	1000 ppm	
Isobutene	[2-Methyl-1-propene]	C_4H_8	115-11-7	250 ppm	
Isobutyl acetate	[2-Methylpropyl acetate]	$C_6H_{12}O_2$	110-19-0	150 ppm	
Isobutyl nitrite		$C_4H_9NO_2$	542-56-3		C 1 ppm
Isopentane	[2-Methylbutane]	C_5H_{12}	78-78-4	600 ppm	
Isopentyl acetate	[Isoamyl acetate]	$C_7H_{14}O_2$	123-92-2	50 ppm	100 ppm
Isophorone	[3,5,5-Trimethyl-2-cyclohexen-1-one]	$C_9H_{14}O$	78-59-1		C 5 ppm
Isophorone diisocyanate		$C_{12}H_{18}N_2O_2$	4098-71-9	0.005 ppm	
Isopropenylbenzene	[α-Methyl styrene]	C_9H_{10}	98-83-9	50 ppm	100 ppm
2-Isopropoxyethanol		$C_5H_{12}O_2$	109-59-1	25 ppm	
Isopropyl acetate	[1-Methylethyl acetate]	$C_5H_{10}O_2$	108-21-4	100 ppm	200 ppm
Isopropylamine	[2-Propanamine]	C_3H_9N	75-31-0	5 ppm	10 ppm
N-Isopropylaniline		$C_9H_{13}N$	768-52-5	2 ppm	
Isopropylbenzene	[Cumene]	C_9H_{12}	98-82-8	50 ppm	
Isopropyl glycidyl ether	[IGE]	$C_6H_{12}O_2$	4016-14-2	50 ppm	75 ppm
Kaolin			1332-58-7	2 mg/m³	
Kerosene			8008-20-6	200 mg/m³	
Ketene		C_2H_2O	463-51-4	0.5 ppm	1.5 ppm
Lead	metal and inorganic compounds, as Pb	Pb	7439-92-1	0.05 mg/m³	
Lead(II) arsenate		$Pb_3(AsO_4)_2$	3687-31-8	0.15 mg/m³	
Lead(II) chromate	as Pb	$PbCrO_4$	7758-97-6	0.05 mg/m³	
Lithium hydride		LiH	7580-67-8	0.025 mg/m³	
Magnesium oxide	[Magnesia]	MgO	1309-48-4	10 mg/m³	
Malathion		$C_{10}H_{19}O_6PS_2$	121-75-5	1 mg/m³	
Maleic anhydride		$C_4H_2O_3$	108-31-6	0.1 ppm	

Substance	Notes	Formula	CAS Reg. No.	Time-weighted average	Short-term exposure limit
Manganese	metal and inorganic compounds, as Mn	Mn	7439-96-5	0.2 mg/m^3	
Manganese cyclopentadienyl tricarbonyl	as Mn	$C_8H_5MnO_3$	12079-65-1	0.1 mg/m^3	
Manganese 2-methylcyclopentadienyl tricarbonyl	as Mn	$C_9H_7MnO_3$	12108-13-3	0.2 mg/m^3	
Mercury	metal and inorganic compounds, as Hg	Hg	7439-97-6	0.025 mg/m^3	
Mercury	alkyl compounds, as Hg	Hg	7439-97-6	0.01 mg/m^3	0.03 mg/m^3
Mercury	aryl compounds, as Hg	Hg	7439-97-6	0.1 mg/m^3	
Mesityl oxide	[Isobutenyl methyl ketone]	$C_6H_{10}O$	141-79-7	15 ppm	25 ppm
Methacrylic acid	[2-Methylpropenoic acid]	$C_4H_6O_2$	79-41-4	20 ppm	
Methane		CH_4	74-82-8	1000 ppm	
Methanethiol	[Methyl mercaptan]	CH_4S	74-93-1	0.5 ppm	
Methanol	[Methyl alcohol]	CH_4O	67-56-1	200 ppm	250 ppm
Methomyl	[Acetimidic acid, *N*-[(methylcarbamoyl) oxy]thio-, methyl ester]	$C_5H_{10}N_2O_2S$	16752-77-5	2.5 mg/m^3	
2-Methoxyaniline	[*o*-Anisidine]	C_7H_9NO	90-04-0	0.5 mg/m^3	
4-Methoxyaniline	[*p*-Anisidine]	C_7H_9NO	104-94-9	0.5 mg/m^3	
Methoxychlor		$C_{16}H_{15}Cl_3O_2$	72-43-5	10 mg/m^3	
2-Methoxyethanol	[Ethylene glycol monomethyl ether; EGME]	$C_3H_8O_2$	109-86-4	0.1 ppm	
2-Methoxyethyl acetate	[Ethylene glycol monomethyl ether acetate; EGMEA]	$C_5H_{10}O_3$	110-49-6	0.1 ppm	
2-Methoxy-2-methylbutane	[Methyl *tert*-pentyl ether; TAME]	$C_6H_{14}O$	994-05-8	20 ppm	
4-Methoxyphenol		$C_7H_8O_2$	150-76-5	5 mg/m^3	
1-Methoxy-2-propanol	[1,2-Propylene glycol monomethyl ether; PGME]	$C_4H_{10}O_2$	107-98-2	100 ppm	150 ppm
Methyl acetate		$C_3H_6O_2$	79-20-9	200 ppm	250 ppm
Methyl acrylate	[Methyl propenoate]	$C_4H_6O_2$	96-33-3	2 ppm	
2-Methylacrylonitrile	[2-Methylpropenenitrile]	C_4H_5N	126-98-7	1 ppm	
Methylamine	[Methanamine]	CH_5N	74-89-5	5 ppm	15 ppm
2-Methylaniline	[*o*-Toluidine]	C_7H_9N	95-53-4	2 ppm	
3-Methylaniline	[*m*-Toluidine]	C_7H_9N	108-44-1	2 ppm	
4-Methylaniline	[*p*-Toluidine]	C_7H_9N	106-49-0	2 ppm	
N-Methylaniline	[*N*-Methylbenzenamine]	C_7H_9N	100-61-8	0.5 ppm	
3-Methyl-1-butanol	[Isoamyl alcohol]	$C_5H_{12}O$	123-51-3	100 ppm	125 ppm
2-Methyl-1-butanol acetate		$C_7H_{14}O_2$	624-41-9	50 ppm	100 ppm
3-Methyl-2-butanol acetate		$C_7H_{14}O_2$	5343-96-4	50 ppm	100 ppm
3-Methyl-2-butanone	[Methyl isopropyl ketone]	$C_5H_{10}O$	563-80-4	200 ppm	
Methyl *tert*-butyl ether	[MTBE]	$C_5H_{12}O$	1634-04-4	50 ppm	
Methyl 2-cyanoacrylate	[Mecrylate]	$C_5H_5NO_2$	137-05-3	0.2 ppm	
Methylcyclohexane		C_7H_{14}	108-87-2	400 ppm	
Methylcyclohexanol	all isomers	$C_7H_{14}O$	25639-42-3	50 ppm	
2-Methylcyclohexanone		$C_7H_{12}O$	583-60-8	50 ppm	75 ppm
Methyl demeton		$C_6H_{15}O_3PS_2$	8022-00-2	0.05 mg/m^3	
2-Methyl-3,5-dinitrobenzamide	[3,5-Dinitro-*o*-toluamide; Dinitolmide]	$C_8H_7N_3O_5$	148-01-6	1 mg/m^3	
2-Methyl-4,6-dinitrophenol	[Dinitro-*o*-cresol]	$C_7H_6N_2O_5$	534-52-1	0.2 mg/m^3	
4,4'-Methylenebis[2-chloroaniline]	[MBOCA]	$C_{13}H_{12}Cl_2N_2$	101-14-4	0.01 ppm	
Methylenebis(4-cyclohexylisocyanate)		$C_{15}H_{22}N_2O_2$	5124-30-1	0.005 ppm	
Methyl formate		$C_2H_4O_2$	107-31-3	100 ppm	150 ppm
6-Methyl-1-heptanol	[Isooctyl alcohol]	$C_8H_{18}O$	26952-21-6	50 ppm	
5-Methyl-3-heptanone	[Ethyl amyl ketone]	$C_8H_{16}O$	541-85-5	10 ppm	
5-Methyl-3-heptanone	[Ethyl 2-methylbutyl ketone]	$C_8H_{16}O$	541-85-5	10 ppm	
5-Methyl-2-hexanone	[Methyl isopentyl ketone]	$C_7H_{14}O$	110-12-3	50 ppm	
Methylhydrazine		CH_6N_2	60-34-4	0.01 ppm	

Substance	Notes	Formula	CAS Reg. No.	Time-weighted average	Short-term exposure limit
Methyl isocyanate		C_2H_3NO	624-83-9	0.02 ppm	
Methyl methacrylate	[Methyl 2-methyl-2-propenoate]	$C_5H_8O_2$	80-62-6	50 ppm	100 ppm
1-Methylnaphthalene		$C_{11}H_{10}$	90-12-0	0.5 ppm	
2-Methylnaphthalene		$C_{11}H_{10}$	91-57-6	0.5 ppm	
2-Methyl-5-nitroaniline		$C_7H_8N_2O_2$	99-55-8	1 mg/m³	
Methyloxirane	[1,2-Propylene oxide]	C_3H_6O	75-56-9	2 ppm	
Methyl parathion		$C_8H_{10}NO_5PS$	298-00-0	0.2 mg/m³	
2-Methylpentane	[Isohexane]	C_6H_{14}	107-83-5	500 ppm	1000 ppm
3-Methylpentane		C_6H_{14}	96-14-0	500 ppm	1000 ppm
2-Methyl-2,4-pentanediol	[Hexylene glycol]	$C_6H_{14}O_2$	107-41-5		C 25 ppm
4-Methyl-2-pentanol	[Methyl isobutyl carbinol]	$C_6H_{14}O$	108-11-2	25 ppm	40 ppm
4-Methyl-2-pentanone	[Isobutyl methyl ketone]	$C_6H_{12}O$	108-10-1	50 ppm	75 ppm
2-Methyl-1-propanol	[Isobutyl alcohol]	$C_4H_{10}O$	78-83-1	50 ppm	
2-Methyl-2-propanol	[*tert*-Butyl alcohol]	$C_4H_{10}O$	75-65-0	100 ppm	
Methylstyrene	all isomers	C_9H_{10}	25013-15-4	50 ppm	100 ppm
N-Methyl-*N*,2,4,6-tetranitroaniline	[Tetryl]	$C_7H_5N_5O_8$	479-45-8	1.5 mg/m³	
Metribuzin		$C_8H_{14}N_4OS$	21087-64-9	5 mg/m³	
Mevinphos		$C_7H_{13}O_6P$	7786-34-7	0.01 ppm	
Mica			12001-26-2	3 mg/m³	
Mineral oil	mist			5 mg/m³	10 mg/m³
Molybdenum	metal and insoluble compounds, as Mo	Mo	7439-98-7	10 mg/m³	
Molybdenum	soluble compounds, as Mo	Mo	7439-98-7	0.5 mg/m³	
Monocrotophos	[*trans*-Dimethyl 1-methyl-3-(methylamino)-3-oxo-1-propenyl phosphate]	$C_7H_{14}NO_5P$	6923-22-4	0.05 mg/m³	
Morpholine	[Tetrahydro-1,4-oxazine]	C_4H_9NO	110-91-8	20 ppm	
Naled	[1,2-Dibromo-2,2-dichloroethylphosphoric acid, dimethyl ester]	$C_4H_7Br_2Cl_2O_4P$	300-76-5	0.1 mg/m³	
Naphtha			8030-30-6	400 ppm	
Naphthalene		$C_{10}H_8$	91-20-3	10 ppm	15 ppm
1-Naphthalenylthiourea	[ANTU]	$C_{11}H_{10}N_2S$	86-88-4	0.3 mg/m³	
2-Naphthylamine	levels as low as possible	$C_{10}H_9N$	91-59-8		
Neopentane	[2,2-Dimethylpropane]	C_5H_{12}	463-82-1	600 ppm	
Nickel	metal	Ni	7440-02-0	1.5 mg/m³	
Nickel	soluble compounds, as Ni	Ni	7440-02-0	0.1 mg/m³	
Nickel	insoluble compounds, as Ni	Ni	7440-02-0	0.2 mg/m³	
Nickel carbonyl	as Ni	C_4NiO_4	13463-39-3	0.05 ppm	
Nickel subsulfide	as Ni	Ni_3S_2	12035-72-2	0.1 mg/m³	
Nicotine	[3-(1-Methyl-2-pyrrolidinyl)pyridine, (*S*)-]	$C_{10}H_{14}N_2$	54-11-5	0.5 mg/m³	
Nitrapyrin	[Pyridine, 2-chloro-6-(trichloromethyl)-]	$C_6H_3Cl_4N$	1929-82-4	10 mg/m³	
Nitric acid		HNO_3	7697-37-2	2 ppm	4 ppm
Nitric oxide		NO	10102-43-9	25 ppm	
4-Nitroaniline		$C_6H_6N_2O_2$	100-01-6	3 mg/m³	
Nitrobenzene		$C_6H_5NO_2$	98-95-3	1 ppm	
4-Nitrobiphenyl	levels as low as possible	$C_{12}H_9NO_2$	92-93-3		
Nitroethane		$C_2H_5NO_2$	79-24-3	100 ppm	
Nitrogen dioxide		NO_2	10102-44-0	3 ppm	5 ppm
Nitrogen trifluoride		NF_3	7783-54-2	10 ppm	
Nitromethane	[Nitrocarbol]	CH_3NO_2	75-52-5	20 ppm	
1-Nitropropane		$C_3H_7NO_2$	108-03-2	25 ppm	
2-Nitropropane	[Isonitropropane]	$C_3H_7NO_2$	79-46-9	10 ppm	
N-Nitrosodimethylamine	levels as low as possible	$C_2H_6N_2O$	62-75-9		
Nitrotoluene	all isomers	$C_7H_7NO_2$	1321-12-6	2 ppm	

Substance	Notes	Formula	CAS Reg. No.	Time-weighted average	Short-term exposure limit
Nitrous oxide		N_2O	10024-97-2	50 ppm	
3,3,4,4,5,5,6,6,6,-Nonafluoro-1-hexene	[Perfluorobutylethene]	$C_6H_3F_9$	19430-93-4	100 ppm	
Nonane	all isomers	C_9H_{20}	111-84-2	200 ppm	
Octachloronaphthalene	[Perchloronaphthalene]	$C_{10}Cl_8$	2234-13-1	0.1 mg/m^3	0.3 mg/m^3
Octane	all isomers	C_8H_{18}	111-65-9	300 ppm	375 ppm
Osmium(VIII) oxide	[Osmium tetroxide]	OsO_4	20816-12-0	0.0002 ppm	0.0006 ppm
Oxalic acid		$C_2H_2O_4$	144-62-7	1 mg/m^3	2 mg/m^3
2-Oxetanone	[β-Propiolactone]	$C_3H_4O_2$	57-57-8	0.5 ppm	
Oxirane	[Ethylene oxide]	C_2H_4O	75-21-8	1 ppm	
Oxiranemethanol	[Glycidol]	$C_3H_6O_2$	556-52-5	2 ppm	
4,4′-Oxybis(benzenesulfonyl hydrazide)		$C_{12}H_{14}N_4O_5S_2$	80-51-3	0.1 mg/m^3	
Ozone	depends on workload	O_3	10028-15-6	0.1 ppm	
Paraquat		$C_{12}H_{14}N_2$	4685-14-7	0.5 mg/m^3	
Parathion		$C_{10}H_{14}NO_5PS$	56-38-2	0.05 mg/m^3	
Pentaborane(9)		B_5H_9	19624-22-7	0.005 ppm	0.015 ppm
Pentachloronaphthalene	all isomers	$C_{10}H_3Cl_5$	1321-64-8	0.5 mg/m^3	
Pentachloronitrobenzene	[Quintozene]	$C_6Cl_5NO_2$	82-68-8	0.5 mg/m^3	
Pentachlorophenol		C_6HCl_5O	87-86-5	0.5 mg/m^3	
Pentaerythritol		$C_5H_{12}O_4$	115-77-5	10 mg/m^3	
Pentanal	[Valeraldehyde]	$C_5H_{10}O$	110-62-3	50 ppm	
Pentane	all isomers	C_5H_{12}	109-66-0	600 ppm	750 ppm
Pentanedial	[Glutaraldehyde]	$C_5H_8O_2$	111-30-8		C 0.05 ppm
2-Pentanone	[Methyl propyl ketone]	$C_5H_{10}O$	107-87-9		150 ppm
3-Pentanone	[Diethyl ketone]	$C_5H_{10}O$	96-22-0	200 ppm	300 ppm
Pentyl acetate	[Amyl acetate]; all isomers	$C_7H_{14}O_2$	628-63-7	50 ppm	100 ppm
Perchloryl fluoride	[Chlorine trioxide fluoride]	ClO_3F	7616-94-6	3 ppm	6 ppm
Perfluoroacetone	[Hexafluoroacetone]	C_3F_6O	684-16-2	0.1 ppm	
Perfluoroisobutene	[Perfluoroisobutylene]	C_4F_8	382-21-8		C 0.01 ppm
Perfluoropropene	[Hexafluoropropene]	C_3F_6	116-15-4	0.1 mg/m^3	
Phenol	[Hydroxybenzene]	C_6H_6O	108-95-2	5 ppm	
10*H*-Phenothiazine	[Thiodiphenylamine]	$C_{12}H_9NS$	92-84-2	5 mg/m^3	
Phenyl glycidyl ether	[PGE]	$C_9H_{10}O_2$	122-60-1	0.1 ppm	
Phenylhydrazine		$C_6H_8N_2$	100-63-0	0.1 ppm	
N-Phenyl-2-naphthalenamine	levels as low as possible	$C_{16}H_{13}N$	135-88-6		
Phenylphosphine	[Monophenylphosphine]	C_6H_7P	638-21-1		C 0.05 ppm
Phorate	[Phosphorodithioic acid, *O,O*-diethyl *S*-[(ethylthio)methyl] ester]	$C_7H_{17}O_2PS_3$	298-02-2	0.05 mg/m^3	0.2 mg/m^3
Phosphine	[Phosphorus hydride]	PH_3	7803-51-2	0.3 ppm	1 ppm
Phosphoric acid	[Orthophosphoric acid]	H_3PO_4	7664-38-2	1 mg/m^3	3 mg/m^3
Phosphorus (yellow)	[White phosphorus]	P	7723-14-0	0.1 mg/m^3	
Phosphorus(III) chloride	[Phosphorus trichloride]	PCl_3	7719-12-2	0.2 ppm	0.5 ppm
Phosphorus(V) chloride	[Phosphorus pentachloride]	PCl_5	10026-13-8	0.1 ppm	
Phosphorus(V) sulfide	[Phosphorus pentasulfide]	P_2S_5	1314-80-3	1 mg/m^3	3 mg/m^3
Phosphoryl chloride	[Phosphorus(V) oxychloride]	$POCl_3$	10025-87-3	0.1 ppm	
Phthalic anhydride		$C_8H_4O_3$	85-44-9	1 ppm	
α-Pinene	[2-Pinene]	$C_{10}H_{16}$	80-56-8	20 ppm	
β-Pinene	[Nopinene]	$C_{10}H_{16}$	127-91-3	20 ppm	
Piperazine dihydrochloride	[Diethylenediamine dihydrochloride]	$C_4H_{12}Cl_2N_2$	142-64-3	5 mg/m^3	
2-Pivaloyl-1,3-indandione	[Pindone]	$C_{14}H_{14}O_3$	83-26-1	0.1 mg/m^3	
Platinum	metal dust	Pt	7440-06-4	1 mg/m^3	
Platinum	soluble salts, as Pt	Pt	7440-06-4	0.002 mg/m^3	
Polychlorinated biphenyls (42% chlorine)	[PCBs]		53469-21-9	1 mg/m^3	

Substance	Notes	Formula	CAS Reg. No.	Time-weighted average	Short-term exposure limit
Polychlorinated biphenyls (54% chlorine)	[PCBs]		11097-69-1	0.5 mg/m³	
Poly(vinyl chloride)	[PVC]		9002-86-2	1 mg/m³	
Portland cement			65997-15-1	10 mg/m³	
Potassium hydroxide		KOH	1310-58-3		C 2 mg/m³
Propanal	[Propionaldehyde]	C_3H_6O	123-38-6	20 ppm	
Propane	[LPG]	C_3H_8	74-98-6	1000 ppm	
1,3-Propane sultone	levels as low as possible	$C_3H_6O_3S$	1120-71-4		
Propanoic acid	[Propionic acid]	$C_3H_6O_2$	79-09-4	10 ppm	
1-Propanol	[Propyl alcohol]	C_3H_8O	71-23-8	100 ppm	400 ppm
2-Propanol	[Isopropyl alcohol]	C_3H_8O	67-63-0	200 ppm	400 ppm
Propargyl alcohol	[2-Propyn-1-ol]	C_3H_4O	107-19-7	1 ppm	
Propene	[Propylene]	C_3H_6	115-07-1	500 ppm	
Propoxur	[Phenol, 2-(1-methylethoxy)-, methylcarbamate]	$C_{11}H_{15}NO_3$	114-26-1	0.5 mg/m³	
Propyl acetate		$C_5H_{10}O_2$	109-60-4	200 ppm	250 ppm
1,2-Propylene glycol dinitrate		$C_3H_6N_2O_6$	6423-43-4	0.05 ppm	
Propyleneimine	[2-Methylaziridine]	C_3H_7N	75-55-8	2 ppm	
Propyl nitrate		$C_3H_7NO_3$	627-13-4	25 ppm	40 ppm
Propyne	[Methylacetylene]	C_3H_4	74-99-7	1000 ppm	
Pyrethrin I	[Pyrethrum]	$C_{21}H_{28}O_3$	8003-34-7	5 mg/m³	
2-Pyridinamine	[2-Aminopyridine]	$C_5H_6N_2$	504-29-0	0.5 ppm	
Pyridine	[Azine]	C_5H_5N	110-86-1	1 ppm	
Pyrocatechol	[Catechol]	$C_6H_6O_2$	120-80-9	5 ppm	
Resorcinol	[1,3-Benzenediol]	$C_6H_6O_2$	108-46-3	10 ppm	20 ppm
Rhodium	metal and insoluble compounds, as Rh	Rh	7440-16-6	1 mg/m³	
Rhodium	soluble compounds, as Rh	Rh	7440-16-6	0.01 mg/m³	
Ronnel		$C_8H_8Cl_3O_3PS$	299-84-3	5 mg/m³	
Rotenone		$C_{23}H_{22}O_6$	83-79-4	5 mg/m³	
Rubber	natural latex, as inhalable proteins			0.0001 mg/m³	
Selenium	element and compounds, as Se	Se	7782-49-2	0.2 mg/m³	
Selenium hexafluoride		SeF_6	7783-79-1	0.05 ppm	
Sesone	[Sodium 2-(2,4-dichlorophenoxy)ethyl sulfate]	$C_8H_7Cl_2NaO_5S$	136-78-7	10 mg/m³	
Silane	[Silicon tetrahydride]	SiH_4	7803-62-5	5 ppm	
Silicon carbide	lower limits for fibrous SiC	SiC	409-21-2	10 mg/m³	
Silicon dioxide (α-quartz)	[Quartz]	SiO_2	14808-60-7	0.025 mg/m³	
Silicon dioxide (cristobalite)	[Cristobalite]	SiO_2	14464-46-1	0.025 mg/m³	
Silver	metal	Ag	7440-22-4	0.1 mg/m³	
Silver	soluble compounds, as Ag	Ag	7440-22-4	0.01 mg/m³	
Sodium azide	[Smite]	NaN_3	26628-22-8		C 0.29 mg/m³
Sodium fluoroacetate		$C_2H_2FNaO_2$	62-74-8	0.05 mg/m³	
Sodium hydrogen sulfite	[Sodium bisulfite]	$NaHSO_3$	7631-90-5	5 mg/m³	
Sodium hydroxide	[Caustic soda]	NaOH	1310-73-2		C 2 mg/m³
Sodium metabisulfite	[Sodium pyrosulfite]	$Na_2S_2O_5$	7681-57-4	5 mg/m³	
Sodium tetraborate decahydrate	[Borax]	$Na_2B_4O_7 10H_2O$	1303-96-4	2 mg/m³	6 mg/m³
Starch			9005-25-8	10 mg/m³	
Stibine	[Antimony hydride]	SbH_3	7803-52-3	0.1 ppm	
Stoddard solvent			8052-41-3	100 ppm	
Strontium chromate	as Cr	$SrCrO_4$	7789-06-2	0.0005 mg/m³	
Strychnine		$C_{21}H_{22}N_2O_2$	57-24-9	0.15 mg/m³	
Styrene	[Vinylbenzene]	C_8H_8	100-42-5	20 ppm	40 ppm
Subtilisins	as crystalline active enzyme		9014-01-1		C 0.00006 mg/m³

Substance	Notes	Formula	CAS Reg. No.	Time-weighted average	Short-term exposure limit
Sucrose		$C_{12}H_{22}O_{11}$	57-50-1	10 mg/m³	
Sulfometuron methyl		$C_{15}H_{16}N_4O_5S$	74222-97-2	5 mg/m³	
Sulfotep	[Tetraethyl thiodiphosphate; TEDP]	$C_8H_{20}O_5P_2S_2$	3689-24-5	0.2 mg/m³	
Sulfur chloride	[Disulfur dichloride]	S_2Cl_2	10025-67-9		C 1 ppm
Sulfur decafluoride	[Sulfur pentafluoride]	S_2F_{10}	5714-22-7		C 0.01 ppm
Sulfur dioxide		SO_2	7446-09-5	2 ppm	5 ppm
Sulfur hexafluoride		SF_6	2551-62-4	1000 ppm	
Sulfuric acid	[Oil of vitriol]	H_2SO_4	7664-93-9	0.2 mg/m³	
Sulfur tetrafluoride		SF_4	7783-60-0		C 0.1 ppm
Sulfuryl fluoride		SO_2F_2	2699-79-8	5 ppm	10 ppm
Sulprofos		$C_{12}H_{19}O_2PS_3$	35400-43-2	1 mg/m³	
Talc		$3MgO4SiO_2H_2O$	14807-96-6	2 mg/m³	
Tantalum	dust	Ta	7440-25-7	5 mg/m³	
Tantalum(V) oxide	dust, as Ta	Ta_2O_5	1314-61-0	5 mg/m³	
Tellurium	and compounds, as Te (except H_2Te)	Te	13494-80-9	0.1 mg/m³	
Tellurium hexafluoride		TeF_6	7783-80-4	0.02 ppm	
Terbufos		$C_9H_{21}O_2PS_3$	13071-79-9	0.01 mg/m³	
Terephthalic acid	[1,4-Benzenedicarboxylic acid]	$C_8H_6O_4$	100-21-0	10 mg/m³	
Terphenyl	all isomers	$C_{18}H_{14}$	26140-60-3		5 mg/m³
1,1,2,2-Tetrabromoethane	[Acetylene tetrabromide]	$C_2H_2Br_4$	79-27-6	0.1 ppm	
Tetrabromomethane	[Carbon tetrabromide]	CBr_4	558-13-4	0.1 ppm	0.3 ppm
1,1,1,2-Tetrachloro-2,2-difluoroethane	[Tetrachloro-1,1-difluoroethane]	$C_2Cl_4F_2$	76-11-9	100 ppm	
1,1,2,2-Tetrachloro-1,2-difluoroethane	[Tetrachloro-1,2-difluoroethane]	$C_2Cl_4F_2$	76-12-0	50 ppm	
1,1,2,2-Tetrachloroethane	[Acetylene tetrachloride]	$C_2H_2Cl_4$	79-34-5	1 ppm	
Tetrachloroethene	[Perchloroethylene]	C_2Cl_4	127-18-4	25 ppm	100 ppm
Tetrachloromethane	[Carbon tetrachloride]	CCl_4	56-23-5	5 ppm	10 ppm
Tetrachloronaphthalene	all isomers	$C_{10}H_4Cl_4$	1335-88-2	2 mg/m³	
Tetraethyl lead	as Pb	$C_8H_{20}Pb$	78-00-2	0.1 mg/m³	
Tetraethyl pyrophosphate	[TEPP]	$C_8H_{20}O_7P_2$	107-49-3	0.01 mg/m³	
Tetrafluoroethene	[Tetrafluoroethylene]	C_2F_4	116-14-3	2 ppm	
Tetrahydrofuran	[Oxolane]	C_4H_8O	109-99-9	50 ppm	100 ppm
Tetramethyl lead	as Pb	$C_4H_{12}Pb$	75-74-1	0.15 mg/m³	
Tetramethyl silicate	[Methyl silicate]	$C_4H_{12}O_4Si$	681-84-5	1 ppm	
Tetramethylsuccinonitrile	[Tetramethylbutanedinitrile]	$C_8H_{12}N_2$	3333-52-6	0.5 ppm	
Tetranitromethane		CN_4O_8	509-14-8	0.005 ppm	
Thallium	and soluble compounds, as Tl	Tl	7440-28-0	0.1 mg/m³	
4,4′-Thiobis(6-*tert*-butyl-*m*-cresol)	[Bis(5-*tert*-butyl-4-hydroxy-2-methylphenyl) sulfide]	$C_{22}H_{30}O_2S$	96-69-5	10 mg/m³	
Thioglycolic acid		$C_2H_4O_2S$	68-11-1	1 ppm	
Thionyl chloride	[Sulfinyl dichloride]	$SOCl_2$	7719-09-7		C 1 ppm
Thiram		$C_6H_{12}N_2S_4$	137-26-8	0.05 mg/m³	
Tin	metal	Sn	7440-31-5	2 mg/m³	
Tin	inorganic compounds, as Sn	Sn	7440-31-5	2 mg/m³	
Tin	organic compounds, as Sn	Sn	7440-31-5	0.1 mg/m³	
Titanium(IV) oxide	[Titanium dioxide]	TiO_2	13463-67-7	10 mg/m³	
Toluene	[Methylbenzene]	C_7H_8	108-88-3	20 ppm	
Toluene-2,4-diisocyanate		$C_9H_6N_2O_2$	584-84-9	0.005 ppm	0.02 ppm
Toluene-2,6-diisocyanate		$C_9H_6N_2O_2$	91-08-7	0.005 mg/m³	0.02 ppm
Toxaphene	[Polychlorocamphene]	$C_{10}H_{10}Cl_8$	8001-35-2	0.5 mg/m³	
1*H*-1,2,4-Triazol-3-amine	[Amitrole]	$C_2H_4N_4$	61-82-5	0.2 mg/m³	
Tribromomethane	[Bromoform]	$CHBr_3$	75-25-2	0.5 ppm	
Tributyl phosphate	[Butyl phosphate]	$C_{12}H_{27}O_4P$	126-73-8	0.2 ppm	
Trichlorfon	[Trichlorphon]	$C_4H_8Cl_3O_4P$	52-68-6	1 mg/m³	

Substance	Notes	Formula	CAS Reg. No.	Time-weighted average	Short-term exposure limit
Trichloroacetic acid		$C_2HCl_3O_2$	76-03-9	1 ppm	
1,2,4-Trichlorobenzene		$C_6H_3Cl_3$	120-82-1		C 5 ppm
1,1,1-Trichloroethane	[Methyl chloroform]	$C_2H_3Cl_3$	71-55-6	350 ppm	450 ppm
1,1,2-Trichloroethane	[Vinyl trichloride]	$C_2H_3Cl_3$	79-00-5	10 ppm	
Trichloroethene	[Trichloroethylene]	C_2HCl_3	79-01-6	10 ppm	25 ppm
Trichlorofluoromethane	[CFC-11]	CCl_3F	75-69-4		C 1000 ppm
Trichloromethane	[Chloroform]	$CHCl_3$	67-66-3	10 ppm	
Trichloromethanesulfenyl chloride	[Perchloromethyl mercaptan]	CCl_4S	594-42-3	0.1 ppm	
(Trichloromethyl)benzene	[Benzotrichloride]	$C_7H_5Cl_3$	98-07-7		C 0.1 ppm
Trichloronaphthalene	all isomers	$C_{10}H_5Cl_3$	1321-65-9	5 mg/m³	
Trichloronitromethane	[Chloropicrin]	CCl_3NO_2	76-06-2	0.1 ppm	
2,4,5-Trichlorophenoxyacetic acid	[2,4,5-T]	$C_8H_5Cl_3O_3$	93-76-5	10 mg/m³	
1,2,3-Trichloropropane	[Allyl trichloride]	$C_3H_5Cl_3$	96-18-4	10 ppm	
1,1,2-Trichloro-1,2,2-trifluoroethane	[CFC-113]	$C_2Cl_3F_3$	76-13-1	1000 ppm	1250 ppm
Tri-*o*-cresyl phosphate	[Tri-*o*-tolyl phosphate]	$C_{21}H_{21}O_4P$	78-30-8	0.1 mg/m³	
Triethanolamine	[Tris(2-hydroxyethyl)amine]	$C_6H_{15}NO_3$	102-71-6	5 mg/m³	
Triethylamine	[*N*,*N*-Diethylethanamine]	$C_6H_{15}N$	121-44-8	1 ppm	3 ppm
1,3,5-Triglycidyl-*s*-triazinetrione	[1,3,5-Tris(oxiranemethyl)-1,3,5-triazine-2,4,6(1*H*,3*H*,5*H*)-trione]	$C_{12}H_{15}N_3O_6$	2451-62-9	0.05 mg/m³	
Triiodomethane	[Iodoform]	CHI_3	75-47-8	0.6 ppm	
Trimellitic anhydride	[1,2,4-Benzenetricarboxylic anhydride]	$C_9H_4O_5$	552-30-7	0.0005 mg/m³	C 0.002 mg/m³
Trimethylamine	[*N*,*N*-Dimethylmethanamine]	C_3H_9N	75-50-3	5 ppm	15 ppm
Trimethylbenzene	all isomers	C_9H_{12}	25551-13-7	25 ppm	
Trimethyl phosphite		$C_3H_9O_3P$	121-45-9	2 ppm	
Trinitroglycerol	[Nitroglycerin; NG]	$C_3H_5N_3O_9$	55-63-0	0.05 ppm	
2,4,6-Trinitrophenol	[Picric acid]	$C_6H_3N_3O_7$	88-89-1	0.1 mg/m³	
2,4,6-Trinitrotoluene	[TNT]	$C_7H_5N_3O_6$	118-96-7	0.1 mg/m³	
Triphenyl phosphate		$C_{18}H_{15}O_4P$	115-86-6	3 mg/m³	
Tungsten	metal and insoluble compounds, as W	W	7440-33-7	5 mg/m³	10 mg/m³
Tungsten	soluble compounds, as W	W	7440-33-7	1 mg/m³	3 mg/m³
Turpentine			8006-64-2	20 ppm	
Uranium	metal and compounds, as U	U	7440-61-1	0.2 mg/m³	0.6 mg/m³
Vanadium(V) oxide	dust or fume; [Vanadium pentoxide]	V_2O_5	1314-62-1	0.05 mg/m³	
Vinyl acetate		$C_4H_6O_2$	108-05-4	10 ppm	15 ppm
4-Vinylcyclohexene		C_8H_{12}	100-40-3	0.1 ppm	
1-Vinyl-2-pyrrolidinone		C_6H_9NO	88-12-0	0.05 ppm	
Warfarin	[Coumadin]	$C_{19}H_{16}O_4$	81-81-2	0.1 mg/m³	
Xylene	[Dimethylbenzene], all isomers	C_8H_{10}	1330-20-7	100 ppm	150 ppm
Xylidine	[Dimethylaniline], all isomers	$C_8H_{11}N$	1300-73-8	0.5 ppm	
Yttrium	metal and compounds, as Y	Y	7440-65-5	1 mg/m³	
Zinc chloride	fume	$ZnCl_2$	7646-85-7	1 mg/m³	2 mg/m³
Zinc chromate	as Cr	$ZnCrO_4$	13530-65-9	0.01 mg/m³	
Zinc oxide	[Zincite]	ZnO	1314-13-2	2 mg/m³	10 mg/m³
Zirconium	metal and compounds, as Zr	Zr	7440-67-7	5 mg/m³	10 mg/m³

OCTANOL–WATER PARTITION COEFFICIENTS

The octanol–water partition coefficient, *P*, is a widely used parameter for correlating biological effects of organic substances. It is a property of the two-phase system in which water and 1-octanol are in equilibrium at a fixed temperature and the substance is distributed between the water-rich and octanol-rich phases. *P* is defined as the ratio of the equilibrium concentration of the substance in the octanol-rich phase to that in the water-rich phase, in the limit of zero concentration. In general, *P* tends to be large for compounds with extended non-polar structures (such as long chain or multi-ring hydrocarbons) and small for compounds with highly polar groups. Thus *P* (or, in its more common form of expression, log *P*) provides a measure of the lipophilic vs. hydrophilic nature of a compound, which is an important consideration in assessing the potential toxicity. A discussion of methods of measurement and accuracy considerations for log *P* may be found in Reference 1.

This table gives selected values of log *P* for about 450 organic compounds, including many of environmental importance. All values refer to a nominal temperature of 25 °C. The source of each value is indicated in the last column. These references contain data on many more compounds than are included here.

Compounds are listed by molecular formula following the Hill convention. To locate a compound by name or CAS Registry Number when the molecular formula is not known, use the table "Physical Constants of Organic Compounds" in Section 3 and its indexes to determine the molecular formula.

References

1. Sangster, J., *J. Phys. Chem. Ref. Data*, 18, 1111, 1989.
2. Mackay, D., Shiu, W. Y., and Ma, K. C., *Illustrated Handbook of Physical-Chemical Properties and Environmental Fate for Organic Chemicals*, Lewis Publishers/CRC Press, Boca Raton, FL, 1992.
3. Shiu, W. Y., and Mackay, D., *J. Phys. Chem. Ref. Data*, 15, 911, 1986.
4. Pinsuwan, S., Li, L., and Yalkowsky, S. H., *J. Chem. Eng. Data*, 40, 623, 1995.
5. *Solubility Data Series, International Union of Pure and Applied Chemistry, Vol. 20*, Pergamon Press, Oxford, 1985.
6. *Solubility Data Series, International Union of Pure and Applied Chemistry, Vol. 38*, Pergamon Press, Oxford, 1985.
7. Miller, M. M., Ghodbane, S., Wasik, S. P., Tewari, Y. B., and Martire, D. E., *J. Chem. Eng. Data*, 29, 184, 1984.

Mol. form.	Name	log *P*	Ref.
CCl_2F_2	Dichlorodifluoromethane	2.16	2
CCl_3F	Trichlorofluoromethane	2.53	2
CCl_4	Tetrachloromethane	2.64	2
$CHBr_3$	Tribromomethane	2.38	2
$CHCl_3$	Trichloromethane	1.97	2
CH_2BrCl	Bromochloromethane	1.41	2
CH_2Br_2	Dibromomethane	2.3	2
CH_2Cl_2	Dichloromethane	1.25	2
CH_2F_2	Difluoromethane	0.20	1
CH_2I_2	Diiodomethane	2.5	2
CH_2O	Formaldehyde	0.35	1
CH_2O_2	Formic acid	−0.54	1
CH_3Br	Bromomethane	1.19	2
CH_3Cl	Chloromethane	0.91	2
CH_3F	Fluoromethane	0.51	1
CH_3I	Iodomethane	1.5	2
CH_3NO	Formamide	−1.51	1
CH_3NO_2	Nitromethane	−0.33	1
CH_4O	Methanol	−0.74	1
CH_5N	Methylamine	−0.57	1
$C_2Cl_3F_3$	1,1,2-Trichlorotrifluoroethane	3.16	2
C_2Cl_4	Tetrachloroethylene	2.88	2
C_2Cl_6	Hexachloroethane	4.00	4
C_2HCl_3	Trichloroethylene	2.53	2
C_2HCl_5	Pentachloroethane	2.89	2
$C_2H_2Cl_2$	1,1-Dichloroethylene	2.13	2
$C_2H_2Cl_2$	*cis*-1,2-Dichloroethylene	1.86	2
$C_2H_2Cl_2$	*trans*-1,2-Dichloroethylene	1.93	2
$C_2H_2Cl_4$	1,1,2,2-Tetrachloroethane	2.39	2
C_2H_3Cl	Chloroethylene	1.38	2
$C_2H_3Cl_3$	1,1,1-Trichloroethane	2.49	2
$C_2H_3Cl_3$	1,1,2-Trichloroethane	2.38	2
C_2H_3N	Acetonitrile	−0.34	1
$C_2H_4Cl_2$	1,1-Dichloroethane	1.79	2
$C_2H_4Cl_2$	1,2-Dichloroethane	1.48	2
C_2H_4O	Acetaldehyde	0.45	1
C_2H_4O	Ethylene oxide	−0.30	1
$C_2H_4O_2$	Acetic acid	−0.17	1
C_2H_5Br	Bromoethane	1.6	2
C_2H_5Cl	Chloroethane	1.43	2
C_2H_5I	Iodoethane	2	2
C_2H_5NO	Acetamide	−1.26	1
$C_2H_5NO_2$	Nitroethane	0.18	1
C_2H_6O	Ethanol	−0.30	1
C_2H_6O	Dimethyl ether	0.10	1
C_2H_6OS	Dimethyl sulfoxide	−1.35	1
$C_2H_6O_2S$	Dimethyl sulfone	−1.41	1
C_2H_7N	Ethylamine	−0.13	1
C_2H_7N	Dimethylamine	−0.38	1
C_3H_3N	2-Propenenitrile	0.25	1
$C_3H_4Cl_2$	*cis*-1,3-Dichloropropene	2.03	2
C_3H_4O	Propargyl alcohol	−0.38	1
C_3H_4O	Acrolein	0.01	1
C_3H_5Br	3-Bromopropene	1.79	1
C_3H_5ClO	Epichlorohydrin	0.30	2
$C_3H_5Cl_3$	1,2,3-Trichloropropane	2.63	2
C_3H_5N	Propanenitrile	0.16	1
C_3H_5NO	Acrylamide	−0.78	1
$C_3H_6Cl_2$	1,2-Dichloropropane	2.0	2
C_3H_6O	Allyl alcohol	0.17	1
C_3H_6O	Propanal	0.59	1
C_3H_6O	Acetone	−0.24	1
C_3H_6O	Methyloxirane	0.03	1
$C_3H_6O_2$	Propanoic acid	0.33	1
$C_3H_6O_2$	Methyl acetate	0.18	1
C_3H_7Br	1-Bromopropane	2.1	2
C_3H_7Br	2-Bromopropane	1.9	2
C_3H_7Cl	1-Chloropropane	2.04	1
C_3H_7Cl	2-Chloropropane	1.90	1
C_3H_7I	1-Iodopropane	2.5	2
C_3H_7N	Allylamine	0.03	1
C_3H_7NO	*N,N*-Dimethylformamide	−1.01	1
C_3H_7NO	*N*-Methylacetamide	−1.05	1
$C_3H_7NO_2$	1-Nitropropane	0.87	1

Mol. form.	Name	log *P*	Ref.
C_3H_8O	1-Propanol	0.25	1
C_3H_8O	2-Propanol	0.05	1
C_3H_8S	1-Propanethiol	1.81	1
C_3H_9N	Propylamine	0.48	1
C_3H_9N	Isopropylamine	0.26	1
C_3H_9N	Ethylmethylamine	0.15	1
C_3H_9N	Trimethylamine	0.16	1
C_4H_4O	Furan	1.34	1
C_4H_4S	Thiophene	1.81	1
C_4H_5N	Pyrrole	0.75	1
C_4H_6	1,3-Butadiene	1.99	1
C_4H_6	2-Butyne	1.46	1
C_4H_6O	2,5-Dihydrofuran	0.46	1
$C_4H_6O_2$	Methacrylic acid	0.93	1
$C_4H_6O_2$	Vinyl acetate	0.73	1
$C_4H_6O_2$	Methyl acrylate	0.80	1
C_4H_7N	Butanenitrile	0.60	1
C_4H_8	*cis*-2-Butene	2.33	1
C_4H_8	*trans*-2-Butene	2.31	1
C_4H_8	Isobutene	2.35	1
$C_4H_8Cl_2O$	Bis(2-chloroethyl) ether	1.12	2
C_4H_8O	Ethyl vinyl ether	1.04	1
C_4H_8O	Butanal	0.88	1
C_4H_8O	2-Butanone	0.29	1
C_4H_8O	Tetrahydrofuran	0.46	1
$C_4H_8O_2$	Butanoic acid	0.79	1
$C_4H_8O_2$	Propyl formate	0.83	1
$C_4H_8O_2$	Ethyl acetate	0.73	1
C_4H_9Br	1-Bromobutane	2.75	1
C_4H_9Cl	1-Chlorobutane	2.64	2
C_4H_9F	1-Fluorobutane	2.58	1
C_4H_9I	1-Iodobutane	3	2
C_4H_9N	Pyrrolidine	0.46	1
C_4H_9NO	Butanamide	−0.21	1
C_4H_9NO	*N,N*-Dimethylacetamide	−0.77	1
$C_4H_9NO_2$	1-Nitrobutane	1.47	1
C_4H_{10}	Isobutane	2.8	2
$C_4H_{10}O$	1-Butanol	0.84	1
$C_4H_{10}O$	2-Butanol	0.65	1
$C_4H_{10}O$	2-Methyl-1-propanol	0.76	1
$C_4H_{10}O$	2-Methyl-2-propanol	0.35	1
$C_4H_{10}O$	Diethyl ether	0.89	1
$C_4H_{10}S$	1-Butanethiol	2.28	1
$C_4H_{10}S$	Diethyl sulfide	1.95	1
$C_4H_{11}N$	Butylamine	0.86	1
$C_4H_{11}N$	*tert*-Butylamine	0.40	1
$C_4H_{11}N$	Diethylamine	0.58	1
C_5H_5N	Pyridine	0.65	1
C_5H_6O	2-Methylfuran	1.85	1
C_5H_7N	1-Methylpyrrole	1.21	1
C_5H_8	1,4-Pentadiene	2.48	1
C_5H_8	1-Pentyne	1.98	1
$C_5H_8O_2$	Methyl methacrylate	1.38	1
$C_5H_8O_2$	Ethyl acrylate	1.32	1
C_5H_9N	Pentanenitrile	0.94	1
C_5H_{10}	1-Pentene	2.2	2
C_5H_{10}	Cyclopentane	3.00	1
$C_5H_{10}O$	2-Pentanone	0.84	1
$C_5H_{10}O$	3-Pentanone	0.82	1
$C_5H_{10}O$	3-Methyl-2-butanone	0.56	1
$C_5H_{10}O$	Tetrahydropyran	0.82	1
$C_5H_{10}O$	2-Methyltetrahydrofuran	1.85	2
$C_5H_{10}O_2$	Pentanoic acid	1.39	1
$C_5H_{10}O_2$	Propyl acetate	1.24	1
$C_5H_{10}O_2$	Ethyl propanoate	1.21	1
$C_5H_{10}O_3$	Diethyl carbonate	1.21	1
$C_5H_{11}Br$	1-Bromopentane	3.37	1
$C_5H_{11}F$	1-Fluoropentane	2.33	1
$C_5H_{11}N$	Piperidine	0.84	1
$C_5H_{11}NO_2$	1-Nitropentane	2.01	1
C_5H_{12}	Pentane	3.45	1
C_5H_{12}	Neopentane	3.11	1
$C_5H_{12}O$	1-Pentanol	1.51	1
$C_5H_{12}O$	2-Pentanol	1.25	1
$C_5H_{12}O$	3-Pentanol	1.21	1
$C_5H_{12}O$	3-Methyl-1-butanol	1.28	1
$C_5H_{12}O$	2-Methyl-2-butanol	0.89	1
$C_5H_{12}O$	3-Methyl-2-butanol	1.28	1
$C_5H_{12}O$	2,2-Dimethyl-1-propanol	1.31	1
$C_5H_{12}O$	Methyl *tert*-butyl ether	0.94	1
$C_5H_{13}N$	Pentylamine	1.49	1
C_6Cl_6	Hexachlorobenzene	5.47	5
C_6HCl_5	Pentachlorobenzene	5.03	5
C_6HCl_5O	Pentachlorophenol	5.07	4
$C_6H_2Cl_4$	1,2,3,4-Tetrachlorobenzene	4.55	5
$C_6H_2Cl_4$	1,2,3,5-Tetrachlorobenzene	4.65	5
$C_6H_2Cl_4$	1,2,4,5-Tetrachlorobenzene	4.51	5
$C_6H_3Cl_3$	1,2,3-Trichlorobenzene	4.04	5
$C_6H_3Cl_3$	1,2,4-Trichlorobenzene	3.98	5
$C_6H_3Cl_3$	1,3,5-Trichlorobenzene	4.02	5
$C_6H_4Cl_2$	*o*-Dichlorobenzene	3.38	5
$C_6H_4Cl_2$	*m*-Dichlorobenzene	3.48	5
$C_6H_4Cl_2$	*p*-Dichlorobenzene	3.38	5
$C_6H_4Cl_2O$	2,4-Dichlorophenol	3.23	4
C_6H_5Br	Bromobenzene	2.99	2
C_6H_5Cl	Chlorobenzene	2.84	1
C_6H_5F	Fluorobenzene	2.27	2
C_6H_5I	Iodobenzene	3.28	2
$C_6H_5NO_2$	Nitrobenzene	1.85	1
C_6H_6	Benzene	2.13	1
C_6H_6O	Phenol	1.48	4
C_6H_6S	Benzenethiol	2.52	1
C_6H_7N	Aniline	0.90	1
C_6H_7N	2-Methylpyridine	1.11	1
C_6H_7N	3-Methylpyridine	1.20	1
C_6H_7N	4-Methylpyridine	1.22	1
C_6H_8	1,4-Cyclohexadiene	2.3	2
C_6H_8O	5-Hexyn-2-one	0.58	1
C_6H_8O	2-Cyclohexen-1-one	0.61	1
C_6H_8O	2-Ethylfuran	2.40	1
C_6H_{10}	1,5-Hexadiene	2.8	2
C_6H_{10}	1-Hexyne	2.73	2
C_6H_{10}	Cyclohexene	2.86	1
$C_6H_{10}O$	5-Hexen-2-one	1.02	1
$C_6H_{10}O$	Cyclohexanone	0.81	1
$C_6H_{10}O_2$	Ethyl methacrylate	1.94	1
$C_6H_{11}Br$	Bromocyclohexane	3.20	1
$C_6H_{11}N$	Hexanenitrile	1.66	1
C_6H_{12}	1-Hexene	3.40	1
C_6H_{12}	4-Methyl-1-pentene	2.5	2
C_6H_{12}	Cyclohexane	3.44	1
C_6H_{12}	Methylcyclopentane	3.37	2

Mol. form.	Name	log *P*	Ref.
$C_6H_{12}O$	Cyclohexanol	1.23	1
$C_6H_{12}O$	Hexanal	1.78	1
$C_6H_{12}O$	2-Hexanone	1.38	1
$C_6H_{12}O$	4-Methyl-2-pentanone	1.31	1
$C_6H_{12}O_2$	Hexanoic acid	1.92	1
$C_6H_{12}O_2$	Butyl acetate	1.82	1
$C_6H_{13}Br$	1-Bromohexane	3.80	1
$C_6H_{13}N$	Cyclohexylamine	1.49	1
C_6H_{14}	Hexane	4.00	1
C_6H_{14}	3-Methylpentane	3.60	2
C_6H_{14}	2,2-Dimethylbutane	3.82	1
C_6H_{14}	2,3-Dimethylbutane	3.85	2
$C_6H_{14}O$	1-Hexanol	2.03	1
$C_6H_{14}O$	2-Hexanol	1.76	1
$C_6H_{14}O$	3-Hexanol	1.65	1
$C_6H_{14}O$	3,3-Dimethyl-2-butanol	1.48	1
$C_6H_{14}O$	Dipropyl ether	2.03	1
$C_6H_{14}O$	Diisopropyl ether	1.52	1
$C_6H_{15}N$	Hexylamine	2.06	1
$C_6H_{15}N$	Dipropylamine	1.67	1
$C_6H_{15}N$	Triethylamine	1.45	1
$C_7H_5BrO_2$	2-Bromobenzoic acid	2.20	4
$C_7H_5BrO_2$	3-Bromobenzoic acid	2.87	4
$C_7H_5BrO_2$	4-Bromobenzoic acid	2.86	4
C_7H_5N	Benzonitrile	1.56	1
C_7H_6O	Benzaldehyde	1.48	1
$C_7H_6O_2$	Benzoic acid	1.88	4
$C_7H_6O_2$	Phenyl formate	1.26	1
$C_7H_6O_3$	Salicylic acid	2.20	4
C_7H_7Br	(Bromomethyl)benzene	2.92	1
C_7H_7Cl	*o*-Chlorotoluene	3.42	1
C_7H_7Cl	*m*-Chlorotoluene	3.28	1
C_7H_7Cl	*p*-Chlorotoluene	3.33	1
C_7H_7Cl	(Chloromethyl)benzene	2.30	1
$C_7H_7NO_2$	*p*-Nitrotoluene	2.42	1
C_7H_8	Toluene	2.73	1
C_7H_8	1,3,5-Cycloheptatriene	2.63	2
C_7H_8O	*o*-Cresol	1.98	1
C_7H_8O	*m*-Cresol	1.98	1
C_7H_8O	*p*-Cresol	1.97	1
C_7H_8O	Benzyl alcohol	1.05	1
C_7H_8O	Anisole	2.11	1
C_7H_9N	Benzylamine	1.09	1
C_7H_9N	*o*-Methylaniline	1.32	1
C_7H_9N	*m*-Methylaniline	1.40	1
C_7H_9N	*p*-Methylaniline	1.39	1
C_7H_9N	*N*-Methylaniline	1.66	1
C_7H_{14}	1-Heptene	3.99	1
C_7H_{14}	Methylcyclohexane	3.88	1
$C_7H_{14}O$	2-Heptanone	1.98	1
$C_7H_{14}O$	5-Methyl-2-hexanone	1.88	1
$C_7H_{15}Br$	1-Bromoheptane	4.36	1
$C_7H_{15}Cl$	1-Chloroheptane	4.15	1
$C_7H_{15}I$	1-Iodoheptane	4.70	1
C_7H_{16}	Heptane	4.50	1
$C_7H_{16}O$	1-Heptanol	2.62	1
$C_7H_{16}O$	2-Heptanol	2.31	1
$C_7H_{16}O$	3-Heptanol	2.24	1
$C_7H_{16}O$	4-Heptanol	2.22	1
$C_7H_{17}N$	Heptylamine	2.57	1
C_8H_6	Phenylacetylene	2.40	1
C_8H_6O	Benzofuran	2.67	1
C_8H_6S	Benzo[b]thiophene	3.12	1
C_8H_7N	Benzeneacetonitrile	1.56	1
C_8H_7N	Indole	2.14	1
C_8H_8	Styrene	3.05	1
C_8H_8O	Acetophenone	1.63	1
C_8H_8O	2-Methylbenzaldehyde	2.26	1
C_8H_8O	Benzeneacetaldehyde	1.78	1
C_8H_8O	2,3-Dihydrobenzofuran	2.14	1
C_8H_8O	Phenyloxirane	1.61	1
$C_8H_8O_2$	*o*-Toluic acid	2.32	4
$C_8H_8O_2$	*m*-Toluic acid	2.37	1
$C_8H_8O_2$	*p*-Toluic acid	2.34	1
$C_8H_8O_2$	Benzeneacetic acid	1.41	1
$C_8H_8O_2$	Phenyl acetate	1.49	1
$C_8H_8O_2$	Methyl benzoate	2.20	1
C_8H_{10}	Ethylbenzene	3.15	1
C_8H_{10}	*o*-Xylene	3.12	1
C_8H_{10}	*m*-Xylene	3.20	1
C_8H_{10}	*p*-Xylene	3.15	1
$C_8H_{10}O$	*o*-Ethylphenol	2.47	1
$C_8H_{10}O$	*m*-Ethylphenol	2.50	1
$C_8H_{10}O$	*p*-Ethylphenol	2.50	1
$C_8H_{10}O$	2,4-Xylenol	2.35	1
$C_8H_{10}O$	2,5-Xylenol	2.34	1
$C_8H_{10}O$	2,6-Xylenol	2.36	1
$C_8H_{10}O$	3,4 Xylenol	3.23	1
$C_8H_{10}O$	3,5-Xylenol	2.35	1
$C_8H_{10}O$	Benzeneethanol	1.36	1
$C_8H_{10}O$	α-Methylbenzyl alcohol	1.42	1
$C_8H_{10}O$	3-Methylbenzenemethanol	1.60	1
$C_8H_{10}O$	4-Methylbenzenemethanol	1.58	1
$C_8H_{10}O$	Phenetole	2.51	1
$C_8H_{10}O$	Benzyl methyl ether	1.35	1
$C_8H_{10}O$	2-Methylanisole	2.74	1
$C_8H_{10}O$	3-Methylanisole	2.66	1
$C_8H_{10}O$	4-Methylanisole	2.81	1
$C_8H_{11}N$	*p*-Ethylaniline	1.96	1
$C_8H_{11}N$	*N,N*-Dimethylaniline	2.31	1
$C_8H_{11}N$	Benzeneethanamine	1.41	1
$C_8H_{14}O_2$	Butyl methacrylate	2.88	1
$C_8H_{15}N$	Octanenitrile	2.75	1
C_8H_{16}	1-Octene	4.57	1
C_8H_{16}	Cyclooctane	4.45	2
$C_8H_{16}O$	2-Octanone	2.37	1
$C_8H_{16}O_2$	Octanoic acid	3.05	1
$C_8H_{17}Br$	1-Bromooctane	4.89	1
C_8H_{18}	Octane	5.15	1
$C_8H_{18}O$	1-Octanol	3.07	1
$C_8H_{18}O$	2-Octanol	2.90	1
$C_8H_{18}O$	4-Octanol	2.68	1
$C_8H_{18}O$	Dibutyl ether	3.21	1
C_9H_7N	Quinoline	2.03	1
C_9H_7N	Isoquinoline	2.08	1
C_9H_8	Indene	2.92	1
$C_9H_8O_2$	*trans*-Cinnamic acid	2.13	1
C_9H_9N	Benzenepropanenitrile	1.72	1
C_9H_{10}	Indan	3.33	1
$C_9H_{10}O$	1-Phenyl-1-propanone	2.19	1
$C_9H_{10}O$	1-Phenyl-2-propanone	1.44	1
$C_9H_{10}O$	4-Methylacetophenone	2.19	1

Mol. form.	Name	log *P*	Ref.
$C_9H_{10}O_2$	2-Phenylpropanoic acid	1.80	1
$C_9H_{10}O_2$	Benzyl acetate	1.96	1
$C_9H_{10}O_2$	4-Methylphenyl acetate	2.11	1
$C_9H_{10}O_2$	Ethyl benzoate	2.64	1
C_9H_{12}	Propylbenzene	3.69	1
C_9H_{12}	Isopropylbenzene	3.66	1
C_9H_{12}	*o*-Ethyltoluene	3.53	1
C_9H_{12}	*p*-Ethyltoluene	3.63	2
C_9H_{12}	1,2,3-Trimethylbenzene	3.60	1
C_9H_{12}	1,2,4-Trimethylbenzene	3.63	1
C_9H_{12}	1,3,5-Trimethylbenzene	3.42	1
$C_9H_{12}O$	2-Propylphenol	2.93	1
$C_9H_{12}O$	4-Propylphenol	3.20	1
$C_9H_{12}O$	2,3,6-Trimethylphenol	2.67	1
$C_9H_{12}O$	2,4,6-Trimethylphenol	2.46	1
$C_9H_{12}O$	Benzenepropanol	1.88	1
$C_9H_{13}N$	*N,N*-Dimethylbenzylamine	1.98	1
$C_9H_{13}N$	Amphetamine	1.76	1
C_9H_{18}	1-Nonene	5.15	1
$C_9H_{18}O$	2-Nonanone	3.16	1
$C_9H_{18}O$	5-Methyl-2-octanone	2.92	1
C_9H_{20}	Nonane	5.65	1
$C_9H_{20}O$	1-Nonanol	4.02	1
$C_9H_{21}N$	Tripropylamine	2.79	1
$C_{10}H_7Cl$	1-Chloronaphthalene	3.90	1
$C_{10}H_7Cl$	2-Chloronaphthalene	3.98	1
$C_{10}H_8$	Naphthalene	3.34	4
$C_{10}H_8$	Azulene	3.22	1
$C_{10}H_8O$	1-Naphthol	2.84	1
$C_{10}H_8O$	2-Naphthol	2.70	1
$C_{10}H_{12}O_2$	Isopropyl benzoate	3.18	1
$C_{10}H_{14}$	Butylbenzene	4.26	1
$C_{10}H_{14}$	*tert*-Butylbenzene	4.11	1
$C_{10}H_{14}$	Isobutylbenzene	4.01	2
$C_{10}H_{14}$	*p*-Cymene	4.10	1
$C_{10}H_{14}$	1,2,4,5-Tetramethylbenzene	4.10	2
$C_{10}H_{14}$	1,2,3,4-Tetramethylbenzene	4.00	1
$C_{10}H_{14}$	1,2,3,5-Tetramethylbenzene	4.10	1
$C_{10}H_{14}O$	4-Butylphenol	3.65	1
$C_{10}H_{20}O$	2-Decanone	3.77	1
$C_{10}H_{20}O_2$	Decanoic acid	4.09	1
$C_{10}H_{22}$	Decane	6.25	1
$C_{10}H_{22}O$	1-Decanol	4.57	1
$C_{11}H_9N$	4-Phenylpyridine	2.59	1
$C_{11}H_{10}$	1-Methylnaphthalene	3.87	1
$C_{11}H_{10}$	2-Methylnaphthalene	4.00	1
$C_{11}H_{16}$	Pentylbenzene	4.90	1
$C_{11}H_{16}$	Pentamethylbenzene	4.56	1
$C_{11}H_{22}O$	2-Undecanone	4.09	1
$C_{11}H_{22}O_2$	Methyl decanoate	4.41	1
$C_{12}Cl_{10}$	Decachlorobiphenyl	8.26	3
$C_{12}HCl_9$	2,2′,3,3′,4,5,5′,6,6′-Nonachlorobiphenyl	8.16	3
$C_{12}H_2Cl_8$	2,2′,3,3′,5,5′,6,6′-Octachlorobiphenyl	7.10	3
$C_{12}H_3Cl_7$	2,2′,3,3′,4,4′,6-Heptachlorobiphenyl	6.70	3
$C_{12}H_4Cl_6$	2,2′,3,3′,4,4′-Hexachlorobiphenyl	7.00	3
$C_{12}H_4Cl_6$	2,2′,4,4′,6,6′-Hexachlorobiphenyl	7.00	3
$C_{12}H_4Cl_6$	2,2′,3,3′,6,6′-Hexachlorobiphenyl	6.70	3
$C_{12}H_5Cl_5$	2,3,4,5,6-Pentachlorobiphenyl	6.30	3
$C_{12}H_5Cl_5$	2,2′,4,5,5′-Pentachlorobiphenyl	6.40	3
$C_{12}H_6Cl_4$	2,3,4,5-Tetrachlorobiphenyl	5.72	3
$C_{12}H_6Cl_4$	2,2′,4′,5-Tetrachlorobiphenyl	5.73	7
$C_{12}H_7Cl_3$	2,4,5-Trichlorobiphenyl	5.60	3
$C_{12}H_7Cl_3$	2,4,6-Trichlorobiphenyl	5.47	3
$C_{12}H_8Cl_2$	2,5-Dichlorobiphenyl	5.10	3
$C_{12}H_8Cl_2$	2,6-Dichlorobiphenyl	5.00	3
$C_{12}H_8O$	Dibenzofuran	4.12	1
$C_{12}H_9Cl$	2-Chlorobiphenyl	4.52	1
$C_{12}H_9Cl$	3-Chlorobiphenyl	4.58	1
$C_{12}H_9Cl$	4-Chlorobiphenyl	4.61	1
$C_{12}H_9N$	Carbazole	3.72	1
$C_{12}H_{10}$	Acenaphthene	3.96	4
$C_{12}H_{10}$	Biphenyl	3.76	6
$C_{12}H_{10}N_2$	Azobenzene	3.82	1
$C_{12}H_{10}O$	Diphenyl ether	4.21	1
$C_{12}H_{10}S$	Diphenyl sulfide	4.45	1
$C_{12}H_{11}N$	Diphenylamine	3.44	4
$C_{12}H_{12}$	1-Ethylnaphthalene	4.40	1
$C_{12}H_{12}$	1,2-Dimethylnaphthalene	4.31	1
$C_{12}H_{12}$	1,4-Dimethylnaphthalene	4.37	1
$C_{12}H_{14}O$	4-Phenylcyclohexanone	2.45	1
$C_{12}H_{18}$	Hexylbenzene	5.52	1
$C_{12}H_{18}$	Hexamethylbenzene	4.69	4
$C_{12}H_{22}O$	Cyclododecanone	4.10	1
$C_{12}H_{24}O_2$	Dodecanoic acid	4.6	1
$C_{12}H_{26}O$	1-Dodecanol	5.13	1
$C_{13}H_8O$	9H-Fluoren-9-one	3.58	1
$C_{13}H_9N$	Acridine	3.40	1
$C_{13}H_{10}$	9H-Fluorene	4.20	4
$C_{13}H_{10}O$	Benzophenone	3.18	1
$C_{13}H_{10}O_2$	Phenyl benzoate	3.59	1
$C_{13}H_{11}NO$	*N*-Phenylbenzamide	2.62	1
$C_{13}H_{12}$	Diphenylmethane	4.14	1
$C_{13}H_{12}$	4-Methylbiphenyl	4.63	1
$C_{13}H_{12}O$	Diphenylmethanol	2.67	1
$C_{13}H_{12}O$	Benzyl phenyl ether	3.79	1
$C_{14}H_{10}$	Anthracene	4.56	4
$C_{14}H_{10}$	Phenanthrene	4.52	4
$C_{14}H_{12}$	*trans*-Stilbene	4.81	1
$C_{14}H_{12}$	1-Methylfluorene	4.97	1
$C_{14}H_{12}O$	2-Phenylacetophenone	3.18	1
$C_{14}H_{12}O_2$	Benzyl benzoate	3.97	1
$C_{14}H_{14}$	1,2-Diphenylethane	4.70	1
$C_{14}H_{14}$	4,4′-Dimethylbiphenyl	5.09	1
$C_{14}H_{22}$	Octylbenzene	6.30	1
$C_{14}H_{28}O_2$	Tetradecanoic acid	6.1	1
$C_{15}H_{12}$	2-Methylanthracene	5.15	2
$C_{15}H_{12}$	9-Methylanthracene	5.07	1
$C_{15}H_{12}$	1-Methylphenanthrene	5.14	2
$C_{16}H_{10}$	Fluoranthene	5.07	4
$C_{16}H_{10}$	Pyrene	5.08	4
$C_{16}H_{14}$	9,10-Dimethylanthracene	5.69	1
$C_{16}H_{32}O_2$	Hexadecanoic acid	7.17	1
$C_{17}H_{12}$	11H-Benzo[a]fluorene	5.40	1
$C_{17}H_{12}$	11H-Benzo[b]fluorene	5.75	1
$C_{18}H_{12}$	Benz[a]anthracene	5.91	1
$C_{18}H_{12}$	Chrysene	5.73	4
$C_{18}H_{12}$	Naphthacene	5.76	1
$C_{18}H_{12}$	Triphenylene	5.49	4
$C_{18}H_{15}N$	Triphenylamine	5.74	1
$C_{18}H_{30}O_2$	Linolenic acid	6.46	1
$C_{18}H_{32}O_2$	Linoleic acid	7.05	1

Mol. form.	Name	log *P*	Ref.
$C_{18}H_{34}O_2$	Oleic acid	7.64	1
$C_{18}H_{36}O_2$	Stearic acid	8.23	1
$C_{19}H_{16}O$	Triphenylmethanol	3.68	1
$C_{20}H_{12}$	Perylene	6.25	1
$C_{20}H_{12}$	Benzo[a]pyrene	6.20	4
$C_{20}H_{32}O_2$	Arachidonic acid	6.98	1
$C_{20}H_{40}O_2$	Arachidic acid	9.29	1
$C_{21}H_{16}$	1,2-Dihydro-3-methylbenz[j]aceanthrylene	6.75	1
$C_{22}H_{12}$	Benzo[ghi]perylene	6.90	1
$C_{24}H_{12}$	Coronene	6.05	4

PROTECTION AGAINST IONIZING RADIATION

The following data and rules of thumb are helpful in estimating the penetrating capability of and danger of exposure to various types of ionizing radiation. More precise data should be used for critical applications.

Alpha Particles

Alpha particles of at least 7.5 MeV are required to penetrate the epidermis, the protective layer of skin, which is about 0.07 mm thick.

Electrons

Electrons of at least 70 keV are required to penetrate the epidermis, the protective layer of skin, which is about 0.07 mm thick.

The range of electrons in g/cm^2 is approximately equal to the maximum energy (E) in MeV divided by 2.

The range of electrons in air is about 3.65 m per MeV; for example, a 3 MeV electron has a range of about 11 m in air.

A chamber wall thickness of 30 mg/cm^2 will transmit 70% of the initial fluence of 1 MeV electrons and 20% of that of 0.4 MeV electrons.

When electrons of 1 to 2 MeV pass through light materials such as water, aluminum, or glass, less than 1% of their energy is dissipated as bremsstrahlung.

The bremsstrahlung from 1 Ci of ^{32}P aqueous solution in a glass bottle is about 1 mR/h at 1 meter distance.

When electrons from a 1 Ci source of ^{90}Sr – ^{90}Y are absorbed, the bremsstrahlung hazard is approximately equal to that presented by the gamma radiation from 12 mg of radium. The average energy of the bremsstrahlung is about 300 keV.

Gamma Rays

The air-scattered radiation (sky-shine) from a 100 Ci ^{60}Co source placed 1 ft behind a 4 ft high shield is about 100 mrad/h at 6 ft from the outside of the shield.

Within ±20% for point source gamma emitters with energies between 0.07 and 4 MeV, the exposure rate (R/h) at 1 ft is $6C{\cdot}E{\cdot}n$ where C is the activity in curies, E is the energy in MeV, and n is the number of gammas per disintegration.

Neutrons

An approximate HVL (thickness of absorber for which the neutron flux falls to half its initial value) for 1 MeV neutrons is 3.2 cm of paraffin; that for 5 MeV neutrons is 6.9 cm of paraffin.

Miscellaneous

The activity of any radionuclide is reduced to less than 1% after 7 half-lives (i.e., $2^{-7} = 0.8\%$).

For nuclides with a half-life greater than 6 days, the change in activity in 24 hours will be less than 10%.

10 HVL (half-value layers) attenuates approximately by 10^{-3}.

There is 0.64 mm^3 of radon gas at STP in transient equilibrium with 1 Ci of radium.

The natural background from all sources in most parts of the world leads to an equivalent dose rate of about 0.04 to 4 mSv per year for the average person. About 84% of this comes from terrestrial sources, the remainder from cosmic rays. The U. S. average is about 3.6 mSv/yr but can range up to 50 mSv/yr in some areas. A passenger in a plane flying at 12,000 meters receives 5 μSv/hr from cosmic rays (as compared to about 0.03 μSv/hr at sea level).

The ICRP recommended exposure limit to man-made sources of ionizing radiation (Reference 2) is 20 mSv/yr averaged over 5 years, with the dose in any one year not to exceed 50 mSv.

A whole-body dose of about 3 Gy over a short time interval will typically lead to 50% mortality in 30 days assuming no medical treatment.

Units

The gray (Gy) is the SI unit of absorbed dose; it is a measure of the mean energy imparted to a sample of irradiated matter, divided by the mass of the sample. Gy is a special name for the SI unit J/kg.

The sievert (Sv) is the SI unit of equivalent dose, which is defined as the absorbed dose multiplied by a weighting factor that expresses the long-term biological risk from low-level chronic exposure to a specified type of radiation. The Sv is another special name for J/kg.

1 curie (Ci) = $3.7{\cdot}10^{10}$ becquerel (Bq); i.e., $3.7{\cdot}10^{10}$ disintegrations per second.

1 roentgen (R) = $2.58{\cdot}10^{-4}$ coulomb per kilogram (C/kg); a measure of the charge (positive or negative) liberated by x-ray or gamma radiation in air, divided by the mass of air.

1 rad = 0.01 Gy

1 rem = 0.01 Sv

References

1. Padikal, T. N., and Fivozinsky, S. P., *Medical Physics Data Book, National Bureau of Standards Handbook 138*, U.S. Government Printing Office, Washington, D.C., 1981.
2. *1990 Recommendations of the International Commission on Radiological Protection*, ICRP Publication 60, *Annals of the ICRP*, Pergamon Press, Oxford, 1991.
3. *Radiation: Doses, Effects, Risks*, United Nations Sales No. E.86.III.D.4, 1985.
4. Eidelman, S., et al., *Physics Letters*, B592, 1, 2004.

ANNUAL LIMITS ON INTAKES OF RADIONUCLIDES

K. F. Eckerman

The following table lists, for workers, the annual limits on oral and inhalation intakes (ALI) for selected radionuclides based on the occupational radiation protection guidance of the International Commission on Radiological Protection (References 1 and 2). An intake of one ALI corresponds to an annual whole body dose of 0.02 Sv (2 rem).

The ALI is expressed in the SI unit of activity, the becquerel (Bq), and in the conventional unit, the microcurie (µCi); 1 µCi = $3.7 \cdot 10^4$ Bq. The chemical form of inhaled radionuclides is, in most instances, stated in terms of the rate of absorption to blood from the lungs and the fractional absorption from the small intestine. Types F, M, and S denote chemical forms that are absorbed from the lungs at rates characterized as fast, moderate, and slow, respectively. The time to absorb 90% of the deposited radionuclide, in the absence of radioactive decay, corresponds to about 10 minutes, 150 days, and 7000 days for Type F, M, and S compounds, respectively. Type F compounds can be considered to be more soluble than M or S, S being the most insoluble. Chemical form consideration for ingestion is specified by the fractional absorption from the small intestine, denoted as f_1. The f_1 values range from 10^{-5} to 1. Higher fractional absorption is associated with greater solubility of the compound.

References

1. *1990 Recommendations of the International Commission on Radiological Protection, ICRP Publication 60, Annals of the ICRP 21,* (1–3), Pergamon Press, Oxford, 1991.
2. *Dose Coefficients for Intakes of Radionuclides by Workers, ICRP Publication 68, Annals of the ICRP,* 24(4), Pergamon Press, Oxford, 1995.

		Inhalation intakes			Oral intakes		
	Physical half-life	Chemical form type/f_1	ALI Bq	ALI µCi	Chemical form f_1	ALI Bq	ALI µCi
^{3}H	12.3 y	HT gas	1.1E+13	3.0E+08	1.000	1.1E+13	3.0E+08
		HTO vapor	1.1E+09	3.0E+04			
^{11}C	0.340 h	CO	1.7E+10	4.5E+05	1.000	8.3E+08	2.3E+04
		CO_2	9.1E+09	2.5E+05			
		Organic compounds	6.2E+09	1.7E+05			
^{14}C	5730 y	CO	2.5E+10	6.8E+05	1.000	3.4E+07	9.3E+02
		CO_2	3.1E+09	8.3E+04			
		Organic compounds	3.4E+07	9.3E+02			
^{18}F	1.83 h	F 1.000	3.7E+08	1.0E+04	1.000	4.1E+08	1.1E+04
		M 1.000	2.2E+08	6.1E+03			
		S 1.000	2.2E+08	5.8E+03			
^{22}Na	2.60 y	F 1.000	1.0E+07	2.7E+02	1.000	6.3E+06	1.7E+02
^{24}Na	15.0 h	F 1.000	3.8E+07	1.0E+03	1.000	4.7E+07	1.3E+03
^{32}P	14.3 d	F 0.800	1.8E+07	4.9E+02	0.800	8.3E+06	2.3E+02
		M 0.800	6.9E+06	1.9E+02			
^{35}S	87.4 d	Inorganic compounds					
		F 0.800	2.5E+08	6.8E+03	0.800	1.4E+08	3.9E+03
		M 0.800	1.8E+07	4.9E+02	0.100	1.1E+08	2.8E+03
		Vapor	1.7E+08	4.5E+03			
		Organic compounds			1.000	2.6E+07	7.0E+02
^{42}K	12.4 h	F 1.000	1.0E+08	2.7E+03	1.000	4.7E+07	1.3E+03
^{43}K	22.6 h	F 1.000	7.7E+07	2.1E+03	1.000	8.0E+07	2.2E+03
^{45}Ca	163 d	M 0.300	8.7E+06	2.4E+02	0.300	2.6E+07	7.1E+02
^{47}Ca	4.53 d	M 0.300	9.5E+06	2.6E+02	0.300	1.3E+07	3.4E+02
^{51}Cr	27.7 d	F 0.100	6.7E+08	1.8E+04	0.100	5.3E+08	1.4E+04
		M 0.100	5.9E+08	1.6E+04	0.010	5.4E+08	1.5E+04
		S 0.100	5.6E+08	1.5E+04			
^{54}Mn	312 d	F 0.100	1.8E+07	4.9E+02	0.100	2.8E+07	7.6E+02
		M 0.100	1.7E+07	4.5E+02			
^{52}Fe	8.28 h	F 0.100	2.9E+07	7.8E+02	0.100	1.4E+07	3.9E+02
		M 0.100	2.1E+07	5.7E+02			
^{55}Fe	2.70 y	F 0.100	2.2E+07	5.9E+02	0.100	6.1E+07	1.6E+03
		M 0.100	6.1E+07	1.6E+03			

		Inhalation intakes			Oral intakes		
		Chemical form	ALI		Chemical form	ALI	
	Physical half-life	type/f_1	Bq	μCi	f_1	Bq	μCi
^{59}Fe	44.5 d	F 0.100	6.7E+06	1.8E+02	0.100	1.1E+07	3.0E+02
		M 0.100	6.3E+06	1.7E+02			
^{57}Co	271 d	M 0.100	5.1E+07	1.4E+03	0.100	9.5E+07	2.6E+03
		S 0.050	3.3E+07	9.0E+02	0.050	1.1E+08	2.8E+03
^{58}Co	70.8 d	M 0.100	1.4E+07	3.9E+02	0.100	2.7E+07	7.3E+02
		S 0.050	1.2E+07	3.2E+02	0.050	2.9E+07	7.7E+02
^{60}Co	5.27 y	M 0.100	2.8E+06	7.6E+01	0.100	5.9E+06	1.6E+02
		S 0.050	1.2E+06	3.2E+01	0.050	8.0E+06	2.2E+02
^{64}Cu	12.7 h	F 0.500	2.9E+08	7.9E+03	0.500	1.7E+08	4.5E+03
		M 0.500	1.3E+08	3.6E+03			
		S 0.500	1.3E+08	3.6E+03			
^{59}Ni	75000 y	F 0.050	9.1E+07	2.5E+03	0.050	3.2E+08	8.6E+03
		M 0.050	2.1E+08	5.8E+03			
		Vapor	2.4E+07	6.5E+02			
^{63}Ni	96.0 y	F 0.050	3.8E+07	1.0E+03	0.050	1.3E+08	3.6E+03
		M 0.050	6.5E+07	1.7E+03			
		Vapor	1.0E+07	2.7E+02			
^{65}Zn	244 d	S 0.500	7.1E+06	1.9E+02	0.500	5.1E+06	1.4E+02
^{67}Ga	3.26 d	F 0.001	1.8E+08	4.9E+03	0.001	1.1E+08	2.8E+03
		M 0.001	7.1E+07	1.9E+03			
^{68}Ga	1.13 h	F 0.001	4.1E+08	1.1E+04	0.001	2.0E+08	5.4E+03
		M 0.001	2.5E+08	6.7E+03			
^{68}Ge	288 d	F 1.000	2.4E+07	6.5E+02	1.000	1.5E+07	4.2E+02
		M 1.000	2.5E+06	6.8E+01			
^{75}Se	120 d	F 0.800	1.4E+07	3.9E+02	0.800	7.7E+06	2.1E+02
		M 0.800	1.2E+07	3.2E+02	0.050	4.9E+07	1.3E+03
^{79}Se	65000 y	F 0.800	1.3E+07	3.4E+02	0.800	6.9E+06	1.9E+02
		M 0.800	6.5E+06	1.7E+02	0.050	5.1E+07	1.4E+03
^{86}Rb	18.6 d	F 1.000	1.5E+07	4.2E+02	1.000	7.1E+06	1.9E+02
^{85}Sr	64.8 d	F 0.300	3.6E+07	9.7E+02	0.300	3.6E+07	9.7E+02
		S 0.010	3.1E+07	8.4E+02	0.010	6.1E+07	1.6E+03
^{87m}Sr	2.80 h	F 0.300	9.1E+08	2.5E+04	0.300	6.7E+08	1.8E+04
		S 0.010	5.7E+08	1.5E+04	0.010	6.1E+08	1.6E+04
^{89}Sr	50.5 d	F 0.300	1.4E+07	3.9E+02	0.300	7.7E+06	2.1E+02
		S 0.010	3.6E+06	9.7E+01	0.010	8.7E+06	2.4E+02
^{90}Sr	29.1 y	F 0.300	6.7E+05	1.8E+01	0.300	7.1E+05	1.9E+01
		S 0.010	2.6E+05	7.0E+00	0.010	7.4E+06	2.0E+02
^{99}Mo	2.75 d	F 0.800	5.6E+07	1.5E+03	0.800	2.7E+07	7.3E+02
		S 0.050	1.8E+07	4.9E+02	0.050	1.7E+07	4.5E+02
^{99m}Tc	6.02 h	F 0.800	1.0E+09	2.7E+04	0.800	9.1E+08	2.5E+04
		M 0.800	6.9E+08	1.9E+04			
^{99}Tc	213000 y	F 0.800	5.0E+07	1.4E+03	0.800	2.6E+07	6.9E+02
		M 0.800	6.3E+06	1.7E+02			
^{106}Ru	1.01 y	F 0.050	2.0E+06	5.5E+01	0.050	2.9E+06	7.7E+01
		M 0.050	1.2E+06	3.2E+01			
		S 0.050	5.7E+05	1.5E+01			
^{111}In	2.83 d	F 0.020	9.1E+07	2.5E+03	0.020	6.9E+07	1.9E+03
		M 0.020	6.5E+07	1.7E+03			
^{113m}In	1.66 h	F 0.020	1.1E+09	2.8E+04	0.020	7.1E+08	1.9E+04
		M 0.020	6.3E+08	1.7E+04			
^{113}Sn	115 d	F 0.020	2.5E+07	6.8E+02	0.020	2.7E+07	7.4E+02
		M 0.020	1.1E+07	2.8E+02			
^{123}I	13.2 h	F 1.000	1.8E+08	4.9E+03	1.000	9.5E+07	2.6E+03
		Vapor	9.5E+07	2.6E+03			
^{125}I	60.1 d	F 1.000	2.7E+06	7.4E+01	1.000	1.3E+06	3.6E+01
		Vapor	1.4E+06	3.9E+01			
^{129}I	$1.57 \cdot 10^7$ y	F 1.000	3.9E+05	1.1E+01	1.000	1.8E+05	4.9E+00
		Vapor	2.1E+05	5.6E+00			

		Inhalation intakes			Oral intakes		
	Physical half-life	Chemical form type/f_1	ALI Bq	ALI µCi	Chemical form f_1	ALI Bq	ALI µCi
^{131}I	8.04 d	F 1.000	1.8E+06	4.9E+01	1.000	9.1E+05	2.5E+01
		Vapor	1.0E+06	2.7E+01			
^{129}Cs	1.34 d	F 1.000	2.5E+08	6.7E+03	1.000	3.3E+08	9.0E+03
^{134}Cs	2.06 y	F 1.000	2.1E+06	5.6E+01	1.000	1.1E+06	2.8E+01
^{136}Cs	13.1 d	F 1.000	1.1E+07	2.8E+02	1.000	6.7E+06	1.8E+02
^{137}Cs	30.0 y	F 1.000	3.0E+06	8.1E+01	1.000	1.5E+06	4.2E+01
^{141}Ce	32.5 d	M 5.0E-04	7.4E+06	2.0E+02	5.0E-04	2.8E+07	7.6E+02
		S 5.0E-04	6.5E+06	1.7E+02			
^{144}Ce	284 d	M 5.0E-04	8.7E+05	2.4E+01	5.0E-04	3.8E+06	1.0E+02
		S 5.0E-04	6.9E+05	1.9E+01			
^{133}Ba	10.7 y	F 0.100	1.1E+07	3.0E+02	0.100	2.0E+07	5.4E+02
^{140}Ba	12.7 d	F 0.100	1.3E+07	3.4E+02	0.100	8.0E+06	2.2E+02
^{169}Yb	32.0 d	M 5.0E-04	9.5E+06	2.6E+02	5.0E-04	2.8E+07	7.6E+02
		S 5.0E-04	8.3E+06	2.3E+02			
^{198}Au	2.69 d	F 0.100	5.1E+07	1.4E+03	0.100	2.0E+07	5.4E+02
		M 0.100	2.0E+07	5.5E+02			
		S 0.100	1.8E+07	4.9E+02			
^{198m}Au	2.30 d	F 0.100	3.4E+07	9.2E+02	0.100	1.5E+07	4.2E+02
		M 0.100	1.0E+07	2.7E+02			
		S 0.100	1.1E+07	2.8E+02			
^{197}Hg	2.67 d	Inorganic compounds					
		F 0.400	2.4E+08	6.4E+03	1.000	2.0E+08	5.5E+03
					0.400	1.2E+08	3.2E+03
		Vapor	4.5E+06	1.2E+02			
		Organic compounds					
		F 0.020	2.0E+08	5.4E+03	0.020	8.7E+07	2.4E+03
		M 0.020	7.1E+07	1.9E+03			
^{203}Hg	46.6 d	Inorganic compounds					
		F 0.400	2.7E+07	7.2E+02	1.000	1.1E+07	2.8E+02
					0.400	1.8E+07	4.9E+02
		Vapor	2.9E+06	7.7E+01			
		Organic compounds					
		F 0.020	3.4E+07	9.2E+02	0.020	3.7E+07	1.0E+03
		M 0.020	1.1E+07	2.8E+02			
^{201}Tl	3.04 d	F 1.000	2.6E+08	7.1E+03	1.000	2.1E+08	5.7E+03
^{210}Pb	22.3 y	F 0.200	1.8E+04	4.9E-01	0.200	2.9E+04	7.9E-01
^{207}Bi	38.0 y	F 0.050	2.4E+07	6.4E+02	0.050	1.5E+07	4.2E+02
		M 0.050	6.3E+06	1.7E+02			
^{210}Po	138 d	F 0.100	2.8E+04	7.6E-01	0.100	8.3E+04	2.3E+00
		M 0.100	9.1E+03	2.5E-01			
^{224}Ra	3.66 d	M 0.200	8.3E+03	2.3E-01	0.200	3.1E+05	8.3E+00
^{226}Ra	1600 y	M 0.200	1.7E+03	4.5E-02	0.200	7.1E+04	1.9E+00
^{228}Ra	5.75 y	M 0.200	1.2E+04	3.2E-01	0.200	3.0E+04	8.1E-01
^{228}Th	1.91 y	M 5.0E-04	8.7E+02	2.4E-02	5.0E-04	2.9E+05	7.7E+00
		S 2.0E-04	6.3E+02	1.7E-02	2.0E-04	5.7E+05	1.5E+01
^{230}Th	77000 y	M 5.0E-04	7.1E+02	1.9E-02	5.0E-04	9.5E+04	2.6E+00
		S 2.0E-04	2.8E+03	7.5E-02	2.0E-04	2.3E+05	6.2E+00
^{232}Th	$1.40 \cdot 10^{10}$ y	M 5.0E-04	6.9E+02	1.9E-02	5.0E-04	9.1E+04	2.5E+00
		S 2.0E-04	1.7E+03	4.5E-02	2.0E-04	2.2E+05	5.9E+00
^{234}U	$2.44 \cdot 10^{5}$ y	F 0.020	3.1E+04	8.4E-01	0.020	4.1E+05	1.1E+01
		M 0.020	9.5E+03	2.6E-01	0.002	2.4E+06	6.5E+01
		S 0.002	2.9E+03	7.9E-02			
^{235}U	$7.04 \cdot 10^{8}$ y	F 0.020	3.3E+04	9.0E-01	0.020	4.3E+05	1.2E+01
		M 0.020	1.1E+04	3.0E-01	0.002	2.4E+06	6.5E+01

	Physical half-life	Inhalation intakes: Chemical form type/f_1	ALI Bq	ALI µCi	Oral intakes: Chemical form f_1	ALI Bq	ALI µCi
		S 0.002	3.3E+03	8.9E-02			
^{238}U	4.47·10^9 y	F 0.020	3.4E+04	9.3E-01	0.020	4.5E+05	1.2E+01
		M 0.020	1.3E+04	3.4E-01	0.002	2.6E+06	7.1E+01
		S 0.002	3.5E+03	9.5E-02			
^{237}Np	2.14·10^6 y	M 5.0E-04	1.3E+03	3.6E-02	5.0E-04	1.8E+05	4.9E+00
^{239}Np	2.36 d	M 5.0E-04	1.8E+07	4.9E+02	5.0E-04	2.5E+07	6.8E+02
^{238}Pu	87.7 y	M 5.0E-04	6.7E+02	1.8E-02	5.0E-04	8.7E+04	2.4E+00
		S 1.0E-05	1.8E+03	4.9E-02	1.0E-05	2.3E+06	6.1E+01
					1.0E-04	4.1E+05	1.1E+01
^{239}Pu	24100 y	M 5.0E-04	6.3E+02	1.7E-02	5.0E-04	8.0E+04	2.2E+00
		S 1.0E-05	2.4E+03	6.5E-02	1.0E-05	2.2E+06	6.0E+01
					1.0E-04	3.8E+05	1.0E+01
^{241}Pu	14.4 y	M 5.0E-04	3.4E+04	9.3E-01	5.0E-04	4.3E+06	1.2E+02
		S 1.0E-05	2.4E+05	6.4E+00	1.0E-05	1.8E+08	4.9E+03
					1.0E-04	2.1E+07	5.6E+02
^{241}Am	432 y	M 5.0E-04	7.4E+02	2.0E-02	5.0E-04	1.0E+05	2.7E+00
^{244}Cm	18.1 y	M 5.0E-04	1.2E+03	3.2E-02	5.0E-04	1.7E+05	4.5E+00
^{252}Cf	2.64 y	M 5.0E-04	1.5E+03	4.2E-02	5.0E-04	2.2E+05	6.0E+00

CHEMICAL CARCINOGENS

The following substances are listed in the 11th Report on Carcinogens 2004 released by the National Institute of Environmental Health Sciences (NIEHS) under the National Toxicology Program (NTP). Substances are grouped in two classes:

- **Known to be human carcinogens**: There is sufficient evidence of carcinogenicity from studies in humans which indicates a causal relationship between exposure to the substance and human cancer.
- **Reasonably anticipated to be human carcinogens:** There is limited evidence of carcinogenicity from studies in humans which indicates that causal interpretation is credible, but that alternative explanations, such as chance, bias, or confounding factors, could not be adequately excluded; or there is sufficient evidence of carcinogenicity from studies in experimental animals.

The NTP report also lists many poorly defined materials such as soots, tars, mineral oils, coke oven emissions, etc. These materials are not included here.

The table lists the name normally used in the *Handbook of Chemistry and Physics*, followed by additional names and acronyms by which the substance is known. In many cases the primary name given here is different from that used in the NTP report; however, names used in the NTP report appear in the Other Names column. The Chemical Abstracts Service Registry Number (CAS RN), given in the last column, is taken from the NTP report. Extensive details on each substance are given in the reference.

Reference

Public Health Service, National Toxicology Program, 11th Report on Carcinogens, available on the Internet at <http://ntp.niehs.nih.gov/>.

Substance	Other Names	CAS RN
Known to be Human Carcinogens		
Aflatoxins		1402 68-2
4-Aminobiphenyl	*p*-Biphenylamine	92-67-1
Arsenic compounds, inorganic		
Asbestos		1332-21-4
Azathioprine	6-[(1-Methyl-4-nitro-1*H*-imidazol-5-yl)thio]-1*H*-purine	446-86-6
Benzene		71-43-2
p-Benzidine (includes dyes metabolized to benzidene)	[1,1'-Biphenyl]-4,4'-diamine	92-87-5
Beryllium and beryllium compounds		7440-41-7
Bis(2-chloroethyl) sulfide	Mustard gas	505-60-2
Bis(chloromethyl) ether		542-88-1
1,3-Butadiene		106-99-0
1,4-Butanediol dimethylsulfonate	Myleran; Busulfan	55-98-1
Cadmium and cadmium compounds		7440-43-9
Chlorambucil		305-03-3
Chloroethene	Vinyl chloride; Chloroethylene	75-01-4
1-(2-Chloroethyl)-3-(4-methylcyclohexyl)-1-nitrosourea	MeCCNU	13909-09-6
Chloromethyl methyl ether		107-30-2
Chromium hexavalent compounds		
Cyclophosphamide	Cyclophosphane; 2*H*-1,3,2-Oxazaphosphorin-2-amine, *N,N*-bis(2-chloroethyl)tetrahydro-, 2-oxide	50-18-0
Cyclosporin A	Cyclosporine	59865-13-3
Diethylstilbestrol		56-53-1
Erionite		66733-21-9
Estrogens, steroidal		
N-(4-Ethoxyphenyl)acetamide	Phenacetin	62-44-2
Melphalan	*L*-Phenylalanine, 4-[bis(2-chloroethyl)amino]-	148-82-3
Methoxsalen (with UV therapy)	PUVA; 9-Methoxy-7*H*-furo[3,2-g][1]benzopyran-7-one	298-81-7
2-Naphthylamine	2-Aminonaphthalene; β-Naphthylamine	91-59-8
Nickel compounds		
Oxirane	Ethylene oxide	75-21-8

Substance	Other Names	CAS RN
Radon		10043-92-2
Silicon dioxide (respirable size)	Quartz; Silica	14808-60-7
Silicon dioxide (respirable size)	Cristobalite; Silica	14464-46-1
Silicon dioxide (respirable size)	Tridymite; Silica	15468-32-3
Sulfuric acid (strong acid mists)	Oil of vitriol	7664-93-9
Tamoxifen		10540-29-1
2,3,7,8-Tetrachlorodibenzo-*p*-dioxin	TCDD; Dioxin	1746-01-6
Thorium(IV) oxide	Thorium dioxide	1314-20-1
Triethylenethiophosphoramide	Thiotepa; Tris(1-aziridinyl)phosphine, sulfide	52-24-4
Reasonably Anticipated to be Human Carcinogens		
Acetaldehyde	Ethanal	75-07-0
2-(Acetylamino)fluorene		53-96-3
Acrylamide	2-Propenamide	79-06-1
Acrylonitrile	Propenenitrile	107-13-1
4-Allyl-1,2-dimethoxybenzene	Methyleugenol; 1,2-Dimethoxy-4-allylbenzene	93-15-2
2-Amino-9,10-anthracenedione	2-Aminoanthraquinone	117-79-3
1-Amino-2,4-dibromo-9,10-anthracenedione	1-Amino-2,4-dibromoanthraquinone	81-49-2
2-Amino-3,4-dimethylimidazo[4,5-f]quinoline	MeIQ	77094-11-2
2-Amino-3,8-dimethylimidazo[4,5-f]quinoxaline	MeIQx	77500-04-0
1-Amino-2-methyl-9,10-anthracenedione	1-Amino-2-methylanthraquinone	82-28-0
2-Amino-3-methyl-3*H*-imidazo[4,5-f]quinoline	IQ	76180-96-6
2-Amino-1-methyl-6-phenylimidazo[4,5-b]pyridine	PhIP	105650-23-5
Azacitidine	5-Azacytidine; 4-Amino-1-β-*D*-ribofuranosyl-1,3,5-triazine-2(1*H*)-one	320-67-2
Benz[a]anthracene		56-55-3
Benzo[b]fluoranthene	Benz[e]acephenanthrylene	205-99-2
Benzo[j]fluoranthene		205-82-3
Benzo[k]fluoranthene	2,3,1′,8′-Binaphthylene	207-08-9
Benzo[a]pyrene		50-32-8
2,2′-Bioxirane	Diepoxybutane	1464-53-5
Bis(4-amino-3-chlorophenyl)methane	4,4-Methylene-bis(2-chloraniline); MBOCA	101-14-4
2,2-Bis(bromomethyl)-1,3-propanediol	BBMP; Pentaerythritol dibromide	3296-90-0
Bis(2-chloroethyl)methylamine	Nitrogen mustard hydrochloride	55-86-7
N,N′-Bis(2-chloroethyl)-*N*-nitrosourea	BCNU; Carmustine	154-93-8
Bis[4-(dimethylamino)phenyl]methane	Michler's Base; 4,4-Methylenebis(*N,N*-dimethylbenzenamine)	101-61-1
1,3-Bis(2,3-epoxypropoxy)benzene	Diglycidyl resorcinol ether	101-90-6
Bis(2-ethylhexyl) phthalate	DEHP; Di(2-ethylhexyl) phthalate	117-81-7
Bromodichloromethane		75-27-4
Bromoethene	Vinyl bromide	593-60-2
tert-Butyl-4-hydroxyanisole	BHA; Butylated hydroxyanisole	25013-16-5
Chloramphenicol		56-75-7
Chlorendic acid	1,4,5,6,7,7-Hexachloro-5-norbornene-2,3-dicarboxylic acid	115-28-6
Chlorinated paraffins (C_{12}, 60% Cl)		108171-26-2
4-Chloro-1,2-benzenediamine	4-Chloro-*o*-phenylenediamine	95-83-0
2-Chloro-1,3-butadiene	Chloroprene	126-99-8
1-(2-Chloroethyl)-3-cyclohexyl-1-nitrosourea	CCNU; Lomustine; Belustine	13010-47-4
4-Chloro-2-methylaniline	*p*-Chloro-*o*-toluidine	95-69-2
4-Chloro-2-methylaniline hydrochloride	*p*-Chloro-*o*-toluidine hydrochloride	3165-93-3

Substance	Other Names	CAS RN
1-Chloro-2-methylpropene	Dimethylvinyl chloride	513-37-1
3-Chloro-2-methylpropene		563-47-3
Chlorozotocin	2-[[[(2-Chloroethyl)nitrosoamino]carbonyl]amino]-2-deoxy-*D*-glucose	54749-90-5
Fuchsin	C.I. Basic Red 9, monohydrochloride	569-61-9
Cobalt(II) sulfate	Cobaltous sulfate	10124-43-3
Cupferron		135-20-6
Dacarbazine	5-(3,3-Dimethyl-1-triazenyl)-1*H*-imidazole-4-carboxamide	4342-03-4
Decabromobiphenyl		13654-09-6
*c*is-Diaminedichloroplatinum	Cisplatin	15663-27-1
2,4-Diaminoanisole sulfate	1,3-Benzenediamine, 4-methoxy, sulfate	39156-41-7
4,4'-Diaminodiphenyl ether	4,4-Oxydianiline	101-80-4
4,4'-Diaminodiphenylmethane	4,4'-Methylenedianiline	101-77-9
4,4'-Diaminodiphenylmethane dihydrochloride	4,4'-Methylenedianiline dihydrochloride	13552-44-8
4,4'-Diaminodiphenyl sulfide	4,4'-Thiodianiline	139-65-1
Dibenz[a,h]acridine		226-36-8
Dibenz[a,j]acridine	7-Azadibenz[a,j]anthracene	224-42-0
Dibenz[a,h]anthracene	1,2:5,6-Dibenzanthracene	53-70-3
7*H*-Dibenzo[c,g]carbazole		194-59-2
Dibenzo[a,e]pyrene	Naphtho[1,2,3,4-def]chrysene	192-65-4
Dibenzo[a,h]pyrene	Dibenzo[b,def]chrysene	189-64-0
Dibenzo[a,i]pyrene	Benzo[rst]pentaphene	189-55-9
Dibenzo[a,l]pyrene	Dibenzo[def,p]chrysene	191-30-0
1,2-Dibromo-3-chloropropane		96-12-8
1,2-Dibromoethane	Ethylene dibromide; EDB	106-93-4
2,3-Dibromo-1-propanol	DBP	96-13-9
2,3-Dibromo-1-propanol, phosphate (3:1)	Tris(2,3-dibromopropyl) phosphate	126-72-7
p-Dichlorobenzene	1,4-Dichlorobenzene	106-46-7
3,3'-Dichloro-*p*-benzidine	3,3'-Dichloro[1,1'-biphenyl]-4,4'-diamine	91-94-1
3,3'-Dichloro-*p*-benzidine dihydrochloride	3,3'-Dichloro-[1,1'-biphenyl]-4,4'-diamine dihydrochloride	612-83-9
1,2-Dichloroethane	Ethylene dichloride	107-06-2
Dichloromethane	Methylene chloride	75-09-2
1,3-Dichloropropene (unspecified isomer)		542-75-6
Diethyl sulfate		64-67-5
2,3-Dihydro-6-propyl-2-thioxo-4(1H)-pyrimidinone	Propylthiouracil	51-52-5
1,8-Dihydroxy-9,10-anthracenedione	Danthron; 1,8-Dihydroxyanthraquinone	117-10-2
3,3'-Dimethoxybenzidine (and dyes metabolized to 3,3'-Dimethoxybenzidine)	Dianisidine	119-90-4
p-(Dimethylamino)azobenzene		60-11-7
2',3-Dimethyl-4-aminoazobenzene	*o*-Aminoazotoluene; 4-*o*-Tolylazo-*o*-toluidine	97-56-3
Dimethylcarbamic chloride	Dimethylcarbamoyl chloride	79-44-7
1,1-Dimethylhydrazine	UDMH	57-14-7
Dimethyl sulfate		77-78-1
1,6-Dinitropyrene		42397-64-8
1,8-Dinitropyrene		42397-65-9
1,4-Dioxane		123-91-1
1,2-Diphenylhydrazine	Hydrazobenzene	122-66-7
1,3-Diphenyl-1-triazene	Diazoaminobenzene	136-35-6

Substance	Other Names	CAS RN
Disperse Blue No. 1	1,4,5,8-Tetraamino-9,10-anthracenedione	2475-45-8
Doxorubicin hydrochloride	Adriamycin	23214-92-8
Epichlorohydrin	(Chloromethyl)oxirane	106-89-8
1,2-Epoxy-4-(epoxyethyl)cyclohexane	4-Vinyl-1-cyclohexene dioxide	106-87-6
Ethyl carbamate	Urethane	51-79-6
Ethyl methanesulfonate		62-50-0
N-Ethyl-*N*-nitrosourea	ENU; *N*-Nitroso-*N*-ethylurea	759-73-9
Fluoroethene	Vinyl fluoride	75-02-5
Formaldehyde (gas)	Methanal	50-00-0
Furan		110-00-9
Hexabromobiphenyl isomers	Firemaster FF-1	67774-32-7
Hexachlorobenzene	Perchlorobenzene	118-74-1
Hexachlorocyclohexane isomers		608-73-1
1,2,3,4,5,6-Hexachlorocyclohexane, (1α,2α,3β,4α,5β,6β)	α-Hexachlorocyclohexane	319-84-6
1,2,3,4,5,6-Hexachlorocyclohexane, (1α,2β,3α,4β,5α,6β)	β-Hexachlorocyclohexane	319-85-7
1,2,3,4,5,6-Hexachlorocyclohexane, (1α,2α,3β,4α,5α,6β)	Lindane; γ-Hexachlorocyclohexane	58-89-9
Hexachloroethane	Perchloroethane	67-72-1
Hexamethylphosphoric triamide	Hexamethylphosphoramide; Tris(dimethylamino)phosphine oxide	680-31-9
Hydrazine		302-01-2
Hydrazine sulfate		10034-93-2
2-Imidazolidinethione	Ethylene thiourea	96-45-7
Indeno[1,2,3-cd]pyrene	1,10-(1,2-Phenylene)pyrene	193-39-5
Kepone	Chlordecone	143-50-0
Lead and lead compounds	Includes all lead compounds	7439-92-1
o-Methoxyaniline hydrochloride	*o*-Anisidine hydrochloride	134-29-2
2-Methoxy-5-methylaniline	*p*-Cresidine; 5-Methyl-*o*-anisidine	120-71-8
o-Methylaniline	*o*-Toluidine	95-53-4
o-Methylaniline hydrochloride	*o*-Toluidine hydrochloride	636-21-5
2-Methyl-1,3-butadiene	Isoprene	78-79-5
5-Methylchrysene		3697-24-3
4,4-Methylenedianiline dihydrochloride	Benzenamine, 4,4'-methylenedi-, dihydrochloride	13552-44-8
Methyl methanesulfonate		66-27-3
N-Methyl-*N*'-nitro-*N*-nitrosoguanidine		70-25-7
N-Methyl-*N*-nitrosourea	*N*-Nitroso-*N*-methylurea	684-93-5
Methyloxirane	1,2-Propylene oxide	75-56-9
Metronidazole	2-Methyl-5-nitro-1*H*-imidazole-1-ethanol	443-48-1
Mirex	Hexachloropentadiene dimer	2385-85-5
Naphthalene		91-20-3
Nickel (metallic)		7440-02-0
Nitrilotriacetic acid	*N,N*-Bis(carboxymethyl)glycine	139-13-9
2-Nitroanisole	1-Methoxy-2-nitrobenzene	91-23-6
Nitrobenzene		98-95-3
6-Nitrochrysene		7496-02-8
Nitrofen	2,4-Dichloro-1-(4-nitrophenoxy)benzene	1836-75-5
Nitromethane		75-52-5
2-Nitropropane		79-46-9
1-Nitropyrene		5522-43-0

Substance	Other Names	CAS RN
4-Nitropyrene		57835-92-4
N-Nitrosodibutylamine	Dibutylnitrosamine	924-16-3
N-Nitrosodiethanolamine	2,2'-(Nitrosoimino)ethanol	1116-54-7
N-Nitrosodiethylamine	DEN; Diethylnitrosamine	55-18-5
N-Nitrosodimethylamine	DMN; Dimethylnitrosamine	62-75-9
4-(*N*-Nitrosomethylamino)-1-(3-pyridyl)-1-butanone	NNK; Ketone, 3-pyridyl-3-(*N*-methyl-*N*-nitrosamino)propyl	64091-91-4
N-Nitroso-*N*-methylvinylamine	*N*-Methyl-*N*-nitrosoethenamine	4549-40-0
4-Nitrosomorpholine	*N*-Nitrosomorpholine	59-89-2
N-Nitrosonornicotine	*N'*-Nitroso-3-(2-pyrrolidinyl)pyridine	16543-55-8
N-Nitrosopiperidine	1-Nitrosopiperidine	100-75-4
N-Nitroso-*N*-propyl-1-propanamine	*N*-Nitrosodipropylamine	621-64-7
N-Nitrosopyrrolidine		930-55-2
N-Nitrososarcosine	*N*-Methyl-*N*-nitrosoglycine	13256-22-9
Norethisterone	19-Norpregn-4-en-20-yn-3-one, 17-hydroxy-, (17 α)-	68-22-4
Ochratoxin A		303-47-9
Octabromobiphenyl isomers		61288-13-9
2-Oxetanone	β-Propiolactone	57-57-8
Oxiranemethanol	Glycidol	556-52-5
Oxymetholone	Androstan-3-one, 17-hydroxy-2-(hydroxymethylene)-17-methyl-	434-07-1
Phenazopyridine hydrochloride	3-(Phenylazo)-2,6-pyridinediamine, monohydrochloride	136-40-3
Phenolphthalein	3,3-Bis(4-hydroxyphenyl)-1(3*H*)-isobenzofuranone	77-09-8
Phenoxybenzamine hydrochloride		63-92-3
Phenyloxirane	Styrene-7,8-oxide	96-09-3
Phenytoin	5,5-Diphenyl-2,4-imidazolidinedione	57-41-0
Polybrominated biphenyls	PBBs	
Polychlorinated biphenyls	PCBs	1336-36-3
Procarbazine hydrochloride		366-70-1
Progesterone	Pregn-4-ene-3,20-dione	57-83-0
1,3-Propane sultone	1,2-Oxathiolane, 2,2-dioxide	1120-71-4
Propyleneimine	2-Methylaziridine	75-55-8
Reserpine		50-55-5
Safrole	5-(2-Propenyl)-1,3-benzodioxole	94-59-7
Selenium sulfide		7446-34-6
Streptozotocin	*D*-Glucopyranose, 2-deoxy-2-[[(methylnitrosoamino)carbon yl]amino]-	18883-66-4
Sulfallate	*N,N*-Diethyldithiocarbamic acid, 2-chloroallyl ester	95-06-7
Tetrachloroethene	Tetrachloroethylene; Perchloroethylene	127-18-4
Tetrachloromethane	Carbon tetrachloride	56-23-5
Tetrafluoroethene	Tetrafluoroethylene	116-14-3
N,N,N',N'-Tetramethyl-4,4'-diaminobenzophenone	Bis(dimethylamino)benzophenone; Michler's Ketone	90-94-8
Tetranitromethane		509-14-8
Thioacetamide	Ethanethioamide	62-55-5
Thiourea	Thiocarbamide	62-56-6
o-Tolidine	3,3'-Dimethylbenzidine	119-93-7
Toluene-2,4-diamine	2,4-Diaminotoluene	95-80-7
Toluene diisocyanate (unspecified isomer)		26471-62-5
Toxaphene	Polychlorocamphene	8001-35-2

Substance	Other Names	CAS RN
1*H*-1,2,4-Triazol-3-amine	Amitrole	61-82-5
1,1,1-Trichloro-2,2-bis(4-chlorophenyl)ethane	DDT; Dichlorodiphenyltrichloroethane	50-29-3
Trichloroethene	Trichloroethylene	79-01-6
Trichloromethane	Chloroform	67-66-3
(Trichloromethyl)benzene	Benzotrichloride	98-07-7
2,4,6-Trichlorophenol		88-06-2
1,2,3-Trichloropropane		96-18-4

Appendix A
Mathematical Tables

Miscellaneous Mathematical Constants A-1
Decimal Equivalents of Common Fractions A-2
Quadratic Formula A-2
Exponential and Hyperbolic Functions and Their Common Logarithms A-3
Natural Trigonometric Functions to Four Places A-6
Relation of Angular Functions in Terms of One Another A-8
Derivatives A-9
Integration A-11
Integrals A-15
Differential Equations A-46
Fourier Series A-57
Fourier Expansions for Basic Periodic Functions A-59
The Fourier Transforms A-61
Series Expansion A-65
Vector Analysis A-68
Orthogonal Curvilinear Coordinates A-75
Transformation of Integrals A-77
Bessel Functions A-78
The Factorial Function A-80
The Gamma Function A-81
The Beta Function A-82
The Error Function A-83
Orthogonal Polynomials A-83
Tables of Orthogonal Polynomials A-86
Clebsch–Gordan Coefficients A-87
Normal Probability Function A-88
Percentage Points, Student's *t*-Distribution A-91
Percentage Points, Chi-Square Distribution A-91
Percentage Points, *F*-Distribution A-93
Moment of Inertia for Various Bodies of Mass A-97

MISCELLANEOUS MATHEMATICAL CONSTANTS

π CONSTANTS

π = 3.14159 26535 89793 23846 26433 83279 50288 41971 69399 37511

$1/\pi$ = 0.31830 98861 83790 67153 77675 26745 02872 40689 19291 48091

π^2 = 9.86960 44010 89358 61883 44909 99876 15113 53136 99407 24079

$\log_e \pi$ = 1.14472 98858 49400 17414 34273 51353 05871 16472 94812 91531

$\log_{10} \pi$ = 0.49714 98726 94133 85435 12682 88290 89887 36516 78324 38044

$\log_{10} \sqrt{2\pi}$ = 0.39908 99341 79057 52478 25035 91507 69595 02099 34102 92128

CONSTANTS INVOLVING e

e = 2.71828 18284 59045 23536 02874 71352 66249 77572 47093 69996

$1/e$ = 0.36787 94411 71442 32159 55237 70161 46086 74458 11131 03177

e^2 = 7.38905 60989 30650 22723 04274 60575 00781 31803 15570 55185

M = $\log_{10} e$ = 0.43429 44819 03251 82765 11289 18916 60508 22943 97005 80367

$1/M$ = $\log_e 10$ = 2.30258 50929 94045 68401 79914 54684 36420 76011 01488 62877

$\log_{10} M$ = 9.63778 43113 00536 78912 29674 98645 − 10

π^e AND e^π CONSTANTS

π^e = 22.45915 77183 61045 47342 71522

e^π = 23.14069 26327 79269 00572 90864

$e^{-\pi}$ = 0.04321 39182 63772 24977 44177

$e^{\pi/2}$ = 4.81047 73809 65351 65547 30357

i^i = $e^{-\pi/2}$ = 0.20787 95763 50761 90854 69556

NUMERICAL CONSTANTS

$\sqrt{2}$ = 1.41421 35623 73095 04880 16887 24209 69807 85696 71875 37695

$\sqrt[3]{2}$ = 1.25992 10498 94873 16476 72106 07278 22835 05702 51464 70151

$\log_e 2$ = 0.69314 71805 59945 30941 72321 21458 17656 80755 00134 36026

$\log_{10} 2$ = 0.30102 99956 63981 19521 37388 94724 49302 67881 89881 46211

$\sqrt{3}$ = 1.73205 08075 68877 29352 74463 41505 87236 69428 05253 81039

$\sqrt[3]{3}$ = 1.44224 95703 07408 38232 16383 10780 10958 83918 69253 49935

$\log_e 3$ = 1.09861 22886 68109 69139 52452 36922 52570 46474 90557 82275

$\log_{10} 3$ = 0.47712 12547 19662 43729 50279 03255 11530 92001 28864 19070

OTHER CONSTANTS

Euler's Constant γ = 0.57721 56649 01532 86061

$\log_e \gamma$ = −0.54953 93129 81644 82234

Golden Ratio ϕ = 1.61803 39887 49894 84820 45868 34365 63811 77203 09180

DECIMAL EQUIVALENTS OF COMMON FRACTIONS

		1/64	0.015625			33/64	0.515625
	1/32	2/64	0.03125		17/32	34/64	0.53125
		3/64	0.046875			35/64	0.546875
1/16	2/32	4/64	0.0625	9/16	18/32	36/64	0.5625
		5/64	0.078125			37/64	0.578125
	3/32	6/64	0.09375		19/32	38/64	0.59375
		7/64	0.109375			39/64	0.609375
1/8	4/32	8/64	0.125	5/8	20/32	40/64	0.625
		9/64	0.140625			41/64	0.640625
	5/32	10/64	0.15625		21/32	42/64	0.65625
		11/64	0.171875			43/64	0.671875
3/16	6/32	12/64	0.1875	11/16	22/32	44/64	0.6875
		13/64	0.203125			45/64	0.703125
	7/32	14/64	0.21875		23/32	46/64	0.71875
		15/64	0.234375			47/64	0.734375
1/4	8/32	16/64	0.25	3/4	24/32	48/64	0.75
		17/64	0.265625			49/64	0.765625
	9/32	18/64	0.28125		25/32	50/64	0.78125
		19/64	0.296875			51/64	0.796875
5/16	10/32	20/64	0.3125	13/16	26/32	52/64	0.8125
		21/64	0.328125			53/64	0.828125
	11/32	22/64	0.34375		27/32	54/64	0.84375
		23/64	0.359375			55/64	0.859375
3/8	12/32	24/64	0.375	7/8	28/32	56/64	0.875
		25/64	0.390625			57/64	0.890625
	13/32	26/64	0.40625		29/32	58/64	0.90625
		27/64	0.421875			59/64	0.921875
7/16	14/32	28/64	0.4375	15/16	30/32	60/64	0.9375
		29/64	0.453125			61/64	0.953125
	15/32	30/64	0.46875		31/32	62/64	0.96875
		31/64	0.484375			63/64	0.984375
1/2	16/32	32/64	0.5	1/1	32/32	64/64	1

QUADRATIC FORMULA

The solutions of the equation $ax^2 + bx + c = 0$, where $a \neq 0$, are given by:

$$x = \frac{-b \pm \sqrt{b^2 - 4ac}}{2a}.$$

EXPONENTIAL AND HYPERBOLIC FUNCTIONS AND THEIR COMMON LOGARITHMS

	e^x		e^{-x}	$\sinh x$		$\cosh x$		$\tanh x$
x	Value	$\log_{10}$	Value	Value	$\log_{10}$	Value	$\log_{10}$	Value
0.00	1.0000	0.00000	1.00000	0.0000	$-\infty$	1.0000	0.00000	0.00000
0.01	1.0101	.00434	0.99005	.0100	−2.00001	1.0001	.00002	.01000
0.02	1.0202	.00869	.98020	.0200	−2.30106	1.0002	.00009	.02000
0.03	1.0305	.01303	.97045	.0300	−2.47719	1.0005	.00020	.02999
0.04	1.0408	.01737	.96079	.0400	−2.60218	1.0008	.00035	.03998
0.05	1.0513	.02171	.95123	.0500	−2.69915	1.0013	.00054	.04996
0.06	1.0618	.02606	.94176	.0600	−2.77841	1.0018	.00078	.05993
0.07	1.0725	.03040	.93239	.0701	−2.84545	1.0025	.00106	.06989
0.08	1.0833	.03474	.92312	.0801	−2.90355	1.0032	.00139	.07983
0.09	1.0942	.03909	.91393	.0901	−2.95483	1.0041	.00176	.08976
0.10	1.1052	.04343	.90484	.1002	−1.00072	1.0050	.00217	.09967
0.11	1.1163	.04777	.89583	.1102	−1.04227	1.0061	.00262	.10956
0.12	1.1275	.05212	.88692	.1203	−1.08022	1.0072	.00312	.11943
0.13	1.1388	.05646	.87809	.1304	−1.11517	1.0085	.00366	.12927
0.14	1.1503	.06080	.86936	.1405	−1.14755	1.0098	.00424	.13909
0.15	1.1618	.06514	.86071	.1506	−1.17772	1.0113	.00487	.14889
0.16	1.1735	.06949	.85214	.1607	−1.20597	1.0128	.00554	.15865
0.17	1.1853	.07383	.84366	.1708	−1.23254	1.0145	.00625	.16838
0.18	1.1972	.07817	.83527	.1810	−1.25762	1.0162	.00700	.17808
0.19	1.2092	.08252	.82696	.1911	−1.28136	1.0181	.00779	.18775
0.20	1.2214	.08686	.81873	.2013	−1.30392	1.0201	.00863	.19738
0.21	1.2337	.09120	.81058	.2115	−1.32541	1.0221	.00951	.20697
0.22	1.2461	.09554	.80252	.2218	−1.34592	1.0243	.01043	.21652
0.23	1.2586	.09989	.79453	.2320	−1.36555	1.0266	.01139	.22603
0.24	1.2712	.10423	.78663	.2423	−1.38437	1.0289	.01239	.23550
0.25	1.2840	.10857	.77880	.2526	−1.40245	1.0314	.01343	.24492
0.26	1.2969	.11292	.77105	.2629	−1.41986	1.0340	.01452	.25430
0.27	1.3100	.11726	.76338	.2733	−1.43663	1.0367	.01564	.26362
0.28	1.3231	.12160	.75578	.2837	−1.45282	1.0395	.01681	.27291
0.29	1.3364	.12595	.74826	.2941	−1.46847	1.0423	.01801	.28213
0.30	1.3499	.13029	.74082	.3045	−1.48362	1.0453	.01926	.29131
0.31	1.3634	.13463	.73345	.3150	−1.49830	1.0484	.02054	.30044
0.32	1.3771	.13897	.72615	.3255	−1.51254	1.0516	.02187	.30951
0.33	1.3910	.14332	.71892	.3360	−1.52637	1.0549	.02323	.31852
0.34	1.4049	.14766	.71177	.3466	−1.53981	1.0584	.02463	.32748
0.35	1.4191	.15200	.70469	.3572	−1.55290	1.0619	.02607	.33638
0.36	1.4333	.15635	.69768	.3678	−1.56564	1.0655	.02755	.34521
0.37	1.4477	.16069	.69073	.3785	−1.57807	1.0692	.02907	.35399
0.38	1.4623	.16503	.68386	.3892	−1.59019	1.0731	.03063	.36271
0.39	1.4770	.16937	.67706	.4000	−1.60202	1.0770	.03222	.37136
0.40	1.4918	.17372	.67032	.4108	−1.61358	1.0811	.03385	.37995
0.41	1.5063	.17806	.66365	.4216	−1.62488	1.0852	.03552	.33847
0.42	1.5220	.18240	.65705	.4325	−1.63594	1.0895	.03723	.39693
0.43	1.5373	.18675	.65051	.4434	−1.64677	1.0939	.03897	.40532
0.44	1.5527	.19109	.64404	.4543	−1.65738	1.0984	.04075	.41364
0.45	1.5683	.19543	.63763	.4653	−1.66777	1.1030	.04256	.42190
0.46	1.5841	.19978	.63128	.4764	−1.67797	1.1077	.04441	.43008
0.47	1.6000	.20412	.62500	.4875	−1.68797	1.1125	.04630	.43820
0.48	1.6161	.20846	.61878	.4986	−1.69779	1.1174	.04822	.44624
0.49	1.6323	.21280	.61263	.5098	−1.70744	1.1225	.05018	.45422
0.50	1.6487	.21715	.60653	.5211	−1.71692	1.1276	.05217	.46212
0.51	1.6653	.22149	.60050	.5324	−1.72624	1.1329	.05419	.46995
0.52	1.6820	.22583	.59452	.5438	−1.73540	1.1383	.05625	.47770
0.53	1.6989	.23018	.58860	.5552	−1.74442	1.1438	.05834	.48538

	e^x		e^{-x}	sinh x		cosh x		tanh x
x	Value	$\log_{10}$	Value	Value	$\log_{10}$	Value	$\log_{10}$	Value
0.54	1.7160	.23452	.58275	.5666	−1.75330	1.1494	.06046	.49299
0.55	1.7333	.23886	.57695	.5782	−1.76204	1.1551	.06262	.50052
0.56	1.7507	.24320	.57121	.5897	−1.77065	1.1609	.06481	.50798
0.57	1.7683	.24755	.56553	.6014	−1.77914	1.1669	.06703	.51536
0.58	1.7860	.25189	.55990	.6131	−1.78751	1.1730	.06929	.52267
0.59	1.8040	.25623	.55433	.6248	−1.79576	1.1792	.07157	.52990
0.60	1.8221	.26058	.54881	.6367	−1.80390	1.1855	.07389	.53705
0.61	1.8404	.26492	.54335	.6485	−1.81194	1.1919	.07624	.54413
0.62	1.8589	.26926	.53794	.6605	−1.81987	1.1984	.07861	.55113
0.63	1.8776	.27361	.53259	.6725	−1.82770	1.2051	.08102	.55805
0.64	1.8965	.27795	.52729	.6846	−1.83543	1.2119	.08346	.56490
0.65	1.9155	.28229	.52205	.6967	−1.84308	1.2188	.08593	.57167
0.66	1.9348	.28664	.51685	.7090	−1.85063	1.2258	.08843	.57836
0.67	1.9542	.29098	.51171	.7213	−1.85809	1.2330	.09095	.58498
0.68	1.9739	.29532	.50662	.7336	−1.86548	1.2402	.09351	.59152
0.69	1.9937	.29966	.50158	.7461	−1.87278	1.2476	.09609	.59798
0.70	2.0138	.30401	.49659	.7586	−1.88000	1.2552	.09870	.60437
0.71	2.0340	.30835	.49164	.7712	−1.88715	1.2628	.10134	.61068
0.72	2.0544	.31269	.48675	.7838	−1.89423	1.2706	.10401	.61691
0.73	2.0751	.31703	.48191	.7966	−1.90123	1.2785	.10670	.62307
0.74	2.0959	.32138	.47711	.8094	−1.90817	1.2865	.10942	.62915
0.75	2.1170	.32572	.47237	.8223	−1.91504	1.2947	.11216	.63515
0.76	2.1383	.33006	.46767	.8353	−1.92185	1.3030	.11493	.64108
0.77	2.1598	.33441	.46301	.8484	−1.92859	1.3114	.11773	.64693
0.78	2.1815	.33875	.45841	.8615	−1.93527	1.3199	.12055	.65721
0.79	2.2034	.34309	.45384	.8748	−1.94190	1.3286	.12340	.65841
0.80	2.2255	.34744	.44933	.8881	−1.94846	1.3374	.12627	.66404
0.81	2.2479	.35178	.44486	.9015	−1.95498	1.3464	.12917	.66959
0.82	2.2705	.35612	.44043	.9150	−1.96144	1.3555	.13209	.67507
0.83	2.2933	.36046	.43605	.9286	−1.96784	1.3647	.13503	.68048
0.84	2.3164	.36481	.43171	.9423	−1.97420	1.3740	.13800	.68581
0.85	2.3396	.36915	.42741	.9561	−1.98051	1.3835	.14099	.69107
0.86	2.3632	.37349	.42316	.9700	−1.98677	1.3932	.14400	.69626
0.87	2.3869	.37784	.41895	.9840	−1.99299	1.4029	.14704	.70137
0.88	2.4100	.38218	.41478	.9981	−1.99916	1.4128	.15009	.70642
0.89	2.4351	.38652	.41066	1.0122	0.00528	1.4229	.15317	.71139
0.90	2.4596	.39087	.40657	1.0265	.01137	1.4331	.15627	.21630
0.91	2.4843	.39521	.40242	1.0409	.01741	1.4434	.15939	.72113
0.92	2.5093	.39955	.39852	1.0554	.02341	1.4539	.16254	.72590
0.93	2.5345	.40389	.39455	1.0700	.02937	1.4645	.16570	.73059
0.94	2.5600	.40824	.39063	1.0847	.03530	1.4753	.16888	.73522
0.95	2.5857	.41258	.38674	1.0995	.04119	1.4862	.17208	.73978
0.96	2.6117	.41692	.38289	1.1144	.04704	1.4973	.17531	.74428
0.97	2.6379	.42127	.37908	1.1294	.05286	1.5085	.17855	.74870
0.98	2.6645	.42561	.37531	1.1446	.05864	1.5199	.18181	.75307
0.99	2.6912	.42995	.37158	1.1598	.06439	1.5314	.18509	.75736
1.00	2.7183	.43429	.36788	1.1752	.07011	1.5431	.18839	.76159
1.10	3.0042	.47772	.33287	1.3356	.12569	1.6685	.22233	.80050
1.20	3.3201	.52115	.30119	1.5095	.17882	1.8107	.25784	.83365
1.30	3.6693	.56458	.27253	1.6984	.23004	1.9709	.29467	.86172
1.40	4.0552	.60801	.24660	1.9043	.27974	2.1509	.33262	.88535
1.50	4.4817	.65144	.22313	2.1293	.32823	2.3524	.37151	.90515
1.60	4.9530	.69487	.20190	2.3756	.37577	2.5775	.41119	.92167
1.70	5.4739	.73830	.18268	2.6456	.42253	2.8283	.45153	.93541
1.80	6.0496	.78173	.16530	2.9422	.46867	3.1075	.49241	.94681
1.90	6.6859	.82516	.14957	3.2682	.51430	3.4177	.53374	.95624
2.00	7.3891	.86859	.13534	3.6269	.55953	3.7622	.57544	.96403

	e^x		e^{-x}	sinh x		cosh x		tanh x
x	Value	$\log_{10}$	Value	Value	$\log_{10}$	Value	$\log_{10}$	Value
2.10	8.1662	.91202	.12246	4.0219	.60443	4.1443	.61745	.97045
2.20	9.0250	.95545	.11080	4.4571	.64905	4.5679	.65972	.97574
2.30	9.9742	.99888	.10026	4.9370	.69346	5.0372	.70219	.98010
2.40	11.023	1.04231	.09072	5.4662	.73769	5.5569	.74484	.98367
2.50	12.182	1.08574	.08208	6.0502	.78177	6.1323	.78762	.98661
2.60	13.464	1.12917	.07427	6.6947	.82573	6.7690	.83052	.98903
2.70	14.880	1.17260	.06721	7.4063	.86960	7.4735	.87352	.99101
2.80	16.445	1.21602	.06081	8.1919	.91339	8.2527	.91660	.99263
2.90	18.174	1.25945	.05502	9.0596	.95711	9.1146	.95974	.99396
3.00	20.086	1.30288	.04979	10.018	1.00078	10.068	1.00293	0.99505
3.50	33.115	1.52003	.03020	16.543	1.21860	16.573	1.21940	0.99818
4.00	54.598	1.73718	.01832	27.290	1.43600	27.308	1.43629	0.99933
4.50	90.017	1.95433	.01111	45.003	1.65324	45.014	1.65335	0.99975
5.00	148.41	2.17147	.00674	74.203	1.87042	74.210	1.87046	0.99991
5.50	244.69	2.38862	.00409	122.34	2.08758	122.35	2.08760	0.99997
6.00	403.43	2.60577	.00248	201.71	2.30473	201.72	2.30474	0.99999
6.50	665.14	2.82291	.00150	332.57	2.52188	332.57	2.52189	1.00000
7.00	1096.6	3.04006	.00091	548.32	2.73904	548.32	2.73903	1.00000
7.50	1808.0	3.25721	.00055	904.02	2.95618	904.02	2.95618	1.00000
8.00	2981.0	3.47436	.00034	1490.5	3.17333	1490.5	3.17333	1.00000
8.50	4914.8	3.69150	.00020	2457.4	3.39047	2457.4	3.39047	1.00000
9.00	8103.1	3.90865	.00012	4051.5	3.60762	4051.5	3.60762	1.00000
9.50	13360.	4.12580	.00007	6679.9	3.82477	6679.9	3.82477	1.00000
10.00	22026.	4.34294	.00005	11013.	4.04191	11013.	4.04191	1.00000

NATURAL TRIGONOMETRIC FUNCTIONS TO FOUR PLACES

x radians	*x* degrees	sin *x*	cos *x*	tan *x*	cot *x*	sec *x*	csc *x*		
.0000	0° 00′	.000	1.0000	.0000	–	1.000	–	90° 00′	1.5708
.0029	10	.0029	1.0000	.0029	343.8	1.000	343.8	50	1.5679
.0058	20	.0058	1.0000	.0058	171.9	1.000	171.9	40	1.5650
.0087	30	.0087	1.0000	.0087	114.6	1.000	114.6	30	1.5621
.0116	40	.0116	.9999	.0116	85.94	1.000	85.95	20	1.5592
.0145	50	.0145	.9999	.0145	68.75	1.000	68.76	10	1.5563
.0175	1° 00′	.0175	.9998	.0175	57.29	1.000	57.30	89° 00′	1.5533
.0262	30	.0262	.9997	.0262	38.19	1.000	38.20	30	1.5446
.0349	2° 00′	.0349	.9994	.0349	28.64	1.001	28.65	88° 00′	1.5359
.0436	30	.0436	.9990	.0437	22.90	1.001	22.93	30	1.5272
.0524	3° 00′	.0523	.9986	.0524	19.08	1.001	19.11	87° 00′	1.5184
.0611	30	.0610	.9981	.0612	16.35	1.002	16.38	30	1.5097
.0698	4° 00′	.0698	.9976	.0699	14.30	1.002	14.34	86° 00′	1.5010
.0785	30	.0785	.9969	.0787	12.71	1.003	12.75	30	1.4923
.0873	5° 00′	.0872	.9962	.0875	11.43	1.004	11.47	85° 00′	1.4835
.0960	30	.0958	.9954	.0963	10.39	1.005	10.43	30	1.4748
.1047	6° 00′	.1045	.9945	.1051	9.514	1.006	9.597	84° 00′	1.4661
.1134	30	.1132	.9936	.1139	8.777	1.006	8.834	30	1.4573
.1222	7° 00′	.1219	.9925	.1228	8.144	1.008	8.206	83° 00′	1.4486
.1309	30	.1305	.9914	.1317	7.596	1.009	7.661	30	1.4399
.1396	8° 00′	.1392	.9903	.1405	7.115	1.010	7.185	82° 00′	1.4312
.1484	30	.1478	.9890	.1495	6.691	1.011	6.765	30	1.4224
.1571	9° 00′	.1564	.9877	.1584	6.314	1.012	6.392	81° 00′	1.4137
.1658	30	.1650	.9863	.1673	5.976	1.014	6.059	30	1.4050
.1745	10° 00′	.1736	.9848	.1763	5.671	1.015	5.759	80° 00′	1.3963
.1833	30	.1822	.9833	.1853	5.396	1.017	5.487	30	1.3875
.1920	11° 00′	.1908	.9816	.1944	5.145	1.019	5.241	79° 00′	1.3788
.2007	30	.1994	.9799	.2035	4.915	1.020	5.016	30	1.3701
.2094	12° 00′	.2079	.9781	.2126	4.705	1.022	4.810	78° 00′	1.3614
.2182	30	.2164	.9763	.2217	4.511	1.025	4.620	30	1.3526
.2269	13° 00′	.2250	.9744	.2309	4.331	1.026	4.445	77° 00′	1.3439
.2356	30	.2334	.9724	.2401	4.165	1.028	4.284	30	1.3352
.2443	14° 00′	.2419	.9703	.2493	4.011	1.031	4.134	76° 00′	1.3265
.2531	30	.2404	.9681	.2586	3.867	1.033	3.994	30	1.3177
.2618	15° 00′	.2588	.9659	.2679	3.732	1.035	3.864	75° 00′	1.3090
.2705	30	.2672	.9636	.2773	3.606	1.038	3.742	30	1.3003
.2793	16° 00′	.2756	.9613	.2867	3.487	1.040	3.628	74° 00′	1.2915
.2880	30	.2840	.9588	.2962	3.376	1.043	3.521	30	1.2828
.2967	17° 00′	.2924	.9563	.3057	3.271	1.046	3.420	73° 00′	1.2741
.3054	30	.3007	.9537	.3153	3.172	1.049	3.326	30	1.2654
.3142	18° 00′	.3090	.9511	.3249	3.078	1.051	3.236	72° 00′	1.2566
.3229	30	.3173	.9483	.3346	2.989	1.054	3.152	30	1.2479
.3316	19° 00′	.3256	.9455	.3443	2.904	1.058	3.072	71° 00′	1.2392
.3403	30	.3338	.9426	.3541	2.824	1.061	2.996	30	1.2305
.3491	20° 00′	.3420	.9397	.3640	2.747	1.064	2.924	70° 00′	1.2217
.3578	30	.3502	.9367	.3739	2.675	1.068	2.855	30	1.2130
.3665	21° 00′	.3584	.9336	.3839	2.605	1.071	2.790	69° 00′	1.2043
.3752	30	.3665	.9304	.3939	2.539	1.075	2.729	30	1.1956
.3840	22° 00′	.3746	.9272	.4040	2.475	1.079	2.669	68° 00′	1.1868
.3927	30	.3827	.9239	.4142	2.414	1.082	2.613	30	1.1781
.4014	23° 00′	.3907	.9205	.4245	2.356	1.086	2.559	67° 00′	1.1694
.4102	30	.3987	.9171	.4348	2.300	1.090	2.508	30	1.1606
.4189	24° 00′	.4067	.9135	.4452	2.246	1.095	2.459	66° 00′	1.1519
		cos *y*	sin *y*	cot *y*	tan *y*	csc *y*	sec *y*	*y* degrees	*y* radians

x radians	x degrees	sin x	cos x	tan x	cot x	sec x	csc x		
.4276	30	.4147	.9100	.4557	2.194	1.099	2.411	30	1.1432
.4363	25° 00′	.4226	.9063	.4663	2.145	1.103	2.366	65° 00′	1.1345
.4451	30	.4305	.9026	.4770	2.097	1.108	2.323	30	1.1257
.4538	26° 00′	.4384	.8988	.4877	2.050	1.113	2.281	64° 00′	1.1170
.4625	30	.4462	.8949	.4986	2.006	1.117	2.241	30	1.1083
.4712	27° 00′	.4540	.8910	.5095	1.963	1.122	2.203	63° 00′	1.0996
.4800	30	.4617	.8870	.5206	1.921	1.127	2.166	30	1.0908
.4887	28° 00′	.4695	.8829	.5317	1.881	1.133	2.130	62° 00′	1.0821
.4974	30	.4772	.8788	.5430	1.842	1.138	2.096	30	1.0734
.5061	29° 00′	.4848	.8746	.5543	1.804	1.143	2.063	61° 00′	1.0647
.5149	30	.4924	.8704	.5658	1.767	1.149	2.031	30	1.0559
.5236	30° 00′	.5000	.8660	.5774	1.732	1.155	2.000	60° 00′	1.0472
.5323	30	.5075	.8616	.5890	1.698	1.161	1.970	30	1.0385
.5411	31° 00′	.5150	.8572	.6009	1.664	1.167	1.942	59° 00′	1.0297
.5498	30	.5225	.8526	.6128	1.632	1.173	1.914	30	1.0210
.5585	32° 00′	.5299	.8480	.6249	1.600	1.179	1.887	58° 00′	1.0123
.5672	30	.5373	.8434	.6371	1.570	1.186	1.861	30	1.0036
.5760	33° 00′	.5446	.8397	.6494	1.540	1.192	1.836	57° 00′	.9948
.5847	30	.5519	.8339	.6619	1.511	1.199	1.812	30	.9861
.5934	34° 00′	.5592	.8290	.6745	1.483	1.206	1.788	56° 00′	.9774
.6021	30	.5664	.8241	.6873	1.455	1.213	1.766	30	.9687
.6109	35° 00′	.5736	.8192	.7002	1.428	1.221	1.743	55° 00′	.9599
.6196	30	.5807	.8141	.7133	1.402	1.228	1.722	30	.9512
.6283	36° 00′	.5878	.8090	.7265	1.376	1.236	1.701	54° 00′	.9425
.6370	30	.5948	.8039	.7400	1.351	1.244	1.681	30	.9338
.6458	37° 00′	.6018	.7986	.7536	1.327	1.252	1.662	53° 00′	.9250
.6545	30	.6088	.7934	.7673	1.303	1.260	1.643	30	.9163
.6632	38° 00′	.6157	.7880	.7813	1.280	1.269	1.624	52° 00′	.9076
.6720	30	.6225	.7826	.7954	1.257	1.278	1.606	30	.8988
.6807	39° 00′	.6293	.7771	.8098	1.235	1.287	1.589	51° 00′	.8901
.6894	30	.6361	.7716	.8243	1.213	1.296	1.572	30	.8814
.6981	40° 00′	.6428	.7660	.8391	1.192	1.305	1.556	50° 00′	.8727
.7069	30	.6494	.7604	.8541	1.171	1.315	1.540	30	.8639
.7156	41° 00′	.6561	.7547	.8693	1.150	1.325	1.524	49° 00′	.8552
.7243	30	.6626	.7490	.8847	1.130	1.335	1.509	30	.8465
.7330	42° 00′	.6691	.7431	.9004	1.111	1.346	1.494	48° 00′	.8378
.7418	30	.6756	.7373	.9163	1.091	1.356	1.480	30	.8290
.7505	43° 00′	.6820	.7314	.9325	1.072	1.367	1.466	47° 00′	.8203
.7592	30	.6884	.7254	.9490	1.054	1.379	1.453	30	.8116
.7679	44° 00′	.6947	.7193	.9657	1.036	1.390	1.440	46° 00′	.8029
.7767	30	.7009	.7133	.9827	1.018	1.402	1.427	30	.7941
.7854	45° 00′	.7071	.7071	1.0000	1.0000	1.414	1.414	45° 00′	.7854
		cos y	sin y	cot y	tan y	csc y	sec y	y degrees	y radians

RELATION OF ANGULAR FUNCTIONS IN TERMS OF ONE ANOTHER

Trigonometric Functions

Function	$\sin\alpha$	$\cos\alpha$	$\tan\alpha$	$\cot\alpha$	$\sec\alpha$	$\csc\alpha$
$\sin\alpha$	$\sin\alpha$	$\pm\sqrt{1-\cos^2\alpha}$	$\frac{\tan\alpha}{\pm\sqrt{1+\tan^2\alpha}}$	$\frac{1}{\pm\sqrt{1+\cot^2\alpha}}$	$\frac{\pm\sqrt{\sec^2\alpha-1}}{\sec\alpha}$	$\frac{1}{\csc\alpha}$
$\cos\alpha$	$\pm\sqrt{1-\sin^2\alpha}$	$\cos\alpha$	$\frac{1}{\pm\sqrt{1+\tan^2\alpha}}$	$\frac{\cot\alpha}{\pm\sqrt{1+\cot^2\alpha}}$	$\frac{1}{\sec\alpha}$	$\frac{\pm\sqrt{\csc^2\alpha-1}}{\csc\alpha}$
$\tan\alpha$	$\frac{\sin\alpha}{\pm\sqrt{1-\sin^2\alpha}}$	$\frac{\pm\sqrt{1-\cos^2\alpha}}{\cos\alpha}$	$\tan\alpha$	$\frac{1}{\cot\alpha}$	$\pm\sqrt{\sec^2\alpha-1}$	$\frac{1}{\pm\sqrt{\csc^2\alpha-1}}$
$\cot\alpha$	$\frac{\pm\sqrt{1-\sin^2\alpha}}{\sin\alpha}$	$\frac{\cos\alpha}{\pm\sqrt{1-\cos^2\alpha}}$	$\frac{1}{\tan\alpha}$	$\cot\alpha$	$\frac{1}{\pm\sqrt{\sec^2\alpha-1}}$	$\pm\sqrt{\csc^2\alpha-1}$
$\sec\alpha$	$\frac{1}{\pm\sqrt{1-\sin^2\alpha}}$	$\frac{1}{\cos\alpha}$	$\pm\sqrt{1+\tan^2\alpha}$	$\frac{\pm\sqrt{1+\cot^2\alpha}}{\cot\alpha}$	$\sec\alpha$	$\frac{\csc\alpha}{\pm\sqrt{\csc^2\alpha-1}}$
$\csc\alpha$	$\frac{1}{\sin\alpha}$	$\frac{1}{\pm\sqrt{1-\cos^2\alpha}}$	$\frac{\pm\sqrt{1+\tan^2\alpha}}{\tan\alpha}$	$\pm\sqrt{1+\cot^2\alpha}$	$\frac{\sec\alpha}{\pm\sqrt{\sec^2\alpha-1}}$	$\csc\alpha$

Note: The choice of sign depends upon the quadrant in which the angle terminates.

Hyperbolic Functions

Function	$\sinh x$	$\cosh x$	$\tanh x$
$\sinh x =$	$\sinh x$	$\pm\sqrt{\cosh^2 x-1}$	$\frac{\tanh x}{\sqrt{1-\tanh^2 x}}$
$\cosh x =$	$\sqrt{1+\sinh^2 x}$	$\cosh x$	$\frac{1}{\sqrt{1-\tanh^2 x}}$
$\tanh x =$	$\frac{\sinh x}{\sqrt{1+\sinh^2 x}}$	$\pm\frac{\sqrt{\cosh^2 x-1}}{\cosh x}$	$\tanh x$
$\mathrm{cosech}\, x =$	$\frac{1}{\sinh x}$	$\pm\frac{1}{\sqrt{\cosh^2 x-1}}$	$\frac{\sqrt{1-\tanh^2 x}}{\tanh x}$
$\mathrm{sech}\, x =$	$\frac{1}{\sqrt{1+\sinh^2 x}}$	$\frac{1}{\cosh x}$	$\sqrt{1-\tanh^2 x}$
$\coth x =$	$\frac{\sqrt{1+\sinh^2 x}}{\sinh x}$	$\frac{\pm\cosh x}{\sqrt{\cosh^2 x-1}}$	$\frac{1}{\tanh x}$

Function	$\mathrm{cosech}\, x$	$\mathrm{sech}\, x$	$\coth x$
$\sinh x =$	$\frac{1}{\mathrm{cosech}\, x}$	$\pm\frac{\sqrt{1-\mathrm{sech}^2 x}}{\mathrm{sech}\, x}$	$\frac{\pm 1}{\sqrt{\coth^2 x-1}}$
$\cosh x =$	$\pm\frac{\sqrt{\mathrm{cosech}^2 x+1}}{\mathrm{cosech}\, x}$	$\frac{1}{\mathrm{sech}\, x}$	$\pm\frac{\coth x}{\sqrt{\coth^2 x-1}}$
$\tanh x =$	$\frac{1}{\sqrt{\mathrm{cosech}^2 x+1}}$	$\pm\sqrt{1+\mathrm{sech}^2 x}$	$\frac{1}{\coth x}$
$\mathrm{cosech}\, x =$	$\mathrm{cosech}\, x$	$\pm\frac{\mathrm{sech}\, x}{\sqrt{1-\mathrm{sech}^2 x}}$	$\pm\frac{\sqrt{\coth^2 x-1}}{1}$
$\mathrm{sech}\, x =$	$\pm\frac{\mathrm{cosech}\, x}{\sqrt{\mathrm{cosech}^2 x+1}}$	$\mathrm{sech}\, x$	$\pm\frac{\sqrt{\coth^2 x-1}}{\coth x}$
$\coth x =$	$\sqrt{\mathrm{cosech}^2 x+1}$	$\pm\frac{1}{\sqrt{1-\mathrm{sech}^2 x}}$	$\coth x$

Whenever two signs are shown, choose + sign if x is positive, − sign if x is negative.

DERIVATIVES

In the following formulas u, v, w represent functions of x, while a, c, n represent fixed real numbers. All arguments in the trigonometric functions are measured in radians, and all inverse trigonometric and hyperbolic functions represent principal values *Let $y = f(x)$ and $\frac{dy}{dx} = \frac{d[f(x)]}{dx} = f'(x)$ define, respectively, a function and its derivative for any value x in their common domain. The differential for the function at such a value x is accordingly defined as

$$dy = d[f(x)] = \frac{dy}{dx}\,dx = \frac{d[f(x)]}{dx}\,dx = f'(x)\,dx$$

Each derivative formula has an associated differential formula. For example, formula 6 below has the differential formula

$$d(uvw) = uv\,dw + vw\,du + uw\,dv$$

1. $\frac{d}{dx}(a) = 0$

2. $\frac{d}{dx}(x) = 1$

3. $\frac{d}{dx}(au) = a\frac{du}{dx}$

4. $\frac{d}{dx}(u + v - w) = \frac{du}{dx} + \frac{dv}{dx} - \frac{dw}{dx}$

5. $\frac{d}{dx}(uv) = u\frac{dv}{dx} + v\frac{du}{dx}$

6. $\frac{d}{dx}(uvw) = uv\frac{dw}{dx} + vw\frac{du}{dx} + uw\frac{dv}{dx}$

7. $\frac{d}{dx}\left(\frac{u}{v}\right) = \frac{v\frac{du}{dx} - u\frac{dv}{dx}}{v^2} = \frac{1}{v}\frac{du}{dx} - \frac{u}{v^2}\frac{dv}{dx}$

8. $\frac{d}{dx}(u^n) = nu^{n-1}\frac{du}{dx}$

9. $\frac{d}{dx}(\sqrt{u}) = \frac{1}{2\sqrt{u}}\frac{du}{dx}$

10. $\frac{d}{dx}\left(\frac{1}{u}\right) = -\frac{1}{u^2}\frac{du}{dx}$

11. $\frac{d}{dx}\left(\frac{1}{u^n}\right) = -\frac{n}{u^{n+1}}\frac{du}{dx}$

12. $\frac{d}{dx}\left(\frac{u^n}{v^m}\right) = \frac{u^{n-1}}{v^{m+1}}\left(nv\frac{du}{dx} - mu\frac{dv}{dx}\right)$

13. $\frac{d}{dx}(u^n v^m) = u^{n-1}v^{m-1}\left(nv\frac{du}{dx} + mu\frac{dv}{dx}\right)$

14. $\frac{d}{dx}[f(u)] = \frac{d}{du}[f(u)] \cdot \frac{du}{dx}$

15. $\frac{d^2}{dx^2}[f(u)] = \frac{df(u)}{du} \cdot \frac{d^2u}{dx^2} + \frac{d^2 f(u)}{du^2} \cdot \left(\frac{du}{dx}\right)^2$

16. $\frac{d^n}{dx^n}[uv] = \binom{n}{0}v\frac{d^nu}{dx^n} + \binom{n}{1}\frac{dv}{dx}\frac{d^{n-1}u}{dx^{n-1}} + \binom{n}{2}\frac{d^2v}{dx^2}\frac{d^{n-2}u}{dx^{n-2}}$

$$+ \cdots + \binom{n}{k}\frac{d^kv}{dx^k}\frac{d^{n-k}u}{dx^{n-k}} + \cdots + \binom{n}{n}u\frac{d^nv}{dx^n}$$

where $\binom{n}{r} = \frac{n!}{r!(n-r)!}$ is the binomial coefficient, n non-negative integer, and $\binom{n}{0} = 1$.

17. $\frac{du}{dx} = \frac{1}{\frac{dx}{du}}$ if $\frac{dx}{du} \neq 0$

18. $\frac{d}{dx}(\log_a u) = (\log_a e)\frac{1}{u}\frac{du}{dx}$

19. $\frac{d}{dx}(\log_e u) = \frac{1}{u}\frac{du}{dx}$

20. $\frac{d}{dx}(a^u) = a^u(\log_e a)\frac{du}{dx}$

21. $\frac{d}{dx}(e^u) = e^u\frac{du}{dx}$

22. $\frac{d}{dx}(u^v) = vu^{v-1}\frac{du}{dx} + (\log_e u)\, u^v\frac{dv}{dx}$

23. $\frac{d}{dx}(\sin u) = (\cos u)\frac{du}{dx}$

24. $\frac{d}{dx}(\cos u) = -(\sin u)\frac{du}{dx}$

25. $\frac{d}{dx}(\tan u) = (\sec^2 u)\frac{du}{dx}$

26. $\frac{d}{dx}(\cot u) = -(\csc^2 u)\frac{du}{dx}$

27. $\frac{d}{dx}(\sec u) = \sec u \cdot \tan u\frac{du}{dx}$

28. $\frac{d}{dx}(\csc u) = -\csc u \cdot \cot u\frac{du}{dx}$

29. $\frac{d}{dx}(\text{vers}\, u) = \sin u\frac{du}{dx}$

30. $\frac{d}{dx}(\arcsin u) = \frac{1}{\sqrt{1-u^2}}\frac{du}{dx}, \quad \left(-\frac{\pi}{2} \le \arcsin u \le \frac{\pi}{2}\right)$

31. $\frac{d}{dx}(\arccos u) = -\frac{1}{\sqrt{1-u^2}}\frac{du}{dx}, \quad (0 \le \arccos u \le \pi)$

32. $\frac{d}{dx}(\arctan u) = \frac{1}{1+u^2}\frac{du}{dx}, \quad \left(-\frac{\pi}{2} < \arctan u < \frac{\pi}{2}\right)$

33. $\frac{d}{dx}(\text{arc}\cot u) = -\frac{1}{1+u^2}\frac{du}{dx}, \quad (0 \le \text{arc}\cot u \le \pi)$

34. $\frac{d}{dx}(\text{arc}\sec u) = \frac{1}{u\sqrt{u^2-1}}\frac{du}{dx}, \quad \left(0 \le \text{arc}\sec u < \frac{\pi}{2}, -\pi \le \text{arc}\sec u < -\frac{\pi}{2}\right)$

35. $\frac{d}{dx}(\text{arc}\csc u) = -\frac{1}{u\sqrt{u^2-1}}\frac{du}{dx}, \quad \left(0 < \text{arc}\csc u \le \frac{\pi}{2}, -\pi < \text{arc}\csc u \le -\frac{\pi}{2}\right)$

36. $\frac{d}{dx}(\text{arc vers}\, u) = \frac{1}{\sqrt{2u-u^2}}\frac{du}{dx}, \quad (0 \le \text{arc vers}\, u \le \pi)$

37. $\frac{d}{dx}(\sinh u) = (\cosh u)\frac{du}{dx}$

38. $\frac{d}{dx}(\cosh u) = (\sinh u)\frac{du}{dx}$

39. $\frac{d}{dx}(\tanh u) = (\text{sech}^2 u)\frac{du}{dx}$

40. $\frac{d}{dx}(\coth u) = -(\text{csch}^2 u)\frac{du}{dx}$

41. $\frac{d}{dx}(\text{sech}\, u) = -(\text{sech}\, u \cdot \tanh u)\frac{du}{dx}$

42. $\frac{d}{dx}(\text{csch}\, u) = -(\text{csch}\, u \cdot \coth u)\frac{du}{dx}$

43. $\frac{d}{dx}(\sinh^{-1} u) = \frac{d}{dx}[\log(u + \sqrt{u^2+1})] = \frac{1}{\sqrt{u^2+1}}\frac{du}{dx}$

44. $\frac{d}{dx}(\cosh^{-1} u) = \frac{d}{dx}[\log(u + \sqrt{u^2-1})] = \frac{1}{\sqrt{u^2-1}}\frac{du}{dx}, \quad (u > 1, \cosh^{-1} u > 0)$

45. $\frac{d}{dx}(\tanh^{-1} u) = \frac{d}{dx}\left[\frac{1}{2}\log\frac{1+u}{1-u}\right] = \frac{1}{1-u^2}\frac{du}{dx}, \quad (u^2 < 1)$

46. $\frac{d}{dx}(\coth^{-1} u) = \frac{d}{dx}\left[\frac{1}{2}\log\frac{u+1}{u-1}\right] = \frac{1}{1-u^2}\frac{du}{dx}, \quad (u^2 > 1)$

47. $\frac{d}{dx}(\mathrm{sech}^{-1}\, u) = \frac{d}{dx}\left[\log\frac{1+\sqrt{1-u^2}}{u}\right] = -\frac{1}{u\sqrt{1-u^2}}\frac{du}{dx}, \quad (0 < u < 1,\ \mathrm{sech}^{-1}\, u > 0)$

48. $\frac{d}{dx}(\mathrm{csch}^{-1}\, u) = \frac{d}{dx}\left[\log\frac{1+\sqrt{1+u^2}}{u}\right] = -\frac{1}{|u|\sqrt{1+u^2}}\frac{du}{dx}$

49. $\frac{d}{dq}\int_p^q f(x)\,dx = f(q), \quad [p \text{ constant}]$

50. $\frac{d}{dp}\int_p^q f(x)\,dx = -f(p), \quad [q \text{ constant}]$

51. $\frac{d}{da}\int_p^q f(x,a)\,dx = \int_p^q \frac{\partial}{\partial a}[f(x,a)]\,dx + f(q,a)\frac{dq}{da} - f(p,a)\frac{dp}{da}$

INTEGRATION

The following is a brief discussion of some integration techniques. A more complete discussion can be found in a number of good textbooks. However, the purpose of this introduction is simply to discuss a few of the important techniques which may be used, in conjunction with the integral table which follows, to integrate particular functions.

No matter how extensive the integral table, it is a fairly uncommon occurrence to find in the table the exact integral desired. Usually some form of transformation will have to be made. The simplest type of transformation, and yet the most general, is substitution. Simple forms of substitution, such as $y = ax$, are employed almost unconsciously by experienced users of integral tables. Other substitutions may require more thought. In some sections of the tables, appropriate substitutions are suggested for integrals that are similar to, but not exactly like, integrals in the table. Finding the right substitution is largely a matter of intuition and experience.

Several precautions must be observed when using substitutions:

1. Be sure to make the substitution in the dx term, as well as everywhere else in the integral.

2. Be sure that the function substituted is one-to-one and continuous. If this is not the case, the integral must be restricted in such a way as to make it true. See the example following.

3. With definite integrals, the limits should also be expressed in terms of the new dependent variable. With indefinite integrals, it is necessary to perform the reverse substitution to obtain the answer in terms of the original independent variable. This may also be done for definite integrals, but it is usually easier to change the limits.

Example:

$$\int \frac{x^4}{\sqrt{a^2 - x^2}}\,dx$$

Here we make the substitution $x = |a|\sin\theta$. Then $dx = |a|\cos\theta\,d\theta$, and

$$\sqrt{a^2 - x^2} = \sqrt{a^2 - a^2\sin^2\theta} = |a|\sqrt{1 - \sin^2\theta} = |a\cos\theta|$$

Notice the absolute value signs. It is very important to keep in mind that a square root radical always denotes the positive square root, and to assure the sign is always kept positive. Thus $\sqrt{x^2} = |x|$. Failure to observe this is a common cause of errors in integration.

Notice also that the indicated substitution is not a one-to-one function; that is, it does not have a unique inverse. Thus we must restrict the range of θ in such a way as to make the function one-to-one. Fortunately, this is easily done by solving for θ

$$\theta = \sin^{-1}\frac{x}{|a|}$$

and restricting the inverse sine to the principal values, $-\frac{\pi}{2} \le \theta \le \frac{\pi}{2}$.

Thus the integral becomes

$$\int \frac{a^4 \sin^4 \theta |a| \cos \theta \, d\theta}{|a| \, |\cos \theta|}$$

Now, however, in the range of values chosen for θ, $\cos \theta$ is always positive. Thus we may remove the absolute value signs from $\cos \theta$ in the denominator. (This is one of the reasons that the principal values of the inverse trigonometric functions are defined as they are.) Then the $\cos \theta$ terms cancel, and the integral becomes

$$a^4 \int \sin^4 \theta \, d\theta$$

By application of integral formulas 299 and 296, we integrate this to

$$-a^4 \frac{\sin^3 \theta \cos \theta}{4} - \frac{3a^4}{8} \cos \theta \sin \theta + \frac{3a^4}{8} \theta + C$$

We now must perform the inverse substitution to get the result in terms of x. We have

$$\theta = \sin^{-1} \frac{x}{|a|}$$

$$\sin \theta = \frac{x}{|a|}$$

Then

$$\cos \theta = \pm \sqrt{1 - \sin^2 \theta} = \pm \sqrt{1 - \frac{x^2}{a^2}} = \pm \frac{\sqrt{a^2 - x^2}}{|a|}$$

Because of the previously mentioned fact that $\cos \theta$ is positive, we may omit the $\pm$ sign. The reverse substitution then produces the final answer

$$\int \frac{x^4}{\sqrt{a^2 - x^2}} dx = -\frac{1}{4} x^3 \sqrt{a^2 - x^2} - \frac{3}{8} a^2 x \sqrt{a^2 - x^2} + \frac{3}{8} a^4 \sin^{-1} \frac{x}{|a|} + C.$$

Any rational function of x may be integrated, if the denominator is factored into linear and irreducible quadratic factors. The function may then be broken into partial fractions, and the individual partial fractions integrated by use of the appropriate formula from the integral table. See the section on partial fractions for further information.

Many integrals may be reduced to rational functions by proper substitutions. For example,

$$z = \tan \frac{x}{2}$$

will reduce any rational function of the six trigonometric functions of x to a rational function of z. (Frequently there are other substitutions that are simpler to use, but this one will always work. See integral formula number 484.)

Any rational function of x and $\sqrt{ax + b}$ may be reduced to a rational function of z by making the substitution

$$z = \sqrt{ax + b}.$$

Other likely substitutions will be suggested by looking at the form of the integrand.

The other main method of transforming integrals is integration by parts. This involves applying formula number 5 or 6 in the accompanying integral table. The critical factor in this method is the choice of the functions u and v. In order for the method to be successful, $v = \int dv$ and $\int v \, du$ must be easier to integrate than the original integral. Again, this choice is largely a matter of intuition and experience.

Example:

$$\int x \sin x \, dx$$

Two obvious choices are $u = x$, $dv = \sin x \, dx$, or $u = \sin x$, $dv = x \, dx$. Since a preliminary mental calculation indicates that $\int v \, du$ in the second choice would be more, rather than less, complicated than the original integral (it would contain x^2), we use the first choice.

$$u = x \qquad du = dx$$

$$dv = \sin x \, dx \qquad v = -\cos x$$

$$\int x \sin x \, dx = \int u \, dv = uv - \int v \, du = -x \cos x + \int \cos x \, dx$$

$$= \sin x - x \cos x$$

Of course, this result could have been obtained directly from the integral table, but it provides a simple example of the method. In more complicated examples the choice of u and v may not be so obvious, and several different choices may have to be tried. Of course, there is no guarantee that any of them will work.

Integration by parts may be applied more than once, or combined with substitution. A fairly common case is illustrated by the following example.

Example:

$$\int e^x \sin x\, dx$$

Let

$$u = e^x \qquad \text{Then } du = e^x\, dx$$
$$dv = \sin x\, dx \qquad v = -\cos x$$

$$\int e^x \sin x\, dx = \int u\, dv = uv - \int v\, du = -e^x \cos x + \int e^x \cos x\, dx$$

In this latter integral,

$$\text{Let } u = e^x \qquad \text{Then } du = e^x\, dx$$
$$dv = \cos x\, dx \qquad v = \sin x$$

$$\begin{aligned}\int e^x \sin x\, dx = -e^x \cos x + \int e^x \cos x\, dx &= -e^x \cos x + \int u\, dv \\ &= -e^x \cos x + uv - \int v\, du \\ &= -e^x \cos x + e^x \sin x - \int e^x \sin x\, dx\end{aligned}$$

This looks as if a circular transformation has taken place, since we are back at the same integral we started from. However, the above equation can be solved algebraically for the required integral:

$$\int e^x \sin x\, dx = \frac{1}{2} e^x \sin x - \frac{1}{2} e^x \cos x$$

In the second integration by parts, if the parts had been chosen as $u = \cos x$, $dv = e^x\, dx$, we would indeed have made a circular transformation, and returned to the starting place.

In general, when doing repeated integration by parts, one should never choose the function u at any stage to be the same as the function v at the previous stage, or a constant times the previous v.

The following rule is called the extended rule for integration by parts. It is the result of $n+1$ successive applications of integration by parts. If

$$\begin{aligned} g_1(x) &= \int g(x)\, dx, \qquad g_2(x) = \int g_1(x)\, dx, \\ g_3(x) &= \int g_2(x)\, dx, \ldots, g_m(x) = \int g_{m-1}(x)\, dx, \ldots, \end{aligned}$$

then

$$\begin{aligned} \int f(x) \cdot g(x)\, dx &= f(x) \cdot g_1(x) - f'(x) \cdot g_2(x) + f''(x) \cdot g_3(x) - + \cdots \\ &\quad + (-1)^n f^{(n)}(x) g_{n+1}(x) + (-1)^{n+1} \int f^{(n+1)}(x) g_{n+1}(x)\, dx. \end{aligned}$$

A useful special case of the above rule is when $f(x)$ is a polynomial of degree n. Then $f^{(n+1)}(x) = 0$, and

$$\int f(x) \cdot g(x)\, dx = f(x) \cdot g_1(x) - f'(x) \cdot g_2(x) + f''(x) \cdot g_3(x) - + \cdots + (-1)^n f^{(n)}(x) g_{n+1}(x) + C.$$

Example: If $f(x) = x^2$, $g(x) = \sin x$

$$\int x^2 \sin x \, dx = -x^2 \cos x + 2x \sin x + 2 \cos x + C.$$

Another application of this formula occurs if

$$f''(x) = a f(x) \qquad \text{and} \qquad g''(x) = bg(x),$$

where a and b are unequal constants. In this case, by a process similar to that used in the above example for $\int e^x \sin x \, dx$, we get the formula

$$\int f(x)g(x)\, dx = \frac{f(x) \cdot g'(x) - f'(x) \cdot g(x)}{b - a} + C.$$

This formula could have been used in the example mentioned. Here is another example.

Example: If $f(x) = e^{2x}$, $g(x) = \sin 3x$, then $a = 4$, $b = -9$, and

$$\int e^{2x} \sin 3x \, dx = \frac{3\, e^{2x} \cos 3x - 2\, e^{2x} \sin 3x}{-9 - 4} + C = \frac{e^{2x}}{13}(2 \sin 3x - 3 \cos 3x) + C$$

The following additional points should be observed when using this table.

1. A constant of integration is to be supplied with the answers for indefinite integrals.
2. Logarithmic expressions are to base $e = 2.71828\ldots$, unless otherwise specified, and are to be evaluated for the absolute value of the arguments involved therein.
3. All angles are measured in radians, and inverse trigonometric and hyperbolic functions represent principal values, unless otherwise indicated.
4. If the application of a formula produces either a zero denominator or the square root of a negative number in the result, there is usually available another form of the answer which avoids this difficulty. In many of the results, the excluded values are specified, but when such are omitted it is presumed that one can tell what these should be, especially when difficulties of the type herein mentioned are obtained.
5. When inverse trigonometric functions occur in the integrals, be sure that any replacements made for them are strictly in accordance with the rules for such functions. This causes little difficulty when the argument of the inverse trigonometric function is positive, since then all angles involved are in the first quadrant. However, if the argument is negative, special care must be used. Thus if $u > 0$,

$$\sin^{-1} u = \cos^{-1} \sqrt{1 - u^2} = \csc^{-1} \frac{1}{u}, \text{ etc.}$$

However, if $u < 0$,

$$\sin^{-1} u = -\cos^{-1} \sqrt{1 - u^2} = -\pi - \csc^{-1} \frac{1}{u}, \text{ etc.}$$

See the section on inverse trigonometric functions for a full treatment of the allowable substitutions.

6. In integrals 340–345 and some others, the right side includes expressions of the form

$$A \tan^{-1}[B + C \tan f(x)].$$

In these formulas, the $\tan^{-1}$ does not necessarily represent the principal value. Instead of always employing the principal branch of the inverse tangent function, one must instead use that branch of the inverse tangent function upon which $f(x)$ lies for any particular choice of x. (This is not an issue when the antiderivative is continuous.)

Example:

$$\begin{aligned}
\int_0^{4\pi} \frac{dx}{2 + \sin x} &= \frac{2}{\sqrt{3}} \tan^{-1} \frac{2 \tan(x/2 + 1)}{\sqrt{3}} \Bigg]_0^{4\pi} \\
&= \frac{2}{\sqrt{3}} \left[\tan^{-1} \left(\frac{2 \tan 2\pi + 1}{\sqrt{3}} \right) - \tan^{-1} \left(\frac{2 \tan 0 + 1}{\sqrt{3}} \right) \right] \\
&= \frac{2}{\sqrt{3}} \left[\frac{13\pi}{6} - \frac{\pi}{6} \right] = \frac{4\pi}{\sqrt{3}} = \frac{4\sqrt{3}\pi}{3}
\end{aligned}$$

Here

$$\tan^{-1}\frac{2\tan 2\pi+1}{\sqrt{3}}=\tan^{-1}\frac{1}{\sqrt{3}}=\frac{13\pi}{6},$$

since $f(x)=2\pi$; and

$$\tan^{-1}\frac{2\tan 0+1}{\sqrt{3}}=\tan^{-1}\frac{1}{\sqrt{3}}=\frac{\pi}{6},$$

since $f(x)=0$.

7. B_n and E_n where used in integrals represents the Bernoulli and Euler numbers as defined in tables of Bernoulli and Euler polynomials contained in certain mathematics reference and handbooks.

INTEGRALS

ELEMENTARY FORMS

1. $\int a\,dx = ax$

2. $\int a\cdot f(x)\,dx = a\int f(x)\,dx$

3. $\int \phi(y)\,dx = \int \frac{\phi(y)}{y'}\,dy, \quad \text{where } y'=\frac{dy}{dx}$

4 $\int (u+v)\,dx = \int u\,dx + \int v\,dx, \quad \text{where } u \text{ and } v \text{ are any functions of } x$

5. $\int u\,dv = u\int dv - \int v\,du = uv - \int v\,du$

6. $\int u\frac{dv}{dx}\,dx = uv - \int v\frac{du}{dx}\,dx$

7. $\int x^n\,dx = \frac{x^{n+1}}{n+1}, \quad \text{except } n=-1$

8. $\int \frac{f'(x)\,dx}{f(x)} = \log f(x), \quad (df(x)=f'(x)\,dx)$

9. $\int \frac{dx}{x} = \log x$

10. $\int \frac{f'(x)\,dx}{2\sqrt{f(x)}} = \sqrt{f(x)}, \quad (df(x)=f'(x)\,dx)$

11. $\int e^x\,dx = e^x$

12. $\int e^{ax}\,dx = e^{ax}/a$

13. $\int b^{ax}\,dx = \frac{b^{ax}}{a\log b}, \quad (b>0)$

14. $\int \log x\,dx = x\log x - x$

15. $\int a^x \log a\,dx = a^x, \quad (a>0)$

16. $\int \frac{dx}{a^2+x^2} = \frac{1}{a}\tan^{-1}\frac{x}{a}$

17. $\int \frac{dx}{a^2-x^2} = \begin{cases} \frac{1}{a}\tanh^{-1}\frac{x}{a} \\ \text{or} \\ \frac{1}{2a}\log\frac{a+x}{a-x}, \quad (a^2>x^2) \end{cases}$

18. $\int \frac{dx}{x^2-a^2} = \begin{cases} -\frac{1}{a}\coth^{-1}\frac{x}{a} \\ \text{or} \\ \frac{1}{2a}\log\frac{x-a}{x+a}, \quad (x^2>a^2) \end{cases}$

19. $$\int \frac{dx}{\sqrt{a^2 - x^2}} = \begin{cases} \sin^{-1} \frac{x}{|a|} \\ \text{or} \\ -\cos^{-1} \frac{x}{|a|}, \quad (a^2 > x^2) \end{cases}$$

20. $$\int \frac{dx}{\sqrt{x^2 \pm a^2}} = \log(x + \sqrt{x^2 \pm a^2})$$

21. $$\int \frac{dx}{x\sqrt{x^2 - a^2}} = \frac{1}{|a|} \sec^{-1} \frac{x}{a}$$

22. $$\int \frac{dx}{x\sqrt{a^2 \pm x^2}} = -\frac{1}{a} \log \left(\frac{a + \sqrt{a^2 \pm x^2}}{x} \right)$$

FORMS CONTAINING $(a + bx)$

For forms containing $a + bx$, but not listed in the table, the substitution $u = \frac{a+bx}{x}$ may prove helpful.

23. $$\int (a + bx)^n \, dx = \frac{(a + bx)^{n+1}}{(n+1)b}, \quad (n \neq -1)$$

24. $$\int x(a + bx)^n \, dx = \frac{1}{b^2(n+2)}(a + bx)^{n+2} - \frac{a}{b^2(n+1)}(a + bx)^{n+1}, \quad (n \neq -1, -2)$$

25. $$\int x^2(a + bx)^n \, dx = \frac{1}{b^3} \left[\frac{(a + bx)^{n+3}}{n+3} - 2a\frac{(a + bx)^{n+2}}{n+2} + a^2 \frac{(a + bx)^{n+1}}{n+1} \right]$$

26. $$\int x^m(a + bx)^n \, dx = \begin{cases} \frac{x^{m+1}(a+bx)^n}{m+n+1} + \frac{an}{m+n+1} \int x^m(a + bx)^{n-1} \, dx \\ \text{or} \\ \frac{1}{a(n+1)} \left[-x^{m+1}(a + bx)^{n+1} + (m + n + 2) \int x^m(a + bx)^{n+1} \, dx \right] \\ \text{or} \\ \frac{1}{b(m+n+1)} \left[x^m(a + bx)^{n+1} - ma \int x^{m-1}(a + bx)^n \, dx \right] \end{cases}$$

27. $$\int \frac{dx}{a + bx} = \frac{1}{b} \log (a + bx)$$

28. $$\int \frac{dx}{(a + bx)^2} = -\frac{1}{b(a + bx)}$$

29. $$\int \frac{dx}{(a + bx)^3} = -\frac{1}{2b(a + bx)^2}$$

30. $$\int \frac{x \, dx}{a + bx} = \begin{cases} \frac{1}{b^2}[a + bx - a \log(a + bx)] \\ \text{or} \\ \frac{x}{b} - \frac{a}{b^2} \log(a + bx) \end{cases}$$

31. $$\int \frac{x \, dx}{(a + bx)^2} = \frac{1}{b^2} \left[\log (a + bx) + \frac{a}{a + bx} \right]$$

32. $$\int \frac{x \, dx}{(a + bx)^n} = \frac{1}{b^2} \left[\frac{-1}{(n-2)\,(a + bx)^{n-2}} + \frac{a}{(n-1)(a + bx)^{n-1}} \right], \quad n \neq 1, 2$$

33. $$\int \frac{x^2 \, dx}{a + bx} = \frac{1}{b^3} \left[\frac{1}{2}(a + bx)^2 - 2a(a + bx) + a^2 \log (a + bx) \right]$$

34. $$\int \frac{x^2 \, dx}{(a + bx)^2} = \frac{1}{b^3} \left[a + bx - 2a \log (a + bx) - \frac{a^2}{a + bx} \right]$$

35. $$\int \frac{x^2 \, dx}{(a + bx)^3} = \frac{1}{b^3} \left[\log (a + bx) + \frac{2a}{a + bx} - \frac{a^2}{2(a + bx)^2} \right]$$

36. $$\int \frac{x^2 \, dx}{(a + bx)^n} = \frac{1}{b^3} \left[\frac{-1}{(n-3)\,(a + bx)^{n-3}} + \frac{2a}{(n-2)\,(a + bx)^{n-2}} - \frac{a^2}{(n-1)\,(a + bx)^{n-1}} \right], \quad n \neq 1, 2, 3$$

37. $$\int \frac{dx}{x(a + bx)} = -\frac{1}{a} \log \frac{a + bx}{x}$$

38. $$\int \frac{dx}{x(a + bx)^2} = \frac{1}{a(a + bx)} - \frac{1}{a^2} \log \frac{a + bx}{x}$$

39. $$\int \frac{dx}{x(a+bx)^3} = \frac{1}{a^3}\left[\frac{1}{2}\left(\frac{2a+bx}{a+bx}\right)^2 + \log\frac{x}{a+bx}\right]$$

40. $$\int \frac{dx}{x^2(a+bx)} = -\frac{1}{ax} + \frac{b}{a^2}\log\frac{a+bx}{x}$$

41. $$\int \frac{dx}{x^3(a+bx)} = \frac{2bx-a}{2a^2x^2} + \frac{b^2}{a^3}\log\frac{x}{a+bx}$$

42. $$\int \frac{dx}{x^2(a+bx)^2} = -\frac{a+2bx}{a^2x(a+bx)} + \frac{2b}{a^3}\log\frac{a+bx}{x}$$

FORMS CONTAINING $c^2 \pm x^2$ or $x^2 - c^2$

43. $$\int \frac{dx}{c^2+x^2} = \frac{1}{c}\tan^{-1}\frac{x}{c}$$

44. $$\int \frac{dx}{c^2-x^2} = \frac{1}{2c}\log\frac{c+x}{c-x}, \quad (c^2 > x^2)$$

45. $$\int \frac{dx}{x^2-c^2} = \frac{1}{2c}\log\frac{x-c}{x+c}, \quad (x^2 > c^2)$$

46. $$\int \frac{x\,dx}{c^2 \pm x^2} = \pm\frac{1}{2}\log(c^2 \pm x^2)$$

47. $$\int \frac{x\,dx}{(c^2 \pm x^2)^{n+1}} = \mp\frac{1}{2n(c^2 \pm x^2)^n}$$

48. $$\int \frac{dx}{(c^2 \pm x^2)^n} = \frac{1}{2c^2(n-1)}\left[\frac{x}{(c^2 \pm x^2)^{n-1}} + (2n-3)\int\frac{dx}{(c^2 \pm x^2)^{n-1}}\right]$$

49. $$\int \frac{dx}{(x^2 - c^2)^n} = \frac{1}{2c^2(n-1)}\left[-\frac{x}{(x^2 - c^2)^{n-1}} - (2n-3)\int\frac{dx}{(x^2 - c^2)^{n-1}}\right]$$

50. $$\int \frac{x\,dx}{x^2-c^2} = \frac{1}{2}\log(x^2-c^2)$$

51. $$\int \frac{x\,dx}{(x^2-c^2)^{n+1}} = -\frac{1}{2n(x^2-c^2)^n}$$

FORMS CONTAINING $a + bx$ AND $c + dx$

Define $u = a + bx$, $v = c + dx$, and $k = ad - bc$. If $k = 0$, then $v = \frac{c}{a}u$.

52. $$\int \frac{dx}{u \cdot v} = \frac{1}{k}\cdot\log\left(\frac{v}{u}\right)$$

53. $$\int \frac{x\,dx}{u \cdot v} = \frac{1}{k}\left[\frac{a}{b}\log(u) - \frac{c}{d}\log(v)\right]$$

54. $$\int \frac{dx}{u^2 \cdot v} = \frac{1}{k}\left(\frac{1}{u} + \frac{d}{k}\log\frac{v}{u}\right)$$

55. $$\int \frac{x\,dx}{u^2 \cdot v} = \frac{-a}{bku} - \frac{c}{k^2}\log\frac{v}{u}$$

56. $$\int \frac{x^2\,dx}{u^2 \cdot v} = \frac{a^2}{b^2ku} + \frac{1}{k^2}\left[\frac{c^2}{d}\log(v) + \frac{a(k-bc)}{b^2}\log(u)\right]$$

57. $$\int \frac{dx}{u^n \cdot v^m} = \frac{1}{k(m-1)}\left[\frac{-1}{u^{n-1}\cdot v^{m-1}} - (m+n-2)b\int\frac{dx}{u^n \cdot v^{m-1}}\right]$$

58. $$\int \frac{u}{v}\,dx = \frac{bx}{d} + \frac{k}{d^2}\log(v)$$

59. $$\int \frac{u^m\,dx}{v^n} = \begin{cases} \frac{-1}{k(n-1)}\left[\frac{u^{m+1}}{v^{n-1}} + b(n-m-2)\int\frac{u^m}{v^{n-1}}\,dx\right] \\ \text{or} \\ \frac{-1}{d(n-m-1)}\left[\frac{u^m}{v^{n-1}} + mk\int\frac{u^{m-1}}{v^n}\,dx\right] \\ \text{or} \\ \frac{-1}{d(n-1)}\left[\frac{u^m}{v^{n-1}} - mb\int\frac{u^{m-1}}{v^{n-1}}\,dx\right] \end{cases}$$

FORMS CONTAINING $(a + bx^n)$

60. $$\int \frac{dx}{a+bx^2} = \frac{1}{\sqrt{ab}} \tan^{-1} \frac{x\sqrt{ab}}{a}, \quad (ab > 0)$$

61. $$\int \frac{dx}{a+bx^2} = \begin{cases} \frac{1}{2\sqrt{-ab}} \log \frac{a+x\sqrt{-ab}}{a-x\sqrt{-ab}}, & (ab < 0) \\ \text{or} \\ \frac{1}{\sqrt{-ab}} \tanh^{-1} \frac{x\sqrt{-ab}}{a}, & (ab < 0) \end{cases}$$

62. $$\int \frac{dx}{a^2+b^2x^2} = \frac{1}{ab} \tan^{-1} \frac{bx}{a}$$

63. $$\int \frac{x\,dx}{a+bx^2} = \frac{1}{2b} \log(a+bx^2)$$

64. $$\int \frac{x^2\,dx}{a+bx^2} = \frac{x}{b} - \frac{a}{b} \int \frac{dx}{a+bx^2}$$

65. $$\int \frac{dx}{(a+bx^2)^2} = \frac{x}{2a(a+bx^2)} + \frac{1}{2a} \int \frac{dx}{a+bx^2}$$

66. $$\int \frac{dx}{a^2-b^2x^2} = \frac{1}{2ab} \log \frac{a+bx}{a-bx}$$

67. $$\int \frac{dx}{(a+bx^2)^{m+1}} = \begin{cases} \frac{1}{2ma} \frac{x}{(a+bx^2)^m} + \frac{2m-1}{2ma} \int \frac{dx}{(a+bx^2)^m} \\ \text{or} \\ \frac{(2m)!}{(m!)^2} \left[\frac{x}{2a} \sum_{r=1}^{m} \frac{r!(r-1)!}{(4a)^{m-r}(2r)!(a+bx^2)^r} + \frac{1}{(4a)^m} \int \frac{dx}{a+bx^2} \right] \end{cases}$$

68. $$\int \frac{x\,dx}{(a+bx^2)^{m+1}} = -\frac{1}{2bm(a+bx^2)^m}$$

69. $$\int \frac{x^2\,dx}{(a+bx^2)^{m+1}} = \frac{-x}{2mb(a+bx^2)^m} + \frac{1}{2mb} \int \frac{dx}{(a+bx^2)^m}$$

70. $$\int \frac{dx}{x(a+bx^2)} = \frac{1}{2a} \log \frac{x^2}{a+bx^2}$$

71. $$\int \frac{dx}{x^2(a+bx^2)} = -\frac{1}{ax} - \frac{b}{a} \int \frac{dx}{a+bx^2}$$

72. $$\int \frac{dx}{x(a+bx^2)^{m+1}} = \begin{cases} \frac{1}{2am(a+bx^2)^m} + \frac{1}{a} \int \frac{dx}{x(a+bx^2)^m} \\ \text{or} \\ \frac{1}{2a^{m+1}} \left[\sum_{r=1}^{m} \frac{a^r}{r(a+bx^2)^r} + \log \frac{x^2}{a+bx^2} \right] \end{cases}$$

73. $$\int \frac{dx}{x^2(a+bx^2)^{m+1}} = \frac{1}{a} \int \frac{dx}{x^2(a+bx^2)^m} - \frac{b}{a} \int \frac{dx}{(a+bx^2)^{m+1}}$$

74. $$\int \frac{dx}{a+bx^3} = \frac{k}{3a} \left[\frac{1}{2} \log \frac{(k+x)^3}{a+bx^3} + \sqrt{3} \tan^{-1} \frac{2x-k}{k\sqrt{3}} \right], \quad \left(k = \sqrt[3]{\frac{a}{b}} \right)$$

75. $$\int \frac{x\,dx}{a+bx^3} = \frac{1}{3bk} \left[\frac{1}{2} \log \frac{a+bx^3}{(k+x)^3} + \sqrt{3} \tan^{-1} \frac{2x-k}{k\sqrt{3}} \right], \quad \left(k = \sqrt[3]{\frac{a}{b}} \right)$$

76. $$\int \frac{x^2\,dx}{a+bx^3} = \frac{1}{3b} \log(a+bx^3)$$

77. $$\int \frac{dx}{a+bx^4} = \frac{k}{2a} \left[\frac{1}{2} \log \frac{x^2+2kx+2k^2}{x^2-2kx+2k^2} + \tan^{-1} \frac{2kx}{2k^2-x^2} \right], \quad \left(ab > 0,\ k = \sqrt[4]{\frac{a}{4b}} \right)$$

78. $$\int \frac{dx}{a+bx^4} = \frac{k}{2a} \left[\frac{1}{2} \log \frac{x+k}{x-k} + \tan^{-1} \frac{x}{k} \right], \quad \left(ab < 0,\ k = \sqrt[4]{-\frac{a}{b}} \right)$$

79. $$\int \frac{x\,dx}{a+bx^4} = \frac{1}{2bk} \tan^{-1} \frac{x^2}{k}, \quad \left(ab > 0,\ k = \sqrt{\frac{a}{b}} \right)$$

80. $$\int \frac{x\,dx}{a+bx^4} = \frac{1}{4bk} \log \frac{x^2-k}{x^2+k}, \quad \left(ab < 0,\ k = \sqrt{-\frac{a}{b}} \right)$$

81. $$\int \frac{x^2\,dx}{a+bx^4} = \frac{1}{4bk} \left[\frac{1}{2} \log \frac{x^2-2kx+2k^2}{x^2+2kx+2k^2} + \tan^{-1} \frac{2kx}{2k^2-x^2} \right], \quad \left(ab > 0,\ k = \sqrt[4]{\frac{a}{4b}} \right)$$

82. $$\int \frac{x^2\,dx}{a+bx^4} = \frac{1}{4bk} \left[\log \frac{x-k}{x+k} + 2\tan^{-1} \frac{x}{k} \right], \quad \left(ab < 0,\ k = \sqrt[4]{-\frac{a}{b}} \right)$$

83. $$\int \frac{x^3\,dx}{a+bx^4} = \frac{1}{4b} \log(a+bx^4)$$

84. $$\int \frac{dx}{x(a+bx^n)} = \frac{1}{an} \log \frac{x^n}{a+bx^n}$$

85. $$\int \frac{dx}{(a+bx^n)^{m+1}} = \frac{1}{a}\int \frac{dx}{(a+bx^n)^m} - \frac{b}{a}\int \frac{x^n\,dx}{(a+bx^n)^{m+1}}$$

86. $$\int \frac{x^m\,dx}{(a+bx^n)^{p+1}} = \frac{1}{b}\int \frac{x^{m-n}\,dx}{(a+bx^n)^p} - \frac{a}{b}\int \frac{x^{m-n}\,dx}{(a+bx^n)^{p+1}}$$

87. $$\int \frac{dx}{x^m(a+bx^n)^{p+1}} = \frac{1}{a}\int \frac{dx}{x^m(a+bx^n)^p} - \frac{b}{a}\int \frac{dx}{x^{m-n}(a+bx^n)^{p+1}}$$

88. $$\int x^m(a+bx^n)^p\,dx = \begin{cases} \frac{1}{b(np+m+1)}\left[x^{m-n+1}(a+bx^n)^{p+1} - a(m-n+1)\int x^{m-n}(a+bx^n)^p\,dx\right] \\ \text{or} \\ \frac{1}{np+m+1}\left[x^{m+1}(a+bx^n)^p + anp\int x^m(a+bx^n)^{p-1}\,dx\right] \\ \text{or} \\ \frac{1}{a(m+1)}\left[x^{m+1}(a+bx^n)^{p+1} - (m+1+np+n)b\int x^{m+n}(a+bx^n)^p\,dx\right] \\ \text{or} \\ \frac{1}{an(p+1)}\left[-x^{m+1}(a+bx^n)^{p+1} + (m+1+np+n)\int x^m(a+bx^n)^{p+1}\,dx\right] \end{cases}$$

FORMS CONTAINING $c^3 \pm x^3$

89. $$\int \frac{dx}{c^3 \pm x^3} = \pm\frac{1}{6c^2}\log\frac{(c\pm x)^3}{c^3\pm x^3} + \frac{1}{c^2\sqrt{3}}\tan^{-1}\frac{2x\mp c}{c\sqrt{3}}$$

90. $$\int \frac{dx}{(c^3\pm x^3)^2} = \frac{x}{3c^3(c^3\pm x^3)} + \frac{2}{3c^3}\int \frac{dx}{c^3\pm x^3}$$

91. $$\int \frac{dx}{(c^3\pm x^3)^{n+1}} = \frac{1}{3nc^3}\left[\frac{x}{(c^3\pm x^3)^n} + (3n-1)\int \frac{dx}{(c^3\pm x^3)^n}\right]$$

92. $$\int \frac{x\,dx}{c^3\pm x^3} = \frac{1}{6c}\log\frac{c^3\pm x^3}{(c\pm x)^3} \pm \frac{1}{c\sqrt{3}}\tan^{-1}\frac{2x\mp c}{c\sqrt{3}}$$

93. $$\int \frac{x\,dx}{(c^3\pm x^3)^2} = \frac{x^2}{3c^3(c^3\pm x^3)} + \frac{1}{3c^3}\int \frac{x\,dx}{c^3\pm x^3}$$

94. $$\int \frac{x\,dx}{(c^3\pm x^3)^{n+1}} = \frac{1}{3nc^3}\left[\frac{x^2}{(c^3\pm x^3)^n} + (3n-2)\int \frac{x\,dx}{(c^3\pm x^3)^n}\right]$$

95. $$\int \frac{x^2\,dx}{c^3\pm x^3} = \pm\frac{1}{3}\log(c^3\pm x^3)$$

96. $$\int \frac{x^2\,dx}{(c^3\pm x^3)^{n+1}} = \mp\frac{1}{3n(c^3\pm x^3)^n}$$

97. $$\int \frac{dx}{x(c^3\pm x^3)} = \frac{1}{3c^3}\log\frac{x^3}{c^3\pm x^3}$$

98. $$\int \frac{dx}{x(c^3\pm x^3)^2} = \frac{1}{3c^3(c^3\pm x^3)} + \frac{1}{3c^6}\log\frac{x^3}{c^3\pm x^3}$$

99. $$\int \frac{dx}{x(c^3\pm x^3)^{n+1}} = \frac{1}{3nc^3(c^3\pm x^3)^n} + \frac{1}{c^3}\int \frac{dx}{x(c^3\pm x^3)^n}$$

100. $$\int \frac{dx}{x^2(c^3\pm x^3)} = -\frac{1}{c^3x} \mp \frac{1}{c^3}\int \frac{x\,dx}{c^3\pm x^3}$$

101. $$\int \frac{dx}{x^2(c^3\pm x^3)^{n+1}} = \frac{1}{c^3}\int \frac{dx}{x^2(c^3\pm x^3)^n} \mp \frac{1}{c^3}\int \frac{x\,dx}{(c^3\pm x^3)^{n+1}}$$

FORMS CONTAINING $c^4 \pm x^4$

102. $$\int \frac{dx}{c^4+x^4} = \frac{1}{2c^3\sqrt{2}}\left[\frac{1}{2}\log\frac{x^2+cx\sqrt{2}+c^2}{x^2-cx\sqrt{2}+c^2} + \tan^{-1}\frac{cx\sqrt{2}}{c^2-x^2}\right]$$

103. $$\int \frac{dx}{c^4-x^4} = \frac{1}{2c^3}\left[\frac{1}{2}\log\frac{c+x}{c-x} + \tan^{-1}\frac{x}{c}\right]$$

104. $$\int \frac{x\,dx}{c^4+x^4} = \frac{1}{2c^2}\tan^{-1}\frac{x^2}{c^2}$$

105. $$\int \frac{x\,dx}{c^4-x^4} = \frac{1}{4c^2}\log\frac{c^2+x^2}{c^2-x^2}$$

106. $$\int \frac{x^2\,dx}{c^4+x^4} = \frac{1}{2c\sqrt{2}}\left[\frac{1}{2}\log\frac{x^2-cx\sqrt{2}+c^2}{x^2+cx\sqrt{2}+c^2} + \tan^{-1}\frac{cx\sqrt{2}}{c^2-x^2}\right]$$

107. $$\int \frac{x^2\,dx}{c^4 - x^4} = \frac{1}{2c}\left[\frac{1}{2}\log\frac{c+x}{c-x} - \tan^{-1}\frac{x}{c}\right]$$

108. $$\int \frac{x^3\,dx}{c^4 \pm x^4} = \pm\frac{1}{4}\log(c^4 \pm x^4)$$

FORMS CONTAINING $(a + bx + cx^2)$

Define $X = a + bx + cx^2$ and $q = 4ac - b^2$. If $q = 0$, then $X = c\left(x + \frac{b}{2c}\right)^2$, and formulas starting with 23 should be used in place of these.

109. $$\int \frac{dx}{X} = \frac{2}{\sqrt{q}}\tan^{-1}\frac{2cx+b}{\sqrt{q}}, \quad (q > 0)$$

110. $$\int \frac{dx}{X} = \begin{cases} \frac{-2}{\sqrt{-q}}\tanh^{-1}\frac{2cx+b}{\sqrt{-q}} \\ \text{or} \\ \frac{1}{\sqrt{-q}}\log\frac{2cx+b-\sqrt{-q}}{2cx+b+\sqrt{-q}}, \quad (q < 0) \end{cases}$$

111. $$\int \frac{dx}{X^2} = \frac{2cx+b}{qX} + \frac{2c}{q}\int\frac{dx}{X}$$

112. $$\int \frac{dx}{X^3} = \frac{2cx+b}{q}\left(\frac{1}{2X^2} + \frac{3c}{qX}\right) + \frac{6c^2}{q^2}\int\frac{dx}{X}$$

113. $$\int \frac{dx}{X^{n+1}} = \begin{cases} \frac{2cx+b}{nqX^n} + \frac{2(2n-1)c}{qn}\int\frac{dx}{X^n} \\ \text{or} \\ \frac{(2n)!}{(n!)^2}\left(\frac{c}{q}\right)^n\left[\frac{2cx+b}{q}\sum_{r=1}^{n}\left(\frac{q}{cX}\right)^r\left(\frac{(r-1)!r!}{(2r)!}\right) + \int\frac{dx}{X}\right] \end{cases}$$

114. $$\int \frac{x\,dx}{X} = \frac{1}{2c}\log X - \frac{b}{2c}\int\frac{dx}{X}$$

115. $$\int \frac{x\,dx}{X^2} = \frac{bx+2a}{qX} - \frac{b}{q}\int\frac{dx}{X}$$

116. $$\int \frac{x\,dx}{X^{n+1}} = -\frac{2a+bx}{nqX^n} - \frac{b(2n-1)}{nq}\int\frac{dx}{X^n}$$

117. $$\int \frac{x^2}{X}\,dx = \frac{x}{c} - \frac{b}{2c^2}\log X + \frac{b^2-2ac}{2c^2}\int\frac{dx}{X}$$

118. $$\int \frac{x^2}{X^2}\,dx = \frac{(b^2-2ac)x+ab}{cqX} + \frac{2a}{q}\int\frac{dx}{X}$$

119. $$\int \frac{x^m\,dx}{X^{n+1}} = -\frac{x^{m-1}}{(2n-m+1)cX^n} - \frac{n-m+1}{2n-m+1}\cdot\frac{b}{c}\int\frac{x^{m-1}\,dx}{X^{n+1}} + \frac{m-1}{2n-m+1}\cdot\frac{a}{c}\int\frac{x^{m-2}\,dx}{X^{n+1}}$$

120. $$\int \frac{dx}{xX} = \frac{1}{2a}\log\frac{x^2}{X} - \frac{b}{2a}\int\frac{dx}{X}$$

121. $$\int \frac{dx}{x^2X} = \frac{b}{2a^2}\log\frac{X}{x^2} - \frac{1}{ax} + \left(\frac{b^2}{2a^2} - \frac{c}{a}\right)\int\frac{dx}{X}$$

122. $$\int \frac{dx}{xX^n} = \frac{1}{2a(n-1)X^{n-1}} - \frac{b}{2a}\int\frac{dx}{X^n} + \frac{1}{a}\int\frac{dx}{xX^{n-1}}$$

123. $$\int \frac{dx}{x^mX^{n+1}} = -\frac{1}{(m-1)ax^{m-1}X^n} - \frac{n+m-1}{m-1}\cdot\frac{b}{a}\int\frac{dx}{x^{m-1}X^{n+1}} - \frac{2n+m-1}{m-1}\cdot\frac{c}{a}\int\frac{dx}{x^{m-2}X^{n+1}}$$

FORMS CONTAINING $\sqrt{a + bx}$

124. $$\int \sqrt{a+bx}\,dx = \frac{2}{3b}\sqrt{(a+bx)^3}$$

125. $$\int x\sqrt{a+bx}\,dx = -\frac{2(2a-3bx)\sqrt{(a+bx)^3}}{15b^2}$$

126. $$\int x^2\sqrt{a+bx}\,dx = \frac{2(8a^2-12abx+15b^2x^2)\sqrt{(a+bx)^3}}{105b^3}$$

127. $$\int x^m\sqrt{a+bx}\,dx = \begin{cases} \frac{2}{b(2m+3)}\left[x^m\sqrt{(a+bx)^3} - ma\int x^{m-1}\sqrt{a+bx}\,dx\right] \\ \text{or} \\ \frac{2}{b^{m+1}}\sqrt{a+bx}\sum_{r=0}^{m}\frac{m!(-a)^{m-r}}{r!(m-r)!(2r+3)}(a+bx)^{r+1} \end{cases}$$

128. $$\int \frac{\sqrt{a+bx}}{x}\,dx = 2\sqrt{a+bx} + a\int \frac{dx}{x\sqrt{a+bx}}$$

129. $$\int \frac{\sqrt{a+bx}}{x^2}\,dx = \frac{\sqrt{a+bx}}{x} + \frac{b}{2}\int \frac{dx}{x\sqrt{a+bx}}$$

130. $$\int \frac{\sqrt{a+bx}}{x^m}\,dx = -\frac{1}{(m-1)a}\left|\frac{\sqrt{(a+bx)^3}}{x^{m-1}} + \frac{(2m-5)b}{2}\int \frac{\sqrt{a+bx}}{x^{m-1}}\,dx\right|$$

131. $$\int \frac{dx}{\sqrt{a+bx}} = \frac{2\sqrt{a+bx}}{b}$$

132. $$\int \frac{x\,dx}{\sqrt{a+bx}} = -\frac{2(2a-bx)}{3b^2}\sqrt{a+bx}$$

133. $$\int \frac{x^2\,dx}{\sqrt{a+bx}} = \frac{2(8a^2-4abx-3b^2x^2)}{15b^3}\sqrt{a+bx}$$

134. $$\int \frac{x^m\,dx}{\sqrt{a+bx}} = \begin{cases} \frac{2}{(2m+1)b}\left[x^m\sqrt{a+bx} - ma\int \frac{x^{m-1}\,dx}{\sqrt{a+bx}}\right] \\ \text{or} \\ \frac{2(-a)^m\sqrt{a+bx}}{b^{m+1}}\sum_{r=0}^{m} \frac{(-1)^r m!(a+bx)^r}{(2r+1)r!(m-r)!a^r} \end{cases}$$

135. $$\int \frac{dx}{x\sqrt{a+bx}} = \frac{1}{\sqrt{a}}\log\left(\frac{\sqrt{a+bx}-\sqrt{a}}{\sqrt{a+bx}+\sqrt{a}}\right), \quad (a>0)$$

136. $$\int \frac{dx}{x\sqrt{a+bx}} = \frac{2}{\sqrt{-a}}\tan^{-1}\sqrt{\frac{a+bx}{-a}}, \quad (a<0)$$

137. $$\int \frac{dx}{x^2\sqrt{a+bx}} = -\frac{\sqrt{a+bx}}{ax} - \frac{b}{2a}\int \frac{dx}{x\sqrt{a+bx}}$$

138. $$\int \frac{dx}{x^n\sqrt{a+bx}} = \begin{cases} -\frac{\sqrt{a+bx}}{(n-1)ax^{n-1}} \quad \frac{(2n-3)b}{(2n-2)a}\int \frac{dx}{x^{n-1}\sqrt{a+bx}} \\ \text{or} \\ \frac{(2n-2)!}{[(n-1)!]^2}\left[-\frac{\sqrt{a+bx}}{a}\sum_{r=1}^{n-1}\frac{r!(r-1)!}{x^r 2(r)!}\left(-\frac{b}{4a}\right)^{n-r-1} + \left(-\frac{b}{4a}\right)^{n-1}\int \frac{dx}{x\sqrt{a+bx}}\right] \end{cases}$$

139. $$\int (a+bx)^{\pm\frac{n}{2}}\,dx = \frac{2(a+bx)^{\frac{2\pm n}{2}}}{b(2\pm n)}$$

140. $$\int x(a+bx)^{\pm\frac{n}{2}}\,dx = \frac{2}{b^2}\left[\frac{(a+bx)^{\frac{4\pm n}{2}}}{4\pm n} - \frac{a(a+bx)^{\frac{2\pm n}{2}}}{2\pm n}\right]$$

141. $$\int \frac{dx}{x(a+bx)^{\frac{m}{2}}} = \frac{1}{a}\int \frac{dx}{x(a+bx)^{\frac{m-2}{2}}} - \frac{b}{a}\int \frac{dx}{(a+bx)^{\frac{m}{2}}}$$

142. $$\int \frac{(a+bx)^{n/2}\,dx}{x} = b\int (a+bx)^{(n-2)/2}\,dx + a\int \frac{(a+bx)^{(n-2)/2}}{x}\,dx$$

143. $$\int f(x,\sqrt{a+bx})\,dx = \frac{2}{b}\int f\left(\frac{z^2-a}{b}, z\right) z\,dz, \quad (z=\sqrt{a+bx})$$

FORMS CONTAINING $\sqrt{a+bx}$ and $\sqrt{c+dx}$

Define $u = a+bx$, $v = c+dx$, and $k = ad-bc$. If $k=0$, then, $v = (\frac{c}{a})u$, and formulas starting with 124 should be used in place of these.

144. $$\int \frac{dx}{\sqrt{uv}} = \begin{cases} \frac{2}{\sqrt{bd}}\tanh^{-1}\frac{\sqrt{bduv}}{bv}, & bd>0, \quad k<0 \\ \text{or} \\ \frac{2}{\sqrt{bd}}\tanh^{-1}\frac{\sqrt{bduv}}{du}, & bd>0, \quad k>0 \\ \text{or} \\ \frac{1}{\sqrt{bd}}\log\frac{(bv+\sqrt{bduv})^2}{v}, & (bd>0) \end{cases}$$

145. $$\int \frac{dx}{\sqrt{uv}} = \begin{cases} \frac{2}{\sqrt{-bd}}\tan^{-1}\frac{\sqrt{-bduv}}{bv} \\ \text{or} \\ -\frac{1}{\sqrt{-bd}}\sin^{-1}\left(\frac{2bdx+ad+bc}{|k|}\right), & (bd<0) \end{cases}$$

146. $$\int \sqrt{uv}\,dx = \frac{k+2bv}{4bd}\sqrt{uv} - \frac{k^2}{8bd}\int \frac{dx}{\sqrt{uv}}$$

147. $$\int \frac{dx}{v\sqrt{u}} = \begin{cases} \frac{1}{\sqrt{kd}} \log \frac{d\sqrt{u}-\sqrt{kd}}{d\sqrt{u}+\sqrt{kd}} \\ \text{or} \\ \frac{1}{\sqrt{kd}} \log \frac{(d\sqrt{u}-\sqrt{kd})^2}{v}, \quad (kd > 0) \end{cases}$$

148. $$\int \frac{dx}{v\sqrt{u}} = \frac{2}{\sqrt{-kd}} \tan^{-1} \frac{d\sqrt{u}}{\sqrt{-kd}}, \quad (kd < 0)$$

149. $$\int \frac{x\,dx}{\sqrt{uv}} = \frac{\sqrt{uv}}{bd} - \frac{ad+bc}{2bd}\int \frac{dx}{\sqrt{uv}}$$

150. $$\int \frac{dx}{v\sqrt{uv}} = \frac{-2\sqrt{uv}}{kv}$$

151. $$\int \frac{v\,dx}{\sqrt{uv}} = \frac{\sqrt{uv}}{b} - \frac{k}{2b}\int \frac{dx}{\sqrt{uv}}$$

152. $$\int \sqrt{\frac{v}{u}}\,dx = \frac{v}{|v|}\int \frac{v\,dx}{\sqrt{uv}}$$

153. $$\int v^m \sqrt{u}\,dx = \frac{1}{(2m+3)d}\left(2v^{m+1}\sqrt{u} + k\int \frac{v^m\,dx}{\sqrt{u}}\right)$$

154. $$\int \frac{dx}{v^m\sqrt{u}} = -\frac{1}{(m-1)k}\left(\frac{\sqrt{u}}{v^{m-1}} + \left(m-\frac{3}{2}\right) b\int \frac{dx}{v^{m-1}\sqrt{u}}\right)$$

155. $$\int \frac{v^m\,dx}{\sqrt{u}} = \begin{cases} \frac{2}{b(2m+1)}\left[v^m\sqrt{u} - mk\int \frac{v^{m-1}}{\sqrt{u}}\,dx\right] \\ \text{or} \\ \frac{2(m!)^2\sqrt{u}}{b(2m+1)!}\sum_{r=0}^{m}\left(-\frac{4k}{b}\right)^{m-r}\frac{(2r)!}{(r!)^2}v^r \end{cases}$$

FORMS CONTAINING $\sqrt{x^2 \pm a^2}$

156. $$\int \sqrt{x^2 \pm a^2}\,dx = \frac{1}{2}\left[x\sqrt{x^2 \pm a^2} \pm a^2 \log(x + \sqrt{x^2 \pm a^2})\right]$$

157. $$\int \frac{dx}{\sqrt{x^2 \pm a^2}} = \log(x + \sqrt{x^2 \pm a^2})$$

158. $$\int \frac{dx}{x\sqrt{x^2 - a^2}} = \frac{1}{|a|}\sec^{-1}\frac{x}{a}$$

159. $$\int \frac{dx}{x\sqrt{x^2 + a^2}} = -\frac{1}{a}\log\left(\frac{a + \sqrt{x^2 + a^2}}{x}\right)$$

160. $$\int \frac{\sqrt{x^2 + a^2}}{x}\,dx = \sqrt{x^2 + a^2} - a\log\left(\frac{a + \sqrt{x^2 + a^2}}{x}\right)$$

161. $$\int \frac{\sqrt{x^2 - a^2}}{x}\,dx = \sqrt{x^2 - a^2} - |a|\sec^{-1}\frac{x}{a}$$

162. $$\int \frac{x\,dx}{\sqrt{x^2 \pm a^2}} = \sqrt{x^2 \pm a^2}$$

163. $$\int x\sqrt{x^2 \pm a^2}\,dx = \frac{1}{3}\sqrt{(x^2 \pm a^2)^3}$$

164. $$\int \sqrt{(x^2 \pm a^2)^3}\,dx = \frac{1}{4}\left[x\sqrt{(x^2 \pm a^2)^3} \pm \frac{3a^2x}{2}\sqrt{x^2 \pm a^2} + \frac{3a^4}{2}\log(x + \sqrt{x^2 \pm a^2})\right]$$

165. $$\int \frac{dx}{\sqrt{(x^2 \pm a^2)^3}} = \frac{\pm x}{a^2\sqrt{x^2 \pm a^2}}$$

166. $$\int \frac{x\,dx}{\sqrt{(x^2 \pm a^2)^3}} = \frac{-1}{\sqrt{x^2 \pm a^2}}$$

167. $$\int x\sqrt{(x^2 \pm a^2)^3}\,dx = \frac{1}{5}\sqrt{(x^2 \pm a^2)^5}$$

168. $$\int x^2\sqrt{x^2 \pm a^2}\,dx = \frac{x}{4}\sqrt{(x^2 \pm a^2)^3} \mp \frac{a^2}{8}x\sqrt{x^2 \pm a^2} - \frac{a^4}{8}\log(x + \sqrt{x^2 \pm a^2})$$

169. $$\int x^3\sqrt{x^2+a^2}\,dx = (\frac{1}{5}x^2 - \frac{2}{15}a^2)\sqrt{(a^2+x^2)^3}$$

170. $$\int x^3\sqrt{x^2-a^2}\,dx = \frac{1}{5}\sqrt{(x^2-a^2)^5} + \frac{a^2}{3}\sqrt{(x^2-a^2)^3}$$

171. $$\int \frac{x^2\,dx}{\sqrt{x^2\pm a^2}} = \frac{x}{2}\sqrt{x^2\pm a^2} \mp \frac{a^2}{2}\log(x+\sqrt{x^2\pm a^2})$$

172. $$\int \frac{x^3\,dx}{\sqrt{x^2\pm a^2}} = \frac{1}{3}\sqrt{(x^2\pm a^2)^3} \mp a^2\sqrt{x^2\pm a^2}$$

173. $$\int \frac{dx}{x^2\sqrt{x^2\pm a^2}} = \mp\frac{\sqrt{x^2\pm a^2}}{a^2x}$$

174. $$\int \frac{dx}{x^3\sqrt{x^2+a^2}} = \frac{\sqrt{x^2+a^2}}{2a^2x^2} + \frac{1}{2a^3}\log\frac{a+\sqrt{x^2+a^2}}{x}$$

175. $$\int \frac{dx}{x^3\sqrt{x^2-a^2}} = \frac{\sqrt{x^2-a^2}}{2a^2x^2} + \frac{1}{2|a^3|}\sec^{-1}\frac{x}{a}$$

176. $$\int x^2\sqrt{(x^2\pm a^2)^3}\,dx = \frac{x}{6}\sqrt{(x^2\pm a^2)^5} \mp \frac{a^2x}{24}\sqrt{(x^2\pm a^2)^3} - \frac{a^4x}{16}\sqrt{x^2\pm a^2} \mp \frac{a^6}{16}\log(x+\sqrt{x^2\pm a^2})$$

177. $$\int x^3\sqrt{(x^2\pm a^2)^3}\,dx = \frac{1}{7}\sqrt{(x^2\pm a^2)^7} \mp \frac{a^2}{5}\sqrt{(x^2\pm a^2)^5}$$

178. $$\int \frac{\sqrt{x^2\pm a^2}\,dx}{x^2} = -\frac{\sqrt{x^2\pm a^2}}{x} + \log(x+\sqrt{x^2\pm a^2})$$

179. $$\int \frac{\sqrt{x^2+a^2}}{x^3}\,dx = -\frac{\sqrt{x^2+a^2}}{2x^2} - \frac{1}{2a}\log\frac{a+\sqrt{x^2+a^2}}{x}$$

180. $$\int \frac{\sqrt{x^2-a^2}}{x^3}\,dx = -\frac{\sqrt{x^2-a^2}}{2x^2} + \frac{1}{2|a|}\sec^{-1}\frac{x}{a}$$

181. $$\int \frac{\sqrt{x^2\pm a^2}}{x^4}\,dx = \mp\frac{\sqrt{(x^2\pm a^2)^3}}{3a^2x^3}$$

182. $$\int \frac{x^2\,dx}{\sqrt{(x^2\pm a^2)^3}} = \frac{-x}{\sqrt{x^2\pm a^2}} + \log(x+\sqrt{x^2\pm a^2})$$

183. $$\int \frac{x^3\,dx}{\sqrt{(x^2\pm a^2)^3}} = \sqrt{x^2\pm a^2} \pm \frac{a^2}{\sqrt{x^2\pm a^2}}$$

184. $$\int \frac{dx}{x\sqrt{(x^2+a^2)^3}} = \frac{1}{a^2\sqrt{x^2+a^2}} - \frac{1}{a^3}\log\frac{a+\sqrt{x^2+a^2}}{x}$$

185. $$\int \frac{dx}{x\sqrt{(x^2-a^2)^3}} = -\frac{1}{a^2\sqrt{x^2-a^2}} - \frac{1}{|a^3|}\sec^{-1}\frac{x}{a}$$

186. $$\int \frac{dx}{x^2\sqrt{(x^2\pm a^2)^3}} = -\frac{1}{a^4}\left[\frac{\sqrt{x^2\pm a^2}}{x} + \frac{x}{\sqrt{x^2\pm a^2}}\right]$$

187. $$\int \frac{dx}{x^3\sqrt{(x^2+a^2)^3}} = -\frac{1}{2a^2x^2\sqrt{x^2+a^2}} - \frac{3}{2a^4\sqrt{x^2+a^2}} + \frac{3}{2a^5}\log\frac{a+\sqrt{x^2+a^2}}{x}$$

188. $$\int \frac{dx}{x^3\sqrt{(x^2-a^2)^3}} = \frac{1}{2a^2x^2\sqrt{x^2-a^2}} - \frac{3}{2a^4\sqrt{x^2-a^2}} - \frac{3}{2|a^5|}\sec^{-1}\frac{x}{a}$$

189. $$\int \frac{x^m}{\sqrt{x^2\pm a^2}}\,dx = \frac{1}{m}x^{m-1}\sqrt{x^2\pm a^2} \mp \frac{m-1}{m}a^2\int\frac{x^{m-2}}{\sqrt{x^2\pm a^2}}\,dx$$

190. $$\int \frac{x^{2m}}{\sqrt{x^2\pm a^2}}\,dx = \frac{(2m)!}{2^{2m}(m!)^2}\left[\sqrt{x^2\pm a^2}\sum_{r=1}^{m}\frac{r!(r-1)!}{(2r)!}(\mp a^2)^{m-r}(2x)^{2r-1} + (\mp a^2)^m\log(x+\sqrt{x^2\pm a^2})\right]$$

191. $$\int \frac{x^{2m+1}}{\sqrt{x^2\pm a^2}}\,dx = \sqrt{x^2\pm a^2}\sum_{r=0}^{m}\frac{(2r)!(m!)^2}{(2m+1)!(r!)^2}(\mp 4a^2)^{m-r}x^{2r}$$

192. $$\int \frac{dx}{x^m\sqrt{x^2\pm a^2}} = \mp\frac{\sqrt{x^2\pm a^2}}{(m-1)a^2x^{m-1}} \mp \frac{(m-2)}{(m-1)a^2}\int\frac{dx}{x^{m-2}\sqrt{x^2\pm a^2}}$$

193. $$\int \frac{dx}{x^{2m}\sqrt{x^2 \pm a^2}} = \sqrt{x^2 \pm a^2} \sum_{r=0}^{m-1} \frac{(m-1)!m!(2r)!2^{2m-2r-1}}{(r!)^2(2m)!(\mp a^2)^{m-r}x^{2r+1}}$$

194. $$\int \frac{dx}{x^{2m+1}\sqrt{x^2+a^2}} = \frac{(2m)!}{(m!)^2}\left[\frac{\sqrt{x^2+a^2}}{a^2}\sum_{r=1}^{m}(-1)^{m-r+1}\frac{r!(r-1)!}{2(2r)!(4a^2)^{m-r}x^{2r}} + \frac{(-1)^{m+1}}{2^{2m}a^{2m+1}}\log\frac{\sqrt{x^2+a^2}+a}{x}\right]$$

195. $$\int \frac{dx}{x^{2m+1}\sqrt{x^2-a^2}} = \frac{(2m)!}{(m!)^2}\left[\frac{\sqrt{x^2-a^2}}{a^2}\sum_{r=1}^{m}\frac{r!(r-1)!}{2(2r)!(4a^2)^{m-r}x^{2r}} + \frac{1}{2^{2m}|a|^{2m+1}}\sec^{-1}\frac{x}{a}\right]$$

196. $$\int \frac{dx}{(x-a)\sqrt{x^2-a^2}} = -\frac{\sqrt{x^2-a^2}}{a(x-a)}$$

197. $$\int \frac{dx}{(x+a)\sqrt{x^2-a^2}} = \frac{\sqrt{x^2-a^2}}{a(x+a)}$$

198. $$\int f(x, \sqrt{x^2+a^2})\,dx = a\int f(a\tan u, a\sec u)\sec^2 u\,du, \quad \left(u = \tan^{-1}\frac{x}{a},\ a > 0\right)$$

199. $$\int f(x, \sqrt{x^2-a^2})\,dx = a\int f(a\sec u, a\tan u)\sec u\tan u\,du, \quad \left(u = \sec^{-1}\frac{x}{a},\ a > 0\right)$$

FORMS CONTAINING $\sqrt{a^2 - x^2}$

200. $$\int \sqrt{a^2-x^2}\,dx = \frac{1}{2}\left[x\sqrt{a^2-x^2} + a^2\sin^{-1}\frac{x}{|a|}\right]$$

201. $$\int \frac{dx}{\sqrt{a^2-x^2}} = \begin{cases} \sin^{-1}\frac{x}{|a|} \\ \text{or} \\ -\cos^{-1}\frac{x}{|a|} \end{cases}$$

202. $$\int \frac{dx}{x\sqrt{a^2-x^2}} = -\frac{1}{a}\log\left(\frac{a+\sqrt{a^2-x^2}}{x}\right)$$

203. $$\int \frac{\sqrt{a^2-x^2}}{x}\,dx = \sqrt{a^2-x^2} - a\log\left(\frac{a+\sqrt{a^2-x^2}}{x}\right)$$

204. $$\int \frac{x\,dx}{\sqrt{a^2-x^2}} = -\sqrt{a^2-x^2}$$

205. $$\int x\sqrt{a^2-x^2}\,dx = -\frac{1}{3}\sqrt{(a^2-x^2)^3}$$

206. $$\int \sqrt{(a^2-x^2)^3}\,dx = \frac{1}{4}\left[x\sqrt{(a^2-x^2)^3} + \frac{3a^2x}{2}\sqrt{a^2-x^2} + \frac{3a^4}{2}\sin^{-1}\frac{x}{|a|}\right]$$

207. $$\int \frac{dx}{\sqrt{(a^2-x^2)^3}} = \frac{x}{a^2\sqrt{a^2-x^2}}$$

208. $$\int \frac{x\,dx}{\sqrt{(a^2-x^2)^3}} = \frac{1}{\sqrt{a^2-x^2}}$$

209. $$\int x\sqrt{(a^2-x^2)^3}\,dx = -\frac{1}{5}\sqrt{(a^2-x^2)^5}$$

210. $$\int x^2\sqrt{a^2-x^2}\,dx = -\frac{x}{4}\sqrt{(a^2-x^2)^3} + \frac{a^2}{8}\left(x\sqrt{a^2-x^2} + a^2\sin^{-1}\frac{x}{|a|}\right)$$

211. $$\int x^3\sqrt{a^2-x^2}\,dx = (-\frac{1}{5}x^2 - \frac{2}{15}a^2)\sqrt{(a^2-x^2)^3}$$

212. $$\int x^2\sqrt{(a^2-x^2)^3}\,dx = -\frac{1}{6}x\sqrt{(a^2-x^2)^5} + \frac{a^2x}{24}\sqrt{(a^2-x^2)^3} + \frac{a^4x}{16}\sqrt{a^2-x^2} + \frac{a^6}{16}\sin^{-1}\frac{x}{|a|}$$

213. $$\int x^3\sqrt{(a^2-x^2)^3}\,dx = \frac{1}{7}\sqrt{(a^2-x^2)^7} - \frac{a^2}{5}\sqrt{(a^2-x^2)^5}$$

214. $$\int \frac{x^2\,dx}{\sqrt{a^2-x^2}} = -\frac{x}{2}\sqrt{a^2-x^2} + \frac{a^2}{2}\sin^{-1}\frac{x}{|a|}$$

215. $$\int \frac{dx}{x^2\sqrt{a^2-x^2}} = -\frac{\sqrt{a^2-x^2}}{a^2x}$$

216. $$\int \frac{\sqrt{a^2-x^2}}{x^2}\,dx = -\frac{\sqrt{a^2-x^2}}{x} - \sin^{-1}\frac{x}{|a|}$$

217. $$\int \frac{\sqrt{a^2-x^2}}{x^3}\,dx = -\frac{\sqrt{a^2-x^2}}{2x^2} + \frac{1}{2a}\log\frac{a+\sqrt{a^2-x^2}}{x}$$

218. $$\int \frac{\sqrt{a^2-x^2}}{x^4}\,dx = -\frac{\sqrt{(a^2-x^2)^3}}{3a^2x^3}$$

219. $$\int \frac{x^2\,dx}{\sqrt{(a^2-x^2)^3}} = \frac{x}{\sqrt{a^2-x^2}} - \sin^{-1}\frac{x}{|a|}$$

220. $$\int \frac{x^3\,dx}{\sqrt{a^2-x^2}} = -\frac{2}{3}(a^2-x^2)^{3/2} - x^2(a^2-x^2)^{1/2} = -\frac{1}{3}\sqrt{a^2-x^2}(x^2+2a^2)$$

221. $$\int \frac{x^3\,dx}{\sqrt{(a^2-x^2)^3}} = 2(a^2-x^2)^{1/2} + \frac{x^2}{(a^2-x^2)^{1/2}} = -\frac{a^2}{\sqrt{a^2-x^2}} + \sqrt{a^2-x^2}$$

222. $$\int \frac{dx}{x^3\sqrt{a^2-x^2}} = -\frac{\sqrt{a^2-x^2}}{2a^2x^2} - \frac{1}{2a^3}\log\frac{a+\sqrt{a^2-x^2}}{x}$$

223. $$\int \frac{dx}{x\sqrt{(a^2-x^2)^3}} = \frac{1}{a^2\sqrt{a^2-x^2}} - \frac{1}{a^3}\log\frac{a+\sqrt{a^2-x^2}}{x}$$

224. $$\int \frac{dx}{x^2\sqrt{(a^2-x^2)^3}} = \frac{1}{a^4}\left[-\frac{\sqrt{a^2-x^2}}{x} + \frac{x}{\sqrt{a^2-x^2}}\right]$$

225. $$\int \frac{dx}{x^3\sqrt{(a^2-x^2)^3}} = -\frac{1}{2a^2x^2\sqrt{a^2-x^2}} + \frac{3}{2a^4\sqrt{a^2-x^2}} - \frac{3}{2a^5}\log\frac{a+\sqrt{a^2-x^2}}{x}$$

226. $$\int \frac{x^m}{\sqrt{a^2-x^2}}\,dx = -\frac{x^{m-1}\sqrt{a^2-x^2}}{m} + \frac{(m-1)a^2}{m}\int\frac{x^{m-2}}{\sqrt{a^2-x^2}}\,dx$$

227. $$\int \frac{x^{2m}}{\sqrt{a^2-x^2}}\,dx = \frac{(2m)!}{(m!)^2}\left[-\sqrt{a^2-x^2}\sum_{r=1}^{m}\frac{r!(r-1)!}{2^{2m-2r+1}(2r)!}a^{2m-2r}x^{2r-1} + \frac{a^{2m}}{2^{2m}}\sin^{-1}\frac{x}{|a|}\right]$$

228. $$\int \frac{x^{2m+1}}{\sqrt{a^2-x^2}}\,dx = -\sqrt{a^2-x^2}\sum_{r=0}^{m}\frac{(2r)!(m!)^2}{(2m+1)!(r!)^2}(4a^2)^{m-r}x^{2r}$$

229. $$\int \frac{dx}{x^m\sqrt{a^2-x^2}} = -\frac{\sqrt{a^2-x^2}}{(m-1)a^2x^{m-1}} + \frac{m-2}{(m-1)a^2}\int\frac{dx}{x^{m-2}\sqrt{a^2-x^2}}$$

230. $$\int \frac{ax}{x^{2m}\sqrt{a^2-x^2}} = -\sqrt{a^2-x^2}\sum_{r=0}^{m-1}\frac{(m-1)!m!(2r)!2^{2m-2r-1}}{(r!)^2(2m)!a^{2m-2r}x^{2r+1}}$$

231. $$\int \frac{dx}{x^{2m+1}\sqrt{a^2-x^2}} = \frac{(2m)!}{(m!)^2}\left[-\frac{\sqrt{a^2-x^2}}{a^2}\sum_{r=1}^{m}\frac{r!(r-1)!}{2(2r)!(4a^2)^{m-r}x^{2r}} + \frac{1}{2^{2m}a^{2m+1}}\log\frac{a-\sqrt{a^2-x^2}}{x}\right]$$

232. $$\int \frac{dx}{(b^2-x^2)\sqrt{a^2-x^2}} = \frac{1}{2b\sqrt{a^2-b^2}}\log\frac{(b\sqrt{a^2-x^2}+x\sqrt{a^2-b^2})^2}{b^2-x^2}, \quad (a^2>b^2)$$

233. $$\int \frac{dx}{(b^2-x^2)\sqrt{a^2-x^2}} = \frac{1}{b\sqrt{b^2-a^2}}\tan^{-1}\frac{x\sqrt{b^2-a^2}}{b\sqrt{a^2-x^2}}, \quad (b^2>a^2)$$

234. $$\int \frac{dx}{(b^2+x^2)\sqrt{a^2-x^2}} = \frac{1}{b\sqrt{a^2+b^2}}\tan^{-1}\frac{x\sqrt{a^2+b^2}}{b\sqrt{a^2-x^2}}$$

235. $$\int \frac{\sqrt{a^2-x^2}}{b^2+x^2}\,dx = \frac{\sqrt{a^2+b^2}}{|b|}\sin^{-1}\frac{x\sqrt{a^2+b^2}}{|a|\sqrt{x^2+b^2}} - \sin^{-1}\frac{x}{|a|}$$

236. $$\int f(x,\sqrt{a^2-x^2})\,dx = a\int f(a\sin u, a\cos u)\cos u\,du, \quad \left(u=\sin^{-1}\frac{x}{a}, a>0\right)$$

FORMS CONTAINING $\sqrt{a+bx+cx^2}$

Define $X = a+bx+cx^2$, $q = 4ac-b^2$, and $k = \frac{4c}{q}$. If $q=0$, then $\sqrt{X} = \sqrt{c}\,\left|x+\frac{b}{2c}\right|$.

237. $$\int \frac{dx}{\sqrt{X}} = \begin{cases} \frac{1}{\sqrt{c}}\log(2\sqrt{cX}+2cx+b) \\ \text{or} \\ \frac{1}{\sqrt{c}}\sinh^{-1}\frac{2cx+b}{\sqrt{q}}, \quad (c>0) \end{cases}$$

238. $$\int \frac{dx}{\sqrt{X}} = -\frac{1}{\sqrt{-c}}\sin^{-1}\frac{2cx+b}{\sqrt{-q}}, \quad (c<0)$$

239. $$\int \frac{dx}{X\sqrt{x}} = \frac{2(2cx+b)}{q\sqrt{x}}$$

240. $$\int \frac{dx}{X^2\sqrt{x}} = \frac{2(2cx+b)}{3q\sqrt{x}}\left(\frac{1}{X}+2k\right)$$

241. $$\int \frac{dx}{X^n\sqrt{x}} = \begin{cases} \frac{2(2cx+b)\sqrt{x}}{(2n-1)qX^n} + \frac{2k(n-1)}{2n-1}\int \frac{dx}{X^{n-1}\sqrt{x}} \\ \text{or} \\ \frac{(2cx+b)(n!)(n-1)!4^n k^{n-1}}{q[(2n)!]\sqrt{x}} \sum_{r=0}^{n-1} \frac{(2r)!}{(4kX)^r (r!)^2} \end{cases}$$

242. $$\int \sqrt{x}\,dx = \frac{(2cx+b)\sqrt{x}}{4c} + \frac{1}{2k}\int \frac{dx}{\sqrt{x}}$$

243. $$\int X\sqrt{x}\,dx = \frac{(2cx+b)\sqrt{x}}{8c}\left(X+\frac{3}{2k}\right) + \frac{3}{8k^2}\int \frac{dx}{\sqrt{x}}$$

244. $$\int X^2\sqrt{x}\,dx = \frac{(2cx+b)\sqrt{x}}{12c}\left(X^2+\frac{5X}{4k}+\frac{15}{8k^2}\right) + \frac{5}{16k^3}\int \frac{dx}{\sqrt{x}}$$

245. $$\int X^n\sqrt{x}\,dx = \begin{cases} \frac{(2cx+b)X^n\sqrt{x}}{4(n+1)c} + \frac{2n+1}{2(n+1)k}\int X^{n-1}\sqrt{x}\,dx \\ \text{or} \\ \frac{(2n+2)!}{[(n+1)!]^2(4k)^{n+1}} \\ \left[\frac{k(2cx+b)\sqrt{x}}{c}\sum_{r=0}^{n} \frac{r!(r+1)!(4kX)^r}{(2r+2)!} + \int \frac{dx}{\sqrt{x}}\right] \end{cases}$$

246. $$\int \frac{x\,dx}{\sqrt{x}} = \frac{\sqrt{x}}{c} - \frac{b}{2c}\int \frac{dx}{\sqrt{x}}$$

247. $$\int \frac{x\,dx}{X\sqrt{x}} = -\frac{2(bx+2a)}{q\sqrt{x}}$$

248. $$\int \frac{x\,dx}{X^n\sqrt{x}} = -\frac{\sqrt{x}}{(2n-1)cX^n} - \frac{b}{2c}\int \frac{dx}{X^n\sqrt{x}}$$

249. $$\int \frac{x^2\,dx}{\sqrt{x}} = \left(\frac{x}{2c}-\frac{3b}{4c^2}\right)\sqrt{x} + \frac{3b^2-4ac}{8c^2}\int \frac{dx}{\sqrt{x}}$$

250. $$\int \frac{x^2\,dx}{X\sqrt{x}} = \frac{(2b^2-4ac)x+2ab}{cq\sqrt{x}} + \frac{1}{c}\int \frac{dx}{\sqrt{x}}$$

251. $$\int \frac{x^2\,dx}{X^n\sqrt{x}} = \frac{(2b^2-4ac)x+2ab}{(2n-1)cq\,X^{n-1}\sqrt{x}} + \frac{4ac+(2n-3)b^2}{(2n-1)cq}\int \frac{dx}{X^{n-1}\sqrt{x}}$$

252. $$\int \frac{x^3\,dx}{\sqrt{x}} = \left(\frac{x^2}{3c}-\frac{5bx}{12c^2}+\frac{5b^2}{8c^3}-\frac{2a}{3c^2}\right)\sqrt{x} + \left(\frac{3ab}{4c^2}-\frac{5b^3}{16c^3}\right)\int \frac{dx}{\sqrt{x}}$$

253. $$\int \frac{x^n\,dx}{\sqrt{x}} = \frac{1}{nc}x^{n-1}\sqrt{x} - \frac{(2n-1)b}{2nc}\int \frac{x^{n-1}\,dx}{\sqrt{x}} - \frac{(n-1)a}{nc}\int \frac{x^{n-2}\,dx}{\sqrt{x}}$$

254. $$\int x\sqrt{x}\,dx = \frac{X\sqrt{x}}{3c} - \frac{b(2cx+b)}{8c^2}\sqrt{x} - \frac{b}{4ck}\int \frac{dx}{\sqrt{x}}$$

255. $$\int xX\sqrt{x}\,dx = \frac{X^2\sqrt{x}}{5c} - \frac{b}{2c}\int X\sqrt{x}\,dx$$

256. $$\int xX^n\sqrt{x}\,dx = \frac{X^{n+1}\sqrt{x}}{(2n+3)c} - \frac{b}{2c}\int X^n\sqrt{x}\,dx$$

257. $$\int x^2\sqrt{x}\,dx = \left(x-\frac{5b}{6c}\right)\frac{X\sqrt{x}}{4c} + \frac{5b^2-4ac}{16c^2}\int \sqrt{x}\,dx$$

258. $$\int \frac{dx}{x\sqrt{x}} = -\frac{1}{\sqrt{a}}\log\frac{2\sqrt{aX}+bx+2a}{x}, \quad (a>0)$$

259. $$\int \frac{dx}{x\sqrt{x}} = \frac{1}{\sqrt{-a}}\sin^{-1}\left(\frac{bx+2a}{|x|\sqrt{-q}}\right), \quad (a<0)$$

260. $$\int \frac{dx}{x\sqrt{x}} = -\frac{2\sqrt{x}}{bx}, \quad (a=0)$$

261. $$\int \frac{dx}{x^2\sqrt{x}} = -\frac{\sqrt{x}}{ax} - \frac{b}{2a}\int \frac{dx}{x\sqrt{x}}$$

262. $$\int \frac{\sqrt{x}\,dx}{x} = \sqrt{x} + \frac{b}{2}\int \frac{dx}{\sqrt{x}} + a\int \frac{dx}{x\sqrt{x}}$$

263. $$\int \frac{\sqrt{x}\,dx}{x^2} = -\frac{\sqrt{x}}{x} + \frac{b}{2}\int \frac{dx}{x\sqrt{x}} + c\int \frac{dx}{\sqrt{x}}$$

FORMS INVOLVING $\sqrt{2ax - x^2}$

264. $\displaystyle\int \sqrt{2ax - x^2}\,dx = \frac{1}{2}\left[(x-a)\sqrt{2ax - x^2} + a^2 \sin^{-1}\frac{x-a}{|a|}\right]$

265. $\displaystyle\int \frac{dx}{\sqrt{2ax - x^2}} = \begin{cases} \cos^{-1}\frac{a-x}{|a|} \\ \text{or} \\ \sin^{-1}\frac{x-a}{|a|} \end{cases}$

266. $\displaystyle\int x^n\sqrt{2ax - x^2}\,dx = \begin{cases} -\frac{x^{n-1}(2ax-x^2)^{3/2}}{n+2} + \frac{(2n+1)a}{n+2}\int x^{n-1}\sqrt{2ax - x^2}\,dx \\ \text{or} \\ \sqrt{2ax - x^2}\left[\frac{x^{n+1}}{n+2} - \sum_{r=0}^{n} \frac{(2n+1)!(r!)^2 a^{n-r+1}}{2^{n-r}(2r+1)!(n+2)!n!}x^r\right] \\ \quad + \frac{(2n+1)!a^{n+2}}{2^n n!(n+2)!}\sin^{-1}\frac{x-a}{|a|} \end{cases}$

267. $\displaystyle\int \frac{\sqrt{2ax - x^2}}{x^n}\,dx = \frac{(2ax - x^2)^{1/2}}{(3-2n)ax^n} + \frac{n-3}{(2n-3)a}\int \frac{\sqrt{2ax - x^2}}{x^{n-1}}\,dx$

268. $\displaystyle\int \frac{x^n\,dx}{\sqrt{2ax - x^2}} = \begin{cases} \frac{-x^{n-1}\sqrt{2ax-x^2}}{n} + \frac{a(2n-1)}{n}\int \frac{x^{n-1}}{\sqrt{2ax-x^2}}\,dx \\ \text{or} \\ -\sqrt{2ax - x^2}\sum_{r=1}^{n} \frac{(2n)!r!(r-1)!a^{n-r}}{2^{n-r}(2r)!(n!)^2}x^{r-1} + \frac{(2n)!a^n}{2^n(n!)^2}\sin^{-1}\frac{x-a}{|a|} \end{cases}$

269. $\displaystyle\int \frac{dx}{x^n\sqrt{2ax - x^2}} = \begin{cases} \frac{\sqrt{2ax - x^2}}{a(1-2n)x^n} + \frac{n-1}{(2n-1)a}\int \frac{dx}{x^{n-1}\sqrt{2ax-x^2}} \\ \text{or} \\ -\sqrt{2ax - x^2}\sum_{r=0}^{n-1} \frac{2^{n-r}(n-1)!n!(2r)!}{(2n)!(r!)^2 a^{n-r} x^{r+1}} \end{cases}$

270. $\displaystyle\int \frac{dx}{(2ax - x^2)^{3/2}} = \frac{x-a}{a^2\sqrt{2ax - x^2}}$

271. $\displaystyle\int \frac{x\,dx}{(2ax - x^2)^{3/2}} = \frac{x}{a\sqrt{2ax - x^2}}$

MISCELLANEOUS ALGEBRAIC FORMS

272. $\displaystyle\int \frac{dx}{\sqrt{2ax + x^2}} = \log(x + a + \sqrt{2ax + x^2})$

273. $\displaystyle\int \sqrt{ax^2 + c}\,dx = \frac{x}{2}\sqrt{ax^2 + c} + \frac{c}{2\sqrt{a}}\log\left(x\sqrt{a} + \sqrt{ax^2 + c}\right), \quad (a > 0)$

274. $\displaystyle\int \sqrt{ax^2 + c}\,dx = \frac{x}{2}\sqrt{ax^2 + c} + \frac{c}{2\sqrt{-a}}\sin^{-1}\left(x\sqrt{-\frac{a}{c}}\right), \quad (a < 0)$

275. $\displaystyle\int \sqrt{\frac{1+x}{1-x}}\,dx = \sin^{-1}x - \sqrt{1 - x^2}$

276. $\displaystyle\int \frac{dx}{x\sqrt{ax^n + c}} = \begin{cases} \frac{1}{n\sqrt{c}}\log\frac{\sqrt{ax^n+c}-\sqrt{c}}{\sqrt{ax^n+c}+\sqrt{c}} \\ \text{or} \\ \frac{2}{n\sqrt{c}}\log\frac{\sqrt{ax^n+c}-\sqrt{c}}{\sqrt{x^n}}, \quad (c > 0) \end{cases}$

277. $\displaystyle\int \frac{dx}{x\sqrt{ax^n + c}} = \frac{2}{n\sqrt{-c}}\sec^{-1}\sqrt{-\frac{ax^n}{c}}, \quad (c < 0)$

278. $\displaystyle\int \frac{dx}{\sqrt{ax^2 + c}} = \frac{1}{\sqrt{a}}\log(x\sqrt{a} + \sqrt{ax^2 + c}), \quad (a > 0)$

279. $\displaystyle\int \frac{dx}{\sqrt{ax^2 + c}} = \frac{1}{\sqrt{-a}}\sin^{-1}\left(x\sqrt{-\frac{a}{c}}\right), \quad (a < 0)$

280. $\displaystyle\int (ax^2 + c)^{m+1/2}\,dx = \begin{cases} \frac{x(ax^2+c)^{m+1/2}}{2(m+1)} + \frac{(2m+1)c}{2(m+1)}\int (ax^2 + c)^{m-1/2}\,dx \\ \text{or} \\ x\sqrt{ax^2 + c}\sum_{r=0}^{m} \frac{(2m+1)!(r!)^2 c^{m-r}}{2^{2m-2r+1}m!(m+1)!(2r+1)!}(ax^2 + c)^r \\ \quad + \frac{(2m+1)!c^{m+1}}{2^{2m+1}m!(m+1)!}\int \frac{dx}{\sqrt{ax^2+c}} \end{cases}$

281. $\displaystyle\int x(ax^2 + c)^{m+\frac{1}{2}}\,dx = \frac{(ax^2 + c)^{m+\frac{3}{2}}}{(2m+3)a}$

282. $$\int \frac{(ax^2+c)^{m+1/2}}{x}\,dx = \begin{cases} \frac{(ax^2+c)^{m+1/2}}{2m+1} + c\int \frac{(ax^2+c)^{m-1/2}}{x}\,dx \\ \text{or} \\ \sqrt{ax^2+c}\;\sum_{r=0}^{m} \frac{c^{m-r}(ax^2+c)^r}{2r+1} + c^{m+1}\int \frac{dx}{x\sqrt{ax^2+c}} \end{cases}$$

283. $$\int \frac{dx}{(ax^2+c)^{m+1/2}} = \begin{cases} \frac{x}{(2m-1)c(ax^2+c)^{m-1/2}} + \frac{2m-2}{(2m-1)c}\int \frac{dx}{(ax^2+c)^{m-1/2}} \\ \text{or} \\ \frac{x}{\sqrt{ax^2+c}}\sum_{r=0}^{m-1} \frac{2^{2m-2r-1}(m-1)!m!(2r)!}{(2m)!(r!)^2c^{m-r}(ax^2+c)^r} \end{cases}$$

284. $$\int \frac{dx}{x^m\sqrt{ax^2+c}} = -\frac{\sqrt{ax^2+c}}{(m-1)cx^{m-1}} - \frac{(m-2)a}{(m-1)c}\int \frac{dx}{x^{m-2}\sqrt{ax^2+c}}$$

285. $$\int \frac{1+x^2}{(1-x^2)\sqrt{1+x^4}}\,dx = \frac{1}{\sqrt{2}}\log \frac{x\sqrt{2}+\sqrt{1+x^4}}{1-x^2}$$

286. $$\int \frac{1-x^2}{(1+x^2)\sqrt{1+x^4}}\,dx = \frac{1}{\sqrt{2}}\tan^{-1}\frac{x\sqrt{2}}{\sqrt{1+x^4}}$$

287. $$\int \frac{dx}{x\sqrt{x^n+a^2}} = -\frac{2}{na}\log \frac{a+\sqrt{x^n+a^2}}{\sqrt{x^n}}$$

288. $$\int \frac{dx}{x\sqrt{x^n-a^2}} = -\frac{2}{na}\sin^{-1}\frac{a}{\sqrt{x^n}}$$

289. $$\int \sqrt{\frac{x}{a^3-x^3}}\,dx = \frac{2}{3}\sin^{-1}\left(\frac{x}{a}\right)^{3/2}$$

FORMS INVOLVING TRIGONOMETRIC FUNCTIONS

290. $$\int (\sin ax)\,dx = -\frac{1}{a}\cos ax$$

291. $$\int (\cos ax)\,dx = \frac{1}{a}\sin ax$$

292. $$\int (\tan ax)\,dx = -\frac{1}{a}\log\cos ax = \frac{1}{a}\log\sec ax$$

293. $$\int (\cot ax)\,dx = \frac{1}{a}\log\sin ax = -\frac{1}{a}\log\csc ax$$

294. $$\int (\sec ax)\,dx = \frac{1}{a}\log(\sec ax + \tan ax) = \frac{1}{a}\log\tan\left(\frac{\pi}{4}+\frac{ax}{2}\right)$$

295. $$\int (\csc ax)\,dx = \frac{1}{a}\log(\csc ax - \cot ax) = \frac{1}{a}\log\tan\frac{ax}{2}$$

296. $$\int (\sin^2 ax)\,dx = -\frac{1}{2a}\cos ax\sin ax + \frac{1}{2}x = \frac{1}{2}x - \frac{1}{4a}\sin 2ax$$

297. $$\int (\sin^3 ax)\,dx = -\frac{1}{3a}(\cos ax)(\sin^2 ax + 2)$$

298. $$\int (\sin^4 ax)\,dx = \frac{3x}{8} - \frac{\sin 2ax}{4a} + \frac{\sin 4ax}{32a}$$

299. $$\int (\sin^n ax)\,dx = -\frac{\sin^{n-1}ax\cos ax}{na} + \frac{n-1}{n}\int (\sin^{n-2}ax)\,dx$$

300. $$\int (\sin^{2m} ax)\,dx = -\frac{\cos ax}{a}\sum_{r=0}^{m-1}\frac{(2m)!(r!)^2}{2^{2m-2r}(2r+1)!(m!)^2}\sin^{2r+1}ax + \frac{(2m)!}{2^{2m}(m!)^2}x$$

301. $$\int (\sin^{2m+1} ax)\,dx = -\frac{\cos ax}{a}\sum_{r=0}^{m}\frac{2^{2m-2r}(m!)^2(2r)!}{(2m+1)!(r!)^2}\sin^{2r}ax$$

302. $$\int (\cos^2 ax)\,dx = \frac{1}{2a}\sin ax\cos ax + \frac{1}{2}x = \frac{1}{2}x + \frac{1}{4a}\sin 2ax$$

303. $$\int (\cos^3 ax)\,dx = \frac{1}{3a}(\sin ax)(\cos^2 ax + 2)$$

304. $$\int (\cos^4 ax)\,dx = \frac{3x}{8} + \frac{\sin 2ax}{4a} + \frac{\sin 4ax}{32a}$$

305. $$\int (\cos^n ax)\,dx = \frac{1}{na}\cos^{n-1} ax \sin ax + \frac{n-1}{n}\int (\cos^{n-2} ax)\,dx$$

306. $$\int (\cos^{2m} ax)\,dx = \frac{\sin ax}{a}\sum_{r=0}^{m-1}\frac{(2m)!(r!)^2}{2^{2m-2r}(2r+1)!(m!)^2}\cos^{2r+1} ax + \frac{(2m)!}{2^{2m}(m!)^2}x$$

307. $$\int (\cos^{2m+1} ax)\,dx = \frac{\sin ax}{a}\sum_{r=0}^{m}\frac{2^{2m-2r}(m!)^2(2r)!}{(2m+1)!(r!)^2}\cos^{2r} ax$$

308. $$\int \frac{dx}{\sin^2 ax} = \int (\csc^2 ax)\,dx = -\frac{1}{a}\cot ax$$

309. $$\int \frac{dx}{\sin^m ax} = \int (\csc^m ax)\,dx = -\frac{1}{(m-1)a}\cdot\frac{\cos ax}{\sin^{m-1} ax} + \frac{m-2}{m-1}\int \frac{dx}{\sin^{m-2} ax}$$

310. $$\int \frac{dx}{\sin^{2m} ax} = \int (\csc^{2m} ax)\,dx = -\frac{1}{a}\cos ax\sum_{r=0}^{m-1}\frac{2^{2m-2r-1}(m-1)!m!(2r)!}{(2m)!(r!)^2\sin^{2r+1} ax}$$

311. $$\int \frac{dx}{\sin^{2m+1} ax} = \int (\csc^{2m+1} ax)\,dx = -\frac{1}{a}\cos ax\sum_{r=0}^{m-1}\frac{(2m)!(r!)^2}{2^{2m-2r}(m!)^2(2r+1)!\sin^{2r+2} ax} + \frac{1}{a}\cdot\frac{(2m)!}{2^{2m}(m!)^2}\log\tan\frac{ax}{2}$$

312. $$\int \frac{dx}{\cos^2 ax} = \int (\sec^2 ax)\,dx = \frac{1}{a}\tan ax$$

313. $$\int \frac{dx}{\cos^n ax} = \int (\sec^n ax)\,dx = \frac{1}{(n-1)a}\cdot\frac{\sin ax}{\cos^{n-1} ax} + \frac{n-2}{n-1}\int \frac{dx}{\cos^{n-2} ax}$$

314. $$\int \frac{dx}{\cos^{2m} ax} = \int (\sec^{2m} ax)\,dx = \frac{1}{a}\sin ax\sum_{r=0}^{m-1}\frac{2^{2m-2r-1}(m-1)!m!(2r)!}{(2m)!(r!)^2\cos^{2r+1} ax}$$

315. $$\int \frac{dx}{\cos^{2m+1} ax} = \int (\sec^{2m+1} ax)\,dx = \frac{1}{a}\sin ax\sum_{r=0}^{m-1}\frac{(2m)!(r!)^2}{2^{2m-2r}(m!)^2(2r+1)!\cos^{2r+2} ax} + \frac{1}{a}\cdot\frac{(2m)!}{2^{2m}(m!)^2}\log(\sec ax + \tan ax)$$

316. $$\int (\sin mx)(\sin nx)\,dx = \frac{\sin(m-n)x}{2(m-n)} - \frac{\sin(m+n)x}{2(m+n)}, \quad (m^2 \neq n^2)$$

317. $$\int (\cos mx)(\cos nx)\,dx = \frac{\sin(m-n)x}{2(m-n)} + \frac{\sin(m+n)x}{2(m+n)}, \quad (m^2 \neq n^2)$$

318. $$\int (\sin ax)(\cos ax)\,dx = \frac{1}{2a}\sin^2 ax$$

319. $$\int (\sin mx)(\cos nx)\,dx = -\frac{\cos(m-n)x}{2(m-n)} - \frac{\cos(m+n)x}{2(m+n)}, \quad (m^2 \neq n^2)$$

320. $$\int (\sin^2 ax)(\cos^2 ax)\,dx = -\frac{1}{32a}\sin 4ax + \frac{x}{8}$$

321. $$\int (\sin ax)(\cos^m ax)\,dx = -\frac{\cos^{m+1} ax}{(m+1)a}$$

322. $$\int (\sin^m ax)(\cos ax)\,dx = \frac{\sin^{m+1} ax}{(m+1)a}$$

323. $$\int (\cos^m ax)(\sin^n ax)\,dx = \begin{cases} \frac{\cos^{m-1} ax \sin^{n+1} ax}{(m+n)a} + \frac{m-1}{m+n}\int (\cos^{m-2} ax)(\sin^n ax)\,dx \\ \text{or} \\ -\frac{\sin^{n-1} ax \cos^{m+1} ax}{(m+n)a} + \frac{n-1}{m+n}\int (\cos^m ax)(\sin^{n-2} ax)\,dx \end{cases}$$

324. $$\int \frac{\cos^m ax}{\sin^n ax}\,dx = \begin{cases} -\frac{\cos^{m+1} ax}{(n-1)a\sin^{n-1} ax} - \frac{m-n+2}{n-1}\int \frac{\cos^m ax}{\sin^{n-2} ax}\,dx \\ \text{or} \\ \frac{\cos^{m-1} ax}{a(m-n)\sin^{n-1} ax} + \frac{m-1}{m-n}\int \frac{\cos^{m-2} ax}{\sin^n ax}\,dx \end{cases}$$

325. $$\int \frac{\sin^m ax}{\cos^n ax}\,dx = \begin{cases} \frac{\sin^{m+1} ax}{a(n-1)\cos^{n-1} ax} - \frac{m-n+2}{n-1}\int \frac{\sin^m ax}{\cos^{n-2} ax}\,dx \\ \text{or} \\ -\frac{\sin^{m-1} ax}{a(m-n)\cos^{n-1} ax} + \frac{m-1}{m-n}\int \frac{\sin^{m-2} ax}{\cos^n ax}\,dx \end{cases}$$

326. $$\int \frac{\sin ax}{\cos^2 ax}\,dx = \frac{1}{a\cos ax} = \frac{\sec ax}{a}$$

327. $$\int \frac{\sin^2 ax}{\cos ax}\,dx = -\frac{1}{a}\sin ax + \frac{1}{a}\log\tan\left(\frac{\pi}{4}+\frac{ax}{2}\right)$$

328. $$\int \frac{\cos ax}{\sin^2 ax}\,dx = -\frac{1}{a\sin ax} = -\frac{\csc ax}{a}$$

329. $$\int \frac{dx}{(\sin ax)(\cos ax)} = \frac{1}{a}\log\tan ax$$

330. $$\int \frac{dx}{(\sin ax)(\cos^2 ax)} = \frac{1}{a}\left(\sec ax + \log\tan\frac{ax}{2}\right)$$

331. $$\int \frac{dx}{(\sin ax)(\cos^n ax)} = \frac{1}{a(n-1)\cos^{n-1} ax} + \int \frac{dx}{(\sin ax)(\cos^{n-2} ax)}$$

332. $$\int \frac{dx}{(\sin^2 ax)(\cos ax)} = -\frac{1}{a}\csc ax + \frac{1}{a}\log\tan\left(\frac{\pi}{4}+\frac{ax}{2}\right)$$

333. $$\int \frac{dx}{(\sin^2 ax)(\cos^2 ax)} = -\frac{2}{a}\cot 2ax$$

334. $$\int \frac{dx}{\sin^m ax\cos^n ax} = \begin{cases} -\dfrac{1}{a(m-1)(\sin^{m-1} ax)(\cos^{n-1} ax)} \\ \quad + \dfrac{m+n-2}{m-1}\displaystyle\int \frac{dx}{(\sin^{m-2} ax)(\cos^n ax)} \\ \text{or} \\ \dfrac{1}{a(n-1)\sin^{m-1} ax\cos^{n-1} ax} + \dfrac{m+n-2}{n-1}\displaystyle\int \frac{dx}{\sin^m ax\cos^{n-2} ax} \end{cases}$$

335. $$\int \sin(a+bx)\,dx = -\frac{1}{b}\cos(a+bx)$$

336. $$\int \cos(a+bx)\,dx = \frac{1}{b}\sin(a+bx)$$

337. $$\int \frac{dx}{1\pm\sin ax} = \mp\frac{1}{a}\tan\left(\frac{\pi}{4}\mp\frac{ax}{2}\right)$$

338. $$\int \frac{dx}{1+\cos ax} = \frac{1}{a}\tan\frac{ax}{2}$$

339. $$\int \frac{dx}{1-\cos ax} = -\frac{1}{a}\cot\frac{ax}{2}$$

340. $$\int \frac{dx}{a+b\sin x} = \begin{cases} \dfrac{2}{\sqrt{a^2-b^2}}\tan^{-1}\dfrac{a\tan\frac{x}{2}+b}{\sqrt{a^2-b^2}} \\ \text{or} \\ \dfrac{1}{\sqrt{b^2-a^2}}\log\dfrac{a\tan\frac{x}{2}+b-\sqrt{b^2-a^2}}{a\tan\frac{x}{2}+b+\sqrt{b^2-a^2}} \end{cases}$$

341. $$\int \frac{dx}{a+b\cos x} = \begin{cases} \dfrac{2}{\sqrt{a^2-b^2}}\tan^{-1}\dfrac{\sqrt{a^2-b^2}\tan\frac{x}{2}}{a+b} \\ \text{or} \\ \dfrac{1}{\sqrt{b^2-a^2}}\log\left(\dfrac{\sqrt{b^2-a^2}\tan\frac{x}{2}+a+b}{\sqrt{b^2-a^2}\tan\frac{x}{2}-a-b}\right) \end{cases}$$

342. $$\int \frac{dx}{a+b\sin x+c\cos x}$$
$$= \begin{cases} \dfrac{1}{\sqrt{b^2+c^2-a^2}}\log\left(\dfrac{b-\sqrt{b^2+c^2-a^2}+(a-c)\tan\frac{x}{2}}{b+\sqrt{b^2+c^2-a^2}+(a-c)\tan\frac{x}{2}}\right) & (\text{if } a^2<b^2+c^2,\ a\neq c), \\ \dfrac{2}{\sqrt{a^2-b^2-c^2}}\tan^{-1}\left(\dfrac{b+(a-c)\tan\frac{x}{2}}{\sqrt{a^2-b^2-c^2}}\right) & (\text{if } a^2>b^2+c^2), \\ \dfrac{1}{a}\left[\dfrac{a-(b+c)\cos x-(b-c)\sin x}{a-(b-c)\cos x+(b+c)\sin x}\right] & (\text{if } a^2=b^2+c^2,\ a\neq c). \end{cases}$$

343. $$\int \frac{\sin^2 x\,dx}{a+b\cos^2 x} = \frac{1}{b}\sqrt{\frac{a+b}{a}}\tan^{-1}\left(\sqrt{\frac{a}{a+b}}\tan x\right) - \frac{x}{b}, \quad (ab>0,\ \text{or } |a|>|b|)$$

344. $$\int \frac{dx}{a^2\cos^2 x+b^2\sin^2 x} = \frac{1}{ab}\tan^{-1}\left(\frac{b\tan x}{a}\right)$$

345. $$\int \frac{\cos^2 cx}{a^2+b^2\sin^2 cx}\,dx = \frac{\sqrt{a^2+b^2}}{ab^2c}\tan^{-1}\frac{\sqrt{a^2+b^2}\tan cx}{a} - \frac{x}{b^2}$$

346. $$\int \frac{\sin cx\cos cx}{a\cos^2 cx+b\sin^2 cx}\,dx = \frac{1}{2c(b-a)}\log(a\cos^2 cx+b\sin^2 cx)$$

347. $\displaystyle\int \frac{\cos cx}{a\cos cx + b\sin cx}\,dx = \int \frac{dx}{a + b\tan cx} = \frac{1}{c(a^2+b^2)}[acx + b\log(a\cos cx + b\sin cx)]$

348. $\displaystyle\int \frac{\sin cx}{a\sin cx + b\cos cx}\,dx = \int \frac{dx}{a + b\cot cx} = \frac{1}{c(a^2+b^2)}[acx - b\log(a\sin cx + b\cos cx)]$

349. $\displaystyle\int \frac{dx}{a\cos^2 x + 2b\cos x\sin x + c\sin^2 x} = \begin{cases} \dfrac{1}{2\sqrt{b^2-ac}}\log\dfrac{c\tan x + b - \sqrt{b^2-ac}}{c\tan x + b + \sqrt{b^2-ac}}, & (b^2 > ac) \\ \text{or} & \\ \dfrac{1}{\sqrt{ac-b^2}}\tan^{-1}\dfrac{c\tan x + b}{\sqrt{ac-b^2}}, & (b^2 < ac) \\ \text{or} & \\ -\dfrac{1}{c\tan x + b}, & (b^2 = ac) \end{cases}$

350. $\displaystyle\int \frac{\sin ax}{1 \pm \sin ax}\,dx = \pm x + \frac{1}{a}\tan\left(\frac{\pi}{4} \mp \frac{ax}{2}\right)$

351. $\displaystyle\int \frac{dx}{(\sin ax)(1 \pm \sin ax)} = \frac{1}{a}\tan\left(\frac{\pi}{4} \mp \frac{ax}{2}\right) + \frac{1}{a}\log\tan\frac{ax}{2}$

352. $\displaystyle\int \frac{dx}{(1+\sin ax)^2} = -\frac{1}{2a}\tan\left(\frac{\pi}{4} - \frac{ax}{2}\right) - \frac{1}{6a}\tan^3\left(\frac{\pi}{4} - \frac{ax}{2}\right)$

353. $\displaystyle\int \frac{dx}{(1-\sin ax)^2} = \frac{1}{2a}\cot\left(\frac{\pi}{4} - \frac{ax}{2}\right) + \frac{1}{6a}\cot^3\left(\frac{\pi}{4} - \frac{ax}{2}\right)$

354. $\displaystyle\int \frac{\sin ax}{(1+\sin ax)^2}\,dx = -\frac{1}{2a}\tan\left(\frac{\pi}{4} - \frac{ax}{2}\right) + \frac{1}{6a}\tan^3\left(\frac{\pi}{4} - \frac{ax}{2}\right)$

355. $\displaystyle\int \frac{\sin ax}{(1-\sin ax)^2}\,dx = -\frac{1}{2a}\cot\left(\frac{\pi}{4} - \frac{ax}{2}\right) + \frac{1}{6a}\cot^3\left(\frac{\pi}{4} - \frac{ax}{2}\right)$

356. $\displaystyle\int \frac{\sin x\,dx}{a + b\sin x} = \frac{x}{b} - \frac{a}{b}\int \frac{dx}{a + b\sin x}$

357. $\displaystyle\int \frac{dx}{(\sin x)(a + b\sin x)} = \frac{1}{a}\log\tan\frac{x}{2} - \frac{b}{a}\int \frac{dx}{a + b\sin x}$

358. $\displaystyle\int \frac{dx}{(a + b\sin x)^2} = \frac{b\cos x}{(a^2 - b^2)(a + b\sin x)} + \frac{a}{a^2 - b^2}\int \frac{dx}{a + b\sin x}$

359. $\displaystyle\int \frac{\sin x dx}{(a + b\sin x)^2} = \frac{a\cos x}{(b^2 - a^2)(a + b\sin x)} + \frac{b}{b^2 - a^2}\int \frac{dx}{a + b\sin x}$

360. $\displaystyle\int \frac{dx}{a^2 + b^2\sin^2 cx} = \frac{1}{ac\sqrt{a^2+b^2}}\tan^{-1}\frac{\sqrt{a^2+b^2}\tan cx}{a}$

361. $\displaystyle\int \frac{dx}{a^2 - b^2\sin^2 cx} = \begin{cases} \dfrac{1}{ac\sqrt{a^2-b^2}}\tan^{-1}\dfrac{\sqrt{a^2-b^2}\tan cx}{a}, & (a^2 > b^2) \\ \text{or} & \\ \dfrac{1}{2ac\sqrt{b^2-a^2}}\log\dfrac{\sqrt{b^2-a^2}\tan cx + a}{\sqrt{b^2-a^2}\tan cx - a}, & (a^2 < b^2) \end{cases}$

362. $\displaystyle\int \frac{\cos ax}{1 + \cos ax}\,dx = x - \frac{1}{a}\tan\frac{ax}{2}$

363. $\displaystyle\int \frac{\cos ax}{1 - \cos ax}\,dx = -x - \frac{1}{a}\cot\frac{ax}{2}$

364. $\displaystyle\int \frac{dx}{(\cos ax)(1 + \cos ax)} = \frac{1}{a}\log\tan\left(\frac{\pi}{4} + \frac{ax}{2}\right) - \frac{1}{a}\tan\frac{ax}{2}$

365. $\displaystyle\int \frac{dx}{(\cos ax)(1 - \cos ax)} = \frac{1}{a}\log\tan\left(\frac{\pi}{4} + \frac{ax}{2}\right) - \frac{1}{a}\cot\frac{ax}{2}$

366. $\displaystyle\int \frac{dx}{(1 + \cos ax)^2} = \frac{1}{2a}\tan\frac{ax}{2} + \frac{1}{6a}\tan^3\frac{ax}{2}$

367. $\displaystyle\int \frac{dx}{(1 - \cos ax)^2} = -\frac{1}{2a}\cot\frac{ax}{2} - \frac{1}{6a}\cot^3\frac{ax}{2}$

368. $\displaystyle\int \frac{\cos ax}{(1 + \cos ax)^2}\,dx = \frac{1}{2a}\tan\frac{ax}{2} - \frac{1}{6a}\tan^3\frac{ax}{2}$

369. $\displaystyle\int \frac{\cos ax}{(1 - \cos ax)^2}\,dx = \frac{1}{2a}\cot\frac{ax}{2} - \frac{1}{6a}\cot^3\frac{ax}{2}$

370. $\displaystyle\int \frac{\cos x\,dx}{a + b\cos x} = \frac{x}{b} - \frac{a}{b}\int \frac{dx}{a + b\cos x}$

371. $\displaystyle\int \frac{dx}{(\cos x)(a + b\cos x)} = \frac{1}{a}\log\tan\left(\frac{x}{2} + \frac{\pi}{4}\right) - \frac{b}{a}\int \frac{dx}{a + b\cos x}$

372. $\displaystyle\int \frac{dx}{(a + b\cos x)^2} = \frac{b\sin x}{(b^2 - a^2)(a + b\cos x)} - \frac{a}{b^2 - a^2}\int \frac{dx}{a + b\cos x}$

373. $\int \frac{\cos x}{(a+b\cos x)^2}\,dx = \frac{a\sin x}{(a^2-b^2)(a+b\cos x)} - \frac{b}{a^2-b^2}\int \frac{dx}{a+b\cos x}$

374. $\int \frac{dx}{a^2+b^2-2ab\cos cx} = \frac{2}{c(a^2-b^2)}\tan^{-1}\left(\frac{a+b}{a-b}\tan\frac{cx}{2}\right)$

375. $\int \frac{dx}{a^2+b^2\cos^2 cx} = \frac{1}{ac\sqrt{a^2+b^2}}\tan^{-1}\frac{a\tan cx}{\sqrt{a^2+b^2}}$

376. $\int \frac{dx}{a^2-b^2\cos^2 cx} = \begin{cases} \frac{1}{ac\sqrt{a^2-b^2}}\tan^{-1}\frac{a\tan cx}{\sqrt{a^2-b^2}}, & (a^2>b^2) \\ \text{or} & \\ \frac{1}{2ac\sqrt{b^2-a^2}}\log\frac{a\tan cx-\sqrt{b^2-a^2}}{a\tan cx+\sqrt{b^2-a^2}}, & (b^2>a^2) \end{cases}$

377. $\int \frac{\sin ax}{1\pm\cos ax}\,dx = \mp\frac{1}{a}\log(1\pm\cos ax)$

378. $\int \frac{\cos ax}{1\pm\sin ax}\,dx = \pm\frac{1}{a}\log(1\pm\sin ax)$

379. $\int \frac{dx}{(\sin ax)(1\pm\cos ax)} = \pm\frac{1}{2a(1\pm\cos ax)} + \frac{1}{2a}\log\tan\frac{ax}{2}$

380. $\int \frac{dx}{(\cos ax)(1\pm\sin ax)} = \mp\frac{1}{2a(1\pm\sin ax)} + \frac{1}{2a}\log\tan\left(\frac{\pi}{4}+\frac{ax}{2}\right)$

381. $\int \frac{\sin ax}{(\cos ax)(1\pm\cos ax)}\,dx = \frac{1}{a}\log(\sec ax\pm 1)$

382. $\int \frac{\cos ax}{(\sin ax)(1\pm\sin ax)}\,dx = -\frac{1}{a}\log(\csc ax\pm 1)$

383. $\int \frac{\sin ax}{(\cos ax)(1\pm\sin ax)}\,dx = \frac{1}{2a(1\pm\sin ax)} \pm \frac{1}{2a}\log\tan\left(\frac{\pi}{4}+\frac{ax}{2}\right)$

384. $\int \frac{\cos ax}{(\sin ax)(1\pm\cos ax)}\,dx = -\frac{1}{2a(1\pm\cos ax)} \pm \frac{1}{2a}\log\tan\frac{ax}{2}$

385. $\int \frac{dx}{\sin ax\pm\cos ax} = \frac{1}{a\sqrt{2}}\log\tan\left(\frac{ax}{2}\pm\frac{\pi}{8}\right)$

386. $\int \frac{dx}{(\sin ax\pm\cos ax)^2} = \frac{1}{2a}\tan\left(ax\mp\frac{\pi}{4}\right)$

387. $\int \frac{dx}{1+\cos ax\pm\sin ax} = \pm\frac{1}{a}\log\left(1\pm\tan\frac{ax}{2}\right)$

388. $\int \frac{dx}{a^2\cos^2 cx - b^2\sin^2 cx} = \frac{1}{2abc}\log\frac{b\tan cx+a}{b\tan cx-a}$

389. $\int x(\sin ax)\,dx = \frac{1}{a^2}\sin ax - \frac{x}{a}\cos ax$

390. $\int x^2(\sin ax)\,dx = \frac{2x}{a^2}\sin ax - \frac{a^2x^2-2}{a^3}\cos ax$

391. $\int x^3(\sin ax)\,dx = \frac{3a^2x^2-6}{a^4}\sin ax - \frac{a^2x^3-6x}{a^3}\cos ax$

392. $\int x^m \sin ax\,dx = \begin{cases} -\frac{1}{a}x^m\cos ax + \frac{m}{a}\int x^{m-1}\cos ax\,dx \\ \text{or} \\ \cos ax\sum_{r=0}^{\left[\frac{m}{2}\right]}(-1)^{r+1}\frac{m!}{(m-2r)!}\cdot\frac{x^{m-2r}}{a^{2r+1}} \\ \quad +\sin ax\sum_{r=0}^{\left[\frac{m-1}{2}\right]}(-1)^r\frac{m!}{(m-2r-1)!}\cdot\frac{x^{m-2r-1}}{a^{2r+2}} \end{cases}$

Note: $[s]$ means greatest integer $\le s$; Thus $[3.5]$ means 3; $[5]=5$, $\left[\frac{1}{2}\right]=0$.

393. $\int x(\cos ax)\,dx = \frac{1}{a^2}\cos ax + \frac{x}{a}\sin ax$

394. $\int x^2(\cos ax)\,dx = \frac{2x\cos ax}{a^2} + \frac{a^2x^2-2}{a^3}\sin ax$

395. $\int x^3(\cos ax)\,dx = \frac{3a^2x^2-6}{a^4}\cos ax + \frac{a^2x^3-6x}{a^3}\sin ax$

396. $\int x^m(\cos ax)\,dx = \begin{cases} \frac{x^m\sin ax}{a} - \frac{m}{a}\int x^{m-1}\sin ax\,dx \\ \text{or} \\ \sin ax\sum_{r=0}^{|m/2|}(-1)^r\frac{m!}{(m-2r)!}\cdot\frac{x^{m-2r}}{a^{2r+1}} \\ \quad +\cos ax\sum_{r=0}^{|(m-1)/2|}(-1)^r\frac{m!}{(m-2r-1)!}\cdot\frac{x^{m-2r-1}}{a^{2r+2}} \end{cases}$

Note: $[s]$ means greatest integer $\le s$; Thus $[3.5]$ means 3; $[5]=5$, $\left[\frac{1}{2}\right]=0$.

397. $\displaystyle\int \frac{\sin ax}{x}\,dx = \sum_{n=0}^{r}(-1)^n \frac{(ax)^{2n+1}}{(2n+1)(2n+1)!}$

398. $\displaystyle\int \frac{\cos ax}{x}\,dx = \log x + \sum_{n=1}^{r}(-1)^n \frac{(ax)^{2n}}{2n(2n)!}$

399. $\displaystyle\int x(\sin^2 ax)\,dx = \frac{x^2}{4} - \frac{x\sin 2ax}{4a} - \frac{\cos 2ax}{8a^2}$

400. $\displaystyle\int x^2(\sin^2 ax)\,dx = \frac{x^3}{6} - \left(\frac{x^2}{4a} - \frac{1}{8a^3}\right)\sin 2ax - \frac{x\cos 2ax}{4a^2}$

401. $\displaystyle\int x(\sin^3 ax)\,dx = \frac{x\cos 3ax}{12a} - \frac{\sin 3ax}{36a^2} - \frac{3x\cos ax}{4a} + \frac{3\sin ax}{4a^2}$

402. $\displaystyle\int x(\cos^2 ax)\,dx = \frac{x^2}{4} + \frac{x\sin 2ax}{4a} + \frac{\cos 2ax}{8a^2}$

403. $\displaystyle\int x^2(\cos^2 ax)\,dx = \frac{x^3}{6} + \left(\frac{x^2}{4a} - \frac{1}{8a^3}\right)\sin 2ax + \frac{x\cos 2ax}{4a^2}$

404. $\displaystyle\int x(\cos^3 ax)\,dx = \frac{x\sin 3ax}{12a} + \frac{\cos 3ax}{36a^2} + \frac{3x\sin ax}{4a} + \frac{3\cos ax}{4a^2}$

405. $\displaystyle\int \frac{\sin ax}{x^m}\,dx = -\frac{\sin ax}{(m-1)x^{m-1}} + \frac{a}{m-1}\int \frac{\cos ax}{x^{m-1}}\,dx$

406. $\displaystyle\int \frac{\cos ax}{x^m}\,dx = -\frac{\cos ax}{(m-1)x^{m-1}} - \frac{a}{m-1}\int \frac{\sin ax}{x^{m-1}}\,dx$

407. $\displaystyle\int \frac{x}{1\pm\sin ax}\,dx = \mp\frac{x\cos ax}{a(1\pm\sin ax)} + \frac{1}{a^2}\log(1\pm\sin ax)$

408. $\displaystyle\int \frac{x}{1+\cos ax}\,dx = \frac{x}{a}\tan\frac{ax}{2} + \frac{2}{a^2}\log\cos\frac{ax}{2}$

409. $\displaystyle\int \frac{x}{1-\cos ax}\,dx = -\frac{x}{a}\cot\frac{ax}{2} + \frac{2}{a^2}\log\sin\frac{ax}{2}$

410. $\displaystyle\int \frac{x+\sin x}{1+\cos x}\,dx = x\tan\frac{x}{2}$

411. $\displaystyle\int \frac{x-\sin x}{1-\cos x}\,dx = -x\cot\frac{x}{2}$

412. $\displaystyle\int \sqrt{1-\cos ax}\,dx = -\frac{2\sin ax}{a\sqrt{1-\cos ax}} = -\frac{2\sqrt{2}}{a}\cos(\frac{ax}{2})$

413. $\displaystyle\int \sqrt{1+\cos ax}\,dx = \frac{2\sin ax}{a\sqrt{1+\cos ax}} = \frac{2\sqrt{2}}{a}\sin(\frac{ax}{2})$

414. $\displaystyle\int \sqrt{1+\sin x}\,dx = \pm 2\left(\sin\frac{x}{2} - \cos\frac{x}{2}\right),$
[use + if $(8k-1)\frac{\pi}{2} < x \leq (8k+3)\frac{\pi}{2}$, otherwise − ; k an integer]

415. $\displaystyle\int \sqrt{1-\sin x}\,dx = \pm 2\left(\sin\frac{x}{2} + \cos\frac{x}{2}\right),$
[use + if $(8k-3)\frac{\pi}{2} < x \leq (8k+1)\frac{\pi}{2}$, otherwise −; k an integer]

416. $\displaystyle\int \frac{dx}{\sqrt{1-\cos x}} = \pm\sqrt{2}\,\log\tan\frac{x}{4},$
[use + if $4k\pi < x < (4k+2)\pi$, otherwise −; k an integer]

417. $\displaystyle\int \frac{dx}{\sqrt{1+\cos x}} = \pm\sqrt{2}\,\log\tan\left(\frac{x+\pi}{4}\right),$
[use + if $(4k-1)\pi < x < (4k+1)\pi$, otherwise −; k an integer]

418. $\displaystyle\int \frac{dx}{\sqrt{1-\sin x}} = \pm\sqrt{2}\,\log\tan\left(\frac{x}{4} - \frac{\pi}{8}\right),$
[use + if $(8k+1)\frac{\pi}{2} < x < (8k+5)\frac{\pi}{2}$, otherwise −; k an integer]

419. $\displaystyle\int \frac{dx}{\sqrt{1+\sin x}} = \pm\sqrt{2}\log\tan\left(\frac{x}{4} + \frac{\pi}{8}\right),$
[use + if $(8k-1)\frac{\pi}{2} < x < (8k+3)\frac{\pi}{2}$, otherwise −; k an integer]

420. $\displaystyle\int \tan^2(ax)\,dx = \frac{1}{a}\tan ax - x$

421. $\displaystyle\int \tan^3(ax)\,dx = \frac{1}{2a}\tan^2 ax + \frac{1}{a}\log\cos ax$

422. $\int \tan^4(ax)\,dx = \frac{\tan^3 ax}{3a} - \frac{1}{a}\tan ax + x$

423. $\int \tan^n(ax)\,dx = \frac{\tan^{n-1} ax}{a(n-1)} - \int (\tan^{n-2} ax)\,dx$

424. $\int \cot^2(ax)\,dx = -\frac{1}{a}\cot ax - x$

425. $\int \cot^3(ax)\,dx = -\frac{1}{2a}\cot^2 ax - \frac{1}{a}\log\sin ax$

426. $\int \cot^4(ax)\,dx = -\frac{1}{3a}\cot^3 ax + \frac{1}{a}\cot ax + x$

427. $\int \cot^n(ax)\,dx = -\frac{\cot^{n-1} ax}{a(n-1)} - \int (\cot^{n-2} ax)\,dx$

428. $\int \frac{x}{\sin^2 ax}\,dx = \int x(\csc^2 ax)\,dx = -\frac{x\cot ax}{a} + \frac{1}{a^2}\log\sin ax$

429. $\int \frac{x}{\sin^n ax}\,dx = \int x(\csc^n ax)\,dx = -\frac{x\cos ax}{a(n-1)\sin^{n-1} ax} - \frac{1}{a^2(n-1)(n-2)\sin^{n-2} ax} + \frac{(n-2)}{(n-1)}\int \frac{x}{\sin^{n-2} ax}\,dx$

430. $\int \frac{x}{\cos^2 ax}\,dx = \int x(\sec^2 ax)\,dx = \frac{1}{a}x\tan ax + \frac{1}{a^2}\log\cos ax$

431. $\int \frac{x}{\cos^n(ax)}\,dx = \int x(\sec^n ax)\,dx = \frac{x\sin ax}{a(n-1)\cos^{n-1} ax} - \frac{1}{a^2(n-1)(n-2)\cos^{n-2} ax} + \frac{n-2}{n-1}\int \frac{x}{\cos^{n-2} ax}\,dx$

432. $\int \frac{\sin ax}{\sqrt{1+b^2\sin^2 ax}}\,dx = -\frac{1}{ab}\sin^{-1}\frac{b\cos ax}{\sqrt{1+b^2}}$

433. $\int \frac{\sin ax}{\sqrt{1-b^2\sin^2 ax}}\,dx = -\frac{1}{ab}\log(b\cos ax + \sqrt{1-b^2\sin^2 ax})$

434. $\int \sin(ax)\sqrt{1+b^2\sin^2 ax}\,dx = -\frac{\cos ax}{2a}\sqrt{1+b^2\sin^2 ax} - \frac{1+b^2}{2ab}\sin^{-1}\frac{b\cos ax}{\sqrt{1+b^2}}$

435. $\int \sin(ax)\sqrt{1-b^2\sin^2 ax}\,dx = -\frac{\cos ax}{2a}\sqrt{1-b^2\sin^2 ax} - \frac{1-b^2}{2ab}\log(b\cos ax + \sqrt{1-b^2\sin^2 ax})$

436. $\int \frac{\cos ax}{\sqrt{1+b^2\sin^2 ax}}\,dx = \frac{1}{ab}\log(b\sin ax + \sqrt{1+b^2\sin^2 ax})$

437. $\int \frac{\cos ax}{\sqrt{1-b^2\sin^2 ax}}\,dx = \frac{1}{ab}\sin^{-1}(b\sin ax)$

438. $\int \cos(ax)\sqrt{1+b^2\sin^2 ax}\,dx = \frac{\sin ax}{2a}\sqrt{1+b^2\sin^2 ax} + \frac{1}{2ab}\log(b\sin ax + \sqrt{1+b^2\sin^2 ax})$

439. $\int \cos(ax)\sqrt{1-b^2\sin^2 ax}\,dx = \frac{\sin ax}{2a}\sqrt{1-b^2\sin^2 ax} + \frac{1}{2ab}\sin^{-1}(b\sin ax)$

440. $\int \frac{dx}{\sqrt{a+b\tan^2 cx}} = \frac{\pm 1}{c\sqrt{a-b}}\sin^{-1}\left(\sqrt{\frac{a-b}{a}}\sin cx\right), \quad (a > |b|)$

[use + if $(2k-1)\frac{\pi}{2} < x \leq (2k+1)\frac{\pi}{2}$, otherwise −; k an integer]

FORMS INVOLVING INVERSE TRIGONOMETRIC FUNCTIONS

441. $\int \sin^{-1}(ax)\,dx = x\sin^{-1} ax + \frac{\sqrt{1-a^2x^2}}{a}$

442. $\int \cos^{-1}(ax)\,dx = x\cos^{-1} ax - \frac{\sqrt{1-a^2x^2}}{a}$

443. $\int \tan^{-1}(ax)\,dx = x\tan^{-1} ax - \frac{1}{2a}\log(1+a^2x^2)$

444. $\int \cot^{-1}(ax)\,dx = x\cot^{-1} ax + \frac{1}{2a}\log(1+a^2x^2)$

445. $\int \sec^{-1}(ax)\,dx = x\sec^{-1}ax - \frac{1}{a}\log\left(ax + \sqrt{a^2x^2-1}\right)$

446. $\int \csc^{-1}(ax)\,dx = x\csc^{-1}ax + \frac{1}{a}\log\left(ax + \sqrt{a^2x^2-1}\right)$

447. $\int \sin^{-1}\frac{x}{a}\,dx = x\sin^{-1}\frac{x}{a} + \sqrt{a^2-x^2}, \qquad (a>0)$

448. $\int \cos^{-1}\frac{x}{a}\,dx = x\cos^{-1}\frac{x}{a} - \sqrt{a^2-x^2}, \qquad (a>0)$

449. $\int \tan^{-1}\frac{x}{a}\,dx = x\tan^{-1}\frac{x}{a} - \frac{a}{2}\log(a^2+x^2)$

450. $\int \cot^{-1}\frac{x}{a}\,dx = x\cot^{-1}\frac{x}{a} + \frac{a}{2}\log(a^2+x^2)$

451. $\int x\sin^{-1}(ax)\,dx = \frac{1}{4a^2}\left[(2a^2x^2-1)\sin^{-1}(ax) + ax\sqrt{1-a^2x^2}\right]$

452. $\int x\cos^{-1}(ax)\,dx = \frac{1}{4a^2}\left[(2a^2x^2-1)\cos^{-1}(ax) - ax\sqrt{1-a^2x^2}\right]$

453. $\int x^n\sin^{-1}(ax)\,dx = \frac{x^{n+1}}{n+1}\sin^{-1}(ax) - \frac{a}{n+1}\int \frac{x^{n+1}\,dx}{\sqrt{1-a^2x^2}}, \ (n \neq -1)$

454. $\int x^n\cos^{-1}(ax)\,dx = \frac{x^{n+1}}{n+1}\cos^{-1}(ax) + \frac{a}{n+1}\int \frac{x^{n+1}\,dx}{\sqrt{1-a^2x^2}}, \ (n \neq -1)$

455. $\int x\tan^{-1}(ax)\,dx = \frac{1+a^2x^2}{2a^2}\tan^{-1}ax - \frac{x}{2a}$

456. $\int x^n\tan^{-1}(ax)\,dx = \frac{x^{n+1}}{n+1}\tan^{-1}ax - \frac{a}{n+1}\int \frac{x^{n+1}}{1+a^2x^2}\,dx$

457. $\int x(\cot^{-1}ax)\,dx = \frac{1+a^2x^2}{2a^2}\cot^{-1}ax + \frac{x}{2a}$

458. $\int x^n\cot^{-1}(ax)\,dx = \frac{x^{n+1}}{n+1}\cot^{-1}ax + \frac{a}{n+1}\int \frac{x^{n+1}}{1+a^2x^2}\,dx$

459. $\int \frac{\sin^{-1}(ax)}{x^2}\,dx = a\log\left(\frac{1-\sqrt{1-a^2x^2}}{x}\right) - \frac{\sin^{-1}(ax)}{x}$

460. $\int \frac{\cos^{-1}(ax)\,dx}{x^2} = -\frac{1}{x}\cos^{-1}(ax) + a\log\frac{1+\sqrt{1-a^2x^2}}{x}$

461. $\int \frac{\tan^{-1}(ax)\,dx}{x^2} = -\frac{1}{x}\tan^{-1}(ax) - \frac{a}{2}\log\frac{1+a^2x^2}{x^2}$

462. $\int \frac{\cot^{-1}(ax)}{x^2}\,dx = -\frac{1}{x}\cot^{-1}ax - \frac{a}{2}\log\frac{x^2}{a^2x^2+1}$

463. $\int \sin^{-1}(ax)^2\,dx = x(\sin^{-1}ax)^2 - 2x + \frac{2\sqrt{1-a^2x^2}}{a}\sin^{-1}ax$

464. $\int \cos^{-1}(ax)^2\,dx = x(\cos^{-1}ax)^2 - 2x - \frac{2\sqrt{1-a^2x^2}}{a}\cos^{-1}ax$

465. $$\int (\sin^{-1}ax)^n\,dx = \begin{cases} x(\sin^{-1}ax)^n + \dfrac{n\sqrt{1-a^2x^2}}{a}(\sin^{-1}ax)^{n-1} - n(n-1)\displaystyle\int(\sin^{-1}ax)^{n-2}\,dx \\ \text{or} \\ \displaystyle\sum_{r=0}^{[n/2]}(-1)^r\frac{n!}{(n-2r)!}x(\sin^{-1}ax)^{n-2r} + \sum_{r=0}^{[n-1/2]}(-1)^r\frac{n!\sqrt{1-a^2x^2}}{(n-2r-1)!a}(\sin^{-1}ax)^{n-2r-1} \end{cases}$$

Note: $[s]$ means greatest integer $\leq s$. Thus $[3.5]$ means 3; $[5] = 5$, $[\frac{1}{2}] = 0$.

466. $$\int (\cos^{-1}ax)^n\,dx = \begin{cases} x(\cos^{-1}ax)^n - \dfrac{n\sqrt{1-a^2x^2}}{a}(\cos^{-1}ax)^{n-1} - n(n-1)\displaystyle\int(\cos^{-1}ax)^{n-2}\,dx \\ \text{or} \\ \displaystyle\sum_{r=0}^{[n/2]}(-1)^r\frac{n!}{(n-2r)!}x(\cos^{-1}ax)^{n-2r} \times \sum_{r=0}^{[n-1/2]}(-1)^r\frac{n!\sqrt{1-a^2x^2}}{(n-2r-1)!a}(\cos^{-1}ax)^{n-2r-1} \end{cases}$$

467. $\int \frac{1}{\sqrt{1-a^2x^2}}(\sin^{-1}ax)\,dx = \frac{1}{2a}(\sin^{-1}ax)^2$

468. $\int \frac{x^n}{\sqrt{1-a^2x^2}}(\sin^{-1}ax)\,dx = -\frac{x^{n-1}}{na^2}\sqrt{1-a^2x^2}\sin^{-1}ax + \frac{x^n}{n^2a} + \frac{n-1}{na^2}\int \frac{x^{n-2}}{\sqrt{1-a^2x^2}}\sin^{-1}ax\,dx$

469. $\int \frac{1}{\sqrt{1-a^2x^2}}(\cos^{-1} ax)\,dx = -\frac{1}{2a}(\cos^{-1} ax)^2$

470. $\int \frac{x^n}{\sqrt{1-a^2x^2}}(\cos^{-1} ax)\,dx = -\frac{x^{n-1}}{na^2}\sqrt{1-a^2x^2}\cos^{-1} ax - \frac{x^n}{n^2 a} + \frac{n-1}{na^2}\int \frac{x^{n-2}}{\sqrt{1-a^2x^2}}\cos^{-1} ax\,dx$

471. $\int \frac{\tan^{-1} ax}{a^2x^2+1}\,dx = \frac{1}{2a}(\tan^{-1} ax)^2$

472. $\int \frac{\cot^{-1} ax}{a^2x^2+1}\,dx = -\frac{1}{2a}(\cot^{-1} ax)^2$

473. $\int x\sec^{-1} ax\,dx = \frac{x^2}{2}\sec^{-1} ax - \frac{1}{2a^2}\sqrt{a^2x^2-1}$

474. $\int x^n \sec^{-1} ax\,dx = \frac{x^{n+1}}{n+1}\sec^{-1} ax - \frac{1}{n+1}\int \frac{x^n\,dx}{\sqrt{a^2x^2-1}}$

475. $\int \frac{\sec^{-1} ax}{x^2}\,dx = -\frac{\sec^{-1} ax}{x} + \frac{\sqrt{a^2x^2-1}}{x}$

476. $\int x\csc^{-1} ax\,dx = \frac{x^2}{2}\csc^{-1} ax + \frac{1}{2a^2}\sqrt{a^2x^2-1}$

477. $\int x^n \csc^{-1} ax\,dx = \frac{x^{n+1}}{n+1}\csc^{-1} ax + \frac{1}{n+1}\int \frac{x^n\,dx}{\sqrt{a^2x^2-1}}$

478. $\int \frac{\csc^{-1} ax}{x^2}\,dx = -\frac{\csc^{-1} ax}{x} - \frac{\sqrt{a^2x^2-1}}{x}$

FORMS INVOLVING TRIGONOMETRIC SUBSTITUTIONS

479. $\int f(\sin x)\,dx = 2\int f\left(\frac{2z}{1+z^2}\right)\frac{dz}{1+z^2}, \quad \left(z = \tan\frac{x}{2}\right)$

480. $\int f(\cos x)\,dx = 2\int f\left(\frac{1-z^2}{1+z^2}\right)\frac{dz}{1+z^2}, \quad \left(z = \tan\frac{x}{2}\right)$

481. $\int f(\sin x)\,dx = \int f(u)\frac{du}{\sqrt{1-u^2}}, \quad (u = \sin x)$

482. $\int f(\cos x)\,dx = -\int f(u)\frac{du}{\sqrt{1-u^2}}, \quad (u = \cos x)$

483. $\int f(\sin x, \cos x)\,dx = \int f\left(u, \sqrt{1-u^2}\right)\frac{du}{\sqrt{1-u^2}}, \quad (u = \sin x)$

484. $\int f(\sin x, \cos x)\,dx = 2\int f\left(\frac{2z}{1+z^2}, \frac{1-z^2}{1+z^2}\right)\frac{dz}{1+z^2}, \quad \left(z = \tan\frac{x}{2}\right)$

LOGARITHMIC FORMS

485. $\int (\log x)\,dx = x\log x - x$

486. $\int x(\log x)\,dx = \frac{x^2}{2}\log x - \frac{x^2}{4}$

487. $\int x^2(\log x)\,dx = \frac{x^3}{3}\log x - \frac{x^3}{9}$

488. $\int x^n(\log ax)\,dx = \frac{x^{n+1}}{n+1}\log ax - \frac{x^{n+1}}{(n+1)^2}$

489. $\int (\log x)^2\,dx = x(\log x)^2 - 2x\log x + 2x$

490. $\int (\log x)^n\,dx = \begin{cases} x(\log x)^n - n\int (\log x)^{n-1}\,dx, & (n \neq -1) \\ \text{or} \\ (-1)^n n!\,x\sum_{r=0}^{n} \frac{(-\log x)^r}{r!} \end{cases}$

491. $\int \frac{(\log x)^n}{x}\,dx = \frac{1}{n+1}(\log x)^{n+1}$

492. $\int \frac{dx}{\log x} = \log(\log x) + \log x + \frac{(\log x)^2}{2\cdot 2!} + \frac{(\log x)^3}{3\cdot 3!} + \cdots$

493. $\int \frac{dx}{x\log x} = \log(\log x)$

494. $$\int \frac{dx}{x(\log x)^n} = -\frac{1}{(n-1)(\log x)^{n-1}}$$

495. $$\int \frac{x^m\,dx}{(\log x)^n} = -\frac{x^{m+1}}{(n-1)(\log x)^{n-1}} + \frac{m+1}{n-1}\int \frac{x^m\,dx}{(\log x)^{n-1}}$$

496. $$\int x^m(\log x)^n\,dx = \begin{cases} \frac{x^{m+1}(\log x)^n}{m+1} - \frac{n}{m+1}\int x^m(\log x)^{n-1}\,dx \\ \text{or} \\ (-1)^n \frac{n!}{m+1} x^{m+1} \sum_{r=0}^{n} \frac{(-\log x)^r}{r!(m+1)^{n-r}} \end{cases}$$

497. $$\int x^p \cos(b\ln x)\,dx = \frac{x^{p+1}}{(p+1)^2+b^2}\,[b\sin(b\ln x) + (p+1)\cos(b\ln x)] + c$$

498. $$\int x^p \sin(b\ln x)\,dx = \frac{x^{p+1}}{(p+1)^2+b^2}\,[(p+1)\sin(b\ln x) - b\cos(b\ln x)] + c$$

499. $$\int [\log(ax+b)]\,dx = \frac{ax+b}{a}\log(ax+b) - x$$

500. $$\int \frac{\log(ax+b)}{x^2}\,dx = \frac{a}{b}\log x - \frac{ax+b}{bx}\log(ax+b)$$

501. $$\int x^m[\log(ax+b)]\,dx = \frac{1}{m+1}\left[x^{m+1} - \left(-\frac{b}{a}\right)^{m+1}\right]\log(ax+b) - \frac{1}{m+1}\left(-\frac{b}{a}\right)^{m+1}\sum_{r=1}^{m+1}\frac{1}{r}\left(-\frac{ax}{b}\right)^r$$

502. $$\int \frac{\log(ax+b)}{x^m}\,dx = -\frac{1}{m-1}\frac{\log(ax+b)}{x^{m-1}} + \frac{1}{m-1}\left(-\frac{a}{b}\right)^{m-1}\log\frac{ax+b}{x} + \frac{1}{m-1}\left(-\frac{a}{b}\right)^{m-1}\sum_{r=1}^{m-2}\frac{1}{r}\left(-\frac{b}{ax}\right)^r, \quad (m>2)$$

503. $$\int \left[\log\frac{x+a}{x-a}\right]dx = (x+a)\log(x+a) - (x-a)\log(x-a)$$

504. $$\int x^m\left[\log\frac{x+a}{x-a}\right]dx = \frac{x^{m+1} - (-a)^{m+1}}{m+1}\log(x+a) - \frac{x^{m+1} - a^{m+1}}{m+1}\log(x-a) + \frac{2a^{m+1}}{m+1}\sum_{r=1}^{\left[\frac{m+1}{2}\right]}\frac{1}{m-2r+2}\left(\frac{x}{a}\right)^{m-2r+2}$$

Note: $[s]$ means greatest integer $\leq s$; Thus $[3.5]$ means 3; $\lfloor 5 \rfloor = 5$, $\left[\frac{1}{2}\right] = 0$.

505. $$\int \frac{1}{x^2}\left[\log\frac{x+a}{x-a}\right]dx = \frac{1}{x}\log\frac{x-a}{x+a} - \frac{1}{a}\log\frac{x^2-a^2}{x^2}$$

506. $$\int (\log X)\,dx = \begin{cases} \left(x+\frac{b}{2c}\right)\log X - 2x + \frac{\sqrt{4ac-b^2}}{c}\tan^{-1}\frac{2cx+b}{\sqrt{4ac-b^2}}, & (b^2-4ac<0) \\ \text{or} \\ \left(x+\frac{b}{2c}\right)\log X - 2x + \frac{\sqrt{b^2-4ac}}{c}\tanh^{-1}\frac{2cx+b}{\sqrt{b^2-4ac}}, & (b^2-4ac>0) \\ \text{where} \\ X = a+bx+cx^2 \end{cases}$$

507. $$\int x^n(\log(a+bx+cx^2))\,dx = \frac{x^{n+1}}{n+1}\log X - \frac{2c}{n+1}\int\frac{x^{n+2}}{X}\,dx - \frac{b}{n+1}\int\frac{x^{n+1}}{X}\,dx$$

508. $$\int \log(x^2+a^2)\,dx = x\log(x^2+a^2) - 2x + 2a\tan^{-1}\frac{x}{a}$$

509. $$\int \log(x^2-a^2)\,dx = x\log(x^2-a^2) - 2x + a\log\frac{x+a}{x-a}$$

510. $$\int x\log(x^2\pm a^2)\,dx = \frac{1}{2}(x^2\pm a^2)\log(x^2\pm a^2) - \frac{1}{2}x^2$$

511. $$\int \log(x+\sqrt{x^2\pm a^2})\,dx = x\log(x+\sqrt{x^2\pm a^2}) - \sqrt{x^2\pm a^2}$$

512. $$\int x\log(x+\sqrt{x^2\pm a^2})\,dx = \left(\frac{x^2}{2}\pm\frac{a^2}{4}\right)\log(x+\sqrt{x^2\pm a^2}) - \frac{x\sqrt{x^2\pm a^2}}{4}$$

513. $$\int x^m\log(x+\sqrt{x^2\pm a^2})\,dx = \frac{x^{m+1}}{m+1}\log(x+\sqrt{x^2\pm a^2}) - \frac{1}{m+1}\int\frac{x^{m+1}}{\sqrt{x^2\pm a^2}}\,dx$$

514. $$\int \frac{\log(x+\sqrt{x^2+a^2})}{x^2}\,dx = -\frac{\log(x+\sqrt{x^2+a^2})}{x} - \frac{1}{a}\log\frac{a+\sqrt{x^2+a^2}}{x}$$

515. $$\int \frac{\log(x+\sqrt{x^2-a^2})}{x^2}\,dx = -\frac{\log(x+\sqrt{x^2-a^2})}{x} + \frac{1}{|a|}\sec^{-1}\frac{x}{a}$$

516. $$\int x^n\log(x^2-a^2)\,dx = \frac{1}{n+1}\left[x^{n+1}\log(x^2-a^2) - a^{n+1}\log(x-a) - (-a)^{n+1}\log(x+a) - 2\sum_{r=0}^{[n/2]}\frac{a^{2r}x^{n-2r+1}}{n-2r+1}\right]$$

Note: $[s]$ means greatest integer $\leq s$; Thus $[3.5]$ means 3; $[5] = 5$, $\left[\frac{1}{2}\right] = 0$.

EXPONENTIAL FORMS

517. $\int e^x\,dx = e^x$

518. $\int e^{-x}\,dx = -e^{-x}$

519. $\int e^{ax}\,dx = \dfrac{e^{ax}}{a}$

520. $\int x\,e^{ax}\,dx = \dfrac{e^{ax}}{a^2}(ax-1)$

521. $$\int x^m e^{ax}\,dx = \begin{cases} \frac{x^m e^{ax}}{a} - \frac{m}{a}\int x^{m-1}e^{ax}\,dx \\ \text{or} \\ e^{ax}\sum_{r=0}^{m}(-1)^r \frac{m!x^{m-r}}{(m-r)!a^{r+1}} \end{cases}$$

522. $\int \dfrac{e^{ax}\,dx}{x} = \log x + \dfrac{ax}{1!} + \dfrac{a^2x^2}{2\cdot 2!} + \dfrac{a^3+x^3}{3\cdot 3!} + \cdots$

523. $\int \dfrac{e^{ax}}{x^m}\,dx = -\dfrac{1}{m-1}\dfrac{e^{ax}}{x^{m-1}} + \dfrac{a}{m-1}\int \dfrac{e^{ax}}{x^{m-1}}\,dx$

524. $\int e^{ax}\log x\,dx = \dfrac{e^{ax}\log x}{a} - \dfrac{1}{a}\int \dfrac{e^{ax}}{x}\,dx$

525. $\int \dfrac{dx}{1+e^x} = x - \log(1+e^x) = \log\dfrac{e^x}{1+e^x}$

526. $\int \dfrac{dx}{a+be^{px}} = \dfrac{x}{a} - \dfrac{1}{ap}\log(a+be^{px})$

527. $\int \dfrac{dx}{ae^{mx}+be^{-mx}} = \dfrac{1}{m\sqrt{ab}}\tan^{-1}\left(e^{mx}\sqrt{\dfrac{a}{b}}\right), \quad (a>0, b>0)$

528. $$\int \frac{dx}{ae^{mx}-be^{-mx}} = \begin{cases} \frac{1}{2m\sqrt{ab}}\log\frac{\sqrt{a}\,e^{mx}-\sqrt{b}}{\sqrt{a}e^{mx}+\sqrt{b}} \\ \text{or} \\ \frac{-1}{m\sqrt{ab}}\tanh^{-1}\left(\sqrt{\frac{a}{b}}e^{mx}\right), \quad (a>0, b>0) \end{cases}$$

529. $\int (a^x - a^{-x})\,dx = \dfrac{a^x + a^{-x}}{\log a}$

530. $\int \dfrac{e^{ax}}{b+ce^{ax}}\,dx = \dfrac{1}{ac}\log(b+ce^{ax})$

531. $\int \dfrac{x\,e^{ax}}{(1+ax)^2}\,dx = \dfrac{e^{ax}}{a^2(1+ax)}$

532. $\int x\,e^{-x^2}\,dx = -\dfrac{1}{2}e^{-x^2}$

533. $\int e^{ax}\sin(bx)\,dx = \dfrac{e^{ax}[a\sin(bx)-b\cos(bx)]}{a^2+b^2}$

534. $\int e^{ax}\sin(bx)\sin(cx)\,dx = \dfrac{e^{ax}[(b-c)\sin(b-c)x + a\cos(b-c)x]}{2[a^2+(b-c)^2]} - \dfrac{e^{ax}[(b+c)\sin(b+c)x + a\cos(b+c)x]}{2[a^2+(b+c)^2]}$

535. $$\int e^{ax}\sin(bx)\cos(cx)\,dx = \begin{cases} \frac{e^{ax}[a\sin(b-c)x-(b-c)\cos(b-c)x]}{2[a^2+(b-c)^2]} + \frac{e^{ax}[a\sin(b+c)x-(b+c)\cos(b+c)x]}{2[a^2+(b+c)^2]} \\ \text{or} \\ \frac{e^{ax}}{\rho}[(a\sin bx - b\cos bx)[\cos(cx-\alpha)] - c(\sin bx)\sin(cx-\alpha)] \\ \text{where} \\ \rho = \sqrt{(a^2+b^2-c^2)^2+4a^2c^2}, \\ \rho\cos\alpha = a^2+b^2-c^2, \quad \rho\sin\alpha = 2ac \end{cases}$$

536. $\int e^{ax}\sin(bx)\sin(bx+c)\,dx = \dfrac{e^{ax}\cos c}{2a} - \dfrac{e^{ax}[a\cos(2bx+c)+2b\sin(2bx+c)]}{2(a^2+4b^2)}$

537. $\int e^{ax}\sin(bx)\cos(bx+c)\,dx = -\dfrac{e^{ax}\sin c}{2a} + \dfrac{e^{ax}[a\sin(2bx+c)-2b\cos(2bx+c)]}{2(a^2+4b^2)}$

538. $\int e^{ax}\cos(bx)\,dx = \dfrac{e^{ax}}{a^2+b^2}[a\cos(bx)+b\sin(bx)]$

539. $\int e^{ax}\cos(bx)\cos(cx)\,dx = \dfrac{e^{ax}[(b-c)\sin(b-c)x + a\cos(b-c)x]}{2[a^2+(b-c)^2]} + \dfrac{e^{ax}[(b+c)\sin(b+c)x + a\cos(b+c)x]}{2[a^2+(b+c)^2]}$

540. $\int e^{ax}\cos(bx)\cos(bx+c)\,dx = \dfrac{e^{ax}\cos c}{2a} + \dfrac{e^{ax}[a\cos(2bx+c)+2b\sin(2bx+c)]}{2(a^2+4b^2)}$

541. $\int e^{ax}\cos(bx)\sin(bx+c)\,dx = \dfrac{e^{ax}\sin c}{2a} + \dfrac{e^{ax}[a\sin(2bx+c)-2b\cos(2bx+c)]}{2(a^2+4b^2)}$

542. $$\int e^{ax}\sin^n(bx)\,dx = \frac{1}{a^2+n^2b^2}\left[(a\sin bx - nb\cos bx)e^{ax}\sin^{n-1}bx + n(n-1)b^2\int e^{ax}[\sin^{n-2}bx]\,dx\right]$$

543. $$\int e^{ax}\cos^n(bx)\,dx = \frac{1}{a^2+n^2b^2}\left[(a\cos bx + nb\sin bx)e^{ax}\cos^{n-1}bx + n(n-1)b^2\int e^{ax}[\cos^{n-2}bx]\,dx\right]$$

544. $$\int x^m e^x\sin x\,dx = \frac{1}{2}x^m e^x(\sin x - \cos x) - \frac{m}{2}\int x^{m-1}e^x\sin x\,dx + \frac{m}{2}\int x^{m-1}e^x\cos x\,dx$$

545. $$\int x^m e^{ax}\sin(bx)\,dx = \begin{cases} x^m e^{ax}\frac{a\sin bx - b\cos bx}{a^2+b^2} - \frac{m}{a^2+b^2}\int x^{m-1}e^{ax}(a\sin bx - b\cos bx)\,dx \\ \text{or} \\ e^{ax}\sum_{r=0}^{m}\frac{(-1)^r m! x^{m-r}}{\rho^{r+1}(m-r)!}\sin[bx-(r+1)\alpha] \\ \text{where} \\ \rho = \sqrt{a^2+b^2}, \quad \rho\cos\alpha = a, \quad \rho\sin\alpha = b \end{cases}$$

546. $$\int x^m e^x\cos x\,dx = \frac{1}{2}x^m e^x(\sin x + \cos x) - \frac{m}{2}\int x^{m-1}e^x\sin x\,dx - \frac{m}{2}\int x^{m-1}e^x\cos x\,dx$$

547. $$\int x^m e^{ax}\cos(bx)\,dx = \begin{cases} x^m e^{ax}\frac{a\cos bx + b\sin bx}{a^2+b^2} - \frac{m}{a^2+b^2}\int x^{m-1}e^{ax}(a\cos bx + b\sin bx)\,dx \\ \text{or} \\ e^{ax}\sum_{r=0}^{m}\frac{(-1)^r m! x^{m-r}}{\rho^{r+1}(m-r)!}\cos[bx-(r+1)\alpha] \\ \rho = \sqrt{a^2+b^2}, \quad \rho\cos\alpha = a, \quad \rho\sin\alpha = b \end{cases}$$

548. $$\int e^{ax}(\cos^m x)(\sin^n x)\,dx = \begin{cases} \frac{e^{ax}\cos^{m-1}x\sin^n x[a\cos x + (m+n)\sin x]}{(m+n)^2+a^2} \\ \quad -\frac{na}{(m+n)^2+a^2}\int e^{ax}(\cos^{m-1}x)(\sin^{n-1}x)\,dx \\ \quad +\frac{(m-1)(m+n)}{(m+n)^2+a^2}\int e^{ax}(\cos^{m-2}x)(\sin^n x)\,dx \\ \text{or} \\ \frac{e^{ax}\cos^m x\,sin^{n-1}x[a\sin x - (m+n)\cos x]}{(m+n)^2+a^2} \\ \quad +\frac{ma}{(m+n)^2+a^2}\int e^{ax}(\cos^{m-1}x)(\sin^{n-1}x)\,dx \\ \quad +\frac{(n-1)(m+n)}{(m+n)^2+a^2}\int e^{ax}(\cos^m x)(\sin^{n-2}x)\,dx \\ \text{or} \\ \frac{e^{ax}(\cos^{m-1}x)(\sin^{n-1}x)(a\sin x\cos x + m\sin^2 x - n\cos^2 x)}{(m+n)^2+a^2} \\ \quad +\frac{m(m-1)}{(m+n)^2+a^2}\int e^{ax}(\cos^{m-2}x)(\sin^n x)\,dx \\ \quad +\frac{n(n-1)}{(m+n)^2+a^2}\int e^{ax}(\cos^m x)(\sin^{n-2}x)\,dx \\ \text{or} \\ \frac{e^{ax}(\cos^{m-1}x)(\sin^{n-1}x)(a\cos x\sin x + m\sin^2 x - n\cos^2 x)}{(m+n)^2+a^2} \\ \quad +\frac{m(m-1)}{(m+n)^2+a^2}\int e^{ax}(\cos^{m-2}x)(\sin^{n-2}x)\,dx \\ \quad +\frac{(n-m)(n+m-1)}{(m+n)^2+a^2}\int e^{ax}(\cos^m x)(\sin^{n-2}x)\,dx \end{cases}$$

549. $$\int xe^{ax}\sin(bx)\,dx = \frac{xe^{ax}}{a^2+b^2}(a\sin bx - b\cos bx) - \frac{e^{ax}}{(a^2+b^2)^2}[(a^2-b^2)\sin bx - 2ab\cos bx]$$

550. $$\int xe^{ax}\cos(bx)\,dx = \frac{xe^{ax}}{a^2+b^2}(a\cos bx - b\sin bx) - \frac{e^{ax}}{(a^2+b^2)^2}[(a^2-b^2)\cos bx - 2ab\sin bx]$$

551. $$\int \frac{e^{ax}}{\sin^n x}\,dx = -\frac{e^{ax}[a\sin x + (n-2)\cos x]}{(n-1)(n-2)\sin^{n-1}x} + \frac{a^2+(n-2)^2}{(n-1)(n-2)}\int \frac{e^{ax}}{\sin^{n-2}x}\,dx$$

552. $$\int \frac{e^{ax}}{\cos^n x}\,dx = -\frac{e^{ax}[a\cos x - (n-2)\sin x]}{(n-1)(n-2)\cos^{n-1}x} + \frac{a^2+(n-2)^2}{(n-1)(n-2)}\int \frac{e^{ax}}{\cos^{n-2}x}\,dx$$

553. $$\int e^{ax}\tan^n x\,dx = e^{ax}\frac{\tan^{n-1}x}{n-1} - \frac{a}{n-1}\int e^{ax}\tan^{n-1}x\,dx - \int e^{ax}\tan^{n-2}x\,dx$$

HYPERBOLIC FORMS

554. $$\int \sinh x\,dx = \cosh x$$

555. $$\int \cosh x\,dx = \sinh x$$

556. $$\int \tanh x\,dx = \log\cosh x$$

557. $$\int \coth x\,dx = \log \sinh x$$

558. $$\int \operatorname{sech} x\,dx = \tan^{-1}(\sinh x)$$

559. $$\int \operatorname{csch} x\,dx = \log \tanh\left(\frac{x}{2}\right)$$

560. $$\int x \sinh x\,dx = x\cosh x - \sinh x$$

561. $$\int x^n \sinh x\,dx = x^n \cosh x - n\int x^{n-1}(\cosh x)\,dx$$

562. $$\int x\cosh x\,dx = x\sinh x - \cosh x$$

563. $$\int x^n \cosh x\,dx - x^n \sinh x - n\int x^{n-1}(\sinh x)\,dx$$

564. $$\int \operatorname{sech} x \tanh x\,dx = -\operatorname{sech} x$$

565. $$\int \operatorname{csch} x \coth x\,dx = -\operatorname{csch} x$$

566. $$\int \sinh^2 x\,dx = \frac{\sinh 2x}{4} - \frac{x}{2}$$

567. $$\int (\sinh^m x)(\cosh^n x)\,dx = \begin{cases} \frac{1}{m+n}(\sinh^{m+1} x)(\cosh^{n-1} x) + \frac{n-1}{m+n}\int (\sinh^m x)(\cosh^{n-2} x)\,dx \\ \text{or} \\ \frac{1}{m+n}\sinh^{m-1} x \cosh^{n+1} x - \frac{m-1}{m+n}\int (\sinh^{m-2} x)(\cosh^n x)\,dx, \quad (m+n \neq 0) \end{cases}$$

568. $$\int \frac{dx}{(\sinh^m x)(\cosh^n x)} \begin{cases} -\frac{1}{(m-n)(\sinh^{m-1} x)(\cosh^{n-1} x)} - \frac{m+n-2}{m-1}\int \frac{dx}{(\sinh^{m-2} x)(\cosh^n x)}, \quad (m \neq 1) \\ \text{or} \\ \frac{1}{(n-1)\sinh^{m-1} x \cosh^{n-1} x} + \frac{m+n-2}{n-1}\int \frac{dx}{(\sinh^m x)(\cosh^{n-2} x)}, \quad (n \neq 1) \end{cases}$$

569. $$\int \tanh^2 x\,dx = x - \tanh x$$

570. $$\int \tanh^n x\,dx = -\frac{\tanh^{n-1} x}{n-1} + \int (\tanh^{n-2} x)\,dx, \quad (n \neq 1)$$

571. $$\int \operatorname{sech}^2 x\,dx = \tanh x$$

572. $$\int \cosh^2 x\,dx = \frac{\sinh 2x}{4} + \frac{x}{2}$$

573. $$\int \coth^2 x\,dx = x - \coth x$$

574. $$\int \coth^n x\,dx = -\frac{\coth^{n-1} x}{n-1} + \int \coth^{n-2} x\,dx, \quad (n \neq 1)$$

575. $$\int \operatorname{csch}^2 x\,dx = -\operatorname{ctnh} x$$

576. $$\int \sinh(mx)\sinh(nx)\,dx = \frac{\sinh(m+n)x}{2(m+n)} - \frac{\sinh(m-n)x}{2(m-n)}, \quad (m^2 \neq n^2)$$

577. $$\int \cosh(mx)\cosh(nx)\,dx = \frac{\sinh(m+n)x}{2(m+n)} + \frac{\sinh(m-n)x}{2(m-n)}, \quad (m^2 \neq n^2)$$

578. $$\int \sinh(mx)\cosh(nx)\,dx = \frac{\cosh(m+n)x}{2(m+n)} + \frac{\cosh(m-n)x}{2(m-n)}, \quad (m^2 \neq n^2)$$

579. $$\int \sinh^{-1}\frac{x}{a}\,dx = x\sinh^{-1}\frac{x}{a} - \sqrt{x^2+a^2}, \quad (a > 0)$$

580. $$\int x\sinh^{-1}\frac{x}{a}\,dx = \left(\frac{x^2}{2} + \frac{a^2}{4}\right)\sinh^{-1}\frac{x}{a} - \frac{x}{4}\sqrt{x^2+a^2}, \quad (a > 0)$$

581. $$\int x^n \sinh^{-1} x\,dx = \left(\frac{x^{n+1}}{n+1}\right)\sinh^{-1} x - \frac{1}{n+1}\int \frac{x^{n+1}}{(1+x^2)^{\frac{1}{2}}}\,dx, \quad (n \neq -1)$$

582. $$\int \cosh^{-1}\frac{x}{a}\,dx = \begin{cases} x\cosh^{-1}\frac{x}{a} - \sqrt{x^2-a^2}, \ \left(\cosh^{-1}\frac{x}{a} > 0\right) \\ \text{or} \\ x\cosh^{-1}\frac{x}{a} + \sqrt{x^2-a^2}, \ \left(\cosh^{-1}\frac{x}{a} < 0\right), \quad (a > 0) \end{cases}$$

583. $\int x\cosh^{-1}\frac{x}{a}\,dx = \frac{2x^2-a^2}{4}\cosh^{-1}\frac{x}{a} - \frac{x}{4}(x^2-a^2)^{\frac{1}{2}}$

584. $\int x^n(\cosh^{-1}x)\,dx = \frac{x^{n+1}}{n+1}\cosh^{-1}x - \frac{1}{n+1}\int \frac{x^{n+1}}{(x^2-1)^{\frac{1}{2}}}dx, \quad (n \neq -1)$

585. $\int \tanh^{-1}\frac{x}{a}\,dx = x\tanh^{-1}\frac{x}{a} + \frac{a}{2}\log(a^2-x^2), \quad \left(\left|\frac{x}{a}\right| < 1\right)$

586. $\int \coth^{-1}\frac{x}{a}\,dx = x\coth^{-1}\frac{x}{a} + \frac{a}{2}\log(x^2-a^2), \quad \left(\left|\frac{x}{a}\right| > 1\right)$

587. $\int x\tanh^{-1}\frac{x}{a}\,dx = \frac{x^2-a^2}{2}\tanh^{-1}\frac{x}{a} + \frac{ax}{2}, \quad \left(\left|\frac{x}{a}\right| < 1\right)$

588. $\int x^n\tanh^{-1}x\,dx = \frac{x^{n+1}}{n+1}\tanh^{-1}x - \frac{1}{n+1}\int \frac{x^{n+1}}{1-x^2}dx, \quad (n \neq -1)$

589. $\int x\coth^{-1}\frac{x}{a}\,dx = \frac{x^2-a^2}{2}\coth^{-1}\frac{x}{a} + \frac{ax}{2}, \left(\left|\frac{x}{a}\right| > 1\right)$

590. $\int x^n\coth^{-1}x\,dx = \frac{x^{n+1}}{n+1}\coth^{-1}x + \frac{1}{n+1}\int \frac{x^{n+1}}{x^2-1}dx, \quad (n \neq -1)$

591. $\int \operatorname{sech}^{-1}x\,dx = x\operatorname{sech}^{-1}x + \sin^{-1}x$

592. $\int x\operatorname{sech}^{-1}x\,dx = \frac{x^2}{2}\operatorname{sech}^{-1}x - \frac{1}{2}\sqrt{1-x^2}$

593. $\int x^n\operatorname{sech}^{-1}x\,dx = \frac{x^{n+1}}{n+1}\operatorname{sech}^{-1}x + \frac{1}{n+1}\int \frac{x^n}{\sqrt{1-x^2}}dx, \quad (n \neq -1)$

594. $\int \operatorname{csch}^{-1}x\,dx = x\operatorname{csch}^{-1}x + \frac{x}{|x|}\sinh^{-1}x$

595. $\int x\operatorname{csch}^{-1}x\,dx = \frac{x^2}{2}\operatorname{csch}^{-1}x + \frac{1}{2}\frac{x}{|x|}\sqrt{1+x^2}$

596. $\int x^n\operatorname{csch}^{-1}x\,dx = \frac{x^{n+1}}{n+1}\operatorname{csch}^{-1}x + \frac{1}{n+1}\frac{x}{|x|}\int \frac{x^n}{\sqrt{x^2+1}}dx, \quad (n \neq -1)$

DEFINITE INTEGRALS

597. $\int_0^\infty x^{n-1}e^{-x}\,dx = \int_0^1 \left(\log\frac{1}{x}\right)^{n-1}dx = \frac{1}{n}\prod_{m=1}^{\infty}\frac{\left(1+\frac{1}{m}\right)^n}{1+\frac{n}{m}} = \Gamma(n)$

for $n \neq 0, -1, -2, -3, \ldots$ (This is the Gamma function)

598. $\int_0^\infty t^n p^{-t}dt = \frac{n!}{(\log p)^{n+1}}, \quad (n = 0, 1, 2, 3, \ldots \text{ and } p > 0)$

599. $\int_0^\infty t^{n-1}e^{-(a+1)t}\,dt = \frac{\Gamma(n)}{(a+1)^n}, \quad (n > 0, a > -1)$

600. $\int_0^1 x^m\left(\log\frac{1}{x}\right)^n dx = \frac{\Gamma(n+1)}{(m+1)^{n+1}}, \quad (m > -1, n > -1)$

601. $\Gamma(n)$ is finite if $n > 0$; $\Gamma(n+1) = n\Gamma(n)$

602. $\Gamma(n)\cdot\Gamma(1-n) = \frac{\pi}{\sin n\pi}$

603. $\Gamma(n) = (n-1)!$ if $n =$ integer > 0

604. $\Gamma(\frac{1}{2}) = 2\int_0^\infty e^{-t^2}dt = \sqrt{\pi} = 1.7724538509\cdots = \left(-\frac{1}{2}\right)!$

605. $\Gamma(n+\frac{1}{2}) = \frac{1\cdot 3\cdot 5\ldots(2n-1)}{2^n}\sqrt{\pi} \quad n = 1, 2, 3, \ldots$

606. $\Gamma(-n+\frac{1}{2}) = \frac{(-1)^n 2^n\sqrt{\pi}}{1\cdot 3\cdot 5\ldots(2n-1)} \quad n = 1, 2, 3, \ldots$

607. $\int_0^1 x^{m-1}(1-x)^{n-1}\,dx = \int_0^\infty \frac{x^{m-1}}{(1+x)^{m+n}}dx = \frac{\Gamma(m)\Gamma(n)}{\Gamma(m+n)} = B(m,n)$

(This is the Beta function)

608. $B(m,n) = B(n,m) = \frac{\Gamma(m)\Gamma(n)}{\Gamma(m+n)}$, where m and n are any positive real numbers.

609. $\int_a^b (x-a)^m(b-x)^n\,dx = (b-a)^{m+n+1}\frac{\Gamma(m+1)\cdot\Gamma(n+1)}{\Gamma(m+n+2)}, \quad (m > -1, n > -1, b > a)$

610. $\int_1^\infty \frac{dx}{x^m} = \frac{1}{m-1}, \quad [m > 1]$

611. $\int_0^\infty \frac{dx}{(1+x)x^p} = \pi\csc p\pi, \quad [0 < p < 1]$

612. $\int_0^\infty \frac{dx}{(1-x)x^p} = -\pi \cot p\pi, \quad [0 < p < 1]$

613. $\int_0^\infty \frac{x^{p-1}\,dx}{(1+x)} = \frac{\pi}{\sin p\pi} = B(p, 1-p) = \Gamma(p)\Gamma(1-p), \quad [0 < p < 1]$

614. $\int_0^\infty \frac{x^{m-1}\,dx}{1+x^n} = \frac{\pi}{n \sin \frac{m\pi}{n}}, \quad [0 < m < n]$

615. $\int_0^\infty \frac{x^a\,dx}{(m+x^b)^c} = \frac{m^{\frac{a+1-bc}{b}}}{b}\left[\frac{\Gamma\left(\frac{a+1}{b}\right)\Gamma\left(c-\frac{a+1}{b}\right)}{\Gamma(c)}\right] \quad \left(a > -1,\ b > 0,\ m > 0,\ c > \frac{a+1}{b}\right)$

616. $\int_0^\infty \frac{dx}{(1+x)\sqrt{x}} = \pi$

617. $\int_0^\infty \frac{a\,dx}{a^2+x^2} = \begin{cases} \frac{\pi}{2} & (\text{if } a > 0), \\ 0 & (\text{if } a = 0), \\ -\frac{\pi}{2} & (\text{if } a < 0) \end{cases}$

618. $\int_0^a (a^2-x^2)^{n/2}\,dx = \frac{1}{2}\int_{-a}^a (a^2-x^2)^{n/2}\,dx = \frac{1\cdot 3\cdot 5\ldots n}{2\cdot 4\cdot 6\ldots(n+1)}\cdot\frac{\pi}{2}\cdot a^{n+1} \quad (n \text{ odd},\ a > 0)$

619. $\int_0^a x^m (a^2-x^2)^{n/2}\,dx = \begin{cases} \frac{1}{2}a^{m+n+1} B\left(\frac{m+1}{2}, \frac{n+2}{2}\right) & (a > 0,\ m > -1,\ n > -2) \\ \text{or} & \\ \frac{1}{2}a^{m+n+1}\frac{\Gamma\left(\frac{m+1}{2}\right)\Gamma\left(\frac{n+2}{2}\right)}{\Gamma\left(\frac{m+n+3}{2}\right)} & (a > 0,\ m > -1,\ n > -2) \end{cases}$

620. $\int_0^{\pi/2} \sin^n x\,dx = \begin{cases} \int_0^{\pi/2} (\cos^n x)\,dx & \\ \frac{1\cdot 3\cdot 5\cdot 7\ldots(n-1)}{2\cdot 4\cdot 6\cdot 8\ldots(n)}\frac{\pi}{2}, & (n \text{ an even integer, } n \neq 0), \\ \frac{1\cdot 3\cdot 5\cdot 7\ldots(n-1)}{2\cdot 4\cdot 6\cdot 8\ldots(n)}, & (n \text{ an odd integer, } n \neq 0), \\ \frac{\sqrt{\pi}}{2}\frac{\Gamma\left(\frac{n+1}{2}\right)}{\Gamma\left(\frac{n}{2}+1\right)} & (n > -1) \end{cases}$

621. $\int_0^\infty \frac{\sin mx\,dx}{x} = \frac{\pi}{2};\ \text{if } m > 0;\ 0,\ \text{if } m = 0;\ -\frac{\pi}{2},\ \text{if } m < 0$

622. $\int_0^\infty \frac{\cos x\,dx}{x} = \infty$

623. $\int_0^\infty \frac{\tan x\,dx}{x} = \frac{\pi}{2}$

624. $\int_0^\pi \sin ax \cdot \sin bx\,dx = \int_0^\pi \cos ax \cdot \cos bx\,dx = 0, \quad (a \neq b;\ a,\ b \text{ integers})$

625. $\int_0^{\pi/a} [\sin(ax)][\cos(ax)]\,dx = \int_0^\pi [\sin(ax)][\cos(ax)]dx = 0$

626. $\int_0^\pi [\sin(ax)][\cos(bx)]\,dx = \frac{2a}{a^2-b^2},\ \text{if } a-b \text{ is odd, or } 0 \text{ if } a-b \text{ is even}$

627. $\int_0^\infty \frac{\sin x \cos mx\,dx}{x} = 0, \quad \text{if } m < -1 \text{ or } m > 1;\ \frac{\pi}{4},\ \text{if } m = \pm 1;\ \frac{\pi}{2},\ \text{if } m^2 < 1$

628. $\int_0^\infty \frac{\sin ax \sin bx}{x^2}\,dx = \frac{\pi a}{2}, \quad (a \leq b)$

629. $\int_0^\pi \sin^2 mx\,dx = \int_0^\pi \cos^2 mx\,dx = \frac{\pi}{2} \quad (m \text{ is a non-zero integer})$

630. $\int_0^\infty \frac{\sin^2(px)}{x^2}\,dx = \frac{\pi|p|}{2}$

631. $\int_0^\infty \frac{\sin x}{x^p}\,dx = \frac{\pi}{2\Gamma(p)\sin(p\pi/2)}, \quad 0 < p < 1$

632. $\int_0^\infty \frac{\cos x}{x^p}\,dx = \frac{\pi}{2\Gamma(p)\cos(p\pi/2)}, \quad 0 < p < 1$

633. $\int_0^\infty \frac{1-\cos px}{x^2}\,dx = \frac{\pi|p|}{2}$

634. $\int_0^\infty \frac{\sin px \cos qx}{x}\,dx = \left\{0,\ q > p > 0;\ \frac{\pi}{2},\ p > q > 0;\ \frac{\pi}{4},\ p = q > 0\right\}$

635. $\int_0^\infty \frac{\cos(mx)}{x^2+a^2}\,dx = \frac{\pi}{2|a|}e^{-|ma|}$

636. $\int_0^\infty \cos(x^2)\,dx = \int_0^\infty \sin(x^2)\,dx = \frac{1}{2}\sqrt{\frac{\pi}{2}}$

637. $\int_0^\infty \sin ax^n\,dx = \frac{1}{na^{1/n}}\Gamma(1/n)\sin\frac{\pi}{2n},$ if $n > 1$

638. $\int_0^\infty \cos ax^n\,dx = \frac{1}{na^{1/n}}\Gamma(1/n)\cos\frac{\pi}{2n},$ if $n > 1$

639. $\int_0^\infty \frac{\sin x}{\sqrt{x}}\,dx = \int_0^\infty \frac{\cos x}{\sqrt{x}}\,dx = \sqrt{\frac{\pi}{2}}$

640. (a) $\int_0^\infty \frac{\sin^3 x}{x}\,dx = \frac{\pi}{4}$ (b) $\int_0^\infty \frac{\sin^3 x}{x^2}\,dx = \frac{3}{4}\log 3$

641. $\int_0^\infty \frac{\sin^3 x}{x^3}\,dx = \frac{3\pi}{8}$

642. $\int_0^\infty \frac{\sin^4 x}{x^4}\,dx = \frac{\pi}{3}$

643. $\int_0^{\pi/2} \frac{dx}{1 + a\cos x} = \frac{\cos^{-1} a}{\sqrt{1-a^2}},$ $(|a| < 1)$

644. $\int_0^\pi \frac{dx}{a + b\cos x} = \frac{\pi}{\sqrt{a^2 - b^2}},$ $(a > b \geq 0)$

645. $\int_0^{2\pi} \frac{dx}{1 + a\cos x} = \frac{2\pi}{\sqrt{1-a^2}},$ $(a^2 < 1)$

646. $\int_0^\infty \frac{\cos ax - \cos bx}{x}\,dx = \log\left|\frac{b}{a}\right|$

647. $\int_0^{\pi/2} \frac{dx}{a^2\sin^2 x + b^2\cos^2 x} = \frac{\pi}{2|ab|}$

648. $\int_0^{\pi/2} \frac{dx}{(a^2\sin^2 x + b^2\cos^2 x)^2} = \frac{\pi(a^2+b^2)}{4a^3b^3},$ $(a, b > 0)$

649. $\int_0^{\pi/2} \sin^{n-1} x\cos^{m-1} x\,dx = \frac{1}{2}\mathrm{B}\left(\frac{n}{2}, \frac{m}{2}\right),$ (if m and n are positive integers)

650. $\int_0^{\pi/2} (\sin^{2n+1}\theta)\,d\theta = \frac{2\cdot 4\cdot 6\ldots(2n)}{1\cdot 3\cdot 5\ldots(2n+1)},$ $(n = 1, 2, 3, \ldots)$

651. $\int_0^{\pi/2} (\sin^{2n}\theta)\,d\theta = \frac{1\cdot 3\cdot 5\ldots(2n-1)}{2\cdot 4\ldots(2n)}\left(\frac{\pi}{2}\right),$ $(n = 1, 2, 3, \ldots)$

652. $\int_0^{\pi/2} \frac{x}{\sin x}\,dx = 2\left\{\frac{1}{1^2} - \frac{1}{3^2} + \frac{1}{5^2} - \frac{1}{7^2} + \cdots\right\}$

653. $\int_0^{\pi/2} \frac{dx}{1 + \tan^m x} = \frac{\pi}{4}$

654. $\int_0^{\pi/2} \sqrt{\cos\theta}\,d\theta = \frac{(2\pi)^{\frac{3}{2}}}{\left[\Gamma\left(\frac{1}{4}\right)\right]^2}$

655. $\int_0^{\pi/2} (\tan^h\theta)\,d\theta = \frac{\pi}{2\cos\left(\frac{h\pi}{2}\right)},$ $(0 < h < 1)$

656. $\int_0^\infty \frac{\tan^{-1}(ax) - \tan^{-1}(bx)}{x}\,dx = \frac{\pi}{2}\log\frac{a}{b},$ $(a, b > 0)$

657. The area enclosed by a curve defined through the equation $x^{\frac{b}{c}} + y^{\frac{b}{c}} = a^{\frac{b}{c}}$ where $a > 0$, c a positive odd integer and b a positive even integer is given by $\frac{\left[\Gamma\left(\frac{c}{b}\right)\right]^2}{\Gamma\left(\frac{2c}{b}\right)}\left(\frac{2ca^2}{b}\right)$

658. $I = \iiint\limits_R x^{h-1}y^{m-1}z^{n-1}\,dv$, where R denotes the region of space bounded by the co-ordinate planes and that portion of the surface $\left(\frac{x}{a}\right)^p + \left(\frac{y}{b}\right)^q + \left(\frac{z}{c}\right)^k = 1$, which lies in the first octant and where $h, m, n, p, q, k, a, b, c$, denote positive real numbers is given by

$$\int_0^a x^{h-1}\,dx \int_0^b \left[1 - \left(\frac{x}{a}\right)^p\right]^{\frac{1}{q}} y^m dy \int_0^c \left[1 - \left(\frac{x}{a}\right)^p - \left(\frac{y}{b}\right)^q\right]^{\frac{1}{k}} z^{n-1} dz = \frac{a^h b^m c^n}{pqk}\,\frac{\Gamma\left(\frac{h}{p}\right)\Gamma\left(\frac{m}{q}\right)\Gamma\left(\frac{n}{k}\right)}{\Gamma\left(\frac{h}{p} + \frac{m}{q} + \frac{n}{k} + 1\right)}$$

659. $\int_0^\infty e^{-ax}\,dx = \frac{1}{a}, \quad (a > 0)$

660. $\int_0^\infty \frac{e^{-ax} - e^{-bx}}{x}\,dx = \log\frac{b}{a}, \quad (a, b > 0)$

661. $\int_0^\infty x^n e^{-ax}\,dx = \begin{cases} \frac{\Gamma(n+1)}{a^{n+1}} & (\text{if } n > -1 \text{ and } a > 0) \\ \text{or} & \\ \frac{n!}{a^{n+1}} & (\text{if } a > 0 \text{ and } n \text{ is a positive integer}) \end{cases}$

662. $\int_0^\infty x^n \exp(-ax^p)\,dx = \frac{\Gamma(k)}{pa^k}, \quad \left(n > -1,\ p > 0,\ a > 0,\ k = \frac{n+1}{p}\right)$

663. $\int_0^\infty e^{-a^2x^2}\,dx = \frac{1}{2a}\sqrt{\pi} = \frac{1}{2a}\Gamma\left(\frac{1}{2}\right), \quad (a > 0)$

664. $\int_0^\infty xe^{-x^2}\,dx = \frac{1}{2}$

665. $\int_0^\infty x^2e^{-x^2}\,dx = \frac{\sqrt{\pi}}{4}$

666. $\int_0^\infty x^{2n}e^{-ax^2}\,dx = \frac{1\cdot 3\cdot 5\ldots(2n-1)}{2^{n+1}a^n}\sqrt{\frac{\pi}{a}} \quad (a > 0, n > -\tfrac{1}{2})$

667. $\int_0^\infty x^{2n+1}e^{-ax^2}\,dx = \frac{n!}{2a^{n+1}}, \quad (a > 0, n > -1)$

668. $\int_0^1 x^m e^{-ax}\,dx = \frac{m!}{a^{m+1}}\left[1 - e^{-a}\sum_{r=0}^{m}\frac{a^r}{r!}\right]$

669. $\int_0^\infty e^{\left(-x^2 - \frac{a^2}{x^2}\right)}\,dx = \frac{e^{-2a}\sqrt{\pi}}{2}, \quad (a \geq 0)$

670. $\int_0^\infty e^{-nx}\sqrt{x}\,dx = \frac{1}{2n}\sqrt{\frac{\pi}{n}} \quad (n > 0)$

671. $\int_0^\infty \frac{e^{-nx}}{\sqrt{x}}\,dx = \sqrt{\frac{\pi}{n}} \quad (n > 0)$

672. $\int_0^\infty e^{-ax}(\cos mx)\,dx = \frac{a}{a^2 + m^2}, \quad (a > 0)$

673. $\int_0^\infty e^{-ax}(\sin mx)\,dx = \frac{m}{a^2 + m^2}, \quad (a > 0)$

674. $\int_0^\infty xe^{-ax}[\sin(bx)]\,dx = \frac{2ab}{(a^2 + b^2)^2}, \quad (a > 0)$

675. $\int_0^\infty xe^{-ax}[\cos(bx)]\,dx = \frac{a^2 - b^2}{(a^2 + b^2)^2}, \quad (a > 0)$

676. $\int_0^\infty x^n e^{-ax}[\sin(bx)]\,dx = \frac{n![(a + ib)^{n+1} - (a - ib)^{n+1}]}{2i(a^2 + b^2)^{n+1}}, \quad (i^2 = -1, a > 0)$

677. $\int_0^\infty x^n e^{-ax}[\cos(bx)]\,dx = \frac{n![(a - ib)^{n+1} + (a + ib)^{n+1}]}{2(a^2 + b^2)^{n+1}}, \quad (i^2 = -1, a > 0, n > -1)$

678. $\int_0^\infty \frac{e^{-ax}\sin x}{x}\,dx = \cot^{-1} a, \quad (a > 0)$

679. $\int_0^\infty e^{-a^2x^2}\cos bx\,dx = \frac{\sqrt{\pi}}{2|a|}\exp\left(-\frac{b^2}{4a^2}\right), \quad (ab \neq 0)$

680. $\int_0^\infty e^{-t\cos\phi}t^{b-1}[\sin(t\sin\phi)]\,dt - [\Gamma(b)]\sin(b\phi), \quad \left(b > 0, -\frac{\pi}{2} < \phi < \frac{\pi}{2}\right)$

681. $\int_0^\infty e^{-t\cos\phi}t^{b-1}[\cos(t\sin\phi)]\,dt - [\Gamma(b)]\cos(b\phi), \quad \left(b > 0, -\frac{\pi}{2} < \phi < \frac{\pi}{2}\right)$

682. $\int_0^\infty t^{b-1}\cos t\,dt = [\Gamma(b)]\cos\left(\frac{b\pi}{2}\right), \quad (0 < b < 1)$

683. $\int_0^\infty t^{b-1}(\sin t)\,dt = [\Gamma(b)]\sin\left(\frac{b\pi}{2}\right), \quad (0 < b < 1)$

684. $\int_0^1 (\log x)^n\,dx = (-1)^n \cdot n! \quad (n > -1)$

685. $\int_0^1 \left(\log\frac{1}{x}\right)^{\frac{1}{2}}\,dx = \frac{\sqrt{\pi}}{2}$

686. $\int_0^1 \left(\log\frac{1}{x}\right)^{-\frac{1}{2}}\,dx = \sqrt{\pi}$

687. $\int_0^1 \left(\log \frac{1}{x}\right)^n dx = n!$

688. $\int_0^1 x \log(1-x)\, dx = -\frac{3}{4}$

689. $\int_0^1 x \log(1+x)\, dx = \frac{1}{4}$

690. $\int_0^1 x^m (\log x)^n\, dx = \frac{(-1)^n n!}{(m+1)^{n+1}}, \qquad (m > -1,\ n = 0, 1, 2, \ldots)$
If $n \neq 0, 1, 2, \ldots$ replace $n!$ by $\Gamma(n+1)$.

691. $\int_0^1 \frac{\log x}{1+x}\, dx = -\frac{\pi^2}{12}$

692. $\int_0^1 \frac{\log x}{1-x}\, dx = -\frac{\pi^2}{6}$

693. $\int_0^1 \frac{\log(1+x)}{x}\, dx = \frac{\pi^2}{12}$

694. $\int_0^1 \frac{\log(1-x)}{x}\, dx = -\frac{\pi^2}{6}$

695. $\int_0^1 \log(x)\log(1+x)\, dx = 2 - 2\log 2 - \frac{\pi^2}{12}$

696. $\int_0^1 \log(x)\log(1-x)\, dx = 2 - \frac{\pi^2}{6}$

697. $\int_0^1 \frac{\log x}{1-x^2}\, dx = -\frac{\pi^2}{8}$

698. $\int_0^1 \log\left(\frac{1+x}{1-x}\right) \cdot \frac{dx}{x} = \frac{\pi^2}{4}$

699. $\int_0^1 \frac{\log x\, dx}{\sqrt{1-x^2}} = -\frac{\pi}{2}\log 2$

700. $\int_0^1 x^m \left[\log\left(\frac{1}{x}\right)\right]^n dx = \frac{\Gamma(n+1)}{(m+1)^{n+1}}, \qquad (\text{if } m+1 > 0 \text{ and } n+1 > 0)$

701. $\int_0^1 \frac{(x^p - x^q)\, dx}{\log x} = \log\left(\frac{p+1}{q+1}\right), \qquad (p+1 > 0,\ q+1 > 0)$

702. $\int_0^1 \frac{dx}{\sqrt{\log\left(\frac{1}{x}\right)}} = \sqrt{\pi}, \qquad$ (same as integral 686)

703. $\int_0^\infty \log\left(\frac{e^x+1}{e^x-1}\right) dx = \frac{\pi^2}{4}$

704. $\int_0^{\pi/2} \log(\sin x)\, dx = \int_0^{\pi/2} \log\cos x\, dx = -\frac{\pi}{2}\log 2$

705. $\int_0^{\pi/2} \log(\sec x)\, dx = \int_0^{\pi/2} \log\csc x\, dx = \frac{\pi}{2}\log 2$

706. $\int_0^{\pi} x \log(\sin x)\, dx = -\frac{\pi^2}{2}\log 2$

707. $\int_0^{\pi/2} \sin x \log(\sin x)\, dx = \log 2 - 1$

708. $\int_0^{\pi/2} \log\tan x\, dx = 0$

709. $\int_0^{\pi} \log(a \pm b\cos x)\, dx = \pi \log\left(\frac{a + \sqrt{a^2-b^2}}{2}\right), \qquad (a \geq b)$

710. $\int_0^{\pi} \log(a^2 - 2ab\cos x + b^2)\, dx = \begin{cases} 2\pi \log a & a \geq b > 0 \\ 2\pi \log b & b \geq a > 0 \end{cases}$

711. $\int_0^\infty \frac{\sin ax}{\sinh bx}\, dx = \frac{\pi}{2|b|}\tanh\frac{a\pi}{2b}$

712. $\int_0^\infty \frac{\cos ax}{\cosh bx}\, dx = \frac{\pi}{2|b|}\operatorname{sech}\frac{a\pi}{2b}$

713. $\int_0^\infty \frac{dx}{\cosh ax} = \frac{\pi}{2|a|}$

714. $\int_0^\infty \frac{x\,dx}{\sinh ax} = \frac{\pi^2}{4a^2} \quad (a > 0)$

715. $\int_0^\infty e^{-ax}\cosh bx\,dx = \frac{a}{a^2 - b^2}, \quad (0 \le |b| \quad < a)$

716. $\int_0^\infty e^{-ax}\sinh bx\,dx = \frac{b}{a^2 - b^2}, \quad (0 \le |b| \quad < a)$

717. $\int_0^\infty \frac{\sinh ax}{e^{bx} + 1}\,dx = \frac{\pi}{2b}\csc\frac{a\pi}{b} - \frac{1}{2a} \quad (b > 0)$

718. $\int_0^\infty \frac{\sinh ax}{e^{bx} - 1}\,dx = \frac{1}{2a} - \frac{\pi}{2b}\cot\frac{a\pi}{b} \quad (b > 0)$

719. $\int_0^{\pi/2} \frac{dx}{\sqrt{1 - k^2\sin^2 x}} = \frac{\pi}{2}\left[1 + \left(\frac{1}{2}\right)^2 k^2 + \left(\frac{1\cdot 3}{2\cdot 4}\right)^2 k^4 + \left(\frac{1\cdot 3\cdot 5}{2\cdot 4\cdot 6}\right)^2 k^6 + \cdots\right], \quad \text{if } k^2 < 1$

720. $\int_0^{\pi/2} \sqrt{1 - k^2\sin^2 x}\,dx = \frac{\pi}{2}\left[1 - \left(\frac{1}{2}\right)^2 k^2 - \left(\frac{1\cdot 3}{2\cdot 4}\right)^2 \frac{k^4}{3} - \left(\frac{1\cdot 3\cdot 5}{2\cdot 4\cdot 6}\right) 2\frac{k^6}{5} - \cdots\right], \quad \text{if } k^2 < 1$

721. $\int_0^\infty e^{-x}\log x\,dx = -\gamma = -0.5772157\ldots$

722. $\int_0^\infty e^{-x^2}\log x\,dx = -\frac{\sqrt{\pi}}{4}(\gamma + 2\log 2)$

723. $\int_0^\infty \left(\frac{1}{1 - e^{-x}} - \frac{1}{x}\right)e^{-x}\,dx = \gamma = 0.5772157\ldots$ [Euler's Constant]

724. $\int_0^\infty \frac{1}{x}\left(\frac{1}{1+x} - e^{-x}\right)dx = \gamma = 0.5772157\ldots$

For n even :

725. $\int \cos^n x\,dx = \frac{1}{2^{n-1}}\sum_{k=0}^{n/2-1}\binom{n}{k}\frac{\sin(n-2k)x}{(n-2k)} + \frac{1}{2^n}\binom{n}{n/2}x$

726. $\int \sin^n x\,dx = \frac{1}{2^{n-1}}\sum_{k=0}^{n/2-1}\binom{n}{k}\frac{\sin[(n-2k)(\frac{\pi}{2}-x)]}{2k-n} + \frac{1}{2^n}\binom{n}{n/2}x$

For n odd:

727. $\int \cos^n x\,dx = \frac{1}{2^{n-1}}\sum_{k=0}^{(n-1)/2}\binom{n}{k}\frac{\sin(n-2k)x}{n-2k}$

728. $\int \sin^n x\,dx = \frac{1}{2^{n-1}}\sum_{k=0}^{(n-1)/2}\binom{n}{k}\frac{\sin\left[(n-2k)\left(\frac{\pi}{2}-x\right)\right]}{2k-n}$

DIFFERENTIAL EQUATIONS

Certain types of differential equations occur sufficiently often to justify the use of formulas for the corresponding particular solutions. The following set of Tables I to XIV covers all first, second, and nth order ordinary linear differential equations with constant coefficients for which the right members are of the form $P(x)e^{rx}\sin sx$ or $P(x)e^{rx}\cos sx$, where r and s are constants and $P(x)$ is a polynomial of degree n.

When the right member of a reducible linear partial differential equation with constant coefficients is not zero, particular solutions for certain types of right members are contained in Tables XV to XXI. In these tables both F and P are used to denote polynomials, and it is assumed that no denominator is zero. In any formula the roles of x and y may be reversed throughout, changing a formula in which x dominates to one in which y dominates. Tables XIX, XX, XXI are applicable whether the equations are reducible or not. The symbol $\binom{m}{n}$ stands for $\frac{m!}{(m-n)!n!}$ and is the $(n+1)^{\text{st}}$ coefficient in the expansion of $(a+b)^m$. Also $0! = 1$ by definition.

The tables as herewith given are those contained in the text *Differential Equations* by Ginn and Company (1955) and are published with their kind permission and that of the author, Professor Frederick H. Steen.

SOLUTION OF LINEAR DIFFERENTIAL EQUATIONS WITH CONSTANT COEFFICIENTS

Any linear differential equation with constant coefficients may be written in the form

$$p(D)y = R(x)$$

where

- D is the differential operation: $Dy = \frac{dy}{dx}$
- $p(D)$ is a polynomial in D,
- y is the dependent variable,
- x is the independent variable,
- $R(x)$ is an arbitrary function of x.

A power of D represents repeated differentiation, that is

$$D^n y = \frac{d^n y}{dx^n}$$

For such an equation, the general solution may be written in the form

$$y = y_c + y_p$$

where y_p is any particular solution, and y_c is called the *complementary function*. This complementary function is defined as the general solution of the *homogeneous equation*, which is the original differential equation with the right side replaced by zero, i.e.,

$$p(D)y = 0$$

The complementary function y_c may be determined as follows:

1. Factor the polynomial $p(D)$ into real and complex linear factors, just as if D were a variable instead of an operator.

2. For each nonrepeated linear factor of the form $(D - a)$, where a is real, write down a term of the form

$$ce^{ax}$$

where c is an arbitrary constant.

3. For each repeated real linear factor of the form $(D - a)^n$, write down n terms of the form

$$c_1 e^{ax} + c_2 x e^{ax} + c_3 x^2 e^{ax} + \cdots + c_n x^{n-1} e^{ax}$$

where the c_i's are arbitrary constants.

4. For each non-repeated conjugate complex pair of factors of the form $(D - a + ib)(D - a - ib)$, write down two terms of the form

$$c_1 e^{ax} \cos bx + c_2 e^{ax} \sin bx$$

5. For each repeated conjugate complex pair of factors of the form $(D - a + ib)^n (D - a - ib)^n$, write down $2n$ terms of the form

$$c_1 e^{ax} \cos bx + c_2 e^{ax} \sin bx + c_3 x e^{ax} \cos bx + c_4 x e^{ax} \sin bx$$
$$+ \cdots + c_{2n-1} x^{n-1} e^{ax} \cos bx + c_{2n} x^{n-1} e^{ax} \sin bx$$

6. The sum of all the terms thus written down is the complementary function y_c.

To find the particular solution y_p, use the following tables, as shown in the examples. For cases not shown in the tables, there are various methods of finding y_p. The most general method is called *variation of parameters*. The following example illustrates the method:

Example: Find y_p for $(D^2 - 4)y = e^x$.
This example can be solved most easily by use of equation 63 in the tables following. However, it is given here as an example of the method of variation of parameters.
The complementary function is

$$y_c = c_1 e^{2x} + c_2 e^{-2x}$$

To find y_p, replace the constants in the complementary function with unknown functions,

$$y_p = u e^{2x} + v e^{-2x}$$

We now prepare to substitute this assumed solution into the original equation. We begin by taking all the necessary derivatives:

$$y_p = ue^{2x} + ve^{-2x}$$
$$y'_p = 2ue^{2x} - 2ve^{-2x} + u'e^{2x} + v'e^{-2x}$$

For each derivative of y_p except the highest, we set the sum of all the terms containing u' and v' to 0. Thus the above equation becomes

$$u'e^{2x} + v'e^{-2x} = 0 \quad \text{and} \quad y'_p = 2ue^{2x} - 2ve^{-2x}$$

Continuing to differentiate, we have

$$y''_p = 4ue^{2x} + 4ve^{-2x} + 2u'e^{2x} - 2v'e^{-2x}$$

When we substitute into the original equation, all the terms not containing u' or v' cancel out. This is a consequence of the method by which y_p was set up.

Thus all that is necessary is to write down the terms containing u' or v' in the highest order derivative of y_p, multiply by the constant coefficient of the highest power of D in $p(D)$, and set it equal to $R(x)$. Together with the previous terms in u' and v' which were set equal to 0, this gives us as many linear equations in the first derivatives of the unknown functions as there are unknown functions. The first derivatives may then be solved for by algebra, and the unknown functions found by integration. In the present example, this becomes

$$u'e^{2x} + v'e^{-2x} = 0$$
$$2u'e^{2x} - 2v'e^{-2x} = e^x$$

We eliminate v' and u' separately, getting

$$4u'e^{2x} = e^x$$
$$4v'e^{-2x} = -e^x$$

Thus

$$u' = \tfrac{1}{4}e^{-x}$$
$$v' = -\tfrac{1}{4}e^{3x}$$

Therefore, by integrating

$$u = -\tfrac{1}{4}e^{-x}$$
$$v = -\tfrac{1}{12}e^{3x}$$

A constant of integration is not needed, since we need only one particular solution. Thus

$$y_p = ue^{2x} + ve^{-2x} = -\frac{1}{4}e^{-x}e^{2x} - \frac{1}{12}e^{3x}e^{-2x}$$
$$= -\frac{1}{4}e^x - \frac{1}{12}e^x = -\frac{1}{3}e^x$$

and the general solution is

$$y = y_c + y_p = c_1e^{2x} + c_2e^{-2x} - \frac{1}{3}e^x$$

The following samples illustrate the use of the tables.

Example 1: Solve $(D^2 - 4)y = \sin 3x$. Substitution of $q = -4$, $s = 3$ in formula 24 gives

$$y_p = \frac{\sin 3x}{-9 - 4}$$

wherefore the general solution is

$$y = c_1e^{2x} + c_2e^{-2x} - \frac{\sin 3x}{13}$$

Example 2: Obtain a particular solution of $(D^2 - 4D + 5)y = x^2e^{3x}\sin x$.

Applying formula 40 with $a = 2$, $b = 1$, $r = 3$, $s = 1$, $P(x) = x^2$, $s + b = 2$, $s - b = 0$, $a - r = -1$, $(a - r)^2 + (s + b)^2 = 5$, $(a - r)^2 + (s - b)^2 = 1$, we have

$$\begin{aligned} y_p &= \frac{e^{3x}\sin x}{2}\left[\left(\frac{2}{5} - \frac{0}{1}\right)x^2 + \left(\frac{2(-1)2}{25} - \frac{2(-1)0}{1}\right)2x + \left(\frac{3\cdot 1\cdot 2 - 2^3}{125} - \frac{3\cdot 1\cdot 0 - 0}{1}\right)2\right] \\ &\quad - \frac{e^{3x}\cos x}{2}\left[\left(\frac{-1}{5} - \frac{-1}{1}\right)x^2 + \left(\frac{1-4}{25} - \frac{1-0}{1}\right)2x + \left(\frac{-1-3(-1)4}{125} - \frac{-1-3(-1)0}{1}\right)2\right] \\ &= \left(\frac{1}{5}x^2 - \frac{4}{25}x - \frac{2}{125}\right)e^{3x}\sin x + \left(-\frac{2}{5}x^2 + \frac{28}{25} - \frac{136}{125}\right)e^{3x}\cos x \end{aligned}$$

The special formulas effect a very considerable saving of time in problems of this type.

Example 3: Obtain a particular solution of $(D^2 - 4D + 5)y = x^2e^{2x}\cos x$. (Compare with Example 2.)

Formula 40 is not applicable here since for this equation $r = a$, $s = b$, wherefore the denominator $(a - r)^2 + (s - b)^2 = 0$. We turn instead to formula 44. Substituting $a = 2$, $b = 1$, $P(x) = x^2$ and replacing sin by cos, cos by – sin, we obtain

$$\begin{aligned} y_p &= \frac{e^{2x}\cos x}{4}\left(x^2 - \frac{2}{4}\right) + \frac{e^{2x}\sin x}{2}\int\left(x^2 - \frac{1}{2}\right)dx \\ &= \left(\frac{x^2}{4} - \frac{1}{8}\right)e^{2x}\cos x + \left(\frac{x^3}{6} - \frac{x}{4}\right)e^{2x}\sin x \end{aligned}$$

which is the required solution.

Example 4: Find z_p for $(D_x - 3D_y)\,z = \ln(y + 3x)$. Referring to Table XV we note that formula 69 (not 68) is applicable. This gives

$$z_p = x\ln(y + 3x)$$

It is easily seen that $-y/3\ln(y + 3x)$ would serve equally well.

Example 5: Solve $(D_x + 2D_y - 4)\,z = y\cos(y - 2x)$.

Since R in formula 76 contains a polynomial in x, not y, we rewrite the given equation in the form $(D_y + \frac{1}{2}D_x - 2)\,z = \frac{1}{2}\,y\cos(y - 2x)$. Then

$$z_c = e^{2y}F\left(x - \frac{1}{2}y\right) = e^{2x}f(2x - y)$$

and by the formula

$$z_p = -\frac{1}{2}\cos(y - 2x)\cdot\left(\frac{y}{2} + \frac{\frac{1}{2}}{2}\right) = -\frac{1}{8}(2y + 1)\cos(y - 2x)$$

Example 6: Find z_p for $(D_x + 4D_y)^3\,z = (2x - y)^2$.

Using formula 79, we obtain

$$z_p = \frac{\iiint u^2\,du^3}{[2 + 4(-1)]^3} = \frac{u^5}{5\cdot 4\cdot 3\cdot(-8)} = -\frac{(2x - y)^5}{480}$$

Example 7: Find z_p for $(D_x^3 + 5D_x^2D_y - 7D_x + 4)z = e^{2x+3y}$. By formula 87

$$z_p = \frac{e^{2x+3y}}{2^3 + 5\cdot 2^2\cdot 3 - 7\cdot 2 + 4} = \frac{e^{2x+3y}}{58}$$

Example 8: Find z_p for

$$(D_x^4 + 6D_x^3D_y + D_xD_y + D_y^2 + 9)z = \sin(3x + 4y)$$

Since every term in the left number is of even degree in the two operators D_x and D_y, formula 90 is applicable. It gives

$$\begin{aligned} z_p &= \frac{\sin(3x + 4y)}{(-9)^2 + 6(-9)(-12) + (-12) + (-16) + 9} \\ &= \frac{\sin(3x + 4y)}{710} \end{aligned}$$

Table I: $(D-a)y = R$

R	y_p
1. e^{rx}	$\frac{e^{rx}}{r-a}$
2. $\sin sx$*	$-\frac{a\sin sx + s\cos sx}{a^2+s^2} = \frac{1}{\sqrt{a^2+s^2}}\sin\left(sx + \tan^{-1}\frac{s}{a}\right)$
3. $P(x)$	$-\frac{1}{a}\left[P(x) + \frac{P'(x)}{a} + \frac{P''(x)}{a^2} + \cdots + \frac{P^{(n)}(x)}{a^n}\right]$
4. $e^{rx}\sin sx$*	Replace a by $a-r$ in formula 2 and multiply by e^{rx}.
5. $P(x) \quad e^{rx}$	Replace a by $a-r$ in formula 3 and multiply by e^{rx}.
6. $P(x)\sin sx$*	$-\sin sx\left[\frac{a}{a^2+s^2}P(x) + \frac{a^2-s^2}{(a^2+s^2)^2}P'(x) + \frac{a^3-3as^2}{(a^2+s^2)^3}P''(x) + \cdots + \frac{a^k - \binom{k}{2}a^{k-2}s^2 + \binom{k}{4}a^{k-4}s^4 - \cdots}{(a^2+s^2)^k}P^{(k-1)}(x) + \cdots\right]$ $-\cos sx\left[\frac{s}{a^2+s^2}P(x) + \frac{2as}{(a^2+s^2)^2}P'(x) + \frac{3a^2s-s^3}{(a^2+s^2)^3}P''(x) + \cdots + \frac{\binom{k}{1}a^{k-1}s - \binom{k}{3}a^{k-3}s^3 + \cdots}{(a^2+s^2)^k}P^{(k-1)}(x) + \cdots\right]$
7. $P(x)e^{rx}\sin sx$*	Replace a by $a-r$ in formula 6 and multiply by e^{rx}.
8. e^{ax}	xe^{ax}
9. $e^{ax}\sin sx$*	$-\frac{e^{ax}\cos sx}{s}$
10. $P(x)e^{ax}$	$e^{ax}\int P(x)\,dx$
11. $P(x)e^{ax}\sin sx$	$\frac{e^{ax}\sin sx}{s}\left[\frac{P'(x)}{s^3} - \frac{P'''(x)}{s^3} + \frac{P^{v}(x)}{s^5} - \cdots\right] - \frac{e^{ax}\cos sx}{s}\left[P(x) - \frac{P''(x)}{s^2} + \frac{P^{iv}(x)}{s^4} - \cdots\right]$

* For $\cos sx$ in R replace "sin" by "cos" and "cos" by "$-$sin" in y_p.

$$D^n = \frac{d^n}{dx^n} \qquad \binom{m}{n} = \frac{m!}{(m-n)!n!} \qquad 0! = 1$$

Table II: $(D-a)^2y = R$

R	y_p
12. e^{rx}	$\frac{e^{rz}}{(r-a)^2}$
13. $\sin sx$*	$\frac{1}{(a^2+s^2)}[(a^2-s^2)\sin sx + 2as\cos sx] = \frac{1}{a^2+s^2}\sin\left(sx + \tan^{-1}\frac{2as}{a^2-s^2}\right)$
14. $P(x)$	$\frac{1}{a^2}\left[P(x) + \frac{2P'(x)}{a} + \frac{3P''(x)}{a^2} + \cdots + \frac{(n+1)P^{(n)}(x)}{a^n}\right]$
15. $e^{rx}\sin sx$*	Replace a by $a-r$ in formula 13 and multiply by e^{rx}.
16. $P(x)e^{rx}$	Replace a by $a-r$ in formula 14 and multiply by e^{rx}.
17. $P(x)\sin sx$*	$\sin sx\left[\frac{a^2-s^2}{(a^2+s^2)^2}P(x) + 2\frac{a^3-3as^2}{(a^2+s^2)^3}P'(x) + 3\frac{a^4-6a^2s^2+s^4}{(a^2+s^2)^4}P''(x) + \cdots + (k-1)\frac{a^k - \binom{k}{2}a^{k-2}s^2 + \binom{k}{4}a^{k-4}s^4 - \cdots}{(a^2+s^2)^k}P^{(k-2)}(x) + \cdots\right]$ $+\cos sx\left[\frac{2as}{(a^2+s^2)^2}P(x) + 2\frac{3a^2s-s^3}{(a^2+s^2)^3}P'(x) + 3\frac{4a^3s-4as^3}{(a^2+s^2)^4}P''(x) + \cdots + (k-1)\frac{\binom{k}{1}a^{k-1}s - \binom{k}{3}a^{k-3}s^3 + \cdots}{(a^2+s^2)^k}P^{(k-2)}(x) + \cdots\right]$
18. $P(x)e^{rx}\sin sx$*	Replace a by $a-r$ in formula 17 and multiply by e^{rx}.
19. e^{ax}	$\frac{1}{2}x^2e^{ax}$
20. $e^{ax}\sin sx$*	$-\frac{e^{ax}\sin sx}{s^2}$
21. $P(x)e^{ax}$	$e^{ax}\int\int P(x)\,dx\,dx$
22. $P(x)e^{ax}\sin sx$*	$-\frac{e^{ax}\sin sx}{s^2}\left[P(x) - \frac{3P''(x)}{s^2} + \frac{5P^{iv}(x)}{s^4} - \frac{7P^{vi}(x)}{s^6} + \cdots\right]$ $-\frac{e^{ax}\cos sx}{s^2}\left[\frac{2P'(x)}{s} + \frac{4P'''(x)}{s^3} - \frac{6P^{v}(x)}{s^5} - \cdots\right]$

* For $\cos sx$ in R replace "sin" by "cos" and "cos" by "$-$ sin" in y_p.

Table III: $(D^2 + q)y = R$

R	y_p
23. e^{rx}	$\frac{e^{rx}}{r^2+q}$
24. $\sin sx*$	$\frac{\sin sx}{-s^2+q}$
25. $P(x)$	$\frac{1}{q}\left[P(x) - \frac{P''(x)}{q} + \frac{P^{iv}(x)}{q^2} - \cdots + (-1)^k \frac{P^{(2k)}(x)}{qk} \cdots\right]$
26. $e^{rx}\sin sx$	$\frac{(r^2-s^2+q)e^{rx}\sin sx - 2rse^{rx}\cos sx}{(r^2-s^2+q)^2+(2rs)^2} = \frac{e^{rx}}{\sqrt{(r^2-s^2+q)^2+(2rs)^2}}\sin\left[sx - \tan^{-1}\frac{2rs}{r^2-s^2+q}\right]$
27. $P(x)e^{rx}$	$\frac{e^{rx}}{r^2+q}\left[P(x) - \frac{2r}{r^2+q}P'(x) + \frac{3r^2-q}{(r2+q)^2}P''(x) - \frac{4r^3-4qr}{(r^2+q)^3}P'''(x) + \cdots + \cdots + (-1)^{k-1}\frac{\binom{k}{1}r^{k-1}-\binom{k}{3}r^{k-3}q+\binom{k}{5}r^{k-5}q^2-\cdots}{(r^2+q)^{k-1}}P^{(k-1)}(x) + \cdots\right]$
28. $P(x)\sin sx*$	$\frac{\sin sx}{(-s^2+q)}\left[P(x) - \frac{3s^2+q}{(-s^2+q)^2}P''(x) + \frac{5s^4+10s^2q+q^2}{(-s^2+q)^4}P^{iv}(x) + \cdots + (-1)^k\frac{\binom{2k+1}{1}s^{2k}+\binom{2k+1}{3}s^{2k-2}q+\binom{2k+1}{5}s^{2k-4}q^2+\cdots}{(-s^2+q)^{2k}}P^{(2k)}(x) + \cdots\right] - \frac{s\cos sx}{(-s^2+q)}\left[\frac{2P'(x)}{(-s^2+q)} - \frac{4s^2+4q}{(-s^2+q)^3}P'''(x) + \cdots + (-1)^{k+1}\frac{\binom{2k}{1}s^{2k-2}+\binom{2k}{3}s^{2k-4}q+\cdots}{(-s^2+q)^{2k-1}}P^{(2k-1)}(x) + \cdots\right]$

Table IV: $(D^2 + b^2)y = R$

R	y_p
29. $\sin bx*$	$-\frac{x\cos bx}{2b}$
30. $P(x)\sin bx*$	$\frac{\sin bx}{(2b)^2}\left[P(x) - \frac{P''(x)}{(2b)^2} + \frac{P^{iv}(x)}{(2b)^4} - \cdots\right] - \frac{\cos bx}{2b}\int\left[P(x) - \frac{P''(x)}{(2b)^2} + \cdots\right]dx$

* For $\cos sx$ in R replace "sin" by "cos" and "cos" by "– sin" in y_p.

Table V: $(D^2 pD + q)y = R$

R	y_p
31. e^{rx}	$\frac{e^{rx}}{r^2+pr+q}$
32. $\sin sx*$	$\frac{(q-s^2)\sin sx - ps\cos sx}{(q-s^2)^2+(ps)^2} = \frac{1}{\sqrt{(q-s^2)^2+(ps)^2}}\sin\left(sx - \tan^{-1}\frac{ps}{q-s^2}\right)$
33. $P(x)$	$\frac{1}{q}\left[P(x) - \frac{p}{q}P'(x) + \frac{p^2-q}{q^2}P''(x) - \frac{p^3-2pq}{q^3}P'''(x) + \cdots + (-1)^n\frac{p^n-\binom{n-1}{1}p^{n-2}q+\binom{n-2}{2}p^{n-4}q^2-\cdots}{q^n}P^{(n)}(x)\right]$
34. $e^{rx}\sin sx*$	Replace p by $p + 2r$, q by $q+pr+r^2$ in formula 32 and multiply by e^{rx}.
35. $P(x)e^{rx}$	Replace p by $p + 2r$, q by $q+pr+r^2$ in formula 33 and multiply by e^{rx}.

Table VI: $(D - b)(D - a)y = R$

R	y_p
36. $P(x)\sin sx*$	$\frac{\sin sx}{b-a}\left[\left(\frac{a}{a^2+s^2} - \frac{b}{b^2+s^2}\right)P(x) + \left(\frac{a^2-s^2}{(a^2+s^2)^2} - \frac{b^2-s^2}{(b^2+s^2)^2}\right)P'(x) + \left(\frac{a^3-3as^2}{(a^2+s^2)^3} - \frac{b^3-3bs^2}{(b^2+s^2)^3}\right)P''(x) + \cdots\right] + \frac{\cos sx}{b-a}\left(\frac{s}{a^2+s^2} - \frac{s}{b^2+s^2}\right)P(x) + \left(\frac{2as}{(a^2+s^2)^2} - \frac{2bs}{(b^2+s^2)^2}\right)P'(x) + \left(\frac{3a^2s-s^2}{(a^2+s^2)^3} - \frac{3b^2s-s^3}{(b^2+s^2)^3}\right)P''(x) + \cdots\Big]^{\dagger}$
37. $P(x)e^{rx}\sin sx*$	Replace a by $a–r$, b by $b - r$ in formula 36 and multiply by e^{rx}.
38. $P(x)e^{ax}$	$\frac{e^{ax}}{a-b}\left[\int P(x)\,dx + \frac{P(x)}{(b-a)} + \frac{P'(x)}{(b-a)^2} + \frac{P''(x)}{(b-a)^3} + \cdots + \frac{P^{(n)}(x)}{(b-a)^{n+1}}\right]$

* For $\cos sx$ in R replace "sin" by "cos" and "cos" by "– sin" in y_p.

† For additional terms, compare with formula 6.

Table VII: $(D^2 - 2aD + a^2 + b^2)y = R$

R	y_p
39. $P(x)\sin sx^*$	$\frac{\sin sx}{2b}\left[\left(\frac{s+b}{a^2+(s+b)^2} - \frac{s-b}{a^2+(s-b)^2}\right)P(x) + \left(\frac{2a(s+b)}{[a^2+(s+b)^2]^2} - \frac{2a(s-b)}{[a^2+(s-b)^2]^2}\right)P'(x) + \left(\frac{3a^2(s+b)-(s+b)^3}{[a^2+(s+b)^2]^3} - \frac{3a^2(s-b)-(s-b)^3}{[a^2+(s-b)^2]^3}\right)P''(x) + \cdots\right]$ $- \frac{\cos sx}{2b}\left[\left(\frac{a}{a^2+(s+b)^2} - \frac{a}{a^2+(s-b)^2}\right)P(x) + \left(\frac{a^2-(s+b)^2}{[a^2+(s+b)^2]^2} - \frac{a^2-(s-b)^2}{[a^2+(s-b)^2]^2}\right)P'(x) + \left(\frac{a^2-3a(s+b)^2}{[a^2+(s+b)^2]^3} - \frac{a^3-3a(s-b)^2}{[a^2+(s-b)^2]^3}\right)P''(x) + \cdots\right]^{\dagger}$
40. $P(x)e^{rx}\sin sx*$	Replace a by $a-r$ in formula 39 and multiply by e^{rx}.
41. $P(x)e^{ax}$	$\frac{e^{ax}}{b^2}\left[P(x) - \frac{P''(x)}{b^2} + \frac{P^{\text{iv}}(x)}{b^4} - \cdots\right]$
42. $e^{ax}\sin sx*$	$\frac{e^{ax}\sin sx}{-s^2+b^2}$
43. $e^{ax}\sin bx*$	$-\frac{xe^{ax}\cos bx}{2b}$
44. $P(x)e^{ax}\sin bx*$	$\frac{e^{ax}\sin bx}{(2b)^2}\left[P(x) - \frac{P''(x)}{(2b)^2} + \frac{P^{\text{iv}}(x)}{(2b)^4} - \cdots\right]$ $-\frac{e^{ax}\cos bx}{2b}\int\left[P(x) - \frac{P''(x)}{(2b)^2} + \frac{P^{\text{iv}}(x)}{(2b)^4} - \cdots\right]dx$

* For $\cos sx$ in R replace "sin' by "cos' and "cos" by "– sin" in y_p.

† For additional terms, compare with formula 6.

Table VIII: $f(D)y = [D^n + a_{n-1}D^{n-1} + \cdots + a_1D + a_0]y = R$

R	y_p
45. e^{rx}	$\frac{e^{rx}}{f(r)}$
46. $\sin sx*$	$\frac{[a_0-a_2s^2+a_4s^4-\cdots]\sin sx-[a_1s-a_3s^3+a_5s^5+\cdots]\cos sx}{[a_0-a_2s^2+a_4s^4-\cdots]^2+[a_1s-a_3s^3+a_5s^5-\cdots]^2}$

Table IX: $f(D^2)y = R$

R	y_p
47. $\sin sx*$	$\frac{\sin sx}{f(-s^2)} = \frac{\sin sx}{a_0-a_2s^2+\cdots\pm s^{2n}}$

Table X: $(D-a)^n y = R$

R	y_p
48. e^{rx}	$\frac{e^{rx}}{(r-a)^n}$
49. $\sin sx*$	$\frac{(-1)^n}{(a^2+s^2)^2}\{[a^n - \binom{n}{2}a^{n-2}s^2 + \binom{4}{n}a^{n-4}s^4 - \cdots]\sin sx$ $+[\binom{n}{1}a^{n-1}s - \binom{n}{3}a^{n-3}s^3 + \cdots]\cos sx\}$
50. $P(x)$	$\frac{(-1)^n}{a^n}\left[P(x) + \binom{n}{1}\frac{P'(x)}{a} + \binom{n+1}{2}\frac{P'(x)}{a^2} + \binom{n+2}{3}\frac{P'''(x)}{a^2} + \cdots\right]$
51. $e^{rx}\sin sx*$	Replace a by $a-r$ in formula 49 and multiply by e^{rx}.
52. $e^{rx}P(x)$	Replace a by $a-r$ in formula 50 and multiply by e^{rx}.
53. $P(x)\sin sx*$	$(-1)^n\sin sx[A_nP(x) + \binom{n}{1}A_{n+1}P'(x) + \binom{n+1}{2}A_{n+2}P''(x) + \binom{n+2}{3}A_{n+3}P'''(x) + \cdots]$ $+(-1)^n\cos sx[B_nP(x) + \binom{n}{1}B_{n+1}P'(x) + \binom{n+1}{2}B_{n+2}P''(x) + \binom{n+2}{3}B_{n+3}P'''(x) + \cdots]$ $A_1 = \frac{a}{a^2+s^2},\ A_2 = \frac{a^2-s^2}{(a^2+s^2)^2},\ \ldots,\ A_k = \frac{a^k - \binom{k}{2}a^{k-2}s^2 + \binom{k}{4}a^{k-4}s^4 - \cdots}{(a^2+s^2)^k}$ $B_1 = \frac{a}{a^2+s^2},\ B_2 = \frac{2as}{(a^2+s^2)^2},\ \ldots,\ B_k = \frac{\binom{k}{1}a^{k-1}s - \binom{k}{3}a^{k-3}s^3 + \cdots}{(a^2+s^2)^k}$
54. $e^{rx}\sin sx*$	Replace a by $a-r$ in formula 53 and multiply by e^{rx}.
55. $e^{ax}P(x)$	$e^{ax}\int\int\cdots\int P(x)\,dx^n$
56. $P(x)e^{ax}\sin sx*$	$\frac{(-1)^{\frac{n-1}{2}}e^{ax}\sin sx}{s^n}\left[\binom{n}{n-1}\frac{P'(x)}{s} - \binom{n+2}{n-1}\frac{P'''(x)}{s^3} + \binom{n+4}{n-1}\frac{P^{\text{v}}(x)}{s^5} - \cdots\right]$ $+\frac{(-1)^{\frac{n+1}{2}}e^{ax}\cos sx}{s^n}\left[\binom{n-1}{n-1}P(x) - \binom{n+1}{n-1}\frac{P''(x)}{s^2} + \binom{n+3}{n-1}\frac{P^{\text{iv}}(x)}{s^4} - \cdots\right]$ (n odd) $\frac{(-1)^{\frac{n}{2}}e^{ax}\sin sx}{s^n}\left[\binom{n-1}{n-1}P(x) - \binom{n+1}{n-1}\frac{P''(x)}{s^2} + \binom{n+3}{n-1}\frac{P^{\text{iv}}(x)}{s^4} - \cdots\right]$ $+\frac{(-1)^{\frac{n}{2}}e^{ax}\cos sx}{s^n}\left[\binom{n}{n-1}\frac{P'(x)}{s} - \binom{n+2}{n-1}\frac{P'''(x)}{s^3} + \binom{n+4}{n-1}\frac{P^{\text{v}}(x)}{s^5} - \cdots\right]$ (n even)

* For $\cos sx$ in R replace "sin" by "cos" and "cos" by "– sin" in y_p.

Table XI: $(D-a)^n f(D)y = R$

57. e^{ax} $\quad \frac{x^n}{n!} \cdot \frac{e^{ax}}{f(a)}$

* For $\cos sx$ in R replace "sin" by "cos" and "cos" by "– sin" in y_p.

Table XII: $(D^2+q)^n y = R$

R	y_p
58. e^{rx}	$e^{rx}/(r^2+q)^n$
59. $\sin sx*$	$\sin sx/(q-s^2)^n$
60. $P(x)$	$\frac{1}{q^n}\left[P(x) - \binom{n}{1}\frac{P''(x)}{q^2} + \binom{n+1}{2}\frac{P^{\text{iv}}(x)}{q^2} - \binom{n+2}{3}\frac{P^{\text{vi}}(x)}{q^3} + \cdots\right]$
61. $e^{rx}\sin sx*$	$\frac{e^{rx}}{(A^2+B^2)^n}\left\{\left[A^n - \binom{n}{2}A^{n-2}B^2 + \binom{n}{4}A^{n-4}B^4 - \cdots\right]\sin sx - \left[\binom{n}{1}A^{n-1}B - \binom{n}{3}A^{n-3}B^3 + \cdots\right]\cos sx\right\}$ $A = r^2 - s^2 + q, \quad B = 2rs$

Table XIII: $(D^2+b^2)^n y = R$

62. $\sin bx*$ $\quad (-1)^{n+1/2}\frac{x^n \cos bx}{n!(2b)^n}$ $(n$ odd), $(-1)^{n/2}\frac{x^n \sin bx}{n!(2b)^n}$ $(n$ even)

Table XIV: $(D^n - q)y = R$

63. e^{rx}	$e^{rx}/(r^n \quad q)$
64. $P(x)$	$-\frac{1}{q}\left[P(x)\frac{P^{(n)}(x)}{q} + \frac{P^{(2n)}(x)}{q^2} + \cdots\right]$
65. $\sin sx*$	$-\frac{q\sin sx + (-1)^{\frac{n-1}{2}} s^n \cos sx}{q^2+s^{2n}}$ $(n$ odd), $\quad \frac{\sin sx}{(-s^2)^{n/2}-q}$ $(n$ even)
66. $e^{rx}\sin sx*$	$\frac{Ae^{rx}\sin sr - Be^{rx}\cos sx}{A^2+B^2} = \frac{e^{rx}}{\sqrt{A^2+B^2}}\sin\left(sx - \tan^{-1}\frac{B}{A}\right)$ $A = \left[r^n - \binom{n}{2}r^{n-2}s^2 + \binom{n}{4}r^{n-4}s^4 - \cdots\right] - q,$ $B = \left[\binom{n}{1}r^{n-1}s - \binom{n}{3}r^{n-3}s^3 + \cdots\right]$

* For $\cos sx$ in R replace "sin" by "cos" and cos by "– sin" in y_p.

Table XV: $(D_x + mD_y)z = R$

R	z_p
67. e^{ax+by}	$\frac{e^{ax+by}}{a+mb}$
68. $f(ax+by)$	$\frac{\int f(u)du}{a+mb}, \quad u = ax+by$
69. $f(y-mx)$	$xf(y-mx)$
70. $\phi(x,y)\,f(y-mx)$	$f(y-mx)\int \phi(x, a+mx)\,dx \quad (a = y - mx$ after integration)

Table XVI: $(D_x + mD_y - k)z = R$

71. e^{ax+by}	$\frac{e^{ax+by}}{a+mb-k}$
72. $\sin(ax+by)*$	$-\frac{(a+bm)\cos(ax+by)+k\sin(ax+by)}{(a+bm)^2+k^2}$
73. $e^{\alpha x+\beta y}\sin(ax+by)*$	Replace k in 72 by $k-\alpha-m\beta$ and multiply by $e^{\alpha x+\beta y}$
74. $e^{xk}f(ax+by)$	$\frac{e^{kx}\int f(u)du}{a+mb}, \quad u = ax+by$
75. $f(y-mx)$	$-\frac{f(y-mx)}{k}$
76. $p(x)\,f(y-mx)$	$-\frac{1}{k}f(y-mx)\left[p(x) + \frac{p'(x)}{k} + \frac{p''(x)}{k^2} + \cdots + \frac{p^{(n)}(x)}{k^n}\right]$
77. $e^{kx}f(y-mx)$	$xe^{kx}f(y-mx)$

* For $\cos(ax+by)$ replace "sin" by "cos" and "cos" by "– sin" in z_p.

$D_x = \frac{\partial}{\partial x}; \quad D_y = \frac{\partial}{\partial y}; \quad D_{x^k}D_{y^r} = \frac{\partial^{k+r}}{\partial_{x^k}\partial_{y^r}}$

Table XVII: $(D_z + mD_y)^n z = R$

R	z_p
78. e^{ax+by}	$\frac{e^{ax+by}}{(a+mb)^n}$
79. $f(ax+by)$	$\frac{\int\int\cdots\int f(u)du^n}{(a+mb)^n}$, $u = ax+by$
80. $f(y-mx)$	$\frac{x^n}{n!} f(y-mx)$
81. $\phi(x, y) f(y+mx)$	$f(y-mx)\int\int\cdots\int \phi(x, a+mx)\, dx^n (a = y - mx$ after integration)

Table XVIII: $(D_x + mD_y - k)^n z = R$

R	z_p
82. e^{ax+by}	$\frac{e^{ax+by}}{(a+mb-k)^n}$
83. $f(y-mx)$	$\frac{(-1)^n f(y-mx)}{k^n}$
84. $P(x) f(y-mx)$	$\frac{(-1)^n}{k^n} f(y-mx)\left[p(x) + \binom{n}{1}\frac{p'(x)}{k} + \binom{n+1}{2}\frac{p''(x)}{k^3} + \binom{n+2}{3}\frac{p'''(x)}{k^3} + \cdots\right]$
85. $e^{kz} f(ax+by)$	$\frac{e^{kx}\int\int\cdots\int f(u)du^n}{(a+mb)^n}$, $u = ax+by$
86. $e^{kx} f(y-mx)$	$\frac{x^n}{n!} e^{kx} f(y-mx)$

Table XIX: $[D_x^n + a_1 D_x^{n-1} D_y + a_2 D_x^{n-2} D_y^2 + \cdots + a^n D_y^n]z = R$

R	z_p
87. e^{ax+by}	$\frac{e^{ax+by}}{a + a_1 a^{n-1} b + a_2 a^{n-2} b^2 + \cdots + a_n b^n}$
88. $f(ax+by)$	$\frac{\int\int\cdots\int\int f(u)du^n}{a^n + a_1 a^{n-1} b + a_2 a^{n-2} b^2 + \cdots + a^n b^n}$, $(u = ax+by)$

Table XX: $F(D_x, D_y)z = R$

R	z_p
89. e^{ax+by}	$\frac{e^{ax+by}}{F(a,b)}$

Table XXI: $F(D_x^2, D_x D_y, D_y^2)z = R$

R	z_p
90. $\sin(ax+by)*$	$\frac{\sin(ax+by)}{F(-a^2, -ab, -b^2)}$

* For $\cos(ax+by)$ replace "sin " by "cos", and "cos" by "– sin" in z_p.

Differential equation	Method of solution
$yF(xy)\,dx + x\,G(xy)\,dy = 0$	$\ln x = \int \frac{G(v)\,dv}{v\{G(v) - F(v)\}} + c$ where $v = xy$. If $G(v) = F(v)$, then the solution is $xy = c$.
Linear, homogeneous, second order equation $\frac{d^2y}{dx^2} + b\frac{dy}{dx} + cy = 0$ where b and c are real constants	Let m_1, m_2 be the roots of $m^2 + bm + c = 0$. Then there are 3 cases: Case 1. m_1, m_2 real and distinct: $y = c_1e^{m_1x} + c_2e^{m_2x}$ Case 2. m_1, m_2 real and equal: $y = c_1e^{m_1x} + c_2xe^{m_1x}$ Case 3. $m_1 = p + qi,\ m_2 = p - qi$: $y = e^{px}(c_1 \cos qx + c_2 \sin qx)$ where $p = -b/2,\ q = \sqrt{4c - b^2}/2$
Linear, nonhomogeneous, second order equation $\frac{d^2y}{dx^2} + b\frac{dy}{dx} + cy = R(x)$ where b and c are real constants	There are 3 cases corresponding to those above: Case 1. $y = c_1e^{m_1x} + c_2e^{m_2x}$ $+ \frac{e^{m_1x}}{m_1 - m_2}\int e^{-m_1x}R(x)\,dx$ $+ \frac{e^{m_2x}}{m_2 - m_1}\int e^{-m_2x}R(x)\,dx$ Case 2. $y = c_1e^{m_1x} + c_2xe^{m_1x}$ $+ xe^{m_1x}\int e^{-m_1x}R(x)\,dx$ $- e^{m_1x}\int xe^{\;m_1x}R(x)\,dx$ Case 3. $y = e^{px}(c_1 \cos qx + c_2 \sin qx)$ $+ \frac{e^{px}\sin qx}{q}\int e^{-px}R(x)\cos qx\,dx$ $- \frac{e^{px}\cos qx}{q}\int e^{-px}R(x)\sin qx\,dx$

Differential equation	Method of solution
Euler or Cauchy equation $x^2\frac{d^2y}{dx^2}+bx\frac{dy}{dx}+cy=S(x)$	Putting $x=e^t$, the equation becomes $\frac{d^2y}{dt^2}+(b-1)\frac{dy}{dt}+cy=S(e^t)$ and can then be solved as a linear second order equation.
Bessel's equation $x^2\frac{d^2y}{dx^2}+x\frac{dy}{dx}+(\lambda^2x^2-n^2)y=0$	$y=c_1J_n(\lambda x)+c_2Y_n(\lambda x)$
Transformed Bessel's equation $x^2\frac{d^2y}{dx^2}+(2p+1)x\frac{dy}{dx}+$ $(\alpha^2x^{2r}+\beta^2)\,y=0$	$y=x^{-p}\left\{c_1J_{q/r}\left(\frac{\alpha}{r}x^r\right)+c_2Y_{q/r}\left(\frac{\alpha}{r}x^r\right)\right\}$ where $q=\sqrt{p^2-\beta^2}$.
Legendre's equation $(1-x^2)\frac{d^2y}{dx^2}-2x\frac{dy}{dx}+n(n+1)y=0$	$y=c_1P_n(x)+c_2Q_n(x)$
Separation of variables $f_1(x)g_1(y)\,dx+f_2(x)g_2(y)\,dy=0$	$\int\frac{f_1(x)}{f_2(x)}\,dx+\int\frac{g_2(y)}{g_1(y)}dy=c$
Exact equation $M(x,y)\,dx+N(x,y)\,dy=0$ where $\partial M/\partial y=\partial N/\partial x$	$\int M\partial x+\int\left(n-\frac{\partial}{\partial y}\int M\partial x\right)dy=c$ where ∂x indicates that the integration is to be performed with respect to x keeping y constant.
Linear first order equation $\frac{dy}{dx}+P(x)y=Q(x)$	$ye^{\int P\,dx}=\int Qe^{\int P\,dx}\,dx+c$
Bernoulli's equation $\frac{dy}{dx}+P(x)y=Q(x)y^n$	$ve^{(1-n)}\int P\,dx=\int Qe^{(1-n)\int P\,dx}\,dx+c$ where $v=y^{1-n}$. If $n=1$, then the solution is $\ln y=\int(Q-P)\,dx+c$.
Homogeneous equation $\frac{dy}{dx}=F\left(\frac{y}{x}\right)$	$\ln x=\int\frac{dv}{F(v)-v}+c$ where $v=y/x$. If $F(v)=v$, then the solution is $y=cx$.
Reducible to homogeneous $(a_1x+b_1y+c_1)\,dx$ $+(a_2x+b_2y+c_2)\,dy=0$ with $\frac{a_1}{a_2}\neq\frac{b_1}{b_2}$	Set $u=a_1x+b_1y+c_1$ and $v=a_2x+b_2y+c_2$. Then eliminate x and y and the equation becomes homogenous.
Reducible to separable $(a_1x+b_1y+c_1)\,dx$ $+(a_2x+b_2y+c_2)\,dy=0$ with $\frac{a_1}{a_2}=\frac{b_1}{b_2}$	Set $u=a_1x+b_1y$. Then eliminate x or y and the equation becomes separable.

FOURIER SERIES

1. If $f(x)$ is a bounded periodic function of period 2L (i.e., $f(x+2L) = f(x)$), and satisfies the *Dirichlet conditions*:

   (a) In any period $f(x)$ is continuous, except possibly for a finite number of jump discontinuities.

   (b) In any period $f(x)$ has only a finite number of maxima and minima.

   Then $f(x)$ may be represented by the *Fourier series*

$$\frac{a_0}{2} + \sum_{n=1}^{\infty}\left(a_n \cos\frac{n\pi x}{L} + b_n \sin\frac{n\pi x}{L}\right)$$

where a_n and b_n are as determined below. This series will converge to $f(x)$ at every point where $f(x)$ is continuous, and to

$$\frac{f(x^{+}) + f(x^{\ })}{2}$$

(i.e., the average of the left-hand and right-hand limits) at every point where $f(x)$ has a jump discontinuity.

$$a_n = \frac{1}{L}\int_{-L}^{L} f(x)\cos\frac{n\pi x}{L}\,dx, \quad n = 0, 1, 2, 3, \ldots,$$

$$b_n = \frac{1}{L}\int_{-L}^{L} f(x)\sin\frac{n\pi x}{L}\,dx, \quad n = 1, 2, 3, \ldots$$

We may also write

$$a_n = \frac{1}{L}\int_{\alpha}^{\alpha+2L} f(x)\cos\frac{n\pi x}{L}\,dx \text{ and } b_n = \frac{1}{L}\int_{\alpha}^{\alpha+2L} f(x)\sin\frac{n\pi x}{L}\,dx$$

where α is any real number. Thus if $\alpha = 0$,

$$a_n = \frac{1}{L}\int_{0}^{2L} f(x)\cos\frac{n\pi x}{L}\,dx, \quad n = 0, 1, 2, 3, \ldots,$$

$$b_n = \frac{1}{L}\int_{0}^{2L} f(x)\sin\frac{n\pi x}{L}\,d, \quad n = 1, 2, 3, \ldots$$

2. If in addition to the restrictions in (1), $f(x)$ is an even function (i.e., $f(-x) = f(x)$), then the Fourier series reduces to

$$\frac{a_0}{2} + \sum_{n=1}^{\infty} a_n \cos\frac{n\pi x}{L}$$

That is, $b_n = 0$. In this case, a simpler formula for a_n is

$$a_n = \frac{2}{L}\int_{0}^{L} f(x)\cos\frac{n\pi x}{L}\,dx, \quad n = 0, 1, 2, 3, \ldots$$

3. If in addition to the restrictions in (1), $f(x)$ is an odd function (i.e., $f(-x) = -f(x)$), then the Fourier series reduces to

$$\sum_{n=1}^{\infty} b_n \sin\frac{n\pi x}{L}$$

That is, $a_n = 0$. In this case, a simpler formula for the b_n is

$$b_n = \frac{2}{L}\int_{0}^{L} f(x)\sin\frac{n\pi x}{L}\,dx, \quad n = 1, 2, 3, \ldots$$

4. If in addition to the restrictions in (2) above, $f(x) = -f(L-x)$, then a_n will be 0 for all even values of n, including $n = 0$. Thus in this case, the expansion reduces to

$$\sum_{m=1}^{\infty} a_{2m-1}\cos\frac{(2m-1)\pi x}{L}$$

5. If in addition to the restrictions in (3) above, $f(x) = f(L - x)$, then b_n will be 0 for all even values of n. Thus in this case, the expansion reduces to

$$\sum_{m=1}^{\infty} b_{2m-1} \sin \frac{(2m-1)\pi x}{L}$$

(The series in (4) and (5) are known as *odd-harmonic series*, since only the odd harmonics appear. Similar rules may be stated for even-harmonic series, but when a series appears in the even-harmonic form, it means that $2L$ has not been taken as the smallest period of $f(x)$. Since any integral multiple of a period is also a period, series obtained in this way will also work, but in general computation is simplified if $2L$ is taken to be the smallest period.)

6. If we write the Euler definitions for $\cos\theta$ and $\sin\theta$, we obtain the complex form of the Fourier series known either as the "Complex Fourier Series" or the "Exponential Fourier Series" of $f(x)$. It is represented as

$$f(x) = \frac{1}{2} \sum_{n=-\infty}^{n=+\infty} c_n e^{i\omega_n x}$$

where

$$c_n = \frac{1}{L} \int_{-L}^{L} f(x)\, e^{-i\omega_n x}\, dx, \quad n = 0, \pm 1, \pm 2, \pm 3, \ldots$$

with $\omega_n = \frac{n\pi}{L}$ for $n = 0, \pm 1, \pm 2, \ldots$ The set of coefficients c_n is often referred to as the Fourier spectrum.

7. If both sine and cosine terms are present and if $f(x)$ is of period $2L$ and expandable by a Fourier series, it can be represented as

$$f(x) = \frac{a_0}{2} + \sum_{n=1}^{\infty} c_n \sin\left(\frac{n\pi x}{L} + \phi_n\right), \quad \text{where}$$

$$a_n = c_n \sin\phi_n, \quad b_n = c_n \cos\phi_n, \quad c_n = \sqrt{a_n^2 + b_n^2}, \quad \phi_n = \arctan\left(\frac{a_n}{b_n}\right)$$

It can also be represented as

$$f(x) = \frac{a_0}{2} + \sum_{n=1}^{\infty} c_n \cos\left(\frac{n\pi x}{L} + \phi_n\right), \quad \text{where}$$

$$a_n = c_n \cos\phi_n, \quad b_n = -c_n \sin\phi_n, \quad c_n = \sqrt{a_n^2 + b_n^2}, \quad \phi_n = \arctan\left(-\frac{b_n}{a_n}\right)$$

where ϕ_n is chosen so as to make a_n, b_n, and c_n hold.

8. The following table of trigonometric identities should be helpful for developing Fourier series.

	n	n even	$n$odd	$n/2$ odd	$n/2$ even
$\sin n\pi$	0	0	0	0	0
$\cos n\pi$	$(-1)^n$	+1	−1	+1	+1
$*\sin\frac{n\pi}{2}$		0	$(-1)^{(n-1)/2}$	0	0
$*\cos\frac{n\pi}{2}$		$(-1)^{n/2}$	0	−1	+1
$\sin\frac{n\pi}{4}$			$\frac{\sqrt{2}}{2}(-1)^{(n^2+4n+11)/8}$	$(-1)^{(n-2)/4}$	0

*A useful formula for $\sin\frac{n\pi}{2}$ and $\cos\frac{n\pi}{2}$ is given by

$$\sin\frac{n\pi}{2} = \frac{(i)^{n+1}}{2}[(-1)^n - 1] \quad \text{and} \quad \cos\frac{n\pi}{2} = \frac{(i)^n}{2}[(-1)^n + 1], \quad \text{where } i^2 = -1.$$

Auxiliary Formulas for Fourier Series

$$1 = \frac{4}{\pi}\left[\sin\frac{\pi x}{k} + \frac{1}{3}\sin\frac{3\pi x}{k} + \frac{1}{5}\sin\frac{5\pi x}{k} + \cdots\right] \qquad [0 < x < k]$$

$$x = \frac{2k}{\pi}\left[\sin\frac{\pi x}{k} - \frac{1}{2}\sin\frac{2\pi x}{k} + \frac{1}{3}\sin\frac{3\pi x}{k} - \cdots\right] \qquad [-k < x < k]$$

$$x = \frac{k}{2} - \frac{4k}{\pi^2}\left[\cos\frac{\pi x}{k} + \frac{1}{3^2}\cos\frac{3\pi x}{k} + \frac{1}{5^2}\cos\frac{5\pi x}{k} + \cdots\right] \qquad [0 < x < k]$$

$$x^2 = \frac{2k^2}{\pi^3}\left[\left(\frac{\pi^2}{1} - \frac{4}{1}\right)\sin\frac{\pi x}{k} - \frac{\pi^2}{2}\sin\frac{2\pi x}{k} + \left(\frac{\pi^2}{3} - \frac{4}{3^3}\right)\sin\frac{3\pi x}{k}\right.$$
$$\left. - \frac{\pi^2}{4}\sin\frac{4\pi x}{k} + \left(\frac{\pi^2}{5} - \frac{4}{5^3}\right)\sin\frac{5\pi x}{k} + \cdots\right] \qquad [0 < x < k]$$

$$x^2 = \frac{k^2}{3} - \frac{4k^2}{\pi^2}\left[\cos\frac{\pi x}{k} - \frac{1}{2^2}\cos\frac{2\pi x}{k} + \frac{1}{3^2}\cos\frac{3\pi x}{k} - \frac{1}{4^2}\cos\frac{4\pi x}{k} + \cdots\right]$$
$$[-k < x < k]$$

$$1 - \frac{1}{3} + \frac{1}{5} - \frac{1}{7} + \cdots = \frac{\pi}{4}$$

$$1 - \frac{1}{2^2} + \frac{1}{3^2} + \frac{1}{4^2} + \cdots = \frac{\pi^2}{6}$$

$$1 - \frac{1}{2^2} + \frac{1}{3^2} - \frac{1}{4^2} + \cdots = \frac{\pi^2}{12}$$

$$1 + \frac{1}{3^2} + \frac{1}{5^2} - \frac{1}{7^2} + \cdots = \frac{\pi^2}{8}$$

$$\frac{1}{2^2} + \frac{1}{4^2} + \frac{1}{6^2} + \frac{1}{8^2} + \cdots = \frac{\pi^2}{24}$$

FOURIER EXPANSIONS FOR BASIC PERIODIC FUNCTIONS

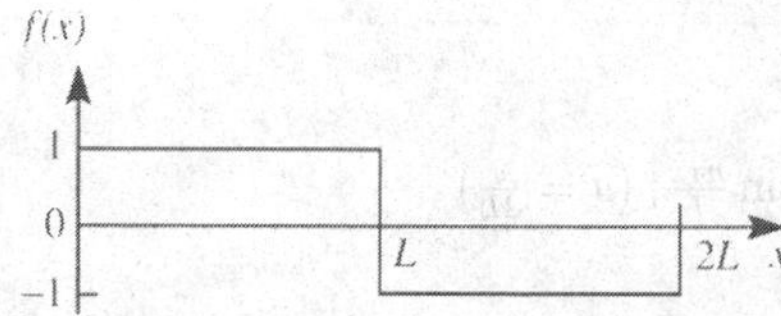

$$f(x) = \frac{4}{\pi}\sum_{n=1,3,5\ldots}\frac{1}{n}\sin\frac{n\pi x}{L}$$

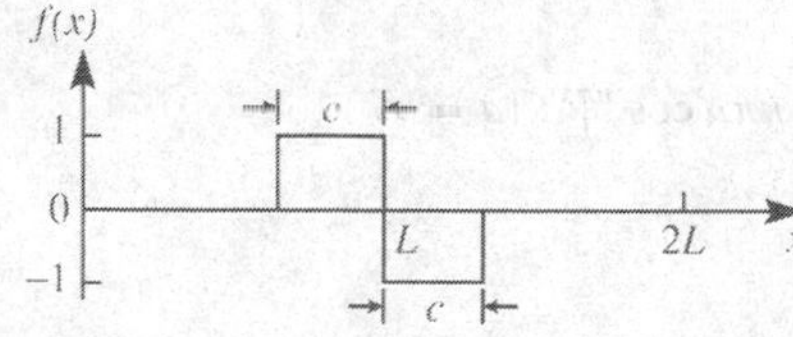

$$f(x) = \frac{2}{\pi}\sum_{n=1}^{\infty}\frac{(-1)^n}{n}\left(\cos\frac{n\pi c}{L} - 1\right)\sin\frac{n\pi x}{L}$$

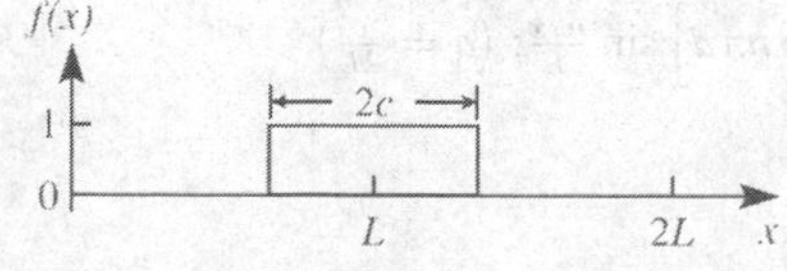

$$f(x) = \frac{c}{L} + \frac{2}{\pi}\sum_{n=1}^{\infty}\frac{(-1)^n}{n}\sin\frac{n\pi c}{L}\cos\frac{n\pi x}{L}$$

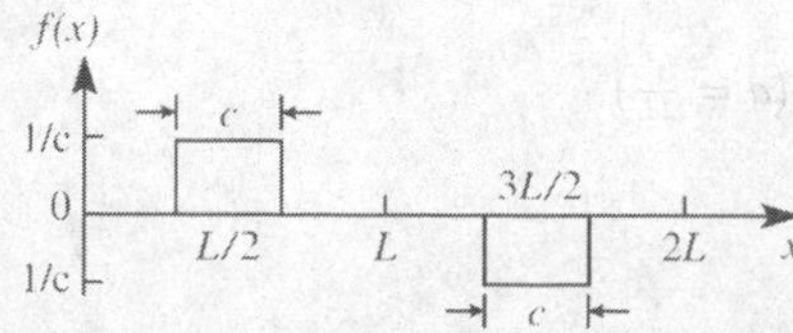

$$f(x) = \frac{2}{L}\sum_{n=1}^{\infty}\sin\frac{n\pi}{2}\,\frac{\sin(\frac{1}{2}n\pi c/L)}{\frac{1}{2}n\pi c/L}\sin\frac{n\pi x}{L}$$

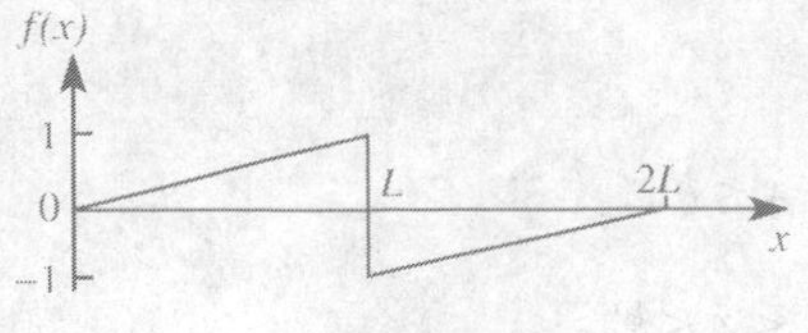

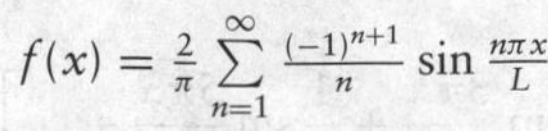

$$f(x)=\frac{2}{\pi}\sum_{n=1}^{\infty}\frac{(-1)^{n+1}}{n}\sin\frac{n\pi x}{L}$$

$$f(x)=\frac{1}{2}-\frac{4}{\pi^2}\sum_{n=1,3,5,\ldots}\frac{1}{n^2}\cos\frac{n\pi x}{L}$$

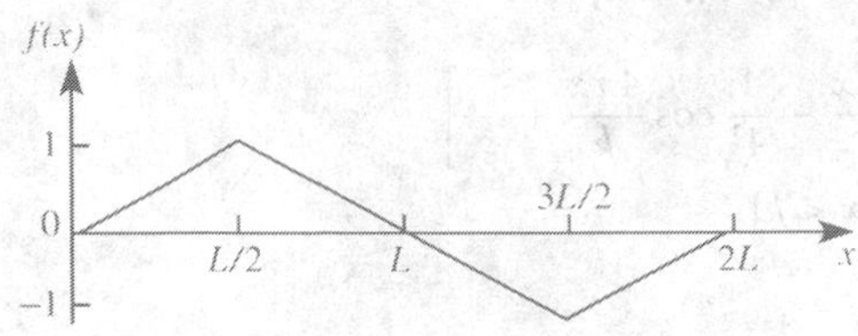

$$f(x)=\frac{8}{\pi^2}\sum_{n=1,3,5,\ldots}\frac{(-1)^{(n-1)/2}}{n^2}\sin\frac{n\pi x}{L}$$

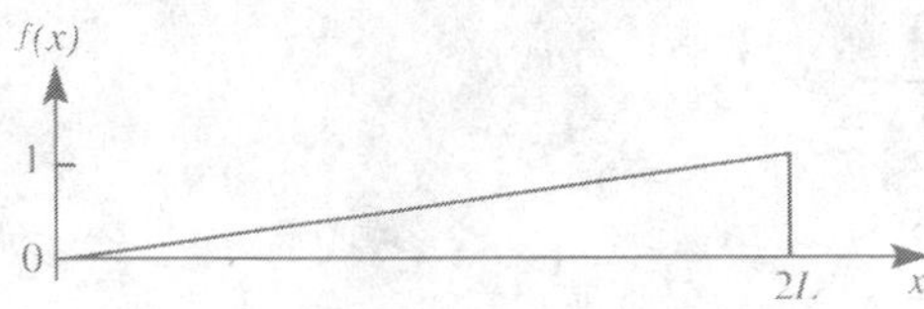

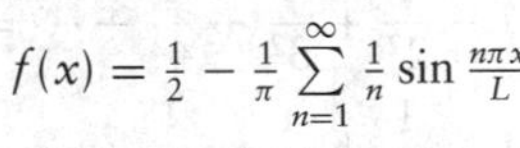

$$f(x)=\frac{1}{2}-\frac{1}{\pi}\sum_{n=1}^{\infty}\frac{1}{n}\sin\frac{n\pi x}{L}$$

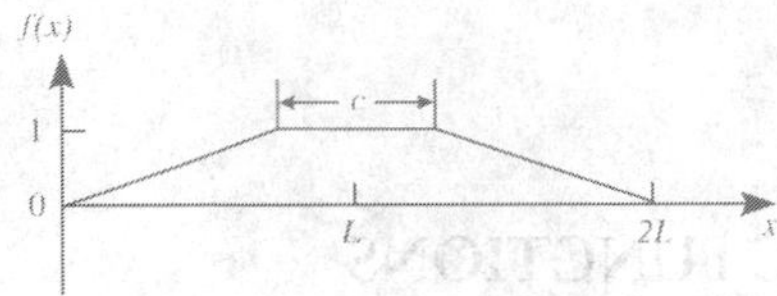

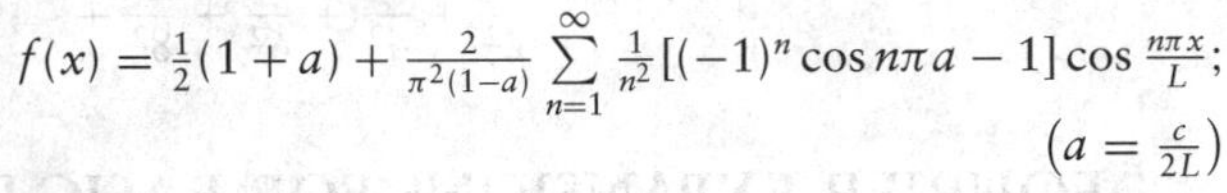

$$f(x)=\frac{1}{2}(1+a)+\frac{2}{\pi^2(1-a)}\sum_{n=1}^{\infty}\frac{1}{n^2}[(-1)^n\cos n\pi a-1]\cos\frac{n\pi x}{L};\quad\left(a=\frac{c}{2L}\right)$$

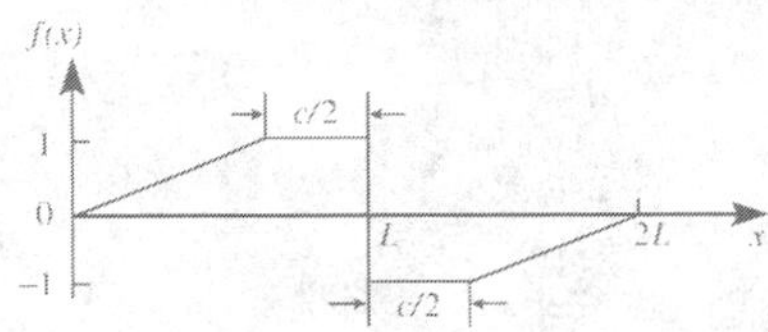

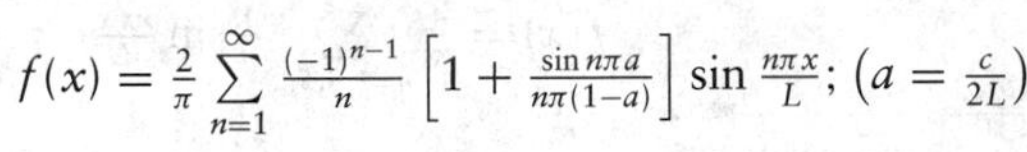

$$f(x)=\frac{2}{\pi}\sum_{n=1}^{\infty}\frac{(-1)^{n-1}}{n}\left[1+\frac{\sin n\pi a}{n\pi(1-a)}\right]\sin\frac{n\pi x}{L};\ \left(a=\frac{c}{2L}\right)$$

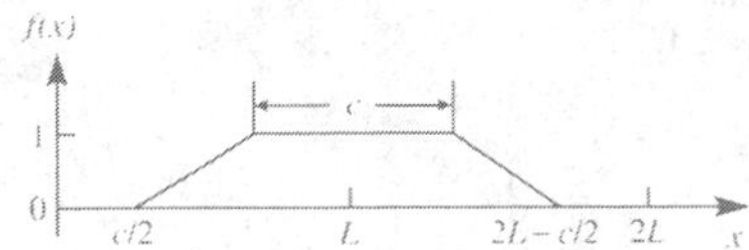

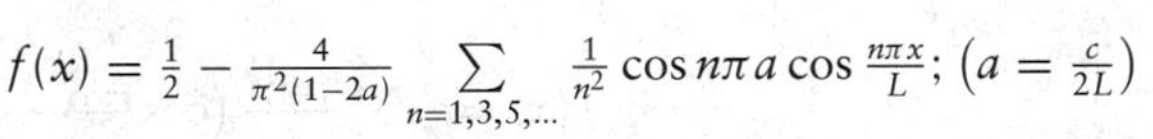

$$f(x)=\frac{1}{2}-\frac{4}{\pi^2(1-2a)}\sum_{n=1,3,5,\ldots}\frac{1}{n^2}\cos n\pi a\cos\frac{n\pi x}{L};\ \left(a=\frac{c}{2L}\right)$$

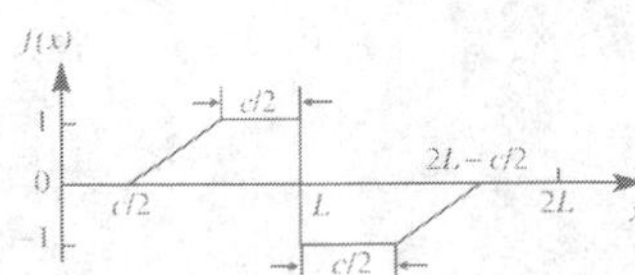

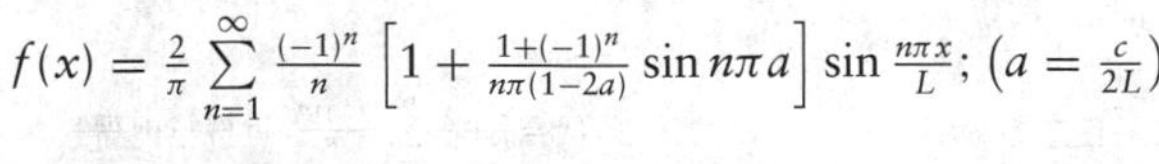

$$f(x)=\frac{2}{\pi}\sum_{n=1}^{\infty}\frac{(-1)^n}{n}\left[1+\frac{1+(-1)^n}{n\pi(1-2a)}\sin n\pi a\right]\sin\frac{n\pi x}{L};\ \left(a=\frac{c}{2L}\right)$$

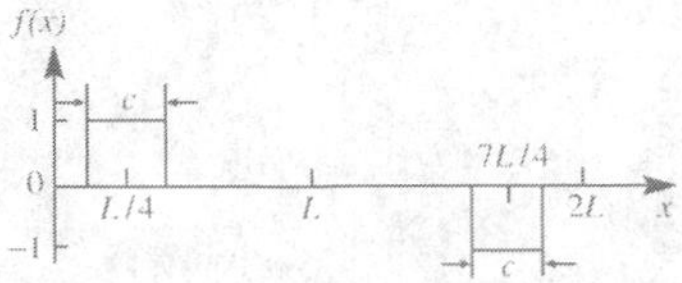

$$f(x)\frac{4}{\pi}\sum_{n=1}^{\infty}\frac{1}{n}\sin\frac{n\pi}{4}\sin n\pi a\sin\frac{n\pi x}{L};\ \left(a=\frac{c}{2L}\right)$$

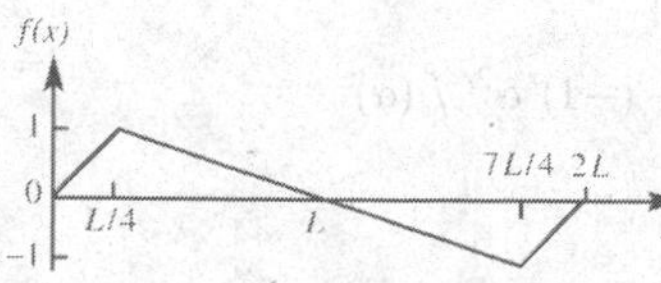

$$f(x) = \tfrac{9}{\pi^2} \sum_{n=1}^{\infty} \tfrac{1}{n^2} \sin \tfrac{n\pi}{3} \sin \tfrac{n\pi x}{L};\ (a = \tfrac{c}{2L})$$

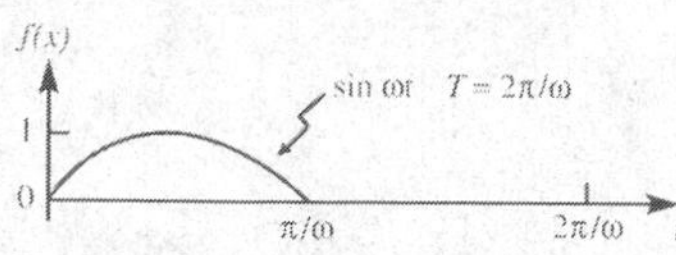

$$f(x) = \tfrac{32}{3\pi^2} \sum_{n=1}^{\infty} \tfrac{1}{n^2} \sin \tfrac{n\pi}{4} \sin \tfrac{n\pi x}{L};\ (a = \tfrac{c}{2L})$$

$$f(x) = \tfrac{1}{\pi} + \tfrac{1}{2} \sin \omega t - \tfrac{2}{\pi} \sum_{n=2,4,6,\ldots} \tfrac{1}{n^2-1} \cos n\omega t$$

Extracted from graphs and formulas, pages 372, 373, *Differential Equations in Engineering Problems*, Salvadori and Schwarz, published by Prentice-Hall, Inc., 1954.

THE FOURIER TRANSFORMS

For a piecewise continuous function $F(x)$ over a finite interval $0 \leq x \leq \pi$; the *finite Fourier cosine transform* of $F(x)$ is

$$f_c(n) = \int_0^{\pi} F(x) \cos nx\, dx \quad (n = 0, 1, 2, \ldots)$$

If x ranges over the interval $0 \leq x \leq L$, the substitution $x' = \pi x/L$ allows the use of this definition, also. The inverse transform is written.

$$\overline{F}(x) = \frac{1}{\pi} f_c(0) - \frac{2}{\pi} \sum_{n=1}^{x} f_c(n) \cos nx \quad (0 < x < \pi)$$

where $F(x) = \frac{F(x+\epsilon)+F(x-\epsilon)}{2}$. We observe that $F(x+) = F(x-) = F(x)$ at points of continuity. The formula

$$\begin{aligned} f_c^{(2)}(n) &= \int_0^{\pi} F''(x) \cos nx\, dx \\ &= -n^2 f_c(n) - F'(0) + (-1)^n F'(\pi) \end{aligned} \tag{1}$$

makes the finite Fourier cosine transform useful in certain boundary value problems. Analogously, the *finite Fourier sine transform* of $F(x)$ is

$$f_s(n) = \int_0^{\pi} F(x) \sin nx\, dx \quad (n = 1, 2, 3, \ldots)$$

and

$$\overline{F}(x) = \frac{2}{\pi} \sum_{n=1}^{\infty} f_s(n) \sin nx \quad (0 < x < \pi)$$

Corresponding to (1) we have

$$\begin{aligned} f_s^{(2)}(n) &= \int_0^{\pi} F''(x) \sin nx\, dx \\ &= -n^2 f_s(n) - n F(0) - n(-1)^n F(\pi) \end{aligned} \tag{2}$$

If $F(x)$ is defined for $x \leq 0$ and is piecewise continuous over any finite interval, and if $\int_0^x F(x)\, dx$ is absolutely convergent, then

$$f_c(\alpha) = \sqrt{\frac{2}{\pi}} \int_0^x F(x) \cos(\alpha x)\, dx$$

is the *Fourier cosine transform* of $F(x)$. Furthermore,

$$\overline{F}(x) = \sqrt{\frac{2}{\pi}} \int_0^x f_c(\alpha) \cos(\alpha x)\, d\alpha.$$

If $\lim_{x\to\infty} d^n F/dx^n = 0$, then an important property of the Fourier cosine transform is

$$f_c^{(2r)}(\alpha) = \sqrt{\frac{2}{\pi}} \int_0^x \left(\frac{d^{2r} F}{dx^{2r}}\right) \cos(\alpha x)\, dx = -\sqrt{\frac{2}{\pi}} \sum_{n=0}^{r-1} (-1)^n a_{2r-2n-1} \alpha^{2n} + (-1)^r \alpha^{2r} f_c(\alpha) \qquad (3)$$

where $\lim_{x\to\infty} d^r F/dx^r = a_r$, makes it useful in the solution of many problems.

Under the same conditions.

$$f_s(\alpha) = \sqrt{\frac{2}{\pi}} \int_0^x F(x) \sin(\alpha x)\, dx$$

defines the *Fourier sine transform* of $F(x)$, and

$$\overline{F}(x) = \sqrt{\frac{2}{\pi}} \int_0^x f_s(\alpha) \sin(\alpha x)\, d\alpha$$

Corresponding to (3) we have

$$f_s^{(2r)}(\alpha) = \sqrt{\frac{2}{\pi}} \int_0^\infty \frac{d^{2r} F}{dx^{2r}} \sin(\alpha x)\, dx = -\sqrt{\frac{2}{\pi}} \sum_{n=1}^{r} (-1)^n \alpha^{2n-1} a_{2r-2n} + (-1)^{r-1} \alpha^{2r} f_s(\alpha)$$

Similarly, if $F(x)$ is defined for $-\infty < x < \infty$, and if $\int_{-\infty}^{\infty} F(x)\, dx$ is absolutely convergent, then

$$f(\alpha) = \frac{1}{\sqrt{2\pi}} \int_{-\infty}^{\infty} F(x) e^{i\alpha x}\, dx$$

is the *Fourier transform* of $F(x)$, and

$$\overline{F}(x) = \frac{1}{\sqrt{2\pi}} \int_{-\infty}^{\infty} f(\alpha) e^{-i\alpha x}\, d\alpha$$

Also, if

$$\lim_{|x|\to\infty} \left|\frac{d^n F}{dx^n}\right| = 0 \quad (n = 1, 2, \ldots, r-1)$$

then

$$f^{(r)}(\alpha) = \frac{1}{\sqrt{2\pi}} \int_{-\infty}^{\infty} F^{(r)}(x) e^{i\alpha x}\, dx = (-i\alpha)^r f(\alpha)$$

Finite Sine Transforms

$f_s(n)$	$F(x)$
1. $f_s(n) = \int_0^\pi F(x) \sin nx\, dx \ (n = 1, 2, \ldots)$	$F(x)$
2. $(-1)^{n+1} f_s(n)$	$F(\pi - x)$
3. $\frac{1}{n}$	$\frac{\pi - x}{\pi}$
4. $\frac{(-1)^{n+1}}{n}$	$\frac{x}{\pi}$
5. $\frac{1-(-1)^n}{n}$	1
6. $\frac{2}{n^2} \sin \frac{n\pi}{2}$	$\begin{cases} x & \text{when } 0 < x < \pi/2 \\ \pi - x & \text{when } \pi/2 < x < \pi \end{cases}$
7. $\frac{(-1)^{n+1}}{n^3}$	$\frac{x(\pi^2 - x^2)}{6\pi}$
8. $\frac{1-(-1)^n}{n^3}$	$\frac{x(\pi - x)}{2}$
9. $\frac{\pi^2(-1)^{n-1}}{n} - \frac{2[1-(-1)^n]}{n^3}$	x^2
10. $\pi(-1)^n \left(\frac{6}{n^3} - \frac{\pi^2}{n}\right)$	x^3
11. $\frac{n}{n^2+c^2}[1 - (-1)^n e^{c\pi}]$	e^{cx}
12. $\frac{n}{n^2+c^2}$	$\frac{\sinh c(\pi - x)}{\sinh c\pi}$

$f_s(n)$	$F(x)$
13. $\frac{n}{n^2-k^2} \quad (k \neq 0, 1, 2, \ldots)$	$\frac{\sin k(\pi - x)}{\sin k\pi}$
14. $\begin{cases} \frac{\pi}{2} & \text{when } n = m \\ 0 & \text{when } n \neq m \end{cases} \quad (m = 1, 2, \ldots)$	$\sin mx$
15. $\frac{n}{n^2 - k^2}[1 - (-1)^n \cos k\pi] \quad (k \neq 1, 2, \ldots)$	$\cos kx$
16. $\begin{cases} \frac{n}{n^2-m^2}[1 - (-1)^{n+m}] & \text{when } n \neq m = 1, 2, \ldots \\ 0 & \text{when } n = m \end{cases}$	$\cos mx$
17. $\frac{n}{(n^2-k^2)^2} (k \neq 0, 1, 2, \ldots)$	$\frac{\pi \sin kx}{2k \sin^2 k\pi} - \frac{x \cos k(\pi - x)}{2k \sin k\pi}$
18. $\frac{b^n}{n} \ (\lvert b \rvert \leq 1)$	$\frac{2}{\pi} \arctan \frac{b \sin x}{1 - b \cos x}$
19. $\frac{1-(-1)^n}{n} b^n \quad (\lvert b \rvert \leq 1)$	$\frac{2}{\pi} \arctan \frac{2b \sin x}{1 - b^2}$

Finite Cosine Transforms

	$f_c(n)$	$F(x)$
1.	$f_c(n) = \int_0^\pi F(x)\cos nx\,dx \quad (n = 0, 1, 2, \ldots)$	$F(x)$
2.	$(-1)^n f_c(n)$	$F(\pi - x)$
3.	0 when $n = 1, 2, \cdots; \quad f_c(0) = \pi$	1
4.	$\frac{2}{n}\sin\frac{n\pi}{2}; \quad f_c(0) = 0$	$\begin{cases} 1 & \text{when } 0 < x < \pi/2 \\ -1 & \text{when } \pi/2 < x < \pi \end{cases}$
5.	$-\frac{1-(-1)^n}{n^2}; \quad f_c(0) = \frac{\pi^2}{2}$	x
6.	$\frac{(-1)^n}{n^2}; \quad f_c(0) = \frac{\pi^2}{6}$	$\frac{x^2}{2\pi}$
7.	$\frac{1}{n^2}; \quad f_c(0) = 0$	$\frac{(\pi-x)^2}{2\pi} - \frac{\pi}{6}$
8.	$3\pi^2\frac{(-1)^n}{n^2} - 6\frac{1-(-1)^n}{n^4}; \quad f_c(0) = \frac{\pi^4}{4}$	x^3
9.	$\frac{(-1)^n e^c\pi - 1}{n^2+c^2}$	$\frac{1}{c}e^{cx}$
10.	$\frac{1}{n^2+c^2}$	$\frac{\cosh c(\pi-x)}{c\sinh c\pi}$
11.	$\frac{k}{n^2-k^2}[(-1)^n\cos\pi k - 1]$ $(k \neq 0, 1, 2, \cdots)$	$\sin kx$
12.	$\frac{(-1)^{n+m}-1}{n^2-m^2}; \; f_c(m) = 0 \quad (m = 1, 2, \cdots)$	$\frac{1}{m}\sin mx$
13.	$\frac{1}{n^2-k^2} \quad (k \neq 0, 1, 2, \ldots)$	$-\frac{\cos k(\pi-x)}{k\sin k\pi}$
14.	$\begin{cases} 0 & \text{for } n = 1, 2, \cdots;\ n \neq m \\ \frac{\pi}{2} & \text{for } n = m \end{cases}$	$\cos mx \quad$ for $m = 1, 2, 3, \ldots$

Fourier Sine Transforms

	$F(x)$	$f_s(\alpha)$
1.	$\begin{cases} 1 & (0 < x < a) \\ 0 & (x > a) \end{cases}$	$\sqrt{\frac{2}{\pi}}\left[\frac{1-\cos\alpha}{\alpha}\right]$
2.	$x^{p-1}(0 < p < 1)$	$\sqrt{\frac{2}{\pi}}\frac{\Gamma(p)}{\alpha^p}\sin\frac{p\pi}{2}$
3.	$\begin{cases} \sin x & (0 < x < a) \\ 0 & (x > a) \end{cases}$	$\frac{1}{\sqrt{2\pi}}\left[\frac{\sin[a(1-\alpha)]}{1-\alpha} - \frac{\sin[a(1+\alpha)]}{1+\alpha}\right]$
4.	e^{-x}	$\sqrt{\frac{2}{\pi}}\left[\frac{\alpha}{1+\alpha^2}\right]$
5.	$xe^{-x^2/2}$	$\alpha e^{-\alpha^2/2}$
6.	$\cos\frac{x^2}{2}$	$\sqrt{2}\left[\sin\frac{\alpha^2}{2}C\left(\frac{\alpha^2}{2}\right) - \cos\frac{\alpha^2}{2}S\left(\frac{\alpha^2}{2}\right)\right]^*$
7.	$\sin\frac{x^2}{2}$	$\sqrt{2}\left[\cos\frac{\alpha^2}{2}C\left(\frac{\alpha^2}{2}\right) + \sin\frac{\alpha^2}{2}S\left(\frac{\alpha^2}{2}\right)\right]^*$

Here $C(y)$ and $S(y)$ are the Fresnel integrals:

$$C(y) = \frac{1}{\sqrt{2\pi}}\int_0^y \frac{1}{\sqrt{t}}\cos t\,dt, \qquad S(y) = \frac{1}{\sqrt{2\pi}}\int_0^y \frac{1}{\sqrt{t}}\sin t\,dt$$

*More extensive tables of the Fourier sine and cosine transforms can be found in Fritz Oberhettinger, *Tabellen zur-Fourier Transformation*, Springer, 1957.

Fourier Cosine Transforms

	$F(x)$	$f_c(\alpha)$
1.	$\begin{cases} 1 & (0 < x < a) \\ 0 & (x > a) \end{cases}$	$\sqrt{\frac{2}{\pi}}\frac{\sin a\alpha}{\alpha}$
2.	$x^{p-1} \quad (0 < p < 1)$	$\sqrt{\frac{2}{\pi}}\frac{\Gamma(p)}{\alpha^p}\cos\frac{p\pi}{2}$
3.	$\begin{cases} \cos x & (0 < x < a) \\ 0 & (x > a) \end{cases}$	$\frac{1}{\sqrt{2\pi}}\left[\frac{\sin[a(1-\alpha)]}{1-\alpha} + \frac{\sin[a(1+\alpha)]}{1+\alpha}\right]$
4.	e^{-x}	$\sqrt{\frac{2}{\pi}}\left(\frac{1}{1+\alpha^2}\right)$
5.	$e^{-x^2/2}$	$e^{-\alpha^1/2}$
6.	$\cos\frac{x^2}{2}$	$\cos\left(\frac{\alpha^2}{2} - \frac{\pi}{4}\right)$
7.	$\sin\frac{x^2}{2}$	$\cos\left(\frac{\alpha^2}{2} + \frac{\pi}{4}\right)$

Fourier Transforms

	$F(x)$	$f(\alpha)$
1.	$\frac{\sin ax}{x}$	$\begin{cases} \sqrt{\frac{\pi}{2}} & \lvert\alpha\rvert < a \\ 0 & \lvert\alpha\rvert > a \end{cases}$
2.	$\begin{cases} e^{iwx} & (p < x < q) \\ 0 & (x < p,\ x > q) \end{cases}$	$\frac{i}{\sqrt{2\pi}} \frac{e^{ip(w+\alpha)} - e^{iq(w+\alpha)}}{(w+\alpha)}$
3.	$\begin{cases} e^{-cx+iwx} & (x > 0) \\ 0 & (x < 0) \end{cases} \quad (c > 0)$	$\frac{i}{\sqrt{2\pi}(w+\alpha+ic)}$
4.	$e^{-px^2}\ R(p) > 0$	$\frac{1}{\sqrt{2p}} e^{-\alpha^2/4p}$
5.	$\cos px^2$	$\frac{1}{\sqrt{2p}} \cos\left[\frac{\alpha^2}{4p} - \frac{\pi}{4}\right]$
6.	$\sin px^2$	$\frac{1}{\sqrt{2p}} \cos\left[\frac{\alpha^2}{4p} + \frac{\pi}{4}\right]$
7.	$\lvert x\rvert^{-p} \quad (0 < p < 1)$	$\sqrt{\frac{2}{\pi}} \frac{\Gamma(1-p) \sin \frac{p\pi}{2}}{\lvert\alpha\rvert^{(1-p)}}$
8.	$\frac{e^{-a\lvert x\rvert}}{\sqrt{\lvert x\rvert}}$	$\frac{\sqrt{\sqrt{(a^2+\alpha^2)}+a}}{\sqrt{a^2+\alpha^2}}$
9.	$\frac{\cosh ax}{\cosh \pi x} \quad (-\pi < a < \pi)$	$\sqrt{\frac{2}{\pi}} \frac{\cos \frac{a}{2} \cosh \frac{\alpha}{2}}{\cosh \alpha + \cos a}$
10.	$\frac{\sinh ax}{\sinh \pi x} \quad (-\pi < a < \pi)$	$\frac{1}{\sqrt{2\pi}} \frac{\sin a}{\cosh \alpha + \cos a}$
11.	$\begin{cases} \frac{1}{\sqrt{a^2-x^2}} & (\lvert x\rvert < a) \\ 0 & (\lvert x\rvert > a) \end{cases}$	$\sqrt{\frac{\pi}{2}} J_0(a\alpha)$
12.	$\frac{\sin[b\sqrt{a^2+x^2}]}{\sqrt{a^2+x^2}}$	$\begin{cases} 0 & (\lvert\alpha\rvert > b) \\ \sqrt{\frac{\pi}{2}} J_0(a\sqrt{b^2-\alpha^2}) & (\lvert\alpha\rvert < b) \end{cases}$
13.	$\begin{cases} p_{n(x)} & (\lvert x\rvert < 1) \\ 0 & (\lvert x\rvert > 1) \end{cases}$	$\frac{i^n}{\sqrt{\alpha}} J_{n+\frac{1}{2}}(\alpha)$
14.	$\begin{cases} \frac{\cos[b\sqrt{a^2-x^2}]}{\sqrt{a^2-x^2}} & (\lvert x\rvert < a) \\ 0 & (\lvert x\rvert > a) \end{cases}$	$\sqrt{\frac{\pi}{2}} J_0(a\sqrt{a^2+b^2})$
15.	$\begin{cases} \frac{\cosh[b\sqrt{a^2-x^2}]}{\sqrt{a^2-x^2}} & (\lvert x\rvert < a) \\ 0 & (\lvert x\rvert > a) \end{cases}$	$\sqrt{\frac{\pi}{2}} J_0(a\sqrt{\alpha^2-b^2})$

*More extensive tables of Fourier transforms can be found in *W.* Magnus and *F.* Oberhettinger, *Formulas and Theorems of the Special Functions of Mathematical Physics*. Chelsea, 1949, 116–120.

SERIES EXPANSION

The expression in parentheses following certain of the series indicates the region of convergence. If not otherwise indicated it is to be understood that the series converges for all finite values of x.

Binomial Series

$$(x+y)^n = x^n + nx^{n-1}y + \frac{n(n-1)}{2!}x^{n-2}y^2 + \frac{n(n-1)(n-2)}{3!}x^{n-3}y^3 + \cdots (y^2 < x^2)$$

$$(1 \pm x)^n = 1 \pm nx + \frac{n(n-1)x^2}{2!} \pm \frac{n(n-1)(n-2)x^3}{3!} + \cdots (x^2 < 1)$$

$$(1 \pm x)^{-n} = 1 \mp nx + \frac{n(n+1)x^2}{2!} \mp \frac{n(n+1)(n+2)x^3}{3!} + \cdots (x^2 < 1)$$

$$(1 \pm x)^{-1} = 1 \mp x + x^2 \mp x^3 + x^4 \mp x^5 + \cdots \quad (x^2 < 1)$$

$$(1 \pm x)^{-2} = 1 \mp 2x + 3x^2 \mp 4x^3 + 5x^4 \mp 6x^5 + \cdots \quad (x^2 < 1)$$

Reversion of Series

Let a series be represented by

$$y = a_1x + a_2x^2 + a_3x^3 + a_4x^4 + a_5x^5 + a_6x^6 + \cdots$$

with $a_1 \neq 0$. The coefficients of the series

$$x = A_1y + A_2y^2 + A_3y^3 + A_4y^4 + \cdots$$

are

$$A_1 = \frac{1}{a_1} \qquad A_2 = -\frac{a_2}{a_1^3} \qquad A_3 = \frac{1}{a_1^5}(2a_2^2 - a_1a_3)$$

$$A_4 = \frac{1}{a_1^7}(5a_1a_2a_3 - a_1^2a_4 - 5a_2^3)$$

$$A_5 = \frac{1}{a_1^9}(6a_1^2a_2a_4 + 3a_1^2a_3^2 + 14a_2^4 - a_1^3a_5 - 21a_1a_2^2a_3)$$

$$A_6 = \frac{1}{a_1^{11}}(7a_1^3a_2a_5 + 7a_1^3a_3a_4 + 84a_1a_2^3a_3 - a_1^4a_6 - 28a_1^2a_2^2a_4 - 28a_1^2a_2a_3^2 - 42a_2^5)$$

$$A_7 = \frac{1}{a_1^{13}}(8a_1^4a_2a_6 + 8a_1^4a_3a_5 + 4a_1^4a_4^2 + 120a_1^2a_2^3a_4 + 180a_1^2a_2^2a_3^2 + 132a_2^6 - a_1^5a_7$$
$$-36a_1^3a_2^2a_5 - 72a_1^3a_2a_3a_4 - 12a_1^3a_3^3 - 330a_1a_2^4a_3)$$

Taylor Series

1. $f(x) = f(a) + (x-a)f'(a) + \frac{(x-a)^2}{2!}f''(a) + \frac{(x-a)^3}{3!}f'''(a)$
$+ \cdots + \frac{(x-a)^n}{n!}f^{(n)}(a) + \cdots$ (Taylor Series)

(Increment form)

2. $f(x+h) = f(x) + hf'(x) + \frac{h^2}{2!}f''(x) + \frac{h^3}{3!}f'''(x) + \cdots$
$= f(h) + xf'(h) + \frac{x^2}{2!}f''(h) + \frac{x^3}{3!}f'''(h) + \cdots$

3. If $f(x)$ is a function possessing derivatives of all orders throughout the interval $a \le x \le b$, then there is a value X, with $a < X < b$, such that

$$f(b) = f(a) + (b-a)f'(a) + \frac{(b-a)^2}{2!}f''(a) + \cdots + \frac{(b-a)^{n-1}}{(n-1)!}f^{(n-1)}(a) + \frac{(b-a)^n}{n!}f^{(n)}(X)$$

$$f(a+h) = f(a) + hf'(a) + \frac{h^2}{2!}f''(a) + \cdots + \frac{h^{n-1}}{(n-1)!}f^{(n-1)}(a) + \frac{h^n}{n!}f^{(n)}(a+\theta h)$$

where $b = a + h$ and $0 < \theta < 1$. Or

$$f(x) = f(a) + (x-a)f'(a) + \frac{(x-a)^2}{2!}f''(a) + \cdots + (x-a)^{n-1}\frac{f^{(n-1)}(a)}{(n-1)!} + R_n,$$

where

$$R_n = \frac{f^{(n)}[a + \theta \cdot (x-a)]}{n!}(x-a)^n, 0 < \theta < 1.$$

The above forms are known as Taylor series with the remainder term.

4. Taylor series for a function of two variables

If $\left(h\frac{\partial}{\partial x} + k\frac{\partial}{\partial y}\right) f(x,y) = h\frac{\partial f(x,y)}{\partial x} + k\frac{\partial f(x,y)}{\partial y}$;

$$\left(h\frac{\partial}{\partial x} + k\frac{\partial}{\partial y}\right)^2 f(x,y) = h^2\frac{\partial^2 f(x,y)}{\partial x^2} + 2hk\frac{\partial^2 f(x,y)}{\partial x \partial y} + k^2\frac{\partial^2 f(x,y)}{\partial y^2}$$

etc., and if $\left(h\frac{\partial}{\partial x} + k\frac{\partial}{\partial y}\right)^n f(x,y)\Big|_{x=a}^{y=b}$ where the bar and subscripts mean that after differentiation we are to replace x by a and y by b, then

$$f(a+h, b+k) = f(a,b) + \left(h\frac{\partial}{\partial x} + k\frac{\partial}{\partial y}\right) f(x,y)\Big|_{x=a}^{y=b} + \cdots + \frac{1}{n!}\left(h\frac{\partial}{\partial x} + k\frac{\partial}{\partial y}\right)^n f(x,y)\Big|_{x=a}^{y=b} + \cdots$$

Maclaurin Series

$$f(x) = f(0) + xf'(0) + \frac{x^2}{2!}f''(0) + \frac{x^3}{3!}f'''(0) + \cdots + x^{n-1}\frac{f^{(n-1)}(0)}{(n-1)!} + R_n,$$

where

$$R_n = \frac{x^n f^{(n)}(\theta x)}{n!}, \quad 0 < \theta < 1.$$

Exponential Series

$$e = 1 + \frac{1}{1!} + \frac{1}{2!} + \frac{1}{3!} + \frac{1}{4!} + \cdots$$

$$e^x = 1 + x + \frac{x^2}{2!} + \frac{x^3}{3!} + \frac{x^4}{4!} + \cdots$$

$$a^x = 1 + x\log_e a + \frac{(x\log_e a)^2}{2!} + \frac{(x\log_e a)^3}{3!} + \cdots$$

$$e^x = e^a\left[1 + (x-a) + \frac{(x-a)^2}{2!} + \frac{(x-a)^3}{3!} + \cdots\right]$$

Logarithmic Series

$$\log_e x = \tfrac{x-1}{x} + \tfrac{1}{2}\left(\tfrac{x-1}{x}\right)^2 + \tfrac{1}{3}\left(\tfrac{x-1}{x}\right)^3 + \cdots \qquad (x > \tfrac{1}{2})$$

$$\log_e x = (x-1) - \tfrac{1}{2}(x-1)^2 + \tfrac{1}{3}(x-1)^3 - \cdots \qquad (2 \geq x > 0)$$

$$\log_e x = 2\left[\tfrac{x-1}{x+1} + \tfrac{1}{3}\left(\tfrac{x-1}{x+1}\right)^3 + \tfrac{1}{5}\left(\tfrac{x-1}{x+1}\right)^5 + \cdots\right] \qquad (x > 0)$$

$$\log_e(1+x) = x - \tfrac{1}{2}x^2 + \tfrac{1}{3}x^3 - \tfrac{1}{4}x^4 + \cdots \qquad (-1 < x \leq 1)$$

$$\log_e(n+1) - \log_e(n-1) = 2\left[\tfrac{1}{n} + \tfrac{1}{3n^3} + \tfrac{1}{5n^5} + \cdots\right]$$

$$\log_e(a+x) = \log_e a + 2\left[\tfrac{x}{2a+x} + \tfrac{1}{3}\left(\tfrac{x}{2a+x}\right)^3 + \tfrac{1}{5}\left(\tfrac{x}{2a+x}\right)^5 + \cdots\right] \qquad (a > 0, -a < x < +\infty)$$

$$\log_e \tfrac{1+x}{1-x} = 2\left[x + \tfrac{x^3}{3} + \tfrac{x^5}{5} + \cdots + \tfrac{x^{2n-1}}{2n-1} + \cdots\right] \qquad -1 < x < 1$$

$$\log_e x = \log_e a + \tfrac{(x-a)}{a} - \tfrac{(x-a)^2}{2a^2} + \tfrac{(x-a)^3}{3a^3} - + \cdots \qquad 0 < x \leq 2a$$

Trigonometric Series

$$\sin x = x - \tfrac{x^3}{3!} + \tfrac{x^5}{5!} - \tfrac{x^7}{7!} + \cdots \quad \text{(all real values of } x\text{)}$$

$$\cos x = 1 - \tfrac{x^2}{2!} + \tfrac{x^4}{4!} - \tfrac{x^6}{6!} + \cdots \quad \text{(all real values of } x\text{)}$$

$$\tan x = x + \tfrac{x^3}{3} + \tfrac{2x^5}{15} + \tfrac{17x^7}{315} + \tfrac{62x^9}{2835} + \cdots + \tfrac{(-1)^{n-1}2^{2n}(2^{2n}-1)B_{2n}}{(2n)!}x^{2n-1} + \cdots,$$
$$\left[x^2 < \tfrac{\pi^2}{4} \quad \text{and } B_n \text{ represents the } n^{\text{th}} \text{ Bernoulli number}\right]$$

$$\cot x = \tfrac{1}{x} - \tfrac{x}{3} - \tfrac{x^3}{45} - \tfrac{2x^5}{945} - \tfrac{x^7}{4725} - \cdots - \tfrac{(-1)^{n+1}2^{2n}}{(2n)!}B_{2n}x^{2n-1} - \cdots,$$
$$\left[x^2 < \pi^2 \quad \text{and } B_n \text{ represents the } n^{\text{th}} \text{ Bernoulli number}\right]$$

$$\sec x = 1 + \tfrac{x^2}{2} + \tfrac{5}{24}x^4 + \tfrac{61}{720}x^6 + \tfrac{277}{8064}x^8 + \cdots + \tfrac{(-1)^n}{(2n)!}E_{2n}x^{2n} + \cdots,$$
$$\left[x^2 < \tfrac{\pi^2}{4} \quad \text{and } E_n \text{ represents the } n^{\text{th}} \text{ Euler number}\right]$$

$$\csc x = \tfrac{1}{x} + \tfrac{x}{6} + \tfrac{7}{360}x^3 + \tfrac{31}{15{,}120}x^5 + \tfrac{127}{604{,}800}x^7 + \cdots + \tfrac{(-1)^{n+1}2(2^{2n-1}-1)}{(2n)!}B_{2n}x^{2n-1} + \cdots,$$
$$\left[x^2 < \pi^2 \quad \text{and } B_n \text{ represents the } n^{\text{th}} \text{ Bernoulli number}\right]$$

$$\sin x = x\left(1 - \tfrac{x^2}{\pi^2}\right)\left(1 - \tfrac{x^2}{2^2\pi^2}\right)\left(1 - \tfrac{x^2}{3^2\pi^2}\right)\cdots \qquad (x^2 < \infty)$$

[2pt] $$\cos x = \left(1 - \tfrac{4x^2}{\pi^2}\right)\left(1 - \tfrac{4x^2}{3^2\pi^2}\right)\left(1 - \tfrac{4x^2}{5^2\pi^2}\right)\cdots \qquad (x^2 < \infty)$$

[2pt] $$\sin^{-1} x = x + \tfrac{x^3}{2\cdot 3} + \tfrac{1\cdot 3}{2\cdot 4\cdot 5}x^5 + \tfrac{1\cdot 3\cdot 5}{2\cdot 4\cdot 6\cdot 7}x^7 + \cdots \qquad \left(x^2 < 1, -\tfrac{\pi}{2} < \sin^{-1} x < \tfrac{\pi}{2}\right)$$

[2pt] $$\cos^{-1} x = \tfrac{\pi}{2} - \left(x + \tfrac{x^3}{2\cdot 3} + \tfrac{1\cdot 3}{2\cdot 4\cdot 5}x^5 + \tfrac{1\cdot 3\cdot 5x^7}{2\cdot 4\cdot 6\cdot 7} + \cdots\right) \qquad (x^2 < 1,\ 0 < \cos^{-1} x < \pi)$$

[2pt] $$\tan^{-1} x = x - \tfrac{x^3}{3} + \tfrac{x^5}{5} - \tfrac{x^7}{7} + \cdots \qquad (x^2 < 1)$$

[2pt] $$\tan^{-1} x = \tfrac{\pi}{2} - \tfrac{1}{x} + \tfrac{1}{3x^3} - \tfrac{1}{5x^5} + \tfrac{1}{7x^7} - \cdots \qquad (x > 1)$$

[2pt] $$\tan^{-1} x = -\tfrac{\pi}{2} - \tfrac{1}{x} + \tfrac{1}{3x^3} - \tfrac{1}{5x^5} + \tfrac{1}{7x^7} - \cdots \qquad (x < -1)$$

[2pt] $$\cot^{-1} x = \tfrac{\pi}{2} - x + \tfrac{x^3}{3} - \tfrac{x^5}{5} + \tfrac{x^7}{7} - \cdots \qquad (x^2 < 1)$$

$$\log_e \sin x = \log_e x - \frac{x^2}{6} - \frac{x^4}{180} - \frac{x^6}{2835} - \cdots \qquad (x^2 < \pi^2)$$

$$\log_e \cos x = -\frac{x^2}{2} - \frac{x^4}{12} - \frac{x^6}{45} - \frac{17x^8}{2520} - \cdots \qquad \left(x^2 < \frac{\pi^2}{4}\right)$$

$$\log_e \tan x = \log_e x + \frac{x^2}{3} + \frac{7x^4}{90} + \frac{62x^6}{2835} + \cdots \qquad \left(x^2 < \frac{\pi^2}{4}\right)$$

$$e^{\sin x} = 1 + x + \frac{x^2}{2!} - \frac{3x^4}{4!} - \frac{8x^5}{5!} - \frac{3x^6}{6!} + \frac{56x^7}{7!} + \cdots$$

$$e^{\cos x} = e\left(1 - \frac{x^2}{2!} + \frac{4x^4}{4!} - \frac{31x^6}{6!} + \cdots\right)$$

$$e^{\tan x} = 1 + x + \frac{x^2}{2!} + \frac{3x^3}{3!} + \frac{9x^4}{4!} + \frac{37x^5}{5!} + \cdots \qquad \left(x^2 < \frac{\pi^2}{4}\right)$$

$$\sin x = \sin a + (x-a)\cos a - \frac{(x-a)^2}{2!}\sin a - \frac{(x-a)^3}{3!}\cos a + \frac{(x-a)^4}{4!}\sin a + \cdots$$

VECTOR ANALYSIS

Definitions

Any quantity that is completely determined by its magnitude is called a *scalar*. Examples of such are mass, density, temperature, etc. Any quantity that is completely determined by its magnitude and direction is called a *vector*. Examples of such are velocity, acceleration, force, etc. A vector quantity is represented by a directed line segment, the length of which represents the magnitude of the vector. A vector quantity is usually represented by a boldfaced letter such as $\mathbf{V}$. Two vectors $\mathbf{V}_1$ and $\mathbf{V}_2$ are equal to one another if they have equal magnitudes and are acting in the same directions. A negative vector, written as $-\mathbf{V}$, is one that acts in the opposite direction to $\mathbf{V}$, but is of equal magnitude to it. If we represent the magnitude of $\mathbf{V}$ by v, we write $|\mathbf{V}| = v$. A vector parallel to $\mathbf{V}$, but equal to the reciprocal of its magnitude is written as $\mathbf{V}^{-1}$ or as $1/\mathbf{V}$.

The *unit vector* $\mathbf{V}/v$ (when $v \neq 0$) is that vector which has the same direction as $\mathbf{V}$, but has a magnitude of unity (sometimes represented as $\mathbf{V}_0$ or $\hat{\mathbf{v}}$).

Vector Algebra

The vector sum of $\mathbf{V}_1$ and $\mathbf{V}_2$ is represented by $\mathbf{V}_1+\mathbf{V}_2$. The vector sum of $\mathbf{V}_1$ and $-\mathbf{V}_2$, or the difference of the vector $\mathbf{V}_2$ from $\mathbf{V}_1$ is represented by $\mathbf{V}_1 - \mathbf{V}_2$.

If r is a scalar, then $r\mathbf{V}=\mathbf{V}r$, and represents a vector r times the magnitude of $\mathbf{V}$, in the same direction as $\mathbf{V}$ if r is positive, and in the opposite direction if r is negative. If r and s are scalars, $\mathbf{V}_1$, $\mathbf{V}_2$, $\mathbf{V}_3$, vectors, then the following rules of scalars and vectors hold:

$$\mathbf{V}_1 + \mathbf{V}_2 = \mathbf{V}_2 + \mathbf{V}_1$$
$$(r+s)\mathbf{V}_1 = r\mathbf{V}_1 + s\mathbf{V}_1; \quad r(\mathbf{V}_1 + \mathbf{V}_2) = r\mathbf{V}_1 + r\mathbf{V}_2$$
$$\mathbf{V}_1 + (\mathbf{V}_2 + \mathbf{V}_3) = (\mathbf{V}_1 + \mathbf{V}_2) + \mathbf{V}_3 = \mathbf{V}_1 + \mathbf{V}_2 + \mathbf{V}_3$$

Vectors in Space

A plane is described by two distinct vectors $\mathbf{V}_1$ and $\mathbf{V}_2$. Should these vectors not intersect each other, then one is displaced parallel to itself until they do (Figure 1). Any other vector $\mathbf{V}$ lying in this plane is given by

$$\mathbf{V} = r\mathbf{V}_1 + s\mathbf{V}_2$$

A *position vector* specifies the position in space of a point relative to a fixed origin. If therefore $\mathbf{V}_1$ and $\mathbf{V}_2$ are the position vectors of the points A and B, relative to the origin O, then any point P on the line AB has a position vector $\mathbf{V}$ given by

$$\mathbf{V} = r\mathbf{V}_1 + (1-r)\mathbf{V}_2$$

The scalar "r" can be taken as the metric representation of P since $r = 0$ implies $P = B$ and $r = 1$ implies $P = A$ (Figure 2). If P divides the line AB in the ratio $r{:}s$ then

$$\mathbf{V} = \left(\frac{r}{r+s}\right)\mathbf{V}_1 + \left(\frac{s}{r+s}\right)\mathbf{V}_2$$

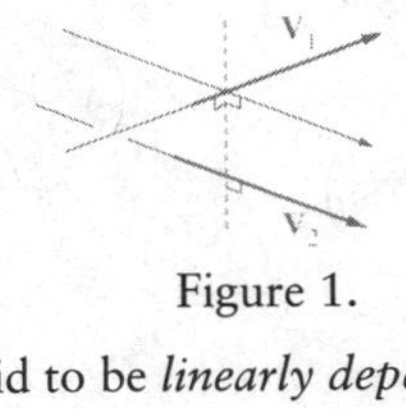

Figure 1.

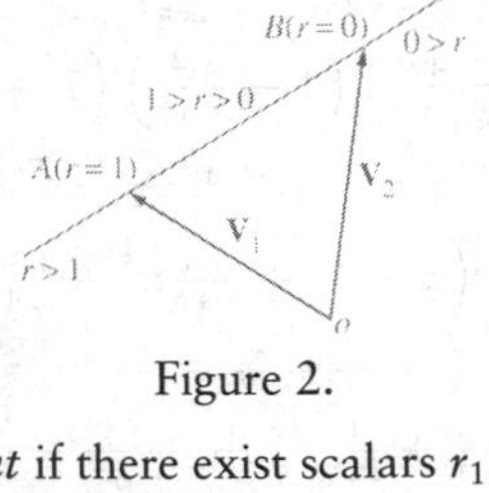

Figure 2.

The vectors $\mathbf{V}_1$, $\mathbf{V}_2$, $\mathbf{V}_3, \ldots, \mathbf{V}_n$ are said to be *linearly dependent* if there exist scalars $r_1, r_2, r_3, \ldots, r_n$, not all zero, such that

$$r_1\mathbf{V}_1 + r_2\mathbf{V}_2 + \cdots + r_n\mathbf{V}_n = 0$$

A vector $\mathbf{V}$ is linearly dependent upon the set of vectors $\mathbf{V}_1, \mathbf{V}_2, \mathbf{V}_3, \ldots, \mathbf{V}_n$ if

$$\mathbf{V} = r_1\mathbf{V}_1 + r_2\mathbf{V}_2 + r_3\mathbf{V}_3 + \cdots + r_n\mathbf{V}_n$$

Three vectors are linearly dependent if and only if they are co-planar.

All points in space can be uniquely determined by linear dependence upon three *base vectors*, i.e., three vectors any one of which is linearly independent of the other two. The simplest set of base vectors is the unit vectors along the coordinate Ox, Oy and Oz axes. These are usually designated by **i, j** and **k**, respectively.

If **V** is a vector in space, and a, b and c are the respective magnitudes of the projections of the vector along the axes then

$$\mathbf{V} = a\mathbf{i} + b\mathbf{j} + c\mathbf{k}$$

and

$$v = \sqrt{a^2 + b^2 + c^2}$$

and the direction cosines of **V** are

$$\cos\alpha = a/v, \quad \cos\beta = b/v, \quad \cos\gamma = c/v.$$

The law of addition yields

$$\mathbf{V}_1 + \mathbf{V}_2 = (a_1 + a_2)\mathbf{i} + (b_1 + b_2)\mathbf{j} + (c_1 + c_2)\mathbf{k}$$

The Scalar, Dot, or Inner Product of Two Vectors

This product is represented as $\mathbf{V}_1 \cdot \mathbf{V}_2$ and is defined to be equal to $v_1 v_2 \cos\theta$, where θ is the angle from $\mathbf{V}_1$ to $\mathbf{V}_2$, i.e.,

$$\mathbf{V}_1 \cdot \mathbf{V}_2 = v_1 v_2 \cos\theta$$

The following rules apply for this product:

$$\mathbf{V}_1 \cdot \mathbf{V}_2 = a_1 a_2 + b_1 b_2 + c_1 c_2 = \mathbf{V}_2 \cdot \mathbf{V}_1$$

It should be noted that this verifies that scalar multiplication is commutative.

$$(\mathbf{V}_1 + \mathbf{V}_2) \cdot \mathbf{V}_3 = \mathbf{V}_1 \cdot \mathbf{V}_3 + \mathbf{V}_2 \cdot \mathbf{V}_3$$
$$\mathbf{V}_1 \cdot (\mathbf{V}_2 + \mathbf{V}_3) = \mathbf{V}_1 \cdot \mathbf{V}_2 + \mathbf{V}_1 \cdot \mathbf{V}_3$$

If $\mathbf{V}_1$ is perpendicular to $\mathbf{V}_2$ then $\mathbf{V}_1 \cdot \mathbf{V}_2 = 0$, and if $\mathbf{V}_1$ is parallel to $\mathbf{V}_2$, then $\mathbf{V}_1 \cdot \mathbf{V}_2 = v_1 v_2 = r w_1^2$. In particular

$$\mathbf{i} \cdot \mathbf{i} = \mathbf{j} \cdot \mathbf{j} = \mathbf{k} \cdot \mathbf{k} = 1,$$

and

$$\mathbf{i} \cdot \mathbf{j} = \mathbf{j} \cdot \mathbf{k} = \mathbf{k} \cdot \mathbf{i} = 0$$

The Vector or Cross Product of Two Vectors

This product is represented as $\mathbf{V}_1 \times \mathbf{V}_2$ and is defined to be equal to $v_1 v_2 (\sin\theta)\mathbf{1}$, where θ is the angle from $\mathbf{V}_1$ to $\mathbf{V}_2$ and **1** is a unit vector perpendicular to the plane of $\mathbf{V}_1$ and $\mathbf{V}_2$ and so directed that a right-handed screw driven in the direction of **1** would carry $\mathbf{V}_1$ into $\mathbf{V}_2$, i.e.,

$$\mathbf{V}_1 \times \mathbf{V}_2 = v_1 v_2 (\sin\theta)\mathbf{1}$$

and $\tan\theta = \dfrac{|\mathbf{V}_1 \times \mathbf{V}_2|}{\mathbf{V}_1 \cdot \mathbf{V}_2}$

The following rules apply for vector products:

$$\begin{aligned}
\mathbf{V}_1 \times \mathbf{V}_2 &= -\mathbf{V}_2 \times \mathbf{V}_1 \\
\mathbf{V}_1 \times (\mathbf{V}_2 + \mathbf{V}_3) &= \mathbf{V}_1 \times \mathbf{V}_2 + \mathbf{V}_1 \times \mathbf{V}_3 \\
(\mathbf{V}_1 + \mathbf{V}_2) \times \mathbf{V}_3 &= \mathbf{V}_1 \times \mathbf{V}_3 + \mathbf{V}_2 \times \mathbf{V}_3 \\
\mathbf{V}_1 \times (\mathbf{V}_2 \times \mathbf{V}_3) &= \mathbf{V}_2(\mathbf{V}_3 \cdot \mathbf{V}_1) - \mathbf{V}_3(\mathbf{V}_1 \cdot \mathbf{V}_2) \\
\mathbf{i} \times \mathbf{i} &= \mathbf{j} \times \mathbf{j} = \mathbf{k} \times \mathbf{k} = 0 \quad \text{(the zero vector)} \\
\mathbf{i} \times \mathbf{j} = \mathbf{k}, \quad \mathbf{j} \times \mathbf{k} = \mathbf{i}, \quad \mathbf{k} \times \mathbf{i} &= \mathbf{j}
\end{aligned}$$

If $\mathbf{V}_1 = a_1\mathbf{i} + b_1\mathbf{j} + c_1\mathbf{k}$, $\mathbf{V}_2 = a_2\mathbf{i} + b_2\mathbf{j} + c_2\mathbf{k}$, and $\mathbf{V}_3 = a_3\mathbf{i} + b_3\mathbf{j} + c_3\mathbf{k}$, then

$$\mathbf{V}_1 \times \mathbf{V}_2 = \begin{vmatrix} \mathbf{i} & \mathbf{j} & \mathbf{k} \\ a_1 & b_1 & c_1 \\ a_2 & b_2 & c_2 \end{vmatrix} = (b_1 c_2 - b_2 c_1)\mathbf{i} + (c_1 a_2 - c_2 a_1)\mathbf{j} + (a_1 b_2 - a_2 b_1)\mathbf{k}$$

It should be noted that, since $\mathbf{V}_1 \times \mathbf{V}_2 = -\mathbf{V}_2 \times \mathbf{V}_1$, the vector product is not commutative.

Scalar Triple Product

There is only one possible interpretation of the expression $\mathbf{V}_1 \cdot \mathbf{V}_2 \times \mathbf{V}_3$ and that is $\mathbf{V}_1 \cdot (\mathbf{V}_2 \times \mathbf{V}_3)$ which is obviously a scalar. Further

$$\mathbf{V}_1 \cdot (\mathbf{V}_2 \times \mathbf{V}_3) = (\mathbf{V}_1 \times \mathbf{V}_2) \cdot \mathbf{V}_3 = \mathbf{V}_2 \cdot (\mathbf{V}_3 \times \mathbf{V}_1)$$

$$= \begin{vmatrix} a_1 & b_1 & c_1 \\ a_2 & b_2 & c_2 \\ a_3 & b_3 & c_3 \end{vmatrix}$$

$$= r_1 r_2 r_3 \cos\phi \sin\theta,$$

Where θ is the angle between $\mathbf{V}_2$ and $\mathbf{V}_3$ and ϕ is the angle between $\mathbf{V}_1$ and the normal to the plane of $\mathbf{V}_2$ and $\mathbf{V}_3$.

This product is called the *scalar triple product* and is written as $[\mathbf{V}_1\mathbf{V}_2\mathbf{V}_3]$.

The determinant indicates that it can be considered as the volume of the parallelepiped whose three determining edges are $\mathbf{V}_1$, $\mathbf{V}_2$ and $\mathbf{V}_3$.

It also follows that cyclic permutation of the subscripts does not change the value of the scalar triple product so that

$$[\mathbf{V}_1\mathbf{V}_2\mathbf{V}_3] = [\mathbf{V}_2\mathbf{V}_3\mathbf{V}_1] = [\mathbf{V}_3\mathbf{V}_1\mathbf{V}_2]$$

$$\text{but} \quad [\mathbf{V}_1\mathbf{V}_2\mathbf{V}_3] = -[\mathbf{V}_2\mathbf{V}_1\mathbf{V}_3] \quad \text{etc.} \qquad \text{and} \qquad [\mathbf{V}_1\mathbf{V}_1\mathbf{V}_2] \equiv 0 \quad \text{etc.}$$

Given three non-coplanar reference vectors $\mathbf{V}_1$, $\mathbf{V}_2$ and $\mathbf{V}_3$, the *reciprocal system* is given by $\mathbf{V}_1^*$, $\mathbf{V}_2^*$ and $\mathbf{V}_3^*$, where

$$1 = v_1 v_1^* = v_2 v_2^* = v_3 v_3^*$$

$$0 = v_1 v_2^* = v_1 v_3^* = v_2 v_1^* \quad \text{etc.}$$

$$\mathbf{V}_1^* = \frac{\mathbf{V}_2 \times \mathbf{V}_3}{[\mathbf{V}_1\mathbf{V}_2\mathbf{V}_3]}, \quad \mathbf{V}_2^* = \frac{\mathbf{V}_3 \times \mathbf{V}_1}{[\mathbf{V}_1\mathbf{V}_2\mathbf{V}_3]}, \quad \mathbf{V}_3^* = \frac{\mathbf{V}_1 \times \mathbf{V}_2}{[\mathbf{V}_1\mathbf{V}_2\mathbf{V}_3]}$$

The system **i**, **j**, **k** is its own reciprocal.

Vector Triple Product

The product $\mathbf{V}_1 \times (\mathbf{V}_2 \times \mathbf{V}_3)$ defines the *vector triple product*. Obviously, in this case, the brackets are vital to the definition.

$$\mathbf{V}_1 \times (\mathbf{V}_2 \times \mathbf{V}_3) = (\mathbf{V}_1 \cdot \mathbf{V}_3)\mathbf{V}_2 - (\mathbf{V}_1 \cdot \mathbf{V}_2)\mathbf{V}_3$$

$$= \begin{vmatrix} \mathbf{i} & \mathbf{j} & \mathbf{k} \\ a_1 & b_1 & c_1 \\ \begin{vmatrix} b_2 & c_2 \\ b_3 & c_3 \end{vmatrix} & \begin{vmatrix} c_2 & a_2 \\ c_3 & a_3 \end{vmatrix} & \begin{vmatrix} a_2 & b_2 \\ a_3 & b_3 \end{vmatrix} \end{vmatrix}$$

i.e., it is a vector, perpendicular to $\mathbf{V}_1$, lying in the plane of $\mathbf{V}_2$, $\mathbf{V}_3$. Similarly

$$(\mathbf{V}_1 \times \mathbf{V}_2) \times \mathbf{V}_3 = \begin{vmatrix} i & j & k \\ \begin{vmatrix} b_1 & c_1 \\ b_2 & c_2 \end{vmatrix} & \begin{vmatrix} c_1 & a_1 \\ c_2 & a_2 \end{vmatrix} & \begin{vmatrix} a_1 & b_1 \\ a_2 & b_2 \end{vmatrix} \\ a_3 & b_3 & c_3 \end{vmatrix}$$

$$\mathbf{V}_1 \times (\mathbf{V}_2 \times \mathbf{V}_3) + \mathbf{V}_2 \times (\mathbf{V}_3 \times \mathbf{V}_1) + \mathbf{V}_3 \times (\mathbf{V}_1 \times \mathbf{V}_2) \equiv 0$$

If $\mathbf{V}_1 \times (\mathbf{V}_2 \times \mathbf{V}_3) = (\mathbf{V}_1 \times \mathbf{V}_2) \times \mathbf{V}_3$, then $\mathbf{V}_1$, $\mathbf{V}_2$, $\mathbf{V}_3$ form an *orthogonal set*. Thus **i**, **j**, **k** form an orthogonal set.

Geometry of the Plane, Straight Line and Sphere

The position vectors of the fixed points A, B, C, D relative to O are $\mathbf{V}_1$, $\mathbf{V}_2$, $\mathbf{V}_3$, $\mathbf{V}_4$ and the position vector of the variable point P is $\mathbf{V}$.

The vector form of the equation of the straight line through A parallel to $\mathbf{V}_2$ is

$$\mathbf{V} = \mathbf{V}_1 + r\mathbf{V}_2$$

$$\text{or} \quad (\mathbf{V} - \mathbf{V}_1) = r\mathbf{V}_2$$

$$\text{or} \quad (\mathbf{V} - \mathbf{V}_1) \times \mathbf{V}_2 = 0$$

while that of the plane through A perpendicular to $\mathbf{V}_2$ is

$$(\mathbf{V} - \mathbf{V}_1) \cdot \mathbf{V}_2 = 0$$

The equation of the line AB is

$$\mathbf{V} = r\mathbf{V}_1 + (1 - r)\mathbf{V}_2$$

and those of the bisectors of the angles between $\mathbf{V}_1$ and $\mathbf{V}_2$ are

$$\mathbf{V} = r\left(\frac{\mathbf{V}_1}{v_1} \pm \frac{\mathbf{V}_2}{v_2}\right) \qquad \text{or}$$

$$\mathbf{V} = r(\hat{\mathbf{v}}_1 \pm \hat{\mathbf{v}}_2)$$

The perpendicular from C to the line through A parallel to $\mathbf{V}_2$ has as its equation

$$\mathbf{V} = \mathbf{V}_1 - \mathbf{V}_3 - \hat{\mathbf{v}}_2 \cdot (\mathbf{V}_1 - \mathbf{V}_3)\hat{\mathbf{v}}_2.$$

The condition for the intersection of the two lines, $\mathbf{V} = \mathbf{V}_1 + r\mathbf{V}_3$ and $\mathbf{V} = \mathbf{V}_2 + s\mathbf{V}_4$, is

$$[(\mathbf{V}_1 - \mathbf{V}_2)\mathbf{V}_3\mathbf{V}_4] = 0.$$

The common perpendicular to the above two lines is the line of intersection of the two planes

$$[(\mathbf{V} - \mathbf{V}_1)\mathbf{V}_3(\mathbf{V}_3 \times \mathbf{V}_4)] = 0 \quad \text{and} \quad [(\mathbf{V} - \mathbf{V}_2)\mathbf{V}_4(\mathbf{V}_3 \times \mathbf{V}_4)] = 0$$

and the length of this perpendicular is

$$\frac{[(\mathbf{V}_1 - \mathbf{V}_2)\mathbf{V}_3\mathbf{V}_4]}{|\mathbf{V}_3 \times \mathbf{V}_4|}.$$

The equation of the line perpendicular to the plane ABC is

$$\mathbf{V} = \mathbf{V}_1 \times \mathbf{V}_2 + \mathbf{V}_2 \times \mathbf{V}_3 + \mathbf{V}_3 \times \mathbf{V}_1$$

and the distance of the plane from the origin is

$$\frac{[\mathbf{V}_1\mathbf{V}_2\mathbf{V}_3]}{|(\mathbf{V}_2 - \mathbf{V}_1) \times (\mathbf{V}_3 - \mathbf{V}_1)|}.$$

In general the vector equation

$$\mathbf{V} \cdot \mathbf{V}_2 = r$$

defines the plane which is perpendicular to $\mathbf{V}_2$, and the perpendicular distance from A to this plane is

$$\frac{r - \mathbf{V}_1 \cdot \mathbf{V}_2}{v_2}.$$

The distance from A, measured along a line parallel to $\mathbf{V}_3$, is

$$\frac{r - \mathbf{V}_1 \cdot \mathbf{V}_2}{\mathbf{V}_2 \cdot \hat{\mathbf{v}}_3} \quad \text{or} \quad \frac{r - \mathbf{V}_1 \cdot \mathbf{V}_2}{v_2 \cos\theta}$$

where θ is the angle between $\mathbf{V}_2$ and $\mathbf{V}_3$. (If this plane contains the point C then $r = \mathbf{V}_3 \cdot \mathbf{V}_2$ and if it passes through the origin, then $r = 0$.) Given two planes

$$\mathbf{V} \cdot \mathbf{V}_1 = r$$
$$\mathbf{V} \cdot \mathbf{V}_2 = s$$

then any plane through the line of intersection of these two planes is given by

$$\mathbf{V} \cdot (\mathbf{V}_1 + \lambda\mathbf{V}_2) = r + \lambda s$$

where λ is a scalar parameter. In particular $\lambda = \pm v_1/v_2$ yields the equation of the two planes bisecting the angle between the given planes.

The plane through A parallel to the plane of $\mathbf{V}_2$, $\mathbf{V}_3$ is

$$\mathbf{V} = \mathbf{V}_1 + r\mathbf{V}_2 + s\mathbf{V}_3$$
$$\text{or} \quad (\mathbf{V} - \mathbf{V}_1) \cdot \mathbf{V}_2 \times \mathbf{V}_3 = 0$$
$$\text{or} \quad [\mathbf{V}\mathbf{V}_2\mathbf{V}_3] - [\mathbf{V}_1\mathbf{V}_2\mathbf{V}_3] = 0$$

so that the expansion in rectangular Cartesian coordinates yields (where $\mathbf{V} \equiv x\mathbf{i} + y\mathbf{j} + z\mathbf{k}$):

$$\begin{vmatrix} (x - a_1) & (y - b_1) & (z - c_1) \\ a_2 & b_2 & c_2 \\ a_3 & b_3 & c_3 \end{vmatrix} = 0$$

which is obviously the usual linear equation in x, y, and z.

The plane through AB parallel to $\mathbf{V}_3$ is given by

$$[(\mathbf{V} - \mathbf{V}_1)(\mathbf{V}_1 - \mathbf{V}_2)\mathbf{V}_3] = 0$$
$$\text{or} \quad [\mathbf{V}\mathbf{V}_2\mathbf{V}_3] - [\mathbf{V}\mathbf{V}_1\mathbf{V}_3] - [\mathbf{V}_1\mathbf{V}_2\mathbf{V}_3] = 0$$

The plane through the three points A, B and C is

$$\mathbf{V} = \mathbf{V}_1 + s(\mathbf{V}_2 - \mathbf{V}_1) + t(\mathbf{V}_3 - \mathbf{V}_1)$$
$$\text{or} \quad \mathbf{V} = r\mathbf{V}_1 + s\mathbf{V}_2 + t\mathbf{V}_3 \qquad (r + s + t \equiv 1)$$
$$\text{or} \quad [(\mathbf{V} - \mathbf{V}_1)(\mathbf{V}_1 - \mathbf{V}_2)(\mathbf{V}_2 - \mathbf{V}_3)] = 0$$
$$\text{or} \quad [\mathbf{V}\mathbf{V}_1\mathbf{V}_2] + [\mathbf{V}\mathbf{V}_2\mathbf{V}_3] + [\mathbf{V}\mathbf{V}_3\mathbf{V}_1] - [\mathbf{V}_1\mathbf{V}_2\mathbf{V}_3] = 0$$

For four points A, B, C, D to be coplanar, then

$$r\mathbf{V}_1 + s\mathbf{V}_2 + t\mathbf{V}_3 + u\mathbf{V}_4 \equiv 0 \equiv r + s + t + u$$

The following formulas relate to a sphere when the vectors are taken to lie in three-dimensional space and to a circle when the space is two dimensional. For a circle in three dimensions, take the intersection of the sphere with a plane.

The equation of a sphere with center O and radius OA is

$$\mathbf{V} \cdot \mathbf{V} = v_1^2 \quad (\text{not} \mathbf{V} = \mathbf{V}_1)$$
$$\text{or} \quad (\mathbf{V} - \mathbf{V}_1) \cdot (\mathbf{V} + \mathbf{V}_1) = 0$$

while that of a sphere with center B radius v_1 is

$$(\mathbf{V} - \mathbf{V}_2) \cdot (\mathbf{V} - \mathbf{V}_2) = v_1^2$$
$$\text{or} \quad \mathbf{V} \cdot (\mathbf{V} - 2\mathbf{V}_2) = v_1^2 - v_2^2$$

If the above sphere passes through the origin, then

$$\mathbf{V} \cdot (\mathbf{V} - 2\mathbf{V}_2) = 0$$

Note that in two-dimensional polar coordinates this is simply

$$r = 2a \cdot \cos\theta$$

while in three-dimensional Cartesian coordinates it is

$$x^2 + y^2 + z^2 - 2(a_2x + b_2y + c_2x) = 0.$$

The equation of a sphere having the points A and B as the extremities of a diameter is

$$(\mathbf{V} - \mathbf{V}_1) \cdot (\mathbf{V} - \mathbf{V}_2) = 0.$$

The square of the length of the tangent from C to the sphere with center B and radius v_1 is given by

$$(\mathbf{V}_3 - \mathbf{V}_2) \cdot (\mathbf{V}_3 - \mathbf{V}_2) = v_1^2$$

The condition that the plane $\mathbf{V} \cdot \mathbf{V}_3 = s$ is tangential to the sphere $(\mathbf{V} - \mathbf{V}_2) \cdot (\mathbf{V} - \mathbf{V}_2) = v_1^2$ is

$$(s - \mathbf{V}_3 \cdot \mathbf{V}_2) \cdot (s - \mathbf{V}_3 \cdot \mathbf{V}_2) = v_1^2 v_3^2.$$

The equation of the tangent plane at D, on the surface of sphere $(\mathbf{V} - \mathbf{V}_2) \cdot (\mathbf{V} - \mathbf{V}_2) = v_1^2$, is

$$(\mathbf{V} - \mathbf{V}_4) \cdot (\mathbf{V}_4 - \mathbf{V}_2) = 0$$
$$\text{or} \quad \mathbf{V} \cdot \mathbf{V}_4 - \mathbf{V}_2 \cdot (\mathbf{V} + \mathbf{V}_4) = v_1^2 - v_2^2$$

The condition that the two circles $(\mathbf{V} - \mathbf{V}_2) \cdot (\mathbf{V} - \mathbf{V}_2) = v_1^2$ and $(\mathbf{V} - \mathbf{V}_4) \cdot (\mathbf{V} - \mathbf{V}_4) = v_3^2$ intersect orthogonally is clearly

$$(\mathbf{V}_2 - \mathbf{V}_4) \cdot (\mathbf{V}_2 - \mathbf{V}_4) = v_1^2 + v_3^2$$

The polar plane of D with respect to the circle

$$(\mathbf{V} - \mathbf{V}_2) \cdot (\mathbf{V} - \mathbf{V}_2) = v_1^2 \quad \text{is}$$
$$\mathbf{V} \cdot \mathbf{V}_4 - \mathbf{V}_2 \cdot (\mathbf{V} + \mathbf{V}_4) = v_1^2 - v_2^2$$

Any sphere through the intersection of the two spheres $(\mathbf{V} - \mathbf{V}_2) \cdot (\mathbf{V} - \mathbf{V}_2) = v_1^2$ and $(\mathbf{V} - \mathbf{V}_4) \cdot (\mathbf{V} - \mathbf{V}_4) = v_3^2$ is given by

$$(\mathbf{V} - \mathbf{V}_2) \cdot (\mathbf{V} - \mathbf{V}_2) + \lambda(\mathbf{V} - \mathbf{V}_4) \cdot (\mathbf{V} - \mathbf{V}_4) = v_1^2 + \lambda v_3^2$$

while the radical plane of two such spheres is

$$\mathbf{V} \cdot (\mathbf{V}_2 - \mathbf{V}_4) = -\frac{1}{2}(v_1^2 - v_2^2 - v_3^2 + v_4^2)$$

Differentiation of Vectors

If $V_1 = a_1 i + b_1 j + c_1 k$, and $V_2 = a_2 i + b_2 j + c_2 k$, and if V_1 and V_2 are functions of the scalar t, then

$$\frac{d}{dt}(V_1 + V_2 + \cdots) = \frac{dV_1}{dt} + \frac{dV_2}{dt} + \cdots$$

$$\frac{dV_1}{dt} = \frac{da_1}{dt}i + \frac{db_1}{dt}j + \frac{dc_1}{dt}k, \quad \text{etc}$$

$$\frac{d}{dt}(V_1 \cdot V_2) = \frac{dV_1}{dt} \cdot V_2 + V_1 \cdot \frac{dV_2}{dt}$$

$$\frac{d}{dt}(V_1 \times V_2) = \frac{dV_1}{dt} \times V_2 + V_1 \times \frac{dV_2}{dt}$$

$$V \cdot \frac{dV}{dt} = v \cdot \frac{dv}{dt}$$

In particular, if V is a vector of constant length, then the right-hand side of the last equation is identically zero showing that V is perpendicular to its derivative.

The derivatives of the triple products are

$$\frac{d}{dt}[V_1V_2V_3] = \left[\left(\frac{dV_1}{dt}\right)V_2V_3\right] + \left[V_1\left(\frac{dV_2}{dt}\right)V_3\right] + \left[V_1V_2\left(\frac{dV_3}{dt}\right)\right] \quad \text{and}$$

$$\frac{d}{dt}\{V_1 \times (V_2 \times V_3)\} = \left(\frac{dV_1}{dt}\right) \times (V_2 \times V_3) + V_1 \times \left(\left(\frac{dV_2}{dt}\right) \times V_3\right) + V_1 \times \left(V_2 \times \left(\frac{dV_3}{dt}\right)\right)$$

Geometry of Curves in Space

s = the *length of arc*, measured from some fixed point on the curve (Figure 3).

V_1 = the position vector of the point A on the curve.

$V_1 + \delta V_1$ = the position vector of the point P in the neighborhood of A.

$\hat{t}$ = the *unit tangent* to the curve at the point A, measured in the direction of s increasing.

The *normal plane* is that plane which is perpendicular to the unit tangent. The principal normal is defined as the intersection of the normal plane with the plane defined by V_1 and $V_1 + \delta V_1$ in the limit as $\delta V_1 - 0$.

$\hat{n}$ = the *unit normal* (principal) at the point A. The plane defined by $\hat{t}$ and $\hat{n}$ is called the *osculating plane* (alternatively plane of curvature or local plane).

ρ = the radius of curvature at A.

$\delta\theta$ = the angle subtended at the origin by δV_1.

$$\kappa = \frac{d\theta}{ds} = \frac{1}{\rho}$$

$\hat{b}$ = the *unit binormal* i.e., the unit vector which is parallel to $\hat{t} \times \hat{n}$ at the point A

λ = the *torsion* of the curve at A.

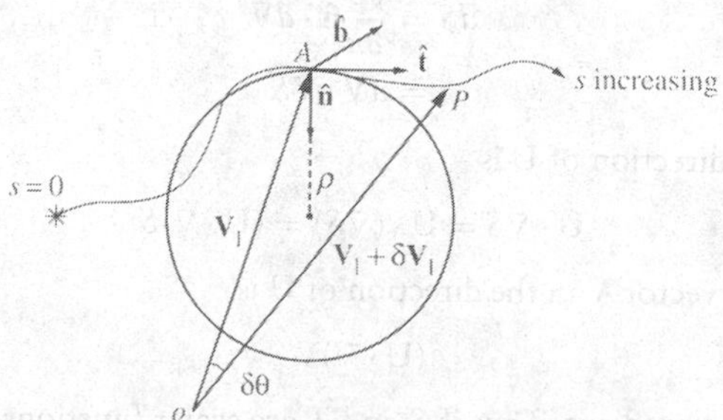

Figure 3.

Frenet's Formulas:

$$\frac{d\hat{t}}{ds} = \kappa\hat{n}$$

$$\frac{d\hat{n}}{ds} = -\kappa\hat{t} + \lambda\hat{b}$$

$$\frac{d\hat{b}}{ds} = -\lambda\hat{n}$$

The following formulas are also applicable:

Unit tangent	$\hat{\mathbf{t}} = \frac{d\mathbf{V}_1}{ds}$
Equation of the tangent	$(\mathbf{V} - \mathbf{V}_1) \times \hat{\mathbf{t}} = 0$ or $\mathbf{V} = \mathbf{V}_1 + q\hat{\mathbf{t}}$
Unit normal	$\hat{\mathbf{n}} = \frac{1 d^2\mathbf{V}_1}{\kappa ds^2}$
Equation of the normal plane	$(\mathbf{V} - \mathbf{V}_1) \cdot \hat{\mathbf{t}} = 0$
Equation of the normal	$(\mathbf{V} - \mathbf{V}_1) \times \hat{\mathbf{n}} = 0$ or $\mathbf{V} = \mathbf{V}_1 + r\hat{\mathbf{n}}$
Unit binormal	$\hat{\mathbf{b}} = \hat{\mathbf{t}} \times \hat{\mathbf{n}}$
Equation of the binormal	$(\mathbf{V} - \mathbf{V}_1) \times \hat{\mathbf{b}} = 0$
	or $\mathbf{V} = \mathbf{V}_1 + u\hat{\mathbf{b}}$
	or $\mathbf{V} = \mathbf{V}_1 + w\frac{d\mathbf{V}_1}{ds} \times \frac{d^2\mathbf{V}_1}{ds^2}$
Equation of the osculating plane	$[(\mathbf{V} - \mathbf{V}_1)\hat{\mathbf{t}}\hat{\mathbf{n}}] = 0$
	or $\left[(\mathbf{V} - \mathbf{V}_1)\left(\frac{d\mathbf{V}_1}{ds}\right)\left(\frac{d^2\mathbf{V}_1}{ds^2}\right)\right] = 0$

Differential Operators—Rectangular Coordinates

$$dS = \frac{\partial S}{\partial x} \cdot dx + \frac{\partial S}{\partial y} \cdot dy + \frac{\partial S}{\partial z} \cdot dz$$

By definition

$$\nabla \equiv \text{del} \equiv \mathbf{i}\frac{\partial}{\partial x} + \mathbf{j}\frac{\partial}{\partial y} + \mathbf{k}\frac{\partial}{\partial z}$$

$$\nabla^2 \equiv \text{Laplacian} \equiv \frac{\partial^2}{\partial x^2} + \frac{\partial^2}{\partial y^2} + \frac{\partial^2}{\partial z^2}$$

If S is a scalar function, then $\nabla S \equiv \text{grad}\, S \equiv \frac{\partial S}{dx}\mathbf{i} + \frac{\partial S}{dy}\mathbf{j} + \frac{\partial S}{dz}\mathbf{k}$.

Grad S defines both the direction and magnitude of the maximum rate of increase of S at any point. Hence the name *gradient* and also its vectorial nature. ∇S is independent of the choice of rectangular coordinates.

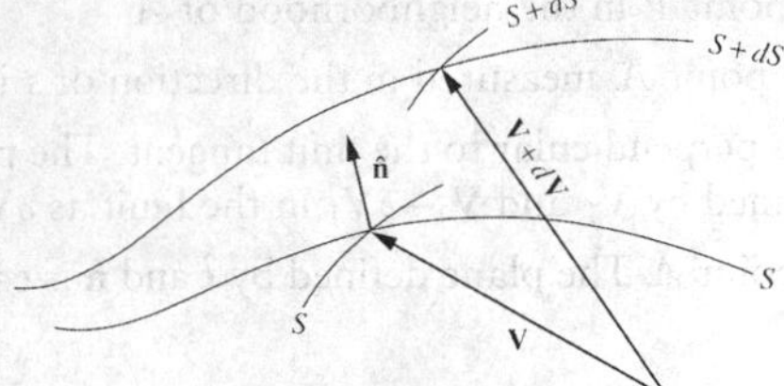

Figure 4.

$$\nabla S = \frac{\partial S}{\partial n}\hat{\mathbf{n}} \tag{4}$$

where $\hat{\mathbf{n}}$ is the unit normal to the surface $S =$ constant, in the direction of S increasing. The total derivative of S at a point having the position vector $\mathbf{V}$ is given by (Figure 4)

$$dS = \frac{\partial S}{\partial n}\hat{\mathbf{n}} \cdot d\mathbf{V}$$
$$= d\mathbf{V} \cdot \nabla S$$

and the directional derivative of S in the direction of $\mathbf{U}$ is

$$\mathbf{U} \cdot \nabla S = \mathbf{U} \cdot (\nabla S) = (\mathbf{U} \cdot \nabla)S$$

Similarly the directional derivative of the vector $\mathbf{V}$ in the direction of $\mathbf{U}$ is

$$(\mathbf{U} \cdot \nabla)\mathbf{V}$$

The *distributive* law holds for finding a gradient. Thus if S and T are scalar functions

$$\nabla(S + T) = \nabla S + \nabla T$$

The *associative* law becomes the rule for differentiating a product:

$$\nabla(ST) = S\nabla T + T\nabla S$$

If $\mathbf{V}$ is a vector function with the magnitudes of the components parallel to the three coordinate axes V_x, V_y, V_z, then

$$\nabla \cdot \mathbf{V} \equiv \text{div}\, \mathbf{V} \equiv \frac{\partial V_x}{\partial x} + \frac{\partial V_y}{\partial y} + \frac{\partial V_z}{\partial z}$$

The divergence obeys the distributive law. Thus, if **V** and **U** are vector functions, then

$$\nabla \cdot (\mathbf{V}+\mathbf{U}) = \nabla \cdot \mathbf{V} + \nabla \cdot \mathbf{U}$$
$$\nabla \cdot (S\mathbf{V}) = (\nabla S) \cdot \mathbf{V} + S(\nabla \cdot \mathbf{V})$$
$$\nabla \cdot (\mathbf{U} \times \mathbf{V}) = \mathbf{V} \cdot (\nabla \times \mathbf{U}) - \mathbf{U} \cdot (\nabla \times \mathbf{V})$$

As with the gradient of a scalar, the divergence of a vector is invariant under a transformation from one set of rectangular coordinates to another.

$$\nabla \times \mathbf{V} \equiv \text{curl}\,\mathbf{V} \qquad (\text{ sometimes } \nabla \wedge \mathbf{V} \text{ or rot}\,\mathbf{V})$$
$$\equiv \left(\frac{\partial \mathbf{V}_x}{\partial y} - \frac{\partial \mathbf{V}_y}{\partial z}\right)\mathbf{i} + \left(\frac{\partial \mathbf{V}_x}{\partial z} - \frac{\partial \mathbf{V}_z}{\partial x}\right)\mathbf{j} + \left(\frac{\partial \mathbf{V}_y}{\partial x} - \frac{\partial \mathbf{V}_x}{\partial y}\right)\mathbf{k}$$
$$= \begin{vmatrix} \mathbf{i} & \mathbf{j} & \mathbf{k} \\ \frac{\partial}{\partial x} & \frac{\partial}{\partial y} & \frac{\partial}{\partial z} \\ V_x & V_y & V_z \end{vmatrix}$$

The *curl* (or *rotation*) of a vector is a vector that is invariant under a transformation from one set of rectangular coordinates to another.

$$\nabla \times (\mathbf{U}+\mathbf{V}) = \nabla \times \mathbf{U} + \nabla \times \mathbf{V}$$
$$\nabla \times (S\mathbf{V}) = (\nabla S) \times \mathbf{V} + S(\nabla \times \mathbf{V})$$
$$\nabla \times (\mathbf{U} \times \mathbf{V}) = (\mathbf{V} \cdot \nabla)\mathbf{U} - (\mathbf{U} \cdot \nabla)\mathbf{V} + \mathbf{U}(\nabla \cdot \mathbf{V}) - \mathbf{V}(\nabla \cdot \mathbf{U})$$

If $\mathbf{V} = V_x\mathbf{i} + V_y\mathbf{j} + V_z\mathbf{k}$, then

$$\nabla \cdot \mathbf{V} = \nabla V_x \cdot \mathbf{i} + \nabla V_y \cdot \mathbf{j} + \nabla V_z \cdot \mathbf{k}$$
$$\text{and} \quad \nabla \times \mathbf{V} = \nabla V_x \times \mathbf{i} + \nabla V_y \times \mathbf{j} + \nabla V_z \times \mathbf{k}$$

The operator ∇ can be used more than once. The possibilities where ∇ is used twice are:

$$\nabla \cdot (\nabla \theta) \equiv \text{div}\,\text{grad}\,\theta$$
$$\nabla \times (\nabla \theta) \equiv \text{curl}\,\text{grad}\,\theta$$
$$\nabla(\nabla \cdot \mathbf{V}) = \text{grad}\,\text{div}\,\mathbf{V}$$
$$\nabla \cdot (\nabla \times \mathbf{V}) \equiv \text{div}\,\text{curl}\,\mathbf{V}$$
$$\nabla \times (\nabla \times \mathbf{V}) \equiv \text{curl}\,\text{curl}\,\mathbf{V}$$

Thus, if S is a scalar and **V** is a vector:

$$\text{div}\,\text{grad}\,S \equiv \nabla \cdot (\nabla S) \equiv \text{ Laplacian } S \equiv \nabla^2 S \equiv \frac{\partial^2 S}{\partial x^2} + \frac{\partial^2 S}{\partial y^2} + \frac{\partial^2 S}{\partial z^2}$$
$$\text{curl}\,\text{grad}\,S \equiv 0$$
$$\text{curl}\,\text{curl}\,\mathbf{V} \equiv \text{grad}\,\text{div}\,\mathbf{V} - \nabla^2\mathbf{V};$$
$$\text{div}\,\text{curl}\,\mathbf{V} \equiv 0$$

Taylor expansion in three dimensions can be written

$$f(\mathbf{V}+\varepsilon) = e^{\varepsilon \cdot \nabla} f(\mathbf{V}) \qquad \text{where } \mathbf{V} = x\mathbf{i} + y\mathbf{j} + z\mathbf{k}$$
$$\text{and } \varepsilon = h\mathbf{i} + l\mathbf{j} + m\mathbf{k}$$

ORTHOGONAL CURVILINEAR COORDINATES

If at a point P there exist three uniform point functions u, v and w so that the surfaces $u =$ const., $v =$ const., and $w =$ const., intersect in three distinct curves through P, then the surfaces are called the *coordinate surfaces* through P. The three lines of intersection are referred to as the *coordinate lines* and their tangents a, b, and c as the *coordinate axes*. When the coordinate axes form an orthogonal set the system is said to define *orthogonal curvilinear coordinates* at P.

Consider an infinitesimal volume enclosed by the surfaces u, v, w, $u + du$, $v + dv$, and $w + dw$ (Figure 5).

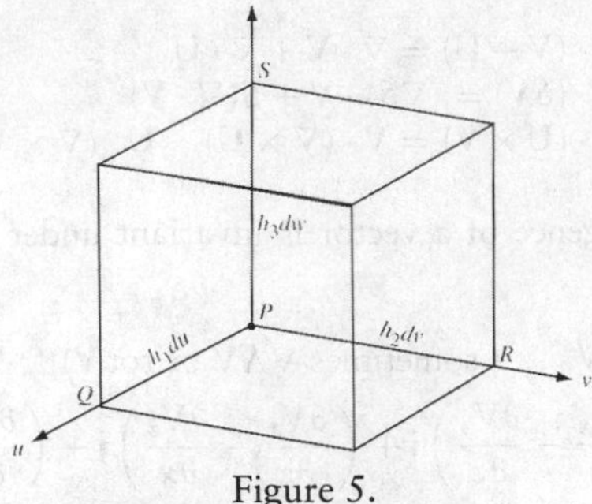

Figure 5.

The surface $PRS \equiv u =$ constant, and the face of the curvilinear figure immediately opposite this is $u + du =$ constant, etc. In terms of these surface constants

$$\begin{aligned} P &= P(u, v, w) \\ Q &= Q(u + du, v, w) \quad \text{and} \quad PQ = h_1\, du \\ R &= R(u, v + dv, w) \quad \text{and} \quad PR = h_2\, dv \\ S &= S(u, v, w + dw) \quad \text{and} \quad PS = h_3\, dw \end{aligned}$$

where h_1, h_2, and h_3 are functions of u, v, and w.

- In rectangular Cartesians **i**, **j**, **k**

$$h_1 = 1, \qquad h_2 = 1, \qquad h_3 = 1.$$

$$\frac{\hat{\mathbf{a}}}{h_1}\frac{\partial}{\partial u} = \hat{\mathbf{i}}\frac{\partial}{\partial x}, \qquad \frac{\hat{\mathbf{b}}}{h_2}\frac{\partial}{\partial v} = \hat{\mathbf{j}}\frac{\partial}{\partial y}, \qquad \frac{\hat{\mathbf{c}}}{h_3}\frac{\partial}{\partial w} = \hat{\mathbf{k}}\frac{\partial}{\partial z}.$$

- In cylindrical Cartesians $\hat{\mathbf{r}}, \hat{\theta}, \hat{\mathbf{z}}$

$$h_1 = 1, \qquad h_2 = r, \qquad h_3 = 1.$$

$$\frac{\hat{\mathbf{a}}}{h_1}\frac{\partial}{\partial u} = \hat{r}\frac{\partial}{\partial r}, \qquad \frac{\hat{\mathbf{b}}}{h_2}\frac{\partial}{\partial v} = \frac{\hat{\boldsymbol{\theta}}}{r}\frac{\partial}{\partial \theta}, \qquad \frac{\hat{\mathbf{c}}}{h_3}\frac{\partial}{\partial w} = \hat{\mathbf{z}}\frac{\partial}{\partial z}.$$

- In spherical coordinates $\hat{\mathbf{r}}, \hat{\theta}, \hat{\psi}$

$$h_1 = 1, \qquad h_2 = r, \qquad h_3 = r\sin\theta$$

$$\frac{\hat{\mathbf{a}}}{h_1}\frac{\partial}{\partial u} = \hat{\mathbf{r}}\frac{\partial}{\partial r}, \qquad \frac{\hat{\mathbf{b}}}{h_2}\frac{\partial}{\partial v} = \frac{\hat{\boldsymbol{\theta}}}{r}\frac{\partial}{\partial \theta}, \qquad \frac{\hat{\mathbf{c}}}{h_3}\frac{\partial}{\partial w} = \frac{\hat{\boldsymbol{\psi}}}{r\sin\theta}\frac{\partial}{\partial \psi}$$

The general expressions for grad, div and curl together with those for ∇^2 and the directional derivative are, in orthogonal curvilinear coordinates, given by:

$$\begin{aligned} \nabla S &= \frac{\hat{\mathbf{a}}}{h_1}\frac{\partial S}{\partial u} + \frac{\hat{\mathbf{b}}}{h_2}\frac{\partial S}{\partial v} + \frac{\hat{\mathbf{c}}}{h_3}\frac{\partial S}{\partial w} \\ (\mathbf{V}\cdot\nabla)S &= \frac{V_1}{h_1}\frac{\partial S}{\partial u} + \frac{V_2}{h_2}\frac{\partial S}{\partial v} + \frac{V_3}{h_3}\frac{\partial S}{\partial w} \\ \nabla\cdot\mathbf{V} &= \frac{1}{h_1h_2h_3}\left\{\frac{\partial}{\partial u}(h_2h_3V_1) + \frac{\partial}{\partial v}(h_3h_1V_2) + \frac{\partial}{\partial w}(h_1h_2V_3)\right\}. \\ \nabla\times\mathbf{V} &= \frac{\hat{\mathbf{a}}}{h_2h_3}\left\{\frac{\partial}{\partial v}(h_3V_3) - \frac{\partial}{\partial w}(h_2V_2)\right\} + \frac{\hat{\mathbf{b}}}{h_3h_1}\left\{\frac{\partial}{\partial w}(h_1V_1) - \frac{\partial}{\partial u}(h_3V_3)\right\} \\ &\quad + \frac{\hat{\mathbf{c}}}{h_1h_2}\left\{\frac{\partial}{\partial u}(h_2V_2) - \frac{\partial}{\partial v}(h_1V_1)\right\} \\ \nabla^2 S &= \frac{1}{h_1h_2h_3}\left\{\frac{\partial}{\partial u}\left(\frac{h_2h_3}{h_1}\frac{\partial S}{\partial u}\right) + \frac{\partial}{\partial v}\left(\frac{h_3h_1}{h_2}\frac{\partial S}{\partial v}\right) + \frac{\partial}{\partial w}\left(\frac{h_1h_2}{h_3}\frac{\partial S}{\partial w}\right)\right\} \end{aligned}$$

Formulas of Vector Analysis

	Rectangular coordinates	Cylindrical coordinates	Spherical coordinates
Conversion to rectangular coordinates		$x = r\cos\varphi \quad y = r\sin\varphi \quad z = z$	$x = r\cos\varphi\sin\theta \quad y = r\sin\varphi\sin\theta \quad z = r\cos\theta$
Gradient...	$\nabla\phi = \frac{\partial\phi}{\partial x}\mathbf{i} + \frac{\partial\phi}{\partial y}\mathbf{j} + \frac{\partial\phi}{\partial z}\mathbf{k}$	$\nabla\phi = \frac{\partial\phi}{\partial r}\mathbf{r} + \frac{1}{r}\frac{\partial\phi}{\partial\varphi}\Phi + \frac{\partial\phi}{\partial z}\mathbf{k}$	$\nabla\phi = \frac{\partial\phi}{\partial r}\mathbf{r} + \frac{1}{r}\frac{\partial\phi}{\partial\theta}\theta + \frac{1}{r\sin\theta}\frac{\partial\phi}{\partial\varphi}\Phi$
Divergence...	$\nabla\cdot\mathbf{A} = \frac{\partial A_x}{\partial x} + \frac{\partial A_y}{\partial y} + \frac{\partial A_z}{\partial z}$	$\nabla\cdot\mathbf{A} = \frac{1}{r}\frac{\partial(rA_r)}{\partial r} + \frac{1}{r}\frac{\partial A_\varphi}{\partial\varphi} + \frac{\partial A_z}{\partial z}$	$\nabla\cdot\mathbf{A} = \frac{1}{r^2}\frac{\partial(r^2 A_r)}{\partial r} + \frac{1}{r\sin\theta}\frac{\partial(A_\theta\sin\theta)}{\partial\theta} + \frac{1}{r\sin\theta}\frac{\partial A_\varphi}{\partial\varphi}$
Curl...	$\nabla\times\mathbf{A} = \begin{vmatrix} \mathbf{i} & \mathbf{j} & \mathbf{k} \\ \frac{\partial}{\partial x} & \frac{\partial}{\partial y} & \frac{\partial}{\partial z} \\ A_x & A_y & A_z \end{vmatrix}$	$\nabla\times\mathbf{A} = \begin{vmatrix} \frac{1}{r}\mathbf{r} & \Phi & \frac{1}{r}\mathbf{k} \\ \frac{\partial}{\partial r} & \frac{\partial}{\partial\varphi} & \frac{\partial}{\partial z} \\ A_r & rA_\varphi & A_z \end{vmatrix}$	$\nabla\times\mathbf{A} = \begin{vmatrix} \frac{\mathbf{r}}{r^2\sin\theta} & \frac{\theta}{r\sin\theta} & \frac{\Phi}{r} \\ \frac{\partial}{\partial r} & \frac{\partial}{\partial\theta} & \frac{\partial}{\partial\varphi} \\ A_r & rA_\theta & rA_\varphi\sin\theta \end{vmatrix}$
Laplacian...	$\nabla^2\phi = \frac{\partial^2\phi}{\partial x^2} + \frac{\partial^2\phi}{\partial y^2} + \frac{\partial^2\phi}{\partial z^2}$	$\nabla^2\phi = \frac{1}{r}\frac{\partial}{\partial r}\left(r\frac{\partial\phi}{\partial r}\right) + \frac{1}{r^2}\frac{\partial^2\phi}{\partial\varphi^2} + \frac{\partial^2\phi}{\partial z^2}$	$\nabla^2\phi = \frac{1}{r^2}\frac{\partial}{\partial r}\left(r^2\frac{\partial\phi}{\partial r}\right) + \frac{1}{r^2\sin\theta}\frac{\partial}{\partial\theta}\left(\sin\theta\frac{\partial\phi}{\partial\theta}\right) + \frac{1}{r^2\sin^2\theta}\frac{\partial^2\phi}{\partial\varphi^2}$

TRANSFORMATION OF INTEGRALS

If

1. s is the distance along a curve "C" in space and is measured from some fixed point.
2. S is a surface area
3. V is a volume contained by a specified surface
4. $\hat{\mathbf{t}}$ = the unit tangent to C at the point
5. P $\hat{\mathbf{n}}$ = the unit outward pointing normal
6. F is some vector function
7. ds is the vector element of curve ($= \hat{\mathbf{t}}\, ds$)
8. dS is the vector element of surface ($= \hat{\mathbf{n}}\, dS$)

then

$$\int_{(c)} \mathbf{F}\cdot\hat{\mathbf{t}}\, ds = \int_{(c)} \mathbf{F}$$

and when $\mathbf{F} = \nabla\phi$

$$\int_{C} (\nabla\phi)\cdot\hat{\mathbf{t}}\, ds = \int_{C} d\phi$$

Gauss' Theorem

When S defines a closed region having a volume V:

$$\iiint_V (\nabla\cdot\mathbf{F})\, dV = \iint_S \mathbf{F}\cdot(\hat{\mathbf{n}})\, dS = \iint_S \mathbf{F}\cdot dS$$

also

$$\iiint_V (\nabla\phi)\, dV = \iint_S \phi\,\hat{\mathbf{n}}\, dS$$

and

$$\iiint_V (\nabla\times\mathbf{F})\, dV = \iint_S (\hat{\mathbf{n}}\times\mathbf{F})\, dS$$

Stokes' Theorem

When C is closed and bounds the open surface S:

$$\iint_S \hat{\mathbf{n}}\cdot(\nabla\times\mathbf{F})\, dS = \int_C \mathbf{F}\cdot ds$$

also

$$\iint_S (\hat{\mathbf{n}}\times\nabla\phi)\, dS = \int_{(c)} \phi\, ds$$

Green's Theorem

$$\iint_S (\nabla\phi\cdot\nabla\theta)\, dS = \iint_S \phi\,\hat{\mathbf{n}}\cdot(\nabla\theta)\, dS = \iiint_V \phi(\nabla^2\theta)\, dV$$

$$= \iint_S \theta\cdot\hat{\mathbf{n}}(\nabla\phi)\, dS = \iiint_V \phi(\nabla^2\theta)\, dV$$

BESSEL FUNCTIONS

1. Bessel's differential equation for a real variable x is

$$x^2\frac{d^2y}{dx^2}+x\frac{dy}{dx}+(x^2-n^2)y=0$$

2. When n is not an integer, two independent solutions of the equation are $J_n(x)$ and $J_{-n}(x)$ where

$$J_n(x)=\sum_{k=0}^{\infty}\frac{(-1)^k}{k!\Gamma(n+k+1)}\left(\frac{x}{2}\right)^{n+2k}$$

3. If n is an integer, then $J_n(x)=(-1)^nJ_n(x)$, where

$$J_n(x)=\frac{x^n}{2^nn!}\left\{1-\frac{x^2}{2^2\cdot 1!(n+1)}+\frac{x^4}{2^4\cdot 2!(n+1)\,(n+2)}+\frac{x^6}{2^6\cdot 3!(n+1)\,(n+2)\,(n+3)}+\ldots\right\}$$

4. For $n=0$ and $n=1$, this formula becomes

$$J_0(x)=1-\frac{x^2}{2^2(1!)^2}+\frac{x^4}{2^4(2!)^2}-\frac{x^6}{2^6(3!)^2}+\frac{x^8}{2^8(4!)^2}-\cdots$$

$$J_1(x)=\frac{x}{2}-\frac{x^3}{2^3\cdot 1!2!}+\frac{x^5}{2^5\cdot 2!3!}-\frac{x^7}{2^7\cdot 3!4!}+\frac{x^9}{2^9\cdot 4!5!}-\cdots$$

5. When x is large and positive, the following asymptotic series may be used

$$J_0(x)=\left(\frac{2}{\pi x}\right)^{\frac{1}{2}}\left\{P_0(x)\cos\left(x-\frac{\pi}{4}\right)-Q_0(x)\sin\left(x-\frac{\pi}{4}\right)\right\}$$

$$J_1(x)=\left(\frac{2}{\pi x}\right)^{\frac{1}{2}}\left\{P_1(x)\cos\left(x-\frac{3\pi}{4}\right)-Q_1(x)\sin\left(x-\frac{3\pi}{4}\right)\right\}$$

where

$$P_0(x)\sim 1-\frac{1^2\cdot 3^2}{2!(8x)^2}+\frac{1^2\cdot 3^2\cdot 5^2\cdot 7^2}{4!(8x)^4}-\frac{1^2\cdot 3^2\cdot 5^2\cdot 7^2\cdot 9^2\cdot 11^2}{6!(8x)^6}+\cdots$$

$$Q_0(x)\sim -\frac{1^2}{1!8x}+\frac{1^2\cdot 3^2\cdot 5^2}{3!(8x)^3}-\frac{1^2\cdot 3^2\cdot 5^2\cdot 7^2\cdot 9^2}{5!(8x)^5}+-\cdots$$

$$P_1(x)\sim 1+\frac{1^2\cdot 3\cdot 5}{2!(8x)^2}-\frac{1^2\cdot 3^2\cdot 5^2\cdot 7\cdot 9}{4!(8x)^4}+\frac{1^2\cdot 3^2\cdot 5^2\cdot 7^2\cdot 9^2\cdot 11\cdot 13}{6!(8x)^6}-+\cdots$$

$$Q_1(x)\sim \frac{1\cdot 3}{1!8x}-\frac{1^2\cdot 3^2\cdot 5\cdot 7}{3!(8x)^3}+\frac{1^2\cdot 3^2\cdot 5^2\cdot 7^2\cdot 9\cdot 11}{5!(8x)^5}-\cdots$$

[In $P_1(x)$ the signs alternate from + to − after the first term]

6. The zeros of $J_0(x)$ and $J_1(x)$.

If j_{0s} and j_{1s} are the sth zeros of $J_0(x)$ and $J_1(x)$, respectively, and if $a=4_s-1$, $b=4_s+1$

$$j_{0,s}\sim\frac{1}{4}\pi a\left\{1+\frac{2}{\pi^2a^2}-\frac{62}{3\pi^4a^4}+\frac{15,116}{15\pi^6a^6}-\frac{12,554,474}{105\pi^8a^8}+\frac{8,368,654,292}{315\pi^{10}a^{10}}-+\cdots\right\}$$

$$j_{1,s}\sim\frac{1}{4}\pi b\left\{1-\frac{6}{\pi^2b^2}+\frac{6}{\pi^4b^4}-\frac{4716}{5\pi^6b^6}+\frac{3,902,418}{35\pi^8b^8}-\frac{895,167,324}{35\pi^{10}b^{10}}+\cdots\right\}$$

$$J_1(j_{0,s})\sim\frac{(-1)^{s+1}2^{\frac{3}{2}}}{\pi a^{\frac{1}{2}}}\left\{1-\frac{56}{3\pi^4a^4}+\frac{9664}{5\pi^6a^6}-\frac{7,381,280}{21\pi^8a^8}+\cdots\right\}$$

$$J_0(j_{1,s})\sim\frac{(-1)^s2^{\frac{3}{2}}}{\pi b^{\frac{1}{2}}}\left\{1+\frac{24}{\pi^4b^4}-\frac{19,584}{10\pi^6b^6}+\frac{2,466,720}{7\pi^8b^8}-\cdots\right\}$$

7. Table of zeros for $J_0(x)$ and $J_1(x)$

Define $\{\alpha_n, \beta_n\}$ by $J_0(\alpha_n) = 0$ and $J_1(\beta_n) = 0$.

Roots α_n	$J_1(\alpha_n)$	Roots β_n	$J_0(\beta_n)$
2.4048	0.5191	0.0000	1.0000
5.5201	−0.3403	3.8317	−0.4028
8.6537	0.2715	7.0156	0.3001
11.7915	−0.2325	10.1735	−0.2497
14.9309	0.2065	13.3237	0.2184
18.0711	−0.1877	16.4706	−0.1965
21.2116	0.1733	19.6159	0.1801

8. Recurrence formulas

$$J_{n-1}(x) + J_{n+1}(x) = \frac{2n}{x} J_n(x) \qquad nJ_n(x) + xJ_n'(x) = xJ_{n-1}(x)$$
$$J_{n-1}(x) - J_{n+1}(x) = 2J_n'(x) \qquad nJ_n(x) - xJ_n'(x) = xJ_{n+1}(x)$$

9. If J_n is written for $J_n(x)$ and $J_n^{(k)}$ is written for $\frac{d^k}{dx^k}\{J_n(x)\}$, then the following derivative relationships are important

$$J_0^{(r)} = -J_1^{(r-1)}$$
$$J_0^{(2)} = -J_0 + \tfrac{1}{x}J_1 = \tfrac{1}{2}(J_2 - J_0)$$
$$J_0^{(3)} = \tfrac{1}{x}J_0 + \left(1 - \tfrac{2}{x^2}\right)J_1 = \tfrac{1}{4}(-J_3 + 3J_1)$$
$$J_0^{(4)} = \left(1 - \tfrac{3}{x^2}\right)J_0 - \left(\tfrac{2}{x} - \tfrac{6}{x^3}\right)J_1 = \tfrac{1}{8}(J_4 - 4J_2 + 3J_0), \text{ etc.}$$

10. Half-order Bessel functions

$$J_{\frac{1}{2}}(x) = \sqrt{\tfrac{2}{\pi x}}\sin x$$
$$J_{-\frac{1}{2}}(x) = \sqrt{\tfrac{2}{\pi x}}\cos x$$
$$J_{n+\frac{3}{2}}(x) = -x^{n+\frac{1}{2}}\tfrac{d}{dx}\{x^{-(n+\frac{1}{2})}J_{n+\frac{1}{2}}(x)\}$$
$$J_{n-\frac{1}{2}}(x) = x^{-(n+\frac{1}{2})}\tfrac{d}{dx}\{x^{n+\frac{1}{2}}J_{n+\frac{1}{2}}(x)\}$$

n	$\left(\frac{\pi x}{2}\right)^{\frac{1}{2}} J_{n+\frac{1}{2}}(x)$	$\left(\frac{\pi x}{2}\right)^{\frac{1}{2}} J_{-(n+\frac{1}{2})}(x)$
0	$\sin x$	$\cos x$
1	$\frac{\sin x}{x} - \cos x$	$-\frac{\cos x}{x} - \sin x$
2	$\left(\frac{3}{x^2} - 1\right)\sin x - \frac{3}{x}\cos x$	$\left(\frac{3}{x^2} - 1\right)\cos x + \frac{3}{x}\sin x$
3	$\left(\frac{15}{x^3} - \frac{6}{x}\right)\sin x - \left(\frac{15}{x^2} - 1\right)\cos x$ etc.	$-\left(\frac{15}{x^3} - \frac{6}{x}\right)\cos x - \left(\frac{15}{x^2} - 1\right)\sin x$

11. Additional solutions to Bessel's equation are

$Y_n(x)$ (also called Weber's function, and sometimes denoted by $N_n(x)$)

$H_n^{(1)}(x)$ and $H_n^{(2)}(x)$ (also called Hankel functions)

These solutions are defined as follows

$$Y_n(x) = \begin{cases} \dfrac{J_n(x)\cos(n\pi) - J_{-n}(x)}{\sin(n\pi)} & n \text{ not an integer} \\ \lim\limits_{\nu \to n} \frac{J_\nu(x)\cos(\nu\pi) - J_{-\nu}(x)}{\sin(\nu\pi)} & n \text{ an integer} \end{cases} \qquad \begin{aligned} H_n^{(1)}(x) &= J_n(x) + iY_n(x) \\ H_n^{(2)}(x) &= J_n(x) - iY_n(x) \end{aligned}$$

The additional properties of these functions may all be derived from the above relations and the known properties of $J_n(x)$.

12. Complete solutions to Bessel's equation may be written as

$$c_1 J_n(x) + c_2 J_{-n}(x) \qquad \text{if } n \text{ is not an integer}$$

or, for any value of n,

$$c_1 J_n(x) + c_2 Y_n(x) \qquad \text{or} \qquad c_1 H_n^{(1)} x + c_2 H_n^{(2)}(x)$$

13. The modified (or hyperbolic) Bessel's differential equation is

$$x^2\frac{d^2y}{dx^2}+x\frac{dy}{dx}-(x^2+n^2)y=0$$

14. When n is not an integer, two independent solutions of the equation are $I_n(x)$ and $I_{-n}(x)$, where

$$I_n(x)=\sum_{k=0}^{\infty}\frac{1}{k!\Gamma(n+k+1)}\left(\frac{x}{2}\right)^{n+2k}$$

15. If n is an integer,

$$I_n(x)=I_{-n}(x)=\frac{x^n}{2^n n!}\left\{1+\frac{x^2}{2^2\cdot 1!(n+1)}+\frac{x^4}{2^4\cdot 2!(n+1)(n+2)}+\frac{x^6}{2^6\cdot 3!(n+1)(n+2)(n+3)}+\cdots\right\}$$

16. For $n=0$ and $n=1$, this formula becomes

$$I_0(x)=1+\frac{x^2}{2^2(1!)^2}+\frac{x^4}{2^4(2!)^2}+\frac{x^6}{2^6(3!)^2}+\frac{x^8}{2^8(4!)^2}+\cdots$$

$$I_1(x)=\frac{x}{2}+\frac{x^3}{2^3\cdot 1!2!}+\frac{x^5}{2^5\cdot 2!3!}+\frac{x^7}{2^7\cdot 3!4!}+\frac{x^9}{2^9\cdot 4!5!}+\cdots$$

17. Another solution to the modified Bessel's equation is

$$K_n(x)=\begin{cases}\frac{1}{2}\pi\frac{I_{-n}(x)-I_n(x)}{\sin(n\pi)} & n \text{ not an integer}\\ \lim\limits_{\nu\to n}\frac{1}{2}\pi\frac{I_{-\nu}(x)-I_\nu(x)}{\sin(\nu\pi)} & n \text{ an integer}\end{cases}$$

This function is linearly independent of $I_n(x)$ for all values of n. Thus the complete solution to the modified Bessel's equation may be written as

$$c_1I_n(x)+c_2I_{-n}(x)\quad n \text{ not an integer}$$

or

$$c_1I_n(x)+c_2K_n(x)\quad \text{any } n$$

18. The following relations hold among the various Bessel functions:

$$I_n(z)=i^{-m}J_m(iz)$$
$$Y_n(iz)=(i)^{n+1}I_n(z)-\tfrac{2}{\pi}i^{-n}K_n(z)$$

Most of the properties of the modified Bessel function may be deduced from the known properties of $J_n(x)$ by use of these relations and those previously given.

19. Recurrence formulas

$$I_{n-1}(x)-I_{n+1}(x)=\tfrac{2n}{x}I_n(x)\qquad I_{n-1}(x)+I_{n+1}(x)=2I_n'(x)$$
$$I_{n-1}(x)-\tfrac{n}{x}I_n(x)=I_n'(x)\qquad I_n'(x)=I_{n+1}(x)+\tfrac{n}{x}I_n(z)$$

THE FACTORIAL FUNCTION

For non-negative integers n, the factorial of n, denoted $n!$, is the product of all positive integers less than or equal to n; $n! = n\cdot(n-1)\cdot(n-2)\cdots 2\cdot 1$. If n is a negative integer ($n=-1,-2,\ldots$), then $n!=\pm\infty$.

Approximations to $n!$ for large n include Stirling's formula

$$n!\approx\sqrt{2\pi e}\left(\frac{n}{e}\right)^{n+\frac{1}{2}},$$

and Burnsides's formula

$$n!\approx\sqrt{2\pi}\left(\frac{n+\frac{1}{2}}{e}\right)^{n+\frac{1}{2}}.$$

n	$n!$	$\log_{10} n!$	n	$n!$	$\log_{10} n!$
0	1	0.00000	1	1	0.00000
2	2	0.30103	3	6	0.77815
4	24	1.38021	5	120	2.07918
6	720	2.85733	7	5040	3.70243
8	40320	4.60552	9	3.6288×10^{5}	5.55976
10	3.6288×10^{6}	6.55976	11	3.9917×10^{7}	7.60116
12	4.7900×10^{8}	8.68034	13	6.2270×10^{9}	9.79428
14	8.7178×10^{10}	10.94041	15	1.3077×10^{12}	12.11650
16	2.0923×10^{13}	13.32062	17	3.5569×10^{14}	14.55107
18	6.4024×10^{15}	15.80634	19	1.2165×10^{17}	17.08509
20	2.4329×10^{18}	18.38612	25	1.5511×10^{25}	25.19065
30	2.6525×10^{32}	32.42366	40	8.1592×10^{47}	47.91165
50	3.0414×10^{64}	64.48307	60	8.3210×10^{81}	81.92017
70	1.1979×10^{100}	100.07841	80	7.1569×10^{118}	118.85473
90	1.4857×10^{138}	138.17194	100	9.3326×10^{157}	157.97000
110	1.5882×10^{178}	178.20092	120	6.6895×10^{198}	198.82539
130	6.4669×10^{219}	219.81069	150	5.7134×10^{262}	262.75689
500	1.2201×10^{1134}	1134.0864	1000	4.0239×10^{2567}	2567.6046

THE GAMMA FUNCTION

Definition: $\Gamma(n) = \int_0^{\infty} t^{n-1} e^{-t}\, dt \quad n > 0$

Recursion Formula: $\Gamma(n+1) = n\Gamma(n)$

$\Gamma(n+1) = n! \quad$ if $n = 0, 1, 2, \ldots$ where $0! = 1$

For $n < 0$ the gamma function can be defined by using

$\Gamma(n) = \frac{\Gamma(n+1)}{n}$

Graph:

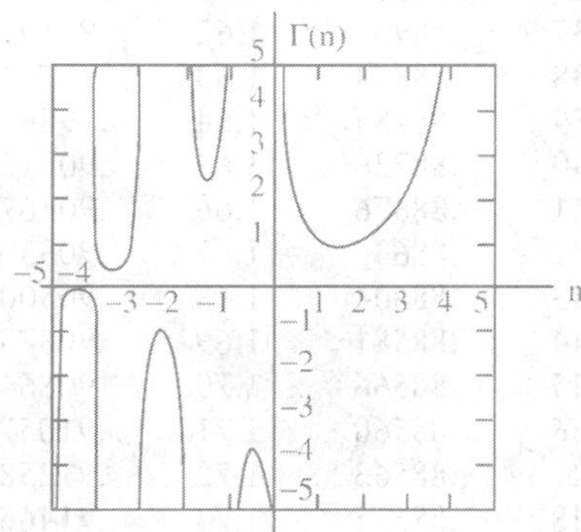

Special Values: $\Gamma(1/2) = \sqrt{\pi}$

$$\Gamma(m + 1/2) = \frac{1 \cdot 3 \cdot 5 \cdots (2m-1)}{2^m}\sqrt{\pi} \quad m = 1, 2, 3, \ldots$$

$$\Gamma(-m + 1/2) = \frac{(-1)^m 2^m \sqrt{\pi}}{1 \cdot 3 \cdot 5 \cdots (2m-1)} \quad m = 1, 2, 3, \ldots$$

Definition:

$$\Gamma(x+1) = \lim_{k \to \infty} \frac{1 \cdot 2 \cdot 3 \cdots k}{(x+1)(x+2)\cdots(x+k)} k^x$$

$$\frac{1}{\Gamma(x)} = x e^{\gamma x} \prod_{m=1}^{\infty} \left\{ \left(1 + \frac{x}{m}\right) e^{-x/m} \right\}$$

This is an infinite product representation for the gamma function where γ is Euler's constant.

Properties:

$$\Gamma'(1) = \int_0^\infty e^{\gamma x} \ln x \, dx = -\gamma$$

$$\frac{\Gamma'(x)}{\Gamma(x)} = -\gamma + \left(\frac{1}{1} - \frac{1}{x}\right) + \left(\frac{1}{2} - \frac{1}{x+1}\right) + \cdots + \left(\frac{1}{n} - \frac{1}{x+n-1}\right) + \cdots$$

$$\Gamma(x+1) = \sqrt{2\pi x}\, x^x e^{-x} \left\{1 + \frac{1}{12x} + \frac{1}{288x^2} - \frac{139}{51,840x^3} + \cdots\right\}$$

This is called *Stirling's asymptotic series*.

Values of $\Gamma(n) = \int_0^\infty e^{-x} x^{n-1}\, dx$; $\Gamma(n+1) = n\Gamma(n)$

n	$\Gamma(n)$	n	$\Gamma(n)$	n	$\Gamma(n)$	n	$\Gamma(n)$
1.00	1.00000	1.25	.90640	1.50	.88623	1.75	.91906
1.01	.99433	1.26	.90440	1.51	.88659	1.76	.92137
1.02	.98884	1.27	.90250	1.52	.88704	1.77	.92376
1.03	.98355	1.28	.90072	1.53	.88757	1.78	.92623
1.04	.97844	1.29	.89904	1.54	.88818	1.79	.92877
1.05	.97350	1.30	.89747	1.55	.88887	1.80	.93138
1.06	.96874	1.31	.89600	1.56	.88964	1.81	.93408
1.07	.96415	1.32	.89464	1.57	.89049	1.82	.93685
1.08	.95973	1.33	.89338	1.58	.89142	1.83	.93969
1.09	.95546	1.34	.89222	1.59	.89243	1.84	.94261
1.10	.95135	1.35	.89115	1.60	.89352	1.85	.94561
1.11	.94740	1.36	.89018	1.61	.89468	1.86	.94869
1.12	.94359	1.37	.88931	1.62	.89592	1.87	.95184
1.13	.93993	1.38	.88854	1.63	.89724	1.88	.95507
1.14	.93642	1.39	.88785	1.64	.89864	1.89	.95838
1.15	.93304	1.40	.88726	1.65	.90012	1.90	.96177
1.16	.92980	1.41	.88676	1.66	.90167	1.91	.96523
1.17	.92670	1.42	.88636	1.67	.90330	1.92	.96877
1.18	.92373	1.43	.88604	1.68	.90500	1.93	.97240
1.19	.92089	1.44	.88581	1.69	.90678	1.94	.97610
1.20	.91817	1.45	.88566	1.70	.90864	1.95	.97988
1.21	.91558	1.46	.88560	1.71	.91057	1.96	.98374
1.22	.91311	1.47	.88563	1.72	.91258	1.97	.98768
1.23	.91075	1.48	.88575	1.73	.91466	1.98	.99171
1.24	.90852	1.49	.88595	1.74	.91683	1.99	.99581
						2.00	1.00000

THE BETA FUNCTION

Definition: $B(m,n) = \int_0^1 t^{m-1}(1-t)^{n-1} dt \quad m > 0, n > 0$

Relationship with Gamma function: $B(m,n) = \dfrac{\Gamma(m)\Gamma(n)}{\Gamma(m+n)}$

Properties:

$$B(m,n) = B(n,m)$$

$$B(m,n) = 2\int_0^{\pi/2} \sin^{2m-1}\theta \cos^{2n-1}\theta \, d\theta$$

$$B(m,n) = \int_0^\infty \frac{t^{m-1}}{(1+t)^{m+n}} dt$$

$$B(m,n) = r^n(r+1)^m \int_0^1 \frac{t^{m-1}(1-t)^{n-1}}{(r+t)^{m+n}} dt$$

Inequality: $\max_{-1\le x\le 1} |P_n^{(\alpha,\beta)}(x)| = \begin{cases} \binom{n+q}{n} \sim n^q \text{if } q = \max(\alpha,\beta) \ge -\frac{1}{2} \\ |P_n^{(\alpha,\beta)}(x')| \sim n^{-1/2} \text{if } q < -\frac{1}{2} \\ x' \text{ is one of the two maximum points nearest} \\ \frac{\beta-\alpha}{\alpha+\beta+1} \end{cases}$

V: Generalized Laguerre

Name: Generalized Laguerre *Symbol*: $L_n^{(\alpha)}(x)$ *Interval*: $[0,\infty]$

Differential Equation: $xy'' + (\alpha + 1 - x)y' + ny = 0$

Explicit Expression: $L_n^{(\alpha)}(x) = \sum_{m=0}^{n} (-1)^m \binom{n+\alpha}{n-m} \frac{1}{m!} x^m$

Recurrence Relation: $(n+1)L_n^{(\alpha)} + 1(x) = [(2n+\alpha+1) - x]L_n^{(\alpha)}(x) - (n+\alpha)L_n^{(\alpha)} - 1(x)$

Weight: $x^\alpha e^{-x}, \alpha > -1$ *Standardization*: $L_n^{(\alpha)}(x) = \frac{(-1)^n}{n!} x^n + \cdots$

Norm: $\int_0^\infty x^\alpha e^{-x} [L_n^{(\alpha)}(x)]^2 \, dx = \frac{\Gamma(n+\alpha+1)}{n!}$

Rodrigues' Formula: $L_n^{(\alpha)}(x) = \frac{1}{n! x^\alpha e^{-x}} \frac{d^n}{dx^n} \{x^{n+\alpha} e^{-x}\}$

Generating Function: $(1-z)^{-\alpha-1} \exp\left(\frac{xz}{z-1}\right) = \sum_{n=0}^{\infty} L_n^{(\alpha)}(x) z^n$

Inequality: $|L_n^{(\alpha)}(x) \le \frac{\Gamma(n+\alpha+1)}{n!\Gamma(\alpha+1)} e^{x/2}; \quad \begin{matrix} x \ge 0 \\ \alpha > 0 \end{matrix}$

$|L_n^{(a)}(x)| \le \left[2 - \frac{\Gamma(\alpha+n+1)}{n!\Gamma(\alpha+1)}\right] e^{x/2}; \quad \begin{matrix} x \ge 0 \\ -1 < \alpha < 0 \end{matrix}$

VI: Hermite

Name: Hermite *Symbol*: $H_n(x)$ *Interval*: $[-\infty,\infty]$

Differential Equation: $y'' - 2xy' + 2ny = 0$

Explicit Expression: $H_n(x) = \sum_{m=0}^{[n/2]} \frac{(-1)^m n!(2x)^{n-2m}}{m!(n-2m)!}$

Recurrence Relation: $H_{n+1}(x) = 2xH_n(x) - 2nH_{n-1}(x)$

Weight: e^{-x^2} *Standardization*: $H_n(1) = 2^n x^n + \cdots$

Norm: $\int_{-\infty}^{\infty} e^{-x^2} [H_n(x)]^2 \, dx = 2^n n! \sqrt{\pi}$

Rodrigues' Formula: $H_n(x) = (-1)^n e^{x^2} \frac{d^n}{dx^n}(e^{-x^2})$

Generating Function: $e^{-x^2+2zx} = \sum_{n=0}^{\infty} H_n(x) \frac{z^n}{n!}$

Inequality: $|H_n(x)| e^{x^2/2} k 2^{n/2} \sqrt{n!} \; k \approx 1.086435$

TABLES OF ORTHOGONAL POLYNOMIALS

$H_0 = 1$
$H_1 = 2x$
$H_2 = 4x^2 - 2$
$H_3 = 8x^3 - 12x$
$H_4 = 16x^4 - 48x^2 + 12$
$H_5 = 32x^5 - 160x^3 + 120x$
$H_6 = 64x^6 - 480x^4 + 720x^2 - 120$
$H_7 = 128x^7 - 1344x^5 + 3360x^3 - 1680x$
$H_8 = 256x^8 - 3584x^6 + 13440x^4 - 13440x^2 + 1680$
$H_9 = 512x^9 - 9216x^7 + 48384x^5 - 80640x^3 + 30240x$
$H_{10} = 1024x^{10} - 23040x^8 + 161280x^6 - 403200x^4 + 302400x^2 - 30240$

$x^{10} = (30240H_0 + 75600H_2 + 25200H_4 + 2520H_6 + 90H_8 + H_{10})/1024$
$x^9 = (15120H_1 + 10080H_3 + 1512H_5 + 72H_7 + H_9)/512$
$x^8 = (1680H_0 + 3360H_2 + 840H_4 + 56H_6 + H_8)/256$
$x^7 = (840H_1 + 420H_3 + 42H_5 + H_7)/128$
$x^6 = (120H_0 + 180H_2 + 30H_4 + H_6)/64$
$x^5 = (60H_1 + 20H_3 + H_5)/32$
$x^4 = (12H_0 + 12H_2 + H_4)/16$
$x^3 = (6H_1 + H_3)/8$
$x^2 = (2H_0 + H_2)/4$
$x = (H_1)/2$
$1 = H_0$

$L_0 = 1$
$L_1 = -x + 1$
$L_2 = (x^2 - 4x + 2)/2$
$L_3 = (-x^3 + 9x^2 - 18x + 6)/6$
$L_4 = (x^4 - 16x^3 + 72x^2 - 96x + 24)/24$
$L_5 = (-x^5 + 25x^4 - 200x^3 + 600x^2 - 600x + 120)/120$
$L_6 = (x^6 - 36x^5 + 450x^4 - 2400x^3 + 5400x^2 - 4320x + 720)/720$

$x^6 = 720L_0 - 4320L_1 + 10800L_2 - 14400L_3 + 10800L_4 - 4320L_5 + 720L_6$
$x^5 = 120L_0 - 600L_1 + 1200L_2 - 1200L_3 + 600L_4 - 120L_5$
$x^4 = 24L_0 - 96L_1 + 144L_2 - 96L_3 + 24L_4$
$x^3 = 6L_0 - 18L_1 + 18L_2 - 6L_3$
$x^2 = 2L_0 - 4L_1 + 2L_2$
$x = L_0 - L_1$
$1 = L_0$

$P_0 = 1$
$P_1 = x$
$P_2 = (3x^2 - 1)/2$
$P_3 = (5x^3 - 3x)/2$
$P_4 = (35x^4 - 30x^2 + 3)/8$
$P_5 = (63x^5 - 70x^3 + 15x)/8$
$P_6 = (231x^6 - 315x^4 + 105x^2 - 5)/16$
$P_7 = (429x^7 - 693x^5 + 315x^3 - 35x)/16$
$P_8 = (6435x^8 - 12012x^6 + 6930x^4 - 1260x^2 + 35)/128$
$P_9 = (12155x^9 - 25740x^7 + 18018x^5 - 4620x^3 + 315x)/128$
$P_{10} = (46189x^{10} - 109395x^8 + 90090x^6 - 30030x^4 + 3465x^2 - 63)/256$

$x^{10} = (4199P_0 + 16150P_2 + 15504P_4 + 7904P_6 + 2176P_8 + 256P_{10})/46189$
$x^9 = (3315P_1 + 4760P_3 + 2992P_5 + 960P_7 + 128P_9)/12155$
$x^8 = (715P_0 + 2600P_2 + 2160P_4 + 832P_6 + 128P_8)/6435$
$x^7 = (143P_1 + 182P_3 + 88P_5 + 16P_7)/429$
$x^6 = (33P_0 + 110P_2 + 72P_4 + 16P_6)/231$
$x^5 = (27P_1 + 28P_3 + 8P_5)/63$
$x^4 = (7P_0 + 20P_2 + 8P_4)/35$
$x^3 = (3P_1 + 2P_3)/5$
$x^2 = (P_0 + 2P_2)/3$
$x = P_1$
$1 = P_0$

$T_0 = 1$
$T_1 = x$
$T_2 = 2x^2 - 1$
$T_3 = 4x^3 - 3x$
$T_4 = 8x^4 - 8x^2 + 1$
$T_5 = 16x^5 - 20x^3 + 5x$
$T_6 = 32x^6 - 48x^4 + 18x^2 - 1$
$T_7 = 64x^7 - 112x^5 + 56x^3 - 7x$
$T_8 = 128x^8 - 256x^6 + 160x^4 - 32x^2 + 1$
$T_9 = 256x^9 - 576x^7 + 432x^5 - 120x^3 + 9x$
$T_{10} = 512x^{10} - 1280x^8 + 1120x^6 - 400x^4 + 50x^2 - 1$

$x^{10} = (126T_0 + 210T_2 + 120T_4 + 45T_6 + 10T_8 + T_{10})/512$
$x^9 = (126T_1 + 84T_3 + 36T_5 + 9T_7 + T_9)/256$
$x^8 = (35T_0 + 56T_2 + 28T_4 + 8T_6 + T_8)/128$
$x^7 = (35T_1 + 21T_3 + 7T_5 + T_7)/64$
$x^6 = (10T_0 + 15T_2 + 6T_4 + T_6)/32$
$x^5 = (10T_1 + 5T_3 + T_5)/16$
$x^4 = (3T_0 + 4T_2 + T_4)/8$
$x^3 = (3T_1 + T_3)/4$
$x^2 = (T_0 + T_2)/2$
$x = T_1$
$1 = T_0$

$U_0 = 1$
$U_1 = 2x$
$U_2 = 4x^2 - 1$
$U_3 = 8x^3 - 4x$
$U_4 = 16x^4 - 12x^2 + 1$
$U_5 = 32x^5 - 32x^3 + 6x$
$U_6 = 64x^6 - 80x^4 + 24x^2 - 1$
$U_7 = 128x^7 - 192x^5 + 80x^3 - 8x$
$U_8 = 256x^8 - 448x^6 + 240x^4 - 40x^2 + 1$
$U_9 = 512x^9 - 1024x^7 + 672x^5 - 160x^3 + 10x$
$U_{10} = 1024x^{10} - 2304x^8 + 1792x^6 - 560x^4 + 60x^2 - 1$

$x^{10} = (42U_0 + 90U_2 + 75U_4 + 35U_6 + 9U_8 + U_{10})/1024$
$x^9 = (42U_1 + 48U_3 + 27U_5 + 8U_7 + U_9)/512$
$x^8 = (14U_0 + 28U_2 + 20U_4 + 7U_6 + U_8)/256$
$x^7 = (14U_1 + 14U_3 + 6U_5 + U_7)/128$
$x^6 = (5U_0 + 9U_2 + 5U_4 + U_6)/64$
$x^5 = (5U_1 + 4U_3 + U_5)/32$
$x^4 = (2U_0 + 3U_2 + U_4)/16$
$x^3 = (2U_1 + U_3)/8$
$x^2 = (U_0 + U_2)/4$
$x = (U_1)/2$
$1 = U_0$

CLEBSCH–GORDAN COEFFICIENTS

$$\begin{pmatrix} j_1 & j_2 & j \\ m_1 & m_2 & m \end{pmatrix} = \delta_{m,m_1+m_2} \sqrt{\frac{(j_1+j_2-j)!(j+j_1-j_2)!(j+j_2-j_1)!(2j+1)}{(j+j_1+j_2+1)!}}$$

$$\times \sum_k \frac{(-1)^k \sqrt{(j_1+m_1)!(j_1-m_1)!(j_2+m_2)!(j_2-m_2)!(j+m)!(j-m)!}}{k!(j_1+j_2-j-k)!(j_1-m_1-k)!(j_2+m_2-k)!(j-j_2+m_1+k)!(j-j_1-m_2+k)!}.$$

1. Conditions:

(a) Each of $\{j_1, j_2, j, m_1, m_2, m\}$ may be an integer, or half an integer. Additionally: $j > 0$, $j_1 > 0$, $j_2 > 0$ and $j + j_1 + j_2$ is an integer.
(b) $j_1 + j_2 - j \geq 0$.
(c) $j_1 - j_2 + j \geq 0$.
(d) $-j_1 + j_2 + j \geq 0$.
(e) $|m_1| \leq j_1$, $|m_2| \leq j_2$, $|m| \leq j$.

2. Special values:

(a) $\begin{pmatrix} j_1 & j_2 & j \\ m_1 & m_2 & m \end{pmatrix} = 0$ if $m_1 + m_2 \neq m$.
(b) $\begin{pmatrix} j_1 & 0 & j \\ m_1 & 0 & m \end{pmatrix} = \delta_{j_1,j}\delta_{m_1,m}$.
(c) $\begin{pmatrix} j_1 & j_2 & j \\ 0 & 0 & 0 \end{pmatrix} = 0$ when $j_1 + j_2 + j$ is an odd integer.
(d) $\begin{pmatrix} j_1 & j_1 & j \\ m_1 & m_1 & m \end{pmatrix} = 0$ when $2j_1 + j$ is an odd integer.

3. Symmetry relations: all of the following are equal to $\begin{pmatrix} j_1 & j_2 & j \\ m_1 & m_2 & m \end{pmatrix}$:

(a) $\begin{pmatrix} j_2 & j_1 & j \\ -m_2 & -m_1 & -m \end{pmatrix}$,
(b) $(-1)^{j_1+j_2-j} \begin{pmatrix} j_2 & j_1 & j \\ m_1 & m_2 & m \end{pmatrix}$,
(c) $(-1)^{j_1+j_2-j} \begin{pmatrix} j_1 & j_2 & j \\ -m_1 & -m_2 & -m \end{pmatrix}$,
(d) $\sqrt{\frac{2j+1}{2j_1+1}}(-1)^{j_2+m_2} \begin{pmatrix} j & j_2 & j_1 \\ -m & m_2 & -m_1 \end{pmatrix}$,
(e) $\sqrt{\frac{2j+1}{2j_1+1}}(-1)^{j_1-m_1+j-m} \begin{pmatrix} j & j_2 & j_1 \\ m & -m_2 & m_1 \end{pmatrix}$,
(f) $\sqrt{\frac{2j+1}{2j_1+1}}(-1)^{j-m+j_1-m_1} \begin{pmatrix} j_2 & j & j_1 \\ m_2 & -m & -m_1 \end{pmatrix}$,
(g) $\sqrt{\frac{2j+1}{2j_2+1}}(-1)^{j_1-m_1} \begin{pmatrix} j_1 & j & j_2 \\ m_1 & -m & -m_2 \end{pmatrix}$,
(h) $\sqrt{\frac{2j+1}{2j_2+1}}(-1)^{j_1-m_1} \begin{pmatrix} j & j_1 & j_2 \\ m & -m_1 & m_2 \end{pmatrix}$.

By use of the symmetry relations, Clebsch–Gordan coefficients may be put in the standard form $j_1 \leq j_2 \leq j$ and $m \geq 0$.

m_2	m	j_1	j	$\left(\begin{array}{cc\|c} j_1 & \frac{1}{2} & j \\ m_1 & m_2 & m \end{array}\right)$
$-\frac{1}{2}$	0	$\frac{1}{2}$	1	$\frac{\sqrt{2}}{2} \approx 0.707107$
0	$\frac{1}{2}$	$\frac{1}{2}$	1	$\frac{\sqrt{3}}{2} \approx 0.866025$
$\frac{1}{2}$	0	$\frac{1}{2}$	1	$\frac{\sqrt{2}}{2} \approx 0.707107$
$\frac{1}{2}$	$\frac{1}{2}$	$\frac{1}{2}$	1	$\frac{\sqrt{3}}{2} \approx 0.866025$
$\frac{1}{2}$	1	$\frac{1}{2}$	1	$1 \approx 1.000000$

m_2	m	j_1	j	$\left(\begin{array}{cc\|c} j_1 & 1 & j \\ m_1 & m_2 & m \end{array}\right)$
-1	0	1	1	$\frac{\sqrt{2}}{2} \approx 0.707107$
-1	0	1	2	$\frac{\sqrt{6}}{6} \approx 0.408248$
$-\frac{1}{2}$	0	$\frac{1}{2}$	$\frac{3}{2}$	$\frac{\sqrt{2}}{2} \approx 0.707107$
$-\frac{1}{2}$	$\frac{1}{2}$	1	1	$\frac{3}{4} \approx 0.750000$
$-\frac{1}{2}$	$\frac{1}{2}$	1	2	$\frac{\sqrt{5}}{4} \approx 0.559017$
0	0	1	2	$\frac{\sqrt{6}}{3} \approx 0.816496$
0	0	$\frac{1}{2}$	$\frac{3}{2}$	$\frac{\sqrt{3}}{2} \approx 0.866025$
0	$\frac{1}{2}$	$\frac{1}{2}$	$\frac{3}{2}$	$\frac{\sqrt{6}}{3} \approx 0.8164967$
0	$\frac{1}{2}$	1	1	$\frac{\sqrt{2}}{4} \approx 0.353553$
0	$\frac{1}{2}$	1	2	$\frac{\sqrt{10}}{4} \approx 0.790569$
0	1	1	1	$\frac{\sqrt{2}}{2} \approx 0.707107$

m_2	m	j_1	j	$\left(\begin{array}{cc\|c} j_1 & 1 & j \\ m_1 & m_2 & m \end{array}\right)$
0	1	1	2	$\frac{\sqrt{2}}{2} \approx 0.707107$
$\frac{1}{2}$	0	$\frac{1}{2}$	$\frac{3}{2}$	$\frac{\sqrt{2}}{2} \approx 0.707107$
$\frac{1}{2}$	$\frac{1}{2}$	1	1	$-\frac{\sqrt{2}}{4} \approx -0.353553$
$\frac{1}{2}$	$\frac{1}{2}$	1	2	$\frac{\sqrt{10}}{4} \approx 0.790569$
$\frac{1}{2}$	1	$\frac{1}{2}$	$\frac{3}{2}$	$\frac{\sqrt{30}}{6} \approx 0.912871$
$\frac{1}{2}$	$\frac{3}{2}$	1	2	$\frac{\sqrt{105}}{12} \approx 0.853913$
1	0	1	1	$-\frac{\sqrt{2}}{2} \approx -0.707107$
1	0	1	2	$\frac{\sqrt{6}}{6} \approx 0.408248$
1	$\frac{1}{2}$	$\frac{1}{2}$	$\frac{3}{2}$	$\frac{\sqrt{3}}{3} \approx 0.577350$
1	$\frac{1}{2}$	1	1	$-\frac{3}{4} \approx -0.750000$
1	$\frac{1}{2}$	1	2	$\frac{\sqrt{5}}{4} \approx 0.559017$
1	1	$\frac{1}{2}$	$\frac{3}{2}$	$\frac{\sqrt{10}}{4} \approx 0.790569$
1	1	1	1	$-\frac{\sqrt{2}}{2} \approx -0.707107$
1	1	1	2	$\frac{\sqrt{2}}{2} \approx 0.707107$
1	$\frac{3}{2}$	$\frac{1}{2}$	$\frac{3}{2}$	$1 \approx 1.000000$
1	$\frac{3}{2}$	1	2	$\frac{\sqrt{105}}{12} \approx 0.853913$
1	2	1	2	$1 \approx 1.000000$

NORMAL PROBABILITY FUNCTION

Table of the Normal Distribution

For a standard normal random variable ($\Phi(z)$ is the area under the Standard Normal Curve from $-\infty$ to z).

Limits		Proportion of the total area	Remaining area
$\mu - \lambda\sigma$	$\mu + \lambda\sigma$	(%)	(%)
$\mu - \sigma$	$\mu + \sigma$	68.27	31.73
$\mu - 1.65\sigma$	$\mu + 1.65\sigma$	90	10
$\mu - 1.96\sigma$	$\mu + 1.96\sigma$	95	5
$\mu - 2\sigma$	$\mu + 2\sigma$	95.45	4.55
$\mu - 2.58\sigma$	$\mu + 2.58\sigma$	99.0	0.99
$\mu - 3\sigma$	$\mu + 3\sigma$	99.73	0.27
$\mu - 3.09\sigma$	$\mu + 3.09\sigma$	99.8	0.2
$\mu - 3.29\sigma$	$\mu + 3.29\sigma$	99.9	0.1

x	1.282	1.645	1.960	2.326	2.576	3.090
$\Phi(x)$	0.90	0.95	0.975	0.99	0.995	0.999
$2[1 - \Phi(x)]$	0.20	0.10	0.05	0.02	0.01	0.002

x	3.09	3.72	4.26	4.75	5.20	5.61	6.00	6.36
$1 - \Phi(x)$	10^{-3}	10^{-4}	10^{-5}	10^{-6}	10^{-7}	10^{-8}	10^{-9}	10^{-10}

Areas under the Standard Normal Curve from 0 to z

z	0	1	2	3	4	5	6	7	8	9
0.0	.0000	.0040	.0080	.0120	.0160	.0199	.0239	.0279	.0319	.0359
0.1	.0398	.0438	.0478	.0517	.0557	.0596	.0636	.0675	.0714	.0754
0.2	.0793	.0832	.0871	.0910	.0948	.0987	.1026	.1064	.1103	.1141
0.3	.1179	.1217	.1255	.1293	.1331	.1368	.1406	.1443	.1480	.1517
0.4	.1554	.1591	.1628	.1664	.1700	.1736	.1772	.1808	.1844	.1879
0.5	.1915	.1950	.1985	.2019	.2054	.2088	.2123	.2157	.2190	.2224
0.6	.2258	.2291	.2324	.2357	.2389	.2422	.2454	.2486	.2518	.2549
0.7	.2580	.2612	.2652	.2673	.2704	.2734	.2764	.2794	.2823	.2852
0.8	.2881	.2910	.2939	.2967	.2996	.3023	.3051	.3078	.3106	.3133
0.9	.3159	.3186	.3212	.3238	.3264	.3289	.3315	.3340	.3365	.3389
1.0	.3413	.3438	.3461	.3485	.3508	.3531	.3554	.3577	.3599	.3621
1.1	.3643	.3665	.3686	.3708	.3729	.3749	.3770	.3790	.3810	.3830
1.2	.3849	.3869	.3888	.3907	.3925	.3944	.3962	.3980	.3997	.4015
1.3	.4032	.4049	.4066	.4082	.4099	.4115	.4131	.4147	.4162	.4177
1.4	.4192	.4207	.4222	.4236	.4251	.4265	.4279	.4292	.4306	.4319
1.5	.4332	.4345	.4357	.4370	.4382	.4394	.4406	.4418	.4429	.4441
1.6	.4452	.4463	.4474	.4484	.4495	.4505	.4515	.4525	.4535	.4545
1.7	.4554	.4564	.4573	.4582	.4591	.4599	.4608	.4616	.4625	.4633
1.8	.4641	.4649	.4656	.4664	.4671	.4678	.4686	.4693	.4699	.4706
1.9	.4713	.4719	.4726	.4732	.4738	.4744	.4750	.4756	.4761	.4767
2.0	.4772	.4778	.4783	.4788	.4793	.4798	.4803	.4808	.4812	.4817
2.1	.4821	.4826	.4830	.4834	.4838	.4842	.4846	.4850	.4854	.4857
2.2	.4861	.4864	.4868	.4871	.4875	.4878	.4881	.4884	.4887	.4890
2.3	.4893	.4896	.4898	.4901	.4904	.4906	.4909	.4911	.4913	.4916
2.4	.4918	.4920	.4922	.4925	.4927	.4929	.4931	.4932	.4934	.4936
2.5	.4938	.4940	.4941	.4943	.4945	.4946	.4948	.4949	.4951	.4952
2.6	.4953	.4955	.4956	.4957	.4959	.4960	.4961	.4962	.4963	.4964
2.7	.4965	.4966	.4967	.4968	.4969	.4970	.4971	.4972	.4973	.4974
2.8	.4974	.4975	.4976	.4977	.4977	.4978	.4979	.4979	.4980	.4981
2.9	.4981	.4982	.4982	.4983	.4984	.4984	.4985	.4985	.4986	.4986
3.0	.4987	.4987	.4987	.4988	.4988	.4989	.4989	.4989	.4990	.4990
3.1	4990	.4991	.4991	.4991	.4992	.4992	.4992	.4992	.4993	.4993
3.2	4993	.4993	.4994	.4994	.4994	.4994	.4994	.4995	.4995	.4995
3.3	4995	.4995	.4995	.4996	.4996	.4996	.4996	.4996	.4996	.4997
3.4	4997	.4997	.4997	.4997	.4997	.4997	.4997	.4997	.4997	.4998
3.5	4998	.4998	.4998	.4998	.4998	.4998	.4998	.4998	.4998	.4998
3.6	4998	.4998	.4999	.4999	.4999	.4999	.4999	.4999	.4999	.4999
3.7	4999	.4999	.4999	.4999	.4999	.4999	.4999	.4999	.4999	.4999
3.8	4999	.4999	.4999	.4999	.4999	.4999	.4999	.4999	.4999	.4999
3.9	5000	.5000	.5000	.5000	.5000	.5000	.5000	.5000	.5000	.5000

Common sample size calculations

Parameter	Estimate	Sample size
μ	$\bar{x}$	$n = \left(\frac{z_{\alpha/2} \cdot \sigma}{E}\right)^2$
p	$\hat{p}$	$n = \frac{(z_{\alpha/2})^2 \cdot pq}{E^2}$
$\mu_2 - \mu_2$	$\bar{x}_1 - \bar{x}_2$	$n_1 = n_2 = \frac{(z_{\alpha/2})^2(\sigma_1^2 + \sigma_2^2)}{E^2}$
$p_1 - p_2$	$\hat{p}_1 - \hat{p}_2$	$n_1 = n_2 = \frac{(z_{\alpha/2})^2(p_1 q_1 + p_2 q_2)}{E^2}$

Common one-sample confidence intervals

Parameter	Assumptions	$100(1-\alpha)\%$ Confidence interval
μ	n large, σ^2 known, or normality, σ^2 known	$\bar{x} \pm z_{\alpha/2} \cdot \frac{\sigma}{\sqrt{n}}$
μ	normality, σ^2 unknown	$\bar{x} \pm t_{\alpha/2,n-1} \cdot \frac{s}{\sqrt{n}}$
σ^2	normality	$\left(\frac{(n-1)s^2}{\chi^2_{\alpha/2,n-1}}, \frac{(n-1)s^2}{\chi^2_{1-\alpha/2,n-1}}\right)$
p	binomial experiment, n large	$\hat{p} \pm z_{\alpha/2} \cdot \sqrt{\frac{\hat{p}(1-\hat{p})}{n}}$

Common two-sample confidence intervals

Parameter	Assumptions	$100(1-\alpha)\%$ Confidence interval
$\mu_1 - \mu_2$	normality, independence, σ_1^2, σ_2^2 known or n_1, n_2 large, independence, σ_1^2, σ_2^2 known	$(\bar{x}_1 - \bar{x}_2) \pm z_{\alpha/2} \cdot \sqrt{\frac{\sigma_1^2}{n_1} + \frac{\sigma_2^2}{n_2}}$
$\mu_1 - \mu_2$	normality, independence, $\sigma_1^2 = \sigma_2^2$ unknown	$(\bar{x}_1 - \bar{x}_2) \pm$ $t_{\frac{\alpha}{2},n_1+n_2-2} \cdot s_p\sqrt{\frac{1}{n_1} + \frac{1}{n_2}}$ $s_p^2 = \frac{(n_1-1)s_1^2 + (n_2-1)s_2^2}{n_1+n_2-2}$
$\mu_1 - \mu_2$	normality, independence, $\sigma_1^2 \neq \sigma_2^2$ unknown	$(\bar{x}_1 - \bar{x}_2) \pm t_{\alpha/2,\nu} \cdot \sqrt{\frac{s_1^2}{n_1} + \frac{s_2^2}{n_2}}$ $\nu \approx \frac{\left(\frac{s_1^2}{n_1} + \frac{s_2^2}{n_2}\right)^2}{\frac{(s_1^2/n_1)^2}{n_1-1} + \frac{(s_2^2/n_2)^2}{n_2-1}}$
$\mu_1 - \mu_2$	normality, n pairs, dependence	$\bar{d} \pm t_{\alpha/2,n-1} \cdot \frac{s_d}{\sqrt{n}}$
$p_1 - p_2$	binomial experiments, n_1, n_2 large, independence	$(\hat{p}_1 - \hat{p}_2) \pm$ $z_{\alpha/2} \cdot \sqrt{\frac{\hat{p}_1(1-\hat{p}_1)}{n_1} + \frac{\hat{p}_2(1-\hat{p}_2)}{n_2}}$

PERCENTAGE POINTS, STUDENT'S *T*-DISTRIBUTION

This table gives values of t such that

$$F(t) = \int_{-\infty}^{t} \frac{\Gamma\left(\frac{n+1}{2}\right)}{\sqrt{n\pi}\,\Gamma\left(\frac{n}{2}\right)} \left(1 + \frac{x^2}{n}\right) - \frac{n+1}{2}\, dx$$

for n, the number of degrees of freedom, equal to 1, 2, . . . , 30, 40, 60, 120, ∞; and for $F(t) = 0.60, 0.75, 0.90, 0.95, 0.975, 0.99, 0.995$, and 0.9995. The t-distribution is symmetrical, so that $F(-t) = 1 - F(t)$

n/F	.60	.75	.90	.95	.975	.99	.995	.9995
1	.325	1.000	3.078	6.314	12.706	31.821	63.657	636.619
2	.289	.816	1.886	2.920	4.303	6.965	9.925	31.598
3	.277	.765	1.638	2.353	3.182	4.541	5.841	12.924
4	.271	.741	1.533	2.132	2.776	3.747	4.604	8.610
5	.267	.727	1.476	2.015	2.571	3.365	4.032	6.869
6	.265	.718	1.440	1.943	2.447	3.143	3.707	5.959
7	.263	.711	1.415	1.895	2.365	2.998	3.499	5.408
8	.262	.706	1.397	1.860	2.306	2.896	3.355	5.041
9	.261	.703	1.383	1.833	2.262	2.821	3.250	4.781
10	.260	.700	1.372	1.812	2.228	2.764	3.169	4.587
11	.260	.697	1.363	1.796	2.201	2.718	3.106	4.437
12	.259	.695	1.356	1.782	2.179	2.681	3.055	4.318
13	.259	.694	1.350	1.771	2.160	2.650	3.012	4.221
14	.258	.692	1.345	1.761	2.145	2.624	2.977	4.140
15	.258	.691	1.341	1.753	2.131	2.602	2.947	4.073
16	.258	.690	1.337	1.746	2.120	2.583	2.921	4.015
17	.257	.689	1.333	1.740	2.110	2.567	2.898	3.965
18	.257	.688	1.330	1.734	2.101	2.552	2.878	3.922
19	.257	.688	1.328	1.729	2.093	2.539	2.861	3.883
20	.257	.687	1.325	1.725	2.086	2.528	2.845	3.850
21	.257	.686	1.323	1.721	2.080	2.518	2.831	3.819
22	.256	.686	1.321	1.717	2.074	2.508	2.819	3.792
23	.256	.685	1.319	1.714	2.069	2.500	2.807	3.767
24	.256	.685	1.318	1.711	2.064	2.492	2.797	3.745
25	.256	.684	1.316	1.708	2.060	2.485	2.787	3.725
26	.256	.684	1.315	1.706	2.056	2.479	2.779	3.707
27	.256	.684	1.314	1.703	2.052	2.473	2.771	3.690
28	.256	.683	1.313	1.701	2.048	2.467	2.763	3.674
29	.256	.683	1.311	1.699	2.045	2.462	2.756	3.659
30	.256	.683	1.310	1.697	2.042	2.457	2.750	3.646
40	.255	.681	1.303	1.684	2.021	2.423	2.704	3.551
60	.254	.679	1.296	1.671	2.000	2.390	2.660	3.460
120	.254	.677	1.289	1.658	1.980	2.358	2.617	3.373
∞	.253	.674	1.282	1.645	1.960	2.326	2.576	3.291

*This table is abridged from the *Statistical Tables* by R. A. Fisher and Frank Yates published by Oliver & Boyd. Ltd., Edinburgh and London, 1938. It is published here with the kind permission of the authors and their publishers.

PERCENTAGE POINTS, CHI-SQUARE DISTRIBUTION

This table gives values of χ^2 such that

$$F(\chi)^2 = \int_0^{\chi^2} \frac{1}{2^{n/2}\Gamma\left(\frac{n}{2}\right)} x^{(n-2)/2} e^{-x/2}\, dx$$

for n, the number of degrees of freedom, equal to 1, 2, . . ., 30. For $n > 30$, a normal approximation is quite accurate. The expression $\sqrt{2x^2} - \sqrt{2n-1}$ is approximately normally distributed as the standard normal distribution. Thus χ_α^2, the α-point of the distribution, may be computed by the formula

$$\chi_\alpha^2 = \frac{1}{2}[x_\alpha + \sqrt{2n-1}]^2,$$

where x_α is the α-point of the cumulative normal distribution. For even values of n, $F(\chi^2)$ can be written as

$$1 - F(\chi^2) = \sum_{x=0}^{x'-1} \frac{e^{-\lambda}\lambda^x}{x!}$$

with $\lambda = \frac{1}{2}\chi^2$ and $x' = \frac{1}{2}n$. Thus the cumulative chi-square distribution is related to the cumulative Poisson distribution.
Another approximate formula for large n

$$\chi_\alpha^2 = n\left(1 - \frac{2}{9n} + z_\alpha\sqrt{\frac{2}{9n}}\right)^3$$

n = degrees of freedom
z_α = the normal deviate (the value of x for which $F(x)$ = the desired percentile).

x	1.282	1.645	1.960	2.326	2.576	3.090
$F(x)$	.90	.95	.975	.99	.995	.999

$\chi^2_{.99} = 60[1 - 0.00370 + 2.326(0.06086)]^3 = 88.4$ is the 99th percentile for 60 degrees of freedom.

$$F(\chi^2) = \int_0^{\chi^2} \frac{1}{2^{n/2}\Gamma\left(\frac{n}{2}\right)} x^{n-2/2} e^{-x/2}\, dx$$

$n \backslash F$	.005	.010	.025	.050	.100	.250	.500	.750	.900	.950	.975	.990	.995
1	.0000393	.000157	.000982	.00393	.0158	.102	.455	1.32	2.71	3.84	5.02	6.63	7.88
2	.0100	.0201	.0506	.103	.211	.575	1.39	2.77	4.61	5.99	7.38	9.21	10.6
3	.0717	.115	.216	.352	.584	1.21	2.37	4.11	6.25	7.81	9.35	11.3	12.8
4	.207	.297	.484	.711	1.06	1.92	3.36	5.39	7.78	9.49	11.1	13.3	14.9
5	.412	.554	.831	1.15	1.61	2.67	4.35	6.63	9.24	11.1	12.8	15.1	16.7
6	.676	.872	1.24	1.64	2.20	3.45	5.35	7.84	10.6	12.6	14.4	16.8	18.5
7	.989	1.24	1.69	2.17	2.83	4.25	6.35	9.04	12.0	14.1	16.0	18.5	20.3
8	1.34	1.65	2.18	2.73	3.49	5.07	7.34	10.2	13.4	15.5	17.5	20.1	22.0
9	1.73	2.09	2.70	3.33	4.17	5.90	8.34	11.4	14.7	16.9	19.0	21.7	23.6
10	2.16	2.56	3.25	3.94	4.87	6.74	9.34	12.5	16.0	18.3	20.5	23.2	25.2
11	2.60	3.05	3.82	4.57	5.58	7.58	10.3	13.7	17.3	19.7	21.9	24.7	26.8
12	3.07	3.57	4.40	5.23	6.30	8.44	11.3	14.8	18.5	21.0	23.3	26.2	28.3
13	3.57	4.11	5.01	5.89	7.04	9.30	12.3	16.0	19.8	22.4	24.7	27.7	29.8
14	4.07	4.66	5.63	6.57	7.79	10.2	13.3	17.1	21.1	23.7	26.1	29.1	31.3
15	4.60	5.23	6.26	7.26	8.55	11.0	14.3	18.2	22.3	25.0	27.5	30.6	32.8
16	5.14	5.81	6.91	7.96	9.31	11.9	15.3	19.4	23.5	26.3	28.8	32.0	34.3
17	5.70	6.41	7.56	8.67	10.1	12.8	16.3	20.5	24.8	27.6	30.2	33.4	35.7
18	6.26	7.01	8.23	9.39	10.9	13.7	17.3	21.6	26.0	28.9	31.5	34.8	37.2
19	6.84	7.63	8.91	10.1	11.7	14.6	18.3	22.7	27.2	30.1	32.9	36.2	38.6
20	7.43	8.26	9.59	10.9	12.4	15.5	19.3	23.8	28.4	31.4	34.2	37.6	40.0
21	8.03	8.90	10.3	11.6	13.2	16.3	20.3	24.9	29.6	32.7	35.5	38.9	41.4
22	8.64	9.54	11.0	12.3	14.0	17.2	21.3	26.0	30.8	33.9	36.8	40.3	42.8
23	9.26	10.2	11.7	13.1	14.8	18.1	22.3	27.1	32.0	35.2	38.1	41.6	44.2
24	9.89	10.9	12.4	13.8	15.7	19.0	23.3	28.2	33.2	36.4	39.4	43.0	45.6
25	10.5	11.5	13.1	14.6	16.5	19.9	24.3	29.3	34.4	37.7	40.6	44.3	46.9
26	11.2	12.2	13.8	15.4	17.3	20.8	25.3	30.4	35.6	38.9	41.9	45.6	48.3
27	11.8	12.9	14.6	16.2	18.1	21.7	26.3	31.5	36.7	40.1	43.2	47.0	49.6
28	12.5	12.6	15.3	16.9	18.9	22.7	27.3	32.6	37.9	41.3	44.5	48.3	51.0
29	13.1	14.3	16.0	17.7	19.8	23.6	28.3	33.7	39.1	42.6	45.7	49.6	52.3
30	13.8	15.0	16.8	18.5	20.6	24.5	29.3	34.8	40.3	43.8	47.0	50.9	53.7

PERCENTAGE POINTS, F-DISTRIBUTION

This table gives values of F such that

$$F(F) = \int_0^F \frac{\Gamma\left(\frac{m+n}{2}\right)}{\Gamma\left(\frac{m}{2}\right)\Gamma\left(\frac{n}{2}\right)} m^{m/2} n^{n/2} x^{m-2/2} (n+mx)^{-(m+n)/2}\, dx$$

for selected values of m, the number of degrees of freedom of the numerator of F; and for selected values of n, the number of degrees freedom of the denominator of F. The table also provides values corresponding to $F(F) = .10, .05, .025, .01, .005, .001$ since $F_{1-\alpha}$ for m and n degrees of freedom is the reciprocal of F_α for n and m degrees of freedom. Thus

$$F_{.05}(4,7) = \frac{1}{F_{.95}(7,4)} = \frac{1}{6.09} = .164$$

$$F(F) - \int_0^F \frac{\Gamma\left(\frac{m+n}{2}\right)}{\Gamma\left(\frac{m}{2}\right)\Gamma\left(\frac{n}{2}\right)} m^{m/2} n^{n/2} x^{(m/2)-1} (n+mx)^{-(m+n)/2}\, dx = .90$$

$n \backslash m$	1	2	3	4	5	6	7	8	9	10	12	15	20	24	30	40	60	120	∞
1	39.86	49.50	53.59	55.83	57.24	58.20	58.91	59.44	59.86	60.19	60.71	61.22	61.74	62.00	62.26	62.53	62.79	63.06	63.33
2	8.53	9.00	9.16	9.24	9.29	9.33	9.35	9.37	9.38	9.39	9.41	9.42	9.44	9.45	9.46	9.47	9.47	9.48	9.49
3	5.54	5.46	5.39	5.34	5.31	5.28	5.27	5.25	5.24	5.23	5.22	5.20	5.18	5.18	5.17	5.16	5.15	5.14	5.13
4	4.54	4.32	4.19	4.11	4.05	4.01	3.98	3.95	3.94	3.92	3.90	3.87	3.84	3.83	3.82	3.80	3.79	3.78	3.76
5	4.06	3.78	3.62	3.52	3.45	3.40	3.37	3.34	3.32	3.30	3.27	3.24	3.21	3.19	3.17	3.16	3.14	3.12	3.10
6	3.78	3.46	3.29	3.18	3.11	3.05	3.01	2.98	2.96	2.94	2.90	2.87	2.84	2.82	2.80	2.78	2.76	2.74	2.72
7	3.59	3.26	3.07	2.96	2.88	2.83	2.78	2.75	2.72	2.70	2.67	2.63	2.59	2.58	2.56	2.54	2.51	2.49	2.47
8	3.46	3.11	2.92	2.81	2.73	2.67	2.62	2.59	2.56	2.54	2.50	2.46	2.42	2.40	2.38	2.36	2.34	2.32	2.29
9	3.36	3.01	2.81	2.69	2.61	2.55	2.51	2.47	2.44	2.42	2.38	2.34	2.30	2.28	2.25	2.23	2.21	2.18	2.16
10	3.29	2.92	2.73	2.61	2.52	2.46	2.41	2.38	2.35	2.32	2.28	2.24	2.20	2.18	2.16	2.13	2.11	2.08	2.06
11	3.23	2.86	2.66	2.54	2.45	2.39	2.34	2.30	2.27	2.25	2.21	2.17	2.12	2.10	2.08	2.05	2.03	2.00	1.97
12	3.18	2.81	2.61	2.48	2.39	2.33	2.28	2.24	2.21	2.19	2.15	2.10	2.06	2.04	2.01	1.99	1.96	1.93	1.90
13	3.14	2.76	2.56	2.43	2.35	2.28	2.23	2.20	2.16	2.14	2.10	2.05	2.01	1.98	1.96	1.93	1.90	1.88	1.85
14	3.10	2.73	2.52	2.39	2.31	2.24	2.19	2.15	2.12	2.10	2.05	2.01	1.96	1.94	1.91	1.89	1.86	1.83	1.80
15	3.07	2.70	2.49	2.36	2.27	2.21	2.16	2.12	2.09	2.06	2.02	1.97	1.92	1.90	1.87	1.85	1.82	1.79	1.76
16	3.05	2.67	2.46	2.33	2.24	2.18	2.13	2.09	2.06	2.03	1.99	1.94	1.89	1.87	1.84	1.81	1.78	1.75	1.72
17	3.03	2.64	2.44	2.31	2.22	2.15	2.10	2.06	2.03	2.00	1.96	1.91	1.86	1.84	1.81	1.78	1.75	1.72	1.69
18	3.01	2.62	2.42	2.29	2.20	2.13	2.08	2.04	2.00	1.98	1.93	1.89	1.84	1.81	1.78	1.75	1.72	1.69	1.66
19	2.99	2.61	2.40	2.27	2.18	2.11	2.06	2.02	1.98	1.96	1.91	1.86	1.81	1.79	1.76	1.73	1.70	1.67	1.63
20	2.97	2.59	2.38	2.25	2.16	2.09	2.04	2.00	1.96	1.94	1.89	1.84	1.79	1.77	1.74	1.71	1.68	1.64	1.61
21	2.96	2.57	2.36	2.23	2.14	2.08	2.02	1.98	1.95	1.92	1.87	1.83	1.78	1.75	1.72	1.69	1.66	1.62	1.59
22	2.95	2.56	2.35	2.22	2.13	2.06	2.01	1.97	1.93	1.90	1.86	1.81	1.76	1.73	1.70	1.67	1.64	1.60	1.57
23	2.94	2.55	2.34	2.21	2.11	2.05	1.99	1.95	1.92	1.89	1.84	1.80	1.74	1.72	1.69	1.66	1.62	1.59	1.55
24	2.93	2.54	2.33	2.19	2.10	2.04	1.98	1.94	1.91	1.88	1.83	1.78	1.73	1.70	1.67	1.64	1.61	1.57	1.53
25	2.92	2.53	2.32	2.18	2.09	2.02	1.97	1.93	1.89	1.87	1.82	1.77	1.72	1.69	1.66	1.63	1.59	1.56	1.52
26	2.91	2.52	2.31	2.17	2.08	2.01	1.96	1.92	1.88	1.86	1.81	1.76	1.71	1.68	1.65	1.61	1.58	1.54	1.50
27	2.90	2.51	2.30	2.17	2.07	2.00	1.95	1.91	1.87	1.85	1.80	1.75	1.70	1.67	1.64	1.60	1.57	1.53	1.49
28	2.89	2.50	2.29	2.16	2.06	2.00	1.94	1.90	1.87	1.84	1.79	1.74	1.69	1.66	1.63	1.59	1.56	1.52	1.48
29	2.89	2.50	2.28	2.15	2.06	1.99	1.93	1.89	1.86	1.83	1.78	1.73	1.68	1.65	1.62	1.58	1.55	1.51	1.47
30	2.88	2.49	2.28	2.14	2.05	1.98	1.93	1.88	1.85	1.82	1.77	1.72	1.67	1.64	1.61	1.57	1.54	1.50	1.46
40	2.84	2.44	2.23	2.09	2.00	1.93	1.87	1.83	1.79	1.76	1.71	1.66	1.61	1.57	1.54	1.51	1.47	1.42	1.38
60	2.79	2.39	2.18	2.04	1.95	1.87	1.82	1.77	1.74	1.71	1.66	1.60	1.54	1.51	1.48	1.44	1.40	1.35	1.29
120	2.75	2.35	2.13	1.99	1.90	1.82	1.77	1.72	1.68	1.65	1.60	1.55	1.48	1.45	1.41	1.37	1.32	1.26	1.19
∞	2.71	2.30	2.08	1.94	1.85	1.77	1.72	1.67	1.63	1.60	1.55	1.49	1.42	1.38	1.34	1.30	1.24	1.17	1.00

$F = \frac{s_1^2}{s_2^2} = \frac{S_1}{m} / \frac{S_2}{n}$, where $s_1^2 = S_1/m$ and $s_2^2 = S_2/n$ are independent mean squares estimating a common variance σ^2 and based on m and n degrees of freedom, respectively.

$$F(F) = \int_0^F \frac{\Gamma\left(\frac{m+n}{2}\right)}{\Gamma\left(\frac{m}{2}\right)\Gamma\left(\frac{n}{2}\right)} m^{m/2} n^{n/2} x^{(m/2)-1}(n+mx)^{-(m+n)/2}\, dx = .95$$

$n \backslash m$	1	2	3	4	5	6	7	8	9	10	12	15	20	24	30	40	60	120	∞
1	161.4	199.5	215.7	224.6	230.2	234.0	236.8	238.9	240.5	241.9	243.9	245.9	248.0	249.1	250.1	251.1	252.2	253.3	254.3
2	18.51	19.00	19.16	19.25	19.30	19.33	19.35	19.37	19.38	19.40	19.41	19.43	19.45	19.45	19.46	19.47	19.48	19.49	19.50
3	10.13	9.55	9.28	9.12	9.01	8.94	8.89	8.85	8.81	8.79	8.74	8.70	8.66	8.64	8.62	8.59	8.57	8.55	8.53
4	7.71	6.94	6.59	6.39	6.26	6.16	6.09	6.04	6.00	5.96	5.91	5.86	5.80	5.77	5.75	5.72	5.69	5.66	5.63
5	6.61	5.79	5.41	5.19	5.05	4.95	4.88	4.82	4.77	4.74	4.68	4.62	4.56	4.53	4.50	4.46	4.43	4.40	4.36
6	5.99	5.14	4.76	4.53	4.39	4.28	4.21	4.15	4.10	4.06	4.00	3.94	3.87	3.84	3.81	3.77	3.74	3.70	3.67
7	5.59	4.74	4.35	4.12	3.97	3.87	3.79	3.73	3.68	3.64	3.57	3.51	3.44	3.41	3.38	3.34	3.30	3.27	3.23
8	5.32	4.46	4.07	3.84	3.69	3.58	3.50	3.44	3.39	3.35	3.28	3.22	3.15	3.12	3.08	3.04	3.01	2.97	2.93
9	5.12	4.26	3.86	3.63	3.48	3.37	3.29	3.23	3.18	3.14	3.07	3.01	2.94	2.90	2.86	2.83	2.79	2.75	2.71
10	4.96	4.10	3.71	3.48	3.33	3.22	3.14	3.07	3.02	2.98	2.91	2.85	2.77	2.74	2.70	2.66	2.62	2.58	2.54
11	4.84	3.98	3.59	3.36	3.20	3.09	3.01	2.95	2.90	2.85	2.79	2.72	2.65	2.61	2.57	2.53	2.49	2.45	2.40
12	4.75	3.89	3.49	3.26	3.11	3.00	2.91	2.85	2.80	2.75	2.69	2.62	2.54	2.51	2.47	2.43	2.38	2.34	2.30
13	4.67	3.81	3.41	3.18	3.03	2.92	2.83	2.77	2.71	2.67	2.60	2.53	2.46	2.42	2.38	2.34	2.30	2.25	2.21
14	4.60	3.74	3.34	3.11	2.96	2.85	2.76	2.70	2.65	2.60	2.53	2.46	2.39	2.35	2.31	2.27	2.22	2.18	2.13
15	4.54	3.68	3.29	3.06	2.90	2.79	2.71	2.64	2.59	2.54	2.48	2.40	2.33	2.29	2.25	2.20	2.16	2.11	2.07
16	4.49	3.63	3.24	3.01	2.85	2.74	2.66	2.59	2.54	2.49	2.42	2.35	2.28	2.24	2.19	2.15	2.11	2.06	2.01
17	4.45	3.59	3.20	2.96	2.81	2.70	2.61	2.55	2.49	2.45	2.38	2.31	2.23	2.19	2.15	2.10	2.06	2.01	1.96
18	4.41	3.55	3.16	2.93	2.77	2.66	2.58	2.51	2.46	2.41	2.34	2.27	2.19	2.15	2.11	2.06	2.02	1.97	1.92
19	4.38	3.52	3.13	2.90	2.74	2.63	2.54	2.48	2.42	2.38	2.31	2.23	2.16	2.11	2.07	2.03	1.98	1.93	1.88
20	4.35	3.49	3.10	2.87	2.71	2.60	2.51	2.45	2.39	2.35	2.28	2.20	2.12	2.08	2.04	1.99	1.95	1.90	1.84
21	4.32	3.47	3.07	2.84	2.68	2.57	2.49	2.42	2.37	2.32	2.25	2.18	2.10	2.05	2.01	1.96	1.92	1.87	1.81
22	4.30	3.44	3.05	2.82	2.66	2.55	2.46	2.40	2.34	2.30	2.23	2.15	2.07	2.03	1.98	1.94	1.89	1.84	1.78
23	4.28	3.42	3.03	2.80	2.64	2.53	2.44	2.37	2.32	2.27	2.20	2.13	2.05	2.01	1.96	1.91	1.86	1.81	1.76
24	4.26	3.40	3.01	2.78	2.62	2.51	2.42	2.36	2.30	2.25	2.18	2.11	2.03	1.98	1.94	1.89	1.84	1.79	1.73
25	4.24	3.39	2.99	2.76	2.60	2.49	2.40	2.34	2.28	2.24	2.16	2.09	2.01	1.96	1.92	1.87	1.82	1.77	1.71
26	4.23	3.37	2.98	2.74	2.59	2.47	2.39	2.32	2.27	2.22	2.15	2.07	1.99	1.95	1.90	1.85	1.80	1.75	1.69
27	4.21	3.35	2.96	2.73	2.57	2.46	2.37	2.31	2.25	2.20	2.13	2.06	1.97	1.93	1.88	1.84	1.79	1.73	1.67
28	4.20	3.34	2.95	2.71	2.56	2.45	2.36	2.29	2.24	2.19	2.12	2.04	1.96	1.91	1.87	1.82	1.77	1.71	1.65
29	4.18	3.33	2.93	2.70	2.55	2.43	2.35	2.28	2.22	2.18	2.10	2.03	1.94	1.90	1.85	1.81	1.75	1.70	1.64
30	4.17	3.32	2.92	2.69	2.53	2.42	2.33	2.27	2.21	2.16	2.09	2.01	1.93	1.89	1.84	1.79	1.74	1.68	1.62
40	4.08	3.23	2.84	2.61	2.45	2.34	2.25	2.18	2.12	2.08	2.00	1.92	1.84	1.79	1.74	1.69	1.64	1.58	1.51
60	4.00	3.15	2.76	2.53	2.37	2.25	2.17	2.10	2.04	1.99	1.92	1.84	1.75	1.70	1.65	1.59	1.53	1.47	139
120	3.92	3.07	2.68	2.45	2.29	2.17	2.09	2.02.	1.96	1.91	1.83	1.75	1.66	1.61	1.55	1.50	1.43	1.35	1.25
∞	3.84	3.00	2.60	2.37	2.21	2.10	2.01	1.94	1.88	1.83	1.75	1.67	1.57	1.52	1.46	1.39	1.32	1.22	1.00

$F = \frac{s_1^2}{s_2^2} = \frac{S_1}{m} / \frac{S_2}{n}$, where $s_1^2 = S_1/m$ and $s_2^2 = S_2/n$ are independent mean squares estimating a common variance σ^2 and based on m and n degrees of freedom, respectively.

$$F(F)=\int_0^F \frac{\Gamma\left(\frac{m+n}{2}\right)}{\Gamma\left(\frac{m}{2}\right)\Gamma\left(\frac{n}{2}\right)} m^{m/2} n^{n/2} x^{(m/2)-1}(n+mx)^{-(m+n)/2}\,dx = .975$$

$n\backslash m$	1	2	3	4	5	6	7	8	9	10	12	15	20	24	30	40	60	120	∞
1	647.8	799.5	864.2	899.6	921.8	937.1	948.2	956.7	963.3	968.6	976.7	984.9	993.1	997.2	1001	1006	1010	1014	1018
2	38.51	39.00	39.17	39.25	39.30	39.33	39.36	39.37	39.39	39.40	39.41	39.43	39.45	39.46	39.46	39.47	39.48	39.49	39.50
3	17.44	16.04	15.44	15.10	14.88	14.73	14.62	14.54	14.47	14.42	14.34	14.25	14.17	14.12	14.08	14.04	13.99	13.95	13.90
4	12.22	10.65	9.98	9.60	9.36	9.20	9.07	8.98	8.90	8.84	8.75	8.66	8.56	8.51	8.46	8.41	8.36	8.31	8.26
5	10.01	8.43	7.76	7.39	7.15	6.98	6.85	6.76	6.68	6.62	6.52	6.43	6.33	6.28	6.23	6.18	6.12	6.07	6.02
6	8.81	7.26	6.60	6.23	5.99	5.82	5.70	5.60	5.52	5.46	5.37	5.27	5.17	5.12	5.07	5.01	4.96	4.90	4.85
7	8.07	6.54	5.89	5.52	5.29	5.12	4.99	4.90	4.82	4.76	4.67	4.57	4.47	4.42	4.36	4.31	4.25	4.20	4.14
8	7.57	6.06	5.42	5.05	4.82	4.65	4.53	4.43	4.36	4.30	4.20	4.10	4.00	3.95	3.89	3.84	3.78	3.73	3.67
9	7.21	5.71	5.08	4.72	4.48	4.32	4.20	4.10	4.03	3.96	3.87	3.77	3.67	3.61	3.56	3.51	3.45	3.39	3.33
10	6.94	5.46	4.83	4.47	4.24	4.07	3.95	3.85	3.78	3.72	3.62	3.52	3.42	3.37	3.31	3.26	3.20	3.14	3.08
11	6.72	5.26	4.63	4.28	4.04	3.88	3.76	3.66	3.59	3.53	3.43	3.33	3.23	3.17	3.12	3.06	3.00	2.94	2.88
12	6.55	5.10	4.47	4.12	3.89	3.73	3.61	3.51	3.44	3.37	3.28	3.18	3.07	3.02	2.96	2.91	2.85	2.79	2.72
13	6.41	4.97	4.35	4.00	3.77	3.60	3.48	3.39	3.31	3.25	3.15	3.05	2.95	2.89	2.84	2.78	2.72	2.66	2.60
14	6.30	4.86	4.24	3.89	3.66	3.50	3.38	3.29	3.21	3.15	3.05	2.95	2.84	2.79	2.73	2.67	2.61	2.55	2.49
15	6.20	4.77	4.15	3.80	3.58	3.41	3.29	3.20	3.12	3.06	2.96	2.86	2.76	2.70	2.64	2.59	2.52	2.46	2.40
16	6.12	4.69	4.08	3.73	3.50	3.34	3.22	3.12	3.05	2.99	2.89	2.79	2.68	2.63	2.57	2.51	2.45	2.38	2.32
17	6.04	4.62	4.01	3.66	3.44	3.28	3.16	3.06	2.98	2.92	2.82	2.72	2.62	2.56	2.50	2.44	2.38	2.32	2.25
18	5.98	4.56	3.95	3.61	3.38	3.22	3.10	3.01	2.93	2.87	2.77	2.67	2.56	2.50	2.44	2.38	2.32	2.26	2.19
19	5.92	4.51	3.90	3.56	3.33	3.17	3.05	2.96	2.88	2.82	2.72	2.62	2.51	2.45	2.39	2.33	2.27	2.20	2.13
20	5.87	4.46	3.86	3.51	3.29	3.13	3.01	2.91	2.84	2.77	2.68	2.57	2.46	2.41	2.35	2.29	2.22	2.16	2.09
21	5.83	4.42	3.82	3.48	3.25	3.09	2.97	2.87	2.80	2.73	2.64	2.53	2.42	2.37	2.31	2.25	2.18	2.11	2.04
22	5.79	4.38	3.78	3.44	3.22	3.05	2.93	2.84	2.76	2.70	2.60	2.50	2.39	2.33	2.27	2.21	2.14	2.08	2.00
23	5.75	4.35	3.75	3.41	3.18	3.02	2.90	2.81	2.73	2.67	2.57	2.47	2.36	2.30	2.24	2.18	2.11	2.04	1.97
24	5.72	4.32	3.72	3.38	3.15	2.99	2.87	2.78	2.70	2.64	2.54	2.44	2.33	2.27	2.21	2.15	2.08	2.01	1.94
25	5.69	4.29	3.69	3.35	3.13	2.97	2.85	2.75	2.68	2.61	2.51	2.41	2.30	2.24	2.18	2.12	2.05	1.98	1.91
26	5.66	4.27	3.67	3.33	3.10	2.94	2.82	2.73	2.65	2.59	2.49	2.39	2.28	2.22	2.16	2.09	2.03	1.95	1.88
27	5.63	4.24	3.65	3.31	3.08	2.92	2.80	2.71	2.63	2.57	2.47	2.36	2.25	2.19	2.13	2.03	2.00	1.93	1.85
28	5.61	4.22	3.63	3.29	3.06	2.90	2.78	2.69	2.61	2.55	2.45	2.34	2.23	2.17	2.11	2.05	1.98	1.91	1.83
29	5.59	4.20	3.61	3.27	3.04	2.88	2.76	2.67	2.59	2.53	2.43	2.32	2.21	2.15	2.09	2.03	1.96	1.89	1.81
30	5.57	4.18	3.59	3.25	3.03	2.87	2.75	2.65	2.57	2.51	2.41	2.31	2.20	2.14	2.07	2.01	1.94	1.87	1.79
40	5.42	4.05	3.46	3.13	2.90	2.74	2.62	2.53	2.45	2.39	2.29	2.18	2.07	2.01	1.94	1.88	1.80	1.72	1.64
60	5.29	3.93	3.34	3.01	2.79	2.63	2.51	2.41	2.33	2.27	2.17	2.06	1.94	1.88	1.82	1.74	1.67	1.58	1.48
120	5.15	3.80	3.23	2.89	2.67	2.52	2.39	2.30	2.22	2.16	2.05	1.94	1.82	1.76	1.69	1.61	1.53	1.43	1.31
∞	5.02	3.69	3.12	2.79	2.57	2.41	2.29	2.19	2.11	2.05	1.94	1.83	1.71	1.64	1.57	1.48	1.39	1.27	1.00

$F=\frac{s_1^2}{s_2^2}=\frac{S_1}{m}/\frac{S_2}{n}$, where $s_1^2=S_1/m$ and $s_2^2=S_2/n$ are independent mean squares estimating a common variance σ^2 and based on m and n degrees of freedom, respectively.

$$F(F)=\int_0^F \frac{\Gamma\left(\frac{m+n}{2}\right)}{\Gamma\left(\frac{m}{2}\right)\Gamma\left(\frac{n}{2}\right)} m^{m/2} n^{n/2} x^{(m/2)-1}(n+mx)^{-(m+n)/2}\,dx=.99$$

$n \backslash m$	1	2	3	4	5	6	7	8	9	10	12	15	20	24	30	40	60	120	∞
1	4052	4999.5	5403	5625	5764	5859	5928	5982	6022	6056	6106	6157	6209	6235	6261	6287	6313	6339	6366
2	98.50	99.00	99.17	99.25	99.30	99.33	99.36	99.37	99.39	99.40	99.42	99.43	99.45	99.46	99.47	99.47	99.48	99.49	99.50
3	34.12	30.82	29.46	28.71	28.24	27.91	27.67	27.49	27.35	27.23	27.05	26.87	26.69	26.60	26.50	26.41	26.32	26.22	26.13
4	21.20	18.00	16.69	15.98	15.52	15.21	14.98	14.80	14.66	14.55	14.37	14.20	14.02	13.93	13.84	13.75	13.65	13.56	13.46
5	16.26	13.27	12.06	11.39	10.97	10.67	10.46	10.29	10.16	10.05	9.89	9.72	9.55	9.47	9.38	9.29	9.20	9.11	9.02
6	13.75	10.92	9.78	9.15	8.75	8.47	8.26	8.10	7.98	7.87	7.72	7.56	7.40	7.31	7.23	7.14	7.06	6.97	6.88
7	12.25	9.55	8.45	7.85	7.46	7.19	6.99	6.84	6.72	6.62	6.47	6.31	6.16	6.07	5.99	5.91	5.82	5.74	5.65
8	11.26	8.65	7.59	7.01	6.63	6.37	6.18	6.03	5.91	5.81	5.67	5.52	5.36	5.28	5.20	5.12	5.03	4.95	4.86
9	10.56	8.02	6.99	6.42	6.06	5.80	5.61	5.47	5.35	5.26	5.11	4.96	4.81	4.73	4.65	4.57	4.48	4.40	4.31
10	10.04	7.56	6.55	5.99	5.64	5.39	5.20	5.06	4.94	4.85	4.71	4.56	4.41	4.33	4.25	4.17	4.08	4.00	3.91
11	9.65	7.21	6.22	5.67	5.32	5.07	4.89	4.74	4.63	4.54	4.40	4.25	4.10	4.02	3.94	3.86	3.78	3.69	3.60
12	9.33	6.93	5.95	5.41	5.06	4.82	4.64	4.50	4.39	4.30	4.16	4.01	3.86	3.78	3.70	3.62	3.54	3.45	3.36
13	9.07	6.70	5.74	5.21	4.86	4.62	4.44	4.30	4.19	4.10	3.96	3.82	3.66	3.59	3.51	3.43	3.34	3.25	3.17
14	8.86	6.51	5.56	5.04	4.69	4.46	4.28	4.14	4.03	3.94	3.80	3.66	3.51	3.43	3.35	3.27	3.18	3.09	3.00
15	8.68	6.36	5.42	4.89	4.56	4.32	4.14	4.00	3.89	3.80	3.67	3.52	3.37	3.29	3.21	3.13	3.05	2.96	2.87
16	8.53	6.23	5.29	4.77	4.44	4.20	4.03	3.89	3.78	3.69	3.55	3.41	3.26	3.18	3.10	3.02	2.93	2.84	2.75
17	8.40	6.11	5.18	4.67	4.34	4.10	3.93	3.79	3.68	3.59	3.46	3.31	3.16	3.08	3.00	2.92	2.83	2.75	2.65
18	8.29	6.01	5.09	4.58	4.25	4.01	3.84	3.71	3.60	3.51	3.37	3.23	3.08	3.00	2.92	2.84	2.75	2.66	2.57
19	8.18	5.93	5.01	4.50	4.17	3.94	3.77	3.63	3.52	3.43	3.30	3.15	3.00	2.92	2.84	2.76	2.67	2.58	2.49
20	8.10	5.85	4.94	4.43	4.10	3.87	3.70	3.56	3.46	3.37	3.23	3.09	2.94	2.86	2.78	2.69	2.61	2.52	2.42
21	8.02	5.78	4.87	4.37	4.04	3.81	3.64	3.51	3.40	3.31	3.17	3.03	2.88	2.80	2.72	2.64	2.55	2.46	2.36
22	7.95	5.72	4.82	4.31	3.99	3.76	3.59	3.45	3.35	3.26	3.12	2.98	2.83	2.75	2.67	2.58	2.50	2.40	2.31
23	7.88	5.66	4.76	4.26	3.94	3.71	3.54	3.41	3.30	3.21	3.07	2.93	2.78	2.70	2.62	2.54	2.45	2.35	2.26
24	7.82	5.61	4.72	4.22	3.90	3.67	3.50	3.36	3.26	3.17	3.03	2.89	2.74	2.66	2.58	2.49	2.40	2.31	2.21
25	7.77	5.57	4.68	4.18	3.85	3.63	3.46	3.32	3.22	3.13	2.99	2.85	2.70	2.62	2.54	2.45	2.36	2.27	2.17
26	7.72	5.53	4.64	4.14	3.82	3.59	3.42	3.29	3.18	3.09	2.96	2.81	2.66	2.58	2.50	2.42	2.33	2.23	2.13
27	7.68	5.49	4.60	4.11	3.78	3.56	3.39	3.26	3.15	3.06	2.93	2.78	2.63	2.55	2.47	2.38	2.29	2.20	2.10
28	7.64	5.45	4.57	4.07	3.75	3.53	3.36	3.23	3.12	3.03	2.90	2.75	2.60	2.52	2.44	2.35	2.26	2.17	2.06
29	7.60	5.42	4.54	4.04	3.73	3.50	3.33	3.20	3.09	3.00	2.87	2.73	2.57	2.49	2.41	2.33	2.23	2.14	2.03
30	7.56	5.39	4.51	4.02	3.70	3.47	3.30	3.17	3.07	2.98	2.84	2.70	2.55	2.47	2.39	2.30	2.21	2.11	2.01
40	7.31	5.18	4.31	3.83	3.51	3.29	3.12	2.99	2.89	2.80	2.66	2.52	2.37	2.29	2.20	2.11	2.02	1.92	1.80
60	7.08	4.98	4.13	3.65	3.34	3.12	2.95	2.82	2.72	2.63	2.50	2.35	2.20	2.12	2.03	1.94	1.84	1.73	1.60
120	6.85	4.79	3.95	3.48	3.17	2.96	2.79	2.66	2.56	2.47	2.34	2.19	2.03	1.95	1.86	1.76	1.66	1.53	1.38
∞	6.63	4.61	3.78	3.32	3.02	2.80	2.64	2.51	2.41	2.32	2.18	2.04	1.88	1.79	1.70	1.59	1.47	1.32	1.00

$F=\frac{s_1^2}{s_2^2}=\frac{S_1}{m}/\frac{S_2}{n}$, where $s_1^2=S_1/m$ and $s_2^2=S_2/n$ are independent mean squares estimating a common variance σ^2 and based on m and n degrees of freedom, respectively.

MOMENT OF INERTIA FOR VARIOUS BODIES OF MASS

The mass of the body is indicated by m

Body	Axis	Moment of inertia
Uniform thin rod of length l	Normal to the length, at one end	$m\frac{1}{3}l^2$
Uniform thin rod of length l	Normal to the length, at the center	$m\frac{1}{12}l^2$
Thin rectangular sheet, sides a and b	Through the center parallel to b	$m\frac{1}{12}a^2$
Thin rectangular sheet, sides a and b	Through the center perpendicular to the sheet	$m\frac{1}{12}(a^2+b^2)$
Thin circular sheet of radius r	Normal to the plate through the center	$m\frac{1}{2}r^2$
Thin circular sheet of radius r	Along any diameter	$m\frac{1}{4}r^2$
Thin circular ring. Radii r_1 and r_2	Through center normal to plane of ring	$m\frac{1}{2}(r_1^2+r_2^2)$
Thin circular ring. Radii r_1 and r_2	Any diameter	$m\frac{1}{4}(r_1^2+r_2^2)$
Rectangular parallelepiped, edges a, b, and c	Through center perpendicular to face ab, (parallel to edge c)	$m\frac{1}{12}(a^2+b^2)$
Sphere, radius r	Any diameter	$m\frac{2}{5}r^2$
Spherical shell, external radius, r_1, internal radius r_2	Any diameter	$m\frac{2}{5}\frac{(r_1^5-r_2^5)}{(r_1^3-r_2^3)}$
Spherical shell, very thin, mean radius, r	Any diameter	$m\frac{2}{3}r^2$
Right circular cylinder of radius r, length l	The longitudinal axis of the solid	$m\frac{1}{2}r^2$
Right circular cylinder of radius r, length l	Transverse diameter	$m\left(\frac{r^2}{4}+\frac{l^2}{12}\right)$
Hollow circular cylinder, length l, radii r_1 and r_2	The longitudinal axis of the figure	$m\frac{1}{2}(r_1^2+r_2^2)$
Thin cylindrical shell, length l, mean radius, r	The longitudinal axis of the figure	mr^2
Hollow circular cylinder, length l, radii r_1 and r_2	Transverse diameter	$m\left(\frac{r_1^2+r_2^2}{4}+\frac{l^2}{12}\right)$
Hollow circular cylinder, length l, very thin, mean radius r	Transverse diameter	$m\left(\frac{r^2}{2}+\frac{l^2}{12}\right)$
Elliptic cylinder, length l, transverse semiaxes a and b	Longitudinal axis	$m\frac{1}{4}(a^2+b^2)$
Right cone, altitude h, radius of base r	Axis of the figure	$m\frac{3}{10}r^2$
Spheroid of revolution, equatorial radius r	Polar axis	$m\frac{2}{5}r^2$
Ellipsoid, axes $2a$, $2b$, $2c$	Axis $2a$	$m\frac{1}{5}(b^2+c^2)$

Appendix B
Sources of Physical and Chemical Data

Sources of Physical and Chemical Data B-1